Y0-CDR-970

HANDBOOK OF PHYSICS

OTHER McGRAW-HILL HANDBOOKS OF INTEREST

AMERICAN INSTITUTE OF PHYSICS · American Institute of Physics Handbook
AMERICAN SOCIETY OF MECHANICAL ENGINEERS · ASME Handbooks:

Engineering Tables	Metals Engineering—Processes
Metals Engineering—Design	Metals Properties

BAUMEISTER AND MARKS · Standard Handbook for Mechanical Engineers
BERRY, BOLLAY, AND BEERS · Handbook of Meteorology
BLATZ · Radiation Hygiene Handbook
BRADY · Materials Handbook
BURINGTON · Handbook of Mathematical Tables and Formulas
BURINGTON AND MAY · Handbook of Probability and Statistics with Tables
CALLENDER · Time-Saver Standards
CHOW · Handbook of Applied Hydrology
CONSIDINE · Process Instruments and Controls Handbook
CONSIDINE AND ROSS · Handbook of Applied Instrumentation
ETHERINGTON · Nuclear Engineering Handbook
FLÜGGE · Handbook of Engineering Mechanics
GRANT · Hackh's Chemical Dictionary
HAMSHER · Communication System Engineering Handbook
HARRIS AND CREDE · Shock and Vibration Handbook
HENNEY · Radio Engineering Handbook
HUNTER · Handbook of Semiconductor Electronics
HUSKEY AND KORN · Computer Handbook
IRESON · Reliability Handbook
JURAN · Quality Control Handbook
KAELBLE · Handbook of X-rays
KALLEN · Handbook of Instrumentation and Controls
KING AND BRATER · Handbook of Hydraulics
KLERER AND KORN · Digital Computer User's Handbook
KNOWLTON · Standard Handbook for Electrical Engineers
KOELLE · Handbook of Astronautical Engineering
KORN AND KORN · Mathematical Handbook for Scientists and Engineers
LANDEE, DAVIS, AND ALBRECHT · Electronic Designer's Handbook
MACHOL · System Engineering Handbook
MANTELL · Engineering Materials Handbook
MARKUS · Electronics and Nucleonics Dictionary
MEITES · Handbook of Analytical Chemistry
PERRY · Engineering Manual
RICHEY · Agricultural Engineers' Handbook
ROTHBART · Mechanical Design and Systems Handbook
STREETER · Handbook of Fluid Dynamics
TERMAN · Radio Engineers' Handbook
TOULOUKIAN · Retrieval Guide to Thermophysical Properties Research Literature
TRUXAL · Control Engineers' Handbook
URQUHART · Civil Engineering Handbook
WOLMAN · Handbook of Clinical Psychology

HANDBOOK OF PHYSICS

EDITED BY

E. U. CONDON, Ph.D.

Professor of Physics
and
Fellow of the Joint Institute for Laboratory Astrophysics
University of Colorado

HUGH ODISHAW, D.Sc.

Executive Secretary
Division of Physical Sciences
National Academy of Sciences

SECOND EDITION

McGRAW-HILL BOOK COMPANY

New York St. Louis San Francisco Dusseldorf London

Mexico Panama Sydney Toronto

HANDBOOK OF PHYSICS

 Library of Congress Catalog Card Number 66-20002

ISBN 07-012403-5

90 MPM 75

Contributors

Lawrence Aller
University of California
Los Angeles

Franz L. Alt
American Institute of Physics
New York

D. P. Ames
McDonnell Douglas Corp.
St. Louis

K. R. Atkins
University of Pennsylvania
Philadelphia

John Bardeen
University of Illinois
Urbana

Richard A. Beth
Brookhaven National Laboratory
Long Island

R. B. Bird
University of Wisconsin
Madison

John P. Blewett
Brookhaven National Laboratory
Long Island

Sanborn C. Brown
Massachusetts Institute of Technology
Cambridge

William Fuller Brown, Jr.
University of Minnesota
Minneapolis

Stephan Brunauer
Clarkson College of Technology
Potsdam

Herbert B. Callen
University of Pennsylvania
Philadelphia

G. M. Clemence
Yale University
New Haven

Bernard L. Cohen
University of Pittsburgh
Pittsburgh

E. Richard Cohen
North American Aviation Science Center
Thousand Oaks

E. U. Condon
University of Colorado
Boulder

J. L. B. Cooper
University of Toronto
Toronto

L. E. Copeland
Portland Cement Association
Chicago

C. F. Curtis
University of Wisconsin
Madison

Jesse W. M. DuMond
California Institute of Technology
Pasadena

Leonard Eisenbud
State University of New York
Long Island

Churchill Eisenhart
National Bureau of Standards
Washington, D.C.

William M. Fairbank
Stanford University
Stanford

Alexander L. Fetter
Stanford University
Stanford

William E. Forsythe
General Electric Company
Cleveland

M. M. Frocht
Illinois Institute of Technology
Chicago

G. T. Garvey
Princeton University
Princeton

D. T. Goldman
National Bureau of Standards
Washington, D.C.

Walter Gordy
Duke University
Durham

R. W. Gurney (*deceased*)
Bristol University
Bristol

Andrew Guthrie
California State College
Hayward

David Halliday
University of Pittsburgh
Pittsburgh

Walter J. Hamer
National Bureau of Standards
Washington, D.C.

R. W. Hayward
National Bureau of Standards
Washington, D.C.

Max Herzberger
Eastman Kodak Company
Rochester

E. L. Hill
University of Minnesota
Minneapolis

R. D. Hill
University of Illinois
Urbana

John A. Hipple
Philips Laboratories
Briarcliff Manor

Joseph G. Hirschberg
University of Miami
Coral Gables

J. O. Hirschfelder
University of Wisconsin
Madison

A. J. Hoffman
International Business Machines Corporation
Yorktown Heights

Theodor Hurlimann
Federal Institute for Reactor Research
Wuerenlingen, Switzerland

Uno Ingard
Massachusetts Institute of Technology
Cambridge

John C. Ingraham
Massachusetts Institute of Technology
Cambridge

Fritz John
New York University
New York

Walter H. Johnson, Jr.
University of Minnesota
Minneapolis

Deane B. Judd
National Bureau of Standards
Washington, D.C.

C. Lanczos
Ford Motor Company
Dearborn

Gerald L. Landsman
Douglas Aircraft Company
Los Angeles

Julian E. Mack (*deceased*)
University of Wisconsin
Madison

J. Rand McNally, Jr.
Oak Ridge National Laboratory
Oak Ridge

R. D. Maurer
Corning Glass Works
Corning

Robert J. Maurer
University of Illinois
Urbana

Walter C. Michels
Bryn Mawr College
Bryn Mawr

Elliott W. Montroll
University of Rochester
Rochester

Philip M. Morse
Massachusetts Institute of Technology
Cambridge

C. O. Muehlhouse
National Bureau of Standards
Washington, D.C.

Harold H. Nielson
Ohio State University
Columbus

A. O. Nier
University of Minnesota
Minneapolis

Jan S. Nilsson
Chalmers University of Technology
Göteborg, Sweden

Richard M. Noyes
University of Oregon
Eugene

Hugh Odishaw
National Academy of Sciences
Washington, D.C.

Chester H. Page
National Bureau of Standards
Washington, D.C.

Yash Pal
Tata Institute of Fundamental Research
Bombay

Ray Pepinsky
Nova University
Fort Lauderdale

Louis A. Pipes
University of California
Los Angeles

Norman F. Ramsey
Harvard University
Cambridge

M. Reiner
Technion—Israel Institute of Technology
Haifa, Israel

F. L. Roesler
University of Wisconsin
Madison

M. E. Rose
University of Virginia
Charlottesville

Frederick D. Rossini
University of Notre Dame
Notre Dame

Ivan G. Schröder
National Bureau of Standards and Brookhaven National Laboratory
Long Island

R. J. Seeger
National Science Foundation
Washington, D.C.

E. J. Seldin
Union Carbide Corporation
Cleveland

Harold K. Skramstad
Naval Ordnance Laboratory
Corona

Lloyd P. Smith
Stanford Research Institute
Menlo Park

R. Smoluchowski
Princeton University
Princeton

E. S. Steeb, Jr.
General Electric Company
Cleveland

William E. Stephens
University of Pennsylvania
Philadelphia

S. D. Stookey
Corning Glass Works
Corning

James F. Swindells
National Bureau of Standards
Washington, D.C.

A. H. Taub
University of California
Berkeley

Olga Taussky
California Institute of Technology
Pasadena

John Todd
California Institute of Technology
Pasadena

C. Tompkins
University of California
Los Angeles

V. Vand
Pennsylvania State University
University Park

A. R. von Hippel
Massachusetts Institute of Technology
Cambridge

John A. Wheeler
Princeton University
Princeton

Eugene P. Wigner
Princeton University
Princeton

J. Gibson Winans
State University of New York at Buffalo
Buffalo

Hugh C. Wolfe
American Institute of Physics
New York

Reuben E. Wood
George Washington University
Washington, D.C.

W. J. Youden
National Bureau of Standards
Washington, D.C.

Marvin Zelen
National Cancer Institute
Bethesda

Preface to the Second Edition

To the preface of the first edition we need add only an indication of the changes. That volume, which seems to have served needs of an appreciable number of scientists and engineers, ran some 1500 large pages. By condensation and elimination, we got this down to about 1400 pages, but then revisions and the inclusion of some new topics produced another 300 pages, and so this edition totals some 1700 pages.

The extent of revision of material in the first edition can be summarized as follows. About one-third remains almost as it was. Another third received fair to considerable revision. A final third called for total revision, enough having happened in those areas to require and warrant a fresh approach. Useful expansion of some chapters and inclusion of new topics accounts for the increase in size.

New topics include linear spaces and operators, numerical analysis, theory of relativity, plasma physics, superconductivity, and magnetic resonance. Parts 3 and 6 required the least revision; Parts 1, 2, 4, 5, 7, and 8 averaged out to about better than a third; Part 9 on nuclear physics called for an essentially new and extended treatment, as might be expected. The index has been revised and expanded: even while the first edition was in print, its index was expanded at the suggestion of colleagues; that for this edition is much fuller, consisting of about 12,000 entries.

We again hope that scientists and engineers will find this one-volume compendium useful and would appreciate suggestions from its readers.

E. U. CONDON
HUGH ODISHAW

Preface to the First Edition

This book was first planned nearly ten years ago when we were closely associated at the National Bureau of Standards. We set ourselves the problem of making a judicious selection from the vast literature of physics of materials which might reasonably be called "What every physicist should know."

As the planning went forward we became increasingly aware of what a difficult task we had undertaken. The literature of physics has become so great, and is growing at such a rate, that it is very difficult for a physicist to be really well-informed on more than a relatively narrow specialty within the subject. Nevertheless the unity of the science is such that much research progress depends considerably on utilization of advances in one part to provide the means for solving problems in another. Therefore it is necessary for physicists to make strong efforts to resist tendencies toward over-specialization.

One way in which the rapidity of progress has complicated our task is the tendency for parts of the book to become out-of-date while being set up in type. We have made efforts to avoid this by making more than the usual number of additions and corrections while the book was going through galley proof. Our thanks are due the contributing specialists for their willingness to go to the extra trouble of making their chapters as up-to-date as possible in spite of this difficulty.

By the very nature of the preparation and publication process, a handbook cannot be completely current with journal literature, and there is variation even emong the chapters, as revealed by their references. Within this restriction, we believe that the Handbook fulfills its function as a one-volume compendium.

It is our sincere hope that physicists the world over will find this selection of materials to be a useful one. We think that there is considerable economy of effort to be gained in a one-volume synthesis of the principal parts of the science in that so many techniques find use again and again in different parts of the subject and only need to be explained once in a work of this kind. We will appreciate receiving suggestions from readers as to how the book's usefulness may be improved in future editions.

E. U. CONDON
HUGH ODISHAW

Contents

HANDBOOK OF PHYSICS

Part 1 · Mathematics

Chapter 1

Fundamentals

By FRANZ L. ALT, National Bureau of Standards

1. Numbers and Arithmetic Operations

Numbers. It is possible to define numbers and prove statements about them without specific recourse to experience or intuition, using only a few simple concepts of logic. For this reason mathematical statements are considered infallible except for demonstrable errors in reasoning, while statements in the physical sciences are subject to empirical verification or revision. Such a definition of numbers, together with a list of their fundamental properties which can be proved logically on the basis of this definition, is given in Sec. 2.

For convenience, numbers are written in the *decimal system*. In this system, ten integers are represented by special symbols, called digits: 0,1,2,3,4,5,6,7,8,9. Any positive integer (*natural number*) greater than 9 is represented by a group of digits

$$d_n d_{n-1} \cdots d_2 d_1 d_0$$

where $0 \leq d_i \leq 9$ for $i = 0,1,2, \ldots ,n$. This group is understood to represent the number

$$d_n 10^n + d_{n-1} 10^{n-1} + \cdots + d_2 10^2 + d_1 10 + d_0$$

Negative integers are written in the same way with the minus sign prefixed. *Rational numbers* (see Sec. 2) are represented in either of two ways: as fractions (a/b), where a and b are integers, or in decimal notation

$$d_n d_{n-1} \cdots d_2 d_1 d_0 \cdot d_{-1} d_{-2} \cdots d_{-m} \cdots$$

Such a decimal number represents a rational number if, and only if, it either terminates or becomes periodic after a finite number of decimal places; for example, $0.1875 = 3/16$, $12.50\,675\,675\,675 \ldots = 1851/148$. Terminating decimal numbers may alternatively be written with periods of 9s, for example,

$$0.1875 = 0.1874999 \ldots$$

Decimal numbers which are neither periodic nor terminating represent irrational numbers (see Sec. 2). In actual computations *irrational numbers* are commonly approximated by rational numbers, usually by terminating decimal numbers. Examples of irrational numbers are:

$$\begin{aligned} \sqrt{2} &= 1.4142136 \ldots \\ \sqrt{3} &= 1.7320508 \ldots \\ \pi &= 3.1415927 \ldots \\ e &= 2.7182818 \ldots \\ \log 2 &= 0.3010300 \ldots \end{aligned}$$

Instead of the decimal system, in which powers of 10 play a fundamental role, systems based on other integers are occasionally used. Thus a system using the base b contains b digits, whose values are $0,1,2, \ldots ,b - 1$. In particular, the *binary* system, with base 2, has only 0 and 1 for digits. A number like 101.1001 is understood to mean

$$\begin{aligned} 1 \times 2^2 + 0 \times 2^1 + 1 \times 2^0 + 1 \times 2^{-1} \\ + 0 \times 2^{-2} + 0 \times 2^{-3} + 1 \times 2^{-4} \\ = 4 + 1 + \tfrac{1}{2} + \tfrac{1}{16} = 5.5625 \end{aligned}$$

The binary system is used in some types of computing machines, as are the systems based on 4, 8, and 16. The duodecimal system (base 12) has a certain historical importance. Other number systems are used only infrequently.

Addition and Subtraction. The simplest arithmetic operation is *addition*, $a + b$. The two numbers on which the operation is carried out are called *terms;* if they are to be distinguished, the first is called *augend*, the second *addend* (or occasionally *auctor*). The result of the operation is called the *sum*. Because of the associative law of addition, according to which $a + (b + c) = (a + b) + c$ (see Sec. 2), it is permissible to write sums of more than two terms:

$$a_1 + a_2 + \cdots + a_n$$

without indicating which of the $(n - 1)$ additions is to be carried out first, second, etc. Also, because of the commutative law (Sec. 2) the terms of such a sum may be rearranged at will without changing the result.

Subtraction is the inverse of addition. By this we mean the problem of finding a number c such that

$$c + b = a$$

i.e., of finding the augend if the addend and sum are known. (Because of the commutativity of addition, the problem of finding the addend leads to the same operation.) The result is denoted by $a - b$. Here the number a is called the *minuend*, b the *subtrahend*, and the result of the subtraction the *difference*.

If a is any real number $\neq 0$, exactly one of the two numbers a, $-a$ is positive; it is denoted by $|a|$, the absolute value or *modulus* of a. We define $|0| = 0$. For complex numbers $a = a' + ia''$ (see Sec. 2), the absolute value is defined by

$$|a| = \sqrt{a'^2 + a''^2}$$

Thus we have, for both real and complex numbers,

$$|a| \geq 0$$

as well as the *triangle inequality*

$$|a + b| \leq |a| + |b|$$

Multiplication and Division. Multiplication of any number by a natural number may be defined by repeated addition of the first number to itself. This definition can be extended to cover multiplication of any two numbers. Two numbers to be multiplied, as ab, are called *factors;* if they are to be distinguished, the first is called *multiplicand,* the second *multiplier.* The result of the multiplication is called the *product.*

The preferred notations for multiplication are the symbol $\times$ for specific numbers and a dot or mere juxtaposition for algebraic symbols. Thus 5×7 is preferable to $5 \cdot 7$ and ab or $a \cdot b$ is preferable to $a \times b$. These recommendations are not, however, generally accepted.

As in the case of addition, the commutative and associative laws (Sec. 2) cause products of more than two factors, like $a_1a_2 \ldots a_n$, to have the same value without regard to arrangement of factors or the order in which the multiplications are carried out.

Because of its commutativity, multiplication has only one inverse operation, division. This is denoted by $\frac{a}{b}$ or $a \div b$, or sometimes a/b or $a:b$. The number a is called the *dividend,* b the *divisor,* and the result of the division is called the *quotient* or *ratio.*

The division $a/b = c$ may be regarded as the solution of the problem of finding that number c for which $bc = a$. If a and b are integers, then sometimes division is understood to mean finding the absolutely largest *integer* c such that

$$|bc| \leq |a|$$

in other words, solving the equation

$$a = bc + d$$

where c and d are integers, d is of the same sign as a, and $|d| < |b|$. The number d is called the *remainder.*

The following properties of multiplication are often used: a product is zero if and only if at least one of its factors is zero. The product of a real number a by itself (denoted by a^2) is not negative.

When several arithmetic operations occur in succession, multiplications and divisions are performed before additions and subtractions, and in the order in which they occur. If any other order is desired, it is indicated by parentheses, brackets, or braces. For example:

$$\begin{aligned} 3 + 6 \div 2 \times 4 - 7 &= 3 + 3 \times 4 - 7 \\ &= 3 + 12 - 7 = 15 - 7 = 8 \end{aligned}$$

$$\begin{aligned} 6 \div (3 \times (5 - (1 + 2))) &= 6 \div \{3 \times [5 - (1 + 2)]\} \\ &= 6 \div \{3 \times [5 - 3]\} \\ &= 6 \div \{3 \times 2\} = 6 \div 6 = 1 \end{aligned}$$

Powers, Roots, and Logarithms. Just as the repeated addition of a number to itself was used as a definition of multiplication by a natural number, so the operation of multiplying a number by itself n times, where n is a natural number, leads to the operation a^n, *the nth power of a.* This may be generalized for any numbers, a^b. Here a is called the *base,* b the *exponent.* This operation is neither commutative nor associative; that is, propositions like $a^b = b^a$ or $(a^b)^c = a^{(b^c)}$ are in general not true. Because of the noncommutativity of the operation $a^b = c$ there are two operations which are inverse to it,

$$a = \sqrt[b]{c} \qquad \text{or} \qquad a = c^{1/b}$$

the bth *root* of c, and

$$b = \log_a c$$

the *logarithm* of c to the base a. Logarithms to the base 10 are called *common* or *Briggs* logarithms. Logarithms to the base $e = 2.718281828 \ldots$ are of particular importance; they are called natural logarithms. The symbol log without indication of the base is used in the literature sometimes for common, sometimes for natural logarithms; some authors use ln for natural logarithms. For powers with base e, as e^a, the symbol exp a is often used (read: "exponential a"), especially if the exponent is not a specific number or single letter, as a, but a longer algebraic expression.

Logarithms, especially common logarithms, are used to simplify numerical computations, by means of the relations

$$\begin{aligned} \log(ab) &= \log a + \log b \\ \log \frac{a}{b} &= \log a - \log b \\ \log a^b &= b \log a \\ \log \sqrt[b]{a} &= \frac{1}{b} \log a \end{aligned}$$

Thus the computer is able to replace multiplications and divisions by additions and subtractions, and the computing of powers (and roots) by multiplications, provided that he can efficiently obtain $\log x$ from a given x, and vice versa. This is done by use of tables of logarithms.

Rounding. Irrational numbers are usually replaced by rational numbers for numerical computation. In addition, periodic decimal numbers are frequently replaced by terminating ones, and long terminating ones by shorter ones. Such replacement of a number by another which is approximately, but not exactly, equal is called *rounding;* the difference between the true number and the approximate one, the *rounding error.* The process of rounding is usually carried out in such a way that the rounding error does not exceed ½ unit of the last retained decimal place. This is accomplished by the following rule: if the first omitted digit has one of the values from 5 to 9, then one unit is added to the last retained decimal place (*rounding up*); thus $1.375 \sim 1.4$; $0.9965 \sim 1.00$. If the first omitted digit has one of the values from 0 to 4, no change is made in the retained digits (*rounding off* or *down*); thus $1.425 \sim 1.4$. Sometimes the following further rule is adopted: if the sequence of omitted digits is $5000 \ldots$ or $4999 \ldots$, round up or down in such a way as to make the last retained digit even; thus $1.375 \sim 1.38$, $0.9965 \sim 0.996$.

When numbers represent the result of physical measurements, the extent of rounding is usually determined by the precision with which those measurements can be performed.

If arithmetic operations are carried out with rounded numbers the results are affected by rounding errors. An exact evaluation of these errors is cumbersome. The following rules serve as guide lines for computation with rounded numbers, by determining the number of digits that may validly be retained in the result of an operation.

The number of valid decimal places (i.e., places to the right of the decimal point) in a sum or difference is not greater than the number of such places in either term.

The number of places to the left and right of the decimal point in a number, beginning with the first *nonzero* digit and ending with the last validly retained digit, is called the number of significant digits. It is usually denoted by S, while the number of valid decimal places (to the right of the point) is denoted by D. Thus

3.1416 has $5S$, $4D$
0.0167 has $3S$, $4D$
1.3010 has $5S$, $4D$

The number of significant digits of a product or quotient is not greater than the number of such digits in either factor.

In long computations, to reduce the effect of accumulation of rounding errors, it is advisable to carry one or more digits beyond the last valid one in all intermediate results, remembering that they are invalid and rounding the final result to the valid number of digits.

2. Logical Foundation of Arithmetic

Natural Numbers. Consider various classes (sets) of objects. Two classes A, B are called *equivalent* if it is possible to establish a one-to-one correspondence between their elements (i.e., to associate with each element of A one element of B in such a way that with different elements of A different elements of B are associated and that all elements of B are used). Having thus defined equivalence of two classes, we start with an arbitrary class A of objects and consider the *class of all classes* which are *equivalent to* A. This class of classes is called a (cardinal) number; more specifically, it is called the *cardinal number of the class* A or *the number of elements of* A. In particular, if a class A has the property that whenever a and b are elements of A, it is true that a and b are identical, then the cardinal number of A is called 1. The cardinal number of the "void" class (the class containing no elements) is called 0. Starting from these definitions, other *natural numbers* (2,3,4, . . .) can be defined successively.

This method of introducing numbers becomes more accessible to intuitive understanding (but also to misunderstanding) if we note that two classes are equivalent, by the above definitions, if and only if they contain the same number of elements; and if we note further that defining a cardinal number as a *class of classes* amounts to the same as defining it as *that characteristic which is common to all equivalent classes.*

Properties of Natural Numbers. The *principle of complete induction* states that if $S(n)$ is some statement about the natural number n of which it can be shown that $S(n)$ implies $S(n + 1)$ and that $S(n)$ is true for some number n_0, then $S(n)$ is true for every natural number $> n_0$.

It can be proved, again by purely logical reasoning and without recourse to intuition, that the natural numbers possess the elementary properties usually associated with them. Among these are:

Theorem 1. The natural numbers are *ordered.* This means that there exists a relation between numbers, $a > b$ ("a is greater than b"), such that:

a. If $a > b$ and $b > c$, then $a > c$ (the relation is transitive).
b. If $a > b$ is true, then $b > a$ is not true (the relation is asymmetric).
c. For any two numbers, a,b, at least one of the statements $a > b$, $a = b$, $b > a$ is true (it follows that *exactly* one is true).

We write $a \geq b$ instead of "$a > b$ or $a = b$," $a \neq b$ instead of "$a > b$ or $a < b$," $a < b$ instead of $b > a$.

d. $1 > 0$.

Theorem 2. Addition and multiplication are defined; the sum, $a + b$, and product, ab, of two natural numbers are again natural numbers. They satisfy these relations:

a. $a + b = b + a$ (commutative law of addition)
b. $(a + b) + c = a + (b + c)$ (associative law of addition)
c. $a + 0 = a$
d. $ab = ba$ (commutative law of multiplication)
e. $(ab)c = a(bc)$ (associative law of multiplication)
f. $a \cdot 1 = a$
g. $a(b + c) = ab + ac$ (distributive law of multiplication)
h. If $a > b$, then $a + c > b + c$
i. If $a > b$, then $ac > bc$

Other Types of Numbers. It is possible in principle, but cumbersome in practice, to express statements about physical measurements in terms of natural numbers alone. Such statements are made briefer by the introduction of other types of numbers: negative numbers, fractions, and irrational and complex numbers.

In order to introduce *fractions* (*rational numbers*), we consider ordered pairs of natural numbers (a,b) with $b \neq 0$, calling the first number of such a pair *numerator* and the second *denominator;* we call two pairs (a,b) and (a',b') equivalent if $ab' = a'b$; we define the class of all pairs which are equivalent with a given pair (a,b) as the rational number a/b. We then define the relation $a/b > a'/b'$ as meaning that $ab' > a'b$; the sum $a/b + a'/b'$ as $(ab' + a'b)/(bb')$; the product $(a/b)(a'/b')$ as $(aa')/(bb')$. Again, as in the case of natural numbers, we can prove the usual elementary properties of rational numbers. Among these are most, but not all, of the theorems about natural numbers.

By the above definitions, fractions with denominator 1 have properties exactly analogous to those of the natural numbers. Thus, $(a/1) + (b/1) = (a + b)/1$; $(a/1)(b/1) = (ab/1)$; and $a/1 > b/1$ if and only if $a > b$. For brevity, we shall write a instead of $a/1$, etc., although there is a logical difference between the natural number a and the fraction $a/1$.

Customarily, the introduction of rational numbers, just given, is preceded by a similarly simple introduction of *relative* (positive or negative) integers. After this is done, rational numbers are defined as classes of pairs of integers, with slight modifications in the procedure shown above. The introduction of *irrational* numbers is somewhat more complicated and will not be explained here. Finally, *complex* numbers are defined as ordered pairs of *real* (i.e., rational or irrational) numbers (a,b), where a is called the *real part* and b the *imaginary part*, with the conventional definitions

$$(a, b) + (a', b') = (a + a', b + b')$$
$$(a, b)(a', b') = (aa' - bb', ab' + a'b)$$

For pairs whose imaginary part is 0, these operations give the same result as those with real numbers, that is, $(a, 0) + (a', 0) = (a + a', 0)$; $(a, 0)(a', 0) = (aa', 0)$; for abbreviation we write a for $(a,0)$, etc. If we further set $i = (0,1)$, we obtain

$$(a,b) = (a,0) + (0,1)(b,0) = a + ib$$

From the definition of multiplication,

$$i^2 = (0,1)(0,1) = (-1,0) = -1$$

or

$$i = \sqrt{-1}$$

The two numbers $(a + ib)$, $(a - ib)$ are called *complex conjugates*. For hypercomplex numbers, see Part 1, Chap. 2, Sec. 13.

Every time we widen the domain of numbers, some theorems lose their validity. For instance, when we go from natural to (positive and negative) integer numbers, Theorem 2*i* is no longer true; in its place is Theorem 2*i'*:

Theorem 2*i'*. If $a < b$ and $c < 0$, then $ac > bc$. On the other hand, new theorems can be proved for the enlarged domain which do not hold for the narrower one. For example, for the domain of all integers (positive, negative, and zero):

Theorem 3*a*. For any two integers x,y, there is an integer z such that $x = y + z$. By way of definition we set $z = x - y$.

For the domain of all rational numbers:

Theorem 3*b*. For any two rational numbers p,q, where $q \neq 0$, there exists a rational number r such that $p = rq$.

The following fundamental property of rational numbers can be derived from the simple theorems listed so far: If x,y are two rational numbers such that $y < x$, then there exists a rational number z such that $y < z < x$. This property is expressed briefly by saying that the rational numbers are *dense*.

The real numbers have the important fundamental property of being *continuous*. In order to formulate this property, we introduce two definitions: a set M of real numbers is said to precede another set N if any number in M is smaller than every number in N; and a number z is said to separate the sets M and N if $z \geq x$ for every number x in M and $z \leq y$ for every number y in N. The continuity theorem states:

Theorem 3*c*. Whenever two sets M and N of real numbers have the property that M precedes N, there exists a real number z which separates M and N.

It should be noted that this theorem is false as long as only rational numbers are considered. Once it is established, a number of frequently applied statements about real numbers can be proved. One of these states that each infinite sequence $r_1, r_2, \ldots r_n, \ldots$ of real numbers, all of which lie between two fixed real numbers, has at least one "point of accumulation," that is, a real number r such that each neighborhood of r, no matter how small, contains infinitely many of the numbers r_n. If, moreover, the sequence r_k is monotonic (for example, monotonically increasing, that is, each number of the sequence is greater than all preceding numbers), then it has exactly one point of accumulation (which is then called the *limit* of the sequence). A special case of the last statement is the fact that each decimal number, whether terminating or not, represents a real number.

While Theorems 1*a* to 1*d*, 2*a* to 2*i'*, and 3*a* to 3*c* must be proved from the definition of real numbers, all other statements about such numbers can instead be derived from these few numbered theorems, without further reference to the definition of real numbers. That is to say, these theorems form a complete set of axioms for the system of real numbers (Part 1, Chap. 3, Sec. 1).

3. Algebra of Sets and Logic

Propositional Calculus. Because of the crucial importance of logical deduction in mathematics and science, it is convenient to express logical operations by formulas, both for brevity and to achieve a rigorous formulation of the rules of inference.

We say (postponing a more formal treatment until later) that a proposition (statement) is either true or false. If p is a proposition, its *negation* is denoted by $\neg p$ ("non p"). This proposition is true if p is false, and vice versa. If p and q are propositions, their *disjunction* is denoted by $p \vee q$ ("p or q"). This proposition is true if at least one of p and q is true and is false if both p and q are false ("nonexclusive or"). The *conjunction*, $p \wedge q$ ("p and q"), is true if and only if both p and q are true. The *implication*, $p_1 \rightarrow q$ ("p implies q" or "if p then q"), is false if and only if p is true and q false. (By this definition, a false statement p implies any statement q, whether true or false and regardless of any connection between the meanings of the two statements; similarly, a true statement q is implied by any p.) The *equivalence* $p \leftrightarrow q$ ("if and only if p then q") is true if and only if p and q are either both true or both false. The *incompatibility* $p|q$ is false if and only if p and q are both true.

The last four of these operations can be expressed in terms of the first two. For instance,

$$p \wedge q = \neg(\neg p \vee \neg q) \qquad p \rightarrow q = \neg p \vee q$$
$$p \leftrightarrow q = (p \rightarrow q) \wedge (q \rightarrow p) \qquad p|q = \neg p \vee \neg q \tag{1.1}$$

From these operations, compound *expressions* can be built up. The right-hand sides of the preceding definitions are examples of such expressions. An expression consisting of only a single symbol, such as p, is called a *propositional variable*. Parentheses are used to indicate the order in which operations are performed, as in algebra. Unless otherwise indicated by parentheses, the operations $\neg$ are performed first, $\vee$ and $\wedge$ next, and all others afterward. In the literature one sometimes finds dots in place of parentheses; also, other operations symbols are used, such as $\sim$ in place of $\neg$, $\cdot$ or & in place of $\wedge$, $\supset$ in place of $\rightarrow$, $\equiv$ in place of $\leftrightarrow$.

The use of parentheses is made superfluous by notations of the type introduced by J. Łukasiewicz (so-called Polish notations), in which operations symbols are put before the operands. For instance, if we agree to write Np, Mpq, Dpq, Cpq, instead of $\neg p$, $p \wedge q$, $p \vee q$, $p \rightarrow q$, respectively, then $(p \rightarrow q) \wedge (q \rightarrow p)$ becomes $MCpqCqp$, and there is no confusion about the order of performing the operations. Similar notations can be used in ordinary algebra.

As stated before, a proposition p is either true or false; we shall write 1 for true, 0 for false. Expressions formed by means of the operations $\neg$, $\vee$, etc., are also propositions and have the *truth values* 0 or 1. We have indicated above, in words, how the truth value of a composite expression depends on the truth values of its components. These assignments of truth values can be represented with clarity in the form of a truth table:

p	q	$\neg p$	$p \vee q$	$p \wedge q$	$p \rightarrow q$	$p \leftrightarrow q$	$p \mid q$
1	1	0	1	1	1	1	0
1	0	0	1	0	0	0	1
0	1	1	1	0	1	0	1
0	0	1	0	0	1	1	1

It is obvious how truth values are obtained for more complicated expressions. As an aid to memory, note that the truth values of $p \wedge q$ are the arithmetic products of the truth values of p and q.

Of particular importance are expressions that have a truth value of 1 regardless of the truth values of their components. Examples, which are easily verified, are $p \rightarrow (p \vee q)$, or $\neg(p \vee q) \rightarrow (\neg p \wedge \neg q)$. Such expressions are called *tautologies*. They constitute the valid statements of logic.

The following four tautologies are often chosen as a basis for symbolic logic:

(*a*) $(p \vee p) \rightarrow p$ (tautology)
(*b*) $p \rightarrow (q \vee p)$ (addition)
(*c*) $(p \vee q) \rightarrow (q \vee p)$ (permutation)
(*d*) $(q \rightarrow r) \rightarrow [(p \vee q) \rightarrow (p \vee r)]$ (summation)
(1.2)

Their role is similar to that of *axioms* in other formal theories. From known tautologies, other tautologies can be obtained by applying the following *rules of inference:*

a. Substitution Rule. If in a tautology a propositional variable is replaced by some expression (the same expression every time the variable occurs) the result is a tautology.

b. Detachment Rule. If P, Q are expressions such that P and $P \rightarrow Q$ are tautologies, then Q is a tautology.

All tautologies can be obtained by repeated application of rules a and b, starting from (1.2). Consequently, by merely filling in its truth table, one can *decide* for any given expression whether it can be derived from the axioms by means of the rules of inference.

To stress the formal character of the symbolism, let us for a moment forget the meanings of all the foregoing symbols and consider a system S of undefined elements denoted by $p, q, \ldots$, and assume that two operations $\neg$, $\vee$ are defined within the system; i.e., if p is an element of S then $\neg p$ is also an element of S, and if p, q are elements of S then so is $p \vee q$. Define $p \rightarrow q$ as an abbreviation for $\neg p \vee q$. Assume further that some of the elements of S are distinguished; they form a subsystem T of S. Assume that the four elements (1.2) belong to T and any elements derived from them by means of procedures a and b also belong to T. Under these conditions the system S is called a *formal logic.*

Any set of objects that satisfies all these conditions constitutes an *interpretation* of the system. For instance, propositions in the ordinary sense of the word form an interpretation of S, if $\neg$, $\vee$ are defined as ordinary negation and disjunction and T is defined as the set of all tautologies (composite expressions that are always true regardless of the truth of their elementary components). A related but much simpler interpretation is that in which S consists of only two elements, namely, the numbers 0 and 1, and T consists of the number 1 alone. The operation $\neg p$ is defined as $1 - p$, and $p \vee q$ as $p + q - pq$. (In these two definitions, addition, subtraction, and multiplication are understood in the ordinary algebraic sense.) One can verify that, if p and q are 0 or 1, so are $\neg p$ and $p \vee q$ and, therefore, all other expressions; also, that the expressions (1.2) are always $= 1$ and that rules a and b preserve this value. Therefore the system (0,1) is indeed an interpretation of S.

One can choose other sets of axioms in place of (1.2). Of the systems so obtained, some lead to the same set T of tautologies; others do not. Łukasiewicz has given a single axiom from which the entire propositional calculus can be derived. The limitation to two truth values is not essential; *m-valued logics* have been studied extensively.

The operations $\wedge$, $\rightarrow$, $\leftrightarrow$, $\mid$ can all be replaced by combinations of $\neg$ and $\vee$. Alternatively, one can start with the single operation $\mid$ and define $\neg p = p|p$, $p \vee q = \neg p | \neg q$, etc. Representations in terms of $\neg$, $\vee$, and $\wedge$ enjoy important properties of simplicity and symmetry. For instance, from every tautology expressed in terms of $\neg$, $\vee$, $\wedge$, another one can be obtained by interchanging $\vee$ and $\wedge$ and negating the whole expression. This fact is known as the *principle of duality.* For example, to $\neg p \vee (q \vee p)$ corresponds $\neg[\neg p \wedge (q \wedge p)]$, which is also a tautology.

An important unsolved problem is that of *simplification,* i.e., expressing a proposition in terms of $\neg$, $\vee$, $\wedge$ so as to use the smallest possible number of operations $\vee$, $\wedge$. This problem has practical applications, e.g., to the design of electric circuits and to the formulation of questions for information retrieval.

Predicate Calculus. The propositional calculus deals with propositions (statements) as a whole, without regard to the concepts occurring in them. To deal with the latter, we consider an enlarged system, which includes as undefined elements not only the propositional variables $p, q, \ldots$ of the propositional calculus but also two other kinds of symbols. These are called *object variables*, denoted by the letters $x, y, z, \ldots,$ and *predicate variables* (sometimes called *functional variables*), denoted by the letters $f, g, \ldots$. A predicate variable is followed by one or more object variables, as in $f(x)$, $f(x,y)$.

The usual interpretation of these symbols in terms of conventional logic is that object variables represent objects or other concepts, and predicate variables represent statements about them. For example, $x, y, \ldots$ might stand for human beings, and the predicate $f(x,y)$ for "are brothers," or $x, y, \ldots$ might represent numbers, and $f(x)$ might mean "x is rational."

For operations we use first of all the propositional operations $\neg$, $\vee$ as before; these, as well as others defined from them ($\wedge$, $\rightarrow$, etc.), are applicable not only to propositional but also to predicate variables (but not to object variables). Furthermore, we introduce an operation denoted by $(\forall x)$, read "for all x," which operates only on predicate variables followed by x (possibly in addition to other object variables), for example, $(\forall x)f(x)$, $(\forall x)f(x,y)$, $(\forall y)f(x,y,z)$. We define another operation by $(\exists x)f(x) = \neg(\forall x)\neg f(x)$ (read "for some x" or "there exists an x such that"). We agree that $(\forall x)f(x)$ and $(\forall y)f(y)$ denote the same element. The operations $(\forall x)$ and $(\exists x)$ are called *quantifiers*.

An object variable x is called *bound* if it follows a predicate variable preceded by $(\forall x)$ or $(\exists x)$; otherwise it is called *free*. The operators $(\forall x)$, $(\exists x)$ may be applied not only to predicate variables but to any expression containing x as a free variable. For example, in $(\exists x)f(x,y,z)$, x is bound while y, z are free; therefore we may form $(\forall y)(\exists x)f(x,y,z)$. In the latter expression only z is free.

We use the axioms (1.2) of the propositional calculus and in addition

$$\begin{array}{ll}(a) & ((\forall x)f(x)) \rightarrow f(y) \\ (b) & f(y) \rightarrow ((\exists x)f(x))\end{array} \qquad (1.3)$$

We define as "true statements" the axioms (1.2), (1.3), and all expressions derivable from them by means of the following rules of inference. (This definition includes, in particular, all tautologies of the propositional calculus.)

The rules of inference are as follows:

a. Substitution. For $p, q, \ldots$ we may substitute any expression formed from propositional or predicate variables; for $f(x)$, any expression containing x as a free variable; etc.*

b. Detachment. As in the propositional calculus.

c. Passage. Let F be an expression not containing x as a free variable, and $G(x)$ an expression containing x as a free variable. If $F \rightarrow G(x)$ is a true statement, so is $F \rightarrow (\forall x)G(x)$, and if $G(x) \rightarrow F$ is a true statement, so is $(\exists x)G(x) \rightarrow F$.

The principle of duality holds just as in the propositional calculus if, in addition to interchanging $\vee$ and $\wedge$, we interchange $\forall$ and $\exists$. For instance, the expression

$$[(\forall x)f(x) \vee (\forall x)g(x)] \rightarrow [(\forall x)(f(x) \vee g(x))]$$

is a true statement in our system. (The converse implication, incidentally, is not true.) Its dual is

$$\neg\{\neg[(\exists x)f(x) \wedge (\exists x)g(x)] \wedge [(\exists x)(f(x) \wedge g(x))]\}$$

which is indeed also true; it can be transformed into the more transparent equivalent form

$$[(\exists x)(f(x) \wedge g(x))] \rightarrow [(\exists x)f(x) \wedge (\exists x)g(x)]$$

Calculus of Classes. The entities with which we deal in this section have been called "classes" by some authors and "sets" by others. The latter term, which we shall use here, has tended to displace the former, especially in mathematical contexts.

We denote by $x \in A$ the statement that the object x is an *element* of the *set* A. We write $x \notin A$ if x is not an element of A. If sets A and B have the same elements, we declare A and B to be the same set: $A = B$ if and only if $(x \in A) \leftrightarrow (x \in B)$. This is called the (intuitive) *principle of extension.* The elements of a set may be objects of any kind; they may even themselves be sets. Thus we may form a set S, say, whose elements are the sets $A, B, \ldots$; we write $A \in S$, $B \in S$, etc.

If a set A has only a finite number of elements, say $x_1, x_2, \ldots, x_n$, it is sometimes denoted by $A = \{x_1, x_2 \ldots, x_n\}$. In particular, $\{x\}$ is the set having the object x for its only element. A distinction is drawn between the object x and the set $\{x\}$.

A set, whether finite or infinite, can also be defined by a property common to all its elements. This kind of definition establishes a connection between the predicate calculus of the preceding section and the algebra of sets. (More precisely, we need only a portion of predicate calculus, namely, that concerned with predicates of a single variable.) If $f(x)$ is a predicate, we can form the set A of all objects x for which $f(x)$ is true. This set is denoted by $A = \{x|f(x)\}$ (*principle of abstraction*).*

If a set B has the property that every element of B is also an element of A (but not necessarily vice versa), then B is called a *subset* of A; we write $B \subseteq A$ or $A \supseteq B$ (read "A *includes* B"). This inclusion relation between sets has the following properties:

$$\begin{array}{lll}(a) & A \subseteq A & (reflexivity) \\ (b) & [(A \subseteq B) \wedge (B \subseteq C)] \rightarrow (A \subseteq C) & (transitivity) \\ (c) & [(A \subseteq B) \wedge (B \subseteq A)] \rightarrow (A = B) & \end{array} \qquad (1.4)$$

If $A \subseteq B$ but not $B \subseteq A$, that is, if there is at least one element in B which is not in A, then A is called a

* By "expressions" we mean those obtained by repeated application of the operations $\neg$, $\vee$, $\forall$ in accordance with their definitions (usually called "well-formed expressions"). Substitution introducing additional variables, e.g., replacing $f(x)$ by $q(x,y)$, is not permitted in some cases if y also occurs outside of f in the same formula. We omit here the detailed conditions for this case. (See [12, especially p. 218].†)

† Numbers in brackets refer to References at end of chapter.

* Our presentation is limited to the so-called intuitive set theory. This disregards certain difficulties such as Russell's paradox of the set $R = \{X|X \notin X\}$, for which each of the statements $R \in R$, $R \notin R$ leads to a contradiction.

proper subset of B; in symbols, $A \subset B$ or $B \supset A$ ("*B properly includes A*"). This relation has the property of transitivity (*b*), that is, $(A \subset B) \wedge (B \subset C) \rightarrow (A \subset C)$, but not of reflexivity; on the contrary, $\neg(A \subset A)$ is true for all A (*irreflexivity*).

We define a set V, called the *empty* (void) set, which has no element at all. For every set A, $V \subset A$. In some contexts it is desirable also to define a *universal set* U, for which $A \subset U$ for all sets A, and $x \in U$ for every x.

The *complement* of a set A is the set of all objects that are not elements of A. We shall denote it by $\sim A$. Obviously

$$\sim(\sim A) = A \tag{1.5}$$

$$(x \in \sim A) \leftrightarrow (x \notin A) \tag{1.6}$$

Another way of stating the latter relation is

$$[A = \{x|f(x)\}] \leftrightarrow [\sim A = \{x|\neg f(x)\}] \tag{1.7}$$

If A is any set, the set of all subsets of A is called the *power* set of A and is denoted by $\mathcal{P}(A)$. We have

$$\begin{gathered} A \in \mathcal{P}(A) \\ V \in \mathcal{P}(A) \\ (B \subseteq A) \leftrightarrow (B \in \mathcal{P}(A)) \end{gathered} \tag{1.8}$$

If sets are defined by the principle of abstraction,

$$[f(x) \rightarrow g(x)] \leftrightarrow [\{x|f(x)\} \subseteq \{x|g(x)\}] \tag{1.9}$$

That is to say, the inclusion relation $\subseteq$ for sets is the counterpart of the operation of implication for propositions.

The set of elements common to two sets A and B is called the *intersection* (also the *product* or *meet*) of A and B and is denoted by $A \cap B$. Obviously

$$[(x \in A) \wedge (x \in B)] \leftrightarrow [x \in (A \cap B)] \tag{1.10}$$

The set of elements that occur in at least one of the sets A and B is called the *union* (also *sum* or *join*), denoted by $A \cup B$. Obviously

$$[(x \in A) \vee (x \in B)] \leftrightarrow [x \in (A \cup B)] \tag{1.11}$$

Also, if $A = \{x|f(x)\}$ and $B = \{x|g(x)\}$, then

$$A \cap B = \{x|f(x) \wedge g(x)\}$$

and

$$A \cup B = \{x|f(x) \vee g(x)\}$$

Thus the operations $\cap$ and $\cup$ for sets are counterparts of the operations $\wedge$ and $\vee$, respectively, for propositions.

These operations have the following properties, among others;

(*a*) $A \cap A = A$	$A \cup A = A$ (*idempotent law*)	
(*b*) $A \cap B = B \cap A$	$A \cup B = B \cup A$ (*commutative law*)	
(*c*) $A \cap (B \cap C) = (A \cap B) \cap C$	$A \cup (B \cup C) = (A \cup B) \cup C$ (*associative law*)	
(*d*) $A \cap (B \cup C) = (A \cap B) \cup (A \cap C)$	$A \cup (B \cap C) = (A \cup B) \cap (A \cup C)$ (*distributive law*)	
(*e*) $A \cap U = A$	$A \cup V = A$	
(*f*) $A \cap V = V$	$A \cup U = U$	
(*g*) $\sim A \cap A = V$	$\sim A \cup A = U$	(1.12)

All these relations have their counterparts in the propositional calculus. All can be derived logically from the definitions of $\cup$, $\cap$, etc.; in particular, one way to prove them is by means of their propositional counterparts.

Because of (*b*) and (*c*) of (1.12) the union or intersection of any finite number of sets can be formed without regard to order. We denote by $\sum_i A_i$ the union and by $\prod_i A_i$ the intersection of several sets A_i.

Two sets A, B are called *disjoint* if $A \cap B = V$, that is, if they have no element in common.

A *partition* of the set A is a set P of subsets of A, which is mutually disjoint and whose union equals A, that is, a set of sets $P \subseteq \mathcal{P}(A)$ such that

$$[(X \in P) \wedge (Y \in P)] \rightarrow (X \cap Y = V)$$

and that $\sum_{X \in P} X = A$.

The notation $\sum_{X \in P}$ causes no difficulty as long as P is finite. The concepts of partition and more generally of union and intersection of several sets can be extended to infinite collections of sets if certain safeguards are observed.

A duality principle holds for the calculus of classes, similar to the one for the propositional calculus. The two columns of (1.12) are duals of each other.

In a looser way there are also analogies between the operations $\cup$, $\cap$ for sets and the operations of addition and multiplication in algebra. The sets V, U have as counterparts the numbers 0, 1. Lines (*b*), (*c*), and (*e*) of (1.12) as well as the left halves of (*d*) and (*f*) are true in algebra; remarkably, the idempotent laws and the distributive law for addition, among others, have no algebraic counterpart. On the other hand, in this analogy there is nothing in the calculus of classes closely comparable to the inverse operations in algebra, subtraction and division. There is still another correspondence between ordinary algebra and the calculus of classes, in which algebraic multiplication corresponds to $X \cap Y$ as before, while addition corresponds to the operation $(X \cup Y) \cap \sim(X \cap Y)$ for sets. This operation has a unique inverse.

Among the many other frequently applied relations in the calculus of classes, we mention

$$A \cap B \subseteq A \subseteq A \cup B \tag{1.13}$$

and the *De Morgan laws*

$$\begin{aligned} \sim(A \cap B) &= \sim A \cup \sim B \\ \sim(A \cup B) &= \sim A \cap \sim B \end{aligned} \tag{1.14}$$

Relations and Functions. If S is a set with elements $a, b, c, \ldots$, we define a (*binary*) *relation* on S as a set of ordered pairs of elements of S. An ordered pair, $<a,b>$, is defined as

$$<a,b> = \{\{a\},\{a,b\}\} \tag{1.15}$$

Thus, if a, b are elements of S, $<a,b>$ is a subset of the power set $\mathcal{P}(S)$. While $\{x,y\}$ and $\{y,x\}$ denote the same set ("unordered pair"), $<x,y>$ is different from $<y,x>$. We call a the left member, b the right member of $<a,b>$.

If R is a set of such ordered pairs, the elements a, b of S are said to "stand in the relation R to each other" if and only if $<a,b>$ is an element of R. To each relation R corresponds a predicate of two variables, say $r(x,y)$, which is defined as true or false, depending on whether $<x,y>$ is or is not an element of R, that is, whether the objects x, y do or do not stand in the given relation to each other; that is to say, $r(x,y) \leftrightarrow <x,y> \in R$. This correspondence between binary relations and predicates of two variables is analogous to the correspondence between sets and predicates of one variable which was used for the definition of sets by abstraction. In obvious generalization of the latter, we can write

$$R = \{<x,y> | r(x,y)\} \tag{1.16}$$

For a binary relation R the *domain* of R, $\mathfrak{D}(R)$, is defined as the set of those elements that occur as the left member in an ordered pair of R:

$$\mathfrak{D}(R) = \{x | (\exists y) r(x,y)\}$$

Similarly the *range* of the relation R, denoted by $\mathfrak{R}(R)$, is defined as the set of all elements that occur as the right member in an ordered pair of R:

$$\mathfrak{R}(R) = \{y | (\exists x) r(x,y)\}$$

The pair $<b,a>$ is defined as the *converse pair* of any ordered pair $<a,b>$. If R is a binary relation, the *converse relation* R is defined as the set of all ordered pairs whose converses are in R:

$$\hat{R} = \{<x,y> | r(y,x)\}$$

A special binary relation is the *identity relation* I, consisting of all (ordered) pairs whose two elements are identical: $I = \{<x,y> | x = y\}$.

Another important relation is the set of all ordered pairs whose left member is an element of a set A and whose right member is an element of a set B. This set of ordered pairs, which may properly be considered as a binary relation on the set $A \cup B$, is called the *cartesian product* of A and B and is denoted by $A \times B$:

$$A \times B = \{<x,y> | (x \in A) \wedge (y \in B)\}$$

For example, if A is the set of points x lying on a straight line (axis) X, and B the set of all points y lying on another (intersecting) straight line Y, $A \times B$ is the set of all pairs $<x,y>$; the name "cartesian" recalls the fact that it was Descartes who devised the mapping of points of the (Euclidean) plane onto the set of pairs now called the cartesian product of the two axes.

Just as we have defined binary relations as sets of ordered pairs, so ternary, quaternary, . . . relations can be defined as sets of ordered triplets, quadruplets, etc. However, in the following we shall discuss mainly binary relations.

Instead of $<x,y> \in R$ one often writes xRy; this is because many of the relations used in mathematics are customarily denoted by symbols placed between the two members, as in $x = y$, $x > y$, etc.

A relation is called:

$$
\begin{array}{ll}
(a) & \textit{Transitive} \text{ if } [(xRy) \wedge (yRz)] \rightarrow (xRz) \\
(b) & \textit{Symmetric} \text{ if } (xRy) \rightarrow (yRx) \\
(c) & \textit{Asymmetric} \text{ if } (xRy) \rightarrow \neg(yRx) \\
(d) & \textit{Reflexive} \text{ if } (\forall x) xRx \\
(e) & \textit{Irreflexive} \text{ if } \neg(\exists x) xRx
\end{array}
\tag{1.17}
$$

Let S be a set and E a relation defined on S (i.e., a relation whose domain and range coincide with S) and having the properties (a), (b), and (d) of (1.17). Such an E is called an *equivalence relation.* One can prove the fundamental fact that E induces a *partition* $P(E)$ of S into disjoint subsets such that two elements x, y of S are in the same subset if and only if xEy; conversely, every partition of S defines an equivalence relation. Examples of equivalence relations are the relations of equality $(=)$ in arithmetic or algebra, of equivalence $(\leftrightarrow)$ in the propositional or predicate calculus, of "having the same cardinal number" between sets (Sec. 2), and of isomorphism (Part 1, Chap. 2, Secs. 13 and 14).

A relation R on the set S is called a *partial ordering* of S if it is transitive and asymmetric. Examples are the relations $\subset$ between sets and $<$ between real numbers. A *simple* (or *linear*) *ordering* of S is a partial ordering such that, for any two elements x, y of R, there is either xRy or yRx. For example, the real numbers are simply ordered by the relation $<$. If the set S is simply ordered by the relation R, so is every subset of S. [More precisely speaking, the subset $T \subset S$ is simply ordered by the relation $(T \times T) \cap R$, the "restriction" (see below) of R to T.]

In a partial ordering R of the set S, an element y of S is called a "first" member if yRx holds for all x in S. Such a first member need not exist. The set S is said to be *well ordered* by R if R is a simple ordering and if every subset of S contains a first member. For example, the natural numbers are well ordered by the relation $<$ whereas rational or real numbers are not.

A *function* (more precisely, a single-valued function of one variable) is a relation in which no two of the ordered pairs have the same first element, in other words, a relation F for which

$$[(xFy) \wedge (xFz)] \rightarrow (y = z) \tag{1.18}$$

Thus, for a given x there is at most one y for which xFy holds; if there is one, we customarily write $y = F(x)$. The words *mapping, transformation,* and *operator* are synonyms for the word "function" in this general sense. The predicates of one variable defined above are a special case of the definition given here; so are the customary functions of algebra and analysis.

The domain and range of a function are defined as for relations in general. The domain $\mathfrak{D}(F)$ is the set of all elements x that occur as left members, and the range $\mathfrak{R}(F)$ is the set of all y that occur as right members, of the relation F. It is said that F maps $\mathfrak{D}(F)$ *onto* $\mathfrak{R}(F)$ or *into* any set of which $\mathfrak{R}(F)$ is a subset.

Let F, G be two functions for which $\mathfrak{D}(F) \subset \mathfrak{D}(G)$. Then if, for any x in $\mathfrak{D}(F)$, $G(x) = F(x)$, we call F the *restriction* of G to the domain $\mathfrak{D}(F)$, and G an *extension* of F to the domain $\mathfrak{D}(G)$. For example, most operations on real numbers are extensions of the corresponding operations on rationals or integers; the analytic continuations described in Part 1, Chap. 3, Sec. 8, are extensions in our sense.

If F, G are two functions such that $\mathfrak{R}(F) \subseteq \mathfrak{D}(G)$, the *composite function*, denoted by $G \circ F$, may be obtained by applying the function F to any element x in $\mathfrak{D}(F)$, resulting in an element y of $\mathfrak{R}(F)$, and then applying G to y. (Often the notation GF is used in place of $G \circ F$.) It is easy to prove that the operation $\circ$ is associative: $H \circ (G \circ F) = (H \circ G) \circ F$. It is in general not commutative; frequently $G \circ F \neq F \circ G$.

The function F is called *one-to-one* if

$$[F(x_1) = F(x_2)] \rightarrow (x_1 = x_2)$$

that is, no two different arguments give the same value for F. For a one-to-one function F it is possible to form the *inverse* function, denoted by F^{-1}. In terms of the definition of F as a relation, i.e., a certain set of ordered pairs $<x,y>$, F^{-1} is the set of corresponding converse pairs $<y,x>$. Obviously,

$$\mathfrak{R}(F^{-1}) = \mathfrak{D}(F) \qquad \mathfrak{D}(F^{-1}) = \mathfrak{R}(F)$$

Also, F^{-1} is itself a one-to-one function, and

$$(F^{-1})^{-1} = F$$

If F, G are one-to-one functions such that

$$\mathfrak{R}(F) = \mathfrak{D}(G)$$

then the composite $G \circ F$ is itself a one-to-one function. Also, it has an inverse, and it is easy to see that $(G \circ F)^{-1} = F^{-1} \circ G^{-1}$.

An important class of functions are those for which the domain and the range are ordered and are such that the ordering is preserved by F. They are called *isotone* functions. The monotonic functions of real variables are examples. If F is isotone, it is necessarily one-to-one and therefore has an inverse, and F^{-1} is also isotone.

We have limited ourselves to functions of one variable, corresponding to binary relations. Many of the definitions and theorems given here have obvious generalizations to functions of several variables.

Numbers. Consider a set U and the set of all its subsets, $\mathcal{P}(U)$. We define a relation E among sets of $\mathcal{P}(U)$ as follows: Two sets A, B of $\mathcal{P}(U)$ (that is, two subsets of U) are in the relation E (that is, $<A,B> \in E$) if and only if there exists a one-to-one function F that maps A onto B. We have seen that, in this case, F^{-1} is a one-to-one function that maps B onto A; that is, if $<A,B> \in E$, then $<B,A> \in E$. In other words, the relation E is symmetric. If $<A,B> \in E$ and $<B,C> \in E$, that is, if there exist one-to-one functions F, G that map A onto B and B onto C, respectively, then $G \circ F$ is a one-to-one mapping of A onto C, and therefore E is transitive. Finally, the identity relation I, defined above, maps every A one-to-one onto itself, so that E is reflexive. Thus, E is an equivalence relation. It therefore defines equivalence classes in $\mathcal{P}(U)$. Any equivalence class under the relation E is called a *cardinal number*.

The natural numbers introduced in Sec. 2 are cardinal numbers in this sense. In fact, the reasoning we have just presented is merely a more rigorous formulation of the explanation of Sec. 2. The cardinal number 1 is defined as the class

$$1 = \{A|(\exists x)[(x \in A) \wedge ((\forall y)((y \in A) \rightarrow (y = x)))]\}$$

for which one can show that it is an equivalence class under E. The cardinals 2, 3, . . . are similarly defined. These *natural numbers* do not exhaust all cardinals; in general (unless U is chosen too narrow) there exist *transfinite cardinals* corresponding to subsets of U with infinitely many elements.

Boolean Algebra. The close analogy between the operations of the class calculus and the propositional calculus suggests that both may be considered as special cases of the same general theory. In order to present such a theory, we could use the operation symbols of either of these calculi or a third set of symbols. For simplicity we shall use those of the propositional calculus. Just as before, then, we consider a set S of elements p, q, . . . with operations $\neg$, $\vee$ and a subset $T \subset S$ which contains the four elements (1.2) and all other elements obtained from them by repeated application of the operations of substitution and detachment. One interpretation of such a system is to consider p, q, . . . as variables representing propositions in the sense of ordinary logic and $\neg$, $\vee$ as ordinary logical negation and disjunction. Then T is the set of all tautologies. Another interpretation is to consider p, q, . . . as sets, most simply, as subsets of some given set U. That is to say, the set S is identical with the power set $\mathcal{P}(U)$. The operations $\neg$, $\vee$ are interpreted as complement and union. Then T contains only the single element U, that is, $T = \{U\}$. Indeed one verifies easily that all four of the expressions (1.2) are U, no matter what are the sets p, q, r. The 14 equations (1.12) translate into expressions of the propositional calculus. [The symbol $=$ translates into $\leftrightarrow$, in line with the fact that $=$ is tantamount to "$\subseteq$ and $\supseteq$," just as $p \leftrightarrow q$ is defined as $(p \rightarrow q) \wedge (q \rightarrow p)$; in (*c*), (*f*), and (*g*) of (1.12), any tautology may be substituted for U, the negation of any tautology for V.] These expressions are all tautologies; each of them can be derived from (1.2) by repeated substitution and detachment.

Vice versa, we may choose the counterparts of (1.12) as the axioms of our theory; the statements (1.2) can be derived from them.

It is also possible to interpret $\vee$ as the intersection of sets, in which case $\wedge$ becomes the union, $\rightarrow$ is translated into $\supseteq$ rather than $\subseteq$, and the set of tautologies becomes $\{V\}$. Because of the duality principle, all derivations are analogous.

Another interpretation of Boolean algebra, alluded to before, is that by a set of only two elements, such as the numbers 0 and 1, with $\neg p$ represented by $1 - p$, $p \vee q$ represented by $p + q - pq$. The set T consists of the number 1 alone.

Apart from these interpretations in logic, however, various interpretations of Boolean algebra are of immediate practical applicability. The best known is to electric circuits, specifically switching circuits, where the two truth values 0 and 1 may be translated either into the absence and presence of a signal (electric pulse) or into the state—open or closed—of a circuit or more generally into the states of any bistable circuit element. This kind of application has acquired prominence because of the importance of bistable elements and switching circuits in computer technology.

In computer programming it is often convenient to use Boolean notation for the logical operations involved in information handling. Many computers admit instructions for performing logical operations such as $\wedge$, $\vee$ on individual binary digits.

We have already mentioned the simplification of propositional expressions, i.e., the problem of finding equivalent expressions that minimize either the number of operations or the number of variables. This problem assumes importance for economy in designing electric circuits.

This application has also led to an important generalization of Boolean algebra. Since the states of a circuit change with time, it is suggestive to consider Boolean variables that are functions of time. In the sense in which functions have been described in the foregoing, we consider a domain representing either a number of discrete time instants or a continuous time variable and a range consisting only of the Boolean values 0 and 1. Thus we have functions $p(t)$, $q(t)$, . . . , with Boolean operations performed on them, and we can, for example, form difference equations such as $p(t + h) = p(t) \vee q(t)$ or the like. This has become an important tool in the theory of automata.

Formal Theories. One of the purposes in exhibiting the foregoing theories was to illustrate the concept of a formal theory. The propositional calculus and its equivalent, Boolean algebra, as well as the predicate calculus, have been presented as formal theories; the class calculus obviously could have been, although we have chosen a looser approach. Elsewhere in this volume (Part 1, Secs. 13 and 14, Chap. 2, as well as much of Chap. 8) are examples of formally introduced mathematical theories; the reader will realize that many of the fundamental physical disciplines—notably mechanics of rigid bodies, fluid mechanics, electromagnetism—could be completely formalized without major changes in the body of equations describing them. Last but not least, the logical foundations of arithmetic have been presented, in Sec. 2, in a form which, while falling short of complete formalization, points the way to it. Historically, this problem of logical foundations of mathematics led to the modern development of symbolic logic and the methodology of deductive sciences.

A formal theory begins with three kinds of listings: a set of *undefined concepts*, a set of *axioms*, and *rules of inference*. For example, in the propositional calculus one can choose as undefined concepts the symbols for propositional variables p, q, . . . and the operations $\neg$, $\vee$. The axioms are the four formulas (1.2); the rules of inference are substitution and detachment. One can make other choices and arrive at the same theory. It is indeed typical of most formal theories that the choice of the fundamental elements is not unique.

The undefined concepts are frequently of two kinds, one of which we may call, for want of a better name, "objects," while the other consists of operations and/or relations. There can be several types of objects. The propositional calculus has only one type, but in the predicate calculus there are propositional, predicate, and "object" (in a narrower sense) variables. Group theory has only one kind of elements, but geometry uses points, lines, and planes. Some theories use certain constants, in addition to variables, among their undefined objects, for instance, U and V in the calculus of classes. (Constants differ formally from variables by not admitting substitution as a rule of inference.) The distinction between operations and relations is fluid; for instance, the binary operation $a + b$ may be replaced by the ternary relation $c = a + b$, or vice versa.

The undefined objects and operations together define the set of *meaningful expressions*. Some of these are selected as axioms. The set of axioms is the most characteristic part of a formal theory and the one that attracts the most attention. For instance, Euclidean and non-Euclidean geometries differ in the choice of axioms. It is well known that concern with axioms far antedates the modern development of formalization of theories.

As rules of inference, most theories use the entire apparatus of logic. More rigorously speaking, valid inferences are obtained by substitution into tautologies of the propositional or predicate calculus or other formal logical systems. When, however, logic itself is the subject of a formal theory, it is necessary to define the rules of inference more explicitly and more narrowly, for instance, the rules of substitution, detachment, and passage for the predicate calculus. Repeated application of the rules of inference to the axioms results in a set of expressions called the "*true statements*" of the theory. The word "true" in this context is understood relative to the chosen set of axioms.

Several properties of sets of axioms are of greatest importance in assaying their value or the value of the theories they engender. They are independence, consistency, completeness, and decidability. A set of axioms is called independent if none of them can be derived from the others; if a set is not independent, then some of the axioms can be omitted without diminishing the usefulness of the theory. This is only a matter of economy or aesthetics; there is no harm in retaining dependent axioms. It can be shown, for example, that axioms (1.2) for the propositional calculus are independent. A set of axioms or the theory based on it is called *consistent* if no two contradictory expressions (that is, P and $\neg P$) are in the set T. If one such contradictory pair appears, then *every* meaningful expression is in T; such a theory would be considered uninteresting. A set of axioms, or the theory based on it, is called *complete* if every meaningful expression P can (in principle) be either proved or disproved; i.e., either P or $\neg P$ is in T. This may be considered desirable but is not essential; incomplete theories can be interesting and useful, and indeed most theories of mathematics and science are incomplete. Finally, a theory is called *decidable* if there is a general procedure for deciding whether or not a given meaningful expression can be proved or not. The propositional calculus has been shown to be consistent, complete, and decidable. The predicate calculus is consistent but it is neither complete nor decidable. More than that, it has been proved that no formal theory can be either complete or decidable if it is comprehensive enough to include simple statements about the natural numbers among its meaningful expressions. This is the essence of Gödel's famous proof.

References

Foundations of Mathematics:

1. Bourbaki, N.: "Eléments d'histoire des mathématiques," Hermann & Cie, Paris, 1960.
2. Denbow, C. H., and V. Goedicke: "Foundations of Mathematics," Harper & Row, New York, 1959.
3. Eves, H., and C. V. Newsom: "An Introduction to the Foundations and Fundamental Concepts of Mathematics," Rinehart, New York, 1958.
4. Fujii, J. N.: "An Introduction to the Elements of Mathematics," Wiley, New York, 1961.
5. Kleene, S. C.: "Introduction to Metamathematics," Van Nostrand, Princeton, N.J., 1952.
6. Kneebone, G. T.: "Mathematical Logic and the Foundations of Mathematics," Van Nostrand, Princeton, N.J., 1963.
7. Mostow, G. D., J. H. Sampson, and J. P. Meyer: "Fundamental Structures of Algebra," McGraw-Hill, New York, 1963.
8. Russell, B.: "Introduction to Mathematical Philosophy," G. Allen, London, 1919.
9. Singh, J.: "Great Ideas of Modern Mathematics: Their Nature and Use," Dover, New York, 1959.
10. Wajsmann, F.: "Introduction to Mathematical Thinking," Ungar, New York, 1951.

Mathematical Logic:

11. Carnap, R.: "Introduction to Symbolic Logic and Its Applications," Dover, New York, 1958.
12. Church, A.: "Introduction to Mathematical Logic," vol. 1, Princeton University Press, Princeton, N.J., 1956.
13. Lewis, C. I., and C. H. Langford: "Symbolic Logic," Dover, New York, 1951.
14. Carrol, Lewis (C. C. Dodgson): "Symbolic Logic," 4th ed., Dover, New York.
15. Hilbert, D., and W. Ackerman: "Principles of Mathematical Logic," Chelsea Publishing Company, New York, 1950.
16. Nagel, E., and J. R. Newman: "Gödel's Proof," New York University Press, New York, 1958.
17. Rosenbloom, P. C.: "Elements of Mathematical Logic," Dover, New York.
18. Stoll, R. R.: "Sets, Logic and Axiomatic Theories," Freeman, San Francisco, 1961.
19. Tarski, A.: "Introduction to Logic," Oxford University Press, Fair Lawn, N.J., 1946.

Theory of Sets:

20. Bernays, P.: "Axiomatic Set Theory," North Holland Publishing Company, Amsterdam, 1958.
21. Committee on the Undergraduate Program: "Elementary Mathematics of Sets," Mathematical Association of America, 1958.
22. Fraenkel, A. A.: "Abstract Set Theory," North Holland Publishing Company, Amsterdam, 1953.
23. Kamke, E.: "Theory of Sets," Dover, New York, 1950.

Chapter 2

Algebra

By OLGA TAUSSKY, California Institute of Technology

1. Polynomials

The main task of algebra is the solution of algebraic equations in one or more unknowns and of systems of such equations.

An *algebraic equation* in n unknowns $x_1, \ldots, x_n$ is an equation which can be brought into the form

$$\Sigma c_{e_1,\ldots,e_n} x_1^{e_1} x_2^{e_2} \cdots x_n^{e_n} = 0$$

where $c_{e_1,\ldots,e_n}$ are numbers, called *coefficients;* the e_i are integers. The sum on the left-hand side of this equation is called a *polynomial* in the n unknowns. The single element $c_{e_1,\ldots,e_n} x_1^{e_1} \cdots x_n^{e_n}$ is called a *term.* The sum Σe_i is its *degree;* the largest actually occurring degree in a polynomial is the *degree of the polynomial.* If all terms have the same degree, the polynomial is called *homogeneous* or a *form.* The best known forms are the *quadratic forms*

$$\sum_{i,k=1}^{n} a_{ik} x_i x_k$$

(see under symmetric matrices).

A polynomial which has only one term is called a *monomial,* if two a *binomial.*

The binomial theorem gives the polynomial expansion for the nth power of the sum of two quantities, when n is a positive integer:

$$(a + b)^n = \sum_{r=0}^{n} \binom{n}{r} a^r b^{n-r}$$

The numbers (also for nonintegral n)

$$\binom{n}{r} = \frac{n(n-1)\cdots(n-r+1)}{1 \cdot 2 \cdots r}$$

are called *binomial coefficients.* The number $1 \cdot 2 \cdots r$ is denoted by $r!$ (factorial r) and $0! = 1$ (Part 1, Chap 3, Sec. 11). The binomial coefficients have the properties that (for integral n)

$$(1) \quad \binom{n}{r} = \binom{n}{n-r} = \frac{n!}{r!(n-r)!} = \frac{n(n-1)\cdots(n-r+1)}{r!}$$

$$(2) \quad \binom{n}{r} + \binom{n}{r-1} = \binom{n+1}{r}$$

By definition $\binom{n}{0} = 1.$

These numbers can be arranged in the *Pascal triangle:*

n	$\binom{n}{r}$
0	1
1	1 1
2	1 2 1
3	1 3 3 1
4	1 4 6 4 1
.	. . . .

From relation (2), each value is the sum of the two that are above it.

A generalization of the binomial theorem is the *multinomial theorem:*

$$(a_1 + a_2 \cdots + a_t)^n = \sum_{r_1 + \cdots + r_t = n} \frac{n!}{r_1! r_2! \cdots r_t!} a_1^{r_1} a_2^{r_2} \cdots a_t^{r_t}$$

When $t = 3$, the coefficients can be arranged in a pyramid and multidimensional analogues exist in the higher cases.

Even when n is not an integer, the number $\binom{n}{r}$ has meaning and the binomial theorem is true in many cases. It leads, however, to infinite series (Part 1, Chap. 3, Sec. 1).

2. Algebraic Equations in One Unknown, Complex Numbers

These equations are of the form

$$a_0 x^n + a_1 x^{n-1} + \cdots + a_{n-1} x + a_n = 0 \quad (2.1)$$

If the a_i are rational numbers, then the *solutions* or *roots* of such equations are called *algebraic numbers.* All other numbers are called *transcendental,* for example, e and π. Many algebraic equations would not have any roots unless the idea of number is extended to include quantities more general than rational numbers and numbers composed of surds or limits of sequences of such numbers (Part 1, Chap. 3, Sec. 1).

All these numbers form what is called the set of *real* numbers. The simplest example of an equation without real roots is the equation $x^2 + 1 = 0$. It has been found essential to define the imaginary unit $i = \sqrt{-1}$ and so the roots are $x = \pm i$. Generally, numbers of the form $a + ib$ are called *complex numbers,* where a and b are real numbers—called the real and imaginary parts of $a + ib$. No other numbers need to be introduced as roots for algebraic equations. The following *fundamental theorem of algebra* was proved by Gauss: In the domain of complex numbers every algebraic equation has a root. Thus an equation (2.1) of degree n has exactly n roots $\alpha_1, \ldots, \alpha_n$ and the corresponding polynomial can be written in the form

$$a_0(x - \alpha_1)(x - \alpha_2) \cdots (x - \alpha_n) \qquad (2.2)$$

These last two facts are equally true if the coefficients a_i are themselves complex numbers. If they are, however, real and the complex number $\alpha = a + ib$ is a root, then the *conjugate complex* number $\bar{\alpha} = a - ib$ is a root too. This implies that equations of odd degree have at least one real root.

The principal laws which govern the use of complex numbers are:

(1) $$a + ib = c + id$$

if and only if $a = c$, $b = d$.

(2) $(a_1 + ib_1) + (a_2 + ib_2) = a_1 + a_2 + i(b_1 + b_2)$

(3) $(a_1 + ib_1)(a_2 + ib_2) = a_1a_2 - b_1b_2 + i(a_1b_2 + a_2b_1)$

Property (1) is used very frequently, for example, when the square root of a complex number is to be found.

Other number systems (so-called *hypercomplex numbers*) have been introduced (see Sec. 13 below). However, no other system preserves the fundamental properties of complex numbers, in particular the fact that $ab = ba$ for any two numbers and the existence of a reciprocal to every number $\neq 0$. (For further remarks on complex numbers, see Part 1, Chap. 3, Sec. 6.)

3. Equations of Degree 2 (Quadratic Equations)

The equation

$$a_0x^2 + a_1x + a_2 = 0 \qquad (2.3)$$

has the roots

$$x = \frac{-a_1 \pm \sqrt{a_1^2 - 4a_0a_2}}{2a_0}$$

If the coefficients are real, then the roots are real, identical, or conjugate complex according as the discriminant $D = a_1^2 - 4a_0a_2 \gtreqless 0$.

4. Equations of Degree 3 (Cubic Equations)

The equation

$$a_0x^3 + a_1x^2 + a_2x + a_3 = 0 \qquad (2.4)$$

is usually transformed into

$$y^3 + ay + b = 0 \qquad (2.5)$$

by putting $x = y - a_1/3a_0$.

Denote by Δ the *discriminant* $-108D$, where $D = b^2/4 + a^3/27$, and by ϵ a complex root of $x^3 = 1$; then the roots of Eq. (2.5) can be written in the form (*Cardan's formula*):

$$y_1 = \left[-\frac{b}{2} + \sqrt{D}\right]^{1/3} + \left[-\frac{b}{2} - \sqrt{D}\right]^{1/3}$$

$$y_2 = \epsilon\left[-\frac{b}{2} + \sqrt{D}\right]^{1/3} + \epsilon^2\left[-\frac{b}{2} - \sqrt{D}\right]^{1/3}$$

$$y_3 = \epsilon^2\left[-\frac{b}{2} + \sqrt{D}\right]^{1/3} + \epsilon\left[-\frac{b}{2} - \sqrt{D}\right]^{1/3}$$

If the cubic equation is real and $D < 0$ then there are three real roots; if $D > 0$, there are one real and two complex roots; if $D = 0$, there are at least two identical real roots.

This formula is inconvenient for computational purposes. A solution in terms of trigonometric or hyperbolic expressions is more useful. For instance, in the case $a < 0$, $D < 0$, the roots are, with

$$\cos 3\phi = -(b/2)/\sqrt{-a^3/27}$$

$$y_1 = 2\sqrt{-a/3}\cos\phi$$
$$y_2 = -y_1/2 + \sqrt{-a}\sin\phi$$
$$y^3 = -y_1/2 - \sqrt{-a}\sin\phi$$

5. Equations of Degree 4 (Biquadratic Equations)

The equation

$$a_0x^4 + a_1x^3 + a_2x^2 + a_3x + a_4 = 0 \qquad (2.6)$$

is transformed to

$$y^4 + ay^2 + by + c = 0 \qquad (2.7)$$

by putting $x = y - a_1/4a_0$. Denote the three roots of the equation

$$z^3 + az^2 + (a^2 - 4c)z - b^2 = 0 \qquad (2.8)$$

by $\alpha_1, \alpha_2, \alpha_3$; the roots of Eq. (2.7) are

$$y_1 = \tfrac{1}{2}(+\sqrt{\alpha_1} + \sqrt{\alpha_2} + \sqrt{\alpha_3})$$
$$y_2 = \tfrac{1}{2}(+\sqrt{\alpha_1} - \sqrt{\alpha_2} - \sqrt{\alpha_3})$$
$$y_3 = \tfrac{1}{2}(-\sqrt{\alpha_1} + \sqrt{\alpha_2} - \sqrt{\alpha_3})$$
$$y_4 = \tfrac{1}{2}(-\sqrt{\alpha_1} - \sqrt{\alpha_2} + \sqrt{\alpha_3})$$

The discriminant Δ (see Sec. 4) of Eq. (2.8) is also the discriminant of Eq. (2.6). If the equation is real and $\Delta < 0$, there are a pair of conjugate complex roots and a pair of real roots; if $\Delta > 0$, all roots are real, provided that $a < 0$ and $a^2 - 4c > 0$; otherwise there are four complex roots. If $\Delta = 0$, the roots are not distinct.

6. Equations of Degree n

Investigations using the theory of permutation groups show that no formulas of a similar nature can be obtained for equations of degree $n > 4$, although special equations can be solved in terms of radicals.

7. Discriminants and General Symmetric Functions

The *discriminant* of a general polynomial of the form (2.1) is defined as

$$\prod_{i>k} (\alpha_i - \alpha_k)^2 \tag{2.9}$$

where α_i are the zeros of the polynomial. It vanishes if and only if two roots are identical. The discriminant is unaltered if the roots α are subjected to a permutation; such a function of the roots is called a *symmetric function.* The best-known symmetric functions are the *elementary symmetric* functions:

$$\sum_i \alpha_i, \quad \sum_{i<k} \alpha_i\alpha_k, \quad \sum_{i<k<l} \alpha_i\alpha_k\alpha_l, \quad . \; . \; . \tag{2.10}$$

From (2.1) and (2.2) the ith elementary symmetric function coincides with $(-1)^i \, a_i/a_0$. Every symmetric polynomial in $\alpha_1, \ldots, \alpha_n$ is a polynomial in the elementary symmetric functions.

8. Matrices

An array of $n \times m$ complex numbers, a_{ik}, when $i = 1, 2, \ldots, n$ and $k = 1, 2, \ldots, m$, is called an $n \times m$ matrix. Each a_{ik} is called an element of the matrix. They are commonly represented with a_{ik} written in the ith row and in the kth column of a rectangular array:

$$A = a_{ik} = \begin{bmatrix} a_{11} & a_{12} & \cdots & a_{1m} \\ a_{21} & a_{22} & \cdots & a_{2m} \\ \cdot & \cdot & \cdots & \cdot \\ a_{n1} & a_{n2} & \cdots & a_{nm} \end{bmatrix} \tag{2.11}$$

The element in the ith row and the kth column is also denoted as the (i,k) element. Matrices are also used quite often when elements are not numbers but are other mathematical objects; e.g., the elements can be matrices themselves. Operations may be defined for matrices in which they are regarded as single entities.

For $m = 1$ the matrix consists of a single column and is called a *column vector;* for $n = 1$ the matrix consists of a single row and is called a *row vector.* For $n = m$ the matrix is called *square.*

Addition is defined for two matrices having the same $n \times m$ structure; the sum is an $n \times m$ matrix obtained by adding corresponding elements:

$$A + B = C \qquad \text{where } c_{ik} = a_{ik} + b_{ik}$$

Multiplication of A by an ordinary number r is defined, giving the matrix whose elements are ra_{ik}.

Multiplication of two matrices A, B is defined only in case A is an $n \times m$ matrix and B is an $m \times p$ matrix; in this case the product $C = AB$ is an $n \times p$ matrix whose elements are

$$c_{ik} = \sum_{s=1}^{m} a_{is}b_{sk} \tag{2.12}$$

Only when A and B are square is the product BA also defined as the $n \times n$ matrix whose (i,k) element is $\sum_{s=1}^{n} b_{is}a_{sk}$, and so, in general, $AB \neq BA$. If $AB = BA$, the matrices are said to *commute.*

Associated with each $n \times m$ matrix A with elements a_{ik} is an $m \times n$ matrix with elements a_{ki}, obtained by interchanging rows and columns in the rectangular array. It is denoted by A' and is called the *transpose* of A. The transpose of a product is the product of the transposes in reverse order: $(AB)' = B'A'$.

Also associated with A is the complex-conjugate matrix A^* whose elements are $\bar{a}_{ik}$.‡

The matrix derived from A by taking first its transpose and then the complex conjugate (or vice versa) is called the *Hermitian conjugate* $A\dagger = \bar{A}'$. The (i,k) element of $A\dagger$ is $\bar{a}_{ki}$.

For a column vector x, represented by the $n \times 1$ matrix with elements x_k, the transpose x' is the $1 \times n$ matrix (row vector) having the same elements, while the Hermitian conjugate $x\dagger$ is a $1 \times n$ matrix having elements $\bar{x}_k$.

The *scalar* (or *inner*) *product* of two vectors x and y is defined as the ordinary number (1×1 matrix or *scalar*)

$$xy = yx = \Sigma x_k y_k \tag{2.13}$$

Two vectors are said to be *orthogonal* if $xy = 0$.

For square $n \times n$ matrices, the elements a_{ii} are called *diagonal elements.* Their sum, Σa_{ii}, is called the *trace* of A, often written $\mathrm{Tr}(A)$ or $\mathrm{tr}(A)$. In all cases, $\mathrm{Tr}\,(AB) = \mathrm{Tr}\,(BA)$.

The $n \times n$ matrix whose elements are δ_{ik}, that is, 1 for $i = k$, and 0 for $i \neq k$, is called the *unit matrix* I_n or simply I. The quantity δ_{ik} is often called the *Kronecker delta.*

The matrix A is called *nonsingular* if a matrix B exists such that $AB = I_n$; otherwise it is called *singular.* If B exists, it is unique and is called the *inverse* of A; it is usually written A^{-1} so that

$$AA^{-1} = I_n = A^{-1}A$$

Square matrices are described by various adjectives according to the following properties:

Diagonal	if	$a_{ik} = 0$	$i \neq k$	
Upper (lower) triangular	if	$a_{ik} = 0$	$i > k$	$(i < k)$
Symmetric	if	$A = A'$		
Skew or antisymmetric	if	$A = -A'$		
Real	if	$A = A^*$		
Pure imaginary	if	$A = -A^*$		
Hermitian	if	$A = A\dagger$		
Skew Hermitian	if	$A = -A\dagger$		
Orthogonal	if	$A^{-1} = A'$		
Unitary	if	$A^{-1} = A\dagger$		

Matrices, in one aspect, are a special case of *linear operators* (see Part 1, Chap. 6). This point of view also motivates the product definition. In the theory of linear operators the rules of matrix algebra are extended to infinite matrices. In this case the product

‡ This notation is used by physicists. In mathematics the notation $\bar{A}$ is used while A^* is the matrix denoted by $A\dagger$.

AB need not exist; the rules Tr (AB) = Tr (BA) and the uniqueness of A^{-1} and the rule $(A^{-1})^{-1} = A$ no longer hold in general; thus care must be exercised in taking over results established in finite matrix algebra.

A system of linear equations

$$\begin{aligned} a_{11}x_1 + \cdots + a_{1m}x_m &= u_1 \\ a_{21}x_1 + \cdots + a_{2m}x_m &= u_2 \\ \cdots\cdots\cdots\cdots\cdots \\ a_{n1}x_1 + \cdots + a_{nm}x_m &= u_n \end{aligned}$$

can be written as

$$\sum_k a_{ik}x_k = u_i \qquad i = 1, \ldots, n$$

which becomes, in matrix notation,

$$Ax = u$$

where A is the $n \times m$ matrix with elements a_{ik}, x is the $m \times 1$ column matrix with elements x_k, and u is the $n \times 1$ column matrix with elements u_k. If $n = m$ and A is nonsingular, the solution of the system takes the form

$$x = A^{-1}u \tag{2.14}$$

If, in this case, a change of variables is made by the linear substitution $y = Bx$, where B is a nonsingular $n \times n$ matrix, then the linear equations imply that the y_i are related by the equations

$$(BAB^{-1})y = Bu$$

The matrix A is said to have been *transformed* by B, and the matrix BAB^{-1} is called *similar* to A. Similar matrices have equal trace. Other important concepts which are invariant under similarity transformations are discussed below in Sec. 12.

If A is symmetric and B is orthogonal, BAB^{-1} is also symmetric. Likewise, if A is Hermitian and B is unitary, then BAB^{-1} is also Hermitian.

For certain applications (e.g., in group representations, Sec. 14) another product between two matrices is used. Here the matrices can be rectangular and of arbitrary dimensions. The product $A \times B$ is called the *Kronecker product* or *tensor product*. Let $A = (a_{ik})$ be $n \times m$ and $B = (b_{ik})$ be $p \times q$. Then $A \times B$ is the $np \times mq$ matrix

$$\begin{bmatrix} a_{11}B & a_{12}B & \cdots & a_{1m}B \\ a_{21}B & a_{22}B & \cdots & a_{2m}B \\ \cdot & \cdot & \cdots & \cdot \\ a_{n1}B & a_{n2}B & \cdots & a_{nm}B \end{bmatrix}$$

9. Determinants

A determinant associated with a square matrix (a_{ik}) is defined as the sum

$$|a_{ik}| = \Sigma \pm a_{1i_1}a_{2i_2} \cdots a_{ni_n}$$

where $i_1, \ldots, i_n$ is one of the $n!$ permutations of the numbers $1, \ldots, n$ and the sign $\pm$ is chosen accordingly as the permutation is even or odd. A permutation has the same parity as the number of transpositions needed to change $(i_1,i_2, \ldots ,i_n)$ into $(1,2,3, \ldots ,n)$. The value of the determinant is rarely computed from this expression, but rather by using some of its properties: (1) A matrix and its transpose have equal determinants. (2) If all numbers in a fixed row (or column) are multiplied by the same number, the determinant is multiplied by that number. (3) If a multiple (by the same number) of the elements of a row (or column) is added to another row (or column), the value of the determinant is unchanged. (4) Denote by A_{ik} the value of the subdeterminant of the matrix obtained from A by omitting the ith row and the kth column, multiplied by $(-1)^{i+k}$. A subdeterminant is called a *minor;* A_{ik} is called the *cofactor* of the element a_{ik}. The determinant can be expressed in the form

$$|a_{ik}| = \sum_k a_{ik}A_{ik}$$

On the other hand

$$\sum_k a_{ik}A_{jk} = 0 \qquad \text{if } i \neq j$$

Using (4), the inverse matrix of a nonsingular matrix A is

$$A^{-1} = \frac{(A_{ik})'}{|A|}$$

This is, however, not generally of much use for the computation of the inverse of a numerical matrix (Part 1, Chap. 6, Sec. 4). (5) The determinant of the product of two matrices is equal to the product of the determinants of the two matrices. (6) From the definition of the determinant it is evident that a matrix which has a row or a column of zeros has a vanishing determinant. Further, the determinant vanishes if there is a *linear dependence* between the rows or the columns of the matrix, that is, if numbers $\alpha_1, \ldots, \alpha_n$ exist such that not all $\alpha_k = 0$ and

$$\sum_k \alpha_k a_{kj} = 0 \qquad \text{or} \qquad \sum_k a_{ik}\alpha_k = 0$$

A square matrix is singular if and only if its determinant vanishes.

Three special determinants will be mentioned.

1. Let $x_1, \ldots, x_n$ be n quantities; then the corresponding *Vandermonde determinant* is $|a_{ik}|$, where $a_{ik} = x_i^{k-1}$. Its value is

$$(-1)^{n(n-1)/2} \prod_{i<k} (x_i - x_k)$$

and its square is the discriminant (see Sec. 7).

2. The *Wronskian* concerns n functions $f_i(x)$. It is defined as $|f_i^{(k)}|$, where

$$f_i^{(k)} = \frac{d^k f_i}{dx^k}$$

$k = 0,1, \ldots, n-1$. The vanishing of this determinant means that there is a relation

$$c_1f_1 + \cdots + c_nf_n = 0$$

where c_i are constants, not all of which are zero.

3. The *Jacobian* concerns n functions f_i of n variables x_i and their partial derivatives. It is the determinant whose elements are $\partial f_i/\partial x_k$. The vanishing of this determinant signifies that there is a functional dependency between the f_i, that is, there exists a function $F(f_1, \ldots, f_n)$ whose value is 0 for all values of the variables x_i.

10. Systems of Linear Equations

Using determinants, an explicit solution for any linear system with square nonsingular matrix

$$Ax = u \tag{2.15}$$

can be obtained by means of *Cramer's rule;* it is, however, not convenient for numerical computation:

$$x_i = \frac{|U_i|}{|A|} \qquad i = 1, 2, \ldots, n$$

where U_i is the matrix obtained from A by replacing the ith column by the vector u. The solution is unique.

If u is the vector $(0,0, \ldots, 0)$—the *homogeneous* case—the system (2.15) thus has only the trivial solution

$$x_1 = \cdots = x_n = 0$$

If A is singular and the system is homogeneous, solutions other than the trivial one are always possible. If the system is inhomogeneous, then in general there is no solution at all. If a solution is possible, it means—since the homogeneous system has a solution—that there is a linear dependency (Sec. 9) between the n equations so that at least one of them can be discarded. Consider then, generally, a system of linear equations with rectangular matrix A, with n rows and m columns. If $n > m$, such a system has a solution only if there is a linear dependency between the equations so that certain of them are a consequence of the others. The case $n = m$ has been discussed. If $n < m$ and the system is homogeneous, it will always have solutions; if it is inhomogeneous, nontrivial solutions exist only if other conditions are fulfilled. Any two solutions of an inhomogeneous system differ by a solution of the corresponding homogeneous system.

11. Characteristic Roots of Matrices and Quadratic Forms

Like the trace of a matrix, its determinant is invariant under transformation (Sec. 8); both facts follow from the invariance of the characteristic roots. These numbers $\lambda_1, \ldots, \lambda_n$ have the property that

$$Ax = \lambda_i x \tag{2.16}$$

for a suitable vector X (different from the null vector), called the modal vector or characteristic vector that belongs to λ_i.

It follows that the vector x is a solution of the homogeneous system of equations

$$(A - \lambda I_n)x = 0$$

Hence the determinant $|A - \lambda I_n| = 0$. This is an algebraic equation of degree n in λ, the *characteristic equation* of A, and this implies that there are n (not necessarily different) values λ_i (as stated above). If the coefficients of the powers of λ are investigated, λ^{n-1} has as coefficient the trace of A and the constant term is $|a_{ik}|$. Hence trace $A = \Sigma\lambda_i$ and $|A| = \Pi\lambda_i$.

If B is any nonsingular $n \times n$ matrix, then the determinant of $B^{-1}(A - \lambda I_n)B = B^{-1}AB - \lambda I_n$, is equal to $|B^{-1}|\,|A - \lambda I_n|\,|B| = |A - \lambda I_n|$. This shows that the similar matrices A and $B^{-1}AB$ have the same characteristic equation and therefore the same characteristic roots. The characteristic equation of a product AB is equal to the characteristic equation of BA. However, it is not true in general that the characteristic roots of AB are products of the characteristic roots of A and those of B.

The characteristic roots of a symmetric real matrix are real; so are the characteristic roots of a Hermitian matrix; those of a skew symmetric real matrix are purely imaginary; those of an orthogonal or unitary matrix have absolute value 1 (Sec. 8).

The characteristic vectors which correspond to two distinct roots of a symmetric matrix are orthogonal. This is a special instance of the fact that to every Hermitian matrix A a unitary matrix S can be found such that

$$S^{-1}AS = \begin{bmatrix} \lambda_1 & & & \\ & \cdot & & \\ & & \cdot & \\ & & & \cdot \\ & & & & \lambda_n \end{bmatrix}$$

The matrix S is sometimes called the *modal matrix of* A since it is composed of the characteristic vectors which correspond to the λ_i.

These results concerning Hermitian matrices can be translated into results concerning the corresponding Hermitian forms $\Sigma a_{ik}x_i x_k^*$. Every such form can be transformed into the form

$$\sum_{i=1}^{n} \lambda_i |y_i|^2$$

(*transformation to principal axes*) by means of a linear transformation which can further be chosen in such a way that it transforms the form $\Sigma|x_i|^2$ into $\Sigma|y_i|^2$. A Hermitian form can be transformed into another by means of a unitary transformation if and only if their corresponding matrices have the same characteristic roots. If all $\lambda_i > 0$, the form and the matrix are called *positive definite;* if all $\lambda_i \geqq 0$, then they are called *positive semidefinite.*

If the variables of a Hermitian form are subjected to any linear transformation X, a new Hermitian form results. Unless X is unitary, the corresponding matrix will not necessarily have the same characteristic roots as the original one; however, the distribution of the signs of the characteristic roots remains unaltered (*theorem of inertia* of Hermitian forms).

If A is a positive definite real symmetric matrix and B arbitrary real symmetric, the determinantal equation $|B - \lambda A| = 0$ has real roots; they coincide with the characteristic roots of $A^{-1}B$. Although A^{-1} is again symmetric, the matrix $A^{-1}B$ is in general not symmetric; in fact, any real matrix can be expressed as the product of two symmetric matrices.

Corresponding results for skew-symmetric matrices follow. Every skew-symmetric real matrix A can be transformed by means of an orthogonal transformation into

$$\begin{bmatrix} E_1 & & & & & \\ & E_2 & & & & \\ & & \cdot & & & \\ & & & \cdot & & \\ & & & & E_s & \\ & & & & & 0 \end{bmatrix}$$

where

$$E_i = \begin{bmatrix} 0 & \alpha_i \\ -\alpha_i & 0 \end{bmatrix}$$

and 0 is a zero matrix of suitable size (possibly absent). The characteristic roots of such a matrix are $\pm i\alpha_1$, $\pm i\alpha_2$, . . . or $\pm i\alpha_1$, $\pm i\alpha_2$, . . . , 0.

Arbitrary matrices cannot be transformed to diagonal form. Generally a matrix A is called *normal* if it can be transformed to diagonal form by a unitary matrix. These matrices are also characterized by the fact that $AA\dagger = A\dagger A$. A normal matrix with real characteristic roots is Hermitian. Every matrix, however, can be transformed into the *Jordan normal* form

$$\begin{bmatrix} A_1 & & & & \\ & A_2 & & & \\ & & \cdot & & \\ & & & \cdot & \\ & & & & A_r \end{bmatrix}$$

where the A_i are themselves square matrices which are either one-dimensional or Jordan blocks:

$$\begin{bmatrix} \alpha & 1 & & & \\ & \alpha & 1 & & \\ & & \cdot & \cdot & \\ & & & \alpha & 1 \\ & & & & \alpha \end{bmatrix}$$

It can further be shown that to every matrix A two square matrices P and Q can be found such that PAQ is of diagonal form; however, P and Q will not in general be inverse matrices. Also, the elements in the diagonal of PAQ are not, in general, the characteristic roots of A.

If in the characteristic equation of A

$$x^n + a_1x^{n-1} + \cdots + a_n = 0$$

the unknown x is replaced by A, the constant a_n by a_nI_n, and the number 0 by the n-dimensional zero matrix, an identity is obtained. This fact constitutes the *Cayley-Hamilton theorem.* It deals with polynomials in matrices and is thus closely connected with hypercomplex systems. These properties find much application in quantum mechanics (Part 2, Chap. 6).

Every matrix can be transformed to triangular form by a unitary similarity. Special sets of matrices can be transformed to triangular form simultaneously by a unitary similarity; e.g., commuting matrices have this property (see also Sec. 12).

Two Hermitian matrices A, B, one of which is positive definite, can be transformed to diagonal form simultaneously by a so-called congruence transformation $XAX\dagger$, $XBX\dagger$.

The practical computation of characteristic roots is discussed in Part 1, Chap. 7. A number of useful bounds for the characteristic roots of a finite matrix are available. Some are invariant under unitary transformations, e.g., the fact that all characteristic roots lie inside or on the boundary of the region in the complex plane determined by all complex numbers $\Sigma a_{ik}x_ix_k{}^*$ for $\Sigma|x_i|^2 = 1$. In the case that A is a Hermitian matrix, it follows that the largest characteristic root of A is max $\Sigma a_{ik}x_ix_k{}^*$ for $\Sigma|x_i|^2 = 1$ and that the smallest characteristic root of A is min $\Sigma a_{ik}x_ix_k{}^*$ for $\Sigma|x_i|^2 = 1$.

A very useful bound is $\max_i \sum_k |a_{ik}|$. An even better one is given by the fact that the characteristic roots of A lie inside or on the boundary of the n circles with centers a_{ii} and radii $\sum_{k\neq i} |a_{ik}|$.

Any polynomial equation of degree n

$$x^n + a_1x^{n-1} + \cdots + a_{n-1}x + a_n = 0$$

can be interpreted as the characteristic equation of an $n \times n$ matrix, for example, of the so-called companion matrix

$$\begin{bmatrix} 0 & 1 & 0 & \cdot & \cdot & \cdot & 0 \\ 0 & 0 & 1 & 0 & \cdot & \cdot & \cdot \\ \cdot & \cdot & \cdot & \cdot & \cdot & \cdot & \cdot \\ 0 & \cdot & \cdot & \cdot & \cdot & \cdot & 1 \\ -a_n & -a_{n-1} & \cdot & \cdot & \cdot & \cdot & -a_1 \end{bmatrix}$$

Characteristic roots of special matrices:

1. *Matrices with non-negative elements.* Such matrices have a non-negative characteristic root. In particular, matrices with positive elements have a positive characteristic root that is larger than the modulus of any other root. The corresponding characteristic vector can be chosen to have positive components only. The same facts are true if the matrix involves zero elements but is irreducible (or indecomposable, i.e., cannot be transformed to the form $\begin{pmatrix} P & Q \\ O & R \end{pmatrix}$ by the same permutation in rows and columns, where P, R are square matrices and O consists of zeros only). If the matrix is irreducible and has only one characteristic root of maximal absolute value, a finite power of it is positive. Such a matrix is called *primitive.* The converse is also true. If the matrix is irreducible and has k roots of maximal absolute value, it can be transformed by the same permutations of rows and columns to the form

$$\begin{bmatrix} 0 & P_1 & & & & \\ & & P_2 & & & \\ & & & \cdot & & \\ & & & & \cdot & \\ & & & & & P_{k-1} \\ P_k & \cdot & \cdot & \cdot & \cdot & 0 \end{bmatrix}$$

where the P_i are block matrices and all other elements are zero.

If all the minors of all dimensions of a matrix are positive, then all its characteristic roots are positive. Such matrices play a role in dynamics. They are called totally positive.

2. *Stochastic matrices.* These have non-negative elements, and the sum of the elements in each row is 1. If also the sum of the elements in each column is 1, they are called *doubly stochastic.* They play a role in the theory of stochastic processes. Every irreducible non-negative matrix is similar to a stochastic matrix via a diagonal matrix. A permutation matrix is a special case of a doubly stochastic matrix; it contains a 1 in each row and column and 0's elsewhere. It is also a special unitary matrix.

3. *Diagonally dominant matrices.* The $n \times n$ matrix $A = (a_{ik})$ is called diagonally dominant if $a_{ii} > \sum_{\substack{k=1 \\ k \neq i}}^{n} |a_{ik}|$, $i = 1, \ldots, n$. Such a matrix is nonsingular.

4. *M matrices.* These have been introduced by Ostrowski. They have non-negative inverses and their off-diagonal elements are nonpositive. Because of the latter property, such a matrix A has the property that $\lambda I - A$ is positive for a sufficiently large real number λ. Hence A has a real characteristic root.

5. *Stable matrices.* A matrix is called stable if its characteristic roots have negative real parts. Let $a_0x^n + a_1x^{n-1} + \cdots + a_{n-1}x + a_n$ be the characteristic polynomial of such a matrix. It can be tested for stability by the Routh-Hurwitz inequalities:

$$a_0 > 0, \ |a_1| > 0, \ \begin{vmatrix} a_1 & a_3 \\ a_0 & a_2 \end{vmatrix} > 0, \ \begin{vmatrix} a_1 & a_3 & a_5 \\ a_0 & a_2 & a_4 \\ 0 & a_1 & a_3 \end{vmatrix} > 0,$$

$$\ldots, \begin{vmatrix} a_1 & a_3 & a_5 & \cdots & 0 \\ a_0 & a_2 & a_4 & & \\ 0 & a_1 & a_3 & & \\ 0 & a_2 & a_4 & & \\ \cdot & \cdot & \cdot & & a_n \end{vmatrix} > 0$$

Another important test was formed by Lyapunov: A is stable if and only if a positive definite Hermitian G exists such that

$$AG + GA^* = -I$$

6. *Kronecker products.* If A is a square matrix with characteristic roots α_i and B is a square matrix with characteristic roots β_j then $A \times B$ has the characteristic roots $\alpha_i\beta_j$.

7. *Nilpotent matrix.* This is a matrix such that some power of it is the zero matrix. All the characteristic roots are zero.

8. *Idempotent matrices.* They coincide with their square. Their characteristic roots are all 0 or 1.

9. *Integral matrices.* These are matrices whose elements are rational integers. They are of importance in crystallography. Their characteristic roots are algebraic integers.

12. Functions of Matrices and Infinite Sequences

Any two polynomials in the same matrix commute, but not every two commutative matrices are polynomials in the same matrix. A rational function $p(A)/q(A)$ can always be expressed as a polynomial in A, provided $q(A)$ is nonsingular.

From polynomials we are led to power series in a matrix A:

$$c_0I + c_1A + c_2A^2 + \cdots$$

where the c_i are numbers. Generally, an infinite sequence $A_n = (a_{ik}^{(n)})$ of matrices is said to converge to a matrix $A = (a_{ik})$, denoted as $\lim_{n\to\infty} A_n = A$, if $\lim_{n\to\infty} a_{ik}^{(n)} = a_{ik}$ for all values of i and k. From this it is evident that an infinite series ΣA_n of matrices converges if the infinite series corresponding to each element converges. The convergence of a power series in A is linked up with the characteristic roots of A. In particular the geometric series $I + A + A^2 + \cdots$ converges if and only if the characteristic roots of A all lie inside the unit circle. The exponential series

$$e^A = \sum \frac{1}{n!} A^n$$

converges for all matrices A. Not all properties of the exponential function remain if matrices are introduced as exponents, e.g., in general

$$e^Ae^B \neq e^{A+B}$$

If, however, A and B commute, $e^Ae^B = e^{A+B}$.

The converse is not true. The relation

$$e^Ae^B = e^{A+B} + f(A,B)$$

when $f(A,B)$ is an infinite expansion in expressions of the form $XY - YX$ is known as the Baker-Campbell-Hausdorff formula.

The characteristic roots of a polynomial $p(A)$ in a matrix A are of the form $p(\alpha_i)$ if α_i are the characteristic roots of A. This follows, for example, from the fact that the matrix is similar to a triangular matrix. The characteristic roots of polynomials in several matrix variables A_i cannot be expressed easily in terms of the characteristic roots of the A_i except in special cases, e.g., when the matrices can be transformed to triangular form by a simultaneous similarity transformation. This is the case, for example, when the A_i commute pairwise. Various bounds are known for the difference between the characteristic roots of $A + B$ and $\alpha_i + \beta_i$, where α_i are the characteristic roots of A and β_i the roots of B; and similarly for products. In particular, this has been studied where A and B are real and symmetric or Hermitian or even normal. For example, if A and B are real and symmetric with characteristic roots $\alpha_1 \geq \alpha_2 \cdots \geq \alpha_n$, and $\beta_1 \geq \beta_2 \cdots \geq \beta_n$, and if $\gamma_1 \geq \gamma_2 \cdots \geq \gamma_n$ are the characteristic roots of $A + B$, then

$$\gamma_{i+j-1} \leq \alpha_i + \beta_j$$

if $i + j \leq n + 1$. This was proved by H. Weyl.

Another important fact concerning the characteristic roots of $A + B$ was proved by Lidskii and Wielandt: The γ_i are in the convex closure of the $n!$ points of $\alpha_i + P\beta_i$ when P runs through all $n!$ permutations.

In order to proceed from Hermitian matrices to general ones, the *singular values* λ_i of a matrix A are studied. They are the positive square roots of the characteristic roots of AA^*. The following inequalities between the $|\alpha_i|$ and λ_i were found by E. T.

Browne:

$$\min \lambda_i \leq |\alpha_i| \leq \max \lambda_i$$

Many other inequalities are known for the AB case.

13. Hypercomplex Systems or Algebras

A hypercomplex system with respect to the real (or complex) numbers has a finite set of *base elements* or *units* $e_1, \ldots, e_n$ such that every element of the system is of the form $a_1e_1 + \cdots + a_ne_n$, where a_i are real (or complex) numbers. The base elements have multiplication rules $e_ie_k = \Sigma a_{ikj}e_j$, where a_{ikj} are again real (or complex) numbers which are arbitrary as long as $(e_ie_k)e_j = e_i(e_ke_j)$. If this last condition is not fulfilled, the system is called *nonassociative*. From the multiplication of the base elements a multiplication of any two elements $a_1e_1 + \cdots + a_ne_n$ and $b_1e_1 + \cdots + b_ne_n$ is given by putting

$$(a_1e_1 + \cdots + a_ne_n)(b_1e_1 + \cdots + b_ne_n) = \Sigma a_ib_ke_ie_k$$

Addition of these two elements is defined by

$$(a_1 + b_1)e_1 + \cdots + (a_n + b_n)e_n$$

The element $0 \cdot e_1 + \cdots + 0 \cdot e_n$ plays the role of the 0 among ordinary numbers. An element corresponding to 1 among ordinary numbers does not always exist. The following are several examples of hypercomplex systems, the first four of associative systems and the last three of nonassociative systems:

1. *The complete matrix algebra with real or complex coefficients*, i.e., the set of all $n \times n$ matrices with real (or complex) coefficients. The n^2 matrices which have one element 1 and 0 elsewhere may be taken as base elements. This set has a 1 element namely, I_n. However, in many other respects, as already mentioned earlier, it does not behave like ordinary numbers, for example, $AB \neq BA$ in general; further, the product of two matrices may be the zero matrix, without either factor being the zero matrix. It may even happen that $AB = 0$ and $BA \neq 0$. The nonsingular matrices have reciprocals; the singular matrices do not.

2. *Quaternions.* This system has four units, usually called $1,i,j,k$, or $1,i_1,i_2,i_3$, with the multiplication rules

$$\begin{gathered} 1 \cdot i_\alpha = i_\alpha \cdot 1 = i_\alpha \\ i_\alpha i_\beta = -i_\beta i_\alpha \quad (\alpha \neq \beta),\ i_\alpha{}^2 = -1 \\ i_1i_2 = i_3 \quad i_2i_3 = i_1 \quad i_3i_1 = i_2 \end{gathered}$$

If the coefficients are assumed real, then the quaternions have no *divisors of zero*, i.e., no product

$$(a_0 + a_1i_1 + a_2i_2 + a_3i_3)(b_0 + b_1i_1 + b_2i_2 + b_3i_3) = 0$$

without either all $a_i = 0$ or all $b_i = 0$. In this case every quaternion $\neq 0$ has an inverse, namely,

$$\frac{a_0 - a_1i_1 - a_2i_2 - a_3i_3}{a_0{}^2 + a_1{}^2 + a_2{}^2 + a_3{}^2}$$

The quaternions are a most important discovery by Hamilton, who used them for expressing the rotations of the sphere as ordinary complex numbers describe the rotations of the circle. Apart from the reals and the complex numbers they are the only associative hypercomplex systems with real coefficients and no divisors of zero. One of the applications of quaternions is to furnish a proof that every integer is a sum of four squares.

3. *Clifford algebras* have 2^n base elements which are generated by n elements e_i with the relations

$$e_i{}^2 = -1 \qquad e_ie_j = -e_je_i\ (i \neq j)$$

Sets of matrices which are *anticommuting* have been studied in various connections. The *Pauli spin matrices* form a set of anticommuting 2×2 matrices each of which has as square the unit matrix

$$\begin{bmatrix} 0 & 1 \\ 1 & 0 \end{bmatrix}, \begin{bmatrix} 0 & -i \\ i & 0 \end{bmatrix}, \begin{bmatrix} 1 & 0 \\ 0 & -1 \end{bmatrix}$$

(See Part 2, Chap. 1, Sec. 3.) Dirac obtains four 4×4 matrices with the same properties:

$$\alpha_1 = \begin{bmatrix} 0 & 0 & 0 & 1 \\ 0 & 0 & 1 & 0 \\ 0 & 1 & 0 & 0 \\ 1 & 0 & 0 & 0 \end{bmatrix}$$

$$\alpha_2 = \begin{bmatrix} 0 & 0 & 0 & -i \\ 0 & 0 & i & 0 \\ 0 & -i & 0 & 0 \\ i & 0 & 0 & 0 \end{bmatrix}$$

$$\alpha_3 = \begin{bmatrix} 0 & 0 & 1 & 0 \\ 0 & 0 & 0 & -1 \\ 1 & 0 & 0 & 0 \\ 0 & -1 & 0 & 0 \end{bmatrix}$$

$$\alpha_4 = \begin{bmatrix} 1 & 0 & 0 & 0 \\ 0 & 1 & 0 & 0 \\ 0 & 0 & -1 & 0 \\ 0 & 0 & 0 & -1 \end{bmatrix}$$

Eddington proved that there cannot be more than five anticommuting matrices E_i in four dimensions, such that $E_i{}^2 = -1$. If all are real or pure imaginary, then two are real and three are pure imaginary.

4. *Dual numbers* have two base elements 1 and e with $e^2 = 0$.

5. *Cayley numbers* have eight base elements: $1,e_1, \ldots, e_7$ with $e_i{}^2 = -1$, $e_ie_j = -e_je_i$ and $e_1e_2 = e_3$, $e_1e_4 = e_5$, $e_1e_6 = e_7$, $e_2e_5 = e_7$, $e_2e_4 = -e_6$, $e_3e_4 = e_7$, $e_3e_5 = e_6$.

All other products of base elements are obtained from the further rule that $e_ie_k = e_j$ implies $e_ke_j = e_i$, $e_je_i = e_k$.

Like quaternions, the Cayley numbers with real coefficients have no zero divisors, and every element $\neq 0$ has a unique inverse. The conjecture that no hypercomplex system with more than eight base elements and real coefficients is without divisors of zero has been established recently.

6. *Matrix algebra with* $\frac{1}{2}(AB + BA)$ *as composition instead of* AB. This composition defines a Jordan algebra. The expression $AB + BA$ is called an *anticommutator*. This is a commutative (even if $AB \neq BA$) but not associative composition. It plays a role in quantum mechanics.

7. *Matrix algebra with* $AB - BA$ *as composition instead of* AB. This composition defines a *Lie algebra*. It is an anticommuting and nonassociative composi-

tion. Lie algebras are much used in physics because of their close connection with Lie groups (see Sec. 14). The expression $(A,B) = AB - BA$ is called a *commutator*. It has the property $(A,B) = -(B,A)$ and satisfies the *Jacobi identity*

$$(A,(B,C)) + (B,(C,A)) + (C,(A,B)) = 0$$

While complete matrix algebras are hypercomplex systems, a certain converse is true too: To every associative hypercomplex system S there corresponds an *isomorphic* set of matrices, i.e., a set which is in a one-to-one correspondence with S such that the sum (or product) of two elements of S corresponds to the sum (or product) of the corresponding matrices. For the complex numbers this isomorphism is established through the correspondence

$$a + ib \rightleftarrows \begin{bmatrix} a & b \\ -b & a \end{bmatrix}$$

for the quaternions through the correspondence

$$a + bi + cj + dk \rightleftarrows \begin{bmatrix} a & b & c & d \\ -b & a & -d & c \\ -c & d & a & -b \\ -d & -c & b & a \end{bmatrix}$$

14. Theory of Groups

A group is a set of elements (finite or infinite) with a composition law (frequently called multiplication) which satisfies the following axioms:

1. The multiplication is associative.
2. The product of two elements of a group belongs to the group.
3. There is a unit element e such that $ex = xe = x$ for all x of the group.
4. To each element x of the group there is an inverse x^{-1} with respect to e, that is,

$$xx^{-1} = x^{-1}x = e$$

The number of elements of the group is called its *order*. A subset of a group which forms a group in itself is called a *subgroup*. If C is a subset of a group and g a fixed element, the set gC (respectively Cg) denotes the set of all elements gc (respectively cg) when c runs through C. If C is a subgroup, such a set is called a *coset* with respect to C. If C is a subgroup, the set $g^{-1}Cg$ is called a *conjugate* subgroup. If a subgroup C coincides with all its conjugate subgroups, it is called *self-conjugate* or *normal* or *invariant*. The cosets with respect to a self-conjugate subgroup form a group again: the *quotient group* or *factor group* with respect to the subgroup. A group without a proper subgroup that is normal is called *simple*. The long-standing conjecture that all nontrivial simple groups are of even order was proved in 1963.

If c is a fixed element, the elements $g^{-1}cg$ are called conjugates of c. All the elements of a group can be decomposed into disjoint *classes* such that any two elements of the same class are conjugate, but no two elements belonging to different classes are conjugate.

If any two elements x,y of the group commute, that is, $xy = yx$, then the group is called *Abelian*. The set of all elements c which commute with all the elements of a group form a subgroup, the *center*. If $xy \neq yx$, then $x^{-1}y^{-1}xy$ differs from the unit element; this element is called the *commutator* of x and y. The smallest subgroup which contains all the commutators is called the *commutator subgroup;* it is self-conjugate and has an Abelian factor group.

Isomorphisms for groups are defined analogously as in Sec. 13. An isomorphism of a group with itself is called *automorphism*. A special type of the latter is the inner automorphism defined by $c^{-1}xc$, where c is a fixed element of the group and x runs through all the elements of the group.

Examples of Groups. *Example* 1. The permutations of n elements form a group S_n of order $n!$ (the *symmetric* group of n elements). Every finite group of order m is isomorphic with a subgroup of the S_m. The even permutations of n elements form a subgroup of S_n, the *alternating* group of n elements.

Example 2. All nonsingular real $n \times n$ matrices form the general linear group $GL(n)$ under the operation of multiplication. A subgroup is formed by the matrices of determinant ± 1, called the unimodular or special linear group $SL(n)$; a subgroup of the latter is formed by the orthogonal matrices $O(n)$. This group has as subgroup the rotation group $O^+(n)$ consisting of matrices with determinant $+1$ only. The $n \times n$ unitary matrices form the group $U(n)$ with the subgroup $SU(n)$ of matrices of determinant $+1$. The group $O(n)$ can also be characterized as the matrices of the linear transformations under which the quadratic form $x_1^2 + \cdots + x_n^2$ stays invariant. Transformations that leave other quadratic forms invariant play a role too, in particular the *Lorentz transformations* which leave $-x_1^2 + x_2^2 + x_3^2 + x_4^2$ invariant. They form the *Lorentz group*.

Example 3. *The Space Groups.* These groups are connected with the n-dimensional Euclidean space R_n, that is, the set of n-tuples $(x, \ldots, x_n)$. The n-dimensional *lattices* are the sets of points which are linear combinations with *integral* coefficients of n linearly independent points. By the sum of two points is understood the point whose coordinates are the sums of the corresponding coordinates of the two points. For example, in R_2 the set of all points with integral coefficients form a lattice; another lattice is found if also the centers of the squares of the previous lattice are admitted as lattice points (see Figs. 2.1 and 2.2).

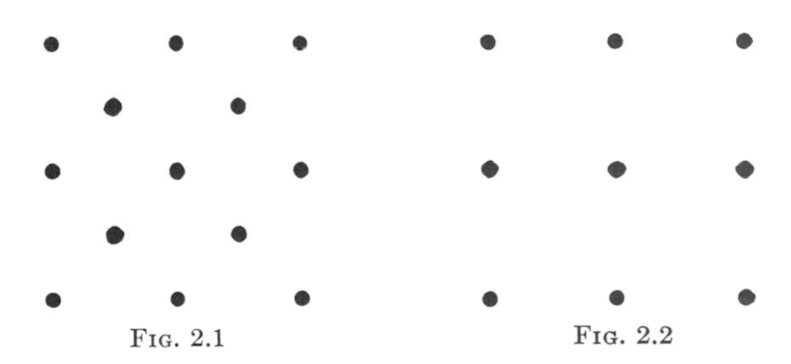

FIG. 2.1 FIG. 2.2

In modern crystallography the lattices in R_3 are used as the main tool for the classifications; there are seven different types of lattices, called *triclinic, monoclinic, tetragonal, hexagonal, rhombohedric, rhombic,* and *cubic*. (Cf. Part 8, Chap. 1.)

A *symmetry* is a motion of the R_n which transforms

lattice points into lattice points. The symmetries of a lattice form a group. It is easy to see that in R_2 symmetries which leave (0,0) invariant must be of period 2, 3, 4, or 6. Generally, a motion in R_n is given by the transformation $y = Ax + a$, where A is an n-dimensional orthogonal matrix and a a column of n real numbers. In the case where $|A| = 1$ the matrix A alone defines a rotation, the column a alone a translation. Among the groups whose elements are rigid motions the *discrete groups* are characterized by the property that the orbit of any given point (i.e., the images of the point under the action of the group) is a discrete point set. *Space groups* are discrete groups containing n linearly independent translations. For example, the symmetry group of a lattice is a space group. There are 230 space groups in R_3 among which 14 are lattice symmetry groups; also there are 17 space groups in R_2 of which 3 are lattice symmetry groups.

Example 4. *Continuous Groups*. These are infinite groups whose elements are functions of the points of an r-dimensional Euclidean space E_r and for which the group composition is a continuous function of both factors. By multiplying every element in the group by a fixed element, a transformation of the space E_r into or onto itself results. More generally, continuous groups of transformations of an E_n $(n \neq r)$ depending on r real parameters have been studied for a long time. The best known are discussed in the next example.

Example 5. *Lie Groups*. These are continuous groups depending on r parameters in which the composition is an analytic function of both factors. If $r = 1$, it is a case of a one-parameter group, which is commutative. It is isomorphic with the translations of the E_n. For $r > 1$ the group is in general not commutative, but a certain additive commutative group is associated with it, called the *infinitesimal group*. This group has r generators called the infinitesimal generators or transformations or operators. They completely determine the Lie group. However, multiplicatively they are also generators of a nonassociative hypercomplex system, a *Lie algebra*, because the additive commutators of the infinitesimal generators are linear expressions in the commutators. The coefficients are called the *structure constants* of the Lie group. For $r = 1$ the infinitesimal operator is obtained as follows: Let α_1 be any fixed value of the parameter and α_2 the parameter of the corresponding inverse transformation. Let the elements of the group be the transformations

$$x'_i = f_i(x_1, \ldots, x_n; \alpha) \qquad i = 1, \ldots, n$$

We then consider

$$\begin{aligned} x'_i &= f_i(x_1, \ldots, x_n; \alpha_1) \\ x_i &= f_i(x'_1, \ldots, x'_n; \alpha_2) \end{aligned}$$

Let δt be a small quantity and consider the transformation

$$f_i(f_1(x_1, \ldots, x_n; \alpha_1), \ldots, f_n(x_1, \ldots, x_n; \alpha_1); \alpha_2 + \delta t)$$

By the mean-value theorem this can be expressed as

$$x_i + \xi_i(x_1, \ldots, x_n; \alpha_1)\delta t$$

where the ξ_i are independent of δt if terms of the second or higher order are neglected. This is then the infinitesimal operator.

A Lie group that has been studied particularly well is the rotation group. Its infinitesimal generators are the components of the "angular momentum."

A set of matrices that is homomorphic with a group is called a *representation* of the group. Every finite group has representations; many infinite groups, although not all of them, can be represented by finite matrices. Every infinite group can be represented by infinite matrices. While every finite group can be represented by unitary matrices, this is not true for infinite groups.

The dimension of the matrices of a representation is called the *degree* of the representation. The sets of traces of the matrices which occur in the representation are called its *character*. If all the matrices of a representation are transformed by the same nonsingular matrix, an *equivalent* representation is obtained. If a representation is equivalent to one in which all the matrices have the form

$$\begin{bmatrix} A_1 & Q \\ P & A_2 \end{bmatrix}$$

where A_1, A_2 are square matrices and P,Q consist of zeros only, then it is called *reducible;* otherwise it is called *irreducible*.

All irreducible representations of a finite group are equivalent to representations by matrices whose elements are algebraic numbers. It was shown by R. Brauer that these can be chosen as rational functions with rational coefficients of the hth root of unity if h is the order of the group.

All representations of a finite group are equivalent to representations by unitary matrices. The number of nonequivalent irreducible representations is equal to the sum of classes in the group.

Every representation of a finite group can be reduced completely, i.e., an equivalent representation of the form

$$\begin{bmatrix} A_1 & 0 & \cdot & \cdot & \cdot & 0 \\ 0 & A_2 & 0 & \cdot & \cdot & 0 \\ \cdot & \cdot & \cdot & \cdot & \cdot & \cdot \\ 0 & \cdot & & \cdot & \cdot & A_r \end{bmatrix}$$

can be found where the A_i are square matrices and where 0 means a zero matrix of appropriate size, and where the representation A_i is irreducible. The irreducible representations of a commutative group are all one-dimensional. The reduction is always unique.

Two irreducible representations which have the same character are equivalent. The degrees n_i of the irreducible representations divide the order of the group. The sum Σn_i^2 is equal to the order of the group.

The only matrices which commute with all matrices of an irreducible representation are the *scalar matrices*, i.e., the matrices which have the same constant along the main diagonal and zeros elsewhere.

Of great use is the Schur lemma: Let $D_1(g)$, $D_2(g)$ be two irreducible representations of degrees n_1, n_2 and let

$$AD_1(g) = D_2(g)A$$

for some rectangular matrix A and for all elements g in

the group. If $n_1 \neq n_2$, then $A = 0$; if $n_1 = n_2$ and $A \neq 0$, then A is nonsingular and the representations are equivalent.

The orthogonality relations: Let h be the order of the group; let $D^1, D^2, \ldots$ be the inequivalent irreducible representations of dimensions $n_1, n_2, \ldots$ and let D^i have the elements $D_{\alpha\beta}{}^i(g)$ where g runs through all group elements; then

$$\sum_g D_{\alpha\beta}{}^{(i)}(g) D_{\mu\nu}{}^j(g^{-1}) = \frac{h}{n_i} \delta_{ij} \delta_{\alpha\nu} \delta_{\beta\mu}$$

Let $\chi^1, \chi^2, \ldots$ be the characters of the irreducible representations. They obey the following orthogonality relations:

$$\sum_g \chi^i(g^{-1}) \chi^j(g) = h\delta_{ij}$$

The regular representation is obtained by multiplying the group elements on the left by a fixed element and associating with this element the permutation of the group generated in this way. The regular representation contains n_i times every irreducible representation of dimension n_i.

The representations of the symmetric group have been studied in great detail.

Among infinite groups the representations of the groups $O(n)$ and $SU(n)$ have received particular attention.

Spin representations play a big role in applications. They are homomorphisms; e.g., the group $SU(2)$ is homomorphic to $O^+(3)$ and thus gives a spin representation of $O^+(3)$.

The group U_3 is contained in O_6. The representations of SU_3 have received particular attention recently, especially the eight-dimensional representation of SU_3 over its center, called the *eightfold way*.

15. Rings and Fields

A hypercomplex system is a special case of a ring. This is a system with two compositions, called addition and multiplication. The addition is usually assumed commutative, but not the multiplication. The two compositions are linked by the distributive laws

$$a(b + c) = ab + ac$$
$$(a + b)c = ac + bc$$

With respect to addition, a ring forms a group. The unit element of this group is usually called "zero" and denoted by 0. An element e for which $ex = x$ for all x in the ring is called a *left unit*. A similar definition holds for *right units* and *two-sided units*. If the elements $\neq 0$ form a group with respect to multiplication, the ring is called a *skew field*. If, further, multiplication is commutative, then it is called a *field*. (However, some authors use the term "field" instead of "skew field" and then speak of commutative fields; also the term "noncommutative field" is used instead of "skew field.") A hypercomplex system that is a skew field is called a *division algebra*. If a skew field has only a finite number of elements, multiplication is necessarily commutative. The number of elements in a finite field is a power of a prime number p^n. The field is then denoted by $GF(p^n)$, where GF stands for *Galois field*. Apart from isomorphisms there is exactly one field with p^n elements. Even if the field is infinite, every element can have a finite order with respect to the additive group. This order is then a prime number.

References

Algebra, Theory of Equations:

1. Albert, A. A.: "Modern Higher Algebra," University of Chicago Press, Chicago, 1937.
2. Birkhoff, G., and S. MacLane: "A Survey of Modern Algebra," Macmillan, New York, 1965.
3. Bôcher, M.: "Introduction to Higher Algebra," Macmillan, New York, 1907; Dover, New York, 1964.
4. Burnside, W. S., and A. W. Panton: "Theory of Equations," 7th ed., Longmans, London, 1912; Dover, New York, 1960.
5. Chrystal, G.: "Algebra, an Elementary Textbook," 2 vols., Chelsea Publishing Company, New York, 1952.
6. Dickson, L. E.: "Algebras and Their Arithmetics," University of Chicago Press, Chicago, 1923.
7. Herstein, I. N.: "Topics in Algebra," Blaisdell Publishing Co., New York, 1964.
8. Jacobson, N.: "Lectures in Abstract Algebra," 2 vols., Van Nostrand, Princeton, N.J., 1953, 1955.
9. Sawyer, W. W.: "A Concise Approach to Abstract Algebra," Freeman, San Francisco, 1959.
10. Van der Waerden, B. L.: "Modern Algebra," 2 vols., Springer, Berlin, 1931; Ungar, New York, 1943.
11. Weber, H.: "Lehrbuch der Algebra," 3 vols., Chelsea Publishing Company, New York.

Linear Algebra and Matrices:

12. Amir-Moez, A. R.: Extreme Properties of Eigenvalues of a Hermitian Transformation and Singular Values of the Sum and Product of Linear Transformations, *Duke Math.*, **23**: 463–476 (1956).
13. Bellman, R.: "Introduction to Matrix Analysis," McGraw-Hill, New York, 1960.
14. Browne, E. T.: The Characteristic Equation of a Matrix, *Bull. Am. Math. Soc.*, **34**: 363–368 (1928).
15. Dickson, L. E.: "Algebraic Theories," Dover, New York, 1959.
16. Faddeev, D. K., and V. N. Faddeeva: "Computational Methods of Linear Algebra," Freeman, San Francisco, 1963.
17. Frazer, R. A., W. J. Duncan, and A. R. Collar: "Elementary Matrices and Some Applications to Dynamics and Differential Equations," Macmillan, New York, 1947.
18. Gantmacher, F. R.: "The Theory of Matrices," 2 vols., Chelsea Publishing Company, New York, 1959.
19. Greub, W. H.: "Linear Algebra," Academic, New York, 1963.
20. Halmos, P. R.: "Finite-dimensional Vector Spaces," 2d ed., Van Nostrand, Princeton, N.J., 1958.
21. Hausdorff, F.: Die Symbolische Exponentialformel in der Gruppentheorie, *Ber. Sächs. Akad. Wiss., Leipzig*, **58**: 19–48 (1906).
22. Higman, B.: "Applied Group-theoretic and Matrix Methods," Oxford University Press, Fair Lawn, N.J., 1955; Dover, New York, 1964.
23. Lidskii, V. C.: The Proper Values of the Sum and Products of Symmetric Matrices, *Dokl. Akad. Nauk, SSSR*, **75**: 769–772 (1950).
24. MacDuffee, C. C.: The Theory of Matrices, "Ergebnisse der Mathematik und ihrer Grenzgebiete," vol. 2, Springer, Berlin, 1933.

25. MacDuffee, C. C.: Vectors and Matrices, *Carus Math. Monograph* 7, Open Court, LaSalle, Ill., 1949.
26. Mal'cev, A. I.: "Foundations of Linear Algebra," Freeman, San Francisco, 1963.
27. Mirsky, L.: "An Introduction to Linear Algebra," Oxford University Press, Fair Lawn, N.J., 1955.
28. Muir, T.: "The Theory of Determinants in the Historical Order of Development," 4 vols. bound as 2, Dover, New York, 1960.
29. Ostrowski, A.: Uber die Determinanten mit Überwiegender Hauptdiagonale, *Comment. Math. Helv.*, **10**: 69–96 (1937).
30. Schwerdtfeger, H.: "Introduction to Linear Algebra and the Theory of Matrices," Erven P. Noordhoff, NV, Groningen, Netherlands, 1950.
31. Shilov, G. E.: "An Introduction to the Theory of Linear Spaces," Prentice-Hall, Englewood Cliffs, N.J., 1961.
32. Silverman, R. A. (ed).: "Academician V. I. Smirnov's Linear Algebra and Group Theory, McGraw-Hill, New York, 1961.
33. Stoll, R. R.: "Linear Algebra and Matrix Theory," McGraw-Hill, New York, 1952.
34. Taussky, O.: Some Topics Concerning Bounds for Eigen Values of Finite Matrices, in J. Todd (ed.), "Survey of Numerical Analysis," chap. 8, McGraw-Hill, New York, 1962.
35. Taussky, O., and J. Todd: Systems of Equations, Matrices and Determinants, in "The Tree of Mathematics," pp. 305–337, The Digest Press, Pacoima, Calif., 1957.
36. Turnbull, H. W., and A. C. Aitken: "Theory of Canonical Matrices," Blackie, Glasgow, 1932; Dover, New York, 1961.
37. Turnbull, H. W.: "The Theory of Determinants, Matrices and Invariants," Dover, New York, 1960.
38. Wedderburn, J. H. M.: "Lectures on Matrices," American Mathematical Society Colloquium Publications, XVII, 1934.
39. Weyl, H.: Das asymptotische Verteilungsgesetz der Eigenwerte linearer partieller Differentialgleichungen, *Math. Ann.*, **71**: 441–479 (1912).
40. Wielandt, H.: An Extremum Property of Sums of Eigen Values, *Proc. Am. Math. Soc.*, **6**: 106–110 (1955).

Hypercomplex Systems, Theory of Groups:

41. Boerner, H.: "Representations of Groups," North Holland Publishing Company, Amsterdam, 1963.
42. Burckhardt, J. J.: "Die Bewegungsgruppen der Kristallographie," Birkhäuser Verlag, 1966.
43. Burnside, W.: "Theory of Groups of Finite Order," Cambridge University Press, New York, 1911.
44. Carmichael, R. D.: "Introduction to the Theory of Groups of Finite Order," Ginn, Boston, 1937.
45. Curtis, C. W., and I. Reiner: "Representation Theory of Finite Groups and Associative Algebras," Wiley, New York, 1962.
46. Dade, E. C.: The Maximal Finite Groups of 4 × 4 Integral Matrices, *Illinois J. Math.*, **9**: 99–122 (1965).
47. Dickson, L. E.: "Linear Groups with an Exposition of the Galois Field Theory," Dover, New York, 1958.
48. Hall, M., Jr.: "The Theory of Groups," Macmillan, New York, 1959.
49. Hamermesh, M.: "Group Theory and Its Applications to Physical Problems," Addison-Wesley, Reading, Mass., 1962.
50. Heine, V.: "Group Theory in Quantum Mechanics," Pergamon Press, New York, 1960.
51. Jacobson, N.: "Lie Algebras," Interscience, New York, 1962.
52. Jordan, P., J. von Neumann, and E. Wigner: On an Algebraic Generalization of the Quantum Mechanical Formalism, *Ann. Math.*, **35**: 29–64 (1934).
53. Kurosh, A. G.: "The Theory of Groups," 2 vols., Chelsea Publishing Company, New York, 1960.
54. Littlewood, D. E.: "The Theory of Group Characters and Matrix Representations of Groups," Oxford University Press, Fair Lawn, N.J., 1950.
55. Lomont, J. S.: "Applications of Finite Groups," Academic, New York, 1959.
56. Lyubarskii, G. Ya.: "Group Theory and Its Applications to Physics," Pergamon Press, New York, 1960.
57. Mathews, J., and R. L. Walker: "Mathematical Methods of Physics," Benjamin, New York, 1964.
58. Miller, G. A., H. F. Blichfeldt, and L. E. Dickson: "Theory and Applications of Finite Groups," Dover, New York, 1961.
59. Murnaghan, F. D.: "The Theory of Group Representations," Dover, New York, 1963.
60. Pontrjagin, L.: "Topological Groups," Princeton University Press, Princeton, N.J., 1959.
61. Racah, G.: Lectures on Lie Groups, Gordon and Breach, Science Publishers, Inc. New York, 1964.
62. Scott, W. R.: "Group Theory," Prentice-Hall, Englewood Cliffs, N.J., 1964.
63. Speiser, A.: "Die Theorie der Gruppen von endlicher Ordnung," Dover, New York, 1954.
64. Van der Waerden, B. L.: "Gruppen von linearen Transformationen," Chelsea Publishing Company, New York, 1948.
65. Van der Waerden, B. L.: "Die Gruppentheoretische Methode in der Quantenmechanik," Springer, Berlin, 1932.
66. Weyl, H.: "Gruppentheorie und Quantenmechanik," Hirzel, Leipzig, 1928.
67. Wigner, E.: "Group Theory and Its Application to the Quantum Mechanics of Atomic Spectra," Academic, New York, 1959.
68. Zassenhaus, H.: On an Algorithm for the Determination of Space Groups, *Comment. Math. Helv.*, **21**: 117 (1948).
69. Zassenhaus, H.: "The Theory of Groups," Chelsea Publishing Company, New York, 1958.

Chapter 3

Analysis

By JOHN TODD, California Institute of Technology

1. Real Numbers, Limits

Real Numbers, Convergence of Sequences. From the non-negative integers 0,1,2, . . . we first generate the negative integers, -1, -2, . . . and then the fractions or rational numbers, $\pm p/q$ (p,q integers, $q \neq 0$) (see Part 1, Chap. 1, Sec. 2). We may then generate the real numbers as limits of convergent sequences of rational numbers. We then prove that the limit of a convergent sequence of real numbers is a real number, so that the real numbers have a completeness property which is not possessed by the rational numbers. For example, the sequence

$$1,\ 14/10,\ 141/100,\ 1{,}414/1{,}000,\ \ldots$$

converging to the real number $\sqrt{2}$, does not have a rational limit.

The set of real numbers (or points) x such that $a \leq x \leq b$ is called a closed interval and is denoted by $[a,b]$. An open interval is the set of real numbers (or points) x such that $a < x < b$; it is denoted by (a,b). A neighborhood of a point a is an open interval containing it: $(a - \epsilon, a + \eta)$ is a neighborhood of a for any $\epsilon > 0$, $\eta > 0$. We shall speak simply of intervals when the question of their closure or openness is not relevant. The sequence $\{x_n\}$ converges to x if any neighborhood of x, no matter how small, contains all but a finite number of terms of the sequence. Arithmetically this means that corresponding to any $\epsilon > 0$ (no matter how small) there is an integer n_0 (depending on ϵ) such that if $n \geq n_0$ then $|x_n - x| < \epsilon$. We denote this situation by the symbols $x_n \to x$ or by $\lim_{n\to\infty} x_n = x$ or simply by $\lim x_n = x$ if there is no doubt about the current variable. Examples of convergent sequences, with their limits follow. $\lim a^n = 0$ if $|a| < 1$; $\lim na^n = 0$, if $|a| < 1$; $\lim (1 + n^{-1})^n = e = 2.71828$ . . . .

A sequence $\{x_n\}$ is said to be *bounded* if there is a number B such that, for all n, $|x_n| \leq B$. A bounded sequence always contains a convergent sub-sequence, i.e., there are numbers $n_1 < n_2 < n_3 < \cdots$ such that

$$\lim_{i\to\infty} x_{n_i}$$

exists. This is the *Bolzano-Weierstrass theorem*. For instance, the sequence $\{x_n\}$, where $x_n = (-1)^n$, is bounded but is not convergent; the sub-sequence $\{x_{2n}\}$ is convergent to 1.

A sequence is called increasing if $x_1 \leq x_2 \leq \cdots$ and decreasing if $x_1 \geq x_2 \geq \cdots$; the term *monotone* is used to cover both these cases. The following result is fundamental: a monotone sequence is convergent if and only if it is bounded.

This is proved by use of the *Dedekind section theorem:* if all the real numbers are divided into two classes L,R such that any member of L is less than any member of R, and if each class L,R contains at least one member, then there is a real number ξ such that if $x < \xi$, then x belongs to L; and if $x > \xi$, then x belongs to R. Other applications of this theorem establish the existence of exact bounds of a bounded sequence $\{x_n\}$, the least upper bound, $\overline{\text{bound}}\, x_n$, and the greatest lower bound, $\underline{\text{bound}}\, x_n$.

If a sequence $\{x_n\}$ is such that to every A, however large, there corresponds an $n_0 = n_0(A)$ such that if $n \geq n_0$ then $x_n \geq A$, we say that the sequence tends to plus infinity and write $x_n \to +\infty$ or $\lim x_n = +\infty$. Similarly we define $x_n \to -\infty$, $\lim x_n = -\infty$. If $x_n = n$, then $x_n \to +\infty$.

The following results, which cannot be extended to the infinite case in general, are valid when the limits (on the right) are finite:

$$\begin{aligned} \lim\,(ax_n + by_n) &= a \lim x_n + b \lim y_n \\ \lim\,(x_n y_n) &= \lim x_n \lim y_n \\ \lim x_n^{-1} &= (\lim x_n)^{-1} \qquad \text{if } \lim x_n \neq 0 \end{aligned}$$

A sequence need have no limit, finite or infinite. Let us confine our attention to bounded sequences—for example, $\{x_n\}$ where $x_n = (-1)^n$. Here there are sub-sequences, $\{x_{2n}\}$ for instance, convergent to $+1$ and sub-sequences $\{x_{2n-1}\}$ convergent to -1. No convergent sub-sequence has a limit greater than $+1$ or less than -1. This holds in general: There is (1) a number L, denoted by $\overline{\lim}\, x_n$, such that, for all $\epsilon > 0$, there is an infinite number of the terms of the sequence greater than $L - \epsilon$ and only a finite number greater than $L + \epsilon$ and (2) a number l, denoted by $\underline{\lim}\, x_n$, such that, for all $\epsilon > 0$, there is an infinite number of terms of the sequence less than $l + \epsilon$ and only a finite number less than $l - \epsilon$. A necessary and sufficient condition for the convergence of the sequence is the equality of L and l: if $L = l$, then $\lim x_n$ exists and has this value and, conversely, if $\lim x_n$ exists, then $L = l = \lim x_n$. The numbers $\overline{\lim}\, x_n$ and $\underline{\lim}\, x_n$ are called the upper and lower limits of the sequence; they are to be carefully distinguished from the upper and lower bounds of the sequence.

All rational numbers between 0 and 1 can be arranged in the form of a sequence. They can be ordered as follows

$$0/1;\ 1/1;\ 1/2;\ 1/3;\ 1/4,2/3;\ 1/5;\ 1/6,2/5,3/4;\ 1/7,3/5;\ 1/8,2/7,4/5;\ \cdots$$

In the nth group we include numbers p/q (in their lowest terms) such that $p + q = n$ and we arrange these in order of magnitude. This group is called the Farey series of order n. This sequence has $l = 0$, $L = 1$ and there are sub-sequences in it which converge to any *real* number between 0,1. Cantor has proved that it is not possible to arrange *all* the real numbers between 0,1 in the form of a sequence.

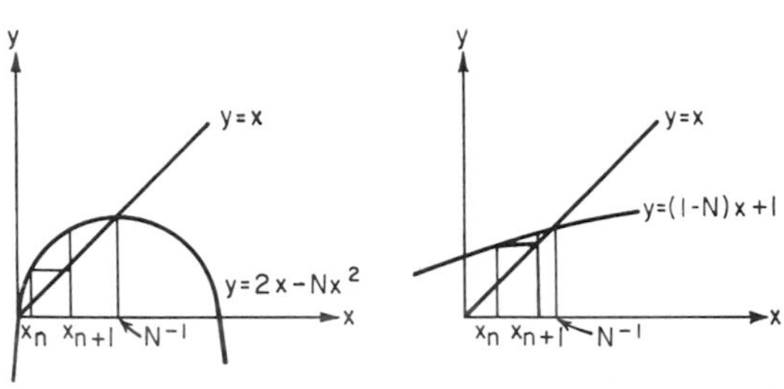

FIG. 3.1

The convergence properties of the following sequences are of interest. The geometrical interpretation is indicated in Fig. 3.1. The first sequence is

$$x_{n+1} = x_n(2 - Nx_n) \tag{3.1}$$

If this sequence has a limit, it is zero or N^{-1}. Suppose $0 < x_0 < N^{-1}$, so that $N^{-1} - x_0 > 0$. Then, since $x_{n+1} - x_n = Nx_n(N^{-1} - x_n)$, we have $x_{n+1} > x_n$. On the other hand

$$x_{n+1} - N^{-1} = -N\{x_n - N^{-1}\}^2 < 0$$

that is, $$x_{n+1} < N^{-1}$$

Thus $\{x_n\}$ is a bounded increasing sequence: it must have a limit and this can only be N^{-1}. Let δ_n denote the error $x_n - N^{-1}$. We have $\delta_{n+1} = -N\delta_n^2$. Convergence of this type is called quadratic; roughly speaking, this implies that if x_n and N^{-1} coincide to a certain number of decimals, x_{n+1} and N^{-1} will coincide to about twice that number. As an example take $N = \frac{1}{2}$ and $x_0 = 1$. Then the sequence is 1, 1.5, 1.875, 1.9921875, ...

The second sequence is

$$x_{n+1} = (1 - N)x_n + 1 \tag{3.2}$$

Here we have linear convergence to N^{-1}:

$$\eta_{n+1} = (x_{n+1} - N^{-1}) = (1 - N)(x_n - N^{-1}) = (1 - N)\eta_n$$

With the same $x_0 = 1$ we obtain the sequence 1,1.5, 1.75,1.875, ...

Infinite Series. An infinite series

$$\sum_{n=1}^{\infty} x_n = x_1 + x_2 + \cdots + x_n + \cdots$$

is said to be *convergent* if the sequence $\{S_n\}$ of its partial sums

$$S_n = x_1 + x_2 + \cdots + x_n$$

is convergent. If $\lim_{n\to\infty} S_n = S$, we write

$$\sum_{n=1}^{\infty} x_n = S$$

and call S the sum of the series.

The geometric series $a + ar + ar^2 + \cdots$ is convergent if $|r| < 1$. An arithmetic series

$$a + (a + d) + (a + 2d) + \cdots$$

can only be convergent in the trivial case $a = 0 = d$.

The terms of a convergent series themselves form a sequence which converges to zero; such a sequence is called a *null sequence.* This condition, while necessary for convergence, is not sufficient, as is shown by the case of the series $1 + 1/2 + 1/3 + \cdots + 1/n + \cdots$ for which the partial sums are unbounded. A series which does not converge is called *divergent.* Several types of divergence are possible, as indicated by the following examples: $1 + 1 + 1 + 1 + \cdots$; $1 - 1 + 1 - 1 + \cdots$; $1 - 2 + 3 - 4 + \cdots$.

The theory of sequences and the theory of series are coextensive: given a series Σx_n we consider the sequence of its partial sums while, on the other hand, given any sequence S_n we can construct a series Σx_i, which has $\{S_n\}$ for its partial sums by putting $x_i = S_i - S_{i-1}$, $i > 1$, $S_1 = x_1$.

Efforts have been made, since the time of Euler, to assign conventional sums to series divergent in the sense just defined. This theory is of considerable interest mathematically and has occasional physical significance. Two examples follow. The series $1 - x + x^2 + \cdots$ has sum $(1 + x)^{-1}$ for $|x| < 1$; it is reasonable to consider assigning the sum

$$(1 + 1)^{-1} = \tfrac{1}{2}$$

to this series when $x = 1$. This is the basis of the Abel or Poisson summation method. An alternative approach is the method of arithmetic means due to Cesàro. We associate with the series Σx_i having partial sums S_n the value

$$\lim_{n\to\infty} \frac{S_1 + S_2 + \cdots + S_n}{n}$$

if this exists. In the case of $1 - 1 + 1 - 1 + \cdots$ the partial sums are 1,0,1,0,1, . . . and we reach the same value $\frac{1}{2}$. Both these methods are *consistent*, i.e., when applied to convergent series they produce the sum in the ordinary sense.

The behavior of series all (but a finite number) of whose terms are of one sign is particularly simple. We begin with this case, then consider alternating series, and conclude with series of real terms whose signs are arbitrary.

Series with Positive Terms. A series of positive terms must either converge or diverge to $+\infty$. This follows from the remark on monotone sequences. To prove that such a series converges, it is sufficient to show that its partial sums are bounded above. Thus since $n! \geq 2^{n-1}$ for $n = 1, 2, \ldots$ it follows that the partial sums of $\Sigma(n!)^{-1}$ do not exceed those of $\Sigma 2^{-n}$; since those of the latter series are obviously bounded by 2, so are the former. It follows that $\Sigma(n!)^{-1}$ is convergent. This criterion is not always convenient. Many other practically useful tests for the convergence or divergence of series have been devised.

Cauchy's Test. Σx_n is convergent if $\lim (x_n)^{1/n} < 1$ and divergent if $\lim (x_n)^{1/n} > 1$.

D'Alembert's Test. Σx_n is convergent if $\lim (x_n/x_{n+1}) > 1$, and divergent if $\lim (x_n/x_{n+1}) < 1$.

When the limits are actually unity, more refined tests are necessary. Convergence follows from $\overline{\lim}\ (x_n)^{1/n} < 1$; similar extensions are available for the other cases.

It is often more convenient to operate with integrals than with sums. For this reason the following test is useful. For definitions of infinite integrals, see Sec. 3.

Integral Test. If $\phi(x) \geq 0$, $x \geq 1$ then the series $\Sigma\phi(n)$ and

$$\int_1^\infty \phi(x)\, dx$$

converge or diverge together. For instance, since

$$\int_1^\infty t^{-\alpha}\, dt$$

is convergent for $\alpha > 1$ and divergent for $\alpha \leq 1$, it follows that $\Sigma n^{-\alpha}$ is convergent for $\alpha > 1$ and divergent for $\alpha \leq 1$. Even in the divergent case the difference between the partial sums of the series and the corresponding definite integral has a finite limit, for example, in the case $\alpha = 1$ using

$$\int_1^n t^{-1}\, dt = \ln n$$

we can infer the existence of

$$\lim \left(1 + \frac{1}{2} + \cdots + \frac{1}{n} - \ln n\right)$$

This limit is *Euler's constant* $\gamma = 0.5772 \ldots$.

Alternating Series. A series $x_1 - x_2 + x_3 - \cdots$ where $x_1 \geq x_2 \geq \cdots \geq 0$ *and* $x_n \to 0$ is always convergent. This is established by first showing that the sequences of even partial sums and odd partial sums are each monotone and bounded and therefore convergent. Using the hypothesis $x_n \to 0$, it can be concluded that the limits of these two sequences are the same and that the series itself is therefore convergent. For instance, $1 - 1/2 + 1/3 - 1/4 - \cdots$ is convergent (to the sum ln 2).

Series with Arbitrary Real Terms. If $\Sigma|x_n|$ is convergent so is Σx_n, but the converse of this is not true, as shown by the example of $\Sigma(-1)^n n^{-1}$. Series Σx_n such that $\Sigma|x_n|$ is convergent are called *unconditionally* or *absolutely* convergent and have a particularly simple behavior.

For instance, no matter how the terms are rearranged the series still converges to the same sum. This is not so even for $\Sigma(-1)^n n^{-1}$, for it can be rearranged in the form

$$1 - \frac{1}{2} - \frac{1}{4} + \frac{1}{3} - \frac{1}{6} - \frac{1}{8} + \frac{1}{5} - \frac{1}{10} - \frac{1}{12} + \cdots + \frac{1}{2n-1} - \frac{1}{4n-2} - \frac{1}{4n} + \cdots$$

which is convergent to the sum ½ log 2. Any convergent series which is not absolutely convergent can be rearranged in such a way that the resulting series has any prescribed behavior, for example, converges to any assigned sum or diverges to $+\infty$ or to $-\infty$, or oscillates between any assigned limits (possibly infinite).

Convergent series can be added and subtracted term by term. In fact for any α, β, if $\Sigma a_n = A$, $\Sigma b_n = B$, then the series Σc_n, where $c_n = \alpha a_n + \beta b_n$, is convergent to the sum $\alpha A + \beta B$.

The question of *multiplication* of two series is more complicated. The product of two series Σa_n, Σb_n can be arranged in the form Σc_n, where

$$c_n = a_0 b_n + a_1 b_{n-1} + \cdots + a_{n-1} b_1 + a_n b_0$$

The series Σc_n is called the *Cauchy product*, and its terms are obtained by taking the sums of successive diagonals of the doubly infinite array of terms $a_i b_j$. If at least one of two convergent series Σa_n, Σb_n is absolutely convergent, then the Cauchy product is convergent and has the *correct sum*.

Other Infinite Algorithms. In addition to representation of numbers (or functions) as the limits of sequences or as infinite series, representations as infinite products and occasionally as infinite continued fractions appear.

We say that the infinite product

$$\prod_{n=1}^\infty (1 + a_n)$$

is convergent if $a_n \neq -1$ and if

$$\lim_{m\to\infty} \prod_{n=1}^m (1 + a_n)$$

exists and is different from zero (this last condition is introduced for convenience). When $a_n \geq 0$, then $\Pi(1 + a_n)$ and Σa_n converge or diverge together. The product $\Pi(1 + a_n)$ is said to be absolutely convergent if $\Pi(1 + |a_n|)$ is convergent; in which case the order of the factors can be changed in any fashion.

Examples of infinite products follow.

$$\sin z = \pi z \prod_{n=1}^\infty \left(1 - \frac{z^2}{n^2}\right)$$

$$\frac{1}{\Gamma(z)} = ze^\gamma \prod_{n=1}^\infty \left(1 + \frac{z}{n}\right) e^{-z/n}$$

A simple infinite continued fraction is of the form

$$a_1 + \frac{1}{a_2 +}\frac{1}{a_3 +} \cdots \frac{1}{a_n +} \cdots$$

If the a_i are positive integers, this is always convergent in the sense of the existence of

$$\lim_{n\to\infty}\left(a_1+\frac{1}{a_2+}\cdots+\frac{1}{a_{n-1}+}\frac{1}{a_n}\right)$$

Periodic continued fractions such as

$$1+\frac{1}{1+}\frac{1}{2+}\frac{1}{1+}\frac{1}{2+}\frac{1}{1+}\cdots$$

represent quadratic surds, in this case $\sqrt{3}$, and conversely. Elaborate theories have been developed. One important application is in the representation of the characteristic values of Mathieu functions (Sec. 11). Another nontrivial example is

$$\frac{\tan x}{x}=\frac{1}{1-}\frac{x^2}{3-}\frac{x^2}{5-}\frac{x^2}{7-}\cdots$$

2. Real Functions

Definitions and Examples. A law which defines a correspondence between the individuals of two given sets of numbers is called a *function.* Denote a typical number of one of the sets by x, the whole set by $X = \{x\}$, the numbers that correspond to x in the second set Y by y. If to every x, there corresponds exactly one value of y, then y is called a *single-valued function of* x, frequently denoted by $y = f(x)$. The function maps X on Y. A many-valued function of x is one which assumes more than one value for one value of x. The set of all values of x for which y is defined is called the *domain* (of definition) of the function. Since x stands for any element in the set $X = \{x\}$, it is called a *variable.* So is y, which stands for any element in the range (of values) of f: x is called the *independent* variable (or argument) and y the *dependent* variable. Analogously functions $f(x_1,x_2, \ldots ,x_n)$ *of several independent* variables are defined.

A function may be given through a mathematical formula or through a table. If the mathematical relation which defines the function is of the form $y = f(x)$, the function is said to be given *explicitly.* If the relation is defined by an expression of the form $f(x,y) = 0$, then the function is said to be given *implicitly.*

If y is a function of x, then x is some function of y—sometimes denoted by $x = f^{-1}(y)$, the so-called inverse function with respect to the original one. If $x = F(y)$ is inverse to $y = \phi(x)$, then $y = \phi(x)$ is inverse to $x = F(y)$. An inverse function need not be unique; for example, the inverse of $y = \tan x$, which is denoted by $x = \arctan y$, is infinitely many-valued. Given any particular determination of arctan y, say the so-called principal one denoted by Arctan y which satisfies $-\frac{1}{2}\pi < \text{Arctan } y \leq \frac{1}{2}\pi$, then $n\pi + \text{Arctan } y$ is also an inverse function of $\tan x$.

Some functions satisfy *functional equations,* for example, the function $f(x) = ax$, where a is a constant, satisfies the functional equation:

$$f(x+y) = f(x) + f(y)$$

The function $f(x) = a^x$ satisfies $f(x)f(y) = f(x + y)$. If $f(x + y)$ can be expressed as a rational function of $f(x)$ and $f(y)$, the function is said to have an *addition theorem,* for example,

$$\tan(x+y) = \frac{\tan x + \tan y}{1 - \tan x \tan y}$$

Functions are often represented graphically. For this purpose various coordinate systems can be used. The most usual are the rectangular cartesian. The graph of the inverse function is then obtained by interchanging the x and y coordinate axes.

Limits of Functions of a Continuous Variable. We have already considered the limits of functions of a positive integral variable, n. Similar definitions and theorems are available for the limits of functions of a real variable, x. For example,

$$\lim_{x\to 0}\frac{\sin x}{x} = 1 \qquad \lim_{x\to 1}(x-1)^{-2} = +\infty$$

$$\lim_{x\to+\infty} x^n e^{-x} = 0 \qquad \text{for any } n$$

In certain cases we must specify in what way the argument approaches its limit: for example, x^{-1} has no limit as $x \to 0$ unrestrictedly, but if it approaches x through positive values it has a limit $+\infty$, while if x approaches through negative values it has a limit $-\infty$. The following notation is used to indicate these situations:

$$\lim_{x\to 0+} x^{-1} = +\infty \qquad \lim_{x\to 0-} x^{-1} = -\infty$$

An important case is that of infinite integrals discussed in Sec. 3.

The following notations are convenient. We shall say that $f(x)$ is of smaller order than $\phi(x)$, for example, as $x \to \infty$, if

$$\lim \frac{f(x)}{\phi(x)} = 0$$

In these circumstances we write

$$f = o(\phi) \qquad x \to \infty$$

For instance, $x^2 = o(e^x)$, $x \to \infty$ and $x^{-2} = o(1)$, $x \to \infty$. We also write $\{x/(x - 1)\} = 1 + o(1)$, $x \to \infty$.

We shall say that $f(x)$ is of the same order as $\phi(x)$, for example, as $x \to \infty$, if the ratio $f(x)/\phi(x)$ is bounded as $x \to \infty$; we denote this by $f = O(\phi)$, $x \to \infty$. Thus we write $[x/(x - 1)] = O(1)$ as $x \to \infty$, $\sin x = O(x)$ as $x \to 0$.

The standard function ϕ with which we compare f is usually a monotone function tending to ∞, or to 0.

The notation is also applicable to functions of a positive integral variable:

$$1 + \frac{1}{2} + \frac{1}{3} + \cdots + \frac{1}{n} = O(\log n) \qquad n \to \infty$$

or to functions of a complex variable.

Continuous Functions. Let $f(x)$ be defined in an interval. It is continuous at a point c in this interval if whenever $x_n \to c$ then $f(x_n) \to f(c)$, or, expressed arithmetically, when corresponding to any $\epsilon > 0$, no matter how small, there is a $\delta = \delta(\epsilon) > 0$ such that when $|x - c| < \delta$ then $|f(x) - f(c)| < \epsilon$. If a function is continuous at all points of a bounded closed interval, it is bounded in that interval and it is also uniformly continuous there, that is, we can choose one δ, *independent of* x, such that if $|\xi - x| < \delta$ then $|f(\xi) - f(x)| < \epsilon$ for all x, ξ in the interval. It is easy to see that these properties do not necessarily hold in an interval which is not closed, for example, x^{-1} in the interval $0 < x \leq 1$, or in an infinite interval.

As examples of points where a function is not continuous we mention the origin in the case of the functions $f(x)$ and $g(x)$ defined by

$$f(x) = \frac{|x|}{x} \qquad x \neq 0, f(0) = 0$$

and

$$g(x) = \sin x^{-1} \qquad x \neq 0, g(0) = 0$$

This concept of continuity is far from being as strong as the intuitive idea of a function whose graph can be drawn; there are, indeed, continuous functions which are nowhere differentiable. Weierstrass showed that the function

$$f(x) = \sum_{0}^{\infty} 2^{-n} \cos \{(13)^n \pi x\}$$

is everywhere continuous but nowhere differentiable.

The theorems of Sec. 1 above about limits of sums, products, and quotients of sequences give rise to similar results about continuous functions. Using them, and the obvious fact that $f(x) = x$ is a continuous function of x, we can deduce that x^r is continuous for any integral value of n, that any polynomial $a_0x^r + a_1x^{r-1} + \cdots + a_{r-1}x + a_r$ is continuous, and that any rational function (i.e., quotient of two polynomials) is continuous except at points where the denominator vanishes.

The following theorem asserts the existence of an inverse function in certain circumstances: If $F(y)$ is a function of y, continuous and strictly increasing in the neighborhood of $y = b$, and $F(b) = a$, then there is a unique continuous function $y = \phi(x)$, such that $\phi(a) = b$, which satisfies $F(y) = x$ identically in the neighborhood of $x = a$. By strictly increasing in the neighborhood of $y = b$, we mean that there is a $\delta > 0$ such that if $b - \delta < y_1 < y_2 < b + \delta$ then $F(y_1) > F(y_2)$. It is important to note this merely asserts the existence of an inverse function locally. See also the special case of a power series later in this section.

If a function is not defined for all points in an interval (e.g., if its values are only given by a table), it is of interest to find a continuous function, preferably a polynomial, which coincides with the function wherever it is known. The construction of such a function is called (polynomial) *interpolation* (see Part 1, Chap. 7).

Differentiation. Suppose $f(x)$ defined in the neighborhood of $x = x_0$. If the limit

$$\lim_{h \to 0} \frac{f(x_0 + h) - f(x_0)}{h}$$

exists, it is called the derivative or differential coefficient of $f(x)$ at x_0; the limit is variously denoted by $f'(x_0)$, $(df/dx)_{x_0}$, $Df(x_0)$. Unless otherwise stated, it is understood that the variable h is unrestricted (except by $h \neq 0$). In special cases one can distinguish between right and left derivatives, for example, $f(x) = |x|$, where $f'_+(0) = 1$, $f'_-(0) = -1$.

A necessary condition for the existence of a derivative at a point is the continuity of the function at that point, but this is not sufficient. For instance, $f(x) = |x|$ is continuous but not differentiable at $x = 0$. See also the Weierstrass example given above.

If $f(x)$ is differentiable at all points in a neighborhood of x_0, we can ask whether $f'(x)$ is differentiable at x_0. If it is, the derivative of $f'(x)$ at x_0 is called the second derivative of $f(x)$ at x_0 and denoted by $f''(x)$, $(d^2f/dx^2)_{x_0}$, $D^2f(x_0)$. In the same way higher derivatives can be defined. The nth derivative will be denoted by $f^{(n)}$ or D^nf. When $n = 0$, these are interpreted as f itself. It is not difficult to construct functions (of a real variable) which are differentiable n times at a point x_0 but which cannot be differentiated $n + 1$ times.

The following general results are available, where f and g are differentiable functions of x and α, β are constants:

$$(\alpha f + \beta g)' = \alpha f' + \beta g'$$
$$(fg)' = f'g + fg'$$

If f is a differentiable function of g and g is a differentiable function of x,

$$\frac{d}{dx} f(g(x)) = \frac{df(g)}{dg} \frac{dg(x)}{dx} \tag{3.3}$$

For instance, let $f = \sin g$ and $g = x^2$; then

$$\frac{d}{dx} \sin x^2 = \cos x^2 \cdot 2x$$

From the rule for differentiating a product, using induction, we find the Leibniz formula

$$\frac{d^n}{dx^n} [fg] = \sum_{r=0}^{n} \binom{n}{r} f^{(r)} g^{(n-r)}$$

a result which is especially convenient if f (or g) is a polynomial.

The evaluation of differential coefficients is an exercise in evaluation of limits. The proof that, when n is a positive integer,

$$(x^n)' = nx^{n-1} \tag{3.4}$$

depends on the identity

$$(x^n - a^n) = (x - a)(x^{n-1} + x^{n-2}a + \cdots + a^{n-1})$$

and the continuity of x^r. The relation (3.4) is true for all values of n. From (3.3) and (3.4) for all n,

$$\frac{d}{dx} [f(x)]^n{}_{x=x_0} = n[f(x_0)]^{n-1} f'(x_0)$$

provided that f is differentiable at x_0 and that $f(x_0) \neq 0$ when $n < 1$.

The proof that $(\sin x)' = \cos x$, or

$$(\sinh x)' = \cosh x$$

depends on the addition theorem for $\sin x$, or $\sinh x$,

and the existence of

$$\lim_{x \to 0} \frac{\sin x}{x} = 1 \qquad \text{or} \qquad \lim_{x \to 0} \frac{\sinh x}{x} = 1$$

Again, the fact that $(e^x)' = e^x$ (from which follows $(a^x)' = \log a \cdot a^x$) depends on the existence of $\lim_{h \to 0} (e^h - 1)/h = 1$. As a last example we note that $(\log x)' = x^{-1}$.

Geometrical Interpretations. The geometrical interpretation of the existence of $df(x)/dx$ at x_0 is that of the existence of a tangent to the graph of $f(x)$ at $(x_0, f(x_0))$ and the slope (or gradient) of this tangent is the value of the derivative. The equation of the tangent is in fact

$$y - y_0 = (x - x_0)f'(x_0)$$

The slope of the chord PQ of the graph (Fig. 3.2) of $y = f(x)$ is $h^{-1}[f(x_0 + h) - f(x_0)]$ and on the assumption that $f(x)$ is differentiable at all points between x_0 and $x_0 + h$ and continuous at x_0 and $x_0 + h$, there is a point ξ, $x_0 \leq \xi \leq x_0 + h$ such that the tangent RS at $R[\xi, f(\xi)]$ is parallel to the chord PQ. Arithmetically, this is expressed by

$$f(x_0 + h) - f(x_0) = hf'(\xi)$$

a result known as the *first mean-value theorem.*

If $f'(x_0) \geq 0$, the function is increasing at x_0, while if $f'(x_0) \leq 0$, $f(x)$ is decreasing at x_0. As in the case of sequences, the adjective monotone is used to cover both cases. If $f'(x_0) = 0$, there will be a maximum if $f''(x_0) < 0$ and a minimum if $f''(x_0) > 0$. If $f''(x_0) = 0$ and if $f''(x)$ changes sign as the curve passes through x_0, the curve is said to have an inflexion at x_0, as in the case of $y = x^3$ at the origin.

Differential Geometry. Differentiation provides a new tool for the study of curves, e.g., the angle between two curves at a point of intersection is defined as the angle between their tangents at that point. If they have a common tangent at that point, i.e., if their differential coefficients coincide, the curves are said to *touch* and the intersection point is called a *point of contact.* If the second differential coefficients also coincide, the curves have *contact of second order.* Contacts of higher order are defined similarly (Part 1, Chap. 8).

A circle which has contact of second order with a curve at a point is called the *circle of curvature.* The reciprocal of the *radius of curvature* is the *curvature* at the point. The study of differential geometry is carried out conveniently using the techniques of vector and tensor analysis (Part 1, Chaps. 9 and 10).

Taylor's Theorem. The mean-value theorem (Fig. 3.2) is a special case of Taylor's theorem. If $f(x)$ is differentiable n times in the closed interval $(x_0, x_0 + h)$, then

$$f(x_0 + h) = f(x_0) + \frac{h}{1!}f'(x_0) + \frac{h^2}{2!}f''(x_0) + \cdots + \frac{h^{n-1}}{(n-1)!}f^{(n-1)}(x_0) + \frac{h^n(1-\theta)^{n-p}}{(n-1)!p}f^{(n)}(x_0 + \theta h)$$

where θ is a number between 0 and 1. The last term is called the remainder term R_n. Specializing this expression for R_n by taking $p = n$ and $p = 1$, respectively, we get the forms of the remainder due, respectively, to Lagrange and Cauchy:

$$R_n = \frac{h^n}{n!}f^{(n)}(x_0 + \theta_1 h)$$

$$R_n = \frac{h^n(1-\theta_2)^{n-1}}{(n-1)!}f^{(n)}(x_0 + \theta_2 h)$$

where θ_1, θ_2 are numbers between 0 and 1.

A special case of Taylor's theorem is *Maclaurin's,* where $x_0 = 0$, $h = x$:

$$f(x) = f(0) + \frac{x}{1!}f'(0) + \frac{x^2}{2!}f''(0) + \cdots + \frac{x^{n-1}}{(n-1)!}f^{(n-1)}(0) + R_n$$

where $R_n = x^n f^{(n)}(\theta x)/n!$, $0 \leq \theta \leq 1$.

If $f(x)$ is indefinitely differentiable and if it can be proved that $R_n \to 0$, then (1) the series $\Sigma x^n f^{(n)}(0)/n!$ is convergent and (2) it converges to the correct sum $f(x)$. Statement (1) does not imply statement (2).

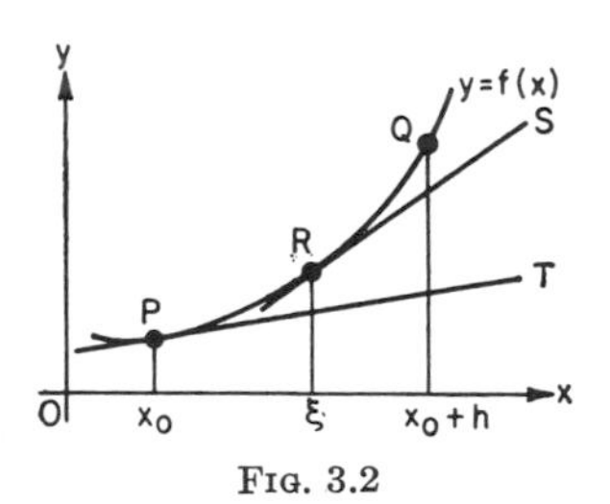

FIG. 3.2

This is shown by $f(x) = \exp(-x^{-1})$, $x \neq 0$, $f(x) = 0$, for which $f^{(n)}(0) = 0$ for all $n = 0,1,2, \ldots$ so that the series $\Sigma x^n f^n(0)/n!$ is identically zero, whereas $f(0) \neq 0$ except at $x = 0$.

Power Series for Elementary Functions

(1) $e^x = 1 + \dfrac{x}{1!} + \dfrac{x^2}{2!} + \cdots + \dfrac{x^n}{n!} + \cdots$ all x

(2) $\ln(1 + x) = x - \dfrac{x^2}{2} + \dfrac{x^3}{3} - \cdots + (-1)^{n-1}\dfrac{x^n}{n} + \cdots$ $|x| < 1$

(3) $\ln\left(\dfrac{1+x}{1-x}\right) = 2\left(x + \dfrac{x^3}{3} + \dfrac{x^5}{5} + \cdots + \dfrac{x^{2n+1}}{2n+1} + \cdots\right)$ $|x| < 1$

(4) $(1 + x)^\mu = 1 + \dbinom{\mu}{1}x + \dbinom{\mu}{2}x^2 + \cdots + \dbinom{\mu}{r}x^r + \cdots$ $|x| < 1$

The binomial coefficients $\binom{\mu}{r}$ were defined in Part 1, Chap. 2, Sec. 1 for the case of the positive integral μ, but the definition and the functional equations given there apply for arbitrary real or complex μ. Note that in all cases the right-hand side is one-valued but the left-hand side may be many-valued; the value given by the series is the principal value.

(5) $\sin x = \frac{x}{1!} - \frac{x^3}{3!} + \cdots + (-1)^n \frac{x^{2n+1}}{(2n+1)!} + \cdots$ all x

(6) $\cos x = 1 - \frac{x^2}{2!} + \cdots + (-1)^n \frac{x^{2n}}{(2n)!} + \cdots$ all x

(7) $\tan x = x + \frac{1}{3}x^3 + \frac{2}{15}x^5 + \frac{17}{315}x^7 + \cdots$ $|x| < \pi/2$

The coefficients in this series are expressible in terms of the Bernoulli numbers B_n. The coefficient of x^{2n-1} is

$$\frac{(-1)^{n-1}2^{2n}(2^{2n}-1)B_{2n}}{(2n)!}$$

The Bernoulli numbers are defined as follows:

$$B_0 = 1 \qquad B_1 = -\frac{1}{2} \qquad B_{2n+1} = 0 \qquad (n < 0)$$

$$(-1)^{n+1}B_{2n} = \frac{2(2n)!}{(2\pi)^{2n}} \sum_{k=1}^{\infty} \frac{1}{k^{2n}} \qquad (n \geq 1)$$

(8) It is not possible to differentiate $\cot x$ at the origin and thus it is not possible to express $\cot x$ as a power series in x. However, $\cot x - x^{-1}$ can be expressed in this way:

$$\cot x - \frac{1}{x} = -\frac{1}{3}x - \frac{1}{45}x^3 - \frac{2}{945}x^5 - \cdots$$

In another form:

$$x \cot x = \sum_{n=0}^{\infty} (-1)^n \frac{B_{2n}(2x)^{2n}}{(2n)!} \qquad |x| < \tfrac{1}{2}\pi$$

where the B_n are the Bernoulli numbers.

(9) $\arcsin x = x + \frac{1}{2}\frac{x^3}{3} + \frac{1 \cdot 3}{2 \cdot 4}\frac{x^5}{5} + \cdots$ $|x| < 1$

$$\arccos x = \pi/2 - \arcsin x$$

(10) $\arctan x = \frac{x}{1} - \frac{x^3}{3} + \frac{x^5}{5} - \frac{x^7}{7} + \cdots$ $|x| < 1$

$$\text{arccot}\, x = \arctan x^{-1} = \tfrac{1}{2}\pi - \arctan x$$

(11) $\sinh x = \frac{x}{1!} + \frac{x^3}{3!} + \cdots + \frac{x^{2n+1}}{(2n+1)!} + \cdots$ all x

(12) $\cosh x = 1 + \frac{x^2}{2!} + \cdots + \frac{x^{2n}}{(2n)!} + \cdots$ all x

(13) $\tanh x = x - \frac{1}{3}x^3 + \frac{2}{15}x^{15} - \frac{17}{315}x^7 + \cdots$ $|x| < \pi/2$

The coefficients of this last series are the same as those for $\tan x$ except for the signs.

The evaluation of $f^{(n)}(0)$ is generally not the simplest way to obtain the formal power series representing a given function. Two other methods are the following:

(*a*) $y = \arctan x$

We have $(1 + x^2)^{-1} = 1 - x^2 + x^4 - \cdots$. Integrating this over the interval $[0,x]$, we obtain (see Sec. 4)

$$\arctan x = x - \tfrac{1}{3}x^3 + \tfrac{1}{5}x^5 - \cdots$$

(*b*) $y = \sin(m \arcsin x)$

Since $y' = \cos(m \arcsin x) \times m(1 - x^2)^{-1/2}$, we have $(1 - x^2)(y^{(1)})^2 = m^2(1 - y)^2$. If we differentiate this $n + 1$ times by Leibniz' formula,

$$(1 - x^2)y^{(n+2)} - (2n+1)xy^{(n+1)} + (m^2 - n^2)y^{(n)} = 0$$

Putting $x = 0$,

$$y^{(n+2)}(0) = -(m^2 - n^2)y^{(n)}(0)$$

By successive use of this, beginning with $y(0) = 0$, $y^{(1)}(0) = m$, we obtain

$$\sin(m \arcsin x) = mx - \frac{m(m^2 - 1^2)}{3!}x^3 + \frac{m(m^2 - 1^2)(m^2 - 3^2)}{5!}x^5 - \cdots$$

If we put $x = \sin\theta$ in this, we obtain

$$\sin m\theta = m \sin\theta - \frac{m(m^2 - 1^2)}{3!}\sin^3\theta + \frac{m(m^2 - 1^2)(m^2 - 3^2)}{5!}\sin^5\theta - \cdots$$

The formal manipulations in these methods require justification in each case.

Power series provide a convenient method of evaluating functions for small values of the argument, and estimates of the remainder used in establishing the convergence of the series to the function can be used to give guarantees on the accuracy of the values. They can be evaluated by using tables of powers, or by the following recurrence process for the evaluation of a polynomial, if the number of terms required has been decided. If

$$p_0 = a_0 \quad \text{and} \quad p_{i+1} = xp_i + a_i \quad \text{for } i = 0, 1, \ldots, n-1$$

then

$$p_n = a_0x^n + a_1x^{n-1} + \cdots + a_{n-1}x + a_n$$

This process involves repeated multiplications by x.

Various properties of the special functions concerned can be used to obtain series more rapidly convergent than the one obtained in the most straightforward fashion. For instance the series for $\arctan x$ while convergent for $-1 < x < 1$, is not satisfactory in practice near the values $x = 1$. The following series may be used near $x = \pm 1$:

$$\arctan x = \frac{x}{1 + x^2}\left[1 + \frac{2}{3}\left(\frac{x^2}{1 + x^2}\right) + \frac{2 \cdot 4}{3 \cdot 5}\left(\frac{x^2}{1 + x^2}\right)^2 + \cdots\right]$$

3. Integration

The Indefinite Integral. This is introduced as an operation inverse to differentiation. If

$$f(x) = \frac{d}{dx} F(x)$$

we say that $F(x)$ is an indefinite integral of $f(x)$ and write

$$F(x) = \int f(x)\,dx$$

This function is not unique. If $F(x)$ is an indefinite integral of $f(x)$, so is $F(x) + C$, for any constant C. This arbitrariness is often indicated by such means as

$$\int \cos x\,dx = \sin x + C$$

$$\int x^n\,dx = \frac{x^{n+1}}{n+1} + C \qquad n \neq -1$$

The following rules hold for the indefinite integral:

$$\int af(x)\,dx = a \int f(x)\,dx \qquad (a \text{ constant})$$
$$\int [f_1(x) + f_2(x)]\,dx = \int f_1(x)\,dx + \int f_2(x)\,dx$$
$$\int f'(x)g(x)\,dx = f(x)g(x) - \int f(x)g'(x)\,dx$$

The last rule for integration by parts follows from the rules for differentiating a product. An example of its use follows:

$$\begin{aligned} I &= \int \sin^2 x\,dx \\ &= \int \sin x \frac{d}{dx}(-\cos x)\,dx \\ &= -\sin x \cos x + \int \cos^2 x\,dx \\ &= -\sin x \cos x + \int (1 - \sin^2 x)\,dx \end{aligned}$$

whence $I = -\sin x \cos x + x - I$, giving

$$I = \tfrac{1}{2}(x - \sin x \cos x)$$

There is a rule for change of variable in an indefinite integral, corresponding to the (chain) rule for the differentiation of a composite function:

$$\int f(x)\,dx = \int f(u)\phi'(u)\,du$$

where $x = \phi(u)$.

The formal process is valid on the assumption that ϕ is monotone and that ϕ' exists. From this rule for change of variable, or otherwise, we have

$$\int \frac{f'(x)}{f(x)}\,dx = \log f(x)$$

As an example:

$$\int \tan x\,dx = -\int \frac{-\sin x}{\cos x}\,dx = -\log \cos x$$
$$= \log \sec x$$

Elaborate tables of indefinite integrals have been compiled. A convenient brief listing is given in Table 3.1. (See Sec. *e* of the References at the end of this chapter.)

The Definite Integral. This is introduced initially as an area. The symbol

$$\int_a^b f(x)\,dx$$

for suitable $f(x)$ represents the area between the curve $y = f(x)$ and the x axis, and the ordinates $x = a$, $x = b$. A more arithmetic approach is the following definition as a limit of a sum: Assume $[a,b]$ a finite and closed interval and $f(x)$ bounded in $a \leq x \leq b$. Consider a subdivision of the interval $[a,b]$ thus:

$$a = x_0 < x_1 < x_2 \cdots < x_n = b$$

Let

$$m_i = \underline{\text{bound}}_{x_{i-1} \leq x \leq x_i} f(x) \qquad M_i = \overline{\text{bound}}_{x_{i-1} \leq x \leq x_i} f(x)$$

Consider the sums

$$s = \Sigma(x_i - x_{i-1})m_i \qquad S = \Sigma(x_i - x_{i-1})M_i$$

Table 3.1

$f(x)$	$\int f(t)\,dt$	$f(x)$	$\int f(t)\,dt$
x^n	$x^n/(n+1) \quad n \neq -1$	$(a^2 + x^2)^{-1}$	$a^{-1} \arctan (x/a) \quad a \neq 0$
x^{-1}	$\ln \|x\|$	$(a^2 - x^2)^{-1}$	$a^{-1} \operatorname{arctanh} (x/a) = (2a)^{-1} \ln [(a+x)/(a-x)]$
e^{ax}	$a^{-1}e^{ax}$		
a^x	$a^x/\ln a$	$(a^2 - x^2)^{-\frac{1}{2}}$	$\arcsin (x/a)$
$\sin x$	$-\cos x$	$(a + x^2)^{-\frac{1}{2}}$	$\operatorname{arcsinh} (x/a) = \ln \{[x + (x^2 + a^2)^{\frac{1}{2}}]/a\}$
$\cos x$	$\sin x$		
$\tan x$	$\ln \sec x$	$(a^2 - x^2)^{\frac{1}{2}}$	$\frac{1}{2}[x(a^2 - x^2)^{\frac{1}{2}} + a^2 \arcsin (x/a)]$
$\cot x$	$\ln \sin x$		
$\operatorname{cosec} x$	$\ln (\tan \frac{1}{2}x)$	$(x^2 \pm a^2)^{\frac{1}{2}}$	$\frac{1}{2}\{x(x^2 \pm a^2)^{\frac{1}{2}} \pm a^2 \ln [x + (x^2 \pm a^2)^{\frac{1}{2}}]\}$
$\sinh x$	$\cosh x$		
$\cosh x$	$\sinh x$		
$\tanh x$	$\ln (\cosh x)$		
$\arcsin x$	$x \arcsin x + (1 - x^2)^{\frac{1}{2}}$	$e^{ax} \sin bx$	$e^{ax}(a \sin bx - b \cos bx)/(a^2 + b^2)$
$\ln x$	$x \ln x - x$	$\cosh x \cos x$	$\frac{1}{2}(\sinh x \cos x + \cosh x \sin x)$
$(x \ln x)^{-1}$	$\ln (\ln x)$		
$(1 + \cos x)^{-1}$	$\tan (x/2)$		

NOTE: Some of these results are valid only for certain values of the parameters involved. We always assume $a > 0$ and in those involving $a^2 - x^2$ we require $a > x > 0$. In the thirteenth relation on the left we require $|x| < 1$.

The sums s are bounded above and the sums S are bounded below, no matter what subdivision of $[a,b]$ we take. We denote the corresponding bounds by

$$\underline{\int} f(x)\,dx = \overline{\text{bound}}\ s \qquad \overline{\int} f(x)\,dx = \underline{\text{bound}}\ S$$

It can be shown that

$$\underline{\int} f(x)\,dx \leq \overline{\int} f(x)\,dx$$

If these bounds are equal, we say that $f(x)$ is integrable (in the Riemann sense) over the interval $[a,b]$ and denote their common value by

$$\int_a^b f(x)\,dx$$

If $f(x)$ is continuous in $[a,b]$, then it is integrable; this is also the case if $f(x)$ is monotone, in which case it can have a (perhaps) infinite set of discontinuities (simple jumps), that is, points d_i $(i = 1,2, \ldots)$ such that

$$\lim_{x \to d_i -} f(x) \neq \lim_{x \to d_i +} f(x)$$

A necessary and sufficient condition for integrability (in the sense of Riemann) is that the discontinuities of $f(x)$ in $[a,b]$ should form a set of *zero measure*, i.e., one which can be enclosed in a finite number of intervals whose total length is arbitrarily small. The function which is equal to 1 when x is rational and zero when x is irrational is not integrable; it has a discontinuity at every point.

If $f(x)$ is integrable, the value of the definite integral can be obtained by considering the limit of the sum s (or S or indeed $\Sigma(x_i - x_{i-1})f(\xi_i)$, where ξ_i lies between x_{i-1} and x_i) for any sequence of subdivisions of $[a,b]$, for which the length of the largest of the intervals (x_{i-1},x_i) tends to zero. We can conveniently consider the limit of sums for equal intervals, n in number, as $n \to \infty$. For example, consider

$$\int_a^b x\,dx$$

Write $h = (b - a)/n$. Then an approximating sum is

$$\begin{aligned} h[a + (a + h) + \cdots + (a + (n - 1)h)] &= h[na + \tfrac{1}{2}(n - 1)nh] \\ &= (b - a)a + \tfrac{1}{2}(b - a)^2 + \tfrac{1}{2}(b - a)h \\ &= \tfrac{1}{2}(b^2 - a^2) + \tfrac{1}{2}(b - a)h \end{aligned}$$

As $n \to \infty$, a and b remaining fixed, $h \to 0$ and the above expression for the approximating sum has the limit $\frac{1}{2}(b^2 - a^2)$.

This definition is of little practical value in the evaluation analytically of particular definite integrals. We usually rely on the theorem that if $F(x)$ is an indefinite integral of $f(x)$ then

$$\int_a^b f(x)\,dx = F(b) - F(a)$$

(The choice of the indefinite integral is immaterial because the constant involved cancels out.)

Various devices (integration by parts, or integration by substitution, for instance) introduced in connection with indefinite integrals can be used in the manipulation of definite integrals. That care must, however, be taken is indicated by the following example:

$$I = \int_0^3 (x^2 - 4x + 5)\,dx = \left|\tfrac{1}{3}x^3 - 2x^2 + 5x\right|_0^3 = 6$$

Put $y = x^2 - 4x + 5$ so that $x = 2 \pm \sqrt{y - 1}$, giving $y = 5$ for $x = 0$ and $y = 2$ for $x = 3$. Then the integral becomes

$$\begin{aligned} \int_5^2 y\,dx = \int_5^2 y \frac{dx}{dy}\,dy &= \int_5^2 y\left[\pm\frac{1}{2}(y - 1)^{-1/2}\right] dy \\ &= \pm\tfrac{1}{2}\int_5^2 [(y - 1)^{1/2} + (y - 1)^{-1/2}]\,dy \\ &= \pm[\tfrac{1}{3}(y - 1)^{3/2} + (y - 1)^{1/2}] = 1\tfrac{2}{3} \end{aligned}$$

which is certainly wrong. A more careful examination of the transformation from the variable x to y shows that as x increases from 0 to 3 the variable y decreases from 5 to 1 and then increases to 2. The correct integration with respect to y should be

$$I = \int_5^1 y[-\tfrac{1}{2}(y - 1)^{-1/2}]\,dy + \int_1^2 y[+\tfrac{1}{2}(y - 1)^{-1/2}]\,dy = 6$$

Elaborate tables of definite integrals have been compiled. See Sec. *e* of the References at the end of the chapter.

In the other direction, if $f(x)$ is continuous at b, the derivative with respect to b of $R(b)$

$$R(b) = \int_a^b f(x)\,dx$$

is $f(b)$. Its derivative with respect to a is $-f(a)$.

Infinite Integrals. The definitions of a definite integral do not apply when the interval extends to infinity in one or both directions or when the function is unbounded in the interval. As examples:

$$\int_0^\infty e^{-x}\,dx \qquad \text{or} \qquad \int_0^1 x^{-1/2}\,dx$$

Suppose

$$\int_a^X f(x)$$

exists for all $X \geq a$; then if

$$\lim_{X \to \infty} \int_a^X f(x)\,dx$$

exists, it is denoted by

$$\int_a^\infty f(x)\,dx$$

This is often called a Cauchy-Riemann integral. We define

$$\int_{-\infty}^a f(x)\,dx$$

in a similar way. Further,

$$\int_{-\infty}^{+\infty} f(x)\,dx$$

is defined as

$$\int_{-\infty}^{a} f(x)\,dx + \int_{a}^{\infty} f(x)\,dx$$

if both these exist for some a. In particular

$$\int_0^\infty e^{-x}\,dx = \lim_{x\to\infty} [1 - e^{-x}] = 1$$

since $\lim_{x\to\infty} e^{-x} = 0$. However,

$$\int_0^\infty \sin x\,dx$$

does not exist since

$$\int_0^X \sin x\,dx = 1 - \cos X$$

and this has no limit as $X \to \infty$.

The integrals just discussed are *improper* because the range is infinite. It can happen that the usual definitions do not apply because the functions become unbounded. Suppose that $f(x) \to \infty$ as $x \to a+$ and suppose that

$$\int_{a+\epsilon}^{b} f(x)\,dx$$

exists for all $\epsilon > 0$. If

$$\lim_{\epsilon\to+0} \int_{a+\epsilon}^{b} f(x)\,dx$$

exists it is denoted by

$$\int_a^b f(x)\,dx$$

This definition can be extended in an obvious way to cover singularities at the upper end of the range. For instance, since

$$\int_\epsilon^1 x^{-1/2}\,dx = \left| 2x^{1/2} \right|_\epsilon^1 = 2 - 2\epsilon^{1/2} \to 2$$

as $\epsilon \to 0$, we write

$$\int_0^1 x^{-1/2}\,dx = 2$$

A more complicated situation can arise at a singularity in the interior of the range of integration, (a,b). It can happen that

$$\int_a^c f(x)\,dx \qquad \text{and} \qquad \int_c^b f(x)\,dx$$

exist independently as improper integrals in the sense just discussed. However, if we consider

$$\int_a^b (x-c)^{-1}\,dx$$

where $a \le c \le b$, this is not the case. Nevertheless

$$\begin{aligned}\int_a^{c-\delta} (x-c)^{-1}\,dx &+ \int_{c+\delta}^{b} (x-c)^{-1}\,dx \\ &= \log \delta - \log (c-a) + \log (b-c) - \log \delta' \\ &= \log \frac{b-c}{c-a} + \log \frac{\delta}{\delta'}\end{aligned}$$

has a definite limit if we put $\delta = \delta'$ and let $\delta \to 0$. This is known as a *principal value integral:*

$$PV \int_a^b (x-c)^{-1}\,dx = \log \frac{b-c}{c-a}$$

Generalizations of the Concept of Integral. In addition to the immediate extensions of the concept of integral already introduced there are many of interest. The *Stieltjes integral* of a function $f(x)$ with respect to a weight function $w(x)$ is defined as

$$\int_a^b f(x)\,dw(x) = \lim \sum f(\xi_{i-1})[w(x_i) - w(x_{i-1})]$$

where the limit is taken as the maximum length of the subintervals (x_{i-1},x_i) tends to zero, and where, as before, ξ_{i-1} is a point in this interval (x_{i-1},x_i). This integral exists whenever f is continuous and $w(x)$ is a monotone function. This type of integral is often used in statistics. It reduces to the Riemann integral when $w(x) \equiv x$ and when $w(x)$ has a continuous derivative, it can be evaluated as the Riemann integral

$$\int_a^b f(x)w'(x)\,dx$$

Another important type of integral is the *Lebesgue integral.* Without this a satisfactory account of the theory of Fourier series, for instance, is impossible. To introduce this we essentially subdivide the range of values assumed by $f(x)$ for $a \le x \le b$, instead of the interval $[a,b]$. We then consider sums of the form

$$s = \Sigma y_{i-1}\mu_i \qquad S = \Sigma y_i \mu_i$$

where μ_i is the measure of the set E_i of points x for which the $f(x)$ assumes values between y_{i-1} and y_i. In the case when $f(x)$ is continuous, E_i consists of a set of intervals on the x axis and then the corresponding measure is the sum of the lengths of the intervals. The definition of (Lebesgue) measure applies to much more general cases, but not every set is measurable, i.e., has a measure in this sense.

If f is bounded and is such that all the sets E_i are measurable, then the upper bound of the sums s and the lower bound of the sums S coincide. This common bound is called the Lebesgue integral of f.

In considering the integration of a function f in this sense, we can therefore always disregard the values of $f(x)$ on a set of measure zero, since these make no contribution to the sums s,S. Thus we can integrate such functions as the one equal to 1 for x rational and 0 for x irrational; for if we disregard the values at the rational points, which form a set of measure zero, the function becomes (essentially) constant and is certainly integrable in the Lebesgue sense.

If f is integrable in the Riemann sense, it is integrable in the Lebesgue sense, and the two integrals have the same value. This is not true for the extended type Riemann integral as is shown by the case of

$$\int_0^\infty \frac{\sin x}{x}\,dx$$

which exists as an improper Riemann integral with the value $\pi/2$, but this does not exist in the Lebesgue sense, for it is a fundamental property of the Lebesgue integral that if $\int f(x)\,dx$ exists so does $\int |f(x)|\,dx$, which is easily seen to be impossible in the present case.

Double and Multiple Integrals. Multiple integrals can be defined as limits of double sums in an obvious extension of the definition introduced above. A double integral

$$\iint_S f(x,y)\,dx\,dy$$

can be interpreted as a volume: the volume bounded by the surface $z = f(x,y)$, the plane $z = 0$, and the lines parallel to the z axis, through the boundary of the region S over which integration takes place. It can be shown that (usually) when a double integral exists so do the repeated integrals

$$\int dx[\int f(x,y)\,dy],\ \int dy[\int f(x,y)\,dx]$$

and all three are equal. This enables calculation of double integrals to be reduced to that of simple ones. The main difficulty involved is usually the determination of the appropriate limits of the simple integral.

As an example of the evaluation of multiple integrals consider

$$I = \int_{x=0}^{a} \int_{y=0}^{a} \exp(-x^2)\exp(-y^2)\,dx\,dy$$

where the integration is over the square with vertices (0,0), (0,a), (a,0), (a,a). The limits in the corresponding repeated integral are constants, and it follows from the definition that the integral can be expressed as the product of simple integrals:

$$I = \int_{x=0}^{x=a} \exp(-x^2)\,dx \int_{y=0}^{y=a} \exp(-y^2)\,dy = I_a{}^2$$

where $$I_a = \int_0^a \exp(-x^2)\,dx$$

Consider

$$J_a = \int\int \exp(-x^2)\exp(-y^2)\,dx\,dy$$

where the integration is to be over the first quadrant of the circle of radius a. To handle this, it is convenient to change the variable to polar coordinates.

$$x = r\cos\theta \qquad y = r\sin\theta$$

The element of area is now enclosed between two circles, radii r, $r + dr$, and two radii at angles θ, $\theta + d\theta$. It has area $r\,d\theta\,dr$ and

$$\begin{aligned} J_a &= \int_{\theta=0}^{\pi/2} \int_{r=0}^{a} \exp(-r^2)\,rdr\,d\theta \\ &= \int_{\theta=0}^{\pi/2} \left[\int_{r=0}^{a} \exp(-r^2)\,rdr\right] d\theta \\ &= \int_{\theta=0}^{\pi/2} \{\tfrac{1}{2}[1 - \exp(-a^2)]\}\,d\theta \\ &= \frac{\pi}{4}[1 - \exp(-a^2)] \end{aligned}$$

Since the integrand is everywhere positive

$$J_a < I_a{}^2 < J_{2a}$$

the quadrant being included between the two squares. However, if $a \to \infty$, J_a and J_{2a} have the same limit, and so the limit of I_a is $\sqrt{\pi}/2$. Hence

$$\int_0^\infty e^{-x^2}\,dx = \sqrt{\pi}/2$$

Transformations of the following type are often helpful:

$$\int_0^a dx \int_0^x f(x,y)\,dy = \int_0^a dy \int_y^a f(x,y)\,dx$$

which correspond to the interpretation of a double integral over the triangle bounded by the lines $y = 0$, $y = x$, $x = a$, as a repeated integral in two ways.

The general formulas for the transformation of double integrals depend on the *Jacobian* of the transformation (see Sec. 5):

$$\iint f(x,y)\,dx\,dy = \iint \phi(u,v)\,\frac{\partial(u,v)}{\partial(x,y)}\,du\,dv$$

where $\phi(u,v) = f(x(u,v),\ y(u,v))$, and the regions of integration correspond.

Green's theorem, expressing a double integral over a region in terms of an integral around its boundary, plays an important role in applied mathematics: If $\frac{\partial G}{\partial x}, \frac{\partial F}{\partial y}$ are continuous within a region A and on its boundary C, then

$$\iint_A \left(\frac{\partial G}{\partial x} - \frac{\partial F}{\partial y}\right) dx\,dy = \int_C (F\,dx + G\,dy)$$

The curvilinear integral on the right is defined as

$$\int_0^T \left(F\frac{dx}{dt} + G\frac{dy}{dt}\right) dt$$

where t is a parameter such that as t runs from 0 to T the point $x = x(t)$, $y = y(t)$ describes the curve once in the positive direction.

A consequence of this theorem is that if

$$\frac{\partial G}{\partial x} = \frac{\partial F}{\partial y}$$

then the curvilinear integral $\int_{\widehat{PQ}} (F\,dx + G\,dy)$ does not depend on the path of integration but only on the end points P,Q.

Stokes' theorem expresses a curvilinear integral in terms of a surface integral bounded by the curve. It is discussed most conveniently in vector notation (Part 1, Chap. 9).

Improper Multiple Integrals. In the same way as we proceeded from integrals of the form

$$\int_a^b f(x)\,dx$$

to integrals of the form

$$\int_0^\infty f(x)\,dx$$

we can introduce improper multiple integrals. We shall not give the definitions in detail, but all manipulations of infinite integrals must be handled with care.

As an example of an integral, improper because of

unboundedness of the integrand, consider

$$I = \int_0^1 \int_0^1 \frac{dx\,dy}{1 - xy}$$

For $0 < \epsilon < 1$, $0 < \eta < 1$,

$$I_{\epsilon,\eta} = \int_0^{1-\epsilon} \int_0^{1-\eta} \frac{dx\,dy}{1 - xy}$$
$$= X + \frac{X^2}{2^2} + \frac{X^3}{3^2} + \cdots$$

where $X = (1 - \epsilon)(1 - \eta)$. It follows from *Abel's theorem on power series* that

$$\lim_{\epsilon \to 0, \eta \to 0} I_{\epsilon,\eta} = 1 + \frac{1}{2^2} + \frac{1}{3^2} + \cdots = \frac{\pi^2}{6}$$

Hence the improper integral I exists and has this value.

As an example of formulas involving integrals, improper because of the infinite ranges, we mention the convolution theorem for Laplace transforms (see Sec. 4 and Chap. 6 of Part 1).

$$\int_0^\infty e^{-pt} f(t)\,dt \int_0^\infty e^{-p\tau} g(\tau)\,d\tau = \int_0^\infty e^{-pt} \int_0^t g(\tau) f(t - \tau)\,d\tau\,dt \quad (3.5)$$

If the integrals on the left are absolutely convergent (for a particular p, which we keep fixed), their product can be written as

$$\int_0^\infty \int_0^\infty e^{-p(t+\tau)} f(t) g(\tau)\,dt\,d\tau$$

which can be transformed by

$$t + \tau = u \qquad \tau = v$$

into

$$\int_0^\infty e^{-pu} \left[\int_0^u f(u - v) g(v)\,dv \right] du$$

which is the result required.

In the language of transforms, (3.5) can be stated as the product of the transforms of two functions f, g, which is the transform of their convolution or resultant $h = f * g$ defined by

$$h(t) = \int_0^t g(\tau) f(t - \tau)\,d\tau$$

4. Integral Transforms

For an additional account of this subject and its application, see Chap. 6 of Part 1.

Definitions and Examples. With a function $f(t)$ it is often convenient to associate certain transforms: in particular the *Laplace transform* defined by

$$\mathfrak{L}[f(t),p] = \int_0^\infty e^{-pt} f(t)\,dt = \phi(p)$$

and the *Fourier cosine and sine transforms*

$$\mathfrak{F}_c[f(t),u] = \sqrt{\frac{2}{\pi}} \int_0^\infty f(t) \cos ut\,dt = F_c(u)$$

$$\mathfrak{F}_s[f(t),u] = \sqrt{\frac{2}{\pi}} \int_0^\infty f(t) \sin ut\,dt = F_s(u)$$

It is also often convenient to use the complex *Fourier transform* in the form

$$\mathfrak{F}\{f(t),u\} = \frac{1}{\sqrt{2\pi}} \int_{-\infty}^{+\infty} f(t)\, e^{iut}\,dt = F(u)$$

These, of course, exist only for certain functions f, and then only for certain values of p, or u. The calculation of the transforms are exercises in the integral calculus. Of these we give two examples, one trivial, the other reasonably complicated.

Example 1. $I_n = \int_0^\infty e^{-pt} t^n\,dt = \mathfrak{L}\{t^n,p\}$

where n is a positive integer, can be evaluated by successive integrations by parts. Thus

$$I_n = \left[\frac{e^{-pt}}{-p} t^n \right]_0^\infty + n \int_0^\infty \frac{e^{-pt}}{p} t^{n-1}\,dt = \frac{n}{p} I_{n-1}$$

so that, since $I_0 = p^{-1}$, we have

$$I_n = \frac{n!}{p^{n+1}}$$

Example 2. As a further example consider the evaluation of the Laplace transform of

$$J_0(t) = \sum_{n=0}^{\infty} \frac{(-1)^n}{(n!)^2} \left(\frac{t}{2} \right)^{2n}$$

We have

$$\mathfrak{L}\{J_0(t),p\} = \sum_0^\infty \frac{(-1)^n}{(n!)^2} \mathfrak{L}\left\{ \left(\frac{t}{2} \right)^{2n}, p \right\}$$
$$= \sum_0^\infty \frac{(-1)^n}{(n!)^2 \cdot 2^{2n}} \frac{(2n)!}{p^{2n+1}}$$

Now

$$\frac{(-1)^n 2n!}{n! 2^n \cdot 2^n \cdot n!} = \frac{[-\tfrac{1}{2}][-\tfrac{3}{2}] \cdots [-(2n-1)/2]}{n!} = \binom{-1/2}{n}$$

Hence

$$\mathfrak{L}\{J_0(t),p\} = \frac{1}{p} \sum_0^\infty \binom{-1/2}{n} (p^{-2})^n = \frac{1}{p} (1 + p^{-2})^{-1/2}$$
$$= \frac{1}{\sqrt{1 + p^2}} \quad (3.6)$$

The inversion of summation and integration in (3.5) and the summation of the binomial series in (3.6) are valid if $\Re p > 1$. The final result is true of $\Re p > 0$ in virtue of the permanence of functional relations, since $\mathfrak{L}\{J_0(t), p\}$ exists for these values of p (see Sec. 8). By letting $p \to 0$ in the relation (3.6), we obtain

$$\int_0^\infty J_0(t)\,dt = 1$$

a result not conveniently obtained directly.

These transforms have certain reciprocity properties, for example, if $F_s(u)$ is the sine transform of $f(t)$,

TABLE 3.2. LAPLACE TRANSFORMS

$f(t)$		$\phi(p)$
t^n	$n > 0$	$\frac{n!}{p^{n+1}}$
$e^{\alpha t}$		$\frac{1}{p-\alpha}$
$\cos \alpha t$		$\frac{p}{p^2+\alpha^2}$
$\sin \alpha t$		$\frac{\alpha}{p^2+\alpha^2}$
$t \cos \alpha t$		$\frac{p^2-\alpha^2}{(p^2+\alpha^2)^2}$
$t \sin \alpha t$		$\frac{2\alpha p}{(p^2+\alpha^2)^2}$
$J_0(t)$		$\frac{1}{(1+p^2)^{1/2}}$
$\frac{e^{at}}{\sqrt{t}}$		$\left(\frac{\pi}{p-a}\right)^{1/2}$
$J_0(2\sqrt{at})$		$\frac{1}{p}e^{-a/p}$
$e^{bt}\frac{\sin at}{t}$		$\arctan \frac{a}{p-b}$

then $f(u)$ is the sine transform of $F_s(t)$. In the complex case we have the following:

$$F(u) = \frac{1}{\sqrt{2\pi}} \int_{-\infty}^{+\infty} f(t)e^{iut}\, dt$$

then
$$f(x) = \frac{1}{\sqrt{2\pi}} \int_{-\infty}^{+\infty} F(u)e^{-ixu}\, du$$

This depends on the Fourier double integral formula

$$f(x) = \frac{1}{2\pi} \int_{-\infty}^{+\infty} e^{-ixu}\, du \int_{-\infty}^{+\infty} f(t)e^{iut}\, dt$$

where $f(x)$ on the left must be replaced by the average of its limiting values at x if it has a simple discontinuity there.

Tables 3.2 and 3.3 present some tables of Laplace and Fourier transforms. More extensive tables of Laplace and Fourier transforms are listed in Sec. *d* of the References at the end of this chapter.

5. Functions of Several Real Variables

Functions of Two Real Variables. *Partial Derivatives.* A real function of two real variables x,y is defined by a correspondence between certain pairs of real numbers x,y (which may be interpreted as points in a plane) and certain real numbers. It can be represented graphically as (part of) a surface $z = f(x,y)$. Functions may be given explicitly as $z = \sin (x^2 + y^2)$, or $z = (1 - x^2y^2)$, or implicitly as

$$x^2 + y^2 + z^2 = 1$$

or numerically by a double-entry table.

We can define continuity of $z = f(x,y)$ at (x_0,y_0) requiring that $f(x,y)$ shall be near $f(x_0,y_0)$ when (x,y) is near (x_0,y_0), that is, when $(x - x_0)^2 + (y - y_0)^2$ or when $|x - x_0| + |y - y_0|$ is small. This condition is much more restrictive than the condition of being continuous with respect to x for $y = y_0$ and being continuous with respect to y for $x = x_0$. For example, $f(x,y) = 2xy/(x^2 + y^2)$ is constant for $x = 0$ or for $y = 0$ and therefore continuous with respect to

TABLE 3.3. FOURIER TRANSFORMS

f		F_c
e^{-x}		$\sqrt{\frac{2}{\pi}}\frac{1}{1+u^2}$
$f(x) = 1$ $f(x) = 0$	$0 \le x < a$ $a \le x < \infty$	$\sqrt{\frac{2}{\pi}}\frac{\sin au}{u}$
$\frac{1}{\cosh \pi x}$		$\frac{1}{\sqrt{2\pi}\cosh \frac{1}{2}u}$
$\exp(-\frac{1}{2}x^2)$		$\exp(-\frac{1}{2}u^2)$
$\cos \frac{1}{2}x^2$		$\sqrt{2}(\cos \frac{1}{2}u^2 + \sin \frac{1}{2}u^2)$

f	F_s
e^{-x}	$\sqrt{\frac{2}{\pi}}\frac{u}{1+u^2}$
$\frac{\sin x}{x}$	$\frac{1}{\sqrt{2\pi}}\log\left\|\frac{1+u}{1-u}\right\|$
$x \exp(-\frac{1}{2}x^2)$	$u \exp(-\frac{1}{2}u^2)$

f	F
e^{-x}	$\sqrt{\frac{2}{\pi}}\frac{1}{1+x^2}$

each variable separately at (0,0) but, if we take $y = mx$ we have $f(x,mx) = 2m/(1 + m^2)$ and so there are points as near as we please to (0,0) at which $f(x,y)$ has any value between -1 and 1.

We say that the partial derivative of $f(x,y)$ with respect to x at (x_0,y_0) exists if

$$\lim_{x \to x_0} \frac{f(x,y_0) - f(x_0,y_0)}{x - x_0}$$

exists. We denote this limit by

$$\frac{\partial f}{\partial x} \quad \text{or} \quad f_x$$

Similarly we define the partial derivatives f_y with respect to y. Repeated derivatives $(f_x)_x = f_{xx}$, $(f_x)_y = f_{xy}$, $(f_y)_x = f_{yx}$, $(f_y)_y = f_{yy}$, . . . are defined in the manner indicated. There seems to be disagreement on this notation. Some authors define

$$\frac{\partial}{\partial y}\left(\frac{\partial f}{\partial x}\right) = \frac{\partial^2 f}{\partial x\, \partial y} = f_{xy}$$

others define

$$\frac{\partial}{\partial y}\left(\frac{\partial f}{\partial x}\right) = \frac{\partial^2 f}{\partial y\, \partial x} = f_{yx}$$

The question of the equality of f_{xy} and f_{yx} arises. Without going into details irrelevant here we give the following result: if f, f_x, f_y, and f_{xy} exist in a neighborhood of (x_0,y_0) and are continuous at that point then f_{yx} exists at that point and $f_{xy} = f_{yx}$.

Mean-value Theorems. Corresponding to the result

$$f(x + h) - f(x) = hf'(x) + o(h)$$

which is true on the hypothesis of the existence of $f'(x)$, we might expect the result

$$f(x + h, y + k) - f(x, y) = hf_x + kf_y + o(\sqrt{h^2 + k^2})$$

on the hypothesis of the existence of the partial derivatives at (x,y). This result is false. We can, however, obtain it under the assumption that f_x and f_y exist and are *continuous* in the neighborhood of x,y.

Corresponding to the mean-value theorem

$$f(a + h) - f(a) = hf'(a + \theta h) \qquad 0 \leq \theta \leq 1$$

valid on the hypothesis of the existence of $f'(x)$ in $a \leq x \leq a + h$ we have

$$(a + h, b + k) - f(a, b) = hf_x(a + \theta h, b + \theta k) + kf_y(a + \theta h, b + \theta k) \qquad 0 \leq \theta \leq 1$$

valid when f_x, f_y are continuous in the neighborhood of (a,b).

More generally, assuming that f and all its partial derivatives of order $1,2, \ldots ,n + 1$ are continuous, we have

$$f(a + h, b + k) = f(a, b) + \sum_{r=1}^{n} \frac{1}{r!} \left[h \frac{\partial}{\partial x} + k \frac{\partial}{\partial y} \right]^r f(a, b) + R_{n+1}$$

where the powers of the operators are interpreted as partial derivatives, evaluated at (a,b) and where

$$R_{n+1} = \frac{1}{(n + 1)!} \left[h \frac{\partial}{\partial x} + k \frac{\partial}{\partial y} \right]^{n+1} f(a + \theta h, b + \theta k) \qquad 0 \leq \theta \leq 1$$

From this, for $n = 2$, we can deduce sufficient conditions for the occurrence of maxima and minima of $f(x,y)$. We have, in fact, the following: if $f_x = 0$, $f_y = 0$ and if f_{xx}, f_{yy} and $f_{xy} = f_{yx}$ exist at (x_0,y_0), then if $f_{xx} < 0$ there is a relative maximum if

$$\Delta = f_{xy}^2 - f_{xx}f_{yy} < 0$$

while if $f_{xx} > 0$ there is a relative minimum if $\Delta < 0$; if $\Delta > 0$ there is neither a maximum nor minimum.

It is important to investigate how a relation of the form $F(x,y) = 0$ defines y as a function of x. Consideration of examples like $x^2 + y^2 + 1 = 0$, $x - y^2 = 0$ shows that y need not exist (x,y being restricted to real values) or that y may have more than one value. The following result is sufficiently general for most applications. Let $F(x,y)$ be continuous in a neighborhood of $(0,0)$ and let $F(0,0) = 0$. Let $F'_y(x,y)$ exist at $(0,0)$ and be different from zero. Then for any $\epsilon > 0$ there corresponds a $\delta = \delta(\epsilon) > 0$ such that to all x, $|x| < \delta$, there corresponds $ay = y(x)$ such that $F(x,y) = 0$ and such that $|y| < \epsilon$.

If we take $F(x,y) = x - f(y)$ we obtain results about the function inverse to f.

Functions of n Real Variables. Results similar to those mentioned on the two-variable case are available in the general case. Among the important types of functions of n variables are homogeneous functions. We say that $f(x_1,x_2, \ldots ,x_n)$ is homogeneous of degree r (which may be positive, negative, or zero) if

$$f(\lambda x_1,\lambda x_2, \ldots ,\lambda x_n) = \lambda^r f(x_1,x_2, \ldots ,x_n)$$

For instance,

$$x^r,\ x^\alpha y^{r-\alpha} + x^{r-\alpha} y^\alpha, (x^r + y^r + z^r) \arctan \frac{xyz}{x^3 + y^3 + z^3}$$

are homogeneous functions of degree r in one, two, and three variables.

A characteristic property of homogeneous functions is the following: if $f(x_1,x_2, \ldots ,x_n)$ is homogeneous of degree r and if it is continuously differentiable, we have Euler's formula:

$$x_1 f_{x_1} + x_2 f_{x_2} + \cdots + x_n f_{x_n} = rf$$

Jacobians. Consider n functions $f_1,f_2, \ldots ,f_n$ of n real variables $x_1,x_2, \ldots ,x_n$; suppose each partial derivative $\partial/\partial x_i(f_j) = f_{ij}$ exists at a point in the common domain of definition of $f_1,f_2, \ldots ,f_n$. The determinant $|f_{ij}|$ is called the Jacobian of $f_1,f_2, \ldots ,f_n$ at the point. It is also denoted by

$$|f_{ij}| = \frac{\partial(f_1,f_2, \ldots ,f_n)}{\partial(x_1,x_2, \ldots ,x_n)}$$

and many of the results suggested by this notation can be established.

For instance, if $f_i = x_i$, then $f_{ij} = \delta_{ij}$; that is, $f_{ij} = 1$, $i = j$, $f_{ij} = 0$, $i \neq j$. We have

$$\frac{\partial(x_1,x_2, \ldots ,x_n)}{\partial(x_1,x_2, \ldots ,x_n)} = 1$$

If $F_i = F_i(f_1,f_2, \ldots ,f_n)$ and the Jacobian

$$\frac{\partial(F_1,F_2, \ldots ,F_n)}{\partial(f_1,f_2, \ldots ,f_n)}$$

is defined at values of $f_1,f_2, \ldots ,f_n$ corresponding to $x_1,x_2, \ldots ,x_n$, then we have

$$\frac{\partial(F_1,F_2, \ldots ,F_n)}{\partial(x_1,x_2, \ldots ,x_n)} = \frac{\partial(F_1,F_2, \ldots ,F_n)}{\partial(f_1,f_2, \ldots ,f_n)} \frac{\partial(f_1,f_2, \ldots ,f_n)}{\partial(x_1,x_2, \ldots ,x_n)}$$

An important application of Jacobians is in connection with the functional independence of functions. (This idea is to be compared and contrasted with the idea of *linear* independence already introduced in Part 1, Chap. 2.) For simplicity in notation, we take the two-dimensional case. Let $u = u(x,y)$, $v = v(x,y)$ be defined and have continuous partial derivatives u_x, u_y, v_x, v_y in a domain‡ D; suppose that $\partial(u,v)/\partial(x,y)$ is not identically zero in D. Let $\mathfrak{D}$ be the domain described by (u,v) as (x,y) describes D. Let $F(u,v)$ be a continuous function not vanishing throughout any domain $\mathfrak{D}_1 \subset \mathfrak{D}$. Then $F(u,v) = 0$ cannot hold throughout D. Roughly, there is no functional relation connecting u,v if the Jacobian does not vanish. Conversely, too, if the functions satisfy a relation of the type $F(u,v) = 0$, then $\partial(u,v)/\partial(x,y) = 0$. A sec-

‡ The term *domain* is defined in Sec. 6.

ond application has already been given in connection with transformation of multiple integrals (Sec. 3).

6. Complex Numbers

Definitions. Many facts concerning functions can only be properly understood through the introduction of complex numbers. These are of the form $a + bi$, where a and b are real and i is a symbol representing $\sqrt{-1}$. The following rules are observed:

$$a + bi = c + di \quad \text{if and only if } a = c \text{ and } b = d \tag{3.7}$$

$$(a + bi) \pm (c + di) = (a \pm c) + (b \pm d)i \tag{3.8}$$

$$(a + bi) \times (c + di) = (ac - bd) + (ad + bc)i \tag{3.9}$$

From this follows the existence of a unique quotient $(a + bi)/(c + di)$, provided not both $c = 0$ and $d = 0$. It is not necessary to interpret the symbol $a + bi$ as a sum although this is very convenient; the sum could equally well be replaced by the ordered pair of real numbers a,b with the above rules of composition (see Part 1, Chap. 1). This idea is also helpful for the geometric interpretation of complex numbers suggested by Gauss.

The conjugate $\bar{z}$ or z^* of $z = a + bi$ is the number $a - bi$. The *norm* of $a + bi$ is $a^2 + b^2$. It has the property that

$$\text{norm } (a + bi) \times \text{norm } (c + di) = \text{norm } (a + bi)(c + di)$$

The *modulus* or *absolute* value of $a + bi$ is the positive square root of the norm, denoted by $a + bi$. The absolute value of a product is again equal to the product of the absolute values of the factors; for the absolute value of a sum, however, only an inequality can be obtained. We have $|z_1 + z_2| \leq |z_1| + |z_2|$, that is, the absolute value of a sum of two complex numbers is less than or equal to the sum of the absolute value of the terms, a fact which corresponds to one side of a triangle being less than or equal to the sum of the other two.

Using the absolute value $r = |a + bi|$ of the complex number $a + ib \neq 0$, we can write

$$a + bi = r\left[\left(\frac{a}{r}\right) + \left(\frac{b}{r}\right) i\right]$$

In this way the number is expressed as the product of a positive real number and a complex number with norm 1. Since $\left|\frac{a}{r}\right| \leq 1$ and $\left|\frac{b}{r}\right| \leq 1$, but

$$\frac{a^2}{r^2} + \frac{b^2}{r^2} = 1$$

it is clear that a uniquely defined angle θ, $0 \leq \theta < 2\pi$ exists such that

$$\frac{a}{r} = \cos\theta \qquad \frac{b}{r} = \sin\theta$$

We then obtain

$$a + ib = r(\cos\theta + i\sin\theta)$$

The angle θ is called the argument of the complex number. For real numbers we have $b = 0$ so that $\theta = 0$ for $a > 0$, $\theta = \pi$ for $a < 0$. The product and quotient of two complex numbers are easily obtained in this trigonometric representation. Let

$$a_j + ib_j = r_j(\cos\theta_j + i\sin\theta_j) \text{ for } j = 1,2$$

Then

$$(a_1 + ib_1) \times (a_2 + ib_2) = r_1r_2[\cos(\theta_1 + \theta_2) + i\sin(\theta_1 + \theta_2)]$$

$$\frac{a_1 + ib_1}{a_2 + ib_2} = r_1r_2[\cos(\theta_1 - \theta_2) + i\sin(\theta_1 - \theta_2)]$$

In particular *De Moivre's theorem* holds for the powers of complex numbers with any real number as exponent:

$$(a + ib)^n = r^n(\cos n\theta + i\sin n\theta)$$

If $n = m^{-1}$, where m is a positive integer, that is, if the mth root of $a + ib$ has to be found, we obtain in this way a representation for the m different roots that exist:

$$(a + ib)^{1/m} = r^{1/m}\left(\cos\frac{\theta + 2k\pi}{m} + i\sin\frac{\theta + 2k\pi}{m}\right)$$

$$k = 0,1, \ldots ,m - 1$$

If $a = 1$, $b = 0$, the mth *roots of unity* are obtained; they are of the form

$$\xi_k = \cos\frac{2\pi k}{m} + i\sin\frac{2\pi k}{m} \qquad k = 0,1, \ldots ,m - 1$$

The complex number $a + ib$ may be interpreted as the point (a,b) in the plane. This point P has distance $r = |a + ib|$ from O and the straight line OP makes the angle θ with the positive direction of the x axis. The sum and difference of two complex numbers are found geometrically in the same way as the sum and difference of two vectors (Part 1, Chap. 9).

In order to generalize the definition of interval to complex numbers, circles are used: if all the points of the boundary are included, they are called closed circles; if none are included, we have open circles. A *neighborhood* of the (complex) point α is the set of all numbers z such that $|\alpha - z| < r$ when r is a fixed but arbitrary (positive) number. The numbers z with $|z| < 1$ form a neighborhood of O, the so-called *unit circle*. The distance between the (complex) points α,β is $|\alpha - \beta|$.

An open set is one which can be represented as the set of points belonging to one or more of a system of open circles. In particular an open circle is itself an open set; an annulus, e.g., the set of points z such that $1 < |z| < 3$ is open, for it consists of all the points in the open circles $|z - 2e^{i\theta}| < 1$, $0 \leq \theta < 2\pi$. The open sets most used are the sets bounded by one or more closed curves.

Among the open sets we distinguish regions (or domains): these are the connected open sets and are such that any two points in them can be joined by a polygonal path consisting entirely of points of the set. An open circle or an annulus is a connected open set, but the open set which consists of two open circles with no points in common is not connected.

Among the regions, we distinguish those which are simply connected from those which are multiply connected. This can be done by distinguishing between the number of components or (pieces) of the boundary: one in the case of an open circle, two in the case of an annulus. Alternatively, we can define a simply connected region as one in which any closed curve can be shrunk to a point without going outside the region, which is intuitively obvious not to be the case, e.g., in the case of the circle $|z| = 2$ in the annulus $1 < |z| < 3$.

Limits of Sequences of Complex Numbers. The sequence $\alpha_1, \alpha_2, \ldots$ has the limit α if every neighborhood of α contains all but a finite number of elements of the sequence or, alternatively, if the distance between α and α_i becomes arbitrarily small for sufficiently large i. A necessary, but not sufficient condition for $\lim \alpha_n = \alpha$ is that $\lim |\alpha_n| = |\alpha|$. Let $\alpha_n = a_n + ib_n$ and $\alpha = a + ib$; then a necessary and sufficient condition for $\lim \alpha_n = \alpha$ is that $\lim a_n = a$, $\lim b_n = b$.

As for real numbers the following rules hold for convergent sequences:

$$\begin{aligned} \lim (\alpha_n + \beta_n) &= \lim \alpha_n \pm \lim \beta_n \\ \lim \alpha_n \beta_n &= \lim \alpha_n \lim \beta_n \\ \lim (\alpha_n)^{-1} &= (\lim \alpha_n)^{-1} \qquad \text{if } \lim \alpha_n \neq 0 \end{aligned}$$

The definition of the sum of a series of complex terms is formally the same as that for real terms. The rules for the addition and subtraction are the same. The concept of absolute convergence is introduced as before and we show that an absolutely convergent series can be rearranged without altering the sum. There is a slight difference in the behavior of the possible sums of a series which is not absolutely convergent: the points representing these either fill a line, or fill the whole complex plane. The multiplication of two convergent complex series is again permissible if at least one of the two factors converges absolutely.

7. Series of Functions

Power Series. Special Cases. We have already discussed the convergence of series (and sequences) whose terms were numbers, real or complex. We now consider series $\Sigma u_n(x)$, where each term $u_n(x)$ is a function defined for certain values of a variable x. Special cases are those in which $u_n(x) = a_n x^n$, $u_n(x) = a_n e^{nix}$. The questions of convergence for each value of x can be discussed as before, but there are often general criteria available as in the following case.

A series of the form $\Sigma a_n x^n$ is called a power series. It can be shown that there exists a *radius of convergence* ρ, $0 \leq \rho \leq \infty$ with the following property: $\Sigma a_n x^n$ is convergent and indeed absolutely convergent if $|x| < \rho$; $\Sigma a_n x^n$ is divergent if $|x| > \rho$. Moreover, $\rho = 1/\overline{\lim}\, (|a_n|)^{1/n}$; and $\rho = \lim (|a_n|/|a_{n+1}|)$ if this limit exists. In virtue of this absolute convergence and the result on multiplication noted earlier, it follows that two power series can be multiplied term by term and rearranged to give a new power series convergent to the correct product in the interior of the smaller of the two circles of convergence.

The familiar series for e^x, $\sin x$, . . . have radii of convergence $\rho = \infty$, that for $\log (1 + x)$ has radius of convergence 1, while the series $\Sigma n! x^n$ has radius of convergence zero (it only converges for $x = 0$).

A power series with $\rho > 0$ defines a function which has an inverse function which can itself under certain circumstances be represented as a power series. Specifically, suppose $w = f(z) = a_1 z + a_2 z^2 + \cdots$ is convergent in $|z| < r$ and that $a_1 \neq 0$. Then there exists a unique function $z = \phi(w)$, expressible as a power series

$$z = b_1 w + b_2 w^2 + \cdots$$

convergent in a certain circle $|w| < s$ and satisfying

$$f[\phi(w)] = w$$

The coefficients $b_1, b_2, \ldots$ can be expressed in terms of the $a_1, a_2, \ldots$; in particular,

$$b_1 = a_1^{-1} \qquad b_2 = -a_2 \qquad b_3 = 2a_2^2 - a_3 \cdots$$

The following result, due to Abel, is often useful. Suppose that Σa_n is convergent. Then $\Sigma a_n x^n$ is convergent for $|x| < 1$. Denote its sum by $A(x)$. Abel's theorem then states that $\lim_{x \to 1} A(x)$ exists and has the value Σa_n.

Uniform Convergence. The new questions which arise in this section are ones concerning the continuity, differentiability, or integrability of the sum function assuming that individual terms are continuous, differentiable, or integrable. The following examples indicate some of the possibilities.

Example 1.

$$\begin{aligned} \sum_0^\infty x^2(1 + x^2)^{-n} &= 1 + x^2 \qquad x \neq 0 \\ &= 0 \qquad\qquad x = 0 \end{aligned}$$

Here each term is a continuous function, but the sum function is not continuous.

Example 2.

$$\Sigma(n^{-1}x^n - (n + 1)^{-1}x^{n+1}) = x \qquad 0 < x < 1$$

The differentiated series is $\Sigma(x^{n-1} - x^n)$ which has sum 1 for $x \neq 0$ and sum 0 for $x = 0$.

Example 3.

$$S_n(x) = n^2 x e^{-nx} \qquad S_n(s) \to S(x) \equiv 0 \qquad 0 \leq x \leq 1$$

$$\begin{aligned} \int_0^1 S_n(x)\, dx &= \int_0^1 nxe^{-nx} n\, dx \\ &= \int_0^{nx} te^{-t}\, dt \to \int_0^\infty te^{-t}\, dt = 1 \neq \int_0^1 S(x)\, dx \end{aligned}$$

The main new concept required is that of uniform convergence. The series $\Sigma u_n(z)$, convergent to $u(z)$ for z in a set A, is said to be uniformly convergent in A if given any $\epsilon > 0$ there is a number N *independent* of z (but dependent on ϵ) such that

$$\left| u(z) - \sum_{n=1}^{N} u_n(z) \right| < \epsilon$$

for all z in A. A sufficient condition for this is given by the following criterion.

M Test. $\Sigma u_n(z)$ is uniformly convergent for z in A if there exists a convergent series ΣM_n of positive constants such that $|u_n(z)| \leq M_n$ for all z in A, $n = 1, 2, \ldots$. From this $\Sigma n^{-2} z^n$ is uniformly convergent for $|z| \leq 1$ and $\Sigma n^{-2}e^{ni\theta}$ is uniformly convergent for all θ.

A more detailed study of Examples 1, 2, and 3 shows that neither of the series $\Sigma x^2(1 + x^2)^{-n}$, $\Sigma(x^{n-1} - x^n)$ nor the sequence $n^n x e^{-nx}$ is uniformly convergent in any interval including the origin. The number of terms necessary to get a given accuracy $|u(x) - \Sigma u_n(x)| < \epsilon$ increases indefinitely as $x \to 0$ (for sufficiently small ϵ).

The following theorems are often sufficient to justify manipulations with series.

Theorem 1. $u(x) = \Sigma u_n(x)$ is continuous for x in A if $\Sigma u_n(x)$ is uniformly convergent in A and each $u_n(x)$ is continuous in A.

Theorem 2. $\Sigma u_n(x)$ is uniformly convergent in A and may be differentiated term by term if each term $u_n(x)$ is differentiable in A, if the *differentiated* series $\Sigma u'_n(x)$ is uniformly convergent in A, and if $\Sigma u_n(x)$ is convergent at a single point in A.

Theorem 3. $u(x) = \Sigma u_n(x)$ may be integrated term by term, that is,

$$\int_a^b u(x)\,dx = \sum \int_a^b u_n(x)\,dx$$

if the series $\Sigma u_n(x)$ is uniformly convergent in the interval $[a,b]$.

A power series $\Sigma a_n z^n$ with a radius of convergence $\rho \neq 0$ can be differentiated at any point inside its circle of convergence and integrated over any range inside this circle.

Sufficient conditions for the convergence or uniform convergence of Fourier series, or of other orthogonal series can be found in the literature. The only general result we mention is the fact that a Fourier series can always be integrated term by term.

Multiple Series. A doubly infinite sequence $S_{m,n}$ of complex numbers (possibly depending on a parameter) is said to be convergent, as $m,n \to \infty$, to a limit S if, given any $\epsilon > 0$, there are numbers $m_0 = m_0(\epsilon)$, $n_0 = n_0(\epsilon)$ such that

$$|S_{m,n} - S| < \epsilon$$

if $m \geq m_0$, $n \geq n_0$. If the choice of m_0,n_0 can be made independently of a parameter involved the convergence is said to be uniform with respect to that parameter. For instance,

$$(m + n)^{-1} \to 0$$

and $(m + n + x)^{-1} \to 0$, uniformly with respect to x, when x is restricted to the positive real axis.

The situation just described is indicated by

$$\lim_{m,n \to \infty} S_{m,n} = S$$

The existence of this double limit implies the existence and equality of the two repeated limits

$$\lim_{m \to \infty} (\lim_{n \to \infty} S_{m,n}) \qquad \lim_{n \to \infty} (\lim_{m \to \infty} S_{m,n})$$

but the converse is not true. If

$$S_{m,n} = \frac{(m - n)}{(m + n)}$$

then $\lim_{m,n \to \infty} S_{m,n}$ does not exist but

$$\lim_{m \to \infty} \left(\lim_{n \to \infty} \frac{m - n}{m + n} \right) = -1$$

$$\lim_{n \to \infty} \left(\lim_{m \to \infty} \frac{m - n}{m - n} \right) = 1$$

The definition of the sum of a double series is reduced to that of the limit of a double sequence thus:

$$\lim_{m,n \to \infty} \left(\sum_{\lambda=1}^{m} \sum_{\nu=1}^{n} a_{\lambda,\nu} \right) = \sum_{\lambda=1}^{\infty} \sum_{\nu=1}^{\infty} a_{\lambda,\nu}$$

Corresponding to the repeated limits we have the sums by rows and the sums by columns of the double series. The example given above can be modified to show that the sum by rows and the sum by columns can be different; other examples show that even when they are equal the (proper) double sum may not exist. However, the following result is available: if all terms of the series are positive and if one method of summation (by rows, or by columns, or by diagonals) gives a finite sum, the series has this value for its proper double sum. The same remains true if the series is absolutely convergent, i.e., if the series whose terms are $|a_{\lambda,\nu}|$ is convergent.

The questions of analytical manipulation, e.g., passage to a limit, differentiation, integration of a double series of terms depending on one or more parameters, are often quite delicate. Very often, however, these processes can be justified by appeals to obvious extensions of the M test (see above).

An important example of a double series is the case where $a_{\lambda,\nu} = (z + \lambda\omega_1 + \nu\omega_2)^\alpha$. It can be proved that $\Sigma\Sigma a_{\lambda,\nu}$ is convergent if $-\alpha > 2$, provided $\mathfrak{g}(\omega_1/\omega_2) \neq 0$ and that $z + \lambda\omega_1 + \nu\omega_2 \neq 0$.

Asymptotic Series. Another type of series of considerable interest is asymptotic series. These can be introduced as follows. Consider that solution $y = u(x)$ of $xy' - xy + 1 = 0$ which vanishes for $x = +\infty$. By using an integrating factor, choosing the constant of integration appropriately and integrating by parts, we find

$$\begin{aligned} y &= \int_x^\infty \frac{e^{x-t}}{t}\,dt \\ &= \frac{1}{x} - \frac{1}{x^2} + \cdots + (-1)^{n-1}\frac{(n-1)!}{x^n} \\ &\qquad + (-1)^n n! \cdot \int_x^\infty \frac{e^{x-t}}{t^{n+1}}\,dt \end{aligned}$$

The series $\Sigma u_n(x)$, where $u_n(x) = (-1)^n n!/x^{n+1}$ does not converge for any x since $|u_n(x)/u_{n+1}(x)| \to 0$. Nevertheless

$$\left| y - \sum_{r=1}^{n} u_r(x) \right| = \left| (n+1)! \int_x^\infty \frac{e^{x-t}}{t^{n+1}}\,dt \right|$$

$$< (n+1)! \left| \int_x^\infty \frac{dt}{t^{n+1}} \right| = \frac{(n+1)!}{nx^n}$$

and this tends to zero, *for fixed n*, as $x \to +\infty$.

This statement is to be contrasted with that about the remainder of a convergent series, which tends to zero as $n \to \infty$, for all x for which the series is convergent. In other words we get any desired accuracy for any x by taking a sufficient number of terms, while in the present case the accuracy attainable is limited but increases as $x \to \infty$. Since $n!/x^{n+1}$ decreases as n increases from 1 to x and then increases, the best we can do is to take about x terms of the series.

Series with the above property are called *asymptotic*. The formal definition is as follows:

$$F(x) \sim a_0 + a_1x^{-1} + a_2x^{-2} + \cdots$$

if, for each fixed n,

$$\lim_{x\to\infty} x^n[F(x) - a_0 - a_1x^{-1} - \cdots - a_nx^{-n}] = 0$$

The term asymptotic is often used more loosely to signify, for example, that $\lim f(x)/g(x)$ is unity.

One of the more important asymptotic expansions is *Stirling's formula* for $n!$, or for $\Gamma(x)$. In the above sense it can be shown that

$$\ln \Gamma(x) - \left(x - \frac{1}{2}\right) \ln x + x \sim \frac{1}{2} \ln 2\pi + \frac{B_2}{1 \cdot 2x} + \frac{B_4}{3 \cdot 4x^3} + \cdots$$

where the $B_2, B_4, \ldots$ are the *Bernoulli* numbers. Using the first term only,

$$n! \doteqdot \left(\frac{n}{e}\right)^n \sqrt{2\pi n}$$

Other examples of asymptotic expansions occur in the section on special functions. As for their general properties, it is permissible to add, subtract, and multiply them term by term, to integrate them, and usually to differentiate them. While a function has a unique asymptotic expansion, many functions can have the same expansion. For example, $e^{-x} \sim 0 + 0 \cdot x^{-1} + 0 \cdot x^{-2} + \cdots$ and thus $f(x) + Ae^{-x}$ will have the same asymptotic expansion as $f(x)$.

8. Functions of a Complex Variable

Definitions and Examples. Let $z = x + iy$ be an independent variable whose values lie in the complex plane and $f(z)$ a complex number uniquely associated with it under some correspondence. Then $w = f(z)$ is called a one-valued function of z; it is the dependent variable. All definitions used for functions of a real variable apply here too; in particular, *continuity* can be introduced: $f(z)$ is continuous at z if

$$f(\lim_{n\to\infty} z_n) = \lim_{n\to\infty} f(z_n)$$

for all sequences z_n such that $z_n \to z$.

Every complex function $w = f(z)$ can be written in the form

$$w = u(x,y) + iv(x,y)$$

when $u(x,y)$ and $v(x,y)$ are real functions of the real variables x and y.

The definition of differentiability of a function of a complex variable is the same as that for a real variable. We require the existence of

$$\lim_{h\to 0} \frac{f(z + h) - f(z)}{h}$$

where the approach to the limit may be in any manner. More precisely, given any $\epsilon > 0$ there must be an $\delta > 0$, such that for some complex constant A, which we denote by $f'(z)$, $|h| < \delta$ implies

$$\left| \frac{f(z + h) - f(z)}{h} - A \right| < \epsilon$$

In particular if we let $h \to 0$ along the x and y axes the limits of the incrementary ratio must exist and be equal. From this follows the fact that if

$$f(z) = u(x,y) + iv(x,y)$$

then u and v satisfy the Cauchy-Riemann differential equations

$$u_x = v_y \qquad u_y = -v_x$$

From this it follows that if f is differentiable at all points of a neighborhood of z then u and v are harmonic, that is, satisfy the (potential) equation

$$w_{xx} + w_{yy} = 0$$

in that neighborhood. It follows, furthermore, that the curves u = const and v = const cut orthogonally. These curves correspond to equipotentials and streamlines or lines of force in certain realizations. A function $f(z)$ that is differentiable at all points of a region is said to be regular in it.

In certain respects the theory of harmonic functions and the theory of regular functions of a complex variable are equivalent: any harmonic function can be regarded as the real (or imaginary) part of a regular function. Indeed if u is harmonic in the interior of a curve C, then

$$v = \int_{(x_0,y_0)}^{(x,y)} (-u_y\, dx + u_x\, dy)$$

is a one-valued harmonic function, the conjugate of u and $u + iv$ is regular in D. The path of integration can be any curve within C joining a fixed point to the current point.

The fundamental example of a function regular in a domain is a power series Σa_nz^n in its circle of convergence. This can be established directly or a proof can be based on the fact that z^n is differentiable and that a series of functions, each regular in domain, which is uniformly convergent there, is itself regular and can be differentiated term by term.

Properties of Regular Functions. It is essential to realize the following difference between a real function $F(x)$ differentiable in an interval and a complex function $f(z)$ regular in a region. In the latter case the existence of $f'(z)$ implies the existence of all succeeding derivatives $f''(z)$, $f'''(z)$, . . . , while the existence of $F'(x)$ has no such implication, for example, if $F(x) = x^2$, $x < 0$ and $F(x) = 2x^2$, $x > 0$, then $F'(0)$ exists but $F''(x)$ does not exist at $x = 0$. One method of establishing this fundamental property of regular functions is the following. One first shows that if $f(z)$ is a (complex) function continuous on a curve C, then the function

$$F(\xi) = \int_C \frac{f(z)\,dz}{z - \xi}$$

is a regular function of ξ for all ξ not on C, and its derivatives can be obtained by successive differentiation under the sign of integration:

$$F^{(n)}(\xi) = n! \int_C f(z)(z - \xi)^{-n-1}\,dz$$

Next one establishes *Cauchy's theorem:* if $F(z)$ is regular within and on a closed curve C, then

$$\int_C F(z)\,dz = 0$$

In other words the definite integral of a regular function is independent of the path whenever the paths form a closed curve within which the function is regular.

The last condition is essential for, for example,

$$\int_{ACB} z^{-1}\,dz = \pi i$$

while

$$\int_{ADB} z^{-1}\,dz = -\pi i$$

In fact, if C is the circle $|z| = r$, $r > 0$ (Fig. 3.3), then

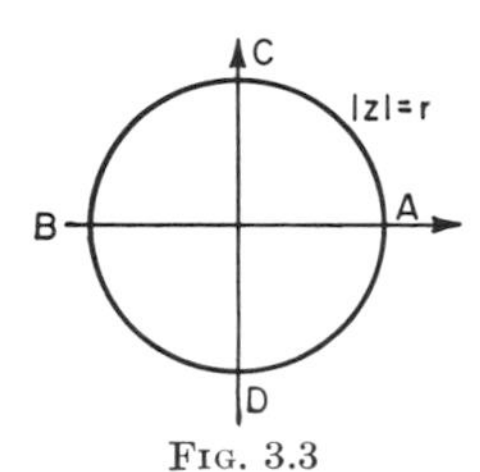

FIG. 3.3

$\int_C z^{-1}\,dz = 2\pi i$. Using this last result, one establishes *Cauchy's formula:* If ξ is a point inside a closed curve within and on which $f(z)$ is regular, then

$$\int_C f(z)(z - \xi)^{-1}\,dz = 2\pi i f(\xi)$$

From this it follows that $f(z)$ is indefinitely differentiable.

It can then be shown that if $f(z)$ is regular in a region D, then $f(z)$ can be represented as a Taylor series about each point of D, convergent in some circle, the radius of which depends on the position of the point,

$$f(z) = \sum \frac{f^{(n)}(z_0)(z - z_0)^n}{n!}$$

This is in contrast with the real case when the existence of one derivative, or even all and the convergence of the Taylor series, does not suffice to ensure that the Taylor series converges to the right sum. For example, consider the function $C(x) = e^{-x^2}$, $x \neq 0$, $C(0) = 0$. All derivatives of $C(x)$ at $x = 0$ exist and are zero so that the formal Taylor series is identically zero. We notice also, as another consequence of Cauchy's formula, that a regular function is completely determined inside a closed curve, by its boundary values on the curve. This is also true for harmonic functions. In particular, if C is the circle $|z - a| = r$, we have the *Poisson integral formula*

$$P(a + \rho e^{i\phi}) = \frac{1}{2\pi} \int_0^{2\pi} P(a + re^{i\theta}) \frac{r^2 - \rho^2}{r^2 + \rho^2 - 2r\rho \cos(\theta - \phi)}\,d\theta$$

Another consequence of Cauchy's formula is *Liouville's theorem:* if $f(z)$ is regular in the whole plane and bounded, then it is a constant. We prove $f(a) = f(b)$ for any a,b. We have

$$f(a) - f(b) = \frac{1}{2\pi i} \oint dz \left(\frac{1}{z - a} - \frac{1}{z - b}\right) f(z) = \frac{a - b}{2\pi i} \oint \frac{f(z)\,dz}{(z - a)(z - b)}$$

where the contour integral is taken round a large circle, center $\frac{1}{2}(a + b)$, for instance, and radius R. Since $f(z)$ is bounded, it is clear that the integrand is $O(R^{-2})$ and thus the integral is $O(R^{-1})$. Letting $R \to \infty$, we deduce $f(a) = f(b)$.

Singularities. Among the types of points at which a function is not regular are poles. If $f(z)$ is regular in the neighborhood of a, but not at a, and if there is a positive integer n such that the function $g(z)$, defined by $g(z) = (z - a)^n f(z)$ for $z \neq a$ and by continuity at a, is regular at a, then $f(z)$ is said to have a pole at a. The least n for which this assertion is true is called the *order* of the pole. Thus z^{-n} has a pole of order n at the origin ($n = 1, 2, \ldots$). The behavior of $f(z)$ near a pole a is simple: $f(z) \to \infty$ no matter how $z \to a$. There are, however, other types of singularity; for example, $z = 0$ for $f(z) = e^{1/z}$ at which the behavior of the function is violent; in this case we can find $z_n \to 0$ such that $\{f(z_n)\}$ has any assigned limit and we can find points as close as we please to $z = 0$ for which $f(z)$ assumes any complex value different from zero. This point is called an *isolated essential singularity:* it is not a pole but there are no other singularities in the neighborhood of $z = 0$. The point $z = 0$ is an essential singularity for cosec z^{-1}, for there are poles at $z = (n\pi)^{-1}$, $n = 1, 2, \ldots$.

While the Taylor expansion of a function is not available in the neighborhood of a point at which it is not regular, there is a generalization, the *Laurent expansion,* valid in annular regions. We have, at a pole of order r,

$$f(z) = A_{-r}(z - a)^{-r} + \cdots + A_{-1}(z - a)^{-1} + \phi(z)$$

where $\phi(z)$ is regular at a and can therefore be expanded as a Taylor series

$$\phi(z) = \sum_{n=0}^{\infty} A_n(z - a)^n$$

The sum

$$\sum_{s=1}^{r} A_{-s}(z - a)^{-s}$$

is called the *principal part* of $f(z)$ at a. The A_{-s} are uniquely determined: A_{-1} is called the *residue* of

$f(z)$ at A. The following result, which is an extension of Cauchy's formula, is of considerable practical importance. If $f(z)$ is regular inside (and on) a closed curve C except at a finite number of poles $a_1, \ldots, a_n$,

$$\int_C f(z)\,dz = 2\pi i \sum_{i=1}^{n} \{\text{residue of } f(z) \text{ at } a_i\}$$

A consequence of this is the following:

$$\frac{1}{2\pi i}\int_C \frac{f'(z)\,dz}{f(z)} = Z - P$$

where Z is the number of zeros inside C and P the number of poles there (multiple poles or zeros being counted according to their multiplicity). From this we deduce in the case when there are no poles inside C, since

$$\int \frac{f'(z)\,dz}{f(z)} = \log f(z) = \log |f(z)| + i \arg f(z)$$

that the increment in $\arg f(z)$ as f describes a closed curve is $2\pi Z$.

The following results are of use in the evaluation of residues. If $f(z)$ has a *simple* pole at $z = p$, then

(1) residue of $f(z)$ at $p = \lim_{z \to p} (z - p)f(z)$

If $f(z) = g(z)/h(z)$, where $h(z)$ has a simple zero at p and $g(h) \neq 0$ and both are regular in the neighborhood of $z = p$, then

(2) residue of $f(z)$ at $p = \dfrac{g(p)}{h'(p)}$

Examples of Contour Integrations

$$\int_0^{2\pi} \frac{\sin^2\theta\,d\theta}{a + b\cos\theta} = \frac{2\pi(a - \sqrt{a^2 - b^2})}{b^2} \qquad a > b > 0$$

Put $z = e^{i\theta}$, then the integrand can be expressed as a rational function of z,

$$I = \int_0^{2\pi} \frac{\sin^2\theta\,d\theta}{a + b\cos\theta} = \oint \frac{(z^2 - 1)^2\,dz}{-2iz^2(bz^2 + 2az + b)}$$

where the contour integral is taken round the unit circle C. Using the residue theorem,

$$I = 2\pi i \left\{\text{sum of residues of } \frac{(z^2 - i)^2}{-2iz^2(bz + 2az + b)} \text{ at its poles inside } C\right\}$$

When $a > b > 0$ the poles inside C are at 0,

$$[-a + \sqrt{(a^2 - b^2)}]/b$$

the former being a double pole, the latter a simple one.

We use relation (1) above to calculate the residue at $p_1 = (-a + \sqrt{(z^2 - b^2)})/b$. It is

$$\lim_{z \to p_1} \frac{(z - p_1)(z^2 - 1)^2}{-2iz^2b(z - p_1)(z - p_1^{-1})} = \frac{(p_1^2 - 1)}{-2ip_1^2b(p_1 - p_1^{-1})} = \frac{(p_1^2 - 1)}{-2ibp_1} = \frac{p_1 - p_1^{-1}}{-2ib} = \frac{2\sqrt{a^2 - b^2}}{-2ib^2}$$

To handle the double pole at the origin, we proceed as follows. We obtain the Laurent expansion from the elementary theory of partial fractions which gives

$$\frac{(z^2 - 1)^2}{-2iz^2(bz^2 + 2az + b)} = \frac{B}{z^2} + \frac{A}{z} + \frac{C}{z - p_1} + \frac{D}{b(z - p_1^{-1})}$$

so that

$$\frac{(z^2 - 1)^2}{-2i(bz^2 + 2az + b)} = B + Az + z^2[C(z - p_1^{-1})^{-1} + Db^{-1}(z - p_1^{-1})^{-1}]$$

Therefore if we expand the left-hand side in powers of z, the coefficient of z is the residue required. This can be done, for instance, by long division:

$$\begin{array}{r} i(2b)^{-1} - iab^{-2}z + \cdots \\ -2i(bz^2 + 2az + b)\overline{)1 - 2z^2 + z^4} \\ 1 + 2ab^{-1}z + z^2 \\ \hline -2ab^{-1}z - 3z^2 + z^4 \\ \cdots\cdots\cdots \end{array}$$

The residue required is therefore $-iab^{-2}$. Hence

$$I = 2\pi i\left(\frac{2\sqrt{a^2 - b^2}}{-2ib^2} - \frac{ia}{b^2}\right) = \frac{2\pi(a - \sqrt{a^2 - b^2})}{b^2}$$

As a second example we choose the evaluation of

$$I = \int_0^\infty \frac{x^6\,dx}{x^8 + 1}$$

Here we consider $\oint \dfrac{z^6\,dz}{z^8 + 1}$, where the contour is the real axis from $-R$ to R, and the upper half of the semicircle of radius R, where R is a (large) parameter. The absolute value of the integral along the semicircle does not exceed:

$$\pi R \frac{R^6}{R^8 - 1}$$

and this tends to zero as R increases indefinitely. We therefore have, as $R \to \infty$,

$$\oint \frac{z^6\,dz}{z^8 + 1} = \int_{-R}^{+R} \frac{x^6}{x^8 + 1}\,dx + o(1)$$

Hence,

$$\int_0^\infty \frac{x^6\,dx}{x^8 + 1} = \frac{1}{2}2\pi i\left(\text{sum of residues of } \frac{z^6}{z^8 + 1} \text{ at its poles within upper half of the circle } |z| = R\right)$$

The poles in question are simple ones at

$$p_r = \exp\left[\frac{\pi(2r + 1)i}{8}\right]$$

where $r = 0,1,2,3$. Using relation (2) above, we see that the residue at p_r is

$$(8p_r)^{-1} = \frac{1}{8}\exp\left[\frac{-\pi(2r + 1)i}{8}\right]$$

and the sum, which must be purely imaginary, is

$$-\frac{i}{8}\left(\sin\frac{\pi}{8} + \sin\frac{3\pi}{8} + \sin\frac{5\pi}{8} + \sin\frac{7\pi}{8}\right) = \frac{-i}{4}\left(\sqrt{1 + \frac{1}{\sqrt{2}}}\right)$$

Hence $$I = \frac{\pi}{4}\sqrt{1 + \frac{1}{\sqrt{2}}}$$

A third example: integration of $\exp(i\pi z^2)\operatorname{cosec}\pi z$ around the parallelogram having two sides through the points $\pm\frac{1}{2}$, making an angle of $\frac{1}{4}\pi$ with the x axis, the other two sides being part of the lines $y = \pm R$, R being large. This integral has the value $2i$, since the only pole of the integral is a simple one at the origin where the residue is $-\pi^{-1}$. Combining the contributions from the larger sides, and observing that the contributions from the smaller sides become negligible as $R \to \infty$, we obtain

$$\int_0^\infty \exp(-x^2)\,dx = \tfrac{1}{2}\sqrt{\pi}$$

Analytic Continuation. An important concept in the general theory of functions of a complex variable is that of analytic continuation. Through this a proper understanding of many-valued functions and Riemann surfaces is obtained. Consider the series

$$f(z) = (1 + z)^{-1} = 1 - z + z^2 - \cdots$$

This is regular for $|z| < 1$ and in particular at $z = \frac{1}{2}$. Since $f^{(m)}(z) = (-1)^m m!(1 + z)^{-m-1}$ the Taylor expansion of $f(z)$ about $z = \frac{1}{2}$ is

$$f(\tfrac{1}{2} + w) = \Sigma(-1)^m(\tfrac{2}{3})^m w^m$$

where $w = z - \frac{1}{2}$, which is convergent for $|w| < \frac{3}{2}$. The series

$$F(w) = \Sigma(-1)^m(\tfrac{2}{3})^m w^m$$

coincides with $f(z)$ in the whole circle $|z| < 1$ but has a meaning in a larger region. In such circumstances F is said to be a direct analytic continuation of f. The situation just described is not typical; in general, the new series converges in a circle which overlaps but does not include the original circle.

This process of direct continuation can be repeated and, in certain circumstances, we can obtain an analytic continuation of a function into a region where the original function was defined, but having values distinct from those of the original function. For instance, it can be proved, without use of the properties of the logarithmic function, that the series

$$P(z) = z + \tfrac{1}{2}z + \tfrac{1}{3}z^3 + \cdots$$

can be continued by means of Taylor series at points on the circle $|z - 1| = 1$ in such a way as to obtain, as a continuation of this series,

$$2\pi i + z + \tfrac{1}{2}z^2 + \tfrac{1}{3}z^3 + \cdots$$

Repetition of this process gives all the determinations

$$2n\pi i + z + \tfrac{1}{2}z^2 + \tfrac{1}{3}z^3 + \cdots \qquad n = 0, \pm1, \pm2, \ldots$$

of the continuations of $P(z)$.

Whereas the representation by power series is the basic theoretical method, it is often convenient practically to contemplate analytic continuations expressed in different forms, e.g., the following representations of $\Gamma(z)$:

$$\Gamma(z) = \int_0^\infty e^{-t}t^{z-1}\,dt$$

valid for $\Re z > 0$ and

$$\Gamma(z) = \int_1^\infty e^{-t}t^{z-1}\,dt + \sum_{n=0}^{\infty}\frac{(-1)^n}{n!(z+n)}$$

for $z \neq -1, -2, \ldots$.

It is easy to construct functions which cannot be continued beyond their original domain of definition. For instance, Σz^{2^n} has the unit circle as a natural boundary.

The Principle of the Permanence of Functional Relations. This principle is important. Assume that the relation

$$\sin 2x = 2\sin x\cos x$$

has been established for x real. Assume further that the functions

$$s(z) = z - \frac{z^3}{3!} + \frac{z^5}{5!} - \cdots$$

$$c(z) = 1 - \frac{z^2}{2!} + \frac{z^4}{4!} - \cdots$$

are known to coincide with $\sin x$ and $\cos x$ for $z = x$ real. We wish to conclude that

$$s(2z) = 2s(z)c(z)$$

for all z. This is obtained as a special case of the principle, deduced from the fact that if two functions are regular in a simply connected domain G and coincide at a set of (infinitely many distinct) points in G which have a limit point in G, then they coincide throughout G. The condition that the limit point is in G is essential.

Consider $f(z) = \sin z^{-1}$, $g(z) \equiv 0$ in the circle $|z - 1| < 1$. They coincide at

$$z_n = (n\pi)^{-1} \qquad n = 1,2, \ldots$$

but are not identical: the limit 0 of the sequence $\{z_n\}$ does not lie within the circle $|z - 1| < 1$.

A very simple method of analytic continuation is given by the following reflection method due to Schwarz. If $f(z)$ is regular in the upper half of the unit circle (that is, $|z| < 1$, $\mathcal{I}(z) > 0$) and continuous in $|z| < 1$, $\mathcal{I}(z) \geq 0$, then the function $F(z)$ defined

$$\begin{aligned} F(z) &= f(z) & |z| &< 1 & \mathcal{I}(z) &\geq 0 \\ F(z) &= [f(z^*)]^* & |z| &< 1 & \mathcal{I}(z) &< 0 \end{aligned}$$

is regular in the whole unit circle $|z| < 1$. There are many extensions of this.

Special Classes of Functions. Functions such as e^z, $\sin z$ which have no singularities are called *integral* or *entire functions,* and are represented by power series with an infinite radius of convergence. Such functions can be factored, in somewhat the same way as polynomials, but the analogy is not complete: e^z has no zeros, $(a_n z^n + \cdots + a_n)e^z$ has

n zeros, $\sin z$ has infinite number of zeros. Typical of the representation in the product form are the following:

$$\sin x = x \prod_{n=1}^{\infty} \left(1 - \frac{x^2}{n^2\pi^2}\right)$$

$$\frac{1}{\Gamma(z)} = ze^{\gamma z} \prod_{n=1}^{\infty} \left[\left(1 + \frac{z}{n}\right) e^{-z/n}\right]$$

Functions which are regular except at poles of which only a finite number lie in any finite part of the plane have properties somewhat similar to those of rational functions, e.g., expansions in partial functions typified by

$$\operatorname{cosec}^2 z = \sum_{m=-\infty}^{\infty} (z - m\pi)^{-2}$$

$$\zeta(z) = \frac{1}{2} + \sum\sum{}' \left[\frac{1}{z - \omega(m_1,m_2)} + \frac{1}{\omega(m_1,m_2)} + \frac{z}{\omega(m_1,m_2)^2}\right]$$

where the prime indicates summation over all pairs of (integral) indices m_1, m_2 except $m_1 = 0$, $m_2 = 0$, where $\omega(m_1,m_2) = 2m_1\omega_1 + 2m_2\omega_2$ and where $\mathscr{I}(\omega_1/\omega_2) > 0$. Such functions are called *meromorphic*.

A function $f(z)$, such that for some constant ω we have $f(z) = f(z + \omega)$ for all z, is said to have period ω. For instance $e^{2\pi iz}$ has period 1. It can be shown that, in general, any function with period 1, is a rational function of $e^{2\pi iz} = u$, for example,

$$\sin 2\pi z = -\tfrac{1}{2}i(u - u^{-1})$$

A regular function with two periods whose ratio is real and irrational is necessarily a constant. If we allow the periods ω_1,ω_2 to have a complex ratio, then we can have nontrivial examples, but they cannot be integral functions: they must be at least meromorphic. A simple doubly periodic function is the *Weierstrass elliptic function*

$$\wp(z) = \frac{1}{z^2} + \sum\sum{}' \left[\frac{1}{(z - \omega(m_1,m_2))^2} - \frac{1}{\omega(m_1,m_2)^2}\right]$$

The double periodicity is formally obvious and can be established by showing that the double series is absolutely convergent and can therefore be rearranged at will. In addition to having $2\omega_1$, $2\omega_2$, as periods, $\wp(z)$ has also all the numbers $\omega(m_1,m_2)$ for $m_1 = 0, \pm 1, \pm 2, \ldots, m_2 = 0, \pm 1, \pm 2, \ldots$ as periods. The periods $2\omega_1$, $2\omega_2$, are called primitive periods. $\wp(z)$ has double poles at all the points $\omega(m_1,m_2)$ and no other finite singularities.

9. Conformal Mapping

Introduction. If $w = f(z) = u + iv$ is a regular function of z, then u and v satisfy the Cauchy-Riemann equations

$$u_x = v_y \qquad u_y = -v_x$$

from which it follows that u, v are harmonic functions, satisfying $h_{xx} + h_{yy} = 0$ and that the curves $u = \alpha$, $v = \beta$ (α,β any constants) cut orthogonally. The transformation from z to w is (locally) conformal; configurations in the z plane near a point z_0 go over into configurations in the w plane which are similar to them (in the sense of elementary geometry); in fact they are subjected to a rotation of amount $\arg f'(z_0)$ and a magnification of amount $|f'(z_0)|$, provided that $|f'(z_0)| \neq 0$.

The subject of conformal mapping develops these two ideas. Extensive treatments, from the practical viewpoint, are given in the books cited in Sec. *c* of the References at the end of the chapter.

Examples. The mapping generated by $w = z^2$ and by the exponential functions which we take in the form $w = e^{2\pi iz}$, are indicated in Fig. 3.4*a* to *h* where corresponding curves in the two planes are marked similarly (e.g., nature of lines).

In the former case, the conformal property breaks down at the origin (see Fig. 3.4*a* and *b*) where

$$f'(z) = 0$$

To $u = c$, $v = c'$ correspond (parts of) the rectangular hyperbolas $x^2 - y^2 = c$, $2xy = c'$ (see Fig. 3.4*c* and *d*). To (parts of) the curves $x = d$, $y = d'$ correspond the parabolas $v = 4d(d - u)$, and

$$v = 4d'(d' + u)$$

(Fig. 3.4*e* and *f*). The hyperbolas cut orthogonally except for $c = c' = 0$, and similarly the parabolas cut orthogonally except for $d = d' = 0$.

Since $w = e^{2\pi iz}$ has period 1, we consider the correspondence between the strip $0 < |\Re z| < 1$ and the w plane cut along the real axis from 0 to ∞. The lines $x = c$, $y = c'$ transform into lines through the origin and circles center the origin, respectively (see Fig. 3.4*g* and *h*). We have, in fact,

$$z = x + iy_0 \qquad 0 \le x \le 1 \qquad u = e^{-2\pi y_0}\cos 2\pi x, \quad v = e^{-2\pi y_0}\sin 2\pi x$$

$$z = x_0 + iy \qquad -\infty < y < \infty \qquad w = Re^{i\phi}, \quad R = e^{-2\pi y}, \quad \phi = 2\pi x_0$$

The transformation $w = e^z$ is obtained from this by a rotation through a right angle and a change of scale in the z plane.

The bilinear transformation $w = (az + b)/(cz + d)$ where $ad - bc \neq 0$, is of considerable importance. Important special cases are

(1) $w = z^{-1}$

(2) $w = e^{i\delta}\dfrac{z - \alpha}{\bar{\alpha}z - 1} \qquad \alpha = a + ib \qquad |\alpha| < 1,\ \delta$ real

(3) $w = e^{i\delta}\dfrac{z - \alpha}{z - \bar{\alpha}} \qquad \alpha = a + ib \qquad b > 0,\ \delta$ real

(4) $w = \dfrac{az + b}{cz + d} \qquad a,b,c,d$ proportional to real numbers, $ad - bc > 0$

The first of these corresponds to inversion in the unit circle: it maps $0 < |z| < 1$ conformally onto $1 < |w| < \infty$. The second is the most general mapping of $|z| < 1$ onto $|w| < 1$; the third is the most general mapping of $\mathscr{I}(z) > 0$ onto $|w| < 1$ and the fourth the most general mapping of $\mathscr{I}(z) > 0$ onto $\mathscr{I}(w) > 0$. Most of these results can be established geometrically.

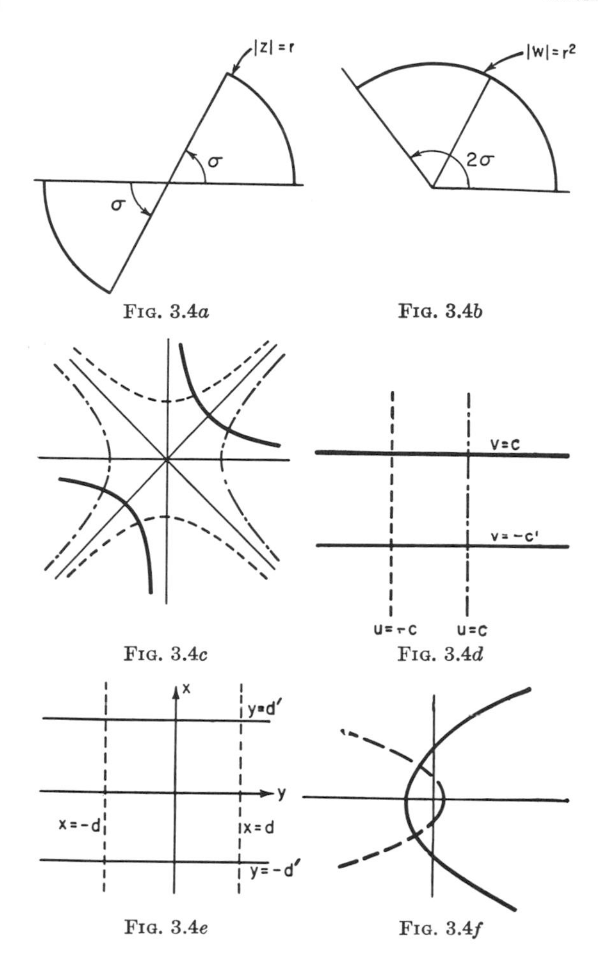

Fig. 3.4a

Fig. 3.4b

Fig. 3.4c

Fig. 3.4d

Fig. 3.4e

Fig. 3.4f

The case $w = \frac{1}{2}(z + z^{-1})$ is of interest in classical aerodynamics. The fact that w has a pole at the origin and that w' has zeros at ± 1 make this case more complicated. If we write $z = re^{i\theta}$, we find $u = \frac{1}{2}(r + r^{-1}) \cos \theta$, $v = \frac{1}{2}(r - r^{-1}) \sin \theta$ so that to a circle in z plane there corresponds an ellipse on the w plane, and the same ellipse corresponds to reciprocal values of r. Confining attention to $|z| < 1$, we see that as $r \to 0$ the ellipse expands indefinitely becoming more circular, while as $r \to 1$ the ellipse shrinks to the line $-1 < u < 1$. The behavior of $|z| > 1$ is similar.

For $0 < d < \frac{1}{2}$, the map of the interior of the circle $|z + d| = 1 - d$ or the exterior of the circle $|z - d(1 - 2d)^{-1}| = (1 - d)(1 - 2d)^{-1}$ is a symmetric "airfoil" having a zero angle at $w = d$. If we choose to map a circle with center not on the real axis, we obtain an unsymmetrical airfoil. Generalization of this *Joukowski transformation*, and the specialization of the various parameters introduced, make possible the conformal mapping of airfoils of realistic shape.

Suppose $f(z)$ is regular in D and that f never assumes the same value more than once in D. Then it can be shown that $f'(z)$ does not vanish in D. Let $\mathfrak{D}$ denote the set of points $w = f(z)$ for z in D. Then f generates a one-to-one mapping of $\mathfrak{D}$ onto D. Let $\phi(w)$ denote the inverse function of this mapping. This is a regu-

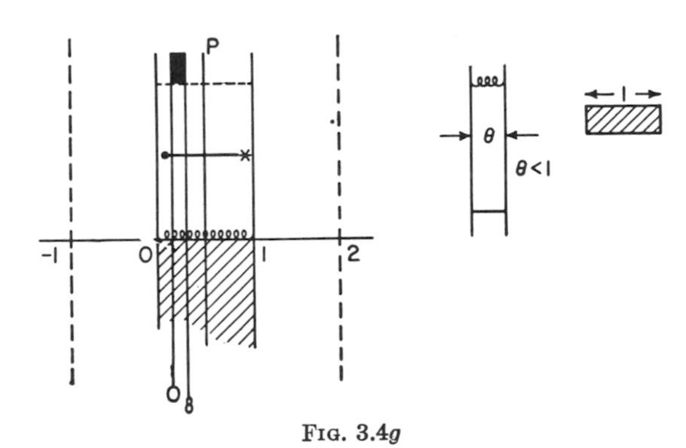

Fig. 3.4g

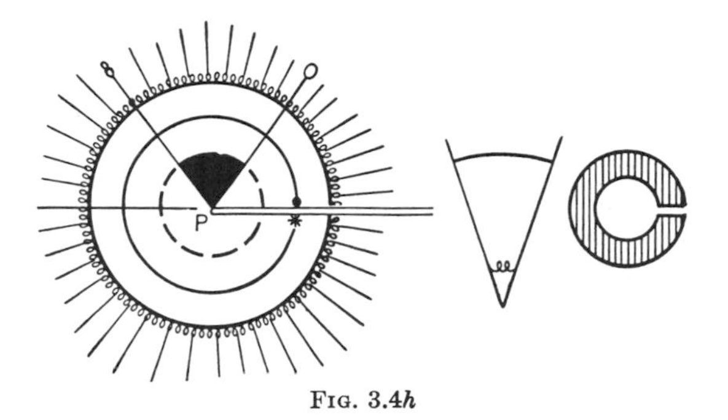

Fig. 3.4h

lar function of w in virtue of the existence of

$$\lim_{\omega \to \omega_0} \frac{\phi(w) - \phi(w_0)}{w - w_0} = \lim \frac{z - z_0}{w - w_0} = \left(\lim \frac{w - w_0}{z - z_0}\right)^{-1} = [f'(z_0)]^{-1}$$

the critical step following since $f'(z_0) \neq 0$. There is therefore complete symmetry between the behavior of f in D and that of $\phi = f^{-1}$ in $\mathfrak{D}$. It follows that it is unnecessary to study the problem of the mapping of one general domain onto another; it is sufficient to study that of a general domain onto a standard domain, e.g., a circle or a half plane or a circular annulus.

The study of functions $f(z)$ which are regular and one-valued in $|z| < 1$ has been carried a considerable distance. Such functions can be expressed as a power series with radius of convergence unity, and it is convenient to normalize them by setting $f(0) = 0$, $f'(0) = 1$. The power series is therefore

$$f(z) = z + a_2 z^2 + a_3 z^3 + \cdots$$

The question arises as to what restrictions on the coefficients are implied by our hypothesis; it has been shown that $|a_2| \le 2$, $|a_3| \le 3$, $|a_4| \le 4$, and it has been conjectured that $|a_n| \le n$. The critical function would be

$$f(z) = z(1 - z)^{-2} = z + 2z^2 + 3z^3 + \cdots$$

Restrictions on the distortion produced by mappings of this kind have been investigated: it can be shown that $\mathfrak{D}$ must always include the circle $|w| < \frac{1}{4}$ and this result is best possible, since in the case of

$f(z) = z(1 - z)^{-2}$ the point $z = 1$ transforms into $w = -\frac{1}{4}$.

A general existence theorem is available: if D is a simply connected domain, the interior of a closed curve C, there is a function $f(z)$, regular in D which maps D conformally on the interior of the unit circle. This function $f(z)$ is not unique; in fact,

$$F(z) = e^{i\delta} \frac{f(z) - \alpha}{\bar{\alpha} f(z) - 1}$$

has the same properties for any real δ, and any α with $|\alpha| < 1$. This is, however, the most general function of its kind. There are essentially three real parameters available in F and these can be chosen, for example, to make three points on C correspond to three given points on the circle $|z| = 1$ or to make a point, and a direction through it correspond to a given point and direction.

This existence theorem is of little use in practice, although it is conceivable that some of the proofs are of such a constructive character that they can be adapted for use on high-speed computing machines. For the present, however, reference to dictionaries, together with a knowledge of some of the grammar of the subject, is recommended.

This result cannot be extended to any simply connected domain whatsoever. The whole plane is such a domain, and if $w = f(z)$ mapped this on $|w| < 1$, we could have a function regular in the whole plane and bounded, which is impossible by Liouville's theorem. However, the theorem is true if the domain has more than one boundary point.

The mapping function for any polygon can be written down *implicitly*. This result is due to Schwarz and Christoffel. For instance, a triangle with angles $\alpha\pi$, $\beta\pi$, $\gamma\pi$ is mapped on the half plane $\mathfrak{I}(w) > 0$ by an expression of the form

$$z = A \int (w - a)^{\alpha - 1}(w - b)^{\beta - 1}(w - c)^{\gamma - 1}\, dw + B$$

where a,b,c, are arbitrary and A,B are then determined by the triangle.

The case of the mapping of a rectangle on $\mathfrak{I}(w) > 0$ is accomplished by

$$z = A \int [(w - a)(w - b)(w - c)(w - d)]^{-1/2}\, dw + B$$

Consider the special case

$$z = \int_0^w \frac{d\xi}{\sqrt{(1 - \xi^2)(1 - k^2\xi^2)}}$$

where $0 < k^2 < 1$. Without reference to the theory of elliptic functions it can be shown that this maps the interior of the rectangle with vertices $K, K + iK'$, $-K + iK'$, $-K$ on $\mathfrak{I}(w) > 0$. Here K, K' are constants depending on k (see Sec. 11).

Example. The practical problem of conformal mapping is greatly simplified by the following general theorem, which states, roughly, if the boundaries match, so do the interiors. Precisely: if $w = f(z)$ is regular in D, the interior of a simple closed curve C (and continuous on $C + D$) and, if as z describes C positively, $w = f(z)$ describes a simple closed curve $\mathfrak{C}$ exactly once, then $\mathfrak{C}$ is described positively and $w = f(z)$ maps D conformally on the interior $\mathfrak{D}$ of $\mathfrak{C}$. This result cannot, however, be extended in a natural way to the case of infinite domains.

As an example of the combination of basic mappings, let us determine the mapping which carries $|z| < 1$ on the exterior of $v^2 = -4u$ in such a way that $z = 0$ becomes $w = 3$ and that the positive real direction at these points correspond.

Examination of $w = z^2$ shows that $\mathfrak{R}(z) > \alpha$ becomes the exterior of $v^2 = -4\alpha^2(u - \alpha^2)$. Take $\alpha = 1$; then $\mathfrak{R}(z) > 1$ goes onto the exterior of $v^2 = -4(u - 1)$, and therefore onto the exterior of $v^2 = -4u$ under $w = z^2 + 1$. Further, with

$$w = z^2 + 1$$

we shall let $z = \sqrt{2}$ correspond to $w = 3$ and, since $w' = 2\sqrt{2}$ there, the positive real directions at the points correspond. We are left therefore with the determination of a (bilinear) transformation which maps $|z| < 1$ onto $\mathfrak{R}(z) > 1$ in such a way that $z = 0$ goes into $z = \sqrt{2}$, positive real directions being preserved at these points. Or, with the change $z = \zeta + 1$ we have to map $|z| < 1$ onto $\mathfrak{R}(\zeta) > 0$ in such a way that $z = 0$ becomes $\zeta = \sqrt{2} - 1$. This is accomplished by

$$z = e^{i\delta}\left[\frac{\zeta - (\sqrt{2} - 1)}{\zeta + (\sqrt{2} - 1)}\right]$$

and, in order to preserve the positive direction, we take $\delta = 0$.

Hence, we have

$$w = z^2 + 1 = (\zeta + 1)^2 + 1 = \left[\frac{z(\sqrt{2} - 2) + \sqrt{2}}{1 - z}\right]^2 + 1$$

10. Orthogonality

General. Two (possibly) complex-valued functions f, ϕ defined on a real interval (a,b) (possibly infinite) are said to be *orthogonal* if

$$(\phi, f) = \int_a^b \phi^* f\, dx = 0$$

where ϕ^* denotes the complex conjugate of ϕ. A system of functions $\{\phi_n\}$ is said to be *orthogonal* if every pair of different functions are orthogonal; it is said to be *normal*, if, in addition

$$(\phi_n, \phi_n) = \int_a^b \phi_n^* \phi_n\, dx = 1$$

for all n. If we define the Kronecker delta $\delta_{m,n}$ by

$$\delta_{m,n} = 0 \qquad \text{if } m \neq n \qquad \delta_{n,n} = 1$$

we can write the conditions for a normal orthogonal system very compactly as $(\phi_m, \phi_n) = \delta_{n,n}$. The word orthonormal is often used in place of normal and orthogonal. The classical example is that of the set of functions

$$e^{nix} \qquad n = 0, \pm 1, \pm 2, \ldots$$

in the interval $(0, 2\pi)$. These form an orthogonal, but

not a normal set. The set $e^{inx}/\sqrt{2\pi}$, $n = 0, \pm 1, \pm 2, \ldots$, is normal and orthogonal.

Given a normal orthogonal system ϕ_n, we define the *Fourier coefficients* a_n of a function f with respect to $\{\phi_n\}$ and the interval (a,b) by the equations

$$a_n = (\phi_n, f) = \int_a^b \phi_n^* f\,dx$$

In order that these should exist, f and ϕ_n must be suitably restricted, e.g., it is sufficient to assume f and each ϕ_n are of integrable square. This follows from the *Schwarz inequality*

$$|(f,g)| \leq \{(f,f)(g,g)\}^{1/2}$$

We call the series

$$\sum_{n=1}^{\infty} a_n \phi_n(x)$$

the *Fourier series* of f (with respect to ϕ_n) and the successive partial sums f_r of this series are called the *Fourier polynomials* of f. We indicate this formal relation by

$$f(x) \sim \sum_{n=1}^{\infty} a_n \phi_n(x)$$

(There will be no confusion with the symbol for asymptotic equality.) Examples in the classical case are the following: If $f(x)$ has period 2π and $f(x) = [(\pi - x)/2]^2$, $0 \leq x < 2\pi$, then

$$f(x) \sim \frac{\pi^2}{12} + \sum_{k=1}^{\infty} \frac{\cos kx}{k^2}$$

If $f(x)$ has period 2π and $f(x) = (\pi - x)/2$,

$$0 < x < 2\pi$$

then

$$f(x) \sim \sum_{k=1}^{\infty} \frac{\sin kx}{k}$$

When does the series $\Sigma a_n \phi_n$ converge to the sum $f(x)$? If $f(x)$ and the system $\{\phi_n\}$ are reasonably behaved, the series will converge to the correct sum. The second example above indicates an exceptional case which occurs frequently: when $f(x)$ has a simple jump at a point ($x = 0$, in this case) the series converges to the average of the two limiting values $-\pi/2$, $+\pi/2$ in this case.

Even if the Fourier series does not converge, the *Fourier* polynomials give the best mean-square approximation to $f(x)$ among all the "trigonometric" polynomials, i.e., expressions

$$s_n = \sum_{r=1}^{n} c_r \phi_r$$

where the c_r are any constants. Expanding and using the definitions of the a_r, we find

$$\int_a^b |f - s_n|^2\,dx = \int_a^b |f|^2\,dx + \sum_1^n |c_r - a_r|^2 - \sum_1^n |a_r|^2$$

In particular, taking $c_r = a_r$ so that $s_r = f_r$,

$$\int_a^b |f - f_n|^2\,dx = \int_a^b |f|^2\,dx - \sum_1^n a_r^2$$

and thus $\displaystyle\int_a^b |f - f_n|^2\,dx \leq \int_a^b |f - s_n|^2\,dx$

with equality occurring if and only if each $a_r = c_r$ ($r = 1, 2, \ldots, n$), that is, the Fourier polynomial gives the best mean-square approximation. It is clear that the left-hand side is non-negative and this means

$$\sum_{r=0}^{n} |a_r|^2 \leq \int_a^b |f|^2\,dx$$

so that, if f is of integrable square, Σa_r^2 is convergent. This is known as *Bessel's inequality*.

An important property of an orthogonal system is that of completeness: $\{\phi_n\}$ is said to be *complete* if there is no (nontrivial) function orthogonal to each ϕ_n. It can be shown that if $\{\phi_n\}$ is a complete normal orthogonal system, then if f is of integrable square,

$$\int_a^b |f - f_n|^2\,dx \to 0 \qquad \text{as } n \to \infty$$

and

$$\sum |a_n|^2 = \int_a^b |f|^2\,dx$$

The latter result is known as *Parseval's theorem*. Application of it to $f + g$ and to $f - g$ gives the following more general form:

$$\sum a_n b_n = \int_a^b f g^*\,dx$$

where b_n are the Fourier coefficients of g.

The following converse of Parseval's theorem is true. If the series $\Sigma |c_r|^2$ is convergent, there exists a function f of integrable square which has c_r for its Fourier coefficients with respect to $\{\phi_n\}$. This is the *Riesz-Fischer theorem.*

Given any system of functions $\{\phi_n\}$, it is possible to construct a system of finite linear combinations of them which constitute an orthogonal system $\{f_n\}$. The process can be made clear by discussing the case when the given sequence is $1, x, x^2, \ldots$ and the interval is $-1, +1$. We choose $f_1 = 1$. We try $f_2 = x$ and find that this is satisfactory since

$$(f_1, f_2) = \int_{-1}^{+1} x\,dx = 0$$

We try $f_3 = x^2$ and find this is unsatisfactory since $(f_1, f_3) = \int_{-1}^{+1} x^2\,dx \neq 0$. We therefore try

$$f_3 = \alpha f_1 + \beta f_2 + x^2$$

and choose α, β to satisfy $(f_1, f_3) = 0$, $(f_2, f_3) = 0$; that is,

$$(f_1, f_1) + \alpha \cdot 0 + (f_1, x^2) = 0$$

and

$$0 + \beta(f_2, f_2) + (f_2, x^2) = 0$$

This gives $\alpha = -\frac{1}{3}$, $\beta = 0$ and thus $f_3 = x^2 - \frac{1}{3}$. Continuing, we try $f = \alpha f_1 + \beta f_2 + \gamma f_3 + x^3$ and find $f_4 = x^3 - \frac{3}{4}x$. This process can be continued and we obtain an orthogonal system $\{f_n\}$ which is essentially the Legendre system $\{P_n(x)\}$. We obtain a normal orthogonal system by taking the $\{f_n\}$ just constructed and dividing each f_n by the constant $(f_n,f_n)^{1/2}$. This orthonormalization process is called the Gram-Schmidt algorithm.

The system

$$\sin x, \sin 2x, \ldots$$

is not complete in $(0,\pi)$ since the Fourier coefficients of $f = \cos x$ with respect to this system are all zero. The system is, however, complete in $(0,2\pi)$.

The system $\{P_n(x)\}$ is complete in $(-1,1)$. The proof of this can be made to depend on the *Weierstrass approximation theorem:* if $f(x)$ is a function continuous in a finite closed interval, there exists a sequence of polynomials $P_n(x)$ such that $P_n(x)$ converges uniformly to $f(x)$ in the interval. The completeness of $\{P_n(x)\}$ is essentially equivalent to the *finite-moment theorem:* if all the moments

$$\mu_n = \int_a^b f(x)x^n\,dx$$

of a continuous function $f(x)$ over a *finite* interval a,b, are zero, then $f(x)$ is identically zero. This result cannot be extended to the infinite case, as is shown by

$$f(x) = \exp[-x^{1/4}]\sin[x^{1/4}]$$

A generalization of the concept of orthogonality: two functions f,ϕ are *orthogonal* with respect to a weight function $w(x)$ (usually assumed to be nonnegative in the range) if

$$\int_a^b f(x)\phi(x)w(x)\,dx = 0$$

In this circumstance it is clear that $f(x)\sqrt{w(x)}$ and $\phi(x)\sqrt{w(x)}$ are orthogonal in the usual sense.

Particularly important are the systems of polynomials obtained by orthogonalizing with respect to certain $w(x)$, in the manner just described, the set of polynomials $1, x, x^2, \ldots$. We shall discuss the following cases in some detail.

Legendre:

$P_n(x)$	$w(x) = 1$	$a = -1,$	$b = 1$

Chebyshev:

$T_n(x)$	$w(x) = \dfrac{1}{\sqrt{1-x^2}}$	$a = -1,$	$b = 1$

Laguerre:

$L_n(x)$	$w(x) = e^{-x}$	$a = 0,$	$b = \infty$

Hermite:

$H_n(x)$	$w(x) = e^{-x^2}$	$a = -\infty,$	$b = \infty$

Many properties are common to any system of orthogonal polynomials, e.g., the fact that all roots of such polynomials are real and separated by roots of the preceding polynomial.

Legendre Polynomials

Representations

$$P_n(x) = \frac{1}{2^n n!}\frac{d^n}{dx^n}[(x^2-1)^n]$$
$$= \frac{(2n)!}{2^n(n!)^2}\left[x^n - \frac{n(n-1)}{2(2n-1)}x^{n-2} + \frac{n(n-1)(n-2)(n-3)}{2\cdot 4(2n-1)(2n-3)}x^{n-4} - \cdots\right]$$

$$P_n(\cos\theta) = \frac{1}{\pi}\int_0^\pi (\cos\theta \pm i\sin\theta\cos\phi)^n\,d\phi$$
$$= \frac{2}{\pi}\int_0^\pi \frac{\cos(n+\frac{1}{2})\phi\,d\phi}{\sqrt{2(\cos\phi - \cos\theta)}}$$

The corresponding (complete) normal orthogonal system is

$$\phi_n(x) = \sqrt{n+\tfrac{1}{2}}\,P_n(x)$$

Generating Functions

$$\frac{1}{\sqrt{1-2hx+h^2}} = \sum_{n=1}^{\infty} h^n P_n(x) \qquad |h| < 1$$

Recurrence Relations

$$(n+1)P_{n+1} - (2n+1)xP_n + nP_{n-1} = 0$$
$$(x^2-1)P_n = nxP_n - nP_{n-1}$$
$$= (n+1)P_{n+1} - (n+1)xP_n$$

Differential Equation $y = P_n(x)$

$$\text{satisfies } (1-x^2)y'' - 2xy' + n(n+1)y = 0$$

Asymptotic Behavior

$$P_n(\cos\theta) \sim \sqrt{\frac{2}{n\pi\sin\theta}}\cos\left[\left(n+\frac{1}{2}\right)\theta - \frac{\pi}{4}\right] + o(n^{-3/2})$$

Bounds

$$|P_n(x)| \le 1 \qquad -1 < x < 1$$
$$(n\sin\theta)^{1/2}|P_n(\cos\theta)| \le 1 \qquad \theta \le \theta \le \pi$$

Special Series

$$\frac{1}{\sqrt{2(1-x)}} \sim P_0(x) + P_1(x) + P_2(x) + \cdots \quad |x| < 1$$

Observe that the Parseval result is false: the function $1/\sqrt{2(1-x)}$ is not of integrable square in $(-1,1)$.

Chebyschev Polynomials

Representation

$$T_n(x) = 2^{1-n}\cos[n\arccos x]$$

The corresponding complete normal orthogonal system is

$$\phi_n(x) = \frac{2^{(2n-1)/2}T_n(x)}{\sqrt{\pi}\,(1-x^2)^{1/4}}$$

Generating Function

$$\frac{1-t^2}{1-2tx+t^2} = \sum_{n=0}^{\infty} T_n(x)(2t)^n$$

Recurrence Relations

$$T_{n+1} - xT_n + \tfrac{1}{4}T_{n-1} = 0 \qquad n > 2$$
$$T_2 - xT_1 + \tfrac{1}{4}T_n = -\tfrac{1}{4} \qquad T_1 - xT_0 = 0$$

Differential Equation $y = T_n(x)$

$$\text{satisfies } (1 - x^2)y'' - xy' + n^2y = 0$$

Special Property of Minimum Deviation. Among all the polynomials $w_n(x) = x^n + \cdots$ of degree n and leading coefficient unity, $T_n(x)$ is the one for which

$$\max_{-1 \leq x \leq 1} |w_n(x)|$$

is least. The value in question is 2^{1-n}. This property of the $T_n(x)$ is exploited in numerical analysis.

The $T_n(x)$ are more properly called the Chebyshev polynomials of the first kind. The Chebyshev polynomials of the second kind, defined by

$$U_n(x) = 2^{-n} \sin \{(n + 1) \arccos x\}/\sin \{\arccos x\}$$

which are orthogonal with respect to $(1 - x^2)^{1/2}$ in $(-1,1)$ are also of importance.

Laguerre Polynomials

Representations

$$n!\, L_n(x) = e^x D^n[e^{-x}x^n] = (-)^n \left[x^n - \frac{n^2}{1} x^{n-1} + \frac{n^2(n-1)^2 x^{n-2}}{1 \cdot 2} - \cdots (-)^n n! \right]$$

$$n!\, L_n(x) = \int_0^\infty e^{x-t} t^n J_0(2\sqrt{xt})\, dt$$

The corresponding (complete) normal orthogonal system is

$$\phi_n(x) = e^{-x/2} L_n(x)$$

Generating Function

$$\frac{1}{1-t} \exp\left[\frac{-xt}{1-t}\right] = \sum_{n=0}^{\infty} L_n(x)\, t^n$$

Recurrence Relations

$$(n+1)L_{n+1} - (2n + 1 - x)L_n + nL_{n-1} = 0$$
$$xL'_n = nL_n - nL_{n-1} = L_{n+1} - (n + 1 - x)L_n$$

Differential Equation $y = L_n(x)$

$$\text{satisfies } xy'' + (1 - n)y' + ny = 0$$

Asymptotic Behavior

$$\phi_n(x) \sim \pi^{-1/2} x^{-1/4} n^{-1/4} \cos\left(2\sqrt{\pi x} - \frac{\pi}{4}\right)$$

as $n \to \infty$, x fixed

Bounds

$$|\phi_n(x)| \leq 1$$

Generalized Laguerre Polynomials. If we consider orthogonality with respect to the weight function $e^{-x}x^\alpha$ $(\alpha \geq 0)$ in the interval $(0, \infty)$, we obtain a system denoted by $L_n{}^\alpha(x)$. We have

$$L_n{}^{(\alpha)}(x) = \sum_{r=0}^{n} \binom{n+\alpha}{n-r} \frac{(-x)^r}{r!}$$

The corresponding normal orthogonal functions are

$$\phi_n(x) = \frac{e^{-x/2} x^{\alpha/2}}{\sqrt{\Gamma(1+\alpha)\binom{n+\alpha}{n}}} L_n{}^{(\alpha)}(x)$$

and the generating function is

$$(1 - t)^{-\alpha-1} e^{-xt/(1-t)} = \Sigma L_n{}^\alpha(x) t^n \qquad |t| < 1$$

Hermite Polynomials

Representations

$$H_n(x) = e^{x^2}(-D)^n e^{-x^2}$$
$$= (2x)^n - \frac{n \cdot n-1}{1!}(2x)^{n-2} + \frac{n \cdot n-1 \cdot n-2 \cdot n-3}{2!}(2x)^{n-4} - \cdots$$

The corresponding complete normal orthogonal system is

$$\phi_n(x) = e^{\;x^2/2} \frac{H_n(x)}{2^{n/2}\pi^{1/4}(n!)^{1/2}}$$

Generating Function

$$\exp(-t^2 + 2xt) = \sum_0^\infty \frac{H_n(x)t^n}{n!}$$

Recurrence Relations

$$H_{n+1} = 2xH_n - 2nH_{n-1} \qquad H'_n = 2nH_{n-1}$$

Differential Equation $y = H_n(x)$

$$\text{satisfies } y'' - 2xy' + 2ny = 0$$

Sturm-Liouville Orthogonal Systems. These are systems associated with certain two-point boundary-value problems for a differential equation of the following form:

$$\frac{d}{dx}[py'] + (q\rho - r)y = 0$$

where p,q,r are positive functions (with continuous first and second derivatives) in $[a,b]$ and where ρ is a parameter. After making the substitutions

$$\xi = \int_a^x \sqrt{q}\,\frac{dt}{\rho} \qquad \eta = y\sqrt[4]{pq} \qquad \sigma^2 = \rho$$

and assuming that

$$\int_a^b \sqrt{\frac{q}{p}}\, dt = \pi$$

the equations become

$$\frac{d^2\eta}{d\xi^2} + (\sigma^2 - l)\eta = 0$$

where l is a function of ξ, depending on p,q,r. We

seek solutions of this equation which satisfy

$$\eta' - k\eta = 0 \text{ at } \xi = 0 \qquad \eta' + K\eta = 0 \text{ at } \xi = \pi$$

It can be shown that such solutions exist only for a discrete set of values of ρ, the so-called characteristic values ρ_n. These are distinct and $\rho_n \to \infty$. The solutions, u_n, called characteristic functions, form an orthogonal system (in the simple sense).

The case $l = 0$, $k = 0$, $K = 0$ gives the characteristic values $\sigma^2 = n^2$ ($n = 1,2, \ldots$) with characteristic functions $u_n = \cos n\xi$. Less trivially consider the case when $k \neq 0$, $K \neq 0$ and to fix the scale take $y(0) = 1$. The general solution of the differential equation is

$$y = \cos \sigma x + c \sin \sigma x$$

and, in order that the boundary conditions be satisfied we must have $k = c\sigma$

$$K(\cos \sigma\pi + c \sin \sigma\pi) = (\sigma \sin \sigma\pi - c\sigma \cos \sigma\pi)$$

This means that solutions are possible only if σ is a root of the transcendental equation

$$\tan \sigma\pi = \frac{\sigma(k + K)}{\sigma^2 - kK}$$

an equation first studied by Cauchy. Graphical consideration indicates that this has an infinite set of roots $\sigma_n \to \infty$ and that $\sigma_n \sim n\pi$ as $n \to \infty$.

11. Special Functions

Classification. There are various classifications of the special functions which arise in mathematics. We can, for instance, begin with polynomials and rational functions (the ratio of two polynomials). We have already discussed certain classes of (orthogonal) polynomials. Algebraic functions, in general many-valued, arise as solutions of algebraic equations with polynomial coefficients, for example, $w = \sqrt{z}$ satisfies $w^2 - z = 0$. A transcendental function is one which is not algebraic. The so-called elementary transcendental functions include e^x, $\log x$, and the trigonometrical and hyperbolic functions and their inverses. Among the higher transcendental functions of particular interest are the Bessel (and related) functions, elliptic functions, and the gamma function. Detailed accounts of these properties will be found in special monographs; many of the functions have been tabulated.

A general survey of the theory of special functions is found in [9]‡; a more recent account is available in [29]. A useful collection of formulas is given in [40, 46]. Comprehensive information about the tabulation of special functions is available in [39] and in [16]. Useful selections of tables of special functions are found in [40] and [45].

Conventions

x = real variable
z = complex variable
n = integral index, non-negative, unless otherwise mentioned
ν = arbitrary complex index
$x = \cos\theta \qquad -1 < x < 1$

‡ Numbers in brackets refer to References at end of chapter.

Gamma Function

$$\Gamma(z) = \int_0^\infty e^{-t} t^{z-1}\, dt \qquad \Re(z) > 0$$
$$= z^{-1} e^{-\gamma z} \prod \left[\left(1 + \frac{z}{n}\right)^{-1} e^{z/n} \right] \qquad \text{all } z$$

This function is regular for all z except $z = 0, -1, -2, \ldots$ where it has simple poles with residue $(-1)^l/l!$ at $z = -l$. It is easy to verify that

$$\Gamma(\tfrac{1}{2}) = \sqrt{\pi} \quad \text{and} \quad \Gamma(n+1) = n! \quad n = 1,2,3, \ldots$$

It satisfies the following functional equations:

$$\Gamma(z + 1) = z\Gamma(z) \qquad \Gamma(z)\Gamma(1 - z) = \pi \operatorname{cosec} \pi z$$

Asymptotic Behavior: Stirling's Formula

$$\ln \Gamma(z) \sim (z - \tfrac{1}{2}) \ln z - z + \tfrac{1}{2} \ln 2\pi$$
$$n! \sim \left(\frac{n}{e}\right)^n \sqrt{2\pi n}$$

Related Functions: Beta Functions

$$B(z,w) = \int_0^1 t^{z-1}(1 - t)^{w-1}\, dt$$
$$= \frac{\Gamma(z)\Gamma(w)}{\Gamma(z + w)} \qquad \Re(z) > 0,\ \Re(w) > 0$$

Bessel Functions

$$J_\nu(z) = \left(\frac{z}{2}\right)^\nu \sum_{r=0}^\infty \left(\frac{iz}{2}\right)^{2r} \frac{1}{r!\Gamma(\nu + r + 1)}$$

$$Y_\nu(z) = \frac{1}{\sin \nu\pi} [\cos \nu\pi J_\nu(z) - J_{-\nu}(z)]$$
$$\nu \neq 0, \pm 1, \pm 2, \ldots$$

$$Y_n(z) = \frac{2}{\pi} J_n(z) \left[\gamma + \ln \frac{1}{2} z\right]$$
$$- \frac{1}{\pi} \sum_{r=0}^{n-1} \frac{(n - r - 1)!}{r!} \left(\frac{z}{2}\right)^{-n+2r}$$
$$- \frac{1}{\pi} \sum_{r=0}^{n-1} \frac{(-1)^r}{r!(n + r)!} \left(\frac{z}{2}\right)^{n+2r} \left(\frac{1}{1} + \frac{1}{2} + \cdots + \frac{1}{r}\right.$$
$$\left. + \frac{1}{1} + \frac{1}{2} + \cdots + \frac{1}{r + n}\right) \qquad n = 0,1,2, \ldots$$

are independent solutions of the differential equation

$$z^2 \mathcal{C}'' + \mathcal{C}' + (z^2 - \nu^2)\mathcal{C} = 0$$

These functions are one-valued integral functions for $n = 0,1,2, \ldots$, but for other indices are many-valued and have a branch point at $z = 0$, in view of the factor z^ν.

The *Hankel functions*

$$H_\nu^{(1)}(z) = J_\nu(z) + iY_\nu(z)$$
$$H_\nu^{(2)}(z) = J_\nu(z) - iY_\nu(z)$$

are also independent solutions of the Bessel equation.

Generating Function

$$\exp\left[\tfrac{1}{2}z(t - t^{-1})\right] = J_n(z) + \sum_{n=1}^{\infty} [t^n + (-t)^{-n}]J_n(z)$$

Recurrence Relations

$$z\mathcal{C}_{\nu+1} = 2\nu C_\nu - z\mathcal{C}_{\nu-1}$$
$$\mathcal{C}'_\nu = \mathcal{C}_{\nu-1} - \mathcal{C}_{\nu+1}$$

where $\mathcal{C}$ denotes J or Y or indeed any solution of the Bessel equation.

Asymptotic Behavior

$$J_\nu(z) \sim \sqrt{\frac{2}{\pi z}}\left\{\cos\left(z - \frac{1}{2}\nu\pi - \frac{1}{4}\pi\right)\left[1 - \frac{(4\nu^2 - 1^2)(4\nu^2 - 3^2)}{2!(8z)^2} + \cdots\right] - \sin\left(z - \frac{1}{2}\nu\pi - \frac{1}{4}\pi\right)\left[\frac{(4\nu^2 - 1^2)}{1!(8z)} - \cdots\right]\right\}$$

$$Y_\nu(z) \sim \sqrt{\frac{2}{\pi z}}\left\{\sin\left(z - \frac{1}{2}\nu\pi - \frac{1}{4}\pi\right)\left[1 - \frac{(4\nu^2 - 1^2)(4\nu^2 - 3^2)}{2!(8z)^2} + \cdots\right] + \cos\left(z - \frac{1}{2}\nu\pi - \frac{1}{4}\pi\right)\left[\frac{(4\nu^2 - 1^2)}{1!(8z)} - \cdots\right]\right\}$$

provided $|z| \gg |\nu|$, $|z| \gg 1$.

Zeros. $J_n(z)$ has an infinite set of simple zeros. We denote by $j_{n,r}$ the rth positive zero of $J_n(x)$. Since $J_n(-x) = (-1)^n J_n(x)$, it follows that $-j_{n,r}$ is also a zero of $J_n(x)$; also, except for $n = 0$, $J_n(0) = 0$. It is known that $j_{n,r} \to \infty$ and that $j_{n,1} > n$.

Functions Related to Bessel Functions. Modified Bessel Functions

$$I_\nu(x) = e^{-\nu\pi i/2}J_\nu(ix) \qquad K_\nu(x) = \tfrac{1}{2}\pi i e^{\nu\pi i/2}H_\nu^{(1)}(ix)$$

These are independent solutions of the equation

$$x^2y'' + xy' - (x^2 + \nu^2)y = 0$$

Kelvin Functions. The Bessel functions with argument $r\sqrt{i}$ or $r\sqrt{i^3}$ occur in problems in electrical engineering. The following are the usual definitions:

$$\operatorname{ber}_\nu x + i\operatorname{bei}_\nu x = J_\nu(xe^{3\pi i/4})$$
$$\operatorname{ker}_\nu x + i\operatorname{kei}_\nu x = \tfrac{1}{2}\pi i H_\nu^{(1)}(xe^{3\pi i/4})$$
$$\operatorname{her}_\nu x = (2/\pi)\operatorname{kei}_\nu x \qquad \operatorname{hei}_\nu x = -(2/\pi)\operatorname{ker}_\nu x$$

In particular, we note that for $x > 0$

$$J_0(x\sqrt{i}) = \operatorname{ber} x - i\operatorname{bei} x$$

Spherical Bessel Functions. The functions

$$j_n(x) = \sqrt{\pi/2x}\,J_\nu(x)$$

$\nu = \pm(n + \tfrac{1}{2})$, sometimes called Stokes functions, are independent solutions of the equation

$$x^2y'' + 2xy' + [x^2 - n(n+1)]y = 0$$

These are elementary functions, for example

$$j_0(x) = \sqrt{\frac{\pi}{2x}}J_{1/2}(x) = \frac{\sin x}{x}$$

$$j_{-1}(x) = \sqrt{\frac{\pi}{2x}}J_{-1/2}(x) = \frac{\cos x}{x}$$

$$j_1(x) = \sqrt{\frac{\pi}{2x}}J_{3/2}(x) = \left(\frac{-\cos x}{x} + \frac{\sin x}{x^2}\right)$$

In general,

$$j_n(x) = \sqrt{\frac{\pi}{2x}}J_{n+1/2}(x) = x^n\left(-\frac{1}{x}\frac{d}{dx}\right)^n\left(\frac{\sin x}{x}\right)$$

In this case the asymptotic formulas reduce to an expression in finite terms.

Bessel-Clifford Functions

$$C_\nu(x) = x^{-\nu/2}J_\nu(2\sqrt{x}) \qquad D_\nu(x) = x^{-\nu/2}Y_\nu(2\sqrt{x})$$

These are independent solutions of the equation

$$xy'' + (\nu + 1)y' + y = 0$$

The solutions of the equation $xy'' + y = 0$ are of the form

$$y = cxC_1(x) + dx\,D_1(x) \qquad c,d \text{ constants}$$

Airy Integrals

$$\operatorname{Ai}(x) = \frac{1}{\pi}\int_0^\infty \cos\left(\frac{1}{3}x^3 + xt\right)dt$$

$$\operatorname{Bi}(x) = \frac{1}{\pi}\int_0^\infty \left[\exp\left(-\frac{1}{3}t^3 + xt\right) + \sin\left(\frac{1}{3}t^3 + xt\right)\right]dt$$

are independent solutions of the equation

$$y'' + xy = 0$$

They arise, for instance, in the theory of diffraction. They can be expressed in terms of Bessel functions of order $\pm\tfrac{1}{3}$ and $\pm\tfrac{2}{3}$ and of argument $\tfrac{2}{3}x^{3/2}$.

Struve Functions

$$H_\nu(z) = \sum_{m=0}^{\infty}\frac{(-1)^m(\tfrac{1}{2}z)^{\nu+2m+1}}{\Gamma(m + \tfrac{3}{2})\Gamma(\nu + m + \tfrac{3}{2})}$$

is a solution of

$$z^2w'' + zw' + (z^2 - \nu^2)w = \frac{4(\tfrac{1}{2}z)^{\nu+1}}{\Gamma(\nu + \tfrac{1}{2})\Gamma(\tfrac{1}{2})}$$

It can be represented as

$$H_\nu(z) = \frac{2(\tfrac{1}{2}z)^\nu}{\Gamma(\nu + \tfrac{1}{2})\Gamma(\tfrac{1}{2})}\int_0^{\pi/2}\sin(z\cos\theta)\sin^{2\nu}\theta\,d\theta$$

As in the case of Bessel functions, the Struve functions are elementary when $\nu = n + \frac{1}{2}$, for example,

$$\boldsymbol{H}_{1/2}(z) = \sqrt{\frac{2}{\pi z}}\,(1 - \cos z)$$

and

$$\boldsymbol{H}_{3/2}(z) = \sqrt{\frac{2}{\pi z}}\left[\frac{z}{2} + \frac{1}{z} - \sin z - \frac{\cos z}{z}\right]$$

Riccati-Bessel Functions. These are defined as

$$S_n(x) = \sqrt{\pi/2x}\,J_{n+1/2}(x)$$
$$C_n(x) = (-1)^n \sqrt{\pi/2x}\,J_{-n-1/2}(x)$$

and are independent solutions of the equation

$$x^2y'' + x^2(n^2 - \tfrac{1}{4})y = 0$$

They occur in the theory of scattering.

Legendre Functions. Properties of the Legendre polynomials have been already given. It was pointed out, in particular, that $P_n(x)$ satisfies the differential equation

$$(1 - x^2)y'' - 2xy' + [n(n+1)]y = 0$$

An independent solution of this is

$$Q_n(x) = \frac{1}{2}P_n(x)\ln\frac{1+x}{1-x} - \omega_{n-1}(x) \qquad -1 < x < 1$$
$$= \frac{1}{2}P_n(x)\ln\frac{1+x}{1-x} + \frac{1}{2^n n!}\frac{d^n}{dx^n}\left[(x^2-1)^n \ln\frac{1+x}{1-x}\right]$$

where

$$\omega_{n-1}(x) = \sum_{r=1}^{n} r^{-1}P_{n-1}(x)P_{n-r}(x)$$

Wronskian

$$P_n(x)Q'_n(x) - P'_n(x)Q_n(x) = (1 - x^2)^{-1} \qquad -1 < x < 1$$

The functions $Q_n(x)$ satisfy the same recurrence relations as the $P_n(x)$.

Asymptotic Behavior

$$Q_n(\cos\theta) \sim \sqrt{\frac{\pi}{2n\sin\theta}}\cos\left[\left(n + \frac{1}{2}\right)\theta + \frac{1}{4}\pi\right]$$
$$n \gg \epsilon^{-1}, \qquad 0 < \epsilon < \theta < \pi - \epsilon$$

Associated Legendre Functions

$$P_n{}^m(x) = (-1)^m(1 - x^2)^{m/2}D^mP_n(x)$$

for $m = 0,1,2,\ldots,n$ satisfies the differential equations

$$(1 - x^2)y'' - 2xy' + \left[n(n+1) - \frac{m^2}{1-x^2}\right]y = 0$$

These functions are polynomials in $\sin n\theta$, $\cos n\theta$, if we write $x = \cos\theta$.

Orthogonality

$$\int_{-1}^{+1} P_n{}^mP_{n'}{}^m\,dx = \frac{2}{2n+1}\frac{(n+m)!}{(n-m)!} \qquad n = n'$$
$$= 0 \qquad n \neq n'$$

Recurrence Relations

Fixed m, $0 < n - 1$,

$$(2n+1)xP_n{}^m - (n - m + 1)P_{n+1}{}^m - (n + m)P_{n-1}{}^m = 0$$

Fixed n, $0 < m < n - 2$,

$$P_n{}^{m+2} - 2(m+1)\cot\theta P_n{}^{m+1} + (n - m)(n + m + 1)P_n{}^m = 0$$

General Legendre Functions. We have so far confined our attention mainly to real arguments, x arbitrary in the case of $P_n{}^m$ and $-1 < x < 1$ in the case of Q_n. The question of the definition of these quantities for general complex arguments and indices is rather intricate [29, 56].

Error Function. It seems convenient to use the notations

$$\operatorname{erf} x = \frac{2}{\sqrt{\pi}}\int_0^x \exp(-t^2)\,dt = H(x)$$

This function plays an important role in probability theory, where the alternative normalization

$$\alpha(x) = \sqrt{\frac{\pi}{2}}\int_0^x \exp\left(-\frac{t^2}{2}\right)dt$$

is often used. In both cases $H(x) \to 1$, $\alpha(x) \to 1$ as $x \to +\infty$.

The following asymptotic expansion is valid for large x:

$$H(x) \sim 1 - \frac{\exp(-x^2)}{x\sqrt{\pi}}\left[1 - \frac{1}{2x^2} + \frac{1\cdot 3}{(2x^2)^2} - \frac{1\cdot 3\cdot 5}{(2x^2)^3} + \cdots\right]$$

Related to this function are the Fresnel integrals which are also expressible as indefinite integrals of spherical Bessel functions:

$$C(x) = \int_0^x \cos(\pi t^2/2)\,dt = \tfrac{1}{2}\int_0^u J_{-1/2}(t)\,dt$$
$$S(x) = \int_0^x \sin(\pi t^2/2)\,dt = \tfrac{1}{2}\int_0^u J_{1/2}(t)\,dt$$

where $x = \frac{1}{2}\pi u^2$. We have

$$H[\tfrac{1}{2}x(1+i)\sqrt{\pi}] = (1+i)[C(x) - iS(x)]$$

Hypergeometric Function. Many of the functions discussed earlier satisfy the Gaussian hypergeometric equation

$$(1) \qquad z(1-z)w'' + [c - (a+b+1)]zw' - abw = 0$$

for appropriate values of the parameters a,b,c, or limiting (confluent) cases of this such as the Kummer equation

$$(2) \qquad zw'' + (c - z)w' - aw = 0$$

or the Whittaker equation

$$(3) \qquad w'' + \left(-\frac{1}{4} + \frac{k}{z} + \frac{1}{4} - \frac{\mu^2}{z^2}\right)w = 0$$

A solution of (1) regular for $|z| < 1$ is given by

$$F(a,b;c;z) = 1 + \frac{a \cdot b}{c \cdot 1} z + \frac{a(a+1)b(b+1)}{c(c+1) \cdot 1 \cdot 2} z^2 + \cdots$$

As a special case we observe that

$$\ln(1+z) = z F(1,1;2;-x)$$

A solution of (2) regular for $|z| < 1$ is

$${}_1F_1(a;c;z) = 1 + \frac{a}{c} \cdot \frac{z}{1!} + \frac{a(a+1)}{c(c+1)} \cdot \frac{z^2}{2!} + \cdots$$

As a special case we observe

$$\int_0^x \exp(-t^2)\, dt = x\, {}_1F_1(\tfrac{1}{2};\tfrac{3}{2};-x^2)$$

Detailed discussion of these functions is given in [6].

Mathieu Functions. There are various functions of this type. We take the solutions of

$$(1) \qquad y'' + (b - s\cos^2 x)y = 0$$

as standard Mathieu (or elliptic cylinder) functions. The solutions of

$$(2) \qquad y'' - (b - s\cosh^2 x)y = 0$$

can be called modified Mathieu (or hyperbolic cylinder) functions, (2) being obtained from (1) by the change from x to ix.

In many cases we require periodic solutions, in particular, solutions with period π or 2π. In order that such exist, the parameter b must have one of a countably infinite set of characteristic values. In the case $s = 0$, the characteristic values are $b = n^2$ and the corresponding characteristic functions are $\sin nx$, $\cos nx$. For $s \neq 0$, there corresponds to a given characteristic number exactly one periodic solution, which is either even or odd. The characteristic numbers corresponding to these are denoted by $be_n(s)$, $bo_n(s)$, respectively. The subscript n indicates those which approach n^2 as $s \to 0$. The solutions are denoted by:

Even, period π: $Se_{2r}(s,x)$; Odd, period π: $So_{2r}(s,x)$
Even, period 2π: $Se_{2r+1}(s,x)$; Odd, period 2π: $So_{2r+1}(s,x)$

They can be expressed as Fourier series of appropriate form, for example,

$$Se_{2r}(s,x) \sim \sum_{k=0}^{\infty} De_{2k}^{(2r)} \cos 2kx$$

Tables of the characteristic values and the Fourier coefficients have been prepared; these enable the periodic Mathieu function themselves to be calculated without great labor [59]. The following normalization has been used:

$$Se_r(s,0) = 1 \qquad \left[\frac{d}{dx} So_r(s,x)\right]_{x=1} = 1$$

Elliptic Functions—Weierstrassian. Let ω_1, ω_2, be two (complex) numbers whose ratio has a positive imaginary part. It can be shown that the function

$$\wp(z) = \frac{1}{z^2} + \sum_{m_1,m_2}{}' \left[\frac{1}{(z - 2m_1\omega_1 - 2m_2\omega_2)^2} - \frac{1}{(2m_1\omega_1 + 2m_2\omega_2)^2}\right]$$

(where the summation is over all pairs of m_1,m_2

$$m_1 = 0, \quad \pm 1, \quad \pm 2, \quad \ldots \quad m_2 = 0, \quad \pm 1, \quad \pm 2, \quad \ldots$$

except $m_1 = 0 = m_2$) is doubly periodic, that is,

$$\wp(z + 2m_1\omega_1 + 2m_2\omega_2) = \wp(z)$$

for any integral m_1,m_2. $\wp(z)$ has a double pole, with zero residue at each point

$$z = \Omega_{m_1,m_2} = 2m_1\omega_1 + 2m_1\omega_2$$

and it satisfies the differential equation

$$\left(\frac{d\wp}{dz}\right)^2 = 4[\wp(z) - e_1][\wp(z) - e_2][\wp(z) - e_3]$$

where $e_1 = \wp(\omega_1)$, $e_2 = \wp(\omega_2)$, $e_3 = \wp(\omega_1 + \omega_2)$. It can be shown that $e_1 + e_2 + e_3 = 0$. Writing

$$g_2(\omega_1,\omega_2) = g_2 = -4(e_1e_2 + e_2e_3 + e_3e_1) = 60 \sum_{m_1,m_2,}{}' \Omega^{-4}{}_{m_1,m_2},$$

$$g_3(\omega_1,\omega_2) = g_3 = 4e_1e_2e_3 = 140 \sum_{m_1,m_2}{}' \Omega^{-6}{}_{m_1,m_2}$$

the above equation reduces to

$$\left(\frac{d\wp}{dz}\right)^2 = 4\wp^3 - g_2\wp - g_3$$

Thus $\wp$ is the inverse function associated with

$$\int \frac{dt}{\sqrt{4t^3 - g_2t - g_3}}$$

and can be defined by

$$u = \int_0^{\wp(u)} \frac{dt}{\sqrt{4t^3 - g_2t - g_3}}$$

Given an equation of the form $y'^2 = 4y^3 - a_2y - a_3$, it can be shown that numbers ω_1, ω_2 exist for which

$$g_2(\omega_1,\omega_2) = a_2 \qquad g_3(\omega_1,\omega_2) = a_3$$

Thus any equation of this type can be solved in terms of an elliptic function with appropriate periods. We note that $\wp(u)$ has an algebraic addition theorem similar to that for $\sin x$ which can be written as

$$\sin(x + y) = \sin x(\sin y)' + \sin y(\sin x)'$$

That for $\wp(z)$ is

$$\wp(u + v) = -\wp(u) - \wp(v) + \frac{1}{4}\left[\frac{\wp'(u) - \wp'(v)}{\wp(u) - \wp(v)}\right]^2$$

The Weierstrassian elliptic function is a basic one, for it can be shown that any doubly periodic function whose only singularities are poles can be expressed in the form

$$R_1(\wp) + \wp' R_2(\wp)$$

where R_1, R_2 are rational functions and $\wp$ is the Weierstrassian function with the periods of the given function.

Elliptic Functions—Jacobian Type. These are denoted by $\operatorname{sn} u = \operatorname{sn}(u,k^2)$; $\operatorname{cn} u = \operatorname{cn}(u,k^2)$,

$$\operatorname{dn} u = dn\,(u,k^2)$$

and can be introduced in various ways. They depend on a parameter, the modulus k^2, where in many cases $0 < k^2 < 1$. They satisfy the equations

$$[(\operatorname{sn} u)']^2 = (1 - \operatorname{sn}^2 u)(1 - k^2 \operatorname{sn}^2 u)$$
$$[(\operatorname{cn} u)']^2 = (1 - \operatorname{cn}^2 u)(k'^2 + k^2 \operatorname{cn}^2 u)$$
$$[(\operatorname{dn} u)']^2 = -(1 - \operatorname{dn}^2 u)(k'^2 - \operatorname{dn}^2 u)$$

where $k'^2 = 1 - k^2$ and $\operatorname{sn} 0 = 0$, $\operatorname{cn} 0 = 1$, $\operatorname{dn} 0 = 1$. It follows that

$$\operatorname{sn}^2 u + \operatorname{cn}^2 u = 1 \qquad \operatorname{dn}^2 u + k^2 \operatorname{sn}^2 u = 1$$

and that

$$\operatorname{sn} u = u - \frac{(1 + k^2)u^3}{3!} + \frac{(1 + 14k^2 + k^4)u^5}{5!} + \cdots$$
$$\operatorname{cn} u = 1 - \frac{u^2}{2!} + \frac{(1 + 4k^2)u^4}{4!} + \cdots$$
$$\operatorname{dn} u = 1 - \frac{k^2u^2}{2!} + \frac{k^2(4 + k^2)u^4}{4!} - \cdots$$

The range of validity of these series can be deduced from the fact that they hold up to the nearest singularity. It can be shown as $k \to 0$, $\operatorname{sn} u \to \sin u$, while as $k \to 1$, $\operatorname{sn} u \to \tanh u$. These functions are best regarded as functions of a complex variable, with two periods, one of which is purely real and the other purely imaginary, in the case $0 < k^2 < 1$. The only singularities of the functions are simple poles. The periodicity of the functions, and their values at the points $m_1K + m_2iK'$ are indicated in the diagrams (Fig. 3.5) where we have written

$$K = K(k^2) = \int_0^1 \frac{dt}{\sqrt{(1 - t^2)(1 - k^2t^2)}}$$
$$K' = K'(k^2) = \int_0^1 \frac{dt}{\sqrt{(1 - t^2)(1 - k'^2t^2)}} = K(k'^2)$$

These functions satisfy algebraic addition theorems:

$$\operatorname{sn}(u + v) = \frac{\operatorname{sn} u \operatorname{cn} v \operatorname{dn} v + \operatorname{sn} v \operatorname{cn} u \operatorname{dn} u}{1 - k^2 \operatorname{sn}^2 u \operatorname{sn}^2 v}$$
$$\operatorname{cn}(u + v) = \frac{\operatorname{cn} u \operatorname{cn} v - \operatorname{sn} u \operatorname{dn} u \operatorname{sn} v \operatorname{dn} v}{1 - k^2 \operatorname{sn}^2 u \operatorname{sn}^2 v}$$
$$\operatorname{dn}(u + v) = \frac{\operatorname{dn} n \operatorname{dn} v - k^2 \operatorname{sn} u \operatorname{cn} u \operatorname{sn} v \operatorname{cn} v}{1 - k^2 \operatorname{sn}^2 u \operatorname{sn}^2 v}$$

Use of these formulas, together with the values of the functions at the points m_1K, m_2iK', will indicate the effect of adding $m_1K + m_2iK'$ to the argument of an elliptic function. For instance, using

$$\operatorname{sn}(K + iK') = k^{-1}$$

$\operatorname{cn}(K + iK') = -ik'k^{-1}$ and $\operatorname{dn}(K + iK') = 0$, we find

$$\operatorname{dn}(u + K + iK') = \frac{\operatorname{dn} u \cdot 0 - k^2 \operatorname{sn} u \operatorname{cn} u \cdot k^{-1} \cdot -ik'k^{-1}}{1 - k^2 \operatorname{sn}^2 u \cdot k^2} = ik' \operatorname{sc} u$$

Use of these formulas together with the results of the Jacobi imaginary transformation:

$$\operatorname{sn}(iu,k) = i \operatorname{sc}(u,k') \qquad \operatorname{cn}(iu,k) = \operatorname{nc}(u,k')$$
$$\operatorname{dn}(iu,k) = \operatorname{dc}(u,k')$$

enables one to compute the elliptic functions of a

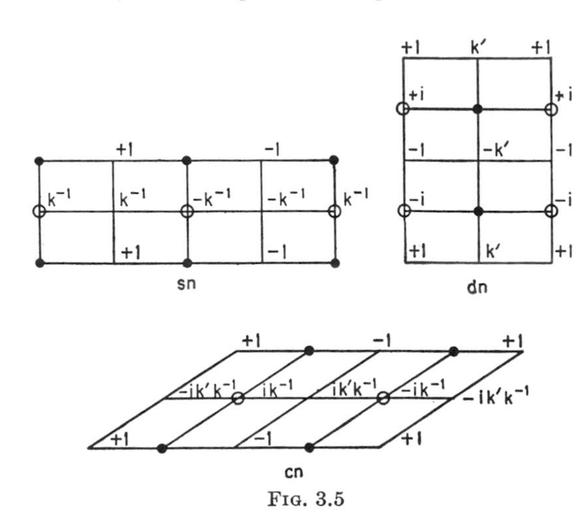

Fig. 3.5

complex argument in terms of those of real argument. Here we use the notation of Glaisher:

$$\operatorname{sc} u = \frac{\operatorname{sn} u}{\operatorname{cn} u} \qquad \operatorname{nc} u = \frac{1}{\operatorname{cn} u} \qquad \operatorname{dc} u = \frac{\operatorname{dn} u}{\operatorname{cn} u}$$

These are various relations between the Jacobian and Weierstrassian elliptic functions. A typical one is

$$\wp(u/\sqrt{e_1 - e_3}) = e_3 + (e_1 - e_3) \operatorname{ns}^2 u$$

Elliptic Integrals. It is well known that any integral of the form

$$\int R(x, \sqrt{ax^2 + bx + c})\, dx$$

can be evaluated in terms of elementary functions, no matter what rational function R may be. The enumeration of all cases, and the manipulations necessary to work in the real field, are tedious but elementary. The problem of the evaluation of integrals of the form

$$\int R(x, \sqrt{ax^4 + bx^3 + cx^2 + dx + e})\, dx$$

is of a similar type, but more complicated. We give here a few simple examples.

Consider first

$$\int_0^x \frac{dt}{\sqrt{(a^2 - t^2)(b^2 - t^2)}}$$

We have noted that sn z satisfies the differential equation

$$(w')^2 = (1 - w^2)(1 - k^2w^2)$$

This suggests the transformation

$$t = b \text{ sn } (ax) \qquad k^2 = \frac{b^2}{a^2}$$

which gives, formally,

$$\int_0^x \frac{dt}{\sqrt{(a^2 - t^2)(b^2 - t^2)}}$$
$$= \int_0^{a^{-1}\text{sn}^{-1}(xb^{-1})} \frac{ab \text{ cn } (a\tau) \text{ dn } (a\tau)\, d\tau}{\sqrt{\{a^2 - b^2 \text{ sn}^2 (a\tau)\}\{b^2 - b^2 \text{ sn}^2 (a\tau)\}}}$$
$$= \int_0^{a^{-1}\text{sn}^{-1}(xb^{-1})} d\tau = \frac{1}{a} \text{ sn}^{-1} \left(\frac{x}{b}\right)$$

This result is actually correct if $a > b > x$ and if we choose that value of sn^{-1} which lies between 0 and $K(k^2)$.

If the quartic form is a general one, it is necessary first to reduce it, by means of a real linear (fractional) transform to the form $(\pm t^2 \pm a^2)(\pm t^2 \pm b^2)$ and then to choose an appropriate transformation, according to the distribution of the ambiguous signs.

Consider next, for simplicity, an integral of the form

$$I = \int \frac{t^2\, dt}{\sqrt{(1 - t^2)(1 - k^2t^2)}}$$

Putting $t = \text{sn } \tau$, we find

$$I = \int \text{sn}^2 \tau \, d\tau = k^{-2}\tau - k^{-2} \int \text{dn}^2 \tau \, d\tau$$

It has been found convenient to introduce the *Jacobian Zeta function:*

$$Z(z) = \int_0^z \text{dn}^2 \tau \, d\tau - zEK^{-1}$$

where $E = \int_0^1 (1 - k^2x^2)^{1/2}(1 - x^2)^{-1/2}\, dx$

It is clear that I can be expressed in terms of Z and thus tables of $Z(z)$, together with table of sn (or rather sn^{-1}), enable an integral of this kind to be evaluated.

The remaining third kind of elliptic integral is more awkward to handle. An algebraic integral of the form

$$\int \frac{dt}{(1 + At^2)\sqrt{(\pm k^2 + a^2)(\pm t^2 + b^2)}}$$

can be expressed, by an appropriate substitution, in terms of known functions and

$$\int \frac{\text{sn}^2 \tau \, d\tau}{1 + \nu \text{ sn}^2 \tau} \quad \text{or} \quad \int \frac{k^2 \text{ sn } \alpha \text{ cn } \alpha \text{ dn } \alpha \text{ sn}^2 \tau \, d\tau}{1 - k^2 \text{ sn}^2 \alpha \text{ sn}^2 \tau} = \Pi(x,\alpha)$$

It can be shown that

$$\Pi(x,\alpha) = xZ(\alpha) + \frac{1}{2} \ln \frac{\Theta(x - \alpha)}{\Theta(x + \alpha)}$$

where the Jacobian Θ function is defined in terms of $x = \pi\Pi/2k$ and $q = \exp(-\pi K'/K)$ by

$$\Theta(u) = 1 - 2q \cos 2x + 2q^4 \cos 4x - 2q^9 \cos 6x + \cdots$$

a series which is rapidly convergent when $|q| < 1$. The following series is a convenient one for the evaluation of q:

$$q = \frac{k^2}{16}\left[1 + 2\left(\frac{k}{4}\right)^2 + 15\left(\frac{k}{4}\right)^4 + 150\left(\frac{k}{4}\right)^6 + 1{,}707\left(\frac{k}{4}\right)^8 + \cdots\right]^4$$
$$= \frac{1}{2}\epsilon + \frac{2}{2^5}\epsilon^5 + \frac{15}{2^9}\epsilon^9 + \frac{150}{2^{13}}\epsilon^{13} + \frac{1{,}707}{2^{17}}\epsilon^{17} + \cdots$$

where $\epsilon = \dfrac{1 - (1 - k^2)^{1/4}}{1 + (1 - k^2)^{1/4}}$

References

a. Calculus, Analysis:

1. Apostol, T. M.: "Calculus," 2 vols., Blaisdell Publishing Co., New York, 1961.
2. Apostol, T. M.: "Mathematical Analysis," Addison-Wesley, Reading, Mass., 1957.
3. Courant, R.: "Differential and Integral Calculus," 2 vols., Blackie, Glasgow, 1936–8.
4. Edwards, J.: "An Elementary Treatise in the Differential Calculus," MacMillan, New York, 1886.
5. Edwards, J.: "A Treatise on the Integral Calculus," 2 vols., Chelsea Publishing Company, New York, 1954.
6. Hardy, G. H.: "A Course of Pure Mathematics," Cambridge University Press, New York, 1964.
7. Knopp, K.: "Theory and Application of Infinite Series," Blackie, Glasgow, 1928.
8. Protter, M. H., and C. B. Morrey: "Modern Mathematical Analysis," Addison-Wesley, Reading, Mass., 1964.
9. Whittaker, E. T., and G. N. Watson: "A Course of Modern Analysis," 4th ed., Cambridge University Press, New York, 1927.

b. Functions of a Complex Variable:

10. Ahlfors, L. V.: "Complex Analysis," 2d ed., McGraw-Hill, New York, 1966.
11. Caratheodory, C.: "Theory of Functions of a Complex Variable," 2 vols., Chelsea Publishing Company, New York, 1958, 1960.
12. Copson, E. T.: "Functions of a Complex Variable," Oxford University Press, Fair Lawn, N.J., 1935.
13. Nehari, Z.: "Introduction to Complex Analysis," Allyn and Bacon, Boston, 1961.
14. Hurwitz, A., and R. Courant: "Funktionentheorie," 4th ed., Springer, Berlin, 1964.
15. Knopp, K.: "Elements of the Theory of Functions," Dover, New York, 1952.
16. Knopp, K.: "Theory of Functions," 2 vols., Dover, New York, 1945.
17. Knopp, K.: "Problem Book in the Theory of Functions," 2 vols., Dover, New York, 1948.
18. Rothe, R., F. Ollendorff, and K. Pohlhausen: "Theory of Functions as Applied to Engineering Problems," Technology Press, Cambridge, Mass., 1933; Dover, New York, 1961.
19. Titchmarsh, E. C.: "Theory of Functions," Oxford University Press, Fair Lawn, N.J., 1932.

c. Conformal Representation:

20. Betz, A.: "Konforme Abbildung," Springer, Berlin, 1949.
21. Von Koppenfels, W., and F. Stallman: "Praxis der konformen Abbildung," Springer, Berlin, 1959.

22. Bieberbach, L.: "Conformal Mapping," Chelsea Publishing Company, New York, 1953.
23. Kober, H.: "Conformal Representations," Dover, New York, 1952.
24. Gaier, D.: "Konstruktive Methoden der konformen Abbildung," Springer, Berlin, 1964.

d. Fourier Series and Integrals; Orthogonal Functions:

25. Campbell, G. A., and R. M. Foster: "Fourier Integrals for Practical Applications," Van Nostrand, Princeton, N.J., 1948.
26. Carslaw, H. S.: "Introduction to the Theory of Fourier's Series and Integrals," Macmillan, New York, 1921.
27. Franklin, P.: "An Introduction to Fourier Methods and the Laplace Transformation," Dover, New York, 1958.
28. Erdélyi, A.: "Operational Calculus and Generalized Functions," Holt, Rinehart and Winston, New York, 1962.
29. Erdélyi, A., W. Magnus, F. Oberhettinger, and F. G. Tricomi: "Higher Transcendental Functions," 3 vols., McGraw-Hill, New York, 1953, 1955. "Tables of Integral Transforms," 2 vols., McGraw-Hill, New York, 1954.
30. Jackson, D.: Fourier Series and Orthogonal Polynomials, *Carus Math. Monograph* 6, Open Court, LaSalle, Ill., 1941.
31. Kaczmarz, S., and H. Steinhaus: "Theorie der Orthogonalreihen," Chelsea Publishing Company, New York, 1951.
32. Lighthill, M. J.: "An Introduction to Fourier Analysis and Generalised Functions," Cambridge University Press, New York, 1958.
33. Oberhettinger, F.: "Tabellen zur Fourier Transformation," Springer, Berlin, 1937.
34. Papoulis, A.: "The Fourier Integral and Its Applications," McGraw-Hill, New York, 1962.
35. Szegö, G.: "Orthogonal Polynomials," American Mathematical Society, Providence, R. I., 1939.
36. Tranter, C. J.: "Integral Transforms in Mathematical Physics," Wiley, New York, 1959.
37. Wiener, N.: "The Fourier Integral and Certain of Its Applications," Cambridge University Press, New York, 1933.
38. Rogosinski, W.: "Fourier Series," Chelsea Publishing Company, New York, 1959.

e. Tables of Integrals and Special Functions:

39. Fletcher, A., J. C. P. Miller, L. Rosenhead, and L. J. Comrie: "An Index of Mathematical Tables," 2 vols., Addison-Wesley, Reading, Mass., 1962.
40. Abramowitz, M., and I. A. Stegun: Handbook of Mathematical Functions, Natl. Bur. Std. (U.S.), *Appl. Math. Ser.* 55, 1964.
41. Bois, G. Petit: "Tables of Indefinite Integrals," Dover, New York, 1961.
42. Carmichael, R. D., and E. R. Smith: "Mathematical Tables and Formulas," Dover, New York, 1962.
43. Dwight, H. B.: "Tables of Integrals and Other Mathematical Data," Macmillan, New York, 1934.
44. Dwight, H. B.: "Mathematical Tables of Elementary and Some Higher Mathematical Functions," 2d ed., Dover, New York, 1958.
45. Jahnke, E., F. Emde, and F. Lösch: "Tables of Higher Functions," 6th ed., McGraw-Hill, New York, 1960.
46. Magnus, W., F. Oberhettinger, and D. Sonti: "Formulas and Theorems for the Special Functions of Mathematical Physics," Springer, New York, 1966.
47. Peirce, B. O.: "A Short Table of Integrals," Ginn, Boston, 1929.
48. Gröbner, W., and N. Hofreiter: "Integraltafel," 2 vols., Springer, Berlin, 1949, 1950.
49. Gradshteyn, I. S., and I. M. Ryzhik: "Table of Integrals, Series, and Products," Academic Press, New York, 1965.

f. Elliptic Functions:

50. Bowman, F.: "Introduction to Elliptic Functions with Applications," Dover, New York, 1961.
51. Milne-Thomson, L. M.: "Jacobian Elliptic Function Tables," Dover, New York, 1950.
52. Oberhettinger, F., and W. Magnus: "Anwendung der elliptischen Funktionen in Physik und Technik," Springer, Berlin, 1949.
53. Byrd, P. F., and M. D. Friedman: "Handbook of Elliptic Integrals for Engineers and Physicists," Springer, Berlin, 1954.

g. Other Special Functions:

54. Watson, G. N.: "Bessel Functions," Cambridge University Press, New York, 1944.
55. Gray, A., G. B. Matthews, and T. M. MacRobert: "A Treatise on Bessel Functions and Their Applications to Physics," Macmillan, New York, 1952.
56. Hobson, E. W.: "The Theory of Spherical and Ellipsoidal Harmonics," Cambridge University Press, New York, 1931.
57. MacRobert, T. M.: "Spherical Harmonics," Dover, New York, 1947.
58. McLachlan, N. W.: "Theory and Application of Mathieu Functions," Oxford University Press, Fair Lawn, N.J., 1947.
59. Meixner, J., and F. W. Schäfke: "Mathieusche Funktionen und Sphäroid-Funktionen," Springer, Berlin, 1954.
60. Sneddon, I. N.: "Special Functions of Mathematical Physics and Chemistry," Oliver & Boyd, Edinburgh, 1956.
61. Schäfke, F. W.: "Einführung in die Theorie der Speziellen Funktionen der mathematischen Physik," Springer, Berlin, 1963.

h. General:

62. Courant, R., and D. Hilbert: "Methods of Mathematical Physics," vols. 1 and 2, Interscience, New York, vol. 3 to appear.
63. Morse, P. M., and H. Feshbach: "Methods of Mathematical Physics," 2 vols., McGraw-Hill, New York, 1953.
64. Jeffreys, H., and B. S. Jeffreys: "Methods of Mathematical Physics," Cambridge University Press, New York, 1956.
65. Mathews, J., and R. L. Walker: "Mathematical Methods of Physics," Benjamin, New York, 1964.

Chapter 4

Ordinary Differential Equations

By OLGA TAUSSKY, California Institute of Technology

1. Introduction

Ordinary differential equations are equations involving functions and their derivatives, without being identities. They are an important source for the discovery of new functions. They are classified (1) according to the number of (independent) variables, (2) according as the derivatives are total or partial into *ordinary* or *partial* differential equations, (3) according to the *order* which is the order of the highest differential coefficient involved, (4) according to the *degree* which is the power to which the highest differential coefficient is raised, and (5) into *linear* and *nonlinear*. A linear differential equation is one in which all differential coefficients and the dependent variable are multiplied by functions of the independent variable only. The theory of linear equations is more highly developed than that of nonlinear equations. Among the important nonlinear differential equations is *van der Pol's equation:*

$$x'' - k(1 - x^2)x' + x = bk\lambda \cos (\lambda t + \alpha)$$

If a nonlinear equation is too difficult to solve, it is sometimes made linear by replacing the dependent variables in the coefficients of the derivatives by constants, a process called *linearizing*.

If several dependent variables occur, we have *systems of differential equations*.

Every indefinite integral is the solution of a differential equation. The solutions of a differential equation are also called its *integrals;* even identities which imply solutions are called integrals. Like indefinite integrals, differential equations have an infinite number of solutions unless further conditions are imposed, e.g., in the case of a definite integral. These conditions are usually called *boundary conditions* or *initial conditions*.

In an ordinary differential equation of order n, the solution which contains n arbitrary parameters is called the *complete primitive* or *general solution*. For instance, the differential equation $y'' + \lambda^2 y = 0$ has the complete primitive $y = A \cos \lambda t + B \sin \lambda t$, where A,B are arbitrary constants. A solution free of parameters is called a *particular* integral. A partial differential equation of order n has as its most general solution an expression which contains n arbitrary functions, for example,

$$\frac{\partial^2 z}{\partial x^2} - c^2 \frac{\partial^2 z}{\partial t^2} = 0$$

has as the most general solution

$$z = f(x - ct) + g(x + ct)$$

where f and g are arbitrary functions. A *singular* solution of a differential equation is one which is not included in the complete primitive, for example, $y = x(dy/dx) + c/(dy/dx)$ has as complete primitive $y = mx + c/m$, while $y^2 = 4cx$ is a singular solution. As in this case, a singular solution often represents the envelope of the curves given by the complete primitive.

If a boundary-value problem has to be solved for a linear differential equation with y as the dependent variable which involves a parameter λ in such a way that λ appears as a factor of y but in no other way, then a so-called *eigenvalue* problem has to be solved. The differential equation may not have solutions satisfying the required boundary conditions for all values of λ but only for special ones, the so-called *eigenvalues* or *characteristic values*. The solution corresponding to such a value of λ is called an *eigenfunction, eigenvector*, or *characteristic function*. The set of all eigenvalues of a differential equation is called its *spectrum*. It depends also on the boundary conditions. Consider, for example, the differential equation $y'' + \lambda^2 y = 0$ subject to the conditions

$$y(0) = 0,\ y(\pi) = 0 \qquad \text{and} \qquad y(0) = 0,\ y'(0) = 1$$

In the first case solutions nontrivial exist if and only if $\lambda^2 = 1, 4, 9, \ldots$ and the corresponding solutions are $y = A \sin \lambda t$. In the second case solutions exist for all values of λ.

2. Simple Cases

There is no general method to solve all differential equations; however, some types yield to special methods. Some differential equations can be made *exact* by multiplication by an *integrating factor*, i.e., after multiplication both sides become the differential coefficients of a function of x and y with respect to x. An example of an exact differential equation is

$$\sin y \cos x + \sin x \cos y \frac{dy}{dx} = 0$$

with the solution $\sin y \sin x = c$. The equation $\tan y + \tan x(dy/dx) = 0$ is not exact, but reduces to the one above after multiplication by $\cos x \cos y$.

All linear equations of the first order,

$$\frac{dy}{dx} + Py = Q$$

where P,Q are functions of x only, can be solved by applying the integrating factor $\exp (\int P\, dx)$.

Bernoulli Equation. $(dy/dx) + Py = Qy^n$, though not linear for $n \neq 0,1$, can be reduced to the linear case by first dividing by y^n and then changing the dependent variable through the substitution $y^{-(n-1)} = z$.

Linear Equations of Higher Order

$$\frac{d^n y}{dx^n} + p_1(x) \frac{d^{n-1}y}{dx^{n-1}} + \cdots + p_{n-1}(x) \frac{dy}{dx} + p_n(x)y = q(x)$$

If $q(x) \equiv 0$, the differential equation is called *homogeneous;* otherwise it is *nonhomogeneous.* For this case the *superposition principle* holds: the sum of two solutions is a solution again. Hence in the case $n = 2$ all solutions can be obtained in the form $y = ay_1 + by_2$, where y_1,y_2 are two linearly independent solutions and a,b are arbitrary constants. Two solutions are called linearly independent if no relation $c_1y_1 + c_2y_2 = 0$ exists where c_1,c_2 are constants which are not both zero (Part 1, Chap. 2, Sec. 9). Any function $ay_1 + by_2$ is called a *complementary function* if a,b are constants. If $q(x) \neq 0$, all solutions may be obtained in the form $y = \bar{y} + ay_1 + by_2$ where $\bar{y}$ is a particular integral and y_1,y_2 are linearly independent solutions of the corresponding homogeneous equation.

If the coefficients $p_i(x)$ are constants a_i and $q(x) = 0$, a routine method is available. Let α_i, $i = 1, \ldots ,n$, be the roots of the algebraic equation

$$x^n + a_1x^{n-1} + \cdots + a_{n-1}x + a_n = 0$$

—the *characteristic equation* of the differential equation—then the complete primitive is of the form

$$y = \sum_{i=1}^{n} A_i \exp (\alpha_i x)$$

A_i arbitrary constants, if all α_i are different. If, say, α_1 is of multiplicity r (exactly) so that

$$\alpha_1 = \alpha_2 = \cdots = \alpha_r$$

then the part of the solution involving these roots takes the form

$$\left(\sum_{i=1}^{r} A_i x^{i-1} \right) \exp (\alpha_1 x)$$

If the equation is inhomogeneous, the solution can be obtained, for example, by the *Laplace transform method:* Consider, for simplicity, the solution of

$$y'' + ay' + by = f(t) \qquad y(0) = c \qquad y'(0) = d \qquad (4.1)$$

where a and b are constants and primes denote differentiation with respect to t. Let $\eta(p)$ and $\phi(p)$ be the Laplace transforms of y and f. Then, multiplying (4.1) by e^{-pt} and integrating with respect to t from 0 to ∞,

$$\int_0^\infty y''e^{-pt}\, dt + a \int_0^\infty y'e^{-pt}\, dt + b\eta(p) = \phi(p) \qquad (4.2)$$

Integrating by parts and assuming ye^{-pt}, $y'e^{-pt} \to 0$ as $t \to \infty$,

$$\begin{aligned} \int_0^\infty y''e^{-pt}\, dt &= \Big[y'e^{-pt} \Big]_0^\infty + p \int_0^\infty y'e^{-pt}\, dt \\ &= -d + p \left\{ \Big[ye^{-pt} \Big]_0^\infty + p \int_0^\infty ye^{-pt}\, dt \right\} \\ &= -d + p[-c + p\eta(p)] \end{aligned}$$

and

$$\begin{aligned} \int_0^\infty y'e^{-pt}\, dt &= \Big[ye^{-pt} \Big]_0^\infty + p \int_0^\infty ye^{-pt}\, dt \\ &= -c + p\eta(p) \end{aligned}$$

Hence (4.2) becomes

$$-d - pc + p^2\eta(p) - ac + ap\eta(p) + b\eta(p) = \phi(p)$$

that is,

$$\eta(p) = \frac{\phi(p)}{p^2 + ap + b} + \frac{d + ac + pc}{p^2 + ap + b} \qquad (4.3)$$

If we can find functions $g(t)$ and $h(t)$ whose Laplace transforms are the two terms on the right of (4.3), then, using the fact that the Laplace transforms are unique, $y(t) = g(t) + h(t)$. This is the required solution of (4.1) and is presented as the sum of a complementary function $g(t)$ and a particular integral $h(t)$.

This method will obviously generalize to the case of equations of higher order, with constant coefficients. The solution in each case can be obtained by reference to a sufficiently extensive dictionary of Laplace transforms.

In the case $f(t) = 0$, only the second term on the right of (4.3) is involved. This can be expressed in the form of partial fractions; when the quantities a,b,c,d are general, the form is $a_1/(p - \alpha_1) + a_2/(p - \alpha_2)$ and this is clearly the transform of

$$a_1 \exp (\alpha_1 t) + a_2 \exp (\alpha_2 t)$$

3. Existence Theorems

A special case of a general existence theorem ensures that a very large class of differential equations has solutions. Consider the differential equation $dy/dx = f(x,y)$ with the condition that the solution reduces to $y = y_0$ when $x = x_0$. Such a solution will exist in some interval about x_0 if the function $f(x,y)$ is one-valued and continuous inside a rectangle which surrounds the point (x_0,y_0) and if inside the rectangle it satisfies the so-called *Lipschitz condition:*

$$|f(x,Y) - f(x,y)| < K|Y - y|$$

where K is a constant. The solution is, moreover, unique and continuous with respect to the initial point. This theorem refers to the *one-point boundary-value problem.* An analogous theorem covers the case where the solution is studied in the complex plane and can be extended to linear equations of order n:

$$\frac{d^n y}{dx^n} = f\left(x, y, \frac{dy}{dx}, \ldots, \frac{d^{n-1}y}{dx^{n-1}}\right)$$

in particular

$$\frac{d^n y}{dx^n} + p_1(x)\frac{d^{n-1}y}{dx^{n-1}} + \cdots + p_{n-1}(x)\frac{dy}{dx} + p_n(x)y = q(x)$$

where the values of y, (dy/dx), . . . , $(d^{n-1}y/dx^{n-1})$ are prescribed for a certain value x_0 of x. The point x_0 is called an *ordinary point* if the value of $d^n y/dx^n$ at x_0 is finite whatever choice was made for $d^i y/dx^i$, $i < n$; otherwise it is called *singular*. If x_0 is an ordinary point and all functions $p_i(x)$ are analytic, then there exists a unique and analytic solution. There is no nonanalytic solution satisfying the same boundary conditions.

4. Methods for Solution

The linear case $n = 2$ has received particular attention. As a practical method *expansion of the solution into a power series* is one of the most useful devices. To obtain the power series, the solution is written tentatively as a power series with unknown coefficients and introduced into the differential equation. In this way recurrence relations for the coefficients are obtained. If these relations involve two coefficients only, the method is very practical. If the solution is studied near a pole of the coefficients, say 0, then an expansion of the form

$$y_1 = x^\alpha(a_0 + a_1 x + \cdots)$$

can be tried. By comparison an equation of second degree for the index α, the *indicial equation* and recurrence relations for the coefficients a_i are found. If such a solution exists and if the indicial equation is of degree 2 (in the general case of degree n), then $x = 0$ is called a *regular singular point;* otherwise it is called an *irregular singular point.*

A second solution can sometimes be found easily by putting $y = uy_1$. The second solution is frequently of the form $y = y_1 \log x + \Sigma b_r x^{\alpha_2 + r}$, where α_2 is the other root of the indicial equation. The *method of Frobenius* consists in taking the solution in the form $y = x^\alpha(1 + a_1 x + a_2 x^2 + \cdots)$ and obtaining $\partial y/\partial\alpha$ as a second solution in the case that the indicial equation has a double root. A similar treatment is necessary when the roots of the indicial equation differ by an integer.

Certain differential equations with variable coefficients yield to attack by Laplace transforms. One of the methods for finding a particular integral of a nonhomogeneous equation is called *variation of parameters.* This means that certain constants are replaced by functions. Let y_1 and y_2 be two independent solutions of

$$y'' + P(x)y' + Q(x)y = 0 \tag{4.4}$$

so that the Wronskian determinant

$$W = y_1 y'_2 - y'_1 y_2 \neq 0$$

The general solution of this equation is

$$y = A_1 y_1 + A_2 y_2 \tag{4.5}$$

where A_1 and A_2 are constants. To solve

$$y'' + P(x)y' + Q(x)y = R(x) \tag{4.6}$$

assume a solution in the form (4.5) where A_1 and A_2 are assumed to be functions of x. Then

$$y' = A'_1 y_1 + A'_2 y_2 + A_1 y'_1 + A_2 y'_2$$

Assume that

$$A'_1 y_1 + A'_2 y_2 = 0 \tag{4.7}$$

Then

$$y'' = A_1 y''_1 + A_2 y''_2 + A'_1 y'_1 + A'_2 y'_2$$

If $y = A_1(x)y_1 + A_2(x)y_2$ is a solution to (4.6), then

$$A'_1 y'_1 + A'_2 y'_2 = R \tag{4.8}$$

Hence (4.7) and (4.8) are a pair of simultaneous equations from which we may determine A'_1 and A'_2:

$$A'_1 = \frac{Ry_2}{W} \qquad A'_2 = \frac{-Ry_1}{W}$$

giving A_1 and A_2 by a quadrature. We have

$$y = y_1(x)\int_{x_1}^{x} \frac{R(t)y_2(t)\,dt}{W(t)} - y_2(x)\int_{x_2}^{x} \frac{R(t)y_1(t)\,dt}{W(t)}$$

The limits of integration x_1, x_2 provide the two arbitrary constants.

An important special case is the solution of

$$y'' + n^2 y = f(x)$$

which can be written as

$$y = -\frac{1}{n}\int_0^x f(t) \sin n(t - x)\,dt + A \cos nx + B \sin nx$$

The function $\sin n(t - x)$ is *Green's function* of the differential expression (d^2y/dx^2) (Sec. 6). A number of examples of differential equations that occur in physics are found in Part 1, Chap. 3, Secs. 9 and 10.

5. Some General Theorems

These theorems are about linear differential equations of second order whose coefficients are continuous functions of a parameter. Let the differential equation be of the form

$$p_0(x)y''(x) + p_1(x)y'(x) + p_2(x)y(x) = 0 \tag{4.9}$$

The functions $p_0(x)$, $p_1(x)$, $p_2(x)$ are assumed continuous in an interval (a,b); they are further continuous functions of a parameter λ. Any solution of (4.9) is then defined in (a,b) and is a continuous function of x and λ. It is uniquely determined by the conditions

$$y(a) = \alpha \qquad y'(a) = \beta$$

Hence, the solution for which

$$y(a) = 0 \qquad y'(a) = 0$$

must be the function $y = 0$. Hence, a solution which does not vanish identically can only have simple zeros. There is only a finite number of zeros in a finite interval (a,b) and the zeros of two independent solutions separate each other. Any solution which has a zero in (a,b) is said to *oscillate* there.

Consider the case when Eq. (4.9) can be written in the form

$$\frac{d}{dx}[K(x)y'] - G(x)y = 0 \qquad K(x) > 0 \qquad (4.10)$$

Let $G(x)$ be a monotonically decreasing function of λ. Then the solutions of the equation which correspond to the larger value of the parameter oscillate more rapidly than the ones with smaller parameters.

In certain problems in physics it is necessary to investigate those solutions of (4.10) which assume given values at the two end points a,b of the interval; these are usually taken as 0 and π. If the boundary-value conditions are of the form

$$\begin{aligned} a_{11}y(a) + a_{12}y'(a) &= 0 \\ a_{21}y(b) + a_{22}y'(b) &= 0 \end{aligned}$$

the problem is called a *Sturm-Liouville boundary-value problem.* A special case arises when the parameter λ occurs simply as a factor which leads to an eigenvalue problem, and the differential equation is of the form

$$(pv')' + \lambda\rho\nu = 0 \qquad (4.11)$$

where ρ is a known function. A more general problem is the differential equation

$$(pv')' - qv + \lambda\rho v = 0 \qquad (4.12)$$

The most important boundary-value conditions usually imposed on (4.11) are:

$$\begin{aligned} v(0) &= 0 & v(\pi) &= 0 \\ h_0v(0) &= v'(0) & -h_1v(\pi) &= v'(\pi) \\ v'(0) &= 0 & v'(\pi) &= 0 \\ v(0) &= v(\pi) & p(0)v'(0) &= p(\pi)v'(\pi) \end{aligned}$$

(condition for periodicity)

The main results concerning the solution of (4.11) are the following. There is an infinite sequence of values $\lambda = \lambda_1, \lambda_2, \ldots$ for which the differential equation can be solved; they are called the *eigenvalues* of the differential equation. The corresponding solutions $v_i(x)$, the so-called *eigenfunctions,* form a system of orthogonal functions with ρ as weight function:

$$\int_0^\pi v_n v_m \rho\, dx = 0 \qquad n \neq m$$

Every function f which satisfies the boundary-value conditions can be expanded into a series

$$f = \sum_{n=1}^{\infty} c_n v_n, \qquad c_n = \int_0^\pi \rho f v_n\, dx$$

With the exception of the periodicity condition the eigenvalues of these problems are simple, i.e., there are not two linearly independent solutions of the problems. If $h_0,h_1 > 0$, then all $\lambda_i \geqq 0$. This means that the solution is of oscillatory character. The simple case $v'' + \lambda v = 0$, with the boundary-value conditions $v(0) = v(\pi) = 0$, illustrates all these results. The general solution of this equation is

$$c_1 \exp(i\lambda^{1/2}x) + c_2 \exp(-i\lambda^{1/2}x)$$

The boundary-value conditions can be satisfied if and only if $\lambda = n^2$. The corresponding solutions are $v_n = \sin nx$.

6. Nonhomogeneous Equations, Green's Function

Let the equation be of the form

$$p_0(x)y''(x) + p_1(x)y'(x) + p_2(x)y(x) = f(x) \qquad (4.13)$$

where $p_0(x)$, $p_1(x)$, $p_2(x)$, $f(x)$ are continuous functions in (a,b) and $p_0(x) \neq 0$. Let $y_1(x)$, $y_2(x)$ be two independent solutions of the homogeneous equations. It then follows that

$$\gamma(x,\xi) = \frac{y_2(\xi)y_1(x)}{p_0(\xi)[y_2'(\xi)y_1(\xi) - y_1'(\xi)y_2(\xi)]} \qquad \text{for } a \leq x \leq \xi$$

and

$$= \frac{y_1(\xi)y_2(x)}{p_0(\xi)[y_2'(\xi)y_1(\xi) - y_1'(\xi)y_2(\xi)]} \qquad \text{for } \xi \leq x \leq b$$

is a solution of the homogeneous equation for all x in (a,b) apart from $x = \xi$. This function is continuous at $x = \xi$, but its derivative makes a jump of length 1 at $x = \xi$. For any function γ with these properties the function

$$\int_a^b \gamma(x,\xi)f(\xi)\, d\xi$$

is a solution of the inhomogeneous equation.

Consider now a differential equation

$$L(y) + \lambda y = f(x) \qquad (4.14)$$

with the boundary condition,

$$\alpha_1 y(0) - \alpha y'(0) = \beta_1 y(\pi) - \beta y'(\pi) = 0$$

The symbol $L(y)$ denotes a differential operator, of the form used in (4.13).

This problem can be solved at once if a function $\gamma(x,\xi)$ as defined above is known which also satisfies the boundary-value conditions. Such a function is called the *Green's function* $\Gamma(x,\xi)$ of the problem. The solution is then

$$y = \int_0^\pi \Gamma(x,\xi)f(\xi)\, d\xi$$

In order to find Γ, take a solution $u_1(x,\lambda)$ of the homogeneous equation for which

$$u_1(0,\lambda) = \alpha, \qquad u'_1(0,\lambda) = \alpha_1$$

and a solution $u_2(x,\lambda)$ of the homogeneous equation for which $u_2(\pi,\lambda) = \beta$, $u'_2(\pi,\lambda) = \beta_1$. Put

$$\begin{aligned} \gamma(x,\xi,\lambda) &= u_1(x,\lambda)u_2(\xi,\lambda) & \text{for } x \leqq \xi \\ &= u_1(\xi,\lambda)u_2(x,\lambda) & \text{for } x \geqq \xi \end{aligned}$$

Then

$$\Gamma(x,\xi,\lambda) = \frac{\gamma(x,\xi,\lambda)}{p_0(\xi)u'_2(\xi,\lambda)u_1(\xi,\lambda) - u'_1(\xi,\lambda)u_2(\xi,\lambda)}$$

If u_1,u_2 are linearly independent, the denominator will not vanish (Part 1, Chap. 2, Sec. 9). Linear dependency of u_1,u_2 for certain values of λ means that $u_1(x,\lambda)$ is an eigenfunction and λ the corresponding eigenvalue. In this case the problem can in general not be solved. If λ is not an eigenvalue exactly one solution exists. Assume that $\lambda = 0$ is not an eigenvalue (that is not a restriction) and put

$$\Gamma(x,\xi,0) = G(x,\xi)$$

Then

$$\int_0^\pi G(x,\xi)f(\xi)\,d\xi$$

is the solution of $L(y) = f(x)$.

If $y(x)$ is the solution of (4.14) and we put

$$L[y(x)] = u(x)$$

then $y(x)$ can be regarded as the solution of

$$L(y) = u(x)$$

hence

$$y = \int_0^\pi G(x,\xi)u(\xi)\,d\xi$$

Thus the problem of solving (4.14) is identical with the problem of finding a function u that has

$$u(x) + \lambda \int_0^\pi G(x,\xi)u(\xi)\,d\xi = f(x)$$

For the case $f(x) = 0$ the *integral equation* (Part 1, Chap. 6, Sec. 3)

$$u(x) + \lambda \int_0^\pi G(x,\xi)u(\xi)\,d\xi = 0$$

is obtained.

7. Systems of Simultaneous Differential Equations

Here we consider a set of more than one dependent variables $x_1, \ldots, x_n$, but only one independent variable t. If the system has constant coefficients, it is of the form

$$\begin{aligned} f_{11}(D)x_1 + \cdots + f_{1n}(D)x_n - \xi_1(t) &= 0 \\ f_{21}(D)x_1 + \cdots + f_{2n}(D)x_n - \xi_2(t) &= 0 \\ \cdots\cdots\cdots\cdots\cdots\cdots\cdots\cdots \\ f_{nn}(D)x_1 + \cdots + f_{nn}(D)x_n - \xi_n(t) &= 0 \end{aligned} \tag{4.15}$$

where $D = d/dt$ and f_{ik} are polynomials with constant coefficients in D; $\xi_i(t)$ are given functions of t. If the highest degree in D among the polynomials f_{ik} is m, the system is said to be of order m.

The study of such systems is closely connected with matrix theory. An alternative form of expressing (4.14) is

$$(A_0D^m + A_1D_{m-1} + \cdots + A_{m-1}D + A_m)x = \xi$$

where the A_i are square $n \times n$ matrices with constant elements, x is the column vector $(x_1, \ldots, x_n)$, and ξ the column vector $(\xi_1, \ldots, \xi_n)$.

If $\xi = 0$ the system is called *homogeneous;* otherwise it is called *nonhomogeneous.* The general solution of a nonhomogeneous system is the sum of the general solution of the corresponding homogeneous system and a particular integral of the nonhomogeneous system.

Replace the operator D in the matrix $(f_{ik}(D))$ by an indeterminate λ, and consider the determinant $|f_{ik}(\lambda)|$. It is called the characteristic determinant of the system. If the roots $\lambda_1, \ldots, \lambda_n$ of the determinantal equation $|f_{ik}(\lambda)| = 0$ are all different, then

$$x = \exp(\lambda_i k_i t)$$

is a solution where k_i is a constant solution vector of $f_{ik}(\lambda_i)k_i = 0$. In this way the complementary function is obtained. More complicated expressions are obtained if multiple roots occur.

The general solution contains m arbitrary constants and particular integrals can be determined by supplementary conditions.

Some well-known examples are

$$(AD^2 + E)x = 0$$

where A and E are both symmetric matrices. Systems of first order have received particular attention. Such a system in the homogeneous case can be written in the form

$$\begin{aligned} \frac{dx_1}{dt} &= a_{11}x_1 + \cdots + a_{1n}x_n \\ &\vdots \\ \frac{dx_n}{dt} &= a_{n1}x_1 + \cdots + a_{nn}x_n \end{aligned}$$

or briefly,

$$\frac{dx}{dt} = Ax$$

where $A = (a_{ik})$ and $dx/dt = dx_1/dt, \ldots, dx_n/dt$.

The solution of this system can be expressed in the form

$$x = e^{At}x_0$$

where $x_0 = x(0)$ is the initial vector.

In the nonhomogeneous case when we are concerned with the system

$$\frac{dx}{dt} = Ax + f(t)$$

where $f(t)$ is a vector $(f_1(t), \ldots, f_n(t))$, we use variation of the constants, i.e., we put

$$x = e^{At}v$$

and obtain $dv/dt = e^{-At}f(t)$.

The solution of the inhomogeneous system is therefore of the form

$$x = e^{At}\left[\int_0^t e^{-At}f(t)\,dt + x(0)\right]$$

Consider next systems with variable coefficients

$$\frac{dx}{dt} = A(t)x$$

The solution can again be expressed in terms of an infinite series of matrices: Assume that $x = x_0$ for $t = t_0$. The solution can then be expressed in the form

$$x(t) = \Omega_0{}^t(A)x(t_0)$$

where $\Omega_0{}^t(A)$ is the matrix series.

$$I + \int_{t_0}^{t} A(\tau_1)\,d\tau_1 + \int_{t_0}^{t} A(\tau_1) \int_{t_0}^{\tau_1} A(\tau_2)\,d\tau_2\,d\tau_1 + \int_{t_0}^{t} A(\tau_1) \int_{t_0}^{\tau_1} A(\tau_2) \int_{t_0}^{t} A(\tau_3)\,d\tau_3\,d\tau_2\,d\tau_1 + \cdots$$

which converges for every matrix $A(t)$ with continuous functions of t as elements. This series is called the *matrizant* of A. It has the property that $(d/dt)\Omega(A) = A\Omega(A)$. If the elements of A are constants the matrizant is e^{At}. A nonhomogeneous system can be treated correspondingly. Of special interest is the case when the elements of $A(t)$ are periodic functions of t. In this case the system can be reduced to one with a constant coefficient matrix A.

References

1. Bateman, H.: "Differential Equations," Longmans, London, 1926.
2. Bellman, R.: "Stability Theory of Differential Equations," McGraw-Hill, New York, 1953.
3. Bieberbach, L.: "Einjühorung in die Theorie der Differential gleichungen in reellen Gebiet," Springer, Berlin, 1956.
4. Coddington, E. A., and N. Levinson: "Theory of Ordinary Differential Equations," McGraw-Hill, New York, 1955.
5. Courant, R., and D. Hilbert: "Methods of Mathematical Physics," Interscience, New York, 1953.
6. Frazer, R. A., A. W. J. Duncan, and A. R. Collar: "Elementary Matrices," Cambridge, New York and London, 1938.
7. Hoheisel, G.: "Gewöhnliche Differentialgleichungen," Sammlung Göschen, De Gruyter, Berlin, 1926.
8. Ince, E. L.: "Ordinary Differential Equations," Longmans, London, 1927.
9. Kamke, E.: "Differentialgleichungen, Lösungsmethoden und Lösungen," Vol. 1, Akademische Verlagsgesellschaft, Leipzig, 1944.
10. Lefschetz, S.: "Lectures on Differential Equations," Princeton University Press, Princeton, N.J., 1948.
11. McLachlen, N. W.: "Bessel Functions for Engineers," Oxford, New York and London, 1934.
12. Nemickii, V. V., and V. V. Stepanov: "Qualitative Theory of Differential Equations," Moscow, 1949.
13. Birkhoff, G., and G.-C. Rota: "Ordinary Differential Equations," Ginn & Co., New York, 1962.
14. Wasow, W.: "Asymptotic Expansions for Ordinary Differential Equations," Interscience, New York, 1965.

Chapter 5

Partial Differential Equations

By FRITZ JOHN, New York University

1. General Properties

Definitions. A system of partial differential equations for a set of functions $u_1, \ldots, u_N$ of the independent variables $x_1, x_2, \ldots, x_n$ has the form $F_1 = 0, \ldots, F_p = 0$, where the F_i are functions of the x_k, u_j and of a finite number of partial derivatives of the u_j. The *order* m of the system is the order of the highest derivative that occurs in any of the equations. The system is called *linear* if the F_i are linear in the u_j and all their derivatives. It is called *quasi-linear* if the F_i are linear in the derivatives of the u_j of order m. Ordinarily we are concerned with the case where the number N of unknown functions equals the number p of differential equations. An important special case is that of a single equation for a single unknown function u.

Vector Notation. The independent variables $x_1, \ldots, x_n$ are combined conveniently into a vector $x = (x_1, \ldots, x_n)$; similarly the vector $u = (u_1, \ldots, u_N)$ describes the system of unknown functions. The symbols $D_1 = \partial/\partial x_1, \ldots, D_n = \partial/\partial x_n$ for the partial-differentiation operations also can be combined into a vector $D = (D_1, \ldots, D_n)$. For a scalar $\phi = \phi(x) = \phi(x_1, \ldots, x_n)$ the vector $D\phi = (D_1\phi, \ldots, D_n\phi)$ is the *gradient* of ϕ.

Very useful is the abbreviated notation of Laurent Schwartz for higher partial derivatives. Let $\eta = (\eta_1, \ldots, \eta_n)$ be a vector and $\alpha = (\alpha_1, \ldots, \alpha_n)$ be a *multi-index*, i.e., a vector all of whose components $\alpha_1, \ldots, \alpha_n$ are non-negative integers. The scalar ("monomial")

$$(\eta_1)^{\alpha_1}(\eta_2)^{\alpha_2} \cdots (\eta_n)^{\alpha_n}$$

is then denoted simply η^α. Accordingly D^α stands for the partial-derivative operator

$$(D_1)^{\alpha_1} \cdots (D_n)^{\alpha_n} = \frac{\partial^{\alpha_1 + \cdots + \alpha_n}}{\partial x_1^{\alpha_1} \cdots \partial x_n^{\alpha_n}}$$

For the order $\alpha_1 + \cdots + \alpha_n$ of the partial derivative the symbol $|\alpha|$ is used.

In this notation the most general linear mth-order differential equation for a single function

$$u = u(x) = u(x_1, \ldots, x_n)$$

takes the simple form

$$\sum_{|\alpha| \leq m} A_\alpha(x) D^\alpha u = B(x) \tag{5.1}$$

Here the summation is supposed to be extended over all multi-indices $\alpha = (\alpha_1, \ldots, \alpha_n)$ with

$$|\alpha| = \alpha_1 + \cdots + \alpha_n$$

not exceeding m, and the $A_\alpha = A_{\alpha_1 \ldots \alpha_n}$ and B are given scalar functions of $x = (x_1, \ldots, x_n)$. Equation (5.1) also serves to represent the most general mth-order *system* of linear partial differential equations, if we interpret u and B as vectors with m components, or, more precisely, as matrices with one column and N rows, and the A_α as square matrices with N rows and columns. The general mth-order quasi-linear system is of the form

$$\sum_{|\alpha| = m} A_\alpha D^\alpha u = B \tag{5.2}$$

where the A_α are square matrices and B a vector with components that are given functions of the u_j and of their derivatives of orders $< m$.

General nonlinear systems of partial differential equations can formally be reduced to quasi-linear systems of first order. One has only to introduce all derivatives of the u_j that occur in the equations as new unknown functions and to differentiate each differential equation once.

Characteristic Equation. With a single linear partial differential equation (5.1) for a single function u (with the A_α, B, u taken as scalars) we can associate the expression

$$Q = \sum_{|\alpha| = m} A_\alpha(x) \xi^\alpha \tag{5.3}$$

which is a homogeneous polynomial ("form") in the components of the vector $\xi = (\xi_1, \ldots, \xi_n)$ with coefficients depending on x. This is the *characteristic form* of Eq. (5.1). It is formed from the terms in the differential equation involving the derivatives of highest order (the "principal part" of the differential equation). If (5.1) is interpreted as a system with square matrices $A_\alpha(x)$ the characteristic form Q is defined as

$$Q = \text{determinant} \left(\sum_{|\alpha| = m} A_\alpha(x) \xi^\alpha \right) \tag{5.4}$$

Thus for the equation of sound waves (see Part 3, Chap. 8, Sec. 2)

$$D_4^2 u - c^2(D_1^2 + D_2^2 + D_3^2)u = 0 \tag{5.5}$$

(with x_4 denoting the time) the characteristic form is

$$Q = \xi_4^2 - c^2(\xi_1^2 + \xi_2^2 + \xi_3^2) \quad (5.6)$$

For waves in an isotropic elastic body (see Part 3, Chap. 7, Sec. 1) the differential equations are

$$\rho D_4^2 u_i = (\lambda + \mu) D_i(D_1 u_1 + D_2 u_2 + D_3 u_3) + \mu(D_1^2 + D_2^2 + D_3^2) u_i \quad (5.7)$$

for $i = 1, 2, 3$. Here the characteristic form is

$$Q = [\rho\xi_4^2 - \mu(\xi_1^2 + \xi_2^2 + \xi_3^2)]^2 \times [\rho\xi_4^2 - (\lambda + 2\mu)(\xi_1^2 + \xi_2^2 + \xi_3^2)] \quad (5.8)$$

A characteristic form Q also can be associated with a nonlinear system of equations $F_1 = 0, F_2 = 0, \ldots, F_N = 0$ for functions $u_1, \ldots, u_N$. Here we take

$$Q = \text{determinant}\left(\sum_{|\alpha|=m} \frac{\partial F_i}{\partial D^\alpha u_k} \xi^\alpha\right) \quad (5.9)$$

An $(n-1)$-dimensional surface S in n-dimensional x-space can be described by an equation $\phi(x) = \phi(x_1, \ldots, x_n) = 0$. The gradient vector $D\phi = (D_1\phi, \ldots, D_n\phi)$ points in the direction of the normal of S. The surface S is *characteristic* at a point x if the characteristic form has the value 0 when $D\phi$ is substituted for ξ. In the case of a linear equation or system of equations of the form (5.3) the partial differential equation alone determines if a surface S is characteristic. For example, in Eq. (5.7) for elastic waves the surface $\phi = 0$ is characteristic if either

$$\rho(D_4\phi)^2 - \mu((D_1\phi)^2 + (D_2\phi)^2 + (D_3\phi)^2) = 0 \quad (5.10a)$$

or

$$\rho(D_4\phi)^2 = (\lambda + 2\mu)((D_1\phi)^2 + (D_2\phi)^2 + (D_3\phi)^2) \quad (5.10b)$$

In the case of a nonlinear equation the coefficients of Q depend on u and its derivatives; hence knowledge of the special solution u is needed here to decide whether a surface S is characteristic or not.

The Cauchy Problem. The characteristic equation is of importance in connection with the *Cauchy problem* for the system of differential equations. In this problem one has to determine a solution u of the mth-order system from prescribed *Cauchy data* on a surface S, that is, from prescribed values of u and of its derivatives of order $<m$ on S. (Actually only the values of u and its first $m - 1$ *normal* derivatives have to be prescribed on S, since all other derivatives are obtainable from them.) If S is noncharacteristic and the differential equations are linear or quasilinear, the differential equations can be used to obtain algebraically the values of all mth-order derivatives of u along S. Similarly, by differentiating the differential equations, it is possible to obtain the values of any higher derivatives of u on S, provided u and the coefficients of the differential equations are sufficiently often differentiable.

It follows that two solutions of a linear system of partial differential equations that agree with their derivatives of orders $<m$ along a surface S also will agree in all higher-order derivatives along S, provided S is not characteristic. This shows that discontinuities of solutions in the derivatives of order m can "propagate" only along characteristic surfaces and suggests the identification of such surfaces with possible *wavefronts*. In this connection, if the variable x_n is considered the time, it is better to interpret the equation $\phi(x_1, \ldots, x_n) = 0$ as that of an $(n-2)$-dimensional surface in $(n-1)$-dimensional $x_1 \cdots x_{n-1}$-space that moves with the time x_n.

In general, the Cauchy problem with data prescribed on a characteristic surface S has no solution. The fact that the characteristic form vanishes, in general, leads to an inconsistency between the Cauchy data and the differential equations. If the surface S is not characteristic, the Cauchy data on S together with the differential equations permit one formally to calculate all higher derivatives of u on S and to construct a formal power-series solution of the Cauchy problem. But in general no actual solution exists, unless the "initial" surface S is "spacelike" (see below). Only if one is restricted to *analytic* differential equations, initial surfaces S, and Cauchy data (i.e., those that can be represented locally in terms of convergent power series expansions) will there exist for noncharacteristic S an analytic solution u of the Cauchy problem in a sufficiently small neighborhood of S (theorem of Cauchy and Kowalewski).

Classification of Partial Differential Equations. Different types of partial differential equations can be distinguished according to the algebraic properties of the characteristic form. The behavior of the solution of the differential equations depends to a large extent on the "type" of the equation. In particular, the kind of additional information needed to determine a solution varies from type to type. Following Hadamard, a problem can be considered as *well posed* if its solution exists and is unique for all sufficiently regular data and if the solution depends continuously on the data. But what constitutes a well-posed problem for a partial differential equation depends on the type of the equation. For partial differential equations occurring in applications, physical considerations often suggest the mathematical problems that are well posed for those equations, with data and corresponding solutions somewhat in the relation of causes and effects. However, mathematical problems that are not well posed in the sense of Hadamard occur as well, for example, wherever the state of a system is to be deducted from observations.

No complete classification of equations is attempted here, but a few types will be mentioned.

A differential equation or system of equations is called *elliptic* at a point x if the characteristic form Q is definite, that is, $Q \neq 0$ for all vectors $\xi \neq 0$. (In a nonlinear equation elliptic character may depend on the special solution considered.) For an elliptic equation no real characteristic surfaces exist. Solutions are very smooth except possibly at the boundary of the region of definition of the solution. (Solutions are even analytic in case of analytic elliptic equations.) In the common well-posed problems for such equations "suitable" data are prescribed over the whole boundary of the region of definition. The Cauchy problem would not be well posed for elliptic equations; in the usual situations Cauchy data on one portion of the boundary would be inconsistent with the

Cauchy data in any other portion. Elliptic equations occur in connection with problems from which dependence on time has been eliminated, for example, in cases of equilibrium or of steady state or of harmonic dependence on time.

The other most important type, that of the *hyperbolic* equations, consists of those equations (or systems of equations) for which there exist surfaces $\phi(x_1, \ldots, x_n) = 0$ (so-called *spacelike* surfaces) on which the Cauchy data of a solution can be prescribed arbitrarily. If we substitute in the characteristic form Q for the vector ξ the expression $\eta + \lambda D\phi$, where ξ is a vector and λ a scalar, then Q will go over into a polynomial $P(\lambda)$ which still depends on the choice of the vector η. The surface $\phi = 0$ is spacelike if for each real nonvanishing vector η independent of $D\phi$ the polynomial $P(\lambda)$ has real and distinct roots. The surface is certainly not spacelike if for some real η one of the roots λ is imaginary. The intermediate case where all roots λ are real for real $\eta \neq 0$ but not necessarily distinct requires special considerations. For a geometric interpretation we consider the *normal cone* with vertex x in n-dimensional Euclidean space; it consists of the points $y = (y_1, \ldots, y_n)$ of the form $y = x + \xi$ for which ξ satisfies the characteristic equation $Q = 0$ belonging to the point x. The surface $\phi = 0$ is certain to be spacelike at one of its points x if every real two-dimensional plane through the normal of the surface at x intersects the normal cone with vertex x in real distinct lines. If we identify the independent variable x_n with the time, we can prescribe the Cauchy data of a solution arbitrarily at the initial time provided the surfaces $x_n = \text{const}$ are spacelike.

These notions are best illustrated by the case of a single linear second-order equation for a scalar function u:

$$\sum_{i,k=1}^{n} a_{ik}(x) D_i D_k u + \sum_{i=1}^{n} b_i(x) D_i u + c(x) u = d(x) \tag{5.11}$$

Here the characteristic form is given by

$$Q = \sum_{i,k=1}^{n} a_{ik}(x) \xi_1 \xi_k \tag{5.12}$$

In the *elliptic* case Q is a definite quadratic form in the ξ_1. The standard example is the *Laplace equation*

$$\Delta u = \sum_{i=1}^{n} \frac{\partial^2 u}{\partial x_i^2} = 0 \tag{5.13}$$

Equation (5.11) is *hyperbolic* if there exist spacelike surfaces $\phi(x_1, \ldots, x_n) = 0$, that is, surfaces with the property that every two-dimensional plane through the surface normal at a point x intersects the normal cone

$$\sum_{i,k=1}^{n} a_{ik}(x)(y_i - x_i)(y_k - x_k) = 0 \tag{5.14}$$

in two distinct real straight lines. For a hyperbolic equation the characteristic form (5.12) can always be reduced to

$$\pm(\xi_1^2 + \xi_2^2 + \cdots + \xi_{n-1}^2 - \xi_n^2)$$

by a real linear substitution for the variables ξ. The standard example for a hyperbolic equation is the equation of waves (5.5). For that equation the surface $\phi(x_1,x_2,x_3,x_4) = 0$ is spacelike if

$$\left(\frac{\partial \phi}{\partial x_4}\right)^2 - c^2 \left[\left(\frac{\partial \phi}{\partial x_1}\right)^2 + \left(\frac{\partial \phi}{\partial x_2}\right)^2 + \left(\frac{\partial \phi}{\partial x_3}\right)^2\right] > 0 \tag{5.15}$$

In particular, the planes $x_4 = \text{const}$ are spacelike.

The only other type of equations of importance in mathematical physics is presented by the *parabolic* equations. The second-order equation (5.11) is parabolic when the characteristic form (5.12) is semidefinite. To this type belongs the equation of heat conduction (see Part 5, Chap. 5, Sec. 2)

$$\frac{\partial u}{\partial x_4} = \alpha \left(\frac{\partial^2 u}{\partial x_1^2} + \frac{\partial^2 u}{\partial x_2^2} + \frac{\partial^2 u}{\partial x_3^2}\right) \tag{5.16}$$

This equation is distinguished from the hyperbolic ones, when x_4 is interpreted as the time, by the fact that disturbances propagate with *infinite* speed.

A nonlinear equation may be elliptic for some solutions and hyperbolic for others. Even in the case of a linear equation the type may vary with the region considered. An example is furnished by *Tricomi's equation*

$$x_2 \frac{\partial^2 u}{\partial x_1^2} + \frac{\partial^2 u}{\partial x_2^2} = 0 \tag{5.17}$$

which is hyperbolic for $x_2 < 0$ and elliptic for $x_2 > 0$.

Green's Identity. Every linear partial differential equation or system of such equations gives rise to integral identities that are derived by applying the formula for *integration by parts* to a multiple integral. This formula (the *divergence theorem*) asserts that, under suitable regularity assumptions for two functions u, v in a domain R with boundary B,

$$\int_R u(D_i v)\, dx = -\int_R v(D_i u)\, dx + \int_B uv\xi_i\, dS$$

Here $dx = dx_1 \cdots dx_n$ denotes the n-dimensional element of volume, dS the surface element of B, and $\xi = (\xi_1, \ldots, \xi_n)$ the unit vector in the direction of the exterior normal. Let L be a scalar differential operator defined by

$$Lu = \sum_{|\alpha| \leqq m} A_\alpha(x) D^\alpha u \tag{5.18}$$

We associate with L the *adjoint operator* $\bar{L}$ defined by

$$\bar{L}v = \sum_{|\alpha| \leq m} (-1)^{|\alpha|} D^\alpha (A_\alpha v) \tag{5.19}$$

Repeated application of integration by parts to the expression vLu leads to *Green's identity*

$$\int_R (vLu - u\bar{L}v)\, dx = \int_B M(u,v,\xi)\, dS \tag{5.20}$$

Here the expression M is linear in the functions u, v and their derivatives and in the components of ξ (the direction cosines of the exterior normal).

For the Laplace operator in n dimensions

$$L = \sum_{i=1}^{n} D_i^2 = \Delta \tag{5.21}$$

we have $\bar{L} = L$, and Green's identity takes the well-known form

$$\int_R (v\,\Delta u - u\,\Delta v)\,dx = \int_B \left(v\frac{\partial u}{\partial n} - u\frac{\partial v}{\partial n}\right) dS \tag{5.22}$$

Here $\partial u/\partial n$ is the *normal derivative* of u defined by

$$\frac{\partial u}{\partial n} = \sum_{i=1}^{n} \xi_i D_i u \tag{5.23}$$

Green's identity for systems of equations has the same form. If now in the definition (5.18) of L the A_α denote square matrices, we have to replace the A_α in the definition (5.19) of $\bar{L}$ by the transposed matrices A_α^T. Instead of the expression $vLu - u\bar{L}v$ in identity (5.20) we have to take $v^T(Lu) - (\bar{L}v)^T u$.

If u is a solution of $Lu = 0$ and v any function vanishing in a neighborhood of the boundary B of R, identity (5.20) becomes

$$\int_R u\bar{L}v\,dx = 0 \tag{5.24}$$

This relation can be used to characterize the solutions u of $Lu = 0$ in a way that does not involve any derivatives of u at all and applies to functions that are not necessarily differentiable. We call u a *weak solution* of $Lu = 0$ if Eq. (5.24) holds for all infinitely often differentiable functions v that vanish near the boundary of R.

2. First-order Equations

The General First-order Equation for a Single Unknown Function. If $u = u(x_1, \ldots, x_n)$ is an unknown function and $p_i = u_{x_i} = \partial u/\partial x_i$ denotes its first partial derivative with respect to x_i,

$$F(x_1, \ldots, x_n, u, p_1, \ldots, p_n) = 0 \tag{5.25}$$

is the most general first-order partial differential equation for u. Geometrically a solution

$$u = u(x_1, \ldots, x_n)$$

can be interpreted as an n-dimensional surface ("integral surface") in $(n+1)$-dimensional $x_1 \cdots x_n u$-space. The Cauchy problem for Eq. (5.25) consists in finding an integral surface passing through a given $(n-1)$-dimensional manifold S in $x_1 \cdots x_n u$-space. The most important fact that sets equations of the form (5.25) apart from higher-order equations and from systems of equations is that *the solution of the Cauchy problem for* (5.10) *is equivalent to the solution of a system of ordinary differential equations.*

We associate with (5.10) the system of ordinary differential equations ("characteristic equations")

$$\frac{dx_1}{F_{p_1}} = \cdots = \frac{dx_n}{F_{p_n}} = \frac{du}{x_1 F_{p_1} + \cdots x_n F_{p_n}} = \frac{dp_1}{-F_{x_1} - p_1 F_u} = \cdots = \frac{dp_n}{-F_{x_n} - p_n F_u} = dt \tag{5.26}$$

for $x_1, \ldots, x_n, u, p_1, \ldots, p_n$ as functions of a single independent variable t. The expression $F(x_1, \ldots, x_n, u, p_1, \ldots, p_n)$ is constant for a solution of (5.26). A solution $x_1(t), \ldots, x_n(t), u(t), p_1(t), \ldots, p_n(t)$ of the characteristic equations (5.26) for which $F = 0$ is called a *characteristic strip.* Geometrically each set of $2n+1$ variables $x_1, \ldots, x_n, u, p_1, \ldots, p_n$ can be interpreted as an *element* in $(n+1)$-dimensional space, consisting of a point $(x_1, \ldots, x_n, u)$ and an n-dimensional plane through that point with the equation

$$v - u = p_1(y_1 - x_1) + \cdots + p_n(y_n - x_n)$$

in running coordinates $y_1, \ldots, y_n, v$. A *strip* is a one-dimensional set of elements consisting of the points of a curve with a tangent plane selected in each point. A *characteristic element* is an element satisfying $F = 0$. Through each characteristic element $x_1^0, \ldots, x_n^0, u^0, p_1^0, \ldots, p_n^0$ there passes exactly one characteristic strip which can be found by solving the system of ordinary differential equations (5.26) with initial values $x_i = x_i^0$, $u = u^0$, $p_i = p_i^0$.

The integral surfaces $u = u(x_1, \ldots, x_n)$ of the partial differential equation (5.25) are made up of characteristic strips in the sense that they are covered by an $(n-1)$-parameter family of curves, each of which together with the tangent planes of the integral surface along the curve forms a characteristic strip. Integral surfaces through a given $(n-1)$-dimensional manifold S are found by selecting in each point of S a characteristic element tangent to S and passing a characteristic strip through each of those elements. Of special interest are the *singular integral surfaces* ("conoids") that are obtained when the initial manifold S degenerates into a point $(x_1^0, \ldots, x_n^0, u^0)$. They are formed by passing characteristic strips through all characteristic elements passing through that point, giving rise to an integral surface with a conical singularity.

Alternatively all these notions can be interpreted geometrically in n-dimensional $x_1 \cdots x_n$-space with u as the time. The level surfaces

$$u(x_1, \ldots, x_n) = \text{const}$$

of a solution u of (5.25) are then the different positions taken by a moving $(n-1)$-dimensional surface in x-space. An "element" now consists of a point $(x_1^0, \ldots, x_n^0)$ and a "hyperplane" (i.e., plane of dimension $n-1$) through that point with direction numbers $(p_1^0, \ldots, p_n^0)$ for its normal. For a geometrically given element and time u^0 there may be several ways of adjusting the arbitrary factor of proportionality in the p_j^0 in such a way that the condition $F(x_1^0, \ldots, x_n^0, u^0, p_1^0, \ldots, p_n^0) = 0$ is satisfied. Forming the solution of the characteristic differential equations (5.26) with initial conditions $x_i = x_i^0$,

$p_i = p_i^0$ for $u = u^0$, in which now u is considered as the independent variable, we obtain a definite way in which the original geometrically given element has moved from its position at the time u^0. The different possible factors of proportionality for the direction numbers of the normal give rise to different possible modes of propagation of the element along certain curves ("rays").

The Cauchy problem in this interpretation consists in finding for a solution $u(x_1, \ldots, x_n)$ of (5.25) the level surfaces S_λ given by $u = \text{const} = \lambda$ if S_0 is prescribed. [In the important case where the first-order partial differential equation (5.25) itself is just the characteristic equation for a higher-order system the S_λ represent the *wavefronts* in the propagation of discontinuities in the highest-order derivatives for the solutions of that system.] To construct S_λ we break up S_0 into elements consisting of the points of S_0 and the corresponding tangent planes. Each of these elements at the time $u = 0$ can be propagated by means of the characteristic equations along some ray and arrives at a certain position at the time $u = \lambda$. The elements at that time then form the points and corresponding tangent planes of the surface S_λ. An alternative solution of the Cauchy problem ("Huygens' principle") makes use of the conoid solutions, obtained by propagating all elements that pass through a given point P at the time $u = 0$. Forming the level surfaces at the time λ for these singular solutions "originating" from all the various points of S_0 and taking their envelope also yield the surface S_λ.

The simplest illustration with $n = 3$ is furnished by the differential equation

$$F = p_4^2 - c^2(p_1^2 + p_2^2 + p_3^2) = 0 \tag{5.27}$$

This is the equation for the characteristic surfaces $\phi = x_4 - u(x_1,x_2,x_3) = 0$ of the wave equation (5.5) and hence describes the propagation of discontinuities in the waves, as in geometrical optics of homogeneous isotropic media (see Part 6, Chap. 2, Sec. 3). Here the propagation of elements is determined by the characteristic equations

$$\frac{dx_i}{du} = c^2 p_i \qquad \frac{dp_i}{du} = 0 \qquad i = 1, 2, 3$$

[see Eq. (5.25)], which have to be combined with (5.27). Here the rays along which an element propagates with the constant speed c are straight lines perpendicular to the element. Given the initial surface S_0, we find S_λ by shifting each surface element of S_0 parallel to itself a distance $c\lambda$ along the normal. Alternatively we could also take the envelope of the spheres of radius $c\lambda$ about the points of S_0.

An example of an equation of the type (5.25) in dynamics is furnished by Hamilton's partial differential equation for a function $W(q_1, \ldots, q_n)$:

$$\frac{\partial W}{\partial t} + H\left(q_1, \ldots, q_n, \frac{\partial W}{\partial q_1}, \ldots, \frac{\partial W}{\partial q_n}, t\right) = 0$$

The corresponding ordinary characteristic differential equations are Hamilton's equations of motion [see Part 2, Chap. 2, Eqs. (2.24)].

3. Elliptic Equations

The Laplace Equations. The prime example of an elliptic equation is the "equation of Laplace" or "potential equation"

$$\Delta u = \sum_{i=1}^{n} u_{x_i x_i} = 0 \tag{5.28}$$

The solutions are called *harmonic* or *potential* functions.

The theory of the harmonic functions of two independent variables x, y is essentially equivalent to that of the *analytic* ("regular," "holomorphic") functions of one complex variable $z = x + iy$. Indeed, an analytic function $f(z) = f(x + iy)$ defined in a domain D of the complex-number plane can always be written in the form $f(x + iy) = u(x,y) + iv(x,y)$, where $u(x,y)$ and $v(x,y)$ are real-valued functions that satisfy the *Cauchy-Riemann equations*

$$u_x = v_y \qquad u_y = -v_x \tag{5.29}$$

Equations (5.29) imply that u and v are harmonic: $u_{xx} + u_{yy} = 0$, $v_{xx} + v_{yy} = 0$. Conversely we can find for any function $u(x,y)$ that is harmonic in a simply connected domain D a "conjugate" harmonic function $v(x,y)$ so that (5.29) holds and $u + iv$ is analytic in $x + iy$. A pair of conjugate harmonic functions, i.e., of solutions of the Cauchy-Riemann equations, defines a *conformal mapping* $(x,y) \to (u,v)$ (see Part 1, Chap. 3, Sec. 9). Of importance is the *invariance of the Laplace equation under conformal mappings*. Thus a function $w(x,y)$ for which $w_{xx} + w_{yy} = 0$ also satisfies the equation

$$w_{uu} + w_{vv} = 0$$

if we apply the substitution $u = u(x,y)$, $v = v(x,y)$ involving two harmonic conjugate functions u, v. Many two-dimensional problems in potential theory can be simplified by a suitable conformal mapping of the domain D. In particular, any simply connected domain with at least two boundary points can always be mapped conformally onto a half plane or onto a circular disk ("Riemann's mapping theorem").

If the dimension n of the space exceeds 2, the group of transformations leaving (5.28) invariant is relatively small. It consists of trivial linear similarity transformations combined with the *transformation by reciprocal radii* (also called *inversion*). The latter is given by the formulas

$$x'_i = x_i/r^2 \tag{5.30}$$

where

$$r^2 = x_1^2 + \cdots + x_n^2 \tag{5.31}$$

More precisely, if $u(x_1, \ldots, x_n)$ is harmonic, so is

$$r^{2-n}u(x_1/r^2, \ldots, x_n/r^2)$$

Of great importance for the study of the Laplace equation (as for that of linear partial differential equations in general) are the *singular solutions*. They can be described conveniently by using the generalized functions ("distributions") introduced by Laurent Schwartz. A *distribution* w in x-space does not necessarily have a value at each point $x = (x_1, \ldots, x_n)$, but the volume integral

$$\int wf\, dx_1 \cdots dx_n = \int wf\, dx$$

is defined for every *test function* $f(x)$, that is, every infinitely differentiable function $f(x)$ that has the value zero outside some finite set. Any first derivative u_{x_i} of a continuous function $u(x)$ is then defined by integration by parts as that distribution w for which for every test function $f(x)$

$$\int wf\, dx = -\int uf_{x_i}\, dx \tag{5.32}$$

More generally, any higher-order derivative $D^\alpha u = w$ of a continuous function $u(x)$ is defined as the distribution for which

$$\int wf\, dx = (-1)^{|\alpha|} \int uD^\alpha f\, dx \tag{5.33}$$

The most important distribution is the "Dirac function" $\delta(x) = \delta(x_1, \ldots, x_n)$ defined by the condition that

$$\int \delta(x) f(x)\, dx = f(0) \tag{5.34}$$

for every test function f.

A function $v(x)$ is called a *fundamental solution* of the linear equation

$$Lu = \sum_{|\alpha| \leqq m} A_\alpha(x) D^\alpha u = 0$$

with pole y, if it satisfies the symbolic equation

$$Lv = \delta(x - y) \tag{5.35}$$

i.e., if the identity for volume integrals

$$\int v(x) \sum_{|\alpha| \leqq m} (-1)^{|\alpha|} A_\alpha(x) D^\alpha f\, dx = \int v\bar{L}f\, dx = f(y)$$

holds for every test function f. For an elliptic operator L the function v is a regular solution of the equation $Lv = 0$ for $x \neq y$; near $x = y$ the mth-order derivatives of v become infinite like r^{-n}, where r is the distance between x and y. Special fundamental solutions with pole y for the Laplace equation in n dimensions are the expressions $v(x) = K(x,y)$ defined by

$$K(x,y) = \frac{r^{2-n}}{(2-n)\omega_n} \quad \text{for } n > 2 \tag{5.36a}$$

$$K(x,y) = \frac{\log r}{2\pi} \quad \text{for } n = 2 \tag{5.36b}$$

Here r is the distance of the points x and y, and ω_n denotes the surface area of the unit sphere in n-dimensional Euclidean space; thus $\omega_2 = 2\pi$, $\omega_3 = 4\pi$, $\omega_4 = 2\pi^2$, and generally

$$\omega_n = \frac{2\pi^{n/2}}{\Gamma(n/2)} \tag{5.36c}$$

(See Part 1, Chap. 3, Sec. 11, for the definition of the Γ function.)

The most general solution $v(x)$ of the equation

$$\Delta v = \delta(x - y) \tag{5.37}$$

has the form $v(x) = K(x,y) + w(x)$, where $w(x)$ is a regular harmonic function. Let R denote a region with boundary B in n-space. Under suitable regularity assumptions, it follows from Green's identity (5.22) and (5.37) that the value of an arbitrary function $u(x)$ at the point y can be represented in the form

$$u(y) = \int_B \left(u \frac{\partial v}{\partial n} - v \frac{\partial u}{\partial n}\right) dS + \int_R v\, \Delta u\, dx \tag{5.38}$$

where $\partial/\partial n$ indicates differentiation in the direction of the exterior normal of B. Here v is any fundamental solution with pole y. For $v = K(x,y)$ Eq. (5.38) is the representation of an arbitrary function u in R as the *Newtonian potential* of a mass distribution in R combined with the potentials of a *single* and of a *double layer* on B.

The *Green's function* $G(x,y)$ of the region R is that fundamental solution with pole y that vanishes on the boundary B of R. It has the form

$$G(x,y) = K(x,y) + w(x,y)$$

where for fixed y $w(x,y)$ is a harmonic function of x with boundary values $w = -K(x,y)$ on B. If $u(x)$ is harmonic in R and $u = f$ on B, the formula (5.38) with $v(x) = K(x,y)$ gives the representation

$$u(y) = \int_B f \frac{\partial G}{\partial n}\, dS \tag{5.39}$$

The *Dirichlet problem* ("first boundary-value problem") consists in finding a function $u(x)$ that is harmonic in R and takes prescribed values f on the boundary B of R. Formula (5.39) gives the solution of the Dirichlet problem if the Green's function for the region R is known.

Only in special cases can one obtain an explicit representation of Green's function. For $n = 2$ determination of $G(x,y) = G(x_1,x_2,y_1,y_2)$ for a simply connected domain R can be reduced to the problem of finding a complex function $f(z) = f(x_1 + ix_2)$ which maps R conformally onto the interior of the unit circle. For known f, then G is given by

$$G = \text{real part of } \frac{1}{2\pi} \log \frac{f(x_1 + ix_2) - f(y_1 + iy_2)}{1 - f(x_1 + ix_2)\overline{f(y_1 + iy_2)}} \tag{5.40}$$

For any dimension n a Green's function can be constructed explicitly for the case where R is the spherical region $x_1^2 + \cdots + x_n^2 < a^2$. The corresponding explicit solution of the Dirichlet problem for the sphere constitutes *Poisson's formula*

$$u(y) = \frac{1}{a\omega_n} \int_B u(x) \frac{a^2 - \rho^2}{r^n}\, dS \tag{5.41}$$

where r is again the distance of the point x on the boundary B from y and ρ is the distance of y from the origin. [The constant ω_n is defined by (5.36*c*).] Formula (5.41) for $\rho = 0$ yields *Gauss' law of the arithmetic mean:* A harmonic function u can assume its maximum only on the boundary of its domain of definition, unless u is a constant.

The solution of the Dirichlet problem also can be characterized by a minimum property (*Dirichlet's principle*): Among all functions u in R having the prescribed boundary values f on the boundary B

the one satisfying Laplace's equation yields the smallest value for the *Dirichlet integral*

$$D(u) = \int_R \sum_{i=1}^{n} \left(\frac{\partial u}{\partial x_i}\right)^2 dx \qquad (5.42)$$

This variational formulation lends itself to an approximate solution of the Dirichlet problem by the *Rayleigh-Ritz method:* We restrict u to the linear combinations $u = u_0 + a_1u_1 + \cdots + a_ku_k$ of a finite number of suitable selected functions u_0, $u_1, \ldots, u_k$ satisfying the boundary conditions $u_0 = f$, $u_1 = \cdots = u_k = 0$ on B. Minimizing $D(u)$ yields a simple finite system of algebraic linear equations for the constants $a_1, \ldots, a_k$ that can be solved explicitly.

The Dirichlet problem can also be reduced to an *integral equation.* For this purpose we assume that the unknown harmonic function in the domain R is given by the potential of a double layer

$$u(y) = \int_B w \frac{\partial v}{\partial n} dS \qquad (5.43)$$

where $v = K(x,y)$ is the fundamental solution of Eqs. (5.36*a*) and (5.36*b*) and the density w is still to be determined from the boundary condition $u = f$ on B. As the point y moves into the boundary, Eq. (5.43) must be replaced by

$$f(y) = \tfrac{1}{2}w(y) + \int_B w \frac{\partial v}{\partial n} dS \qquad (5.44)$$

which constitutes an integral equation (see Part 1, Chap. 6) of the Fredholm type for w. Once this equation has been solved for w we find u in the interior points from Eq. (5.43).

In the *Neumann problem* for the Laplace equation we ask for a solution u with prescribed normal derivative $\partial u/\partial n = g$ on the boundary B. Necessary for existence is the *compatibility* condition

$$\int_B g\, dS = 0 \qquad (5.45)$$

In analogy to the Green's function for the Dirichlet problem, the *Neumann function* $N = N(x,y)$ is introduced as a fundamental solution with pole y whose normal derivative $\partial N/\partial n$ is constant on B. The solution of the Neumann problem can then be written [see Eq. (5.38)] in the form

$$u(y) = -\int_B gN\, dS \qquad (5.46)$$

Other Elliptic Equations with Constant Coefficients. The inhomogeneous equation

$$\Delta u = f(x) \qquad \Delta = \sum_{h=1}^{n} \frac{\partial^2}{\partial x_k^2} \qquad (5.47)$$

is known as *Poisson's differential equation.* It is satisfied by the potentials of a mass distribution of density f

$$u(y) = \int f(x)K(x,y)\, dx \qquad (5.48)$$

with K defined by Eqs. (5.36*a*) and (5.36*b*), provided f vanishes outside a bounded set and is sufficiently regular.

The equation

$$\Delta u + \lambda u = 0 \qquad (5.49)$$

(*reduced wave equation*) arises naturally in connection with the *wave equation*

$$v_{tt} = c^2 \Delta v \qquad (5.50)$$

For a solution of the hyperbolic equation (5.50) of the special form

$$v(x,t) = u(x)e^{-i\omega t} \qquad (5.51)$$

the factor u has to satisfy the elliptic equation (5.49) with $\lambda = (\omega/c)^2$. A fundamental solution of Eq. (5.49) for $n = 2$ is given by

$$K(x,y) = \frac{1}{4i} H_0^{(1)}(\sqrt{\lambda}\, r) \qquad (5.52)$$

where r is the distance of the points x and y, and $H_0^{(1)}$ denotes the *Hankel function.* For large r this solution behaves like

$$\pi^{-\frac{1}{2}}\lambda^{-\frac{1}{4}}r^{-\frac{1}{2}} \exp(ir\lambda^{\frac{1}{2}} - 3\pi i/4) \qquad (5.53)$$

and satisfies the *radiation condition*

$$\lim_{r\to\infty} r^{\frac{1}{2}}(\partial/\partial r - i\lambda^{\frac{1}{2}})K = 0 \qquad (5.54)$$

The corresponding solution of Eq. (5.50) represents an *outgoing wave.* The analogous fundamental solution for $n = 3$ is given by

$$K(x,y) = -\frac{\exp(ir\lambda^{\frac{1}{2}})}{4\pi r} \qquad (5.55)$$

Many of the properties of the potential equation apply with suitable modifications to the solutions of (5.49). There are also, however, certain characteristic differences: for example, the point at infinity plays a different role from the other points. A solution of (5.49), which is defined in a neighborhood of infinity and tends to 0 faster than $r^{(1-n)/2}$ for $r \to \infty$, must vanish identically.

Another essential difference between (5.49) and the potential equation is the fact that for the solution of (5.49) the maximum principle does not apply. In fact, a solution can be positive in the interior of a region while vanishing identically on its boundary. A simple example for the case $n = 3$ is given by the function $(\sin \sqrt{\lambda}\, r)/r$, a regular solution of (5.49) which vanishes for $r = \pi/\sqrt{\lambda}$ and is positive in the interior of the sphere of radius $\pi/\sqrt{\lambda}$. In this case the Dirichlet problem for the sphere of radius $\pi/\sqrt{\lambda}$ can have several solutions. In general, there exists for a given region R and a given value of λ no solution of (5.49) that vanishes on the boundary of R except the trivial solution $u = 0$. If there exists a nontrivial solution u of (5.49) vanishing on the boundary B of R, the function u is called an *eigenfunction* [with respect to the Dirichlet problem for Eq. (5.49) in the region R] belonging to the *eigenvalue* λ. For a given region R there exist infinitely many eigenfunctions and corresponding eigenvalues. The eigenvalues can be shown to be all real and positive and to form a discrete set. They can be arranged into a sequence $\lambda_1, \lambda_2, \lambda_3, \ldots$, where $\lambda_n \leqq \lambda_{n+1}$ and $\lim_{n\to\infty} \lambda_n = \infty$. Here one and the same eigenvalue λ appears as often in the sequence as there are linearly independent eigenfunctions belonging to λ. The corresponding eigenfunctions form a sequence $u_1, u_2, \ldots$ with the *orthogonality* property:

$$\int_R u_n u_m \, dx = \begin{cases} 1 & \text{for } n = m \\ 0 & \text{for } n \neq m \end{cases} \tag{5.56}$$

if for each eigenvalue a suitable system of linearly independent eigenfunctions has been chosen. This system of functions is "complete." It can even be shown that every function vanishing on B and possessing a sufficient number of continuous derivatives in R can be represented by an absolutely and uniformly convergent series of the form

$$u = \sum_{n=1}^{\infty} a_n u_n \tag{5.57}$$

The coefficients in this expansion are given by the formula

$$a_n = \int_R u u_n \, dx \tag{5.58}$$

Expansions of the form (5.57) can be used to solve a great number of boundary-value problems. Thus a solution u of the Poisson equation (5.47) with vanishing boundary values can be found by expanding f into eigenfunctions

$$f = \sum c_n u_n \qquad c_n = \int_R f u_n \, dx \tag{5.59}$$

and taking the u given by (5.57) with $a_n = -c_n/\lambda_n$. Similarly a solution v of the wave equation (5.50) for $(x_1, \ldots, x_n)$ in R and $t > 0$ can be found from given initial data v and v_t for $t = 0$ and boundary data $v = 0$ on the boundary of R by taking for v a series of the form

$$v = \sum_n [a_n \cos (c\sqrt{\lambda_n}\, t) + b_n \sin (c\sqrt{\lambda_n}\, t)] u_n(x_1, \ldots, x_n) \tag{5.60}$$

Here the coefficients a_n, b_n have to be determined from the initial conditions.

The eigenvalues λ_n can also be characterized by *minimum properties*. Introducing the Dirichlet integral

$$D(u) = \int_R \left(\sum_i u_{x_i}{}^2\right) dx \tag{5.61}$$

and the expression

$$H(u) = \int_R u^2 \, dx \tag{5.62}$$

we find that, among all functions u that vanish on the boundary of R and for which $H(u) = 1$, the function $u = u_1$ gives the smallest value of $D(u)$ and that $D(u_1) = \lambda_1$. Similarly

$$\lambda_k = D(u_k) = \min D(u)$$

among all functions u satisfying the conditions

$$H(u) = 1, \qquad \int_R u u_m \, dx = 0 \qquad \text{for } m = 1, \ldots, k-1$$

and vanishing on the boundary of R. The value λ_k can also be characterized directly by a *maximum-minimum* principle. For any $k - 1$ functions $v_1, \ldots, v_{k-1}$ we form the greatest lower bound of the expression $D(u)$ among all functions u that vanish on the boundary of u and satisfy

$$H(u) = 1, \qquad \int_R u v_i \, dx = 0 \qquad \text{for } i = 1, \ldots, k-1$$

If $d(v_1, \ldots, v_{k-1})$ denotes the value of this greatest lower bound, then λ_k = maximum $d(v_1, \ldots, v_{k-1})$ among all systems of functions $(v_1, \ldots, v_{k-1})$. Here all functions u and v_i admitted are assumed to be continuous in the closure of R and to have continuous first derivatives in the interior of R.

Other eigenvalue problems are associated with different boundary conditions for the differential equation (5.49) and also with different self-adjoint elliptic differential equations. For many cases, results similar to the ones given in the preceding example have been established.

As an example of an elliptic equation of higher order we consider the *plate equation*

$$\frac{\partial^4 u}{\partial x^4} + 2\frac{\partial^4 u}{\partial x^2 \, \partial y^2} + \frac{\partial^4 u}{\partial y^4} = \Delta^2 u = 0 \tag{5.63}$$

A fundamental solution of this equation is given by

$$u = \frac{1}{8\pi} r^2 \log r \tag{5.64}$$

The *Dirichlet problem* for Eq. (5.63) would consist in determining a solution u in a region R if the values of u and of its normal derivative u_n are prescribed on the boundary B of R. The solution of this problem is unique. The solution u yields the smallest value for the integral

$$\int_R (u_{xx} + u_{yy})^2 \, dx \, dy$$

among all functions u with the same prescribed values of u and u_n on B.

Linear Elliptic Equations with Variable Coefficients. A linear equation of order m

$$Lu = \sum_{|\alpha| \leqq m} A_\alpha(x) D^\alpha u = f(x) \tag{5.65}$$

is *elliptic* if the characteristic form

$$Q = \sum_{|\alpha| = m} A_\alpha(x) \xi^\alpha \tag{5.66}$$

is a definite form of degree m in ξ for each point x in question. The solution of (5.65) can be expected to be determined by $m/2$ data on the boundary. Thus the *Dirichlet problem* would consist in finding a solution of $Lu = f$ if u and its first $(m - 2)/2$ normal derivatives are prescribed on the boundary B. The Cauchy data are not independent from each other on B and cannot be prescribed arbitrarily.

The *maximum principle* cannot be expected to hold for equations of order $m > 2$. It does hold for the solutions u of the equation $Lu = 0$ in the case $m = 2$, provided that form Q is positive definite and the coefficient of the undifferentiated term u in L is nonpositive.

In the case $n = 2$, if the independent variables are denoted by x, y, the general linear second-order equation has the form

$$Lu = au_{xx} + 2bu_{xy} + cu_{yy} + 2du_x + 2eu_y + fu = g \tag{5.67}$$

with coefficients $a, b, \ldots, g$ that are known functions of x, y. Equation (5.67) is elliptic if the condition

$$ac - b^2 > 0 \tag{5.68}$$

is satisfied for all x, y in question. By introducing suitable new independent variables, the principal part $au_{xx} + 2bu_{yy} + cu_{yy}$ can become the Laplace expression Δu. An equation of the form

$$\Delta u + 2du_x + 2eu_y = 0 \tag{5.69a}$$

is equivalent to the first-order system

$$\sigma u_x + \tau u_y = v_y \qquad \sigma u_y - \tau u_x = -v_x \tag{5.69b}$$

in analogy to the Cauchy-Riemann equations (5.29); we only have to determine the coefficients σ, τ from the relations

$$\sigma_x - \tau_y = 2d\sigma \qquad \sigma_y + \tau_x = 2e\sigma \tag{5.69c}$$

The solution functions u, v of (5.69*b*) can be combined into a *pseudoanalytic* function $u + iv$.

Nonlinear Elliptic Equations. Any system of nonlinear equations can be reduced by differentiation to a *quasi-linear* system of differential equations for the original unknown functions and a certain number of their derivatives. Most of the results that have been obtained concern rather special types of second-order equations in two independent variables. Such equations arise in differential geometry and in the study of subsonic flows.

The equations of the special form

$$A(u_x,u_y)u_{xx} + 2B(u_x,u_y)u_{xy} + C(u_x,u_y)u_{yy} = 0$$

can be reduced to the *linear* equations

$$A(\xi,\eta)w_{\eta\eta} - 2B(\xi,\eta)w_{\xi\eta} + C(\xi,\eta)w_{\xi\xi} = 0$$

in the independent variables ξ, η by means of the *Legendre transformation*

$$\xi = u_x \qquad \eta = u_y \qquad w = x\xi + y\eta - u$$

In many cases problems involving a quasi-linear equation can be reduced to integral equations or integrodifferential equations by considering the coefficients of the highest derivatives occurring as known and by solving formally the corresponding problem for the resulting linear equation. The integral equations obtained in this way are often solvable by *iteration* methods.

4. Parabolic Equations of Second Order

The Equation of Heat. In a suitable system of units conduction of heat in a homogeneous and isotropic medium is described by the differential equation

$$u_t = \Delta u = u_{xx} + u_{yy} + u_{zz} \tag{5.70}$$

In the situation ordinarily considered the point (x,y,z) is restricted to a region R, which is independent of the time t, and t to values > 0. The standard problem consists in finding a solution of (5.70) for which there are prescribed the initial values

$$u(x,y,z,0) = f(x,y,z) \qquad \text{for } (x,y,z) \text{ in } R \tag{5.71}$$

and in addition a boundary condition of the form

$$Lu = g \qquad \text{for } (x,y,z) \text{ on } B \text{ and } t > 0 \tag{5.72}$$

is imposed. Here B is the boundary of R, and L is a differential operator acting on u. In case R, the whole infinite space, condition (5.72) is replaced by certain boundedness assumptions on u and its derivatives.

The only *characteristic surfaces* of Eq. (5.70) are the manifolds t = const. Interpreting t as the time, this implies infinite speed of propagation of discontinuities. Connected with this feature is the fact that the solutions of (5.70) are analytic functions of (x,y,z) for fixed t. Another point in which the behavior of Eq. (5.70) resembles that of elliptic equations is the existence of a *maximum principle*. If R is closed and bounded and u is continuous in the four-dimensional region D in $xyzt$-space formed by the points with (x,y,z) in R and $0 \leqq t \leqq t_1$, and if u is a solution of (5.70) in the interior of R, then u assumes its maximum value for D either in a point for which (x,y,z) lies on the boundary B of R or in a point for which $t = 0$. If R is the whole infinite space and u,u_t,u_x,u_{xx} are continuous and bounded for $t \geqq 0$, then the values of u for $t > 0$ will not exceed the least upper bound of the values of u for $t = 0$.

The maximum principle implies the *uniqueness* of the boundary-value problem, Eqs. (5.70) to (5.72), for the case where the boundary operator Lu reduces to the function u itself. Other uniqueness proofs can be obtained from the *energy integral* relation

$$\frac{d}{dt}\iiint_R \tfrac{1}{2}(u_x^2 + u_y^2 + u_z^2)\,dx\,dy\,dz = -\iiint_R u_t^2\,dx\,dy\,dz + \iint_B u_t\frac{du}{dn}\,dS \tag{5.73}$$

which is valid for a $u(x,y,z,t)$ that is continuous together with its first derivatives for (x,y,z) in the closed and bounded region R for $t \geqq 0$, and satisfies (5.70) in the interior of R for $t > 0$. Relation (5.73) shows that u vanishes identically for $t > 0$, if u vanishes in R for $t = 0$, and if u or du/dn vanish on B for $t > 0$. This shows that the solution of Eqs. (5.70) to (5.72) is determined uniquely for $Lu = u$ or $Lu = du/dn$. More generally it will be unique for expressions L of the form

$$Lu = u + \lambda \frac{du}{dn} \tag{5.74}$$

where λ is non-negative and independent of t.

Formal solutions of boundary problems for (5.70) of the type mentioned can often be obtained by expansions in terms of eigenfunctions for the potential equation. For example, let Eq. (5.49) have nontrivial solutions $u_n(x,y,z)$ which vanish on the boundary B of R for $\lambda = \lambda_n$. Then

$$u(x,y,z,t) = \sum_n c_n \exp(-\lambda_n t) u_n(x,y,z) \tag{5.75}$$

represents a solution of (5.70) vanishing on B. The coefficients c_n can be determined by means of the initial condition (5.71). They are given by the formula

$$c_n = \iiint_R f u_n \, dx \, dy \, dz \tag{5.76}$$

according to (5.58) if the u_n are normalized so as to satisfy the orthogonality relations (5.56). In the particular case, where f is the Dirac function

$$f(x,y,z) = \delta(x - \xi, y - \eta, z - \zeta) \tag{5.77}$$

we obtain from (5.75) the formal expansion

$$u(x,y,z,t) = \sum_n \exp(-\lambda_n t) u_n(\xi,\eta,\zeta) u_n(x,y,z) \tag{5.78}$$

Formula (5.78) represents essentially the *Green's function* for Eq. (5.70), the region R and the boundary operator $Lu = u$. This function can be defined for any linear boundary operator Lu as the solution $u = G(x,y,z,t,\xi,\eta,\zeta)$ of Eqs. (5.70) to (5.72) with boundary function $g = 0$ and initial values f given by Eq. (5.77). The solution with more general initial values f is then given by

$$u(x,y,z,t) = \iiint_R G(x,y,z,t,\xi,\eta,\zeta) f(\xi,\eta,\zeta) \, d\xi \, d\eta \, d\zeta \tag{5.79}$$

Simple explicit expressions for G can be obtained for the limiting case, where R is the whole xyz-space, and the boundary conditions (5.72) become superfluous. A solution of (5.70) with initial conditions (5.77) in the whole xyz-space is easily obtained by Fourier transformation. One finds

$$G(x, y, z, t, \xi, \eta, \zeta) = K(x - \xi, y - \eta, z - \zeta, t) \tag{5.80}$$

where

$$K(x,y,z,t) = (2\pi)^{-3} \iiint_{-\infty}^{\infty} \exp[i(\alpha x + \beta y + \gamma z) - (\alpha^2 + \beta^2 + \gamma^2)t] \, d\alpha \, d\beta \, d\gamma$$
$$= (4\pi t)^{-3/2} \exp\left(-\frac{x^2 + y^2 + z^2}{4t}\right) \tag{5.81}$$

The corresponding formula in the one-dimensional case for a solution of

$$u_t - u_{xx} = 0 \tag{5.82}$$

with initial values

$$u(x,0) = \delta(x - \xi) \tag{5.83}$$

is given by

$$G(x,t,\xi) = (2\pi)^{-1} \int_{-\infty}^{+\infty} \exp[i\alpha(x - \xi) - \alpha^2 t] \, d\alpha = (2\pi t)^{-1/2} \exp\left[-\frac{(x - \xi)^2}{4t}\right]$$
$$= K(x - \xi, t) \tag{5.84}$$

The fundamental solution represented by (5.81) or (5.84) is continuous and has continuous derivatives of all orders with respect to its arguments for $t > 0$, except at the origin of space. In a neighborhood of the origin, K becomes unbounded for $t \to 0$. Considered as an analytic function, K has an essential singularity at all points of space for $t = 0$.

Formula (5.81) provides the solution of the initial value problem (5.70) and (5.71) for the case that R is the whole space, in the form

$$u(x, y, z, t) = (4\pi t)^{-3/2} \iiint_{-\infty}^{+\infty} \exp\left(-\frac{\xi^2 + \eta^2 + \zeta^2}{4t}\right) f(x + \xi, y + \eta, z + \zeta) \, d\xi \, d\eta \, d\zeta \tag{5.85}$$

provided u, u_x, u_{xx}, u_t are bounded at ∞.

Explicit expressions for solutions of *mixed* boundary-initial problems can be obtained by expansions in terms of eigenfunctions of the potential equation, in case the boundary conditions are homogeneous or at least time-independent. Thus a solution of the *one-dimensional* problem

$$u_t = u_{xx} \quad \text{for } t > 0,\ 0 < x < L \tag{5.86}$$
$$u(x,0) = f(x) \quad \text{for } 0 < x < L \tag{5.87}$$
$$u(0,t) = u(L,t) = 0 \quad \text{for } t > 0$$

can be obtained on the basis of (5.75) and (5.76) in the form

$$u(x,t) = \sum_n c_n \exp(-\lambda_n t) u_n(x) t \tag{5.88}$$

with

$$c_n = \int_0^L f(\xi) u_n(\xi) \, d\xi \tag{5.89}$$

Here the λ_n and u_n are given by

$$\lambda_n = \frac{n^2\pi^2}{L^2} \qquad u_n(x) = \frac{1}{\sqrt{L/2}} \sin \frac{n\pi x}{L} \qquad n = 1,2, \ldots \tag{5.90}$$

For the special initial function $f(x) = \delta(x - \xi)$, we obtain for u the *Green's function for a finite interval:*

$$G(x,t,\xi) = \sum_{n=1}^{\infty} \frac{2}{L} \sin \frac{n\pi\xi}{L} \sin \frac{n\pi x}{L} \exp\left(-\frac{n^2\pi^2 t}{L^2}\right) \tag{5.91}$$

Results of this type are also obtainable by application of the *Laplace transformation.*

The General Second-order Linear Parabolic Equation. The notions developed in connection with the equation of heat can be extended to more general equations of the form

$$u_t - \Lambda u = u_t - \sum_{i,k} a_{ik} u_{x_i x_k} - 2 \sum_i b_i u_{x_i} - cu = w \tag{5.92}$$

where Λ is an elliptic differential operator in the independent variables $x_1, \ldots, x_n$. The coefficients a_{ik}, b_i, c, w are given functions of $x_1, \ldots, x_n, t$. The ellipticity of Λ is equivalent to the requirement that the quadratic form $\Sigma_{i,k} a_{ik} \xi_i \xi_k$ is positive definite. Equation (5.92) is assumed to hold for $t > 0$ and $(x_1, \ldots, x_n)$ confined to a region R with boundary B. Usually there is imposed an initial condition

$$u(x_1, \ldots, x_n, 0) = f(x_1, \ldots, x_n) \text{ in } R \tag{5.93}$$

and a boundary condition of the form

$$Lu = h \quad \text{for } (x_1, \ldots, x_n) \text{ on } B \text{ and } t > 0 \tag{5.94}$$

where L is a differential operator containing derivatives with respect to the x_i.

Linear equations of the type (5.92) often can be simplified by transformations. As an example we can consider an equation of the form

$$u_t = a(x,t) u_{xx} + 2b(x,t) u_x + c(x,t) u \qquad a > 0 \tag{5.95}$$

Introducing instead of t and x the new independent variables t and

$$\int^x \frac{dx}{\sqrt{a(x,t)}}$$

we obtain a differential equation of the same form (5.95) in which, however, the coefficient a is replaced by the constant 1. Introducing, moreover, a new dependent variable v by

$$v = u \exp \left(\int^x b \, dx \right)$$

we obtain for v an equation of the type

$$v_t = v_{xx} + \gamma(x,t) v \tag{5.96}$$

The initial value problem for equations of the form (5.96) and even for the more general quasi-linear equations of the form

$$v_t = v_{xx} + \gamma(x,t,v) \tag{5.97}$$

can be reduced conveniently to integral equations. Making use of the fundamental solution (5.84) of the one-dimensional heat equation, we find that the solution v of (5.97) with initial values $v(x,0) = f(x)$ for $-\infty < x < +\infty$ satisfies

$$v(x,t) = \int_{-\infty}^{+\infty} \frac{\exp[-(x-\xi)^2/4t]}{\sqrt{4\pi t}} f(\xi) \, d\xi + \int_0^t d\tau \int_{-\infty}^{+\infty} \frac{\exp[-(x-\xi)^2/4(t-\tau)]}{[4\pi(t-\tau)]^{1/2}} \gamma[\xi,\tau,v(\xi,\tau)] \, d\xi$$

This integral equation, which is essentially an equation of the type of *Volterra* (see Part 1, Chap. 6, Sec. 3), can be solved by iteration.

5. Hyperbolic Equations in Two Independent Variables

Linear Equation of Second Order. These are equations of the form

$$Lu = au_{xx} + 2bu_{xy} + cu_{yy} + 2du_x + 2eu_y + fu = g \tag{5.98}$$

where the coefficients a,b,c,d,e,f,g are known functions of x and y satisfying the *hyperbolicity condition*

$$ac - b^2 < 0 \tag{5.99}$$

The characteristics of (5.98) are the curves satisfying the ordinary first-order differential equation

$$a\,dy^2 - 2b\,dx\,dy + c\,dx^2 = 0 \tag{5.100}$$

obtained from (5.3) by taking ϕ in the form

$$\phi = y - y(x)$$

On the basis of (5.99) this equation can be split into the two real first-order equations

$$a\,dy - (b + \sqrt{b^2 - ac})\,dx = 0 \quad \text{and} \quad a\,dy - (b - \sqrt{b^2 - ac})\,dx = 0 \tag{5.101}$$

Thus there exist two one-parameter families of characteristic curves. If these curves are used as coordinate curves α = const and β = const in a curvilinear $\alpha\beta$ coordinate system, Eq. (5.98) will take the form

$$u_{\alpha\beta} + 2\bar{d}u_\alpha + 2\bar{e}u_\beta + \bar{f}u = \bar{g} \tag{5.102}$$

If the coefficients of L are constants, the transformation from (5.98) into the form (5.102) can be accomplished by a linear substitution with constant coefficients, and the new coefficients $\bar{d},\bar{e},\bar{f}$ will again be constants. Introducing a new dependent variable v by the substitution

$$v = \exp[2(\bar{d}\beta + \bar{e}\alpha)]u$$

the equation with constant coefficients can be reduced to the even simpler form

$$v_{\alpha\beta} + kv = h \tag{5.103}$$

with a suitable constant k. Putting $\alpha + \beta = \lambda$, $\alpha - \beta = \mu$, we obtain the alternative canonical form

$$v_{\lambda\lambda} - v_{\mu\mu} + kv = h \tag{5.104}$$

Green's identity (5.20) for an operator L of the form (5.98) can be given the form

$$\iint_R (vLu - u\bar{L}v)\,dx\,dy = \int_B \left[\sigma \left(v \frac{\partial u}{\partial \nu} - u \frac{\partial v}{\partial \nu} \right) + Auv \right] ds \tag{5.105}$$

Here R is a region with boundary curve B, for which the exterior normal has direction cosines ξ, η. The adjoint operator $\bar{L}$ is determined by

$$\bar{L}v = (av)_{xx} + (2bv)_{xy} + (cv)_{yy} - (2dv)_x - (2ev)_y + fv \tag{5.106}$$

The functions σ and A are given by

$$\sigma = \sqrt{(a\xi + b\eta)^2 + (b\xi + c\eta)^2} \qquad (5.107)$$
$$A = (2d - a_x - b_y)\xi + (2e - b_x - c_y)\eta \qquad (5.108)$$

Finally, $\partial/\partial\nu$ is defined as the directional derivative in the *binormal* direction which shall be the direction with direction cosines

$$\xi' = \frac{a\xi + b\eta}{\sigma} \qquad \eta' = \frac{b\xi + c\eta}{\sigma} \qquad (5.109)$$

Thus $\partial/\partial\nu$ denotes the operator defined by

$$\frac{\partial}{\partial\nu} = \xi'\frac{\partial}{\partial x} + \eta'\frac{\partial}{\partial y} \qquad (5.110)$$

The characteristic curves have normals, whose direction cosines satisfy

$$0 = a\xi^2 + 2b\xi\eta + c\eta^2 = \sigma(\xi\xi' + \eta\eta') = Q(\xi,\eta)$$

Hence a characteristic curve can be defined as one for which the binormal direction becomes tangential. Along such a curve

$$\frac{\partial}{\partial\nu} = \pm\frac{d}{ds} \qquad \text{and} \qquad \sigma = \sqrt{b^2 - ac} \qquad (5.111)$$

The binormal direction is essentially the normal of B taken in the Riemann metric associated with the operator L.

Riemann's method of solving the *Cauchy problem* for Eq. (5.98) is based on an application of (5.105) for a suitable region R and auxiliary function v. Let Cauchy data for a solution u of (5.98) be prescribed on a curve C. Let $P = (X,Y)$ be a point of the xy plane, such that the characteristics from P together with the curve C bound a curvilinear triangle $PST = R$, as shown in Fig. 5.1.

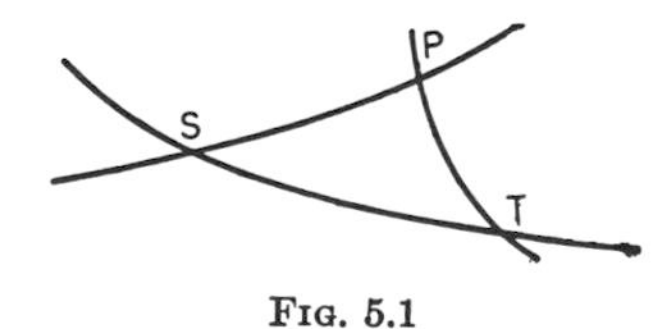

FIG. 5.1

Without restriction of generality one can assume that on TP we have

$$\frac{\partial}{\partial\nu} = +\frac{d}{ds}$$

and on PS we have

$$\frac{\partial}{\partial\nu} = -\frac{d}{ds}$$

where s is measured counterclockwise. The Riemann function $v = v(x,y;X,Y)$ is then defined as the solution of the equation $\bar{L}v = 0$, satisfying the boundary conditions

$$\sigma\frac{dv}{d\gamma} + \frac{d\sigma v}{d\gamma} = Av \qquad \text{on } PT \text{ and on } PS \qquad (5.112)$$
$$\sigma v = 1 \text{ at } P$$

Then

$$2u(X,Y) = (v\sigma u)_T + (v\sigma u)_S - \int_S^T \left[\sigma\left(v\frac{\partial u}{\partial\nu} - \frac{\partial v}{\partial\nu}\right) + Auv\right] ds + \iint_R vg\,dx\,dy \qquad (5.113)$$

The solution of the Cauchy problem is necessarily given by this formula. However, there may not exist such a solution if the curve C becomes characteristic in some of its points.

The preceding relations can be illustrated by the example of the equation

$$u_{yy} - c^2u_{xx} + ku = g \qquad (5.114)$$

to which every hyperbolic equation (5.98) with constant coefficients can be reduced. The characteristics are here the lines $x \pm cy = \text{const.}$ The Riemann function becomes

$$v(x,y;X,Y) = \frac{1}{c}J_0\left\{\left[k(y - Y)^2 - \frac{k}{c^2}(x - X)^2\right]\right\}^{1/2} \qquad (5.115)$$

(J_0 is the Bessel function defined in Part 1, Chap. 3, Sec. 11.) The solution of the Cauchy problem for (5.114) with data prescribed on the x axis becomes in a point (X,Y) with $Y > 0$

$$2u(X, Y) = u(X + cY, 0) + u(X - cY, 0) + \int_{X-cY}^{X+cY} (vu_y - uv_y)_{y=0}\,dx + \iint_R vg\,dx\,dy \qquad (5.116)$$

where the triangle R has vertices (X, Y), $(X - cY, 0)$, and $(X + cY, 0)$. In the special case where the constant k in Eq. (5.114) reduces to 0, the Riemann function v has the constant value $1/c$. Formula (5.116) goes over into the formula

$$u(X, Y) = \frac{1}{2}[u(X + cY, 0) + u(X - cY, 0)] + \frac{1}{2c}\int_{X-cY}^{X+cY} u_y(\xi, 0)\,d\xi + \frac{1}{2c}\int_0^Y \int_{X-c(Y-y)}^{X+c(Y-y)} g(x, y)\,dx\,dy \qquad (5.117)$$

giving the solution of the initial value problem for the *one-dimensional wave equation*. As indicated by (5.117) the general solution $u(x,y)$ of the homogeneous wave equation

$$u_{yy} - c^2u_{xx} = 0 \qquad (5.118)$$

can always be written in the form

$$u(x,y) = f(x + cy) + g(x - cy) \qquad (5.119)$$

with suitable functions f and g of one argument.

Formula (5.113) shows that the solution of the Cauchy problem for the linear hyperbolic equation does not depend on the totality of all "data" (meaning the function g and the values of u and its first derivatives on the initial curve C), but only on the values of these data in a certain region, the *domain of dependence* of the point (X,Y). This domain of dependence of (X,Y) is bounded by the initial curve C and by the characteristics issuing from the point $P = (X,Y)$. The values of the data in other points have no influence on the value of the solution at P. Conversely the *domain of influence* of data at a point is an angular region bounded by characteristic rays with the opposite sense issuing from that point. In the case of the equation with constant coefficients (5.114) the domain of dependence of a point (X,Y) consists of the points (x,y) with

$$c^2(Y - y)^2 - (x - X)^2 > 0 \qquad y < Y \qquad (5.120)$$

The same relation describes the points (X,Y) belonging to the domain of influence of the point (x,y) (Fig. 5.2).

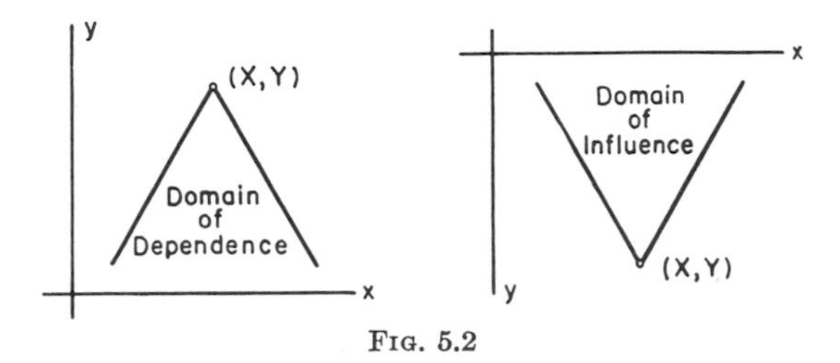

FIG. 5.2

Another class of equations of the type (5.98) frequently encountered in applications are the equations of Euler and Poisson which have the form

$$u_{xy} - \frac{\beta'}{x - y} u_x - \frac{\beta}{y - x} u_y = 0$$

with constants β, β'. Here the Riemann function has the explicit expression

$$v(x,y;X,Y) = \left(\frac{y - x}{Y - x}\right)^{\beta} \left(\frac{y - x}{y - X}\right)^{\beta'} F(\beta,\beta';1;z)$$

where

$$z = \frac{(x - X)(y - Y)}{(x - Y)(y - X)}$$

and F denotes the *hypergeometric* function (Part 1, Chaps. 3 and 12).

Hyperbolic Systems of First-order Equations. The general system of n linear first-order equations for n functions $u_1, \ldots, u_n$ of two independent variables x,y can be written in the form

$$\sum_{k=1}^{n} a_{ik} \frac{\partial u_k}{\partial y} + \sum_{k=1}^{n} b_{ik} \frac{\partial u_k}{\partial x} = \sum_{k=1}^{n} c_{ik} u_k + g_i \qquad i = 1, \ldots, n \qquad (5.121)$$

where the $a_{ik}, b_{ik}, c_{ik}, g_i$ are given functions of x,y. Assuming that the determinant of the a_{ik} does not vanish in the region considered, we can solve for the y derivatives and put the system into the *normal form*

$$\frac{\partial u_i}{\partial y} = \sum_{k=1}^{n} \alpha_{ik} \frac{\partial u_k}{\partial x} + \sum_{k=1}^{n} \beta_{ik} u_k + \gamma_i \qquad i = 1, \ldots, n \qquad (5.122)$$

The *initial value problem* for (5.122) or Cauchy problem with the x axis as initial line consists then in finding a solution such that

$$u_i(x,0) = f_i(x) \qquad i = 1, \ldots, n \qquad (5.123)$$

Cauchy problems for linear equations of higher order can be reduced in general to Cauchy problems for first-order systems by introducing suitable derivatives of the unknown functions as additional dependent variables. For example, the Cauchy problem for the second-order equation (5.98) consists in finding a solution u such that

$$u = f(x) \qquad u_y = \phi(x) \qquad \text{for } y = 0 \qquad (5.124)$$

We can convert this problem into a similar one involving the three unknown functions u_1,u_2,u_3 defined by

$$u_1 = u \qquad u_2 = \frac{\partial u}{\partial x} \qquad u_3 = \frac{\partial u}{\partial y}$$

for which we have the three first-order differential equations

$$\frac{\partial u_1}{\partial y} = u_3, \qquad \frac{\partial u_2}{\partial y} = \frac{\partial u_3}{\partial x},$$

$$a\frac{\partial u_2}{\partial x} + 2b\frac{\partial u_3}{\partial x} + c\frac{\partial u_3}{\partial y} + 2du_2 + 2eu_3 + fu_1 = g \qquad (5.125)$$

If here $c \neq 0$ (that is, if the x axis is not a characteristic), we can solve the third equation for $\partial u_3/\partial y$ and obtain a system in the normal form (5.122). The initial values of the u_i are given by

$$u_1(x,0) = f(x) \qquad u_2(x,0) = f'(x) \qquad u_3(x,0) = \phi(x) \qquad (5.126)$$

The differential equation (5.98) is not *equivalent* to the system (5.125), for not *every* solution of (5.125) leads to one of (5.98). However, (5.98) plus initial conditions (5.124) is *equivalent* to the system (5.125) plus initial conditions (5.126).

By (5.4) the equation

$$\det(\eta\delta_{ik} - \alpha_{ik}\xi) = 0 \qquad (5.127)$$

represents the condition that a curve with direction cosines ξ, η for its normal is a characteristic curve. If $\lambda_1, \ldots, \lambda_n$ are the characteristic roots of the matrix α_{ik} (see Part 1, Chap. 2, Sec. 11), Eq. (5.127) can be factored into the form

$$(\eta - \lambda_1\xi)(\eta - \lambda_2\xi) \cdots (\eta - \lambda_n\xi) = 0 \qquad (5.128)$$

Here the λ_i are in general functions of x,y, which can be found from the given functions α_i by solving an nth-order algebraic equation. If we disregard the degenerate case of multiple eigenvalues, system (5.122) is *hyperbolic* if all the λ_i are real and distinct from each other.

Equation (5.128) shows that there exist in the hyperbolic case n distinct real families of characteristic curves; each family is given by a first-order ordinary differential equation

$$\frac{dx}{dy} + \lambda_i(x,y) = 0 \qquad (5.129)$$

In the hyperbolic case the matrix of α_{ik} is *similar* to a diagonal matrix with the λ_i as diagonal elements. There exists a real nonsingular matrix e_{ik} such that

$$(e_{ik})^{-1}(\alpha_{ik})(e_{ik}) = (\lambda_i\delta_{ik})$$

The e_{ik} are functions of x,y which can be determined by solving algebraic equations. If we introduce a new set of dependent variables U_i in place of the u_i, defined by the linear equations

$$u_i = \sum_{k=1}^{n} e_{ik}(x,y)U_k \tag{5.130}$$

the U_i will satisfy a system of differential equations of the canonical form

$$\frac{\partial U_i}{\partial y} = \lambda_i(x,y)\frac{\partial U_i}{\partial x} + \sum_{k=1}^{n} h_{ik}(x,y)U_k + l_i(x,y) \qquad i = 1, \ldots, n \tag{5.131}$$

If C_i is a characteristic curve belonging to the ith family given by (5.129), (5.131) gives rise to the relation

$$dU_i = \left(\sum_{k=1}^{n} h_{ik}U_k + l_i\right) dy \tag{5.132}$$

along C_i. This observation permits the conversion of equations in the canonical form into integral equations. Let $P = (X,Y)$ be a point such that there exist n characteristic arcs $PQ_1, PQ_2, \ldots, PQ_n$ connecting P to points of the x axis. This situation will arise if the x axis is noncharacteristic, the equations hyperbolic, and P close enough to the x axis. Integrating (5.132) along the arc $C_i = PQ_i$, we obtain integral relations of the form

$$U_i(P) = U_i(Q_i) + \int_{Q_i}^{P} \left(\sum_k h_{ik}U_k + l_i\right) dy \tag{5.133}$$

As the $U_i(Q_i)$ are known through the given initial data, we have here again a system of integral equations similar to Volterra equations (see Part 1, Chap. 6, Sec. 3). They are of the general form

$$U = TU + V \tag{5.134}$$

where U stands for the vector $(U_1, \ldots, U_n)$. T is a linear integral operator, and V is a given vector field. Because the domain of integration shrinks into a point as P approaches the x axis (and hence in abstract terminology: the *norm* of the operator T is arbitrarily small for P close to the x axis), Eq. (5.134) can be solved by iteration:

$$U = \lim_{n\to\infty} U^{(n)} \qquad U^{(n+1)} = TU^{(n)} + V \qquad U^{(1)} = 0$$

In the more general case of a quasi-linear system (5.121) all the coefficients depend on the u_i in addition to the independent variables x,y. In this case the solution of the Cauchy problem is more difficult, as the characteristic curves cannot be determined a priori by solving ordinary differential equations, but depend on the solution u_i one is attempting to find. It is, however, possible to bring the system into a canonical form analogous to (5.131) with the λ_i and l_i still depending on the U_k and to devise an iteration scheme solving this system. Here the iteration is complicated by the fact that integrations are performed over curves C_i that depend on the approximation U^n to U already obtained.

The importance of quasi-linear equations lies in the fact that general nonlinear systems of equations can be reduced to quasi-linear ones by the processes of differentiating the original equations and of adding derivatives of the original unknown functions as further unknown functions.

6. Hyperbolic Equations with More than Two Independent Variables

The Linear Equation with Constant Coefficients. We shall restrict ourselves to the case of a single equation for a single scalar unknown function. (Most of the results carry over with proper modifications to the case of linear systems with constant coefficients.) The spacelike independent variables $x_1, \ldots, x_n$ will be combined into a vector x; the independent variable playing the role of time is t. The corresponding differentiation symbols are $D = (\partial/\partial x_1, \ldots, \partial/\partial x_n)$ and $\tau = \partial/\partial t$. The differential equations considered have the form

$$P(D,\tau)u = 0 \tag{5.135}$$

where $u = u(x,t)$ and where P is a polynomial of degree m with constant coefficients. The *characteristic form* of the equation is

$$Q(\xi,\lambda) = Q(\xi_1, \ldots, \xi_n,\lambda) = P_m(\xi,\lambda)$$

where P_m, the *principal part* of P, consists of the mth-order terms in P. The *initial value problem* or Cauchy problem for the differential equation (5.135) consists in finding a solution with prescribed initial data

$$\tau^k u(x,t) = f_k(x) \qquad \text{for } t = 0 \text{ and } k = 0, \ldots, m-1 \tag{5.136}$$

The existence of a solution for sufficiently regular data f_k is assured when the initial surface $t = 0$ is *spacelike*, i.e., when the roots $\lambda = \lambda(\xi)$ of the equation

$$P(i\xi,i\lambda) = 0 \tag{5.137}$$

have their imaginary parts bounded uniformly in ξ for real ξ. This condition is certainly satisfied when Eq. (5.135) is *strictly hyperbolic* in the sense that the roots λ of the characteristic equation

$$P_m(\xi,\lambda) = 0 \tag{5.138}$$

are real and distinct for every real $\xi \neq 0$.

It is sufficient to be able to solve the Cauchy problem for initial data of the *standard form*

$$\tau^k u = 0 \qquad \text{for } k = 0, \ldots, m-2 \qquad \tau^{m-1}u = F(x) \qquad \text{for } t = 0 \tag{5.139}$$

Indeed every solution u of the Cauchy problem with the more general initial data (5.136) can be written in the form

$$u = \sum_{j=0}^{m-1} \tau^j u_j$$

where u_j is a solution of a Cauchy problem with initial data of the standard form (5.139) with a suitable $F = F_j(x)$.

A formal solution of Eq. (5.135) for the standard initial values (5.139) is easily obtained by Fourier transformation. Let

$$F(x) = \int \exp (ix \cdot \xi) g(\xi)\, d\xi \qquad (5.140)$$

where $x \cdot \xi = x_1\xi_1 + \cdots + x_n\xi_n$, $d\xi = d\xi_1 \cdots d\xi_n$, and the integration is extended over the whole real ξ-space. Then

$$u(x,t) = \int K(\xi,t) \exp (ix \cdot \xi) g(\xi)\, d\xi \qquad (5.141)$$

Here the kernel $K(\xi,t)$ is given by the contour integral

$$K(\xi,t) = \frac{1}{2\pi} P_m(0,1) \oint \frac{\exp i\lambda t}{P(i\xi,i\lambda)}\, d\lambda$$

extended over a path in the λ plane that includes all roots λ of Eq. (5.137). In the strictly hyperbolic case the roots λ_k of Eq. (5.137) are all distinct, and for $\xi \neq 0$ there is the alternative expression

$$K(\xi,t) = P_m(0,1) \sum_{k=1}^{m} \frac{\exp i\lambda_k t}{P'(i\xi,i\lambda_k)}$$

where $P'(i\xi,i\lambda) = \partial P(i\xi,i\lambda)/\partial(i\lambda)$.

Since the Fourier transform $f(\xi)$ of F can be obtained from the well-known formula

$$g(\xi) = (2\pi)^{-n} \int \exp (-iy \cdot \xi) F(y)\, dy \qquad (5.142)$$

(5.141) is an explicit formal solution of the initial value problem, which for F sufficiently regular and vanishing of sufficiently high order at infinity is an actual solution. In some cases simpler expressions for the solution can be obtained by substituting for g in (5.141) its expression (5.142) and interchanging the integrations with respect to ξ and y. The convergence difficulties encountered are overcome best by using the symbolism of the theory of distributions.

To solve the initial value problem for the *inhomogeneous* equation

$$P(D,\tau)u = v(x,t) \qquad (5.143)$$

it is sufficient to be able to solve the corresponding problem for the homogeneous equation and, in addition, to find a special solution u of (5.143) with initial values of 0. Such a special solution is obtained by *Duhamel's principle* in the form

$$u(x,t) = \int_0^t U(x, t - s, s)\, ds$$

Here $U(x,t,s)$ denotes for every value of the parameter s the solution of the homogeneous equation $P(D,\tau)U = 0$ with initial values of the standard form

$$\tau^k U(x,t,s) = 0 \quad \text{for } k = 0, \ldots, m-2 \text{ and } t = s$$

$$\tau^{m-1} U(x,t,s) = \frac{v(x,s)}{P_m(0,1)} \quad \text{for } t = s$$

Hence the solution of the general initial value problem for Eq. (5.143) is reduced to that for (5.135) with standard initial values.

The Equation of Waves and Related Equations. The most important special case is the solution of the equation

$$u_{tt} = c^2(u_{xx} + u_{yy} + u_{zz}) + v(x,y,z,t) \qquad (5.144)$$

with initial values

$$u(x,y,z,0) = f(x,y,z) \qquad u_t(x,y,z,0) = g(x,y,z) \qquad (5.145)$$

The solution of this problem, which can be obtained by Fourier analysis, is given by the formula

$$\begin{aligned} 4\pi c^2 u(x, y, z, t) &= \frac{\partial}{\partial t}\left[\frac{1}{t} \iint_{r=ct} f(x + \xi, y + \eta, z + \zeta)\, dS\right] \\ &+ \frac{1}{t} \iint_{r=ct} g(x + \xi, y + \eta, z + \zeta)\, dS \\ &+ \iiint_{r<ct} \frac{1}{r} v\left(x + \xi, y + \eta, z + \zeta, t - \frac{r}{c}\right) d\xi\, d\eta\, d\zeta \end{aligned} \qquad (5.146)$$

where ξ, η, ζ denote the variables of integration and $r^2 = \xi^2 + \eta^2 + \zeta^2$. Equation (5.146) shows that the value of the solution at a point (x,y,z,t) depends only on the values of the initial data and of their derivatives on the sphere of radius ct about the point (x,y,z). The wave equation (5.144) has this unusual feature, sometimes called the *strong form of Huygen's principle,* that the domain of dependence is a *surface* in the initial space. In general, the values of the solution of a hyperbolic equation depend on the data in a full domain of the initial space. The exceptional behavior is encountered for the wave equation if the number of space dimensions has one of the values 3,5,7, ...

The corresponding formula for the equation

$$u_{tt} = c^2(u_{xx} + u_{yy}) + v(x,y,t) \qquad (5.147)$$

in two space dimensions, can be obtained from (5.146) by assuming all quantities to be independent of z:

$$\begin{aligned} 2\pi c^2 u(x, y, t) &= c \frac{\partial}{\partial t} \iint_{r<ct} \frac{f(x + \xi, y + \eta)}{\sqrt{c^2t^2 - r^2}}\, d\xi\, d\eta \\ &+ c \iint_{tr<c} \frac{g(x + \xi, y + \eta)}{\sqrt{c^2t^2 - r^2}}\, d\xi\, d\eta \\ &+ \int_0^{ct} d\zeta \iint_{r<\zeta} \frac{v\left(x + \xi, y + \eta, t - \frac{\zeta}{c}\right)}{\sqrt{\zeta^2 - r^2}}\, d\xi\, d\eta \end{aligned}$$

where $r^2 = \xi^2 + \eta^2$.

The general equation

$$u_{tt} = c^2(u_{xx} + u_{yy} + u_{zz} + au_x + bu_y + cu_z + du)$$

can be reduced to the form

$$u_{tt} = c^2(u_{xx} + u_{yy} + u_{zz} + \lambda^2 u) \qquad (5.148)$$

by a substitution of the form

$$u = \bar{u} \exp(\alpha x + \beta y + \gamma z + \delta t)$$

with suitable constants α, β, γ, δ. The solution of (5.148) with initial values (5.145) is given by

$$4\pi c^2 u(x, y, z, t) = \frac{1}{c}\frac{\partial}{\partial t}\frac{1}{t}\frac{\partial}{\partial t}\iiint\limits_{r<ct} f(x+\xi, y+\eta, z+\zeta)\cdot J_0(i\lambda\sqrt{c^2t^2 - r^2})\,d\xi\,d\eta\,d\zeta + \frac{1}{c}\frac{1}{t}\frac{\partial}{\partial t}\iiint\limits_{r<ct} g(x+\xi, y+\eta, z+\zeta)\cdot J_0(i\lambda\sqrt{c^2t^2 - r^2})\,d\xi\,d\eta\,d\zeta$$

The corresponding formula for two space dimensions, which can be obtained from (5.146) by the substitution $u(x,y,z) = e^{\lambda z}\bar{u}(x,y)$, is given by

$$2\pi c u(x, y, t) = \frac{\partial}{\partial t}\iint\limits_{r<ct}\frac{f(x+\xi, y+\eta)\cosh(\lambda\sqrt{c^2t^2 - r^2})}{\sqrt{c^2t^2 - r^2}}\,d\xi\,d\eta + \iint\limits_{r<ct}\frac{g(x+\xi, y+\eta)\cosh(\lambda\sqrt{c^2t^2 - r^2})}{\sqrt{c^2t^2 - r^2}}\,d\xi\,d\eta$$

References

General:

1. Courant, R., and D. Hilbert: "Methods of Mathematical Physics," vols. 1 and 2, Interscience, New York, 1953, 1962.
2. Garabedian, Paul R.: "Partial Differential Equations," Wiley, New York, 1964.
3. Hellwig, G.: "Partial Differential Equations," Blaisdell Pub. Co., New York, 1964.
4. Bers, L., F. John, and M. Schechter: "Partial Differential Equations," Interscience, New York, 1964.
5. John, F.: "Plane Waves and Spherical Means Applied to Partial Differential Equations," Interscience, New York, 1955.
6. Hörmander, Lars: "Linear Partial Differential Operators," Springer, Berlin, 1963.
7. Friedman, Avner: "Generalized Functions and Partial Differential Equations," Prentice-Hall, Englewood Cliffs, N.J., 1963.
8. Schwartz, L.: "Theorie des distributions," 2 vols., Paris, 1950–1951.
9. Sommerfeld, A.: "Partial Differential Equations in Physics," Academic, New York, 1949.
10. Churchill, R. V.: "Fourier Series and Boundary Value Problems," 2d ed., McGraw-Hill, New York, 1963.
11. Frank, P., and R. von Mises: "Die Differential und Integralgleichungen der Mechanik und Physik," Mary S. Rosenberg, 1943.
12. Bateman, H.: "Partial Differential Equations of Mathematical Physics," Cambridge University Press, New York, 1933.
13. Webster, A. G.: "Partial Differential Equations of Mathematical Physics," Stechert-Hafner, New York, 1933.
14. Petrovskii, T. G.: "Lectures on Partial Differential Equations," Interscience, New York, 1954.

Elliptic Equations:

15. Kellogg, O. D.: "Foundations of Potential Theory," Ungar, New York, 1954.
16. Miranda, C.: "Equazioni alle derivate parziali di tipo ellitico," Springer, Berlin, 1955.
17. Hellwig, G.: "Differentialoperatoren der Mathematischen Physik," Springer, Berlin, 1964.
18. Bergman, S., and M. M. Schiffer: "Kernel Functions and Elliptic Differential Equations in Mathematical Physics," Academic, New York, 1953.

Parabolic Equations:

19. Friedman, Avner: "Partial Differential Equations of Parabolic Type," Prentice-Hall, Englewood Cliffs, N.J., 1964.
20. Carslaw, H. S.: "Mathematical Theory of the Conduction of Heat in Solids," Dover, New York, 1945.
21. Doetsch, G.: "Theorie und Anwendung der Laplace-transformation," Dover, New York, 1943.
22. Hille, E., and R. Phillips: "Functional Analysis and Semi-groups," American Mathematical Society, Providence, R.I., 1957.

Hyperbolic Equations:

23. Hadamard, J.: "Lectures on Cauchy's Problem," New Haven, Conn., 1923.
24. Baker, S. G., and E. T. Copson: "The Mathematical Theory of Huygens' Principle," Oxford University Press, Fair Lawn, N.J., 1939.
25. Courant, R., and K. O. Friedrichs: "Supersonic Flows and Shock Waves," Interscience, New York, 1948.
26. Sauer, R.: "Anfangswertprobleme bei partiellen Differentialgleichungen," Springer, Berlin, 1958.
27. Leray, J.: "Lectures on Hyperbolic Equations with Variable Coefficients," Institute for Advanced Study, Princeton, N.J., 1952.

Chapter 6

Linear Spaces and Operators; Integral Equations

By J. L. B. COOPER, University of Toronto

The theory of linear spaces and of operators on them is the most highly developed part of functional analysis. Its characteristic feature is that classes of functions are considered as entities whose elements, the functions, are subject to algebraic operations such as addition or multiplication and to topological processes such as convergence to limits. This establishes a link between the theory of differential and integral operators and the theory of transformations of finite-dimensional spaces.

1. Linear Spaces

A *linear space* is a set E of elements $x, y, \ldots$ with which is associated a set of scalars $\lambda, \mu, \ldots$ such that for any elements x, y in the set and any scalars λ, μ an element $\lambda x + \mu y$ exists, and the following laws hold:

(*a*) $x + y = y + x$

(*b*) $\lambda(x + y) = \lambda x + \lambda y$

(*c*) $(\lambda + \mu)x = \lambda x + \mu x$

(*d*) $1x = x$

(*e*) $\lambda(\mu x) = (\lambda\mu)x$

The zero element of E is also written 0;

$$0x = x - x = 0 \qquad \text{for all } x$$

In almost all physical applications the set of scalars is either the set of real or the set of complex numbers; quaternions occur in a few applications. In what follows we shall generally suppose the scalars to be the complex field.

Elements $x_1, \ldots, x_r$ of a space are linearly independent if whenever $\lambda_1 x_1 + \cdots + \lambda_r x_r = 0$ then $\lambda_1 = \cdots = \lambda_r = 0$. If $x = \lambda_1 x_1 + \cdots + \lambda_p x_p$, x is said to be linearly dependent on $x_1 \ldots x_p$. The rank or dimension of a linear space is the maximum number of any set of linearly independent elements that can be found in it. If this is finite, the space is finite-dimensional; any set of linearly independent elements equal in number to the rank has the property that any element of the space is linearly dependent on that set. A subset of a linear space that obeys the axioms for a linear space (i.e., contains $\lambda x + \mu y$ whenever it contains x, y) is called a *linear subspace*. The elements linearly dependent on a given set of elements are said to be spanned by that set and constitute a linear subspace whose rank is not greater than the number of elements in the set.

Norms and Seminorms. A function $p(x)$ defined on a linear space and taking real non-negative values is called a *seminorm* if for all scalars λ, μ and all x, y.

$$p(\lambda x) = |\lambda| p(x), \qquad p(x + y) \leq p(x) + p(y)$$

If $p(x) = 0$ and if and only if $x = 0$, p is called a *norm*. A norm is written $\|x\|$ rather than $p(x)$.

With any set of seminorms, $\{p_\alpha\}$ say, there is associated a topology defined by the set of seminorms; a point x_0 of the space has as neighborhoods the sets of the form $\{x : p_{\alpha i}(x - x_0) < \epsilon, i = 1,2, \ldots, r\}$, where ϵ is any positive number, and $p_{\alpha i} \cdots p_{\alpha r}$ is any finite collection of seminorms. In particular, it is said that $x_n \to x_0$ as $n \to \infty$ if $p_\alpha(x_n - x_0) \to 0$ for each α as $n \to \infty$. For limits to be unique it is necessary and sufficient that $p_\alpha(x) = 0$ for all α implies $x = 0$.

If the topology is defined by a single norm, the space is called a *normed space*. If in addition it is complete, that is, if whenever (x_n) is a sequence such that $\|x_m - x_n\| \to 0$ as $m, n \to \infty$ then there is an x such that $x_n \to x$ as $n \to \infty$, the space is called a *Banach space*.

Hilbert Space. A *Hilbert space* is a linear space H with properties 1 and 2:

1. There is defined in the space H a Hermitian scalar product; i.e., with each pair of vectors x, y is associated a complex number (x,y) so that for all scalars λ, μ and all x, y

(i) $(\lambda x, y) = \lambda(x,y)$

(ii) $(x,y) = \overline{(y,x)}$

(iii) $(x,x) > 0 \qquad$ unless $x = 0$

It follows that, if we write $(x,x)^{1/2} = \|x\|$, then $\|x\|$ is a norm, and $|(x,y)| \leq \|x\| \, \|y\|$.

2. The space H is complete for the norm $\|x\|$.

It is said that x is *orthogonal* or *perpendicular* to y if $(x,y) = 0$. Hilbert space has many of the properties of (complex) Euclidean space; thus Pythagoras' theorem holds: If x is perpendicular to y,

$$\|x + y\|^2 = \|x\|^2 + \|y\|^2$$

If L is a closed *linear manifold*, then for any element x of the space there is an element y in L such that $x = y + z$, where z is perpendicular to all of L. y is called the projection of x on L and is written $P_L x$.

A set of elements is called *orthonormal* if all elements of the set have norm 1 and any two different elements are orthogonal to one another. Such a set is called complete if there is no nonzero x in the space such that $(x,x_\alpha) = 0$ for all α. In that case there is for every x in the space a unique expression

$$x = \sum_{1}^{\infty} c_n x_{\alpha n}$$

where $(x_{\alpha n})$ is a countable set of the complete orthonormal system and $c_n = (x,x_{\alpha n})$. A Hilbert space containing a countable everywhere-dense set, that is, a sequence of points (z_n) such that every point of the space is a limit of some sub-sequence of (z_n), is called *separable;* such a space has a countable complete orthonormal set. A general Hilbert space contains complete orthonormal sets, but they are not countable unless the space is separable.

Example 1. The space of all bounded continuous functions over a set G in n-dimensional space, with the norm $\|x\| = \sup \{|x(t)|:t\epsilon G\}$ is a *Banach space.*

Example 2. The set D of all infinitely differentiable functions on the real line, or in n-dimensional space, such that each function vanishes outside a bounded set depending on the function, is a linear space. This can be given a topology in which $x_n \to x$ if and only if there is a bounded set outside which all the $x_n(t)$ and $x(t)$ vanish, and $x_n(t) \to x(t)$ as $n \to \infty$ uniformly inside the bounded set for each p.

Example 3. The set S of all infinitely differentiable functions on the real line which have the property that for all r, k

$$\sup_t |t^k D^r x(t)| = p_{k,r}(x)$$

is a linear space in which the $p_{k,r}$ are a set of seminorms. The same applies to the similar set of functions in several variables with D^r denoting an arbitrary differential operator

$$D^r = \frac{\partial r}{\partial x_1^{r_1} \cdots \partial x_n^{r_n}}$$

Example 4. The set l^2 of all complex sequences $x = (x_n)$ such that $\Sigma|x_n|^2 < \infty$ is a separable Hilbert space when the scalar product of (x_n) and (y_n) is taken to be $(x,y) = \Sigma x_n\overline{y_n}$. If e_n is the vector with 1 as nth component and all other components 0, (e_n) is a complete orthonormal set.

Example 5. The set of all functions defined and measurable on a set G in n-dimensional space, and with the property that

$$\|x\|^2 = \int_G |x(t)|^2 \, dt < \infty$$

is a separable Hilbert space, which is designated $L^2(G)$, when the scalar product is taken to be

$$(x,y) = \int_G x(t)\overline{y(t)} \, dt$$

If, for example, G is the interval $(0,2\pi)$ of the real line, the space is written $L^2(0,2\pi)$, and the set of functions $\{e^{inx}/\sqrt{(2\pi)}\}$ with n running through all integers constitutes a complete orthonormal set; the corresponding expansion is the usual Fourier series.

2. Linear Functionals and Operators

Let E and F be linear spaces, and let L be a linear subspace of E. A function T defined in L and such that Tx has its values in F and satisfies

$$T(\lambda_1 x_1 + \lambda_2 x_2) = \lambda_1 T x_1 + \lambda_2 T x_2$$

for all scalars λ_1, λ_2 and x_1, x_2 in L is called a *linear operator* in E to F.

If T is defined throughout E, and if $Tx \to 0$ as $x \to 0$, the operator is *continuous.* If the spaces E and F are normed, T is continuous if and only if there is a constant M such that $\|Tx\| \le M\|x\|$ for all x in E; the smallest number M for which this holds is called the *norm,* $\|T\|$, of T. When a constant M exists, the operator is called *bounded.*

Examples of linear operators are matrix transformations of finite-dimensional spaces; integral operators, considered in Sec. 3; and differential operators. Thus the operator d/dt, considered as an operator in the space $C(a,b)$ of continuous functions on an interval (a,b) to itself, is an *unbounded operator,* defined only for those functions that have continuous derivatives. Since the function e^{int} is multiplied by (in) by this operator, it is clearly unbounded.

If the values of Tx are scalars and if T is continuous, T is called a *linear functional.* The set of all linear functionals on a space E is called the *dual space* of E, written E'. If E is a Hilbert space, then to every linear functional g there corresponds an element a of the Hilbert space such that $g(x) = (x,a)$ for all a: Hilbert space coincides with its dual. If E is the space $C(a,b)$ of continuous functions on (a,b), every continuous functional on E takes the form

$$g(x) = \int x(t) \, d\mu(t)$$

where μ is a function of bounded variation on (a,b); a similar formula holds for functions of several variables.

An operator T to a normed space is said to be *compact* if, whenever (x_n) is a bounded sequence of elements of the space, a sub-sequence of the numbers n, say n_i, exists such that (Tx_{n_i}) is convergent. Important examples of compact integral operators are mentioned below. In the Hilbert space l^2 of sequences (x_n), an operator A defined by $Ax = y$ when

$$y_r = \Sigma a_{rs} x_s$$

is compact if $\sum_{r,s} |a_{rs}|^2$ is finite. Operators with this property are called *Hilbert-Schmidt operators.*

Not all continuous operators are compact; the operator taking x into itself is not compact in a space of infinite dimensions. The operator taking the sequence (x_n) into $(n^{-1/2}x_n)$ is compact but not a Hilbert-Schmidt operator.

Hermitian and Self-adjoint Operators. Let A be an operator in a Hilbert space defined on a linear subset whose closure is the entire space. Then there is defined a linear operator A^* with the property that the domain of definition of A^* consists of all those x in the space such that there is an element A^*x satisfying $(x,Ay) = (A^*x,y)$ for all y in the domain of definition of A. If $A = A^*$, A is called *self-adjoint.* If the domain of definition D_A of A is contained in that of A^* and the two operators agree in their common domain, i.e., if $(x,Ay) = (Ax,y)$ for all x, y in D_A, then A is called *symmetric*, or *Hermitian*. Any bounded Hermitian operator is self-adjoint, but this is not true of unbounded operators, for example, the operator $i(d/dt)$ on $L^2(0,\infty)$ with domain consisting of those functions $x(t)$ that have derivatives in $L^2(0,\infty)$ and are such that $x(0) = 0$ is Hermitian but its adjoint is the formally identical operator without the restriction on the value of $x(0)$.

Spectral Resolution. If A is a compact self-adjoint operator, there exist a complete orthonormal set (x_n) and a set of real numbers (λ_n) with $\lambda_n \to 0$ as $n \to \infty$ and such that $Ax_n = \lambda_n x_n$ for all n. The (x_n) are called the *proper vectors* or *eigenvectors* of A, and the (λ_n) the *proper values* or *eigenvalues* of A. If A is a Hilbert-Schmidt operator, $\Sigma\lambda_n^2$ is finite.

For noncompact operators the spectral theory is more complicated because the spectrum need not consist only of discrete points. The general situation is as follows: Given a *self-adjoint operator* A there exists, for each real number, a projector $E(\lambda)$; the spaces onto which the $E(\lambda)$ project increase with λ; that is, $E(\lambda)E(\mu) = E(\min(\lambda,\mu))$ for all λ, μ, and $E(-\infty) = 0$, $E(\infty) = I$. An element x is in the domain of A if and only if $\int\lambda^2\|E(d\lambda)x\|^2$ is finite; then

$$Ax = \int\lambda E(d\lambda)x \tag{6.1}$$

and $E(\lambda)$ is called the *resolution of the identity* for A.

Unitary Operators. An operator U is said to be unitary if it maps the Hilbert space onto itself and preserves scalar products, so that for each x there is a y such that $Uy = x$, and $(Ux,Uy) = (x,y)$ for all x, y. If $U(t)$ is a set of unitary operators such that $U(0) = 1$ and $U(t + s) = U(t)U(s)$ for all real numbers t and s, there exists a resolution of the identity $E(\lambda)$, as described above, such that for all x

$$U(t)x = \int e^{i\lambda t}E(d\lambda)x \tag{6.2}$$

The self-adjoint operator A corresponding to this resolution of the identity in the manner of the previous paragraph satisfies

$$AU(t) = \frac{1}{i}\frac{dU(t)}{dt} \tag{6.3}$$

For example, let A be the operator $\frac{1}{i}\frac{d}{dt}$ over $L^2(-\infty,\infty)$. The spectral resolution for A is determined by the Fourier transformation formula. If

$$x(t) = \int_{-\infty}^{\infty} X(u)e^{iut}\,dt$$

then

$$[E(\lambda)x](t) = \int_{-\infty}^{\lambda} X(u)e^{iut}\,dt$$

The corresponding operators $U(t)$ given by formula (6.2) are such that $U(t)x(s) = x(t + s)$.

For A compact, with proper values (λ_n) and proper vectors (x_n), $E(\lambda)$ is the projector onto the space spanned by all x_n such that $\lambda_n \le \lambda$.

Quantum-mechanical Applications. In nonrelativistic quantum theory the states of a physical system are represented by points (actually rays) of unit norm in a Hilbert space; to each physically measurable quantity A (observable) there corresponds a self-adjoint operator A. The correspondence is such that in a state x the expectation value of any function $f(A)$ of A is $(f(A)x,x)$. Thus for a particle in three dimensions the state is represented by a function $\phi(x,y,z)$ (*wave function*), the x coordinate of position by the operator taking $\phi(x,y,z)$ into $x\phi(x,y,z)$, the x coordinate of momentum by the operator. Proper vectors, if they exist, correspond to states in which the operator has an exact value. That the operators corresponding to observables are Hermitian can be seen to correspond to the requirement that expectation values must be real. That the operators are self-adjoint has less evident physical significance; it can be shown to correspond to a condition that the states can be subjected to a unitary transformation through all time in which expectation values of all functions of the observable remain constant.

If H is the *Hamiltonian operator*, which represents the energy of the system, the evolution of the system through time is given by the *Schroedinger equation:*

$$H\psi(t) = \frac{1}{i}\frac{d\psi}{dt}$$

Here $\psi(t) = U(t)\psi(0)$; the $U(t)$ form a group of unitary operators as discussed in connection with a resolution (6.2) above.

Differential Operators. An operator L on the space $L^2(a,b)$, where

$$Ly = \frac{d}{dt}\left[p(t)\frac{dy}{dt}\right] + q(t)y$$

and where p and q are real-valued functions, is a *Hermitian operator* if the domain of definition is restricted to those functions y for which the operator has a value in the space and for which suitable boundary conditions are obeyed at a and b. If the domain (a,b) is finite and if p is continuous over the domain, the operator can be extended by imposing boundary conditions of a suitable type to a self-adjoint operator, which then has a pure point spectrum. If p is discontinuous or if the domain is infinite, the spectrum may be made up of either continuous or point spectrum or both.

Similarly, the operator $(\nabla^2 + V(x,y,z))$ applied to functions defined over all three-dimensional space, with V a real function, gives rise to a self-adjoint operator in L^2. The nature of the spectrum depends on the behavior of the function V, in particular, on its growth at infinity. Thus if $V = 1/r$, so that the operator is essentially the Hamiltonian operator for the hydrogen atom, the spectrum consists of a continuous spectrum over the entire set of positive numbers with a discrete spectrum consisting of points with negative values. If $V = -r^2$, essentially the case of the harmonic oscillator, the spectrum is a pure point spectrum.

3. Integral Equations

A linear operator K on the spaces $L^2(G)$ or $C(G)$ defined in (6.2) and having the form

$$(Kf)(s) = \int_G K(s,t)f(t)\,dt \tag{6.4}$$

where $s = (s_1, \ldots, s_n)$, is called an *integral operator;* $K(s,t)$ is called the *kernel of the operator.* The operator K^* given by

$$(K^*f)(s) = \int \overline{K(t,s)}f(t)\,dt \qquad (6.5)$$

is the *adjoint operator,* or *conjugate operator.* An equation

$$\int_G K(s,t)\phi(t)\,dt = g(s) \qquad \text{that is, } K\phi = g \qquad (6.6)$$

is called an *integral equation of the first kind* for ϕ; an equation

$$\phi(s) - \lambda\int K(s,t)\phi(t)\,dt = g(s) \qquad \text{that is, } \phi - \lambda K\phi = g \qquad (6.7)$$

is called an *integral equation of the second kind.*

Such equations arise, for instance, in potential theory, where the potential due to a mass is expressed as an integral operator on the mass distribution and so the solution of the Laplace-Poisson equation is related to that of an integral equation. It is frequently convenient to transform differential equations into integral equations.

If $\iint|K(s,t)|^2\,ds\,dt = \|K\|^2 < \infty$, the operator is a compact operator on $L^2(G)$ to itself and, indeed, is a Hilbert-Schmidt operator. Equation (6.7) is then called a *Fredholm equation,* and the following statements hold:

For any complex number λ, either (1) Eq. (6.7) has a unique solution ϕ in $L^2(G)$ for each g in $L^2(G)$ or (2) the corresponding homogeneous equation

$$\phi(0) - \lambda\int K(s,t)\phi(t)\,dt = 0 \qquad (6.8)$$

has a nonzero solution.

If statement 1 holds, λ is called a regular point of the equation. The solution of (6.7) is then given by an integral operator

$$\phi(s) = \int K(s,t;\lambda)g(t)\,dt \qquad (6.9)$$

$K(s,t;\lambda)$ is called the *resolvent kernel.* If λ is regular for (6.7) λ is also regular for the adjoint equation $\phi - \lambda K^*\phi = g$, and the resolvent kernel for this is $K(t,s,\lambda)$.

Every λ such that $|\lambda|\ \|K\| < 1$ is regular, and the solution of the equation is given by the *Neumann series:*

$$\phi = g + \lambda Kg + \lambda^2K^2g + \cdots + \lambda^nK^ng + \cdots \qquad (6.10)$$

For general regular values of λ the resolvent kernel has the form $K(s,t;\lambda) = D(s,t;\lambda)/D(\lambda)$, where

$$D(\lambda) = \sum_0^\infty (-1)^n \frac{\lambda^n}{n!} K_n$$

$$D(s,t;\lambda) = \sum_0^\infty (-1)^n \frac{\lambda^n}{n!} K_n(s,t)$$

In this, K_n is the integral of the $n \times n$ determinant with $K(s_j,t_k)$ in the j, k place as all variables s_j, t_k range over G, and $K(s,t;\lambda)$ is a similar integral but with the determinant modified by bordering it with the terms $K(s,t)$, $K(s,t_k)$, $K(s_j,t)$ in its first row and column. $D(s,t;\lambda)$ and $D(\lambda)$ are integral functions, and $D(\lambda)$ is nonzero at regular points.

In case 2, λ is called a characteristic value of the equation. There are at most a countable set of characteristic values, $\{\lambda_n\}$, and there are at most a finite number in any bounded region. For each, Eq. (6.8) has a finite number of linearly independent solutions, called *characteristic functions;* the conjugate equation has exactly the same number of linearly independent solutions. Equation (6.7) has a solution if and only if g is orthogonal to all solutions of $\psi - \bar{\lambda}K^*\psi = 0$.

An equation need have no characteristic values. In the important case of the *Volterra equation*

$$\phi(s) - \lambda \int_0^s K(s,t)\phi(t)\,dt = g(s)$$

there is a unique solution for every value of λ. This solution is given by the Neumann series (6.10), which is always convergent.

Symmetric Equations. If $K(t,s) = \overline{K(s,t)}$, the kernel is said to be symmetric, and K is then a Hermitian operator in L^2. The theory of Hermitian operators discussed in Sec. 2 applies. If the kernel is a Hilbert-Schmidt kernel—in particular, if the interval of integration is bounded and $K(s,t)$ is continuous or bounded—the kernel has a nonempty set of real characteristic values $\{\lambda_n\}$; each characteristic value has a finite set of characteristic functions associated with it. There are at most a finite number of characteristic values in any bounded part of the plane. The equation $\int K(s,t)\phi(t) = 0$ may have a nonzero solution. In this case, ∞ must be included among the characteristic values; an infinite number of characteristic functions may correspond to the value ∞. This is the case, in particular, if the kernel is *degenerate,* that is, of the form $\sum_1^n k_m(s)\overline{k_m(t)}$. The characteristic functions can be chosen so that they have norm 1; they then form a complete normal orthogonal set. Every function g can be expressed as a series $g = \Sigma g_n\phi_n$, with $\int g(t)\overline{\phi_n(t)}\,dt = g_n$, and where the series converges in the sense of the L^2 norm. The expression for $K(s,t)$ itself is $\sum \frac{\phi_n(s)\overline{\phi_n(t)}}{\lambda_n}$. If λ is a regular value of K, the solution of (6.7) is

$$\phi = \sum \frac{\lambda_n g_n \phi_n}{\lambda - \lambda_n}$$

If $K(s,t)$ is a Hilbert-Schmidt kernel such that

$$\int K(s,u)\overline{K(t,u)}\,du = \int \overline{K(u,s)}K(u,t)\,du \qquad (6.11)$$

the same results apply, save that the characteristic values λ_n need not be real. A kernel obeying condition (6.11) is said to be normal.

For symmetric kernels that are not compact kernels, the spectrum need not be purely a point spectrum but is still entirely real, so that every nonreal λ is a regular point. Normal kernels of general type have a spectrum, which may be both point spectrum and continuous spectrum, distributed in the complex plane.

4. Integral Transforms

Some of the most important integral equations of the first kind occur in connection with integral transforms. The most widely applied of these involve kernels not of the types discussed in Sec. 3. With $K(s,t) = e^{-ist}$ we obtain the Fourier transform

$$\hat{g}(s) = \frac{1}{\sqrt{(2\pi)}} \int_{-\infty}^{\infty} g(t)e^{-ist}\,dt \qquad (6.12)$$

For functions in $L^2(-\infty, \infty)$, the integral for $\hat{g}(s)$ converges in the mean-square sense, i.e., the sense of the norm in L^2, and g is given by

$$g(t) = \frac{1}{\sqrt{(2\pi)}} \int_{-\infty}^{\infty} \hat{g}(s)e^{ist}\,ds \qquad (6.13)$$

This formula is also valid in the mean-square sense. If g is summable over $(-\infty, \infty)$, the integral for g converges in the ordinary sense everywhere; the inversion formula (6.13) holds at all points t such that g is of bounded variation in the neighborhood of t and $g(t) = \frac{1}{2}\{g(t+0) + g(t-0)\}$. More generally the formula

$$g(t) = \lim_{\lambda\to\infty} \frac{1}{\sqrt{(2\pi)}} \int_{-\lambda}^{\lambda} \left(1 - \frac{|s|}{\lambda}\right) \hat{g}(s)e^{ist}\,ds \qquad (6.14)$$

holds for almost every t, in particular, wherever g is continuous.

Associated integral transforms are the Fourier cosine and sine transforms

$$g_c(t) = \sqrt{\frac{2}{\pi}} \int_0^{\infty} g(s)\cos st\,ds \qquad (6.15)$$

$$g_s(t) = \sqrt{\frac{2}{\pi}} \int_0^{\infty} g(s)\sin st\,ds \qquad (6.16)$$

to which (6.12) reduces in the case of even and odd g, respectively, and for which the following inversion formulas hold:

$$g(s) = \sqrt{\frac{2}{\pi}} \int_0^{\infty} g_c(t)\cos st\,dt \qquad (6.17)$$

$$g(s) = \sqrt{\frac{2}{\pi}} \int_0^{\infty} g_s(t)\sin st\,dt \qquad (6.18)$$

under similar conditions of validity to those for (6.13).

The variable s in (6.12) may be allowed to assume complex values. The most common form for dealing with this transform is through the Laplace transform, given by

$$(Lg)(w) = G(w) = \int_0^{\infty} g(t)e^{-wt}\,dt \qquad (6.19)$$

obtained from (6.12) by putting $s = -iw$ when $g(t) = 0$ for $t < 0$. If $\int_0^{\infty} |g(t)|e^{-at}\,dt < \infty$, $G(w)$ is holomorphic throughout the half plane $\operatorname{Re} w > a$. The inversion formula

$$g(t) = \frac{1}{2\pi i} \int_{c-i\infty}^{c+i\infty} G(w)e^{wt}\,dw \qquad (6.20)$$

holds for $c > a$ if g satisfies the conditions under which (6.13) is valid.

The Fourier transformation is important physically in that it involves an expression for a nonperiodic function in terms of periodic components. Other important applications arise in the theory of differential and integral equations from the behavior of these transforms in relation to certain functional operators.

If $y(t)e^{-at}$ and $y'(t)e^{-at}$ are summable over $(0, \infty)$ then integration by parts of the Laplace transform Ly' shows that, if $\operatorname{Re} w > a$,

$$(Ly')(w) = -y(0) - w(Ly)(w) \qquad (6.21)$$

and generally, if the higher derivatives of y exist and $y^{(n)}(t)e^{-at}$ is summable,

$$(Ly^{(n)})(w) = -y^{(n-1)}(0) - wy^{(n-2)}(0) - \cdots - w^{n-1}y(0) - w^n(Ly)(w) \qquad (6.22)$$

A second important property of these transforms relates to the operation of convolution. If f and g are two functions, their convolution h is

$$h(t) = (f * g)(t) = \int_{-\infty}^{\infty} f(t-s)g(s)\,ds \qquad (6.23)$$

or

$$h(t) = \int_0^t f(t-s)g(s)\,ds \qquad (6.24)$$

in case f and g are defined only for positive values of the variables. Convolution of functions corresponds to multiplication of their Fourier or Laplace transforms:

$$\hat{h}(u) = \sqrt{(2\pi)}\hat{f}(u)\hat{g}(u) \qquad (6.25)$$

or

$$(Lh)(w) = (Lf)(w)(Lg)(w) \qquad (6.26)$$

in the case of Eq. (6.24).

Consider the differential equation

$$P(D)y = D^n y + a_{n-1}D^{n-1}y + \cdots + a_0 y = g(t) \qquad (6.27)$$

where $D = d/dt$ and the coefficients a_r are constant. Application of the Laplace transformation gives

$$P(w)Y(w) = G(w) + y^{(n-1)}(0) + (w + a_{n-1})y^{(n-2)}(0) + \cdots \qquad (6.28)$$

On solving (6.28) for $Y(w)$ and using (6.20), we find $y(t)$. If

$$p(t) = \frac{1}{2\pi i} \int_{c-i\infty}^{c+i\infty} \frac{e^{wt}}{P(w)}\,dw$$

$p(t)$ is that solution of the differential equation $P(D)y = 0$ with $y(0) = y'(0) = \cdots = y^{(n-2)}(0) = 0$, $y^{(n-1)}(0) = 1$. The solution of (6.27) is

$$y(t) = \int_0^t p(t-s)g(s)\,ds + y^{(n-1)}(0)p(t) + y^{(n-2)}(0)[D + a_{n-1}]p(t) + \cdots$$

Similar techniques applied to partial differential equations enable them to be reduced to equations involving derivatives with respect to fewer variables or to ordinary equations. For example, the equation of heat conduction

$$\frac{\partial^2 y}{\partial x^2} = \frac{\partial y}{\partial t} \qquad t > 0 \qquad -\infty < x < \infty$$

reduces, on applying a Laplace transformation with respect to x, to the equation $\partial Y/\partial t = -w^2 Y$, whose solution is $Y(w,t) = Y(w,0)e^{-w^2 t}$. Application of the inversion formula (6.19) gives the solution

$$y(x,t) = \frac{1}{2\sqrt{(\pi t)}} \int_{-\infty}^{\infty} y(u,0) \exp\left[-\frac{(x-u)^2}{4t}\right] du$$

Application to Integral Equations. The integral equation of the second kind

$$\phi(s) - \lambda \int_{-\infty}^{\infty} K(s-t)\phi(t)\,dt = g(s)$$

becomes, after application of a Fourier transformation,

$$[1 - \lambda\sqrt{(2\pi)}\hat{K}(u)]\hat{\phi}(u) = g(u)$$

In all cases where the Fourier technique is legitimate, this gives $\hat{\phi}(u)$ and so $\phi(t)$. By more elaborate applications of these techniques, other equations, for example, the *Wiener-Hopf equation,*

$$g(s) = \int_{-\infty}^{\infty} K(|s-t|)\phi(t)\,dt$$

can be solved.

A classical singular equation of this type is the *Abel equation* which arises in the determination of a curve $y(s)$ such that a particle falls along it at a given rate. A generalized form of this equation is

$$f(t) = \int_0^t (t-s)^{-\alpha}\phi(s)\,ds \qquad 0 < \alpha < 1$$

Application of Laplace-transform methods gives as solution

$$\phi(t) = \frac{\sin \pi\alpha}{\pi} \frac{d}{dt} \int_0^t (t-s)^{\alpha-1} f(s)\,ds$$

Nonlinear Equations. Nonlinear equations arise in a variety of problems and have little systematic theory; some may be solved by iterative or other numerical methods. A typical form is

$$\phi(s) = g(s) + \int_0^s K(s,t,\phi(t))\,dt$$

If K satisfies the conditions

$$K(s,t,u_1) - K(s,t,u_2) \leq N|u_1 - u_2|$$

for some N and all s, t, u_1, u_2, the equation can be solved by iteration, putting

$$\phi_0(s) = g(s)$$

$$\phi_{n+1}(s) = g(s) + \int_0^s K(s,t,\phi_n(t))\,dt \qquad n = 0, 1, \ldots$$

5. Generalized Functions

Improper functions, with properties not possessed by ordinary functions, are needed in a number of applications of mathematical physics. The *Dirac delta function* is an example. A systematic and rigorous treatment can be given by regarding them as functionals, in the sense explained below. The typical problem is that of finding a construct to act in the place of the derivative of a function that is not differentiable.

If $f(x)$ is any integrable function, it defines a functional T_f on the space D of Example 2, Sec. 1, or, if it is not too large at infinity, on the space S of Example 3, according to the formula

$$T_f(\phi) = \int_{-\infty}^{\infty} f(x)\phi(x)\,dx \qquad (6.29)$$

If f is differentiable

$$T_{f'}(\phi) = \int_{-\infty}^{\infty} f'(x)\phi(x)\,dx = -\int_{-\infty}^{\infty} f(x)\phi'(x)\,dx$$
$$= -T_f(\phi') \qquad (6.30)$$

The functional $T'(\phi) = -T_f(\phi')$ exists even if f is not differentiable, and both it and T_f are continuous as functionals on the spaces D or S, respectively. We therefore consider, in place of functions, the functionals that represent them and are members of the spaces D' or S' conjugate to D or S. Within this class of functionals, differentiation is always possible.

For example, the *Heaviside unit function* $H(t)$, which is 1 for positive t and 0 for negative t, corresponds to the functional which takes ϕ into $\int_0^\infty \phi(t)\,dt$; and its derivative, the Dirac delta function, is the functional that takes ϕ into $-\int_0^\infty \phi'(t)\,dt$, that is, into $\phi(0)$.

Generalized functions are often called *distributions*. Operations of multiplication by functions and of convolution, as well as other algebraic operations, can be performed on distributions.

Distributions have important applications in the theory of differential equations, in which it is important to consider solutions which may not have the requisite number of derivatives but which are solutions in a sense of the theory of distributions. For example, the solution $\frac{1}{2}[f(x+t) + f(x-t)]$ of the wave equation obeys that equation only in the sense of a theory of generalized functions when f is not twice differentiable. Generalized solutions of hydrodynamical equations are held to have an application to the theory of turbulent flow. However, it can be shown that in the case of the elliptic equation—in particular, in potential theory—solutions in the distribution sense are solutions in the ordinary sense.

The theory also enables a generalization of the theory of *Fourier transforms* to be carried out. A distribution in S' has a Fourier transform that is also in S'. This allows one, for example, to consider Fourier transforms of functions that are of the order of a polynomial at infinity, since they are in S'. For example, if $f(x)x^{-k}$ is integrable over $(-\infty, \infty)$ for some integer k, the function

$$F_k(x) = \frac{1}{\sqrt{(2\pi)}} \int_{|x|>1} f(x) \frac{e^{-iux}}{(-ix)^k}\,dx$$
$$+ \frac{1}{\sqrt{(2\pi)}} \int_{-1}^{1} f(x) \frac{e^{-iux} - \cdots - (iux)^k/k!}{(-ix)^k}\,dx$$

exists in the ordinary sense, and its kth derivative, which is in general a distribution, is the Fourier transform of f. By considering functionals on various classes of analytic functions it is possible to define Fourier transforms for arbitrarily large functions.

References

Linear Spaces and Operators:

1. Banach, S.: "Théorie des opérations linéares," Monografje Matematyczne, Warsaw, 1932.
2. Von Neumann, J.: "Mathematical Foundations of Quantum Mechanics," Princeton University Press, Princeton, N.J. (trans. from German, Springer, Berlin, 1932).
3. Stone, M. H.: "Linear Transformations in Hilbert Space," American Mathematical Society, New York, 1932.
4. Riesz, F., and B. Sz. Nagy: "Functional Analysis," Ungar, New York, 1955.
5. Taylor, A. E.: "Introduction to Functional Analysis," Wiley, New York, 1958.
6. Zaanen, A. C.: "Linear Analysis," North-Holland Publishing Co., Amsterdam, Groningen, 1953.

Integral Equations:

7. Hilbert, D.: "Grundzüge einer allgemeinen Theorie der linearen Integralgleichungen," Teubner, Leipzig, 1904.
8. Hellinger, E., and O. Toeplitz: Integralgleichungen und Gleichungen mit unendlichvielen Unbekannten, Enzyklopadie der Math. Wiss. II, C. 13, Band II, Teil 2, Teubner Leipzig, 1927.
9. Buckner, H.: "Die praktische Behandlung von Integralgleichungen," Springer, Berlin, 1952.
10. Smithies, F.: "Integral Equations," Cambridge University Press, New York, 1958.
11. Mikhlin, A.: "Integral Equations and Their Applications to Certain Problems in Mechanics, Mathematical Physics and Technology," Pergamon Press, New York, 1957.

Integral Transforms:

12. Titchmarsh, E. C.: "Introduction to the Theory of Fourier Integrals," Oxford University Press, Fair Lawn, N.J., 1948.
13. Doetsch, G.: "Handbuch der Laplace Transformation," vols. I, II, III, Birkhaüser Verlag, Basel, 1950–1956.
14. Van der Pol, B., and H. Bremmer: "Operational Calculus Based on the Two-sided Laplace Integral," Cambridge University Press, New York, 1950.

Generalized Functions

15. Schwartz, L.: "Théorie des distributions," vols. I, II, Hermann et cie, Paris, 1950, 1951.
16. Gel'fand, I. M., et al.: "Generalized Functions," vols. 1–4, Academic, New York, 1964–
17. Garding, L., and J. Lions: Functional Analysis, *Nuovo Cimento Suppl.*, **14**: 9 (1959).
18. Streater, R. F., and A. S. Wightman: "PCT, Spin and Statistics and all that," Benjamin, New York, 1963.

Chapter 7

Numerical Analysis

By JOHN TODD, California Institute of Technology

The automatic computer is now an indispensable tool for every scientist. One must, however, be always aware of the existence of arithmetic traps of all kinds which are ready for the unwary. After a few remarks on computers and computer languages we begin this chapter with a series of "bad examples" which can be worked out using hand calculation; it takes little imagination to realize how such singularities can explode when we move to current practice on microsecond machines. The discovery of such examples is one of the more important activities of the numerical analyst. On the constructive side, the numerical analyst must consider algorithms for the standard computational processes, analyzing them theoretically and practically and pointing out those which are satisfactory in various situations and indicating their limitations. Such considerations complete this chapter.

1. Computers and Computer Languages

It is not appropriate to describe here in any detail the structure of automatic computers or that of the various languages used to communicate with them. Suffice it to say that once a computation has been planned in every detail it is in many cases relatively easy to describe it formally in a "problem-oriented language" such as ALGOL or FORTRAN. This description can then be transcribed onto a deck of cards that is an acceptable input to the computer. The computer then translates this deck into a machine-language program which, when executed by the computer, carries out the planned computation. (This machine-language program may never appear outside the machine, being executed at once or stored internally; it may, however, be realized externally as another deck of cards which can be reused directly without going through the translation process again.) The translation process usually includes searches for more or less clerical errors, and the computer emits a list of such errors; if any exist, they are corrected, for example, in the original language, and the corrected program is resubmitted to the computer. It must not be assumed that a program is correct merely because it is translatable and can be executed; in all cases it should be used to repeat the computation of pilot programs that have been thoroughly checked previously.

We point out that, whereas standard function symbols such as sin, exp, sqrt are translatable, more elaborate ones such as J_n or sn or processes such as evaluating a determinant or evaluating a definite integral are not available in current languages, and use should be made of tested subroutines or procedures which may be available either in the internal store of the computer or externally in the library of the computer (as card decks or on magnetic tape) or in the open literature (e.g., *Journal of Association for Computing Machinery* or *Numerische Mathematik*). In all cases the user should be aware of the scope and limitations of the material he is using. Unfortunately this is often difficult, as documentation is usually poor and hearsay evidence may be all that is available.

In some cases the automatic translation may not provide a sufficiently efficient machine-language program. In such cases, and in cases where automatic translation of the conceived program is not possible, the scientist must make use of an "assembly language" which allows him to take full advantage of the computer's capabilities.

For further information on the matters discussed in this section the reader should refer to general books and to the detailed manuals appropriate for the equipment and languages used.

2. The Problem of Error Analysis

The general problem of numerical analysis can be stated in the form "given x, find $f(x)$," if we interpret x and f liberally, for instance, with $x = 2, f = \sqrt{}$ or with x a positive definite matrix and f its largest eigenvalue. Naturally, the problem is stated in a certain context. The nature and precision of x are given, and so is the equipment available; so should be the desired accuracy. Thus we may want to find $\sqrt{2}$ to $20D$, using a desk machine, or the largest eigenvalue, to within $5S$, of a 100×100 matrix whose elements are given to $6S$. Whereas the first of these problems is readily soluble, the second may very well not be.

Ideally, with each solution should be given an upper bound on the error. It is rarely feasible in significant calculations to obtain realistic error bounds analytically, owing to the complexity of current algorithms. Suggestions have been made to carry out the error analysis automatically, in parallel with the main calculation. Thus with each quantity q calculated there would be associated an interval $[q - \epsilon', q + \epsilon'']$ within which it must lie, where the ϵ', ϵ'' are calculated by the auxiliary error program. An alternative cruder scheme is to carry along with each quantity q a significance index which indicates the number of digits in q that are significant; this can be done by complicating the hardware or by auxiliary programs. These solutions to the error problem may be too costly in time, memory space, or equipment to be generally acceptable. The practical solution seems to be a change in the standards of acceptability. This does not mean that a computing center should be a home of mathematical delinquency but rather a place where mathematical and computational evidence is stored and where fine judgments are made on its basis. The evidence based on controlled computational experiments, when examined in collaboration with the customer, who has technical knowledge in the area of his problem, often permits giving error estimates that have some authority but not that of classical mathematics. We shall base our later treatment on this point of view, discussing convergence, etc., for academic problems, e.g., the solution of a Dirichlet problem in a square, in the hope that conclusions drawn will be applicable in more realistic circumstances. To make up for this encouragement to extrapolate, we shall point out "bad examples." We hope that this will preserve a critical attitude, which seems often to become atrophied in the neighborhood of computers.

We discuss briefly the nature of error analyses. In the case of the classical recurrence relation for computing N^{-1}, say when $0 < N < 1$,

$$x_0 = 1 \qquad x_{n+1} = x_n(2 - Nx_n)$$

theoretical results are available. If $\epsilon_n = x_n - N^{-1}$ then

$$\epsilon_{n+1} = -N\epsilon_n^2$$

and $x_{n+1} > x_n$ so that x_n strictly increases to the limit N^{-1}. Upper estimates for ϵ_n are readily obtained. In actual practice, for simplicity with a $2D$ machine, it is no longer true that x_n steadily increases. For example, with $N = 0.99$ we have

$$x_0 = 1.00 \qquad x_1 = 1.01 \qquad x_2 = 1.01$$

The behavior of the algorithm and the error estimates depend on the detailed arrangement of the computation and the structure of the arithmetic unit and are bound to be lengthy even in such simple cases as that above.

The results of error analyses may be of the form

$$\|\text{output} - f(\text{input})\| < \epsilon \qquad f(x) = \sqrt{x}$$

for example, and are called "forward" by Wilkinson. They may be of the form

$$\|f^{-1}(\text{output}) - (\text{input})\| < \epsilon \qquad f^{-1}(x) = x^2$$

for example, which are "backward" in the terminology of Wilkinson. The significance of this type of analysis from the point of view of the experimental scientist, whose data are subject to error, and from that of the numerical analyst, who cannot introduce most numbers exactly into his machine, is evident. One of the first to point out this importance was Givens. In addition, especially in the case of floating-point operations, it appears that this type of error analysis is easier to carry out.

We observe that if we have an estimate of the form

$$\|f^{-1\prime}(x)\| < K$$

we can recover a result of the forward kind from one of the backward form. Here the norm symbol is to be interpreted appropriately: It may be a discrete one or a continuous one, as in the case when we discretize a continuous problem and produce a solution consisting of the solution to the discrete problem and an interpolation process for obtaining nontabular values.

It is clear that error analysis in digital computation is essentially analytic number theory and so it is not surprising that some of the notable contributions have come from mathematicians who have been expert in this theory.

We give examples of the results available for the square-root process.

1. Assume that we are using an s-place binary machine working in fixed point. Then a program can be found for which the difference between the alleged square root and the actual square never exceeds $.76 \times 2^{-s}$. This is the result of a forward analysis due to Goldstine, Murray, and von Neumann.

2. Suppose we use a floating-point binary machine of the IBM 7090 type and that the input is

$$\bar{a} = 2^{\eta} \times A \qquad \tfrac{1}{2} \leq A \leq 1$$

A program can be found such that the square of the output differs from $\bar{a}$ by at most $2^{\eta-27} \times 4.38$; this is a backward analysis due to Casey.

We conclude with some comments on the practical use of published algorithm and error analyses. We contemplate a scientist with a computing problem for which he finds a published algorithm, complete with an error analysis and test data by a reputable authority. The author will have devised the algorithm, made the error analysis, and, finding it satisfactory, translated the algorithm into ALGOL, for instance, making use of his intimate knowledge of his own processor to ensure that it produces an object program that carries out precisely the arithmetic he has analyzed. He will also have carried out some applications of his algorithm to typical cases to check his ALGOL program and to indicate the realism of his error estimates. The user, typically, will have to translate the published program into another language (FORTRAN or PL/1, for instance) or at any rate into another dialect, submit it to a processor (with the details of which he may not be familiar), and obtain his own object program. To decide whether this will carry out exactly the same arithmetic as originally analyzed is likely to require a prohibitively detailed study of both sets of hardware and software. In these circumstances the best the user can do is to run the test cases, study the results obtained in comparison with the published ones, and, using this and any additional experience that may be available, make some necessarily tentative judgments on the errors to be expected in his own problem. It is clear that standardization of equipment and languages is desirable.

3. Some Bad Examples

Example 1. We are conditioned to accept associativity of multiplication and its consequence that abc^{-1} can be evaluated either as

First ab, then $(ab)c^{-1}$

or as First bc^{-1}, then $a(bc^{-1})$

If we imagine the operations of multiplication and division as those used by a computer, with rounding to a fixed precision, for example, to $2D$, this is no longer the case. If we take

$$a = .12 \qquad b = .11 \qquad c = .13$$

the two calculations are

.0132 rounded to .01 divided by .13 giving .076 which rounds to .08, and

.11 divided by .13 giving .846 which rounds to .85 which multiplied by .12 gives .104 which rounds to .10

The correct answer is .10.

Example 2. Consider the solution of the system of equations

$$\begin{aligned} 10x + 7y + 8z + 7w &= 32 \\ 7x + 5y + 6z + 5w &= 23 \\ 8x + 6y + 10z + 9w &= 33 \\ 7x + 5y + 9z + 10w &= 31 \end{aligned}$$

By inspection, the exact solution is

$$x = 1 \qquad y = 1 \qquad z = 1 \qquad w = 1$$

It is also easy to verify that

$$x = 9.2 \qquad y = -12.6 \qquad z = 4.5 \qquad w = -1.1$$

nearly solves the system, indeed up to errors of .1, −.1, .1, −.1 in the right-hand sides. If we perturb the right-hand sides to

$$32 + \epsilon,\ 23 - \epsilon,\ 33 + \epsilon,\ 31 - \epsilon$$

we can verify that the exact solution of this new problem is

$$x = 1 + 82\epsilon \qquad y = 1 - 136\epsilon \qquad z = 1 + 35\epsilon \qquad w = 1 - 21\epsilon$$

Relative error in the data is magnified by several thousand in the solution. This example is due to T. S. Wilson.

Example 3. We have just seen the effect of small changes in *linear-equation* problems. We now consider the same effect in *polynomial equations.*

$$z^4 - 4z^3 + 6z^2 - 4z + 1 = 0$$

has four roots 1, 1, 1, 1

If we change the middle coefficient from 6 to $6 - 49 \times 10^{-8}$, i.e., by less than 1 in 10^7, the roots change by about 3 in 100, being

$$1.02681,\ 0.97389,\ 0.99965 \pm i0.026455$$

Similarly, if we change from the equation $z^{10} = 0$ to $z^{10} = 10^{-10}$ the roots change from 0 to numbers of modulus 10^{-1}.

The above examples suggest, and rightly, that multiple roots are especially sensitive. However, if we take an equation with roots 1, 2, . . . , 20, say

$$\prod_{r=1}^{20} (z - r) \equiv z^{20} - 210z^{19} + \cdots + (20)! = 0$$

and change the coefficient of z^{19} from 210 to $210 + 2^{-23}$ (i.e., we make a relative error of about 10^{-9}), the roots of the new equation are

1.00000 0000	10.09526 6145 ± 0.64340 0904i
2.00000 0000	11.79363 3881 ± 1.65232 9728i
3.00000 0000	13.99235 8137 ± 2.51883 0070i
4.00000 0000	16.73073 7466 ± 2.81262 4894i
4.99999 9928	19.50243 9700 ± 1.94033 0347i
6.00000 6944	
6.99969 7234	
8.00726 7603	
8.91725 0249	
20.84690 8101	

showing some major discrepancies despite the separation of the roots. This example is due to J. H. Wilkinson.

Example 4. Again we consider calculating $\sin(\pi/6)$ on our $2D$ computer.

We have to replace $\pi/6$ by .53, replace $\sin x$ by $x - \frac{1}{6}x^3$, and obtain after several roundings

$$\begin{aligned} \sin(\pi/6) &= .52[1.00 - .17 \times (.52 \times .52)] \\ &= .52[1. - .05] \\ &= .49 \end{aligned}$$

Example 5. Newton's method for obtaining square roots is motivated as follows: If x_n is an approximation to $\sqrt{N}$, say, too low (high), then N/x_n will be an approximation which is too high (low). Hence their average

$$x_{n+1} = \tfrac{1}{2}\{x_n + (N/x_n)\}$$

should be a better approximation.

For instance, with $N = .25$, $x_0 = 1$ we obtain

$$x_1 = 0.625 \qquad x_2 = .5125 \qquad x_3 = .500152 \qquad x_3 = .500000 \ldots$$

It can be proved that if $\epsilon_n = x_n - \sqrt{N}$ then

$$\epsilon_{n+1} = \tfrac{1}{2}\epsilon_n^2 x_n^{-1}$$

The error at one stage is of the order of the square at the preceding stage, which means that, roughly speaking, the number of correct decimals in the answer is doubled at each application.

It can also be proved that

$$x_{n+1} - x_n = \tfrac{1}{2}(N - x_n^2)x_n^{-1} < 0$$

provided $x_0 > 0$, $n > 0$ so that

$$x_{n+1} < x_n$$

and that x_n does converge to $\sqrt{N}$.

All this is *theoretical arithmetic*. In *practical computation* the infinite descent guaranteed by the strictness of the last inequality just cannot happen. The question arises: At what stage does one stop and take the current x_n as the required square root? It can be shown that the appropriate x_n is the one that first satisfies $x_n \geq x_{n-1}$, that is, when the sequence becomes stationary or reverses its direction. Let us look at two examples, the first on our $2D$ machine.

If $N = .01$, $x_0 = .11$, we find $N \div x_0 = .09$ and then

$$x_1 = .50x_0 + .50(N \div x_0) = .06 + .05 = .11 = x_0$$

This example shows that strict inequality need not happen.

If $N = -\frac{1}{2}$, $x_0 = 1$, we find $x_1 = \frac{1}{4}$, $x_2 = -\frac{7}{8}$, and $x_3 = -\frac{17}{112}$. Hence

$$\sqrt{-\tfrac{1}{2}} = -\tfrac{17}{112}$$

This example shows the necessity to check that $N \geq 0$. Naturally if one is asked directly to find $\sqrt{-\frac{1}{2}}$ one takes appropriate action; in practice, however, intermediate steps of calculation are rarely monitored by human beings, and safeguards must be incorporated in the subroutines.

Example 6. The Bessel function

$$J_n(x) = \frac{1}{\pi} \int_0^\pi \cos (x \sin \theta - n\theta)\, d\theta$$

satisfies the recurrence relation:

$$J_{n+1}(x) = 2nxJ_n(x) - J_{n-1}(x)$$

Tables of J_0 and J_1 are readily available but if, for instance, $J_{20}(1)$ is required, the use of the above relation seems indicated. We proceed from the values of J_0 and J_1 correct to $10D$ to the others, on the left below; the underlining indicates the digits in the generated values that are not correct.

$J_0 =$	$.76519\ 76866$	$\tilde{J}_{40} =$	$0.$
$J_1 =$	$.44005\ 05857$	$\tilde{J}_{39} =$	$.00000\ 001$
$J_2 =$	$.11490\ 3484\underline{8}$	$\tilde{J}_{38} =$	$.00000\ 078$
$J_3 =$	$.01956\ 335\underline{35}$	$\tilde{J}_{37} =$	$.00005\ 927$
$J_4 =$	$.00247\ 663\underline{62}$	$\tilde{J}_{36} =$	$.00438\ 520$
$J_5 =$	$.00024\ 97\underline{381}$	$\tilde{J}_{35} =$	$.31567\ 512$
$J_6 =$	$.00002\ \underline{07248}$		$\cdots$
$J_7 =$	$-.00000\ \underline{10385}$	$\tilde{J}_{20} =$	$.43716\ 256 \times 10^{26}$

It is clear that this seemingly attractive approach is completely unsuitable. The trouble is caused by the "numerical instability" of the process; the unavoidable rounding errors in the data are amplified by the factor $2nx$, and there is also some cancellation.

However, something can be saved. The fact that the ascending recurrence is unstable suggests that the descending one might be stable. There is a difficulty here since we have no initial values but it turns out that arbitrary starting values will do.

For instance, if we take $\tilde{J}_{40} = 0$, $\tilde{J}_{39} = 1 \times 10^{-8}$, and use

$$\tilde{J}_{n-1} = 2n\tilde{J}_n - \tilde{J}_{n+1}$$

we obtain the values on the right above.

This, of course, is not likely to be the correct result, for the difference equation is homogeneous and we can multiply by any scale factor. There are several ways of finding the appropriate multiplier; the simplest is to carry on the above recurrence to obtain

$$\tilde{J}_0 = .86360016 \times 10^{50}$$

Comparing this with the correct value of J_0, we see that the appropriate scale factor is

$$k = J_0/\tilde{J}_0 = .88605550 \times 10^{-50}$$

Applying this to $\tilde{J}_{20}$, we find our estimate

$$J_{20} = k\tilde{J}_{20} = .38735029 \times 10^{-24}$$

which is correct to within a unit in the last place.

The choice of the starting points 40, 39 is largely dictated by experience. If, instead, we start at

$$\tilde{J}_{25} = 0 \qquad \tilde{J}_{24} = 10^{-8}$$

we obtain

$$\begin{aligned} \tilde{J}_0 &= .80486703 \times 10^{23} \\ k &= .95071315 \times 10^{-23} \\ J_{20} &= .38735029 \times 10^{-24} \end{aligned}$$

This method is applicable to compute many of the important special functions that arise in various branches of science.

Example 7. The examples above have been mostly carried out in fixed-point arithmetic. We conclude with an example in the floating-point mode. Consider the solution of the quadratic $ax^2 + 2bx + c = 0$ by the formula

$$x = \frac{-b \pm \sqrt{b^2 - ac}}{a}$$

If x_1, x_2 are the two roots and if $|x_1| \geq |x_2|$, the above formula is satisfactory for evaluating x_1. However if $|ac| \ll b^2$ in the evaluation of x_2 we are subtracting two large, almost equal numbers and there is a consequent loss of significance. It is easy to devise a satisfactory method of calculating x_2; indeed it is sufficient to use the formula

$$x_2 = c/(ax_1)$$

The reader is advised to carry out these calculations either by programming and running them on an actual computer or by simulating the floating-point calculations on a desk calculator, for instance, in the case of $a = 1$, $b = 10^5$, $c = -1$, carrying eight significant decimal digits.

4. The Use of Tables

There is no doubt that the use of tables is the most reliable way to find spot values of special functions, either for direct use or for checking programs. Every computing center should have available a library of tables and indices of tables, and every scientist ought to know efficient ways of interpolation.

Lagrangian Methods for Interpolation. The usual methods of interpolation are based on the assumption that the function f in question behaves like a polynomial. In order to calculate f approximately at a point x, we obtain a polynomial approximation F for f which is good in the neighborhood of the point in question, and we evaluate $F(x)$ as an approximation to $f(x)$. The basis for this is the following theorem due to Lagrange.

There is a unique polynomial of degree n assuming $n + 1$ arbitrary values f_i at any $n + 1$ distinct points $x_0, x_1, \ldots, x_n$. This polynomial is

$$L_n(x) = L_n(f,x) = \sum_{i=0}^{n} f_i l_i^{(n)}(x)$$

where $\quad l_i^{(n)}(x) = \Pi'[(x - x_j)/(x_i - x_j)]$

and the product is over all $j = 0, 1, \ldots, n; j \neq i$.

For simplicity, let us discuss the case $n = 3$, the 4-point case. We change the notation and ask for the existence of a cubic $\alpha x^3 + \beta x^2 + \gamma x + \delta$ that assumes the values A, B, C, D at distinct points a, b, c, d. This assumption requires

$$\begin{aligned} A &= \alpha a^3 + \beta a^2 + \gamma a + \delta \\ B &= \alpha b^3 + \beta b^2 + \gamma b + \delta \\ C &= \alpha c^3 + \beta c^2 + \gamma c + \delta \\ D &= \alpha d^3 + \beta d^2 + \gamma d + \delta \end{aligned}$$

This is a set of linear equations for $\alpha, \beta, \gamma, \delta$ which can be solved uniquely, since the determinant of the system is a Vandermondian which does not vanish, a, b, c, d being assumed distinct. This establishes the existence of an interpolating cubic. That it is unique follows essentially from the fundamental theorem of algebra, for, if there were two, their difference would be a polynomial of degree 3 at most, which would vanish at a, b, c, d.

The general case of $(n + 1)$-point interpolation can be treated similarly. However, the result can be established directly by observing that $l_i(x_j) = \delta_{ij}$, so that

$$L_n(f,x_j) = \Sigma f_i \delta_{ij} = f_j \qquad j = 0, 1, \ldots, n$$

The uniqueness follows by the argument used above. We note that, for all x, $\Sigma l_i(x) = 1$.

It is clear that the $\alpha, \beta, \gamma, \delta$ are linear functions of A, B, C, D, and the same is true for the interpolating cubic $\alpha x^3 + \beta x^2 + \gamma x + \delta$, which we can therefore write as

$$A\mathcal{A}(x) + B\mathcal{B}(x) + C\mathcal{C}(x) + D\mathcal{D}(x)$$

where the polynomials $\mathcal{A}, \mathcal{B}, \mathcal{C}, \mathcal{D}$ are cubics. This shows that tables of $\mathcal{A}, \mathcal{B}, \mathcal{C}, \mathcal{D}$ would greatly facilitate interpolation in tables of any function that allows 4-point interpolation. It is clearly not possible to contemplate tables covering arbitrary a, b, c, d, but the problem becomes practicable if we restrict ourselves to the case when $a - b = b - c = c - d = l$, particularly if we notice that we can assume that $l = -1$ and $a = -1$, $b = 0$, $c = 1$, $d = 2$. Using p as the nondimensional variable, we see that

$$\mathcal{A}(p) = L_{-1}(p) = -\frac{p(p-1)(p-2)}{6}$$

$$\mathcal{B}(p) = L_0(p) = \frac{(p+1)(p-1)(p-2)}{2}$$

$$\mathcal{C}(p) = L_1(p) = -\frac{(p+1)p(p-2)}{2}$$

$$\mathcal{D}(p) = L_2(p) = \frac{(p+1)p(p-1)}{6}$$

Among the many tables of these Lagrange interpolation coefficients that of the National Bureau of Standards (1948) covers the cases of from 3 to 11 points, at varying intervals in p, down to .0001.

The question of the error in interpolation by the Lagrangian formula is significant only when some information about the general behavior of the function is given. The usual remainder formula is

$$f(x) - L_n(f,x) = \frac{f^{(n+1)}(\xi)}{(n+1)!} \prod_{i=0}^{n} (x - x_i)$$

where f is assumed to have an $(n + 1)$st derivative in an interval including $x, x_0, x_1, \ldots, x_n$ and where $\xi = \xi(x)$ is a point in this interval. Note that we do not assume that the nodes $x_0, x_1, \ldots, x_n$ are equally spaced. If we assume the nodes are equally spaced at an interval h, then the above formula reduces to

$$f(x) - L_n(f,x) = \binom{p}{n+1} h^{n+1} f^{(n+1)}(\xi)$$

where $x = a + ph$ and ξ is as before.

It is clear that this error can be made as small as we please by taking h small enough and/or by increasing n, it being assumed that the derivatives of f are bounded. It is also clear that the error is smaller in the center of the range, because of the first factor on the right.

An elegant and practical method for reducing an $(n + 1)$-point interpolation to a sequence of $(\frac{1}{2})n(n - 1)$ linear interpolations has been given by Aitken. It is convenient for both desk calculators and automatic computers and makes the use of interpolation tables unnecessary. We describe the 4-point case for simplicity.

Given $f(a) = A$, $f(b) = B$, $f(c) = C$, $f(d) = D$, we show how to find $f(p)$. Interpolate linearly between (a,A), (b,B) to find (p,B_1); then interpolate linearly between (a,A), (c,C) to find (p,C_1), and then between (a,A), (d,D) to find (p,D_1). The next stage is to interpolate linearly between (b,B_1), (c,C_1) to find (p,C_2) and then between (b,B_1), (d,D_1) to find (p,D_2). Finally, interpolate linearly between (c,C_2) and (d,D_2) to find (p,P).

The scheme is illustrated graphically in Fig. 7.1. We note that there is no assumption that the a, b, c, d are equally spaced.

We give, without comment, two examples which show how the scheme can be carried out and how labor can be saved by dropping common initial figures. Many extensions of the method have been given by Aitken and Neville.

Given $f(1) = 1, f(2) = 125, f(3) = 729, f(4) = 2157$, find $f(2.5)$.

$1 = a$	$1 = A$			
$2 = b$	$125 = B$	$187 = B_1$		
$3 = c$	$729 = C$	$547 = C_1$	$367 = C_2$	
$4 = d$	$2197 = D$	$1099 = D_1$	$415 = D_2$	$343 = P$

Here $f(x) = (4x - 3)^3$, and the interpolation is exact, as it should be.

Given $f(0) = 47.434165$, $f(1) = 47.539457$,

$$f(2) = 47.644517, \quad f(3) = 47.749346,$$

find $f(1.4321)$.

0	47.434165			
1	.539457	.584954		
2	.644517	682	837	
3	.749346	517	60	24

We find $f(1.4321) = 47.584824$, which agrees with the fact that

$$f(x) = \sqrt{2250 + 10x}$$

We shall now show generally that the Aitken algorithm leads to the Lagrangian interpolant. We follow a proof of Feller. We want to evaluate $f(p)$,

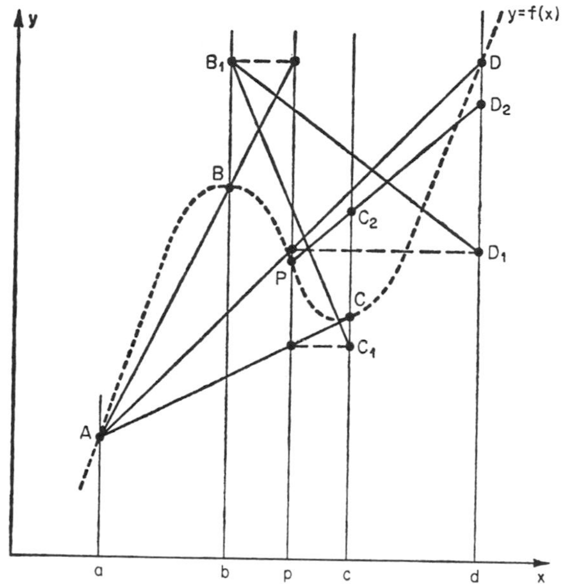

FIG. 7.1. Aitken's algorithm.

where f is a polynomial of degree n determined by its values at the distinct points $x_0, x_1, \ldots, x_n$. Consider

$$f^{(1)}(x) = \frac{\begin{vmatrix} f(x_0) & x_0 - p \\ f(x) & x - p \end{vmatrix}}{x - x_0} \qquad x \neq x_0$$

We observe that $f^{(1)}(x)$ is a polynomial of degree $n - 1$ and that $f^{(1)}(p) = f(p)$. Hence the problem is equivalent to that of evaluating $f^{(1)}(p)$, where $f^{(1)}$ is determined by its values at $x_1, x_2, \ldots, x_n$. Repetition of this process according to the scheme

$$\begin{array}{lllll}
x_0 & f(x_0) & & & \\
x_1 & f(x_1) & f^{(1)}(x_1) & & \\
x_2 & f(x_2) & f^{(1)}(x_2) & f^{(2)}(x_2) & \\
\cdot & \cdot & \cdot & \cdot & \\
\cdot & \cdot & \cdot & \cdot & \\
\cdot & \cdot & \cdot & \cdot & \\
x_n & f(x_n) & f^{(1)}(x_n) & f^{(2)}(x_n) & \cdots \quad f^{(n)}(x_n) = f(p)
\end{array}$$

leads to the determination of $f(p)$.

We have already noted that the Aitken algorithm does not require the abscissas to be equally spaced. It can therefore be applied to carry out inverse interpolation. However, a word of caution is necessary here. Unless $n = 2$, it is not true that, if n-point interpolation is permissible, then n-point *inverse* interpolation is permissible. This is best illustrated by an example. Consider the determination of the zero of $f(x)$, where

$$f(0) = -342 \quad f(1) = -218 \quad f(2) = 386 \quad f(3) = 1854$$

Actually $f(x) = (4x + 1)^3 - 343$, so that 4-point direct interpolation is exact, and the zero is 1.5. But a 4-point inverse interpolation gives 1.9926.

−342	0				−342				
−218	1	2.7581			−218	124			
386	2	.9396	2.1018		386	728	604		
1854	3	.4672	.5171	1.9926	1854	2196	2072	1468	

The numbers on the right are the divisors used. In many cases of direct interpolation it is not necessary to record them explicitly.

In general, the safest way to do inverse interpolation is to subtabulate the original table, for example, by Lagrangian methods, in the neighborhood of the point in question until linear interpolation is possible.

We conclude this section with a remark on the problem of the determination of the position $\bar{x}$, of the minimum or maximum of a tabulated function, or, more generally, where a tabulated function has an assigned derivative; this has been discussed thoroughly by Salzer. We discuss directly a 3-point method. We fit a parabola $y = a + bx + cx^2$ to $f(x)$ at $x = -1$, $x = 0$, $x = 1$. Then

$$a = f(0) \qquad b = \tfrac{1}{2}[f(1) - f(-1)] \qquad c = \tfrac{1}{2}[f(1) + f(-1)]$$

We obtain for the abscissa and ordinate of its vertex the expressions

$$\bar{x} \doteqdot \frac{-b}{2c} = -\frac{1}{2}\frac{f(1) - f(-1)}{f(1) - 2f(0) + f(-1)}$$

and

$$f(\bar{x}) \doteqdot \frac{4ac - b^2}{4c} = f(0) - \frac{1}{8}\frac{[f(1) - f(-1)]^2}{f(1) - 2f(0) + f(-1)}$$

Finite Differences. We now turn to the consideration of finite-difference methods.

The formation of a table of successive differences of a function (tabulated at equal intervals in the argument) is indicated below in the case of a table of cubes. The notation is as follows: If we write $f(n) = f_n = F(a + nh)$, then we define $\Delta^r F = \Delta_n{}^r F$ to be $\Delta^r f$, where

$$\Delta f(n) = f(n + 1) - f(n) \qquad \Delta^r f(n) = \Delta\Delta^{r-1} f \qquad r > 1$$

This gives

$$\begin{aligned}\Delta^2 f(n) = \Delta\Delta f(n) &= [f(n + 2) - f(n + 1)] \\ &\qquad - [f(n + 1) + f(n)] \\ &= f(n + 2) - 2f(n + 1) + f(n)\end{aligned}$$

We find, using the standard notation for binomial coefficients,

$$\begin{aligned}\Delta^r f(n) &= f(n + r) \\ &\quad - \binom{r}{1} f(n + r - 1) + \binom{r}{2} f(n + r - 2) \\ &\quad + \cdots + (-1)^{r-1}\binom{r}{1} f(n + 1) + (-1)^r f(n)\end{aligned}$$

x	$f(x)$	Δf	$\Delta^2 f$	$\Delta^3 f$	$\Delta^4 f$
1	1				
		7			
2	8		12		
		19		6	
3	27		18		0
		37		6	
4	64		24		0
		61		6	
5	125		30		0
		91		6	
6	216		36		0
		127		6	
7	343		42		0
		169		6	
8	512		48		0
		217		6	
9	729		54		
		271			
10	1000				

We shall also find it convenient to use the central-difference notation. This is indicated in the following diagram:

$$
\begin{array}{lllll}
f_{-2} & & & & \\
 & \Delta_{-2} = \delta_{-3/2} & & & \\
f_{-1} & \mu_{-1} & \Delta_{-2}{}^2 = \delta_{-1}{}^2 & & \\
 & \Delta_{-1} = \delta_{-1/2} & \mu_{-1/2}{}^2 & \Delta_{-2}{}^3 = \delta_{-1/2}{}^3 & \\
f_0 & \mu_0 & \Delta_{-1}{}^2 = \delta_0{}^2 & \mu_0{}^3 & \Delta_{-3}{}^4 = \delta_0{}^4 \\
 & \Delta_0 = \delta_{1/2} & \mu_{1/2}{}^2 & \Delta_{-1}{}^3 = \delta_{1/2}{}^3 & \\
f_1 & \mu_1 & \Delta_0{}^2 = \delta_1{}^2 & & \\
 & \Delta_1 = \delta_{3/2} & & & \\
f_2 & & & &
\end{array}
$$

The averaging operator μ is defined by

$$\mu f(r) = \tfrac{1}{2}[f(r - \tfrac{1}{2}) + f(r + \tfrac{1}{2})]$$

and the central-difference operator by

$$\delta f(r) = f(r + \tfrac{1}{2}) - f(r - \tfrac{1}{2})$$

We shall have occasion to use the differentiation operator D, the promotor operator E which is defined by

$$Ef(r) = f(r + 1)$$

and the backward difference operator ∇ which is defined by

$$\nabla f(r) = f(r) - f(r - 1)$$

An operator form of Taylor's theorem can be written as

$$Ef_n = [1 + hD + h^2D^2/2! + h^3D^3/3! + \cdots]f_n$$

which suggests the relations

$$E = \exp hD \qquad hD = \ln E = \log(1 + \Delta)$$

The table above illustrates the following fundamental fact:

The nth differences (at any constant interval) of a polynomial of degree n are constant, and all succeeding differences vanish.

This is readily established by noting that the operation of differencing reduces the degree of a polynomial by unity:

$$\Delta\left(\sum_{r=0}^{n} a_r x^r\right) = \sum_{r=0}^{n} a_r\,\Delta x^r = \sum_{r=0}^{n} a_r[(x + h)^r - x^r]$$

$$= \sum_{r=0}^{n} a_r(rhx^{r-1} + \cdots) = \sum_{r=0}^{n-1} b_r x^r$$

Only rarely in practice do we have to deal with exact polynomials. The effect of rounding off in a difference table is shown by the following table of $.1x^3$ rounded to the nearest integer. The difference that we expect to be constant is oscillatory (and higher differences will get larger).

x	$f(x)$	Δf	$\Delta^2 f$	$\Delta^3 f$	$\Delta^4 f$
1	0				
		1			
2	1		1		
		2		0	
3	3		1		2
		3		2	
4	6		3		−1
		6		1	
5	12		4		−3
		10		−2	
6	22		2		5
		12		3	
7	34		5		−3
		17		0	
8	51		5		0
		22		0	
9	73		5		
		27			
10	100				

Let us see what happens in a bad case—for example,

$$f(x) = \tfrac{1}{2} - (-)^n a$$

for a small. The difference tables for $f(x)$ and for $f(x)$ rounded to the nearest integer are as follows:

x	$f(x)$	Δf	$\Delta^2 f$	$\Delta^3 f$	$\Delta^4 f$	x	$f(x)$	Δf	$\Delta^2 f$	$\Delta^3 f$	$\Delta^4 f$
0	$\frac{1}{2} - a$					0	0				
		$2a$						1			
1	$\frac{1}{2} + a$		$-4a$			1	1		−2		
		$-2a$		$8a$				−1		+4	
2	$\frac{1}{2} - a$		$+4a$		$-16a$	2	0		+2		−8
		$2a$		$-8a$				1		−4	
3	$\frac{1}{2} + a$		$-4a$			3	1		−2		
		$-2a$						−1			
4	$\frac{1}{2} - a$					4	0				

This shows that there may be a spurious contribution of up to 2^{n-1} units in the nth difference, because of rounding off of the function values to the nearest unit.

The effect of a change in the length of the (constant) interval of differencing is easily determined: The constant nth difference of x^n at unit interval is $n!$, whereas for an interval h, the nth difference is $h^n n!$.

The formation of a table of differences is useful in checking a table or locating errors. The way in which an error propagates is indicated in a repetition of the above table of cubes, with a deliberate error at $x = 6$, where we have written 217 instead of 216. The pattern of $(1 - 1)^4$ in the fourth column is characteristic. In practice, however, the exact binomial pattern is more or less obscured by rounding errors, but it is usually easy to pick out errors. even when several are present and interfering.

1	1				
		7			
2	8		12		
		19		6	
3	27		18		0
		37		6	
4	64		24		+1
		61		7	
5	125		31		−4
		92		3	
6	217		34		+6
		126		9	
7	343		43		−4
		169		5	
8	512		48		+1
		217		6	
9	729		54		0
		271		6	
10	1000		60		
		331			
11	1331				

The binomial pattern can also be established by noting that differencing is a linear operation, that the effect of the error is just that of the differences of 0, 0, 0, 0, 0, 1, 0, 0, 0, 0, 0, and that these come immediately from the explicit formula for Δ^n.

So far we have mainly discussed the differences of polynomials. This is almost enough, for we are concerned with polynomial interpolation, which is applicable only when the function in question is approximately representable by a polynomial. If $f(x)$ is differentiable, we have

$$\Delta f(x) = f(x + h) - f(x) = hf'(x + \theta h) \qquad 0 \leq \theta \leq 1$$

Thus, if f' is continuous and h is small, we have

$$\Delta f(x) \doteqdot hf'(x)$$

More generally, we can show that, if $f(x)$ has a continuous nth derivative and h is sufficiently small,

$$\Delta^n f(x) \doteqdot h^n f^{(n)}(x)$$

More precisely, if $f^{(n)}(x)$ is continuous then

$$\Delta^n f(a) = h^n f^{(n)}(\xi) \qquad a \leq \xi \leq a + nh$$

This can be proved by induction.

Interpolation Using Differences. The literature discussing methods of interpolation using finite differences is enormous. We shall confine our discussion to what is probably the most useful one, due to Everett. The 4-point Lagrangian interpolant

$$f(p) = L_{-1}(p)f(-1) + L_0(p)f(0) + L_1(p)f(1) + L_2(p)f(2)$$

can be rearranged as

$$f(p) = \{(1 - p)f(0) + pf(1)\} + \{E_2\delta^2 f(0) + F_2\delta^2 f(1)\}$$

where

$$E_2(p) = F_2(1 - p) = \frac{-p(p - 1)(p - 2)}{3!}$$

If we have tables of E_2 and F_2 and a table of f and its second differences, interpolation is easy. For instance, we evaluate sin 1.234, given the following data:

x	$\sin x$	δ^2
1.2	.9320	−371
1.4	.9854	−392

The Everett coefficients for $p = .17$ are $E_2 = -.0430$, $F_2 = -.0275$, and we find

$$\begin{aligned}\sin 1.234 &= (.83 \times .9320 + .17 \times .9854) \\ &\quad + (.0430 \times .0371 + .0275 \times .0392) \\ &= .9411 + .0027 \\ &= .9438\end{aligned}$$

There are extensions of the above formula to give an Everett form of a 6-point Lagrangian interpolant:

$$f(p) = \{(1 - p)f(0) + pf(1)\} + \{E_2\delta^2 f(0) + F_2\delta^2 f(1)\} + \{E_4\delta^4 f(0) + F_4\delta^4 f(1)\}$$

and so on. One of the advantages of this method over the Lagrangian method is that the size of the successive contributions in the braces is a usually reliable indication of the appropriate formula to use. We consider the same problem with the data on the left.

x	$\sin x$	δ^2	δ^4
.0	.0000	0	0
.5	.4794	−1173	285
1.0	.8415	−2061	507
1.5	.9975	−2442	597
2.0	.9093	−2226	544

x	$\sin x$	δ_m^2
.0	.0000	0
.5	.4794	−1225
1.0	.8415	−2154
1.5	.9975	−2552
2.0	.9093	−2326

The Everett coefficients for $p = .468$ are $E_2 = -.0636$, $F_2 = -.0609$, $E_4 = .0118$, $F_4 = .0115$, and the calculation gives

$$\sin 1.234 = .9145 + .0280 + .0013 = .9438$$

We shall now discuss the "throwback." It was observed that the ratio

$$\frac{E_4}{E_2} = \frac{p^2 - 2p - 3}{20}$$

is approximately constant for $0 \leq p \leq 1$. (The reader should draw a rough graph of the quadratic.) If, therefore, a reasonable mean value k of this ratio is chosen, we can include the fourth-difference contribution by "modifying" the second difference:

$$\delta_m{}^2 f = \delta^2 f + k\delta^4 f$$

Various ways of choosing k have been discussed, and the preferred value is $k = -.18393$.

In practice, the use of this idea is along the following lines: The tablemaker computes f, $\delta^2 f$, $\delta^4 f$ but prints only f, $\delta_m{}^2 f$. If the user treats the $\delta_m{}^2 f$ just as he treats $\delta^2 f$, he receives a bonus, the major part of the contribution of $\delta^4 f$, without having to obtain the E_4, F_4 or to do the multiplications and additions. The actual error incurred by the modification is less than half a unit if the fourth differences are less than 1000 and the fifth difference less than 70.

If we return to the case of a $4D$ table of sin x, we now see that an interval of $h = .5$ will practically suffice. That is to say, the table on the right above will be adequate for interpolation to an accuracy of the order of the rounding error in the data. It is to be compared with the table at interval .2 which is required when the straight second-order Everett process is used or to the interval .02 which is required when only linear interpolation is permitted.

The calculation now reads

$$\begin{aligned} \sin 1.234 &= \{.532 \times .8415 + .468 \times .9975\} \\ &\quad + \{.2154 \times .0636 + .2552 \times .0609\} \\ &= .9145 + .0292 \\ &= .9437 \end{aligned}$$

Actually, $\sin 1.234 = .943818$, and the error can be traced to the marginal choice of h and/or roundings.

Comparison of Methods. There is a violent transatlantic controversy about the methods discussed above. On the whole, Americans favor the Lagrangian methods and Europeans the difference methods. We shall mention some of the arguments.

There is no doubt that the Lagrangian method should be used only for a guaranteed table. The calculation of the differences required for the other methods provides a check on the reliability of a new or doubtful table and, the differences having been obtained, the use of the difference methods is usually less laborious. There is, however, the question of what order of interpolation to use. Only in a few tables is this given, and so, if one does not want to compute differences, one is forced to carry out more than one interpolation and to observe the behavior of the interpolant; for this the Aitken scheme is efficient. Since the sum of the Lagrangian coefficients is necessarily unity, one can provide a good check in desk computations by accumulating the multiplier as well as the product. This check is not available in the finite-difference methods (except in the linear case), where the varying orders of magnitude of the multipliers and the differences encourage wrong settings in the calculator.

Auxiliary Functions. The error estimate in the Lagrangian interpolation shows that there will be difficulty in the neighborhood of a pole of the function or of one of its derivatives. This is illustrated simply by the case of cosec x near $x = 0$, the values of which are given below. However, the functions cosec $x - x^{-1}$ and x cosec x are well behaved near $x = 0$, both analytically and numerically, as indicated by their differences. To interpolate for cosec x near $x = 0$ we interpolate for either of the auxiliary functions and then, using tables of x^{-1} or dividing, we can readily obtain the required value of cosec x.

x	cosec x	cosec $x - x^{-1}$			x cosec x			
0	∞	0			1.000000			
			833			4		
.005	200.001	.000833		1	1.000004		7	
			834			13		1
.010	100.002	.001667		−1	1.000017		8	
			833			21		0
.015	66.6692	.002500		0	1.000038		8	
			833			29		0
.020	50.0033	.003333		1	1.000067		8	
			834			37		1
.025	40.0042	.004167		0	1.000104		9	
			834			46		
.030	33.3383	.005001			1.000150			

The choice of appropriate auxiliary functions is usually determined by the analytic nature of the original function or by asymptotic expansions for it. We give a few examples, referring for details to the better tables, in particular to the National Bureau of Standards "Handbook of Functions."

For arcsin x near $x = 1$, it proves convenient to tabulate arcsin $x + \sqrt{2(1 - x)}$, so that the desired function is obtained by reference to a square-root table.

If we define $A_0(x)$, $B_0(x)$ by

$$J_0(x) = A_0(x) \sin x + B_0(x) \cos x$$

it will be found that $A_0(x)$, $B_0(x)$ are much easier to tabulate than $J_0(x)$ for x large, and $J_0(x)$ can be recovered by reference to tables of sin x and cos x.

Consideration of the Stirling asymptotic expansions (cf. Part 1, Chap. 3) suggests that

$$f_1(x) = \Gamma(x)/\{x^{x-\frac{1}{2}}e^{-x}(2\pi)^{\frac{1}{2}}\}$$

would be nearly linear in x^{-1}. It is found that a table of $f_1(x)$ for $x^{-1} = 0(.001).01$, together with appropriate auxiliary tables, enables $\Gamma(x)$ to be readily obtained to about $8D$ for $x \geq 100$.

Finally (cf. Part 1, Chap. 3), we find that

$$\mathrm{Ei}(x) = \int_{-\infty}^{x} t^{-1}e^{-t}\, dt$$

can be evaluated for large x by way of

$$T(x) = e^{-x}\mathrm{Ei}(x) - x^{-1}$$

Multivariate Interpolation. Much of the preceding material on univariate interpolation can be extended to the multivariate case. The naïve approach to the problem of finding $f(p,q)$ in a double-entry table is to do successive univariate interpolation, that is, to find $f(0,q)$, $f(1,q)$, . . . by interpolation in the y direction and obtain $f(p,q)$ by interpolation in the x direction among these values. If much interpolation is to be done, more efficient methods must be sought. In good tables, advice for interpolation is given, and where interpolation is thought likely to be necessary, the table is planned to make it as convenient as possible.

An important special case of multivariate interpolation is in tables of regular functions of a complex variable. Here use of the Cauchy-Riemann equations can greatly facilitate matters. We refer to the introductions of the many recent tables of this character for various approaches.

Special methods for functions of particular form are often convenient. For instance, in the case of functions defined by integrals, Gaussian quadratures may be applied.

The Theory of Approximation. An automatic (digital) computer cannot deal directly with the usual special functions. These are usually introduced into the computer by means of approximations, either as a polynomial or as a rational function (the ratio of two polynomials). (In some cases it is preferable to introduce a skeleton table of the function and efficient means for interpolation in this table.)

A classical theorem of Weierstrass states that a continuous function $f(x)$ of a real variable x, $0 \leq x \leq 1$, can be represented as the limit of a uniformly convergent sequence of polynomials. A constructive version of this was given by S. Bernstein: If

$$B_n(f,x) = \sum_{k=0}^{n} f\left(\frac{k}{n}\right)\binom{k}{n} x^k(1-x)^{n-k}$$

then $\quad B_n(f,x) \rightarrow f(x)$, uniformly in $[0,1]$

Unfortunately these Bernstein polynomials do not converge very rapidly. Indeed we have, in general, if $f''(x)$ exists,

$$|B_n(f,x) - f(x)| = O(n^{-1})$$

The question arises whether it is possible to find readily, for a given function $f(x)$, in $[0,1]$ say, and a given integer n, a polynomial $\hat{p}_n(x)$ that is a best approximation to $f(x)$ in the sense that

$$\max_{0 \leq x \leq 1} |f(x) - \hat{p}_n(x)|$$

is least over all polynomial $p_n(x)$ of degree $\leq n$. Chebyschev showed that there is a unique such polynomial, but no finite algorithm is yet known that will generate this for general $f(x)$. Similar results are available for rational functions and for functions of several real variables, but in this case uniqueness is not guaranteed.

There are two classical cases in which the best approximation is known.

1. If $f(x) = x^{n+1}$ and the interval is $[-1,1]$, then

$$\hat{p}_n(x) = 2^{-n}T_n(x)_{+1} - x^{n+1}$$

where $T_n(x)$ is the Chebyshev polynomial defined in Part 1, Chap. 3; the error is 2^{-n}.

2. If $f(x) = (1+x)^{-1}$ and the interval is $[0,1]$, then

$$\hat{p}_n(x) = \sqrt{2}\,\{[\tfrac{1}{2} - cT_1(2x-1) + \cdots + (-1)^{n-1}c^{n-1}T_{n-1}(2x-1)] + (-1)^n c^n(1-c^2)^{-1}T_n(2x-1)\}$$

where $c = 3 - 2\sqrt{2}$, and the error is $c^n/4$.

Chebyshev showed that $\hat{p}_n(x)$ was characterized by the equal-ripple property: $e_n(x) = f(x) - \hat{p}_n(x)$ assumes its extreme values in $[-1,1]$ at least $n+2$ times there, with consecutive extrema having opposite sign. This is easily verified in case 1 above for

$$e_n(x) = 2^{-n}T_{n+1}(x) = 2^{-n}\cos[(n+1)\operatorname{arc}\cos x]$$

and $\max |e_n(x)| = 2^{-n}$ and

$$e_n\left(\cos\frac{r\pi}{n+1}\right) = (-1)^r 2^{-n} \qquad r = 0, 1, \ldots, n+1$$

It is not true that if we can obtain a valid expansion of the form

$$f(x) = \sum_{0}^{\infty} a_n T_n(x)$$

the segments of this give the best approximations $\hat{p}_n(x)$.

This can be verified in case 2 above when

$$(1+x)^{-1} = \sqrt{2}\,\{\tfrac{1}{2} + \sum_{n=1}^{\infty}(-1)^n c^n T_n(2x-1)\}$$

but truncating this at $n = 1$ we get

$$\tfrac{1}{2}(7\sqrt{2} - 8) - (6\sqrt{2} - 8)x$$

whereas $\quad \hat{p}_1(x) = \tfrac{1}{4} + \tfrac{1}{2}\sqrt{2} - \tfrac{1}{2}x$

The errors are .0503 and .0429, respectively.

Nevertheless truncated expansions of the form above which are really Fourier expansions

$$f(\cos\theta) = \Sigma a_n \cos n\theta$$

are usually good approximations and can be used as a base for various improvement processes.

There is a growing literature on the theory of approximation, from both the theoretical and practical standpoints; it includes lists of approximations of various functions, to various precisions over various ranges. There are also available various computer algorithms to generate good approximations.

Special Devices. The construction of processes that increase the speed of convergence of sequences and series has been a favorite topic for many numerical analysts. We shall discuss the δ^2 process which has been popularized in numerical analysis by Aitken [it dates back at least to Kummer (1837)] and the Euler summation process.

If $x_n \rightarrow x$ and $x_n - x \doteqdot A\lambda^n$ for some λ, $|\lambda| < 1$, then

$$\frac{x_{n+2} - x}{x_{n+1} - x} \doteqdot \frac{x_{n+1} - x}{x_n - x} \doteqdot \lambda$$

We find

$$x \doteqdot x_{n+2} - \frac{(x_{n+2} - x_{n+1})^2}{x_{n+2} - 2x_{n+1} + x_n}$$

This suggests that the sequence $\{\bar{x}_{n+2}\}$ defined by

$$\bar{x}_{n+2} = x_{n+2} - \frac{(x_{n+2} - x_{n+1})^2}{x_{n+2} - 2x_{n+1} + x_n} \qquad n = 0, 1, 2, \ldots$$

converges more rapidly to x than the original sequence. This is indeed the case; for if

$$x_n - x = A\lambda^n + O(\lambda^n) \qquad |\lambda| < 1$$

then it follows that

$$\bar{x}_n - x = O(\lambda^n)$$

Several remarks are in order. First, this process can be iterated to remove successively components in the remainder of the form

$$A\lambda^n,\ B\mu^n,\ C\nu^n,\ \ldots$$

where $1 > |\lambda| > |\mu| > |\nu| > \cdots$. The cases in which there are equalities such as $|\lambda| = |\mu|$ can be handled by simple modifications. Second, it is important to note that this process can make things worse if the Aitken hypothesis is not satisfied. Here is a simple example involving two of the standard iterative processes for determining the reciprocal of a number N. Consider the sequences

$$y_{n+1} = (1 - N)y_n + 1 \qquad z_{n+1} = z_n(2 - Nz_n)$$

In the case $N = \frac{1}{2}$ with $y_0 = 1$, $z_0 = 1$, we obtain the following table:

y_n			$\bar{y}_n$	z_n			$\bar{z}_n$
1				1			
	.5				.5		
1.5		−.25		1.5		−.125	
	.25				.375		
1.75		−.125	2	1.875		−.2578	3.0000
	.125				.1172		
1.875			2	1.9922			2.0455

The sequence $\{\bar{y}_n\}$ surely converges more rapidly than $\{y_n\}$ while the sequence $\{\bar{z}_n\}$ appears to converge less rapidly than $\{z_n\}$. These results can be easily established. First of all, each sequence converges to N^{-1} if $0 < N < 1$ for

$$y_n - N^{-1} = (1 - N)^n(y_0 - N^{-1})$$
$$z_n - N^{-1} = -N^{2^n-1}(z_0 - N^{-1})^{2^n}$$

Thus $\{y_n\}$ satisfies the Aitken hypothesis while $\{z_n\}$ does not, converging too rapidly.

There has been much recent activity in this area, e.g., by Shanks, Lubkin, Wynn, Henrici, and Bauer.

The Euler summation method is essentially a transformation of one infinite series into another. It is most convenient to exhibit it in the case of an alternating series. We proceed formally, using the finite-difference operators:

$$\begin{aligned} u_0 - u_1 + u_2 - \cdots &= (1 - E + E^2 - \cdots)u_0 = (1 + E)^{-1}u_0 \\ &= (2 + \Delta)^{-1}u_0 = \tfrac{1}{2}(1 + \tfrac{1}{2}\Delta)^{-1}u_0 \\ &= \tfrac{1}{2}u_0 - \tfrac{1}{4}\Delta u_0 + \tfrac{1}{8}\Delta^2 u_0 - \tfrac{1}{16}\Delta^3 u_0 + \cdots \end{aligned}$$

We show the efficacy of this by considering the evaluation of ln 2 from the power series:

$$\ln 2 = 1 - \tfrac{1}{2} + \tfrac{1}{3} - \cdots$$

It is possible to apply the Euler transformation directly to this series, but it is more convenient to apply it to the tail. We obtain

$$1 - \tfrac{1}{2} + \cdots - \tfrac{1}{8} = .63452\,3809$$

and we difference the sequence $\frac{1}{9}, \frac{1}{10}, \frac{1}{11}, \frac{1}{12}, \ldots$ thus:

.11111 1111					
	−.01111 1111				
.10000 0000		+.00202 0202			
	−.00909 0909		−.00050 5051		
.09090 9091		+.00151 5151		+.00015 5402	
	−.00757 5758		−.00034 9649		−.00005 5505
.08333 3333					
⋮					

The next leading differences are $+.00002\,2212$, $-.00000\,9740, \ldots$. Using all of these, we find

$$\sum_{9}^{\infty} (-1)^{n-1}/n = .05555\,5556 + .00277\,7778 + .00025\,2525 + .00003\,1566 + .00000\,4856 + .00000\,0867 + .00000\,0174 + .00000\,0038$$

giving $\qquad \ln 2 = .69314\,7169$

which we compare with the true value

$$\ln 2 = .69314\,7181$$

We remark that the transformed series always converges if the original one does, but not necessarily more rapidly.

We introduced the concept of asymptotic series in Part 1, Chap. 3 and indicated earlier in this section how they can be used to suggest auxiliary functions and so facilitate the tabulation problem. We now illustrate, by a numerical example, the direct use of asymptotic series in the evaluation of functions. We continue with the example

$$f(x) = -e^{-x}\text{Ei}(-x) = \int_x^\infty t^{-1}e^{x-t}\,dt \sim \sum_0^\infty (-1)^n n!\,x^{-n-1}$$

discussed in Chap. 3. We have seen that for a fixed x the estimate for $f(x)$ given by segments of the series improves until about x terms have been used and then worsens. If x is such that the limited accuracy obtainable by an asymptotic series is adequate, this method is usually the most convenient one. We have also noted that this accuracy increases as x does. We shall examine the situation for $f(x)$, with $x = 15$. We record on the left the multiplier $n/x = n/15$, which produces the $(n + 1)$st term from the nth. In the center the actual terms are recorded, and on the right the partial sums.

.06	.06666 66667	.06666 66667
.13	−.00444 44444	.06222 22223
.2	+.00059 25926	.06281 48149
.26	−.00011 85185	.06269 62964
.3	+.00003 16049	.06272 79013
.4	−.00001 05350	.06271 73663
.46	+.00000 42140	.06272 15803
.53	−.00000 19665	.06271 96138
.6	+.00000 10488	.06272 06626
.6	−.00000 06293	.06272 00333
.73	+.00000 04195	.06272 04528
.8	−.00000 03077	.06272 01451
.86	+.00000 02461	.06272 03912
.93	−.00000 02133	.06272 01779
1.	+.00000 01991	.06272 03770
1.06	−.00000 01991	.06272 01779
1.13	+.00000 02124	.06272 03903
	−.00000 02407	.06272 01496

Our arguments suggest that the best estimate for $f(x)$ is that given on the fifteenth line.

The value given by Coulson and Duncanson, differing from ours in the seventh place, is

$$.00000\ 00191\ 8628 \ldots \times 3269017.372 \ldots = .06272\ 028 \ldots$$

Experiments showed that the Euler process applied to asymptotic series gave sensible results. We find, for instance, that the sum of the first four terms in $f(15)$ is 0.06269 62966 and that the Euler sum of the tail is .00002 39226, which gives

$$f(15) = .06272\ 02190$$

If we sum the first eight terms before applying the Euler process, we find

$$f(15) = .06272\ 02790$$

These processes were justified by Rosser.

Another classical device, Richardson's h^2 extrapolation, or deferred approach to the limit, is discussed in Sec. 8 in connection with the Romberg process.

5. The Solution of Polynomial and Transcendental Equations

From the point of view of the computer, there is little to distinguish between these two classes of equations. On the one hand, transcendental equations are usually replaced by the polynomial ones; on the other, classical properties of polynomial equations may no longer hold in the computer context, for example, on a 32-bit machine, working in fixed-point mode, the Chebyshev polynomial $T_{34}(x)$ should be theoretically zero for *all* x in $[-1,1]$ since $|T_{34}(x)| \leq 2^{-3}$. What values actually turn up depend on details of the hardware and the computing algorithm. It is, however, convenient to separate the two cases. We deal first, and mainly, with the polynomial case.

The problems encountered can be separated into the following:

1. To find *the* zero of $f(z)$ near z_0; this assumes that the zero in question has been "isolated."
2. To find *all* zeros of $f(z)$.
3. To find the number of zeros of $f(z)$ in some region of the complex plane.

Clearly a solution of problem 2 implies solutions of problems 1 and 3 but not necessarily efficient ones.

It is necessary in the beginning to decide whether the problems have any significance in the round-off context. This is settled by a theorem on the continuity of zeros of a polynomial as functions of the coefficients.

If
$$f(z) = \sum_{j=0}^{n} a_j z^j,$$
and
$$g(z) = \sum_{j=0}^{n} b_j z^j$$
where $a_0, a_n \neq 0$, are such that there is a $\tau \neq 0$ such that

$$|b_j - a_j| \leq \tau |a_j| \qquad j = 0, 1, 2, \ldots, n$$

then the zeros α_i of f and β_i of g can be so ordered that

$$|(y_j|x_j) - 1| < 8n\tau^{1/n} \qquad j = 1, 2, \ldots, n$$

Two remarks are called for here. First, if we take $f(z) = (z - 1)^n$ and $g(z) = f(z) - \epsilon$, we find that the estimate is near the truth in this case. The example given in Sec. 2 shows that the situation can be bad in the case of well-separated zeros as well as in this extreme case. On the other hand, determination of the relative error in the coefficients which will ensure a relative error of at most 1 per cent in the zeros, say in the case $n = 10$, shows how unpractical this is. The purport of these remarks is to indicate the desirability of examining experimentally, in each particular case on the computer, the stability of the position of the zeros as functions of the coefficients.

1. *Iterative determination of the zero near* z_0. Instead of solving $f(z) = 0$ we may solve

$$z = g(z) \qquad \text{where } g(z) = z - h(z)f(z)$$

Here $h(z)$ is some analytic function not vanishing in the neighborhood of z_0. We then consider the iteration based on the given z_0:

$$z_{i+1} = g(z_i) \qquad i = 0, 1, 2, \ldots$$

It is easy to verify that the sequence $\{z_i\}$ will converge to the desired zero z^* if $|g'(z)| < 1$ for all z in some neighborhood of z^*; indeed, if $g'(z^*) = 0$ the condition $|g'(z)| < 1$ is not needed, and convergence is assured if z_0 is sufficiently close to z^*.

Two classical methods can be exhibited in this form. We begin with Newton's method which is illustrated geometrically in the real domain in Fig. 7.2. Clearly

$$z_{n+1} = z_n - f(z_n)/f'(z_n)$$

so that in the general framework $g(z) = z - f(z)/f'(z)$. We therefore have

$$g'(z) = \frac{f(z)f''(z)}{[f'(z)]^2}$$

which vanishes at z^*. Hence we always have convergence if z_0 is sufficiently close to z^*. Observe that convergence is quadratic:

$$z_{i+1} - z^* = O[(z_i - z^*)^2]$$

This is satisfactory in practice. However, there is a drawback to the Newton process: If $f(z)$ has real coefficients and z_0 is real, all succeeding z_i will be real and we cannot get to a complex zero.

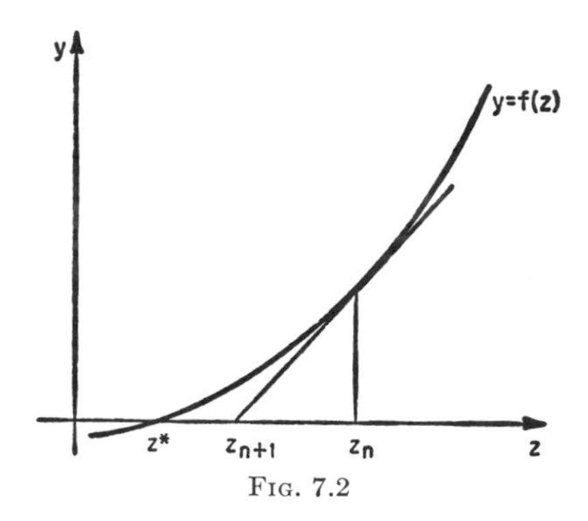

FIG. 7.2

The second method, due to Laguerre, does not have this drawback and has cubic convergence but is rather more complicated. In it

$$g(z) = z - \frac{nf(z)}{f'(z) \pm \sqrt{(n-1)[(n-1)f'(z)^2 - nf(z)f''(z)]}}$$

The ambiguous sign is chosen so that the denominator has the larger of its two values. This method is sometimes incorporated in eigenvalue algorithms as a rootfinder for the "concentrated" matrix.

Another method which has proved satisfactory in practice and has been analyzed thoroughly by Ostrowski is due to Muller. The geometric interpretation of this is that from any three approximations z_0, z_1, z_2 to the zero z^* we obtain another, z_3, by approximating the function $f(z)$ by the quadratic interpolating $f(z)$ at z_0, z_1, z_2, and finding the appropriate zero of this quadratic.

[This method is a generalization of a 2-point method—the "secant" method or rule of false position (Newton's method is the "tangent" method)—and can clearly be generalized to an m-point method.]

This method has convergence of order of about 1.84. Formally, it appears somewhat simpler than, for example, the Newton method since one works with functional values alone and the derivatives are not required; this is not a significant advantage in good practice.

2. *Determination of all the zeros.* When a zero has been obtained, we can "divide out" and obtain a "deflated" polynomial of degree $n - 1$. Several remarks are required. First, the zero obtained might be a multiple one, and appropriate action must be taken. It is possible, by finding the highest common factor of $f(z)$ and $f'(z)$—which is a rational process—to obtain a polynomial that has only simple zeros at the zeros of $f(z)$. There is, of course, the practical problem of deciding between, for example, a genuine double zero and two close simple zeros. Second, each time we deflate we are likely to contaminate the residual polynomial more and more. In general, each problem must be treated individually.

3. *The number of zeros of $f(z)$ in a region R.* This can, in principle, be evaluated by determining

$$\frac{1}{2\pi i} \int_C \frac{f'(z)}{f(z)}\, dz$$

numerically, where C is the boundary of R; only a very rough approximation to the integral is needed since we know it must be an integer.

In aspects of stability investigations, one is concerned with the cases when R is the interior of a unit circle or the left-hand half plane and when it is hoped to find all zeros in R. These two problems are essentially equivalent in view of the Cayley transformation

$$w = \frac{1-z}{1+z}$$

which maps $|z| < 1$ into $w < 0$; accordingly we deal with the second one only. Here the classical result is the Routh-Hurwitz criterion: A necessary and sufficient condition for all the zeros of

$$a_0 z^n + \cdots + a_n$$

to have negative real parts is that a_0 and all the n determinants

$$|a_1|, \quad \begin{vmatrix} a_1 & a_3 \\ a_0 & a_2 \end{vmatrix}, \quad \begin{vmatrix} a_1 & a_3 & a_5 \\ a_0 & a_2 & a_4 \\ 0 & a_1 & a_3 \end{vmatrix}, \quad \ldots, \quad \begin{vmatrix} a_1 & a_3 & a_5 & \cdots & 0 \\ a_0 & a_2 & a_4 & \cdots & 0 \\ 0 & a_1 & a_3 & \cdots & 0 \\ 0 & a_0 & a_2 & \cdots & 0 \\ & & & & \vdots \\ & & \cdots & & a_n \end{vmatrix}$$

should be positive.

In many cases R is an interval on the real axis; for example, if $f(z)$ is the characteristic polynomial of a symmetric matrix, all its zeros are necessarily real. In this case we can make use of the Sturmian theory. In the simplest case this theory gives the following result: The number of zeros in $[a,b]$ is the excess of the number of changes of sign in the sequence

$$f(a),\ f'(a),\ f_2(a),\ \ldots,\ f_n(a)$$

over the number of changes of sign in the sequence

$$f(b),\ f'(b),\ f_2(b),\ \ldots,\ f_n(b)$$

In the theoretical version the sequence of polynomials (each having degree one less than its predecessor) is computed by a highest common-factor algorithm applied to its two predecessors. In many practical cases only the actual values displayed above are computed, the polynomials never appearing explicitly. Obviously, difficulties can occur in practice: The signs of the numbers may be doubtful, but this can be because, for example, a was chosen to be near a zero. Again calculations of this type need careful monitoring.

Efficient algorithms of this kind can lead to solutions of problem 2. Granted that we know that all the real roots of $f(z)$ lie in $[a,b]$, we proceed along following lines. We next apply the algorithm to $[a, \frac{1}{2}(a + b)]$ and to $[\frac{1}{2}(a + b), b]$. If there are no zeros in one of these we proceed to bisect the other. If zeros are in both, we bisect both. Continuing in this way, we can rapidly isolate the zeros and then obtain good approximations to each by use of some of the methods described in 1 above.

6. Systems of Linear Algebraic Equations and Related Problems

The solution of a system of equations, which is written in matrix vector form as

$$A\mathbf{x} = \mathbf{b}$$

is a fundamental one in numerical analysis. The determinantal method, known as Cramer's rule, is easily seen to be impracticable for large systems, even with efficient methods for evaluating the determinant. However, a practical method for the solution of a system of linear equations is the elimination process.

Gaussian Elimination Process. Consider the system

$$\begin{array}{rr} & \text{check sum} \\ x + 2y + 3z + 4w = 30 & -20 \\ 2x + 3y + 4z + 5w = 40 & -26 \\ 3x + 4y + 4z + 5w = 43 & -27 \\ 4x + 5y + 5z + 8w = 61 & -39 \end{array}$$

Subtract multiples of the first equation from the other three so as to eliminate x in each of these, giving

$$\begin{array}{rr} -y - 2z - 3w = -20 & 14 \\ -2y - 5z - 7w = -47 & 33 \\ -3y - 7z - 8w = -59 & 41 \end{array}$$

Now subtract multiples of the first of these from the other two so as to eliminate y:

$$\begin{array}{rr} -z - w = -7 & 5 \\ -z + w = 1 & -1 \end{array}$$

Finally eliminate z by subtracting the first of these from the second to get

$$\begin{array}{rr} 2w = 8 & -6 \\ w = 4 & -3 \end{array}$$

We now back-substitute as follows to get z, y, and x:

$$\begin{array}{rr} -z - 4 = -7 & z = 3 \\ -y - 2 \times 3 - 3 \times 4 = -20 & y = 2 \\ x + 2 \times 2 + 3 \times 3 + 4 \times 4 = 30 & x = 1 \end{array}$$

This method applies whether the system of equations is symmetric or not. It is important, in practice, to reorder the equations and the variables, if necessary, in order to ensure that the "pivotal coefficients" are not too small. This process is called "pivoting" or "positioning for size." An extreme case of this is a system such as

$$\begin{aligned} 0x + y &= 2 \\ x + y &= 3 \end{aligned}$$

If we have a system with a relatively small coefficient of x in the first equation, we have to take a large multiple of this from the second; consequently any uncertainty in the coefficient of y will be amplified and affect the determination of y.

A useful check in processes of this kind is provided by carrying along, as indicated in the extra column on the right, the sum of the coefficients in the equations, operating on these numbers as on the equations, and then checking that the derived equations have the correct sum.

We shall now outline a matrix presentation of the elimination method. It will be convenient throughout the rest of this chapter to reserve the letter L for lower triangular matrices and U for upper triangular matrices. We have the following result of Cholesky. If all the leading submatrices of a matrix A are nonsingular, then $A = LU$. This is readily proved by induction, using partitioning of A. That the condition of nonsingularity is essential is shown by the case of the matrix $\begin{bmatrix} 0 & 1 \\ 1 & 0 \end{bmatrix}$; however, the required situation can be brought about by a simultaneous permutation of the rows and columns of A if $\det A \neq 0$. The proof of the result shows that L and U are essentially unique; specifying the diagonal elements of L (or of U) will make them so. In some cases it is convenient to state the result in the form

$$A = LDU$$

where L, U have unit diagonals and D is a diagonal matrix. In the case when A is symmetric, $L = U'$ and there is uniqueness up to the sign of the diagonal elements; this decomposition is the essence of the "square-root" method.

An important feature of this decomposition is the fact that it can be obtained without solving any simultaneous equations. We illustrate this in a symmetric case. Assuming

$$A_8 \equiv \begin{bmatrix} 1 & 2 & 3 \\ 2 & 3 & 4 \\ 3 & 4 & 4 \end{bmatrix} = \begin{bmatrix} a & 0 & 0 \\ b & c & 0 \\ d & e & f \end{bmatrix} \begin{bmatrix} a & b & d \\ 0 & c & e \\ 0 & 0 & f \end{bmatrix}$$

we equate elements in the first row and obtain $a^2 = 1$, $ab = 2$, $ad = 3$ from which, assuming $a = +1$, we find $b = 2$, $d = 3$. On equating elements in the second row, $b^2 + c^2 = 3$ and $bd + ce = 4$, which gives $c = +i$, say, and $e = 2i$. Finally, we obtain

$$d^2 + e^2 + f^2 = 4$$

giving $f = +1$, say.

If A is positive definite, L is real, and conversely. This remark leads to a convenient test of the definiteness of a matrix. An operation count shows that this decomposition involves n square roots and about $n^3/6$ multiplications; if the LDL' decomposition is organized carefully it can again be done with about $n^3/6$ multiplications. In the general unsymmetric cases about $n^3/3$ multiplications are required.

We observe that since

$$\det A = \deg L \det D \det U = \det D$$
$$\det A = \det L \det U = \Pi l_{ii} \times \Pi d_{ii}$$

the determinant can be evaluated at negligible additional expense. This appears to be the most efficient way of evaluating det A for general A.

We shall now indicate how this decomposition leads to the Gaussian elimination process. Multiplying across

$$LU\mathbf{x} = \mathbf{b}$$

by L^{-1}, we find

$$U\mathbf{x} = \mathbf{y} \qquad \text{where } L\mathbf{y} = \mathbf{b}$$

which is in the Gaussian form. In practice, the determination of $\mathbf{x}$ is obtained at negligible additional expense; about n^2 multiplications are needed to find $\mathbf{y}$ and then about n^2 more to find $\mathbf{x}$.

Using the decomposition above can solve

$$A_3(x_1,x_2,x_3)' = (14,20,23)'$$

by first solving $L\mathbf{y} = \mathbf{b}$, that is,

$$\begin{bmatrix} 1 & 0 & 0 \\ 2 & i & 0 \\ 3 & 2i & 2 \end{bmatrix} \begin{bmatrix} y_1 \\ y_2 \\ y_3 \end{bmatrix} = \begin{bmatrix} 14 \\ 20 \\ 23 \end{bmatrix}$$

to get successively $y_1 = 14$, $y_2 = 8i$, $y_3 = 3i$. Then we solve $L'(x_1,x_2,x_3)' = (14,8i,3i)'$ to find successively $x_3 = 3$, $x_2 = 2$, $x_1 = 1$.

We now take up the inversion problem. We note that the inverse of a triangular matrix is a triangular matrix of the same species which again can be obtained without the solution of simultaneous equations. We then observe that

$$A^{-1} = (LU)^{-1} = U^{-1}L^{-1}$$

and so A^{-1} can be obtained at a cost of about n^3 multiplications in all in the general case.

The Triple Diagonal Case. This case (which will be seen to be of great importance in several contexts later) can be dealt with very efficiently. We take the linear equation case:

$$\begin{cases} b_1x_1 + c_1x_2 = d_1 & \\ a_ix_{i-1} + b_ix_i + c_ix_{i+1} = d_i & i = 2, 3, \ldots, n-1 \\ a_nx_{n-1} + b_nx_n = d_n & \end{cases}$$

The solution can be found by $O(n)$ operations in contrast to the $O(n^3)$ needed in the general case; thus,

$$\beta_1 = \frac{c_1}{b_1} \quad \beta_i = \frac{c_i}{b_i - a_i\beta_{i-1}} \qquad i = 2, 3, \ldots, n-1$$

$$q_1 = \frac{d_1}{b_1} \quad q_i = \frac{d_i - A_iq_{i-1}}{b_i - A_i\beta_{i-1}} \qquad i = 2, 3, \ldots, n$$

$$x_n = q_n \quad x_i = q_i - \beta_ix_{i+1} \qquad i = n-1, n-2, \ldots, 1$$

Iterative Processes. We have just described, in broad outline, efficient methods of handling the solution of $Ax = b$ for general A. If A is of a special form, e.g., if it is a "band" matrix, i.e., if $a_{ij} = 0$ for $|i - j| > m$, where $m \ll n$, or, more generally, if A is "sparse" in the sense that $a_{ij} \neq 0$ for a small proportion of the subscripts, then other "iterative" methods may be more efficient. We discuss some of these briefly here and will return to them in connection with the solution of differential equations (Secs. 9 and 10).

A basic iterative process is that of Newton. If X_n is an approximate inverse of A then

$$X_{n+i} = X_n(2I - AX_n)$$

is usually a better one. Observe that each iteration involves two matrix multiplications (indeed one fore and one aft) and so unless these are cheap (e.g., many zeros in A) it would be better to use the methods already described which give A^{-1} at the expense of about one matrix multiplication. (In special cases the A^{-1} computed directly might be inaccurate and one might want to use the Newton formula to improve the approximate inverse.) Use of this method is specially indicated when a good approximation to A^{-1} can be obtained cheaply, e.g., if A has a dominant diagonal.

Another classical process is the so-called Gauss-Seidel one. In the solution of the system

$$\begin{aligned} 12x - 3y + 2z &= 96 \\ -3x - 8y + z &= 68 \\ x + 2y + 6z &= 3 \end{aligned}$$

it runs as follows: We guess a solution, say $x_0 = 1$, $y_0 = 1$, $z_0 = 1$ and "improve" it as follows:

$$\begin{bmatrix} 1 \\ 1 \\ 1 \end{bmatrix}; \quad \begin{aligned} 12x - 3 + 2 &= 96: \\ -24.24 - 8y + 1 &= 68: \\ 8.08 - 22.80 + 6z &= 3: \end{aligned} \quad \begin{bmatrix} 8.08 \\ -11.40 \\ 2.95 \end{bmatrix};$$

$$\begin{aligned} 12x + 34.20 + 5.90 &= 96: \\ -13.98 - 8y + 2.95 &= 68: \\ 4.66 - 19.76 + 6z &= 3: \end{aligned} \quad \begin{bmatrix} 4.66 \\ -9.88 \\ 3.02 \end{bmatrix}; \quad \ldots$$

$$\begin{bmatrix} 5.03 \\ -10.01 \\ 3.00 \end{bmatrix}; \quad \ldots$$

From (x_n,y_n,z_n) we obtain x_{n+1} by substituting y_n, z_n in the first equation and solving for x; we obtain y_{n+1} by substituting x_{n+1}, z_n in the second equation and solving for y; we obtain z_{n+1} by substituting for x_{n+1}, y_{n+1} in the third equation and solving for z.

This method can be described in matrix notation as follows: We assume that diag $A = I$ and write

$$A = -L + I - U$$

where $L(U)$ is *strictly* lower (upper) triangular. If $\mathbf{x}^{(0)}$ is the initial guess for the solution $\mathbf{x}$ then successive approximations are given by

$$\mathbf{x}^{(n+1)} = (I - L)^{-1}(\mathbf{b} + U\mathbf{x}^{(r)})$$

If $\epsilon^{(n)} = x - x^{(r)}$ is the rth error vector, then

$$\boldsymbol{\varepsilon}^{(r+1)} = (I - L)^{-1}U\boldsymbol{\varepsilon}^{(r)}$$

and convergence will take place for all $\mathbf{x}^{(0)}$ if and only if the spectral radius of $B = (I - L)^{-1}U$, that is, the maximum of the moduli of the eigenvalues of B, is less than unity.

This fact can be established as follows in the case when the iteration matrix B has a full set of characteristic vectors, $\mathbf{c}_s$, $s = 1, 2, \ldots, n$. Suppose β_s is the characteristic value corresponding to c_s. We can write

$$\boldsymbol{\varepsilon}^{(0)} = \Sigma\gamma_s\mathbf{c}_s$$

and then

$$\boldsymbol{\varepsilon}^{(j)} = \Sigma\gamma_s\beta_s{}^j\mathbf{c}_s$$

It is clear that $\boldsymbol{\varepsilon}^{(j)} \to 0$ if $|\beta_s| < 1$ for all s, that is, if the spectral radius $\rho(B) = \max |\beta_s| < 1$. Conversely, if $\rho \geq 1$ then for certain $\boldsymbol{\varepsilon}^{(0)}$ convergence will not take place. We note that the spectral radius gives the attenuation per iteration of the component of the dominant characteristic vector.

This is not a very convenient criterion; it can be shown that convergence always takes place if A is a positive definite symmetric matrix.

A more naïve method of solution, usually ascribed to Jacobi, which is largely of theoretical interest (Sec. 10) consists in *not* using the improved values $x_{n+1}, y_{n+1}, \ldots$ in the determination of $y_{n+1}, z_{n+1}, \ldots$. The analysis of convergence is simpler in this case for, again assuming diag $A = I$, we find

$$\mathbf{x}^{(r+1)} = \mathbf{b} - (A - I)\mathbf{x}^{(r)}$$

and then

$$\boldsymbol{\varepsilon}^{(r+1)} = (I - A)\boldsymbol{\varepsilon}^{(r)} \qquad \boldsymbol{\varepsilon}^{(r)} = (I - A)^r\boldsymbol{\varepsilon}^{(0)}$$

Convergence will take place for all $\mathbf{x}^{(0)}$ if and only if the spectral radius of $I - A$ is less than unity.

It is plausible to think that the Gauss-Seidel method would be better than the Jacobi method; actually the methods are not comparable. In some cases one is better (i.e., the corresponding spectral radius is smaller) and in some, the other.

Condition of Matrices. So far we have been sketching representative direct and iterative methods for handling the linear equations and related problems, and our point of view has been largely formal. We have been concerned with the duration of the solution rather than its numerical stability, i.e., its sensitivity to the fact that our arithmetic is subject to round-off errors. The full examination of any algorithms from this aspect is a very extensive undertaking. We can only outline the sorts of results that are available, but before doing so we discuss some numerical examples. We refer to the example due to Wilson discussed in Sec. 3. It may be expected that if we carry out the solution to this system, working, for instance, to 2 or $3D$, we may expect to get into trouble, and indeed we do. It is easy to construct larger systems with similar behavior when we are working to the usual $8D$ precision. A notorious example is the Hilbert matrix

$$H_n = \begin{bmatrix} 1 & \frac{1}{2} & \frac{1}{3} & \cdots & 1/n \\ \frac{1}{2} & \frac{1}{3} & \frac{1}{4} & \cdots & 1/n+1 \\ \cdot & \cdot & \cdot & & \cdot \\ \cdot & \cdot & \cdot & \cdots & \cdot \\ \cdot & \cdot & \cdot & & \cdot \\ 1/n & 1/n+1 & 1/n+2 & \cdots & 1/2n-1 \end{bmatrix}$$

for which standard inversion programs fail for, say, $n \geq 10$. It will be shown later that this matrix is not a purely academic curiosity but that comparable matrices occur in curve-fitting problems.

It might be thought that the phenomenon of peculiar sensitivity would disappear if other methods of solution were used. This need not be the case. Applying the Gauss-Seidel method to the Wilson example, starting with the vector of zeros we obtain

$$\begin{array}{l} (3.20, .12, .67, .20)' \\ (2.44, .18, 1.06, .35)' \\ (1.98, .21, 1.28, .46)' \\ (1.55, .21, 1.44, .61)' \\ \cdots\cdots\cdots \end{array}$$

which is far from rapidly convergent to the solution.

It is necessary to introduce the concept of the norm $\mathfrak{N}(\mathbf{x})$ of a vector and the norm $\mathfrak{N}(A)$ of a matrix. These quantities measure size. We require that they satisfy the following axioms:

Vectors. $\mathfrak{N}(\mathbf{x}) \geq 0$ and $\mathfrak{N}(\mathbf{x}) = 0$ if and only if $\mathbf{x} = 0$; $\mathfrak{N}(\mathbf{x} + \mathbf{y}) \leq \mathfrak{N}(\mathbf{x}) + \mathfrak{N}(\mathbf{y})$; $\mathfrak{N}(\alpha\mathbf{x}) = |\alpha|\mathfrak{N}(\mathbf{x})$, α any scalar.

Matrices. $\mathfrak{N}(A) \geq 0$ and $\mathfrak{N}(A) = 0$ if and only if $A = 0$; $\mathfrak{N}(A + B) \leq \mathfrak{N}(A) + \mathfrak{N}(B)$;

$$\mathfrak{N}(\alpha A) = |\alpha|\mathfrak{N}(A)$$

α any scalar; $\mathfrak{N}(AB) \leq \mathfrak{N}(A)\mathfrak{N}(B)$.

As examples of vector norms we give

$$\|\mathbf{x}\|_2 = (\Sigma|x_i|^2)^{1/2} \qquad \|\mathbf{x}\|_\infty = \max |x_i|$$

A vector norm $\|\cdot\|$ induces a matrix norm by defining

$$\|A\| = \max_{\mathbf{x}\neq 0} \frac{\|A\mathbf{x}\|}{\|\mathbf{x}\|}$$

The norms induced by $\| \quad \|_2$ and $\| \quad \|_\infty$ are

$$\|A\| = (\rho(AA'))^{1/2} \qquad \|A\| = \max_i \sum_k |a_{ik}|$$

When A is symmetric, the first of these reduces to the spectral radius of A; this norm is often called the spectral norm.

We note that the spectral radius $\rho(A)$ is not a norm, for if $A = \begin{bmatrix} 0 & 1 \\ 0 & 0 \end{bmatrix}$ and $B = \begin{bmatrix} 0 & 0 \\ 1 & 0 \end{bmatrix}$ we have

$$\rho(A + B) = 1$$

while $\rho(A) = \rho(B) = 0$, so that the triangle axiom is not satisfied.

Theoretically, in the finite-dimensional case, all norms are equivalent in the sense that there are *constants* p_{12}, p_{21} such that

$$p_{12} \leq \frac{\mathfrak{N}_1(\mathbf{x})}{\mathfrak{N}_2(\mathbf{x})} \leq p_{21}$$

is true for all $\mathbf{x} \neq 0$.

The problems with which we have to deal are of the form: Given completely specified algorithms for finding A^{-1}, say, find

$$\|A^{-1} - \mathfrak{A}\| \text{ or } \|A\mathfrak{A} - I\|$$

where $\mathfrak{A}$ is the machine output. This error often depends on

$$\|A\| \, \|A^{-1}\|$$

This quantity has been called a *condition number* of the matrix A. We quote a classical result in this area:

If A is positive definite and of order n and $\mathcal{A}$ is the output of a certain fixed-point Gaussian algorithm,

$$\|A\mathcal{A} - I\| \leq c(\lambda/\mu)n^2\epsilon$$

where λ, μ are the maximum and minimum eigenvalues of A, ϵ is the smallest number recognized by the machine, and c is a moderate constant.

Here the ratio λ/μ is the product of the spectral norms of A and A^{-1} and has been called the P-condition number. The norm symbol on the left indicates the spectral norm. This result is of little practical use, for it is an upper bound and far from realistic (see below), and the determination of the P-condition number is more arduous than the inversion problem. However, the result serves many useful purposes, and, in the circumstances, rough estimates of the P-condition number are adequate.

For the case of the system arising from a 5-point discretization of the Dirichlet problem for a square, with n^2 mesh points, the P-condition number can be calculated exactly; it is $O(n^2)$. Thus the error should be $O(n^4)$. The following are typical results:

$n =$	10	20	30	40	50
Error:	2×10^{-7}	6×10^{-7}	1×10^{-6}	2×10^{-6}	3×10^{-6}

These results were obtained with a floating-point program on a machine using 8 bits for the characteristic and 27 bits for the mantissa. The "error" tabulated is the Euclidean norm $\|A\mathcal{R} - I\|_2$, where $\mathcal{R}$ is the "machine inverse" of A. The corresponding errors for a double precision program were:

Error:	3×10^{-14}	3×10^{-13}	1×10^{-12}	4×10^{-12}	6×10^{-12}

For the similar discretization of a biharmonic problem, in which the P-condition number is $O(n^4)$, the corresponding results were

$n =$	10	20	30	40	50
Error:	3×10^{-5}	4×10^{-4}	2×10^{-3}	9×10^{-3}	2×10^{-2}

Least Squares. We conclude this section by a matrix presentation of the curve-fitting problem, using the least-squares criterion. We shall indicate the arithmetic dangers of the process but we must first note that, in the beginning, serious consideration must be given to the question of whether the problem should be attacked at all, i.e., whether polynomial fitting by least squares is appropriate. For instance, exponential fitting with a different norm might be more suitable.

The problem is to find a polynomial of degree k

$$y(x) = c_0 + c_1x + \cdots + c_kx^k$$

such that for an assigned set of (distinct) abscissas $x_1, x_2, \ldots, x_m$ and of data $f_1, f_2, \ldots, f_m$, the discrepancy

$$\sum_{i=1}^{m} \{y(x_i) - f_i\}^2$$

is least. We have a new problem only if $m \geq k + 1$; we assume this is the case. If we introduce the column vectors

$$\mathbf{f} = (f_1, f_2, \ldots, f_m)' \qquad \mathbf{c} = (c_0, c_1, \ldots, c_k)'$$
$$\mathbf{q}^{(i)} = (x_1{}^i, x_2{}^i, \ldots, x_m{}^i)'$$

and the $m \times (k + 1)$ matrix

$$Q = (\mathbf{q}^{(0)}, \mathbf{q}^{(1)}, \ldots, \mathbf{q}^{(k)})$$

then our problem is to find

$$\min_{\mathbf{c}} \|\mathbf{f} - Q\mathbf{c}\|^2$$

The unique solution to this is

$$\mathbf{c} = (Q'Q)^{-1}Q'\mathbf{f}$$

In other words, $\mathbf{c}$ is given by the "normal equation"

$$(Q'Q)\mathbf{c} = Q'\mathbf{f}$$

On the one hand, the matrix of this system is symmetric and positive definite and so the Cholesky method is very satisfactory for solving it; indeed pivoting is not usually required.

On another hand, the matrix $Q'Q$ is often badly conditioned, and multiple precision arithmetic may be required. We have

$$(Q'Q)_{ij} = \sum_{k=1}^{m} x_k{}^i x_k{}^j \doteqdot m \int_0^1 x^i x^j \, dx = \frac{m}{i + j + 1}$$

if we assume the $\{x_k\}$ evenly distributed in (0,1). Thus the matrix $Q'Q$ is approximately the $(k + 1)$-dimensional Hilbert matrix which we have already discussed.

Further objections and remedies have been discussed, for instance, by Forsythe.

7. The Characteristic-value Problem

The basic problem here is to determine the characteristic values and vectors of a finite matrix. In principle this is covered in Secs. 5 and 6; we have merely to calculate the characteristic equation and apply the methods of Sec. 5 and, having got a characteristic value, obtain the corresponding characteristic vector by using the methods of Sec. 6. However, it is usually much better to take advantage of the special nature of the problem. There are several versions of the basic problem:

1. We want one (or a few) characteristic values, e.g., the dominant one.
2. We want them all.
3. We want rough estimates (inclusion or exclusion regions); e.g., we might want to know if all the characteristic values are inside the unit circle or that all are in the left half plane.

The problem naturally splits into two cases: (1) when A is real and symmetric (or Hermitian) and then all the characteristic values are real and (2) the unsymmetric case. We may also have to contend with a generalized problem, the solution of det $(A - \lambda B) = 0$. Finally we may be concerned, as in various vibration problems, with the case when the elements are functions of a parameter (frequency), and we may want to know, for instance, for what range of values of the parameter the characteristic values have a negative real part. The last two problems will not be discussed.

Dominant Characteristic Value and Vector. The dominant characteristic value of a matrix is the one (or ones) with maximum absolute value. In many physical situations it is its value that is important. In the case of positive matrices [i.e., if $A = (a_{ij})$ has $a_{ij} > 0$ for all i, j] the Perron-Frobenius theory tells us that there is a single dominant characteristic value that is positive and the corresponding characteristic vector can be chosen to have all its components positive. If the matrix is non-negative the results have to be modified slightly, as a consideration of the matrices

$$\begin{bmatrix} 0 & 1 \\ 1 & 0 \end{bmatrix} \quad \begin{bmatrix} 1 & 0 \\ 0 & 0 \end{bmatrix}$$

shows. There may be several characteristic values of maximum absolute value but one always is positive (and simple) and the corresponding characteristic vector can be chosen to have all its components positive if the matrix is irreducible [i.e., incapable of transformation into the form

$$\begin{bmatrix} A_{11} & A_{12} \\ 0 & A_{22} \end{bmatrix} \qquad A_{11}, A_{22} \text{ square}$$

by a simultaneous permutation of rows and columns]. The theory of non-negative matrices plays an extensive part in current numerical analysis.

When A has a single dominant characteristic value, the "power" method is an efficient way to find it and the corresponding characteristic vector. Suppose $\alpha_1, \alpha_2, \ldots, \alpha_n$ are the characteristic values of A, that $|\alpha_1| > |\alpha_2| \geq |\alpha_3| \geq \cdots \geq |\alpha_n|$, that $\mathbf{u}_1, \mathbf{u}_2, \ldots, \mathbf{u}_n$ are the corresponding characteristic vectors, and that this is a complete system so that any vector can be expressed as a linear combination of them.

Take any vector $\mathbf{u} = \Sigma a_i \mathbf{u}_i$. Consider the vectors $A^r\mathbf{u}$ obtained by repeatedly multiplying by A. Then, if $a_1 \neq 0$,

$$A^r\mathbf{u} = \Sigma a_i A^r \mathbf{u}_i = \Sigma a_i \alpha_i{}^r \mathbf{u}_i \doteqdot a_1 \alpha_1{}^r \mathbf{u}_1$$

This shows that the directions of the vectors $A^r\mathbf{u}$ converge and if they are normalized (e.g., by making them of unit length or making a particular component unity) the corresponding $\mathbf{v}_r$ satisfy

$$A\mathbf{v}_{r+1} \doteqdot \alpha_1 \mathbf{v}_r$$

As an example, we take

$$A = \begin{bmatrix} 0.2 & 0.9 & 1.32 \\ -11.2 & 22.28 & -10.72 \\ -5.8 & 9.45 & -1.94 \end{bmatrix}$$

We choose an arbitrary vector $\mathbf{u} = (1,0,0)$. Then we find successively

$$A\mathbf{u} = (0.2, -11.2, 15.8) = \lambda^{(1)}\mathbf{v}_1$$

where $\lambda^{(1)} = 0.2$, $\mathbf{v}_1 = (1, -56, -29)$;

$$A\mathbf{v}_1 = (-88.48, -948, -478.74) = \lambda^{(2)}\mathbf{v}_2$$

where $\lambda^{(2)} = -88.48$, $\mathbf{v}_2 = (1,10.7143,5.4107)$;

$$A\mathbf{v}_2 = (16.9850,169.5119,84.9534) = \lambda^{(3)}\mathbf{v}_3$$

where $\lambda^{(3)} = 16.9850$, $\mathbf{v}_3 = (1,9.9801,5.0017)$;

$$A\mathbf{v}_3 = (15.7834,157.5384,78.8086) = \lambda^{(4)}\mathbf{v}_4$$

where $\lambda^{(4)} = 15.7834$, $\mathbf{v}_4 = (1,9.9807,4.9928)$;

$$A\mathbf{v}_4 = (15.7731,157.6472,78.8316) = \lambda^{(5)}\mathbf{v}_5$$

where $\lambda^{(5)} = 15.7731$, $\mathbf{v}_5 = (1,9.9947,4.9979)$.
It can be verified that the largest characteristic root is 15.8 and that the corresponding characteristic vector is (1,10,5).

It is clear that the success of this method depends on the separation of the α's, that is, we would like to have $|\lambda_1| \gg |\lambda_2| \geq \cdots$, and on the goodness of our first approximation, that is, we want $|a_1| \gg |a_i|, i \neq 1$. However, in practice, the method may work even if $a_1 = 0$, for the roundings off will soon introduce a component of $\mathbf{u}_1$. Various acceleration techniques are available for dealing with awkward cases, e.g., with close characteristic values or even characteristic values of equal absolute value.

A troublesome example (given by Bodewig) is the following

$$\begin{bmatrix} 2 & 1 & 3 & 4 \\ 1 & -3 & 1 & 5 \\ 3 & 1 & 6 & -2 \\ 4 & 5 & -2 & -1 \end{bmatrix}$$

which has characteristic values -8.029, 7.932, 5.669, -1.573.

As soon as a characteristic value and vector of A have been determined, it is possible to "deflate" A, that is, obtain an $(n-1) \times (n-1)$ matrix whose characteristic values and vectors are simply related to the remaining ones of A. A simple process is the following: Suppose $A\mathbf{x} = \alpha_1\mathbf{x}$ and that the first component of $\mathbf{x}$ is the unity. Let A_1 be the first *row* of A and define

$$\tilde{A} = A - \mathbf{x}A_1$$

We note that the characteristic values of $\tilde{A}$ are $0, \alpha_2, \alpha_3, \ldots, \alpha_n$ and the characteristic vector corresponding to an $\alpha \neq \alpha_1$ is $\mathbf{u} - \mathbf{u}_1$. Further, the first row of $\tilde{A}$ is zero and the first component of all its characteristic vectors are zero; hence the succeeding calculations can be done on the first minor of $\tilde{A}$.

In the case of the 3×3 matrix above, the deflated matrix is

$$\tilde{A} = A - \begin{bmatrix} 1 \\ 10 \\ 5 \end{bmatrix} (.2,.9,1.32)$$

$$= \begin{bmatrix} 0 & 0 & 0 \\ -13.2 & 13.28 & -23.92 \\ -6.8 & 4.96 & -7.54 \end{bmatrix}$$

In principle, if all the characteristic values of A were distinct, all the α's and $\mathbf{u}$'s could be obtained by successive applications of the power method followed by deflation. In practice, however, α_1 is found only approximately and the deflation process is done only approximately. This means that there will be a progressive contamination. Very soon the information obtained may be of little value.

Complete Characteristic-value Problem. We now consider the determination of all the characteristic values of a real symmetric matrix A. The main devices used are orthogonal transformations of the matrix into more concentrated form. If we can find an orthogonal R such that

$$R'AR = D$$

where D is diagonal and therefore necessarily diag $(\alpha_1, \alpha_2, \ldots, \alpha_n)$ we would be finished, for then the columns of R give the characteristic vectors of A. Following Jacobi, it is possible to approach this situation as follows: If a_{rs} is (one of) the largest (in absolute value) off-diagonal elements of A, consider the matrix

$$R'_{rs}AR_{rs} = B$$

where R_{rs} is the orthogonal matrix with all elements zero except the following:

Diagonal elements not in columns r and s: 1
(r,r) element: $\cos\theta$, (r,s) element: $\sin\theta$
(s,r) element: $-\sin\theta$, (s,s) element: $\cos\theta$

where $\tan 2\theta = -2a_{rs}/(a_{rr} - a_{ss})$. The (r,s) and (s,r) elements of the matrix B are zero, and the sum J of the squares of its off-diagonal elements has been reduced by $2a_{rs}^2$ or by a factor of at least $(1 - 2(n^2 - n)^{-2})$. Repetition of this process, which involves only $O(n)$ arithmetic operations, n^2 times will reduce J by a factor of at least $(1 - 2(n^2 - n)^{-2})^{n^2} \sim e^{-2}$. Thus $O(n^3)$ operations will produce a matrix with arbitrarily small off-diagonal elements and so give arbitrarily good approximations to the α_i.

Two variations on this method have proved more satisfactory in general practice. In each the matrix is reduced to a similar triple diagonal form; the further handling of this is the same in the two methods.

Givens again uses the two-dimensional Jacobi rotations but annihilates, in order, the elements in positions

$$\begin{array}{llr} (1,3)(1,4)\cdots(1,n) & & R_{2,3}, R_{2,4}, \ldots, R_{2,n} \\ \quad(2,4)\cdots(2,n) & \text{by rotations} & R_{3,4}, \ldots, R_{3,n} \\ \quad\ldots\ldots & & \ldots\ldots \\ \quad(n-2,)n & & R_{n-1,n} \end{array}$$

This arrangement ensures that the annihilated off-diagonals *remain* zero—which need not be the case in the Jacobi scheme. Here, then, after $(n-2)(n-1)/2$ rotations we get a triple diagonal matrix. In a minor variation on this, due to Rutishauser, the elements are annihilated from the right in each row; this has certain advantages, e.g., when the matrix is not stored in the main memory.

If we apply this method to the Wilson matrix we find, working to $6D$,

$$\begin{bmatrix} 5 & 10.488089 & 0 & 0 \\ 10.488089 & 25.472729 & 3.521903 & 0 \\ 0 & 3.521898 & 3.680571 & -.185813 \\ 0 & 0 & -.185813 & .846701 \end{bmatrix}$$

Theoretically the triple diagonal matrix is symmetric; it is not necessary to calculate both a_{ij} and a_{ji} but doing so, as above, gives some indication of the amount of error incurred. The characteristic values of this matrix are .0105, .8431, 3.858, 30.29.

Householder annihilates the off-diagonal elements a row at a time but has to use transforming matrices more complicated than the Jacobi rotations—specifically, matrices of the form

$$P = I - 2\mathbf{w}\mathbf{w}'$$

where $\mathbf{w}$ is a column vector of length 1; that is, $\mathbf{w}'\mathbf{w} = 1$. It is easy to verify that $P'P = I$, that is, that P is orthogonal. We write

$$\mathbf{w}_r = (0, \ldots, 0, x_r^{(r)}, \ldots, x_n^{(r)})$$

We can show that, provided the $x_j^{(i)}$ are chosen appropriately,

$$P'_{n-1} \cdots P'_3P'_2AP_2P_3 \cdots P_{n-1}$$

is a symmetric triple diagonal matrix.

An operation count for these two reductions can be made and gives about $\frac{4}{3}n^3$ multiplications for the Givens method and about $\frac{2}{3}n^3$ for the Householder method. Formally, therefore, the Householder method is twice as fast; there remains, however, the question of the relative numerical stability. Theoretical estimates, e.g., by Wilkinson, and experience indicate the use of the Householder reduction in general.

As far as the symmetric characteristic-value problem is concerned, we have still to show how to deal with the triple diagonal case to which we have reduced the general case. We outline two methods. Suppose we have to deal with

$$C = C_n = \begin{bmatrix} b_1 & c_1 & & & & \\ c_1 & b_2 & c_2 & & & \\ & & \cdot & & & \\ & & & \cdot & & \\ & & & \cdot & & \\ & & & c_{n-2} & b_{n-1} & c_{n-1} \\ & & & & c_{n-1} & b_n \end{bmatrix}$$

Then $f_n = \det(xI - C_n)$ can be calculated by the recurrence

$$f_0 = 1 \qquad f_1 = x - b_1$$
$$f_r = (x - b_r)f_{r-1} - c_{r-1}^2 f_{r-2} \qquad r \geq 2$$

Observe that the sequence

$$f_n, f_{n-1}, \ldots, f_0$$

is a Sturm sequence (provided no c_i vanishes—in which case the problem has broken up into two simpler ones). If a and b are bounds for the characteristic values we compute the above sequence for a, b and $(a + b)/2$. Comparison of the number of sign changes indicates whether there are characteristic values in one or both halves. We then compute the Sturm sequence at the mid-point of any half that contains characteristic values, and so on. This method rapidly isolates the characteristic values, and the expense is $O(n)$ operations for each bisection.

An alternative is the QR method, due to Francis and Kublanowskaja. This depends on the fact that any real matrix can be expressed in the form $A = QR$, where Q is orthogonal and R upper triangular; when A is triple diagonal, this can be done at an expense of $O(n)$ operations.

From $A = A_1$ we obtain $A_1 = Q_1R_1$ and then define $A_2 = R_1Q_1$; since the matrix $A_2 = Q_1^{-1}A_1Q_1$ is similar to A_1 it has the same characteristic values. We deal with A_2 in the same way, and so on, getting a sequence of similar matrices A_k. In the special case of A symmetric and triple diagonal, all the matrices A_k have the same two properties. It can be shown that the sequence $\{A_k\}$ converges to an upper triangular matrix.

Analysis of this process shows that it is numerically stable and quite competitive with the Sturm sequence method.

The problem of determining the characteristic vectors is simply discussed. We take the case of the Householder method. We obtain the α's by the Sturm sequence or QR method. It is tempting to try to find the $\mathbf{u}$'s by solving the first $n - 1$ equations of the singular system

$$C\mathbf{x} = \alpha\mathbf{u}$$

but this can be dangerous since α will not be exact. An acceptable method is based on the Wielandt or inverse iteration, which consists in the generation of a sequence of vectors, from an arbitrary initial vector $\mathbf{b}_0$, by

$$(C - \alpha I)\mathbf{b}_{r-1} = \mathbf{b}_r$$

If α is a close approximation to α_1, say, then $\{\mathbf{b}_r\}$ rapidly approaches the corresponding vector (indeed two iterations often suffice). This is seen as follows:

We write $\mathbf{b}_0 = \Sigma\beta_i\mathbf{u}_i$, where the $\mathbf{u}_i$ are the characteristic vectors of C. Then

$$\begin{aligned}\mathbf{b}_r &= \Sigma\beta_i(\alpha_i - \alpha)^{-r}\mathbf{u}_i \\ &\doteqdot \beta_1(\alpha_1 - \alpha)^{-r}\mathbf{u}_1\end{aligned}$$

As before, we need to have $\beta_1 \neq 0$ and assume that α is closer to α_1 than any other characteristic value.

Application of this device means that we have to solve the triple diagonal system twice. We know that this can be accomplished cheaply in the general case, and it is even simpler and safe in the special symmetric case.

The methods useful in the symmetric case are generally satisfactory in the unsymmetric case. The power method does not use symmetry, and if we apply the Householder reduction, choosing the vectors $\mathbf{w}_r$ to kill the elements in the rth row in columns $r + 2, r + 3, \ldots, n$, we do not necessarily kill the symmetric elements; instead of a triple diagonal matrix we are left with a lower Hessenberg matrix, i.e., one with zeros in position (i,j), where $i = 1, 2, \ldots, n - 1; j = i + 2, i + 3, \ldots, n$.

We can deal with a Hessenberg matrix conveniently by use of the QR algorithm, for each decomposition requires only $O(n^2)$ operations and the successive matrices A_k retain the Hessenberg form. An alternative scheme is to observe that it is easy to evaluate $\det(C - xI)$ when C is in Hessenberg form; we eliminate all elements in the first column but the last by adding multiples of the ith column to the first to kill its $(i - 1)$st element and then the expansion of the determinant contains only one term. We can therefore apply the Muller method to determine the characteristic values.

Bounds for Characteristic Values. There is now a vast literature on this subject. From this we pick three theorems, each of which can be of considerable use.

If A is any (complex) matrix then the α_i are included in the union of the n Gerschgorin circles

$$\Gamma_i\colon |z - a_{ii}| \leq \sum_{i \neq j} |a_{ij}|$$

If A is a real symmetric matrix and if for any vector $\mathbf{x}$ we define the Rayleigh quotient

$$R(\mathbf{x}) = \mathbf{x}'A\mathbf{x}/\mathbf{x}'\mathbf{x}$$

we have

$$\min_i \alpha_i \leq R(\mathbf{x}) \leq \max_i \alpha_i$$

and there is equality when $\mathbf{x}$ is the appropriate characteristic vector.

If A is any matrix, then

$$\sum_i |\alpha_i|^2 \leq \sum_{i,j} |a_{ij}|^2$$

with equality if and only if A is normal. This gives

$$|\alpha| \leq \left[\sum_{i,j} |a_{ij}|^2\right]^{1/2}$$

8. Quadrature and Differentiation

The problem with which we are mainly concerned is the numerical evaluation of

$$F(x) = \int_a^x f(t)\,dt$$

for one or more values of x, where $f(t)$ is given analytically or by a table.

Occasionally the simplest solution to this is provided by obtaining the indefinite integral analytically and then referring to tables of the functions involved. Among the more elaborate tables of integrals in common use are those of Gröbner and Hofreiter, Bierens de Haan, Byrd and Friedman, and Ryghik and Gradshteyn. In many cases the explicit analytical form of the indefinite integral $F(x)$ may not be very helpful. However, even if this solution is too unwieldy for general use, it may be valuable for checking —for example, the final value.

Lagrangian Formulas. The idea now to be exploited is the following: In order to evaluate

$$I = \int_a^b f(x)\,dx$$

approximately, we shall evaluate the integral of an approximation to $f(x)$. We make use of the results already available about the approximations to functions by polynomials.

For instance, if

$$a \leq x_0 < x_1 < \cdots < x_n = b$$

and if

$$L_n(f,x) = \Sigma f(x_i)l_i^{(n)}(x)$$

is the corresponding Lagrangian polynomial, we can consider

$$Q = Q_n = \int_a^b L_n(f,x)\,dx = \sum f(x_i) \int_a^b l_i^{(n)}(x)\,dx = \sum A_i f(x_i)$$

as an approximation to I. We note that the coefficients A_i do not depend on the particular function being integrated, only on the nodes $x_0, x_1, \ldots, x_n$. If we take special cases, such as the one in which the nodes are equally spaced, the tabulation of the A_i is feasible, and a convenient solution to the problem is available.

Since $L_n(f,x) \equiv f(x)$ if $f(x)$ is a polynomial of degree at most n, it follows that $Q = I$ in this case. This result clearly remains true in the weighted case when we approximate

$$I = \int_a^b f(x)w(x)\,dx$$

by $Q = \Sigma A_i f(x_i)$, where now

$$A_i = \int_a^b l_i^{(n)}(x)w(x)\,dx$$

We show how to obtain a crude error estimate in the general $(n+1)$-point weightless case. Suppose the $(n+1)$st derivative of $f(x)$ is continuous in $[a,b]$ and bounded there by M_{n+1}, say. We know that

$$f(x) = L_n(x) + \frac{f^{(n+1)}(\xi)}{(n+1)!}(x - x_0)(x - x_1) \cdots (x - x_n)$$

If we consider integrating this with respect to x, we must remember that in general ξ depends on x. About the best we can do is to replace $f^{(n+1)}(\xi)$ by M_{n+1}. We then find

$$|I - Q_n| \leq \frac{M_{n+1}}{(n+1)!} \int_a^b |x - x_0|\,|x - x_1| \cdots |x - x_n|\,dx$$

A crude estimate of the integral is $(b-a)^{n-2}$, so that

$$|I - Q_n| \leq \frac{M_{n+1}(b-a)^{n-2}}{(n+1)!}$$

We now discuss some special cases beginning with the *trapezoidal rule*. We take $n = 1$ and $x_0 = a$, $x_1 = b$ and integrate the linear approximation to $f(x)$

$$f(x) \doteqdot f(a) + \frac{x-a}{b-a}[f(b) - f(a)]$$

to get $$Q = (b-a)\{\tfrac{1}{2}[f(b) + f(a)]\}$$

with an error which does not exceed $\frac{1}{12}M_2(b-a)^3$. If we subdivide the interval $[a,b]$ into $r = (b-a)/h$ parts and apply the above formula to each subinterval we get

$$Q = h\{\tfrac{1}{2}f(x_0) + f(x_1) + \cdots + f(x_{r-1}) + \tfrac{1}{2}f(x_r)\}$$

with an error which does not exceed

$$r \times \tfrac{1}{12}M_2h^3 = \tfrac{1}{12}M_2(b-a)h^2$$

We shall discuss below, in a more sophisticated context, the economies of subdivision. Here the error is reduced by a factor of r^2, but we must know $r+1$ values of f rather than 2.

We obtain the *three-eighths rule* by integrating a cubic agreeing with $f(x)$ at the points a, $a + (b-a)/3$, $a + 2(b-a)/3$, b. Taking $a = -1$, $b = 2$, we get

$$Q = \tfrac{3}{8}(f(-1) + 3f(0) + 3f(1) + f(2))$$

If we take $a = -2$, $b = 2$ and $x_0 = -1$, $x_1 = 0$, $x_2 = 1$, we get

$$Q = \tfrac{4}{3}[2f(-1) - f(0) + 2f(1)]$$

a result due to Milne. Since this estimates $\int_{-2}^{2}$ without using the ordinates at the end points (such estimates are said to be of *open* type), it can be used as a predictor formula in the solution of differential equations (see Sec. 9).

The familiar Simpson's rule is obtained by integrating over $[-1,1]$ a cubic which coincides with $f(x)$ at $-1, 0, 1$ and which in addition has the same derivative as $f(x)$ at 0:

$$\int_{-1}^{-1} f(x)\,dx \doteqdot \tfrac{1}{3}[f(-1) + 4f(0) + f(1)]$$

An error estimate for this is $|I - Q| \leq M_4/90$.

Quadratures Using Differences. Just as in the case of interpolation, there are methods based on differences.

The most efficient method is due to Gauss and uses the following central-difference formula:

$$h^{-1}\int_0^h f(x)\,dx = (1 - \tfrac{1}{12}\delta^2 + \tfrac{11}{720}\delta^4 - \cdots)\mu f(\tfrac{1}{2})$$

How this is used in practice is indicated by the following beginning of an evaluation of

$$I = \int_0^{1/2} (1 - x^2)^{1/2}\,dx$$

using an interval $h = .05$. The integrand is tabulated and differenced.

$I(x)$	δI		x	$(1-x^2)^{1/2}$		δ^2		δ^4	
0		50010	0	1.000000		−2502*		−18*	
	49979				−1251		−9*		−1*
49979		49948	.05	998749		2511		19*	
	49854				3762		28		3*
99833		49760	.10	994987		2539		22	
	49602				6301		50		
149435		49945	.15	988686		2589			
	49223				8890				
198658		49001	.20	979796					

The starred entries are not officially available if we restrict ourselves to the given values of the integrand. However, they may be estimated in various ways; for instance, we may assume two zero fifth differences and then the starred values are replaced by

−2505		−22
	−6	0
		22
		0

This being done, the first column to the left of the argument column is computed; it is

$$h(1 - \tfrac{1}{12}\delta^2 + \tfrac{11}{720}\delta^4 - \cdots)$$

From this column, that labeled δI is obtained by averaging, and from the δI column, that for I is obtained by addition.

It is clear that the errors in our estimates of the missing differences are obliterated by the multiplying factors $-h/12$, $+11h/720$, ...

The correct value, obtained analytically, is

$$I = \frac{1}{8}\sqrt{3} + \frac{\pi}{12} = .478306$$

Gregory's formula for quadrature makes use only of differences that are actually available. It has the disadvantage of involving differences of all orders, not merely those of even order.

$$\int_0^{rh} f(x)\,dx = h(\tfrac{1}{2}f_0 + f_1 + \cdots + f_{r-1} + \tfrac{1}{2}f_r)$$
$$+ h(\tfrac{1}{12}\Delta - \tfrac{1}{24}\Delta^2 + \tfrac{19}{720}\Delta^3 - \cdots)f_0$$
$$- h(\tfrac{1}{12}\nabla + \tfrac{1}{24}\nabla^2 + \tfrac{19}{720}\nabla^3 + \cdots)f_r$$

The succeeding coefficients are $\frac{3}{160}$, $\frac{863}{60480}$, $\frac{275}{24192}$, ... Here the first line on the right is

x	$(1 - x^2)^{1/2}$					
.0	1.000000					
		−1251				
0.05	998749		−2511			
		−3762		−28		
0.10	994987		−2539		−22	
		−6301		−50		+1
0.15	988686		−2589		−21	
		−8890		−71		−5
0.20	979796		−2660		−26	
		−11550		−97		−2
0.25	968246		−2757		−28	
		−14307		−125		−1
0.30	953939		−2882		−29	
		−17189		−164		−12
0.35	936750		−3046		−41	
		−20235		−205		−21
0.40	916516		−3251		−62	
		−23486		−267		
0.45	893029		−3518			
		−27004				
0.50	866025					

Hence

$$\int_0^{0.5} (1 - x^2)^{1/2}\,dx = 0.05[0.500000 + 0.998749 + \cdots$$
$$+ 0.893029 + 0.433012$$
$$+ \tfrac{1}{12}\ (-0.001251) - \tfrac{1}{24}\ (-0.002511)$$
$$+ \tfrac{19}{720}\ (-0.000028) - \tfrac{3}{160}(-0.000022) + \cdots$$
$$- \tfrac{1}{12}\ (-0.027004) - \tfrac{1}{24}\ (-0.003518)$$
$$- \tfrac{19}{720}\ (-0.000267) - \tfrac{3}{160}(-0.000062) - \cdots]$$
$$= 0.478306$$

the first approximation to the integral—the trapezium expression—which is corrected by the second line which involves the first available forward differences and by the third line which involves the last available backward differences. We repeat the example just discussed. We take $h = .05$ and work to $6D$. The differences involved in the correction terms are included in the boxes.

The *Euler-Maclaurin sum formula* is often convenient, either for the evaluation of a sum or of an integral. It is

$$h^{-1}\int_0^{nh} f(x)\,dx = \tfrac{1}{2}f_0 + f_1 + \cdots + f_{n-1} + \tfrac{1}{2}f_n$$
$$- \tfrac{1}{12}h(f'_n - f'_0)$$
$$+ \tfrac{1}{720}h^3(f'''_n - f'''_0)$$
$$- \tfrac{1}{30{,}240}h^5(f_n^{(5)} - f_0^{(5)}) + \cdots$$

If we take $h = 1$ and let $n \to \infty$ we obtain formally, on the assumption that each $f^{(2p+1)}(n) \to 0$,

$$\sum_1^\infty f(n) = \int_0^\infty f(x)\,dx - \tfrac{1}{2}f(0) - \tfrac{1}{12}f'(0)$$
$$+ \tfrac{1}{720}f'''(0) - \cdots$$

This relation can be justified in the case

$$f(n) = (10 + n)^{-2}$$

when we find

$$\sum_1^\infty n^{-2} = (1 + 2^{-2} + \cdots + 10^{-2}) + \sum_1^\infty f(n)$$
$$= 1.54976\ 77312 + \int_{10}^\infty x^{-2}\,dx$$
$$- \tfrac{1}{2}(10)^{-2} + \tfrac{1}{12}(2!)(10)^{-3}$$
$$- \tfrac{1}{720}(4!)(10)^{-5} - \cdots$$
$$= 1.54976\ 77312 + .10000\ 00000$$
$$- .00500\ 00000 + .00016\ 66667$$
$$- .00000\ 03333 + .00000\ 00024 - \cdots$$
$$= 1.64493\ 40670$$

This is to be compared with

$$\pi^2/6 = 1.64493\ 40668$$

A naive approach to the evaluation of $\sum_1^\infty n^{-2}$ shows that the remainder after n terms is about n^{-1}, so that a direct summation is not feasible.

Gaussian Quadratures. We have noted that a Lagrangian $(n + 1)$-point quadrature, with arbitrary nodes, is exact for polynomials of degree at most n. Can we do better than this if we choose the x_i cleverly?

If we consider the equality

$$\int f(x)\,dx = \Sigma A_i f(x_i)$$

or, more generally,

$$\int f(x)p(x)\,dx = \Sigma A_i f(x_i)$$

where $p(x)$ is a fixed, positive weight function, we see that there are $2n + 2$ constants on the right. One might expect to evaluate them by requiring that this relation be satisfied for $f(x) = x^r$, $r = 0, 1, \ldots, 2n + 1$. In this case, since the operations are linear, we would have equality when f is a polynomial of degree $2n + 1$ at most. We shall show that this is indeed possible, relying on the general theory of orthogonal polynomials.

Let $f_0, f_1, \ldots, f_n$ be the normal orthogonal system constructed from $1, x, \ldots, x^n$, where the interval is $[a,b]$ and the weight function is $p(x)$. Then f_n has n real distinct zeros, all in $[a,b]$; denote these by $x_i^{(n)}$, $i = 1, 2, \ldots, n$. Any polynomial $f(x)$ of degree $2n + 1$ at most can be written as

$$f(x) = q(x)f_{n+1}(x) + r(x)$$

where the quotient $q(x)$ and the remainder $r(x)$ are of degree n at most. Integrating this relation, we find

$$\int_a^b f(x)p(x)\,dx = \int_a^b q(x)f_{n+1}(x)p(x)\,dx + \int_a^b r(x)p(x)\,dx$$

The first integral on the right vanishes by orthogonality, and the second is exactly $\Sigma A_i r(x_i)$, where the x_i are the $n + 1$ zeros of $f_{n+1}(x)$ and the A_i are the corresponding Lagrangian weights. Now since

$$f_{n+1}(x_i) = 0$$

it follows from the above representation of $f(x)$ that $f(x_i) = r(x_i)$. Thus we have

$$\int_a^b f(x)p(x)\,dx = \Sigma A_i f(x_i)$$

so that the Lagrangian quadrature based on the $x_i = x_i^{(n+1)}$ is exact for $f(x)$. The coefficients A_i are often called the Christoffel numbers and are always positive. They have been tabulated for many of the usual cases.

It might be expected that the error in such a Gaussian quadrature, in the case when $f(x)$ has a continuous $(2n + 2)$nd derivative, would be bounded by a multiple of the bound of this derivative. It can be shown

$$\int_a^b f(x)p(x)\,dx - \sum A_i f(x_i) = \frac{f^{(2n+2)}(\xi)}{(2n+2)!}\int_a^b \prod_{i=0}^{n} (x - x_i)^2 p(x)\,dx$$

where $a \le \xi \le b$. The integral on the right does not depend on f and can be evaluated once for all.

Comparison of Quadrature Methods. The arguments for and against the Lagrangian as compared with the finite-difference quadratures are essentially the same as those mentioned in the corresponding discussion of interpolation. We have also to evaluate the Gaussian methods. It is clear that they are not likely to be very practical when the integrand is tabulated at equal intervals, for preliminary interpolations are necessary to evaluate the $f(x_i)$, and these may outbalance any gain due to the smaller error estimates. On the other hand, if the integrand is not tabulated, the Gaussian type may be very convenient. This is certainly so in two cases: (1) when automatic computers are used and the evaluation of $f(x_i)$ does not depend on the number of decimals in the argument and (2) when $f(x_i)$ is being evaluated experimentally and the x_i are set once for all.

Aside from the above general considerations, we shall compare the efficiencies of the three-eighths rule, Simpson's rule, and the 2-point Gauss-Legendre quadrature for a function $f(x)$ such that $|f^{(4)}(x)| \le 1$ in $a \le x \le b$.

The absolute values of the errors in Simpson's rule and in the three-eighths rule are bounded by

$$\frac{N[(b-a)/2N]^5}{90} \quad \text{and} \quad N\left(\frac{b-a}{2N}\right)^5\left(\frac{3}{80}\right)$$

where N is the number of panels used. If the error is to be less than $\epsilon(b - a)^5$ we must have

$$N = \sqrt[4]{1/2880\epsilon} \quad \text{and} \quad N = \sqrt[4]{1/6480\epsilon}$$

respectively. The costs, in terms of evaluations of the integrand, are, respectively,

$$\sqrt[9]{1/180\epsilon} \qquad \sqrt[9]{1/80\epsilon}$$

The basic error estimate in the Gauss-Legendre case for the interval $[-1,1]$ is

$$|I - Q| \le \frac{|f^{(2n)}(\xi)|2^{2n+1}(n!)^4}{[(2n)!]^3(2n+1)}$$

When the interval is $[\alpha,\beta]$ and $n = 2$, this becomes

$$|I - Q| \le \frac{(\beta - \alpha)^5}{4320}$$

If we use N panels for $[a,b]$, the total error is

$$\frac{N[(b-a)/N]^5}{4320}$$

To make this equal to $\epsilon(b - a)^5$, we must take

$$N = \sqrt[4]{1/4320\epsilon}$$

and so we require

$$\sqrt[4]{1/270\epsilon}$$

evaluations.

Summing up, the relative efficiencies of the methods are in the ratios

1.36 (Gauss-Legendre):1.22 (Simpson):1 (three-eighths)

The Romberg Process. This quadrature process is based on the idea of computing a linear transformation of a sequence of trapezoidal quadratures, e.g., each at half the interval of the preceding. We define:

$$T_0^{(0)} = (b - a)[\tfrac{1}{2}f(a) + \tfrac{1}{2}f(b)]$$
$$T_0^{(1)} = 2^{-1}(b - a)[\tfrac{1}{2}f(a) + f(\tfrac{1}{2}(a + b)) + \tfrac{1}{2}f(b)]$$
$$T_0^{(2)} = 2^{-2}(b - a)[\tfrac{1}{2}f(a) + f(\tfrac{1}{4}(3a + b)) + f(\tfrac{1}{2}(a + b)) + f(\tfrac{1}{4}(a + 3b)) + \tfrac{1}{2}f(b)]$$

It is clear that the $T_0^{(k)}$, as $k \to \infty$, form a sequence of Riemann sums for $I = \int_a^b f(x)\,dx$, with uniform mesh size, tending to zero. If, for example, f is continuous in $[a,b]$, then $T_0^{(k)} \to I$.

We now recall the fact that the error in a trapezoidal quadrature at interval h is $O(h^2)$; in fact, it is $M_2(b - a)h^2/12$, where M_2 is a mean of the second derivative. This remark suggests the use of the *Richardson extrapolation* which we now explain in a general context.

Let $\phi(x)$ be the solution to a continuous problem and $\phi(x,h)$ be the solution to a discrete version of this, where h indicates the mesh size. In some circumstances we may have

$$\phi(x,h) = \phi(x) + h^2\phi_2(x) + R_2$$

and so, if R_2 is negligible and we obtain $\phi(x,h)$ for two values h_1, h_2 so that

$$\phi(x,h_1) = \phi(x) + h_1^2\phi_2(x) \qquad \phi(x,h_2) = \phi(x) + h_2^2\phi_2(x)$$

we can eliminate ϕ_2 to get (an approximate value of)

$$\phi(x) = \frac{h_2^2\phi(x,h_1) - h_1^2\phi(x,h_2)}{h_2^2 - h_1^2}$$

This plausible device can be applied, or modified, in many situations, e.g., when ϕ is a constant, when ϕ is a function of several variables, and when the representation of $\phi(x,h)$ has a different form, such as

$$\phi(x,h) = \phi(x) + h^4\phi_4(x) + R_4$$

We give one example. Consider the vibrations of a string of length l, and density ρ under a tension T. The fundamental frequency is given by

$$\sigma = \frac{\pi}{l}\sqrt{\frac{T}{\rho}}$$

If we approximate the string by a discrete system of masses $\frac{1}{2}m$, m, . . . , m, $\frac{1}{2}m$ placed at interval h the corresponding frequency is

$$\sigma(h) = 2\sqrt{\frac{T}{mh}}\,\frac{\sin mh}{2l}$$

We can now make the following table:

h^{-1}	2	3	4	5	10	20	40	∞
$\sigma(h)/\sigma$	.9003	.9548	.9745	.9836	.9959	.9990	.9997	1

If we extrapolate from 2 and 4 we get .9992, and from 5 and 10 we get 1.0000.

We now return to the integration problem and use the h^2 extrapolation on the $T_0^{(k)}$ to derive a new column by

$$T_1^{(k)} = \frac{4T_0^{(k+1)} - T_0^{(k)}}{3}$$

It is easy to verify that the numbers $T_1^{(k)}$ are just those obtained from Simpson's rule. We combine the patterns

$$\frac{4}{3}\times\frac{h}{2}\{\tfrac{1}{2}\quad 1\quad 1\quad 1\quad \cdots\quad 1\quad 1\quad 1\quad \tfrac{1}{2}\}$$

$$-\frac{1}{3}\times h\,\{\tfrac{1}{2}\quad 1\quad \cdots\quad 1\quad \tfrac{1}{2}\}$$

to get $\frac{1}{3}\times h\{\frac{1}{2}\quad 2\quad 1\quad 2\quad \cdots\quad 2\quad 1\quad 2\quad \frac{1}{2}\}$

The next stage is clearly to use the fact that Simpson's rule is an h^4 process and extrapolate to find a new column according to

$$T_2^{(k)} = \frac{16T_1^{(k+1)} - T_1^{(k)}}{15}$$

and to carry on in this way to obtain a triangular array

$$\begin{array}{llllll}
T_0^{(0)} & & & & & \\
T_0^{(1)} & T_1^{(0)} & & & & \\
T_0^{(2)} & T_1^{(1)} & T_2^{(0)} & & & \\
 & & \cdots & & & \\
 & & T_{m-1}^{(k)} & & & \\
T_0^{(m+k)} & \cdots & T_{m-1}^{(k+1)} & T_m^{(k)} & \cdots & T_{m+k}^{(0)} \\
 & & \cdots & & &
\end{array}$$

This table is developed from the first column by means of the generic recurrence relation

$$T_m^{(k)} = \frac{4^mT_{m-1}^{(k+1)} - T_{m-1}^{(k)}}{4^m - 1}$$

The following table was obtained in this way for the integral $I = \int_1^2 dx/x = \ln 2 = 0.693147180$:

.750 000 000				
.708 333 333	.694 444 444			
.697 023 809	.693 253 967	.693 174 603		
.694 121 851	.693 154 532	.693 147 901	.693 147 479	
.693 391 202	.693 147 652	.693 147 193	.693 147 182	.693 147 181

The corresponding errors are, in units of the ninth decimal,

56852820				
15186153	1297264			
3876629	106787	27423		
974671	7352	721	299	
244022	472	13	2	1

As we have pointed out before, the main labor in numerical quadrature is the computation of the values of the integrand. In the preparation of the first column of the above table we note that all values obtained are used in succeeding entries. The determination of the entries in the second and later columns is done by the use of the generic recurrence relation at a negligible additional expense.

The remarkable improvement in convergence of the columns and of the diagonals can be established rigorously under quite general conditions. There are, however, some exceptions, e.g., when the original column is "too rapidly" convergent to allow the assumed representation. A case in point is $\int_0^5 e^{-x^2}\,dx$. In all cases we recommend that the whole Romberg table be printed out and its behavior examined, rather than merely printing out a single entry thought to be adequate.

We shall now outline the general theoretical results. If the convergence of the first column only is assumed, it can be shown that all later columns and all diagonals converge to the same limit. In order to establish results about the rate of convergence, we must make some assumptions about the smoothness of the integrand. If we assume that $f^{(2m+2)}(x)$ is continuous in $[a,b]$, we can show that, as $k \to \infty$,

$$T_m^{(k)} - I = O(4^{-k(m+1)})$$

i.e., any entry is 4^{m+1} times as accurate as the entry immediately above. This behavior can be checked in the case of $f(x) = x^{-1}$ discussed above. We tabulate here the ratios of consecutive vertical entries in the error array:

3.74			
3.92	12.1		
3.98	14.5	38.0	
3.99	15.6	55.5	150

If we assume that the integrand is an analytic function in some domain including $[a,b]$, it can be shown that the diagonals of the Romberg table converge faster than any geometric series.

We have mentioned from time to time the importance of algorithms being numerically stable, i.e., insensitive to round off, as well as being efficient theoretically. The Romberg algorithm is very satisfactory from this point of view. Any particular entry $T_m^{(k)}$, say, in the Romberg table is clearly a linear combination of the values of the integrand at points of the form $a + jh$, where $h = (b - a) \times 2^{-m-k}$, specifically

$$T_m^{(k)} = h \sum_{j=0}^{2^{m+k}}{}'' d_j^{(m)} f(a + jh)$$

where the $''$ indicates that the first and last terms are to be taken with weight $\frac{1}{2}$. The weights $d_j^{(m)}$ can be calculated; it can be shown that they are all positive and indeed satisfy

$$.4841 \leq d_j^{(m)} \leq 1.4524$$

Thus it is clear that there is no exceptional weighting of the rounding errors and, moreover, there can be no exceptional building up of errors, as would be the case in the expression $10a - 9b$ if the errors in a and b were of opposite sign.

There have been various modifications and new applications of the Romberg idea. For example, it is not essential to bisect continually, but in this case different recurrence relations are required. Also, the idea has been applied for the numerical solution of differential equations.

Multiple Integrals. Probably the most satisfactory method for occasional use in straightforward repeated quadratures by some of the methods discussed above. However, when much multiple integration has to be done, more powerful methods should be considered. There is a considerable body of recent literature, e.g., suggesting efficient "sampling" points especially where the domain of integration has some regularity.

Monte Carlo Methods. This method has been used as a tool in the evaluation of multiple integrals but is, on the whole, a last resort. Its use is specially indicated when the integral in question has arisen in the analysis of a physical process for which a stochastic model can be made. Attention should be paid to the various statistical refinements of the crude Monte Carlo.

We discuss only the most primitive problem, the estimation of $I = \int_0^1 f(x)\,dx$, where $0 \leq f(x) \leq 1$. We suppose we have a source of random numbers uniformly distributed in $[0,1]$. One method consists in drawing pairs of independent random numbers ξ_i, η_i, observing whether (ξ_i,η_i) lies below the curve $y = f(x)$, and scoring 1 if it does, i.e., if $f(\xi_i) > \eta_i$, and 0 otherwise. The relative score is taken as an estimate for I.

The second method consists of taking single random numbers ξ_i and using

$$N^{-1} \sum_{i=1}^{N} f(\xi_i)$$

as an estimate for I.

It is easy to verify that, in general, the second method is at least as efficient as the first in the sense that the estimator has a smaller variance; in the special case of $f(x) = x$ the variance of the second method is one-third of that of the first. The computational labor involved in the two methods is about the same: In the first method one must compute $f(\xi_i)$ and compare it with η_i and accumulate a 1 or 0, and in the second one has to accumulate $f(\xi_i)$. As we are about to see, the generation of the random numbers can be done at a negligible expense.

The generation of a sequence of numbers that behave as if they were uniformly distributed on $[0,1]$ has been discussed at length, but practically all the results are experimental. We shall merely state that the sequence

$$r_n = 2^{-42} x_n$$

where

$$x_0 = 1 \qquad x_{n+1} \equiv \rho x_n (\text{mod } 2^{42}) \qquad \rho = 5^{2r+1}$$

(for example, $r = 8$)

seems to pass all the tests of randomness devised by statisticians; we note that the sequence has a long period, 2^{40}. Successive numbers are generated by a single "low-order" multiplication on a 42-bit machine. For other machines, different multipliers ρ and different recurrence relations, e.g.,

$$x_{n+1} \equiv \rho x_n + y$$

may be appropriate.

We conclude with some representative results (due to Davis and Rabinowitz) for the evaluation of the volume of a three- and a nine-dimensional sphere; they indicate the sort of accuracy attainable.

Number of points	Three dimensions	Nine dimensions
256	.562500	.015625
1024	.534180	.012695
4096	.521973	.007812
16384	.529968	.005676
Exact answer	.523599	.006442

Numerical Differentiation. Numerical differentiation is notoriously delicate; this is intuitively obvious for one is normally subtracting two large quantities and dividing by a small one. We discuss this process briefly here.

Using formal power series expansions,

$$\begin{aligned} hD &= \Delta - (1/2)\Delta^2 + (1/3)\Delta^3 - \cdots \\ &= \mu\delta - (1/6)\mu\delta^3 + (1/30)\mu\delta^5 - \cdots \\ h^2D^2 &= \Delta^2 - \Delta^3 + (11/12)\Delta^4 - (5/6)\Delta^5 + \cdots \\ &= \delta^2 - (1/12)\delta^4 + (1/90)\delta^6 - \cdots \end{aligned}$$

All orders of differences appear in the formulas involving Δ but only alternate ones appear in those involving δ; further, the coefficients in the latter case are smaller. It is generally true that central-difference formulas are the most satisfactory in practice.

It is often sufficient to tabulate the first derivative of a function at points halfway between the original points of tabulation. For this we have the very convenient formula

$$hf'_{1/2} = \delta f_{1/2} - (1/24)\delta^3 f_{1/2} + (3/640)\delta^5 f_{1/2} - \cdots$$

The following is an abstract from a table of $J_0(x)$. From it we verify that the differential equation $xy'' + y' + xy = 0$ is satisfied at $x = 20$.

x	$J_0(x)$		$\delta^2 J_0(x)$		
19.99	0.16768 47990		−164422		
		[−66 01347]		[741]	
20.00	0.16702 46643	{−66 83188}	−163681	{748}	[14]
		[−67 65028]		[755]	
20.01	0.16565 53661		−162926		

The entries in brackets are computed first; then the mean differences in braces are obtained as the mean of the differences above and below. We then use formulas above to obtain

$$\begin{aligned} J_0(20) &= 100[-6683187 - 1/6(748) + \cdots]10^{-10} \\ &= -0.06683\,312 \\ J''_0(20) &= 10000[-163681 - 1/12(14) + \cdots]10^{-10} \\ &= -0.16368\,2 \end{aligned}$$

We find

$$20J''_0(20) + J'_0(20) + 20J_0(20) = 0.00003$$

Four figures have been lost in determination of the second derivative. Some loss is inevitable in differentiation, whether done graphically or numerically. It is desirable, therefore, to use as efficient formulas as possible and to use as large an interval as is convenient.

9. Difference Equations and Ordinary Differential Equations

Equations of these types occur in many branches of science. We begin with a brief account of linear difference equations with constant coefficients.

Difference Equations. The unique solution of the system

$$u_{n+1} = \alpha u_n \qquad \alpha \neq 0 \qquad u_0 \text{ given}$$

is manifestly

$$u_n = u_0\alpha^n$$

The solution of the system

$$u_{n+2} - (\alpha + \beta)u_{n+1} + \alpha\beta u_n = 0 \qquad \alpha \neq 0,\ \beta \neq 0 \quad u_0,\ u_1 \text{ given}$$

is obtained as follows: Writing it first as

$$(u_{n+2} - \beta u_{n+1}) - \alpha(u_{n+1} - \beta u_n) = 0$$

and then as

$$(u_{n+2} - \alpha u_{n+1}) - \beta(u_{n+1} - \alpha u_n) = 0$$

and applying the previous result, we find

$$u_{n+1} - \beta u_n = (u_1 - \beta u_0)\alpha^n$$

and

$$u_{n+1} - \alpha u_n = (u_1 - \alpha u_0)\beta^n$$

If $\alpha \neq \beta$, we can eliminate u_{n+1} from this to obtain

$$u_n = A\alpha^n + B\beta^n$$

where A, B are certain constants. If $\alpha = \beta$, we find, instead,

$$u_n = (a + bn)\alpha^n$$

where a, b are certain constants. Any second-order linear difference equation with constant coefficients can be put in the above form (after solving a quadratic equation) and so our result is quite general. The result can be generalized to cover recurrence relations or difference equations of higher order. We note that, in general, unless $|\alpha| \leq 1$ and $|\beta| \leq 1$ the solution will be unbounded.

Very few nonlinear difference equations have been discussed. An example of a second-order equation, written for convenience as a pair of first-order equations, which has been studied in detail, is the following:

$$\begin{aligned} y_{n+1} &= e^{x - x_n}y_n \\ x_{n+1} &= e^x y_n(1 - e^{-x_n}) \end{aligned}$$

A classical example is

$$a_{n+1} = \tfrac{1}{2}(a_n + b_n) \qquad b_{n+1} = \sqrt{a_n b_n}$$

for which, if $a_0 \leq b_0$, we have

$$\lim a_n = \lim b_n = M(a_0,b_0)$$

where $M(a,b)$ is the arithmetic-geometric mean of a, b defined by

$$M = \pi b/2K'$$

$$K' = \int_0^1 \{(1 - x^2)(1 - (1 - k^2)x^2)\}^{-1/2}\,dx \qquad k = a/b$$

Differential Equations. We now discuss ordinary differential equations. Occasionally the simplest solution is provided by evaluating the analytical solution, either directly or by use of tables; the compendium of Kamke is invaluable in this connection, followed by reference to indices of tables. But it is easy to construct examples where this process is applicable but not very successful; for instance, the solution of $y' = x^2 - y^2$, $y(0) = 1$ is

$$y(x) = x\,\frac{\Gamma(1/4)I_{3/4}(\tfrac{1}{2}x^2) + 2\Gamma(3/4)I_{-3/4}(\tfrac{1}{2}x^2)}{\Gamma(1/4)I_{-1/4}(\tfrac{1}{2}x^2) + 2\Gamma(3/4)I_{1/4}(\tfrac{1}{2}x^2)}$$

Picard Method. Some of the existence theorems in theoretical analysis are constructive in character and can be used to provide numerical solutions. For instance, the *Picard method* for the initial value problem

$$y' = f(x,y) \qquad y(a) = b$$

suggests the definition of a sequence of function $y_n(x)$ by

$$y_0(x) \equiv b$$

$$y_{n+1}(x) = b + \int_a^x f[t,y_n(t)]\,dt$$

Under mild assumptions on $f(x,y)$, it can be proved that the sequence $\{y_n(x)\}$ has a limit that is the unique solution of the problem. Since we have quadrature methods at our disposal, it would be possible to evaluate each y_n in turn until a satisfactory approximate solution is obtained.

A little experience shows that this method is not a very practical one, except in the neighborhood of a. We find that we have essentially a two-dimensional solution to a one-dimensional problem; we have to tabulate *each* $y_n(x)$. It will be seen that modifications of the quadrature process enable us to traverse the interval (a,x) once only.

Local Taylor Series. Whenever the differential equation is such that it is easy to obtain the nth derivative of its solution—for example, by recurrence relations—the method of *local Taylor series* is very suitable, especially for hand calculation, as it has checks at every stage and large intervals can be used. Two cases where this is applicable are as follows:

$$(1) \qquad y' = y - x^2 \qquad y(0) = 3$$

for which

$$y' = y - x^2 \qquad y'' = y - x^2 - 2x$$
$$y^{(3)} = y^{(4)} = \cdots = y - x^2 - 2x - 2$$

so that the *reduced derivative* $\tau_n = h^n y^{(n)}/n!$ satisfies

$$\tau_{n+1} = \frac{h\tau_n}{n+1}$$

The solution is $y = e^x + x^2 + 2x + 2$.

$$(2) \qquad (x^2 - 1)y'' + 2xy' - 6y = 0 \qquad y,\ y' \text{ given}$$

for which the reduced derivative satisfies

$$(x^2 - 1)\tau_{n+2} + \frac{2hx(n+1)}{n+2}\tau_{n+1} + \frac{h^2(n-2)(n+3)}{(n+1)(n+2)}\tau_n = 0$$

The solution is

$$y = Q_2(x) = [\tfrac{3}{4}x^2 - \tfrac{1}{4}] \log \left[\frac{x+1}{x-1}\right] - \frac{3}{2}x$$

We discuss the first example only, and we produce the following table:

x	$x = 0$	$x = 0.3$	$x = 0.6$
$\tau_0 = y$	3.00000	4.03986	5.38212
$\tau_1 = hy^{(1)}$	0.90000	1.18496	1.50664
$\tau_2 = h^2y^{(2)}/2!$	0.13500	0.15074	0.17200
$\tau_3 = h^3y^{(3)}/3!$	0.00450	0.00607	0.00820
$\tau_4 = h^4y^{(4)}/4!$	0.00034	0.00046	0.00061
$\tau_5 = h^5y^{(5)}/5!$	0.00002	0.00003	0.00004
$y(x + h)$	4.03986	5.38212	7.06961
$y(x - h)$	2.23082	3.00000	4.03985

Beginning in the column $x = 0$ with $\tau_0 = 3$, we compute τ_1, τ_2, and τ_3 directly and then obtain the remainder by the recurrence relation, stopping when the terms become negligible. Using the Taylor series centered at $x = 0$,

$$y(\pm 0.3) = \tau_0 \pm \tau_1 + \tau_2 \pm \tau_3 + \tau_4 \pm \tau_5 + \cdots$$

and summing $\tau_0 + \tau_2 + \tau_4 + \cdots$ and $\tau_1 + \tau_3 + \tau_5 + \cdots$ and adding and subtracting, we obtain $y(\pm 0.3)$. The value at $x = -0.3$ can be used to check a previous value; the value $x = 0.3$ is a new value which is entered at the top of the second column. The second column is completed as was the first, and we obtain $y(0.3 \pm 0.3)$, that is, $y(0)$, which we check with our initial value, and $y(0.6)$, which we use to start the third column.

This method is a very powerful one and can be used for tabulation to high accuracy of pivotal values (at a wide interval) of a function which might be subtabulated. If we attempt to use a larger interval $h = 1.5$ we find that 11 terms are needed instead of 6.

Predictor-Corrector Schemes. We now describe a simple case of *predictor-corrector* schemes. They rely on *open* quadrature formulas to predict and on *closed* quadrature formulas to correct. We consider only scalar equations of the form

$$y'(x) = f(x,y) \qquad y(0) \text{ given}$$

but the schemes can be developed for equations of higher order or for vector equations. We write

$$I = \int_0^{nh} f(t)\,dt$$

and use results from Sec. 8 which we restate in the form

Milne:

$$I_4 - I_0 = 4h[2f_1 - f_2 + 2f_3]/3, \text{ with an error of } 14M'_4h^5/45$$

Simpson:

$$I_4 - I_2 = h[f_2 + 4f_3 + f_4]/3, \text{ with an error of } -M''_4h^5/90$$

where M'_4, M''_4 are values of $f^{(4)}(t)$ at some intermediate points. Note that the error in the second estimate is about $\tfrac{1}{28}$ of that in the first. This fact can be used in comparison of the estimates and gives some idea of the errors being committed. Suppose that we have obtained (adequate approximations to) $f_1 = f(h,y(h))$, f_2, f_3. Then, by means of the first formula, we can predict $I_4 = y_4$. Using this, we can compute f_4 and then use the second formula to obtain another estimate for y_4. If the two values of y_4 do not disagree too violently, we proceed. In the event of violent disagreement, a change to a smaller h is indicated.

We discuss the solution of $y' = x + y$, $y(0) = 1$ for $0 \le x \le 1$, to $4D$. The correct solution is

$y = 2e^x - 1 - x$, so that

$$y(1) = 2e - 2 = 3.43656 \ldots$$

Using an interval $h = .1$ and assuming that $y(.1)$, $y(.2)$, $y(.3)$ are known, our computation begins as follows:

x	y		$y' = f(x,y) = x + y$
0	1.0000		1.0000
.1	1.1103		1.2103
.2	1.2428		1.4428
.3	1.3997		1.6997
.4	1.5836	1.5836	1.9836
.5	1.7974	1.7974	2.2974
.6	2.0442		2.6442
.7			

Stability. The predictor-corrector scheme is a very convenient one for hand calculation but, from the point of view of stability, it has certain weaknesses that imply that care should be taken when it is used extensively, e.g., on automatic equipment. The question of stability is a very delicate one from the theoretical point of view but it can produce catastrophic results, and the computer must always remain on the alert. We give a rather rough account of the problem in the present context.

The trouble occurs in the use of the Simpson corrector. We must now be careful about our notation: y_n, y'_n will denote the theoretical values of y, y' at $x_n = nh$, and Y_n, Y'_n will denote the computed values there (in principle, computed to infinite precision; in practice, the round-off should be negligible compared with the "truncation errors" in the Simpson formula). For the error ϵ' in y' we have

$$\epsilon_n = f(x_n,y_n) - f(x_n,Y_n) = (y_n - Y_n)f_y(x_n,\theta_n)$$

where θ_n is between y_n, Y_n; we assume f_y continuous in y. The error $\epsilon_n = y_n - Y_n$ in y satisfies

$$\epsilon_{n+1} = \epsilon_{n-1} + h[\epsilon'_{n-1} + 4\epsilon'_n + \epsilon'_{n+1}]$$

We now suppose that f_y has a constant value k, say; presumably the behavior when f_y is varying slowly will be similar. We have $\epsilon'_n = k\epsilon_n$; substituting this in the last equation, we get a recurrence relation for ϵ_n:

$$\epsilon_{n+1} = \frac{4hk}{3 - hk}\epsilon_n + \frac{3 + hk}{3 - hk}\epsilon_{n-1}$$

The solution of this difference equation is

$$\epsilon_n = A\alpha^n + B\beta^n$$

when α, β are the roots of

$$\lambda^2 - \frac{4hk}{3 - hk}\lambda - \frac{3 + hk}{3 - hk} = 0$$

Since h is small, we have, approximately,

$$\alpha \doteqdot 1 + hk \qquad \beta \doteqdot -1 + hk/3$$

one or the other of which exceeds 1 in absolute value.

The differential equation has the solution

$$y(x) = Ce^{kx}$$

while the difference equation gives, for $x = nh$,

$$y(x) = A(1 + (kx/n))^n + B(-1)^n[1 - (kx/3n)]^n$$
$$\sim Ae^{kx} + (-1)^nBe^{-kx/3}$$

Suppose $k < 0$ so that the solution to the differential equation is a decaying exponential; then the solution to the difference equation contains that solution contaminated by an oscillating positive exponential term which will swamp the true solution if $B \neq 0$. In practice, components of the unwanted solution are brought in by the roundings in the computation.

We now wish to illustrate one aspect of the phenomenon of instability in numerical detail, in a simple case.

We consider the *numerical* solution of the differential equation

$$y'' = -y \qquad y'(0) = 1 \qquad y(0) = 0$$

We discretize this, reducing the problem to a difference equation. If we use an interval h and write

$$y_n = y(x_n) \qquad x_n = nh$$

one discretization of the differential equation is

$$y_{n+1} = (2 - h^2)y_n - y_{n-1}$$

This amounts to approximating y'' by $(y_{n+1} - 2y_n + y_{n-1})/h^2$; we take as initial conditions $y_0 = 0$, $y_1 = h$. Another, much better approximation to y'' is

$$(-y_{n+2} + 16y_{n+1} - 30y_n + 16y_{n-1} - y_{n-2})/12h^2$$

The comparative strength of these, in approximating $f''(1) = -f(1)$, is indicated by the following table:

x	$\sin x = f(x)$	$f''(1)$
.6	.564642	
.8	.717356	3-point: −.838675
1.0	.841471	
1.2	.932039	5-point: −.841562
1.4	.985450	

For $h = .1$ the difference equations become

$$y_{n+1} = 1.99y_n - y_{n-1}$$

and

$$y_{n+2} = 16y_{n+1} - 29.88y_n + 16y_{n-1} - y_{n-2}$$

and the second is, plausibly, much better. What actually happens, however, is shown in the following table. The column labeled (1) gives the correct values of $\sin x$, that labeled (2) is the result of using the second-order difference equation, while the results in (3) and (4) are obtained by using the fourth-order difference equation and working to $10D$ and to $5D$, respectively.

x	(1)	(2)	(3)	(4)
0	0.00000 00000	0.00000	0.00000 00000	0.00000
0.1	0.09983 34166	0.09983	0.09983 34166	0.09983
0.2	0.19866 93308	0.19866	0.19866 93308	0.19867
0.3	0.29552 02067	0.29550	0.29552 02067	0.29552
0.4	0.38941 83423	0.38939	0.38941 83685	0.38934
0.5	0.47942 55386	0.47939	0.47942 59960	0.47819
0.6	0.56464 24734	0.56460	0.56464 90616	0.54721
0.7	0.64421 76872	0.64416	0.64430 99144	0.40096
0.8	0.71735 60909	0.71728	0.71864 22373	−2.67357
0.9	0.78332 69096	0.78323	0.80125 45441	
1.0	0.84147 09848	0.84135	1.09135 22239	
1.1	0.89120 73601	0.89106	4.37411 56871	
1.2	0.93203 90860	0.93186		
1.3	0.96355 81854	0.96334		
1.4	0.98544 97300	0.98519		
1.5	0.99749 49866	0.99719		
1.6	0.99957 36030	0.99922		

This is a simple example of "instability." The initial errors in (3) and (4) are positive and negative, respectively; they are magnified by a factor of about 14 at each stage and soon swamp the desired solution. The reason for this is that the solution of the fourth-order difference equation is of the form

$$y(n) = A\alpha^n + B\alpha^{-n} + C \cos n\theta + D \sin n\theta$$

where, to four figures,

$$\alpha = 13.94 \qquad \theta = .1000$$

Euler, Heun, and Runge-Kutta Methods. One of the most naïve approaches to the numerical solution of $y' = f(x,y)$, $y(x_0) = y_0$ is associated with Euler. It consists in defining the solution by the relation

$$y(n + 1) = y(n) + hf(nh,y(n)) \qquad n \geq 0 \qquad y(0) = y_0$$

If we assume that y can be expanded as a power series in h, it is clear that the local error is $O(h^2)$. This means that we might expect a total error of about $O(h)$, for over a fixed range we would have $O(h^{-1})$ steps, and if we assume the errors additive, we have $O(h^{-1})O(h^2) = O(h)$. This is therefore not a very practical method; we note, however, that it requires no special starting devices.

A slightly more complicated method is due to Heun; it has a smaller error but retains the advantages of requiring no special starting devices. It consists of the following scheme, for $n \geq 0$:

$$\begin{aligned} y^*(n + 1) &= y(n) + hf(nh,y(n)) \\ y^{**}(n + 1) &= y(n) + hf((n + 1)h, y^*(n + 1)) \\ y(n + 1) &= \tfrac{1}{2}(y^*(n + 1) + y^{**}(n + 1)) \end{aligned}$$

Again, assuming the existence of a power series expansion for y, we find that the local error is $O(h^3)$. This means that the total error, in favorable cases, is $O(h^2)$, which makes the scheme a feasible one.

The above error estimates do not hold whenever the solution cannot be expanded as a power series. If we take a trivial case

$$y' = x^{1/2}$$

we see that the local error is $O(h^{3/2})$, not $O(h^3)$. It is instructive to study the following classical example in the range $[0,1]$:

$$y' = \sqrt{x} + \sqrt{y} \qquad y(0) = 0$$

which has a solution of the form

$$y = \tfrac{2}{3}x^{3/2} + \tfrac{4}{7}(\tfrac{2}{3})^{1/2}x^{7/4} + \tfrac{1}{7}x^2 + \tfrac{1}{49}(\tfrac{2}{3})^{1/2}x^{9/4} - \tfrac{2}{1715}x^{5/2} + \cdots$$

Should the Heun method be unsatisfactory, it is natural to try to get a similar method with a smaller local error. One such method is the Runge-Kutta method, which consists in writing

$$y(n + 1) = y(n) + \tfrac{1}{6}[K(1,n) + 2K(2,n) + 2K(3,n) + K(4,n)]$$

where

$$\begin{aligned} K(1,n) &= hf(nh,y(n)) \\ K(2,n) &= hf((n + \tfrac{1}{2})h, y(n) + \tfrac{1}{2}K(1,n)) \\ K(3,n) &= hf((n + \tfrac{1}{2})h, y(n) + \tfrac{1}{2}K(2,n)) \\ K(4,n) &= hf((n + 1)h, y(n) + K(3,n)) \end{aligned}$$

The local error in this case is $O(h^5)$.

A detailed comparison of the methods discussed is rather difficult. The following remarks apply in many representative cases. The Runge-Kutta scheme requires four evaluations of $f(x,y)$ per step whereas the Heun method requires but two; however, the step size in the Heun method is often much less than half that required in the Runge-Kutta, so that the latter is generally to be preferred.

Comparison of the local errors in the Runge-Kutta and Simpson-Milne methods gives a factor of 32 in favor of Runge-Kutta, which can be balanced by a factor of 2 in the step size. This means that the amount of calculation is comparable.

Summarizing, we may say that, although the Runge-Kutta method has no checks, it is probably to be preferred for automatic calculators, whereas the Simpson-Milne method is to be preferred for desk computers because of its checks, despite the difficulties at the beginning (which reappear if the interval has to be decreased in the course of the calculation), and provided circumstances do not involve instabilities.

We note that it is possible to try the h^2 extrapolation discussed in Sec. 8. We can carry out, for instance, a Heun integration at interval h, then one at interval $\frac{1}{2}h$, say, and then improve each ordinate by eliminating the h^2 component in the error.

Boundary-value Problems. So far we have discussed initial value problems only. However, various boundary problems for ordinary differential equations are of practical importance. Many of them are discussed in detail in the books of Collatz and Fox. We mention briefly two problems.

1. Find a value of λ such that there is a nontrivial solution of the equation

$$y'' + \lambda xy = 0 \qquad y(0) = 0 = y(1)$$

One method of trial and error for this, sometimes called the "shooting method," is the following: We guess a value of λ, say λ_1, and solve the resulting initial value problem, obtaining a value f_1 at $x = 1$. We need to choose some initial slope, which we may take to be $y_1 = 1$.

A reasonable guess for λ_1 can be obtained in the following way. Discretize the problem at interval $h = \frac{1}{4}$ to get

$$\begin{aligned} y(0) - 2y(\tfrac{1}{4}) + y(\tfrac{1}{2}) &= -\tfrac{1}{16} \cdot \tfrac{1}{4} y(\tfrac{1}{4})\lambda \\ y(\tfrac{1}{4}) - 2y(\tfrac{1}{2}) + y(\tfrac{3}{4}) &= -\tfrac{1}{16} \cdot \tfrac{1}{2} y(\tfrac{1}{2})\lambda \\ y(\tfrac{1}{2}) - 2y(\tfrac{3}{4}) + y(1) &= -\tfrac{1}{16} \cdot \tfrac{3}{4} y(\tfrac{3}{4})\lambda \end{aligned}$$

where we put $y(0) = y(1) = 0$. This gives a cubic for λ, of which the relevant root is 17.87.

Having obtained f_1 and λ_1, we take another λ_2 and compute the corresponding f_2. Using these two values, we interpolate and get a new approximation λ_3 for λ. We then obtain f_3 and carry on in this manner.

The exact solution to the problem satisfying the left-hand boundary condition only is

$$y = x^{\frac{1}{2}} J_{\frac{1}{3}}(\tfrac{2}{3}\sqrt{\lambda}\, x^{\frac{3}{2}})$$

The least value of λ for which this vanishes is given by

$$\tfrac{2}{3}\sqrt{\lambda} = 2.9025 \qquad \text{that is, } \lambda = 18.9563$$

2. Find the least value of λ for which there is a nontrivial solution of the relations

$$y'' + \lambda y = 0 \qquad y(0) = 0 = y(\pi)$$

It can be shown that, if $u(x)$ is a function satisfying the boundary conditions, then the required value of $\lambda^{\frac{1}{2}}$ is the minimum value of the Rayleigh quotient,

$$R(u) = -\int_0^\pi uu''\,dx \Big/ \int_0^\pi u^2\,dx$$

and the function u which minimizes $R(u)$ is the required solution.

Thus, if $u_1 = \sin x$,
we get $R(u_1) = 1$
If we take $u_2 = x(x - \pi)$,
we get $R(u_2) = 10/\pi^2 > 1$
and if we take $u_3 = x(x + \pi)(x - \pi)$,
we get $R(u_3) = 21/2\pi^2 > 1$
The development of this idea into a practical method requires considerable experience. Advances have been made by the Aronszajn-Weinstein school.

10. Partial Differential Equations

For clarity we confine our discussion to special cases of three classical problems:

Potential equation: $u_{xx} + u_{yy} = 0$ in a domain D with $u(x,y) = f(x,y)$ on the boundary of D, f given.

Heat equation: $u_t = u_{xx}$ in $t \geq 0$, $0 \leq x \leq 1$ with $u(x,0) = f(x)$ and $u(0,t) \equiv 0$, $u(1,t) \equiv 0$, f given.

Wave equation: $u_{tt} = u_{xx}$, $t \geq 0$ with $u(x,0) = f(x)$ and $u_t(x,0) = g(x)$, f, g given.

These are linear equations with constant coefficients. For computer use, there is little additional formal complication if the coefficients are not constant, but the theoretical complications are considerable.

We note that numerical solutions often can be obtained, in principle, as appropriate linear combinations of particular solutions known analytically. But in practice this method (which has been studied in the context of automatic computers by, for example, the Bergman school) is not generally as successful as the method of finite differences, and we shall restrict our discussion to the latter method. There are many ways of approximating the differential operators by finite-difference operators on a rectangular mesh; this variety makes the theory richer and the pitfalls deeper.

The Potential Equation. Let us suppose D is a square with vertices (0,0), (1,0), (1,1), (0,1). A simple discretization of the problem is obtained as follows: Denote by (r,s) the point (rh,sh), where $h = 1/(n + 1)$, n integral. We shall obtain a function U defined at the points (r,s) which is an approximation to $u(rh,sh)$. The Laplacian operator is approximated by

$$h^{-2}[U(r-1,s) + U(r,s-1) - 4U(r,s) + U(r,s+1) + U(r+1,s)]$$

and the differential system

$$u_{xx} + u_{yy} = 0 \text{ inside } D \qquad u(x,y) = f(x,y)$$

on the boundary of D by a system of n^2 linear equations

$$\mathcal{A}U = b$$

where $\mathcal{A}$ is an $n \times n$ block triple diagonal matrix of $n \times n$ triple diagonal matrices: In fact,

$$\mathcal{A} = \begin{bmatrix} A & I & & & \\ I & A & I & & \\ & & \cdot & & \\ & & \cdot & & \\ & & \cdot & & \\ & & I & A & I \\ & & & I & A \end{bmatrix}$$

with
$$A = \begin{bmatrix} -4 & 1 & & & \\ 1 & -4 & 1 & & \\ & & \cdot & & \\ & & \cdot & & \\ & & \cdot & & \\ & & 1 & -4 & 1 \\ & & & 1 & -4 \end{bmatrix}$$

where U is the n^2-dimensional vector of unknown $U(r,s)$, enumerated in the order

$$(1,1), (2,1), (3,1), \ldots, (n,1); \; (1,2), (2,2), (3,2), \ldots, (n,2); \; \ldots; (1,n), (2,n), (3,n), \ldots, (n,n)$$

and b is an n^2-dimensional vector (of which at most $4n$ components are nonzero). The reader is invited to write the matrix $\mathfrak{A}$ and the vector b in detail in simple cases $n = 3, 4$, for instance.

A more complicated discretization is obtained if, instead of the above 5-point approximation which can be represented by the left-hand stencil below, we use a 9-point approximation represented by the right-hand stencil

$$\begin{matrix} & 1 & \\ 1 & -4 & 1 \\ & 1 & \end{matrix} \qquad\qquad \begin{matrix} 1 & 4 & 1 \\ 4 & -20 & 4 \\ 1 & 4 & 1 \end{matrix}$$

Our first remark is that $\mathfrak{A}$ is nonsingular. This can be established using a refinement of Gerschgorin's theorem of Sec. 7 or, alternatively, using the following results which are of fundamental importance in our account of this subject.

Lemma. Let $A = (a_{ij})$ be a triple diagonal matrix where

$$\begin{aligned} a_{ii} &= a \qquad & i &= 1, 2, \ldots, n \\ a_{i,i+1} &= b \qquad & i &= 1, 2, \ldots, n-1 \\ a_{i,i-1} &= c \qquad & i &= 2, 3, \ldots, n \end{aligned}$$

all other elements being zero. Then the characteristic values of A are

$$\alpha_k = a - 2\sqrt{bc}\cos(k\pi/(n+1)) \qquad k = 1, 2, \ldots, n$$

Theorem. Let $\mathfrak{A} = (A_{ij})$ be an $mn \times mn$ partitioned matrix where each of the m^2 blocks is a rational function of a fixed $n \times n$ matrix, say $\mathfrak{A}_{ij} = f_{ij}(A)$. Then the mn characteristic values of $\mathfrak{A}$ are obtainable as the characteristic values of the n matrices

$$(f_{ij}(\alpha_k)) \qquad i, j = 1, 2, \ldots, m$$

where α_k, $k = 1, 2, \ldots, n$ are the characteristic values of A.

Various generalizations of this theorem are available but it is sufficient for the present purposes to have it in the above original form of Williamson.

Applying these results, we find the characteristic values of $\mathfrak{A}$ to be

$$\alpha_{r,s} = -4 - 2\cos\frac{r\pi}{n+1} - 2\cos\frac{s\pi}{n+1} \qquad r, s = 1, 2, \ldots, n$$

and all are different from zero.

Because of the simple geometry of this problem we can discuss it completely explicitly. It therefore provides a good example on which to carry out controlled computational experiments. When the geometry is more complicated, the structure of the matrix will be less regular but it will nevertheless remain "sparse"; this is true if we leave the constant-coefficient case. We begin by considering the applications of the Jacobi and Gauss-Seidel methods discussed in Sec. 6.

In the Jacobi case, after normalizing the matrix $\mathfrak{A}$ by multiplying by $-\frac{1}{4}$ we find the iteration matrix to be

$$\begin{bmatrix} B & \frac{1}{4}I & & & & \\ \frac{1}{4}I & B & \frac{1}{4}I & & & \\ & & \cdot & & & \\ & & & \cdot & & \\ & & & & \cdot & \\ & & & \frac{1}{4}I & B & \frac{1}{4}I \\ & & & & \frac{1}{4}I & B \end{bmatrix}$$

where

$$B = \begin{bmatrix} 0 & \frac{1}{4} & & & & \\ \frac{1}{4} & 0 & \frac{1}{4} & & & \\ & & \cdot & & & \\ & & & \cdot & & \\ & & & & \cdot & \\ & & & \frac{1}{4} & 0 & \frac{1}{4} \\ & & & & \frac{1}{4} & 0 \end{bmatrix}$$

Williamson's theorem obviously applies, and the characteristic values of the iteration matrix are

$$\tfrac{1}{2}\cos\frac{r\pi}{n+1} + \tfrac{1}{2}\cos\frac{s\pi}{n+1} \qquad r, s = 1, 2, \ldots, n$$

and its spectral radius is

$$\text{Jacobi:} \qquad \cos\frac{\pi}{n+1} \doteq 1 - \frac{\pi^2}{2n^2}$$

It can be shown that the spectral radius for the Gauss-Seidel process is the square of the Jacobi value:

$$\text{Gauss-Seidel:} \qquad \cos^2\frac{\pi}{n+1} \doteq 1 - \frac{\pi^2}{n^2}$$

This comes as the special case $\omega = 1$ of the successive overrelaxation method introduced by D. M. Young in 1950 which we now discuss.

In an application of the Gauss-Seidel scheme we "improve" the value $U(r,s)$ by adding

$$\tfrac{1}{4}[U(r-1, s) + U(r, s-1) - 4U(r,s) + U(r, s+1) + U(r+1, s)]$$

Manual computers of the relaxation era found that, if one "overcorrected" appropriately at each stage, convergence was faster. We now consider overrelaxation with parameter ω, adding ω times the above correction instead, and try to find an optimal value of this parameter.

We discuss this in the general case where the matrix $\mathfrak{A}$ is normalized to have a unit diagonal. We write

$$\mathfrak{A} = -L + I - R$$

where L (and R) coincide with $\mathfrak{A}$ on the left (and right) of the diagonal and have zero elements everywhere else. The iteration equation now becomes

$$\boldsymbol{\varepsilon}^{(j+1)} = (I - \omega L)^{-1}((1-\omega)I + \omega R)\boldsymbol{\varepsilon}^{(j)}$$

and so we are concerned with the spectral radius of

$$\mathfrak{L}_\omega = (I - \omega L)^{-1}((1-\omega)I + \omega R)$$

and with the choice of ω for which this is least.

For simplicity, we take the case when n is even, and we may then assume a reordering of our equations and variables so that the basic equation is in the form

$$\begin{bmatrix} I & -M \\ -M & I \end{bmatrix} \mathbf{x} = \mathbf{b}$$

The corresponding iteration matrix now is of the form

$$\mathcal{L}_\omega = \begin{bmatrix} I & 0 \\ -\omega M & I \end{bmatrix}^{-1} \begin{bmatrix} (1-\omega)I & \omega M \\ 0 & (1-\omega)I \end{bmatrix}$$
$$= \begin{bmatrix} I & 0 \\ \omega M & I \end{bmatrix} \begin{bmatrix} (1-\omega)I & \omega M \\ 0 & (1-\omega)I \end{bmatrix}$$
$$= \begin{bmatrix} (1-\omega)I & \omega M \\ \omega(1-\omega)M & \omega^2 M^2 + (1-\omega)I \end{bmatrix}$$

Hence the characteristic values λ of $\mathcal{L}_\omega$ are those of

$$\begin{bmatrix} (1-\omega), & \omega\mu \\ \omega(1-\omega)\mu, & \omega^2\mu^2 + (1-\omega) \end{bmatrix}$$

where the μ are the characteristic values of M. Thus we have

$$(\lambda + (\omega - 1))^2 = \lambda\omega^2\mu^2$$

and, in particular for the case $\omega = 1$, we have

$$\lambda = \mu^2$$

establishing the announced estimate for the Gauss-Seidel situation.

An elementary investigation of the behavior of (the two values of) λ as μ runs through the characteristic values of M, that is, from $\cos n\pi/(n+1)$ to $\cos \pi/(n+1)$, establishes the optimal value of ω as

$$\omega_b = 2\{1 + \sin \pi/(n+1)\}^{-1}$$

and the corresponding spectral radius as

Young SOR: $\qquad \omega_b - 1 \doteqdot 1 - \dfrac{2\pi}{n}$

The last development which we shall consider is the ADI method, introduced by Peaceman, Rachford, and Douglas, beginning in 1956. Formally, it proceeds as follows: We split $\mathcal{A}$ in some convenient way as

$$\mathcal{A} = H + V$$

and replace the basic equation

$$u^{(j+1)} = b - ((H+V) - I)u^{(j)}$$

by the sequence of pairs

$$\begin{aligned} (H + r_{j+1}I)U^{(j+\frac{1}{2})} &= (r_{j+1}I - V)U^{(j)} + b \\ (V + r_{j+1}I)U^{(j+1)} &= (r_{j+1}I - H)U^{(j+\frac{1}{2})} + b \end{aligned}$$

Here the r_j are parameters to be assigned. In principle, they may be all different, or periodic, with period m, say. Observe that each stage of this iteration involves the inversion of $H + r_{j+1}I$ to obtain $U^{(j+\frac{1}{2})}$ and then that of $V + r_{j+1}I$ to obtain $U^{(j+1)}$. This is the reason for the adjective "implicit." To make this method practical, H and V must be chosen so that this process is easy, e.g., so that they are triple diagonal matrices. A little algebra shows that the iteration relation for the error vectors is

$$\varepsilon^{(j+1)} = [(V + r_{j+1}I)^{-1}(H - r_{j+1}I)(H + r_{j+1}I)^{-1} (V - r_{j+1}I)]\varepsilon^{(j)}$$

Assume H and V have the same characteristic-vector system. We also assume we are dealing with the periodic case. Then the attenuation of the ith characteristic vector per *period* is

$$\mu_i = \left| \prod_{j=1}^{m} \left\{ \frac{\sigma_i - r_j}{\sigma_i + r_j} \cdot \frac{\tau_i - r_j}{\tau_i + r_j} \right\} \right|$$

where σ_i, τ_i are the ith characteristic values of H and V. It is now clear that we ought to choose the parameters (r_j) so that max μ_i is least.

An appropriate decomposition of the matrix A is $A = H + V$ where H is the triple diagonal matrix whose general row is

$$\ldots \; -\tfrac{1}{4}, \tfrac{1}{2}, -\tfrac{1}{4}, \; \ldots$$

and V is a triple diagonal *block* matrix whose general row is

$$\ldots, \; -\tfrac{1}{4}I, \tfrac{1}{2}I, -\tfrac{1}{4}I, \; \ldots$$

It is easy to verify that H and V are positive definite symmetric matrices and that they commute and that their characteristic values are each

$$\sigma_l = \cos^2 l\pi/2(n+1) \qquad l = 1, 2, \ldots, n$$

each n-fold.

The spectral radius of the iteration matrix is max μ_i, where

$$\mu_i = \prod_{j=1}^{m} \left(\frac{\sigma_i - r_j}{\sigma_i + r_j}\right)^2$$

and we can approximate this by

$$\max_{\sigma_n \le x \le \sigma_1} \prod_{j=1}^{m} \left(\frac{x - r_j}{x + r_j}\right)^2$$

This is close to

$$\max_{k' \le x \le 1} \prod_{j=1}^{m} \left(\frac{x - r_j}{x + r_j}\right)^2$$

when $k' = \pi^2/4n^2$.

The determination of the optimal choice of the $\{r_j\}$, that is, those making the last expression minimum, was carried out by W. B. Jordan in 1964. He found the optimal r_j were given by

$$r_j = \operatorname{dn}\left(\frac{(2j-1)K}{2m}, k^2\right) \qquad j = 1, 2, \ldots, m$$

where dn is the Jacobian elliptic function defined in Part 1, Chap. 3. From these r_j we can estimate the corresponding spectral radius. We now note that the value of m is still at our disposal and although it appears, after making due allowance for the complications of our iteration, that the efficiency of the scheme increases indefinitely with m, a choice of m about ln n gives an effective spectral radius of

ADI optimal: $\qquad 1 - c/\ln n$

where c is a moderate constant, which is significantly better than that for the Young process.

We now state our present position: We have available a series of iterative processes for the solution of the discretized Dirichlet problem, for a particular h.

The rate of convergence is acceptable, e.g., in the Young process, in which each iteration involves $O(n)$ operations; we de-e-fold the error in $O(n)$ steps for $(1 - 1/n)^n \sim e^{-1}$, and so we can attain a satisfactory approximation to the solution in $O(n^2)$ operations. There remains the question of the discretization error: Does U approach u? It was only recently (1963) that a satisfactory account of this was given: Bramble and Hubbard gave an a priori estimate of the error in terms of the smoothness of the boundary values and of the boundary itself. A detailed discussion of the convergence of the discrete solution, in the heat-conduction case, is given below.

Our theoretical discussion of the ADI process depended on the commutativity of the matrices H, V. It is remarkable that in the case of the 5-point discretization only for the Dirichlet (or Helmholtz) problem (with constant coefficients), when the domain is a rectangle, is this condition satisfied. In practice, however, satisfactory results have been obtained in more general conditions.

The Heat Equation. In the case of this equation and of the wave equation a new phenomenon arises. We shall discuss this in general terms before proceeding to a detailed account of the special case. In seeking numerical solutions of a partial differential equation, it is natural to replace it by a consistent finite-difference approximation, i.e., one which goes over formally into the continuous problem as the mesh of the lattice shrinks to a point. One very important fact is that this formal *consistency* between difference and differential equation problems gives no guarantee that the solution of the difference equation will converge toward that of the differential equation or will have any tendency to converge at all. It turns out that in addition to formal consistency the difference scheme has to satisfy certain *inequalities* if useful approximations are to be obtained. These inequalities ensure that the scheme is *stable*, where stability of the scheme can be defined as the property that the solution of the problem changes only by a small amount for small changes of the data independently of the mesh size used. This amounts to the assumption that the solution depends continuously on the data, uniformly for all mesh sizes. (Here the words "small" and "continuous" refer to suitable *norms* used for measuring changes in functions.) For a stable scheme the solution of the difference equation will converge toward that of the differential equation, provided a sufficiently regular solution of the differential equation problem exists. In addition the solution of the difference equation will change by little if the difference equation itself is solved only approximately, as long as the errors committed in this approximation are sufficiently small.

The exact solution to the problem stated at the beginning of this section can be given, provided we can expand $f(x)$ in an absolutely convergent Fourier series

$$f(x) = \sum a_i \sin i\pi x \qquad a_i = 2\int_0^1 f(x) \sin i\pi x \, dx$$

In this case it is

$$u(x,t) = \Sigma a_i \sin i\pi x \exp(-i^2\pi^2 t)$$

A simple discretization of this problem, which we can indicate by

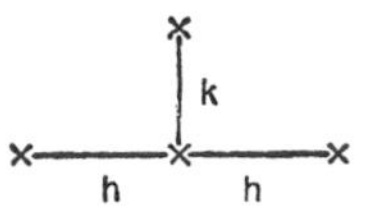

leads to the difference equation

$$U(m, n+1) = rU(m-1, n) + (1-2r)U(m,n) + rU(m+1, n)$$
$$U(m,0) = f(mh) \qquad U(0,n) = U(M+1, n) \equiv 0$$

where $U(m,n)$ is written for $U(mh,nk)$ and r for kh^{-2}, $h = (M+1)^{-1}$. This partial difference system can be solved explicitly, and it can be shown that if $r \le \frac{1}{2}$ then $U \to u$ as $h \to 0$. We shall discuss numerically the case where

$$f(x) = 2x \qquad 0 \le x \le \tfrac{1}{2}$$
$$f(x) = 2(1-x) \qquad \tfrac{1}{2} \le x \le 1$$

and concentrate on the point $(\frac{1}{2}, \frac{3}{64})$. Using the solution

$$u(x,t) = 8\pi^{-2}[\sin \pi x \exp(-\pi^2 t) - (\tfrac{1}{3}) \sin 3\pi x \exp(-9\pi^2 t) + \cdots]$$

we find

$$U(\tfrac{1}{2}, \tfrac{3}{64}) = .5117520442$$

We now fix $r = \frac{1}{4}$, take M odd, and observe the points on the line $x = \frac{1}{2}$; we find, where

$$\theta = \frac{\pi}{2(M+1)}$$

$$U(\tfrac{1}{2}(M+1), n) = 2(M+1)^{-2} \sum_{i \text{ odd}} \operatorname{cosec}^2 i\theta \cos^{2n} i\theta$$

Evaluating these, we find

$M = 3$	.53125	
$M = 7$	.51661 72981	
$M = 15$	.51296 84502 . . .	last three figures doubtful
$M = 31$	.51205 61854 . . .	last three figures doubtful
$M = \infty$	.51175 20442	

Convergence is apparent, and the differences $U_M - u$, rounded to five decimals, are

1950, 487, 122, 31

which makes it clear that the error is $O(h^2)$ (and that Richardson h^2 extrapolation is permissible).

We now note that the restriction $k/h^2 \le \frac{1}{2}$ is rather serious: A reduction in h implies that k is reduced to rh^2 and the number of time steps required to reach a particular point (x,t) will be correspondingly increased. It will be seen that if we keep

$$kh^{-2} = r > \tfrac{1}{2}$$

no sensible results can be obtained. Accordingly we look for more satisfactory discretizations of the problem. Among those considered is the one indicated by the stencil:

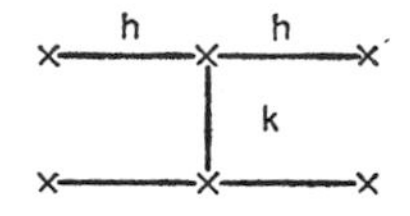

$$U(m, n+1) - U(m,n) = \tfrac{1}{2}r \begin{bmatrix} U(m-1, n) - 2U(m,n) \\ \quad + U(m+1, n) \\ +U(m-1, n+1) - 2U(m, n+1) \\ \quad + U(m+1, n+1) \end{bmatrix}$$

We want to discuss the behavior of the solutions of the finite-difference system in the two cases. If we write

$$\mathbf{l}_n = (U(1,n), U(2,n), \ldots, U(M,n))$$

then the first system has a solution

$$\mathbf{l}_n = A^n \mathbf{l}_0$$

where $A = I + rT_n$ is an $n \times n$ matrix, with the $n \times n$ matrix T_n defined by

$$T_n = \begin{bmatrix} -2, & 1 & & & \\ 1, & -2, & 1 & & \\ & & \cdot & & \\ & & \cdot & & \\ & & \cdot & & \\ & & 1, & -2, & 1 \\ & & & 1, & -2 \end{bmatrix}$$

The recurrence relation for $\mathbf{l}_n$ in the second case is

$$B\mathbf{l}_{n+1} = (-B + 4I)\mathbf{l}_n$$

where $B = 2I - rT_n$; this leads to

$$\mathbf{l}_n = C^n \mathbf{l}_0$$

where $C = 4B^{-1} - I$.

The matrices A, C have the same characteristic vectors $\{v_i\}$, say; let the characteristic values be $\{\alpha_i\}$, $\{\gamma_i\}$. We may expand our data $\mathbf{l}_0$ in terms of the $\{v_i\}$ as

$$\mathbf{l}_0 = \Sigma c_i \mathbf{v}_i$$

where

$$c_i = \{2/(M+1)\} \sum_{s=1}^{M} f(sh) \sin is\pi/(M+1)$$

Then the solutions are given explicitly as

$$\mathbf{l}_n = \Sigma c_i \alpha_i^n \mathbf{v}_i$$

and

$$\mathbf{l}_n = \Sigma c_i \gamma_i^n \mathbf{v}_i$$

Using the lemma stated earlier, we have

$$\alpha_i = 1 - 4r \sin^2 i\theta \qquad \theta = \tfrac{1}{2}\pi/(M+1)$$

$$\gamma_i = \frac{1 - 2r \sin^2 i\theta}{1 + 2r \sin^2 i\theta}$$

From these explicit solutions convergence can be established. We discuss the first case formally. We have

$$U(m,n) = \sum_{i=1}^{M} c_i \sin 2mi\theta (1 - 4r \sin^2 i\theta)^n$$

If f is integrable in the sense of Riemann, it is clear that

$$c_i = c_i(M) \to a_i \qquad M \to \infty$$

Since we are observing a particular point the factor $\sin 2mi\theta$ above and the factor $\sin i\pi x$ in

$$u(mh,nh) = \Sigma a_i \sin i\pi x \exp(-i^2\pi^2 t)$$

are the same. As for the remaining factor, we have

$$\left[1 - 4r \sin^2 \frac{\cdot\, i\pi}{2(M+1)}\right]^n \doteqdot \exp\left[-4rn \frac{i^2\pi^2}{4(M+1)^2}\right] = \exp(-i^2\pi^2 t)$$

All this can be established rigorously if $0 \leq r \leq \tfrac{1}{2}$. The explicit forms of the solutions show that they are uniformly bounded for $0 \leq r \leq \tfrac{1}{2}$ in the first case and for all r in the second. Hence small perturbations in the initial data (or small errors at any time) remain small under the stated circumstances. This is, essentially, the meaning of stability.

We have stated that a condition sufficient for convergence in the first case is that $r \leq \tfrac{1}{2}$, which is the same as that for stability. The equivalence of stability and convergence in a quite general setting has been established by P. Lax.

Many other discretizations of the heat equation have been discussed, some relating values at three time steps and therefore requiring initially the preparation of an additional line of data by other means.

The Wave Equation. We shall assume that we have a periodic problem. Using the simple discretization

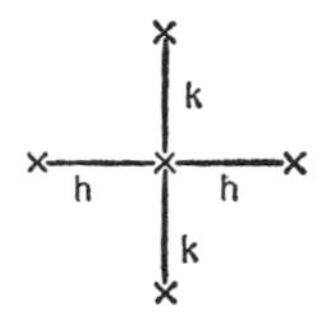

leads to the difference equation

$$U(m, n+1) - 2U(m,n) + U(m, n-1) = k^2h^{-2}\{U(m-1, n) - 2U(m,n) + U(m+1, n)\}$$

where the first coordinate m is to be understood mod $M + 1$.

We can write this equation in matrix form as

$$\begin{bmatrix} \mathbf{l}_{n+1} \\ \mathbf{l}_n \end{bmatrix} = \begin{bmatrix} C & -I \\ I & 0 \end{bmatrix} \begin{bmatrix} \mathbf{l}_n \\ \mathbf{l}_{n-1} \end{bmatrix}$$

where C is the circulant matrix

$$\begin{bmatrix} 2 - 2r^2, & r^2, & & & r^2 \\ r^2, & 2 - 2r^2, & r^2, & & \\ & & \cdot & & \\ & & \cdot & & \\ & & \cdot & & \\ & & r^2, & 2 - 2r^2, & r^2 \\ r^2 & & & r^2, & 2 - 2r^2 \end{bmatrix}$$

with $r^2 = k^2h^{-2}$. We find, as before,

$$\begin{bmatrix} \mathbf{l}_{n+1} \\ \mathbf{l}_n \end{bmatrix} = \begin{bmatrix} C & -I \\ I & 0 \end{bmatrix}^n \begin{bmatrix} \mathbf{l}_1 \\ \mathbf{l}_0 \end{bmatrix}$$

and stability depends on the characteristic values μ of the $(2M \times 2M)$ matrix, which can be found explicitly by using Williamson's theorem and the fact that the characteristic values of C are

$$\gamma_i = 2 - 2r^2 + 2r^2 \cos 2i\pi/(M + 1) \qquad i = 1, 2, \ldots, M$$

We see that

$$\mu^2 - \gamma_i\mu + 1 = 0$$

so that $|\mu| \leq 1$ if and only if $\gamma_i^2 \leq 4$. Now γ_i ranges from 2 to almost $2 - 4r^2$ and stability obtains if $r^2 \leq 1$.

This condition for uniform boundedness of the solutions of the difference equation is also the classical condition of Courant, Friedrichs, and Lewy for the convergence of the solutions of the difference equation to that of the differential equation.

Another way of interpreting this condition is the following: It is known, from the theory of partial differential equations (Part 1, Chap. 5), that the value of the solution u at a point (X,T) can depend only on the values of the data in the range $(X - T \leq x \leq X + T)$. On the other hand, the solution of our difference equation depends only on the data for $(X - (h/k)T \leq x \leq X + (h/k)T)$. It is clear that the solution of the difference equation can, in general, only converge toward that of the differential equation if $h \geq k$.

There are many other discretizations possible for the wave equation. The discussion of their stability both in the case of one and of higher dimensions can be carried out by intensive use of Williamson's theorem, as exemplified here in the simplest cases.

References

This selection is, on the whole, confined to material which considers appropriate use of automatic digital computing equipment.

General and Introductory

Alt, F. L.: "Electronic Digital Computers," Academic, New York, 1958.

Bauer, F. L., J. Heinhold, K. Samelson, and R. Sauer: "Moderne Rechenanlagen," Teubner, Stuttgart, 1964.

Ekman, T., and C.-E. Fröberg: "Introduction to ALGOL Programming," Studentlitteratur, Lund, 1965.

Gruenberger, F., and G. Jaffray: "Problems for Computer Solution," Wiley, New York, 1965.

Hastings, C.: "Approximation for Digital Computers," Princeton University Press, Princeton, N.J., 1955.

Prager, W.: "Introduction to Basic FORTRAN Programming and Numerical Methods," Blaisdell Publishing Co., New York, 1965.

Ralston, A., and H. S. Wilf: "Mathematical Methods for Digital Computers," Wiley, New York, 1960.

Textbooks

"Modern Computing Methods," H. M. Stationery Office, London, 1961.

Berezin, I. S., and N. P. Zhidkov: "Computing Methods," vols. I, II, Pergamon Press, New York, 1965.

Conte, S. D.: "Elementary Numerical Analysis," McGraw-Hill, New York, 1965.

Fox, L.: "Introduction to Numerical Linear Algebra," Clarendon Press, Oxford, 1964.

Fröberg, C.-E.: "Introduction to Numerical Analysis," Addison-Wesley, Reading, Mass., 1965.

Henrici, P.: "Elements of Numerical Analysis," Wiley, New York, 1964.

Noble, B.: "Numerical Methods," vols. I, II, Interscience, New York, 1964.

Ralston, A.: "A First Course in Numerical Analysis," McGraw-Hill, New York, 1965.

Stiefel, E.: "An Introduction to Numerical Analysis," Academic, New York, 1963.

Todd, J. (ed.): "Survey of Numerical Analysis," McGraw-Hill, New York, 1962.

Todd, J.: "Introduction to the Constructive Theory of Functions," Academic, New York, 1963.

Treatises and Monographs

Foundations:

Davis, M.: "Computability and Unsolvability," McGraw-Hill, New York, 1958.

Markov, A. A.: "Theory of Algorithms," Department of Commerce, Office of Technical Services, Washington D.C., 1961.

Languages:

Iverson, K. E.: "A Programming Language," Wiley, New York, 1962.

Knuth, D. E.: "The Art of Computer Programming," Addison-Wesley, Reading, Mass., 1967.

Error Analysis:

Rall, L. B. (ed.): "Error in Digital Computation," vols. I, II, Wiley, New York, 1965.

Wilkinson, J. H.: "Rounding Errors in Algebraic Processes," Prentice-Hall, Englewood Cliffs, N.J., 1963.

Approximation Theory:

Davis, P. J.: "Interpolation and Approximation," Blaisdell Publishing Co., New York, 1963.

Langer, R. E. (ed.): "On Numerical Approximation," University of Wisconsin Press, Madison, Wis., 1959.

Meinardus, G.: "Approximationen von Funktionen und ihre numerische Behandlung," Springer, Berlin, 1964.

Rice, J. R.: "The Approximation of Functions," vols. I, II, Addison-Wesley, Reading, Mass., 1964.

Polynomial and Transcendental Equations:

Ostrowski, A. M.: "Solution of Equations and Systems of Equations," Academic, New York, 1960.

Traub, J. F.: "Iterative Methods for the Solution of Equations," Prentice-Hall, Englewood Cliffs, N. J., 1964.

Matrix Problems:

Faddeev, D. K., and V. N. Faddeeva: "Computational Methods of Linear Algebra," Freeman, San Francisco, 1963.

Householder, A. S.: "The Theory of Matrices in Numerical Analysis," Blaisdell Publishing Co., New York, 1964.

Paige, L. J., and O. Taussky (eds.): "Simultaneous Linear Equations and the Determination of Eigenvalues," Natl. Bur. of Standards, Applied Math. Series 29, Washington, D.C., 1953.

Wilkinson, J. H.: "The Algebraic Eigenvalue Problem," Clarendon Press, Oxford, 1965.

Functional Analysis:

Collatz, L.: "Funktionalanalysis und numerische Mathematik," Springer, Berlin, 1964.

Quadrature:

Krylov, V. I.: "Approximate Calculation of Integrals," (trans. A. H. Stroud), Macmillan, New York, 1962.

Differential Equations:

Collatz, L.: "The Numerical Treatment of Differential Equations," Springer, Berlin, 1960.
Forsythe, G. E., and W. Wasow: "Finite-difference Methods for Partial Differential Equations," Wiley, New York, 1960.
Fox, L.: "Numerical Solution of Two-point Boundary Problems," Clarendon Press, Oxford, 1957.
Henrici, P.: "Discrete Variable Methods in Ordinary Differential Equations," Wiley, New York, 1962.
Henrici, P.: "Error Propagation for Difference Methods," Wiley, New York, 1963.
Richtmyer, R. D.: "Difference Methods for Initial Value Problems," Interscience, New York, 1957.
Varga, R. S.: "Iterative Numerical Analysis," Prentice-Hall, Englewood Cliffs, N.J., 1962.
Wachspress, E. L.: "Iterative Solution of Elliptic Systems and Applications to the Neutron Diffusion Equations of Reactor Physics," Prentice-Hall, Englewood Cliffs, N.J., 1966.

Integral Equations:

Bückner, H.: "Die praktische Behandlung von Integralgleichungen," Springer, Berlin, 1952.

Conformal Mapping:

Gaier, D.: "Konstruktive Methoden der konformen Abbildung," Springer, Berlin, 1964.

Monte Carlo:

Hammersley, J. M., and O. C. Handscomb: "Monte Carlo Methods," Wiley, New York, 1964.
Meyer, H. A. (ed.): "Symposium on Monte Carlo Methods," Wiley, New York, 1956.

Tables

The following should be immediately accessible:

"Interpolation and Allied Tables," H. M. Stationery Office, London, 1956.
"Barlow's Tables of Squares, Cubes, etc.," 4th ed. (extended to 12,500), Chemical Publishing, New York, 1962.
Jahnke, E., F. Emde, and F. Lösch: "Tables of Higher Functions," 6th ed., McGraw-Hill, New York, 1960.
Abramowitz, M., and I. A. Stegun (eds.): "Handbook of Mathematical Functions," National Bureau of Standards, Applied Mathematics Series 55, Washington, D.C., 1964.
Fletcher, A., J. C. P. Miller, L. Rosenhead, and L. J. Comrie: "An Index of Mathematical Tables," vols. I, II, Addison-Wesley, Reading, Mass., 1962.

Standard Collections of Tables

British Association for the Advancement of Science (1931–1952)
continued by the
Royal Society 1950– . . .
Both series published by the Cambridge University Press, London.
U. S. National Bureau of Standards:
MT Series, 1939–1946,
Applied Mathematics Series, 1948– . . .
Both published by Government Printing Office, Washington, D.C., with 13 volumes issued by Columbia University Press, New York, 1943–1951.
U.S.S.R. Academy of Sciences Computing Center (Many volumes are available in the Pergamon Mathematical Tables Series, Macmillan, New York).
Harvard University, Annals Computation Laboratory, Harvard University Press, Cambridge, Mass., 1945– . . . .
National Physical Laboratory, Math. Tables, H. M. Stationery Office, London, 1945– . . . .
"Tracts for Computers," Cambridge University Press, London, 1919– . . . .

The numerical analyst should be familiar with collections of formulas relating to integrals, special functions, differential equations, such as the following:

Gröbner, W., and N. Hofreiter: "Integraltafel," vols. I, II, Springer, Vienna, 1949, 1950.
Byrd, P. F., and M. D. Friedman: "Handbook of Elliptic Integrals for Engineers and Physicists," Springer, Berlin, 1954.
Ryghik, Gradshteyn, I. S., and I. M.: "Tables of Integrals, Series and Products," Academic, New York, 1935.
Erdélyi, A., W. Magnus, F. Oberhettinger, and F. G. Tricomi: "Higher Transcendental Functions," vols. 1, 2, 3, McGraw-Hill, New York, 1953.
Kamke, E.: "Differentialgleichungen, Lösungsmethoden und Lösungen," vols. I, II, Akademische Verlagsgesellschaft Geest & Portig KG, Leipzig, 1944–1950.

In addition to this literature, we mention the existence of many Proceedings of Symposia on various topics, held in various places, and several quasi periodicals:

Alt, F. L. (ed.): "Advances in Computers," Academic, New York, 1960– . . . .
Alder, B., S. Fernbach, and M. Rotenberg (eds.): "Methods in Computational Physics," Academic, New York, 1963.

In view of the rapid development of this subject, attention must be paid to the periodical literature of the various societies in the United States and abroad, such as the following:

Association for Computing Machinery
Society for Industrial and Applied Mathematics
Computer Society

As well, there are various international journals such as *Numerische Mathematik Computing.*

Chapter 8

Geometry

By A. J. HOFFMAN, Thomas J. Watson Research Center

1. Definition and Assumptions

Geometry is the mathematical study of the properties of space. There are various "geometries," which may be roughly classified by the assumptions made about space.

Although geometry began in Egypt as an adjunct to the tasks of surveying—so that a point was a mark or position and a straight line was a taut string—the empirical science of geometry became for the Greeks an abstract science toward which they held a point of view that, sympathetically interpreted, does not differ markedly from that of the contemporary mathematician. We believe today that a geometry consists of assumptions (called *axioms* or *postulates*) about various undefined terms, to which no physical interpretation is necessarily intended, and all conclusions implied by these assumptions. The usefulness and interest of a geometry may depend on how closely the assumptions are fulfilled in some physical interpretation of the undefined terms.

Although Euclidean geometry (the body of geometrical knowledge contained in Euclid's "Elements," which is essentially a summary of Greek geometry) was the earliest known geometry, an understanding of its place in the hierarchy of geometries is best obtained by considering first projective geometry. The other geometries will appear as "subgeometries" of projective geometry.

In projective geometry the notion of measurement plays no role. If we shall not speak of distance or area, what facts of interest are left? As an example, we offer Desargues' theorem, famous in the history of projective geometry: let A, B, C, A', B', C' be points such that the lines AA', BB', CC' meet in a point O. Let C'' be the intersection of AB and $A'B'$, A'' be the intersection of BC and $B'C'$, and B'' be the intersection of AC and $A'C'$. Then A'', B'', and C'' are on a line. Both the hypothesis and the conclusion of the theorem deal only with the relation of incidence of points and lines. This is the characteristic feature of projective geometry, since incidence is the only relation considered in the postulates.

It is more economical to describe our subject analytically rather than to develop it *ab initio* from its postulates. For the sake of simplicity, we shall consider only plane geometry over the real numbers.

2. Projective Plane

Let (x_0,x_1,x_2) be any triple of numbers, not all 0. Two such triples, (x_0,x_1,x_2) and (x'_0,x'_1,x'_2) are said to be equivalent if there is a $\lambda \neq 0$ such that $x'_i = \lambda x_i$, $i = 0, 1, 2$. A class of equivalent triples is a *projective point* (briefly, a point), and any one of the triples of the class is said to be a set of *homogeneous coordinates* (briefly, coordinates) of the point. The set of all points satisfying a linear homogeneous equation

$$y_0x_0 + y_1x_1 + y_2x_2 = 0 \tag{8.1}$$

where the coefficients y_i are not all 0, form a (projective) line, and the triple y_0, y_1, y_2 are (homogeneous) coordinates of the line. Note that λy_0, λy_1, λy_2 describes the same line as y_0, y_1, y_2.

The point $x = (x_0,x_1,x_2)$ is on or incident with the line $y = y_0, y_1, y_2$ if Eq. (8.1) is satisfied. Similarly, we say the line is on the point. A good interpretation of these definitions is to conceive of a projective point as a line through the origin in Euclidean 3-space, and a projective line as a plane through the origin in Euclidean 3-space.

The definitions of point and line in the projective plane are so similar that the following *principle of duality* is true: any theorem of projective plane geometry remains valid if one interchanges the words "point" and "line" wherever they appear. For example, if (x_0,x_1,x_2) and (x'_0,x'_1,x'_2) are coordinates of distinct points, there is one and only one line on them, namely, the line whose coordinates are

$$\left[\begin{vmatrix} x_3 & x'_3 \\ x_2 & x'_2 \end{vmatrix} \quad \begin{vmatrix} x_2 & x'_2 \\ x_0 & x'_0 \end{vmatrix} \quad \begin{vmatrix} x_0 & x'_0 \\ x_1 & x'_1 \end{vmatrix} \right]$$

(see Part 1, Chap. 2, Sec. 9). The dual of this statement is: given any two lines, there is one and only one point on them. Thus, the famous parallel line postulate does not hold in the projective plane: every pair of lines intersects.

3. Projective Group

A one-to-one mapping of the set of points of the projective plane onto itself is called a *collineation* if, whenever three points are on a line, their images are on a line. Every collineation can be described in terms of a matrix (see Part 1, Chap. 2, Sec. 8):

$$\begin{aligned} \lambda x'_0 &= a_{00}x_0 + a_{01}x_1 + a_{02}x_2 \\ \lambda x'_2 &= a_{10}x_0 + a_{11}x_1 + a_{12}x_2 \\ \lambda x'_2 &= a_{20}x_0 + a_{21}x_1 + a_{22}x_2 \end{aligned} \tag{8.2}$$

where $A = (a_{ij})$ is regular. Conversely, for any regular A, the transformations (8.2) describe a collineation. It is clear that ρA, $\rho \neq 0$, describes the same collineation. Further, if a collineation σ

is represented by A, and τ is represented by B, then $\tau\sigma$ (the collineation σ followed by the collineation τ) is a collineation represented by the matrix BA. Thus the group (Part 1, Chap. 2, Sec. 13) of collineations of the projective plane is isomorphic to the group of homogeneous regular matrices of order 3.

The algebra of matrices affords an easy proof of the fact that any collineation admits a fixed point W, that is, a point that is taken into itself by the collineation. Let A be the matrix associated with the collineation: we see from (8.2) that we wish to find a nonzero such that there exist x_0, x_1, x_2, not all zero, satisfying

$$\begin{aligned} 0 &= (a_{00} - \lambda)x_0 + a_{01}x_1 + a_{02}x_2 \\ 0 &= a_{10}x_0 + (a_{11} - \lambda)x_1 + a_{12}x_2 \\ 0 &= a_{20}x_0 + a_{21}x_2 + (a_{22} - \lambda)x_2 \end{aligned}$$

The theory of matrices (Part 1, Chap. 2, Sec. 11) indicates that such x_0, x_1, x_2 will exist if we find a $\lambda \neq 0$ such that

$$\begin{vmatrix} a_{00} - \lambda & a_{01} & a_{02} \\ a_{10} & a_{11} - \lambda & a_{12} \\ a_{20} & a_{21} & a_{22} - \lambda \end{vmatrix} = 0$$

But this determinant is a polynomial of degree 3 in λ; hence it has at least one real root (Part 1, Chap. 2, Sec. 1); further, such a root is not zero, for that would contradict the fact that A is regular.

The definition of a collineation reveals that it can also be regarded as a one-to-one mapping of the set of lines onto itself such that whenever three lines are on a point, their images are on a point. If A is the matrix associated with the point transformation, of a given collineation, then $A^{-1\prime}$ is the matrix associated with the line transformation of the same collineation.

4. Correlations, Polarities, and Conics

A correlation is a one-to-one mapping of the set of points onto the set of lines, together with a one-to-one mapping of the set of lines onto the set of points, such that a point x is on a line y if and only if the line which is the image of x is on the point which is the image of y. With a given correlation, we associate matrices A and $A^{-1\prime}$ such that A describes the mapping of points onto lines:

$$\begin{aligned} \lambda y'_0 &= a_{00}x_0 + a_{01}x_1 + a_{02}x_2 \\ \lambda y'_1 &= a_{10}x_0 + a_{11}x_1 + a_{12}x_2 \\ \lambda y' &= a_{20}x_0 + a_{21}x_1 + a_{22}x_2 \end{aligned} \tag{8.3}$$

and $A^{-1\prime}$ describes the mapping of lines onto points. The correlations do not form a group; indeed, the composition of two correlations is a collineation. Of special interest are those correlations ρ such that ρ^2 is the identity collineation—i.e., the collineation that takes every point onto itself. Such a correlation is called a *polarity*. A point and line corresponding under a polarity are said to be *pole* and *polar*, respectively. A pole (polar) which lies on its polar (pole) is said to be *self-conjugate under the polarity*. If A is the matrix associated with a polarity, it follows from (8.3) that $A = A'$. If the polarity admits a self-conjugate point, it is called a *hyperbolic polarity*. A hyperbolic polarity admits an infinite number of self-conjugate points, namely, all nontrivial zeros of the quadratic form (Part 1, Chap. 2, Sec. 11):

$$x'Ax = a_{00}x_0^2 + a_{11}x_1^2 + a_{22}x_2^2 + 2a_{01}x_0x_1 + 2a_{12}x_1x_2 + 2a_{20}x_2x_0 \tag{8.4}$$

These self-conjugate points form a *conic* (one of several possible definitions of a conic). A self-conjugate line is tangent to the conic at its pole. A polarity that does not admit any self-conjugate points is said to be *elliptic*. It is sometimes convenient to think of an imaginary conic, the locus of complex nontrivial zeros of (8.4), as being associated with an elliptic polarity.

5. Projective Line

The projective line has already been defined as the locus of all points satisfying (8.1). If we wish to isolate a line for special study, it is convenient to choose the line $x_0 = 0$. Any homogeneous linear transformation

$$\begin{aligned} \lambda x'_1 &= a_{11}x_1 + a_{12}x_2 \\ \lambda x'_2 &= a_{21}x_1 + a_{22}x_2 \end{aligned} \qquad \begin{vmatrix} a_{11} & a_{12} \\ a_{21} & a_{22} \end{vmatrix} \neq 0 \tag{8.5}$$

is called a projective transformation of this line, and the group of all such transformations is called the projective group of the line. Any collineation of the plane that leaves $x_0 = 0$ invariant (i.e., takes this line onto itself) (but not uniquely) is a collineation of the plane (8.3).

Let x^0, x^1, x^2, x^3 be four distinct points on some line of the projective plane. Let τ be any collineation such that τ: $x^0 \to (0,0,1)$, $x^1 \to (0,1,0)$, $x^2 \to (0,1,1)$. Then τ: $x^3 \to (0,x_1,x_2)$ (some point on $x = 0$). The number x_2/x_1 is defined as the cross ratio of x^0, x^1, x^2, x^3—written $R(x^0,x^1,x^2,x^3)$. It can be shown that $R(x^0,x^1,x^2,x^3)$ does not depend on the collineation τ, and that two sets, each of four collinear points, correspond under some collineation if and only if they have the same cross ratio. Of special interest is the case $R(x^0,x^1,x^2,x^3) = -1$; then we say that x^0, x^1, x^2, x^3 form a *harmonic* set, or x^3 is the *harmonic conjugate* of x^2 with respect to x^0 and x^1.

If $x^i = (0,x_1^i,x_2^i)$, $i = 0, 1, 2, 3$, then

$$R(x^0,x^1,x^2,x^3) = \frac{\begin{vmatrix} x_1^0 & x_2^0 \\ x_1^2 & x_2^2 \end{vmatrix} \cdot \begin{vmatrix} x_1^1 & x_2^1 \\ x_1^3 & x_2^3 \end{vmatrix}}{\begin{vmatrix} x_1^0 & x_2^0 \\ x_1^3 & x_2^3 \end{vmatrix} \cdot \begin{vmatrix} x_1^3 & x_2^1 \\ x_1^2 & x_2^2 \end{vmatrix}} \tag{8.6}$$

A projective transformation that, when composed with itself, yields the identity transformation is called an *involution*, and this occurs if and only if $a_{22} = -a_{11}$ in (8.5).

6. Subgroups of the Projective Group

Of all the subgroups of the projective group, we shall concentrate especially on four: (1) Affine group, the subgroup of all collineations that leave $x_0 = 0$ invariant. (2) Euclidean group, the subgroup consisting of all "affine" collineations that, on $x_0 = 0$, commute with the involution $\lambda x'_1 = -x_2$, $\lambda x'_2 = x_1$. (3) Hyperbolic group, the subgroup consisting of all collineations that commute with a given hyperbolic polarity. (4) Elliptic group, the subgroup of all collineations that commute with a given elliptic polarity.

7. Affine Group and Plane

Consider the set of points of the projective plane not on $x_0 = 0$. Since for any such point (x_0,x_1,x_2), $x_0 \neq 0$, we may always express its coordinates in the form $(1,x_1,x_2)$. If the 1 in the x_0 position is understood, we can express the coordinates of our affine points in the unique, nonhomogeneous form (x_1,x_2). The general equation of an affine line is

$$a_0 + a_1x_1 + a_2x_2 = 0$$

Whenever $x_0 = 0$ enters the discussion, it will be called the line at infinity. Many of the familiar properties of the Euclidean plane hold in the affine plane, for example, the following: Two distinct lines are called parallel if and only if their intersection is on the line at infinity (hence, their intersection is not an affine point). It follows that on a point not on a given line there is one and only one line parallel to the given line.

We can define all points x^3 lying *between* two points x^1 and x^2. Let x^0 be the intersection of the line on x^1 and x^2 with the line at infinity; then the points between x^1 and x^2 are all points x^3 such that $R(x^1,x^2,x^0,x^3) < 0$. In particular, the mid-point of x^1 and x^2 is the harmonic conjugate of x^0 with respect to x^1 and x^2. The line segment x^1x^2 consists of all points between x^1 and x^2.

A conic which does not intersect the line at infinity is an *ellipse;* a conic which is tangent to the line at infinity is a *parabola;* and a conic which intersects the line at infinity in two points is a hyperbola. The tangents to a hyperbola at these points of intersection are called *asymptotes.* The center of a conic is the pole of the line at infinity under the (unique) polarity which admits the given conic as the locus of self-conjugate points. The general equation for a conic in the affine plane may be obtained from (8.4) by setting $x_0 = 1$. The equation becomes (relabeling the coefficients)

$$0 = Ax_1^2 - Bx_1x_2 + Cx_1^2 + Dx_2 + Ex_2 + F \quad (8.7)$$

If $B^2 - 4AC < 0$, the conic is an ellipse; if

$$B^2 - 4AC = 0$$

the conic is a parabola; if $B^2 - 4AC > 0$, the conic is a hyperbola.

The equations describing the affine collineations are

$$\begin{aligned} x_2^1 &= a_{11}x_1 + a_{12}x_2 + b_1 \\ x_2^1 &= a_{21}x_1 + a_{22}x_2 + b_2 \end{aligned} \qquad \begin{vmatrix} a_{11} & a_{12} \\ a_{21} & a_{22} \end{vmatrix} \neq 0 \quad (8.8)$$

Any two figures in the affine plane are *equivalent* (with respect to the affine group) if there is a collineation (8.8) mapping one figure onto the other. Thus, if we define a line segment as the set of points between two given points (called the *end points* of the segment), any two line segments are equivalent. Similarly, defining a triangle as the set of three line segments whose end points are any two of three given noncollinear points, any two triangles are equivalent. On the other hand, two parallel lines are not equivalent to two intersecting lines.

8. Euclidean Group and Plane

A definition of the Euclidean group equivalent to that given above (Sec. 6) is that it is the group of all transformations (8.8) such that the matrix

$$A = \begin{bmatrix} a_{11} & a_{12} \\ a_{21} & a_{22} \end{bmatrix}$$

has the property that for some number $\rho \neq 0$ the matrix ρA is orthogonal. It operates on the same set of points and lines as the affine group; hence pictorially the Euclidean plane is the same as the affine plane. But since the Euclidean group is smaller than the affine group, fewer figures are equivalent under the Euclidean group to a given figure than are equivalent under the affine group. Two figures are said to be similar if they are equivalent under a Euclidean transformation. Two figures are said to be congruent if A itself is orthogonal.

If (x_1,x_2) and (y_1,y_2) are two points, the distance between them (or the length of the line segment joining them) is $[(x_1 - y_1)^2 + (x_2 - y_2)^2]^{1/2}$. Every congruent transformation preserves distance, and conversely any mapping of the Euclidean plane onto itself that preserves distance is a congruent transformation.

Two lines $a_1x_1 + a_2x_2 + d = 0$ and

$$b_1x_1 + b_2x_2 + e = 0$$

are said to be perpendicular if $a_1b_1 + a_2b_2 = 0$. Every Euclidean transformation preserves perpendicularity; conversely any mapping of the Euclidean plane onto itself that preserves perpendicularity is a Euclidean transformation. The last statement follows from the fact that two lines are perpendicular if and only if their respective intersections with the line at infinity are points which correspond under the involution given in Sec. 6, (2).

The distance from a point (x_1,x_2) to a line

$$a_1z_1 + a_2z_2 + a_0 = 0$$

is the minimum of the distances from (x_1,x_2) to points on the line. This minimum is attained at some point (y_1,y_2) on the line, and the line joining (x_1,x_2) to (y_1,y_2) is perpendicular to the given line. A formula for the distance is

$$\frac{|a_1x_1 + a_2x_2 + a_0|}{\sqrt{a_1^2 - a_2^2}}$$

9. Conics

Since the Euclidean group is smaller than the affine group, we can distinguish additional properties of conics. These are most simply stated in terms of the notion of distance.

Let d be a given line (called the *directrix*), F a given point (called the *focus*), and e a given positive number (called the *eccentricity.*) Let C be the locus of a point P which moves so that e equals the distance from P to F divided by the distance from P to d. If $e < 1$, C is an ellipse; if $e = 1$, C is a parabola:

if $e > 1$, C is a hyperbola. Further, every conic may be described in this way, except for the circles. A circle is an ellipse such that every point has the same distance, called the radius, from the center; indeed, it is the locus of all such points.

A given ellipse (hyperbola) has another focus and directrix, which, with the same eccentricity, prescribes the same ellipse (hyperbola).

Another set of definitions for the ellipse and hyperbola is the following (if P and Q are points, let $\overline{PQ}$ be the distance from P to Q).

Let F_1 and F_2 be two points (called foci, and which are foci in the above sense as well), and let $2a > 0$ be a given number. The locus of all points P such that $\overline{PF_1} + \overline{PF_2} = 2a$ is an ellipse; the locus of all points P such that $\overline{PF_1 + PF_2} = 2a$ is a hyperbola. Of course, setting $\overline{F_1F_2} = 2c$, we must have $c < a$ $(a < c)$ if the ellipse (hyperbola) is to exist. (This follows from the triangle inequality: the sum of the lengths of two sides of a triangle is greater than the length of the third side.)

The line joining the foci contains the center of the ellipse (hyperbola). The line segment of length $2a$ on that line whose mid-point is the center is an axis (called the *major axis* in the case of the ellipse, the *transverse axis* in the case of the hyperbola); the line segment with center as mid-point perpendicular to the major (transverse) axis, of length $2b$, where $b = \sqrt{a^2 - c^2}$ $(b = \sqrt{c^2 - a^2})$, is called the *minor axis* (*conjugate axis*). If (0,0) is taken as the center and $(\pm c,0)$ as the foci, the equation for the ellipse is

$$\frac{x_1^2}{a^2} + \frac{x_2^2}{b^2} = 1 \tag{8.9}$$

For the hyperbola it is

$$\frac{x_1^2}{a^2} - \frac{x_2^2}{b^2} = 1 \tag{8.10}$$

The eccentricity is c/a. The asymptotes of the hyperbola (8.10) are

$$x = \pm \frac{b}{A} x_1$$

10. Angles

We first define a *ray* (or half line) with end point P as the set of all points on some line l containing P that are on one *side* of P (that is, there is a point Q on l such that the ray consists of all points X on l with P between Q and X). Two rays with a common end point divide the plane in general into two parts; either of these parts is called an angle. The common end point of the two rays is called the vertex of the angle; the rays are called the sides of the angle.

We measure an angle as follows: let C be any circle of radius r with center at the vertex and let c be the portion of the circle contained in the angle. Let d be the length of c (for required concept of limit, see Part 1, Chap. 3, Sec. 1). It can be proved that d/r does not depend on C, and further that two angles are congruent if and only if this ratio is the same. Hence d/r is taken as the measure of the angle (in radians). The largest value that d/r can assume is called 2π. The smallest value is, of course, 0. An angle formed by two opposite rays of the same line is called a straight angle; the smaller of the angles formed by two perpendicular rays is called a right angle. An angle whose size is between 0 and $\pi/2$ is called *acute;* an angle whose size is between $\pi/2$ and π is called *obtuse;* an angle whose size exceeds π is called *reflex.*

11. Triangles

As we have seen (Sec. 7 above), three noncollinear points (called *vertices*) determine a triangle. If A is a vertex of triangle ABC, then by $\measuredangle A$ we mean the angle whose vertex is A, whose sides are the pair of rays each having A as end point and including the segments AB and AC, respectively, and containing the segment BC. The sum of the angles of a triangle is π (fundamental theorem of Euclidean geometry, equivalent in a certain sense to the parallel postulate).

Two triangles are similar if and only if the angles of one are, respectively, congruent to the angles of the other; further, two triangles are similar if and only if corresponding sides have proportional lengths.

A triangle ABC, in which C is a right angle, is called a *right triangle.* The side opposite C is called the *hypotenuse;* the other sides are called legs. The square of the length of the hypotenuse equals the sum of the squares of the lengths of the legs (*theorem of Pythagoras*).

Several important functions of a real number have a geometric interpretation if the argument is the size of an angle, considering here angles between 0 and π. Assume A is an acute angle of the right triangle ABC, where C is the right angle. Then sine $A = \overline{BC}/\overline{AB}$, cosine $A = \overline{AC}/\overline{AB}$, and tangent $A = \overline{BC}/\overline{AC}$.

Facts about these functions and their uses are given in ref. 1.

Sine, cosine, and tangent are usually abbreviated to sin, cos, and tan, respectively. By the theorem on similar triangles quoted above these functions depend only on the size of angle A, not on the particular right triangle in which an angle of this size may be located. If A is an obtuse angle and A' is its supplement (that is, $A + A'$ is a straight angle), then $\sin A = \sin A'$, $\cos A = -\cos A'$, $\tan A = -\tan A'$. Also,

$$\sin \frac{\pi}{2} = 1$$

$\cos \pi/2 = 0$, $\tan \pi/2$ is undefined (angles whose size is near $\pi/2$ have tangents with large absolute values). Some useful relations connecting these functions of angles and the sizes of any triangle (not necessarily a right triangle) ABC are:

Law of sines:

$$\frac{\sin A}{\overline{BC}} = \frac{\sin B}{\overline{AC}}$$

Law of cosines:

$$\overline{BC}^2 = \overline{AC}^2 + \overline{AB}^2 - 2\overline{AC}\,\overline{AB} \cos A$$

Law of tangents:

$$\frac{\overline{BC} - \overline{AC}}{\overline{BC} + \overline{AC}} = \frac{\tan \frac{1}{2}(A - B)}{\tan \frac{1}{2}(A + B)}$$

If any three of the six quantities (sides and angles) are given (including at least one side), the remaining three can in general be computed using the above formulas and tables for the values of the functions.

The size of a triangle is measured by its area, which may be defined in various ways: Let a_d be the *altitude* from A to BC (the line segment perpendicular to BC with A as end point and a point on the line BC as other end point). Then the area of ABC is $\frac{1}{2}a_d \cdot \overline{BC}$ ($\frac{1}{2}$ the altitude times the base); also area

$$ABC = \tfrac{1}{2}\overline{AB}\,\overline{AC} \sin A$$

If the sides are denoted by a, b, c, then area ABC is $\sqrt{S(S-a)(S-b)(S-c)}$, where $S = \frac{1}{2}(a+b+c)$. If the vertices are (x_1,x_2), (y_1,y_2), (z_1,z_2), then the area is

$$\tfrac{1}{2}\begin{vmatrix} x_1 & x_2 & 1 \\ y_1 & y_2 & 1 \\ z_1 & z_2 & 1 \end{vmatrix}$$

Two triangles are congruent if and only if they are similar and have the same area.

12. Polygons

A set of line segments $l_1, \ldots, l_n$ together with a set of points $A_1, \ldots, A_n$ are said to form a polygon if the end points of l_i $(i = 1, \ldots, n-1)$ are A_i and A_{i+1} and the end points of l_n are A_n and A_0. The segments l_i are called the sides of the polygon. The points A_1 are called the vertices of the polygon. A polygon with the property that the lengths of all sides are the same is called *equilateral;* if in addition all A_i are on a circle, the polygon is *regular*. The area of the region enclosed by a regular polygon of n sides whose vertices are on a circle of radius r is $(n/2)r^2 \sin (2\pi/n)$ (by adding the area of the triangles which makes up the region). As n increases, this number approaches πr^2, which is therefore defined to be the area enclosed by a circle of radius r.

A polygon with four sides is called a *quadrilateral.* If at least one pair of alternate sides is parallel, the quadrilateral is called a *trapezoid,* and the area of the region it encloses is $(\frac{1}{2})h(a+b)$, where a and b are the lengths of the parallel sides and h is the distance from any point of one of the parallel sides to the other. If both pairs of alternate sides are parallel, the trapezoid is a *parallelogram.* A special case is the regular quadrilateral called a *square.* A square of side a encloses an area of a^2.

13. Hyperbolic Group and Plane

The given hyperbolic polarity ρ is called the absolute polarity. Let C be the (absolute) conic of self-conjugate points of ρ. C divides the plane into two regions, one of which is called the interior of C (every point in the interior has the property that every line containing it intersects C in two points). The points of the hyperbolic plane are the interior of C; the lines of the hyperbolic plane are those portions of the projective lines which consist of hyperbolic points (if any exist).

If l is a (hyperbolic) line and P is a (hyperbolic) point not on l, there are an infinite number of (hyperbolic) lines on P which do not intersect l (in a hyperbolic point). Hence the parallel postulate does not hold. It is customary to call two nonintersecting lines parallel if the intersection of the projective lines which contain them is a point on the absolute conic. Hence, for P and l as above, there are exactly two lines through P parallel to l.

Two lines (or points) are said to be perpendicular if the pole (polar) of one is on the other. Two figures in the hyperbolic plane are said to be congruent if there is a hyperbolic transformation mapping one onto the other. All the axioms of the Euclidean plane are valid in the hyperbolic plane except the axiom of parallels.

The formulas for distance, etc., are most conveniently stated by assuming that the absolute is the conic $x_1^2 + x_2^2 - x_0^2 = 0$ [the Euclidean picture of our plane is then, of course, the interior of the circle of radius 1 with center at (0,0)].

The distance between points $x = (x_0,x_1,x_2)$ and $y = (y_0,y_1,y_2)$ is

$$\cosh^{-1} \frac{|x_0y_0 - x_1y_1 - x_2y_2|}{\sqrt{(x_0^2 - x_1^2 - x_2^2)(y_0^2 - y_1^2 - y_2^2)}}$$

The distance from a point $x = (x_0,x_1,x_2)$ to a line $y = [y_0,y_1,y_2]$ is

$$\sinh^{-1} \frac{|x_0y_0 + x_1y_1 + x_2y_2|}{\sqrt{(x_0^2 - x_1^2 - x_2^2)(-y_0^2 + y_1^2 + y_2^2)}}$$

The acute angle formed by rays on intersecting lines $x = [x_0,x_1,x_2]$ and $y = [y_0,y_1,y_2]$ is

$$\cos^{-1} \frac{-x_0y_0 + x_1y_1 + x_2y_2}{\sqrt{(-x_0^2 + x_1^2 + x_2^2)(-y_0^2 + y_1^2 + y_2^2)}}$$

Some formulas connecting the sides a, b, c and angles A, B, C of a triangle are

$$\frac{\sinh a}{\sin A} = \frac{\sinh b}{\sin B}$$

$$\sinh b \sinh c \cos A = \cosh b \cosh c - \cosh a$$
$$\sin B \sin C \cosh a = \cos B \cos C + \cos A$$

The formula for the area of any triangle is $\pi - A - B - C$. An immediate consequence is that the sum of the angles of a triangle is always less than π.

14. Elliptic Group and Plane

The elliptic polarity mentioned in (4), Sec. 6 above, is called the absolute polarity, and we associate with it an imaginary (absolute) conic. The points and lines of the elliptic plane are the same as those of the projective plane. Hence the parallel postulate does not hold in the elliptic plane. The remarks about perpendicularity and congruence made about the hyperbolic plane apply without change here. An excellent model of the elliptic plane is to conceive of an elliptic point as a pair of antipodal points on the Euclidean sphere and an elliptic line as a great circle on that sphere. Then the congruent transformations correspond to the rotations of this sphere.

The formulas for distance, etc., are most conveniently stated by assuming that the absolute is the conic $x_1^2 + x_2^2 + x_0^2 = 0$. Then the distance between a point $x = (x_0,x_1,x_2)$ and a point

$$y = (y_0,y_1,y_2)$$

is

$$\cos^{-1} \frac{|x_0y_0 + x_1y_1 + x_2y_2|}{\sqrt{(x_0^2 + x_1^2 + x_2^2)(y_0^2 + y_1^2 + y_2^2)}}$$

This implies that every line segment has length less than π. The distance from a point $x = (x_0,x_1,x_2)$ to a line $y = y_0,y_1,y_2$ is

$$\sin^{-1} \frac{|x_0y_0 + x_1y_1 + x_2y_2|}{\sqrt{(x_0^2 + x_1^2 + x_2^2)(y_0^2 + y_1^2 + y_2^2)}}$$

The acute angle formed by rays on intersecting lines $x = x_0, x_1, x_2$ and $y = y_0, y_1, y_2$ is

$$\cos^{-1} \frac{|x_0y_0 + x_1y_1 + x_2y_2|}{\sqrt{(x_0^2 + x_1^2 + x_2^2)(y_0^2 + y_1^2 + y_2^2)}}$$

Some formulas connecting the sides and angles of a triangle are

$$\frac{\sin a}{\sin A} = \frac{\sin B}{\sin b}$$

$$\sin b \sin c \cos A = \cos a - \cos b \cos c$$

$$\sin B \sin C \cos a = \cos B \cos C + \cos A$$

By a proper choice of the unit of area, the formula for the area of any triangle is $A + B + C - \pi$. An immediate consequence is that the sum of the angles of a triangle is greater than π.

References

1. Adams, E. P.: "Smithsonian Mathematical Formulae," Washington, 1922.
2. Coxeter, H. S. M.: "Non-Euclidean Geometry," Toronto, 1942.
3. Veblen, O., and J. W. Young: "Projective Geometry," 2 vols., Boston, 1910.

Chapter 9

Vector Analysis

By E. U. CONDON, University of Colorado

Vector and tensor analysis provide a calculus for certain types of geometric quantities of frequent occurrence in physics. The simplest example of these is the vector displacement, or directed line, by which a particle is moved from a point A to another point D in space. Vector analysis develops an algebra of such quantities, independently of any reference to a coordinate system. In tensor analysis (Part 1, Chap. 10) results are obtained by describing such quantities with reference to a coordinate system, taking care to use methods which ensure and emphasize the invariance of the results to transformations of the coordinate system. For most purposes, where only vectors and second-order tensors are involved, the methods of vector analysis are usually preferred. But for problems involving tensors of higher order, or space of more than three dimensions, or curvilinear and nonorthogonal coordinate systems, the methods of tensor analysis are used almost exclusively in the literature of physics.

1. Addition of Vectors

A vector in Euclidean 3-space is represented geometrically by a directed line and will be represented as an algebraic quantity by a letter in boldface type, as **A**.

For most purposes vectors will be regarded as nonlocalized, that is, two vectors are regarded as equal if they have the same magnitude and direction, without regard to their location in space. Associated with two vectors **A** and **B** is a single vector $(\mathbf{A} + \mathbf{B}) = \mathbf{C}$ called their sum, which is also the same as $(\mathbf{A} + \mathbf{B})$ obtained by *addition* by the *parallelogram* rule: the tail of the second vector is placed on the head of the first. The sum is the vector extending from the tail of the first vector to the head of the second.

Subtraction is defined in terms of addition as with ordinary numbers. If $\mathbf{D} = \mathbf{A} - \mathbf{B}$, then **D** is the vector which must be added to **B** in order to give **A**. Subtraction may also be reduced to addition by first defining the vector $-\mathbf{B}$ as the vector represented by the same line as the vector **B**, but with head and tail interchanged. Then $(\mathbf{A} - \mathbf{B})$ is the same as $\mathbf{A} + (-\mathbf{B})$; thereby subtraction is reduced to addition of a negative quantity as with ordinary numbers.

Multiplication of a vector **A** by a positive scalar x gives a vector $x\mathbf{A} = \mathbf{A}x$ whose direction is the same as that of **A** but whose length is x times greater than that of **A**.

Two or more vectors are said to be *collinear* if there is a line parallel to each of them. Three or more vectors are said to be *coplanar* if there is a plane parallel to each of them.

For making the bridge between vector algebra and analytic geometry it is convenient to introduce three unit vectors **i**,**j**,**k** which are mutually orthogonal and directed, respectively, along the x,y,z axes of a cartesian coordinate system. Such a coordinate system is called right-handed, if, as in Fig. 9.1, a right hand grasping the z axis with the thumb pointing along its positive direction would find that the fingers curled naturally in the sense of turning the short way from the positive x axis to the positive y axis.

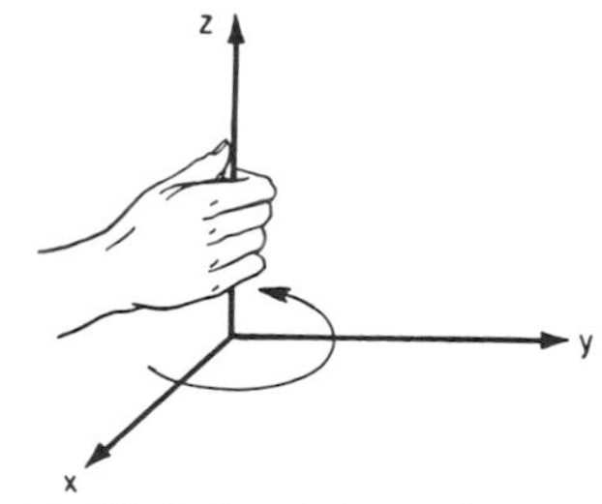

FIG. 9.1. Right-handed coordinate system.

In terms of such a coordinate system any vector **A** is completely specified by giving its three *components* A_1,A_2,A_3, in the relation

$$\mathbf{A} = A_1\mathbf{i} + A_2\mathbf{j} + A_3\mathbf{k} \tag{9.1}$$

The components of $\mathbf{A} + \mathbf{B}$ are then given by $A_1 + B_1$, $A_2 + B_2$, $A_3 + B_3$.

If **P** is the position vector of the point P relative to the origin O and **u** is a vector, then the *equation of the line* through **P** in the direction of **u** is

$$\mathbf{r} = \mathbf{P} + s\mathbf{u}$$

where **r** is a variable vector to any point on the line and s is a real variable, $-\infty < s < +\infty$. Likewise

$$\mathbf{r} = \mathbf{P} + s\mathbf{u} + t\mathbf{v}$$

where s and t are two independent real variables, is the equation of the plane through P which contains the directions of **u** and **v**.

2. Scalar and Vector Products

Two quantities are invariantly related to the vectors **A** and **B** which are called products of **A** and **B** and are known as the scalar and vector products.

The *scalar product* is written $\mathbf{A} \cdot \mathbf{B}$ and is therefore sometimes called the dot product. It is defined as

$$\mathbf{A} \cdot \mathbf{B} = AB \cos \alpha \tag{9.2}$$

that is, as the product of the magnitudes of **A** and **B** by the cosine of the angle α between **A** and **B**. From this it follows that scalar multiplication is commutative, $\mathbf{A} \cdot \mathbf{B} = \mathbf{B} \cdot \mathbf{A}$, and associative,

$$\mathbf{A} \cdot (\mathbf{B} + \mathbf{C}) = \mathbf{A} \cdot \mathbf{B} + \mathbf{A} \cdot \mathbf{C}$$

Whereas in ordinary multiplication $AB = 0$ implies either $A = 0$ or $B = 0$, the corresponding scalar product equation in vectors $\mathbf{A} \cdot \mathbf{B} = 0$ implies either $\mathbf{A} = 0$ or $\mathbf{B} = 0$ or that the two vectors are orthogonal.

The length squared of a vector is $A^2 = \mathbf{A} \cdot \mathbf{A}$. The *law of cosines* in trigonometry may be established as follows: $\mathbf{c} = \mathbf{a} - \mathbf{b}$; therefore

$$\mathbf{c} \cdot \mathbf{c} = c^2 = (\mathbf{a} - \mathbf{b}) \cdot (\mathbf{a} - \mathbf{b}) = a^2 - 2ab \cos C + b^2$$

If **i**,**j**,**k** are the unit vectors of a rectangular cartesian coordinate system, then

$$\begin{aligned} \mathbf{i} \cdot \mathbf{i} = \mathbf{j} \cdot \mathbf{j} = \mathbf{k} \cdot \mathbf{k} = 1 \\ \mathbf{i} \cdot \mathbf{j} = \mathbf{j} \cdot \mathbf{k} = \mathbf{k} \cdot \mathbf{i} = 0 \end{aligned} \tag{9.3}$$

and therefore the scalar product of two vectors is expressible in terms of its components:

$$\begin{aligned} \mathbf{a} \cdot \mathbf{b} &= (a_1\mathbf{i} + a_2\mathbf{j} + a_3\mathbf{k}) \cdot (b_1\mathbf{i} + b_2\mathbf{j} + b_3\mathbf{k}) \\ &= a_1b_1 + a_2b_2 + a_3b_3 \end{aligned} \tag{9.4}$$

The components of **A** along the three axes are $\mathbf{A} \cdot \mathbf{i}$, $\mathbf{A} \cdot \mathbf{j}$, and $\mathbf{A} \cdot \mathbf{k}$; therefore the vector can be written as $\mathbf{A} = (\mathbf{A} \cdot \mathbf{i})\mathbf{i} + (\mathbf{A} \cdot \mathbf{j})\mathbf{j} + (\mathbf{A} \cdot \mathbf{k})\mathbf{k}$.

The *vector product* of **A** and **B**, sometimes called the *cross product* or the skew product, is a vector written $\mathbf{A} \times \mathbf{B}$. It is defined as being a vector perpendicular to the plane determined by **A** and **B** in the sense in which the thumb of the right hand points if the fingers curl around in the sense of the smaller rotation which would carry **A** into **B**, the magnitude of $\mathbf{A} \times \mathbf{B}$ being $AB \sin \alpha$.

The vector product is associative, but is anticommutative, that is,

$$\mathbf{A} \times \mathbf{B} = -\mathbf{B} \times \mathbf{A} \tag{9.5}$$

The equation $\mathbf{A} \times \mathbf{B} = 0$ implies either $\mathbf{A} = 0$, or $\mathbf{B} = 0$, or that **A** and **B** are parallel or antiparallel; hence $\mathbf{A} \times \mathbf{A} = 0$ for any vector.

If **A** and **B** represent the sides of a parallelogram, then $\mathbf{A} \times \mathbf{B}$ has a magnitude equal to the area of the parallelogram and a direction that is normal to the plane of the parallelogram; thus $\mathbf{A} \times \mathbf{B}$ specifies the magnitude and orientation of such a plane element of area.

The vector products of the unit vectors of a right-hand coordinate system are

$$\begin{array}{lll} \mathbf{i} \times \mathbf{j} = \mathbf{k} & \mathbf{j} \times \mathbf{k} = \mathbf{i} & \mathbf{k} \times \mathbf{i} = \mathbf{j} \\ \mathbf{j} \times \mathbf{i} = -\mathbf{k} & \mathbf{k} \times \mathbf{j} = -\mathbf{i} & \mathbf{i} \times \mathbf{k} = -\mathbf{j} \\ \mathbf{i} \times \mathbf{i} = 0 & \mathbf{j} \times \mathbf{j} = 0 & \mathbf{k} \times \mathbf{k} = 0 \end{array} \tag{9.6}$$

Therefore the vector product of two vectors is

$$\begin{aligned} \mathbf{a} \times \mathbf{b} &= (a_1b_2 - a_2b_1)\mathbf{k} + (a_2b_3 - a_3b_2)\mathbf{i} \\ &\qquad + (a_3b_1 - a_1b_3)\mathbf{j} \\ &= \begin{vmatrix} \mathbf{i} & \mathbf{j} & \mathbf{k} \\ a_1 & a_2 & a_3 \\ b_1 & b_2 & b_3 \end{vmatrix} = \begin{vmatrix} \mathbf{i} & a_1 & b_1 \\ \mathbf{j} & a_2 & b_2 \\ \mathbf{k} & a_3 & b_3 \end{vmatrix} \end{aligned} \tag{9.7}$$

The *mixed triple product* of three vectors $(\mathbf{a} \times \mathbf{b}) \cdot \mathbf{c}$, in terms of components, is given by

$$(\mathbf{a} \times \mathbf{b}) \cdot \mathbf{c} = \begin{vmatrix} a_1 & a_2 & a_3 \\ b_1 & b_2 & b_3 \\ c_1 & c_2 & c_3 \end{vmatrix} \tag{9.8}$$

which follows from (9.7) and (9.4). This triple product has the geometrical significance that its magnitude is equal to the volume of the parallelepiped defined by the vectors **a**, **b**, and **c**. It is *equal* to the volume if **c** makes an acute angle with $\mathbf{a} \times \mathbf{b}$; otherwise $(\mathbf{a} \times \mathbf{b}) \cdot \mathbf{c}$ is equal to the negative of the volume. If **a**, **b**, and **c** are three nonvanishing vectors, then $(\mathbf{a} \times \mathbf{b}) \cdot \mathbf{c} = 0$ is the necessary and sufficient condition that the three vectors be coplanar.

Such a triple product is unaltered in value, as follows from its geometric interpretation, if the dot and cross are interchanged, $(\mathbf{a} \times \mathbf{b}) \cdot \mathbf{c} = \mathbf{a} \cdot (\mathbf{b} \times \mathbf{c})$, or if the three factors are cyclically permuted, $\mathbf{a} \cdot (\mathbf{b} \times \mathbf{c}) = \mathbf{b} \cdot (\mathbf{c} \times \mathbf{a}) = \mathbf{c} \cdot (\mathbf{a} \times \mathbf{b})$, but reverses sign if the cyclic order is changed. It is not necessary to show parentheses since $(\mathbf{a} \cdot \mathbf{b}) \times \mathbf{c}$ would be meaningless and thus $\mathbf{a} \cdot \mathbf{b} \times \mathbf{c}$ admits only of the interpretation $\mathbf{a} \cdot (\mathbf{b} \times \mathbf{c})$. Some books use the notation [**abc**] for the mixed triple product.

The vector triple product of three vectors **a**, **b**, and **c** may be written $\mathbf{a} \times (\mathbf{b} \times \mathbf{c})$. This is a vector orthogonal to $\mathbf{b} \times \mathbf{c}$ and therefore lies in the plane determined by **b** and **c**. It is given by

$$\mathbf{a} \times (\mathbf{b} \times \mathbf{c}) = (\mathbf{a} \cdot \mathbf{c})\mathbf{b} - (\mathbf{a} \cdot \mathbf{b})\mathbf{c} \tag{9.9}$$

and is therefore not the same as

$$(\mathbf{a} \times \mathbf{b}) \times \mathbf{c} = (\mathbf{a} \cdot \mathbf{c})\mathbf{b} - (\mathbf{c} \cdot \mathbf{b})\mathbf{a}$$

The component of a vector **b** which is perpendicular to a unit vector **a** is given by $-\mathbf{a} \times (\mathbf{a} \times \mathbf{b})$, as follows from direct application of (9.9).

The following formulas for transforming the products of four vectors are useful:

$$(\mathbf{a} \times \mathbf{b}) \cdot (\mathbf{c} \times \mathbf{d}) = (\mathbf{a} \cdot \mathbf{c})(\mathbf{b} \cdot \mathbf{d}) - (\mathbf{a} \cdot \mathbf{d})(\mathbf{b} \cdot \mathbf{c}) \tag{9.10}$$

$$\begin{aligned} (\mathbf{a} \times \mathbf{b}) \times (\mathbf{c} \times \mathbf{d}) &= (\mathbf{a} \times \mathbf{b} \cdot \mathbf{d})\mathbf{c} - (\mathbf{a} \times \mathbf{b} \cdot \mathbf{c})\mathbf{d} \\ &= (\mathbf{a} \cdot \mathbf{c} \times \mathbf{d})\mathbf{b} - (\mathbf{b} \cdot \mathbf{c} \times \mathbf{d})\mathbf{a} \end{aligned} \tag{9.11}$$

Equating the last two equivalent forms gives a relation by which any vector **d** may be explicitly expressed in terms of the other three vectors **a**,**b**,**c**, provided **a**, **b**, and **c** are not coplanar:

$$\begin{aligned} (\mathbf{a} \cdot \mathbf{b} \times \mathbf{c})\mathbf{d} &= (\mathbf{b} \cdot \mathbf{c} \times \mathbf{d})\mathbf{a} + (\mathbf{c} \cdot \mathbf{a} \times \mathbf{d})\mathbf{b} \\ &\qquad + (\mathbf{a} \cdot \mathbf{b} \times \mathbf{d})\mathbf{c} \end{aligned} \tag{9.12}$$

3. Vectors and Tensors in Oblique Coordinates

If the coordinate axes are not rectangular and the basis vectors, denoted here as $\mathbf{e}_1,\mathbf{e}_2,\mathbf{e}_3$, are not necessarily of unit length, then the description of a vector is most conveniently carried out by introducing two distinct sets of vector components.

The notation for the components of **X** with respect to the $\mathbf{e}_1,\mathbf{e}_2,\mathbf{e}_3$ basis which will be used is

$$\mathbf{X} = x^1\mathbf{e}_1 + x^2\mathbf{e}_2 + x^3\mathbf{e}_3 = x^s\mathbf{e}_s \tag{9.13}$$

In the latter form the *summation convention* is used: an index occurring twice is to be summed over the values 1,2, . . . ,N, where N is the number of dimensions of the space. The components $x^1\mathbf{e}_1, x^2\mathbf{e}_2$, etc., have the geometrical significance that if added by the parallelogram rule they give **X**.

However, if the coordinate axes are oblique, the separate components, as x^1, are no longer given by the simple formula $\mathbf{X} \cdot \mathbf{e}_1$, etc. The scalar products of the basic vectors are not equal to δ_{rs} but to a general symmetric matrix which will be designated g_{rs}, defined by

$$g_{rs} = \mathbf{e}_r \cdot \mathbf{e}_s \tag{9.14}$$

Accordingly the expression for the scalar product of two vectors **X** and **Y** is

$$\mathbf{X} \cdot \mathbf{Y} = x^r g_{rs} y^s \tag{9.15}$$

This suggests that **X** can also be represented by a second set of components, denoted by subscripts instead of superscripts, which are defined by the equation

$$x_s = x^r g_{rs} \tag{9.16}$$

In terms of these components the scalar product $\mathbf{X} \cdot \mathbf{Y}$ takes on the simple form $x_s y^s$ analogous to that in orthogonal coordinates. Alternatively, one could use the original kind of components of **X** and combine them with the new kind for **Y** defined by $y_r = g_{rs} y^s$ to give $\mathbf{X} \cdot \mathbf{Y} = x^r y_r$.

The matrix reciprocal to g_{rs} will be denoted by g^{st},

$$g_{rs} g^{st} = \delta_r^t = \begin{cases} 1 & t = r \\ 0 & t \neq r \end{cases}$$

This matrix provides the explicit solution of (9.15) for the x^r in terms of the x_s in the form $x^r = g^{rs} x_s$. Accordingly the vector expression for **X** in terms of the x_s becomes

$$\mathbf{X} = x_s g^{sr} \mathbf{e}_r = x_s \mathbf{e}^s \tag{9.17}$$

The second form here involves the introduction of the associated set of vectors $\mathbf{e}^s$ through the relation

$$\mathbf{e}^s = g^{sr} \mathbf{e}_r \tag{9.18}$$

Since g^{rs} and g_{st} are reciprocal matrices, it follows that

$$\mathbf{e}^s \cdot \mathbf{e}_t = \delta_t^s \tag{9.19}$$

The two sets of basis vectors related in this way are called *reciprocal systems* of vectors. Any vector **X** can be equally well expressed in terms of its components x^s or x_s with respect to either set,

$$\mathbf{X} = x^r \mathbf{e}_r = x_s \mathbf{e}^s$$

and in view of (9.19) the components with respect to one set are given by forming scalar components with regard to the corresponding vector of the other set:

$$x^r = \mathbf{X} \cdot \mathbf{e}^r \quad \text{and} \quad x_s = \mathbf{X} \cdot \mathbf{e}_s \tag{9.20}$$

In the case of three dimensions the reciprocal set to $\mathbf{e}_1,\mathbf{e}_2,\mathbf{e}_3$ is easily expressed explicitly:

$$\mathbf{e}^1 = \frac{\mathbf{e}_2 \times \mathbf{e}_3}{[\mathbf{e}_1\mathbf{e}_2\mathbf{e}_3]}, \text{ etc.}$$

Hence $\mathbf{e}^1$ is normal to the plane of $\mathbf{e}_2$ and $\mathbf{e}_3$ as it must be to satisfy (9.18).

In the special case in which $\mathbf{e}_1,\mathbf{e}_2,\mathbf{e}_3$ are orthogonal unit vectors, then the basis $\mathbf{e}^1,\mathbf{e}^2,\mathbf{e}^3$ are the same as the $\mathbf{e}_1,\mathbf{e}_2,\mathbf{e}_3$, respectively; so the two kinds of components are equal, and there is no need to distinguish subscripts and superscripts in the notation.

Transformation of Axes. If the coordinate basis is transformed from one oblique set $\mathbf{e}_1,\mathbf{e}_2$ . . . to another $\mathbf{f}_1,\mathbf{f}_2$ . . . the relation of one of these to the other will be described by a nonsingular matrix a_t^s:

$$\mathbf{f}_t = a_t^s \mathbf{e}_s \tag{9.21}$$

Components of vectors with respect to the **f**'s will be denoted by barred letters; thus $\mathbf{X} = \bar{x}^t\mathbf{f}_t = x^s\mathbf{e}_s$. Because of the invariance of form of these two expressions for **X**, it follows that the components x^t do not transform by the same transformation as the basis vectors $\mathbf{e}_t$: $\mathbf{X} = \bar{x}^t\mathbf{f}_t = \bar{x}^t a_t^s \mathbf{e}_s = x^s\mathbf{e}_s$, and therefore $\bar{x}^t a_t^s = x^s$, giving

$$\bar{x}^t = A_s^t x^s \tag{9.22}$$

where A_s^t is the matrix reciprocal to a_t^s in the sense $A_s^t a_u^s = \delta_u^t$. Because the components x^s transform in this different way than the basis vectors $\mathbf{e}_s$, they are called the *contravariant* components of **X**.

The vectors $\mathbf{f}^t$ that are reciprocal to $\mathbf{f}_t$ also transform contravariantly. This is necessary to preserve the validity of $\mathbf{f}^t \cdot \mathbf{f}_s = \delta_s^t$ in the new system of coordinates; hence $\mathbf{f}^t = A_s^t \mathbf{e}^s$.

In the coordinate system defined by the $\mathbf{f}_s$ and $\mathbf{f}^t$, the quantities $\bar{g}_{tu}$ defined in (9.14) as $\mathbf{f}_t \cdot \mathbf{f}_u$ will play the role of the g_{rs} in the system defined by the $\mathbf{e}_s$. These form a set of N^2 quantities [really only $\frac{1}{2}N(N+1)$ in view of $g_{rs} = g_{sr}$] which transform according to the equation

$$\bar{g}_{tu} = (a_t^s \mathbf{e}_s) \cdot (a_u^r \mathbf{e}_r) = (a_t^s a_u^r) g_{sr} \tag{9.23}$$

The transformation matrix for the g_{sr} is related to that for the $\mathbf{e}_s$ by a double use of the same set of coefficients a_t^s.

The set of quantities, g_{sr}, are said to be the covariant components of a second-order symmetric tensor: second-order because there are two indices relating to coordinate axes, symmetric because $g_{rs} = g_{sr}$; covariant because the transformation is based on the a_t^s rather than on the reciprocal matrix A_t^s. Similarly, the quantities $\bar{g}^{rs} = \mathbf{f}^r \cdot \mathbf{f}^s$ form the *contravariant components* of the same tensor while the quantities $g_s^r = \mathbf{f}^r \cdot \mathbf{f}_s = g_s^r = \delta_s^r$ may be called the invariant components of the same tensor.

The tensor is thus describable by any one or all of three sets of components. All of these become identical if the axes are orthogonal and the basis vectors are unit vectors.

In general a tensor of the mth order in N-dimensional space has N^m components. One way of describing it is by means of its *covariant components*, which transform according to the formula

$$\bar{S}_{tuv\ldots} = (a_t^i a_u^j a_v^k \cdots) S_{ijk\ldots} \qquad (9.24)$$

or by its *contravariant components*, which transform as follows:

$$\bar{S}^{tuv} = (A_i^t A_j^u A_k^v \cdots) S^{ijk}\ldots \qquad (9.25)$$

For some purposes it is convenient to introduce the idea of basis tensors. Thus the basis for a second-order tensor is the set of symbolic units $\mathbf{e}_s\mathbf{e}_t$; and a second-order tensor $\mathbf{S}$, having the contravariant components S^{st}, would then be written

$$\mathbf{S} = S^{st}\mathbf{e}_s\mathbf{e}_t \qquad (9.26)$$

in which it is understood that $\mathbf{e}_s\mathbf{e}_t \neq \mathbf{e}_t\mathbf{e}_s$ in order that the formalism will be able to deal with nonsymmetric tensors. The same tensor may also be expressed with relation to the reciprocal basis as $\mathbf{S} = S_{st}\mathbf{e}^s\mathbf{e}^t$, and since $\mathbf{e}_s = g_{sr}\mathbf{e}^r$ and $\mathbf{e}^t = g_{tu}\mathbf{e}^u$ it follows that for *any tensor* the passage from contravariant to covariant components is by means of the g_{sr} tensor

$$S_{ru} = S^{st}g_{sr}g_{tu} \qquad (9.27)$$

Tensors can be added only if they are of the same order and by adding corresponding components of like covariance or contravariance.

The ensemble of N^{m+n} quantities obtained by multiplying together in all possible combinations the N^m components of an mth-order tensor with the N^n components of an Nth order gives a set of quantities which form the components of a tensor of order $(n + m)$ because they transform in this manner when the axes are changed. This process whereby an $(n + m)$th-order tensor is formed from two tensors of order n and m is called *outer multiplication* of the tensors.

In particular the *outer product* of two vectors $\mathbf{X}$ and $\mathbf{Y}$ defines a second-order tensor: $\mathbf{XY} = x^s y^t \mathbf{e}_s\mathbf{e}_t$, when written in terms of contravariant components. The scalar product is derived from this by evaluating the scalar products $\mathbf{e}_s \cdot \mathbf{e}_t = g_{st}$. The vector product is related to the antisymmetric second-order tensor $\frac{1}{2}(\mathbf{XY} - \mathbf{YX})$.

4. Gradient of Scalar and Vector Fields

A scalar function of position in space is called a scalar field. With each scalar field $\phi(\mathbf{r})$ there is associated a vector function of position called the gradient of ϕ, written grad ϕ or $\nabla\phi$.

The gradient of ϕ is defined at a point as the vector whose direction is that of the maximum rate of increase of ϕ and whose magnitude is equal to this maximum rate of increase. Using cartesian coordinates (x,y,z) and Taylor's theorem, the change in $\phi(\mathbf{r})$ to the first order in the vector displacement $\mathbf{s}$ is

$$\phi(\mathbf{r} + \mathbf{s}) = \phi(\mathbf{r}) + s_1\frac{\partial\phi}{\partial x} + s_2\frac{\partial\phi}{\partial y} + s_3\frac{\partial\phi}{\partial z} + \cdots$$
$$= \phi(\mathbf{r}) + \mathbf{s}\cdot\nabla\phi + \cdots$$

where $$\nabla\phi = \mathbf{i}\frac{\partial\phi}{\partial x} + \mathbf{j}\frac{\partial\phi}{\partial y} + \mathbf{k}\frac{\partial\phi}{\partial z} \qquad (9.28)$$

Keeping s constant, the maximum value of $\mathbf{s}\cdot\nabla\phi$ occurs when $\mathbf{s}$ has the same direction as the vector $\nabla\phi$, and when $\mathbf{s}$ has this direction the rate of change of ϕ with displacement in this direction is given by the magnitude of $\nabla\phi$. Hence $\nabla\phi$, given by (9.28), is the gradient in the sense of the preceding paragraph.

The term $\mathbf{s}\cdot\nabla\phi$ is called the *directional derivative* in the direction of $\mathbf{s}$, if $\mathbf{s}$ is restricted to being a unit vector.

For various values of the constant C the equation $\phi(x,y,z) = C$ defines a family of surfaces. The vector, grad ϕ, is normal to the surface of this family through the point for which it is calculated.

The gradient may be regarded as resulting from the application to the function ϕ of the vector differential operator, del,

$$\nabla = \left(\mathbf{i}\frac{\partial}{\partial x} + \mathbf{j}\frac{\partial}{\partial y} + \mathbf{k}\frac{\partial}{\partial z}\right) \qquad (9.29)$$

in a cartesian coordinate system.

If an oblique coordinate system $\mathbf{e}_1,\mathbf{e}_2,\mathbf{e}_3$ is used and the coordinates of a point in space are given by the contravariant components of the position vector $\mathbf{r} = x^s\mathbf{e}_s$, then the scalar ϕ will be expressed as a function of x^1, x^2, and x^3. If s^1, s^2, s^3 are now the contravariant components of the displacement, then the expression for the first-order change in ϕ is, by Taylor's theorem,

$$s^1\frac{\partial\phi}{\partial x^1} + s^2\frac{\partial\phi}{\partial x^2} + s^3\frac{\partial\phi}{\partial x^3}$$

an expression which is invariant to change of axes, and therefore the three quantities $\partial\phi/\partial x^s$ must form the *covariant* components of a vector. Hence in oblique coordinates the expression for the gradiant operator is

$$\nabla = \mathbf{e}^1\frac{\partial}{\partial x^1} + \mathbf{e}^2\frac{\partial}{\partial x^2} + \mathbf{e}\frac{\partial}{\partial x^3} \qquad (9.30)$$

Alternatively by a similar argument ∇ can be expressed in terms of $\mathbf{e}_1$, etc., if ϕ is given as a function of the covariant components of the position vector.

By a *vector field* is meant a vector function of position in space. The *gradient* of a vector field $\mathbf{F}$ is defined as the second-order tensor which gives the leading term in the Taylor's expansion of $\mathbf{F}$:

$$\mathbf{F}(\mathbf{r} + \mathbf{s}) - \mathbf{F}(\mathbf{r}) = s_1\frac{\partial\mathbf{F}}{\partial x} + s_2\frac{\partial\mathbf{F}}{\partial y} + s_3\frac{\partial\mathbf{F}}{\partial z} + \cdots$$
$$= \mathbf{s}\cdot\nabla\mathbf{F}$$

where $\nabla\mathbf{F}$ is the second-order tensor

$$\begin{aligned}\nabla\mathbf{F} &= \left(\mathbf{i}\frac{\partial}{\partial x} + \mathbf{j}\frac{\partial}{\partial y} + \mathbf{k}\frac{\partial}{\partial z}\right)\mathbf{F} \\ &= \mathbf{ii}\frac{\partial F_x}{\partial x} + \mathbf{ij}\frac{\partial F_y}{\partial x} + \mathbf{ik}\frac{\partial F_z}{\partial x} \\ &+ \mathbf{ji}\frac{\partial F_x}{\partial y} + \mathbf{jj}\frac{\partial F_y}{\partial y} + \mathbf{jk}\frac{\partial F_z}{\partial y} \\ &+ \mathbf{ki}\frac{\partial F_x}{\partial z} + \mathbf{kj}\frac{\partial F_y}{\partial z} + \mathbf{kk}\frac{\partial F_z}{\partial z}\end{aligned} \qquad (9.31)$$

In the same way the gradient of a tensor of any order may be defined as a tensor of one higher order.

5. Divergence of a Vector Field

The divergence of a vector field **F** is a scalar field which is obtained formally in cartesian coordinates as the scalar product of ∇ with **F**,

$$\text{div } \mathbf{F} = \nabla \cdot \mathbf{F} = \frac{\partial F_x}{\partial x} + \frac{\partial F_y}{\partial y} + \frac{\partial F_z}{\partial z} \qquad (9.32)$$

In oblique coordinates, it follows from (9.30) that this same simple form will be applicable if F_x, F_y, F_z are the contravariant components of **F** expressed as functions of the contravariant components (x,y,z) of the position vector.

A definition of divergence that is more related to its physical significance may be given as follows. Consider any closed surface S in the field and let $\mathbf{dS}$ be a vector element of area reckoned positively in the sense of the *outward normal* to the element of area. Then $\mathbf{F} \cdot \mathbf{dS}$ is called the *flux* of F across the element of area. The name comes from analogy with the case in which **F** is the velocity of a fluid and $\mathbf{F} \cdot \mathbf{dS}$ measures the amount of fluid crossing the area element in unit time. The total outward flux of **F** across the closed surface S is then the surface integral $\iint \mathbf{F} \cdot \mathbf{dS}$ extended over the closed surface.

If the region bounded by S is cut into parts by any number of internal partitioning surfaces, then the total flux across S is equal to the sum of the total fluxes across each of the parts, for in calculating the sum the flux across any internal bounding surface will be counted twice in opposite senses with respect to two adjoining regions. Hence, if the region internal to S is partitioned into very small volume elements dV, it will be possible to represent the total flux $\iint \mathbf{F} \cdot \mathbf{dS}$ as the volume integral of the flux per unit volume from each volume element within S. This limiting value of the flux out of a small volume element in the neighborhood of a point, divided by the volume of the volume element, is equal to the *divergence* of **F**.

The relation between the flux through S and the volume integral of the divergence is called the *divergence theorem:*

$$\iint \mathbf{F} \cdot \mathbf{dS} = \iiint \text{div } \mathbf{F} \, dV \qquad (9.33)$$

where the volume integral is extended over the region enclosed by the closed surface S.

To show that (9.32) represents the divergence in the sense of the divergence theorem, one may consider a volume element from x to $x + dx$, y to $y + dy$, z to $z + dz$. The flux outward across the face at $x + dx$ is $F_x(x + dx) \, dy \, dz$ and that outward across the face at x is $-F_x(x) \, dy \, dz$ and therefore the total outward flux across these two faces is

$$\frac{\partial F_x}{\partial x} \, dx \, dy \, dz$$

Closely related to the divergence theorem are the following:

1. For a scalar field ϕ, the vector surface integral of ϕ is equal to the volume integral of the gradient of ϕ:

$$\iint \phi \, \mathbf{dS} = \iiint \text{grad } \phi \, dV \qquad (9.34)$$

2. For a vector field **F**, the surface integral of the cross product of **F** with the vector element of area is equal to the negative volume integral of the curl of F:

$$\iint \mathbf{F} \times \mathbf{dS} = -\iiint \text{curl } \mathbf{F} \, dV \qquad (9.35)$$

where curl **F** is defined in (9.37).

3. For a vector field **F**, the tensor surface integral of **F** over a closed surface is equal to the volume integral of the tensor gradient of **F**:

$$\iint \mathbf{F} \, d\mathbf{S} = \iiint \text{grad } \mathbf{F} \, dV \qquad (9.36)$$

6. Curl of a Vector Field

The curl of a vector field **F** is a vector field which is obtained formally, in cartesian coordinates, as the vector product of ∇ and **F**,

$$\text{curl } \mathbf{F} = \nabla \times \mathbf{F} = \mathbf{i}\left(\frac{\partial Fz}{\partial y} - \frac{\partial F_y}{\partial z}\right) + \mathbf{j}\left(\frac{\partial F_x}{\partial z} - \frac{\partial F_z}{\partial x}\right) + \mathbf{k}\left(\frac{\partial F_y}{\partial x} - \frac{\partial F_x}{\partial y}\right) \qquad (9.37)$$

The field, curl **F**, may also be considered in relation to a theorem for transformation of integrals known as *Stokes' theorem.*

If C is a closed curve in the field, then the line integral $\int \mathbf{F} \cdot d\mathbf{s}$ extended over the curve, where $d\mathbf{s}$ is a vector element of length, is called the *circulation* of **F** with respect to the curve C.

If now S is *any* open surface which is bounded by C, then on S one can partition the region within C by laying out a fine mesh of closed circuits with a double system of curves. The total circulation of **F** around C will be the sum of the circulation around these little closed circuits, for in calculating the sum each internal boundary is traversed twice in opposite senses. This discussion indicates that the circulation is also given by the surface integral of an appropriate quantity over S, the appropriate quantity being defined as curl **F**:

$$\int_C \mathbf{F} \cdot d\mathbf{s} = \iint_S \text{curl } \mathbf{F} \cdot \mathbf{dS} \qquad (9.38)$$

The positive sense of $d\mathbf{S}$ is related to the positive sense of traversing C by the following *right-hand rule:* if the thumb of the right hand points along the curve in the sense of transversal, then the fingers will point toward the sense of $\mathbf{dS}$ to be taken as positive.

The fact that the surface integral may be calculated over *any* surface bounded by C requires that the surface integral have the same value over all such surfaces. Any two of them together form a closed surface, but for one of these $\mathbf{dS}$, as defined for Stokes' theorem, will be directed inward. Using outward normals, equality of the two integrals requires that $\iint \text{curl } \mathbf{F} \cdot \mathbf{dS} = 0$ over any closed surface bounded by any two open surfaces which each have C as a boundary. This, by the divergence theorem, requires

$$\text{div curl } \mathbf{F} = 0 \qquad (9.39)$$

which also follows from the definition in terms of ∇ in (9.37) and (9.32).

Likewise, on applying Stokes' theorem to a field **F** which is the gradient of a scalar ϕ, the line integral $\int_C \text{grad}\ \phi \cdot d\mathbf{s}$ will be zero since it is the difference in value of ϕ between the initial and final points, here the same. Hence it follows that

$$\text{curl grad}\ \phi = 0 \tag{9.40}$$

which also follows from the definitions in terms of components.

Any vector field **F** whose curl vanishes identically is said to be *lamellar* or *irrotational*. For such a vector field there exists a scalar field ϕ such that $\mathbf{F} = \text{grad}\ \phi$.

Likewise a vector field whose divergence vanishes identically is said to be *solenoidal*. For such a vector field there exists another vector field **A** such that $\mathbf{F} = \text{curl}\ \mathbf{A}$.

Closely related to Stokes' theorem are the following relations between line and surface integrals:

1. For a scalar field ϕ,

$$\int_C \phi\, d\mathbf{s} = \iint_S d\mathbf{S} \times \text{grad}\ \phi \tag{9.41}$$

2. For a vector field **F**, the tensor line integral is related to the tensor gradient in the same way:

$$\int_C \mathbf{F}\, d\mathbf{s} = \iint_S d\mathbf{S} \times \text{grad}\ \mathbf{F} \tag{9.42}$$

7. Expansion Formulas

The following formulas are useful in calculations of the applications of the operator ∇ to fields which are themselves at each point the product of two fields. Here ϕ is a scalar field and **F** and **G** are two vector fields:

$$\text{div}\ (\phi\mathbf{F}) = (\nabla\phi) \cdot \mathbf{F} + \phi\ \text{div}\ \mathbf{F} \tag{9.43}$$

$$\text{curl}\ (\phi\mathbf{F}) = (\nabla\phi) \times \mathbf{F} + \phi\ \text{curl}\ \mathbf{F} \tag{9.44}$$

$$\text{div}\ (\mathbf{F} \times \mathbf{G}) = \mathbf{G} \cdot \text{curl}\ \mathbf{F} - \mathbf{F} \cdot \text{curl}\ \mathbf{G} \tag{9.45}$$

$$\text{curl}\ (\mathbf{F} \times \mathbf{G}) = \mathbf{G} \cdot \nabla\mathbf{F} - \mathbf{F} \cdot \nabla\mathbf{G} + \mathbf{F}\ \text{div}\ \mathbf{G} - \mathbf{G}\ \text{div}\ \mathbf{F} \tag{9.46}$$

$$\text{grad}\ (\mathbf{F} \cdot \mathbf{G}) = \mathbf{G} \cdot \nabla\mathbf{F} + \mathbf{F} \cdot \nabla\mathbf{G} + \mathbf{G} \times \text{curl}\ \mathbf{F} + \mathbf{F} \times \text{curl}\ \mathbf{G} \tag{9.47}$$

The second-order differential functions are

$$\text{curl grad}\ \phi = 0 \tag{9.48}$$

$$\text{div curl}\ \mathbf{F} = 0 \tag{9.49}$$

$$\text{div grad}\ \phi = \frac{\partial^2\phi}{\partial x^2} + \frac{\partial^2\phi}{\partial y^2} + \frac{\partial^2\phi}{\partial z^2} = \nabla^2\phi \tag{9.50}$$

$$\text{curl curl}\ \mathbf{F} = \text{grad div}\ \mathbf{F} - \nabla^2\mathbf{F} \tag{9.51}$$

The operator div grad, or ∇^2, is called the *Laplace operator* and the quantity $\nabla^2\phi$ is called the *Laplacian* of ϕ.

8. Orthogonal Curvilinear Coordinates

Consider three scalar functions of position u, v, and w. In general, one and only one of each of the family of level surfaces for these will pass through a point P. These values of u, v, and w for P are called its curvilinear coordinates in this system.

At each point in space it is possible to introduce three unit vectors which will be denoted $\mathbf{u}_1,\mathbf{v}_1,\mathbf{w}_1$, whose directions are the directions of grad **u**, grad **v**, and grad **w**, respectively. It will be supposed that these unit vectors in this order form a right-handed coordinate system. Since the surfaces are curved, the basic unit vectors vary in direction from point to point.

It will be supposed in this section that the surfaces intersect orthogonally, that is, that at each point three conditions of the form $\text{grad}\ \mathbf{u} \cdot \text{grad}\ \mathbf{v} = 0$ are satisfied.

Consider the rectangular curvilinear volume element extending from u to $u + du$, v to $v + dv$, and w to $w + dw$. The length of the edge PA can be written

$$h_1\, du = \frac{du}{|\text{grad}\ \mathbf{u}|}$$

and similarly for the two other edges. Then the vector displacement from P to P' is given by

$$h_1\, du\, \mathbf{u}_1 + h_2\, dv\, \mathbf{v}_1 + h_3\, dw\, \mathbf{w}_1$$

If a scalar field ϕ is given as a function of its curvilinear coordinates, then the change in ϕ on going from P to P' is

$$\begin{aligned}\phi(P') - \phi(P) &= du\frac{\partial\phi}{\partial u} + dv\frac{\partial\phi}{\partial v} + dw\frac{\partial\phi}{\partial w}\\ &= (h_1\, du\, \mathbf{u}_1 + \cdots + \cdots)\\ &\quad \cdot \left(\frac{1}{h_1}\frac{\partial\phi}{\partial u}\mathbf{u}_1 + \cdots + \cdots\right)\end{aligned}$$

and therefore the curvilinear expression for the gradient of ϕ is

$$\text{grad}\ \phi = \frac{1}{h_1}\frac{\partial\phi}{\partial u}\mathbf{u}_1 + \frac{1}{h_2}\frac{\partial\phi}{\partial v}\mathbf{v}_1 + \frac{1}{h_3}\frac{\partial\phi}{\partial w}\mathbf{w}_1 \tag{9.52}$$

The curvilinear expression for the divergence of **F** in terms of its components on the curvilinear basis $\mathbf{F} = F_1\mathbf{u}_1 + F_2\mathbf{v}_1 + F_3\mathbf{w}_1$ is obtained by calculating the total flux over the basic volume element u to $u + du$, etc., and dividing by the volume $h_1h_2h_3\, du\, dv\, dw$. This result is

$$\text{div}\ \mathbf{F} = \frac{1}{h_1h_2h_3}\left[\frac{\partial}{\partial u}(h_2h_3F_1) + \frac{\partial}{\partial v}(h_3h_1F_2) + \frac{\partial}{\partial w}(h_1h_2F_3)\right] \tag{9.53}$$

Combining these results, the curvilinear expression for the *Laplacian* of a scalar ϕ is found to be

$$\nabla^2\phi = \text{div grad}\ \phi = \frac{1}{h_1h_2h_3}\left[\frac{\partial}{\partial u}\left(\frac{h_2h_3}{h_1}\frac{\partial\phi}{\partial u}\right) + \frac{\partial}{\partial v}\left(\frac{h_3h_1}{h_2}\frac{\partial\phi}{\partial v}\right) + \frac{\partial}{\partial w}\left(\frac{h_1h_2}{h_3}\frac{\partial\phi}{\partial w}\right)\right] \tag{9.54}$$

The expression for curl **F** is

$$\text{curl } \mathbf{F} = \frac{\mathbf{u}_1}{h_2h_3}\left[\frac{\partial}{\partial v}(h_3F_3) - \frac{\partial}{\partial w}(h_2F_2)\right] + \frac{\mathbf{v}_1}{h_3h_1}\left[\frac{\partial}{\partial w}(h_1F_1) - \frac{\partial}{\partial u}(h_3F_3)\right] + \frac{\mathbf{w}_1}{h_1h_2}\left[\frac{\partial}{\partial u}(h_2F_2) - \frac{\partial}{\partial v}(h_1F_1)\right] \quad (9.55)$$

The two curvilinear coordinate systems most used in physics are *spherical polar coordinates* r,θ,ϕ given by

$$x = r \sin \theta \cos \phi$$
$$y = r \sin \theta \sin \phi$$
$$z = r \cos \theta$$

and *cylindrical polar coordinates* (r,ϕ,z) in which the z is the same as the cartesian and

$$x = r \cos \phi$$
$$y = r \sin \phi$$

as in plane polar coordinates.

The explicit forms assumed by the vector operators in these two systems are:

1. *Spherical Polar Coordinates*

$$h_1 = 1 \qquad h_2 = r \qquad h_3 = r \sin \theta$$

$$\text{grad } V = \mathbf{r}_1 \frac{\partial V}{\partial r} + \boldsymbol{\theta}_1 \frac{1}{r}\frac{\partial V}{\partial \theta} + \frac{\boldsymbol{\phi}_1}{r \sin \theta}\frac{\partial V}{\partial \phi} \quad (9.56)$$

$$\text{div } \mathbf{F} = \frac{1}{r^2}\frac{\partial}{\partial r}(r^2F_1) + \frac{1}{r \sin \theta}\left[\frac{\partial}{\partial \theta}(\sin \theta \cdot F_2) + \frac{\partial F_3}{\partial \phi}\right] \quad (9.57)$$

$$\nabla^2 V = \frac{1}{r^2}\frac{\partial}{\partial r}\left(r^2 \frac{\partial V}{\partial r}\right) + \frac{1}{r^2}\left[\frac{1}{\sin \theta}\frac{\partial}{\partial \theta}\left(\sin \theta \frac{\partial V}{\partial \theta}\right) + \frac{1}{\sin^2 \theta}\frac{\partial^2 V}{\partial \phi^2}\right] \quad (9.58)$$

$$\text{curl } \mathbf{F} = \frac{\mathbf{r}_1}{r \sin \theta}\left[\frac{\partial}{\partial \theta}(F_3 \sin \theta) - \frac{\partial F_2}{\partial \phi}\right] + \frac{\boldsymbol{\theta}_1}{r}\left[\frac{1}{\sin \theta}\frac{\partial F_1}{\partial \phi} - \frac{\partial}{\partial r}(rF_3)\right] + \frac{\boldsymbol{\phi}_1}{r}\left[\frac{\partial}{\partial r}(rF_2) - \frac{\partial F_1}{\partial \theta}\right] \quad (9.59)$$

2. *Cylindrical Polar Coordinates*

$$h_1 = 1 \qquad h_2 = r \qquad h_3 = 1$$

$$\text{grad } V = \mathbf{r}_1 \frac{\partial V}{\partial r} + \frac{\boldsymbol{\phi}_1}{r}\frac{\partial V}{\partial \phi} + \mathbf{k}\frac{\partial V}{\partial z} \quad (9.60)$$

$$\text{div } \mathbf{F} = \frac{1}{r}\frac{\partial}{\partial r}(rF_1) + \frac{1}{r}\frac{\partial F_2}{\partial \phi} + \frac{\partial F_3}{\partial z} \quad (9.61)$$

$$\nabla^2 V = \frac{1}{r}\frac{\partial}{\partial r}\left(r\frac{\partial V}{\partial r}\right) + \frac{1}{r^2}\frac{\partial^2 V}{\partial \phi^2} + \frac{\partial^2 V}{\partial z^2} \quad (9.62)$$

$$\text{curl } \mathbf{F} = \mathbf{r}_1\left[\frac{1}{r}\frac{\partial F_3}{\partial \phi} - \frac{\partial F_2}{\partial z}\right] + \boldsymbol{\phi}_1\left[\frac{\partial F_1}{\partial z} - \frac{\partial F_3}{\partial r}\right] + \mathbf{k}\left[\frac{1}{r}\frac{\partial}{\partial r}(rF_2) - \frac{1}{r}\frac{\partial F_1}{\partial \phi}\right] \quad (9.63)$$

9. Transformation of Curvilinear Coordinates

In this section it will be supposed that a point in space is specified by three curvilinear (not necessarily orthogonal) coordinates u^1, u^2, u^3. At each point in space a set of basic vectors $\mathbf{e}_1$, $\mathbf{e}_2$, $\mathbf{e}_3$ may be defined from the relation $\mathbf{r}(u^1,u^2,u^3)$, which expresses the position vector in terms of these coordinates.

$$\mathbf{e}_1 = \frac{\partial \mathbf{r}}{\partial u^1} \qquad \mathbf{e}_2 = \frac{\partial \mathbf{r}}{\partial u^2} \qquad \mathbf{e}_3 = \frac{\partial \mathbf{r}}{\partial u^3} \quad (9.64)$$

$\mathbf{e}_1$ is tangent to the curve along which u^1 increases while holding u_2 and u_3 constant. The reciprocal set of vectors (10.18) will be denoted $\mathbf{e}^1,\mathbf{e}^2,\mathbf{e}^3$. The vector expression for $\mathbf{dr}$ is

$$\mathbf{dr} = \mathbf{e}_1\, du^1 + \mathbf{e}_2\, du^2 + \mathbf{e}_3\, du^3$$
$$= \mathbf{e}^1\, du_1 + \mathbf{e}^2\, du_2 + \mathbf{e}^3\, du_3$$

where, however, the differentials du_1, etc., are in general not perfect differentials of functions of the original coordinates u^1,u^2,u^3.

Consideration of the length of $\mathbf{dr}$ leads to the introduction of the g_{rs} tensor at each point in space, $g_{rs} = \mathbf{e}_r \cdot \mathbf{e}_s$ and $ds^2 = \mathbf{dr} \cdot \mathbf{dr} = g_{rs}\, du^r\, du^s g_{rs}$ is called the differential line element.

An arbitrary vector **A** can be expressed in terms of either its contravariant components A^r or its covariant components A_r,

$$\mathbf{A} = A^r\mathbf{e}_r = A_r\mathbf{e}^r \quad (9.65)$$

with $\quad A^r = g^{rs}A_s \quad$ or $\quad A_s = g_{sr}A^r$

Similarly, a tensor of the second order, **A**, can be written in terms of various sets of components and corresponding tensor units:

$$\mathbf{A} = A^{rs}\mathbf{e}_r\mathbf{e}_s = A_{rs}\mathbf{e}^r\mathbf{e}^s = A_r{}^s\mathbf{e}^r\mathbf{e}_s = A^r{}_s\mathbf{e}_r\mathbf{e}^s \quad (9.66)$$

where the corresponding relations are

$$A_r{}^s = g_{ru}A^{us} \qquad A_{rs} = g_{sv}A_r{}^v$$
$$A^r{}_s = g_{su}A^{ru}$$

which indicates the general way in which the components of the metric tensor act to raise and lower indices.

The tensor **I**, which when multiplied into any vector leaves it unaltered, is called the *idemfactor*. It has these forms:

$$\mathbf{I} = \mathbf{e}_r\mathbf{e}^r = \mathbf{e}^r\mathbf{e}_r = g^{rs}\mathbf{e}_r\mathbf{e}_s = g_{rs}\mathbf{e}^r\mathbf{e}^s \quad (9.67)$$

Hence the metric tensor may also be regarded as giving the tensor components of the idemfactor.

Consider the small parallelepiped whose edges are $\mathbf{dr}_{(1)} = \mathbf{e}_1\, du^1$, $\mathbf{dr}_{(2)} = \mathbf{e}_2\, du^2$, $\mathbf{dr}_{(3)} = \mathbf{e}_3\, du^3$. The lengths of these edges are $\sqrt{g_{11}}\, du^1$, $\sqrt{g_{22}}\, du^2$, $\sqrt{g_{33}}\, du^3$. Hence $\sqrt{g_{11}}$ is the h_1 of the previous section in case the system is orthogonal.

Likewise the vector elements of area are

$\mathbf{ds}_{(1)} = \mathbf{dr}_{(2)} \times \mathbf{dr}_{(3)}$, etc., and

$$ds_{(1)} = \sqrt{g_{22}g_{33} - g_{23}^2}\, du^2\, du^3 \quad (9.68)$$

The volume dV of the elementary parallelepiped is $dV = \pm\mathbf{dr}_{(1)} \cdot (\mathbf{dr}_{(2)} \times \mathbf{dr}_{(3)})$, the + sign occurring if $\mathbf{dr}_{(1)}$, $\mathbf{dr}_{(2)}$, and $\mathbf{dr}_{(3)}$ form a right-handed system. This gives $dV = \sqrt{g}\, du^1\, du^2\, du^3$, where g is written for the determinant of the g_{rs} matrix.

A second curvilinear system of curvilinear coordinates will now be introduced, denoted by v^1,v^2,v^3, where these are related to the u^1,u^2,u^3 by some set of functional relations:

$$v^1 = v^1(u^1,u^2,u^3)$$
$$v^2 = v^2(u^1,u^2,u^3)$$
$$v^3 = v^3(u^1,u^2,u^3)$$

and the second system will have its sets of reciprocal vectors $\mathbf{f}_r$ and $\mathbf{f}^r$ corresponding to the $\mathbf{e}_r$ and $\mathbf{e}^r$ of the original coordinates. If $\mathbf{dr}$ is a vector displacement from a point P to a nearby Q,

$$\mathbf{dr} = \mathbf{f}_r\, dv^r = \mathbf{e}_s\, du^s$$

and the differentials transform linearly according to the relations

$$dv^r = \frac{\partial v^r}{\partial u^s}\, du^s \qquad du^s = \frac{\partial u^s}{\partial v^r}\, dv^r \tag{9.69}$$

Let J and K be the Jacobians of these two transformations and suppose neither vanishes. The two matrices are inverses in the sense

$$\frac{\partial u^r}{\partial v^s}\frac{\partial v^s}{\partial u^t} = \delta^r_t$$

and likewise $JK = 1$.

The basic vectors transform in a way determined by invariance of the form of $\mathbf{dr}$:

$$\mathbf{f}_r = \frac{\partial u^s}{\partial v^r}\,\mathbf{e}_s \qquad \mathbf{e}_r = \frac{\partial v^s}{\partial u^r}\,\mathbf{f}_s \tag{9.70}$$

that is, by the matrix inverse to that used for transforming the coordinate differentials. These results indicate a geometric interpretation of the partial derivatives in terms of scalar products of the basic vectors:

$$\frac{\partial u^s}{\partial v^r} = \mathbf{f}^r \cdot \mathbf{e}_s \qquad \frac{\partial v^s}{\partial u^r} = \mathbf{e}^r \cdot \mathbf{f}_s$$

Writing h_{rs} for the metric tensor in the new coordinate system, $ds^2 = h_{rs}\, dv^r\, dv^s$, it follows that the transformation law for the h_{rs} is

$$h_{rs} = \frac{\partial u^t}{\partial v^r}\frac{\partial u^u}{\partial v^s}\, g_{tu} \qquad g_{rs} = \frac{\partial v^t}{\partial u^r}\frac{\partial v^u}{\partial u^s}\, h_{tu} \tag{9.71}$$
$$h^{rs} = \frac{\partial v^r}{\partial u^t}\frac{\partial v^s}{\partial u^u}\, g^{tu} \qquad g^{rs} = \frac{\partial u^r}{\partial v^t}\frac{\partial u^s}{\partial v^u}\, h^{tu}$$

The transformation is such that

$$g^r_s = h^r_s = \delta^r_s$$

thus the mixed components have the same values in all systems.

The transformation of the determinant of the g_{rs} and of that of g^{rs} is

$$h = K^2 g \qquad g = J^2 h$$
$$h^1 = J^2 g^1 \qquad g^1 = K^2 h \tag{9.72}$$

where g^1 is the determinant of the g^{rs}.

The gradient of a scalar field function $U(u^1,u^2,u^3)$ will be defined such that the change in U in passing from P to Q is

$$dU = \nabla U \cdot \mathbf{dr} = \frac{\partial u}{\partial u^r}\, du^r$$

and therefore

$$\nabla U = \mathbf{e}^r \frac{\partial U}{\partial u^r} \tag{9.73}$$

which is the appropriate generalization to oblique curvilinear coordinates of (8.1). This gives the form

$$\nabla = \mathbf{e}^r \frac{\partial}{\partial u^r}$$

for the basic vector differential operator.

The *divergence* of a vector field $\mathbf{A}$ is obtained by calculating the flux out of the elementary parallelepiped whose edges lie along the curves of increasing u^1, u^2, and u^3. It takes the form

$$\text{div } \mathbf{A} = \frac{1}{\sqrt{g}}\frac{\partial}{\partial u^r}(\sqrt{g}\, A^r) \tag{9.74}$$

The generalized form of the *Laplacian*, sometimes called the *Lamé operator*, is obtained by combining the two preceding results:

$$\text{div grad } U = \frac{1}{\sqrt{g}}\frac{\partial}{\partial u^r}\left(\sqrt{g}\, g^{rs}\frac{\partial U}{\partial u^s}\right) \tag{9.75}$$

The *curl* of a vector field $\mathbf{A}$ closely resembles its form in cartesian coordinates:

$$\text{curl } \mathbf{A} = \frac{1}{\sqrt{g}}\left[\left(\frac{\partial A_3}{\partial u^2} - \frac{\partial A_2}{\partial u^1}\right)\mathbf{e}_1 + \cdots + \cdots\right] \tag{9.76}$$

where it is to be noted that covariant components of $\mathbf{A}$ are used to get the contravariant components of curl $\mathbf{A}$.

References

1. Craig, H. V.: "Vector and Tensor Analysis," McGraw-Hill, New York, 1943.
2. Gibbs, J. W., and E. B. Wilson: "Vector Analysis," Yale University Press, New Haven, 1901.
3. Jeffreys, H.: "Cartesian Tensors," Cambridge University Press, New York, 1935.
4. Weatherburn, C. E.: "Elementary Vector Analysis," G. Bell, London, 1924.
5. Weatherburn, C. E.: "Advanced Vector Analysis," G. Bell, London, 1924.
6. Wills, A. P.: "Vector and Tensor Analysis," Prentice-Hall, Englewood Cliffs, N.J., 1931.

Chapter 10

Tensor Calculus

By C. LANCZOS, Dublin Institute for Advanced Studies

1. Scalars, Vectors, Tensors

Certain quantities in physics seem to have absolute significance while other quantities can be defined only *relative to a certain frame of reference.* Mass, density, temperature, and specific heat are represented by *pure numbers*, assigned to certain physical categories. Such quantities are called *scalars.* Other quantities, however, involve the dimensions of space. In a one-dimensional world all measurements would be reducible to scalars, but in a two- or higher-dimensional manifold quantities occur which cannot be measured by pure numbers. They involve magnitude and direction and require a definite frame of reference for analytical characterization, e.g., a vector may be visualized as an arrow put in space The invariant description of such directed quantities gave rise to a relatively recently developed branch of mathematical physics, called *absolute calculus* or *tensor calculus.* In it vectors are a special case of a more general class of directed quantities, called tensors, which play a fundamental role in the functional relations of the physical universe.

2. Analytic Operations with Vectors

A vector may be visualized as an arrow which has magnitude and direction. In vector analysis, Chap. 9, such a quantity is represented by an algebraic symbol with suitable properties. Certain geometrical operations on two vectors are denoted by $\mathbf{A} + \mathbf{B}$, or $\mathbf{A} - \mathbf{B}$, or $\mathbf{A} \cdot \mathbf{B}$, or $\mathbf{A} \times \mathbf{B}$. The tools of ordinary algebra are thus put into the service of directed quantities. Differentiation and integration are also applicable to certain operations with vectors.

In this procedure some of the basic postulates of ordinary algebra have to be sacrificed. Vector algebra is less simple than ordinary algebra by requiring two kinds of multiplications: the *scalar* product $\mathbf{A} \cdot \mathbf{B}$ and the *vector* product $\mathbf{A} \times \mathbf{B}$. This complication can be avoided by the use of *Hamilton's quaternions* which combine the two kinds of multiplications into one single operation: the product $\mathbf{AB}$ of the two quaternions $\mathbf{A}$ and $\mathbf{B}$. Even so the commutative law of ordinary multiplication $\mathbf{AB} = \mathbf{BA}$ has to be abandoned, although the other postulates of algebra are retained.

3. Unit Vectors; Components

A different and more far-reaching approach is obtained by introducing a system of mutually perpendicular unit vectors for the analysis of vectors. In space of three dimensions three such vectors are necessary and sufficient for the description of an arbitrary vector $\mathbf{A}$. For analytical purposes the three-dimensional nature of physical space is of accidental significance and can be replaced by the concept of an *n-dimensional* space in which n mutually perpendicular unit vectors of the length 1:

$$\mathbf{U}_1, \mathbf{U}_2, \ldots, \mathbf{U}_n \tag{10.1}$$

are sufficient for the representation of an arbitrary vector $\mathbf{A}$. The vector $\mathbf{A}$ is now obtained as a linear superposition of the vectors (10.1):

$$\mathbf{A} = a_1\mathbf{U}_1 + a_2\mathbf{U}_2 + \cdots + a_n\mathbf{U}_n \tag{10.2}$$

The quantities

$$a_1, a_2, \ldots, a_n \tag{10.3}$$

called the *components* of the vector $\mathbf{A}$ are obtained by projecting $\mathbf{A}$ on the unit vectors:

$$a_i = \mathbf{A} \cdot \mathbf{U}_i \tag{10.4}$$

While these a_i are ordinary real numbers which satisfy all the postulates of ordinary algebra, they cannot be conceived as an aggregate of scalars since they have additional significance because of their association with the frame of axes (10.1) to which they belong. These components are comparable to the digits of the decimal number 3425. The given digits characterize this number only if the base 10 is given to which they belong. The same digits, if associated with the base 8, belong to an entirely different number; on the other hand, the same number appears in the new form 6541 if the base 8 is adopted. The number remained the same but its components have changed in the new reference system.

Thus a vector is defined by a set of n real numbers (10.3) in relation to a particular frame of axes. An important property of these numbers is the rule of transformation to find the components on changing to some other frame of n unit vectors

$$\bar{\mathbf{U}}_1, \bar{\mathbf{U}}_2, \ldots, \bar{\mathbf{U}}_n \tag{10.5}$$

These rules are developed in the *absolute calculus*

or *tensor calculus*, which falls into two main chapters: the algebraic operations with tensors, *tensor algebra*, and the infinitesimal operations with tensor fields, *tensor analysis*.

4. Adjoint Set of Axes

In a more general reference frame the basic vectors need not be mutually perpendicular or of length 1, even though for practical purposes we usually prefer such a system. Tensor calculus uses an arbitrary skew-angular set of basic vectors,

$$\mathbf{V}_1, \mathbf{V}_2, \ldots, \mathbf{V}_n \tag{10.6}$$

not restricted in length and mutual positions, except that they be *linearly independent*, i.e., the volume included by them shall not be zero.

Operation with such a system of basic vectors is greatly facilitated by associating with it a second set of basic vectors, called the *adjoint set*. For this new set of vectors the same notation **V** is used with the subscripts in an upper position

$$\mathbf{V}^1, \mathbf{V}^2, \ldots, \mathbf{V}^n \tag{10.7}$$

An orthogonal set of axes is characterized by the customary orthogonality conditions:

$$\mathbf{U}_i \cdot \mathbf{U}_k = 0 \qquad i \neq k \tag{10.8}$$

while the normalization of the length of axes to 1 adds the further condition

$$\mathbf{U}_i^2 = 1 \tag{10.9}$$

Although the general set of vectors (10.6) satisfies neither of these two conditions, we can always define a new set of vectors (10.7) by the conditions

$$\begin{aligned} \mathbf{V}_i \cdot \mathbf{V}^k &= 0 \qquad i \neq k \\ \text{and} \qquad \mathbf{V}_i \cdot \mathbf{V}^i &= 1 \end{aligned} \tag{10.10}$$

To any given $\mathbf{V}_i$ these equations are solvable and the solution is unique, provided that the given $\mathbf{V}_i$ are linearly independent.

The original and the adjoint set of vectors are in a dual relation to each other: the adjoint of the adjoint set leads back to the original set. The conditions (10.10) express the mutual orthogonality or *biorthogonality* of the two vector sets $\mathbf{V}_i$ and $\mathbf{V}^i$ and their mutual normalization.

The special advantage of the orthogonal and normalized set of unit vectors $\mathbf{U}_i$ can now be seen in the fact that here the adjoint set $\mathbf{V}^i$ coincides with the original set $\mathbf{V}_i$. Hence an orthogonal and normalized (*orthonormal*) set of unit vectors is *self-adjoint*, thus avoiding the doubling of the fundamental set of vectors.

The adjoint set $\mathbf{V}^i$ can be generated as a linear superposition of the given vectors $\mathbf{V}_i$:

$$\mathbf{V}^i = g^{i1}\mathbf{V}_1 + g^{i2}\mathbf{V}_2 + \cdots + g^{in}\mathbf{V}_n = \sum_{\alpha=1}^{n} g^{i\alpha}\mathbf{V}_\alpha \tag{10.11}$$

Since the $\mathbf{V}_i$ are given, we have the following dot products:

$$\mathbf{V}_i \cdot \mathbf{V}_k = \mathbf{V}_k \cdot \mathbf{V}_i = g_{ik} \tag{10.12}$$

These $g_{ik} = g_{ki}$ form the elements of a symmetric matrix. The conditions (10.10) now demand:

$$\sum_{\alpha=1}^{n} g^{i\alpha} g_{\alpha k} = \delta_{ik} \tag{10.13}$$

(The *Kronecker symbol* δ_{ik} is defined as follows: its value is 1 for $i = k$, and 0 for $i \neq k$.) The matrix of the g^{ik} is the *reciprocal* of the g_{ik} matrix. The existence of the reciprocal matrix demands that the determinant

$$g = \|g_{ik}\| \tag{10.14}$$

shall not be zero. The geometrical significance of this determinant is the square of the volume included by the n base vectors $\mathbf{V}_i$. Since the $\mathbf{V}_i$ are linearly independent, according to our basic assumption, this volume cannot vanish, and the existence (and uniqueness) of the $g^{ik} = g^{ki}$ is guaranteed.

The duality of the adjoint sets permits us to complete (10.11) by the analogous dual equation

$$\mathbf{V}_i = \sum_{\alpha=1}^{n} g_{i\alpha}\mathbf{V}^\alpha \tag{10.15}$$

with

$$\mathbf{V}^i \cdot \mathbf{V}^k = \mathbf{V}^k \cdot \mathbf{V}^i = g^{ik} \tag{10.16}$$

The symmetric matrices g_{ik} and g^{ik} are fundamental for the general theory of tensors and for Einstein's theory of general relativity.

In the special case of an orthonormal set of axes (10.8) and (10.9) the g_{ik} are reduced to the elements of the unit matrix:

$$g_{ik} = g^{ik} = \delta_{ik} \tag{10.17}$$

and we obtain $\mathbf{V}^i = \mathbf{V}_i$

5. Covariant and Contravariant Components of a Vector

In view of the complete duality of the vectors $\mathbf{V}_i$ and $\mathbf{V}^i$, each set can equally be used for the analysis of a given vector **A**. We can put

$$\begin{aligned} \mathbf{A} &= a^1\mathbf{V}_1 + a^2\mathbf{V}_2 + \cdots + a^n\mathbf{V}_n \\ &= \sum_{\alpha=1}^{n} a^\alpha \mathbf{V}_\alpha \end{aligned} \tag{10.18}$$

with

$$a^i = \mathbf{A} \cdot \mathbf{V}^i \tag{10.19}$$

and likewise

$$\mathbf{A} = a_1\mathbf{V}^1 + a_2\mathbf{V}^2 + \cdots + a_n\mathbf{V}^n = \sum_{\alpha=1}^{n} a_\alpha \mathbf{V}^\alpha \tag{10.20}$$

with

$$a_i = \mathbf{A} \cdot \mathbf{V}_i \tag{10.21}$$

The a^i and the a_i are two independent sets of components, associated with the same vector **A**, but expressing that vector in the reference system of the $\mathbf{V}_i$ and in the adjoint reference system of the $\mathbf{V}^i$. The a^i are called the *contravariant*, the a_i the *covariant* components of the same vector **A**. The relation between these two sets of components can be found with the help of (10.11) and (10.15).

$$a^i = \sum_{\alpha=1}^{n} g^{i\alpha} a_\alpha \tag{10.22}$$

$$a_i = \sum_{\alpha=1}^{n} g_{i\alpha} a^\alpha \tag{10.23}$$

If the axes are self-adjoint (orthonormal), the g_{ik} assume the normal values δ_{ik}, and a^i and a_i become identical.

6. Transformation of the Basic Vectors $\mathbf{V}_i$

A different set of basic vectors $\bar{\mathbf{V}}_i$ can be expressed in the reference systems of the original $\mathbf{V}_i$, giving rise to relations of the form:

$$\bar{\mathbf{V}}_i = \sum_{\alpha=1}^{n} \beta_i^\alpha \mathbf{V}_\alpha \tag{10.24}$$

while the inverse transformation takes the form:

$$\mathbf{V}_i = \sum_{\alpha=1}^{n} \bar{\beta}_i^\alpha \bar{\mathbf{V}}_\alpha \tag{10.25}$$

The matrices β_i^k and $\bar{\beta}_i^k$ are reciprocal to each other:

$$\sum_{\alpha=1}^{n} \beta_i^\alpha \bar{\beta}_\alpha^k = \sum_{\alpha=1}^{n} \beta_\alpha^i \bar{\beta}_k^\alpha = \delta_{ik} \tag{10.26}$$

Existence of the inverse matrix $\bar{\beta}_i^k$ is guaranteed by the demand that the vectors $\bar{\mathbf{V}}_i$ shall also be linearly independent.

The definition of the adjoint set of vectors gives

$$\begin{aligned} \beta_i^k &= \bar{\mathbf{V}}_i \cdot \mathbf{V}^k \\ \bar{\beta}_i^k &= \mathbf{V}_i \cdot \bar{\mathbf{V}}^k \end{aligned} \tag{10.27}$$

and the transformation of the adjoint vectors is given by the following equations, dual to (10.24) and (10.25):

$$\bar{\mathbf{V}}^i = \sum_{\alpha=1}^{n} \bar{\beta}_\alpha^i \mathbf{V}^\alpha$$

$$\mathbf{V}^i = \sum_{\alpha=1}^{n} \beta_\alpha^i \bar{\mathbf{V}}^\alpha \tag{10.28}$$

7. Transformation of Vector Components

The vector $\mathbf{A}$ can be analyzed in the new set of axes, obtaining

$$\bar{a}_i = \mathbf{A} \cdot \bar{\mathbf{V}}_i = \sum_{\alpha=1}^{n} \beta_i^\alpha a_\alpha$$

$$\bar{a}^i = \mathbf{A} \cdot \bar{\mathbf{V}}^i = \sum_{\alpha=1}^{n} \bar{\beta}_\alpha^i a^\alpha \tag{10.29}$$

The covariant components follow the transformation law of the $\mathbf{V}_i$ (*are covariant with* the $\mathbf{V}_i$), while the contravariant components follow the transformation law of the $\mathbf{V}^i$ (are *contravariant to* the $\mathbf{V}_i$).

8. Radius Vector R

The position of an ordinary point P in space can be characterized by a set of contravariant coordinates

$$x^1, x^2, \ldots, x^n \tag{10.30}$$

defined as the contravariant components of the radius vector $\mathbf{R}$:

$$\mathbf{R} = x^1\mathbf{V}_1 + x^2\mathbf{V}_2 + \cdots + x^n\mathbf{V}_n = \sum_{\alpha=1}^{n} x^\alpha \mathbf{V}_\alpha \tag{10.31}$$

The same point P can likewise be characterized in terms of the covariant coordinates

$$x_1, x_2, \ldots, x_n \tag{10.32}$$

defined by the covariant components of the radius vector $\mathbf{R}$:

$$\mathbf{R} = x_1\mathbf{V}^1 + x_2\mathbf{V}^2 + \cdots + x_n\mathbf{V}^n = \sum_{\alpha=1}^{n} x_\alpha \mathbf{V}^\alpha \tag{10.33}$$

The square of the radius vector R has an important geometrical significance. It expresses the square of the distance of the point P from the origin in terms of the coordinates of P. Making use of the definition of the g_{ik} and g^{ik} according to (10.12) and (10.16):

$$\mathbf{R}^2 = \sum_{i=1}^{n} \sum_{k=1}^{n} g_{ik} x^i x^k = \sum_{i,k=1}^{n} g_{ik} x^i x^k \tag{10.34}$$

Similarly

$$\mathbf{R}^2 = \sum_{i,k=1}^{n} g^{ik} x_i x_k \tag{10.35}$$

An expression of the form (10.34) is called a *quadratic form* of the variables x^i. The particular quadratic form which defines the square of the distance of the point x^i from the origin is called the *fundamental metrical form.*

The relation between the x_i and the x^i is established on the basis of (10.22) and (10.23):

$$x^i = \sum_{\alpha=1}^{n} g^{i\alpha} x_\alpha \tag{10.36}$$

$$x_i = \sum_{\alpha=1}^{n} g_{i\alpha} x^\alpha \tag{10.37}$$

9. Abstract Definition of a Vector

A more abstract definition of a vector may be given which brings the central principle of tensor calculus, *the principle of invariance,* into sharp focus:

We start with the variables $x^1, x^2, \ldots, x^n$, which characterize the position of an arbitrary point P in space, and assume that the square of the distance s of that point from the origin of the reference system is given by the quadratic form

$$s^2 = \sum_{i,k=1}^{n} g_{ik}x^i x^k \tag{10.38}$$

We then introduce the covariant x_i by the definition

$$x_i = \sum_{\alpha=1}^{n} g_{i\alpha}x^\alpha \tag{10.39}$$

Hence

$$s^2 = \sum_{\alpha=1}^{n} x_\alpha x^\alpha \tag{10.40}$$

We now consider the *linear form* of the variables x^i

$$A = a_1x^1 + \cdots + a_nx^n = \sum_{\alpha=1}^{n} a_\alpha x^\alpha \tag{10.41}$$

and define the coefficients $a_1, \ldots, a_n$ of this linear form as the covariant components of a vector **A**. The same form A can also be written in terms of the x_i:

$$A = a^1x_1 + \cdots + a^nx_n = \sum_{\alpha=1}^{n} a^\alpha x_\alpha \tag{10.42}$$

with

$$a^i = \sum_{\alpha=1}^{n} g^{i\alpha}a_\alpha \tag{10.43}$$

thus defining the *contravariant* components of the same vector **A**.

If the vector **A** is regarded as constant force, then the physical significance of the scalar A is the *work* done by the force during the displacement OP = **R**. In the abstract definition of a vector the justification of calling **A** a vector is taken from the fact that the work of the force **A** for arbitrary positions of the radius vector **R** appears as a linear form of the coordinates x^i. The coefficients of this form define the covariant components of the force **A**.

A mere set of numbers $a_i, \ldots, a_n$ does not establish a vector since these coefficients have significance only in connection with a given set of coordinate axes. The abstract definition of a vector takes this property of the vector components into account since the linear form A is established solely in connection with the variables $x^1, x^2, \ldots, x^n$. In particular we consider first purely *rectilinear* systems, i.e., coordinate systems whose parameter lines are parallel straight lines. This means in terms of the x^i that we consider arbitrary *linear* transformations of the variables x^i:

$$\bar{x}^i = \sum_{\alpha=1}^{n} \bar{\beta}^i_\alpha x^\alpha \tag{10.44}$$

with non-vanishing determinant $\|\bar{\beta}^i_\alpha\|$. The inverse transformation is then given by

$$x^i = \sum_{\alpha=1}^{n} \beta^i_\alpha \bar{x}^\alpha \tag{10.45}$$

where the matrix β^i_k is the reciprocal of the matrix $\bar{\beta}^i_k$.

The transformation of the covariant x_i is established by the principle that the *bilinear form*

$$s^2 = \sum_{\alpha=1}^{n} x_\alpha x^\alpha \tag{10.46}$$

shall be an invariant of the transformation:

$$\sum_{\alpha=1}^{n} x_\alpha x^\alpha = \sum_{\alpha=1}^{n} \bar{x}_\alpha \bar{x}^\alpha \tag{10.47}$$

This principle establishes the transformation of the x_i as the reciprocal of the transformation of the x^i:

$$\bar{x}_i = \sum_{\alpha=1}^{n} \beta^\alpha_i x_\alpha \tag{10.48}$$

$$x_i = \sum_{\alpha=1}^{n} \bar{\beta}^\alpha_i \bar{x}_\alpha \tag{10.49}$$

Transformation of vector components is established by the principle that the linear form A shall be an invariant of the transformation:

$$\sum_{\alpha=1}^{n} a_\alpha x^\alpha = \sum_{\alpha=1}^{n} \bar{a}_\alpha \bar{x}^\alpha \tag{10.50}$$

The individual coefficients a_i change their values if the frame of axes is changed. The value of the *entire linear form A*, however, must *not* be influenced by the transformation, no matter what the position of the point P is. This principle establishes the transformation law of the a_i in the form

$$\bar{a}_i = \sum_{\alpha=1}^{n} \beta^\alpha_i a_\alpha \tag{10.51}$$

The transformation of the contravariant a^i is similarly established by the invariance of the linear form (10.41):

$$a^i = \sum_{\alpha=1}^{n} \bar{\beta}^i_\alpha a^\alpha \tag{10.52}$$

The duality of the components a_i and a^i and their transformation laws *without any reference to unit vectors* have been developed by using the following tools: (1) The definition of a vector on the basis of an invariant linear form. (2) The existence of a distance square defined by an invariant form of second order.

10. Invariants and Covariants

In theory of relativity the distinction between quantities which change with the reference system (*covariants*) and quantities which do not change with the reference system (*invariants*) is of paramount importance. The coefficients of a linear form are covariants since they depend on the reference system employed. The *entire linear form*, however, is an *invariant* of the transformation which does *not* change its value in any rectilinear reference system and thus has absolute significance. In the prerelativistic phase of physics certain quantities which belong to the realm of covariants were treated as invariants, and vice versa. In particular, the time t was considered as an absolute, unchangeable variable which does not participate in any transformations, while in fact nature forms a four-dimensional manifold of space *and* time. This relegates the time t into the realm of a fourth coordinate which is transformed together with the three space variables x,y,z. The orthogonal transformations of 3-space, characterized by the invariance of the quadratic form $x^2 + y^2 + z^2$, were enlarged to the orthogonal transformations of 4-space, characterized by the invariance of the quadratic form

$$x^2 + y^2 + z^2 - c^2t^2 = \bar{x}^2 + \bar{y}^2 + \bar{z}^2 - c^2\bar{t}^2 \qquad (10.53)$$

where c is the velocity of light which plays the role of a scale factor of the fourth dimension and becomes thus one of the basic constants of nature. The physical significance of the invariance principle (10.53) is that light travels in *any* nonaccelerated reference system with the same velocity c, in every direction, irrespective of the motion of the observer.

In general relativity the expression (10.53), which introduces a Euclidean geometry into the four-dimensional space-time world, is recognized as a macroscopic approximation to reality since the actual metric of the universe is of the Riemannian type and has to be developed on the basis of general tensor calculus.

11. Abstract Definition of a Tensor

Invariant algebraic forms of first order are only a special example of the much wider class of invariant algebraic forms of *any* order. This gives a natural introduction of the tensor concept: a tensor of mth order is defined with the help of an *invariant algebraic form of order m*. The coefficients of this form define the components of the tensor, covariant if the variables are the contravariant coordinates x^i, and contravariant if the variables are the covariant coordinates x_i. A vector is thus a special case appearing as a tensor of first order.

12. Tensors of Second Order

Tensors of second order occur particularly often in the mathematical description of natural phenomena. For example, the elastic stress tensor is a symmetric tensor of second order, as well as the Maxwellian electric stress tensor. In the theory of relativity the electromagnetic field strength is an antisymmetric tensor of second order in space-time.

All these tensors represent mathematically the coefficients of an invariant algebraic form of second order:

$$A = \sum_{\alpha,\beta=1}^{n} a_{\alpha\beta}x^\alpha x^\beta \qquad (10.54)$$

This quadratic form has only $n(n + 1)/2$ independent elements, since the terms $a_{ik}x^ix^k$ and $a_{ki}x^kx^i$ combine into one. We make the tensor unique by adding the symmetry condition

$$a_{ki} = a_{ik} \qquad (10.55)$$

and speak of a *symmetric tensor*.

A general tensor of second order is defined by the following device: consider two *different* positions of the radius vector **R**, say x^i and y^i, and define the form

$$A = \sum_{\alpha,\beta=1}^{n} a_{\alpha\beta}x^\alpha y^\beta \qquad (10.56)$$

The terms with a_{ik} and a_{ki} are now independent and the form defines n^2 separate elements.

The symmetry pattern (10.55) of the coefficients can be augmented by the pattern

$$a_{ki} = -a_{ik} \qquad (10.57)$$

which defines an antisymmetric tensor of second order having $n(n - 1)/2$ independent components. The six independent components of such a tensor in 4-space combine the electric and magnetic field strength into one entity.

The same tensor of second order may be given in covariant or contravariant or mixed form, according to the nature of the variables employed:

$$\begin{aligned} A &= \sum_{\alpha,\beta=1}^{n} a_{\alpha\beta}x^\alpha y^\beta \\ &= \sum_{\alpha,\beta=1}^{n} a^{\alpha\beta}x_\alpha y_\beta \\ &= \sum_{\alpha,\beta=1}^{n} a^\alpha{}_\beta x_\alpha y^\beta \\ &= \sum_{\alpha,\beta=1}^{n} a_\alpha{}^\beta x^\alpha y_\beta \end{aligned} \qquad (10.58)$$

13. Einstein Sum Convention

The homogeneous notation of the variables and the consistent use of lower and upper indices for the distinction of covariant and contravariant components and variables contributed greatly to the systematic development of tensor calculus. An additional operational simplification was introduced by Einstein. In all previous formulas the position of the indices is such that the summation occurs over an index which in one factor is in the upper position and in the other factor in the lower position. Now whenever the same index appears twice in opposite positions in a formula, we shall *automatically sum over that index.* Hence the notation

$$a_\alpha x^\alpha$$

shall mean

$$\sum_{\alpha=1}^{n} a_\alpha x^\alpha$$

The same convention holds if a product contains more than one pair of equal indices, e.g., the double sum

$$\sum_{i,k=1}^{n} a_{ik}x^i y^k$$

is now written in the form

$$a_{ik}x^i y^k$$

This simplification greatly facilitates the symbolic manipulations with tensors.

14. Tensor Algebra

Algebraic operations with tensors are an immediate consequence of the general definition of a tensor as the coefficients of an invariant algebraic form of the order m. The general definition makes use of m independent positions of the radius vector **R**:

$$x^i,\ y^k,\ \ldots,\ z^m$$

Moreover, any of these sets of variables may be put in covariant form

$$x_i,\ y_k,\ \ldots,\ z_m$$

Every set of variables is associated with one subscript or superscript in the coefficients, corresponding indices being always in juxtaposition, for example,

$$A = a_{ik\cdots m}x^i y^k \cdots z^m$$
$$A = a_{k}^{i}{\cdots}^{m} x_i y^k \cdots z_m$$

Addition of Tensors. The sum of two invariants is again an invariant. The sum of the two forms

$$a_i x^i + b_i x^i = (a_i + b_i)x^i \tag{10.59}$$

defines a new invariant form of first order. The quantities

$$c_i = a_i + b_i \tag{10.60}$$

form the covariant components of a new tensor of first order.

Generally two tensors of the same order whose components are in homologous positions can be added; the sum defines a new tensor of the same order, with the same distribution of covariant and contravariant components:

$$a_i{}^{km} + b_i{}^{km} = c_i{}^{km} \tag{10.61}$$

Multiplication by a Constant. The multiplication of all tensor components by the same constant defines a new tensor of the same order and same distribution of indices:

$$\alpha a_i{}^{km} = c_i{}^{km} \tag{10.62}$$

Multiplication of Two Tensors. The product of two invariant algebraic forms gives once more an invariant algebraic form. The order of the new form is the sum of the orders of the composing factors:

$$(a_i x^i)(b_k y^k) = a_i b_k x^i y^k \tag{10.63}$$

Hence

$$a_i b_k = c_{ik} \tag{10.64}$$

defines a tensor of second order, covariant in both indices. Generally the product of any two tensors, with any distribution of covariant and contravariant components, yields a tensor whose order is the sum of the order of the composing factors and whose indices exactly repeat the entire set of composing indices, e.g., the product

$$a_i{}^m b_k{}^p = c_i{}^m{}_k{}^p \tag{10.65}$$

yields a tensor of fourth order, covariant in i,k, contravariant in m,p.

Transposition of Indices. If in the definition of an invariant algebraic form the positions of the radius vector are exchanged, we once more obtain an invariant algebraic form of the same order. For example, if

$$A = a_{ik}x^i y^k \tag{10.66}$$

is an invariant,

$$B = a_{ik}y^i x^k = a_{ki}x^i y^k \tag{10.67}$$

is also an invariant. This shows that if a_{ik} is a covariant tensor of second order,

$$b_{ik} = a_{ki} \tag{10.68}$$

is also a covariant tensor of second order. Generally it is permissible to exchange any two indices which are both in the upper or both in the lower position.

With the help of this operation we can always decompose a covariant or a contravariant tensor of even order into the sum of two tensors; the one symmetric, the other antisymmetric in one pair of indices. For example, the covariant tensor of second order a_{ik} may be written in the form:

$$a_{ik} = \tfrac{1}{2}(a_{ik} + a_{ki}) + \tfrac{1}{2}(a_{ik} - a_{ki}) \tag{10.69}$$

The first tensor on the right side is symmetric [see (10.55)], the second antisymmetric [see (10.57)] in i, k.

By the same operation two covariant vectors a_i and b_k give rise to a symmetric tensor of second order

$$c_{ik} = a_i b_k + a_k b_i \tag{10.70}$$

and an antisymmetric tensor of second order

$$d_{ik} = a_i b_k - a_k b_i \tag{10.71}$$

Raising and Lowering Indices. The general definition of a tensor of mth order involved m independent positions of the radius vector R, which could be given in either covariant or contravariant form. However, the general relations (10.36) and (10.37) between covariant and contravariant coordinates make it possible to change any covariant index to a contravariant index, and vice versa. This involves a homologous change in the position of the corresponding index of the associated coefficient.

The raising of a covariant index occurs by the process

$$a^i \cdots = g^{i\alpha} a_\alpha \cdots \qquad (10.72)$$

while the *lowering* of a contravariant index occurs by the process

$$a_i{}^{\cdots} = g_{i\alpha} a^{\alpha \cdots} \qquad (10.73)$$

The dots indicate any combination of covariant or contravariant indices which do not participate in the operation and which are carried along without any change.

Contraction of a Tensor. The general definition of a tensor as an invariant algebraic form of a certain order includes the transformation law of the components if the variables are subjected to an arbitrary nonsingular linear transformation. All contravariant variables x^i, y^i, . . . follow the same transformation matrix β^i_α, while all covariant variables x_i, y_i · · · follow the same transformation matrix $\bar{\beta}^\alpha_i$ which is the reciprocal of β^i_α; [see (10.45) and (10.49)]. The transformation law of the tensor components $a_{ik}{}^m$ · · · is identical with the transformation law of the product

$$x_i y_k z^m \cdots \qquad (10.74)$$

Now perform the following operation: Equate one covariant and one contravariant index of a tensor and perform the summation over this index. We might choose, for example, the indices k and m in the example (10.74) and form the quantities

$$a_{i\alpha}{}^\alpha \cdots \qquad (10.75)$$

which follow the transformation law of the product

$$x_i (y_\alpha z^\alpha) \cdots \qquad (10.76)$$

The bilinear form

$$y_\alpha z^\alpha \qquad (10.77)$$

is an *invariant* of a linear transformation. Hence the factor in parentheses in (10.76) behaves like a *constant* during the transformation, and therefore the transformation of the quantities (10.75) follows the transformation law of the product (10.73) omitting the indices k and m. This is equivalent to the statement that the quantities (10.75) form the components of a tensor which has the same indices as the original tensor but omitting the two indices k and m. Thus the operation

$$b_i \cdots = a_{i\alpha}{}^\alpha \cdots \qquad (10.78)$$

called *contraction*, generates a new tensor whose order is *lowered by 2* compared with the original tensor.

In the case of a tensor of second order, contraction results in a tensor of zeroth order, giving a scalar or invariant:

$$a = a^\alpha{}_\alpha \qquad (10.79)$$

If in particular the tensor $a^i{}_k$ is defined as the product of the two vectors b^i and c_k, we obtain the invariant

$$a = bc \qquad (10.80)$$

This invariant is the scalar or dot product **B · C** of the two vectors **B** and **C**.

15. Determinant Tensor

Consider n independent positions of the radius vector **R** and form the product of the following two determinants, composing rows by rows:

$$\begin{vmatrix} x^1 & x^2 & \cdots & x^n \\ y^1 & y^2 & \cdots & y^n \\ \cdot & & & \\ \cdot & & & \\ \cdot & & & \\ z^1 & z^2 & \cdots & z^n \end{vmatrix} \begin{vmatrix} x_1 & x_2 & \cdots & x_n \\ y_1 & y_2 & \cdots & y_n \\ \cdot & & & \\ \cdot & & & \\ \cdot & & & \\ z_1 & z_2 & \cdots & z_n \end{vmatrix} \qquad (10.81)$$

The product is a determinant whose elements are invariants, hence it is an invariant. Moreover:

$$\begin{vmatrix} x_1 & x_2 & \cdots & x_n \\ \cdot & & & \\ \cdot & & & \\ \cdot & & & \\ z_1 & z_2 & \cdots & z_n \end{vmatrix} = \begin{vmatrix} x^1 & x^2 & \cdots & x^n \\ \cdot & & & \\ \cdot & & & \\ \cdot & & & \\ z^1 & z^2 & \cdots & z^n \end{vmatrix} \cdot \begin{vmatrix} g_{11} & g_{12} & \cdots & g_{1n} \\ \cdot & & & \\ \vdots & & & \\ \cdot & & & \\ g_{n1} & g_{n2} & \cdots & g_{nn} \end{vmatrix} \qquad (10.82)$$

Substituting (10.82) in (10.81) and taking the square root, we obtain

$$\sqrt{g} \begin{vmatrix} x^1 & x^2 & \cdots & x^n \\ \cdot & & & \\ \cdot & & & \\ \cdot & & & \\ z^1 & z^2 & \cdots & z^n \end{vmatrix} = \text{invariant} \qquad (10.83)$$

Similarly

$$\frac{1}{\sqrt{g}} \begin{vmatrix} x_1 & x_2 & \cdots & x_n \\ \cdot & & & \\ \cdot & & & \\ \cdot & & & \\ z_1 & z_2 & \cdots & z_n \end{vmatrix} = \text{invariant} \qquad (10.84)$$

From the theory, a determinant of the form (10.83) or (10.84) may be written as an algebraic form of nth order:

$$\begin{vmatrix} x^1 & x^2 & \cdots & x^n \\ \cdot & & & \\ \cdot & & & \\ \cdot & & & \\ z^1 & z^2 & \cdots & z^n \end{vmatrix} = \epsilon_{ik \cdots m} x^i y^k \cdots z^m \qquad (10.85)$$

where $\epsilon_{ik \cdots m}$ vanishes for any combination of indices which are not all different from each other, while the nonvanishing $\epsilon_{ik \cdots m}$ are defined as $+1$ if $ik \cdots m$ represents an *even* permutation of the numbers 1,2, . . . ,n, and -1 if $ik \cdots m$ represents an *odd* permutation of the numbers 1,2, . . . ,n.

Hence in any n-dimensional manifold there exists a tensor of nth order, antisymmetric in any pair of indices which has the covariant components

$$\delta_{ik \cdots m} = \sqrt{g}\, \epsilon_{ik \cdots m} \qquad (10.86)$$

or the contravariant components

$$\delta^{ik \cdots m} = \frac{1}{\sqrt{g}} \epsilon_{ik \cdots m} \qquad (10.87)$$

This tensor is called the *determinant tensor* or *permutation tensor*.

16. Dual Tensor

Multiply a given tensor of mth order, covariant in all its indices, by the contravariant tensor (10.87), contracting over all the indices of the given tensor. The result is a completely contravariant tensor of order $n - m$, antisymmetric in any pair of indices:

$$a^{*pq\cdots s} = a_{\alpha\beta\cdots\gamma}\delta^{\alpha\beta\cdots\gamma pq\cdots s} \qquad (10.88)$$

where a^* is called the *dual* of the tensor a. A similar construction applies to the completely covariant tensor if the original tensor is completely contravariant.

Of particular interest is the application of this operation to 3-space and to 4-space. In 3-space the antisymmetric tensor

$$c_{ik} = \tfrac{1}{2}(a_i b_k - a_k b_i) \qquad (10.89)$$

is associated with the vectors **A** and **B**. The dual of this tensor becomes a tensor of the order $3 - 2 = 1$, that is, a vector. The contravariant components of this vector are

$$\begin{aligned} c^{*1} &= \frac{1}{\sqrt{g}}(a_2b_3 - a_3b_2) \\ c^{*2} &= \frac{1}{\sqrt{g}}(a_3b_1 - a_1b_3) \\ c^{*3} &= \frac{1}{\sqrt{g}}(a_1b_2 - a_2b_1) \end{aligned} \qquad (10.90)$$

which is the customary cross product $\mathbf{A} \times \mathbf{B}$ of vector algebra. This method of associating a third vector to two given vectors is restricted to 3-space because the cross product of two vectors is basically an antisymmetric tensor of second order associated with two vectors according to (10.89). In 3-space the dual of this tensor is a vector, giving rise to the vector components (10.90).

In 4-space the dual of an antisymmetric tensor of second order is again an antisymmetric tensor of second order. This relation is fundamental for the relativistic interpretation of the duality of the Maxwell electromagnetic equations (see Sec. 24).

17. Tensor Fields

The linear algebraic form (10.41) could be interpreted as the work of the force **A** during the displacement $\mathbf{R} = OP$. This required that **A** be a constant force. If a *field* of force is given which changes its magnitude and direction continuously from point to point, we have to think of the infinitesimal displacement x^i which remains in the neighborhood of the point P, the displacement being taken between the points $P = x^i$ and $P' = x^i + dx^i$. The work of the force **A** is then given by the *differential form*

$$A = a_\alpha\, dx^\alpha \qquad (10.91)$$

The coefficients of this differential form are no longer constants but continuous functions of the coordinates $x^1, x^2, \ldots, x^n$.

By changing from algebraic forms to differential forms it is possible to extend the realm of tensor operations from constant tensors to tensor fields. Everything remains valid as before with the understanding that all operations of tensor algebra are now performed *at a definite point* x^i of the field. The differentials dx^i can be interpreted as local coordinates of the point P', measured from the center P. The infinitesimal displacement from P to P' eliminates the variable character of the field, since for such displacements the tensor field assumes the behavior of a constant tensor.

For present discussions the field concept will not be extended to the coordinates x^i themselves. These will still be assumed to be *rectilinear coordinates* which extend to the entire space. Hence the transformation from the x^i to the $\bar{x}^i$ is still a *linear* transformation, and the transformation matrix of the differentials dx^i is the same as the transformation matrix of the coordinates x^i themselves.

The transition from algebraic to differential forms does not modify any of the previous results. The only difference is that components of vectors and tensors are now functions of the point P.

The field concept does *not* extend, however, to the metrical tensor g_{ik}. Since the coordinates are rectilinear, the expression (10.34) for the finite distance $S = OP$ is still valid. The differential form of this equation:

$$(dR)^2 = ds^2 = g_{ik}\, dx^i\, dx^k \qquad (10.92)$$

defines the square of the line element ds. This *line element* ds is associated with the two neighboring points $P = x^i$ and $P' = x^i + dx^i$ and defines the infinitesimal distance between these two points. The g_{ik} coefficients of this quadratic differential form are constants throughout the field.

18. Differentiation of a Tensor

The abstract definition of a tensor of mth order involves m independent positions $dy^i, dz^k \cdots du^p$ of the *infinitesimal* radius vector $d\mathbf{R}$. The definition occurs with the help of the invariant differential form

$$A = a_{ik\cdots p}\, dy^i\, dz^k \cdots du^p \qquad (10.93)$$

where the coefficients $a_{ik\cdots p}$ are functions of the coordinates $x^1, x^2, \ldots, x^n$.

Since the differential of an invariant is again an invariant, we can form the infinitesimal change of A between two neighboring points P and P' of the field. This gives the new invariant

$$dA = \frac{\partial a_{ik\cdots p}}{\partial x^\alpha}\, dy^i\, dz^k \cdots du^p\, dx^\alpha \qquad (10.94)$$

which is a differential form of the order $m + 1$. Hence by definition we have obtained a new tensor of the order $m + 1$:

$$b_{ik\cdots pq} = \frac{\partial a_{ik\cdots p}}{\partial x^q} \qquad (10.95)$$

In order to indicate that this new tensor, called the *covariant derivative* of the tensor $a_{ik\cdots p}$, originated from that tensor, we do not use a new letter for its designation but adopt the following method of notation:

$$a_{ik\cdots p,q} = \frac{\partial a_{ik\cdots p}}{\partial x^q} \qquad (10.96)$$

The same procedure holds if some or all the indices of the given tensor are contravariant. The differentiation of a tensor is the *only typical operation* of tensor analysis. All operations of tensor analysis are a combination of the differentiation of a tensor discussed in this paragraph, and the previous algebraic operations, discussed in Sec. 14.

19. Covariant Derivative of the Metrical Tensor

The metrical tensor g_{ik} forms a symmetric tensor of second order. Since in a rectilinear reference system the g_{ik} are *constants*, the covariant derivative of the tensor g_{ik} *vanishes* at every point of the n-dimensional manifold:

$$g_{ik,m} = \frac{\partial g_{ik}}{\partial x^m} = 0 \qquad (10.97)$$

20. Special and General Relativity

Einstein formulated the principle of special relativity which requires that all reference systems in uniform motion relative to each other shall be equivalent for the formulation of the laws of nature. This requires that the equations of mathematical physics shall have invariance with respect to linear transformations of x,y,z,t under which $c^2t^2 - x^2 - y^2 - z^2$ is invariant.

In 1916 Einstein formulated the principle of general relativity (based on the equivalence of heavy mass and inertial mass) which required that arbitrary reference systems in arbitrary motion relative to each other shall be equivalent for the formulation of the laws of nature. This requires that the equations of mathematical physics shall have variance with respect to *arbitrary curvilinear transformations* of the four variables x,y,z,t.

The tools of tensor calculus were in harmony with the principle of general relativity. These tools are in intimate relation to the concepts of Riemannian geometry and brought the importance of the geometry into sharp focus. For detailed discussion of the physical ideas of relativity theorem, see Part 2, Chap. 6, and Part 6, Chap. 8.

21. Curvilinear Transformations

In place of rectilinear coordinates x^i a more general class of curvilinear coordinates $\bar{x}^i$ will now be used characterized by an arbitrary *point transformation*

$$\bar{x}^i = \bar{f}^i(x^1,x^2, \ldots ,x^n) \qquad (10.98)$$

where the $\bar{f}^i(x^1, \ldots ,x^n)$ are given as arbitrary continuous and twice differentiable functions of the old variables x^i, with nonvanishing Jacobian. The inverse of the transformation (10.98) takes the form

$$x^i = f^i(\bar{x}^1,\bar{x}^2, \ldots ,\bar{x}^n) \qquad (10.99)$$

The relation between the *differentials* of the variables remains *linear:*

$$d\bar{x}^i = \frac{\partial \bar{f}^i}{\partial x^\alpha}\, dx^\alpha \qquad (10.100)$$

This suffices for tensor analysis since invariance properties may be established from differential forms.

The new feature associated with the use of curvilinear coordinates is the fact that the matrix of the transformation

$$\bar{\beta}^i_k = \frac{\partial \bar{f}^i}{\partial x^k} \qquad (10.101)$$

now changes from point to point and hence is a field quantity. The geometrical significance of this change is that the local reference systems, characterized by the dx^i, are no longer in parallel orientation to each other. This, however, is irrevelant for the operations of tensor algebra which are restricted to one definite point of the manifold. The only operation which becomes essentially modified by the introduction of curvilinear coordinates is the differentiation of a tensor, since this operation involves relations between field quantities at neighboring points of the manifold.

22. Covariant Derivative of a Tensor

Consider an arbitrary position dy^i of the infinitesimal radius vector dR. The corresponding differentials $d\bar{y}^i$ in the curvilinear system become

$$d\bar{y}^i = \bar{\beta}^i_\alpha\, dy^\alpha \qquad (10.102)$$

Considering the dy^α as constants, the $d\bar{y}^i$ are *not* constants since the factor $\bar{\beta}^i_\alpha$ changes from point to point. This gives:

$$\begin{aligned} d^2\bar{y}^i &= \frac{\partial \bar{\beta}^i_\alpha}{\partial \bar{x}^\nu}\, dy^\alpha\, d\bar{x}^\nu = \beta^\alpha_\mu \frac{\partial \bar{\beta}^i_\alpha}{\partial \bar{x}^\nu}\, d\bar{y}^\mu\, d\bar{x}^\nu \\ &= -\bar{\beta}^i_\alpha \frac{\partial \beta^\alpha_\mu}{\partial \bar{x}^\nu}\, d\bar{y}^\mu\, d\bar{x}^\nu \end{aligned} \qquad (10.103)$$

But

$$\frac{\partial \beta^\alpha_\mu}{\partial \bar{x}^\nu} = \frac{\partial^2 f^\alpha}{\partial \bar{x}^\mu\, \partial \bar{x}^\nu} \qquad (10.104)$$

and hence, introducing the auxiliary quantities

$$\Gamma^m_{ik} = \bar{\beta}^m_\alpha \frac{\partial \beta^\alpha_i}{\partial \bar{x}^k} = \beta^m_\alpha \frac{\partial^2 f}{\partial \bar{x}^i\, \partial \bar{x}^k} \qquad (10.105)$$

we notice that these quantities (which do *not* form a tensor of third order, in spite of the analogous notation) are symmetric in i,k:

$$\Gamma^m_{ik} = \Gamma^m_{ki} \qquad (10.106)$$

With the help of these quantities the relation (10.103) becomes

$$d^2\bar{y}^i = -\Gamma^i_{\alpha\beta}\, d\bar{y}^\alpha\, d\bar{x}^\beta \qquad (10.107)$$

The corresponding transformation law of the covariant differentials

$$d\bar{y}_i = \beta^\alpha_i\, dy_\alpha \qquad (10.108)$$

yields

$$d^2\bar{y}_i = \Gamma^\alpha_{i\beta}\, d\bar{y}_\alpha\, d\bar{x}^\beta \qquad (10.109)$$

We now consider the invariant differential form

$$A = \bar{a}_\alpha \, d\bar{y}^\alpha \tag{10.110}$$

written down in an arbitrary curvilinear system. From the differential of this invariant, we derive the covariant derivative of the vector a_i. Now we have to differentiate the second factor too, replacing $d^2\bar{y}^i$ by (10.107). Thus the covariant derivative of the vector a_i in an arbitrary curvilinear system is

$$a_{i,k} = \frac{\partial a_i}{\partial x^k} - \Gamma^\alpha_{ik} a_\alpha \tag{10.111}$$

Similarly the invariant form $a^\alpha \, dy$ yields:

$$a^i{}_{,k} = \frac{\partial a^i}{\partial x^k} + \Gamma^i_{\alpha k} a^\alpha \tag{10.112}$$

Generally, applying the same principle to a differential form of arbitrary order, we obtain the result that *every index of the tensor gives rise to a correction term.* If the index is covariant, the correction term follows the pattern of Eq. (10.111), if contravariant, the pattern of Eq. (10.112). The remaining indices are carried along unchanged, e.g.,

$$a^i{}_{k,m} = \frac{\partial a^i{}_k}{\partial x^m} + \Gamma^i_{\alpha m} a^\alpha{}_k - \Gamma^\alpha_{km} a^i{}_\alpha \tag{10.113}$$

23. Covariant Derivative of the Metrical Tensor

If curvilinear coordinates are introduced, the g_{ik} become field quantities. The transformation of Eq. (10.92) to curvilinear coordinates gives

$$\bar{g}_{ik} = g_{\mu\nu}\beta^\mu_i \beta^\nu_k = g_{\mu\nu} \frac{\partial f^\mu}{\partial \bar{x}^i} \frac{\partial f^\nu}{\partial \bar{x}^k} \tag{10.114}$$

which reveals the field character of the new g_{ik}. Nevertheless, the covariant derivative of g_{ik} must vanish since a tensor which is zero in any rectilinear reference system remains zero in *every* reference system. This gives the important relation

$$\frac{\partial g_{ik}}{\partial x^m} - \Gamma^\alpha_{ik} g_{\alpha k} - \Gamma^\alpha_{km} g_{i\alpha} = 0 \tag{10.115}$$

We introduce the so-called *Christoffel symbols of the first kind*:

$$\Gamma^\alpha_{im} g_{\alpha k} = [{}^{ik}_{m}] \tag{10.116}$$

and rewrite (10.115) with these symbols. We also know the symmetry of the Christoffel symbols in the two upper indices, in view of (10.106). We thus deduce by a simple algebraic manipulation:

$$[{}^{ik}_{m}] = \frac{1}{2}\left(\frac{\partial g_{im}}{\partial x^k} + \frac{\partial g_{km}}{\partial x^i} - \frac{\partial g_{ik}}{\partial x^m}\right) \tag{10.117}$$

and obtain the important result that the *auxiliary quantities* Γ^m_{ik}, originally defined in terms of the transformation equations to curvilinear coordinates (cf. (10.105)), *are expressible in terms of the metric associated with that curvilinear system.*

We can thus completely abandon the transformation equations which originally gave rise to a curvilinear system. If we possess the metrical tensor g_{ik} associated with that reference system, we can immediately form the quantities

$$\Gamma^m_{ik} = [{}^{ik}_{\alpha}] g^{\alpha m} \tag{10.118}$$

and thus obtain all the tools for the formation of covariant derivatives. The entire edifice of absolute calculus for arbitrary curvilinear coordinates can thus be erected on the basis of the invariance of differential forms, plus the existence of the metrical tensor g_{ik}.

24. Fundamental Differential Invariants and Covariants of Mathematical Physics

Apart from the fundamental importance of general relativity, the study of absolute calculus has also a purely practical value. The differential equations of mathematical physics have to be solved frequently under boundary conditions which demand the introduction of the proper kind of curvilinear coordinates, such as spherical, cylindrical, parabolic, or other coordinates. The tools of absolute calculus put us in the position to write down the basic differential equations of physics in *any* reference system.

The differential operators of absolute calculus are complicated by the appearance of Γ-quantities. However, in many of the fundamental differential operators of mathematical physics these quantities enter in a highly simplified manner. We list below the most important differential invariants and covariants of mathematical physics. In deducing these expressions, the following relation is of great usefulness:

$$\Gamma^\alpha_{\alpha i} = \frac{1}{2}\frac{\partial g_{\mu\nu}}{\partial x^i} g^{\mu\nu} = \frac{1}{2g}\frac{\partial g}{\partial x^i} = \frac{1}{\sqrt{g}}\frac{\partial \sqrt{g}}{\partial x^i} \tag{10.119}$$

where g is defined by (10.14).

Divergence of a Vector. The following scalar can be associated with a vector field, called the *divergence* of the vector **A**:

$$a^\alpha{}_{,\alpha} = \operatorname{div} a = \frac{\partial \sqrt{g}\, a^\alpha}{\sqrt{g}\, \partial x^\alpha} \tag{10.120}$$

Laplacian Operator $\Delta\phi$: Let the vector a_i of (10.120) be the gradient of the scalar function ϕ:

$$a_i = \frac{\partial \phi}{\partial x^i} \tag{10.121}$$

The divergence of this vector gives the invariant Laplacian operator

$$\Delta\phi = \operatorname{div} \operatorname{grad} \phi = \frac{\partial \sqrt{g}\, g^{\alpha\beta}(\partial\phi/\partial x^\alpha)}{\sqrt{g}\, \partial x^\beta} \tag{10.122}$$

Divergence of a Symmetric Tensor T^{ik}. Let $T^{ik} = T^{ki}$ be a symmetric tensor of second order. We form the following vector, called the divergence of the tensor T^{ik}:

$$T^{i\alpha}{}_{,\alpha} = \frac{\partial \sqrt{g}\, T^{i\alpha}}{\sqrt{g}\, \partial x^\alpha} - \frac{1}{2}\frac{\partial g_{\mu\nu}}{\partial x^i} T^{\mu\nu} \tag{10.123}$$

This differential covariant is of fundamental importance since the conservation law of momentum and energy appears in field physics in the form that the divergence of the symmetric matter tensor T^{ik} vanishes.

Divergence of an Antisymmetric Tensor F^{ik}. Let $F^{ik} = -F^{ki}$ be an antisymmetric tensor of second order (e.g., electromagnetic field strength). Then

$$F^{i\alpha}{}_{,\alpha} = \frac{\partial \sqrt{g}\, F^{i\alpha}}{\sqrt{g}\, \partial x^\alpha} \tag{10.124}$$

25. Maxwell Electromagnetic Equations

The Maxwellian equations of the electromagnetic field, considered relativistically, split into two vector equations. The first equation states that the divergence of the electromagnetic field strength $F^{ik} = -F^{ki}$ is equal to the current density vector:

$$\frac{\partial \sqrt{g}\, F^{i\alpha}}{\sqrt{g}\, \partial x^\alpha} = S^i \tag{10.125a}$$

The second equation states that the divergence of the dual tensor vanishes:

$$\frac{\partial \sqrt{g}\, F^{*i\alpha}}{\partial x^\alpha} = 0 \tag{10.125b}$$

Considering the definition (10.88) of the dual tensor, we obtain

$$\frac{\partial F_{\mu\nu}}{\partial x^\alpha} \epsilon_{\mu\nu\alpha i} = 0 \tag{10.126}$$

The remarkable feature of this equation is that it does not contain any metrical quantity explicitly. It is solvable by putting

$$F_{ik} = \frac{\partial \phi_i}{\partial x^k} - \frac{\partial \phi_k}{\partial x^i} \tag{10.127}$$

which is likewise free of metrical quantities. ϕ_i is called the *vector potential.* It is subject to the conservation law of electricity which has the consequence that the divergence of ϕ_i vanishes:

$$\frac{\partial \sqrt{g}\, \phi_\alpha g^{\alpha\beta}}{\partial x^\beta} = 0 \tag{10.128}$$

26. Curvature Tensor of Riemann

We consider the second covariant derivative of a vector a_i and differentiate once in the sequence jk and once in the sequence kj. If this operation is performed in a rectilinear reference system, we find that the result is in both cases the same. Hence we would expect that also in a curvilinear system the two tensors $a_{i,jk}$ and $a_{i,kj}$ will agree. In fact the result comes out as follows:

$$a_{i,jk} - a_{i,kj} = -R^\alpha{}_{ijk} a_\alpha \tag{10.129}$$

where

$$R^\alpha{}_{ijk} = \frac{\partial \Gamma^\alpha_{ij}}{\partial x^k} - \frac{\partial \Gamma^\alpha_{ik}}{\partial x^j} + \Gamma^\mu_{ij}\Gamma^\alpha_{k\mu} - \Gamma^\mu_{ik}\Gamma^\alpha_{j\mu} \tag{10.130}$$

Since the left side of (10.129) is a tensor of third order, covariant in i,j,k, the factor of a_α must be a tensor of fourth order, contravariant in α and covariant in i,j,k. It is completely composed of the Γ-quantities and thus of a completely *metrical* character.

The apparent paradox that the tensor (10.129) vanishes in a rectilinear system but does not vanish in a curvilinear system is caused by the fact that the assumption of a universally rectilinear reference system strongly prejudices the metrical character of a manifold. The metrical tensor of a manifold *may* have the form (10.114), in which case it is derived from an originally *constant* g_{ik}. In this case we have a metrical geometry which satisfies the postulates of Euclidean geometry. But it is also possible that the g_{ik} of a curvilinear reference system are prescribed as *some* field quantities, without demanding that they shall be of specific form (10.114). Riemann in 1854 established the far-reaching idea of a metrical manifold which is characterized by a quadratic differential form *without* any further restrictions except for the natural conditions of continuity and differentiability. A geometry of this kind, called *Riemannian geometry,* is Euclidean only in *infinitesimal* portions of space but not in finite portions. The Euclidean type of geometry is a specially simple example of Riemannian geometry in which the tensor (10.130), called the *Riemann-Christoffel curvature tensor* or briefly the *Riemann-tensor,* vanishes identically. The second covariant derivative of a tensor is then independent of the sequence of differentiations, which is not true in the more general metrical pattern of Riemannian geometry.

27. Properties of Riemann Tensor

By lowering the first index of (10.130) we obtain the fully covariant tensor $R_{ijkm} = R^\alpha{}_{jkm} g_{\alpha i}$, for which the following expression holds:

$$R_{ijkm} = \frac{1}{2}\left[\frac{\partial^2 g_{ik}}{\partial x_j\, \partial x_m} + \frac{\partial^2 g_{jm}}{\partial x_i\, \partial x_k} - \frac{\partial^2 g_{im}}{\partial x_j\, \partial x_k} - \frac{\partial^2 g_{jk}}{\partial x_i\, \partial x_m}\right] + \left\{\begin{bmatrix} ik \\ \alpha \end{bmatrix}\begin{bmatrix} jm \\ \beta \end{bmatrix} - \begin{bmatrix} im \\ \alpha \end{bmatrix}\begin{bmatrix} jk \\ \beta \end{bmatrix}\right\} g^{\alpha\beta} \tag{10.130a}$$

This tensor is characterized by the following algebraic symmetry properties.

It is antisymmetric in the first pair of indices:

$$R_{jikm} = -R_{ijkm} \tag{10.131}$$

It is antisymmetric in the second pair of indices:

$$R_{ijmk} = -R_{ijkm} \tag{10.132}$$

It is symmetric with respect to an exchange of the first and second pair of indices:

$$R_{kmij} = R_{ijkm} \tag{10.133}$$

It satisfies the *cyclic identity*

$$R_{ijkm} + R_{ikmj} + R_{imjk} = 0 \tag{10.134}$$

These symmetry properties reduce the number of algebraically independent components to $n^2(n^2 - 1)/12$. Hence the number of independent components is 1 in 2-space, 6 in 3-space, and 20 in 4-space.

Differential properties of the curvature tensor. R_{ijkm} satisfies the following differential identity, called the *Bianchi identity:*

$$R_{ijkm,n} + R_{ijmn,k} + R_{ijnk,m} = 0 \qquad (10.135)$$

28. Contracted Curvature Tensor

Einstein thought that the metrical tensor g_{ik} should be considered as a field quantity and subjected to field equations. These field equations must take the form of some partial differential equations which have invariant significance in all curvilinear reference systems. The curvature tensor of Riemann did not seem suitable for this purpose since it is a tensor of *fourth* order with $n^2(n^2 - 1)/12$ algebraically independent components, while the metrical tensor is only a symmetric tensor of *second* order, with $n(n + 1)/2$ independent components. Any statement in terms of the full Riemann tensor would thus be strongly overdetermined.

A contraction changes a tensor of fourth order to a tensor of second order. The contraction over the first two indices of the Riemann tensor gives zero, on account of the antisymmetric nature of these two indices. However, a contraction over the first and third (or second and fourth) indices generates a new tensor of second order, called the *Einstein tensor,* which is symmetric in i and k. We denote this tensor again by the letter R since the possession of only two indices distinguishes this tensor R_{ik} sufficiently from the full Riemann tensor R_{ijkm}:

$$R_{ik} = R_i{}^\alpha{}_{k\alpha} = \frac{\partial^2 \log \sqrt{g}}{\partial x^i \, \partial x_k} - \frac{\partial \sqrt{g}\, \Gamma^\alpha_{ik}}{\sqrt{g}\, \partial x^\alpha} + \Gamma^\alpha_{i\beta}\Gamma^\beta_{k\alpha} \qquad (10.136)$$

A second contraction generates an invariant, called the *scalar Riemannian curvature* or the *Gaussian curvature.* We denote it by the letter R, without any indices:

$$R = R^\alpha{}_\alpha = \left[\frac{\partial^2 \log \sqrt{g}}{\partial x^\mu \, \partial x^\nu} - \frac{\partial \sqrt{g}\, \Gamma^\alpha_{\mu\nu}}{\sqrt{g}\, \partial x^\alpha} + \Gamma^\alpha_{\mu\beta}\Gamma^\beta_{\nu\alpha}\right] g^{\mu\nu} \qquad (10.137)$$

In two dimensions, where the Riemann tensor has only *one* independent component, the scalar Gaussian curvature R is sufficient for the characterization of a Riemannian manifold.

In three dimensions, where the Riemann tensor has *six* independent components, the contracted curvature tensor R_{ik} is sufficient for a full characterization of a Riemannian manifold.

In four dimensions, where the Riemann tensor has 20 independent components, the contracted tensor R_{ik} with its 10 components does not give a full characterization of a Riemannian manifold. Here the vanishing of R_{ik} does *not* necessitate (as in two and three dimensions) the flattening out of space due to the vanishing of the full Riemann tensor.

Bibliography

Original Papers:

Riemann, B.: "Über die Hypothesen, die der Geometrie zu Grunde liegen." Gesammelte Werke, pp. 254–269, Teubner, Leipzig, 1876.

Christoffel, E. B.: "Über die Transformation der homogenen differentialausdrucke Zweiten Grades," *Crelle's J.*, vol. 70, pp. 46–70, 1869.

Ricci, G., and T. Levi-Civita: "Methodes de calcul differentiel absolu et leurs applications," *Math. Ann.*, vol. 54, pp. 125–201, 1901.

Einstein, A.: "Die Grundlagen der allgemeinen Relativitätstheorie," *Ann. Physik*, vol. 49, pp. 769–822, 1916.

Hessenberg, G.: "Vektorielle Begründung der Differentialgeometrie," *Math. Ann.*, vol. 78, pp. 187–217, 1918.

Textbooks of Recent Date:

Tolman, R. C.: "Relativity, Thermodynamics and Cosmology," Clarendon Press, Oxford, 1934.

Bergmann, P. G.: "Introduction to the Theory of Relativity," Prentice-Hall, New York, 1942.

Michal, A. D.: "Matrix and Tensor Calculus," Wiley, New York, 1947.

Brand, L.: "Vector and Tensor Analysis," Wiley, New York, 1947.

Synge, J. L., and A. Schild: "Tensor Calculus," *Math. Expositions*, 5 University of Toronto Press, Toronto, 1949.

Rainich, G. Y.: "Mathematics of Relativity," Wiley, New York, 1950.

Møller, C.: "The Theory of Relativity," Clarendon Press, Oxford, 1952.

Pauli, W.: "Theory of Relativity," Pergamon Press, New York, 1958.

Fock, V.: "The Theory of Space Time and Gravitation," Pergamon Press, New York, 1959.

Synge, J. L.: "Relativity: The General Theory," North Holland Publishing Company, Amsterdam, and Interscience, New York, 1959.

Chapter 11

Calculus of Variations

By C. B. TOMPKINS, University of California, Los Angeles

1. Maxima and Minima of a Function of a Single Variable

Consider a function $f(e)$ defined for real values of e in some interval $a \leq e \leq b$ (permit the values $a = -\infty$, $b = \infty$). The function $f(e)$ attains a relative minimum at the point $e = e_0$ if $f(e_0)$ is defined and if there exists a positive number d such that $f(e) - f(e_0) \geq 0$ for any value of e in the interval of definition of $f(e)$ which satisfies the restriction $|e - e_0| < d$. A similar definition, with the direction of the major inequality reversed, may be written for relative maxima. The study of relative maxima may be made equivalent to the study of relative minima: The function $f(e)$ attains a relative maximum at the point $e = e_0$ if and only if the function

$$g(e) = -f(e)$$

attains a relative minimum at that point.

In the applications to be made here (e.g., the Weierstrass condition for a minimum in the calculus of variations), it will be necessary to consider minima which are attained at the ends of the interval of definition as well as those attained at interior points. The functions to be minimized usually possess derivatives, and the fundamental theorem is the following: If the function $f(e)$ defined over the interval $a \leq e \leq b$ attains a relative minimum at $e = e_0$, and if the derivative at e_0, $f'(e_0)$, exists, then if $e_0 = a$, it is necessary that $f'(e_0) \geq 0$; if $e_0 = b$, it is necessary that $f'(e_0) \leq 0$; if $a < e_0 < b$, it is necessary that $f'(e_0) = 0$.

The proof of this theorem depends on noticing that for e close enough to e_0 the sign of the difference quotient $[f(e) - f(e_0)]/(e - e_0)$ must agree with the sign of $f'(e_0)$ if $f'(e_0) \neq 0$, for the difference quotient must approach the derivative as a limit. If any case of the theorem were false, these difference quotients would indicate points arbitrarily close to e_0 at which $f(e)$ takes on values less than $f(e_0)$, a contradiction.

The application of this theorem to problems usually depends upon writing a function to be minimized and then finding all points at which the derivative of this function is zero. If there are end points to the interval of definition at which the function could conceivably attain a minimum, these points are checked separately.

2. Minima of a Function of Several Variables

Consider a function $f(u_1,u_2, \ldots ,u_n)$ defined in some n-dimensional region including the point $u_i = \bar{u}_i$. Suppose that $f(u)$ attains a relative minimum at $u_i = \bar{u}_i$ in the sense defined for functions of a single variable, with the extension that neighboring points in the n-dimensional u-space must yield values of $f(u)$ which are never lower than $f(\bar{u})$. If at $(\bar{u})$ the partial derivatives $f_{u_i}(\bar{u})$ exist and if the set of numbers $(v_1,v_2, \ldots ,v_n)$ are direction numbers leading from $(\bar{u})$ into the region of definition [so that the function $f(\bar{u} + ev)$ is defined for positive values of e close to zero], then the theory of the last section may be applied. Indeed, if $(\bar{u})$ is a point at which $f(u)$ attains a relative minimum, then $e = 0$ must be a point at which the function $g(e) = f(\bar{u} + ev)$ attains a relative minimum. Furthermore, under the conditions stated the derivative $g'(0)$ exists, and

$$g'(0) = \sum_{i=1}^{n} f_{u_i}(\bar{u})v_i$$

The theory developed in Sec. 1 may be applied here. If the point $(\bar{u})$ is an interior point of the region of definition, then the point $e = 0$ is an interior point of an interval of definition of the function $g(e)$ and it is necessary that $g'(0) = 0$ for any choice of (v). Under these conditions a necessary condition for a relative minimum is $f_{u_i}(\bar{u}) = 0$, a consequence of setting $g'(0) = 0$ with the choice $v_i = 1$ and $v_j = 0$ for $j \neq i$.

For points where the admissible variations are restricted either by the boundary of the region or because the range of definition of the function $f(u)$ is not an n-dimensional region in the first place, the analysis is somewhat more complicated. However, any admissible variation (v) gives rise to a condition in which the fundamental theorem (Sec. 1) may be applied.

In the simplest cases the theory of relative minima of functions of several variables is concerned with a search for points at which all the partial derivatives of the function are equal to zero. Such points are called critical points. The values

$$(\bar{u}) = (\bar{u}_1,\bar{u}_2, \ldots ,\bar{u}_n)$$

are coordinates of a critical point of the function

$f(u_1,u_2, \ldots ,u_n)$ if $f_{u_i}(u) = 0$ for all i. If a function attains a relative minimum at a point at which the partial derivatives of the function all exist and in a neighborhood of which the function is defined, then this point is a critical point.

3. Minima of a Definite Integral—the Euler Equations

The calculus of variations carries the argument of Sec. 2 into a consideration of quantities which are essentially functions of infinitely many variables. In the simplest classical problem, the admissible independent variables (u) of Sec. 2 are replaced by functions $y_i(x)$ of a single variable subject to some restrictions concerning smoothness. The smoothness restrictions are usually considered to be important mathematically, for it is frequently interesting to allow functions with few derivatives (functions with corners, or discontinuous first derivatives, say) to compete with smoother functions, even for physical applications.

The simplest classical problem of the calculus of variations is the fixed end point nonparametric problem. It is to minimize an integral

$$J = \int_a^b F[x, y(x), y'(x)]\, dx \qquad (11.1)$$

among integrals depending on admissible functions $y_i(x)$ with prescribed end points

$$y_i(a) = \alpha_i \qquad y_i(b) = \beta_i \qquad (11.2)$$

To avoid notational difficulty, the basic function of the integrand will be considered to be a function of $2n + 1$ variables, $F(x;y_1,y_2, \ldots ,y_n; p_1,p_2, \ldots ,p_n)$, where the variables (p) are meaningless variables so far as the application is concerned (they are not restricted to be derivatives of the y functions) and the variables (y) are also considered to be independent variables for the function F. With this notation, *subscripts* x, y_i, and p_i will be used, respectively, to denote differentiation with respect to the corresponding independent variables. In computing the integral, with some choice of the functions $y_i(x)$, each value of x gives n values of $y_i(x)$ and n values of $y_i'(x)$, and these values determine a value of the integrand function $F[x,y,y']$ or of its derivatives $F_{y_i}[x,y,y']$ and $F_{p_i}[x,y,y']$.

Suppose that the functions $\bar{y}_i(x)$ are such that $y_i = \bar{y}_i(x)$ satisfies the prescribed end conditions (11.2) and affords a relative minimum to J. Consider any set of functions $g_i(x)$ with $g_i(a) = g_i(b) = 0$. Then for any fixed value of e, the functions $\bar{y}_i(x) + eg_i(x)$ satisfy the required end conditions, and, for e near zero, these functions are near $\bar{y}_i(x)$. Then, if J does attain a relative minimum at $\bar{y}_i$, it is necessary that the function of a single variable

$$J(e) = \int_a^b F[x, \bar{y}(x) + eg(x), \bar{y}'(x) + eg'(x)]\, dx \qquad (11.3)$$

attain a relative minimum at $e = 0$.

The function of the single variable e can be studied by computing $J'(0)$. Straightforward differentiation leads to

$$J'(e) = \int_a^b \{\Sigma_i g_i(x)F_{y_i}[x, \bar{y}(x) + eg(x), \bar{y}'(x) + eg'(x)] + \Sigma_i g'_i(x)F_{p_i}[x, \bar{y}(x) + eg(x), \bar{y}'(x) + eg'(x)]\}\, dx \qquad (11.4)$$

and

$$J'(0) = \int_a^b \Sigma_i g_i(x)F_{y_i}[x,\bar{y},\bar{y}']\, dx + g'_i(x)F_{p_i}[x,\bar{y},\bar{y}']\, dx \qquad (11.5)$$

A necessary condition for a minimum is that this derivative have the value zero.

If the first term is integrated by parts, a useful equation is obtained. The formula used is

$$\int_a^b u_i(x)\, dv_i(x) = u_i(b)v_i(b) - u_i(a)v_i(a) - \int_a^b v_i(x)\, du_i(x)$$

The choices

$$u_i = g_i(x) \qquad \text{and} \qquad v_i = \int_a^x F_{y_i}[x,\bar{y},y']\, dx$$

are made and use is made of the fact that

$$u(a) = u(b) = 0$$

which follows from $g(a) = g(b) = 0$. The equation obtained is

$$\int_a^b \Sigma_i g'_i(x) \left\{F_{p_i}[\bar{x},\bar{y},\bar{y}'] - \int_a^x F_{y_i}[\bar{x},\bar{y},\bar{y}']\right\} dx = 0 \qquad (11.6)$$

and this must be satisfied for every set of functions $g_i(x)$ of the type admitted above as variations.

If for any choice of i the quantity in the braces $\{\ \}$ is not constant (it may, however, differ from zero), then a set of functions $g_i(x)$ can be found for which Eq. (11.6) is not satisfied (lemma of DuBois-Reymond). One such set of functions is

$$g_i(x) = \int_a^x \left[\{\ \} - \frac{1}{b-a}\int_a^b \{\ \}\right] dx \qquad (11.7)$$

where the braces $\{\ \}$ have been used to denote the quantity in similar braces corresponding to any choice of i in Eq. (11.6). This may be verified by noting that $g_i(x)$ meets the end conditions imposed and that the integrand in (11.6) becomes $[g'_i(x)]^2$ for this choice of $g(x)$.

If J attains a relative minimum at $\bar{y}_i(x)$, it is necessary that $\bar{y}$ satisfy equations obtained by setting the expression in braces $\{\ \}$ equal to a constant for each value of i; this gives a set of equations known as Euler's equations in integral form:

$$F_{p_i}[x,\bar{y},\bar{y}'] - \int_a^x F_{y_i}[x,\bar{y},\bar{y}']\, dx = c_i \qquad (11.8)$$

These equations can be differentiated with respect to x, for two of the three terms are differentiable (the indefinite integral and the constant) and the third, which is their sum, must therefore also be differentiable. Differentiation leads to Euler's equations in differential form:

$$\frac{d}{dx} F_{p_i}[x,\bar{y},\bar{y}'] = F_{y_i}[x,\bar{y},\bar{y}'] \qquad (11.9)$$

If the left members of Eqs. (11.9) are expanded formally, a set of second-order differential equations results:

$$F_{p_ix}[x,\bar{y},\bar{y}'] + \Sigma_j F_{p_iy_j}[x,\bar{y},\bar{y}']\bar{y}'_j + \Sigma_j F_{p_ip_j}[x,\bar{y},\bar{y}']\bar{y}''_j = F_{y_i}[x,\bar{y},\bar{y}'] \qquad (11.10)$$

If these second derivatives $\bar{y}''$ exist, then Eqs. (11.9) and (11.10) are equivalent; thus every solution of Euler's equations in any form is a solution of Eqs. (11.10) at all points where the second derivative exists. The expanded equations, however, are meaningless at corners or at other places where second derivatives do not exist.

[The statements above can be proved only under slight restrictions. The only appearance of $\bar{y}''$ is in the last term of the first member; to study possible solutions of the differential equation, a study of the coefficients of this term is in order. If for all interesting values of x the determinant $|F_{p_ip_j}[x,\bar{y},\bar{y}']|$ differs from zero, then any solution of Eqs. (11.10) near $\bar{y}$ has uniquely determined continuous second derivatives, for the implicit function theorem involving the Jacobian matrix (Part 1, Chap. 3, Sec. 5) guarantees these. If this determinant of the Jacobian matrix becomes zero, however, this recourse to the implicit function theorem is not possible, and statements about solutions are somewhat harder to make; this situation arises in the important class of parametric problems, presented in the Sec. 6.]

Functions satisfying Euler's equations and the end conditions are *critical functions* or *critical points* for the integral J. Just as a major part of the study of minima of functions of a finite number of variables was the search for critical points, so a large part of the solution of calculus of variations problems consists of finding solutions to Euler's equations—usually differentiated and expanded into second-order differential equations.

When the determinant of the coefficients of the second derivatives in these differential equations differs from zero along the solution, then the equations may be (in principle) solved uniquely for $\bar{y}''$, and solutions exist which are defined uniquely and continuously by initial conditions (Part 1, Chap. 4, Sec. 3). It is true in perfect generality (subject to the assumption about the determinant of coefficients) that a solution $\bar{y}_i(x)$ can be found complying with initial conditions $\bar{y}_i(a) = \alpha_i$ and $y'_i(a) = \gamma_i$ and that this solution is unique; however, without further knowledge of the equations, the theory of ordinary differential equations does not guarantee either the existence of solutions subject to the boundary conditions $\bar{y}_i(a) = \alpha_i$ and $\bar{y}_i(b) = \beta_i$ or the uniqueness of solutions if they exist. The quest for such a solution and its examination when found may use any method of differential equations, or in some cases it may be attractive to try variations which decrease the integral J [based, for example, on the explicit functions $g_i(x)$ given in Eqs. (11.7)] in the hope that successive iterations will lead to an approximate solution of the problem. This successive approximation scheme is very nearly the one used by physical systems in approaching their equilibrium positions, except that frictional or other damping in the physical system is replaced by arbitrary steps in the arithmetic iterations.

In summary the problem is defined and the main result stated below:

1. An integrand function $F[x,y,p]$ of $2n + 1$ variables, x, $(y) = (y_1,y_2, \ldots ,y_n)$ and

$$(p) = (p_1,p_2, \ldots ,p_n)$$

is admissible for a connected region R of the $(n + 1)$-dimensional space of variables (x,y) if F and all its first and second derivatives are defined in R with (p) unrestricted and if these derivatives are continuous simultaneously in all the variables (x,y,p).

2. A curve or a set of functions $y_i(x)$, $a \leq x \leq b$, is admissible for the connected region R and the end points (a,α) and (b,β) if $[x,y(x)]$ lies in R for every x, if $y_i(a) = \alpha_i$ and $y_i(b) = \beta_i$, and if the functions are continuous and their first derivatives exist and are continuous except at a finite number of points at which the left and right derivatives exist as limits of the derivatives from these directions.

3. The distance between two admissible curves $y_i(x)$ and $z_i(x)$ is $|y - z| = \max_{i,x}|y_i(x) - z_i(x)|$.

Theorem 1. *If the integral*

$$J = \int_a^b F[x,y(x),y'(x)]\, dx$$

attains a minimum at the curve $\bar{y}_i(x)$ *relative to all admissible curves with end points* (a,α) *and* (b,β) *neighboring* $\bar{y}$ *in the sense of Definition* 3, *then it is necessary that the Euler equations in their forms* (11.8) *and* (11.9) *be satisfied.*

4. An admissible curve satisfying the Euler equations is called a critical curve or a critical point for the associated integrand function and end points.

5. An integrand function F is regular along an admissible curve if at every point of that curve the determinant

$$|F_{p_ip_j}| \neq 0$$

Theorem 2. *Every critical curve without corners for an integrand function regular along it satisfies the expanded Euler equations* (11.10), *and such curves are determined uniquely and continuously near the given curve by their initial points and their initial slopes but with the right end point unspecified.*

4. Examples

The usual example of a regular problem is the brachistochrone problem: finding a curve along which a particle starting with preassigned initial tangential velocity $u > 0$ at a point (a,α_1,α_2) and accelerated by constant gravity while being restrained without friction to move on the curve will reach a preassigned point (b,β_1,β_2) in the shortest time. Assume the coordinate system is chosen so that the gravitational acceleration is in the negative direction along the second coordinate axis, the y_1 axis of the (x,y_1,y_2)

frame of reference. The time of descent along the curve $y_i = y_i(x)$ is

$$J = \int_a^b \frac{\sqrt{1 + y'^2_1 + y'^2_2}}{\sqrt{u^2 + 2g(\alpha_1 - y_1)}}\,dx \qquad (11.11)$$

This is a classical fixed end point calculus of variations problem with integrand function

$$F[x,y_1,y_2,p_1,p_2] = \frac{\sqrt{1 + p_1^2 + p_2^2}}{\sqrt{u^2 + 2g(\alpha_1 - y_1)}} \qquad (11.12)$$

which is independent of x and y_2. The region R of definition is any region in which the denominator is positive; the largest possible leaves x and y_2 unrestricted and limits y_1:

$$y_1 < \frac{u^2}{2g} + \alpha_1$$

The determinant

$$\begin{vmatrix} F_{p_1p_1} & F_{p_1p_2} \\ F_{p_2p_2} & F_{p_2p_2} \end{vmatrix} = \frac{1}{(1 + p_1^2 + p_2^2)[u^2 + 2g(\alpha_1 - y_1)]}$$

and this is never zero in R, so that the integrand function is regular for any curve in R. The critical curves without corners satisfy Euler's second-order differential equations (11.10). These differential equations integrate into functions describing arcs of plane cycloids. (This solution is said to have been found first by Johann Bernoulli in 1696.)

Another problem is the problem of finding geodesics, or arcs of least length, on a surface. If y_1 and y_2 represent latitude and longitude, respectively, on a sphere of unit radius and if x represents some parameter, such as time, the length of the path $y_i = y_i(x)$ leading from (α_1,α_2) at time a to (β_1,β_2) at time b is

$$J = \int_a^b \sqrt{y'^2_1 + \cos^2 y_1 y'^2_2}\,dx \qquad (11.13)$$

For this problem the integrand function is independent of x and y_2:

$$F = \sqrt{p_1^2 + \cos^2 y_1 p_2^2} \qquad (11.14)$$

This function has indeterminate derivatives if $p_1 = p_2 = 0$. Furthermore, the integrand function is nowhere regular, for

$$\begin{vmatrix} F_{p_1p_1} & F_{p_1p_2} \\ F_{p_2p_1} & F_{p_2p_2} \end{vmatrix} = \frac{\cos^4 y_1}{F^6} \begin{vmatrix} p_2^2 & -p_1p_2 \\ -p_1p_2 & p_1^2 \end{vmatrix} = 0$$

These difficulties are directly due to the parametric nature of the problem; the inessential role played by the independent variable x in determining the length of the arc prohibits any condition (such as regularity) which would guarantee unique functions defining the arc. Some additional theory will be developed in Sec. 6 to permit the calculation of great circles in the present example. Without this additional development, however, the Euler equations in their expanded form (11.10) may be used and must be satisfied by curves with continuous second derivatives if the functions defining these curves have derivatives which avoid the singularities of F.

A third example concerns the hanging flexible wire, which is well known to assume a catenary shape. If the problem is attacked by minimizing the potential energy of the wire, assuming horizontal coordinate x and vertical coordinate y, then the energy to be minimized is proportional to

$$J = \int_a^b \sqrt{y(1 + y'^2)}\,dx \qquad (11.15)$$

where the curve $y = y(x)$ describes the shape of the wire. However, not only the end points (a,α) and (b,β) are specified, but also the length of the wire. The class of admissible curves must be further restricted to those of this specified length L; that is,

$$\int_a^b \sqrt{1 + y'^2}\,dx = L \qquad (11.16)$$

This problem contains one too many integrals for the theory thus far developed here. The attack which will be made is due to Lagrange. In this special problem the method is easily apparent. Since work is required to shorten the cable, even though the extra length is allowed to follow its original path without frictional loss of energy, it follows that a parameter μ can be evaluated for a correct solution of the catenary problem attacked so that the rate of change of potential energy from shortening the cable can be made momentarily equal to $-\mu L$. For this choice of the integral, $J + \mu L$ must have zero variation among all neighboring curves, this variation being the sum of the zero variation necessary for the minimum of J and the zero rate of change of potential energy got from changing the length of the wire when this energy rate of change is modified by adding the term μL. Incidentally, the same argument shows that this problem is equivalent to finding the shortest possible wire with given potential energy and fixed end points. This important observation is verified analytically by including another constant multiplier λ which clearly will not affect the problem essentially, and minimizing the combination $I = \lambda J + \mu L$ (λ and μ both constant but not both zero). Solutions of the original problem lie among the solutions of this more general problem; the original problem is solved by a specialized choice of λ and μ not deducible in advance.

The integral to be minimized is, then,

$$I = \int_a^b (\lambda y \sqrt{1 + y'^2} + \mu \sqrt{1 + y'^2})\,dx \qquad (11.17)$$

where λ and μ are constants not now specified but not both zero. Here,

$$F = (\lambda y + \mu)\sqrt{1 + p^2}$$

$$F_y = \lambda\sqrt{1 + p^2} \qquad F_p = \frac{p(\lambda y + \mu)}{\sqrt{1 + p^2}} \qquad (11.18)$$

Only the ratio μ/λ is essential so that λ may be set equal to one and the Euler equations solved immediately:

$$y = \mu + \frac{1}{c}\cosh c(x - d)$$

c, d, and μ are to be chosen to satisfy end conditions and the length restriction imposed. There are, then, three conditions and three parameters which are to be determined by the conditions; the rest of the calculation is straightforward.

This example can be expanded by noting that there is no reason for as tight a restriction on end points as has been made. If the flexible wire is attached without friction to curves or surfaces instead of fixed end points, it must find end points and assume the catenary shape between these points to minimize potential energy. Physically it is clear that if the end points can move in any direction on a smooth curve or surface, the catenary must strike this end curve or surface perpendicularly. This result may be obtained analytically from transversality conditions to be derived later. Such problems normally split into two parts: first, to find all critical curves for any end points and, second, to find curves satisfying the transversality conditions.

By way of final example, assume that the Euler equations are known; the inverse problem of the calculus of variations is that of finding an integrand function which leads to these Euler equations. Hamilton tried this, letting x represent time and y represent cartesian coordinates in space, and choosing Newton's equations of motion as the Euler equations:

$$my''_i = K_i \tag{11.19}$$

where K_i is a force vector function and m is the mass of a particle. If there is only one space dimension and if the following identification is tried

$$\frac{d}{dx}[F_p] = my''$$

then it may be that $F_p = my'$ plus a function of x and y alone, and $F = \frac{1}{2}my'^2$ plus a function of x and y alone. Continuing with $F_y = K$ ties down the undetermined part of F to a function of x alone:

$$F = \tfrac{1}{2}m\dot{y}^2 + \int K\,dy \text{ plus a function of } x \text{ alone} \tag{11.20}$$

If the x function is ignored (under fixed end conditions it will integrate to a constant which does not affect the determination of a minimizing curve), the function $F(y,\dot{y})$ represents an energy, the integral $\int F\,dt$ represents action. (Here, the notation has changed from the x designation of time, which fits the earlier development of the calculus of variations above, to the more usual physical notation with t representing time and a dot representing differentiation with respect to time.) Increasing the number of dimensions and the number of particles leads to similar action integrals, and changing the coordinate reference system preserves the form of these equations because of tensor invariant properties of energies (Part 1, Chap. 10). Thus this attack on the inverse problem leads to the versatile Lagrangian equations of mechanics. These equations and the conditions of generality with which Hamilton's principle of least action is applicable lead to variational methods in physics, which have been used in most aspects of physical calculation.

5. Other First Variations: Weierstrass Condition, Corner Conditions, One-side Variations

Some neighboring curves to a curve $\bar{y}_i(x)$ cannot be expressed by functions $\bar{y}(x) + eg_i(x)$ of the form used in deriving the Euler equations. In particular, curves which are close but with derivatives considerably different from those of $\bar{y}_i(x)$ cannot be expressed in this way. For example, let $C(e)$ be a family of curves approaching $\bar{y}$ as e approaches zero but with the curves departing from $\bar{y}$ in a sharp V for each curve, the size of the V tending to zero with e but with the sharpness of the angle of V held fixed. If the integral of Sec. 3 is evaluated along the curve $C(e)$ for any choice of e, the value obtained may be represented by $J(e)$, just as was done in Sec. 3. The resulting function $J(e)$ must attain a relative minimum at $e = 0$ if J attains a relative minimum at $y = \bar{y}$. With a little care, this function can be arranged to be defined and differentiable for $e \geq 0$ and neighboring zero; but if the V is included in the curves $C(e)$, the definition of $J(e)$ cannot be extended past $e = 0$ if the derivative $J'(0)$ is to exist. Hence, when the basic theorem concerning minima of functions of a single variable is applied, it is necessary that $J'(0) \geq 0$. Weierstrass evaluated $J'(0)$ for general families of curves of the type described and found the following necessary condition for $J'(0) \geq 0$ for all possible families of this type.

A necessary condition of Weierstrass for J to attain a relative minimum at $y = \bar{y}$ is

$$E(x,\bar{y},\bar{y}',q) \geq 0 \tag{11.21}$$

for every value of q, where

$$E(x,y,p,q) = F[x,y,q] - F[x,y,p] - \Sigma_i(q_i - p_i)F_{p_i}[x,y,p] \tag{11.22}$$

With reasonable luck in a problem, it will turn out that $E(x,y,p,q) \geq 0$ for any admissible set of variables not necessarily restricted to

$$(x,y,p) = (x,\bar{y},\bar{y}')$$

This is true for the geodesic problem of Sec. 4, for example. For that problem

$$E(x,y,p,q) = \frac{\sqrt{(p_1^2+\cos^2 y_1 p_2^2)(q_1^2+\cos^2 y_1 q_2^2)} - (p_1q_1+\cos^2 y_1 p_2 q_2)}{\sqrt{p_1^2 + \cos^2 y_1 p_2^2}}$$

This quantity can never be negative (according to an inequality attributed to Schwarz). Thus angular variations of the Weierstrass type may be ignored in this geodesic problem.

Another situation not fully studied in Sec. 3 is the possibility of corners in minimizing curves. These are explicitly allowed in the conditions stated, and any corners which do exist must still leave the Euler equations in integral form satisfied. In many problems this rules out all possibility of corners in minimizing curves.

However, conditions as stated in Sec. 3 do not normally admit the conditions existing at an interface between two mediums which are physically different, for two continuous derivatives of $F(x,y,p)$ are demanded. Thus, reflections, refractions, and other such angular changes usually are not treated by the method now being discussed.

The Weierstrass-Erdmann corner condition which governs corners at points where $F(x,y,p)$ has two continuous derivatives follows directly from Eqs. (11.8) and the discussion immediately following it. It is stated in the following theorem: If J attains a relative minimum at $y = \bar{y}$, then it is necessary that the functions $F_{p_i}[x,\bar{y},\bar{y}']$ be continuous in x, even at corners where $\bar{y}'$ is not continuous.

Again, in the geodesic problem it can be shown that for any admissible sets (x,y,p,q) the only condition under which $F_{p_i}(x,y,p) = F_{p_i}(x,y,q)$ is $p_1/p_2 = q_1/q_2$.

Under this condition the tangent has a continuous direction and the only "corners" are discontinuities in the first derivative caused by rough parametrization.

Slightly different applications of the basic theorems govern minimizing curves which lie partially on the boundary of R, the region of definition of F. If the region R is open, a minimizing curve may not exist even for the smoothest problems. For example, if R is the region $x + y > 0$, there is no brachistochrone connecting points far apart but each close to the line $x + y = 0$; for any curve lying in part close to this line, there exists another of shorter time of descent lying closer to the line.

If the region R is closed, so that it contains the line and is defined by $x + y \geq 0$, there is a brachistochrone, part of which may be on the line $x + y = 0$. Variations from this line upward are allowed, but variations downward are not permitted; hence, the zero derivative leading to the Euler equations is not necessary and is replaced by a non-negative derivative for admissible variations. These unilateral variations may lead easily to inequalities to replace the equalities in the integral form of Euler's equations (11.8), or they may be more easily treated as an additional restriction not radically different from the length restriction in the ordinary problem. Both approaches are found in the literature, but the second is more easily made general.

6. Parametric Problems

Parametric problems were introduced in the geodesic example of Sec. 4. These are problems in which the integral depends only on the geometric arc joining the end points in y-space and not on the particular choice of the functions $y_i(x)$ defining the arc. Such a problem may be described by listing the ways in which it differs from the nonparametric problem already discussed: (1) The integrand function $F[y,p]$ does not depend on x (the parameter) and need not be defined when $p_i = 0$ for all i; otherwise it retains the two continuous derivatives previously required in the region R and all remaining values of p_i. (2) The terminal value b of the parameter is not fixed, the prescribed end values being a, α_i, and β_i (of the initial parameter value and the end points α and β, respectively). (3) The function $F[y,p]$ is further restricted so that if $t = f(x)$, $a \leq x \leq b$, is a function of x with $f'(x) > 0$, and $z_i(t) = y_i[f^{-1}(t)]$, then

$$\int_{f(a)}^{f(b)} F[z(t), z'(t)]\, dt = \int_a^b F[y(x), y'(x)]\, dx \qquad (11.23)$$

Condition (3) is to hold for any admissible curve between any end points. It has the following consequences:

The function $F[y,p]$ is positive homogeneous of weight one in p; that is, if $q_i = kp_i$, $k > 0$, then

$$F[y,q] = kF[y,p] \qquad (11.24)$$

The proof is by taking $k = dx/dt$ and demanding that the two integrands be identical in x [with differential $dx = (dx/dt)\, dt$].

Then, *for every admissible* (y) and (p)

$$F[y,p] = \Sigma_i p_i F_{p_i}[y,p] \qquad (11.25)$$

This is Euler's theorem concerning homogeneous functions. It is derived by differentiating Eq. (11.24) with respect to k and then evaluating for $k = 1$.

Finally, *the determinant*

$$|F_{p_ip_j}| = 0$$

Proof. Differentiate Eq. (11.25) with respect to p_j

$$F_{p_j} = F_{p_j} + \Sigma_i p_i F_{p_ip_j}$$

then

$$\Sigma_i p_i F_{p_ip_j} = 0$$

and since p_i differs from zero for some value of i, it follows (Part 1, Chap. 2, Sec. 9) that the determinant is zero. The immediate result is the conclusion that no parametric problem is regular in the sense used for nonparametric problems. It is an obvious fact, indeed, that there can be no guarantee of two continuous derivatives with respect to any parameter, for the parametrization can be roughened deliberately to destroy the second derivative. This leaves in doubt the completeness of a study based on Euler's second-order differential equations.

This is the only part of the nonparametric theory which does not apply directly to the parametric case. This gap is filled by seeking solutions $\bar{y}_i$ of the integral form of Euler's equations which at every point satisfy the condition

$$\Sigma_i y'^2_i = 1 \qquad (11.25a)$$

This condition avoids the points where F is not defined and guarantees enough smoothness to permit a study of the second-order differential equations. Then the argument concerning nonparametric regular problems can be modified to use a new definition of regularity yielding a satisfactory result. The parametric integral is regular along the curve $\bar{y}_i(x)$ if at each point of the curve the determinant of $n + 1$ rows and columns

$$\begin{vmatrix} F_{p_ip_j}[\bar{y},\bar{y}'] & \bar{y}'_i \\ \bar{y}'_j & 0 \end{vmatrix} \neq 0$$

If a parametric integrand is regular along a curve $\bar{y}_i$ with continuous first derivatives which satisfy Eq. (11.25a) and if $\bar{y}_i$ satisfies the integral form of Euler's equations, then $\bar{y}_i(x)$ has continuous second derivatives and is a solution of Euler's second-order differential equations.

This follows by applying the implicit function theorem to the solutions $(u_i\lambda)$ of the equations

$$F_{p_i}[\bar{y},u] + \lambda u_i - \int_a^x F_{y_i}[\bar{y},\bar{y}']\, dx - c_i = 0$$

$$\Sigma_i u_i^2 - 1 = 0$$

near the solutions

$$u_i = \bar{y}'_i = 0$$

Using these definitions, it is easy to show that the integrand of the geodesic problem is regular along every curve.

7. Problems with Variable End Points

Admissible curves were described in Sec. 3 as arcs with prescribed end points. The definition was slightly modified in Sec. 6, but the effect was to retain arcs, independent of parametrization, with prescribed end points. It was noted in Sec. 4 in connection with the examples that weaker restrictions are sometimes desirable: one or both end points are allowed to move on a geometric manifold of some description.

If the minimizing end points in such a problem are known, the arc joining them must minimize the integral among all admissible curves joining these end points. Hence the variable end-point problem includes (1) finding the proper end points and (2) solving the fixed end-point problem between them. Actual solution is frequently in the reverse order.

The problem to be considered is that of minimizing the integral

$$J = \int_{a(z)}^{b(z)} F[x, y(x), y'(x)]\, dx + \theta(z) \qquad (11.26)$$

where θ is a function (with at least two continuous derivatives) of no more than $2n + 2$ variables $z_1, z_2, \ldots, z_r$, where there are given two r-dimensional manifolds defined by the equations

$$x = a(z) \qquad y_i = \alpha_i(z)$$
$$x = b(z) \qquad y_i = \beta_i(z)$$

respectively (all functions to have at least two continuous derivatives), and where the minimum is with respect to both the variables $z_\rho = (z_1, z_2, \ldots, z_r)$ and all admissible curves which are, generally, defined as in Sec. 3 except their end points are $[a(z),\alpha_i(z)]$ and $[b(z),\beta_i(z)]$.

It is difficult to provide variations which are strictly admissible curves, but analysis shows that the conditions satisfied by the end points depend only on the tangent directions to the end manifolds. Once this is proved the following transversality condition is established: A necessary condition for a minimum in the variable end-point problem is a minimizing curve $\bar{y}_i(x)$ between the end points and, in addition, the following transversality condition at the end manifolds:

$$\begin{aligned}\theta_z(\bar{z}) + F_{p_i}[b,\bar{y}(b),\bar{y}'(b)]\beta_{i z_\rho}(\bar{z}) - F_{p_i}[a,\bar{y}(a),\bar{y}'(a)]\alpha_{i z_\rho}(\bar{z})\\ + \{F[b,\bar{y}(b),\bar{y}'(b)] - F_{p_i}[b,\bar{y}(b),\bar{y}'(b)]\bar{y}'_i(b)\}b_{z_\rho}(\bar{z})\\ - \{F[a,y(a),y'(a)] - F_{p_i}[a,\bar{y}(a),\bar{y}'(a)]y'_i(a)\}a_{z_\rho}(\bar{z}) = 0\end{aligned} \qquad (11.27)$$

here $(\bar{z}) = (\bar{z}_1, \bar{z}_2, \ldots, \bar{z}_r)$ represents the values of $(\bar{z})$ at the minimum, and each term with index i appearing is to be summed over all values of i.

This theorem may be expanded to solve problems from physics in which an end manifold is the boundary between two physically different mediums. Assuming one fixed end point in each region, the problem is solved by considering two problems, one in each region, each with one fixed end point and one variable end point (on the interface). However, in this case the variable end points for the two problems coincide, and Eq. (11.27) must be replaced by equations obtained by replacing the left members for each ρ by the sum of the corresponding left members of the two individual problems. In other words, the variations of the two problems individually need not be zero, but they must annihilate one another. This condition and not the Weierstrass-Erdmann corner condition of Sec. 5 governs such interfaces.

Expression (11.26) is not particularly involved, and probably only remarks concerning $\theta(z)$ and (z) need be made here. Briefly, $\theta(z)$ admits some preference on the end manifold so that motion on this manifold need not be completely free; for free motion, $\theta(z) = 0$. The parameters (z) describe the end manifolds. If the parameters are split so that for no ρ does z_ρ appear both in the set of functions $[a(z),\alpha_i(z)]$ and the set $[b(z),\beta_i(z)]$, then the end points on the two manifolds are not related, except a posteriori by the minimizing curve. If this is not the case, then the end points are not independent, but variation of one parameter may vary both end points. The analytic formulation is, thus, general enough to fit most variable end-point problems which might arise.

8. Isoperimetric Problems—the Problem of Bolza

The catenary problem (Sec. 4) is an example of an important class of problems which imposes implicit restrictions on the admissible functions in a way analogous with the restriction of the following simple problem: Find a rectangle with unit area and minimum perimeter. Here, if explicit reduction to a single variable is avoided (in more complex problems such as the catenary problem this reduction to explicit form with fewer variables is frequently decidedly inconvenient), the problem can be attacked through the implicit function theorem.

Analytically, if x_1 and x_2 represent the length and height of the rectangle, the problem is to minimize

$$J = 2x_1 + 2x_2 \qquad (11.28)$$

subject to the restriction

$$I = x_1x_2 - 1 = 0 \qquad (11.29)$$

If $x_1 = \bar{x}_1$ and $x_2 = \bar{x}_2$ minimizes J subject to Eq. (11.29), then it is clear that it is impossible to find values (x_1,x_2) near $(\bar{x}_1,\bar{x}_2)$ which change I and J (as functions of x_1 and x_2) arbitrarily; in particular, a lower value of J and the same value of I must be impossible. However, the implicit function theorem states that such arbitrary variations are possible unless the Jacobian determinant

$$\begin{vmatrix} J_{x_1}(\bar{x}_1,\bar{x}_2) & J_{x_2}(\bar{x}_1,\bar{x}_2) \\ I_{x_1}(\bar{x}_1,\bar{x}_2) & I_{x_2}(\bar{x}_1,\bar{x}_2) \end{vmatrix} = 0 \qquad (11.30)$$

This zero value of the determinant is a necessary condition for the minimum. For the problem stated. it is

$$\begin{vmatrix} 2 & 2 \\ x_2 & x_1 \end{vmatrix} = 0$$

or $x_1 = x_2$.

The same condition is arrived at by seeking critical points of all the functions

$$K = \lambda J + \mu I \tag{11.31}$$

where λ and μ are constants, unrestricted except for the requirement that at least one differs from zero. These critical points occur where

$$\begin{aligned} \lambda J_{x_1} + \mu I_{x_1} &= 0 \\ \lambda J_{x_2} + \mu I_{x_2} &= 0 \end{aligned} \tag{11.32}$$

However, the existence of numbers λ and μ not both zero and satisfying (11.32) is equivalent to Eq. (11.30).

The same attack on the catenary problem of Sec. 4 is plausible; at least the function K of Eq. (11.31) is analogous to integral I of Eq. (11.17). The aim of the attack here is to show: (1) that the implicit function theorem can be applied to general calculus of variations problems in the manner illustrated above; and (2) that the results can be summarized by writing a new integrand, a linear combination of the contributing integrands with undetermined coefficients (known as Lagrange multipliers).

The proof of statement (1) above is straightforward, becoming complicated only because of the care required for problems of increasing complexity. For the catenary problem, or any isoperimetric problem in which one integral is minimized subject to a restriction that the value of another is fixed, the attack depends on producing two-dimensional sets of variations (one dimension per integral in general), admissible except for the abandonment of the requirement that the second integral remain fixed in value and demanding that the Jacobian determinant of the two integrals (with respect to the two independent parameters of the variations) become zero at the minimizing curve. If this Jacobian determinant were different from zero, variations in the family could be found, changing I and L to arbitrarily chosen neighboring values; in particular I could be decreased and L held fast. Statement (2) follows immediately from this application of the implicit function theorem.

A general result can be proved in this way if sufficient care is exercised. This may be put in terms of the problems already stated, augmented by some additional restrictions: The problem of Bolza is the problem of Sec. 7 of minimizing the quantity J defined by Eq. (11.26) with admissible curves $y_i = y_i(x)$ further restricted to satisfy the differential equations (11.28):

$$\phi_\tau[x,y(x),y'(x)] = 0 \qquad \tau = 1, 2, \ldots, m < n, \text{ for all } x$$

where for each τ the function $\phi_\tau[x,y,p]$ is defined in the region of definition of F and has the derivatives specified for F in Sec. 3 and the functions $\phi_\tau[x,y,p]$ are independent in the sense that for any value of x at least one determinant composed of m full columns of the matrix

$$\phi_{\tau p_i}[x, y, y']$$

differs from zero.

The problem of Bolza, one of the most general single integral problems adequately solved in the calculus of variations, includes the isoperimetric problem. An isoperimetric problem with fixed end points minimizes

$$J = \int_a^b F[x,y,y']\,dx \tag{11.33}$$

subject to the restriction that

$$L = \int_a^b G[x,y,y']\,dx \tag{11.34}$$

has a prescribed value, where G is a function satisfying the properties imposed in F in Definition 1 (p. **1**-154). The reduction is made by introducing a new variable

$$Y_{n+1} = \int_a^x G[x,y,y']\,dx \tag{11.35}$$

Then

$$Y'_{n+1} = G[x,y,y'] \tag{11.36}$$

and the problem can be written in terms of new variables

$$\begin{aligned} (y) &= (y_1,y_2, \ldots ,y_n,y_{n+1}) \\ (p) &= (p_1,p_2, \ldots ,p_n,p_{n+1}) \\ J &= \int_a^b F[x,y,y']\,dx \end{aligned}$$

where of the $2(n+1)$ variables (y,p) only the original $2n$ appear in $F[x,y,p]$, and

$$\phi_1[x,y,p] = G[x,y,p] - p_{n+1} \tag{11.37}$$

Equation (11.36) plus an integration constant which fixes the end value of L is equivalent to (11.34).

In particular, for the catenary problem

$$y_2 = \int_a^x \sqrt{1 + y'^2}\,dx$$

or Eq. (11.36) becomes

$$y'_2 = \sqrt{1 + y'^2}$$

and $y_2(a) = 0$, $y_2(b) = L$.

The governing theorem states: *A necessary condition for a minimum in the problem of Bolza is that the Euler equations in integral form and the transversality conditions are satisfied for an integrand function.*

$$H[x,y,p] = \lambda_0 F[x,y,p] + \Sigma\lambda_\tau\phi_\tau[x,y,p] \tag{11.38}$$

where the λ's *are functions of x not all zero.*

Final determination of the λ's depends on the constants of the problem (as in the catenary problem of Sec. 4). This theorem yields the equations found in Sec. 4 for the catenary problem.

9. Second Variations

A stronger condition for a minimum of a function of a single variable is frequently stated in terms of a second derivative. Consider again the function $F(e)$ defined for $a \le e \le b$, and suppose that at $e = e_0$, $a < e_0 < b$, both the first two derivatives exist, and $f'(e) = 0$. If $f''(e_0) < 0$, then for values of e near e_0, $(e - e_0)f'(e) < 0$, and the law of the mean states that for these values of e, $f(e) < f(e_0)$. [The *law of the mean* states that if $f'(e)$ exists for $e_0 < e < e_1$, then $f(e_1) - f(e_0) = (e_1 - e_0)f'(e_2)$, where e_2 is some value between e_0 and e_1; e_2 is, in fact, a point where the function

$$h(e) = e_1f(e_0) - e_0f(e_1) + ef(e_1) - f(e_0) - (e_1 - e_0)f(e)$$

attains a maximum or minimum: such a point clearly lies between e_0 and e_1 since $h(e_0) = h(e_1) = 0$.] This argument leads to the following two theorems.

A necessary condition for a relative minimum at $e = e_0$ when $a < e_0 < b$ and when $f'(e_0)$ and $f''(e_0)$ both exist is $f'(e_0) = 0$ and $f''(e_0) \geq 0$.

The border case, where $f'(e_0) = f''(e_0) = 0$, may be either maximum or minimum or neither, the functions $-e^4$, e^4, and e^3, respectively, being examples at $e = 0$. The matter may be settled in some cases by considering higher derivatives, the parity of the order of the first nonzero derivative and its sign being crucial, if enough derivatives exist; but for the function $k(e)$ defined

$$\begin{aligned} k(0) &= 0 \\ k(e) &= n^{-1/e^2} \qquad e > 0 \\ k(e) &= bn^{-1/e^2} \qquad e < 0 \end{aligned}$$

where $n > 1$ and $b = \pm 1$ has all derivatives equal to zero at $e = 0$ and has a minimum there if $b = 1$ and neither maximum nor minimum if $b = -1$.

At a critical point $(\bar{u})$ of a function of several variables $f(u_1,u_2, \ldots ,u_n)$ a second derivative may be evaluated for every direction. If (as in Sec. 2) for some $(v) = (v_1,v_2, \ldots ,v_n)$

$$g(e) = f(\bar{u} + ev)$$

then

$$g'(0) = \sum_{i=1}^{n} v_i f_{u_i}(\bar{u})$$

and

$$g''(0) = \sum_{i,j=1}^{n} v_i v_j f_{u_i u_j}(\bar{u})$$

If for any choice of (v), except the degenerate choice where all $v_i = 0$, this second derivative is positive, the critical point is a relative minimum, for $f(u)$ is greater than $f(\bar{u})$ for neighboring (u) in every direction as defined by (v). The indeterminate case, however, where

$$\sum_{i,j=1}^{n} v_i v_j f_{u_i u_j}(\bar{u}) = 0$$

can occur in a more complicated and a more interesting way.

Indeed (see Part 1, Chap. 2, Sec. 11), by choosing a set of variables

$$w_i = \sum_{j=1}^{n} a_{ij} v_j$$

where the determinant $|a_{ij}| \neq 0$, so (v) is uniquely determined by (w), the quadratic form for $g''(0)$ may be made identical with a simple expression in (w):

$$\sum_{i,j=1}^{n} v_i v_j f_{u_i u_j}(\bar{u}) = \sum_{i=1}^{n} b_i w_i^2 \qquad (11.39)$$

where for each i, b_i takes one of the values $1, -1, 0$.

The necessary condition for a minimum now is that none of the b's is negative. A sufficient condition for a relative minimum at a critical point is that all the b's are positive.

The critical points for which some b's are positive and some negative represent unstable equilibria if the function minimized is a physical potential. A mountain pass represents such a critical point for the gravitation potential function.

These points are related, in number, in accordance with the geometry of the region of definition of the function. On the (approximately spherical) earth, for example, it is known that n_0 the number of points of relatively minimal altitude, n_1 the number of passes, and n_2 the number of peaks satisfy the following relations under some conditions of regularity:

$$\begin{aligned} n_0 &\geq 1 \\ n_1 - n_0 &\geq 1 \\ n_2 - n_1 + n_0 &= 0 \end{aligned}$$

Adequate regularity is the absence of b's with value zero in the diagonalization of the second derivative quadratic form as in Eq. (11.39) at each critical point.

Similar theory has been developed for the second variation of any calculus of variation problem. For the simplest problem, as discussed in Sec. 3, the second derivative at $e = 0$ of $J(e)$ is

$$J''(0) = \int_a^b \Sigma_{i,j} \{F p_i p_j [x,\bar{y},\bar{y}'] g'_i g'_j + 2F p_i y_j [x,\bar{y},\bar{y}'] g'_i g_i + F y_i y_j [x,\bar{y},\bar{y}'] g_i g_j\} \, dx$$

The analysis demands for a minimum that $J''(0) \geq 0$. The technique generally used is due to Jacobi. A more sensitive technique has been developed by M. Morse to show that the second variation may be analyzed just as the critical points of a function of several variables may be. Jacobi obtained necessary conditions in terms of differential equations for a minimum. Morse showed that for general problems the nonminimal critical points behave in ways completely analogous with the behavior of critical points of a function of a finite number of variables.

The minimizing solutions usually represent stable physical states; the nonminimizing represent unstable equilibria. For many problems the existence of two stable (minimizing) states implies the existence of an unstable one between them.

Jacobi's development more or less completed the study of the minimum problem: when his condition is added to the ones already stated and when all signs $\leq$ are replaced by $<$ and all signs $\geq$ are replaced by $>$, the total condition stated is sufficient for a minimum.

Morse's development pushed the study of nonstable critical curves far along toward a complete solution. For a large class of problems, he established relations analogous to all the relations known to exist between critical points of finite dimensional geometric objects.

10. Multiple-integral Problems

There is no difficulty in writing the differential form of the Euler equations for multiple integrals.

Consider a nonparametric fixed boundary double-integral problem:

$$J = \iint_S F[x,y,z_i(x,y),z_x,z_y]\, dy\, dx \qquad (11.40)$$

integrated over a fixed region S of the (x,y) plane, with admissible surfaces restricted on the boundary of S, $B(s)$: $z_i(x,y) = \alpha_i(x,y)$ if $(x,y)\epsilon B(S)$. Here no count of derivatives will be attempted either for $B(s)$ or for the function $F[x,y,z,p,q]$.

Variation is achieved by writing

$$J(e) = \iint_S F[x,\, y,\, \bar{z} + eg,\, \bar{z}_x + eg_x,\, \bar{z}_y + eg_y]\, dy\, dx \qquad (11.41)$$

where $\bar{z}$ is supposedly a minimizing surface and g is an admissible variation with value zero on $B(S)$. The development continues easily.

$$J'(0) = \iint_S \Sigma_i[F_{z_k}g_i + F_{p_i}g_{i_x} + F_{q_i}g_{i_y}]\, dy\, dx$$

The integration by parts must go in a direction which works for a single integral but which is extravagant with derivatives (leading to the differential, rather than the integral, form of Euler's equations):

$$\iint F_{p_i}g_{i_x}\, dy\, dx = -\iint g_i \frac{\partial}{\partial x} F_{p_i}\, dy\, dx$$

$$\iint F_{q_i}g_{i_y}\, dy\, dx = 1 \iint g_i \frac{\partial}{\partial y} F_{q_i}\, dy\, dx$$

Hence

$$J'(0) = \iint_S \Sigma_i g_i \left[F_{z_i} - \frac{\partial}{\partial x}F_{p_i} - \frac{\partial}{\partial y}F_{q_i}\right] dy\, dx$$

If the quantity in brackets is not identically zero, the g_i can be chosen zero except for a small region near a nonzero part of the bracketed quantity, where g_i can be chosen to make $J'(0)$ positive, for example. Thus, Euler's equations follow

$$F_{z_i} - \frac{\partial}{\partial x}F_{p_i} - \frac{\partial}{\partial y}F_{q_i} = 0 \qquad (11.42)$$

Haar has avoided the assumption that $(\partial/\partial x)F_{p_i}$ exists by a more complicated development. Weierstrass conditions, edge (corner) conditions, second variations, and sufficient conditions have been developed. They are largely analogous to single-integral problems, but with great additional complexity due to the added dimensionality.

The proof of existence of minimizing surfaces and the solution of numerical problems are both difficult because of the nonlinear nature of the partial differential equations (11.42) of Euler. Nonlinear ordinary equations of the Euler type are easily handled, at least in theory. Nonlinear partial equations except for a few examples are beyond the reach of present-day methods.

Problems involving potential functions and minimal area surfaces are solvable reasonably completely, but not simply. General problems have not been adequately solved.

11. Methods of Computation

Numerical solution of problems from the calculus of variations usually depends on replacing the problem by one depending on a finite number of variables or parameters. This may be done in any of several ways.

If a solution of the Euler differential equations can be found in closed form depending only on initial conditions, these initial conditions are finite in number and the solution of the original variational problem is reduced to the solution of a problem in analysis with this finite number of variables. This is the approach favored by most workers whenever it may be applied.

If direct, closed solution of the Euler equations is not feasible, approximate methods must be tried. Three types of approximation are frequently used.

In the first type of approximate solution the Euler differential equations are replaced by analogous difference equations whose solutions approximate those of the differential equations. This is done by introducing a discrete grid of admissible values for the (formerly continuous) independent variable and by replacing derivatives by analogous difference quotients involving neighboring points of this discrete grid.

In the second type of approximation, the discrete grid of the first type is applied to the integrals of the problem. Thus, these integrals are replaced by sums of finite numbers of terms involving the values of the dependent variables at the points of the discrete grid, and these sums (which are functions of a finite number of variable values of the dependent variables at these grid points) are minimized by methods of ordinary differential calculus.

The third type of approximation restricts the admissible functions to be dependent (usually linearly dependent) on a finite number of functions which are likely to be a truncated part of a complete set of orthogonal functions (such as sine and cosine functions suitable for Fourier series development of any function). Here the aim is to choose the finite basis in a way which will admit a close approximation to any accurate solution of the problem. Frequently, the experience (or the hunch) of the solver is dominant in choosing this basis, and no general rule can be stated. The whole family of functions which can be obtained from this finite basis is admissible, and the integrals involved in the calculus of variations problem may be computed for each. The solution sought in the approximation is the member of this truncated family which furnishes a minimum to the problem relative to other functions of the same family.

Of the three approximate methods, the last is probably the most frequently used by physicists.

Each of the approximation schemes reduces the functional problem to a problem involving functions of finitely many variables. These are minimization problems. It may be possible to solve these auxiliary problems exactly or in closed form by methods of differential calculus, but in many cases the solver will be driven to approximate methods for them also.

The most generally applicable method of approximate solution of minimizing problems of many variables is a method of steep descent. Methods of steep descent are iterative methods which start with some (probably guessed) approximation to the minimum and improve the approximation by displacing the point considered approximately along the negative gradient of the function to be minimized (this gradient is evaluated at the point in question). The effectiveness of this method depends on the nature of the particular problem attacked. Its motivation depends on the fact that the positive gradient (see Part 1, Chap. 10) is the direction of steepest ascent of the function, and the negative gradient is the direction of steepest descent of the function. The method depends on the principle that a relative minimum can be reached by moving downward as rapidly as possible.

12. Conclusion

The basis of the solution of any numerical problem is the solution of the Euler equations. The problem of Bolza gives a theory for generating a problem with easily written Euler equations from any likely single-integral problem. If these equations can be solved in general functions, the problem is reduced to one of fitting constants and satisfying transversality conditions.

The Weierstrass condition represents a phenomenon not found in minimizing functions of a finite number of variables. It does not give a constructive generation of a curve as do the Euler equations. It should be used as a test after the solution to the Euler equations has been found.

If the Euler equations or the transversality conditions are not conveniently solvable in general, successive numerical approximations may be used. For Euler's equations the functions of Eqs. (11.7) or discrete analogues of them may be used to decrease the value of J. Small values of e must be used, the value depending on the state of the approximation; J must be made to decrease.

Similarly useful functions are developed in the routine development of the transversality conditions, so that end points may be adjusted by successive approximation also. Finally, the Lagrange multipliers may be adjusted, sometimes rapidly by successive approximation.

References

Expository

1. Bolza, O.: "Lectures on the Calculus of Variations," Chicago, 1946.
2. Bliss, G. A.: "Calculus of Variations," Chicago, 1925.
3. Seifert-Threlfall, "Variationsrechnung im Grossen (Morse-sche Theorie)," Leipzig, 1948.

Advanced

4. Bliss, G. A.: "Lectures on the Calculus of Variations," Chicago, 1946.
5. Bolza, O.: "Vorlesungen über Variationsrechnung," Leipzig, 1949.
6. Caratheodory, C.: "Variationsrechnung und partielle Differential-gleichungen erster Ordnung," Leipzig and Berlin, 1935.
7. Courant, R.: "Dirichlet's Principle," New York, 1950.
8. Courant, R., and D. Hilbert: "Methods of Mathematical Physics," vol. 2, New York, 1957.
9. Morse, M.: "Calculus of Variations in the Large," Ann Arbor, 1947.

Chapter 12

Elements of Probability

By CHURCHILL EISENHART, National Bureau of Standards
and MARVIN ZELEN, National Cancer Institute

1. Probability

Elementary Probability. In many situations in the physical, biological, and social sciences, cases occur where an experiment may be repeated a large number of times under "essentially the same conditions"; yet the outcomes vary in an irregular manner that defies all attempts at prediction. Such situations give rise to sequences of *random experiments*.

Any outcome of a random experiment is termed an *event*. The event E consists of any outcome with the property E. Associated with a long sequence of random experiments will be a number $P(E)$, termed a *probability*, which gives the relative frequency of the occurrence of E. This meaning of the term probability is often referred to as the frequency interpretation of probability.

For example, in tossing a coin, each toss can be considered as a random experiment that must result in one or the other of the two outcomes "heads" or "tails." The outcome of any one toss cannot be predicted, but in a long sequence of trials the relative frequency of heads will tend to equal $P(H)$, say.

The term probability thus used is associated with the outcome of a collection or sequence of random trials and is called *statistical* or *mathematical probability*. Another type of probability termed *intuitive* or *subjective probability* is characterized by being associated with the outcome of a unique event or with the credibility of statements. This use of probability may be associated with expressing a degree of belief, relative to an outcome or proposition; see Good [6]‡ and Savage [46].

Axiomatic Definition. The frequency interpretation of probabilities serves as a useful intuitive definition. However, certain difficulties are inherent in this definition which make difficult a rigorous mathematical theory of probability. Just as geometry is based on a few basic axioms, the modern mathematical theory of probability is also based on axioms from which all properties can be logically derived.

Before giving the basic axioms, it will be convenient to introduce the following terminology. Let an *elementary event* or *sample point* be defined as the indecomposable outcome of a random experiment. An *event* E is composed of one or more sample points.

‡ Numbers in brackets refer to References at end of chapter.

The *sample space* (denoted by S) is defined as the aggregate of all the sample points. An event containing no points of the sample space (denoted by O) is called the *null event*. In point-set terminology, the sample space S is a *set*, a sample point is an *element* of the set, an event E is a *subset* of S, and O is an *empty set*.

Let E_1 and E_2 be two events in S. Then adopting the following operations from point-set theory:

1. *Sum:* $E_1 + E_2$ is the occurrence of *at least one* of the events E_1 or E_2.
2. *Product:* E_1E_2 is the simultaneous occurrence of the events E_1 *and* E_2.
3. *Difference:* $E_1 - E_2$ is the occurrence of elementary events in E_1 not common to E_2.

These operations are both associative and distributive; however, only the sum and product operations are commutative. In particular, if two events contain no elementary points in common, they are said to be *mutually exclusive events* and $E_1E_2 = 0$.

Let $X_1, X_2, \ldots$ be the points of a (denumerable) sample space S, and let $E_1, E_2, \ldots$ be a collection of events which are subsets of S. Then the basic axioms of probability state that associated with every event E_i is a real non-negative number termed a probability and denoted by $P(E_i)$ such that

$$0 \leq P(E_i) \leq 1 \tag{12.1}$$

$$P(S) = 1 \qquad P(\text{O}) = 0 \tag{12.2}$$

If the events E_i are all mutually exclusive events, then

$$P\left(\sum_{i=1}^{\infty} E_i\right) = \sum_{i=1}^{\infty} P(E_i) \tag{12.3}$$

From the above axioms it follows that for any two events E_1 and E_2

$$P(E_1 + E_2) = P(E_1) + P(E_2) - P(E_1E_2) \tag{12.4}$$

If an event E can never occur in a sequence of random experiments, then E is called an *impossible event* and has the probability $P(E) = 0$; similarly, if an event E will always occur at every trial of a random experiment, E is called a *certain event* and has the probability $P(E) = 1$. On the other hand, if an event E has a probability $P(E) = 0$, this does not mean that the event E will never occur. All this means is that in a long sequence of random experiments the relative frequency of E will be close to zero. Similarly, if it is known that $P(E) = 1$ for an event E, this does not mean that E will occur in every random experiment, but only that in a long sequence of trials the relative frequency will be close to 1.

For example, consider all numbers included within the interval (0,1). If one number is drawn at random within this interval, the probability of drawing a rational number is equal to zero. Alternatively, the probability of drawing an irrational number will be equal to one.

The preceding definitions and axioms suffice to define mathematical probability for discrete sample spaces. See Mosteller et al. [13] for an elementary exposition and Feller [3] for a comprehensive advanced discussion. Nondiscrete sample spaces are discussed in detail by Cramér [2], Feller [3*a*], Fisz [4], Loève [11], Parzen [15], and Kolmogorov [33].

Independence and Conditional Probability. Two events E_1 and E_2 are said to be *independent* if the probability of the simultaneous occurrence of E_1 *and* E_2 is equal to the product of the individual probabilities; that is, if

$$P(E_1E_2) = P(E_1)P(E_2) \qquad (12.5)$$

If the event E_2 does not exclude the simultaneous occurrence of the event E_1, then some cases of E_2 will also be realizations of E_1. The "relative frequency" of E_1 in such cases is termed the *conditional probability* of E_1 given E_2, written as $P(E_1|E_2)$. If $P(E_2) > 0$, then

$$P(E_1|E_2) = \frac{P(E_1E_2)}{P(E_2)} \qquad (12.6)$$

A rearrangement yields

$$P(E_1E_2) = P(E_2)P(E_1|E_2) = P(E_1)P(E_2|E_1) \qquad (12.7)$$

and if E_1 and E_2 are independent, then

$$P(E_1|E_2) = P(E_1) \qquad \text{and} \qquad P(E_2|E_1) = P(E_2) \qquad (12.8)$$

which is an equivalent definition for independence, i.e., the probability of the event E_1 does not depend upon the occurrence of E_2, and vice versa.

More generally the simultaneous occurrence of the events $E_1, E_2, \ldots, E_n$ is given by

$$P(E_1E_2, \ldots, E_n) = P(E_1)P(E_2|E_1)P(E_3|E_1E_2) \cdots P(E_n|E_1E_2, \ldots, E_{n-1}) \qquad (12.9)$$

and the events $E_1, E_2, \ldots, E_n$ are said to be *mutually independent* if and only if

$$\begin{aligned} P(E_iE_j) &= P(E_i)P(E_j), & i \neq j \\ P(E_iE_jE_k) &= P(E_i)P(E_j)P(E_k), & i \neq j \neq k \\ &\vdots \\ P(E_1E_2 \cdots E_n) &= P(E_1)P(E_2) \cdots P(E_n) \end{aligned} \qquad (12.10)$$

Bayes' Theorem and Inverse Probability. Let $H_1, H_2, \ldots, H_k$ be a set of mutually exclusive events that are also exhaustive, and let E be an arbitrary event for which $P(E) \neq 0$. Then, by (12.7),

$$P(EH_i) = P(E)P(H_i|E) = P(H_i)P(E|H_i)$$

Solving for $P(H_i|E)$ yields

$$P(H_i|E) = \frac{P(H_i)P(E|H_i)}{P(E)}$$

Since the $\{H_i\}$ are exhaustive,

$$P(E) = \sum_{i=1}^{k} P(H_i)P(E|H_i)$$

and substitution for $P(E)$ in the preceding equation yields

$$P(H_i|E) = \frac{P(H_i)P(E|H_i)}{\sum_{i=1}^{k} P(H_i)P(E|H_i)} \qquad (12.11)$$

Formula (12.11) is called Bayes' theorem, after its discoverer, Rev. Thomas Bayes (1702–1761). It says that the conditional probability of H_i, given E, is proportional to the (unconditional) probability of H_i multiplied by the conditional probability of E, given H_i.

If it is known that an unspecified one of the events $H_1, H_2, \ldots, H_k$ has occurred coincident with, or followed by, the event E, then Bayes' theorem gives the (conditional) probability that the unspecified event was H_i, $i = 1, 2, \ldots, k$. Consequently the conditional probability $P(H_i|E)$ can be regarded as the "posterior probability" of H_i after E is known to have occurred.

Bayes' theorem is the cornerstone of *inverse probability*, a precise form of inductive inference that seeks to reason from observed events to the probabilities of the various hypotheses that may explain them. In such applications, E is taken to represent a given datum, and the $\{H_i\}$, to represent distinct hypotheses (e.g., alternative hypotheses that could give rise to E). The probability $P(H_i)$ is termed the *initial* or *prior probability* associated with H_i; the expression $P(E|H_i)$ is called the *likelihood* of E on H_i; and the probability $P(H_i|R)$ is termed the *posterior probability* of H_i. Bayes' theorem is regarded as giving the probability of a particular hypothesis, say H_i, being true, conditional on the observed datum. The intriguing feature about Bayes' theorem is that it provides a formal way of specifying "how one learns by experience." That is, the prior probabilities are modified by the likelihood function to give the posterior probability associated with a hypothesis.

To use Bayes' theorem, it is necessary to know $P(E|H_i)$, the probability of the datum conditional on a particular hypothesis being true, and $P(H_i)$, the prior probability of the hypothesis. Both the prior and posterior probabilities are taken to refer to the probability of a hypothesis being true and thus interpret "probability" as a degree of belief. Much controversy has surrounded the use of Bayes' theorem since it was first published in 1763 [19]. The controversy stems from the difficulty of specifying the prior probabilities if absolutely nothing is known about them. When the prior probabilities are known, there is no dispute about the applicability of Bayes' theorem. One of the suggestions made is that in the absence of information we should take the prior probabilities to be equal, because "to say that the probabilities are equal is a precise way of saying that we have no ground for choosing between the alternatives." This does not lead to any technical difficulties so long as k remains finite, but when $k \to \infty$ and the set of hypotheses $\{H_i\}$ becomes a continuum in some parameter, say θ, then difficulties arise: If θ has a uniform prior probability distribution over the interval (a,b), then θ^2 cannot have a uniform prior probability distribution over (a^2,b^2), at least not if "prior probability distributions" are to obey the same calculus for "change of variables" as do probability distributions of random variables; see the last portion of Sec. 3. However, there are other rules for choosing the prior probabilities in the absence of information; see, for example, Jeffreys [9] and Savage [46]. Jeffreys' rules are such that the prior probabilities contribute less than "one observation of information," so that the importance of an arbitrary prior probability diminishes as the number of observations becomes large. On the other hand, Savage advocates that the prior probabilities be chosen by using subjective probability to reflect one's degree of belief in a hypothesis.

2. Random Variables and Distribution Functions

Random Variables. A variable whose value depends on the outcome of a random experiment is termed a *random* (or *stochastic*) *variable.* For example, in coin tossing, a random variable Y may be defined by

$$Y = \begin{cases} 1 & \text{if coin shows a head} \\ 0 & \text{if coin shows a tail} \end{cases}$$

Here Y serves as a symbolic *indicator* of the outcome of a random toss, with $Y = 1$ signifying "heads" and $Y = 0$, "not heads." Numbers obtained by direct observation of natural processes or as the results of physical measurement are often regarded as observed values of random variables. Thus N, the number of radioactive atoms of a given type that decay in a fixed interval of time, and T, the elapsed time between any two successive disintegrations, are usually regarded as random variables; and the randomness of their variation taken to be an intrinsic property of radioactivity phenomena. The distance between two lines of a photographic plate under prescribed conditions is usually considered to be a constant, say δ; but a measured value of this distance, say d, may be regarded as an observed value of a random variable D that varies about δ as a consequence of random variations in conditions that cannot be controlled strictly from one measurement to another. (Throughout this chapter, the modern convention of designating *random variables* by *capital letters,* and nonrandom variables and quantities by lower-case letters, is followed wherever practicable.)

Y and N defined as above are one-dimensional *discrete* random variables; T and D, one-dimensional *continuous* random variables. If n tosses of a coin are made and Y_i is defined as Y above for the ith toss, then the vector quantity $(Y_1,Y_2, \ldots ,Y_n)$ is an n-dimensional discrete random variable. Similarly, the elapsed times $T_1,T_2, \ldots ,T_n$ between any $n + 1$ successive radioactive disintegrations, and n-fold repetition of the measurement operation D, define n-dimensional continuous random variables $(T_1,T_2, \ldots ,T_n)$ and $(D_1,D_2, \ldots ,D_n)$.

Any function of a random variable is also a random variable. Thus,

$$S = \sum_{i=1}^{n} Y_i$$

which signifies the number of "heads" in n tosses of the coin, is a random variable; and so are S^2, $\arcsin \sqrt{S/n}$, and e^{itS}, where $i = \sqrt{-1}$ and t is a real variable, $-\infty < t < \infty$.

Distribution Functions. Associated with any one-dimensional random variable X is a unique *distribution function* (d.f.), $F(x)$, defined as

$$F(x) = P\{X \leq x\} \tag{12.12}$$

where $P\{X \leq x\}$ signifies, for $-\infty < x < +\infty$, the probability of the event "$X \leq x$." The probability that X takes on a value within the interval $a < X \leq b$ is given by

$$P\{a < X \leq b\} = F(b) - F(a) \tag{12.13}$$

From the mathematical definition of probability it follows that

(1) $F(x)$ is a nondecreasing function of x, that is,

$$F(x_1) \leq F(x_2) \text{ for } x_1 \leq x_2$$

(2) $F(x)$ is everywhere continuous on the right,

$$F(x) = \lim_{\epsilon \to 0} F(x + \epsilon)$$

(3) $$F(-\infty) = 0 \qquad F(\infty) = 1$$

Discrete Distributions. A random variable X is of the *discrete type,* and is said to have a *discrete distribution,* if it can take on at most an enumerable infinity of values

$$\ldots\ x_{-1},\ x_0,\ x_1,\ \ldots\ ,$$

with finite *point probabilities*

$$p_s = P\{X = x_s\} \tag{12.14}$$

subject only to the restrictions $p_s > 0$ for all s and $\Sigma_s p_s = 1$. When this is the case, the distribution function of X is given by

$$F(x) = P\{X \leq x\} = \sum_{x_s \leq x} p_s \tag{12.15}$$

where the summation is over all values of $x_s \leq x$; and is a step function that is constant over every interval not containing any point of the set $\{x_s\}$, but has a *step* of height p_s at the point x_s ($s = 0, \pm 1, \pm 2, \ldots$). The probability that X takes on any particular value x' different from every x_s is zero, that is, $P\{X = x'\} = 0$. The set $\{x_s\}$ of values for which $P\{X = x\} > 0$ is termed the *spectrum* of the random variable X. Some important one-dimensional discrete distributions are characterized in Table 12.1.‡

A random variable X having a *causal* or *unitary distribution* has its probability distribution concentrated at a single point, that is,

$$P\{X = c\} = 1 \qquad (12.16)$$

where c is any finite constant. Its distribution function is

$$F(x) = \epsilon(x - c) \qquad (12.17)$$

where $\epsilon(z)$ is the step function

$$\epsilon(z) = \begin{cases} 0 & \text{for } z < 0 \\ 1 & \text{for } z \geq 0 \end{cases} \qquad (12.18)$$

The distribution function of any discrete distribution (12.15) can be written

$$F(x) = \sum_{s=1}^{\infty} p_s \epsilon(x - x_s) \qquad (12.19)$$

so that the distribution of any discrete random variable X can be regarded as being composed of a linear function of causal distributions.

Continuous Distributions. A random variable X is of the *continuous type*, and is said to have a *continuous distribution*, if its distribution function $F(x)$ is everywhere continuous and possesses a derivative $F'(x) = f(x)$ that is finite and continuous for all values of x, except possibly at certain individual points, of which any finite interval contains at most a finite number. The distribution function is then *absolutely continuous*, and can be written

$$F(x) = P\{X \leq x\} = \int_{-\infty}^{x} f(t)\,dt \qquad (12.20)$$

with

$$F(\infty) = \int_{-\infty}^{+\infty} f(t)\,dt = 1 \qquad (12.21)$$

Its derivative, $f(x)$, termed the *probability density* (or *frequency*) function, represents the probability density at each point at which it is finite. The differential of $F(x)$,

$$dF(x) = P\{x < X \leq x + dx\} = f(x)\,dx \qquad (12.22)$$

is termed the *probability element* of the distribution. The values of x for which $f(x) > 0$ constitute the *spectrum* of the random variable X. The probability that X takes on any particular value x is zero, that is, $P\{X = x\} = 0$ for all x. Consequently, the probabilities $P\{a < X \leq b\}$, $P\{a < X < b\}$, $P\{a \leq X \leq b\}$, and $P\{a \leq X < b\}$ are all equal, and are given by

$$F(b) - F(a) = \int_a^b f(t)\,dt \qquad (12.23)$$

A number of important one-dimensional continuous distributions are characterized in Table 12.2, where the probability density function is in each case, except for the error function, given in canonical form, $g[(x - \lambda)/\beta]$, where λ is the *location parameter* and β the *scale parameter* of the distribution.

‡ Haight [31] gives explicit formulas for a large number of discrete and continuous probability distributions, together with detailed summaries of their properties.

Stieltjes Integrals. In order to avoid the necessity of repeating for discrete distributions, in terms of summations, various definitions, theorems, proofs, etc., expressed in terms of integrals for continuous distributions, or vice versa, it is often convenient to employ a formalism that enables discrete and continuous distributions to be treated in a unified manner. This can be accomplished rigorously by the use of *Stieltjes integrals*.

If $g(x)$ is continuous for $a \leq x \leq b$, then the *Stieltjes integral* of $g(x)$ with respect to any distribution function $F(x)$, over the half-open interval $a < x \leq b$, written

$$\int_a^b g(x)\,dF(x) \qquad (12.24)$$

is defined as previously described (Part 1, Chap. 3, Sec. 3) and reduces to an ordinary integral,

$$\int_a^b g(x)f(x)\,dx, \qquad f(x) = F'(x) \qquad (12.25)$$

when $F(x)$ is of the continuous type; and to a simple summation,

$$\sum_{a<x_s\leq b} g(x_s)p_s \qquad (12.26)$$

with

$$p_s = P\{X = x_s\} = F(x_s) - F(x_{s-1}) \qquad (12.27)$$

when $F(x)$ is of the discrete type. Consequently, (12.24) serves as a generic representation of both (12.25) and (12.26).

The familiar properties of ordinary integrals extend immediately to Stieltjes integrals of a continuous function $g(x)$ with respect to a distribution function $F(x)$. Thus, for example,

$$\int_a^b dF(x) = F(b) - F(a); \qquad (12.28)$$

$$\int_a^b g(x)\,dF(x) = g(b)F(b) - g(a)F(a) - \int_a^b F(x)\,dg(x); \qquad (12.29)$$

$$\int_a^b g(x)\,dF(x) = \int_a^c g(x)\,dF(x) + \int_c^b g(x)\,dF(x) \qquad a < c < b; \qquad (12.30)$$

$$\sum_{n=0}^{\infty} \int_a^b g_n(x)\,dF(x) = \int_a^b g(x)\,dF(x) \qquad (12.31)$$

if

$$\sum_{n=0}^{\infty} g_n(x)$$

converges uniformly to $g(x)$ on $a \leq x \leq b$; and

$$\int_a^b g(x)\,dF(x) = \int_c^d g(\beta(x))\,dF(\beta(x)) \qquad (12.32)$$

with $a = \beta(c)$ and $d = \beta(d)$, if $\beta(x)$ is continuous and monotonically increasing on $a \leq x \leq b$. It follows that the theorems for ordinary integrals on continuity, differentiation, and integration with respect to a parameter extend to integrals of the form

$$\int_a^b g(x,t)\,dF(x)$$

TABLE 12.1. SOME ONE-DIMENSIONAL DISCRETE DISTRIBUTIONS

Name	Spectrum $\{x_s\}$	Point probabilities p_s	Restrictions on parameters	Mean	Variance	Skewness γ_1	Kurtosis γ_2
Causal, or unitary	$x_0 = c$	$p_0 = 1$	$-\infty < c < +\infty$	c	0		
Binomial, or Bernoulli	$x_s = s$, for $s = 0, 1, 2, \ldots, n$	$\binom{n}{s} p^s(1-p)^{n-s}$	$0 < p < 1$ (Let $q = 1 - p$)	np	npq	$\frac{q-p}{\sqrt{npq}}$	$\frac{1-6pq}{npq}$
Hypergeometric	$x_s = s$, for $s = 0, 1, 2, \ldots, n$	$\frac{\binom{N_1}{s}\binom{N_2}{n-s}}{\binom{N_1+N_2}{n}}$	N_1 and N_2 integers, $n \le N_1 + N_2$ $\left(\text{Let } N = N_1 + N_2,\ p = \frac{N_1}{N},\ q = 1 - p = \frac{N_2}{N}\right)$	np	$npq\frac{N-n}{N-1}$	$\frac{q-p}{\sqrt{npq}}\left(\frac{N-1}{N-n}\right)^{1/2}\left(\frac{N-2n}{N-2}\right)$	Complicated
Poisson	$x = s$, for $s = 0, 1, 2, \ldots, \infty$	$e^{-m}m^s/s!$	$0 < m < \infty$	m	m	$\frac{1}{\sqrt{m}}$	$\frac{1}{m}$
Negative binomial	$x = s$, for $s = 0, 1, 2, \ldots, \infty$	$\binom{r+s-1}{s} p^r(1-p)^s$	$r \ge 0$ and $0 < p < 1$ $\left(\text{Let } p = \frac{1}{Q} \text{ and } 1 - p = \frac{P}{Q}\right)$	rP	rPQ	$\frac{Q+P}{\sqrt{rPQ}}$	$\frac{1+6PQ}{rPQ}$

TABLE 12.2. SOME ONE-DIMENSIONAL CONTINUOUS DISTRIBUTIONS

Name	Spectrum	Probability density function, $f(x)$	Restrictions on parameters	Mean	Variance	Skewness γ_1	Kurtosis γ_2
Error function	$-\infty < x < +\infty$	$\frac{h}{\sqrt{\pi}} \exp(-h^2x^2)$	$0 < h < \infty$	0	$\frac{1}{2h^2}$	0	0
Normal	$-\infty < x < +\infty$	$\frac{1}{\sigma\sqrt{2\pi}} \exp\left[-\frac{1}{2}\left(\frac{x-m}{\sigma}\right)^2\right]$	$-\infty < m < \infty$, $0 < \sigma < \infty$	m	σ^2	0	0
Cauchy	$-\infty < x < +\infty$	$\frac{1}{\pi\beta} \frac{1}{1 + \left(\frac{x-\lambda}{\beta}\right)^2}$	$-\infty < \lambda < \infty$, $0 < \beta < \infty$	Not defined	Not defined	Not defined	Not defined
Laplace, or double exponential	$-\infty < x < +\infty$	$\frac{1}{2\beta} \exp\left(-\left\lvert\frac{x-\lambda}{\beta}\right\rvert\right)$	$-\infty < \lambda < \infty$, $0 < \beta < \infty$	λ	$2\beta^2$	0	3
Extreme-value, or Fisher-Tippett Type I (or doubly exponential)	$-\infty < x < +\infty$	$\frac{1}{\beta} \exp(-y - e^{-y})$ with $y = \frac{x-\lambda}{\beta}$	$-\infty < \lambda < \infty$, $0 < \beta < \infty$	$\lambda + \gamma\beta$, where γ is Euler's constant	$\frac{(\pi\beta)^2}{6}$	1.3	2.4
Exponential	$\lambda \le x < \infty$	$\frac{1}{\beta} \exp\left[-\left(\frac{x-\lambda}{\beta}\right)\right]$	$-\infty < \lambda < \infty$, $0 < \beta < \infty$	$\lambda + \beta$	β^2	2	6
Pearson Type III (or gamma distribution for $\lambda = 0$)	$\lambda \le x < \infty$	$\frac{1}{\beta\Gamma(p)} \left(\frac{x-\lambda}{\beta}\right)^{p-1} \exp\left[-\left(\frac{x-\lambda}{\beta}\right)\right]$	$-\infty < \lambda < \infty$, $0 < \beta < \infty$, $0 < p < \infty$	$\lambda + p\beta$	$p\beta^2$	$\frac{2}{\sqrt{p}}$	$\frac{6}{p}$
Rectangular, or uniform	$\theta - \frac{w}{2} \le x \le \theta + \frac{w}{2}$	$\frac{1}{w}$	$-\infty < \theta < \infty$, $0 < w < \infty$	θ	$\frac{w^2}{12}$	0	$-6/5$

Finally, as in the case of ordinary integrals, by definition

$$\int_{-\infty}^{+\infty} g(x)\, dF(x) = \lim_{\substack{a \to -\infty \\ b \to +\infty}} \int_a^b g(x)\, dF(x) \qquad (12.33)$$

if the limit on the right is finite and unique for $a \to -\infty$ and $b \to +\infty$ independently; otherwise the left-hand integral is undefined.

Using the Stieltjes integral, any distribution function can be written as

$$F(x) = \int_{-\infty}^{x} dF(t) \qquad (12.34)$$

which, for a distribution of the continuous type, becomes

$$F(x) = \int_{-\infty}^{x} f(t)\, dt, \qquad f(t) = F'(t); \qquad (12.35)$$

and, for a distribution of the discrete type,

$$F(x) = \sum_{x_s \leq x} [F(x_s) - F(x_{s-1})] = \sum_{x_s \leq x} p_s \qquad (12.36)$$

where p_s is the point probability (12.27) at x_s. In particular, we may express the distribution function of the causal distribution, $\epsilon(x - c)$, defined by (12.17) and (12.18) in the form

$$\epsilon(x - c) = \int_{-\infty}^{x} d\epsilon(t - c) \qquad (12.37)$$

and applying (12.28) and (12.29), deduce immediately

$$\int_{-\infty}^{+\infty} d\epsilon(x - c) = 1 \qquad (12.38)$$

$$\int_{-\infty}^{+\infty} g(x)\, d\epsilon(x - c) = g(c) \qquad (12.39)$$

Dirac δ Function. In mathematical analyses of physical phenomena it may sometimes be more meaningful to regard some physical quantity not as a strict constant in the mathematical sense, but rather as a random variable X whose values are concentrated in a small neighborhood of c in accordance with a distribution function of the continuous type having a very small dispersion. A physical "constant" equal to c can then be regarded as the limit of the foregoing as the dispersion tends to zero, i.e., as a random variable with distribution function $\epsilon(x - c)$.

In order to apply the formalism of continuous distributions to such limiting distributions without resort to the Stieltjes integral, Dirac [24, p. 72] introduced an improper function, the *Dirac δ function*, "defined" by

$$\delta(x - c) = \frac{\text{"}d\text{"}}{dx}\, \epsilon(x - c) = \begin{cases} \infty & \text{for } x = c \\ 0 & \text{for } x \neq c \end{cases} \qquad (12.40)$$

$$\int_{-\infty}^{+\infty} \delta(x - c)\, dx = 1$$

By analogy with (12.39) he wrote

$$\int_{-\infty}^{+\infty} g(x)\delta(x - c)\, dx = g(c) \qquad (12.41)$$

as a formal expression of "the most important property of the δ function"; and by formal differentiation of both sides of (12.41) with respect to c, obtained another important property

$$\int_{-\infty}^{+\infty} g(x)\delta'(x - c)\, dx = -g(c) \qquad (12.42)$$

where $\delta'(x - c)$ signifies the "derivative" of the δ function. In this manner Dirac arrived at a host of rules for algebraic work involving δ functions [24, p. 75].

From a strictly mathematical viewpoint the "δ function" is not a "function" because the integral of a function that is zero everywhere except at one point must necessarily vanish. Nevertheless, the δ function has become an accepted tool in theoretical physics (see, for example, Morse and Feshbach [40]) and the customary operations with the δ function can be made mathematically rigorous by using the general theory of distributions, as explained in Schwartz [47] and Halperin [32]. In this chapter continuous and discrete distributions are treated by use of the Stieltjes integral, rather than the δ function.

3. Distributions in n Dimensions

The joint probability of the simultaneous events $\{X_i \leq x_i\}$ $(i = 1,2, \ldots ,n)$ is denoted by

$$P\{X_1 \leq x_1, \ldots ,X_n \leq x_n\} = F(x_1,x_2, \ldots ,x_n)$$

and is called the cumulative distribution function of the n random variables X_i, $i = 1,2, \ldots ,n$. The two main types of distributions of importance are the *discrete* and *continuous types*.

Multivariate distributions of the *discrete type* are characterized by random variables which can assume values for an enumerable set of points $(x_1,x_2, \ldots ,x_n)$ such that

$$P\{X_1 = x_{1i_1}, \ldots , X_n = x_{ni_n}\} = p_{i_1 \ldots i_n}$$

$$\text{and} \sum_{i_1 \ldots i_n} p_{i_1 \ldots i_n} = 1$$

(summation is over entire spectrum) (12.43)

Distributions of the *continuous type* are characterized by being absolutely continuous such that the probability density function is given by

$$f(x_1, \ldots ,x_n) = \frac{\partial^n F(x_1, \ldots ,x_n)}{\partial x_1 \cdots \partial x_n} \qquad (12.44)$$

The *k-dimensional marginal distribution* is defined as the distribution obtained by putting $(n - k)$ of the x_i equal to infinity in $F(x_1, \ldots ,x_n)$.

Thus the marginal distribution of X_i, say, is

$$F(\infty, \ldots , \infty, x_i, \infty, \ldots , \infty) = F_i(x_i)$$

and a necessary and sufficient condition for the statistical independence of the random variables $\{X_i\}$ is that their joint distribution be expressible in the form

$$\begin{aligned} F(x_1, \ldots ,x_n) &= F(x_1, \infty, \ldots , \infty) \cdots F(\infty, \ldots , \infty,x_n) \\ &= F_1(x_1)F_2(x_2) \cdots F_n(x_n) \qquad (12.45) \end{aligned}$$

The *conditional probability density function* resulting from putting

$$X_{k+1} = x'_{k+1}, \ldots, X_n = x'_n$$

is given by

$$f(x_1, \ldots, x_k | x'_{k+1}, \ldots, x'_n) = \frac{f(x_1, \ldots, x_n)}{\int_{-\infty}^{\infty} \cdots \int_{-\infty}^{\infty} f(x_1, \ldots, x_k, x'_{k+1}, \ldots, x'_n)\, dx_1 \ldots dx_k} \tag{12.46}$$

Example. The bivariate normal density function

$$f(x,y) = \frac{1}{2\pi\sigma_x\sigma_y\sqrt{1-\rho^2}} \times \exp -\frac{1}{2(1-\rho^2)}\left[\frac{(x-m_x)^2}{\sigma_x^2} - \frac{2\rho(x-m_x)(y-m_y)}{\sigma_x\sigma_y} + \frac{(y-m_y)^2}{\sigma_y^2}\right] \tag{12.47}$$

is an example of a multivariate distribution with two random variables. The probability density function of the marginal distribution of X is then

$$f_1(x) = \int_{-\infty}^{\infty} f(x,y)\, dy = \frac{1}{\sqrt{2\pi}\,\sigma_x} \exp -\frac{1}{2}\left(\frac{x-m_x}{\sigma_x}\right)^2 \tag{12.48}$$

which is the univariate normal density function. The density function of the conditional distribution of Y given that $X = x$ is then, by (12.46),

$$f(y|x) = \frac{f(x,y)}{\int_{-\infty}^{\infty} f(x,y)\, dy} = \frac{f(x,y)}{f_1(x)} = \frac{1}{\sqrt{2\pi}\,\sigma_y\sqrt{1-\rho^2}} \times \exp -\frac{1}{2\sigma_y^2(1-\rho^2)}\left[y - m_y - \frac{\rho\sigma_y}{\sigma_x}(x - m_x)\right]^2 \tag{12.49}$$

Finally if $\rho = 0$, the density function (12.47) can be written as the product of the two marginal distributions, i.e.,

$$f(x,y) = \left[\frac{1}{\sigma_x\sqrt{2\pi}} \exp -\frac{1}{2}\left(\frac{x-m_x}{\sigma_x}\right)^2\right] \times \left[\frac{1}{\sigma_y\sqrt{2\pi}} \exp -\frac{1}{2}\left(\frac{y-m_y}{\sigma_y}\right)^2\right] \tag{12.50}$$

and the random variables X and Y are then statistically independent.

Change of Variables for Continuous Distributions. For a distribution of the continuous type the *probability element* is defined by

$$P\{x_1 < X_1 \le x_1 + dx_1, \ldots, x_n < X_n \le x_n + dx_n\} = f(x_1, \ldots, x_n)\, dx_1 \ldots dx_n \tag{12.51}$$

provided $f(x_1, \ldots, x_n)$ is finite at the point $(x_1, \ldots, x_n)$.

If a new set of random variables $Y_1, \ldots, Y_n$ is introduced by the transformation $y_i = y_i(x_1, \ldots, x_n)$, $(i = 1, \ldots, n)$, and if

a) $y_i(x_1, \ldots, x_n)$ is everywhere unique and continuous (12.52)

b) $\dfrac{\partial y_i}{\partial x_j}$ is everywhere finite and continuous except possibly in certain points on an enumerable number of hypersurfaces (12.52)

c) $y_i(x_1, \ldots, x_n)$ is a one-to-one transformation such that $x_i = x_i(y_1, \ldots, y_n)$ and the x_i are unique (12.52)

d) the Jacobian $J = \dfrac{\partial(x_1, \ldots, x_n)}{\partial(y_1, \ldots, y_n)}$ is different from zero and finite except for points on the exceptional hypersurfaces (12.52)

then $f(x_1, \ldots, x_n)\, dx_1 \cdots dx_n$, the probability element, is transformed into

$$f(x_1, \ldots, x_n)\, dx_1 \cdots dx_n = f(x_1, \ldots, x_n)|J|\, dy_1 \cdots dy_n \tag{12.53}$$

where $f(x_1, \ldots, x_n)$ is to be expressed as a function of $y_1, \ldots, y_n$ on the right-hand side.

When a one-to-one transformation between the original and the transformed variates does not exist over the entire spectrum, then the probability density function of the transformed variates must be found "piecewise" by subdividing the space of the original variables and finding the corresponding contribution to the probability density function of the transformed variates for each subdivision of the original space. This technique is illustrated in Example 3 below.

Example 1. Suppose $Y = (X - a)/b$, where X is a univariate random variable having a frequency function $f(x)$ and a and b are given constants. Then for the case $n = 1$, (12.53) becomes

$$f(x)\, dx = f[x(y)]\left|\frac{dx}{dy}\right| dy \tag{12.54}$$

and the frequency function for Y is

$$g(y) = f[x(y)]\left|\frac{dx}{dy}\right| = f(by + a)|b| \tag{12.55}$$

Example 2. Let the random variables X and Y have the joint probability density function $f(x,y)$, and one requires the probability density of $W = Y/X$. Then on making the change of variables

$$x = x \qquad y = wx$$

the Jacobian is $|J| = |x|$ and thus the probability density of W is

$$g(w) = \int_{-\infty}^{\infty} f(x,wx)|x|\, dx \tag{12.56}$$

If X and Y are independent, then $f(x,y) = f_1(x)f_2(y)$ and (12.56) can be written as

$$g(w) = \int_{-\infty}^{\infty} f_1(x)f_2(wx)|x|\, dx \tag{12.57}$$

Example 3. Let the random variable X have the probability density function $f(x)$ and one requires the density function of $W = X^2$. Since $x = \pm\sqrt{w}$, the density function will be made up of two parts corresponding to $+\sqrt{w}$ and $-\sqrt{w}$. Thus the probability that W is within an interval (α,β) is

$$\begin{aligned} P\{\alpha < W \le \beta\} &= P\{\sqrt{\alpha} < X \le \sqrt{\beta}\} \\ &\quad + P\{-\sqrt{\beta} < X \le -\sqrt{\alpha}\} \\ &= \int_{\sqrt{\alpha}}^{\sqrt{\beta}} f(x)\,dx + \int_{-\sqrt{\beta}}^{-\sqrt{\alpha}} f(x)\,dx \end{aligned} \quad (12.58)$$

Making the change of variable

$$x = \sqrt{w} \quad \text{and} \quad x = -\sqrt{w}$$

in the first and second integrals in (12.58) gives

$$P\{\alpha < W \le \beta\} = \int_\alpha^\beta g^+(w)\,dw + \int_\alpha^\beta g^-(w)\,dw \quad (12.59)$$

where

$$g^+(w) = \frac{f(\sqrt{w})}{2\sqrt{w}}$$
$$g^-(w) = \frac{f(-\sqrt{w})}{2\sqrt{w}} \quad (12.60)$$

Thus the probability density function of W is

$$\begin{aligned} g(w) &= g^+(w) + g^-(w) \\ &= \frac{1}{2\sqrt{w}}[f(\sqrt{w}) + f(-\sqrt{w})] \end{aligned} \quad (12.61)$$

4. Expected Values, Moments, Correlation, Covariance, and Inequalities on Distributions

Let $g(X_1, \ldots, X_n)$ be a function of the random variables $\{X_i\}$ $(i = 1, \ldots, n)$ whose joint distribution is given by $F(x_1, \ldots, x_n)$. Then the *expected value* of $g(X_1, \ldots, X_n)$, denoted by the operator E, thus $E[g(X_1, \ldots, X_n)]$, is defined by

$$E[g(X_1, \ldots, X_n)] = \int_{-\infty}^{\infty} \cdots \int_{-\infty}^{\infty} g(x_1, \ldots, x_n)\,dF(x_1, \ldots, x_x) \quad (12.62)$$

and is said to "exist" only if the series or integral representing $E[g(X_1, \ldots, X_n)]$ is absolutely convergent.

Expected Values of Sums and Products. Let $X = (X_1, \ldots, X_n)$ and $Y = (Y_1, \ldots, Y_m)$ be vector quantities which are n- and m-dimensional random variables having the joint distribution function $F(x,y)$ where

$$x = (x_1, \ldots, x_n) \quad \text{and} \quad y = (y_1, \ldots, y_n);$$

also let $g_1(X)$ and $g_2(Y)$ be functions of X and Y and c_1 and c_2 be given constants. Then

$$E[c_1g_1(X) + c_2g_2(Y)] = c_1E[g_1(X)] + c_2E[g_2(Y)] \quad (12.63)$$

Furthermore, if $F(x,y) = F_1(x)F_2(y)$, where $F_1(x)$ and $F_2(y)$ are the respective marginal distributions, then X and Y are statistically independent and

$$E[g_1(X)g_2(Y)] = E[g_1(X)]E[g_2(Y)] \quad (12.64)$$

Moments. Let

$$g(X_1, \ldots, X_n) = X_1^{a_1}X_2^{a_2} \cdots X_n^{a_n}$$

then the *moments about the origin* associated with $F(x_1, \ldots, x_n)$ are defined by

$$\alpha_{a_1 \ldots a_n} = E[X_1^{a_1} \cdots X_n^{a_n}] \quad (12.65)$$

where $a_1 + \cdots + a_n$ is termed the order of the moment. Similarly, the *central moments,* obtained by replacing $X_i^{a_i}$ in (12.65) by $[X_i - E(X_i)]^{a_i}$, are defined by

$$\mu_{a_1 \ldots a_n} = E\{[X_1 - E(X_1)]^{a_1} \cdots [X_n - E(X_n)]^{a_n}\} \quad (12.66)$$

In applications the most important moments are those of orders one and two. In order to avoid notation difficulties, the moment of order one (usually termed *the mean*) will be denoted by $m_i = E[X_i]$; those of order two by

$$\begin{aligned} \sigma_{ij} &= E[(X_i - m_i)(X_j - m_j)] \quad \text{for } i \ne j \\ \sigma_{ii} &= \sigma_i^2 = E[(X_i - m_i)^2] \end{aligned} \quad (12.67)$$

where σ_{ij} $(i \ne j)$ is termed the *covariance between* X_i and X_j [sometimes written as cov (X_i,X_j)] and σ_i^2 is termed the *variance.* In many situations it is convenient to represent the covariance term σ_{ij} by

$$\sigma_{ij} = \rho_{ij}\sigma_i\sigma_j \quad (12.68)$$

where ρ_{ij} is called the *correlation between* X_i and X_j, and σ_i, σ_j are the *standard deviations* of X_i and X_j. Using Schwarz's inequality it can be shown that ρ_{ij} can only take on values between -1 and $+1$ inclusive.

If X_i and X_j are statistically independent, then $\sigma_{ij} = 0$; hence the correlation $\rho_{ij} = 0$. However, the converse is not true, that is, if $\rho_{ij} = 0$ it does not necessarily follow that X_i and X_j are independent. For example, if (X,Y) has uniform probability density over a circle centered at the origin, then X and Y are *not* independent; yet $\rho_{xy} = 0$. Only in the case of the bivariate normal distribution does $\rho_{xy} = 0$ imply independence.

The *rank* of a distribution is defined as the rank of the moment matrix

$$\Sigma = \|\sigma_{ij}\| \quad i,j = 1, \ldots, n$$

where Σ is always a positive semidefinite matrix. If the rank $r < n$, the distribution is termed *singular* and the entire mass of the distribution is confined to r independent linear functions composed of the n random variables $\{X_i\}$.

Expected Values for One-dimensional Random Variables. For the case where $n = 1$, the expected value operator is

$$E[g(X)] = \int_{-\infty}^{\infty} g(x)\,dF(x) \quad (12.69)$$

The moments about the origin and the central moments are given by

$$\alpha_i = \int_{-\infty}^{\infty} x^i\,dF(x) \quad (12.70)$$

$$\mu_i = \int_{-\infty}^{\infty} [x - E(x)]^i\,dF(x) \quad (12.71)$$

respectively. The relationships between the moments about the origin and the central moments are

$$\begin{aligned} \mu_0 &= 1 \\ \mu_1 &= 0 \\ \mu_2 &= \alpha_2 - m^2 \\ \mu_3 &= \alpha_3 - 3m\alpha_2 + 2m^3 \\ \mu_4 &= \alpha_4 - 4m\alpha_3 + 6m^2\alpha_2 - 3m^4 \end{aligned} \tag{12.72}$$

and in general

$$\mu_n = \sum_{j=0}^{n} \binom{n}{j} (-1)^{n-j}\alpha_j m^{n-j} \tag{12.73}$$

The moments will uniquely define a distribution function $F(x)$ if

$$M(t) = E[e^{tX}] = \sum_{n=0}^{\infty} \frac{\alpha_n}{n!} t^n \tag{12.74}$$

is absolutely convergent for any $t > 0$, in which case $M(t)$ is the *moment-generating function* of the distribution (see Sec. 6). In particular, if the random variable is bounded, this condition will always be satisfied.

Inequalities on Distributions Which Are Functions of Moments. Knowledge about the moments of a distribution enables one to find bounds on the distribution function without knowledge of the explicit functional form of the distribution. Below is a selection of such inequalities which are applicable under very general conditions. More complete surveys are given by Godwin [29], Savage [45], and Uspensky [17].

Markoff Inequality. Let X be a non-negative random variable such that $m = E(X)$ is finite; then for any number $\lambda > 0$

$$P\{X \geq \lambda\} \leq \frac{m}{\lambda} \tag{12.75}$$

Bienaymé-Chebyshev Inequality. Let X be a random variable with mean m and standard deviation σ. Then for every $\lambda > 0$

$$P\{|X - m| \geq \lambda\sigma\} \leq \frac{1}{\lambda^2} \tag{12.76}$$

If $X_1, \ldots, X_n$ is a sequence of uncorrelated random variables from a distribution with common mean m and standard deviation σ, then for every $\lambda > 0$

$$P\left\{|\bar{X} - m| \geq \frac{\lambda\sigma}{\sqrt{n}}\right\} \leq \frac{1}{\lambda^2} \tag{12.77}$$

where

$$\bar{X} = \frac{1}{n}\sum_{i=1}^{n} X_i$$

Gauss Inequality. Let X be a random variable having a unimodal distribution of the continuous type; then for every $\lambda > 0$

$$P\{|X - x_0| \geq \lambda\tau\} \leq \frac{4}{9\lambda^2} \tag{12.78}$$

where x_0 is the mode and

$$\tau^2 = E[X - x_0]^2 = \sigma^2 + (x_0 - m)^2$$

is the second-order moment about the mode.

5. Measures of Location, Dispersion, Skewness, and Kurtosis

In many applications it is necessary to describe a distribution function by a few simple parameters. Any distribution can roughly be characterized by some "central point" or "typical value" of the distribution, and the concentration of the distribution about this central point. Parameters which describe "central points" of distributions are called *measures of location;* parameters which describe the concentration of distributions about these central points are called *measures of dispersion.* (The smaller the dispersion, the greater the concentration of the distribution about the central point, and vice versa.) It is possible to construct infinitely many measures of location and dispersion, but the following are in most frequent use and refer to a one-dimensional random variable X having the distribution $F(x)$.

Measures of Location. *Mean.* The mean of a distribution is defined as

$$m = E(X) = \int_{-\infty}^{\infty} x\, dF(x) \tag{12.79}$$

and can be regarded as the "center of gravity" of the mass in the distribution.

Median. The median divides the mass of the distribution into two equal parts. Thus any root of the equation

$$F(x) = \tfrac{1}{2} \tag{12.80}$$

determines a median. Every continuous-type distribution has at least one median.

Mode. For a continuous-type distribution, a *mode* is defined as any local maximum point of the probability density function $f(x)$. If $f(x)$ has only one maximum point, then the distribution is termed *unimodal.* For discrete-type distributions having point probabilities $\{p_i\}$, the mode is defined as that value of the random variable, say $X = x_0$, for which

$$p_0 = \max\{p_i\} \tag{12.81}$$

Measures of Dispersion. *Variance and Standard Deviation.* The variance σ^2, which is the central moment of order two, μ_2, that is,

$$\sigma^2 = E(X - m)^2 = \int_{-\infty}^{+\infty} (x - m)^2\, dF(x) \tag{12.82}$$

can be interpreted as representing the *moment of inertia* of the unit mass distribution with respect to a perpendicular axis through the center of gravity. The second moment about any other point, say c, will always be greater than the variance by virtue of the fundamental identity

$$E(X - m)^2 = E(X - c)^2 - (c - m)^2 \tag{12.83}$$

The positive square root of the variance, σ, termed the *standard deviation* is also used as a measure of dispersion.

Mean Deviation about the Mean. The *mean (or average) deviation about the mean* is defined by

$$E|X - m| = \int_{-\infty}^{\infty} |x - m|\, dF(x)$$
$$= 2\int_{m}^{\infty} (x - m)\, dF(x) \quad (12.84)$$

which can be written in the alternative form

$$E|X - m| = 2\left[\int_{m}^{\infty} x\, dF(x) - mP\{X > m\}\right]$$

Mean Deviation about the Median. The mean deviation about the median (denoted by m_e) is defined by

$$E|X - m_e| = \int_{-\infty}^{\infty} |X - m_e|\, dF(x) \quad (12.85)$$

The mean deviation about any other point, say c, can never be less than the mean deviation about the median in view of the identity

$$E|x - m_e| = E|x - c| + (m_e - c)[P\{X < c\} - P\{X > c\}] \quad (12.86)$$

in which the two factors of the second term on the right are always of opposite sign for $c \neq m_e$.

Semi-interquartile Range. The semi-interquartile range is defined by

$$\frac{x_{3/4} - x_{1/4}}{2} \quad (12.87)$$

where $x_{1/4}$ and $x_{3/4}$ are the first and third quartiles, defined by

$$F(x_p) = \int_{-\infty}^{x_p} dF(x) = p$$

The mean and variance are the measures of location and dispersion most often used in applications. Tables 12.1 and 12.2 list these quantities for some of the more important distributions.

Mean and Variance of Linear Functions. In applications the mean and variance are the measures of location and dispersion used most often and it is often necessary to find means and variances of linear functions of random variables. For example, let $\{X_i\}$ $(i = 1, \ldots, n)$ be a sequence of random variables having means and variances m_i and σ_i^2, respectively; let the covariance between X_i and X_j be $\rho_{ij}\sigma_i\sigma_j$ and consider the linear function

$$L = \sum_{i=1}^{n} l_i X_i$$

Then

$$E[L] = \sum_{i=1}^{n} l_i m_i \quad (12.88)$$

and

$$\text{var}\,[L] = \sum_{i=1}^{n}\sum_{j=1}^{n} l_i l_j \rho_{ij}\sigma_i\sigma_j$$
$$= \sum_{i=1}^{n} l_i^2\sigma_i^2 + \sum_{i}\sum_{\substack{j \\ i\neq j}} l_i l_j \rho_{ij}\sigma_i\sigma_j \quad (12.89)$$

If the $\{X_i\}$ are mutually independent, $\rho_{ij} = 0$ and

$$\text{var}\,[L] = \sum_{i=1}^{n} l_i^2\sigma_i^2 \quad (12.90)$$

Let $l_i = 1/n$, then

$$L = \bar{X} = \frac{1}{n}\sum_{i=1}^{n} X_i$$

and if $\sigma_i^2 = \sigma^2$, the variance of the average can be written as

$$\text{var}\,[\bar{X}] = \frac{\sigma^2}{n}\{1 + (n - 1)\bar{\rho}\} \quad (12.91)$$

where $$\bar{\rho} = \frac{2}{n(n-1)}\sum_{i}\sum_{\substack{j \\ (i<j)}} \rho_{ij}$$

Note that if $\bar{\rho}$ is negative the variance is smaller, and when positive the variance larger, than would be the case if the random variables were uncorrelated.

Means and Variances of Nonlinear Functions. Often it is necessary to determine the mean and variance of a nonlinear function of several random variables, that is, $Z = \phi(X_1, \ldots, X_n)$. Exact solutions are often difficult to find. In many cases, however, approximate solutions can be obtained by use of a method sometimes called the *propagation of error* technique.

If the probability mass of the joint distribution of $(X_1, X_2, \ldots, X_n)$ is concentrated in a relatively narrow region about $(m_1, m_2, \ldots, m_n)$, where $m_i = E[X_i]$, and if the function

$$Z = \phi(X_1, X_2, \ldots, X_n)$$

can be represented with sufficient approximation by the linear terms of a Taylor series in the neighborhood of $(m_1, m_2, \ldots, m_n)$, then

$$E[Z] \simeq \phi(m_1, m_2, \ldots, m_n) \quad (12.92)$$

$$\text{Var}\, Z \simeq \sum_{i=1}^{n}\sum_{j=1}^{n} \frac{\partial\phi}{\partial x_i}\frac{\partial\phi}{\partial x_j}\rho_{ij}\sigma_i\sigma_j \quad (12.93)$$

where the partial derivatives are evaluated at the point $(m_1, \ldots, m_n)$, and ρ_{ij} and σ_i are the respective correlations and standard deviations. For the case where the random variables $\{X_i\}$ are mutually uncorrelated, then $\rho_{ij} \equiv 0$, and

$$\text{var}\, Z \simeq \sum_{i=1}^{n}\left(\frac{\partial\phi}{\partial x_i}\right)^2 \sigma_i^2 \quad (12.94)$$

For example, let $m = E(X_i)$, $\sigma^2 = \text{var}\, X_i$ and let

$$\bar{X} = \frac{1}{n}\sum_{i=1}^{n} X_i$$

then, by (12.92),

$$E[\bar{X}^2] \simeq m^2$$

However, we know from (12.83) and (12.91) that

$$E[\bar{X}^2] = m^2 + \frac{\sigma^2}{n}$$

exactly. The above approximate answer will be good only if n is relatively large compared to σ^2.

Measures of Skewness and Kurtosis. A distribution is said to be symmetric about the point a if $F(a + x) + F(a - x) = 1$ for all x for which $a + x$ and $a - x$ are continuity points of the distribution. Specifically, if the distribution is of the continuous type, then $f(a - x) = f(a + x)$.

A distribution which is not symmetric is called *skew*. If a unimodal skew distribution has a frequency function, the frequency function will have a "long" and "short" tail about each side of the mode. The distribution is said to be *positively* or *negatively skew* depending on whether the longer tail is to the right or left of the mode, respectively.

Since all odd central moments of a symmetric distribution are zero, any odd central moment (other than μ_1 which is identically zero) can be used as a measure of skewness, the simplest choice being the third central moment μ_3.

Since this contains units of dimension three, a *coefficient of skewness* devoid of units is defined by

$$\beta_1 = \left(\frac{\mu_3}{\sigma^3}\right)^2 \tag{12.95}$$

In a similar manner the fourth central moment can be taken as a measure of the degree of flattening or excess of a frequency function near the mode. A *coefficient of kurtosis* is defined by

$$\beta_2 = \frac{\mu_4}{\sigma^4} \tag{12.96}$$

and can be taken as a measure of the degree of peakedness of a frequency function near the mode.

If $\{X_i\}$, $(i = 1,2, \ldots ,n)$, are a sequence of independent identically distributed random variables with coefficients of skewness and kurtosis $\beta_1(X)$ and $\beta_2(X)$, then the β_1 and β_2 for their average,

$$\bar{X} = \frac{1}{n} \sum_{i=1}^{n} X_i$$

are

$$\beta_1(\bar{X}) = \frac{\beta_1(X)}{n}$$

and

$$\beta_2(\bar{X}) = 3 + \frac{\beta_2(X) - 3}{n} \tag{12.97}$$

respectively.

Sometimes skewness and kurtosis are expressed in terms of

$$\gamma_1 = \frac{K_3}{(K_2)^{3/2}} = \sqrt{\beta_1}$$
$$\gamma_2 = \frac{K_4}{K_2{}^2} = \beta_2 - 3 \tag{12.98}$$

where the K_r denotes the rth *cumulant* of the distribution as defined by Eq. (12.117), Sec. 6. Both γ_1 and γ_2 are zero for the normal distribution. Tables 12.1 and 12.2 list values of γ_1 and γ_2 for some of the more important distributions.

6. Characteristic Functions and Generating Functions

Characteristic Functions. Let X be a random variable having the distribution function $F(x)$. Then the expected value of the function $g(X) = e^{itX}$ $(i = \sqrt{-1})$ is called the *characteristic function* of the distribution corresponding to $F(x)$ and is written as

$$\phi(t) = E[e^{itX}] = \int_{-\infty}^{\infty} e^{itx}\, dF(x) \tag{12.99}$$

If the moment of order n exists, then the characteristic function may be differentiated n times with respect to t to obtain

$$\phi^{(k)}(0) = i^k\alpha_k \qquad k = 0,1, \ldots ,n \tag{12.100}$$

Conversely, if the nth derivative of $\phi(t)$ exists in a neighborhood of the origin, then all moments to order n exist and are given by (12.100).

Corresponding to the two main classes of distributions, Eq. (12.99) can be written as

$$\phi(t) = \int_{-\infty}^{\infty} e^{itx} f(x)\, dx \tag{12.101}$$

$$\text{and } \phi(t) = \sum_x p(x)e^{itx} \qquad \text{(summation is over all } x\text{)} \tag{12.102}$$

for the distributions of the continuous and discrete types, respectively. Table 12.3 contains a summary of characteristic functions for some of the more important distributions.

Uniqueness, Addition, and Continuity Theorems. The characteristic function plays a fundamental role in probability, mainly due to the following three theorems. Lukacs [35] develops the mathematical theory of characteristic functions, and Lukacs and Laha [36] discuss applications.

Uniqueness Theorem. A characteristic function uniquely determines a distribution function; furthermore if $x - h$ and $x + h$ are continuity points of the distribution function $F(x)$, then

$$F(x + h) - F(x - h) = \lim_{T\to\infty} \frac{1}{\pi} \int_{-T}^{T} \frac{\sin ht}{t} e^{-itx}\phi(t)\, dt \tag{12.103}$$

For distributions of the continuous type, the probability density function $f(x)$ is given by the Fourier transform of $\phi(t)$,

$$f(x) = \lim_{h\to 0} \frac{F(x + h) - F(x - h)}{2h} = \frac{1}{2\pi} \int_{-\infty}^{\infty} e^{-itx}\phi(t)\, dt \tag{12.104}$$

If the distribution is of the discrete type, then the point probability p_j at the point of increase x_j is given by

$$p_j = \frac{1}{2\pi} \int_{-\pi}^{\pi} e^{-itx_j}\phi(t)\, dt \tag{12.105}$$

TABLE 12.3. CHARACTERISTIC FUNCTIONS AND CUMULANTS FOR SELECTED DISTRIBUTIONS

Name	Characteristic function	Cumulants
Normal	$\exp(imt - \frac{1}{2}\sigma^2 t^2)$	$K_1 = m,\ K_2 = \sigma^2$ $K_n = 0 \quad \text{for } n > 2$
Cauchy	$\exp(i\lambda t - \beta\|t\|)$	Not defined
Extreme-value	$\Gamma(1 - i\beta t)\exp(i\lambda t)$	$K_1 = \gamma,^* \ K_2 = \frac{(\pi\beta)^2}{6}$ $K_n = \beta^n\Gamma(n)\sum_{r=1}^{\infty}\frac{1}{r^n} \quad n > 2$
Exponential	$\frac{\exp(i\lambda t)}{1 - it\beta}$	$K_1 = \lambda + \beta$ $K_n = \beta^n\Gamma(n) \quad n > 1$
Pearson Type III	$\frac{\exp(i\lambda t)}{(1 - it\beta)^p}$	$K_1 = \lambda + \beta p$ $K_n = \beta^n p\Gamma(n) \quad n > 1$
Rectangular, or uniform	$\exp i\theta t\left[\frac{\exp\left(i\frac{w}{2}t\right) - \exp\left(-i\frac{w}{2}t\right)}{iwt}\right]$	$K_1 = \theta,\ K_{2n+1} = 0$ $K_{2n} = \frac{(-1)^{n+1}w^{2n}B_n\dagger}{2^n}$
Causal, or unitary	$\exp i\lambda t$	$K_1 = \lambda,\ K_n = 0 \quad \text{for } n > 1$
Binomial, or Bernoulli	$(q + pe^{it})^n \quad q = 1 - p$	$K_1 = np$ $K_{n+1} = pq\frac{dK_n}{dp} \quad \text{for } n \geq 1$
Poisson	$\exp[m(e^{it} - 1)]$	$K_n = m \quad n = 1, 2, \ldots$
Geometric	$\frac{p}{1 - qe^{it}} \quad q = 1 - p$	$K_1 = \frac{q}{p}$ $K_{n+1} = q\frac{dK_n}{dq} \quad \text{for } n \geq 1$
Pascal	$\left(\frac{p}{1 - qe^{it}}\right)^r \quad q = 1 - p$	$K_1 = \frac{rq}{p}$ $K_{n+1} = q\frac{dK_n}{dq} \quad \text{for } n \geq 1$
Negative binomial	$(Q - Pe^{it})^{-r} \quad Q = 1 + P$	$K_1 = rP$ $K_{n+1} = PQ\frac{dK_n}{dQ} \quad \text{for } n \geq 1$

* γ is Euler's constant 0.5772 . . . .
† B_n are the Bernoulli numbers $B_1 = \frac{1}{6}$, $B_2 = \frac{1}{30}$, ...

Addition Theorem. Let X and Y be independent random variables with characteristic functions $\phi_x(t)$ and $\phi_y(t)$ corresponding to the distribution functions $F(x)$ and $G(y)$, respectively. Then the distribution function of their sum $X + Y$ has the characteristic function

$$\phi_{x+y}(t) = \phi_x(t)\phi_y(t) \tag{12.106}$$

The corresponding distribution of their sum

$$S = X + Y$$

is

$$H(s) = \int_{-\infty}^{\infty} F(s - t)\,dG(t) = \int_{-\infty}^{\infty} G(s - t)\,dF(t) \tag{12.107}$$

and if $F(x)$ and $G(y)$ are distributions of the continuous type, the probability density function of S is

$$h(s) = \int_{-\infty}^{\infty} f(s - t)g(t)\,dt = \int_{-\infty}^{\infty} g(s - t)f(t)\,dt \tag{12.108}$$

The integrals obtained in (12.107) or (12.108) define what is termed the *convolution* of F and G; and often (12.107) and (12.108) are written as

$$\begin{aligned} H &= F * G = G * F \\ h &= f * g = g * f \end{aligned} \tag{12.109}$$

respectively. These results may be generalized to any number of random variables.

Continuity Theorem. Let $F_1(x)$, $F_2(x)$, . . . , be a sequence of distribution functions and $\phi_1(t)$, $\phi_2(t)$, . . . , their respective characteristic functions. Then a necessary and sufficient condition for the sequence $\{F_n(x)\}$ to converge to a limiting distribution function $F(x)$ is that for every t, the sequence of characteristic functions $\{\phi_n(t)\}$ converge to a limit $\phi(t)$ which is continuous for $t = 0$; in which case $\phi(t)$ is the characteristic function of the limiting distribution function $F(x)$.

Necessary Conditions for Characteristic Functions. The necessary conditions which every characteristic function must satisfy are:

$$\begin{aligned} &(1) \quad \phi(t) \text{ continuous for all real } t \\ &(2) \quad \phi(0) = 1 \\ &(3) \quad |\phi(t)| \leq 1 \\ &(4) \quad \phi(-t) = \overline{\phi(t)} \end{aligned} \tag{12.110}$$

where $\overline{\phi(t)}$ denotes the complex conjugate of $\phi(t)$.

Necessary and sufficient conditions that a function $\phi(t)$ must satisfy to be a characteristic function are usually difficult to apply in practice; see, for example, Cramér [2, p. 91] and Lukacs [35, p. 61]. Polya has given an easily applied set of sufficient conditions; see Lukacs [35, p. 70].

Characteristic Function of a Function of a Random Variable. Let $Y = k(X)$ be a function of the random variable X. Then the characteristic function of Y is

$$\phi_y(t) = \int_{-\infty}^{\infty} e^{itk(x)}\, dF(x) \tag{12.111}$$

In particular if $Y = a + bX$, where a and b are given constants, we have

$$\phi_y(t) = \int_{-\infty}^{\infty} e^{it(a+bx)}\, dF(x) = e^{iat}\phi_x(bt) \tag{12.112}$$

Characteristic Functions for n-dimensional Random Variables. Let $X_1, X_2, \ldots, X_n$ be a sequence of random variables having the joint distribution $F(x_1, x_2, \ldots, x_n)$. Then the characteristic function associated with $F(x_1, x_2, \ldots, x_n)$ is defined by

$$\begin{aligned} \phi(t_1, t_2, \ldots, t_n) &= E\left[\exp\left(i \sum_{j=1}^{n} t_j X_j\right)\right] \\ &= \int_{-\infty}^{\infty} \int_{-\infty}^{\infty} \cdots \int_{-\infty}^{\infty} \exp\left(i \sum_{j=1}^{n} t_j x_j\right) dF\,(x_1, x_2, \ldots, x_n) \end{aligned} \tag{12.113}$$

If the variables are mutually independent, that is, $F(x_1, x_2, \ldots, x_n) = F_1(x_1)F_2(x_2) \cdots F_n(x_n)$, then (12.113) reduces to

$$\phi(t_1, t_2, \ldots, t_n) = \prod_{j=1}^{n} \phi_j(t_j) \tag{12.114}$$

where $\phi_j(t_j) = \int_{-\infty}^{\infty} \exp\,(it_j x_j)\, dF_j(x_j)$

and conversely, if

$$\phi(t_1, t_2, \ldots, t_n) = \prod_{j=1}^{n} \phi_j(t_j)$$

then the random variables $\{X_j\}(j = 1, 2, \ldots, n)$ are mutually independent. In many cases analogous properties hold for characteristic functions in n dimensions as for the univariate case.

Moment-generating Functions. If the integral

$$M(t) = \int_{-\infty}^{\infty} e^{tx}\, dF(x) \tag{12.115}$$

exists for some fixed neighborhood of $t = 0$, then $M(t)$ is said to define the *moment-generating function* of the distribution $F(x)$, and the kth moment about the origin is given by

$$M^{(k)}(0) = \alpha_k \tag{12.116}$$

The moment-generating function is obtained by replacing it by t in Eq. (12.99). Analogous to the uniqueness, addition, and continuity theorems for characteristic functions, there exist similar theorems for the moment-generating function. Characteristic functions are generally preferred over moment-generating functions in theoretical investigations because characteristic functions will always exist, whereas the moment-generating function may not.

Cumulants and Semi-invariants. Let $\phi(t)$ be the characteristic function associated with the distribution $F(x)$. Then the *cumulant-* or *semi-invariant-generating* function is defined by log $\phi(t)$ and can be written in the form

$$\log \phi(t) = \sum_{n=1}^{\infty} \frac{K_n}{n!} (it)^n \tag{12.117}$$

where K_n is termed the nth *cumulant* or *semi-invariant.*

The cumulants can be written as a polynomial function of the moments. In terms of the central moments, the expressions for the first four cumulants are

$$\begin{aligned} K_1 &= m \\ K_2 &= \sigma^2 \\ K_3 &= \mu_3 \\ K_4 &= \mu_4 - 3\mu_2{}^2 \end{aligned} \tag{12.118}$$

where m is the mean, σ^2 the variance, and μ_3, μ_4 the third and fourth central moments, respectively. Formulas for higher order cumulants expressed as polynomial functions of the moments can be found in Kendall and Stuart [10, p. 68].

Properties of Cumulants. If $\{K_n\}$ are the cumulants of the random variable X, then the cumulants of $Y = a + bX$ (denoted by $\{K'_n\}$) are

$$\begin{aligned} K'_1 &= a + bK_1 \\ K'_n &= b^n K_n \qquad \text{for } n > 1 \end{aligned}$$

Thus all cumulants except for the first are invariant with respect to a change of location under a linear transformation.

The most important property of cumulants is that they are completely additive with respect to linear functions of independent random variables. For example, consider the function

$$L = \sum_{i=1}^{m} C_i X_i$$

where $\{X_i\}$ are independent random variables having distribution functions $\{F_i(x)\}$, respectively, for $i = 1, \ldots, n$ and the C_i are given constants. Then if $K_n{}^{(i)}$ refers to the nth cumulant of X_i, the nth cumulant for the linear function L is

$$K_n = \sum_{i=1}^{m} C_i{}^n K_n{}^{(i)} \tag{12.119}$$

TABLE 12.4. PROBABILITY-GENERATING FUNCTIONS FOR SOME IMPORTANT DISCRETE DISTRIBUTIONS

Name	*Probability-generating Function*
Causal	t^{x_0}
Binomial, or Bernoulli	$(q + pt)^n \quad q = 1 - p$
Poisson	$\exp(-m + mt)$
Geometric	$\frac{p}{1 - qt} \quad q = 1 - p$
Pascal	$\left(\frac{p}{1 - qt}\right)^r \quad q = 1 - p$
Negative binomial	$(Q - Pt)^{-r} \quad Q = 1 + P$
Hypergeometric	$\frac{\binom{N - Np}{n}}{\binom{N}{n}} F(-n, -Np; N - Np - n + 1, t)$
Multinomial	$(p_1t_1 + p_2t_2 + \cdots + p_kt_k)^n$

Table 12.3 lists the cumulants for some of the important distribution functions.

Probability-generating Functions. Let X be a random variable having the distribution function $F(x)$. Then if $E[t^X]$ exists, it is called the *probability-generating function,* that is,

$$P(t) = E[t^X] = \int_{-\infty}^{\infty} t^x \, dF(x) \qquad (12.120)$$

The probability-generating function is of especial importance in the study of discrete random variables taking on only integral values $X = 0, 1, 2, \ldots$. Let

$$P\{X = j\} = p_j \qquad j = 0, 1, 2, \ldots \qquad (12.121)$$

then the probability-generating function is

$$P(t) = \sum_{j=0}^{\infty} p_j t^j \qquad (12.122)$$

which always converges absolutely for $|t| \leq 1$. Thus

$$P(1) = \sum_{j=0}^{\infty} p_j = 1$$

$$P'(1) = \sum_{j=0}^{\infty} jp_j = E[X] \qquad (12.123)$$

$$P''(1) + P'(1) - P'^2(1) = \text{Var}[X]$$

and in general

$$P^{(\nu)}(1) = E[X(X - 1) \cdots (X - \nu + 1)] \qquad (12.124)$$

Table 12.4 lists the probability-generating functions of some discrete distributions of the type (12.121). See Feller [3] for a very complete discussion of probability-generating functions and their applications.

Convolutions. Let X and Y be non-negative independent random variables such that

$$P\{X = j\} = a_j \quad \text{and}$$
$$P\{Y = j\} = b_j \quad \text{for } j = 0, 1, 2, \ldots \qquad (12.125)$$

with associated probability-generating functions $A(t)$ and $B(t)$, respectively. Then $S = X + Y$ is a random variable such that

$$P\{S = s\} = a_0b_s + a_1b_{s-1} + \cdots + a_sb_0 = c_s \qquad (12.126)$$

and the generating function of S is

$$C(t) = A(t)B(t) \qquad (12.127)$$

The sequence of probabilities $\{c_s\}$ is usually written as the convolution of $\{a_s\}$ and $\{b_s\}$, that is,

$$\{c_s\} = \{a_s\} * \{b_s\} \qquad (12.128)$$

Compound Distributions. Let $\{X_k\}$ be a sequence of independent identically distributed random variables such that $P\{X_k = j\} = p_j$ for $j = 0, 1, \ldots$. Consider the random variable

$$S_N = \sum_{k=1}^{N} X_k \qquad (12.129)$$

where N, the number of terms, is a random variable such that $P\{N = n\} = \pi_n$; then

$$P\{S_N = j\} = \sum_{n=1}^{\infty} P\{N = n\} \cdot P\{X_1 + X_2 + \cdots + X_n = j\} \qquad (12.130)$$

Distributions of this type are called *compound distributions.*

If $\pi(t)$ and $P(t)$ are the generating functions of $\{N\}$ and $\{X_k\}$, respectively, then the generating function of the compound distribution (12.130) is $\pi[P(t)]$.

Example. Let X_k be a random variable which can assume only the values 1 or 0 with probabilities p and q; and let N follow the Poisson distribution with parameter m. Then the generating functions of X_k and N are $P(t) = (q + pt)$ and $\pi(t) = e^{-m+mt}$, respectively. Hence the generating function of the compound distribution is

$$\pi[P(t)] = \exp[-m + m(q + pt)] = \exp[-mp + mpt] \qquad (12.131)$$

which is the generating function of a Poisson distribution with parameter (mp).

7. Limit Theorems

This section contains some of the more important limit theorems relating to sums of random variables. More complete discussions can be found in Feller [3, 26], Fisz [4], Gnedenko [5], Loève [11], and Gnedenko and Kolmogorov [28].

Let $\{X_i\}$ be an unending sequence of independent random variables with means $E(X_i)$ and variances $\sigma^2(X_i)$, and let

$$\bar{X}_n = \frac{1}{n} \sum_{i=1}^{n} X_i$$

$$m_n = \frac{1}{n} \sum_{i=1}^{n} E(X_i) \qquad (12.132)$$

$$\sigma_n^2 = \frac{1}{n} \sum_{i=1}^{n} \sigma^2(X_i)$$

Weak Law of Large Numbers. The *(weak) law of large numbers* is said to hold for the sequence of random variables $\{X_i\}$ if for any $\epsilon > 0$,

$$\lim_{n \to \infty} P\{|\bar{X}_n - m_n| < \epsilon\} = 1 \qquad (12.133)$$

If such is the case, this means that the probability of the average $\bar{X}_n$ differing from the expectation m_n by more than an arbitrary number ϵ tends to zero as $n \to \infty$, and $(\bar{X}_n - m_n)$ is said to *converge in probability* to zero.

Two cases where the (weak) law of large numbers holds are as follows:

1. *Khinchin's theorem* (identically distributed variables). If $\{X_i\}$ is a sequence of mutually independent identically distributed random variables, then for the (weak) law of large numbers to hold it is necessary and sufficient that the mean of their common distribution, $m = E[X_i]$, be finite.

2. *Chebyshev's theorem* (nonidentically distributed variables). If $\{X_i\}$ is a sequence of uncorrelated random variables with finite means and finite variances, then a sufficient condition for the (weak) law of large numbers to hold is that

$$\lim_{n \to \infty} \frac{\sigma_n^2}{n} = 0 \qquad (12.134)$$

In particular, both (1) and (2) will be satisfied if the random variables are bounded, $|X_i| < k < \infty$.

The (weak) law of large numbers asserts that for any particular value of n that is sufficiently large, say $n = n_0$, the deviation $|\bar{X}_n - m_n|$ is almost certain to be small. It does not imply, however, that $|\bar{X}_n - m_n|$ remains small continually for $n = n_0, n_0 + 1, \ldots$; it can happen that $|\bar{X}_n - m_n|$ is large for some n's $> n_0$, but the probability of this happening for any particular $n > n_0$ is small.

Strong Law of Large Numbers. A stronger statement about the convergence of the sample mean to the population mean is given by the *strong law of large numbers* which states that

$$P\{\lim_{n \to \infty} |\bar{X}_n - m_n| = 0\} = 1 \qquad (12.135)$$

The strong law asserts that for sufficiently large n, $|\bar{X}_n - m_n|$ will with "overwhelming probability" become and remain small, and, indeed, tend to zero, as n increases. The strong law is equivalent to *convergence almost everywhere.*

Two conditions for the strong law of large numbers to hold (first given by Kolmogorov) are given below.

1. *Identically distributed variables.* The strong law of large numbers holds if $\{X_i\}$ is a sequence of mutually independent identically distributed random variables, such that the mean $E[X_i]$ of their common distribution is finite. This is both a necessary and a sufficient condition.

2. *Nonidentically distributed variables.* If the sequence of mutually independent random variables $\{X_i\}$ is such that $E[X_i]$ and $\sigma^2(X_i)$ are both finite, then a sufficient condition for the strong law of large numbers to hold is that the series

$$\sum_{i=1}^{n} \frac{\sigma^2(X_i)}{i^2}$$

converge as $n \to \infty$.

Central-limit Theorem. The normal distribution plays a fundamental role in the theory of statistics primarily because of the *central limit theorem.* A sequence of random variables $\{X_i\}$ is said to obey the central-limit theorem if for every $a \leq b$

$$\lim_{n \to \infty} P\left\{a \leq \frac{(\bar{X}_n - m_n)\sqrt{n}}{\sigma_n} \leq b\right\} = F(b) - F(a) \qquad (12.136)$$

where $$F(x) = \frac{1}{\sqrt{2\pi}} \int_{-\infty}^{x} e^{-t^2/2}\, dt$$

This simply means that the probability that

$$\frac{(\bar{X}_n - m_n)\sqrt{n}}{\sigma_n}$$

exceeds a but does not exceed b can be approximated by the right-hand side of (12.136) as closely as one desires by taking n sufficiently large; $\bar{X}_n$ is then said to be *asymptotically normally distributed* with parameters m_n and $\sqrt{\sigma_n^2/n}$, or alternatively

$$\frac{(\bar{X}_n - m_n)\sqrt{n}}{\sigma_n}$$

is said to be asymptotically normally distributed with parameters 0 and 1. This does not *necessarily* imply that the mean and variance of $\bar{X}_n$ converge to m_n and σ_n^2/n, respectively, nor that the mean and variance of

$$\frac{(\bar{X}_n - m_n)\sqrt{n}}{\sigma_n}$$

converge to 0 and 1, respectively.

Two different conditions under which the central-limit theorem holds are:

1. The *Lundberg-Lévy theorem* (identically distributed variables). If $\{X_i\}$ is a sequence of independent random variables with identical probability distributions and if $E[X_i]$ and $\sigma^2(X_i)$ are both finite, then the central-limit theorem holds. Thus $\bar{X}_n$ is asymptotically normally distributed with parameters $m = E[X_i]$ and $\sqrt{\sigma^2(X_i)/n}$.

2. The *Liapounoff theorem* (nonidentically distributed variables).‡ If $\{X_i\}$ is a sequence of independent random variables for which the means $E(X_i)$, variances $\sigma^2(X_i)$, and absolute third moments $\tau_i^3 = E[|X_i - E[X_i]|^3]$ all are finite, and if

$$\lim_{n\to\infty} \frac{\sum_{i=1}^{n} \tau_i}{\sqrt{n}\sigma_n} = 0 \tag{12.137}$$

then the central-limit theorem holds.

Both (1) and (2) will be satisfied if the random variables are uniformly bounded, $|X_i| < k < \infty$.

Convergence of Random Variables.

Definition 1. The sequence of random variables $\{X_n\}$ is said to *converge in probability* (*converge in measure*) to the random variable X, if for every $\epsilon > 0$,

$$\lim_{n\to\infty} P\{|X_n - X| < \epsilon\} = 1 \tag{12.138}$$

Definition 2. The sequence of random variables $\{X_n\}$ is said to be *convergent in probability* (*convergent in measure*), if for every $\epsilon > 0$

$$\lim_{\substack{m\to\infty\\ n\to\infty}} P\{|X_m - X_n| < \epsilon\} = 1 \tag{12.139}$$

Definition 3. The sequence of random variables $\{X_n\}$ is said to *converge with probability one* (*converge almost everywhere*) to the random variable X if

$$P\{\lim_{n\to\infty} |X_n - X| = 0\} = 1 \tag{12.140}$$

Definition 4. The sequence of random variables $\{X_n\}$ is said to be *convergent with probability one* (*convergent almost everywhere*) if

$$P\{\lim_{\substack{m\to\infty\\ n\to\infty}} |X_m - X_n| = 0\} = 1 \tag{12.141}$$

Some of the important properties with respect to the above definitions of convergence are:

1. If $\{X_n\}$ converges $\left\{\begin{smallmatrix}\text{i.m.}\\ \text{a.e.}\end{smallmatrix}\right\}$ to X, then $\{X_n\}$ is convergent $\left[\begin{smallmatrix}\text{i.m.}\\ \text{a.e.}\end{smallmatrix}\right]$, that is, convergence in measure to X implies $\{X_n\}$ is convergent in measure, convergence almost everywhere to X implies $\{X_n\}$ is convergent almost everywhere.

2. If $\{X_n\}$ converges $\left[\begin{smallmatrix}\text{i.m.}\\ \text{a.e.}\end{smallmatrix}\right]$ to X and also to Y, then $X = Y$ a.e.

3. If $\{X_n\}$ converges a.e. to X, then $\{X_n\}$ also converges in measure to X.

4. If $\{X_n\}$ is convergent in measure, then there exists a sub-sequence $\{X_{n_i}\}$ and a quantity X such that the sub-sequence $\{X_{n_i}\}$ converges to X almost everywhere.

5. If $\{X_n\}$ is convergent $\left[\begin{smallmatrix}\text{i.m.}\\ \text{a.e.}\end{smallmatrix}\right]$, then there exists a quantity X such that $\{X_n\}$ converges to X $\left[\begin{smallmatrix}\text{i.m.}\\ \text{a.e.}\end{smallmatrix}\right]$.

The following two theorems, dealing with random variables which converge in probability to constants, may be useful in applications, see Cramér [2, p. 254].

Theorem 1. Let $\{X_i\}$ be a sequence of random variables with distribution functions $\{F_i(x)\}$ such that $F_i(x)$ tends to a distribution function $F(x)$ as $i \to \infty$. Furthermore, let $\{Y_i\}$ be another sequence of random variables which converges in probability to a constant c. Then in the limit as $i \to \infty$, the distribution functions of

$$X_i + Y_i, \qquad X_iY_i, \qquad \frac{X_i}{Y_i}$$

are $F(x - c)$, $F(x/c)$, and $F(cx)$ respectively, for $c > 0$. [When $c < 0$, the latter two limiting distributions become $1 - F(x/c)$ and $1 - F(cx)$, respectively.]

Theorem 2. Let $\{U_i\}$, $\{V_i\}$, . . . , $\{Z_i\}$ be sequences of random variables which converge in probability to $u,v, \ldots ,z$, respectively. Then any rational function of the random variables,

$$R(U_i, V_i, \ldots, Z_i)$$

converges in probability to the constant

$$R(u,v, \ldots ,z)$$

provided that the latter quantity is finite.

8. The Normal Distribution

The normal distribution is the most important continuous distribution in mathematical statistics and the theory of probability. Its importance derives principally from the central-limit theorem considered in Sec. 7, which says that under very general conditions the distribution of the arithmetic mean (and hence also of the sum when properly "reduced") of a large number of independent random variables can be represented to a good approximation by a normal distribution. The conditions are essentially that the contributions of the respective variables to the dispersion of this sum be individually negligible and that the probability of any single one of the variables making a large contribution to the sum be small. In practice, measurement data taken under conditions of statistical equilibrium often appear to be approximately normally distributed: if not in the original units, then when transformed by some simple (and intuitively sensible) transformation to other units (e.g., logarithms) in which their random variation may more reasonably be supposed to be the summation of a host of individually negligible effects.

A random variable X distributed between $-\infty$ and $+\infty$ with probability density function

$$\frac{1}{\sigma\sqrt{2\pi}} \exp\left[-\frac{1}{2}\left(\frac{x - m}{\sigma}\right)^2\right] \qquad -\infty < x < +\infty \tag{12.142}$$

is said to be *normally distributed* (or, to have a *normal* or *Gaussian distribution*) *with parameters m and σ*. For brevity, one may say *X is normal* (m,σ).

‡ This is actually a special case of the Liapounoff theorem.

A normal distribution with parameters m and σ is symmetric with respect to its single point of maximum probability density, $x = m$, which is, therefore, both the mode and the median of the distribution. The point $x = m$ is also the mean, but this is not a necessary consequence of symmetry or unimodality (see Sec. 5).

A change in the value of m, the parameter of location, results only in a displacement of the distribution, without modifying its shape. On the other hand, a change in the value of the scale parameter σ changes the shape of the distribution, both in "height" and "breadth," without otherwise affecting its location.

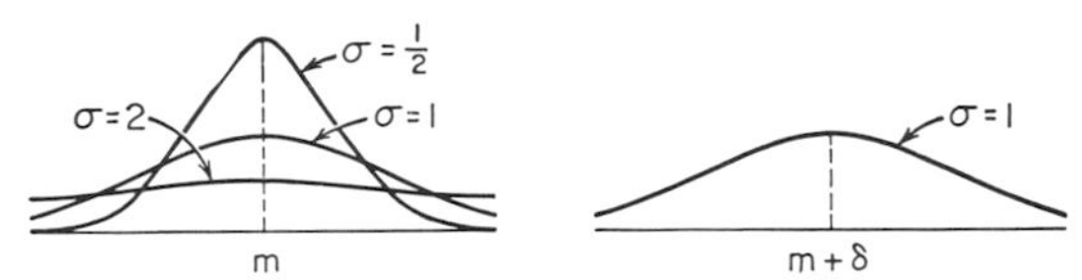

FIG. 12.1. Normal distributions with different parameter values.

The characteristic function of the normal distribution is

$$\phi(t) = \exp(imt - \tfrac{1}{2}\sigma^2 t^2) \qquad (12.143)$$

so that

$$E[X] = m \qquad (12.144)$$

as already noted, and

$$\sigma^2[X] = \sigma^2 \qquad (12.145)$$

The symbols m and σ thus have their usual significance, i.e., the mean and standard deviation, respectively.

All odd central moments of the normal distribution are zero. The central moments of even order are given by

$$\mu_{2r} = \frac{(2r)!}{2^r r!}\sigma^{2r} \qquad r = 0, 1, \ldots \qquad (12.146)$$

and the cumulants by

$$K_1 = m \qquad K_2 = \sigma^2 \qquad (12.147)$$
$$K_r = 0 \qquad \text{for all } r > 2$$

Standardized Normal Distribution. If X is normally distributed with parameters m and σ, then $Y = (X - m)/\sigma$, termed a *standardized normal deviate*, is normally distributed with parameters 0 and 1, and

$$P\{x_1 < X \le x_2\} = P\{y_1 < Y \le y_2\} = \frac{1}{\sqrt{2\pi}}\int_{y_1}^{y_2} \exp(-\tfrac{1}{2}t^2)\,dt \qquad (12.148)$$

for $y_i = (x_i - m)/\sigma$, $(i = 1,2)$. Consequently, the normal distribution need not be tabulated for various combinations of m and σ; it is sufficient to provide tables for the special case of $m = 0$ and $\sigma = 1$, termed the *standardized normal distribution*.

Tables of the Normal Distribution. Discussion of tables of the standardized normal distribution is greatly simplified by the following notation: Let

$$Z(x) = \frac{1}{\sqrt{2\pi}} \exp(-\tfrac{1}{2}x^2)$$
$$P(x) = \int_{-\infty}^{x} Z(t)\,dt = 1 - Q(x) = \tfrac{1}{2}[1 + A(x)] \qquad (12.149)$$
$$Q(x) = \int_{x}^{\infty} Z(t)\,dt = P(-x) = \tfrac{1}{2}[1 - A(x)]$$
$$A(x) = \int_{-x}^{+x} Z(t)\,dt = 2P(x) - 1 = 1 - 2Q(x)$$

Brief tables of $Z(x)$ and $P(X)$ are given in many textbooks and handbooks (for example, [1–5; 10; 12]); some (for example, [54]) give $Z(x)$ and $\tfrac{1}{2}A(x)$; others [8, 15, 17] only $P(x)$ or [18] only $Q(x)$ or [13] $\tfrac{1}{2}A(x)$. The principal extensive tabulations are [58–65]. For references to additional *direct tables*, to *inverse tables* (giving abscissas corresponding to various areas), and to tables of derivatives, see Greenwood and Hartley [52] and the National Bureau of Standards "Guide" [55]. For special methods of direct and inverse interpolation in tables of $P(x)$, $Q(x)$, or $A(x)$, see [55, pt. III]. For power series, asymptotic, and continued fraction expansions, polynomial and rational approximations, and relations of $Z(x)$ and $P(x)$ to other functions, see Zelen and Severo [49, pp. 932–936].

Addition of Independent Normal Variates. Let $\{X_i\}$ $(i = 1, \ldots, n)$ be a sequence of mutually independent normally distributed random variables with parameters (m_i,σ_i); then their weighted sum

$$L = \sum_{i=1}^{n} a_i X_i \qquad (12.150)$$

where the a_i are real and finite and at least one a_i is different from zero, follows a normal distribution with mean and variance given by

$$E[L] = \sum_{i=1}^{n} a_i m_i \qquad (12.151)$$

$$\sigma^2[L] = \sum_{i=1}^{n} a_i^2 \sigma_i^2 \qquad (12.152)$$

In particular, if the X_i are normal and mutually independent, then their sum is also normal (with parameters Σm_i and $\sqrt{\Sigma\sigma_i^2}$). Since this is true whatever the values of the individual m_i and σ_i, the normal distribution is said to be *completely reproductive* (with respect to addition).

The converse to the above general proposition is also true: if L, a linear function of a finite number of independent random variables, is normally distributed, then each component variable X_i is itself normally distributed. (The restriction to a "finite number" is necessary to avoid contradiction by the central-limit theorem.)

Distribution of the Arithmetic Mean. For $a_i = 1/n$, the "sum" (12.150) is simply the arithmetic mean

$$\bar{X} = \frac{1}{n} \sum_{i=1}^{n} X_i \tag{12.153}$$

of the normal random variates X_i. In particular, if the X_i are independent, normal, and identically distributed (that is, $m_i = m$ and $\sigma_i = \sigma$), then $\bar{X}$ is also normal, with parameters m and $\sigma/\sqrt{n}$; and, furthermore, as Daly [23] has shown, $\bar{X}$ and $g(X_1,X_2, \ldots ,X_n)$ are independently distributed when $g(X_1,X_2, \ldots ,X_n) = g(X_1 + \delta,X_2 + \delta, \ldots ,X_n + \delta)$ is any function of the X's [like $\Sigma(X_i - \bar{X})^2$, or $\Sigma|X_i - \bar{X}|$, or their range, $\max(X_i - X_j)$] that is invariant under translation. This is a unique property of the normal distribution. The direct converse is not true.

Distributions of Squares, Products, and Quotients. If X is normally distributed with parameters m and σ, then $[(X - m)/\sigma]^2$ has a *chi-square distribution* (12.239), with one degree of freedom; $(X/\sigma)^2$ has a *noncentral chi-square distribution* (12.244), with one degree of freedom and noncentrality $(m/\sigma)^2$; and $(X - m)^2$ has a *gamma distribution* (Table 12.2) with parameters $\beta = 2\sigma^2$ and $p = \frac{1}{2}$.

If X and Y are independently and normally distributed with means m_x and m_y and variances σ_x^2 and σ_y^2, respectively, then the probability density function of the distribution of their product, $W = XY$, has been expressed by C. C. Craig [21] in a convergent series involving Bessel functions of the first kind with purely imaginary argument, which reduces when $m_x = m_y = 0$ to

$$f(w) = \frac{\sigma_x\sigma_y}{\pi} K_0\left(\frac{w}{\sigma_x\sigma_y}\right) \tag{12.154}$$

where $K_0(u)$ is the Bessel function of the second kind with a purely imaginary argument of zero order.

If X and Y are independent and normal (m_x,σ_x) and (m_y,σ_y), respectively, Marsaglia [38] has shown that, for $m_x \geq 0$ and $m_y \geq 0$, the distribution of the quotient $Z = Y/X$ has the probability density function

$$h(z) = \frac{\exp\left[-\frac{1}{2}(a^2 + b^2)\right]}{\pi(1 + z^2)} \left[1 + \frac{q}{\phi(q)} \int_0^q \phi(t)\, dt\right] \tag{12.155}$$

where $q = (b + az)/\sqrt{1 + z^2}$, $a = m_y/\sigma_y$, $b = m_x/\sigma_x$, and $\phi(t)$ is the standardized normal density, that is, $Z(t)$ of (12.149). For $m_y < \sigma_y$, $h(z)$ is always unimodal; for $m_y > 2.257\sigma_y$, $h(z)$ is always bimodal. Marsaglia [38] presents graphs of $h(z)$ for a great variety of values of a and b. When $m_x = m_y = 0$, $h(z)$ reduces to

$$h(z) = \frac{\sigma_x\sigma_y}{\pi} \frac{1}{\sigma_y^2 + \sigma_x^2 z^2} \qquad -\infty < z < +\infty \tag{12.156}$$

That is, Z has a *Cauchy distribution* (Table 12.2) with $\lambda = 0$ and $\beta = \sigma_y/\sigma_x$. Fieller [27] gives a number of expressions for the distribution of Z in the general case, and from these deduces Geary's result (1930) that

$$\frac{m_x Z - m_y}{(\sigma_x^2 Z^2 + \sigma_y^2)^{1/2}} \tag{12.157}$$

is *approximately* normal (0,1) for m_x/σ_x large (>5, say).

Bivariate Normal Distribution. Two unbounded random variables X and Y whose joint distribution is defined by a probability density function of the form (12.47) are said to have a *bivariate normal distribution with parameters m_X, m_Y, σ_X, σ_Y, and ρ.*

The characteristic function of the distribution is

$$\phi(t_1,t_2) = \exp\{i(m_X t_1 + m_Y t_2) - \tfrac{1}{2}(\sigma_X^2 t_1^2 + 2\rho\sigma_X\sigma_Y t_1 t_2 + \sigma_Y^2 t_2^2)\} \tag{12.158}$$

so that

$$E[X] = K_{10} = m_X \qquad E[Y] = K_{01} = m_Y \tag{12.159}$$

$$\sigma^2[X] = K_{20} = \sigma_X^2 \qquad \sigma^2[Y] = K_{02} = \sigma_Y^2 \qquad \text{cov}\,[X,Y] = K_{11} = \rho\sigma_X\sigma_Y \tag{12.160}$$

and the higher-order cumulants are given by

$$K_{rs} = 0 \qquad \text{for all } r, s > 2 \tag{12.161}$$

The symbols m_X, m_Y, σ_X, σ_Y, and ρ thus have their usual meanings, that is, are the means and standard deviations of X and Y, and their coefficient of correlation [see (12.68)].

If X and Y jointly have a bivariate normal distribution with parameters m_X, m_Y, σ_X, σ_Y, and ρ, then the *marginal distributions* [see Sec. 3, especially Eq. (12.48)] of X and Y are normal (m_X,σ_X) and (m_Y,σ_Y), respectively; the *conditional distribution of Y for $X = x$* [see Eq. (12.49)] is normal with parameters

$$m_{Y|x} = E[Y|X = x] = m_Y + \rho\frac{\sigma_Y}{\sigma_X}(x - m_X) \tag{12.162}$$

and $$\sigma_{Y|x} = \sigma[Y|X = x] = \sigma_Y\sqrt{1 - \rho^2} \tag{12.163}$$

and the *conditional distribution of X for $Y = y$* is normal with parameters $m_{X|y}$ and $\sigma_{X|y}$, the values of which are given by (12.162) and (12.163) with X and Y interchanged and x replaced by y.

When X and Y jointly have a bivariate normal distribution above, the quadratic form

$$Q = \frac{1}{1 - \rho^2}\left[\frac{(X - m_x^2)}{\sigma_x^2} - \frac{2\rho(X - m_x)(Y - m_y)}{\sigma_x\sigma_y} + \frac{(Y - m_y)^2}{\sigma_y^2}\right]$$

has a *chi-square distribution* (12.239), with 2 degrees of freedom, and the integral of the *bivariate normal probability density function* (12.47) over the interior of the ellipse $Q = a^2$ is $1 - \exp(-\frac{1}{2}a^2)$.

When $\rho = 0$ and $\sigma_x = \sigma_y = \sigma$, the random variables X, Y are said to have a *circular normal distribution*, with center (m_x,m_y) and scale parameter σ. Its integral over the interior of a concentric circle of radius r is $1 - \exp[-\frac{1}{2}(r^2/\sigma^2)]$, which when set equal to $\frac{1}{2}$ yields $r = 1.1774\sigma$ as the *circular* or *radial probable error*. The quadratic form

$$Q = \frac{[(X - m_x)^2 + (Y - m_y)^2]}{\sigma^2} = \frac{R^2}{\sigma^2}$$

has a chi-square distribution (12.239), with 2 degrees of freedom, so that $E[R^2] = 2\sigma^2$; R, the length of the radius vector from the center (m_x, m_y) to (X,Y), has a *Rayleigh distribution,* with mean

$$E[R] = \sigma \sqrt{\pi/2} = 1.2533\sigma$$

Tables of the Bivariate Normal Distribution Function. The principal tables for evaluating bivariate normal probabilities are [66–70]. They give values of

$$V(h,k) = \int_0^h dx \int_0^{kx/h} g(x,y,0)\, dy$$

and $$L(h,k,\rho) = \int_h^\infty dx \int_k^\infty g(x,y,\rho)\, dy$$

where

$$g(x,y,\rho) = \frac{1}{2\pi\sqrt{1-\rho^2}} \exp\left[\frac{-(x^2+y^2-2\rho xy)}{2(1-\rho^2)}\right]$$

and of $$T(h,a) = \frac{1}{2\pi} \arctan a - V(h,ah)$$

Owen and Wiesen [71] and Zelen and Severo [72; 49, p. 937] show how to express $L(h,k,\rho)$ in terms of $L(h,0,\rho')$ and $L(h,0,\rho'')$, where ρ' and ρ'' are explicit functions of h, k, and ρ. Owen and Wiesen [71] give charts plotting $L(h,0,\rho)$ versus h with constant contours for ρ. Zelen and Severo [72] give contours of $L(h,0,\rho)$ on h,ρ grids, reprinted in [49, pp. 937–939], from which values of $L(h,0,\rho)$ can be read directly to within $\pm.01$ for $0 \leq h \leq 2.5$, $-1 \leq \rho \leq +1$. Values for $h < 0$ can be obtained from

$$L(-h,0,\rho) = \tfrac{1}{2} - L(h,0,-\rho)$$

If the integral of $g(x,y,\rho)$ is required for a section of the xy plane bounded by any two lines, then a suitable linear transformation will reduce this integral to $L(h,k,\rho')$, with h, k, and ρ' determined by the equations of the particular lines concerned. By a combination of such integrals, the region of integration can be extended to any region bounded by lines. Alternatively, the integral of $g(x,y,\rho)$ over any polygonal region can be explained in terms of $T(h,a)$ values, as explained by Owen [68], or in terms of linear combinations of $V(h,k)$ values and the univariate normal integrals $P(x)$ and $A(x)$ of (12.149), as illustrated by examples in Owen's introduction to the National Bureau of Standards Tables [66, sec. 2.3]. Both of these techniques require a preliminary linear circularizing transformation to new variables $U = U(X,Y)$ and $V(X,Y)$ having a joint circular normal distribution. Some general transformations for this purpose are given in [49, sec. 26.3.22; 66, sec. 2.3; 67, pp. 61–62].

Tables exist for evaluating bivariate normal probabilities corresponding to offset circles, offset ellipses, and various other regions defined by quadratic forms; see Greenwood and Hartley [52] and Owen [56].

Distributions of Products and Quotients. If X and Y have a bivariate normal distribution with means m_x and m_y, variances σ_x^2 and σ_y^2, and correlation ρ, the distribution of their product $W = XY$ has been expressed by Craig [21] as a convergent series involving sums of Bessel functions.

Marsaglia [38] has pointed out that it is always possible to determine two constants C_1 and C_2 such that the linear function $(C_1 + C_2 Z)$ of their quotient $Z = Y/X$ has a distribution of the same form as does Z in the uncorrelated case ($\rho = 0$) considered above. When $m_x = m_y = 0$, Z has the probability density function

$$h(z) = \frac{\sigma_x\sigma_y\sqrt{1-\rho^2}}{\pi} \frac{1}{\sigma_y^2(1-\rho^2) + \sigma_x^2\left(z - \rho\dfrac{\sigma_y}{\sigma_x}\right)^2} \qquad (12.164)$$

a *Cauchy distribution* with mode at the point

$$z = \rho\frac{\sigma_y}{\sigma_x}$$

Corresponding to (12.157) is Geary's result that

$$\frac{m_x Z - m_y}{(\sigma_y^2 - 2\rho\sigma_x\sigma_y Z + \sigma_x^2 Z^2)^{1/2}} \qquad (12.165)$$

is *approximately* normally distributed for $m_x/\sigma_x > 5$, say.

9. Discrete Distributions

Indicator of an Event. Let E be an event (denoted as a "success") which occurs with probability p, and X a random variable that takes on the values 1 or 0, according as E does or does not occur, that is,

$$P\{X = 1\} = p \qquad P\{X = 0\} = 1 - p = q \qquad (12.166)$$

The random variable X is called the *indicator* of the event E. The corresponding probability-generating function [see (12.122)] is

$$P(t) = q \cdot t^0 + p \cdot t^1 = q + pt = 1 + p(t-1) \qquad (12.167)$$

The characteristic function of the distribution is

$$\phi(t) = (q + pe^{it}) = 1 + (e^{it} - 1)p \qquad (12.168)$$

and the distribution function is given by

$$F(x) = q\epsilon(x) + p\epsilon(x-1) \qquad (12.169)$$

where $\epsilon(z)$ is the causal distribution function [see (12.17)].

The mean and variance of X are

$$E[X] = p \qquad \sigma^2[X] = pq \qquad (12.170)$$

Bernoulli Trials and the Binomial Distribution. A sequence of independent trials such that the probability of a "success" is the same for each trial is termed a series of *Bernoulli trials.* Let X_i be the indicator for the ith trial and consider the random variable

$$S = \sum_{i=1}^{n} X_i \qquad (2.171)$$

which is the number of "successes" in a series of n repeat trials.

The probability of obtaining exactly s "successes"

in n independent trials is given by the probabilities of the *binomial distribution*

$$b(s;n,p) = P\{S = s\} = \binom{n}{s} p^s q^{n-s}, \qquad q = 1 - p \tag{12.172}$$

The distribution derives its name from the fact that (12.172) is the general term in the binomial expansion of $(q + p)^n$.

Properties. The corresponding characteristic function, mean, and central moments to order four are

$$\phi(t) = (q + pe^{it})^n \tag{12.173}$$

$$E[S] = np \qquad \sigma^2[S] = npq \tag{12.174}$$

$$\mu_3 = npq(q - p) \qquad \mu_4 = npq[1 + 3(n - 2)pq]$$

Useful recursive expressions for the $(r + 1)$st central moments and cumulants are

$$\mu_{r+1} = pq\left[nr\mu_{r-1} + \frac{d\mu_r}{dp}\right] \tag{12.175}$$

$$K_{r+1} = pq\frac{dK_r}{dp} \tag{12.176}$$

The cumulative distribution function can be written in terms of the incomplete beta function, that is,

$$\begin{aligned} F(x) = P\{S \leq x\} &= \sum_{s=0}^{x} b(s;n,p) \\ &= \frac{1}{B(x+1, n-x)} \int_0^q t^{n-x-1}(1-t)^x \, dt \\ &= I_q(n - x, x + 1) = 1 - I_p(x + 1, n - x) \end{aligned} \tag{12.177}$$

Finally, if S_1 and S_2 are independently and binomially distributed with parameters n_1,p and n_2,p, respectively, then their sum $S = S_1 + S_2$ is binomially distributed with parameters $n_1 + n_2,p$.

Limit Theorems. There are two important limit theorems associated with the binomial distribution which are actually special cases of the *law of large numbers* and the *central-limit theorem*, respectively (see Sec. 7).

Bernoulli's theorem states that the probability of the relative frequency of "successes" in n Bernoulli trials, S/n, differing from p by more than ϵ tends to zero as $n \to \infty$ for any $\epsilon > 0$, that is,

$$\lim_{n\to\infty} P\left\{\left|\frac{S}{n} - p\right| > \epsilon\right\} = 0 \qquad (\epsilon > 0) \tag{12.178}$$

De Moivre's theorem states that in a series of Bernoulli trials

$$\lim_{n\to\infty} P\left\{a \leq \frac{S - np}{\sqrt{npq}} \leq b\right\} \simeq \frac{1}{\sqrt{2\pi}} \int_a^b \exp\left(-\tfrac{1}{2}t^2\right) dt \tag{12.179}$$

Since the right side of (12.179) is the integral from a to b of the normal distribution with parameters $m = 0$ and $\sigma = 1$, S is said to be asymptotically normally distributed with parameters $m = np$ and $\sigma = \sqrt{npq}$ [see Eq. (12.136) and sequel].

Tables of the Binomial Probability Distribution. Discussion of tables of the binomial probability distribution is greatly simplified by the following notation: Let $b(s;n,p)$ denote the term (12.172) giving the probability of *exactly* s successes, and let

$$\begin{aligned} P(s;n,p) &= \sum_{r=0}^{s} b(r;n,p) \\ Q(s-1;\, n,p) &= \sum_{r=s}^{n} b(r;n,p) \end{aligned} \tag{12.180}$$

so that $P(s;n,p)$ gives the probability of s *or less* successes (i.e., *at most* s successes), and $Q(s - 1;\, n,p)$, the probability of *more than* $s - 1$ successes (i.e., *at least* s successes).

Some elementary textbooks and handbooks provide brief tables of $b(s;n,p)$ and $Q(s - 1;\, n,p)$, as exemplified by [73] and [75]; others give only $b(s;n,p)$ (for example, [74]) or only $P(s;n,p)$ (for example, [76]). The principal extensive tabulations of binomial probabilities are [77–82]. For other tables of binomial probabilities, see Greenwood and Hartley [52, secs. 3.11 and 3.22].

When Eq. (12.177) is written in the form

$$P(s;n,p) = I_{1-p}(n - s, s + 1) = 1 - I_p(s + 1, n - s)$$

it is evident that

$$\begin{aligned} Q(s-1;\, n,p) &= I_p(s, n - s + 1) \\ &= 1 - I_{1-p}(n - s + 1, s) \end{aligned}$$

Consequently, values of $P(s;n,p)$ and $Q(s - 1;\, n,p)$ can be obtained directly from tables of the incomplete-beta-function ratio $I_x(a,b)$, for example, from Pearson's table [83]. (For additional tables of $I_x(a,b)$, see Greenwood and Hartley [52, sec. 3.12].) For fixed values of n and s, values of p for which

$$Q(s - 1;\, n,p) = \alpha$$

can be read, for various values of α, from tables of the "Percentage Points of the Beta Distributions," [84–86]. (For additional relevant tables, see Greenwood and Hartley [52, sec. 3.3].)

Approximations to the Binomial Distribution.

NORMAL APPROXIMATION. Although De Moivre's theorem can be used as an approximation to the binomial distribution, more accurate results can be obtained by using a modified form of (12.179). If it is desired to find the normal approximation to $P\{a \leq S_n \leq b\}$, then a better approximation is

$$P\{a \leq S_n \leq b\} \simeq \frac{1}{\sqrt{2\pi}} \int_{t_1}^{t_2} \exp\left(-\tfrac{1}{2}t^2\right) dt \tag{12.181}$$

with $t_1 = \dfrac{a - np - \frac{1}{2}}{\sqrt{npq}}, \qquad t_2 = \dfrac{b - np + \frac{1}{2}}{\sqrt{npq}}$

This approximation may be satisfactory if npq is "large."

The error decreases as n increases and as p approaches $\frac{1}{2}$. An excellent bound for the error when $npq \geq 25$ and t_1 and t_2 are small is given by Uspensky [17, p. 129]. A more refined general analysis of the error has been given by Feller [3].

Raff [44, Table 1] gives values of the maximum error of (12.181) for any combination a, b $(0 \leq a < b \leq n)$, for a large number of combinations of p and n, $0.002 \leq p \leq 0.5$ and $5 \leq n \leq 500$, and concludes that the maximum error of (12.181) is always less than $0.140/\sqrt{npq}$ and is less than 0.05 when $np^{3/2} > 1.07$. Smith [65, Fig. 7] gives contours in the (n,p) plane that bound the regions within which the maximum error of (12.181) *for* $b = n$ and any a $(0 \leq a < n)$ is less than ϵ, for sixteen values of ϵ, from 0.0005 to 0.1, and finds that, *when* $b = n$, the maximum error of (12.181) is less than 0.005 if $np > 37q^3$.

Poisson Exponential Approximation. If npq is small, then a more accurate approximation is

$$P\{a \leq S_n \leq b\} \simeq \sum_{s=a}^{b} \frac{e^{-m}m^s}{s!} \qquad (12.182)$$

with $m = np$.

Bounds to the error of this approximation are given by Uspensky [17, pp. 135–137].

Raff [44, Table 4] gives values of the maximum error of (12.182) for any combination a, b $(0 \leq a < b \leq n)$, for a large number of combinations of p and n, $0.002 \leq p \leq 0.5$ and $5 \leq n \leq 500$, and also a chart [44, Fig. 2] separating the (n,p) combinations for which the maximum error of the Poisson approximation (12.182) is less than that of the normal approximation (12.181). He finds that for $p < 0.075$ the maximum error of (12.182) is always less than that of (12.181). Smith [48, Fig. 11] gives contours bounding regions in the (n,p) plane where, for $b = n$, the maximum error of (12.182) is less than ϵ, for twelve values of ϵ, from 0.0005 to 0.035, and finds that, *when* $b = n$, the maximum error of the Poisson approximation (12.182) is less than 0.01 when $p < 0.07$ for all n and in less than 0.001 when $p < 0.01$ and $n > 25$, and that, *when* $b = n$, the maximum error of the Poisson approximation (12.182) is less than that of the normal approximation when $2n < (q^3/p)^{2.9}$.

Other Approximations. Raff [44] and Smith [48] also examine two-term Gram-Charlier type A series [10, sec. 6.17] approximations to $P\{a \leq S_n \leq b\}$ and to $P\{a \leq S_n \leq n\}$, respectively. Raff [44, p. 296] finds that, for all a, b $(0 \leq a < b \leq n)$ and all n and p, the maximum error is always less than $0.056/\sqrt{npq}$; Smith [48, p. 23] finds that, *when* $b = n$, the maximum error is less than 0.001 for $np^{1.24} > 12.7$. They also examine the corresponding two-term Gram-Charlier type B series [10, sec. 6.24] approximations, finding [44, p. 301] the maximum error to be less than 0.005 for all a, b $(0 \leq a < b \leq n)$ when $np \leq 0.8$ and $n \geq 5$; and [48, p. 31] that, for $b = n$, the maximum error is less than 0.001 when $p < 0.1$ *and* $n \geq 10$. Raff also examines Camp's approximation [20], finding [44, p. 301] that its maximum error for all a, b $(0 \leq a < b \leq n)$ never exceeds $0.007/\sqrt{npq}$ and has an absolute maximum of 0.0122, for any combination of n and p. Raff [44, pp. 300–301] observes that, *for* $n \geq 5$, the maximum error in approximating $P\{a \leq S_n \leq b\}$ can be kept below 0.005 by using the two-term "Poisson Gram-Charlier" (type B) approximation when $np \leq 0.8$ and the "Camp-Paulson" approximation when $np \geq 0.8$.

Observed Proportions, Arcsine Transformation, and Two-way Square-root Paper. If $P = S/n$ denotes the *observed proportion* of successes in n Bernoulli trials, then

$$E[P] = p \qquad 0 \leq p \leq 1$$
$$\sigma^2[P] = \frac{p(1-p)}{n} \qquad (12.183)$$

Thus, the standard deviation, $\sigma[P]$, of an observed proportion P depends upon the magnitude of the "true proportion" p which P serves to estimate; is a maximum $(1/4n)$ when $p = \frac{1}{2}$; and tends to zero as $p \to 0$ or 1.

The fact that $\sigma[P]$ depends on p as well as upon n is a source of inconvenience in some theoretical and applied work. R. A. Fisher has noted (1922) that, when n is large, a separation of variables can be effected by conducting the analysis in terms of *angles* (T) related to the observed proportions (P) by the *arcsine transformation*

$$T = \arcsin \sqrt{P} = \tfrac{1}{2} \arccos (1 - 2P) \qquad (12.184)$$

with $0 \leq T \leq \pi$ for $0 \leq P \leq 1$, since, by (12.92) and (12.94),

$$E[T] \simeq \arcsin \sqrt{p} = \theta, \text{ say}$$
$$\sigma^2[T] \simeq \frac{1}{4n} \qquad \text{for } T \text{ in radians} \qquad (12.185)$$

For further details, and modifications, see [14, Chap. 20].

From (12.185) and the normal approximation (12.181) it follows that, for large n, the observational point $(\sqrt{S}, \sqrt{n - S})$ at a distance n from the origin in the direction T is approximately normally distributed about the point $(\sqrt{np}, \sqrt{n(1 - p)})$, with variance $\frac{1}{4}$, along the arc of a circle of radius n with center at the origin. Mosteller and Tukey have designed a special graph paper [42] with square-root scales for both coordinates and have explained its various uses in considerable detail in a lengthy paper [41], of which Wallis and Roberts [18, sec. 19.6] give a convenient synopsis.

Poisson Trials. Consider a sequence of independent trials such that the probability of a "success" on the jth trial is p_j, $(j = 1,2, \ldots)$. Trials of this kind are called *Poisson trials*—not to be confused with the "Poisson exponential distribution" considered later in this section.

Let X_j be the indicator for the jth trial, and let

$$S = \sum_{j=1}^{n} X_j \qquad (12.186)$$

signify the total number of "successes" in a sequence of n such trials. The trials being independent by assumption, the probability-generating function for S is given [see (12.127)] by

$$P_S(t) = \prod_{j=1}^{n} P_{X_j}(t) = \prod_{j=1}^{n} [1 + p_j(t - 1)] \qquad (12.187)$$

Hence $P\{S = s\}$ can be found by evaluating the

coefficient of t^s in the product on the extreme right of (12.187).

The corresponding characteristic function, mean, and variance are

$$\phi_S(t) = \prod_{j=1}^{n} (q_j + p_j e^{it})$$
$$E[S] = \sum_{j=1}^{n} p_j = n\bar{p} \tag{12.188}$$
$$\sigma^2[S] = \sum_{j=1}^{n} p_j q_j = n\overline{pq} - \sum_{j=1}^{n} (p_j - \bar{p})^2$$

where $\bar{p} = \frac{1}{n}\sum_{j=1}^{n} p_j$ and $\bar{q} = 1 - \bar{p}$

Note that, if the constant probability, p, associated with a series of n Bernoulli trials is equal to the average, $\bar{p}$, of unequal probabilities p_j, $(j = 1,2, \ldots ,n)$, associated with a series of n Poisson trials, then the mean number of successes, $E[S]$, will be the same for both series, but the variance of S for the Poisson series [see (12.188)] will be smaller by an amount proportional to the dispersion of the p_j about their average. Thus, in a succession of simultaneous independent tosses of n different coins, the numbers of "heads" obtained on successive occasions will tend to vary *more* if the probability of "heads" is the same for all the coins, than if this probability varies from coin to coin; and will be a maximum if the probabilities are all equal to 1/2. In this sense, we may say that uniformity *increases* chance fluctuations.

The Hypergeometric Distribution.

Sampling with, and without, Replacement. Consider an urn containing W white and B black balls, respectively. Let a trial consist of selecting a ball at random and noting its color. Repeated trials of this kind define *sampling with replacement*, or *sampling without replacement*, according as the ball drawn in each trial *is*, or *is not*, returned to the urn before initiating the next trial.

Sampling *with replacement* is equivalent to a sequence of Bernoulli trials, with the probability of drawing a white ball equal to $W/(W + B)$ in every trial, and the total number of "successes," S, in n such trials has a *binomial distribution* (12.172).

In sampling *without replacement*, the probability of drawing a white ball varies from trial to trial, being $W/(W + B)$ for the first trial, but either $(W - 1)/(W + B - 1)$ or $W/(W + B - 1)$ on the second trial, according as the first trial yields a white or a black ball, respectively; and so forth. When the sampling is *without replacement*, the number of "successes" in (the first) n trials has a *hypergeometric distribution*.

Hypergeometric Distribution. Consider a finite population of N distinct objects of which pN (an integer) are "defective," and the remaining $(N - pN)$ are "nondefective." If a sample of n objects is drawn from this population, either *en bloc* or *in succession without replacement*, in such a manner that the probability of drawing any particular set of n objects is $1/\binom{N}{n}$, then the probability of drawing a sample containing *exactly s defectives* is given by

$$h(s;n,p,N) = \frac{\binom{Np}{s}\binom{N-Np}{n-s}}{\binom{N}{n}} = \frac{\binom{n}{s}\binom{N-n}{Np-s}}{\binom{N}{Np}}$$
$$\text{for } s = 0,1,2, \ldots ,\min(Np,n) \tag{12.189}$$

The middle expression is the coefficient of $t^s u^n$ in the expansion of the probability-generating function

$$P_S(t) = \frac{(1 + tu)^{N_p}(1 + u)^{N-N_p}}{\binom{N}{n}} \tag{12.190}$$

and the right-hand expression is derived from it by algebraic manipulation. The letter u in (12.190) serves to count the objects assigned to the "sample" and t counts the number of *defectives* so assigned. (The terminology "defective" and "nondefective" is used here in recognition of the fact that many important applications of the hypergeometric distribution deal with sampling from lots or batches of materials composed of items which are individually either "defective" or "nondefective.")

The distribution derives its name from the fact that $h(s;n,p,N)$ is equal to the coefficient of z^s in the series expansion of

$$\frac{\binom{N-Np}{n}}{\binom{N}{n}} F(-n,-Np,N - Np - n + 1;z) \tag{12.191}$$

where $F(\alpha,\beta,\gamma;z)$ is the hypergeometric function (see page 1–57 [Part 1, Chap. 3, Sec. 11]). The characteristic function of the distribution can be obtained by substituting e^{it} for t in (12.190), or for z in (12.191).

The mean and variance of the distribution (12.189) are

$$E[S] = np, \qquad \sigma^2[S] = npq\left(\frac{N-n}{N-1}\right), \qquad (q = 1 - p) \tag{12.192}$$

Note that the variance is less than the variance of the binomial distribution (12.174) having the same mean, which corresponds to sampling *with replacement*.

Tables of the individual terms $h(s;n,p,N)$ and of the partial sums

$$P(s;n,p,N) = \sum_{r=0}^{s} h(r;n,p,N)$$
$$Q(s - 1;\, n,p,N) = \sum_{r=s}^{n} h(r;n,p,N) = 1 - P(s - 1;\, n,p,N) \tag{12.193}$$

of terms of the hypergeometric distribution (12.189) can be held to manageable size by taking advantage of the following identities:

$$\begin{aligned} h(s;n,p,N) &= h(s;Np,n/N,N) \\ &= h(n - s; n, 1 - p, N) \\ &= h(Np - s; N - n, p, N) \\ &= h(N(1 - p) - (n - s); \\ &\qquad N - n, 1 - p, N) \end{aligned} \qquad (12.194)$$

$$\begin{aligned} P(s;n,p,N) &= P(s;Np,n/N,N) \\ &= 1 - P(n - s - 1; n, \\ &\qquad N(1 - p), N) \\ &= P(N(1 - p) - (n - s); \\ &\qquad N - n, N(1 - p), N) \\ &= 1 - P(Np - s - 1; N - n, \\ &\qquad Np, N) \end{aligned} \qquad (12.195)$$

These identities show that every hypergeometric probability can be expressed in terms of a hypergeometric probability with $0 \le s \le Np \le n \le \frac{1}{2}N$. The first line of each set expresses the most important feature, i.e., that n and Np can always be interchanged, as is evident in (12.189); and the second lines, that s can be interchanged with $n - s$, *provided* that p and $1 - p$ are interchanged simultaneously (and the terms summed accordingly).

Tables of the Hypergeometric Probability Distribution. The principal tables of the hypergeometric probability distribution are those of Lieberman and Owen [87]. An excerpt sufficient for all situations with $N \le 20$ is given by Owen [56]; see [88] for details.

Approximations to the Hypergeometric Distribution.

BINOMIAL APPROXIMATIONS. When N is large and n is small compared to N, the terms $h(s;n,p,N)$ of the hypergeometric distribution (12.189) can be approximated by the terms $b(s;n,p)$ of the *corresponding binomial distribution* (12.172), to which the hypergeometric distribution (12.189) tends in the limit as $N \to \infty$.

When N is large but $Np < n$, a somewhat better approximation to $h(s;n,p,N)$ is $b(s;n^*,p^*)$, with $n^* = Np$ and $p^* = n/N$, which is the corresponding term of the binomial distribution to which the hypergeometric distribution (12.189) tends in the limit as $N \to \infty$, $n \to \infty$, and $p \to 0$, with $Np \to n^*$ and $n/N \to p^*$.

Either of these approximations will ordinarily be satisfactory if $n/N < 0.1$.

More generally, the terms of hypergeometric distribution (12.189) can often be approximated quite accurately by the corresponding terms of the binomial distribution which has the same mean and, as closely as possible, the same variance [77]. To find this binomial distribution, first evaluate

$$n' = \frac{E[S]}{p'} \qquad \text{for } p' = 1 - \frac{\sigma^2[S]}{E[S]}$$

where $E[S]$ and $\sigma^2[S]$ are given by (12.192); then, take n'', the nearest integer to n', and evaluate p'' from

$$p'' = \frac{E[S]}{n''}$$

obtaining $b(s,n'',p'')$ as the binomial distribution of "closest fit" to the hypergeometric distribution (12.189).

POISSON APPROXIMATION. If

$$p < 0.1 \qquad \text{and} \qquad n/N < 0.1$$

then the hypergeometric probabilities can be approximated by the Poisson probabilities, that is,

$$h(s;n,p,N) \simeq e^{-np} \frac{(np)^s}{s!} \qquad (12.196)$$

NORMAL APPROXIMATION. If $npq(N - n)/N > 9$ and $n/N < 0.1$, then

$$\begin{aligned} P\{a \le S \le b\} &= \sum_{s=a}^{b} h(s;n,p,N) \\ &\simeq \frac{1}{\sqrt{2\pi}} \int_{t_1}^{t_2} \exp(-\tfrac{1}{2}t^2)\, dt \end{aligned} \qquad (12.197)$$

for

$$t_1 = \frac{a - np - \frac{1}{2}}{\sqrt{npq\left(\dfrac{N - n}{N - 1}\right)}} \qquad t_2 = \frac{b - np + \frac{1}{2}}{\sqrt{npq\left(\dfrac{N - n}{N - 1}\right)}}$$

The Poisson Exponential Distribution. The discrete distribution defined by

$$p(s;m) = \frac{e^{-m}m^s}{s!} \qquad m > 0 \qquad (12.198)$$

for $s = 0(1)\infty$, termed the *Poisson exponential distribution*, is (with the normal, and binomial distributions) one of the three principal distributions of probability theory. It has wide applications, not only as the limiting form of many discrete distributions, but also, in its own right, as an exact distribution fundamental to many physical and biological phenomena. Examples are radioactive disintegrations per unit of time; number of telephone calls per unit time; microorganisms, or other "particles," per unit volume of fluid, or per unit area in the field of a microscope; chromosome interchanges per cell produced by a fixed dosage of X radiation; and "defects" per item (or per unit of length or area) of manufactured product.

Properties. The characteristic function, mean, and variance are given by

$$\phi(t) = \exp m(e^{it} - 1) \qquad (12.199)$$

$$E[S] = m \qquad \sigma^2[S] = m \qquad (12.200)$$

and all cumulants are identically equal to m, i.e.,

$$K_r = m \qquad r = 1, 2, \ldots \qquad (12.201)$$

The cumulative distribution function can be written in terms of the incomplete-gamma-function ratio $I(u,p)$ tabulated by Harter [101] and Pearson [102] and in terms of the integral $Q(\chi^2|\nu)$ of the chi-square distribution for ν degrees of freedom [see (12.239)] tabulated by Pearson and Hartley [57], as follows:

$$F(x) = P\{S \le x\} = \sum_{s=0}^{x} p(s;m) \qquad (12.202)$$

$$\begin{aligned} F(x) &= \frac{1}{\Gamma(x + 1)} \int_m^\infty e^{-t}t^x\, dt = 1 - I(u,p) \\ &= Q(\chi^2|\nu) \end{aligned}$$

for $\quad u = m/\sqrt{s + 1} \qquad p = s$
$\qquad \chi^2 = 2m \qquad \nu = 2(s + 1) \qquad (12.203)$

Furthermore, if S_1 and S_2 are independently and Poisson distributed with parameters m_1 and m_2, respectively, then their sum $S_1 + S_2$ has a Poisson distribution with parameter $m_1 + m_2$.

Poisson Distribution as an Exact Distribution. In order that "events" per unit interval (area, or volume) be distributed *strictly* in accordance with a Poisson distribution it is necessary and sufficient that

(a) $P_s(h)$, the probability of exactly s occurrences in *any* interval (or, region) of length (area, or volume) h, be independent of the number of occurrences in any other nonoverlapping interval (or, region);

(b) $$\sum_{s=0}^{\infty} P_s(h) = 1$$

(c) $P_1(h) = mh + o(h)$, with $m > 0$, where $o(h)$ denotes a quantity of smaller order of magnitude than h as $h \to 0$; and

(d) $$\sum_{s=2}^{\infty} P_s(h) = o(h) \qquad \text{as } h \to 0$$

These assumptions lead directly to the system of differential equations

$$P'_0(h) = -mP_0(h)$$
$$P'_s(h) = -mP_s(h) + mP_{s-1}(h) \qquad s = 1(1)\infty$$

which have as their solution

$$P_s(h) = \frac{e^{-mh}(mh)^s}{s!} \qquad s = 0(1)\infty$$

For $h = 1$, this reduces to (12.198).

Besides establishing the Poisson distribution as a "law" in its own right, this derivation ties in closely with the theory of *stochastic processes,* and lends itself to many generalizations. See Feller [3, p. 364ff].

Poisson Distribution as a Limiting Form of Various Discrete Distributions. Traditionally, the Poisson distribution is best known as an approximation, (12.182), to the *binomial distribution* (12.172), for n "large" and p "small"; and (12.198) is "derived" as the limit of (12.172) as $n \to \infty$ and $p \to 0$ with $np \to m$. More generally, (12.198) is the limiting form of the distribution of the number of successes S in n *Poisson trials* as $n \to \infty$, $p_j \to 0$ (for every j), and

$$\sum_{j=1}^{n} p_j \to m \qquad 0 < m < \infty$$

which may be shown most easily by noting that the corresponding limit of the probability-generating function $P_S(t)$ given by (12.187) is $e^{m(t-1)}$, which is clearly the generating function of (12.198).

The Poisson distribution, (12.198), is also the limiting form of the following:

1. The *hypergeometric distribution,* (12.189), as $N \to \infty$, $p \to 0$, and $n \to \infty$, with $Np \to \infty$ and either $np \to m$ or $n/M \to m$.

2. The *Pascal distribution,* (12.221), as $r \to \infty$ and $p \to 1$, with $r(1 - p) \to m$.

3. The *Polya distribution,* (12.232), as $r \to \infty$ and $d \to 0$, with $rd \to m$.

4. The *distribution of the number of empty cells* when r things are allocated independently at random among n cells (see Feller [3, p. 69]), as $r \to \infty$ and $n \to \infty$, with $ne^{-(r/n)} \to m$.

5. The *distribution of the number of "runs" of exactly r consecutive "successes"* in a series of n *Bernoulli trials* (see Feller [3, p. 278]), as $n \to \infty$, $r \to \infty$, and $np^r(1 - p) \to m$, where p is the probability of "success" in any single trial.

Tables of the Poisson Exponential Distribution. Discussion of tables for evaluating cumulative sums of successive terms of the Poisson exponential distribution (12.198) is greatly simplified by the following notation: Let

$$P(s;m) = \sum_{r=0}^{s} p(r;m) = \sum_{r=0}^{s} \frac{e^{-m}m^r}{r!}$$
$$\text{and} \quad Q(s-1;m) = \sum_{r=s}^{\infty} p(r;m) = \sum_{r=s}^{\infty} \frac{e^{-m}m^r}{r!} \qquad (12.204)$$

The principal tabulations of Poisson probabilities are those of Kitagawa [109] and Molina [110] and Table 7 of Pearson and Hartley [57]; see [98] for details. Brief tables are given in the textbooks of Fisz [4], Gnedenko [5], and Parzen [15]. Somewhat fuller tables are found in the handbooks of Owen [56] and Weast et al. [54]; see [103–107] for details. Values of $m = m(s,\alpha)$ such that

$$Q(s-1; m(s,\alpha)) = 1 - P(s-1; m(s,\alpha)) = \alpha \qquad (12.205)$$

for $s = 0, 1, \ldots$ are given by Campbell [108] for α's ranging from .000001 to .999999.

Since, for $x = s$, the third expression in Eqs. (12.202) is identically $P(s;m)$, these equations serve to express $P(s;m)$ in terms of the incomplete-gamma-function ratio $I(u,p)$ and the integral $Q(\chi^2|\nu)$ of the chi-square distribution for ν degrees of freedom (12.239). The corresponding expressions for $Q(s-1; m)$ are

$$Q(s-1;m) = I(u,p) = 1 - Q(\chi^2|\nu) = P(\chi^2|\nu) \qquad (12.206)$$

for $$u = m/\sqrt{s} \qquad p = s - 1$$
$$\chi^2 = 2m \qquad \nu = 2s$$

Consequently, values of $P(s;m)$ and $Q(s-1;m)$ can be obtained from the Harter [101] and Pearson [102] tables of $I(u,p)$ and read directly from the tables of $Q(\chi^2|\nu)$ given by Fisz [4], Abramowitz and Stegun [50], and Pearson and Hartley [57]; see [91, 93, and 98] for details.

Similarly, values of $m = m(s,\alpha)$ defined by (12.205) can be expressed in terms of corresponding percentage points of the chi-square distribution $\chi^2(\nu,\alpha)$, defined by (12.242), as follows:

$$m(s,\alpha) = \tfrac{1}{2}\chi^2(\nu, 1-\alpha) \qquad \text{for } \nu = 2s$$
$$m(s, 1-\alpha) = \tfrac{1}{2}\chi^2(\nu,\alpha) \qquad \text{for } \nu = 2(s+1) \qquad (12.207)$$

Hence values of $m(s,\alpha)$ can be read directly from the many available tables of $\chi^2(\nu,\alpha)$; see, for example, [89, 90, 92–98, and 101].

Relation to Multinomial Distribution. Let S_i $(i = 1,2, \ldots ,n)$ be independent random variables following Poisson distributions with parameters m_i, respectively, *then the conditional distribution of the* $\{S_i\}$ *for a fixed total,*

$$\sum_{i=1}^{n} S_i = s$$

is a multinomial distribution (12.234) *with parameters* $\{m_i/m\}$ *and* n, that is,

$$P\left\{S_1 = s_1, \ldots ,S_n = s_n \middle| \sum_{i=1}^{n} S_i = s\right\} = \frac{n!}{s_1! \cdots s_n!} \prod_{i=1}^{n} \left(\frac{m_i}{m}\right)^{s_i} \qquad (12.208)$$

where $m = \sum_{i=1}^{n} m_i$ and $\sum_{i=1}^{n} s_i = s$

This is a unique property of the Poisson distribution, serving to characterize the Poisson distribution among all discrete distributions on non-negative integers [39].

Alternatively, let S_j, $(j = 1,2, \ldots ,n)$, be random variables whose joint distribution is the multinomial distribution (12.234) with parameters $\{p_j\}$ and N, and with the "sample size" N itself a *random variable* having a Poisson distribution with parameter m, then the joint over-all "compound distribution" [see (12.129)] of the $\{S_j\}$ is

$$P\{S_1 = s_1, \ldots ,S_n = s_n|\{p_j\},m\} = \prod_{j=1}^{n} \frac{e_j^{-mp}(mp_j)^{s_j}}{s_j!} \qquad (12.209)$$

Since the right-hand side of (12.209) is simply the *product,*

$$\prod_{j=1}^{n} p(s_j;mp_j)$$

of the appropriate terms of n separate Poisson distributions, it follows that the $\{S_j\}$ are individually distributed according to Poisson distributions with parameters mp_j and are *mutually independent.*

Finally, if the S_j are jointly multinomially distributed, as above, but with N *fixed,* then

$$P\{S_1 = s_1, \ldots ,S_n = s_n|\{p_j\},N\} \to \prod_{j=1}^{n} \frac{e^{-m_j}(m_j)^{s_j}}{s_j!} \qquad (12.210)$$

in the limit as $N \to \infty$ with $Np_j \to m_j$, $(0 < m_j < \infty)$, for $j = 1, 2, \ldots , n$. Since the right-hand side of (12.210) is simply the product,

$$\prod_{j=1}^{n} p(s_j;m_j)$$

of the appropriate terms of n separate Poisson distribution, it is seen that, as $N \to \infty$ with $Np_j \to m_j$, the S_j are *asymptotically* mutually independent and individually Poisson distributed, with parameters m_j, respectively.

Relation to the Exponential Distribution. When the Poisson distribution, in the form (12.204), gives the exact distribution of the number of events per linear interval of length h, then, for *any point* a, the probability that the event point nearest to a in a given direction is at a distance of x *or less* from a is simply $1 - e^{-mx}$. This is the probability of at least one event point in the interval $(a, a + x)$, which depends upon x (the length of the interval) and upon m (the mean number of events per unit interval) but *not* upon its start, a. Consequently, if X denotes the distance from an arbitrary point a to the "next" event point, then X is a random variable, termed the *waiting* (*or holding*) *time,* with distribution function

$$F(x) = 1 - e^{-mx} \qquad 0 < x < \infty \qquad (12.211)$$

regardless of a. The corresponding density function is

$$f(x) = me^{-mx} \qquad 0 < x < \infty \qquad (12.212)$$

and waiting times are thus seen to be *exponentially distributed,* i.e., to have an *exponential distribution* (Table 12.2), with parameters $m = 0$ and $\beta = 1/m$.

If a is chosen to be an event point, then X is the waiting time to the "next" event point. Consequently, if the *number* (S) of events per unit interval is *Poisson distributed* according to (12.198), then the *distance* between successive event points is *exponentially distributed* according to (12.211) and (12.212); and conversely.

Finally, it should be noted that under these circumstances the *conditional probability* that $X > x$ *given* that $X > x_0$ reduces to

$$P\{X > x|m;X > x_0\} = \frac{\exp(-mx)}{\exp(-mx_0)} = \exp[-m(x - x_0)], \qquad 0 < x_0 < x < \infty \qquad (12.213)$$

so that the *residual waiting time,* $T = X - x_0$, has the distribution function $F(t) = 1 - e^{mt}$, which is the same as (12.211). In other words, the *elapsed time,* x_0, since the last event does not affect the probability that the next event will occur within time t from now.

Negative Binomial Distributions.

The Geometric Distribution. A discrete random variable X is said to have a *geometric distribution* with parameter p, if

$$P\{X = x\} = q^x p \qquad \text{for } x = 0, 1, \ldots \qquad (12.214)$$

where $q = 1 - p$ and $0 < p < 1$. The corresponding probability-generating function [see (12.122)] is

$$P(t) = p \sum_{x=0}^{\infty} (qt)^x = \frac{p}{1 - qt} \qquad (12.215)$$

and the mean and variance of X are

$$E[X] = \frac{q}{p} \qquad \sigma^2[X] = \frac{q}{p^2} \tag{12.216}$$

Note that the right-hand side of (12.214) is equal to the probability of exactly x "failures" before the first "success" in a (conceptually infinite) sequence of Bernoulli trials (see Bernoulli Trials and the Binomial Distribution above). Hence (12.214) gives the *probability of the first success occurring on the* $(x + 1)$st *trial* in a sequence of Bernoulli trials; and X may be interpreted as the *number of failures preceding the first success.* Furthermore, the probability that the first success occurs *after the* xth *trial* takes the simpler form

$$P\{X > x\} = q^x \qquad x = 0, 1, 2, \ldots \tag{12.217}$$

When X has the geometric distribution (12.214), the *conditional distribution of* X *given that* $X > x_0$ is

$$P\{X = x|X > x_0\} = \frac{q^x p}{q^{x_0}} = P\{X = x - x_0\} \tag{12.218}$$

for $x = x_0 + 1, x_0 + 2, \ldots$; and (12.214) is the only probability distribution on the non-negative integers, 0, 1, . . . having this property (see Feller [3, p. 219]).

Equation (12.218), in the light of the preceding discussion, implies that, given x_0 consecutive failures in a sequence of Bernoulli trials, the probability that the first success occurs on the $(x + 1)$st *trial thereafter* is equal to the right-hand side of (12.214), regardless of x_0. Consequently, taking $x_0 = 0$, (12.214) can be interpreted either as the distribution of the number of successive failures between an *arbitrary trial and the first success thereafter,* or as the distribution of the *gap* (i.e., the number of successive failures) between *consecutive successes* [compare (12.211) and sequel].

The relation of the *discrete geometric distribution* (12.214) to the *continuous exponential distribution* (12.211) is as follows: If in a sequence of Bernoulli trials each trial takes time Δt, then (12.217) gives the probability that the first success occurs *after* time $x \cdot \Delta t$. Letting $T = X \cdot \Delta t$, taking the limit of the right-hand side of (12.217) as $\Delta t \to 0$, $x \to \infty$, and $p \to 0$, with $x \cdot \Delta t \to t$ and $(p/\Delta t) \to m$, yields

$$\Pr\{T > t\} = e^{-mt} \tag{12.219}$$

as the limiting value of the *probability that the first success occurs after time t;* that is, in the limit the *waiting time,* T, is exponentially distributed according to (12.211).

The Pascal Distribution. Consider the random variable $Y = X_1 + X_2 + \cdots + X_r$, where the $X_j (j = 1,2, \ldots ,r)$ are mutually independent and identically distributed according to the geometric distribution (12.214), with probability-generating function, $G(t)$, as in (12.215). From (12.127) it follows that the probability-generating function associated with the distribution of Y is

$$[G(t)]^r = \left(\frac{p}{1 - qt}\right)^r \tag{12.220}$$

so that the distribution of Y is given by

$$\begin{aligned} P\{Y = y\} = f(y;r,p) &= \binom{r + y - 1}{y} p^r q^y \\ &= \binom{-r}{y} p^r(-q)^y \end{aligned} \tag{12.221}$$

for $y = 0, 1, 2, \ldots$, where $q = 1 - p$ and $0 < p < 1$.

The distribution $f(y;r,p)$ defined by (12.221) for r *any positive integer* is known as the *Pascal distribution,* in honor of Blaise Pascal (1623–1662).

The associated characteristic function, mean, and variance are

$$\begin{aligned} \phi(t) &= \left(\frac{p}{1 - qe^{it}}\right)^r \qquad q = 1 - p \\ E[Y] &= \frac{rq}{p} \\ \text{var}\,[Y] &= \frac{rq}{p^2} \end{aligned} \tag{12.222}$$

Moreover, it is evident from the foregoing derivation that if Y_1 and Y_2 are independently and *Pascal distributed* with probabilities $f(y_i;r_i,p)$ $(i = 1,2)$, then their sum $Y = Y_1 + Y_2$ follows a Pascal distribution with probabilities $f(y_1 + y_2;\, r_1 + r_2,\, p)$.

It should be noticed that the random variable Y may be interpreted as the *number of failures preceding the* rth *success* in a sequence of Bernoulli trials; and (12.221), as giving the *probability that the* rth *success occurs on the* $(r + y)$th *trial,* from which viewpoint (12.221) may be derived directly by elementary considerations.

Individual terms, $f(y;r,p)$, of the Pascal distribution (12.221) can be evaluated from tables of *binomial probabilities* (12.172), $b(s;n,p)$, by virtue of the identities

$$\begin{aligned} f(y;r,p) &= p \cdot b(y;\, r + y - 1,\, q) \\ &= p \cdot b(r - 1;\, r + y - 1,\, p) \\ &= \frac{r}{r + y}\, b(r;\, r + y,\, p) \\ &= \frac{r}{r + y}\, b(y;\, r + y,\, q) \end{aligned} \tag{12.223}$$

For *fixed* r, when y is large and p is small, the approximation

$$f(y;r,p) \simeq \frac{r}{y}\frac{e^{-m}m^r}{r!} = \frac{r}{y}\,p(r;m) \tag{12.224}$$

with $m = py$, may be useful in evaluating probabilities in the "tail" of the distribution, $p(r;m)$ being the general term of the *Poisson exponential distribution* (12.198). For any y, as $r \to \infty$ and $p \to 1$, with $r(1 - p) \to m$,

$$f(y;r,p) \to p(y,m)$$

providing another "Poisson approximation," useful alike for evaluating individual terms and sums of consecutive terms. When $r = 1$, then (12.217) applies. For $r > 1$, but small, a sum of consecutive terms of (12.221) can be approximated by the sum of the corresponding terms of a binomial distribution, $b(y;n,$"p"$)$, with mean and variance equated to that of (12.222).

Extensive tables of the Pascal negative binomial distribution have been published by Williamson and Bretherton [111].

The General Negative Binomial Distribution, Inhomogeneity, and "Contagion." The equivalent expressions on the right-hand side of (12.221) serve to define the probability distribution of a discrete random variable when r is *not an integer*, provided that $r \geq 0$. If p and $q = 1 - p$ in (12.221) are replaced by P and $Q = 1 + P$ defined by

$$p = \frac{1}{Q} \qquad q = \frac{P}{Q} \tag{12.225}$$

and y is replaced by x, then, for every

$$r \geq 0, \; 0 \leq P < \infty$$

and $Q = 1 + P$, the sequence

$$f(x;r,P) = \binom{r + x - 1}{x} Q^{-r} \left(\frac{P}{Q}\right)^x \qquad x = 0,1, \ldots \tag{12.226}$$

defines the probability distribution of a discrete random variable X. Since the right-hand side of (12.226) is the general term of the binomial expression of $(Q - P)^{-r}$, X is said to have (or to follow) a *negative binomial distribution* with parameters r and P. The corresponding characteristic function, mean, variance, μ_3 and μ_4 are

$$\begin{aligned} \phi(t) &= (Q - Pe^{it})^{-r}, \qquad Q = 1 + P \\ E[X] &= rP \\ \sigma^2[X] &= rPQ \\ \mu_3 &= rPQ(Q + P) \\ \mu_4 &= rPQ[1 + 3(r + 2)PQ] \end{aligned} \tag{12.227}$$

respectively, and can be obtained from the corresponding expressions for the *binomial distribution*, (12.173) and (12.174), by substituting $-r$ for n, $-P$ for p, and Q for q.

Finally, from the form of the characteristic function $\phi(t)$ in (12.227), it is evident that, if X_1 and X_2 are independently distributed according to (12.226) with index parameters r_1 and r_2, respectively, and a common P, then their sum $X = X_1 + X_2$ has a negative binomial distribution with parameters $(r_1 + r_2)$ and P.

It is a remarkable fact that the *negative binomial distribution* (12.226) can be derived from two distinctly different—and, indeed, mutually contradictory!—sets of assumptions. *First,* following Greenwood and Yule [30], the negative binomial can be derived as an extension of the Poisson distribution (12.198) to *inhomogeneous* situations where the *outcomes in nonoverlapping intervals ARE mutually INDEPENDENT* but the *expected number per interval, m, is a random variable,* varying from interval to interval in accordance with a *gamma distribution*

$$f(m) = \frac{1}{\beta\Gamma(r)} \left(\frac{m}{\beta}\right)^{r-1} e^{-m/\beta} \qquad 0 < m < \infty \tag{12.228}$$

with $0 < r < \infty$ and $0 < \beta < \infty$. Under these circumstances the over-all distribution of S, the number of events per interval, is given by the *compound Poisson distribution* corresponding to (12.228), that is, by

$$\begin{aligned} P\{S = s|r,\beta\} &= \int_0^\infty f(m) \frac{e^{-m}m^s}{s!} dm \\ &= \binom{r + s - 1}{s} \frac{\beta^s}{(1 + \beta)^{r+s}} \\ & \qquad s = 0,1, \ldots \end{aligned} \tag{12.229}$$

which for $s = x$ and $\beta = P$ is the same as (12.226). *Second,* following Polya and Eggenberger (1923) [3, p. 82], (12.226) can be derived as the limiting form of a distribution appropriate to a situation where *successive outcomes ARE NOT INDEPENDENT,* each "favorable" outcome increasing (or decreasing) the probability of future "favorable" outcomes. To be specific, let $p_1 = p$ and $q_1 = 1 - p$ denote the probabilities of "success" and "failure," respectively, on the *first trial,* and let (see Lundberg [37])

$$\begin{aligned} p_{m+1} &= \frac{p + k\gamma}{1 + m\gamma} \\ q_{m+1} &= \frac{q + (m - k)\gamma}{1 + m\gamma} \\ & -1 < \gamma < \infty, \; n = 1,2, \ldots \end{aligned} \tag{12.230}$$

denote the conditional probabilities of "success" and "failure," respectively, on the $(m + 1)$st trial, given that the first m trials resulted in exactly k "successes" and $(m - k)$ "failures." Then the probability of exactly x successes in the first n trials is given by the *generalized Polya-Eggenberger distribution*

$$\Pi(x;n,p,\gamma) = \binom{n}{x} \frac{p(p+\gamma) \cdots [p+(x-1)\gamma]q(q+\gamma) \cdots [q+(n-x-1)\gamma]}{1(1 + \gamma) \cdots [1 + (n - 1)\gamma]} \tag{12.231}$$

for $x = 1(1)n$, with $0 < p < 1$ and $-1 < \gamma < +\infty$. Letting $n \to \infty$, $p \to 0$, and $\gamma \to 0$, with $np \to \lambda$ and $n\gamma \to d$, $(0 < \lambda < \infty, \; -\frac{1}{2} < d < \infty)$, yields the limiting form *(Polya's distribution),*

$$\Pi(x;\lambda,d) = \binom{\lambda/d + x - 1}{x} \left(\frac{1}{1 + d}\right)^{\lambda/d} \left(\frac{d}{1 + d}\right)^x \qquad x = 0,1, \ldots \tag{12.232}$$

which for $r = \lambda/d$ and $P = d$ is the same as (12.226).

In summary, any *compound Poisson distribution* always appears "contagious" in the sense that, compared with the simple Poisson (12.198), there will be too many intervals (area, volumes, etc.) with "no event" and, in comparison with these, too few with "one event"; however, such contagion may not be inherent in the mechanism underlying the observations, but simply in our method of observation. On the other hand, an excellent fit of Polya's distribution to observations is not necessarily indicative of any phenomena of contagion in the mechanism behind the observed distribution. In order to decide whether or not there is a true contagion, it is not sufficient to consider the over-all distribution of events per unit interval (area, volumes, etc.). Rather a study of the pattern of correlations between adjacent, nearby, and distant intervals is necessary.

Finally, it should be noticed that if two distinct types of events are independently Poisson distributed on a common line with parameters m_1 and m_2 respectively then in view of (12.212), the number of events of the first type lying between an arbitrary pair of events of the second type will be Poisson distributed with parameter $m = m_1X$, itself a random variable distributed according to (12.228) with $r = 1$ and $\beta = m_1/m_2$; S_1, the number of events of the first kind per interval between successive events of the second kind, will have a negative binomial distribution (12.226) with parameters $r = 1$ and $P = m_1/m_2$, that is, the *geometric distribution* (12.214) with parameter $p = m_1/(m_1 + m_2)$; and S_2, the number of events of the second kind per interval between successive events of the first kind, will have the geometric distribution with parameter

$$p = \frac{m_2}{m_1 + m_2}$$

The Multinomial Distribution. As a generalization of Bernoulli trials and the binomial distribution (see above), consider a sequence of repeated *independent* trials, such that the outcome of any single trial can (and must) be one of the (at most enumerably infinite set of) mutually exclusive events $E_1, E_2, \ldots, E_j \ldots$, and let $p_j (j = 1,2, \ldots)$ denote the probability that E_j is realized in any single trial, with

$$\sum_{j=1}^{\infty} p_j = 1$$

The probability-generating function [see (12.122)] for a single trial is then given by

$$G(t_1,t_2, \ldots) = (p_1t_1 + p_2t_2 + \cdots + p_jt_j + \cdots) \qquad 0 < p_j < 1 \quad (12.233)$$

with

$$\sum_{j=1}^{\infty} p_j = 1$$

Let $S_j = S_j(n)$ $(j = 1,2, \ldots)$ denote the number of times that the event E_j occurs in n such trials, then $\{S_j\}$ is a set of discrete random variables subject to the restriction

$$\sum_{j=1}^{\infty} S_j = n$$

and the probability that $S_1 = s_1$, $S_2 = s_2, \ldots, S_j = s_j, \ldots$ is given [see (12.127)] by the coefficient of $t_1^{s_1}t_2^{s_2} \cdots t_j^{s_j} \cdots$ in the multinomial expansion of $[G(t_1,t_2, \ldots)]^n$, that is, by the corresponding term of the *multinomial distribution,*

$$P\{S_1 = s_1, \ldots, S_j = s_j, \ldots\} = \frac{n!}{\prod_{j=1}^{\infty} s_j!} \prod_{j=1}^{\infty} p_j^{s_j} \qquad (12.234)$$

$$\text{with } 0 < p_j < 1, \qquad \sum_{j=1}^{\infty} p_j = 1,$$

for all *non-negative integer* values of s_j $(j = 1,2, \ldots)$ subject to the restriction

$$\sum_{j=1}^{\infty} s_j = n$$

The corresponding characteristic function is

$$\phi(t_1,t_2, \ldots, t_j, \ldots) = [p_1 \exp(it_1) + \cdots + p_j \exp(it_j) + \cdots]^n \qquad (12.235)$$

Hence

$$E[S_j] = np_j \quad \sigma^2[S_j] = np_jq_j \quad q_j = 1 - p_j \qquad (12.236)$$
$$\operatorname{cov}[S_j,S_k] = -np_jp_k \quad \text{for } j \neq k,\ j,k = 0,1,2, \ldots$$

The probabilities of the multinomial distribution (12.234) can always be written in the form

$$\frac{\prod_{j=1}^{\infty} e^{-np_j}(np_j)^{s_j}/s_j!}{e^{-n}(n^n/n!)}$$

$$0 < p_j < 1, \quad \sum_{j=1}^{\infty} p_j = 1, \qquad \sum_{j=1}^{\infty} s_j = n \qquad (12.237)$$

that is, as the *conditional distribution* of *independent* random variables S_j that are *Poisson distributed* with parameters $m_j = np_j$, *given that*

$$\sum_{j=1}^{\infty} S_j = n$$

For the converse, and a related result, see (12.208) and (12.209) respectively.

10. Sampling Distributions

Closely connected with the normal distribution are the sampling distributions of functions of random variables having an underlying normal distribution. This section summarizes the properties of the *chi-square distribution, Student's t distribution,* the *variance ratio or F distribution,* and their associated noncentral distributions. (Table 12.5 summarizes the means and variances of the above-mentioned distributions.)

The Chi-square Distribution. Let $\{X_i\}$ $i = 1, 2, \ldots, \nu$ be a sequence of random variables having a normal distribution with parameters m_i and σ_i, respectively. Then the random variable

$$\chi^2 = \sum_{i=1}^{\nu} \left(\frac{X_i - m_i}{\sigma_i}\right)^2 \qquad (12.238)$$

is said to have the *chi-square* distribution with parameter ν. The parameter ν is termed the *degrees of freedom* of the *chi-square distribution* and the probability density function is given by

$$p_\nu(\chi^2) = \begin{cases} \dfrac{1}{2^{\nu/2}\Gamma(\frac{1}{2}\nu)} (\chi^2)^{\frac{1}{2}\nu - 1} \exp(-\frac{1}{2}\chi^2) & \text{for } \chi^2 \geq 0 \\ 0 & \text{for } \chi^2 < 0 \end{cases} \qquad (12.239)$$

TABLE 12.5. MEANS AND VARIANCES OF SAMPLING DISTRIBUTIONS

Distribution	Mean		Variance	
Chi square	ν		2ν	
Student's t	0		$\frac{\nu}{\nu - 2}$	$\nu > 2$
Variance ratio (F)	$\frac{\nu_2}{\nu_2 - 2}$	$\nu_2 > 2$	$\frac{2\nu_2^2(\nu_1 + \nu_2 - 2)}{\nu_1(\nu_2 - 2)^2(\nu_2 - 4)}$	$\nu_2 > 4$
Noncentral chi square	$\nu + \lambda$		$2(\nu + 2\lambda)$	
Noncentral Student's t	$\left(\frac{\nu}{2}\lambda\right)^{1/2} \frac{\Gamma\left(\frac{\nu-1}{2}\right)}{\Gamma\left(\frac{\nu}{2}\right)}$	$\nu > 1$	$\frac{\nu}{\nu - 2} + \nu\lambda\left\{\frac{1}{\nu - 2} - \frac{1}{2}\left[\frac{\Gamma\left(\frac{\nu-1}{2}\right)}{\Gamma\left(\frac{\nu}{2}\right)}\right]^2\right\}$	$\nu > 2$
Noncentral variance ratio (F)	$\frac{\nu_2(\nu_1 + \lambda)}{(\nu_2 - 2)\nu_1}$	$\nu_2 > 2$	$\frac{2\nu_2^2}{(\nu_2 - 2)^2\nu_1^2(\nu_2 - 4)}[(\nu_1 + \lambda)^2 + (\nu_1 + 2\lambda)(\nu_2 - 2)]$	$\nu_2 > 4$

The characteristic function is

$$\phi(t) = (1 - 2it)^{-\nu/2} \tag{12.240}$$

from which the general expression for the cumulants can be obtained, for example,

$$K_n = 2^{n-1}(n - 1)!\nu$$

Tables of the *chi-square distribution* (12.239) are usually expressed in terms of its "right-tail" integral

$$Q(\chi^2|\nu) = \int_{\chi^2}^{\infty} p_\nu(u)\, du = 1 - P(\chi^2|\nu) \tag{12.241}$$

The principal tabulation of $Q(\chi^2|\nu)$ is Table 7 of Pearson and Hartley [57], which is reproduced in part as Table 26.7 in Abramowitz and Stegun [50]; for details, see [98 and 93], respectively. Statistical applications usually require "upper percentage points," i.e., values of $\chi^2 = \chi^2(\nu,\alpha)$ such that

$$Q(\chi^2(\nu,\alpha)|\nu) = \alpha \tag{12.242}$$

for selected values of α and given values of ν. The principal tabulations of values of $\chi^2(\nu,\alpha)$ are Table IV of Fisher and Yates [51], Table 8 of Pearson and Hartley [57], and Table 2 of Harter [101]; see [94, 98, and 101], respectively, for details. Less complete tables of $\chi^2(\nu,\alpha)$ and of $Q(\chi^2|\nu)$ or $P(\chi^2|\nu)$ are given in various textbooks and handbooks; see, for example, [89–93, 95–97].

As $\nu \to \infty$, $Q(\chi^2|\nu) \to Q(x)$ defined by (12.149), for $x = (\chi^2 - \nu)/\sqrt{2\nu}$. For approximations to $Q(\chi^2|\nu)$, and to $\chi^2(\nu,\alpha)$ when ν is large, see Zelen and Severo [49, p. 941].

The sum of two independent χ^2 distributions with ν_1 and ν_2 degrees of freedom will also have a chi-square distribution with $\nu_1 + \nu_2$ degrees of freedom; i.e., the distribution of $\chi_{\nu_1}{}^2 + \chi_{\nu_2}{}^2$ is $p_{\nu_1+\nu_2}(\chi^2)$. Furthermore, if

$$\chi_\nu{}^2 = \sum_{i=1}^{\nu} \left(\frac{X_i - m_i}{\sigma_i}\right)^2$$

can be written as the sum of k quadratic forms Q_i ($i = 1,2, \ldots ,k$), that is

$$\chi_\nu{}^2 = \sum_{i=1}^{k} Q_i$$

such that Q_i has rank ν_i, then the necessary and sufficient condition that Q_i be independently distributed as chi square with ν_i degrees of freedom is that

$$\nu = \sum_{i=1}^{k} \nu_i$$

This last result forms the basis of R. A. Fisher's analysis of variance procedures [7, 16].

Noncentral Chi-square Distribution. Let $\{X_i\}$ be a sequence of independent random variables having a normal distribution with parameters (m_i,σ_i) for $i = 1, 2, \ldots, \nu$. Then

$$\chi'^2{}_\nu = \sum_{i=1}^{\nu} \left(\frac{X_i}{\sigma_i}\right)^2 \tag{12.243}$$

is said to follow the *noncentral chi-square distribution* having the probability density function

$$p_\nu(\chi'^2) = \frac{\exp(-\chi'^2/2 - \lambda/2)}{2^{\nu/2}} \sum_{j=0}^{\infty} \frac{(\chi'^2)^{\frac{1}{2}\nu+j-1}}{2^{2j}\Gamma(\frac{1}{2}\nu + j)} \frac{\lambda^j}{j!} \tag{12.244}$$

where

$$\lambda = \sum_{i=1}^{\nu} \frac{m_i{}^2}{\sigma_i{}^2}$$

is called the noncentral parameter and ν is the number of degrees of freedom.

The characteristic function is given by

$$\phi(t) = \frac{\exp[\lambda it/(1 - 2it)]}{(1 - 2it)^{\nu/2}} \tag{12.245}$$

and the general expression for the cumulants is

$$K_n = 2^{n-1}(n - 1)!(\nu + n\lambda) \tag{12.246}$$

Tables of the distribution of noncentral chi square have been given by Fix [99, 100]. Also Patnaik [43] has shown that χ'^2/C has approximately a χ^2 distribution with m degrees of freedom, that is,

$$\int_0^{\chi'^2} p_\nu(\chi'^2)\, d\chi'^2 \sim \int_0^{\chi'^2/C} p_m(\chi^2)\, d\chi^2 \tag{12.247}$$

where $$C = \frac{\nu + 2\lambda}{\nu + \lambda} \qquad m = \frac{(\nu + \lambda)^2}{\nu + 2\lambda} \qquad (12.248)$$

Student's t Distribution. Let X be a random variable from a normal population with parameters $m = 0$, $\sigma = 1$, and let χ^2 be independent of X and distributed like (12.239). Then

$$t = \frac{X}{\sqrt{\chi_\nu^2/\nu}} \qquad (12.249)$$

follows the *Student's t distribution* with ν *degrees of freedom.* The probability density function is

$$p_\nu(t) = \frac{1}{B\left(\frac{\nu}{2}, \frac{1}{2}\right)\sqrt{\nu}}\left(1 + \frac{t^2}{\nu}\right)^{-(\nu+1)/2} \qquad (12.250)$$

The distribution is unimodal and symmetric about $t = 0$. The ith moment is finite for $i < \nu$. Since the distribution is symmetric all existing odd-order moments are zero and the general expression for even-order moments (about the origin) is given by

$$\alpha_{2n} = \mu_{2n} = \frac{1 \cdot 3 \cdot \cdot \cdot (2n-1)\nu^n}{(\nu-2)(\nu-4) \cdot \cdot \cdot (\nu-2n)} \qquad \text{for } 2n < \nu \qquad (12.251)$$

As the parameter ν increases, Student's distribution approaches a normal distribution with parameters (0,1), that is,

$$\lim_{\nu \to \infty} p_\nu(t) = \frac{1}{\sqrt{2\pi}} \exp -\tfrac{1}{2}t^2$$

If $\nu = 1$, the distribution reduces to

$$p_1(t) = \frac{1}{\pi}(1 + t^2)^{-1} \qquad (12.252)$$

which is the *Cauchy distribution.*

In keeping with (12.149), let

$$P(t|\nu) = \int_{-\infty}^{t} p_\nu(u)\,du = 1 - Q(t|\nu)$$
$$A(t|\nu) = \int_{-t}^{+t} p_\nu(u)\,du = 1 - 2Q(t|\nu) \qquad (12.253)$$

where $p_\nu(u)$ is given by (12.250) with u in place of t. The principal tabulation of $P(t|\nu)$ is Table 9 of Pearson and Hartley [57]; see [122] for details. Statistical applications often require "two-tail percentage points" of t, that is, values of $t = t(\nu,\alpha)$ such that

$$2Q(t(\nu,\alpha)|\nu) = \alpha \qquad (12.254)$$

for various values of α corresponding to given values of ν. The principal tabulations of values of $t(\nu,\alpha)$ are Table III of Fisher and Yates [51], Table 12 of Pearson and Hartley [57], Table 26.10 in Abramowitz and Stegun [50], Table A.2 in Graybill [7], and Table I of Owen [119]; for details, see [114, 117, 113, 112, and 119], respectively. Portions of these tables of $t(\nu,\alpha)$ are given in many textbooks and handbooks; see, for example, [1, 2, 4, 10, 12, 14, and 54].

Noncentral Student's t Distribution. Let X be a random variable from a normal population with mean $m = \Delta$ and variance σ^2, and let χ_ν^2 follow an independent chi-square distribution with ν degrees of freedom. Then

$$t' = \frac{X/\sigma}{\sqrt{\chi_\nu^2/\nu}} \qquad (12.255)$$

is said to follow the *noncentral Student's t distribution* having the probability density function

$$p_\nu(t') = \frac{\exp(-\frac{1}{2}\lambda)}{\sqrt{\nu}}\left(1 + \frac{t'^2}{\nu}\right)^{-(\nu+1)/2} \sum_{j=0}^{\infty} \frac{\lambda^j}{2^j j!\,B\left(\frac{1}{2} + j, \frac{\nu}{2}\right)}\left(\frac{\nu t'^2}{\nu + t'^2}\right)^j \qquad (12.256)$$

with ν degrees of freedom and noncentral parameter $\lambda = \Delta^2/\sigma^2 = \delta^2$.

The *noncentral t distribution* has been tabulated and graphed in two specialized forms appropriate to particular applications. Owen [119, Table II], Croarkin [120], Neyman and Tokarska [121], and Pearson and Hartley [122] have provided tables and graphs of the power function of Student's t test, which is discussed in many textbooks, for example, Graybill [7], Hald [8], Mood and Graybill [12], and Natrella [14]. Johnson and Welch [123], Resnikoff and Lieberman [124], and Scheuer and Spurgeon [125] give tables designed for industrial sampling inspection applications but useful for various other applications which they describe.

Variance Ratio (F) and Fisher's z Distribution. Let χ_1^2, χ_2^2 be two independent random variables following a χ^2 distribution with degrees of freedom ν_1 and ν_2, respectively, and consider the ratio

$$F = \frac{\chi_1^2/\nu_1}{\chi_2^2/\nu_2} \qquad (12.257)$$

The distribution of F is said to follow the *variance ratio* or *F distribution* with ν_1 and ν_2 degrees of freedom. The probability density function is given by

$$p_{\nu_1,\nu_2}(F) = \left(\frac{\nu_1}{\nu_2}\right)^{\nu_1/2} \frac{1}{B\left(\frac{\nu_1}{2}, \frac{\nu_2}{2}\right)} \frac{F^{\frac{1}{2}(\nu_1-2)}}{\left(1 + \frac{\nu_1}{\nu_2}F\right)^{\frac{1}{2}(\nu_1+\nu_2)}} \qquad \text{for } 0 \le F < \infty \qquad (12.258)$$

When $\nu_1 = 1$, the distribution of F is the same as that of t^2, the square of Student's t. The moments up to order $\nu_2/2$ exist and the moments about the origin are

$$\alpha_n = \frac{\Gamma(\frac{1}{2}\nu_1 + n)\,\Gamma(\frac{1}{2}\nu_2 - n)}{\Gamma(\frac{1}{2}\nu_1)\Gamma(\frac{1}{2}\nu_2)}\left(\frac{\nu_2}{\nu_1}\right)^n \qquad \text{for } n < \frac{\nu_2}{2} \qquad (12.259)$$

Many statistical textbooks and handbooks (e.g., [1, 7, 12, 14, 16, 50, 51, 53, 54, 56, 57]) give tables of values

$$Q(F(\nu_1,\nu_2,\alpha)|\nu_1,\nu_2) = \int_{F(\nu_1,\nu_2,\alpha)}^{\infty} p_{\nu_1,\nu_2}(u)\,du = \alpha \qquad (12.260)$$

for selected values of $\alpha < 0.5$ and for various values of ν_1 and ν_2. The principal source tabulations of $F(\nu_1,\nu_2,\alpha)$ are Table V of Fisher and Yates [51] and Table 18 of Pearson and Hartley [57]; for details, see [126 and 127]. Some additional values are given in Table A.3 of Graybill [7]; see [128] for details. Values of $F(\nu_1,\nu_2,\alpha)$ for $\alpha > 0.5$ can be obtained from the relationship

$$F(\nu_1, \nu_2, 1 - \alpha) = \frac{1}{F(\nu_1,\nu_2,\alpha)} \tag{12.261}$$

The above tables have been derived from tables of the incomplete-beta-function ratio $I_x(a,b)$ [see Eq. (12.177) and refs. 83–85] by virtue of the relationship

$$Q(F|\nu_1,\nu_2) = I_x\left(\frac{\nu_2}{2}, \frac{\nu_1}{2}\right) \tag{12.262}$$

where $$x = \frac{\nu_1}{\nu_2 + \nu_1 F}$$

R. A. Fisher in 1924 first derived the distribution of the related quantity

$$z = \tfrac{1}{2} \ln F$$

However, in practice it is more convenient to use the variance ratio F.

Noncentral F Distribution. Let $\chi_1'^2$ follow a noncentral chi-square distribution with ν_1 degrees of freedom and parameter λ and let $\chi_2{}^2$ follow a chi-square distribution with ν_2 degrees of freedom. Then the ratio

$$F' = \frac{\chi_1'^2/\nu_1}{\chi_2{}^2/\nu_2} \tag{12.263}$$

has the *noncentral F distribution* with ν_1 and ν_2 degrees of freedom and noncentral parameter λ.

The probability density function is

$$p_{\nu_1,\nu_2}(F') = \left(\frac{\nu_1}{\nu_2}\right)^{\nu_1/2} \frac{F'^{\frac{1}{2}(\nu_1-2)}}{\left(1 + \frac{\nu_1}{\nu_2}F'\right)^{\frac{1}{2}(\nu_1+\nu_2)}} e^{-\lambda/2} \sum_{j=0}^{\infty} \frac{\lambda^j}{2^j j! B\left(\frac{\nu_1}{2} + j, \frac{\nu_2}{2}\right)} \left(\frac{F'}{1 + \frac{\nu_1}{\nu_2}F'}\right)^j \tag{12.264}$$

where $$\lambda = \sum_i \frac{m_i{}^2}{\sigma_i{}^2}$$

If $\nu_1 = 1$, then (12.264) reduces to the noncentral t'^2 distribution.

The noncentral F distribution has been tabulated and graphed only in the form of power functions of F tests for fixed effects (model I) in analysis of variance by Fox [130], Lehmer [131], and Pearson and Hartley [132]. Fox's graphs are reproduced in Scheffé [16]. Graybill [7] reprints an earlier table of Tang (1938); see [129] for details. F tests in analysis of variance are discussed in the above-mentioned books and also by Hald [8], Mood and Graybill [12], and Natrella [14].

Approximations to the noncentral F distribution are given by Zelen and Severo [49, p. 948].

References

Textbooks and Treatises on Probability and Statistics:

1. Arley, N., and K. R. Buch: "Introduction to the Theory of Probability and Statistics," Wiley, New York, 1950.
2. Cramér, H.: "Mathematical Methods of Statistics," Princeton University Press, Princeton, N.J., 1946.
3. Feller, W.: "An Introduction to Probability Theory and Its Applications," 2d ed., vol. 1, Wiley, New York, 1957.

3*a*. *Ibid.*, vol. 2, 1966.

4. Fisz, M.: "Probability Theory and Mathematical Statistics," 3d ed., Wiley, New York, 1963.
5. Gnedenko, B. V.: "The Theory of Probability," 2d ed., Chelsea Publishing Company, New York, 1963.
6. Good, I. J.: "Probability and the Weighing of Evidence," Hafner, New York, 1950.
7. Graybill, F. A.: "An Introduction to Linear Statistical Models," vol. 1, McGraw-Hill, New York, 1961.
8. Hald, A.: "Statistical Theory with Engineering Applications," Wiley, New York, 1952.
9. Jeffreys, H.: "Theory of Probability," 3d ed., Clarendon Press, Oxford, 1961.
10. Kendall, M. G., and A. Stuart: "The Advanced Theory of Statistics,"vol. 1 ("Distribution Theory"), Griffin, London, 1958.
11. Loève, M.: "Probability Theory," 3d ed., Van Nostrand, Princeton, N.J., 1963.
12. Mood, A. M., and F. A. Graybill: "Introduction to the Theory of Statistics," 2d ed., McGraw-Hill, New York, 1963.
13. Mosteller, F., R. E. K. Rourke, and G. B. Thomas, Jr.: "Probability with Statistical Applications," Addison-Wesley, Reading, Mass., 1961.
14. Natrella, Mary G.: Experimental Statistics, *Natl. Bur. Std. (U.S.), Handbook* 91, 1963.
15. Parzen, E.: "Modern Probability Theory and Its Applications," Wiley, New York, 1960.
16. Scheffé, H.: "The Analysis of Variance," Wiley, New York, 1959.
17. Uspensky, J. V.: "Introduction to Mathematical Probability," McGraw-Hill, New York, 1937.
18. Wallis, W. A., and H. V. Roberts: "Statistics, A New Approach," Free Press, Glencoe, Ill., 1956.

Monographs, Research Papers, Tracts, Etc.:

19. Bayes, Thomas: An Essay Towards Solving a Problem in the Doctrine of Chances, *Phil. Trans.*, **53**: 370–418 (1763). Reprinted in modernized notation, with editorial corrections and a biographical note by G. A. Barnard, *Biometrika*, **45**: 293–315 (1958).
20. Camp, Burton H.: Approximation to the Point Binomial, *Ann. Math. Statist.*, **22**: 130–131 (1951).
21. Craig, C. C.: On the Frequency Function *xy*, *Ann. Math. Statist.*, **7**: 1–15 (1936).
22. Cramér, H.: "Random Variables and Probability Distributions," 2d ed., Cambridge University Press, New York, 1962.
23. Daly, J.: On the Use of the Sample Range in an Analogue of Student's *t*-test, *Ann. Math. Statist.*, **17**: 71–74 (1946).
24. Dirac, P. A. M.: "The Principles of Quantum Mechanics," 3d ed., Clarendon Press, Oxford, 1947.
25. Feller, W.: On the Normal Approximation to the Binomial Distribution, *Ann. Math. Statist.*, **16**: 319–329 (1945).
26. Feller, W.: The Fundamental Limit Theorems in Probability, *Bull. Am. Math. Soc.*, **51**: 800–832 (1945).

27. Fieller, E. C.: The Distribution of the Index in a Normal Bivariate Population, *Biometrika*, **24**: 428–440 (1932).
28. Gnedenko, B. V., and A. N. Kolmogorov: "Limit Distributions for Sums of Independent Random Variables" (trans. by K. L. Chung), Addison-Wesley, Cambridge, Mass., 1954.
29. Godwin, H. J.: On Generalizations of Tchebychef's Inequality, *J. Am. Statist. Assoc.*, **50**: 923–943 (1955).
30. Greenwood, M., and G. U. Yule: An Inquiry into the Nature of Frequency Distributions Representative of Multiple Happenings with Particular Reference to the Occurrence of Multiple Attacks of Disease or of Repeated Accidents, *J. Roy. Statist. Soc.*, **83**: 255–279 (1920).
31. Haight, F. A.: Index to the Distributions of Mathematical Statistics, *J. Res. Natl. Bur. Std.*, **65 B**: 23–60 (1961).
32. Halperin, I.: "Introduction to the Theory of Distributions, Based on the Lectures Given by Laurent Schwartz," University of Toronto Press, Toronto, 1952.
33. Kolmogorov, A. N.: "Foundations of the Theory of Probability," 2d ed., Chelsea Publishing Company, New York, 1956.
34. Loève, M.: Fundamental Limit Theorems of Probability Theory, *Ann. Math. Statist.*, **21**: 321–338 (1950).
35. Lukacs, E.: Characteristic Functions, Griffin's Statistical Monographs and Courses No. 5, Hafner, New York, 1960.
36. Lukacs, E., and R. G. Laha: "Applications of Characteristic Functions," Griffin's Statistical Monographs and Courses No. 14, Hafner, New York, 1964.
37. Lundberg, O.: "On Random Processes and Their Application to Sickness and Accident Statistics," Almquist and Wiksells, Uppsala, Sweden, 1940.
38. Marsaglia, G.: Ratios of Normal Variables and Ratios of Sums of Uniform Variables, *J. Am. Statist. Assoc.*, **60**: 193–204, 1965.
39. Moran, P. A. P.: A Characteristic Property of the Poisson Distribution, *Proc. Cambridge Phil. Soc.*, **48**: 206–207 (1952).
40. Morse, P. M., and H. Feshbach: "Methods of Theoretical Physics," Pt. I, McGraw-Hill, New York, 1953.
41. Mosteller, F., and J. W. Tukey: The Uses and Usefulness of Binomial Probability Paper, *J. Am. Statist. Assoc.*, **44**: 174–212 (1949).
42. Mosteller, F., and J. W. Tukey: "Binomial Probability Graph Paper," Codex Book Co., Norwood, Mass.
43. Patnaik, P. B.: The Non-central χ^2- and F-distributions and Their Applications, *Biometrika*, **36**: 202–232 (1949).
44. Raff, M. S.: On Approximating the Point Binomial, *J. Am. Statist. Assoc.*, **51**: 293–303 (1956).
45. Savage, I. R.: Probability Inequalities of the Tchebycheff Type, *J. Res. Natl. Bur. Std.* **65B**: 211–222 (1961).
46. Savage, L. J., et al.: "The Foundations of Statistical Inference," Methuen, London (Wiley, New York), 1962.
47. Schwartz, L.: "Théorie des distributions," vols. I, II, Hermann & Cie, Paris, 1950, 1951.
48. Smith, E. S.: "Binomial, Normal, and Poisson Probabilities," published by the author, Bel Air, Md., 1953.
49. Zelen, M., and N. C. Severo: Probability Functions, chap. 26 of [50].

Tables

General Collections and Guides:

50. Abramowitz, M., and Irene A. Stegun (eds.): Handbook of Mathematical Functions, with Formulas, Graphs, and Mathematical Tables, *Natl. Bur. Std. (U.S.) Appl. Math. Ser.* 55, 1964.
51. Fisher, R. A., and F. Yates: "Statistical Tables for Biological, Agricultural and Medical Research," 6th ed., Oliver & Boyd, Edinburgh and London (Hafner, New York), 1963.
52. Greenwood, J. A., and H. O. Hartley: "Guide to Tables in Mathematical Statistics," Princeton University Press, Princeton, N.J., 1962. (Catalogues a large selection of tables used in mathematical statistics.)
53. Hald, A.: "Statistical Tables and Formulas," Wiley, New York, 1952.
54. Weast, R. C., S. M. Selby, and C. D. Hodgman (eds.): "Handbook of Mathematical Tables," 2d ed. (supplement to "Handbook of Chemistry and Physics"), Chemical Rubber Co., Cleveland, Ohio, 1964.
55. National Bureau of Standards: A Guide to Tables of the Normal Probability Integral, *Appl. Math. Ser.* 21, 1952.
56. Owen, D. B.: "Handbook of Statistical Tables," Addison-Wesley, Reading, Mass., 1962.
57. Pearson, E. S., and H. O. Hartley (eds.): "Biometrika Tables for Statisticians," vol. 1, Cambridge University Press, London, 1954.

Normal Probability Distribution, Ordinates, and Integral:

58. Tables 26.1, 26.2, 26.4, and 26.5 of [50]:
$Z(x)$ and $P(x)$ for $x = 0(.02)3.00$, $15D$; $Z(x)$ and $P(x)$ for $x = 3.00(.05)5.00$, $10S$. $-\log Q(x)$ for $x = 5(1)50(10)100(50)500$, $5D$. $Z(x)$ for $P(x) = .5(.001)1$, $5D$. x for $Q(x) = 0(.001).5$, $5D$; x for $Q(x) = 0(.0001).025$, $5D$; x for $Q(x) = 10^{-n}$, $n = 4(1)23$, $5D$.
59. Tables I, II, and IX of [51]:
x for $2Q(x) = 0(.01).99$, $6D$. Z for $x = 0(.01)3(.1)3.9$, $4D$. $x + 5$ for $P = .001(.001).980(.0001).9999$, $4D$.
60. Tables I, II, and III of [53]:
$Z(x)$ for $x = 0(.01)4.99$, $4S$. $P(x)$ for $x = 0(.01)1.29$, $4D$; $x = 1.30(.01)2.32$, $5D$; $x = 2.33(.01)3.09$, $6D$; $x = 3.10(.01)3.71$, $7D$; $x = 3.72(.01)4.26$, $8D$; $x = 4.27(.01)4.75$, $9D$; $x = 4.76(.01)4.99$, $10D$. $x + 5$ for $P(x) = .0001(.0001).025(.001).975(.0001).9999$, $3D$.
61. Sections 1.1 and 1.2 of [56]:
$Z(x)$ and $P(x)$ for $x = 0(.01)3.99$, $6D$. x and $Z(x)$ for $Q(x) = .5(.01).10(.005).010(.001).001(.0001).0001$ and for $Q(x) = 5 \times 10^{-n}$ and 1×10^{-n}, $n = 5(1)9$, $5D$.
62. Tables 1 to 5 of [57]:
$Z(x)$ and $P(x)$ for $x = 0(.01)4.50$, $7D$; $x = 4.50(.01)6.00$, $10D$. $-\log Q(x)$ for $x = 5(1)50(10)100(50)500$, $5D$. x for $Q(x) = .02(.0001).0001$, $4D$. x for $Q(x) = .5(.001).001$ and 10^{-n}, $n = 4(1)9$, $4D$. $Z(x)$ for $Q(x) = 0(.001).5$, $5D$.
63. Harvard University: "Tables of the Error Function and Its First Twenty Derivatives," Harvard University Press, Cambridge, Mass., 1952.
$Z(x)$ for $x = 0(.004)5.216$, and $\frac{1}{2}A(x) = P(x) - \frac{1}{2}$ for $x = 0(.004)4.892$, $6D$.
64. Kelley, T. L.: "The Kelley Statistical Tables," Harvard University Press, Cambridge, Mass., 1948.
x and $Z(x)$ corresponding to $P(x) = .5(.0001).9999$, $8D$. x and $Z(x)$ for $Q(x) = 5 \times 10^{-n}$ and 1×10^{-n}, $n = 5(1)9$, $8S$–$6S$.

65. National Bureau of Standards: Tables of Normal Probability Functions, *Appl. Math. Ser.* 23, 1953.
$Z(x)$ and $A(x)$ for $x = 0(.0001)1(.001)7.800$(various)8.285, 15D; $Z(x)$ and $2Q(x)$ for $x = 6(.01)10$, 7S.

Bivariate Normal Probability Integral:

66. National Bureau of Standards: Tables of the Bivariate Normal Distribution Function and Related Functions, with Introduction by Gertrude Blanche and Applications by D. B. Owen, *Appl. Math. Ser.* 50, 1959.
$L(h,k,\rho)$ for h, $k = 0(.1)4$, $\rho = 0(.05).95(.01)1$, 6D; $L(h,k,-\rho)$ for h, $k = 0(.1)A$, $\rho = 0(.05).95(.01)1$, where A is such that $L < .5 \times 10^{-7}$, 7D; $V(h,ah)$ for $h = 0(.01)4(.02)4.6(.1)5.6$, ∞, 7D; $V(ah,h)$ for $a = .1(.1)1$, $h = 0(.01)4(.02)5.6$, ∞, 7D.
67. Nicholson, C.: The Probability Integral for Two Variables, *Biometrika*, **33**: 59–72 (1943).
$V(h,ah)$ for $h = .1(.1)3$, ∞, 6D.
68. Owen, D. B.: Tables for Computing Bivariate Normal Probabilities, *Ann. Math. Statist.*, **27**: 1075–1090 (1956).
$T(h,a)$ for $a = .25(.25)1$, $h = 0(.01)2(.02)3$; $a = 0(.01)1$, ∞, $h = 0(.25)3$; $a = .1$, $.2(.05).5(.1).8$, 1, ∞, $h = 3(.05)3.5(.1)4.7$, 6D.
69. Owen, D. B.: "The Bivariate Normal Probability Function," U.S. Department of Commerce, Office of Technical Services, Washington, D.C., 1957.
$T(h,a)$ for $a = 0(.025)1$, ∞; $h = 0(.01)3.5(.05)4.75$, 6D.
70. Section 8.5 of [56]:
$T(h,a)$ for $a = .25(.25)1$, $h = 0(.01)3.14$; $a = 0(.01)1$, ∞, $h = 0(.25)3.25$; $a = .1$, $.2(.05).5(.1)1$, ∞, $h = 3(.05)3.5(.1)4.0(.2)4.6$, 4.76, 6$D$.
71. Owen, D. B., and J. M. Wiesen: A Method of Computing Bivariate Normal Probabilities, *Bell System Tech. J.*, **38**: 1–20 (1959).
72. Zelen, M., and N. C. Severo: Graphs for Bivariate Normal Probabilities, *Ann. Math. Statist.*, **31**: 619–624 (1960).
$L(h,o,\rho)$ for $0 \le h \le 1$, $-1 \le \rho \le 0$; $0 \le h \le 1$, $0 \le \rho \le 1$; $1 \le h \le 2.50$, $-1 \le \rho \le 1$.

Binomial Distribution and the Incomplete Beta Function:

73. Tables IVA and IVB of [13]:
$b(s;n,p)$ and $Q(s-1;\ n,\ p)$ for $p = .01$, .05, .1(.1).9, .95, .99, $n = 2(1)25$, $s = 0(1)n$, 3D.
74. Table II of [15]:
$b(s;n,p)$ for $p = .01$, .05(.05).30, $\frac{1}{3}$, .35(.05).50 and $p = .49$, $n = 1(1)10$, $s = 0(1)n$, 4D.
75. Pages 537–544 of [54]:
$b(s;n,p)$ for $p = .05(.05).50$, $n = 1(1)20$, $A = 0(1)n$, 4D. $Q(s-1;\ n,p)$ for same p and n, $A = 1(1)n$, 4D.
76. Section 9.5 of [56]:
$P(s;n,p)$ for $p = \frac{1}{16}(\frac{1}{16})\frac{8}{16}$, $n = 2(1)25$, $s = 0(1)n$, 4D.
77. Harvard University Computation Laboratory: "Tables of the Cumulative Binomial Probability Distribution," Harvard University Press, Cambridge, Mass., 1955.
$Q(s-1;\ n,p)$ for $p = .01(.01).5$, $\frac{1}{16}$, $\frac{1}{12}$, $\frac{1}{8}$, $\frac{1}{6}$, $\frac{3}{16}$, $\frac{5}{16}$, $\frac{1}{3}$, $\frac{3}{8}$, $\frac{5}{12}$, $\frac{7}{16}$, $n = 1(1)50(2)100(10)200(20)500(50)1000$, $s = 1(1)n$, 5D.
78. National Bureau of Standards: Tables of the Binomial Probability Distribution, *Appl. Math. Ser.* 6, 1950.
$b(s;n,p)$ for $p = .01(.01).50$, $n = 2(1)49$, $s = 0(1)n-1$, 7D. $Q(s-1;\ n,p)$ for $p = .01(.01).50$, $n = 2(1)49$, $s = 1(1)n$, 7D.
79. Robertson, W. H.: Tables of the Binomial Distribution Function for Small Values of p, *Sandia Corp. Monograph SCR*-143, U.S. Department of Commerce, Office of Technical Services, 1960.
$P(s;n,p)$ for $p = .001(.001).02$, $n = 2(1)100(2)200(10)500(20)1000$, $s = 0(1)$ various, 5D; and for $p = .021(.001).05$, $n = 2(1)50(2)100(5)200(10)300(20)600(50)1000$, $s = 0(1)$ various, 5D.
80. Romig, H. G.: "50–100 Binomial Tables," Wiley, New York, 1953.
$b(s;n,p)$ and $P(s;n,p)$ for $p = .01(.01).5$, $n = 50(5)100$, $s = 0(1)$various, 6D.
81. U.S. Army Ordnance Corps: Tables of Cumulative Binomial Probabilities, *Ordnance Corps Pamph. ORDP* 20-1, U.S. Department of Commerce, Office of Technical Services, 1952.
$Q(s-1;n,p)$ for $p = .01(.01).5$, $n = 1(1)150$, $s = 1(1)n$, 7D.
82. Weintraub, S.: "Tables of the Cumulative Binomial Probability Distribution for Small Values of p," Free Press, New York, 1963.
$Q(s-1;\ n,\ p)$ for $p = .00001$, .0001(.0001).001(.001).1, $n = 1(1)100$, $s = 1(1)n$, 10D.
83. Pearson, K. (ed.): "Tables of the Incomplete Beta Function," Cambridge University Press, London, 1934.
$I_x(a,b)$ for $x = .01(.01)1$, $a = b(.5)11(1)50$, $b = .5(.5)10.5$ and $a = b(1)50$, $b = 11(1)50$, 7D.
84. Table 16 of [57]:
Values for x for which $I_x(a,b) = Q(a-1;\ a+b-1,\ x) = .005$, .01, .025, .05, .1, .25, .5; $2a = 1(1)30$, 40, 60, 120, ∞, $2b = 1(1)10$, 12, 15, 20, 24, 30, 40, 60, 120, ∞, 5D.
85. Table 3 of [101]:
Values of x for which $I_x(a,b) = .0001$, .0005, .001, .005, .01, .025, .05, .1(.1).5, and values of $x < .1$ for which $I_x(a,b) = .6(.1).9$, .95, .975, $a = 1(1)40$, $b = 1(1)40$, 7S.
86. Vogler, L. E.: Percentage Points of the Beta Distribution, *Natl. Bur. Std. (U.S.), Tech. Note* 215, 1964.
Values of x for which $I_x(a,b) = .0001$, .001, .005, .01, .025, .05, .1, .5, $2a = 1(.1)2$, 2.2, 2.5(.5)5(1)10, 12, 15, 20, 24, 30, 40, 60, 120, ∞, $2b = 1(.1)10$, 12, 20, 24, 30, 40, 60, 120, 6S.

Hypergeometric Probability Distribution:

87. Lieberman, G. J., and D. B. Owen: "Tables of the Hypergeometric Probability Distribution," Stanford University Press, Stanford, Calif., 1961.
$h(s;n,p,N)$ and $P(s;n,p,N)$ for $N = 2(1)25$, $n = 1(1)N-1$, $Np = 1(1)n$, $s = 0(1)Np$, 6D; and for $N = 26(1)50(10)100$, $n = 1(1)\dfrac{N}{2}$ for N *even* and $n = 1(1)\dfrac{N-1}{2}$ for N *odd*, $Np = 1(1)n$, $s = 0(1)Np$. $h(s;500,p,1{,}000)$ and $P(s;500,p,1{,}000)$ for $1{,}000p = 1(1)500$, $s = 0(1)\dfrac{1{,}000p}{2}$ or $\dfrac{1{,}000p-1}{2}$, 6D. $h(s;n,p,N)$ and $P(s;n,p,N)$ for $N = 100(100)2{,}000$, $n = \frac{1}{2}N$, $Np = n-1$, n, and $s = 0(1)\dfrac{n}{2}$ or $\dfrac{n-1}{2}$, 6D.
88. Section 18.1 of [56]:
$h(s;n,p,N)$ and $P(s;n,p,N)$ for $N = 2(1)20$, $n = 1(1)\dfrac{N}{2}$ for N *even* and $n = 1(1)\dfrac{N-1}{2}$ for N *odd*, $Np = 1(1)n$, $s = 0(1)n$, 6D.

Chi-square, Noncentral Chi-square, Incomplete-gamma-function Ratio, Poisson Exponential Distribution:

89. Table 3 of [2]:
 Values of χ^2 for which $Q(\chi^2|\nu) = .99, .98, .95, .90, .80, .70, .50, .30, .20, .10, .05, .02, .01, .001$, when $\nu = 1(1)30$, $3D$.
90. Table IV of [4]:
 Values of χ^2 for which $Q(\chi^2|\nu) = .80, .70, .50, .30, .20, .10, .05, .02, .01$, when $\nu = 1(1)30$, $3D$.
91. Pages 435–438 of [5]:
 $Q(\chi^2|\nu)$ for $\chi^2 = 1(1)30$, $\nu = 1(1)29$, $4D$.
92. Table A.1 of [7]:
 Values χ^2 for which $Q(\chi^2|\nu) = .0001, .001, .005, .01, .025, .05, .1, .25, .50, .75, .9, .95, .975, .99, .995, .999, .9999$, $\nu = 1(1)10, 12, 15, 20, 24, 30, 40, 60, 120$, $5S$.
93. Tables 26.7 and 26.8 of [50]:
 $Q(\chi^2|\nu)$ for $\nu = 1(1)30$, $\chi^2 = .001(.001).01(.01).1(.1)2(.2)10(.5)20(1)40(2)76$, $5D$, with auxiliary scale showing $m = \chi^2/2$, to aid in reading the cumulative Poisson probability $P(s-1;\ m) = Q(\chi^2|\nu)$ for $\nu = 2s$. Values of χ^2 for which $Q(\chi^2|\nu) = .995, .99, .975, .95, .9, .75, .5, .25, .1, .05, .025, .01, .005, .001, .0005, .0001$ when $\nu = 1(1)30(10)100$, $5\text{–}6S$.
94. Table IV of [51]:
 Values of χ^2 for which $Q(\chi^2|\nu) = .001, .01, .02, .05, .1, .2, .3, .5, .7, .8, .9, .95, .98, .99$ when $\nu = 1(1)30$, $3D$ or $3S$.
95. Tables 5 and 6 of [53]:
 Values of χ^2 for which $P(\chi^2|\nu) = .0005, .001, .005, .01, .025, .05, .1(.1).9, .95, .975, .99, .995, .999, .9995$ when $\nu = 1(1)100$, $3S$. Values of $\chi^2/\nu = s^2/\sigma^2$ for which $P(\chi^2|\nu)$ = above values when $\nu = 1(1)100(5)200(10)300(50)1{,}000(1{,}000)5{,}000, 10{,}000$, $4D$.
96. Pages 280 and 533 of [54]:
 Values of χ^2 for which $Q(\chi^2|\nu) = .99, .98, .95, .90, .80, .70, .50, .30, .20, .10, .05, .02, .01$ when $\nu = 1(1)30$. $Q(\chi^2|\nu)$ for $\chi^2 = 1(1)4(2)10(5)20$, $\nu = 2(1)11$; and for $\chi^2 = 8(2)20(5)30$, $\nu = 9(1)19$, $3D$.
97. Section 3.1 of [56]:
 Values of χ^2 for which $P(\chi^2|\nu) = .005, .01, .025, .05, .10, .25, .75, .90, .95, .975, .99, .995$, when $\nu = 1(1)100(2)200(50)300(100)1{,}000$, $3D$.
98. Tables 7 and 8 of [57]:
 $Q(\chi^2|\nu)$ for $\nu = 1(1)20(2)70$, $\chi^2 = .001(.001).01(.01).1(.1)2(.2)10(.5)20(1)40(2)134$, $5D$, with auxiliary scale showing $m = \chi^2/2$, to aid in reading the cumulative Poisson probability $P(s-1;\ m) = Q(\chi^2|\nu)$ for $\nu = 2s$. Values of χ^2 for which $Q(\chi^2|\nu) = .995, .99, .975, .95, .90, .75, .5, .25, .1, .05, .025, .01, .005, .001$ when $\nu = 1(1)30(10)100$, $6S$.
99. Fix, E.: Tables of Non-central χ^2, *Univ. of Calif. Publ. in Statist.*, **1**: 15–19 (1949).
 Values of λ for which $P\{\chi'^2 > \chi^2(\alpha,\nu)|\nu,\lambda\} = .1(.1).9$, where the $\chi^2(\alpha,\nu)$ are the values of χ^2 for which $Q(\chi^2|\nu) = \alpha = .01, .05$, when $\nu = 1(1)20(2)40(5)60(10)100$, $3D$ or $3S$.
100. Fix, Evelyn, J. L. Hodges, Jr., and E. L. Lehmann: The Restricted Chi-square Test, in Ulf Grenander, "Probability and Statistics: The Harold Cramér Volume," pp. 92–107, Wiley, New York, 1959.
 Values of λ for which $P\{\chi'^2 > \chi^2(\alpha,\nu)|\nu,\lambda\} = .5(.1), .9\ .95$, where the $\chi^2(\alpha,\nu)$ are the values of χ^2 for which $Q(\chi^2|\nu) = \alpha = .001, .005, .01(.05).3(.1)$, when $\nu = 1(1)6$.
101. Harter, H. Leon: "New Tables of the Incomplete Gamma-function Ratio, and of the Percentage Points of the Chi-square and Beta Distributions," Government Printing Office, Washington, D.C., 1964.
 $I(u,p)$ for $u = .1(.1)25$, $p = -.5(.5)74(1)164$, $9D$.
 Values of χ^2 for which $P(\chi^2|\nu) = .0001, .0005, .001, .005, .01, .025, .05, .1(.1).9, .95, .975, .99, .995, .999, .9995, .9999$ when $\nu = 1(1)150(2)330$, $6S$. (See [85] also.)
102. Pearson, K. (ed.): "Tables of the Incomplete Γ-function," Cambridge University Press, London, 1934.
 $I(u,p)$ for $p = -1(.05)0(.1)5(.2)50$, $u = 0(.1)$ until $I(u,p) = 1$, to $7D$. $I(u,p)$ for $p = -1(.01) - .75$, $u = 0(.1)6$, $5D$.
103. Table I of [4]:
 $p(s;m)$ for $m = .1(.1)1.0(.5)5(1)10$, $s = 0(1)29$, $6D$.
104. Pages 431–434 of [5]:
 $p(s;m)$ and $P(s;m)$ for $m = .1(.1)1(1)9$, $s = 0(1)28$, $6D$.
105. Table III of [15]:
 $p(s;m)$ for $m = .1(.1)2.0(.2)4(1)10$, $s = 0(1)24$, $4D$.
106. Pages 545–552 of [54]:
 $p(s;m)$ and $Q(s-1;\ m)$ for $m = .1(.1)10(1)20$, $s = 0(1)39$, $4D$.
107. Sections 9.3 and 9.4 of [56]:
 $P(s;m)$ for $m = .001, .005(.005).1(.1)1.0(.2)2.0(.5)5(1)10$, $s = 0(1)24$, $4D$. Values of m for which $P(s;m) = .000005, .00001(.00001).0001(.0001).001(.001).01(.01).10(.05).3(.1).5$, $s = 0(1)4$, $3D$.
108. Campbell, G. A.: Probability Curves Showing Poisson's Exponential Summation, *Bell System Tech. J.*, **1**: 95–113 (1923).
 Values of $m = \chi^2/2$ for which $Q(s-1;\ m) = P(\chi^2|\nu) = .000001$, $2D$; $.0001, .01$, $3D$; $.1, .25, .5, .75, .9$, $4D$; $.99, .9999$, $3D$; $.999999$, $2D$ for $s = \nu/2 = 1(1)101$.
109. Kitagawa, T.: "Tables of Poisson Distribution," Baifukan, Tokyo, Japan, 1951.
 $p(s;m)$ for $m = .001(.001)1(.01)5$, $s = 0(1)11$, $8D$; $m = 5(.01)10$, $s = 0(1)31$, $7D$.
110. Molina, E. C.: "Poisson's Exponential Binomial Limit," Van Nostrand, Princeton, N.J., 1947.
 $p(s;m)$ and $Q(s-1;\ m)$ for $m = .001(.001).01(.01).3, .4$, $s = 0(1)7$, $7D$; for $m = .5(.1)15(1)100$, $s = 0(1)150$, $6D$.

Negative Binomial Probability Distribution:

111. Williamson, E., and M. H. Bretherton: "Tables of the Negative Binomial Probability Distribution," Wiley, New York, 1963.

$$f(y;r,p) \text{ and } \sum_{x=0}^{y} f(x;r,p) \text{ for } p = 0.05,\ r = 0.1(0.1)$$

 0.5; $p = 0.10$, $r = 0.1(0.1)1.0$; $p = 0.12(0.02)0.20$, $r = 0.1(0.1)2.5$; $p = 0.22(0.02)0.40$, $r = 0.1(0.1)2.5(0.5)5.0$; $p = 0.42(0.02)0.60$, $r = 0.1(0.1)2.5(0.5)10.0$; $p = 0.62(0.02)0.80$, $r = 0.2(0.2)5.0(1)20$; $p = 0.82(0.02)0.90$, $r = 0.5(0.5)10.0(2)50$; $p = 0.95$, $r = 2(2)50(10)200$; $6D$.

Student's t Distribution and Noncentral t Distribution:

112. Table A.2 of [7]:
 Values of t for which $2Q(t|\nu) = .0001, .001, .005, .01, .025, .05, .1, .25, .5, .75, .9, .95, .975, .99, .995, .999, .9999$, when $\nu = 1(1.2)1.5, 2(1)10, 12, 15, 24, 30, 40, 60, 120, \infty$, $5S$.
113. Table 26.10 of [50]:
 Values of t for which $A(t|\nu) = .2, .5, .8, .9, .95, .99, .95, .995, .998, .999, .9999, .99999, .999999$, when $\nu = 1(1)30, 40, 60, 120, \infty$, $3D$.
114. Table III of [51]; reprinted on p. 279 of [54]: Values of t for which $2Q(t|\nu) = .9(.1), .05, .02, .01, .001$, when $\nu = 1(1)30$, $3D$.
115. Table IV of [53]: Values of t for which $2Q(t|\nu) = .6(.1).9, .95, .975, .99, .995, .999, .9995$ when $\nu = 1(1)30, 40, 50, 60, 80, 100, 200, 500, \infty$, $3D$.

116. Section 2.1 of [56]:
Values of t for which $P(t|\nu) = .75, .90, .95, .975, .99, .995$, when $\nu = 1(1)30$, 4D.

117. Tables 9 and 12, and p. 133 of [57]:
$P(t|\nu)$ for $t = 0.0(.1)4(.2)8$, $\nu = 1(1)20$; 5D; for $t = 0.00(.05)2(.1)4, 5$, $\nu = 20(1)24, 30, 40, 60, 120, \infty$, 5D. Values of t for which $2Q(t|\nu) = .8, .5, .2, .1, .05, .02, .01, .005, .002, .001$, when $\nu = 1(1)30, 40, 60, 120, \infty$, 3D. Values of t for which $Q(t|\nu) = .001, .0001, .00001, .000005$ when $\nu = 1(1)10$, at least 3S.

118. Federighi, E. T.: Extended Tables of Percentage Points of Student's t-Distribution, *J. Am. Statist. Assoc.*, **54**: 683–688 (1959).
Values of t for which $2Q(t|\nu) = .25, .10, .05, .025, .01, .005, .0025, .001, .0005, .00025, .0001, .00005, .000025, .00001$ when $\nu = 1(1)30(5)60(10)100, 200, 500, 1{,}000, 2{,}000, 10{,}000, \infty$, 3D.

119. Owen, D. B.: The Power of Student's t-Test, *J. Am. Statist. Assoc.*, **60**: 320–333 (1965).
Values of $t = t(\nu,Q)$ which $Q(t|\nu) = .05, .01, .025, .005$, when $\nu = 1(1)30(5)100(10)200, \infty$, 5D.
Values of $\delta = \sqrt{\lambda}$ for which $P\{t' > t(\nu,Q)|\nu,\delta\} = 1 - \beta$ for $\beta = .01, .05, .1(.1).9$, $\alpha = Q(t|\nu) = .05, .01, .025, .005$, when $\nu = 1(1)30(5)100(10)200, \infty$, 5D.

120. Croarkin, Mary C.: Graphs for Determining the Power of Student's t-Test, *J. Res. Natl. Bur. Std.*, **66B**: 59–70 (1962).
Figures 1–5‡ show, for $\alpha = 2Q(t|\nu) = .01, .02, .05, .10, .20$, respectively, values of "δ" $= \delta/\sqrt{n} = \sqrt{\lambda/n}$, to 2S, for which $P\{|t'| > t(\nu,\alpha)|\nu$, "$\delta$"$\} = \beta = .10, .5, .9, .99$ ($\beta > \alpha$ only), when $n = \nu + 1 = 3(.1)10(1)61$. Figures 6–10‡ show, for the same α and β combinations, values of "δ" $= \delta\sqrt{2/n} = \sqrt{2\lambda/n}$, to 2S, for which $P\{|t'| > t(\nu,\alpha)|\nu$, "$\delta$"$\} = \beta$, when $n = (\nu + 2)/2 = 2(.1)10(1)32$.

121. Neyman, J., and B. Tokarska: Errors of the Second Kind in Testing Student's Hypothesis, *J. Am. Statist. Assoc.*, **31**: 318–326 (1936).
Values of $\delta = \sqrt{\lambda}$ to 2D for which $P\{t' \le t(\nu,Q|\nu,\delta\} = .01, .05, .1(.1).9$, for $\alpha = Q(t|\nu) = .05, .01$.

122. Table 10 of [57]:
Curves showing $P\{|t'| > t(\nu,\alpha)|\nu,\phi\}$ to 2D, where $\phi = \sqrt{\lambda/2} = \delta/\sqrt{2}$, for $\nu = 6(1)10, 12, 15, 20, 30, 60, \infty$, when $\alpha = 2Q(t|\nu) = .05$ and $1.5 \le \phi \le 3.5$, and when $\alpha = 2Q(t|\nu) = .01$ and $2 \le \phi \le 5$.

123. Johnson, N. L., and B. L. Welch: Applications of the Non-central t-distribution, *Biometrika*, **31**: 362–389 (1939).
Values are given for auxiliary functions that enable calculation of values of $\delta = \sqrt{\lambda}$ for given t' and $P(t'|\nu,\delta)$, or values of t' for given δ and $P(t'|\nu,\delta)$, when $P(t|\nu,\delta) = .005, .01, .025, .05, .1(.1).9, .95, .975, .99, .995$, and $\nu = 4(1)9, 16, 36, .44, \infty$.

124. Resnikoff, G. S., and G. J. Lieberman: "Tables of the Non-central t-Distribution," Stanford University Press, Stanford, Calif., 1957.
$p_\nu(t'|\nu,\delta)$ and $P(t'|\nu,\delta)$ for $\nu = 2(1)24(5)49$, $\delta = \sqrt{\lambda} = x_\alpha\sqrt{\nu + 1}$, and $Q(x_\alpha) = \alpha = .25, .15, .1, .065, .04, .025, .01, .004, .0025, .001$, for $t'/\sqrt{\nu}$ in steps of .05, 4D.
Values of $t'/\sqrt{\nu} = t'(\nu,\delta,\epsilon)/\sqrt{\nu}$ for the above δ for which $Q(t'(\nu,\delta,\epsilon)|\nu,\delta) = \epsilon$, for $\epsilon = .995, .99, .95, .90, .75, .50, .25, .10, .05$, when $\nu = 2(1)4$; and for the foregoing ϵ and $\epsilon = .005$ when $\nu = 5(1)24(5)49$.

125. Scheuer, E. M., and R. A. Spurgeon: Some Percentage Points of the Non-central t-Distribution, *J. Am. Statist. Assoc.*, **58**: 176–182 (1963).
Values of $t'/\sqrt{\nu} = t'(\nu,\delta,\epsilon)/\sqrt{\nu}$ for which $Q(t'(\nu,\delta,\epsilon)|\nu,\delta) = .975, .025$, for $\delta = \sqrt{\nu + 1}\, x_p$, where $Q(x_p) = .25, .15, .10, .065, .04, .025, .01, .004, .0025, .001$, for $\nu = 2(1)24(5)49$, 3D.

F Distribution and Noncentral F Distribution:

126. Table V of [51]:
Values of F and $z = \frac{1}{2} \ln F$ for which $Q(F|\nu_1,\nu_2) = .2, .1, .05, .01, .001$, when $\nu_1 = 1(1)6, 8, 12, 24, \infty$ and $\nu_2 = 1(1)30, 40, 60, 120, \infty$; 2D for F; 4D for z.

127. Table 18 of [57]:
Values of F for which $Q(F|\nu_1,\nu_2) = .25, .10, .05, .025, .01, .005, .001$ when $\nu_1 = 1(1)10, 12, 15, 20, 24, 30, 40, 60, 120, \infty$ and $\nu_2 = 1(1)30, 40, 60, 120, \infty$, 4S or 2D.

128. Table A.3 of [7]:
Values of F for which $Q(F|\nu_1,\nu_2) = .0001, .001, .005, .01, .025, .05, .1, .25, .50, .75, .9, .95, .975, .995, .999, .9999$, when $\nu_1 = 1(1)10, 12, 15, 20, 24, 30, 40, 60, 120, \infty$, and $\nu_2 = 1, 1.2, 1.5, 2(1)10, 12, 15, 20, 24, 30, 40, 60, 120, \infty$, 5S.

129. Table A.4 of [7]:
Values of $P\{F' \le F(\nu_1,\nu_2,\alpha)|\nu_1,\nu_2,\phi\}$ for $\nu_1 = 1(1)8$, $\nu_2 = 2(2)6(1)30, \infty$ and $\phi = \sqrt{\lambda/(\nu_1 + 1)} = 1(.5)3(1)8$, when $\alpha = Q(F|\nu_1,\nu_2) = .01, .05$, 3D.

130. Fox, M.: Charts of the Power of the F-test, *Ann. Math. Statist.*, **27**: 484–497 (1956).
Graphs of values of $\phi = \sqrt{\lambda/(\nu_1 + 1)}$ for which $P\{F' > F(\nu_1,\nu_2,\alpha)|\nu_1,\nu_2,\phi\} = \beta$ for $\nu_1 = 3(1)10(2)20, 40, 60, 80, 100, 200, \infty$, $\nu_2 = 4(1)10(2)20, 40, 60, 80, 100, 200, \infty$, $\alpha = .01, .05$, and $\beta = .5(.1).9$; together with nomograms to facilitate β-wise interpolation.

131. Lehmer, E.: Inverse Tables of Probabilities of Errors of the Second Kind, *Ann. Math. Statist.*, **15**: 388–398 (1944).
Values of $\phi = \sqrt{\lambda/(\nu_1 + 1)}$ for which $P\{F' \le F(\nu_1,\nu_2,\alpha)|\nu_1,\nu_2,\phi\} = .2, .3$ for $\nu_1 = 1(1)10, 12, 15, 20, 24, 30, 40, 60, 120, \infty$, $\nu_2 = 2(2)20, 24, 30, 40, 60, 80, 120, 240, \infty$, and $\alpha = .01, .05$; 3D or 3S.

132. Pearson, E. S., and H. O. Hartley: Charts of the Power Function for the Analysis of Variance Tests, Derived from the Non-central F-Distribution, *Biometrika*, **38**: 112–130 (1951).
Graphs showing $P\{F' > F(\nu_1,\nu_2,\alpha)|\nu_1,\nu_2,\phi\}$ for $\phi = \sqrt{\lambda/(\nu_1 + 1)}$, $\nu_1 = 1(1)8$, $\nu_2 = 6(1)10, 12, 15, 30, 60, \infty$, when $1 \le \phi \le 5$ and $\alpha = .05, .01$.

‡ *Note:* The graph and abscissa index "$\delta = |\mu_1 - \mu_2|/\sigma$" appearing on p. 64 above the caption for Fig. 5 should be interchanged with the graph and abscissa index "$\delta = |\mu_0 - \mu_1|/\sigma$" appearing on p. 69 above the caption for Fig. 10.

Chapter 13

Statistical Design of Experiments

By W. J. YOUDEN, National Bureau of Standards

Statistical design in experimentation has two chief functions. The first has to do with achieving the best possible precision with a given experimental procedure. The second function has to do with the detection of sources of constant error in the procedure. Statistical design does not involve any changes in the equipment or customary measurement operations, but does specify the order of making the measurements. If the measurements are taken in certain groups or patterns, it is often possible to compensate automatically for environmental conditions which have somehow eluded control and which influence the observed results.

The basic idea is not new. Nearly everyone, at some time or other, has made use of the sandwich arrangement *ABBA*. Two objects, *A* and *B*, are to be compared. Possibly the instrument drifts, or the temperature is imperfectly controlled, or some unknown factor is shifting. If the time to take the sequence of readings is not too long, it is reasonable to presume that the shift will be proportional to time; that is, repeated measurements on the same object would be subject to a succession of equal increments. It is apparent that this arrangement provides an automatic compensation for a linear drift. The difference between the average for the two readings for *A* and the average of the *B* readings is just what it would be, assuming no drift at all.

The measurement procedure has not been changed. The order *ABBA* should give better results than *AABB*, because, in the latter case, the difference between the averages is more vulnerable to a drift. Statistical design starts with this primitive sandwich arrangement. There now exist arrangements especially suited for drifts that are nonlinear. The drift may reverse direction repeatedly. Points of discontinuity in the drift also may be taken care of, provided the points of discontinuity are known. In fact, it is these points of discontinuity that determine the pattern or grouping of the measurements into blocks.

When an arrangement like $A_1B_1B_2A_2$ has been used, the averages are taken and presumed to be better than they otherwise would be. Very often, there the matter rests. If the readings have been taken at equal time intervals, the four readings also make it possible to estimate the drift. Clearly $A_2 - A_1$ is the drift over three time intervals, and $B_2 - B_1$ the drift over one interval of time. The proper weighted average of these two estimates is

$$\text{Drift per unit interval} = \frac{3(A_2 - A_1) + (B_2 - B_1)}{10}$$

Even when the experimenter has glanced at the readings for confirmation regarding the drift he rarely realizes that these same four readings also furnish an estimate of the experimental error appropriate to the better precision that has been achieved by the compensation for drift error. This estimate of the error is based on just one degree of freedom. It is easily pooled with the estimates from as many sandwiches as are available.

The mean square for error or error variance for the pattern $A_1B_1B_2A_2$ is

$$\frac{(3B_1 - 3B_2 + A_2 - A_1)^2}{20}$$

Clearly the coefficients cancel the drift effect, and the objects drop out as well. Only experimental error remains. It can be shown that the three squares

$\frac{(A_2 + A_1 - B_1 - B_2)^2}{4}$	measure of difference between A and B
$\frac{(3A_2 - 3A_1 - B_1 + B_2)^2}{20}$	measure of drift
$\frac{(A_2 - A_1 + 3B_1 - 3B_2)^2}{20}$	measure of error variance

add up to the total of the squares of the deviations of the four measurements about the mean m of all four measurements: $(A_1 - m)^2 + (A_2 - m)^2 + (B_1 - m)^2 + (B_2 - m)^2$. All the variation is thus accounted for. The estimate used for the slope has in fact been the least-squares estimate; one that makes the error variance $(A_2 - A_1 + 3B_1 - 3B_2)^2/20$ a minimum.

There are a number of standard patterns or designs for which the least-square solutions are either obvious or already available in textbooks [1, 2, 4].* The more important designs will be mentioned briefly.

The most common and useful is the *randomized block design*. Each block contains a complete set of the objects in a random order. The number of blocks or replications is at the discretion of the experimenter. The blocks may correspond to time periods which are often discontinuous from one another. Because all the objects in any one block or time period undergo the environmental conditions peculiar to

* Numbers in brackets refer to References at end of chapter.

that block the averages are automatically compensated for the block effects. Equally the block averages directly reveal whether or not the environment did change from block to block. The assumption is made that all measurements in a block are subject to a common increment characteristic of that block. Inevitably this assumption is imperfectly realized. However, there is usually a large common component and this component is compensated for by the arrangement. The amounts by which the objects are not equally affected go into the estimate of the error variance, as they should.

If the blocks are equal in number to the objects, there is a possible double grouping which is called a *Latin square.*

	Blocks (Columns)	5	6	7	8
Blocks (Rows)	1	*A*	*B*	*C*	*D*
	2	*D*	*C*	*B*	*A*
	3	*B*	*A*	*D*	*C*
	4	*C*	*D*	*A*	*B*

Again it is assumed that each row and column makes a characteristic contribution to the measurements recorded for the objects *A*, *B*, *C*, and *D*. Subject to this condition the rows and columns may both represent objects and the letters *A*, *B*, *C*, and *D* then represent four groupings. Thus, four resistance thermometers (rows 1, 2, 3, and 4) have been used to calibrate four temperature reservoirs (columns 5, 6, 7, and 8). The four combinations indicated by letter *A* were used on one day, the four marked by *B* on the next day, and so on. If, as was the case, all four reservoirs were nearly identical in temperature, it follows that the differences between the four thermometers persist regardless of the reservoir. A similar statement holds for reservoirs. The important point in this arrangement is that thermometers and the reservoirs are equally exposed to each of the environmental situations or blocks. The day-to-day differences are automatically compensated. The averages for the measurements on days *A*, *B*, *C*, and *D* are used to reveal whether there are day-to-day effects. In one instance the elimination of these day-to-day effects cut the error variance in half compared to that obtained in prior work with exactly the same equipment.

Often the block, whether it be a time period or a space unit, will not accommodate a complete set of the objects. The blocks then consist of incomplete sets of objects. A considerable group of *balanced incomplete block designs* is available. They have been devised to facilitate the computation work. In these designs any given object will be found paired in one block or another with all other objects the same number of times. Many of these designs also permit a grouping across the blocks. The following example shows 9 objects arranged in 12 blocks of 3, each object appearing in 4 blocks and having as its 8 neighbors the other 8 objects.

1	2	3	4	5	6	7	8	9	10	11	12
A	*D*	*G*	*A*	*B*	*C*	*A*	*G*	*D*	*G*	*A*	*D*
B	*E*	*H*	*D*	*E*	*F*	*E*	*B*	*H*	*E*	*H*	*B*
C	*F*	*I*	*G*	*H*	*I*	*I*	*F*	*C*	*C*	*F*	*I*

A second example shows 7 objects arranged in 7 blocks of 4 objects. Each object appears in 4 blocks and has for its 12 neighbors the other 6 objects, each appearing twice. This pattern also shows a cross grouping. The rows are complete blocks.

1	2	3	4	5	6	7
A	*B*	*C*	*D*	*E*	*F*	*G*
C	*D*	*E*	*F*	*G*	*A*	*B*
D	*E*	*F*	*G*	*A*	*B*	*C*
E	*F*	*G*	*A*	*B*	*C*	*D*

Estimates based on incomplete block designs always involve some adjustment to the row observations. The best estimate for any letter in the 4×7 design shown above is given by taking 4 times the total of the 4 observations for that letter and subtracting the block totals for the 4 blocks in which this letter appears. The remainder, after dividing by 14, is added or subtracted as required to the grand average of all 28 observations.

If k is the number of objects per block, b the number of blocks, v the number of objects, and r the number of repetitions of each object, then $bk = vr$. Furthermore, the number of pairings within a block, $k(k - 1)/2$, multiplied by the number of blocks must be an integral multiple of the total pairings that can be formed from the v objects. These constraints limit the solutions to certain combinations of k, b, v, and r. Nevertheless, the inherent symmetry of these designs brings the advantage that the least-square solutions for the best estimates of the objects are easily obtained.

The gain in precision for the comparisons arises from the fact that the estimates or adjusted averages for the objects involve only differences between measurements occurring in the same block. One of the oldest examples of this balanced design is afforded by the comparison of meter bars by pairs. The comparison equipment accommodates just two bars. The very nature of the arrangement tends to provide excellent identity of environment for the two bars thus paired and consequently the difference in length between these two bars will be precisely determined. Later another pair of bars may be compared. It is not so important to attempt to reproduce exactly the environment used for the preceding pair because the difference in length is relatively insensitive to slight shifts in the environment, provided both bars in the pair undergo the same shift in environment (see ref. 6).

A design in blocks of 2 is always possible but will require $v - 1$ repetitions of each object. The number of repetitions ($v - 1$) may be quite unnecessary to give the required precision of the averages. A recent development in statistical design is concerned with the selection of certain subsets of the pairs with a view to retaining, as far as possible, simplicity in the calculations (see ref. 12). For example 9 bars, ordinarily requiring 36 pairings, may be scheduled for just 18 pairs. Furthermore, these 18 pairs may be grouped in 2 squads of 9 pairs. In each squad the pairs themselves may be ordered so that a complete set of 9 bars appears in each column. Thus every bar would be assigned once to the left side of the equipment and once to the right side. The column totals for the left and the right will then provide a searching test of the equality of the two opposing sides of the equipment. The second squad of 9 pairs provides a similar check on the apparatus.

Pair	Left	Right	Pair	Left	Right
1	*A*	*B*	10	*A*	*C*
2	*B*	*C*	11	*B*	*E*
3	*C*	*I*	12	*C*	*F*
4	*D*	*A*	13	*D*	*G*
5	*E*	*F*	14	*E*	*D*
6	*F*	*D*	15	*F*	*I*
7	*G*	*H*	16	*G*	*A*
8	*H*	*E*	17	*H*	*B*
9	*I*	*G*	18	*I*	*H*

The pairs chosen link any letter, such as *A*, once to each of four other letters (*B*, *C*, *D*, *G*) and through these by double links to the remaining four letters.

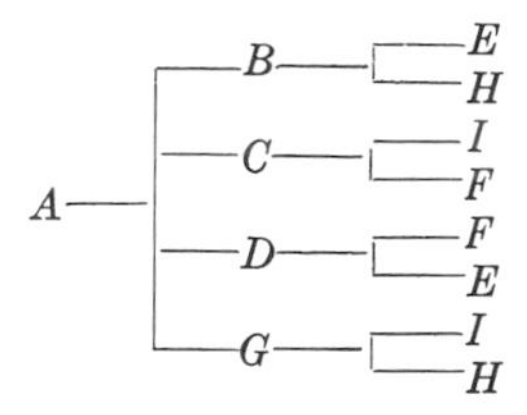

These incomplete sets are called *partially balanced incomplete blocks*. The block size is not limited to pairs; for example, 10 objects may be assigned to 5 blocks, 4 in each block as follows:

1	2	3	4	5
A	*F*	*H*	*C*	*D*
B	*E*	*I*	*J*	*G*
C	*A*	*B*	*H*	*I*
D	*G*	*E*	*F*	*J*

The top two rows contain a complete set of the 10 letters, and so do the two bottom rows.

The successful application of one or another of these statistical designs depends a great deal upon familiarity with the experimental situation. Some applications have already been mentioned. The pairs were also used in a study of Weston standard cells and the left-right grouping brought to light a very small bias between the binding posts of this equipment. The Latin square has been used in Geiger counter measurements on radioactive isotopes (see ref. 7). The pairing technique was also used in a critical comparison of the Canadian, British, and American radium standards (see ref. 3). A recent paper shows the application of statistical design to the balancing out of errors arising from instrumental drift even when the drift reverses during the course of the measurements (see ref. 8).

There is another broad class of experimental studies which offer large opportunities for economy of experimental effort. These are characterized as *factorial experiments*. As the name suggests, such experiments involve several factors. The factor may be a continuous variable such as temperature. The experimenter selects two or more values of the factor for study. The factor may also be an item of equipment such as a thermocouple. The experiment will then be directed to the comparison of two or more thermocouples. Usually there are a number of factors. The experiment has the broad purpose of ascertaining the direct effects of the several factors and whether the factors operate independently of one another in the effects they produce.

A study may involve equipment in which a vital component is a tube for which there is a choice of two materials, two lengths, two diameters, and two wall thicknesses. Let the letters m, l, d, and t refer to these features of the tube. Further suppose that a bridge and a resistance box are required and that two bridges (b) and two resistance boxes (r) are available. Finally, there are two assistants (a), either of whom can make the measurements unassisted. Now if each assistant made one measurement on each one of all possible combinations of tube material, diameter, length, wall thickness, bridge, and resistance box there would be a total of 128 readings.

A standard notation is used to indicate the appropriate combinations. The symbol m for example, will be assigned to one of the materials, and if the tube used in the combination is of this material then m becomes a part of the label. If the other material is used, the symbol for material is simply omitted. This rule is adopted for all factors. Consequently, the combination label may have all seven symbols, or one of seven sets of six symbols, and so on down to a label with no symbols. By convention this combination is designated (1).

It is presumed that each reading enters into a formula which explicitly provides for the various tube characteristics and the readings have been corrected for any bridge and resistance calibration. There will result 128 computed values which should be identical but which will in fact reflect random errors plus any constant errors arising from inexact values assigned to tube diameter, etc., and any residuals for bridge and resistance not completely removed by the calibration. Enormous power now resides in the experimenter's hands. For example, the 128 results may be separated into two groups of 64 readings, one group made with the tube of diameter d_0, and the other group made with the tube of diameter d_1. Furthermore, these 64 readings in each group may be paired into a one-to-one correspondence based on the identity of the other six characteristics including assistants.

Paired Comparison Available to Contrast the Two Diameters

(1)	d	a	da
t	td	ta	tda
l	ld	la	lda
tl	tld	tla	$tlda$
m	md	ma	mda
mt	mtd	mta	$mtda$
ml	mld	mla	$mlda$
mtl	$mtld$	$mtla$	$mtlda$
etc.		etc.	

The averages for the 2 tube diameters are each based on 64 results which are directly comparable because the pairing has provided matched comparisons. Indeed the difference, if any, between these two averages will properly be judged by the variation shown among the 64 individual differences furnished by the 64 pairs.

Seven times the same data may be dichotomized in this manner, each contrast providing a mass of evidence on each of the seven experimentally varied elements and for which it was believed proper and adequate adjustments were made by the formula and the calibrations. These contrasts therefore serve to show whether diameter or other characteristic is, in fact, known with adequate precision. Great efficiency results from this use of the same data for all features of the program that the experimenter wishes assurance about. The direct effects of the factors and whether or not they are independent of one another in their actions may be readily examined. Factors that do not act independently are said to exhibit an *interaction*.

The matter does not end here. Even with seven factors the number of combinations becomes large. The opportunity exists of choosing a subset of the whole number of combinations without sacrificing any important information. In general it will be desired to examine the data fairly closely. Not only is it important to determine whether the formulas hold for both tube lengths, for example, but if there does appear to be a discrepancy depending on tube length, it is important to know whether this persists over both tube diameters, or both materials. That is, do these factors interact? Subsets can be chosen which preserve these comparisons.

The following 32 combinations may be selected, these in turn are subdivided into two sets of 16, one set assigned to one assistant and one to the other assistant.

First Assistant		Second Assistant	
1	*tb*	*tl*	*bl*
tbld	*ld*	*bd*	*td*
blmr	*tlmr*	*tbmr*	*mr*
tdmr	*bdmr*	*ldmr*	*tbldmr*
tblm	*lm*	*bm*	*tm*
dm	*tbdm*	*tldm*	*bldm*
tr	*br*	*lr*	*tblr*
bldr	*tldr*	*tbdr*	*dr*

The symbol *a* for one of the assistants has been omitted as unnecessary. The total number of measurements has now been reduced to 16 by each assistant. More commonly there may be just one operator. In that event these subsets may be used to divide the work into two days or other suitable time periods. It will then be possible to see if some unsuspected factors associated with the two time periods are having any effects.

Inspection will show that the 16 combinations always contain any one symbol 8 times and omit it 8 times. The contrast, using the data from both assistants, involves a comparison of averages based upon 16 observations. Often small effects which otherwise might easily escape notice are revealed by this scheme.

There remains the problem of evaluating any observed effects as to whether or not they exceed effects that might be expected in view of the random errors of measurement. The experiment is self-contained, in that the data also provide an estimate of this experimental error. Seven dichotomies of the data are needed to exhibit possible direct effects of the seven factors (one of these factors may be a time grouping of the measurements). The six other factors must be considered in all possible pairs to determine whether or not they act independently. Tests for these possible interactions will consume another 15 dichotomies of the data. There are, in all, 31 independent comparisons available from the 32 observations, and here 10 of these contrasts remain to estimate the differences that arise from random experimental errors. Full descriptions of these *fractional factorial experiments* are available (see refs. 2 and 4).

References

1. Bose, R. C., W. H. Clatworthy, and S. S. Shrekhande: Tables of Partially Balanced Designs with Two Associate Classes, *N. Carolina Agri. Expt. Sta., Tech. Bull.* 107, 1954.
2. Cochran, W. G., and G. M. Cox: "Experimental Designs," Wiley, New York, 1950.
3. Davenport, W. B., and others: Comparison of Four National Radium Standards, *J. Res. Natl. Bur. Std.*, **53**: 267 (1954).
4. Kempthorne, Oscar: "The Design and Analysis of Experiments," Wiley, New York, 1952.
5. Mandel, John: "The Statistical Analysis of Experimental Data," Wiley, New York, 1964.
6. Page, B. L.: Calibration of Meter Line Standards of Length at The National Bureau of Standards, *J. Res. Natl. Bur. Std.*, **54**: 1 (1955).
7. Seliger, H. H.: A New Method of Radioactive Standard Calibration, *J. Res. Natl. Bur. Std.*, **45**: 496 (1950).
8. Youden, W. J.: Instrumental Drift, *Science*, **120**: 627 (1954).
9. Youden, W. J.: Systematic Errors in Physical Constants, *Phys. Today*, **14**(9): (September, 1961).
10. Youden, W. J.: Measurement Agreement Comparisons, Proceedings 1962 Standards Laboratory Conference, *Natl. Bur. Std.* (U.S.), *Publ.* 248.
11. Youden, W. J.: Uncertainties in Calibration, *IRE Trans. Instrumentation*, **I-11**(3, 4): (December, 1962).
12. Youden, W. J., and W. S. Connor: New Experimental Designs for Paired Observations, *J. Res. Natl. Bur Std.*, **53**: 191 (1954).

Part 2 · Mechanics of Particles and Rigid Bodies

Chapter 1

Kinematics

By E. U. CONDON, University of Colorado

1. Velocity and Acceleration

If the position vector of a moving particle relative to an arbitrary, unaccelerated origin at time t is $\mathbf{r}(t)$, then the *velocity* $\mathbf{v}(t)$ is defined as the vector which is the first time derivative of $\mathbf{r}(t)$ and the *acceleration* $\mathbf{a}(t)$ is defined as the first time derivative of $\mathbf{v}(t)$ and therefore is the second time derivative of $\mathbf{r}(t)$.

$$\mathbf{v}(t) = \dot{\mathbf{r}}(t) \qquad \mathbf{a}(t) = \dot{\mathbf{v}}(t) = \ddot{\mathbf{r}}(t) \tag{1.1}$$

The velocity vector (Fig. 1.1) lies in the direction of the tangent to the path. It can therefore be written $\mathbf{v} = v\mathbf{t}_0$, where $\mathbf{t}_0$ is the unit tangent to the path at the place of the particle and v is the magnitude of the velocity, sometimes called the *speed*.

The acceleration may be regarded as the sum of two parts, one arising from the change of speed, the other from the changing direction of motion

$$\mathbf{a} = \dot{v}\mathbf{t}_0 + v\dot{\mathbf{t}}_0$$

The first term, $\dot{v}\mathbf{t}_0$, is called the *tangential component* of acceleration. Since $\dot{\mathbf{t}}_0 = v\, d\mathbf{t}_0/ds$, where s is the

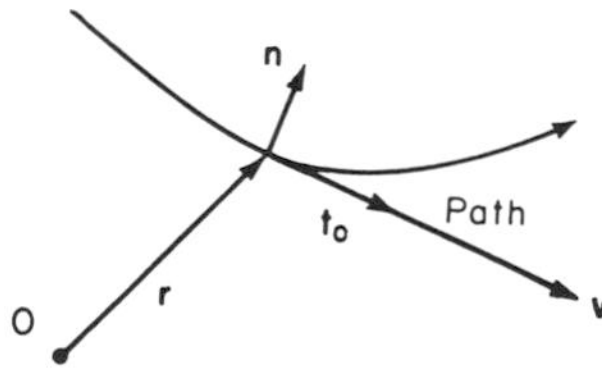

Fig. 1.1

length of arc along the path, and since $d\mathbf{t}_0/ds = \mathbf{n}/R$, where $\mathbf{n}$ is the unit vector drawn toward the center of curvature, and hence normal to $\mathbf{t}_0$, and R is the radius of curvature,

$$\mathbf{a} = \dot{v}\mathbf{t}_0 + \left(\frac{v^2}{R}\right)\mathbf{n} \tag{1.2}$$

These relations are shown in Fig. 1.1. The term involving $\mathbf{n}$ is known as the *centripetal component* of acceleration.

In nonrotating rectangular coordinates

$$\mathbf{r} = x\mathbf{i} + y\mathbf{j} + z\mathbf{k} \tag{1.3}$$

and the components of velocity and acceleration are simply the first and second derivatives of the coordinates, $v_x = \dot{x}$, etc., $a_x = \ddot{x}$, etc.

Rotating Axes. If the motion is referred to a rectangular coordinate system whose axes are rotating, expression (1.3) holds at each instant, but in differentiating it is now necessary to take account of the time variation of the unit vectors **i**, **j**, and **k**. This allows for the physical fact that a particle with constant coordinates in the rotating system has velocity and acceleration.

For example, suppose the axes are rotating with angular velocity ω about the **k** axis (Fig. 1.2). Then

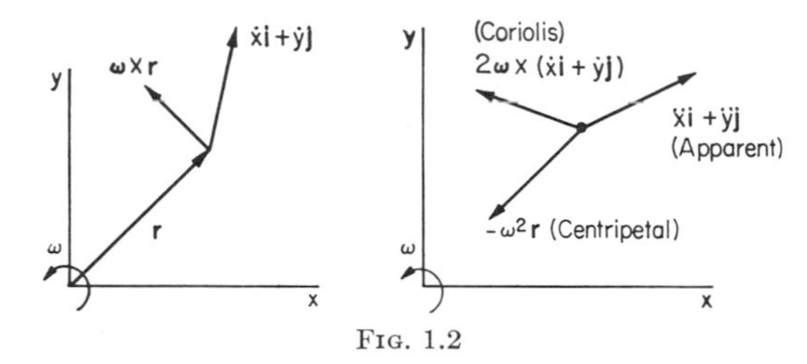

Fig. 1.2

the **k** axis itself does not change with time but $d\mathbf{i}/dt = \omega\mathbf{j}$ and $d\mathbf{j}/dt = -\omega\mathbf{i}$. Hence the velocity vector is

$$\begin{aligned}\mathbf{v} &= (\dot{x}\mathbf{i} + \dot{y}\mathbf{j} + \dot{z}\mathbf{k}) + \omega(-y\mathbf{i} + x\mathbf{j}) \\ &= (\dot{x}\mathbf{i} + \dot{y}\mathbf{j} + \dot{z}\mathbf{k}) + \boldsymbol{\omega} \times \mathbf{r}\end{aligned} \tag{1.4}$$

where $\boldsymbol{\omega} = \omega\mathbf{k}$ is the vector representing the angular velocity of the coordinate axes. The first term is called the apparent velocity; there is no established usage as to a name for the second term.

The acceleration vector is

$$\begin{aligned}\mathbf{a} &= (\ddot{x}\mathbf{i} + \ddot{y}\mathbf{j} + \ddot{z}\mathbf{k}) \\ &\quad + 2\boldsymbol{\omega} \times (\dot{x}\mathbf{i} + \dot{y}\mathbf{j} + \dot{z}\mathbf{k}) \\ &\quad - \omega^2(1 - \mathbf{kk}) \cdot \mathbf{r}\end{aligned} \tag{1.5}$$

or, fully expressed in terms of components,

$$\begin{aligned}\mathbf{a} &= (\ddot{x} - 2\omega\dot{y} - \omega^2 x)\mathbf{i} \\ &\quad + (\ddot{y} + 2\omega\dot{x} - \omega^2 y)\mathbf{j} \\ &\quad + \ddot{z}\mathbf{k}\end{aligned}$$

In (1.5) the first term is known as the *apparent* acceleration, the second as the *Coriolis* acceleration, and the third as the *centripetal* acceleration. The Coriolis acceleration is normal both to the axis of rotation and the direction of the apparent velocity. The centripetal acceleration is proportional to the particle's distance from the origin and is directed toward the axis of rotation.

A system of coordinates which preserves constant orientation with regard to the "fixed" stars is nonrotating, for ordinary purposes. The precise problem from the viewpoint of the astronomer is discussed in Chap. 8, Sec. 7.

A coordinate system fixed with regard to points on the earth's surface is a rotating coordinate system, and since there are 86,164 mean solar seconds in a sidereal day the value of ω is $2\pi/86{,}164 = 7.292 \times 10^{-5}$ and the direction of $\boldsymbol{\omega}$ is toward the North Pole of the heavens. The centripetal acceleration term in (1.5) is indistinguishable dynamically from gravitational acceleration and thus serves merely to modify the latter slightly. But the *Coriolis acceleration* gives rise to distinctive dynamical effects which permit the measurement of the earth's angular velocity *as a vector* and hence provides means of finding the direction of north, the latitude of a place, and the length of a day by means of experiments performed wholly within a closed laboratory.

Units. Velocity (LT^{-1}) is measured in derived units, based on the use of any unit of length for measuring (x,y,z) combined with any unit of time for t. Unit speed is that in which unit displacement occurs in unit time.

Acceleration (LT^{-2}) is likewise expressible with a derived unit based on any length unit and any time unit. For some purposes it is convenient to use two different time units. Thus 1 mile/hr sec is either an acceleration in which the velocity increases at the rate of 1 mile/hr in a second, or an acceleration in which the velocity increases at the rate of 1 mile/sec in an hour. Both are equivalent. (See Appendix.)

Curvilinear Coordinates. The motion may be referred to an orthogonal curvilinear coordinate system. The position vector $\mathbf{r}$ is then a function of three curvilinear coordinates u, v, w which vary with time.

The velocity vector is then

$$\mathbf{v} = \dot{u}\mathbf{u}_1 + \dot{v}\mathbf{v}_1 + \dot{w}\mathbf{w}_1 \tag{1.6}$$

where

$$\mathbf{u}_1 = \frac{\partial \mathbf{r}}{\partial u}, \text{ etc.}$$

The vectors $\mathbf{u}_1$, $\mathbf{v}_1$, $\mathbf{w}_1$ are in general not unit vectors. The vector $\mathbf{u}_1$ lies in the direction of displacement produced by varying u while holding v and w constant and similarly for $\mathbf{v}_1$ and $\mathbf{w}_1$.

The acceleration vector is obtained by differentiating (1.6):

$$\begin{aligned}\mathbf{a} = {} & \ddot{u}\mathbf{u}_1 + \ddot{v}\mathbf{v}_1 + \ddot{w}\mathbf{w}_1 \\ & + \dot{u}^2 \frac{\partial^2 \mathbf{r}}{\partial u^2} + 2\dot{u}\dot{v} \frac{\partial^2 \mathbf{r}}{\partial u\, \partial v} + 2\dot{u}\dot{w} \frac{\partial^2 \mathbf{r}}{\partial u\, \partial w} \\ & + \dot{v}^2 \frac{\partial^2 \mathbf{r}}{\partial v^2} + 2\dot{v}\dot{w} \frac{\partial^2 \mathbf{r}}{\partial v\, \partial w} + \dot{w}^2 \frac{\partial^2 \mathbf{r}}{\partial w^2}\end{aligned} \tag{1.7}$$

The special case of two-dimensional motion referred to plane polar coordinates (r,θ) occurs often in practice. In this case

$$\frac{\partial \mathbf{r}}{\partial r} = \mathbf{r}_0 \qquad \frac{\partial \mathbf{r}}{\partial \theta} = r\boldsymbol{\theta}_0$$

where $\mathbf{r}_0$ and $\boldsymbol{\theta}_0$ are unit vectors in the directions of increasing r and θ, respectively. Therefore the velocity vector is $\mathbf{v} = \dot{r}\mathbf{r}_0 + r\dot{\theta}\boldsymbol{\theta}_0$. The basic second derivatives are $\partial^2\mathbf{r}/\partial r^2 = \mathbf{r}_0$, $\partial^2\mathbf{r}/\partial r\, \partial\theta = \boldsymbol{\theta}_0$,

$$\frac{\partial^2 \mathbf{r}}{\partial \theta^2} = -r\mathbf{r}_0$$

and therefore the acceleration in terms of polar coordinates is

$$\mathbf{a} = (\ddot{r} - r\dot{\theta}^2)\mathbf{r}_0 + (r\ddot{\theta} + 2\dot{r}\dot{\theta})\boldsymbol{\theta}_0 \tag{1.8}$$

The second component can be written in the form $r\ddot{\theta} + 2\dot{r}\dot{\theta} = \dfrac{1}{r}\dfrac{d}{dt}(r^2\dot{\theta})$, a result which finds frequent application in central force problems.

2. Kinematics of a Rigid Body

A rigid body is one for which the relative motion (if any) of its parts is treated as negligible so that the distance apart of any two points in the body remains constant in time. It is convenient to choose a point O', which will move with the body, as the origin of a cartesian coordinate system whose axes **lmn** are fixed in the moving body. Let **ijk** be the unit vectors of a cartesian coordinate system which has the same origin but which moves by parallel displacement so that **ijk** are constant vectors. The kinematics of the motion of the rigid body is reduced to that of the turning of the coordinate system **lmn** relative to **ijk**.

Since **l**, **m**, and **n** are unit vectors, it follows that $d\mathbf{l}/dt$ must be perpendicular to **l** and therefore expressible as a linear combination of **m** and **n**, and similarly for $d\mathbf{m}/dt$ and $d\mathbf{n}/dt$, so that $d\mathbf{l}/dt = A\mathbf{m} + B\mathbf{n}$, $d\mathbf{m}/dt = C\mathbf{n} + D\mathbf{l}$, $d\mathbf{n}/dt = E\mathbf{l} + F\mathbf{m}$. The six coefficients, $A, \ldots, F$, are not independent because the motion of **lmn** must be such as to preserve their mutual orthogonality, hence there are three relations of the form

$$\mathbf{l} \cdot \frac{d\mathbf{m}}{dt} + \frac{d\mathbf{l}}{dt} \cdot \mathbf{m} = 0$$

which require that $D + A = 0$, etc.

Therefore the relations assume the form

$$\begin{aligned}\frac{d\mathbf{l}}{dt} &= \omega_3\mathbf{m} - \omega_2\mathbf{n} \\ \frac{d\mathbf{m}}{dt} &= \omega_1\mathbf{n} - \omega_3\mathbf{l} \\ \frac{d\mathbf{n}}{dt} &= \omega_2\mathbf{l} - \omega_1\mathbf{m}\end{aligned} \tag{1.9}$$

in which the quantities ω_1, ω_2, and ω_3 are the components of the instantaneous angular velocity of the body about the **l**, **m**, and **n** axes, respectively.

If $\mathbf{r} = u\mathbf{l} + v\mathbf{m} + w\mathbf{n}$ so that (u,v,w) are the coordinates of a point P in the rigid body with respect to the moving axes and $\mathbf{r}_0$ is the vector from a fixed point in space to the moving origin of the system **lmn**, then the velocity of P is

$$\mathbf{v} = \mathbf{v}_0 + u\frac{d\mathbf{l}}{dt} + \cdots + \cdots$$

where $\mathbf{v}_0$ is the velocity of the origin. Therefore

$$\mathbf{v} = \mathbf{v}_0 + \boldsymbol{\omega} \times \mathbf{r} \qquad (1.10)$$

where $\boldsymbol{\omega} = \omega_1\mathbf{l} + \omega_2\mathbf{m} + \omega_3\mathbf{n}$.

The vector $\boldsymbol{\omega}$ is the instantaneous *angular velocity* of the body. In the most general motion of a rigid body it can vary both in magnitude and direction and may not have a simple relation either to axes fixed in space or to axes fixed in the body. The succession of values of $\boldsymbol{\omega}(t)$ drawn from the origin in the moving axes develops a cone (in general, not circular) in the moving system. This, after Poinsot ("Théorie Nouvelle de la Rotation des Corps," Paris, 1834), is known as the *polhode*. The same succession of values of $\boldsymbol{\omega}(t)$ drawn in the fixed axes develops another cone in the fixed system known as the *herpolhode*. The most general rotational motion of a rigid body is thus that in which the polhode rolls on the herpolhode, being in contact along the common vector $\boldsymbol{\omega}(t)$ at the time t.

3. Euler's Angles

The orientation of the vectors **lmn** in the moving body relative to the nonrotating reference axes **ijk** involves nine coefficients which are the scalar products of the type $(\mathbf{l} \cdot \mathbf{i})$ of any one of the moving set relative to any one of the fixed set. As there are six equations which express the conditions that **lmn** are unit orthogonal vectors, it follows that only three of these nine scalar products are independent so that all nine of them are expressible in terms of three independent parameters.

For this purpose three independent angles were introduced by Euler. These are known as Euler's angles. These may be introduced in various ways of which the following is one convenient representation (Fig. 1.3).

The angle θ is defined as that between **n** and **k**, so that $\mathbf{n} \cdot \mathbf{k} = \cos\theta$. The vector $\mathbf{k} \times \mathbf{n}$ being perpendicular both to **n** and to **k** lies in the intersection of the (**i**,**j**) plane and the (**l**,**m**) plane which line of intersection is known as the *line of nodes*. The angle between the vector **i** and the line of nodes reckoned from **i** to $(\mathbf{k} \times \mathbf{n})$ in the direction toward **j** is called ϕ. Since the magnitude of $\mathbf{k} \times \mathbf{n}$ is $\sin\theta$, it follows that ϕ is defined by the relations $\mathbf{i} \cdot (\mathbf{k} \times \mathbf{n}) = \sin\theta\cos\phi$, $\mathbf{j} \cdot (\mathbf{k} \times \mathbf{n}) = \sin\theta\sin\phi$. The angles θ and ϕ thus serve to fix the orientation of **n** where, however, ϕ is reckoned from a different origin from the one usually used in introducing a spherical polar coordinate system.

The third angle ψ serves to give the orientation of the **l** axis relative to the line of nodes, being the angle from $\mathbf{k} \times \mathbf{n}$ toward **l** according to the relations $\mathbf{l} \cdot (\mathbf{k} \times \mathbf{n}) = \sin\theta\cos\psi$, $\mathbf{m} \cdot (\mathbf{k} \times \mathbf{n}) = -\sin\theta\sin\psi$.

The range of possible values of the angles corresponding to distinct orientations of the rigid body are

$$0 \le \theta \le \pi \qquad 0 \le \phi \le 2\pi \qquad 0 \le \psi \le 2\pi$$

From the definitions, the expressions for the basic scalar products connecting the two sets of vectors are:

	l	m	n
i	$\alpha_1 = \cos\phi\cos\psi - \sin\phi\sin\psi\cos\theta$	$\alpha_2 = -\cos\phi\sin\psi - \sin\phi\cos\psi\cos\theta$	$\alpha_3 = \sin\theta\sin\phi$
j	$\beta_1 = \sin\phi\cos\psi + \cos\phi\sin\psi\cos\theta$	$\beta_2 = -\sin\phi\sin\psi + \cos\phi\cos\psi\cos\theta$	$\beta_3 = -\sin\theta\cos\phi$
k	$\gamma_1 = \sin\theta\sin\psi$	$\gamma_2 = \sin\theta\cos\psi$	$\gamma_3 = \cos\theta$

The angular velocity $\boldsymbol{\omega}$ of the body may be expressed in terms of the rate of change of Euler's angles. The expressions for the components of angular velocity are:

Fixed Axes	*Moving Axes*
$\boldsymbol{\omega} \cdot \mathbf{i} = \dot\theta\cos\phi + \dot\psi\sin\phi\sin\theta$	$\boldsymbol{\omega} \cdot \mathbf{l} = \dot\phi\sin\theta\sin\psi + \dot\theta\cos\psi$
$\boldsymbol{\omega} \cdot \mathbf{j} = \dot\theta\sin\phi - \dot\psi\cos\phi\sin\theta$	$\boldsymbol{\omega} \cdot \mathbf{m} = \dot\phi\sin\theta\cos\psi - \dot\theta\sin\psi$
$\boldsymbol{\omega} \cdot \mathbf{k} = \dot\phi + \dot\psi\cos\theta$	$\boldsymbol{\omega} \cdot \mathbf{n} = \dot\psi + \dot\phi\cos\theta$

In the special case that $\dot\theta$ and $\dot\phi$ are zero, the body merely rotates with angular velocity $\dot\psi$ about the **n** axis, which has a fixed orientation in space as well as in the body. If $\dot\theta$ and $\dot\psi$ are zero, then the body rotates about the **k** axis in a way that is called a motion of precession. For example, the motion of the earth would be described by giving $\dot\psi = 2\pi/T$, where T is the length of the day, but owing to the precession of the equinoxes $\dot\phi = 2\pi/T'$, where $T' = 26{,}000$ years.

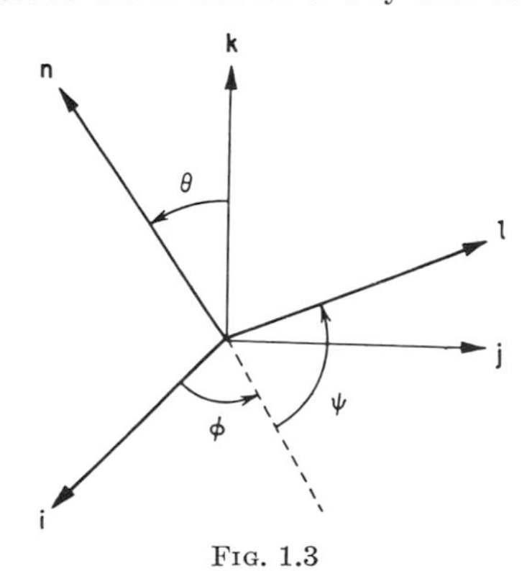

Fig. 1.3

Another way of describing rotations of a rigid body is based on a study of the matrix representation of an orthogonal transformation. If $x_1x_2x_3$ are the coordinates, relative to fixed ones, of any point in the body in its initial position and $x'_1x'_2x'_3$ the coordinates, relative to the same fixed axes of the same point after any rotation of the body, then $(x'_1x'_2x'_3)$ are given as an orthogonal linear transformation $x' = Ax$ in matrix notation. Here A must satisfy the orthogonality condition $A\tilde{A} = 1$, where $\tilde{A}$ is the transpose of A (see Part 1, Chap. 2).

A rotation that is specified by the Eulerian angles may be regarded as, first, a rotation about the **k** axis through the angle ψ, second, a rotation through the angle θ about the **k** $\times$ **n** axis, and, finally, a rotation through the angle ψ about the **n** axis. The combined rotation is represented by the product of the three matrices representing these rotations:

$$A = \begin{bmatrix} \cos\psi & \sin\psi & 0 \\ -\sin\psi & \cos\psi & 0 \\ 0 & 0 & 1 \end{bmatrix} \begin{bmatrix} 1 & 0 & 0 \\ 0 & \cos\theta & \sin\theta \\ 0 & -\sin\theta & \cos\theta \end{bmatrix} \begin{bmatrix} \cos\phi & \sin\phi & 0 \\ -\sin\phi & \cos\phi & 0 \\ 0 & 0 & 1 \end{bmatrix}$$

On multiplying out, this gives the form

$$A = \begin{bmatrix} \alpha_1 & \alpha_2 & \alpha_3 \\ \beta_1 & \beta_2 & \beta_3 \\ \gamma_1 & \gamma_2 & \gamma_3 \end{bmatrix}$$

where the components are the direction cosines in the table of this section.

Cayley-Klein Parameters. The group of three-dimensional rotations of a rigid body is isomorphic with the group of unitary transformations, whose determinant is $+1$, of two complex variables (u,v). Let $u' = \alpha u + \beta v$, $v' = \gamma u + \delta v$ and write

$$Q = \begin{bmatrix} \alpha & \beta \\ \gamma & \delta \end{bmatrix}$$

Then the conditions that Q be unitary and of determinant $+1$ require that Q be of the form

$$Q = \begin{bmatrix} \alpha & \beta \\ -\beta^* & \alpha^* \end{bmatrix} \qquad Q^{-1} = \begin{bmatrix} \alpha^* & -\beta^* \\ \beta & \alpha \end{bmatrix}$$

where α and β are two complex numbers satisfying the condition $\alpha^* + \beta^* = 1$ so that Q contains three independent parameters. The four complex numbers, $\alpha, \beta, \gamma, \delta$, are known as the Cayley-Klein parameters. With each transformation matrix Q one can associate a transformation of the cartesian coordinates (x,y,z) of points in the rigid body as follows,

$$\begin{bmatrix} z' & x' - iy' \\ x' + iy' & -z' \end{bmatrix} = Q \begin{bmatrix} z & x - iy \\ x + iy & -z \end{bmatrix} Q^{-1}$$

This is an orthogonal transformation since this general form of transformation leaves invariant the value of the determinant, which is $-(x^2 + y^2 + z^2)$. In terms of the components $\alpha, \beta, \gamma, \delta$ of Q the transformation matrix A is

$$A = \begin{bmatrix} \tfrac{1}{2}(\alpha^2 - \gamma^2 + \delta^2 - \beta^2) & \frac{i}{2}(\gamma^2 - \alpha^2 + \delta^2 - \beta^2) & \gamma\delta - \alpha\beta \\ \frac{i}{2}(\alpha^2 + \gamma^2 - \beta^2 - \delta^2) & \tfrac{1}{2}(\alpha^2 + \gamma^2 + \beta^2 + \delta^2) & -i(\alpha\beta + \gamma\delta) \\ (\beta\delta - \alpha\gamma) & i(\alpha\gamma + \beta\delta) & (\alpha\delta + \beta\gamma) \end{bmatrix}$$

The Q matrix associated with a rotation about the **k** axis through an angle ϕ is

$$\begin{bmatrix} e^{i\phi/2} & 0 \\ 0 & e^{-i\phi/2} \end{bmatrix}$$

and the Q matrix for the general finite rotation specified by the three Eulerian angles is given by

$$\alpha = e^{i(\psi+\phi)/2}\cos\frac{\theta}{2} \qquad \beta = ie^{i(\psi-\phi)/2}\sin\frac{\theta}{2}$$

$$\gamma = -ie^{-i(\psi-\phi)/2}\sin\frac{\theta}{2} \qquad \delta = e^{-i(\psi+\phi)/2}\cos\frac{\theta}{2}$$

The Cayley-Klein parameters have a close relation to the Pauli matrices used in the description of electron spin. These are:

$$\sigma_x = \begin{bmatrix} 0 & 1 \\ 1 & 0 \end{bmatrix} \qquad \sigma_y = \begin{bmatrix} 0 & -i \\ i & 0 \end{bmatrix} \qquad \sigma_z = \begin{bmatrix} 1 & 0 \\ 0 & -1 \end{bmatrix}$$

hence $\begin{bmatrix} z & x - iy \\ x + iy & -z \end{bmatrix} = x\sigma_x + y\sigma_y + z\sigma_z = \mathbf{r} \cdot \boldsymbol{\sigma}$

The three Pauli matrices together with the unit matrix form a set of four units in terms of which any Hermitian second-order matrix can be expressed. The Q matrix for rotation about the **k** axis through an angle ϕ is

$$\mathbf{Q} = \mathbf{1}\cos\frac{\phi}{2} + i\sigma_z \sin\frac{\phi}{2}$$

where **1** is written for

$$\begin{bmatrix} 1 & 0 \\ 0 & 1 \end{bmatrix}$$

In general the Q matrix for a rotation through an angle ϕ about any axis is of this same form where in place of σ_z one uses the Pauli spin matrix for the component of the vector $\boldsymbol{\sigma}$ along the direction of the axis of rotation.

4. Relativistic Kinematics

Classical kinematics is founded on the idea that it is possible to refer positions to a rigid cartesian framework, which in an absolute sense may be said to be nonrotating, and that time is a single independent variable which is the same for all observers, at all places in space no matter what is their relative motion. On this view if the observer O' (Fig. 1.4) moves with velocity V along the X axis with respect to O, the relation between the coordinates and time used by the two observers for description of motions is

$$\begin{aligned} x &= x' + Vt' \\ y &= y' \\ z &= z' \\ t &= t' \end{aligned} \qquad (1.11)$$

This transformation of coordinates is called a *Galilean transformation.* Experience has shown that this relation is correct for ordinary mechanical phenomena but that it has to be modified if V becomes comparable with the velocity of light, $c = 3 \times 10^{10}$ cm/sec. It will be convenient in what follows to write $\beta = V/c$ and to choose units so that the numerical value of the velocity of light is unity.

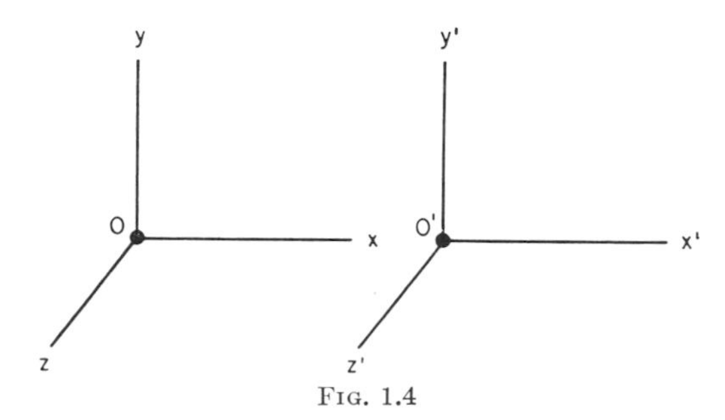

Fig. 1.4

The necessary modification springs from the experimental generalizations from electrodynamic and optical phenomena that (1) the velocity of light as measured by any observer moving in an unaccelerated way will be independent of the motion of the light source and (2) that all such observers will find the same numerical value for the velocity of light. These together are known as the *principle of relativity* (see Part 2, Chap. 6).

There will be an instant when O' coincides with O. For convenience suppose both observers reckon time from this instant so that both $t = 0$ and $t' = 0$ correspond to this instant. If at this instant a pulse of light is emitted from the common origin, the observer O' will find, by any experiment he can devise, that the expanding wavefront of the pulse will be represented by the equation

$$t'^2 - (x'^2 + y'^2 + z'^2) = 0 \qquad (1.12)$$

Likewise, according to the principle of relativity, observer O will find the same expanding wavefront represented by the equation

$$t^2 - (x^2 + y^2 + z^2) = 0 \qquad (1.13)$$

This is in contradiction with the Galilean transformation according to which, if O' finds the expanding light wave to be spherical with a center moving with O', then the observer O would find it represented by $t^2 - [(x - \beta t)^2 + y^2 + z^2] = 0$. That is, it would appear to O to be a sphere with moving center going along with the motion of O'.

In the language of the older wave theory of light it would be said that observer O' was the one who was "at rest with respect to the ether" or "at absolute rest" and that the observer O was moving with velocity $-\beta$ in the x direction "with respect to the ether." It was presumed that optical experiments could be devised which would reveal this motion of the center of the expanding light wave and thus the observer might, by optical experiments carried out in his own laboratory, determine β, his absolute velocity with respect to the ether.

However, such attempts always gave the result that the light wave behaved as if the center of the sphere was at rest in the particular frame of reference employed. Adopting this result as a general principle, it follows that the relation between the time and space measurements of O' and O must be such that Eq. (1.12) transforms into Eq. (1.13).

The fact that the transformation and its inverse must be single-valued requires the transformation to be linear. Assuming the corresponding axes to be parallel in the two systems, one may put $y' = y$ and $z' = z$. The transformation reduces to the form that x' and t' are linear expressions in x and t in which the coefficients depend on β.

Two events appear simultaneous to an observer if they are associated with the same instant of time. If the Galilean transformation were correct, there would be a single time variable for all observers and events noted as simultaneous in one reference frame would be so noted in all other reference frames for observers in uniform motion relative to each other. But if now the time is to be transformed according to a formula like $t' = at + bx$, then it follows that two events that are separated in space $(x_1 \neq x_2)$ and are *simultaneous* to O, that is, $t_1 = t_2$, will appear as *not simultaneous* to O', for then $t'_1 \neq t'_2$. This is the sense in which the principle of relativity implies that space and time are not absolutes in themselves. The events of the physical world take place in a four-dimensional space-time manifold, and the breakdown of the relation between two events into spatial and temporal separation is not of absolute significance, but an accidental attribute of the situation, like the arbitrarily chosen orientation of a cartesian coordinate system in a geometrical problem.

It is convenient to graph the (t,x) coordinates of events on a two-dimensional graph as in Fig. 1.5. The points reached by light pulses traveling to the right and left are given by $x = \pm t$. In four-dimensional space these points whose equation is (1.13) are said to lie on the same light cone. The sequence of events corresponding to successive positions of observer O' is given by the line $x = \beta t$. This is labeled the t' axis in the figure since it corresponds to points for which $x' = 0$.

The transformation is such that $t'^2 - x'^2 = t^2 - x^2$, since units have been chosen so as to assign unit value to the velocity of light. In particular then the value of the O' time corresponding to (t,x) and $x' = 0$ is $t'^2 = t^2 - x^2$. Since $x = \beta t$, this can be written $t'^2 = t^2 - x^2 = t^2(1 - \beta^2)$. The fact that t'^2 is not $t^2 + x^2$ shows that the space of the graph possesses a non-Euclidean geometry. The relation of t' and t for $x' = 0$ is thus $t' = t(1 - \beta^2)^{1/2}$.

It follows that the clock of the moving system appears to run at a slow rate since the interval t' is measured by a smaller number than t, as $(1 - \beta^2)^{1/2}$ is always less than unity. This is a symmetrical situation in that one can consider the appearance to O' of the lapse of time in some apparatus moving with O. In this case $x = 0$, and $t^2 = t'^2 - x'^2$, with $x' = -\beta t$ and thus $t = t'(1 - \beta^2)^{1/2}$. Therefore it is a property of the transformation that the interval of time between two events occurring at the same place in the system O' will appear as a longer interval of time when observed by an observer O moving relative to the system O'.

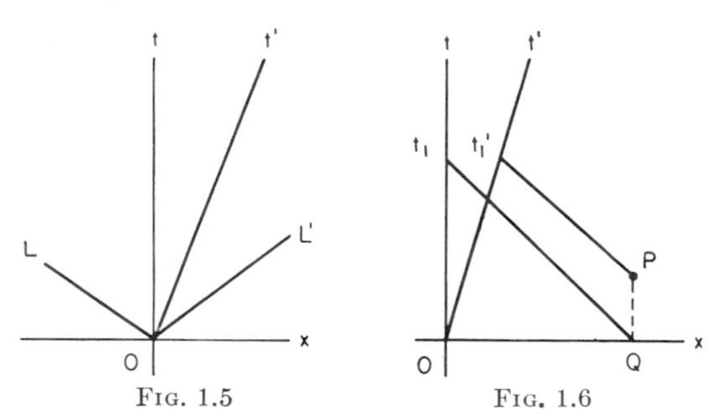

FIG. 1.5 FIG. 1.6

It is next important to see the relation (Fig. 1.6) between the (t,x) coordinates of events that appear as simultaneous to the observer O'. Let it be supposed that P is O' simultaneous with the common origin event and therefore is associated with the value $t' = 0$. If at the instant $t' = 0$ a light signal is sent from P, it will be observed to arrive at $x' = 0$ at the time $t' = t'_1$. By the argument just given this event is characterized by $t = t_1$, $x = \beta t_1$, where $t_1 = t'_1/(1 - \beta^2)^{1/2}$. Since the velocity of light is unity, the quantity t'_1 is also equal to the x' coordinate of P. Likewise a light signal from Q ($t = 0$, $x = x_p$) will arrive at $x = 0$ at the time t_1 and therefore for P, $x = x'/(1 - \beta^2)^{1/2}$ ($t' = 0$). Since the coordinates of P in the two systems must be related by $t'^2 - x'^2 = t^2 - x^2$, $-x^2(1 - \beta^2) = t^2 - x^2$ and therefore $t = \beta x$ is the relation connecting events for which $t' = 0$ and which are therefore simultaneous to observer O'.

The result permits recognition of the way in which distances undergo a *Lorentz contraction* when measured by an observer moving relative to the object. Suppose a bar of length L, as measured by an observer at rest with it, is at rest in the O system with one end at $x = 0$ and the other at $x = L$, at all values of t. The observer O' will measure as its length the separation in x' of two ends of the bar at what *he* regards as simultaneous times, for example, at $t' = 0$. He will therefore find that one end of the bar is at ($x' = 0$, $t' = 0$) and the other at ($x' = L\sqrt{1 - \beta^2}$, $t' = 0$). Observer O' will thus obtain a shorter value for the length of the bar than observer O who rides with the bar.

The foregoing discussion covers the three main points in which the relativity principle leads to results at variance with the classical, naïve "common-sense" view as embodied in the Galilean transformation, namely, (1) the interval of time between two events at the same place in a certain system seems greater to any observer moving with respect to the first system; (2) events at different places which are simultaneous in one system will not appear simultaneous to an observer moving with respect to the system in which the events appear to be simultaneous; and (3) any length in the direction of relative motion appears shorter to an observer moving with respect to the length than to an observer moving with the length.

To summarize, the preceding discussion shows that (1) the relation of t' and t for $x' = 0$, that is, $x = \beta t$, is $t' = t(1 - \beta^2)^{1/2}$; (2) the relation of x' and x for an event simultaneous with the origin, $t' = 0$, is $t = \beta x$ and $x' = x(1 - \beta^2)^{1/2}$.

From these results it follows that the general transformation must be of the form $x' = K(x - \beta t)$, $t' = K(t - \beta x)$, where K is a function of the velocity and must be $1/(1 - \beta^2)^{1/2}$ in order to be consistent with the special results obtained. The Lorentz transformation is therefore

$$x' = \frac{x - \beta t}{\sqrt{1 - \beta^2}} \qquad t' = \frac{t - \beta x}{\sqrt{1 - \beta^2}}$$

As mentioned earlier, the (t,x) space-time has a non-Euclidean geometry. The formulas for a Lorentz transformation are expressible in terms of hyperbolic functions in analogy with the formulas for rotation of axes involving circular functions in ordinary space: $x' = x \cosh \theta - t \sinh \theta$, $t' = -x \sinh \theta + t \cosh \theta$, where $\beta = \tanh \theta$.

This mode of writing the transformation makes it easy to get the relativistic formula for composition of velocities. If O'' moves with speed relative to O' given by β_2 as measured in O', and O' moves with speed β_1 relative to O, then one has $(x'',t'') = L_2(x',t')$, where this notation indicates the Lorentz transformation with parameter β_2. Also $(x',t') = L_1(x,t)$. Thus by elimination $(x'',t'') = L_3(x,t)$, where

$$\theta_3 = \theta_1 + \theta_2$$

and therefore by the addition formula for hyperbolic tangents, calling β_3 the velocity of O'' relative to O, $\beta_3 = (\beta_1 + \beta_2)/(1 + \beta_1\beta_2)$. An important property of these results is that the only real transformations are those for $\beta < 1$ and that the composition of velocities is such that the resultant of two velocities always gives $\beta_3 < 1$ no matter how near unity are the summands β_1 and β_2.

If the velocity of O' relative to O is given by $\boldsymbol{\beta}$, (in any direction, with c as unit) and if $(\mathbf{r}',t')$ are the vectorial coordinates and time in the O' frame and $(\mathbf{r},t)$ are the corresponding quantities in the O frame, then the general Lorentz transformation is

$$\mathbf{r}' = \mathbf{r} + \left[\frac{1}{\sqrt{1 - \beta^2}} - 1\right](\mathbf{r} \cdot \boldsymbol{\beta}_0)\,\boldsymbol{\beta}_0 - \frac{\boldsymbol{\beta} t}{\sqrt{1 - \beta^2}}$$

$$t' = \frac{t - \mathbf{r} \cdot \boldsymbol{\beta}}{\sqrt{1 - \beta^2}}$$

where $\boldsymbol{\beta}_0$ is a unit vector in the direction of the relative velocity $\boldsymbol{\beta}$.

According to the principle of relativity all reference frames which are related to each other by Lorentz transformations are equally valid; in other words, the fundamental laws of physics must be expressible in a form that is invariant under any Lorentz transformation. As the Lorentz transformations together with ordinary rotations of the spatial reference frame form a group, this means that fundamental physical laws must be expressible in forms that are invariant under the transformations of this group.

This calls for an important revision of the kinematical concepts of particle dynamics since in the classical treatment the time t occupies a role as independent variable that is unsymmetrical with respect to the space coordinates. If $(x + dx, y + dy, z + dz, t + dt)$ is an event infinitesimally separated from (x,y,z,t), then $ds^2 = dt^2 - (dx^2 + dy^2 + dz^2)$ is a characteristic of the separation that is invariant under Lorentz transformations. The quantity ds is called the *interval* between the events. It is real or timelike if $(dx^2 + dy^2 + dz^2) < dt^2$; otherwise it is imaginary or spacelike.

If, relative to a particular reference frame, the motion of a particle is given in the usual way by three functions of the time, $x(t)$, $y(t)$, $z(t)$, then $ds^2 = dt^2\ (1 - \beta^2)$, where $\beta^2 = \dot{x}^2 + \dot{y}^2 + \dot{z}^2$, which makes it possible to express the total lapse of interval as observed by a clock which moves with the body $s = \int^t (1 - \beta^2)^{1/2}\, dt$. For measurement of s one would need an idealized clock which is not affected by acceleration in any other way than to adjust its rate always to the time coordinate of a reference frame that is instantaneously at rest relative to the body. The parameter s is called the *proper time* of the moving body and is a suitable invariant parameter in terms of which to describe the motion of a particle. Instead of $x(t)$, $y(t)$, $z(t)$ one has $X(s)$, $Y(s)$, $Z(s)$, $T(s)$, where these four functions satisfy the relation, $\dot{T}^2 - (\dot{X}^2 + \dot{Y}^2 + \dot{Z}^2) = 1$, where dots indicate differentiation with respect to s.

In the four-dimensional space geometry the velocity is replaced by a 4-vector which is tangent to the invariant curve made up of the ensemble of events in the space-time motion of the particle. The fact that the *velocity vector is always a unit vector* is the analogue of the fact that the unit vector tangent to a curve in 3-space, given by parametric equations $x(s)$, $y(s)$, $z(s)$, where s is the arc length of the curve, has the components $\dot{x}(s)$, $\dot{y}(s)$, $\dot{z}(s)$.

Similarly the appropriate invariant definition of acceleration is a 4-vector having the components $\ddot{X}(s)$, $\ddot{Y}(s)$, $\ddot{Z}(s)$, $\ddot{T}(s)$. By differentiation of the equation expressing the constancy of length of the velocity 4-vector, it follows that the acceleration 4-vector is orthogonal to the velocity 4-vector in the sense that $\dot{T}\ddot{T} - \dot{X}\ddot{X} - \dot{Y}\ddot{Y} - \dot{Z}\ddot{Z} = 0$.

In relativistic kinematics a particle is said to be *uniformly accelerated* if in each increment of time, measured in a system with respect to which the particle is instantaneously at rest, the additional velocity gained is the same. Referring back to the discussion of the Lorentz transformation in terms of hyperbolic functions, this implies that the parameter θ increases at a constant rate.

5. Vector Algebra of Space-Time

Ordinary vector algebra in three dimensions provides a means of dealing with directed quantities which expresses the lack of dependence of the intrinsic geometrical relationships on the choice of coordinate system by which they are expressed. To give similar expression to the independence of physical relations to the particular choice of a reference frame from among those related by Lorentz transformations, it is desirable to extend vector algebra to the four-dimensional space-time manifold (Part 1, Chap. 10).

This extension involves not merely passing from three to four dimensions but also taking account of the fact that the manifold has a non-Euclidean geometry, in that the invariant "length" of a space-time vector is $x^2 + y^2 + z^2 - c^2t^2$ instead of $x^2 + y^2 + z^2 + c^2t^2$ as it would be in a Euclidean 4-space. One way is to introduce an imaginary fourth coordinate, $u = ict$, in which case the invariant length takes the form $x^2 + y^2 + z^2 + u^2$. This method will be used in what follows, the discussion being limited to transformations involving orthogonal coordinate systems.

For convenience the four coordinates will be designated by x_α according to the scheme $x_1 = x$, $x_2 = y$, $x_3 = z$, $x_4 = ict$.

A general *Lorentz transformation* to a new set of coordinates x'_α is defined by the fourth-order matrix $a_{\alpha\beta}$, $x'_\alpha = a_{\alpha\beta}x_\beta$. Throughout this section it is to be understood that a repeated index in a term (as β here) implies summation over the values 1, 2, 3, 4 of that index. The orthogonality condition which carries $\Sigma x'_\alpha{}^2$ into $\Sigma x_\alpha{}^2$ requires that $a_{\alpha\beta}$ be such that $a_{\beta\alpha}a_{\beta\gamma} = \delta_{\alpha\gamma}$ (1 if $\alpha = \gamma$, 0 if $\alpha \neq \gamma$). It is convenient to define $\tilde{a}_{\alpha\beta}$ as the matrix which is the transpose of $a_{\alpha\beta}$ (interchange of rows and columns) by $\tilde{a}_{\alpha\beta} = a_{\beta\alpha}$ in order to put this in the form of matrix multiplication, $\tilde{a}_{\alpha\beta}a_{\beta\gamma} = \delta_{\alpha\gamma}$. The inverse transformation to $a_{\alpha\beta}$ is the one which carries x'_α back to x_α. Denoting its matrix by $A_{\alpha\beta}$, $x_\alpha = A_{\alpha\beta}x'_\beta$. It has to be related to $a_{\alpha\beta}$ by the equations

$$A_{\alpha\beta}a_{\beta\gamma} = \delta_{\alpha\gamma}$$

Comparing with the preceding equation, it is seen that the *inverse matrix to an orthogonal matrix is its own transpose:* $A_{\alpha\beta} = \tilde{a}_{\alpha\beta}$.

For a real Lorentz transformation, real x_1, x_2, x_3 and pure imaginary x_4 must transform into real $x'_1x'_2x'_3$ and pure imaginary x'_4. Therefore, $a_{\alpha\beta}$ is pure imaginary if α or $\beta = 4$, but is real if neither α nor $\beta = 4$ or if both α and $\beta = 4$.

Corresponding to the notation $\mathbf{r} = x\mathbf{i} + y\mathbf{j} + z\mathbf{k}$ for a 3-vector, one may introduce 4 unit vectors $\mathbf{e}_\alpha$ in the directions of increasing x_α and write $\mathbf{r} = x_\alpha\mathbf{e}_\alpha$, where it is understood that the unit vectors are orthogonal so that $\mathbf{e}_\alpha \cdot \mathbf{e}_\beta = \delta_{\alpha\beta}$. For any physically real vector the first three components are real numbers and the fourth component is pure imaginary; thus $\mathbf{r}^2 = x_1{}^2 + x_2{}^2 + x_3{}^2 - |x_4|^2$ is the fundamental scalar invariant derived from a real 4-vector.

Likewise if A and B are two real vectors, $\mathbf{A} = A_\alpha\mathbf{e}_\alpha$, $\mathbf{B} = B_\alpha\mathbf{e}_\alpha$, their scalar product

$$\mathbf{A} \cdot \mathbf{B} = A_1B_1 + A_2B_2 + A_3B_3 - |A_4|\,|B_4|$$

is a scalar invariant under any Lorentz transformation.

Second-order tensors are defined as being 4×4 component quantities $C_{\alpha\beta}$ whose components transform on changing to a new coordinate system by the rule $C'_{\alpha\beta} = a_{\alpha\gamma}a_{\beta\delta}C_{\gamma\delta}$. A real second-order tensor is one for which $C_{\alpha\beta}$ is real if α and β are not 4 or are both 4, but is pure imaginary if α or β equal 4. This property is preserved on transformation to new coordinates since $a_{\alpha\beta}$ has the same property.

Any second-order tensor can be split into symmetric and antisymmetric parts $\frac{1}{2}(C_{\alpha\beta} + C_{\beta\alpha})$ and $\frac{1}{2}(C_{\alpha\beta} - C_{\beta\alpha})$, a decomposition which is invariant under Lorentz transformations.

In three-dimensional vector analysis there are three components to a vector and three components to an antisymmetric second-order tensor. This fact makes possible the association of a vector as a geometric object to represent an antisymmetric second-order tensor. But in four dimensions an antisymmetric second-order tensor has six components; therefore no such association is possible.

Nevertheless cross-product multiplication can be defined formally by using the units

$$\mathbf{e}_\beta \times \mathbf{e}_\alpha = -\mathbf{e}_\alpha \times \mathbf{e}_\beta$$

as the basic units of the antisymmetric tensor of cross multiplication. The cross product $\mathbf{A} \times \mathbf{B}$ becomes

$$\begin{aligned}\mathbf{A} \times \mathbf{B} &= (A_\alpha \mathbf{e}_\alpha \times B_\beta \mathbf{e}_\beta) \\ &= (A_\alpha B_\beta - A_\beta B_\alpha)\mathbf{e}_\alpha \times \mathbf{e}_\beta\end{aligned}$$

where $\beta < \alpha$ in the summation over β.

The relativistic formulation of physical theories which are already known in nonrelativistic form consists in the transcription of the classical equations into the 4-vector algebra of Lorentz transformations, making such modifications as are necessary to adapt the classical approximate forms to this requirement.

The motion of a particle in classical mechanics is given by specifying $x_1(t)$, $x_2(t)$, $x_3(t)$ relative to a particular reference frame. This is not in relativistically invariant form as it stands. Instead of t one needs to introduce the proper time τ,

$$\tau = \int^t \sqrt{1 - \beta^2}\, dt$$

as the independent variable, this being a negligible change when $\beta^2 \ll 1$. Then the 4-vector suitable for describing the velocity of a particle is

$$\begin{aligned}v_\alpha &= \left(\frac{dx_1}{d\tau}, \frac{dx_2}{d\tau}, \frac{dx_3}{d\tau}, ic\,\frac{dt}{d\tau}\right) \\ &= \frac{1}{\sqrt{1 - \beta^2}}(v_x, v_y, v_z, ic)\end{aligned}$$

where $v_x = dx/dt$ in the classical way. By direct calculation, $\Sigma v_\alpha{}^2 = -c^2$; thus v_α is a vector of constant length. When the particle is at rest in a particular reference frame, its velocity vector points exactly along the time axis of that frame.

Similarly the *acceleration 4-vector* is defined as $a_\alpha = dv_\alpha/d\tau$. Since v_α is a vector of constant length, it follows that the acceleration is always orthogonal to the velocity. In terms of the usual velocity components, with time as independent variable,

$$a_1 = \frac{1}{\sqrt{1 - \beta^2}} \frac{d}{dt}\left(\frac{v_1}{\sqrt{1 - \beta^2}}\right), \ldots, \ldots,$$

$$a_4 = \frac{ic}{\sqrt{1 - \beta^2}} \frac{d}{dt}\left(\frac{1}{\sqrt{1 - \beta^2}}\right)$$

The moment of the velocity of a particle moving with velocity $\mathbf{v}$ when at position $\mathbf{r}$ (these are 3-vectors) is defined as the vector product $\mathbf{r} \times \mathbf{v}$. In relativistic kinematics the cross product of the corresponding 4-vectors gives an antisymmetric second-order tensor, which has six components. The scheme is

$$\begin{aligned}(\mathbf{r}_\alpha) \times (\mathbf{v}_\alpha) = &(x_1v_2 - x_2v_1)\mathbf{e}_1 \times \mathbf{e}_2 \\ &+ (x_2v_3 - x_3v_2)\mathbf{e}_2 \times \mathbf{e}_3 \\ &+ (x_3v_1 - x_1v_3)\mathbf{e}_3 \times \mathbf{e}_1 \\ &+ (x_1v_4 - x_4v_1)\mathbf{e}_1 \times \mathbf{e}_4 \\ &+ (x_2v_4 - x_4v_2)\mathbf{e}_2 \times \mathbf{e}_4 \\ &+ (x_3v_4 - x_4v_3)\mathbf{e}_3 \times \mathbf{e}_4\end{aligned}$$

The first three lines, the "spacelike" components of the tensor, reduce at low velocities to the usual moment of velocity with the components

$$\frac{(\mathbf{r} \times \mathbf{v})}{\sqrt{1 - \beta^2}}$$

where $\mathbf{r}$ and $\mathbf{v}$ are 3-vectors. The three "timelike" components together form the 3-vector:

$$\frac{ic}{\sqrt{1 - \beta^2}}[(x - v_xt)\mathbf{i} + (y - v_yt)\mathbf{j} + (z - v_zt)\mathbf{k}]$$

In general it is found that the six components of an antisymmetrical second-order tensor group themselves into three spacelike and three timelike components each of which is simply related to a vector in 3-space.

References

Mechanics is the oldest branch of physics and has the largest literature. Kinematics is usually treated in the introductory chapters of any general treatise on mechanics. The following books are a selection from the many available which may be usefully consulted for fuller treatment of the classical mechanics of particles and rigid bodies:

1. Ames, J. S., and F. D. Murnaghan: "Theoretica Mechanics," Ginn, Boston, 1929. Reprint: Dover New York.
2. Goldstein, H.: "Classical Mechanics," Addison-Wesley, Reading, Mass., 1950.
3. Lanczos, C.: "The Variational Principles of Mechanics," University of Toronto Press, Toronto, 1949.
4. Mach, Ernst: "The Science of Mechanics," 5th English ed., Open Court, Chicago, 1942.
5. MacMillan, W. D.: "Theoretical Mechanics," McGraw-Hill, New York, vol. 1 Statics and Dynamics of a Particle (1927) and vol. 3 Dynamics of Rigid Bodies (1936). Reprint: Dover, New York.
6. Pauli, W.: "Theory of Relativity," Pergamon Press, New York, 1958.
7. Poincaré, H.: Les Méthodes nouvelles de la mécanique céleste, 3 vols., Gauthier-Villars, Paris, 1892–1899. Reprint: Dover, New York.
8. Synge, J. L., and B. A. Griffith: "Principles of Mechanics," 2d ed., McGraw-Hill, New York, 1949.
9. Webster, A. G.: "The Dynamics of Particles and of Rigid, Elastic and Fluid Bodies," Stechert, New York, 1920.
10. Whittaker, E. T.: "A Treatise on the Analytical Dynamics of Particles and Rigid Bodies," 4th ed., Cambridge University Press, New York, 1937.

Chapter 2

Dynamical Principles

By E. U. CONDON, University of Colorado

The basis of classical mechanics is contained in Newton's three laws of motion (1687). According to the first of these, a body moves uniformly in a straight line if not acted on by a force. According to the second, the rate of change of momentum of a particle is proportional to the resultant force acting. According to the third, the force acting on particle i due to any kind of interaction with particle j is equal and opposite to that acting on j due to interaction with i and acts in the direction of the line joining i and j.

1. Mass

The property of a body by which it requires force to change its state of motion is called its *inertia,* and mass is the numerical measure of this property. Since it is found experimentally that the acceleration of gravity is the same for all bodies, large or small and of whatever material, it follows that the *weight* (which is the *force* of gravitational attraction) of a body is proportional to its *mass.* This fact justifies the use of a balance to compare the masses of bodies.

The fundamental unit of mass is the international *kilogram,* made of platinum and kept at the International Bureau of Weights and Measures at Sèvres, France. Copies of it have been issued to the national standardizing laboratories of various countries of the world and are checked against the fundamental standard at Sèvres from time to time. Another basic unit of mass is the *gram,* defined as 0.001 kg. In America the avoirdupois pound is defined as 453.59237 g.

2. Momentum

The *momentum* of a particle is a vector defined as the product of its mass m by its vector velocity $\mathbf{v}$. It will be denoted by $\mathbf{p}$ so that $\mathbf{p} = m\mathbf{v}$. The unit of momentum is therefore that of unit mass traveling at unit velocity; in the cgs system that is the momentum of 1 g moving at a speed of 1 cm/sec. The *angular momentum* of a particle about the origin is defined as $\mathbf{L} = \mathbf{r} \times \mathbf{p}$, where $\mathbf{r}$ is its position vector and $\mathbf{p}$ is its linear momentum.

3. Force

Because force is *proportional* to the rate of change of momentum the unit of force may be chosen in such a way that the force is *equal* to the rate of change of momentum. When this is done the force is said to be measured in absolute units.

For a particle of constant mass the rate of change of momentum is equal to the product of mass times acceleration. Unit force is thus defined as the force which gives unit acceleration to unit mass. In the cgs system this is called 1 *dyne:* 1 dyne gives an acceleration of 1 cm/sec^2 to a mass of 1 g, that is, $\mathbf{F} = m\mathbf{a}$ where $\mathbf{F}$ is expressed in dynes, m in grams, and $\mathbf{a}$ in centimeters per second per second.

In the meter-kilogram-second (mks) system, the absolute unit of force is called 1 *newton,* the force necessary to give an acceleration of 1 meter/sec^2 to a mass of 1 kg (1 newton = 10^5 dynes).

In the English system the absolute unit of force is known as the *poundal,* the force needed to give an acceleration of 1 ft/sec^2 to a mass of 1 lb.

At a place where the acceleration of gravity is g, the weight expressed as a force in absolute units is equal to mg, for when it acts alone on a body of mass m it is able to give the body an acceleration g.

The absolute system of force units is used universally in fundamental physics. In the older branches of engineering, another system is in use, based on taking the unit of force to be numerically equal to the weight of 1 gram (g) or 1 pound (lb). When this is done, and with the pound as the unit of mass, it is no longer possible to write $\mathbf{F} = m\mathbf{a}$. Instead an appropriate proportionality factor must be introduced. This factor is g^{-1} because unit force gives unit mass an acceleration equal to g on this system; therefore the equation appropriate to the second law of motion in these units is $\mathbf{F} = m\mathbf{a}/g$.

Another procedure used in engineering amounts to absorbing the g^{-1} factor into the m by defining a new unit of mass. In the English system this is called the *slug,* which is defined as the mass which is given an acceleration of 1 ft/sec^2 by a force of 1 lb weight. One slug is therefore g lb of mass so that the mass of the body expressed in slugs is m/g, where m is its mass expressed in pounds.

4. Impulse

Since the time rate of change of momentum of a body is equal to the force acting on it, the change in momentum in any finite interval is equal to the time integral of the force acting. The time integral of force acting on a body is known in mechanics as the *impulse.* In the absolute system of units it is expressed in the same units as momentum.

If a particle of mass m moving with velocity $\mathbf{v}_0$ is acted on by an impulse $\mathbf{I}$, the velocity will be changed to $\mathbf{v}$ such that

$$m(\mathbf{v} - \mathbf{v}_0) = \mathbf{I}$$

Multiplying through by $\frac{1}{2}(\mathbf{v} + \mathbf{v}_0)$, this gives

$$\tfrac{1}{2}m\mathbf{v}^2 - \tfrac{1}{2}m\mathbf{v}_0^2 = \mathbf{I} \cdot \tfrac{1}{2}(\mathbf{v} + \mathbf{v}_0)$$

Hence the increase in kinetic energy of the particle due to the action of the impulse is the scalar product of the impulse and the arithmetic mean of the velocity before and after the impulse. The idea of impulse is often convenient in describing collision phenomena where the total change of momentum is known but the exact time variation of the force producing it during the collision is not known.

5. Work and Energy: Power

The *work* done on a body when it moves through a displacement $d\mathbf{r}$ under the action of a force $\mathbf{F}$ is defined as being $\mathbf{F} \cdot d\mathbf{r} = dW$. The absolute cgs unit of work is called the *erg* and is the work done when a force of 1 dyne acts through a displacement of 1 cm in the same direction as the force. The absolute mks unit is the *joule* (1 joule = 10^7 ergs), the work done by 1 newton acting through a distance of 1 meter. Similarly, the absolute English unit is the *foot-poundal,* the work done by a force of 1 poundal in moving 1 ft in the direction of the force. In engineering it is often convenient to measure work in gravitational units as the foot-pound or the kilogram-meter.

In the equation of motion for a particle of mass m,

$$\mathbf{F} = \frac{d}{dt}(m\mathbf{v})$$

If both sides are multiplied by $d\mathbf{r} = \mathbf{v}\,dt$,

$$\mathbf{F} \cdot d\mathbf{r} = d(\tfrac{1}{2}m\mathbf{v}^2)$$

so that the work done on the body by the acting force is equal to the increase of the quantity $T = \frac{1}{2}mv^2$. This quantity is defined as the *kinetic energy* of the body and therefore the statement just made is that the increase in kinetic energy of a body is equal to the work done on the body by the forces acting on it. The whole kinetic energy of the body is equal to the whole work done on it by the resultant force in being accelerated from zero speed up to the speed v. Conversely, the kinetic energy is the work which the body can do against external forces in being brought to rest.

In mechanics, *power* means the time rate of doing work. In the mks system it is

$$1 \text{ joule/sec} = 1 \text{ watt} = 10^7 \text{ ergs/sec}$$

For units of energy, see Table 2.1.

TABLE 2.1. UNITS OF ENERGY

Units	*Value in Ergs*
British thermal unit (mean)	1.0548×10^{10}
Gram-calorie (mean)	4.186×10^{7}
15°C	4.185×10^{7}
20°C	4.181×10^{7}
Cubic centimeter-atmosphere	1.01325×10^{6}
Cubic foot-atmosphere	2.8694×10^{10}
Foot-pound	1.35582×10^{7}
Foot-poundal	4.21402×10^{5}
Horsepower-hour	2.6845×10^{13}
Volt-electron	1.60210×10^{-12}
Volt-Faraday	9.64870×10^{11}
Kilogram-meter	9.80665×10^{7}
Kilowatthour	3.6000×10^{13}
Liter-atmosphere	1.01328×10^{9}

Gravitational units are evaluated on basis of a standard acceleration of gravity of $g = 980.665$ cm/sec².

The conventional engineering horsepower is 33,000 ft-lb/min = 550 ft-lb/sec = 746 watts.

6. Potential Energy

Let a force acting on a particle be a function of that particle's position in space so that $\mathbf{F} = \mathbf{F}(x,y,z)$. Then the work done by the force on the body in a finite displacement from A to B is

$$\int_A^B \mathbf{F} \cdot d\mathbf{r}$$

In general this will be dependent on the path followed in going from A to B, but in a very important class of cases the force field is such that the value of the integral is independent of the path. Such a force field is said to be *conservative.*

For such a conservative force field there exists a scalar function of position $V(x,y,z)$ such that

$$\int_A^B \mathbf{F} \cdot d\mathbf{r} = V(x_A,y_A,z_A) - V(x_B,y_B,z_B) \tag{2.1}$$

that is, such that the work done on the body by the force is equal to the *decrease* in the function $V(x,y,z)$ in the change in question. This function is called the *potential energy* of the body in the force field in question. Conversely, the force acting on the body is equal to the negative gradient of the potential energy function (Part 1, Chap. 9, Sec. 4),

$$\mathbf{F} = -\operatorname{grad} V \tag{2.2}$$

In general it may happen that forces act on the body other than those which are a function of position alone or are derivable from a potential energy function. Let $\mathbf{F}$ stand for these so that the whole force is equal to $\mathbf{F} - \operatorname{grad} V$. Then the equation of motion is $\mathbf{F} - \operatorname{grad} V = d(m\mathbf{v})/dt$, giving $\mathbf{F} \cdot d\mathbf{r} = d(T + V)$. The work done by the forces other than those derived from V is thus equal to the increase of the sum of the kinetic energy ($T = \frac{1}{2}mv^2$) and the potential energy.

Conversely, if all the force acting on the body is derivable from the potential energy function, then $\mathbf{F} = 0$ and the preceding equation can be integrated to show that throughout the motion the sum of the kinetic and potential energies is constant. This is known as the *energy integral.*

7. Central Force: Collision Problems

If the force acting on a particle is, at each point, directed either toward or away from a center, which may be chosen as the origin, then $\mathbf{r} \times \mathbf{F} = 0$ at all times.

In the equation of motion $m\, d\mathbf{v}/dt = \mathbf{F}$, one may multiply through by $\mathbf{r}$ and find that $m\mathbf{r} \times d\mathbf{v}/dt = 0$, which gives

$$m \frac{d}{dt}(\mathbf{r} \times \mathbf{v}) = 0 \qquad \text{or} \qquad \frac{d}{dt}\mathbf{L} = 0 \tag{2.3}$$

where $\mathbf{L} = m\mathbf{r} \times \mathbf{v}$, is defined as the *orbital angular momentum*. Therefore in the case of central force the orbital angular momentum of the particle is constant. If the initial position and velocity of the particle are $\mathbf{r}_0$ and $\mathbf{v}_0$, then the orbital angular momentum is equal to $\mathbf{L}_0 = m\mathbf{r}_0 \times \mathbf{v}_0$ throughout the motion. It follows that the motion takes place in a plane through the origin that is normal to $\mathbf{L}_0$; thus the problem is reducible to a two-dimensional one of motion in this plane.

Collision problems are important in atomic physics as a means of studying the laws of interaction between two kinds of colliding particles. The theoretical interpretation is simplest in a coordinate system in which the center of mass is at rest. Denoting the particles by 1 and 2, their momenta in such a system will be represented by equal and opposite vectors, $\mathbf{P}$ and $-\mathbf{P}$, as in Fig. 2.1. The initial lines of motion will be parallel and separated by a distance, a. This is known as the *impact parameter*. After the particles have interacted and are receding from each other, their momenta will still be equal and opposite vectors, because of conservation of momentum (Sec. 8 below). Let the momenta of 1 and 2 after collision be $\mathbf{P}'$ and $-\mathbf{P}'$, respectively. The angle of scattering Θ is the angle between $\mathbf{P}$ and $\mathbf{P}'$, as in Fig. 2.2.

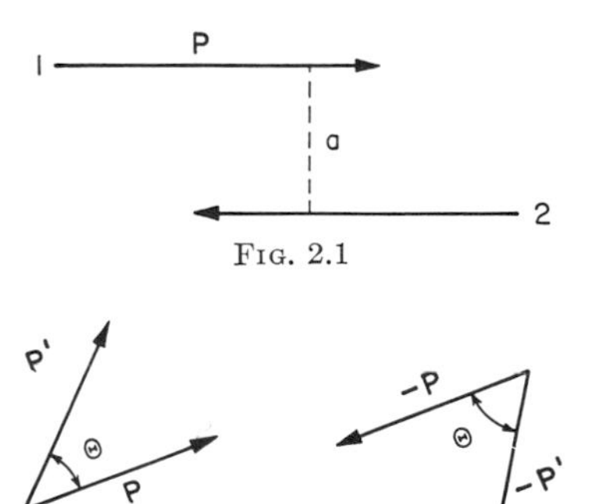

FIG. 2.1

FIG. 2.2

The dynamics of the collision for a particular law of force determines a functional relation between Θ and a and P which may be designated as $a(\Theta,P)$; this gives the impact parameter at a given P needed for a given angle of scattering.

In many problems of atomic physics one is concerned with the scattering of a beam of particles of type 1 directed against a target of particles of type 2. In such cases the lines of motion are distributed at random; thus the relative probability of different values of a is the area of the annular ring, $2\pi a\, da$, between a and $a + da$. It is customary to measure the effectiveness of the scattering by a *collision* cross section for scatter, $\sigma(\Theta)$ into *unit solid angle* between Θ and $\Theta + d\Theta$. This is defined in such a way that if a beam of N particles of type 1 per square centimeter per second is directed against a particle of type 2, the number scattered per second into the solid angle $2\pi \sin \Theta\, d\Theta$ between Θ and $\Theta + d\Theta$ is $N\sigma(\Theta) \cdot 2\pi \sin \Theta\, d\Theta$. The collision cross section is related to the impact parameter $a(\Theta,P)$ as follows

$$\delta(\Theta) = \frac{a}{\sin \Theta}\frac{da}{d\Theta} = \frac{1}{2}\frac{d(a^2)}{d(\cos \Theta)} \tag{2.4}$$

In experimental work involving collisions, a stream of particles of type 1 is allowed to fall on a fixed target of particles of type 2 and the measurements are therefore made in the laboratory in which the particles of type 2 are at rest rather than in a frame in which the center of mass is at rest. Classical mechanics is applicable if all velocities are small compared with that of light; then the laboratory frame moves relative to the center of mass frame in the direction of $\mathbf{P}$ with a velocity P/m_2. Hence in the laboratory frame, target particles (type 2) have no momentum and incident particles have a momentum $\mathbf{p}_1 = (1 + m_1/m_2)\mathbf{P}$. In the laboratory the angle of scattering of the incident particles will be called θ, an angle which is smaller than Θ, as indicated in Fig. 2.3.

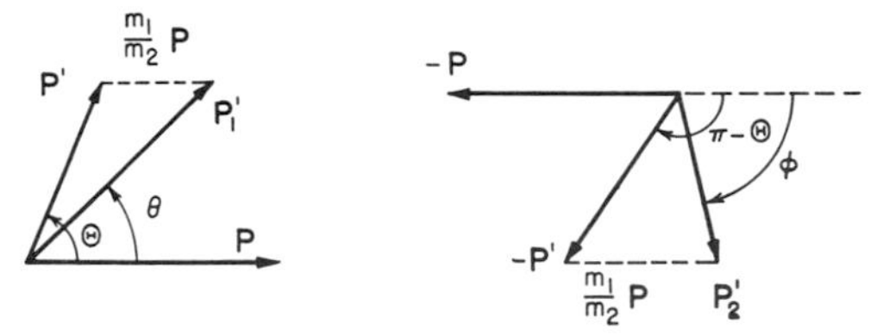

FIG. 2.3

The direction of motion of the target particle after impact relative to the direction of the incident beam is called the angle of recoil, ϕ, in the laboratory frame. It is $(\pi - \Theta)$ in the center-of-mass frame.

Collisions are said to be *elastic* if the total kinetic energy of the particles is unchanged in the collision. For such collisions $|\mathbf{P}'| = |\mathbf{P}|$ since the kinetic energy is

$$T = \frac{1}{2}\left(\frac{1}{m_1} + \frac{1}{m_2}\right)\mathbf{P}^2$$

The relation between angle of scatter, Θ, in the center-of-mass frame and the angle θ in the laboratory frame is therefore

$$\tan \theta = \frac{\sin \Theta}{(m_1/m_2) + \cos \Theta} \tag{2.5}$$

This relation is shown graphically in Fig. 2.4 for three cases, $m_1/m_2 < 1, = 1$, and > 1. If the incident particle is light compared to the struck particle, there is little difference between θ and Θ. If the masses are equal, then at all angles, $\theta = \frac{1}{2}\Theta$. If the incident particle is heavy compared to the struck particle, then there is a maximum value of θ given by $\sin \theta = m_2/m_1$.

Similarly the angle of recoil ϕ in the laboratory system is given by

$$\tan \phi = \frac{\sin \Theta}{(m_1/m_2) - \cos \Theta} \tag{2.6}$$

the general behavior of which is also shown in Fig. 2.4.

The dependence of the scattering cross section on θ in the laboratory system is different from that on Θ

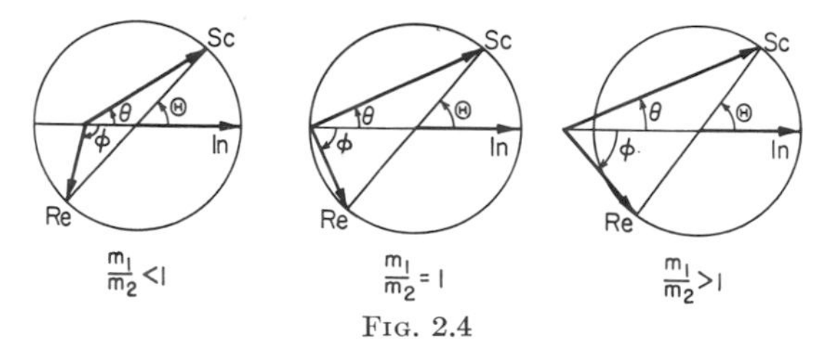

FIG. 2.4

in the center-of-mass system. If $\sigma_l(\theta)$ is written for the laboratory system scattering cross section, then $\sigma_l(\theta) \sin \theta \, d\theta = \sigma(\Theta) \sin \Theta \, d\Theta$; thus

$$\sigma_l(\theta) = \sigma(\Theta) \frac{\sin \Theta \, d\Theta}{\sin \theta \, d\theta} \tag{2.7}$$

Hence, from (2.4),

$$\sigma_l(\theta) = \frac{1}{2} \frac{d(a^2)}{d(\cos \theta)}$$

Inelastic collisions may result when the bodies involved in the collision have structure, so that energy can be taken up or given out from the translatory motion by being exchanged with internal degrees of freedom of one or both of the colliding bodies. With ordinary macroscopic bodies inelastic collisions result from an internal dissipation of some energy, so that the translational kinetic energy after the collision is less than that before.

In atomic systems (and artificially constructed macroscopic systems which might be devised to illustrate the point) *superelastic collisions* may also occur in which the total translational kinetic energy is greater after the collision than before, being gained as a result of some internal change in one or both bodies.

In the case of ordinary inelastic collisions, $P' < P$, and in the case of superelastic collisions, $P' > P$, but in either case the appropriate generalizations of the formulas giving θ and ϕ in terms of Θ are

$$\tan \theta = \frac{\sin \Theta}{(m_1P/m_2P') + \cos \Theta}$$

$$\tan \phi = \frac{\sin \Theta}{(m_1P/m_2P') - \cos \Theta}$$

Therefore if in all the collisions P' bears a constant ratio to P, the effect on the angular distribution is the same as if the collisions were elastic but with a different mass ratio of the particles.

In collisions between macroscopic objects, such as spheres of various hard materials, it is found that P'/P is a constant characteristic of the materials over a wide range of values of P, so that a certain *fraction of the relative momentum is lost* at each collision. This ratio $e = P'/P$ is known as the *coefficient of restitution*. It is unity for perfectly elastic collisions. Collisions in which $e = 0$ are called *perfectly inelastic*. In these the two bodies stick together after impact and the combined body moves in the same direction as the initial motion of the bombarding particle.

In applying quantum mechanics to the behavior of two colliding atomic systems, the only differences result in the mode of calculation of the relative collision cross section for various angles of scatter Θ in the center-of-mass system. The relations of θ and ϕ to Θ involved in passing from the laboratory reference frame to that of the center of mass are the same in quantum mechanics as in classical theory.

8. System of Particles

If the system involves n particles whose position vectors are $\mathbf{r}_i$ $(i = 1,2, \ldots ,n)$, whose masses are m_i, and which are acted on by forces $\mathbf{F}_i + \mathbf{F}_{ij}$, then the equations of motion are $m_i\ddot{\mathbf{r}}_i = \mathbf{F}_i + \Sigma_j\mathbf{F}_{ij}$. Here $\mathbf{F}_{ij}$ is the force acting on the ith particle due to its interaction with the jth particle. From Newton's third law of motion it follows that

$$\mathbf{F}_{ij} = -\mathbf{F}_{ji} \qquad \text{and} \qquad \mathbf{r}_i \times \mathbf{F}_{ij} + \mathbf{r}_j \times \mathbf{F}_{ji} = 0$$

Making use of this property and summing all the equations,

$$\Sigma m_i\ddot{\mathbf{r}}_i = \Sigma \mathbf{F}_i \tag{2.8}$$

The *center of mass* of the system is defined as being a point whose position vector $\mathbf{R}$ is $M\mathbf{R} = \Sigma m_i\mathbf{r}_i$, where $M = \Sigma m_i$. Therefore $M\ddot{\mathbf{R}} = \Sigma \mathbf{F}_i$ so that the *center of mass moves as if all the mass were concentrated there and acted on by the resultant of the external forces.*

In particular, if the resultant of the external forces vanishes, then the acceleration of the center of mass vanishes. This gives an integral of the equations of motion

$$\Sigma m_i\mathbf{v}_i = \text{const} \qquad \text{or} \qquad \Sigma \mathbf{p}_i = \text{const} \tag{2.9}$$

according to which the vector resultant linear momentum of all the particles is constant. This is known as *conservation of momentum*.

If the equations of motion are multiplied through by $\mathbf{r}_i \times$ and summed,

$$\sum_i m\mathbf{r}_i \times \ddot{\mathbf{r}}_i = \sum_i \mathbf{r}_i \times \mathbf{F}_i + \sum_{ij} \mathbf{r}_i \times \mathbf{F}_{ij} \tag{2.10}$$

the double sum vanishes since $\mathbf{F}_{ij}$ acts in the direction of $(\mathbf{r}_i - \mathbf{r}_j)$. The quantity $\Sigma(\mathbf{r}_i \times \mathbf{F}_i)$ gives the vector resultant *torque* of all the external forces acting on the system. The quantity on the left is the rate of increase of the vector sum of the orbital angular momenta of the particles, so that the rate of increase of the resultant angular momentum is equal to the resultant torque of the external forces.

In particular, if the resultant torque of the external forces vanishes, then the resultant angular momentum vector is constant. This is known as *conservation of angular momentum*.

The *virial theorem* of Clausius gives a result concerning the time average of the kinetic energy of the particles in a system which has important applications in the theory of gases.

Let

$$G = \sum_i \mathbf{p}_i \cdot \mathbf{r}_i \tag{2.11}$$

From the equations of motion, writing now $\mathbf{F}_i$ in place of $\mathbf{F}_i + \sum_j \mathbf{F}_{ij}$ for the total force acting on the ith particle,

$$dG/dt = 2T + \sum_i \mathbf{F}_i \cdot \mathbf{r}_i$$

where T is the total kinetic energy of all the particles. If this is averaged over a long time by integrating from 0 to τ and dividing by τ, then in closed systems where all $\mathbf{p}_i$ and $\mathbf{r}_i$ remain finite, the value of G remains finite so that the quantity $[G(\tau) - G(0)]/\tau$ tends to zero, giving

$$\bar{T} = -\tfrac{1}{2}\overline{\sum_i \mathbf{F}_i \cdot \mathbf{r}_i} \tag{2.12}$$

The quantity on the right is known as the *virial*. The virial theorem is the statement that the time average of the kinetic energy is equal to the virial. In case the only forces are those of interaction between pairs of particles described by a potential energy $V(r)$ when the separation is r, the virial reduces to

$$\frac{1}{2}\overline{\sum_{i<j} \frac{\partial V(r_{ij})}{\partial r_{ij}} r_{ij}}$$

9. Lagrange's Equations

In dealing with systems involving a number of particles or rigid bodies which are constrained to move along certain curves or surfaces, it is often convenient to use more general coordinates than the cartesian position coordinates of the particles.

The number of independent coordinates needed to express the positions of all its parts, consistently with the constraints, is known as the number of *degrees of freedom* of the system. The constraints are said to be *holonomic* if they can be expressed in integrated form as a functional relation among the coordinates themselves. Nonholonomic constraints are those requiring nonintegrable relations between differentials of the coordinates for their expression. In what follows the system will be supposed subject only to holonomic constraints. If a generalized set of coordinates for a system of n degrees of freedom is designated by $(q_1, q_2, \ldots, q_n)$, this means that there exist a set of functional relations

$$\mathbf{r}_i = \mathbf{f}_i(q_1, \ldots, q_n, t) \tag{2.13}$$

giving the vector position of each particle in the system in terms of the q's and the time.

If $\mathbf{F}_i$ denotes the resultant force acting on the ith particle, both external and internal, then the equations of motion are $m_i\ddot{\mathbf{r}}_i = \mathbf{F}_i$ for each particle. If each such equation is multiplied through by $\partial\mathbf{f}_i/\partial q_r$ and summed over all the particles, the result is

$$\sum m_i\ddot{\mathbf{r}}_i \cdot \frac{\partial \mathbf{f}_i}{\partial q_r} = \sum \mathbf{F}_i \cdot \frac{\partial \mathbf{f}_i}{\partial q_r}$$

The velocity of the ith particle is given by

$$\mathbf{v}_i = \sum_s \frac{\partial \mathbf{f}_i}{\partial q_s}\dot{q}_s + \frac{\partial \mathbf{f}_i}{\partial t} \tag{2.14}$$

Regarding this as a function of the q's and $\dot{q}$'s and t, it follows that

$$\frac{\partial \mathbf{v}_i}{\partial \dot{q}_s} = \frac{\partial \mathbf{f}_i}{\partial q_s}$$

The quantity

$$\sum \mathbf{F}_i \cdot \frac{\partial \mathbf{f}_i}{\partial q_r}$$

is the coefficient of δq_r in the expression for the total work done by all the forces $\mathbf{F}_i$ on all the particles in a displacement in which the other q's and the time are constant while q_r is changed to $q_r + \delta q_r$. This will be denoted by Q_r. The general expression for the work done by the forces in an arbitrary variation of the q's is

$$\Sigma\, Q_r\, \delta q_r \tag{2.15}$$

The quantity Q_r is known as the *generalized force* associated with the generalized coordinate q_r. Since the q_r are not in general measured in length units, it follows that the Q_r will not in general be measured in force units, but the units of Q_r will always be such that $Q_r\,\delta q_r$ is measured in energy units.

In terms of the generalized coordinates the kinetic energy can be written

$$T = \tfrac{1}{2}\Sigma m_i \mathbf{v}_i^2 \tag{2.16}$$

where (2.14) is to be used to express T in terms of the q's, $\dot{q}$'s, and t. Since

$$\frac{\partial T}{\partial \dot{q}_r} = \sum m_i \mathbf{v}_i \cdot \frac{\partial \mathbf{v}_i}{\partial \dot{q}_r}$$

it follows that

$$\frac{d}{dt}\left(\frac{\partial T}{\partial \dot{q}_r}\right) = \sum m_i \frac{d\mathbf{v}_i}{dt} \cdot \frac{\partial \mathbf{v}_i}{\partial \dot{q}_r} + m_i \mathbf{v}_i \cdot \frac{d}{dt}\left(\frac{\partial \mathbf{f}_i}{\partial q_r}\right)$$

so that

$$\sum m_i \frac{d\mathbf{v}_i}{dt} \frac{\partial \mathbf{v}_i}{\partial \dot{q}_r} = \frac{d}{dt}\left(\frac{\partial T}{\partial \dot{q}_r}\right) - \frac{\partial T}{\partial q_r}$$

Therefore in terms of the generalized coordinates the *Lagrange equations of motion* are

$$\frac{d}{dt}\left(\frac{\partial T}{\partial \dot{q}_r}\right) - \frac{\partial T}{\partial q_r} = Q_r \tag{2.17}$$

In setting up the Lagrange equations for a particular system, one needs only to find the expression for the kinetic energy from the kinematics of the system and, second, to have an expression for the generalized force components Q_r. In setting up the expressions for Q_r, no notice need be taken of forces of internal interaction between parts of a rigid body or the forces involved in constrained motion where a particle slides without friction on a curved surface or one solid body slides on another.

If the connection of the coordinates q with the position coordinates $\mathbf{r}_i$ of the particles is such that the functions $\mathbf{f}_i$ do not involve the time explicitly, the coordinates are said to be *skleronomic*. This is the case in most applications. In such cases the kinetic energy is homogeneous quadratic form in the $\dot{q}$'s, which may be written $T = \frac{1}{2}\Sigma a_{rs}\dot{q}_r\dot{q}_s$, where the a_{rs} are functions of the q's. Then

$$\frac{\partial T}{\partial \dot{q}_r} = \sum a_{rs}\dot{q}_s \qquad \frac{\partial T}{\partial q_r} = \frac{1}{2}\sum_{t,s} \frac{\partial a_{ts}}{\partial q_r}\dot{q}_t\dot{q}_s$$

$$\frac{d}{dt}\left(\frac{\partial T}{\partial \dot{q}_r}\right) = \sum_s a_{rs}\ddot{q}_s + \sum_{t,s} \frac{\partial a_{rs}}{\partial q_t}\dot{q}_t\dot{q}_s$$

and therefore the Lagrange equations of motion can be written

$$\sum_{t,s} a_{rs}\ddot{q}_s + \sum_{t,s} [{}^{t}{}_{r}{}^{s}]\dot{q}_t\dot{q}_s = Q_r \tag{2.18}$$

where $[{}^{t}{}_{r}{}^{s}]$ is a *Christoffel symbol* defined by

$$[{}^{t}{}_{r}{}^{s}] = \frac{1}{2}\left[\frac{\partial a_{rs}}{\partial q_t} + \frac{\partial a_{rt}}{\partial q_s} - \frac{\partial a_{ts}}{\partial q_r}\right]$$

(Part 1, Chap. 10, Sec. 23).

The coefficients a_{rs} form a matrix of the nth order. If A_{rs} is the reciprocal matrix to a_{rs} in the sense that $\Sigma_r A_{ur}a_{rs} = \delta_{us}$, then these equations may be solved explicitly for the q's to give

$$\ddot{q}_u + \sum_{r,t,s} A_{ur}[{}^{t}{}_{r}{}^{s}]\dot{q}_t\dot{q}_s = \sum_r A_{ur}Q_r \tag{2.19}$$

In case the forces which contribute to the Q_r are derivable from a potential, *a potential energy* function can be introduced such that $Q_r = -\partial V/\partial q_r$, where V is a function of the q's not involving the $\dot{q}$'s. In this case the equations of motion depend only on the Lagrangian function of the dynamical system defined as

$$L = T - V \tag{2.20}$$

having the form

$$\frac{d}{dt}\left(\frac{\partial L}{\partial \dot{q}_r}\right) - \frac{\partial L}{\partial q_r} = 0$$

Velocity dependent forces also permit of the introduction of a potential energy function which depends on the $\dot{q}_r$ as well as the q_r, provided the forces are related to the V in such a way that

$$Q_r = -\frac{\partial V}{\partial q_r} + \frac{d}{dt}\left(\frac{\partial V}{\partial \dot{q}_r}\right) \tag{2.21}$$

An important example is the dynamics of motion of a charged particle in an electromagnetic field described by the scalar and vector potentials, ϕ and $\mathbf{A}$, for which the velocity dependent potential energy, $V = e(\phi - \mathbf{A}\cdot\mathbf{v}/c)$ gives the correct equation of motion

$$m\ddot{\mathbf{r}} = e\left(-\operatorname{grad}\phi - \frac{1}{c}\dot{\mathbf{A}} + \frac{1}{c}\mathbf{v}\times\operatorname{curl}\mathbf{A}\right)$$

In case the forces are derivable from a Lagrangian function an integral of energy exists, although it is equal to the sum of kinetic and potential energy only in special cases,

$$\frac{dL}{dt} = \sum \ddot{q}_r\frac{\partial L}{\partial \dot{q}_r} + \sum \dot{q}_r\frac{\partial L}{\partial q_r} = \frac{d}{dt}\sum \dot{q}_r\frac{\partial L}{\partial \dot{q}_r}$$

Therefore the equations of motion always possess an integral in the form

$$\sum \dot{q}_r\frac{\partial L}{\partial \dot{q}_r} - L = \text{const} \tag{2.22}$$

In the special case that the kinetic energy is homogeneous quadratic in the $\dot{q}$, the first term is $2T$ so that the integral takes the form $T + V = \text{const}$, the usual conservation of energy theorem.

10. Ignorable Coordinates

The quantity $p_r = \partial L/\partial \dot{q}_r$ is called the *momentum conjugate* to the coordinate q_r; thus the Lagrange equations can be written $dp_r/dt = \partial L/\partial q_r$. If certain of the coordinates are not explicitly contained in the Lagrangian function, the generalized momenta associated with such coordinates are constant. Coordinates of this kind are called *ignorable* or *cyclic*. Examples are the cartesian coordinates of the center of mass of a system not acted on by forces external to itself, or the coordinate giving the angular position of a gyroscope which is dynamically balanced and not acted on by any torque.

Suppose that the k coordinates $q_1, q_2, \ldots, q_k$ are ignorable. Then the k conjugate momenta $p_1, p_2, \ldots, p_k$, which are known expressions in the $q_{k+1}, \ldots, q_n$ and all the $\dot{q}$'s, are constant. By solving the equations $p_r = \partial L/\partial \dot{q}_r$ $(r = 1,2, \ldots, k)$, one can express the $\dot{q}_1, \ldots, \dot{q}_k$ in terms of the $p_1, \ldots, p_k$, which are constants, and the $q_{k+1}, \ldots, q_n$.

Let R be defined by

$$R = L - \sum_1^k \dot{q}_r\frac{\partial L}{\partial \dot{q}_r}$$

and suppose that the $q_1 \cdots q_k$ have been eliminated from R by use of the cyclic integrals of motion, then R depends only on the q_r, $\dot{q}_r$ for $r = k + 1 \ldots, n$ and parametrically on the $p_1 \cdots p_k$. The equations of motion for the coordinates $q_{k+1} \cdots q_n$ have Lagrangian form with R as the Lagrangian of the reduced system. Since R depends on the constant $p_1, p_2 \cdots p_k$, the dynamical behavior of the other coordinates will depend on the motions in the ignorable coordinates.

11. Hamilton's Equations

In a conservative system of n degrees of freedom characterized by a Lagrangian function L, the momenta are defined by $p_r = \partial L/\partial \dot{q}_r$ $(r = 1,2, \ldots, n)$. These n equations can be regarded as a means of expressing the $\dot{q}_r$ in terms of the p_r so that the independent variables are the $2n$ quantities p_r, q_r instead of $\dot{q}_r$, q_r as in Lagrange's equations.

The function

$$H = \Sigma p_r \dot{q}_r - L \tag{2.23}$$

when expressed as a function of the p_r, q_r by using these relations to eliminate the $\dot{q}_r$ is known as the *Hamiltonian function* for the system. Its total differential is

$$dH = \sum \dot{q}_r\, dp_r + \sum p_r\, d\dot{q}_r - \sum \frac{\partial L}{\partial q_r} dq_r - \sum \frac{\partial L}{\partial \dot{q}_r} d\dot{q}_r - \frac{\partial L}{\partial t} dt$$

Since $p_r = \partial L/\partial \dot{q}_r$, the terms in $d\dot{q}_r$ go out and therefore the partial derivatives of $H(p_r,q_r,t)$ with respect to p_r and q_r are $\partial H/\partial p_r = \dot{q}_r$ and

$$\left(\frac{\partial H}{\partial q_r}\right)_p = -\left(\frac{\partial L}{\partial q_r}\right)_q$$

Therefore the equations of motion can be expressed in the form of $2n$ first-order differential equations, known as *Hamilton's equations:*

$$\dot{q}_r = \frac{\partial H}{\partial p_r} \quad \text{and} \quad \dot{p}_r = -\frac{\partial H}{\partial q_r} \tag{2.24}$$

If $F(p_r,q_r)$ is any function of the coordinates and momenta, then its time derivative is given by

$$\frac{dF}{dt} = \sum \left(\frac{\partial F}{\partial p_r} \dot{p}_r + \frac{\partial F}{\partial q_r} \dot{q}_r\right) = \sum \left(\frac{\partial F}{\partial q_r}\frac{\partial H}{\partial p_r} - \frac{\partial F}{\partial p_r}\frac{\partial H}{\partial q_r}\right)$$

The expression on the right is known as the *Poisson bracket* of F and H and is denoted by (F,H) so that the equation for the rate of change of F is

$$\frac{dF}{dt} = (F,H) \tag{2.25}$$

Hence any function F whose Poisson bracket with the Hamiltonian vanishes is a constant of the motion.

Since this equation is quite general, it follows that the Hamiltonian equations themselves can be written in Poisson bracket form:

$$\dot{q}_r = (q_r,H) \qquad \dot{p}_r = (p_r,H) \tag{2.26}$$

In particular the Poisson bracket (H,H) vanishes identically and therefore an integral of the motion is that $H =$ const.

In terms of Hamilton's equations, a cyclic coordinate is one which does not appear explicitly in the Hamiltonian function so that $\partial H/\partial q_r = 0$, if q_r is cyclic. Hence $p_r =$ const for a cyclic coordinate.

Any transformation from the $2n$ variables p_r, q_r to a new set P_r, Q_r which has the property of preserving the form of the equations is called a *canonical transformation.* Such transformations are more general than coordinate transformations in which each Q_r is a function of the q's alone. Such transformations can be derived from a function V according to the four following schemes:

1. From a function of old and new coordinates, $V(q_r,Q_r,t)$,

$$p_r = \frac{\partial V}{\partial q_r} \qquad P_r = -\frac{\partial V}{\partial Q_r} \tag{2.27}$$

2. From a function of the old coordinates and new momenta, $V(q_r,P_r,t)$,

$$p_r = \frac{\partial V}{\partial q_r} \qquad Q_r = \frac{\partial V}{\partial P_r}$$

3. From a function of old momenta and new coordinates $V(p_r,Q_r,t)$,

$$P_r = -\frac{\partial V}{\partial Q_r} \qquad q_r = -\frac{\partial V}{\partial p_r}$$

4. From a function of the old and new momenta, $V(p_r,P_r,t)$,

$$q_r = -\frac{\partial V}{\partial p_r} \qquad Q_r = \frac{\partial V}{\partial P_r}$$

In each of the foregoing forms the Hamiltonian in the new system becomes

$$\mathcal{H}(P_r,Q_r,t) = H(p_r,q_r,t) + \frac{\partial V}{\partial t} \tag{2.28}$$

The ordinary change of coordinate variables is a special case in which $V(q_r,P_r,t)$ is linear in the P_r:

$$V(q_r,P_r,t) = \Sigma f_r(q_r)P_r + g(q_r)$$

This gives

$$Q_r = f_r(q_1q_2 \cdots q_n) \qquad P_r = \sum \frac{\partial f_s}{\partial q_r} P_s + \frac{\partial g}{\partial q_r}$$

If a canonical transformation can be found such that in the new system of coordinates the Hamiltonian function depends only on the $P_1P_2 \ldots$ and does not depend on the $Q_1Q_2 \ldots$, then the new P's are all constants of the motion. The n values of the P's are n of the $2n$ integration constants of the general solution. Since $\dot{Q}_r = \partial H/\partial P_r$, it follows that the velocities are also constant and therefore $Q_r = \omega_r t + \beta_r$, where ω_r is written for the constant $\partial H/\partial P_r$ and the β_r form another set of n constants. This is the basic idea of the *Hamilton-Jacobi theory of dynamics:* to integrate the equations of motion by seeking a canonical transformation to new coordinates in which the Hamiltonian does not depend on the Q_r.

For example, the Hamiltonian for a *simple harmonic oscillator of* mass m and frequency $\omega/2\pi$ is $H = p^2/2m + \frac{1}{2}m\omega^2q^2$. The canonical transformation generated by $V = \frac{1}{2}m\omega q^2 \cot Q$ gives

$$p = m\omega q \cot Q \qquad P = \frac{1}{2} m\omega q^2 \frac{1}{\sin^2 Q}$$

or

$$\mathcal{H} = \omega P$$

Therefore, by Hamilton's equations, P is a constant of the motion and $\dot{Q} = \omega$ so that $Q = \omega t + \beta$. Since $q = \sqrt{2P/m\omega} \sin Q$ and $p = \sqrt{2Pm\omega} \cos Q$, this gives the usual solution for simple harmonic motion.

12. Relativistic Particle Mechanics

The principles of relativistic kinematics are given in Chap. 1, Secs. 4 and 5. There it is shown that the proper time $\tau = \int^t \sqrt{1 - \beta^2}\, dt$ is the most natural invariant scalar to take as the independent variable in defining the velocity 4-vector, $v_\alpha = dx_\alpha/d\tau$.

For a particle of mass m the Newtonian classical equations of motion are $m\, d\mathbf{v}/dt = \mathbf{F}$, where $\mathbf{F}$ is the force acting (3-vector) and $\mathbf{v} = d\mathbf{r}/dt$ is the ordinary velocity vector. The present problem is to find a modification of this which is equivalent to it at low velocities and is in Lorentz-invariant form.

It is convenient to start by introducing the 4-vector momentum, p_α, whose spacelike components are $p_1 = mv_1$, $p_2 = mv_2$, $p_3 = mv_3$, where $v_1 = dx_1/d\tau$. The timelike component is

$$p_4 = mv_4 = \frac{imc}{\sqrt{1 - \beta^2}} \tag{2.29}$$

An appropriate 4-vector whose spacelike part is related in a simple way to the usual force vector must also be introduced. Multiplying the classical equation through by $dt/d\tau$, this is

$$m \frac{dv_\alpha}{d\tau} = \frac{F_x}{\sqrt{1 - \beta^2}}, \cdots \qquad \alpha = 1, 2, 3$$

This indicates that the spacelike components of the vector should be the ordinary classical force expression divided by $\sqrt{1 - \beta^2}$. The quantity $F_x\, dt/d\tau$ is the rate of change of the x component of momentum due to this component reckoned with respect to the proper time. The fourth component takes the form

$$i \frac{d}{dt} \left(\frac{mc^2}{\sqrt{1 - \beta^2}} \right) = \frac{d\tau}{dt} cF_4$$

For $\beta \ll 1$, the term on the left becomes $i\, d/dt\, (\frac{1}{2}mv^2)$ and therefore the term on the right must be such that it reduces to the classical rate at which work is done by the force which is $\mathbf{F} \cdot \mathbf{v}$. Therefore the fourth-component of the force 4-vector is

$$F_4 = i \frac{\mathbf{F} \cdot \boldsymbol{\beta}}{\sqrt{1 - \beta^2}}$$

where $\boldsymbol{\beta} = (v_x\mathbf{i} + v_y\mathbf{j} + v_z\mathbf{k})/c$ is the classical velocity with time as independent variable. The force 4-vector is therefore

$$F_\alpha = \frac{1}{\sqrt{1 - \beta^2}} (F_x, F_y, F_z, i\mathbf{F} \cdot \boldsymbol{\beta}) \tag{2.30}$$

and the relativistic form of Newton's second law of motion is

$$\frac{dp_\alpha}{d\tau} = F_\alpha \tag{2.31}$$

where the momentum 4-vector is $p_\alpha = mv_\alpha$ and m is the scalar *invariant* mass of the particle.

The so-called *relativistic variation of mass with velocity* arises if the time t of a particular Lorentz frame is used as the independent variable, instead of the proper time τ. The equations of motion for $\alpha = 1, 2, 3$ become

$$\frac{d}{dt} \left(\frac{m\mathbf{v}}{\sqrt{1 - \beta^2}} \right) = \mathbf{F} \tag{2.32}$$

This is usually described by saying that the mass of the particle is variable with velocity so that the mass at velocity v is $m/\sqrt{1 - \beta^2}$. In the 4-vector form $p_\alpha = mv_\alpha$. As the vector v_α is of constant length $\Sigma v_\alpha^2 = -c^2$, it follows that $\Sigma p_\alpha^2 = -m^2c^2$. Therefore, to be consistent, $\Sigma F_\alpha v_\alpha = 0$, and a calculation from the equations shows this to be so with the definition of F_α adopted.

The effect of a force acting on a particle is thus solely to produce a curvature in its world line.

An important example is the equation of motion of a charged particle in an electromagnetic field specified by $\mathbf{E}$ and $\mathbf{H}$. The classical form of the equation of motion is $m\, d\mathbf{v}/dt = e(\mathbf{E} + \frac{1}{c}\mathbf{v} \times \mathbf{H})$, where e is the charge, $\mathbf{E}$ is the electric field, and $\mathbf{H}$ the magnetic field.

Hence the F_1 component of F_α is

$$F_1 = \frac{e/c}{\sqrt{1 - \beta^2}} (v_yH_z - v_zH_y + cE_x)$$

In the relativistic treatment of the electromagnetic field (see Part 4, Chap. 1) it is shown that $\mathbf{E}$ and $\mathbf{H}$ are parts of an antisymmetric tensor $H_{\alpha\beta}$ where

$$H_{\alpha\beta} = \begin{bmatrix} 0 & H_3 & -H_2 & -iE_1 \\ -H_3 & 0 & H_1 & -iE_2 \\ H_2 & -H_1 & 0 & -iE_3 \\ iE_1 & iE_2 & iE_3 & 0 \end{bmatrix}$$

and hence F_1 can be written $F_1 = \Sigma_\alpha (e/c) H_{1\alpha} v_\alpha$ or more generally $F_\alpha = \Sigma_\beta (e/c) H_{\alpha\beta} v_\beta$. The relativistic equation of motion takes the form

$$\frac{dv_\alpha}{d\tau} = \frac{e}{mc} \sum_\beta H_{\alpha\beta} v_\beta \tag{2.33}$$

In the case of motion of charged particles in uniform and time constant electromagnetic fields, these equations provide a starting point for a unified treatment of all possible cases which is simple since they are then a set of linear differential equations with constant coefficients. The equations are consistent with the general constancy of Σv_α^2 because H is antisymmetric.

Case 1. Uniform Electric Field E along x Axis. The only nonvanishing components of $H_{\alpha\beta}$ are H_{41} and H_{14}. Therefore the y and z components of velocity $dy/d\tau$ and $dz/d\tau$ are constant. Note that this implies, however, that the nonrelativistic components, dy/dt and dz/dt, are not constant.

The first and fourth equations of motion are $dv_1/d\tau = -ikv_4$, $dv_4/d\tau = ikv_1$, where $k = eE/mc$.

Hence $v_1 = A \sinh k\tau + B \cosh k\tau$. If, in particular, the particle starts from rest at $\tau = 0$, then $B = 0$ and $v_4 = iA \cosh k\tau$ and $A = c$ is determined by the relation $v_4^2 + v_1^2 = -c^2$. Therefore $dx/d\tau = c \sinh k\tau$, $dt/d\tau = \cosh k\tau_1$, which integrate to give

$$x = \frac{c}{k} (\cosh k\tau - 1)$$

$$t = \frac{1}{k} \sinh k\tau$$

Since $dx/dt = c \tanh k\tau$, the particle is uniformly accelerated as long as $k\tau \ll 1$, and for $k\tau \gg 1$ the classical velocity dx/dt approaches the velocity of light.

Case 2. Uniform Magnetic Field H along z Axis. The only nonvanishing components of $H_{\alpha\beta}$ are H_{12} and H_{21}. Therefore v_3 and v_4 are constant. The constancy of v_4 implies a constant ratio between classical time t and relativistic proper time τ. The equations for v_1 and v_2 are, with $k = eH/mc$,

$$\frac{dv_1}{d\tau} = kv_2 \qquad \text{and} \qquad \frac{dv_2}{d\tau} = -kv_1$$

which with proper initial conditions have the solution $v_1 = u \cos k\tau$, $v_2 = u \sin k\tau$ so that the motion in x and y is uniformly in a circle with angular velocity k, measured relative *to the proper time.* Let v be the constant velocity along the z axis; then $\Sigma v_\alpha^2 = -c^2$ gives the relation

$$t = \left(1 + \frac{u^2 + v^2}{c^2}\right)^{1/2} \tau$$

As the value of $dt/d\tau$ is greater than unity, this means that the frequency with regard to laboratory time of describing the circular orbit becomes smaller at high energies. This plays an important role in design of high energy cyclotrons.

13. Variation Principles

The general equations of dynamics are derivable from a number of variation principles which provide another starting point for the development. One of these is known as *Hamilton's principle.*

Considering first a system whose generalized coordinates are $q_1 \cdots q_n$, having a Lagrangian function L, as defined in Sec. 9, Hamilton's principle may be stated as follows:

The motion of the system from its given configuration at time t_1 to its final configuration at time t_2 is such that

$$I = \int_{t_1}^{t_2} L\, dt \tag{2.34}$$

is an extremum (maximum or minimum) relative to small variations from the actual motion.

If this principle is assumed as a starting point, then the Lagrangian equations of motion (2.20) follow from the calculus of variations (Part 1, Chap 11) as the expression of the condition on the time variations of the q's which makes I an extremum. The principle can be generalized to the case in which nonconservative forces appear in a way which is also applicable to the determination of the forces exerted by constraints whether the forces are conservative or not.

If $\mathbf{F}_i$ is the total force acting on the ith particle whose position is $\mathbf{r}_i$, then the extended form of Hamilton's principle is that the actual motion between the given initial configuration at t_1 to the given final configuration at t_2 is such that

$$I = \int_{t_1}^{t_2} \left(T + \sum_i \mathbf{F}_i \cdot \mathbf{r}_i\right) dt \tag{2.35}$$

is an extreme. The variation of $\Sigma_i \mathbf{F}_i \cdot \mathbf{r}_i$ gives

$$\delta \sum_i \mathbf{F}_i \cdot \mathbf{r}_i = \sum_i \mathbf{F}_i \cdot \delta\mathbf{r}_i = \sum_r Q_r\, \delta q_r$$

where the Q_r are the generalized forces defined in (2.15). If the Q's are derivable from a generalized potential as in (2.21), then (2.35) takes the same form as (2.34) and thus leads to Lagrange's equations.

If the system is nonholonomic with constraints that are expressible in differential form, these will appear as, say, m equations of the form

$$\sum_s A_{rs}\, dq_s + A_r\, dt = 0 \tag{2.36}$$

with $r = 1, 2, \ldots, m$. This means that the variations in δq_r to be used in varying the integral for I are no longer independent but must satisfy these conditions. The variations are taken without varying the time, and the conditions can be taken into account by the method of Lagrange's undetermined multipliers. Let λ_r be the multiplier for the rth equation, $\lambda_r \Sigma_s A_{rs}\, \delta q_s = 0$. Following the usual procedure this gives the variation condition:

$$\int_{t_1}^{t_2} \sum_{r=1}^{n} \left(\frac{\partial L}{\partial q_r} - \frac{d}{dt}\frac{\partial L}{\partial \dot{q}_r} + \sum_s \lambda_s A_{sr}\right) \delta q_r = 0 \tag{2.37}$$

The δq's are not independent but, if the values of λ_s are chosen so that

$$\frac{\partial L}{\partial q_r} - \frac{d}{dt}\left(\frac{\partial L}{\partial \dot{q}_r}\right) = \sum_s \lambda_s A_{sr} \tag{2.38}$$

for $r = (n - m) \cdots n$, then the only δq's left in (2.37) are the independent ones and therefore must hold for the other values of r, namely, $r = 1, 2, \ldots, m$. Therefore in total there are n equations of motion of the form (2.38) which together with the m equations of constraint (2.36) in the form

$$\sum_s A_{rs}\dot{q}_s + A_r = 0 \qquad r = 1, 2, \ldots, m \tag{2.39}$$

suffice to determine the $n + m$ unknown quantities $q_1 \cdots q_n$ and $\lambda_1 \cdots \lambda_m$.

The undetermined multipliers, $\lambda_1 \cdots \lambda_m$, have a simple physical relation with the forces exerted in the operation of the constraints (2.36). If Q_r are the set of forces involved in the operation of the constraints, then since the constraints are only operative in virtue of the forces they bring into play they must in fact be given by

$$Q_r = \sum_s \lambda_s A_{sr} \qquad (2.40)$$

The foregoing discussion, while applicable to nonholonomic constraints expressible in differential form, is also applicable to an ordinary holonomic constraint of the form $f_r(q_1 \cdots q_n,t) = 0$, for this corresponds to a differential constraint of the form (2.39) in which

$$A_{rs} = \frac{\partial f}{\partial q_s} \quad \text{and} \quad A_r = \frac{\partial f}{\partial t} \qquad (2.41)$$

and thus the Lagrange multiplier method may be used in dealing with such a constraint when it is desired to be able to calculate the forces of constraint.

The equations of motion in Hamiltonian form (2.24) follow from the use of Hamilton's principle (2.34) on using (2.23) to express the integrand in terms of the Hamiltonian function. This gives $\delta I = 0$, where

$$I = \int_1^2 \sum_r p_r\, dq_r - \int_1^2 H\, dt \qquad (2.42)$$

where the integration extends from the initial configuration at time t_1 to the final configuration at time t_2. In carrying out the variation of this integral, the procedure is to subject the p_r and q_r to independent variations at times corresponding to the unvaried path. This gives

$$\delta I = \int_1^2 \sum_r \left[\delta p_1 \left(\dot{q}_r - \frac{\partial H}{\partial p_r} \right) - \delta q_1 \left(\dot{p}_r + \frac{\partial H}{\partial q_r} \right) \right] d\tau$$

Since the variations of the p's and q's are independent, the vanishing of δI requires that the coefficients of each δp and δq vanish, giving the Hamiltonian equations of motion.

References

Historical and Philosophical:

1. Newton, Isaac: "Mathematical Principles of Natural Philosophy," Motte's trans. revised by F. Cajori, University of California Press, Berkeley, Calif., 1946.
2. Galileo, Galilei: "Dialogues Concerning Two New Sciences," trans. by Henry Crew and Alfonsio de Salvio, Northwestern University Press, Evanston and Chicago, 1950.
3. Galileo, Galilei: "Dialogues on the Great World Systems," trans. T. Salusbury, edited by Giorgia de Santillana, University of Chicago Press, Chicago, 1953.
4. Clagett, Marshall: "The Science of Mechanics in the Middle Ages," University of Wisconsin Press, Madison, Wis., 1959.
5. Mach, Ernst: "The Science of Mechanics: A Critical and Historical Account of Its Development," 6th ed., Open Court, La Salle, Ill., 1960.
6. Hertz, Heinrich: "The Principles of Mechanics, Presented in a New Form," Dover, New York, 1956.
7. Whitrow, G. J.: "The Natural Philosophy of Time," Nelson, London, 1961. Reprint: Harper & Row, New York, 1963.

Older Expositions:

8. Thomson, William, and P. G. Tait: "Principles of Mechanics and Dynamics," 2 vols., Dover, New York, 1962; originally published as "Treatise on Natural Philosophy," Cambridge University Press, 1879.
9. Ames, J. S., and F. D. Murnaghan: "Theoretical Mechanics, an Introduction to Mathematical Physics," Ginn, Boston, 1929; Dover, New York, 1958.
10. Osgood, W. F.: "Mechanics," Macmillan, New York, 1937.
11. Webster, A. G.: "The Dynamics of Particles and of Rigid, Elastic and Fluid Bodies," 3d ed., Stechert, New York, 1942.
12. MacMillan, W. D.: "Statics and the Dynamics of a Particle," McGraw-Hill, New York, 1927; Dover, New York, 1958.

Special Theory of Relativity:

13. Born, Max: "Einstein's Theory of Relativity," rev. ed., Dover, New York, 1962.
14. Pauli, W.: "Theory of Relativity," Pergamon Press, New York, 1958.
15. Møller, C.: "The Theory of Relativity," Clarendon Press, Oxford, 1952.
16. Aharoni, J.: "The Special Theory of Relativity," Clarendon Press, Oxford, 1959.
17. Synge, J. L.: "Relativity: the Special Theory," Interscience, New York, 1956.

Recent Expositions:

18. Whittaker, E. T.: "A Treatise on the Analytical Dynamics of Particles and Rigid Bodies," 4th ed., Cambridge University Press, New York, 1927.
19. Synge, J. L., and B. A. Griffith: "Principles of Mechanics," 3d ed., McGraw-Hill, New York, 1959.
20. Goldstein, H.: "Classical Mechanics," Addison-Wesley, Reading, Mass., 1959.
21. Corben, H. C., and P. Stehle: "Classical Mechanics," 2d ed., Wiley, New York, 1960.
22. Sommerfeld, A.: "Mechanics, Lectures on Theoretical Physics," vol. 1, Academic, New York, 1952.
23. Landau, L. D., and E. M. Lifshitz: "Mechanics," vol. 1 of "Course of Theoretical Physics," Addison-Wesley, Reading, Mass., 1960.
24. Kilmister, C. W.: "Hamiltonian Dynamics," Wiley, New York, 1964.
25. Ter Haar, D.: "Elements of Hamiltonian Mechanics," North Holland Publishing Company, Amsteram, 1961.
26. Lanczos, C.: "The Variational Principles of Mechanics," University of Toronto Press, Toronto, 1949.

Chapter 3

Theory of Vibrations

By E. U. CONDON, University of Colorado

1. Simple Harmonic Motion

If a particle of mass m is attracted toward the origin by a force kx, proportional to the distance from the origin, its equation of motion is

$$m\ddot{x} + kx = 0 \tag{3.1}$$

If x is a generalized coordinate of any kind, and the kinetic energy is $\frac{1}{2}m\dot{x}^2$ and the potential energy is $\frac{1}{2}kx^2$, independently of the geometrical character of the coordinate x, the equation of motion is of the same form. For instance, if x is the angle of turning of a rigid body about a fixed axis, then m has to be interpreted as its moment of inertia about that axis and k becomes the torque per unit angular displacement of the system of forces which provide a torque proportional to the angle of displacement from the equilibrium position.

The most general solution for the motion is the real part of the expression

$$x = Ae^{i\omega t}$$

where $\omega = (k/m)^{1/2}$ and A is an arbitrary complex constant.

The real motion is given by

$$x = \tfrac{1}{2}(Ae^{i\omega t} + A^*e^{-i\omega t}) \tag{3.2}$$

where A^* is the complex conjugate of A. In terms of the initial values of position and velocity, x_0 and v_0, the complex amplitude A is $A = (x_0 - iv_0/\omega)$. This is known as the simple harmonic motion, of frequency $\omega/2\pi$. The velocity is given by

$$\dot{x} = \frac{i\omega}{2}(Ae^{i\omega t} - A^*e^{-i\omega t}) \tag{3.3}$$

The complex number $x = Ae^{i\omega t}$ describes a circle of radius $|A|$ with angular velocity ω, as indicated in Fig. 3.1. The complex number representing the velocity, $i\omega Ae^{i\omega t}$, describes a circle of radius $\omega|A|$ with phase advanced over that of x by $\pi/2$. The velocity is said to lead the position.

The total energy of the motion,

$$W = \tfrac{1}{2}m\dot{x}^2 + \tfrac{1}{2}kx^2$$

is constant and equal to $\frac{1}{2}kAA^*$.

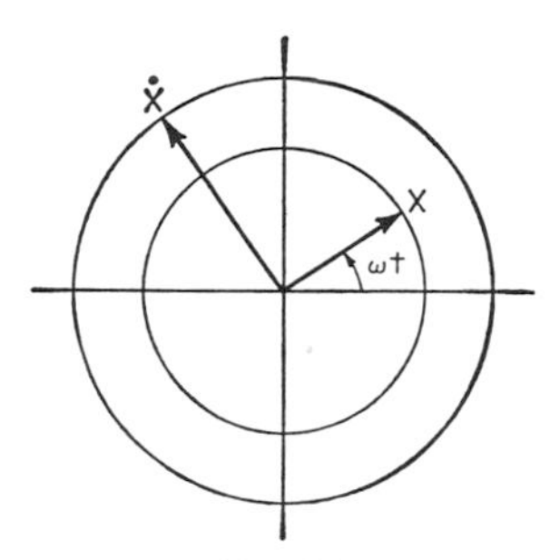

FIG. 3.1

2. Damped Harmonic Motion

If in addition to the elastic restoring force kx, the particle is acted on by a frictional force, $-f\dot{x}$, which acts in a sense opposite to the velocity, the equation of free motion is

$$m\ddot{x} + f\dot{x} + kx = 0 \tag{3.4}$$

Writing $\omega_0 = (k/m)^{1/2}$ and $\lambda = f/2m$, the equation of motion becomes $\ddot{x} + 2\lambda\dot{x} + \omega_0^2 x = 0$. This has solutions of the form $x = Ae^{kt}$, provided that k has one of the two values $k = -\lambda \pm i\sqrt{\omega_0^2 - \lambda^2}$, so that the general solution can be written as the real part of a complex expression

$$x = A \exp(-\lambda t) \exp(i\omega_1 t) \tag{3.5}$$

where $\omega_1 = (\omega_0^2 - \lambda^2)^{1/2}$. This is the general expression for a *damped harmonic motion*. The complex number x moves in a logarithmic spiral (assuming $\lambda < \omega_0$) and the real motion is a simple harmonic motion with exponentially decreasing amplitude.

The natural logarithm of the ratio of the amplitudes before and after one complete period of the harmonic factor is called the *logarithmic decrement* δ of the damped harmonic motion,

$$\delta = \frac{2\pi\lambda}{\omega_1} \tag{3.6}$$

Another measure of the damping is designated by Q, without giving it a name. The *Q of the system* is defined as

$$Q = \frac{m\omega_0}{f} = \frac{\sqrt{mk}}{f} = \frac{\omega_0}{2\lambda} \tag{3.7}$$

Thus Q is large for systems of small damping. In terms of Q the damping factor is $\exp(-2\omega_0 t/Q)$. Therefore $Q/4\pi$ is the number of cycles of the *undamped* frequency needed to reduce the *amplitude* to e^{-1} of its original value. Since the energy of the motion varies as the square of the amplitude, it follows that $Q/2\pi$ is the number of cycles of the undamped frequency in which the *energy* of the motion is reduced to e^{-1} of its original value.

In terms of Q the altered frequency is

$$\omega_1 = \omega_0 \sqrt{1 - \tfrac{1}{4}Q^{-2}} \tag{3.8}$$

thus the change is small for systems of high Q or little damping. The exact relation between δ and Q is

$$\delta = \frac{\pi/Q}{\sqrt{1 - \tfrac{1}{4}Q^{-2}}}$$

which for systems of high Q is approximately $\delta = \pi/Q$.

If the initial values of position and velocity are x_0 and v_0, then the complex amplitude is

$$A = \left(1 + \frac{\lambda}{i\omega_1}\right) x_0 - \frac{iv_0}{\omega_1}$$

which reduces to the formula of Sec. 1 as $\lambda \to 0$.

As the magnitude of the damping is increased, the rate of decay of the oscillations becomes greater and their frequency, $\omega_1/2\pi$, is reduced until when $\lambda = \omega_0$ (that is, when Q has been reduced to $\frac{1}{2}$), the motion becomes aperiodic and is said to be *critically damped.* For larger values of λ and therefore smaller values of Q, the motion remains aperiodic and is said to be *overdamped.* In the overdamped case, ω_1 becomes pure imaginary but the formulas of this section retain their validity. In the critically damped case, the general solution for the motion is of the form $x = (A + Bt)e^{-\lambda t}$, or in terms of initial velocity and displacement

$$x = [x_0 + (v_0 + \lambda x_0)t]e^{-\lambda t}$$

3. Forced Harmonic Motion

If the particle is acted on by an external driving force which can be represented by the real part of $Fe^{i\omega t}$, in addition to the elastic restoring force and the linear frictional term, the equation of motion for complex x whose real part gives the actual motion is

$$m\ddot{x} + f\dot{x} + kx = Fe^{i\omega t} \tag{3.9}$$

The general solution of this consists of the general solution for the free motion, as in (3.5), plus a particular solution, $Be^{i\omega t}$, which has the same frequency as the driving force. The solution for the free motion is called the *transient* term since it becomes negligible after times such that $\lambda t \gg 1$. The $Be^{i\omega t}$ is called the steady state because it is the state of motion approached after the transient dies away.

The complete expression for the motion is therefore the real part of

$$x = A \exp(-\lambda + i\omega_1)t + B \exp(i\omega t) \tag{3.10}$$

In fitting such a solution to the initial conditions, B is determined by the equation of motion and complex A is to be adjusted. The calculation of A is to be made as in Sec. 2 except that $[x_0 - \frac{1}{2}(B + B^*)]$ and $v_0 - \frac{1}{2}i\omega(B - B^*)$ are to be used in place of x_0 and v_0 in determining the complex amplitude of the transient.

The complex amplitude of the steady motion is given by

$$B = \frac{F}{-m\omega^2 + if\omega + k} \tag{3.11}$$

which, in the notation of (3.10), is

$$B = \frac{F}{if\omega[1 + iQ(\omega/\omega_0 - \omega_0/\omega)]} \tag{3.12}$$

In the limit of very low driving frequency, $\omega \ll \omega_0$, this reduces approximately to $B = F/k$, so that in this limit the amplitude of the motion is essentially controlled by the stiffness of the restoring force. At the other extreme of high driving frequency, $\omega \gg \omega_0$, this reduces approximately to $B = -F/m\omega^2$ so that the amplitude of forced oscillation is governed mainly by the inertia. At the resonant frequency, $\omega = \omega_0$, the amplitude is $B = F/if\omega_0$ and is therefore inversely proportional to the magnitude of the frictional term.

The maximum amplitude $|B|$ of forced oscillation as a function of driving frequency comes neither at ω_0 nor ω_1 but at $\omega_2 = (\omega_0^2 - 2\lambda^2)^{1/2}$.

The maximum amplitude of velocity (that is, the maximum of ωB) comes at $\omega = \omega_0$.

4. Mechanical Impedance

The relations involved in the forced oscillation of a damped harmonic oscillator are most conveniently discussed in terms of the relation between the complex amplitude of the velocity and that of the force. Writing $v = Ce^{i\omega t}$, one has from the equation of motion $F = CZ$, where Z is a complex factor known as the *impedance* of the mechanical system,

$$Z = \left(\frac{k}{i\omega} + i\omega m + f\right) \tag{3.13}$$

The real part of the impedance is known as the *resistance;* it is just equal to the coefficient of the frictional resistance. For a particle having no restoring force and negligible inertia the motion is such that the velocity is in phase with the driving force. The imaginary part of the impedance is known as the *reactance.* The reactance is the sum of two contributions: $-ik/\omega$, the *elastic reactance,*

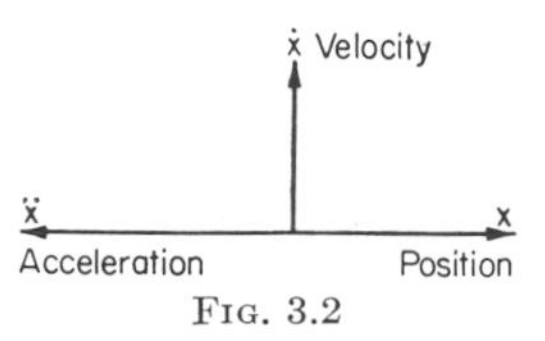

FIG. 3.2

and $i\omega m$, the *inertial reactance.* Since the phase relations of position, velocity, and acceleration are as shown in Fig. 3.2, elastic reactance requires a driving force component in phase with the displacement, while inertial reactance gives rise to a driving force component in phase with acceleration.

A convenient form for the impedance is

$$Z = f[1 + iQ(\omega/\omega_0 - \omega_0/\omega)] \tag{3.14}$$

which brings out clearly the fact that the resonance is sharper, the higher the Q of the system. The mean rate at which power is supplied and dissipated by the resistance is $\frac{1}{2}f|C|^2 = \frac{1}{2}fF^2/|Z|^2$. Hence the power excitation of the system is reduced to half when $|Z|^2$ is doubled. This occurs at the frequency such that $(\omega/\omega_0 - \omega_0/\omega) = \pm Q^{-1}$; if Q is large, this gives $\Delta\omega/\omega_0 = \pm\frac{1}{2}Q$. Hence the full range of frequency from half-power excitation above to that below resonance is $\omega_0 Q^{-1}$; thus Q^{-1} is directly a measure of the sharpness of resonance.

For a given magnitude of driving force, *resonance* in the velocity amplitude occurs at minimum impedance which occurs when the reactance vanishes, i.e., when the inertial reactance just balances the elastic reactance, making the impedance be real so that the velocity is in phase with the driving force.

For specific computations involving amplitude and phase relations, it is convenient to note from (3.14) that if α is the angle of Z, and hence the angle by which the phase of the driving force leads the velocity,

$$\tan\alpha = Q\left(\frac{\omega}{\omega_0} - \frac{\omega_0}{\omega}\right)$$
$$\sim 2\left(\frac{\Delta\omega}{\omega_0}\right)Q$$

and the magnitude $|Z|$ which governs the ratio of driving force to velocity amplitude is given by $|Z| = f/\cos\alpha$. These two formulas permit the use of trigonometric tables for rapid solution of resonance problems.

It is instructive to consider the energy changes taking place. If the velocity is given by $v = C\cos\omega t$ with C real, then the position is given by

$$x = +\frac{C}{\omega}\sin\omega t$$

The driving force needed to overcome the frictional resistance is equal to $F_1 = fc\cos\omega t$ and the instantaneous power delivered by this part of the driving force is $F_1 v = fC^2\cos^2\omega t$ which is always positive, although it drops to zero when the particle is at rest and is at maximum when the speed is greatest.

The driving force needed because of inertial reactance is $F_2 = -\omega mC\sin\omega t$ and the instantaneous power delivered by this part is $F_2 v = -\omega mC^2\sin\omega t\cos\omega t$. This is alternately positive and negative and has zero average value. The kinetic energy of the particle is $T = \frac{1}{2}mC^2\cos^2\omega t$ and the instantaneous power associated with the inertial reactance is just that required to increase and diminish the kinetic energy.

Similarly the driving force associated with elastic reactance is $F_3 = +(k/\omega)C\sin\omega t$ and the instantaneous power associated with it is $(k/\omega)C^2\sin\omega t\cos\omega t$. This is involved in producing the time variations of the potential energy $V = \frac{1}{2}kx^2$ which is stored elastically.

The total instantaneous power supplied by the driving force is thus

$$Fv = fC^2\cos^2\omega t + [(k/\omega) - \omega m]C^2\sin\omega t\cos\omega t$$

The first term has the average value $\frac{1}{2}fC^2$ which is needed to supply the frictional power losses. The second term is spoken of as *reactive power:* during each cycle the oscillating body reacts back on the driver in such a way as to restore to it the same amount of work as the driver does on the body so that the net power transfer is zero.

Far from resonance it is necessary that the driver be able to supply a relatively large amount of reactive power to maintain a prescribed velocity amplitude. But at resonance the inertial and elastic reactive powers just cancel so that the driver needs only to supply the frictional power losses.

5. Two Coupled Oscillators

There are many instances in which two particles of masses m_1 and m_2 are acted on by elastic restoring forces and also by elastic interaction terms so that, if x_1 and x_2 are the two coordinates, then the potential energy of the system is

$$V(x_1,x_2) = \tfrac{1}{2}k_{11}x_1^2 + k_{12}x_1x_2 + \tfrac{1}{2}k_{22}x_2^2 \tag{3.15}$$

while the kinetic energy is

$$T = \tfrac{1}{2}m_1\dot{x}_1^2 + \tfrac{1}{2}m_2\dot{x}_2^2$$

The equations of motion are

$$\begin{aligned} m_1\ddot{x}_1 + k_{11}x_1 + k_{12}x_2 &= 0 \\ m_2\ddot{x}_2 + k_{12}x_1 + k_{22}x_2 &= 0 \end{aligned} \tag{3.16}$$

Such a pair of particles are called *coupled oscillators.* This includes of course the special case when x_1 and x_2 refer to two degrees of freedom of motion of the same particle.

A simple example is that of two masses suspended from springs in series as indicated in Fig. 3.3. The

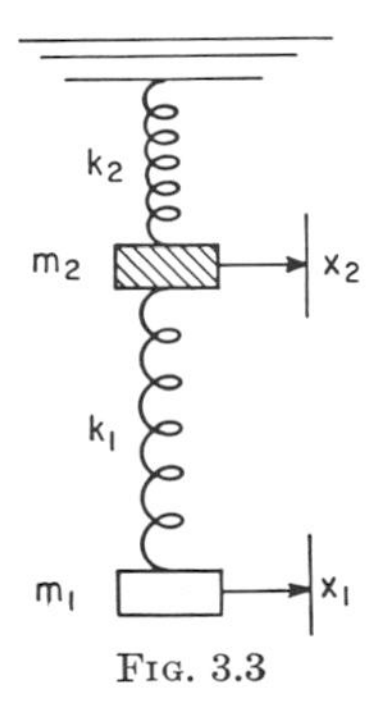

Fig. 3.3

coordinates x_1 and x_2 may be measured positive downward from the equilibrium positions as origin, at which the upward pull of the stretched springs just equals the downward pull of gravity on the weights. The net upward force on the lower mass is $k_1(x_1 - x_2)$ so that its equation of motion is $m_1\ddot{x}_1 + k_1x_1 - k_1x_2 = 0$. Likewise the net upward force on the upper mass is $k_2x_2 - k_1(x_1 - x_2)$ so that the equation of motion is

$$m_2\ddot{x}_2 - k_1x_1 + (k_1 + k_2)x_2 = 0$$

This is of the form of (3.16) with appropriate identification of the k's.

Solutions of these are obtained by assuming $x_1 = a_1e^{i\omega t}$, $x_2 = a_2e^{i\omega t}$ which leads to

$$\begin{aligned}(k_{11} - m_1\omega^2)a_1 + \qquad\qquad k_{12}a_2 &= 0\\ k_{12}a_1 + (k_{22} - m_2\omega^2)a_2 &= 0\end{aligned}$$

These equations have solutions only for two values of ω^2, namely, those given by the roots of the quadratic equation,

$$(k_{11} - m_1\omega^2)(k_{22} - m_2\omega^2) = k_{12}^2$$

If the coupling were absent ($k_{12} = 0$), the two roots would be (1) $\omega_1^2 = k_{11}/m_1$ and the corresponding solution would be $a_2/a_1 = 0$; (2) $\omega_2^2 = k_{22}/m_2$ with $a_1/a_2 = 0$ and in this case the two particles oscillate independently. The equation can be written

$$(\omega_1^2 - \omega^2)(\omega_2^2 - \omega^2) = K^4$$

where $K^4 = k_{12}^2/m_1m_2$ is a coupling parameter having the dimensions of frequency. The roots are now

$$\omega^2 = \tfrac{1}{2}(\omega_1^2 + \omega_2^2) \pm \sqrt{\tfrac{1}{4}(\omega_1^2 - \omega_2^2)^2 + K^4} \qquad (3.17)$$

Supposing $\omega_1 > \omega_2$, one of the frequencies of the coupled system is higher than ω_1 and one is lower than ω_2. If $K^2 \ll \tfrac{1}{2}(\omega_1^2 + \omega_2^2)$, the coupling is said to be *weak* and the roots are, approximately

$$\begin{aligned}\omega_a^2 &= \omega_1^2 + \frac{K^4}{\omega_1^2 - \omega_2^2}\\ \omega_b^2 &= \omega_2^2 - \frac{K^4}{\omega_1^2 - \omega_2^2}\end{aligned} \qquad (3.18)$$

The ratio of amplitudes associated with the upper frequency is

$$\left(\frac{\sqrt{m_1}a_1}{\sqrt{m_2}a_2}\right)_a = \frac{\omega_1^2 - \omega_2^2}{K^2}$$

so that in the limit of small coupling the frequency approaches the uncoupled value ω_1 and the associated motion is one in which the second particle has a vanishingly small amplitude compared to the first. Similarly, associated with the lower frequency the ratio of amplitudes is

$$\left(\frac{\sqrt{m_2}a_2}{\sqrt{m_1}a_1}\right)_b = -\frac{\omega_1^2 - \omega_2^2}{K^2}$$

so that for this mode the motion is mostly that of the second particle.

Hence there are two basic *normal modes* of vibration in which each particle executes a pure simple harmonic motion. If $k_{12} > 0$, then the particles oscillate *in phase* in the mode of higher frequency and *out of phase* in the mode of lower frequency.

In the example of two masses suspended from springs in series, the coupling parameter k_{12} is intrinsically negative; thus it is the low-frequency mode in which the masses oscillate in phase.

If the two uncoupled frequencies are equal, $\omega_1 = \omega_2$, then the system is said to be *degenerate* in the uncoupled condition and the effect on the frequencies of a small amount of coupling is much greater than when the system is initially nondegenerate.

The general motion of the system results from the superposition of the two normal modes with arbitrary amplitudes and relative phases, which have to be determined from the initial conditions.

6. Small Oscillations about Equilibrium

The ideas of Sec. 5 may be generalized to small oscillations of a system of N degrees of freedom. Let the Lagrangian coordinates of a system of N degrees of freedom be $q_1 \cdots q_n$ and suppose these have been so chosen that $q_1 = q_2 \cdots q_n = 0$ corresponds to a minimum of the potential energy function, and let the value $V = 0$ be assigned to this minimum. Then for small values of the q's the potential energy will be approximated by the positive definite quadratic form

$$V = \tfrac{1}{2}\sum_{r,s} k_{rs}q_rq_s \qquad (3.19)$$

The kinetic energy is usually a quadratic form in the $\dot{q}$'s whose coefficients are functions of the q's. But if the q's are small throughout the motion, then the kinetic energy may be well approximated by assigning the values $q_1 \cdots q_n = 0$ in calculating the coefficients in which case the kinetic energy is a positive definite quadratic form in the $\dot{q}$'s with constant coefficients,

$$T = \tfrac{1}{2}\Sigma m_{rs}\dot{q}_r\dot{q}_s \qquad (3.20)$$

The equations of motion are

$$\sum_s (m_{rs}\ddot{q}_s + k_{rs}q_s) = 0 \qquad r = 1, 2, \ldots, N \qquad (3.21)$$

Let m_{tr}^{-1} be the matrix reciprocal to m_{rs}:

$$\sum_r m_{tr}^{-1}m_{rs} = \delta_{ts}$$

Multiplying the equations of motion through by m_{tr}^{-1} and summing with regard to r gives

$$\ddot{q}_t + \sum_s\left(\sum_r m_{tr}^{-1}k_{rs}\right)q_s = 0 \qquad (3.22)$$

Thus a matrix $(\Sigma_r m_{tr}^{-1}k_{rs})$ takes the place of the simple k/m of the one-dimensional oscillator.

These equations can be formally solved by introducing a suitably chosen linear transformation of the q's,

$$q_s = \sum_\alpha S_{s\alpha}Q_\alpha \qquad (3.23)$$

The matrix reciprocal to $S_{s\alpha}$ will be designated $S_{\beta s}^{-1}$ so that $\Sigma_s S_{\beta s}^{-1}S_{s\alpha} = \delta_{\beta\alpha}$. Making this substitution in the equations of motion

$$\sum_\alpha S_{t\alpha}\ddot{Q}_\alpha + \sum_{s,\alpha}\left(\sum_r m_{tr}^{-1}k_{rs}\right)S_{s\alpha}Q_\alpha = 0$$

Multiplying by $S_{\beta t}^{-1}$ and summing over t,

$$\ddot{Q}_\beta + \sum_\alpha\sum_{t,r,s} S_{\beta t}^{-1}m_{tr}^{-1}k_{rs}S_{s\alpha}Q_\alpha = 0$$

Provided now the S matrix can be chosen in such a way that

$$\sum_{t,r,s} S_{\beta t}{}^{-1}m_{tr}{}^{-1}k_{rs}S_{s\alpha} = \omega_\beta{}^2\,\delta_{\beta\alpha} \tag{3.24}$$

this will reduce to $\ddot{Q}_\beta + \omega_\beta{}^2 Q_\beta = 0$ and therefore the motion of each coordinate Q_β will be simple harmonic with frequency $\omega_\beta/(2\pi)$.

The Q's are then said to be *normal coordinates.* A state of motion of the system in which all but one of the normal coordinates have zero amplitude is called a *normal mode of vibration* of the system. From the form of the equations of motion it follows that in terms of the normal coordinates the kinetic and potential energies are in the simple form of sum of squares:

$$T = \tfrac{1}{2}\sum_\alpha \dot{Q}_\alpha{}^2 \qquad V = \tfrac{1}{2}\sum_\alpha \omega_\alpha{}^2 Q_\alpha{}^2$$

If the kinetic and potential energies are both positive definite quadratic forms, then all the quantities ω_α will be real and also the transformation matrix $S_{s\alpha}$ will be real.

In case several of the ω_α are equal, then the normal coordinates are not uniquely determined. For if, say, Q_1, Q_2, Q_3 all have the same frequency, then any linear combination of these as $\lambda Q_1 + \mu Q_2 + \nu Q_3$ has all the properties of a normal coordinate with this same frequency.

The condition imposed on the S matrix in order that the Q's be normal coordinates provides the system of equations for determining S. Multiplying through by $S_{\omega\beta}$ and summing over β this gives $\Sigma_s(\Sigma_r m_{tr}{}^{-1}k_{rs})S_{s\alpha} = \omega_\alpha{}^2 S_{t\alpha}$. This is a set of N linear homogeneous equations for the N quantities $S_{s\alpha}$ with $s = 1, 2, \ldots, N$, for fixed α. The condition that they be solvable is therefore the vanishing of the determinant of their coefficients

$$\left|\sum_r m_{tr}{}^{-1}k_{rs} - \omega^2\delta_{ts}\right| = 0 \tag{3.25}$$

This is an algebraic equation of the Nth degree for ω^2 whose roots give the frequencies of the normal modes. With each root a solution can be found for the ratios of the $S_{s\alpha}$ associated with the frequency $\omega_\alpha/2\pi$. The magnitudes of the $S_{s\alpha}$ are arbitrary, as different choices merely amount to a change of scale for the corresponding Q_α. In practice it will usually be found to be simpler to work directly with the equations of motion in the form (3.21) to find explicit solutions of a particular problem.

7. Oscillations with Dissipation

If there are frictional forces which are proportional to the velocities, these can be formally represented by another positive definite quadratic form in the q's known as *Rayleigh's dissipation function.*

$$F = \tfrac{1}{2}\sum_{ij} f_{ij}\dot{q}_i\dot{q}_j \tag{3.26}$$

and the equations of motion of free oscillation take the form

$$\frac{d}{dt}\left(\frac{\partial T}{\partial \dot{q}_r}\right) - \frac{\partial F}{\partial \dot{q}_r} + \frac{\partial V}{\partial q_r} = G_r$$

if G_r is the generalized component of applied force on the coordinate q_r in the sense that $G_r\,dq_r$ is the work done by this force on the system in the displacement dq_r.

In this case there are three quadratic forms to deal with. In general the introduction of normal coordinates as in Sec. 6 above will not also reduce the dissipation term to a constant multiple of a single $\dot{Q}$. After introduction of the normal coordinates the equations of motion for free oscillation will thus be of the form

$$\ddot{Q}_r + \sum_s F_{rs}\dot{Q}_s + \omega_r{}^2 Q_r = 0 \tag{3.27}$$

in the general case. If the damping is small, each normal mode will be damped with its own characteristic damping constant λ_r so that

$$Q_r = Q_{r0}\exp\left(-\lambda_r t + i\omega_r t\right)$$

Neglecting λ^2 and terms of the form $F_{rs}\lambda_s$, an approximate form for λ_r is $\lambda_r = \tfrac{1}{2}\omega_r^{-1}\,\Sigma_s F_{rs}\omega_s$. At the other extreme, if the damping is very great, the velocities will be slow enough so that the inertia of the parts will be of small effect—in such a case a kind of *damping normal coordinates* can be introduced by making a transformation of coordinates which simultaneously reduces the dissipation function and the potential energy to a sum of squares. In this case the equations of motion assume the form

$$\sum_s M_{rs}\ddot{Q}_s + \dot{Q}_r + \lambda_r Q_r = 0 \tag{3.28}$$

(These Q's have a different relation to the q's than the normal coordinates.) Neglecting the inertial terms, each Q_r varies with the time with a factor $\exp\,(-\lambda_r t)$. If the inertial effects are small, one can assume that Q_r varies as $\exp\,[-(\lambda_r + \delta_r)t]$, where δ_r is the small change in the damping constant which the inertia produces. Approximately $\delta_r = \Sigma_s M_{rs}\lambda_s{}^2$.

8. Forced Oscillations of Coupled Systems

Following the notation of Sec. 7 let it be supposed that the system is driven by an applied force of the form $g_r e^{i\omega t}$, where g_r is a constant for a particular value of r, and the other applied forces vanish. This gives rise to a forced oscillation of all the coordinates q_s. In the more general case in which there are applied forces acting on several of the coordinates, the resulting forced oscillation will be the linear superposition of what is produced by each of the applied forces acting alone, so that the general case is covered by dealing with the case in which the applied force acts on but one coordinate.

By analogy with the results of Secs. 3 and 4 above, the various q's will all undergo harmonic motion with the same frequency as the driving force after the transient motion has died away. Replacing each q_s by $(-iv_s/\omega)e^{i\omega t}$, so that the constant v_s will have the physical meaning of the velocity amplitude of the sth coordinate, the equations of motion assume the form

$$\sum_j (m_{ij}\ddot{q}_j + f_{ij}\dot{q}_j + k_{ij}q_j) = g_i e^{i\omega t}$$

$$\sum_j \left(f_{ij} + i\omega m_{ij} - \frac{ik_{ij}}{\omega}\right) v_j = g_i$$

This is a system of linear equations for the N quantities v_j in terms of the amplitudes g_i of the harmonic applied force.

Each coefficient is of the same form as the mechanical impedance of a single resonator, as introduced above in Sec. 4:

$$Z_{ij} = f_{ij} + i\left(\omega m_{ij} - \frac{k_{ij}}{\omega}\right) \tag{3.29}$$

The coefficients of the type Z_{ii} are sometimes called *blocked impedances* because they give the response of the ith coordinate to a driving force applied to that coordinate on the supposition that all the other coordinates are constrained not to oscillate. The coefficient Z_{ij} is called a *transfer impedance* since it gives the response of the ith coordinate to a driving force in the jth coordinate.

The ensemble of all the coefficients Z_{ij} forms the *impedance matrix* for the system. It is a symmetrical matrix, $Z_{ij} = Z_{ji}$. The matrix that is reciprocal to Z_{ij} will be called the *admittance matrix* and denoted by A_{ki}. It is also a symmetrical matrix defined by the relation $\Sigma_i A_{ki} Z_{ij} = \delta_{kj}$ and therefore the formal solution of the equations of forced oscillation is

$$v_k = \sum_i A_{ki} g_i \tag{3.30}$$

A_{ki} is not the reciprocal of Z_{ki} as in a system of one degree of freedom, but is a component of the matrix that is the reciprocal of the matrix Z_{ki}.

9. General Driving Force

Let an oscillator be acted on by an exciting force which depends in any way on the time. The equation of motion is then of the form

$$\ddot{x} + \omega^2 x = F(t) \tag{3.31}$$

where $F(t)$ is the force acting at time t divided by the mass of the particle. The free motion is

$$x = \xi e^{i\omega t} + \eta e^{-i\omega t}$$

where ξ and η are constants. The forced motion may be expressed in this same form, but with ξ and η appropriately chosen functions of the time.

The essential point is to require that ξ and η have at each instant the values they would have in the ensuing harmonic motion if at that instant $F(t)$ were to become and remain zero. This requires not only that $x(t)$ be given as above but also

$$\dot{x}(t) = i\omega(\xi e^{i\omega t} - \eta e^{-i\omega t})$$

With ξ and η variable

$$\dot{x}(t) = \dot{\xi} e^{i\omega t} + \dot{\eta} e^{-i\omega t} + i\omega(\xi e^{i\omega t} - \eta e^{i\omega t})$$

hence, to fulfill the condition that ξ and η give the correct velocity at any instant, the condition

$$\dot{\xi} e^{i\omega t} + \dot{\eta} e^{-i\omega t} = 0$$

must be imposed. In this case the equation for $\ddot{x}(t)$ in terms of ξ and η becomes

$$\ddot{x}(t) = -\omega^2(\xi e^{i\omega t} + \eta e^{-i\omega t}) + i\omega(\dot{\xi} e^{i\omega t} - \dot{\eta} e^{-i\omega t})$$

Substitution in the equation of motion gives

$$\begin{aligned} \xi &= \xi_0 + \frac{1}{2i\omega}\int_0^t F(t)e^{-i\omega t}\,dt \\ \eta &= \eta_0 - \frac{1}{2i\omega}\int_0^t F(t)e^{+i\omega t}\,dt \end{aligned} \tag{3.32}$$

In particular if the oscillator is at rest at $t = -\infty$, then the limiting effect of the disturbance toward $t \to \infty$ is given by

$$\begin{aligned} \xi &= \frac{1}{2i\omega}\int_{-\infty}^{+\infty} F(t)e^{-i\omega t}\,dt \\ \eta &= -\frac{1}{2i\omega}\int_{-\infty}^{+\infty} F(t)e^{i\omega t}\,dt \end{aligned}$$

which shows that the part of $F(t)$ effective in exciting the oscillator is the component in its Fourier integral representation associated with the natural frequency of the oscillator.

For example, if $F(t) = (K/\tau\sqrt{\pi}) \exp(-t^2/\tau^2)$, the motion produced for $t \gg \tau$ is

$$x(t) = \frac{K}{\omega}\exp(-\tfrac{1}{4}\omega^2 t^2)\sin \omega t \tag{3.33}$$

or if $F(t) = K\tau/\pi(t^2 + \tau^2)$ the motion produced for $t \gg \tau$ is given by $x(t) = (K/\omega)e^{-\omega\tau}\sin \omega t$.

In the general case of N degrees of freedom let $f_r(t)$ be the generalized disturbing force acting on the rth coordinate (notation of Sec. 3.6); then the equations of motion are

$$\sum_s (m_{rs}\ddot{q}_s + k_{rs}q_s) = f_r(t)$$

or

$$\ddot{q}_t + \sum_s \left(\sum_r m_{tr}^{-1}k_{rs}\right) q = \sum_r m_{tr}^{-1}f_r(t)$$

Making the substitution for the transformation to normal coordinates, $q_s = \Sigma S_{s\alpha}Q_\alpha$, the equation of motion for the Q_β becomes

$$\ddot{Q}_\beta + \omega_\beta^2 Q_\beta = \sum_{tr} S_{\beta t}^{-1} m_{tr}^{-1} f_r(t)$$

so the term on the right gives the measure of the effectiveness of the set of disturbing forces $f_r(t)$ in exciting the βth normal mode of vibration. This quantity will be denoted by $F_\beta(t)$.

10. Physical Pendulum

If a rigid body of mass M is free to rotate about a horizontal axis fixed both in space and in the body and if Ma^2 is its moment of inertia about this axis and h is the distance of its center of mass from the axis of rotation, then the equation of motion is

$$\ddot{\theta} + \frac{gh}{a^2}\sin\theta = 0 \tag{3.34}$$

where θ is measured from the equilibrium position, and therefore for amplitudes sufficiently small that $\sin\theta$ can be approximated by θ, the motion is approximately simple harmonic with frequency $\omega_0/2\pi$ where $\omega_0 = \sqrt{gh/a^2}$. The quantity $L = a^2/h$ is therefore called the length of the equivalent simple pendulum.

The energy integral is $\dot{\theta}^2 + 2\omega_0^2(1 - \cos\theta) = \omega^2$, where ω is the angular velocity possessed by the pendulum when $\theta = 0$. If $\omega^2 < 4\omega_0^2$, the energy of motion will not be sufficient to make the body keep on rotating in the same sense so the motion will be oscillatory.

The exact solution for the motion can be expressed in terms of elliptic functions. If $y = \sin\theta/2$, then the energy integral becomes

$$\dot{y}^2 = \omega_0^2(1 - y^2)\left(\frac{\omega^2}{4\omega_0^2} - y^2\right)$$

The solution of this is

$$y = \frac{\omega}{2\omega_0}\,\mathrm{sn}\,[\omega_0(t - t_0), k] \tag{3.35}$$

where the modulus is $k = \omega/2\omega_0$. The period of the oscillatory motion is $T = 4K/\omega_0$, where $K(k)$ is the complete elliptic integral

$$K(k) = \int_0^1 (1 - t^2)^{-1/2}(1 - k^2t^2)^{-1/2}\,dt$$

If α is the maximum amplitude of swing, the modulus is $k = \sin\alpha/2$ and the period is given by

$$T = 2\pi\sqrt{\frac{L}{g}}\left[1 + \left(\frac{1}{2}\right)^2\sin^2\frac{\alpha}{2} + \left(\frac{1\cdot 3}{2\cdot 4}\right)^2\sin^4\frac{\alpha}{2} + \left(\frac{1\cdot 3\cdot 5}{2\cdot 4\cdot 6}\right)^2\sin^6\frac{\alpha}{2} + \cdots\right]$$

If $\alpha = 1°$, the correction is less than one part in fifty thousand.

11. Nonharmonic Vibrations

The physical pendulum is a particular case of a nonharmonic vibration, i.e., of a motion of a particle about equilibrium in which it is necessary to consider the departure of the force from strict proportionality with the displacement from equilibrium. Denoting the coordinate by x and the mass by m (recognizing that x may be an angle in which case m is a moment of inertia), the kinetic energy is given by $\frac{1}{2}m\dot{x}^2$ and the potential energy is $V(x)$, where it may be assumed that $x = 0$ is the equilibrium position so that $V(0) = 0$, and also it may be assumed for convenience that $V(0) = 0$.

The energy integral takes the form

$$\tfrac{1}{2}m\dot{x}^2 + V(x) = W \tag{3.36}$$

where W is the constant total energy. The motion is formally described by the integral

$$t = \int_{x_0}^{x} [(m/2)(W - V(x))]^{-1/2}\,dx \tag{3.37}$$

where x_0 is the displacement at $t = 0$, and the initial velocity $\dot{x}_0$ enters through its effect on the value of W,

$$\tfrac{1}{2}m\dot{x}_0^2 + V(x_0) = W$$

This completes the formal integration of the problem. Not many analytic forms for $V(x)$ lead to integrable forms in (3.37). In case $V(x)$ is simply $\frac{1}{2}kx^2$, this leads to simple harmonic motion; for $V(x)$ given by $V(x) = \frac{1}{2}kx^2 + \alpha x^3 + \beta x^4$, the integral can be expressed in terms of elliptic functions but usually, in a particular case, it is simpler to make a direct numerical integration than to express the results as an elliptic function and use tables of such functions.

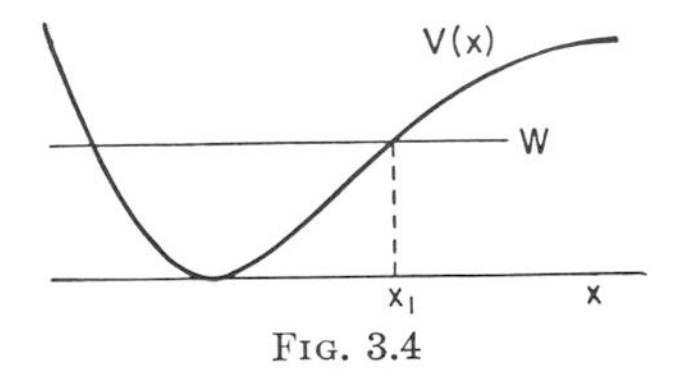

FIG. 3.4

The following qualitative discussion of the motion is instructive. The integrand in (3.37) is, of course, the reciprocal of the velocity, $v(x)$, defined as a function of position through the energy integral (3.36). As long as $v(x)$ is not zero, the direction of motion does not change so the physical interest attaches to the behavior in the neighborhood of the *turning points*, x_1, or places where $V(x_1) = W$, at which the velocity vanishes and the integrand in (3.37) becomes infinite. The simplest case is that in which x_1 is a simple root: in this case $V(x)$ is approximately of the form $V(x) = W + mg_1(x - x_1)$ for values of x near to x_1 and the integral gives a finite time of approach to $x = x_1$ with a reversal of the sign of the velocity at the turning point, so that the motion in the neighborhood of the turning point is like that of a freely falling body under uniform acceleration g_1. If there are two such turning points, x_1 and x_2, there will be a finite time of travel from x_1 to x_2 and back to x_1 after which the motion repeats itself. The period T of the motion is

$$T = \sqrt{2m}\int_{x_1}^{x_2}\frac{dx}{\sqrt{W - V(x)}} \tag{3.38}$$

If, however, $W - V(x) = 0$ has a quadratic root at $x = x_1$, that is, if $x = x_1$ is a maximum of $V(x)$ having the value W, then the integral in (3.38) diverges, and the time of approach to such an equilibrium point is logarithmically infinite.

References

1. Andronow, A. A., and C. E. Chaikin: "Theory of Oscillations," Princeton University Press, Princeton, N.J., 1949.
2. Goldstein, H.: "Classical Mechanics," Chap. 10, Addison-Wesley, Reading, Mass., 1950.
3. Den Hartog, J. P.: "Mechanical Vibrations," 3d ed., McGraw-Hill, New York, 1947.
4. Morse, P. M.: "Vibration and Sound," 2d ed., Chap. 2, McGraw-Hill, New York, 1948.
5. Lord Rayleigh: "The Theory of Sound," Chaps. 2 to 5, Dover, New York, 1945.
6. Timoshenko, S., and D. H. Young: "Advanced Dynamics," McGraw-Hill, New York, 1948.
7. Whittaker, E. T.: "Analytical Dynamics," Chap. 7, Cambridge University Press, New York, 1927.

Chapter 4

Orbital Motion

By E. U. CONDON, University of Colorado

1. Motion under Constant Gravity

The motion of a particle under the constant downward acceleration of gravity is very simply described. Let the z axis be positive upward; then if g is the acceleration of gravity, the force downward on a particle of mass m is mg and the potential energy at height z is $V = mgz$.

The equation of motion is

$$\ddot{\mathbf{r}} = -g\mathbf{k}$$

and thence the vector velocity at time t is

$$\mathbf{v} = \mathbf{v}_0 - gt\mathbf{k} \tag{4.1}$$

where $\mathbf{v}_0$ is the initial velocity. The horizontal component of velocity remains constant while the vertical component decreases linearly with the time.

The vector position at time t is given by

$$\mathbf{r} = \mathbf{r}_0 + \mathbf{v}_0 t - \tfrac{1}{2}gt^2\mathbf{k} \tag{4.2}$$

The entire motion takes place in the vertical plane determined by $\mathbf{r}_0$ and $\mathbf{v}_0$. Choosing the x axis in this plane, so that $v_{0y} = 0$, the parametric equations of the path are

$$z = v_{0z}t - \tfrac{1}{2}gt^2$$
$$x = v_{0x}t$$

showing that the path is a parabola. The particle reaches maximum altitude at the time $t_1 = v_{0z}/g$, at which the vertical component of velocity vanishes. The value of the maximum height reached is

$$z_1 = \frac{v_{0z}^2}{2g}$$

The *range* is the horizontal distance traveled before returning to the plane $z = 0$. For a particle projected at an angle of elevation α with the horizontal at a speed V, the range is

$$R = \frac{V^2}{g}\sin 2\alpha \tag{4.3}$$

Hence for a given speed V the maximum range is V^2/g obtained at an angle of projection of $\alpha = 45°$. Ranges less than the maximum can be reached by projecting at two different angles of elevation, one greater and one less than 45°, giving a high and a low trajectory for the same range.

2. Effect of Earth's Rotation

Coordinate axes fixed relative to the earth's surface are rotating about the north pole of the heavens with one revolution in a sidereal day or 86,169 mean solar seconds. Hence the angular velocity is

$$\boldsymbol{\omega} = 7.292 \times 10^{-5}\mathbf{n}/\text{sec}^{-1}$$

where $\mathbf{n}$ is a unit vector pointing in the direction of the earth's north polar axis.

From Chap. 1, Sec. 1, it follows that the acceleration contains extra terms due to the time rate of change of the coordinate basis; thus the acceleration is

$$(\ddot{x}\mathbf{i} + \ddot{y}\mathbf{j} + \ddot{z}\mathbf{k}) + 2\boldsymbol{\omega} \times (\dot{x}\mathbf{i} + \dot{y}\mathbf{j} + \dot{z}\mathbf{k}) + \boldsymbol{\omega} \times (\boldsymbol{\omega} \times \mathbf{r})$$

A person who likes to think of his coordinate system as if it were nonrotating will regard the first term alone as giving the acceleration and transpose the other two terms to the other side of the equation and regard them as constituting additional forces acting on the body. In this way the equation of motion takes the form

$$m\ddot{\mathbf{r}} = m\mathbf{g}_0 - m\boldsymbol{\omega} \times (\boldsymbol{\omega} \times \mathbf{r}) - 2m\boldsymbol{\omega} \times \mathbf{v} \tag{4.4}$$

in which it is to be understood that $\mathbf{v}$ means $(\dot{x}\mathbf{i} + \dot{y}\mathbf{j} + \dot{z}\mathbf{k}$ and $\ddot{\mathbf{r}} = \ddot{x}\mathbf{i} + \ddot{y}\mathbf{j} + \ddot{z}\mathbf{k}$ consistent with the pretense that the coordinate axes are not rotating. In (4.4) $\mathbf{g}_0$ is the part of the acceleration of gravity due to gravitational attraction of the actual mass distribution of the earth. In the second term on the right, it is important to remember that the origin of $\mathbf{r}$ is on the axis of rotation of the earth, say its center, so that for motions near the earth's surface for short ranges $\mathbf{r}$ is nearly a constant. The direction of this term is *centrifugal*, that is, directed away from the earth's axis of rotation. Its magnitude is $\omega^2 a \cos\psi$ at a place whose latitude is ψ, the radius of the earth being a. This term combines vectorially with the first to give an apparent local acceleration of gravity

$$\mathbf{g} = \mathbf{g}_0 - \boldsymbol{\omega} \times (\boldsymbol{\omega} \times \mathbf{r})$$

The direction of $\mathbf{g}$ is the direction of the vertical at the place in question. Since $a = 6.36 \times 10^8$ cm, $\omega^2 a = 3.40$ cm/sec². The term $\omega^2 a \cos\psi$ does not give the entire variation of effective g with latitude: another source of variation is the fact that the earth's rotation gives rise to a nonspherical shape for the earth which gives a latitude variation for $\mathbf{g}_0$, as discussed in Chap. 7, Sec. 3.

The last term on the right in (4.4) gives rise to an apparent force due to the earth's rotation which is known as the *Coriolis force*. It acts at right angles to the apparent velocity $\mathbf{v}$ and to the axis of rotation. Introducing the effective local gravity $\mathbf{g}$, the equation of motion for a particle moving near the earth's surface takes the form

$$\ddot{\mathbf{r}} = \mathbf{g} + 2\mathbf{v} \times \boldsymbol{\omega} \tag{4.5}$$

Since $g = 980$ cm/sec², it follows that the linear speed, when at right angles to $\boldsymbol{\omega}$, would have to be $v = 67$ km/sec in order for the Coriolis force to be equal to the gravitational. Therefore under most circumstances the Coriolis force is small compared to the gravitational. Nevertheless, it produces important observable effects.

To discuss these explicitly it will be supposed that $\mathbf{i}$ points to the east, $\mathbf{j}$ to the north, and $\mathbf{k}$ vertically upward, in which case the direction of the North Pole is

$$\mathbf{n} = (\mathbf{j} \cos \psi + \mathbf{k} \sin \psi)$$

For a freely falling body, released from rest, the velocity will be at first in the direction $-\mathbf{k}$ so that the Coriolis acceleration is directed initially to the east and is equal to $2(gt)\ \omega \cos \psi$. As a result of its action the body after falling a time t acquires an easterly component of velocity

$$\dot{x} = gt^2 \cdot \omega \cdot \cos \psi$$

and hence an easterly displacement

$$x = \tfrac{1}{3} g t^3\ \omega \cos \psi$$

Therefore the easterly displacement accompanying a free fall through a distance h is

$$x = \frac{1}{2} (\omega \cos \psi) \left(\frac{8h^3}{g}\right)^{1/2}$$

The deflection is small but definitely measurable: a drop of 1,000 ft at the equator gives an easterly deflection of 4.6 in.

As the body falls, it acquires an easterly component of velocity which gives rise to a Coriolis force in the direction $\mathbf{i} \times \mathbf{n}$ which may be regarded as consisting of an upward component which acts to reduce the effective value of gravity and make the vertical motion a little slower than otherwise, and a north-south component which produces a second-order deflection of a freely falling body, that is toward the equator in either hemisphere, whose amount is

$$\tfrac{1}{6} g t^4 \omega^2 \sin \psi \cos \psi$$

Both the easterly and the southerly deflections were experimentally observed in 1832 by Reich at Freiburg.

If the particle is constrained to move essentially in a horizontal plane, then only the vertical component of $\boldsymbol{\omega}$ is effective in the Coriolis force, and the equation of motion takes the form

$$\ddot{\mathbf{r}} = 2\omega \sin \psi (\mathbf{v} \times \mathbf{k})$$

This gives rise to an apparent force tending to deflect the particle to the right in the Northern Hemisphere and to the left in the Southern Hemisphere. A particle not acted on by any other forces moves in a circular path whose radius is $v/2\omega \sin \psi$, where v is the linear speed.

This Coriolis deflection of a high-speed projectile can have relatively large effects which must be taken into account in ballistics. For example, a projectile whose maximum range is 10 km must have a muzzle speed of $V = 3.15$ km/sec with a horizontal component of 2.22 km/sec and therefore the horizontal projection of its path will not be straight but will be an arc of a circle whose radius is 3.10^4 km at a place of latitude 30°.

The *Foucault pendulum* was devised in 1851 to exhibit the earth's rotation in the laboratory. It is a pendulum with a suspension carefully made to be symmetrical about a vertical axis so that its natural frequency is the same for motion in any vertical plane and also with a bob sufficiently massive that it will swing for some hours in spite of damping due to air resistance. The observed effect is that the plane of oscillation does not remain fixed, but veers around in the clockwise sense (in the Northern Hemisphere) with an angular velocity equal to $\omega \sin \psi$.

The simplest way to derive this result is to observe that if a frame of reference is introduced which is rotating about the vertical axis with an angular velocity $-\omega \sin \psi\ \mathbf{k}$ relative to the axes fixed on the earth, then the combined effect of this rotation and the earth's rotation is to give a coordinate frame in which the resultant vertical component of angular velocity vanishes. Therefore in this reference frame no Coriolis force acts and the motion will take place in a fixed vertical plane, that is, one which rotates with angular velocity $\omega \sin \psi$ relative to coordinate axes fixed on the earth. This is the observed result.

The problem may also be considered as one of coupled oscillations. If the effective length of the pendulum is l and $k^2 = g/l$, then the equations of motion are

$$\begin{aligned} \ddot{x} - 2\omega \sin \psi\ \dot{y} + k^2 x &= 0 \\ \ddot{y} + 2\omega \sin \psi\ \dot{x} + k^2 y &= 0 \end{aligned} \tag{4.6}$$

There are two normal modes, corresponding to motions in which the x and y amplitudes are equal but $\pi/2$ out of time phase, giving a resultant circular motion. One mode corresponds to motion in the clockwise sense, the other to motion in the counterclockwise sense. The circular frequencies of these are $k \pm \omega \sin \psi$, the clockwise mode being the one of higher frequency. Motion in a plane has to be regarded as arising from superposition of these two modes with equal amplitudes, and the gradual precession of the plane as due to the difference in frequency of the two modes.

The dynamic effects of the earth's rotation also find application in the *gyrocompass*, in which a rotating rigid body is used to determine the direction of north If such a body is carefully suspended in such a way that it can spin about a figure axis, with that axis free to turn in any direction, then the vector angular momentum of the body, $\boldsymbol{\Omega}$, will remain constant relative to axes fixed in space.

3. General Integrals of Central-force Problem

The problem of the motion of two particles which interact with forces of attraction or repulsion which are a function of the distance between them is of fundamental importance in physics. It lies at the basis of Newton's dynamical interpretation of Kepler's empirical laws of planetary motion in the solar system, and also provides the explanation of motions of binary stars. It provided the basis for the pre-quantum-mechanical Bohr theory of atomic structure, and governs the study of atomic phenomena by collision processes.

Let $\mathbf{r}_1$ and $\mathbf{r}_2$ be the position vectors of the two particles whose masses are m_1 and m_2, and let $\mathbf{F}_{12}$ be the force on 1 due to 2. The equations of motion are

$$m_1\ddot{\mathbf{r}}_1 = \mathbf{F}_{12} \qquad m_2\ddot{\mathbf{r}}_2 = \mathbf{F}_{21} = -\mathbf{F}_{12} \tag{4.7}$$

These are a set of six differential equations, each of the second order, so that twelve constants of integration must appear in the general solution.

It is convenient to introduce the relative coordinate $\mathbf{r} = \mathbf{r}_2 - \mathbf{r}_1$ and the position of the center of mass

$$\mathbf{R} = \frac{m_1\mathbf{r}_1 + m_2\mathbf{r}_2}{m_1 + m_2}$$

In terms of these the equations of motion become

$$\ddot{\mathbf{R}} = 0 \qquad \mu\ddot{\mathbf{r}} = \mathbf{F}_{21}$$

where μ is called the *reduced mass,*

$$\mu = \frac{m_1 m_2}{m_1 + m_2}$$

The first equation integrates to give

$$\mathbf{R} = \mathbf{R}_0 + \mathbf{V}_0 t$$

indicating that the center of mass moves uniformly in a straight line.

It will be supposed that $\mathbf{F}_{21}$ lies in the direction of $\mathbf{r}$. Multiplying the equation of relative motion through by $\mathbf{r}\times$ gives

$$\mu\mathbf{r} \times \ddot{\mathbf{r}} = 0$$

which can be written

$$\dot{\mathbf{L}} = 0 \qquad \text{hence} \qquad \mathbf{L} = \text{const}$$

where $\mathbf{L}$ is the vector orbital angular momentum of the system, $\mathbf{L} = \mu\mathbf{r} \times \mathbf{v}$. Therefore the vector orbital angular momentum of the system $\mathbf{L}$ is constant throughout the motion.

The path of the motion lies in a plane normal to $\mathbf{L}$. It is convenient now to use polar coordinates r and θ in the plane of the motion. Also it will be supposed that the forces of interaction are conservative so that there exists a potential energy function $V(r)$ such that $\mathbf{F}_{21} = -(\partial V/\partial r)\mathbf{r}_0$, where $\mathbf{r}_0$ is the unit vector in the direction of $\mathbf{r}$.

The equations of motion in the plane are now

$$\mu(\ddot{r} - r\dot{\theta}^2) = -\frac{\partial V}{\partial r}$$

$$\mu r^2\dot{\theta} = L$$

Use of the angular momentum integral permits the elimination of $\dot{\theta}^2$ from the radial equation,

$$\mu\ddot{r} = -\frac{\partial}{\partial r}\left(V + \frac{L^2}{2\mu r^2}\right)$$

Therefore the existence of orbital angular momentum affects the radial motion just as if there were an additional repulsive force of interaction between the two particles which varies as the inverse cube of the distance between them.

The radial equation possesses an integral of energy in the form

$$\frac{1}{2}\mu\dot{r}^2 + \frac{L^2}{2\mu r^2} + V(r) = E \tag{4.8}$$

where E is the constant total energy. The first term on the left is the kinetic energy of the radial motion, the second is the kinetic energy of the motion in θ, since it can be written $\frac{1}{2}\mu r^2\dot{\theta}^2$, and the third is the potential energy. It is convenient to use the abbreviation

$$U(r) = V(r) + \frac{L^2}{2\mu r^2}$$

regarding $U(r)$ as the effective potential energy for the radial motion.

The energy integral may be further integrated by separation of variables to give

$$t = \int_{r_0}^{r} \frac{dr}{\sqrt{(2/\mu)[E - U(r)]}} \tag{4.9}$$

where r_0 is the distance apart of the two particles at $t = 0$.

In principle this integral determines r as a function of the time, thus permitting one further integration to determine θ as a function of the time,

$$\theta - \theta_0 = \frac{L}{\mu}\int_0^t \frac{dt}{r^2} \tag{4.10}$$

Thus the general integration of the two-body problem is completely reduced to quadratures. The twelve integration constants are

$$\mathbf{R}_0,\ \mathbf{V}_0,\ \mathbf{L},\ E,\ r_0,\ \theta_0,$$

in which the three vector integration constants each count as three.

4. Differential Equation for Orbit

The procedure of Sec. 3 may be modified in a way that permits determination first of the geometrical shape of the orbit, after which the time dependence of position in the orbit may be determined.

This is done by using $\mu r^2\dot{\theta} = L$ to replace t by θ as the independent variable in the radial equation of motion. The resulting equation takes a simpler form if $u = 1/r$ is used as the dependent variable instead of r itself.

The equation for the orbit is

$$\frac{d^2u}{d\theta^2} + u + \frac{\mu}{L^2}\frac{\partial V}{\partial u} = 0 \tag{4.11}$$

In particular, if no force acts between the particles, then this equation possesses the solution

$$u = A^{-1} \cos (\theta - \theta_0)$$

which is the equation of a straight line as it should be.

Suppose that initially the particles are at a great distance apart and that coordinates are chosen so that $\theta_0 = 0$. The speed of approach is $(2E/\mu)^{1/2}$ on a straight line which comes within a minimum distance a, where a is called the *impact parameter*. Then $L = a(2\mu E)^{1/2}$ and $\dot{r} = -(L/\mu)\, du/d\theta$. Writing $v = a/r = au$, the equation for the orbit takes the form

$$\frac{d^2v}{d\theta^2} + v + \frac{1}{2E}\frac{\partial V}{\partial v} = 0 \qquad (4.12)$$

and the initial conditions are $v = 0$ and $dv/d\theta = 1$ at $\theta = 0$.

One asymptote of the orbit is the line $\theta = 0$ and the other is the next value of θ for which $v = 0$.

5. Motion under Inverse-square-law Attraction

If the two particles have unlike electrostatic charges the product of whose magnitude is e^2, then the potential energy of interaction is

$$V(r) = -\frac{e^2}{r} = -e^2u$$

In the case of gravitational attraction of two particles of masses m_1 and m_2 the potential energy is

$$V(r) = -\frac{Gm_1m_2}{r} = -Gm_1m_2u$$

where G is the universal gravitation constant,

$$G = (6.670 \pm 0.005) \times 10^{-8} \text{ dyne-cm}^2\text{-gr}^{-2}$$

Equation (4.11) for the orbit takes the form

$$\frac{d^2u}{d\theta^2} + u - \frac{\mu e^2}{L^2} = 0$$

the solution of which is

$$u = \frac{\mu e^2}{L^2}[1 + \epsilon \cos (\theta - \theta_0)]$$

where ϵ and θ_0 are constants of integration.

In terms of r and θ the equation of the orbit is

$$r = \frac{p}{1 + \epsilon \cos (\theta - \theta_0)} \qquad p = \frac{L^2}{\mu e^2} \qquad (4.13)$$

This is the polar form of the equation of a conic section having one focus at the origin, with the line $\theta = \theta_0$ as its major axis, p as its semilatus rectum and ϵ as its eccentricity. The orbit is:

Circle	for $\epsilon = 0$
Ellipse	for $0 < \epsilon < 1$
Parabola	for $\epsilon = 1$
Hyperbola	for $\epsilon > 1$

To calculate the value of the total energy, one may evaluate $L^2/2\mu r^2 - e^2/r$ at the minimum value of r which is $p/(1 + \epsilon)$, when the radial velocity is zero:

$$E = -\frac{e^2}{2p}(1 - \epsilon^2) = -\frac{e^2}{2a} \qquad (4.14)$$

Therefore circular and elliptical orbits have negative energy, parabolic orbits have zero energy, and hyperbolic orbits positive energy.

For a given value of p the circular orbits are those of least energy. The total energy of a circular orbit is $-e^2/2p$, and since the potential energy is $-e^2/p$, it follows that the kinetic energy is $+e^2/2p$.

In the case of circular or elliptic motion the period T of complete revolution follows from the fact that $\frac{1}{2}r^2\dot{\theta} = L/2\mu$ is the rate at which the radius vector sweeps out area and the total area of an ellipse of semiaxes a and b is πab or, in terms of a and ϵ, $\pi a^2[(1 - \epsilon^2)^{1/2}]$, so that

$$T = \frac{\mu}{L} 2\pi a^2 \sqrt{1 - \epsilon^2}$$
$$= 2\pi \sqrt{\frac{\mu a^3}{e^2}} \qquad (4.15)$$

These results contain all three of *Kepler's laws of planetary motion:*

1. Orbits of the planets are ellipses with the sun in one focus.
2. Motion in the orbit is such that the rate of sweeping out area by the radius vector is constant.
3. Squares of the periods are proportional to cubes of semimajor axes.

The result (4.15) in the gravitational case (Gm_1m_2 in place of e^2) becomes

$$T = 2\pi \sqrt{\frac{a^3}{G(m_1 + m_2)}} \qquad (4.16)$$

Kepler's third law is thus based on an approximation which neglects the mass of the planet relative to that of the sun.

6. Motion in Elliptic Orbit

The angular momentum integral, $L = \text{const}$, is sometimes called the *integral of areas* because $r^2\dot{\theta}/2$ has the geometrical interpretation of being the rate at which area is swept over by the moving radius vector r. The rate of sweeping out area is $L/2\mu$.

The quantity $n = 2\pi/T$, where T is the period of revolution, is the mean angular velocity of the radius vector, and is called the *mean motion* in the astronomical literature; thus

$$n = \left(\frac{e^2}{\mu a^3}\right)^{1/2}$$

The problem of describing the position in an ellipse at any instant is complicated by the lack of a simple formula for $A(\theta)$, the area of an elliptical sector of angular opening θ measured from the major axis. In astronomical literature, θ is called the *true anomaly* of the moving body. Computation of the motion is carried out in relation to an auxiliary angle ϕ which is called the *eccentric anomaly*. Referring to the figure, the moving particle is at P and S is the focus of the elliptical orbit, and A is the position of nearest

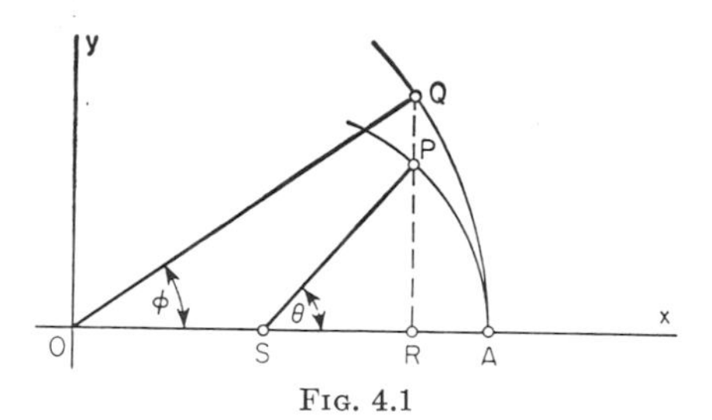

FIG. 4.1

approach or perihelion, so that θ is the angle ASP. O is the center of the ellipse and Q is a point on a circle of radius OA, or a, which has the same abscissa as P. The eccentric anomaly is defined as the angle AOQ. One has

$$x_P = a\cos\phi \qquad x_Q = a\cos\phi$$
$$y_P = b\sin\phi \qquad y_Q = a\sin\phi$$

Also $y_P = r\sin\theta$; thus the relation between the true anomaly θ and eccentric anomaly ϕ is

$$\sin\phi = \frac{(1-\epsilon^2)^{1/2}\sin\theta}{1+\epsilon\cos\theta}$$

The desired elliptical sector area $A(\theta)$ is the sum of the triangle RPS and the elliptical triangle APR. This latter is (b/a) times the circular triangle AQR, which in turn is the circular sector AQO minus the triangle RQO. Therefore $A(\theta)$ is simply expressed in terms of the eccentric anomaly,

$$A(\theta) = \tfrac{1}{2}ab(\phi - \epsilon\sin\phi)$$

Since $A(\theta) = \pi abt/T$, the time dependence of ϕ is given by

$$\phi - \epsilon\sin\phi = nt$$

which is known as *Kepler's equation.*

To compute the position in the orbit at time t, one has first to calculate ϕ from Kepler's equation, after which the coordinates x_P and y_P are easily calculated. Kepler's equation defines ϕ as a function of nt which depends parametrically on ϵ, which functional relation is the subject of a large mathematical literature.

Bibliography

Introductory orbit theory is dealt with in most of the general works on dynamics in the references of Part 2, Chap. 2.

Baker, R. M. L., Jr., and M. W. Makemson: "An Introduction to Astrodynamics," Academic, New York, 1960.

Brouwer, D., and G. M. Clemence: "Methods of Celestial Mechanics," Academic, New York, 1961.

Clarke, A. C.: "Interplanetary Flight, an Introduction to Astronautics," Harper & Row, New York, 1962.

Crawford, R. T.: "Determination of Orbits of Comets and Asteroids," McGraw-Hill, New York, 1930.

Herrick, Samuel: "Astrodynamics," Van Nostrand, Princeton, N.J., 1961.

Moulton, F. R.: "Celestial Mechanics," Macmillan, New York, 1928.

Smart, W. M.: "Celestial Mechanics," Longmans, London, 1960.

Van de Kamp, Peter: "Elements of Astromechanics," Freeman, San Francisco, 1964.

Watson, James C.: "Theoretical Astronomy, Relating to the Motions of the Heavenly Bodies," Dover, New York, 1964. (Originally, Lippincott, Philadelphia, 1868.)

Chapter 5

Dynamics of Rigid Bodies

By E. U. CONDON, University of Colorado

1. Angular Momentum

The angular momentum of a particle of mass m with respect to an origin is defined (Chap. 2, Sec. 7) as $m\mathbf{r} \times \mathbf{v}$, where $\mathbf{r}$ is its position and $\mathbf{v}$ its velocity. The total angular momentum of a system of particles is defined as the vector sum of such quantities for all the particles in the system.

It is convenient to write $\mathbf{R} + \mathbf{r}$ for $\mathbf{r}$ and $\mathbf{V} + \mathbf{v}$ for $\mathbf{v}$, where $\mathbf{R}$ and $\mathbf{V}$ are the position and velocity of the center of mass of the system of particles and therefore $\mathbf{r}$ and $\mathbf{v}$ are the position and velocity of the particular particle relative to the center of mass. Writing M for the total mass ($M = \Sigma m$), the total angular momentum for the system becomes

$$\begin{aligned} \mathbf{L} &= \Sigma m(\mathbf{R} + \mathbf{r}) \times (\mathbf{V} + \mathbf{v}) \\ &= M\mathbf{R} \times \mathbf{V} + \Sigma m\mathbf{r} \times \mathbf{v} \end{aligned}$$

since $\Sigma m\mathbf{r} = 0$ and $\Sigma m\mathbf{v} = 0$ by definition of the center of mass.

If the system of particles is a rigid body turning with instantaneous angular velocity $\boldsymbol{\omega}$ about an axis through the center of mass, then

$$\mathbf{v} = \boldsymbol{\omega} \times \mathbf{r}$$

(Chap. 1, Sec. 2) and therefore the *internal* angular momentum (that is, omitting $M\mathbf{R} \times \mathbf{V}$) is

$$\begin{aligned} \mathbf{L} &= \Sigma m\mathbf{r} \times (\boldsymbol{\omega} \times \mathbf{r}) \\ &= \Sigma m(r^2\mathbf{I} - \mathbf{rr}) \cdot \boldsymbol{\omega} \end{aligned} \tag{5.1}$$

The quantity which multiplies into $\boldsymbol{\omega}$ to give the angular momentum is a symmetric second-order tensor which depends on the distribution of mass in the rigid body. For a spherically symmetric body, the tensor reduces to a scalar, in which case the angular momentum has the same direction as the angular velocity. In general they are not parallel.

The tensor

$$\boldsymbol{\Phi} = \Sigma m(r^2\mathbf{I} - \mathbf{rr}) \tag{5.2}$$

is called the *inertia tensor* for the body. If the components of $\mathbf{r}$ with respect to axes (1,m,n) fixed in the body are (x,y,z), then the tensor components are

$$\begin{bmatrix} A & -H & -G \\ -H & B & -F \\ -G & -F & C \end{bmatrix} = \begin{bmatrix} \Sigma m(y^2 + z^2) & -\Sigma mxy & -\Sigma mxz \\ -\Sigma mxy & \Sigma m(z^2 + x^2) & -\Sigma myz \\ -\Sigma mxz & -\Sigma myz & \Sigma m(x^2 + y^2) \end{bmatrix} \tag{5.3}$$

The diagonal components are known in mechanics as the *moments of inertia* of the body with respect to the particular axes used. Thus $\Sigma m(y^2 + z^2)$ is the moment of inertia for rotation of the body about an axis through its center of mass and parallel to the 1 axis. The three components of the form Σmxy are called *products of inertia.*

Because the inertia tensor is symmetric, there will exist a choice of axes such that when the inertia tensor is calculated for them, the products of inertia will vanish and the inertia tensor will be represented in diagonal form. These directions in the body are known as its *principal axes of inertia.* For rotation about a principal axis the angular momentum is in the same direction as the angular velocity.

If the body rotates with angular velocity $\boldsymbol{\omega}$ about an axis which does not pass through the center of mass, then if $\mathbf{R}$ is the position vector of the center of mass relative to an origin located anywhere on the axis of rotation, the velocity of the center of mass will be $\boldsymbol{\omega} \times \mathbf{R}$; thus the angular momentum due to motion of the center of mass will be (5.1)

$$\begin{aligned} \mathbf{L}_c &= M\mathbf{R} \times (\boldsymbol{\omega} \times \mathbf{R}) \\ &= M(R^2\mathbf{I} - \mathbf{RR}) \cdot \boldsymbol{\omega} \end{aligned}$$

and therefore the whole angular momentum is given by (5.1) by adding the tensor $M(R^2\mathbf{I} - \mathbf{RR})$ to inertia tensor $\boldsymbol{\Phi}$ defined in (5.2). If a, b, and c are the coordinates of the center of mass with respect to the origin, this additional tensor is

$$M \begin{bmatrix} (b^2 + c^2) & -ab & -ac \\ -ab & (c^2 + a^2) & -bc \\ -ac & -bc & (a^2 + b^2) \end{bmatrix} \tag{5.4}$$

In general the sum of (5.3) and (5.4) have different principal axes from (5.3) alone; but if the new origin itself lies along a principal axis through the center of mass, then two of the quantities (a,b,c) vanish so that (5.4) has the same principal axes as (5.3).

If $\boldsymbol{\Phi}$ is the inertia tensor and $\mathbf{r}$ is the position vector of a variable point then

$$\mathbf{r} \cdot \boldsymbol{\Phi} \cdot \mathbf{r} = 1$$

is the equation of an ellipsoid that forms a convenient way of visualizing the inertial properties of the body. The semiaxes of this ellipsoid are $A^{-1/2}$, $B^{-1/2}$, and $C^{-1/2}$, where A, B, and C are the principal moments of inertia of the body. This ellipsoid is sometimes called the *Poinsot ellipsoid of inertia.*

The moment of inertia about any axis is defined as $K = \Sigma mp^2$, where p is the perpendicular distance of a particle from the axis. The Poinsot ellipsoid has the property that $K = 1/r^2$, where r is the radius vector out to the ellipsoid along the axis in question even when this is not a principal axis.

In terms of the tensor components the angular momentum components are related to the angular velocity components in this way,

$$\begin{aligned} L_1 &= A\omega_1 - H\omega_2 - G\omega_3 \\ L_2 &= -H\omega_1 + B\omega_2 - F\omega_3 \\ L_3 &= -G\omega_1 - F\omega_2 + G\omega_3 \end{aligned} \tag{5.5}$$

The time rate of change of angular momentum when referred to axes which rotate with the body is made up of two contributions. One is that due to time variation of the components with respect to the rotating axes; the other is due to the fact that the axes themselves are rotating,

$$\frac{d\mathbf{L}}{dt} = \dot{L}_1\mathbf{l} + \dot{L}_2\mathbf{m} + \dot{L}_3\mathbf{n} + \boldsymbol{\omega} \times \mathbf{L} \tag{5.6}$$

2. Kinetic Energy

The velocity of a particle being $\mathbf{V} + \mathbf{v}$ where $\mathbf{V}$ is the velocity of the system's center of mass and $\mathbf{v}$ is the particle's own velocity relative to the center of mass, the total kinetic energy of the body is

$$\begin{aligned} T &= \tfrac{1}{2}\Sigma m(\mathbf{V} + \mathbf{v})^2 \\ &= \tfrac{1}{2}M\mathbf{V}^2 + \tfrac{1}{2}\Sigma m\mathbf{v}^2 \end{aligned} \tag{5.7}$$

Since $\mathbf{v} = \boldsymbol{\omega} \times \mathbf{r}$ for a rigid body rotating with instantaneous angular velocity $\boldsymbol{\omega}$ and for a particle whose position vector is $\mathbf{r}$, the internal kinetic energy (that is, omitting $\frac{1}{2}M\mathbf{V}^2$) is

$$T = \tfrac{1}{2}\Sigma m(\boldsymbol{\omega} \times \mathbf{r})^2 = \tfrac{1}{2}\Sigma m[\mathbf{r}^2\boldsymbol{\omega}^2 - (\mathbf{r} \cdot \boldsymbol{\omega})^2] \tag{5.8}$$

If K_ω is the moment of inertia about the instantaneous axis of rotation, then

$$T = \tfrac{1}{2}K_\omega\omega^2 \tag{5.9}$$

and in terms of the angular momentum

$$T = \tfrac{1}{2}\mathbf{L} \cdot \boldsymbol{\omega} \tag{5.10}$$

and therefore the kinetic energy is expressible in terms of the angular velocity components as

$$T = \tfrac{1}{2}(A\omega_1^2 + B\omega_2^2 + C\omega_3^2 - 2F\omega_2\omega_3 - 2G\omega_3\omega_1 - 2H\omega_1\omega_2) \tag{5.11}$$

3. Equations of Motion

Let $\mathbf{F}_i$ be the external force acting on the ith particle and F_{ij} be the internal force on the ith particle arising from its interaction with the jth particle. By Newton's third law of motion,

$$\mathbf{F}_{ij} = -\mathbf{F}_{ji} \quad \text{and} \quad (\mathbf{r}_i - \mathbf{r}_j) \times \mathbf{F}_{ij} = 0 \tag{5.12}$$

that is, the forces are equal and opposite and act in the line joining the interacting particles.

The equation of motion of the ith particle is then

$$m_i(\dot{\mathbf{V}} + \dot{\mathbf{v}}_i) = \mathbf{F}_i + \sum_j \mathbf{F}_{ij}$$

Summing for all particles of the system,

$$M\dot{\mathbf{V}} = \mathbf{F}_i \tag{5.13}$$

so that the center of mass moves as if the internal forces were not acting and as if the external forces acted directly on the center of mass.

Multiplying each equation by $(\mathbf{R} + \mathbf{r}_i)$ and summing, one has

$$\frac{d}{dt}(\mathbf{L}_0 + \mathbf{L}) = \mathbf{R} \times \sum \mathbf{F}_i + \sum \mathbf{r}_i \times \mathbf{F}_i \tag{5.14}$$

where $\mathbf{L}_0 = M\mathbf{R} \times \mathbf{V}$. Now $d\mathbf{L}_0/dt$ follows from (5.13) by multiplying by $\mathbf{R}\times$; therefore the equation for the rate of change of the internal angular momentum is

$$\frac{d\mathbf{L}}{dt} = \sum \mathbf{r}_i \times \mathbf{F}_i$$

The quantity $\mathbf{r}_i \times \mathbf{F}_i$ is the *torque* due to the external force $\mathbf{F}_i$ on the ith particle calculated with respect to the center of mass.

If $(M_1\mathbf{l} + M_2\mathbf{m} + M_3\mathbf{n})$ is the resultant torque on the body, $(\Sigma\mathbf{r}_i \times \mathbf{F}_i)$, resolved along axes fixed in the body, then the equations of motion for the angular momentum components resolved along these axes are

$$\begin{aligned} \frac{dL_1}{dt} + (\omega_2L_3 - \omega_3L_2) &= M_1 \\ \frac{dL_2}{dt} + (\omega_3L_1 - \omega_1L_3) &= M_2 \\ \frac{dL_3}{dt} + (\omega_1L_2 - \omega_2L_1) &= M_3 \end{aligned} \tag{5.15}$$

These take their simplest form if the moving axes at all times coincide with the principal axes of inertia of the rigid body; therefore

$$L_1 = \omega_1A \qquad L_2 = \omega_2B \qquad L_3 = \omega_3C$$

Then the equations for the components of angular momentum become

$$\begin{aligned} \frac{dL_1}{dt} + (B^{-1} - C^{-1})L_2L_3 &= M_1 \\ \frac{dL_2}{dt} + (C^{-1} - A^{-1})L_3L_1 &= M_2 \\ \frac{dL_3}{dt} + (A^{-1} - B^{-1})L_1L_2 &= M_3 \end{aligned} \tag{5.16}$$

These are known as *Euler's equations of motion* for a rotating rigid body.

Sometimes these are used as equations for the components of angular velocity $(\omega_1,\omega_2,\omega_3)$ with regard to axes parallel to the principal axes of inertia of the body. Since, by (5.5), $L_1 = A\omega_1$, etc., these become

$$\begin{aligned} A\dot{\omega}_1 + (C - B)\omega_2\omega_3 &= M_1 \\ B\dot{\omega}_2 + (A - C)\omega_3\omega_1 &= M_2 \\ C\dot{\omega}_3 + (B - A)\omega_1\omega_2 &= M_3 \end{aligned} \tag{5.17}$$

These are also called Euler's equations.

4. Rotation about a Fixed Axis

The simplest case of rotation of a rigid body is that of rotation about an axis that is fixed in space and fixed in the body. In general the axis will neither pass through the center of mass nor be one of the principal axes of inertia.

Let (x,y,z) be the coordinates of a particle with regard to a coordinate system fixed in the body whose origin is on the axis of rotation and whose z axis is on the axis of rotation, and x', y', z' be the coordinates with regard to a coordinate system fixed in space such that the instantaneous positions of the two systems coincide at $t = 0$; then

$$\begin{aligned} x' &= x \cos \omega t - y \sin \omega t \\ y' &= x \sin \omega t + y \cos \omega t \\ z' &= z \end{aligned}$$

so that the velocity of such a particle with regard to fixed axes is

$$\begin{aligned} v'_x &= -\omega(x \sin \omega t + y \cos \omega t) \\ v'_y &= +\omega(x \cos \omega t - y \sin \omega t) \\ v'_z &= 0 \end{aligned}$$

and the acceleration of such a particle is

$$\begin{aligned} a'_x &= -\omega^2(x \cos \omega t - y \sin \omega t) \\ a'_y &= -\omega^2(x \sin \omega t + y \cos \omega t) \\ a'_z &= 0 \end{aligned}$$

The total *force* which needs to act on the body to keep it rotating is the sum of mass times acceleration for each particle:

$$\begin{aligned} F'_x &= -\omega^2[(\Sigma mx) \cos \omega t - (\Sigma my) \sin \omega t] \\ F'_y &= -\omega^2[(\Sigma mx) \sin \omega t + (\Sigma my) \cos \omega t] \\ F'_z &= 0 \end{aligned}$$

This force will be supplied by forces of reaction which develop at the bearings which hold the axis of rotation fixed in space. It follows that these forces of reaction vanish if the axis of rotation passes through the body's center of mass. If this is the case, the body is said to be *statically balanced*, the term arising from the fact that if the axis has a horizontal component then the body will be in neutral equilibrium with regard to the action of gravity, which affords a convenient static way of adjusting the mass distribution of a piece of rotating machinery so as to put the center of mass on the axis and avoid these reactions of unbalance on the bearings.

Let $\mathbf{l}$, $\mathbf{m}$, $\mathbf{n}$ be the unit vectors of the coordinate system fixed in the body. Then the angular momentum $\mathbf{L}$ if the angular velocity is $\omega\mathbf{n}$ is, from (5.5),

$$\mathbf{L} = [(-G\mathbf{l} - F\mathbf{m}) + C\mathbf{n}]\omega$$

The component $C\omega\mathbf{n}$ is also constant in space, but the off-axis component $-\omega(G\mathbf{l} + F\mathbf{m})$ varies with time because $\mathbf{l}$ and $\mathbf{m}$ rotate with the body.

The bearings therefore must supply a *torque* of reaction equal to the time rate of change of $\mathbf{L}$ which is $\boldsymbol{\omega} \times \mathbf{L}$ or

$$\boldsymbol{\omega} \times \mathbf{L} = \omega^2(F\mathbf{l} - G\mathbf{m})$$

The magnitude of this remains constant but its direction in space rotates with the changing direction of the vectors $\mathbf{l}$ and $\mathbf{m}$. If the axis of rotation is fixed by means of bearings at two ends of a shaft, the torque will be supplied by means of forces of reaction provided at those bearings. These reactions are called *reactions of dynamic unbalance*. In the case of static unbalance the forces at the bearings at opposite ends of the shaft are in the same direction at each instant, whereas in the case of dynamic unbalance the forces are in opposite directions at opposite ends of the shaft.

The forces due to dynamic unbalance vanish if the axis of rotation is a principal axis of inertia of the body so that the products of inertia F and G vanish. To avoid excessive reaction on the bearings, the rotating parts in high-speed machinery have to be dynamically balanced as well as statically balanced. The design of the rotating part is made as symmetric as possible. Then the piece is put in a balancing machine in which it is rotated in bearings which are equipped with electric transducers to measure the varying forces of reaction at the bearings. Some such machines are fully automatic, using the transducer output to solve for the amount of the unbalance and feeding the information to an automatic drill which removes by drilling the right amount of metal to leave the piece both statically and dynamically balanced.

5. Rotation about a Fixed Point with No External Forces

If the fixed point is not the center of mass, then the inertia tensor has to be calculated with regard to axes having the fixed point as origin [sum of (5.3) and (5.4)], but this does not alter the form of the Euler equations of motion, (5.16) and (5.17). Of course when the fixed point is not the center of mass there will be a dynamic reaction to be supplied by the point of support as discussed in Sec. 4.

If no external forces act, then the equations of motion are (5.16) with $(M_1,M_2,M_3) = 0$. Multiplying by L_1, L_2, and L_3 and adding, a first integral is obtained which says that

$$\mathbf{L}^2 = \text{const} \tag{5.18}$$

Similarly on multiplying by $A^{-1}L_1$, $B^{-1}L_2$, and $C^{-1}L_3$ and adding, another first integral is obtained expressing the constancy of total kinetic energy,

$$T = \tfrac{1}{2}(A^{-1}L^2_1 + B^{-1}L^2_2 + C^{-1}L^2_3) \tag{5.19}$$

The individual components of angular momentum referred to axes fixed in space must also be constant since no external torque acts.

In the coordinates fixed in the body, using L_1, L_2, and L_3 as coordinates of a point, the integral (5.19) represents an ellipsoid whose principal semiaxes are $(2AT)^{1/2}$, $(2BT)^{1/2}$, and $(2CT)^{1/2}$. The equation expressing constancy of $\mathbf{L}^2$ in terms of angular momentum components is a sphere of radius L. The variation of $\mathbf{L}$ in the body must therefore be along a curve represented by the intersection of the ellipsoid and the sphere (Fig. 5.1). This makes possible a visualization of the main qualitative properties of the motion. The construction also shows that $\mathbf{L}$ must lie between the smallest and the largest semiaxes of the ellipsoid. Otherwise the sphere does not intersect the ellipsoid; thus (5.18) and (5.19) cannot be simultaneously satisfied.

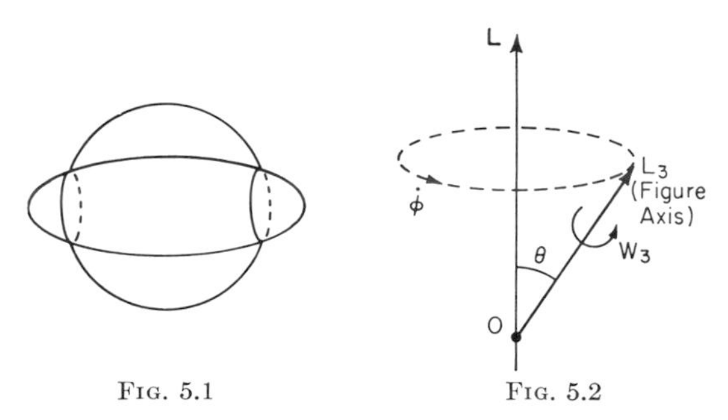

FIG. 5.1 FIG. 5.2

Symmetrical Top. If two of the principal moments of inertia are equal, say $A = B$, then the C axis is an axis of symmetry and the rotating body is called a symmetrical top. The exact solution is easily obtained. The equation for $\dot{L}_3$ shows that this component is constant.

This fact makes the equations for L_1 and L_2 easily solvable. Let $\lambda = (A^{-1} - C^{-1})$ and $\mathfrak{L} = (\mathbf{L}^2 - L_3^2)^{1/2}$ be the constant magnitude of the angular momentum vector perpendicular to the symmetry axis. Then the equations of motion give

$$L_1 = \mathfrak{L} \cos \lambda L_3 t$$
$$L_2 = \mathfrak{L} \sin \lambda L_3 t$$

Choose now a system of axes fixed in space whose $\mathbf{k}$ axis is along the direction of the constant resultant angular momentum $\mathbf{L}$. Suppose the moving axes $(\mathbf{l},\mathbf{m},\mathbf{n})$ are specified by Eulerian angles (θ,ϕ,ψ). The component of angular momentum along the figure axis is $L_3 = L \cos \theta$ and since both L_3 and L are constant, it follows that θ is constant, so that the figure axis of the body preserves a constant angle to the vector of total angular momentum.

Similarly, $L_1 = L \sin \theta \sin \psi$, $L_2 = L \sin \theta \cos \psi$ from which

$$\psi = \frac{\pi}{2} - \lambda L_3 t$$

so that the rate of change of the Eulerian angle ψ is constant and equal to

$$\dot{\psi} = \left(\frac{1}{C} - \frac{1}{A}\right) L_3$$

Since the angular velocity about the figure axis is $\omega_3 = L_3/C = \dot{\psi} + \dot{\phi} \cos \theta$, it follows that $\dot{\phi}$ is also constant and equal to

$$\dot{\phi} = \frac{L}{A}$$

The geometrical situation is shown in Fig. 5.2. The total angular momentum $\mathbf{L}$ is fixed in space. The body turns about its figure axis with an angular velocity ω_3 and the figure axis in turn precesses around $\mathbf{L}$ with the angular velocity $\dot{\phi}$. Although the sense of rotation of ω_3 and $\dot{\phi}$ are related to the resultant angular momentum $\mathbf{L}$ and the position of the figure axis as indicated, the sense of rotation for $\dot{\psi}$ depends on whether the body is prolate $(C < A)$ or oblate $(C > A)$.

6. Asymmetrical Top

If the three principal moments of inertia are unequal, the body is called an asymmetrical top. For convenience suppose $A < B < C$.

The kinetic energy ellipsoid is fixed in the body. The angular momentum vector is fixed in space. The body moves in such a way that the end point of $\mathbf{L}$ always remains in the ellipsoid, tracing a wavy curve in its surface, which reduces to a circle in the case of a symmetrical top (Fig. 5.3).

Suppose $L > \sqrt{2BT}$. The motion of the end of $\mathbf{L}$ along the ellipsoid is confined to the region near either end, as indicated in Fig. 5.3. The possible variation of the ellipsoid relative to $\mathbf{L}$ is along one or the other of two separated closed curves, and the initial conditions will determine along which of them the actual motion takes place.

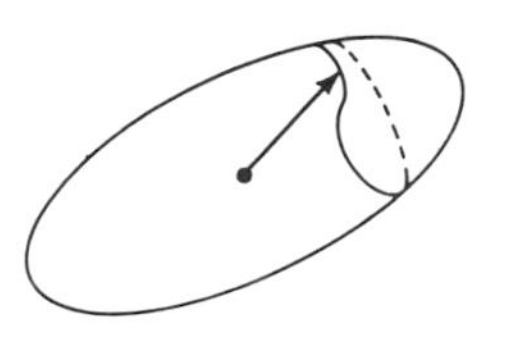

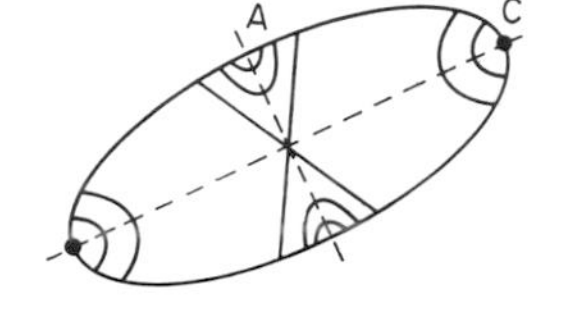

FIG. 5.3 FIG. 5.4

The family of possible curves for various values of L for a fixed value of the total energy are indicated qualitatively in Fig. 5.4. Here it is supposed that the axis for the intermediate value of the moment of inertia B is perpendicular to the paper. For L nearly equal to (a little less than) $\sqrt{2CT}$ or nearly equal to (a little more than) $\sqrt{2AT}$, the curves become small closed loops around the A and C axes, respectively.

This indicates that *stable rotation* can occur about these axes in which the angular momentum, if initially close to one or the other of these axes in the body, remains close to it throughout the motion.

However, rotation initially about an axis near to the intermediate moment of inertia is unstable. The curves for the value $L = \sqrt{2BT}$ do not remain close to this axis, but loop around the ellipsoid. This indicates that if the angular momentum is slightly displaced from the intermediate axis, the axis of rotation will move from this axis in the ensuing motion.

The third of the equations of motion gives

$$\dot{L}^2_3 = \left(\frac{1}{A} - \frac{1}{B}\right)^2 L^2_1 L^2_2$$

By use of the two integrals

$$L^2{}_1 + L^2{}_2 + L^2{}_3 = L^2$$

and

$$\frac{L^2{}_1}{A} + \frac{L^2{}_2}{B} + \frac{L^2{}_3}{C} = 2T$$

one can eliminate L_1 and L_2 obtaining

$$\dot{L}^2{}_3 = \left[\left(\frac{L_2}{A} - 2T\right) - \left(\frac{1}{A} - \frac{1}{C}\right) L^2{}_3\right] \left[\left(2T - \frac{L^2}{B}\right) + \left(\frac{1}{B} - \frac{1}{C}\right) L^2{}_3\right]$$

This equation shows that L_3 is an elliptic function of the time. It carries out a libration in values between the roots of the right-hand side of the equation.

References

1. Whittaker, E. T.: "Analytical Dynamics," chap. 6, Cambridge University Press, New York, 1927.
2. Webster, A. G.: "The Dynamics of Particles and of Rigid, Elastic and Fluid Bodies," chaps. 6, 7, Stechert, New York, 1942.
3. Klein, F., and A. Sommerfeld: "Theorie des Kreisels," 4 vols., Leipzig, 1897 to 1910.
4. Gray, A.: "A Treatise on Gyrostatics and Rotational Motion," Macmillan, London, 1918. Reprint: Dover, New York, 1959.
5. MacMillan, W. D.: "Dynamics of Rigid Bodies," McGraw-Hill, New York, 1936.

Chapter 6

The Theory of Relativity

By E. L. HILL, University of Minnesota

1. Introduction

The term *theory of relativity* is used to denote the two theories advanced by Albert Einstein (1879–1955) in 1905 [1]† and in 1916 [2]. Einstein's concepts were formulated as an extension of classical mechanics and electromagnetic-field theory. Their connections with quantum mechanics and the modern theory of quantized fields unquestionably are fundamental but as yet are subject to major uncertainties both of physical interpretation and of mathematical formulation [3, 4].

The *special theory of relativity* is concerned with phenomena in which the gravitational field is absent or plays only a minor role. The *general theory of relativity* deals with gravitation and its influence on other physical phenomena. The special theory has been indispensable in the formulation of a theoretical basis for modern high-energy physics, while to date the general theory has found its major applications in the study of astronomical and cosmological phenomena.

The theory arose from the study of certain incompatibilities that exist between the theories of kinematics associated with classical mechanics and with the electromagnetic field and optical theory. Limitations of space do not admit discussion of the historical background here but excellent treatments are readily accessible in the literature [5–8].

2. Space and Time in Classical Physics

In the theory of mechanics proposed by Isaac Newton (1642–1727) physical forces are supposed to be uniquely determinate and to be clearly distinguishable from the kinetic reactions ("centrifugal force," "Coriolis force," etc.) that arise from the use of moving reference systems. In applications of Newton's third law of motion, it is assumed that all forces representing actions at a distance between particles, such as the force of gravitation, depend only on the relative positions of the particles in space at each instant of time.

The kinematic assumptions of Newton's theory are adapted to this view of the nature of force. In particular, it is essential to Newton's third law that time must have a universal character, the same for all observers however they may be moving. This interpretation of time leads at once to the result that, while absolute motion may not have a clear significance relative motion is defined uniquely by the vector law of addition of velocities. Any given object can be reduced to apparent permanent rest by the introduction of a co-moving reference system attached to it [8*a*].

The forces that arise in the study of electromagnetic fields and their actions on electrically charged matter and current systems do not obey Newton's third law of motion. When at rest electric charges give rise to electrostatic fields but when in motion they produce combined electric and magnetic fields. Similarly, steady electric currents generate magnetic fields, but when they are oscillatory, or when they are set in motion, they give rise to magnetic and electric fields.

More importantly, transient electric and magnetic fields are propagated in free space with the speed of light, as was first shown on theoretical grounds by J. C. Maxwell (1831–1879) in 1864. It follows that, if the principles of conservation of energy, momentum, and angular momentum, which are derivable from Newton's third law for systems of uncharged mass particles, are to hold when electrodynamic forces come into play, the electromagnetic field must be supposed to act as a carrier for these quantities. The discovery of the dynamic nature of the electromagnetic field was a major turning point in the history of physical theory [8].

The failure of many attempts to detect an absolute motion of the earth in space by various optical and electromagnetic experiments led H. Poincaré (1854–1912) in 1904 to suggest that a new system of mechanics might be developed on the principle of invariance of the speed of light. The theory thus adumbrated by Poincaré was worked out independently by Einstein as a theory of the electrodynamic properties of moving media [1, 9].

3. The Minkowski Space-Time Manifold

The major mathematical accomplishment of Einstein's special theory of relativity was the construction of a system of kinematics that is compatible with the principle of invariance of the speed of light in free space. Einstein accompanied his formal derivation of the transformation equations to moving reference systems with a detailed examination of the new interpretation of space and time that it implied. The principal idea used in his discussion was the measurement of space and time quantities by means of light signals. This specific formulation of Einstein's argu-

† Numbers in brackets refer to References at end of chapter.

ment has been the subject of much discussion in the literature [10, 11].

As a formal basis for his analysis, Einstein took as fundamental the combination of (1) *a point of space* and (2) *an instant of time,* which he called an *event.* In his view, the primary result of empirical observation is the establishment of causal relationships among physical events which can be described as occurring at given points in space and time. The set of all possible Einsteinian events constitute the four-dimensional space-time manifold.

In his first exposition of the special theory Einstein used the concept of space-time primarily as an economical way of describing real physical phenomena as they occur in space and time. It was emphasized by H. Minkowski (1864–1909) [9, 12] that the axiomatic statement of the theory can be simplified materially if one considers the space-time manifold itself to be a definite entity having the geometrical character of a pseudo-Riemannian manifold with a line element of the special form (*Minkowski line element*)

$$\phi_0 \equiv c^2(dt)^2 - (dx)^2 - (dy)^2 - (dz)^2 \qquad (6.1)$$

where c is the speed of light in free space. Since the quadratic form (6.1) is not positive-definite it does not constitute a Riemannian metric in the strict sense [13]. It is called a *pseudo-metric* or simply a *line element.* When the space-time manifold is assumed to have one coordinate system in which the line element everywhere has the Minkowski form (6.1) it is referred to as *Minkowski space-time.*

Minkowski's interpretation of the special theory of relativity was accepted by Einstein, after some initial hesitation, and later was made the starting point of his general theory of relativity. In this way the theory of relativity as a whole has come to be understood as a study of the geometry of the space-time manifold and its influence on physical phenomena.

In a strict sense, Minkowski's introduction of a geometrical structure of physical character for the space-time manifold is not demanded by the empirical evidence which underlies the special theory of relativity [14]. However, since it simplifies the statement of the theory, particularly in the development of a relativistic theory of particle mechanics, and since it also prepares the way for the study of the general theory, it has been adopted widely by theoretical physicists as the most suitable axiomatic starting point for the theory. In the interest of brevity and clarity of statement we adopt Minkowski's formulation as the guiding principle of our discussion (cf. Sec. 13).

It is convenient for mathematical purposes to introduce the notation

$$x^0 = ct \qquad x^1 = x \qquad x^2 = y \qquad x^3 = z \qquad (6.2)$$

with which the Minkowski line element can be written in the condensed form

$$\phi_0 \equiv \eta_{\lambda\mu}\, dx^\lambda\, dx^\mu \qquad (6.3)$$

Here

$$\eta_{00} = +1 \qquad \eta_{11} = \eta_{22} = \eta_{33} = -1 \qquad \eta_{\lambda\mu} = 0 \text{ if } \mu \neq \lambda \qquad (6.4)$$

with the matrix

$$\eta = \left[\begin{array}{c|ccc} 1 & 0 & 0 & 0 \\ \hline 0 & -1 & 0 & 0 \\ 0 & 0 & -1 & 0 \\ 0 & 0 & 0 & -1 \end{array}\right] \qquad (6.5)$$

Here and subsequently Greek letters are used to indicate indices running over the values 0, 1, 2, 3 and Latin letters for values 1, 2, 3. When a letter is repeated as a subscript and a superscript it is to be summed over the appropriate values (Einstein summation convention).

4. The Einstein Space-Time Manifold

In developing his general theory of relativity, Einstein assumed that in the presence of a gravitational field the space-time manifold has a geometrical structure that is characterized by a pseudo-Riemannian line element of the general form

$$\phi(x) \equiv g_{\lambda\mu}(x)\, dx^\lambda\, dx^\mu \qquad (6.6)$$

where the 10 functions $\{g_{\lambda\mu}(x) = g_{\mu\lambda}(x)\}$, called the *metric functions,* depend on the distribution of matter and radiation in the universe. The metric functions play the role of potential functions for the gravitational field resulting from the physical action of the matter and radiation.

It is assumed further that at every point of space-time there is a choice of coordinate system that reduces (6.6) to the Minkowski form. This is the mathematical formulation of the *equivalence principle.* It is not necessary that, even in the absence of an external gravitational field, there exist a reference frame such that (6.6) takes the Minkowski form in the large. The establishment of a suitable set of differential equations and boundary conditions for the metric functions is the first major problem of the theory (Sec. 30).

The introduction of the line element (6.6) for the space-time manifold permits the use of the analytical machinery of Riemannian geometry and its associated tensor analysis. The reader is referred to Part 1, Chap. 10, and to the extensive literature on tensor calculus for the required material [14*a*].

5. Einstein's Principle of Causality

The assumption that relations of cause and effect exist among the observed events of the material universe imposes certain restrictions on the forms into which physical theories can be cast. Among the most demanding of these conditions is the dictum that *events must not precede their causes in time.*

It is logically conceivable that causal relations might hold throughout all space at each instant of time. Relations of this type, of which Newton's law of gravitation is the most important example, are said to be propagated with infinite speed.

The theory of relativity is based on the assumption that all causal physical influences travel with finite speed. Stated more completely, this principle takes the following form:

1. *At each point of space-time there exists a finite*

upper bound for the local speeds of propagation of all physical influences that are capable of producing causal results.

2. *The upper limiting speed is attained by the local speed of propagation of light.*

A mathematical formulation of this principle is stated later (Sec. 23). In the special theory of relativity it is assumed that the speed of light in free space is an absolute constant, while in the general theory it is admitted that a gravitational field can influence the local value of the speed of light. In every case the local speed of light has the significance of being maximal for the speeds of propagation of all causal influences.

6. Moving Reference Systems

Much of the mathematical argumentation associated with the theory of relativity is concerned with transformations of reference systems. This feature of the discussion not infrequently leads to serious misstatements concerning the meaning of empirical results since such transformations never are actually carried out in the course of experimental measurements. As a consequence, it sometimes is difficult to arrive at definitive interpretations of empirical observations that are free from the theoretical ambiguities involved in the definition of moving reference systems.

This source of misunderstanding becomes most severe when it is desired to think of a co-moving reference system that is attached to an object observed to be moving in some general manner. *In the special theory of relativity this problem has an agreed solution only for objects that are observed to move with uniform velocity in the standard reference system.* In the general theory the choice of reference systems that is allowed is so great that there is no generally accepted solution, and each case must be discussed on its own merits.

This interpretational ambiguity is compounded when the theory is applied to quantum-mechanical systems. It is a fundamental tenet of quantum mechanics that it is impossible to use the methods of classical kinematics for such particles as the electron, proton, meson, etc., although it usually is argued also that when these particles are combined into macroscopic matter the procedures of classical kinematics become applicable. However, the considerations that underlie this assertion are nonrelativistic in nature and do not follow logically from the principles of quantum mechanics, so that they are subject to serious question [15, 16]. There seems to be no escape from the conclusion that any combination of nonlocalizable particles itself, in principle, is nonlocalizable.

The consequence is that in relativistic forms of quantum mechanics the concept of moving reference systems is restricted to the role of reference frames for observers who themselves stand outside the physical context of the problem but the concept cannot be used in the sense of classical kinematic theory for the definition of co-moving reference frames attached to arbitrarily moving physical objects. In view of this difficulty, it is best to minimize, so far as is possible, the introduction of arguments that make essential use of the concept of co-moving reference systems in the interpretation of experimental data [8*a*].

7. Physical Quantities as Tensors

The adoption of the space-time manifold as the basic mathematical structure of the theory of relativity permits the use of four-dimensional tensor analysis as the fundamental tool in the construction of differential equations and other symbolic relations. Its use in physical terms depends on the ability to identify physical quantities as tensors. In practice this identification is accomplished by reformulation of the equations of classical mathematical physics to a suitable form for use in the special theory, with subsequent extension to the general theory. Any ambiguity arising in this process must be adjudicated on grounds of physical analysis [6, 17, 18, 18*a*].

THE SPECIAL THEORY OF RELATIVITY

8. The Inhomogeneous Lorentz Group

The kinematical structure of the special theory of relativity is based on the condition that under allowable transformations of the space-time frame from the standard reference system the Minkowski line element is invariant both in numerical magnitude and in algebraic form; i.e.,

$$\eta_{\lambda\mu}\, dx'^{\lambda}\, dx'^{\mu} = \eta_{\alpha\beta}\, dx^{\alpha}\, dx^{\beta}$$

Expressed more fully, it is required that

$$c^2(dt')^2 - (dx')^2 - (dy')^2 - (dz')^2 = c^2(dt)^2 - (dx)^2 - (dy)^2 - (dz)^2 \quad (6.7)$$

This is called *the condition of form invariance of the Minkowski line element.* It is included in, but is more restrictive than, the condition of invariance of the speed of light, which demands only that

$$\begin{aligned}(dx'/dt')^2 + (dy'/dt')^2 + (dz'/dt')^2 &= (dx/dt)^2 + (dy/dt)^2 + (dz/dt)^2 \\ &= c^2 \end{aligned} \quad (6.8)$$

The group of transformations that satisfy (6.7) is called the *inhomogeneous Lorentz group* $\mathfrak{L}^*$. It consists of the following generating subgroups:

1. T_4: *translations in space-time* (4-*parameters*)

$$x'^{\lambda} = x^{\lambda} + a^{\lambda} \quad (6.9)$$

2. $\mathfrak{L}$: *homogeneous Lorentz transformations* (6-*parameters*)

$$x'^{\lambda} = \ell^{\lambda}_{\mu} x^{\mu} \quad (6.10)$$

where the real 4×4 matrix $L = [\ell^{\lambda}_{\mu}]$ satisfies the condition

$$L^{-1} = \eta L^t \eta \quad (6.11)$$

Here L^t and L^{-1} are the transpose and inverse matrices of L, while η is the Minkowski matrix (6.5). Matrices of this type are called *Lorentz matrices.*

Analysis of (6.11) shows that for any Lorentz matrix

$$\begin{aligned}(\det L)^2 &= 1 \\ (\ell^0_0)^2 - (\ell^0_1)^2 - (\ell^0_2)^2 - (\ell^0_3)^2 &= 1\end{aligned} \quad (6.12)$$

from which it follows that

$$\det(L) = \pm 1 \qquad (\ell^0_0)^2 \geq 1 \quad (6.13)$$

These results lead to a classification of Lorentz matrices which is shown in the following table [19, 20]:

		det(L)	sgn (ℓ_0^0)	Cosets of $\mathcal{E}$
(a) Einstein group, $\mathcal{E}$	$\mathcal{E}'$, $\mathcal{L}_+$, $\mathcal{L}$	+	+	$\mathcal{E}$
(b)	$\mathcal{E}'$, $\mathcal{L}$	−	+	$\eta\mathcal{E}$
(c)	$\mathcal{L}_+$, $\mathcal{L}$	+	−	$-\mathcal{E}$
(d)	$\mathcal{L}$	−	−	$-\eta\mathcal{E}$

The Lorentz matrices for which $\det(L) = +1$, $\ell_0^0 \geq 1$ form an invariant subgroup of $\mathfrak{L}$ which is called the *homogeneous Einstein group* $\mathcal{E}$. When the translation group T_4 is adjoined to $\mathcal{E}$ it becomes the *inhomogeneous Einstein group* $\mathcal{E}^*$. This group provides the kinematical basis of the original formulation of the special theory of relativity.

The Minkowski matrix η is a Lorentz matrix which defines the *space-inversion transformation*

$$x' = -x \qquad y' = -y \qquad z' = -z \qquad t' = t \tag{6.14}$$

By adjoining this transformation to the Einstein group one obtains the *extended Einstein group* $\mathcal{E}'$, which consists of all Lorentz matrices for which $\ell_0^0 \geq 1$.

The family of Lorentz matrices for which

$$\det(L) = +1$$

is called the *improper Lorentz group* $\mathfrak{L}_+$.

The Lorentz matrix $-\eta$ defines the *time-inversion transformation*

$$x' = x \qquad y' = y \qquad z' = z \qquad t' = -t \tag{6.15}$$

By adjoining this transformation to either $\mathcal{E}'$ or $\mathfrak{L}_+$ one obtains the full homogeneous Lorentz group $\mathfrak{L}$.

While the kinematical theory underlying Einstein's formulation of the special theory of relativity is provided by the inhomogeneous Einstein group $\mathcal{E}^*$, the formulation of current quantum-mechanical field theory requires use of all transformations of tne inhomogeneous Lorentz group $\mathfrak{L}^*$.

9. The Einstein Group

The particular features of the special theory of relativity which distinguish it from classical theory arise from the use of the homogeneous Einstein group of transformations for the definition of moving reference systems. Every transformation matrix L_e of the Einstein group can be expressed as a product of space rotations and special transformations to moving frames. These can be regarded symbolically as transformations in various coordinate planes of the Minkowski space-time manifold.

The general matrix form for space rotations is

$$L(R) = \left[\begin{array}{c|ccc} 1 & 0 & 0 & 0 \\ \hline 0 & & & \\ 0 & & R & \\ 0 & & & \end{array}\right] \tag{6.16}$$

where R is an arbitrary 3×3 real unimodular orthogonal matrix $[R^{-1} = R^t, \det(R) = +1]$.

As a typical case of the special transformations to moving reference systems we consider a frame S' which moves along the positive z axis of the standard frame S with speed V. The appropriate transformation from S to S' is ($\beta = V/c$)

$$x' = x \qquad y' = y \qquad z' = \frac{z - \beta ct}{(1 - \beta^2)^{1/2}}$$
$$ct' = \frac{ct - \beta z}{(1 - \beta^2)^{1/2}} \tag{6.17}$$†

with the inverse transformation from S' to S

$$x = x' \qquad y = y' \qquad z = \frac{z' + \beta ct'}{(1 - \beta^2)^{1/2}}$$
$$ct = \frac{ct' + \beta z'}{(1 - \beta^2)^{1/2}} \tag{6.18}$$

The matrix corresponding to transformation (6.17) is

$$L_{(03)}(\beta) = \left[\begin{array}{c|ccc} \gamma & 0 & 0 & -\beta\gamma \\ \hline 0 & 1 & 0 & 0 \\ 0 & 0 & 1 & 0 \\ -\beta\gamma & 0 & 0 & \gamma \end{array}\right]$$
$$\gamma = 1/(1 - \beta^2)^{1/2} \tag{6.19}$$

where the subscript (03) identifies the z axis as the distinguished direction of motion of S' with respect to S. Comparison with (6.18) shows that

$$[L_{(03)}(\beta)]^{-1} = L_{(03)}(-\beta) \tag{6.20}$$

Every transformation of the homogeneous Einstein group $\mathcal{E}$ can be factorized in the form

$$L_e = L(R') \cdot L_{(03)}(\beta) \cdot L(R'') \tag{6.21}$$

where $L(R')$ and $L(R'')$ are suitably chosen space rotations. This shows that for the analysis of the physical content of the special theory of relativity it is sufficient to study the special Lorentz transformations of type (6.19) together with the space rotations.

Frequently it is convenient to have available the transformation equations to a system S' which moves in an arbitrary direction with velocity $\mathbf{V}$ with respect to the standard system S. In the following formulas it is assumed that the space axes of S and S' are similarly oriented and that the origins of the two systems coincide at the instant $t = t' = 0$. The formulas as given are not true vector relations but are only a condensed form of the transformation equation (Fig. 6.1)

$$\begin{aligned} \mathbf{r}' &= (\gamma - 1)(\mathbf{r} \times \mathbf{n}_0) \times \mathbf{n}_0 + \gamma(\mathbf{r} - \mathbf{n}_0 Vt) \\ ct' &= \gamma(ct - (\mathbf{r} \cdot \mathbf{n}_0)V/c) \end{aligned} \tag{6.22}$$

with
$$\mathbf{n}_0 = \mathbf{V}/V$$

Any reference system which can be obtained from the standard frame S by a transformation of the inhomogeneous Einstein group $\mathcal{E}^*$ is called an *Einstein reference system*.

† We follow here current practice in quantum-mechanical theory by using the z axis as the axis of space symmetry.

10. The Addition Law for Velocities

Let

$$\mathbf{u} = d\mathbf{r}/dt \qquad \mathbf{u}' = d\mathbf{r}'/dt' \tag{6.23}$$

be the apparent velocity vectors of a moving particle in the standard frame S and in the moving frame S' defined by (6.22). By differentiation of (6.22) the transformation law for velocities is found to be

$$\mathbf{u}' = \frac{(\gamma - 1)(\mathbf{u} \times \mathbf{n}_0) \times \mathbf{n}_0 + \gamma(\mathbf{u} - \mathbf{V})}{\gamma(1 - \mathbf{u} \cdot \mathbf{V}/c^2)} \tag{6.24}$$

If two particles are moving with velocities $\mathbf{u}_1$ and $\mathbf{u}_2$ in S, the relative velocity $\mathbf{u}_{12}$ of particle 1 with respect to particle 2 is defined to be the velocity of the former in any Einstein frame in which the latter is at rest. It can be found by substitution of the formulas

$$\mathbf{u} = \mathbf{u}_1 \qquad \mathbf{n}_0 V = \mathbf{u}_2 \qquad \gamma_2 = \frac{1}{(1 - u_2{}^2/c^2)^{1/2}}$$

in (6.24), from which we find

$$\mathbf{u}_{12} = \frac{(\gamma_2 - 1)(\mathbf{u}_1 \times \mathbf{u}_2/u_2) \times \mathbf{u}_2/u_2 + \gamma_2(\mathbf{u}_1 - \mathbf{u}_2)}{\gamma_2(1 - \mathbf{u}_1 \cdot \mathbf{u}_2/c^2)} \tag{6.25}$$

It is apparent that this is not symmetric in the two particles, so that in general $\mathbf{u}_{21} \neq \mathbf{u}_{12}$. Calculation of the magnitude of the vector (6.25) shows that

$$|\mathbf{u}_{12}| = |\mathbf{u}_{21}| = \frac{(|\mathbf{u}_1 - \mathbf{u}_2|^2 - |\mathbf{u}_1 \times \mathbf{u}_2|^2/c^2)^{1/2}}{(1 - \mathbf{u}_1 \cdot \mathbf{u}_2/c^2)} \tag{6.26}$$

It follows from (6.26) that, if $|\mathbf{u}_1|$, $|\mathbf{u}_2| < c$, then $|\mathbf{u}_{12}| < c$ for any orientation of the velocity vectors of the particles.

11. FitzGerald-Lorentz Contraction of Moving Bodies

Consider a body of any form which is permanently at rest in the Einstein frame S' defined by (6.17). Let its surface be described by a functional equation $F(x',y',z') = 0$. On use of (6.17) it is seen that the apparent form of this surface, as referred to S, is

$$F\left(x, y, \frac{z - Vt}{(1 - V^2/c^2)^{1/2}}\right) = 0 \tag{6.27}$$

The form of this equation indicates that in S the body appears to be transported with uniform speed V and to be contracted along the direction of its motion by a factor $(1 - V^2/c^2)^{1/2}$. For example, if it is a sphere of radius a, with center at O' as seen in S', the equation of its surface in S is

$$\frac{x^2}{a^2} + \frac{y^2}{a^2} + \frac{(z - Vt)^2}{a^2(1 - V^2/c^2)} = 1 \tag{6.28}$$

which is the equation of an oblate spheroid of semi-axes $[a, a, a(1 - V^2/c^2)^{1/2}]$, with its center at the moving point O'.

This apparent foreshortening of moving objects in the direction of motion is known as the *FitzGerald-Lorentz contraction*. It was suggested independently by G. F. FitzGerald (1851–1901) and by H. A. Lorentz (1853–1928) prior to the development of Einstein's theory as an explanation of the null result of the Michelson-Morley experiment.

It should be noted that our discussion does not imply that this contraction can be observed directly by examination of light emitted from, or reflected by, a moving body. For this purpose it is necessary to take into account the aberration and Doppler effect of the light [21, 22].

12. The Time Dilatation

A second novel result which follows from the relativistic transformation formulas is the *time-dilatation effect*. If two events take place at a fixed point (x',y',z') in S' but at different times t'_1, $t'_2 > t'_1$, it follows from (6.18) that the times at which these events appear to occur in S are

$$t_1 = \frac{t'_1 + Vz'/c^2}{(1 - V^2/c^2)^{1/2}} \qquad t_2 = \frac{t'_2 + Vz'/c^2}{(1 - V^2/c^2)^{1/2}}$$

from which it is found that the apparent time lapses between them, as related to the two reference frames, are connected by the formula

$$t_2 - t_1 = \frac{t'_2 - t'_1}{(1 - V^2/c^2)^{1/2}} \geq t'_2 - t'_1 \tag{6.29}$$

This result shows that the time interval between the events appears to be *longer*, as measured in system

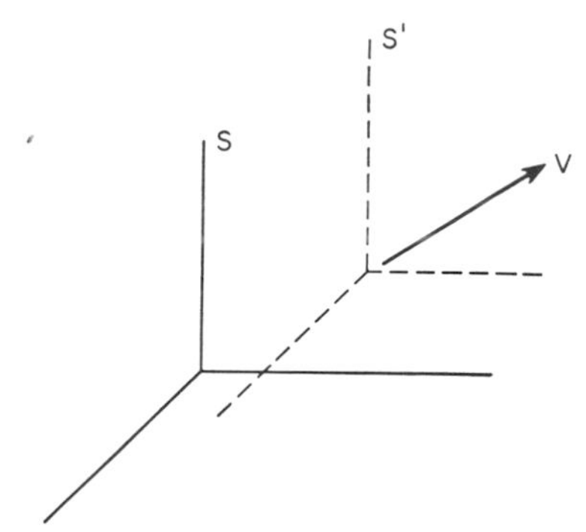

FIG. 6.1. Geometric relation of the moving system s' to the standard system s.

S, than it is in the moving system S'. This result is often described by the phrase *moving clocks run slowly*. It is evident that in the theory of relativity the universality of time for all observers fails to hold.

13. Einstein's Kinematical Theory

These results as derived are of restricted generality since the relative motion of the two reference frames S and S' must be unaccelerated. An object at rest in one system can then never be at rest in the other. Einstein [1, 9] proposed a generalization of the theory which aims at lifting this restriction to uniform motion.

For this purpose we observe first that, if a particle has a motion in system S which is such that its speed always is less than the speed of light, it is possible to find an infinity of uniformly moving systems in which the particle will appear to be at rest at one particular instant. For example, if in S at time t_0 the particle is at the space point $\mathbf{r}_0 = x_0\mathbf{i} + y_0\mathbf{j} + z_0\mathbf{k}$ and has velocity $\mathbf{V}_0$, use of (6.22) shows that we can find an Einstein frame S' such that

$$\mathbf{r}' = (\gamma_0 - 1)[(\mathbf{r} - \mathbf{r}_0) \times \mathbf{n}_0] \times \mathbf{n}_0 + \gamma_0[(\mathbf{r} - \mathbf{r}_0) - \mathbf{V}_0(t - t_0)]$$

$$ct' = \gamma_0[c(t - t_0) - (\mathbf{r} - \mathbf{r}_0) \cdot \mathbf{V}_0/c]$$

$$\mathbf{n}_0 = \mathbf{V}_0/V_0 \qquad \gamma_0 = \frac{1}{(1 - V_0^2/c^2)^{1/2}}$$

Inspection of these relations shows that in S' the particle is at rest at the origin ($\mathbf{r}' = 0$) at time $t' = 0$.

In order to make use of such families of uniformly moving reference system for the discussion of *accelerated* motions, we state the hypothesis which was used implicitly by Einstein:

Einstein's Kinematical Assumption. The apparent rate of a moving ideal clock at any instant in the standard system S is equal to the apparent rate of a similar ideal clock which is attached permanently to any Einstein reference frame in which the moving clock is instantaneously at rest.

This assumption implies that the rates of the two ideal clocks can be synchronized whenever they are instantaneously at rest with respect to each other, even if their relative motion is accelerated at that instant. It is tantamount to the assertion that the Minkowski line element is a physically real property of the space-time manifold (Sec. 3).

Einstein's hypothesis allows us to interpret the space coordinates that appear in the Minkowski line element as the coordinates of a moving particle (clock), so that the latter can be written as

$$c^2(dt)^2 - (dx)^2 - (dy)^2 - (dz)^2 = c^2(dt)^2(1 - v(t)^2/c^2) \tag{6.30}$$

where $v(t)$ is the instantaneous speed of the particle in S. By use of this result the natural rate of running of the moving clock can be compared with its apparent rate in S through the formula

$$d\tau = dt \cdot (1 - v(t)^2/c^2)^{1/2} \tag{6.31}$$

If dt represents a time interval in S, $d\tau$ measures the corresponding interval as recorded on the clock. Since the two rates agree when the clock is at rest in S, time as recorded on the clock often is called the *proper time of the clock* [23].

According to this formula, if an ideal clock whose rate has been synchronized with clocks in S leaves a space point A at time t_1 in S and arrives at a point B at time $t_2 > t_1$ in this system, the proper time interval as recorded by the clock is given by the formula [24]

$$\Delta\tau = \int_{t_1}^{t_2} (1 - v(t)^2/c^2)^{1/2}\, dt \leq t_2 - t_1 \tag{6.32}$$

The possible implications of this formula for moving biological systems (*twin paradox*) were first considered in detail by P. Langevin (1872–1946) [25] and since that time have received much attention in the literature [26, 27]. We need insist here only on the fact that the validity of (6.32) depends on that of Einstein's hypothesis, which presumably could fail to hold empirically for accelerated biological systems without prejudicing the correctness of the special theory of relativity.

14. Differential Geometry in Minkowski Space-Time [28]

In this section we develop a description of moving reference axes attached to the world line of a particle in Minkowski space-time which is the analog of the Frenet-Serret triad of three-dimensional Euclidean geometry [29]. The associated reference frame of a moving particle which is provided by this analysis, in general, is not to be construed as a co-moving frame that is suitable for physical interpretation in the sense of the special theory of relativity, since it is not necessarily an Einstein reference frame. Further comments on this point appear at the end of this section.

We introduce a notation for 4-vectors,

$$A = \{A_t, \mathbf{A}_s\}$$

where A_t is the time component of A and $\mathbf{A}_s$ is its set of space components expressed as an ordinary 3-vector. The scalar product of two 4-vectors is defined by

$$(A,B) = (B,A) = A_tB_t - \mathbf{A}_s \cdot \mathbf{B}_s \tag{6.33}$$

This permits the following classification of types of 4-vectors:

$$\begin{aligned} (A,A) &> 0 \qquad A \text{ is } \textit{timelike} \\ &= 0 \qquad \textit{null} \\ &< 0 \qquad \textit{spacelike} \end{aligned} \tag{6.34}$$

The magnitude of A is $|(A,A)|^{1/2}$.

In the standard reference system S the *displacement 4-vector* of a moving material particle is $X = \{ct, \mathbf{r}(t)\}$. With the proper-time interval defined by

$$d\tau = dt\{1 - [v(t)]^2/c^2\}^{1/2},$$

we let

$$\gamma = 1/\{1 - [v(t)]^2/c^2\}^{1/2}$$

so that in taking derivatives with respect to time

$$\frac{d}{d\tau} = \gamma \frac{d}{dt}$$

A superposed dot indicates differentiation with respect to proper time.

The *velocity 4-vector* of the particle is defined by

$$u \equiv \dot{X} = \{\gamma c, \gamma\mathbf{v}(t)\} \tag{6.35}$$

The velocity 4-vector is timelike with $(u, u) = c^2$.

The *acceleration 4-vector* is defined similarly as

$$a \equiv \dot{u} = \{\gamma^4(\mathbf{a} \cdot \mathbf{v})/c, \gamma^2\mathbf{a} + \gamma^4(\mathbf{v} \cdot \mathbf{a})\mathbf{v}/c^2\} \tag{6.36}$$

where we use the relation $\dot{\gamma} = \gamma^4(\mathbf{v} \cdot \mathbf{a})/c^2$.

We next introduce a vector frame $\{t^{(0)}, n^{(1)}, n^{(2)}, n^{(3)}\}$ attached to the world line of the particle by the relations

$$u = ct^{(0)} \tag{6.37a}$$

$$a = \frac{c}{\rho} n^{(1)} \tag{6.37b}$$

$$\dot{a} = \frac{c}{\rho\sigma} n^{(2)} - \frac{c\dot{\rho}}{\rho^2} n^{(1)} + \frac{c}{\rho^2} t^{(0)} \tag{6.37c}$$

$$\ddot{a} = \frac{c}{\rho\sigma\zeta} n^{(3)} - \frac{c}{\rho^2\sigma^2}(\rho\dot{\sigma} + 2\sigma\dot{\rho})n^{(2)} + \frac{c}{\rho^3}\left(1 - \frac{\rho^2}{\sigma^2} + 2\dot{\rho}^2 - \rho\ddot{\rho}\right) n^{(1)} - \frac{3c\dot{\rho}}{\rho^3} t^{(0)} \tag{6.37d}$$

It is not difficult to verify that this frame satisfies the conditions

$$(t^{(0)},t^{(0)}) = 1 \qquad (n^{(k)},t^{(0)}) = 0 \qquad (n^{(k)},n^{(\ell)}) = -\delta^{k\ell} \tag{6.38}$$

The change in the vector frame with motion along the world line of the particle is determined by the matrix relation

$$\begin{bmatrix} \dot{t}^{(0)} \\ \dot{n}^{(1)} \\ \dot{n}^{(2)} \\ \dot{n}^{(3)} \end{bmatrix} = \begin{bmatrix} 0 & 1/\rho & 0 & 0 \\ 1/\rho & 0 & 1/\sigma & 0 \\ 0 & -1/\sigma & 0 & 1/\zeta \\ 0 & 0 & -1/\zeta & 0 \end{bmatrix} \cdot \begin{bmatrix} t^{(0)} \\ n^{(1)} \\ n^{(2)} \\ n^{(3)} \end{bmatrix} \tag{6.39}$$

The timelike vector $t^{(0)}$ is the unit tangent to the world line in the direction of motion of the particle, while the spacelike unit vectors $n^{(1)}$, $n^{(2)}$, $n^{(3)}$ are the normal vectors. The three scalar quantities ρ, σ, ζ determine the three kinds of curvature of the world line.

Some special 4-vectors associated with the world line which have received attention in the literature in the study of accelerated motion are the *Abraham 4-vector* and the *Haantjes 4-vector* [30, 31]:

1. *Abraham 4-vector:*

$$\begin{aligned} \Gamma &\equiv \dot{a}/c + (a,a)u/c^3 \qquad (6.40) \\ &= \frac{1}{\rho\sigma} n^{(2)} - \frac{\dot{\rho}}{\rho^2} n^{(1)} \end{aligned}$$

2. *Haantjes 4-vector:*

$$\Xi \equiv \dot{K} - (\dot{K},u)u/c^2 \tag{6.41}$$

where $K = \dfrac{\Gamma}{|\Gamma|} = n^{(2)} \operatorname{sech} \alpha + n^{(1)} \tanh \alpha$

$$\cosh \alpha = \rho\sigma|\Gamma| \qquad \tanh \alpha = \frac{1}{|\Gamma|} \frac{d}{d\tau} \left(\frac{1}{\rho}\right)$$

A short computation yields the result

$$\begin{aligned} \Xi = \operatorname{sech} \alpha \Bigg[&\frac{n^{(3)}}{\zeta} \\ &+ \left(\dot{\alpha} \operatorname{sech} \alpha - \frac{1}{\sigma}\right) (-n^{(2)} \sinh \alpha + n^{(1)}) \Bigg] \qquad (6.42) \end{aligned}$$

These quantities are of importance in the replacement of condition (6.7) by (6.8) and in the development of a theory of relativistic uniformly accelerated coordinate systems. The results have been widely discussed in the literature but have not received general acceptance among theoretical physicists [32, 33].

15. Relativistic Mechanics of a Particle

In order to arrive at a dynamical equation of motion of a mass particle one must adopt a suitable definition of force. This was first done by Einstein [1] by consideration of the motion of a particle of electric charge ϵ moving in an external electromagnetic field. This allows one to take advantage of the accepted definition of force in this case. Einstein's results were modified by M. Planck (1858–1947) [34] to take the following form:

Einstein-Planck Equation of Motion

$$\frac{d\mathbf{p}}{dt} = \epsilon \left(\mathbf{E} + \frac{\mathbf{v}}{c} \times \mathbf{H}\right) \tag{6.43}$$

where
$$\mathbf{p} = \frac{m\mathbf{v}}{(1 - v^2/c^2)^{1/2}} \tag{6.44}$$

is the *kinematic* 3-*momentum* of the particle.

This equation can be cast into four-dimensional form if we define the *kinematic* 4-*momentum* of the particle in terms of the 4-velocity of (6.35) by the formula

$$p \equiv mu = \left\{\frac{mc}{(1 - v^2/c^2)^{1/2}}, \frac{m\mathbf{v}}{(1 - v^2/c^2)^{1/2}}\right\} \tag{6.45}$$

The *electromagnetic* 4-*force* is

$$f = \left\{\frac{\epsilon(\mathbf{E} \cdot \mathbf{v})/c}{(1 - v^2/c^2)^{1/2}}, \frac{\epsilon(\mathbf{E} + \mathbf{v}/c \times \mathbf{H})}{(1 - v^2/c^2)^{1/2}}\right\} \tag{6.46}$$

With this definition, the equation of motion (6.43) takes the four-dimensional form

$$dp/d\tau = f \tag{6.47}$$

A suitable definition of kinetic energy is found by application of the classical equation of activity of the forces, writing

$$dT/dt = \epsilon(\mathbf{E} \cdot \mathbf{v}) \tag{6.48}$$

On requiring the kinetic energy to vanish when the particle is at rest, one finds the solution

$$T = mc^2 \left[\frac{1}{(1 - v^2/c^2)^{1/2}} - 1\right] \tag{6.49}$$

These relations can be extended in a formal way to motion under more general forces if it is required that every three-dimensional force $\mathbf{F}$ be associated with a 4-force of the form

$$f = \left\{\frac{(\mathbf{F} \cdot \mathbf{v})/c}{(1 - v^2/c^2)^{1/2}}, \frac{\mathbf{F}}{(1 - v^2/c^2)^{1/2}}\right\} \tag{6.50}$$

This complex of ideas led Poincaré and Einstein independently to the conclusion that the total energy of a moving particle is given by the expression

$$w = \frac{mc^2}{(1 - v^2/c^2)^{1/2}} = \gamma mc^2 \tag{6.51}$$

This concept was deepened by G. N. Lewis (1875–1946) and Einstein by the assumption that the quantity

$$w_0 = mc^2 \tag{6.52}$$

is the *internal* (*rest*) *energy of the particle*. The extension of this idea to arbitrary physical masses, indicating a complete equivalence of mass and energy, played an important part in Einstein's later interpretation of the gravitational field.

With some modifications, these formulas can be applied to photons, the latter being considered as particles of zero rest mass which move with the speed of light. The energy of a photon is given by Einstein's quantum formula

$$w = h\nu = \hbar\omega \tag{6.53}$$

where $h = 2\pi\hbar = 6.625 \times 10^{-27}$ erg-sec is Planck's constant, while $\nu = \omega/2\pi = c/\lambda$ is the frequency of the radiation. The relation between energy and momentum for a photon is simply $w = pc$. The equation of motion (6.47) has no analogue for photons within the context of the special theory of relativity.

16. Relativistic Center-of-mass Systems

The concept of center of mass of a system of particles, which is so useful in classical mechanics, has no simple generalization in relativistic mechanics. Its nearest equivalent for an arbitrary system of free particles is that of an Einstein reference frame in which the total momentum vanishes.

Since the 3-momentum and total energy of a particle form a 4-vector (6.45) in relativity theory, their transformation equations under Einstein transformations follow from (6.22):

$$\mathbf{p}' = (\gamma - 1)(\mathbf{p} \times \mathbf{n}_0) \times \mathbf{n}_0 + \gamma(\mathbf{p} - \mathbf{n}_0 Vw/c^2)$$
$$w' = \gamma(w - \mathbf{p}\cdot\mathbf{n}_0 V) \qquad \gamma = \frac{1}{(1 - V^2/c^2)^{1/2}} \tag{6.54}$$

The total momentum and energy of a system of free particles are defined by the sums

$$\mathbf{P} = \Sigma \mathbf{p}_i \qquad W = \Sigma w_i \tag{6.55}$$

The linearity of relations (6.54) in momentum and energy leads at once to the transformation equations

$$\mathbf{P}' = (\gamma - 1)(\mathbf{P} \times \mathbf{n}_0) \times \mathbf{n}_0 + \gamma(\mathbf{P} - \mathbf{n}_0 VW/c^2)$$
$$W' = \gamma(W - \mathbf{P}\cdot\mathbf{n}_0 V) \tag{6.56}$$

Inspection of (6.56) reveals that if we make the identification

$$\mathbf{V} = \mathbf{n}_0 V = \mathbf{P}c^2/W \tag{6.57}$$

the total momentum of the set of particles is zero in the moving Einstein frame, while the total energy is

$$W' = (W^2 - P^2c^2)^{1/2} \tag{6.58}$$

This moving reference system, or any Einstein frame obtained from it by translations and space rotations, can be taken to be an appropriate relativistic center-of-mass (CM) system for the set of particles. The principal advantage of such a system is that it can be used even when the particles interact through close-contact collisions which may alter the total number of particles, the principles of conservation of momentum and energy being assumed to hold for such collisions. The consistency of this assumption with the relativistic definitions of energy and momentum has been demonstrated by G. N. Lewis and R. C. Tolman (1881–1948) [35, 36].

17. Collision Processes

Consider a collision between two particles of rest masses m_1 and m_2, as a result of which two particles of rest masses m_3 and m_4 are ejected. For convenience we assume that particle 2 initially is at rest at the origin of the standard (laboratory) frame S and that particle 1 approaches it with momentum p_1 along

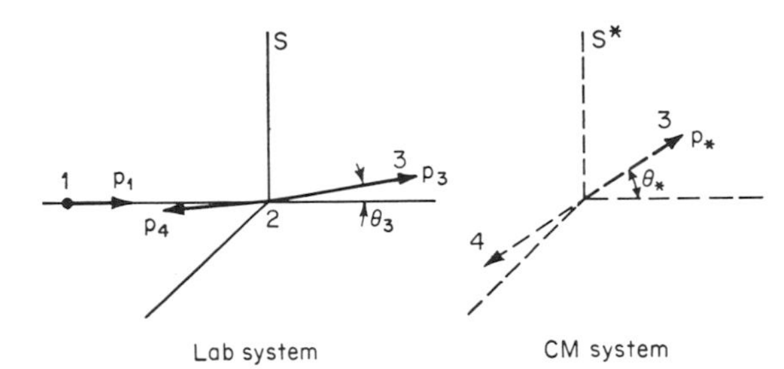

FIG. 6.2. Geometry of a two-body collision in laboratory and center-of-mass (CM) coordinate systems.

the negative z axis (Fig. 6.2). The initial momentum and energy in S are

$$\mathbf{P} = p_1\mathbf{k}$$
$$W = w_1 + m_2c^2 = (m_1^2c^2 + p_1^2)^{1/2}c + m_2c^2 \tag{6.59}$$

We introduce the CM system S^* defined in Sec. 16, which moves with the velocity

$$\mathbf{V} = \frac{p_1c^2}{w_1 + m_2c^2}\mathbf{k} \tag{6.60}$$

In the latter frame the total momentum is zero, while the energy is

$$W_* = [(w_1 + m_2c^2)^2 - p_1^2c^2]^{1/2}$$
$$= [(m_1^2 + m_2^2)c^4 + 2w_1m_2c^2]^{1/2} \tag{6.61}$$

As viewed in the CM system, the ejected particles move with equal, but oppositely directed, momenta $\pm\mathbf{p}_*$ and total energy W_*. The energy equation in this system is

$$W_* = (m_3^2c^4 + p_*^2c^2)^{1/2} + (m_4^2c^4 + p_*^2c^2)^{1/2} \tag{6.62}$$

On solution between (6.61) and (6.62) one finds that

$$p_* = \frac{1}{2}\left\{\frac{[(m_1^2 + m_2^2 - m_3^2 - m_4^2)c^2 + 2w_1m_2]^2 - 4m_3^2m_4^2c^4}{(m_1^2 + m_2^2)c^2 + 2w_1m_2}\right\}^{1/2} \tag{6.63}$$

Let us select one of the outgoing particles for investigation, say particle 3. Its momentum and energy in the laboratory system are

$$\mathbf{p}_3 = (\gamma - 1)(\mathbf{p}_* \times \mathbf{k}) \times \mathbf{k} + \gamma(\mathbf{p}_* + \mathbf{k}w_{3*}V/c^2) \tag{6.64}$$
$$w_3 = \gamma(w_{3*} + \mathbf{p}_*\cdot\mathbf{k}V)$$

with

$$w_{3*} = (m_3^2c^4 + p_*^2c^2)^{1/2}$$

It is convenient to separate the momentum vectors into components parallel and perpendicular to the direction of the initial motion of particle 1, writing

$$\mathbf{p}_* = \mathbf{p}_{*\parallel} + \mathbf{p}_{*\perp} \qquad \mathbf{p}_3 = \mathbf{p}_{3\parallel} + \mathbf{p}_{3\perp} \tag{6.65}$$

It is found from (6.64) that

$$\mathbf{p}_{3\perp} = \mathbf{p}_{*\perp} \qquad \mathbf{p}_{3\parallel} = \gamma(\mathbf{p}_{*\parallel} + w_{3*}\mathbf{V}/c^2) \tag{6.66}$$

Introducing the angles of ejection of particle 3 we have

$$p_3 \sin\theta_3 = p_* \sin\theta_*$$
$$p_3 \cos\theta_3 = \gamma(p_* \cos\theta_* + w_{3*}V/c^2) \tag{6.67}$$

This yields our final transformation formulas

$$\tan\theta_3 = \frac{\sin\theta_*}{\gamma(\cos\theta_* + w_{3*}V/p_*c^2)} \tag{6.68}$$

$$p_3 = p_*\left[\sin^2\theta_* + \gamma^2\left(\cos\theta_* + \frac{w_{3*}V}{p_*c^2}\right)^2\right]^{1/2} \tag{6.69}$$

with $$\gamma = \frac{1}{1 - V^2/c^2}$$

where V is defined in (6.60).

All the kinematic properties of the collision process can be determined by use of these formulas. In applications to low-energy nuclear processes it is sufficiently accurate to set $\gamma = 1$. For a more detailed treatment of collision processes in classical mechanics see Part 2, Chap. 2, Sec. 7. A useful graphical method for the study of low-energy nuclear disintegration processes has been devised by J. L. McKibben [37].

18. The Electromagnetic Field

Maxwell's equations for the electromagnetic field in the usual differential form are (Part 4, Chap. 1)

$$\begin{aligned} &\operatorname{div}\mathbf{E} = 4\pi\rho \qquad \operatorname{curl}\mathbf{E} + \frac{1}{c}\frac{\partial\mathbf{H}}{\partial t} = 0 \\ &\operatorname{div}\mathbf{H} = 0 \qquad \operatorname{curl}\mathbf{H} - \frac{1}{c}\frac{\partial\mathbf{E}}{\partial t} = \frac{4\pi}{c}\mathbf{j} \end{aligned} \tag{6.70}$$

$$\frac{\partial\rho}{\partial t} + \operatorname{div}\mathbf{j} = 0 \qquad \text{(conservation of charge)} \tag{6.71a}$$

$$\mathbf{f} = \rho\mathbf{E} + \frac{\mathbf{j}}{c}\times\mathbf{H} \qquad \text{(Lorentz force density)} \tag{6.71b}$$

It has been shown by R. Hargreaves (1853–1939) and H. Bateman (1882–1946) that for some purposes it is advantageous to formulate the field equations as integral equations in Minkowski space-time [38]. For this purpose we construct the following differential forms [39]:

$$\begin{aligned} \omega_1 \equiv\ & H_x[dy\,dz] + H_y[dz\,dx] + H_z[dx\,dy] + cE_x[dx\,dt] \\ & + cE_y[dy\,dt] + cE_z[dz\,dt] \qquad (6.72a) \\ \omega_2 \equiv\ & E_x[dy\,dz] + E_y[dz\,dx] + E_z[dx\,dy] - cH_x[dx\,dt] \\ & - cH_y[dy\,dt] - cH_z[dz\,dt] \qquad (6.72b) \\ \omega_3 \equiv\ & 4\pi(\rho[dx\,dy\,dz] - j_x[dy\,dz\,dt] - j_y[dz\,dx\,dt] \\ & - j_z[dx\,dy\,dt]) \qquad (6.72c) \end{aligned}$$

If Σ is an arbitrary open domain in Minkowski space-time, with closed boundary $\partial\Sigma$, the generalized form of Stokes' theorem shows that for any differential form ω

$$\int_{\partial\Sigma}\omega = \int_\Sigma d\omega \tag{6.73}$$

where $d\omega$ is the exterior derivative of ω. We have, for example,

$$\begin{aligned} d\omega_1 = & \left(\frac{\partial E_y}{\partial x} - \frac{\partial E_x}{\partial y} + \frac{1}{c}\frac{\partial H_z}{\partial t}\right)c[dx\,dy\,dt] \\ & + \left(\frac{\partial E_z}{\partial y} - \frac{\partial E_y}{\partial z} + \frac{1}{c}\frac{\partial H_x}{\partial t}\right)c[dy\,dz\,dt] \\ & + \left(\frac{\partial E_x}{\partial z} - \frac{\partial E_z}{\partial x} + \frac{1}{c}\frac{\partial H_y}{\partial t}\right)c[dz\,dx\,dt] \\ & + \left(\frac{\partial H_x}{\partial x} + \frac{\partial H_y}{\partial y} + \frac{\partial H_z}{\partial z}\right)[dx\,dy\,dz] \end{aligned}$$

which vanishes identically by Maxwell's equations.

It is found that Maxwell's equations are expressed by the relations

$$d\omega_1 = 0 \qquad d\omega_2 = \omega_3 \qquad d\omega_3 = 0 \tag{6.74}$$

This formulation of the field equations expresses the fact that Maxwell's equations involve only skew-symmetric tensors (p-vectors) and so do not have any metric implications for space-time [40]. The transformation equations for the field quantities are readily written explicitly from the invariance of the differential forms (6.72) under Einstein transformations.

If one prefers to use the more customary tensor notation, the electromagnetic field is expressed by the (real) skew-symmetric matrix

$$[F_{\lambda\mu}] \equiv \left[\begin{array}{c|ccc} 0 & E_x & E_y & E_z \\ \hline -E_x & 0 & -H_z & H_y \\ -E_y & H_z & 0 & -H_x \\ -E_z & -H_y & H_x & 0 \end{array}\right] \tag{6.75}$$

with the transformation formulas

$$F'_{\lambda\mu}(x') = \frac{1}{2}\frac{\partial(x^\alpha,x^\beta)}{\partial(x'^\lambda,x'^\mu)}F_{\alpha\beta}(x) \tag{6.76}$$

Maxwell's equations take the form

$$\partial_\lambda F_{\mu\nu} + \partial_\mu F_{\nu\lambda} + \partial_\nu F_{\lambda\mu} = 0 \tag{6.77}$$

$$\partial_\lambda F^{\lambda\mu} = \frac{4\pi}{c}j^\mu \tag{6.78}$$

$$\partial_\lambda j^\lambda = 0 \tag{6.79}$$

with $$\{j^\lambda\} = \{c\rho,j_x,j_y,j_z\} \tag{6.80}$$

It can be seen from (6.72c) that the charge and current functions are density functions on the space-time manifold, but when the Minkowski line element is assigned to the manifold they can be formulated as the 4-vector (6.80).

The transformation equations of the field vectors corresponding to the Einstein transformation (6.22) are

$$\begin{aligned} \mathbf{H}' &= (1-\gamma)(\mathbf{H}\cdot\mathbf{n}_0)\mathbf{n}_0 + \gamma\left(\mathbf{H} - \frac{\mathbf{V}}{c}\times\mathbf{E}\right) \\ \mathbf{E}' &= (1-\gamma)(\mathbf{E}\cdot\mathbf{n}_0)\mathbf{n}_0 + \gamma\left(\mathbf{E} + \frac{\mathbf{V}}{c}\times\mathbf{H}\right) \end{aligned} \tag{6.81}$$

There exist two invariant density functions

$$\omega_1^2 = -\omega_2^2 = c(\mathbf{E}\cdot\mathbf{H})[dx\,dy\,dz\,dt] \tag{6.82a}$$

$$\omega_1\omega_2 = -c(\mathbf{H}^2 - \mathbf{E}^2)[dx\,dy\,dz\,dt] \tag{6.82b}$$

Under all transformations of the Einstein group the differential form $[dx\,dy\,dz\,dt]$ itself is invariant and serves as a volume element on the Minkowski manifold. The quantities

$$\Theta_1 = \mathbf{E}\cdot\mathbf{H} \qquad \Theta_2 = \mathbf{H}^2 - \mathbf{E}^2 \tag{6.83}$$

are then invariants. This result is used in Part 6, Chap. 8.

Bateman's classic paper [38] gives a full study of the invariance of Maxwell's equations under the conformal group in Minkowski space-time.

19. Macroscopic Electrodynamics

The theory of relativity does not offer a panacea for the difficulties encountered in developing atomic theories of the condensed states of matter. However, it does offer some useful help in the construction of a framework for such theories. The following notes are intended primarily only as a general guide in some points of the macroscopic theory of electrodynamics which have application in solid-state and plasma physics. For a more detailed treatment of the formal theory the reader is referred to the book by J. Frenkel (1894–1952) [41].

The phenomenological theory is based on the idea that the charge and current functions, when averaged over the high-frequency atomic properties, can be separated into the two categories of "free" (conduction) and "bound" (polarization) charges and currents

$$\rho = \rho_f + \rho_b \qquad \mathbf{j} = \mathbf{j}_f + \mathbf{j}_b \tag{6.84}$$

The bound charges and currents then are replaced by equivalent electric and magnetic polarization densities, **P** and **M**, in the medium, by use of the relations

$$\rho_b = -\operatorname{div}\mathbf{P} \qquad \mathbf{j}_b = \frac{\partial \mathbf{P}}{\partial t} + c\operatorname{curl} M \tag{6.85}$$

It is evident that these definitions admit the use of separate conservation relations for the free and bound charges:

$$\frac{\partial \rho_f}{\partial t} + \operatorname{div}\mathbf{j}_f = 0 \qquad \frac{\partial \rho_b}{\partial t} + \operatorname{div}\mathbf{j}_b = 0 \tag{6.86}$$

If these relations are assumed to hold in a reference system that is at rest with respect to the medium, they can be transformed to an Einstein moving frame by formal considerations. For the general Einstein transformation (6.22) the formulas are

$$\begin{aligned} \mathbf{P}' &= (1-\gamma)(\mathbf{P}\cdot\mathbf{n}_0)\mathbf{n}_0 + \gamma(\mathbf{P} - \mathbf{n}_0 \times \mathbf{M}V/c) \\ \mathbf{M}' &= (1-\gamma)(\mathbf{M}\cdot\mathbf{n}_0)\mathbf{n}_0 \\ &\qquad + \gamma(\mathbf{M} + \mathbf{n}_0 \times \mathbf{P}V/c) \end{aligned} \tag{6.87}$$

These formulas show that the electric- and magnetic-dipole polarizations are not completely separable from each other but appear to be intermixed by motion of the medium, just as are the electric and magnetic fields themselves.

The theory given so far does not depend on any assumption of constitutive relations for the medium, according to which the polarization densities are assumed to be determinable from the averaged fields in the medium. For the development of this theory (Minkowski electrodynamics) reference is made to the literature [18, 18*a*, 42].

The relationship between optics and relativity theory is discussed in Part 6, Chap. 8.

THE GENERAL THEORY OF RELATIVITY

20. Gravitational and Inertial Mass

It has been remarked (Sec. 2) that in Newtonian mechanics the distinction between physical forces and kinetic reactions is absolute. A characteristic property of the latter is that they always are proportional to the *inertial mass* of the system on which they appear to act. The only known physical force that is proportional to mass is that of the gravitational field. When the term mass is used in this sense, it is called *gravitational mass.*

This striking resemblance between kinetic reactions and gravitational forces accounts for Galileo's statement that, apart from resistive forces of aerodynamic origin, all bodies fall in the earth's gravitational field with the same acceleration. A possible source of error in this conclusion is that it might be necessary to draw a distinction between the two types of mass in the sense that their ratio might depend on the body concerned. Galileo's observation, if universally valid, implies that this is not the case. In the cgs system of units this ratio is normalized to unity so that one can speak simply of equality or inequality of inertial and gravitational mass for different bodies.

Very precise measurements of this ratio were instituted in 1889 by R. v. Eötvös (1848–1919) [43]. The aim of his experiment was to balance on a given body the kinetic reaction arising from the rotation of the earth against the earth's gravitational attraction. Although the idea is simple, high accuracy can be achieved only by taking great care in the measurements. The experiment was repeated by P. Zeeman (1868–1943) [44] and by Eötvös, Pekar, and Fekete [45] to an accuracy of about 1 part in 10^9. In 1910 a pendulum method was used by L. Southerns to test the possibility that the radioactive substance uranium oxide might behave differently from lead oxide in a gravitational field. No difference was observed to an accuracy of 1 part in 10^4 [46]. This result has been confirmed since then to a higher order of accuracy.

More recently, R. H. Dicke and his associates have modified the experiment in order to measure the gravitational attraction of the sun on masses of aluminum and gold. They find equality of inertial and gravitational mass for these substances, to an accuracy of 1 part in 10^{11}, despite their appreciable differences in nuclear structure [47]. The identity of the two types of mass thus seems to be among the best-demonstrated facts of empirical physics.

21. Equations of Motion of a Particle

Newton's first law of motion for a freely moving particle holds equally well in classical mechanics and in the special theory of relativity. In this case both Newton's equation of motion and the Einstein-Planck equation (6.43) reduce to the form

$$d^2x/dt^2 = 0 \qquad d^2y/dt^2 = 0 \qquad d^2z/dt^2 = 0 \tag{6.88}$$

We wish to replace these equations by a formulation which is adapted to analysis in four-dimensional space-time.

Let the world line of the particle be expressed in parametric form

$$x = F(\sigma) \qquad y = G(\sigma) \qquad z = H(\sigma) \qquad t = T(\sigma)$$

We find on substitution in (6.88) a set of equations of the type

$$\frac{dx}{dt} = \frac{dx}{d\sigma}\frac{d\sigma}{dt} \qquad \frac{d^2x}{dt^2} = \frac{d^2x}{d\sigma^2}\left(\frac{d\sigma}{dt}\right)^2 + \frac{dx}{d\sigma}\frac{d^2\sigma}{dt^2}$$

Equations (6.88) will be satisfied identically if we set

$$\frac{d^2x}{d\sigma^2} = 0 \qquad \frac{d^2y}{d\sigma^2} = 0 \qquad \frac{d^2z}{d\sigma^2} = 0 \qquad \frac{d^2\sigma}{dt^2} = 0$$

The last of these equations implies that σ is a linear function of t and conversely that t is a linear function of σ, so that

$$d^2t/d\sigma^2 = 0 \qquad (6.89)$$

Equations (6.88), together with the statement (6.89), are thus expressible in four-dimensional form by the relations

$$d^2x^\lambda/d\sigma^2 = 0 \qquad \lambda = 0, 1, 2, 3 \qquad (6.90)$$

These relations are not form-invariant under general changes of coordinates, however, so that it is necessary to convert them to tensor form before they can be used in all coordinate systems.

Let $\{\Omega_{\mu\nu}{}^\lambda(x) = \Omega_{\nu\mu}{}^\lambda(x)\}$ be any set of functions of the coordinates required to transform under change of coordinates by the formula [48]

$$\Omega'_{\beta\gamma}{}^\alpha(x') = \Omega_{\mu\nu}{}^\lambda(x)\frac{\partial x'^\alpha}{\partial x^\lambda}\frac{\partial x^\mu}{\partial x'^\beta}\frac{\partial x^\nu}{\partial x'^\gamma} + \frac{\partial x'^\alpha}{\partial x^\theta}\frac{\partial^2 x^\theta}{\partial x'^\beta\,\partial x'^\gamma} \qquad (6.91)$$

Direct computation shows that

$$\frac{d^2x'^\alpha}{d\sigma^2} + \Omega'_{\beta\gamma}{}^\alpha(x')\frac{dx'^\beta}{d\sigma}\frac{dx'^\gamma}{d\sigma} = \frac{\partial x'^\alpha}{\partial x^\lambda}\cdot\left\{\frac{d^2x^\lambda}{d\sigma^2} + \Omega_{\mu\nu}{}^\lambda(x)\frac{dx^\mu}{d\sigma}\frac{dx^\nu}{d\sigma}\right\}$$

which expresses the fact that the quantities

$$\frac{d^2x^\lambda}{d\sigma^2} + \Omega_{\mu\nu}{}^\lambda\frac{dx^\mu}{d\sigma}\frac{dx^\nu}{d\sigma}$$

transform as contravariant components of a 4-vector.

This result allows (6.90) to be generalized to the tensor form

$$\frac{d^2x^\lambda}{d\sigma^2} + \Omega_{\mu\nu}{}^\lambda\frac{dx^\mu}{d\sigma}\frac{dx^\nu}{d\sigma} = 0 \qquad (6.92)$$

with the proviso that, in the coordinate system S in which (6.88) hold, we are to set

$$\Omega_{\mu\nu}{}^\lambda(x) = 0 \qquad (\text{in } S) \qquad (6.93)$$

This latter condition can be accepted since transformation equations (6.91) show that the functions $\{\Omega_{\mu\nu}{}^\lambda(x)\}$ do not form components of a tensor. Their vanishing in one coordinate system therefore does not imply vanishing in every coordinate system.

Any set of functions that transform by law (6.91) provides a *linear connection* on the Einstein space-time manifold. The particular linear connection that satisfies (6.93) gives a covariant generalization of the Newton-Einstein first law of motion for which the world lines of freely moving particles form a distinguished family of curves in space-time.

22. The Christoffel Linear Connection

One particular linear connection is related in a unique manner to the geometric structure to be attributed to the Einstein space-time manifold. To find it, we first define the *Christoffel symbols of first and second kinds* by the formulas [49]

$$[\lambda\mu,\nu] = [\mu\lambda,\nu] = \frac{1}{2}\left(\frac{\partial g_{\mu\nu}}{\partial x^\lambda} + \frac{\partial g_{\lambda\nu}}{\partial x^\mu} - \frac{\partial g_{\lambda\mu}}{\partial x^\nu}\right) \qquad (6.94)$$

$$\left\{{\nu \atop \lambda\mu}\right\} = \left\{{\nu \atop \mu\lambda}\right\} = g^{\nu\theta}[\lambda\mu,\theta] \qquad (6.95)$$

Calculation shows that under coordinate transformations the latter set of functions transform by the rule

$$\left\{{\gamma \atop \alpha\beta}\right\}' = \left\{{\nu \atop \lambda\mu}\right\}\frac{\partial x'^\gamma}{\partial x^\nu}\frac{\partial x^\lambda}{\partial x'^\alpha}\frac{\partial x^\mu}{\partial x'^\beta} + \frac{\partial x'^\gamma}{\partial x^\theta}\frac{\partial^2 x^\theta}{\partial x'^\alpha\,\partial x'^\beta}$$

which agrees with (6.91). This is called the *Christoffel linear connection* since it was studied in 1869 by E. B. Christoffel (1829–1900).

The curves defined by (6.92) with this linear connection are the *geodesics* of the Einstein space-time manifold. Any geodesic for which, in addition, $ds = 0$ is called a *null geodesic* (cf. Sec. 4).

23. Einstein's Geodesic Principle

The accepted equality of inertial and gravitational mass for all bodies suggests that the world lines of matter on the Einstein space-time manifold are determined by some universal condition that is independent of the nature of the body. Considerations of this nature led Einstein to introduce the following important principle [2]:

1. *Under the action of the gravitational field determined by the geometric structure of the space-time manifold the motion of a particle is a non-null geodesic satisfying the differential equations*

$$\frac{d^2x^\lambda}{ds^2} + \left\{{\lambda \atop \mu\nu}\right\}\frac{dx^\mu}{ds}\frac{dx^\nu}{ds} = 0 \qquad (6.96)$$

2. *Photons move like particles along null geodesics.*

Einstein's geodesic principle can be interpreted as the statement that, so far as the influence of gravitational forces on a test particle is concerned, they are completely equivalent to free motion under the influence of the geometric structure of the space-time manifold. This is the most positive interpretation of the equivalence of gravitation and four-dimensional geometry provided by the theory of relativity.

The geodesic principle is used normally in the sense that it determines the world lines of small test particles and photons that do not themselves contribute to the field in which they move. For larger masses that contribute to the total field it is necessary to extend the principle by a proof that Einstein's field equations (Sec. 30) determine the equations of motion of matter. This deeper interpretation of the theory was made the subject of extensive investigation by Einstein and his associates [50, 51].

The null geodesics passing through an event P of space-time form the *characteristic null cone* of the manifold at P [52, 53]. This surface separates the non-null geodesics through P into the two classes of timelike and spacelike geodesics. The fact that the timelike geodesics, which can be the paths of material particles, lie in the interior of the null cone is the geometric expression of Einstein's principle of causality (Sec. 5).

24. Geodesic Coordinates

At a given event P of space-time we consider a particular non-null geodesic G which starts from P with an initial direction specified by the tangent values

$$p_0{}^{\lambda} = (dx^{\lambda}/ds)_P \qquad g_{\lambda\mu}(P)p_0{}^{\lambda}p_0{}^{\mu} \neq 0 \tag{6.97}$$

The coordinates of points on G can be parametrized by the geodesic distance s and in some open neighborhood of P can be expressed as the power series

$$x^{\lambda}(s) = x^{\lambda}(P) + p_0{}^{\lambda}s - \tfrac{1}{2}\left\{{\lambda \atop \mu\nu}\right\}_P p_0{}^{\mu}p_0{}^{\nu}s^2 + O(s^3) \tag{6.98}$$

For points on G we define new coordinates $\{y^{\lambda}\}$ by the formula

$$y^{\lambda} = p_0{}^{\lambda}s \tag{6.99}$$

When this operation is carried out for each geodesic through P, the new variables so obtained are called *geodesic coordinates* in the neighborhood of P.

On substitution into (6.98) we have the coordinate transformations

$$x^{\lambda} - x^{\lambda}(P) = y^{\lambda} - \tfrac{1}{2}\left\{{\lambda \atop \mu\nu}\right\}_P y^{\mu}y^{\nu} + O(y^3) \tag{6.100}$$

Inversion of these relations (by series expansions) in an open region about P yields the transformation equations from the original coordinates to the geodesic coordinates.

Since (6.99) can be regarded as the series expansion of the geodesic coordinates in terms of the geodesic distance on each geodesic, the linearity of these relations shows by comparison with (6.98) that *when geodesic coordinates are employed the Christoffel symbols vanish at P.* However, in general, not all the derivatives of these symbols can be made to vanish at P by any choice of coordinates since this would imply automatically the vanishing of the Riemann-Christoffel tensor.

25. Differentiation of Tensor Fields

The process of differentiation of the components of vector and tensor fields, which is indispensable in the formulation of differential equations, normally does not yield new tensor fields. This is the case even in Euclidean space when curvilinear coordinates are employed. We have had an important example of this fact already in the circumstance that the tangent vector on a geodesic

$$p^{\lambda} \equiv dx^{\lambda}/ds \tag{6.101}$$

satisfies the differential equation

$$\frac{dp^{\lambda}}{ds} + \left\{{\lambda \atop \mu\nu}\right\} p^{\mu}p^{\nu} = 0 \tag{6.102}$$

which is identical with (6.96). Since (6.102) is a tensor equation it holds in any coordinate system.

Tensor equations are said to be *covariant* under transformations of coordinates. *It is to be noted carefully that this is a weaker requirement than that of strict form invariance of equations which characterized our discussion of the special theory of relativity.* For example, if we use the geodesic coordinates of Sec. 24, (6.102) takes the simpler form

$$(dp^{\lambda}/ds)_P = 0 \qquad \text{(geodesic coordinates)} \tag{6.103}$$

owing to the vanishing of the Christoffel symbols in this coordinate system. However, under coordinate transformations, (6.103) is not form-invariant but reverts to the covariant formulation (6.102).

This special property of geodesic coordinates can be employed for the introduction of differentiation processes which exhibit tensor covariance under transformation of coordinates. By analogy with (6.102), we take an arbitrary covariant vector field $\{\xi^{\lambda}(x)\}$ and replace the usual partial-differentiation procedure by *covariant differentiation*

$$\xi^{\lambda}{}_{;\mu} \equiv \frac{\partial \xi^{\lambda}}{\partial x^{\mu}} + \left\{{\lambda \atop \mu\nu}\right\}\xi^{\nu} \tag{6.104}$$

Covariant differentiation of covariant vectors is defined by the formula

$$\eta_{\lambda;\mu} \equiv \frac{\partial \eta_{\lambda}}{\partial x^{\mu}} - \left\{{\theta \atop \mu\lambda}\right\}\eta_{\theta} \tag{6.105}$$

It follows from these definitions that

$$(\xi^{\lambda}\eta_{\lambda})_{;\mu} \equiv \frac{\partial}{\partial x^{\mu}}(\xi^{\lambda}\eta_{\lambda}) \tag{6.106}$$

In order to continue this process to higher-ordered derivatives it is best to proceed by formal definition, since derivatives of the Christoffel symbols enter and the reduction to geodesic coordinates loses its simplicity. Indices of covariant differentiation always appear in the subscript position, following the semicolon.

The general formula for covariant differentiation of a mixed tensor is

$$\begin{aligned} K^{\lambda_1,\ldots,\lambda_n}_{\mu_1,\ldots,\mu_m;\nu} \equiv{} & \frac{\partial}{\partial x^{\nu}}(K^{\lambda_1,\ldots,\lambda_n}_{\mu_1,\ldots,\mu_m}) \\ & + \sum_{t=1}^{n} \left\{{\lambda_t \atop \nu\sigma}\right\} K^{\lambda_1,\ldots,\lambda_{t-1}\,\sigma\lambda_{t+1},\ldots,\lambda_n}_{\mu_1,\ldots,\mu_m} \\ & - \sum_{r=1}^{m} \left\{{\theta \atop \nu\mu_r}\right\} K^{\lambda_1,\ldots,\lambda_n}_{\mu_1,\ldots,\mu_{r-1}\,\theta\mu_{r+1},\ldots,\mu_m} \end{aligned} \tag{6.107}$$

This rule leads at once to the result

$$g_{\lambda\mu;\nu} = 0 \tag{6.108}$$

showing that the covariant derivative of the metric tensor is identically zero.

26. The Riemann-Christoffel and Ricci Tensors

Covariant differentiation is more complicated than is the usual process of partial differentiation in that, in general, it is not commutative. One has by direct computation

$$\xi^{\lambda}_{;\mu\nu} - \xi^{\lambda}_{;\nu\mu} = -R^{\lambda}_{\sigma\mu\nu}\xi^{\sigma} \tag{6.109}$$

$$\eta_{\lambda;\mu\nu} - \eta_{\lambda;\nu\mu} = R^{\sigma}_{\lambda\mu\nu}\eta_{\sigma} \tag{6.110}$$

where

$$R^{\lambda}_{\sigma\mu\nu} \equiv \frac{\partial}{\partial x^{\mu}}\left\{{\lambda \atop \sigma\nu}\right\} - \frac{\partial}{\partial x^{\nu}}\left\{{\lambda \atop \sigma\mu}\right\} + \left\{{\theta \atop \sigma\nu}\right\}\left\{{\lambda \atop \theta\mu}\right\} - \left\{{\theta \atop \sigma\nu}\right\}\left\{{\lambda \atop \theta\mu}\right\} \tag{6.111}$$

is called the *Riemann-Christoffel tensor.*†

† It is to be noted that there is no standard sign convention for this tensor in the literature.

It is convenient at times to express the Riemann-Christoffel tensor with all indices in covariant positions by the formula

$$R_{\lambda\sigma\mu\nu} = g_{\lambda\kappa}R^{\kappa}_{\sigma\mu\nu} \qquad (6.112)$$

The $4^4 = 256$ components of this tensor satisfy the following set of identities which reduce the number of independent components to 20:

$$(a) \qquad R_{\lambda\sigma\nu\mu} = -R_{\lambda\sigma\mu\nu} \qquad (6.113)$$

$$(b) \qquad R_{\sigma\lambda\mu\nu} = -R_{\lambda\sigma\mu\nu} \qquad (6.114)$$

$$(c) \qquad R_{\mu\nu\lambda\sigma} = R_{\lambda\sigma\mu\nu} \qquad (6.115)$$

$$(d) \quad R^{\lambda}_{\sigma\mu\nu} + R^{\lambda}_{\mu\nu\sigma} + R^{\lambda}_{\nu\sigma\mu} = 0 \qquad (6.116)$$

The Riemann-Christoffel tensor often is called the *curvature tensor.* It can be shown that a necessary condition that Einstein space-time be Minkowskian in the large is that this tensor vanishes identically.

If the covariant derivative of the Riemann-Christoffel tensor is computed from (6.109) the following formula can be shown to hold:

$$R^{\lambda}_{\sigma\mu\nu;\tau} + R^{\lambda}_{\sigma\nu\tau;\mu} + R^{\lambda}_{\sigma\tau\mu;\nu} = 0 \qquad (6.117)$$

This important condition is known as the *Bianchi identity.*

The *Ricci tensor* $\{R_{\lambda\mu} = R_{\mu\lambda}\}$ is defined by the relation [54]

$$R_{\lambda\mu} = R^{\sigma}_{\lambda\mu\sigma} \qquad (6.118)$$

This tensor has 10 independent components. The vanishing of the Ricci tensor does not imply the vanishing of the Riemann-Christoffel tensor and consequently does not insure the Minkowski character of Einstein space-time in the large.

The *curvature invariant* is

$$R = R^{\lambda}_{\lambda} = g^{\lambda\mu}R_{\lambda\mu} \qquad (6.119)$$

The Bianchi identity (6.117) leads to the following formula satisfied by the Ricci tensor:

$$R^{\lambda}_{\mu;\lambda} = \tfrac{1}{2}R^{\lambda}_{\lambda;\mu} = \frac{1}{2}\frac{\partial R}{\partial x^{\mu}} \qquad (6.120)$$

which is called the *Ricci identity.*

27. The Einstein Tensor

The Einstein tensor $\{E_{\lambda\mu}(x) = E_{\mu\lambda}(x)\}$ is defined by the relation

$$E_{\lambda\mu} \equiv R_{\lambda\mu} - \tfrac{1}{2}g_{\lambda\mu}(R - 2\Lambda) \qquad (6.121)$$

where Λ is an arbitrary real constant which is called the *cosmical constant.* It was first introduced into relativity theory by Einstein.

On raising the first index of the Einstein tensor one has the relation equivalent to (6.121):

$$E^{\lambda}_{\mu} \equiv R^{\lambda}_{\mu} - \tfrac{1}{2}\delta^{\lambda}_{\mu}(R - 2\Lambda) \qquad (6.122)$$

Application of the Ricci identity (6.120) shows that

$$E^{\lambda}_{\mu;\lambda} = 0 \qquad (6.123)$$

This identity for the Einstein tensor, which holds for all values of the cosmical constant, is of great importance in relativity theory.

28. The Physical Model of Space-Time

The mathematical apparatus that has been constructed suffices to determine the paths of material particles (timelike geodesics) and of photons (null geodesics) once the 10 functions $\{g_{\lambda\mu}(x)\}$ of the line element have been specified. Conversely, it has been shown by H. Weyl (1885–1955) that the geodesics and null geodesics determine the geometry of the manifold [55].

More significant and more difficult is the problem of studying the structure of the space-time manifold under the assumption that this is determined by the physical content of the universe. To this end, it is necessary to have a physical model on which the mathematical structure can be based.

The obvious complexity of the observed astronomical universe makes it evident that one can proceed only by the introduction of an appropriately smoothed description of matter that eliminates the strong discontinuities produced by the existence of local concentrations of matter from the atoms to the stars and galaxies. The physical model starts from the hypothesis that for the study of the general characteristics of space-time structure it is sufficient to consider matter and radiation as smoothed into a kind of perfect fluid whose most important characteristic is its energy distribution. This approach resulted from Einstein's conviction that the gravitational behavior of any body or substance is determined in the first instance by its total energy content; however, that energy might be subdivided (as atomic energy, nuclear energy, radiation energy, etc.) for the purposes of more detailed physical theories.

While there is no particular reason to doubt the cogency of Einstein's hypothesis, we must note that it poses the intriguing and difficult practical problem of showing by experiment that such diverse physical entities as photons, neutrons, atomic beams, mesons, and so on, all are subject to gravitational forces.

29. The Perfect Fluid

The theory used to describe the perfect fluid in the general theory of relativity is a direct extension of that of classical hydrodynamics and its reformulation in the special theory. The fluid is assumed to be characterized by the physical properties of pressure p and mass density ρ. The energy density is defined by the symmetric tensor ($T^{\lambda\mu} = T^{\mu\lambda}$) [17]

$$T^{\lambda\mu} = \left(\rho + \frac{p}{c^2}\right)u^{\lambda}u^{\mu} - \frac{g^{\lambda\mu}p}{c^2} \qquad \text{g/cm}^3 \qquad (6.124)$$

where $\{u^{\lambda} = dx^{\lambda}/ds\}$ is the 4-velocity vector at any point in the fluid. (This definition of the 4-velocity makes it dimensionless, in contrast to the notation used in Sec. 14.)

The equations of motion and the conservation laws of momentum and energy of the fluid in the special theory of relativity are contained in the relations

$$\partial T^{\lambda\mu}/\partial x^{\mu} = 0 \qquad \text{(special theory)} \qquad (6.125)$$

The extension of these relations to the general theory is obtained simply by replacing partial differentiation by covariant differentiation, so that (6.125) is replaced by the formula

$$T^{\lambda\mu}{}_{;\mu} = 0 \qquad (6.126)$$

30. Einstein's Field Equations

The field equations for the determination of the metric functions $\{g_{\lambda\mu}(x)\}$, considered as potential functions determined by a particular distribution of momentum and energy of the fluid representing the distribution of matter and radiation in the universe, are assumed to be

$$E^{\lambda\mu} = -\kappa T^{\lambda\mu} \qquad \text{cm}^{-2} \qquad (6.127)$$

where $G = 6.670 \times 10^{-8}$ dyne-cm$^2/g^2$
$\kappa = 8\pi G/c^2 = 1.864 \times 10^{-27}$ cm/g

Writing (6.127) more completely, lowering the tensor indices for convenience, we have Einstein's field equations in the form

$$R_{\lambda\mu} - \tfrac{1}{2}g_{\lambda\mu}(R - 2\Lambda) = -\kappa T_{\lambda\mu} \qquad (6.128)$$

Inspection of (6.128) shows that the Einstein equations are quasi-linear; that is, they are linear functions of the second derivatives of the potential functions $\{g_{\lambda\mu}(x)\}$ but are nonlinear in these functions and their first derivatives. This property is of considerable importance in their study. Detailed examinations of the mathematical structure of Einstein's equations have been made by E. Cartan (1869–1951) [56], H. C. Levinson and E. B. Zeisler [57], and A. Lichnerowicz [58].

31. The Cosmical Constant

The cosmical constant Λ enters Einstein's field equations as an arbitrary constant of integration, so that its significance and magnitude can be determined only by physical considerations. There has been a strong tendency among theorists [59], following Einstein's views, to require this constant to be zero on theoretical grounds. However, the cosmological applications of the theory [60, 61] indicate that, while the numerical value to be attributed to Λ is small (and negative), it cannot be zero if the available data on the Hubble red shift of the expanding universe and on the density of matter as measured are accepted. The question of giving a definite physical theory of the cosmical constant is entirely open.

32. The Choice of Coordinate System

The choice of coordinate system in space-time in which one is to seek a solution of the Einstein field equations always presents a major problem. This was recognized by Einstein [9, 62] shortly after the proposal of his general theory of relativity. He pointed out that the necessity of using boundary conditions in order to obtain complete solutions of the field equations presents unexpectedly difficult questions of principle. This provided the major motivation for his turning his attention to cosmological applications of the theory.

These questions can be settled only in part by physical considerations alone, even in astronomical and cosmological applications, so that the resulting mathematical models are incomplete without further considerations of a topological nature. This has raised interesting questions concerning the mathematical nature of the Einstein space-time manifold in the large [63]. The vital importance of topological considerations for the physical models has been exhibited in a dramatic manner by Wheeler's theory of geometrodynamics [64, 65].

33. The Schwarzschild Solution

The most important exact solution of Einstein's field equations was given by K. Schwarzschild (1873–1916) in 1916 [66]. In form it describes the spherically symmetric static field about a point mass or the exterior field of a bounded spherical distribution. It has been shown by G. D. Birkhoff (1884–1944) [67] that any spherically symmetric solution, static or not, that satisfies Einstein's equations for free space outside a given spherical surface can be put in the Schwarzschild form by a suitable transformation of coordinates. In this sense the Schwarzschild solution is a canonical form for all spherically symmetric solutions [68].

If the central system at $r = 0$ has mass M, the Schwarzschild line element has the standard form ($\Lambda = 0$)

$$(ds)^2 = \left(1 - \frac{r_0}{r}\right) c^2(dt)^2 - \frac{(dr)^2}{(1 - r_0/r)} - r^2(d\theta)^2 - r^2 \sin^2 \theta (d\varphi)^2 \qquad (6.129)$$

where $r_0 = 2GM/c^2$ is the *Schwarzschild radius*. Since the whole mass is here supposed to be concentrated at the origin, the field equations take the form

$$R_{\lambda\mu} - \tfrac{1}{2}g_{\lambda\mu}R = 0 \qquad r \neq 0$$

It follows from this equation that $R = 0$, so that we have the simpler form

$$R_{\lambda\mu} = 0 \qquad r \neq 0 \qquad (6.130)$$

Direct calculation shows that relations (6.130) are satisfied by (6.129). The line element (6.129) then yields the equations of motion of test particles and photons in the study of the non-null and null geodesics.

34. Einstein's Tests of the General Theory

During the development of his theory of gravitation Einstein devised three tests suitable for experimental examination. The detailed mathematical theory of these tests follows from the study of the geodesics of (6.129) when applied to the gravitational field of the sun. This analysis is given in many places in the literature [17, 69, 70] and so we content ourselves with a brief statement of the results.

a. Perihelion Advance of a Planet. A study of the geodesic motion of a test particle in a planetary orbit from the line element (6.129) shows that the motion is almost elliptical, as in Keplerian motion, but that there is a slow direct precession of the perihelion given by the formula [70]

$$\langle d\varphi/dt \rangle = \left[\frac{3''.84}{\eta^{5/2}(1 - e)^2}\right]$$

seconds of arc per century

where e is the eccentricity and η is the semimajor axis in astronomical units of the planetary orbit. The formula gives the residue after subtraction of known perturbations from other planets. For Mercury and the Earth the formula yields values of $43''.03$ and $3''.84$ while the measured values are $43''.11$ and $5''.0$.

b. Bending of Starlight. A study of the null geodesics of (6.129) shows that a beam of light passing close to the edge of the sun is bent toward the latter. This can be interpreted as a gravitational attraction of the sun on the photons in the beam. This effect was first measured at the eclipse of May 29, 1919, and has been studied systematically by many eclipse expeditions since that time. The attendant publicity was very important in bringing the theory of relativity to the attention of the general public [71, 72, 72*a*].

The theoretically predicted deviation is $1''.745$. The measured values have been both larger and smaller than this figure. The data require troublesome corrections that make it difficult to arrive at definitive results. Measurements made in 1947 average $2''.01$, while those of 1952 yield $1''.70$ [73].

c. Einstein Red Shift. Light emitted from the surface of a massive star should have a shift of frequency toward the red (gravitational red shift). This can be regarded as a loss of energy of the photons in escaping from the gravitational attraction of the star. The theoretical value for the wavelength increase is

$$\delta\lambda/\lambda = 2.12 \times 10^{-6} M/r$$

where M and r are the mass and radius of the star measured in units of those of the sun. This is equivalent to the first-order Doppler shift for a speed of recession of 0.6 km/sec in the case of the sun ($M = 1$, $r = 1$).

Wavelength displacements of the predicted order of magnitude have been measured in the sun's spectrum, but recent discussion of the observational data has emphasized the difficulties of interpretation arising from effects of atomic collisions in the sun's atmosphere. The situation is complicated by the fact that the observed shift varies with the part of the sun's surface from which the light is received.

In the case of the star Sirius B (companion of Sirius) the measured shift agrees with the expected value, being about 31 times larger than for the sun. Similar good agreement is obtained for o_2Eridani B. For certain other white dwarfs for which large values of the Einstein red shift would be expected the measurements indicate only small values at best [74, 75].

35. Other Tests

Recent improvements in the accuracy of laboratory techniques and interest in problems of space physics have led to the suggestion of a number of possible new tests of the theory of relativity. Since most of these have not been proved as yet, we confine our remarks to a few cases which appear to be of particular interest and illustrate some of the more unusual problems involved.

a. Gravitational Energy of γ Rays. It has been remarked (Sec. 34*c*) that the Einstein (gravitational) red shift can be regarded as a measurement of the energy of a photon in the gravitational field of a star. The change in energy of photons moving vertically in the earth's gravitational field over a distance of 22.5 m has been measured for the 14-kev γ rays of Fe^{57} by use of the Mössbauer effect by Pound and Rebka and with improved accuracy by Pound and Snider [76]. They have verified the red-shift formula to an estimated accuracy of 1 per cent.

b. Gravitational Deflection of Slow Neutrons. The vertical displacement of a beam of slow neutrons moving horizontally in the earth's gravitational field has been measured by Reynolds [77]. Both thermal and filtered neutron beams were used. The observed drop corresponded to a value of the gravitational acceleration $g = 935 \pm 70$ cm/sec².

c. Gravitational Deflection of Electrons and Positrons. Preliminary experiments have been reported on the vertical motion of free electrons and positrons [78].

36. Cosmological Applications

The first attempt to apply the general theory of relativity to the study of the physical universe in the large was made by Einstein [9, 62]. Important extensions were made shortly after by De Sitter and by Friedman [5]. These theories go beyond the basic postulates of the theory of gravitation, owing to the necessity of introducing new hypotheses concerning the large-scale distribution of matter in the universe. Their most interesting consequence has been the suggestion that the universe might be finite in size and that it might be expanding or contracting with time.

These ideas received great impetus from the discovery by Hubble and Humason [79] that there is a red shift in the light received from distant nebulae (Hubble red shift) by an amount that increases roughly linearly with the nebular distance. When this shift is interpreted as a Doppler shift it indicates a mean recessional speed amounting to about $H = 100$ (km/sec)/megaparsec [80].

The postulates of the relativistic theory of the expanding universe have been examined in detail by Weyl [81] and Robertson [82], and extensive analyses of the various models have been given by McVittie [83] and by Heckmann and Schücking [84].

The introduction of the new techniques of radio astronomy and the discovery of galactic sources having very large red shifts have made this one of the most active fields of observational and theoretical astrophysics. One of the most pressing questions arises from the observation that some of the most distant objects emit energy at an unexpectedly large rate (quasars). The general theory of relativity plays an indispensable role in current attempts to understand the physical mechanisms that may be involved [85, 86].

References

1. Einstein, A.: *Ann. Physik*, **17**(4): 891 (1905).
2. Einstein, A.: *Ann. Physik*, **49**(4): 769 (1916).
3. Bergmann, P.: The Special Theory of Relativity, in "Encyclopedia of Physics," 2d ed., vol. 4, p. 109, Springer, Berlin, 1962.
4. Bergmann, P.: The General Theory of Relativity, in "Encyclopedia of Physics," 2d ed., vol. 4, p. 203, Springer, Berlin, 1962.
5. Einstein, A.: "The Meaning of Relativity," 5th ed., Princeton, Princeton, N.J., 1955.

6. Pauli, W.: "The Theory of Relativity," Pergamon Press, New York, 1958.
7. Von Laue, M.: "La Théorie de la relativité," 2 vols., Gauthier-Villars, Paris, 1922, 1926.
8. Whittaker, E. T.: "History of the Theories of Aether and Electricity," 2 vols., Nelson, London, 1951, 1953.
8*a*. Hill, E. L., *Phys. Rev.*, **84**: 1165 (1951).
9. Lorentz, H., et al.: "The Principle of Relativity," Dover, New York, 1951.
10. Reichenbach, H.: "The Philosophy of Space and Time," Dover, New York, 1958.
11. Grünbaum, A.: "Philosophical Problems of Space and Time," Knopf, New York, 1963.
12. Minkowski, H.: *Physik. Z.*, **10**: 104 (1909).
13. Cartan, E.: "Géométrie des espaces de Riemann," 2d ed., Gauthier-Villars, Paris, 1946.
14. Murnaghan, F.: *Phys. Rev.*, **17**: 73 (1921).
14*a*. Langwitz, D.: "Differential and Riemannian geometry," Academic, New York, 1965.
15. Rosen, N.: *Am. J. Phys.*, **32**: 597 (1964).
16. Hill, E. L.: Classical Mechanics as a Limiting Form of Quantum Mechanics, in P. Feyerabend (ed.), volume of essays dedicated to Herbert Feigl, University of Minnesota Press, Minneapolis, 1966.
17. Tolman, R. C.: "Relativity, Thermodynamics, and Cosmology," Oxford University Press, London, 1932.
18. Von Laue, M.: "Die Relativitätstheorie," 2 vols., Friedr. Vieweg & Sohn, Brunswick, Germany, 1952, 1953.
18*a*. Cunningham, E.: "The Principle of Relativity," Cambridge University Press, New York, 1921.
19. Murnaghan, F.: "The Theory of Group Representations," chap. 12, Dover, New York, 1963.
20. Gelfand, I. M., R. A. Minlos, and Z. Ya. Shapiro: "Representations of the Rotation and Lorentz Groups and Their Applications," Pergamon Press, New York, 1963.
21. Terrell, J.: *Phys. Rev.*, **116**: 1041 (1959).
22. Ney, E. P.: "Electromagnetism and Relativity," p. 59, Harper & Row, New York, 1962.
23. Crawford, F. S.: *Nature*, **179**: 35 (1957).
24. MacDuffee, C. C.: *Proc. Cambridge Phil. Soc.*, **56**: 176 (1960).
25. Langevin, P.: *Scientia*, **10**: 31 (1911).
26. Von Hoerner, S.: *Science*, **137**: 18 (1962).
27. Pilgeram, L.: *Science*, **138**: 1180 (1962).
28. Gursey, F.: *Rev. Fac. Sci. Univ. Istanbul*, **21(A)**: 129 (1956).
29. Smirnov, V. I.: "A Course of Higher Mathematics," vol. 2, p. 362, Addison-Wesley, Reading, Mass., 1964.
30. Haantjes, J.: *Proc. Ned. Akad. Wet.*, **44**: 814 (1941).
31. Hill, E. L.: *Phys. Rev.*, **72**: 143 (1947).
32. Schouten, J. A.: *Revs. Mod. Phys.*, **21**: 421 (1949).
33. Fulton, T., F. Rohrlich, and L. Witten: *Nuovo Cimento*, **26**: 652 (1962).
34. Planck, M.: *Ann. Physik*, **26**(4): 1 (1908).
35. Lewis, G. N., and R. C. Tolman: *Phil. Mag.*, **18**: 510 (1909).
36. Tolman, R. C.: "Relativity, Thermodynamics, and Cosmology," chap. 3, Oxford University Press, London, 1932.
37. McKibben, J. L.: *Phys. Rev.*, **70**: 101 (1946).
38. Bateman, H.: *Proc. London Math. Soc.*, **8**(2): 223 (1910).
39. Flanders, H.: "Differential Forms with Applications to the Physical Sciences," Academic, New York, 1963.
40. Murnaghan, F.: "Vector Analysis and the Theory of Relativity," p. 37, Johns Hopkins, Baltimore, 1922.
41. Frenkel, J.: "Lehrbuch der Elektrodynamik," vol. 2, Springer, Berlin, 1928.
42. Becker, R., and F. Sauter: "Electromagnetic Fields and Interactions," vol. 1, Blaisdell Publishing Co., New York, 1964.
43. Eötvös, R. v.: *Math. Naturwissen. Ber. Ungarn*, **8**: 65 (1890).
44. Zeeman, P.: *Proc. Ned. Akad. Wet.*, **20**: 542 (1917).
45. Eötvös, R. v., D. Pekar, and E. Fekete: *Ann. Physik*, **68**(4): 11 (1922).
46. Southerns, L.: *Proc. Roy. Soc. (London)*, **A84**: 325 (1910).
47. Dicke, R. H.: Experimental Relativity, in "Relativity, Groups, and Topology" (Les Houches Lectures, 1963), p. 165, Gordon and Breach, Science Publishers, New York, 1964.
48. Eisenhart, L. P.: "Non-Riemannian Geometry," vol. 8, American Mathematical Society Colloquium Publications, American Mathematical Society, Providence, R. I.
49. Eisenhart, L. P.: "Riemannian Geometry," Princeton, Princeton, N.J., 1926.
50. Einstein, A., and L. Infeld: *Can. J. Math.*, **1**: 209 (1949).
51. Infeld, L., and J. Plebanski: "Motion and Relativity," Pergamon Press, New York, 1960.
52. Adler, R., M. Bazin, and M. Schiffer: "Introduction to General Relativity," chap. 7, McGraw-Hill, New York, 1965.
53. Duff, G. F. D.: "Partial Differential Equations," chap. 4, University of Toronto Press, Toronto, Canada, 1956.
54. McVittie, G. C.: "General Relativity and Cosmology," 2d ed., p. 32, The University of Illinois Press, Urbana, Ill., 1965.
55. Weyl, H.: *Göttingen Nachr.*, p. 99, 1921: *Selecta*, p. 249, 1956, Birkhäuser Verlag, Basel.
56. Cartan, E.: *J. Math. Pures Appl.*, **1**: 141 (1922). "Oeuvres complètes," Pt. III, vol. 1, p. 549, Gauthier-Villars, Paris, 1955.
57. Levinson, H. C., and E. B. Zeisler: "The Law of Gravitation in Relativity," University of Chicago Press, Chicago, 1929.
58. Lichnerowicz, A.: "Théories relativistes de la gravitation et de l'Électromagnétisme," Masson & Cie, Paris, 1955.
59. Pauli, W.: "The Theory of Relativity," p. 220, Pergamon Press, New York, 1958.
60. McVittie, G. C.: "Fact and Theory in Cosmology," chap. 6, Eyre and Spottiswoode, London, 1961.
61. McVittie, G. C.: Galaxies as Members of the Universe, in "Problems of Extra-Galactic Research," p. 441, Macmillan, New York, 1962.
62. Einstein, A.: *Sitzber. Preuss. Akad. Wissen.*, p. 142, 1917.
63. Calabi, E., and L. Markus: *Ann. Math.*, **75**: 63 (1962).
64. Wheeler, J. A.: "Geometrodynamics," Academic, New York, 1962.
65. Marzke, R. F., and J. A. Wheeler: Gravitation as Geometry, in "Gravitation and Relativity," p. 40, W. A. Benjamin, New York, 1964.
66. Schwarzschild, K.: *Sitzber. Preuss. Akad. Wissen.*, p. 189, 1916.
67. Birkhoff, G. D.: "Relativity and Modern Physics," 2d ed., chap. 15, Harvard, Cambridge, Mass., 1927.
68. Bonnor, W.: On Birkhoff's Theorem, in "Recent Developments in General Relativity," p. 167, Pergamon Press, New York, 1962.
69. Kottler, Fr.: Gravitation und Relativitätstheorie, in "Encyklopädie der Mathematischen Wissenschaften," vol. 6:2:2, p. 159, B. G. Teubner, Leipzig, 1922.
70. McVittie, G. C.: "General Relativity and Cosmology," 2d ed., The University of Illinois Press, Urbana, Ill., 1965.
71. Ginzburg, V. L.: Experimental Verifications of the General Theory of Relativity, in "Recent Developments in General Relativity," p. 57, Pergamon Press, New York, 1962.

72. Bertotti, B., D. Brill, and R. Krotkov: Experiments on Gravitation, in "Gravitation: an Introduction to Current Research," p. 1, Wiley, New York, 1962.

72*a*. Von Klüber, H.: The Determination of Einstein's Light Reflection in the Gravitational Field of the Sun, "Vistas in Astronomy," vol. 3, p. 47, Pergamon Press, New York, 1960.

73. Van Biesbroek, G.: *Astron. J.*, **55**: 49 (1950); **58**: 87 (1953).

74. Luyten, W. J.: White Dwarfs and Degenerate Stars, in "Vistas in Astronomy," vol. 2, p. 1048, Pergamon Press, New York, 1956.

75. Luyten, W. J.: White Dwarfs, in "Advances in Astronomy and Astrophysics," vol. 2, p. 199, Academic, New York, 1963.

76. Pound, R. V., and J. L. Snider: *Phys. Rev. Letters*, **13**: 539 (1964).

77. Reynolds, A. W.: *Phys. Rev.*, **82**: 172 (1951).

78. Fairbank, W. M., F. C. Wittleborn, and L. V. Knight: *Science*, **144**: 562 (1964).

79. Hubble, E.: "The Realm of the Nebulae," Yale, New Haven, Conn., 1936.

80. McVittie, G. C. (ed.): "Problems of Extra-Galactic Research," Macmillan, New York, 1962.

81. Weyl, H.: *Physik. Z.*, **24**: 230 (1923): *Phil. Mag.*, **9**(7): 936 (1930).

82. Robertson, H. P.: *Proc. Natl. Acad. Sci. U.S.*, **15**: 822 (1929); *Revs. Mod. Phys.*, **5**:62 (1933).

83. McVittie, G. Ċ.: "General Relativity and Cosmology," 2d ed., chaps. 8 and 9, The University of Illinois Press, Urbana, Ill., 1965.

84. Heckmann, O., and E. Schücking: Newtonsche und Einsteinsche Kosmologie, and Andere Kosmologische Theorien, in "Encyclopedia of Physics," 2d ed., vol. 53, pp. 489, 520, Springer, Berlin, 1959.

85. McVittie, G. C.: *Astrophys. J.*, **140**: 401 (1964); **141**: 333 (1965).

86. Robinson, I., A. Schild, and E. L. Schücking (eds.): "Quasistellar Sources and Gravitational Collapse," The University of Chicago Press, Chicago, 1965.

Chapter 7

Gravitation

By HUGH C. WOLFE, American Institute of Physics

1. Inverse-square Law

Kepler's laws of planetary motion were derived by Newton on the hypothesis of a universal law of gravitational attraction (for derivation, see Chap. 4, Secs. 3 to 5). He postulated that every particle of matter attracts every other particle of matter in the universe with a force proportional to the product of their masses and inversely proportional to the square of the distance between them, the direction of the force being along the straight line between the particles:

$$F = G\frac{m_1 m_2}{r^2} \tag{7.1}$$

where G is known as the gravitational constant. Superposition is assumed, the resultant force on any one particle being the vector sum of the forces exerted on it by all other particles.

The gravitational field described by (7.1) is conservative, i.e., the work done against the mutual gravitational force between two particles in changing their separation is independent of the path and of the rate of motion and depends only on the magnitudes of the initial and final distances between them. Since equal displacements of the two particles with equal but opposite gravitational forces of mutual attraction give no net work, only the motion of m_2 relative to m_1 need be considered.

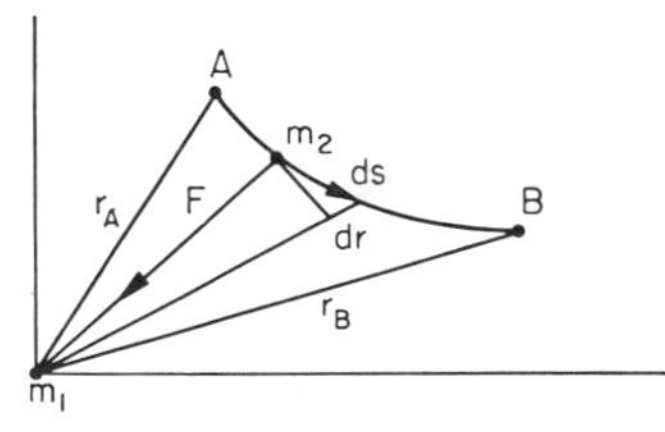

FIG. 7.1

The work done by an external agency against the gravitational force as m_2 moves from A to B relative to m_1 is

$$W = \int_A^B -\mathbf{F}\cdot\mathbf{ds} = \int_{r_A}^{r_B} \frac{Gm_1m_2}{r^2}\,dr$$

$$= Gm_1m_2\left[-\frac{1}{r}\right]_{r_A}^{r_B} = Gm_1m_2\left(\frac{1}{r_A} - \frac{1}{r_B}\right) \tag{7.2}$$

Let $r_A = \infty$ and replace r_B by r; then the mutual potential energy of the two particles, normalized to zero at infinite separation, is

$$W = -\frac{Gm_1m_2}{r} \tag{7.3}$$

The potential energy of interaction of m_1 with all other particles is

$$W = -Gm_1\int\frac{dm}{r} \tag{7.4}$$

where the integration is over all particles except m_1; and

$$V = \frac{W}{m_1} = -G\int\frac{dm}{r} \tag{7.5}$$

is the gravitational potential at the position of m_1, due to all other matter.

The motion of m_1 is determined by the resultant force on it,

$$\mathbf{F}_1 = -m_1G\int\frac{dm}{r^2}\mathbf{r}_0 \tag{7.6}$$

where r is the radial distance from dm to m_1 and $\mathbf{r}_0$ is a unit vector in the direction of r. This force is m_1 times the negative gradient at the position of m_1 of the potential V,

$$\mathbf{F}_1 = -m_1 \operatorname{grad} V \tag{7.7}$$

The quantity $\mathbf{E} = \mathbf{F}_1/m_1 = -\operatorname{grad} V$ is known as the intensity of the gravitational field at the position of m_1. Its value at any point in space is the gravitational force per unit mass which would act on a test particle if it were placed at that point without displacing the other masses.

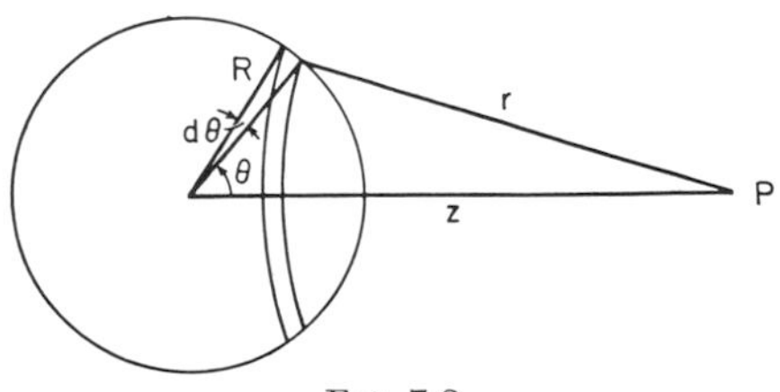

FIG. 7.2

The earth and other heavenly bodies being roughly spherical, it is important to determine the potential and field due to spherically symmetrical distribution of mass. The potential at point P at distance z

from the center of a spherical shell of radius R and thickness δR with uniform density ρ is

$$\delta V = -G \int_0^\pi \frac{\rho 2\pi R^2\, \delta R \sin\theta\, d\theta}{r} \tag{7.8}$$

Since $r^2 = R^2 + z^2 - 2zR\cos\theta$,

$$\frac{R\sin\theta\, d\theta}{r} = \frac{dr}{z} \tag{7.9}$$

and, hence, if $z > R$,

$$\delta V = -G\frac{\rho 2\pi R\,\delta R}{z}\int_{z-R}^{z+R} dr = -G\frac{\rho 4\pi R^2\,\delta R}{z}$$

$$= -G\frac{\delta M}{z} \tag{7.10}$$

At exterior points the potential (and hence the field also) due to a spherical shell of uniform density is the same as though all of the mass were concentrated at the center. This result holds for any aggregate of concentric spherical shells, i.e., for any spherically symmetrical body, even if the density varies with radius. Likewise, the mutual potential energy of any aggregate of spherically symmetric bodies, such as the solar system, is the same as though the mass of each body were concentrated at its center.

Considering a point inside a spherical shell, where $z < R$, the lower limit of integration is $R - z$, since r is an essentially positive quantity. Instead of (7.10)

$$\delta V = -G\frac{\rho 2\pi R\,\delta R}{z}\int_{R-z}^{R+z} dr = -G\rho 4\pi R\,\delta R$$

$$= -G\frac{\delta M}{R} \tag{7.11}$$

This potential is constant throughout the interior of the spherical shell and equal to the value the potential would have at the surface if all the mass were concentrated at the center. The gradient of this potential vanishes so that there is no force on a mass particle due to a spherical shell of mass surrounding it.

For a solid sphere of radius R_0 with uniform density ρ, integration of (7.10) and (7.11) yields

$$z > R_0 \qquad V = -G\frac{M}{z} \tag{7.12}$$

$$z < R_0 \qquad V = -G\rho\left(2\pi R_0^2 - \frac{2\pi}{3}z^2\right) \tag{7.13}$$

The corresponding gravitational field intensities are

$$z > R \qquad \mathbf{E} = -\text{grad } V = -G\frac{M}{z^2}\mathbf{r}_0 \tag{7.14}$$

$$z < R \qquad \mathbf{E} = -\text{grad } V = -G\frac{4\pi\rho z}{3}\mathbf{r}_0 \tag{7.15}$$

These results for spherical mass distributions may also be obtained from a general theorem of Gauss relative to inverse-square fields. Consider any closed mathematical surface surrounding a particle of mass m. An infinitesimal cone of solid angle $d\Omega$ cuts out of this surface a vector element of area $\mathbf{dS}$ in the direction of the outward normal. Forming the scalar product, $\mathbf{E}\cdot\mathbf{dS}$, where E is the field due to m at the position of dS,

$$\mathbf{E}\cdot\mathbf{dS} = -G\frac{m}{r^2}dS\cos\theta \tag{7.16}$$

where θ is the angle between the radius and the outward normal. But $dS\cos\theta$ is the projection of dS on a sphere of radius r and

$$\frac{dS\cos\theta}{r^2} = d\Omega \tag{7.17}$$

Hence $\quad \iint \mathbf{E}\cdot\mathbf{dS} = -Gm\iint d\Omega = -4\pi Gm \quad (7.18)$

A cone $d\Omega$ may cut the surface an odd number of times but each additional pair of cuttings will give a pair of equal and opposite contributions to the integral. If m were located outside the surface, every cone would cut the surface an even number of

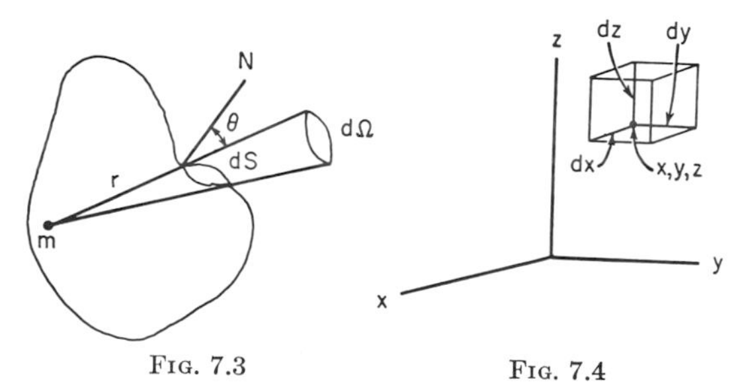

FIG. 7.3 FIG. 7.4

times and would give no net contribution to the integral. Hence, *Gauss' theorem:* If E is the resultant field due to a distribution of particles, both within and without a closed surface,

$$\iint \mathbf{E}\cdot\mathbf{dS} = -4\pi G\Sigma m \tag{7.19}$$

where Σm is the sum of the masses of all the particles enclosed by the surface.

An equivalent differential equation may be found by applying (7.19) to a differential volume element in the form of a parallelepiped of sides dx, dy, dz with one corner at x, y, z.

$$\begin{aligned}\iint \mathbf{E}\cdot\mathbf{dS} &= \left(\frac{\partial E_x}{\partial x} + \frac{\partial E_y}{\partial y} + \frac{\partial E_z}{\partial z}\right)dx\,dy\,dz \\ &= \text{div }\mathbf{E}\,dv \\ &= -\left(\frac{\partial^2 V}{\partial x^2} + \frac{\partial^2 V}{\partial y^2} + \frac{\partial^2 V}{\partial z^2}\right)dv = -\nabla^2 V\,dv\end{aligned} \tag{7.20}$$

By *Gauss' theorem* (7.19), this is equal to $-4\pi G\Sigma m$ or $-4\pi G\rho\,dv$, where ρ is the density at x, y, z. Hence

$$\text{div }\mathbf{E} = -\nabla^2 V = -4\pi G\rho \tag{7.21}$$

known as *Poisson's equation.* In empty space, where $\rho = 0$,

$$\nabla^2 V = 0 \tag{7.22}$$

known as *Laplace's equation.* All the equations of this section are equally valid for electrostatics, with charge substituted for mass, since they depend specifically on the inverse-square law.

2. Gravitational Constant, G

Direct determination of the constant G by measurement of the force between known masses in the laboratory was carried out by Cavendish in 1798. Lead balls 2 in. in diameter were hung by short wires from the ends of a thin wooden bar 6 ft long suspended at its center by a torsion wire 39 in. long. The torsion constant of the wire was found from the period and moment of inertia. If T is the period, I the moment of inertia, and $k\theta$ the restoring torque corresponding to an angular deflection θ from the equilibrium position,

$$T = 2\pi\sqrt{\frac{I}{k}} \qquad \text{or} \qquad k = \frac{4\pi^2 I}{T^2} \tag{7.23}$$

The equilibrium position was taken as the center of swing for small oscillations. The shift in equilibrium position, $\Delta\theta$, was observed when a pair of 12-in. lead spheres were turned on their supporting suspension from a position where their attractions for the pair of small spheres gave additive clockwise torques to a position where they gave counterclockwise torques. The mean distance d between centers in each position was about 9 in. Then

$$k\,\Delta\theta = 2G\frac{Mm}{d^2}a \tag{7.24}$$

where a is the distance from the axis to the centers of the small spheres. Making a number of corrections for the oversimplifications in this equation, Cavendish found

$$G = 6.754 \times 10^{-11}\ \text{N m}^2\ \text{kg}^{-2} \tag{7.25}$$

A more refined repetition of the experiment by P. R. Heyl,* using small spheres of gold, platinum, and optical glass on the torsion balance *in vacuo* and using steel cylinders of 66 kg as the attracting masses, yields the currently most reliable value,

$$G = 6.670 \times 10^{-11}\ \text{N m}^2\ \text{kg}^{-2} \tag{7.26}$$

Equation (7.14) leads to

$$E = \frac{GM}{R_0^2} = G\frac{4}{3}\pi R_0\rho$$

or

$$\rho = \frac{3E}{4\pi R_0 G} \tag{7.27}$$

Taking the mean radius of the earth as about 6.371×10^8 cm and the gravitational field intensity at the earth's surface as approximately 981 cm/sec² yields for the mean density of the earth

$$\rho = 5.51\ \text{g/cm}^3 \tag{7.28}$$

A more refined calculation gives for the mass of the earth

$$M = 5.976 \times 10^{27}\ \text{g} \tag{7.29}$$

* P. R. Heyl: *Natl. Bur. Standards J. Research*, vol. 5, p. 1243, 1930.

3. Acceleration of Gravity g

The force on a mass m at the earth's surface due to the earth's gravitational field **E** is

$$\mathbf{F} = m\mathbf{E} \tag{7.30}$$

In the absence of other forces, according to Newtonian mechanics, the mass will have a gravitational acceleration

$$\mathbf{a} = \frac{\mathbf{F}}{m} = \mathbf{E} \tag{7.31}$$

If the earth were a nonrotating, spherically symmetric body, remote from all other bodies, the gravitational acceleration of a mass at its surface would be given by (7.14)

$$E = G\frac{M}{R_0^2} \tag{7.32}$$

with direction toward the center.

What is generally called the acceleration of gravity, g, differs from (7.32) in several important respects. As pointed out in Chap. 4, Sec. 2, **g** is measured relative to coordinates attached to the rotating

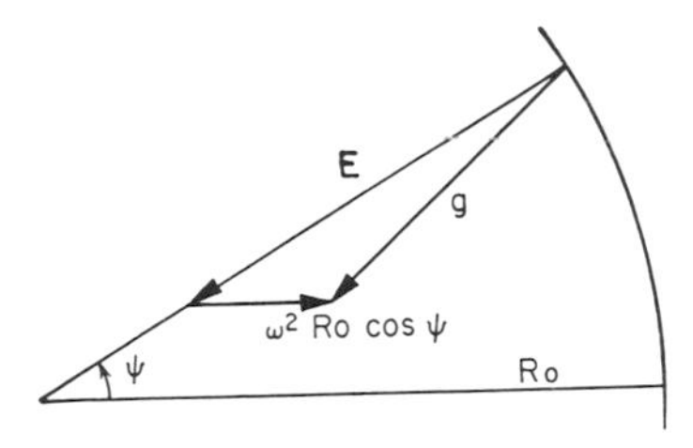

FIG. 7.5

earth and differs from **E** by the centrifugal term $\omega^2 R_0 \cos\psi$, where ω is the earth's angular velocity of rotation, R_0 is the earth's radius, and ψ is the latitude at the place where g is measured. The centrifugal term amounts to 3.392 cm/sec² at the equator and zero at the poles. In measurements of g by means of pendulums or gravimeters, the *Coriolis term*, $2\omega \times \mathbf{v}$, which is responsible for rotation of the plane of oscillation of a Foucault pendulum, plays no role.

The earth is not spherical. Its shape, undoubtedly associated with rotation, is most nearly described as a triaxial spheroid, bulging at the equator and flattened at the poles. Its shape is moderately well represented by an ellipsoid of revolution: polar radius = 6,356,912 meters; equatorial radius = 6,378,388 meters.

If (7.32) were directly applicable, this difference in R_0 would mean an increase in E from the equator to the pole by 6.63 cm/sec² and the two effects together would give an excess in g of about 10 cm/sec² at the poles as compared with the equator. The nonspherical distribution of mass of the earth compensates for about half of this difference.* The sea-level values of g are 978.0524 cm/sec² at the equator, and 983.2329 cm/sec² at the poles.

* Sigmund Hammer, *Geophysics*, vol. 8, pp. 57–60, 1943.

The *figure of the earth* is a name sometimes associated with an equation describing the variation of g with position on the reference spheriod. With values of the constants determined by Heiskanen in 1938, the equation is

$$g = 978.0524[1 + 0.005297 \sin^2 \psi - 0.0000059 \sin^2 2\psi + 0.0000276 \cos^2 \psi \cos 2(\lambda + 25°)] \quad (7.33)$$

where ψ is latitude and λ is longitude (positive east of Greenwich).

Actual values of g at any location differ from those on an ideal reference spheroid. The three effects taken into account by geophysicists are:

1. The *free-air correction for elevation.* Measured g would be reduced by $0.000094h$ cm/sec², where h is elevation above sea level in feet, as a simple inverse-square-law effect.

2. The *Bouguer correction for elevation.* The free-air correction is too large by $0.000034h$ cm/sec², which is the added gravitational intensity due to a uniform slab of material lying between sea level and the elevation h and having a density of 2.67 g/cm³, the average density of crustal rocks.

3. The *topographic correction.* The Bouguer correction is too large by the gravitational effect of nearby land masses above the elevation h and also of masses that would be required to fill nearby depressions up to the elevation h.

Deviations not accounted for by these three corrections are called *Bouguer anomalies* and their occurrence indicates abnormal variation of density with distance from the earth's center. Bouguer anomalies extend over large areas of the earth's surface. After the corrections (Bouguer corrections are made in the case of the oceans by computing the change in observed g due to replacing the ocean waters with material of the average density of crustal rocks), observed values of g are in general too low on high plateaus and too high over the oceans, being about right over land near sea level. These anomalies give evidence for *isostasy*, the hypothesis that the earth's crust "floats" on a denser liquid core with the lower density crust penetrating more deeply below mountain chains and with the base of the crust nearer to sea level under the oceans.* Isostatic anomalies are differences between measured g and the theoretical value found by combining the other three corrections with an isostatic correction—the computed effect on g of the variation in thickness of the earth's crust required for isostatic equilibrium. Their existence indicates departure from isostatic equilibrium with consequent mechanical stresses in the earth's crust. The continents are close to isostatic equilibrium, but some islands have positive anomalies as large as 0.1 cm/sec², and some ocean deeps have negative anomalies as large as 0.2 cm/sec².

Measurement of g. The best absolute measurements of g have been made by determination of the periods of reversible pendulums. The period (for infinitesimal amplitude) of a physical pendulum is

$$T = 2\pi \sqrt{\frac{I}{mgh}} \quad (7.34)$$

* G. B. Airy, *Trans. Roy. Soc. (London)*, *A*, vol. 145, pp. 101–104, 1855.

where I is the moment of inertia about the axis of rotation and h is the distance from the axis to the center of gravity. For each axis there is another parallel axis, with the same period, through a point called the center of oscillation. This axis lies at a distance h' below the center of gravity, and the plane through the two axes passes through the center of gravity. By the parallel-axis theorem,

$$I = m(k^2 + h^2) \quad (7.35)$$

where mk^2 is the moment of inertia about a parallel axis through the center of gravity. Equality of the two periods requires that

$$\frac{k^2 + h^2}{h} = \frac{k^2 + h'^2}{h'} \quad (7.36)$$

or

$$k^2 = hh' \quad (7.37)$$

Putting this value into either expression for T yields

$$T = 2\pi \sqrt{\frac{h + h'}{g}} \quad (7.38)$$

or

$$g = \frac{4\pi^2(h + h')}{T^2} \quad (7.39)$$

Kater's convertible pendulum (1817) was designed with two knife-edges turned inward on opposite sides of the center of gravity so that it could be swung from each in turn and their separation adjusted until the two periods were equal. The pendulum is so designed that the period comes out extremely close to 2 sec, and the period is determined by measuring the time interval between consecutive coincidences of the pendulum with a chronometer. From the period and the measured distance between the knife-edges the value of g is obtained. For precise results further refinements are necessary. The pendulum is made symmetrical in geometrical shape, but not in mass distribution, with the knife-edges equidistant from the center of figure. The two periods are adjusted to be nearly equal; g is now given by

$$g = 8\pi^2 \left(\frac{T_1^2 + T_2^2}{h_1 + h_2} + \frac{T_1^2 - T_2^2}{h_1 - h_2} \right) \quad (7.40)$$

The buoyant force of air on the pendulum and the added moment of inertia due to air dragged along affect the individual periods T_1 and T_2 but the symmetry of the pendulum causes these effects to cancel when the two periods are combined as in Eq. (7.40). To obtain g with precision of 1 part in 10^6, the periods, T_1 and T_2, and the distance between the knife-edges, $h_1 + h_2$, must be measured with slightly greater precision. If T_1 and T_2 differ by less than 0.1 per cent, it is not necessary to measure h_1 and h_2 separately to better than 0.1 per cent when the unsymmetrical mass distribution makes h_1 and h_2 significantly different.

In evaluating g by Eq. (7.40), the axes of rotation are assumed to be at the knife-edges, neglecting the finite curvature of the knife-edges. If the radii of curvature of the two knife-edges are equal, the effect cancels out exactly when the two periods are combined as in Eq. (7.40). In practice, this is accomplished by having planes instead of knife-edges on the pendulum and swinging it from both planes on the same knife-edge. An alternative procedure is to take the average of two values of g as given by Eq. (7.40), the second value being based on measurements with the two knife-edges demounted and interchanged.

Yielding of the support on which the pendulum swings has the effect of raising the axis of rotation above the center of curvature of the knife-edge. In computing g from Eq. (7.40), one must add to $h_1 + h_2$ a length equal to the horizontal displacement of the supports at the axis of rotation under a horizontal force equal to the weight of the pendulum.

One further correction is necessary. The equations used here are those for the period of the pendulum in the limit of zero amplitude. Since measurements are made at finite amplitude, the periods must be adjusted to zero amplitude by use of the equation

$$T = T_{\mathrm{lim}} \left[1 + \frac{1^2}{2^2} \sin^2 \frac{\alpha}{2} + \frac{1^2 \cdot 3^2}{2^2 \cdot 4^2} \sin^4 \frac{\alpha}{2} + \cdots \right] \quad (7.41)$$

where α is the amplitude of swing (Chap. 3, Sec. 10).

Taking into account all these refinements and using six different reversible pendulums, some with planes on the pendulums and some with interchangeable knife-edges, an absolute determination of g was made at Potsdam in 1906.* The value obtained at the Potsdam laboratory was 981.274 cm/sec^2. Values of g at a network of stations over the world have been determined from the Potsdam value as a base by comparison of the periods of the same pendulum or group of pendulums at the various locations. Everything else being the same or reduced to equivalence by corrections for pressure, temperature, yielding of support, and amplitude of swing, the period for a given pendulum varies inversely as the square root of g.

Newer absolute determinations of g at the National Bureau of Standards in Washington† and at the National Physical Laboratory, Teddington, England,‡ plus re-evaluation of the correction for systematic errors in the original Potsdam data* lead to the conclusion that g values in the internationally accepted system, based on Potsdam, are too high by about 14 parts per million.†

* F. Kuhnen and P. Furtwangler, "Bestimmung der absoluten Grosse der Schwerkraft zu Potsdam mit Reversionspendelm," Königliche Preussiche Geodatische Institut, 1906.

† P. R. Heyl and G. S. Cook, *Natl. Bur. Standards J. Research*, vol. 17, p. 805, 1936.

‡ J. S. Clark, *Trans. Roy. Soc. (London)*, A, vol. 238, p. 65, 1939.

4. Equivalence of Inertial and Gravitational Mass

The Einstein interpretation of gravitational acceleration as a geometrical property of space would be impossible if this acceleration were different for different bodies. In Eq. (7.32) the expression for gravitational acceleration, independent of the body accelerated, was obtained by canceling out the "gravitational" mass in the numerator and the "inertial" mass in the denominator. Hence equality of gravitational acceleration can be described as equality of inertial and gravitational mass. In view of the great differences in ratio of protons to neutrons, in binding energies associated with strong nuclear forces, etc., it is particularly important to investigate this question by comparing materials of high and low atomic mass. The experiments of Eötvös early in this century showed the equivalence of inertial and gravitational mass for various materials with a claimed accuracy of 1 part in 10^9. New experiments by Dicke‡ and his collaborators at Princeton over the past several years have established to 1 part in 10^{11} the equality of gravitational acceleration for lead and copper.

The experiment involved a suspended triangular system of two copper masses and one lead chloride mass. The earth, carrying this system, moves around the sun with its centripetal acceleration determined by the gravitational pull of the sun. The orientation of the suspended body relative to the sun changes with earth's rotation, and any failure of the expected equality would lead to an oscillation with a 24-hr period, which was not observed. Elaborate precautions were taken to eliminate disturbing phenomena and to observe very small effects.

* H. L. Dryden, *Natl. Bur. Standards J. Research*, vol. 29, p. 303, 1942.

† A full treatment is given by A. H. Cook, *Metrologia* I, **3**: 84–114 (July, 1965).

‡ R. H. Dicke, in H. Y. Chiu and W. F. Hoffmann (eds.), "Gravitation and Relativity," chap. 1, W. A. Benjamin, New York, 1964. R. H. Dicke, *Sci. Am.*, December, 1961, p. 84. See also E. L. Hill, The Theory of Relativity, Part 2, Chap. 6 in this Handbook.

Chapter 8

Dynamics of the Solar System

By G. M. CLEMENCE, Yale University

1. Introduction

The dynamics of the solar system takes its peculiar form because of the distinctive characteristics of astronomical observations, as compared with those of ordinary physics. These characteristics are:

1. Both the observer and all the objects observed are in constant motion relative to one another. This requires that the reference frames be dynamically defined and means that in the systems commonly used the coordinates of any body are continuously changing from the motion of the coordinate system, in addition to the changes caused by the body's own motion.

2. Distances cannot be measured directly; the only directly measurable quantity is the angle subtended at the observer by the lines of sight to two celestial objects or by one line of sight and the local direction of gravity. Either of these measurements can be made with a precision of a tenth of a second of arc for a single observation, and the precision can be much increased by repetition of the observations; the amplitude of a periodic change can be measured with a precision of a thousandth of a second or even better in favorable cases. Relative distances in the solar system may be inferred to eight or nine significant figures from the periods of revolution by means of the law of gravitation. It is customary to express them in astronomical units, the astronomical unit being defined by a in the equation

$$n^2a^3 = k^2(1 + m) \tag{8.1}$$

If n is the angular mean motion expressed in radians per mean solar day, m the mass of a planet expressed in terms of the sun's mass as the unit, and k the Gaussian constant of gravitation (0.017 202 098 95 exactly), then a (which is a close approximation to the mean distance of a planet from the sun) comes out in astronomical units. In the case of the earth the value of a determined by (8.1) is 1.000 000 03 a.u., but due to perturbations by the other planets this is not identical with the mean value of the distance between the earth and the sun.

Natural units of length and mass (the astronomical unit and the mass of the sun) are used in preference to the meter and gram for convenience; there is no astronomical requirement for any other units of length and mass. Moreover, the meter and gram cannot be employed with sufficient precision for astronomical purposes. To express the astronomical unit in centimeters, it is necessary first to measure the radius of the earth in centimeters, then to employ direct triangulation with the radius of the earth as the base line, the angular displacement of a planet against the background of stars being observed from opposite sides of the earth. The result, 1.497×10^{13} cm, is uncertain in the last figure. Similarly, masses in grams can be determined to only three significant figures, while relative masses are known to four or five in favorable cases.

3. Experiments cannot be performed. All that can be done is to observe what is taking place. In consequence the observational material is great in quantity, and much ingenuity is required in interpretation. Long periods of time are dealt with; observations made a century ago are still in daily use, and for some purposes observations made twenty centuries ago are of value. Doubtless for this reason, indefinite integrals have been more frequently employed in the dynamics of the solar system than in other branches of physics.

The limitations imposed by the impossibility of performing experiments can be in part removed by refinement of observational techniques; precise observations continued for a short time may yield as much as imprecise ones continued for a long time. Hence astronomers continually strive for greater accuracy of observation; the techniques are more sophisticated and more precise than in any other branch of science. The interpretation of such precise observations requires much arithmetic, many significant figures, and facility and ingenuity in computational techniques.

Some of the foregoing statements now (1965) require modification since radar echoes have been received from Mercury and Venus. The high precision of such observations makes it practicable (assuming the velocity of propagation of radar radiation to be known in meters per second) to introduce the meter as a measure of astronomical distances.

2. Equations of Motion

Let $\xi_a, \eta_a, \zeta_a, \xi_b, \eta_b, \zeta_b$ be the coordinates of two point masses m_a and m_b in a right-handed cartesian system of coordinates, which is supposed to be an inertial system, not rotating or subject to any acceleration. If the distance between m_a and m_b is r, then by Newton's law the force acting between them is

$$F = \frac{k^2 m_a m_b}{r^2}$$

The constant k depends on the units of mass, time, and length chosen. If the sun's mass is taken as unity, the mean solar day as unit of time, and the astronomical unit as unit of length, then k is the Gaussian constant. (In theoretical work it is sometimes convenient to choose the units so as to make k unity.)

The vector from m_a to m_b has the direction cosines

$$\frac{\xi_b - \xi_a}{r} \qquad \frac{\eta_b - \eta_a}{r} \qquad \frac{\zeta_b - \zeta_a}{r}$$

Hence the ξ component of the force acting upon m_a is

$$F_{a\xi} = \frac{F(\xi_b - \xi_a)}{r} = k^2 m_a m_b \frac{\xi_b - \xi_a}{r^3}$$

with similar expressions for the η and ζ components; and the ξ component of the force acting upon m_b is

$$F_{b\xi} = k^2 m_a m_b \frac{\xi_a - \xi_b}{r^3}$$

with similar expressions for the η and ζ components.

Since the rate of change of momentum is proportional to the force and is in the direction in which the force acts,

$$F_{a\xi} = m_a \frac{d^2\xi_a}{dt^2} \qquad F_{b\xi} = m_b \frac{d^2\xi_b}{dt^2}$$

whence

$$m_a \frac{d^2\xi_a}{dt^2} = k^2 m_a m_b \frac{\xi_b - \xi_a}{r^3}$$

$$m_b \frac{d^2\xi_b}{dt^2} = k^2 m_a m_b \frac{\xi_a - \xi_b}{r^3}$$

These are the ξ equations of motion of m_a and m_b, with

$$r^2 = (\xi_a - \xi_b)^2 + (\eta_a - \eta_b)^2 + (\zeta_a - \zeta_b)^2$$

Introduce additional point masses m_j ($j = 1,2,3, \ldots$) into the system. The attractions of m_j on m_a and m_b are given by equations similar to the preceding and the total accelerations of m_a and m_b may be obtained by summing all these accelerations, giving

$$m_a \frac{d^2\xi_a}{dt^2} = k^2 m_a m_b \frac{\xi_b - \xi_a}{r^3} + \Sigma_j k^2 m_a m_j \frac{\xi_j - \xi_a}{\rho_{j,a}^3} \tag{8.2}$$

$$m_b \frac{d^2\xi}{dt^2} = k^2 m_a m_b \frac{\xi_a - \xi_b}{r^3} + \Sigma_j k^2 m_b m_j \frac{\xi_j - \xi_b}{\rho_{j,b}^3} \tag{8.3}$$

with similar equations for η and ζ, and

$$\rho_{j,a}^2 = (\xi_a - \xi_j)^2 + (\eta_a - \eta_j)^2 + (\zeta_a - \zeta_j)^2$$
$$\rho_{j,b}^2 = (\xi_b - \xi_j)^2 + (\eta_b - \eta_j)^2 + (\zeta_b - \zeta_j)^2$$

Let the origin of coordinates be taken at m_a, which is equivalent to the linear transformation

$$\xi_b - \xi_a = x \qquad \xi_j - \xi_a = x_j$$

from which follows

$$\xi_j - \xi_b = x_j - x$$

and put

$$r_j^2 = x_j^2 + y_j^2 + z_j^2$$
$$\rho_j^2 = (x_j - x)^2 + (y_j - y)^2 + (z_j - z)^2$$

Divide (8.2) by m_a and (8.3) by m_b, and subtract the first from the second. The result is the equation of motion of m_b relative to m_a,

$$\frac{d^2x}{dt^2} = -k^2(m_a + m_b)\frac{x}{r^3} - \Sigma_j k^2 m_j \frac{x_j}{r_j^3} + \Sigma k^2 m_j \frac{x_j - x}{\rho_j^3} \tag{8.4}$$

with similar equations for y and z.

Suppose that m_a is taken as the unit of mass. Then $m_a = 1$ and the subscript may be dropped from m_b, (8.4) being written

$$\frac{d^2x}{dt^2} = -k^2(1 + m)\frac{x}{r^3} + \Sigma_j k^2 m_j \left[\frac{x_j - x}{\rho_j^3} - \frac{x_j}{r_j^3}\right] \tag{8.5}$$

This is the general equation of motion of a planet or comet m with the sun as origin of coordinates, under the action of disturbing planets m_j, on the assumption (true for all practical purposes) that the mass of a planet or the sun may be regarded as a point mass. The first term represents the action of the sun on m, the first term in the square brackets the action of m_j on m, and the second the action of m_j on the sun (this is often called the *indirect term*).

With slight modification the same equation may be used for the motion of a satellite around its primary.

3. Method of Solution

If all the m_j vanish, then (8.5) and the similar equations for y and z are easily solved. The six constants of integration are the values of x, y, z, $\dot{x}$, $\dot{y}$, $\dot{z}$ at the origin from which time is reckoned, and they determine the position of m in a conic section as a function of the time.

The *problem of three bodies* is the case where one of the m_j (say m_1) is different from zero. The practical methods of solution all depend on successive approximations, which are possible because m_1 and m are small; for Jupiter, the most massive planet, $m = 10^{-3}$. In the first approximation m_1 and m are put equal to zero. In the second approximation the right-hand sides of the equations may be evaluated on the assumption that m_1 and m move in known conic sections. A double integration then gives improved coordinates of m. The second approximation for m_1 (m being different from zero) may be obtained by interchanging the roles of m and m_1 in the equations. The coordinates resulting from the second approximation may then be substituted in the right-hand sides of the equations, leading to yet better values for the coordinates.

An additional planet m_2 may be introduced by first considering the case of m and m_2 alone, then m_1 and m_2 alone, in the first two approximations, introducing all three planets in the third approximation. For the principal planets three or four approximations are necessary if the results are to be comparable in precision with the observations. The masses of the minor planets and comets are negligible, and this leads to simplifications; the motions of the principal planets (m_j) being assumed already known, the successive approximations are necessary only for the particular m under investigation.

4. Form of Solution

When the interval of time involved is relatively short, as for a newly discovered object, or when the motion is relatively slow, as for the outer planets, the method of *special perturbations* may be used. In this method the equations of motion are integrated numerically step by step. The quantities given by the integration may be the rectangular coordinates and velocity components or any other set of six so-called *elements* that define the position and velocity vector of the object. In any case the results appear in the form of numbers tabulated at intervals of time which are chosen small enough to obtain the requisite convergence of the differences, but not much smaller; in practice it is usually desired to neglect the differences (or numerical derivatives with respect to the time) of higher order than the eighth. A tabular interval as long as 200 days may be used for the outer planets, while for objects passing close to the sun or to Jupiter an interval shorter than a day may be necessary. No matter what choice of elements is made, at least one double integration is necessary in effect if not in form. It follows that the effect of errors of rounding increases proportionally to the $\frac{3}{2}$ power of the number of steps, whence many steps require many significant figures. In the most extensive application of this method yet made the motions of the five outer planets were simultaneously traced for 400 years at intervals of 40 days, and 14 significant figures were required in order to prevent the errors of rounding from exceeding the errors of observation. With the method of special perturbations the six constants of integration for each body must be known with entire precision at the start; and hence the computer must integrate, compare with observation, revise the constants of integration, then integrate again, and repeat this cycle of operations once or twice before the integration can be extended far into the future with confidence that it can be used for accurate prediction. Besides this drawback, the method is very laborious when long intervals of time are involved. On the other hand, it is the only known method that can always be applied, no matter what the character of the motion may be.

When a body is relatively close to its primary and when at the same time its departure from circular motion is not too pronounced, as is the case with the moon disturbed by the sun, the reciprocal of the distance between it and a disturbing body may be developed in convergent multiple Fourier series on the assumptions that the motions are elliptic. The development may be completely literal, all the elements being denoted by symbols, or partly literal and partly numerical, according to the exigencies of the case. In any case, the time, or one or more monotonic functions of it, are left as literal quantities. Similar series representing the disturbing forces can be obtained and integrated term by term to give series expressions for the coordinates, which, however, are only approximate. These series are said to be accurate to the first power of the disturbing masses. They may be used to calculate increments to the disturbing forces that were used, which then yield improved expressions for the coordinates; and this cycle may be repeated as often as is desired. This is known as the method of *general perturbations*, because it yields general expressions for the coordinates as functions of the time. Formally, the position of the object can be found for any time whatever by simple substitution into these series; but owing to the limited precision of the constants of integration, which must always be determined from observations, the errors of the results unavoidably increase with the time from the initial epoch. Furthermore, it is not known whether the series for the coordinates converge or not in the cases actually arising; the question of convergence is at once of great interest and great difficulty.

In the case of bodies relatively more distant from their primaries, such as the principal planets, the method of general perturbations has not yet been applied in its pure form, owing to the excessively slow numerical convergence of the series developments in their most general form. By allowing the time and its integral positive powers to appear as factors of certain periodic terms, the practical range of application of the method can be greatly extended. In this form it has been applied successfully to all the principal planets except Pluto and to many minor planets. While it necessarily fails to give accurate results for very remote epochs, it seems to be adequate for a few centuries in most cases where it has been applied.

Nothing has been definitely established regarding the configuration of the solar system a million years ago. On the basis of plausible assumptions it appears that the character of the motions of the more massive planets and satellites has not greatly changed since they were originally formed. On the other hand, it is highly probable that many minor planets are fragments of what were once larger bodies.

Any mathematical expression yielding the coordinates of a body as functions of the time is known as a *theory of the motion*, or briefly a *theory*. If the coordinates are explicitly tabulated for special values of the time as with the method of special perturbations, the theory is said to be a *special theory*. If the time reckoned from a *fundamental epoch* is expressed by a symbol, the theory is called a *general theory*. General theories may be *numerical*, in which the time is the only symbolic quantity, or *literal*, in which everything is symbolic, or a combination of the two. At first sight it would seem that a literal theory could be constructed without knowledge of the numerical values of the constants of integration, but this is true only in a trivial sense. General series expansions in powers of the constants are needed, and their approximate values must be known in order to determine the degree of approximation required; a literal theory in so general a form that it could be directly applied to many different bodies is not to be thought of, owing to the prohibitive amount of labor required.

5. Precession and Nutation

By reason of its rotation, the earth is nearly an oblate spheroid, with equatorial radius about 6.378 400

$\times 10^8$ cm and polar radius about 2.15×10^6 cm shorter. Owing to the asphericity the moon and sun produce important perturbations of the direction of the earth's axis of rotation in inertial space. The perturbations may be developed analytically by the method of general perturbations applied to Poisson's equations. When this is done, it is found that the motion of the axis can be expressed as the sum of two components. One of these components, called the *lunisolar precession*, is the continuous description of a cone by the axis of rotation, the angle of the cone being about 47° and its axis at any time being approximately perpendicular to the plane of the earth's orbit around the sun. The period is about 26,000 years; but inasmuch as the axis of the cone is also in motion, the period is not exactly defined. The other component, called the *nutation*, is a superposed somewhat irregular motion, approximating to a cone of angle about 18 seconds of arc and period about 18.6 years.

The foregoing description may be made more precise by introducing the fundamental concepts of spherical astronomy. Imagine a sphere of radius very large by comparison with distances in the solar system and center at the center of the earth. Suppose further that this sphere, the *celestial sphere*, is absolutely fixed in inertial space. The axis of rotation of the earth at any instant intersects the celestial sphere in two points called the *true poles of the equator*, and the plane of the earth's equator intersects the celestial sphere in a great circle 90° from the poles, called the *true equator*. Either pole approximately describes a small circle on the celestial sphere of radius 23.5°, in about 26,000 years, by virtue of the lunisolar precession, and a smaller circle of radius 9 seconds of arc, in about 18.6 years, by virtue of the nutation. The nutation having been accurately calculated at any time, it can be applied to the true pole, yielding a hypothetical *mean pole* which is affected only by precession, and a *mean equator*. It is useful also to introduce the notion of the *mean pole of date*, by which is meant the instantaneous position of the moving mean pole on the celestial sphere, and the *fixed pole*, which is fixed on the celestial sphere and coincides with the mean pole at a specified epoch, as Jan. 1, 1950, Greenwich mean noon. The mean pole moves on the celestial sphere at a rate of about 20 seconds of arc per year.

6. Frames of Reference

The most natural coordinate system to use on the celestial sphere is one analogous to latitude and longitude on the earth. In principle the true pole is conveniently established on any night as the center of the circular arcs traced in the sky by stars near the pole, and the true equator may be found by means of a divided circle placed so that its plane contains the earth's axis of rotation. Angular distances north or south of the true equator are called *true declinations*. They may by calculation be freed from the effects of nutation, then becoming *mean declinations*.

To define the origin of *right ascension* (which corresponds to longitude on the earth), it is necessary to examine the orbital motion of the earth. By developing the theory of the earth around the sun it is found that a plane of reference can be defined which passes through the center of the sun and is such that the departures of the earth from this plane can be expressed as the sum of a number of strictly periodic terms of very small total amplitude (about 0.5 second of arc as seen from the sun). These small terms are called perturbations of the sun's latitude, and the plane is called the plane of the moving ecliptic or the *ecliptic of date*. The theory of the earth's motion shows that the ecliptic is slowly rotating about an axis through the sun at a rate of 47 seconds of arc per century. Knowing the rate of motion and the orientation of the axis, it is possible to introduce the concept of a *fixed ecliptic*, defined similarly to the fixed equator. With the exception of the perturbations of the sun's latitude the moving ecliptic is the path on the celestial sphere in which the sun appears to revolve once a year. The intersections of the moving ecliptic with the true equator are called the *true equinoxes*, and the one of these that the sun passes in March is the *vernal* equinox or simply equinox. The vernal equinox is the origin of right ascensions, which are reckoned eastward, that is, in the direction that makes the sun's right ascension always increase. The angle between the plane of the true equator and the plane of the moving ecliptic is the *true obliquity*, which differs from the *mean obliquity* by the nutation in obliquity. It is now easy to introduce the concept of a *mean equinox of date* and a fixed equinox.

Enough has been said to indicate that true right ascensions and declinations are observable coordinates and that a correct theory of the nutation permits the derivation of mean right ascensions and declinations. The mean right ascension and declination of a star are slowly varying in a progressive manner; and the greater part of the variation is due to precession, a small remainder being due to the star's own motion in space and to the translational motion of the solar system. The part of the change in the mean coordinates of a star that is not due to precession is called *proper motion*. Only a few stars have proper motions as large as a second of arc per year, while the changes due to precession average about 50 seconds over the sky.

The laws of motion are referred to an inertial frame of reference, as are the theories of motion derived from these laws. The celestial sphere was defined as fixed with reference to an inertial frame, but the mean right ascensions and declinations of all celestial bodies are continuously changed by precession. Hence in order to refer astronomical observations to an inertial frame it is necessary to know the precession. In fact the practical realization of an inertial frame is precisely equivalent to the determination of the precession.

7. Determination of the Precession

The precession is a precisely known function of the masses of the moon and earth, the elements of their orbits, and the dynamical flattening of the earth. If the equatorial and polar moments of inertia of the earth could be directly measured with sufficient precision, the precession might be calculated by a formula, but this is not the case; and in practice the dynamical flattening must be derived from the observed value of the precession.

From the dynamical point of view the most obvious method of determining the precession is by comparing the theoretical and observed motions of the planets. This method failed before the advent of the theory of general relativity because the theories were not sufficiently precise, and it is difficult to apply because of the very large amount of calculation required. The difficulty is that the principal effect of the precession is of precisely the same form as one of the constants of integration in the theories; in order to separate the two it is necessary to measure the amplitude of small periodic terms in the planetary motions. Notwithstanding this the method shows greater promise than any other, at least for the immediate future, since the necessary observations are already available.

Historically the method most used has been the analysis of proper motions of stars. The stars were assumed as a whole to constitute an inertial system (except, in recent years, for galactic rotation, which in this method must be determined simultaneously with the precession), and on this assumption it was attempted to impose the condition that the average value of all the proper motions must be zero. No difficulty is encountered with the first four significant figures, but beyond these the method fails for two reasons. First, it is necessary to determine at least six quantities simultaneously: the precession, the two constants of galactic rotation, and the three components of the velocity of the solar system. The only available method is that of least squares; but the motions of the stars do not have a Gaussian distribution, and hence the method is not strictly valid. The presence of a few stars with motions much greater than the average constitutes an embarrassment. The second difficulty is that the observational material contains systematic errors of obscure origin, so that different choices of stars and different hypotheses of errors lead to different results.

Recently it has been proposed to use the extragalactic nebulas as an inertial frame, and extensive observational programs have been commenced with this object in view. These programs will require 30 to 50 years to reach the fifth significant figure, owing to the limited precision of the observations, but if pursued with sufficient vigor may eventually prove more precise than any other method.

8. Perturbations of Planets and Satellites

When the disturbing forces exerted by one planet on another are developed in trigonometric series, on the assumption that the planets move in fixed ellipses of known shape, relative orientation, and relative size, the forces may be expressed as the sum of terms of the form

$$C_{j,k} \cos (jg' + kg) + S_{j,k} \sin (jg' + kg)$$

C and S being numbers, g' and g being known periodic and linear functions of the time, and j and k being integers such that for any non-negative value of j, k takes on all values positive and negative including zero. It is found that large values of either j or k are associated with small values of C and S; this property makes the series practically useful. When a term of this form is integrated, the quantity $jg' + kg$ appears as a divisor of $C_{j,k}$ and $S_{j,k}$. The functions g' and g relate one to one planet and the other to the other; they are closely connected with the rate of angular revolution around the sun, and are determined by observation. When j and k are both positive, C and S will be diminished in size by the integration, and in this case the perturbation is said to be one of short period, the period being shorter than the period of revolution of either planet. When j and k differ in sign, it may happen that for small values of j and k the function $jg' + kg$ becomes small by comparison with either g' or g; in such cases the perturbation is one of long period and C and S are augmented by the integration. Occasionally it may happen that $jg' + kg$ vanishes within the errors of observation for small values of j and k; such a case is said to be one of *resonance*, and a modification of the usual technique must be made.

Evidently, values of j and k may always be found that will make $jg' + kg$ as small as we please, but it is not found in practice that any difficulty is introduced in this way if the smaller of j and k is greater than about 15; in such cases C and S are so small that they do not become appreciable after the integration.

Cases of resonance and of perturbations of long period may occur among three bodies if the motion of one is commensurable, or nearly so, with a near commensurability between two others.

The earliest-known perturbation of long period, and the first to receive a gravitational explanation, is the great inequality between Jupiter and Saturn. In the time required for five revolutions of Jupiter, Saturn makes nearly two. The period of the perturbation is about 900 years, Jupiter being disturbed to the extent of 0.3° and Saturn 0.7°.

An interesting case of resonance is that of the Trojan minor planets. Nearly a score of minor planets have mean periods of revolution equal to that of Jupiter. Their mean positions are nearly in Jupiter's orbit, some remaining 60° in advance of Jupiter and the rest 60° behind. But they exhibit quite large oscillations, or *librations*, about their mean positions. Another case of resonance is the inner three of the four great satellites of Jupiter, which preserve such a relation that the positions of any two can be immediately calculated from knowledge of the third.

Among the nearly 1,600 well-known minor planets between the orbits of Mars and Jupiter not one is known to move exactly in resonance with either large planet. At the distances from the sun where such resonances would occur there are gaps containing no planets, and the gaps are most pronounced in the simplest cases, where the ratio of the period of Jupiter to that of the planet would be 2/1, 3/2, 3/1, 4/3, 5/2, etc. Similar gaps exist in the rings of Saturn, where the periods would be commensurable with those of the larger satellites. The existence of such gaps is a necessary consequence of a smooth distribution of the energy of the moving bodies; a smooth distribution of orbits in space would have been accompanied by anomalies in the energy distribution.

9. Determination of Time

Any recurring phenomenon, the recurrences of which can be counted, is a measure of time. What physicists require is an invariable measure of time, suitable for use as a standard, which is to say as an independent variable in the study of phenomena. Until now at least, the only recurring phenomena that can be supposed invariable, and that can be counted, are the periods of rotation of celestial bodies on their axes, and the periods of revolution about their primaries; the most convenient one of these, and the one universally used until the present century, is the period of rotation of the earth.

In practice the period used is not the true period of rotation (with respect to an inertial frame) but the period with respect to the fictitious mean sun, called the *mean solar day,* which is defined by means of gravitational theory in such a way that its ratio to the true period of rotation is absolutely fixed; the true period of rotation is approximately 23 hr 56 min 4.0988 sec of mean solar time. Mean solar time is actually determined from night to night by observations of transits of stars across the meridian, using stars whose mean solar times of transit can be very precisely predicted on the basis of past observations.

During the first half of the present century it became possible to compare the rotation of the earth with the orbital motions of the moon, Mercury, Venus, the earth, and the four great satellites of Jupiter with sufficient precision to show that the mean solar day is not an invariable unit of time. The principal variation is an irregular one, such as would be caused by the accumulation of small random changes in the derivative of the rotational velocity with respect to the time. The observed changes in the length of the day have amounted to a part in 3×10^7, and the cause has not yet been discovered. Besides these random changes there is a gradual increase in the length of the day due to tidal friction, amounting to a part in 10^8 in a century, and a periodic change in the course of a year of the same order of magnitude.

In recent years a better unit of time than the mean solar day has been required, and the mean period of the earth's revolution has come into use. The actual period is not strictly invariable, but the gravitational theory of the earth's motion permits the mean period to be defined. The mean period has a slight progressive change, and the period actually used is the value of the mean period at 1900.

It is desirable to give a rigorous statement of what is meant by an invariable measure of time. The only definition at present possible to use in practice, and one that seems adequate for all purposes, is: an invariable measure of time is the one defined by the accepted laws of motion, that is, Newton's laws as slightly modified by general relativity; in other words, an invariable measure of time is the one according to which observations of celestial bodies continuously agree with the theories of their motions. This definition has always been used, although the fact may not have been appreciated. The substitution of the year for the day as a standard of time has been brought about by the realization that the theory of the rotation of the earth is inadequate.

During the past few years a radically new sort of clock has appeared, generally known as an atomic clock. The essential feature of the most common variety consists of a resonant cavity containing cesium, the atoms of which are excited electrically to change their state with a very high frequency, which in turn is used to monitor an ordinary quartz-crystal clock. Such clocks are now used to control the frequency of standard radio-frequency broadcasting stations. They are not truly invariable because, in accordance with a well-known principle of relativity, their running depends on the intensity of the gravitational field in which they are placed; the largest effect is an annual variation of about 3.3 parts in 10×10^9. Probably atomic time is to be identified with the well-known proper time s of a moving relativistic observer, while astronomical time is the coordinate (sometimes called cosmic) time t, but this supposition has yet to be demonstrated by experiment. Furthermore, the question whether or not the ratio of the year 1900 to the atomic second is changing progressively, as has been supposed by some theoretical physicists, must remain undecided for at least some years while observational data are being collected.

10. Relativity

The theory of relativity shows that in the most general sense time and space are not mathematically distinguishable from each other. A few remarks on this subject are necessary in order to show that the definitions of Sec. 9 are not ambiguous.

In the equations of motion given by relativity, the three rectangular coordinates are not expressed as functions of the time as an independent variable. Equations so expressed could not possibly be true simultaneously for all observers. The general equations of motion are of a type that preserves the invariance of a certain relation between the coordinates and the time. However, it is found that the general equations may be particularized for a single observer and that they then may be made to take the form of Newton's equations with the addition of very small corrective terms, expressing the so-called relativistic motion of the perihelia. In these particularized equations the time may properly be regarded as the independent variable *for the observer in question.* The statements of Sec. 9 must be understood in this restricted sense. As long as physical observations are made from the earth (or in fact from any other place in the solar system, unless the present precision is greatly increased), all observers have sensibly the same time scale; and all may use the same particularized equations without fear of sensible ambiguity.

11. National Ephemerides

The labor and difficulty of calculating precise right ascensions and declinations of celestial objects, as well as of precise predictions of phenomena depending on them such as the rising and setting of the sun and moon, are so great that it must be done once and for all; no observing astronomer could possibly provide himself with the necessary results. In consequence the governments of the larger nations publish annual volumes of ephemerides and astronomical phenomena, sharing the calculations to avoid duplication of labor. An ephemeris is simply a table in which the argument is the time. The annual volume published in the United States is the *American Ephemeris and Nautical Almanac.* Such a volume serves two distinct purposes. First, it enables observations to be made by indicating when events will occur as seen from different parts of the earth. Second, it permits the comparison of observations with theory; the analysis of such comparisons is the only means by which the theories can be tested and improved and new discoveries made in the field.

12. Celestial Navigation

The governments of the larger nations also publish navigational almanacs, which supply the astronomical data needed by navigators or surveyors in determining latitude and longitude from observations of celestial bodies. The principle of such determinations is as follows: If a straight line be conceived to be projected from the center of the earth to a celestial body at any instant of time, this line intersects the surface of the earth in a single point. An observer at this point will see the object directly overhead, at an angular altitude of 90° above the horizon. His latitude is equal to the declination of the object, and his longitude is equal to its Greenwich hour angle. (The Greenwich hour angle of an object is the angle between two planes, one being the plane of the Greenwich meridian and the other the plane through the true poles and the object; it is very simply calculated when the right ascension and the Greenwich mean time are known.) The navigator need only be supplied with tables giving the declinations and Greenwich hour angles of various objects as functions of the Greenwich time and with a timepiece showing Greenwich time so that he can use the tables. A navigational almanac consists principally of such tables.

The navigator measures altitudes of celestial bodies directly with a sextant. In practice it is easier not to observe objects precisely in the zenith but to make observations at moderate altitudes. The principle just described can be easily extended to this case. Suppose an object is to be observed at an altitude h. The observer is then necessarily located somewhere on a small circle of radius 90° minus h, and center at the point described in the preceding paragraph. Suppose now that another object is observed, locating the observer on a second small circle. If certain simple rules are followed in selecting the objects to be observed, the two circles will intersect, placing the observer definitely at one of the two points of intersection; and in general the two points are so far apart that the observer finds it easy to decide which of the two he is at. In practice it is not necessary to draw the complete circles; the navigator is likely to know his position nearly enough so that he need draw only two short arcs near the point of intersection where he is, and he is further able to regard the arcs as straight lines on his chart without sensible error. The lines are called *lines of position,* and their intersection is a *fix.*

At night there is no dearth of celestial objects to be observed, but by day only the sun, moon, and Venus can be observed, and often only the sun is available. Using only the sun, the navigator can still locate himself at any time on a line of position. Observations of the sun at noon give a line nearly perpendicular to those obtained in the early morning or late afternoon, and knowledge of his approximate speed and direction of motion enables the navigator to establish a *running fix* from such lines.

13. Astronomical Constants

The constants of importance in the dynamics of the solar system may be divided into several classes: (1) the elements of the orbits of the several bodies; (2) their masses, the mass of the sun being taken as unity; (3) the constants specifying their size, shape, rotation, and inner constitution, and the positions of their axes of rotation (the constants relating to the earth are of special importance); and (4) the velocity of light, which is needed in order to deduce the actual position of a body at any instant from the observed position, which is valid for the earlier instant when the light seen by the observer left the body.

Orbital Elements. Any set of six constants that serve for the calculation of the position and velocity vector at a specified instant are properly regarded as the elements of an orbit. Among the many useful sets are the rectangular coordinates themselves and the rectangular components of the velocity. This set is convenient when the coordinates are obtained by numerical integration. It is customary to express the coordinates in astronomical units, and the velocity components in astronomical units per day, or per any other convenient unit of time.

In the first approximation to the orbit, which for most bodies may be regarded as a fixed ellipse, a set of elements more directly connected with the geometry of an ellipse is more convenient than rectangular coordinates. Such a set is the following:

1. The inclination J of the plane of the ellipse to the fixed equator of 1950.

2. The right ascension N of the ascending node on the fixed equator, this node being on the line of intersection of the plane of the orbit and the plane of the equator, at the point on the celestial sphere where the body crosses the equator from south to north.

3. The angular distance ω measured in the plane of the orbit from the node to the perihelion in the direction of motion of the body. The perihelion is the point of the ellipse nearest the sun and is at one end of the major axis of the ellipse. The elements J, N, ω define the orientation of the orbit in space.

4. The period of revolution P, or indifferently the angular mean motion n, connected with P by the relation $Pn = 2\pi$. Instead of either P or n, half the major axis of the ellipse may be used, this quantity a being found by (8.1).

5. The time T of perihelion passage, that is, one of the times when the body is at its perihelion, or alternatively, the angular distance f from the perihelion to the body at a specified instant of time.

6. The eccentricity e of the ellipse, which defines its shape. If r is the distance from the sun to the object, then at perihelion $r = a(1 - e)$, and at aphelion $r = a(1 + e)$.

If t is the time reckoned from the instant T, then the mean anomaly g is given by $g = nt$. If the eccentric anomaly ϵ is defined by Kepler's equation

$$g = \epsilon - e \sin \epsilon$$

then the rectangular coordinates may be found from the equations

$$x = A_x(\cos \epsilon - e) + B_x \sin \epsilon$$
$$y = A_y(\cos \epsilon - e) + B_y \sin \epsilon$$
$$z = A_z(\cos \epsilon - e) + B_z \sin \epsilon$$

where the A and B are constants given by

$$A_x = a(\cos \omega \cos N - \sin \omega \sin N \cos J)$$
$$B_x = a \cos \phi \,(-\sin \omega \cos N - \cos \omega \sin N \cos J)$$
$$A_y = a(\cos \omega \sin N + \sin \omega \cos N \cos J)$$
$$B_y = a \cos \phi \,(-\sin \omega \sin N + \cos \omega \cos N \cos J)$$
$$A_z = a(\sin \omega \sin J)$$
$$B_z = a \cos \phi \cos \omega \sin J$$

where $\cos \phi = (1 - e^2)^{1/2}$.

The velocity components $\dot{x}$, $\dot{y}$, $\dot{z}$ are given by

$$\dot{x} = (-A_x \sin \epsilon + B_x \cos \epsilon)\frac{an}{r}$$

$$\dot{y} = (-A_y \sin \epsilon + B_y \cos \epsilon)\frac{an}{r}$$

$$\dot{z} = (-A_z \sin \epsilon + B_z \cos \epsilon)\frac{an}{r}$$

where $r^2 = x^2 + y^2 + z^2$.

The above formulas give the rectangular coordinates and velocity components at any instant in terms of the elements J, N, ω, a, T, and e.

When we pass beyond elliptic motion and consider the actual motion of a body disturbed by the principal planets, it is still possible and useful to assign precise meanings to the six elements just mentioned; having given the actual values of x, y, z, $\dot{x}$, $\dot{y}$, $\dot{z}$ at a particular instant, the six elements may be determined in such a way as to preserve the relations given above. Suppose now that the same process is carried through for some other instant; it will be found that the elements obtained are not quite the same as before. Elements determined in this way are called *osculating elements*, and in order to have precise meaning any set of them must always be associated with the appropriate instant of time as *epoch of osculation*. The changes that they undergo with the time may be regarded as perturbations of the osculating elements.

Osculating elements are almost invariably used with the method of special perturbations, but with analytic theories they are inconvenient and it is customary then to define the elements rigorously in connection with certain terms of the perturbations. The elements then have analytic definitions instead of simple geometrical ones; they are called *mean elements*.

In Table 8.1 are given the elements of the nine principal planets. To the accuracy given, no distinction need be made between osculating elements and mean elements. For convenience, they are given in the usual form, which is slightly different from that described above. The plane of reference is the fixed ecliptic of the stated epoch instead of the equator. The inclination to this plane is denoted by i, and the longitude of the node (the distance from the equinox to the node on the ecliptic) by Ω. The angle π is the sum of Ω and the arc from the node to the perihelion, called the longitude of the perihelion. The mean motion n is expressed in seconds of arc per mean solar day, and the other four angular elements in degrees.

TABLE 8.1. ELEMENTS OF THE ORBITS OF THE PRINCIPAL PLANETS
Epoch Jan. 1, 1950

Planet	a	P, years	n	e	i	Ω	π	Longitude
Mercury...	0.39	0.24	14,732	0.206	7	48	77	16
Venus......	0.72	0.62	5,768	0.007	3	76	131	82
Earth......	1.00	1.00	3,548	0.017	0	...	102	100
Mars.......	1.52	1.88	1,887	0.093	2	49	335	146
Jupiter.....	5.20	11.86	299	0.048	1	100	14	311
Saturn.....	9.54	29.46	120	0.056	2	113	92	164
Uranus.....	19.19	84.02	42	0.047	1	74	170	93
Neptune...	30.07	164.79	22	0.008	2	131	44	195
Pluto......	39.46	247.70	14	0.249	17	110	224	166

Masses. In Table 8.2 are given the reciprocals of the masses of the principal planets, the mass of the sum being taken as unity. They are the conventionally adopted values, and they are not to be regarded as strictly correct. Owing to the very complicated way in which the masses affect the ephemerides and to the great confusion that would be caused by frequent changes as well as the great labor involved, it is necessary and desirable to retain the same values for long periods of time; it is much more important that the theories of the planets be gravitationally *consistent* with one another than to have any one theory as accurate as it is possible to make it at the moment. The values given here are sufficiently accurate so that only the most exhaustive investigations can discover any error in them; they have been in general use since 1900.

TABLE 8.2. RECIPROCALS OF MASSES

Planet	Reciprocal mass
Mercury..................	6,000,000
Venus....................	408,000
Earth....................	333,432
Earth + moon...........	329,390
Mars.....................	3,093,500
Jupiter..................	1,047.355
Saturn...................	3,501.6
Uranus...................	22,869
Neptune..................	19,314
Pluto....................	360,000

Other Constants. Among the constants describing the size, shape, and other physical features of bodies of the solar system, those relating to the earth are of special practical importance since many of them must be used daily in the prediction of astronomical phenomena. In Table 8.3 are given the more important of them. More or less rigorous relations are known among some of these constants, and some are connected with those in Tables 8.1 and 8.2. The solar parallax, for example, is connected with the mass of the earth-moon system by a relation that is known more accurately than either of the constants themselves. The constants entering into these various relations have come to be known as the *system of astronomical constants;* among them it is

Table 8.3. Astronomical Constants

Solar parallax, angle subtended by earth's equatorial radius at distance of one astronomical unit (seconds of arc)	8.80
Constant of nutation (seconds of arc)	9.21
Precessional motion of the pole (seconds of arc per tropical year at 1900)	20.0468
Speed of rotation of the ecliptic (seconds of arc per tropical year at 1900)	0.4711
Obliquity of the ecliptic at 1900	23°27′8.26″
Time for light to travel one astronomical unit (seconds)	498.580
Tropical year (days at 1900)	365.242 198.79
Mass of earth/mass of moon	81.53
Mean synodic month (mean period of phases of the moon), days	29.530 588

possible to choose a number of independent ones, the *fundamental constants*, from which all the others can be derived by theoretical relations. Different choices are possible. The questions of which of these constants can be determined most accurately by direct observation and which can be most accurately derived from the others are very complicated and difficult and cannot be discussed here. The values given are, as with the masses, conventional ones; they are not all correct in the last figure given, and not all the theoretical relations among them are rigorously satisfied. But normally they will serve as the basis of any theoretical investigation; and, taken as a whole, no set could be chosen that would have the practical advantages of these. They are properly regarded as parts of a system of reference, to which observations are referred.

References

1. Brown, E. W.: "Introductory Treatise on the Lunar Theory," Cambridge, New York and London, 1896.
2. Brown, E. W., and C. A. Shook: "Planetary Theory," Cambridge, New York and London, 1896.
3. Chauvenet, W.: "Manual of Spherical and Practical Astronomy," 5th ed., vols. I, II, Lippincott, Philadelphia, 1889, 1887.
4. Eckert, W. J., D. Brouwer, and G. M. Clemence: "Coordinates of the Five Outer Planets 1653–2060," *Astron. Papers of the American Ephemeris*, vol. XII, 1951.
5. Hill, G. W.: "A New Theory of Jupiter and Saturn," *Astron. Papers of the American Ephemeris*, vol. IV, 1890.
6. Moulton, F. R.: "Introduction to Celestial Mechanics," 2d rev. ed., Macmillan, New York, 1928.
7. Newcomb, S.: "The Elements of the Four Inner Planets and the Fundamental Constants of Astronomy," *Supplement to the American Ephemeris and Nautical Almanac* for 1897, Washington, 1895.
8. Newcomb, S.: "Theories of the Four Inner Planets," *Astron. Papers of the American Ephemeris*, vol. III, part V, Washington, 1891.
9. Tisserand, F.: "Traité de mécanique céleste," vols. 1889–1896, Gauthier-Villars, Paris.

Chapter 9

Control Mechanisms

By HAROLD K. SKRAMSTAD, Naval Ordnance Laboratory
and GERALD L. LANDSMAN, Douglas Aircraft Company

1. Introduction

A *control mechanism* is a device in which a controlled quantity (or output) is made to be a desired function of a controlling quantity (or input).

Control mechanisms can be divided into two classes: (1) *closed loop*, in which a function of the output is compared with the input, with their difference being used to control the output to be the desired function in the input; and (2) *open loop*, in which no function of the output is used in controlling the output to be the desired function of the input. Figure 9.1 is a block diagram of an open-loop control mechanism and Fig. 9.2 of a closed-loop control mechanism.

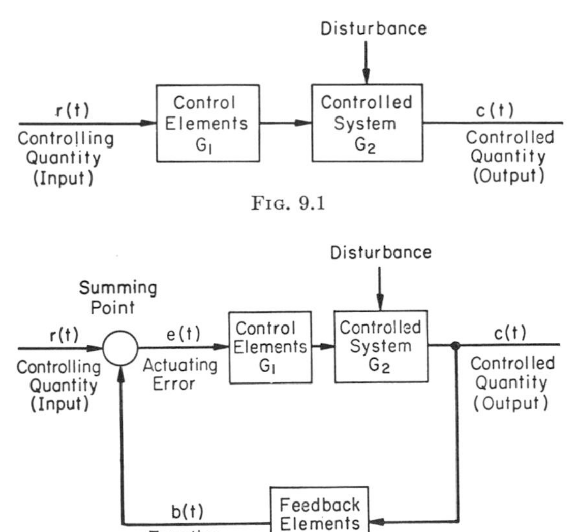

FIG. 9.1

FIG. 9.2

Control mechanisms take many forms. Consider a control mechanism consisting of a variable voltage source applied to the input of a power amplifier, which is used to control the armature current and thus the speed of an electric motor. This is an open-loop system (Fig. 9.1). The controlling quantity, or input, is the variable voltage source; the control elements are the power amplifier and control windings on the motor, the controlled system is the motor armature and connected load, and the controlled quantity is the speed of the motor.

If, in addition, a tachometer is connected to the output shaft of the motor and the difference between the output voltage of the tachometer (function of the output) and the voltage source (input) is applied to the amplifier, the system becomes closed loop (Fig. 9.2). The controlling quantity, control elements, and controlled system are the same as before. The tachometer is the feedback element, the voltage output from the tachometer is the function of the output, and the difference between this voltage and the input voltage is the actuating error.

If, in the motor control system described above, a potentiometer instead of a tachometer is connected to the output shaft, and a voltage proportional to shaft position is combined with the input to form the error, the control mechanism becomes a position control instead of a speed control.

In general, closed-loop mechanisms give much more accurate control than the corresponding open-loop mechanisms. When the motor control system above is connected open loop, variation of the characteristics of the amplifier and motor, load on the motor, etc., would cause differences between the desired and actual output. In the closed-loop connection with tachometer or potentiometer feedback, these same component variations would have much smaller effects, since the output speed (or position) is corrected by the difference between actual and desired speeds (or position).

A closed-loop control mechanism in which it is desired to maintain the output at an arbitrary (but adjustable) value is sometimes called a *regulator*. A common example is an automatic furnace control system for maintaining the temperature in a room at a designated value. The controlled quantity (or output) is the temperature in the room, and the difference between the thermostat setting (input) and the measured temperature (function of the output) is the error, which by turning the furnace on and off controls the room temperature.

A closed-loop control mechanism in which the output, usually a position, is made continuously a desired function of the input over a wide range of input values, and where external power is supplied, is often called a *servomechanism*. The distinction between a regulator and a servomechanism is primarily in the way they are used; if emphasis is on keeping a quantity at a constant value, it is called a regulator. If emphasis is on forcing the controlled quantity to follow a changing quantity, it is called a servomechanism.

Another example of a servomechanism is a star-tracking telescope. The system consists of a motor-driven telescope, a photocell arrangement, and an electronic power amplifier. The input to the servomechanism is the angular position of the star being tracked; the output is the angular position of the telescope. The difference between the input and the output is measured by a photocell arrangement and is applied to the power amplifier, which, in turn, drives the motors to position the telescope in the direction of the star.

2. Differential Equation Analysis

The performance of a control mechanism can be described by the solution of the differential equations which represent its characteristics. For example, consider the motor control system with potentiometer feedback shown in Fig. 9.3.

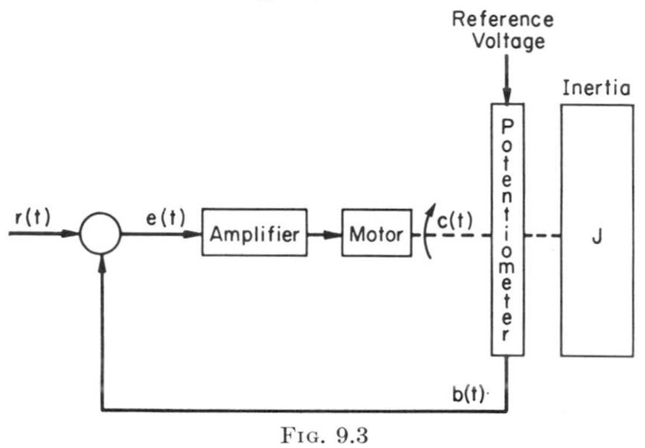

Fig. 9.3

For simplicity, assume that the torque applied to the motor output shaft is proportional to the input voltage to the power amplifier.

Let $c(t)$ = position of motor shaft (controlled variable)
J = moment of inertia of motor shaft and connected load
$r(t)$ = variable input voltage (controlling quantity)
$b(t)$ = potentiometer output voltage
$e(t)$ = error voltage to power amplifier = $r(t) - b(t)$
L = torque applied to motor shaft
K_1 = proportionality constant between motor torque and input voltage to power amplifier
K_2 = proportionality constant between potentiometer output voltage and shaft position

$$L = J\frac{d^2c}{dt^2} = K_1e = K_1(r - b) \tag{9.1}$$

$$b = K_2c \tag{9.2}$$

$$J\frac{d^2c}{dt^2} + K_1K_2c = K_1r \tag{9.3}$$

This has the solution:

$$c = \frac{K_1}{Jp^2 + K}r(t) + A\sin\left(\sqrt{\frac{K}{J}}t + \phi\right) \tag{9.4}$$

where $K = K_1K_2$, A and ϕ are constants of integration, and p is the operator d/dt. The solution consists of a particular integral $K_1/(Jp^2 + K)r(t)$, which depends upon the form of $r(t)$, and a complementary function consisting of an undamped oscillation term of angular frequency $\omega = (K/J)^{1/2}$, having an amplitude A and phase ϕ determined by the initial conditions.

To stabilize this control mechanism a viscous damper may be connected to the output shaft in addition to the potentiometer so as to produce a torque proportional to the angular velocity of the output shaft. Such a system is shown in Fig. 9.4.

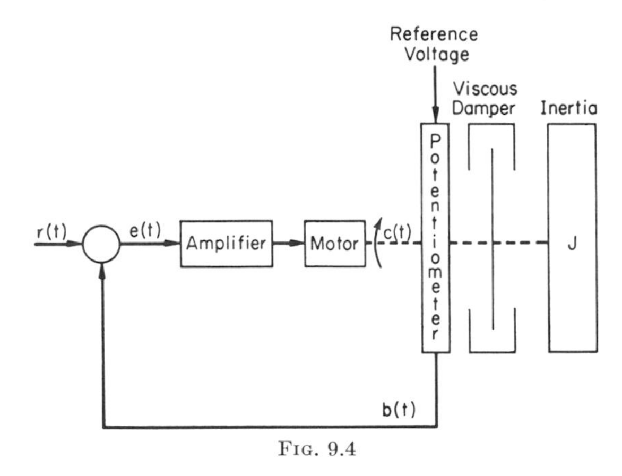

Fig. 9.4

Let F = proportionality constant between torque of viscous damper and angular velocity of shaft. The differential equation of this system is

$$J\frac{d^2c}{dt^2} = K_1e - F\frac{dc}{dt} \tag{9.5}$$

$$J\frac{d^2c}{dt^2} + F\frac{dc}{dt} + K_1K_2c = K_1r \tag{9.6}$$

This has as its solution:

$$c = \frac{K_1}{Jp^2 + Fp + K}r(t) + A\exp(-\zeta_n t)\sin(\omega_n\sqrt{1 - \zeta^2}t + \phi) \tag{9.7}$$

ζ and ω_n are defined by the relations

$$\zeta = \frac{F}{2\sqrt{JK}} \qquad \text{and} \qquad \omega_n = \sqrt{\frac{K}{J}}$$

The solution consists of a particular integral depending upon the input signal $r(t)$ and a complementary function which is a damped oscillation of angular frequency $\omega_n(1 - \zeta^2)^{1/2}$ and coefficient of damping $\zeta\omega_n$, having an amplitude A and a phase ϕ determined by the initial conditions. The quantity K, the torque produced on the output shaft by unit angular displacement of the shaft, is commonly called the *loop gain.* If r is a step function applied at time $t = 0$, ($r = r_1$, $t < 0$ and $r = 0$, $t > 0$), Eq. (9.7) becomes for $\zeta < 1$ (less than critical damping)

$$c = \frac{K_1 r_1}{K\sqrt{1-\zeta^2}} \exp(-\zeta\omega_n t) \sin\left(\omega_n \sqrt{1-\zeta^2} t + \tan^{-1}\frac{\sqrt{1-\zeta^2}}{\zeta}\right) \quad (9.8)$$

When $\zeta = 1$ (critically damped system),

$$c = \frac{K_1 r_1}{K}(1 + \omega_n t)\exp(-\omega_n t) \quad (9.9)$$

When $\zeta > 1$ (overdamped system),

$$c = \frac{K_1 r_1}{K}\exp(-\zeta\omega_n t)\left(\frac{\zeta}{\sqrt{\zeta^2-1}}\sinh\sqrt{\zeta^2-1}\,\omega_n t + \cosh\sqrt{\zeta^2-1}\,\omega_n t\right) \quad (9.10)$$

Curves of cK/K_1r_1 plotted as a function of $\omega_n t$ for various values of ζ are shown in Fig. 9.5.

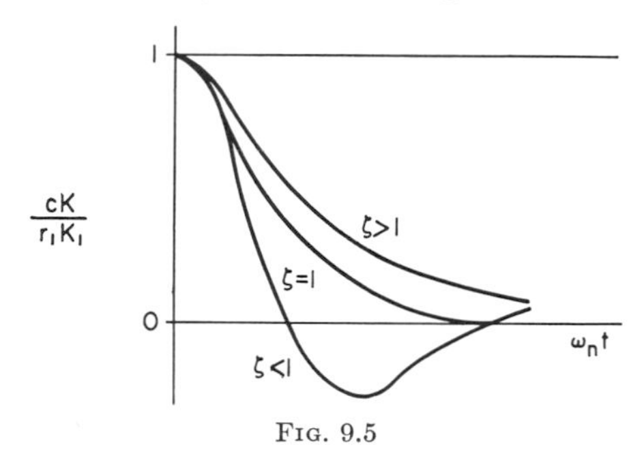

FIG. 9.5

These curves represent the response obtained from any dynamical system described by a second-order differential equation, such as a system consisting of a mass in a viscous damping fluid supported by a spring.

By examination of the relationship $\zeta^2 = F^2/4JK$ and $\omega_n^2 t = K/J$, it is possible to determine the effect of variation of K (the loop gain) and F (the damping gain) on ω_n (the natural frequency) and ζ (the coefficient of damping) of the control mechanism.

Useful information is obtained by examination of the steady-state response of such a second-order system to a sinusoidal input. This corresponds to the particular integral of Eq. (9.7) for $r = r_0 \cos \omega t$, which is of the form

$$c = c_0 \cos(\omega t + \theta) \quad (9.11)$$

where

$$c_0 = \frac{K_1 r_0}{K\sqrt{(1-\omega^2/\omega_n^2)^2 + 4\zeta^2\,\omega^2/\omega_n^2}} \quad (9.12)$$

and

$$\theta = \tan^{-1}\left(\frac{2\zeta\,\omega/\omega_n}{1-\omega^2/\omega_n^2}\right) \quad (9.13)$$

Figure 9.6 is a plot of c_0 and θ as a function of frequency.

We have so far assumed that certain simple control systems may be represented by linear second-order differential equations with constant coefficients. Although all physical systems are nonlinear to varying degrees, the simplification of linearity yields valuable information in their analysis and synthesis.

All but the simplest control mechanisms require differential equations of higher order than the second to represent adequately their behavior. The classical method of solving these higher-order equations for the particular integral and complementary function and determining their constants of integration from initial conditions may be used. However, a more systematic method is that of using the Laplace transform.

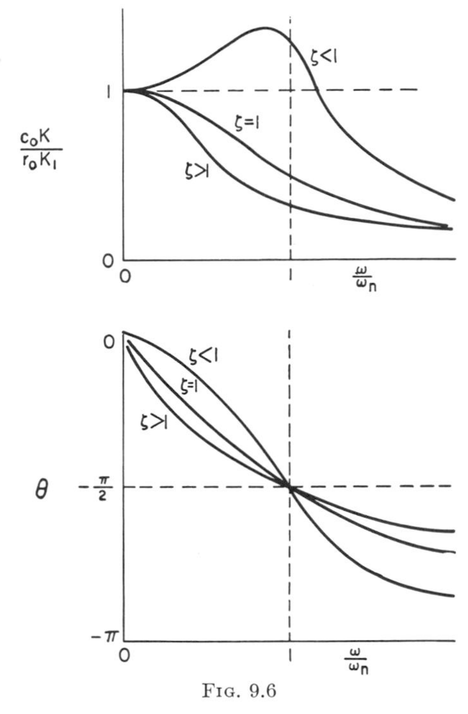

FIG. 9.6

The Laplace transform is a mathematical tool for facilitating the solution of linear differential equations. It enables differential equations to be transformed into relatively simple algebraic equations which can be manipulated until a desired form is obtained, and then transformed back into complete solutions of the original differential equations.

The Laplace transform of a function $f(t)$, denoted by $\mathcal{L}f(t)$, is defined by the following integral:

$$\mathcal{L}f(t) = F(s) = \int_0^\infty e^{-st}f(t)\,dt$$

This transformation replaces $f(t)$ with a new function $F(s)$, which is a function of a complex operator s in place of the time t. All variables are assumed zero for negative values of t.

In practice, it is seldom necessary to use the transformation integral, since extensive tables of the transforms of functions are available in standard textbooks.

By use of the Laplace transforms, transient and steady-state solutions are simultaneously obtained. Generally it is not necessary to change the Laplace transform back into a complete solution of the original differential equation, since much may be determined about the system from the transform itself. Details in the use of this method are found in most recent textbooks, and more completely in refs. 11 and 20.

3. Frequency-response Analysis

The solution of differential equations representing a control mechanism gives a clear picture of its performance. However, the equations must be solved for each set of system parameters, which is time-consuming; and interpretation of the solution into design criteria is difficult.

The transient response of any linear system can be determined from its response to sinusoidal inputs from zero to infinite frequency, thus making it possible to study the dynamics of control mechanisms by examination of the steady-state frequency response alone, eliminating the need for solving its differential equations. For example, the response of a second-order system to a step input shown in Fig. 9.5 is derivable from the frequency-response curves shown in Fig. 9.6. The frequency response often provides all the information that is required, including relative stability and resonant frequencies.

Nyquist developed a frequency-response method for determining the stability of feedback amplifiers which has been applied to control mechanisms and is commonly referred to as a Nyquist diagram. To apply the method to single-loop systems, the closed loop is broken and sinusoidal inputs of various frequencies are applied to the input of the resulting open-loop system. The amplitude ratio of output to input and phase difference between output and input (phase of output minus phase of input) are plotted on a polar diagram in the complex plane for values of frequency from $-\infty$ to $+\infty$. The plot for "negative" frequencies is considered to be the plot for positive frequencies mirrored in the real axis. If a vector drawn from the $(-1,0)$ point to the curve as the frequency varies from $-\infty$ to $+\infty$ makes one or more revolutions in the clockwise direction, the system is unstable; otherwise it is stable. If, as the frequency approaches zero on the negative frequency curve, the amplitude ratio approaches infinity and the phase difference approaches $n\pi/2$ radians, the curves for negative and positive frequency are connected by making n clockwise semicircles of infinite radius about the origin as the frequency goes from negative to positive values.

For multiple-loop systems each closed loop may be opened in turn, and the method may be applied to the resulting open loop. A detailed discussion of stability of multiple-loop systems is given in most textbooks, for example, in refs. 9, 15, 26, and 27.

An example of the application of the method is given in curves A and B of Fig. 9.7. In each case the amplitude ratio approaches infinity and the phase difference approaches $\pi/2$ as the frequency approaches zero from the negative direction, and semicircles of infinite radii are drawn clockwise from $\pi/2$ to $-\pi/2$ to connect the loci for negative and positive frequencies. In curve A a vector drawn from the $(-1,0)$ point to the curve makes no revolutions as the frequency varies from $-\infty$ to $+\infty$, and the locus represents a stable system. In curve B the vector makes two revolutions in the clockwise sense, and the system is unstable.

In practice, the Nyquist diagram is plotted for only positive values of frequency and stability or instability determined by inspection. This convention is used in Fig. 9.8, which shows Nyquist diagrams of three types of stable (A) and three types of unstable (B) control mechanisms.

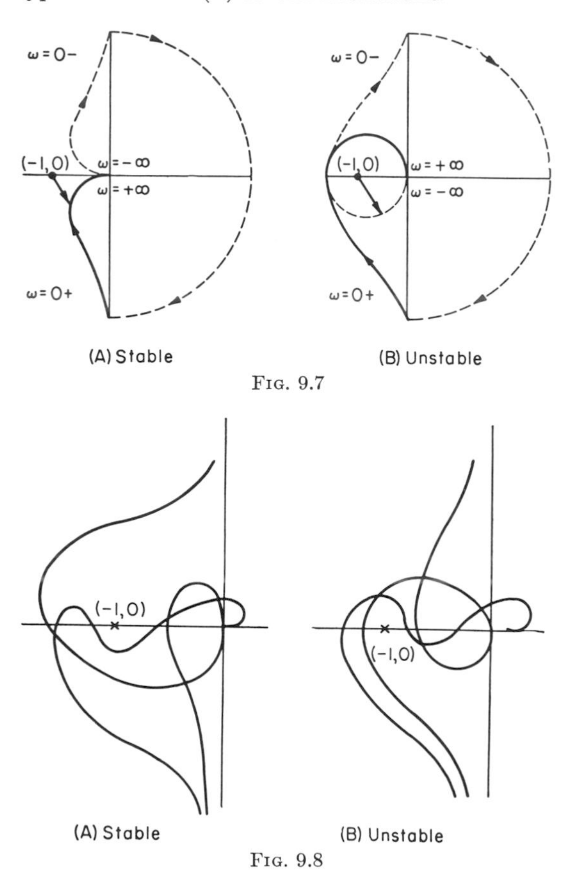

FIG. 9.7

FIG. 9.8

As an example of the application of the frequency method to a control mechanism, let us again consider the motor control system with potentiometer feedback and no viscous damping (Fig. 9.3). If the loop at the output of the potentiometer is opened and a sinusoidal input is applied to the amplifier, the output voltage b is related to the input voltage e (or r) by the equation

$$J\frac{d^2b}{dt^2} = K_2J\frac{d^2c}{dt^2} = K_1K_2e = Ke \tag{9.14}$$

Since the particular integral of a linear differential equation with constant coefficients with the right-hand member a sinusoidal function is found by replacing the operator d/dt by $i\omega$, we obtain

$$\frac{B}{E} = -\frac{K}{J\omega^2} \tag{9.15}$$

where B and E represent the complex amplitudes of the potentiometer output and actuating error voltages, respectively, as functions of $i\omega$. It should be noted that the form of the equation is the same as for the Laplace transform of b/e, with $i\omega$ substituted for s and with the initial values of b and e and their derivatives equated to zero. The ratio of output to input of a system as a function of frequency is designated as the *transfer function* of the system.

The Nyquist diagram for the system of Fig. 9.3 is a straight line along the negative real axis and, of course, passes through the point $(-1,0)$, as shown in curve A of Fig. 9.9. This represents an oscillating system whose natural frequency corresponds to the value of ω at point $(-1,0)$. It should be noted that varying the loop gain K only varies the natural frequency of the oscillation.

Now consider the motor control system with potentiometer feedback and viscous damping shown in Fig. 9.4. Opening the loop at the output of the potentiometer, the differential equation of the system becomes

$$J\frac{d^2b}{dt^2} = K_2J\frac{d^2c}{dt^2} = Ke - F\frac{db}{dt} \tag{9.16}$$

For this case, we have

$$\frac{B}{E} = \frac{K}{i\omega(i\omega J + F)} \tag{9.17}$$

The Nyquist diagram in this case is shown as curve B of Fig. 9.9. It approaches asymptotically to the negative imaginary axis as ω approaches zero. Since the curve approaches the negative real axis only for $\omega \to \infty$ and amplitude ratio equal to zero, the system is stable.

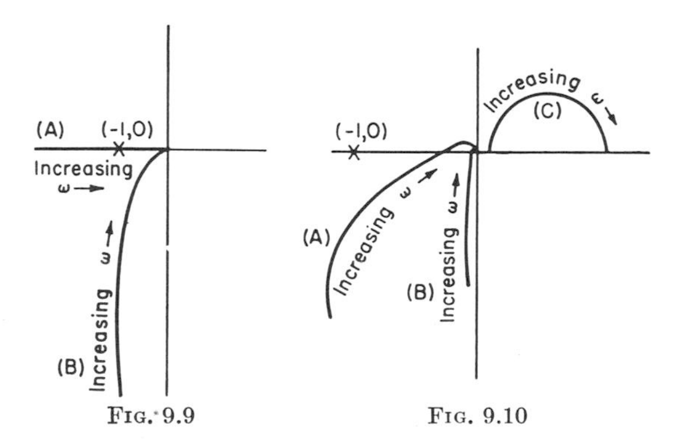

FIG. 9.9   FIG. 9.10

Consider now the motor control system with potentiometer feedback and viscous damping shown in Fig. 9.4, taking into account exponential (transfer) lag between the motor torque and the input to the amplifier. Such a lag is usually present in an actual system due to the inductance of the motor winding or other causes. For this case, Eq. (9.1) becomes

$$L + T\frac{dL}{dt} = K_1e \tag{9.18}$$

where T is the time constant of build-up of torque on the motor shaft. The differential equation of the system becomes

$$TJ\frac{d^3b}{dt^3} + (J + FT)\frac{d^2b}{dt^2} + F\frac{db}{dt} = Ke \tag{9.19}$$

For that case, we have

$$\frac{B}{E} = \frac{K}{i\omega(i\omega J + F)(i + i\omega T)} \tag{9.20}$$

The effect of the exponential lag is to multiply the transfer function B/E of Eq. (9.17) by the factor $1/(1 + i\omega T)$. The Nyquist diagram in this case is shown as curve A of Fig. 9.10. The curve now intersects the real axis for finite ω, and there will be some critical value of K that will cause the curve to pass through the $(-1,0)$ point. The system is stable only for values of K smaller than that critical value.

4. System Improvement by Compensation

By addition of frequency-sensitive electrical networks or other elements, it is often possible to change the shape of the Nyquist diagram in such a way as to improve system performance. For example, in the motor control system considered above, it is possible to compensate for the time lag between the motor torque and the input to the power amplifier by inserting a lead network, as shown in Fig. 9.11, ahead of the power amplifier.

For this network, the following relation holds:

$$\frac{E_{\text{out}}}{E_{\text{in}}} = \frac{\alpha(1 + i\omega T_1)}{1 + i\alpha\omega T_1} \tag{9.21}$$

where

$$\alpha = \frac{R_2}{R_1 + R_2} \qquad \text{and} \qquad T_1 = R_1C$$

The transfer function for this case, from Eqs. (9.20) and (9.21), is given by

$$\frac{B}{E} = \frac{K}{i\omega(i\omega J + F)(1 + i\omega T)}\,\frac{\alpha(1 + i\omega T_1)}{(1 + i\omega\alpha T_1)} \tag{9.22}$$

The effect of the lead network is to add the quantity $\alpha(1 + i\omega T_1)/(1 + i\omega\alpha T_1)$ as a factor in the new expression for the transfer function. The Nyquist diagram for this case is shown in Fig. 9.10, curve B. Curve C shows the diagram for the lead network of Fig. 9.11 alone. If the time constant T_1 of the lead network is equal to the time constant in the lag of the motor control, the effect of introduction of the lead network is to decrease the time constant from T to αT, and decrease the gain from K to $K\alpha$. However, the reduction in time lag permits increasing the gain of the amplifier by a factor greater than $1/\alpha$ with increased stability.

The determination of the proper constants for a lead network to increase the stability of any system, which may involve many lag terms, is not usually based on compensating the lag terms individually but is done by using constants which will move the

Nyquist curve farther from the $(-1,0)$ point. A limitation in the use of such lead networks results because their insertion magnifies the noise in the error signal by an amount that varies inversely with the magnitude of α. The possible resulting saturation of the system may introduce lagging phase shifts that completely nullify their advantages.

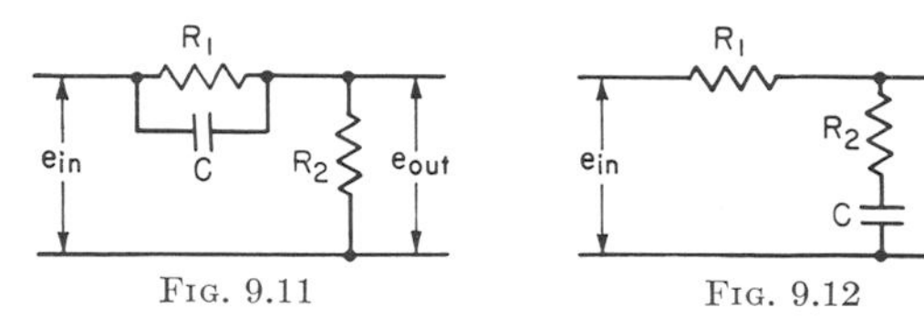

FIG. 9.11 FIG. 9.12

The use of the frequency-response diagram has thus far been concerned with the establishment of the criteria for absolute stability. It is more often necessary to analyze or design a closed-loop control system with respect to some selected degree of relative stability. Thus, in the example of the motor control system, the effect of gain variation on the damping ratio ζ was examined, ζ representing a measure of relative system stability. Figure 9.5 shows the transient response of the motor control system for different values of ζ, and Fig. 9.6 shows the sinusoidal magnitude and phase shift response of C/R for different values of ζ. It is thus possible to relate the relative stability, as measured by ζ, of this system with the system's over-all frequency response.

This determination of relative system stability can be determined in general by drawing loci of constant C/R on the C/E plane and examining the relation of a given system's transfer function C/E to these loci. They are circles of radii $|M/(M^2 - 1)|$ and with center at $x = -M^2/(M^2 - 1)$, where $M = C/R$. These circles of constant M provide a means of quantitatively evaluating the performance of a control mechanism as a function of frequency. Two figures of merit that are used to describe the performance of a control system are values of the maximum value of M and the frequency at which this maximum value occurs. Values of M_m (maximum value of M) representing good design practice are in the neighborhood of 1.3 but vary, depending upon the type of system. A high value of ω_n (value of ω for $M = M_m$) corresponds to a high speed of response of the system. The use of the lead network illustrates improvement in the high-frequency response. The introduction of an integral or lag network is often used to improve low-frequency response. Such a network is shown in Fig. 9.12. For this network, the following relation holds:

$$\frac{E_{\text{out}}}{E_{\text{in}}} = \frac{1 + i\alpha\omega T_1}{1 + i\omega T_1}$$

where $\alpha = \dfrac{R_2}{R_1 + R_2}$ and $T_1 = (R_1 + R_2)C$

For example, a control mechanism with a Nyquist diagram, as shown in Fig. 9.13, curve A, is not amenable to compensation by the use of lead networks, since their introduction would tend to shift the points at or near maximum M to greater values of M for equal values of steady-state gain, bringing the locus closer to the $(-1,0)$ point. Introduction of a lag network, however, does not appreciably change the frequency at which the maximum M occurs, but it does allow the use of a higher steady-state gain. This is true because the low-frequency parts of the locus have low values of M; the lagging phase shift of the integral network cannot then increase the low-frequency values of M to values too much in excess of 1. At the high frequencies the attenuation characteristic of the integral network more than offsets the lagging phase-shift characteristic, bringing the locus farther from the $(-1,0)$ point, and thus enabling an over-all increase in gain without increasing the M of the high-frequency part of the locus. Curve B of Fig. 9.13 shows the effect of adding a lag network to the control mechanism whose locus is shown in curve A. The frequency locus of the lag network of Fig. 9.12 is shown in curve C of Fig. 9.13.

Combinations of lead and lag networks are sometimes useful in simultaneously improving the high- and low-frequency response of a control mechanism. Such a combination network is shown in Fig. 9.14. This network will act as a lag network at low frequencies and as a lead network at high frequencies. By proper choice of parameters, the frequency locus of a control mechanism may be modified by the addition of the network so as to be more remote from the point $(-1,0)$ and thus permit higher gains with the same degree of stability.

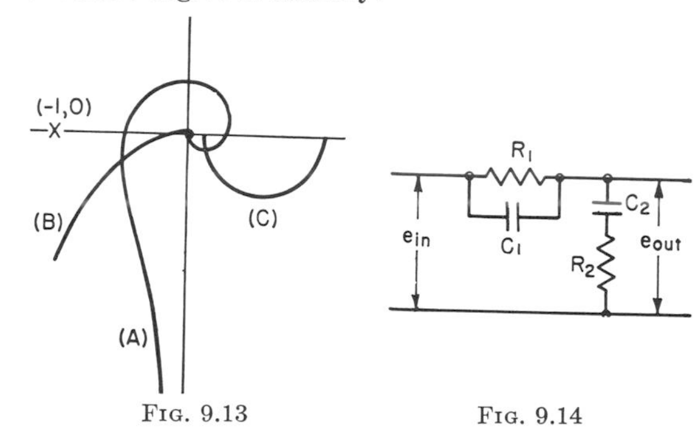

FIG. 9.13 FIG. 9.14

The stabilization of a control system by tachometer feedback illustrates the general class of feedback stabilization methods as contrasted with series stabilization by lead and lag networks. Dynamically, the feeding back of a tachometric signal in parallel with the existing feedback is equivalent to the insertion of an "ideal" lead network in which the phase lead of the output with respect to the input increases continually as the frequency increases. For example, in the case of the motor control system, shown in Fig. 9.15, which includes both potentiometer and tachometer feedback, viscous friction, and motor torque time delay, B will be given by

$$B = (K_2 + i\omega D)C \tag{9.23}$$

where D is the proportionality constant between the voltage output of the tachometer and the angular rate of rotation of the motor shaft. Then

$$\frac{B}{E} = \frac{(1 + i\omega D/K_2)K}{i\omega(i\omega J + F)(1 + i\omega T)} \tag{9.24}$$

If D/K_2 is equal to T, the effect of the lag in the motor torque will be canceled by the tachometric feedback. However, as in the case of lead networks, the determination of the proper tachometric feedback gain is not usually based on compensating the lag terms individually but on using a gain which will move the Nyquist curve farther from the $(-1,0)$ point.

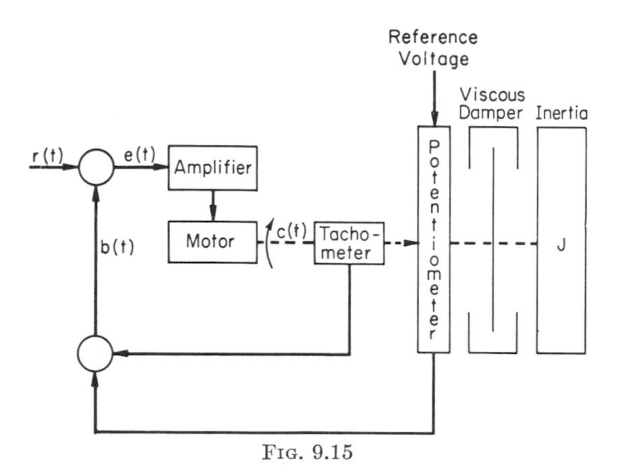

FIG. 9.15

One of the advantages of the tachometric feedback over a lead network is that the noise in the error signal is not magnified. However, the tachometer itself may introduce noise; and the choice of either method of stabilization depends upon the particular system being stabilized.

5. Steady-state Error

The steady-state error of a control mechanism is defined as the value of the error signal after all transients due to disturbances to the system have died out. The types of steady-state error of most interest are those due to step functions in position, velocity, and acceleration and to steady load disturbances. For example, let us consider the effect of step function in position, velocity, and acceleration applied to the motor control system of Fig. 9.4, with potentiometer feedback and viscous damping described by Eq. (9.5). Writing Eq. (9.5) in terms of E and R, we have

$$J\frac{d^2e}{dt^2} + F\frac{de}{dt} + Ke = J\frac{d^2r}{dt^2} + F\frac{dr}{dt} \tag{9.25}$$

If r is a step function in position ($r = 0$, $t < 0$ and $r = a$, $t > 0$), the particular integral of this equation is $e = 0$. If r is a ramp function ($r = 0$, $t < 0$ and $r = at$, $t > 0$), the particular integral is $e = aF/K$. The system has a steady-state error which is proportional to the slope of the input ramp function and the viscous friction and inversely proportional to the loop gain. This shows the advantage of large loop gains in reducing steady-state errors. If r is a parabolic function (step function in acceleration), it is easily shown that the steady-state error would continuously increase with time.

Suppose a constant load torque L_0 is applied to the output shaft of the motor; Eq. (9.5) becomes

$$J\frac{d^2c}{dt^2} = K_1e - F\frac{dc}{dt} - L_0 \tag{9.26}$$

Rewriting in terms of e and r,

$$J\frac{d^2e}{dt^2} + F\frac{de}{dt} + Ke = J\frac{d^2r}{dt^2} + F\frac{dr}{dt} + K_2L_0 \tag{9.27}$$

For constant r, the particular integral is

$$e = \frac{L_0K_2}{K} = \frac{L_0}{K_1} \tag{9.28}$$

This again points out the importance of high loop gain in minimizing steady-state errors.

In general, if the lowest order of derivative of r on the right side of the equation is one, the steady-state error for a step position in position is zero; for a step function in velocity, a constant; and for a step function in acceleration, a quantity increasing with time. The Nyquist diagram for this case [see Eq. (9.25)] becomes asymptotic to the negative imaginary axis as ω approaches zero. If the lowest derivative on the right side of the equation is two, the steady-state error in both position and velocity is zero; and for a step acceleration input, it is a constant. The Nyquist diagram for this case approaches the negative real axis as ω approaches zero.

6. Other Methods of Analysis

Methods other than the plotting of Nyquist diagrams are commonly used in the analysis and synthesis of control mechanisms, and offer particular advantages in many cases. By plotting curves of the reciprocal of the amplitude ratio vs. phase angle, it is easier to analyze systems with parallel feedback loops. By plotting the logarithms of the amplitude ratio and the phase angle, it is easier to identify the specific effects of lead or lag components of the over-all system. Detailed explanations of how these methods are used are contained in the References, especially refs. 3, 5, and 26.

A method of analysis, called the *root-locus* method, uses the roots of the open-loop function to find the roots of the closed-loop function. Stability exists in any linear system if, and only if, there are no roots with positive real parts in the characteristic equation. In the root-locus method, the values of the roots of the open-loop function are plotted in the s plane. The locus of points where the phase shift is 180° is determined graphically. All closed-loop denominator roots are located on this locus, but the exact position depends upon the value of the gain constant K. As K is varied, the closed-loop root locations shift along the 180° phase locus. K is selected so that the closed-loop denominator roots have acceptable values. Descriptions of the root-locus method are given in refs. 8, 12, 16, 21, 23, 26, 27, and 29.

Most of the previous analysis of control mechanisms is based on the assumption of linearity in the operation of all components in the system. In general, control mechanisms contain such components as valves, amplifiers, motors, gears, etc., and exhibit nonlinear characteristics to varying degrees. Although methods of analysis are available for certain types of nonlinear differential equations, there are no simple methods of solution of general application. In many cases where methods are available, so much labor is involved that the effort is not justified. Some of these methods are detailed in refs. 1, 9, 12, 16, 17, 23, 27, 29, and 30. Systems which operate by error data supplied intermittently are discussed in refs. 16 and 26–29.

An expedient is to ignore the nonlinearities and apply linear theory to the nonlinear problems. The accuracy depends, of course, on the type and degree of nonlinearity. Since mathematics is used only as a guide to the design of any real system, final adjustments being made on the actual equipment, linear analysis in most cases gives results which are sufficiently close for engineering purposes.

The use of automatic computers, both digital and analogue, makes possible the analysis and synthesis of linear and nonlinear control mechanisms that cannot be handled adequately by analytical means. The modern electronic analogue computer is especially useful in the solution of control problems. Variables in the problem are represented by electrical voltages. Problems may be solved by interconnection of computer components, such as summers, integrators, multipliers, function generators, etc., to solve the differential equations describing the system. Alternately, components of the computer may be arranged to possess the same transfer functions as corresponding components of the control mechanism, and interconnected in the same open- or closed-loop arrangements as are the components of the control mechanism under study. In this way the control mechanism is "simulated" by the computer, and the response of the computer to any disturbance will yield the response of the control mechanism to the same type of disturbance. If the problem is solved in "real time," that is, if the computer solution proceeds at the same rate as the control system being simulated, it is often convenient to use the analogue computer to simulate part of the system, with actual system "hardware" used in the remainder. Details on the use of analogue computers in the solution of control problems are discussed in refs. 17, 20, 24, and 28.

References

1. Ahrendt, William R., and John F. Taplin: "Automatic Feedback Control," McGraw-Hill, New York, 1951.
2. Bode, H. W.: "Network Analysis and Feedback Amplifier Design," Van Nostrand, Princeton, N.J., 1945.
3. Brown, G. S., and D. P. Campbell: "Principles of Servomechanisms," Wiley, New York, 1948.
4. Brown, Robert Grover, and James Wilson Nilsson: "Introduction to Linear Systems Analysis," Wiley, New York, 1962.
5. Chestnut, Harold, and Robert W. Mayer: "Servomechanisms and Regulating System Design," vols. 1 and 2, Wiley, New York, 1951, 1955.
6. Clark, Robert N.: "Introduction to Automatic Control Systems," Wiley, New York, 1962.
7. D'Azzo, John J., and Constantine H. Houpis: "Feedback Control System Analysis and Synthesis," 2d ed., McGraw-Hill, New York, 1965.
8. Evans, Walter R.: "Control-system Dynamics," McGraw-Hill, New York, 1954.
9. Fett, Gilbert H.: "Feedback Control Systems," Prentice-Hall, Englewood Cliffs, N.J., 1954.
10. Flugge-Lotz, Irmgard: "Discontinuous Automatic Control," Princeton, N.J., 1953.
11. Gardner, M. F., and J. L. Barnes: "Transients in Linear Systems," Wiley, New York, 1942.
12. Gille, J. C., M. J. Pélegrin, and P. Decaulne: "Feedback Control Systems," McGraw-Hill, New York, 1959.
13. Greenwood, Ivan A., Jr., J. Vance Holdam, Jr., and Duncan MacRae, Jr.: "Electronic Instruments," McGraw-Hill, New York, 1948.
14. Hall, Albert C.: "The Analysis and Synthesis of Linear Servomechanisms," Technology Press, Cambridge, Mass., 1943.
15. James, Hubert M., Nathaniel B. Nichols, and Ralph S. Phillips: "Theory of Servomechanisms," McGraw-Hill, New York, 1947.
16. Kuo, Benjamin C.: "Automatic Control Systems," Prentice-Hall, Englewood Cliffs, N.J., 1962.
17. Lauer, Henry, Robert Lesnick, and Leslie E. Matson: "Servomechanism Fundamentals," 2d ed., McGraw-Hill, New York, 1960.
18. MacColl, Lero A.: "Fundamental Theory of Servomechanisms," Van Nostrand, Princeton, N.J., 1945.
19. Macmillan, R. H.: "An Introduction to the Theory of Control in Mechanical Engineering," Cambridge University Press, London, 1951.
20. Murphy, Gordon J.: "Control Engineering," Van Nostrand, Princeton, N.J., 1959.
21. Nixon, Floyd E.: "Principles of Automatic Controls," Prentice-Hall, Englewood Cliffs, N.J.
22. Oldenbourg, R. C., and H. Sartorius: "Dynamik der selbstatiger Regelungen," trans. H. L. Mason, American Society of Mechanical Engineers, 1948.
23. Savant, C. J., Jr.: "Control System Design," 2d ed., McGraw-Hill, New York, 1964.
24. Soroka, Walter W.: "Analog Methods in Computation and Simulation," McGraw-Hill, New York, 1954.
25. Smith, Ed Sinclair: "Automatic Control Engineering," McGraw-Hill, New York, 1944.
26. Thaler, G. J., and R. G. Brown: "Analysis and Design of Feedback Control Systems," 2d ed., McGraw-Hill, New York, 1960.
27. Truxal, John G.: "Automatic Feedback Control System Synthesis," McGraw-Hill, New York, 1955.
28. Truxal, John G. (ed.): "Control Engineers' Handbook," McGraw-Hill, New York, 1958.
29. Tsien, H. S.: "Engineering Cybernetics," McGraw-Hill, New York, 1954.
30. Tustin, A.: "Automatic and Manual Control," Academic, New York, 1952.

Part 3 · *Mechanics of Deformable Bodies*

Chapter 1

Kinematics and Dynamics

By E. U. CONDON, University of Colorado

In the mechanics of deformable bodies Newton's laws of motion govern the movements of each particle. In addition to external field forces, such as the gravitational field acting on each particle, there are internal molecular forces whereby contiguous elements interact with equal and opposite forces. These internal forces are not available to direct observation. A wide variety of special motions is described phenomenologically in terms of a few simple assumptions about such internal forces. Molecular theories then relate these assumed forces to models of the particular solid or fluid in question [1].†

The entire subject falls into three parts, solid mechanics, fluid mechanics, and, for intermediate states, rheology. The basic distinction is that solid bodies show rigidity; i.e., when they are deformed by application of external force, internal forces are brought into play which oppose the deformation and tend to restore the original shape, whereas fluids do not have this property. Solids, moreover, are capable of holding together only for some finite total amount of deformation; otherwise they rupture or fracture into parts between which the forces of interaction are quite negligible. For most solids rupture occurs before the fractional change in linear dimensions is more than about 1 per cent, and therefore a large part of solid mechanics can be developed with a mathematical formalism in which relative displacements of neighboring elements are small throughout the motion. For fluids, on the other hand, the relative motions of parts are quite large except in some special cases, like the oscillatory motions involved in propagation of a sound wave. Therefore fluid mechanics must use mathematical methods capable of dealing with large relative deformations.

Because of these distinctions the classical developments of solid mechanics and fluid mechanics took place along rather different lines, each with its own literature. There is actually a much closer unity to these subjects than this historical development suggests, which is recognized in the presentation which follows. This unity is important not only from the mathematical point of view but also in connection with certain extreme conditions, such as the plastic flow of metals at very high rates of strain and the quasi-solid behavior of fluids of very great viscosity (Part 3, Chap. 3).

1. Kinematics of Continuous Media

Two general methods are in use for describing the motions of the parts of a deformable body. These are usually called Lagrangian and Eulerian (though both are due to Euler). In the Lagrangian method a vector function of time $\mathbf{r}(t,\mathbf{s})$ depends also parametrically on a vector argument $\mathbf{s}$ and is such that for $t = 0$, $\mathbf{r}(0,\mathbf{s}) = \mathbf{s}$. The physical meaning is that the function $\mathbf{r}(t,\mathbf{s})$ gives the vector position at time t of the element of fluid that is initially at $\mathbf{r} = \mathbf{s}$. In this case the velocity and acceleration of the element originally at $\mathbf{s}$ are given by $\dot{\mathbf{r}}(t,\mathbf{s})$ and $\ddot{\mathbf{r}}(t,\mathbf{s})$, where the dots refer to partial differentiation with respect to time.

The classical analysis of strain, used in solid mechanics for small deformations, is based essentially on the Lagrangian method, as will appear in Sec. 2.

In the Eulerian method, attention is fastened on a volume element at a fixed location in space. As the motion continues, material flows across the walls of such a fixed volume element, so the identity of the material contained within it is changing with time.

In a macroscopically small volume around the point whose position vector is $\mathbf{r}$ the average velocity of all the material particles (weighted proportionally to the relative masses) contained in it will be a vector field $\mathbf{v}(t,\mathbf{r})$. This is the velocity field of the fluid flow. Likewise if $\rho(t,\mathbf{r})$ is the average mass per unit volume in such a volume element, this scalar field is called the density. Since $\mathbf{v}$ is defined as the mass-weighted average, the total momentum $\Sigma_i m_i \mathbf{v}_i$ of all the particles in a volume element is given by $\rho\mathbf{v}\, dV$.

The mass of material crossing a surface element $\mathbf{dS}$ in time dt toward the positive direction of $\mathbf{dS}$ is given by $\rho\mathbf{v} \cdot \mathbf{dS}\, dt$, and therefore the total rate of outflow from a region bounded by a fixed closed surface S is the closed-surface integral $\iint \rho\mathbf{v} \cdot \mathbf{dS}$, where $\mathbf{dS}$ is the outward-drawn normal to the closed surface. By the divergence theorem this is equal to $\iiint \operatorname{div}\,(\rho\mathbf{v}) \cdot dV$ extended over the volume enclosed by S. (Part 1, Chap. 9, Sec. 5.)

The total amount of material contained inside the same volume is $\iiint \rho\, dV$, and therefore the time rate at which the material inside S is increasing is $\iiint \dot{\rho}\, dV$. If matter is not created or destroyed inside the region enclosed by S, these two rates must balance, giving $\iiint [\dot{\rho} + \operatorname{div}\,(\rho\mathbf{v})]\, dV = 0$. As this must be true for any bounding surface S, the integrand must vanish everywhere.

This gives a differential relation between the density and velocity fields

$$\dot{\rho} + \operatorname{div}\,(\rho\mathbf{v}) = 0 \tag{1.1}$$

known as the *equation of continuity*.

† Numbers in brackets refer to References at end of chapter.

In considering the time variation of any field quantity such as ρ it is important to consider two kinds of time differentiation. Ordinary $\dot{\rho}$, as in (1.1), refers to variation with time of the density of material at a fixed point in space.

But it is also of interest in fluid mechanics to consider the variation with time of ρ or any other field quantity as it would appear to an observer who is instantaneously moving with the field, i.e., moving with velocity $\mathbf{v}$. In the time dt the motion is from $\mathbf{r}$ to $\mathbf{r} + \mathbf{v}\,dt$, and therefore the total change in the quantity ρ in unit time, which will be denoted by $d\rho/dt$, is given by

$$\frac{d\rho}{dt} = \dot{\rho} + \mathbf{v} \cdot \nabla\rho \tag{1.2}$$

This kind of time derivative is sometimes called a *convected derivative*.

In terms of the convected derivative the equation of continuity becomes

$$\frac{d\rho}{dt} + \rho \operatorname{div} \mathbf{v} = 0 \tag{1.3}$$

A fluid is said to be *incompressible* if the changes in density of the fluid in its motion are considered to be negligible for the phenomena in question. In this approximation $d\rho/dt = 0$, and therefore the velocity field must satisfy the condition $\operatorname{div} \mathbf{v} = 0$.

2. Stress

In analyzing the motion of deformable bodies it is necessary to suppose that internal forces act between contiguous parts of the material. Forces of this kind act across the bounding surface of any given element of the material. Let $\mathbf{T}$ be the *force per unit area*, or the traction, acting across the surface element $\mathbf{dS}$, counted as positive if it acts like a tension, i.e., having a positive component in the direction of the outward-drawn normal. As $\mathbf{T}$ is the force per unit area acting on the material inside the bounding surface due to its interaction with the material outside this bounding surface, so also $-\mathbf{T}$ is the force on the material outside due to its interaction with the material inside.

In general the value of $\mathbf{T}$ will depend on the orientation of the surface element at a given point across which the interaction takes place. The ensemble of all values of $\mathbf{T}$ for all possible orientations of the surface element describes the state of *stress* in the body. If $\mathbf{T}$ is normal to the surface element across which it acts, the stress on such a surface element is said to be a *normal traction*. As defined here, a positive value corresponds to a tension or pull, but the opposite sign convention, in which positive $\mathbf{T}$ represents a pressure, is also found in the literature. If $\mathbf{T}$ is parallel to the surface element across which it acts, the stress is said to be a *tangential* or *shearing stress*. The terms normal and tangential refer to aspects of the state of stress with regard to particular surface elements: the same over-all state of stress may appear as a normal stress across certain surface elements and as a simple shear across others. It is now important to consider the relations which exist between the values of $\mathbf{T}$ corresponding to all possible orientations of the surface element.

The whole force on the material inside a region bounded by a closed surface S due to the surface tractions of the stress is $\iint \mathbf{T}\,dS$, and the torque about the origin acting on the body as a result of these forces is $\iint \mathbf{r} \times \mathbf{T}\,dS$. If the surface S is considered to be variable in size and a is a linear dimension, then the surface area is of the order a^2, and the volume enclosed is of order a^3. The system of stresses must be such that it is possible for each small volume element to be in equilibrium without regard to the body forces which are proportional to a^3. This argument provides the basis for determining the dependence of $\mathbf{T}$ on the direction of dS.

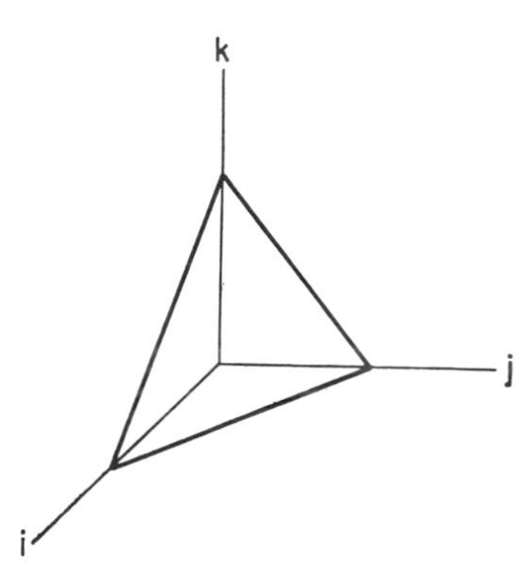

Fig. 1.1

Considering the equilibrium of the tetrahedron (Fig. 1.1) enclosed by a plane normal to the vector $\mathbf{n}$, let $\mathbf{T}_n$ be the traction acting across this plane and $\mathbf{T}_i$, $\mathbf{T}_j$, $\mathbf{T}_k$ be the tractions acting across the planes normal to $\mathbf{i}$, $\mathbf{j}$, and $\mathbf{k}$, respectively. If S is the area of the triangular face normal to $\mathbf{n}$, then the areas of the triangular faces normal to $\mathbf{i}$, $\mathbf{j}$, and $\mathbf{k}$ are $S\mathbf{n} \cdot \mathbf{i}$, $S\mathbf{n} \cdot \mathbf{j}$, and $S\mathbf{n} \cdot \mathbf{k}$, respectively. For equilibration of the forces one must have

$$\mathbf{T}_n = (\mathbf{T}_i\mathbf{i} + \mathbf{T}_j\mathbf{j} + \mathbf{T}_k\mathbf{k}) \cdot \mathbf{n} \tag{1.4}$$

The quantity in parentheses is called the *stress tensor* (Part 1, Chap. 9, Sec. 3). It is made up of the three vector stresses which act across unit area of the coordinate planes and permits the calculation of the stress acting across a plane of arbitrary orientation.

Considering now the equilibration of the torque of the stresses acting over the surface of a small element, it may be supposed, if the element is sufficiently small, that the stress tensor can be regarded as constant at all points on the bounding surface. In this case the condition $\iint \mathbf{r} \times \mathbf{T}\,dS = 0$ becomes

$$(\mathbf{i} \times \mathbf{T}_i)\iint x\mathbf{i} \cdot \mathbf{n}\,dS + (\mathbf{i} \times \mathbf{T}_j)\iint x\mathbf{j} \cdot \mathbf{n}\,dS + \cdots + \cdots = 0$$

which reduces to

$$V[(\mathbf{i} \times \mathbf{T}_i) + (\mathbf{j} \times \mathbf{T}_j) + (\mathbf{k} \times \mathbf{T}_k)] = 0$$

where V is the volume of the element over which the torque must vanish. In terms of components

$$\begin{aligned} \mathbf{T}_i &= \mathbf{i}T_{xx} + \mathbf{j}T_{yx} + \mathbf{k}T_{zx} \\ \mathbf{T}_j &= \mathbf{i}T_{xy} + \mathbf{j}T_{yy} + \mathbf{k}T_{zy} \\ \mathbf{T}_k &= \mathbf{i}T_{xz} + \mathbf{j}T_{yz} + \mathbf{k}T_{zz} \end{aligned}$$

this vanishing of the torques requires that the components form a symmetric tensor, $T_{xy} = T_{yx}$, etc.

Therefore the general state of stress inside a body is represented by a *symmetric second-order tensor.* Such a tensor has six independent components. The state of stress at a point is most easily visualized by regarding it as resulting from an isotropic part, $p = -\frac{1}{3}(T_{xx} + T_{yy} + T_{zz})$, which is called the mean *pressure* and may be negative as well as positive. In tensor notation this part is written as $p(\mathbf{ii} + \mathbf{jj} + \mathbf{kk})$ and represents a force per unit area that acts normally across a surface $\mathbf{dS}$ of any orientation whose magnitude is the same whatever the orientation of $\mathbf{dS}$.

For convenience, the notation T_{xx}, etc., will continue to be used for the stress tensor from which the isotropic part has been removed, so that now T represents a second-order tensor of zero trace and is therefore a five-component quantity. Being a symmetric tensor, it can be referred to principal axes with respect to which the shearing stresses vanish, which amounts to a specification of three of the five components. Since the tensor has zero trace, if $\mathbf{l}$, $\mathbf{m}$, $\mathbf{n}$ are unit vectors along its principal axes, it must have the form $A(\mathbf{ll} - \mathbf{nn}) + B(\mathbf{mm} - \mathbf{nn})$, where A and B are two quantities required to specify the magnitude of the stress. In the special case in which $A = B$, the state of stress has the $\mathbf{n}$ axis as an axis of symmetry. It is convenient to regard these two terms separately and to consider the entire state of stress as being made up of three terms: the isotropic, or pressure, part and the two terms measured by A and B, which are obviously alike in general character.

A stress of the form $A(\mathbf{ll} - \mathbf{nn})$ will be called a *pure shear.* It corresponds (for $A > 0$) to a normal tension across surface elements normal to the $\mathbf{l}$ axis and a normal pressure across surface elements normal to the $\mathbf{n}$ axis. For a surface element oriented normal to $\mathbf{l} \pm \mathbf{n}$ the stress becomes a pure shear, since

$$A(\mathbf{ll} - \mathbf{nn}) \cdot (\mathbf{l} \pm \mathbf{n}) = A(\mathbf{l} \mp \mathbf{n})$$

which is normal to $\mathbf{l} \pm \mathbf{n}$, and therefore the stress acts in the plane of these surface elements.

3. Equations of Motion

The equations of motion for a deformable body are obtained by equating the density ρ times the convective derivative of velocity with respect to time to the forces acting on unit volume. The forces acting on a volume are partly due to body forces, such as those acting on the volume elements due to a gravitational field. Let $\mathbf{g}$ be the acceleration of gravity; then the force on the volume dV due to gravity is $\rho\mathbf{g}\,dV$. The other part of the force acting on a volume element is equal to the resultant of the surface tractions acting because of the state of stress of the body.

If $\mathbf{dS}$ is the outward-drawn vector surface element of the bounding surface of a small volume of the body and $\mathfrak{T}$ is the stress tensor, then by definition of the stress tensor the force acting on the body across this surface element is $\mathfrak{T} \cdot \mathbf{dS}$; so the whole force is $\iint \mathfrak{T} \cdot \mathbf{dS}$ extended over the bounding surface. This is equal to $\iiint \operatorname{Div} \mathfrak{T}\, dV$ extended over the volume, and therefore the force on unit volume due to the state of stress is equal to the divergence of the stress tensor, i.e.,

$$\operatorname{Div} \mathfrak{T} = \mathbf{i}\left(\frac{\partial T_{11}}{\partial x_1} + \frac{\partial T_{12}}{\partial x_2} + \frac{\partial T_{13}}{\partial x_3}\right) + \cdots + \cdots \tag{1.5}$$

The equation of motion for the deformable material is thus given by

$$\rho \frac{d\mathbf{v}}{dt} = \rho\mathbf{g} + \operatorname{Div} \mathfrak{T} \tag{1.6}$$

This basic equation is applicable to solids or to fluids, the distinction between these arising according to the nature of the internal stresses $\mathfrak{T}$. Thus a solid has rigidity, meaning that shearing stresses which are brought into play by a shearing deformation oppose the deformation, whereas a fluid has no rigidity. On the other hand, fluids have viscosity, meaning that shearing stresses which are brought into play by a shearing rate of deformation oppose the continuance of the shearing motion. In order to proceed with further analysis of (1.6) it is necessary to have a supplementary equation expressing a characteristic equation of the material which expresses the stress $\mathfrak{T}$ in terms of the deformation, or the rate of deformation, or the past history of deformation. This is called a rheological equation of state of the material (Part 3, Chap. 3).

A good deal of the complication in dealing with (1.6) arises from the fact that terms like $v_x(\partial v_x/\partial x)$ occur in the convective derivative for the velocity components, and therefore the equations are nonlinear, so the *superposition principle cannot be used* to derive general solutions as linear combinations of special solutions.

In nearly every real problem the body force per unit mass $\mathbf{g}$ is derivable from a potential function, $\mathbf{g} = -\operatorname{grad} U$. Moreover, the convective derivative terms can be written as

$$\frac{d\mathbf{v}}{dt} = \frac{\partial \mathbf{v}}{\partial t} + \tfrac{1}{2} \operatorname{grad} \mathbf{v}^2 - \mathbf{v} \times \operatorname{curl} \mathbf{v} \tag{1.7}$$

It will be convenient to separate out from $\mathfrak{T}$ the isotropic component called pressure so that $\mathfrak{T}$ will be the part of the stress having a zero diagonal sum, or a combination of simple shearing stresses. For an isotropic stress

$$T_{11} = T_{22} = T_{33} = -p$$

and the other stress components are zero, and so for such a stress $\operatorname{Div} \mathfrak{T} = -\operatorname{grad} p$. With these substitutions the equation of motion takes the form

$$\frac{\partial \mathbf{v}}{\partial t} - \mathbf{v} \times \operatorname{curl} \mathbf{v} + \operatorname{grad}(P + U + \tfrac{1}{2}\mathbf{v}^2) = \rho^{-1} \operatorname{Div} \mathfrak{T} \tag{1.8}$$

in which $P = \int^p \rho^{-1}\, dp$, where $\mathfrak{T}$ is that part of the internal stress which is due to stress differences, the isotropic part being represented by the pressure p.

With the equations of motion in this form, the classification of various fields of special interest in mechanics of deformable bodies becomes evident:

Perfect fluids are those which can support no shearing stress (so $\mathfrak{T} = 0$). *Viscous* fluids are those in which there can be a shearing stress. *Newtonian* fluids are those in which the shearing stress is proportional to the rate of shearing deformation, including also the possibility of linear volume viscosity in which there is an isotropic pressure term that is proportional to the rate of change of volume. *Non-Newtonian* fluids are those in which the stress depends on deformation rates in more complicated ways, including the past history as well as the presently existent state and rate of deformation. An *incompressible* fluid is one which undergoes negligible change of density with changes of pressure occurring in the given situation. *Irrotational flow* is a kind of fluid motion in which curl $\mathbf{v} = 0$ throughout the region occupied.

Elastic solids are generally considered to be those in which Hooke's law is obeyed, i.e., in which stress is linearly related to strain, for the range of stresses occurring in the given situation. Many solids obey Hooke's law for small enough stresses but with increasing departure from a linear relation after a certain characteristic stress is reached. This is an inexactly defined quantity known as the *elastic limit*. Usually for applied stresses less than the elastic limit the elasticity is quite reversible, but for higher stresses the elasticity shows *hysteresis*, i.e., the stresses diminish more rapidly to zero with reduced strain after such a high stress has been applied, and there is a *permanent set* or change of form after the stress is entirely removed. If such a permanent set was produced by the action of nonisotropic stresses, then the material will generally show nonisotropic properties after such a deformation. Actually, such a change of form is often not permanent; the body slowly returns to its original form with lapse of time, a behavior which is called *elastic aftereffect*. Likewise when some solid bodies are kept under stress, they may show a flow or creep.

These various kinds of relations of stress to deformation exist in varying degree in different classes of materials, as metals, polymers, and so on. The properties are in general strongly temperature-dependent and are accompanied by absorption or release of heat, so important differences in behavior occur according to whether the motions take place under isothermal or adiabatic conditions. In general, such thermal effects will give rise to nonuniform temperature distributions, resulting in irreversible heat flow and thus providing an important mechanism for the degradation of mechanical energy into thermal energy in such materials. Also many materials of technical interest have a microscopically (or even coarser) heterogeneous structure, as with polycrystalline metals, many rocks, or concrete.

The whole subject is therefore complicated, not only by mathematical difficulties, but also by a wide range of different types of flow behavior exhibited by different materials under different conditions. But it is of importance because of the wide range of problems of applied physics involved, ranging from the high-speed flow of air around the wings of an airplane traveling at supersonic speeds to the rolling of metals and the molding of plastics.

4. Molecular Standpoint

A fluid or solid body is made up of molecules; therefore the observed macroscopic behavior of such a body is related to certain statistical averages over the behavior of the individual molecules. Considerations of this kind are more fully developed in Part 5 in connection with analysis of the thermal properties of matter. It is instructive in connection with solid and fluid mechanics to consider the equations of motion from this viewpoint separately from the fuller consideration of the thermal properties.

For simplicity of notation it will be supposed that the material consists of but a single kind of molecule; the formal generalization to the case in which several kinds are present is easily made. Let $\mathbf{r}$ be the position of the volume element $dx\,dy\,dz$ of an individual molecule whose velocity components lie in the range $du_1\,du_2\,du_3$ at $\mathbf{u} = u_1\mathbf{i} + u_2\mathbf{j} + u_3\mathbf{k}$. The statistical distribution of velocities of the molecules in this volume element at any instant is given by a distribution function $f(\mathbf{u},\mathbf{r},t)$ such that the number of molecules in the volume element $dx\,dy\,dz$ whose velocity components lie in the ranges u_1 to $u_1 + du_1$ and so on are given by $f(\mathbf{u},\mathbf{r},t)\,du_1\,du_2\,du_3 \cdot dx\,dy\,dz$. The total number of molecules of all speeds in unit volume at $\mathbf{r}$ and $\mathbf{t}$ is given by

$$N(\mathbf{r},t) = \iiint f(\mathbf{u},\mathbf{r},t)\,du_1\,du_2\,du_3 \tag{1.9}$$

The mean speed of the molecules in the volume element surrounding $\mathbf{r}$ is given by the vector $\mathbf{v}(\mathbf{r},t)$, where

$$N(\mathbf{r},t)\mathbf{v}(\mathbf{r},t) = \iiint \mathbf{u}f(\mathbf{u},\mathbf{r},t)\,du_1\,du_2\,du_3 \tag{1.10}$$

Writing m for the mass of a molecule, the density is

$$\rho(\mathbf{r},t) = mN(\mathbf{r},t) \tag{1.11}$$

As the assumption of a statistical velocity distribution implies, the individual molecules move with a variety of speeds which deviate from the mean. Therefore, even if there were no interaction of any kind between the individual molecules, the actual motion that ensues is considerably different from what it would be if all the molecules moved with the same speed $\mathbf{v}$. In consequence the change with time of the velocity field $\mathbf{v}$ is different from what it would be in the absence of such effects. A simple example is provided by considering a gas in a small finite region initially, say a sphere, whose molecules are moving in such a way that each cartesian component is distributed according to the Gauss error function but whose density is so low that the effects of collisions and other molecular interactions can be neglected. If it is in a container at rest, $\mathbf{v} = 0$ at all points inside the container. Now suppose the container is suddenly taken away. Molecules which would have collided with the wall will now go on out into space. After a time there will be molecules at a considerable distance away from the finite initial region in all directions. The molecules in a volume element away from the initial region will have an average velocity directed away from the initial region: this is not due to the action of any "forces" in the usual sense of the word, for it has been supposed that the density is so low that molecular collisions are negligible. It is due simply to the kinematics of the individual motion: the particular molecules which turn up in the volume element under consideration are those which in the initial statistical distribution had the proper velocities to lead them to reach this volume element. Thus the average velocity of *these* molecules will not be zero but a finite vector directed away from the initial region. Remaining near the center of the region which initially contained the gas will be only those molecules which were initially the slow ones of the statistical distribution. If the situation is spherically symmetrical, the mean velocity of these molecules will still be zero, but the rms deviation of the speed from the mean for the molecules left behind will be much less than initially, for the fast-moving molecules have gone away.

How does this situation appear to a phenomenological theory which does not look closely enough into the real situation to be aware of these statistical effects? Such a theory tacitly pretends that the only motion it needs to consider is the mean velocity $\mathbf{v}$ in each volume element. It sees that in the initial situation this $\mathbf{v}$ equaled zero throughout the region. Then it sees that $\mathbf{v}$ develops nonzero values directed away from the initial region, and in order for this to happen it has to postulate the existence of unbalanced stresses (simply pressures in the case of a gas) which produce these observed accelerations. It is clear that the mere existence of a velocity distribution of molecules departing from the mean gives rise to behavior which a theory that works only with the dynamics of the mean velocity must regard as having resulted from the existence of stresses in the material.

Such stresses are called *kinetic stresses*. In addition to the kinetic stress the statistical effects of the actual forces of molecular interaction also affect the change with time of the mean velocity field $\mathbf{v}$. These actions are also describable in terms of a stress which is due to the composite effect of real molecular interactions and so is more obviously like the elementary notion of stress as a force per unit area.

These general considerations will now be made explicit by a calculation of the hydrodynamic equation of continuity and the equation of motion based on detailed consideration of the statistical distribution of molecular velocities.

By arguments like those in Sec. 1 the net rate at which molecules leave the volume element from x to $x + dx$, y to $y + dy$, z to $z + dz$ is

$$\left[\iiint \left(u_1 \frac{\partial f}{\partial x} + u_2 \frac{\partial f}{\partial y} + u_3 \frac{\partial f}{\partial z} \right) du_1\, du_2\, du_3 \right] dx\, dy\, dz$$

which is easily expressed in terms of the mean velocity $\mathbf{v}$ and molecular density N, already defined, as $\operatorname{div}(N\mathbf{v})\, dx\, dy\, dz$. This must equal the rate of loss of the number in this volume element expressed as a diminution with time of N, i.e., $(-\partial N/\partial t)\, dx\, dy\, dz$. Therefore the equation of continuity takes the form

$$\frac{\partial N}{\partial t} + \operatorname{div}(N\mathbf{v}) = 0 \qquad (1.12)$$

which is the same as (1.1) since $\rho = Nm$.

The dynamical equations of motion are obtained by analysis of the time rate of change of the total momentum of the molecules contained within such a volume element.

This total momentum changes in three ways: (*a*) because of long-range forces such as gravitational and electromagnetic fields interacting with the molecules, (*b*) because of short-range forces of molecular interaction between pairs of molecules one of which is within and the other without the volume element, and (*c*) because of the actual change in population of molecules within the volume element due to the statistical spread in their velocities.

The long-range forces (*a*) are usually called body forces in the literature, whereas (*b*) and (*c*) together make up what the phenomenological theory regards as stresses due to the internal constitution of the material. Letting $\mathbf{F}_i$ stand for the long-range force acting on the ith molecule and $\mathbf{G}_{ij}$ stand for the short-range force on the ith molecule due to its interaction with the jth molecule (where j refers to a molecule outside the volume element), the total rate of change of momentum of the molecules i within the volume element is

$$\sum_i \mathbf{F}_i + \sum_{i,j} \mathbf{G}_{ij}$$

The net rate of loss of momentum in the volume element due to molecular transport of momentum is

$$\left[\iiint m\mathbf{u} \left(u_1 \frac{\partial f}{\partial x} + u_2 \frac{\partial f}{\partial y} + u_3 \frac{\partial f}{\partial z} \right) du_1\, du_2\, du_3 \right] \cdot dx\, dy\, dz$$

This is the divergence of the tensor

$$m \iiint f\, \mathbf{u}\, \mathbf{u}\, du_1\, du_2\, du_3 \qquad (1.13)$$

of the second-order moments of the velocity distribution. It is convenient to write $\mathbf{u} = \mathbf{v} + \mathbf{w}$ and to transform the variables of integration to $dw_1\, dw_2\, dw_3$, where $\mathbf{w}$ is the velocity of an individual molecule relative to the average velocity, and therefore the mean value of $\mathbf{w}$ is zero. This gives for this tensor $\rho\mathbf{v}\mathbf{v} + \rho\mathfrak{C}$, in which $\mathfrak{C}$ is written for the second-order tensor giving the second-order moments of the velocities of the molecules relative to their average velocity:

$$\mathfrak{C} = \iiint f\, \mathbf{w}\, \mathbf{w}\, dw_1\, dw_2\, dw_3 \qquad (1.14)$$

so, for example,

$$\begin{aligned} C_{11} &= \iiint w_1^2 f\, dw_1\, dw_2\, dw_3 \\ C_{12} &= \iiint w_1 w_2 f\, dw_1\, dw_2\, dw_3 \qquad \text{etc.} \end{aligned}$$

The total amount of momentum in the volume element also changes with time because of the time dependence of f, the rate of increase per unit volume being

$$\iiint m\mathbf{u} \frac{\partial f}{\partial t}\, du_1\, du_2\, du_3 = \frac{\partial}{\partial t}(\rho\mathbf{v})$$

By the principle of conservation of momentum this rate of increase must equal the rates of increase due to (*a*) action of forces and (*b*) transport of molecules across boundaries (equation of continuity for momentum); therefore,

$$\frac{\partial}{\partial t}(\rho\mathbf{v}) = \sum_i \mathbf{F}_i + \sum_{i,j} \mathbf{G}_{ij} - \text{Div}(\rho\mathbf{vv}) - \text{Div}(\rho\mathfrak{C})$$

Transposing the term in Div ($\rho\mathbf{vv}$) and using the equation of continuity, this becomes

$$\rho\left(\frac{\partial\mathbf{v}}{\partial t} + \mathbf{v}\cdot\nabla\mathbf{v}\right) = \sum_i \mathbf{F}_i + \sum_{i,j} \mathbf{G}_{ij} - \text{Div}(\rho\mathfrak{C}) \quad (1.15)$$

The left side is recognized as ρ times the convected derivative of the velocity, as in Sec. 3. On the right, the first term gives the sum of the long-range forces acting on the molecules in unit volume. This is the term denoted by $\rho\mathbf{g}$ in (1.6). The third term on the right is already in the form of the divergence of a second-order tensor: this is the part of the total stress which is called the *kinetic stress* arising from the momentum transport due to dispersion of the molecular speeds, as already discussed.

It remains to see how $\Sigma_{ij}\mathbf{G}_{ij}$ can be reduced to the form of the divergence of a tensor describing the statistical average of all the forces of molecular interaction. The physical dimensions of the differential volume of phenomenological theory need to be large enough to include a large number of molecules, and hence the linear dimensions of such a volume element will be large compared to a molecular diameter, which is also large compared with the range of interaction of intermolecular forces. For this reason the summation over all the molecules i contained in the volume element will be, in effect, only a summation over those lying within one or two molecular diameters of the surface of the volume element, and the summation over all the outside molecules j for a given i will be really a summation over just those molecules which lie near enough to the ith molecule to be within range of the intermolecular forces. In this way $\Sigma_{ij}\mathbf{G}_{ij}$ gives rise to a term acting across each bounding surface of the volume element whose magnitude is proportional to the area of that bounding surface. A fuller discussion of the equilibrium of a small tetrahedron along the lines of that given in Sec. 2 can be given to show that this term can be written in the form $\mathfrak{U}\cdot d\mathbf{S}$, where $\mathfrak{U}$ is a symmetrical second-order tensor representing the summation $\Sigma_{ij}\mathbf{G}_{ij}$ of the interactions of pairs of molecules.

This completes the reduction of the equation of motion for the velocity field to the form (1.6) and shows that the over-all stress $\mathfrak{T}$ of the phenomenological theory is made up of the two parts $\mathfrak{U}$, representing actual intermolecular forces, and $-\rho\mathfrak{C}$, representing the kinetic effects arising from the statistical dispersion of the molecular speeds,

$$\mathfrak{T} = \mathfrak{U} - \rho\mathfrak{C} \quad (1.16)$$

The simplest instance of kinetic stress is that considered in the kinetic theory of gases (Part 5, Chap. 2), in which all forms of molecular interaction which would give rise to $\mathfrak{U}$ are neglected. If the gas is essentially in equilibrium, the distribution of velocities is essentially spherically symmetrical, so the kinetic stress tensor reduces to its isotropic part, which is simply a hydrostatic pressure p equal to

$$p = \langle\rho w_1^2\rangle_{\text{av}} = \langle\rho w_2^2\rangle_{\text{av}} = \langle\rho w_3^2\rangle_{\text{av}} = \tfrac{1}{3}\rho c^2 \quad (1.17)$$

where c is written for the rms velocity of the molecules.

5. Energy Relations for Fluid

If the equation of motion (1.8) is multiplied through by $\mathbf{v}\,\cdot$, the term in $\mathbf{v}\times\text{curl}\,\mathbf{v}$ will go out, since this vector is orthogonal to $\mathbf{v}$, giving

$$\left(\frac{\partial}{\partial t} + \mathbf{v}\cdot\text{grad}\right)(\tfrac{1}{2}\rho\mathbf{v}^2) + \mathbf{v}\cdot\text{grad}(p + U) = \mathbf{v}\cdot\text{Div}\,\mathfrak{T} \quad (1.18)$$

The quantity $\frac{1}{2}\rho\mathbf{v}^2$ is the kinetic energy in unit volume reckoned as if all the material moved with the average velocity. It is therefore not the same as the total kinetic energy of motion in unit volume, which is $\frac{1}{2}\rho(\mathbf{v}^2 + c^2)$, where c is the rms speed of the molecules relative to the mean velocity.

In many cases the potential-energy function U is independent of time, $\partial U/\partial t = 0$. It is convenient for the moment to consider that $\mathfrak{T}$ includes all the stress including the isotropic pressure, absorbing the $\mathbf{v}\cdot\text{grad}\,p$ term into $\mathbf{v}\cdot\text{Div}\,\mathfrak{T}$. Then (1.18) takes the form

$$\left(\frac{\partial}{\partial t} + \mathbf{v}\cdot\text{grad}\right)(\tfrac{1}{2}\rho\mathbf{v}^2 + U) = \mathbf{v}\cdot\text{Div}\,\mathfrak{T} \quad (1.19)$$

so the convected rate of increase of the sum of kinetic and potential energy is given by the term $\mathbf{v}\cdot\text{Div}\,\mathfrak{T}$.

From the definition of the stress tensor the force acting across the bounding surface element $d\mathbf{S}$ is $\mathfrak{T}\cdot d\mathbf{S}$, and therefore the rate at which this force does work on the material inside the region of which $d\mathbf{S}$ is part of the bounding surface is $\mathbf{v}\cdot\mathfrak{T}\cdot d\mathbf{S}$; so the total rate at which work is done on the material inside is

$$\begin{aligned}\iint \mathbf{v}\cdot\mathfrak{T}\cdot d\mathbf{S} &= \iiint \text{div}(\mathbf{v}\cdot\mathfrak{T})\,dV \\ &= \iiint \mathbf{v}\cdot\text{Div}\,\mathfrak{T}\,dV + \iiint(\nabla\mathbf{v}:\mathfrak{T})\,dV \quad (1.20)\end{aligned}$$

Here the integrand of the second integral is the scalar product of the rate of strain tensor whose components are $\partial v_x/\partial x$, $\partial v_y/\partial y$ with the tensor $\mathfrak{T}$, so

$$\begin{aligned}\nabla\mathbf{v}:\mathfrak{T} = {}& \frac{\partial v_x}{\partial x}T_{11} + \left(\frac{\partial v_x}{\partial y} + \frac{\partial v_v}{\partial x}\right)T_{12} + \frac{\partial v_y}{\partial y}T_{22} \\ &+ \left(\frac{\partial v_y}{\partial z} + \frac{\partial v_z}{\partial y}\right)T_{23} + \frac{\partial v_z}{\partial z}T_{33} + \left(\frac{\partial v_z}{\partial x} + \frac{\partial v_x}{\partial z}\right)T_{31}\end{aligned}$$

The first term on the right of (1.20) is equal to the convected rate of increase per unit volume of the sum of the kinetic and potential energies of the fluid in unit volume. The presence of the second term on the right in (1.20) shows that not all of the work done by the internal stresses goes into increasing the kinetic and potential energies but that part of it goes to the quantity $\nabla \mathbf{v}:\mathfrak{T}$, which is also a rate of change of energy per unit volume with time.

It is not possible to give a brief general discussion of $\nabla \mathbf{v}:\mathfrak{T}$ because such a variety of special cases arise in different physical situations. In the case of an elastic material obeying Hooke's law the stresses $\mathfrak{T}$ are proportional to certain strain components. In this case it is possible to introduce a quadratic form in the strain components which can be interpreted as internal strain energy of the material, i.e., energy that is reversibly stored as a kind of potential energy of elastic deformation. In particular if the only stress is an isotropic pressure, $\nabla \mathbf{v}:\mathfrak{T}$ becomes $-p$ div $\mathbf{v}$, and if the material can be characterized by some functional relation $p(\rho)$ by which the pressure is dependent on ρ, then from the equation of continuity

$$-p \operatorname{div} \mathbf{v} = -\frac{p}{\rho}\frac{d\rho}{dt} = \rho \frac{d}{dt}\int^{\rho} p d\frac{1}{\rho}$$

which suggests that this term can be taken into account by regarding it as the convected rate of change of an intrinsic energy per unit mass defined as

$$E = -\int^{\rho} p d\frac{1}{\rho} \tag{1.21}$$

the lower limit being arbitrary.

In the case of a viscous fluid the stress includes terms which are proportional to the symmetrical part of the tensor $\nabla \mathbf{v}$ of rate of relative deformation. The term $\nabla \mathbf{v}:\mathfrak{T}$ is then expressible as a quadratic form in the rate of strain components which is positive definite. This quadratic form is known as a dissipation function and gives the rate at which the work done on unit volume by the stresses at the boundaries is dissipated into heat by the irreversible frictional processes which give rise to the viscous stresses.

6. Strain

In dealing with elastic solids, it is convenient to describe the deformation in terms of a symmetrical second-order tensor known as the *strain*. Suppose the deformation is such that the particle whose initial position is $\mathbf{s} = x\mathbf{i} + y\mathbf{j} + z\mathbf{k}$ is carried over into the final position $\mathbf{r}$, where the vector $\mathbf{r}$ is a function of $\mathbf{s}$, i.e., of x, y, z. In the following it is assumed not that $\mathbf{r} - \mathbf{s}$, the resultant displacement of the particle initially at $\mathbf{s}$, is small but that the relative displacements of parts initially close together are nearly the same. Writing $\mathbf{r} = r_1\mathbf{i} + r_2\mathbf{j} + r_3\mathbf{k}$, the displacement gradient is defined as the second-order tensor $\{\nabla \mathbf{r}\}$, where (Part 1, Chap. 9, Sec. 4)

$$\begin{aligned}\{\nabla \mathbf{r}\} &= \left(\frac{\partial}{\partial x}\mathbf{i} + \frac{\partial}{\partial y}\mathbf{j} + \frac{\partial}{\partial z}\mathbf{k}\right)(r_1\mathbf{i} + r_2\mathbf{j} + r_3\mathbf{k}) \\ &= \mathbf{ii}\frac{\partial r_1}{\partial x} + \mathbf{ij}\frac{\partial r_2}{\partial x} + \mathbf{ik}\frac{\partial r_3}{\partial x} \\ &\quad + \mathbf{ji}\frac{\partial r_1}{\partial y} + \mathbf{jj}\frac{\partial r_2}{\partial y} + \mathbf{jk}\frac{\partial r_3}{\partial y} \\ &\quad + \mathbf{ki}\frac{\partial r_1}{\partial z} + \mathbf{kj}\frac{\partial r_2}{\partial z} + \mathbf{kk}\frac{\partial r_3}{\partial z}\end{aligned} \tag{1.22}$$

The individual components are known as relative displacements. They are physically dimensionless and usually less than 10^{-2} in elastic solids.

This nine-component tensor can be regarded as the sum of three parts. First, it can be represented as the sum of a symmetric and antisymmetric part in the usual way. Then the symmetric part can be presented as the sum of an isotropic part and a part which is composed of pure shears, analogous to the corresponding decomposition of the symmetric stress tensor. Using the notation

$$\begin{aligned} e_{ij} &= \frac{1}{2}\left(\frac{\partial r_j}{\partial x_i} + \frac{\partial r_i}{\partial x_j}\right) \\ w_{ij} &= \frac{1}{2}\left(\frac{\partial r_j}{\partial x_i} - \frac{\partial r_i}{\partial x_j}\right)\end{aligned} \tag{1.23}$$

the three parts are (*a*) the rotation, (*b*) the cubical dilatation, and (*c*) the distortion.

The rotation can be written

$$w_{12}(\mathbf{ij} - \mathbf{ji}) + w_{23}(\mathbf{jk} - \mathbf{kj}) + w_{31}(\mathbf{ki} - \mathbf{ik}) \tag{1.24}$$

This gives rise to relative displacements

$$\delta\mathbf{r} = \tfrac{1}{2}(\operatorname{curl} \mathbf{r}) \times \delta\mathbf{s}$$

which is the relative displacement which results from a rigid rotation of the material about the direction of curl $\mathbf{r}$ through an angle equal to half the magnitude of curl $\mathbf{r}$.

The cubical dilatation is given by

$$\Delta(\mathbf{ii} + \mathbf{jj} + \mathbf{kk}) \quad \text{where} \quad \Delta = \left(\frac{\partial r_1}{\partial x_1} + \frac{\partial r_2}{\partial x_2} + \frac{\partial r_3}{\partial x_3}\right) \tag{1.25}$$

and is equal to the fractional increase of volume.

The distortion is given by

$$\left(\frac{\partial r_1}{\partial x_1} - \tfrac{1}{3}\Delta\right)\mathbf{ii} + e_{12}(\mathbf{ij} + \mathbf{ji}) + \cdots + \cdots \tag{1.26}$$

and corresponds to a change in shape without change in volume, and so it is called a distortional strain. It is a five-component quantity which takes the form

$$e_{s1}(\mathbf{ll} - \mathbf{nn}) + e_{s2}(\mathbf{mm} - \mathbf{nn})$$

when referred to its principal axes.

This decomposition into three parts is the most satisfactory from the point of view of general theory, as it is invariant to choice of axes. However, it has the property that some physically simple types of deformation are regarded as made up of two parts. Thus a *simple extension* in the ratio $(1 + a):1$ in the z direction without change in the x and y directions is a sum of a cubical dilatation and a pure distortion in accord with the matrix array of the components:

$$\begin{bmatrix} 0 & 0 & 0 \\ 0 & 0 & 0 \\ 0 & 0 & a \end{bmatrix} = \frac{a}{3}\begin{bmatrix} 1 & 0 & 0 \\ 0 & 1 & 0 \\ 0 & 0 & 1 \end{bmatrix} + \frac{a}{3}\begin{bmatrix} -1 & 0 & 0 \\ 0 & -1 & 0 \\ 0 & 0 & 2 \end{bmatrix}$$

the first of these being a dilatation, and the second being composed of two simple shears. A *simple shear* is defined as one which produces relative displacements

$$\delta r_1 = a\,\delta y \qquad \delta r_2 = 0 \qquad \delta r_3 = 0$$

and, in terms of the invariant decomposition into rotation, dilatation, and shear, the simple shear is really a resultant of a rotation and a pure shear, according to the matrix scheme

$$\begin{bmatrix} 0 & a & 0 \\ 0 & 0 & 0 \\ 0 & 0 & 0 \end{bmatrix} = \frac{a}{2}\begin{bmatrix} 0 & 1 & 0 \\ -1 & 0 & 0 \\ 0 & 0 & 0 \end{bmatrix} + \frac{a}{2}\begin{bmatrix} 0 & 1 & 0 \\ 1 & 0 & 0 \\ 0 & 0 & 0 \end{bmatrix}$$

The first matrix represents a rotation about the z axis through an angle $a/2$, and the second represents a pure shear whose principal axes are in the diagonal directions $\mathbf{i} + \mathbf{j}$.

The *strain quadric* is the surface whose equation is

$$e_{11}x^2 + e_{22}y^2 + e_{33}z^2 + (e_{12} + e_{21})xy + \cdots = 1 \tag{1.27}$$

It has the property that the reciprocal of the square of the radius vector to it from the center is proportional to the extension of a line in that direction. If the strain is one in which the material is expanded in all directions, the quadric is an ellipsoid, but if there is expansion in some directions and contraction in others, it is a hyperboloid, and the lines which undergo neither expansion nor contraction are generators of the cone which is asymptotic to the hyperboloid.

By a *plane strain* is meant one in which there are no displacements in, say, the z direction and in which the displacements in the x and y directions are independent of z. Such a strain is represented by a matrix of the form

$$\begin{bmatrix} e_{xx} & e_{yx} & 0 \\ e_{xy} & e_{yy} & 0 \\ 0 & 0 & 0 \end{bmatrix}$$

in which the nonvanishing components do not depend on z.

Because the strain components are related to the derivatives of $\mathbf{r}$, they are related to each other. They therefore are not independent functions of position but must satisfy certain *conditions of compatibility:*

$$\begin{aligned}
\frac{\partial^2 e_{yy}}{\partial z^2} + \frac{\partial^2 e_{zz}}{\partial y^2} &= \frac{\partial^2}{\partial y\,\partial z}(e_{yz} + e_{zy}) \\
\frac{\partial^2 e_{zz}}{\partial x^2} + \frac{\partial^2 e_{xx}}{\partial z^2} &= \frac{\partial^2}{\partial z\,\partial x}(e_{zx} + e_{xz}) \\
\frac{\partial^2 e_{xx}}{\partial y^2} + \frac{\partial^2 e_{yy}}{\partial x^2} &= \frac{\partial^2}{\partial x\,\partial y}(e_{xy} + e_{yx}) \\
\frac{\partial^2 e_{xx}}{\partial y\,\partial z} &= \frac{\partial}{\partial x}\left(-\frac{\partial e_{yz}}{\partial x} + \frac{\partial e_{zx}}{\partial y} + \frac{\partial e_{xy}}{\partial z}\right) \\
\frac{\partial^2 e_{yy}}{\partial z\,\partial x} &= \frac{\partial}{\partial y}\left(+\frac{\partial e_{yz}}{\partial x} - \frac{\partial e_{zx}}{\partial y} + \frac{\partial e_{xy}}{\partial z}\right) \\
\frac{\partial^2 e_{zz}}{\partial x\,\partial y} &= \frac{\partial}{\partial z}\left(+\frac{\partial e_{yz}}{\partial x} + \frac{\partial e_{zx}}{\partial y} - \frac{\partial e_{xy}}{\partial z}\right)
\end{aligned} \tag{1.28}$$

In the last three of these e_{xy}, etc., is to be understood to mean the symmetrized component $\frac{1}{2}(e_{xy} + e_{yx})$.

Rate of strain is obtained from strain by differentiation with respect to time. If $\mathbf{r}(\mathbf{s},t)$ is the position at time t of the particle initially at $\mathbf{s}$, then the velocity is $\dot{\mathbf{r}}$, and the rate of strain is defined as the tensor $\{\nabla\dot{\mathbf{r}}\}$, which can also be analyzed into rate of rotation, dilatation, and shear as with strain itself.

7. Hooke's Law

Elastic materials are characterized by relations between stress and recoverable strain. The classical theory of elasticity is limited almost entirely to situations in which the strains are small. For small strains the elastic stresses in a material are linear functions of the strain. In recent years an increasing amount of attention has been given to more general cases in which large strains and nonlinear relations between stress and strain are required for the description of the observed phenomena (Part 3, Chap. 3) [2].

The general relation for elastic materials that stress is a linear function of strain for small strains is known as *Hooke's law.* It was published by its discoverer, Robert Hooke, in 1676 in the form *ceiiinosssttuu,* which is an anagram for the Latin *ut tensio, sic vis.* This form of publication today would hardly be considered an adequate basis for a claim to recognition as discoverer of an important law of physics.

The real situation in many cases of practical importance is more complicated than an elementary presentation often seems to imply. Most elastic materials have a microscopic heterogeneity, being actually aggregates of small crystals which are individually anisotropic but oriented more or less at random, perhaps showing an over-all average behavior that is anisotropic because of the existence of some measure of preferred orientation of the individual crystals. Then, too, some materials of interest are quite grossly heterogeneous, e.g., structural concrete, containing sand and coarse gravel as well as portland cement. Moreover, the technical processes by which many solids are prepared may result in variously complicated states of imperfect equilibrium, giving rise to situations in which parts of the material are held in a state of stress in their relations to other parts even when the entire body is in the "unstressed" state in the sense that no external forces are acting on it. The possibility of such complications needs to be constantly borne in mind in applying the theory of elasticity to actual physical situations. In spite of this, simple forms of theory which ignore such complications and work with Hooke's law using empirically determined over-all elastic constants have a wide field of usefulness and are an appropriate starting point for more elaborate analyses.

The general form of Hooke's law is that the six-component symmetrical stress tensor is a linear function of the components of the six-component symmetrical strain tensor,

$$T_{ij} = \sum_{k,l} C_{ijkl}\,e_{kl} \tag{1.29}$$

Since the e's are dimensionless, the c's are measured in the same stress units as the T_{ij}. For most solids the nonvanishing elastic constants have experimental values of the order of 10^{12} dynes/cm^2 $\sim 10^6$ atm.

For a gas the isothermal modulus of compression is just equal to the pressure itself, and so the stress change due to a given cubical dilatation is at ordinary atmospheric pressure of the order of one-millionth what it is in most solids.

The coefficients C_{ijkl} form a fourth-order tensor. A general tensor of fourth order in three dimensions has $3^4 = 81$ components. However, because of the symmetry of T_{ij} and e_{kl}, the tensor C_{ijkl} satisfies symmetry relations of the form

$$C_{ijkl} = C_{jikl} \quad \text{and} \quad C_{ijkl} = C_{ijlk} \tag{1.30}$$

which reduces the number of independent coefficients from 81 to 36.

Further reduction in the number of independent coefficients is obtained if the material has the property that a strain-energy function exists; i.e., the elastic energy stored in unit volume represented by the time integral of $\iiint \nabla \mathbf{v} : \mathfrak{T}\, dV$ in (1.20) is independent of the path of variation of the strain by which a given final state of strain is produced, starting from the unstrained condition. This gives rise to the further symmetry condition on C_{ijkl} that

$$C_{ijkl} = C_{klij} \tag{1.31}$$

which further reduces the number of independent components from 36 to 21. This can be seen by recognizing that the relative velocity components $\partial v_x/\partial x$, etc., in (1.20) can also be considered as the time derivative of the strain components, so $\partial v_x/\partial x$ is the same as $\partial e_{11}/\partial t$, and $\partial v_x/\partial y + \partial v_y/\partial x$ is the same as $\partial(e_{21} + e_{12})/\partial t$, and therefore the rate of doing work on unit volume considered in (1.20) is

$$\sum_{i,j} T_{ij} \frac{\partial}{\partial t} e_{ij} = \sum_{i,j,k,l} C_{ijkl}\, e_{kl} \frac{\partial}{\partial t} e_{ij}$$

The condition stated, $C_{ijkl} = C_{klij}$, is the condition that this can be written as the time derivative of a quadratic form in the strain components, which quadratic form is known as the strain-energy function. Denoting the strain-energy function per unit volume by W, it takes the form

$$\begin{aligned} W = \tfrac{1}{2}[&C_{1111}e_{11}^2 + 2C_{1122}e_{11}e_{22} + 2C_{1133}e_{11}e_{33} \\ &+ C_{2222}e_{22}^2 + 2C_{2233}e_{22}e_{33} + C_{3333}e_{33}^2 \\ &+ 2C_{1123}e_{11}(e_{23} + e_{32}) + 2C_{1131}e_{11}(e_{31} + e_{13}) \\ &+ 2C_{1112}e_{11}(e_{12} + e_{21}) + 2C_{2223}e_{22}(e_{23} + e_{32}) \\ &+ 2C_{2231}e_{22}(e_{31} + e_{13}) + 2C_{2212}e_{22}(e_{12} + e_{21}) \\ &+ 2C_{3323}e_{33}(e_{23} + e_{32}) + 2C_{3331}e_{33}(e_{31} + e_{13}) \\ &+ 2C_{3312}e_{33}(e_{12} + e_{21}) + C_{2323}(e_{23} + e_{32})^2 \\ &+ 2C_{2331}(e_{23} + e_{32})(e_{31} + e_{13}) \\ &+ 2C_{2312}(e_{23} + e_{32})(e_{12} + e_{21}) \\ &+ C_{3131}(e_{31} + e_{13})^2 + 2C_{3112}(e_{31} + e_{13})(e_{12} + e_{21}) \\ &+ C_{1212}(e_{12} + e_{21})^2] \end{aligned} \tag{1.32}$$

The full fourth-order-tensor notation is cumbersome, and it is customary in the literature of elasticity theory to use a two-index notation for the tensor of elastic constants based on a one-index numbering of the stress and strain components. In this scheme

Single:	1	2	3	4	5	6
Double:	11	22	33	23,32	31,13	12,21

In this notation the elastic constant C_{1123} is written C_{14}, and so on. In comparing results in other books it is important to watch carefully for minor differences in notation. For example, in Love's treatise on the "Theory of Elasticity" he writes e_{xx}, e_{xy}, etc., for the strain components, except that, for the nondiagonal components, he writes

$$e_{xy} = \frac{\partial r_2}{\partial x} + \frac{\partial r_1}{\partial y}$$

in the notation used here, i.e., $e_{12} + e_{21}$. As such his e_{xy} is equal to *twice* the 12 component of the symmetric strain tensor.

The number of independent components of C_{ijkl} is reduced still further by taking account of the symmetry properties of the different crystal classes [3].

In the case of an isotropic solid there are no preferred axes of any kind applying to the material. This has the consequence of introducing one further relation between the three independent nonvanishing constants of the cubic system in the form

$$C_{11} = C_{12} + 2C_{44} \tag{1.33}$$

so that for an isotropic solid there are two independent elastic constants.

Aside from restrictions imposed by crystal symmetry, if it were supposed that all the interactions between molecules giving rise to the elastic stresses were ultimately due to *central* forces, there would be a number of additional restrictions imposed on the c's. These, known as *Cauchy relations*, are

$$\begin{matrix} C_{23} = C_{44} & C_{31} = C_{55} & C_{12} = C_{66} \\ C_{14} = C_{56} & C_{25} = C_{46} & C_{45} = C_{36} \end{matrix}$$

In the case of an isotropic body this would imply that $\lambda = \mu$ and therefore $\sigma = \frac{1}{4}$ [see (1.34) and (1.35) for definitions of λ and μ]. In general the experimental values of elastic constants do *not* satisfy the Cauchy relations, indicating that the stresses are not due entirely to central forces.

The case of an isotropic body may be simply considered from first principles. If the strain and stress are each regarded as invariantly decomposed into the isotropic and the shear parts, there will be one elastic constant characterizing the proportionality between the isotropic (hydrostatic) stress and the isotropic strain (dilatation). When the shear strain is referred to principal axes, it consists of two parts of the form $\mathbf{ll} - \mathbf{nn}$ and $\mathbf{mm} - \mathbf{nn}$. It is assumed that a strain of either of these forms sets up stresses of the same form, and isotropy requires that the coefficient be the same for each form. Hence there is one other elastic constant relating shear stress to shear strain.

A variety of ways of expressing the two-constant stress-strain relations for isotropic elastic materials is used in the literature. The two constants, due to Lamé, are written λ and μ and relate stress to strain as follows:

$$T_{ij} = \lambda \Delta \delta_{ij} + 2\mu e_{ij} \tag{1.34}$$

where $\delta_{ij} = 1$ for $i = j$, and 0 for $i \neq j$. The quantity μ is called the *rigidity* or the *modulus of rigidity*, but the quantity λ does not have a name other than *Lamé's lambda*. Adding the first three equations, it is found that the isotropic part of the stress

$$T = \tfrac{1}{3}(T_{11} + T_{22} + T_{33})$$

is proportional to the dilatation Δ,

$$T = (\lambda + \tfrac{2}{3}\mu)\Delta = k\Delta \tag{1.35}$$

The quantity $k = \lambda + \frac{2}{3}\mu$ is called the *modulus of compression*, and its reciprocal is called the compressibility. For fluids $\mu = 0$.

If a body is subjected to a uniform tension along the x axis so that all values of T_{ij} except T_{11} vanish, then

$$T_{11} = Ee_{11} \qquad \text{where } E = \frac{\mu(3\lambda + 2\mu)}{\lambda + \mu} \tag{1.36}$$

$$\text{and } e_{22} = e_{33} = -\sigma e_{11} \qquad \text{where } \sigma = \frac{\lambda}{2(\lambda + \mu)}$$

The coefficient E, which relates the fractional increase in length to the stress producing it, is known as *Young's modulus*. The second part of (1.36) shows that the body contracts laterally while it expands in the direction of the applied tension. The ratio of the two strains σ is known as *Poisson's ratio*. Some illustrative values for common materials are given in Table 1.1.

TABLE 1.1. ELASTIC COEFFICIENTS OF COMMON MATERIALS (Unit, 10^{11} dyne/cm²)

Material	k	μ	E	σ	Limit, %*
Gelatine (80% water)	......	......	0.0002	0.50	10
Natural rubber	0.0019	......	0.0001	0.49	100–200
Polyethylene	......	......	0.001	......	2
Nylon	......	......	0.03	......	3
Silk thread	......	......	0.6	......	1
Lead	3	0.4–0.6	1–1.8	0.40	
Wood (par. to fiber)	......	......	1–1.5		
Sandstone	......	......	1.5		
Concrete	......	......	2–3	0.08–0.18	
Granite	3	1	3–6	0.2–0.3	0.5
Tin	5.2	1.7	4–5.3	0.33	
Glass	3.5–5.5	2–3	5–7	0.24–0.40	0.02
Aluminum	......	......	7	0.13	
Zinc	7–11	4	8–13	0.20	
Copper	13	6	10–12	0.32	
Wrought iron	15	7.5–8	18–20	0.27	
Steel	16–18	8–8.5	22–25	0.28–0.30	2.5

* The last column gives the approximate limit (per cent) of the strain for which a linear relation between stress and strain may be used.

With respect to coordinates in which both T_{ij} and e_{ij} are referred to principal axes, the strains can be expressed in terms of the stresses as follows:

$$\begin{aligned} e_{11} &= E^{-1}[T_{11} - \sigma(T_{22} + T_{33})] \\ e_{22} &= E^{-1}[T_{22} - \sigma(T_{33} + T_{11})] \\ e_{33} &= E^{-1}[T_{33} - \sigma(T_{11} + T_{22})] \end{aligned} \tag{1.37}$$

In the case of an isotropic solid the strain-energy function takes the form

$$\begin{aligned} W = {} & \tfrac{1}{2}(\lambda + 2\mu)\Delta^2 \\ & + \tfrac{1}{2}\mu[(e_{23} + e_{32})^2 + (e_{31} + e_{13})^2 + (e_{12} + e_{21})^2 \\ & \qquad - 4(e_{22}e_{33} + e_{33}e_{11} + e_{11}e_{22})] \end{aligned}$$

The term in Δ is the elastic energy stored in producing an isotropic cubical dilatation. The term proportional to μ is the elastic energy stored in virtue of the rigidity of the material opposing the shearing strain. In the simple case of a strain of the form $a(\mathbf{ii} - \mathbf{kk})$, one has $e_{11} = a = -e_{33}$, and other strain components vanish, giving $2\mu a^2$ for this part of the elastic energy of unit volume.

8. Viscosity

A fluid has a vanishing modulus of rigidity, and so is characterized by a single elastic constant k, modulus of compression, as defined in (1.35), whose reciprocal is called the compressibility. The modulus of compression is the ratio of the increase in hydrostatic pressure to the fractional diminution of volume which it produces.

When a fluid flows in such a way as to undergo a deformation, internal stresses are brought into play which are proportional to the *rate* of deformation instead of to the shearing deformation itself, as in the rigidity of an elastic solid. Such stresses are called *viscous stresses*, and the property of possessing such stresses is called viscosity.

Some special fluids show more complicated relationships between the stress and the rate of shear strain. Such fluids are called *non-Newtonian* (Part 3, Chap. 3), whereas those in which simple proportionality exists are called *Newtonian*.

The tensor formalism is exactly analogous to that leading to (1.34) in the case of an isotropic elastic solid. This leads one to expect that a fluid has two viscosity coefficients just as it has two elastic coefficients. In place of the elastic strains in (1.34) there appears the time rates of change of these, so, in terms of velocity components, these take the form

$$\begin{aligned} T_{ii} &= \lambda_v \operatorname{div} \mathbf{v} + 2\mu_v \frac{\partial v_i}{\partial x_i} \\ T_{ij} &= 2\mu_v \left(\frac{\partial v_i}{\partial x_j} + \frac{\partial v_j}{\partial x_i}\right) \qquad i \neq j \end{aligned} \tag{1.38}$$

Adding the first three, the isotropic part of the viscous stress $T = T_{11} + T_{22} + T_{33}$, is given by

$$T = (\lambda_v + \tfrac{2}{3}\mu_v) \operatorname{div} \mathbf{v} \tag{1.39}$$

in exact analogy with (1.35) in the elastic case.

A nonvanishing value of $\lambda_v + \frac{2}{3}\mu_v$ corresponds to the phenomenological possibility that the pressure in a fluid which is expanding or contracting may depart from the instantaneous static equilibrium value. This property is called volume viscosity (Part 3, Chap. 3). Most of the literature of fluid flow, while recognizing this formal possibility, proceeds to assume that this part of the viscous stress vanishes. This is equivalent to assuming that $\lambda_v + \frac{2}{3}\mu_v = 0$. Recently, study of the absorption of ultrasonic waves in fluids has given evidence which indicates that $\lambda_v + \frac{2}{3}\mu_v \neq 0$ for some fluids.

Except where the contrary is explicitly stated, it will be supposed that $\lambda_v + \frac{2}{3}\mu_v = 0$ and that a fluid is characterized by a single viscosity coefficient $\eta = \mu_v$. It is the shearing stress per unit rate of shearing strain. Its cgs unit therefore corresponds to a stress of 1 dyne/cm² with a shear velocity gradient of 1 cm/sec/cm. It is called the *poise*. It is 1 dyne-sec/cm², and its dimensions are $ML^{-1}T^{-1}$. Some illustrative values of viscosity at 20°C are given in Table 1.2.

TABLE 1.2. VISCOSITY AND MODULUS OF COMPRESSION OF SOME FLUIDS

Material	Viscosity at 25°C, 10^{-3} poise	k, 10^{11} dynes/cm²
Acetone	3.31	1,080
Benzene	6.47	970
Water	8.94	500
Mercury	15.5	38
40% sucrose in water	62	330
60% sucrose in water	565	260
Glycerol	5,450	220
Castor oil	10,300	470

TABLE 1.3. TEMPERATURE VARIATION OF VISCOSITY OF WATER

Temperature, °C	*Viscosity*, 10^{-2} *poise*
0	1.792
20	1.002
40	0.656
60	0.469
80	0.357
100	0.284

The ratio μ_v/ρ, or η/ρ, which occurs in the equations of motion, is usually denoted by ν and, following Maxwell, is called the *kinematic viscosity*. Its dimensions are L^2T^{-1}, so it is measured in square centimeters per second in the cgs system, the same as diffusivity and thermal diffusivity.

In the equations of motion (1.6) or (1.8) the divergence of the viscous stress tensor (1.39) appears. Assuming $\lambda_v + \frac{2}{3}\mu_v = 0$, this becomes

$$\text{Div}\,\mathfrak{T} = \eta(\tfrac{4}{3}\,\text{grad div}\,\mathbf{v} - \text{curl curl}\,\mathbf{v}) \quad (1.40)$$

Using this, assuming a static equation of state $p = p(\rho)$ and a characteristic function $P(\rho) = \int dp/\rho$ and assuming that the external forces are derivable from a potential energy per unit mass U, the equation of motion for a viscous fluid takes the form

$$\frac{d\mathbf{v}}{dt} + \text{grad}\,(P + U - \tfrac{4}{3}\nu\,\text{div}\,\mathbf{v}) + \nu\,\text{curl curl}\,\mathbf{v} = 0 \quad (1.41)$$

This equation of motion is called the *Navier-Stokes equation* because it was first derived in 1822 by C. L. M. H. Navier and was extensively studied by G. G. Stokes in a memoir published in 1845.

Most of the hydrodynamic literature is confined to the case of an incompressible fluid in which case the term in (1.41) involving div **v** vanishes. Likewise, of course, if this assumption is made, there is no need to assume $\lambda_v + \frac{2}{3}\mu_v = 0$. In dealing with incompressible fluids, using a cartesian coordinate system, curl curl $\mathbf{v} = -\nabla^2\mathbf{v}$, and likewise the transformation (1.7) may be used to obtain the dynamical equation for an incompressible viscous fluid

$$\frac{\partial\mathbf{v}}{\partial t} - \mathbf{v}\times\text{curl}\,\mathbf{v} + \text{grad}\,(P + U + \tfrac{1}{2}\mathbf{v}^2) = \nu\,\nabla^2\mathbf{v} \quad (1.42)$$

which is a special form of (1.8).

References

1. Principal Books:

Fluid Mechanics:

1. Lamb, Horace: "Hydrodynamics," 5th ed., Cambridge University Press, New York, 1924.
2. Basset, A. B.: "A Treatise on Hydrodynamics," 2 vols. (originally published, 1888), Dover, New York, 1961.
3. Milne-Thomson, L. M.: "Theoretical Hydrodynamics," Macmillan, New York, 1950.
4. Goldstein, S.: "Modern Developments in Fluid Dynamics," 2 vols., Oxford University Press, Fair Lawn, N.J., 1938. Reprint: Dover, New York, 1965.

Elasticity Theory:

1. Love, A. E. H.: "A Treatise on the Mathematical Theory of Elasticity," 4th ed., Cambridge University Press, New York, 1934.
2. Timoshenko, S., and J. N. Goodier: "Theory of Elasticity," 2d ed., McGraw-Hill, New York, 1951.
3. Prager, William: "Introduction to Mechanics of Continua," Ginn, Boston, 1961.
4. Sokolnikoff, I. S.: "Mathematical Theory of Elasticity," 2d ed., McGraw-Hill, New York, 1956.
5. Jaeger, J. C.: "Elasticity, Fracture and Flow," Wiley, New York, 1956.

2. Eringen, A. Cemal: "Nonlinear Theory of Continuous Media," McGraw-Hill, New York, 1962.
3. Sommerfeld, A.: "Mechanics of Deformable Bodies," p. 288, Academic, New York, 1950.

Chapter 2

Fluid Mechanics

By R. J. SEEGER, National Science Foundation

1. Statics of Fluids

According to Eq. (1.6) of Chap. 1, the equation of motion for a deformable material in equilibrium becomes

$$\mathbf{g} + \operatorname{div} \mathfrak{T} = 0$$

In the case of a fluid with isotropic stress

$$\operatorname{div} \mathfrak{T} = -\operatorname{grad} p$$

so

$$\rho \mathbf{g} = \operatorname{grad} p$$

We have also the thermal equation of state, viz., $p = p(\rho,T)$, where T is the temperature. In mechanics per se the temperature is considered constant, so a physical process can be regarded as isothermal. In adiabatic processes, involving reversible thermodynamic changes, the fact that the entropy remains constant makes it possible for the temperature to be eliminated. In either case, we can write $p(\rho)$. *Barotropic flows* (V. Bjerknes, 1898) are those for which lines of constant pressure coincide everywhere with those of constant density; *baroclinic flows* are those for which these lines do not coincide (owing to the dependence of p also on temperature, humidity of the atmosphere, salinity of the ocean, etc.). Introducing an auxiliary pressure function $P(\rho) = \int^p dp/\rho$, we obtain for barotropic flows

$$\mathbf{g} = \operatorname{grad} P$$

Hence curl $\mathbf{g} = 0$, so equilibrium is possible only if the body-force field is a conservative one.

In most practical applications $\mathbf{g}$ is the gravitational acceleration field, i.e., $\mathbf{g} = -\operatorname{grad} V$, where V is the gravitational potential. In this instance

$$V + P = \text{const}$$

The surfaces of constant gravitational potential, accordingly, are surfaces of constant P and hence of constant pressure; they are called *level surfaces.* Near the earth's surface there is a uniform gravitational field, so $V = gz$, and

$$gz + P = \text{const}$$

If the surface of an incompressible field is taken at $z = 0$, p_0 the atmospheric pressure there, and the positive z axis vertically downward, then

$$-gz + \frac{p}{\rho} = \frac{p_0}{\rho}$$

A change in p_0 at $z = 0$ will result in the same change in p at an arbitrary point z. This transmissibility of pressure is a specific instance of *Pascal's principle* (1647). If p_a is the overpressure (gauge pressure) relative to the atmospheric pressure at $z = 0$, then

$$p_a = \rho g z$$

This relation contains all the elementary rules of hydrostatics about pressure at different levels in communicating vessels. The equilibrium condition for a free surface between a stationary liquid with pressure p_1 and the atmosphere above it with pressure p_2 is

$$p_1 - p_2 = T_s\left(\frac{1}{R_1} + \frac{1}{R_2}\right)$$

where T_s is the surface tension of the liquid and R_1, R_2 the radii of curvature of the surface. For a plane surface, $1/R_1 = 1/R_2 = 0$, the pressures on each side of the surface are equal.

The general equation for $V + P$ is useful also for measurement of atmospheric height; it can be reduced to a *formula for barometric height* (Laplace). For an ideal gas the thermal equation of state is $p = (R/M)\rho T$, where R is the universal gas constant and M the molecular weight of the gas. If the atmosphere is considered isothermal, then

$$P = \frac{RT}{M} \log \rho \qquad \text{and} \qquad \frac{p}{p_0} = e^{-Mgz/RT}$$

hence the atmospheric pressure decreases exponentially with the height. The temperature, however, is not actually constant throughout the atmosphere. It is customary to regard the atmosphere as being divided into layers of constant mean temperatures when applying the above barometric formula.

A *polytropic change* (G. Zeuner, 1887) is a reversible change of state during which any particularly defined specific heat remains constant. In the case of convective equilibrium (Kelvin, 1862; V. R. Emden, 1907) of an ideal gas $p = \text{constant } \rho^n$, where n is the *polytropic exponent.* For $n = 1$ a polytropic gas suffers an isothermal change. In the case of adiabatic changes n is equal to γ, ratio of the specific heat at constant pressure to that at constant volume; then

$$P = \frac{\gamma}{\gamma - 1}\frac{p}{\rho}$$

and

$$-gz + \frac{\gamma}{\gamma - 1}\frac{p}{\rho} = \frac{\gamma}{\gamma - 1}\frac{p_0}{\rho_0}$$

For $p = 0$ there is a limiting height of the atmosphere, which is about 48 km for $\gamma = 1.2$ (semihumid, adiabatic). This height is 28 km for the adiabatic case with $\gamma = 1.4$; for isothermal conditions it is infinite. The troposphere, which extends to an average height of 12 km, is approximately characterized by adiabatic changes with $\gamma = 1.2$. The stratosphere above it, however, is more nearly isothermal (−50°C).

The polytropic distribution of air masses can be interpreted as a sedimentation equilibrium both in a gravity field and in a centrifugal field. For a dilute emulsion behaving like an ideal gas the ratio between pressures at two different levels can be replaced by the ratio between the numbers of particles at the same places. The *molecular weight* is given by the Avogadro number times the mass of a particle (corrected for buoyancy). With these considerations the formula for barometric height can be used to determine experimentally the Avogadro number. If the external-force field is due to a rotation of the fluid with constant angular velocity ω about a vertical axis at a distance r from it, the equivalent centrifugal potential is given by

$$V = \tfrac{1}{2}r^2\omega^2$$

and, therefore, $\frac{1}{2}r^2\omega^2 + P = \text{constant}$. Hence

$$\frac{p}{p_0} = e^{(M/2)(r^2\omega^2/RT)}$$

At a given distance the greater the partial pressure of a gas, the greater its molecular weight. Utilizing this differentiation, ultracentrifuges up to 100,000 rpm are employed for separating protein molecules in incompressible mixtures.

When a body is immersed in a fluid at rest, the total force acting on it by the fluid is $-\iint p\,\mathbf{dS}$, where $\mathbf{dS}$ is a vector element of area directed outward from the body. But $\iint p\,\mathbf{dS} = \iiint \text{grad}\ p\,dV$. Therefore $-\iint p\,\mathbf{dS} + \iiint \rho\mathbf{g}\,dV = 0$. The resultant force on the immersed body is equal and opposite to the total weight of the displaced fluid. This relationship, known as *Archimedes' principle*, holds for compressible fluids in nonuniform fields as well as for incompressible liquids in uniform fields.

This principle is utilized for determining specific gravity experimentally. The loss of weight of an immersed body of mass M and density ρ is $\rho_0 Vg$, where ρ_0 is the density of the liquid. Hence ρ/ρ_0 is equal to the ratio of the weight in air to the loss of weight in the liquid. If either ρ or ρ_0 is known, the other can be found. More accurately, such weighings have to be corrected for the difference in buoyancy of the body and of the weights used.

2. Inviscid-fluid Dynamics

The term *hydrodynamics* (D. Bernoulli, 1738) applies strictly to the dynamics of liquids but is sometimes used generally to include all fluid dynamics. *Aerodynamics* relates primarily to the dynamics of gases associated with aeronautics; in the incompressible approximation it becomes identical with hydrodynamics.

The equation of motion (3.4) of Chap. 1 for a perfectly inviscid fluid ($\mathfrak{T} = 0$) becomes

$$\left(\frac{\partial \mathbf{v}}{\partial t} - \mathbf{v} \times \text{curl}\ \mathbf{v}\right) + \text{grad}\ (\tfrac{1}{2}v^2 + P + U) = 0 \tag{2.1}$$

The convective terms in this vector equation (Euler's equations, 1755) are nonlinear, so the useful principle of linear superposition of solutions does not apply.

Such fluid motion can be formulated in terms of the concept of circulation. The *circulation* $\Gamma = \oint_C \mathbf{v} \cdot \mathbf{dr}$ of a fluid is given by the line integral of the velocity along any *fluid contour* C. For example, motion of a fluid along circular paths, with speed varying inversely as the distance from the center, has circulation (it is irrotational, except at the center). If all closed fluid contours can be contracted to a point, the space is said to be *simply connected;* the circulation is then invariant (the velocity potential is single-valued; a *multiply connected* space, e.g., an annulus, may have a many-valued potential). The material (substantial) derivative of the circulation equals zero, i.e., $(d/dt)\oint_C \mathbf{v} \cdot \mathbf{dr} = 0$ (Kelvin's theorem, 1869) if $\mathbf{v}$ is a continuous function of the coordinates. An initial circulation in a fluid will never change in magnitude. For example, initial irrotational motion (zero circulation) will persist indefinitely. Conversely, if $\oint_C \mathbf{v} \cdot \mathbf{dr}$ is constant for each circuit path in a fluid, Euler's equation of motion follows.

In a uniform flow of velocity $\mathbf{V}$ there is a force on a body transverse to $\mathbf{V}$ if there is a circulation about the body; this *lift* L is given by $L = \rho\Gamma V$. The lift coefficient C_L of a body is given by

$$C_L = \frac{L}{(\rho/2)V^2A}$$

where A is an average cross-sectional area of the body. A two-dimensional body extending to infinity in both directions (like an infinite wing) will have exponentially a definite circulation lift only if it has a shape with a cusp or corner (finite angle). Such a point, however, is mathematically singular, so that the speed there would generally be infinite and the pressure negative (cf. Bernoulli's principle). Joukowski showed that finite speed exists at such a singular point for real flow only for one particular value of the circulation. A three-dimensional body has circulation and lift only if it has a knife-edge. For a finite body, such as a wing of finite span, the circulation varies with the cross section; it is zero for a fluid contour pulled away from the body.

The *drag coefficient* C_D of a body is given by

$$C_D = \frac{D}{(\rho/2)V^2A}$$

where D is the retarding force on the body along the direction of flow $\mathbf{V}$. Newton (1687) first proposed a V^2 drag law on the basis of the inertial effect of fluid particles impinging on a body.

The three components of velocity and the pressure $p(\rho)$ can be obtained from the three scalar equations of motion and the *equation of continuity* of flow, which expresses the conservation of mass,

$$\frac{d\rho}{dt} + \rho \operatorname{div} \mathbf{v} = 0$$

For steady motion,

$$\operatorname{div} (\rho \mathbf{v}) = 0$$

By Gauss' theorem in this case the total flux through any closed surface in the fluid is zero. For steady flows in an incompressible fluid,

$$\operatorname{div} \mathbf{v} = 0$$

The solution must satisfy the boundary conditions, including the possibility of any velocity component being tangent to a surface but not normal to it.

A flow is said to be *steady* if $\partial \mathbf{v}/\partial t = 0$ at each point in space. In this case the fluid particles move along a *streamline* curve, which is defined as one with the tangent at each point having the direction of the instantaneous velocity there. In steady motion the streamlines coincide with the *streaklines,* which are the loci of all particles passing through definite points in space at some time or other. In unsteady flow, the streamlines are not the Lagrangian paths of the particles.

If the equations of motion are integrated for steady flow along a streamline, for which the path element $\mathbf{dr}$ is normal to $\mathbf{v} \times \operatorname{curl} \mathbf{v}$, there results along the streamline

$$\tfrac{1}{2}\mathbf{v}^2 + P + V = \text{const} \tag{2.2}$$

This relation is sometimes called the *weak form of Bernoulli's principle* (1738). The constant may be different for each streamline. For an incompressible fluid Eq. (2.2) becomes

$$\tfrac{1}{2}\mathbf{v}^2 + \frac{p}{\rho} + gz = \text{const} \tag{2.3}$$

whereas for a compressible fluid it becomes

$$\tfrac{1}{2}\mathbf{v}^2 + \int \frac{dp}{\rho} + gz = \text{const} \tag{2.4}$$

At each point of the fluid there is a definite state of thermodynamic equilibrium specified by two variables such as the density ρ and the specific entropy S. The first law of thermodynamics (conservation of mechanical and thermal energy) states that for reversible changes of volume

$$dE = T\,dS - p\,d\frac{1}{\rho}$$

where E is the specific internal energy (per unit mass).

Therefore, for reversible adiabatic changes ($dS = 0$) along a streamline

$$dE = -p\,d\frac{1}{\rho}$$

In the case of isentropic flow for an ideal gas with negligible viscosity and thermal conductivity the caloric equation of state is

$$E = \frac{1}{\gamma - 1}\frac{p}{\rho}$$

along a streamline p = constant ρ^γ. A theoretically convenient $p(\rho)$ relation is $p = A + k\rho^\gamma$; particularly useful for approximate computations, though physically meaningless, is the *Chaplygin-Kármán-Tsien relation* (1904) $p = A - 1/\rho$. Equation (2.4) reduces to

$$\tfrac{1}{2}\mathbf{v}^2 + \frac{\gamma}{\gamma - 1}\frac{p}{\rho} + gz = \text{const} \tag{2.5}$$

For an ideal gas the acoustic speed c is given by $\sqrt{\gamma(p/\rho)}$. Hence

$$\tfrac{1}{2}\mathbf{v}^2 + \frac{c^2}{\gamma - 1} + gz = \text{const}$$

where $c^2/(\gamma - 1)$ is the specific enthalpy $E + p/\rho$. If the gravitational variation is negligible in a flow, as along a horizontal level,

$$\tfrac{1}{2}\mathbf{v}^2 + \frac{c^2}{\gamma - 1} = \text{const}$$

As the density, and hence the specific enthalpy, decreases to zero, there is a *limiting speed* v_L of steady flow determined by the Bernoulli constant:

$$\tfrac{1}{2}\mathbf{v}^2 + \frac{c^2}{\gamma - 1} = \tfrac{1}{2}v_L^2$$

The speed of flow equal to the local acoustic speed is a *critical speed* in that greater speeds of flow are *supersonic* and lesser speeds *subsonic. Transonic flows* occur in the neighborhood of the acoustic speed. A flow that is not everywhere subsonic or supersonic is said to be a *mixed flow.* The transition regions are difficult to ascertain. The local *Mach number* $M = v/c$ can increase indefinitely (on the assumption of an ideal-gas continuum); M^2 measures the kinetic energy of the flow relative to the thermal energy of the molecules.

The line integral of $\mathbf{v} \times \operatorname{curl} \mathbf{v}$ vanishes also along a vortex line, i.e., a curve whose tangent at each point is in the direction of the angular velocity $\boldsymbol{\omega}$ of the fluid (equal to $\frac{1}{2}$ curl $\mathbf{v}$). Hence Eq. (2.2) is true also along a vortex line. The integration constant along a vortex line is the same constant for each streamline passing through that vortex line. In particular, if $\boldsymbol{\omega} = 0$ throughout the fluid, so the flow is *irrotational,* the constant for an ideal gas has the same value throughout the isentropic flow. Bernoulli's principle is then said to take on a *strong form.*

For flow along a horizontal level, Eqs. (2.3) and (2.5) become, respectively,

$$\tfrac{1}{2}\mathbf{v}^2 + \frac{p}{\rho} = \text{const} \tag{2.6}$$

and

$$\tfrac{1}{2}\mathbf{v}^2 + \frac{\gamma}{\gamma - 1}\frac{p}{\rho} = \text{const} \tag{2.7}$$

The smaller the pressure, the greater the speed of flow. This fact may seem strange at first glance. If the fluid speed increases, however, an unbalanced force is required to produce this acceleration. In other words, the pressure must decrease. Whereas in particle mechanics one is accustomed to regarding motions as the consequence of given forces, in fluid dynamics the velocity of flow along a pipe of variable cross section is determined by the equation of continuity. The resulting pressure distribution is such as to produce forces to maintain this flow.

Bernoulli's principle is the most important one of elementary fluid dynamics; it has numerous applications in aerodynamics. For example, it is used in the *venturi meter*, by which velocity of flow is determined by measuring the pressure change at a constriction in a pipe. The equation of continuity for mean velocities v and v_0 across constricted cross sections of area A and A_0, respectively, is

$$vA = v_0A_0$$

If gravitational effects are negligible, then substitution for v in Eq. (2.2) gives

$$v_0^2 = \frac{2(P_0 - P)}{(A_0/A)^2 - 1}$$

A *pitot tube* can be used to measure within 10 per cent both the static pressure and the so-called dynamical pressure. If a closed-end tube with a side opening is placed parallel to the flow, the observed pressure within the tube (fluid at rest) equals the static pressure p outside. If an open-end tube is placed facing upstream, the observed pressure p_t within the tube is equal to the fluid pressure, i.e., by Eq. (2.6)

$$p_t = p + \tfrac{1}{2}\rho v^2$$

The difference between these two measurements is the *dynamical pressure* $\frac{1}{2}\rho \mathbf{v}^2$; it can be determined with a single apparatus.

If p_0 and ρ_0 are the pressure and density for zero velocity, then Eq. (2.7) becomes

$$\tfrac{1}{2}\mathbf{v}^2 + \frac{\gamma}{\gamma - 1}\frac{p}{\rho} = \frac{\gamma}{\gamma - 1}\frac{p_0}{\rho_0}$$

For adiabatic changes

$$\frac{p}{p_0} = \left(\frac{\rho}{\rho_0}\right)^\gamma$$

so

$$\frac{p}{p_0} = \left(1 - \frac{\gamma - 1}{2}\frac{\mathbf{v}^2}{C_0^2}\right)^{\gamma/(\gamma-1)}$$

For small $\mathbf{v}^2/c_0^2$

$$\frac{p}{p_0} = 1 - \frac{1}{2}\frac{\rho_0\mathbf{v}^2}{p_0} + \frac{\gamma}{8}\frac{\mathbf{v}^2}{c_0^2} + \cdots$$

or

$$\tfrac{1}{2}\mathbf{v}^2 + \frac{p}{\rho_0} = \frac{p_0}{\rho_0}$$

This equation for compressible fluids is the same as (2.6) for incompressible fluids. For airspeeds up to 300 mph the use of this relation results in only about 2 per cent error in speed measurements with a pitot tube.

Bernoulli's principle can be applied also to the speed of efflux of a fluid from a large vessel. Let the free surface of the fluid be in the plane $z = 0$, the atmospheric pressure there be p_0, and the velocity there negligible; then Eq. (2.2) becomes

$$\tfrac{1}{2}\mathbf{v}^2 + P + gz = P_0'$$

If the hole for the efflux is at $z = -h$, then the pressure at the *vena contracta* outside, to which the streamlines converge, is also p_0, so that $\mathbf{v}^2 = 2gh$. This result, observed originally by Torricelli (1643), states that the speed of efflux is the same as that acquired by fluid falling freely through the height h; it is commonly known as *Torricelli's theorem*.

Euler's equation for unsteady flow can be integrated for irrotational flow:

$$\frac{\partial \phi}{\partial t} + \tfrac{1}{2}\mathbf{v}^2 + P + U = \text{const} \tag{2.8}$$

where ϕ is the scalar velocity potential. This relation, too, may be called a form of Bernoulli's principle. Still another form is discussed in Sec. 6 in connection with discontinuous flows involving shock waves.

3. Irrotational, Continuous Flows of Inviscid Fluids

Since a continuous vector field can be represented as the sum of a solenoidal field and of an irrotational field, the velocity field can be expressed in terms of a scalar potential ϕ and of a vector potential $\mathbf{A}$,

$$\mathbf{v} = -\operatorname{grad}\phi - \operatorname{curl}\mathbf{A}$$

In the case of general *irrotational motion*, or potential flow,

$$\mathbf{v} = -\operatorname{grad}\phi$$

A flow initially irrotational will remain so if not disturbed externally.

Substituting in the equation of continuity,

$$\nabla^2\phi = \frac{d \ln \rho}{dt} \tag{2.9}$$

Accordingly, if a fluid is incompressible, or at least if the fractional changes in density are negligible, *Laplace's equation* must be satisfied:

$$\nabla^2\phi = 0 \tag{2.10}$$

The general solution of Eq. (2.10) can be expressed simply in the case of two dimensions, x, y; it is given by the real part of an arbitrary analytic function f of the complex variable $z = x + iy$. The *complex flow potential* can be written as

$$f(z) = \phi(x,y) + i\psi(x,y)$$

The imaginary part ψ, which is the complex conjugate of ϕ, is called the *stream function* (Lagrange, 1781; Stokes, 1847) because the curves ψ = constant are the streamlines of flow. The necessary and sufficient conditions that $f(z)$ should be analytic are the *Cauchy-Riemann equations*,

$$\frac{\partial \phi}{\partial x} = \frac{\partial \psi}{\partial y} \qquad \frac{\partial \psi}{\partial x} = -\frac{\partial \phi}{\partial y}$$

The flow for a line vortex, with no radial component of velocity but with total speed inversely proportional to the distance from it, is everywhere irrotational except at the center, which is a singular point where the representative complex function becomes infinite. These relations are also sufficient if the first partial derivatives are all continuous. Eliminating ψ,

$$\frac{\partial^2 \phi}{\partial x^2} + \frac{\partial^2 \phi}{\partial y^2} = 0$$

Eliminating ϕ, on the other hand,

$$\frac{\partial^2 \psi}{\partial x^2} + \frac{\partial^2 \psi}{\partial y^2} = 0$$

Both the real part and the imaginary part of an analytic function are potential functions. The two-dimensional theory of potential flow is identical with the powerful theory of analytic functions of a complex variable (see section on conformal mapping in Part 1, Chap. 3). Also

$$\frac{\partial \phi}{\partial x}\frac{\partial \psi}{\partial x} + \frac{\partial \phi}{\partial y}\frac{\partial \psi}{\partial y} = 0$$

so families of curves ϕ = constant and ψ = constant are orthogonal to each other and isometric; i.e., an infinitesimal area in the $\phi + i\psi$ plane is a square, and the mapping of the $x + iy$ plane on it is conformal. Conformal mapping of boundaries can be used to transform them into simple boundaries like a circle, for which solutions can be obtained. In general, no stream function can be associated with the velocity potential in three-dimensional flow. Even in the case of an axially symmetric flow, where a stream function ψ exists, it does not combine with a velocity potential ϕ to form a complex function $\phi + i\psi$, so the theory of a complex variable is inapplicable to three-dimensional potential flow except for such specialized conditions as axial symmetry.

The classical treatment of hydrodynamics is largely concerned with the determination of flows around various bodies, such as flows associated with different functions $f(z)$. For example, $f = Az^n$ represents the simply connected flow between two radii vectors from the origin for real A and n. The doubly connected region outside a body, however, involves an indeterminate third real constant. The Kutta-Joukowski (1902–1906) physical requirement of smooth flow at the trailing edge of an airfoil makes possible the determination of this constant (a stagnation point occurs at the trailing edge, so the velocity is finite everywhere).

The velocity component normal to the boundary is unchanged across the boundary. The velocity component tangent to the surface may change; hence in ideal fluids a slip is assumed along a solid. Because of viscous effects in real fluids, there is a narrow region about a body, called a *boundary layer* (Prandtl, 1904), in which the velocity along the surface decreases rapidly to zero at the surface itself. **Even though the viscous effects in the main stream may be negligible for small coefficient of viscosity, they become considerable in the boundary layer because of the large velocity gradient there. Classical solutions of hydrodynamics, in so far as they describe natural flows, are valid only outside the boundary layer.**

Laplace's equation for ϕ being linear, in the case of inviscid incompressible fluids, every linear combination of known solutions is also a solution, so that it is comparatively easy to satisfy a given set of boundary conditions. Use is made of this fact in the Rankine *source-sink* method for flows about bodies of revolution. If a liquid is at rest at infinity and the mass m is moving symmetrically outward or inward across a surface surrounding a point A, then for $+m$, A is called a *simple source*, and for $-m$ it is called a *simple sink* ($\phi = \pm m/r$)(m is the *strength of the source*); in each case the pressure gradient is inward. If a source and equivalent sink are close together, the combination is called a *doublet*. Sources and sinks correspond to north and south magnetic poles, respectively.

Potential-flow patterns can be studied experimentally by means of the Taylor-Sharman (1928) electric-current analogy. The electric-flow equations are

$$\frac{\partial V}{\partial x} = \frac{-j_x}{\sigma} \qquad \frac{\partial V}{\partial y} = \frac{j_y}{\sigma}$$
$$\frac{\partial W}{\partial x} = tj_y \qquad \frac{\partial W}{\partial y} = -tj_x$$

where V is the electric potential, W the electric stream function, j the electric-current density, σ the electrical conductivity, and t the variable thickness of a conducting sheet. For flow with circulation one must identify $V = -\psi$, $W = \phi$, $t = 1/\rho\sigma$; for flow without circulation one may take also $V = \phi$, $W = \psi$, $t = \rho/\sigma$.

The motion of an incompressible inviscid fluid (without free or solid surface boundaries and without general acceleration) past a sphere can be simulated by the superposed motions of a uniform flow and of a doublet. The velocity field in the latter case decreases inversely as the cube of the distance from the sphere. The total change in momentum integrated across a so-called control surface far removed from the disturbing sphere turns out to be zero. It is generally true that continuous solutions in this idealized theory lead to the result that there is no force (a couple is possible) on a body of arbitrary, but smooth, shape in such a fluid. This theorem is sometimes called the *paradox of d'Alembert* (1744) *or Dirichlet;* it is not verified observationally, as noted first by d'Alembert himself.

If a body of mass M moves through an incompressible fluid with a constant speed v_0, the total kinetic energy of the body and of the fluid is equal to $\frac{1}{2}(M + M')v_0^2$. The fluid gives the body an additional *virtual* (*induced*) *mass* of M', which must be taken into account for unsteady motion. In general, the induced mass depends upon the direction of motion. An induced mass is important also in the case of impact of a body on the surface of a liquid. In the case of a sphere M' equals half the mass of the displaced fluid.

Differentiating the Bernoulli equation for unsteady flow (2.8) with respect to the time,

$$\frac{\partial^2 \phi}{\partial t^2} = p'(\rho)\frac{\partial \ln \rho}{\partial t} + \frac{1}{2}\frac{\partial \mathbf{v}^2}{\partial t}$$

provided the external field U is not time-dependent. On substitution of (2.9),

$$p'(\rho)\,\nabla^2\phi - \frac{\partial^2\phi}{\partial t^2} = p'(\rho)\mathbf{v}\cdot\text{grad}\,(\ln\rho) - \frac{\partial}{\partial t}\,(\tfrac{1}{2}\mathbf{v}^2)$$

If the disturbances about a mean density ρ_0 are relatively small, so $p'(\rho)$ is approximated by $p'(\rho_0)$, then ϕ satisfies the wave equation, with $\sqrt{p'(\rho_0)}$ being the speed c of phase propagation. In the case of acoustic waves in compressible liquids and gases $p'(\rho_0)$ must be evaluated adiabatically for agreement with observed values. For incompressible fluids the acoustic speed becomes infinite.

In general, for polytropic changes and steady motion

$$\nabla^2\phi = \frac{1}{c_0^2 - \frac{1}{2}(\gamma - 1)(v^2 - v_0^2)}\, v\,\frac{\partial(\frac{1}{2}\mathbf{v}^2)}{\partial s}$$

where c_0 is the value of c at the point along the path s for which v has the value v_0. If the right-hand side is relatively small, an approximate value of ϕ can be obtained by solving $\nabla^2\phi = 0$ and substituting the value of $\mathbf{v}$ in the right-hand side so

$$\nabla^2\phi = f(x,y,z)$$

where $f(x,y,z)$ is known. This method is then repeated; it breaks down as v approaches c in magnitude.

4. Discontinuous Flows of Inviscid Fluids

In the preceding section it was assumed that the velocity varies continuously throughout the fluid. Discontinuities (Helmholtz, 1868), however, have to be allowed to describe the more quiescent wake regions behind sharp-edged obstacles in moving streams and the jets issuing through sharp-edged holes out of a reservoir into a quiet fluid. In these cases the velocity at the edge is not infinite, inasmuch as the fluid does not flow completely around it. The boundary of the free surface is itself a streamline, across which the pressure of the fluid is continuous, but the velocity components along it may be discontinuous.

If the fixed boundaries are straight lines, flows in two dimensions can be determined by the use of conformal transformations. In this manner the flow past a flat plate (Kirchhoff, 1869) at right angles to a stream has been investigated. A definite drag is found, whereas in continuous potential motions the theoretical resistance would always be zero. The drag coefficient obtained theoretically for an incompressible liquid is 0.88, in comparison with the observed value of 2.0. The discrepancy results from the idealization of the wake, which is not infinite, and which involves vortices along the boundary because of the viscous shearing there. Consideration of an infinite trail of vortices in an inviscid fluid gives a theoretical value of 1.6 for the total drag coefficient.

The conformal-mapping method can be applied also in the case of a two-dimensional incompressible jet to determine the degree of contraction of the jet at the vena contracta, i.e., the ratio of the cross-sectional area at a large distance from the orifice to that in its immediate neighborhood. The theoretical relative contraction of such a jet is 0.611, which is approximately that observed. One can treat also the problem of the three-dimensional jet symmetric about an axis. In the case of *Borda's mouthpiece*, which consists of a long straight tube projection into an infinite reservoir, the theoretical contraction ratio is 0.5. Jets, rather than the resistance of bodies, are more properly the domain of such theoretical treatment omitting vortex consideration, particularly if the jet enters a medium of much smaller density, say, water into air.

In the wake behind a flat plate, the boundary has a surface of discontinuity. For equilibrium in an inviscid flow the *streamline of discontinuity* has constant pressure (and by Bernoulli's theorem constant speed). In this case the separation points are known, and the problem can be solved for plane flows (only). In general, however, as in the flows about curved obstacles, there is a mathematical indeterminacy, usually resolved on the basis of physical assumptions that are not necessarily realized. The fact is that a streamline of discontinuity in an inviscid fluid is unstable. First there develops a wavy form, then a turbulent *mixing zone,* then a curling up or even a *vortex street* for lower Reynolds numbers. The resulting physical state is the existence of an underpressure in the wake. Two dimensionless physical parameters have been suggested for inclusion in the description of such phenomena, particularly cavities and jets: the density ratio for the fluids on the two sides of the streamline and the so-called *cavitation number*

$$Q = \frac{p - p_c}{\frac{1}{2}\rho v^2}$$

where p is the free-stream pressure and p_c is the wake pressure on the streamline of discontinuity ($Q = 0$ for infinitely long cavities). For physical cavities of finite length Q is usually positive. These conditions, however, are incompatible with irrotational flow in an ideal fluid. Only in the case in which the relative density is small is the classical streamline theory applicable—and even then in a very limited sense, as is evident in such water-entry phenomena as surface seal and downward refraction.

Cavitation is not peculiar to liquids; it occurs also in gases if the characteristic radius a of the body is large in comparison with the mean free path of the gas and the speed V of the body is large in comparison with the mean molecular speed c.

$$\frac{V(a/c)}{a} = M$$

gives the ratio of the cavity length to its radius—as in the case of a meteorite entering the atmosphere ($M \sim 25 - 175$) or that of the earth entering the neutral ionized stream of solar gas ($M \sim 100$). In the latter instance the predominant physical factor producing cavitation is the dipole component of the earth's magnetic field; the problem itself remains unsolved except for two idealizations.

5. Vortex Flows of Inviscid Fluids

The *vorticity* or vortex vector $\boldsymbol{\Omega}$ is given by curl $\mathbf{v}$ ($\frac{1}{2}\boldsymbol{\Omega}$, the average rotation of an element of volume, is called the vorticity by some authors). For example, a parallel flow with transverse velocity gradient has vorticity; a vortex-free motion, $\boldsymbol{\omega} = 0$, is irrotational. A *vortex line* is a curve for which the tangent at each point has the direction of $\boldsymbol{\Omega}$. A *vortex tube* consists of all the vortex lines passing through a surface element normal to them; the *vortex strength* is the flux of the vortex vector through the tube. Consideration of a closed curve completely on the wall of a vortex tube leads to the fact that vortex lines are material lines; i.e., they consist permanently of the same fluid particles (*Helmholtz's first theorem*). By Stokes' theorem (1847) the vortex flux through all cross-sectional curves of a tube is equal to the circulation around it, which remains unchanged. Hence the vorticity of a vortex tube remains unchanged during the motion (*Helmholtz's second theorem*). Vortices cannot be created or destroyed; a vortex-free motion will remain so. These theorems, due to Helmholtz (1858), are true for steady and unsteady motions for incompressible and compressible fluids [$p(\rho)$]; they are sufficient to give Kelvin's circulation theorem. Hence either Helmholtz's vortex theorems or Kelvin's theorem together with the equation of continuity are equivalent to Euler's equation of motion and the equation of continuity. Water spouts and tornadoes are approximate vortex filaments (modified by the uniform rotation and compressibility of the atmosphere).

For low speeds vortices form behind a body like a cylinder. As the speed increases, the vortices break away and move downstream. A system consisting of an infinite number of similar positive vortices spread evenly along a line and an infinite number of negative vortices in between along a parallel line is called a *Kármán street* (1912); the stability criterion is sinh $(h/a)\pi = 1$, where h is the distance between rows and a is the distance between neighboring vortices in the same row. Stable vortex streets are often found in the wake of an obstacle (Bénard, 1908). The alternate vortices leaving the body produce a periodic force on it, so the body vibrates (Stroubal, 1878).

A circular vortex filament has no radial velocity in the plane of the ring, so the radius of the ring remains constant, while the ring moves forward with constant speed. Smoke rings are such closed (circular) vortices. When two such rings have the same axis and the same sense of rotation, the induced velocity will result in the enlargement of the leading ring and the diminution of the other, which may eventually pass through the first ring, after which the process is repeated for the new leading ring. When two such rings have the same axis but the opposite sense of rotation, the induced velocity will result in enlargement and diminution of velocity of each ring.

There is an analogy between fluid dynamics for solenoidal fields and electrodynamics: the vortex strength corresponds to the current intensity and the vortex vector to the current density. Vortices are surrounded by velocity lines (streamlines for steady flows) just as electric currents are by magnetic lines of force. In these terms the flow velocity is said to be *induced* by the vorticity. The formula for induced velocity corresponds exactly to the law of Biot and Savart for the magnetic effect of an electric current (see Part 4, Chap. 1, Sec. 6).

6. Flows of Compressible, Inviscid Fluids

Liquids and gases differ greatly with respect to compressibility. *Gas dynamics* is concerned with phenomena for which compressibility is physically significant, e.g., pressure differences large relative to the total pressure (gas flow through a granular medium), and rapid oscillations (acoustic frequencies). In Fig. 2.1 isentropic flow of a compressible fluid through a given stream tube is compared with flow of an incompressible fluid through it.

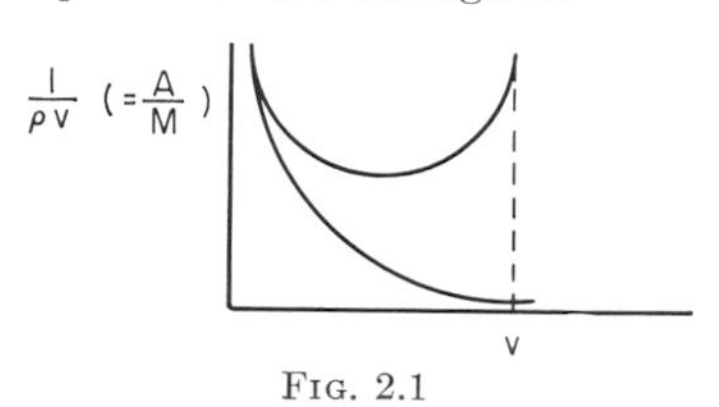

FIG. 2.1

Two solutions of the fundamental equations of continuity and motion, including the pressure-density relation, are said to be *dynamically similar* if one can be transformed into the other by a simple change in the scales of length, time, and density. For example, if $\mathbf{r} = v_0 t_0 \mathbf{r}'$, $t = t_0 t'$, $\rho = \rho_0 \rho'$, and $\mathbf{v} = v_0 \mathbf{v}'$, then these equations become

$$\frac{\partial \rho'}{\partial t'} + \text{div}\,(\rho' \mathbf{v}') = 0$$

and

$$\frac{d\mathbf{v}'}{dt} = -\frac{c^2}{v^2}\frac{v'^2}{\rho'}\,\text{grad}\,\rho'$$

The two flows will be similar only if the dimensionless Mach number $M^2 = v^2/c^2$ is the same for both flows. For adiabatic flows this criterion is satisfied if the adiabatic exponent is the same for the two flows and if the Mach numbers are the same for one set of corresponding points.

If the Mach number for the undisturbed state is very small, then the disturbances are small, and the dimensionless density satisfies the linear wave equation. The speed of propagation c_0 is a constant; it depends upon the medium alone and not upon the magnitude of the disturbance.

For example, where $\delta\rho$ is the change in density, one finds for the *one-dimensional wave equation*

$$\frac{\partial^2 \delta\rho}{\partial x^2} = \frac{1}{c_0^2}\frac{\partial^2 \delta\rho}{\partial t^2} \tag{2.11}$$

and its general solution is

$$\delta\rho = f(x - c_0 t) + g(x + c_0 t)$$

where f and g, both arbitrary functions, represent waves traveling with the velocities c_0 and $-c_0$, respectively. Thus small disturbances travel with acoustic speed in compressible fluids, whereas they are transmitted instantaneously in incompressible fluids. If initially $(t = 0)$ $\delta\rho = f(x)$ and $\partial\delta\rho/\partial t = f_1(x)$, then the disturbance $\delta\rho$ will be completely determined by

$$\delta\rho(x,t) = \tfrac{1}{2}[f(x + c_0t) + f(x - c_0t)] + \tfrac{1}{2}c_0[F(x + c_0t) - F(x - c_0t)]$$

where $F(x) = \int_0^x f_1(y)\, dy$.

The linearized *three-dimensional wave equation* is

$$\frac{\partial^2\delta\rho}{\partial t^2} = c_0^2\, \nabla^2\sigma\rho \qquad (2.12)$$

If initially $\delta\rho = f(x,y,z)$ and $\partial\delta\rho/\partial t = f_1(x,y,z)$, then the solution of this equation is given by Poisson's formula

$$\delta\rho(x,y,z,t) = \frac{\partial}{\partial t}(tF) + tF_1$$

where

$$F \equiv \frac{1}{4\pi R^2}\int f\, dS \qquad \text{and} \qquad F_1 \equiv \frac{1}{4\pi R^2}\int f_1\, dS$$

the surface S being a sphere with radius $R = c_0t$.

If a body itself is traveling with a speed V greater than the acoustic speed c, its disturbances generate a conical envelope. From Fig. 2.2 it follows that $\alpha = \sin^{-1}(1/M)$, where α is called the *Mach angle*. Optical observations, however, show a larger angle and a more sharply defined envelope than might be expected.

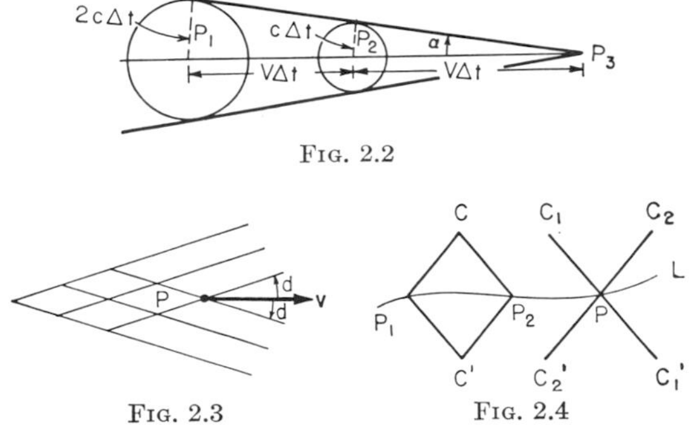

FIG. 2.2

FIG. 2.3 FIG. 2.4

In the case of nonuniform two-dimensional flow it is convenient to draw a network of physical *Mach lines* (mathematical-characteristic curves) in terms of $\mathbf{v}$, c_0, and α at each point P (see Fig. 2.3). Consider an arbitrary curve L (not a characteristic) passing through two points P_1 and P_2 (see Fig. 2.4). Let characteristics be drawn through these two points, as well as through a point P on L. A disturbance at P is restricted between PC_1 and PC_2; it will not affect the region within P_1CP_2C', which is determined by physical quantities along P_1P_2. In this respect supersonic flow differs radically from subsonic flow.

Flow of a gas through a channel with slightly varying cross-sectional area can be treated hydraulically as a one-dimensional problem. The equation of continuity can be written simply

$$\rho q A = m$$

where m is the rate of mass flow, which is constant throughout the channel. From this equation and Bernoulli's principle

$$\frac{dM}{M} = \frac{dq}{q}\left[1 + \frac{M^2}{2}(\gamma - 1)\right]$$

Subsonic flow in a compressible fluid behaves qualitatively like that in an incompressible fluid, but supersonic flow in a compressible fluid is quite different. For example, in subsonic flow the speed increases when the cross-sectional area decreases, and in supersonic flow it increases when the area increases. In the case of a convergent-divergent (de Laval) nozzle subsonic flow changes to supersonic at the minimum section, where $M = 1$. For a given geometry (nozzle) smoothly supersonic flow is possible at only one Mach number, depending upon the exit pressure.

Consider two-dimensional irrotational steady flow in compressible fluids, where it is assumed that $\mathbf{v} = \mathbf{v}(\rho)$ as a special type of solution. Put $v_x \equiv v_x(\rho)$ and $v_y \equiv v_y(\rho)$. Then from the equation of continuity and the condition for irrotational flow

$$u'(\rho)\frac{\partial\rho}{\partial y} - v'(\rho)\frac{\partial\rho}{\partial x} = 0$$

we obtain

$$\left(\frac{dv}{d\theta}\right)^2 = \frac{v^2}{M^2 - 1} \qquad (2.13)$$

The solution of this equation is a one-parameter family of two-branched curves (characteristics) in the (v,θ) hodograph plane. The condition for irrotational flow gives straight lines in the physical plane for any given ρ; these constant-density lines are perpendicular to the hodograph curve.

Applying these results to simple flows at a smooth corner, for flow into a corner the constant density lines form an envelope with a cusp, beyond which a shock is physically formed. For $f(\rho) = 0$ the constant lines intersect at a point, and one has a so-called *Prandtl-Meyer flow* (1908) around a sharp corner. In this case Bernoulli's principle in (2.13) gives

$$\frac{dp}{\rho v^2} = \frac{\pm d\theta}{\sqrt{M^2 - 1}}$$

This equation gives the maximum deflection for $dp = 0$ or $M = \infty$. For example, if $M = 1$ for the initial uniform stream, the maximum deflection will be 129.3° for $\gamma = 1.4$. For greater corner angles, separation from the wall will occur.

The partial differential equation for two-dimensional irrotational flow may be written

$$(c^2 - v_x^2)\phi_{xx} + 2v_xv_y\phi_{xy} + (c^2 - v_y^2)\phi_{yy} = 0$$

where c is a function of ϕ_x and ϕ_y. The characteristic equation in this case is

$$(c^2 - v_x^2)_0\left(\frac{dy_0}{dt}\right)^2 - (2v_xv_y)_0\frac{dy_0}{dt}\frac{dx_0}{dt} - (c^2 - v_y^2)_0\left(\frac{dx_0}{dt}\right)^2 = 0$$

where the initial values (potential equation) are given along the curve $x_0(\sigma)$, $y_0(\sigma)$ in the xy plane, σ being a parameter. These curves in the physical xy plane may have second derivatives of ϕ discontinuous on them. They are Mach lines and are perpendicular to the characteristics in the hodograph plane, which (epicycloids) are given by

$$\left(\frac{dv}{d\theta}\right)^2 = \frac{v^2}{M^2 - 1}$$

The method of characteristics fails in the case of *limiting lines*, for which the streamlines would turn back on themselves. It is useful, however, for numerical and graphical solutions of two-dimensional supersonic flows.

In general, any gain in obtaining a solution in the hodograph plane is offset practically by the difficulty of satisfying by a method of successive approximations the boundary conditions, which have been given in the physical plane.

In the case of three-dimensional steady irrotational compressible flows in a compressible fluid the velocity potential ϕ is found to satisfy the following equation derived from the basic equations of continuity and motion:

$$\frac{\partial^2\phi}{\partial x^2}\left[c^2 - \left(\frac{\partial\phi}{\partial x}\right)^2\right] + \frac{\partial^2\phi}{\partial y^2}\left[c^2 - \left(\frac{\partial\phi}{\partial y}\right)^2\right] - \frac{\partial^2\phi}{\partial z^2}\left[c^2 - \left(\frac{\partial\phi}{\partial z}\right)^2\right] - 2\left[\frac{\partial^2\phi}{\partial x\,\partial y}\frac{\partial\phi}{\partial x}\frac{\partial\phi}{\partial y} + \frac{\partial^2\phi}{\partial y\,\partial z}\frac{\partial\phi}{\partial y}\frac{\partial\phi}{\partial z} + \frac{\partial^2\phi}{\partial z\,\partial x}\frac{\partial\phi}{\partial z}\frac{\partial\phi}{\partial x}\right] = 0 \tag{2.14}$$

This quasi-linear partial differential equation has not yet been solved generally. Let $\mathbf{V}$ be the velocity of flow at infinity past a body at rest and M_∞ the Mach number of the flow at infinity. Three approximate methods of evaluation of Eq. (2.13) are used for subsonic flows.

1. As noted above, for $M_\infty \ll 1$ Eq. (2.13) reduces to Laplace's equation $\nabla^2\phi = 0$ for incompressible flow. The *Janzen-Rayleigh perturbation method* (1913, 1916) for compressible fluids assumes

$$\phi = VL(\phi_0 + M_\infty{}^2\phi_1 + M_\infty{}^4\phi_2 - \cdot\cdot\cdot)$$

where L is a characteristic body length. Substituting in (2.14) and equating coefficients of equal powers of M_∞ give as a second approximation in dimensionless variables ($x = Lx^1$)

$$\nabla^2\phi_1 = \left(\frac{\partial\phi_0}{\partial x^1}\right)^2\frac{\partial^2\phi_0}{\partial x^{12}} + \left(\frac{\partial\phi_0}{\partial y^1}\right)^2\frac{\partial^2\phi_0}{\partial y^{12}} + \left(\frac{\partial\phi_0}{\partial z^1}\right)^2\frac{\partial^2\phi_0}{\partial z^{12}} + \frac{\partial\phi_0}{\partial x^1}\frac{\partial\phi_0}{\partial y^1}\frac{\partial^2\phi_0}{\partial x^1\,\partial y^1} + \frac{\partial\phi_0}{\partial y^1}\frac{\partial\phi_0}{\partial z^1}\frac{\partial^2\phi_0}{\partial y^1\,\partial z^1} + \frac{\partial\phi_0}{\partial x^1}\frac{\partial\phi_0}{\partial z^1}\frac{\partial^2\phi_0}{\partial x^1\,\partial z^1} \tag{2.15}$$

In supersonic flow one equates equal powers of $1/M$ to obtain the *hypersonic approximation*.

2. *The Prandtl-Glauert perturbation method* (1927, 1930) is applicable to uniform flows over *thin bodies* with small slopes, which have small thickness in one direction, e.g., a supersonic airfoil. In this case the solution is assumed to take the form

$$\phi = x + \epsilon\phi_1 + \epsilon^2\phi_2 + \cdot\cdot\cdot$$

where ϵ is a small dimensionless parameter associated with the slope. Upon substituting and equating coefficients of equal powers of ϵ we obtain as a first approximation the linear equation

$$\nabla^2\phi_1 = M_\infty{}^2\frac{\partial^2\phi_1}{\partial x^2} \tag{2.16}$$

Put $x' = x/\sqrt{1 - M_\infty{}^2}$, $y' = y$, $z' = z$. As a result, we obtain Laplace's equation for incompressible fluids

$$\frac{\partial^2\phi_1}{\partial x'^2} + \frac{\partial^2\phi_1}{\partial y'^2} + \frac{\partial^2\phi_1}{\partial z'^2} = 0$$

Hence a solution for a body in an incompressible fluid can be translated into a solution for the same body $x'(y',z')$ in a compressible fluid by a contraction of $\sqrt{1 - M_\infty{}^2}$ in the direction of flow.

In the case of supersonic flow the elliptic partial differential equation becomes hyperbolic. Equation (2.15) is then identical with the two-dimensional wave equation. For example, in two dimensions the solution is

$$\phi_1(x^1,y^1) = f_1(x^1 - \sqrt{M_\infty{}^2 - 1}\,y^1) + f_2(x^1 + \sqrt{M_\infty{}^2 - 1}\,y^1)$$

The straight lines, $x^1 = \sqrt{M_\infty{}^2 - 1} =$ constant, are the Mach lines, inclined at an angle to $\mathbf{V}$ of

$$\alpha = \sin^{-1}(1/M_\infty)$$

The arbitrary functions $f_1\,f_2$ are constant along their respective Mach lines. For a thin body with small variations of slope along the direction of the main stream, there is a lift that is proportional to the angle of attack and a so-called *wave drag*, due to the energy transmitted to the medium in the form of waves. This theory can be applied also to *slender bodies*, e.g., a pointed body of revolution, which have small thicknesses in two directions; it fails for low supersonic Mach numbers and large angles of attack, when the wave is found to be detached from the leading edge (say, of a wedge).

3. It is possible sometimes to use variational principles in order to solve boundary-value problems of two-dimensional flows in compressible fluids either by solving an approximate problem exactly or by using the *Rayleigh-Ritz approximate method*. The method is not practicable for numerical solution of supersonic flow.

The nonlinear partial differential equation for steady compressible two-dimensional potential flow can be written

$$\left[c^2 - \left(\frac{\partial\phi}{\partial x}\right)^2\right]\frac{\partial^2\phi}{\partial x^2} - 2\frac{\partial\phi}{\partial x}\frac{\partial\phi}{\partial y}\frac{\partial^2\phi}{\partial x\,\partial y} + \left[c^2 - \left(\frac{\partial\phi}{\partial y}\right)^2\right]\frac{\partial^2\phi}{\partial y^2} = 0$$

It can be linearized by a transformation from the physical xy plane to the hodograph $v_x v_y$ plane—provided the Jacobian of the transformation does not vanish. For example, consider the Legendre contact transformation

$$\psi = \phi - v_x x - v_y y$$

Then the above equation becomes

$$(c^2 - v_x^2)\frac{\partial^2\psi}{\partial v_y^2} + 2v_x v_y \frac{\partial^2\psi}{\partial v_x\,\partial v_y} + (c^2 - v_y^2)\frac{\partial^2\psi}{\partial v_x^2} = 0 \tag{2.17}$$

where c^2 is a function of $v_x^2 + v_y^2$ alone. When the flow is subsonic, this partial differential equation is of the elliptic type; when the flow is supersonic, it is hyperbolic.

The *Molenbroek-Chaplygin transformation* in terms of polar coordinates (v,θ) is more useful. In this case the basic equations for the potential function ϕ and the stream function ψ become

$$\frac{\rho}{v}\frac{\partial\phi}{\partial\theta} = \frac{\partial\psi}{\partial v}$$

and

$$\frac{1 - M^2}{\rho v}\frac{\partial\psi}{\partial\theta} = -\frac{\partial\phi}{\partial v} \tag{2.18}$$

These equations take the place of the Cauchy-Riemann relations for incompressible flows. The linear partial differential equations for ϕ and ψ are

$$\frac{v}{\rho}\frac{\partial}{\partial v}\left(\frac{\rho v}{1 - M^2}\frac{\partial\phi}{\partial v}\right) + \frac{\partial^2\phi}{\partial\theta^2} = 0$$

and

$$\frac{\rho v}{1 - M^2}\frac{\partial}{\partial v}\left(\frac{v}{\rho}\frac{\partial\psi}{\partial v}\right) - \frac{\partial^2\psi}{\partial\theta^2} = 0 \tag{2.19}$$

Putting

$$\lambda \equiv \int_0^v \frac{\sqrt{1 - M^2}}{v}\,dv \qquad \phi^* \equiv \left(\frac{\rho}{\sqrt{1 - M^2}}\right)^{1/2}\phi$$

$$\psi^* \equiv \left(\frac{\sqrt{1 - M^2}}{\rho}\right)^{1/2}\psi$$

we obtain

$$\frac{\partial\phi^*}{\partial\theta} = \frac{\partial\psi^*}{\partial\lambda} - \psi^*\frac{d}{d\lambda}\ln\left(\frac{\sqrt{1 - M^2}}{\rho}\right)^{\frac{1}{2}}$$

and

$$-\frac{\partial\psi^*}{\partial\theta} = \frac{\partial\psi^*}{\partial\lambda} - \phi^*\frac{d}{d\lambda}\ln\left(\frac{\rho}{\sqrt{1 - M^2}}\right)^{\frac{1}{2}} \tag{2.20}$$

These so-called *canonical equations* of S. Bergman (1945) can be solved by the method of successive approximations,

$$\psi^* = \psi_0(\lambda,\theta) + \sum_{n=1}^{\infty} p_n(\lambda)\psi_n(\lambda,\theta)$$

where ψ_0 is the stream function for an incompressible fluid.

The above equation for the stream function ψ in the hodograph plane can be written in terms of the dimensionless variable $z \equiv v/v_{\max}$,

$$\frac{\partial^2\psi}{\partial z^2} + A(z)\frac{\partial\psi}{\partial z} + B(z)\frac{\partial^2\psi}{\partial\theta^2} = 0$$

where

$$A(z) \equiv \frac{1}{z}\left(1 - \frac{2}{\gamma - 1}\frac{z^2}{1 - z^2}\right)$$

$$B(z) \equiv \frac{1}{z^2}\left(1 - \frac{2}{\gamma - 1}\frac{z^2}{1 - z^2}\right)$$

Put $\psi(z,\theta) = f_n(z)\sin n\theta$. Then

$$f_n'' + A(z)f_n' - n^2B(z)f_n = 0$$

These equations can be reduced to a hypergeometric equation. The general solution can be written

$$\psi(z,\theta) = a_0(\theta) + \sum_{n=1}^{\infty} A_n f_n(z)\sin(n\theta + \alpha_n)$$

Analytical solutions of the equation for the stream function can be obtained also by S. Bergman's *operator method*, as well as by the above method proposed by Chaplygin. Analogously to the incompressible operation by which the stream function is derived from a complex potential, an integral operator is sought to obtain $\psi(v,\theta)$ from $aw(z)$.

For large disturbances and unsteady flow even the one-dimensional case is difficult to solve. An instructive, complete solution was found by Riemann (1858–1859) with the aid of the function

$$l(\rho) \equiv \int_{\rho_0}^{\rho}\frac{c\,d\rho}{\rho}$$

for barotropic flows. The basic equations become

$$\left[\frac{\partial}{\partial t} + (v_x + c)\right]r = 0$$

and

$$\left[\frac{\partial}{\partial t} + (v_x - c)\right]s = 0$$

where $2r \equiv v_x + l$ and $2s \equiv v_x - l$. The *Riemann invariants* r, s are constant for observers moving with speeds $v_x + c$ and $v_x - c$, respectively, in the xt plane; these curves are the characteristics. A *simple wave* is a progressive one for which one set of characteristics consists of a family of straight lines with constant v and c along each; this is an analogue of a Prandtl-Meyer wave. For a *compound wave* neither r nor s is constant. If $dp/d\rho > 0$, the speed of propagation increases with ρ, so the shape of the wave becomes distorted. As the wavefront becomes vertical, a mathematical discontinuity develops. Physically, of course, the neglected viscosity does not permit this limit to be attained. Instead, there occurs a shock wave, across which there is a sudden increase of physical quantities within a very short space.

The observed discrepancy of the Mach angle for a body in supersonic flow is then ascribed to the envelope, being a shock wave of finite disturbance with greater than acoustic speed and with larger density change. Even though the main flow can be regarded as that of an inviscid fluid, the shock itself is fundamentally a viscous phenomenon inasmuch as viscosity is proportional to the space rate of change of velocity, which takes on larger values the smaller the space in which the changes take place. The apparent mathematical discontinuity, however, serves as a convenient model of a shock wave for many approximate discussions. Mathematicians have given considerable attention to the solutions of nonlinear partial differential equations involving discontinuities in ideal inviscid fluids arising from different initial conditions, say, of a moving piston. Unique analytic solutions may be found. It is convenient to consider the flow relative to the shock front (Fig. 2.5) as stationary. The over-all conservation laws of mass, momentum, and energy now take the following algebraic forms, where M is the mass of fluid crossing unit area in unit time:

$$\rho_1(v_1 - V) = \rho_2(v_2 - V) \equiv M \tag{2.21}$$

$$\begin{aligned} M(v_2 - v_1) &= p_1 - p_2 \\ M(E_2 + \tfrac{1}{2}\mathbf{v}_2^2) - (E_1 + \tfrac{1}{2}\mathbf{v}_1^2) &= p_1v_1 - p_2v_2 \end{aligned} \tag{2.22}$$

These are known as the *Rankine-Hugoniot equations* (1870, 1889). Any one of the four quantities p_2, ρ_2, v_2, V is sufficient to determine all properties of such a shock if the initial state of the gas is given. The energy equation for an ideal gas can be written

$$\int_0^{p_1} \frac{dp}{\rho} + \tfrac{1}{2}(v_1 - V)^2 = \int_0^{p_2} \frac{dp}{\rho} + \tfrac{1}{2}(v_2 - V)^2$$

Thus Bernoulli's principle holds across a shock. From the above equations

$$\frac{p_2 - p_1}{\rho_2 - \rho_1} = \gamma \frac{p_2 - p_1}{\rho_2 + \rho_1} \tag{2.23}$$

so,

$$\frac{\Delta p}{\Delta \rho} = \gamma \frac{p_{av}}{\rho_{av}} \to \frac{dp}{d\rho} = \gamma \frac{p}{\rho}, \quad \lim_{\Delta\rho \to 0}$$

Equation (2.23) can be written

$$\frac{\rho_2}{\rho_1} = \frac{(\gamma + 1)(p_2/p_1) + (\gamma - 1)}{(\gamma - 1)(p_2/p_1) + (\gamma + 1)}$$

This *Hugoniot adiabatic* is shown in Fig. 2.5 in comparison with the usual isentropic adiabatic, $p = k\rho^\gamma$; they agree in the acoustic limit. For a given pressure change there is always a greater change in density, temperature, and material speed associated with a shock wave than in adiabatic compression. From the basic equations also one obtains *Prandtl's relation,*

$$(v_1 - V)(v_2 - V) = c_*^2 \tag{2.24}$$

where c_* is the *critical speed.* The constant c_* can be shown to be the acoustic speed when

$$\frac{v - V}{c} = 1$$

Bernoulli's principle then takes the form

$$\frac{c^2}{\gamma - 1} + \tfrac{1}{2}q^2 = \frac{c_*^2}{2}\frac{\gamma + 1}{\gamma - 1} = \text{const}$$

where $q = v - V$. From this equation and (2.24) the material velocity on one side of the shock must be supersonic and that on the other side subsonic. There is a change in entropy across the shock; for weak shocks it is of the third order in the pressure change (and density change). From a study of the change in entropy the flow must always be from supersonic to subsonic under adiabatic conditions, provided no other energy change occurs. Only these theoretically possible *compression shocks* have been observed, i.e., no expansion shocks.

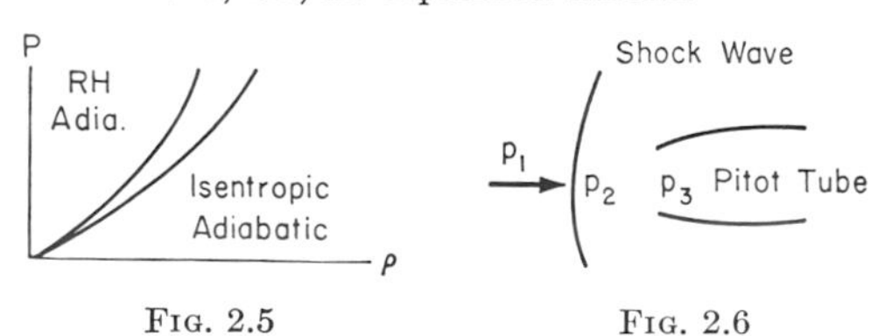

FIG. 2.5 FIG. 2.6

For a curved shock the entropy history along each streamline will be different. Isentropic flow ahead of the shock will become nonisentropic and rotational behind it. For general *rotational flow* with variable Bernoulli constant the vorticity (L. Crocco, 1937) is a linear function of the temperature (also p and ρ),

$$-\frac{2\omega}{\rho} = \frac{dB}{d\psi} - T\frac{dS}{d\psi}$$

where ψ is the stream function. The differential equation for two-dimensional steady adiabatic flow is

$$\left(\frac{v_x^2}{c^2} - 1\right)\frac{\partial^2\psi}{\partial x^2} + 2\frac{v_xv_y}{c^2}\frac{\partial^2\psi}{\partial x\,\partial y} + \left(\frac{v_y^2}{c^2} - 1\right)\frac{\partial^2\psi}{\partial y^2} = \rho^2\left[-\frac{dB}{d\psi} + \frac{c^{12} + (\gamma - 1)v^2}{\gamma R}\frac{dS}{d\psi}\right]$$

it reduces to the potential equation for B and S constant.

A detached curved shock wave forms in front of a pitot tube in a supersonic stream; but the central portion can be regarded as plane.

$$\frac{p_2}{p_1} = \frac{2\gamma}{\gamma + 1}M_1^2 - \frac{\gamma - 1}{\gamma + 1}$$

and

$$\left(\frac{p_3}{p_1}\right)^{(\gamma-1)/\gamma} = \left(\frac{p_2}{p_1}\right)^{(\gamma-1)/\gamma}\left[\frac{(\gamma + 1)^2}{4\gamma} + \frac{\gamma^2 - 1}{4\gamma}\frac{p_1}{p_2}\right]$$

The pitot tube gives p_3; M_1 can be determined from the observed Mach angle; p_1 can then be obtained.

Since a shock leaves the tangential velocity component unchanged, the streamline is found to bend toward the shock according to the equation

$$\frac{\tan \beta_2}{\tan \beta_1} = \frac{\rho_2}{\rho_1}$$

where β_1 is the angle between the normal to the wave and the streamline upstream and β_2 the angle between the normal to the wave and the streamline downstream. The angle of deviation varies for a given β_1 and pressure ratio across the shock; it has a maximum value in each case. There is a highest value of β_1 beyond which no steady state exists. These facts can be used to ascertain the flow about a symmetric wedge with a shock wave attached at the vertex. Two positions are theoretically possible for such a shock wave, but only the weaker (less entropy change) shock wave with the larger β_1 is observed. For a given wedge angle there is a limiting M_∞ beyond which no steady flow with attached shock wave is possible; a detached shock wave results.

An axially symmetric three-dimensional irrotational flow can be shown to exist such that the flow parameters are constant on conical surfaces through a vertex. For example, such a *conical flow* (A. Busemann, 1929) is observed behind the conical shock wave attached at an angle depending on the Mach number to a cone placed in a uniform supersonic stream. As in the case of a wedge there is a physically extraneous solution and a "limiting" cone angle beyond which the shock wave is detached. Numerical solutions of this problem for air were calculated first by G. I. Taylor and J. W. Maccoll and later by Z. Kopal; they agree well with actual measurements.

Two-dimensional shock-wave relations can be conveniently represented by *Busemann's shock polar* (1929) in the hodograph plane. In Fig. 2.7 $\overline{OF}$ is $\mathbf{V}$; $\overline{OP}$ is the material velocity behind the shock. The shock wave is perpendicular to $\overline{FP}$ (the flow represented by $\overline{FPI}$ does not occur physically). The equation of the third-order shock polar (a cartesian leaf) is

$$q^2\left(\frac{c_*^2}{V}+\frac{2}{\gamma+1}V-q\right)=(V-q)^2\left(q-\frac{c_*}{V}\right)$$

where c_* is the critical speed for oblique flow (velocity component of flow along the shock wave), i.e.,

$$q_{n1}q_{n2}=c_*^2-\frac{\gamma-1}{\gamma+1}q_t^2$$

where q_{n1}, q_{n2} are the normal components of velocities of flow and q_t the tangential component. Thus the shock polar is determined by V and c_*. The shock polar between F and A is not physically real, inasmuch as it involves a decrease in entropy across the shock wave. For angles greater than the largest to the shock polar from O no attached shock wave exists.

One-dimensional interactions of shocks and rarefactions or other shocks can be analyzed in the *vp* plane. From the shock conditions one can express

$$v_2=v_1\pm\phi_1(p_2)$$

where ϕ_1 is a monotonic function and the signs (plus, minus) signify the directions of motion (right, left, respectively). Likewise, for rarefactions

$$v_2=v_1\pm\psi_1(p_2)$$

where ψ also is monotonic and the signs have the same significance. The locus L of all physical states that from a given state r can exist on the right of a right-moving shock or rarefaction wave is shown in Fig. 2.8, while in Fig. 2.9 is shown the locus of all

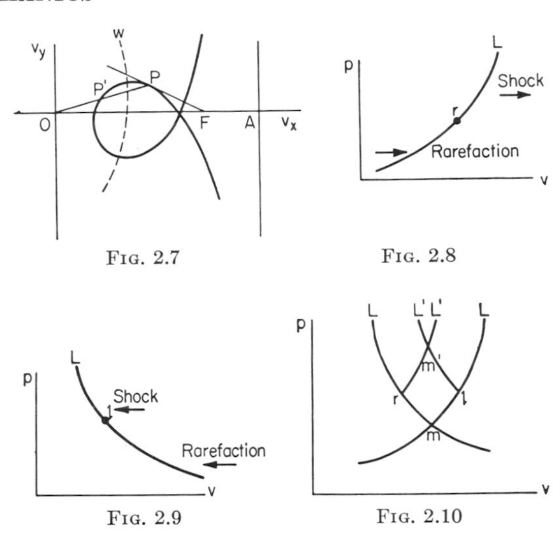

FIG. 2.7

FIG. 2.8

FIG. 2.9

FIG. 2.10

states which can be reached from a given state L on the left of a left-moving wave. Consider two waves separated by a middle zone m before interaction. If the two states l and r are known, then the two L curves drawn through them intersect initially at the common state m. After interaction the right-moving wave moves into the right r state and the left-moving wave into the left l state. New L' curves through l and r will intersect at a new state of separation m' (Fig. 2.10) for head-on collision of two shock waves. Numerical analysis is needed to supplement this qualitative result. Thus, in principle, the interaction of shock waves is soluble.

The two-dimensional interaction of shocks and rarefactions is much more complicated. Significant results, however, have been obtained by the study of the *regular reflection* of shocks from a rigid wall (equivalent to the oblique interaction of similar shocks). In Fig. 2.11 the shock S incident on the wall and its reflected wave R are considered stationary. One asks whether stationary solutions can be found for the supersonic flow from the right. From the diagram

$$c=V\sin\omega$$

and

$$c'=V'\sin\omega'$$

The total change in the material velocity along the wall is given by

$$V-V'=\Delta u\sin\omega+\Delta u'\sin\omega'$$

where Δu and $\Delta u'$ are the changes of the normal material speeds across the respective shocks. Substituting for V and V', we obtain

$$\frac{c-\Delta u\sin^2\omega}{\sin\omega}=\frac{c'+\Delta u'\sin^2\omega'}{\sin\omega'}\tag{2.25}$$

The condition for no flow normal to the wall is

$$\Delta u\cos\omega=\Delta u'\cos\omega'\tag{2.26}$$

For weak shocks $\Delta u/c$ and $\Delta u'/c'\ll 1$ and $c=c'$ we find from (2.25) that $\omega=\omega'$ and from (2.26) that $\Delta u=\Delta u'$. In other words, the angle of incidence equals the angle of reflection, and the overpressure is

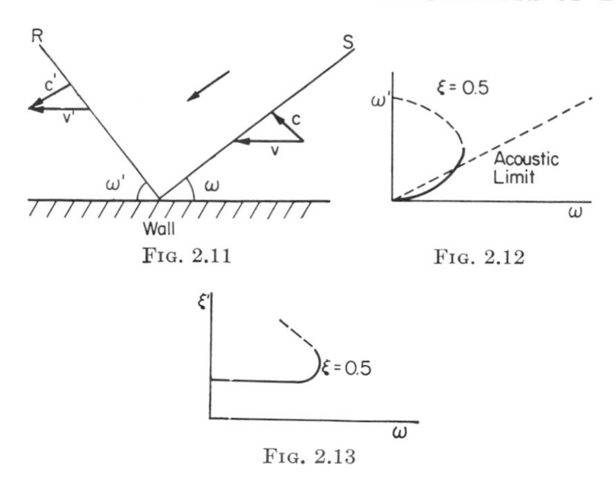

FIG. 2.11

FIG. 2.12

FIG. 2.13

doubled upon reflection. These results are true for the acoustic approximation of infinitesimal disturbances. For shocks (2.25) and (2.26) can be solved numerically for given initial shock strength $\xi \equiv p_1/p_2$ and angle of incidence ω. The results for air are shown for the angle of reflection ω' in Fig. 2.12 and for the reflected shock strength $\xi' = p_2'/p_2$ in Fig. 2.13. For normal incidence the ratio of the reflected pressure to the incident pressure is given by

$$\frac{p'}{p} = \frac{(3\gamma - 1)(p/p_0) - (\gamma - 1)}{(\gamma - 1)(p/p_0) + (\gamma + 1)}$$

The comparable expression for a reflected rarefaction is

$$\left(\frac{p'}{p_0}\right)^{(\gamma-1)/2\gamma} = 2\left(\frac{p}{p_0}\right)^{(\gamma-1)/2\gamma} - 1$$

The angle of reflection may be greater than, less than, or equal to, the angle of incidence. The reflection pressure at oblique incidence may be greater than that at normal incidence. Most significantly, however, for large angles of incidence there are no solutions whatever. This simple theory of regular reflection is well verified experimentally. Only one solution, however, is observed: the one with the weaker shock strength and smaller angle of reflection. In this case of so-called *Mach reflection* one observes that the intersection point moves away from the wall and is connected by a third shock to it. To date no completely satisfactory explanation of this three-shock intersection has been forthcoming.

Most precise measurements on such shock phenomena are made possible through the use of a *shock tube*, which is essentially a tube with a partition separating a high-pressure gaseous region from one of lower pressure. If the separating diaphragm is punctured, a uniform shock wave rapidly develops. The shock-wave strength can be determined either from interferometric measurements of density or from velocity measurements. Each method gives a nominal value within 0.1 per cent, but together they give a value within 2 per cent for shock strengths $\xi = 0.3$ to 0.9. This experimental procedure affords the most accurate verification of the Rankine-Hugoniot equations. Transient phenomena, such as boundary-layer growth, can be studied precisely in the material afterflow of the shock.

Regular refraction of shock waves has been studied at a gaseous interface. As in the case of regular reflection, one sets down the equations for stationary flow and then looks for solutions of the resulting twelfth-degree equation numerically. To eliminate some of the physically extraneous solutions, use is made of limiting solutions, e.g., the analogue to Snell's law of refraction for infinitesimal shocks, and interaction of finite shocks at normal incidence. Another special solution is the *transition angle*, for which no reflection exists, and for which the reflected shock changes to rarefaction or vice versa for a change of angle of incidence—a phenomenon unknown in acoustics. Regular refraction, as theoretically predicted, has not been observed for glancing incidence.

Shock-wave phenomena in gases still present many enigmas; they occur in novae (astrophysics), supersonic ballistics (aeroballistics), boundary-layer interactions (aerodynamics), explosions (blast waves), air jets, etc.

Shock-wave phenomena in liquids have been less investigated, primarily in connection with explosions in water. A *waterlike substance* is one for which intrinsic energy is separable into one part that depends upon the density only and a second part that depends upon the entropy only. In this case $p = p(\rho)$ under all conditions; e.g., explosion processes in water are characterized over a range of high pressures by $p = A\rho^{7.15}$ (A constant). Interactions of shock waves in waterlike substances can be studied similarly to those in gases. For water the angle of reflection for *regular reflection* is always greater than the angle of incidence, and the reflected pressure is always greater than at normal incidence; Mach reflection has been observed.

The phenomenon of the *hydraulic jump* (Jouguet, 1920) for a liquid flowing in an open shallow channel is described by the waterlike relation $p_h = A\rho_h^2$, where p_h and ρ_h are, respectively, the values of the pressure and of the density integrated over the depth. Regular reflection and Mach reflection have both been observed. It has become customary to regard the hydraulic jump as an analogue of a weak shock in a fictitious ideal gas with $\gamma = 2$ inasmuch as the Hugoniot relation for weak shocks reduces to $p = k\rho^2$. The coefficient k, however, is constant only up to third-order variations in entropy. A weak shock is strictly a polytropic process rather than an analogue of the hydraulic jump. A critical examination of the mathematical limit of a weak shock indicates that the interactions of shock waves in waterlike substances exhibit, both qualitatively and quantitatively, differences from those in ideal gases undergoing polytropic changes.

Optical techniques are particularly useful in investigating compressible flows, inasmuch as the latter are not disturbed. *Interferograms* yield the density distribution directly; *schlieren analysis* depends on the density gradient; *shadowgrams* give the gradient of the density gradient.

7. Flows of Viscous Fluids

A *viscous fluid* has internal tangential stresses and adheres to a wall, so the fluid in contact with the wall has no velocity relative to it. Slip occurs for highly rarefied gases and for organic liquids with very large molecules. In general, the viscosity will be a function of the thermodynamic variables of state, say, the pressure and temperature. *Rheology* deals with abnormal liquids, such as colloids and emulsions, for which the viscosity depends upon the velocity distribution of the granular material. A *Newtonian fluid* has constant coefficient of viscosity.

The general *Navier-Stokes equation* (1847) of motion is

$$\frac{d\mathbf{v}}{dt} + \text{grad } (P + U - \tfrac{4}{3}\nu \text{ div } \mathbf{v}) + \nu \text{ curl curl } \mathbf{v} = 0 \tag{2.27}$$

where the *kinematic coefficient of viscosity* $\nu = \mu/\rho$ measures the effect of viscosity relative to that of inertia; thus ν for air is 15 times more than ν for water. The addition of the viscosity term alters the essential nature of the partial differential equation in that it has the highest derivative. Compressibility, on the other hand, adds only low-order derivatives. Hence compressibility adds at most a correction to viscous incompressible flows. Comparatively few solutions of the Navier-Stokes equations for stipulated boundary conditions are known. A fruitful approach is the *inverse method,* in which one seeks solutions having certain specific properties (analytic, geometric, etc.). For incompressible fluids Eq. (2.27) reduces to

$$\frac{\partial \mathbf{v}}{\partial t} - \mathbf{v} \times \text{curl } \mathbf{v} + \text{grad } (P + U + \tfrac{1}{2}\mathbf{v}^2) = \nu \nabla^2 \mathbf{v} \tag{2.28}$$

Consider an incompressible liquid flowing steadily under gravity between two parallel fixed planes ($z = 0, d$) at a small distance d apart. The equations of motion are

$$v_x \frac{\partial v_x}{\partial x} + v_y \frac{\partial v_x}{\partial y} = -\frac{1}{\rho}\frac{\partial p}{\partial x} + \nu \nabla^2 v_x$$

and

$$v_x \frac{\partial v_v}{\partial x} + v_y \frac{\partial v_v}{\partial y} = -\frac{1}{\rho}\frac{\partial p}{\partial y} + \nu \nabla^2 v_y$$

and

$$0 = g - \frac{1}{\rho}\frac{\partial p}{\partial z}$$

Integration with respect to the proper boundary conditions results in the following mean speeds:

$$\overline{v_x} = -\frac{d^2}{12\mu}\frac{\partial p}{\partial x}$$

and

$$\overline{v_y} = -\frac{d^2}{12\mu}\frac{\partial p}{\partial y}$$

Hence $\overline{v_x}$ and $\overline{v_y}$ can be regarded as velocity components derived from a virtual potential $\phi = d^2p/12\mu$. The flow of a perfectly inviscid fluid around an obstacle can be thus simulated for demonstration purposes (Hele-Shaw, 1898).

If each term in (2.28) is divided by V^2L^{-1}, where L is a characteristic length of the flow and V a characteristic speed, then for no external forces each term becomes invariant for a kinematically similar transformation from one problem to another. The flows will be *dynamically similar* if the dimensionless quantity $R \equiv VL/\nu$, the *Reynolds number,* in the term on the right remains the same. For example, the drag coefficient will be a function of the Reynolds number. In the case of a free surface of a heavy liquid another number, the *Froude number* $= V^2/Lg$, must be kept constant, too; this is not possible for a scaled model in water. Physical data for full-size bodies can be obtained from model (small L) experiments if the density of the fluid is proportionately increased so as to keep the Reynolds number the same (M. M. Munk, 1925).

Physically, the Reynolds number represents the ratio of the inertial reaction to the viscous force. In the kinetic theory of gases the kinematic viscosity is shown to be proportional to the mean free path λ of the molecules and to their mean speed, which, in turn, is proportional to the acoustic speed c (see Part 5, Chap. 2). Hence R can be expressed also in terms of two factors

$$R \sim \frac{L}{\lambda}\frac{V}{c}$$

In ordinary low-altitude aerodynamics $L/\lambda \gg 1$, and $V/c \ll 1$ (*subsonic* flow). For modern high-speed flows $V/c \sim 1$ (*transonic*), $V/c > 1$ (*supersonic*), and $V/c \gg 1$ (*hypersonic*). For high-altitude aerodynamics $L/\lambda \sim 1$. If either factor for R becomes different from the usual values, the Reynolds number becomes inadequate, and additional parameters are needed to ensure similarity, e.g., the Mach number $\equiv V/c$ for compressible flows and the *Knudsen number* λ/L for low-density (molecular) flows.

There are two cases where the solution of the Navier-Stokes equation itself does not depend upon R, so kinematically similar flows are valid for kinematically similar problems:

1. $\nabla^2\mathbf{v} = 0$, or curl $\boldsymbol{\omega} = 0$, i.e., irrotational flow with $\boldsymbol{\omega} = 0$;
2. $d\mathbf{v}/dt = 0$, for which the flow does not depend upon R, but the pressure does.

An example of the first type of flow (with $\nabla^2\mathbf{v} = 0$) is plane Couette flow, i.e., flow between two infinite plane parallel walls that are at a distance d apart and move at constant relative velocity v. For a steady state the velocity distribution between the walls is linear. The drag D per unit breadth is given by $D = \mu V(L/d)$ along a distance L on the plates. Another example is *Couette flow* (M. Couette, 1890) itself, i.e., flow between two coaxial cylinders having different radii and different angular velocities of rotation—also two coaxial cylinders having different radii and different velocities along the axis. When the outer cylinder rotates faster than the inner one, the motion tends to become stabilized; whereas when the inner one rotates faster, there is a tendency to instability. At a critical speed the laminar flow breaks down, and ring-shaped vortices develop. The solution (G. I. Taylor, 1923) of the complete Navier-Stokes equation for this case has been verified experimentally—one of the few instances.

The second kind of flow is illustrated by plane Poiseuille flow, i.e., a fluid moving between two fixed plane parallel walls at a distance d apart. In this instance the velocity distribution between the walls for a steady state is a parabolic distribution with the maximum velocity midway. The drag D per unit breadth is given by

$$D = 3\mu\bar{v} L/d$$

where $\bar{v}$ is the mean speed. *Poiseuille flow* (J. L. M. Poiseuille, 1840) itself is steady flow along the axis in an infinitely long cylindrical pipe with circular cross section of radius a. The velocity distribution v_x satisfies Poisson's equation

$$\nabla^2 v_x = \text{const}$$

In this case the velocity distribution is a paraboloid of revolution. The drag D for a pipe of length L is given by

$$D = 8\pi\mu\bar{v}L$$

Plane Couette flow and the above flow between coaxial cylinders moving axially are also illustrations of this general kind.

Approximations to such flows are those with small Reynolds numbers, as in the case of fluids with large viscosity. Such flows are called "slow" motion, inasmuch as the acceleration term may be neglected as a first approximation. By taking the curl of the Navier-Stokes equation,

$$\nabla^2 \boldsymbol{\omega} = 0$$

For flow around an infinite circular cylinder this reduces to the biharmonic equation for the stream function

$$\nabla^4 \psi = 0$$

A solution does not exist for this specific two-dimensional problem with appropriate boundary conditions. Physically the problem has a solution only if the acceleration is not neglected at large distances from the body. A mathematical solution does exist for the three-dimensional case; it gives an approximate value for the observed resistance.

For three-dimensional flow about a sphere of radius a

$$\mu \nabla^2 \mathbf{v} = \text{grad } p \qquad \text{and} \qquad \text{div } \mathbf{v} = 0$$

The solution for appropriate boundary conditions gives Stokes' (1843) formula for the drag,

$$D = 6\pi\mu V_a$$

where V is the flow velocity at infinity. In other words $C_D = 3/2R$. The acceleration, however, cannot be neglected at large distances from the sphere. In Oseen's (1910) treatment the acceleration is approximated by $V(\partial y/\partial x)$. Stokes' solution still holds near the sphere. The drag can be expressed as a power series of the Reynolds number, which converges to the exact solution of the Navier-Stokes equations (S. Goldstein, 1929).

Another limiting type of flow is that for which the Reynolds number R approaches infinity, as in the case that $\nu \to 0$. Many aeronautical problems have Reynolds numbers in this range. For two-dimensional flow of a viscous fluid, in general, the stream function satisfies the fourth-order equation

$$\psi_x \nabla^2 \psi_y - \psi_y \nabla^2 \psi_x = \nu \nabla^4 \psi$$

In two dimensions when $\nu \to 0$, the stream function is found to satisfy the third-order equation

$$\psi_y \nabla^2 \psi_x - \psi_x \nabla^2 \psi_y = 0$$

Hence the flow of a perfectly inviscid fluid cannot satisfy all the boundary conditions (first derivatives of ψ given) for the flow of a viscous fluid, inasmuch as a perfectly inviscid fluid slides along walls and does not adhere to them. Hence as $\nu \to 0$, the solution for $\nu \neq 0$ does not necessarily converge to the solution $\nu = 0$ at the walls, although it may in the interior of the fluid.

The thin region of the fluid along the wall in which the tangential velocity decreases to zero at the wall is called the *boundary layer* (Prandtl, 1904); it occurs in wakes and jets, as well as along walls. The thickness of the boundary layer approaches zero as $\nu \to 0$. The tangential force of the fluid in the boundary layer, integrated over the surface, is responsible for the so-called *skin friction*. The boundary layer on a rotating cylinder in a uniform stream produces a circulation flow about the cylinder and a resultant cross force, the *Magnus effect* (1851).

The vorticity $\boldsymbol{\omega}$ satisfies the equation

$$\frac{D\boldsymbol{\omega}}{Dt} = \nu \nabla^2 \boldsymbol{\omega} + \boldsymbol{\omega} \text{ grad } \mathbf{v}$$

relative to the moving fluid. For a uniform stream this equation is analogous to the equation for the conduction of heat and affords a picture of the diffusion of the vorticity constituting the boundary layer. For $\nu = 0$, the differential equation reduces to Helmholtz's theorem of conservation of vorticity.

Consider two-dimensional flow which takes on the values $V(x)$ and $P(x)$ for v and p, respectively, on the boundary-layer edge that borders on the flow of the perfectly inviscid fluid in the main stream. The equation of motion on this edge is

$$\rho V V_x = -P_x$$

Assume $\nu\sigma^{-2}$ = constant. Then the Navier-Stokes equation within the boundary layer becomes in the limit (with the proper regard in use for merging with the main-stream flow)

$$v_x v_{xx} + v_y v_{xy} = \frac{-p_x}{\rho} + \nu v_{yy}$$

and

$$\frac{P_y}{\rho} = 0$$

The second equation says that the pressure is constant across the boundary layer. Using the edge equation, the equation within the boundary layer is

$$v_x v_{xx} + v_y v_{xy} = V V_x + \nu v_{xy}$$

Integration of the boundary-layer equation by use of the continuity equation from $y = 0$ to y_0 along x_0 gives the so-called *momentum equation* for the shearing stress at the wall

$$\mu \left.\frac{\partial v_x}{\partial y}\right|_{y=0} = \rho \int_0^\infty \left(V \frac{\partial V}{\partial x} - V \frac{\partial v}{\partial x} - 2v \frac{\partial v}{\partial x}\right) dy$$

The *momentum thickness* σ_1^* is defined by

$$\sigma_1^* = \frac{1}{V^2}\int_0^\sigma v(V - v)\,dy$$

The momentum equation is used primarily for obtaining approximate solutions of the boundary-layer equations. In the *Kármán-Pohlhausen method* one assumes a kind of similarity of spatial velocity distribution given by

$$v = Vf\left(\frac{y}{\delta(x)}\right)$$

where the coefficients of the quartic polynomial f are to be determined. The assumption is not always correct, e.g., in the neighborhoods of a separation point, but it gives fairly good predictions about boundary-layer phenomena.

The simplest boundary-layer problem is two-dimensional flow along a flat plate; the thickness grows parabolically. If V = constant and $P = 0$ are the values for the main-stream flow, the partial differential equation for the boundary layer has been shown by Prandtl to reduce to an ordinary differential equation in the variable $\theta \equiv y(\lambda x)^{-1/2}$, where λ is a constant length, say, $2\nu/V$ (λx is the Reynolds number):

$$h'''(\theta) + h(\theta)h''(\theta) = 0$$

where $\psi = \sqrt{\lambda x}\, Vh(\theta)$. The drag coefficient, obtained by H. Blasius (1908), is

$$C_D = 2\frac{\alpha}{R^{1/2}}$$

where $\alpha = 1.3283$.

The effect of the retardation at the wall causes the streamlines to be displaced through a distance σ^*, the so-called *displacement thickness*, defined by $\sigma^*(x) = 1/V(x)\int_0^x[V(x) - v(x,y)]\,dy$. The actual boundary-layer thickness may be taken as $3\sigma^*$. A convenient definition is the distance at which the velocity becomes 1 per cent of that in the undisturbed stream. For the case of a flat plate

$$\sigma^* = 1.72\sqrt{\frac{\nu x}{V}}$$

Separation of flow from a curved surface will occur if the fluid outside the boundary layer is slowed down, as in a diverging channel. Beyond this point the streamlines curl backward. In the separated region, vortices may form and produce a region of higher pressure. In the case of separation on the upper side of an aircraft wing, where it is likely to occur, the lift is thereby reduced.

The Poisson equation for the Poiseuille flow in a long cylindrical pipe is identical with that for the Boussinesq-Prandtl stress function of the torsion of a cylindrical rod of the same cross section, so analogies for the latter are equally applicable to the former. For example, the deformation of a thin membrane stretched across a cylindrical vessel of corresponding cross section and with excess air pressure is indicative of the flow velocity $v(x,y)$.

If a shock wave interacts with a laminar boundary, a disturbance spreads upstream through the subsonic boundary layer and produces a λ *shock*. This phenomenon does not occur for a turbulent boundary layer.

8. Turbulence

Flows such that $v_x = v_z = 0$ and $v(x,t)$ are said to be *laminar*, because all fluid particles in a sheet parallel to the plane $z = 0$ have the same velocity. Such flows occur in straight tubes, provided the Reynolds number does not exceed a certain critical value. In the case of low Reynolds numbers, e.g., large viscosity, lateral components of velocity are smoothed out, so laminar flow is favored. For higher Reynolds numbers the laminar motion is unstable, so that a slight disturbance produces a state of irregular eddying motion called *turbulence*. Between the turbulent boundary layer and a smooth solid a thin *laminar sublayer* with a very steep velocity gradient occurs. Another, even thinner layer sometimes occurs between this sublayer and the turbulent boundary layer; the layer is characterized by fluctuations along it but not across it. In the case of a high Reynolds number, e.g., large inertia, lateral components of motion persist, so turbulence is favored. The whirling fluid produces a greater velocity gradient along the walls (the parabolic distribution is flattened), so that the resistance is increased. Experiment shows a minimum Reynolds number below which turbulence cannot occur and a maximum value above which laminar flow does not seem to occur. Turbulent motion may involve such homogeneous dispersoid systems as a river carrying sand or silt and the atmosphere carrying sand or dust.

Turbulence has defied theoretical description ever since the introduction of the statistical concept by Reynolds, who regarded a fluid as consisting of molecules having solely translatory motion (see Chap. 1, Sec. 4). Mean quantities averaged over the molecular motions for intervals 10^{-15} cm-sec $\ll$ space time $\ll 10^{-8}$ cm-sec are regarded as satisfying the Navier-Stokes equation. (Some investigators have believed that the phenomenon of turbulence cannot be explained on the assumption that the flow is governed by the Navier-Stokes equation. Nowadays it is generally accepted that these equations may permit sufficiently "wild" solutions corresponding to the turbulent flow.) Observed quantities, however, are averaged over space-time intervals of the order of 1 cm-sec. Comparable mean quantities (O. Reynolds, 1895) are obtained from the equations of motion by substituting and averaging with respect to the time

$$v_i = \bar{v}_i + v_i'$$
$$p = \bar{p} + p^1$$

where the velocity component v_i is given as the sum of the mean velocity component $\bar{v}_i$ (stochastic average or space average or time average) and the random fluctuating turbulent velocity component v_i'. The resultant equation for an incompressible inviscid fluid is

$$\frac{\partial \bar{v}_i}{\partial t} + \frac{\partial}{\partial x_j}(\bar{v}_i\bar{v}_j + \sigma_{ij}p) = \nu\left(\frac{\partial}{\partial x_j}\right)^2 \bar{v}_1 - \frac{\partial}{\partial x_j}(v_i'v_j')_{\text{av}}$$

The new eddy stresses, *Reynolds stresses*, form a tensor with components $\tau_{ij}/\rho = -(v_i'v_j')_{av}$; they represent the net transfer of momentum across the respective surfaces because of the velocity fluctuations.

Semiempirical *mixing-length theories* have been used primarily to obtain approximate answers to engineering problems, e.g., the flow resistance in pipes and channels. They are based upon a somewhat ill-defined analogy with the kinetic theory of gases owing to the lack of individuality for eddies and the variation of turbulence intensity. In the *momentum-transfer theory* (Prandtl, 1925),

$$\frac{\tau}{\rho} = \nu' \frac{\partial \bar{v}_i}{\partial x}$$

where $\nu' = l((v_i'^2)_{av})^{1/2}$ is the *coefficient of kinematic turbulent (eddy) viscosity* (J. Boussinesq) and l is the *mixing length.* One assumes

$$((v_i'^2)_{av})^{1/2} = l\left(\frac{\partial \bar{v}_i}{\partial x}\right)$$

In the *vorticity-transfer theory* (Taylor, 1915) one arrives at a similar relation for two-dimensional motion if ν' is constant. The mixing length can be visualized in the spreading of a turbulant jet as it emerges from a slit into still air. The eddy viscosity has been determined for atmospheric turbulence (from distribution of wind speed according to height) and for ocean currents (from distribution of salinity according to depth).

The more recent statistical theory (Taylor, 1935) of turbulence starts with the concept of measurable correlation between two fluctuating quantities in the turbulent flow. For example, the double *time-correlation coefficient* between the fluctuating velocity component v_i^1 at a point P' and v_j'' at a point P'' is given by

$$R_{ij}(v_i',v_j'') = \frac{(v_i'v_j'')_{av}}{\sqrt{(v_i'^2)_{av}}\ \sqrt{(v_j''^2)_{av}}}$$

where the time average is over a time long with respect to the time of a single turbulent fluctuation but short relative to the over-all decay (viscous) time of the turbulence. If P' coincides with P'', R_i reduces to a tensor which is the Reynolds stress tensor. For statistically uniform turbulence (averages independent of the coordinates) with two points on the y axis a length λ, *microscale of turbulence,* can be defined in terms of the correlation coefficient $R_y = R(v_i',v_j'')$, i.e.,

$$R_y = \frac{v_i',v_i''}{\sqrt{(v_i'^2)_{av}}\ \sqrt{(v_i''^2)_{av}}}$$

and

$$\frac{1}{\lambda^2} = \lim_{y\to 0} \frac{1 - R_y}{y^2} = \frac{1}{2\sqrt{(v_i'^2)_{av}}}\left(\frac{(\partial v_i')_{av}}{\partial y}\right)^2$$

The simplest type of turbulence is *statistically isotropic turbulence,* for which mean values of functions of velocity components and their space derivatives are independent of rotation and reflection of the axes; hence $(v_x'^2)_{av} = (v_y'^2)_{av} = (v_z'^2)_{av}$ and $(v_i'v_j')_{av} = 0$. Therefore

$$R_{ij} = \frac{(v_i'v_j'')_{av}}{(v_i'^2)_{av}}$$

where $\sqrt{(v_i'^2)_{av}/U}$ is the *intensity of turbulence* (U is the mean velocity along the axis).

The time rate of *decay of turbulence* kinetic energy per unit volume for isotropic turbulence is given by $-\frac{3}{2}\rho((dv_i'^2)_{av}/dt)$; it is equal to the corresponding rate of dissipation $\bar{W}$, where

$$\bar{W} = \mu\left(2\frac{(\partial v_i')_{av}}{\partial x_k}\frac{(\partial v_j')_{av}}{\partial x_1}\right) \qquad i, j, k, l = 1, 2, 3$$

It turns out that

$$\bar{W} = \frac{15\mu(v_i'^2)_{av}}{\lambda^2}$$

and

$$\frac{d}{dt}\sqrt{(v_i'^2)_{av}} = -\frac{5\nu\sqrt{(v_i'^2)_{av}}}{\lambda^2}$$

The spectrum of turbulence is the distribution of energy among eddies of different sizes, e.g., with respect to the frequency or wave number k. (A small $|k|$ means a large eddy and a large $|k|$ means a small eddy.) The *spectral density* $(v_i'^2)_{av}\ F(k)$ of the component v_i' is related to the correlation coefficient $R(t_1)$ between $v_i'(t)$ and $v_i'(t + t_1)$ at the same point P' in a statistically steady stream:

$$R(t_1) = \frac{[v_i'(t)v_i'(t + t_1)]_{av}}{v_i'^2}$$

$$F(k) = \frac{2}{\pi}\int_0^\infty R(t_1)\cos kt\,dt$$

[The spectral density of the kinetic energy per unit volume for isotropic turbulence is $\frac{3}{2}\rho F(k)$.] In a low-turbulence wind tunnel with a mean velocity U along the x axis, the observations in the moving stream can be related to a theoretical model without mean motion for a time $t_1 = x/U$. For R_x approximately equal to $R(t_1)$, then,

$$R_x = \int_0^\infty F(k)\cos\frac{kx}{U}\,dx$$

Both R_x and $F(k)$ can be measured; the observed values fit the above relation for x not too large and for k not too small. Correlation between velocity fluctuations is measured by two hot-wire anemometers, each of which has a wire 0.001 to 0.005 in. in diameter that loses electrically maintained heat by fluid convection dependent upon the speed of flow (H. L. Dryden and A. M. Kuethe, 1929).

The correlation between the turbulent-velocity component v_i' at the point P' and the component v_j'' at P'' can be represented by a second-order tensor with components $R_{\alpha\beta} = (v_\alpha'v_\beta'')_{av}$ (double-suffix-summation convention). In statistically homogeneous turbulence this tensor is a function of the vector distance $\mathbf{r}$ between the two points and of the time; in statistically isotropic turbulence it is a function of the scalar distance r and of the time. A useful *triple correlation* between the turbulent velocity components at two points is given by the tensor

$$T_{\alpha\beta\gamma} = (v_\alpha' v_\beta' v_\gamma'')_{av}$$

For isotropic turbulence these two tensors each involve only one scalar function: f an even function of r and h an odd function of r, respectively,

$$R_{\alpha\beta} = -\left[-\frac{f'}{2r}\xi_\alpha\xi_\beta + (f + \tfrac{1}{2}rf')\sigma_{\alpha\beta}\right] v_i'^2$$

where $\xi_\alpha \equiv x_\alpha'' - x_\alpha'$, and

$$T_{\alpha\beta\gamma} = -\left[\frac{rh' - h}{r^3}\xi_\alpha\xi_\beta\xi_\gamma + \frac{h}{r}\sigma_{\alpha\beta}\xi_\gamma - \left(\frac{h}{r} + \tfrac{1}{2}h'\right)(\sigma_{\gamma\alpha}\xi_\beta + \sigma_{\beta\gamma}\xi_\alpha)\right]((v_i^{12})_{av})^{3/2}$$

By the equation of continuity the *longitudinal correlation coefficient* f is related to the *transverse correlation coefficient* g,

$$g = f + \frac{r}{2}\frac{\partial f}{\partial r}$$

From the Navier-Stokes equation one obtains the following *Kármán-Howarth equation* for isotropic turbulence:

$$\frac{\partial}{\partial t}((v_i'^2)_{av}f) + 2((v_i'^2)_{av})^{3/2}\left(h' + \frac{4h}{r}\right) = 2\nu(v_i'^2)_{av}\left(f'' + \frac{4}{r}f'\right) \qquad (2.29)$$

Equation (2.29) for the second correlation f cannot be solved completely because it involves the unknown third correlation h. Calculation of h, in turn, involves a still higher correlation and unknown scalar function—ad infinitum in an ever-increasing number of such unknown functions for each higher correlation. The equation, however, does afford significant information on isotropic turbulence.

Substituting power expansions of f and h (both assumed to be analytic functions) in Eq. (2.29) and equating the first powers of r yield the equation for *rate of decay of isotropic turbulence,*

$$\frac{\partial(v_i'^2)_{av}}{\partial t} = 10\nu(v_i'^2)_{av}\frac{\partial^2 f(0)}{\partial r^2}$$

From *Taylor's microscale of turbulence* λ given by

$$\frac{1}{\lambda^2} = -\frac{\partial^2 f(0)}{\partial r^2}$$

we have

$$\frac{d(v_i'^2)_{av}}{dt} = 10\frac{\nu}{\lambda^2}(v_i'^2)_{av} \qquad (2.30)$$

Equating second powers of r gives an equation for the rate of change of the mean square vorticity; in addition to viscous decay there is a production of vorticity by random diffusive extension of vortex lines.

On the assumption that $r^m h \to 0$ and $r^m h' \to 0$ as $r \to \infty$, one obtains by integration of (2.29) for $m = 4$

$$\frac{d}{dt}\left((v_i'^2)_{av}\int_0^\infty r^4 f\,dr\right) = 0 \qquad (2.31)$$

The expression $(v_i'^2)_{av}\int_0^\infty r^4 f\,dr$ is known as *L. G. Loitsianski's invariant* Λ (1945).

A physical-space spectrum is essentially three-dimensional (W. Heisenberg, 1948), although spectral measurements are necessarily made for a one-dimensional spectrum. The following Fourier transforms are introduced:

$$\Gamma_{\alpha\beta}(\mathbf{k}) = \frac{1}{8\pi^3}\int_{-\infty}^{\infty} R_{\alpha\beta}(\mathbf{r})e^{-i\mathbf{k}\cdot\mathbf{r}}\,d\tau(\mathbf{r})$$

$$\chi_{\alpha\beta}(\mathbf{k}) = \frac{1}{8\pi^3}\int_{-\infty}^{\infty} T_{\alpha\beta}(\mathbf{r})e^{-i\mathbf{k}\cdot\mathbf{r}}\,d\tau(\mathbf{r})$$

$$\Pi_{\alpha\beta}(\mathbf{k}) = \frac{1}{8\pi^3}\int_{-\infty}^{\infty} P_{\alpha\beta}(\mathbf{r})e^{-i\mathbf{k}\cdot\mathbf{r}}\,d\tau(\mathbf{r})$$

where

$$T_{\alpha\beta}(\mathbf{r}) = \frac{\partial}{\partial\xi_2}[(v_\alpha' v_\gamma' v_\beta''{}_{av}) - (v_\alpha' v_\beta'' v_\gamma'')_{av}]$$

and

$$P_{\alpha\beta}(\mathbf{r}) = \frac{\partial}{\partial\xi_\alpha}\frac{(p'' v_\beta')_{av}}{\rho} - \frac{\partial}{\partial\xi_\beta}\frac{(p'' v_\alpha')_{av}}{\rho}$$

The *spectral density of energy* $E(k)$ per unit volume (contribution from wave numbers between k and $k + dk$) is given by

$$E(k) = \tfrac{1}{2}F_{\alpha\alpha}(k)$$

where $F_{\alpha\beta}(k) = \iint \Gamma_{\alpha\beta}(k)\,dS(k)$

and $d\tau(\mathbf{k}) = dk\,dS(\mathbf{k})$

In powers of k,

$$E(k) = k^4 \text{ const} + \cdots$$

for isotropic turbulence,

$$E(k) = \frac{\Lambda}{3\pi}k^4 + \cdots$$

so Loitsianski's invariant is associated with the energies of the smallest wave numbers.

From the Navier-Stokes equation,

$$\frac{\partial}{\partial t}R_{\alpha\beta} - T_{\alpha\beta} = P_{\alpha\beta} + 2\nu\nabla^2 R_{\alpha\beta} \qquad (2.32)$$

the Fourier transform of (2.32) is

$$\frac{\partial}{\partial t}\Gamma_{\alpha\beta}(\mathbf{k}) = \chi_{\alpha\beta}(\mathbf{k}) + \pi_{\alpha\beta}(\mathbf{k}) - 2\nu k^2\Gamma_{\alpha\beta}(\mathbf{k}) \qquad (2.33)$$

The pressure term vanishes for isotropic turbulence.

If the viscosity term is predominant on the right side of (2.33), one finds $E_\alpha k_\epsilon^4 - 2\nu k^2 t$, so each wave will be damped independently, the small eddies more rapidly. The energy of the smallest wave numbers remains unchanged throughout the course of the turbulent motion. In the final stage of the decay of isotropic turbulence the Reynolds number is sufficiently small for the slow motion to allow the quadratic terms of the basic equations to be neglected. The energy decay for this strictly viscous dissipation is given by

$$(v_i'^2)_{av} = \frac{\Lambda}{48\sqrt{2\pi}} [\nu(t - t_0)]^{-5/2}$$

In the case of the inertial term $\chi_{\alpha\beta}(\mathbf{k})$ in (8.5) $T_{\alpha\beta}(0) = 0$ so inertia makes no contribution to $(\partial/\partial t)\int_0^\infty F_{\alpha\beta}(\mathbf{k})\, d\mathbf{k}\, [= (\partial/\partial t)\int_0^\infty \Gamma_{\alpha\beta}\, d\tau(\mathbf{k})]$. The inertia term itself does not vanish; its effect is merely to transfer the energy associated with one component from one range of wave numbers to another—physically assumed to be a net transfer from small wave numbers to large ones.

In the intermediate, or transfer, stage of turbulence development, where the quadratic inertia forces are appreciable, the decrease of energy in a wave-number band is the sum of a viscous-dissipation term and a transfer term (algebraic sum of the energy entering the band from all smaller wave numbers and the energy leaving the band toward all larger wave numbers)—an eddy cascade. Quantitative laws have been derived on the basis of some hypothesis of similarity (A. M. Kolmogoroff, 1941; L. Onsager, 1945; C. F. von Weizsäcker, 1948; W. Heisenberg, 1948) to describe this region of statistical equilibrium. For example, consider *locally isotropic turbulence*, for which the physical variables are isotropic relative to a specified point. On *Kolmogoroff's first similarity hypothesis* (viscous and inertia forces comparable) $E(k,t)$ is a function only of k and two parameters ν and $\epsilon = -(\partial/\partial t)\int_0^\infty E(k,t)\, dk$ (rate of energy dissipation per unit volume). On the *second similarity postulate* (viscous forces negligible with respect to the inertia forces) $E(k,t)$ is independent of ν also, so $E(k,\epsilon)$ can be determined from dimensional reasoning. In this case

$$E(k) \sim \frac{\epsilon^{2/3}}{k^{5/3}}$$

which is true neither for too large wave numbers (viscosity appreciable) nor for too small wave numbers (dimensions comparable with those of the apparatus, so a regular eddy cascade does not exist). More general spectral functions $E(k)$ can be derived from the spectral equation of motion on the assumption of special forms for the inertia term as a function of $E(k)$ and k in terms of the preceding physical picture.

Physically, then, the turbulent-energy spectrum for sufficiently high Reynolds numbers has primarily a high energy density at wave numbers of the order of magnitude of the reciprocal of the appropriate apparatus dimensions. These turbulent eddies, being unstable, immediately break up, so that the energy becomes distributed over most of the energy range. At the low end of the range the spectral energy density is of the order k^4; at the high end, of the form $k^4\epsilon^{-k^2\nu t}$, and in the intermediate, Kolmogoroff range $E(k) \sim k^{-5/3}$. As time increases, the region of large density continuously shifts to larger wave numbers, while decreasing in magnitude owing to viscous dissipation. The Kolmogoroff range shifts to larger wave numbers and progressively narrows, until it disappears, and the final stage develops. The smallest Kolmogoroff eddy η is given by

$$\eta = \left(\frac{\nu^3}{\epsilon}\right)^{1/4}$$

and the largest eddy size H by

$$H = \frac{(v_i^{3\prime})_{av}}{\epsilon}$$

The following formula has been proposed (Kármán, 1948) to connect the k^4 region of the spectrum with the $k^{-5/3}$ region

$$E(k) \alpha \Lambda (v_i'^2)_{av} 1 \frac{(k1)^4}{(1 + k^2 1^2)^{17/6}}$$

The similarity assumptions for locally isotropic turbulence have been used (C. C. Lin, 1948) to obtain

$$v_x'^2 = (t - t_0)^{-1}\beta$$

and $$\lambda^2 = 10\nu(t - t_0)\left[1 + \frac{\beta}{\alpha}(t - t_0)\right]$$

where α, β, and t_0 are constants.

Measurements show satisfactory agreement between experiment and theory. The early statistical theory of turbulence was based on the assumption that both double and triple correlation functions of the distance and time could be expressed as functions of a single dimensionless variable combining both distance and time. This somewhat artificial physical assumption has a mathematical advantage in yielding the turbulent-intensity variation and the scale relating the so-called self-preserving double and triple correlations.

The frequent occurrence of turbulence is practically important in the boundary layer, particularly in aeronautics, where high Reynolds numbers are found. As the local Reynolds number (based on the mean velocity outside the boundary layer and the boundary-layer thickness) increases for subsonic speeds, a *critical Reynolds number* is reached, at which there is a transition from laminar to turbulent flow. For a bluff body like a cylinder this *transition point* is associated with, but not usually identical with, the *separation point*, where the fluid leaves the body contour. The most celebrated example is the discrepancy in the drag of a sphere as measured in different wind tunnels, first by Eiffel at Paris and by Föppl at Göttingen. As Prandtl noted, the turbulence levels of the tunnels differed. The strange fact, however, was that the more turbulent tunnel (France) gave less drag. The reason is that the turbulent tunnel produces a turbulent boundary layer, which separates from the body farther upstream than the laminar one. Hence the lower pressure in the resultant wake counteracts to a less degree the pressure on the forward part of the sphere. This drag is called *form or body drag*. Streamlining a body (well-rounded, slender, pointed at rear) decreases the wake and hence this type of drag. Increase of turbulence beyond the critical Reynolds number thickens the boundary layer, so the separation points move forward, and the drag coefficient, in turn, increases.

The *total drag* on a body includes this type and skin friction, which also is modified by a turbulent boundary layer. Of course, in supersonic flow the *wave drag*, too, has to be added; it may amount to 90 per cent, so that changes in the Reynolds number, resulting in laminar or turbulent conditions, may have comparatively little effect on the total drag.

The turbulence limitation of Poiseuille's law for laminar flow in pipes was first noted by Reynolds. Poiseuille's law holds for Reynolds numbers up to about 1,000 (except in the case of specially smooth, rounded tubes for which $R \sim 75{,}000$ can be attained). It is found experimentally that the maximum speed across a section of the tube is $\frac{5}{4}\bar{u}$, where $\bar{u}$ is the mean speed across the section. The mean speed $\bar{u}_y$ at a distance y from the wall is given by *Prandtl's seventh-power law*:

$$\frac{\bar{u}_y}{V} = 8.74 \left(\frac{yV}{\nu}\right)^{1/7}$$

where $V = l(d\bar{u}/dy)$. If the steady flow past a flat plate is turbulent, the drag coefficient is found to be

$$C_D = 0.72 \frac{l}{R^{1/5}}$$

on the assumption that the mean speed in the boundary layer is given by the seventh-power law.

Although the nature of turbulence is not known, the physical origin of the phenomenon is becoming understood. One looks for unstable flows. Specifically, one considers the effect of sinusoidal perturbations upon a steady-state solution. An eigenvalue problem results. If instability is found for these small disturbances, it will exist also for large ones. Stability of small disturbances, however, allows no inference to be made about large ones. It has been shown (H. B. Squire, 1933) that if a two-dimensional parallel flow admits an unstable three-dimensional periodic disturbance for certain Reynolds numbers, then it admits an unstable two-dimensional disturbance at a lower Reynolds number, so that consideration of a two-dimensional problem is sufficient.

A study of the development of the boundary layer in Couette flows shows that the decisive factor for the stability of the initial laminar motion from rest is the velocity profile of the boundary layer. *W. Tollmien's criterion* (1930) is that boundary layers with velocity profiles having inflection points, e.g., in boundary layers with positive pressure gradient, have a great probability of becoming unstable. Flows that remain laminar have velocity profiles that approach asymptotically a straight line. On the other hand, early transient states may have velocity profiles that lead to instability. Hence the decisive factor is not the stability of the final state but rather that of each yearly state of flow development. A fruitful experimental method is the imposing of a small simple harmonic motion on a steady flow about an object (G. B. Schubauer and H. K. Skramstad, 1947). Above a minimum Reynolds number for any given Reynolds number there is a critical region of frequencies where superimposed frequencies become amplified and hence are potential sources of turbulence. There are many flow patterns that show unstable changes, e.g., intermittent turbulence in a straight conduit. Unsteady flows are largely in the observational stage; but they may well turn out to be the ones prevalent in nature.

The turbulence observed in a laboratory, say in a wind tunnel, is quite different from that of the atmosphere, owing primarily to its nonuniform temperature (the Coriolis force of the earth's rotation also becomes significant). In laboratory turbulence, characterized by large velocity gradients, inertia is the predominant factor together with pressure gradients and viscous forces. There are two layers in the atmosphere, separated at about 2,500 m above the ground. In the upper layer the temperature decreases monotonically with altitude, whereas in the lower *thermal boundary layer* the temperature alternately changes with time from a monotonically increasing to a monotonically decreasing function with respect to the altitude. In atmospheric turbulence fluctuations of density are important, so that the usual Reynolds stress has to be modified for compressible flow. The equations of motion (x south, y east) in the boundary layer are

$$2\omega \sin \phi \, \bar{v}_y = \frac{1}{\rho}\frac{\partial \bar{p}}{\partial x} + \frac{1}{\rho}\frac{d}{dz}\left[\rho(v_x{}^1 v_z{}^1)_{\text{av}}\right]$$

$$-2\omega \sin \phi \, \bar{v}_x = \frac{1}{\rho}\frac{\partial p}{\partial y} + \frac{1}{\rho}\frac{d}{dz}\left[\rho(v_y{}^1 v_z{}^1)_{\text{av}}\right]$$

where ω is the angular velocity of the earth, ϕ the latitude, and Reynolds stresses are assumed to depend on z only. In order to solve these equations, hypotheses must be made about the Reynolds stresses. In this connection, the *coefficient of turbulent viscosity* ν' is useful:

$$(v_x{}^1 v_z{}^1)_{\text{av}} = -\nu' \frac{d\bar{v}_x}{dz}$$

$$(v_y{}^1 v_z{}^1)_{\text{av}} = -\nu' \frac{d\bar{v}_y}{dz}$$

Outside the boundary layer both the viscous stresses and the Reynolds stresses are negligible. In the atmospheric boundary layer one can use fine instruments such as the wind-tunnel hot-wire anemometer for measuring the small eddies there, i.e., *microturbulence*. There are still many unknown factors in the general field of atmospheric turbulence.

The scale at which turbulence is investigated is a concept of paramount importance in discussing atmospheric turbulence. Evidently the description of phenomena will be quite different if the size of measurable eddies is greater than 100 or 0.01 mm. For example, the coefficient of molecular diffusion of a very small puff of air in a small box is of the order of 10^{-1} cm^2/sec, whereas for a large meteorological puff of polar air the coefficient of diffusion is of the order of 10^8 cm^2/sec. In general, the larger the scale, the larger the coefficient of diffusion.

Turbulent diffusion differs from molecular diffusion in that it concerns a continuous medium rather than discrete particles. The essential feature of purely turbulent diffusion is that the fluid particles retain their initial properties throughout the history of the motion but undergo a redistribution in space. The complete solution of this problem, therefore, requires the joint probability distribution of "marked" particles—a complicated task.

The basic equation for *turbulent diffusion* (from the equation of continuity) is

$$\frac{\partial \bar{S}}{\partial t} + \operatorname{div}(\bar{S}v) = -\operatorname{div}(\langle S'\mathbf{v}'\rangle_{av})$$

where $S(x,y,z,t) = \bar{S} + S'$ is the concentration of certain material in a fluid in x, y, z at t. The solution of this equation for $\bar{S}$ requires knowledge of $\langle S'\mathbf{v}'\rangle_{av}$. It has been customary to use the *diffusion tensor* ν_{ij}^* with these assumptions: steady state, v_x constant $(\equiv v)$, $\langle v_y\rangle_{av} = \langle v_z\rangle_{av} = 0$; x, y, z axes same as principal axes of tensor; each $\nu_{ij}^* =$ constant, $\nu_{ij}^* = 0$ for $i \neq j$. Then

$$v_0 \frac{\partial \bar{S}}{\partial x} = \nu_{xx}^* \frac{\partial^2 \bar{S}}{\partial x^2} + \nu_{yy}^* \frac{\partial^2 S}{\partial y^2} + \nu_{zz}^* \frac{\partial^2 S}{\partial z^2}$$

Solutions have been found in terms of elementary solutions, $S_n = (1/r)\omega_n(r)P_n \cos\theta\, \epsilon^{r\cos\theta - r}$, where $\omega_n(r)$ is a polynomial and $P_n \cos\theta$ is the Legendre polynomial. In recent years the theory of random functions, e.g., $f(x,y,z,t,w)$ dependent on a parameter w and chosen at random according to some probability law in a space r, has been developed to solve the diffusion equation. For example, a stationary (average invariant with respect to the time) random function can approximate velocity fluctuations in a particular turbulence. A large class of turbulent-diffusion problems can be solved by relating the Lagrangian and Eulerian points of view. In Eulerian language *homogeneous turbulence* is such that the three random functions describing in Lagrangian language the trajectory of a particle in a turbulent flow are stationary random functions with respect to x, y, and z. The system (not yet solved) of differential equations that have to be solved to find the trajectory is

$$\frac{dx}{dt} = v_x(x,y,z,t,w)$$
$$\frac{dy}{dt} = v_y(x,y,z,t,w)$$
$$\frac{dz}{dt} = v_z(x,y,z,t,w)$$

where v_x, v_y, and v_z are stationary random functions of x, y, z for all t.

9. Fluids with Heat

If heat is important, as in the case of high-speed flows, then the first law of thermodynamics requires inclusion of all heat exchanges to or from the fluid, such as thermal conduction both external and internal to the fluid, convection, diffusion, radiation, and chemical or nuclear reactions. In the case of the first mentioned the molecular-momentum transfer, interpreted as a viscous stress, necessarily involves energy transfer.

In *forced convection* the flow is determined by the hydrodynamic equations of motion and the boundary conditions, whereas in *natural convection* it is the result of density gradients established by heat transfer and gravitation. The latter problem is very difficult but has been solved for some special problems, e.g., the Clusius-Dickel thermal-diffusion column (annular region between two vertical tubes at different temperatures) and cellular convection between two large horizontal plates at different temperatures. In the energy equation the energy flux due to radiation must be included as in the case of flames. It can be obtained explicitly for two limiting cases, viz., opaque materials, for which thermodynamic equilibrium is established due to large absorption, and transparent materials with little absorption. For astrophysical phenomena (and also atomic bombs) one must add the radiation energy to the internal energy and the radiation pressure to the hydrodynamic pressure tensor. In a mixture an energy term has to be included for the diffusion processes, although thermal diffusion (due to temperature gradients; in liquids, the Soret effect) and its reciprocal, the Dufour effect (direct effect of concentration and pressure gradients) can usually be neglected.

For a binary mixture in the absence of external forces the *equation of diffusion* is

$$w = -\frac{n_0^2}{n_1 n_2}\left[D_{12}\frac{\partial(n_1/n_0)}{\partial z} + D_T \frac{1}{T}\frac{\partial T}{\partial z}\right]$$

where w is the difference between the mean velocities of the two gases with n_1, n_2 the respective number of molecules per unit volume, $n_0 = n_1 + n_2$, D_{12} Maxwell's coefficient of diffusion, and D_T the *coefficient of thermal diffusion*. Hence for a steady state ($w = 0$) a concentration gradient $\partial(n_1/n_0)/\partial z$ will be set up in opposition to a thermal gradient $\partial T/\partial z$. For a completely ionized gas the heavier gas (the ions) tends to move toward the central, hotter region. In the upper levels of the earth's atmosphere the rate of diffusive separation (increases inversely as the density) may overcome the mixing tendency of turbulence. The so-called *thermal-diffusion ratio* ($k_T = D_T/D_{12}$) depends upon many factors. One result is the practical application of a thermal-diffusion column (with horizontal temperature gradient) for the separation of isotopes. Thermal diffusion in liquids, the *Ludwig-Soret effect*, may be a factor on a small scale in volcanic dikes. An inverse effect to thermal diffusion is the *diffusion thermo-effect*, for which, if a temperature gradient is generated, the resulting thermal diffusion will decrease the primary diffusion.

The heat transfer q per unit area and time by conduction is given by

$$q = -k \operatorname{grad} T$$

where k is the thermal conductivity. The complete thermodynamic-energy statement as based on the mechanical-energy equation is

$$\left(\frac{\partial}{\partial t} + \mathbf{v}\cdot\operatorname{grad}\right)(\tfrac{1}{2}\rho\mathbf{v}^2) + \mathbf{v}\cdot\operatorname{grad}(p + E) = \mathbf{v}\cdot\operatorname{div}\mathfrak{T} + k\nabla^2 T + Q' \quad (2.34)$$

The first term is the time rate of change of kinetic energy per unit volume (in a volume element considered fixed in space). The second term represents the net convective input of kinetic energy per unit volume. The third and fourth terms, respectively, are the time rate of change and net convective input of the specific energy E ($= c_vT$ for an ideal gas with c_v its specific heat at constant volume). The first term on the right is the time rate of work done on the fluid by the surface stresses. The second term on the right is the net rate of conductive heat input. Q' represents any other external heat.

For steady motions not involving free convection the various terms of the energy equation are of the type, respectively, $\rho Vc_pT/L$, pV/L, kT/L, $\mu V^2/L^2$, where c_p is the specific heat at constant pressure. In addition, the momentum equation gives another independent term $\rho V^3/L$. From these five terms four dimensionless ratios may be obtained: the *pressure coefficient* $C_p = p/\frac{1}{2}\rho V^2$; the Reynolds number $R = \rho VL/\mu$; the Mach number $M = V/c$; and the *Péclet number* $\mathrm{Pe} = \rho Vlc_p/K$. The pressure coefficient is adequate for the ideal incompressible fluid; i.e., a given geometric configuration need be solved only once, say, for the pressure coefficient C_p at any point. The addition of internal friction requires an infinity of solutions for one geometric configuration: $C_p = f_1(R,M)$. The further introduction of compressibility requires a double infinity of solutions: $C_p = f_2(R,M)$. Finally, the inclusion of heat necessitates $C_p = f_3(R,M,\mathrm{Pe})$. Instead of the Péclet number, which measures the convective effect relative to conduction, another dimensionless number, the *Prandtl number* $\mathrm{Pr} = c_p\rho/k$, is used; it measures the relative importance of viscous heat and thermal diffusivities. The Prandtl number depends upon the general properties of the material only as contrasted with the Péclet number, which depends upon the specific characteristics of flow. (For unsteady motion a nondimensional parameter such as Vt_c/L is needed, where t_c may be a characteristic time such as the period of oscillation.) The Prandtl number for gases is of the order of 1, inasmuch as the kinematic viscosity and the kinematic thermal conductivity are both proportional to the mean molecular speed and the mean free path. Hence if viscous effects are included, thermal conduction must be considered too.

The transfer of heat from a solid body to a fluid is analogous to the transfer of momentum. At the critical Reynolds number of flow, however, there is no sharp increase in heat loss as in drag.

Heat transfer is better expressed in terms of the dimensionless *Nusselt number*, $Q = \mathrm{Nu}kST/L$, as a similarity parameter than in terms of the heat-transfer coefficient. In general, for low speeds the Nusselt number is a function of the Reynolds number, the Prandtl (or Péclet) number, and the *Grashof number*, $\mathrm{Gr} = L^3g\alpha T/\nu^2$, where α is the thermal coefficient of expansion. In the case of *forced convection*, for which gravitational effects are insignificant, the Nusselt number is a function only of the Reynolds number and of the Prandtl number. In *natural* (*free*) *convection* the Nusselt number is a function only of the Grashof number and of the Prandtl number, inasmuch as there is no representative speed, so that the Reynolds number is not significant.

For an incompressible fluid Eq. (2.34) becomes the equation for thermal potential flow:

$$v_x \frac{\partial T}{\partial x} + v_y \frac{\partial T}{\partial y} = \frac{K}{c_v} \nabla^2 T \tag{2.35}$$

where K/c_v is the *thermal diffusivity*. A *thermal boundary layer* is observed close to bodies; it often coincides with the viscous boundary layer. In this case the same mechanism is responsible both for heat transfer and for momentum transfer. The *dust-free space* (Tyndall, 1870) surrounding a heated cylinder in a dusty or misty atmosphere is closely associated with the thermal boundary layer.

In the case of compressible fluids the variability of density and temperature in the boundary layer must be considered (μ and k, too, must be regarded as variable in that they may be functions of the temperature). Hence from the two-dimensional force equation

$$\rho \left(v_x \frac{\partial v_x}{\partial x} + v_y \frac{\partial v_x}{\partial y} \right) = -\frac{\partial p}{\partial x} + \frac{\partial}{\partial y} \left(\mu \frac{\partial v_x}{\partial y} \right)$$

and

$$\frac{\partial p}{\partial y} = 0$$

and from the enthalpy equation

$$\rho \left[v_x \frac{\partial (c_pT)}{\partial x} + v_y \frac{\partial (c_pT)}{\partial y} \right] - v_x \frac{\partial p}{\partial x} = \frac{\partial}{\partial y} \left(K \frac{\partial T}{\partial y} \right) + \mu \left(\frac{\partial v_x}{\partial y} \right)^2$$

Assuming c_p is constant and $\mathrm{Pr} \sim 1$,

$$\frac{dW}{dt} = \frac{\partial}{\partial y} \left(\mu \frac{\partial W}{\partial y} \right)$$

where $W = c_pT + \frac{1}{2}\mathrm{v}^2$ is the total energy. This equation is analogous to the one for vorticity. For typical boundary conditions, viz., T_w at wall $y = 0$ with $v_x = 0$, and T_0 in free stream $y = \infty$ with $v_x = V$,

$$c_pT + \tfrac{1}{2}v_x^2 = c_pT_w + [c_p(T_0 - T) + \tfrac{1}{2}V^2] \frac{v_x}{V}$$

If the wall is thermally insulated, the total energy of each particle is preserved as if the flow in the boundary layer were devoid of viscosity and heat conduction.

The heat transfer at the wall is $q_w = k_w(\partial T/\partial y)_w$ and the skin friction is $f_w = \mu_w(\partial v_x/\partial y)$. Hence the above equation becomes

$$\frac{q_w}{f_w} = \frac{\lambda_w}{\mu_w} \frac{T_0}{V} \left(1 - \frac{T_w}{T_0} + \frac{\nu - 1}{2} M^2 \right)$$

where M is the free-stream Mach number (cf. Reynolds's low-speed analogy between skin friction and heat transfer for laminar layers). Hence as $M \to 0$, the cooling of a surface by a moving fluid is more efficient than at high speeds. The heat transfer is zero if $T_w = T_0\{1 + [(\nu - 1)/2]M^2\}$. Accordingly, if M increases so that this relation holds, the cooling will cease and heating (from the boundary-layer friction) will then take place.

A *flame* is a thermal wave, accompanied by exothermic chemical reactions, which travels with subsonic velocity dependent upon chemical kinetics and the coefficients of thermal conduction and diffusion. A *detonation*, on the other hand, travels with supersonic velocity determined by the conservation laws of mass, momentum, and energy. A steady-state flame must be stabilized in practice by the presence of a flame holder, as in a bunsen-burner flame. The solution of the basic equations describing a real flame involves difficult numerical procedures. A simple case is one described kinetically by a single reversible unimolecular reaction: $A \rightleftharpoons B + Q$ (Q the heat of reaction). The equation of continuity can be written in dimensionless variables and the distance z from the flame holder:

$$\frac{dG}{d\xi} = \frac{1}{\mu^2} f(x,\tau) \tag{2.36}$$

where $G = mn(v + \bar{V})/\mathfrak{M}$, n being the concentration, $\bar{V}$ the diffusion velocity, $\mathfrak{M} = \rho v$; $\xi = \mathfrak{M} c_p \int_0^z dz/\kappa$; $\mu = M\sqrt{C_p/k\kappa\rho}$, k being the steric factor; and $f(x,t) \equiv \epsilon^{-1/\tau}[-xt(1-x)\epsilon^{-\beta/\tau}]$, x being the mole fraction, $\beta \equiv Q/E^*$ (E^* the activation energy per molecule for forward reaction), $\tau = kT/E^*$. The equation of motion is

$$\frac{dv}{d\xi} = \frac{1}{\omega} h(v,T) \tag{2.37}$$

where

$$v = \frac{v}{v_\infty} \qquad h \equiv (v - 1) + \frac{1}{M_\infty^2 \gamma'}\left(\frac{T}{vT_\infty} - 1\right)$$

$$M_\infty \equiv \left(\frac{v}{c}\right)_\infty$$

at hot boundary,

$$\gamma' \equiv 1 + 1\Big/\left[\frac{1}{\gamma - 1} + \left(\frac{Q}{kT}\right)^2 \frac{\epsilon^{+Q/kT}}{(1 + \epsilon^{+Q/kT})^2}\right]$$

$\omega \equiv 4(c_p n/x)D_{AB}$, D_{AB} being the diffusion coefficient. The energy equation is

$$\frac{d\tau}{d\xi} = g(v,\tau,G) \tag{2.38}$$

where

$$g = \frac{\gamma - 1}{\gamma}\beta(G - G_\infty) - (\tau_\infty - \tau) + \frac{\gamma - 1}{\gamma}\left[(v\tau_\infty - \tau) - \frac{M_\infty^2}{2}\gamma'\tau_\infty(v - 1)^2\right]$$

Finally, the diffusion equation is

$$\frac{dx}{df} = \frac{1}{\sigma}(x - g)$$

where $\sigma \equiv c_p\rho D_{AB}/K$. Solutions of these equations have been made with proper hot and cold boundary conditions with neglected viscosity. For ordinary flames both kinetic energy and diffusion can often be neglected, but the inclusion of diffusion affects the results. On the other hand, the neglect of diffusion, but not of kinetic energy, is important for fast flames. In this case $M = 1$ is the highest Mach number for steady-state propagation of flames.

In a detonation the chemical composition changes, so that energy is released and chemical equilibrium is attained after the gas passes through the wave, i.e., reaction rates are all zero. The Hugoniot curve for a detonation wave differs from that of a shock wave in that the initial point is not on the curve of all possible final states. The only steady-state detonation that occurs is the one for which the line between the initial and final states is tangent to the Hugoniot curve. This solution with the lowest possible detonation velocity is called the *Chapman-Jouguet condition;* the velocity of the hot gases with respect to the detonation front is the local acoustic speed. It can be shown that a detonation is physically a combustion process initiated by a shock wave. The pressure increases abruptly within the shock thickness, while the temperature also increases; by the end of the reaction zone (of the order of 1 cm) the pressure has decreased to half its maximum, whereas the temperature has increased to twice its value. The velocity of detonation is subsonic with respect to the material in the reaction zone, so that the energy released by the chemical reactions in the form of heat can flow toward the front and thus maintain the steady state.

The structure of a detonation wave in a gas is described by the equations used in flame propagation. Diffusion can be neglected, but viscosity is important, as in the case of a shock. Consider p/p_∞, $v = \rho_\infty/\rho$, and x (mole fraction of $A \rightarrow B$ for unimolecular decomposition) with $a = 1$. Then, subject to the Chapman-Jouguet condition $M_\infty = 1$,

$$\frac{dx}{d\zeta} = \frac{1}{v^2} f\left(x, v, \frac{p}{p_\infty}\right)$$

$$\frac{dv}{d\zeta} = \frac{1}{\omega} h\left(v, \frac{p}{p_\infty}\right)$$

$$\frac{d(p/p_\infty)}{d\zeta} = j\left(x, v, \frac{p}{p_\infty}\right)$$

where $$j\left(x, v, \frac{p}{p_\infty}\right) = \frac{1}{v\tau_\infty} g - \frac{1}{\omega v}\frac{p}{p_\infty} h$$

A simple mathematical model of a steady *aerothermodynamic shock wave* is a step shock wave with an addition or subtraction of a quantity of heat Q at the surface of discontinuity. Such a model is a good representation of a detonation wave and of a *condensation shock* (formation of droplets at the shock front). The Rankine-Hugoniot equations are modified only in that the energy equation involves Q on its right side. The possible steady states connecting M_1 on one side with M_2 on the other are shown in Fig. 2.14, where the parameter $K = (\gamma/Q)[M_1^2 + 2/(\gamma - 1)]$ is a function of M_1 as well as Q. The 45° straight line is the acoustic limit (no shock); the equilateral hyperbola is the shock-wave limit (no energy Q). Deflagration is represented by $0 \leq M_1 \leq 1$, $0 \leq M_2 \leq 2$; detonation by $1 \leq M_1$, $0 \leq M_2 \leq 1$. The physical existence of such solutions is not determined by the algebraic relations alone; other conditions, such as the reaction rate, must be considered. Note, however, that an aerothermodynamic shock wave with subsonic flow becoming supersonic flow is not excluded by the second law of thermodynamic shocks, as is the case of ordinary shock waves ($Q = 0$). The interaction of aerothermodynamic shock waves can be investigated in a manner similar to that discussed above for $Q = 0$; the results are more complex.

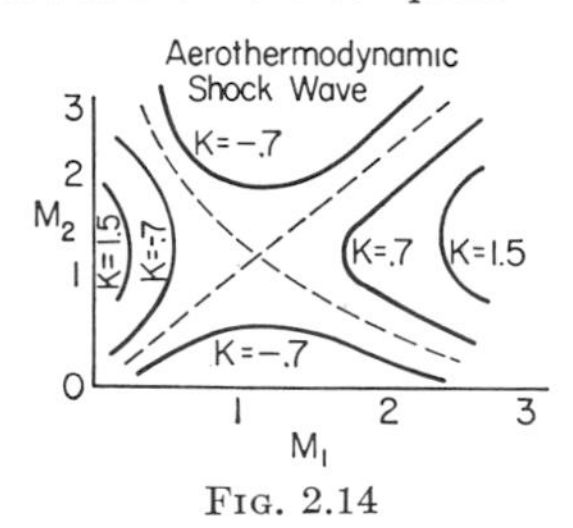

FIG. 2.14

A shock wave, being essentially a viscous phenomenon, must be investigated in terms of the Navier-Stokes equation of motion involving spatial variations. The basic equations for one-dimensional steady-state ideal gas are

$$\frac{d}{dx}(\rho v_x) = 0 \qquad \text{(continuity)}$$

$$\rho v_x \frac{dv_x}{dx} = \frac{\partial T_{xx}}{\partial x} \qquad \text{(motion)}$$

$$\rho v_x \frac{d}{dx}\left(\frac{v_x^2}{2} + E\right) = \frac{d}{dx}(T_{xx}v_x) + \frac{d}{dx}\left(M\frac{dT}{dx}\right) \qquad \text{(energy)}$$

where $T_{xx} = -p + \frac{4}{3}\mu\,(dv_x/dx)$. Integrating,

$$\rho v_x = \mathfrak{M}$$

$$\mathfrak{M}v_x + p - \frac{4}{3}\mu\frac{dv_x}{dx} = \mathfrak{M}B$$

$$\mathfrak{M}\left(\frac{v_x^2}{2} + E + \frac{p}{\rho} - \frac{4}{3}\frac{\mu}{\rho}\frac{dv_x}{dx}\right) - K\frac{dT}{dx} = \mathfrak{M}A$$

where $\mathfrak{M}$, B, A are constants of integration. Far from the shock there is little variation in v_x and T, so the Rankine-Hugoniot equations are valid.

The *structure of the shock wave* depends upon the transition region. Using the dimensionless variables v, ζ, ω of Eqs. (2.36) to (2.38),

$$v\frac{d}{d\zeta}\left(\omega v\frac{dv}{d\zeta}\right) + \frac{\gamma - 1}{2\gamma}v\frac{dv}{d\zeta}\left[1 + v_0 - \frac{4\gamma}{\gamma + 1}\left(1 + \frac{\omega}{2\gamma}\right)v\right] - \frac{\gamma + 1}{\gamma}v(v - 1)(v_0 - v) = 0 \qquad (2.39)$$

where

$$v_0 \equiv \frac{\gamma - 1}{\gamma + 1}v_\infty\left(1 + \frac{2\gamma}{\gamma - 1}\frac{p_\infty}{\rho_\infty v_\infty^2}\right)$$

This equation can be solved for $\omega = 1$ (Prandtl number equals $\frac{3}{4}$). There results

$$\frac{p}{p_\infty} = \left(\frac{\gamma + 1}{\gamma - 1}\frac{v_0}{v} - v\right)\left(\frac{\gamma + 1}{\gamma - 1}v_0 - 1\right)$$

Let the *shock thickness* σ be the distance between the point where $v = v_0 - \epsilon(v_0 - 1)$ and one where $v = v_0 + \epsilon(v_0 - 1)$, ϵ arbitrarily small. Then

$$\Delta\zeta = \frac{2\gamma}{\gamma + 1}\frac{v_0 + 1}{v_0 - 1}\ln\left(\frac{1}{\epsilon} - 1\right) \qquad (2.40)$$

For constant thermal conductivity

$$\sigma = \frac{2(\gamma - 1)}{\gamma + 1}\frac{m\kappa}{Mk}\frac{v_0 + 1}{v_0 - 1}\ln\left(\frac{1}{\epsilon} - 1\right)$$

Introducing kinetic theory relations for K and c in terms of l,

$$\frac{\sigma}{l} = \frac{4}{3(\gamma + 1)}\sqrt{\frac{2}{\gamma\pi}}\frac{v_0 + 1}{v_0 - 1}\ln\left(\frac{1}{\epsilon} - 1\right)$$

Expression (2.40) has been evaluated for $\kappa(T)$ and $\mu(T)$ (L. H. Thomas, 1944) with the result that $1/\sigma$ varies from 0 to 0.7 for Mach numbers up to 4.5. The gradients, therefore, are so large that the Navier-Stokes equation for a continuum cannot be used. One has to consider higher-order Chapman-Enskog approximations (such as the Burnett equations) of the Boltzmann kinetic-theory integral equation. Convergence takes place only slowly. In the case of strong shocks other methods of solving the Boltzmann equation have to be found. One direct approach (H. M. Mott-Smith, 1951) has led to shock thicknesses about five-thirds those of Thomas and more nearly in accord with the experimental values obtained by observing changes in optical reflectivity (D. F. Horning and G. R. Cowan, 1951).

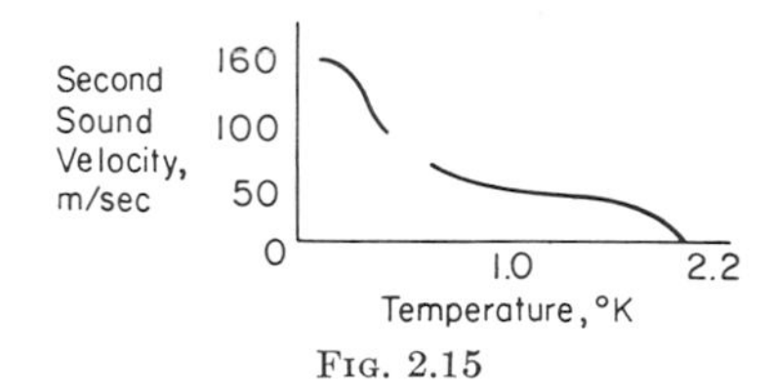

FIG. 2.15

Much information about nonequilibrium fluids is possible through the so-called thermodynamics of irreversible processes. In addition, time effects, such as relaxation phenomena, must be considered in the establishment of steady states, e.g., in a shock wave.

A new field of *quantum hydrodynamics* is that of a quantum liquid like He^{II}, which is dominated largely by its zero-point energy at very low temperatures. He^{II} may be regarded as composed of two fluids below the λ point at 2.19°K, the *normal fluid* and a *superfluid*, which has negligible viscosity (at absolute zero He^{II} consists entirely of superfluid but at the λ point entirely of normal fluid). Some peculiar phenomena due to the superfluid are the *fountain effect* resulting from the flow of the He^{II} superfluid against a temperature gradient, the flow of the He^{II} superfluid through narrow channels impervious to He^{I}, and the propagation of *second sound*, i.e., a temperature wave with no net mass flow of liquid ($\rho_n v_n + \rho_s v_s = 0$). Measured second-sound velocities for He^2 are given in Fig. 2.15. Much remains to be understood about quantum liquids with their high ordering in momentum space. (See Part 5, Chap. 8.)

10. Flows in Electric and Magnetic Fields

If a gas is completely ionized, as in the interior of the sun and stars, it may be considered a binary mixture. Or it may be slightly ionized, as in the earth's ionosphere, which consists of three kinds of particles, neutral (several kinds), ionic (more than one positive kind, and even negative ions with electrons attached to neutral particles), and electronic. The kinetic theory of a binary mixture of neutral particles is complicated and leads to cumbersome formulas, so the general problem, at best, is first simplified and then treated approximately. In ionized gases the relative diffusion of the positive and negative ions constitutes an electric current (slightly ionized gases of low density are virtually neutral electrically, although a distribution of charge may be found near a boundary, say, of a cloud).

In an *electric field* there will be a convection electric current if the gas is not electrically neutral and a conduction current due to the average motion of the charges relative to the gas. Gradients of concentration, pressure, and temperature will also produce an electric current. For weak fields each kind of charged particle makes its own independent contribution to the ohmic current.

In a *magnetic field* each charged particle e is acted on by an electromotive force $e\mathbf{v} \times \mathbf{H}$. If the field $\mathbf{H}$ is uniform, its effect is to cause the particles to describe spiral paths (opposite directions for opposite charges) of radius mv_t/eH, where v_t is the velocity component transverse to the field. If there is also a nonmagnetic force F, it accelerates the charge along the field and causes an average drift $(\mathbf{F} \times \mathbf{H})/eH^2$ from a trochoidal motion due to the force component F_t transverse to the field. If F_t always has the same sign, as in gravity, the opposite drifts will cause a current in the same direction; whereas for drifts in the same direction, as in the case of an electric field, the electric currents will be in opposite directions. Collisions, however, reduce these drift motions. A transverse electric field produces a transverse current and a Hall-effect current perpendicular to $\mathbf{E}$ and $\mathbf{H}$. These phenomena are important for the ionosphere in the earth's magnetic field.

If an electrically conducting fluid is present in a magnetic field, any motion will induce electric fields and hence electric currents. There will be a coupling between the mechanical and electrodynamic forces resulting in a *magnetohydrodynamic wave motion*. The Lagrangian equations of inviscid incompressible motion become

$$\frac{d\mathbf{v}}{dt} = \frac{1}{\rho}\left(\frac{\mathbf{i} \times \mathbf{B}}{c} - \operatorname{grad} p\right) + \mathbf{G}$$

where $\mathbf{i}$ is the conduction current, $\mathbf{B}$ the magnetic induction, and $\mathbf{G}$ the nonelectromagnetic forces. In addition, Maxwell's equations hold for the electromagnetic field.

For a fluid with $i_z = 0$ in a primary homogeneous magnetic field $Hz = H_0$ with $\mathbf{G} = 0$, we find

$$\frac{\partial^2 h_y}{\partial t^2} = \frac{\mu H_0{}^2}{4\pi\rho}\frac{\partial^2 h_y}{\partial z^2} + \frac{c^2}{4\pi\mu\sigma}\frac{\partial^3 h_y}{\partial^2 z\, \partial t}$$

where σ is the electrical conductivity, μ the magnetic permeability, and $\mathbf{H} = \mathbf{H}_0 + \mathbf{h}$. In the case of infinite conductivity

$$\frac{\partial^2 h_y}{\partial t^2} = \frac{\mu H_0{}^2}{4\pi\rho}\frac{\partial^2 h_y}{\partial z^2}$$

The solution of this equation represents a plane wave propagated in the direction of H_0 with a velocity equal to $\pm H_0 \sqrt{\mu/4\pi\rho}$. The magnetic lines of force move with the same velocity as the fluid. If $H_0 = 100$ g, $\mu = \rho = 1$, then the velocity is 28 cm/sec. For finite conductivity the wave experiences damping. The transient cold sunspots with their strong magnetic fields may owe their origin to magnetohydrodynamic waves that have originated deep in the interior of the sun.

References

1. Thompson, W. B.: "An Introduction to Plasma Physics," 1962.
2. Bershader, D. (ed.): "The Magnetohydrodynamics of Conducting Fluids," Stanford University Press, Stanford, Calif., 1959.
3. Seeger, R. J., and G. Temple (eds.): "Research Frontiers in Fluid Dynamics," Wiley, New York, 1965.
4. Seeger, R. J., and A. B. Tayler: *Phys. Letters*, **2**: 339, 1962.
5. Hayes, W. D., and R. F. Probstein: "Hypersonic Flow Theory," Academic, New York, 1959.
6. Von Mises, R., R. H. Geiringer, and G. S. S. Ludford: "Mathematical Theory of Compressible Fluid Flow," Academic, New York, 1958.
7. Wright, J. K.: "Shock Tubes," Methuen, London, 1961.
8. Lin, C. C.: "Liquid Helium," Academic, New York, 1963.
9. Chandrasekhar, S.: "Hydrodynamic and Hydromagnetic Stability," Oxford University Press, Fair Lawn, N.J., 1961.
10. Parker, E. N.: "Interplanetary Dynamical Processes," Wiley, New York, 1963.
11. Stoker, J. J.: "Water Waves," Interscience, New York, 1957.
12. Gurevich, M. I.: "Theory of Jets in an Ideal Fluid," Moscow, 1961.
13. Pasquill, F.: "Atmospheric Diffusion," Von Nostrand, Princeton, N.J., 1961.
14. Birkhoff, G., and E. H. Zaranenello: "Jets, Waves and Cavities," Academic, New York, 1957.
15. Allen, L. H.: "The Atmospheres of the Sun and Stars," Ronald, New York, 1953.
16. Chapman, S., and T. G. Cowling: "Mathematical Theory of Non-uniform Gases," Cambridge University Press, New York, 1949.
17. Reiner, M., and D. Abin (eds.), "Second-order Effects in Elasticity, Plasticity, and Fluid Dynamics," Pergamon Press, New York, 1964.
18. Van Dyke, M.: "Fluid Mechanics Perturbation Methods," Academic, New York, 1964.
19. Sears, W. R.: "Small Perturbation Theory," Princeton University Press, Princeton, N.J., 1960.
20. Eringen, A. C.: "Nonlinear Theory of Continuous Media," McGraw-Hill, New York, 1962.
21. Green, A. E., and J. E. Adkins: "Large Elastic Deformations," Oxford University Press, Fair Lawn, N.J., 1960.
22. Eirich, F. E.: "Rheology: Theory and Applications," Academic, New York, 1956.

23. Nelson, W. C. (ed.): "High Temperature Aspects of Hypersonic Flow," Pergamon Press, New York, 1963.
24. Ferri, A. (ed.): "Fundamental Data Obtained from Shock-tube Experiments," Pergamon Press, New York, 1961.
25. Clauser, F. H. (ed.): "Plasma Dynamics," Addison-Wesley, Reading, Mass., 1960.
26. Riddell, F. R. (ed.): "Hypersonic Flow Research," Academic, New York, 1962.
27. Haltiner, G. J., and F. L. Martin: "Dynamical and Physical Meteorology," McGraw-Hill, New York, 1957.
28. Sutton, O. G.: "Micrometeorology," McGraw-Hill, New York, 1953.
29. Scorer, R. S.: "Natural Aerodynamics," Pergamon Press, New York, 1958.
30. Greene, Edward F., and J. Peter Tolnnies: "Chemical Reactions in Shock Waves," Academic, New York, 1964.

Chapter 3

Rheology

By M. REINER, Israel Institute of Technology

1. Introduction

Rheology is that part of the mechanics of deformable bodies which treats states intermediate between the classical (or Hookian) elastic solid (Part 3, Chaps. 1 and 5) and the classical (or Newtonian) viscous liquid (Part 3, Chap. 2). The basic equation is (1.5) and (1.6) of Part 3. This comprises the three *stress* or *momentum equations* with the six unknown components of the symmetrical stress tensor T_{ij} and is therefore indeterminate. (Cartesian tensors shall be used; for general coordinates, see [10].*)

The solution is made possible through the *rheological equation* of the material under consideration, which expresses T in terms of the displacement vector $\mathbf{u}$ or the velocity vector $\mathbf{v}$ through the tensor of *deformation* D_{ij} or of *flow* f_{ij} (also named *rate of deformation*). When the deformation is infinitesimal (d_{ij}),

$$d_{ij} = \frac{1}{2}\left(\frac{\partial u_i}{\partial x_j} + \frac{\partial u_j}{\partial x_i}\right) \tag{3.1}$$

With

$$v_i = \frac{du_i}{dt} \tag{3.2}$$

we get analogously the *flow tensor*

$$f_{ij} = \frac{1}{2}\left(\frac{\partial v_i}{\partial x_j} + \frac{\partial v_j}{\partial x_i}\right) \tag{3.3}$$

The finite deformation D_{ij} can be defined in different ways.

Equation (3.1) is not suitable because it is not applicable in the case of finite rotations. The question of the measure of finite deformation is treated in Sec. 2.

When the external forces are removed, part of the deformation will always be recovered. This part is *elastic* and is called *strain* (e_{ij}). The measure of infinitesimal strain is given by (3.1) when the displacement u_i is such that it is recovered upon the release of the load.

Having expressed the six components of the strain tensor through the three components of the displacement vector, the rheological equation of the Hooke solid serves to make Eqs. (1.5) and (1.6) of Part 3 determinate. The result is

$$(\lambda + \mu)\frac{\partial \Delta}{\partial x_i} + \mu \nabla^2 u_i + \rho\left(B_i - \frac{dv_i}{dt}\right) = 0 \tag{3.4}$$

* Numbers in brackets refer to references at end of chapter.

The analogous rheological equation of the Newtonian liquid leads to the Navier-Stokes equation, which is the viscous analogue of (3.4).

When the stress in a Hookian solid reaches a certain limit ϑ_{ij}, which determines the strength of the material, the body either fractures or is plastically deformed. In ideal, or *Saint-Venant, plasticity* it is assumed that the plastic deformation proceeds at constant yield stress ϑ.

When acting upon a body, the external forces P perform *work* w_P per unit volume, the specific *power input* being $\dot{w}_P$. Part of this is used to impart to the unit volume in unit time kinetic energy, and the rest is the *stress power.*

$$\dot{w}_s = T_{\alpha\beta}\dot{d}_{\beta\alpha} \tag{3.5}$$

In isotropic materials this can be expressed as the sum of two independent terms, one (the cubical) related to change of volume, the other (the distortional) to change of shape.

The stress power must comply with the two thermodynamical laws. For isothermal processes both are combined in the Gibbs-Helmholtz equation

$$\delta w = \rho\,\delta\phi + \rho\,\delta\Psi \qquad \delta\Psi \geq 0 \tag{3.6}$$

where ϕ is the *intrinsic free energy density,* i.e., after deduction of the kinetic energy produced (effected in viscometry through the *kinetic-energy correction*), and Ψ the *bound energy density.* Equation (3.6) expresses the condition that, while in the case of thermodynamic equilibrium $\delta\Psi$ vanishes, in other cases it can only increase. In the ideal Hooke body, $\delta\Psi$ is supposed to vanish identically, and the equals sign in Eq. (3.6) refers to this ideal case. In the Newtonian liquid the cubical stress power is entirely conserved and the distortional stress power entirely dissipated. Generally, in a rheological process of every *real* material *both* functions ϕ and Ψ will be involved.

The elastic properties of an isotropic material are defined by any two of the following elastic moduli: K, μ, E, σ, λ. The analogous viscous coefficients are ζ, η, λ_T, δ_v, λ_v, where ζ is the coefficient of volume viscosity, defined in Sec. 5, and λ_T is *Trouton's coefficient of viscous traction,* used when the viscosity of a very viscous material, such as bitumen, is determined by extending a prismatic rod and relating rate of extension to the tension. The reciprocal of η is the *fluidity* φ.

Rheology transcends the classical concepts in different directions (see Reiner in [4]) by considering, roughly in chronological order of development: (*a*) a variability of the classical rheological coefficients η and E (or μ) (see Sec. 7), (*b*) materials which exhibit combinations of the fundamental properties of elasticity, viscosity, and plasticity (see Sec. 4), (*c*) volume changes which are not elastic (see Sec. 5), (*d*) second-order effects in elastic solids and viscous liquids (see Sec. 2). In these developments its approach is *phenomenological*, i.e., it treats the materials as they appear to our senses, without taking account of the discrete constitution of matter. Three approximations can be distinguished. In *macrorheology* the materials are treated as homogeneous or quasi-homogeneous and the processes as isothermal. In *microrheology* the heterogeneity of dispersed systems is taken into account. In *metarheology* certain processes are taken into consideration which transcend isothermal mechanics, such as kinetic (gas and rubbery) elasticity, surface tension, rate processes, the theory of which has been developed by Glasstone et al. [25], and psychophysics of rheological measurement, as developed by Scott Blair et al. [9]. The present chapter is confined mainly to macrorheology with a short reference to microrheology in Sec. 7.

2. Second-order Effects in Elasticity and Viscosity

This recent development is treated first because of its intimate connection with classical elasticity and viscosity. It finds its place in rheology because of the nature of the materials which exhibit the effects conspicuously. These materials are mostly high-molecular sub-substances in the form of dispersed systems. Four kinds of second-order effects must be distinguished: (*i*) in Hookian elasticity, (*ii*) cross elasticity, (*iii*) cross viscosity, and (*iv*) in elastic liquids.

i. Keeping Eq. (1.5) of Part 3, Chap. 1, as the definition of Hookian elasticity, this linear tensor equation involves second-order effects as soon as the strain is not infinitesimal, in which case it depends upon the kind of measure adopted. Considering an elementary sphere of unit radius around a particle of the body, this will in general be deformed into a three-axial ellipsoid of half axes $\lambda(\lambda_1,\lambda_2,\lambda_3)$. Any function of λ can be adopted as a measure of deformation which will vanish for $\lambda = 1$ and give the classical Cauchy measure

C: $$e = \lambda - 1 \tag{3.7}$$

as a first approximation. The following other measures have so far been used,

G: $$e = \frac{\lambda^2 - 1}{2} = e + \frac{e^2}{2} \tag{3.8}$$

A: $$e = \frac{1 - 1/\lambda^2}{2} = e - \frac{3e^2}{2} + \cdots \tag{3.9}$$

H: $$e = ln\,\lambda = e - \frac{e^2}{2} + \cdots \tag{3.10}$$

S: $$e = 1 - \frac{1}{\lambda} = e - e^2 + \cdots \tag{3.11}$$

named after Green, Almansi, Hencky, and Swainger, respectively. Braun, Schoenfeld-Reiner, and Traum [17] found that the application of the Swainger measure automatically effects a correction in the Trouton experiment when tensile tractions are referred to the *unstrained* specimen obeying simple Hookian or Newtonian behavior.

If the elastic response of some material can be adequately described by the linear relation referred to above when some such measure is used, the application of any other measure will result in a nonlinear relation with second- and higher-order terms. In addition, two wholly new phenomena not included even approximately in the infinitesimal theory appear: for instance, while the infinitesimal theory yields for simple shear in accordance with the (tangential) displacement gradient,

$$\gamma = \frac{du_x}{dy} \tag{3.12}$$

the shearing stress

$$S = T_{xy} = T_{yx} = \mu\gamma \tag{3.13}$$

with all other stress components vanishing; in the case of finite strain a method indicated by Love [3] shows that normal stress components are necessary as well in order to maintain equilibrium. These are listed in Table 3.1.

As can be seen, in order to maintain simple shear, pressures T_{yy} and tensions T_{xx} resulting in an isotropic pressure $p + T_{zz}$ are necessary in addition to the shearing stress T_{xy}, which, moreover, is not proportional with γ. Contrariwise, should these normal tractions be absent, there will be an extension in the y direction, contraction in the x direction, and cubical dilatation.

ii. The developments of subsection *i* show that the theory embodied in the linear stress equation prejudices experimental results. If *some* definite measure of strain is assumed, this, for instance, carries with it a definite distribution of pressures over the ends of a cylinder in torsion, which may or may not be confirmed by experiment. This drawback can possibly be overcome by searching after a suitable measure, e.g., by combining some of the measures listed above. Alternatively μ need not be a constant but could be a function of the three strain invariants. Such variability is dealt with in Sec. 6. There is, however, another difficulty which cannot be overcome in this manner. All measures result in *pressures* at the ends of the cylinder or *elongations* when the ends are stress-free. While it is true that this is in conformity with the scanty experimental investigations on large strains in steel, copper, and rubber, there is no a priori reason known why some other material should not show *shortening* of a cylinder in torsion. Experiments by Swift [57] have in fact shown that in plastic torsion, rods made of steel or copper lengthen but those of lead shorten. Inasmuch as the plastic deformation results from a freezing in of elastic strains, this points to the fact that the linear stress equation is not general enough. It has been generalized by Reiner [42] to

$$T_{ij} = F_0\delta_{ij} + 2\mu e_{ij} + 4\mu_c e_{i\alpha}e_{\alpha j} \tag{3.14}$$

TABLE 3.1. STRESSES IN FINITE SIMPLE SHEAR

$\alpha = (1 + \gamma^2/4)^{1/2}$

Stress \ Strain	C: $e = \lambda - 1$	G: $e = \frac{\lambda^2 - 1}{2}$	H: $e = \ln \lambda$	A: $e = \frac{1 - 1/\lambda^2}{2}$	S: $e = 1 - \frac{1}{\lambda}$
$\frac{T_{xx} - T_{zz}}{\mu}$	$\frac{1 + \gamma^2/2}{\alpha} - 1$	$\frac{\gamma^2}{2}$	$\frac{\gamma}{4\alpha} \ln \frac{\alpha + \gamma/2}{\alpha - \gamma/2}$	0	$1 - \frac{1}{\alpha}$
$\frac{T_{yy} - T_{zz}}{\mu}$	$\frac{1}{\alpha} - 1$	0	$-\frac{\gamma}{4\alpha} \ln \frac{\alpha + \gamma/2}{\alpha - \gamma/2}$	$-\frac{\gamma^2}{2}$	$1 - \frac{1 + \gamma^2/2}{\alpha}$
$\frac{T_{xy}}{\mu}$	$\frac{\gamma}{2\alpha}$	$\frac{\gamma}{2}$	$\frac{1}{2}\alpha \ln \frac{\alpha + \gamma/2}{\alpha - \gamma/2}$	$\frac{\gamma}{2}$	$\frac{\gamma}{2\alpha}$
$-\frac{(p + T_{zz})}{\mu}$	$\frac{2}{3}(\alpha - 1)$	$\frac{\gamma^2}{6}$	0	$-\frac{\gamma^2}{6}$	$\frac{2}{3}(1 - \alpha)$

with a second-order term in the strain tensor. The coefficient μ_c has been named modulus of *cross elasticity*. *Any* measure of strain can now be used, and as μ_c may have any magnitude and sign, the second-order term may imply either pressures or tensions at the cylinder ends and therefore either extension or shortening in the absence of normal stress [45].

iii. Second-order effects of this kind cannot arise in classical viscosity. In viscous flow the amount of finite deformation reached at any time is of no physical significance. The viscous resistance at any moment t depends upon the gradient of the *infinitesimal* displacement during the time element dt following t. Here the second-order effects result from a generalization of Eqs. (1.5) and (1.38) of Part 3, Chap. 1, in analogy with (3.14), to

$$T_{ij} = F_0\delta_{ij} + 2\eta f_{ij} + 4\eta_c f_{i\alpha} f_{\alpha j} \qquad (3.15)$$

Equation (3.15) was derived by Reiner [41] as the most general law of a viscous liquid. The factor η_c has been named coefficient of *cross viscosity*.

Two second-order effects are present:

a. In order to maintain simple shear, an isotropic tension or compression is necessary, in accordance with the sign of η_c. If this is absent, there will result a cubical dilatation, named *dilatancy* to distinguish it from the cubical dilatation due to an isotropic stress.

b. There are *cross stresses* in the direction of flow x and in the direction of its gradient y, both being algebraically equal. Braun and Reiner [16] have solved the problems of flow of a liquid following Eq. (3.15). In the case of torsional flow between two parallel circular platens of radius R at distance H from each other, one at rest and the other rotating with angular velocity Ω, one finds from the application of Eq. (3.15)

$$T_{rr} = -\Omega^2(R^2 - r^2)\frac{\eta_c}{2H^2}$$
$$T_{zz} = -\Omega^2(R^2 - 3r^2)\frac{\eta_c}{2H^2} \qquad (3.16)$$

If the upper platen has an opening in the center to which a vertical tube is attached, the liquid will be forced by the pressure T_{rr} to flow radially toward the center against the action of centrifugal forces, and, arriving at the center, it will be forced by the pressure T_{zz} to rise in the tube against the action of gravity. The whole arrangement thus forms a centripetal pump.

Note that $T_{zz} = 0$ for $R/\sqrt{3}$ and that for $r = R$, T_{rr} vanishes, while T_{zz} is a tension. This provides a criterion for the occurrence or otherwise of cross viscosity. Greensmith and Rivlin [29] claim to have proved experimentally the existence of a cross-viscosity effect in solutions of polyisobutylene, but Roberts [50] repeated their experiments with contradictory results.

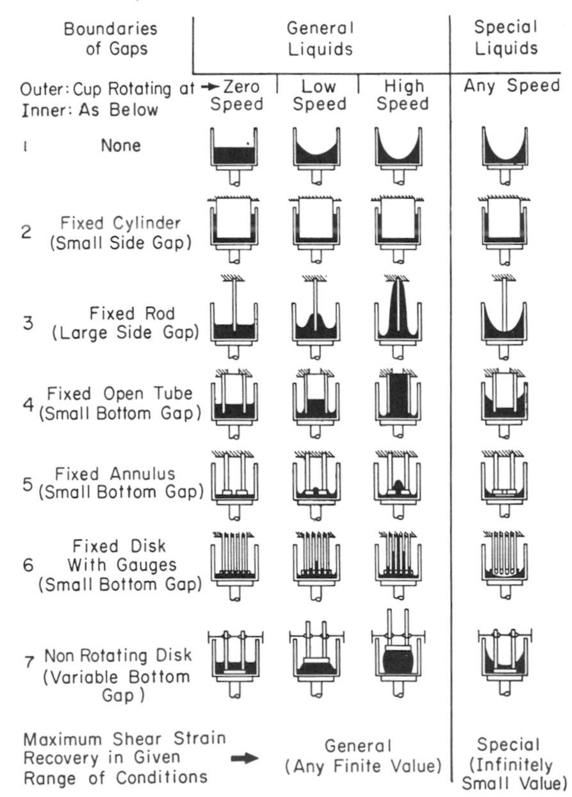

FIG. 3.1. Centripetal pump effects as demonstrated by Weissenberg (1947). Weissenberg calls Newtonian liquids *special*, and elastic liquids *general*. This is justified when considering that the former result from the Maxwell liquid when $\mu_1 = \infty$.

iv. Elastic liquids are treated below in Sec. 4. Here brief mention will be made of a single aspect which enlisted the interest of the rheologists in the second-order effects. Weissenberg [61] demonstrated a series of striking phenomena, shown diagrammatically in Fig. 3.1. They constitute essentially a centripetal-pump effect, as described in subsection *iii*. They could therefore be attributed to cross viscosity. This attitude has been taken by Rivlin [49]. All liquids which Weissenberg mentions (cf. [47, 34]) exhibit elasticity of shape under the action of transient forces, and Weissenberg maintains that the phenomena are due to the elasticity of the dispersed phase. When using the Green measure, this implies that T_{zz} vanishes at the outer edge of torsional flow, a condition in contradiction with cross viscosity [see Eq. (3.16)]. The problem has not yet found its definite solution. There is no reason why the Weissenberg effect, which is a gross phenomenon, should not be due to different micromechanisms in different materials. For instance, any difference in the tension and compression behavior of the dispersed phase partaking in the flow of the liquid medium would give rise to cross stresses in shear. If the dispersed phase were to consist of small threads which can take up tension but collapse under compression, this would constitute such difference. This may be the mechanism in macromolecular solutions.

3. Rheological Properties

Rheological properties are of two kinds:

a. Intrinsic rheological properties are those the investigation of which has reached the stage in which they can be exactly defined as parameters in a postulated rheological equation. Intrinsic properties are either *fundamental* or *complex*. There are four fundamental properties, viz., elasticity, viscosity, plasticity (internal solid friction), and strength. Complex properties result from combinations of the fundamental properties. Only for those combining *two* fundamental properties have special designations been coined, viz., firmoviscosity, elasticoviscosity, and plasticoviscosity. Certain other properties for which designations exist, such as *elastic fore-effect, elastic aftereffect, relaxation*, etc., are not independent properties but are derived properties which can be expressed in terms of intrinsic properties (see Sec. 4).

b. Technological properties are those for which a method of measurement has been devised, the theoretical investigation of which has, however, not yet reached a stage where the property is either shown to be fundamental or can be expressed in terms of known fundamental properties. The result of the measurement is an *index*, which may be of relative significance or which serves only as an identification number. Examples are *penetration, ductility, tack* (for a derivation of tack from a combination of viscosity and yield stress, see [26]), *thixotropy* [28], etc. Investigation of other technological properties has not even reached the stage where they can be measured in any apparatus. Such are *seepage, covering power of a pigment*, and others.

The present chapter deals only with intrinsic properties defined as parameters (coefficients, moduli, etc.) in rheological equations. In establishing a rheological equation from experimental observations, it is of great help to build, even if only in imagination, a mechanical model which will behave qualitatively in a manner similar to that of the actual material in some degree of approximation and to describe that behavior in terms of forces and elongations. If these terms are translated into stresses and deformations, the result easily leads to a rheological equation,

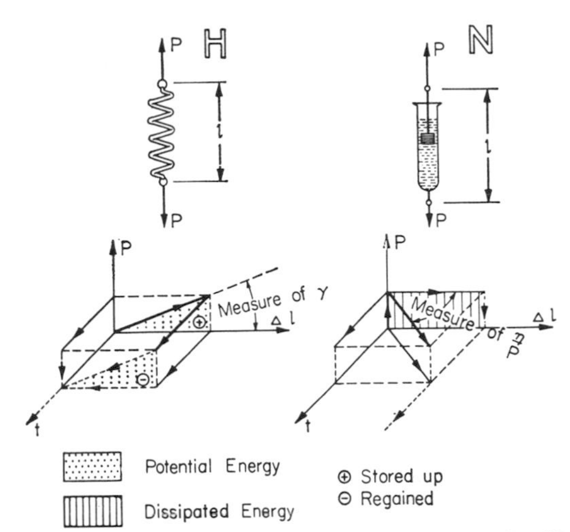

FIG. 3.2. Model of Hooke-solid and Newtonian liquid.

a procedure which constitutes the heuristic value of the model. For the three fundamental bodies the following models suggest themselves:

i. A helical steel spring H (Fig. 3.2) to represent elastic Hookian behavior which in terms of simple shear is defined by

$$S = \mu\gamma \tag{3.17}$$

where S is the shearing (tangential) stress.

ii. A dashpot, e.g., in the form of a test tube filled with a viscous oil in which a stopper is loosely fitted, N (Fig. 3.2), to represent viscous Newtonian behavior defined by

$$S = \eta\dot{\gamma} \tag{3.18}$$

iii. A weight resting on a table top with solid friction between both and pulled via a Hookian spring, StV (Fig. 3.3), to represent plastic Saint-Venant behavior in which the deformation proceeds under constant *yield stress* ϑ, so that

$$S = \vartheta \tag{3.19}$$

while if $S < \vartheta$, the rheological equation (3.17) applies.

These elements can be coupled either in parallel (|) or in series (—). When coupled in parallel, the loads taken by each one of the elements are additive, while the rates of elongation of both are the same. When coupled in series, the rates of elongation are additive, while each takes the same total load. While models work with elongations Δl under pulls P, they can serve to represent not only linear dilatation but also shear or cubical dilatation.

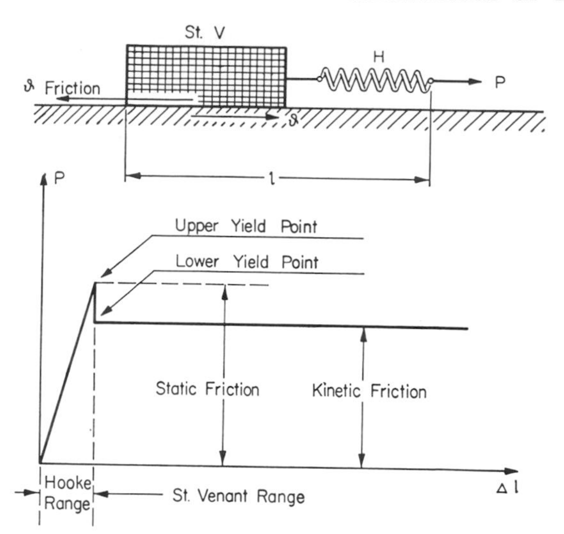

FIG. 3.3. Model for Saint-Venant plastic solid.

4. Complex Bodies

Complex materials are represented by combinations of two or more elements. *Double* bodies are:

iv. The *Kelvin body,* which defines *firmoviscosity.* It is shown in model in Fig. 3.4. This can be written in shorthand as

$$K = H|N \tag{3.20}$$

Such an equation is called a *structural formula.* By combining (3.17) and (3.18), adding stresses, we find the rheological equation

$$S = S_H + S_N = \mu\gamma + \eta_s\dot{\gamma} \tag{3.21}$$

A model for this material was first proposed by Kelvin [33] in a different form, viz., as a sponge imbibed with water. The rheological equation (3.21) was first postulated by Voigt [60]. It has been used to describe the damping of free oscillations of solids, e.g., of torsional oscillations of metal wires. Quantitative agreement can be obtained when it is assumed that the material is composed of a *spectrum* of Kelvin bodies coupled in series. It should be noted that a number of Kelvin bodies coupled in parallel are equivalent to one single Kelvin body.

The materials with which rheology has to deal are mostly dispersed systems composed of solid and liquid phases. In such solid-liquid systems the H element may stand for the solid, the N element for the liquid phase. When explaining solid viscosity, Kelvin considered a gel-like structure, but Voigt postulated Eq. (3.21) for a homogeneous one-phase material. As the designation *Voigt body* is sometimes used for a material represented by the model of Fig. 3.4, this is a case where the symbolic character of the models becomes evident. They should not be taken literally.

Equation (3.21) is a linear differential equation in γ, the integral of which is

$$\gamma = \exp\left(-\frac{\mu}{\eta_s}t\right)\left[\gamma_0 + \frac{1}{\eta}\int S \exp\left(\frac{\mu}{\eta_s}t\right)dt\right] \tag{3.22}$$

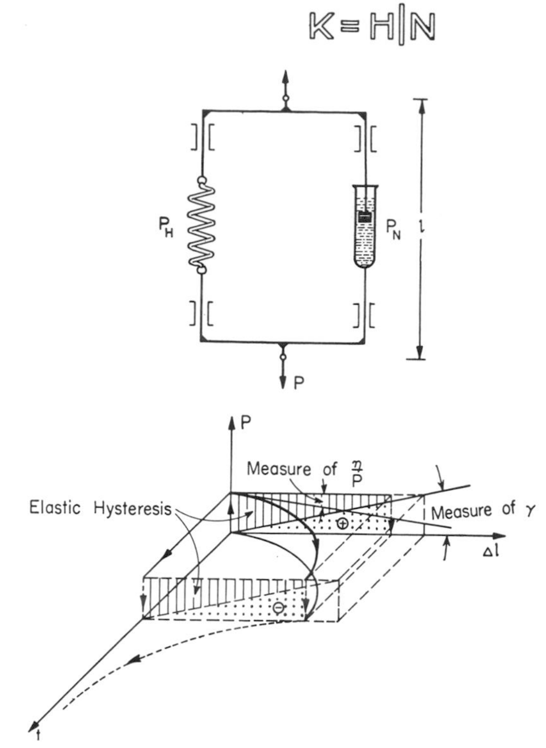

FIG. 3.4. Model for Kelvin solid $P = P_H + P_N$.

where γ_0 is the *initial* strain. Let the stress S be constant $= S_0$. Equation (3.22) then gives

$$\gamma = \frac{S_0}{\mu} + \left(\gamma_0 - \frac{S_0}{\mu}\right)\exp\left(-\frac{t}{t_{\text{ret}}}\right) \tag{3.23}$$

where we have introduced

$$t_{\text{ret}} = \eta_s\mu \tag{3.24}$$

If $\gamma_0 = S_0/\mu$, there will be stationary equilibrium, as with a Hooke body. The Kelvin body is accordingly a solid, and its viscosity is *solid* viscosity, indicated by the subscript s in η_s. On the other hand, if $\gamma_0 = 0$, constant stress will produce a continuous deformation proceeding in time, albeit at a decreasing rate. This, when t_{ret} is very large, may simulate slow flow. It is therefore sometimes spoken of as (primary) *creep.* Finally (for $t = \infty$) the shear S_0/μ is reached. The strain therefore does not appear instantaneously, as in the Hooke body, but is delayed in an *elastic fore-effect* with t_{ret} as time of *retardation.* If the stress is removed ($S_0 = 0$), the strain vanishes in an elastic *aftereffect* (or *creep recovery*), theoretically at $t = \infty$, but if t_{ret} is not too large, practically at finite time. Both elastic fore- and aftereffect constitute delayed or *retarded elasticity.*

v. The *Maxwell body,* which defines *elasticoviscosity,* is shown in model in Fig. 3.5. Its structural formula is

$$M = H\text{—}N \tag{3.25}$$

By taking time derivatives on both sides of (3.17)

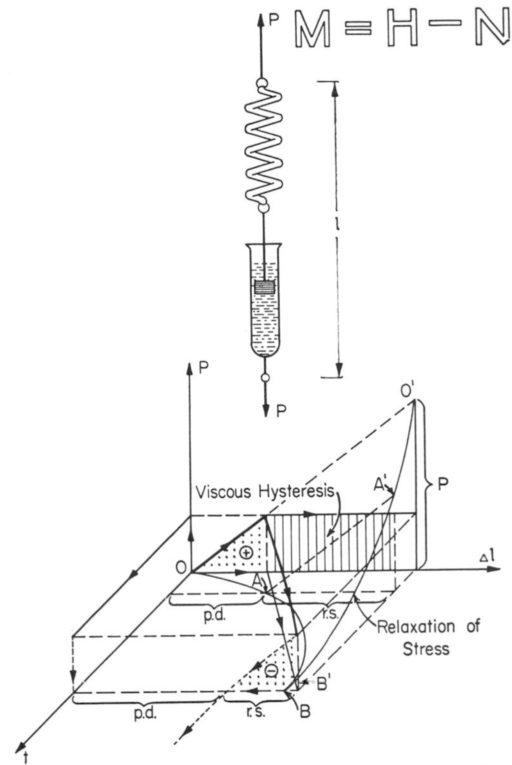

FIG. 3.5. Model for Maxwell liquid $\Delta l = \Delta l_H + \Delta l_N$.
p.d. = permanent deformation
r.s. = recovered strain

and adding rates of deformation, we find as its rheological equation

$$\dot{\gamma} = \dot{\gamma}_H + \dot{\gamma}_N = \frac{S}{\mu_l} + \frac{S}{\eta} \tag{3.26}$$

Equation (3.26) yields on integration

$$S = \exp\left(-\frac{\mu_l}{\eta}t\right)\left[S_0 + \mu_l \int \dot{\gamma} \exp\left(\frac{\mu_l}{\eta}t\right) dt\right] \tag{3.27}$$

where S_0 is the initial stress. Let the rate of shear $\dot{\gamma}$ be constant $= \dot{\gamma}_0$. Equation (3.27) then gives

$$S = \dot{\gamma}_0\eta + (S_0 - \dot{\gamma}_0\eta) \exp\left(-\frac{t}{t_{\text{rel}}}\right) \tag{3.28}$$

where
$$t_{\text{rel}} = \frac{\eta}{\mu_l} \tag{3.29}$$

If $S_0 = \dot{\gamma}_0\eta$, there will be steady flow with the internal stress in equilibrium with the load, as with a Newtonian liquid. The Maxwell body is accordingly a liquid, and its elasticity is liquid elasticity, indicated by the subscript l in μ_l. When the Maxwell body is deformed by some stress S_0 and the deformation is kept constant ($\dot{\gamma}_0 = 0$), the stress diminishes in time with t_{rel} as time of *relaxation*. When the time of relaxation is large, the flow, as seen from $\dot{\gamma}_0 = S_0/\mu_l t_{\text{rel}}$, is very small, and one speaks of (secondary) creep. This creep is not recovered; on the removal of the load the deformation stays put.

The rheological equation of the Maxwell body has been used for the description of the creep of glasses and high-polymer substances. Here also quantitative agreement requires the assumption of a spectrum of Maxwell bodies, coupled in parallel. A number of Maxwell bodies coupled in series are equivalent to another Maxwell body.

Maxwell calculated for air a relaxation time of the order of 10^{-9} sec. Relaxation times for sols (pitch, benzopurpurine, etc.) are of the order of 10^2 sec, of such gels as concrete and glass of the order of 10^7 sec.

vi. The *Bingham body,* which defines *plastico-viscosity,* has the structural formula

$$B = StV\text{—}N \tag{3.30}$$

By combining (3.18) and (3.19) we find

$$\dot{\gamma} = \frac{S - \vartheta}{\eta_{pl}} \tag{3.31}$$

The Bingham body is a plastic solid which resists the plastic flow not only with a *static friction,* as does the Saint-Vénant body, but also with a *plastic* viscosity η_{pl}. It was introduced by Bingham [13] to describe the rheological behavior of concentrated clay suspensions and later used for oil paints [14], flour dough [63], and other soft plastic solids, where the rate of deformation is not negligible (as it is in the plastic deformation of metals).

In this manner intrinsic rheological properties can be systematically expressed in a hierarchy of ideal bodies. Figure 3.6 shows a "tree" of rheological bodies, together with their structural formulas and rheological equations. The latter are written in terms of simple shear; when a general distortion is under consideration, the deviators of the tensors must be used in place of shearing stress and shearing deformation. In the case of linear dilatation, replace S, γ, μ, η, by T, ϵ, E, λT; in the case of cubical dilatation by $-p$, Δ, K, ζ.

The ideal materials listed in Fig. 3.6 were postulated by their authors to describe the rheological behavior of the following materials: Schwedoff [54], a 0.5 per cent gelatine solution; Poynting and Thomson [38], glass fibers (this ideal material has recently been used to represent *anelasticity* in metals); Lethersich [36], elastic bitumen; Jeffreys [31], for the description of the rheological behavior of the earth's crust; Burgers's [19] was found by Reiner [45] to be suitable for concrete; Trouton and Rankine [59], lead wire in torsion; and Schofield and Scott Blair [51], for flour dough.

If the rheological equation of the J body is differentiated in respect of time and the integral eliminated between the original and the differentiated equation, the resulting equation is

$$S + \dot{S}t_{\text{rel}} = \eta(\dot{\gamma} + \ddot{\gamma}t_{\text{ret}}) \tag{3.32}$$

Other equations listed in Fig. 3.6 can also be brought into this sometimes more convenient form, which is the one used by Jeffreys.

Number of Elements	LIQUID	SOLID	
	Viscous Flow	Plastic Flow	Elastic Strain
1	N (Dashpot) 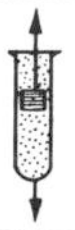 $\dot{\gamma} = s/\eta$ 0	St V (Friction Weight) 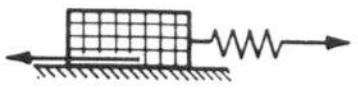 $s - \vartheta = 0$ 0	H (Spring)  $s = \mu\gamma$ 0
2	$M = N - H$ $\dot{\gamma} = s/\eta + s/\mu_1$ $1/\mu_1$	$B = St\ V - N$ $\dot{\gamma} = (s - \vartheta)/\eta_{pl}$ 0 η_{pl}	$K = H \mid N$ $s = \mu\gamma + \eta_s\dot{\gamma}$ η_s
3	$J = N - K$ $(L = N \mid M)$ $\dot{\gamma} = s\frac{\eta + \eta_s}{\eta\eta_s} - \frac{\mu}{\eta_s} e^{-\mu/\eta_s t}\left(\gamma_0 + \frac{1}{\eta_s}\int_0^t s e^{\mu/\eta_s t} dt\right)$ 0	$Schw = B - H$ $\gamma = (s - \vartheta)/\eta_{pl} + s/\mu_1$ $1/\mu_1$	$PTh = H \mid M$ $s = \mu\gamma + e^{-\mu_1/\eta t}\left(t_0 + \mu_1\int_0^t \dot{\gamma} e^{\mu_1/\eta t} dt\right)$ 0
4	$Bu = M - K$ $TR = N - PTh$ $\dot{\gamma} = s\frac{\eta + \eta_s}{\eta\eta_s} - \frac{\mu}{\eta_s} e^{-\mu/\eta_s t}\left(\gamma_0 + \frac{1}{\eta_s}\int_0^t s e^{\mu/\eta_s t} dt\right) + \dot{s}/\mu_1$ $1/\mu_1$	$B - K$ 0	K M H L η_s
5		$Sch\ ScB = Schw - K$ $\dot{\gamma} = (s - \vartheta)\frac{\eta_{pl} + \eta_s}{\eta_{pl}\eta_s} - \frac{\mu}{\eta_s} e^{-\mu/\eta_s t}\left[\gamma_0 + \frac{1}{\eta_s}\int_0^t (s - \vartheta) e^{\mu/\eta_s t} dt\right] + \dot{s}/\mu_1$ $1/\mu_1$	

FIG. 3.6. Tree of rheological bodies.

5. Volume Changes

As a first approximation, the *volumetric rheological equation* for *all materials*, whether solids or liquids, is

$$-p = K\Delta \tag{3.33}$$

where p is the *hydrostatic* or *thermodynamic static* pressure and K is the *bulk modulus*.

The elastic straining of a solid body will in general be accompanied by a viscous resistance, which can be neglected only if the rate of strain is sufficiently small. In analogy with the two moduli of elasticity referring to change of volume K and change of shape μ, there will be two coefficients of *solid viscosity*, one, ζ_s, accompanying volume strain, and the other, η_s, accompanying shearing strain. The latter appears in the rheological equation (3.21) of the Kelvin body. Inasmuch as every liquid has volume elasticity, it will also have solid volume viscosity As a second approximation one therefore replaces (3.33) for both solids and liquids by

$$-p_m = K\Delta + \zeta_s\dot{e}_v \tag{3.34}$$

where p_m, the mean pressure, is now different from p. As mentioned in Sec. 8 of Part 3, Chap. 1, in classical hydrodynamics ζ_s is supposed to vanish. The coefficient ζ_s is known as the second coefficient of viscosity in liquids [32]. Liquids may, however, possess a third coefficient of viscosity (or liquids and solids alike a second coefficient of volume viscosity). It is found that in certain dispersed systems a constant isotropic stress will cause volume flow, i.e., a continuous cubical dilatation or change of density of the material progressing in time. Positive volume flow was found by Lee, Reiner, and Rigden [35] and observed quantitatively by Reiner, Rigden, and Thrower [46], and negative volume flow was found by Glanville and Thomas [24] in concrete. Bosworth [15] has observed volume flow in solidified carbon dioxide. These materials show slow flow or creep, as in a Maxwell body, with viscosities η of the order of 10^{12}, 10^{17}, and 10^{10} poises. Conservation of mass requires, of course, that volume flow be connected with an increase or decrease of voids in the material. As, in accordance with Eyring [23], "a liquid is a binary mixture of molecules and holes," volume flow in liquids may have to be taken into account when pressures are exceedingly high, as in geophysical problems. Gases in explosions may also show a volume-flow effect.

When there is volume flow, a *coefficient of liquid volume viscosity* ζ_l is defined by

$$f_v - f_{\alpha\alpha} = -\frac{p_m}{\zeta_l} \tag{3.35}$$

In analogy with the relations between the moduli of elasticity, there will be the relation

$$\lambda_T = \frac{9\zeta_l\eta}{3\zeta_l + \eta} \tag{3.36}$$

For volume flow to be absent, $\zeta_l = \infty$. In this case from (3.36)

$$\lambda_T = 3\eta \tag{3.37}$$

The relation (3.37) was derived by Trouton [58] and is correct for very dense, highly viscous materials, e.g., bitumen.

When both solid and liquid volume viscosity are present, the volumetric rheological behavior of the material can be represented by the Bu model. Holes or pores in solids give rise also to *volume plasticity.* The whole complex of volume rheology is treated by Reiner in [5].

6. Strength

Strength, one of the fundamental rheological properties, is governed by the maximum energy, also called *resilience* (R), which a volume element of the material of which the body consists can absorb as free energy [48].

In accordance with the resolution of work (see Sec. 1), every material possesses two independent kinds of strength, one referring to a change of volume, the other to distortion. The two criteria for strength are accordingly

$$\phi_v \leq R_v \qquad \phi_{(0)} \leq R_{(0)} \tag{3.38}$$

Failure on volume strength causes brittle *fracture by separation;* failure on distortional strength, either *plastic flow* or *fracture by glide.* The volume strength against isotropic pressure is theoretically infinite. In practice materials will fail through local crushing around pores or holes. In isotropic tension there is a limiting stress corresponding to a limiting dilatation.

For the Hooke body we have for the deviator

$$\dot{e}_{(0)} = \frac{\dot{T}_{(0)}}{2\mu} \tag{3.39}$$

and in accordance with (3.5)

$$\dot{w}_{(0)} = T_{(0)}\dot{e}_{(0)} = \frac{d[T_{(0)}]^2}{4\mu}\,dt \tag{3.40}$$

Since for the Hooke body $\delta\Psi$ vanishes [16], there results the flow *condition*

$$\phi_{(0)} = \frac{\vartheta_{(0)}{}^2}{4\mu} = R_{(0)} \tag{3.41}$$

where $\vartheta_{(0)}$ is the deviator of the *yield stress.* But

$$\vartheta_{(0)}{}^2 = \vartheta_{(0)\alpha\beta}\vartheta_{(0)\beta\alpha} \tag{3.42}$$

and we so get *Hencky's flow condition*

$$\vartheta_{(0)\alpha\beta}\vartheta_{(0)\beta\alpha} = 4\mu R_{(0)} \tag{3.43}$$

If we introduce

$$\vartheta_{(0)\alpha\beta}\vartheta_{(0)\beta\alpha} = 2\text{II}_{\vartheta(0)} \tag{3.44}$$

where II is the second invariant of the tensor, we get *Mises' flow condition.* If the material is brittle, the breaking stress $\bar{T}_{(0)}$ can be introduced for $\vartheta_{(0)}$, and we get *Huber's condition of fracture.*

The Mises-Hencky plastic-flow condition and Huber's fracture condition are expressions of a statical theory, in which neither the rate of deformation nor the rate of stressing appears. When $\delta\Psi$ does not vanish, we find from the application of (3.38) dynamical strength phenomena as follows: (*i*) For the Kelvin body

$$\phi_{(0)} = e_{(0)}{}^2\mu \tag{3.45}$$

and therefore as condition of yield or fracture

$$\bar{e}_{(0)} = \sqrt{\frac{R_{(0)}}{\mu}} \tag{3.46}$$

or

$$\bar{T}_{(0)} = 2\sqrt{\mu R_{(0)}} + 2\eta_s\dot{e}_{(0)} \tag{3.47}$$

A Kelvin body accordingly fails when the *strain* $e_{(0)}$ reaches a definite limit $\bar{e}_{(0)}$. The stress at which the material fails $\bar{T}_{(0)}$ increases with the rate of strain $\dot{e}_{(0)}$. This is the case with mild steel. Note that the tangential component of strain is one-half of the displacement gradius, or $=\gamma/2$. (*ii*) Similarly for the Maxwell body:

$$\bar{T}_{(0)} = 2\sqrt{\mu_l R_{(0)}} \tag{3.48}$$

and

$$\bar{\dot{e}}_{(0)} = \sqrt{R_{(0)}\mu_l\eta^2} + \frac{\dot{T}_{(0)}}{2\mu_l} \tag{3.49}$$

A Maxwell body fails when the stress $T_{(0)}$ reaches a definite limit $\bar{T}_{(0)}$. The rate of strain at which the material fails $\bar{\dot{e}}_{(0)}$ increases with the rate at which the stress is applied $\dot{T}_{(0)}$. This is the case with materials showing (secondary) creep, such as plastics.

7. Microrheological Aspects

In isotropic materials the rheological coefficients must be scalars. They will therefore in general be functions of the three principal invariants of either stress or deformation or of a joint invariant such as work or power. The latter aspect has been elaborated in Sec. 6 in respect of strength. The dependence of the bulk modulus K upon the first invariant is expressed in empirical formulas such as those established by Bridgman [18]. The variability of the coefficient of viscosity η is of importance in colloid physics.

When η varies with the state of stress of flow, the liquid is said to be non-Newtonian possessing *structural* viscosity. This may be genuine, in which case η or its reciprocal, the fluidity φ should be a function of the power input, rising from the *fluidity at rest* φ_0 to the *maximum fluidity* φ more or less rapidly in accordance with a *stability coefficient* χ. Various expressions for the function have been proposed, none sufficiently supported. The variability of φ may, however, also be spurious and due to the rotation of the principal axes of deformation in shear [20] in a viscoelastic liquid in accordance with the phenomenon mentioned in Sec. 2*i*.

When the liquid is a solution, its viscosity can, in principle, be calculated from the hydrodynamics of the flow of a Newtonian liquid past solid bodies of different shapes. Einstein [21, 22], assuming a very dilute suspension of rigid spheres, calculated

$$\eta = \eta_{\text{solv}}(1 + 2.5c_v) \tag{3.50}$$

where η_{solv} is the viscosity of the solvent or continuous phase. His treatment has been generalized to include higher concentrations and other shapes of particles [55].

8. Rheometry

Rheological properties are determined quantitatively in rheometers (plastometers), for which the viscometers are the prototype, either with their *absolute* value or relatively to the magnitude of the same property in a standard material. These instruments are of three types: in type I the material under test is subjected to pure homogeneous deformation; in type II it is subjected to laminar semihomogeneous shear; in type III it is deformed in streamlined flow of more complicated type. Examples for I are the tensile test of mild steel, the compression test of concrete, and Trouton's viscous traction. End effects must be taken into account, but otherwise this is the theoretically most satisfactory method. Examples for II are capillary-tube and rotating-coaxial-cylinder viscometers. Because of nonhomogeneity of deformation, these require either *integration* of a postulated rheological equation or *differentiation* of empirical results [39, 37, 30].

An example for III is the free fall of a small heavy sphere of radius r and density ρ_2 in the material under investigation. This has been solved only for the case of a viscous liquid of density ρ_1. The equation connecting the velocity v, when it has become constant, with the viscosity η of the liquid

$$v = \tfrac{2}{9} r^2(\rho_2 - \rho_1)\frac{g}{\eta} \tag{3.51}$$

is known as *Stokes' law.*

In instruments of types I and II the method of testing may be either *static*, under conditions of equilibrium, or *dynamic*, with either loads or deformations altering in a given manner.

In apparatus of type I, such as the extensometer, the cross section of the cylindrical or prismatic specimen gradually decreases. When the load is kept constant, the tensile traction therefore gradually increases. Different arrangements exist to reduce the load in such a way that the traction remains constant [11, 12, 53]. However, as mentioned in Sec. 2*i*, if the material is a Kelvin body or one of its elements and the extension is plotted in the Swainger measure against time, the resulting curve depends only upon the initial traction even if this changes under constant load. This has been used in measurements on elastic bitumen.

In apparatus of type II, equations must be expressed in cylindrical coordinates: r, θ, z. There is axial symmetry, and as the lamina are hollow cylinders moving as wholes, u_r vanishes, and u_θ and u_z are functions of r only. The displacement can consist of:

a. A translation parallel to the z axis (telescopic movement), where

$$u_\theta = 0 \qquad u = u_z = f_1(r) \qquad \gamma_{zr} = f_1'(r) \tag{3.52}$$

b. A rotation around the z axis, where

$$u_z = 0 \qquad u = u_\theta = f_2(r) \qquad \gamma_{r\theta} = f_2'(r) - \frac{f_2(r)}{r} \tag{3.53}$$

while all other strain components vanish.

If u is replaced by v, the flow components take the place of the strain components. One introduces these into the rheological equations and finds expressions for the stress components. When the latter, in turn, are introduced in the dynamical equations, these can be integrated and yield, for instance, for the Bingham body:

a. The *Buckingham-Reiner equation*

$$\frac{Q}{t} = \frac{\pi R^4\,\Delta p}{8\eta_{pl} l}\left[1 - \frac{4}{3}\frac{2l}{R\,\Delta p} + \frac{1}{3}\left(\frac{2l\vartheta}{R\,\Delta p}\right)^4\right] \tag{3.54}$$

where Q is the quantity of material flowing through a capillary of radius R and length l under the pressure head Δp in unit time. This equation was found applicable by Scott Blair and Crowther [52] to clay and soil pastes and by Wolarowitch, Kulakoff, and Romansky [62] to peat suspensions.

b. The *Reiner and Riwlin equation*

$$\dot{\Omega} = \frac{M}{4\pi l \eta_{pl}}\left(\frac{1}{R_i^2} - \frac{1}{R_e^2}\right) - \vartheta/\eta_{pl} \cdot \ln\frac{R_e}{R_i} \tag{3.55}$$

where $\dot{\Omega}$ is the velocity of rotation of the external cylinder of radius R_e, the internal cylinder of radius R_i being kept stationary through the application of the torque M, while l is the wetted length of the internal cylinder. This equation was found applicable by Wolarowitch and Tolstoi [64] to clay suspensions and by Wolarowitch and Samarina [63] to flour dough. Green [27] investigated different inks in a high-speed apparatus and found complete confirmations. For vanishing ϑ, Eq. (3.54) is reduced to the *Poiseuille equation* and Eq. (3.55) to the *Margules equation* of Newtonian liquids. For non-Newtonian liquids, power developments have been used (see Reiner in [4]).

References

Textbooks and Surveys Containing Original References

1. Committee for the Study of Viscosity of the Academy of Sciences at Amsterdam: "First Report on Viscosity and Plasticity," Amsterdam, 1935; "Second Report," Amsterdam, 1938.
2. Green, H.: "Industrial Rheology and Rheological Structures," New York and London, 1949.
3. Love, A. E. H.: "A Treatise on the Mathematical Theory of Elasticity," 4th ed., Cambridge, 1927.
4. *Proc. Intern. Rheol. Congr.*, 1*st Congr., Amsterdam, 1948*, 1949.
5. *Proc. Intern. Rheol. Congr.*, 2*d Congr., London, 1953*, 1954.
6. Reiner, M.: "Deformation, Strain and Flow," Lewis, London, 1960.
7. ———: A survey of developments in rheology, *Appl. Mechanics Revs.*, **4**: 202 (1951).
8. ———: "Lectures on Theoretical Rheology," Interscience, New York, 1960.
9. Scott Blair, G. W.: "A Survey of General and Applied Rheology," 2d ed., London, 1949.
10. Truesdell, C.: The mechanical foundations of elasticity and fluid dynamics, *J. Rat. Mechanics Anal.*, **1**: 125 (1925).

Original Papers

11. Andrade, E. N. da C.: *Proc. Roy. Soc. (London)*, **A84**: 1 (1911).
12. ——— and B. Chalmers: *Proc. Roy. Soc. (London)*, **A138**: 348 (1932).
13. Bingham, E. C.: *Natl. Bur. Standards Bull.*, **13**: 309 (1916).
14. ——— and H. Green: *Am. Soc. Testing Materials Proc., Pt. II*, **19**: 640 (1919).

15. Bosworth, R. C. L.: *Australian J. Sci. Research*, **A2**: 394 (1949).
16. Braun, I., and M. Reiner: *Quart. J. Mechanics Appl. Math.*, **5**: 42 (1952).
17. ———, R. Schoenfeld-Reiner, and E. Traum: *Bull. Research Council Israel*, **2**: 89, 332 (1952).
18. Bridgman, P. W.: *Proc. Am. Acad. Arts Sci.*, **58**: 166 (1923).
19. Burgers, J. M.: in [1].
20. ———: *Koninkl. Ned. Akad. Wetenschap. Proc.*, **51**: 787 (1948).
21. Einstein, A.: *Ann. Physik*, **19**: 289 (1906).
22. ———: *Ann. Physik*, **34**: 591 (1911).
23. Eyring, H.: *J. Chem. Phys.*, **4**: 283 (1936).
24. Glanville, W. H., and F. C. Thomas: *Bldg. Research Tech. Papers* 21, 1939.
25. Glasstone, S., K. J. Laidler, and H. Eyring: "The Theory of Rate Processes," McGraw-Hill, New York, 1941.
26. Green, H.: *Ind. Eng. Chem.*, **13**: 637 (1941).
27. ———: *Ind. Eng. Chem.*, **14**: 576 (1942).
28. ——— and R. N. Weltmann: *Ind. Eng. Chem.*, **15**: 201 (1943).
29. Greensmith, H. W., and R. S. Rivlin: *Trans. Roy. Soc. (London)*, **A245**: 899 (1953).
30. Hersey, M. D.: *J. Rheol.*, **3**: 196 (1932).
31. Jeffreys, H.: "The Earth," Cambridge, 1929.
32. Karim, M., and L. Rosenhead: *Rev. Modern Phys.*, **24**: 108 (1952).
33. Kelvin, Lord (W. Thomson): "Encyclopaedia Britannica," 9th ed., 1875.
34. Lax-Weiner, K., and R. Schoenfeld-Reiner: *Bull. Research Council Israel*, **2**: 66 (1952).
35. Lee, A. R., M. Reiner, and P. Rigden: *Nature*, **158**: 706 (1946).
36. Lethersich, W.: *Brit. Elec. Allied Ind. Research Assoc. Tech. Rept. and Suppl.* A/T83, 1941.
37. Mooney, M. J.: *Rheology*, **2**: 210 (1931).
38. Poynting, J. H., and J. J. Thomson: "Properties of Matter," London, 1902.
39. Rabinowitsch, B. Z.: *J. Phys. Chem.*, **145A**: 1 (1929).
40. Reiner, M.: *Appl. Sci. Research*, **A1**: 475 (1940).
41. ———: *Am. J. Math.*, **67**: 350 (1945).
42. ———: *Am. J. Math.*, **70**: 433 (1948).
43. ———: *Bull. Research Council Israel*, **1** (3): 3 (1951).
44. ———: *Bull Research Council Israel*, **3**: (1954).
45. ———: "Elasticity," chap. 1, Amsterdam, 1954.
46. ———, P. Rigden, and E. N. Thrower: *J. Soc. Chem. Ind. (London)*, **69**: 257 (1950).
47. ———, G. W. Scott Blair, and H. B. Hawley: *J. Soc. Chem. Ind. (London)*, **68**: 327 (1949).
48. ——— and K. Weissenberg: *Rheol. Leaflet* 10, p. 12, 1939.
49. Rivlin, R. S.: *Proc. Roy. Soc. (London)*, **A193**: 260 (1948).
50. Roberts, J. E.: *Proc. Intern. Congr., 2d Congr., London, 1954*.
51. Schofield, R. K., and G. W. Scott Blair: *Proc. Roy. Soc. (London)*, **139**: 557 (1933); **141**: 72 (1933).
52. Scott Blair, G. W., and E. M. Crowther: *J. Phys. Chem.*, **33**: 321 (1929).
53. ——— and B. C. Veinoglou: *J. Sci. Instr.*, **20**: 58 (1943).
54. Schwedoff, T.: *J. phys.*, **9** (2): 34 (1890).
55. Simha, R.: *Record Chem. Progr. Kresge-Hooker Sci. Lib.*, **1949**: 157.
56. Swainger, K. H.: *Phil. Mag.*, **38** (7): 422 (1947).
57. Swift, H. W.: "Engineering," 1947.
58. Trouton, F.: *Proc. Phys. Soc. (London)*, **19**: 47 (1905); *Proc. Roy. Soc. (London)*, **A77**: 326 (1906).
59. ——— and Rankine, A. D.: *Phil. Mag.*, **8** (6): 555 (1904).
60. Voigt, W.: *Abhandl. Ges. Wiss. Göttingen Mathphysik. Kl.*, **36** (1890).
61. Weissenberg, K.: *Nature*, **159**: 310 (1947).
62. Wolarowitch, M. P., N. N. Kulakoff, and A. N. Romansky: *Kolloid-Z.*, **71**: 267 (1935).
63. ——— and K. J. Samarina: *Kolloid-Z.*, **70**: 280 (1935).
64. ——— and D. M. Tolstoi: *Kolloid-Z.*, **70**: 165 (1935).

Chapter 4

Wave Propagation in Fluids

By A. H. TAUB, University of California

1. Conservation Laws

Study of wave propagation in fluids is concerned with correlation of the state of the fluid macroscopically described by five variables at a point A at a time t with that at a different place B at an earlier time. The five variables describing the macroscopic state of the fluid may be taken to be two thermodynamic ones, such as the pressure p and the density ρ, and three kinematic ones, such as the components of the Eulerian velocity field (velocity of that part of the fluid which is at a certain place at a certain time) or the components of the Lagrange velocity field (velocity of an element of the fluid which was at a certain place at a certain time).

Since changes in state of a given element of a fluid must be in accordance with the laws of conservation of mass, momentum, and energy, and since these laws relate the changes of state of one element of the fluid with those of its neighbors, the study of wave propagation deals with time-dependent solutions of the five mathematical equations describing these conservation laws which satisfy the initial and exterior boundary conditions.

Discussion is here confined to ideal fluids, those without heat conductivity or viscosity. Results from the theory of ideal fluids are closely related to those for a nonideal gas with dissipative mechanisms. Although the theory of ideal fluids contains many complications, it seems more amenable to mathematical treatment than the theory of a fluid with dissipative mechanisms.

For an ideal fluid discontinuities in the state variables may occur even for continuous initial and boundary values of these variables (see Part 3, Chap. 2, Secs. 4, 6). This means one cannot use the partial differential equations embodying the conservation laws [see Eqs. (1.3), (1.6), and (1.19) of Part 3, Chap. 1] throughout the region but must supplement these with various algebraic statements, derived from the conservation laws, relating the jumps in the state variables across the discontinuities. Thus a general propagation problem involves the patching together of various solutions of the partial differential equations, each holding in different domains, across unknown moving surfaces of discontinuity in such a way that the Rankine-Hugoniot equations are satisfied.

The equations which must hold in regions when the variables are differentiable are conservation of mass:

$$\frac{d\rho}{dt} + \rho \operatorname{div} \mathbf{v} = 0 \tag{4.1}$$

conservation of momentum:

$$\rho \frac{d\mathbf{v}}{dt} = \rho \mathbf{g} - \operatorname{grad} p \tag{4.2}$$

and conservation of energy:

$$\rho \frac{d}{dt} (\tfrac{1}{2}v^2 + E) = -\operatorname{div} (p\mathbf{v}) + \rho \mathbf{g} \cdot \mathbf{v} \tag{4.3}$$

where $\mathbf{g}$ is the external force per unit mass, p the pressure, ρ the density, E the internal energy per unit mass, $\mathbf{v}$ the Eulerian velocity, and d/dt the convected derivative.

If there is a surface of discontinuity Σ representating a shock, a slipstream, or a density discontinuity, moving with a velocity $\mathbf{V}$ and having a normal vector $\mathbf{n}$ drawn from the region R_1 to the region R_2, then the law of conservation of mass is contained in the statement

$$\rho_1(V_n - v_{1n}) = \rho_2(V_2 - v_{2n}) = m \tag{4.4}$$

The law of conservation of momentum gives

$$(p_1 - p_2)\mathbf{n} = m(\mathbf{v}_1 - \mathbf{v}_2) \tag{4.5}$$

and the law of conservation of energy becomes

$$m\left(\frac{v_1^2}{2} + E_1\right) - m\left(\frac{v_2^2}{2} + E_2\right) = p_1v_{1n} - p_2v_{2n} \tag{4.6}$$

The variables p, ρ, v, E have the same meaning as in Eqs. (4.3) to (4.5). The subscripts 1 and 2 denote on which side of the surface Σ the variable is to be evaluated, and the subscript n attached to a vector quantity denotes the normal component of that vector. The derivation of Eqs. (4.1) to (4.6) when no external forces are present may be found in [1].*

If the scalar product of each side of Eq. (4.2) with the vector $\mathbf{v}$ is taken and the resulting equation is subtracted from Eq. (4.3), one obtains

* Numbers in brackets refer to references at end of chapter.

$$\rho \frac{dE}{dt} = -p \operatorname{div} \mathbf{v} = +\frac{p}{\rho}\frac{d\rho}{dt}$$

or

$$\rho \left[\frac{dE}{dt} + p \frac{d}{dt}\left(\frac{1}{\rho}\right) \right] = 0$$

From the definition of entropy per unit mass S it follows that

$$T \frac{dS}{dt} = \frac{dE}{dt} + p \frac{d}{dt}\left(\frac{1}{\rho}\right)$$

Hence for continuous fluid motion the conservation of energy is equivalent to

$$\rho T \frac{dS}{dt} = 0 \tag{4.7}$$

That is, along a streamline the energy of a particle of the gas is a constant in regions where the flow is continuous. The entropy of particles is changed in crossing a shock.

For an ideal fluid the entropy per unit mass is a function only of the pressure and density. Conversely, the density may be considered as a function of the pressure and entropy.

$$\frac{d\rho}{dt} = \left(\frac{\partial \rho}{\partial p}\right)_S \left(\frac{dp}{dt}\right) + \left(\frac{\partial \rho}{\partial S}\right)_p \frac{dS}{dt}$$

In view of Eq. (4.7)

$$\frac{d\rho}{dt} = \frac{1}{c^2}\frac{dp}{dt}$$

where

$$c^2 = \left[\left(\frac{\partial \rho}{\partial p}\right)_S \right]^{-1} \tag{4.8}$$

and Eq. (4.1) may be written as

$$\operatorname{div} \mathbf{v} = -\frac{1}{\rho c^2}\frac{dp}{dt} \tag{4.9}$$

If the fluid is incompressible, it has no internal energy ($E = 0$), and Eq. (4.3) is a consequence of (4.2) and (4.1).

If the motion is irrotational, we may write

$$\mathbf{v} = -\operatorname{grad} \phi$$

and Eq. (4.2) becomes

$$\operatorname{grad}\left(-\frac{\partial \phi}{\partial t} + \tfrac{1}{2}v^2\right) = \mathbf{g} - \frac{1}{\rho}\operatorname{grad} p$$

If the external forces are derivable from a potential function, we may write this equation as

$$-\frac{\partial \phi}{\partial t} + \tfrac{1}{2}v^2 + U + \int \frac{dp}{\rho} = C(t) \tag{4.10}$$

where $C(t)$ is a function of time alone. This is called the strong form of Bernoulli's equation (see Part 3, Chap. 2). Equation (4.10) applies to irrotational compressible flow as well as irrotational incompressible flow. It and the condition of irrotational flow are equivalent to the conservation-of-momentum equations (4.2). Only in the case of incompressible irrotational flow are Eqs. (4.10) and (4.3), the conservation-of-energy equation, related.

It follows from Eq. (4.5) that

$$m(v_{1t} - v_{2t}) = 0$$

where v_{1t} and v_{2t} are the components of $\mathbf{v}$ in any direction tangential to the surface Σ. This equation can be satisfied in two entirely different ways: (1) $m = 0$ and (2) $m \neq 0$ and $v_{1t} = v_{2t}$. In the first case there is no flow across the discontinuity surface, and in the second case there is such a flow but the tangential components of the velocity are continuous across the surface. In the second case the discontinuity surface is called a shock, and Eqs. (4.4) to (4.6) reduce to the Rankine-Hugoniot equations discussed in Part 3, Chap. 2, Sec. 6. In the first case ($m = 0$) it follows from (4.5) and (4.6) that there is no discontinuity in pressure or in the normal component of velocity. Hence the tangential component of velocity is discontinuous, or if it is continuous, there must be a discontinuity in density (and hence of temperature and entropy). Both possibilities may ensue. Depending on whether the discontinuous change in tangential velocity or the discontinuous change in density is to be emphasized, a discontinuity with $m = 0$ is called a *slipstream* or a *density discontinuity*.

The problems to be discussed below may then be mathematically described by the system of partial differential equations and algebraic equations (4.1) to (4.6) subject to suitable initial and boundary conditions. If there exists an exterior boundary, fixed or moving, the velocity relative to the boundary of a particle in the boundary must be tangential to the boundary; otherwise there would be a finite flow of fluid across it. If $F(x,y,z,t) = 0$ is the equation of the bounding surface at time t at every point of it, we must have [2]

$$\frac{dF}{dt} = 0 \tag{4.11}$$

Equations (4.1) to (4.6) and the boundary conditions (4.11) are numerically invariant under the Newtonian transformations relating the coordinates x of a fixed observer with the coordinates $x^{\#}$ of an observer moving with a constant velocity $\mathbf{u}$ relative to the first one:

$$\mathbf{x}^{\#} = \mathbf{x} - \mathbf{u}t$$

That is, if we replace all velocity vectors $\mathbf{v}$ by $\mathbf{v} - u$, x by $x^{\#}$, leave p and ρ unaltered, Eqs. (4.1) to (4.6) and (4.11) hold.

2. Small Disturbances

In many problems such as the propagation of sound, called problems in *small* motion or problems in *small disturbances*, it is sufficient to assume that the five state variables of the fluid may be written as

$$p = p_0 + \epsilon p_1 \qquad \rho = \rho_0 + \epsilon \rho_1 \qquad \mathbf{v} = \mathbf{v}_0 + \epsilon \mathbf{v}_1 \tag{4.12}$$

where p_0, ρ_0, and $\mathbf{v}_0$ are constants and terms involving powers of ϵ greater than the first may be neglected. It follows from the observation made at the end of the preceding section that there is no loss in generality in setting $\mathbf{v}_0 = 0$. If this is done and Eqs. (4.12) are substituted into Eqs. (4.9) and (4.2), we obtain

$$\operatorname{div} \mathbf{v}_1 = -\frac{1}{\rho_0 c_0^2}\frac{\partial p_1}{\partial t} \tag{4.13}$$

and

$$\frac{\partial \mathbf{v}_1}{\partial t} = \mathbf{g}_1 - \frac{1}{\rho_0}\operatorname{grad} p_1 \tag{4.14}$$

as the equations of conservation of mass and momentum, respectively, where we have written $g = \epsilon g_1$ and have defined

$$c_0^{-2} = \left[\left(\frac{\partial \rho}{\partial p}\right)_S\right]_{p=p_0,\rho=\rho_0} \tag{4.15}$$

If $\mathbf{g}_1 = 0$, we may eliminate between Eqs. (4.13) and (4.14)

$$\nabla^2 p_1 - \frac{1}{c_0^2}\frac{\partial^2 p_1}{\partial t^2} = 0$$

and

$$\nabla^2 \mathbf{v}_1 - \frac{1}{c_0^2}\frac{\partial^2 \mathbf{v}_1}{\partial t^2} = 0$$

for the quantities v_1 and p_1. If the motion is irrotational, we may write

$$\mathbf{v}_1 = -\operatorname{grad} \phi \tag{4.16}$$

where ϕ is the velocity potential. Then Eq. (4.14) (with $g_1 = 0$) is satisfied by

$$p_1 = \rho_0 \frac{\partial \phi}{\partial t} \tag{4.17}$$

and Eq. (4.13) becomes

$$\nabla^2 \phi - \frac{1}{c_0^2}\frac{\partial^2 \phi}{\partial t^2} = 0 \tag{4.18}$$

In this case the small disturbance is characterized by the velocity potential ϕ, which determines the velocity field through Eq. (4.16) and the pressure through Eq. (4.17). Since ϕ satisfies the wave equation (4.18), the constant c_0 defined by (4.15) is the velocity of propagation of the small disturbance in the medium. For a perfect gas [$p = A(S)\rho^\gamma$, with γ the ratio of specific heats]

$$c_0^2 = \frac{\gamma p_0}{\rho_0}$$

The constant c_0 is called the velocity of sound in the fluid at pressure p_0 and density ρ_0.

Equations (4.17) and (4.18) together with the relation between pressure and density describe the behavior of the state variables during the propagation of sound wave in the fluid. Solutions to these equations for a great variety of problems will be found in [3, 2].

If the fluid is incompressible, ρ is a constant ($\rho_1 = 0$), and $c_0^{-2} = 0$. Equation (4.13) becomes

$$\operatorname{div} \mathbf{v}_1 = 0 \tag{4.19}$$

If the motion is also irrotational, we may use Eq. (4.16), and (4.19) becomes

$$\nabla^2 \phi = 0 \tag{4.20}$$

If it is further assumed that the external force is derivable from a time-independent potential function, we may replace Eq. (4.14) by Bernoulli's equation (4.10), which may be written in this approximation as

$$-\frac{\partial \phi}{\partial t} + U + \frac{p}{\rho} = 0 \tag{4.21}$$

where the arbitrary function of time appearing in Eq. (4.10) has been absorbed into the velocity potential ϕ.

The discussion of surface waves of small amplitude in water of variable depth is concerned with determining solutions of Eqs. (4.20) and (4.21), with U representing the gravitational potential energy, subject to the following boundary conditions: (1)

$$\partial \phi / \partial n = 0$$

at the bottom of the water; i.e., the normal derivative of the velocity potential (the normal component of the velocity of the water) must vanish at the bottom, and (2) the equations derived from an application of Eq. (4.11) along the unknown free boundary of the surface at which the pressure is a constant.

If we introduce a rectangular coordinate system in which the y axis is vertical and the water is assumed to fill the region

$$-h(x,z) \le y \le 0$$

when at rest where the nonnegative quantity $h(x,z)$ is the depth of the water, and if $\eta(x,z,t)$ represents the vertical displacement of the free surface, then the boundary conditions mentioned above become

$$\left(\frac{\partial \phi}{\partial n}\right)_{y=-h} = 0 \tag{4.22}$$

and, as will be shown below,

$$\frac{\partial^2 \phi}{\partial t^2} + g\frac{\partial \phi}{\partial y} = 0 \tag{4.23}$$

at $y = 0$. The derivation of Eq. (4.23) is as follows: Eq. (4.11) may be written as

$$\frac{d}{dt}[y - \eta(x,z,t)] = 0$$

i.e., as

$$v = \frac{\partial \eta}{\partial t} + u\frac{\partial \eta}{\partial x} + w\frac{\partial \eta}{\partial z}$$

at $y = \eta$, where u, v, w are the components of the velocity of the fluid. Assuming the displacement η is of the order ϵ, we may write this as

$$-\frac{\partial \phi}{\partial y} = \frac{\partial \eta}{\partial t} \tag{4.24}$$

and may evaluate both terms at $y = 0$ instead of $y = \eta$. However, Eq. (4.21), when evaluated on the free surface, where $U = g\eta$ and $p = 0$, may be written as

$$\eta = \frac{1}{g}\left(\frac{\partial \phi}{\partial t}\right)_{y=\eta} = \frac{1}{g}\left(\frac{\partial \phi}{\partial t}\right)_{y=0} + \cdots \tag{4.25}$$

where the terms not written explicitly are of order ϵ^2. Substituting (4.25) into (4.24), we obtain (4.23). Equation (4.25) determines $\eta(x,z,t)$ when ϕ is known.

Equations (4.20), (4.22), (4.23), and (4.25) are sometimes referred to as the equations of the *exact linear theory* of surface water waves. Solutions of these equations of the form

$$\phi(x,y,z,t) = \cos \sigma t\, \psi(x,y,z)$$

or

$$\phi(x,y,z,t) = \sin \sigma t\, \psi(x,y,z)$$

are called *standing waves*. For solutions of this type Eqs. (4.20), (4.23), (4.22), and (4.25) become

$$\nabla^2\psi = 0$$

$$\frac{\partial \psi}{\partial y} - \frac{\sigma^2}{g}\psi = 0 \qquad \text{at } y = 0$$

$$\frac{\partial \psi}{\partial y} - \frac{\partial \psi}{\partial x}\frac{\partial h}{\partial x} - \frac{\partial \psi}{\partial z}\frac{\partial h}{\partial z} = 0 \qquad \text{at } y = -h(x,z)$$

and

$$\eta(x,z,t) = -\frac{\sigma}{g}\sin \sigma t\, \psi(x,0,z)$$

or

$$\eta(x,z,t) = \frac{\sigma}{g}\cos \sigma t\, \psi(x,0,z)$$

It is of interest to consider plane waves, i.e., functions ψ which do not depend on z. If the water is of infinite depth, a solution of these equations is then given by

$$\psi = A(m)e^{my}\cos(mx + \alpha)$$

where $A(m)$, m, and α are arbitrary. The frequency σ is then determined from the condition at $y = 0$ by the relation

$$\sigma^2 = gm = g\frac{2\pi}{\lambda} \tag{4.26}$$

between the frequency and wavelength. Linear combinations of solutions of this type are again solutions, and by suitably choosing the $A(m)$ in the expressions

$$\psi = \Sigma A(m)e^{my}\cos(mx + \alpha)$$

solutions may be found in which $\eta(x,0)$ is an arbitrarily prescribed curve.

By suitably combining two of the standing-wave solutions one obtains *progressive waves* of the form

$$\phi(x,y,t) = Ae^{my}\cos(mx \pm \sigma t + \alpha)$$

which represent waves moving in either direction along the x axis. The propagation speed of these waves is

$$c = \frac{\sigma}{m} = \left(\frac{g}{m}\right)^{1/2} = \left(\frac{g\lambda}{2\pi}\right)^{1/2}$$

and depends on the square root of the wavelength.

In case the depth of the water is a constant h, a progressive-plane-wave solution of Eqs. (4.20), (4.22), (4.23), and (4.25) is given by

$$\phi(x,y,t) = A\cosh m(y + h)\cos(mx \pm \sigma t + \alpha)$$

where A and α are constants and σ and m are now related by

$$\sigma^2 = gm\tanh mh$$

The propagation speed is now

$$c = \frac{\sigma}{m} = \left(\frac{g\lambda}{2\pi}\tanh\frac{2\pi h}{\lambda}\right)^{1/2} \tag{4.27}$$

If h is very large compared to λ, the relation between c and λ is practically that given above for progressive waves in water of infinite depth. When λ is large compared to h, $\tanh(2\pi h/\lambda)$ is approximately equal to $2\pi h/\lambda$, and

$$c = (gh)^{1/2} \tag{4.28}$$

approximately and hence is approximately independent of the wavelength.

The problem of the construction of progressive water waves when the depth, given by $h(x)$, instead of being a constant has a constant slope, has been discussed by Hanson [5], Bondi [6], Miche [7], Stoker [4], and Lewy [8]. The first four writers have assumed that the angle the bottom makes with the undisturbed surface is $\pi/2q$ with q an integer. Lewy has dealt with the case where this angle is $p\pi/2q$, where p and q are integers, p is odd, and $p < 2q$, and has given closed expressions for the progressive waves.

Stoker has shown [4] that once the frequency and amplitude at infinity are prescribed, the additional condition that the wave at infinity is a progressing wave moving toward shore leads to a unique solution, which has a logarithmic singularity at the shore line. The solution is also uniquely determined if the singularity at the shore line is prescribed—the behavior at infinity is then determined. This theory furnishes two types of standing-wave solutions from which solutions behaving like arbitrary simple harmonic progressing waves at infinity can be constructed, but it furnishes no criterion by which one can decide what type of wave would actually occur in practice.

By assuming that the waves move from infinity toward shore with no reflection from the shore back to infinity and by prescribing the frequency and amplitude at infinity Stoker finds a unique progressive wave. By numerical calculation of the case of a 6° sloping beach he finds that Eq. (4.27) between the propagation speed and the wavelength holds quite accurately even though the depth h is not a constant but is slowly varying.

Friedrichs [9] has obtained asymptotic expressions for the integrals representing the solutions derived by the methods of Lewy and Stoker and thus is able to give approximate forms for the waves on shallow sloping beaches.

3. Interactions of Waves of Small Amplitude

If there exist boundaries in the body of the fluid other than those considered above, such as a surface of discontinuity separating fluids of different density or a rigid body submerged in the fluid, then combinations of solutions of the type given above must be taken in order to satisfy the additional boundary conditions which must obtain. These conditions are, of course, those already mentioned in Sec. 1: the normal component of the velocity of the fluid relative to the velocity of a boundary other than a shock must vanish, and the difference in pressure across such a boundary must equal the stress in the boundary; the latter quantity is zero when the fluid is a perfect fluid, as has been heretofore assumed, but is not zero when surface tension is taken into account.

The process of forming combinations of solutions varies with the problem under consideration. Thus in some problems (to be discussed below) boundary conditions may be satisfied by assuming that a single progressive wave exists in one region of the fluid and that one of another type exists in another region. In other problems it may be necessary to assume that two progressive waves, one coming from infinity and one going to infinity, exist in one region and only a single progressive wave coming from infinity exists in another region, in order to satisfy the boundary conditions on the surface separating the two regions. The latter type of problem is quite common in the discussion of the reflection of waves from an interface between two fluids. The various waves are then called the incident wave, the reflected wave, and the transmitted wave in the order in which they were introduced.

We illustrate the process of forming combinations of solutions by discussing the problem arising when we have two fluids of density ρ and ρ', one beneath the other, moving parallel to a horizontal x axis with velocities V and V', respectively, the common surface when undisturbed being plane and horizontal. We shall assume that a plane wave disturbs this surface, which is initially $y = 0$. The first fluid will be assumed to be infinite in extent and to fill the space below $y = 0$, and the second will also be assumed to be infinite in extent and to fill the space above $y = 0$. We may then write for the velocity potentials in each of the fluids

$$\phi = -Vx + \epsilon\phi_1 \qquad \phi' = -V'x + \epsilon\phi_1' \tag{4.29}$$

and terms in powers of ϵ higher than the first will be neglected. We denote by $\epsilon\eta$ the ordinate of the displaced surface, and the velocity boundary conditions become at $y = 0$

$$\frac{\partial \eta}{\partial t} + V\frac{\partial \eta}{\partial x} = -\frac{\partial \phi}{\partial y} \qquad \frac{\partial \eta}{\partial t} + V'\frac{\partial \eta}{\partial x} = -\frac{\partial \phi'}{\partial y} \tag{4.30}$$

Equation (4.10) may be written as

$$\begin{aligned} \frac{p}{\rho} &= \epsilon\left(\frac{\partial \phi_1}{\partial t} - V\frac{\partial \phi_1}{\partial x}\right) - gy \\ \frac{p'}{\rho'} &= \epsilon\left(\frac{\partial \phi_1'}{\partial t} - V'\frac{\partial \phi_1'}{\partial x}\right) - gy \end{aligned} \tag{4.31}$$

We assume that the boundary surface cannot withstand tension, and then the pressure boundary condition is $p = p'$. It then follows that at $y = 0$

$$\rho\left(\frac{\partial \phi_1}{\partial t} - V\frac{\partial \phi_1}{\partial x} - g\eta\right) = \rho'\left(\frac{\partial \phi_1'}{\partial t} - V\frac{\partial \phi_1'}{\partial x} - g\eta\right) \tag{4.32}$$

The functions ϕ_1 and ϕ_1' may be taken to be

$$\phi_1 = ce^{ky+i(\sigma t - kx)} \qquad \text{and} \qquad \phi_1' = c_1'e^{-ky+i(\sigma t-kx)}$$

Substituting these expressions into (4.30) and (4.32), we obtain

$$\eta = ae^{i(\sigma t - kx)}$$

$$\text{where} \quad i(\sigma - kV)a = -kc \qquad i(\sigma - kV')a = kc' \tag{4.33}$$

and the propagation velocity is

$$c = \frac{\sigma}{k} = \frac{\rho V + \rho' V'}{\rho + \rho'} \pm \left[\left(\frac{g}{k}\frac{\rho - \rho'}{\rho + \rho'}\right) - \frac{\rho\rho'}{(\rho + \rho')^2}(V - V')^2\right]^{1/2} \tag{4.34}$$

If

$$(V - V')^2 > \frac{g}{k}\frac{\rho^2 - \rho'^2}{\rho\rho'} = \frac{g\lambda}{2\pi}\frac{\rho^2 - \rho'^2}{\rho\rho'}$$

σ is complex, and the common boundary is unstable for sufficiently small wavelengths.

If the stress between two elements of the bounding surface per unit length of the line bounding these elements is not zero but T_1, the pressure boundary condition may be written as

$$p - p' + ET_1\frac{\partial^2 \eta}{\partial x^2} = 0$$

Equation (4.32) is then replaced by

$$\rho\left(\frac{\partial \phi_1}{\partial t} - V\frac{\partial \phi_1}{\partial x} - g\eta\right) - \rho'\left(\frac{\partial \phi_1'}{\partial t} - V'\frac{\partial \phi_1'}{\partial x} - g\eta\right) = -T_1\frac{\partial^2 \eta}{\partial x^2}$$

Assuming the forms used above for ϕ_1, ϕ_1', and η, we find (4.33) holding, and the propagation speed is

$$c = \frac{\sigma}{k} = \frac{V\rho + V'\rho'}{\rho + \rho'} \pm \sqrt{c_0^2 - \frac{\rho\rho'}{(\rho + \rho')^2}(V - V')^2} \tag{4.35}$$

$$\text{where} \qquad c_0^2 = \frac{g}{k}\frac{\rho - \rho'}{\rho + \rho'} + \frac{T_1 k}{\rho + \rho'} \tag{4.36}$$

is the square of the propagation speed when no currents are present. That is, the propagation speed when currents are present is equal to the mean velocity of the streams plus or minus the term under the radical. If the currents are such that the term under the radical vanishes, there is no dispersion, and the wave profile is unaltered as the wave progresses.

The velocity of propagation c_0 has a minimum given by

$$c_m^2 = \frac{2(\rho - \rho')}{\rho + \rho'}\left(\frac{gT_1}{\rho - \rho'}\right)^{1/2}$$

and this minimum is attained for the wavelength

$$\lambda_m = \frac{2\pi}{k_m} = 2\pi\left(\frac{T_1}{g(\rho - \rho')}\right)^{1/2}$$

We may write

$$\frac{c_0^2}{c_m^2} = \frac{1}{2}\left(\frac{\lambda}{\lambda_m} + \frac{\lambda_m}{\lambda}\right)$$

For large λ the first term predominates, and the force governing the motion is mainly gravitational. On the other hand, when λ is small, the second term is important, and the motion is mainly governed by cohesion. The *group velocity* of a train of waves is given by

$$c_g = c - \lambda \frac{dc}{d\lambda} = c\left(1 - \frac{1}{2}\frac{\lambda^2 - \lambda_m^2}{\lambda^2 + \lambda_m^2}\right)$$

and is greater or less than the propagation velocity according as $\lambda \gtrless \lambda_m$.

If currents are present, instability ensues only when the term under the radical in (4.35) is negative. Since c_0 has a minimum c_m, it follows that the equilibrium surface when plane is stable for disturbance of all wavelengths as long as

$$|V - V'| < \frac{\rho + \rho'}{(\rho\rho')^{1/2}} c_m$$

Numbers σ and k satisfying (4.35) determine the functions ϕ_1, ϕ_1', and η given above, provided c, c', and a satisfy (4.34), which are solutions to problems involving an interface between two fluids. Linear combinations of such functions will also be solutions. Various initial conditions may be satisfied by taking appropriate linear combinations (see [2, chap. 9]).

Other illustrations of the process of combining solutions of problems which do not satisfy certain boundary conditions to achieve a solution of a problem with these boundary conditions may be found in [2, 10–12].

4. Small Disturbances in Shallow Water

The equations describing the propagation of a small disturbance on a fluid of variable depth have been derived in Sec. 2. Another set of equations may be derived from the assumption that the depth is so small that the pressure at any point below the surface may be taken to be that given by hydrostatics; i.e., the vertical acceleration may be neglected, and in the coordinate system used above

$$p - p_0 = g\rho(y_0 + \eta - y) \tag{4.37}$$

where p_0 is the uniform external pressure and y_0 is the ordinate of the undisturbed state. If this is substituted into Eq. (4.2), it follows that the horizontal acceleration is independent of y, and hence the particle velocities will be independent of y if they are so initially.

In the case of a plane wave $u = u(x,t)$, $w = 0$, Eq. (4.2) becomes

$$\frac{\partial u}{\partial t} + u\frac{\partial u}{\partial x} = -g\frac{\partial}{\partial x}(y_0 + \eta) + X \tag{4.38}$$

The equation of continuity may be written as

$$\frac{\partial u}{\partial x} + \frac{\partial v}{\partial y} = 0$$

or as

$$v(y,x,t) - v(-h,x,t) = -\int_{h(x)}^{y} \frac{\partial u}{\partial x}\,dy = -(y + h)\frac{\partial u}{\partial x}$$

The boundary conditions [see Eq. (4.11)] at the surface are

$$v(y_0 + \eta, x, t) = \frac{\partial}{\partial t}(y_0 + \eta) + u\frac{\partial}{\partial x}(y_0 + \eta)$$

and at the bottom

$$v(-h,x,t) = -u\frac{\partial h}{\partial x}$$

Hence

$$\frac{\partial}{\partial t}(y_0 + \eta) + u\frac{\partial}{\partial x}(y_0 + \eta) + u\frac{\partial h}{\partial x} = -(y_0 + \eta + h)\frac{\partial u}{\partial x}$$

That is, the equation of continuity may be written as

$$\frac{\partial}{\partial t}(y_0 + \eta + h) + \frac{\partial}{\partial x}[u(y_0 + \eta + h)] = 0 \tag{4.39}$$

Equations (4.38) and (4.39) describe waves of finite amplitude in shallow water. Their integration may be accomplished by methods similar to those discussed in later sections (see [13]).

Under the assumption of small disturbances we may neglect the nonlinear terms in (4.38) and (4.39) and obtain

$$\frac{\partial u}{\partial t} = -g\frac{\partial \eta}{\partial x} + X \qquad \frac{\partial \eta}{\partial t} = -\frac{\partial}{\partial x}(hu) \tag{4.40}$$

where we have set $y_0 = 0$. We may eliminate η from these equations and obtain

$$\frac{\partial^2 u}{\partial t^2} = g\frac{\partial}{\partial x}\left(h\frac{\partial u}{\partial x}\right) + \frac{\partial X}{\partial t} \tag{4.41}$$

or we may eliminate u and obtain

$$\frac{\partial^2 \eta}{\partial t^2} = g\frac{\partial}{\partial x}\left(h\frac{\partial \eta}{\partial x}\right) - \frac{\partial(hX)}{\partial x} \tag{4.42}$$

If we introduce the variable ξ, the displacement of the particle, by the relation

$$\frac{\partial \xi}{\partial t} = u$$

then Eq. (4.41) may be written as

$$\frac{\partial^2 \xi}{\partial t^2} = g\frac{\partial}{\partial x}\left(h\frac{\partial \xi}{\partial x}\right) + X \tag{4.43}$$

and the second of (4.40) becomes

$$\eta = -\frac{\partial}{\partial x}(h\xi) \tag{4.44}$$

Equations (4.40) are often called the equations of shallow-water theory. The above derivation is open to criticism, for it does not indicate clearly the essential role played by the depth in determining the accuracy of the approximation. If we take the case of constant depth h and no external forces acting, Eq. (4.42) admits solutions of the form

$$\eta = F(x + ct) + f(x - ct)$$

i.e., progressive waves with a propagation speed

$$c = (gh)^{1/2}$$

However, the theory of Sec. 2, the exact linear theory, admits progressive waves in which the propagation speed depends on wavelength in accordance with Eq. (4.27). The two theories will, of course, coincide when the wavelength is large compared to the depth.

Friedrichs [14] has shown that, if the product of the depth h times the maximum initial curvature of the free surface is such that its square and higher powers may be neglected, then Eq. (4.38) follows from (4.2). Generalizations of Eq. (4.38) which take into account motions in two dimensions, or even motions of a sheet of fluid on a rotating sphere, may be derived by arguments similar to those used above (see [2, chap. 8]) or by Friedrichs's methods.

The theory of tides [2, chap. 7] is concerned with the solutions of the linear equations describing the motion in shallow water, i.e., Eqs. (4.40) or various generalizations of these equations where the disturbing force is due to the gravitational attraction of a distant body. We shall illustrate the methods used and the type of results obtained by discussing the forced oscillations in a canal of uniform section coincident with the earth's equator.

The disturbing effect of the moon at a point P of the earth's surface may be represented by a potential Ω, whose approximate value is

$$\Omega = \frac{3}{2}\frac{\gamma M a^2}{D^3}\left(\tfrac{1}{3} - \cos^2\theta\right)$$

where M denotes the mass of the moon, D its distance from the earth's center, a the earth's radius, γ the constant of gravitation, and θ the moon's zenith distance at the place P. This gives a horizontal force $-(\partial\Omega/a\,\partial\theta)$ or

$$X = -\frac{gH}{a}\sin 2\theta$$

where E is the mass of the earth,

$$H = \tfrac{3}{2}a\frac{M}{E}\left(\frac{a}{D}\right)^3 \qquad \text{and} \qquad g = \frac{\gamma E}{A^2}$$

The quantity $H = 1.8$ ft if M is the mass of the moon, and $H = 0.79$ ft if M is the mass of the sun. If n denotes the angular velocity of the moon westward relative to a fixed meridian, then

$$\theta = nt + \phi + \epsilon$$

and Eq. (4.43) becomes

$$\frac{\partial^2\xi}{\partial t^2} - \frac{c^2}{a^2}\frac{\partial^2\xi}{\partial\phi^2} = -\frac{gH}{a}\sin 2(nt + \phi + \epsilon)$$

Solutions of these equations consist of solutions of the homogeneous equation obtained by setting the right-hand side equal to zero plus a solution of the nonhomogeneous equation, called a forced oscillation. The former are called free oscillations and are represented by

$$\xi = \sum_{0}^{\infty}[P_r(t)\cos r\phi + Q_r(t)\sin r\phi]$$

where P_r and Q_r satisfy the equation

$$\frac{d^2P_r}{dt^2} + \frac{r^2c^2}{a^2}P_r = 0$$

and therefore

$$P_r = A\sin\left(\frac{rct}{a} + \epsilon\right)$$

The forced oscillations are given by

$$\xi = -\frac{1}{4}\frac{agH}{c^2 - n^2a^2}\sin 2(nt + \phi + \epsilon)$$

and hence

$$\eta = \frac{1}{2}\frac{c^2H}{c^2 - n^2a^2}\cos 2(nt + \phi + \epsilon)$$

The period of the tide is thus half the period of the moon in its orbit, and there is high or low water beneath the moon according as the velocity of a point on the earth's surface which moves so as to be always vertically beneath the moon is less or greater than the velocity of a free wave, i.e., $na < c$ or $na > c$.

5. Plane Waves of Finite Amplitude

The theory of waves of finite amplitude in shallow water is governed by Eqs. (4.38) and (4.39). If in these equations we set $X = 0$, $y_0 = 0$ and assume that h is a constant, we obtain

$$\begin{aligned}\frac{\partial u}{\partial t} + u\frac{\partial u}{\partial x} &= -g\frac{\partial\eta}{\partial x}\\ \frac{\partial\eta}{\partial t} + \frac{\partial}{\partial x}(u\eta) &= 0\end{aligned} \tag{4.45}$$

These equations are mathematically identical with those describing the equations of conservation of momentum and mass, respectively, for one-dimensional motion of a compressible fluid, i.e., Eqs. (4.2) and (4.1), respectively, when we set $\mathbf{g} = 0$, the components of $\mathbf{v}$ equal to $(u,0,0)$, $\rho = \eta$, $p = g(\rho^2/2)$ and assume that u, p, and ρ are functions of x and t alone. Thus the theory of waves of finite amplitude in a shallow uniform canal is identical with the theory of isentropic plane waves of finite amplitude of a fictitious ideal gas with a ratio of specific heats $\gamma = 2$. This subject has been treated in great detail by Stoker [13].

We shall now discuss the general equations (4.1) to (4.3) for the case of one-dimensional motions when no external force is acting. Equation (4.3) or its equivalent (4.7) states that the entropy per unit mass is conserved along streamlines. If we assume that each particle of the fluid comes from the same initial state p_0, ρ_0, we have as an integrated form of Eq. (4.3)

$$p = p_0 \left(\frac{\rho}{\rho_0}\right)^\gamma \tag{4.46}$$

Equations (4.2) and (4.1) become

$$\frac{\partial u}{\partial t} + u \frac{\partial u}{\partial x} = -\frac{1}{\rho}\frac{\partial p}{\partial x} \tag{4.47}$$

and

$$\frac{\partial \rho}{\partial t} + \frac{\partial}{\partial x}(\rho u) = 0 \tag{4.48}$$

respectively.

If we introduce the Lagrange coordinate x^0 and the time t as independent variables through the solution of the equation

$$\frac{dx}{dt} = u(x,t) \tag{4.49}$$

where the value of x at time $t = 0$ is x^0, then Eq. (4.48) may be integrated to give

$$\frac{\partial x}{\partial x^0} = \frac{\rho_0}{\rho} \tag{4.50}$$

which states that the amount of matter between x^0 and $x^0 + dx^0$ at time $t = 0$ is the same as the amount between x and $x + dx$ at time t. In Eq. (4.49) the symbol d/dt represents differentiation keeping x_0 fixed, and in Eq. (4.50) partial differentiation with respect to x^0 means that t is fixed. We shall use these conventions in the sequel.

Equation (4.47) may then be written as

$$\frac{du}{dt} = -\frac{1}{\rho_0}\frac{\partial p}{\partial x^0} \tag{4.51}$$

and the differential form of (4.50) is

$$\frac{\partial u}{\partial x^0} = -\frac{\rho_0}{\rho^2}\frac{d\rho}{dt} \tag{4.52}$$

as follows from (4.50) by differentiation with respect to t. We shall solve these equations for various initial and boundary values by a method due to Riemann [15]. Let

$$\phi = \int_0^\rho \frac{1}{\rho}\sqrt{\frac{dp}{d\rho}}\, d\rho = \phi_0 \left(\frac{\rho}{\rho_0}\right)^{(\gamma-1)/2} = \phi_0 \left(\frac{p}{p_0}\right)^{(\gamma-1)/2\gamma} \tag{4.53}$$

and

$$c = \frac{\rho}{\rho_0}\sqrt{\frac{dp}{d\rho}} = a_0 \left(\frac{\phi}{\phi_0}\right)^{(\gamma+1)/(\gamma-1)} \tag{4.54}$$

where

$$a_0^2 = \frac{\gamma p_0}{\rho_0} \qquad \phi_0 = \frac{2}{\gamma - 1} a_0$$

Then Eqs. (4.51) and (4.52) may be written as

$$\frac{dr}{dt} + c\frac{\partial r}{\partial x^0} = 0 \qquad \frac{ds}{dt} - c\frac{\partial s}{\partial x^0} = 0 \tag{4.55}$$

where

$$r = \tfrac{1}{2}(\phi + u) \qquad s = \tfrac{1}{2}(\phi - u) \tag{4.56}$$

as follows by multiplying both sides of Eq. (4.52) by c, writing $(1/\rho_0)(\partial p/\partial x^0)$ as $c(\partial\phi/\partial x^0)$, and taking the sum and difference of the resulting equations.

Disturbances which are propagated in the direction of positive (or negative) x^0 are called *simple* or progressive waves and are characterized by s = constant (or r = constant, respectively). In case s = constant, the solution of (4.55) is either ϕ = constant or

$$F(\phi) = x^0 - k - c(\phi)(t - t_0)$$

where F is an arbitrary function of ϕ and k and t_0 are constants. In view of (4.54) this may be written as

$$x^0 - k = a_0\left(\frac{\phi}{\phi_0}\right)^{(\gamma+1)/(\gamma-1)}[t - t_0 - T(\phi)] \tag{4.57}$$

The function $T(\phi)$ may be determined from the boundary conditions. Thus if the progressive wave is generated by moving a piston in a tube, and if the piston is initially located at $x_0 = k$ at $t = t_0$, then for this value of x^0 the velocity must be $u = f(t)$ when $f(t)$ is a known function of the time. Since $s = s_0$ = constant, $\phi = u + 2s_0$, and hence ϕ is a known function of the time, $\phi = f(t) + 2s_0$. The function $T(\phi)$ occurring in (4.57) must be the inverse of this function; i.e., $\phi \equiv 2s_0 + f(T(\phi))$.

It follows from (4.57) that a given value of ϕ, hence of u, is propagated in the x_0,t plane along straight lines of slope $a_0(\phi/\phi_0)^{(\gamma+1)/(\gamma-1)}$. These lines are called characteristics. Thus greater values of ϕ and greater velocities u are propagated faster and hence will overtake each other. At a point when this occurs, the solution given by (4.57) cannot satisfy the requirements of a physical theory, viz., that the velocity and pressure be single-valued functions of x^0 and t. Hence if $u = f(t)$ is an increasing function of t, i.e., if the piston motion is a compressive one, such as to decrease the volume of the tube, the multiple-valued solution given by (4.57) must be discarded. In practice it is known that a shock forms at the point when two characteristics intersect, and the strength of this shock changes as the motion progresses. The equations describing the motion after the shock forms are no longer the differential equations derived from (4.1) to (4.3) alone but involve also the system of algebraic equations (4.4) to (4.6). We shall return to this augmented system later.

If the disturbance is a rarefaction caused by an expansive motion of the piston [$f(t)$ is a decreasing function of t], the characteristics will not intersect for $x^0 > k$, and the solution given by (4.57) will be physically acceptable, i.e., continuous and single-valued even if finite discontinuities are present in the function $u = f(t)$.

Solutions of Eqs. (4.55) when neither r nor s is constant are called compound waves. Riemann [15] gave a method for solving (4.55) for compound waves which is particularly simple when

$$\gamma = \frac{2n+1}{2n-1} \tag{4.58}$$

and n is an integer. For a compound wave the quantity

$$J = \frac{\partial r}{\partial x^0}\frac{ds}{dt} - \frac{dr}{dt}\frac{\partial s}{\partial x^0} = 2c\frac{\partial c}{\partial x^0}\frac{\partial s}{\partial x^0} \neq 0$$

Hence we may interchange independent and dependent variables in Eqs. (4.55) and obtain the equivalent equations

$$\frac{\partial x^0}{\partial s} - c\frac{\partial t}{\partial s} = 0 \qquad \frac{\partial x^0}{\partial r} + c\frac{\partial t}{\partial r} = 0 \tag{4.59}$$

If we now let

$$Z = x^0\frac{\partial x}{\partial x^0} + tu - x = x^0\left(\frac{\phi}{\phi_0}\right)^{-(2n-1)} + tu - x$$

then
$$t = \frac{1}{2}\left(\frac{\partial Z}{\partial r} - \frac{\partial Z}{\partial s}\right) = \frac{\partial Z}{\partial u}$$
$$x^0 = -\frac{c}{2}\left(\frac{\partial Z}{\partial r} + \frac{\partial Z}{\partial s}\right) = -c\frac{\partial Z}{\partial \phi} \tag{4.60}$$

and
$$x = -a_0\frac{\phi}{\phi_0}\frac{\partial Z}{\partial \phi} + u\frac{\partial Z}{\partial u} - Z$$

That is, a knowledge of Z as a function of r and s or u and ϕ determines x^0 and t as functions of these variables and hence the particle paths $x(x^0,t)$ after these relations have been inverted. If the first two of Eqs. (4.60) are substituted into (4.59), we obtain the single linear equation

$$\frac{\partial^2 Z}{\partial r\,\partial s} + \frac{n}{r+s}\left(\frac{\partial Z}{\partial r} + \frac{\partial Z}{\partial s}\right) = 0 \tag{4.61}$$

for the determination of Z. The solution of this Darboux equation for integral n may be written as

$$Z = \left[\frac{1}{r+s}\left(\frac{\partial}{\partial r} + \frac{\partial}{\partial s}\right)\right]^{n-1}\left[\frac{F(r) + f(s)}{r+s}\right] \tag{4.62}$$

where $F(r)$ and $f(s)$ are arbitrary functions of their arguments with a suitable number of derivatives.

When a simple wave given by Eq. (4.57) interacts with a boundary such as a rigid wall, a density discontinuity, or a simple wave traveling in the opposite direction, a compound wave is formed. Taub has shown [16] that then the function $F(r)$ may be taken to be

$$F(r) = (r + s_0)^n I_n - \frac{\phi}{\phi_0}\frac{k}{2(n-1)!}\int_{s_0}^{r}(r^2 - s_0{}^2)^{n-1}\,dr \tag{4.63}$$

where
$$I_n = \int \cdots \int T(\phi)\,d\phi \cdots d\phi \tag{4.64}$$

the integration being carried out n times, each time with the lower limit $\phi = 2s_0$ and the upper limit $\phi = r + s_0$. In the reference cited the determination of $f(s)$ in Eq. (4.62) for compound waves due to the interaction of a progressive wave with a density discontinuity is accomplished by solving an ordinary differential equation.

6. Formation and Decay of Shocks in One Dimension

In a one-dimensional flow, when a shock wave of varying strength progresses, the flow behind the shock is not isentropic. Equations (4.1) to (4.3) do not simplify to the form used in Sec. 5 in such a case. However, as has been pointed out in Chap. 2 of Part 3, the change in entropy is of third order in the strength of the shock when this quantity is measured by the $(p_1 - p_0)/p_0$, where p_1 is the pressure behind the shock and p_0 is the pressure ahead of the shock. Hence if the shock is always weak, i.e., if higher powers than the second of $(p_1 - p_0)/p_0$ may be neglected, we may still apply the equations of Sec. 5 to the discussion of flows containing shocks. To this order of accuracy the characteristic parameter s, introduced above, is constant in the flow resulting when a simple wave interacts with a shock. That is, for weak and moderate shocks the flow behind a varying shock which is progressing into a flow represented by a simple wave is a simple wave.

By neglecting the nonisentropic character of the flow behind a varying shock and using the fact that the flow can be represented by a simple wave Chandrasekhar [17] and Friedrichs [18] were able to give approximate solutions for the shock trajectory in the xt plane for the case of a shock progressing into a gas at rest and interacting with a rarefaction wave catching up to the shock. Friedrichs [18, 19] also discusses the formation of a shock.

The method used by these authors will be illustrated by discussing the problem of a shock progressing into a gas at rest followed by a rarefaction (see Fig. 4.1). The characteristics in the xt plane (note x and not x^0) may be written as

$$x = \xi + w(\xi)t \tag{4.65}$$

where
$$w(\xi) = u(\xi) + a(\xi) \tag{4.66}$$

and $a = \sqrt{dp/d\rho}$ is the velocity of sound. The characteristic quantity which is constant along such a characteristic is

$$r = \tfrac{1}{2}\left(u + \frac{2}{\gamma - 1}a\right)$$

The characteristic quantity

$$s = \tfrac{1}{2}\left(u - \frac{2}{\gamma - 1}a\right) = -\frac{1}{\gamma - 1}a_0$$

is constant throughout the simple wave. It follows from the last of these equations and (4.66) that

$$u = \frac{2}{\gamma - 1}(a - a_0) = \frac{2}{\gamma + 1}(w - a_0) \tag{4.67}$$

and
$$a - a_0 = \frac{\gamma - 1}{\gamma + 1}(w - a_0) \tag{4.68}$$

where use has been made of (4.66).

Any curve in the xt plane may be described by giving t as a function of ξ and then determining x from (4.65). If the curve is to represent a shock wave, we must then have

$$\frac{dx}{dt} = V(\xi)$$

where V is the shock velocity. Differentiating (4.65), we obtain as the differential equation for the shock trajectory

$$[V(\xi) - w(\xi)]\frac{dt}{d\xi} - \frac{dw}{d\xi}t = 1 \tag{4.69}$$

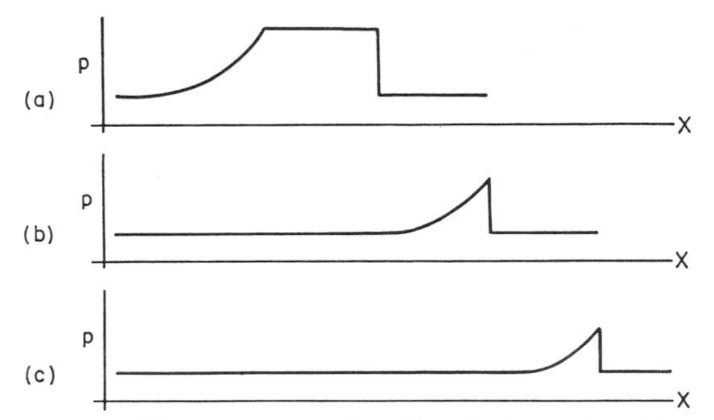

FIG. 4.1. The pressure as a function of x for various times for a shock followed by a rarefaction: (a) At time $t < 0$, before catch-up; (b) At catch-up; (c) After catch-up.

This equation can be solved for t as a function of ξ when $V(\xi)$ is known. However, the Rankine-Hugoniot equations (see Chap. 2 of Part 3) enable us to express the shock velocity in terms of the change in particle velocity as

$$V = u_0 + a_0 + \frac{\gamma + 1}{4}(u - u_0) + \frac{(\gamma + 1)^2}{32}\frac{(u - u_0)^2}{a_0} + \cdots$$

where u_0 and a_0 are the particle and sound velocity in the gas ahead of the shock. In the case under consideration $u_0 = 0$ since the gas ahead of the shock is at rest. In view of (4.67) this equation may be written as

$$V(\xi) = a_0 + \tfrac{1}{2}[w(\xi) - a_0] + \frac{1}{8a_0}[w(\xi) - a_0]^2 \quad (4.70)$$

on writing $w(\xi) = a_0[1 + \sigma(\xi)]$ the differential equation (4.69) becomes

$$\frac{a_0\sigma}{8}(\sigma - 4)\frac{dt}{d\xi} - a_0 t\frac{d\sigma}{d\xi} = 1$$

which may be integrated to give

$$a_0 t = 8\left(\frac{4}{\sigma} - 1\right)^2 \int_\xi^0 \frac{\sigma(\eta)\, d\eta}{(4 - \sigma)^3} \quad (4.71)$$

Since u is positive behind the shock, it follows from (4.67) that $\sigma > 0$ behind the shock. Further, since the shock must be subsonic relative to the gas behind it, we must have

$$w - V = \frac{a_0\sigma}{8}(4 - \sigma) > 0$$

hence $4 > \sigma$. If the simple wave, characterized by s = constant, is compressive, the shock front is accelerated and consequently increases in strength. If it is a rarefaction, the shock front is decelerated and consequently decreases in strength. These statements follow from the fact that $dw/d\xi > 0$ for a rarefaction wave and $dw/d\xi < 0$ for a compressive wave. These latter statements are consequences of the fact that the characteristics do not have an envelope in the region to the right of the piston path for a rarefaction wave and do have one for a compression wave. It then follows from (4.70) that $dV/d\xi > 0$ if $dw/d\xi > 0$ and $dV/d\xi < 0$ if $dw/d\xi < 0$. Moreover $d\xi/dt < 0$, as follows from (4.69). Hence $dV/dt > 0$ for a compression wave, and $dV/dt < 0$ for a rarefaction wave.

When the function $t(\xi)$ given by (4.71) is substituted into (4.65), we obtain $x(\xi)$. These two expressions then determine the parametric equations of the shock front. The functional form of $t(\xi)$ depends on the function $w(\xi)$. This in turn depends on the nature of the rarefaction wave, for it gives the slope of the characteristics (4.65). If the rarefaction is caused by a piston motion, then the velocity of the fluid at the piston is prescribed on the curve representing this motion in the xt plane (see Fig. 4.2). It then

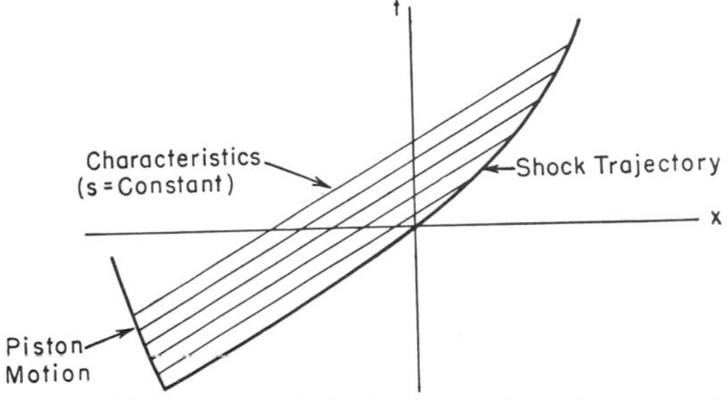

FIG. 4.2. The characteristics in the xt plane for the problem of the decay of a shock.

follows from the first of Eqs. (4.67) that the sound velocity is also prescribed along this curve. Hence w is prescribed along this curve. Thus the family of straight lines (4.65) may be determined, and the parameter ξ is the value of x at which $t = 0$, the time at which the first characteristic intersects the shock trajectory.

The pressure density and particle velocity at any point in the flow behind the shock are determined by the equations

$$p = p_0\left(\frac{a}{a_0}\right)^{2\gamma/(\gamma-1)} \qquad \rho = \rho_0\left(\frac{a}{a_0}\right)^{2/(\gamma-1)}$$

and Eq. (4.67), which, when written in terms of σ, become

$$\begin{aligned} p &= p_0\left(1 + \frac{2\gamma}{\gamma + 1}\sigma + \frac{\gamma}{\gamma + 1}\sigma^2\right) \\ \rho &= \rho_0\left[1 + \frac{2}{\gamma + 1}\sigma + \frac{3 - \gamma}{(\gamma + 1)^2}\sigma^2\right] \end{aligned} \quad (4.72)$$

up to second-order terms in σ.

If we assume that the pressure at the tail of the rarefaction wave interacting with the shock equals the pressure ahead of the shock, then the gas behind the rarefaction is at rest. The tail is then characterized by the parameter ξ_0, and $w(\xi_0) = a_0$, i.e., $\sigma(\xi_0) = 0$. If we now let

$$A = 32\int_{\xi_0}^0 \frac{\sigma(\eta)\, d\eta}{[4 - \sigma(\eta)]^3}$$

Eq. (4.70) becomes for ξ close to ξ_0

$$\sqrt{a_0 t} = \left(\frac{1}{\sigma} - \frac{1}{4}\right)\sqrt{4A} + \cdots$$

or for large values of t

$$\sigma = 2\sqrt{\frac{A}{a_0 t}} - \frac{A}{a_0 t} + \cdots$$

The asymptotic form of the path of the shock front is then

$$x = \xi_0 + a_0 t + 2\sqrt{A a_0 t} - A$$

Since the tail of the shock wave has the path

$$x = \xi_0 + a_0 t$$

the asymptotic expression for the width of the shock front is

$$d(t) = 2\sqrt{A a_0 t} - A$$

If we substitute the asymptotic expression for σ into (4.71), we obtain the expression for the shock strength, $(p - p_0)/p_0$, behind the shock. It decreases in inverse proportion to the square root of the time after the rarefaction has caught up with the shock.

7. Spherical Waves of Finite Amplitude

In a spherically symmetric hydrodynamic flow only the radial component of the particle velocity is nonvanishing, and all quantities characterizing the flow are functions of r, the distance from the origin, and the time t. If we denote by u the particle velocity in the radial direction, and if there are no external forces acting, then the equations of continuity of mass, energy, and momentum become

$$\frac{\partial \rho}{\partial t} + u\frac{\partial \rho}{\partial r} + \rho\left(\frac{\partial u}{\partial r} + \frac{2u}{r}\right) = 0 \qquad (4.73)$$

$$\frac{\partial u}{\partial t} + u\frac{\partial u}{\partial r} + \frac{1}{\rho}\frac{\partial p}{\partial r} = 0 \qquad (4.74)$$

and

$$\frac{\partial}{\partial t}(p\rho^{-\gamma}) + u\frac{\partial}{\partial r}(p\rho^{-\gamma}) = 0 \qquad (4.75)$$

respectively.

For the flow behind a spherical shock wave progressing outward from some center we cannot in general assume that the entropy behind the shock front is constant, for the strength of the spherical shock wave will be varying. This will not be the case if the spherical shock is caused by an expanding sphere where the velocity of the sphere is constant. This special problem has been discussed by Taylor [20] by a method similar to that which will be used below to discuss the problem of spherical motion caused by the release of a finite amount of energy. The latter problem is of interest in the theory of spherical blast waves, and was discussed by Taylor [21] and von Neumann [22]. Other types of approximate treatments of the propagation of shock waves caused by explosions have been given by Brinkley and Kirkwood [23].

When the position of the head of an outward-going spherical wave is given by

$$r = R(t) \qquad (4.76)$$

the total energy in the perfect gas over which the wave has passed is

$$\mathcal{E} = 4\pi \int_0^{R(t)} \left(\tfrac{1}{2}\rho u^2 + \frac{1}{\gamma - 1} p\right) r^2\, dr \qquad (4.77)$$

where the first term in the integrand represents the kinetic energy imparted to the gas particles and the second term represents their internal energy. We seek a solution of Eqs. (4.73) to (4.75) subject to the conditions that $\mathcal{E}$ given by (4.77) is constant and further that across the moving spherical surface given by (4.76) the Rankine-Hugoniot equations (4.4) to (4.6) are satisfied.

We shall assume that the shock wave at the head of the spherically symmetric disturbed region is very strong. Then if ρ_0, p_0, c_0 represent the density, pressure, and sound velocity in the undisturbed region, we may write Rankine-Hugoniot equations as

$$\begin{aligned} \rho_1 &= \frac{\gamma+1}{\gamma-1}\rho_0 \\ p_1 &= \frac{2\gamma}{\gamma+1} p_0 \frac{\dot{R}^2}{c_0^2} = \frac{2}{\gamma+1}\rho_0 \dot{R}^2 \\ u_1 &= \frac{2}{\gamma+1}\dot{R} \\ c_1 &= \frac{\sqrt{2\gamma(\gamma-1)}}{\gamma+1}\dot{R} \end{aligned} \qquad (4.78)$$

where the subscript 1 denotes the values of the variables immediately behind the shock.

If we now assume that

$$\begin{aligned} u &= art^{-1}U(\xi) \qquad & c &= art^{-1}C(\xi) \\ \rho &= \rho_0 \Omega(\xi) \qquad & p &= a^2 r^2 t^{-2} \rho_0 P(\xi) \end{aligned} \qquad (4.79)$$

where a is a dimensionless constant, and

$$\xi = \frac{r}{R(t)} \qquad (4.80)$$

we find, on substituting the first of (4.79) into the third of (4.78), that

$$aRt^{-1}U(1) = \frac{2}{\gamma+1}\dot{R}$$

Hence if we normalize the dimensionless function $U(\xi)$ so that

$$U(1) = \frac{2}{\gamma+1} \qquad (4.81)$$

then we have

$$R(t) = R_0 t^a \qquad (4.82)$$

where R_0 is a constant. The remaining equations of (4.78) then lead to the normalizations

$$P(1) = \frac{2}{\gamma+1} \qquad C(1) = \frac{\sqrt{2\gamma(\gamma-1)}}{\gamma+1}$$

$$\Omega(1) = \frac{\gamma+1}{\gamma-1} \qquad (4.83)$$

In view of (4.82) Eq. (4.80) may be written as

$$R_0\xi = rt^{-a} \tag{4.84}$$

The energy behind the shock at any time t is then given by

$$\mathcal{E} = 4\pi a^2 t^{5a-2} \int_0^1 \left[\tfrac{1}{2}U^2(\xi)\Omega(\xi) + \frac{1}{\gamma - 1} P(\xi) \right] \xi^4 \, d\xi \tag{4.85}$$

This will be a constant independent of the time if, and only if,

$$a = \tfrac{2}{5} \tag{4.86}$$

Hence the motion of the shock front is given in this case by

$$R(t) = R_0 t^{2/5} \tag{4.87}$$

On substituting this expression into (4.78) we may determine the approximate expressions for pressure, particle velocity, and sound velocity behind the shock front as functions of the time or equivalently as functions of the position of the shock front. Thus behind the shock front, $r = R$,

$$p = \frac{A}{R^{3/2}} \tag{4.88}$$

where A is a constant.

The determination of the flow behind the shock front may be reduced to solving ordinary differential equations for the functions $U(\xi)$, $C(\xi)$, and $P(\xi)$. If ρ is eliminated from Eqs. (4.73) to (4.75) by using the relation

$$\rho = \frac{\gamma p}{c^2} \tag{4.89}$$

we obtain the equations

$$\begin{aligned} \frac{\partial u}{\partial r} + \frac{2u}{r} &= -\frac{2}{\gamma - 1}\frac{1}{c}\left(\frac{\partial c}{\partial t} + u\frac{\partial c}{\partial r}\right) \\ \frac{\partial u}{\partial t} + u\frac{\partial u}{\partial r} &= -\frac{c^2}{\gamma p}\frac{\partial p}{\partial r} \\ \frac{1}{p}\left(\frac{\partial p}{\partial t} + u\frac{\partial p}{\partial r}\right) &= \frac{2\gamma}{\gamma - 1}\frac{1}{c}\left(\frac{\partial c}{\partial t} + u\frac{\partial c}{\partial r}\right) \end{aligned} \tag{4.90}$$

When we substitute from Eqs. (4.79) into these equations, we obtain three ordinary differential equations for the three functions $U(\xi)$, $C(\xi)$, $P(\xi)$, which may be written as

$$\begin{aligned} [(1 - U)^2 - C^2]\xi U' &= C^2\left[3U + \frac{2}{\gamma}\left(1 - \frac{1}{a}\right)\right] + U(1 - U)\left(U - \frac{1}{a}\right) \\ [(1 - U)^2 - C^2]\xi C' &= C\left\{\frac{\gamma - 1}{2}U\left(U - \frac{1}{a}\right) + (1 - U)\left(U - \frac{1}{a}\right) + \tfrac{3}{2}(\gamma - 1)U(1 - U) \right. \\ &\qquad \left. - C^2(1 - U)^{-1}\left[U - \frac{1}{a} - \frac{\gamma - 1}{\gamma}\left(1 - \frac{1}{a}\right)\right]\right\} \\ [(1 - U)^2 - C^2]\xi P' &= P\left[2C^2 + 2(1 - U)\left(U - \frac{1}{a}\right) + 3\gamma U(1 - U) + \gamma U\left(U - \frac{1}{a}\right)\right] \end{aligned} \tag{4.91}$$

where the prime denotes differentiation with respect to ξ.

These equations may be solved by first integrating the expression for dU/dC which results on dividing the first of (4.91) by the second, then integrating the second and third of these for C and P as functions of ξ. The integrations were carried out numerically by Taylor [20, 21] both for the case of constant energy in the disturbed region and for the case of the shock wave for an expanding sphere.

Another method for obtaining solutions of the form given by Eqs. (4.79) of Eqs. (4.73) to (4.75) has been described by McVittie [24]. He specializes a general solution of the equations of conservation of mass and momentum which he had previously obtained [25] so that Eq. (4.75) and the shock conditions are satisfied. This general solution obtained by use of the Einstein equations is as follows: let ϕ be an arbitrary function of x, y, z, and t, and define

$$\begin{aligned} \rho &= -\nabla^2\phi \\ \rho\mathbf{v} &= -\operatorname{grad}\frac{\partial\phi}{\partial t} \\ p &= -\frac{\partial^2\phi}{\partial t^2} + \tfrac{1}{3}\nabla^2\left(\sum_{i=1}^{3}\phi_i\right) - \tfrac{1}{3}\sum_{i=1}^{3}\frac{\partial^2\phi_i}{\partial x^{i2}} \\ &\quad + \tfrac{1}{3}\sum_{i=1}^{3}\left(\frac{\partial^2\phi}{\partial x^i\,\partial t}\right)\Big/\nabla^2\phi \end{aligned} \tag{4.92}$$

where ∇^2 is the three-dimensional Laplacian operator and the three functions ϕ_i satisfy the equations

$$\begin{aligned} \nabla^2\phi\,\frac{\partial^2\phi_l}{\partial x^m\,\partial x^n} &= \frac{\partial^2\phi}{\partial x^m\partial t}\,\frac{\partial^2\phi}{\partial x^n\,\partial t} \\ \nabla^2\phi\left(\frac{\partial^2\phi_n}{\partial x^{l2}} - \frac{\partial^2\phi_n}{\partial x^{m2}} + \frac{\partial^2\phi_l}{\partial x^{n2}} - \frac{\partial^2\phi_m}{\partial x^{n2}}\right) &= \frac{\partial^2\phi}{\partial x^l\,\partial t} - \frac{\partial^2\phi}{\partial x^n\partial t} \end{aligned} \tag{4.93}$$

l, m, n being any cyclic permutation of 123. Then the quantities defined by (4.92) and (4.93) satisfy Eqs. (4.1) and (4.2) when no external force is acting, as may be verified by substitution. In the spherically symmetric case Eqs. (4.93) are satisfied by

$$\phi_1 = \phi_2 = \phi_3 = \psi$$

where ψ is defined in terms of ϕ by the equation

$$\frac{\partial\psi}{\partial r} = r\int\frac{I}{r}\,dr + \tfrac{1}{2}rP(t)$$

where $P(t)$ is an arbitrary function of time and

$$I = \frac{\partial^2\phi/\partial r\,\partial t}{\nabla^2\phi}$$

Equations (4.92) become

$$\begin{aligned} \rho &= -\nabla^2\phi \qquad \rho q = -\frac{\partial^2\phi}{\partial r\,\partial t} \\ p &= P - \frac{\partial^2\phi}{\partial t^2} + 2\int\frac{I}{r}\,dr + I \end{aligned} \tag{4.94}$$

where q is the radial velocity of the fluid.

Equations (4.94) are then solution of Eqs. (4.1) and (4.2) in the spherically symmetric case depending on the arbitrary function $\phi(r,t)$ and the arbitrary function $P(t)$. These functions are then to be chosen so that conservation of energy is taken care of and appropriate boundary conditions are satisfied.

8. Effect of Viscosity and Heat Conduction

The discussion of the preceding sections of this chapter assumed that heat conduction and viscosity in the fluid could be neglected and took account of regions where the five state variables describing the fluid underwent abrupt transitions by representing such regions as mathematical discontinuities which were governed by Eqs. (4.4) to (4.6). Such discontinuities do not occur if one takes heat conductivity and viscosity into account in the fundamental conservation laws. However, it can be shown that the results derived on the assumption of the existence of discontinuities in an ideal fluid are closely related to those for a nonideal fluid with dissipative mechanisms such as viscosity and heat conductivity. These dissipative mechanisms are usually neglected in order to obtain more tractable mathematical problems.

However, in many cases the problems involved can be dealt with only by numerical means. When an automatic digital computer is used to obtain numerical solutions, the taking into account of the Rankine-Hugoniot equations (4.4) to (4.6), which must hold across the shocks, is often difficult. Von Neumann and Richtmyer [26] have described a procedure which is convenient for numerical solution of problems involving shocks by a stepwise procedure involving a network of points, with a mesh size Δx, at which the state variables are computed at various times. They introduce artificial dissipative terms into the equations so as to give the shocks a thickness comparable to the spacing of the points of the network. Then the difference equations which approximate the differential equations may be used for the entire calculation just as though there were no shocks at all. In the numerical results obtained the shocks are immediately evident as abrupt transitions that move through the fluid with very nearly the correct speed and across which pressure, density, and particle velocity have nearly the correct jumps.

For a one-dimensional fluid motion the conservation equations become

$$\frac{d\rho}{dt} + \rho \frac{\partial u}{\partial x} = 0 \tag{4.95}$$

$$\rho \frac{du}{dt} = -\frac{\partial}{\partial x}(p + q) \tag{4.96}$$

$$\frac{dE}{dt} + (p + q)\frac{d}{dt}\left(\frac{1}{\rho}\right) = 0 \tag{4.97}$$

if the viscous forces are taken into account and represented by q. Von Neumann and Richtmyer take

$$\begin{aligned} q &= -(\rho_0 c\,\Delta x)^2 \rho \frac{d}{dt}\left(\frac{1}{\rho}\right)\left|\frac{d}{dt}\left(\frac{1}{\rho}\right)\right| \\ &= (\rho_0 c\,\Delta x)^2 \frac{1}{\rho}\frac{\partial u}{\partial x}\left|\frac{\partial u}{\partial x}\right| \end{aligned} \tag{4.98}$$

where c is a dimensionless constant near unity, and replace Eqs. (4.95) to (4.98) by difference equations.

Ludford, Polachek, and Seeger [27] have carried out numerical integrations of the system of equations (4.95) to (4.97) using

$$q = -\tfrac{4}{3}\mu \frac{\partial u}{\partial x} \tag{4.99}$$

where μ is the coefficient of viscosity of the fluid, by giving a particular difference approximation to these equations. They find that, when a low viscosity coefficient (of the order of magnitude expected for a real gas) is used, and if the interval size in x is too large, i.e., larger than the thickness of the shock by an order of magnitude, the effects of viscosity are entirely lost, and the solution is characterized by oscillations of the particles behind the shock front. In order to avoid such difficulties and yet not use a small interval size they have had to resort to using a high value for the viscosity coefficient in their calculations. The methods of von Neumann and Richtmyer avoid this difficulty by using an entirely different type of term for q [see (4.98) and (4.99)].

References

1. Bleakney, W., and A. H. Taub: *Rev. Modern Phys.*, **21**: 584–605 (1949).
2. Lamb, H.: "Hydrodynamics," p. 7, Dover, New York, 1945.
3. Rayleigh: "Theory of Sound."
4. Stoker, J. J.: Surface waves in water of variable depth, *Quart. Appl. Math.*, **5**: 1–54 (1947).
5. Hanson, E. T.: The theory of ship waves, *Proc. Roy. Soc. (London)*, **A111**: 491–529 (1926).
6. Bondi, O.: "On the Problem of Breakers," Admiralty Computing Service, 1943.
7. Miche, A.: Mouvements ondulatoires de la mer en profondeur constante ou décroissante, *Ann. ponts et chaussées*, **114**: 25–78, 131–164, 270–292, 369–406 (1944).
8. Lewy, H.: Water on sloping beaches, *Bull. Am. Math. Soc.*, **52**: 737–775 (1946).
9. Friedrichs, K. O.: Water waves on shallow sloping beaches, *Comm. Appl. Math.*, **1**: 109–134 (1948).
10. ——— and H. Lewy: The dock problem, *Comm. Appl. Math.*, **1**: 135–148 (1948).
11. Heins, A. E.: Water waves over a channel with a dock, *Bull. Am. Math. Soc.*, **53**: 497 (1947).
12. John, F.: Waves in the presence of an inclined barrier, *Comm. Appl. Math.*, **1**: 149–200 (1948).
13. Stoker, J. J.: The formation of breakers and bores, *Comm. Pure Appl. Math. Mechanics*, **1**: 1–86 (1948).
14. Friedrichs, K. O.: On the derivation of shallow water theory: Appendix to the formation of breakers and bores, by J. J. Stoker. *Comm. Pure Appl. Math. Mechanics*, **1**: 1–86 (1948).
15. Riemann, B.: *Abhandl. Ges. Wiss. Göttingen, Math-physik. Kl.*, **8**: 43 (1860) or "Gesamelte Werke," p. 144, 1876.
16. Taub, A. H.: *Ann. Math.*, **47**: 811 (1946).
17. Chandrasekhar, S.: On the Decay of Plane Shock Waves, *Aberdeen Proving Ground, Ballistic Research Laboratory, Report* 423, Nov. 8, 1943.
18. Friedrichs, K. O.: Formation and decay of shock waves, *Comm. Pure Appl. Math. Mechanics*, **1**: 211–245 (1948).
19. Courant, R., and K. O. Friedrichs: "Supersonic Flow and Shock Waves," Interscience, New York, 1948.
20. Taylor, G. I.: The air wave surrounding an expanding sphere, *Proc. Roy. Soc. (London)*, **186**: 273–292 (1946).

21. ———: The Formation of a Blast Wave by a Very Intense Explosion, *Ministry of Home Security R. C.* 210, 1941.
22. von Neumann, J.: Shock Waves started by an infinitesimally short detonation of given (positive and finite) energy, National Defense Research Committee, Div. B, June 30, 1941. 32 pp.
23. Brinkley, S. R., Jr., and J. G. Kirkwood: Theory of propagation of shock waves, *Phys. Rev.*, **71**: 606 (1947).
24. McVittie, G. C.: "Spherically Symmetric Solutions of the Equations of Gas-Dynamics."
25. ———: A method of solution of the equations of classical gas dynamics using Einstein's equations, *Quart. Appl. Math.*, **11**: 327 (1953).
26. von Neumann, J., and R. D. Richtmyer: *J. Appl. Phys.*, **21**: 232 (1950).
27. Ludford, G., H. Polachek, and R. J. Seeger: On unsteady flow of compressible viscous fluids, *J. Appl. Phys.*, **24**: 490 (1953).

Chapter 5

Statics of Elastic Bodies

By RICHARD A. BETH, Brookhaven National Laboratory

1. Elastic Bodies and Structures

Before applying the theory of elasticity in detail to particular cases, such as beams, columns, etc., we review in this section some general ideas and theorems concerning elastic bodies and structures.

Elasticity is the property of recovery of an original size and shape [6, p. 92; 7, p. 141].† Properly speaking, elasticity deals only with reversible stress-strain relations, i.e., those for which the added strain produced by an increment of stress completely vanishes when the stress increment is removed. For a material whose dimensions depend on temperature a change of stress produces a slight change in temperature. Elastic reversibility may still be maintained by specifying either that changes be slow enough to be isothermal or fast enough to be adiabatic; the former is appropriate in considering the statics of elastic bodies and the latter in considering vibrations [6, pp. 94–99, 108–109; 7, p. 159; 16, p. 73; 24, p. 139].

The highest stress that will still permit recovery of the original shape and size of the element when the stress is removed is called the *elastic limit* for the type of stress involved. Some authorities [19, p. 305; 17, p. 141] define, in addition, a *proportional limit* as the greatest stress which a material is capable of developing without a deviation from Hooke's law of proportionality of stress to strain. For practical purposes the distinction is often ignored. Stress beyond the elastic limit results in permanent plastic deformation (set) and, ultimately, rupture of the material.

The definition of elasticity implies that the final displacements of the points of an elastic body produced by certain forces applied simultaneously will vanish if the forces are removed in any order. By reversing the steps of removal we see that the *final displacements are independent of the order of application of the forces.* We must, however, exclude from this statement situations involving stability and buckling, where the stresses are not always uniquely related to the applied forces, and the removal of greater than critical loads is, consequently, not uniquely reversible (Sec. 4).

Saint-Venant's principle states that if the forces acting on a small portion of an elastic body are replaced by another statically equivalent system of forces acting on the same small portion of the body, this redistribution of loading, although it may produce substantial changes in the stresses locally, will have a negligible effect on the stresses at distances which are large in comparison with the linear dimensions of the portion on which the forces are changed [18, pp. 33, 150; 2, pp. 80–83; 6, pp. 21, 131; 15, pp. 95–96 (literature references to proofs); 17, pp. 98–100]. Statically equivalent systems of forces have the same resultant force and moment of force. They are called statically equipollent systems in English books.

At the cost of some rigor,‡ it will save space and make the main ideas easier to follow if, on the basis of Saint-Venant's principle, we replace distributed body and surface forces by equivalent discrete forces acting at selected points of the elastic body or structure.

Since we are interested only in static equilibrium, we imagine the body to be held by constraints that are just sufficient to prevent translation and rotation while permitting elastic deformation. We may use, for example, a rectangular coordinate system determined by three noncollinear points a, b, c of the body at which constraints hold

$$0 = x_a = y_a = z_a = y_b = z_b = z_c$$

at all times. We may then specify any elastic deformation in terms of the *displacement components* s_i, with $i = 1, 2, 3, \ldots$, defined as the changes in x_b, x_c, y_c and in x, y, z at other points. Let P_i be the corresponding *force components* that produce the deformation. The P_i shall include all deforming loads and body forces, e.g., gravity, in as much detail as needed in view of Saint-Venant's principle but not the six constraint reactions required to hold the body in equilibrium at all times.

Even without introducing *Hooke's law* the definition of elasticity implies that an elastic body or structure is a conservative mechanical system in the sense that the work done in applying the forces P_i and producing the displacements s_i is stored as elastic strain energy U, which may be recovered by reducing the forces to zero again. The strain energy stored in the whole or any part of an elastic body cannot be negative, because elastic energy cannot be extracted from a previously unstrained body by applying forces to it, nor can the removal of forces require an input of energy.

† Numbers in brackets refer to references at end of chapter.

‡ Which can be made up by reading [2, 5, 6, 15, 24].

The increment of strain energy associated with a change in the displacements is equal to the work done by the applied forces:

$$dU = \Sigma P_i \, ds_i \tag{5.1}$$

This is a total differential of U regarded as a function of the displacements. Hence

$$P_i = \frac{\partial U}{\partial s_i} \tag{5.2}$$

If the system of external forces is conservative with a potential $V = V(s_i)$, then

$$P_i = -\frac{\partial V}{\partial s_i}$$

which must equal P_i in (5.2) for equilibrium, or

$$\frac{\partial}{\partial s_i}(U + V) = 0 \tag{5.3}$$

Thus, *the total potential energy $U + V$ has a stationary value for equilibrium and a minimum value for stable equilibrium* [17, pp. 17–18].

We denote prescribed or given values by an asterisk and distinguish between component directions, index i, for which the forces $P_i{}^*$ but not the displacements s_i are prescribed, and component directions, index j, for which the displacements $s_j{}^*$ but not the forces P_j are prescribed. Then the potential V of the external forces need not vary with the prescribed displacements, and since the $P_i{}^*$ are given constants,

$$V = -\Sigma P_i{}^* s_i$$

The *minimum principle for displacements* [2, pp. 72–74; 24, pp. 71–72; 27, p. 620] states that *the unprescribed displacements s_i are such that the total potential energy $U - \Sigma P_i{}^* s_i$ is a minimum for stable equilibrium.*

Finally, *if only displacements but no forces are prescribed*, we set $P_i{}^* = 0$ and find that *the unprescribed displacements s_i are such that the strain energy U is a minimum for stable equilibrium.*

Relations (5.1) to (5.3) and the italicized principles apply to all elastic phenomena, including those involving stability and buckling, without regard to any particular law, such as Hooke's law relating forces and displacements [27, p. 620].

The history of the quantitative theory of elasticity† really begins with Hooke's law about 1676. Hooke originally meant his law to apply to any "springy" body as a whole [8, pp. 93–95], and this concept is more useful for our present purpose than the later more general form given in Eq. (1.29) of Part 3, Chap. 1.

In terms of our previous notation, Hooke's law for an elastic body or structure as a whole states that, *within the proportional limits, if a system of forces P_i produces displacements s_i, then forces aP_i will produce displacements as_i, where a is a positive or negative proportionality constant.* Hooke's law in this form covers all first-order elastic phenomena except those involving buckling and stability.

By using Hooke's law we can find *Clapeyron's expression* for the strain energy [2, p. 79; 6, p. 173; 15, p. 90; 17, p. 9; 20, part I, pp. 305–307; 24, p. 118],

$$U = \tfrac{1}{2}\Sigma P_i s_i \tag{5.4}$$

by integrating (5.1) from $a = 0$ to $a = 1$ after inserting aP_i for P_i and $s_i \, da$ for ds_i. We subtract (5.1) from twice the differential of (5.4) and obtain

$$dU = \Sigma s_i \, dP_i \tag{5.5}$$

If we compare this with the total differential of U considered as a function of the P_i, we find *Castigliano's theorem* for the displacements s_i [2, p. 83; 17, p. 15; 19, pp. 283–291; 27, p. 622]:

$$s_i = \frac{\partial U}{\partial P_i} \tag{5.6}$$

If, as before, we denote prescribed values by $P_i{}^*$ and $s_i{}^*$ and the associated unprescribed values by s_i and P_j, then, by (5.5), the variation of U with the unprescribed forces is

$$\delta U = \Sigma s_j{}^* \, \delta P_j = \delta \Sigma s_j{}^* P_j$$

and, hence,

$$\delta(U - \Sigma s_j{}^* P_j) = 0 \tag{5.7}$$

where the quantity $(U - \Sigma s_j{}^* P_j)$ is called the *complementary energy* [15, p. 286] (*Ergänzungsarbeit* [24, p. 72]). The *minimum principle for forces* [2, pp. 74–77; 15, p. 286; 24, p. 72] states that *the unprescribed forces P_j are such that the complementary energy $U - \Sigma s_j{}^* P_j$ is a minimum for stable equilibrium.* The content of this principle, formulated in various ways as *Castigliano's principle, Castigliano's second theorem, Castigliano's principle of least work*, etc., finds many engineering applications [3, vol. II; 10; 25; 2, pp. 85–86; 17, pp. 91–97; 20, part I, pp. 320–325]. For example, the load P_j carried by a redundant member in a strained structure may be found by interpreting $s_j{}^*$ as the "lack of fit" that would result if the redundant member were cut, i.e., as the displacement that is imposed on the elastic structure by considering the redundant member to be intact and uncut.

It can be shown as follows that the principle of superposition holds for an elastic body or structure that obeys Hooke's law. Consider two force systems, aP_i' and bP_i'', that produce displacements as_i' and bs_i'', respectively, when applied alone and displacements as_{ib}' and bs_{ia}'', respectively, when either one is applied *after* the other has been applied and is held constant. We indicate by the added subscripts a and b that each of the latter displacements *may* depend on the previously applied forces. However, as noted before, the definition of elasticity implies that the final displacements s_i due to the application of the forces $P_i = aP_i' + bP_i''$ are independent of the order of application of the two force systems. Hence

$$s_i = as_i' + bs_{ia}'' = bs_i'' + as_{ib}' \tag{5.8}$$

† The history from Galileo to Kelvin is critically discussed in great detail in [23]; see also the Historical Introduction in [6, pp. 1–31]; references to original papers may be found in items marked with an asterisk in the references.

If we "separate the variables,"

$$\frac{s_i' - s_{ib}'}{b} = \frac{s_i'' - s_{ia}''}{a} = k \tag{5.9}$$

the first expression can depend only on b and the second only on a. Thus each is equal to the same constant k, as indicated. Then (5.8) becomes $s_i = as_i' + bs_i'' - kab$, and if we choose $a = b$, we have

$$s_i = a(s_i' + s_i'') - ka^2$$

which contradicts Hooke's law unless $k = 0$. Hence, from (5.9),

$$s_{ia}' = s_i' \quad \text{and} \quad s_{ib}'' = s_i'' \tag{5.10}$$

i.e., within the Hooke's-law range, the displacements produced by any force system are independent of the presence or absence of other force systems, and the *principle of superposition* holds, viz., *the total displacements produced by a combination of force systems acting together is simply the algebraic sum of the displacements that would be produced by each of the force systems acting separately* [17, pp. 4–7].

Kirchhoff's theorem concerning the uniqueness of solution [6, p. 170; 15, pp. 92–94; 17, pp. 13–15; 24, pp. 75–76] asserts that, when some forces P_i^* and other displacements s_j^* are prescribed and *Hooke's law applies, only one configuration of equilibrium is possible.* Suppose that different sets s_i', P_j' and s_i'', P_j'' were possible for the unprescribed quantities. By superposition we can subtract the second system from the first to form a third system, for which the strain energy, by Clapeyron's theorem (5.4), is

$$U = \tfrac{1}{2}\Sigma[(P_i^* - P_i^*)(s_i' - s_i'') + (P_j' - P_j'')(s_j^* - s_j^*)] = 0$$

Since the strain energy stored in any part of the body cannot be negative, it must be zero in every part of the body for this third system, which, therefore, can only represent the original, undistorted body. Hence, contrary to the assumption, $s_i' = s_i''$,

$$P_j' = P_j''$$

and the theorem is proved.

The *reciprocal theorem* [2, p. 80; 6, pp. 173–176 (refers to applications); 15, pp. 297–300; 17, pp. 11–12; 24, pp. 118–121] of Betti and Rayleigh states that, if the force systems P_i' and P_i'' produce displacements s_i' and s_i'', respectively, then

$$\Sigma P_i's_i'' = \Sigma P_i''s_i' \tag{5.11}$$

i.e., *the work done by the forces of one system acting through the displacements of a second is equal to the work done by the forces of the second system acting through the displacements of the first, provided that Hooke's law holds.* The strain energy stored in the body when the system P_i' is applied first and then the system P_i'' is applied while the P_i' are held constant is, by (5.4) and (5.10),

$$U = \tfrac{1}{2}\Sigma P_i's_i' + \Sigma P_i's_i'' + \tfrac{1}{2}\Sigma P_i''s_i''$$

Since this expression must be equal to the similar expression for the strain energy when the force systems are applied in reverse order, (5.11) follows.

The *influence coefficient* a_{jk} is defined as the value of the displacement component s_j when the only deforming force acting on the body or structure is the unit component $P_k = 1$. We define a_{kj} similarly, and then (5.11) gives

$$a_{kj} = a_{jk} \tag{5.12}$$

which is Maxwell's form of the reciprocal theorem.

By Hooke's law and superposition, the displacements produced by any force system P_k are

$$s_j = \Sigma a_{jk}P_k \tag{5.13}$$

where the summation is over the repeated index. By (5.4) the strain energy may be written as a positive definite form in the deforming force components [17, p. 13]

$$U = \tfrac{1}{2}\sum_{j,k} a_{jk}P_jP_k \tag{5.14}$$

Evidently, because of (5.12) and (5.13), the relation (5.6) is satisfied, and

$$a_{jk} = \frac{\partial s_j}{\partial P_k} = \frac{\partial^2 U}{\partial P_j\,\partial P_k} \tag{5.15}$$

If the linear equations (5.13) are solved for

$$P_k = \Sigma A_{kj}s_j \tag{5.16}$$

then A_{kj} is the required value of the force component P_k when we prescribe a unit displacement, $s_j = 1$, and compel all other displacement components to be zero [2, p. 95]. A_{jk} is defined similarly, and by the reciprocal theorem (5.11) we find

$$A_{jk} = A_{kj} \tag{5.17}$$

The strain energy, by (5.4), is a positive definite quadratic form in the displacement components:

$$U = \tfrac{1}{2}\sum_{k,j} A_{kj}s_ks_j \tag{5.18}$$

Because of (5.16) and (5.17), the relation (5.2) is satisfied, and

$$A_{kj} = \frac{\partial P_k}{\partial s_j} = \frac{\partial^2 U}{\partial s_k\,\partial s_j} \tag{5.19}$$

2. The Elastic Moduli

The ideas of Sec. 1 apply to any elastic structure (framework, truss, bridge, etc.) or any structural element (beam, spring, plate, etc.) as well as to any differential element of volume of an elastic body. The general principles do not require that a structure be simply connected, that it be composed of just one material, or that the materials be isotropic. The literature cited shows that the conclusions of Sec. 1 can be derived from the analytical basis given in Part 3, Chap. 1, viz., (*a*) Eqs. (1.34), which constitute Hooke's law for isotropic materials, (*b*) the equations of compatibility (1.28), which ensure that strains be continuous throughout the body, and (*c*) the equation of equilibrium for the stresses [16, p. 60]

$$\text{Div}\,\mathfrak{T} + \rho\mathbf{g} = 0 \tag{5.20}$$

which results from (1.6) when the acceleration is zero. In principle, any problem in isotropic elasticity reduces to the solution of these equations with the prescribed loads and displacements as boundary conditions. However, the analysis becomes quite involved, and it is practically impossible to find exact solutions for even the simplest structural elements (for the simplest soluble cases, see [24, pp. 91–105]). Very good approximate solutions may be obtained on the basis of the minimum principles by suitable choice of the form of the strain-energy expression [24, pp. 82–91].

Instead of using this general analytical approach, applied elasticity and the strength of materials more often attack problems by what is a synthetic method: a solution is built up by visualizing the strain distribution and relating it to the stress distribution and the applied forces by means of the elementary moduli (bulk, k; shear, μ; Young's, E) and Poisson's ratio σ, which are commonly tabulated for various materials. The first-order Hooke's-law theory neglects higher-order effects, and therefore stress is computed in terms of the *unstrained* dimensions and orientation of the element of area on which it acts, although in stability problems (columns, etc.) the net displacement of the element may have to be taken into account in relating stresses to applied forces. Simplicity justifies the assumption that materials are isotropic even when it is probably not strictly true, as with rolled or drawn metals. In design work the deflections and a safe estimate of the maximum stress to which the material will be subjected are of major interest.

In applied elasticity the definitions of the moduli are visualized as follows. Let Fig. 5.1 represent a rectangular element of volume, whose dimensions a, b, c are increased by e_1a, e_2b, and e_3c, respectively, by the action of the principal stresses T_1, T_2, and T_3, which, therefore, give it a *relative* increase in volume, or dilatation,

$$\Delta = (1 + e_1)(1 + e_2)(1 + e_3) - 1 = e_1 + e_2 + e_3 + \cdots \quad (5.21)$$

where higher terms in the strains e_1, e_2, and e_3 may be neglected.

When the stress is the same in all directions, $T = T_1 = T_2 = T_3$, we define the *bulk modulus*, or *modulus of compression*.

$$k = \frac{T}{\Delta} \quad (5.22)$$

as the ratio of stress to volume strain Δ.

When only one principal stress is different from zero, say $T_2 = T_3 = 0$, we define *Young's modulus*

$$E = \frac{T_1}{e_1} \quad (5.23)$$

as the ratio of the uniaxial tension to the extensional strain in the same direction. In this case there is a transverse contraction σe_1 in every direction perpendicular to the uniaxial tension, i.e.,

$$e_2 = e_3 = -\sigma e_1 \quad \text{and} \quad \Delta = e_1(1 - 2\sigma) \quad (5.24)$$

where *Poisson's ratio* σ is a dimensionless constant of the material. Since materials subjected to positive longitudinal tension do not expand laterally, $\sigma \geq 0$, and since their volume does not decrease, $\sigma \leq \frac{1}{2}$. Instead of Poisson's ratio as here defined European writers often use its reciprocal, $m = 1/\sigma$, and call m *Poisson's number* or *Poisson's constant*.

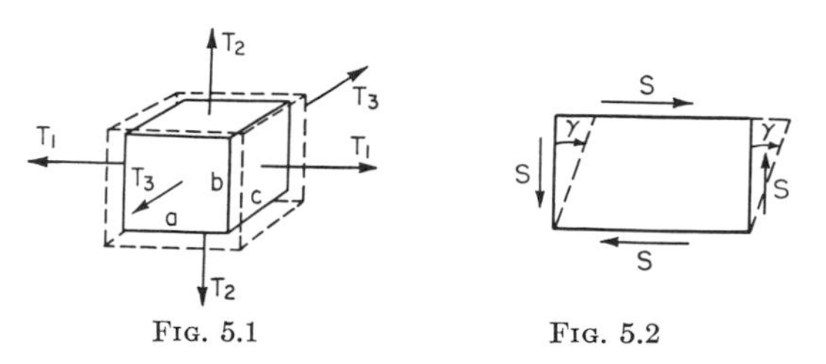

FIG. 5.1 FIG. 5.2

Shear is first introduced without reference to principal stress directions. Let Fig. 5.2 represent an initially rectangular element of volume subjected only to the shearing stresses, S = tangential force per unit area, on the lateral faces as shown. We define the *shear modulus*, or *modulus of rigidity*,

$$\mu = \frac{S}{\gamma} \quad (5.25)$$

as the ratio of shearing stress to the shear strain, γ radians, which is the angular change in inclination (much exaggerated in the figure) between the originally perpendicular sides. That the *shearing stresses on mutually perpendicular planes are always equal in magnitude as shown* [19, p. 59] follows at once from the requirement of equilibrium with respect to rotation for any rectangular element like that of Fig. 5.2.

The definition of μ involves change of shape without change of volume, while the definition of k involves change of volume without change of shape.

In order to obtain the principal strains in the general case of Fig. 5.1 we may, by the principle of superposition, simply add the strains that would be produced by uniaxial stresses T_1, T_2, and T_3 acting separately:

$$e_1 = \frac{T_1 - \sigma(T_2 + T_3)}{E} \qquad e_2 = \frac{T_2 - \sigma(T_3 + T_1)}{E} \qquad e_3 = \frac{T_3 - \sigma(T_1 + T_2)}{E} \quad (5.26)$$

By (5.21) the sum of these equations is the volume strain

$$\Delta = \frac{3T(1 - 2\sigma)}{E} \quad (5.27)$$

where

$$T = \tfrac{1}{3}(T_1 + T_2 + T_3) \quad (5.28)$$

is the *average* stress When $T_1 = T_2 = T_3$,

$$k = \frac{T}{\Delta} = \frac{E}{3(1 - 2\sigma)} \quad (5.29)$$

and it follows that the *bulk modulus k is the ratio of average stress to volume strain even when the principal stresses are unequal.* This result may also be obtained by applying the reciprocal theorem (5.11) to the equal and unequal stress systems.

We may solve for the principal stresses in terms of the strains by writing Eqs. (5.26) in the form

$$Ee_i = (1 + \sigma)T_i - 3\sigma T \qquad i = 1, 2, 3 \tag{5.30}$$

and inserting $T = k\Delta$ from (5.29) to obtain

$$T_i = \frac{3\sigma k}{1 + \sigma}\Delta + \frac{E}{1 + \sigma}e_i \tag{5.31}$$

Equations (5.31) and (5.26), derived here from the elementary definitions of E, σ, and k, correspond to the principal axis forms of Eqs. (1.34) and (1.37), respectively, of Part 3, the coefficient of Δ in (5.31) being *Lamé's modulus*,

$$\lambda = \frac{3\sigma k}{1 + \sigma} \tag{5.32}$$

An elementary physical definition of the modulus λ may be obtained by considering the volume element of Fig. 5.1 subjected to stresses that make $e_i = 0$ while $T_1 \neq 0$; i.e., $T_2 + T_3 = T_1/\sigma$. Then, according to (5.31), $\lambda = T_1/\Delta$; i.e., λ is the ratio of the tension required to prevent contraction in any direction to the volume strain involved.

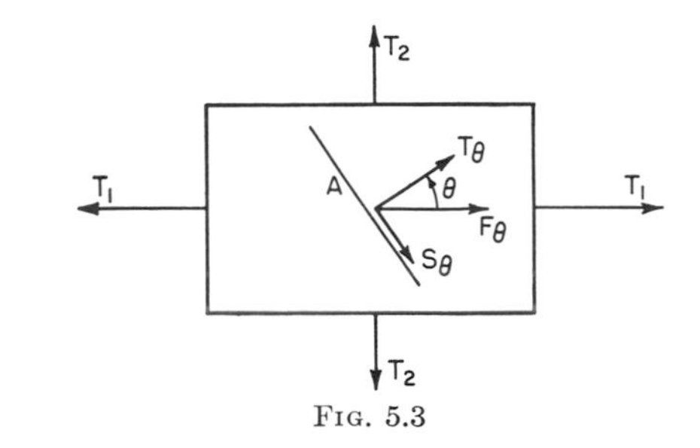

FIG. 5.3

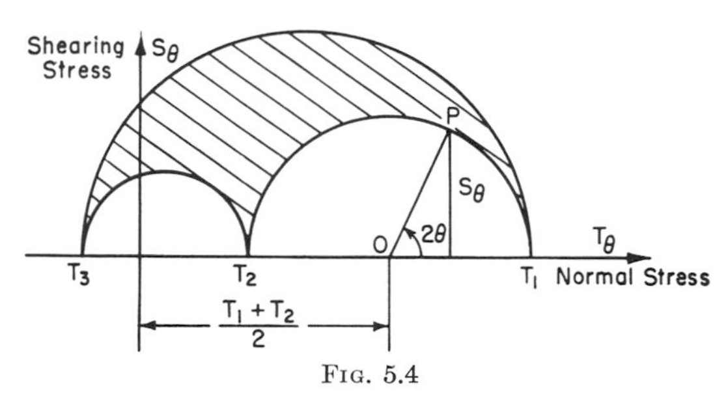

FIG. 5.4

In order to relate shearing stress to the principal stresses we must investigate the directional properties of the stress tensor (Part 3, Chap. 1, Sec. 2). These may be portrayed by *Mohr's circles of stress* [17, pp. 272–276; 19, pp. 59–65; 24, pp. 53–54; 2, pp. 7–12]. Consider first the plane stress system shown in Fig. 5.3. The stress due to T_1 across a unit area A, whose normal lies in the plane of the figure and makes an angle θ with the T_1 direction, is $F_\theta = T_1 \cos\theta$. Thus, owing to T_1 alone, the normal stress across A is $T_\theta = T_1\cos^2\theta$, and the shearing stress across A is $S_\theta = T_1\cos\theta\sin\theta$. If we add to these the similar components due to T_2, we find, for the total normal stress across A,

$$\begin{aligned} T_\theta &= T_1\cos^2\theta + T_2\sin^2\theta \\ &= \tfrac{1}{2}(T_1 + T_2) + \tfrac{1}{2}(T_1 - T_2)\cos 2\theta \end{aligned} \tag{5.33}$$

and for the total shearing stress across A

$$S_\theta = (T_1 - T_2)\sin\theta\cos\theta = \tfrac{1}{2}(T_1 - T_2)\sin 2\theta \tag{5.34}$$

Thus, *the maximum absolute value of the shearing stress,* $|S_\theta|_{\max} = \frac{1}{2}|T_1 - T_2|$, *occurs across planes inclined at 45° to the principal axes,* and, as noted in connection with Fig. 5.2, $|S_\theta| = |S_{\theta \pm 90°}|$.

When S_θ is plotted against T_θ, as in Fig. 5.4, it will be seen from Eqs. (5.33) and (5.34) that corresponding values lie at point P on Mohr's circle of stress, T_1PT_2, such that the radius OP makes an angle 2θ with the normal stress axis.

The third principal stress T_3 is perpendicular to the plane of Fig. 5.3 and therefore contributes nothing to the stress components discussed so far. However, Mohr's circles for the T_1T_3 and T_2T_3 planes may also be put on the diagram of Fig. 5.4. It can then be shown [24, p. 54; 17, pp. 274–275] that *the point representing the normal and shearing stress across any plane, however inclined to the three principal axes, will lie on the shaded area bounded by the three Mohr's circles.* For example, the *maximum shearing stress that occurs at any point of a stressed material is, in magnitude, one-half the difference between the largest and smallest principal stress at that point.* The diagram is also useful in considering theories of elastic failure [17, pp. 282–284].

In order to relate shearing strain to the principal strains we may analyze the directional properties of the strain tensor (see Part 3, Chap. 1, Sec. 6) in a corresponding way. Consider an originally rectangular element of volume with edges a, b, c, lying with edge c parallel to T_3 and edges a and b lying obliquely in the plane of Fig. 5.5, making the angles θ with the principal directions as shown in the unstrained state. Imposition of the principal strains e_1 and e_2 changes the lengths and inclinations of edges a and b as shown in Fig. 5.6, where, neglecting higher powers of e_1 and e_2,

$$e_a = e_1\cos^2\theta + e_2\sin^2\theta \tag{5.35}$$

$$\delta_a = \delta_b = (e_1 - e_2)\sin\theta\cos\theta = \frac{\gamma_\theta}{2} \tag{5.36}$$

$$e_a + e_b = e_1 + e_2 \tag{5.37}$$

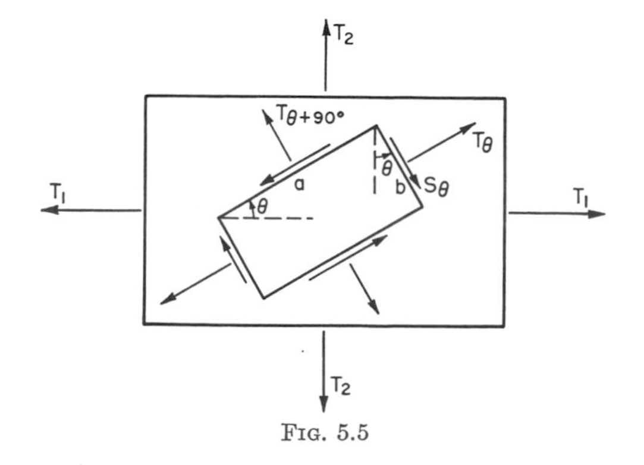

FIG. 5.5

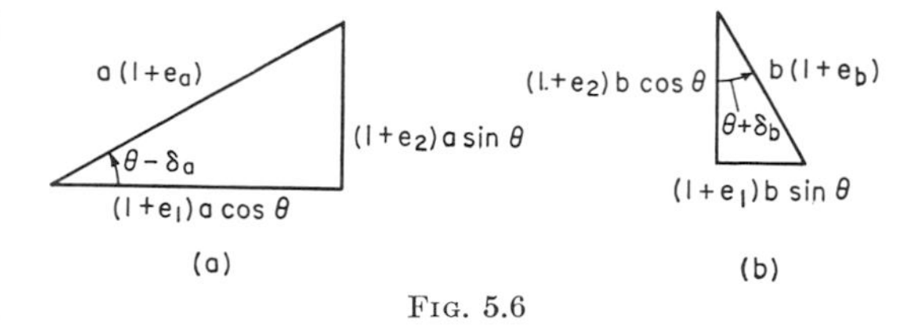

Fig. 5.6

and γ_θ is the shear strain of the inclined rectangle in Fig. 5.5 due to the shearing stress S_θ, which is given by (5.34). We thus find, *independent of* θ

$$\mu = \frac{S_\theta}{\gamma_\theta} = \frac{T_1 - T_2}{2(e_1 - e_2)} = \frac{E}{2(1 + \sigma)} \tag{5.38}$$

where the last expression is obtained by using either (5.26) or (5.31). Thus, the coefficient of e_i in (5.31) is twice the shear modulus μ, as defined in (5.25).

To illustrate the principle of superposition we note that the extensional strain (5.35) is, by (5.26) and (5.33),

$$e_a = \frac{T_\theta - \sigma(T_{\theta+90°} + T_3)}{E}$$

just as would be expected, independent of shear, from the tensions acting in Fig. 5.5, and, by (5.37), the volume strain is independent of the orientation of the element of volume considered.

The strain equations (5.35) and (5.36) have the same form as the stress equations (5.33) and (5.34). The directional properties of the strain tensor at a point can, therefore, be portrayed by *circles of strain*, analogous to Mohr's circles of stress, by plotting a diagram in which the abscissas are extensional strains and the ordinates are *half* the shearing strains [28].

The elastic moduli for isotropic materials, k, E, μ, and λ, each have the dimensions force/area. In Fig. 5.7 dimensionless ratios of these moduli are shown as functions of Poisson's ratio σ, which, as noted with Eqs. (5.24), is a dimensionless constant lying in the range $0 \leq \sigma \leq \frac{1}{2}$ for any material. These ratios have here been derived from the elementary definitions of the moduli. Thus Eqs. (5.29), (5.32), and (5.38) give the ratios

$$\frac{E}{k} = 3(1 - 2\sigma) \qquad \frac{\lambda}{k} = \frac{3\sigma}{1 + \sigma} \qquad \text{and}$$

$$\frac{E}{\mu} = 2(1 + \sigma) \tag{5.39}$$

from which all other ratios plotted may be obtained. Values of σ for typical materials [19, p. 69; 17, p. 140] are indicated at the top of Fig. 5.7. For fluids μ is, by definition, zero. Handbook tables often show poor agreement with the relations (5.31), probably because the moduli were measured on different samples or because the samples were not strictly isotropic.

By using the relations derived in this section we can express the strain energy per unit volume W (see Part 3, Chap. 1, Sec. 7) in various ways involving only *two* of the three kinds of quantities: (*a*) principal stresses, (*b*) principal strains, and (*c*) elastic moduli.

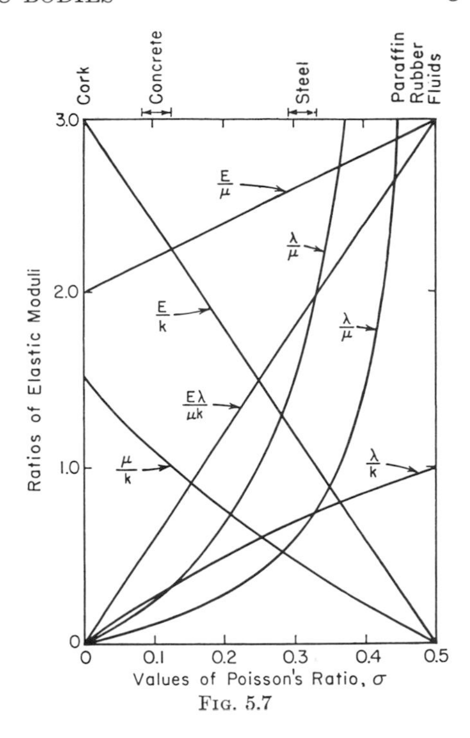

Fig. 5.7

In order to apply the relations of Sec. 1 we consider W to be the strain energy in a unit cube of the material, the T_i being the forces acting on the faces of the cube, and the e_i being the corresponding displacements of the faces. From (5.4) we have W in terms of stresses and strains:

$$\begin{aligned} W_{T,e} &= \tfrac{1}{2}(T_1e_1 + T_2e_2 + T_3e_3) \\ &= \tfrac{1}{2}T\Delta + \tfrac{1}{6}[(T_1 - T_2)(e_1 - e_2) \\ &\quad + (T_2 - T_3)(e_2 - e_3) + (T_3 - T_1)(e_3 - e_1)] \end{aligned} \tag{5.40}$$

In terms of strains and moduli, corresponding to (5.18), W is

$$\begin{aligned} W_e &= \tfrac{1}{2}(\lambda + 2\mu)(e_1^2 + e_2^2 + e_3^2) \\ &\qquad + \lambda(e_1e_2 + e_2e_3 + e_3e_1) \\ &= \tfrac{1}{2}k\Delta^2 + \tfrac{1}{3}\mu[(e_1 - e_2)^2 + (e_2 - e_3)^2 \\ &\qquad + (e_3 - e_1)^2] \end{aligned} \tag{5.41}$$

Finally, in terms of stresses and moduli, corresponding to (5.14), we have

$$\begin{aligned} W_T &= \\ &\frac{\tfrac{1}{2}(T_1^2 + T_2^2 + T_3^2) - \sigma(T_1T_2 + T_2T_3 + T_3T_1)}{E} \\ &= \frac{1}{2}\frac{T^2}{k} + \frac{1}{12\mu}[(T_1 - T_2)^2 + (T_2 - T_3)^2 \\ &\qquad + (T_3 - T_1)^2] \end{aligned} \tag{5.42}$$

The second expression in each pair separates the strain energy per unit volume into two parts, one connected with the change of volume and the other connected with the change of shape, Δ being the volume strain (5.21), and T being the average stress (5.28).

These relations may be used to illustrate the minimum principles of Sec. 1. The minimum principle for displacements tells us that, if T_i^* is prescribed e_i is such that $W_e - T_i^* e_i$ is a minimum. It is easily shown that ·

$$\frac{\partial}{\partial e_i}(W_a - T_i^* e_i) = \frac{\partial W_e}{\partial e_i} - T_i^* = 0$$

yields Eq. (5.31).

Conversely, if e_j^* is prescribed, the minimum principle for forces tells us that P_j is such that $W_T - e_j^* T_j$ is a minimum. Calculation shows that

$$\frac{\partial}{\partial T_j}(W_T - e_j^* T_j) = \frac{\partial W_T}{\partial T_j} - e_j^* = 0$$

yields (5.26).

3. Beams

The primary problems of applied elasticity deal with the elastic behavior of bodies whose extension in one or two dimensions is large relative to that in other directions: beams, columns, bars, rods, plates, and shells. The equilibrium configurations of strings and diaphragms are taken up in connection with vibrations.

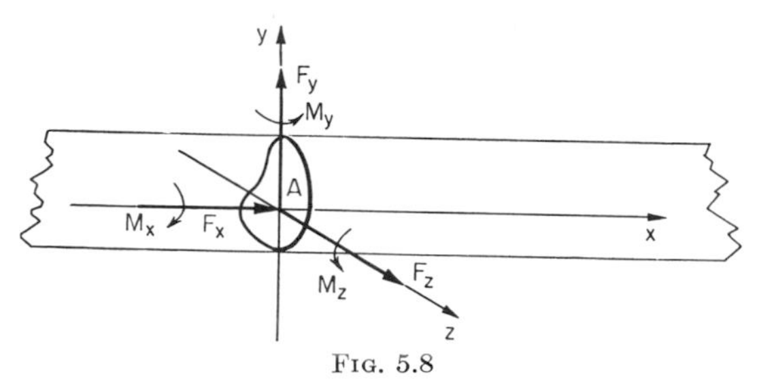

Fig. 5.8

Consider a long, straight bar of uniform cross section, as shown in Fig. 5.8. For reasons that will appear, we choose as x axis the line of centroids of the cross sections and as y and z axes the principal axes of some cross section A, so that the product of inertia of the area A with respect to the axes y and z is zero; i.e.,

$$\int yz\,dA = 0 \tag{5.43}$$

Let the resultant of the stresses acting across A on the part of the bar to the right of A be a force F and a moment of force M. If we resolve F and M into their components along the axes as shown, we can sort out the various possibilities as follows.

Under the heading of *beams* the effect of stresses statically equivalent to F_y and M_z acting alone is analyzed. These tend to bend and shear the rod in the xy plane. The analysis of the effects of adding the components F_z and M_y brings in nothing new, and the resultant total effects on the beam may be obtained by the principle of superposition.

Under the heading of *columns* we deal with the effects of stresses statically equivalent to F_x acting alone. These may give rise to buckling, in which case the principle of superposition does not apply to the bar as a whole. When, in addition, components like F_y and M_z act, we have a beam column or transversely loaded strut.

The effects of stresses equivalent to M_x acting alone are considered under *torsion*. More complicated cases arise when other components are also imposed or when the unstrained member is curved or of nonuniform cross section.

We begin the analysis of a simple beam with the case of *pure bending* in which only a *bending moment*, $M = -M_z$ in Fig. 5.8, acts in the beam and the *shearing force*, $V = F_y$, is zero. This sign convention [19, p. 93] is chosen so that positive M tends to make the beam concave upward.

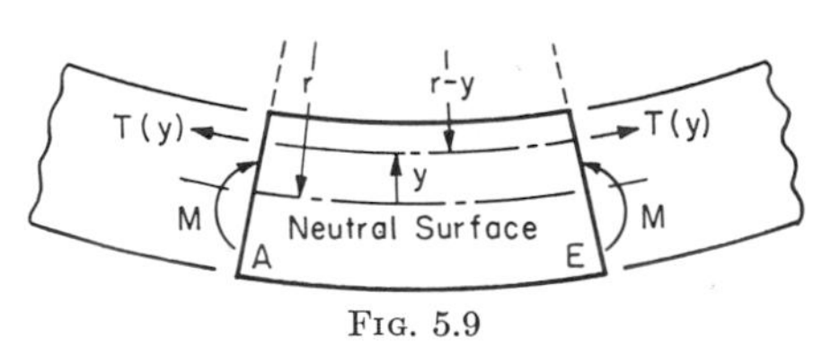

Fig. 5.9

Let Fig. 5.9 be a segment of such a beam sufficiently far from the points at which the external forces are applied so that, by Saint-Venant's principle, the stress distribution no longer depends on the exact manner in which the external forces are applied. (Alternatively, we can imagine Fig. 5.9 to be a segment of a sufficiently long bar bent, without twisting, into a closed loop. When the ends are then smoothly joined, every section will be similar to each other [17, p. 158]). The originally plane cross sections A and B remain plane and perpendicular to the longitudinal filaments, or *fibers*, that compose the beam. The fiber segments become arcs of circles whose centers lie on the line of intersection of planes A and B, which is perpendicular to Fig. 5.9. Unstrained fibers form a cylindrical *neutral surface* of radius r. Fibers which lie a distance y above the neutral surface have a radius of curvature $r - y$ and, therefore, suffer longitudinal strain equal to $-y/r$, provided their distance y from the neutral fibers is not changed by the bending.

Thus, *if we neglect transverse strains* (see Eqs. 5.24) *and stresses*, the fibers must be subject to longitudinal tensions

$$T(y) = -\frac{Ey}{r} \tag{5.44}$$

according to (5.23). By the same assumption the size and shape of the cross sections remain unchanged. Since the net axial tension ($-F_x$ in Fig. 5.8) is zero, we have, for the integral of (5.44) over any cross section A,

$$\int T(y)\,dA = -\frac{E}{r}\int y\,dA = 0 \tag{5.45}$$

i.e., *the centroid of every section lies in the neutral surface.*

The moment M in the beam is the integral of the moments due to the stresses,

$$-yT(y)\,dA = \left(\frac{E}{r}\right) y^2\,dA$$

over the area, which gives the *Bernoulli-Euler law* for the bending of beams

$$M = \frac{EI}{r} = \frac{B}{r} \tag{5.46}$$

where $I = \int y^2 \, dA$ is the moment of inertia of the cross-sectional area about the *neutral axis* in which the cross section intersects the neutral surface, $y = 0$, and $B = EI$ is the *flexural rigidity* [17, p. 45; 20, part I, p. 91] (*Biegungssteifigkeit* [5, p. 165]) of the beam. An *effective flexural rigidity* may be defined for composite beams [17, p. 169].

Unless the integral of the moments of the stresses about the y axis in Fig. 5.9,

$$M_y = - \int zT(y) \, dA = \frac{E}{r} \int yz \, dA$$

is zero, any bending of the beam in the xy plane will induce a moment M_y tending to bend the beam in the yz plane also. Hence we chose principal axes of the cross-sectional area for the y and z axes to begin with by specifying (5.43).

The *maximum fiber stress* T_m (absolute value), which is important for design purposes, will occur for fibers that lie farthest from the neutral surface and for which $y = y_m$. By (5.44) and (5.46),

$$T_m = \frac{My_m}{I} = \frac{M}{Z} \tag{5.47}$$

where $Z = I/y_m$ is called the *section modulus*. Values of Z for various *profile sections* (I beams, channels, angles, etc.) are listed in structural-design handbooks.

The strain energy per unit length u in a bar or beam subjected to pure bending can be calculated by integrating $\frac{1}{2}$ (stress)(strain) over the cross section. Using (5.44) for the stress, $-y/r$ for the strain, and (5.46) yields

$$u = \frac{1}{2}\frac{B}{r^2} = \frac{1}{2}\frac{M}{r} = \frac{1}{2}\frac{M^2}{B} \tag{5.48}$$

The total bending energy stored in a beam can be obtained by integrating (5.48) over the length of the beam. If T_m and u_m are the maximum permissible fiber stress and energy storage per unit length, respectively, then by (5.47)

$$u_m = \frac{1}{2}\frac{Z^2}{B} T_m{}^2 \tag{5.49}$$

which is useful in designing springs for maximum elastic-energy storage [19, pp. 287–288].

The approximation made in neglecting the lateral fiber stresses and strains has practically no effect on the validity and usefulness of the Bernoulli-Euler law (5.46). However, it can be verified experimentally that the tension fibers on the convex side of a bent beam contract laterally, according to (5.24), while the compression fibers on the concave side expand laterally. Thus, an originally rectangular cross section becomes slightly wedge-shaped, and the bent surfaces show hyperbolic double (antielastic) curvature [6, pp. 131, 361; 17, pp. 161–162; 19, p. 114]. The latter can be observed interferometrically to give a measure of Poisson's ratio σ [5, p. 170; 18, p. 254]. (This distortion can be shown qualitatively by bending a rubber eraser with rectangular cross section.) Theories of pure bending that take these effects into account, nevertheless, confirm the Bernoulli-Euler law [5, p. 169; 15, pp. 112–116; 18, pp. 250–254].

So far we have restricted ourselves to pure bending with the shearing force V (F_y in Fig. 5.8) equal to zero. However, even if V is not zero, beam calculations based on (5.46) alone still give very good results and are almost always adequate (see [17, pp. 221–226; 19, pp. 129–139] for the analysis of shearing stress distribution in a beam and [19, pp. 182–184] for the additional deflection due to shear).

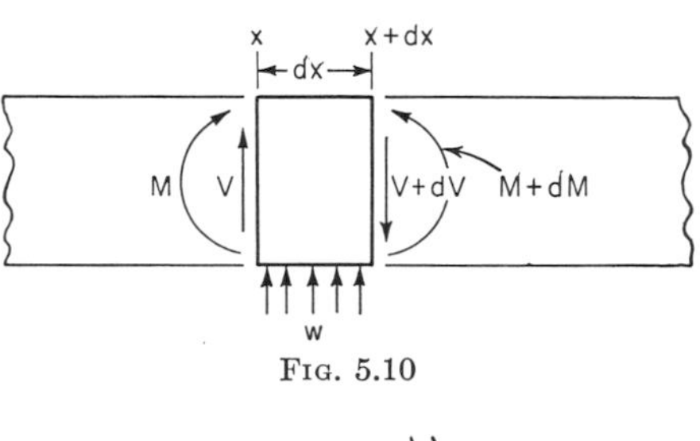

FIG. 5.10

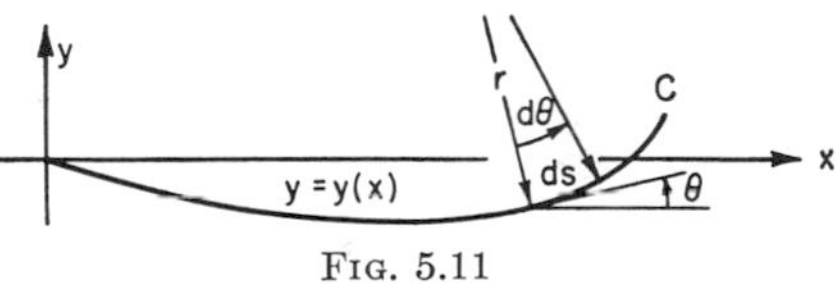

FIG. 5.11

In order to find the curve of deflection of a loaded beam we must first consider the equilibrium of any short segment of the beam. We define the positive directions of the bending moment M, the shearing force V, and the intensity of the distributed load per unit length of the beam w, as shown in Fig. 5.10, acting on the segment of axial length dx between the plane sections at x and $x + dx$. The condition that the total vertical force on the segment be zero gives

$$dV = w \, dx \quad \text{or} \quad w = \frac{dV}{dx} \tag{5.50}$$

and the condition that the total moment of force acting on the segment be zero gives

$$dM = V \, dx \quad \text{or} \quad V = \frac{dM}{dx} \tag{5.51}$$

where the term in dx^2 is, of course, neglected. If a concentrated external force P, positive upward, or a concentrated moment of force L, positive clockwise, acts at any section of the beam, the values of V and M just to the right (+) of that section will be greater than those just to the left (−) by the amounts

$$V_+ - V_- = P \quad \text{and} \quad M_+ - M_- = L \tag{5.52}$$

Let curve C in Fig. 5.11 be the *elastic curve* or *deflection curve* in which the neutral surface of a bent beam or rod intersects the plane of bending. If we measure distance s along the curve and denote by $\theta = \theta(s)$ the angle of inclination of the curve to the x axis, then the curvature of C is $1/r = d\theta/ds$, and

Table 5.1*

	Slope at ends	Deflection at any section in terms of x: δ is positive downward	Maximum and center deflection
1. Cantilever beam—concentrated load P at the free end			
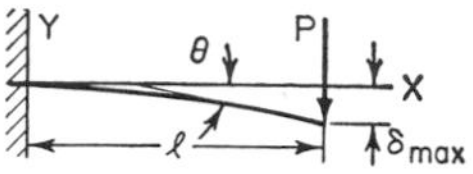	$\theta = \frac{Pl^2}{2EI}$	$\delta = \frac{Px^2}{6EI}(3l - x)$	$\delta_{max} = \frac{Pl^3}{3EI}$
2. Cantilever beam—concentrated load P at any point			
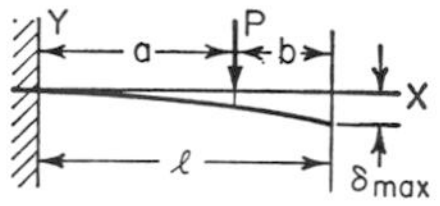	$\theta = \frac{Pa^2}{2EI}$	$\delta = \frac{Px^2}{6EI}(3a - x)$ for $0 < x < a$ $\delta = \frac{Pa^2}{6EI}(3x - a)$ for $a < x < l$	$\delta_{max} = \frac{Pa^2}{6EI}(3l - a)$
3. Cantilever beam—uniformly distributed load of w lb per unit length			
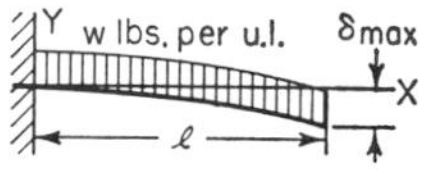	$\theta = \frac{wl^3}{6EI}$	$\delta = \frac{wx^2}{24EI}(x^2 + 6l^2 - 4lx)$	$\delta_{max} = \frac{wl^4}{8EI}$
4. Cantilever beam—uniformly varying load; maximum intensity w lb per unit length			
	$\theta = \frac{wl^3}{24EI}$	$\delta = \frac{wx^2}{120lEI}(10l^3 - 10l^2x + 5lx^2 - x^3)$	$\delta_{max} = \frac{wl^4}{30EI}$
5. Cantilever beam—couple M applied at the free end			
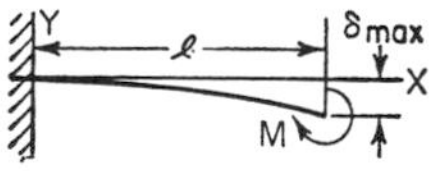	$\theta = \frac{Ml}{EI}$	$\delta = \frac{Mx^2}{2EI}$	$\delta_{max} = \frac{Ml^2}{2EI}$
6. Beam freely supported at ends—concentrated load P at the center			
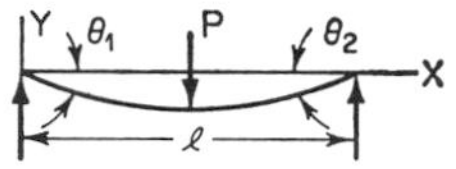	$\theta_1 = \theta_2 = \frac{Pl^2}{16EI}$	$\delta = \frac{Px}{12EI}\left(\frac{3l^2}{4} - x^2\right)$ for $0 < x < \frac{l}{2}$	$\delta_{max} = \frac{Pl^3}{48EI}$
7. Beam freely supported at the ends—concentrated load at any point			
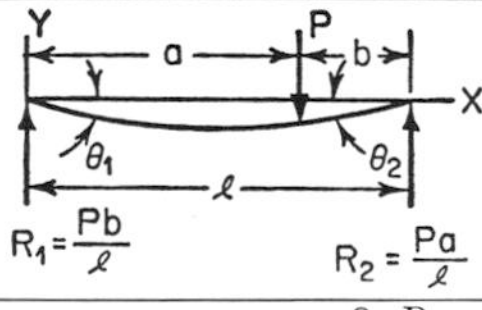 $R_1 = \frac{Pb}{l}$, $R_2 = \frac{Pa}{l}$	Left end $\theta_1 = \frac{Pb(l^2 - b^2)}{6lEI}$ Right end $\theta_2 = \frac{Pab(2l - b)}{6lEI}$	To the left of load P: $\delta = \frac{Pbx}{6lEI}(l^2 - x^2 - b^2)$ To the right of load P: $\delta = \frac{Pb}{6lEI}\left[\frac{l}{b}(x - a)^3 + (l^2 - b^2)x - x^3\right]$	$\delta_{max} = \frac{Pb(l^2 - b^2)^{3/2}}{9\sqrt{3}\,l\,EI}$ at $x = \sqrt{\frac{l^2 - b^2}{3}}$ At center, if $a > b$ $\delta = \frac{Pb}{48EI}(3l^2 - 4b^2)$
8. Beam freely supported at the ends—uniformly distributed load of w lb per unit length			
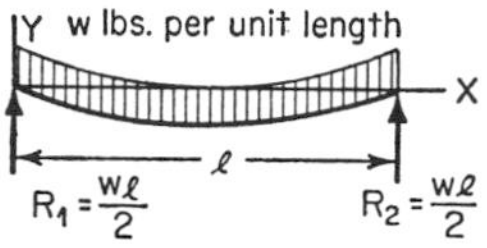 $R_1 = \frac{wl}{2}$, $R_2 = \frac{wl}{2}$	$\theta_1 = \theta_2 = \frac{wl^3}{24EI}$	$\delta = \frac{wx}{24EI}(l^3 - 2lx^2 + x^3)$	$\delta_{max} = \frac{5wl^4}{384EI}$
9. Beam freely supported at the ends—couple M at the right end			
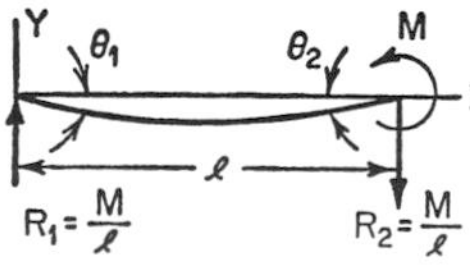 $R_1 = \frac{M}{l}$, $R_2 = \frac{M}{l}$	$\theta_1 = \frac{Ml}{6EI}$ $\theta_2 = \frac{Ml}{3EI}$	$\delta = \frac{Mlx}{6EI}\left(1 - \frac{x^2}{l^2}\right)$	$\delta_{max} = \frac{Ml^2}{9\sqrt{3}\,EI}$ at $x = \frac{l}{\sqrt{3}}$ At center $\delta = \frac{Ml^2}{16EI}$
10. Beam freely supported at the ends—couple M at the left end			
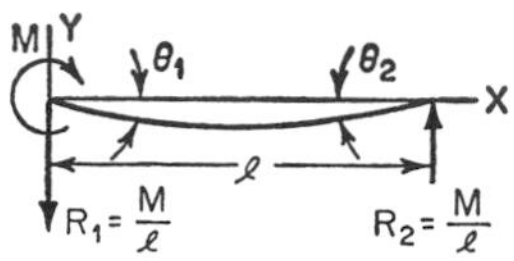 $R_1 = \frac{M}{l}$, $R_2 = \frac{M}{l}$	$\theta_1 = \frac{Ml}{3EI}$ $\theta_2 = \frac{Ml}{6EI}$	$\delta = \frac{Mx}{6lEI}(l - x)(2l - x)$	$\delta_{max} = \frac{Ml^2}{9\sqrt{3}\,EI}$ at $x = \left(1 - \frac{1}{\sqrt{3}}\right)l$ At center $\delta = \frac{Ml^2}{16EI}$

* From Timoshenko and MacCullough, "Elements of Strength of Materials," 3d ed., pp. 182–183, Van Nostrand, Princeton, N.J., 1949, by permission.

the Bernoulli-Euler law (5.46) may be written in the form

$$M = B\frac{d\theta}{ds} \qquad (5.53)$$

where $B = EI$ for homogeneous beams.

In structural design the deflections are usually very small compared to distances along the beam, and it is amply justified to assume that θ (in radians) is so small that θ^2 can be neglected compared to unity. Then $dx/ds \doteq \cos\theta = 1 - \theta^2/2 \cdots \approx 1$, and x may be used for s to measure distance along the beam. In this approximation (5.53) becomes

$$M = B\frac{d\theta}{dx} \qquad (5.54)$$

and

$$\theta = \frac{dy}{dx} \qquad (5.55)$$

gives the slope of the deflection curve, $y = y(x)$. Thus the quantities y, θ, M, V, and w are connected chainwise by the successive derivatives (5.55), (5.54), (5.51), and (5.50), each being a function of x. When a distributed load $w(x)$ is specified and point loads and reactions are put in the form of boundary conditions of the form (5.52), the deflection curve can, therefore, be obtained by successive integration. Results for a number of cases are summarized in Table 5.1 [19, pp. 182–183; 17, p. 191].

Useful results can often be obtained by applying the principle of superposition and the reciprocal theorem (see Sec. 1) to loads and deflections of beams. For engineering work ingenious methods have been developed for evaluating slopes and deflections, such as the area-moment method and the conjugate-beam method [19].

4. Columns

If, in Fig. 5.8, only an axial compressive force $P_x = P$ acts on the end sections of a straight and uniform rod or bar of length L and P is truly axial, i.e., with its line of action along the line of centroids of the cross sections, we have the case of a simple column with "hinged" ends ($M = 0$ at the ends). When the axis of the bar remains straight, it will be shortened by an amount

$$\Delta L = \frac{PL}{AE} \qquad (5.56)$$

where A is the cross section, E is Young's modulus (5.23), and we either assume that P is uniformly distributed across the ends or, on the basis of Saint-Venant's principle, neglect end effects. The strain energy stored is

$$U = \tfrac{1}{2}P\,\Delta L = \frac{P^2L}{2AE} = \frac{AE(\Delta L)^2}{2L} \qquad (5.57)$$

If, however, P is large enough, it turns out that the straight-axis case is one of unstable equilibrium, and stable equilibrium exists only for a configuration in which the axis is curved or buckled. The field of problems of which this is an example is treated under the heading of *elastic stability* [21; 2, pp. 500–620; 5, pp. 277–308; 14, pp. 333–411; 17, pp. 424–464]. It is characteristic of these cases that deflections grow very rapidly with small increases in load after buckling sets in, that stresses beyond the elastic limit and possible failure of the member are imminent, and that the principle of superposition cannot be applied to the body as a whole.

Let the curve of Fig. 5.12 be the deflection curve of such an elastically buckled rod (or column) with hinged ends. We assume that the length L of the elastic curve remains unchanged and that the Bernoulli-Euler law in the form (5.53) applies:

$$-Py = B\frac{d\theta}{ds} \qquad (5.58)$$

or, since $dy/ds = \sin\theta$,

$$\frac{d^2\theta}{ds^2} + \frac{P}{B}\sin\theta = 0 \qquad (5.59)$$

The fact that this relation is of the same mathematical form as the equation of motion of a simple pendulum is an instance of Kirchhoff's kinetic analogue [5, p. 188; 6, p. 399; 17, p. 431; 21, p. 70].

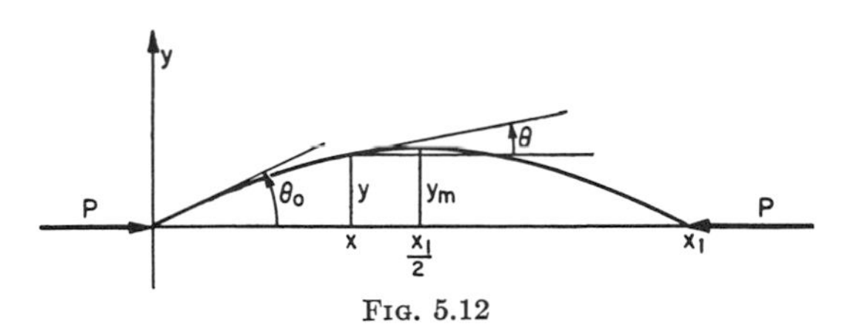

FIG. 5.12

Let P_1 be the lower bound of values of P that will just maintain a slightly buckled form of the bar in equilibrium. P_1 is the Euler, or critical, load for a column. For P just greater than P_1 we may again assume $\theta^2 \ll 1$ and therefore replace $\sin\theta$ by θ in (5.59). The solution for which $\theta = \theta_0$ and $d\theta/ds = 0$ at $s = 0$ is then found to be

$$\theta = \theta_0 \cos\sqrt{\frac{P_1}{B}}\,s \qquad (5.60)$$

so that the deflection curve is, using (5.58),

$$y = -\frac{B}{P_1}\frac{d\theta}{ds} = \theta_0\sqrt{\frac{B}{P_1}}\sin\sqrt{\frac{P_1}{B}}\,s \qquad (5.61)$$

In Fig. 5.12 y is zero for $s = L$; hence, from (5.61), $\sqrt{P_1/B}\,L = \pi$, and the first critical load is

$$P_1 = \frac{\pi^2 B}{L^2} \qquad (5.62)$$

With this value for P_1 the incipient deflection curve (5.61) is

$$y = L\frac{\theta_0}{\pi}\sin\frac{\pi}{L}s \qquad (5.63)$$

Mathematical consideration of the sinusoidal form of (5.63) leads to higher critical values of P as well, viz.,

$$P_n = n^2 P_1 \qquad (5.64)$$

for which $\sqrt{P_n/B}\,L = n\pi$. The corresponding deflection curves have $n - 1$ zeros between $s = 0$ and $s = L$ and are not stable by themselves. However, when θ^2 is not neglected, the actual deflections $y = y(s)$ may be given as a Fourier series of sine terms in $n(\pi/L)s$ for the elastic curve in Fig. 5.12 [9; 5, p. 282].

Obviously θ_0 and the amplitude of the approximate deflection curve (5.63) depend on the amount by which P exceeds P_1. One way to obtain these values is by the use of elliptic integrals, as follows.

If we multiply (5.59) by $(d\theta/ds)\,ds = d\theta$ and integrate from $\theta = \theta_0$ and $d\theta/ds = 0$ at $s = 0$, we obtain, using (5.58) for the last expression,

$$\tfrac{1}{2}B\left(\frac{d\theta}{ds}\right)^2 = P(\cos\theta - \cos\theta_0) = u(s) = \frac{1}{2}\frac{P^2}{B}y^2 \tag{5.65}$$

for the strain energy per unit length (5.48) in the bar. The total strain energy U may be obtained by noting that $\cos\theta = dx/ds$ and integrating over the whole length of the bent bar from $s = 0$ to $s = L$, which gives

$$U = P(x_1 - L\cos\theta_0) \tag{5.66}$$

If we separate variables in the first equation of (5.65) [note from Fig. 5.12 and Eq. (5.58) that $d\theta/ds$ is negative] and introduce k and ϕ by

$$k = \sin\frac{\theta_0}{2} \qquad k\sin\phi = \sin\frac{\theta}{2} \tag{5.67}$$

so that, as s goes from 0 to L, θ goes from θ_0 to $-\theta_0$, and ϕ goes from $\pi/2$ to $-\pi/2$, we find

$$\sqrt{\frac{P}{B}}\,ds = -\frac{d\theta}{\sqrt{2(\cos\theta - \cos\theta_0)}} = -\frac{d\phi}{\sqrt{1 - k^2\sin^2\phi}}$$

and

$$\sqrt{\frac{P}{B}}\,(dx + ds) = \sqrt{\frac{P}{B}}\,(\cos\theta + 1)\,ds = -2\sqrt{1 - k^2\sin^2\phi}\,d\phi$$

Integrating from $s = 0$ to $s = L$ and from $x = 0$ to $x = x_1$ yields

$$\sqrt{\frac{P}{B}}\,L = 2K \qquad \sqrt{\frac{P}{B}}\,(x_1 + L) = 4E \tag{5.68}$$

where K and E are Legendre's complete elliptic integrals of the first and second kinds with modulus k. With $\sqrt{P/B} = \sqrt{P/P_1}\,\pi/L$ from (5.62) and the series expansions in k^2 for K and E, these relations become

$$\sqrt{\frac{P}{P_1}} = \frac{2}{\pi}K = 1 + \tfrac{1}{4}k^2 + \tfrac{9}{64}k^4 + \tfrac{25}{256}k^6 + \cdots \tag{5.69}$$

$$1 + \frac{x_1}{L} = 2\frac{E}{K} = 2\sqrt{\frac{P_1}{P}}\,(1 - \tfrac{1}{4}k^2 - \tfrac{3}{64}k^4 - \tfrac{5}{256}k^6 - \cdots) \tag{5.70}$$

If we square the series (5.69), use

$$p^2 = 2\frac{P - P_1}{P} \tag{5.71}$$

as a measure of the amount by which P exceeds P_1, and solve for k^2, we obtain

$$k^2 = \frac{1 - \cos\theta_0}{2} = p^2(1 - \tfrac{3}{16}p^2 - \tfrac{3}{128}p^4 - \cdots) \tag{5.72}$$

and hence

$$k = \sin\frac{\theta_0}{2} = p(1 - \tfrac{3}{32}p^2 - \tfrac{33}{2048}p^4 - \cdots) \tag{5.73}$$

for the modulus k of the elliptic integrals in terms of the excess of P over P_1.

Using $2\sin(\theta_0/2) \approx 2p$ for θ_0 in (5.63) gives

$$y = \frac{2\sqrt{2}\,L}{\pi}\sqrt{1 - \frac{P_1}{P}}\sin\frac{\pi}{L}s \tag{5.74}$$

as the first approximation to the buckled curve [2, p. 505; 5, p. 282; 6, p. 406; 19, p. 261 (slightly different expressions); 21, p. 74] in terms of the load P.

On the basis of the theory outlined here we can now give series in p for the magnitude of the maximum bending moment $M_m = Py_m$, the total strain energy U, and the other quantities in Fig. 5.12. The inclination θ_0 at the hinged ends can be gotten from (5.72) or (5.73) and

$$\begin{aligned} M_m &= \frac{2L}{\pi}P_1 p(1 + \tfrac{5}{32}p^2 + \tfrac{111}{2048}p^4 + \cdots) \\ U &= LP_1p^2(1 + \tfrac{3}{16}p^2 + \tfrac{7}{128}p^4 + \cdots) \\ y_m &= \frac{2L}{\pi}p(1 - \tfrac{11}{32}p^2 - \tfrac{49}{2048}p^4 - \cdots) \\ x_1 &= L(1 - p^2 + \tfrac{1}{16}p^4 + \tfrac{1}{128}p^6 + \cdots) \end{aligned} \tag{5.75}$$

Neglecting terms in p^3, we therefore have the approximations

$$\begin{aligned} \theta_0 &= 2\sqrt{2}\sqrt{1 - \frac{P_1}{P}} \quad \text{radians} \\ M_m &= \frac{\theta_0}{\pi}L\sqrt{PP_1} \\ U &= 2L(P - P_1) \\ y_m &= \frac{M_m}{P} = \frac{\theta_0}{\pi}L\sqrt{\frac{P_1}{P}} \\ x_1 &= L\left(2\frac{P_1}{P} - 1\right) \end{aligned} \tag{5.76}$$

The buckled rod of Fig. 5.12 is a segment of an *elastica*, which is the elastic curve formed by a uniform originally straight rod bent in a principal plane by forces and couples applied at the ends only. Various forms of the elastica are shown in Fig. 5.13 [5, p. 190; 6, p. 404; 21, p. 74]. The problem of the elastica was important historically in the hands of the Bernoullis and Euler, not only for the theory of stability but for the theory of bending in general [6, pp. 2-3].

Kirchhoff's kinetic analogue is based on the mathematical similarity between the elastic equations of a uniform rod bent and twisted by forces and couples applied at the ends only and the equations of motion of a heavy rigid body turning about a fixed point [5, p. 188; 6, p. 399; 17, p. 431; 21, p. 70]. As was pointed out with Eq. (5.59), the elastica (without twist) corresponds to a plane pendulum—in fact, any one of the forms in Fig. 5.13 can be generated by a point moving with constant speed in a direction continuously determined by the instantaneous position of an oscillating or rotating pendulum; the length of a complete cycle ($2L$) of the elastica corresponds to the period of the pendulum.

The buckled configuration of an end-loaded uniform column of length h is always a segment of an elastica, and may be computed by appropriate adaptation of the formulas developed for the hinged-end case shown in Fig. 5.13. The critical load P_1 is given by (5.62) in terms of the distance L along the equivalent elastica

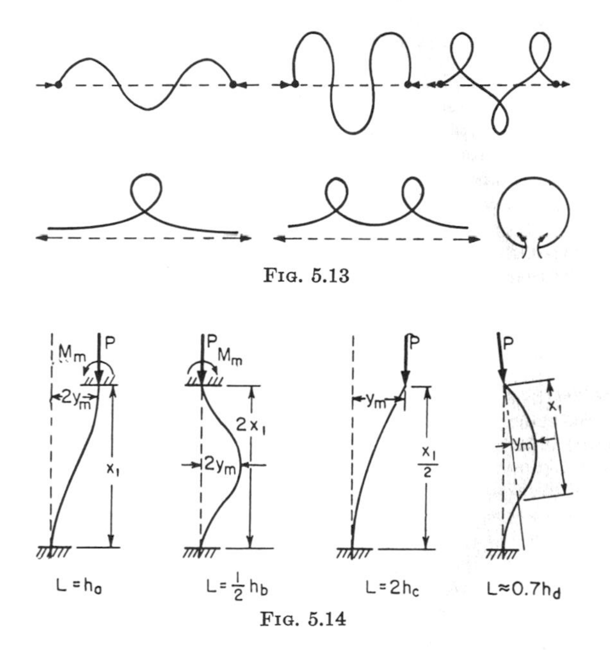

FIG. 5.13

FIG. 5.14

between successive points of inflection where the curve makes the angles θ_0 or $-\theta_0$ with the line of thrust and the bending moments are zero. The relation between L and h depends on the constraints imposed at the ends of the column; various cases are illustrated in Fig. 5.14. With the bottom end fixed in direction (built in) in the initial line of thrust of P, the top end may be built in or hinged and, in each case, may move in a direction normal to the line of thrust or be constrained (guided) to remain in the line tangent to the built-in bottom end. In the last case pictured the resultant line of thrust tilts as shown as soon as P exceeds P_1, a horizontal component being supplied by the guiding constraint at the upper end [2, p. 508; 19, pp. 262–263; 21, pp. 88–90]. Analogous to (5.64), higher critical loads may be computed for each case, but the corresponding configurations are not stable by themselves.

It is characteristic of these buckling situations that the lateral deflections and the magnitude of the maximum bending moment, $M_m = Py_m$, are not proportional to P but grow *very* rapidly with P as soon as buckling sets in, as shown in Fig. 5.15. Complete failure of the member may result soon after the critical load is exceeded. The critical load, and hence such failure, is determined by the elastic properties of the material, and these are, in general, much more reproducible than yield points and ultimate strengths. This fact has been used to design mechanical "fuses."

In the structural design of columns it is necessary to take account of possible eccentric loading and other deviations of actual conditions from the simplified mathematical assumptions underlying the theory given above [19, chap. XI; 21, chaps. II, III]. Buckling phenomena in beams, tubes, plates, shells, etc., are also important in engineering [21; 2, pp. 500–620; 5, pp. 277–308; 14, pp. 333–411; 17, pp. 424–426].

The deflection curve of a straight uniform bar whose lower end is built in at an angle θ_1 to the line of thrust of the load P applied to its upper end is a segment of an elastica, as shown in Fig. 5.16. Such a member may be called a cantilever strut, and series expressions in terms of θ_1 and P have been worked

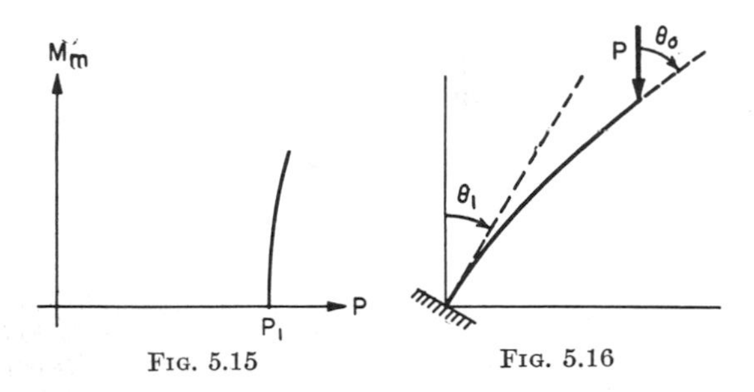

FIG. 5.15

FIG. 5.16

out for the deflected coordinates of the upper end, θ_0, and the strain energy [1].

Axially compressed beams subjected to various concentrated and distributed transverse loadings are sometimes called beam columns [14, chap. 14; 17, p. 203]. While questions of stability are usually not involved, the principle of superposition cannot be directly applied because the axial compression alone would bring in questions of stability. Circle diagrams [17, pp. 205–207] and camptograms [13] are among the graphical methods that have been devised to deal with beam columns.

5. Torsion

If, in Fig. 5.8, only an axial moment M_x acts at any section of a straight uniform bar of length L, we have the case of pure torsion. By Saint-Venant's principle every section sufficiently far from the ends is similar to every other, and, neglecting small end effects, the angular displacement ϕ of one end with respect to the other will be proportional to the length of the bar,

$$\phi = \theta L \tag{5.77}$$

where θ is the twist per unit axial length in the bar.

By Hooke's law (Sec. 1) we may also expect the twist per unit length to be proportional to the torque

M_x in the bar; hence,

$$M_x = C\theta = c\phi \tag{5.78}$$

where C is called the *torsional rigidity* and $c = C/L$ the *torsion constant* of the bar. The torsional strain energy per unit length is

$$u = \tfrac{1}{2}M_x\theta = \tfrac{1}{2}C\theta^2 = \frac{1}{2}\frac{M_x^2}{C} \tag{5.79}$$

and the total torsional strain energy in the bar of length L is

$$U = uL = \tfrac{1}{2}M_x\phi = \tfrac{1}{2}c\phi^2 = \frac{1}{2}\frac{M_x^2}{c} \tag{5.80}$$

If, by means of the reciprocal theorem (Sec. 1), we compare the twist ϕ produced by M_x and the change of volume that would be produced by hydrostatic pressure, we see that, since the tangential stresses on an end section A that compose M_x would do no work during a displacement of A normal to itself, the hydrostatic pressure would do no work

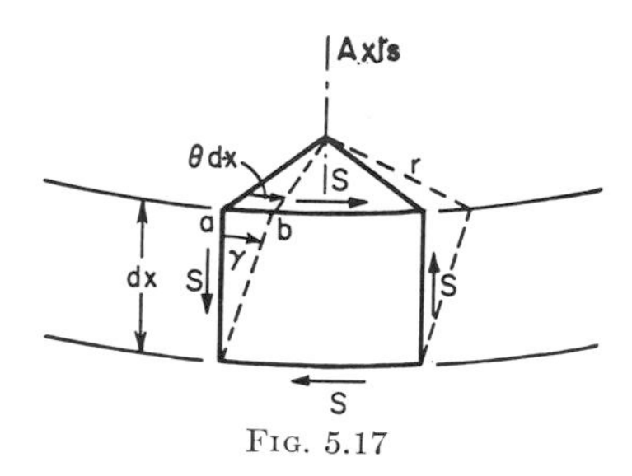

FIG. 5.17

during the setting up of the twist ϕ. Therefore, the volume of the bar does not change while the twist is set up, and C will depend only on the shear modulus μ and not on the bulk modulus k. C is, in fact, proportional to μ.

The simplest case is that of a bar or shaft of circular cross section, which we may picture as composed of circular tubes of radius r and wall thickness dr. Let the rectangle of Fig. 5.17 represent a piece of the wall of such a tube of axial length dx and included between two elements of the cylinder. Because of the circular symmetry all such sections are similar, and a twist of θ radians per unit axial length will shear the rectangle into the parallelogram shown (see Fig. 5.2). The displacement ab along the circle is $r\theta\, dx = \gamma\, dx$, and, by (5.25), the required shearing stresses are $S = \mu\gamma = \mu r\theta$ in the directions shown. Since these are the only stresses acting, it can be shown from (5.33) and (5.34) that the principal stresses are $T_1 = S$ (tension) and $T_2 = -S$ (compression), lying at 45° to the sides of the rectangle in Fig. 5.17, and $T_3 = 0$ in the radial direction. Thus the *stress trajectories* [19, pp. 76, 141] or *lines of stress* [6, p. 88] in this case are right- and left-handed 45° circular helices about the axis of the tube.

The torque in the tube is the resultant moment of force due to the shearing stresses S acting normal to r in the circular-ring cross section:

$$dM_x = r(2\pi r\, dr)S = 2\pi r^3\mu\theta\, dr \tag{5.81}$$

Hence the torsional rigidity of a thin-walled tube of wall thickness t is

$$C = 2\pi r^3 t\mu \tag{5.82}$$

and for a thick-walled tube we obtain, by integrating (5.81) from the inner radius r_1 to the outer radius r_2,

$$C = \tfrac{1}{2}\pi(r_2^4 - r_1^4)\mu \tag{5.83}$$

Similarly, the torsional rigidity for a solid shaft or circular cylinder of radius R is

$$C = \tfrac{1}{2}\pi R^4\mu \tag{5.84}$$

In all these cases of circular symmetry it will be seen that

$$C = I_x\mu \tag{5.85}$$

where I_x is the polar moment of inertia of the cross-sectional area about the longitudinal x axis. For circular sections I_x is twice the moment of inertia I that appears in the definition of the flexural rigidity, $B = EI$, for the same member in pure bending [see (5.46)]. Thus, by (5.38),

$$\frac{B}{C} = \frac{E}{2\mu} = 1 + \sigma \tag{5.86}$$

i.e., *for members with circular symmetry, the ratio of the flexural to the torsional rigidity is* $1 + \sigma$ *and thus depends only on Poisson's ratio* σ.

The first investigations of torsion were made by Coulomb on wires of circular section, and he found the relationships presented by (5.77), (5.78), and R^4 in (5.84) [8, pp. 98–103]. Navier, by erroneously assuming that plane cross sections would remain plane under torsion, deduced the relation (5.85) for cross sections of any shape [5, p. 143; 18, p. 258]. Actually, noncircular cross sections are warped by torsion, as can be illustrated qualitatively by twisting a rubber eraser or any wide flat strip of rectangular section. The warping occurs because the shearing stresses can have no components normal to the force-free lateral surface of the bar on which shearing stresses of equal magnitude would be required to hold them in equilibrium [see (5.34) and Fig. 5.2]. Two principal stress trajectories must lie in any surface to which no shearing forces are applied, and when no pressure is applied either, as in the case of torsion, the third principal stress is zero. Saint-Venant reduced the solution of the general equations (first paragraph, Sec. 2) in the case of torsion of a uniform straight member to the problem of finding one function of the coordinates of the cross section that satisfies Laplace's equation and suitable boundary conditions, and solutions have been worked out for many common sections [5, pp. 144–161; 6, pp. 19–21, chap. XIV; 17, pp. 321–326; 18, chap. 11]. Hydrodynamical and membrane (soap-film) analogies have been helpful in such torsion problems [5, pp. 146–148; 18, pp. 268–272, 289–294].

According to the minimum principle for displacements we may expect that, for a given θ, u in (5.79), and therefore C, will be smaller for the warped cross sections that actually occur than for any other configuration, such as that assumed by Navier in which the cross sections remain plane. Hence we may modify (5.85) to read

$$C = \alpha I_x \mu \tag{5.87}$$

and expect the dimensionless α to be characteristic of the cross section and less than unity for noncircular cross sections. The maximum shearing stress S_m to which the material of the bar is subjected is also important for design purposes. It is given by

$$S_m = \beta \mu \theta \tag{5.88}$$

where the length β is characteristic of the cross section. While for a circular cross section β is equal to the maximum radius, the maximum shear for noncircular sections usually occurs at points of the lateral surface nearest the axis or at sharp reentrant corners, e.g., keyways.

For an elliptical section with semiaxes $a > b$,

$$\alpha = \left(\frac{2ab}{a^2 + b^2}\right)^2 \qquad \beta = \frac{2a^2 b}{a^2 + b^2} \tag{5.89}$$

For an equilateral-triangle section with side a and height $h = \sqrt{3}\,a/2$,

$$\alpha = \tfrac{3}{5} \qquad \beta = \tfrac{1}{2}h \tag{5.90}$$

For a square section of side a,

$$\alpha = 0.844 \qquad \beta = 0.675a \tag{5.91}$$

Various rectangular sections are treated in the literature [5, pp. 150–153; 18, pp. 275–278; 19, pp. 82–83]. For a wide, flat strip of thickness t and width $w > 5t$ it is shown that

$$C = \tfrac{1}{3}(w - 0.63t)t^3\mu \qquad \beta = t \tag{5.92}$$

If a helical spring of N turns is made of a rod or wire of length L and torsional rigidity C so that the center line lies on a cylinder of radius R, then $L = 2\pi NR$. If the spring is stretched by a force P acting along the axis of the helix, P will produce a twisting moment $PR = C\theta$ on each cross section and will therefore extend the spring by an amount

$$\delta = R\theta L = \frac{2\pi NPR^3}{C} \tag{5.93}$$

More elaborate calculations, taking the net shear on each cross section due to P into account, show that the maximum shearing stress occurs on the *inside* of the coils, and experience with railway-car springs shows that cracks usually start there [18, pp. 391–395; 19, pp. 84–86].

References

1. Beth, R. A., and C. P. Wells: Finite deflections of a cantilever-strut, *J. Appl. Phys.*, **22**: 742–746 (1951).
2. Biezeno, C. B., and R. Grammel: "Technische Dynamik," Springer, Berlin, 1939.
3. Ensslin, M.: "Elastizitätslehre für Ingenieure," Sammlung Göschen No. 519, vol. I, 1921; No. 957, vol. II, 1927, De Gruyter, Berlin.
4. Föppl, A., and L. Föppl: "Drang und Zwang," 2d ed., vol. I, 1924, vol. II, 1928, Oldenbourg, Munich.
*5. Geckeler, J. W.: Elastostatik, chap. 3 in "Handbuch der Physik," vol. VI ("Mechanik der Elastischen Körper"), Springer, Berlin, 1928.
*6. Love, A. E. H.: "Mathematical Theory of Elasticity," Dover, New York, 1944, or 4th ed., Cambridge, London, 1927.
7. ———: Elasticity, "Encyclopaedia Britannica," 11th ed., vol. IX, Cambridge, London, 1910.
8. Magie, W. F.: "A Source Book in Physics," McGraw-Hill, New York, 1935.
9. von Mises, R.: *Z. angew. Math. u. Mech.*, **4**: 435 (1924).
10. Pippard, A. J. S.: "Strain Energy Methods of Stress Analysis," Longmans, London, 1928.
11. Pöschl, T., "Über die Minimalprinzipe der Elastizitätstheorie," pp. 160–164, Bauingenieur, 1936.
12. Prescott, J., "Applied Elasticity," Dover, New York, 1946.
13. Rojansky, V., and Beth, R. A.: Camptograms for beams in compression, *J. Appl. Mechanics*, **14**: 202–208 (1947).
14. Sechler, E. E.: "Elasticity in Engineering," Wiley, New York, 1952.
*15. Sokolnikoff, I. S.: "Mathematical Theory of Elasticity," McGraw-Hill, New York, 1946.
16. Sommerfeld, A.: Mechanics of Deformable Bodies in "Lectures on Theoretical Physics," vol. II, Academic, New York, 1950.
17. Southwell, R. V.: "Theory of Elasticity," Clarendon, Oxford, 1936.
18. Timoshenko, S., and J. N. Goodier: "Theory of Elasticity," 2d ed., McGraw-Hill, New York, 1951.
19. ——— and G. H. MacCullough: "Elements of Strength of Materials," 3d ed., Van Nostrand, Princeton, N.J., 1949.
20. ———: "Strength of Materials," 2d ed., parts I, II, Van Nostrand, Princeton, N.J., 1940.
21. ———: "Theory of Elastic Stability," McGraw-Hill, New York, 1936.
22. ———: "Theory of Plates and Shells," McGraw-Hill, New York, 1940.
*23. Todhunter, I., and K. Pearson: "A History of the Theory of Elasticity and the Strength of Materials," vol. I, 1886, vol. II, 1893, Cambridge, London.
*24. Trefftz, E.: Mathematische Elastizitätstheorie, chap. 3 in "Handbuch der Physik," vol. VI ("Mechanik der Elastischen Körper"), Springer, Berlin, 1928.
25. Van den Broek, J. A.: "Elastic Energy Theory," 2d ed., Wiley, New York, 1942.
26. Webster, A. G.: "The Dynamics of Particles and of Rigid, Elastic, and Fluid Bodies," 2d ed., Teubner, Leipzig, 1912.
27. Williams, D.: The relations between the energy theorems applicable in structural theory, *Phil. Mag.*, **26**(7): 617–635 (1938).
28. Wise, J. A.: Circles of strain, *J. Aeronaut. Sci.*, **7**: 438–440 (1940).

* References to original papers.

Chapter 6

Experimental Stress Analysis

By M. M. FROCHT, Illinois Institute of Technology

1. Two-dimensional Stresses and Strains [1, 2]*

Notation. In two-dimensional stress systems there are at each point three independent stress components, which will be denoted by σ_x, σ_y, and τ_{xy} or τ_{yx} (Fig. 6.1*a*). Numerically $\tau_{xy} = \tau_{yx}$. Tensile stresses will be treated as positive, compressive stresses as negative. Shearing stresses will be called positive if their moments about a material point in the immediate vicinity of the stress vector are counterclockwise. Thus in Fig. 6.1*a*, τ_{xy} is positive, and τ_{yx} is negative.

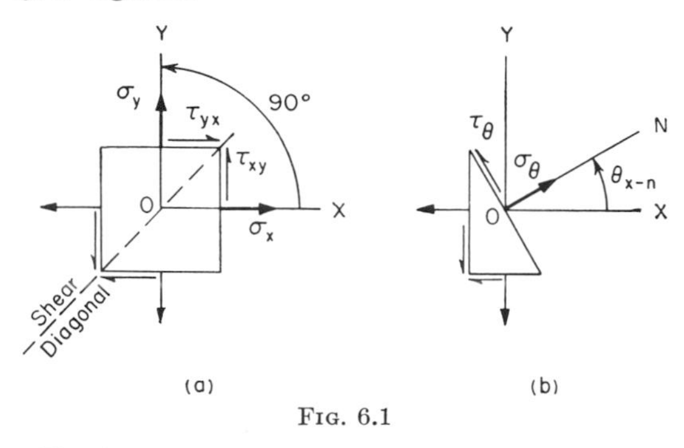

FIG. 6.1

The four shear stresses viewed collectively will be referred to as the *shear system* τ_{xy}, τ_{yx}. The diagonal on which the arrowheads of the shear vectors converge will be called the *shear diagonal.* The shear system τ_{xy}, τ_{yx} is called positive when the shear diagonal passes through the first quadrant; otherwise it is called negative.

The stress components on a plane perpendicular to the XY plane and whose normal makes an angle θ_{x-n} with the X axis (Fig. 6.1*b*) will be denoted by σ_θ and τ_θ. The angle θ_{x-n} will be called *positive* if it is *counterclockwise* and negative if it is clockwise.

From consideration of equilibrium

$$\sigma_\theta = \frac{\sigma_x + \sigma_y}{2} + \frac{\sigma_x - \sigma_y}{2} \cos 2\theta_{x-n} + \tau_{xy} \sin 2\theta_{x-n} \tag{6.1a}$$

$$\tau_\theta = \frac{\sigma_y - \sigma_x}{2} \sin 2\theta_{x-n} + \tau_{xy} \cos 2\theta_{x-n} \tag{6.1b}$$

In Eqs. (6.1*a*) and (6.1*b*), one may treat τ_{xy} either as the shear stress or as the shear system. The result will be the same as long as the positive branch of the Y axis is chosen 90° counterclockwise from the positive branch of the X axis, which is arbitrary (Fig. 6.1).

Strain components will be denoted by ϵ_x, ϵ_y, and γ_{xy}. The linear strains ϵ_x, ϵ_y, parallel to the X and Y axes, respectively, are called positive for extensions and negative for contractions. Shear strain γ_{xy} is called positive when the *shear-strain diagonal,* the diagonal through the acute angles, passes through the first quadrant (Fig. 6.2*a*); otherwise γ_{xy} is negative.

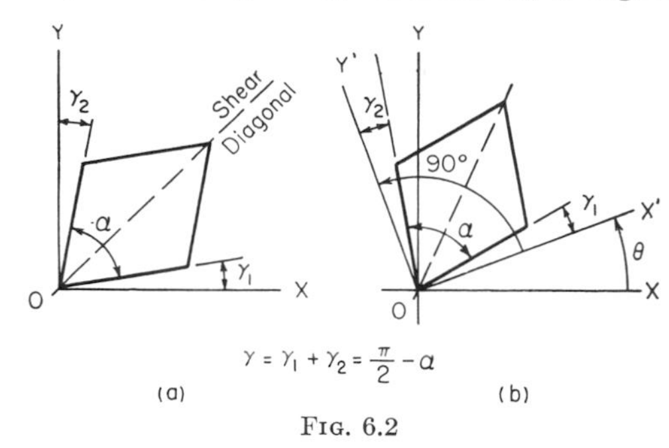

FIG. 6.2

Thus γ_{xy} is positive when the right angle XY gets smaller. With these conventions positive stresses give rise to positive strains. From geometric considerations

$$\epsilon_\theta = \frac{\epsilon_x + \epsilon_y}{2} + \frac{\epsilon_x - \epsilon_y}{2} \cos 2\theta_{x-n} + \frac{\gamma_{xy}}{2} \sin 2\theta_{x-n} \tag{6.2a}$$

$$\frac{\gamma_\theta}{2} = \frac{\epsilon_y - \epsilon_x}{2} \sin 2\theta_{x-n} + \frac{\gamma_{xy}}{2} \cos 2\theta_{x-n} \tag{6.2b}$$

Shear Strains. It is rather difficult to measure shear strains experimentally. The shear strain γ_θ in the right angle $X'OY'$ (Fig. 6.2*b*) is given by

$$\gamma_\theta = \epsilon_{\theta+45°} - \epsilon_{\theta-45°} \tag{6.3}$$

This shows that a shear strain γ_θ can be determined from linear strains along lines making angles ±45° with the directions of OX'.

Mohr's Circles for Stress and Strain. If we eliminate the angle θ from Eqs. (6.1*a*) and (6.1*b*),

$$\left(\sigma_\theta - \frac{\sigma_x + \sigma_y}{2}\right)^2 + \tau_\theta^2 = \left(\frac{\sigma_x - \sigma_y}{2}\right)^2 + \tau_{xy}^2 = B_\sigma^2 \tag{6.4}$$

* Numbers in brackets refer to references at end of chapter.

This defines a *circle* of radius

$$B_\sigma = \sqrt{\left(\frac{\sigma_x - \sigma_y}{2}\right)^2 + \tau_{xy}^2}$$

having a center at $\sigma_\theta = (\sigma_x + \sigma_y)/2 = A_\sigma$ and $\tau_\theta = 0$ (Fig. 6.3). Similarly, Eqs. (6.2*a*) and (6.2*b*) define a circle of radius $B_\epsilon = \sqrt{[(\epsilon_x - \epsilon_y)/2]^2 + (\gamma_{xy}/2)^2}$ with a center at $\epsilon_\theta = (\epsilon_x + \epsilon_y)/2 = A_\epsilon$ and $\gamma_\theta = 0$. Construction of these circles is shown in Fig. 6.3. In both circles the axes are positive to the right and downward. The point K in the stress circle defined by σ_x and τ_{xy}, i.e., by the stress components on the plane for which $\theta_{x-n} = 0$, is called the reference or key point. The key point in the strain circle is defined in an analogous manner.

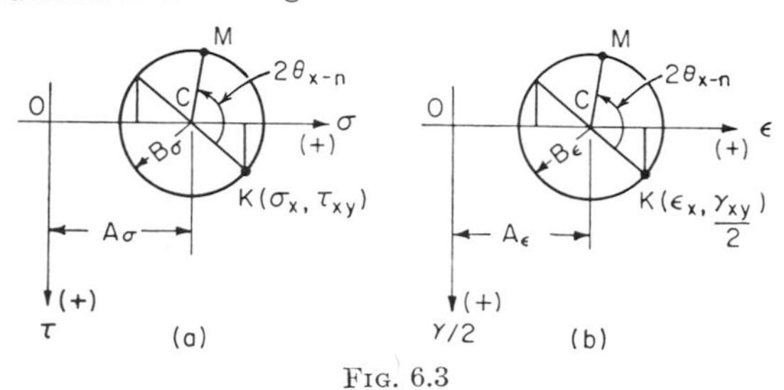

FIG. 6.3

Consider a point M on Mohr's stress circle located so that the angle $KCM = 2\theta_{x-n}$. The coordinates of point M are the stress components σ_θ and τ_θ given by Eqs. (6.1*a*) and (6.1*b*), respectively.

Strain components ϵ_θ and $\gamma_{\theta/2}$ given by Eqs. (6.2*a*) and (6.2*b*) can be obtained in a similar manner from the strain circle (Fig. 6.3*b*). The important point here is to measure the angle $2\theta_{x-n}$ from the *key point* and in the same direction as θ_{x-n}.

All the stress relations at a point follow directly from the stress circle:

1. The principal stresses p and q, with p algebraically greater than q, are given by

$$p, q = \tfrac{1}{2}[\sigma_x + \sigma_y \pm \sqrt{(\sigma_x - \sigma_y)^2 + 4\tau_{xy}^2}] = A_\sigma \pm B_\sigma \qquad (6.5)$$

and
$$\tan 2\theta_{x-p,q} = \frac{2\tau_{xy}}{\sigma_x - \sigma_y} \qquad (6.6)$$

2. The shears are zero on the principal planes, i.e., on planes of maximum and minimum normal stress.

3.
$$\tau_{max} = \frac{p - q}{2} \qquad (6.7)$$

and the corresponding normal stress

$$\sigma_i = \frac{p + q}{2} \qquad (6.8)$$

4. The sum of orthogonal normal stresses is constant:

$$\sigma_x + \sigma_y = p + q = \text{const} \qquad (6.9)$$

5. The algebraically maximum stress p passes through the 45° angle formed by the shear diagonal and the algebraically greater normal stress σ_x or σ_y (Fig. 6.4*a*).

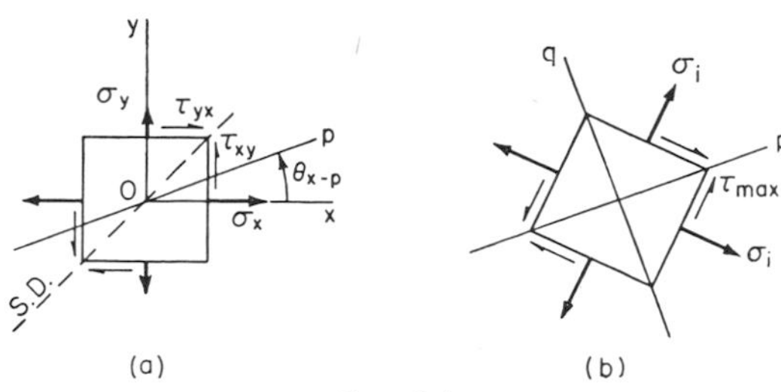

FIG. 6.4

6. The arrowheads of the maximum shear stresses converge on the p line (Fig. 6.4*b*).

Strain relations follow similarly from the strain circle:

1. Principal strains ϵ_p and ϵ_q, $\epsilon_p > \epsilon_q$, are given by

$$\epsilon_p, \epsilon_q = \tfrac{1}{2}[\epsilon_x + \epsilon_y \pm \sqrt{(\epsilon_x - \epsilon_y)^2 + \gamma_{xy}^2}] = A_\epsilon \pm B_\epsilon \qquad (6.10)$$

and
$$\tan 2\theta_{x-p,q} = \frac{\gamma_{xy}}{\epsilon_x - \epsilon_y} \qquad (6.11)$$

2. Shear strains are zero on the principal planes, i.e., on the planes of maximum and minimum normal strain.

3. Maximum shear strain is

$$\gamma_{max} = \epsilon_p - \epsilon_q \qquad (6.12)$$

and the corresponding linear strain is

$$\epsilon_i = \frac{\epsilon_p + \epsilon_q}{2} \qquad (6.13)$$

4.
$$\epsilon_x + \epsilon_y = \epsilon_p + \epsilon_q = \text{const} \qquad (6.14)$$

Stress-Strain Relations. We restrict the discussion to homogeneous and isotropic materials. Within the stress range governed by Hooke's law the principal stresses are collinear with the principal strain, and

$$\epsilon_p = \frac{p}{E} - \nu\frac{q}{E} \qquad (6.15a)$$

$$\epsilon_q = \frac{q}{E} - \nu\frac{p}{E} \qquad (6.15b)$$

Hence
$$p = \frac{E(\epsilon_p + \nu\epsilon_q)}{1 - \nu^2} \qquad (6.16a)$$

$$q = \frac{E(\epsilon_q + \nu\epsilon_p)}{1 - \nu^2} \qquad (6.16b)$$

The main objective of strain-gauge methods in experimental stress analysis is to determine ϵ_p and ϵ_q, from which the principal stresses p and q on free surfaces are found.

Strain Rosettes. A combination of several gauge elements permitting the determination of the linear strains in several different directions through a given point is called a *strain rosette*. The basic type of rosette is one consisting of three elements such as G_1, G_2, G_3 (Fig. 6.5). Such a rosette furnishes necessary and sufficient data for determination of the rectangular strain components ϵ_x, ϵ_y, γ_{xy} at a point which determines the strain circle. Four types of rosettes are used in practice: rectangular rosette,

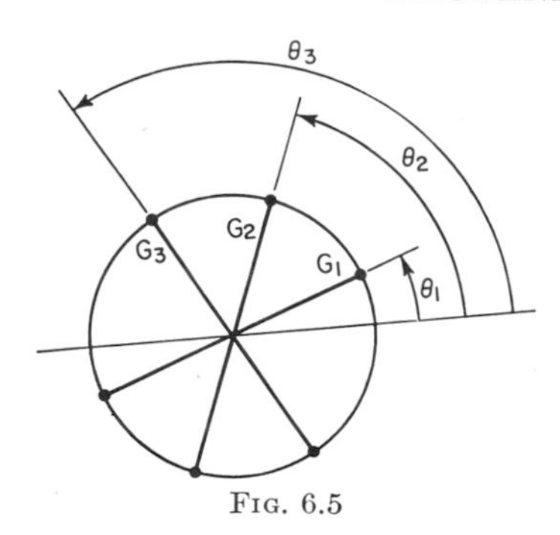

FIG. 6.5

equiangular rosette, four-gauge 45° rosette, and tee-delta rosette.

In the *rectangular rosette* $\theta_1 = 0°$, $\theta_2 = 45°$, and $\theta_3 = 90°$. The resulting principal strains are

$$\epsilon_{p,q} = A_\epsilon \pm B_\epsilon \tag{6.10}$$

where
$$A_\epsilon = \frac{\epsilon_1 + \epsilon_3}{2} \tag{6.17}$$

$$B_\epsilon = \frac{\sqrt{2}}{2}\sqrt{(\epsilon_1 - \epsilon_2)^2 + (\epsilon_2 - \epsilon_3)^2} \tag{6.18}$$

Also
$$\tan 2\theta_{x-p,q} = \frac{2\epsilon_2 - (\epsilon_1 + \epsilon_3)}{\epsilon_1 - \epsilon_3} \tag{6.19}$$

In the *equiangular rosette* $\theta_1 = 0°$, $\theta_2 = 60°$, and $\theta_3 = 120°$.

$$\epsilon_{p,q} = A_\epsilon \pm B_\epsilon \tag{6.10}$$

where

$$A_\epsilon = \tfrac{1}{3}(\epsilon_1 + \epsilon_2 + \epsilon_3) \tag{6.20}$$

$$B_\epsilon = \frac{\sqrt{2}}{3}\sqrt{(\epsilon_1 - \epsilon_2)^2 + (\epsilon_1 - \epsilon_3)^2 + (\epsilon_2 - \epsilon_3)^2} \tag{6.21}$$

and
$$\tan 2\theta_{x-p,q} = \frac{\sqrt{3}\,(\epsilon_2 - \epsilon_3)}{2\epsilon_1 - \epsilon_2 - \epsilon_3} \tag{6.22}$$

The four-gauge 45° rosette is in essence a double rectangular rosette. The redundant reading which is furnished by the fourth gauge serves as a check.

The rectangular rosette is generally used when the approximate directions of the principal strains are known. If the two outside gauges are placed along the expected directions of the principal strains, reliable results can be obtained. The equiangular rosette tends to give greater accuracy, although the evaluation of the results is more complicated.

The rectangular rosette is particularly useful when the shear strain along an arbitrary line is required, as in certain aircraft problems. If the middle gauge coincides with the line along which the shear strain is to be determined, then the required shear strain is given by the difference between the linear strains in the outside gauges, Eq. (6.3). The reading in the middle gauge is not used, so a two-gauge rectangular rosette would serve the same purpose just as well.

Relations between Stress and Strain Circles. The main purpose of strain-gauge measurements is to provide data for determination of principal stresses. The strain circle is an effective aid. Graphical methods are available for the construction of strain circles from all types of strain rosettes. Methods are also available for rapid conversion of strain data into stress data.

By definition (Fig. 6.3),

$$A_\sigma = \frac{\sigma_x + \sigma_y}{2} = \frac{p + q}{2}$$

Introducing the expression from Eq. (6.16),

$$A_\sigma = \frac{E}{1 - \nu}\,\frac{\epsilon_p + \epsilon_q}{2} = \frac{E}{1 - \nu} A_\epsilon \tag{6.23}$$

Similarly

$$B_\sigma = \tfrac{1}{2}\sqrt{(\sigma_x - \sigma_y)^2 + 4\tau_{xy}^2} = \frac{p - q}{2}$$
$$= \frac{E}{1 + \nu}\,\frac{\epsilon_p - \epsilon_q}{2} = \frac{E}{1 + \nu} B_\epsilon \tag{6.24}$$

The circles of stress and strain become concentric when a unit on the stress scale is taken equal to $E/(1 - \nu)$ units of strain. The radius B_σ of the stress circle, in units of the ϵ scale, is then equal to $B_\epsilon(1 - \nu)/(1 + \nu)$. It is thus a simple matter to construct the stress circle from the strain circle.

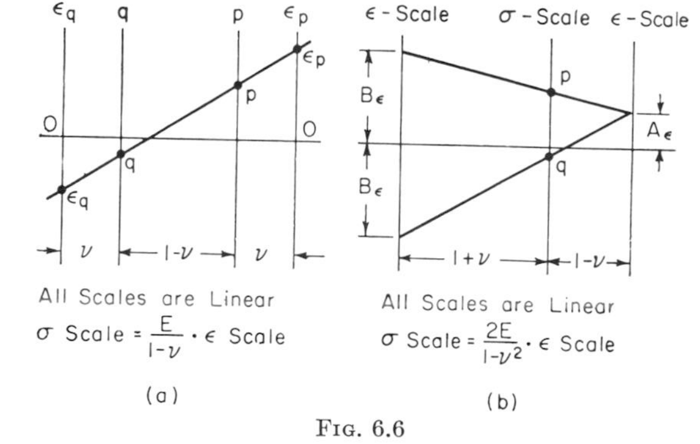

FIG. 6.6

When only the principal stresses are required, the nomograph in Fig. 6.6*a* is convenient. Figure 6.6*b* shows a nomograph for obtaining the principal stresses from the abscissa and radius of the strain circle.

2. Bonded Wire-resistance Strain Gauges [1, 2]*

Introduction. Strains on the surface of complicated bodies are generally determined experimentally by means of strain gauges. These instruments measure the linear deformations δL in prescribed gauge lengths L. The ratio $\delta L/L$ gives the average linear strain. Strain-gauge measurements are inherently confined to free surfaces, although some attempts have been made to insert strain gauges into the interior of such bodies as concrete. Strain gauges yield linear strains only. No instruments are in use to measure shear strains directly.

The basic objective of strain gauges is to magnify small deformations. There are mechanical gauges,

* In collaboration with Dr. John E. Goldberg, Professor of Structural Engineering, Department of Civil Engineering and Engineering Mechanics, Purdue University.

optical gauges, mechanical-optical instruments, acoustical devices, pneumatic gauges, and others. A few of the more widely used gauges are the Huggenberger, Whittemore, Porter-Lipp, and Tuckerman. Discussion is confined to the bonded wire-resistance type of gauge developed during the past fifteen years and today the most widely used and versatile type of strain gauge. Wire-resistance gauges are used to measure strains and also to furnish active elements in such instruments as dynamometers and accelerometers.

Fundamental Principles of Bonded Wire Gauges. When a metallic wire is deformed, its electrical resistance changes by an amount dependent upon the material of the wire and the magnitude of the strain. Consequently, if a suitable wire is cemented to the surface of a test specimen so that the local deformation of the specimen is impressed upon the wire, it is possible to measure the change in resistance of the wire and to interpret this change as the local average strain in the specimen. This is the principle of the electrical-resistance wire strain gauge.

A circular wire of length L and radius r has a resistance

$$R = \frac{\rho L}{\pi r^2} \tag{6.25}$$

in which ρ is the resistivity of the material. If the wire is stretched a small amount δL, the resistance changes by

$$\delta R = \frac{\rho L}{\pi r^2}\left(\frac{\delta\rho}{\rho} + \frac{\delta L}{L} - 2\frac{\delta r}{r}\right) \tag{6.26}$$

Hence

$$\frac{\delta R}{R} = \frac{\delta\rho}{\rho} + \frac{\delta L}{L} - 2\frac{\delta r}{r} \tag{6.27}$$

This may be written

$$\frac{\Delta}{\epsilon} = \frac{\delta\rho/\rho}{\epsilon} + 1 + 2\nu \tag{6.28}$$

in which Δ is the unit change of resistance $\delta R/R$, ϵ denotes the linear strain $\delta L/L$, and ν is Poisson's ratio, which equals $-(\delta r/r)/\epsilon$.

The dimensionless quotient Δ/ϵ is called the *strain sensitivity* of the wire and is denoted by WS. Since the value of Poisson's ratio ν is between 0.25 and 0.35 for most metals and alloys, the last two terms in Eq. (6.28) indicate that dimensional changes alone should contribute between 1.5 and 1.7 to the strain-sensitivity factor. The remaining term, $(\delta\rho/\rho)/\epsilon$, is positive for most materials but is negative in some cases, e.g., nickel.

Classification and Manufacture of Gauges. Strain gauges employing the principle of change in resistance of a strained wire may be divided into two classes, the bonded type and the unbonded type. In the bonded type the strain-sensitive wire is bonded or cemented directly to the specimen or to a thin sheet of paper or other dielectric, which is in turn cemented to the specimen. The strain at the surface of the specimen is thus impressed more or less directly upon the gauge wire. Being attached throughout its length, the gauge wire in the bonded type of gauge will respond to either tensile or compressive strains if the cement is sufficiently strong.

In the unbonded type the strain-sensitive wire may be in the form of a single strand attached to the specimen only at end points of the wire or in the form of a stretched coil attached only at the ends of its loops. Attachment may be through a mechanical-linkage system. Since the wire is unsupported, the compression range of measurement is limited by the amount of initial tension in the wire.

Prefabricated bonded wire-resistance gauges are commercially available in two basic types, flat-wound and helically wound. The usual flat-wound gauge is formed by looping a fine wire (usually 0.001 in. in diameter and 5 in. in length) back and forth over two sets of pins to form a grid of the desired gauge length, as shown in Fig. 6.7. After the desired number of loops have been formed, the cement-coated paper base is placed in contact with the wire grid. The pins are then withdrawn, and the paper is cemented over the loops.

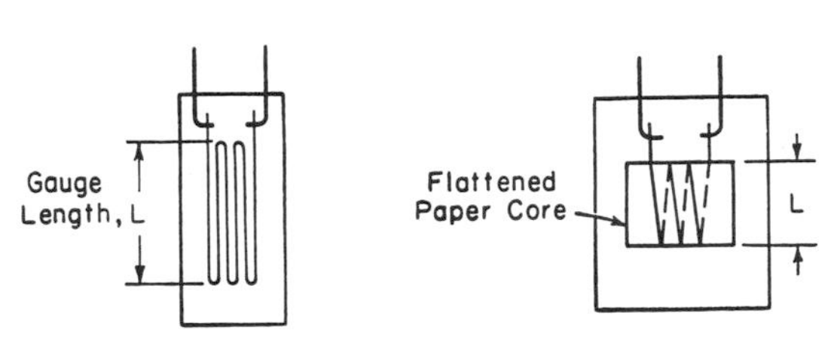

FIG. 6.7

A single gauge element (Fig. 6.7) measures primarily the strain in the direction of the gauge axis. To determine the complete state of strain at a point on the surface of a test specimen the linear strains along three directions are required. For this purpose three gauge elements may be superposed to form a rosette gauge. Four-, three-, and two-element rosette gauges are commercially available.

Calibration and Cross Sensitivity. Owing to the grid structure of the electrical-resistance strain gauges, the strain sensitivity of the wire cannot be used to determine accurate values of the axial strain in the gauge. There are two important differences between the state of strain in the straight wire and the mounted gauge: (1) the strain in the wire is constant, whereas in the loops of the gauge the strain is variable; (2) the straight wire is inherently sensitive only to the axial strain, whereas the loops of the gauge are sensitive also to the transverse strains at a point of a stressed surface. For these reasons manufacturers calibrate gauges in their completed form.

Single-element SR-4 gauges are calibrated in a uniaxial stress field on steel specimens for which $\nu_0 = 0.285$. The strain sensitivity obtained from such calibration is called the *gauge factor* and is denoted by GF. Thus:

$$\mathrm{GF} = \frac{\delta R}{R}\frac{1}{\epsilon_a} = \frac{\Delta}{\epsilon_a} \tag{6.29}$$

in which ϵ_a is the true axial strain in the calibration member. The gauge factor gives accurate results whenever the ratio of the transverse strain ϵ_t to the axial strain ϵ_a is equal to the ratio of these strains in the calibration member. This is clearly not the general case, so strains calculated on the basis of the gauge factor are somewhat in error.

The correction used in practice is based on the assumption that in a biaxial strain field the unit change in resistance Δ is a linear combination of the true strains ϵ_a and ϵ_t,

$$\Delta = \mathrm{GS}_a\epsilon_a + \mathrm{GS}_t\epsilon_t \tag{6.30}$$

in which GS_a is a constant representing the strain sensitivity of the gauge for a uniaxial strain ϵ_a, i.e., when $\epsilon_t = 0$, and GS_t is similarly a constant representing the strain sensitivity of the gauge for a transverse strain ϵ_t with $\epsilon_a = 0$.

The sensitivities GS_a and GS_t can be determined analytically. Figure 6.8 shows an elementary point of view which yields good first approximations. If the loops are replaced by straight lines,

$$\mathrm{GS}_a = \frac{L - W}{L}\,(\mathrm{WS}) \tag{6.31}$$

$$\mathrm{GS}_t = \frac{W}{L}\,(\mathrm{WS}) \tag{6.32}$$

in which L denotes the total length of the wire in the gauge and W is as in Fig. 6.8. In practice the sensitivities GS_a and GS_t are determined experimentally by testing the gauges in unidimensional strain fields.

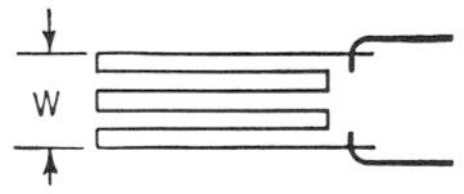

Fig. 6.8

Setting $\mathrm{GS}_t/\mathrm{GS}_a = k$, Eq. (6.30) becomes

$$\Delta = \mathrm{GS}_a(\epsilon_a + k\epsilon_t) \tag{6.33}$$

Using this value of Δ and the gauge factor GF, the approximate or apparent axial strain ϵ_a' becomes

$$\epsilon_a' = \frac{\Delta}{\mathrm{GF}} = \frac{\mathrm{GS}_a}{\mathrm{GF}}\,(\epsilon_a + k\epsilon_t) \tag{6.34}$$

In the calibration member the apparent strain equals the true strain:

$$\epsilon_a' = \epsilon_a \qquad \text{where } \epsilon_t = -\nu_0\epsilon_a \tag{6.35}$$

and therefore

$$\frac{\mathrm{GS}_a}{\mathrm{GF}} = \frac{1}{1 - \nu_0 k} \tag{6.36}$$

so Eq. (6.34) becomes

$$\epsilon_a' = \frac{\epsilon_a + k\epsilon_t}{1 - \nu_0 k} \tag{6.37}$$

The true strain ϵ_a cannot be found from ϵ_a' alone. A second gauge set with its axis parallel to ϵ_t will give an apparent strain

$$\epsilon_t' = \frac{\epsilon_t + k\epsilon_a}{1 - \nu_0 k} \tag{6.38}$$

which together with Eq. (6.37) determines both ϵ_a and ϵ_t:

$$\epsilon_a = \frac{1 - \nu_0 k}{1 - k^2}\,(\epsilon_a' - k\epsilon_t') \tag{6.39}$$

$$\epsilon_t = \frac{1 - \nu_0 k}{1 - k^2}\,(\epsilon_t' - k\epsilon_a') \tag{6.40}$$

The true strains ϵ_a and ϵ_t can thus be obtained from a rectangular rosette gauge or from an orthogonal two-element gauge.

If correction for transverse sensitivity is ignored, the error e in the axial strain expressed in terms of ϵ_p is

$$e = \frac{\epsilon_a' - \epsilon_a}{\epsilon_p} \tag{6.41}$$

From Eq. (6.37),

$$e = \frac{k}{1 - \nu_0 k}\,\frac{\epsilon_t + \nu_0\epsilon_a}{\epsilon_p} \tag{6.42}$$

The maximum error is about 3 per cent of the maximum principal strain.

In rosette gauges two methods can be used to find true strains. (1) With every rosette gauge the manufacturer supplies two constants, a and b, used in appropriate formulas to give the true strain. For the rectangular rosette gauge Baldwin gives

$$\epsilon_1 = \frac{\Delta_1}{a} - \frac{\Delta_3}{ab} \qquad \epsilon_2 = \left(\frac{1}{a} + \frac{1}{ab}\right)\Delta_2 - \frac{1}{ab}\,(\Delta_1 + \Delta_3)$$

$$\epsilon_3 = \frac{\Delta_3}{a} - \frac{\Delta_1}{ab} \tag{6.43}$$

(2) The second method employs apparent strains ϵ' based on the wire sensitivity:

$$\epsilon_1' = \frac{\Delta_1}{\mathrm{WS}_1} \qquad \epsilon_2' = \frac{\Delta_2}{\mathrm{WS}_2} \qquad \epsilon_3' = \frac{\Delta_3}{\mathrm{WS}_3} \tag{6.44}$$

The strain circle constructed from ϵ_1', ϵ_2', and ϵ_3' is concentric with the strain circle of the true strains ϵ_1, ϵ_2, and ϵ_3 as long as the value of $k = \mathrm{GS}_t/\mathrm{GS}_a$ is constant for the gauge elements. Moreover, the radius B_ϵ of the true strain circle is related to B_ϵ' of the apparent circle by the relation

$$B_\epsilon = \left(\frac{1 + k}{1 - k}\right) B_\epsilon' \tag{6.45}$$

The value of k, found by calibration, is furnished by the manufacturer. In general, errors resulting from neglecting the cross sensitivity are small.

High-temperature Gauges, Measurement of Large Strains, and Moistureproofing. At high temperatures special problems arise. Paper base, nitrocellulose cement, and soldering of connections are no longer suitable. Satisfactory gauges for use up to about 500°F have been obtained by using a phenolic resin, such as bakelite, to impregnate the paper base of the gauge and to serve as the bonding cement. The assembly is baked for some 5 hr at around 250°F to produce polymerization. Soldering is replaced by welding or mechanical connections.

No satisfactory techniques are at present available for the production of completely prefabricated gauges to be used at high temperatures. At such temperatures ceramic cements and Sauereisen Cement No. 19 are frequently employed [4, 5].

Difficulties also arise in measuring large strains, say over 2 per cent, where connections between the filament and the lead wires tend to break. For this purpose special gauges are constructed with a small, unbonded loop between filament and lead wire [6].

When gauges have to be used over a long period of time or in wet, humid locations, they must be carefully moistureproofed. At ordinary temperatures silicone grease, water-pump grease, or yellow petroleum jelly may be used. Above 150°F, Petrosene wax is satisfactory. Moistureproofing can be obtained by means of neoprene brushing compound at temperatures around 200°F. Properly constructed shells or coverings well bonded to the test specimen may also be used. (See [7–9] for additional material.)

Basic Strain-gauge Circuits: The Wheatstone Bridge. The most convenient circuit for measuring static strains is the Wheatstone bridge with the strain gauge in one arm (Fig. 6.9). If the bridge is balanced when the gauge is unstrained, the potential difference between the output terminals which develops when the gauge is strained becomes a direct measure of the strain. Two methods are available for measuring the output, the null method and the deflection method. The null method requires that one arm of the bridge, say R_4, be a variable resistance and that the strain be constant for a sufficiently long period of time to permit balancing the bridge. In the deflection method the magnitude of the output is measured directly with a galvanometer or other indicator.

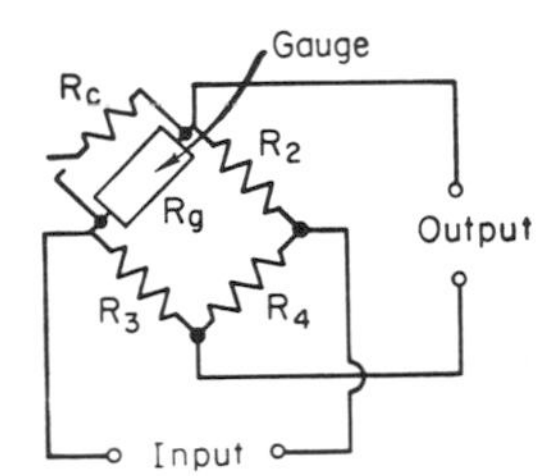

Fig. 6.9. Wheatstone bridge.

Calibration for either method may be obtained by introducing a known resistance change into the gauge arm by means of the shunt resistor R_c. This is equivalent to applying a known strain to the gauge, since $\epsilon = (\delta R_g/R_g)/\text{GF}$. In the null method the variable resistance can be calibrated to read strain directly, and in the deflection method the output can be converted to units of strain.

Temperature Effects and the Dummy Gauge. Changes in temperature during the test will affect readings of the bridge because the resistance of the gauge wire is altered and temperature stresses are set up in the test specimen in addition to those produced by the loads. The effect of the change in the resistance of the wire can be eliminated by mounting a second gauge similar to the active gauge on a free or unloaded specimen of the same material and inserting this second gauge into the bridge to replace R_2 or R_3 (Fig. 6.9). This is called a *dummy* gauge. The dummy gauge does not compensate for the temperature stresses, so the galvanometer reading indicates the sum of the stresses produced by the loads and those induced by the temperature change.

Gauges have been developed which require no dummy to compensate for temperature effects. At present these are available commercially only for steel and dural (see *Baldwin Bull.* 174).

Potentiometer Circuit. The simple potentiometer circuit (Fig. 6.10) may be used for measurement of dynamic strains when thermal effects are negligible. The gauge is connected in series with a battery and a fixed or *ballast* resistor. Calibration is obtained by introducing a known resistance change by means of the shunt resistor R_c.

Useful Results from Two or More Active Gauges. Gauges in adjacent arms unbalance the bridge in proportion to the difference in the strains and gauges in opposite arms unbalance the bridge in proportion to the sum of the strains. This makes it possible to utilize two or more active gauges to increase sensitivity or eliminate undesirable strains. For example, in bending, undesirable axial strains and temperature effects can be eliminated by mounting one gauge on the tension side of the beam and the other on the compression side (Fig. 6.11). If

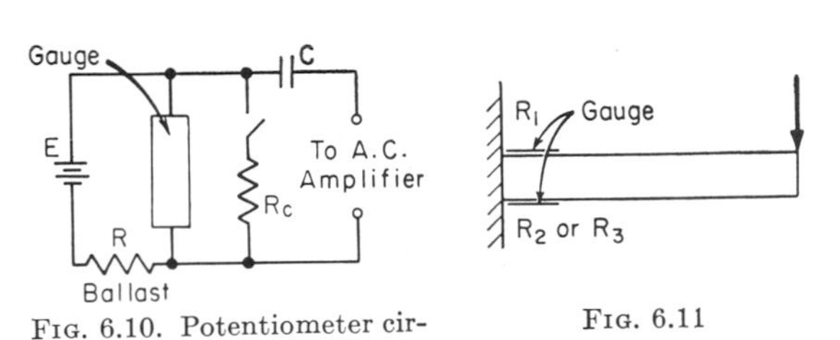

Fig. 6.10. Potentiometer circuit.

Fig. 6.11

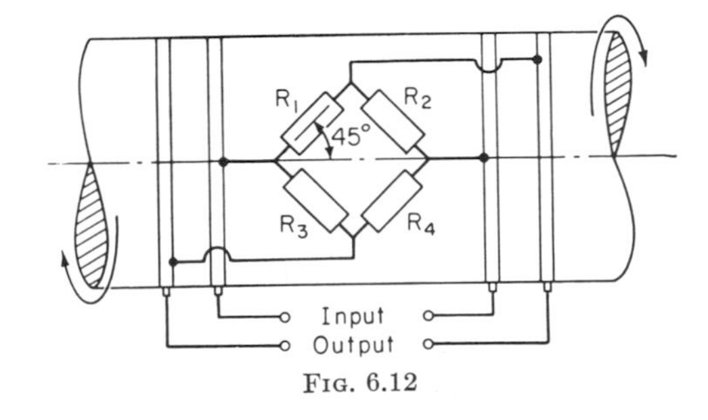

Fig. 6.12

the second gauge replaces R_2 or R_3 of Fig. 6.9, it will also double the bending-strain output of the bridge.

If the second gauge of Fig. 6.11 replaces R_4 of Fig. 6.9, the bending strains are eliminated, and the system is doubly sensitive to axial strain. However, this arrangement is not temperature-compensated.

A temperature-compensated torquemeter of increased sensitivity is obtained by placing four strain gauges on a shaft, as shown in Fig. 6.12. Input and output connections are made by brushes and slip rings at points outside of the strain-gauge loop. Silver-graphite brushes and silver, brass, or monel slip rings have performed satisfactorily. The resistance change in each gauge tends to unbalance the bridge in the same direction, so sensitivity is quadrupled. The arrangement in Fig. 6.12 automatically eliminates strains due to bending.

Instrumentation. For measurement of static strains, instruments are available commercially which provide for the external connection of two strain gauges into an internal bridge circuit. Both gauges may be active, or one may be active while the other is a dummy. The strain is determined in some instruments by manually balancing the bridge and reading the strain from a graduated scale. Others use a pen or pointer and indicate automatically the magnitude of strain. A cantilever with gauges arranged as in Fig. 6.11 is often used to provide the variable resistance R_3 and R_4.

For multiple-gauge installation switching and balancing units are available, so a single indicator may be used.

Dynamic strains up to about 100 cps can be recorded using an amplifier and pen-motor recorder. Oscillograph galvanometers can be made to respond accurately to frequencies up to 5,000 cps. For transient or high-frequency strains the cathode-ray oscilloscope is used.

3. Photoelasticity [10–12; 1, chap. 17]

Temporary Birefringence and the Stress-optic Law. In 1816, Brewster discovered that transparent materials become temporarily birefringent when subjected to mechanical stress. His work and that of Neumann, Wertheim, Kerr, Pockels, and Fresnel established the fundamental principles governing this phenomenon. They proved that a stressed transparent isotropic material exhibits temporary optical properties identical with those found in permanent crystals. (Part 6, Chap. 6, Sec. 8.)

In crystalline materials the permanent birefringence at a point may be described by the index ellipsoid whose principal semiaxes are proportional to the principal refractive indices. In temporary birefringence, due to stresses within the proportional limit, the principal stresses p, q, r are in the directions of the principal axes of the index ellipsoid and are related to the refractive indices as follows:

$$\mu_p - \mu_0 = C_1 p + C_2(q + r) \qquad (6.46)$$
$$\mu_q - \mu_0 = C_1 q + C_2(r + p) \qquad (6.47)$$
$$\mu_r - \mu_0 = C_1 r + C_2(p + q) \qquad (6.48)$$

where μ_0 is the refractive index of the unstressed material, μ_p, μ_q, μ_r are the refractive indices for light vectors along the p, q, and r axes, respectively, and C_1, C_2 are secondary stress-optical coefficients.

Equations (6.46) to (6.48) are the general expressions for the stress-optic law. Brewster's form of this law follows directly:

$$\mu_p - \mu_q = C'(p - q) \qquad (6.49a)$$
$$\mu_p - \mu_r = C'(p - r) \qquad (6.49b)$$
$$\mu_q - \mu_r = C'(q - r) \qquad (6.49c)$$

in which $C' = C_1 - C_2$. If λ is the vacuum wavelength of the light used, then in the emergent beams the relative retardation n in wavelengths, for an optical path t, is

$$n = Ct(p - q) \qquad (6.50a)$$
$$n = Ct(p - r) \qquad (6.50b)$$
$$n = Ct(q - r) \qquad (6.50c)$$

where $C = C'/\lambda$.

These relations are for rays parallel to the principal axes. They can be transformed to show that for a ray of arbitrary direction the relative retardation n is given by

$$n = Ct'(p' - q') \qquad (6.51)$$

where t' is the length of the optical path and p', q' are the *secondary principal stresses* for the given ray. These are the principal stresses obtained by treating a thin stressed plate normal to the direction of propagation of the given ray as if it were in a two-dimensional state of stress. Thus, for a ray parallel to the Z axis

$$p', q' = \tfrac{1}{2}[\sigma_x + \sigma_y \pm \sqrt{(\sigma_x - \sigma_y)^2 + 4\tau_{xy}^2}] \qquad (6.52)$$

This does not contain σ_z, τ_{xz}, or τ_{yz}, i.e., any of the stress components with a subscript z. Equation (6.51) assumes principal planes of constant direction along the optical path. When these directions are not constant, the value of n is altered. However, the rotational effect is generally small, and may be neglected.

4. Two-dimensional Photoelasticity

Stress Pattern, Isoclinics, and Stress Trajectories. When investigating two-dimensional states

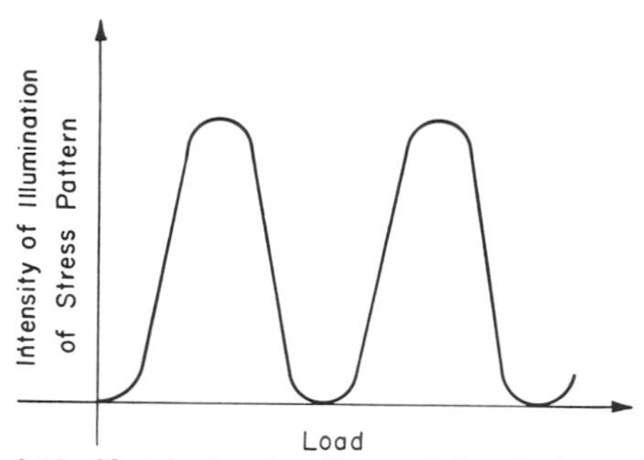

FIG. 6.13. Sketch showing the variation in intensity of light at a point in a model as a function of the applied loads.

of stress, the model, a plate of uniform thickness, is placed in the field of the polariscope normal to a collimated beam of polarized light so the light traverses the model at normal incidence. Resulting retardations are then

$$n = Ct(p - q) \qquad (6.50a)$$

in which p, q are the actual principal stresses in the plate.

Assuming a monochromatic source of light, circular polarization, and a polariscope set for extinction, the intensity I of the transmitted light at each point of the model is

$$I = I_1 \sin^2 \frac{\alpha}{2} \qquad (6.53)$$

where I_1 is the intensity of the incident light and α is the angular phase difference introduced by the stressed model. Thus $I = 0$ for $\alpha = 2n\pi$ ($n = 0$, 1, 2, . . .). The fundamental photoelastic effect of an increasing stress at a point is thus to produce a continuous cyclical variation in the intensity of the transmitted light (Fig. 6.13). Points at which the principal stress differences $p - q$ are equal produce equal retardations and therefore equal light intensities. The locus of such points is called a *fringe*, and the retardation in wavelengths corresponding to a given fringe, i.e., the number of light cycles n, is called the *order* of the fringe. Photoelastic patterns form from such fringes, which originate and multiply in regions of high stresses and move toward regions of lower stresses as the load is applied. A typical stress pattern of a circular disk under diametral compression is given in Fig. 6.14.

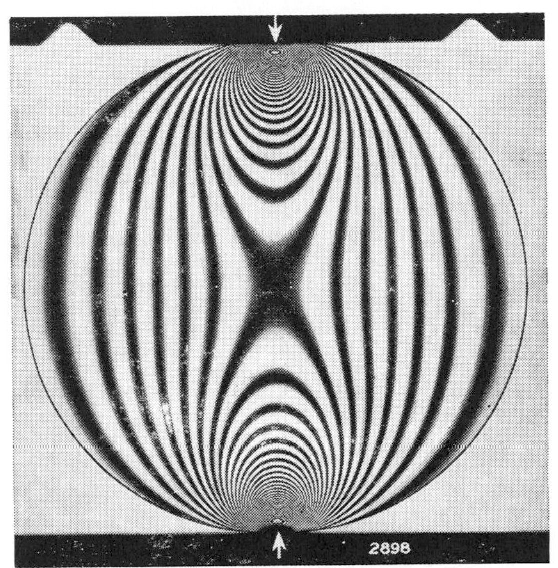

FIG. 6.14. Stress pattern of a circular disk subjected to concentrated loads. Applied loads, $P = 275$ lb; diameter, $D = 1.250$ in.; thickness, $t = 0.200$ in.; material fringe value, $f = 43$ psi shear stress; fringe order at center $= 6\frac{1}{2}$.

If the quarter-wave plates are removed and a *plane* polariscope is used, the transmitted light is

$$I = I_1 \sin^2 2\theta \sin^2 \frac{\alpha}{2} \qquad (6.54)$$

where θ is the angle between the plane of polarization and the nearer principal stress. Now $I = 0$ for either $\alpha = 2n\pi$ or $\theta = 0$. The first condition is the same as in the circular polariscope, so the general stress pattern will be the same in both cases. The plane polariscope gives in addition the lines for which $\theta = 0$, the locus of all points whose principal stress directions coincide with the directions of the polarizer and analyzer.

The locus of points along which the principal stresses have parallel directions is called an *isoclinic*. A plane polariscope can be used to obtain isoclinics for any value of θ. This is done by rotating the polarizer and analyzer as a unit and so keeping them crossed. Curves whose tangents represent directions of one of the principal stresses are known as principal *stress trajectories* or *isostatics*. They are constructed by a graphical process on the basis of isoclinics. A set of isoclinics and stress trajectories for the disk in Fig. 6.14 is given in Fig. 6.15.

Much of the early work in photoelasticity was carried out using white light. With a white light source the stress patterns consist of colored bands, called *isochromatics*, which form in the order of yellow, red, and green followed by similar cycles. For quantitative, precise work a monochromatic source, generally the green line 5461 A, which is readily obtainable from a mercury lamp or an H4-type lamp with suitable filter, is most frequently used. However, isoclinics stand out more sharply against colored backgrounds, and white light is used for that purpose.

Fringe Value and Calibration Members. For purposes of calculation Eq. (6.50a) may be written

$$\frac{p - q}{2} = \frac{n}{2Ct} \qquad (6.55)$$

or by Eq. (6.7)

$$\tau_{\max} = Fn \qquad (6.56)$$

in which $F = 1/2Ct$ is a constant, called the *model* fringe value. From the definition of F,

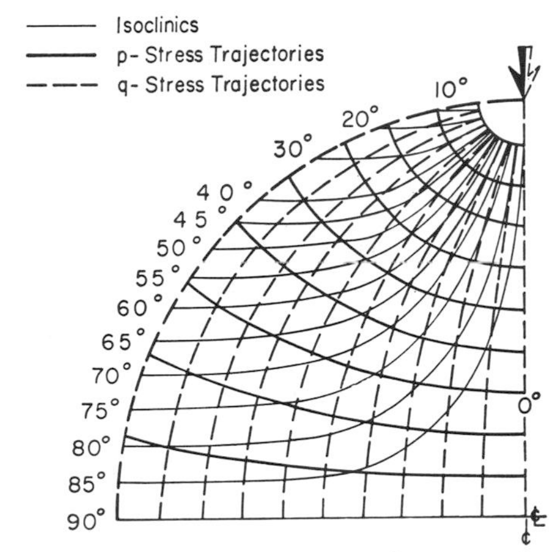

FIG. 6.15. Drawings of isoclinics and stress trajectories for one quadrant of a circular disk subjected to diametral compression.

$$Ft = \frac{1}{2C} = \text{const} \qquad (6.57)$$

The particular value of F for $t = 1$ in. is called the *material* fringe value and is denoted by f:

$$Ft = f \qquad (6.58)$$

The value of F or f can be determined photoelastically from any model for which the principal stresses p, q, or more directly $\tau_{\max}$, can be calculated. Thus

$$F = \frac{\tau_{\max}}{n} = \frac{p - q}{2n} \qquad (6.59)$$

Models used to determine the fringe value are called *calibration members*. A tension bar and the circular disk under diametral compression are most suitable for this purpose. The disk is probably the simplest to machine and to load. Here [11, vol. II, p. 154]

$$F = \frac{2P}{n_c \pi t R} \qquad (6.60)$$

where P is the load on the disk, R its radius, t its thickness, and n_c the fringe order at the center. In addition to ease of machining the disk also yields extremely accurate values of n_c. Bars in compression, beams in pure bending, and wedges under concentrated loads have been used as calibration members.

Determination of Fringe Order. Several methods exist to determine the fringe order, depending on whether its value is relatively high or low. High fringe order can be readily found in patterns similar to that shown in Fig. 6.14. Here a curve of n can be plotted for any path and the high values determined from that curve. Low fringe values are determined by means of the Tardy method or by means of Babinet or Babinet-Soleil compensators. Where extreme accuracy is required and the fringe orders involved are small fractions of a wavelength, compensation is effected with the aid of a sensitive photometer. Retardations can in this way be measured to within 0.002 of a fringe.

Where white light is used, retardations are often determined by means of a tension or compression compensator properly set with respect to the maximum stress at a point. A two-dimensional stress system can be regarded as an isotropic system of stress plus a tension of $p - q$ acting in the p direction, the stress p being algebraically greater than q. The isotropic stresses produce zero retardation. Hence the photoelastic effect is due only to the $p - q$ tensile stress. By placing a tension member close to the point in the model and at right angles to the direction of the p stress it is possible to adjust the tension so that the combined effect of the tension strip and the model is zero. If the compensating strip is of the same material and thickness as the model, the stress in this strip at extinction gives directly the value of $p - q$.

Processing of Direct Photoelastic Data. Combining the data from the stress pattern and isoclinics, it is possible to find $p - q$, τ_{max}, and the directions of the principal stresses. Further, if compensation is used, the directions of p and q may be separated. At a free boundary one of the principal stresses is zero, so that the magnitude and sense of the other stress can be determined. Also, the shear stress τ_{xy} along an arbitrary X axis can be computed. For example, $\tau_{xy} = [(p - q)/2] \sin 2\theta = nF \sin 2\theta$, where n is determined by one of the ways previously outlined and θ is obtained from isoclinic data. However, since isotropic stresses produce no optical effects, it is not possible to obtain the principal stresses themselves from normal photoelastic data except at free boundaries (see d below).

Stress Concentrations. Maximum stresses generally occur on free boundaries, and the optical method is particularly suited for their determination. Among the many practical contributions of photoelasticity, determination of factors of stress concentrations occupies a place of special prominence. These factors, denoted by k, are defined as the ratio of maximum stress divided by nominal stress,

$$k = \frac{\sigma_{max}}{\sigma_{nom}} \tag{6.61}$$

In tension bars containing discontinuities and subjected to axial loads P the nominal stress is generally defined by

$$\sigma_{nom} = \frac{P}{A} \tag{6.62}$$

where A is the minimum area through the discontinuity. The expression for k becomes

$$k = \frac{\sigma_{max}}{P/A} = \frac{2n_{max}F}{P/A} \tag{6.63}$$

In the case of bending, the nominal stress is given by the flexure formula

$$\sigma_{nom} = \frac{Mc}{I} \tag{6.64}$$

Hence

$$k = \frac{2n_{max}F}{Mc/I} \tag{6.65}$$

In more complicated cases the nominal stress must be carefully defined, primarily for simplicity in calculations. The factor k is then calculated from the basic definition, Eq. (6.61), and the maximum stresses are given by

$$\sigma_{max} = k\sigma_{nom} \tag{6.66}$$

In using a particular value of k, the designer must make sure that he follows the same method of calculating the nominal stress as was used in the definition of the factor k. Nominal stresses are often defined in different ways and lead to different values of k.

Separation of Principal Stresses. *a. Lateral Extensometer.* To determine the principal stresses, and thus find the complete state of stress at each point, various experimental, numerical, and graphical methods have been devised. The lateral extensometer is probably the easiest to understand, although in practice it is rather difficult to use. Its operation depends on the fact that the lateral deformation δ, the change in the thickness t of the model at a point, is given by

$$\delta = -\frac{\nu t}{E}(p + q) \tag{6.67}$$

where ν is Poisson's ratio and E is the modulus of elasticity. From a knowledge of $p + q$ and $p - q$ both p and q can be computed. However, at room temperature the lateral deformation is of the same order of magnitude as the natural variations in the thickness of the model, and it is rather difficult to measure these deformations accurately. The lateral extensometer has therefore found only limited use. For an approximate method for $p + q$ using only micrometers see Sec. 5.

b. Isopachics. Curves of constant $p + q$ are called *isopachics*. These are contour lines on the lateral surface of an initially flat model of uniform thickness. Interferometers have been successfully used to obtain isopachic patterns from steel models [13].

When body forces are absent or constant, $p + q$ satisfies Laplace's equation. At the free surface, $p + q = p - q =$ either p or q, so boundary values of $p + q$ are known. By Dirichlet's theorem, values of $p + q$ throughout the region are uniquely determined. Methods for numerical solution of Laplace's equation with prescribed boundary conditions have been perfected and used to obtain isopachic curves in rather complicated cases [11, vol. II, chaps. 8, 9].

c. Shear-difference Method. Numerical or graphical integration of the stress equations of equilibrium may be used to separate the principal stresses. Filon's graphical integration proceeds along a stress trajectory using the appropriate Lamé-Maxwell equilibrium equations. A more effective procedure is found in the shear-difference method, where the integration is carried out along an arbitrary straight line using the equilibrium equations expressed in cartesian form. The shear-difference method is the only method thus far developed which has been successfully used to determine the complete state of stress at an arbitrary interior point in a general three-dimensional problem

d. Oblique Incidence. Principal stresses in two-dimensional problems can also be separated by oblique incidence. However, this procedure holds no advantage over other methods. In three-dimensional problems oblique incidence often simplifies the solution and is of decided practical value.

5. Three-dimensional Photoelasticity

The Frozen Stress Pattern. The practical development of three-dimensional photoelasticity is due largely to the development of the frozen stress pattern. When the loads are removed from a stressed model, the photoelastic effects vanish. However, if a model is made from certain plastics, is kept under loads at a suitably high temperature, and is allowed to cool slowly to room temperature before removing the loads, then, even after the loads are removed, the model will exhibit a photoelastic stress pattern similar in every respect to the pattern at room temperature. Furthermore, the model can be filed, sawed, machined, or cut into pieces without disturbing the fixed stresses in each part. The stresses are then said to be "frozen" into the model.

This phenomenon, first observed in an isinglass gel by Maxwell [14] in 1853, can be explained on the basis that plastics exhibiting frozen patterns are diphase materials: they consist of two distinct molecular networks having different properties. Of particular significance is the difference in fusibilities. One group of bonds, called secondary bonds, becomes soft and fuses at elevated temperature, whereas another, stronger group, called primary bonds, remains infusible and elastic up to a much higher temperature. At the critical temperature the loads are carried only by the primary, infusible network. Upon cooling the model, the soft material becomes hard again. Hence, if the model is cooled and the loads are not removed until room temperature is restored, no external loads are needed to maintain the deformation in the primary elastic network. These deformations are sustained by the hardened secondary bonds with little change in magnitude. The resulting stress patterns have been shown to correspond to the elastic state of the model [15]. For a molecular theory approach to the mechanism of stress freezing, see [15*a*] or [39, p. 62].

The optical and mechanical properties at the critical temperature differ considerably from those at room temperature. For example, the material fringe value of bakelite decreases from 43 to 1.6 psi shear, Young's modulus drops from 615,000 to 1,220 psi, and the tensile strength changes from 17,000 to 400 psi (see Table 6.1).

TABLE 6.1. APPROXIMATE PHYSICAL PROPERTIES OF PHOTOELASTIC MATERIALS

Material	Tensile strength, psi	Modulus of elasticity, E, psi	Poisson's ratio, ν	Fringe value, f, psi shear
Catalin 61-893 (formerly bakelite 61-893):				
At 70°F	17,000	615,000	0.36	43
At 230°F	400–450	1,200	0.50	1.60
Catalin (general-purpose)	4,000	200,000	....	23
Cellulose nitrate (celluloid)	7,000	280,000	0.35	110–190
Fosterite at 185°F	515	2,320	0.48	1.5
Gelatin (13% aqueous solution, no glycerin)	......	6	....	0.07
Glass	10,000	10,000,000	0.4	1,150
Kriston:				
At 70°F	8,000	540,000	....	40.0
At 275°F	680	13,000	....	3.12
Castolite:				
At 70°F	8,100	705,000	0.36	79
At 245°F	350	4,290	0.5	4.4

The great reduction in Young's modulus results in a pronounced increase in the deformation even for moderate loads. These large deformations are undesirable because they tend to alter the geometry excessively. However, they make it possible to determine approximate values of $p + q$ in a two-dimensional plate by measuring the change in thickness with an ordinary micrometer. On free boundaries this method yields the existing boundary stresses.

Stress Data from Frozen Patterns. Slices taken from frozen stress models are examined in the polariscope in much the same way that two-dimensional models are studied. In plane problems the stresses are constant both in magnitude and direction through the depth of the model. This is not the case in slices taken from frozen models, and only the integrated effect is observed. For this reason thin slices are desirable, but this in turn may necessitate the use of compensation and photometric devices for obtaining fringe order and isoclinic data with accuracy.

Photoelastic data furnish at most five independent relations, from which only the principal shears can be determined. To obtain the principal stresses a sixth relation is needed. This limitation does not hold at free boundary surfaces. A free surface can have only two nonvanishing principal stresses. Tangential slices yield directly the difference between the principal surface stresses. If in addition a slice is taken normal to the surface and parallel to one principal stress, it is possible to determine the individual principal stresses on the surface. Oblique incidence may be used in place of the second or normal slice.

Separation of Principal Stresses. *a. Lines of Symmetry.* Jessop [16] has developed an extension of the Lamé-Maxwell equations to three dimensions for determination of principal stresses along stress trajectories in planes of symmetry. In practice this method becomes feasible only for lines of symmetry.

b. Mechanical Measurements of Deformation after Annealing. The slices containing frozen stresses can be annealed and mechanical measurements of deformations made after annealing [17]. Theoretically, this would supply the necessary sixth relation for determination of the complete state of stress. Hooke's law shows that the strains vanish for an isotropic stress system when Poisson's ratio $\nu = 0.5$. This means that two stress systems differing by an isotropic system cannot be distinguished by this method when $\nu = 0.5$, approximately. Unfortunately, the photoelastic materials (such as fosterite, bakelite, and Castolite) have values of ν approximately equal to 0.5 at the stress-freezing temperature, so this method fails [18].

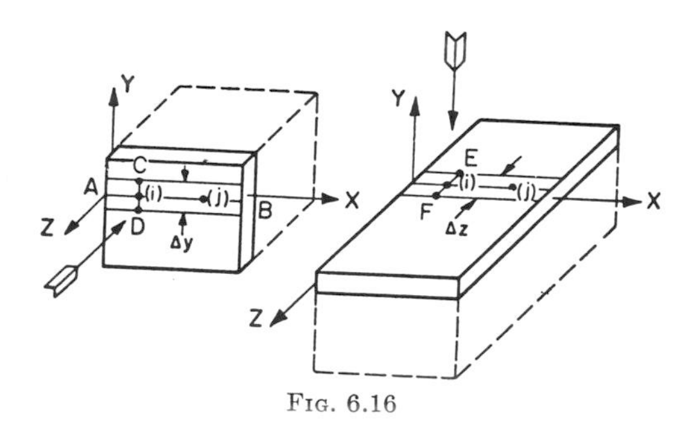

Fig. 6.16

c. Shear-difference Method. The shear-difference method [11, vol. I, chap. 8], which has long been effectively used for determination of separate principal stresses in two-dimensional cases, has been extended to the general three-dimensional problem [18]. Consider a model containing a frozen stress system with the set of axes XYZ, as shown in Fig. 6.16. Let a straight line AB be drawn through i in the interior and let this line be taken as the X axis. At any point along this line the first partial differential equation of equilibrium, with body forces neglected, is

$$\frac{\partial \sigma_x}{\partial x} + \frac{\partial \tau_{yx}}{\partial y} + \frac{\partial \tau_{zx}}{\partial z} = 0 \qquad (6.68)$$

from which, approximately,

$$(\sigma_x)_j = (\sigma_x)_a - \sum_a^j \frac{\Delta \tau_{yx}}{\Delta y} \Delta x - \sum_a^j \frac{\Delta \tau_{zx}}{\Delta z} \Delta x \qquad (6.69)$$

in which $\Delta\tau_{yx}$ is the difference between the shear stresses at points C and D and $\Delta\tau_{zx}$ is the shear difference between points F and E. The partial derivatives in Eq. (6.68) can thus be approximated and $(\sigma_x)_j$ determined by numerical integration.

With σ_x known, the stress components σ_y and σ_z can be determined from

$$(\sigma_y)_j = (\sigma_x)_j - (p' - q') \cos 2\phi_j' \qquad (6.70)$$
$$(\sigma_z)_j = (\sigma_x)_j - (p'' - q'') \cos 2\phi_j'' \qquad (6.71)$$

in which $p' - q'$ and ϕ' are, respectively, the difference between the secondary principal stresses and their isoclinic parameter in the XY plane, and $p'' - q''$ and ϕ'' are similar quantities in the XZ plane, at normal incidence.

In the first problems the necessary experimental quantities were determined from two identically stressed models. One model was sliced parallel to the XY plane and gave $(p' - q')$, ϕ'', and τ_{xz}. However, a technique has now been perfected which makes it possible to obtain the complete solution from one model only. This technique employs a subslice in the form of a slender parallelepiped or square cross section centered around the line of interest, AB. The required quantities are obtained from incidence normal to the oblong faces.

In this way five of the six independent stress components σ_x, σ_y, σ_z, τ_{yx}, τ_{zx} can be found at all points of the line AB. The remaining one unknown stress component τ_{yz} can be found from oblique incidence.

The first problem solved by the shear-difference method was that of a sphere subjected to concentrated diametral loads. The results showed internal consistency and agreed remarkably well with results from a theoretical solution [19]. This procedure opens up vast possibilities for application of photoelasticity to difficult practical problems. If a model can be made and the loads simulated, the stresses in the model can be found. Stresses have been determined in such complex problems as railroad rails and propeller blade retention systems [57, 58].

Other Methods. *a. Convergent Light.* Methods used in crystallography may be applied to slices cut from a frozen stress model. The convergent-light method, using either a petrographic microscope or a modified polariscope, may be used to determine at an arbitrary point the orientation of the principal stress axes, the order of the three principal stresses, and the three principal shearing stresses.

b. Scattered Light. In the methods described thus far the birefringence and isoclinic parameter are determined from observations of transmitted light. It is possible to obtain the necessary data from observations of scattered light looking in directions normal to the axis of propagation [20–22]. Considerable work in this direction has been done. Quite recently this procedure has been successfully employed to obtain the complete data for the shear-difference method for a general three-dimensional problem. The accuracy and practicability of this procedure have been demonstrated by redetermining the stresses in a sphere under diametral loads. This method is nondestructive and, in principle, does not require the use of frozen stresses. The method is somewhat more complicated than that of transmitted light but has a significant theoretical advantage which is mentioned later. See also [75, 76].

c. O'Rourke-Saenz Method. An interesting method suitable for residual stresses in rotationally symmetric bodies was developed by O'Rourke and Saenz [23] and applied to lightning arrestors. This method does not require slicing of the model. See also [77].

Influence of Physical Constants and Reliability of Results. In simply connected bodies subjected only to known boundary forces producing plane stress systems the stress distribution is independent of Young's modulus and Poisson's ratio. The same conclusion holds in multiply connected bodies in which the tractions on the holes reduce to a zero resultant or to a couple. Moreover, even where the stresses are affected by Poisson's ratio, the effect is small and so may usually be neglected. Experiments support these conclusions. Thus, photoelastic results from transparent plastic models have been

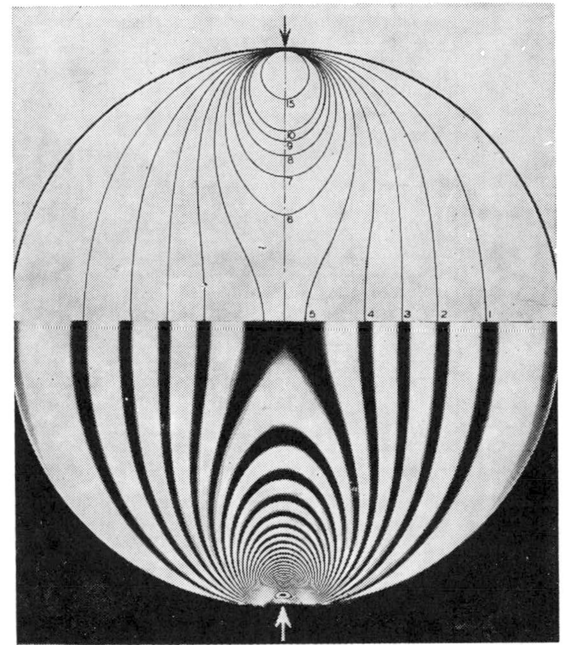

Fig. 6.17. Comparison of theoretical and photoelastic stress patterns for the horizontal half of a disk in diametral compression.

found to be in excellent agreement with stress data from metal specimens [13]. Further corroboration is found in comparisons of theoretical and photoelastic stress patterns, such as in Fig. 6.17.

In three-dimensional problems the stress distribution depends on Poisson's ratio. However, no general theorems on this subject are available. From the solutions which have been obtained to date it would appear that Poisson's ratio has only a small effect on the largest stresses. Its greatest effect is on the smaller stresses [24].

The method of scattered light may help to determine the effect of Poisson's ratio, since it permits stress determinations at room temperature at which Poisson's ratio of the plastic models is approximately the same as for metals.

6. Photoplasticity [25, 26]

The photoplastic method, or more precisely the optical method based on birefringence, which in the main has been confined to the elastic state, is also being extended to the determination of stress distributions in the plastic or mixed states. In order to solve such problems a material must be used which can be stressed plastically, such as celluloid, and its nonlinear stress-optic law established. The shear-difference method can then be used in combination with optical data to separate the principal stresses in two dimensions. The equations which enter do not involve stress-strain relations and are therefore valid in the plastic state, so that no theoretical difficulties are encountered in the determination of the stresses in the model.

However, the conditions for transition from the transparent plastic models to metal prototypes are more severe than those for the purely elastic case. These conditions involve the shape of the stress-strain curve, Poisson's ratio, and the yield criterion. By a fortunate coincidence, the behavior of celluloid with regard to these conditions approximates closely that of aluminum and steel, so that the stresses from celluloid models may be taken as good approximations to those in these metals. Thus far the method has been applied to two-dimensional cases only, such as bars with holes and grooves. Extension to three-dimensional cases can perhaps be effected by the use of scattered light.

7. Dynamic Photoelasticity

The photoelastic method is also being extended to the determination of stresses under dynamic and impact loading. This is done by means of high speed photography with exposure times of one microsecond or less. Streak photography has proved to be a particularly useful tool [27]. At the present time the studies are confined to free boundary stresses in two dimensions. Dynamic problems being studied include stress-wave propagation [28], verification of some basic theoretical solutions [29], and stresses in beams. Considerable attention is being devoted to the study of the dynamic stress-optic law. This work is still in the early stages of development.

8. Brittle Coatings*

Introduction. In two-dimensional states of stress brittle materials fail in tension, cracks forming along directions perpendicular to the maximum local tensile stress. The method of stress analysis by means of brittle coating rests on this property. In addition, the assumption is made that the coating breaks when the maximum tensile strain at a point exceeds a critical value. The basic procedure is to cover the test specimen with a thin brittle coating before loading and to observe the formation of the cracks in this coating under progressively increasing loads. This requires no models, the test being performed on the prototype itself.

Compressive strains are evaluated by applying the brittle coating to the loaded specimen and releasing the loads. The recovery from compressive deformation sets up tensile strains in the coating which cause

* In collaboration with M. M. Leven, Research Engineer, Westinghouse Research Laboratories, East Pittsburgh, Pa.

it to crack. Another method is to apply the coating before loading and to maintain the loads on the structure for several hours. Upon removal of the loads cracks form at right angles to the compressive stresses.

In principle the method is simple and, with ideal coatings which are still to be developed, capable of yielding the complete state of surface stresses under both static and dynamic conditions. However, with the best coating available at present it is possible to locate accurately only the regions of highest stress and the directions of the principal stresses. The assumption that failure of the coating occurs at a critical strain is not generally true, and the quantitative evaluation of stresses is subject to unknown error. Brittle coating is therefore most useful in connection with strain-gauge analysis, where it saves much time and expense. It shows where gauges are to be mounted and along what directions. With proper precautions the method can also be used to determine approximate stresses and their distribution.

Stresscoat. To be applicable to mild steel a practical brittle lacquer must crack at strains below the yield point. The first brittle lacquer to meet this requirement was developed in Germany in 1932 and was called the Maybach lacquer. This was a solution of colophony in benzol.

The lacquer most widely used today is Stresscoat (marketed by the Magnaflux Corporation, developed at the Massachusetts Institute of Technology around 1940). It is a solution of zinc rosinate (wood rosin and 3 per cent zinc oxide) in carbon disulfide with a small amount of dibutyl phthalate as plasticizer. These ingredients are mixed in a variety of proportions to produce various solutions suitable for different conditions of temperature and humidity.

The sensitivity of brittle lacquer is defined as the minimum strain at which it will crack. The sensitivity of Stresscoat and the factors influencing it are determined from steel cantilever strips 12 in. long and 1 by ¼ in. cross section. The calibration strip is sprayed with the lacquer, which is allowed to dry, after which the free end is given a known deflection. A special holder is provided by the Magnaflux Corporation which gives an automatic indication of the strain along the bar. Inspection of the cracks permits determination of the strain sensitivity and the factors influencing it.

In addition, comparison of crack-pattern density on the specimen with crack density of the calibration strip is sometimes used to show the approximate strain level. The crack spacing decreases as the strain increases.

Stresscoat is rather sensitive to changes in temperature. For example, the strain sensitivity of ST 1204 will vary from 0.0004 at 65°F to 0.0006 at 70°F, a 50 per cent change in 5°F. For quantitative results the variation in temperature must be kept to less than 2°F. Variation of strain sensitivity with humidity is appreciable. Variation in sensitivity with film thickness is much less pronounced. The sensitivity varies only by a small percentage for a change in film thickness from 0.004 to 0.008 in., the variation normally encountered in a properly sprayed film. Uniformity of the thickness as well as its magnitude is recognized generally from the color of the film. Stresscoat does not seem to be sensitive to variations in the strain rate, so that static calibrations are considered applicable for dynamic tests.

The sensitivity is affected by many factors [30]. Under normal operating conditions it can be made to vary from 0.0005 to 0.003. It can be increased, by careful chilling or refrigeration, to about 0.0001. Another method of sensitizing the lacquer is by heat-treating it during the drying period.

Selection of the proper coating is made on the basis of the temperature and humidity in accordance with Fig. 6.18.

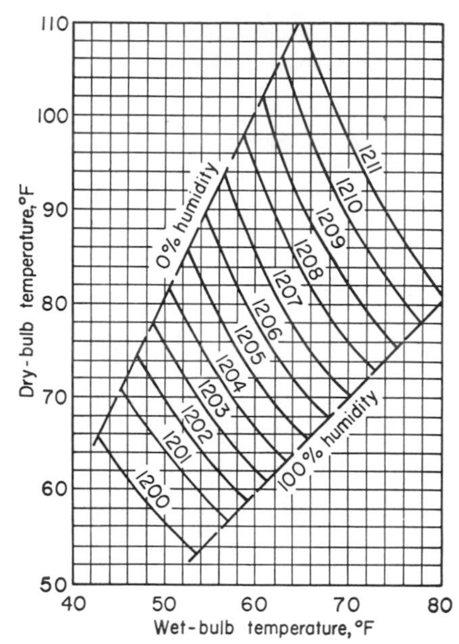

Fig. 6.18. Stresscoat selection chart based on wet- and dry-bulb psychrometer readings. Increasing numbers denote more brittle lacquers.

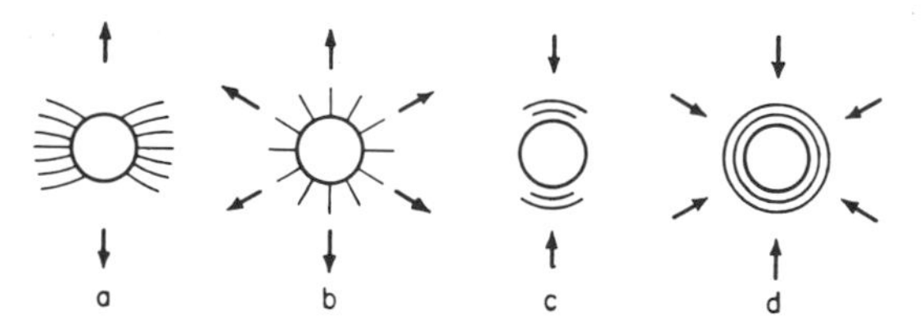

Fig. 6.19. Sketch showing residual-stress patterns obtained in a brittle coated material by the hole-drilling technique: (*a*) one-dimensional tension, (*b*) two-dimensional tension, (*c*) one-dimensional compression, (*d*) two-dimensional compression.

Brittle lacquers, in general, and Stresscoat, in particular, have been applied to a great number of problems. Stresscoat has been used with some degree of success, at least qualitatively, to indicate residual stresses [31]. A small hole, not over ⅛ in. in diameter, is carefully drilled (preferably using a hand drill) into the coated object at the point where the residual stress is to be determined. The depth of the hole is about equal to the diameter. As the hole is drilled, a crack pattern will be developed because of release of residual stress, as in the relaxation method used to indicate compressive strains (Fig. 6.19).

9. X Rays* [32, 33]

Fundamental Principles. The basic data in the X-ray method of stress analysis are obtained from diffraction patterns. Under suitable conditions such patterns provide sufficient data for the calculation of strains along directions approximately perpendicular to the atomic planes.

The spacing d between atomic planes, Fig. 6.20, is altered by elastic deformations and can be determined

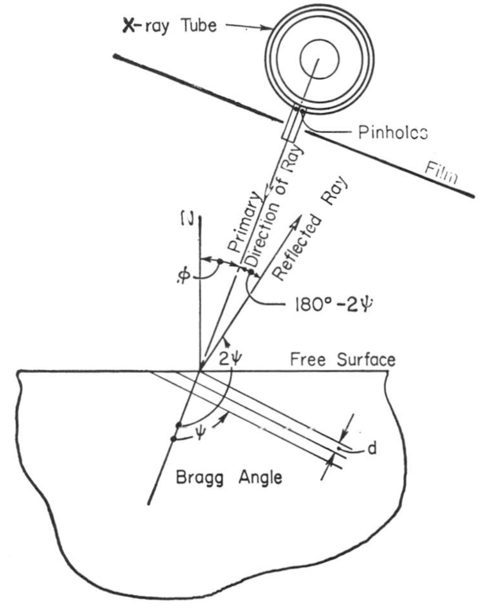

FIG. 6.20

with considerable accuracy from X-ray diffraction patterns by means of Bragg's law

$$d = n\lambda/2 \sin \psi \tag{6.72}$$

Plastic deformations consist essentially of a sliding of crystalline blocks along definite crystallographic surfaces with little or no change in the interatomic spacings within the blocks and affect diffraction patterns only in a qualitative way. They cannot be measured with any degree of accuracy from the angular position of Debye-Scherrer rings.

Two diffraction patterns, one taken in the stress-free state and the other in the stressed state, provide sufficient data to compute the average linear strain ϵ in the direction perpendicular to the atomic planes. Thus, if d_0 and d denote respectively the spacings in the stress-free and stressed states

$$\epsilon = (d - d_0)/d_0 \tag{6.73}$$

Two diffraction patterns may be viewed as a strain gauge with a gauge length equal to the interatomic spacing.

The angle of reflection of a collimated beam of X rays incident on a specimen depends on the material and on the wavelength of radiation. With proper choice of wavelength, a set of reflections arises from atomic planes which are nearly normal to the primary beam. For example, $K\alpha$ radiation from cobalt will reflect from crystallographic planes in steel at an angle $\psi = 80°37.5'$, Fig. 6.20. This is independent of the angle of inclination ϕ of the primary beam. It follows that if a collimated beam of X rays is perpendicular to the surface of the specimen, it can be made to reflect from atomic planes which are nearly parallel to the surface and to provide data for strains which are nearly normal to the surface.

Stress Computation from X-ray Data. In the application of the atomic strain ϵ to the calculation of stress, it is assumed that the small volume of material irradiated by the X-ray beam behaves like an elastic, isotropic, homogeneous material and that the stresses are essentially constant throughout the small volume. The assumption of isotropy is made notwithstanding the fact that the stressed material is known to consist of a large number of crystallites of diverse crystallographic orientation.

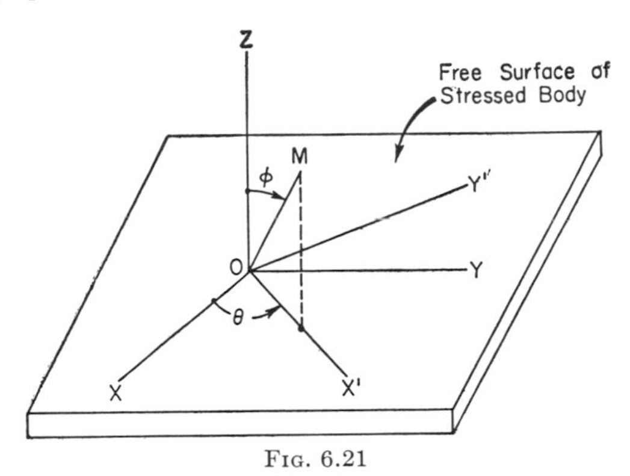

FIG. 6.21

In Fig. 6-21, the X and Y axes lie in a free surface of a stressed body so that the equation of the strain ellipsoid [34] assumes the simplified form

$$\epsilon = \epsilon_x l^2 + \epsilon_y m^2 + \epsilon_z n^2 + \gamma_{xy} lm \tag{6.74}$$

Inserting the values of l, m, n in terms of θ and ϕ we obtain for the strain $\epsilon_{\theta\phi}$ along OM

$$\epsilon_{\theta\phi} = (\epsilon_x \cos^2 \theta + \epsilon_y \sin^2 \theta + \gamma_{xy} \sin \theta \cos \theta) \sin^2 \phi + \epsilon_z \cos^2 \phi \tag{6.75}$$

The quantity in parentheses represents the strain $\epsilon_{x'}$ along OX' so that Eq. (6.75) may be written

$$\epsilon_{\theta\phi} - \epsilon_z = (\epsilon_{x'} - \epsilon_z) \sin^2 \phi \tag{6.76}$$

Since the surface is assumed to be free, i.e., $\sigma_z = 0$, we have from Hooke's law

$$\epsilon_z = -\frac{\nu}{E}(\sigma_{x'} + \sigma_{y'}) \tag{6.77}$$

and

$$\epsilon_{x'} = \frac{\sigma_{x'}}{E} - \frac{\nu\sigma_{y'}}{E} \tag{6.78}$$

From the last three equations

$$\sigma_{x'} = \frac{E(\epsilon_{\theta\phi} - \epsilon_z)}{(1 + \nu) \sin^2 \phi} \tag{6.79}$$

* In collaboration with Dr. Hans Ekstein, Senior Physicist, Argonne National Laboratory.

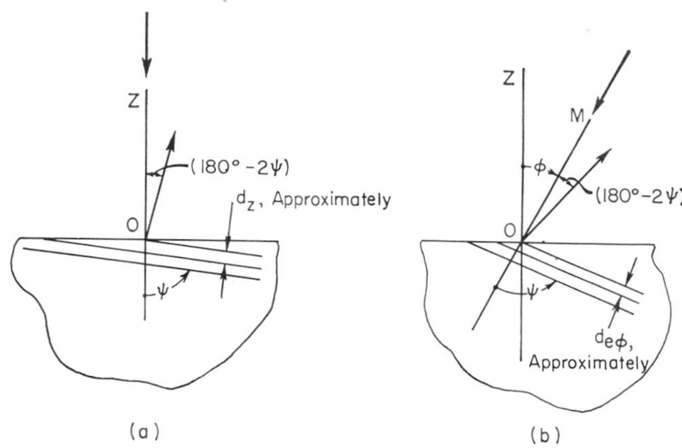

FIG. 6.22

If, now, diffraction patterns are obtained from two primary beam directions, one normal and the other inclined to the surface, and if in both patterns Debye-Scherrer lines with Bragg angles of nearly 90° are used to measure d, then

$$\epsilon_{\theta\phi} - \epsilon_z = \frac{d_{\theta\phi} - d_0}{d_0} - \frac{d_z - d_0}{d_0} = \frac{d_{\theta\phi} - d_z}{d_0} \quad (6.80)$$

in which $d_{\theta\phi}$ and d_z are distances between atomic planes in the stressed state measured along OM and along the Z axis, respectively, Fig. 6.22. Since the angle of reflection has been chosen small, d_0 may be replaced by d_z in the denominator, so

$$\sigma_{x'} = \frac{E}{(1+\nu)\sin^2\phi} \cdot \frac{d_{\theta\phi} - d_z}{d_z}, \text{ approximately} \quad (6.81)$$

Thus the stress $\sigma_1 = \sigma_{x'}$ can be determined from two diffraction patterns in the stressed state. From two additional patterns two more stresses σ_2 and σ_3 can be determined. The equations relating the principal surface stresses p,q to σ_1, σ_2, and σ_3 are identical with those relating the principal strains ϵ_p, ϵ_q to ϵ_1, ϵ_2, and ϵ_3. Hence, the graphical methods of strain analysis (Mohr's circle, 6.1) are also applicable in the X-ray method.

The principal field of application of the X-ray method is in the study of residual stresses where, in theory at least, it provides a means to determine the stresses without destroying the specimen.

The accuracy of the X-ray method has been tested directly for steel specimens subjected to purely elastic stresses by comparing experimental with theoretical values. Under the most favorable circumstances, errors of 2,000 to 3,000 psi have been found.

Under other circumstances, agreement with mechanical measurements was poor. In particular, the values of Young's modulus and Poisson's ratio measured in elastically stressed brass have been found to depend on the wavelength used if the evaluation is made by the conventional formulas. Differences of 60 per cent for different wavelengths have been found. Also, after plastic deformation, residual stresses measured by use of conventional X-ray methods may appear *tensile* or *compressive* according to the choice of wavelength.

These findings raise serious doubts regarding the soundness of the assumptions of quasi-isotropy made in deriving the basic formulas. The reliability of the method is still a subject of controversy.

10. Recent Developments

Five major conferences—in the Soviet Union, East Germany, France, and the United States—have been held since 1958 [35–39]. Several new books have also been published [40–43]. The problems dealt with and the vast scope of technical applications range from underground mines to every conceivable machine and structure.

Free-surface Stresses without Models. Much work has been concerned with new or improved devices and methods to measure free-surface strains directly on metal prototypes.

Foil Gauges. Foil gauges, which came into use in the mid 1950s, are now widely used, probably more than wire gauges. The grid of the foil gauge is made from a thin metal sheet or foil (150 to 200 μin.) by a process of etching which permits the production of precise grid configurations of almost any shape or size. The cross section of the straight portions of a foil gauge is rectangular and that of the wire gauge is circular (0.0005 to 0.001 in. in diameter). The bonding area of a foil gauge is greater than that of a wire gauge. As a result, the shearing stresses in the cement are smaller; this increases the strain range and leads to improved creep and hysteresis characteristics. Also, because the end loops of foil gauges have less resistance than those of wire gauges, cross sensitivity has practically been eliminated.

The precise etching process produces more uniform gauge factors and initial resistance as well as better fatigue characteristics. Foil gauges are better heat dissipators than wire gauges, permitting higher gauge currents with a resulting improvement in the sensitivity of the associated electrical readout devices. One disadvantage of foil gauges is the lower resistance to ground. The method of installation of both wire and foil gauges has been much simplified by the development of Eastman's 910 adhesive.

Semiconductor Gauges. Another type of gauge, which has found limited but important applications, is the semiconductor gauge. These gauges are generally made from silicon crystals doped with a trace of impurities to affect n or p piezoresistive response. The gauge factor may be positive or negative and may have a magnitude of about 130 or 65 times greater than the metallic gauges. The semiconductors can be employed to advantage in dynamic strain measurements and high-signal-output transducers [44].

Method of Replication. The process of replication consists of forming a system of fine (random, straight, or circular) scratches on the surface to be strained and taking replicas or impressions of the scratch pattern before, during, and after loading. Subsequent matching of the replicas in a special microscope comparator permits the determination of the strains with high precision. This method is particularly useful for strain analysis in fillets and other zones difficult to reach with a strain gauge. (See [45] for a description of this method and also [46], which includes a discussion of the pneumatic gauge.)

Birefringent Coatings. The method consists of spraying or cementing a transparent birefringent coating, such as Photostress (Budd Instrument Division), to the free surface of the structure and measuring the induced stresses by means of a portable reflection polariscope. The method has aroused considerable commercial interest and also considerable technical controversy regarding its accuracy, especially about errors introduced by the thickness of the coating and its reinforcing effect [47–49].

Brittle Coatings. An improved coating made of vitreous enamel, called Alltemp, has been developed by Magnaflux [50].

Moiré Fringes. The essential point in this method can be demonstrated by placing two sheets of ruled paper face to face and giving them a relative displacement. Viewing the two sheets against the light, one can see that the points of intersection of the two grids tend to form fringes, called Moiré fringes. The continuity or smoothness of these fringes increases with the density of the grids. Each fringe is the locus of a constant displacement [51].

Free-surface strains can be determined in the following manner. One grid is printed on, or cemented to, the surface to be strained. Loads are applied, and the strained grid is viewed through a master or undeformed grid. The image of the superimposed grids will contain Moiré fringes from which the displacement u and v, and therefore the strains

$$\epsilon_x = \frac{\partial u}{\partial x} \qquad \epsilon_y = \frac{\partial v}{\partial y} \qquad \gamma_{xy} = \frac{\partial u}{\partial v} + \frac{\partial v}{\partial x}$$

can be found.

As far as surface strains go, the method is limited to flat surfaces. However, in 1962, Dantu showed an interesting three-dimensional application [52]. He cut a plexiglass sphere into two identical parts, cemented a grid of 500 lines per inch to one half, and then cemented the two parts together. The sphere was then subjected to diametral loads and the strains and stresses determined from the Moiré pattern. A way is thus opened for the study of three-dimensional strains by means of composite, or sandwich-type, models containing one or more embedded grids. Any plane of interest can be selected for observation by means of a camera with a suitable lens system.

Diffraction-grating Strain Gauge. A strain gauge in which the essential element is a diffraction grating and the change in the diffraction angle is the measure of strain has been developed by Bell. The grating may be ruled directly on the specimen or may be bounded to it. It is claimed that static and dynamic strains up to 10 per cent have been measured at temperatures in excess of 1000°F. In addition to measuring strains, this method may also be used to measure strain gradients, surface angles and their gradients, as well as displacements [53].

Photoelasticity. *Model Material.* For space problems epoxy resins have now generally replaced older model materials. Kriston and Fosterite are no longer available, while bakelite and Castolite have serious shortcomings.

The photoelastic properties of polymerized epoxy resins depend chiefly on the amount and type of the curing agent used, chiefly acid anhydrides and amines. Epoxies are available in liquid and solid forms. Liquid epoxies can be cast to make complicated models of large size and machined when necessary. The fumes of liquid epoxies are toxic. They have a very high abrasive effect and require carbide-tipped tools. Most epoxies exhibit time-edge stresses which can be eliminated by storing for a time at constant humidity or by annealing at a suitable temperature. (For a comprehensive study, see [54].)

For ordinary two- or three-dimensional problems the materials used in the United States are the same as in the Soviet Union—mostly phenolic and epoxy resins. However, unlike the situation in the United States, there exist in the Soviet Union polymer laboratories for the special purpose of developing materials of prescribed mechanical and optical properties. S. I. Sokolov and N. A. Shchegolevskaya have synthesized ordinary polymers and epoxy resins into materials possessing optical and mechanical properties which vary over a wide range. The new types of materials are being utilized in the photoelastic solution of stresses not only in machine parts and structural elements but also in geology and soil mechanics. For example, in order to simulate the rock strata surrounding a mine tunnel or gallery, the photoelastic models for these structures are made of optically sensitive layers having different mechanical properties.

Static Three-dimensional Photoelasticity. Monolithic epoxy models combined with the method of freezing and slicing are generally employed for ordinary three-dimensional problems, and the necessary optical data are taken with a transmitted-light polariscope with or without a photometer.

Interior principal stresses are generally determined by means of the shear-difference method, for which Krol designed a special computer for the automation of the necessary calculations [55–58].

Dynamic Photoelasticity. Numerous difficulties develop when one deals with fast stress waves. Among the first of these is the difficulty of obtaining clear stress patterns without blurring. One approach is to use model materials of low moduli in which the speed of the stress waves is relatively slow. Another approach is to use conventional, more rigid materials and ultra-high-speed photographic techniques [27, 59, 60]. A recent survey [61] lists 79 references which show the great variety of cameras, instrumentation, and techniques employed.

The results obtained by dynamic photoelasticity have been limited mainly to free boundary stresses and to interior secondary maximum shears in two dimensions. For the separation of principal stresses under dynamic conditions, the oblique-incidence method has proved effective for lines of symmetry. A dual-beam polariscope for the simultaneous recording of normal and oblique-incidence stress patterns was developed by P. D. Flynn [62]. Considerable attention has been given to the dynamic stress-optic law [63] and to the properties of photoelastic materials under dynamic conditions [64]. Although substantial progress has been made in dynamic photoelasticity, much remains to be done. Challenging problems lie ahead in the extensions to three-dimensional configurations, to inelastic behavior, and in the transition from model to prototype.

Photoplasticity. It has been established experimentally [65] that, as in elasticity, the general stress-optic law for the plastic state can be written in the form

$$p' - q' = f(n)$$

in which (p',q') are secondary principal stresses and $f(n)$ is a nonlinear function of the birefringence. The value of $p' - q'$ is uniquely determined by the birefringence alone in both the plastic and elastic states. Further, when the directions of the principal stresses and strains do not coincide, the isoclinic parameters in celluloid models give the direction of the principal stresses and not of the principal strains [66, 69]. It has also been shown that factors of stress concentration in the plastic state are in essential agreement with the results from theory and strain-gauge measurements [67, 68]. In the Soviet Union, transparent metals, such as silver chloride and silver bromide, are used for photoplastic studies. Also, in the study of forming processes, anisotropic materials are employed [70].

Photothermoelasticity. To determine the stresses at a point due to temperature differentials the fringe order n at the point and the temperature-dependent fringe value f must be known. In two-dimensional problems the temperatures can be determined from thermocouples attached to the surfaces of the model and the appropriate fringe values from tension calibration specimens at different temperatures.

In three-dimensional problems difficulties develop. The complete temperature distribution must be known, and this requires embedding many thermocouples in the model. Gerard found that for some materials the fringe value is approximately constant for a rather large range of temperature. For example, in Hysol 6000-OP the fringe value remains approximately unchanged from 70 to −60°F. For thermal stresses the transition from model to prototype depends on the value of $f/\alpha E$ of the model material, with α denoting the coefficient of thermal expansion. Leven and Johnson found that while f, α, and E all vary with temperature the value of the ratio $f/\alpha E$ remains virtually constant for epoxy resins over a wide range of temperature [72].

The method of freezing and slicing generally employed at room temperature for three-dimensional problems breaks down for stresses produced by temperature differentials. Two alternative procedures suggest themselves. A model can be constructed in which a thin plate containing the plane of the desired stresses is made of optically sensitive material, and the rest of the model from an optically insensitive material, but otherwise homogeneous with the plate, and the parts cemented to make a monolithic assembly. The observed birefringence and isoclinics from a transmitted-light polariscope would be due to the required stresses in the plate. This procedure seems to be used in the Soviet Union. In the United States, Tramposch and Gerard followed a somewhat different procedure. They removed a thin plate from a model made of one optically sensitive material, attached to its faces sheets of polaroid, and cemented the assembly to form a unit [71]. This construction also gives the necessary data for the plate. In the first procedure the polariscope is outside the model and in the latter it is inside the model.

Frocht and Srinath [73] proposed that the birefringence and isoclinics be determined from scattered-light observations instead of transmitted light. This would eliminate the need for a composite or sandwich-type model. It would also make it possible to determine the complete distribution of birefringence and isoclinic parameters from one and the same model, not always the case in either of the first two procedures.

Photoviscoelasticity. In conventional photoelasticity, optical data are determined at room temperature from elastic models under static conditions when the birefringence and isoclinics do not depend on the rate of loading. However, in viscoelastic models, at temperatures giving rise to the state of transition between the glassy and rubbery states, the optical data are temperature- and rate-dependent. These conditions change the normal mechanical and optical properties of the model material.

The current theoretical view is that the usual stress-optic law is not valid in the transition state and that the principal stress difference $(p - q)$ cannot be determined from the birefringence alone but requires a combination of the histories of the isoclinics and birefringence. Practical situations may arise wherein the optical and mechanical behavior in the viscoelastic transition state must be considered, e.g., in wave-propagation studies. The subject is also of importance to polymer chemists who use birefringence to investigate dynamic phenomena [74].

Scattering, Lasers, and Further Progress. The advantages of scattered light are many. Even with present equipment the scattered-light method is capable of giving accurate reproducible results in general space problems [75]. Its general adoption has been retarded because the present light sources are in many cases too weak to give good stress patterns. The development of lasers at reasonable cost holds great promise for the rapid adoption of the scattering technique. The few laboratories that have already acquired lasers report favorably on their preliminary experiments. Lasers and other improved equipment may open a new era in photoelastic research, and extensions to three-dimensional photoelasticity, photoplasticity, and photothermoelasticity will proceed at an accelerated rate [76].

References

1. Hetenyi, M. (ed.): "Handbook of Experimental Stress Analysis," Wiley, New York, 1950.
2. Frocht, Max M.: "Strength of Materials," chap. 5, Ronald, New York, 1951.
3. Dow, Norris F.: The analysis of strains indicated by multiple-strand resistance-type wire strain gages used as rosettes, NACA advance restricted report (now unclassified), January, 1943.
4. Carpenter, J. E., and L. D. Morris: A wire resistance strain gage for the measurement of static strains at temperatures up to 1600°F, *Proc. Soc. Exp. Stress Anal.*, **9**(1): 191–200 (1951).
5. Gorton, R. E.: Development and use of high-temperature strain gages, *Proc. Soc. Exp. Stress Anal.*, **9**(1): 163–176 (1951).
6. Swainger, K. H.: Electrical resistance wire strain-gauges to measure large strains, *Nature*, **159**(4028): 61–62 (Jan. 11, 1947).
7. Clark, D. S., and G. Datwyler: Stress-strain relations under tension impact loadings, *Am. Soc. Testing Materials Proc.*, **38**(II): 98–111 (1938).
8. Dobie, W. B., and P. C. G. Isaac: "Electrical Resistance Strain Gages," English Universities Press, London, 1948.
9. Day, E. E., and A. H. Sevand: Characteristics of electric strain gages at low temperatures, *Proc. Soc. Exp. Stress Anal.*, **8**(1): 133–142 (1950).
10. Coker, E. G., and L. N. G. Filon: "A Treatise on Photoelasticity," Cambridge, London, 1931; 2d ed., 1953.
11. Frocht, Max M.: "Photoelasticity," vols. I, II, Wiley, New York, 1941, 1948.
12. Jessop, H. T., and F. C. Harris: "Photoelasticity: Principles and Methods," Dover, New York, 1950.
13. Frocht, Max M.: Isopachic stress patterns, *J. Appl. Phys.*, **10**(4): 248–257 (April, 1939).
14. Maxwell, J. C.: On the equilibrium of elastic solids, *Trans. Roy. Soc. Edinburgh*, **20**(1): 87–120 (1853).
15. Hetenyi, M.: The fundamentals of three-dimensional photoelasticity, *J. Appl. Mech.*, **5**(4): 149–155 (December, 1938).

15*a*. See also Treloar, L. R. G., "Physics of Rubber Elasticity," a.a. 2nd ed., Oxford University Press, London, 1958.

16. Jessop, H. T.: The determination of the separate stresses in three-dimensional stress investigations by the frozen stress method, *J. Sci. Instr.*, **26**: 27–31 (January, 1949).
17. Prigorovsky, N. I., and A. K. Preiss: A study of the state of stress in transparent three-dimensional models by means of beams of parallel polarized light, *Izvest. Akad. Nauk SSSR Otdel. Tekh. Nauk*, **1949**(5): 686–700.
18. Frocht, M. M., and R. Guernsey, Jr.: Studies in three-dimensional photoelasticity: The application of the shear difference method to the general space problem, *1st U.S. Natl. Congr. Appl. Mechanics*, December, 1952.
19. Sternberg, E., and F. Rosenthal: The elastic sphere under concentrated loads, *J. Appl. Mech.*, **19**(4): 413–421 (December, 1952).
20. Weller, R.: A new method for photoelasticity in three-dimensions, *J. Appl. Phys.*, **10**(4): 266 (April, 1939).
21. Menges, Hermann Josef: Die experimentelle Ermittlung raümlicher Spannungszustande an durchsichtigen Modellen mit Hilfe des Tyndalleffektes, *Z. angew. Math. u. Mech.*, **20**(4): 210–217 (August, 1940).
22. Jessop, H. T.: The scattered light method of exploration of stresses in two- and three-dimensional models, *Brit. J. Appl. Phys.*, **2**(9): 249–260 (September, 1951).
23. O'Rourke, R. C., and A. W. Saenz: Quenching stresses in transparent isotropic media and the photoelastic method, *Quart. Appl. Math.*, **8**(3): 303–311 (October, 1950).
24. Frocht, M. M.: The growth and present state of three-dimensional photoelasticity, *Appl. Mech. Revs.*, **5**(8): 337–340 (August, 1952). See also The growth and present state of photoelasticity, "Applied Mechanics Surveys," pp. 627–644, Spartan Books, Washington, D.C., 1966.
25. Fried, B.: Some observations on photoelastic materials stressed beyond the elastic limit, *Proc. Soc. Exp. Stress Anal.*, **8**(2): 143–148 (1951).
26. Hiltscher, R.: Theorie und Anwendung der Spannungsoptik im elastoplastischen Gebiet, zD Z. des VDI, **97**(2), (1955).
27. Frocht, M. M., P. D. Flynn, and D. Landsberg: Dynamic photoelasticity by means of streak photography, *Proc. Soc. Exp. Stress Anal.*, **14**(2): 81–90 (1957).
28. Kolsky, H.: "Stress waves in solids," Oxford University Press, London, 1953.
29. Frocht, M. M., and P. D. Flynn: Studies in dynamic photoelasticity, *J. Appl. Mech.*, **23**(1): 116–122, March, 1956.
30. Durelli, A. J., S. Okubo, and R. H. Jacobson: Study of some properties of Stresscoat, *Proc. Soc. Exp. Stress Anal.*, **12**(2): 55–76 (1955).
31. Tokarcik, A. G., and M. H. Polzin: Quantitative evaluation of residual stresses by the Stresscoat drilling technique, *Proc. Soc. Exp. Stress Anal.*, **9**(2): 195 (1952).
32. Barrett, Charles S.: "Structure of Metals," 2d ed., McGraw-Hill, New York, 1952.
33. Glocker, R., B. Hess, and O. Schaaber: *Z. tech. Physik*, Vol. 19, p. 194 (1938).
34. Timoshenko, S., and J. N. Goodier: "Theory of Elasticity," McGraw-Hill, New York, 1951.
35. "Polaryzationno-Optichesky Metod Issledovanya Napryazhenyi" (Russian), "Photoelastic Method for Stress Analysis," S. P. Shikhobalov, Chairman Editorial Board, Publication of Leningrad University, 1960.
36. Internationales Spannungsoptishes Symposium, bearbeit von G. Haberland, *Abhandl. Deut. Akad. Wiss. Berlin*, 1962.
37. Deuxieme conference d'analyse des contraintes à Paris, Memoires GAMAC, *Bull. Soc. Franc. Méchanique* 4, 1962. *Rev. Franc. Méchanique.*
38. First International Congress on Experimental Mechanics, B. E. Rossi (ed.), Pergamon Press, New York, 1963.
39. International Symposium on Photoelasticity, M. M. Frocht (ed.), Pergamon Press, New York, 1963.
40. Durelli, A. J., E. A. Phillips, and C. H. Tsao: "Introduction to the Theoretical and Experimental Analysis of Stress and Strain," McGraw-Hill, New York, 1958.
41. Perry, C. C., and H. R. Lissner: "The Strain Gage Primer," 2d ed., McGraw-Hill, New York, 1962.
42. Dove, R. C., and T. H. Adams: "Experimental Stress Analysis and Motion Measurement," Charles E. Merrill Books, Inc., Columbus, Ohio, 1964.
43. Dally, J. W., and W. F. Riley: "Experimental Stress Analysis," McGraw-Hill, New York, 1965.
44. Sanchez, I. C.: The Semi-Conductor Strain Gauge—A New Tool for Experimental Stress Analysis, Ref. 38, pp. 254–276.
45. Hickson, V. M.: A replica technique for measuring static strains, *J. Mech. Eng. Sci.*, **1**(2): (September, 1959).

46. Hickson, V. M.: A Special Technique in Experimental Stress Analysis, Ref. 38, pp. 221–236.
47. Zandman, F.: Analyse des contraintes pour vernis photoelastiques, groupement pour l'avancement des methode d'analyse de contraintes, *Rev. Franc. Méchanique*, **2**(6): 3–14 (1956).
48. Post, D., and Zandman, F.: Accuracy of birefringent coating method for coatings of arbitrary thickness, *Proc. Soc. Exp. Stress Anal.*, **18**(1): 21–32 (1961). See also Post, D., Isochromatic fringe sharpening and fringe multiplication in photoelasticity, *Proc. Soc. Exp. Stress Anal.*, **12**(2): 143–156 (1955).
49. Duffy, J., and Mylonas, C.: An Experimental Study on the Effects of the Thickness of Birefringent Coating, Ref. 39, pp. 27–42.
50. Singdale, F. N.: Improved brittle coatings for use under widely varying temperature conditions, *Proc. Soc. Stress Anal.*, **11**(2): 173–178.
51. Morse, Stanley, August J. Durelli, and Cesar A. Sciammarella: Geometry of moiré fringes in strain analysis, *Proc. Am. Soc. Civil Engrs. J. Eng. Mech. Div.*, **86**(EM 4): 105–126 (August, 1960).
52. Dantu, P.: Deformation d'une sphere de plexiglass comprimie entre deux planes paralleles rigides en dehors du domain elastique, *Lab. Central des Ponts et Chausses, Publ.* 622, 1962.
53. Bell, J. F.: Diffraction grating strain gauge, *Proc. Soc. Exp. Stress Anal.*, **17**(2): 51–64 (1960).
54. Leven, M. M.: Epoxy Resins for Photoelastic Use, Ref. 39, pp. 145–165.
55. Krol, K. G.: Automatic Process for the Evaluation of Photoelastic Data, Ref. 35, pp. 196–206.
56. Frocht, M. M., and R. Guernsey, Jr.: Further work on the general three-dimensional photoelastic problem, *J. Appl. Mech.*, **22**(2): (June, 1955).
57. Frocht, M. M., and B. C. Wang: A three dimensional photoelastic study of interior stresses in the head of a railroad rail in the region under a wheel, *Proc. 4th U.S. Natl. Congr. Appl. Mech.*, **1**: 603–609 (1962).
58. Frocht, M. M., and B. C. Wang: A Three Dimensional Photoelastic Investigation of a Propeller Blade Retention, "Proceedings of International Symposium on Photoelasticity," pp. 123–140, Pergamon Press, New York, 1963.
59. Wells, A. A., and D. Post: The dynamic stress distribution surrounding a running crack—a photoelastic analysis, *Proc. Soc. Exp. Stress Anal.*, **16**(1): 69–96 (1958).
60. Flynn, P. D., J. T. Gilbert, and A. A. Roll: Some recent developments in dynamic photoelasticity, *SPIE J.*, **2**: 128–131 (1964).
61. Goldsmith, W.: in W. J. Worley (ed.), "Dynamic Photoelasticity, Experimental Techniques in Shock and Vibration," pp. 25–54, American Society of Mechanical Engineers, New York, 1962. See also: NAVWEPS Report 8037, U.S. Naval Ordnance Test Station, China Lake, Calif., November, 1962.
62. Flynn, P. D.: A dual-beam polariscope for oblique incidence, *Exptl. Mech.*, **4**: 182–184 (1964).
63. Frocht, M. M.: Studies in Dynamic Photoelasticity with Special Emphasis on the Stress-Optic Law, in N. Davids (ed.), "International Symposium on Stress Wave Propagation in Materials," pp. 91–118, Interscience, New York, 1960.
64. Clark, A. B. J., and R. J. Sanford: A Comparison of Static and Dynamic Properties of Photoelastic Materials, *Proc. Soc. Exp. Stress Anal.*, **20**(1): 148–151 (1963).
65. Frocht, M. M., and R. A. Thomson: Studies in Photoplasticity, *Proc.* 3d *U.S. Natl. Congr. Appl. Mech.*, June, 1958.
66. Frocht, M. M., and R. A. Thomson: Experiments on mechanical and optical coincidence in photoplasticity, *Exptl. Mech.*, **1**(2): 43–47 (1961).
67. Thomson, R. A., and M. M. Frocht: Further Work on Plane Elasto-Plastic Stress Distribution, Ref. 39, pp. 185–193.
68. Frocht, M. M., and Y. F. Cheng: Foundations for Three-dimensional Photoplasticity, Ref. 39, pp. 195–216.
69. Frocht, M. M., and Y. F. Cheng: On the meaning of isoclinic parameters in the plastic state in cellulose nitrate, *J. Appl. Mech.*, March, 1962, pp. 1–6.
70. Gubkin, S. I., S. I. Dobrovolsky, and B. B. Boiko: "Photoplasticity," Publication of the White Russian Academy of Science at Minsk, 1957 (Russian).
71. Tramposch, Herbert, and George Gerard: An exploratory study of three-dimensional photothermoelasticity, *J. Appl. Mech.*, **28**(1): 35–40 (March, 1961). See also Ref. 39, pp. 81–94.
72. Leven, M. M., and R. L. Johnson: Thermal stresses on the surface of tube sheet plates of 10 and 33⅓ percent ligament efficiency, *Exptl. Mech.*, December, 1964, pp. 356–365.
73. Frocht, M. M., and L. S. Srinath: *J. Appl. Mech.*, **26**(2): 310 (June, 1959) (discussion).
74. Williams, M. L., and R. J. Arenz: The engineering analysis of linear photoviscoelastic materials, *Exptl. Mech.*, September, 1964, pp. 249–262. This paper contains an extensive bibliography in which items 3, 4, 7 are of immediate interest.
75. Srinath, L. S., and M. M. Frocht: The Potentialities of the Method of Scattered Light, Ref. 39, pp. 277–292.
76. Srinath, L. S., and M. M. Frocht: Scattered light photoelasticity, basic equipment and technique, *Proc. 4th Natl. Congr. Appl. Mech.*, 1962.
77. Aben, H. K.: Optical phenomena in photoelastic models by the rotation of principal axes, *Exptl. Mech.*, **6**(1): 13–22 (1966).

Chapter 7

Vibrations of Elastic Bodies; Wave Propagation in Elastic Solids

By PHILIP M. MORSE, Massachusetts Institute of Technology

The theory of the internal motions of solids is far from being thoroughly developed, because of the great complexity of the phenomena. Actual solids exhibit properties which are combinations of elasticity, viscosity, and plasticity; experimental arrangements to measure one of these properties to the exclusion of the others are very difficult to devise. Moreover, the internal constants (elastic, plastic, etc.) of a substance depend strongly on details of the thermal, chemical, and structural state of the substance. For these reasons and others, values of the internal constants are known for only a few materials, and the probable accuracy of those which are known is seldom better than two significant figures.

But even in motions which do allow of separation of various effects, as in small-amplitude vibrations, the corresponding mathematical theory is so complex that exact solutions are possible for only the simplest configurations, which usually correspond very poorly to configurations possible experimentally. Hence correspondence between theoretical and experimental results, to determine elastic moduli and to search for possible inadequacies of theory, is difficult.

This chapter deals with the small-amplitude vibrations in homogeneous solids, where simple elastic forces predominate over the viscous or plastic forces. For such material and for small enough amplitudes the system may be considered conservative. The decrements and attenuations produced by internal friction may be inserted as afterthoughts. A few of the wave motions possible to such systems will be discussed, beginning with the plane wave and going on to some of the more complicated motions which can be analyzed, either approximately or completely, by existing theory.

1. Equation of Motion; Energy and Intensity

Referring to Part 3, Chap. 1, for a solid which is predominantly elastic, rather than plastic, and for displacements from equilibrium $\mathbf{r}(x,t)$ of the point originally at x, which are sufficiently small to make Hooke's law valid, the internal potential energy density produced by the displacement is

$$W = \tfrac{1}{2} \sum_{i,j,k,l} C_{ijkl} e_{ij} e_{kl} = \tfrac{1}{2}\mathbf{e}:\mathbf{C}:\mathbf{e} \qquad (7.1)$$

where $e_{ij} = \frac{1}{2}(\partial r_j/\partial x_i) + \frac{1}{2}(\partial r_i/\partial x_j)$ and $\mathbf{C}$ is the fourth-order tensor representing the elastic moduli [see Eqs. (1.1) and (1.4) of Part 3, Chap. 1]. Symmetry requirements make many of the components of C equal:

$$C_{ijkl} = C_{ijlk} = C_{jikl} = C_{jilk} = C_{klij} = C_{klji} = C_{lkij} = C_{lkji} \qquad (7.2)$$

Thus the 81 components of $\mathbf{C}$ depend on, at most, 21 independent moduli. For solids which are single crystals of the more symmetric type many of the components of $\mathbf{C}$, referred to the principal axes of the crystal, are zero (see Part 3, Chap. 1, Table 1.1). For isotropic solids, there are only two independent moduli, λ and μ:

$$W = \tfrac{1}{2}\lambda\Delta^2 + \mu(e_{11}^2 + e_{22}^2 + e_{33}^2 + 2e_{12}^2 + 2e_{13}^2 + 2e_{23}^2) \qquad \mathbf{(7.3)}$$

$$= \tfrac{1}{2}\lambda\left(\frac{\partial r_1}{\partial x_1} + \frac{\partial r_2}{\partial x_2} + \frac{\partial r_3}{\partial x_3}\right)^2 + \mu\left[\left(\frac{\partial r_1}{\partial x_1}\right)^2 + \left(\frac{\partial r_2}{\partial x_2}\right)^2 + \left(\frac{\partial r_3}{\partial x_3}\right)^2\right]$$
$$+ \tfrac{1}{2}\mu\left[\left(\frac{\partial r_1}{\partial x_2} + \frac{\partial r_2}{\partial x_1}\right)^2 + \left(\frac{\partial r_1}{\partial x_3} + \frac{\partial r_3}{\partial x_1}\right)^2 + \left(\frac{\partial r_2}{\partial x_3} + \frac{\partial r_3}{\partial x_2}\right)^2\right]$$

The kinetic energy density of the material is

$$\mathrm{KE} = \tfrac{1}{2}\rho\left(\frac{\partial \mathbf{r}}{\partial t}\right)^2$$

and the Lagrangian density is $L = \mathrm{KE} - W$. To obtain the Lagrange-Euler equations of motion, Hamilton's principle applied to the integral of L over the whole system gives

$$\frac{\partial}{\partial t}\left[\frac{\partial L}{\partial(\partial r_n/\partial t)}\right] + \sum_{m=1}^{3} \frac{\partial}{\partial x_m}\left[\frac{\partial L}{\partial(\partial r_n/\partial x_m)}\right] = \frac{\partial L}{\partial r_n} \qquad n = 1, 2, 3 \qquad \mathbf{(7.4)}$$

For the most general case (using the two-index coefficients defined in Part 3, Chap. 1),

$$\rho \frac{\partial^2 r_1}{\partial t^2} = \frac{\partial^2}{\partial x_1^2}(C_{11}r_1 + C_{16}r_2 + C_{15}r_3) + \frac{\partial^2}{\partial x_2^2}(C_{66}r_1 + C_{26}r_2 + C_{46}r_3) + \frac{\partial^2}{\partial x_3^2}(C_{55}r_1 + C_{45}r_2 + C_{35}r_3)$$
$$+ \frac{\partial^2}{\partial x_1\,\partial x_2}[2C_{16}r_1 + (C_{12} + C_{66})r_2 + (C_{14} + C_{56})r_3]$$
$$+ \frac{\partial^2}{\partial x_1\,\partial x_3}[2C_{15}r_1 + (C_{14} + C_{56})r_2 + (C_{13} + C_{55})r_3]$$
$$+ \frac{\partial^2}{\partial x_2\,\partial x_3}[2C_{56}r_1 + (C_{46} + C_{25})r_2 + (C_{36} + C_{45})r_3]$$

plus two similar equations. In terms of the four-index moduli these have the form

$$\rho \frac{\partial^2 r_i}{\partial t^2} = \sum_{m,n} \frac{\partial^2}{\partial x_m\,\partial x_n} \sum_j C_{imnj} r_j = \sum_m \frac{\partial}{\partial x_m} T_{im} \tag{7.5}$$

where $\mathfrak{T}$ is the stress tensor. For isotropic solids

$$\rho \frac{\partial^2 r_1}{\partial t^2} = \frac{\partial}{\partial x_1}\left(\lambda \frac{\partial r_1}{\partial x_1} + \lambda \frac{\partial r_2}{\partial x_2} + \lambda \frac{\partial r_3}{\partial x_3} + 2\mu \frac{\partial r_1}{\partial x_1}\right) + \mu \frac{\partial}{\partial x_2}\left(\frac{\partial r_1}{\partial x_2} + \frac{\partial r_2}{\partial x_1}\right) + \mu \frac{\partial}{\partial x_3}\left(\frac{\partial r_1}{\partial x_3} + \frac{\partial r_3}{\partial x_1}\right) \qquad \text{etc.}$$

which is the first component of the vector equation

$$\rho \frac{\partial^2 \mathbf{r}}{\partial t^2} = (\lambda + \mu)\,\text{grad div}\,\mathbf{r} + \mu \nabla^2 \mathbf{r} = (\lambda + 2\mu)\,\text{grad div}\,\mathbf{r} - \mu\,\text{curl curl}\,\mathbf{r} \tag{7.6}$$

The momentum density in the material, caused by the motion, is the vector having components

$$p_n = \frac{\partial L}{\partial(\partial r_n/\partial t)} = \rho \frac{\partial r_n}{\partial t}$$

and the stress tensor is

$$T_{ij} = -\frac{\partial L}{\partial(\partial r_j/\partial x_i)} = \sum_{m,n} C_{ijmn} e_{mn}$$

The energy-momentum tensor is defined as

$$E_{ij} = \sum_{m=1}^{3} \frac{\partial r_m}{\partial x_i} \frac{\partial L}{\partial(\partial r_m/\partial x_j)} - L\delta_{ij} \tag{7.7}$$

where i and j can run from 1 to 4, with $x_4 = t$. This tensor satisfies the divergence condition

$$\sum_{j=1}^{4} \frac{\partial E_{ij}}{\partial x_j} = 0 \qquad i = 1, 2, 3, 4$$

The equation for $i = 4$ is the equation of continuity for energy and intensity, as E_{44} is the Hamiltonian density

$$E_{44} = \sum_{m=1}^{3} p_m \frac{\partial r_m}{\partial t} - L = \frac{1}{2\rho} \sum_{m=1}^{3} p_m^2 + W = H \tag{7.8}$$

and the vector $\mathbf{S}$, with components

$$S_m = E_{4m} = -\sum_{m=1}^{3} T_{mn} \frac{\partial r_n}{\partial t} \tag{7.9}$$

is the intensity vector corresponding to energy flow, which satisfies the energy-continuity equation

$$\frac{\partial H}{\partial t} + \text{div}\,\mathbf{S} = 0$$

For small-amplitude motions the difference between the partial $(\partial/\partial t)$ and the convected derivative (d/dt), the term $\mathbf{v} \cdot \nabla$, is a second-order term, which may be neglected.

2. Plane Waves in Homogeneous Media

When the elastic medium is infinite in extent, plane, simple harmonic waves are possible, for which

$$\mathbf{r} = \mathbf{A} \exp\left(ik \sum_n \gamma_n x_n - i\omega t\right) \tag{7.10}$$

where $\mathbf{A}$ is a constant vector, ω is the angular frequency of the wave motion (supposed specified), γ_n are the direction cosines of the direction of wave propagation (also specified), and k is the wave number of the wave. For each frequency and each direction of propagation there exist a trio of directions of $\mathbf{A}$ and a corresponding trio of values of k, which satisfy this equation.

Inserting the above expression for $\mathbf{r}$ into Eq. (7.5), we obtain

$$\sum_j D_{ij} A_j = \frac{\rho\omega^2 A_i}{k^2} \tag{7.11}$$

where $D_{ij} = \Sigma_{m,n}\,\gamma_m\gamma_n C_{imnj}$, and where $\omega/k = c$ is the velocity of propagation of the wave. The contracted tensor D_{ij} is symmetric, because of the symmetry of C_{imnj}. Consequently the equation has three real eigenvector solutions for $\mathbf{A}$, in three mutually perpendicular directions. Referred to these directions, tensor D_{ij} becomes a diagonal tensor, with principal values D_1, D_2, D_3. Thus for any given value of ω and for each direction of propagation (choice of the γ's) there are three possible plane waves, each corresponding to a definite direction of displacement (direction of $\mathbf{A}$) and each having a distinct wave velocity $c_m = \sqrt{D_m/\rho}$. The three directions of displacement are mutually perpendicular, but none of them is simply related to the direction of propagation (in the most general case).

To illustrate the possibilities in a cubic crystal $C_{1111} = C_{2222} = C_{3333} = C_{11}$;

$$C_{1122} = \cdot \cdot \cdot = C_{3322} = C_{12}$$

$C_{1212} = C_{1221} = \cdot \cdot \cdot = C_{2332} = C_{44}$, and all other C's are zero if the crystal axes are the coordinate axes. Choosing a direction of propagation perpendicular to the z axis and at an angle ϕ with respect to the x axis the tensor $\mathfrak{D}$ has the following form:

$$\mathfrak{D} = \begin{bmatrix} C_{11}\cos^2\phi + C_{44}\sin^2\phi & (C_{12}+C_{44})\sin\phi\cos\phi & 0 \\ (C_{12}+C_{44})\sin\phi\cos\phi & C_{11}\sin^2\phi + C_{44}\cos^2\phi & 0 \\ 0 & 0 & C_{44} \end{bmatrix}$$

One direction of displacement is therefore parallel to the z axis, and the corresponding wave velocity is $c = \sqrt{C_{44}/\rho}$. This wave is *transverse*, i.e., **A** is perpendicular to the direction of propagation.

The other two directions of displacement are in the xy plane and must be obtained by "diagonalizing" the tensor. The results of the calculation are that the angle between the allowed direction of displacement and the x axis is ψ, where

$$\tan\psi = -\frac{(C_{11} - C_{44})\cos 2\phi}{(C_{12} + C_{44})\sin 2\phi} \pm \frac{\sqrt{(C_{11} - C_{44})^2\cos^2 2\phi + (C_{12} + C_{44})^2\sin^2 2\phi}}{(C_{12} + C_{44})\sin 2\phi}$$

with corresponding wave velocities $c = \omega/k$ given by

$$c^2 = \tfrac{1}{2}(C_{11} + C_{44}) \pm \tfrac{1}{2}\sqrt{(C_{11} - C_{44})\cos^2 2\phi + (C_{12} + C_{44})^2\sin^2 2\phi} \qquad (7.12)$$

When $C_{11} - C_{44} = C_{12} + C_{44}$, then $\psi = \phi$, or $\phi + \frac{1}{2}\pi$, i.e., the direction of displacement is either parallel or perpendicular to the direction of propagation, and the wave velocity for the one parallel (longitudinal wave) is $\sqrt{C_{11}/\rho}$; the velocity for the two perpendicular (transverse) waves is then $\sqrt{C_{44}/\rho}$.

In Part 3, Chap. 1, it was pointed out that, when $C_{11} - C_{44}$ equals $C_{12} + C_{44}$, the material is isotropic; waves in isotropic media are discussed in more detail later. For the more general case of the cubic crystal the two waves with directions of displacement in the xy plane are neither transverse nor longitudinal unless $\phi = 0$ or $\pi/4$, i.e., unless the direction of propagation is parallel to one of the crystal axes or bisects a pair of them. In the first case ($\phi = 0$) the velocity of the longitudinal wave is $\sqrt{C_{11}/\rho}$, and that of the two transverse waves is $\sqrt{C_{44}/\rho}$. In the second case the longitudinal velocity is

$$\sqrt{\frac{C_{11} + C_{12} + 2C_{44}}{2\rho}}$$

that of the wave transverse to the xy plane is still $\sqrt{C_{44}/\rho}$, and that of the transverse wave in the xy plane is $\sqrt{(C_{11} - C_{12})/2\rho}$.

Calculations for less symmetric crystals are more tedious. In general the three possible waves for each direction of propagation are neither transverse nor longitudinal, though the three allowed directions of displacement are always mutually perpendicular. As long as the C's are independent of ω, the medium is not dispersive, and solutions of the more general form

$$\mathbf{r} = \mathbf{A}f\left(\sum_n \gamma_n x_n - ct\right)$$

may be obtained for the same directions of **A** and values of c as those obtained for Eq. (7.10).

Turning now to isotropic media, the simpler equation (7.6) gives the longitudinal and transverse waves. Since any vector field can be separated into a part having zero curl (longitudinal or lamellar field) and another part having zero divergence (transverse or solenoidal field), a solution of Eq. (7.6) exists in the form

$$\mathbf{r} = \operatorname{grad}\psi + \operatorname{curl}\mathbf{A} \qquad \operatorname{div}\mathbf{A} = 0 \qquad (7.13)$$

$$\nabla^2\psi - \left(\frac{1}{c_l}\right)^2\frac{\partial^2\psi}{\partial t^2} = 0 \qquad \operatorname{curl}\operatorname{curl}\mathbf{A} + \left(\frac{1}{c_t}\right)^2\frac{\partial^2\mathbf{A}}{\partial t^2} = 0$$

where $c_l = \sqrt{(\lambda + 2\mu)/\rho}$ is the velocity of longitudinal or compressional waves and $c_t = \sqrt{\mu/\rho}$ is the velocity of transverse or shear waves. If $\mathbf{a}_1$, $\mathbf{a}_2$, and $\mathbf{a}_3$ are mutually perpendicular unit vectors, the most general plane wave propagated in the $\mathbf{a}_1$ direction is

$$\mathbf{r} = \mathbf{a}_1 f(\mathbf{a}_1\cdot\mathbf{x} - c_l t) + \mathbf{a}_2 g(\mathbf{a}_1\cdot\mathbf{x} - c_t t) + \mathbf{a}_3 h(\mathbf{a}_1\cdot\mathbf{x} - c_t t)$$

For simple harmonic waves of angular frequency ω,

$$\mathbf{r} = [A_1\mathbf{a}_1 e^{ik_l\mathbf{a}_1\cdot\mathbf{x}} + (A_2\mathbf{a}_2 + A_3\mathbf{a}_3)e^{ik_t\mathbf{a}_1\cdot\mathbf{x}}]e^{-i\omega t} \qquad (7.14)$$

where $k_l = \omega/c_l$ and $k_t = \omega/c_t$. This wave has energy density and intensity vector

$$E_{44} = \rho\omega^2 \sum_n |A_n|^2$$

$$\mathbf{S} = \rho\omega^2\mathbf{a}_1[c_l|A_1|^2 + c_t(|A_2|^2 + |A_3|^2)]$$

respectively.

For isotropic media, therefore, there are only two different kinds of plane waves: the longitudinal, or compressional, wave, where the displacement is in the direction of propagation (as with sound waves), and the transverse, or shear, wave, with displacement orthogonal to the direction of propagation. The two velocities differ, that for compressional waves being larger (usually between 1.5 and 2 times the shear-wave velocity). For example, for steel $c_l \simeq 18{,}000$ fps, and $c_t \simeq 10{,}500$ fps, whereas for granite $c_l \simeq 16{,}000$ fps, and $c_t \simeq 10{,}000$ fps. Two constants (such as amplitude and phase) are needed to specify a compressional wave, in addition to the direction of propagation; four constants (amplitude and phase in each direction normal to the propagation vector) are required to specify a shear wave.

In seismology, the compressional, or longitudinal, waves are called P waves; the shear, or transverse, waves are called S waves. If distinction is needed, the shear waves with motion in a horizontal plane are called SH waves; those with motion in a vertical plane are called SV waves. Geological materials are, of course, far from being homogeneous or isotropic; nevertheless, a rock, such as granite, behaves approximately as a homogeneous, isotropic solid for waves of wavelength long compared to the dimensions of the grains which make up the granite.

3. Spherical Waves, Green's Tensor for Isotropic Media

To describe the elastic waves caused by body forces or internal transients (such as seismic sources) solutions of the inhomogeneous equation

$$\rho \frac{\partial^2 r_i}{\partial t^2} - \sum_{m,n} \frac{\partial^2}{\partial x_m \, \partial x_n} \sum_j C_{imnj} r_j = F_i$$

are needed, or, for isotropic media,

$$\rho \left(\frac{\partial^2 \mathbf{r}}{\partial t^2} - c_l^2 \text{ grad div } \mathbf{r} + c_t^2 \text{ curl curl } \mathbf{r} \right) = \mathbf{F}(\mathbf{x},t) \tag{7.15}$$

where $\mathbf{F}$, with components F_i, is the body force per unit volume on the medium at the point $\mathbf{x}$ at time t.

To solve this equation we find the solution for a "point source" at some point $\mathbf{x}'$, a solution called Green's function. The solution for any given force distribution $\mathbf{F}$ is then the integral of the Green's function, times $\mathbf{F}$, over the region where $\mathbf{F}$ differs from zero. Since, in this case, the solution is a vector and $\mathbf{F}$ is a vector, Green's function $\mathfrak{G}$ must be a tensor, a function of $\mathbf{x}'$, the position of the point source, and of $\mathbf{x}$, the observation point where $\mathbf{r}$ is to be measured. $\mathfrak{G}$ must satisfy a reciprocity condition: the effect at $\mathbf{x}$ caused by a source at $\mathbf{x}'$ must be the same as the effect at $\mathbf{x}'$ caused by a similar source at $\mathbf{x}$.

The effect of a time variation of the driving force $\mathbf{F}$ may be handled in two ways. The more straightforward way (though often more difficult in application) is to find the Green's function for a source at $\mathbf{x}'$ which is just a unit pulse at $t = t'$. The final answer for the wave produced by $\mathbf{F}(\mathbf{x}',t')$ is then an integral of the vector of $\mathbf{F} \cdot \mathfrak{G}$ over $\mathbf{x}'$ and also over t' from the initiation of the force (which might be set at $t = 0$) to time t, when the value of $\mathbf{r}$ is to be observed at $\mathbf{x}$. The other way is to separate $\mathbf{F}(\mathbf{x}',t')$ into its simple harmonic components by means of either the Fourier or the Laplace transform. The Green's tensor is then the disturbance at $\mathbf{x}$ caused by a point source at $\mathbf{x}'$ with simple harmonic dependence on time. The final answer is then the reciprocal transform of the volume integral of the product of the transforms of $\mathbf{F}$ and $\mathfrak{G}$.

If the equation to be solved were

$$\rho \left(\frac{\partial^2 \mathbf{r}}{\partial t^2} - c^2 \nabla^2 \mathbf{r} \right) = \mathbf{F} \tag{7.16}$$

instead of Eq. (7.15), the expression for the Green's function would be simple. For an infinite medium it is

$$\mathfrak{G}_0(c;\mathbf{x},\mathbf{x}';t,t') = \frac{\mathfrak{J}}{4\pi R} \delta \left(t - t' - \frac{R}{c} \right)$$

where $\mathfrak{J}$ is the identity tensor, with components δ_{mn}, such that $\mathbf{F} \cdot \mathfrak{J} = \mathbf{F}$ for any vector $\mathbf{F}$. Function $\delta(z)$ is the Dirac delta function, which is zero for $z \neq 0$ and is infinite at $z = 0$ in such a way that

$$\int_{-\infty}^{\infty} \delta(z) \, dz = 1$$

and

$$\int_{-\infty}^{\infty} \delta(z - z') f(z') \, dz' = f(z)$$

for any bounded function f. Finally $R = |\mathbf{x} - \mathbf{x}'|$ is the distance between source and observation point.

However, Eq. (7.16) corresponds to a case where longitudinal and transverse waves have equal velocities, so the corresponding $\mathfrak{G}_0$ is a combination of longitudinal and transverse waves. To find a Green's tensor for Eq. (7.15) $\mathfrak{G}_0$ must be separated into its longitudinal and transverse parts. This can be done fairly easily *except* for the point $R = 0$, where one has difficulties with orders of infinity. The result, except for $R = 0$, is

$$\begin{aligned}
\mathfrak{P}(c;\mathbf{x},\mathbf{x}';t,t') &= -\frac{\mathfrak{J}}{4\pi R} \left[\left(\frac{c}{R}\right)^2 v(z) + \left(\frac{c}{R}\right) u(z) \right] \\
&\quad + \frac{\mathfrak{R}\mathfrak{R}}{4\pi R^3} \left[3 \left(\frac{c}{R}\right)^2 v(z) + \left(\frac{c}{R}\right) u(z) + \delta(z) \right] \\
\mathfrak{S}(c;\mathbf{x},\mathbf{x}';t,t') &= \frac{\mathfrak{J}}{4\pi R} \left[\left(\frac{c}{R}\right)^2 v(z) + \left(\frac{c}{R}\right) u(z) + \delta(z) \right] \\
&\quad - \frac{\mathfrak{R}\mathfrak{R}}{4\pi R^3} \left[3 \left(\frac{c}{R}\right)^2 v(z) + \left(\frac{c}{R}\right) u(z) + \delta(z) \right]
\end{aligned} \tag{7.17}$$

where $z = t - t' - R/c$, and where

$$u(z) = \int_{-\infty}^{z} \delta(x) \, dx = \begin{cases} 0 & z < 0 \\ 1 & z > 0 \end{cases}$$

$$v(z) = \int_{-\infty}^{z} u(x) \, dx = \begin{cases} 0 & t < 0 \\ z & t > 0 \end{cases}$$

Since $\mathbf{R} = \mathbf{x} - \mathbf{x}'$, quantity $\mathfrak{R}\mathfrak{R}$ is a tensor, with elements $(x_m - x_m')(x_n - x_n')$.

Function $\mathfrak{P}$ is longitudinal, i.e., curl $\mathfrak{P} = 0$, except at $\mathbf{R} = 0$, and function $\mathfrak{S}$ is transverse, i.e., div $\mathfrak{S} = 0$, except at $\mathbf{R} = 0$. Except at $\mathbf{R} = 0$, $\mathfrak{P} + \mathfrak{S} = \mathfrak{G}_0$. For an elastic, isotropic medium, however, satisfying Eq. (7.15), the longitudinal and transverse waves go at different speeds. Consequently, except at $\mathbf{R} = 0$, a Green's tensor for Eq. (7.15), for isotropic elastic media, is

$$\mathfrak{G}_e(\mathbf{x},\mathbf{x}';t,t') = \frac{1}{c_l^2} \mathfrak{P}(c_l;\mathbf{x},\mathbf{x}';t,t') + \frac{1}{c_t^2} \mathfrak{S}(c_t;\mathbf{x},\mathbf{x}';t,t') \tag{7.18}$$

and a solution of Eq. (7.15) outside the region where $\mathbf{F}$ differs from zero is

$$\mathbf{r}(\mathbf{x},t) = \int_0^t dt' \iiint \mathbf{F}(\mathbf{x}',t') \cdot \mathfrak{G}_e(\mathbf{x},\mathbf{x}';t,t') \, dx_1' \, dx_2' \, dx_3' \tag{7.19}$$

Here it is assumed that $\mathbf{F}$ is everywhere zero for $t < 0$ and is always zero outside a bounded region B. The volume integration is over B, and the expression for $\mathbf{r}$ is valid when $\mathbf{x}$ is *outside* of B.

A simple example will indicate how these waves behave. Suppose $\mathbf{F}$ is a unit, point source, pulsed at $t = 0$, directed along the x_1 axis; then the solution is unit vector $\mathbf{i}$ times $\mathfrak{G}_e$ itself. Sufficiently far from the source the $1/R$ terms will predominate, and, for this case

$$\mathbf{r} \to \frac{\cos\vartheta}{4\pi R}\,\mathbf{a}_R\,\delta\left(t - \frac{R}{c_R}\right) + \frac{\sin\vartheta}{4\pi R}\,\mathbf{a}_\vartheta\,\delta\left(t - \frac{R}{c_t}\right)$$

where $\mathbf{a}_R$ is a unit vector pointing along $\mathbf{R} = \mathbf{x} - \mathbf{x}'$ at an angle ϑ with respect to the x_1 axis and $\mathbf{a}_\vartheta$ is a unit vector perpendicular to $\mathbf{a}_R$ in the plane defined by $\mathbf{R}$ and the x_1 axis and pointing toward the x_1 axis rather than away from it ($\mathbf{i} - \mathbf{a}_R\cos\vartheta = \mathbf{a}_\vartheta\sin\vartheta$). The compressional pulse will arrive first ($c_l > c_t$), and the motion will be along the radius vector $\mathbf{R}$; the shear wave will follow, and its motion will be normal to $\mathbf{R}$, pointed toward the axis of the motion of the source (the x_1 axis in the example quoted).

If, on the other hand, the point source is simple harmonic, of angular frequency ω, the corresponding Green's dyadics (for an unbounded medium) are

$$\mathfrak{p}(k;\mathbf{x},\mathbf{x}') = \left(\mathfrak{J}\,\frac{1 - ikR}{k^2R^2} - \frac{\mathbf{RR}}{R^2}\,\frac{3 - ikR - k^2R^2}{k^2R^2}\right)\frac{e^{ikR}}{4\pi R}$$

$$\mathfrak{s}(k;\mathbf{x},\mathbf{x}') = \left(-\mathfrak{J}\,\frac{1 - ikR - k^2R^2}{k^2R^2} + \frac{\mathbf{RR}}{R^2}\,\frac{3 - ikR - k^2R^2}{k^2R^2}\right)\frac{e^{ikR}}{4\pi R} \qquad (7.20)$$

and the steady-state solution of Eq. (7.15), when $\mathbf{F} = \mathbf{f}(\omega,\mathbf{x})e^{-i\omega t}$ is

$$\mathbf{r}(\omega,\mathbf{x}) = \iiint \mathbf{f}(\omega,\mathbf{x}')\cdot\left[\frac{1}{c_l^2}\,\mathfrak{p}\left(\frac{\omega}{c_l};\mathbf{x},\mathbf{x}'\right) + \frac{1}{c_t^2}\,\mathfrak{s}\left(\frac{\omega}{c_t};\mathbf{x},\mathbf{x}'\right)\right]e^{-i\omega t}\,dx_1'\,dx_2'\,dx_3' \qquad (7.21)$$

the integration being carried out over region B, outside of which $\mathbf{f}$ is zero. If $\mathbf{f}$ is a Laplace transform of $\mathbf{F}(\mathbf{x},t)$, then the solution of Eq. (7.15) for $\mathbf{F}$ is the inverse transform of $\mathbf{r}(\omega,x)$ of Eq. (7.21).

4. Reflection from a Plane Interface, Surface Waves

Waves are reflected from a surface of discontinuity in the medium. For example, the discontinuity might be the yz plane, the elastic constants and densities being λ', μ', ρ' for $x > 0$ and λ, μ, ρ for $x < 0$. (We work out only the isotropic case; the nonisotropic case is similar but more tedious.) Suppose a plane compressional wave

$$\mathbf{r}_i = (\mathbf{i}\cos\phi_l + \mathbf{j}\sin\phi_l)\cdot\exp\left[\frac{i\omega}{c_l}(x\cos\phi_l + y\sin\phi_l) - i\omega t\right] \qquad x < 0$$

is incident from the left on this plane, at an angle of incidence ϕ_l. Four plane waves will result: a reflected compressional and shear wave (for $x < 0$)

$$\mathbf{r}_- = A_l^l(-\mathbf{i}\cos\phi_l + \mathbf{j}\sin\phi_l)\cdot\exp\left[\frac{i\omega}{c_l}(-x\cos\phi_l + y\sin\phi_l) - i\omega t\right] + A_t^l(\mathbf{i}\sin\phi_t + \mathbf{j}\cos\phi_t)\cdot\exp\left[\frac{i\omega}{c_t}(-x\cos\phi_t + y\sin\phi_t) - i\omega t\right]$$

and a refracted pair (for $x > 0$)

$$\mathbf{r}_+ = B_l^l(\mathbf{i}\cos\theta_l + \mathbf{j}\sin\theta_l)\cdot\exp\left[\frac{i\omega}{c_l'}(x\cos\theta_l + y\sin\theta_l) - i\omega t\right] + B_t^l(-\mathbf{i}\sin\theta_t + \mathbf{j}\cos\theta_t)\cdot\exp\left[\frac{i\omega}{c_t'}(x\cos\theta_t + y\sin\theta_t) - i\omega t\right]$$

where $c_l = \sqrt{(\lambda + 2\mu)/\rho}$, $c_l' = \sqrt{(\lambda' + 2\mu')/\rho'}$, $c_t = \sqrt{\mu/\rho}$, $c_t' = \sqrt{\mu'/\rho'}$. Simple harmonic waves are used for simplicity; waves of other shape may be built up by the use of the Fourier transform.

To determine the A's and B's, the ϕ's and θ's we must require that $\mathbf{r}_i + \mathbf{r}_-$ equal $\mathbf{r}_+$ at $x = 0$ and that the stress components T_{xm} across the surface $x = 0$ be continuous. From the requirement of continuity of displacement $\mathbf{r}$, for the exponentials all to fit at $x = 0$, for all values of y,

$$\frac{1}{c_l}\sin\phi_l = \frac{1}{c_t}\sin\phi_t = \frac{1}{c_l'}\sin\theta_l = \frac{1}{c_t'}\sin\theta_t$$

thus relating the angles of reflection ϕ_l and ϕ_t and the angles of refraction θ_l and θ_t to the angle of incidence ϕ_l.

In addition, continuity of displacement at $x = 0$ requires that

$$\begin{aligned} A_l^l\cos\phi_l - A_t^l\sin\phi_t + B_l^l\cos\theta_l - B_t^l\sin\theta_t &= \cos\phi_l \\ -A_l^l\sin\phi_l - A_t^l\cos\phi_t + B_l^l\sin\theta_l + B_t^l\cos\theta_t &= \sin\phi_l \end{aligned} \qquad (7.22)$$

Continuity of stress requires that

$$T_{xx} = \lambda\,\operatorname{div}\mathbf{r} + 2\mu\frac{\partial r_x}{\partial x}$$

and
$$T_{xy} = \mu\frac{\partial r_x}{\partial y} + \mu\frac{\partial r_y}{\partial x}$$

be continuous across $x = 0$, which produces two more equations:

$$-\frac{1}{c_l}(\lambda + 2\mu - 2\mu\sin^2\phi_l)A_l^l + \frac{\mu}{c_t}\sin 2\phi\,A_t + \frac{1}{c'}(\lambda' + 2\mu' - 2\mu'\sin^2\theta_l)B_l^l - \frac{\mu'}{c'}\sin 2\phi_t\,B_t^l = \frac{1}{c_l}(\lambda - 2\mu - 2\mu\sin^2\phi_l) \qquad (7.23)$$

$$\frac{\mu}{c_l}\sin 2\phi_l\,A_l^l + \frac{\mu}{c_t}\cos 2\phi_t\,A_t + \frac{\mu'}{c_l'}\sin 2\theta_l\,B_l^l - \frac{\mu'}{c_t'}\cos 2\theta_t\,B_t^l = \frac{\mu}{c_l}\sin 2\phi_l$$

These four simultaneous equations determine values of A_l^l, the reflected compressional amplitude, A_t^l, the reflected shear amplitude, and B_l^l, B_t^l, the refracted compressional and shear amplitudes, for an incident P wave at angle of incidence ϕ_l.

If there is no material in the region $x > 0$, the surface $x = 0$ is a free surface, and T_{xx} and T_{xy} are zero at $x = 0$. The equations for the reflected amplitudes may then be obtained from the second pair ($T_{xx} = T_{xy} = 0$ at $x = 0$) by setting $\rho' = 0$, thus eliminating the unwanted coefficients B_l^l and B_t^l (we cannot use the first pair of equations because $\mathbf{r}$ is not zero at a free surface). The equations for A_l^l and A_t^l become

$$1 + A_l^l - \frac{(c_l/c_t) \sin 2\phi_t}{(c_l/c_t)^2 - 2 \sin^2 \phi_l} A_t^l = 0$$

$$1 - A_l^l - \frac{c_l \cos 2\phi_t}{c_t \sin 2\phi_l} A_t^l = 0$$

from which

$$A_l^l = \frac{\cos (2\phi_t + 2\psi)}{\cos (2\phi_t - 2\psi)};\; A_t^l = 2\frac{c_t}{c_l}\frac{\sin 2\phi_l \cos 2\psi}{\cos (2\phi_t - 2\psi)} \tag{7.24}$$

where

$$\tan 2\psi = \frac{\sin 2\phi_l}{(c_l/c_t)^2 - 2 \sin^2 \phi_l} = \left(\frac{c_t}{c_l}\right)^2 \frac{\sin 2\phi_l}{\cos 2\phi_t}$$

Constants A_l^l and A_t^l are the amplitudes of the reflected longitudinal and transverse reflected waves, respectively, when a unit-amplitude longitudinal wave is incident on a free plane surface at an angle of incidence ϕ_l; the angle of reflection for the P wave is also ϕ_l, and the angle for the reflected S wave is ϕ_t ($c_l \sin \phi_t = c_t \sin \phi_l$).

If the incident wave is a shear wave at angle ϕ_t,

$$(-\mathbf{i} \sin \phi_t + \mathbf{j} \cos \phi_t) \cdot \exp \left[\frac{i\omega}{c_t} (x \cos \phi_t + y \sin \phi_t) - i\omega t\right]$$

the reflected shear wave will be at angle ϕ_t, and the reflected compressional wave will be at angle ϕ_l, where again $c_l \sin \phi_t = c_t \sin \phi_l$. The relative amplitudes of these reflected waves turn out to be

$$A_l^t = 2\frac{c_l}{c_t}\frac{\sin 2\psi \cos 2\phi_t}{\cos (2\phi_t - 2\psi)} \qquad A_t^t = \frac{\cos (2\phi_t - 2\psi)}{\cos (2\phi_t - 2\psi)} \tag{7.25}$$

respectively, where ψ has the same definition as above. By manipulation of the relations between ϕ_l, ϕ_t, and ψ it follows that

$$(A_l^l)^2 + \frac{c_t}{c_l} (A_t^l)^2 = 1 \qquad \frac{c_l}{c_t} (A_l^t)^2 + (A_t^t)^2 = 1 \tag{7.26}$$

indicating that energy is conserved on reflection. When $\phi_t + \psi = \frac{1}{4}\mu$, A_l^l and $A_t^t = 0$; in other words, for the angle of incidence (ϕ_l in one case, ϕ_t in the other) for which $\phi_t + \psi = \frac{1}{4}\mu$ an incident P wave is reflected as a pure S wave and vice versa. At zero and at 90° angle of incidence $A_t^l = A_l^t = 0$; so the incident wave is reflected without change of type.

The other kind of shear wave

$$\mathbf{r} = \mathbf{k} \exp \left[\frac{i\omega}{c_t} (x \cos \phi + y \sin \phi) - i\omega t\right] \quad \text{incident}$$

(if the free surface were the surface of the earth, these would be SH waves) is reflected without change of type and with amplitude -1. In other words, P and SV waves interchange on reflection, but SH waves remain SH waves.

When an SV wave is incident [see Eqs. (7.23)], it is possible for the angle of incidence ϕ_t to be large enough for the angle of reflection of the P wave, ϕ_l, to become complex. If $c_l/c_t \sin \phi_t$ is larger than unity (which can happen for $\sin \phi_t$ less than unity), then $\sin \phi_l$ must be larger than unity. In this case set $\phi_l = \frac{1}{2}\pi + i\alpha$, so that

$$\sin \phi_l = \frac{c_l}{c_t} \quad \sin \phi_t = \cosh \alpha$$

$$\cos \phi_l = -i \sinh \alpha = -i \sqrt{\left(\frac{c_l}{c_t}\right)^2 \sin^2 \phi_t - 1}$$

The reflected wave will then have the form

$$A_l^t(i\mathbf{i} \sinh \alpha + \mathbf{j} \cosh \alpha) \cdot \exp \left(\frac{\omega x}{c_l} \sinh \alpha + \frac{i\omega y}{c_l} \cosh \alpha - i\omega t\right)$$
$$+ A_t^t(\mathbf{i} \sin \phi_t + \mathbf{j} \cos \phi_t) \cdot \exp \left[\frac{i\omega}{c_t} (-x \cos \phi_t + y \sin \phi_t) - i\omega t\right]$$

The fact that the term in the first exponential that is dependent on x is real means that this compressional wave is a surface one, decreasing exponentially as x is increased in the negative direction. Both A_l^t and A_t^t are complex because angle ψ is complex in this case, indicating that the reflected shear wave and the surface compressional wave are each out of phase with the incident wave, only their sum matching the incident phase.

When ϕ_t becomes $\frac{1}{2}\pi$, the reflected shear wave, also traveling parallel to the y axis, just cancels the incident wave, showing that a plane SV wave cannot travel parallel to a free plane surface. A purely surface wave can travel parallel to the free surface, however, if both the SV and the P waves attenuate as $-x$ is increased. Such waves are called *Rayleigh waves.* Because of the interaction between P and SV waves, it is possible to have waves which "cling" to the region near a free surface, traveling along parallel to the surface but damping out exponentially away from the surface. The equivalent angle of incidence for such a wave is $\phi_l = \frac{1}{2}\pi$ plus some imaginary quantity (call it $-i\alpha$), making

$$\cos \phi_l = i \sinh \alpha$$

and $\sin \phi_l = \cosh \alpha$, whereas ϕ_t is $\frac{1}{2}\pi - i\beta$. Since there is no distinction between incident and reflected waves here, we need only include two waves, a P and an S one, each traveling at the same speed parallel to the y axis but penetrating by different amounts into the material in the region $x < 0$;

$$\mathbf{r} = (-i\mathbf{i} \sinh \alpha + \mathbf{j} \cosh \alpha) \cdot \exp\left[\frac{i\omega}{c_l}(-ix \sinh \alpha + y \cosh \alpha - ic_lt)\right]$$
$$+ A(\mathbf{i} \cosh \beta + i\mathbf{j} \sinh \beta) \cdot \exp\left[\frac{i\omega}{c_t}(-ix \sinh \beta + y \cosh \beta - ic_tt)\right] \quad (7.27)$$

In order that the two waves match in phase at $x = 0$

$$c_l \cosh \beta = c_t \cosh \alpha$$

or

$$\lambda - 2\mu \sinh^2 \alpha = -(\lambda + 2\mu) \cosh 2\beta$$

so that α cannot be zero, for example. The attenuation of the longitudinal part of the wave (the first term) in the negative x direction is then by a factor of $1/e$ in a distance $c_l/(\omega \sinh \alpha)$; the corresponding distance for the transverse part is $c_t/(\omega \sinh \beta)$. The velocity of the wave is not c_l or c_t but

$$\frac{c_l}{\cosh \alpha} = \frac{c_t}{\cosh \beta}$$

less than either c_l or c_t.

To find values of A, α, and β and thus of the phase velocity of the surface wave set $T_{xx} = T_{xy} = 0$ at $x = 0$, giving

$$\lambda - 2\mu \sinh^2 \alpha = -(\lambda + 2\mu) \cosh 2\beta = iA\mu \frac{c_l}{c_t} \sinh 2\beta$$

$$\sinh 2\alpha = -iA \frac{c_l}{c_t} \cosh 2\beta$$

or

$$A = i \frac{c_l}{c_l} \cosh 2\beta$$

and

$$\sinh 2\alpha \sinh 2\beta = \left(\frac{c_l}{c_t}\right)^2 \cosh^2 2\beta$$

which last equation serves to determine α or β. It can be recast into

$$(\cosh 2\beta + 1)\left(\cosh 2\beta + \frac{\lambda + \mu}{\lambda + 2\mu}\right)(\cosh^2 2\beta - 1) = \cosh^4 2\beta \quad (7.28)$$

This is a cubic equation for $\cosh 2\beta$, with two complex roots and one real root. The real root represents the solution desired; from it can be obtained values for α and A and thus of the phase velocity of the Rayleigh wave and its penetration beneath the free surface. Values of α, β, A/i, and of wave velocity $c = c_t/\cosh \beta = c_l/\cosh x$ are tabulated for three ratios of μ to λ; other values may be obtained by interpolation:

$\frac{\mu}{\lambda}$	σ*	$\frac{c_l}{c_t}$	α	β	$\frac{A}{i}$	$\frac{c}{c_t}$	$\frac{c}{c_l}$
⅓	0.375	2.236	1.5064	0.3295	3.870	0.9480	0.4226
⅔	0.300	1.871	1.3060	0.3426	3.165	0.9441	0.5048
1	0.250	1.732	1.2192	0.3508	2.861	0.9414	0.5435

* $\sigma = \lambda/2(\lambda + \mu)$ is Poisson's ratio (see Part 3, Chap. 1).

The velocity of the surface wave is about 0.94 times that of free shear waves for reasonable ratios of μ to λ.

5. Waves in a Plate

Waves in a plate of thickness a, of infinite extent, exhibit reflections from both free surfaces. Suppose the material between the planes $z = \frac{1}{2}a$ and $z = -\frac{1}{2}a$ is isotropic, with elastic constants λ and μ. The boundary conditions are that

$$T_{13} = \mu\left(\frac{\partial r_1}{\partial z} + \frac{\partial r_3}{\partial x}\right)$$
$$= T_{23} = \mu\left(\frac{\partial r_2}{\partial z} + \frac{\partial r_3}{\partial y}\right)$$
$$= T_{33} = \lambda \operatorname{div} \mathbf{r} + 2\mu \frac{\partial r_3}{\partial z} = 0 \quad (7.29)$$

at $z = \pm\frac{1}{2}a$.

As before, SH waves can satisfy these conditions by themselves. Solutions for wave motion along the x axis are

$$\mathbf{r} = \mathbf{j} \begin{matrix}\cos\\ \sin\end{matrix} \frac{\pi m z}{a} \left\{\exp\left[i \frac{\omega x}{c_t} \sqrt{1 - \left(\frac{\pi m c_t}{\omega a}\right)^2} - i\omega t\right]\right\} \quad (7.30)$$

where m is zero or an even integer for the cosine factor, an odd integer for the sine. For m larger than $\omega a/\pi c_t$ ($2\pi c_t/\omega$ is the S wavelength in an infinite medium) there is no true wave motion in the x direction; the dependence on x is by a real exponential, which goes to infinity at $x = +\infty$ or $-\infty$, so such waves are not present in an infinite plate.

The wave for $m = 0$ is called the *principal wave;* its velocity is just c_t, the velocity of shear waves in an infinite medium. For $m > 0$ and less than $\omega a/\pi c_t$ wave motion occurs, but the wave velocity is *greater* than c_t [it is $c_t/\sqrt{1 - (\pi m c_t/\omega a)^2}$]. The wave reflects back and forth from one free surface to the other; since the wavefronts are at an angle with respect to the x axis, the distance along the x axis between wavefronts is greater than the wavelength; consequently the velocity along the x axis, which is the ratio between this distance and the frequency, is greater than c_t. We see that for low enough frequencies only the principal wave is propagated along the plate. The principal wave is the only one which is not dispersive.

The SY and the P waves "mix up" on reflection, so the other types of waves are mixtures of the two. To simplify the calculations we first set up the components of the pure compressional and pure shear waves. We can take the gradient of a scalar ψ for our P wave (it thus has zero curl) and the curl of a vector $\mathbf{A}$ for our SV wave (it thus has zero divergence). The functions having the proper symmetry are

$$\psi = \left(\frac{c_l}{i\omega}\right) \begin{matrix}\cos\\ \sin\end{matrix} \left(\frac{\omega z}{c_l} \sin \phi\right) \exp\left(\frac{i\omega x}{c_l} \cos \phi - i\omega t\right)$$
$$\mathbf{A} = \mathbf{j} \frac{c_t}{\omega} A \begin{matrix}\sin\\ \cos\end{matrix} \left(\frac{\omega z}{c_t} \sin \chi\right) \exp\left(\frac{i\omega x}{c_t} \cos \chi - i\omega t\right) \quad (7.31)$$

where $(1/c_l) \cos \phi$ must equal $(1/c_t) \cos \chi$ to have the two waves stay in phase.

Taking the upper choice for the trigonometric functions of z, the function grad ψ + curl $\mathbf{A}$ is

$$\mathbf{r} = \left\{ \mathbf{i} \left[\cos \phi \cos \left(\frac{\omega z}{c_l} \sin \phi \right) - A \sin \chi \cos \left(\frac{\omega z}{c_t} \sin \chi \right) \right] + i\mathbf{k} \left[\sin \phi \sin \left(\frac{\omega z}{c_l} \sin \phi \right) + A \cos \chi \sin \left(\frac{\omega z}{c_t} \sin \chi \right) \right] \right\} \exp \left(\frac{i\omega x}{c_l} \cos \phi - i\omega t \right)$$

Setting $T_{13} = T_{33} = 0$ at $z = \frac{1}{2}a$ (which, by symmetry, also makes them zero at $z = -\frac{1}{2}a$) produces the following equations:

$$\begin{aligned} \sin 2\phi \sin \left(\frac{\omega a}{2c_l} \sin \phi \right) + A \frac{c_l}{c_t} \cos 2\chi \sin \left(\frac{\omega a}{2c_t} \sin \chi \right) &= 0 \\ \cos 2\chi \cos \left(\frac{\omega a}{2c_l} \sin \phi \right) - A \frac{c_t}{c_l} \sin 2\chi \cos \left(\frac{\omega a}{2c_t} \sin \chi \right) &= 0 \end{aligned} \tag{7.32}$$

which are to be solved for A and ϕ (angle χ is related to ϕ by the relation $c_t \cos \phi = c_l \cos \chi$). For example, the equation for χ or ϕ is

$$\sin 2\phi \sin 2\chi \tan \left(\frac{\omega a}{2c_l} \sin \phi \right) = - \frac{\lambda + 2\mu}{\mu} \cos^2 2\chi \tan \left(\frac{\omega a}{2c_t} \sin \chi \right) \tag{7.33}$$

As with the SH wave, there are a number of solutions giving wave propagation when $\omega a/c_t$ is larger than 2π. When $\omega a/2c_t$ is less than 2π, i.e., when the S wavelength in the medium is longer than the plate thickness a, only the principal wave is transmitted. When $\omega a/2c_t$ is small enough so that the tangents may be replaced by their arguments, then the first approximation to the solution becomes

$$\begin{aligned} \cos \chi &= \sqrt{\frac{\lambda + 2\mu}{4(\lambda + \mu)}}; \quad \sin \chi = \sqrt{\frac{3\lambda + 2\mu}{4(\lambda + \mu)}} \\ \cos \phi &= \frac{\lambda + 2\mu}{\sqrt{4\mu(\lambda + \mu)}}; \quad \sin \phi = \frac{i\lambda}{\sqrt{4\mu(\lambda + \mu)}} \end{aligned} \tag{7.34}$$

with the wave velocity

$$c = \frac{c_l}{\cos \phi} = \frac{c_t}{\cos \chi} = \sqrt{\frac{\mu(\lambda + 2\mu)}{4\rho(\lambda + \mu)}}$$

which is larger than c_t but smaller than c_l, e.g., for $\mu = \lambda$, $c = 1.61c_t = 0.95c_l$. Here $\cos \phi > 1$, and $\sin \phi$ is imaginary; so ϕ is imaginary. The result is that for the principal wave (for low frequencies) the largest term in the expression for the displacement is $\mathbf{i} \cosh \alpha \cosh [(\omega z/c_l) \sinh \alpha]$, where $\phi = i\alpha$. This type of wave is predominantly compressional, most of the motion being parallel to the plate surfaces, in the direction of the wave.

In contrast, the wave corresponding to the other choice of trigonometric functions in ψ and A,

$$\mathbf{r} = \left\{ \mathbf{i} \left[\cos \phi \sin \left(\frac{\omega z}{c_l} \sin \phi \right) + A \sin \chi \sin \left(\frac{\omega z}{c_t} \sin \chi \right) \right] + i\mathbf{k} \left[- \sin \phi \cos \left(\frac{\omega z}{c_l} \sin \phi \right) + A \cos \chi \cos \left(\frac{\omega z}{c_t} \sin \chi \right) \right] \right\} \exp \left(\frac{i\omega x}{c_t} \cos \chi - i\omega t \right)$$

has its principal wave primarily transverse to the plate surfaces, at least for low frequencies. Here the equations for A and χ, analogous to Eqs. (7.32), are

$$\begin{aligned} \cos 2\chi \sin \left(\frac{\omega a}{2c_l} \sin \phi \right) &= -A \frac{c_t}{c_l} \sin 2\chi \sin \left(\frac{\omega a}{2c_t} \sin \chi \right) \\ \sin 2\phi \cos \left(\frac{\omega a}{2c_l} \sin \phi \right) &= A \frac{c_l}{c_t} \cos 2\chi \cos \left(\frac{\omega a}{2c_t} \sin \chi \right) \end{aligned} \tag{7.35}$$

and the equation analogous to (7.33) is

$$\cos^2 2\chi \tan \left(\frac{\omega a}{2c_l} \sin \phi \right) = - \frac{\mu}{\lambda + 2\mu} \sin 2\phi \sin 2\chi \tan \left(\frac{\omega a}{2c_t} \sin \chi \right) \tag{7.36}$$

When $\omega a/2c_t$ is quite small, using only the first term in the series expansion for the tangents results in the impossible equation $\cos^2 2\chi = - \sin^2 2\chi$, so we must include the first *two* terms in the expansion. When $\omega a/2c_t \ll 1$ we have, to the third approximation in $\omega a/2c_t$,

$$\tan^2 2\chi \simeq - \left[1 - \frac{1}{12} \frac{\rho(\omega a)^2(\lambda + \mu)}{\mu(\lambda + 2\mu)} \right]$$

This indicates that χ is imaginary and large, so that

$$\chi = i\alpha \qquad \tan 2\chi \simeq 1 - 2e^{-4\alpha}$$
$$-i \sin \chi \simeq \cos \chi \simeq \tfrac{1}{2}e^{\alpha}$$

Consequently, $\sin \chi \simeq i \cos \chi$, and $\sin \phi \simeq i \cos \phi$, and

$$\cos \chi \simeq \left[\frac{3\mu(\lambda + 2\mu)}{\rho(\omega a)^2(\lambda + \mu)} \right]^{1/4}$$
$$\cos \phi \simeq \left[\frac{3(\lambda + 2\mu)^3}{\rho(\omega a)^2\mu(\lambda + \mu)} \right]^{1/4}$$

from which $\cos \chi$, $\sin \chi$, and $\cos \phi$ or $\sin \phi$ are large quantities (if $\omega a/c_l$ is small enough), but $(\omega a/2c_l) \sin \phi$ and $(\omega a/2c_t) \cos \chi$ are small quantities; which justifies expansion of the tangent factors in Eq. (7.36).

To the approximation used here, the expression for the displacement $\mathbf{r}$ is

$$\mathbf{r} \simeq \mathbf{k}C \exp \left\{ \left[\frac{3\rho(\lambda + 2\mu)\omega^2}{\mu(\lambda + \mu)a^2} \right]^{1/4} ix - i\omega t \right\} \tag{7.37}$$

where C is a constant approximately equal to $-i \sin \phi + iA \cos \chi$. This indicates that the motion is almost completely transverse (the x component of $\mathbf{r}$ is vanishingly small compared to the z component) and that the wave velocity is strongly dependent on the frequency. In fact this is a wave of transverse *bending* of the plate. To the approximation used here the transverse displacement of the plate from equilibrium, $\eta = r_z$, obeys the differential equation

$$\frac{\partial^4 \eta}{\partial x^4} + \frac{3\rho(\lambda + 2\mu)}{\mu(\lambda + \mu)a^2} \frac{\partial^2 \eta}{\partial t^2} = 0 \qquad (7.38)$$

which is not the usual wave equation.

Adding plane-wave solutions for different directions to form more general transverse waves of bending, the general equation for the most general two-dimensional wave of bending along the plate approximately satisfies the equation

$$\nabla^4 \eta = \frac{\partial^4 \eta}{\partial x^4} + 2\frac{\partial^4 \eta}{\partial x^2 \partial y^2} + \frac{\partial^4 \eta}{\partial y^4} = -D^4 \eta$$

$$D^4 = \frac{3\rho(\lambda + 2\mu)}{\mu(\lambda + \mu)a^2} = \frac{3\rho(1 - \sigma^2)}{E(\frac{1}{2}a)^2}$$

where $\sigma = \lambda/2(\lambda + \mu)$ is Poisson's ratio and

$$E = \frac{\mu(3\lambda + 2\mu)}{(\lambda + \mu)}$$

is Young's modulus (see Part 3, Chap. 1). Transverse waves of this sort are discussed later in this chapter.

6. Waves along a Cylindrical Rod

Longitudinal and transverse waves of more general form than plane waves are more complex to express in mathematical formulas. For example, cylindrical waves in an isotropic medium can be expressed in terms of the gradient of a scalar solution and the curl of two vector solutions of the wave equation

$$\begin{aligned}
\mathfrak{L}(x,y,z) &= \frac{c_l}{i\omega} \text{grad } [\phi(x,y)e^{ik_\alpha z}]e^{-i\omega t} \\
&= \frac{c_l}{i\omega} (\text{grad } \phi + i\mathbf{k}k_z\phi)e^{ik_\alpha z - i\omega t} \\
\mathfrak{M}(x,y,z) &= \frac{c_t}{i\omega} \text{curl } [\mathbf{k}\psi(x,y)e^{ik_\beta z}]e^{-i\omega t} \\
&= \frac{ic_t}{\omega} (\mathbf{k} \times \text{grad } \psi)e^{ik_\beta z - i\omega t} \\
\mathfrak{N}(x,y,z) &= \left(\frac{c_t}{\omega}\right)^2 \text{curl curl } [\mathbf{k}\chi(x,y)e^{ik_\gamma z}]e^{-i\omega t} \\
&= \left(\frac{c_t}{\omega}\right)^2 (ik_z \text{ grad } \chi + \mathbf{k}\gamma^2\chi)e^{ik_\gamma z - i\omega t}
\end{aligned} \qquad (7.39)$$

where $c_l^2 = (\lambda + 2\mu)/\rho$, $c_t^2 = \mu/\rho$, φ, ψ, χ are scalar solutions of the two-dimensional Helmholtz equations

$$\nabla^2\varphi + \alpha^2\varphi = 0 \qquad \nabla^2\psi + \beta^2\psi = 0 \qquad \nabla^2\chi + \gamma^2\chi = 0$$

and where $k_\alpha^2 = (\omega/c_l)^2 - \alpha^2$, $k_\beta^2 = (\omega/c_t)^2 - \beta^2$, and $k_\gamma^2 = (\omega/c_t)^2 - \gamma^2$. Solution $\mathfrak{L}$ is longitudinal, and solutions $\mathfrak{M}$ and $\mathfrak{N}$ are transverse. Values of α, β, and γ are determined by the boundary conditions at the outer, cylindrical boundary.

For example, for a circular cylindrical rod of radius a, the functions φ, ψ, χ are products of cosine or sine of $m\phi$ and Bessel functions J_m of αr, βr, or γr, where r, ϕ, z are the usual circular cylindrical coordinates. Each of the functions φ, ψ, χ satisfies different boundary conditions at the surface $r = a$. If this surface is a free surface, $\mathbf{a}_r \cdot \mathfrak{T} = 0$, where $\mathbf{a}_r$ is the unit vector in the r direction and $\mathfrak{T}$ is the stress tensor. For the functions $\mathfrak{L}$ in circular cylindrical coordinates

$$\mathbf{a}_r \cdot \mathfrak{T} = \left[-\mathbf{a}_r \frac{2i\mu c_l}{\omega} \frac{\partial^2 \varphi}{\partial r^2} - \mathbf{a}_\phi \frac{2i\mu c_l}{\omega} \frac{\partial}{\partial r}\left(\frac{1}{r}\frac{\partial \varphi}{\partial \phi}\right) + \mathbf{k}2\mu \frac{k_\alpha c_l}{\omega} \frac{\partial \varphi}{\partial r} + \mathbf{a}_r \frac{i\lambda\omega}{c_l}\varphi \right] e^{ik_\alpha z - i\omega t} \qquad (7.40)$$

for functions $\mathfrak{M}$ it is

$$\left\{ -\mathbf{a}_r \frac{2i\mu c_t}{\omega} \frac{\partial}{\partial r}\left(\frac{1}{r}\frac{\partial \psi}{\partial \phi}\right) + \mathbf{a}_\phi \frac{i\mu c_t}{\omega}\left[r\frac{\partial}{\partial r}\left(\frac{1}{r}\frac{\partial \psi}{\partial r}\right) - \frac{1}{r^2}\frac{\partial^2 \psi}{\partial \phi^2}\right] + \mathbf{k}\frac{\mu k_\alpha c_t}{\omega r}\frac{\partial \psi}{\partial \phi} \right\} e^{ik_\beta z - i\omega t}$$

and for the function $\mathfrak{N}$ it is

$$\left[\mathbf{a}_r \frac{2ik_\gamma \mu c_t^2}{\omega^2} \frac{\partial^2 \chi}{\partial r^2} + \mathbf{a}_\phi \frac{2i\mu k_\gamma c_t^2}{\omega^2} \frac{\partial}{\partial r}\left(\frac{1}{r}\frac{\partial \chi}{\partial \phi}\right) + \mathbf{k}\left(\frac{2\gamma^2 c_t^2 - \omega^2}{\omega^2}\right)\frac{\partial \chi}{\partial r}\right] e^{ik_\gamma z - i\omega t}$$

For example, the function $\mathfrak{M}$, for $m = 0$, corresponds to waves of torsion moving along the rod. Here $\psi = AJ_0(\beta r)$ and

$$\mathbf{r} = \mathbf{a}_\phi \frac{\beta c_t}{i\omega} AJ_1(\beta r)e^{ik_\beta z - i\omega t}$$

and

$$\mathbf{a}_r \cdot \mathfrak{T} = \mathbf{a}_\phi \frac{i\mu c_t \beta^2}{\omega} AJ_2(\beta r)e^{ik_\beta z - i\omega t}$$

where β must be a root of the equation $J_2(\beta a) = 0$. The principal torsional wave corresponds to $\beta \to 0$, where $\psi = A'r^2$ and

$$\mathbf{r} = \mathbf{a}_\phi Ar \exp\left(i\frac{\omega z}{c_t} - i\omega t\right) \qquad (7.41)$$

Here $\mathbf{a}_r \cdot \mathfrak{T} = 0$ for any r (though $\mathbf{a}_z \cdot \mathfrak{T}$ is not zero), and the wave velocity is that for S waves in an infinite medium. These waves are purely transverse.

The other types of waves must be a combination of $\mathfrak{L}$ and $\mathfrak{N}$ types, i.e., a combination of P and S waves, to satisfy the requirement that $\mathbf{a}_r \cdot \mathfrak{T}$ be zero at $r = a$. The proper combination to remove the z component of the traction at $r = a$ is

$$\frac{\omega a}{2c_l k_\alpha} \frac{\mathfrak{L}}{\alpha J_1(\alpha a)} + \frac{\omega^2 a}{\omega^2 - 2\gamma^2 c_t^2} \frac{\mathfrak{N}}{\gamma J_1(a\gamma)} \qquad (7.42)$$

where in $\mathfrak{L}$ we use $\varphi = J_0(\alpha r)$, in $\mathfrak{N}$ we use $\chi = J_0(\gamma r)$, and, in order that the phases coincide along z, we must have that $k_\gamma = k_\alpha$, or

$$\gamma = \sqrt{\alpha^2 + \left(\frac{\omega}{c_t}\right)^2 - \left(\frac{\omega}{c_l}\right)^2}$$

This leaves only the parameter α to be adjusted so that the r component of $\mathbf{a}_r \cdot \mathfrak{T}$ is zero at $r = a$.

For the principal wave, for frequencies low enough so that $\omega a/c_t \ll 1$, both αa and γa are small enough so that the first terms alone in the series expansions of the Bessel functions may be used, and, as $r \to a$,

$$\mathbf{a}_r \cdot \mathfrak{T} \to$$

$$\mathbf{a}_r \frac{i\mu}{2ak_\alpha}\left[1 + \frac{\lambda\omega^2 c_l^2}{\mu\alpha^2} - \frac{(\omega/c_l)^2 - \alpha^2}{2(\omega/c_l)^2 - (\omega/c_t)^2 - 2\alpha^2}\right]$$

For this to be zero we must have

$$k_\alpha^2 = k_\gamma^2 \to \rho\omega^2 \frac{\lambda + \mu}{\mu(3\lambda + 2\mu)} \qquad \rho\omega^2 a^2 \ll \mu \qquad (7.43)$$

Inserting this back into the formula for r, we see that the motion is predominantly parallel to the rod axis. The effective velocity, along the z axis, of this principal wave (which is as close to a purely longitudinal wave as can occur in a rod) is not $c_l = \sqrt{(\lambda + 2\mu)/\rho}$ but $c_y = \sqrt{\mu(3\lambda + 2\mu)/\rho(\lambda + \mu)}$. The modulus $E = \mu(3\lambda + 2\mu)/(\lambda + \mu)$ is called Young's modulus (see Part 3, Chap. 1). These waves produce the kind of stretching of the rod (with transverse shrinking) which corresponds to Young's modulus. As soon as the frequency increases so that $\rho\omega^2 a^2$ is no longer small compared to μ, the principal wave is no longer a pure *stretch* wave, and Eq. (7.43) is no longer valid.

7. Standing Waves

Possible steady-state motions of elastic media of finite dimensions correspond to standing waves inside the material. In general these waves are quite complex; it is impossible to calculate exactly their forms or their frequencies except for a few especially symmetric waves and shapes of boundary surfaces or except for a few very simple boundary conditions. A few typical cases, of some practical interest, will be discussed. Only the case of the isotropic medium will be treated here; the results for nonisotropic media can be worked out by similar techniques; numerous cases are discussed in the literature, referred to in the references for this chapter.

First consider the standing waves in a cylinder of finite length l, with axis along the z axis, so that the end surfaces are at $z = 0$ and $z = l$. The cross section, normal to the z axis, is independent of z. For such a case we can use combinations of the general solutions given in Eqs. (7.39). Suppose all the boundary surfaces are free surfaces, where the tractions must be zero; this is a case of considerable practical interest, though other boundary conditions are at times more appropriate.

Perhaps the simplest type of vibration which can be set up in such an elastic solid is the simple *extensional* vibration, where the vibration is predominantly parallel to the z axis. For such vibrations the predominant term in the expression for the displacement must be, approximately,

$$\mathbf{r} \simeq A\mathbf{k} \cos\frac{\pi n z}{l}\, e^{-i\omega t} \qquad n = 1, 2, 3, \ldots \qquad (7.44)$$

for with such a choice of the z factor the tractions $\mathbf{a}_z \cdot \mathbf{T}$ at the ends $z = 0$ and $z = l$ are zero for this largest term. When the wavelength $2l/n$ is very small compared to the transverse dimensions of the cylinder, the boundary conditions are approximately fulfilled by the pure P wave in the z direction, with function ϕ a constant [see Eqs. (7.39)]. In this case $\pi n/l$ must be equal to k_α, which is approximately equal to ω/c_l, and the allowed frequencies of extensional vibration are given approximately by the equation

$$\omega_n \simeq \frac{\pi n}{l}\sqrt{\frac{\lambda + 2\mu}{\rho}} \qquad \frac{2l}{n} \ll \text{transverse dimensions} \qquad (7.45)$$

This formula would be exact if the transverse dimensions were infinite, the vibrations being then the extensional oscillations between the two surfaces of an infinite plate. The elastic modulus here is just the quantity $\lambda + 2\mu$ connected with compressional waves in an infinite medium. On the other hand, when the wavelength is large compared to the transverse dimensions of the bar, the appropriate modulus is Young's modulus, $\mu(3\lambda + 2\mu)/(\lambda + \mu)$, which is to be substituted in Eq. (7.45) instead of $\lambda + 2\mu$. This corresponds to the solution given in Eqs. (7.42) and (7.43) for a circularly cylindrical rod. For the free circular cylinder, of radius a and length l, the displacement $\mathbf{r}$ for extensional vibrations is

$$\mathbf{r} \simeq A\left(\mathbf{k}\cos\frac{\pi n z}{l} + \mathbf{a}_r \frac{\lambda/2}{\lambda + \mu}\frac{\pi n r}{l}\sin\frac{\pi n z}{l}\right) e^{-i\omega t} \qquad (7.46)$$

where $\omega \simeq (\pi n/l)\sqrt{\rho(\lambda + \mu)/\mu(3\lambda + 2\mu)}$. This is valid only when $\pi n a/l \ll 1$, in which case the second term is small compared to the first, even at the outer surface $r = a$. The formula shows clearly the nature of the elongation; where the stretching is greatest, there the transverse shrinking is greatest and vice versa. Equation (7.46) cannot be the exact solution, for the boundary conditions $\mathbf{a}_r \cdot \mathbf{T} = 0$ at $r = a$ and $\mathbf{k} \cdot \mathbf{T} = 0$ at $z = 0$ and $z = l$ are not satisfied exactly.

Thus purely extensional vibrations of the cylinder have frequencies given by Eq. (7.45), when the wavelength $2l/n$ is much smaller than the smallest transverse dimension of the cylinder. When the largest transverse dimension of the cylinder is smaller than the wavelength $2l/n$, then Eq. (7.45) for the frequency is modified by substituting the smaller quantity $\sqrt{\mu(3\lambda + 2\mu)/(\lambda + \mu)}$ for the larger quantity $\sqrt{\lambda + 2\mu}$. For intermediate transverse dimensions, of the same size as the wavelength, the allowed frequencies of extensional vibration will have intermediate values; the correct formula will depend on the cross-sectional shape of the cylinder.

For a circular cylinder, pure torsional standing waves are possible, of the general type [see Eq. (7.41)]

$$\mathbf{r} = \mathbf{a}_\phi A r \cos\frac{\pi n z}{l}\, e^{-i\omega_n t} \qquad n = 1, 2, 3, \ldots \qquad (7.47)$$

where $\omega_n = (\pi n/l)\sqrt{\mu/\rho}$. For less symmetric cross sections a pure torsional wave is not possible, and the standing-wave frequencies are correspondingly affected by the presence of a certain admixture of P wave.

Other types of standing waves can be worked out approximately in some detail for rectangular parallelepipeds, with thickness a in the z direction, length l in the x direction, and width b in the y direction. If the material is much wider than it is long or thick ($b \gg l > a$), then some modes of vibration will involve displacements almost wholly in the xz plane, and the calculations of Sec. 5 may be used.

For example, the extensional modes parallel to the x axis have a displacement given by

$$\mathbf{r} = \Big\{ \mathbf{i} \Big[\frac{\pi n c_l}{\omega_n l} \cos\left(\frac{\omega_n z}{c_l} \sin \phi_n\right) - A_n \sin \chi_n \cos\left(\frac{\omega_n z}{c_t} \sin \chi_n\right)\Big] \cos \frac{\pi n x}{l} - \mathbf{k} \Big[\sin \phi_n \sin\left(\frac{\omega_n z}{c_l} \sin \phi_n\right) + A_n \frac{\pi n c_t}{\omega_n l} \sin\left(\frac{\omega_n z}{c_t} \sin \chi_n\right)\Big] \sin \frac{\pi n x}{l} \Big\} e^{-i\omega_n t}$$

where

$$\sin^2 \phi_n = 1 - \left(\frac{\pi n c_l}{\omega_n l}\right)^2, \quad \sin^2 \chi_n = 1 - \left(\frac{\pi n c_t}{\omega_n l}\right)^2$$

and, in order that the tractions on the surfaces $z = \pm\frac{1}{2}a$ be zero, the quantities A_n and ω_n must be adjusted to satisfy a pair of equations similar to (7.32). Other, higher modes, involving combinations of shear and extension, can also be worked out.

8. Transverse Oscillations of Rods and Plates

Transverse vibrations of plates and rods, similar to those given in Eq. (7.37), may also be worked out. In the case of plates, where the x and y dimensions are large compared to the z dimension, we can use Eq. (7.38); examples will be given later in this section. In the case of rods, where the y and z dimensions are small compared to the x dimension, we can use a simpler, approximate analysis to work out the transverse vibrations.

Suppose the rod is uniform in cross section, density, and elasticity along its principal axis, which is set along the x axis. To be specific, consider rods with cross sections which are symmetrical about both the y and z axes, and suppose the x axis to coincide with its center when the rod is in equilibrium. The displacement of this center line away from equilibrium will be called $\eta(x,t)$ (we assume this displacement to be in the xy plane). In general, if the rod is displaced, some part of it will be bent, and if the displacement is small, the curvature of the rod at point x will be $\partial^2\eta/\partial x^2$. To produce this bending there must be a bending moment $M(x,t)$ acting across the cross section at x.

Suppose the curvature is such that two cross sections, parallel and a distance dx apart at equilibrium, are at an angle $d\theta = (\partial^2\eta/\partial x^2)\,dx$ to each other. All the material in the rod between y and $y + dy$ and between x and $x + dx$ will be compressed (if y is positive) or stretched (if y is negative) by an amount $y\,d\theta$, and the force required to stretch or compress this portion of the rod will be $Ey(d\theta/dx)\omega(y)\,dy$, where $E = \mu(3\lambda + 2\mu)/(\lambda + \mu)$ is Young's modulus and $\omega(y)$ is the width of the strip of rod between y and $y + dy$. The moment of this force about the z axis is then

$$-E\left(\frac{d\theta}{dx}\right) y^2\omega(y)\,dy = -E\left(\frac{\partial^2\eta}{\partial x^2}\right) y^2\omega(y)\,dy$$

in taking the positive moment in a clockwise direction.

Consequently the total moment required to produce the specified bending of the rod is

$$M(x,t) = -E\,\frac{\partial^2 y}{\partial x^2} \int y^2\omega(y)\,dy$$

where the integration is over the area of cross section [$\int\omega(y)\,dy$ equals S, the area of cross section]. This integral is the usual one for the moment of inertia of the cross-sectional area about its center line (parallel to the z axis). The square root of the moment of inertia divided by the area S is called the radius of gyration κ:

$$\kappa^2 = \frac{\int y^2\omega(y)\,dy}{\int\omega(y)\,dy}$$

Consequently the bending moment is

$$M(x,t) = -ES\kappa^2\,\frac{\partial^2\eta}{\partial x^2} \qquad (7.48)$$

This bending moment changes with x; consequently a portion of the rod between x and $x + dx$ will have a net moment $(\partial M/\partial x)\,dx$, which will have to be balanced by a shearing force

$$F(x,t) = \frac{\partial M}{\partial x} = -ES\kappa^2\,\frac{\partial^3\eta}{\partial x^3}$$

Actually the moment of the shear should not be exactly equal to the net moment; when the rod is in motion, it will require a certain moment to rotate the part of the bar as it bends. But if the displacement η is small, the necessary moment balancing the rotation will be small to a higher order.

Finally, the shear F changes with x, and its rate of change represents a net force which accelerates the rod back and forth. In other words, the equation of motion of the rod for transverse motion is

$$\rho S\,\frac{\partial^2\eta}{\partial t^2} = -ES\kappa^2\,\frac{\partial^4\eta}{\partial x^4} \qquad (7.49)$$

This is an approximate expression, valid only for small values of the displacement η and for radii of curvature of the rod axis long compared to the transverse dimensions of the rod. We have neglected the rotational acceleration of the rod as each part of it is bent; we consider only the translational acceleration in the y direction of each part. Furthermore we have used Young's modulus, which assumes that the stretched part of the rod shrinks transversely by a relative amount determined by Poisson's ratio $\sigma = \lambda/2(\lambda + \mu)$ and that the compressed part swells sidewise by a like factor. The corresponding change in shape of the cross section is not considered in this equation, which means that we have tacitly assumed that the dimension of the rod (width) in the z direction (at right angles to the displacement) is not large compared with its size in the y direction (depth). (The case of width large compared with depth is the case of the plate, to be discussed shortly.)

We also note that the equation of motion of the rod is similar to that given in Eq. (7.38) for the transverse vibrations of a plate. The coefficients are different, however. In the case of the bar the coefficient multiplying the time derivative is

$$\frac{\rho}{E\kappa^2} = \frac{12\rho(\lambda + \mu)}{a^2\mu(3\lambda + 2\mu)}$$

(if the bar's cross section is a rectangle of depth a and width b, with b not large compared to a), whereas for the plate ($b \gg a$) it is $3\rho(\lambda + 2\mu)/a^2\mu(\lambda + \mu)$. The difference is closely related to Poisson's ratio σ; in the case of the bar the width b is not so large but what it can swell and contract transversally as it bends; in the case of the plate the width is large enough to prevent it.

Returning to Eq. (7.49), the possible simple harmonic bending motions of the rod are given by

$$\eta = (A \sin \gamma x + B \cos \gamma x + C \sinh \gamma x + D \cosh \gamma x)e^{-i\omega t}$$

where $$\gamma^4 = \frac{\omega^2\rho}{E\kappa^2} = \frac{\omega^2\rho(\lambda + \mu)}{\kappa^2\mu(3\lambda + 2\mu)}$$

The values of A, B, C, and D and the allowed values of ω are determined by the boundary conditions at the two ends of the rod. If the bar end is rigidly clamped, then η and $\partial\eta/\partial x$ are zero at that end; if the end of the bar is free, M and F and thus $\partial^2\eta/\partial x^2$ and $\partial^3\eta/\partial x^3$ are zero at this end.

Two cases will be discussed here: the free-free bar, where both ends of the bar (at $x = 0$ and $x = l$) are free, and the clamped-free bar, where the end at $x = 0$ is clamped and the end at $x = l$ is free. The solution

$$\eta = [(\cosh \gamma x + \cos \gamma x) + a(\sinh \gamma x + \sin \gamma x)]e^{-i\omega t}$$

has its second and third derivatives with respect to x zero at $x = 0$. To have this also true at $x = l$, γ and a must satisfy

$$\cosh \gamma l - \cos \gamma l = -a(\sinh \gamma l - \sin \gamma l)$$
$$\sinh \gamma l + \sin \gamma l = -a(\cosh \gamma l - \cos \gamma l)$$

or

$$(\cosh \gamma l - \cos \gamma l)^2 = \sinh^2 \gamma l - \sin^2 \gamma l$$

or

$$\cos \gamma l \cosh \gamma l = 1$$

Roots of this equation are $\gamma l = 4.7300$; 7.8532; 10.9956; . . . ; $\frac{1}{2}\pi(2n + 1)$ for n large. Consequently the allowed frequencies of a free-free bar are

$$\omega_n = \frac{\kappa\epsilon_n^2}{l^2}\sqrt{\frac{E}{\rho}} \qquad \text{(free-free)} \qquad (7.50)$$

where ϵ_n is the nth root for γl, as given above. From the values of ϵ_n one can obtain values of a_n and thus obtain expressions for the shapes of the various modes of vibration.

For a clamped-free bar take

$$\eta = [\cosh \gamma x - \cos \gamma x + b (\sinh \gamma x - \sin \gamma x)]e^{-i\omega t}$$

where $\cosh \gamma l + \cos \gamma l = -b (\sinh \gamma l + \sin \gamma l)$ and $\cos \gamma l \cosh \gamma l = -1$ serve to determine b and γ and therefore ω. Then

$$\omega_n = \frac{\kappa\delta_n^2}{l^2}\sqrt{\frac{E}{\rho}} \qquad \text{(clamped-free)} \qquad (7.51)$$

where $\delta_1 = 1.8751$, $\delta_2 = 4.6941$, $\delta_3 = 7.8548$,

$$\delta_4 = 10.9955, \ . \ . \ .$$

$\delta_n \simeq \frac{1}{2}\pi(2n - 1)$ are the roots for γl of the equation $\cos \gamma l \cosh \gamma l = -1$.

The preceding applied to uniform bars. To calculate the transverse vibrations of bars with density, cross-sectional area, or elasticity a function of x we must use the approximate methods of the variational principle or of the perturbation series.

The variational method for standing waves usually starts with the expression for the Lagrange density [see Eq. (7.4)]. For the case of a bar, with small transverse displacements from equilibrium and radii of curvature of this displacement long compared with the transverse dimensions of the bar, the Lagrange density (per length of bar along its axis) is

$$L = \tfrac{1}{2}\rho S\left(\frac{\partial\eta}{\partial t}\right)^2 - \tfrac{1}{2}SE\kappa^2\left(\frac{\partial^2\eta}{\partial x^2}\right)^2 \qquad (7.52)$$

with corresponding energy density

$$H = \tfrac{1}{2}\rho S\left[\left(\frac{\partial\eta}{\partial t}\right)^2 + \frac{E\kappa^2}{\rho}\left(\frac{\partial^2\eta}{\partial x^2}\right)^2\right]$$

The variational principle for standing transverse waves is then

$$\delta\mathfrak{L} = 0 \qquad \mathfrak{L} = \int_0^l L\,dx$$

where the function to be varied, η, satisfies the proper boundary conditions at the two ends, corresponding to a free end or a clamped end, for example. The Lagrange-Euler equation corresponding to this principle is Eq. (7.49) if ρ, S, κ^2, and E are independent of x. If any of them are not, one may assume some form for η, say $\phi(x)e^{-i\omega t}$, where $\phi(x)$ is a function which satisfies the specified boundary conditions at $x = 0$ and $x = l$ and which contains several parameters $\alpha_1, \alpha_2, \ . \ . \ . \ , \alpha_n$, which produce appropriate changes in the intermediate shape of the displacement. Then the values of α_m which give a stationary value of $\mathfrak{L}$ are those which make ϕ the nearest possible to the correct solution η for the form chosen.

Since, by integration by parts,

$$\tfrac{1}{2}\int_0^l SE\kappa^2\left(\frac{\partial^2\phi}{\partial x^2}\right)^2 dx = \tfrac{1}{2}\left[SE\kappa^2\frac{\partial\phi}{\partial x}\frac{\partial^2\phi}{\partial x^2} - \phi\frac{\partial}{\partial x}\left(SE\kappa^2\frac{\partial^2\phi}{\partial x^2}\right)\right]_0^l + \tfrac{1}{2}\int_0^l \phi\frac{\partial^2}{\partial x^2}\left(SE\kappa^2\frac{\partial^2\phi}{\partial x^2}\right)dx$$

and since the terms in brackets are zero at the two

ends, if ϕ satisfies the usual boundary conditions there, we can modify $\mathfrak{L}$ to become

$$\mathfrak{L} = \tfrac{1}{2}\omega^2 \int_0^l \rho S\phi^2\,dx - \tfrac{1}{2}\int_0^l \phi\,\frac{\partial^2}{\partial x^2}\left(SE\kappa^2\,\frac{\partial^2\phi}{\partial x^2}\right)dx$$

with the corresponding Lagrange-Euler equation

$$\frac{1}{\rho S}\,\frac{\partial^2}{\partial x^2}\left(SE\kappa^2\,\frac{\partial^2\eta}{\partial x^2}\right) = \omega^2\phi \tag{7.53}$$

which is the equation of motion for a nonuniform bar, to the degree of approximation mentioned above.

Consequently the quantity to be "minimized" may be

$$Q = \frac{\int_0^l \phi\,\frac{\partial^2}{\partial x^2}\left(SE\kappa^2\,\frac{\partial^2\phi}{\partial x^2}\right)dx}{\int_0^l \rho S\phi^2\,dx} = \frac{\int_0^l SE\kappa^2\left(\frac{\partial^2\phi}{\partial x^2}\right)^2 dx}{\int_0^l \rho S\phi^2\,dx} \tag{7.54}$$

Solving for the parameters α_m by setting $\partial Q/\partial\alpha_m = 0$ $(m = 1, 2, \ldots, n)$, the resulting value of Q is then nearest to the correct value of ω^2 and the corresponding shape of ϕ the nearest to the correct η as is possible with the choice of the trial function ϕ.

The perturbation formulas may be derived from this principle by using solutions for the uniform bar to build up the trial function ϕ. Since the combinations $\eta = y_n$ of cos, cosh, sin, sinh of γx corresponding to Eqs. (7.50) or (7.51) (depending on the boundary conditions at $x = 0$ and l) are eigenfunctions for Eq. (7.49), satisfying the usual properties

$$\int_0^l [y_n(x)]^2\,dx = N_n \qquad \int_0^l y_n(x)y_m(x)\,dx = 0 \qquad n \neq m$$

we can use the y's as a complete set to express ϕ,

$$\phi(x) = \sum_n A_n y_n(x) \qquad A_n = \frac{1}{N_n}\int_0^l y_n\phi\,dx$$

If the y's are the ones corresponding to the right set of boundary conditions, ϕ will automatically satisfy these conditions.

To solve for the case in which ρ, S, E, or κ^2 changes with x, set this series in (7.54) and obtain

$$Q = \frac{\sum_{n,m} A_nA_mV_{nm}}{\sum_{n,m} A_nA_mM_{nm}} \qquad V_{nm} = \int_0^l SE\kappa^2\,\frac{\partial^2 y_n}{\partial x^2}\,\frac{\partial^2 y_m}{\partial x^2}\,dx \qquad M_{nm} = \int_0^l \rho S y_n y_m\,dx \tag{7.55}$$

(if ρ and S are independent of x, $M_{nm} = \rho S\,\delta_{nm}N_n$). The best value of Q is then obtained by minimizing with respect to the A's.

If we allow all possible variations of the A's, the minimum value of Q will correspond to the value of ω^2 for the lowest allowed frequency. To obtain the higher modes, we must insist that the coefficient of one, say A_n, be unity and the rest of the A's be small correction terms. For example, the first approximation to $Q \simeq \omega_n^2$ is

$$Q_n^1 = \frac{V_{nn}}{M_{nn}}$$

To the next approximation we set

$$A_m = \frac{V_{nm} - Q_n^1M_{nm}}{V_{mn} - Q_n^1M_{mm}} \qquad A_n = 1$$

in Eq. (7.55) to obtain Q_n^2, and so on. Further tricks of calculation must be worked out for each individual case.

Finally we can calculate the free vibrations for thin plates, for a few simple boundary shapes and boundary conditions. The equation used is Eq. (7.38). As an example we compute the transverse vibrations of the uniform circular plate clamped at its edges (like a telephone-receiver diaphragm). The solutions of Eq. (7.38) in polar coordinates r, ϕ are

$$\eta = {\cos \atop \sin} m\phi\left[J_m\left(\frac{\pi\beta_{mn}r}{R}\right) - \frac{J_m(\pi\beta_{mn})}{I_m(\pi\beta_{mn})}\,I_m\left(\frac{\pi\beta_{mn}r}{R}\right)\right]e^{-i\omega_{mn}t}$$

where $I_m(x)$ is the hyperbolic Bessel function

$$i^{-m}J_m(ix)$$

and where, to satisfy the requirement that $\partial\eta/\partial r = 0$ at the edge of the plate $r = R$, we must have

$$I_m(\pi\beta)\,\frac{d}{d\beta}\,J_m(\pi\beta) - J_m(\pi\beta)\,\frac{d}{d\beta}\,I_m(\pi\beta) = 0$$

which corresponds to the allowed frequencies and eigenvalues:

$$\omega_{mn} = \frac{\pi^2 a}{R^2}\sqrt{\frac{\mu(\lambda+\mu)}{3\rho(\lambda+2\mu)}}\,(\beta_{mn})^2$$

$$\beta_{01} = 1.015 \qquad \beta_{02} = 2.007$$

$$\beta_{11} = 1.468 \qquad \beta_{12} = 2.483 \qquad \beta_{mn} \simeq n + \tfrac{1}{2}m$$

for a circular plate of radius R, thickness a, clamped at its edge.

The vibrations of a clamped rectangular plate may be worked out in a similar way. The boundary conditions for a free edge are not definitely determined, so solutions for free plates are only approximately determined.

9. Scattering of Elastic Waves

It is at times useful to compute the scattering of elastic waves from some object imbedded in an otherwise uniform, isotropic, infinite medium. The Green's tensors, given in Eqs. (7.20), may be used to compute the scattering if the incident wave is simple harmonic of frequency $\omega/2\pi$. Suppose the scattering object has a surface S and is placed so its center of mass is at the origin of coordinates. At the area element dS on its surface the outward-pointing unit vector normal to dS will be called $\mathbf{n}$. Application of the vector Green's theorem shows that on and outside the surface of the scatterer the incident plus scattered elastic wave at the point x, y, z, indicated by vector $\mathbf{x}$, is given by the integral equation

$$\mathbf{F}(\mathbf{x}) = \mathbf{F}_0(\mathbf{x}) + \int \Big\{c_l^2[\mathbf{n}' \cdot \mathfrak{G}(\omega;\mathbf{x},\mathbf{x}')][\text{div}'\,\mathbf{F}(\mathbf{x}')] - \left[\text{div}'\,\mathfrak{p}\left(\frac{\omega}{c_l};\,\mathbf{x},\,\mathbf{x}'\right)\right][\mathbf{n}' \cdot \mathbf{F}(\mathbf{x}')] - c_t^2\mathfrak{G}(\omega;\mathbf{x},\mathbf{x}') \cdot [\mathbf{n}' \times \text{curl}'\,\mathbf{F}(\mathbf{x}')] - \left[\text{curl}'\,\mathfrak{s}\left(\frac{\omega}{c_t};\,\mathbf{x},\,\mathbf{x}'\right)\right] \cdot [\mathbf{n}' \times \mathbf{F}(\mathbf{x}')]\Big\}\,dS' \qquad (7.56)$$

where

$$\mathfrak{G}(\omega;x,x') = \frac{1}{c_l^2}\,\mathfrak{p}\left(\frac{\omega}{c_l};\,\mathbf{x},\,\mathbf{x}'\right) + \frac{1}{c_t^2}\,\mathfrak{s}\left(\frac{\omega}{c_t};\,\mathbf{x},\,\mathbf{x}'\right)$$

is the Green's tensor given in Eqs. (7.20), where $\mathbf{x}'$ is the radius vector to the element dS' on the surface of the scatterer, $\mathbf{n}'$ is the normal to the surface at x', where div′, curl′ are vector operators in the x', y', z' coordinates, and where the integration in the primed coordinates is over the surface of the scatterer. The result of the integration is a vector function of x, y, z; since $\mathfrak{G}$, $\mathfrak{p}$, and $\mathfrak{s}$ are tensors, each term in the integrand is a vector. We assume no sources of waves nearer than infinity and that the source at infinity has produced the incident wave F_0. Exact solution of this integral equation will give an exact solution of the scattering problem; since this is rarely possible, we shall indicate how approximate solutions may be obtained.

As an example, suppose that the scattering "object" is a void in the medium, so that the surface S is a free surface. In this case div $\mathbf{F}$ and $\mathbf{n} \times$ curl $\mathbf{F}$ will be zero at S. Suppose also that the incident wave is a plane P wave in the z direction. Then the elastic displacement $\mathbf{r} = \mathbf{F}$ is a solution of

$$\mathbf{F}(x) = \mathbf{a}_z e^{i\mathbf{k}_i\cdot\mathbf{x}-i\omega t} - \int\{[\mathbf{n}' \cdot \mathbf{F}(\mathbf{x}')]\,\text{div}'\,\mathbf{p} + [\mathbf{n}' \times \mathbf{F}(\mathbf{x}')] \cdot \text{curl}'\,\mathbf{s}\}\,dS' \qquad (7.57)$$

where $\mathbf{k}_i = \mathbf{a}_i(\omega/c_l)$ ($\mathbf{a}_i$ is the unit vector in the direction of the incident wave). At very large distances from the scatterer the integral, which represents the scattered wave, takes on a simpler form. It can be shown that, for $|\mathbf{x}| \gg |\mathbf{x}'|$ and for $(\omega/c_l)|x| \gg 1$,

$$\text{div}'\,\mathfrak{p} \to -\frac{i\omega}{4\pi c_l|\mathbf{x}|}\,\mathbf{a}\,\exp\left(\frac{i\omega|x|}{c_l} - \frac{i\omega\mathbf{a}\cdot\mathbf{x}'}{c_l}\right)$$

$$\text{curl}'\,\mathfrak{s} \to -\frac{i\omega}{4\pi c_t|\mathbf{x}|}\,\mathbf{a} \times \mathfrak{J}\,\exp\left(\frac{i\omega|x|}{c_t} - \frac{i\omega\mathbf{a}\cdot\mathbf{x}'}{c_t}\right)$$

where $\mathbf{a} = \mathbf{x}/|\mathbf{x}|$ is the unit vector in the direction of the observer, i.e., in the direction in which the scattered wave is being measured, and $\mathfrak{J}$ is the unit tensor $\mathbf{ii} + \mathbf{jj} + \mathbf{kk}$.

Inserting all this into the integral, the wave scattered in the direction given by unit vector $\mathbf{a}$ has the asymptotic form

$$\mathbf{F}_s(\mathbf{x}) \to \frac{e^{i\omega|x|/c_l}}{|\mathbf{x}|}\,\mathbf{f}_l(\omega;\mathbf{a}_i,\mathbf{a}) + \frac{e^{i\omega|x|/c_t}}{|x|}\,\mathbf{f}_t(\omega;\mathbf{a}_i,\mathbf{a}) \qquad (7.58)$$

$$\mathbf{f}_l = \frac{i\omega}{4\pi c_l}\,\mathbf{a}\int[\mathbf{n}' \cdot \mathbf{F}(\mathbf{x}')]\exp\left(-\frac{i\omega\mathbf{a}\cdot\mathbf{x}'}{c_l}\right)dS'$$

$$\mathbf{f}_t = \frac{i\omega}{4\pi c_t}\int\{[\mathbf{a} \cdot \mathbf{F}(\mathbf{x}')]\mathbf{n}' - (\mathbf{a} \cdot \mathbf{n}')\mathbf{F}(\mathbf{x}')\}\exp\left(-\frac{i\omega\mathbf{a}\cdot\mathbf{x}'}{c_t}\right)dS'$$

(the vector in the braces is normal to $\mathbf{a}$). The quantities $\mathbf{f}_l$ and $\mathbf{f}_t$ are called the angle-distribution factors for the scattered longitudinal and transverse waves, respectively. If a unit-intensity longitudinal wave is incident on the scatterer, then the scattered intensities in direction $\mathbf{a}$ at distance $|\mathbf{x}|$ from the center of the scatterer are $|\mathbf{f}_l|^2/|\mathbf{x}|^2$ and $c_t|\mathbf{f}_t|^2/c_l|\mathbf{x}|^2$, respectively.

Equation (7.58) is still not a solution, for function $\mathbf{F}$ in the integrand is still unknown. It may be solved by successive approximations, however, or else by a variational principle. For example, if the scattering object is small compared with the wavelength, it is possible that the amount of scattering may be small, and, to a first approximation, the incident wave $\mathbf{F}_0$ may be substituted for $\mathbf{F}$ in the equations above. This approximation is called the *Born approximation;* the expressions for the angle-distribution factors are

$$\mathbf{f}_l \simeq \frac{i\omega}{4\pi c_l}\,\mathbf{a}\int(\mathbf{n}' \cdot \mathbf{a}_i)\exp\left[i\omega(\mathbf{a}_i - \mathbf{a}) \cdot \frac{x}{c_l}\right]ds'$$

$$\mathbf{f}_t \simeq \frac{i\omega}{4\pi c_t}\int\mathbf{a} \times (\mathbf{n} \times \mathbf{a}_i)\exp\left[i\omega\left(\frac{\mathbf{a}_i}{c_l} - \frac{\mathbf{a}}{c_t}\right) \cdot \mathbf{x}'\right]ds' \qquad (7.59)$$

For example, if the scattering "hole" is a sphere of radius b, then, if

$$\mathbf{x}' = \mathbf{i}b\sin\theta'\cos\phi' + \mathbf{j}b\sin\phi'\sin\phi' + \mathbf{k}b\cos\theta'$$

and if the angle of scattering (the angle between $\mathbf{a}$ and $\mathbf{a}_i$) is ϑ ($\mathbf{a}_i$ is along the z axis), then $\mathbf{n}' = \mathbf{x}'/b$. The exponentials may be expanded in terms of spherical harmonics of the angles between x' and the vectors $\mathbf{a}_i - \mathbf{a}$ or $\mathbf{a}_i/c_l - \mathbf{a}/c_t$ and spherical Bessel functions (see Part 1, Chap. 3) of the argument. If $\omega(\mathbf{a}_i - \mathbf{a})/c_l = \mathbf{q}_l$ at angles θ_l,ϕ_l with respect to the spherical axis $\mathbf{a}_i$, and if $\omega\mathbf{a}_i/c_l - \omega\mathbf{a}/c_t = \mathbf{q}_t$ has angles θ_t, ϕ_t, we find that

$$\mathbf{f}_l \simeq -\mathbf{a}\,\frac{\omega b^2}{c_l}\cos\theta_l\,j_1(bq_l)$$

$$\mathbf{f}_t \simeq -\mathbf{a} \times (\mathbf{a}_t \times \mathbf{a}_i)\,\frac{\omega b^2}{c_t}\,j_1(bq_t) \qquad (7.60)$$

where $\mathbf{a}_t = \mathbf{q}_t/q_t$ and q_l and q_t are the magnitudes of the vectors $\mathbf{q}_l$ and $\mathbf{q}_t$ respectively.

More accurate expressions for F may be calculated by inserting the improved form for F back into the integral or by use of a variational principle. The present results, however, show the general behavior of the scattered wave. Vector $\mathbf{q}$ (for incident P wave) vanishes for angle of scattering $\vartheta = 0$ and is nearly normal to $\mathbf{a}_i$ for ϑ small; consequently a P wave scatters very little P wave from a spherical hole in a forward direction. Expressing θ_l and q_l in terms of ϑ (the angle between $\mathbf{a}$ and $\mathbf{a}_i$) shows that the scattered P intensity varies as $\sin^4 \frac{1}{2}\vartheta$ for ϑ small. For larger values of ϑ the factor $j_1(bq_l)$ indicates that diffraction effects will be apparent if $\omega b/c_l$ is large enough. On the other hand, $\mathbf{q}_t$ is parallel to $\mathbf{a}_i$ for $\vartheta = 0$, and $\mathbf{q}_t$ is a minimum there, though not zero. The S wave scattered by the hole, for an incident P wave, is also small in the forward direction, though it is proportional to $\sin^2\vartheta$ for ϑ

small, so it increases more rapidly than $|f_l|^2$ as ϑ is increased from zero.

Other scattered waves may be computed for other types of scatterers and other incident waves. The scattered wave is always a combination of P and S waves, for the waves mix up on reflection.

References

The subject of elastic vibrations has been studied for more than a century, and a complete bibliography would be redundant. The classical results are summarized in:

Brillouin, L.: "Les Tenseurs en méchanique et en elastique," Masson, Paris, 1938.
Love, A. E. H.: "Mathematical Theory of Elasticity," Dover, New York, 1944.
Morse, P. M., and H. Feshbach: "Methods of Theoretical Physics," McGraw-Hill, New York, 1953.
Rayleigh, Lord: "The Theory of Sound," Dover, New York, 1945.
Sommerfeld, A.: "Mechanics of Deformable Bodies," Academic, New York, 1950.

A discussion of the applications of elastic vibrations to seismology and acoustics is given in:

Bergmann, L.: "Ultrasonics," Wiley, New York, 1938.
Bullen, K. E.: "Theory of Seismology," Cambridge, London, 1947.
Morse, P. M.: "Vibration and Sound," McGraw-Hill, New York, 1948.

For a treatment of the elastic vibrations of crystals (including piezoelectric effects) and of the use of such vibrations as a means of measuring elastic constants, see:

Atanasoff, J. V., and P. J. Hart: *Phys. Rev.*, **59**: 85 (1941).
Heising, R. A., et al.: "Quartz Crystals for Electrical Circuits," Van Nostrand, Princeton, N.J., 1946.
Koga, I.: *Physics*, **3**: 70 (1932).
Voigt, W.: "Lehrbuch der Krystalphysik," Teubner Leipzig, 1928.

Chapter 8

Acoustics

By UNO INGARD, Massachusetts Institute of Technology

INTRODUCTION

The field of acoustics as commonly understood deals with the generation and propagation of mechanical vibrations in matter, the application of sound in various fields of science, and its effect on men. Of these branches the general equations of wave motions in solids and fluids have already been treated in Part 3, Chaps. 7 and 4, and we need give only a supplementary summary of some forms of the equations of sound which will be used in the subsequent discussions. Some basic concepts and relations will also be included.

1. Limits of Frequency and Sound Pressure

The range of frequencies encountered in acoustics is quite large; the audible range itself extends over almost 10 octaves (cf. the visual electromagnetic frequency range of about 1 octave). The lower and upper limits of the audible range are approximately 20 and 20,000 cps, respectively, and much of acoustics is concerned with this range.

However, the frequencies in acoustics are by no means limited to the audible range. Frequencies as high as 500 megacycles have been generated, a wavelength of 0.6×10^{-4} cm in air. In liquids and solids the corresponding wavelengths are approximately 2.4×10^{-4} and 8×10^{-4} cm. These wavelengths are of the same order of magnitude as that of visible light.

A gas ceases to behave like a continuum when the wavelength of sound becomes of the order of the mean free path. Strong dispersion and absorption result, and when the sound frequency becomes considerably greater than the collision frequency, the ordered sound motion of the molecules will quickly be transformed into random thermal motion, and no sound propagation can take place [1].† At ordinary atmospheric conditions the mean free path is of the order of 10^{-5} cm, a limit frequency of the order of 10^9 cps.

In solids the assumption of continuum is senseless when the wavelength approaches the intermolecular distance, approximately 10^{-8} cm with corresponding limiting frequency of about 10^{12} cps. The ultimate limit is actually reached when the wavelength is twice the spacing of the unit cell of a crystal. In this region propagation of sound (multiply scattered) resembles diffusion of heat [2].

The range of sound pressure is also considerable. The ear responds to sound pressures from 0.0002 to 2,000 dynes/cm². The lower limit corresponds to an intensity of 10^{-16} watt/cm² at continuous exposure. The least amount of acoustic energy that the ear can detect is of the order of $kT \simeq 10^{-20}$ watt-sec (cf. sensitivity of the eye: about one quantum of light in the middle of the visible region $h\nu \simeq 4 \times 10^{-19}$).

The upper limit of the sound pressure in a pure tone that can be generated in the medium is set approximately by the static pressure. At this pressure the rarefaction part of the sound cycle would create vacuum and breakdown, or cavitation, and the medium could no longer "support" the wave. The intensity in air of a plane sound wave with this limiting pressure equals approximately 1.20×10^3 watts/cm² ($\simeq$191 db). In water at atmospheric pressure the corresponding intensity is 0.36 watt/cm². (The cavitation pressure in a liquid is frequency-dependent and can under certain conditions considerably exceed the static pressure.)

Before the upper pure-tone intensity limit is reached, nonlinearity of the medium causes distortion of the wave: energy is in effect removed from the fundamental frequency and distributed on higher harmonics. Therefore, a large-amplitude wave propagating in air will change waveform and finally, after a certain distance of travel, break into a shock, reaching a stable saw-toothed form. This behavior is somewhat similar to the familiar breaking of waves on a water surface (see Sec. 17).

Sound Pressure and Intensity Levels. In expressing the sound intensity on a logarithmic scale, the reference intensity I_0 is usually taken to be $I_0 = 10^{-16}$ watt/cm², so that

$$\text{Intensity level in decibels} = 10 \log \frac{I}{I_0}$$

Correspondingly, the sound pressure is expressed in decibels by

$$\text{Sound-pressure level in decibels} = 20 \log \frac{p}{p_0}$$

where $p_0 = 0.0002$ dyne/cm², rms value. The reference pressure has been chosen to be approximately the threshold of hearing. The intensity level and the sound-pressure level would be identical if p_0 were the

† Numbers in brackets refer to references at end of chapter.

sound pressure that corresponds to the intensity I_0. This is not exactly true, however, except at one temperature T_0 approximately equal to 300°K (20°C). For temperatures higher than T_0 the intensity level as defined above will be larger than the sound-pressure level by an additional term 10 log $(T/T_0)^{1/2}$ db.

2. General Linear Equations of Sound Propagation [3, 4]

From the linearized equations of sound propagation obtained by keeping only first-order terms of the variations $\mathbf{u}$, δ, p, and σ in the fluid-field variables, velocity $\mathbf{v}$, density ρ, pressure P, and entropy S, respectively, in the general hydrodynamic equations (see Part 3, Chap. 4), one finds the wave equation for a *homogeneous* moving medium

$$\frac{D^2p}{Dt^2} = \left(\frac{\partial}{\partial t} + \mathbf{v}\cdot\nabla\right)^2 p = c^2\,\nabla^2 p \tag{8.1}$$

where $\mathbf{v}$ is the flow velocity of the medium.

If the medium is *inhomogeneous*, the basic equations become considerably more complicated [4]:

$$\frac{\partial\delta}{\partial t} + \mathbf{v}\cdot\nabla\delta + \mathbf{v}\cdot\nabla\rho + \rho\nabla\cdot\mathbf{u} + \delta\nabla\cdot\mathbf{v} = 0 \qquad \text{(conservation of mass)} \tag{8.2}$$

$$\frac{\partial\mathbf{u}}{\partial t} + (\nabla\times\mathbf{v})\times\mathbf{u} + (\nabla\times\mathbf{u})\times\mathbf{v} + \nabla(\mathbf{v}\cdot\mathbf{u}) = -\frac{\nabla p}{\rho} + \frac{(\nabla P)\delta}{\rho^2} \qquad \text{(conservation of momentum)} \tag{8.3}$$

$$\frac{\partial\sigma}{\partial t} + \mathbf{v}\cdot\nabla\sigma + \mathbf{u}\cdot\nabla S = 0 \qquad \text{(conservation of energy)} \tag{8.4}$$

$$p = c^2\delta + h\sigma \qquad \text{(equation of state)} \tag{8.5}$$

where

$$c = \text{velocity of sound} = \left(\frac{\partial P}{\partial\rho}\right)_S \qquad h = \left(\frac{\partial P}{\partial S}\right)_\rho$$

The effect of dissipation in the medium due to heat conduction and viscosity has not been accounted for in these equations.

Irrotational and Isentropic Flow ($\nabla\times\mathbf{v} = 0$, $\nabla S = 0$). In this particular case the sound-particle velocity $\mathbf{u}$ is irrotational, so that, with $\mathbf{u} = -\nabla\phi$, Eqs. (8.2) to (8.5) reduce to [4]

$$\frac{D^2\phi}{Dt^2} = c^2\,\nabla^2\phi + (\nabla\pi_0)\cdot\nabla\phi + \frac{D\phi}{Dt}\,\mathbf{v}\cdot\nabla\ln c^2 \tag{8.6}$$

where $\pi_0 = \int dp/\rho$ is the enthalpy of the original flow. The second term on the right-hand side essentially expresses the effect of a density variation of the medium, and the third term the variation of velocity of propagation (as does the factor c^2 in the first term). For a homogeneous medium Eq. (8.6) reduces to (8.1). If the flow is directed along the x axis, the expanded version of this equation is

$$\nabla^{*2}\phi - \frac{1}{c^2}\frac{\partial^2\phi}{\partial t^2} - \frac{2\beta}{c\sqrt{1-\beta^2}}\frac{\partial^2\phi}{\partial t\,\partial x^*} = 0 \tag{8.7}$$

where

$$\nabla^* = \frac{\partial^2}{\partial x^{*2}} + \frac{\partial^2}{\partial y^2} + \frac{\partial^2}{\partial z^2}$$

$$x^* = \frac{x}{\sqrt{1-\beta^2}} \qquad \beta = \frac{v}{c}$$

Rotational Flow ($\nabla\times\mathbf{v} \neq 0$). The sound particle velocity $\mathbf{u}$ is no longer irrotational but contains a vector potential in addition to the scalar potential. Neglecting terms of second order in v/c and $|\nabla\times\mathbf{v}|/\omega$ and their products, the velocity $\mathbf{u}$ can be expressed in terms of a single quantity ψ, so that [4]

$$\mathbf{u} = -\nabla\psi + \int^t (\nabla\times\mathbf{v})\times\nabla\psi\,dt \tag{8.8}$$

and the Eqs. (8.1) to (8.5) now lead to the wave equation

$$\frac{D^2\psi}{Dt^2} - c^2\,\nabla^2\psi = (\nabla\pi_0)\cdot\nabla\psi + \frac{D\psi}{Dt}\,\mathbf{v}\cdot\nabla\ln c^2 + c^2\int^t(\nabla\psi)\cdot\nabla^2\mathbf{v}\,dt - (\nabla\pi_0)\cdot\int_0^t(\nabla\times\mathbf{v})\times\nabla\psi\,dt \tag{8.9}$$

where

$$\nabla^2\mathbf{v} = -\nabla\times(\nabla\times\mathbf{v})$$

3. Kirchhoff's Formula in a Moving Medium [4]

The velocity potential ϕ at a fixed point, in terms of the values of ϕ and $\partial\phi/\partial n$ on fixed surfaces bounding the region under consideration, becomes, in the case of a moving medium [4],

$$\phi(t) = \frac{1}{4\pi}\int\left\{\frac{1}{R^*}\left[\frac{\partial\phi}{\partial n}\right] - \frac{\partial}{\partial n}\left(\frac{1}{R^*}\right)[\phi] + \frac{1}{c}\frac{1}{R^*}\frac{\partial R}{\partial n}\left[\frac{\partial\phi}{\partial t}\right]\right\}dS + \frac{1}{4\pi}\frac{\beta}{\sqrt{1-\beta^2}}\frac{1}{c}\int\frac{1}{R^*}\left[\frac{\partial\phi}{\partial t}\right]dS_x \tag{8.10}$$

The flow is here assumed to be in the positive x direction. $R^{*2} = x^{*2} + y^2 + z^2$ (see Sec. 2) is approximately the distance from P to the "source" point on the surface. The brackets indicate retarded values, $[\phi] = \phi(t - R/c)$, where

$$R = \frac{-\beta x^* + R^*}{\sqrt{1-\beta^2}}$$

The surface element dS_x in the last integral is the projection of dS upon the direction of the flow (the x axis).

For harmonic time dependence $\phi = \phi_1 e^{-i\omega t}$, the formula reduces to

$$\phi_1 = \frac{1}{4\pi}\int\left\{\frac{\partial\phi}{\partial n}\frac{e^{ikR}}{R^*} - \phi_1\frac{\partial}{\partial n}\left(\frac{e^{ikR}}{R^*}\right)\right\}dS - \frac{1}{4\pi}\frac{2i\beta k}{\sqrt{1-\beta^2}}\int\frac{e^{ikR}}{R^*}\,dS_x \tag{8.10a}$$

If the medium in addition contains a volume-source distribution $Q = Q_1 e^{-i\omega t}$, an additional term

$$\int Q_1 e^{ikR}/R^* \, dv$$

appears in the right-hand side of (8.10*a*).

The generalized Kirchhoff's theorem is applicable not only for surfaces and sources at rest and the medium moving but also for the reversed case (see Secs. 9 and 10).

4. Boundary Conditions. Impedance and Absorption Coefficients

Accompanying the differential equations (8.2) to (8.9) or Kirchhoff's formula are the boundary conditions of *continuity of velocity normal to the boundary* and *continuity of pressure.*

The ratio between the particle velocity at a point in a field and the sound pressure is termed the *specific acoustic admittance.*

$$\frac{\mathbf{n}_i}{\rho c} = \frac{\mathbf{u}_i}{p} \qquad i = 1, 2, 3$$

It is a vector with the same direction as $\mathbf{u}_i$. The inverse of $\mathbf{n}/\rho c$ in a given direction is termed the *specific acoustic impedance*

$$\mathbf{z} = \zeta \rho c = \left(\frac{\mathbf{n}}{\rho c}\right)^{-1} = (\theta - i\chi)\rho c$$

For a plane wave in the x direction the specific acoustic impedance or the *characteristic impedance of the medium* is real and equals

$$\mathbf{Z} = \frac{p}{u} = \rho c$$

(for air $\simeq 41.5$ cgs at 20°C) [5], where ρ = density, c = velocity of sound ($\simeq 340$ m/sec at 20°C in air). The *radiation impedance* of a vibrating surface is the ratio of the pressure at the boundary and the particle velocity of the surface.

The ratio between the pressure and normal velocity at a boundary is referred to as the *normal impedance* of the boundary. In general this quantity is not known a priori and can be determined first after the field has been found, utilizing the boundary conditions mentioned above. The normal impedance will then in general be a function of the angle of incidence. However, for some special material (a "wall" with pores normal to the surface) the particle velocity is always normal to the boundary and will depend only on the local pressure at the point under consideration. For such a *locally reacting* or *point-reacting* boundary the normal impedance will be independent of the sound field and can be specified in advance as a characteristic property of the boundary. Under those conditions the analysis of many field problems, sound waves in rooms, etc., is considerably simplified [6]. Many materials met in practice are approximately locally reacting, e.g., perforated porous tiles, dense porous homogeneous material, cavity-resonator arrangements, etc.

The *absorption coefficient* of a plane boundary exposed to a plane wave with an angle of incidence ϕ is

$$\alpha(\phi) = 1 - \left|\frac{\zeta \cos\phi - 1}{\zeta \cos\phi + 1}\right|^2 = \frac{4\theta \cos\phi}{(\theta\cos\phi + 1)^2 + (\chi\cos\phi)^2} \tag{8.11}$$

In a diffuse sound field the probability $B(\phi)$ that an elementary plane wave has an angle of incidence ϕ is proportional to the solid angle $2\pi \sin\phi \, d\phi$, so that the average absorption coefficient

$$[\int B(\phi)\alpha(\phi) \cos\phi \, d\phi]/[\int B(\phi) \cos\phi \, d\phi]$$

becomes

$$\bar{\alpha} = 2 \int_0^{\pi/2} \alpha(\phi) \sin\phi \cos\phi \, d\phi$$

This coefficient is usually referred to as the *statistical-average absorption coefficient.* With $\alpha(\phi)$ given by (8.11), $\bar{\alpha}$ in general cannot be expressed in closed form. However, if ζ is independent of ϕ, the integral reduces to [7, p. 388]

$$\bar{\alpha} = 8\mu \left[1 + \frac{\mu^2 - \sigma^2}{\sigma} \tan^{-1} \frac{\sigma}{\sigma^2 + \mu^2 + \mu} - \mu \ln \frac{(\mu + 1)^2 + \sigma^2}{\mu^2 + \sigma^2}\right] \tag{8.12}$$

$$\mu + i\sigma = \zeta^{-1} = (\theta - i\chi)^{-1}$$

For relatively hard boundaries, $(\mu,\sigma) \ll 1$, we get $\bar{\alpha} \simeq 8\mu$. Graphical representation of (8.12) is available [8].

5. Second-order Quantities

Any quantity that contains the product of two first-order variables of the field is of second order and hence of the same order of magnitude as the terms originally neglected in linearizing the fundamental equations. In the calculation of such quantities as energy flow, mass flow, and time averages like radiation pressure the contribution of the originally neglected terms has to be evaluated.

Energy Flow. Such an investigation [9] of the energy flow in the wave leads to an expression that involves only first-order quantities, so that the sound intensity equals

$$I(t) = pu \tag{8.13}$$

With $\mathbf{u} = -\nabla\phi$ and $\phi = \phi_1 e^{-i\omega t}$, the time average of the intensity is

$$I = \tfrac{1}{2} \operatorname{Re}(p\tilde{u}_1) = \frac{i\omega\rho_0}{4}(\tilde{\phi}\nabla\phi - \phi\nabla\tilde{\phi}) \tag{8.14}$$

where $\tilde{u}_1$ is the complex conjugate of $\mathbf{u}_1$ (Re = real part of).

Radiation Pressure. Correspondingly, the radiation pressure of a plane wave incident on a perfect absorber equals *exactly* twice the mean kinetic energy density of the wave motion. For *small amplitudes* this expression can be written as the mean *total* energy density,

$$p_{\text{rad}} = \text{energy density} = \frac{\rho_0|u_1|^2}{2} = \frac{I}{c}$$

i.e., the same expression as in the electromagnetic case. For a perfect reflector the radiation pressure is *twice* as large. A thorough discussion of subtle questions regarding the effect of large amplitudes, the infinite extension of the sound beam, the angle of incidence, etc., can be found in Borgnis [10].

6. Electromechanical Analogues

The electrical analogue of a mechanical system is the electrical network (or field) which is described by the same (Lagrangian) equations of motion as the mechanical system. As an illustration consider the simple mechanical system shown in Fig. 8.1*a*, consisting of a horizontal bar, the vertical motion of which is impeded by an attached mass M, a spring with spring constant K, and a friction force Dv.

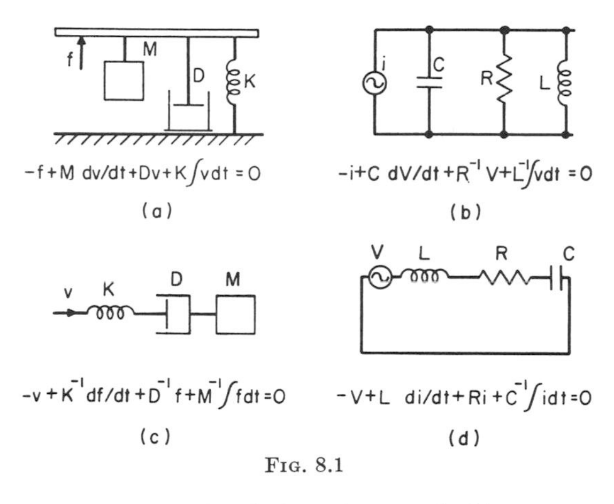

FIG. 8.1

There are two electrical analogues of this system: one, a series circuit shown in Fig. 8.1*d*, and the other, the *dual* of Fig. 8.1*d*, a parallel circuit shown in Fig. 8.1*b*. In the first of these analogues, often called the *classical*, we have the following correspondence between mechanical and electrical quantities: force—voltage, velocity—current, mass—inductance, compressibility (K^{-1})—capacitance, mechanical resistance (D)—electrical resistance. In the second analogue, often called *mobility* analogue, we get correspondingly: force—current, velocity—voltage, mass—capacitance, compressibility—inductance, and mechanical resistance—inverse of electrical resistance. Le Corbeiller and Yeung [11] have pointed out that the picture is not complete without introducing the *mechanical dual* shown in Fig. 8.1*c*. This system is topologically similar to Fig. 8.1*d*. The choice of representation most convenient for a particular problem has been the subject of many studies [12].

SOUND SOURCES AND THEIR FIELDS

7. The "Natural" Sources of Sound [13]

The linear equations governing the propagation of sound in air are

$$\frac{\partial(\rho u_i)}{\partial x_i} + \frac{\partial \rho}{\partial t} = 0 \qquad \text{(conservation of mass)} \qquad (8.15)$$

$$\frac{\partial(\rho u_i)}{\partial t} + c^2 \frac{\partial \rho}{\partial x_i} = 0 \qquad \text{(conservation of momentum)} \qquad (8.16)$$

$$dU = c_v\, dT + P\, dv \qquad \text{(conservation of energy)} \qquad (8.17)$$

$$P = (c_p - c_v)\rho T \qquad \text{(equation of state)} \qquad (8.18)$$

The first of these is the *exact* equation of conservation of mass, and the second an *approximation* of conservation of momentum. The two last equations combine, under isentropic conditions, to the relation $p = (\gamma P_0/\rho_0)\delta = c^2\delta$ between the sound pressure and the density $\delta = \rho - \rho_0$. The three resulting equations lead to the wave equation

$$\nabla^2 p - \frac{1}{c^2}\frac{\partial^2 p}{\partial t^2} = 0$$

with the *source term* on the right-hand side being zero. The simplest physical means required to bring about a source term is found directly by inspection of the equations above, and amounts to introduction of

a. Mass at a rate Q per unit volume in the medium. This means an addition of the term Q on the right-hand side of Eq. (8.15).

b. Force F_i per unit volume in the medium, which enters as an additional term on the right-hand side of Eq. (8.16).

c. An addition $\partial(\rho v_i v_j)/\partial x_j$ per unit volume of the rate of change of momentum. This term, introduced by Lighthill, is due to fluctuations (turbulence), $\rho v_i v_j$ being the Reynolds stress tensor. The term enters in the left-hand side of Eq. (8.16).

d. Heat, at a rate ρH per unit volume. Appears in the left-hand side of Eq. (8.17) (after time differentiation of this equation).

By consideration of these additional terms, the resulting wave equation becomes [13]

$$\nabla^2 p - \frac{1}{c^2}\frac{\partial^2 p}{\partial t^2} = -\left(\dot{Q} + \frac{\gamma - 1}{c^2}\rho \dot{H}\right) + \frac{\partial F_i}{\partial x_i} - \frac{\partial^2 T_{ij}}{\partial x_i\, \partial x_j} \qquad (8.19)$$

where $\dot{Q} = \dfrac{\partial Q}{\partial t} \qquad \dot{H} = \dfrac{\partial H}{\partial t} \qquad T_{ij} = \rho v_i v_j$

The three source terms in this equation are those of a simple source, a dipole, and a quadrupole distribution, respectively. The heat source is equivalent to a mass source of strength $\rho H(\gamma - 1)/c^2$.

The Simple Source. The first term in (8.19) leads to a pressure field given by

$$p = \frac{1}{4\pi}\int_{v'} \frac{\dot{Q}(t - r/c)}{r}\, dv' \qquad (8.20)$$

$$r^2 = (x_1 - x_1')^2 + (x_2 - x_2')^2 + (x_3 - x_3')^2$$

where x_i' = source point, x_i = field point. In the case where the source is concentrated in a point with the flow strength q the field becomes

$$p = \frac{\dot{q}(t - r/c)}{4\pi r} \qquad (8.21)$$

or

$$p = \frac{-i\omega q_0}{4\pi r} e^{ikr} e^{-i\omega t}$$

if $q = q_0e^{-i\omega t}$. The total radiated power is

$$W(t) = \frac{1}{4\pi\rho c}\left[\dot{q}\left(t - \frac{r}{c}\right)\right]^2$$
$$\overline{W(t)} = \frac{\omega^2|q_0|^2}{8\pi\rho c} \quad \text{when } q = q_0e^{-i\omega t} \tag{8.22}$$

The simple-source field (8.21) is realized by means of a pulsating sphere having a radial velocity $U_0e^{-i\omega t}$ such that $q = 4\pi a^2U_0\rho$. At low frequencies the simple-source field represents the major contribution to the far field of *any* finite source with a net flow strength different from zero.

The Dipole. The second source term in (8.19) corresponds to a *dipole* distribution equal to the *vector field* F_i. The source term $\partial f_i/\partial x_i'$ of a concentrated force f_i can be considered as the sum of two simple sources

$$-\frac{1}{\Delta x_i'}f_i(x_i' + \Delta x_i') \quad \text{and} \quad \frac{1}{\Delta x_i'}f_i(x_i')$$

of opposite sign, a distance $\Delta x_i'$ apart, each giving rise to a field (8.21). In the limit $\Delta x_i' = 0$ the resulting field will then be that of a dipole of strength f_i,

$$p = -\frac{1}{4\pi}\frac{\partial}{\partial x_i}\left[\frac{f_i(t - r/c)}{r}\right] \tag{8.23}$$

If the force is in the x_1 direction, the far field becomes

$$p = \frac{1}{4\pi rc}\cos\theta\frac{\partial}{\partial t}\left[f_1\left(t - \frac{r}{c}\right)\right]$$

or

$$p = \frac{-i\omega f_1}{4\pi rc}\cos\theta\, e^{ikr}e^{-i\omega t} \tag{8.24}$$

if $f_i = f_1e^{-i\omega t}$ and $x_1/r = \cos\theta$. The total radiated power is

$$W(t) = \frac{1}{12\pi c^3\rho}\left[\frac{\partial}{\partial t}f_1\left(t - \frac{r}{c}\right)\right]^2$$

or

$$\overline{W(t)} = \frac{\omega^2|f_1|^2}{24\pi c^3\rho} \tag{8.25}$$

if $f_i = f_1e^{-i\omega t}$. In general for a continuous force (or dipole) distribution of strength F_i the field becomes

$$p = \frac{1}{4\pi}\int_{v'}\frac{\partial}{\partial x_i'}\left[\frac{F_i(t - r/c)}{r}\right]dv' \tag{8.26}$$

or in the far field

$$p = \frac{1}{4\pi c}\int\frac{x_i - x_i'}{r^2}\frac{\partial}{\partial t}\left[F_i\left(t - \frac{r}{c}\right)\right]dv' \tag{8.27}$$

The dipole field (8.24) is obtained, for example, from an oscillating sphere having a velocity $U_1e^{-i\omega t}$ corresponding to the dipole strength $f_1 = -i\omega(\pi a^3\rho)U_1$, where a = sphere radius.

It follows from (8.24) that the dipole pressure field will contain one more factor of ω than do the simple-source fields from which it is made up. The total radiated acoustic power, being proportional to ω^2 for the simple source of a given flow strength, will therefore be proportional to ω^4 for the dipole.

The mounting of a loudspeaker or an oscillating piston in an infinite wall or in a closed box will in effect convert the radiator from a dipole to a simple source and hence improve the low-frequency efficiency.

The Quadrupole. In complete analogy with the derivation of the field from the force distribution, it follows that the field caused by the third term in (8.19) is that of a quadrupole distribution equal to the stress *tensor* T_{ij} [13]. The pressure distribution becomes

$$p = \frac{1}{4\pi}\int_{v'}\frac{\partial^2}{\partial x_i'\,\partial x_j'}\frac{T_{ij}(t - r/c)}{r}dv'$$

or in the far field

$$p = \frac{1}{4\pi c^2}\frac{x_ix_j}{r^3}\int_{v'}\frac{\partial^2}{\partial t^2}T_{ij}\left(t - \frac{r}{c}\right)dv' \tag{8.28}$$

It follows from (8.28) that the quadrupole pressure field will contain one more factor of ω than do the dipole fields from which it is composed. If the total acoustic power from the dipole is proportional to ω^4, it will be proportional to ω^6 for the quadrupole.

For a *longitudinal* quadrupole of total strength t_{11} concentrated at $r = 0$ and with both axes in the x_1 direction, the field is

$$p = \frac{1}{4\pi c^2}\frac{1}{r}\cos^2\theta\frac{\partial^2}{\partial t^2}t_{11}\left(t - \frac{r}{c}\right)$$
$$p = -\frac{\omega^2}{4\pi c^2}\frac{1}{r}\cos^2\theta\, t_{11}e^{ikr}e^{-i\omega t} \tag{8.29}$$

if $t_{ii} = t_{11}e^{-i\omega t}$, with the corresponding radiated total power

$$\overline{W} = \frac{\omega^4|t_{11}|^2}{40\pi\rho c^5} \tag{8.30}$$

if $t_{ii} = t_{11}e^{-i\omega t}$.

For a *lateral* quadrupole the expression for the pressure is the same except for a factor $\sin\theta$ replacing one of the factors $\cos\theta$ in (8.29). The total radiated power is

$$\overline{W} = \frac{\omega^4|t_{12}|^2}{120\pi\rho c^5} \tag{8.31}$$

if $t_{ij} = t_{12}e^{-i\omega t}$. Although the quadrupole source has little importance in general, it is the sole contribution in the generation of sound by turbulence in free space.

Sources of higher order can be built and superimposed correspondingly to represent (in a *multipole expansion*) the field from an arbitrary finite source.

8. Generation of Sound by Turbulent Flow [13, 14]

As shown by Lighthill [13], the generation of sound by turbulent flow is due to the (Reynolds) stress-tensor (quadrupole) source distribution discussed above, leading to a sound field given by (8.28). As far as the dependence on the flow parameters is concerned, the radiated power is of the form given by (8.30) or (8.31). Since $t_{ij} \sim \rho v^2l^3$ and $\omega \sim v/l$, it follows that

$$\text{Acoustic power from turbulence} \sim \rho v^8c^{-5}l^2 \tag{8.32}$$

In other words, since the rate of kinetic energy of the flow entering a region is $l^2\rho v^3$, it follows that the *efficiency of sound generation by turbulence goes as the fifth power of the Mach number $M = v/c$.*

9. Radiation from a Simple Source in a Moving Medium [4, 15, 16]

An important question concerns the effect of steady motion of the medium on the field distribution from a stationary simple source of sound located in free space. In the absence of motion of the medium the sound field will be spherically symmetrical, and the surfaces of constant phase will coincide with the surfaces of constant amplitude. Motion of the medium will split this coincidence.

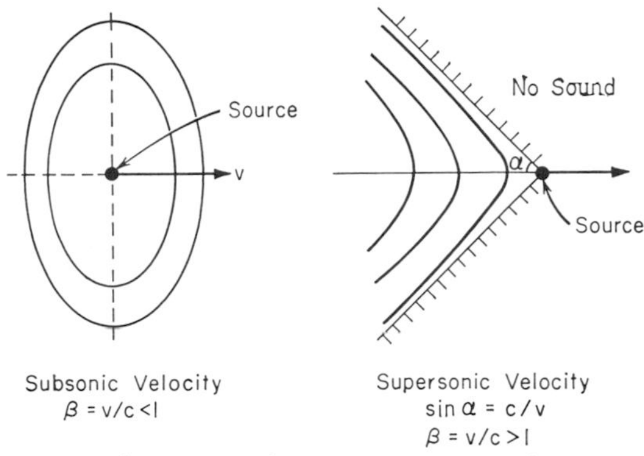

FIG. 8.2. Equal sound pressure contours from sound source in motion or for a stationary source in a moving medium. The contours are in both cases measured in a coordinate system attached to the source.

The field distribution of a stationary source in a moving medium measured in the stationary coordinate system is the same as that of a moving source in a stationary medium measured in the frame connected with the source. The field from a point source located at the origin of the stationary coordinate system xyz in which the medium moves with a constant velocity v in the direction of the x axis is

$$p(x,y,z,t) = \frac{\dot{q}(t - R/c)}{4\pi R^* \sqrt{1 - \beta^2}} \tag{8.33}$$

where $$R = \frac{-\beta x^* + R^*}{\sqrt{1 - \beta^2}}$$

$$R^{*2} = x^{*2} + y^2 + z^2 \qquad x^* = \frac{x}{\sqrt{1 - \beta^2}}$$

which can readily be seen to satisfy Eqs. (8.7).

The surfaces of *constant phase*, given by R = constant, are spheres of radius $R\sqrt{1 + \beta^2}$ with the origin at $x = R\beta$. This can be easily seen in an elementary way by calculating the time it takes for a pulse of sound to reach x, y, z. The surfaces of *constant sound pressure*, on the other hand, are given by R^* = constant, which corresponds to the ellipsoid $x^2/(1 - \beta^2) + y^2 + z^2$ = constant = R^{*2}, as pictured in Fig. 8.2. It is interesting to notice that the field is the same up and down wind and that the intensity is larger in a direction at right angles to the flow. Physically the decrease of sound pressure in the directions with and against the wind can be explained as follows. Down wind the space occupied by a pulse of energy of certain length is "stretched" out, and the energy density is correspondingly decreased. Up wind the wave has effectively to travel further to reach the point of observation, and the effect of spherical divergence will be comparatively larger.

10. Radiation from a Moving Sound Source [4, 15, 16]

Consider a simple source of strength q moving in an arbitrary path defined by the coordinates $X(t)$, $Y(t)$, and $Z(t)$ with respect to the stationary coordinate system xyz. In a way similar to that used in the derivation of the field from a moving charge, it can be shown that the sound-pressure field becomes

$$p(x,y,z,t) = \frac{\dot{q}(t - R/c)}{R(1 - [v_R]/c)4\pi} \tag{8.34}$$

where the distance R is found from the equation

$$f(R) = \left[x - X\left(t - \frac{R}{c}\right)\right]^2 + \left[y - Y\left(t - \frac{R}{c}\right)\right]^2 + \left[z - Z\left(t - \frac{R}{c}\right)\right]^2 - R^2 = 0 \tag{8.35}$$

and $[v_R]$ is the projection of the velocity at time $t - R/c$ upon R.

Subsonic Velocity. For velocities $v < c$ the equation for determination of R has only one root, so that at time t there is effectively only one point on the path of the source which contributes to the field at x, y, z. In the particular case of source motion in the x direction with a constant velocity v ($\beta = v/c$), the distance R becomes

$$R = \frac{\beta\xi^* + R^*}{\sqrt{1 - \beta^2}} \qquad R^{*2} = \xi^{*2} + y^2 + z^2$$

$$\xi^* = \frac{x - vt}{\sqrt{1 - \beta^2}} \tag{8.36}$$

and $$R\left(1 - \frac{[v_R]}{c}\right) = R^*\sqrt{1 - \beta^2}$$

In the coordinate system attached to the source $\xi = x - vt$, $\eta = y$, and $\zeta = z$, the sound-pressure field becomes $p(\xi,\eta,\zeta,t) = \dot{q}(t - R/c)/R^*4\pi\sqrt{1 - \beta^2}$ so that the surfaces of *constant phase* given by

$$t - \frac{R}{c} = \text{constant}$$

become circles with their centers displaced along the ξ axis (see Sec. 9), and the *surfaces of constant pressure* are the ellipsoids $\xi^2/(1 - \beta^2) + \eta^2 + \zeta^2$ = constant. Hence the sound pressure is higher in a direction at right angles to the direction of motion (see Fig. 8.2).

Supersonic Velocity. If we carry through a formal solution for the case of $v > c$, there are two solutions to Eq. (8.35), R_1 and R_2, and the pressure field becomes

$$p(x,y,z,t) = \frac{\dot{q}(t - R_1/c)}{4\pi R_1|1 - [v_{R_1}]/c|} + \frac{\dot{q}(t - R_2/c)}{4\pi R_2|1 - [v_{R_2}]/c|} \quad (8.37)$$

For a rectilinear motion in the x direction the two values of R are

$$R_{1,2} = \frac{\pm R^* - \beta\xi^*}{\beta^2 - 1} \qquad R^{*2} = \xi^{*2} - (y^2 + z^2)$$

$$\xi^* = \frac{x - vt}{\sqrt{\beta^2 - 1}} = \frac{\xi}{\sqrt{\beta^2 - 1}}$$

$$\text{and } R_1\left|1 - \frac{[v_{R_1}]}{c}\right| = R_2\left|1 - \frac{[v_{R_2}]}{c}\right| = R^*\sqrt{\beta^2 - 1}$$

Hence, the surfaces of constant pressure are, in the coordinate system $\xi\eta\zeta$ connected with the source, hyperboloids

$$R^{*2} = \text{constant} = \frac{\xi^2}{\beta^2 - 1} - \eta^2 - \zeta^2 > 0$$

as shown in Fig. 8.2. The limiting curve, for $R^* = 0$, corresponds to the so-called Mach cone

$$\eta^2 + \zeta^2 = \frac{\xi^2}{\beta^2 - 1}$$

with the half angle $\alpha = \sin^{-1}(c/v)$. In this idealized case with a point source of sound the pressure goes to infinity at the origin and hence along the whole Mach cone. In front of the cone there will be no sound.

For a sound source of finite size there will be complicated disturbance in the medium and a shock wave extending from the front (or any discontinuity) on the source. This will change the local properties of the medium, which will affect the sound propagation.

11. The Doppler Effect

When a sound source of frequency ω passes a stationary receiver, the sound observed will in general have a continuous spectrum. At any instance, however, frequency can be defined as the time ratio of change of the phase $\omega = (d/dt)(\omega t - R/c)$. The frequencies thus obtained in the case of subsonic and supersonic velocities are:

Subsonic Velocity.

$$\omega' = \omega\left[\frac{(1 + \beta\xi^*/R^*)}{(1 - \beta^2)}\right] \simeq \omega(1 + \beta\cos\theta)$$

where the distance between the source and observer is $(x - vt)^2 + y^2 + z^2$, $\xi^* = (x - vt)/\sqrt{1 - \beta^2}$, and $R^* = \xi^{*2} + y^2 + z^2$. When the observer is on the x axis in front of the source, $\omega' = \omega/(1 - \beta)$, and when behind the source, $\omega' = \omega/(1 + \beta)$.

Supersonic Velocity. In this case we have two instantaneous frequencies

$$|\omega'| = \omega\frac{\beta\xi^*/R^* - 1}{\beta^2 - 1} \qquad |\omega''| = \omega\frac{\beta\xi^*/R^* + 1}{\beta^2 - 1}$$

which reduce to

$$\omega' = \frac{\omega}{\beta + 1} \quad \text{and} \quad \omega'' = \frac{\omega}{\beta - 1}$$

when the observer is in the x axis. If $1 < \beta < 2$, both ω' and ω'' are less than ω.

12. Radiation and Scattering

In addition to the basic fields from point sources described above, a number of other radiation fields from finite sources are known. With a given velocity distribution on the surface of the source, the field is formally given by the Kirchhoff theorem in Eq. (8.10). Evaluation of the field leads often to great analytical difficulties; some important radiation problems solved in the literature are indicated below. Generally only the "far" field is of interest. However, to find the reactive part of the radiation impedance of the source, the field over the surface of the source must be evaluated. The radiation *resistance* can readily be found directly from the far field by calculation of the radiated power. Among the large number of papers in this field we find solutions of, for example, the following problems: pulsating and oscillating spheres and cylinders [7, p. 244], piston in a sphere or cylinder [7, p. 244], piston in an infinite wall in a medium at rest [7, p. 244] and in a moving medium [18], vibrating piston in free space [19], radiation from open end of a pipe [20].

The scattering problem is closely related to the radiation problem with the additional difficulty that the field distribution on the scatterer is not known a priori. The resulting integral equations must often be solved approximately. Work on scattering includes spheres and cylinders [7, p. 244], spherical aperture in an infinite screen [21], spherical disk [17], straight edge [22], resonators [23], absorbing strips [24], cylindrical vortex [25] using general equations in Sec. 2.

13. Technical Aspects of Sound Generation [5, 26]

The methods used in the generation of sound can essentially be divided into two groups: (1) conversion of electrical oscillations into mechanical ones, and (2) conversion of nonoscillatory mechanical energy (or heat) into oscillatory motion. As a third group one could perhaps specify (3) explosions, electric sparks, and similar effects.

In the first group the means of making the conversion is usually a *linear* system, an electromechanical transducer, whereas in the second group *nonlinear* mechanisms are essential. A sound source like a bell, in which the eigenoscillations are excited, requires for continuous operation an oscillatory driving force and can therefore be considered to belong to the sources in group 1.

In general it is impossible to cover the entire acoustic spectrum by one and the same transducer, and several different mechanisms have been developed for different frequency regions. In the *first group* are membrane and piston vibrators (loudspeakers) driven electromagnetically or electrostatically [27], thermophones [28], magnetostrictive vibrators [27], piezoelectric crystals [29], modulated electric-discharge sources (ionophone, corona). Among the mechanisms of conversion in the *second group* there are modulation of an air stream (rotating-disk siren [30], speech mechanism [31], relaxation oscillators [32] (violin type or corresponding longitudinal oscillations of a bar), vortex tones [33] (edge tones, propeller noise, whistles, musical instruments [33]), heat-maintained oscillators (Rijkes tube [34]).

TABLE 8.1

Frequency range	Generator	Applications
Infrasound, 0–25 cps	Explosions	Seismic exploration; sound ranging, upper-atmosphere research
	Electromagnetic vibrators	Vibration analysis of structures
Audible range, 16 cps–20 kcps	Diaphragm (electromagnetic or electrostatic)	Communication, theaters, public address systems, etc.
	Siren, diaphone	Signaling; smoke coagulation
	Thermophone	Absolute calibration
	Human voice	
	Musical instruments	
	Airplane propeller	Noise source for propagation studies
	Jet engine	
Ultrasound: 20–80 kcps.....	Air whistles	Underwater signaling
	Piezoelectric and magnetostrictive devices	Biological, chemical, and medical applications (destruction of bacteria, emulsification, diathermy)
80 kcps–1 Mcps	Piezoelectric devices (quartz, ADP, barium titanate crystals)	Flaw detection in structures, degassing, etc.
Entire range.....	Modulated electric discharge (ionophone)	Communications, etc.

Table 8.1 gives an idea of the frequency ranges in which the different sources are used and some of their applications. The *acoustic power* generated by the various sources varies over a large range. For example, the average human voice (ordinary speech level) generates about 200 μw of acoustic power, whereas for a jet engine it is of the order of 10 kw.

14. The Human Voice and Speech Mechanism

Physically the sound source represented by the human voice is of the siren type, in which a steady stream of air is modulated. The modulation results from the periodic contraction of the vocal chords. The frequency characteristics of the pulsating air stream on its exit from the mouth have been greatly modified and are largely determined by the transmission characteristics of a tube with constrictions. These characteristics depend essentially on the shape and the setting of the tongue and the mouth opening. For each vowel there is a particular cavity geometry with one or two characteristic resonance frequencies (formants) which carry a large part of the sound energy in that vowel. In the generation of voiceless consonants like *s* the vocal chords are not active, and the modulation of the air stream is now caused by turbulence created when the air is forced with high velocity between the teeth. The speech-power output in ordinary speech is of the order of 200 μw distributed in a frequency range between 100 to 7,000 cps with the major part around 500 to 1,000 cps.

Electrical-filter analogues of the vocal tract have been built by which speech sounds and even sentences can be produced by proper continuous variation of the inductances and the capacitances in the network [35].

PROPAGATION OF SOUND

15. Propagation of Sound in the Atmosphere [3, 4]

The atmosphere is in motion and is inhomogeneous, and the general theory of sound propagation under such conditions, based on the general equations (8.1) to (8.5), is little known. However, when the variation of the properties of the medium is small in a distance of a wavelength, the high-frequency approximation of these equations will give useful information.

a. Ray Acoustics [36, 37]. If a solution of the general equations of sound propagation (8.1) to (8.5) is sought as a series $p = p' + p''/ik + \cdots$ (and correspondingly for other sound variables) of descending powers of $k = \omega/c$, the solution p' in the limit of infinite frequency is the sound field which is referred to as that of ray or geometrical acoustics.

If the field quantities are written in the form $p = p_1 e^{ik_0\psi} e^{-i\omega t}$ (with corresponding expressions for δ, $\mathbf{u}$, and σ) where both p_1 and ψ are functions of the space coordinates, the first-order quantities p_1' satisfy a homogeneous set of linear equations from which result, as Blokhintzev has shown [3, 4]

$$\mathbf{u}_1' = \frac{\nabla\psi}{q}\frac{p_1'}{\rho} \tag{8.38}$$

and the eikonal equation

$$|\nabla\psi|^2 = \frac{q^2}{c^2} \tag{8.39}$$

where
$$q = c_0 - \mathbf{v}\cdot\nabla\psi \tag{8.40}$$

For $\mathbf{v} = 0$ these relations reduce to the well-known $u_1' = p_1'/\rho c$ and $\nabla\psi = c_0/c$, where c_0/c is the index of refraction of the medium. When $\mathbf{v} \neq 0$, the generalized *index of refraction* q/c will depend on the direction of propagation.

The Phase Velocity. Since the phase velocity of the sound wave is $\mathbf{c}_f = \mathbf{c}_0/(\partial\psi/\partial n)$, where n indicates the coordinate along the normal of the phase surface, it follows from Eqs. (8.38) and (8.40) that the phase velocity is the sum of the local sound velocity and the component of the medium velocity in the n direction.

$$\mathbf{c}_f = \mathbf{c} + \mathbf{v}_n$$

The Group Velocity. As will be shown below, the velocity of energy flow is the vector sum of the local sound velocity and the medium velocity

$$\mathbf{c}_g = c\mathbf{n} + \mathbf{v}$$

where $\mathbf{n}$ = unit vector in the direction $\nabla\psi$. Although the direction of the phase normal will *not* be affected by flow orthogonal to it, the ray evidently *will.*

The Energy-transport Equation. The equations (8.38) and (8.39) give no information about the sound pressure or intensity distribution in the field and should be accompanied by the energy-transport equation

$$\frac{\partial E}{\partial t} + \nabla \cdot (E\mathbf{c}_g) = 0$$

to complete in this respect the high-frequency or ray description of the field. The sound energy density is, using (8.38),

$$E = \tfrac{1}{2}(\rho_0 + \delta_1)(v + u_1) - \tfrac{1}{2}\rho v^2 + \frac{1}{2}\frac{p_1{}^2}{\rho c^2} = \frac{c_0}{q}\frac{p_1{}^2}{\rho c^2}$$

and the corresponding intensity vector, $\mathbf{I} = E\mathbf{c}_g$, becomes

$$\mathbf{I} = \left(p_1\mathbf{u}_1 + \frac{p_1{}^2}{\rho c^2}\mathbf{v}\right)\frac{c_0}{q} \qquad (8.41)$$

If the field is stationary, we get $E\mathbf{c}_g S$ = constant (S = area of ray tube). In other words, if the pressure is known at a location x_1, the pressure at an arbitrary point x in the ray tube is obtained from

$$(p_1{}^2 c_g S)_x = \frac{(\rho c^2 q)_x}{(\rho c^2 q)_{x_1}}(p_1{}^2 c_g S)_{x_1}$$

q cannot be evaluated until ψ has been determined from Eq. (8.39).

Vertical Temperature and Wind Gradient. This important case is met in propagation over ground when the temperature and wind velocity are both functions of the height y over ground. The equation for the phase is, from (8.39),

$$\psi_x{}^2 + \psi_y{}^2 = \frac{c^2}{c_0{}^2}\frac{1}{[1 - (v/c_0)\,\nabla\psi]^2} \simeq \frac{c^2}{c_0{}^2}\left(1 + \frac{v}{c_0}\nabla\psi\right)^2$$

Since for $v/c \ll 1$ the right-hand side is approximately a function of y only, $\partial\psi/\partial x$ = constant = $\cos\phi_0$, and

$$\cos\phi = \frac{\psi_x}{|\psi|} \simeq \cos\phi_0\frac{c}{c_0}\frac{1}{1 - v/c\cos\phi_0\cos\psi}$$

$$\simeq \cos\phi_0\left[\frac{c}{c_0} + \frac{v(y)}{c_0}\cos\phi_0\cos\gamma\right]$$

where γ = angle between wind and sound direction, ϕ_0 = elevation angle at origin ($y = 0$ where $c = c_0$), and ϕ = elevation angle of ray at height y [where $c = c(y)$]. Since the curvature of the ray is

$$\frac{1}{R} = -\frac{d\cos\phi}{dy}$$

$$\frac{1}{R} = -\frac{1}{c_0}\left(\frac{dc}{dy} + \frac{dv}{dy}\cos\phi_0\cos\gamma\right)\cos\phi_0$$

In other words, the curvature contributions from wind and temperature are approximately additive. The bending effect of the temperature gradient and the wind gradient cancel each other when

$$-\frac{dc}{dy} = \frac{dv}{dy}\cos\phi_0\cos\gamma$$

As an example, consider a temperature gradient equal to the adiabatic lapse rate $dT/dy = -1°\text{C}$ per 100 m. Then

$$\frac{dc}{dy} \simeq \left(\frac{c_0}{2T_0}\right)\frac{dT}{dy} \simeq 0.57\left(\frac{dT}{dy}\right)$$

$$\simeq -0.57 \text{ m/sec per 100 m}$$

In other words, a sound ray leaving the sound source horizontally ($\phi_0 = 0$) in the downwind direction ($\gamma = 0$) will remain horizontal if the wind gradient equals dv/dy = 0.57 m/sec per 100 m. The variation of the radius of curvature is usually quite small, and the ray path can often be considered to be a circle, at least in a limited region of space.

Shadow Zones [3, 4, 36, 37]. As a result of the wind and temperature gradients, the formation of shadow zones can occur, as illustrated in Fig. 8.3.

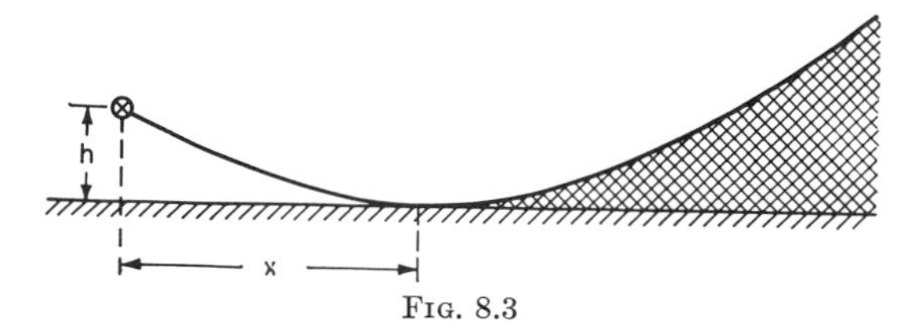

FIG. 8.3

The limiting ray defining the shadow boundary is touching the ground at a distance X from the sound source, which is located a distance h over ground. If the receiver is at the same height, the distance to the shadow zone at this level is $D = 2X$. If only a temperature gradient and no wind is present, the sound field will be symmetrical around the source, and if the temperature decreases with height (lapse rate), the sound rays will be bent upward, and a shadow is formed all around the source. However, with only a wind gradient in the field, the behavior of the sound ray will depend on the direction. Ordinarily the wind strength increases with height, and in that case the sound rays will be bent downward down wind and upward in the up-wind direction. At right angles to the wind there will be no refraction. The distance to the shadow zone is evidently

$$X = \int dx = \int_0^\phi R\cos\phi\,d\phi$$

For a constant gradient of temperature this distance becomes

$$X \simeq \left(\frac{2hc_0}{\alpha + \beta\cos\gamma}\right)^{1/2} \qquad (8.42)$$

where $\alpha = dc/dy$ = constant, $\beta = dv/dy$ = constant, c_0 = velocity of sound at ground $\simeq$ 340 m/sec. In practice the gradients vary with height, so that close to the ground, to a height of 5 m or so, the gradients are nearly inversely proportional to the height.

Therefore, instead of being proportional to $h^{1/2}$ according to (8.42), the distance to the shadow zone, distance X, will be closer to proportional to h [38].

Shadow formation is of great practical importance since it affects the intensity distribution around a source over ground considerably. On an ordinary summer day with a wind of, say, 4 m/sec the distance to the shadow boundary will be approximately 30 times the height of the source over ground. Typical examples of shadow formation around a point source under various weather and ground conditions are given in ref. 38.

b. The Intensity in the Shadow Zone [38]. An actual shadow, as in Fig. 8.3, in the literal sense of the word is obtained on the wave theory only in the limit of infinite frequencies when we have conditions of geometrical acoustics. At finite frequencies some sound energy will be diffracted into the shadow region. The frequency dependence f of the diffracted field intensity in the shadow is given [39] approximately by $e^{-\alpha f^{1/3} r}$, where r is the horizontal distance to the shadow boundary and α is a constant dependent on the gradient. However, the presence of turbulence in the atmosphere will scatter sound into the shadow zone with an intensity which, at least for high frequencies, is proportional to f^2. Hence the frequency dependence of the total intensity in the shadow zone will be $I_S \simeq e^{-\alpha f^{1/3} r} + \text{constant} \times f^2$. For a given position in the shadow there will evidently exist one frequency *at which the intensity is a minimum*, an effect which has been observed experimentally [38]. With a wind of about 4 m/sec this minimum intensity is of the order of 25 db below the level which would be obtained when no shadow is present.

c. Scattering of Sound by Turbulence. Using his general theory of sound generation by turbulence (see Sec. 8), Lighthill has calculated the scattered energy in a turbulent field [40] using the scattering quadrupoles $\mathbf{u}_i\mathbf{v}_j' + \mathbf{u}_j\mathbf{v}_i'$ in the integral (8.28) ($\mathbf{u}$ = particle velocity in the incident sound field, $\mathbf{v}'$ = fluctuating part of the flow velocity in the medium). Scattering is comparatively large when $k_1 = 2k \sin \theta/2$ ($k = \omega/c$, θ = angle between direction of propagation and direction of scattered intensity) is of the same order of magnitude as the wave number of the main energy-bearing eddies: (*i*) When the wavelength of sound λ is larger than the size of these eddies l, there will be little scattering. The maximum occurs in the backward direction $\theta = \pi$. (*ii*) When $\lambda \simeq l$, the scattering will be fairly uniformly distributed in all directions with a minimum in the forward direction. (*iii*) When $\lambda < l$, the scattering will be concentrated in the forward direction close to $\theta \simeq 0$, approaching $\theta = 0$ when the frequency increases. Under the conditions given in (*iii*) the total scattered energy from unit volume of the flow becomes

$$\text{Scattered power/unit volume} = 2k^2 L_1 M^2 I \qquad (8.43)$$

This corresponds to an attenuation of the sound wave per wavelength

$$\sigma\lambda = 4kL_1M^2 \qquad \text{nepers}$$

where L_1 = *macroscale* of turbulence, $k = \omega/c$, I = intensity of incident wave, $v_1'^2 = c^2M^2$ = mean square velocity fluctuation in the *direction* of propagation, and c = velocity of sound. The attenuation due to scattering will exceed the maximum possible molecular absorption $(\sigma\lambda)_m \simeq 0.002$ in air at 20°C if $(v_1'^2)^{1/2} \simeq 0.005c \simeq 1.7$ m/sec, a value often encountered.

In the general case, when no restrictions are put on k, the scattered energy can be evaluated in a simple form when the *turbulence is isotropic*. The scattered energy per unit volume is then [40]

$$W_s = \frac{\pi I k^2}{c^2} \int_0^{2k} \frac{E(k_1)}{k_1} F \frac{k_1}{k} dk_1$$

where $F(x) = (1 - \frac{1}{2}x^2)^2(1 - \frac{1}{4}x^2)$, $k_1 = 2k \sin \theta/2$, $E(k_1)$ = energy spectrum of turbulent energy with respect to wave number k_1, and θ = angle between directions of incident and scattered sound. This expression reduces to (8.43) when $k_1/k \ll 1$. For eddies with $k_1/k > 2$ no energy is scattered; approximately, eddies with $k_1 < k$ give their full contribution to the scattering $\pi I k^2 E(k_1)/c^2 k_1$ [86–88].

d. Ground Absorption. The effect of ground absorption on propagation is analogous to that in electromagnetic theory, and Sommerfeld's solution [41] of the electromagnetic-dipole problem has been carried over to the acoustic case [42]. Also, Weyl's [43] approach has been used [44] in the case of a ground impedance $\zeta\rho c$ independent of angle of incidence, which simplifies the analysis considerably. In the far field the pressure becomes

$$p \simeq \text{const} \left(\frac{|\zeta|}{r}\right)^2 \qquad \left|\frac{\zeta}{r}\right| \ll 1 \qquad \frac{h}{r} \ll 1$$

where h = height of source over ground, i.e., intensity decreases with the fourth power of distance.

16. Propagation in Tubes [7, p. 233]

a. The Plane Wave. Measurement of Acoustic Impedance. The propagation of sound in tubes has wide application in all branches of acoustics and deserves some attention. Ordinarily one is interested only in the plane-wave component, since many experiments performed in tubes are based on the assumption of a plane wave only. The most common of these is the measurement of the acoustic impedance. The material under test is placed as a termination of a tube and exposed to a plane wave. From the location and the maximum and minimum values of the pressure in the standing wave the impedance of the sample can be evaluated.

If the plane wave incident on the sample (at $x = l$ and with impedance z_b) is $p = P_1 e^{ikx} e^{-i\omega t}$, a wave field is set up characterized by the following set of useful relations:

Sound Pressure

$$p = 2P_1 e^{-\psi} \sinh [ik(x - l) + \psi]$$

$$\text{where} \quad \tanh \psi = \frac{z_b}{\rho c} \qquad \psi + i2x/\lambda = \pi(\alpha - i\beta)$$

$$p = 2P_1 e^{-\pi\alpha}(\cosh^2 \pi\alpha - \cos^2 \pi\beta)^{1/2} \qquad (8.44)$$

$$\frac{p_{\max}}{p_{\min}} = \coth \pi\alpha = n \qquad (8.45)$$

Velocity Magnitude

$$|u| = \frac{2P_1}{\rho c} e^{-\pi\alpha} (\cosh^2 \pi\alpha - \sin^2 \pi\beta)^{1/2}$$

Impedance

$$z(x) = \rho c \tanh [ik(x - l) + \psi] = |z|e^{-i\phi}$$

$$\frac{|z(x)|}{\rho c} = \left(\frac{\cosh^2 \pi\alpha - \cos^2 \beta\pi}{\cosh^2 \pi\alpha - \sin^2 \beta\pi}\right)^{1/2}$$

$$\phi = \arg z(x) = -\tan^{-1}\left[\frac{tg\pi\beta}{\tanh \pi\alpha}\left(\frac{1 - \tanh^2 \pi\alpha}{1 - tg^2\pi\beta}\right)\right]$$

Impedance at pressure maximum

$$\zeta_1 = \frac{z_1}{\rho c} = \frac{1}{n} = \coth \pi\alpha$$

Impedance at pressure minimum

$$\zeta_2 = \frac{z_2}{\rho c} = n = \tanh \pi\alpha$$

Distance d to the mth Pressure Minimum from Sample

$$d = (m - \beta)\frac{\lambda}{2} \tag{8.46}$$

It follows from (8.45) and (8.46) that measurement of p_{max}/p_{min} and d determines α and β and hence the impedance z_b of the sample. Tables and graphs of the transformation (8.45) are available [7, p. 233].

b. Higher-order Modes. In addition to the plane-wave component in the tube, there are higher-order modes, which are analogous to the electromagnetic waves in a waveguide. For each of these modes the tube acts as a high-pass filter, allowing propagation above a certain "cutoff" frequency and causing attenuation below this frequency. Although most acoustic measurements are made at frequencies at which the higher modes cannot propagate, there are many cases where the higher-order modes play an important role, e.g., in problems dealing with their propagation characteristics [45], discontinuities in tubes [46], radiation of sound into tubes [47], impedance measurements with higher modes [4], and attenuation [48].

c. Lined Ducts. The propagation of sound in ducts lined with various kinds of sound-absorptive materials is of particular importance in noise abatement and has received considerable attention in the literature, ranging from empirical studies [49], low-frequency approximations [50, 51], wave theory on the assumption of point-reacting boundary [52] to rigorous treatment in the case of a porous boundary [51, 53].

Low-frequency Approximation. At low frequencies when the pressure can be assumed uniform across the duct, the problem is completely analogous to that of propagation of electromagnetic waves on a "lossy" cable. The absorptive boundary makes the average compressibility of the air in the duct complex, which results in a complex propagation constant,

$$k_1 \simeq k\left(1 + \frac{i\eta}{kL}\right)^{1/2} = \epsilon + i\sigma \qquad kL \ll 1$$

where $k = \omega/c$, L = duct area/duct perimeter, $\eta/\rho c$ = average boundary admittance. In the special case of a porous lining of thickness d, the corresponding attenuation constant increases with the square of the frequency

$$\sigma L \simeq 1.5 \frac{rd}{\rho c}(kd)^2 \qquad \text{db}$$

where rd = total flow resistance.

High-frequency Approximation. The attenuation decreases with frequency when $kL \gg 1$ because of the tendency of the wave to recede from the boundary. For a duct of square cross section, the attenuation becomes, on the assumption of a point-reacting boundary with specific impedance

$$\rho c = (\theta - i\chi)\rho c$$
$$\sigma a \simeq 4(n + 1)^2\pi^2\theta(ka)^{-2} \qquad \text{nepers}$$

The rigorous expression for σa in the case of a porous boundary is

$$\sigma a \simeq 4(n + 1)^2\pi^2\left(\frac{rd}{\rho c}\right)^{-1/2}(ka)^{-3/2} \qquad \text{nepers}$$

where n = mode number, $n = 1$ is fundamental mode, a = cross dimension.

17. Propagation of Large-amplitude Waves [54]

Sound waves of large amplitude break into shock after a comparatively short distance of travel and obtain then a stable saw-toothed shape. The reason for this deformation of waveform is, at least in part, that the crest of the sound wave always has a higher temperature than the trough and thus travels faster than the trough. By simple arguments along these lines, the distance of travel after which shock occurs is [55]

$$X = \frac{P_0}{\Delta p}\frac{\lambda}{2\gamma + 1} \qquad \gamma = \frac{c_p}{c_v}$$

For example, with $\Delta p = 0.05P_0$ ($\simeq$165 db) the *shock distance* is only 5.8λ. Since there is an entropy change $\Delta S \simeq (c_p - c_v)(\Delta p/P_0)^3(\gamma + 1)/12\gamma^2$ across the discontinuity in the wave, the attenuation will be different from that of an infinitesimal isentropic sinusoidal wave. The attenuation of the saw-tooth wave or repeated shock wave is

$$\sigma\lambda = \frac{1}{E}\frac{dE}{dx}\lambda = -\frac{\gamma + 1}{\gamma}\frac{\Delta p}{p_0}$$

where E is the mechanical energy of vibration per wavelength. In terms of the pressure amplitude of the wave, the corresponding expression becomes

$$\frac{(\Delta p)_{x_0}}{(\Delta p)_x} = 1 + \frac{(\Delta p)_{x_0}}{p_0}\frac{\gamma + 1}{2\gamma}\frac{x - x_0}{\lambda}$$

The rate of attenuation is proportional to the pressure amplitude. With an original amplitude of the saw-tooth wave $\Delta p = 0.05p_0$, i.e., 165 db, the attenuation in a distance of 10 wavelengths is approximately 3 db.

18. Acoustic Streaming

Large-intensity sound waves set up steady flow in the field because of the effect of viscosity and the nonlinearity properties of the wave. (The loss of momentum of the wave motion due to dissipation is taken up by the medium.) Striking circulating streaming patterns have been observed around cylinders, apertures, etc., and even oscillatory jet formations at very large amplitudes result [56]. The streaming near obstacles has been found to change direction when the incident sound intensity exceeds a certain value. These effects have, at least in part, been explained theoretically [57]. In measurements of radiation pressure the streaming effect is often disturbing, but it can be eliminated, e.g., by means of sound-transparent membranes.

ABSORPTION OF SOUND

Sound is absorbed and transformed into heat, owing to viscosity, heat conduction, and *molecular* absorption or through interaction with elastic bodies in the field (panels, bubbles in water, etc.). Often absorption of sound is intentional, so that highly absorptive materials are desired. We consider the most common of these materials and later the "unavoidable" absorption of sound.

19. Absorption Materials

The different types of common absorption materials are porous structures, cavity resonators, and thin panels. The first of these is particularly efficient at high frequencies, whereas the other two have high absorption around their resonance frequencies, which usually fall in the low-frequency region (below 500 cps). The absorption material is usually applied on boundaries but is also used in the form of *unit (functional) absorbers* placed in the medium.

a. Porous Materials [7, p. 388; 8; 58; 59]. The acoustic properties of a homogeneous porous material with a rigid structure can be described in terms of three parameters: the *flow resistance* r per unit length, the *porosity* P, and the so-called *structure factor* m. The latter factor accounts for the effective increase of the dynamic mass of the air in the porous material, which is caused by at least four different mechanisms; they have been discussed in some detail by Zwikker and Kosten [58] and by Cremer [59]. In terms of these quantities the propagation constant and the characteristic impedance of the porous material are, respectively,

$$k_1 = k\left[\left(m + \frac{ir}{\omega\rho}\right)P\right]^{1/2}$$
$$Z_1 = \rho c\left(\frac{m + ir/\omega\rho}{P}\right)^{1/2} \qquad (8.47)$$

assuming the adiabatic compressibility of the air in the material. At very low frequencies the isothermal value should be used, which in effect can be accounted for by replacing P by $1.4P$. For intermediate conditions the compressibility of the air is complex, leading to additional sound attenuation [59].

Impedance, Penetration Depth, and Absorption. If a plane wave strikes a plane boundary of an infinitely thick porous layer, it will be refracted into the material with an angle of refraction ϕ_r given by $\cos\phi_r = [1 - (k/k_1)^2 \sin^2\phi]^{1/2}$. This angle will be complex except at normal incidence $\phi = 0$, which means that the surfaces of constant phase and constant amplitude do not coincide in the material. The amplitude surfaces are parallel to the boundary, whereas the phase surfaces, from being parallel to the boundary at low frequencies, penetrate the material unchanged in the direction of the incident wave at high frequencies. The normal impedance z of the boundary will then be dependent on the angle of incidence; in fact,

$$z = \frac{Z}{\cos\phi_r} \qquad (8.48)$$

However, at sufficiently low frequencies, when $k_1 \gg k$, the angular dependence is negligible, and the boundary becomes point-reacting. On the other hand, at high frequencies $z \simeq Z/\cos\phi$, and the assumption of angular independence of z may lead to marked errors. For example, the high-frequency value of the attenuation in a duct with a porous boundary becomes proportional to f^{-2}, whereas the correct value is $f^{-3/2}$ (see Sec. 16).

The pressure of the refracted wave will be reduced by a factor e^{-1} after a distance of travel in the material, the *penetration depth,* which equals $2\rho c/r$ and $(2\rho c^2/\omega r)^{1/2}$ at high and low frequencies ($r/\omega\rho \ll 1$ and $r/\omega\rho \gg 1$), respectively. A layer of finite thickness exceeding the penetration depth will therefore have essentially the same boundary impedance and absorption coefficient as an infinite layer. However, the penetration depth varies with frequency, and the absorption coefficient of a layer of finite thickness will vary correspondingly. In Fig. 8.4 is shown the frequency dependence of the average absorption coefficient $\bar{\alpha}$ (see Sec. 4) of a porous layer backed by a rigid wall for some different values of the total flow resistance rd. A value of $rd \simeq 6\rho c$ is ordinarily the most satisfactory. The normal impedance of the layer is given in Table 8.2.

The Effect of Air Space. In order to obtain a relatively high absorption at low frequencies by means of a porous layer with a given total flow resistance, the absorption material is often placed a certain distance out from the wall. The resulting effect on α is noticeable only if the porous layer thickness is smaller than the penetration depth, and is illustrated in simple cases in Table 8.2, which contains the impedances and low-frequency absorption coefficients of (*i*) a homogeneous porous layer of thickness d backed by a hard wall, (*ii*) a thin layer with the same total flow resistance placed at a distance d from the wall, and (*iii*) the same as in (*ii*) but with the air space partitioned. The partitions eliminate the strong angular dependence of z, which is caused by the air layer in (*ii*). In the low-frequency region considered here it is evident that for a given flow resistance R the structure (*iii*) gives the best absorption.

The Effect of Perforated Facing [61]. A perforated hard panel is often placed in front of the porous material. The perforations cause a local constriction

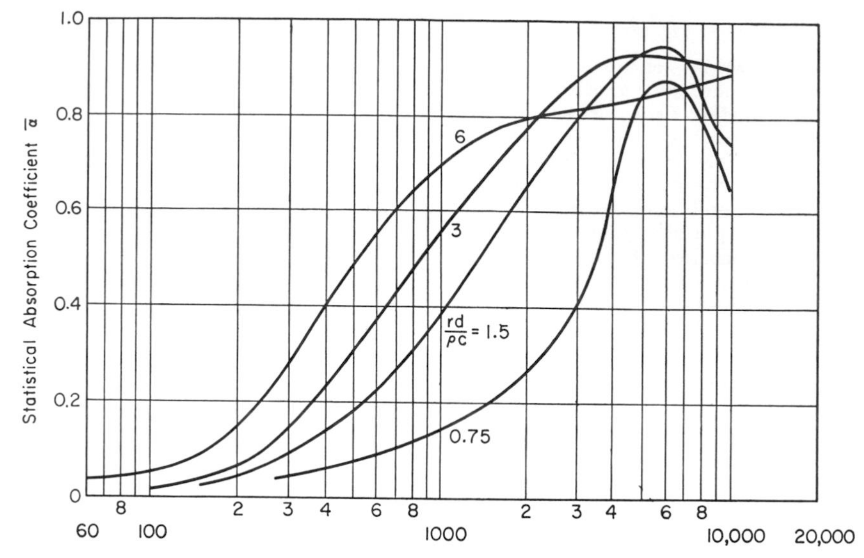

FIG. 8.4

of the flow (near field) that ordinarily extends only about one perforation diameter out from the facing. Acoustically the effect of the facing is an additional impedance, a mass reactance plus a resistance, the former making the system *resonant* (for calculation of the resonance frequency see Sec. 19*b*). The resistive component r_f caused by the near-field dissipation in the porous material is often about twice as large as the resistive component of the porous layer alone. However, r_f decreases rapidly to zero when the separation between the facing and the porous material increases to about one perforation diameter. The effect of a perforated facing on the absorption characteristics of a porous layer is shown in a special case in Fig. 8.5 [61].

Perforated Porous Tiles [62, 63]. Most commercial porous tiles are made of partly perforated, comparatively dense material ($rd/\rho c \simeq 30$) with a surface coating of protective impervious paint. The depth of the perforations is ordinarily three-fourths of the thickness of the tile. The impedance of this material can be calculated by considering the perforation as a lined duct (see Sec. 16) terminated by a cavity filled with porous material. In Fig. 8.6 is shown the

TABLE 8.2

	Structure	Normal impedance	Low-frequency absorption coefficient ($kd \ll 1$)
(*i*)	d	$z = \dfrac{Zi \cot (k_1 d \cos \phi_r)}{\cos \phi_r}$ $\cos \phi_r = [1 - (k/k_1)^2 \sin^2 \phi]^{1/2}$	$\alpha \simeq \dfrac{4}{3} \dfrac{rd}{\rho c} (kd)^2/\cos \phi$ $\bar{\alpha} \simeq \dfrac{8}{3} \dfrac{rd}{\rho c} (kd)^2$
(*ii*)	d	$\dfrac{z}{\rho c} = \dfrac{R}{\rho c} + i \dfrac{\cot (kd \cos \phi)}{\cos \phi}$	$\alpha \simeq 4 \dfrac{R}{\rho c} (kd)^2 \cos^3 \phi$ $\bar{\alpha} \simeq \dfrac{8}{5} \dfrac{R}{\rho c} (kd)^2$
(*iii*)	d	$\dfrac{z}{\rho c} = \dfrac{R}{\rho c} + i \cot kd$ ($R = rd$; ϕ = angle of incidence)	$\alpha \simeq 4 \dfrac{R}{\rho c} (kd)^2/\cos \phi$ $\bar{\alpha} \simeq 8 \dfrac{R}{\rho c} (kd)^2$

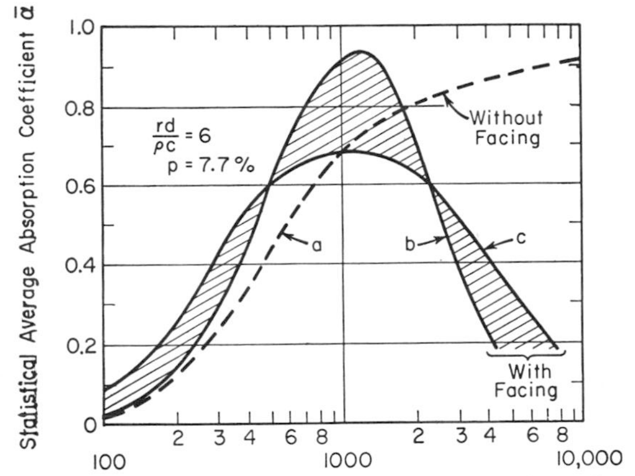

Fig. 8.5. Statistical average absorption coefficient of a one-inch porous layer backed by a rigid wall: (*a*) without perforated facing; (*b*) with perforated facing a distance of about one perforation diameter from the porous layer; (*c*) with perforated facing in close contact with porous layer. (Facing: thickness $\frac{3}{16}$ in., perforation diameter $\frac{5}{32}$ in., holes $\frac{1}{2}$ in. apart.)

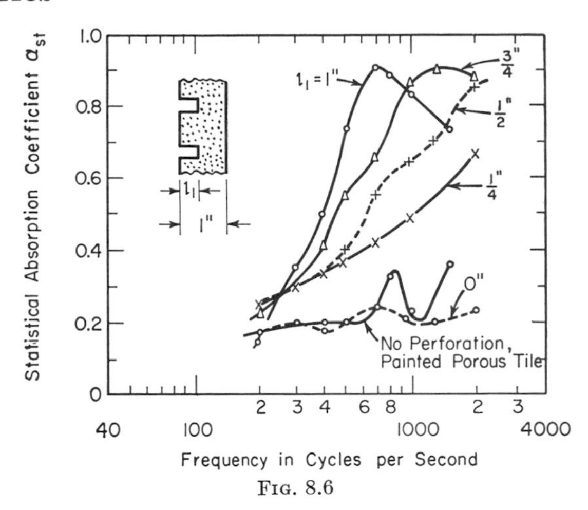

Fig. 8.6

absorption coefficient $\bar{\alpha}$ for different values of the perforation depth in a special case.

Nonrigid Structure [60]. So far the porous structure has been assumed rigid. However, in general the structure will also vibrate, in particular when the material is soft like sponge rubber. The structure wave and the air wave will be coupled through the action of viscosity. The resulting impedance characteristics, both for open and membrane-coated surfaces, have been studied by Kosten and Zwikker [60].

b. Absorption and Scattering by Helmholtz Resonators [58, 25]. A straight tube of length l, open in one end and closed in the other, has a lowest resonance wavelength $\lambda = 4l$. By constricting the open end of the tube, e.g., by adding a plate containing an aperture, the *resonance frequency* will decrease to the value

$$f_0 = \frac{c}{2\pi}\left[\frac{A}{V(t+\delta)}\right]^{1/2} \tag{8.49}$$

in which A = area of aperture, V = cavity volume, t = length of the aperture neck, and δ = the *mass end correction* of the aperture. When the aperture is small compared to the cavity dimensions, δ can be set approximately equal to that of an aperture in an infinite wall $\delta_0 = (16/3\pi)r_0 \simeq 1.7r_0$, where r_0 is the radius of the circular aperture. For an arbitrary aperture one can use $\delta \simeq 0.96A^{1/2}$. The actual end correction is generally smaller, and the calculated resonance frequency based on the above information will be somewhat too large.

A resonator of this constricted type is commonly called a Helmholtz resonator, a classical element in

Table 8.3

A_0 = aperture area $\quad V$ = resonator volume $\quad k_0 = \frac{\omega_0}{c} = \frac{2\pi}{\lambda_0}$

Resonator in Free Field		
Total specific aperture impedance	$\zeta = \theta_i + \theta_r - ik_0(t+\delta)$	in ρc units
	θ_i (see Eq. 8.50)	
	θ_r = radiation resistance $= \frac{A_0\pi}{\lambda_0^2}$	
Absorption cross section	$\sigma_a = \frac{A_0\theta_i}{\lvert\zeta\rvert^2}$	
At resonance	$\sigma_a' = \frac{4x^2}{(1+x^2)^2}(\sigma_a)_{\max}$	$x = \frac{V}{V_1}$
Maximum value of σ_a'	$(\sigma_a)_{\max} = \frac{\lambda_0^2}{4\pi}$	
Corresponding volume	$V_1 \simeq 0.0038 f_0^{1/4}\lambda_0^3$	$\lambda_0 = \frac{c}{f_0}$ where f_0 in cps
Scattering cross section	$\sigma_s = \frac{A_0\theta_r}{\lvert\zeta\rvert^2}$	
At resonance	$\sigma_s' = \sigma_a'$ (see above)	
Q value	$Q = 2\frac{x}{1+x^2}Q_{\max}$	$x = \frac{V}{V_1}$
Maximum value	$Q_{\max} \simeq 270 f_0^{-1/4}$	for $V = V_1$ (see above)
Resonator in an Infinite Wall		
Same relations as above but with	$(\theta_r)_{\text{wall}} = 2\theta_r = \frac{2A_0\pi}{\lambda_0^2}$	
	$[(\sigma_a)_{\max}]_{\text{wall}} = 2\sigma_{\max} = \frac{\lambda_0^2}{2\pi}$	
	$(V_1)_{\text{wall}} = \frac{1}{2}V_1$	V_1 (see above)

TABLE 8.4

μ = kinematic viscosity $\qquad \nu = \dfrac{K_h}{\rho c_p}$ (K_h = heat-conduction coefficient)

f = frequency $\qquad k = \dfrac{\omega}{c} \qquad c_i$ = isothermal velocity $\qquad \gamma = \dfrac{c_p}{c_v}$

Plane-wave attenuation, free space	$\sigma \simeq \dfrac{[\frac{2}{3}\mu + \frac{1}{2}(\gamma - 1)\nu]\omega^2}{c^3}$
	$\sigma \simeq 1.3 \times 10^{-13} f^2$ nepers/cm (air)
	$\sigma \simeq 8.1 \times 10^{-17} f^2$ nepers/cm (water)
Thermal boundary layer:	
Temperature distribution	$\vartheta = \vartheta_0(1 - e^{-(1-i)y/l_\nu})e^{-i\omega t}$
Thickness	$l_\nu = \left(\dfrac{2\nu}{\omega}\right)^{1/2} \simeq 1.16 l_\mu \simeq 0.25 f^{-1/2}$ cm
Compressibility	$K = \dfrac{1}{P_0\gamma}[1 + (\gamma - 1)e^{-(1-i)y/l_\nu}]$ $\quad y$ = distance from boundary
Viscous boundary layer:	
Velocity distribution	$u' = u_0(1 - e^{-(1-i)y/l_\mu})e^{-i\omega t}$
Thickness	$l_\mu = \left(\dfrac{2\mu}{\omega}\right)^{1/2} \simeq 0.22 f^{-1/2}$ cm
Thick tube (area A, perimeter S, $L = A/S$):	
Propagation constant (plane wave)	$k_1 = k\left[1 + (1 + i)\dfrac{l_t}{4L}\right]$
	$l_t = l_\mu + (\gamma - 1)l_\nu \simeq 1.46 l_\mu \simeq 0.32 f^{-1/2}$ cm
Attenuation constant	$\sigma = \dfrac{kl_t}{4L} \simeq \dfrac{1.47}{L} 10^{-5} f^{1/2}$ nepers/cm
Phase velocity	$c = c_0\left(1 - \dfrac{l_t}{4L}\right) \simeq c_0(1 - 0.078 f^{-1/2})$
Narrow channel:	
Flow resistance (p/u)	$r_i \simeq \dfrac{8\mu\rho}{a^2}$ $\quad$ circular tube, radius a
	$r_i \simeq \dfrac{12\mu\rho}{b^2}$ $\quad$ plane parallel plates, b apart when $f < 250 r_i$
	$r_i \simeq \dfrac{2R_s}{a}$ $\quad f_f > 250 r_i$
Propagation constant	$k_1 \simeq k\left(\frac{4}{3} + i\dfrac{r_i}{\omega\rho}\right)^{1/2} \simeq \dfrac{\omega}{c_i}(1 + i)\left(\dfrac{r_i}{2\omega\rho}\right)^{1/2}$
Attenuation constant	$\sigma \simeq \sqrt{8}\,\dfrac{kl_\mu}{a} \simeq 1.1 \times 10^{-4}\dfrac{f^{1/2}}{a}$ $\quad$ circular tube
	$\sigma \simeq \sqrt{12}\,\dfrac{kl_\mu}{b} \simeq 1.4 \times 10^{-4}\dfrac{f^{1/2}}{b}$ $\quad$ plane parallel plates
Hard plane:	
Admittance	$\eta = \dfrac{1 + i}{2\cos\phi}[(\gamma - 1)kl_\nu + kl_\mu \sin^2\phi]$
Absorption coefficient	$\alpha(\phi) = \dfrac{1}{\cos\phi}[(\gamma - 1)kl_\nu + kl_\mu \sin^2\phi]$
Statistical average	$\bar{\alpha} \simeq 1.8 \times 10^{-4} f^{1/2}$ $\quad \phi$ = angle of incidence
Approximate formulas for calculating losses at an arbitrary boundary:	
Surface resistance	$R_s = \dfrac{\rho c k l_u}{2} \simeq 0.83 \times 10^{-3} f^{1/2}$ cgs units
Heat-loss factor	$H = \dfrac{\gamma - 1}{2\rho c} kl_\nu \simeq 10^{-6} f^{1/2}$ cgs units
Boundary dissipation	$W = \frac{1}{2}R_s\int \lvert u_s\rvert^2\,dS + \frac{1}{2}H\int \lvert p\rvert^2\,dS$
	u_s = tangential velocity from loss-free solution
	p = pressure at boundary

acoustics which has been studied extensively ever since Helmholtz. It is of importance for absorption of low-frequency sound and in various other applications.

The *damping of the resonator* is mainly caused by the viscous losses in the region around the aperture. The effect of heat conduction is generally negligible. It can be shown [25] that for a small circular aperture the viscous dissipation corresponds to a specific aperture resistance

$$\theta_i\rho c = \frac{2R_s}{r_0}(t + 2r_0) = 4R_s\left(1 + 0.5\frac{t}{r_0}\right) \tag{8.50}$$

where R_s = viscous *surface resistance* (see Table 8.3). It is implied that r_0 must be larger than the viscous boundary-layer thickness (see Table 8.4).

Absorption and Scattering Cross Sections, etc. [25]. Most properties of an acoustic resonator are well demonstrated by its effect on a plane wave in free field. Regardless of the shape of the resonator, the rates of energy absorbed and scattered by the resonator at sufficiently low frequencies, i.e., wavelength much larger than the dimensions of the resonator, are, respectively,

$$W_a = \frac{I\theta_i A}{|\zeta|^2} \qquad W_s = \frac{I\theta_r A}{|\zeta|^2} \tag{8.51}$$

where I is the incident intensity. The corresponding

absorption and scattering cross section σ_a and σ_s, defined by $W_a = I\sigma_a$ and $W_s = I\sigma_s$, are given in Table 8.3 together with some other useful relations.

A design of a resonator for maximum resonance absorption cross section necessarily means a very high Q value, of the order of 40 to 50, if the only damping in the resonator is that due to viscosity of the boundaries. Therefore, in most applications it is desirable to introduce some additional damping in the resonator. This can be accomplished by means of a thin layer of porous material across the aperture. The Q value can then be brought down to favorable values without sacrificing on the maximum absorption. However, the lower the Q value, the larger the volume must be in order to retain the maximum resonance absorption cross sections.

20. Sound Attenuation [88]

The effect of viscosity and heat conduction in a sound field is in general negligible, except near boundaries and in the study of effects (circulations, etc.) whose existence depends on the presence of losses in the medium. Kirchhoff [64], in his investigations in this field, transformed the general equations of sound, including viscosity and heat conduction, into the form

$$\vartheta = (\gamma - 1)T_0(A_1Q_1 + A_2Q_2) = \text{sound temperature} \tag{8.52}$$

$$\mathbf{u} = A\mathbf{u}' + A_1\left(\frac{i\omega}{k_1^2} - \gamma\nu\right)\nabla Q_1 + A_2\left(\frac{i\omega}{k_2^2} - \gamma\nu\right)\nabla Q_2 \tag{8.53}$$

$$\nabla^2 Q_1 + k_1^2 Q_1 = 0 \qquad \nabla^2 Q_2 + k_2^2 Q_2 = 0 \tag{8.54}$$

$$\mu\,\nabla^2\mathbf{u}' = -i\omega\mathbf{u}' \qquad \nabla\cdot\mathbf{u}' = 0 \tag{8.55}$$

in which

$$k_{1,2}^2 = -\frac{B}{2D}\left[1 \mp \left(1 + \frac{4\omega^2 D}{B^2}\right)^{1/2}\right]$$

$$B = -[c^2 - i\omega(\tfrac{4}{3}\mu + \gamma\nu)]$$

$$D = \frac{\nu}{-i\omega}(c^2 - i\omega\gamma\tfrac{4}{3}\mu)$$

where A, A_1, A_2 = arbitrary constants (for values of μ, ν see Table 8.4) and $\gamma = C_p/C_v$.
At low frequencies $kl_\nu \ll 1$,

$$k_1 \simeq \frac{\omega}{c} + i[\tfrac{2}{3}\mu + \tfrac{1}{2}(\gamma - 1)\nu]\frac{\omega^2}{c^3} \qquad kl_\nu \ll 1 \tag{8.56}$$

$$k_2 \simeq (1 + i)\frac{\omega}{2\nu} = \frac{1 + i}{l_\nu} \qquad l_\nu = \left(\frac{2\nu}{\omega}\right)^{1/2} \tag{8.57}$$

An approximate expression for k_1 can obviously be obtained also when $kl_\nu \gg 1$, but this is *physically meaningless* since the wavelength corresponding to $kl_\nu \simeq 1$ is of the order of the *mean free path* of the gas.

The disturbances Q_2 and u' due to heat conduction and viscosity can be neglected except close to boundaries, where they form a *boundary layer* of thickness of the order of l_ν (see Table 8.4). Superposing these disturbances on the loss-free solution and using (8.52) and (8.53) for the boundary conditions $\vartheta = 0$, $u = 0$, approximate solutions to (8.52) to (8.55) can be readily found [59].

Table 8.4 summarizes some useful results which follow from the general relations above.

Molecular Absorption. More important than the attenuation due to viscosity and heat conduction is the molecular absorption which occurs in polyatomic gases. In simple terms the molecular absorption stems from the difference in the time (the relaxation time) required for the translational and the rotational energy of the molecules to reach equilibrium conditions. The corresponding two components of the specific heat will be out of phase in a periodic change of state. This results in a complex compressibility and hence losses.

Experimental observations by Knudsen [66] and others were essentially explained on this basis by Kneser [67]. He found the following frequency dependence of the absorption coefficient:

$$\sigma_m = 2\pi\frac{\gamma - 1}{\gamma}\frac{c_{vr}}{c_v}\frac{\omega\tau}{1 + (\omega\tau)^2}$$

where c_{vr} = rotational specific heat, c_v = specific heat (translation), and τ = relaxation time. The corresponding frequency dependence of the velocity of propagation is

$$c = c_0\left[1 + \frac{\gamma - 1}{2\gamma}\frac{c_{vr}}{c_v}\frac{(\omega\tau)^2}{1 + (\omega\tau)^2}\right]$$

For oxygen Kneser found from experiments $c_{vr}/c_v \simeq 0.012$. The maximum value of the absorption coefficient of air is then estimated to be

$$\sigma_{\max} = \frac{\pi}{5}\frac{\gamma - 1}{\gamma}0.012 \simeq 2 \times 10^{-3}$$

where the factor ⅕ accounts for the fact that only one-fifth of the air molecules are oxygen.

For pure oxygen the relaxation time is of the order of several seconds, and the effects of molecular absorption are negligible. However, the presence of water vapor reduces the relaxation time considerably, so that in air under ordinary conditions the relaxation time is 10^{-3} sec or less. The relation between the relaxation time τ and the relative humidity h per cent in air can approximately be written

$$\frac{1}{2\pi\tau} = f_{\max} \simeq 30h^2$$

For example, with $\phi = 18$ per cent we get $f_{\max} \simeq$ 10,000 cps. Later measurements by Knudsen [68] are well represented by the empirical relation [59].

$$\sigma \simeq \left(\frac{f}{1{,}000}\right)^{3/2}\frac{0.28}{20 + \phi} \qquad \text{db/m}$$

where $h > 30$.

DETECTION OF SOUND AND BASIC MEASUREMENTS [5, 82]

21. Microphones

Each of the quantities sound pressure, particle velocity, displacement, and density and temperature

TABLE 8.5

Mechanism	Frequency range	Approximate open-circuit sensitivity, db, below 1 volt/dynes/cm²	Approximate impedance, ohms	Application and remarks
Carbon microphone....	Up to 5 kcps	−40	100	Telephones
Electrodynamic........	Up to 20 kcps	−85	10	Field measurements, communications, etc.
Electrostatic...........	Up to 50 kcps	−50	500,000	Precision measurements, standards
Piezoelectric:				
Rochelle............	Audio region	−50	100,000	Temperature-dependent
	Ultrasonic	....	....	Hygroscopic
ADP crystal.........	....	−50	....	Use in underwater sound, solids, etc.
Quartz..............	Mainly ultrasonic	−90−−100	High	
Barium titanate.....	....	−90	Low	High-intensity work in air; water; measurements with small probes
Magnetostrictive......	Mainly ultrasonic	−100	Low	Underwater sound
Ribbon microphone....	Audio region	−100	1	Directive

variations can be used for the detection and measurement of sound. The principles employed in the design of the corresponding detectors or microphones are mostly the same as those used for the generation of sound. However, not all transducers are reversible, a typical example being the carbon microphone.

Pressure-sensitive mechanisms used in the design of microphones include electromagnetic, electrostatic, piezoelectric, and magnetostrictive, all of which are reversible. Examples of nonreversible pressure-sensitive systems are the carbon microphone, in which the pressure dependence of the electrical resistance of carbon powder is utilized, and the electric corona discharge. The former is a rugged and widely used microphone, whereas the latter has only limited applications.

Among *velocity-sensitive mechanisms* is the hot-wire anemometer, in which the cooling effect of the air flow past a thin wire is detected through a change of the resistance of the wire. This is used extensively in turbulence measurements. Absolute measurement of velocity can be made by means of the Rayleigh disk, in which the torque on a disk in the field provides the coupling. The ribbon microphone (thin metal ribbon in a magnetic field), often referred to as a velocity-sensitive instrument, actually responds to the pressure gradient.

Density variations due to the sound wave can be detected and measured optically through the accompanying change of the index of refraction. In the ultrasonic frequency range the diffraction of light from a sound wave can be used also for measuring the wavelength.

Temperature variations can be detected by small (resistance) probes with very low heat capacity.

Some of the properties of the most-used transducers are summarized in the Table 8.5. The different values quoted for the sensitivity of the crystals refer to microphones of approximately the same size. However, the mode of operation and the frequency affect the sensitivity considerably, and the numbers quoted are only orders of magnitude.

22. Microphone Calibration [5]

Reciprocity Calibration. This is the most widely used and most accurate method of microphone calibration. It requires, in addition to the microphone T_1 under test, one reversible transducer T_2 and a sound source Q. The calibration merely involves (1) obtaining ratio R between sensitivity S_1 of T_1 and sensitivity S_2 of T_2 by comparison of T_1 and T_2 in a given field from Q; (2) using T_2 as a sound source. The pressure p generated at T_1 a distance d from the T_2, when the latter is driven by a current i, is $p = CS_2 i$ (C = constant because of reversibility of T_2). The voltage measured on T_1 is $e = S_1 p = CS_2 S_1 i = C(S_1^2/R)i$. Hence

$$S_1 = C^{-1}R\left(\frac{e}{i}\right)^{1/2} \tag{8.58}$$

The success of the method depends on the fact that the constant C is known, $C = 2d\lambda/\rho c$, the same for any reversible transducer. With R measured in step 1 and e/i measured in step 2, the sensitivity of S_1 is thus found from Eq. (8.58). Reciprocity calibrations can also be carried out in a small cavity (*coupler*) [5].

The Rayleigh Disk. The particle velocity in a plane wave can be obtained by measuring the torque which is shown to exist on a small disk inserted in the field. The torque is

$$L = \frac{\frac{4}{3}\rho c^3 u^2 m}{m + \frac{8}{3}\rho c^3} \sin 2\phi$$

where m = mass of disk, c = velocity of sound, ρ = density of medium, u = particle velocity, and ϕ = angle between disk normal and wave normal. Existence of steady circulations around the disk (see Sec. 18) has been accounted for by empirical corrections [5]. Knowing u, the sound pressure is found from $p = \rho c u$.

Pistonphone and Electrostatic Actuator. In the pistonphone the displacement of a piston radiating into a small cavity is measured by a microscope, and the corresponding sound pressure in the cavity is calculated. It is limited to relatively low frequencies. In the electrostatic actuator the membrane of the condenser microphone to be tested is driven electrostatically with a known force.

23. Other Measurements

A large number of special measuring techniques and instruments have been developed in the various branches of acoustics, of which only a few will be mentioned here. A detailed survey of measurements in the audible frequency range has been made by Beranek [5], and further information is found in the treatises on special topics of acoustics, such as Bergmann's "Ultrasonics" [82].

Measurement of Intensity. Acoustic intensity can be measured indirectly by measuring the radiation pressure in a plane wave (see Sec. 5). Such devices, in the form of pressure balances, are used only in water for ultrasonic frequencies [82]. The *streaming* (see Sec. 18) caused by a sound beam in a fluid can be used for determining intensity or absorption in the medium [69].

Frequency and Spectra. The frequency standards in the laboratory are usually quartz-crystal oscillators operating under well-controlled conditions. Precision measurement of the frequency of a signal involves a comparison with the standard, e.g., by observation of the Lissajous figures formed on an oscilloscope by the standard signal and the measured. Frequency measurements by comparison with continuously variable *stroboscopes*, e.g., tuning-fork-controlled, are often used, yielding accuracies of the order of 1 part in 1,000. For less accurate measurements various kinds of bridge circuits or variable filters can be employed.

Spectrum analysis of steady-state sound is usually made by means of variable or fixed band filters (*analyzers*). Special *spectrographs*, applicable also for transient studies, have been developed particularly for use in the study of speech sounds. Transients are also commonly studied by recording them on magnetic tape. By playing back the signal repeatedly, using a continuous loop, it can then be analyzed in the same way as a stationary signal.

Velocity of Sound. For measurement of velocity of sound in gases, various types of interferometers are used in which the wavelength of a standing wave can be measured with an accuracy up to 1 part in 10^5. In liquids, the standing wave is often used as a diffraction grating on light.

Acoustic Impedance. Normal incidence (see Sec. 4): the *angular dependence* of the normal impedance has been measured by essentially three different methods (for survey see [58]): (1) free-field methods [70], (2) decay measurements of normal modes in rooms (for survey see [58]), and (3) impedance tubes with higher-order modes [71].

Vibrations and waves in solids can be measured by means of *strain gauges*, which can be used up to about 2,000 cps (resistance change in a wire due to tension), and various kinds of electromagnetic, electrostatic, and piezoelectric pickups.

24. The Ear and Hearing [5, 26, 72]

From a physical standpoint and the method of classification adopted above, the ear should be referred to as a nonreversible pressure microphone with built-in amplifiers and analyzers and many other features for which physical analogues are hard to find. In a sense, the object of the research on the mechanism of hearing is to find these physical analogues.

a. The Ear. The external part of the ear, with the *auditory canal* of about 3 cm length and 7 mm diameter, is terminated at the inner end by the eardrum, which has the shape of a shallow cone. Behind the drum is the *middle ear*, containing the ossicles, the *hammer*, *anvil*, and *stirrup*, which couple the drum to the oval window, behind which is the liquid-filled, spiral-shaped double channel, the *cochlea*. The cochlea, of length about 3 cm and width 0.1 to 0.5 mm, is divided lengthwise by the *basilar membrane*, containing the more than 10,000 *hair cells*, which form the first elements of the neural system, where pulses of electrical activity are started. The cells are connected to *nerve fibers*, which are grouped along the axis of the cochlea forming part of the *auditory nerve*.

b. Hearing Characteristics [26, 72]. *Sensitivity*. The audible frequency range is approximately 20 to 20,000 cps, and the maximum sensitivity of the ear lies between 2,000 and 4,000 cps, where a pressure of about 0.0002 dyne/cm^2 in the ear canal can be detected. The pressure around the head in a sound field is quite complicated and varies considerably with position and frequency. At about 3,000 cps the pressure just outside the ear can be about 20 db higher than the level of the incident wave.

Loudness of Sound. (The Auditory Sensation of Sound.) A loudness of one unit (called 1 millisone) is defined as the loudness of a 1,000-cps tone of sound-pressure level 0 db (0.0002 dynes/cm^2). The further relationship between loudness and the level of the 1,000-cps tone can be found in most texts on acoustics [1]. The loudness of an *arbitrary sound* is found by measuring the level of an equally loud 1,000-cps tone. This level is called the *loudness level* (in phons) of the sound. The frequency dependence of the loudness level of a pure tone is well known and is presented in the form of equal loudness-level contours (Fletcher and Munson) in most texts and booklets on sound. The upper limit of the loudness level bordering on pain lies in the region of about 130 phons.

Masking. Under normal conditions a tone with a loudness level of 0 phons can just be heard. In the presence of another sound, the threshold level of the tone will be increased. This effect is called masking. The increase of the threshold level of pure tones as a function of frequency is called the *masking spectrum* of the particular masking sound. The masking effect is residual. The time for recovery of the ear is ordinarily only of the order of a few tenths of a second but increases with the sound level and the time of exposure [26].

Pitch. The pitch of a tone of frequency 1,000 cps and 40 db above threshold is chosen to be 1,000 units (mels). The pitch-frequency relation is not linear, and the pitch seems to increase somewhat with sound-pressure level at high frequencies. The pitch of a complex sound is found by comparison with the pitch of a pure tone 40 db above threshold.

ARCHITECTURAL ACOUSTICS [73–77]

From a *physical standpoint* architectural acoustics concerns the propagation of sound in building structures and the study of sound waves in rooms. Problems in *psychoacoustics* must also be considered, particularly in connection with design of auditoriums and sound systems. Sound waves in rooms and in the building structure are, of course, not independent, since the transmission of sound from one room to another is caused by the vibration of the structure. However, the coupling between the room waves and the structure waves is so small that the problems can in general be treated independently.

25. Room Acoustics [8, 59, 88]

Necessary conditions for "good acoustics" in a room are (*a*) uniform and sufficiently high sound intensity, (*b*) freedom from echoes, and (*c*) proper reverberation. These conditions may not always be sufficient, however. For example, such questions as that of *presence* (in a movie theater the sound should appear to come from the screen, etc.) often play a role in the evaluation of a room.

Regarding uniformity of intensity, the shapes of the rooms under consideration are usually not simple enough to make wave-theoretical analysis successful, and geometrical studies are ordinarily made by following the reflections and distribution of bundles of sound rays emanating from the source. Experimentally this is often done on small room models by means of light or shock waves.

Problems of intensity and reverberation are also handled by approximate means, using the elementary (Sabine) theory [73] of a diffuse sound field, which, at least at sufficiently high frequencies, is often satisfactory. Exact solutions of the acoustic performance of a room have been given for the rectangular (and other simple shapes) room [7, chap. 8], by means of which the important shortcomings of the elementary theory have been analyzed.

a. Uniform-field Assumption. If the sound field is assumed to be uniform throughout the room and such that the intensity falling on a test surface in the field is independent of the orientation of the surface, the relation between the *energy density* and the *sound intensity* in the room becomes

$$W = \frac{4I}{c}$$

(For a plane wave the corresponding relation is $W = I/c$.)

The rate of *energy absorbed* at a boundary is

$$W_{\text{abs}} = \bar{\alpha} I$$

where $\bar{\alpha}$ is the average absorption, discussed in Sec. 19.

Reverberation Time. The rate of decay of the sound energy in a room is readily found on the elementary theory from the energy balance

$$V \frac{dW}{dt} = -A\bar{\alpha} I = -A\alpha \frac{c}{4} W \qquad W = W_0 e^{-(\alpha A c/4v)t} \tag{8.59}$$

where V = room volume, c = velocity of sound, and $\bar{\alpha}A = \Sigma \bar{\alpha}_i A_i$, where $\bar{\alpha}_i$ is the statistical average absorption coefficient of the area A_i. The reverberation time is defined as the time required for the energy density to drop 60 db in the room. Hence,

$$T = \frac{4V}{(\bar{\alpha}A)c} 6 \ln 10 \simeq \frac{0.16V}{\bar{\alpha}A} \tag{8.60}$$

where V is in cubic meters and A is in square meters (for modifications of this formula see [8]).

Measurement of Reverberation Time. The decay of sound in a room is generally studied by means of a logarithmic, high-speed graphic level recorder [5]. The recorded decay is then approximately a straight line, the slope of which determines the reverberation time. It is clear from (8.60) that by measuring the reverberation time in a room with and without absorption material the absorption coefficient of the material can be determined.

Steady-state Intensity. If a sound source in a room emits an acoustic power W_0, the average steady-state intensity in the room will be $I_r = W_0/\bar{\alpha}A$. Close to the source, assumed having spherical symmetry, the intensity is $I_f = W_0/4\pi r^2$. A good approximation of the intensity distributed in the room is therefore $I = I_r$ when $r > (\bar{\alpha}A/4\pi)^{1/2}$ and $I = I_f$ when $r < (\bar{\alpha}A/4\pi)^{1/2}$.

b. Exact Solution for Rectangular Room. Rigorous treatment of the theory of sound waves in rooms shows that the elementary theory is inadequate except in the limit of infinite frequency. The quantity determining the decay was found in the diffuse case to be proportional to the statistical average absorption coefficient. The wave theory shows not only that the individual modes generally have different decay rates but also that it is the conductance (real part of the admittance) of the wall rather than its absorption coefficient that is the important factor determining the decay rate. Furthermore, the field in the room generated by a source will be highly irregular (at least at low frequencies) both in regard to space and frequency dependence.

The degree of field uniformity is essentially determined by the number of eigenmodes that are excited in the room. For a rectangular room the corresponding eigenfrequencies are given by

$$f_{lmn} = \frac{c}{2}\left[\left(\frac{l}{l_x}\right)^2 + \left(\frac{m}{l_y}\right)^2 + \left(\frac{n}{l_z}\right)^2\right]^{1/2}$$

where l, m, n are integers and l_x, l_y, l_z are the linear dimensions of the room. The number of modes which have frequencies less than f is [8; 7, chap. 8]

$$N \simeq \frac{4\pi f^3 V}{3c^3} + \frac{\pi \nu^2 A}{4c^2} + \frac{\nu L}{8c}$$

where $V = l_x l_y l_z$ and $L = 4(l_x + l_y + l_z)$. At sufficiently high frequencies it is evident that N becomes independent of the shape, i.e., of A and L, and dependent only on the volume V, a fact that Weyl has shown to be true for a room of arbitrary shape [79]. The number of modes falling into a frequency band Δf wide is in that case

$$N \simeq \frac{4\pi f^2 V}{c^3} \Delta f$$

For example, in the frequency range corresponding to a half tone step [59], $\Delta f/f \simeq 0.06$, the number of modes at 100 cps in an ordinary concert hall ($V \simeq$ 10,000 m^2) would be about 200, which will increase by a factor of 8 for each octave.

The exact solution for the actual pressure distribution in a room containing a sound source has been studied for a rectangular room with wall impedances independent of the angle of incidence [7, chap. 8]. In that case the field will be built up of eigenmodes of the form

$$p = X(x)Y(y)Z(z)$$
$$X(x) = \cosh(ik_x x + \psi_x)$$

(correspondingly for Y and Z) in which k_x is determined from the boundary conditions leading to

$$ik_x l_x = \coth^{-1} \frac{k_x}{k} \zeta_{2x} + \coth^{-1} \frac{k_x}{k} \zeta_{x1}$$

where $\rho c \zeta_{x1}$, $\rho c \zeta_{x2}$ = specific impedances of the x walls, with corresponding expressions for $k_y l_y$ and $k_z l_z$. The complex resonance frequencies ω are then given by

$$\left(\frac{\omega}{c}\right)^2 = k_x^2 + k_y^2 + k_z^2 \tag{8.61}$$

In the case of relatively hard walls ($|\zeta| \gg 1$) the damping coefficients δ_{lmn} resulting from (8.61) for the different modes can also easily be obtained directly from the first-order approximation [80]

$$\delta_{lmn} = 2 \frac{\int \mu |p_{lmn}|^2 \, dS}{\int |p_{lmn}|^2/\rho c \, dV}$$

where p_{lmn} is the hard-wall solution. [See Eq. (8.12).]

26. Transmission of Sound in Building Structures [59, 88]

The transmission of sound from one room to another in a building is a problem of obvious importance and has in some countries even been a subject for consideration in the building codes. The transmission takes place through the excitation of the walls, which in turn radiate sound into adjacent rooms. To estimate the isolating effect of a wall partition it is usually assumed in engineering work that the wall acts as a mass only, so that the normal impedance of the wall becomes

$$z = -i\omega m + \rho c = \rho c \left(1 - \frac{i\omega m}{\rho c}\right)$$

where m is the mass per unit area of the wall. The intensity transmitted through such a wall is then

$$I_t = \frac{4}{4 + (\omega m/\rho c)^2} I = \frac{I_0}{d}$$

where I_0 = incident intensity. The real transmission-loss factor d is usually expressed in decibels so that

$$D = 10 \log d = 10 \log \left[1 + \left(\frac{\omega m}{2\rho c}\right)^2\right] \simeq 20 \log \left(\frac{\omega m}{2\rho c}\right)$$

In other words, according to this formula a doubling of the frequency or the weight of a wall increases the transmission loss by 6 db. The transmission loss of average wall constructions between apartments is usually of the order of 45 db.

The measured values of the transmission loss are usually found to be less than what this formula predicts, in particular at high frequencies. This may in part be explained by the stiffness of the wall that will compensate for the inertia somewhat. In fact, considering the stiffness, the wall impedance becomes [59]

$$z = -i\omega M \left(1 - \frac{c_b^4 \sin^4 \phi}{c^4}\right)$$

where $c_b = (\omega^2 B/M)^{1/4}$ = velocity of free bending waves, B = bending stiffness = $EI/(1 - \sigma^2)$, E = Young's modulus, I = momentum of inertia per unit thickness, σ = Poisson ratio, and ϕ = angle of incidence. For a given $c_b > c$, resonance (*coincidence*) occurs when

$$\frac{c}{\sin \phi} = c_b \tag{8.62}$$

which means that resonance can take place only for frequencies f larger than a limiting value f_g

$$f_g = \frac{c^2}{2\pi} \left(\frac{M}{B}\right)^{1/2}$$

For ordinary walls the frequency f_g falls in the audible region. At resonance, perfect transmission will occur for waves with angle of incidence ϕ given by (8.62), which actually has been observed in experiments on transmission of sound through plates in water [78].

The extension of the considerations above to include multiple-wall structures, internal damping, and *flanking* transmission has been reviewed by Cremer [59].

Transmission Measurements. For measurements of the transmission loss of a wall construction, the so-called two-room method is usually employed. The wall sample is then inserted in an opening (usually 8 by 8 ft) in the heavy partition wall between a source room and a receiver room. The receiver room is ordinarily reverberant. If the sound-pressure levels in the source room and receiver room are measured, L_1 and L_2 respectively, the transmission loss of the panel can be determined from

$$TL = L_1 - L_2 + 10 \log \frac{S}{A\bar{\alpha}}$$

where S = area of test wall, $\bar{\alpha}A$ = absorption area in the receiving room. Measurements of this kind have been standardized [81].

ULTRASONICS [82–85]

Ultrasonics, dealing with sound above the audible range, is in many respects analogous to microwave physics in the field of electromagnetism. Both are characterized by special instrumentation and measuring techniques and a great number of applications in various fields. For example, the work with radar in the atmosphere has its acoustic counterpart, sonar, in the sea.

27. Generation

Use of piezoelectric crystals and magnetostriction oscillators is efficient for generation of sound only in solids and liquids. For efficient generation of sound in air, generators of much lower impedance are required, and whistles (Hartmann) of various kinds and also the modulated electric discharges (ionophone) therefore are more satisfactory.

28. Measurements

In regard to special measuring techniques in ultrasonics, smallness of the wavelengths involved permits successful application of *optical methods* of detection and measurement. Such techniques and others are described in detail by Bergmann [82]. Of fundamental interest is the measurement of the *velocity of sound* in liquids and gases, from which the compressibility of the medium and its specific heat are obtained. Determination of the latter quantity from such measurements requires a priori knowledge of the static compressibility. Velocity measurements have been useful in studies of electrolytes to check various theories concerning the relation between compressibility and concentration of the solution.

Velocity measurements are mostly made by interferometers, in which a standing wave is set up between the generating quartz crystal and a reflector. The wavelength is measured by moving the reflector or by using the standing wave as a grating for diffraction of light. In this way velocity can be measured with an accuracy better than 1 part in 10^5. Elastic constants of solids can also be determined from velocity measurements and elastooptic studies.

29. Applications

The possibility of generating a well-defined beam is the essential feature of ultrasound. Many of the effects ordinarily referred to as ultrasonic, such as *coagulation, emulsification, chemical, biological,* and *thermal effects,* which do not depend on the beaming, are more likely to depend more on intensity than frequency and can be produced also by audible sound of sufficiently high intensity. However, the generation of high intensities is more convenient with ultrasound than with audible sound. In applications which are built upon the beaming of sound, high frequency and not intensity is obviously the essential factor. Such applications include *flaw detection* in metals and sonar.

Underwater Sound. Sonar. An important application of high frequency sound, known as Sound Navigation And Ranging (SONAR), deals with underwater communication, depth sounding, acoustical navigational aids, and the locating and tracking of fish and submarines. Most of the work in this field is done with frequencies around 25 kcps.

The intensity power that can be generated from the transducer is limited by cavitation, which occurs when the sound pressure is about equal to the static pressure. With a transducer close to the surface of the sea the corresponding limiting intensity is 0.36 watt/cm². This value increases with the square of the static pressure. The problem of *propagation of sound in the sea* is analogous to that in the atmosphere discussed in Secs. 15 to 18. The major inhomogeneities in the sea are essentially temperature and pressure gradients, which in the deep layers give rise to a characteristic effect known as *channel or duct propagation* [85]. The velocity of propagation in such a duct is larger both below and above the central layer of the duct, and the resulting refraction effects keep the sound wave in the duct.

Communication by sound signaling in water over ranges of more than 10 km can be made both by code modulation or speech.

Applications involving *detection* and navigation utilize short pulses of the order of a few milliseconds, which are generated and received (after reflection from the target) by the same transducer and displayed on an oscilloscope. In measuring the depth of the sea by sonic means [84] separate loudspeaker and microphones are mostly used. The time of flight of the sound pulse is measured by means of a moving pen, which is started when the pulse goes out and is made to set a mark on the recording paper when the pulse comes back. This sequence is repeated at short intervals so that the contour of the bottom of the sea is displayed on the recording paper.

References

1. Primakoff, H.: *J. Acoust. Soc. Amer.*, **14**: 14 (1942).
2. Slater, J. C.: "Quantum Theory of Matter," p. 316, McGraw-Hill, New York, 1951.
3. Over 100 references are given in the bibliography of J. L. Stone: Acoustic Propagation through Inhomogeneous Media, *Princeton University, Department of Electrical Engineering Tech. Rep.* 7, August, 1951.
4. Blokhintzev, D.: "The Acoustics of an Inhomogeneous Moving Medium," trans. R. T. Beyer and D. Mintzer, Brown University, 1952.
5. Tables for ρc for different media and temperatures are compiled in L. L. Beranek: "Acoustic Measurements," Wiley, New York, 1949.
6. For further discussion on impedance, see P. M. Morse: *J. Acoust. Soc. Amer.*, **11**: 56 (1939).
7. Morse, P. M.: "Vibration and Sound," McGraw-Hill, New York, 1948.
8. ——— and R. H. Bolt: *Rev. Modern Phys.*, **16**: 69 (1944).
9. Shoch, A.: *Acustica*, **3**: 181 (1953); J. Markham: *Phys. Rev.*, **89**: 972 (1953).
10. Borgnis, F. E.: *Rev. Modern Phys.*, **25**: 653 (1953) (includes references to earlier papers).
11. Le Corbeiller, P., and Ying-Wa Yeung: *J. Acoust. Soc. Amer.*, **24**: 643 (1952).
12. Olson, H.: "Dynamical Analogies," Van Nostrand, Princeton, N.J., 1943.

13. Lighthill, M. J.: *Proc. Roy. Soc. (London)*, **A211**: 565 (1952); Baker and Copson: "The Mathematical Theory of Huygens' Principle," p. 11, Oxford, London, 1950.
14. Symposium on aircraft noise, *J. Acoust. Soc. Amer.*, **25**: 363 (1953).
15. Hönl, H.: *Ann. Physik*, **43**: 437 (1943).
16. Walter, A. G.: *Proc. Cambridge Phil. Soc.*, **47**: 109 (1951).
17. Morse, P. M., and H. Feshbach: "Methods of Theoretical Physics," parts I, II, McGraw-Hill, New York, 1953.
18. Chetayev, D. N.: *Doklady Akad. Nauk SSSR*, **90**: 355 (1953).
19. Sommerfeld, A.: *Ann. Physik*, **42**: 389 (1942).
20. Levine, A., and J. Schwinger: *Phys. Rev.*, **73**: 383 (1948).
21. Bouwkamp: Thesis, Groeningen, 1941; A. Levine and J. Schwinger: *Phys. Rev.*, **74**: 958 (1948).
22. Sommerfeld, A., in Frank and Mises: "Differential Gleichungen der Mathematischen Physik," vol. II, p. 844, Rosenberg, New York, 1943.
23. Ingard, U.: *J. Acoust. Soc. Amer.*, **25**: 1037, 1062 (1953).
24. Pellam, J., and R. H. Bolt: *J. Acoust. Soc. Amer.*, **12**: 24 (1940).
25. Dyer, I.: Thesis, M.I.T. Physics Department, 1954.
26. Richardson, E. G.: "Technical Aspects of Sound," Elsevier, New York, 1953.
27. Mason, W. P.: "Electromechanical Transducers and Wave Filters," Van Nostrand, Princeton, N.J., 1942.
28. de Lange, P.: *Proc. Roy. Soc. (London)*, **A91**: 239 (1951).
29. Mason, W. P.: "Piezoelectric Crystals and Their Application to Ultrasonics," Van Nostrand, Princeton, N.J., 1950.
30. King, L. V.: *Trans. Roy. Soc. (London)*, **A218**: 211 (1919); R. C. Jones: *J. Acoust. Soc. Amer.*, **18**: 371 (1946).
31. Fletcher, H.: "Speech and Hearing," Van Nostrand, Princeton, N.J., 1929.
32. Stoker, J. J.: "Nonlinear Vibrations in Mechanical and Electrical Systems," Interscience, New York, 1950.
33. Richardson, E. G.: "Sound," p. 163, Longmans, New York, 1947.
34. Rayleigh: "Theory of Sound," vol. II, p. 224, Dover, New York, 1945; E. Kerwin: Thesis, M.I.T. Electrical Engineering Department, 1954.
35. Dunn, H. K.: *J. Acoust. Soc. Amer.*, **22**: 740 (1950); K. Stevens et al.: *J. Acoust. Soc. Amer.*, **25**: 734 (1953).
36. Emden, R.: *Meteorol. Z.*, **35**: 13, 114 (1918).
37. Stewart and Lindsay: "Acoustics," Van Nostrand, Princeton, N.J., 1930.
38. Ingard, U.: *Proc. Natl. Noise Abatement Symposium, 4th Ann. Symposium, Chicago, 1953*, p. 11.
39. Pekeris, C. L.: *J. Acoust. Soc. Amer.*, **18**: 295 (1946).
40. Lighthill, M. J.: *Proc. Cambridge Phil. Soc.*, **49**(1): 531 (1952).
41. Sommerfeld, A.: *Ann. Physik*, **28**: 665 (1909).
42. Schuster, K.: *Akust. Z.*, **4**: 335 (1939); H. Stenzel: *Ann. Physik*, **43**: 1 (1943); I. Rudnick: *J. Acoust. Soc. Amer.*, **19**: 348 (1947).
43. Weyl, H.: *Ann. Physik*, **60**: 481 (1919).
44. Ingard, U.: *J. Acoust. Soc. Amer.*, **19**: 348 (1947).
45. Hartig, H. E., and C. E. Swanson: *Phys. Rev.*, **54**: 618 (1938); L. Brillouin: *Rev. acoust.*, **8**: 1 (1939).
46. Miles, J. W.: *J. Acoust. Soc. Amer.*, **17**: 259 (1946).
47. Ingard, U.: *Trans. Chalmers Univ. Technol. Gothenburg* 70, 1948.
48. Lambert, R. F.: *J. Acoust. Soc. Amer.*, **25**: 1068 (1953) (contains many references to earlier papers).
49. Sabine, H.: *J. Acoust. Soc. Amer.*, **12**: 53 (1940).
50. Bosquet, I. P.: *Bull. tech. A.I.Br.*, **31**: 12 (1935); L. J. Sivian: *J. Acoust. Soc. Amer.*, **9**: 135 (1937).
51. Willms, W.: *Akust. Z.*, **6**: 150, (1941).
52. Morse, P. M.: *J. Acoust. Soc. Amer.*, **11**: 205 (1939); L. Cremer: *Akust. Z.*, **5**: 57 (1940).
53. Scott, R. A.: *Proc. Phys. Soc. (London)*, **58**: 358 (1946).
54. Fay, R. D.: *J. Acoust. Soc. Amer.*, **3**: 222 (1931).
55. Rudnick, I.: *J. Acoust. Soc. Amer.*, **25**: 1012 (1953); G. C. Werth and L. P. Delsasso: *J. Acoust. Soc. Amer.*, **26**: 59 (1954).
56. Carriere, Z.: *J. phys. radium*, **10**(1): 198 (1929); E. N. Andrade: *Proc. Roy. Soc. (London)*, **A134**: 445 (1939); U. Ingard and S. Labate: *J. Acoust. Soc. Amer.*, **22**: 211 (1950).
57. Westervelt, P.: *J. Acoust. Soc. Amer.*, **25**: 60 (1953); W. Nyborg: *J. Acoust. Soc. Amer.*, **25**: 68 (1953) (both these papers contain several references to earlier work); J. M. Andres and U. Ingard: *J. Acoust. Soc. Amer.*, **25**: 928, 932 (1953); H. Medwin and I. Rudnick: *J. Acoust. Soc. Amer.*, **25**: 538 (1953).
58. Zwikker, E., and C. W. Kosten: "Sound Absorbing Materials," Elsevier, New York, 1949.
59. Cremer, L.: "Die Wissenschaftlichen Grundlagen der Raumakustik" ("The Scientific Foundations of Room Acoustics"), vol. III, Hirzel, Leipzig, 1950.
60. Kosten, C. W., and C. Zwikker: *Physica*, **8**: 968 (1941); L. L. Beranek: *J. Acoust. Soc. Amer.*, **19**: 556 (1947).
61. Ingard, U.: *J. Acoust. Soc. Amer.*, **26**: 151 (1954) (contains references to earlier papers).
62. Jordan, W. L.: *Akust. Z.*, **5**: 79 (1940).
63. Ingard, U.: *J. Acoust. Soc. Amer.*, **26**: 289 (1954).
64. Kirchhoff, G.: *Pogg. Ann. Bd.*, **1868**: 134.
65. Nielsen, A. K.: *Trans. Danish Acad. Tech. Sci.*, **10** (1949).
66. Knudsen, V. O.: *J. Acoust. Soc. Amer.*, **3**: 126 (1931).
67. Kneser, H.: *J. Acoust. Soc. Amer.*, **5**: 122 (1933).
68. Knudsen, V. O.: *J. Acoust. Soc. Amer.*, **6**: 199 (1935)
69. Cady, W. G., and C. E. Gittings: *J. Acoust. Soc. Amer.*, **25**: 892 (1953).
70. Ingard, U., and R. H. Bolt: *J. Acoust. Soc. Amer.*, **23**: 509 (1951).
71. Shaw, E. A. G.: *J. Acoust. Soc. Amer.*, **25**: 224, 231 (1953).
72. Stevens, S. S., and H. Davis: "Hearing," Wiley, New York, 1938.
73. Sabine, W. C.: "Collected Papers on Acoustics," Harvard, Cambridge, 1927.
74. Knudsen, V. O.: *Rev. Modern Phys.*, **6**: 1 (1924).
75. Watson, F. R.: "Acoustics of Buildings," 3d ed., Wiley, New York, 1941.
76. Sabine, P. E.: "Acoustics and Architecture," McGraw-Hill, New York, 1932.
77. Knudsen, V. O., and C. M. Harris: "Acoustical Designing in Architecture," Wiley, New York, 1950.
78. Fay, R. D., and O. V. Fortier: *J. Acoust. Soc. Amer.*, **23**: 339 (1951).
79. Weyl, H.: *Math. Ann.*, **71**: 441 (1912).
80. van den Dungen, F. H.: "Acoustique des Salles," Gauthier-Villars, Paris, 1934.
81. Parkin, P. H.: *Phys. Soc. (London) Acoust. Group Symposium*, 36 (1949); A. London: *J. Acoust. Soc. Amer.*, **23**: 75 (1951).
82. Bergmann, L.: "Der Ultraschall," Hirzel, Zürich, 1949 (very extensive review of entire subject).
83. Hiedemann, E.: "Ultraschall," Springer, Berlin, 1935.
84. Olson, H.: "Elements of Acoustical Engineering," chaps. 13, 14, Van Nostrand, Princeton, N.J., 1947.
85. Kinsler, L., and A. Frey: "Fundamentals of Acoustics," chaps. 15, 16, Wiley, New York, 1950.

86. Chernov, L. A.: "Wave Propagation in a Random Medium," McGraw-Hill, New York, 1960.
87. Tatarski, V. I.: "Wave Propagation in a Turbulent Medium," McGraw-Hill, New York, 1961.
88. Morse, P. M., and K. U. Ingard: Linear Acoustic Theory, in S. Flügge (ed.), "Encyclopedia of Physics," vol. 11, pp. 1–128, Springer, Berlin, 1961.

Part 4 · Electricity and Magnetism

Chapter 1

Basic Electromagnetic Phenomena

By E. U. CONDON, University of Colorado

1. Electric Charge and Coulomb's Law

Physics is based on the concept that all matter is made of atoms and that these atoms in turn are made up of electrically charged particles whose behavior is governed by quantum dynamics (Part 7, Chap. 1). The class of macroscopic phenomena known as electrostatic effects provided the historical basis for development of the concept of electrostatic charge as a measurable physical entity.

The basic law of electrostatics, discovered by C. A. Coulomb (1785), is that electrically charged bodies whose size is small compared with the distance between them interact with equal and opposite forces which vary inversely as the square of the distance between them. The force depends on the intervening medium. The simplest case is that in which the medium is vacuum, although the correction for air at atmospheric pressure is quite small. For this section the medium is assumed to be vacuum.

There are two kinds of charge. Like charges of either kind interact with repulsive forces, whereas unlike charges attract each other. This is in accord with the algebraic behavior of a force which is proportional to the product of the charges on the two interacting bodies, one kind being arbitrarily called positive and the other negative. All these results are included in the formula for the force interaction

$$F = \frac{e_1 e_2}{4\pi\epsilon_0 r^2} \tag{1.1}$$

where repulsive force is positive and e_1 and e_2 are numerical measures of the amount of the two electric charges (including a sign as a way of specifying the kind of electricity), r is the distance apart of the bodies, and ϵ_0 is a constant whose value depends on the choice of units.

The original work of Coulomb was done with a torsion balance and provided direct experimental verification of the r^{-2} variation for values of the order 10 cm. Indirect later work has given no reason to suppose the inverse-square law not to be valid at all distances, and so its general validity over the entire range is usually assumed.

The cgs electrostatic unit of charge e is called the *statcoulomb* and corresponds to the choice $4\pi\epsilon_0 = 1$, with r in centimeters and F in dynes.

The mks system measures r in meters and F in newtons (1 newton $= 10^5$ dynes) and measures the charge in *coulombs*, which is a quantity of charge that is independently defined (Sec. 7). Thus $(4\pi\epsilon_0)^{-1}$ becomes a quantity to be experimentally determined. It is the force in newtons between two charges, each of 1 coulomb, when separated by a distance of 1 m in vacuum. It is most readily measured indirectly through the relation

$$(4\pi\epsilon_0)^{-1} = 10^{-7}c^2 = 3^2 \cdot 10^9$$

where c is the velocity of light in meters per second. In the second form "3" is to be interpreted as $c/10^8$. More accurately, it is 3 = 2.9917925 (1), where (1) means that the standard deviation is one unit of the last significant digit.

The *abcoulomb* is the cgs electromagnetic unit of charge, defined as c statcoulombs, where c is the velocity of light in meters per second. From this relation it follows that

$$1 \text{ abcoulomb} = 10 \text{ coulombs}$$

For calculations in atomic structure it is convenient to introduce atomic electromagnetic units (Part 7, Chap. 1).

2. Electric Field and Potential

Experiment shows that the resultant force on a small body under the simultaneous influence of two or more charged bodies is the vector sum of the forces with which it acts with each of them when they are present singly. Thus the interactions are linearly superposable.

The force acting on a small charged test body at any point in space when it is under the influence of other charged bodies will be proportional to its own charge and to a vector field characteristic of the influence due to the other charged bodies. Thus

$$\mathbf{F} = e\mathbf{E} \tag{1.2}$$

which equation serves to define $\mathbf{E}$, called the *electric field* or *electric vector*. The electric field can, in principle, be determined by measuring the force on a test body at each point in space, even without a knowledge of the location of the charges which give rise to it.

The significance of the stipulation made thus far that the charges giving rise to the field be on bodies small compared with the distances from them is this: All matter contains a vast amount of electric charge, and the electrostatic effects observed in usual experiments result from a very slight unbalance in the normally equal amounts of charge of each sign present in "uncharged" matter. The effects of the charge contained in matter neutralize each other only in case the amounts of positive and negative charge are equal in each volume element of the material.

A body is said to be uncharged if the total algebraic charge on the body as a whole is zero. Such an uncharged body may still modify an electric field by its presence because of effects broadly known as *electrostatic induction.* The charged particles in the matter are not entirely rigidly bound to fixed locations but have varying freedom to move. When a piece of matter is put in an electric field, the positive charges in it tend to move in the direction of **E** and the negative charges to move in the opposite sense. The extent to which they actually do move is a specific property of the kind of matter under consideration and the time allowed. Insofar as any motion occurs, there will result a partial separation of the positive and negative charges within the body, which is then said to be *polarized.* In consequence, the over-all electric field will be altered from what it would have been if the charges within the matter had not had this freedom to move, because **E** has to be calculated with Coulomb-law contributions from *all* the charge in the system, not only the free or unbalanced charges but also the charge distributions arising from displacement of the electric charges within the structure of the material. Such alterations of the field due to polarization of matter tend to be small at distances large compared with the linear dimensions of the bodies, which was the reason for stipulating their smallness in defining the electric field. When a test body having a charge e is introduced into a field, its own field produces alterations in the preexistent electric field $\mathbf{E}_0$ which are of the order e, and therefore the force on such a test body may be more accurately written

$$\mathbf{F} = e\mathbf{E}_0 + e^2\mathbf{E}_1 \tag{1.3}$$

For this reason, a more careful definition of **E** would say that it is the limit of $\mathbf{F}/e$ as the charge on the test body tends to zero. This point can give rise to confusion in atomic theory where charge always occurs in integral multiples of the electronic charge so that it becomes physically meaningless to speak of carrying out the limit of letting e tend to zero.

Returning to the case of a rigidly prescribed charge distribution, with charge e_s located at the position $\mathbf{r}_s$, then the electric field **E** at the location **r** is given by

$$\mathbf{E}(\mathbf{r}) = (4\pi\epsilon_0)^{-1} \sum_s \frac{e_s(\mathbf{r} - \mathbf{r}_s)}{|\mathbf{r} - \mathbf{r}_s|^3} \tag{1.4}$$

Each term in this sum can be written as the negative gradient of $e_s/4\pi\epsilon_0|\mathbf{r} - \mathbf{r}_s|$. Therefore it is possible to define a scalar field by the equation

$$V = (4\pi\epsilon_0)^{-1} \sum_s e_s/|\mathbf{r} - \mathbf{r}_s| \tag{1.5}$$

so that

$$\mathbf{E} = -\operatorname{grad} V \tag{1.6}$$

This scalar field is known as the *electric potential,* or the *electrostatic potential,* or merely the *potential.* With the definition (1.5) it vanishes at infinitely large distances from all the charges that give rise to the field. The force needed to equilibrate a particle of charge e in the field against the electric forces acting on it is $-e\mathbf{E} = e \operatorname{grad} V$, so that the work done by such a force against the forces of the electrostatic field in a displacement $d\mathbf{r}$ is $e \operatorname{grad} V \cdot d\mathbf{r}$ which is $e\,dV$. Therefore the total work needed to bring a charge from infinity, where $V = 0$, to a point by any path is eV, where V is the potential at the final location. This gives a direct physical interpretation of the potential. Similarly, the work done against electric forces in moving charge e from point A to point B is $e(V_B - V_A)$, independent of the path from A to B.

Gauss's Theorem. If S represents any closed surface, then

$$\iint \frac{(\mathbf{r} - \mathbf{r}_s) \cdot d\mathbf{S}}{|\mathbf{r} - \mathbf{r}_s|^3} = \begin{cases} 4\pi \\ 0 \end{cases}$$

The value 4π is obtained if $\mathbf{r}_s$ ends inside the closed surface S, whereas 0 is obtained if it lies outside. If this is applied to the electric field, it follows that $\iint \mathbf{E} \cdot d\mathbf{S}$ over a closed surface equals ϵ_0^{-1} times the algebraic sum of all the charges inside the surface. This is known as Gauss's theorem.

Often it is convenient to think of the charge as continuously distributed with a volume density ρ charge units per unit volume, together with surface distributions σ charge units per unit area over certain surfaces. In this case, the electrostatic potential is given by

$$V = (4\pi\epsilon_0)^{-1} \left(\iint \frac{\sigma\, dS}{R} + \iiint \frac{\rho\, dv}{R} \right) \tag{1.7}$$

where R is written for the magnitude of the vector from the location of the volume or surface element to the point in space for which V is being calculated. This is an obvious generalization of (1.5) for a distribution of point charges.

From Gauss's theorem for a volume distribution of charges, it follows that

$$\operatorname{div} \mathbf{E} = \rho/\epsilon_0 \tag{1.8}$$

where ρ is the volume density of all the charges in the field, not only the free charge but also that bound in polarized matter. At surfaces where there is a surface distribution of charge of amount σ, there is a discontinuity in the normal component of **E**, the amount of which is conveniently expressed in an equation which is the formal analogue of (1.8). Let **n** be the unit normal vector pointing from side 1 to side 2 of the surface of discontinuity and let $\mathbf{E}_1$ and $\mathbf{E}_2$ be the values of **E** at the surface on opposite sides of it. Then it is convenient to define the surface divergence as Div **E** by

$$\operatorname{Div} \mathbf{E} = \mathbf{n} \cdot (\mathbf{E}_2 - \mathbf{E}_1) \tag{1.9}$$

as a quantity which measures the discontinuous change in the normal component of **E** over such a surface. For a surface distribution of surface-charge density σ, Gauss's theorem gives

$$\text{Div } \mathbf{E} = \sigma/\epsilon_0 \qquad (1.10)$$

where σ is the entire surface-charge density including that arising from the state of polarization of the charge present in matter.

Because of the existence of the scalar potential representing $\mathbf{E}$ as its gradient, it follows that the electric field also obeys the equation

$$\text{curl } \mathbf{E} = 0 \qquad (1.11)$$

which also implies the corresponding condition at a *surface* of discontinuity,

$$\text{Curl } \mathbf{E} = 0 \qquad (1.12)$$

where the surface "Curl" is defined by the relation

$$\text{Curl } \mathbf{E} = \mathbf{n} \times (\mathbf{E}_2 - \mathbf{E}_1) \qquad (1.13)$$

andth erefore the condition at surfaces of discontinuity is that the tangential component of $\mathbf{E}$ is continuous.

Combining (1.6) and (1.8), it follows that the potential satisfies *Poisson's equation:*

$$\Delta V + \epsilon_0^{-1}\rho = 0 \qquad (1.14)$$

where Δ is the Laplace operator. The equation reduces to the Laplace equation in regions where the charge density vanishes.

The cgs electrostatic unit of potential is called the statvolt: 1 erg of work is required to move 1 statcoulomb through a potential difference of 1 statvolt. Similarly, that of $\mathbf{E}$ is the statvolt per centimeter: 1 dyne of force acts on 1 statcoulomb of charge at a place where $\mathbf{E}$ is 1 statvolt/cm. These units correspond to the choice $4\pi\epsilon_0 = 1$.

Similarly the cgs electromagnetic units correspond to the choice $4\pi\epsilon_0 = c^{-2}$. That for potential is called the *abvolt,* and so

$$1 \text{ statvolt} = c \text{ abvolts}$$

For the field the corresponding unit is abvolt per centimeter.

The mks unit of potential is the *volt* and corresponds to the use of $4\pi\epsilon_0 = 10^7/c^2$, with c in meters per second and length in meters, force in newtons, and energy in joules. Thus 1 joule of work is required to move a coulomb of charge to a place of 1 volt higher potential. The corresponding unit of field is volt per meter; the force on 1 coulomb is 1 newton in a field of 1 volt/m.

3. Conductors and Dielectrics

Matter can be classified into two main categories with regard to properties in the electrostatic field, namely, *conductors* and *dielectrics.* In conductors there are charges which are relatively free to move over macroscopic distances in the material. In dielectrics there are no such charges, all the electric charge in the material being bound in such a way that it can be displaced within the material only over distances of the order of atomic dimensions (except for sufficiently high values of $\mathbf{E}$ for which a disruptive spark discharge through the material occurs).

When a conductor is placed in an electric field, free positive charges in it move in the direction of $\mathbf{E}$, and free negative charges move in the opposite direction. These accumulate at the surface of the conductor where it is bounded by free space or dielectric material. The separation of charge so resulting contributes to the field in such a way as to reduce $\mathbf{E}$ within the conductor, and the process of flow of charges goes on until the over-all effect is such that $\mathbf{E} = 0$ at all points within the conductor. Therefore, throughout a conductor the potential V is constant; in particular, the surface of a conductor in equilibrium is an equipotential surface. The separation of charges produced by electrostatic induction in a conductor of course produces an external contribution to the electric field as well and therefore will affect the distribution of charges on other material bodies in the field.

In consequence, if given, say, the total charge on each of several conductors and required to find the resulting field, one cannot do this by application of (1.7) because the detailed location of the surface charges on the conductors will not be known. Instead, one must proceed by finding a solution of Laplace's equation for V valid in the region between the conductors which makes each conductor an equipotential and which satisfies other appropriate conditions of the problem. With the distribution of potential so obtained, the surface distribution of charges can be found by

$$E_n = \epsilon_0^{-1}\sigma \qquad (1.15)$$

which relates the normal component of the electric field at the surface of a conductor to the surface density of induced charge on it. This is an appropriate special case of (1.10) which follows from Gauss's theorem, since $\mathbf{E} = 0$ inside the conductor. Thus the mutual readjustments of induced charge which occur when several charged conductors influence each other are reduced to the solution of a boundary-value problem for the field.

In the case of dielectrics, a similar but less complete displacement of positive charge toward $\mathbf{E}$ and of negative charge the opposite way occurs, with corresponding alterations of the initial field. The difference here is that the charges are bound and can move only over atomic dimensions. Within each structural unit, hereinafter called *molecule,* there will be a partial separation of electric charge which can be conveniently described by giving the electric-dipole moment of each molecule. This quantity $\mathbf{p}$ is defined as

$$\mathbf{p} = \iiint \rho \mathbf{r} \, dv \qquad (1.16)$$

where ρ is the charge-density distribution in a molecule and the integration extends over the molecule. In some kinds of material the molecules have permanent dipole moments, but in the absence of an electric field these are oriented at random, and so the vector sum of the $\mathbf{p}$ for all the molecules in a macroscopic volume element is zero unless a preferred statistical orientation is produced by action of an electric field (Part 4, Chap. 7).

In a medium in which the molecules are polarized, either by orientation of preexistent dipole moments in the molecules or by the induction of dipole moments in the molecules by field action, there will be a macroscopic field called the *polarization* $\mathbf{P}$, which is defined as the dipole moment per unit volume, so that

$$\mathbf{P} = v^{-1}\Sigma\mathbf{p} \tag{1.17}$$

where the sum is extended over the molecules contained in the macroscopic volume element v surrounding the place where $\mathbf{P}$ is being calculated.

The potential V at a point $\mathbf{r}_2$ due to a dipole of moment $\mathbf{p}$ at a point $\mathbf{r}_1$, where $r = |\mathbf{r}_2 - \mathbf{r}_1|$ and $\mathbf{r}_{21} = \mathbf{r}_2 - \mathbf{r}_1$, is given by

$$V = (4\pi\epsilon_0)^{-1}\mathbf{p} \cdot \frac{\mathbf{r}_{21}}{r^3} = (4\pi\epsilon_0)^{-1}\mathbf{p} \cdot \text{grad}_1\,(1/r) \tag{1.18}$$

where grad_1 means that the coordinates of the dipole are to be treated as the variable in calculating the gradient.

Therefore, the potential due to a distribution of polarization $\mathbf{P}$ is given by

$$V = (4\pi\epsilon_0)^{-1}\textstyle\iiint \mathbf{P} \cdot \text{grad}_1\,(1/r)\,dv \tag{1.19}$$

in which the gradient is calculated with respect to the location of dv. Applying the divergence theorem, the volume integral transforms into

$$V = (4\pi\epsilon_0)^{-1}\left(\iint \frac{\mathbf{P} \cdot d\mathbf{S}}{r} - \iiint \frac{\text{div}\,\mathbf{P}\,dv}{r}\right) \tag{1.20}$$

in which the surface integral extends over the bounding surface of the polarized region and the second integral extends over its volume. When this is compared with (1.7), the field due to the polarization of the medium is equivalent to that of a surface and volume distribution of charge given by

$$\sigma_p = \mathbf{P} \cdot \mathbf{n} \qquad \text{and} \qquad \rho_p = -\,\text{div}\,\mathbf{P} \tag{1.21}$$

These are known as the equivalent surface and volume densities of the bound charge due to polarization.

As emphasized in Sec. 2, the electric field $\mathbf{E}$ has to be calculated with all the charges, including the bound charges, used in the expression for its divergence. Therefore if σ and ρ now designate the free surface- and volume-charge densities, $\mathbf{E}$ will be determined by

$$\begin{aligned} \text{div}\,\mathbf{E} &= \epsilon_0^{-1}(\rho + \rho_p) \\ \text{Div}\,\mathbf{E} &= \epsilon_0^{-1}(\sigma + \sigma_p) \\ \text{curl}\,\mathbf{E} &= 0 \end{aligned}$$

Using (1.19), one can eliminate the unknown σ_p and ρ_p by introducing a new vector field called the dielectric displacement $\mathbf{D}$, defined by

$$\mathbf{D} = \epsilon_0\mathbf{E} + \mathbf{P} \tag{1.22}$$

so that the vector satisfies divergence equations involving only the densities of free charge:

$$\text{div}\,\mathbf{D} = \rho \qquad \text{Div}\,\mathbf{D} = \sigma \tag{1.23}$$

Further progress can be achieved only if some additional assumption is made about the amount of the polarization $\mathbf{P}$ that is associated with a given electric field $\mathbf{E}$. The assumption made for ideal dielectrics is that $\mathbf{P}$ is linear in $\mathbf{E}$,

$$\mathbf{P} = \epsilon_0 k\mathbf{E} \tag{1.24}$$

where the coefficient k is called the (electric) susceptibility and is to be understood as being a second-order symmetric tensor in the case of crystalline dielectrics. But the relation (1.24) is not universally valid. For example, *electrets* can be made of certain kinds of waxes which show a quasi-permanent polarization $\mathbf{P}$ that is independent of $\mathbf{E}$. Likewise, materials exist which are extremely polarizable in which $\mathbf{P}$ is a nonlinear function of $\mathbf{E}$, showing saturation with increasing $\mathbf{E}$. These materials are called *ferroelectric*, not because they involve iron in any way but because their behavior is the electric analogue to the behavior of ferromagnetic materials like iron.

If (1.24) is valid, as it is for most common dielectrics, then it implies the relation

$$\mathbf{D} = \epsilon_0(1 + k)\mathbf{E} \tag{1.25}$$

where the factor $\epsilon = 1 + k$ is usually called the *dielectric constant*, although some authors assign this name to the product $\epsilon_0\epsilon$. The factor ϵ is also called the *specific inductive capacity*.

Both the $\mathbf{D}$ and the $\mathbf{E}$ fields may be represented geometrically in terms of curves known as *lines of induction* or *lines of force*. These are curves drawn everywhere tangent to the direction of $\mathbf{D}$ or $\mathbf{E}$, respectively. By drawing a definite number of lines originating on each positive charge and ending on each negative charge, the density of lines at each point in space will represent the magnitude of $\mathbf{D}$ or $\mathbf{E}$. Consistently with (1.23), the lines of $\mathbf{D}$ originate and end only at free charges, whereas those of $\mathbf{E}$ originate and end at all charges, both free and bound.

Some physicists prefer to define $\mathbf{D}$ as 4π times the $\mathbf{D}$ defined in (1.22) so that, writing $\mathbf{D}'$ for their $\mathbf{D}$,

$$\mathbf{D}' = 4\pi\mathbf{D} = 4\pi\epsilon_0\mathbf{E} + 4\pi\mathbf{P} \tag{1.22a}$$

and correspondingly (1.23) becomes

$$\text{div}\,\mathbf{D}' = 4\pi p \qquad \text{Div}\,\mathbf{D}' = 4\pi\sigma \tag{1.23a}$$

This choice is attractive in the cgs electrostatic system in that it makes $\mathbf{D}'$ numerically equal to $\mathbf{E}$ in free space.

This matter is related to "rationalization" of the electrical units as advocated by Heaviside. If one regards (1.1) as fundamental, it is natural to define unit charge by setting $4\pi\epsilon_0 = 1$, as is done with the conventional cgs electrostatic units, which also simplifies (1.4) and (1.5). But then, on applying Gauss's theorem, one has $\iint \mathbf{E} \cdot d\mathbf{S} = 4\pi Q$, where Q is the net charge within the surface S. Heaviside preferred to introduce altered units of charge, $\mathbf{E}$ and V, by setting $\epsilon_0 = 1$. This has the effect of leaving 4π in equations such as (1.1), (1.4), and (1.5) but gets rid of the 4π in equations such as (1.8), (1.10), (1.14), and (1.15). A great deal of confusion arises in the literature because authors often fail to state which choice they have made.

4. Forces and Energy in the Electric Field

The work necessary to assemble the charges of a volume distribution ρ from an initial state where all charges are at infinity and ρ and V are zero everywhere may be calculated as follows: Consider a sequence of similar systems for which the density is $\gamma\rho$ and the potential is γV, where γ is a factor constant throughout the field which varies between 0 and 1. To increase the charge from γ to $\gamma + d\gamma$ by bringing up the amount of charge $\rho\, d\gamma$ to each volume element against the potential γV, one must do an amount of work against the electrical forces,

$$dW = \gamma\, d\gamma \iiint \rho(x,y,z)V(x,y,z)\, dv$$

and so the total work done in bringing up all the charge will be obtained on integrating γ from 0 to 1, which gives a factor $\frac{1}{2}$. Therefore the energy needed to establish the system by bringing up the charges is

$$W = \tfrac{1}{2}\iiint \rho V\, dv \qquad (1.26)$$

Thus, for example, if the system consists of a sphere of radius a having a uniform volume distribution of charge in it such that the total charge is q, then inside the sphere

$$V = (q/8\pi\epsilon_0)(3 - x^2) \qquad x < 1$$

where $x = r/a$ and the charge density is $(3/4\pi)(q/a^3)$. The total energy of such a charged sphere is

$$W = (3/20\pi\epsilon_0)(q^2/a) \qquad (1.27)$$

In the case of a uniform surface distribution of the same amount on a sphere of radius a, all the charge is brought to potential $q/4\pi\epsilon_0 a$, and so the energy is $W = q^2/8\pi\epsilon_0 a$. Thus the surface distribution is a state of somewhat lower energy than the uniform volume distribution.

Since either of these expressions becomes infinite as the radius tends to zero, the energy needed to assemble a finite "point" charge is infinite. This kind of consideration indicates that either the Coulomb law must break down at small distances or the fundamental charged particles of atomic physics must have finite size.

If one has a system of point charges and ignores their infinite self-energy by reckoning the state of zero energy to be that in which the finite point charges are held together but located at great distances from each other initially, the extra energy of the system which must be supplied to bring them up to their final positions is

$$W = \tfrac{1}{2}\sum_s e_s V_s \qquad (1.28)$$

in analogy with (1.26).

In case the field involves dielectrics, the density ρ is to be interpreted as the density of free charge, for it is only this part which is brought up from infinity to the bodies which are presumed to be already in place before any charge is brought up. Using (1.23),

$$\begin{aligned} W &= \tfrac{1}{2}\iiint(\text{div }\mathbf{D})V\, dv \\ &= \tfrac{1}{2}\iiint \mathbf{D}\cdot\mathbf{E}\, dv \qquad (1.29)\end{aligned}$$

Comparing (1.26) and (1.29), the former leads naturally to a viewpoint which regards the energy as being localized in space at the particular places where the free charge is located, which leads to ways of speaking that suggest that the energy is "possessed" by the several elements of free charge. This viewpoint also seems a natural one in that the work is done directly on the charges themselves by applying forces to them to work against the electric field. But (1.29) shows that the energy of the system may equally well be represented by regarding the energy as being distributed over the whole field, the density at each place being $\frac{1}{2}\mathbf{D}\cdot\mathbf{E}$ units in unit volume. The two expressions being fully equivalent, there is nothing about these considerations which gives preference to one or the other of these viewpoints. The latter, however, is the one more in accord with the general field-theory development of electrodynamics.

In the electric field the force on any free charge is $e\mathbf{E}$; similarly, the force on unit area of a surface distribution σ is $\sigma\mathbf{E}$, in which $\mathbf{E}$ is interpreted as $(\mathbf{E}_2 + \mathbf{E}_1)/2$, and the force on unit volume of a volume distribution is $\rho\mathbf{E}$. In particular, therefore, at the surface of a conductor where $\mathbf{E}_1 = 0$ and $\mathbf{E}_2$ is normal and equal to $\epsilon_0^{-1}\sigma$, the force on unit area is a tension tending to move the conductor into the nonconducting region, the amount of the tension being

$$\tfrac{1}{2}\epsilon_0^{-1}\sigma^2 = \tfrac{1}{2}\epsilon_0\mathbf{E}^2 \qquad (1.30)$$

Since the maximum electric field in air at atmospheric pressure without sparkover is about 30 kv/cm or 100 statvolts/cm, this gives a maximum tension of about 400 dynes/cm², which is equal to the stress due to about a 4-mm head of hydrostatic pressure in water.

In addition to the forces which act on free-charge distributions, electric forces act on dielectric bodies in an electric field, both at the surfaces of discontinuous change in ϵ and throughout the volume. The surface force acting on unit area in the direction tending to make the boundary move, so that medium 1 encroaches on the space occupied by medium 2, is

$$\mathbf{S} = \sigma\mathbf{E} - \tfrac{1}{2}(\mathbf{E}_1\cdot\mathbf{E}_2)\text{ Grad }\epsilon \qquad (1.31)$$

where $\text{Grad }\epsilon = \mathbf{n}(\epsilon_2 - \epsilon_1)$ and $\mathbf{n}$ is the unit normal drawn from medium 1 toward medium 2. Here also $\mathbf{E}$ is understood as the mean of $\mathbf{E}_1$ and $\mathbf{E}_2$, and σ is the free-charge surface density. If the free surface charge vanishes, the entire surface force is given by the second term; it is normal to the boundary and tends to make the medium of higher dielectric constant move into the space occupied by the medium of lower dielectric constant.

Likewise, if the mass density of the medium is λ and the dielectric constant ϵ depends on the density, there is a volume force acting on unit volume of the dielectric that is given by

$$\mathbf{F} = \rho\mathbf{E} - \frac{\epsilon_0\mathbf{E}^2}{2}\text{ grad }\epsilon + \frac{\epsilon_0}{2}\text{ grad}\left[\mathbf{E}^2\left(\lambda\frac{d\epsilon}{d\lambda}\right)\right] \qquad (1.32)$$

If there is no volume distribution of free charge, the first term vanishes; if there is no inhomogeneity of ϵ, the second term vanishes. With regard to the last term, consider the special case in which the field is uniform. Since $d\epsilon/d\lambda$ is positive for most, if not all, materials, the body force will be in the direction of the gradient of density, and it will tend to make material move toward regions of greater density, thus making the material even more dense. This term is known as the *electrostriction term* for this reason and is the explanation of observed *electrostrictive effects* in liquids. Likewise, if the field is not uniform, this term tends to make the dielectric move into regions where the field is stronger.

According to important results due to Maxwell, the forces **S** and **F** may be regarded as the net effect of unbalance of effective stresses in the field in which the stress across any unit area in the field normal to the unit vector **n** is **K**, where

$$\mathbf{K} = \epsilon_0[(\mathbf{E}\cdot\mathbf{D})\mathbf{n} + \mathbf{D}\times(\mathbf{E}\times\mathbf{n})] \qquad (1.33)$$

This is known as Maxwell's equivalent stress. The net forces given by (1.31) and (1.32) result as the unbalance of such stresses at places where ϵ is changing.

Minimum-energy Theorems. Suppose that the volume- and surface-density distributions of free charge are given and that the total charge q_s on each of the conductors in the field is also given. Then the **D** field must satisfy

$$\operatorname{div}\mathbf{D} = \rho \qquad \operatorname{Div}\mathbf{D} = \sigma \qquad \text{and} \qquad \iint_s \mathbf{D}\cdot d\mathbf{S} = q_s \qquad (1.34)$$

In addition, the electric field in equilibrium satisfies curl **E** = 0 and Curl **E** = 0 (that is, **E** must be normal to the conducting surfaces). It is possible to consider fields which satisfy (1.34) but do not satisfy "curl" equations. Such fields are necessarily nonstatic, as appears in Sec. 9. Thomson proved that the electric-field energy of such nonstatic fields is greater than that of the electrostatic fields, or, in other words, that the electrostatic-field energy is a minimum. Let the nonstatic field be $\mathbf{E} + \delta\mathbf{E}$ and $\mathbf{D} + \delta\mathbf{D}$, where **E** and **D** describe the static field. Then

$$W + \delta W = \tfrac{1}{2}\iiint(\mathbf{E} + \delta\mathbf{E})(\mathbf{D} + \delta\mathbf{D})\,dv$$

$$\delta W = \tfrac{1}{2}\left[\iiint(\mathbf{E}\cdot\delta\mathbf{D} + \delta\mathbf{E}\cdot\mathbf{D})\,dv + \iiint(\delta\mathbf{E}\cdot\delta\mathbf{D})\,dv\right]$$

The first term vanishes because it equals

$$\iiint(\mathbf{E}\cdot\delta\mathbf{D})\,dv$$

and **E** satisfies curl **E** = 0, while $\delta\mathbf{D}$ satisfies div $\delta\mathbf{D} = 0$; therefore, δW equals the last term which is essentially positive.

Another minimum-energy theorem is this: A number of fixed surfaces are given, and the total charge on each surface is given. The electric-field energy is minimum for that distribution of charge on each surface which makes each surface an equipotential. Thus the actual distribution of charge induced on a system of fixed conductors makes the electric-field energy a minimum.

Let V be the potential for the equipotential field and $V + \delta V$ that for another; then

$$\delta W = \iiint(\operatorname{grad} V)\cdot(\operatorname{grad}\delta V)\,dv + \tfrac{1}{2}\iiint(\operatorname{grad}\delta V)^2\,dv$$

The first term may be shown to vanish, and so δW equals the last term, which is an essentially positive quantity.

Suppose that a field due to fixed charges or to conductors with a fixed total charge is given. If a new insulated and uncharged conductor is introduced into the field, the total energy will be lessened, and therefore the new conductor will be acted on by attractive forces on the whole. The change in electric energy is

$$\delta W = -\tfrac{1}{2}\iiint\mathbf{E}^2\,dv \qquad \text{(over the introduced conductor)}$$

$$= -\tfrac{1}{2}\iiint(\operatorname{grad}\delta V)^2\,dv \qquad \text{(over the whole field after the conductor is introduced)} \qquad (1.35)$$

Both terms are essentially negative.

Similarly, in such a field as that just considered, an increase anywhere, $\delta\epsilon$, in the dielectric constant of part of the space lessens the energy of the field, so that a small piece of uncharged dielectric will always be attracted into strong parts of the field. The change δW is

$$\delta W = -\tfrac{1}{2}\iiint\mathbf{E}^2\,\delta\epsilon\,dv \qquad (1.36)$$

5. Ohm's Law and Electromotive Force

Transport of charge within a conductor is measured by current density **i** which gives the amount of charge crossing unit area normal to itself in unit time. Current J is measured as the surface integral of **i** over a surface and measures the net rate at which electric charge crosses that surface.

Corresponding to the three units of charge are three commonly used units of current:

1 statamp = 1 statcoulomb/sec
1 amp = 1 coulomb/sec
1 abamp = 1 abcoulomb/sec

Therefore,

$$1 \text{ abamp} = 10 \text{ amp} = c \quad \text{statamp}$$

where c is in centimeters per second. In every case, actual transport of free charge is involved, and so ρ and **i** are related by the equation of continuity,

$$\operatorname{div}\mathbf{i} + \dot{\rho} = 0 \qquad (1.37)$$

Steady currents are therefore characterized by div **i** = 0, for otherwise the piling up of free charge would set up strong electric fields which would act to modify the currents.

Since div **D** = ρ, this leads to

$$\operatorname{div}(\mathbf{i} + \dot{\mathbf{D}}) = 0 \qquad (1.38)$$

which is of great importance in the theory of magnetic fields excited by nonsteady currents (Sec. 7).

Experimental study of steady currents in linear conductors shows that each segment of such a conductor has an attribute known as its *resistance* R such that the drop in electric potential between two ends is JR when a current J is flowing. In particular, if J is in amperes and the potential drop is in volts, this defines a unit of R called the *ohm*. The proportionality of potential drop to current is known as *Ohm's law*. Other units of resistance called the *statohm* and the *abohm* follow in the other systems, as follows:

$$1 \text{ statohm} = c^2/10^9 \text{ ohms} = c^2 \quad \text{abohms}$$

where c is in centimeters per second.

Ohm's law stated in this way is an integral form of a differential law according to which a conductor is characterized by a resistivity ρ (not to be confused with the other use of ρ to represent charge density) which relates the value of $\mathbf{E}$ needed to produce a current density $\mathbf{i}$:

$$\mathbf{E} = \rho\mathbf{i} \tag{1.39}$$

where ρ is in ohm-centimeters if $\mathbf{i}$ is in amperes per square centimeter and $\mathbf{E}$ in volts per centimeter. A common mistake is to write this unit of ρ as ohms per cubic centimeter, stemming from the fact that ρ numerically is the resistance of a conductor having a cross-sectional area S of 1 cm^2 and a length s of 1 cm (hence, colloquially, it is "the resistance of a cubic centimeter"). The resistance of a conductor of cross section S and length s is

$$R = \rho s/S \tag{1.40}$$

which again shows that the unit of ρ is the resistance unit multiplied by the first power of the length unit. In the mks system the length unit is the meter, and so $\mathbf{E}$ is in volts per meter, $\mathbf{i}$ in amperes per square meter, and ρ is in ohm-meters, where

$$1 \text{ ohm-m} = 100 \text{ ohm-cm}$$

The current transfers charges of each sign to places of lower potential energy (that is, lower eV), in the course of which work is done on them by the electric field. This loss of energy is converted into heat in ohmic conductors, the heat appearing throughout the volume of the conducting material. This quantitative conversion of electrical energy into heat is known as *Joule's law*, and the heat is often called Joule heat.

In integral form, the rate at which heat is generated in a conductor is VJ watts (with V and J in volts and amperes) which is equal to J^2R or V^2/R in view of Ohm's law. In differential form, $\mathbf{E} \cdot \mathbf{i}$ is the heat developed in unit volume (watts per cubic centimeter) with $\mathbf{E}$ in volts per centimeter and $\mathbf{i}$ in amperes per square centimeter. This is also $\rho\mathbf{i}^2$ or $\mathbf{E}^2/\rho$ with ρ in ohm-centimeters. The generation of heat is a consequence of frictional dissipative forces acting on the charges as they move through the conductor.

In the case of acceleration of a beam of free electrons or ions in a high-vacuum device, a current J falling through a potential drop V has electrical power delivered to it in the amount VJ which in the first instance assumes the form of kinetic energy of the accelerated particles and which later may be converted into energy of other forms (principally heat) when the beam of charged particles strikes a target.

Observed values of resistivity of materials cover an enormous range. Superconducting materials behave in many respects as if $\rho = 0$, but their behavior (Part 5, Chap. 11) is not wholly describable merely by supposing their resistivity to vanish. Copper at 20°C (International annealed standard) has $\rho = 1.7241 \times 10^{-6}$ ohm-cm. Many dielectrics that are used as insulators have volume resistivities upwards of 10^{12} ohm-cm and ranging up to 10^{18} ohm-cm. With metals, the resistivity is so low that it does not, as a rule, serve to limit the time rate at which an equilibrium distribution of charge is established, but rather the time rate is set by magnetic inductive limitations. In *insulators* the resistivity is so high that it is the governing factor in determining the rate at which the flow of charges in them nullifies $\mathbf{D}$ and $\mathbf{E}$ in their interior.

Using Ohm's law in (1.38),

$$\text{div}\,(\mathbf{D} + \epsilon\epsilon_0\rho\dot{\mathbf{D}}) = 0$$

which indicates that $\mathbf{D}$ decays exponentially because of conduction, with a time constant

$$T = \epsilon\epsilon_0\rho \qquad \text{sec} \tag{1.41}$$

For example, in a material for which $\epsilon = 10$ and for which $\rho = 10^{14}$ ohm-cm $= 10^{12}$ ohm-m, using

$$4\pi\epsilon_0 = \frac{10^7}{c^2}$$

with c in meters per second, the value of T becomes 88.3 sec.

In dealing with ohmic conduction in thin surface films it is convenient to speak of a surface-current density $\mathbf{i}_s$ (amp/cm) giving the current in the film crossing unit length in the film normal to $\mathbf{i}_s$. Analogously, the surface resistivity ρ_s relates this to the $\mathbf{E}$ (tangential component only) needed to maintain this surface-current density,

$$\mathbf{E} = \frac{\mathbf{i}_s}{\rho_s} \tag{1.42}$$

where now, if $\mathbf{i}_s$ is in amperes per centimeter and $\mathbf{E}$ in volts per centimeter, ρ_s is measured in ohms. The resistance of a film of width w, length s, thickness t, and of material having volume resistivity ρ is

$$R = \frac{\rho s}{tw}$$

which is equivalent to a surface resistivity of $\rho_s = \rho/t$, where t is the film thickness.

The surface layers of insulators are often in a condition very different from the volume of the material, especially in humid atmospheres where some moisture may condense on the surface. Thus there may be surface-leakage currents that are much larger than could possibly occur with true volume conductivity of the material in the actual thickness that is present. For example, typical values reported for sulfur at room temperature are $\rho = 10^{17}$ ohm-cm and $\rho_s = 7 \times 10^{15}$ ohms at 50 per cent humidity, dropping to $\rho_s = 10^{14}$ ohms at 90 per cent humidity. Thus the surface layers carry currents which, if carried by the volume conductivity, would require thicknesses of 14.3 cm and 1,000 cm, respectively.

In the case of a complete circuit in a conductor, since curl $\mathbf{E} = 0$ for the electric field, the line integral of $\mathbf{E}$ around a closed circuit must vanish. Therefore in situations in which steady currents are maintained in a circuit there must be acting on the charges other forces than those described by $\mathbf{E}$.

In order to maintain such currents, some kind of device known as a *generator* must be introduced into the circuit. Many devices of this kind have been developed and are in general use. All have in common that forces other than the electric field act on the charges; therefore there is a conversion of energy in the generator from some other form to energy of the electric field. Simplest conceptually perhaps is the Van de Graaff generator in which the charges are sprayed onto a moving belt and conveyed mechanically from a place of higher potential to a place of lower potential. In d-c rotating generators this is accomplished by electromagnetic inductions. In galvanic cells it is accomplished by electrochemical forces (Part 4, Chap. 9).

All such generators have two electrodes, or terminals, by which they are connected to an external circuit. When no current is flowing between the terminals, they act to place one electrode (the positive terminal) at a higher potential than the other by an amount E, which is called the electromotive force (emf) of the generator. When current flows in them, various energy-dissipative processes occur within the generator because of the flow of current, and this has the effect that the potential difference between the terminals is less than that when the generator is on open circuit. In the simplest case, the potential when current J is flowing will be $E - JR_g$, which can be described by saying that the generator possesses an ideal emf E and an internal resistance R_g. In this case, Ohm's law for the circuit takes the form

$$J = \frac{E}{R_g + R} \tag{1.43}$$

where R is the resistance connected externally to the generator, called the load. The power delivered to the load is J^2R or

$$\frac{(E^2/R_g)x}{(1 + x)^2}$$

where $x = R/R_g$. This has a maximum value of $E^2/4R_g$ when $x = 1$, that is, when $R = R_g$. In this case, equal amounts of power are developed as heat in the generator and in the load, and the load is said to be matched to the generator. The fraction of the total power developed as heat that is developed in the load is $x/(1 + x)$ so that good efficiency of power transfer calls for making x as large as possible.

In more complicated cases, as when the generator involves electromagnetic induction in a device containing iron which shows magnetic saturation, the potential difference available at the terminals is a nonlinear function of the current, $E(J)$, which is known as the *characteristic* of the generator. When attached to a load of resistance R, the current which flows is given by solving $E(J) = JR$; this can be done graphically for a given $E(J)$. Plotting $E(J)/J$ against J gives a curve which shows as ordinate the resistance of the load that is needed to give a circuit current J. The power developed in the load is $JE(J)$, and so the resistance which gives maximum power to the load is that value of $E(J)/J$ for which $JE(J)$ is a maximum.

Galvanic cells may also show complicated polarization effects in which the potential across the terminals at any time depends not only on the current being drawn but on the past history of flow of J.

6. Magnetic Fields Due to Permanent Magnets

Certain special materials such as the magnetic iron ore Fe_3O_4, which is also called lodestone, and various hard steels are capable of being brought into a condition where they are *magnetized*, in which they interact with forces of a type known as *magnetic*. These forces resemble electric forces in many of their mathematical properties. Coulomb used a torsion balance to discover that the force of interaction between two long slender magnets is pretty much localized at the ends, called *poles*, and that the force of interaction between poles is repulsive between like poles and attractive between unlike poles and varies inversely as the square of the distance between them. Magnetic force and electric force are nevertheless distinct and different in that there is no static interaction between an electric charge and a magnet.

Up to 1820 there was no known link, save that of mathematical analogy, between the electrical phenomena and the magnetic phenomena. Then Oersted discovered the magnetic interactions of steady electric currents and between magnets and current-carrying conductors. Ampère studied these phenomena quantitatively and formulated mathematically the laws of interaction between electric currents. Before the discoveries of Oersted and Ampère it was natural to suppose that magnetic forces had their origin in inverse-square-law interactions between magnetic charges or poles and that these were completely analogous to the electric charges which give rise to the electric field. However, one extremely important difference was noted: Under all circumstances magnetized bodies are found to be magnetically neutral; that is, the total strength of its positive poles is equal to the total strength of its negative poles. When the attempt is made to achieve a separation by breaking a magnet in two pieces, it is found that new poles appear at the new surfaces in just such strength that each of the two pieces is magnetically neutral.

This observation can be brought within the framework of the theory of polarized media as developed in Sec. 3 for dielectrics, if one postulates that the only kind of magnetic elements occurring in nature are *magnetic dipoles* and that all magnetic fields due to permanent magnets and all magnetization induced in soft magnetic materials are the fields arising from aggregations of such dipoles. At this stage of development there was no connection between electric and magnetic phenomena; electric fields were set up by or excited by electric charge, and magnetic fields were set up by or excited by material containing magnetic dipoles.

With the discovery of the magnetic effects of electric currents, Ampère recognized that it was not necessary to postulate the independent existence of magnetic dipoles as a new kind of entity in nature and that magnetic effects can be fully accounted for by supposing that the magnetic dipoles are a manifestation of circulating molecular electric currents moving in circuits of subatomic dimensions.

Corresponding to these two different viewpoints are two somewhat different developments of the theory of the magnetic field. The one proceeds to introduce magnetic dipoles as independent elements, and the other starts from the magnetic effects of electric currents. The former is still useful because it emphasizes a mathematical analogy between corresponding problems in electrostatics and magnetostatics. The latter is at present believed to be more fundamental in relating all magnetism to circulating currents in atomic and subatomic units. In this section the magnetic-dipole viewpoint is adopted, while the magnetic effects of currents are considered in the next section.

Unit magnetic-pole strength is defined by Coulomb's law in analogy with (1.1) as

$$F = \frac{p_1 p_2}{4\pi\mu_0 r^2} \tag{1.44}$$

where μ_0 depends on the units used for the other quantities. The cgs unit of pole strength, here called 1 pole, corresponds to $4\pi\mu_0 = 1$, with F in dynes and r in centimeters. Hence this is the analogue of the statcoulomb of charge.

Analogous to (1.2) one defines the magnetic-field vector **H** by

$$\mathbf{F} = p\mathbf{H} \tag{1.45}$$

The unit of **H**, with **F** in dynes and p in poles, is called the *oersted*. The *gamma* (γ) is defined as 10^{-5} oersted. Magnetic-dipole moment is defined by analogy with (1.16):

$$\mathbf{m} = \iiint \rho \mathbf{r}\, dv \tag{1.46}$$

where ρ is understood as volume density of pole strength. The torque on a magnetic dipole **m** in a field **H** is

$$\mathbf{T} = \mathbf{m} \times \mathbf{H} \tag{1.47}$$

where **T** is in dyne-centimeters with **m** in pole-centimeters and **H** in oersteds.

In analogy with (1.5) the magnetic field **H** can be expressed as a gradient of a magnetic scalar potential,

$$\mathbf{H} = -\operatorname{grad} U \qquad \operatorname{curl} \mathbf{H} = 0 \tag{1.48}$$

in which U is expressed in oersted-centimeters, another name for which is *gilberts*.

The intensity of magnetization **M** is defined as the magnetic moment in unit volume, in analogy with (1.17),

$$\mathbf{M} = v^{-1}\Sigma\mathbf{m} \tag{1.49}$$

the summation being over those magnetic dipoles in a macroscopically small volume at the point for which **M** is calculated.

The magnetic potential due to a dipole **m** is

$$U = (4\pi\mu_0)^{-1}\mathbf{m} \cdot \operatorname{grad}_1 \left(\frac{1}{r}\right) \tag{1.50}$$

in analogy with (1.18). Thus the magnetic potential due to a volume distribution of **M** is

$$U = (4\pi\mu_0)^{-1} \iiint \mathbf{M} \cdot \operatorname{grad} \left(\frac{1}{r}\right) dv \tag{1.51}$$

which may be transformed to give the analogue of (1.20). Thus the magnetic field **H** due to a magnetically polarized medium is equivalent to one given by surface and volume distributions of pole strength, analogous to (1.21),

$$\sigma_m = \mathbf{n} \cdot \mathbf{M} \qquad \rho_m = -\operatorname{div} \mathbf{M} \tag{1.52}$$

We can now introduce

$$\mathbf{B} = \mu_0 H + \mathbf{M} \tag{1.53}$$

so that **B** is the analogue of **D** of (1.22). As before, some physicists prefer to work with $\mathbf{B}' = 4\pi\mathbf{B}$

$$\mathbf{B}' = 4\pi\mu_0\mathbf{H} + 4\pi M \tag{1.53a}$$

In the cgs system in which $4\pi\mu_0 = 1$ this is attractive in that it makes **B**′ numerically equal to **H** in free space. The unit of **B**′ in the cgs system is called the *gauss*.

Magnetic materials differ from dielectrics in that for strongly magnetic materials whose behavior resembles iron and steel (ferromagnetic materials) the relation between **M** and **H** is more complicated than is implied by the magnetic analogue of (1.24). This analogue clearly leads to the definition of a magnetic susceptibility γ which is the analogue of the electric susceptibility k:

$$\mathbf{M} = 4\pi\mu_0\gamma\mathbf{H} \tag{1.54}$$

This relation holds for weakly magnetic materials, with a further difference between the dielectric case and the magnetic case. For dielectrics the susceptibility is always positive. In the magnetic case the weakly magnetic materials are of two kinds: *paramagnetic* (having a positive γ) and *diamagnetic* (having a negative γ), which have no electrical analogue.

In ferromagnetic materials the relation between **M** and **H** is much more complicated than (1.54) in two ways: The value of **M** depends not only on the present value of **H** but on the past history of variation of **H** to which the material has been subjected. This behavior is known as *hysteresis*. Materials in which hysteresis is pronounced are called *hard* magnetic materials and those in which it is not pronounced are called *soft*. In both cases the value of **M** increases with increasing **H** but approaches a limiting or saturation value. This is consistent with the picture that the magnetization is the result of the lining up along the field of a definite amount of preexistent dipole moment in unit volume, and so **M** can increase no more after all the dipoles in unit volume are all pointing in the same direction.

Strictly speaking, there are no permanent magnets, but if a hard magnetic material has been magnetized by treatment in a strong field and is later used only in relatively weak fields, it is found that, to a good approximation, the magnetization keeps a constant value $\mathbf{M}_0$ depending on the previous magnetization treatment. More accurately, the weak field **H** will produce small departures from the "permanent" magnetization $\mathbf{M}_0$. To this approximation, therefore, the behavior of permanent magnet materials in weak fields can be described by

$$\begin{aligned} \mathbf{M} &= \mathbf{M}_0 + \mu_0\gamma\mathbf{H} \\ \mathbf{B} &= \mathbf{M}_0 + \mu\mathbf{H} \\ \mu &= \mu_0(1 + \gamma) \end{aligned} \tag{1.55}$$

The potential energy of a preexistent dipole $\mathbf{m}$ at a place where the potential field is U, defined as the work needed to bring the dipole from infinity to its final position against the magnetic forces, is

$$W = \mathbf{m} \cdot \operatorname{grad} U \tag{1.56}$$

Combining this with (1.51) for the potential due to another dipole $\mathbf{m}'$, the energy of interaction of two magnetic dipoles becomes

$$W = (4\pi\mu_0)^{-1}\mathbf{m} \cdot \operatorname{grad} (\mathbf{m}' \cdot \operatorname{grad} r^{-1}) \tag{1.57}$$

The minus sign originates from the fact that in this expression the gradient is taken in both cases for variation of the location of $\mathbf{m}$.

The energy of a system in which the distribution of permanent magnetization is $\mathbf{M}_0$ is

$$W = \tfrac{1}{2}\iiint \mathbf{M}_0 \cdot \operatorname{grad} U \, dv \tag{1.58}$$

by reasoning like that leading to (1.26). This can be written

$$W = -\tfrac{1}{2}\iiint \mathbf{M}_0 \cdot \mathbf{H} \, dv = \tfrac{1}{2}\iiint \mu H^2 \, dv \tag{1.59}$$

since $\iiint \mathbf{B} \cdot \mathbf{H} \, dv = 0$ for any field. This result follows from the fact that div $\mathbf{B} = 0$ and curl $\mathbf{H} = 0$.

The scheme presented in this section is applicable for description of the fields due to magnets and their interaction with magnetizable matter, but it is not suitable for discussing the case of magnetization of material produced by magnetic fields set up by electric currents. For this latter case, which is much more commonly met in practice, the methods described in the next section are to be used.

7. Magnetic Fields Due to Electric Currents

Ampère's experimental investigations of the magnetic forces between two rigid circuits carrying currents J_1 and J_2 led him to the following expression for the total force on circuit 2 due to its interaction with circuit 1;

$$\mathbf{F}_2 = \frac{\mu_0 J_1 J_2}{4\pi} \oint\oint \frac{d\mathbf{r}_2 \times (d\mathbf{r}_1 \times \mathbf{r}_{21})}{r^3} \tag{1.60}$$

where $\mathbf{r}_{21}$ is the vector drawn from the element of length of the first circuit, $d\mathbf{r}_1$, to the element $d\mathbf{r}_2$ and r is the magnitude of the distance between these two elements of length of the circuit. The two integrations are to be performed as line integrals around each circuit. Here μ_0 is a constant which depends on the units used.

If $\mathbf{F}$ is measured in dynes, then in the cgs electromagnetic system Eq. (1.60) implies the fundamental definition of the *abampere* by setting $\mu_0/4\pi = 1$. The abampere is equal to 10 amp, where the ampere is the practical unit. In the mks system, $\mathbf{F}$ is measured in newtons and lengths in meters, and the quantity $\mu_0/4\pi$ is assigned the value 10^{-7}, which has the effect that the unit of current so defined is the *ampere*.

At this point, comparison should be made with the discussion of units for electric charge based on (1.1). There it was stated that the value of ϵ_0 which must be used in (1.1) if the charge is in coulombs (ampere-second), r in meters, and $\mathbf{F}$ in newtons is $4\pi\epsilon_0 = 10^7/c^2$, where c is the velocity of light in meters per second. The 10^7 appearing here is really the value just assigned to $4\pi/\mu_0$, and so the general relation between the basic parameters ϵ_0 and μ_0 is

$$c^2 = (\epsilon_0\mu_0)^{-1} \tag{1.61}$$

Here the connection is stated as a rule of practical convenience. Its full theoretical significance appears later when the equations of the electromagnetic field are fully developed, using ϵ_0 and μ_0, and it is found that they lead to propagation of electromagnetic waves with a velocity c that is given by (1.61).

The expression for $\mathbf{F}_2$ is unsymmetrical in appearance, but on expanding the triple product it becomes

$$\mathbf{F}_2 = \frac{\mu_0 J_1 J_2}{4\pi} \oint\oint \left[\frac{d\mathbf{r}_2 \cdot \mathbf{r}_{21}}{r^3}\, d\mathbf{r}_1 - \frac{(d\mathbf{r}_1 \cdot d\mathbf{r}_2)\mathbf{r}_{21}}{r^3}\right] \tag{1.62}$$

The first integral vanishes, since $\mathbf{r}_{21}/r^3 = \operatorname{grad}(1/r)$, on integration around circuit 2, leaving only the second term. This is clearly antisymmetric in the indices 1 and 2, since $\mathbf{r}_{12} = -\mathbf{r}_{21}$, and therefore $\mathbf{F}_1 = -\mathbf{F}_2$, and so the forces of interaction of the two rigid circuits, as given by Ampère's formula, are in accord with Newton's third law of motion.

In (1.60) the force of interaction is expressed as the sum of contributions from pairs of elements of length of each circuit. It is natural to regard the integrand as being, in some sense, the contribution arising from the interaction of pairs of elements. But the same thing could be said of the nonvanishing integral in (1.62) which gives a different expression for the detailed contributions of specific elements. Since $\mathbf{F}_2$ is unaltered by adding to it any integral expression which vanishes when integrated around either complete circuit, there exists an even wider range of possible choices for the elementary law of interaction between differential elements of the two circuits than just these two.

In electromagnetic theory the form (1.60) is preferred because it lends itself naturally to regarding force 2 as being due to the interaction of current elements in 2 with a field set up by the current in circuit 1. This is done by separating $\mathbf{F}_2$ as follows:

$$\mathbf{F}_2 = J_2 \oint d\mathbf{r}_2 \times \mathbf{B}_2 \tag{1.63}$$

$$\mathbf{B}_2 = \frac{\mu_0 J_1}{4\pi} \oint \frac{d\mathbf{r}_1 \times \mathbf{r}_{21}}{r^3} = \frac{-\mu_0 J_1}{4\pi} \oint d\mathbf{r}_1 \times \operatorname{grad}_1 (1/r) \tag{1.64}$$

In this viewpoint the current in circuit 1 is regarded as setting up a field of *magnetic induction* at the location of circuit 2 with which circuit 2 interacts.

In the case of a volume distribution of current density $\mathbf{i}$, the force acting on it in a field of magnetic induction $\mathbf{B}$ is

$$\mathbf{F}_2 = \iiint \mathbf{i}_2 \times \mathbf{B}_2 \, dv_2 \tag{1.65}$$

Likewise, the field of magnetic induction $\mathbf{B}$ set up by a volume distribution of current **i** is

$$\mathbf{B}_2 = \frac{\mu_0}{4\pi} \iiint \frac{\mathbf{i} \times \mathbf{r}_{21}}{r^3}\, dv_1 \tag{1.66}$$

where $\mathbf{r}_{21}$ is the vector from the volume element dv_1 to the location for which **B** is calculated.

As to the units, the cgs unit of **B** is called the gauss and is obtained on using in (1.66) **i** in abamperes per square centimeter and lengths in centimeter with $\mu_0/4\pi = 1$. The mks unit is called the tesla and is obtained on using **i** in amperes per square meter and lengths in meters, with $\mu_0/4\pi = 10^{-7}$. The tesla is sometimes called the weber per square meter. From this it follows that the mks unit of **B**, the tesla, is equal to 10^4 gauss. This relation may also be inferred from the fact that in (1.65) in the cgs system **F** is given in dynes when **i** is in abamperes per square centimeter and **B** in gauss and dv in cubic centimeters, whereas in the mks system **F** is given in newtons when **i** is in amperes per square meter and **B** is in teslas and dv in cubic meters.

From (1.66),

$$\mathbf{B}_2 = \frac{-\mu_0}{4\pi} \iiint \mathbf{i}_1 \times \operatorname{grad}_2 \frac{1}{r}\, dv_1$$

Since here grad_2 does not operate on the variables of the integrand, this can be written

$$\mathbf{B}_2 = \operatorname{curl} \mathbf{A}_2 \tag{1.67}$$

$\mathbf{A}_2$ is an auxiliary field called the *vector potential*, defined as

$$\mathbf{A}_2 = \frac{\mu_0}{4\pi} \iiint \frac{\mathbf{i}_1\, dv_1}{r} \tag{1.68}$$

from which it follows that the field **B** as defined has the property

$$\operatorname{div} \mathbf{B} = 0 \tag{1.69}$$

Consider the solid angle Ω_2 subtended at any field point 2 by circuit 1. If point 2 is moved by $d\mathbf{r}_2$, the change in this solid angle $d\Omega_2$ will be the same as the change which would be produced if point 2 remained fixed and every part of circuit 1 received the opposite displacement, $-d\mathbf{r}_2$. The solid angle subtended by any surface at point 2 is the surface integral

$$\iint \frac{\mathbf{r}_{21} \cdot d\mathbf{S}_1}{r^3}$$

where $d\mathbf{S}_1$ is the vector element of area of the surface. In this case the surface generating $d\Omega_2$ is the narrow ribbon-shaped surface generated by moving the constant vector $-d\mathbf{r}_2$ around circuit 1. Therefore the change in solid angle $d\Omega_2$ that is under consideration is

$$d\Omega_2 = \frac{\mathbf{r}_{21} \cdot (d\mathbf{r}_2 \times d\mathbf{r}_1)}{r^3}$$

From (1.64) this, when multiplied by $\mu_0 J_1$, is equal to $\mathbf{B}_2 \cdot d\mathbf{r}_2$. Therefore the line integral of **B** in the field from point A to point B is

$$\int_A^B \mathbf{B} \cdot d\mathbf{r} = (\Omega_B - \Omega_A) \frac{\mu_0 J_1}{4\pi} \tag{1.70}$$

where Ω_B is the solid angle subtended by the circuit at point B and Ω_A is that subtended at point A.

The quantity Ω, regarded as a function of position of the point for which it is calculated, is not a single-valued function of position. If for B it has the value Ω_B, then on traversing a closed path in the field which does not loop through circuit 1 it returns to its original value. But on returning to B by a path which loops circuit 1 once, Ω_B will have increased or decreased by 4π, depending on the sign conventions and the sense in which the looping of the circuit occurred.

The line integral of **B** around a closed path is therefore related to the total current enclosed by that path in much the same way as the surface integral of **D** over a closed surface is related to the total charge inside that surface, the result being

$$\oint \mathbf{B} \cdot d\mathbf{r} = \mu_0 J \tag{1.71}$$

where J is the total current flowing through the closed path defined by the path of the line integral of **B**. Here J is to be understood as

$$J = \iint \mathbf{i} \cdot d\mathbf{S}$$

the integral extending over any open surface that is bounded by the path of the line integral and the value of J so obtained being independent of a particular choice of surface, since for steady currents $\operatorname{div} \mathbf{i} = 0$.

Applying Stokes' theorem to (1.71) (since the result is true for any path of integration), one obtains the basic field equation relating **B** to the current density:

$$\operatorname{curl} \mathbf{B} = \mu_0 \mathbf{i} \tag{1.72}$$

which together with (1.69) completely specifies the field of magnetic induction set up by a steady volume distribution of currents. At surfaces of discontinuity, if there is a current $\mathbf{i}_s$ per unit length flowing in a surface, the surface condition is

$$\operatorname{Curl} \mathbf{B} = \mu_0 \mathbf{i}_s$$

where $\operatorname{Curl} \mathbf{B} = \mathbf{n} \times (\mathbf{B}_2 - \mathbf{B}_1)$ and **n** is the unit normal drawn from side 1 to side 2. In particular, if no such surface current flows, the tangential component of **B** is continuous.

As already mentioned, this result is limited in its original experimental basis to steady currents. In case of unsteady currents, (1.72) evidently requires modification, since $\operatorname{div} \operatorname{curl} \mathbf{B} \equiv 0$ for any field, whereas, in general, $\operatorname{div} \mathbf{i}$ does not vanish.

Maxwell was led to the correct generalization of (1.72) by observing that, in the unsteady state,

$$\operatorname{div} \mathbf{i} + \frac{\partial \rho}{\partial t} = 0$$

where ρ is the electromagnetic measure of charge associated with the electromagnetic measure of current density. Therefore, if ρ is an electrostatic measure, one has to write $c^{-1}\, \partial\rho/\partial t$ in place of $\partial\rho/\partial t$. Since $\operatorname{div} \mathbf{D} = \rho$, it follows that the vector field $(\mathbf{i} + \dot{\mathbf{D}}/c)$ has, in all cases, a vanishing divergence. Maxwell named the additional term, $\dot{\mathbf{D}}/c$, *displacement current* (density) and postulated that this complete current is what properly belongs in (1.72), making the complete form of this equation

$$\text{curl } \mathbf{B} - \mu_0 c^{-1} \dot{\mathbf{D}} = \mu_0 \mathbf{i} \qquad (1.73)$$

This supposition lies at the foundation of all electrodynamics and is well-supported by the general verification of the theory of electromagnetic waves (Part 6, Chap. 1).

In this section it is supposed that no magnetic material is present; **M**, the dipole moment of unit volume, is zero everywhere, and therefore $\mathbf{B} = \mu_0 \mathbf{H}$ at all points. In the next section the problem of dealing with additional contributions due to magnetization of matter is considered.

Returning to the expression (1.63) for the force on a current-carrying conductor, since $\mathbf{i} = ne\mathbf{v}/c$ (where n is the number of charges in unit volume and **v** is their mean drift velocity), it appears natural to suppose that **F** is the aggregate effect of the force **f**,

$$\mathbf{f} = (e/c)\mathbf{v} \times \mathbf{B} \qquad (1.74)$$

acting on each one of the moving charges which make up the current. This inference is known to be correct from studies of the deflection of beams of electrons or positive ions in high-vacuum devices. The force given by (1.74) is called the *Lorentz force.*

8. Magnetization and Molecular Currents

Magnetic properties of matter are interpreted by supposing that the volume density of the magnetic-dipole moment, called intensity of magnetization **M** (Sec. 6), is really due to molecular currents. In the electric field the total charge present is the sum of the free charge and the bound charge that is internal to the atomic structure and which gives rise to the electric polarization. Analogously, the total current giving rise to the magnetic field is the free current and additional currents circulating within the structure of atoms and molecules. This interpretation of magnetic-dipole moments is in the interests of simplicity; it might turn out that some magnetic moment exists in nature that is not easily traceable to the flow of electric currents.

The magnetic-dipole moment **m** that corresponds to a volume distribution of current density **i** localized in a small region is

$$\mathbf{m} = \tfrac{1}{2}\iiint \mathbf{r} \times \mathbf{i}\, dv \qquad (1.75)$$

In particular, if the current J flows in a circular circuit of radius a, then the dipole moment is normal to the plane of the circuit, and its magnitude is $\pi a^2 J$, or J times the circuit area.

For a linear circuit of larger area carrying current J, the field will be the same as that of an equivalent magnetic shell of dipole moment J per unit area spread uniformly over any open surface that is bounded by the circuit.

If the intensity of magnetization **m** is not constant throughout the magnetic material, the molecular currents do not balance out, but there is an equivalent volume distribution of current $\mathbf{i}_m$ given by

$$\mu_0 \mathbf{i}_m = \text{curl } \mathbf{M} \qquad (1.76)$$

Correspondingly, a surface-current density

$$\mu_0 \mathbf{i}_{ms} = \text{Curl } \mathbf{M} = \mathbf{n} \times (\mathbf{M}_2 - \mathbf{M}_1)$$

This means that the magnetic field due to this distribution of equivalent currents will be the same as that due to the distribution of magnetization **M**. One postulates that the magnetic induction **B** is the vector field whose curl is given by all current, in analogy with the way that the electric vector **E** is the field whose divergence is given by all the charge. Then (1.72) gives

$$\text{curl } \mathbf{B} - (\mu_0/c)\dot{\mathbf{D}} = \mu_0(\mathbf{i} + \mathbf{i}_m) \qquad (1.77)$$

together with the obvious parallel form for describing conditions at surfaces of discontinuity. Using $\mathbf{B} = \mu_0 \mathbf{H} + \mathbf{M}$ and (1.76),

$$\text{curl } \mathbf{H} = \mathbf{i} + \dot{\mathbf{D}}/c \qquad (1.78)$$

which is the analogue of div $\mathbf{D} = \rho$.

In summary,

Electric: div **E** = total-charge density
div **D** = free-charge density

Magnetic: curl **B** = total-current density
curl **H** = free-current density

Similarly, (1.72) can be written

$$\text{curl } \mathbf{H} - (\epsilon_0/c)\dot{\mathbf{E}} = \mathbf{i} + c^{-1}\dot{\mathbf{P}} \qquad (1.79)$$

To clarify the distinction between the various kinds of current involved, the following summary may be helpful:

Equation (1.71) was developed on the supposition of no magnetization in unit volume. If there is, the equation becomes modified to the form (1.77) that is, curl **B** is generated both by **i**, the macroscopic current density of conduction currents, and by $\mathbf{i}_m$, the current density that is equivalent to the magnetization **M**. In (1.78) the contribution of the magnetization **M** has dropped out, and the quantity $c^{-1}\dot{\mathbf{D}}$ behaves like an additional current density in generating curl **H**. This additional current density is the Maxwell displacement current. The displacement-current density itself is made up of a part that is proportional to $\dot{\mathbf{E}}$ and a part that is proportional to $\dot{\mathbf{P}}$. The part proportional to $\dot{\mathbf{P}}$ is a current in the familiar sense of rate of transport of charge across a surface, since **P** itself owes its origin to the separation of charges in the molecular electric dipoles, and so $\dot{\mathbf{P}}$ implies a current due to the motion of such charges while **P** changes. The part proportional to $\dot{\mathbf{E}}$, however, is a current only in the sense of giving an additional contribution to curl **H** but not in the sense of being recognizably related to the transport of any kind of charge.

Concerning the force on a charged particle moving through a magnetized medium, as in (1.74),

$$\mathbf{f} = (e/c)\mathbf{v} \times \mathbf{b}$$

where **b** is defined as the vector field which gives the correct force. Wannier* has discussed this problem and has shown that, in actual magnetic media, **b**

* *Phys. Rev.*, **72**: 304 (1947).

tends to **B** as the velocity of the charged particle approaches c but that at lower velocities appreciable deviations occur which depend on the details of the magnetic structure of the material.

9. Electromagnetic Induction

Electromagnetic induction was discovered, by Faraday in England and independently by Henry in America, in 1831. The phenomenon is that of the induction of currents in a conducting circuit by a changing relation to a magnetic field.

The simplest case is that of a linear conductor. If the field of magnetic induction is **B**, due either to currents in other conductors, to a current in the same conductor, to permanent magnets, or to a combination of all three, then the magnetic flux through the circuit is defined as

$$\Phi = \iint \mathbf{B} \cdot d\mathbf{S} \tag{1.80}$$

Here the value of the flux is not affected by the particular choice of surface bounded by the circuit, since $\operatorname{div} \mathbf{B} = 0$ for all fields. The positive sense of $d\mathbf{S}$ is associated with a positive sense of description of the circuit by the right-hand rule.

Faraday found that the current J induced in a circuit of resistance R is given by

$$JR = -a\dot{\Phi} \tag{1.81}$$

where a is a constant whose value depends on the units. The minus sign means that increasing flux induces a current in the opposite sense to that given by the right-hand rule. The flux through such a circuit due to the induced current is therefore in a direction to oppose the flux change which produced it. This formulation is known as *Lenz's law*.

As to the units, $a = 1$ in the cgs electromagnetic system. Here **B** is in gauss, and so Φ is in gauss-square centimeters, a unit which is also called the maxwell, and sometimes the line. Then $\dot{\Phi}$ is in maxwells per second or lines per second, a unit of electromotive force which is also called the abvolt. It gives the emf needed when J is measured in abamperes and R in abohms (1 ohm = 10^9 abohms).

One abvolt is 10^{-8} volt; if one uses $\dot{\Phi}$ in abvolts and expresses J in amperes and R in ohms, then one must use $a = 10^{-8}$ in (1.81). The mks system is designed to make $a = 1$, the factor 10^{-8} being absorbed as two factors of 10^{-4}, one arising from the measurement of area in square meters instead of square centimeters, and the other arising from the introduction of a unit for **B**, called the tesla, which is 10^4 gauss. The weber is a unit of flux which is equal to 10^8 maxwells or to 1 volt-sec.

The Faraday law in (1.81) applies whether the changing flux is the result of motion of the circuit in a field or whether the circuit is fixed in position and the magnetic field is changing. The electromagnetic-field description of the two cases is quite different, however. First consider the case in which the circuit moves in a magnetic field which does not change with time. Let $\mathbf{v}_c$ be the velocity of motion in the circuit. If the circuit moves by parallel displacement, $\mathbf{v}_c$ is constant. If the circuit rotates like a rigid body about an axis through the origin, with angular velocity $\boldsymbol{\omega}$, then $\mathbf{v}_c = \mathbf{r} \times \boldsymbol{\omega}$. The discussion which follows applies in either case and to more general cases in which the form of the circuit is distorted in the motion.

The charges in the conductor are compelled to be convected along with the conductor so that they experience a force (1.74). The electromotive force in the circuit is the line integral of this force on unit charge taken around the circuit:

$$\mathcal{E} = \oint (\mathbf{v}_c \times \mathbf{B}) \cdot d\mathbf{s} \tag{1.82}$$

where $\mathcal{E}$ is in abvolts if $\mathbf{v}_c$ is in centimeters per second and **B** is in gauss. The emf arising in this way is often called motional emf. In time δt, the displacement of the element $d\mathbf{s}$ is $\mathbf{v}_c\,\delta t$, and the contribution to the integral is

$$(\mathbf{v}_c \times \mathbf{B}) \cdot d\mathbf{s}\,\delta t = \mathbf{B} \cdot (d\mathbf{s} \times \mathbf{v}_c\,\delta t) = -\mathbf{B} \cdot d\mathbf{S}$$

where $d\mathbf{S}$ is the vector element of area swept out by the motion of the conductor in time δt, and therefore the motional emf given by (1.82) is consistent with (1.81).

The case in which the flux change results from a change in the magnetic field through a circuit fixed in position clearly calls for a change in the electrostatic equation, $\operatorname{curl} \mathbf{E} = 0$. The emf around a static circuit is the line integral of **E** around the closed circuit which vanishes for all electrostatic fields. From (1.81),

$$\oint \mathbf{E} \cdot d\mathbf{s} = -a \iint \dot{\mathbf{B}} \cdot d\mathbf{S}$$

Using Stokes' theorem, the field quantities must therefore be related by

$$\operatorname{curl} \mathbf{E} = -a\dot{\mathbf{B}} \tag{1.83}$$

where $a = 1/c$ if **E** is in statvolts per centimeter and **B** is in gauss, while $a = 10^{-8}$ to give **E** in volts per centimeter with **B** in gauss. In the mks system, $a = 1$, with **E** in volts per meter and **B** in teslas.

Since $\operatorname{curl} \mathbf{E} \neq 0$ in a changing magnetic field, it is no longer possible to represent **E** as the gradient of a scalar potential. Instead, since $\operatorname{div} \mathbf{B} = 0$, it is possible to write $\mathbf{B} = \operatorname{curl} \mathbf{A}$, where **A** is a vector potential, as in (1.68). Therefore, by (1.83),

$$\operatorname{curl} (\mathbf{E} + a\dot{\mathbf{A}}) = 0$$

and it is possible to introduce a scalar potential V to represent this combination, so that

$$\mathbf{E} = -a\dot{\mathbf{A}} - \operatorname{grad} V \tag{1.84}$$

Consider now the alternative descriptions of the same phenomena by two observers using two different frames of reference that are in uniform relative motion with respect to each other. Let **v** be the velocity of observer B's frame of reference relative to that of observer A, and suppose that $|\mathbf{v}| \ll c$ so that relativistic corrections are negligible. Let $\mathbf{E}_A$, $\mathbf{B}_A$ be the fields as they appear to observer A, and $\mathbf{E}_B$, $\mathbf{B}_B$ be the fields as they appear to observer B. Both observers will agree on the total force which acts on a charge, and so

$$\mathbf{E}_A + \mathbf{v}_A \times \mathbf{B}_A = \mathbf{E}_B + \mathbf{v}_B \times \mathbf{B}_B$$

where $\mathbf{v}_A$ and $\mathbf{v}_B$ are the velocities of the same particle with respect to the two reference frames, so that

$\mathbf{v}_B = \mathbf{v}_A + \mathbf{v}$. Therefore,

$$\mathbf{E}_B = \mathbf{E}_A - \mathbf{v} \times \mathbf{B} \qquad (1.85)$$

to the approximation in which **B** may be regarded as the same in both systems. Thus observer B interprets the part of the force which A calls $\mathbf{v} \times \mathbf{B}$ as included in what seems to him to be the electric field. Actually, the $\mathbf{B}_B$ will differ from $\mathbf{B}_A$ also by quantities of the order of **v** because of the difference in motion of charges giving rise to the magnetic field as they will be reckoned in the two different frames of reference. In the relativistic treatment of the electromagnetic field, **E** and **B** are components of a larger physical entity which transforms like an antisymmetric tensor in the four-dimensional space time (Part 2, Chap. 6).

When a current J flows in a closed linear circuit, it sets up a magnetic field which is proportional to the current, if it is assumed that the permeability of the medium is constant. This magnetic field gives rise to a flux through the circuit which is proportional to the current in the circuit. Hence, for any circuit,

$$\Phi = LJ \qquad (1.86)$$

where L is a coefficient whose value depends on the units and on the shape of the circuit and the distribution of magnetizable material in the field. The coefficient is called the *self-inductance* of the circuit. The mks unit is the henry, which gives the flux through the circuit in webers for a unit current of 1 amp. If the current in the circuit increases for any reason, the flux through the circuit must also increase, and this gives rise to an emf of self-induction which acts to oppose the increase in the current by Lenz's law, and the amount of which is $L\dot{J}$. Therefore the form assumed by Ohm's law for a circuit of resistance R and self-inductance L, in which there is an impressed emf E, is

$$\dot{L} + RJ = E \qquad (1.87)$$

This shows, for example, that if the current is initially zero and the emf E is applied at $t = 0$, the current will rise according to the equation

$$J(t) = \frac{E}{R}(1 - e^{-Rt/L}) \qquad (1.88)$$

The current behaves as if it had inertia, the time constant for arriving at the steady Ohm's-law value of the current being L/R, which is in seconds if L is in henrys and R is in ohms.

Multiplying (1.87) through by $J(t)$, one sees that the total work done by the active source of emf in the circuit is equal to the familiar term J^2R, which is converted into heat by Joule's law, and representing the change in an additional term $\frac{1}{2}LJ^2$. This term represents the energy stored in setting up the magnetic field against the back emf of self-induction and is reversibly recoverable when the current decreases and the emf of self-induction then acts in the sense of working to prevent the diminution of the current.

When a number of circuits are in close relation to each other, then current in one circuit, say J_1, sets up a magnetic field which gives rise to a total flux Φ_2 through circuit 2, of the form

$$\Phi_2 = M_{21}J_1 \qquad (1.89)$$

Here the coefficient M_{21} is called the coefficient of mutual induction or the *mutual inductance* between the two circuits, and the two circuits are said to be inductively coupled. From the basic laws of the electromagnetic field, it follows that $M_{21} = M_{12}$, and so the mutual-inductive effects are symmetrical. The mutual inductance gives rise to an emf in one circuit because of changing of the current in another which has many important applications. Mutual inductance is measured in the same units as self-inductance.

Bibliography

General Treatises:

Becker, R.: in F. Sauter (ed.), "Electromagnetic Fields and Interactions," 2 vols., Blaisdell Publishing Co., New York, 1964.

Bleaney, B. I., and B. Bleaney: "Electricity and Magnetism," Oxford University Press, Fair Lawn, N.J., 1957.

Corson, Dale, and Paul Lorrain: "Introduction to Electromagnetic Fields and Waves," Freeman, San Francisco, 1962.

Hallén, Erik: "Electromagnetic Theory," Chapman & Hall, London, 1962.

Harnwell, G. P.: "Principles of Electricity and Electromagnetism," 2d ed., McGraw-Hill, New York, 1949.

Jackson, J. D.: "Classical Electrodynamics," Wiley, New York, 1962.

Jeans, J. H.: "The Mathematical Theory of Electricity and Magnetism," 5th ed., Cambridge University Press, London, 1925.

O'Rahilly, Alfred: "Electromagnetics, A Discussion of Fundamentals," Longmans, London, 1938.

Panofsky, W. K. H., and Melba Phillips: "Classical Electricity and Magnetism," 2d ed., Addison-Wesley, Reading, Mass., 1962.

Rocard, Y.: "Principles of Electricity and Magnetism," Sir Isaac Pitman & Sons, London, 1959.

Smythe, W. R.: "Static and Dynamic Electricity," 2d ed., McGraw-Hill, New York, 1950.

History:

Whittaker, E. T.: "A History of the Theories of Aether and Electricity, from the Age of Descartes to the Close of the Nineteenth Century," rev. ed., Nelson, London, 1951. Paperback reprint: Harper Torchbooks, Harper & Row, New York, 1960.

Units:

Birge, R. T.: On Electric and Magnetic Units and Dimensions, *Am. Phys. Teacher*, **2**: 41 (1934).

Birge, R. T.: On the Establishment of Fundamental and Derived Units, with Special Reference to Electric Units, *Am. Phys. Teacher*, **3**: 102, 171 (1935).

Brown, W. F., Jr.: *Am. J. Phys.*, **8**: 338 (1940).

Coulomb's Law Committee of the American Association of Physics Teachers: The Teaching of Electricity and Magnetism at the College Level, *Am. J. Phys.*, **18**: 1, 69 (1950).

Page, L., and N. I. Adams, Jr.: A Proposed Reformulation of the Electromagnetic Equations and Revision of Units, *J. Franklin Inst.*, **218**: 517 (1934).

Silsbee, F. B.: Systems of Electrical Units, *Natl. Bur. Std. (U.S.), Monograph* 56, 1962.

Chapter 2

Static Electric and Magnetic Fields

By E. U. CONDON, University of Colorado

1. Field Due to Given Charge Distribution

If the charge density at $\mathbf{r}$ is ρ, then, using units such that $\epsilon = 1$, the electrostatic potential due to this charge distribution is

$$V(\mathbf{R}) = \int \frac{\rho(\mathbf{r})\,dv}{|\mathbf{R} - \mathbf{r}|} \tag{2.1}$$

If the charge distribution occupies a limited region of space, then the potential at $\mathbf{R}$ can be expanded in a series of descending powers of R. The basic expansion is

$$\frac{1}{|\mathbf{R} - \mathbf{r}|} = \frac{1}{R} \sum_{k=0}^{\infty} \left(\frac{r}{R}\right)^k P_k(\cos\omega) \qquad r < R \tag{2.2}$$

where ω is the angle between $\mathbf{r}$ and $\mathbf{R}$ and $P_k(\cos\omega)$ is the kth Legendre polynomial of its argument. Accordingly, $V(\mathbf{R})$ can be written $\Sigma_k V_k(\mathbf{R})$, where

$$V_k(\mathbf{R}) = \frac{1}{R^{k+1}} \int \rho r^k P_k(\cos\omega)\,dv$$

Since $\int_0^\pi P_k(\cos\omega)\sin\omega\,d\omega = 0$, for $k \neq 0$, it follows that, if $\rho(\mathbf{r})$ is spherically symmetric, all the $V_k(\mathbf{R})$ vanish except $V_0(\mathbf{R})$. Therefore, for a spherically symmetric distribution, the field is spherically symmetric and the same as if the total charge were concentrated at the origin.

The term for $k = 1$ is called the dipole term and is directly expressible in terms of the *dipole moment* of the given charge distribution. Since

$$P_1(\cos\omega) = \cos\omega$$

this term is

$$V_1(\mathbf{R}) = \frac{\mathbf{M} \cdot \mathbf{R}_0}{R^2} \tag{2.3}$$

where $\mathbf{M} = \int \rho(\mathbf{r})\mathbf{r}\,dv$ and $\mathbf{R}_0$ is the unit vector in the direction of $\mathbf{R}$. Choosing the polar axis along the direction of $\mathbf{M}$, one has at the point R,Θ,Φ

$$V_1(R,\Theta,\Phi) = \frac{M\cos\Theta}{R^2}$$

and therefore the electric vector is

$$\mathbf{E}_1(R,\Theta,\Phi) = \frac{2M\cos\Theta}{R^3}\mathbf{R}_0 + \frac{M\sin\Theta}{R^3}\mathbf{\Theta}_0 \tag{2.4}$$

where $\mathbf{\Theta}_0$ is a unit vector in the direction of increasing Θ.

The term for $k = 2$ is called the quadrupole term and is directly expressible in terms of the *quadrupole moment* of the given charge distribution. The term $k = 3$ is called the octopole term, and so on, the kth term being known as the 2^k*-pole* term.

A quadrupole term is related to the second-order moments of the charge distribution. Let

$$\mathfrak{J} = \mathbf{ii} + \mathbf{jj} + \mathbf{kk}$$

stand for the unit tensor of second order; then the *quadrupole moment* of the charge distribution $\mathfrak{Q}$ can be defined as the tensor

$$\mathfrak{Q} = 3\int\rho(\mathbf{r})\mathbf{rr}\,dv - \mathfrak{J}\int\rho(\mathbf{r})r^2\,dv \tag{2.5}$$

(Various slightly different definitions of the quadrupole moment are to be found in the literature, and so in consulting a particular reference one must attempt to learn from the context which definition is implied.) In particular, if the charge distribution has an axis of symmetry, which is taken to be the z axis, the quadrupole-moment tensor is entirely expressible in terms of a single quantity,

$$Q = \int\rho(\mathbf{r})(3z^2 - r^2)\,dv \tag{2.6}$$

according to the scheme

$$\mathfrak{Q} = Q(-\tfrac{1}{2}\mathbf{ii} - \tfrac{1}{2}\mathbf{jj} + \mathbf{kk}) \tag{2.7}$$

Since $P_2(\cos\omega) = \frac{3}{2}\cos^2\omega - \frac{1}{2}$ and

$$\cos\omega = \frac{xX + yY = zZ}{rR}$$

it follows that the coefficient of, say, XY/R^5 in the expression for $V_2(\mathbf{R})$ is $3\int\rho xy\,dv = Q_{xy}$, and therefore the tensor form of $V_2(\mathbf{R})$ is

$$V_2(\mathbf{R}) = \frac{\mathbf{R}_0 \cdot \mathfrak{Q} \cdot \mathbf{R}_0}{2R^3} \tag{2.8}$$

The quadrupole moment, although written as a second-order tensor, is really a five-component quantity, because it is represented by a symmetric tensor of zero diagonal sum, and so the number of independent components is the same as the number of independent spherical harmonics of the second order.

For higher values of k it is more convenient to work with the expression of the higher-order moments in terms of the $(2k + 1)$-component quantities associated with spherical harmonics of the kth order than to represent the kth term in terms of a tensor of kth order.

The spherical-harmonic addition theorem,

$$P_k(\cos \omega) = \frac{4\pi}{2k+1} \sum_{m=-k}^{+k} F_{km}(\theta) F_{km}(\Theta) e^{im(\phi - \Phi)}$$

where

$$F_{km}(\theta) = (-1)^m \sqrt{\frac{2k+1}{4\pi} \frac{(k-m)!}{(k+m)!}} \sin^m \theta \frac{d^m}{d(\cos \theta)^m} P_k(\cos \theta)$$

for $m \geq 0$ and the same thing with omission of the factor $(-1)^m$ for $m < 0$, permits the writing of $V_k(\mathbf{R})$ in the form

$$V_k(\mathbf{R}) = \frac{1}{R^{k+1}} \sum_{m=-k}^{+k} Q_{km} F_{km}(\Theta) e^{-im\Phi}$$

where the 2^k-pole moment is fully described by the $2k + 1$ quantities,

$$Q_{km} = \frac{4\pi}{2k+1} \int \rho(\mathbf{r}) F_{km}(\theta) e^{im\phi} \, dv \tag{2.9}$$

From the orthogonality properties of the spherical harmonics, it follows that a charge distribution whose angular dependence is that of any spherical harmonic gives rise to a potential field whose angular dependence is given by the same spherical harmonic.

A similar expansion in spherical harmonics is often useful in case the charge distribution *lies wholly outside* a limited region of space. In this case the potential inside this region is expressible in an *ascending* series of positive powers of R by using (2.2) with r and R interchanged, since now $R < r$. This gives $V = \Sigma V_k(\mathbf{R})$, where

$$V_k(\mathbf{R}) = R^k \int \frac{\rho(\mathbf{r}) P_k(\cos \omega)}{r^{k+1}} \, dv \tag{2.10}$$

In particular, the potential at the origin is

$$V_0(\mathbf{R}) = \int \frac{\rho(\mathbf{r}) \, dv}{r}$$

and the first-order term in the space variation for places near the origin is given by

$$V_1(\mathbf{R}) = -\mathbf{R} \cdot \mathbf{E}_1$$

where $\mathbf{E}_1 = -\int [\rho(r)\mathbf{r}/r^3] \, dv$. Here $\mathbf{E}_1$ is the value of the electric field at the origin. Similar expressions are also derivable to represent the terms in R^k in terms of $2k + 1$ integrals involving the $2k + 1$ independent spherical harmonics of order k.

2. Force on a Rigid Charge Distribution

Suppose that the charge distribution $\rho(\mathbf{r})$ is rigidly attached to a frame that is free to move in translation and rotation. If this distribution is placed in a field $\mathbf{E}(\mathbf{r})$ due to other charges, it will experience a force $\mathbf{F}$ and a torque $\mathbf{T}$ given by

$$\begin{aligned} \mathbf{F} &= \int \rho(\mathbf{r}) \mathbf{E}(\mathbf{r}) \, dv \\ \mathbf{T} &= \int \rho(\mathbf{r}) \mathbf{r} \times \mathbf{E}(\mathbf{r}) \, dv \end{aligned} \tag{2.11}$$

If the charge distribution is confined to a small region of space so that the space variations of $\mathbf{E}(\mathbf{r})$ over the region where $\rho \neq 0$ are not very great, then it is often convenient to expand these expressions in such a way as to exhibit which parts come from the action of a uniform field and which from departures from uniformity. The origin will be taken near the center of the charge distribution. Near the origin, $\mathbf{E}(\mathbf{r})$ has the form

$$\mathbf{E}(\mathbf{r}) = \mathbf{E}_1 + \mathfrak{E}_2 \cdot \mathbf{r} + \cdots \tag{2.12}$$

where $\mathfrak{E}_2$ is a constant symmetric (since curl $\mathbf{E} = 0$) tensor of zero diagonal sum (since div $\mathbf{E} = 0$). Each component of $\mathbf{E}_2$ is of the form $(\partial E_x/\partial y)_0$, and so $\mathfrak{E}_2$ is the tensor which describes the inhomogeneity of the field.

The first two terms in the force expansion are therefore $\mathbf{F} = \mathbf{F}_1 + \mathbf{F}_2 + \cdots$, with

$$\begin{aligned} \mathbf{F}_1 &= e\mathbf{E}_1 \qquad e = \int \rho \, dv \\ \mathbf{F}_2 &= \mathfrak{E}_2 \cdot \mathbf{M} \end{aligned} \tag{2.13}$$

where $\mathbf{M}$ is the dipole moment of the charge distribution, defined in (2.3). The next term involves the quadrupole moment of the charge distribution multiplied into a third-order tensor whose components are the second-order space derivatives of $\mathbf{E}$.

Similarly, the first two terms in the torque are $(\mathbf{T} = \mathbf{T}_1 + \mathbf{T}_2 + \cdots)$.

$$\begin{aligned} \mathbf{T}_1 &= \mathbf{M} \times \mathbf{E}_1 \\ \mathbf{T}_2 &= \int \rho(\mathbf{r}) \mathbf{r} \times \mathfrak{E}_2 \cdot \mathbf{r} \, dv \end{aligned} \tag{2.14}$$

The latter is expressible in terms of the quadrupole moment of the charge distribution, defined in (2.5),

$$\mathbf{T}_2 = \left(\sum Q_{yx} \frac{\partial E_z}{\partial x} - \sum Q_{zx} \frac{\partial E_y}{\partial x} \right) \mathbf{i} + \cdots + \cdots$$

where Σ means the summation of the double index x over the three values x,y,z. If a charge distribution has a vanishing dipole moment, it experiences no torque in a *uniform* electric field.

The potential energy of the charge distribution is likewise expressible in a series involving the interaction of higher moments with terms representing higher orders of inhomogeneity of the field. Thus,

$$U = \int \rho(\mathbf{r}) V(\mathbf{r}) \, dv$$

is the potential energy of the charge distribution $\rho(\mathbf{r})$ in the given external field whose potential is $V(\mathbf{r})$. One may write

$$V(r) = V_0 - \mathbf{E}_1 \cdot \mathbf{r} - \tfrac{1}{2}\mathbf{r} \cdot \mathfrak{E}_2 \cdot \mathbf{r}$$

a form which is consistent with (2.12) since

$$\mathbf{E} = -\operatorname{grad} V$$

This gives $U = \Sigma U_k$, where

$$\begin{aligned} U_0 &= eV_0 \\ U_1 &= -\mathbf{M} \cdot \mathbf{E}_1 \\ U_2 &= -\tfrac{1}{2} \int \rho(\mathbf{r}) \mathbf{r} \cdot \mathfrak{E}_2 \cdot \mathbf{r} \, dv \end{aligned} \tag{2.15}$$

Writing out U_2 in full,

$$U_2 = -\tfrac{1}{6}\left(Q_{xx}\frac{\partial E_x}{\partial x} + Q_{yy}\frac{\partial E_y}{\partial y} + Q_{zz}\frac{\partial E_z}{\partial z}\right)$$

In particular, if the charge distribution is symmetric about the z axis, this reduces to

$$U_2 = -\tfrac{1}{4}Q\frac{\partial^2 V}{\partial Z^2} \tag{2.16}$$

with Q as defined in (2.6). This result finds application in the calculation of the energy of interaction of nuclear quadrupole moments with the electric field at the nucleus due to the extranuclear electrons in a molecule (Part 7, Chap. 4).

3. Interaction of Two Rigid Charge Distributions

The energy of interaction of two charge distributions $\rho_1(\mathbf{r})$ and $\rho_2(\mathbf{r})$ which are separated in space may be given in terms of their moments by using the results of Sec. 1 to express the field due to the first distribution in terms of its moments and (2.15) to express the energy of the second charge distribution in this field.

To the second order, the potential at $\mathbf{R}$ due to the distribution $\rho_1(\mathbf{r})$, supposed concentrated near the origin, is

$$V(\mathbf{R}) = \frac{e_1}{R} + \frac{\mathbf{M}_1\cdot\mathbf{R}_0}{R^2} + \frac{\mathbf{R}_0\cdot\mathfrak{Q}_1\cdot\mathbf{R}_0}{2R^3}$$

The terms in the interaction energy may be classified as to magnitudes as (1) charge-charge, (2) charge-dipole, (3) dipole-dipole, (4) charge-quadrupole interactions, etc.

The charge-charge interaction is simply e_1e_2/R. The charge-dipole interaction is

$$\frac{(e_2\mathbf{M}_1 - e_1\mathbf{M}_2)\cdot\mathbf{R}_0}{R^2} \tag{2.17}$$

where $\mathbf{R}_0$ is the unit vector from the first charge distribution to the second charge distribution. The dipole-dipole interaction is

$$\frac{\mathbf{M}_1\cdot\mathbf{M}_2 - 3\,\mathbf{M}_1\cdot\mathbf{R}_0\,\mathbf{M}_2\cdot\mathbf{R}_0}{R^3} \tag{2.18}$$

This result follows from the fact that the field at the second dipole due to the first is

$$\mathbf{E} = \frac{1}{R^3}(3\,\mathbf{M}_1\cdot\mathbf{R}_0\,\mathbf{R}_0 - \mathbf{M}_1)$$

combined with the fact that the energy of the second dipole in this field is $-\mathbf{M}_2\cdot\mathbf{E}_2$. The charge-quadrupole interaction is

$$\frac{1}{2}\frac{\mathbf{R}_0\cdot(e_1\mathfrak{Q}_2 + e_2\mathfrak{Q}_1)\cdot\mathbf{R}_0}{R^3} \tag{2.19}$$

and is therefore of the same order as the dipole-dipole interaction.

Interactions of higher order are usually of consequence only when those of lower order vanish. Thus between two nonpolar neutral atoms the interaction of lowest order would be the quadrupole-quadrupole interaction, which varies as R^{-5}.

4. Conductor in a Given Field

If a conductor is put in a given field which is described by the potential function $V_0(\mathbf{r})$, there will, in general, be a redistribution of charge on the conductor, giving rise to an additional field whose potential function is $V_1(\mathbf{r})$. In equilibrium, a conductor is always an equipotential, and therefore $V_1(\mathbf{r})$ must be such that $V_0(\mathbf{r}) + V_1(\mathbf{r})$ is constant on the surface of the conductor.

Two cases may be distinguished. If the conductor is uncharged, then $V_1(\mathbf{r})$ is subject to the further condition that $\int \text{grad}\,\mathbf{V}_1\cdot d\mathbf{S}$, taken over the closed boundary surface on the conductor, must vanish. In this case, $V_1(\mathbf{r})$ is determined as being a solution of Laplace's equation, satisfying this integral boundary condition and having values equal to $C - V_0(\mathbf{r})$ on the surface of the conductor. If, however, the conductor is connected to earth (zero potential) by a fine wire, then the charge can pass to the conductor in such a way that the function $V_1(\mathbf{r})$ has to be determined by the condition that

$$V_0(\mathbf{r}) + V_1(\mathbf{r}) = 0$$

over the boundary of the conductor. In this case, the total charge on the conductor is

$$q = -\frac{1}{4\pi}\int \text{grad}\,V\cdot d\mathbf{S}$$

The simplest example is that of a conducting sphere (of radius a) under the influence of a given charge distribution external to it. According to (2.10), the potential of the external field will be expandable in a series of spherical harmonics $S_k(\theta,\phi)$ and powers of the distance r from the center of the sphere. Hence

$$V_0(\mathbf{r}) = \sum_k r^k S_k^0(\theta,\phi)$$

where the $S_k^0(\theta,\phi)$ are spherical harmonics determined by the external charge distribution. The field outside the sphere due to any distribution of charge on the conducting sphere will likewise be representable in the form

$$V_1(\mathbf{r}) = \sum_k \frac{1}{r^{k+1}} S_k^1(\theta,\phi)$$

where the $S_k^1(\theta,\phi)$ are spherical harmonics determined by the unknown distribution of charge on the sphere. The condition that $V_0(\mathbf{r}) + V_1(\mathbf{r}) = \text{const}$ on $r = a$ leads to the requirement that

$$S_k^1(\theta,\phi) = -a^{2k+1}S_k^0(\theta,\phi)$$

for $k = 1, 2, \ldots$. For $k = 0$, the spherical harmonic is constant anyway, and this term in $V_0(\mathbf{r})$ reduces to $V_0(0)$, the potential at the location of the center of the sphere before the sphere was introduced. Because of the orthogonality properties of spherical harmonics, none of the terms for $k \neq 0$ contributes to any net charge on the sphere, and therefore the complete expression for the potential of the given field as modified by an *uncharged* conducting sphere of radius a is

$$V_0(\mathbf{r}) = V_0(0) + \sum_{k=1} \left(r^k - \frac{a^{2k+1}}{r^{k+1}}\right) S_k{}^0(\theta,\phi) \qquad (2.20)$$

which reduces to $V(\mathbf{r}) = V_0(0)$ at $|\mathbf{r}| = a$.

If the sphere is connected to the earth by a fine wire so that the additional condition is $V(\mathbf{r}) = 0$ at $|\mathbf{r}| = a$, then a term $-V_0(0)(a/r)$ has to be added to (2.20) to satisfy this condition. This corresponds to a total charge on the sphere of $-V_0(0)a$ units of charge.

In particular, if the given external field is a *uniform field* in the z direction, then the only term in (2.20) is that for $k = 1$, and, if E_1 is the magnitude of the uniform field,

$$V_0(\mathbf{r}) = V_0(0) - E_1 r \cos\theta$$

so that the field, as modified by the presence of the conducting sphere, is

$$V(r) = V_0(0) - E_1\left(r - \frac{a^3}{r^2}\right)\cos\theta$$

The term introduced by the presence of the sphere corresponds to that of a dipole of moment $M = E_1 a^3$ directed parallel to the external electric field. Because of this fact, one speaks of the sphere as being *polarizable,* and the coefficient of proportionality between the induced dipole moment $\mathbf{M}$ and the external field $\mathbf{E}_1$, which in this case is a^3, is called the *polarizability* of the sphere.

In general, if a conductor of arbitrary shape is put in a uniform field, the leading term in the expansion of $V_1(\mathbf{r})$ will be that which corresponds to an induced dipole moment $\mathbf{M}$ that is proportional to the external field $\mathbf{E}_1$ but not, in general, in the same direction as $\mathbf{E}_1$. The general relation is that the conductor is characterized by a symmetric polarizability tensor $\boldsymbol{\alpha}$ such that $\mathbf{M} = \boldsymbol{\alpha} \cdot \mathbf{E}_1$ which will have three principal axes along which $\mathbf{M}$ is parallel to $\mathbf{E}_1$ and three principal values associated with these axes.

5. System of Conductors

If several conductors different as to shape and location are given, the charges on them will come to equilibrium in such a way as to make each conductor an equipotential. The field has to be determined by solving a boundary-value problem for the potential. The potential V satisfies Laplace's equation outside each conductor and assumes constant values on each of them.

Let the several conductors be designated by the subscripts 1, 2, . . . , n. Consider the particular problem of determining the field when the first conductor is at unit potential and all the others are at zero potential. The field associated with this will be denoted by $v_1(\mathbf{r})$, and likewise $v_s(\mathbf{r})$ will be the potential corresponding to the case in which the sth conductor is at unit potential and the others are at zero. Then, owing to the linearity of Laplace's equation and the boundary conditions, the potential representing the field when the potentials of the conductors are $V_1, V_2, \ldots, V_n$ is

$$V(\mathbf{r}) = \sum_{s=1}^{n} V_s v_s(\mathbf{r}) \qquad (2.21)$$

The charge on the tth conductor when the sth conductor is at unit potential and all others are at zero potential will be denoted by C_{ts}. It is found from $v_s(\mathbf{r})$ by calculation of the surface integral of the normal component of the gradient over the closed bounding surface of the tth conductor:

$$c_{ts} = -\frac{1}{4\pi}\int \operatorname{grad} v_s(\mathbf{r}) \cdot d\mathbf{S} \qquad (2.22)$$

where $d\mathbf{S}$ is the outward drawn element of area of the surface bounding the tth conductor.

When the potentials are given, the total charge on the tth conductor will therefore be

$$Q^t = \sum_s c_{ts} V_s \qquad (2.23)$$

The coefficients of like index, as c_{tt}, are known as *coefficients of capacity* and of unlike index as *coefficients of induction.* They are measured in centimeters in the cgs electrostatic system and in farads in the practical system with Q in coulombs and V in volts. The c_{rr} are always positive, but the c_{rs} are negative since the charge induced on a grounded conductor is opposite in sign to that of the inducing charge.

Suppose that $\rho_s(\mathbf{r})$ is written symbolically for the charge distribution associated with the field $v_s(\mathbf{r})$ and $\rho_t(\mathbf{r})$ for that associated with $v_t(\mathbf{r})$. (In actual fact, the charges involved will be surface distributions on the surfaces of the conductors.) Now,

$$\int \rho_s(\mathbf{R}) v_t(\mathbf{R})\, dV = \iint \frac{\rho_s(\mathbf{R})\rho_t(\mathbf{r})}{|\mathbf{R} - \mathbf{r}|}\, dv\, dV = \int \rho_t(\mathbf{R}) v_s(\mathbf{R})\, dV$$

Since the only charges involved are those on the conductors, which are equipotentials, the integral on the left is just equal to the charge on the tth conductor when the sth conductor is at unit potential; all others vanish; that is, the integral on the left is c_{ts}, and that on the right is c_{st}. Hence $c_{ts} = c_{st}$.

For some purposes it is more convenient to express the potentials in terms of the charges. These coefficients are written p_{rs} and are known as the *coefficients of potential:*

$$V_r = \sum_s p_{rs} Q_s \qquad (2.24)$$

The matrix of the p_{rs} is reciprocal to the matrix of the c_{rs}, and so $p_{rs} = p_{sr}$. Since a positive charge on any one conductor when all others are uncharged will make a positive potential field everywhere, it follows that all $p_{rs} \geq 0$.

The energy needed to add the charge dQ_s on to the sth conductor when it is already at potential V_s is $V_s\,dQ_s$, and so the energy needed to add charges $dQ_1, dQ_2, \ldots, dQ_n$ is $dW = \Sigma_s V_s\,dQ_s$. Hence the total energy expended in bringing up the charges from zero is $W = \frac{1}{2}\Sigma_{sr}p_{sr}Q_sQ_r$ or, in terms of the potentials,

$$W = \tfrac{1}{2}\Sigma c_{sr}V_sV_r \tag{2.25}$$

If λ is a generalized coordinate specifying the position of one of the conductors, then all the p_{sr} change with a change in λ. The generalized force F_λ acting to change λ will be such that $F_\lambda\,d\lambda = -dW$, and so

$$F_\lambda = -\frac{\partial W}{\partial \lambda} = -\frac{1}{2}\sum_{sr}\frac{\partial p_{sr}}{\partial \lambda}Q_sQ_r \tag{2.26}$$

This is the mechanical force acting when the *charges* on the conductors remain constant during the variation of λ.

If, however, the *potentials of the conductors* are maintained constant during the variation of λ, then it is necessary to consider that this will require energy from the batteries or other means used to keep the potentials constant. In the change λ to $\lambda + d\lambda$ the change in energy of the system is

$$dW = \frac{1}{2}\sum\left(\frac{\partial c_{sr}}{\partial \lambda}\right)d\lambda\;V_sV_r$$

and the change in the charges needed to keep the potentials constant is $dQ_r = \Sigma(\partial c_{rs}/\partial\lambda)\,d\lambda\;V_s$, so that the *electrical work* done on the system in moving up these charges is $dW_E = \Sigma V_r\,dQ_r = 2dW$. Hence the mechanical work done in the displacement corresponds to $-dW$; the generalized force tending to increase λ is

$$F = +\frac{\partial W}{\partial \lambda} = \frac{1}{2}\sum_{sr}\frac{\partial c_{rs}}{\partial \lambda}V_sV_r \tag{2.27}$$

In this case, the electrical work done by the batteries is twice the increase in energy of the system, and the extra work done by the batteries is given out as work done mechanically by the system.

When two conductors are so arranged as to enhance their electrostatic interaction, they are together regarded as a *capacitor*. It is convenient to write the equations in such a way that the dependence of Q_1 and Q_2 on the potential difference $V = V_1 - V_2$ is made explicit:

$$\begin{aligned} Q_1 &= (c_{11} + c_{12})V_1 - CV \\ Q_2 &= +CV + (c_{22} + c_{12})V_2 \end{aligned} \tag{2.28}$$

where $C = -c_{12} = -c_{21}$. The single quantity C is known as the *capacitance* of the capacitor. The other coefficients $c_{11} + c_{12}$ and $c_{22} + c_{12}$ are called *stray capacitances* or *capacitances to ground*. In practical capacitors an effort is made to keep these small compared with C, and in circuit theory an ideal capacitor is a device for which the stray capacitances are assumed to be zero.

In other words, a capacitor, denoted by the symbol ⊣⊢, is an ideal element for which the exact equations are assumed to be $Q_1 = -CV$, $Q_2 = +CV$.

When conductor 2 is hollow and completely surrounds a conductor 1, then $c_{11} + c_{12} = 0$, and so this stray capacitance vanishes.

The actual relationship of charges on a capacitor, including the stray capacitances, has therefore to be indicated by the circuit diagram of Fig. 2.1 in which conductor 3 is "ground" or the reference conductor of zero potential.

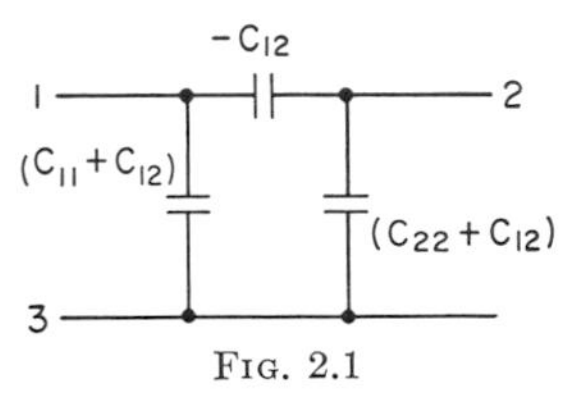

Fig. 2.1

With due regard to stray capacitance, therefore, the capacitance of a capacitor ought to be defined as *the charge induced on either plate when it is grounded and the other plate is brought to minus unit potential.*

In the *parallel-plate capacitor,* the two conductors are flat plates very close together compared with their linear dimensions. The capacitance is

$$C = \epsilon A/4\pi d \tag{2.29}$$

where A is the area of either plate, d is their distance apart, and ϵ is the dielectric constant of the medium between the plates.

In the case of concentric cylinders of radii r_1 and r_2 and of length l, the capacitance is

$$C = \epsilon l/2\ln(r_2/r_1) \tag{2.30}$$

For two *concentric spheres* of radii r_1 and r_2 it is

$$C = \epsilon r_1r_2/(r_2 - r_1) \tag{2.31}$$

In the case of two parallel wires, each of radius a whose centers are separated by distance c, the capacitance per unit length is

$$C = \frac{\epsilon}{4\cosh^{-1}(c/2a)} \to \frac{\epsilon}{4\ln(c/a)} \tag{2.32}$$

for $c \gg a$. In the case of one wire of radius a parallel to an infinite plane at a height h above it, the capacitance per unit length is

$$C = \frac{\epsilon}{2\cosh^{-1}(h/a)} \tag{2.33}$$

6. Magnetic Field Due to a Given Current Distribution

The current distribution is specified by giving the vector current density $\mathbf{I}(\mathbf{r})$ as a function of position. For steady currents, $\operatorname{div}\mathbf{I} = 0$. The magnetic field satisfies the equations

$$\operatorname{div}\mathbf{H} = 0 \qquad \operatorname{curl}\mathbf{H} = 4\pi\mathbf{I} \tag{2.34}$$

The first of (2.34) may be satisfied by introducing the *vector potential* $\mathbf{A}$, according to the equations

$$\operatorname{div}\mathbf{A} = 0 \qquad \mathbf{H} = \operatorname{curl}\mathbf{A} \tag{2.35}$$

The first of (2.35) is not strictly necessary but is introduced to simplify the work. Since

$$\text{curl curl } \mathbf{A} = \text{grad div } \mathbf{A} - \nabla^2 \mathbf{A}$$

it follows that each cartesian component of **A** satisfies Poisson's equation

$$\nabla^2\mathbf{A} + 4\pi\mathbf{I} = 0 \tag{2.36}$$

In consequence, the expansions used in Sec. 1 to express the electric potential in terms of the moments of the charge distribution are also applicable here:

$$\mathbf{A}(\mathbf{R}) = \int \frac{\mathbf{I}(\mathbf{r})\,dv}{|\mathbf{R} - \mathbf{r}|} \tag{2.37}$$

If the current distribution is confined to a limited region of space, then at points *outside that region* the expansion (2.2) may be used to write $\mathbf{A} = \Sigma\mathbf{A}_k(\mathbf{R})$, with

$$\mathbf{A}_k(\mathbf{R}) = \frac{1}{R^{k+1}} \int \mathbf{I}(\mathbf{r}) r^k P_k(\cos\omega)\,dv \tag{2.38}$$

When the spherical harmonics are written out in terms of cartesian coordinates x, y, z and X, Y, Z, the integrals in (2.38) are expressed as polynomials involving integrals of the type $\int I_x x^a y^b z^c\,dv$. Because **I** vanishes outside a limited region and $\text{div } \mathbf{I} = 0$ everywhere, there are a large number of relations between these integrals. If u is any scalar function of position, $\mathbf{I} \cdot \text{grad } u = \text{div } u\mathbf{I}$. Hence

$$\int \mathbf{I} \cdot \text{grad } u\,dv = 0$$

when taken over any volume such that $\mathbf{I} = 0$ on its bounding surface. By assuming all possible monomials of the form $x^r y^s z^t$ for u, this leads to the relations between the moment integrals of the current.

Thus, if u is taken to be x, y, and z in turn, this leads to the result

$$\int \mathbf{I}\,dv = 0 \tag{2.39}$$

Hence there is no $k = 0$ term in the expansion of the vector potential. If u is taken to be x^2, xy, y^2, yz, z^2, zx, this leads to the result that $\int \mathbf{Ir}\,dv$ is antisymmetric, i.e., that integrals of the type $\int I_x x\,dv$ vanish, and that $\int I_x y\,dv + \int I_y x\,dv = 0$. Therefore the $k = 1$ term in the expansion (2.38) can be written

$$\mathbf{A}_1(\mathbf{R}) = \frac{1}{2R^2}\mathbf{R}_0 \cdot \int (\mathbf{Ir} + \mathbf{rI})\,dv = \frac{\mathbf{M} \times \mathbf{R}_0}{R^2} \tag{2.40}$$

where $\mathbf{M} = \frac{1}{2}\int \mathbf{r} \times \mathbf{I}(\mathbf{r})\,dv$. The vector **M** is called the *magnetic-dipole moment* of the current distribution.

Continuing the process, one may introduce magnetic moments of higher order as a means of describing the terms for larger k, but these results find very little use.

Choosing the polar axis along **M** and writing (r,θ,ϕ) for the coordinates of the point at which **A** is calculated,

$$\mathbf{A}_1(r,\theta,\phi) = \frac{M \sin\theta}{r^2}\boldsymbol{\phi}_0 \tag{2.41}$$

where $\boldsymbol{\phi}_0$ is the unit vector in the direction of increasing ϕ. Calculating $\mathbf{H}_1 = \text{curl } \mathbf{A}_1$ (see Part 1, Chap. 9),

$$\mathbf{H}_1(r,\theta,\phi) = \frac{2M\cos\theta}{r^3}\mathbf{r}_0 + \frac{M\sin\theta}{r^3}\boldsymbol{\theta}_0 \tag{2.42}$$

On comparing this with (2.4), it is clear that, at points *outside the current distribution*, the magnetic field due to a magnetic dipole is of exactly the same form as the electric field due to an electric dipole.

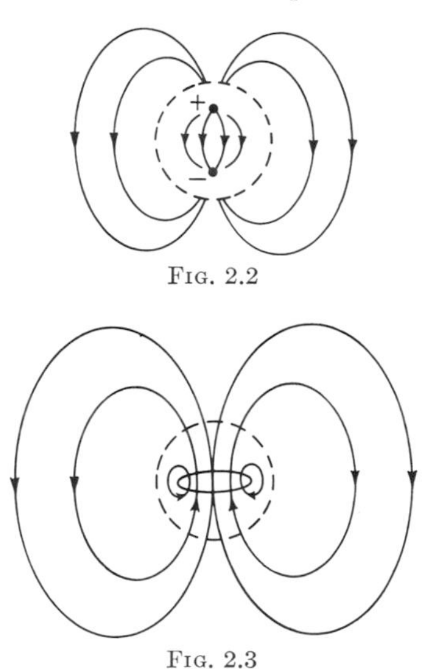

FIG. 2.2

FIG. 2.3

At points *within* the charge and current distributions which give rise to the two kinds of dipole, respectively, the fields are quite different. The simplest prototype of an electric dipole of moment **M** is a pair of charges $+e$ and $-e$ located at a distance a apart such that the line from the minus to the positive charge has the direction of **M** and such that $ea = |\mathbf{M}|$.

The lines of force start at the positive charge and end on the negative charge, and therefore the direction of **E** in the equatorial plane is everywhere opposite to that of M (see Fig. 2.2). Correspondingly, the simplest prototype of a magnetic dipole is a current flowing in a circular loop of radius a lying in the equatorial plane where the magnitude of this current i is given by $\pi a^2 i = |\mathbf{M}|$. In this case, the lines of force of **H** are continuous curves which thread through the current loop (as shown in Fig. 2.3), and therefore in the region inside the loop the direction of **H** is roughly the same as that of **M** and hence opposite to the direction of **H** in the external field.

In the historical development of the theory of magnetism it was at first supposed that the polarized fields needed to describe magnetic materials were of the nature of a magnetic dipole consisting of two close-lying magnetic poles, in exact analogy with the case of electric dipoles. This is a possible alternative when the attention is directed only to the phenomena involving fields external to the dipole. However, when this viewpoint is taken, it requires (1) assumption of the existence of magnetic poles and (2) that these always appear tied together as dipoles, for there has never been any generally accepted evidence for the existence of free magnetic poles (i.e., of magnetically non-neutral material). With the discovery of the magnetic fields due to currents, it was natural to dispense with the hypothesis of independently existent magnetic dipoles, directly analogous to electric dipoles, and to assume that all magnetic-dipole fields are really due to electric currents circulating within atoms or molecules of matter. Because the $k = 0$ term vanishes identically in (2.38), this automatically insures the magnetic neutrality of all matter and gives a basis for the description of all magnetic phenomena without assuming the existence of real magnetic-field sources which are the analogue of electric charges. This view has been completely adequate so far, but there is nothing in electromagnetic theory which rules out the existence of magnetic charges, and it may be that phenomena will be discovered in the future which will require their introduction.

Analogously to (2.10), if the current distribution $\mathbf{I}(r)$ lies wholly *outside a given region*, then an expansion of the same kind can be made. The $k = 0$ term gives rise to a constant $\mathbf{A}_0$ which is without physical significance since $\mathbf{A}$ is used to calculate curl $\mathbf{A}$ only as a means of getting $\mathbf{H}$. The $k = 1$ term gives rise to the parts of $\mathbf{A}$ that are linear in X, Y, Z and which, therefore, on differentiation give rise to the uniform field $\mathbf{H}_1$ near the origin. Interchanging the meaning of $\mathbf{r}$ and $\mathbf{R}$,

$$\mathbf{A}_1(\mathbf{r}) = \mathbf{r} \cdot \int \frac{\mathbf{R}\,\mathbf{I}(\mathbf{R})}{R^3}\,dV \qquad (2.43)$$

In calculating $\mathbf{H}_1 = \text{curl}\,\mathbf{A}_1$, only the antisymmetric part of $\mathbf{RI}$ contributes. This part of $\mathbf{A}_1(\mathbf{r})$ can be written

$$\mathbf{A}_1(\mathbf{r}) = \tfrac{1}{2}\mathbf{H}_1 \times \mathbf{r} \qquad (2.44)$$

where

$$\mathbf{H}_1 = \int \frac{\mathbf{I} \times \mathbf{R}}{R^3}\,dV$$

On taking $\mathbf{H} = \text{curl}\,\mathbf{A}_1$, it is found that $\mathbf{H}_1$ is, in fact, the value of the magnetic field at the origin due to the external current distribution. The discussion has gone through a complete closed loop in that (2.44) agrees with Ampère's law in the form [Eq. (1.51) of Part 4, Chap. 1] from which the field theory of the relation of $\mathbf{H}$ to $\mathbf{I}$ takes its start.

7. Force on a Rigid Current Distribution

According to Eq. (1.47) of Part 4, Chap. 1, the magnetic force on unit volume of a conductor having current density $\mathbf{I}$ is $\mathbf{I} \times \mathbf{H}$. Therefore if $\mathbf{I}(\mathbf{r})$ is a given function of position, placed in an external magnetic field $\mathbf{H}(\mathbf{r})$, the force $\mathbf{F}$ and torque $\mathbf{T}$ acting on the rigid frame in which the current flows are

$$\begin{aligned}\mathbf{F} &= \int \mathbf{I} \times \mathbf{H}\,dv \\ \mathbf{T} &= \int \mathbf{r} \times (\mathbf{I} \times \mathbf{H})\,dv\end{aligned} \qquad (2.45)$$

These expressions are analogous to (2.11) for the electric case.

If the current distribution is confined to a small region of space in which $\mathbf{H}(\mathbf{r})$ is nearly uniform,

$$\mathbf{H}(\mathbf{r}) = \mathbf{H}_1 + \mathfrak{H}_2 \cdot \mathbf{r} + \cdots \qquad (2.46)$$

where $\mathbf{H}_1$ is a constant uniform field and $\mathfrak{H}_2$ is the tensor made up of components like $\partial H_x/\partial y$. Since curl $\mathbf{H} = 0$, because $\mathbf{H}$ is the part of the field due to *external* currents, the tensor is symmetric; since div $\mathbf{H} = 0$, it has a vanishing diagonal sum.

Substituting this expression in that for $\mathbf{F}$, the term involving $\mathbf{H}_1$ vanishes because $\int \mathbf{I}\,dv = 0$, as shown in (2.39). Therefore an arbitrary closed-current distribution experiences no net force in a uniform magnetic field.

The first nonvanishing term is that arising from the inhomogeneity of the field. This is expressible in terms of the magnetic-dipole moment $\mathbf{M}$, defined in (2.40):

$$\mathbf{F} = \mathbf{M} \cdot \mathfrak{H}_2 + \cdots \qquad (2.47)$$

which is analogous to (2.13) for the electric case. Similarly, the first approximation to the torque is $\mathbf{T} = \mathbf{M} \times \mathbf{H}_1 + \cdots$, which is analogous to (2.14) for the electric case. The torque vanishes when the dipole moment is oriented parallel to $\mathbf{H}_1$ (or antiparallel, but this is the unstable orientation).

8. Mutual Inductance and Self-inductance

The current in the sth loop of a set of n closed-loop conductors is written i_s. If unit current flows in the sth loop and no current flows in the other loops, a magnetic field results, which will be denoted $\mathbf{H}_s(\mathbf{r})$ and which is described by a vector potential $\mathbf{A}_s(\mathbf{r})$. The general magnetic field is then $\mathbf{H} = \Sigma_s \mathbf{H}_s(\mathbf{r}) i_s$ derived from $\mathbf{A} = \Sigma_s \mathbf{A}_s(\mathbf{r}) i_s$.

By (2.37), the vector potential at $\mathbf{r}$ due to unit current in the sth loop is

$$\mathbf{A}_s(\mathbf{r}) = \oint \frac{\mathbf{ds}}{|\mathbf{r} - \mathbf{s}|} \qquad (2.48)$$

where the integral is a line integral extended around the sth loop. The $\mathbf{A}_s(\mathbf{r})$ is a dimensionless function of position. The flux through the rth loop due to this field is

$$\iint \mathbf{H}_s(\mathbf{r}) \cdot \mathbf{dS} = \oint \mathbf{A}_s(\mathbf{r}) \cdot \mathbf{dr}$$

where the surface integral extends over a surface bounded by the rth loop and the line integral is extended around the rth loop.

The flux through the rth loop due to unit current in the sth loop is called the *mutual inductance* of the two loops. It is denoted by M_{rs}, where

$$M_{rs} = \oint\oint \frac{\mathbf{dr} \cdot \mathbf{ds}}{|\mathbf{r} - \mathbf{s}|} \qquad r \neq s \qquad (2.49)$$

From the symmetry of this expression, $M_{rs} = M_{sr}$, and so the flux through r due to unit current in s is equal to the flux through s due to unit current in r.

The cgs electromagnetic unit of mutual inductance is the centimeter, and M_{rs} is given in this unit by (2.49) if the lengths are measured in centimeters. In the practical system with unit current being the ampere ($= 10^{-1}$ emu) and unit flux the volt-second ($= 10^8$ emu), the unit of inductance is such that 1 amp produces 1 volt-sec of flux (Part 4, Chap. 1, Sec. 9). This unit is called 1 henry and is equal to 10^9 cm of inductance.

When unit current flows in any loop, it produces flux through that loop. The flux through any loop due to unit current in the same loop is called the *self-inductance* of the loop and is measured in the same units as mutual inductance. For symmetry in general formulas the self-inductance of the rth loop will be designated M_{rr}, but it is more commonly designated L_r, and that notation will also be used.

The self-inductance of a loop cannot be calculated by (2.49), for if the two integrations are extended along the same path the integral diverges. This is correct physically for a loop whose conductor is a mathematical line, since in such a case the magnetic field near such a conductor, because of current in it, becomes infinite and hence the flux through such a loop is infinite. This is analogous to the result that the potential of a point conductor carrying a finite charge is infinite. The finite self-inductance of actual circuits is therefore due to their having conductors of finite cross section.

For a parallel-wire transmission line consisting of two wires of radius a and whose centers are separated by distance d, the self-inductance per unit length of line is given by

$$L = \pi^{-1} \ln (d/a) \tag{2.50}$$

neglecting the flux in the wires themselves. To the same approximation, the self-inductance of unit length of coaxial cable, having the radius a of the inner conductor and inside radius b of the outer conductor, is

$$L = R(\ln (R/a) - 2) + \tfrac{1}{4}R \tag{2.51}$$

where the expression is written this way to exhibit the separate contributions of the flux outside the wire (first term) and inside the wire (second term).

In a system of n conductors the fluxes are given by a symmetric matrix $M_{rs} = M_{sr}$, giving the fluxes in terms of the currents,

$$\Phi_r = \sum_s M_{rs} i_s \tag{2.52}$$

In analogy to the electrostatic case, one could define a matrix reciprocal to M_{rs}, giving the currents in terms of the fluxes:

$$i_r = \Sigma (M^{-1})_{rs} \Phi_s \tag{2.53}$$

9. Magnetic Interaction of Conductors

If a system of n linear conductors has currents $i_1, i_2, \ldots, i_n$, the flux through them is $\Phi_r = \Sigma M_{rs} i_s$, where M_{rs} are the self- and mutual inductances defined in the preceding section. These conductors experience magnetic forces which tend to move them by translation, rotation, and deformation of shape. If these forces result in actual motion of any or all of the conductors, this will, in general, result in changes in the M_{rs}. Hence, even if the i_s remain constant, there will be a change in flux through the circuits. This produces, by Eq. (1.81) of Part 4, Chap. 1, an induced emf $-d\Phi_r/dt$ in the rth circuit.

Therefore, in order to keep the current in the rth circuit constant during such a change, it is necessary to counteract the induced emf by introducing an adjustable battery which at all times gives an emf $\mathcal{E}_r = +d\Phi_r/dt$. The electrical-energy output of this battery is $\int \mathcal{E}_r i_r \, dt = \int i_r \, d\Phi_r = i_r(\Phi_r - \Phi_{r_0})$, where $\Phi_r - \Phi_{r_0}$ is the flux change through the rth circuit during the change.

If the change is a displacement *without deformation* of the first loop, then the self-inductance of the first loop is not changed. The change in flux in the first loop is $\Phi_1 - \Phi_{10} = \sum_{s=2}^{n} \Delta M_{1s} i_s$, where ΔM_{1s} is the change in the mutual inductances resulting from motion of the first loop. The electrical work done by the battery in the first loop is $\Delta W_{e1} = \sum_{s=2}^{n} (\Delta M_{1s}) i_1 i_s$. Likewise, the electrical work done by the batteries which maintain the other currents constant is

$$\Delta W_{es} = \Delta M_{1s} i_1 i_s$$

and so the total electrical work done by all the batteries in maintaining the currents constant is

$$\Delta W_e = 2 \sum_{s=2}^{n} (\Delta M_{1s}) i_1 i_s \tag{2.54}$$

If the first loop is deformed in the process, its self-inductance changes, say by ΔM_{11}. This makes an extra flux change $\Delta M_{11} i_1$ in the first loop, requiring extra electrical work by the battery in this loop to keep i_1 constant, of amount $\Delta M_{11} i_1^2$. Therefore, in this case,

$$\Delta W_e = \Delta M_{11} i_1^2 + 2 \sum_{s=2}^{n} (\Delta M_{1s}) i_1 i_s \tag{2.55}$$

Let x be a parameter which measures length along the first loop and $\mathbf{r}_1(x)$ be the vector initial position of the part whose distance is x from an arbitrary origin. In the displacement it will be supposed that $\mathbf{r}_1(x)$ is displaced to $\mathbf{r}_1(x) + \delta\mathbf{r}(x)$. The magnetic force acting on the length dx of the loop is $i_1 \, d\mathbf{r}_1 \times \mathbf{H}$, where $\mathbf{H}$ is the field due to current in the other loops, and the work done on displacement of this element is therefore $i_1 \, d\mathbf{r}_1 \times \mathbf{H} \cdot \delta\mathbf{r} = i_1 \mathbf{H} \cdot \delta\mathbf{r} \times d\mathbf{r}_1$; consequently, the total work done in the displacement is $i_1 \oint \mathbf{H} \cdot \delta\mathbf{r} \times d\mathbf{r}_1$ integrated around the loop. This is the same as the surface integral of $\mathbf{H}$ over the ribbon-shaped surface bounded by the initial and final positions of the first loop and is therefore equal to $i_1(\Phi'_1 - \Phi'_{10})$, where Φ'_1 is the flux through loop 1 due to currents in the *other* loops. Hence the external mechanical work done by the magnetic forces in this displacement is

$$\Delta W_m = \sum_{s=2}^{n} (\Delta M_{1s}) i_1 i_s \tag{2.56}$$

which is just half the work done by the batteries in maintaining the current constant.

The difference is ascribed to energy stored in the magnetic field. This will be denoted by T, and the change in it during the displacement must be equal to the excess over the electrical work done by the batteries less the external work performed by the magnetic forces; therefore,

$$\Delta T = \sum_{s=2}^{n} (\Delta M_{1s}) i_1 i_s \tag{2.57}$$

The magnetic forces act in a direction which *tends to bring about an increase of* T and are thus opposite in character to electrostatic forces, which act to bring about a decrease in the energy of the electrostatic field. For example, unlike charges attract each other and tend to produce a displacement which would nullify the electric field and hence the energy in the electric field. But currents traversing parallel wires in opposite directions repel each other, thus tending to increase the energy stored in the magnetic field.

If the system of conductors is fixed in form and position, the energy stored in the magnetic field may be calculated as that work which the batteries must do to build up the currents from zero to their final values. Suppose that the currents build up linearly from zero at $t = 0$ to the values $i_1, i_2, \ldots, i_n$ at $t = 1$. Then the counter emf which must be supplied by the battery in the sth loop is

$$\mathcal{E}_s = \frac{d\Phi_s}{dt} = \sum_{s=1}^{n} M_{sr} i_r$$

and the work done by this battery in time dt is $\mathcal{E}_s i_s \, dt$; therefore the total work in the interval from $t = 0$ to $t = 1$ is $\frac{1}{2}\mathcal{E}_s i_s$. The total magnetic energy will be this expression summed over all the loops:

$$T = \tfrac{1}{2} \sum_{s=1}^{n} \mathcal{E}_s i_s = \tfrac{1}{2} \sum M_{sr} i_s i_r \tag{2.58}$$

This expression is consistent with (2.57) in that (2.57) is the change in T resulting from a rigid displacement of the first loop, since such a displacement will not change M_{11} but will change the M_{1s} for $s \neq 1$.

In loops which have no batteries to keep the current constant and are assumed to be of zero resistance, the flux in the first loop due to its own current is $M_{11}i_1$ and that due to interaction with other loops is $\sum_{s=2}^{n} M_{1s} i_s$. If a displacement occurs so that the flux due to external causes changes in time δt by $\delta\Phi'$, the induced emf in the first loop is $-\delta\Phi'/\delta t$; if there is no battery in the circuit to balance this emf, it will produce a rate of increase of current i_1 such that $M_{11}(di_1/dt) = -\delta\Phi'/\delta t$. This is the same as saying that $M_{11}i_1 + \Phi'$ is constant. In case there are no batteries and the loops have zero resistance, therefore, the flux through each loop is constant in changes which alter the mutual inductances of the loops.

Bibliography

Russell, A.: "Alternating Currents," Cambridge University Press, London, 1914.

Smythe, W. R.: "Static and Dynamic Electricity," McGraw-Hill, New York, 1939.

Webster, A. G.: "Electricity and Magnetism," Macmillan, New York, 1897.

Chapter 3

Electric Circuits

By LOUIS A. PIPES, University of California

1. General Considerations

In the majority of electrical circuits encountered in practice, component parameters are constant resistances, inductances, mutual inductances, and capacitances. The problem of determining the response of such circuits to arbitrary impressed potentials and currents leads, in general, to a set of linear differential equations with constant coefficients. These circuits are called *linear* circuits. Another class of practical importance is circuits whose parameters vary only with the time. The theory leads to sets of linear differential equations with variable coefficients. These circuits are termed *linear, time-varying circuits.* Still another class that includes, for example, iron-cored magnetic devices and electronic devices has parameters that vary with the current or with the voltage difference between different parts of the circuit. The mathematical analysis of such circuits leads to nonlinear differential equations, and so they are called *nonlinear circuits.*

Most of this chapter is devoted to the theory of linear electrical circuits with constant parameters, with a brief mention of methods of analysis that have been developed in studying linear time-varying and nonlinear electric circuits.

2. Fundamental Electric-circuit Parameters

The characteristics of an electric circuit may be expressed in terms of four kinds of parameters, defined as follows:

1. *Resistance parameter.* The dissipative parameter is called the resistance. The flow of a current i (amp) through a resistance R (ohms) is accompanied by a drop in electric potential, or voltage drop, e_R (volts) in the direction of this current equal to the product of the resistance and the current:

$$e_R = Ri \tag{3.1}$$

This is commonly known as *Ohm's law.*

2. *Self-inductance parameter.* The inertia parameter is called the self-inductance. The flow of a current i (amp) through a self-inductance L (henrys) is accompanied by a voltage drop e_L (volts) in the direction of the current equal to the product of the inductance and the rate of increase of the current:

$$e_L = L\frac{di}{dt} \tag{3.2}$$

This relation expresses Faraday's law of induction in terms of the self-inductance coefficient L.

3. *Mutual-inductance parameter.* If two neighboring circuits are linked by a magnetic field, a change in the current of one circuit influences the potential difference between the terminals of the other. This effect is expressible in terms of the mutual-inductance parameter M (henrys) of the two circuits. Let the two circuits be A and B, and let (di_B/dt) be the rate of increase of current in circuit B. This change in the current i_B (amp) produces a potential drop v_M (volts) in circuit A:

$$v_M = M\frac{di_B}{dt} \tag{3.3}$$

The polarity of the potential v_M depends on the magnetic orientation of the two windings.

4. *Capacitance parameter.* The capacitance parameter C (farads) may be defined by the equation

$$i = C\frac{de_C}{dt} \tag{3.4}$$

This states that the current i (amp) through a capacitance is proportional to the rate of change of the potential e_C (volts) across its terminals. The proportionality factor is the magnitude of the capacitance C. A capacitor has unit capacitance if unit rate of change of potential across it produces a unit current through it. This may be written alternatively as

$$e_C = \frac{1}{C}\int i\,dt = \frac{q}{C} \tag{3.5}$$

The magnitude of the capacitance C is given by the charge q (coulombs) required to produce unit potential (volts) across its terminals.

3. Kirchhoff's Laws

The descriptive differential equations may be established by applying *Kirchhoff's laws.* These are statements of the principles of conservation of electric charge and conservation of energy. They were first enunciated by Kirchhoff in 1847. The usual statement of the laws is as follows:

1. The algebraic sum of the currents which meet at a junction point in an electric circuit is zero.

2. The algebraic sum of the electromotive forces and potential drops around any closed path or mesh of an electric circuit is zero.

The first law follows from the principle of the conservation of electricity. The second law follows from the principle of the conservation of energy. In applying the second law, all voltages must be included: not only the applied voltages and the voltages due to circuit elements but also the voltages induced by coupling with neighboring circuits.

4. Laws of Combination of Circuit Parameters

By applying the basic principles stated by Kirchhoff's laws, equivalent parameters for various combinations of the fundamental circuit parameters may be obtained, as follows:

1. *Resistances in series.* In a circuit consisting of several resistors in series (Fig. 3.1), as a consequence of Kirchhoff's second law,

$$R_1 i + R_2 i + \cdots + R_n i = (R_1 + R_2 + \cdots + R_n) i = e \qquad (3.6)$$

The composite conductor is equivalent to a single conductor whose resistance is

$$R = \sum_{k=1}^{n} R_k \qquad (3.7)$$

2. *Resistances in parallel.* In the case of several resistors connected in parallel (Fig. 3.2), by Kirchhoff's first law, the total current of the circuit is

$$i = i_1 + i_2 + \cdots + i_n = \frac{e}{R_1} + \frac{e}{R_2} + \cdots + \frac{e}{R_n} = e\left(\sum_{k=1}^{n} \frac{1}{R_k}\right) \qquad (3.8)$$

and the n conductors in parallel are equivalent to a single resistance given by

$$\frac{1}{R} = \sum_{k=1}^{n} \frac{1}{R_k} \qquad (3.9)$$

By introducing the notation

$$\frac{1}{R} = G \qquad \frac{1}{R_k} = G_k$$

where G is the equivalent *conductance* of the circuit and G_k is the *conductance* of R_k, this can be written

$$G = \sum_{k=1}^{n} G_k \qquad (3.10)$$

3. *Inductances in series.* The equivalent inductance L of a circuit composed of n inductances L_k connected in series is given by

$$L = \sum_{k=1}^{n} L_k \qquad (3.11)$$

4. *Inductances in parallel.* The equivalent inductance L of a circuit consisting of n inductances L_k connected in parallel is given by

$$\frac{1}{L} = \sum_{k=1}^{n} \frac{1}{L_k} \qquad (3.12)$$

Inductances combine in the same manner as resistances.

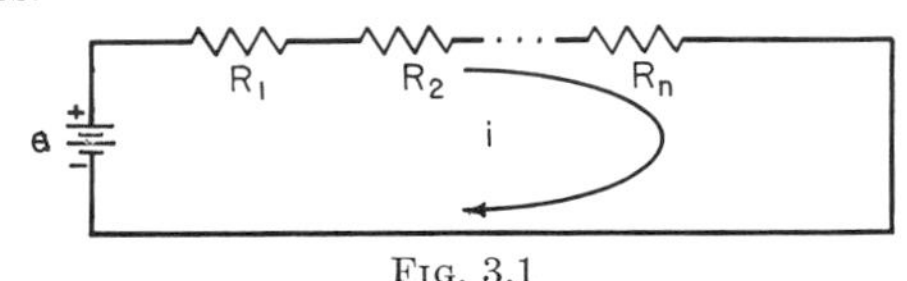

Fig. 3.1

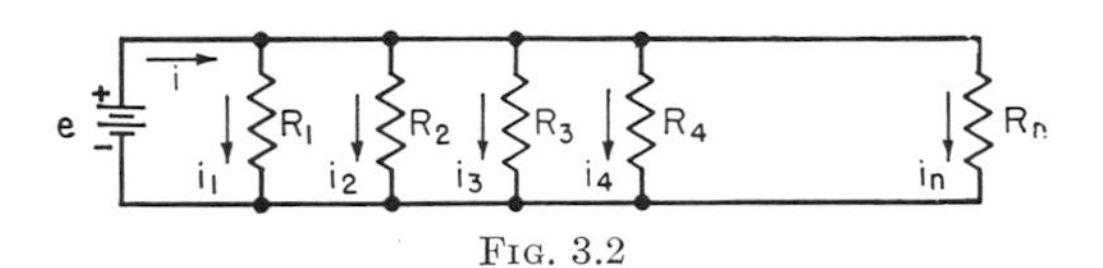

Fig. 3.2

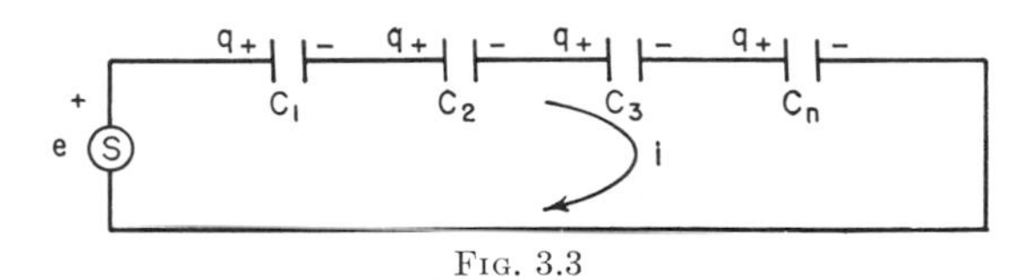

Fig. 3.3

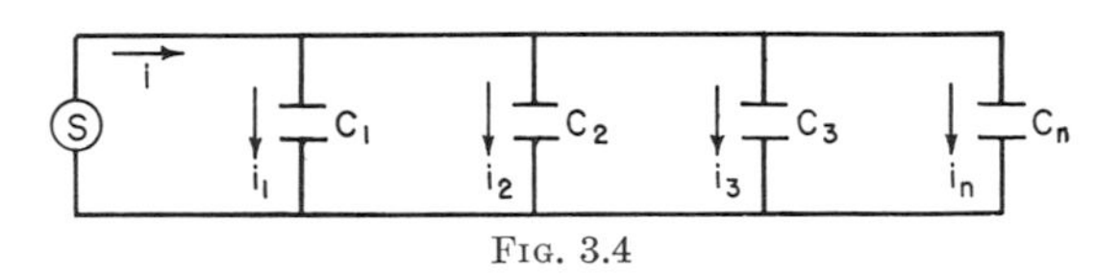

Fig. 3.4

5. *Capacitances in series.* Several capacitances connected in series are shown in Fig. 3.3. The equivalent capacitance C is given by

$$\frac{1}{C} = \sum_{k=1}^{n} \frac{1}{C_k} \qquad (3.13)$$

By introducing the notation

$$\frac{1}{C} = S \qquad \frac{1}{C_k} = S_k$$

where S is the equivalent *elastance* of the circuit and S_k is the *elastance* of the capacitance C_k, this can be written

$$S = \sum_{k=1}^{n} S_k \qquad (3.14)$$

6. *Capacitances in parallel.* For several capacitances connected in parallel (Fig. 3.4), the equivalent capacitance of the circuit is

$$C = \sum_{k=1}^{n} C_k \qquad (3.15)$$

Table 3.1 presents a summary of fundamental circuit parameters and laws.

TABLE 3.1: FUNDAMENTAL CIRCUIT PARAMETERS AND LAWS

Quantity	Symbolic representation	Fundamental relation
Charge = q (coulombs)		$q = \int i\,dt$
Current = i (amperes)	i	$i = \frac{dq}{dt}$ $\sum i = 0$ (at a junction)
Electromotive force = e (volts)	e Voltage Source	$\Sigma e = 0$ (around a closed mesh)
Resistance = R (ohms)	R	$e_R = Ri$ (Ohm's law)
Conductance = G (ohms)$^{-1}$	G	$i_G = Ge$
Self-inductance = L (henrys)	L	$e_L = \frac{L\,di}{dt}$ (Faraday's law of induction)
Mutual-in tance = M (henrys)	A B M	$e_B = M \frac{di_A}{dt}$
Capacitance = C (farads)	C	$\dot{q} = C\dot{e},\ i = C\dot{e}$
Elastance = S (farads)$^{-1}$	S	$e_S = Sq$
Power = $P = ei$ (watts)		$P = ei = \frac{dW}{dt}$
Energy (watt-sec) = $W = \int ei\,dt$		$W = \int P\,dt = \int ei\,dt$

5. Applications of the Fundamental Laws

By the use of Kirchhoff's second law, the electromotive forces around any mesh (completely closed electrical path) may be added to form the differential equation for the mesh. Alternatively, by the use of Kirchhoff's first law, the currents entering any branch point or junction in the network may be added to zero to form the differential equation for the branch point. In Fig. 3.5 a voltage generator or source

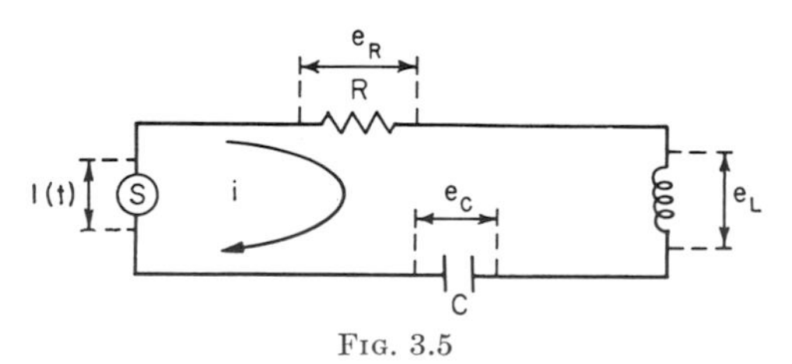

FIG. 3.5

whose potential is $e(t)$ drives a current through the elements R, L, and C in series. If Kirchhoff's second law is applied to this mesh, the following equation is obtained:

$$e_L + e_R + e_C - e(t) = 0 \qquad (3.16)$$

or

$$L\frac{di}{dt} + Ri + \frac{q}{C} = e(t) \qquad (3.17)$$

Since the charge separation q is related to the current by the equation $q = \int i\,dt$, the mesh equation (3.17) may be written in the form

$$L\frac{di}{dt} + Ri + \frac{1}{C}\int i\,dt = e(t) \qquad (3.18)$$

In the circuit of Fig. 3.6, there is a current generator of waveform $i(t)$ supplying the three elements R,

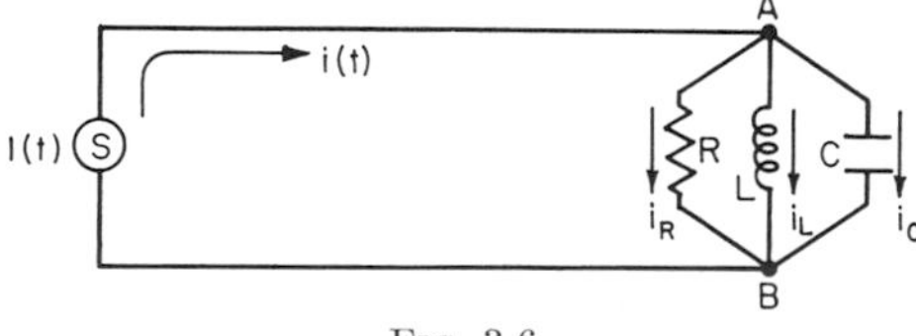

FIG. 3.6

L, and C in parallel. Kirchhoff's first law, applied to the junction or branch point A, gives the equation

$$i_R + i_L + i_C - i = 0 \qquad (3.19)$$

If $e(t)$ denotes the potential difference between the points A and B, the currents i_R, i_L, and i_C may be expressed in terms of this potential by means of the fundamental circuit laws in the form $i_R = e/R$; $i_L = 1/L \int e\,dt$; $i_C = C(de/dt)$. Therefore,

$$C\frac{de}{dt} + \frac{e}{R} + \frac{1}{L}\int e\,dt = i(t) \qquad (3.20)$$

which is the differential equation for the branch point A.

6. Energy Relations

Consider an electromotive source $e(t)$ driving a current $i(t)$ through an electrical device (Fig. 3.7). The instantaneous power input $P(t)$ to the electrical

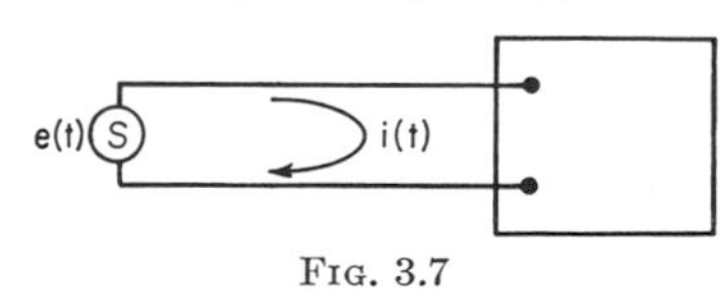

FIG. 3.7

device in watts is given by $P(t) = e(t)i(t)$. The energy input W in watt-seconds or joules in an interval of time $t_2 - t_1$ is given by

$$W = \int_{t_1}^{t_2} P(t)\,dt = \int_{t_1}^{t_2} e(t)i(t)\,dt \qquad (3.21)$$

By the use of these fundamental definitions it is possible to obtain the energy relations of particular circuits. For example, the energy-balance equations for the single-mesh circuit of Fig. 3.5 may be deduced by multiplying both sides of Eq. (3.17) by the current i:

$$Ri^2 + \frac{d}{dt}(\tfrac{1}{2}Li^2) + \frac{d}{dt}\frac{q^2}{2C} = ei \qquad (3.22)$$

The right-hand side represents the power, $P = ei$ (watts), delivered to the circuit by the source potential. The first term, $Ri^2 = P_r$, represents the rate at which the electrical energy is being converted into heat by the resistance element. The second term, $d/dt(\frac{1}{2}Li^2) = P_L$, is the rate of increase of the magnetic energy stored in the inductance element. The third term, $(d/dt)\ (q^2/2C) = P_C$, is the rate of increase of the elastic energy stored in the capacitance element. The potential energy of the circuit is

$$W_C = \tfrac{1}{2}\frac{q^2}{2C} \qquad \text{watt-sec or joules} \qquad (3.23)$$

The kinetic energy of the circuit is

$$W_L = \tfrac{1}{2}Li^2 \qquad \text{watt-sec or joules} \qquad (3.24)$$

7. The Mesh Equations of a General Network

Differential equations governing the relations between the currents and potentials of complicated electrical-circuit structures composed of resistances, inductances, capacitances, and sources of electromotive force may be obtained by application of the Kirchhoff laws. Two formulations are usually given: the mesh-equation formulation (given in this section) and the nodal-equation formulation.

The general network can be regarded as an arrangement of individual branches which may include combinations of resistances, inductances, and capacitances in series connected together at various junction points, or nodes. If the number of junction points of the network is J and the number of branches is B, then the total number n of independent circulating currents, $i_1, i_2, \ldots, i_n$, which may be drawn is given by the equation

$$n = (B - J) + 1 \qquad (3.25)$$

The circulating currents i_k are called the mesh currents of the network. These currents automatically satisfy the first Kirchhoff law. Let external electromotive forces $e_1, e_2, \ldots, e_n$ be applied to the n meshes, or closed circuits, of the network; let L_{kk}, R_{kk}, and S_{kk} denote the total inductance, resistance, and elastance in series in mesh k; and let L_{kr}, R_{kr}, and S_{kr} denote the corresponding mutual elements between the meshes k and r. If Kirchhoff's second law is applied to each independent mesh in turn, the following set of integrodifferential equations is obtained:

$$\begin{aligned}
Z_{11}(D)i_1 + Z_{12}(D)i_2 + Z_{13}(D)i_3 + \cdots \\
+ Z_{1n}(D)i_n &= e_1 \\
Z_{21}(D)i_1 + Z_{22}(D)i_2 + Z_{23}(D)i_3 + \cdots \\
+ Z_{2n}(D)i_n &= e_2 \\
\cdots\cdots\cdots\cdots\cdots\cdots\cdots\cdots \\
Z_{n1}(D)i_1 + Z_{n2}(D)i_2 + Z_{n3}(D)i_3 + \cdots \\
+ Z_{nn}(D)i_n &= e_n
\end{aligned} \qquad (3.26)$$

The coefficients $Z_{kr}(D)$ are *impedance operators*,

$$Z_{kr}(D) = L_{kr}\frac{d}{dt} + R_{kr} + S_{kr}\int dt$$
$$= L_{kr}D + R_{kr} + \frac{S_{kr}}{D} \qquad (3.27)$$

where $D = d/dt$ is the time-derivative operator. The operators $Z_{11}(D)$, $Z_{22}(D)$, etc., are called the *self-impedance operators*, and $Z_{12}(D)$, $Z_{13}(D)$, $Z_{23}(D)$, etc., are called the *mutual-impedance operators*.

The system of Eqs. (3.26) is one of linear integrodifferential equations with constant coefficients. These constitute the *canonical equations* of the network formulated on the mesh basis. Together with a description of the initial conditions, they completely specify its performance.

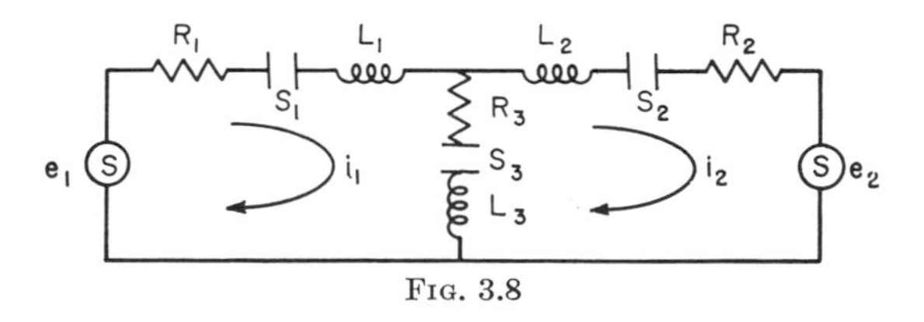

FIG. 3.8

As an example, consider the network of Fig. 3.8, which has three branches ($B = 3$) and two junction points ($J = 2$). Equation (3.25) therefore gives $n = (3 - 2) + 1 = 2$, and two independent circulating currents may be drawn. The performance of the network will be specified in terms of the circulating currents i_1 and i_2. Application of Kirchhoff's second law to the two meshes of the circuit leads to

$$\begin{aligned}
Z_{11}(D)i_1 + Z_{12}(D)i_2 &= e_1 \\
Z_{21}(D)i_1 + Z_{22}(D)i_2 &= e_2
\end{aligned} \qquad (3.28)$$

where

$$Z_{11}(D) = (L_1 + L_3)D + (R_1 + R_3) + \frac{1}{D}(S_1 + S_3)$$

$$Z_{12}(D) = -\left(L_3D + R_3 + \frac{S_3}{D}\right) = Z_{21}(D) \qquad (3.29)$$

$$Z_{22}(D) = (L_2 + L_3)D + (R_2 + R_3) + \frac{1}{D}(S_2 + S_3)$$

The general equations (3.26) possess two important properties responsible for the relative simplicity of classical electric-circuit theory. First, they are linear in both currents and applied electromotive forces; second, the coefficients L_{kr}, R_{kr}, and S_{kr} are constants. Another important property is the reciprocal relation among the coefficients, that is, $L_{kr} = L_{rk}$, $R_{kr} = R_{rk}$, and $S_{kr} = S_{rk}$, so that

$$Z_{kr}(D) = Z_{rk}(D)$$

These reciprocal relations mean that there are no concealed sources or sinks of energy in the network.

8. Energy Relations in a Network

A very important relation involving the energy of the general n-mesh network is obtained by multiplying the first of (3.26) by i_1, the second by i_2, and adding:

$$\frac{d}{dt}\sum_{k=1}^{n}\sum_{r=1}^{n}\tfrac{1}{2}L_{kr}i_ki_r + \frac{d}{dt}\sum_{k=1}^{n}\sum_{r=1}^{n}\tfrac{1}{2}S_{kr}q_kq_r + \sum_{k=1}^{n}\sum_{r=1}^{n}R_{kr}i_ki_r = \sum_{k=1}^{n}e_ki_k = P \qquad (3.30)$$

The quantities $q_k = \int i\,dt$ are the circulating charges of the various meshes. The right-hand member P of (3.30) is the rate at which the applied sources of potential are supplying energy to the network (watts). The magnetic energy W_m of the network in joules is

$$W_m = \tfrac{1}{2}\sum_{k=1}^{n}\sum_{r=1}^{n}L_{kr}i_ki_r \qquad (3.31)$$

The electric energy W_E of the network in joules is

$$W_E = \tfrac{1}{2}\sum_{k=1}^{n}\sum_{r=1}^{n}S_{kr}q_kq_r \qquad (3.32)$$

The rate of dissipation of energy P_R by the resistances of the network in watts or in joules per second is

$$P_R = \sum_{k=1}^{n}\sum_{r=1}^{n}R_{kr}i_ki_r \qquad (3.33)$$

Thus (3.30) expresses the balance of energy in the form

$$\frac{d}{dt}W_m + \frac{d}{dt}W_E + P_R = P \qquad (3.34)$$

The homogeneous quadratic functions for W_m and W_E are the foundations for Maxwell's dynamical theory of electrical networks.

9. General Solution of the Mesh Equations: Transient Phenomena

A concise form of the canonical mesh equations (3.26) may be obtained by the use of matrix notation (Part 1, Chap. 2, Sec. 9). Introduce the following matrices: $[L] = [L_{kr}]$, $[R] = [R_{kr}]$, $[S] = [S_{kr}]$, and construct the square matrix $[Z(D)]$ by addition, in the form $[Z(D)] = [L]D + [R] + [S]D^{-1}$. Introduce the column matrices $(e) = (e_k)$ and $(i) = (i_k)$. In this notation, the canonical mesh equations are

$$[Z(D)](i) = (e) \qquad (3.35)$$

Equation (3.35) together with the initial conditions completely specifies the network performance. In the general problem, the elements of the potential matrix (e) are constants or given functions of time, and the initial mesh currents $i_1^0, i_2^0, \ldots, i_n^0$ and the initial mesh charges $q_1^0, q_2^0, \ldots, q_n^0$ are given at $t = 0$.

Equation (3.35), subject to the given initial conditions at $t = 0$, may be solved by the Laplace-transform method (see Part 1, Chap. 4, Sec. 2). Introduce a column matrix $[E(p)]$ whose elements are the Laplace transforms of the potential matrix (e):

$$[E(p)] = \int_0^\infty e^{-pt}(e)\,dt = \mathfrak{L}(e) \qquad (3.36)$$

Introduce a column matrix $[I(p)]$ whose elements are the transforms of the current matrix (i):

$$[I(p)] = \int_0^\infty e^{-pt}(i)\,dt = \mathfrak{L}(i) \qquad (3.37)$$

The Laplace transform of (3.35) is

$$[Z(p)][I(p)] = [E(p)] + [L](i^0) - \frac{1}{p}[S](q^0) \qquad (3.38)$$

where $$[Z(p)] = p[L] + [R] + \frac{1}{p}[S] \qquad (3.39)$$

is the transform impedance matrix and (i^0) and (q^0) are column matrices whose elements are the initial mesh currents and mesh charges at $t = 0$. The solution of (3.38) for $[I(p)]$ may be expressed in the form

$$[I(p)] = [Z(p)]^{-1}\left\{[E(p)] + [L](i^0) - \frac{1}{p}[S](q^0)\right\} \qquad (3.40)$$

where $[Z(p)]^{-1}$ is the inverse of the matrix $[Z(p)]$. Equation (3.40) enables the transforms of the various mesh currents to be determined by a purely algebraic procedure. The various mesh currents may be found by the equation

$$(i) = \mathfrak{L}^{-1}[I(p)] \qquad (3.41)$$

The mesh currents are determined by a computation of the inverse transforms of the elements of the column matrix $[I(p)]$. If the network has a large number of meshes, the computation of the inverse transforms may be a laborious matter. The labor of the computations may be reduced by consulting a dictionary of Laplace transforms.

10. Examples of Simple Transients

The term *transient* is usually used to designate the result produced by making a sudden change in an electrical circuit, such as switching on or off an electromotive force, the release of the charge on a capacitor, the sudden change in the magnetic flux linking a circuit, etc. A study of the performance of a network produced by these typical changes may be effected by solving the network equations (3.26) by the method of the above section. Special applications of the general theory will now be considered.

1. A constant electromotive force E is suddenly applied at $t = 0$ to the general series circuit of Fig.

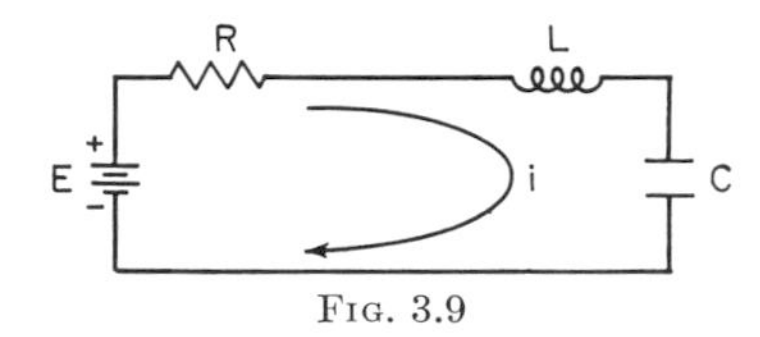

FIG. 3.9

3.9. Initial charge and current are assumed to be zero. The mesh equation is

$$L\frac{di}{dt} + Ri + \int \frac{i}{C}\,dt = E \qquad t > 0 \tag{3.42}$$

The Laplace transform is

$$\left(Lp + R + \frac{1}{Cp}\right) I = \frac{E}{p} \tag{3.43}$$

where $I(p) = \mathfrak{L}i(t)$ is the Laplace transform of the current. Hence,

$$I(p) = \frac{E}{L(p^2 + Rp/L + 1/LC)} = \frac{E}{L[(p + a)^2 + n^2]} \tag{3.44}$$

where $\quad a = \dfrac{R}{2L} \qquad n = \sqrt{\dfrac{1}{LC} - \dfrac{R^2}{4L^2}}$

The current $i(t)$ is the inverse Laplace transform of $I(p)$ and may be obtained by consulting a dictionary of Laplace transforms, thus obtaining

$$\begin{aligned} i(t) &= \frac{E}{nL}e^{-at}\sin nt && \text{if } n^2 > 0 \\ i(t) &= \frac{E}{L}te^{-at} && \text{if } n = 0 \\ i(t) &= \frac{E}{kL}e^{-at}\sinh kt && \text{if } n^2 < 0 \qquad k^2 = -n^2 \end{aligned} \tag{3.45}$$

2. An alternating electromotive force $E_m \sin \omega t$ is suddenly applied to the circuit of Fig. 3.9 at $t = 0$. It is assumed that the initial charge and current of the circuit are zero. The mesh equation is

$$L\frac{di}{dt} + Ri + \frac{i}{C}\,dt = E_m \sin \omega t \qquad t > 0 \tag{3.46}$$

The Laplace transform is

$$\left(Lp + R + \frac{1}{Cp}\right) I = \frac{\omega E_m}{p^2 + \omega^2} \tag{3.47}$$

Hence, $$I(p) = \frac{E_m \omega p}{(Lp^2 + Rp + 1/C)(p^2 + \omega^2)} \tag{3.48}$$

The right-hand member may be expanded into partial fractions with quadratic denominators in the form

$$I(p) = \frac{E}{Z^2(\omega)}\left[\frac{X(p + a) - aX_0}{(p + a)^2 + n^2} - \frac{Xp - R\omega}{p^2 + \omega^2}\right] \tag{3.49}$$

where $\quad a = \dfrac{R}{2L} \qquad n^2 = \dfrac{1}{LC} - \dfrac{R^2}{4L^2}$

$$X = \omega L - \frac{1}{\omega C} \qquad X_0 = \omega L + \frac{1}{\omega C}$$

$$Z^2(\omega) = X^2 + R^2 = \left(\omega L - \frac{1}{\omega C}\right)^2 + R^2$$

If $n^2 > 0$, the inverse transform of $I(p)$ is

$$i(t) = \frac{E_m}{Z(\omega)}\sin(\omega t - \theta) - E_m e^{-at}\frac{\sin(nt - \phi)}{nZ\sqrt{LC}} \tag{3.50}$$

where $\quad \tan\theta = \dfrac{X}{R} \qquad \tan\phi = \dfrac{nX}{aX_0}$

As time increases, the second right-hand term in (3.50) eventually vanishes, and the current assumes a periodic variation given by

$$i_s(t) = \frac{E_m}{Z(\omega)}\sin(\omega t - \theta) \tag{3.51}$$

This is the *steady-state* alternating current produced by the applied harmonic potential. The *transient* term is initiated by the switching operation and rapidly vanishes as a consequence of the resistance of the circuit.

3. A constant electromotive force E is suddenly applied at $t = 0$ to the circuit of Fig. 3.10. It is

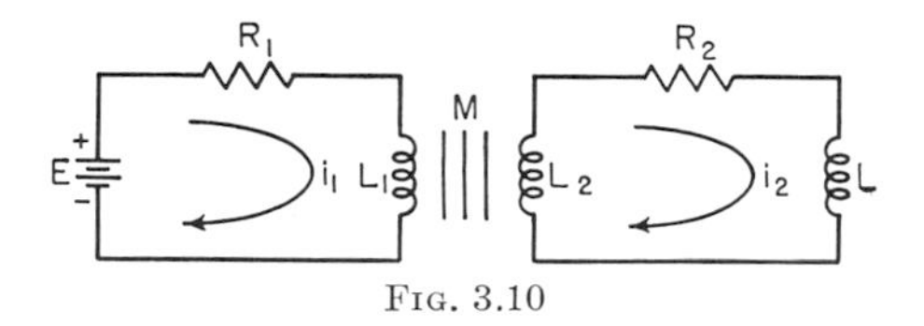

FIG. 3.10

desired to determine the current i_2 if there are no initial currents in the system.

The mesh equations are

$$\begin{aligned} \left(L_1\frac{d}{dt} + R_1\right) i_1 + M\frac{di_2}{dt} &= E \\ M\frac{di_1}{dt} + (L_2 + L)\frac{di_2}{dt} + Ri_2 &= 0 \end{aligned} \tag{3.52}$$

Since the currents are zero at $t = 0$, the Laplace transforms are

$$\begin{aligned} (L_1p + R_1)I_1 + pMI_2 &= \frac{E}{p} \\ MpI_1 + [(L_2 + L)p + R]I_2 &= 0 \end{aligned} \tag{3.53}$$

where I_1 and I_2 are the Laplace transforms of i_1 and i_2. Let the magnetic leakage of the mutual inductance M be negligible so that $M^2 = L_1L_2$. Solving (3.53) for I_2,

$$I_2 = \frac{-En}{L(p+a)(p+b)} \tag{3.54}$$

where $n = \sqrt{\frac{L_2}{L_1}}$ $\qquad A = \frac{R}{L} + \frac{R_1}{L_1} + \frac{R_1L_2}{LL_1}$

and $\left.\begin{matrix} a \\ b \end{matrix}\right\} = \frac{A}{2} \pm \sqrt{\frac{A^2}{4} - B}$ $\qquad B = \frac{RR_1}{LL_1}$

The inverse transform of I_2 is

$$i_2 = \frac{En(e^{-bt} - e^{-at})}{L(b-a)} \tag{3.55}$$

11. Nodal Equations of the General Network: Duality

In the mesh-equation formulation of the general network given in Sec. 7, the potentials applied to the various meshes are regarded as driving sources. The currents in the several closed loops of the network are determined by the condition that all the meshes must be in voltage equilibrium. It is also possible to set up a system of equations that describe the behavior of the network in such a manner that the currents are taken as the activating forces and the response of the network to these currents is the voltages induced between certain points in the network. Consider the circuit of Fig. 3.11.

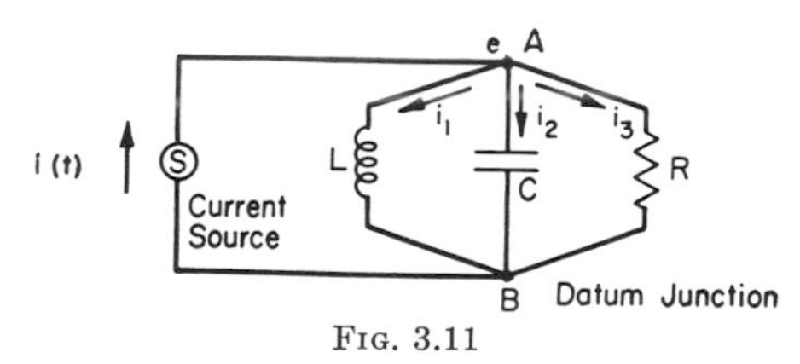

FIG. 3.11

Application of Kirchhoff's first law to the junction point, or node, A gives $i(t) = i_1 + i_2 + i_3$. If the potential difference between the nodal point A and the datum junction B is denoted by $e(t)$, from the fundamental definition of the action of the various circuit parameters, $LDi_1 = e$, $i_2/CD = e$, and $Ri_3 = e$. Hence,

$$i(t) = \frac{e}{LD} + CDe + \frac{e}{R} = [Y_L(D) + Y_C(D) + Y_R]e \tag{3.56}$$

The quantities $Y_L(D)$, $Y_C(D)$, and Y_R are the *operational admittances* of the inductance, capacitance, and resistance elements of Fig. 3.11, defined by

$$Y_L(D) = \frac{1}{LD} = \frac{1}{L}\int (\ \ )\,dt$$

$$Y_C(D) = CD = C\frac{d}{dt}$$

$$Y_R = \frac{1}{R} = G$$

If $Y(D)$ is the sum of the operational admittances of the various elements defined by

$$Y(D) = Y_L(D) + Y_C(D) + Y_R$$

then

$$i(t) = Y(D)e \tag{3.57}$$

This equation is a formulation of the simple circuit of Fig. 3.11 on the basis of the node A.

A general network (Fig. 3.12) may be treated in a similar manner. Consider a network having $n + 1$ nodal points, and choose one nodal point as a datum

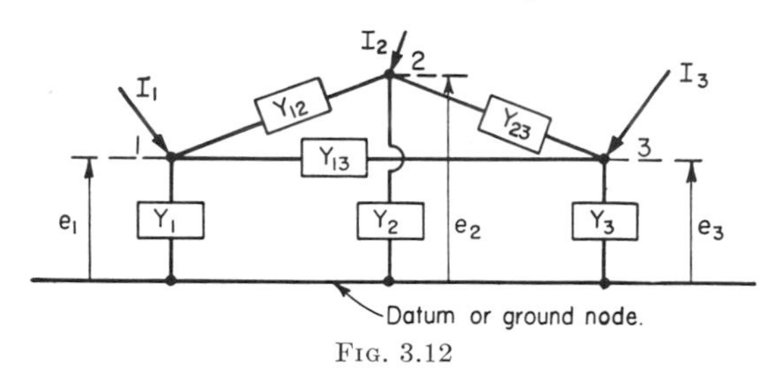

FIG. 3.12

node. The currents $i_1, i_2, \ldots, i_n$ impressed on the nodes 1 to n from some outside sources are regarded as the driving forces of the network. The potentials of the nodal points $1, 2, \ldots, n$ with respect to the datum node are regarded as the responses of the network.

The fundamental equations on the nodal basis are obtained by applying Kirchhoff's first law to each nodal point. The resulting equations express the fact that the driving current flowing into any node from outside the network must equal the total current flowing away from that node into the rest of the network. For example, let i_1 be the current flowing into the first node from the outside and let e_1, e_2, etc., be the potentials of the nodal points 1, 2, etc., with respect to the datum, or ground, node. Let $Y_k(D)$ be the *operational admittance* from the node k to the ground node and let $Y_{rs}(D)$ be the operational-admittance function between the nodes r and s. The current flowing from node 1 to ground is $Y_1(D)e_1$, the current flowing from the first node to the second node is $Y_{12}(e_1 - e_2)$, etc. The equation that expresses the current equilibrium of the first node is

$$Y_1(D)e_1 + Y_{12}(e_1 - e_2) + \cdots + Y_{1n}(e_1 - e_n) = i_1 \tag{3.58}$$

which may be written in the form

$$Y_{11}(D)e_1 - Y_{12}e_2 - Y_{13}e_3 - \cdots - Y_{1n}e_n = i_1 \tag{3.59}$$

where $Y_{11} = Y_1 + Y_{12} + Y_{13} + \cdots + Y_{1n}$. The operator Y_{11} is called the *self-admittance operator* of the first node. The operators Y_{1k} are *mutual-admittance operators*. An equation similar to (3.58) may be written for each node. The complete system of equations for the entire network on the node basis is

$$\begin{array}{l} Y_{11}e_1 - Y_{12}e_2 - Y_{13}e_3 - \cdots - Y_{1n}e_n = i_1 \\ -Y_{21}e_1 + Y_{22}e_2 - Y_{23}e_3 - \cdots - Y_{2n}e_n = i_2 \\ \cdots\cdots\cdots\cdots\cdots\cdots\cdots\cdots \\ -Y_{n1}e_1 - Y_{n2}e_2 - Y_{n3}e_3 - \cdots + Y_{nn}e_n = i_n \end{array} \tag{3.60}$$

In analysis of communication networks, it is usually more convenient to formulate the equations of the circuit on a nodal basis rather than on a mesh basis, because in many cases these circuits contain screen grid tubes that have a very high plate resistance and act as constant-current devices. It also frequently happens that the equations formulated on a nodal basis can be correlated with the physical structure of the network more directly than is possible on a mesh basis.

A certain symmetry exists between the current and voltage relations for a resistance and a conductance, expressed by the equations

$$\begin{aligned} e &= Ri \\ i &= Ge \end{aligned} \tag{3.61}$$

and the corresponding expressions for a capacitance and an inductance,

$$\begin{aligned} e &= L\frac{di}{dt} \\ i &= C\frac{de}{dt} \end{aligned} \tag{3.62}$$

As a consequence of the symmetry of these equations, a set of nodal equations that is formally identical with a given set of mesh equations can be constructed by replacing the general impedance term

$$Z_{ij}(D) = DL_{ij} + R_{ij} + \frac{S_{ij}}{D}$$

on the mesh basis, by $Y_{ij}(D) = DC_{ij} + G_{ij} + 1/DL_{ij}$ on the nodal basis and placing L_{ij} in equivalence with C_{ij}, R_{ij} in equivalence with G_{ij}, and S_{ij} with $1/L_{ij}$.

If the mesh equations for one network correspond to the nodal equations for another one, the two networks are called *inverse structures*. It is not always possible to obtain the inverse of a given structure. The *principle of duality* in network theory is a recognition of the possibilities of realizing inverse electrical structures.

12. Alternating Currents

If all the electromotive forces that are applied to a network are periodic functions of the time, their effect is to cause alternating currents to flow in the various branches of the system. The most common and technically most important of these periodic electromotive forces are those of the harmonic or sinusoidal type. To obtain the amplitudes and phases of the alternating currents produced in a network by application of harmonic potentials, obtain the particular integrals of the network differential equations formulated either on a mesh or a node basis. Consider the general series circuit of Fig. 3.13.

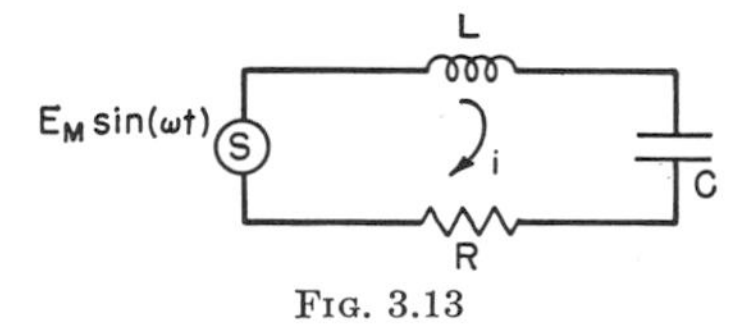

FIG. 3.13

The circuit has a harmonic potential of frequency, $f = \omega/2\pi$ impressed on it. The general solution is given in Sec. 10. After the initial transient disturbances are attenuated, the current settles down to a steady harmonic oscillation. The direct determination of this *steady state* from the mesh equation of the system will now be considered:

$$L\frac{di}{dt} + Ri + \frac{1}{C}\int i\,dt = E_m \sin \omega t = \text{Im}\,(E_m e^{j\omega t}) \tag{3.63}$$

$j = \sqrt{-1}$ and Im = "the imaginary part of"

A simple method is to determine the particular integral of the equation

$$L\frac{di}{dt} + Ri + \frac{1}{C}\int i\,dt = E_m e^{j\omega t}$$

by searching for a periodic solution of $i = Ie^{j\omega t}$. The result is

$$I = \frac{E_m}{R + j(\omega L - 1/\omega C)} = \frac{E_m}{Z} \tag{3.64}$$

The quantity I is the *complex current* of the series circuit. The complex number

$$Z = R + j\left(\omega L - \frac{1}{\omega C}\right) = \sqrt{R^2 + \left(\omega L - \frac{1}{\omega C}\right)^2}\, e^{j\phi}$$

$$\tan \phi = \frac{\omega L - 1/\omega C}{R}$$

is the *complex impedance*, or simply the *impedance*, of the series circuit. The required periodic solution of (3.63) is given by

$$\begin{aligned} i(t) &= \text{Im}\, Ie^{j\omega t} = \text{Im}\left(\frac{E_m}{Z}e^{j\omega t}\right) \\ &= \frac{E_m \sin(\omega t - \phi)}{\sqrt{R^2 + (\omega L - 1)^2/\omega C}} \end{aligned} \tag{3.65}$$

This is the *instantaneous* periodic, or alternating, current that flows in the series circuit in the *steady state.*

Equation (3.64) has exactly the same form as Ohm's law given by (3.1). The impedance Z replaces the resistance, and the complex current I replaces the direct current of Ohm's law. This similarity is of great importance because it reduces the computation of the response of a-c networks to computations similar to those used in computing the response of d-c networks.

The steady-state current (3.65) may be expressed in the alternative forms

$$\begin{aligned} i(t) &= \frac{E_m}{|Z|}\sin(\omega t - \phi) = |I| \sin(\omega t - \phi) \\ &= I_m \sin(\omega t - \phi) \end{aligned} \tag{3.66}$$

where $|Z|$ is the absolute value, or modulus, of the impedance Z and the phase angle ϕ is the argument of Z. $I_m = |I|$ is the modulus of the complex current I. When the impedance Z is written

$$Z = R + j\left(\omega L - \frac{1}{\omega C}\right) = R + jX$$

the real part R of the impedance is called the *resistance* or the *resistive component* of the impedance, and the imaginary part X is called the *reactance* or the *reactive component.* The reciprocal of the impedance is called the *admittance:*

$$Y = \frac{1}{Z} = G - jB$$

It is customary for electrical engineers to write the real and imaginary parts of the admittance in the latter form. The minus sign before jB is introduced so that B will have the same sign as the reactance X. The quantity G is called the *conductance* and B the *susceptance* of the admittance Y.

If the capacitor of the circuit of Fig. 3.13 is short-circuited, this is equivalent to placing $C = \infty$. Therefore, when resistance and inductance only are present in the circuit, (3.65) reduces to

$$i(t) = \frac{E_m}{\sqrt{R^2 + \omega^2 L^2}} \sin\left(\omega t - \tan^{-1}\frac{\omega L}{R}\right) \tag{3.67}$$

In this case, the current is said to *lag* behind the electromotive force. If no inductance is present, $L = 0$, and then the current is given by

$$i(t) = \frac{E_m \omega C}{\sqrt{1 + \omega^2 R^2 C^2}} \sin\left(\omega t + \tan^{-1}\frac{1}{\omega RC}\right) \tag{3.68}$$

In this case the current is said to *lead* the electromotive force.

13. Power, Effective, or Root-mean-square Values; Series Resonance

The power P delivered to the circuit of Fig. 3.13 by the impressed electromotive force at any instant t is

$$\begin{aligned} P = ei &= E_m I_m \sin(\omega t) \sin(\omega t - \phi) \\ &= E_m I_m [\cos\phi \sin^2(\omega t) - \tfrac{1}{2}\sin\phi \sin(2\omega t)] \end{aligned} \tag{3.69}$$

The instantaneous power delivered to the circuit fluctuates during the course of an oscillation of the applied potential. The *mean power* $<P>_{\rm av}$ averaged over a period $T = 2\pi/\omega$ of the harmonic oscillation is

$$\begin{aligned} <P>_{\rm av} &= \frac{1}{T}\int_0^T ei\,dt = E_m I_m \frac{1}{T} \\ &\quad \int_0^T [\cos\phi \sin^2(\omega t) - \tfrac{1}{2}\sin\phi \sin(2\omega t)]\,dt \\ &= \tfrac{1}{2} E_m I_m \cos\phi \end{aligned} \tag{3.70}$$

The mean power is proportional to the cosine of the phase angle ϕ. The quantity $\cos\phi$ is known as the *power factor.* If the circuit does not contain resistance, $\cos\phi = 0$, and there is no average power furnished to the circuit.

The *effective current* I_E is the *steady current* or direct current which will develop the same amount of heat in the resistance R as does the actual current when averaged over a period of the harmonic current oscillation. The average power dissipated in the resistance is

$$\begin{aligned} <P_R>_{\rm av} &= \frac{1}{T}\int_0^T Ri^2\,dt \\ &= RI_m^2 \frac{1}{T}\int_0^T \sin^2(\omega t - \phi)\,dt \\ &= \tfrac{1}{2}RI_m^2 = RI_E^2 \end{aligned} \tag{3.71}$$

Hence the effective or root-mean-square (rms) value of the current I_E is related to the maximum amplitude I_m of the alternating current by the relation

$$I_E = \frac{I_m}{(2)^{1/2}}$$

The effective or root-mean-square electromotive force is $E_e = E_m/(2)^{1/2}$. In terms of the effective values of current and potential, the average power delivered to the circuit by the electromotive force is $<P>_{\rm av} = E_e I_e \cos\phi$.

Series Resonance. At low frequencies the dominating term in the impedance is the one involving the capacitance, whereas at high frequency the term containing the inductance is the important one. The absolute value of the impedance is a minimum, and therefore the current is greatest at the frequency f_r which makes the reactance

$$\left(\omega L - \frac{1}{\omega C}\right)$$

vanish, that is, for

$$f_r = \frac{\omega_0}{2\pi} = \frac{1}{2\pi\sqrt{LC}} \tag{3.72}$$

By analogy with the simple harmonic oscillator, this frequency is known as the *frequency of resonance.* When the circuit is in resonance, the phase angle ϕ vanishes, and the power factor is unity; therefore the power expended in the circuit is a maximum. A coil of large self-inductance and small resistance placed in series with a harmonic potential makes the phase angle approach $\pi/2$ and cuts down the current by increasing the impedance without wasting power as a rheostat does in a d-c circuit. A coil of this type is called a *choke coil.*

It is convenient to write the complex impedance as

$$\begin{aligned} Z = R + j\left(\omega L - \frac{1}{\omega C}\right) &= R + jX_0\left(\frac{\omega}{\omega_0} - \frac{\omega_0}{\omega}\right) \\ &= R\left[1 + jQ\left(\frac{\omega}{\omega_0} - \frac{\omega_0}{\omega}\right)\right] \end{aligned} \tag{3.73}$$

where $\omega_0 = 1/(LC)^{1/2}$, $X_0 = (L/C)^{1/2}$, and $Q = X_0/R$. X_0 is called the *characteristic impedance* of the circuit, and the parameter Q is generally referred to as *the Q of the circuit.* The magnitude of Q is a measure of the sharpness of resonance, since a large Q means that Z will assume large values when ω differs slightly from ω_0.

14. Impedances in Series and Parallel: Parallel Resonance

Use of complex impedances and currents enables a basic principle to be introduced into the theory of alternating currents identical in form with Ohm's law for direct currents. This principle, combined with Kirchhoff's laws, enables the impedances of component branches of a network to be combined in series and parallel in the same manner in which resistances are combined. The equivalent complex impedance Z of a circuit consisting of several impedances in series is $Z = Z_1 + Z_2 + Z_3 + \cdots + Z_n$. The equivalent complex impedance of the impedances $Z_1, Z_2, \ldots, Z_n$ connected in parallel is

$$\frac{1}{Z} = \frac{1}{Z_1} + \frac{1}{Z_2} + \frac{1}{Z_3} + \cdots + \frac{1}{Z_n}$$

This is usually written in terms of the admittances $Y_k = 1/Z_k$ as $Y = Y_1 + Y_2 + Y_3 + \cdots + Y_n$, where Y is the *equivalent admittance* of the entire circuit.

Parallel Resonance. In the circuit of Fig. 3.14, there is a parallel connection of the impedances

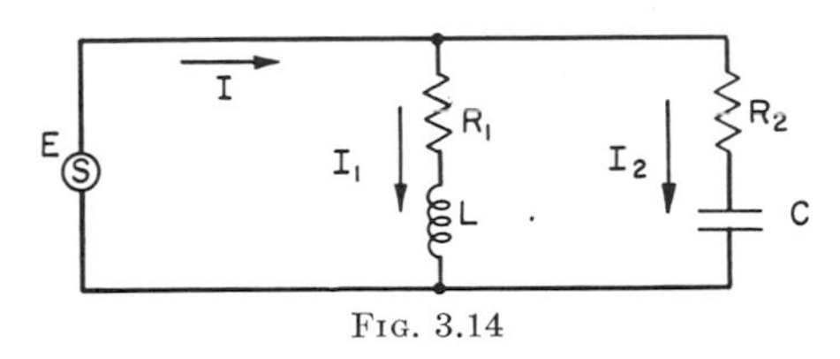

FIG. 3.14

$Z_1 = R_1 + j\omega L$ and $Z_2 = R_2 - j/\omega C$. If the amplitude of the applied harmonic potential is E, the complex currents of the two branches are

$$I_1 = \frac{E}{Z_1} = \frac{E}{R_1 + j\omega L} = \frac{E}{R_1 + jX_L} = \frac{E(R_1 - jX_L)}{|Z_1|}$$
$$I_2 = \frac{E}{Z_2} = \frac{E}{R_2 - j/\omega C} = \frac{E}{R_2 - jX_C} = \frac{E(R_2 + jX_C)}{|Z_2|}$$

where $X_L = \omega L$ is the *inductive reactance* and

$$X_C = \frac{1}{\omega C}$$

is the *capacitive reactance* of the circuit branches.

The resultant complex current of the circuit is $I = I_1 + I_2$. If the circuit parameters are adjusted so that

$$\frac{X_L}{|Z_1|} = \frac{X_C}{|Z_2|}$$

then the resulting complex current is real, and the power factor of the entire circuit is unity. When this condition is attained, the circuit is said to be in *unity-power-factor resonance.* Sometimes this phenomenon is also called *parallel resonance.*

If $R_1 = R$ and $R_2 = 0$, the complex impedance of the entire circuit is

$$Z = \frac{(R + j\omega L)(-j/\omega C)}{R + j(\omega L - 1/\omega C)}$$

If $R^2 \ll \omega^2 L^2$, this impedance has a maximum value of $Z_m = L/RC$ at a frequency for which $\omega = 1/(LC)^{1/2}$. If the resistances R_1 and R_2 are given the value $R_1 = R_2 = (L/C)^{1/2} = R$, the circuit is in unity-power-factor resonance for *all* frequencies, and the resulting equivalent impedance of the circuit is $Z = (L/C)^{1/2} = R$.

15. Transmission of Power

Consider a source of power that generates a harmonic potential of effective amplitude E_e. Let the impedance of the source be $Z_S = R_S + jX_S$ and let

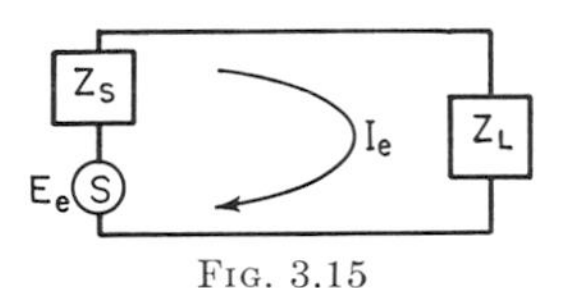

FIG. 3.15

it be connected in series to a load whose impedance is $Z_L = R_L + jX_L$ (Fig. 3.15). The effective value of the current of the circuit is

$$I_e = \frac{E_e}{\sqrt{(R_S + R_L)^2 + (X_S + X_L)^2}} \tag{3.74}$$

The average power consumed by the load is

$$\langle P_L \rangle_{av} = \frac{R_S E_e^2}{(R_S + R_L)^2 + (X_S + X_L)^2} \tag{3.75}$$

The value of Z_L for which the load will absorb maximum power is given by $X_S = -X_L$ and $R_L = R_S$.

The condition for maximum-power transmission is that Z_L should be the conjugate of Z_S. Under this condition, the average power furnished the load is

$$\langle P_L \rangle_{av} = \frac{E_e^2}{4R_L} \tag{3.76}$$

A load impedance is said to be *matched* to its power supply when $Z_L = \bar{Z}_S$, so that the maximum power transfer is effected.

16. General A-C Network: Network Theorems

In Sec. 9 the general mesh equations of a network are written in matrix form: $[Z(D)](i) = (e)$. If the electromotive forces that act in the various meshes of the network are harmonic functions of the same frequency but of different phases, the column electromotive-force matrix is

$$(e) = \begin{Vmatrix} E_1 \sin(\omega t + \theta_1) \\ E_2 \sin(\omega t + \theta_2) \\ \cdot \\ \cdot \\ E_n \sin(\omega t + \theta_n) \end{Vmatrix} = \text{Im} \begin{Vmatrix} E_1 e^{j\theta_1} \\ E_2 e^{j\theta_2} \\ \cdot \\ \cdot \\ \cdot \\ E_n e^{j\theta_n} \end{Vmatrix} \cdot e^{j\omega t} \tag{3.77}$$

and so $(e) = \text{Im}\,(E)e^{j\omega t}$.

The elements of the column matrix (E) are the

complex potentials of the various meshes of the network. The steady-state a-c response of the system has the form $(i) = \text{Im}\,(I)e^{j\omega t}$, where the elements of the column matrix (I) are the complex alternating currents of the network which is related to (E) by $[Z(j\omega)](I) = (E)$.

The complex currents are now determined by $(I) = [Z(j\omega)]^{-1}(E)$. The matrix $[Z(j\omega)]$ is called the *impedance matrix* of the network, and its inverse is $[Z(j\omega)]^{-1}$. The elements of the impedance matrix $Z_{kk} = j\omega L_{kk} + R_{kk} + S_{kk}/j\omega$ are called the *mesh impedances.* The elements

$$Z_{kr} = j\omega L_{kr} + R_{kr} + \frac{S_{kr}}{j\omega} = Z_{rk}$$

are called the *mutual impedances* of the network. As a consequence of the fact that $Z_{kr} = Z_{rk}$, the impedance matrix $[Z(j\omega)]$ is a symmetric matrix.

A number of important network relations or theorems follow:

1. *The superposition theorem.* If a network has two or more sources of electromotive force, the current through any branch of the network is equal to the sum of the currents obtained by considering each source of electromotive force separately while each of the other sources is replaced by its internal impedance. This follows from the linearity of the equations $[Z(j\omega)](I) = (E)$.

2. *The reciprocity theorem.* In any linear network the steady-state current produced in one branch by inserting an alternating electromotive force in a second branch equals, in both magnitude and phase, the current produced in the second branch when the same source of electromotive force is inserted in the first. This is a consequence of the symmetry of the impedance matrix $[Z(j\omega)]$.

3. *The Thévenin-Pollard theorem.* Consider a linear network containing various electromotive-force sources and two output terminals A and B. Let E_0 be the open-circuit potential that appears across these output terminals, and let Z_0 be the input impedance of the network when all the electromotive sources of the network have been replaced by their internal impedances. The current which will flow in an impedance connected across the output terminals AB is $I = E_0/(Z + Z_0)$. This relation follows from the linearity of the equations, and it enables the terminals A and B to be connected into any other circuit as if they were the terminals of a generator of electromotive force E_0 and internal impedance Z_0.

4. *The maximum-power-transfer theorem.* If a source of power of internal impedance Z_s is to supply at its terminals maximum power to a load of impedance Z_L, the load impedance must be the complex conjugate of Z_s. If it is not possible to change the phase angle of the load impedance, maximum power will be supplied to the load if $|Z_L| = |Z_s|$ (see Sec. 15).

5. *The compensation theorem.* If an impedance ΔZ is inserted in a branch of a network, the resulting current increment produced at any point in the network is equal to the current that would be produced at that point by a compensating potential $-I\,\Delta Z$ acting in series with the modified branch, where I is the current in the original branch.

17. Two-terminal Networks; Foster's Reaction Theorem

A general network with two accessible terminals is illustrated schematically by Fig. 3.16. This represents a general linear network having n independent meshes and no internal sources of electromotive force.

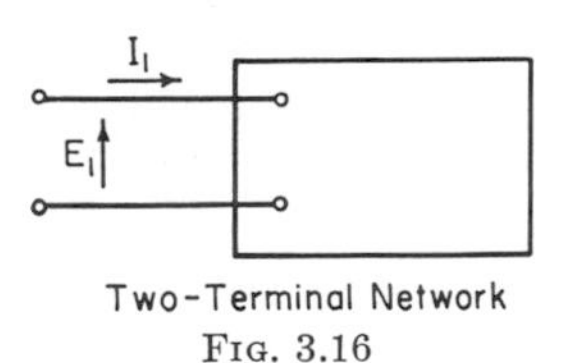

FIG. 3.16

Such a two-terminal network is called a *driving-point impedance* because the two terminals are specified as that point from which the network is excited. If the network is being driven from the first mesh by a complex potential E_1, which produces a complex current I_1, then the impedance looking into the network, Z, is given in the form

$$Z = \frac{E_1}{I_1} = \frac{|Z(j\omega)|}{B_{11}(j\omega)}$$

where $|Z(j\omega)|$ is the determinant of the network impedance matrix $[Z(j\omega)]$ and $B_{11}(j\omega)$ is the cofactor of the element $Z_{11}(j\omega)$ of the impedance matrix. In filters and other important special structures, networks with negligible ohmic resistances play a predominant part. In such nondissipative networks, the elements of the network impedance matrix take the form

$$Z_{rs} = pL_{rs} + \frac{S_{rs}}{p}$$

where p denotes $j\omega$. The impedance Z, looking into such an n-mesh nondissipative network, has the form

$$Z = \frac{|Z(j\omega)|}{B_{11}(j\omega)} = \frac{a_{2n}p^{2n} + a_{2n-2}p^{2n-2} + \cdots + a_2p^2 + a_0}{b_{2n-1}p^{2n-1} + b_{2n-3}p^{2n-3} + \cdots + b_3p^3 + b_1p} \tag{3.78}$$

where the a's and b's are constants. The numerator is a polynomial of even degree and the denominator a polynomial of odd degree. The zeros of these polynomials occur in conjugate-complex pairs, $p = 0$ is a zero of the denominator polynomial, and so the expression may be written

$$Z = H\,\frac{(p^2 - p_1^2)(p^2 - p_3^2)\cdots(p^2 - p^2_{2n-1})}{p(p^2 - p_2^2)(p^2 - p_4)\cdots(p^2 - p^2_{2n-2})} \tag{3.79}$$

where $p_k = \pm j\omega_k$, $k = 1, 3, 5, \ldots, (2n - 1)$, are the roots of the numerator polynomial and $p = 0$, and $p_k = \pm j\omega_k$, $k = 2, 4, 6, \ldots, (2n - 2)$, are the roots of the denominator polynomial. H is a constant, given by

$$H = a_{2n}/b_{2n-1}$$

If the notation $p = j\omega$ and $p_k^2 = -w_k^2$ is introduced, then $Z = jX$, where the *reactance* X has the factored form

$$X = \frac{H(\omega^2 - \omega_1^2)(\omega^2 - \omega_3^2)(\omega^2 - \omega_5^2) \cdots (\omega^2 - \omega^2_{2n-1})}{\omega(\omega^2 - \omega_2^2)(\omega^2 - \omega_4^2)(\omega^2 - \omega_6^2) \cdots (\omega^2 - \omega^2_{2n-2})} \tag{3.80}$$

This expression indicates that the driving point impedance of a general n-mesh, dissipationless network has the form of a reactance which is *zero* at the frequencies $\omega_1, \omega_3, \ldots, \omega_{2n-1}$ and *infinity* at the frequencies $0, \omega_2, \omega_4, \ldots, \omega_{2n-2}$. The reactance $X(\omega)$ has the property that

$$\frac{dX}{d\omega} > 0 \qquad \text{for } 0 < \omega < \infty \tag{3.81}$$

The function $X(\omega)$ must therefore change its sign at a pole. The poles and zeros of $X(\omega)$ alternate, since otherwise two successive poles or zeros would require the slope of the function to be negative over part of the intervening frequency region. This alternation of poles and zeros of $X(\omega)$ is expressed by

$$0 < \omega_1 < \omega_2 < \omega_3 < \cdots < \omega_{2n-2} < \omega_{2n-1} < \infty \tag{3.82}$$

and is called the *separation property* of the zeros and poles of the reactance function. Figure 3.17 illustrates the general appearance of this function.

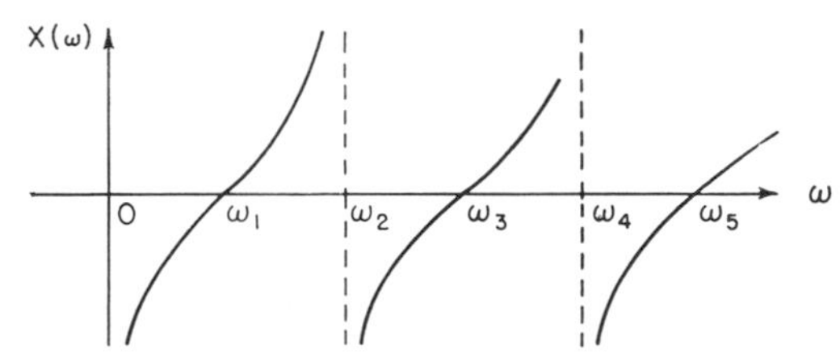

General Appearance of the Reactance Function
FIG. 3.17

The fact that the impedance function of any two-terminal, dissipationless network is a reactance having the form (3.80) and the properties (3.81) and (3.82) is a statement of *Foster's reactance theorem.* This theorem is useful in determining a network which will produce a certain specified current variation for a given source potential. The fact that Foster's reactance theorem applies only to dissipationless networks is not a serious restriction in practical applications, because by design many classes of circuits are constructed with a minimum of resistance so that the reactance theorem describes the behavior of many actual two-terminal networks very well.

18. Four-terminal Networks in the A-C Steady State

A special form of general network of major technical importance is a network with a pair of input terminals and a pair of output terminals. It is customary to refer to such a network as a *four-terminal network* or a *two-terminal-pair network* (Fig. 3.18).

The reference directions for voltage and current will be taken as shown in Fig. 3.18. The network

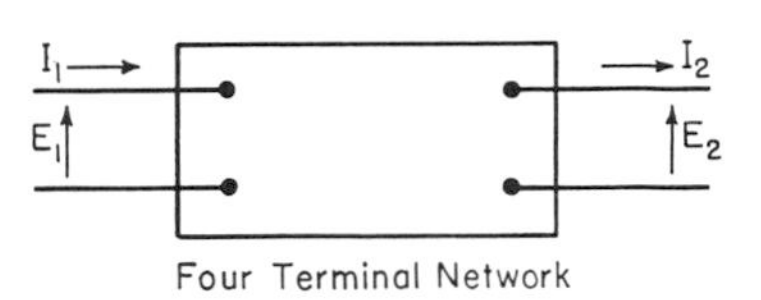

Four Terminal Network
FIG. 3.18

will be supposed to be an n-mesh network having two pairs of accessible terminals. The potentials E_1 and E_2 are complex potentials produced by external sources, and the currents I_1 and I_2 are the complex currents in the corresponding meshes. The internal structure of the network will be assumed to be quite general, and the n meshes will be supposed to have impedances Z_{kk} and complex mutual impedances Z_{ks} between the meshes. The restrictions on the component elements of the network are that they be linear and constant. It is further assumed that the network is passive; i.e., the only sources of electromotive force are E_1 and E_2. The following linear relations (see Sec. 16) exist between the input quantities and the output quantities of the four-terminal network:

$$\begin{aligned} E_1 &= AE_2 + BI_2 \\ I_1 &= CE_2 + DI_2 \end{aligned} \tag{3.83}$$

The coefficients A,B,C,D are called the *general four-terminal-network parameters:*

$$\begin{aligned} A &= (E_1/E_2)_{I_2=0} \qquad & B &= (E_1/I_2)_{E_2=0} \\ C &= (I_1/E_2)_{I_2=0} \qquad & D &= (I_1/I_2)_{E_2=0} \end{aligned} \tag{3.84}$$

A and D are dimensionless transfer ratios, whereas B and C have the dimensions of impedance and of admittance, respectively. As a consequence of the *reciprocity theorem,* the following relation between the network parameters exists:

$$AD - BC = 1 \tag{3.85}$$

Hence only three of the network parameters are independent. If the network contains no resistances and is therefore dissipationless, A and D are real, and B and C are pure imaginary numbers.

For many purposes it is convenient to express the current and potential at the input side of a four-terminal network as a linear transformation of the current and potential on the output side. This may be done by the use of matrix notation (see Part 1, Chap. 2, Sec. 9):

$$\begin{pmatrix} E_1 \\ I_1 \end{pmatrix} = \begin{bmatrix} A & B \\ C & D \end{bmatrix} \begin{pmatrix} E_2 \\ I_2 \end{pmatrix} = [U] \begin{pmatrix} E_2 \\ I_2 \end{pmatrix} \tag{3.86}$$

The square matrix $[U]$ is usually called the *network matrix.* The determinant of the network matrix is equal to unity. The matrix $[U]$ has the inverse

$$[U]^{-1} = \begin{bmatrix} A & B \\ C & D \end{bmatrix}^{-1} = \begin{bmatrix} D & -B \\ -C & A \end{bmatrix} \tag{3.87}$$

If two four-terminal networks of matrices, $[U]_1$ and $[U]_2$, are connected in cascade, as shown in

Fig. 3.19, then the over-all relation between the input and output quantities is given by

$$\begin{pmatrix} E_1 \\ I_1 \end{pmatrix} = [U]_1[U]_2 \begin{pmatrix} E_2 \\ I_2 \end{pmatrix} = [U] \begin{pmatrix} E_2 \\ I_2 \end{pmatrix} \qquad (3.88)$$

where $[U]$ is the matrix of the over-all network:

$$\begin{aligned} [U] &= \begin{bmatrix} A_1 & B_1 \\ C_1 & D_1 \end{bmatrix} \begin{bmatrix} A_2 & B_2 \\ C_2 & D_2 \end{bmatrix} \\ &= \begin{bmatrix} A_1A_2 + B_1C_2 & A_1B_2 + B_1D_2 \\ C_1A_2 + D_1C_2 & C_1B_2 + D_1D_2 \end{bmatrix} \end{aligned} \qquad (3.89)$$

In the case of a cascade connection of several four-terminal networks, the over-all matrix of the new network is the matrix product of the matrices of the individual networks *taken in the order* of connection.

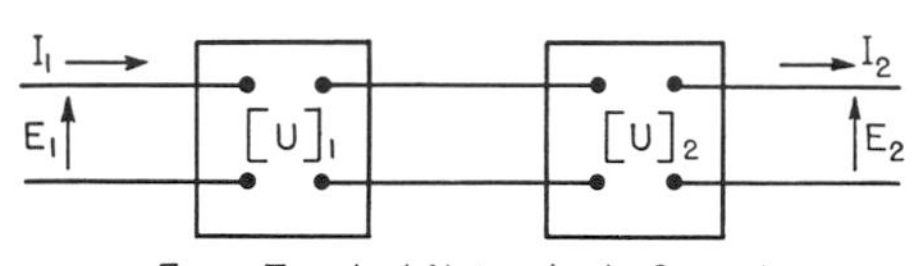

Four-Terminal Networks in Cascade
FIG. 3.19

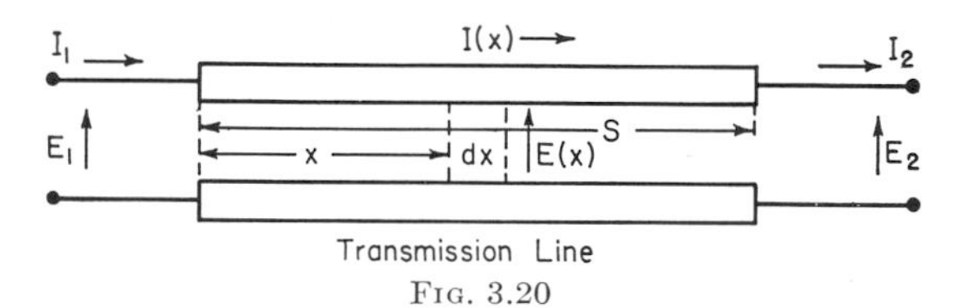

Transmission Line
FIG. 3.20

If two four-terminal networks have the matrices $[U]_1$ and $[U]_2$ and if $[U]_2 = [U]_1^{-1}$, then the second network is said to be the *inverse* of the first one. If a nework whose matrix is $[U]$ is connected to the line in such a manner that the output terminals are connected to the input side and the input terminals are connected to the output side, then $[U]$ is said to be *reversed*. The matrix of the reversed network has the form

$$[U]_R = \begin{bmatrix} A & -B \\ -C & D \end{bmatrix} \qquad (3.90)$$

A four-terminal network is said to be *reversible* if its *inverse* is equal to its *reverse*: $[U]^{-1} = [U]_R$. Then $A = D$, and

$$[U] = \begin{bmatrix} A & B \\ C & A \end{bmatrix} = \begin{bmatrix} \sqrt{1 + BC} & B \\ C & \sqrt{1 + BC} \end{bmatrix} \qquad (3.91)$$

The Transmission Line as a Four-terminal Network. Consider a two-wire transmission line as shown in Fig. 3.20. Let:

R = resistance of line per unit length
L = inductance of line per unit length
G = leakage conductance of line per unit length
C = capacitance of line per unit length
$Z = R + j\omega L$ = series impedance of line per unit length
$Y = G + j\omega C$ = shunt admittance of line per unit length

Let dI be the complex current that flows from one conductor to the other through the admittance $Y\,dx$, and let dE be the change in the potential difference between the conductors caused by the current flowing in the impedance $Z\,dx$. If higher-order infinitesimals are neglected, $dI = -EY\,dx$ and $dE = -IZ\,dx$. The complex current $I(x)$ and the complex potential $E(x)$ therefore satisfy

$$\begin{aligned} \frac{dI}{dx} &= -YE \\ \frac{dE}{dx} &= -ZI \end{aligned} \qquad (3.92)$$

If these are solved subject to the boundary conditions that $E = E_1$ and $I = I_1$ at $x = 0$,

$$\begin{aligned} E(x) &= E_1 \cosh ax - I_1Z_0 \sinh ax \\ I(x) &= -\frac{E_1}{Z_0} \sinh ax + I_1 \cosh ax \end{aligned} \qquad (3.93)$$

where

$$a = (ZY)^{1/2} = [(R + j\omega L)(G + j\omega C)]^{1/2} = \alpha + j\beta$$

and

$$Z_0 = \left(\frac{Z}{Y}\right)^{1/2} = \left[\frac{R + j\omega L}{G + j\omega C}\right]^{1/2}$$

The quantity a is the *propagation function* of the line. The real part of a, α, is the *attenuation function*, and the imaginary part of a, β, is the *phase function* of the line. The impedance Z_0 is the *characteristic impedance* of the line.

If S is the length of the line and I_2 and E_2 are the output complex current and potential of the line, Eqs. (3.93) reduce to

$$\begin{aligned} E(S) &= E_2 = E_1 \cosh aS - I_1Z_0 \sinh aS \\ I(S) &= I_2 = -Z_0^{-1}E_1 \sinh aS + I_1 \cosh aS \end{aligned} \qquad (3.94)$$

Let $\theta = aS$, and write Eqs. (3.94) in matrix form:

$$\begin{pmatrix} E_2 \\ I_2 \end{pmatrix} = \begin{bmatrix} \cosh\theta & -Z_0 \sinh\theta \\ -Z_0^{-1} \sinh\theta & \cosh\theta \end{bmatrix} \begin{pmatrix} E_1 \\ I_1 \end{pmatrix} \qquad (3.95)$$

or

$$\begin{pmatrix} E_1 \\ I_1 \end{pmatrix} = \begin{bmatrix} \cosh\theta & Z_0 \sinh\theta \\ Z_0^{-1} \sinh\theta & \cosh\theta \end{bmatrix} \begin{pmatrix} E_2 \\ I_2 \end{pmatrix} \qquad (3.96)$$

The transmission line is therefore a reversible four-terminal network. The square matrix

$$[U_L] = \begin{bmatrix} \cosh\theta & Z_0 \sinh\theta \\ Z_0^{-1} \sinh\theta & \cosh\theta \end{bmatrix}$$

is the matrix of the transmission line. The quantity θ is called the *angle* of the line.

If the matrices of the fundamental types of four-terminal structures are known, then by matrix multiplication it is easy to obtain many useful properties of more complex structures formed by a cascade connection of basic structures. Table 3.2 is useful for this purpose.

TABLE 3.2. MATRICES OF FUNDAMENTAL FOUR-TERMINAL STRUCTURES

No.	Network	Matrix = $\begin{bmatrix} A & B \\ C & D \end{bmatrix}$
1	Series Impedance	$\begin{bmatrix} 1 & Z \\ 0 & 1 \end{bmatrix}$
2	Shunt Admittance	$\begin{bmatrix} 1 & 0 \\ Y & 1 \end{bmatrix}$
3	Coupled Circuits	$\begin{bmatrix} \frac{L_1}{M} & \frac{jw}{M}(L_1L_2 - M^2) \\ \frac{-j}{wM} & \frac{L_2}{M} \end{bmatrix}$
4	Ideal Transformer $a = \frac{N_2}{N_1}$	$\begin{bmatrix} a^{-1} & 0 \\ 0 & a \end{bmatrix}$
5		$\begin{bmatrix} 1 & Z \\ Y & (1 + YZ) \end{bmatrix}$
6		$\begin{bmatrix} (1 + YZ) & Z \\ Y & 1 \end{bmatrix}$
7	T-Section	$\begin{bmatrix} (1 + YZ_1) & (Z_1 + Z_2 + YZ_1Z_2) \\ Y & (1 + YZ_2) \end{bmatrix}$
8	Pi-Section	$\begin{bmatrix} (1 + ZY_2) & Z \\ (Y_1 + Y_2 + ZY_1Y_2) & (1 + ZY_1) \end{bmatrix}$
9	Transmission Line	$\begin{bmatrix} \cosh(aS) & Z_0 \sinh(aS) \\ Z_0^{-1} \sinh(aS) & \cosh(aS) \end{bmatrix}$ a = Propagation Function Z_0 = Characteristic Impedance S = length of line.
10	Cascade of Symmetrical Identical Networks Matrix of Individual Network $= \begin{bmatrix} A & B \\ C & A \end{bmatrix}$	Over-all matrix of n networks $\begin{bmatrix} \cosh(an) & Z_0 \sinh(an) \\ Z_0^{-1} \sinh(an) & \cosh(an) \end{bmatrix}$ $a = \cosh^{-1}(A)$ = propagation function $Z_0 = \sqrt{\frac{B}{C}}$ = characteristic impedance

19. Wave Propagation along a Cascade of Symmetric Structures

Many important technical problems of electrical-circuit theory involve the determination of the nature of the current and potential distribution along a chain of identical symmetric four-terminal networks, as in Fig. 3.21. Let the matrix of one of the symmetric four-terminal structures of the chain be

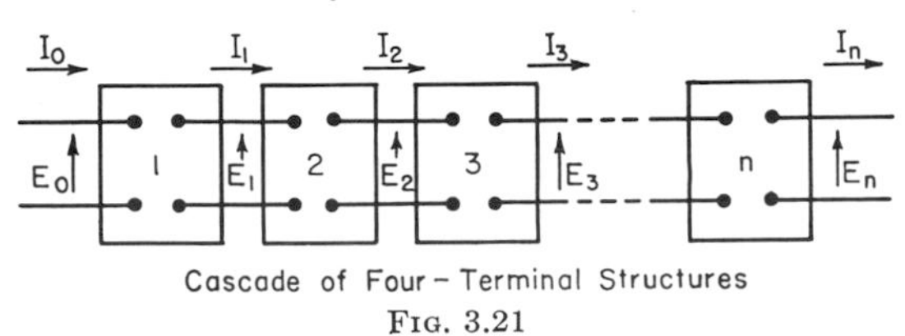

Cascade of Four-Terminal Structures
FIG. 3.21

$$[U] = \begin{bmatrix} A & B \\ C & A \end{bmatrix}$$

Since each of the structures has the same matrix, the potential and the current output of the nth structure of the chain satisfy the equation

$$\begin{pmatrix} E_0 \\ I_0 \end{pmatrix} = \begin{bmatrix} A & B \\ C & A \end{bmatrix}^n \begin{pmatrix} E_n \\ I_n \end{pmatrix} \tag{3.97}$$

By the use of the Cayley-Hamilton theorem of the theory of matrices, the following convenient form for the nth power of the matrix $[U]$ can be obtained (see Part 1, Chap. 2, Sec. 13):

$$\begin{bmatrix} A & B \\ C & A \end{bmatrix}^n = \begin{bmatrix} \cosh an & Z_0 \sinh an \\ Z_0^{-1} \sinh an & \cosh an \end{bmatrix}$$
$$n = 0,1,2,3, \ldots \quad (3.98)$$

where $a = \cosh^{-1} A$ and $Z_0 = (B/C)^{1/2}$.

The quantity a is the *propagation function* of the four-terminal structure of the chain of Fig. 3.21, and Z_0 is the *characteristic impedance.* Equation (3.97) may be written

$$\begin{pmatrix} E_0 \\ I_0 \end{pmatrix} = \begin{bmatrix} \cosh an & Z_0 \sinh an \\ Z_0^{-1} \sinh an & \cosh an \end{bmatrix} \begin{pmatrix} E_n \\ I_n \end{pmatrix} \quad (3.99)$$

or
$$\begin{pmatrix} E_n \\ I_n \end{pmatrix} = \begin{bmatrix} \cosh an & -Z_0 \sinh an \\ -Z_0^{-1} \sinh an & \cosh an \end{bmatrix} \begin{pmatrix} E_0 \\ I_0 \end{pmatrix} \quad (3.100)$$

The potential and current along the chain are

$$\begin{pmatrix} E_k \\ I_k \end{pmatrix} = \begin{bmatrix} \cosh ak & -Z_0 \sinh ak \\ -Z_0^{-1} \sinh ak & \cosh ak \end{bmatrix} \begin{pmatrix} E_0 \\ I_0 \end{pmatrix} \quad (3.101)$$

If the chain is terminated by an impedance equal to the characteristic impedance Z_0, then $I_n = E_n/Z_0$. The impedance Z_i looking into the chain may be obtained from (3.99):

$$Z_i = \frac{E_0}{Z_0} = \frac{E_n \cosh an + Z_0 I_n \sinh an}{(E_n/Z_0) \sinh an + I_n \cosh an} = Z_0 \quad (3.102)$$

Hence, if the cascade of networks is terminated by its own characteristic impedance Z_0, the impedance Z_i looking into the cascade is also equal to the characteristic impedance. In this case, the input current of the chain is $I_0 = E_0/Z_0$. The potential and current distribution along the cascade may be obtained from (3.101) if I_0 is eliminated, giving

$$E_k = E_0 e^{-ak} \qquad k = 0,1,2,3, \ldots ,n$$
$$I_k = \frac{E_0 e^{-ak}}{Z_0} \quad (3.103)$$

Attenuation and Passbands. In general, the propagation function is a complex function of the frequency:

$$a = \alpha + j\beta \quad (3.104)$$

The quantity α is called the *attenuation function,* and β is the *phase function* since it gives the change of phase per section as one progresses along the cascade. If the cascade of four-terminal networks is to pass a certain band of frequencies without attenuation, the real part of the propagation constant must be zero and hence a must be pure imaginary. Thus for a *passband* it is necessary that $\alpha = 0$, $a = j\beta$. Then $\cosh a = \cosh j\beta = \cos \beta = A$.

It follows that the network constant A of the fundamental four-terminal network of the cascade must satisfy the two following conditions in a *passband:* $A = \cos \beta =$ real and, since $\cos (\beta)$ can vary between -1 and $+1$, the inequality $-1 \geq A \geq +1$. A is a real number if the network is dissipationless. For the range of frequencies for which the preceding inequality is *not* satisfied, there will be *attenuation.* In the *attenuation band,*

$$\alpha = \cosh^{-1} A \qquad \text{for } A > +1$$
$$\beta = 0$$

and
$$\alpha = \cosh^{-1} (-A)$$
$$\beta = \pm\pi \qquad \text{for } A < -1$$

In a cascade of dissipationless networks, the characteristic impedance $Z_0 = (B/C)^{1/2}$ is *real* in the *passband* and *imaginary* in the *attenuation band.*

20. Filters

By selection of special structures for the basic four-terminal network of the cascade it is possible to construct recurrent structures that pass certain frequencies freely and attenuate others. Consider the

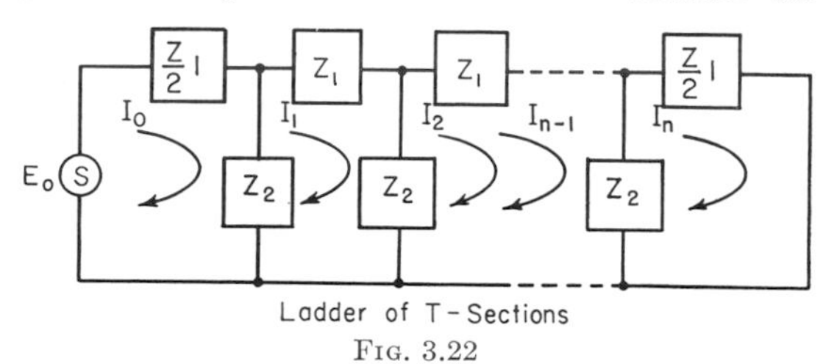

Ladder of T-Sections

FIG. 3.22

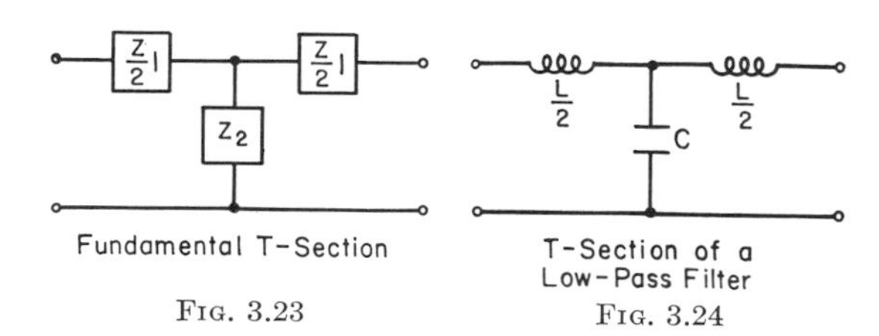

Fundamental T-Section

FIG. 3.23

T-Section of a Low-Pass Filter

FIG. 3.24

ladder network of Fig. 3.22, consisting of a cascade connection of n identical T sections of the form illustrated by Fig. 3.23. The matrix of the fundamental T section is

$$\begin{bmatrix} A & B \\ C & A \end{bmatrix} = \begin{bmatrix} \dfrac{1 + Z_1}{2Z_2} & \dfrac{Z_1 + Z_1{}^2}{4Z_2} \\ \dfrac{1}{Z_2} & \dfrac{1 + Z_1}{2Z_2} \end{bmatrix} \quad (3.105)$$

Hence, $\cosh a = A = 1 + Z_1/2Z_2$ and $Z_0 = Z_2 \sinh a$.

The inequality $-1 \leqq A \leqq +1$ gives the following condition for a *passband:* $-1 \leqq (1 + Z_1/2Z_2) \leqq +1$ or $0 \leqq (-Z_1/Z_2) \leqq 4$.

The current in the nth section may be obtained from (3.99) by placing $E_n = 0$:

$$I_n = \frac{E_0}{Z_0 \sinh an} = \frac{E_0}{Z_2 \sinh (a) \sinh an} \quad (3.106)$$

Constant k Filters. The discriminating property of the ladder network of Fig. 3.22 in passing certain frequencies and attenuating others is employed in the design of a class of filter circuits known as *constant k filters.* In this type of filter, the series and shunt impedances are *inverse reactances* such that $Z_1Z_2 = k^2 =$ const.

A. Low-pass Filter. If the fundamental structure of the ladder network is taken to be that illustrated by Fig. 3.24, then $Z_1 = j\omega L$ and $Z_2 = 1/j\omega C$. Therefore, $Z_1Z_2 = L/C = k^2$ and $-Z_1/Z_2 = \omega^2 LC$.

The limits of the *passband* are $0 \lessgtr \omega^2 LC \lessgtr 4$ or $0 \lessgtr \omega \lessgtr \omega_c$, where $\omega_c = 2/(LC)^{1/2}$.

Hence the filter constructed by the use of the fundamental T structure of Fig. 3.23 passes without attenuation all frequencies between 0 and the *cutoff frequency* $\omega_c = 2/(LC)^{1/2}$. Attenuation sets in at the cutoff frequency and increases steadily as one proceeds to higher frequencies. A filter of this type is called a *low-pass filter*. Its characteristic impedance is

$$Z_0 = \sqrt{\frac{L}{C}\left(1 - \frac{\omega^2}{\omega_c^2}\right)}$$

This impedance is a pure resistance within the passband and becomes a pure reactance in the attenuating band.

B. High-pass Filter. Let the T network of Fig. 3.25 be taken as the fundamental structure of a filter. Then $Z_1 = 1/j\omega C$ and $Z_2 = j\omega L$. Again, in this case, $Z_1 Z_2 = L/C = k^2$. The passband is now determined by $0 \lessgtr 1/LC\omega^2 \lessgtr 4$, that is, $\omega_c \lessgtr \omega \lessgtr \infty$, where $\omega_c = 1/2(LC)^{1/2}$.

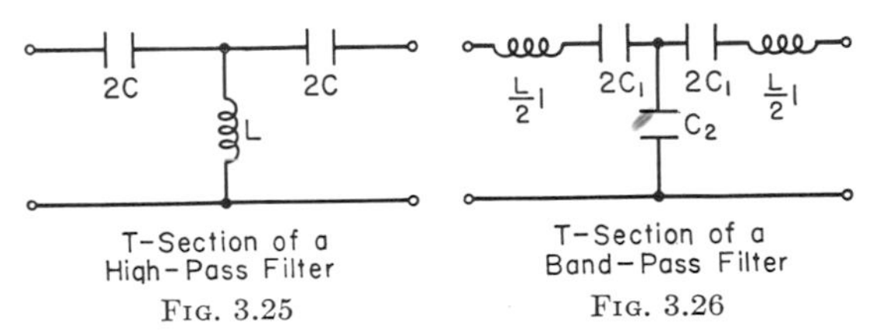

FIG. 3.25 FIG. 3.26

Therefore this filter transmits currents of all frequencies from ∞ down to the cutoff frequency ω_c without attenuation. Such a filter is called a *high-pass filter*. Its characteristic impedance is

$$Z_0 = \sqrt{\frac{L}{C}\left(1 - \frac{\omega_c^2}{\omega^2}\right)}$$

This impedance approaches the value $(L/C)^{1/2}$ at high frequencies.

C. Bandpass Filter. Consider the T network of Fig. 3.26 as the fundamental structure of a filter. For this arrangement, $Z_1 = j(\omega L_1 - 1/\omega C_1)$ and $Z_2 = 1/j\omega C_2$. The passband inequality is

$$0 \lessgtr (\omega^2 L_1 C_2 - C_2/C_1) \lessgtr 4$$

If $\omega_A = \dfrac{1}{\sqrt{L_1 C_1}}$ and $\omega_B = \sqrt{\dfrac{4C_1 + C_2}{L_1 C_1 C_2}}$

the passband range is given by $\omega_A \lessgtr \omega \lessgtr \omega_B$. Hence ω_A is the lowest frequency passed without attenuation and ω_B the highest. A filter of this type is called a *bandpass* filter.

21. Nonlinear Problems in Electric-circuit Theory

Circuits which are *nonlinear* in that their parameters are *not* constant but are functions of the current or potential distribution in their branches are frequently encountered. Examples are those whose components include vacuum or gas-filled tubes, ceramic conductors, nonlinear dielectrics, and iron-cored magnetic devices. These circuits are of great practical importance, and special analytical methods for studying their behavior have been developed.

Nonlinear Resistance. Many semiconductors, such as the black ceramic material formed by a heat-treatment of a mixture of clay and carbon, called *thyrite*, may be considered as resistances which do not obey Ohm's law but in which the relation between the voltage drop V, necessary to develop a current i through them, is given by

$$V = R_0 i^k \tag{3.107}$$

where R_0 depends on the length and cross-sectional area of the specimen and the quantity k is a constant.

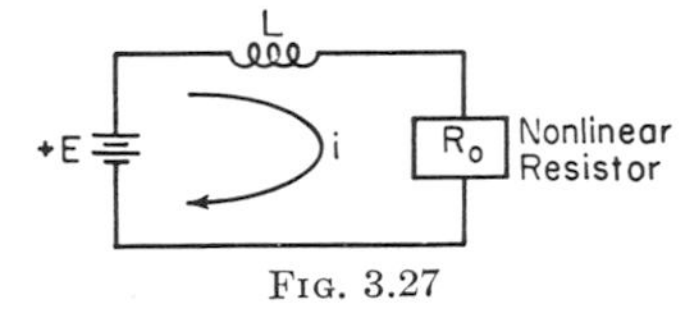

FIG. 3.27

Consider the circuit of Fig. 3.27 which consists of a constant potential E in series with a linear inductance and a nonlinear resistor. If the potential drop through the nonlinear resistance has the form (3.107), the circuit equation is

$$L\frac{di}{dt} + R_0 i^k = E \qquad \text{for } t > 0 \tag{3.108}$$

If it is assumed that the potential is impressed on the circuit at $t = 0$ so that the initial current is zero, it gradually builds up to a final steady-state value. The time is then related to the current by the equation

$$t = L\int_0^i \frac{dx}{E - R_0 x^k} \tag{3.109}$$

The integration may be performed for definite values of the constant k. For example, if $k = 2$,

$$\frac{tR_0}{L} = \int_0^i \frac{dx}{C^2 - x^2} \tag{3.110}$$

where $C = (E/R_0)^{1/2}$, giving

$$i(t) = \sqrt{\frac{E}{R_0}}\tanh\left(t\sqrt{R_0 E}/L\right)$$

The final steady-state current is

$$i_s = \lim_{t\to\infty}\sqrt{\frac{E}{R_0}}\tanh\left(t\sqrt{R_0 E}/L\right) = \sqrt{\frac{E}{R_0}}$$

Nonlinear Inductance. An important problem in the theory of nonlinear circuits is the determination of the steady-state periodic current oscillations in a circuit containing a nonlinear inductor. A circuit that exhibits the curious phenomena that are usually encountered in nonlinear circuits of this type is illustrated by Fig. 3.28. This consists of a capacitor in series with a nonlinear inductor consisting of N turns of wire wound around a core of magnetic material. For simplicity, the circuit will be assumed to be devoid of resistance. The circuit is energized

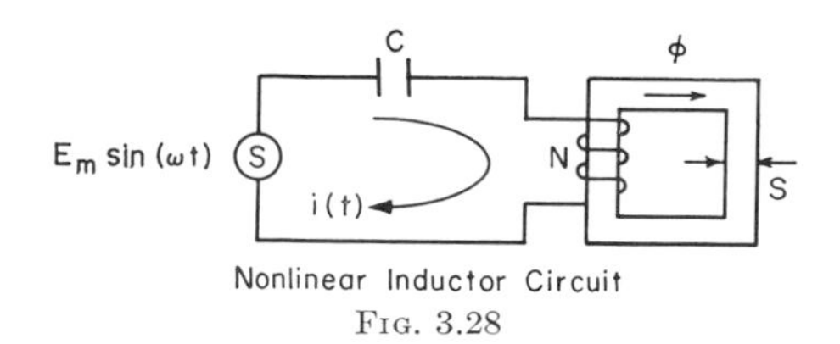

Nonlinear Inductor Circuit
FIG. 3.28

by a harmonic potential $E_m \sin \omega t$, and it is desired to obtain the amplitude of the current oscillation. If q is the separation of charge on the capacitor and ϕ the magnetic flux of the inductor core,

$$N\frac{d\phi}{dt} + \frac{q}{C} = E_m \sin \omega t \tag{3.111}$$

If H is the magnetic intensity of the core and l the average length of the magnetic path of the inductor core, *Ampère's law* expressed in proper units leads to the relation

$$i = \frac{Hl}{N} \tag{3.112}$$

To obtain a relation between the magnetic intensity H and the induction B, the magnetization curve of the material of the core must be expressed in analytical form. The functional relation between H and B is, in general, a many-valued one. However, for certain core materials such as Nicalloy, Permalloy, and certain low-loss steels, it may be approximated by a single-valued odd function of the form

$$H = F(B) = \sum_{n=1,3,5,\ldots} k_n B^n \tag{3.113}$$

where the k_n quantities are the coefficients of the power-series representation of $F(B)$.

The essential features of the problem under consideration may be obtained by using the first two terms of the power series (3.113) and representing the magnetization curve of the material in the form

$$H = \frac{B}{\mu_0} + k_3 B^3 \tag{3.114}$$

The quantity μ_0 is the *initial permeability* of the inductor core material defined by

$$\mu_0 = \left(\frac{dB}{dH}\right)_{H=0} \tag{3.115}$$

and k_3 is a measure of the departure of the magnetization curve from linearity. The core flux ϕ is given by the relation

$$\phi = BS \tag{3.116}$$

where S is the normal cross-sectional area of the magnetic path of the core. The current may be expressed [combining Eqs. (3.112), (3.114), and (3.116)] in terms of the core flux by the equation

$$i = Hl/N = l\phi/NS\mu_0 + lk_3\phi^3/NS^3 \tag{3.117}$$

or

$$i = \frac{N}{L_0}\phi + c_3\phi^3 \tag{3.118}$$

where $L_0 = u_0 N^2 S/l$ and $c_3 = lk_3/NS^3$

The quantity L_0 is called the *initial inductance* of the nonlinear inductor. If (3.111) is differentiated with respect to time, the resulting equation is

$$N\ddot{\phi} + \frac{i}{C} = \omega E_m \cos \omega t \tag{3.119}$$

The current i must be eliminated by substituting (3.118) into (3.119), giving the *nonlinear differential* equation

$$\ddot{\phi} + \omega_0^2\phi + b\phi^3 = \frac{\omega E_m}{N}\cos \omega t \tag{3.120}$$

where $\omega_0 = l/(L_0 C)^{1/2}$ and $b = lk_3/N^2CS^3$. Equation (3.120) is known as *Duffing's differential equation.*

The Theory of the First Approximation. Experimental observations of dynamical systems whose equations of motion are of the form (3.120) indicate that, after some transient disturbances die out, the motion of these systems tends to become periodic as time increases. In the case under consideration, the coefficient of the nonlinear term b is small, and it is reasonable to assume a periodic solution of the form $\phi_0 = A \cos \omega t$ as a first approximation to the solution of (3.120) and to determine its amplitude. Using

$$\cos^3 \omega t = \tfrac{1}{4}[\cos 3\omega t + 3 \cos \omega t] \tag{3.121}$$

the result is

$$\left(\omega_0^2 A - \omega^2 A + \frac{3bA^3}{4}\right)\cos \omega t + \frac{bA^3}{4}\cos 3\omega t = \frac{\omega E_m}{N}\cos \omega t \tag{3.122}$$

If the solution of the nonlinear equation under consideration is to reduce to that of the linear equation for which $b = 0$, the coefficients of the $\cos \omega t$ of both members of (3.121) should equal each other, giving

$$\frac{\omega E_m}{N} + A(\omega^2 - \omega_0^2) = 3b\frac{A^3}{4} \tag{3.123}$$

This cubic equation is used to determine the amplitude A. If we let $y_1 = 3bA^3/4$ and

$$y_2 = (\omega^2 - \omega_0^2)A + \frac{\omega E_m}{N}$$

the solution of (3.123) to determine the possible amplitude of the oscillation A may be effected by plotting the cubical parabola y_1 and the straight line y_2 as functions of A, as is shown in Fig. 3.29.

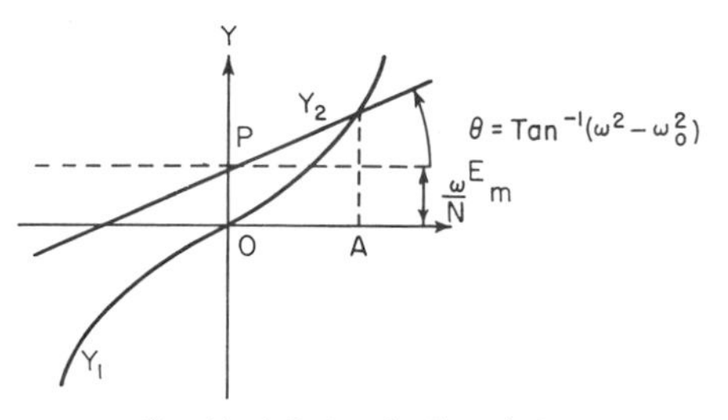

Graphical Determination of A
FIG. 3.29

The possible values of A are the abscissas of the points of intersection of the curves y_1 and y_2. If the angular frequency ω of the applied electromotive force is varied, the distance OP and the slope θ change. For some values of ω, y_1 and y_2 intersect at only *one* point, but for others they intersect at *three* points. This demonstrates the possibility that the amplitude A may change abruptly with a change in the frequency ω.

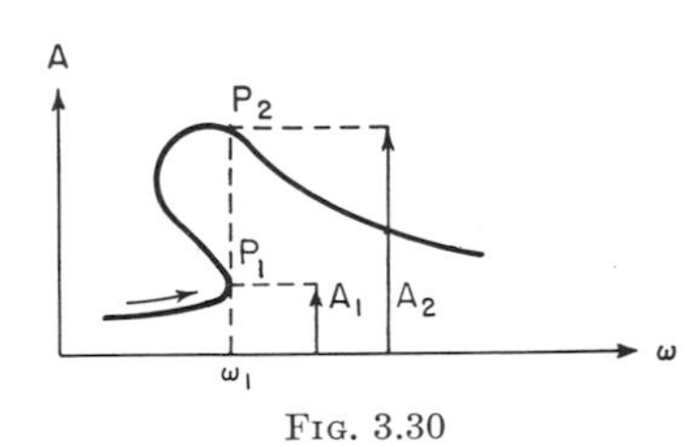

FIG. 3.30

Jump Phenomena. Figure 3.30 illustrates a typical amplitude-vs.-frequency curve. A more precise analysis indicates that, if the angular frequency ω is allowed to increase through low frequencies, the amplitude corresponding to the lower branch of the curve is the stable one. As ω continues to increase, we arrive at the limiting point ω_1. If ω is increased beyond ω_1, the amplitude will suddenly jump from A_1 to A_2. These discontinuities, or *jumps*, in the amplitude of the oscillation are frequently observed in the experimental analysis of nonlinear electrical circuits. If $\omega = \omega_0$, the straight line y_2 is horizontal and the amplitude of the flux oscillation is

$$A = \left(\frac{4\omega E_m}{3bN}\right)^{1/3}$$

The Steady-state Current. The steady-state current may be obtained by substituting $\phi_0 = A \cos \omega t$ into (3.118):

$$i(t) = I_1 \cos \omega t + I_3 \cos 3\omega t \qquad (3.124)$$

where $$I_1 = \frac{NA}{L_0} + 3c_3 \frac{A^3}{4} \qquad I_3 = c_3 \frac{A^3}{4}$$

The first-approximation theory thus accounts for the presence of a *third-harmonic* component in the steady-state current.

The Higher Approximations. The higher approximations to the steady-state solution of Eq. (3.120) may be obtained by writing it in the form

$$\ddot{\phi} = \frac{\omega E_m}{N} \cos \omega t - \omega_0^2 \phi - b\phi^3 \qquad (3.125)$$

If the first approximation ϕ_0 is substituted,

$$\ddot{\phi}_1 = -\omega^2 A \cos \omega t - \tfrac{1}{4} bA^3 \cos 3\omega t \qquad (3.126)$$

Integration gives

$$\ddot{\phi}_1 = A \cos \omega_1 t + \frac{bA^3}{36\omega^2} \cos 3\omega t \qquad (3.127)$$

If the following sequence is established,

$$\ddot{\phi}_{n+1} = \omega E_m \cos \omega t - \omega_0^2 \phi_n - b\phi_n^3 \quad \text{for } n = 0,1,2,3, \ldots$$

then

$$\phi(t) = \lim_{n \to \infty} \phi_n(t)$$

provided that the sequence converges. The sequence of functions, $\phi_0, \phi_1, \phi_2, \ldots, \phi_n$, gives as many terms of the series expansion of $\phi(t)$ as are required. The current may then be determined by (3.118).

Subharmonic Resonance. A method by which the amplitude of the periodic solution of (3.120) may be determined has been given. The fundamental period of the solution obtained in this manner is the same as that of the applied potential of the circuit. Experiments with circuits of this type indicate that, under certain conditions, permanent oscillations whose fundamental frequency is a submultiple of that of the applied potential sometimes occur. This phenomenon is called *subharmonic resonance.* An analysis of the possible conditions for the occurrence of periodic oscillations having a frequency of one-third that of the applied potential will now be investigated. Substitute a solution of the form

$$\phi(t) = A_0 \cos \frac{\omega t}{3} \qquad (3.128)$$

into Eq. (3.120):

$$\left(\omega_0^2 A_0 - \omega^2 \frac{A_0}{9} + \frac{3bA_0^3}{4}\right) \cos \frac{\omega t}{3} + \frac{bA_0^3}{4} \cos \omega t = \frac{\omega E_m}{N} \cos \omega t \qquad (3.129)$$

This equation is satisfied if

$$\frac{bA_0^3}{4} = \frac{\omega E_m}{N} \qquad \text{and} \qquad \omega_0^2 A_0 - \omega^2 \frac{A_0}{9} + 3\frac{bA_0^3}{4} = 0$$

This shows that oscillations of the type (3.128) are *possible* with an amplitude of $A_0 = (4\omega E_m/bN)^{1/3}$, provided that the angular frequency of the applied potential satisfies the equation

$$\omega^2 = 9\omega_0^2 + \frac{27b}{4} \left(\frac{4\omega E_m}{BN}\right)^{2/3} \qquad (3.130)$$

Oscillations of this type lead to a current in the circuit of the form $i = i_0 \cos (\omega t/3) + i_1 \cos \omega t$, where $i_0 = N(A_0/L_0) + \frac{3}{4}c_3 A_0^3$ and $i_1 = c_3(A^3/4)$.

Bibliography

Bartlett, A. C.: "The Theory of Electric Artificial Lines and Filters," Wiley, New York, 1930.

Bode, H. W.: "Network Analysis and Feedback Amplifier Design," Van Nostrand, Princeton, N.J., 1945.

Brenner, E., and M. Javid: "Analysis of Electric Circuits," McGraw-Hill, New York, 1959.

Campbell, G. A.: Cisoidal Oscillations, *Trans. AIEE*, **30**: 873–909 (1911).

Carlin, H. J., and A. B. Giordano: "Network Theory," Prentice-Hall, Englewood Cliffs, N.J., 1964.

Carson, J. R.: "Electric Circuit Theory and Operational Calculus," McGraw-Hill, New York, 1920.

Carson, J. R.: Electromagnetic Theory and the Foundations of Circuit Theory, *Bell System Tech. J.*, **6**: 1 (1927).

Gardner, M. F., and J. L. Barnes: "Transients in Linear Systems," Wiley, New York, 1942.
Guillemin, E. A.: "Communication Networks," vols. 1 and 2, Wiley, New York, 1931, 1935.
Guillemin, E. A.: "Introductory Circuit Theory," Wiley, New York, 1953.
Guillemin, E. A.: "The Mathematics of Circuit Analysis," Wiley, New York, 1949.
Guillemin, E. A.: "Theory of Linear Physical Systems," Wiley, New York, 1963.
Johnson, K. S.: "Transmission Circuits for Telephonic Communication," Van Nostrand, Princeton, N.J., 1925.
Kuo, F.: "Network Analysis and Synthesis," Wiley, New York, 1962.
LePage, W. R., and S. Seely: "General Network Analysis," McGraw-Hill, New York, 1952.
Massachusetts Institute of Technology, Electrical Engineering Staff: "Electric Circuits," Wiley, New York, 1943.
McLachlan, N. W.: "Ordinary Nonlinear Differential Equations in Engineering," Oxford University Press, London, 1950.
Minorsky, N.: "Introduction to Nonlinear Mechanics," Edwards, Ann Arbor, Mich., 1947.
Newstead, Gordon: "General Circuit Theory," Wiley, New York, 1959.
Pierce, G. W.: "Electric Oscillations and Electric Waves," McGraw-Hill, New York, 1920.
Pipes, L. A.: The Matrix Theory of Four-terminal Networks, *Phil. Mag., Ser.* 7, **30**: 370 (1940).
Pipes, L. A.: "Applied Mathematics for Engineers and Physicists," McGraw-Hill, New York, 1946.
Pipes, L. A.: "Matrix Methods for Engineering," Chaps. 9–12, Prentice-Hall, Englewood Cliffs, N.J., 1963.
Russell, A.: "Alternating Currents," Cambridge University Press, London, 1914.
Seshu, S., and N. Balabanian: "Linear Network Analysis," Wiley, New York, 1959.
Skilling, H. H.: "Electrical Engineering Circuits," Wiley, New York, 1957.
Starr, A. T.: "Electric Circuits and Wave Filters," Sir Isaac Pitman & Sons, Ltd., London, 1934.
van der Pol, B.: Nonlinear Theory of Electric Oscillations, *Proc. IRE*, **20**: 1051 (1934).
Van Valkenburg, M. E.: "Network Analysis," Prentice-Hall, Englewood Cliffs, N.J., 1956.
Weber, E.: "Linear Transient Analysis," Wiley, New York, 1954.
Weinberg, L.: "Network Analysis and Synthesis," McGraw-Hill, New York, 1962.

Chapter 4

Electronic Circuits

By CHESTER H. PAGE, National Bureau of Standards

1. General Considerations

The major difference between practical electronic circuits and idealized linear electronic circuits lies in the nonlinearities encountered. Electronic devices have useful properties because they are *nonlinear, unilateral,* or *active.* The *active* elements of electronics can often be idealized to linear behavior, but the devices which are useful only because they are nonlinear cannot be simplified in this way.

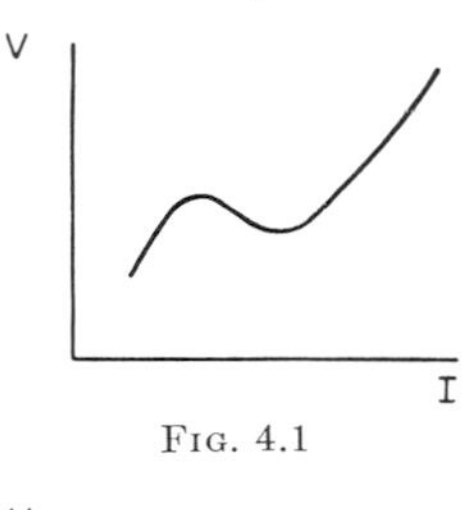

FIG. 4.1

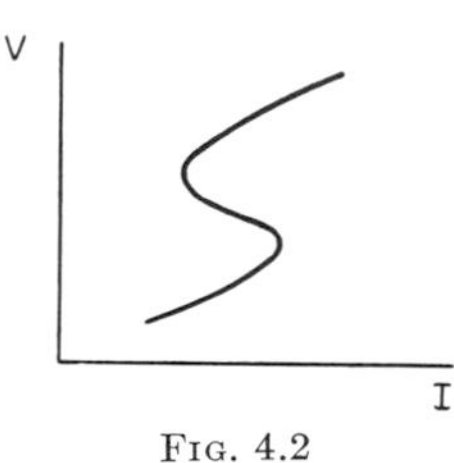

FIG. 4.2

The simplest linear-circuit element is the resistor, described by $V = RI$, and its nonlinear generalization is a device which operates according to the relation $V = V(I)$, an arbitrary function. The *incremental resistance,* or *dynamic resistance,* is defined $R_i = \partial V/\partial I$. If V is a single-valued function of I, and $\partial V/\partial I > 0$ throughout the range of interest, then I is a single-valued function of V, and we can also write

$$R_i = \left(\frac{\partial I}{\partial V}\right)^{-1}$$

Such a nonlinear resistor exhibits positive (dynamic) resistance over its operating range.

In the realm of active circuits one often encounters regions of negative resistance, generally of one or the other of two sample types (Figs. 4.1 and 4.2). In the first, voltage is a single-valued function of the current; the reverse is true in the second. The first exhibits a region of *negative resistance,* often spoken of as *current-controlled,* and is typified by a gas discharge. The negative resistance in the second is often said to be *voltage-controlled* and is typified by a dynatron. It might be better to refer to the second as exhibiting *negative conductance,* since it is the dynamic conductance, $\partial I/\partial V$, that follows directly from the single-valued relationship.

The generalization of capacitance raises new problems. In the linear case, $Q = CV$ or $I = C\dot{V}$, so that a generalized capacitor could be described by an arbitrary $Q(V)$ relation or an arbitrary $I(\dot{V})$. If, however, nonlinear reactance is to retain the property of being nondissipative, we must have $\oint VI\,dt = 0$ for any periodic disturbance. For arbitrary $Q(V)$,

$$\oint VI\,dt = \oint V\dot{Q}(V)\,dt = \oint VQ'(V)\,dV \equiv 0$$

whereas $\oint VI(\dot{V})\,dt \neq 0$ in general. Hence a nonlinear capacitor is characterized by a nonlinear $Q(V)$ relation, and the *incremental capacitance* is given by $C_i = \partial Q/\partial V$ and the current by

$$I = \dot{Q}(V) = C_i(V)\dot{V}$$

Hence I is a function of both V and $\dot{V}$ but *linear* in $\dot{V}$.

Similarly, in the case of linear inductors,

$$V = L\dot{I} \qquad \text{or} \qquad \phi = LI$$

where $\phi = \int E\,dt$ is the total flux linkage and the dual of Q. Restriction to a nondissipative generalization leads to an arbitrary $\phi(I)$ to describe a general inductor. The *incremental inductance* is $L_i = \partial\phi/\partial I$, and the voltage drop is $V = L_i(I)\dot{I}$. Here V is a function of both I and $\dot{I}$ but *linear* in $\dot{I}$.

2. Nonlinear-positive-resistance Elements

The distortion of current or voltage by a circuit containing nonlinear resistance is extremely useful. Thus the distortion products represent the useful output of modulators and demodulators. For small-signal analysis or for demonstrating the basic behavior of modulators, the resistor is assumed to be described by a power series:

$$V = aI + bI^2 + \cdots \tag{4.1}$$

If the current through such a device contains several harmonic components, say

$$I = A \cos pt + B \cos qt \tag{4.2}$$

the voltage appearing across the device will be

$$V = a(A \cos pt + B \cos qt) + \frac{b}{2}(A^2 + B^2 + A^2 \cos 2pt + B^2 \cos 2qt) + bAB[\cos (p+q)t + \cos (p-q)t] + \cdots \quad (4.3)$$

so that *rectification* yields a steady term

$$\frac{b}{2}(A^2 + B^2)$$

and the "beating" effect yields the first-order modulation products of sum and difference frequency.

If V is an infinite power series in I, all modulation products of the type $\cos (rp + mq)t$ will appear.

In an actual circuit, neither the current nor the voltage is known a priori; they will be interrelated through the frequency-selective impedances of the remainder of the circuit. Thus it is necessary to use some approximation scheme to obtain quantitative results.

Demodulation. If the input is a modulated wave,

$$I = (1 + a \cos pt) \cos qt \qquad a < 1 \qquad q \gg p$$

the second-power term yields $(1 + a \cos pt)^2 \cos^2 qt$ which, averaged over a time long compared with $2\pi/q$ but short compared with $2\pi/p$, gives a voltage $1 + 2a \cos pt + a^2 \cos^2 pt \approx 1 + 2a \cos pt$; therefore the modulation, $a \cos pt$, is recovered. The recovered signal is distorted by the $a^2 \cos^2 pt$ term, which yields the second harmonic $\cos 2pt$. This is not satisfactory except for small a.

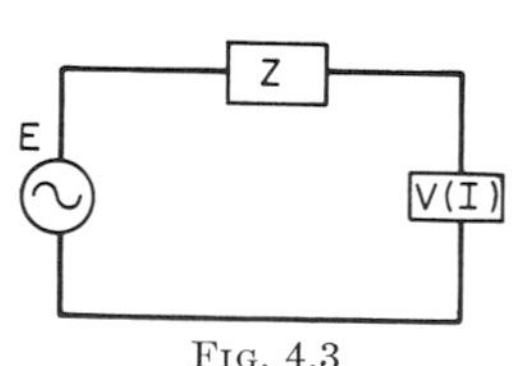

FIG. 4.3

We can, however, use a so-called linear rectifier which is not conveniently described by a power series. For example, let

$$I = \begin{cases} 0 & V < 0 \\ kV & V > 0 \end{cases}$$

and apply a modulated voltage. The current will consist of modulated half waves of the higher frequency q, and the averaging (filtering) will yield a current proportional to the modulation $1 + a \cos pt$.

Consider the circuit of Fig. 4.3 with the equation

$$E(t) = V(I) + Z(p)I \quad (4.4)$$

where $Z(p)$ is an impedance operator, i.e., a frequency-dependent impedance expressed in operational form. Separate $V(I)$ into its linear and nonlinear parts

$$V(I) = RI + f(I) \qquad f(0) = 0$$

so that $E = (R + Z)I + f(I) = Z_0 I + f(I)$

which can be expressed as

$$I = Y[E - f(I)] \quad (4.5)$$

where Y is the inverse of Z_0 and can be found by standard methods. A sequence of approximations to I can be made as follows:

$$\begin{aligned} I_0 &= YE \\ I_1 &= Y[E - f(I_0)] = I_0 - Yf(I_0) \\ I_2 &= I_0 - Yf(I_i) \\ &\text{etc.} \end{aligned} \quad (4.6)$$

This iteration will usually converge. The error at any stage can be treated by an error power concept.

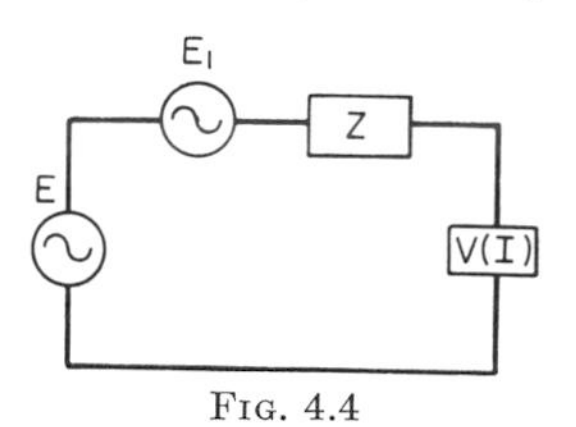

FIG. 4.4

Any particular approximation, say I_2, will be the exact solution of the circuit of Fig. 4.4 if the added generator provides the correcting voltage necessary to make

$$E + E_1 = ZI_2 + V(I_2) = (R + Z)I_2 + f(I_2) \quad (4.7)$$

or

$$E_1 = f(I_2) + (R + Z)I_2 - E$$

Now $(R + Z)I_2 = E - f(I_1)$ by definition; hence $E_1 = f(I_2) - f(I_1)$, and the power drawn from the correcting generator is

$$P_g = \langle [f(I_2) - f(I_1)]I_2 \rangle_{av} \quad (4.8)$$

which can be compared with the power from the real generator

$$\langle EI_2 \rangle_{av}$$

for a measure of the error involved in stopping with any particular approximation.

For large signals, the power-series method is not usually helpful. Then the nonlinear resistor characteristic can frequently be approximated by linear segments, such as

$$I = \begin{cases} 0 & V < 0 \\ kV & V > 0 \end{cases}$$

which is equivalent to an automatic switch in series with a linear resistor. If, in addition, the external circuit resistance is large compared with $1/k$, the nonlinear element reduces to a polarity-sensitive automatic switch. This approximation is customary in the study of copper oxide modulators. A large carrier-frequency voltage is applied, which effectively controls the switching action. If a weak modulating signal is applied, it finds itself switched at carrier frequency, thus generating modulation products. Various current configurations are in use [1].*

Nonlinear reactors can give positive modulation *gain*, i.e., sideband power greater than the power drawn from the signal source, the excess power

* Numbers in brackets refer to References at end of chapter.

being supplied by the carrier generator. This phenomenon cannot be produced by modulation in a passive nonlinear resistance. This fact is frequently assumed, but statements of the theorem are difficult to find in the literature. The theorem is basic.

Consider a passive nonlinear resistance, i.e., one for which $I(V)$ is a nondecreasing function of V. Let the voltage $V(t)$ appearing across the device be almost periodic, representable as a sum of sinusoidal voltages. Consider the voltage $V(t)$ as supplied by a series-connected set of sinusoidal generators, some of which act as power sources, some act as sinks, and some represent pure reactance, depending on the phase difference between each voltage and the corresponding components of the current $I(V)$. Now consider the generators V as comprising two subsets, V_1 and V_2, with V_1 representing the carrier and its harmonics and V_2 the combination of signal, modulation products, and those signal harmonics which are not also carrier harmonics. More precisely, the condition on V_2 is that it contain no frequencies that would be present in the current $I(V_1)$ obtaining with the generators V_2 removed. We have the following theorem:

The sources in the subset V_2 supply at least as much power as that absorbed by the sinks of V_2.

The total power supplied by the generators of V_2 is

$$P_2 = \langle V_2 I(V_1 + V_2) \rangle_t \tag{4.9}$$

where $\langle\ \rangle_t$ indicates the time average. Adding and subtracting $\langle V_2 I(V_1) \rangle_t$,

$$P_2 = \langle V_2[I(V_1 + V_2) - I(V_1)] \rangle_t + \langle V_2 I(V_1) \rangle_t \tag{4.10}$$

where the last term vanishes by the hypothesis on the frequencies present in V_2. Since $I(V)$ is a nondecreasing function of V, the expression in brackets is never opposite in sign to V_2, no matter how complicated the dependence of V_2 on time. Hence the argument of the time average is non-negative, and $P_2 \geq 0$.

The subset of generators V_1 represents a sinusoidal carrier source and sinks for the harmonic distortion or a general periodic source which applies power at some of its harmonics. The set V_2 similarly represents either a sinusoidal signal source, plus harmonic sinks and modulation-product sinks, or a general periodic or almost periodic signal source and modulation-product sinks. The theorem shows that the total available sideband power, even including all the high-order modulation products, is no greater than the power supplied by the signal source.

An application of this theorem to a periodic signal shows that, if only one generator is a source, it must be at fundamental frequency; in other words, a passive nonlinear resistor cannot generate subharmonics.*

3. Negative Resistance

If a negative-resistance device such as is illustrated by Fig. 4.5 is connected in series with a battery of voltage B and a linear resistor, there are two simultaneous equations to be satisfied. The relation

* Further developments in C. H. Page, *J. Research Nat. Bur. Standards,* **56**: 179 (1956).

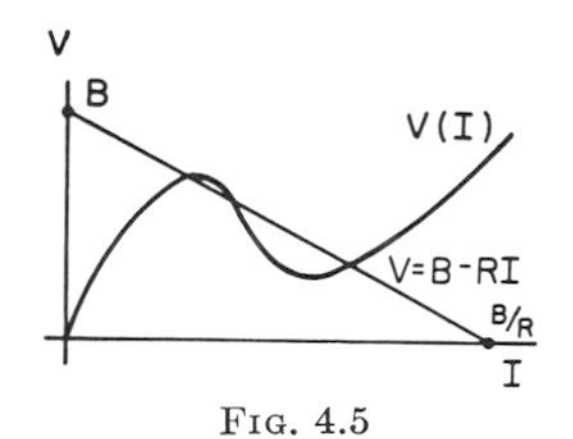

Fig. 4.5

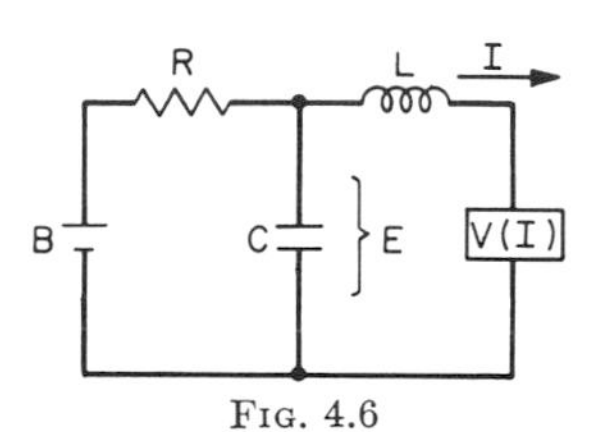

Fig. 4.6

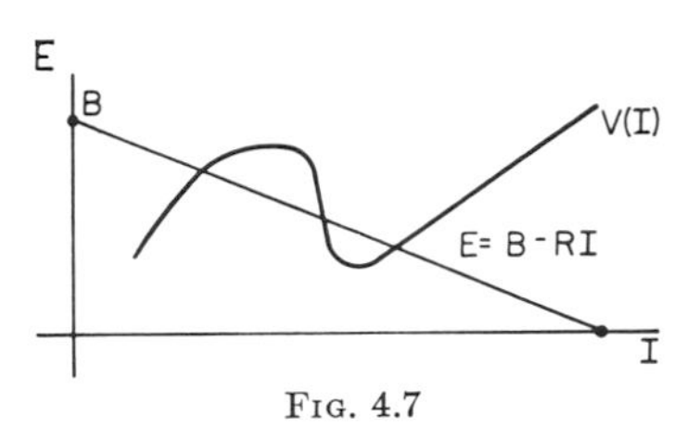

Fig. 4.7

between the current through the device and the voltages across it must satisfy the $V(I)$ relation of Fig. 4.5 and, in addition, must satisfy the relation

$$V = B - RI \tag{4.11}$$

Graphically, the operation of the circuit is indicated by a point in the V,I phase plane, and this point must be on both the lines shown in Fig. 4.5.

These three points of intersection are the only possible operating conditions that satisfy the requirements of both the resistor and the nonlinear device. There are many discussions in the literature as to which of these points are stable or unstable and as to how the operating point can go from one intersection to another. Many of these discussions are confusing, because *for the circuit assumed* there are only the three point solutions, and the operation cannot change. If dynamic changes are to be considered, we *must* include reactance in the circuit.

Series-tuned Circuit. Consider the circuit of Fig. 4.6. Since the current through an inductance and the voltage across a capacitance must be continuous functions of time, we choose these as phase variables. The circuit equations are

$$\begin{aligned} C\dot{E} &= [(B - E)/R] - I \\ L\dot{I} &= E - V(I) \end{aligned} \tag{4.12}$$

The *load line,* $E = B - RI$, of Fig. 4.7 is the locus of operating points such that $\dot{E} = 0$. The curve, $E = V(I)$, is the locus of $\dot{I} = 0$. Hence the intersections now represent the steady-state condition $\dot{E} = 0 = \dot{I}$ and are equilibrium points. The corresponding equilibrium may be either stable or unstable.

The dynamical behavior of the representative point in the phase plane is given by Eqs. (4.12);

these are more convenient when the time parameter is eliminated:

$$\frac{dE}{dI} = \frac{L}{RC} \frac{(B - RI) - E}{E - V(I)} \tag{4.13}$$

If this relation is used, it is easy to construct graphically a dynamic path in the phase plane. Notice that all paths cross the $V(I)$ curve vertically and cross the $B - RI$ line horizontally. The problem of immediate interest is the behavior of such a path in the neighborhood of an equilibrium point, i.e., near a singular point of Eq. (4.13).

If the singular point of interest is E_0, I_0 and if $E = E_0 + e$, $I = I_0 + i$, a power-series expansion yields

$$\frac{de}{di} = \frac{L}{RC} \frac{e + Ri}{-e + \alpha i + \cdots} \tag{4.14}$$

with $\qquad V(I) = V(I_0) + \alpha i + \beta i^2 + \cdots$

Coefficients of the bilinear form (4.14) in e,i determine the behavior of the dynamic path in the neighborhood of the singularity. It may approach the point asymptotically (in time); it may spiral in, spiral out, etc. The various types of behavior are well known [2, 3]; the corresponding points are called nodal points, focal points, saddle points, etc.

In our case, an equilibrium point is unstable if either

$$-\alpha > R \qquad \text{or} \qquad -\alpha > L/RC \tag{4.15}$$

In the first case, the total incremental series resistance is negative, and so instability would be expected regardless of the values of L,C. In the second case, we note that at resonance the circuit of Fig. 4.6,

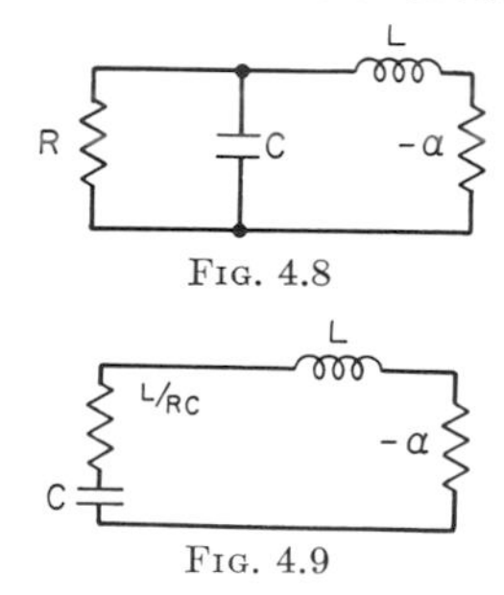

FIG. 4.8

FIG. 4.9

redrawn as Fig. 4.8 omitting the battery, is equivalent to that of Fig. 4.9. Thus, near resonance the total resistance of the circuit is negative, and oscillation occurs.

According to condition (4.15), the singularity is unstable if

$$R < (-\alpha) \qquad \text{or} \qquad R > L/(-\alpha)C \tag{4.16}$$

If these ranges of R overlap, i.e., if $(-\alpha) \geq (L/C)^{1/2}$, the singularity is unstable for all values of R. If Fig. 4.8 is interpreted as a quarter-wave line with characteristic impedance $K = (L/C)^{1/2}$, the input resistance is R at zero frequency and K^2/R at resonance. Hence there is always a frequency at which the input resistance is less than K.

If we let $L \to 0$, approaching the mathematically degenerate case, the oscillations become less and less sinusoidal and are called relaxation oscillations. A typical case occurs when $V(I)$ represents a neon lamp.

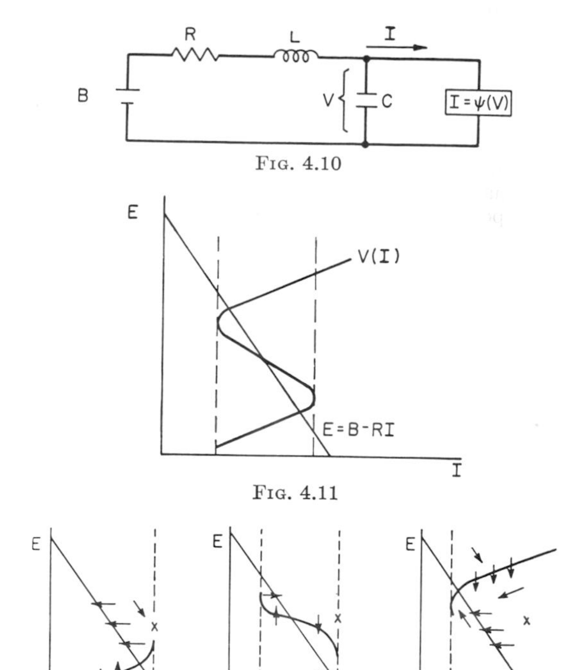

FIG. 4.10

FIG. 4.11

FIG. 4.12

Parallel-tuned Circuit. In this case it is simpler, initially, to assume a negative-resistance device where $I(V)$ is single-valued (Fig. 4.2). Again, we choose for phase variables the current through an inductance and the voltage across a capacitance (Fig. 4.10). The circuit equations are

$$\begin{aligned} C\dot{V} &= I - \psi(V) \\ L\dot{I} &= B - RI - V \end{aligned} \tag{4.17}$$

leading to

$$\frac{de}{di} = \frac{L}{C} \frac{\alpha e - i}{e + Ri} \tag{4.18}$$

where $\alpha = (dI/dV)V_0$.

Equation (4.18) becomes (4.14) on interchanging e and i, L and C, R and $1/R$, so that the instability conditions (4.15) become

$$-\alpha > 1/R \qquad -\alpha > RC/L \tag{4.19}$$

Special Cases. The preceding analyses used a single-valued $V(I)$ device with a series-tuned circuit and a single-valued $I(V)$ device with a parallel-tuned circuit. These choices allowed the independent variable of the device to be one of the necessarily continuous circuit variables, hence a convenient phase variable. If we use a cross combination of circuit type and device type, the situation is awkward.

Consider a series-tuned circuit connected to an $I(V)$ device, say a dynatron. Choose for phase variables the continuous inductor current and capacitor voltage. The dynamic equations are (4.12), but $V(I)$ is multivalued in the central region of the phase plane (Fig. 4.11). The multivalued function

$V(I)$ can be represented on three sheets, so that the motion of the representative point is governed by the appropriate sketch in Fig. 4.12. The arrow shows the direction of motion of the phase point as governed by Eqs. (4.12).

A phase path starting in sheet (1) will spiral in to the equilibrium point if it starts within a certain critical region. Otherwise, the path will reach the sheet boundary (dotted) above the tangent point of the $V(I)$ curve, say at x. Such a path cannot terminate at the boundary; it can terminate only on equilibrium points. The operating voltage V must therefore go to another sheet. This new sheet cannot be sheet (2), for in this region of (2) the velocity is still to the right. The voltage therefore goes to sheet (3), and the velocity will then be to the left. (If the phase path is high enough in the figure, the velocity will be to the right, but the path will circle down and back, eventually reaching the condition assumed.)

The further history of the phase path is similar to that previously discussed, interchanging sheets (1) and (3). The path can never enter sheet (2) and will either terminate on a stable point in sheet (1) or sheet (3) or reach a *limit cycle* in which it continues to oscillate. This limit cycle is indicated in Fig. 4.13, with V alternately in sheets (1) and (3).

A phase path originating in sheet (2) will spiral in to the equilibrium point if it is stable and if the path originates within a suitable region. In all other cases, the path will go into the (1) and (3) routine and never return to sheet (2).

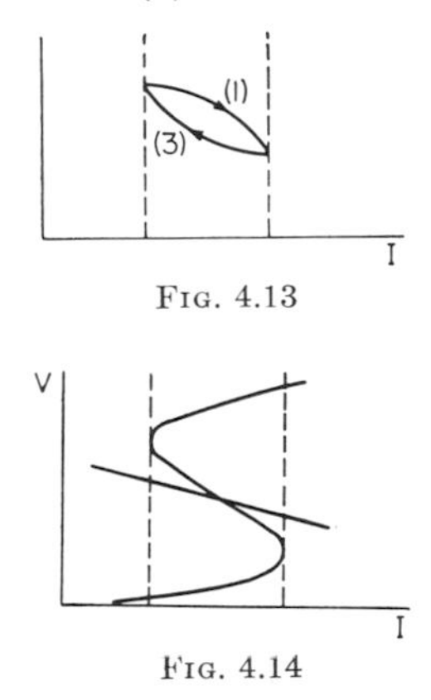

FIG. 4.13

FIG. 4.14

If the load line is shifted as shown in Fig. 4.14, the only singularity is in sheet (2), and this is unstable since $-\alpha > R$. Since there are no equilibrium points in sheets (1) and (3), the *only* solution is that involving the steady-state oscillation of Fig. 4.13.

4. Nonlinear Reactance

The behavior of a single nonlinear reactor can be described in terms of a capacitance having arbitrary $Q(V)$. The nonlinear-inductance case is exactly parallel, interchanging I and V, ϕ and Q throughout the discussion. Series-connected voltage generators must, of course, be replaced by parallel-connected current generators.

If $Q(V)$ is nonlinear, modulation by frequency addition can occur. Let the total voltage comprise three components as follows:

$$V = a \cos(pt + \alpha) + b \cos(qt + \beta) + c \cos[(p + q)t + \gamma] \quad (4.20)$$

where p and q are incommensurate. The power drawn from the generator of pulsatance p is

$$P_p = \langle V_p I \rangle_t = \left\langle V_p \frac{dQ}{dt} \right\rangle_t = -\left\langle Q \frac{dV_p}{dt} \right\rangle_t \quad (4.21)$$

where the *integration by parts* follows from the fact that $\langle d(V_pQ)/dt \rangle_t = 0$ since V_pQ is bounded. Hence

$$P_p = ap \langle Q \sin(pt + \alpha) \rangle_t \quad (4.22)$$

and similar expressions hold for P_q and P_{p+q}.

Since the device is nondissipative,

$$P_p + P_q + P_{p+q} = 0 \quad (4.23)$$

so that, for any $Q(V)$ representable as a power series or sufficiently closely approximated by a polynomial,

$$P_p = pC \qquad P_q = qC \qquad P_{p+q} = -(p + q)C \quad (4.24)$$

with the constant C depending upon the individual component voltages and phases.

Depending upon the sign of C, either the two lower frequencies supply power, which is delivered to a higher-frequency sink, or the sum frequency delivers power, which is absorbed in two lower-frequency sinks [4, 5].

Considering the three radian frequencies, p, q, $p - q$, it is apparent that q and $p - q$ represent generators of the same source or sink nature, whereas p represents the opposite class. We cannot supply power at p and q and absorb it at frequency $p - q$. Hence, the lower sideband cannot be generated alone.

If both upper- and lower-sideband voltages are present, two processes are superposed. The generators of frequencies p and q both act as sources with a sink at $p + q$; the generator at frequency p also acts as a source with sinks at q and $p - q$. Hence, the generator of frequency q is partly source and partly sink; regeneration is present.

Similar relations are found when sources of several frequencies interact to drive a sink at sum frequency and when a single source supplies a set of sinks whose frequencies add up to that of the source.

If a single source is considered as n identical sources, the additive process yields the nth harmonic; the corresponding multiple-sink splitting of power can generate any integral subharmonic. The combination of the two effects, given sufficient nonlinearity, makes possible the conversion of power from one frequency to any rational fraction of that frequency. Hence, in general, if the two frequencies p and q are commensurate, i.e., rationally related, either can represent a sink with the other as source [6]. Thus the modulation process may become regenerative. The general conditions for lack of regeneration are complicated, but it is likely that in the absence of regeneration the relations of (4.24) hold for commensurate p and q.

The implication is that the modulation gain

$$g = \frac{-P_{p+q}}{P_q} = \frac{p+q}{q} \tag{4.25}$$

is equal to the frequency step-up ratio of the modulation process. In the absence of regeneration, this is an upper bound to the modulation gain of any passive device or network [7].

The general subject of frequency-power formulas has been discussed by Penfield [12].

Magnetic and Dielectric Amplifiers. The relation (4.24) shows that the modulation process yields a gain $(p + q)/q$, but the demodulation, or splitting of $p + q$ into the useful q and the by-product p, is accompanied by the inverse gain $q/(q + p)$. We know, however, that resistive demodulators can be operated with relatively small loss. Hence, the combination of a nonlinear reactive modulator and a nonlinear resistive demodulator can comprise an amplifier with a gain greater than unity. This is the basic phenomenon of magnetic and dielectric amplifiers [8].

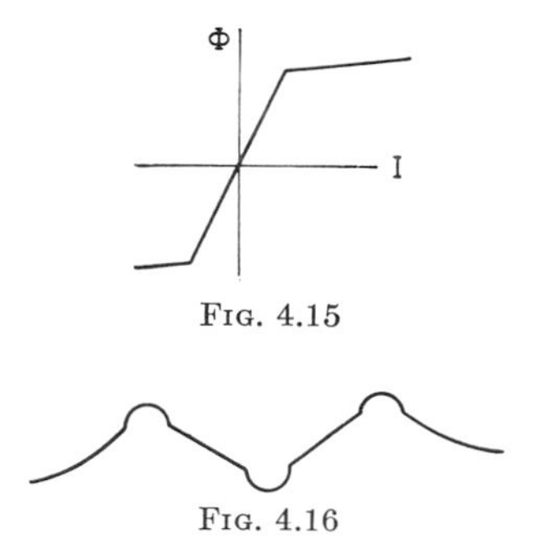

Fig. 4.15

Fig. 4.16

The magnetic amplifier is more flexible and versatile than the dielectric amplifier, because multiple coils can be wound on a magnetic core and interconnect several such assemblies with various polarity combinations. This allows the use of various balanced and bridge circuits for magnetic amplifiers, while nonlinear capacitors are only two-terminal devices.

Large-signal Analysis. Consider the idealized saturable inductor represented by Fig. 4.15, with

$$L \equiv \phi' = \begin{matrix} A \\ a \end{matrix} \qquad \begin{matrix} |I| < 1 \\ |I| > 1 \end{matrix} \qquad A \gg a \tag{4.26}$$

or conversely,

$$\begin{aligned} I &= \phi/A & \phi < A \\ I &= \phi/a - A - a/a & \phi > A \end{aligned} \tag{4.27}$$

If the voltage $E = \alpha \cos \omega t$ is applied to this inductor

$$\phi = (\alpha/\omega) \sin \omega t \tag{4.28}$$

and if $\alpha < \omega A$, the current is sinusoidal, with peak value less than unity. If, however, $\alpha > \omega A$, so that the low-inductance region is entered, the current waveform becomes greatly "amplified" during part of the cycle, appearing as in Fig. 4.16.

Now consider this inductor connected in series with a resistor R, of such value that $R \gg \omega a$. The circuit can be approximated by the resistance alone for $I > 1$, or

$$\begin{aligned} A\dot{I} + RI &= \alpha \cos \omega t & |I| < 1 \\ RI &= \alpha \cos \omega t & |I| > 1 \end{aligned} \tag{4.29}$$

with the solutions

$$I = \frac{\alpha}{R} \cos \omega t \qquad |I| < 1$$

$$I = \frac{\alpha}{\sqrt{R^2 + A^2\omega^2}} \cos\left(\omega t - \cos^{-1} \frac{R}{\sqrt{R^2 + A^2\omega^2}}\right) + Ce^{-(R/A)t} \tag{4.30}$$

where the constant C is determined by the continuity of I at $I = 1$. When $|I| < 1$, most of the voltage appears across the inductance, assuming that $\omega A \gg R$. As the current increases, $I = 1$ is reached, and the voltage drop across the inductance becomes negligible, full power being delivered to the load resistance. The disappearance of the inductive drop allows the current to stay in the saturation region, $I > 1$, until the voltage decreases to a low value. The saturable inductance behaves much like a thyratron, losing control once it has "fired" and regaining control only when the reverse half cycle is approached.

If the applied voltage is reduced so that saturation is never reached, little power will be delivered to the load R, but if a biasing current or flux is applied, the inductance will fire near the appropriate polarity peaks. Thus, a low-power signal can be used to control relatively large power. This is the basis of the usual discussions of magnetic amplifiers [9, 10].

5. Active Circuits

General Considerations. Active elements, e.g., vacuum tubes and transistors, are usually nonlinear, but the nonlinearity is accidental rather than essential. In fact, the ordinary operation of active devices is confined to their approximately linear region. Hence, in this section we consider only idealized linear elements. The various admittances and impedances exhibited at the terminals represent dynamic properties at the operating point, i.e., slopes of the actual E-I relations.

As shown in Part 4, Chap. 3, a linear network can be represented by an admittance matrix or an impedance matrix. At least one of these exists. If a network has an admittance matrix, the external complex voltages and currents are related by

$$I = YE \tag{4.31}$$

The total power input for a given set of applied voltages is

$$P = \tfrac{1}{2}\text{Re}:\tilde{E}_t I = \tfrac{1}{2}\text{Re}:\tilde{E}_t YE \tag{4.32}$$

If $P \geq 0$ for all choices of the vector E, the network can never deliver power and hence is *passive*. Any network that is not *passive* is *active*, but its classification may change with the frequency of the applied voltages.

Equation (4.32) can be expanded and becomes

$$P = \tfrac{1}{2} \sum_{i=1}^{n} \sum_{j=1}^{n} [\text{Re}\,(e_i)\,\text{Re}\,(e_j) + \text{Im}\,(e_i)\,\text{Im}\,(e_j)]\,\text{Re}\,(y_{ij} + y_{ji}) \quad (4.33)$$

which is the sum of two quadratic forms, having the same coefficients Re $(y_{ij} + y_{ji})$. If P is to be non-negative for all choices of e_i, these quadratic forms must be positive definite, or the matrix Re Y must be positive definite.

If the admittance matrix does not exist, the corresponding operation on the impedance matrix yields the expected result that Re Z must be positive definite. The positive-definite nature of a matrix can be examined via its discriminants [11] (Part 1, Chap. 10).

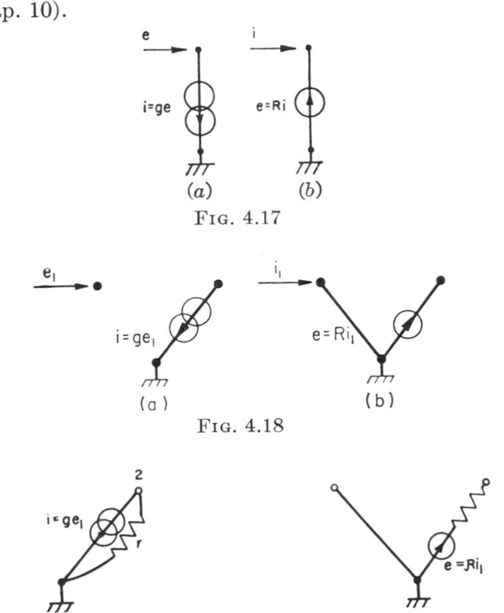

FIG. 4.17

FIG. 4.18

FIG. 4.19

Three-terminal Unilateral Elements. An ordinary conductance or resistance can be thought of as a current generator, controlled by an applied voltage, as shown in Fig. 4.17*a*, or, conversely, as a voltage generator controlled by the applied current (Fig. 4.17*b*). There are two corresponding simple unilateral active elements (Fig. 4.18), wherein the current (voltage) generator is controlled by the voltage (current) applied to a different terminal. Although the representations of Fig. 4.17 are equivalent, those of Fig. 4.18 are not; Fig. 4.18*a* has the admittance matrix

$$Y = \begin{bmatrix} 0 & 0 \\ g & 0 \end{bmatrix}$$

with no inverse, while Fig. 4.18*b* has the singular impedance matrix

$$Z = \begin{bmatrix} 0 & 0 \\ R & 0 \end{bmatrix}$$

As soon as losses are added to the generators of Fig. 4.18, as in Fig. 4.19, we can transform by Thév-

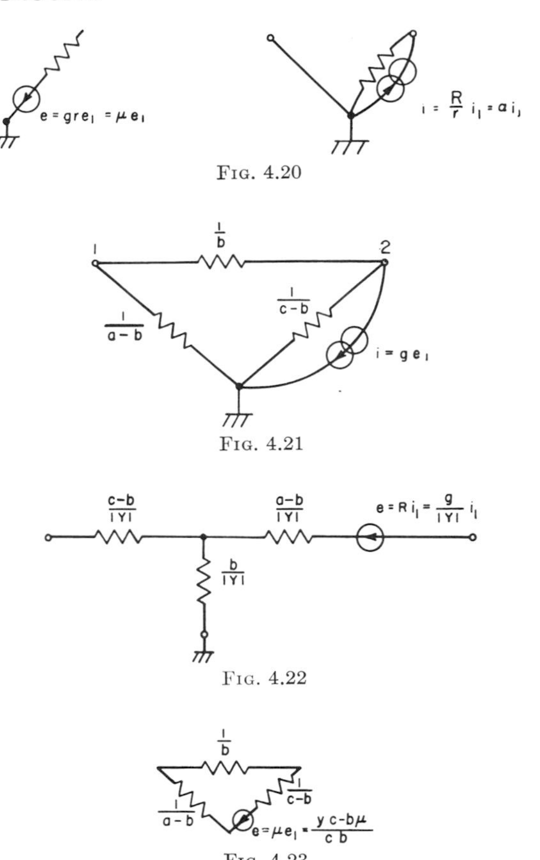

FIG. 4.20

FIG. 4.21

FIG. 4.22

FIG. 4.23

enin's theorem to Fig. 4.20, where the active property is expressed in terms of voltage amplification (μ) and current amplification (α).

Although the μ and α representations are usually used by engineers and are useful from the equivalent-circuit viewpoint, they do not fit the general matrix approach as well as do transconductance and transresistance.

Equivalent Circuits. A resistive π network connected in parallel with the unilateral transconductance element of Fig. 4.18*a* yields a network (Fig. 4.21) with the admittance matrix

$$Y = \begin{bmatrix} a & -b \\ -b & c \end{bmatrix} + \begin{bmatrix} 0 & 0 \\ g & 0 \end{bmatrix} = \begin{bmatrix} a & -b \\ g-b & c \end{bmatrix} \quad (4.34)$$

Its impedance matrix is

$$Z = Y^{-1} = \begin{bmatrix} \frac{c}{|Y|} & \frac{b}{|Y|} \\ \frac{b-g}{|Y|} & \frac{a}{|Y|} \end{bmatrix} = \begin{bmatrix} \frac{c}{|Y|} & \frac{b}{|Y|} \\ \frac{b}{|Y|} & \frac{a}{|Y|} \end{bmatrix} + \begin{bmatrix} 0 & 0 \\ \frac{-g}{|Y|} & 0 \end{bmatrix} \quad (4.35)$$

with the equivalent-circuit representation of Fig. 4.22. By Thévenin's theorem, we have also the equivalent circuits of Figs. 4.23 and 4.24.

If the loads thus applied to the active element are

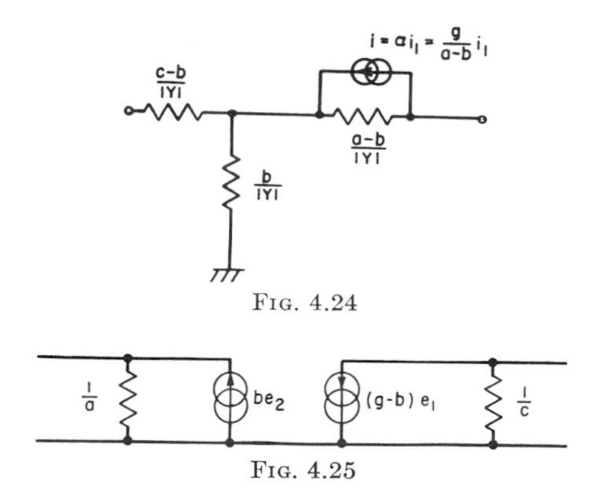

FIG. 4.24

FIG. 4.25

not so heavy as to make the resulting network passive, $g > 4b$, since the discriminant test applied to Y yields

$$0 < ac - \left(\frac{g}{2} - b\right)^2 = ac - b^2 - g^2/4 + gb \tag{4.36}$$

while $ac - b^2 > 0$ for the loading network to be passive.

A less useful equivalent circuit is found by separating Y into its symmetrical and skew-symmetrical parts,

$$Y = \begin{bmatrix} a & -b + g/2 \\ -b + g/2 & c \end{bmatrix} + \begin{bmatrix} 0 & -g/2 \\ g/2 & 0 \end{bmatrix} \tag{4.37}$$

representing the network as the parallel combination of a reciprocal active network containing negative conductance (since $g > 4b > 2b$) and a passive antireciprocal, or gyrostatic, network. The gyrator is not only passive but also lossless, since (4.33) shows that the input power is identically zero.

The admittance matrix can also be split as in Eq. (4.38) below, yielding the circuit of Fig. 4.25,

$$Y = \begin{bmatrix} a & -b \\ 0 & 0 \end{bmatrix} + \begin{bmatrix} 0 & 0 \\ g - b & c \end{bmatrix} \tag{4.38}$$

Either or both of the transconductance current generators in Fig. 4.25 can be converted to a voltage-amplification generator by Thévenin's theorem.

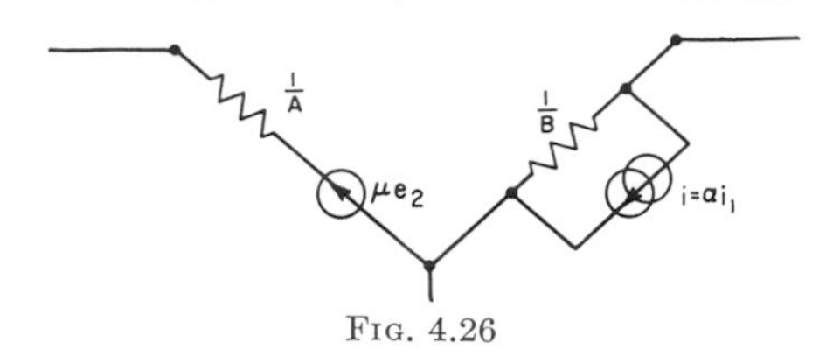

FIG. 4.26

Another equivalent circuit common in transistor work uses two generators, representing forward-current amplification and backward-voltage amplification (Fig. 4.26). Its admittance is

$$Y = \begin{bmatrix} A & -\mu A \\ \alpha A & B - \alpha\mu A \end{bmatrix} \tag{4.39}$$

and its parameters are readily deducible from a given matrix Y by comparison with (4.39).

References

1. Garuthers, R. S.: *Bell System Tech. J.*, **18**: 315 (1939).
2. Stoker, J. J.: "Nonlinear Vibrations," Interscience, New York, 1950.
3. Minorsky, N.: "Introduction to Nonlinear Mechanics," Edwards, Ann Arbor, Mich., 1947.
4. Peterson, E.: *Bell Labs. Record*, **7**: 231 (1929).
5. van der Ziel, A.: *JAP*, **19**: 999 (1948).
6. Compare Fig. 1 of J. M. Manley and E. Peterson: *Trans. AIEE*, **65**: 870 (1946).
7. Page, C. H.: *J. Research Nat. Bur. Standards*, **58**: 227 (1957).
8. Manley, J. M.: *Proc. IRE*, **39**: 242 (1951).
9. Cohen, S. B.: *Proc. IRE*, **39**: 1009 (1951).
10. Ramey, R. A.: *Trans. AIEE*, **70**: 1214 (1951).
11. Frazer, Duncan, and Collar: "Elementary Matrices," Macmillan, New York, 1946.
12. Penfield, P.: "Frequency-power Formulas," The M.I.T. Press and Wiley, 1960.

Chapter 5

Electrical Measurements

By WALTER C. MICHELS, Bryn Mawr College

1. Standards

The art of electrical measurements depends on establishment and maintenance of standards which are adequate in both number and accuracy. The basic electrical unit, *the absolute ampere,* is fixed by international agreement as exactly one-tenth of a cgs electromagnetic unit of current (Part 4, Chap. 1, Sec. 9). It and the other electrical units are based on the international meter, on the international kilogram, on the second, and on an arbitrarily defined magnetic permeability of free space. At the National Bureau of Standards and at other national standardizing laboratories, the ampere can be reproduced to a few parts in a million.

Since the unit of current (I) is not a material or transportable standard, nearly all electrical measurements are made in terms of standards of resistance (R), potential difference (V), capacitance (C), or inductance (L) which have been calibrated, more or less directly, in terms of the absolute ampere. These quantities are connected with the ampere and with the mechanical units of power (P) and time (t) through the relations

$$R = P/I^2 \tag{5.1}$$
$$V = P/I = IR \tag{5.2}$$
$$C = Q/V = \int I\,dt/V \tag{5.3}$$
$$L = V/(dI/dt) \tag{5.4}$$

The units thus defined are known as the *ohm,* the *volt,* the *farad,* and the *henry,* respectively. Precision standards are readily available in decade values from 10^{-3} ohm to 10^6 ohms, from 10^{-10} farad to 10^{-5} farad, and from 10^{-4} henry to 10 henrys. In using resistance and inductance standards, one must be careful to limit the power dissipated in them to values consistent with those at which they have been calibrated, since overheating not only affects the accuracy but also may do permanent damage. Similarly, capacitance standards should never be subjected to potential differences greater than those recommended by the manufacturer.

The only commonly used standard of potential difference is the standard cell. The Weston normal cell* maintains an open-circuit potential difference of 1.018636 volts across its terminals at 20°C. Since this value is quite temperature-sensitive, the unsaturated standard cell is much more commonly used, although it is not perfectly reproducible. Such cells have an emf of 1.0183 to 1.0190 volts, the exact value being determined by comparison with a saturated cell. In using standard cells, care should be taken to avoid abrupt changes in temperature or temperatures outside the range of 4 to 40°C and to insure that currents in excess of 10^{-4} amp are never drawn from the cell. Unsaturated cells should be checked against saturated standards at least once each year.

In recent years digital voltmeters of high precision and good stability have become increasingly available. Many of them employ *breakdown diodes* (*Zener diodes*) which pass nearly constant current over a fairly wide range of potential difference across their terminals. Such devices are more convenient than standard cells, but they should be considered as intermediate standards, because calibration is accomplished by comparison with the latter.

* Vinal, Craig, and Brickwedde, The Weston Cadmium Cell, *Trans. Electrochem. Soc.*, **68**: 139 (1935).

2. Deflection Instruments; the D'Arsonval Galvanometer

The great majority of electrical measurements are made with the help of deflection instruments, i.e., instruments which indicate by the deflection of a pointer the magnitude of an electrical current passing through the instrument. The prototype of many deflection instruments is the *D'Arsonval galvanometer,* the construction of which is shown schematically in Fig. 5.1. A rectangular coil of wire C, usually wound on a bobbin of nonmagnetic material, is suspended in the magnetic field between the poles N-S of a permanent magnet. These pole pieces are ordinarily machined so that their faces are cylindrical, and a soft iron core D is mounted coaxially to insure that the magnetic field in the neighborhood of the coil is radial. The taut suspension F, fixed at its upper end, supports the coil, furnishes a restoring torque to return the coil to its equilibrium position, and acts as a current lead to one end of the coil. The lower suspension G is usually one which furnishes little restoring torque and which serves merely as a second current lead. Both suspensions are usually made of a noncorrosive metal, such as gold or phosphor bronze, in the form of a thin filament or ribbon. The position of the coil is observed by reflection of light from the small mirror M, which is fastened rigidly to the coil.

If the coil consists of N turns, each of area A, and carries a current I and if the magnetic-flux density in the air gap has a magnitude B, the coil is subject to a torque.

$$\tau = -BNAI = KI \tag{5.5}$$

which is independent of position because of the radial direction and constancy of the magnetic field. The rotation of the coil is opposed by the elastic torque of the suspensions. If it is assumed that Hooke's law holds for the torsion of the suspension, the restoring torque is $k\theta$, where k is the torsion constant and θ is the angular deflection of the coil from its equilibrium position (with no current flowing). The new equilibrium position is established when the sum of the two torques is zero, i.e., when

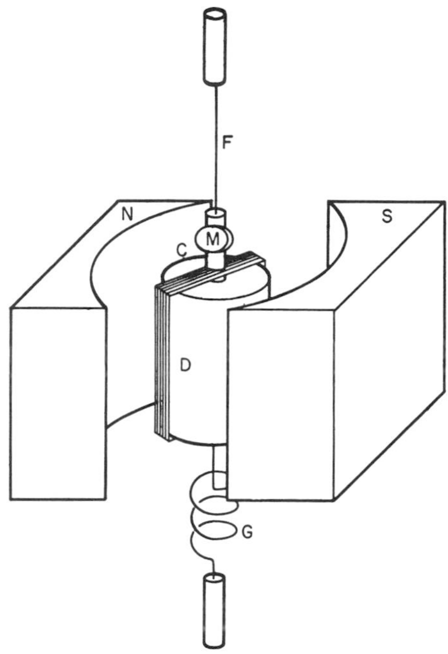

FIG. 5.1. The D'Arsonval galvanometer.

$$\theta = BNAI/k \tag{5.6}$$

The angular deflection of the D'Arsonval galvanometer is therefore proportional to the current flowing in the coil.

Observation of the deflection of the galvanometer is usually accomplished with a lamp-and-scale or a telescope-and-scale arrangement. These two arrangements are illustrated schematically in Fig. 5.2. In either case, the observed deflection is $d = L \tan 2\theta$, where L is the distance from the galvanometer mirror to the scale. This distance is usually between 50 and 200 cm. If, instead of being straight, the scale is curved on a circle centered at the mirror, the deflection is $d = 2L\theta$, and the linearity of deflection with current is preserved.

In portable reflection galvanometers, the lamp and scale are often built into a case with the galvanometer. The light path (L) is sometimes made greater than the length of the case by the use of mirrors which send the beam back and forth several times. In portable needle galvanometers, a needle replaces the mirror and indicates the deflection by its motion over a divided scale.

The relation between deflection and current is expressed either as the current sensitivity or as the current scale factor. The *current sensitivity* is defined as

$$K_I = (dI/dd)_{d=0} = k/2LBNA \tag{5.7}$$

The *current scale factor* is the reciprocal of this, or

$$S_I = (dd/dI)_{I=0} = 2LBNA/k \tag{5.8}$$

Sensitivities and scale factor are commonly expressed in terms of microamperes per millimeter of deflection

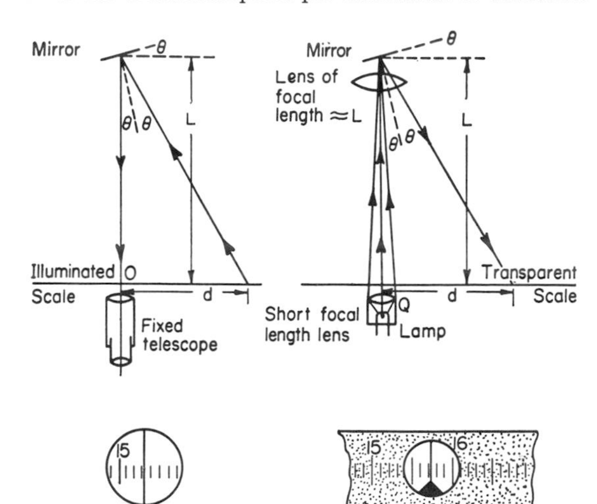

FIG. 5.2. Arrangements for reading galvanometer deflections.

and millimeters per microampere, respectively, both being taken for $L = 1$ m. Commercial galvanometers of the type described above are available with K_I in the range of 10^{-5} to 10^{-1} μa/mm. The more sensitive instruments are characterized by high resistance and long periods (see below). It is sometimes convenient to specify the *voltage sensitivity* or the *voltage scale factor* of a galvanometer which is to be used primarily for measurements of potential differences. These are K_V and S_V, respectively, where

$$K_V = 1/S_V = R_g K_I = kR_g/2LBNA \tag{5.9}$$

R_g being the resistance of the instrument (coil plus suspensions). Available mirror-type commercial galvanometers have values of K_V between 0.05 and 50 μv/mm.

Period and Damping. In practice, the manner in which the galvanometer approaches its equilibrium deflection is often as important as the sensitivity. Suppose that the moving system has a moment of inertia M and that the motion is opposed by a resistive torque proportional to the angular velocity $\dot{\theta}$, the constant of proportionality being D. Then, using the notation of (5.5) and (5.6), the equation of motion of the coil is

$$M\ddot{\theta} + D\dot{\theta} + k\theta = KI \tag{5.10}$$

We may set I equal to zero by shifting the origin of θ. The homogeneous equation thus obtained has two solutions, depending on the relative values of the constants M, D, and k. If

$$\omega^2 = k/M - D^2/4M^2 > 0 \qquad (5.11)$$

the motion is given by

$$\theta = a \cos (\omega t + b) \exp [-(D/2M)t] \qquad (5.12)$$

where a and b are integration constants, depending on the initial conditions. If (5.11) holds, the galvanometer, after a disturbance, follows a damped motion, oscillating about its equilibrium position with constantly decreasing amplitude. Such a galvanometer is said to be *underdamped.* The *period* of the motion is

$$T = 2\pi(k/M - D^2/4M^2)^{-\frac{1}{2}} \qquad (5.13)$$

If the damping is small, this reduces to the free period of the instrument,

$$T_0 = 2\pi \sqrt{M/k} \qquad (5.14)$$

Since the scale factor is inversely proportional to the torsion constant k, it follows from (5.14) that extremely high-sensitivity galvanometers can be constructed only at the cost of long periods, with consequent loss of time in taking readings. If the galvanometer is *overdamped,* i.e., if

$$\omega^2 = k/M - D^2/4M^2 < 0 \qquad (5.15)$$

the solution of (5.10) is

$$\theta = a \exp \left[- \left(\omega + \frac{D}{2M} \right) t \right] \qquad (5.16)$$

The overdamped galvanometer therefore approaches its equilibrium position asymptotically, at a rate depending on the sensitivity and the damping. The transition between the two types of motion, obtained when

$$\omega^2 = k/M - D^2/4M^2 = 0 \qquad (5.17)$$

is known as *critical damping.*

In general, a galvanometer should be slightly underdamped. More specifically, it should overshoot its final reading by an amount about equal to the uncertainty which is allowable in that reading. Usually the air friction and other causes of damping result in an underdamped motion when the galvanometer is on open circuit. The connection of a finite resistance across the galvanometer terminals, however, always increases the damping, as may be seen if we consider the generating action of the moving galvanometer coil. As the coil rotates with an angular velocity in the magnetic field, its sides are cutting magnetic flux at a rate $BNA\dot{\theta}$. In accordance with the law of induction, an emf of magnitude $v = BNA\dot{\theta}$ is set up by this motion. This emf, in turn, causes a current to flow in the coil and in the external circuit. If the galvanometer resistance is R_g and that of the external circuit (as seen from the galvanometer terminals) is R_e, the power dissipated by Joule heating is

$$P = v^2/(R_g + R_e) = \{(BNA)^2/(R_g + R_e)\}\dot{\theta}^2$$

This power can come only from the kinetic energy of the coil. If the coil is subject to a torque τ opposing its motion, the rate of kinetic energy loss is $\tau\dot{\theta}$. Equating this quantity to the electrical power, we find that the torque is

$$\tau = \{(BNA)^2/(R_g + R_e)\}\dot{\theta} \qquad (5.18)$$

This is precisely the type of dependence of torque on velocity that was assumed in (5.10); hence the electrical circuit contributes to the D of that equation. If the damping constant of the open-circuited instrument is D_0,

$$D = D_0 + (BNA)^2/(R_g + R_e) \qquad (5.19)$$

Solving (5.17) for D and substituting into (5.19), we find that the condition for critical damping is

$$D_0 + (BNA)^2/(R_g + R_e) = 2\sqrt{kM} \qquad (5.20)$$

The value of R_e which satisfies (5.20) is known as the *critical damping resistance* (CDRX) and is usually specified by the manufacturer.

When a galvanometer is used to measure current or potential difference from a source whose resistance is high compared with the CDRX, a resistor may be connected in parallel with the terminals to give proper damping. When the source resistance is lower than the CDRX, a series resistor may be inserted. The loss of sensitivity which results from either of these additions is usually a low price to pay for the gain in convenience which results.

Choice of a Galvanometer. It is intrinsic in the design of the D'Arsonval galvanometer that gain in sensitivity can be accomplished only through loss in ruggedness and in portability. As (5.8) and (5.9) show, high scale factors result from large flux densities (B) and small torsion constants (k). The former demand small air gaps between the pole pieces and the core; the latter, fragile suspensions. Most instrument manufacturers produce several distinct types of galvanometer, each of which is characterized by a fixed flux density. Within each type, a wide range of suspensions and coils may be obtained to fit particular requirements.

The most sensitive galvanometer within a given type is obtained by using the thinnest possible suspension, i.e., one which will just support the weight of the coil without danger of elastic yield. Since the coil bobbin and the mirror contribute an appreciable portion of the weight to be supported, there is a lower limit of suspension cross section for any given design. The moment of inertia is also determined, to a large extent, by the size and shape of the mirror and the bobbin. Hence the period, as is shown by (5.14), is a maximum when the finest suspension is used, i.e., when the greatest sensitivity is obtained. It is therefore desirable, for speed in reading, that the least-sensitive galvanometer consistent with the desired accuracy should be used.

With the torsion constant and the moment of inertia fixed, the sensitivity may still be varied by choice of the wire size used for the coil. Within a given weight limitation, the number of turns (N) is inversely proportional to the cross-sectional area of the wire (s). Since the resistance is proportional to

N/s, we may write

$$R_g = CN^2 \qquad (5.21)$$

where C is a constant for a given galvanometer structure. Now suppose that the galvanometer is connected to a source of resistance R, the open-circuit emf of which is V_0. Then the current is

$$I = V_0/(R + R_g)$$

Using (5.7) and (5.21), we find that the deflection $d \propto NV_0/(R + CN^2)$. To find the condition for maximum deflection, we set dd/dN equal to zero and find that

$$CN^2 = R = R_g \qquad (5.22)$$

Hence the ideal galvanometer has a resistance equal to that of the source. This condition is not critical, and departures from it by a factor of 2 are generally not serious.

The above discussion assumes that the galvanometer is connected directly across the terminals of the source, without the addition of series or parallel resistors to adjust the sensitivity or the damping. When such resistors are used, the full circuit may be analyzed, both to determine the best galvanometer to be used and to find the relation between the current through the galvanometer and the current or emf supplied by the source. (For methods of analysis, see Chap. 3.)

3. Direct-current Ammeters and Voltmeters

The instruments used almost invariably for the measurement of potential differences and currents in d-c circuits consist of portable galvanometers of the D'Arsonval type, together with appropriate arrangements of resistors. Because the ruggedness of such meters is far more important than extreme sensitivity, the moving coil is mounted on a shaft supported by hard metal or jeweled bearings, as indicated in Fig. 5.3. The restoring torque is furnished by spiral springs, which also serve as current leads. A pointer moving over a scale along a circular arc, centered at the coil axis, replaces the mirror and light beam of the sensitive galvanometer. Portable galvanometers are usually designed with a free period of the order of 1 sec and are arranged to be somewhat underdamped on open circuit. The sensitivity is commonly such that a current of 0.1 to 10 ma or a potential difference of 50 mv will give a full-scale deflection. Since the considerations of Sec. 2 apply to these instruments, the deflection is approximately linear with the current flowing in the galvanometer. However, nonuniformity in the magnetic field or slight unbalances in the moving system may result in some nonlinearity, and precision instruments must be individually calibrated.

A d-c *ammeter* consists of the movement just described plus a *shunt*, shown schematically by the circuit within the dashed line of Fig. 5.4. It is clear that the ratio of the galvanometer current (I_g) to the current being measured (I) is

$$I_g/I = R_s/(R_g + R_s + r) \qquad (5.23)$$

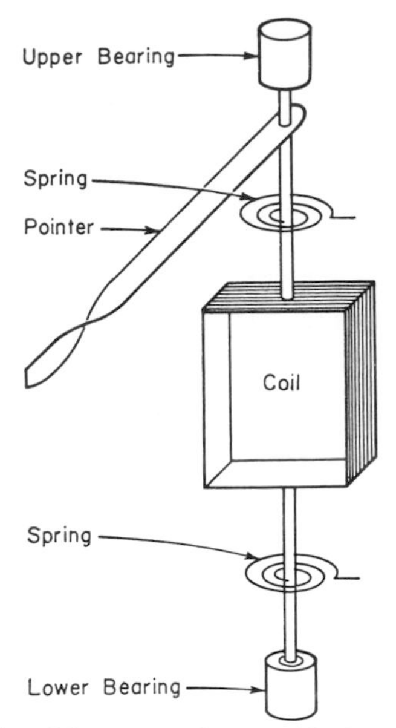

FIG. 5.3. Movement of a portable d-c meter.

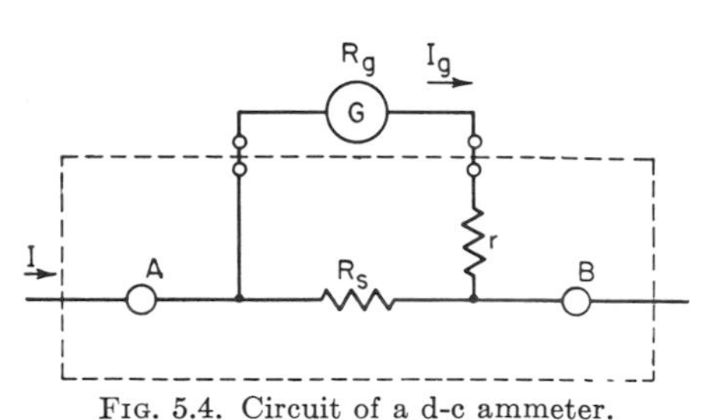

FIG. 5.4. Circuit of a d-c ammeter.

and that the resistance of the meter, as seen from the terminals AB, is

$$R_A = R_s(R_g + r)/(R_g + R_s + r) \qquad (5.24)$$

Since the value of I_g for full-scale deflection is fixed by the galvanometer design, the magnitude of I for full-scale deflection can be made as large as we please by the use of a sufficiently low value of R_s. A low-resistance shunt also insures that the added resistance introduced by placing the meter in series with the circuit elements through which the current is to be measured will not disturb this current appreciably. The resistance r ($R_g > r > R_s$) has only a small effect on the calibration and is included for convenience in adjustment.

Ammeters with ranges up to about 10 amp usually have the shunt built into the same case with the movement. With higher-range instruments this is rendered impractical because of the power (I^2R_A) which must be dissipated, and the shunt is usually external to the instrument. When an external shunt is used, the leads which connect the galvanometer to the shunt should be those with which the instrument was calibrated, since they contribute to r.

A d-c *voltmeter* uses the same galvanometer as an ammeter, connected in series with a *multiplier*, shown as R_m in Fig. 5.5. When the resulting instrument is connected across a source, the potential difference across this source being V, the current through the movement is

$$I_g = V/(R_m + R_g) \tag{5.25}$$

while the resistance of the meter is

$$R_V = R_m + R_g \tag{5.26}$$

A large value of R_m therefore serves both to increase the voltage needed for full-scale deflection and to insure that the connection of the voltmeter in parallel with the circuit being measured will not result in serious disturbance of the potential difference. The inverse of I_g for full-scale deflection is frequently used as a measure in the sensitivity of a voltmeter and is expressed in ohms per volt.

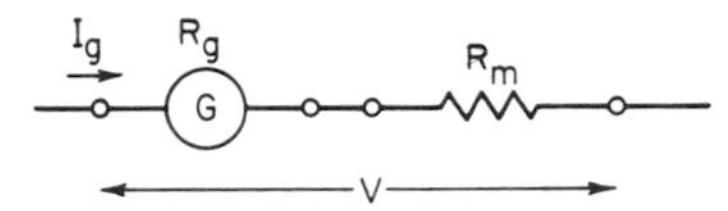

FIG. 5.5. Circuit of a d-c voltmeter.

When ammeters are used in low-resistance circuits or voltmeters are used across high-resistance circuits, the meter may disturb the quantity being measured. If calculation shows that such disturbances are greater than the desired accuracy, it is necessary either that corrections be made for them or that a more sensitive meter be used.

While D'Arsonval-type meters are commonly used for d-c measurements, certain a-c instruments, which are treated in Sec. 4, may also be employed for d-c work.

Vacuum-tube Voltmeters and Electronic Electrometers. The most sensitive portable instruments of the D'Arsonval type have sensitivities of the order of 10,000 ohms/volt. When higher-resistance instruments are needed, *vacuum-tube voltmeters* are ordinarily used. In these instruments, the potential difference to be measured is impressed on the control grid of a vacuum tube, and its effect on the plate current is measured. In the simplified circuit of Fig. 5.6, the various resistance elements

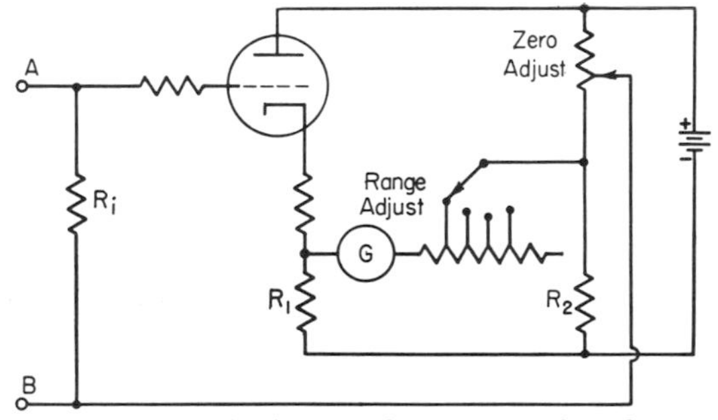

FIG. 5.6. Schematic diagram of a vacuum-tube voltmeter.

are chosen so that the voltage drops across R_1 and R_2 are equal when the terminals AB are at the same potential and the tube is operating well within the linear portion of its transfer characteristic. The "Zero Adjust" is used to satisfy this condition exactly, as indicated by the zero deflection of the portable galvanometer G. A change of the potential of A with respect to B then changes the current through the tube and through R_1, with a resulting deflection of G. With the tubes used in modern vacuum-tube voltmeters, the input resistance (R_i) may be of the order of 1 to 10 megohms, and full-scale deflection may be obtained for potential differences of about 3 volts. Meters of this sort therefore have sensitivities of the order of 10^6 ohms/volt and may be used to measure potentials in circuits with quite high resistance. They are not as accurate as are the meters discussed in Sec. 2, although inverse feedback in the circuit helps to compensate for changes in supply voltage and in tube characteristics.

The vacuum-tube voltmeter may be used to measure currents of the order of 1 μa, if the current is passed through a shunting resistor in parallel with the input resistor. The instrument is limited in this use, however, by the grid current of the tube, which may be of the order of 10^{-7} amp. For the measurement of smaller currents in high-impedance circuits, *electronic electrometers* are now more commonly used than are the older quadrant, binant, or string electrometers. These are actually vacuum-tube voltmeters using special tubes and circuits in order that they may present input resistances of 10^{13} ohms or more and have leakage currents of the order of 10^{-14} amp.

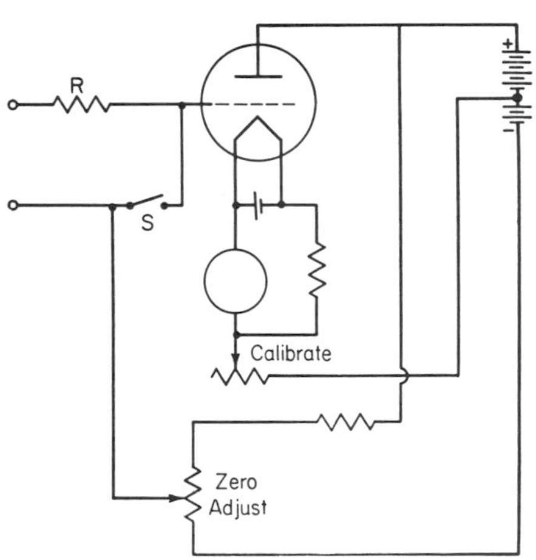

FIG. 5.7. Schematic diagram of an electronic electrometer.

Figure 5.7 shows a schematic diagram of an electronic electrometer which has an input resistance of 10^{14} ohms and an input capacitance of 6 $\mu\mu$f. The similarity between this circuit and that of Fig. 5.6 is apparent. The 1-megohm resistance R serves to protect the vacuum tube from damage if too-high voltages are imposed on the input. The switch S allows the zero of the instrument to be checked.

Ohmmeters. Resistance measurements of the order of 1 to 2 per cent accuracy are ordinarily made by the use of Ohm's law and a *volt-ammeter method* or by means of an ohmmeter, which, in principle, is identical with this method. The two arrangements of the volt-ammeter method are shown in Fig. 5.8. In either case, the resistance is given approximately by the relation

$$R = V/I \qquad (5.27)$$

where V is the potential difference indicated by the voltmeter and I is the current indicated by the ammeter (A). This expression is correct to 1 per cent if the unknown resistance (R) is at least 100 times the resistance of the ammeter and if arrangement (a) is used or if R is not more than 0.01 times the resistance of the voltmeter and arrangement (b) is used.

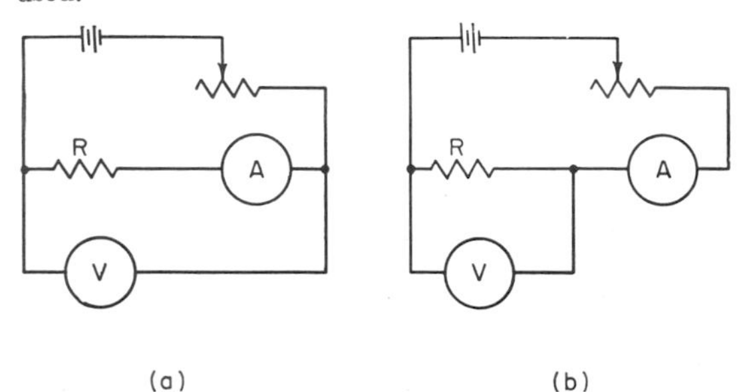

FIG. 5.8. Resistance measurement by the volt-ammeter method.

The volt-ammeter method fails for very high resistances unless a high supply voltage is used or the ammeter is replaced by a sensitive galvanometer. On the other hand, resistances of 10^3 ohms or more can be connected across readily available sources without disturbing the emf of the sources. Advantage is taken of this fact in the use of comparison or substitution methods of measuring resistance. If a high-resistance voltmeter of resistance R_V is connected across the source and indicates a potential difference V_1 and if it is then connected in series with the unknown resistance R across the same source and indicates a potential difference V_2, we have, from Ohm's law,

$$R = R_V(V_1 - V_2)/V_2 \qquad (5.28)$$

This relation obviously gives the greatest percentage accuracy when $V_2 \approx \frac{1}{2}V_1$, that is, when $R_V \approx R$. It is the basis of the comparison method which is used in most electronic *ohmmeters*. These employ vacuum-tube voltmeters as the indicating instruments and may be adapted to a variety of ranges with the help of a number of shunts, which may be placed in parallel with the input resistor to give different values of R_V.

In the *substitution method*, first the unknown resistor and then an adjustable resistance standard are placed in series with the voltmeter. The standard is adjusted until it yields the same reading as the unknown. This method will ordinarily yield somewhat greater accuracy than the comparison method, since it does not depend on the calibration of the deflection instrument.

4. Alternating-current Meters; Electrodynamic Instruments

As is indicated by (5.5), the torque on a D'Arsonval-type instrument reverses with the current through the coil. The solution of (5.10) with a time-variable current substituted for I shows that the deflection is proportional to the instantaneous value of the current if the period of variation is long compared with the period of the meter and that the deflection is very nearly time-independent and is proportional to the average value of the current if the period of variation is short compared with that of the meter. Since the average value of a sinusoidally varying current

$$I = I_0 \sin \omega t \qquad (5.29)$$

is zero, D'Arsonval-type meters cannot measure alternating currents.

The *electrodynamic meter* has a mechanical construction similar to that of the D'Arsonval-type voltmeter or ammeter but uses either an electromagnet or an air-cored coil instead of a permanent magnet. For reasons which will appear later, air-cored coils are used in all precision instruments of this type. In the *electrodynamic ammeter*, this fixed coil is connected in series with the moving coil. (In some cases, the fixed coil carries the full current to be measured, while the moving coil is shunted and carries a definite fraction of this current. This arrangement affects the argument only by a constant factor at commercial frequencies and does not need separate consideration.)

Since the magnetic field of the fixed coil is proportional to the instantaneous value of the current, we may replace B in (5.5) by K_BI, where K_B is a constant of the instrument. Making this substitution and supposing I to have the form of (5.29), we find that the torque on the moving coil is

$$\tau = K_BNAI^2 = K'I^2 = K'I_0^2 \sin^2 \omega t \qquad (5.30)$$

Substitution of this torque into (5.10) shows that, if $\omega \gg 1/T_0$, the deflection is essentially constant and is proportional to the average value of I^2, that is,

$$\theta = K\langle I_0^2 \sin^2 \omega t\rangle_{av} = KI_0^2/2 \qquad (5.31)$$

The electrodynamic ammeter therefore has a squared scale rather than a linear scale; it reads the root-mean-square value of the current passing through it, and it may be calibrated by using direct current ($I = I_0$). It has two great disadvantages as compared with the d-c meter, both due to the fact that the magnetic field produced by an air-cored coil carrying reasonable currents is very much smaller than that obtainable with a permanent magnet:

1. Sensitivities are considerably lower than those of D'Arsonval-type meters.

2. The earth's field or stray magnetic fields due to neighboring currents may appreciably affect the response of the instrument to direct currents.

For these reasons, electrodynamic instruments are seldom used for d-c measurements of current. When they are calibrated against d-c standards, the current should be passed through them first in one direction and then in the other, the meter remaining unmoved, and the average value of the two readings taken as the indication of the meter.

The *electrodynamic voltmeter* has the two coils connected in parallel or the fixed coil connected in parallel with a series combination of the moving coil and a series resistor. If the resistance of the fixed coil is R_f, its inductance is L_f, and a sinusoidally varying potential difference of amplitude V_0 is impressed across it, the current flowing is

$$I_f = \{V_0/(R_f^2 + \omega^2 L_f^2)^{1/2}\} \sin \{\omega t - \tan^{-1} (\omega L_f/R_f)\} \quad (5.32)$$

At frequencies sufficiently low so that $\omega L_f \ll R_f$, the current in the fixed coil, and hence the magnetic field, is proportional to the potential difference and is in phase with it. Since the inductance of the moving coil is usually less than L_f, while its resistance R_m is higher than R_f, the current in it is also proportional to the potential difference, and the deflection of the instrument is

$$\theta = (K/R_f R_m) \langle V_0^2 \sin^2 \omega t \rangle_{av} = K'' V_0^2/2 \quad (5.33)$$

where K'' is a constant, independent of frequency. The electrodynamic voltmeter therefore reads root-mean-square values and may be calibrated on direct current, with the same precautions as are used with the ammeter.

Electrodynamic voltmeters are, in general, not satisfactory for measurements at frequencies exceeding a few hundred cycles per second. At higher frequencies, the terms which were neglected in going from (5.32) to (5.33) assume increased importance, as do the corresponding terms in the expression for the current in the moving coil.

In some applications of electrodynamic instruments (see Secs. 5 and 7), the currents in the two coils are supplied independently, rather than through series of parallel connections. If the moving coil carries a current $I_m \sin \omega_1 t$ and the stationary coil a current $I_f \sin \omega_2 t$, the instantaneous torque is

$$\theta = K I_m I_f \sin \omega_1 t \sin \omega_2 t \quad (5.34)$$

Since the time average of the product of the two sine factors is zero unless $\omega_1 = \omega_2$, the instrument shows zero response except when the frequencies are the same. If the respective currents are $I_m \sin \omega t$ and $I_f \sin (\omega t + \phi)$, the response is

$$\theta = K I_m I_f \langle \sin \omega t \sin (\omega t + \phi) \rangle_{av} = (K I_m I_f/2) \cos \phi \quad (5.35)$$

Power Measurement; Wattmeters. The power consumption of a load operating on direct current is often measured as the product of the current and voltage indicated by two meters used in either of the circuits shown in Fig. 5.8. These measurements are subject to errors related to those in volt-ammeter resistance measurements, since arrangement (*a*) includes in the result the power consumption of the ammeter, while arrangement (*b*) includes the power consumption of the voltmeter. Corrections may be made quite easily if the characteristics of the meters are known, unless the power consumption of the load is small compared with that of either meter.

Electrodynamic instruments are commonly used to measure power consumption in either a-c or d-c loads, one coil replacing the ammeter in either arrangement of Fig. 5.8, the other coil replacing the voltmeter. Considering first the case of alternating current of radian frequency ω, we see from (5.35) that, if the current in one coil is proportional to and in phase with the current in the load, while the current in the other is proportional to and in phase with the potential difference across the load, the deflection

$$\theta \propto I_{rms} V_{rms} \cos \phi \quad (5.36)$$

where I_{rms} and V_{rms} are the root-mean-square values of the current and potential difference, respectively, and ϕ is the phase difference between these two quantities. Since the right-hand side of (5.36) is the power consumed in the circuit, the *electrodynamic wattmeter* has a linear scale. The same errors as occur with the volt-ammeter method of power measurement must, of course, be tolerated or corrected. As with other electrodynamic instruments, errors may become large at frequencies appreciably above the commercial range.

An argument exactly analogous to that used for alternating current shows that the electrodynamic wattmeter may be used for d-c power measurements, with the same scale as that for alternating current. The errors are likely to be somewhat larger than with the volt-ammeter measurement, because of the greater power consumption in electrodynamic instruments, as compared with D'Arsonval-type instruments.

As was indicated above, the errors in a simple wattmeter become intolerable when the power consumption in the load is small compared with that in either coil. This difficulty may be overcome if the coil used in place of V in Fig. 5.8*b* is connected to the output of an amplifier, whose high impedance input replaces V in that circuit. In this way, the power consumption in this coil can be reduced to negligible proportions. The amplifier, of course, must have a linear response and a negligible phase shift.

Moving-vane Instruments. The most commonly used instruments for measurement of low-frequency currents and voltages use a vane of ferromagnetic material (such as soft iron), attached to the rotating shaft and pointer. The current to be measured passes through a rigidly mounted coil, whose magnetic field exerts forces on the vane and thus causes it to move until the torque so produced is equal and opposite to that supplied by the springs. Such instruments are of moderate precision ($\frac{1}{2}$ to 3 per cent) and are cheap and rugged. The relation between current and deflection can be adjusted over wide limits by choice of the shapes of vane and coil. Because the energy stored in the magnetic field of the coil is proportional to the square of the current, the readings of these instruments are functions of the mean-square current to a first approximation. They may therefore be calibrated with direct current subject to the same precautions as with electrodynamic instruments.

Instruments Using Rectifiers. Because of the high cost and high power consumption of electrodynamic instruments, their application is usually limited to precision a-c or power measurements. When accuracies of 1 to 2 per cent are sufficient, a-c measurements are often carried out with meters which combine a D'Arsonval-type movement with a rectifier. The rectifier elements are usually made up of contacts between metals and semiconductors such as copper oxide or selenium. The most commonly used circuit is the *full-wave rectifier bridge* shown in Fig. 5.9. The resistance of each of the rectifier elements at rated voltages is 10 to 100 times as great in the "forward" direction as it is in the "reverse" direction; hence practically the entire current flows over the path $ABCD$ when the left-hand terminal is positive and over the path $DBCA$ when this terminal is negative. At reasonably low frequencies and with the galvanometer carrying more than a quarter of full-scale current, the reverse current (for example, $ACBD$) is negligible, provided only that the resistance of the galvanometer is very low compared with the "reverse" resistance of the rectifier elements. In order that this condition may be satisfied, any multiplying resistance should be connected in series with the bridge, as indicated by R_m, rather than in series with the galvanometer.

Since the current through a rectifier element is not, in general, a linear or other simple function of the applied voltage, an exact general analysis of instruments using rectifiers is difficult. The treatment takes a simple form only if the galvanometer resistance is low and the multiplier resistance high compared with the forward resistance of the rectifier elements. Here the forward current is determined largely by R_m, and the reverse current is negligible. The current through the meter is then proportional to the instantaneous value of the potential difference applied to the terminals, and, in accordance with the discussion of Sec. 3, the meter reading is proportional to the average absolute value of the applied potential. If a sinusoidally varying potential of amplitude V_0 is applied to the meter, the deflection

$$\theta \propto \langle |V_0 \sin \omega t| \rangle_{\text{av}} = 0.636V_0$$

Because the factor of proportionality differs from 0.707, the ratio of the root-mean-square potential to the amplitude, rectifier instruments cannot be calibrated on direct current without correction, nor is their indication independent of waveform. Under the ideal conditions which we have assumed, instruments calibrated on direct current will read high by a factor of $0.707/0.636 = 1.11$ when used on sinusoidal alternating current.

Instruments using copper oxide or other "area-contact" elements are suitable for use at commercial and moderate audio frequencies. At higher frequencies the capacitive reactance of the rectifiers becomes comparable with their resistance, and the bridge is ineffective. On the other hand, "point-contact" rectifiers, such as germanium or silicon elements, have very small capacitances and can be used at radio or even higher frequencies.

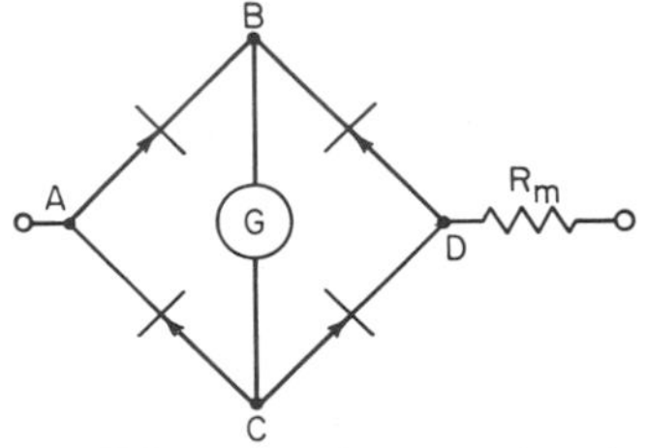

FIG. 5.9. Voltmeter using a rectifier bridge.

Vacuum-tube voltmeters are frequently used with rectifiers to measure a-c potential differences. Because of the intrinsically high input impedance of such meters, the circuit shown in Fig. 5.9 would be impractical with them. More common arrangements

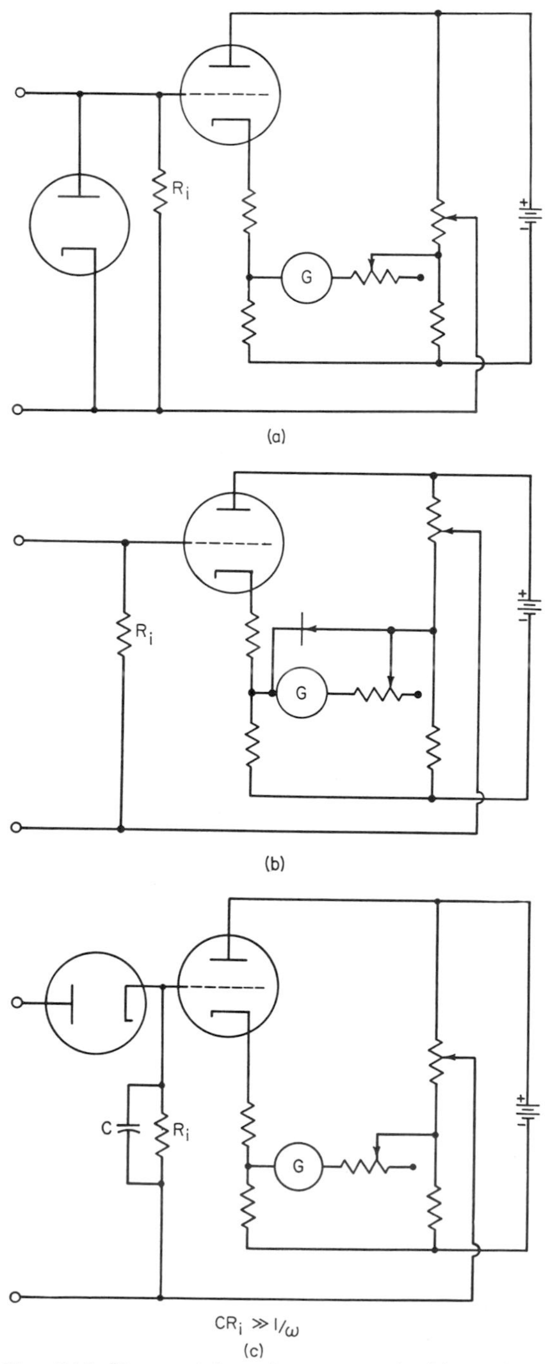

FIG. 5.10. Vacuum-tube voltmeters used with rectifiers (cf. Fig. 5.6).

are of the two types shown in Fig. 5.10. In arrangement (*a*) the rectifier across the input resistor R_i effectively short-circuits the meter during one-half of the cycle, allowing it to read the average value during the other half. In arrangement (*b*) the operation is similar to that described in Sec. 3, but the final indicating meter is effectively short-circuited during one-half of the cycle. With the circuit (*c*) the capacitor acquires a charge which is almost exactly CV_0. This meter therefore reads the peak value of the alternating voltage.

Thermal Instruments. The frequency limitations on electrodynamic and rectifier instruments may be overcome to a large extent by the use of *thermal instruments,* which depend on the Joule heating of a resistor by a current flowing through it. Since this heating effect is proportional to the square of the current, such instruments read root-mean-square values. Almost all modern thermal instruments use a D'Arsonval-type meter to read the emf produced by a thermocouple in contact with a heater through which the current to be measured passes (Fig. 5.11). If the heater has a resistance R and

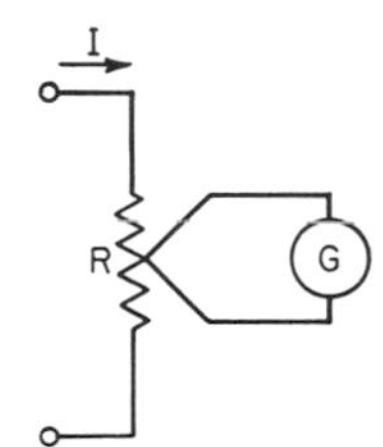

Fig. 5.11. Thermocouple meter.

the current has an rms value of I, the power dissipated in the heater is I^2R. If the equilibrium temperature rise resulting from this power is ΔT, it is found experimentally that

$$\Delta T \propto I^2R$$

The thermocouple attached to the heater develops an emf which is closely proportional to the temperature rise. The deflection of the D'Arsonval-type meter G is proportional to this emf and therefore to the square of the current in the heater. To insure constancy of calibration, the heater and thermocouple are usually mounted in an evacuated glass bulb. Commercial vacuum thermocouples of this sort are available in a wide range of heater and couple resistances, so that they may be adapted to different galvanometers and to voltmeter or ammeter applications. Thermal meters may be calibrated on direct current and used at frequencies up to those at which the inductance and distributed capacitance of the heater become important, as compared with its resistance. If the heater is a straight and fine wire, no appreciable errors are encountered at frequencies in the radio-frequency range.

Electrostatic Instruments. In all the deflection instruments thus far described, the deflection results from the interaction of a current with a magnetic field. *Electrostatic instruments,* however, depend on the interaction between two electric charges. Such instruments vary widely in construction details, but all contain two or more separate conductors, one of which is movable. The potential difference to be measured is applied between the movable conductor and one of the other conductors, causing them to acquire charges. The resulting electrostatic forces cause a motion of the movable conductor against a restoring torque, usually supplied by gravitational forces in relatively insensitive electrostatic voltmeters or by a fine quartz fiber in quadrant electrometers. Because the forces are small, the restoring torques must also be small. The older electrostatic instruments were of low stability, but taut suspensions now have made it possible to produce satisfactory portable instruments of very high impedance.

Electrostatic voltmeters are used principally for the approximate measurement of high potential differences, up to about 10^5 volts. They usually contain only two conductors. Since the charge on each of these conductors is proportional to the applied potential difference and since the force between them is proportional to the product of the two charges, the deflections are, at least approximately, proportional to the square of the voltage. For the same reasons, the meters may be used on alternating current and will read root-mean-square values. They are not satisfactory at high frequencies, since the capacitive current ωVC becomes excessive even for $C \approx 10\ \mu\mu$f.

Quadrant electrometers have been used widely in the past to measure very small currents, either by determination of the potential drop across a high resistance or by observation of the rate at which the potential rises across a small capacitor. They are still unsurpassed for currents of less than 10^{-16} amp but have been almost entirely replaced for larger currents by the electronic instruments described in Sec. 3.

Cathode-ray Oscilloscope. Probably the most versatile of all deflection instruments is the *cathode-ray oscilloscope* (CRO). Its essential element is a *cathode-ray tube,* shown schematically in Fig. 5.12. A block diagram is shown in Fig. 5.13. Electrons in a beam from the hot cathode (K) are accelerated through a series of holes (H_1,H_2) in plates maintained at potentials of the order of 1,000 volts, positive with respect to the cathode. The potentials are so arranged that the beam is focused by the fields around the holes, with the result that nearly all the electrons impinge on the fluorescent screen (S) within a very small area. The point at which the electrons strike the screen may be seen from outside the tube as a bright point of light.* The entire structure is enclosed in an evacuated glass tube.

Shortly after it leaves the last accelerating electrode, the electron beam passes between two deflecting plates (D_1, D_2). The potential difference to be measured (V_y) is applied between these plates. To a first approximation, this potential produces a uniform field $E = V_y/d$ in the region through which the beam is passing. Consequently, the electrons are subject to a force eV_y/d, where e is the electronic charge, and are acclerated transversely with an acceleration eV_y/md, where m is the electronic mass. If the

* The details of the arrangements for focusing and the characteristics of the phosphors used on the screen are discussed in "Television" (Wiley, 1940), by V. K. Zworykin and G. A. Morton.

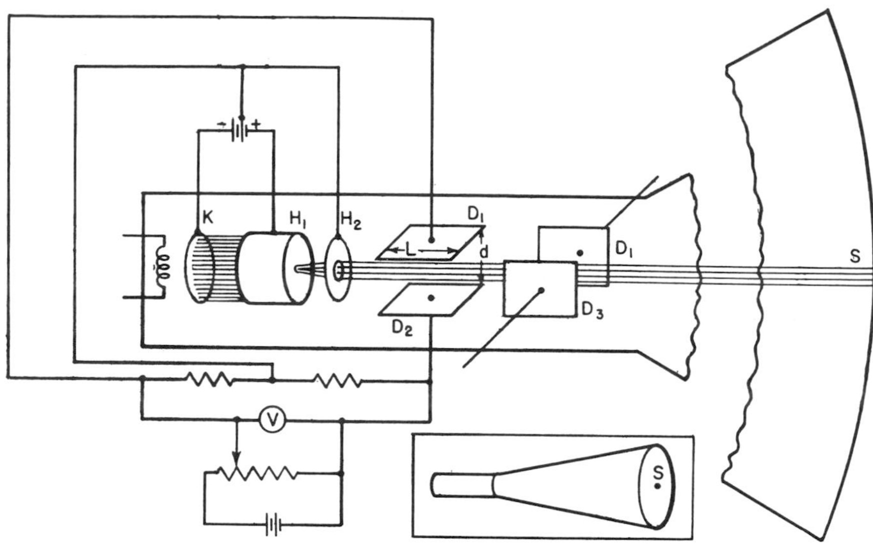

FIG. 5.12. The cathode-ray tube. (The small drawing in the box shows the whole tube approximately to scale; the large drawing is cut away to show the electron gun and the deflecting plates.)

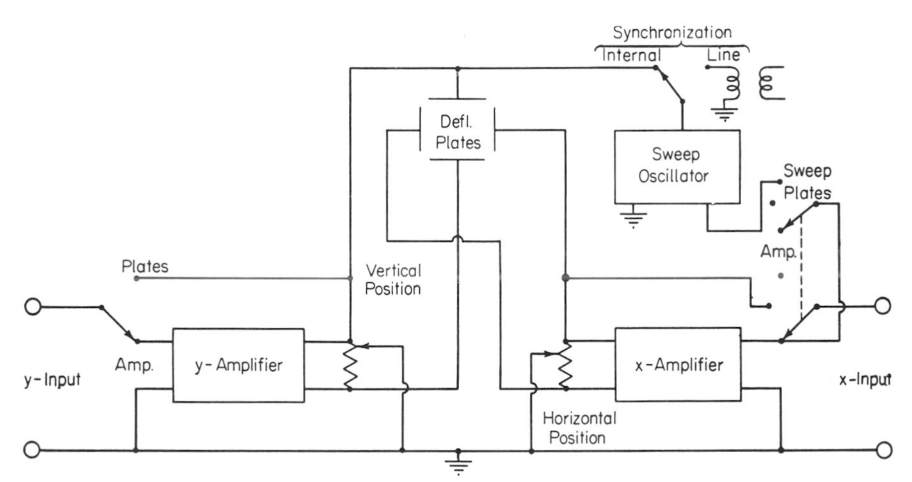

FIG. 5.13. Simplified block diagram of the CRO.

potential of the last accelerating electrode, with respect to the cathode, is V_0, the speed of the electrons is $v_0 = (2eV_0/m)^{1/2}$. Hence the time taken to pass through the deflecting field is $L\,(m/2eV_0)^{1/2}$, and the transverse velocity component acquired during the passage is

$$v_y = (LV_y/d)\sqrt{e/2mV_0}$$

Upon leaving the region of the deflecting plates, the electrons are therefore moving along a trajectory making an angle θ with the axis of the tube, where

$$\theta \approx \tan\theta = v_y/v_0 = LV_y/2dV_0 \qquad (5.37)$$

As (5.37) shows, the deflection of the cathode-ray beam is linear in the applied potential difference. (The first-order theory given here neglects fringing fields, relativistic effects, etc. In good commercial cathode-ray tubes, the fields are shaped so as to compensate for these disturbances, and the linearity of deflection with applied potential difference is maintained to a high degree of accuracy.)

The great advantage of the cathode-ray oscilloscope over competing instruments lies in the speed of its response. The only limitation on this speed is the transit time of the electrons past the deflecting plates. With $V_0 \approx 1{,}000$ volts and $L \approx 2$ cm this time is of the order of 10^{-9} sec, so that the instrument can follow in detail voltage variations at frequencies up to about 10^8 cps. With higher accelerating voltages, faster response is possible but only at the cost of decreased sensitivity, as is shown by (5.37).

Nearly all cathode-ray tubes are fitted with a second pair of deflecting electrodes (D_3, D_4) to which another deflecting voltage (V_x) can be connected to produce a deflection at right angles to that produced by V_y. This potential, known as the *sweep voltage,* may be supplied either by an external independent source or by an oscillator built into the oscilloscope. This oscillator is usually of the relaxation type and is designed to approximate a *"saw-tooth" waveform* of the type shown in Fig. 5.14, the period T being

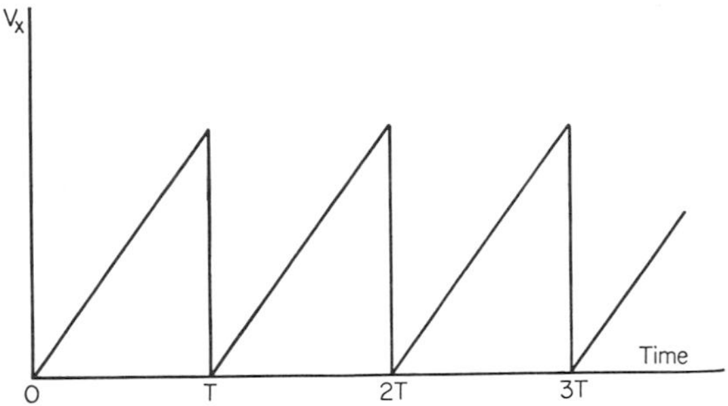

Fig. 5.14. Saw-tooth deflecting potential.

continuously adjustable. If it is desired to maintain this internal oscillator at a precisely determined frequency, it may be "triggered" by the application of a periodic "synchronizing voltage." In most instruments, this synchronizing voltage may be supplied by the power line, by the signal supplied to the y plates, or by an external source of appropriate frequency.

The sensitivity of the usual cathode-ray tube is of the order of 20 volts/in. deflection. The sensitivity at the terminals of the oscilloscope, however, can be increased to the order of 10 mv/in. by the use of amplifiers which are integral parts of the instrument. It is usually the lack of linearity and the phase distortion in these amplifiers, rather than the response of the cathode-ray tube itself, which limit the frequencies at which the cathode-ray oscilloscope may be used. Any interpretation of the deflections obtained with the instrument should be made only after a careful examination of the manufacturer's specifications or after a test of the amplifiers. Cathode-ray oscilloscopes are available with good response over differing ranges extending from direct current to frequencies in the megacycle range.

The amplifiers of most CRO are continuously adjustable but many instruments are also supplied with stepwise variable-gain controls. In the amplifier that carries the signal to be measured, these steps are adjusted to give predetermined sensitivities [v_y/θ in (5.37)]; in the time base amplifier that supplies the saw-tooth wave they give predetermined rates of horizontal movement. When instruments of this sort are used in the continuously variable, "uncalibrated" mode or when CRO without calibrated steps are employed, the instrument should be calibrated immediately before or after the reading is taken, and the amplifier gain should not be adjusted between the reading and the calibration. The calibrating voltage should have as nearly the same amplitude and frequency as the unknown voltage as is practical, in order that errors due to nonlinearity and variations in amplification may be minimized.

A variety of patterns which occur on the screen of the CRO during different measurements of periodic voltages are shown in Fig. 5.15. Patterns such as those shown in (*a*), (*b*), and (*c*) are obtained with a saw-tooth sweep and are used in the measurement of a-c amplitudes, determination of waveform, or comparison of frequencies. The patterns shown in (*d*), (*e*), (*f*), and (*g*) are obtained with a sinusoidal sweep frequency and are used either in frequency comparison or in the measurement of phase difference. If a closed Lissajous figure of the type shown in (*f*) or (*g*) results from the combination of two sinusoidal voltages, the ratio of the number of maxima

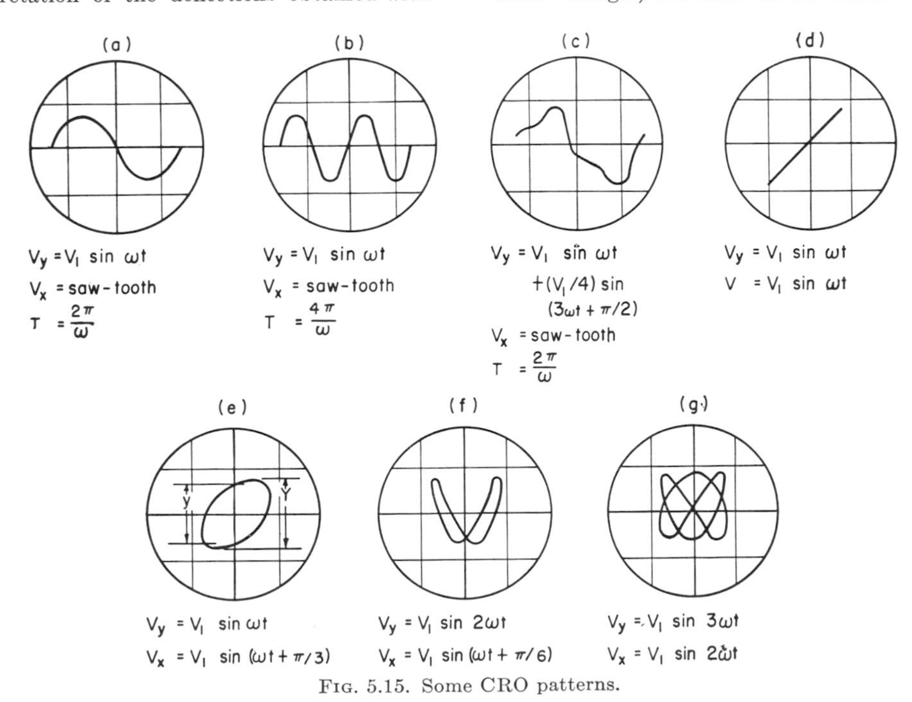

Fig. 5.15. Some CRO patterns.

in the y direction to the number in the x direction is the ratio of the respective frequencies. If the two frequencies are the same, as in (*d*) and (*e*), it is easily shown that the sine of the phase difference between them is y/Y, where the symbols have the meanings indicated in (*e*).

Since the persistence of the phosphor on the screen of a CRO is usually much greater than the period of the phenomenon being studied, nonperiodic voltages cannot be observed clearly with a periodic sweep. Many oscilloscopes are therefore arranged for a single or *nonrecurrent sweep*. To accomplish this, the sawtooth oscillator is biased so that it will not operate until a triggering voltage is applied, after which it will go through one cycle only. The triggering voltage may be either the disturbance being studied or an electrical impulse which is switched on simultaneously with or just prior to that disturbance. Provision is sometimes made for delaying the sweep for a fixed period after the triggering signal is applied.

Instrument Transformers. It was pointed out in Sec. 3 that the ranges of d-c meters may be extended upward by the addition of shunts or multipliers. While these devices are capable of extreme precision, they suffer from the disadvantages that they become bulky when extremely high currents or potential differences are to be measured and that they require the dissipation of appreciable amounts of power. In a-c measurements, particularly at commercial frequencies, these advantages may be overcome by the use of *instrument transformers*. The use of such transformers has become so common that few a-c ammeters are built with ranges greater than 5 amp and few voltmeters with ranges greater than 300 volts.

The *voltage transformer* operates on the principle that a transformer with no flux leakage, with a low-resistance primary, and with a high-impedance secondary load exhibits a voltage across the secondary which is the voltage across the primary divided by the ratio of the number of turns on the primary to the number on the secondary. If an ideal transformer of this sort had its secondary terminals connected to a high-impedance a-c voltmeter and its primary terminals connected across a high-potential a-c source, the potential of the source would be given by the voltmeter reading times the turn ratio. Actually, of course, departures from ideal conditions prevent this simple relation from being true. Compensation for leakage, resistance, etc., is made, in commercial transformers, by the use of a slightly higher turn ratio than that which would be needed in the ideal case. Voltage transformers are usually designed to deliver 100 to 120 volts at the secondary, with ratios between 20:1 and 2,000:1.

The *current transformer* operates on the principle that a transformer with no leakage and with a low-impedance secondary load delivers a current to the secondary which is the current in the primary divided by the ratio of secondary to primary turns. As with the voltage transformer, departures from ideal conditions can be compensated by adjustment of the turn ratio. Current transformers are generally designed for a 5-amp meter connected to the secondary and are made in a wide variety of ratios. Many of the current transformers used in the power industry consist of a magnetic core, on which the secondary is wound, which surrounds a bus bar or other conductor. In this case the primary consists effectively of a single turn.

Since the ratio of secondary to primary turns on a high-ratio current transformer is large, high voltages may be produced across the secondary terminals if these terminals are not connected by an ammeter or other very-low-resistance load. These voltages may constitute a danger both to personnel and to the insulation. If the ammeter is to be disconnected from a current transformer, the secondary terminals should be short-circuited first.

When instrument transformers are used for power or energy measurement, not only their ratios but also the phase shifts between the primary and secondary become important. In well-designed instrument transformers, the phase shift is a small fraction of 1°. For precision work, both phase shift and ratio should be tested under conditions as nearly as possible identical with those under which the transformer will be used.

5. Null Detectors

In many electrical measurements, such as those by the potentiometer and bridge circuits to be discussed in the following sections, circuit parameters are varied until two points in the circuit are brought to the same potential. Instruments used for determining when this "balance" condition has been met are known as *null detectors*. Any of the deflection instruments discussed above can, of course, be used as null detectors, but some have peculiar advantages over others. Furthermore, other instruments, which would be unsuitable for deflection measurements because of nonlinearity or nonconstancy of calibration, are very satisfactory for the determination of a balance.

Null determinations in d-c measurements are almost invariably made with a D'Arsonval-type galvanometer. In some high-resistance bridge circuits, such a galvanometer may effectively short-circuit part of the network, in which case a vacuum-tube voltmeter may be used.

For measurements at commercial frequencies, electrodynamic instruments are commonly used. Such *a-c galvanometers* differ from the deflection instruments described in Sec. 4 in that they use an iron-cored electromagnet in place of the air-cored fixed coil. This increases the sensitivity and decreases the power consumption, at a sacrifice of unimportant linearity. The power consumption may be further reduced and a considerable advantage in convenience gained if the fixed coil is supplied from an exterior source having a frequency identical with that being detected, the latter being connected only to the movable coil. As is indicated by (5.35), the sensitivity of the instrument then depends on the phase angle ϕ between the currents in two coils. In many a-c measurements it is necessary to satisfy two balance conditions simultaneously (see Sec. 7), each of the conditions leading to the elimination of a potential difference which is 90° out of phase with the other. If one of the two coils is supplied through a *phase shifter*, such as that shown in Fig. 5.16, the detector may be made sensitive to one balance condition at a time, and the need of successive approximations to balance may be avoided. In the phase shifter shown, the two contacts are coupled mechanically so that the two resistances are equal at all settings. If the impedance of the circuit to which the output is connected is large compared with both R and $1/\omega C$, it can be shown easily that the output voltage has the same magnitude as the input and that the phase shift is $\tan^{-1} [2RC/(\omega^2 R^2 C^2 - 1)]$.

An electrodynamic instrument used in this way is a type of *synchronous detector,* and it shares with other instruments of this class the advantage that it discriminates against noise or extraneous disturbances, since it responds only to a single frequency.

For measurements in the medium audio-frequency range (500 to 3,000 cps) the telephone receiver is a cheap and convenient null detector, although it is neither frequency- nor phase-selective.

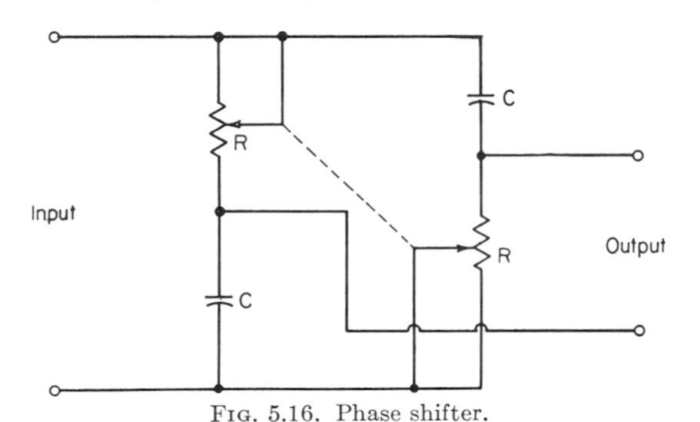

FIG. 5.16. Phase shifter.

In the radio-frequency range the cathode-ray oscilloscope is by far the most effective null detector, and it is also used widely at audio and commercial frequencies. The signal to be detected is usually applied to the vertical deflecting plates, and the horizontal deflection is supplied either by a saw-tooth wave or by a sinusoidal wave of the same frequency as that being detected. In the latter case, the instrument is phase-sensitive (Fig. 5.15*d* and *e*). When the CRO is being used with a bridge, the vertical deflection measuring the bridge output and the horizontal deflection measuring the input to the bridge, it must be remembered that most oscilloscopes have a grounded terminal on both the x and the y inputs. To avoid short-circuiting part of the bridge circuit, it is necessary that one of the two inputs be isolated by the use of a transformer. While the CRO is not frequency-selective, extraneous signals can be distinguished from the input signal, to a fair extent, by visual examination of the pattern on the screen.

6. Potentiometers

Probably the most precise of all present measurements are made with the *potentiometer,* which is a null instrument for the determination of potential differences. The basic circuit of the potentiometer is shown in Fig. 5.17. Between the points A and B there is a fixed and stable resistance (R), made up of coils and slide wires, with adjustable contacts at such points as C, D, E, and F. A current I is supplied by a battery (Ba), which is usually external to the potentiometer. This current may be adjusted with the help of the rheostat (Rheo). Now suppose that the potential difference across a source V_x is to be determined by comparison with a known potential difference supplied by the standard cell V_s. With the contacts C and D set so that a known resistance r_s lies between them, the switch is thrown to the std

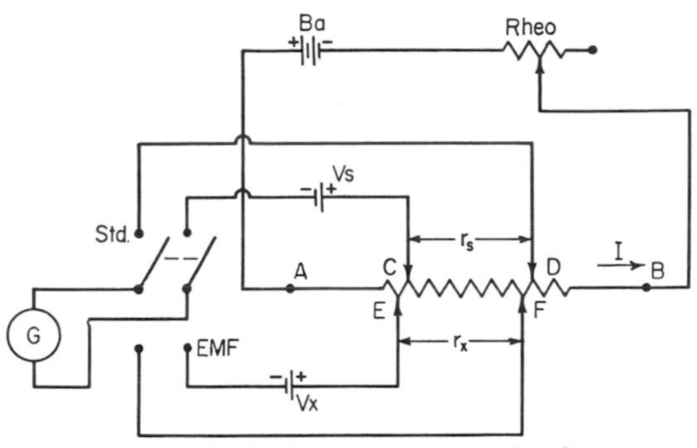

FIG. 5.17. Basic potentiometer circuit.

position, connecting the galvanometer in series with the standard cell, this cell having a polarity opposed to Ba. Rheo is then adjusted until the galvanometer gives a null reading. The potential drop across r_s must then be equal in magnitude to the standard-cell potential difference, and we have

$$Ir_s = V_s$$

After the current has been standardized in this manner, the switch is thrown to the EMF position, so that the unknown V_x replaces the standard cell. The null adjustment is now made by moving the contacts E and F, leaving the setting of Rheo, and therefore of I, unchanged. The condition

$$Ir_x = V_x$$

has now been satisfied. Eliminating I between the two equations, we see that

$$V_x = V_s(r_x/r_s) \tag{5.38}$$

so that the ratio of V_x to V_s is determined with the accuracy with which the relative resistances are known.

The above treatment assumes that I has remained constant throughout the measurement. To check this, it is always advisable that the switch be returned to the std position at the end of the measurement and that the null be checked. If an appreciable departure from a null reading is observed, the entire measurement should be repeated.

In (5.38) only the ratio of two resistances occurs. Hence the unit in which the resistances are calibrated is of no importance. In almost all commercial instruments the positions of the contacts are specified in terms of Ir, where I is a predetermined current. Therefore the potentiometer readings are given directly in volts.

The accuracy with which a potentiometric measurement may be made depends not only on the accuracy with which the resistors have been adjusted but also on the stability of the battery, the sensitivity of the galvanometer, and the resistances at the contacts.

It is usually well to connect Ba to the circuit and to make a preliminary adjustment of I at least an hour before measurements are to be made. The galvanometer should have a resistance which is high compared with R and a high enough voltage sensitivity so that it gives appreciable response to a potential difference equal to the accuracy with which the unknown potential is to be measured. Since this sensitivity would be inconveniently high during the first coarse adjustments, most potentiometers are equipped with a series of keys, some of which, when closed, connect resistances in series with the galvanometer to decrease its voltage sensitivity. In any good instrument, the contacts C,D,E,F have resistances very low compared with that of the galvanometer. If dirt or grease collects on the contacts, their resistances may be greatly increased. This condition is indicated by low and variable sensitivity and may usually be corrected by cleaning the contacts with petroleum ether or a similar highly volatile and noncorrosive solvent.

Commercial "general-purpose" potentiometers are usually built with a range of 0 to about 2 volts, and they may be read to 0.00001 volt. Special-purpose potentiometers have been built to give accuracies of a fraction of a microvolt, but such instruments require very elaborate precautions to eliminate thermal and other parasitic emf's and are therefore much more expensive than are the general-purpose instruments.

While all potentiometers use the basic circuit of Fig. 5.17, there are considerable differences in the details of the circuits used by various manufacturers.

The great advantage of the potentiometer, aside from its high accuracy, is that the current drawn from the unknown is very small, with the result that the internal resistance of the unknown source has little or no effect on the measurement. For some electrochemical measurements, even the small current necessary to produce a noticeable deflection of the galvanometer (10^{-10} to 10^{-7} amp) may be larger than can be tolerated. Under such circumstances, the galvanometer may be replaced by a d-c amplifier of very high input resistance.

While the potentiometer is primarily an instrument for the measurement of potential difference, it may be used for the measurement of currents or resistances. In the former case, the potential drop across a precisely known standard resistor may be measured and (5.2) used. In the latter case, the potential drops across two resistances carrying the same current (i.e., connected in series) are measured, and their ratio is that of the resistances.

7. Bridges; the Four-arm Bridge

The great majority of measurements of resistances, inductances, and capacitances are made by the use of *bridges*, which are, in general, four-terminal networks in which a null indication of a detector shows that a certain precise relation exists among the electrical parameters of the components of the network. Nearly all bridges are equivalent or reducible to the basic circuit of Fig. 5.18, where $\mathbf{Z}_1$, $\mathbf{Z}_2$, $\mathbf{Z}_3$, and $\mathbf{Z}_4$ are impedances made up of combinations of resistances, inductances, and capacitances. The source is either a battery or an oscillator of frequency appropriate to the particular measurement being made. The detector may be any of the null detecting instruments discussed in Sec. 5, provided only that it will respond to the frequency supplied by the source.

If the impedances in the four arms of the bridge are adjusted until the detector shows no deflection, the points C and D must have the same potentials at every instant. The balance conditions are most

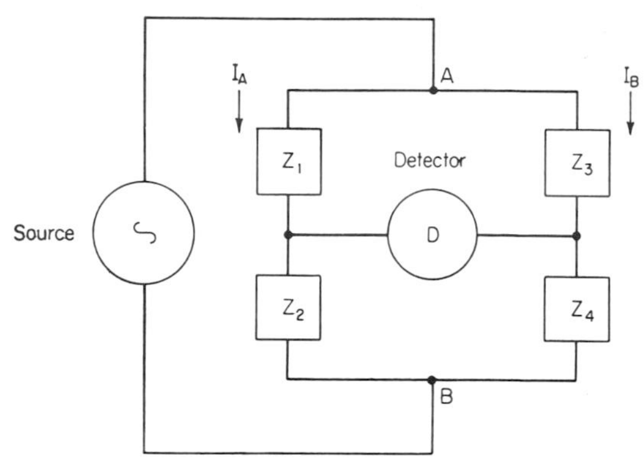

FIG. 5.18. The four-arm bridge.

easily derived by using vectorial or complex representations of the impedances and currents. Representing the impedance $\mathbf{Z}_n$ in the form

$$\mathbf{Z}_n = Z_n e^{j\theta_n} \tag{5.39}$$

and letting the currents in the two sides of the bridge be represented by the complex quantities $\mathbf{I}_A$ and $\mathbf{I}_B$, we have, as the conditions of balance,

$$\mathbf{I}_A Z_1 e^{j\theta_1} = \mathbf{I}_B Z_3 e^{j\theta_3}$$

and

$$\mathbf{I}_A Z_2 e^{j\theta_2} = \mathbf{I}_B Z_4 e^{j\theta_4}$$

[Here Z_1, Z_2, Z_3, and Z_4 are real quantities as are the phase angles θ_1, θ_2, θ_3, and θ_4, and j is used to denote $(-1)^{1/2}$.] Elimination of $\mathbf{I}_A$ and $\mathbf{I}_B$ from these equations yields

$$Z_1 Z_4 e^{j(\theta_1+\theta_4)} = Z_2 Z_3 e^{j(\theta_2+\theta_3)}$$

Separation of the real and imaginary terms in this equation shows that two independent relations among real quantities must be satisfied to give balance:

$$Z_1 Z_4 = Z_2 Z_3 \tag{5.40}$$

$$\theta_1 + \theta_4 = \theta_2 + \theta_3 \tag{5.41}$$

Since these conditions must be satisfied simultaneously, it is clear that at least two parameters in the bridge arms must, in general, be variable in order that a null indication may be obtained. There are two ways of insuring that the two conditions are met simultaneously:

1. One of the parameters is varied until a minimum response of the null detector is obtained; then the other is varied until a further minimum is obtained. One then returns to the first and continues by successive approximations.
2. A phase-sensitive detector and a phase shifter are used, as described in Sec. 5. Certain phase positions will allow the detector to respond to changes

of one parameter and not in the other. When these positions have been found, the adjustments may be made independently.

Variants of the Four-arm Bridge. While only a limited number of choices of the nature of the impedances in the four-arm bridge allow the possibility of balance, there are a considerable number of circuits reducible to that shown in Fig. 5.18. The simplest of these is the *Wheatstone bridge*, in which all four arms are composed of pure resistances, i.e., of real impedances in the notation of Sec. 7 (at all frequencies). In this case (5.41) is automatically satisfied, and the single condition

$$R_1R_4 = R_2R_3 \tag{5.42}$$

is all that is required for balance. The Wheatstone bridge is usually operated from a d-c source and with a D'Arsonval galvanometer as detector.

If inductive or capacitive elements occur in any of the arms, alternating or otherwise variable voltages must be supplied to the bridge to allow a measurement. We shall assume here that the supply is an oscillator, of radian frequency ω. The impedance of a pure inductance L is

$$\mathbf{Z}_L = \omega L e^{-j\pi/2}$$

Similarly, the impedance of a pure capacitance C is

$$\mathbf{Z}_C = (1/\omega C)e^{j\pi/2}$$

The corresponding expressions for more complex impedances may be obtained as combinations of R, $\mathbf{Z}_L$, and $\mathbf{Z}_C$. The substitution of the values of Z and θ obtained from such expressions into (5.40) and (5.41) will show at once whether a proposed bridge configuration can lead to a balance and, if so, what the balance conditions will be. Figure 5.19 shows a number of possible bridges, together with the balance conditions for each. Because of the relatively low cost and ease of calibration of resistors, two of the four arms of a bridge are often pure resistances. Only bridges of this type are shown in Fig. 5.19. Many other circuits, not subject to this limitation, could be shown.

The Wheatstone bridge of Fig. 5.19*a* has already been discussed. Of the others shown, Fig. 5.19*b*, *c*, or *d* may be used for measuring an inductance or a capacitance in terms of a standard of the same general type. Either (*e*) or (*f*) may be used to measure an inductance in terms of a known capacitance, or vice versa. If the bridge (*e*) is used, the frequency of the oscillator must also be known to a fairly high precision, and harmonics cannot be tolerated. This bridge may also be used to measure an unknown frequency in terms of known inductance and capacitance. It should be noted that resistances such as R_3 and R_4 in (*d*) and R_2 in (*f*) occur naturally in capacitors, as the result of leakage, just as resistances such as R_3 in (*c*), (*e*), and (*f*) are inevitably associated with inductances. If the second balance condition given for (*c*) or (*d*) is satisfied by the adjustment of the ratio R_1/R_2 or by variation of one of the capacitors or inductors, it is highly improbable that the first balance condition will be met with the inherent resistances. It will generally, therefore, be necessary

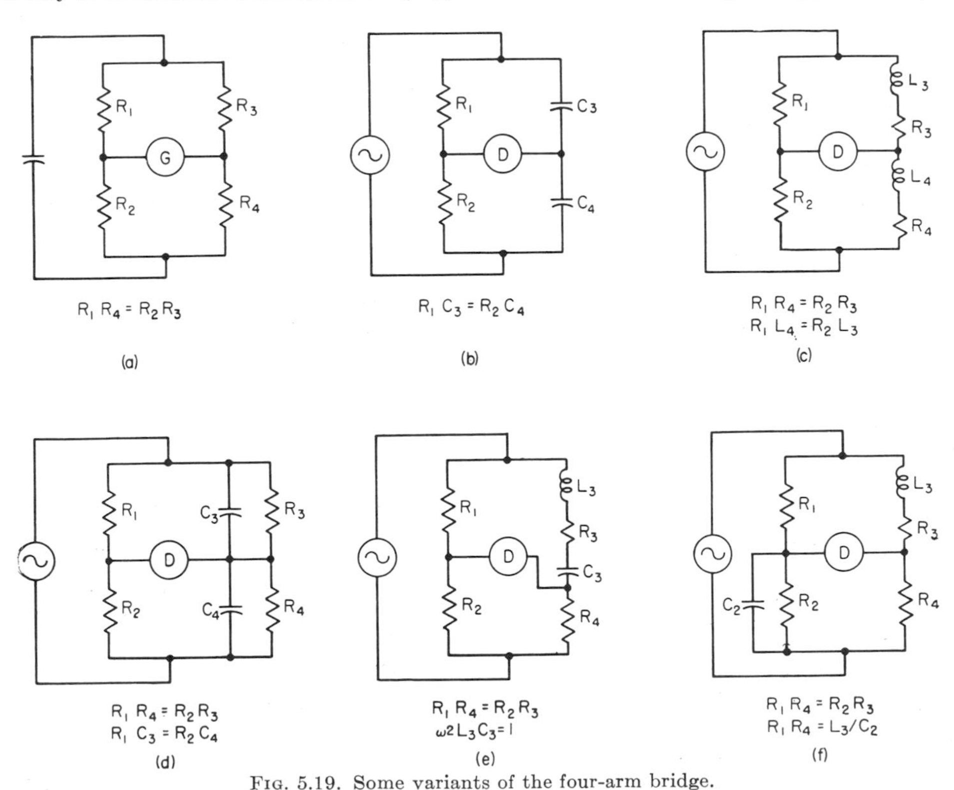

FIG. 5.19. Some variants of the four-arm bridge.

to add series or parallel resistance in one arm until balance is obtained.

Choice of Bridge Elements. It can be shown that the optimum sensitivity of a four-arm bridge is obtained, in general, when all four arms have approximately equal impedances. Similarly, the detector should have an impedance of the same order as that of any one arm. For this reason, the bridges shown in Fig. 5.19 give best results when the ratio R_1/R_2 is made as nearly unity as is possible. If extreme nicety is desired and the value of the unknown is not known approximately, it may be well to set up a bridge with any available standard, to make a preliminary measurement, and then to choose a new standard of the right value to give high sensitivity.

Wheatstone bridges containing three of the four arms and arranged for easy connection of the unknown, the galvanometer, and battery are commercially available. They may also be made up easily of laboratory-type decade resistors of requisite accuracy. If *a-c bridges,* operating at audio frequencies, use such resistance boxes as components, it is essential that the resistance units be wound noninductively. The problems of radio-frequency bridges are treated later.

The most convenient secondary standards of inductance consist of two coils, each of fixed self-inductance, arranged so that they may be moved relatively to each other, thereby varying the mutual inductance between them. If the self-inductances are L_1 and L_2, while the mutual is M, the inductance of the two in series is

$$L = L_1 + L_2 \pm 2M$$

which is variable over as wide a range as M itself. All inductors produce magnetic fields which extend beyond the limits of the instrument. These fields may produce eddy currents in any conductors near the inductor or may cause the magnetization of ferromagnetic materials in the neighborhood. Hence the inductance of the standard may be affected by its surroundings. Such effects may be minimized by removing all large pieces of metal from the vicinity of the inductor. If inaccuracies due to these causes are suspected, their presence may sometimes be detected by changing the geometrical arrangement of the apparatus and testing to see whether the balance of the bridge is affected.

The only capacitors which have sufficiently high stability and sufficiently low leakage to be suitable as standards are those which employ either air or mica as a dielectric. When capacitors of low value are employed, considerable error may be introduced by the capacitance between their plates and their surroundings. If the capacitor is surrounded by a totally enclosed and grounded case, this stray capacitance is increased but is made independent of the surroundings. Hence the best capacitors for audio-frequency measurements are those with metal shields. These should be used with the shields grounded and with the plates bearing the same relation to ground as was used during their calibrations. This relation is usually stated by the manufacturer.

The wires connecting the various components of the bridge exhibit resistance, inductance, and capacitance among themselves and between each and ground. In the Wheatstone bridge only the resistance is important, and it is merely necessary that the leads be short and heavy enough so that they introduce negligible resistance into any arm of the bridge. In bridges operating at audio frequencies, the inductive and capacitive effects may become serious. Hence all leads should be kept as short as possible. Inductive effects may be minimized if wires carrying the same current in opposite directions are kept close together, e.g., as a twisted pair. Capacitive effects are minimized by keeping leads which differ in potential far apart and far removed from other conducting bodies. If the leads are kept short, little trouble will be experienced in the measurement of reasonably large inductances ($>10^{-2}$ henry) or capacitances ($>10^{-9}$ farad) at audio frequencies.

Troubles with stray capacitances and inductances become very serious at radio and higher frequencies. In these frequency ranges resistance standards also become troublesome, both because of their inductance and distributed capacitance and because of skin effect on the resistance itself. Only carefully constructed and shielded bridges, specially designed for high frequencies, will give accurate results, and these may require corrections which are specified by the manufacturer.

Other Bridges. While the four-arm bridge is capable, in principle, of measuring any resistance, inductance, or capacitance, it is sometimes impractical, either because of the effects of leads, contacts, etc., or because other circuits can yield results of the required accuracy more rapidly. A large number of bridges having six or more arms have been devised and are described in the bibliography at the end of the chapter. Their treatment is beyond the scope of a handbook of general physics, and they are mentioned only to indicate their existence.

One arrangement which may make the four-arm bridge more satisfactory than it would otherwise be, particularly for the measurement of small quantities, is the *substitution bridge.* Suppose that any variant of the standard bridge has been balanced with an unknown in one of the arms. If the unknown is then disconnected and replaced by a variable standard, having characteristics as much like the unknown as is practicable, the bridge may again be balanced by adjustment of the standard, leaving everything else in the bridge unchanged. Under these conditions, the equality of the unknown and standard is often insured to a much higher accuracy than could be obtained from the values of the bridge parameters.

8. Measurements Using Resonant Circuits

If a sinusoidally varying potential difference, $V_0 \sin \omega t$, is impressed on a series circuit including a capacitance C, an inductance L, and a resistance R, the amplitude of the current flowing in the circuit is a function of the frequency, i.e., of ω. Explicitly, the current amplitude is

$$I_0 = V_0/[R^2 + (\omega L - 1/\omega C)^2]^{1/2} \qquad (5.43)$$

The admittance of the circuit, i.e., the ratio I_0/V_0, obviously has a maximum at the resonant radian frequency

$$\omega_0 = 1/\sqrt{LC} \tag{5.44}$$

The general shape of the admittance-vs.-frequency curve, and particularly the sharpness of the curve near its maximum, is often expressed in terms of a parameter Q, defined as

$$Q = \omega_0 L/R \tag{5.45}$$

Eliminating L and C from (5.43) with the help of (5.44) and (5.45), we find that the admittance is

$$Y = I_0/V_0 = \frac{1}{R}\left[1 + Q^2\left(\frac{\omega}{\omega_0} - \frac{\omega_0}{\omega}\right)^2\right]^{-1/2} \tag{5.46}$$

Q determines completely the shape of the *resonance curve,* in which YR is plotted as a function of ω/ω_0. Conversely, an experimental determination of the resonance curve allows the Q of the circuit to be measured. Circuits with high Q exhibit sharp resonances, while those with low Q have admittances which vary only slowly with frequency.

Many measurements in the radio-frequency range are made through a utilization of the properties of the resonant circuit. A few of the more common of these will be indicated.

If a coil having an inductance L and a resistance R is connected in series with a good capacitor of capacitance C and with a thermal ammeter and if the resulting circuit is excited by an oscillator having a radian frequency ω_0, maximum reading of the ammeter will be obtained when (5.44) is satisfied. If the frequency of the oscillator at this resonance is known, L can be determined in terms of a known C, or C in terms of a known L. On the other hand, if L and C are known or if their product has previously been determined, the frequency of the oscillator can be obtained. The *wavemeter,* used for the approximate measurement of radio frequencies, operates on this basis, having a fixed inductance and a variable capacitor, which may be adjusted until resonance with a source of unknown frequency is achieved.

In a second application, a coil (L,R) and a capacitor (C) are connected in series with each other and with a source of radio-frequency power. After a source has been tuned to resonance with the circuit, a high-impedance radio-frequency voltmeter is used to measure the voltage V_0 across the source and then the voltage V_L across the coil. If the current in the circuit is I, the former voltage is

$$V_0 = IR$$

in accordance with (5.43) and (5.44). The voltage across the coil is

$$V_L = IR(1 + \omega_0^2 L^2/R^2)^{1/2} = IR(1 + Q^2)^{1/2}$$

and so the ratio

$$V_L/V_0 = (1 + Q^2)^{1/2}$$

The Q of the circuit is thus determined, and (5.45) may be used to give L in terms of R. If C is known, L may also be found in terms of it and the resonant frequency. Hence this measurement gives a determination of resistance at radio frequencies.

9. Measurements at Ultrahigh Frequencies; Distributed Parameters

At all frequencies up to about 3×10^8 cps, the wavelength of electromagnetic radiation (>1 m) is large compared either with the dimensions of the apparatus commonly used in electrical measurements or with the lengths of the wires used for connections. At higher frequencies this condition is not met, with the result that the phase of the wave varies appreciably from point to point within the equipment. This leads to three difficulties:

1. Open wires, coils, capacitors, etc., become efficient radiators of energy. Energy delivered to such elements is largely dissipated through radiation.
2. *Lumped parameters,* such as the inductance of a coil or the capacitance of a capacitor, lose their meaning, since different parts of the coil or different parts of the capacitor are behaving differently. One is forced to abandon such concepts in favor of the idea of *distributed parameters,* such as the inductance or capacitance per unit length of a transmission line.
3. Small changes in the geometry of the apparatus may lead to path differences which are of the order of the wavelength and so may produce large changes in electrical behavior.

These difficulties force one to use, at frequencies above 300 Mc, techniques which are quite different from those employed at lower frequencies.

The methods used for measurements at *ultrahigh frequencies* have expanded so rapidly since 1940 that they now involve a field of specialization comparable in scope with all the rest of electrical measurements. A few of the basic principles involved are presented here as an introduction to more extended treatments.

Coaxial Cables and Waveguides. When an electrical disturbance passes along a conductor, it acts up varying electrical and magnetic fields in the neighborhood of the conductor. Since the boundary conditions developed in Part 4, Chap. 1 must be satisfied at every instant at the interface of the conductor and the surrounding dielectric, the disturbance is propagated at the same speed in the conductor and in the dielectric.

If radiation is to be avoided, the electromagnetic field must be confined to a limited region of space. This is accomplished, in ultrahigh-frequency work, by one of two means. The *coaxial cable* consists of a cylindrical conductor mounted along the axis of a hollow, conducting, cylindrical sheath. The sheath is grounded, and the two conductors act as a transmission line. Since they carry equal currents in opposite directions and equal and opposite charges per unit length at every point along the cable, the field is limited to the region between the two conductors, and it is through this region that the energy passes along the cable. The *waveguide* consists of a hollow metallic duct, usually rectangular in cross section, filled with a dielectric, usually air. As with the coaxial cable, the electromagnetic wave travels in this dielectric, the boundary conditions being satisfied by fluctuating currents which are established in the walls of the guide.

In the ultrahigh-frequency region, waveguides and coaxial cables replace the wires used for connections at low frequencies.

In discussing the properties of high-frequency *lines,* which is a term that may be used for either cables or guides, we may make two simplifications which are valid to the first approximation. First, we neglect all attenuation, or power loss, in the line itself. Such losses are extremely small in well-constructed lines, except at the very highest frequencies. Second, we assume that the only waves which are propagated are those in which the electric field is transverse to the length of the line. This is, in general, true if the linear dimensions of the cross section of the line are small compared with the wavelength.

Suppose that one end of a line is excited by an oscillator supplying a high-frequency electromagnetic wave and that the other end of the line is short-circuited. At the short-circuited end there can be no field in the dielectric. In order that this condition may be satisfied, a reflected wave, 180° out of phase with the incident wave, will be set up at this end. If the length of the line is an odd number of quarter wavelengths,* this wave will combine with the incident wave to form a standing wave along the line. Under these circumstances, no power is transmitted. If we have means to measure the time average value of the electric field at various points along the line (or the time average potential difference between the central axis of the dielectric and ground), we shall find that there are maxima of field (or voltage) at intervals of one-half of a wavelength and that the field (or voltage) at points midway between these maxima is zero.

Now suppose that the termination of the line is changed from a short circuit to a connection to space or by a finite resistance in which Joule heating takes place. In either case, energy must be supplied at the termination, and the reflected wave will not have the same amplitude as the incident wave. Let the amplitude of the incident wave be A and that of the reflected wave be B. Since the power is proportional to the square of the amplitude, we have

$$A^2 - B^2 = kP \tag{5.47}$$

where P is the power delivered at the termination and k is a constant. If the voltages are now measured, there will still be maxima and minima at intervals of one-quarter wavelength, but their values will be $A + B$ and $A - B$, respectively. The ratio of the maximum time average voltage to the minimum time average voltage is known as the *voltage standing-wave ratio* (VSWR). Denoting the VSWR by r, we have

$$r = (A + B)/(A - B) \tag{5.48}$$

Elimination of B from this definition, with the help of (5.47), shows that

$$r = (\sqrt{A^2/kP} + \sqrt{A^2/kP} - 1)^2$$

If the termination is not a pure resistance, the positions of the maxima and minima, as well as the VSWR, will be changed. Thus, suppose that it is a pure reactance X. The average power input is now zero, so that $A = B$, but the incident and reflected waves are not in phase at the end of the line. The end therefore exhibits neither a minimum nor maximum voltage, and the whole interference pattern is shifted from the pattern observed with a short-circuit termination by an amount depending on the magnitude of the reactance. Thus the VSWR measures the resistance of the termination, while the pattern shift measures the reactance.

Measurement of the VSWR. To measure the VSWR we need a device which will extend into the line but which, at the same time, will disturb the line as little as possible. The latter requirement demands that the instrument draw little power, that it cause only negligible reflections, and that it make little change in the characteristic impedance of the line. These conditions are met very well by a small metallic *probe,* inserted through a narrow longitudinal slot in the outer conductor, and connected to a rectifier-type meter. The instrument rectifiers used at low frequencies have such high capacitance that they would be unsatisfactory at these high frequencies. Fortunately, however, the point-contact rectifiers, utilizing a fine metallic "whisker" in contact with a semiconductor such as germanium or silicon, have very low capacitance. In addition, they have the advantage that they pass a current which is almost exactly proportional to the square of the forward-direction voltage. Consequently, the indication of a low-resistance d-c galvanometer connected in series with such a rectifier is quite accurately proportional to the square of the probe voltage.

In precise measurement of the VSWR, corrections need to be made for the departures from perfect square-law behavior in the rectifier and for the effect of the probe on the line.

Power Measurements. Although the probe-rectifier instrument indicates power, it cannot furnish an absolute measure of the power being transmitted in a line. The only satisfactory method that has been found for making this measurement involves the dissipation of the power into heat and the measurement of the rate of heat production. The device most commonly used to accomplish this is a *bead thermistor,* which is a small bead of semiconducting material, having two wire leads extending into it and being covered with a thin layer of glass. The glass merely serves to protect the thermistor from oxidation. The thermistor, enclosed in a glass capsule, is mounted at or near the end of the line, where it, together with its mount, constitutes a line termination. The electromagnetic wave impinging on the thermistor sets up currents within it, and the

* The wavelength in a coaxial cable is the same as that in an infinite dielectric, that is,

$$\lambda = 2\pi/\sqrt{\epsilon\mu\omega^2}$$

where ϵ is the dielectric constant and μ the permeability of the dielectric. In rectangular guides, in which the electric field is perpendicular to one transverse dimension (a), the wavelength is

$$\lambda = 2\pi/\sqrt{\omega^2\epsilon\mu - (\pi/a)^2}$$

Joule heating causes the temperature to rise and the resistance to decrease. If the bead is used as one arm of the Wheatstone bridge, its temperature may be followed, and, after the thermal capacity has been determined, the rate of power input may be computed.

Thin metallic wires, which show an increase of resistance with rising temperature, instead of a decrease, are sometimes used instead of thermistors. These are known as *barretters*. They do not have as high sensitivity as do thermistors, since the absolute value of their temperature coefficient of resistance is low, but they have lower heat capacity and therefore give faster response.

Bibliography

Ginzton, E. L.: "Microwave Measurements," McGraw-Hill, New York, 1957.

Harris, F. K.: "Electrical Measurements," Wiley, New York, 1952.

Hund, A.: "High-frequency Measurements," 2d ed., McGraw-Hill, New York, 1951.

King, D. D.: "Measurements at Centimeter Wavelength," Van Nostrand, Princeton, N.J., 1952.

Kinnard, I. F.: "Applied Electrical Measurements," Wiley, New York, 1956.

Laws, F. A.: "Electrical Measurements," 2d ed., McGraw-Hill, New York, 1938.

Michels, W. C.: "Electrical Measurements and Their Applications," Van Nostrand, Princeton, N.J., 1957.

Montgomery, C. G.: "Technique of Microwave Measurements," McGraw-Hill, New York, 1947.

National Physical Laboratory (England): "Precision Electrical Measurements," Philosophical Library, New York, 1956.

Stout, M. B.: "Basic Electrical Measurements," 2d ed., Prentice-Hall, Englewood Cliffs, N.J., 1960.

Chapter 6

Conduction: Metals and Semiconductors

By JOHN BARDEEN, University of Illinois†

1. General Relations; Metals

Current in metals and semiconductors is carried by electrons which, because of their wave nature, can move through a perfect periodic lattice without being scattered. Resistance results from scattering by lattice imperfections, such as impurities, or by thermal motion of the atoms of the crystal. The current density $\mathbf{j}$ is proportional to the electric intensity (Ohm's law):

$$\mathbf{j} = \sigma \mathbf{\mathcal{E}} \qquad \text{or} \qquad \mathbf{\mathcal{E}} = \rho \mathbf{j} \tag{6.1}$$

In practical units, $\mathbf{j}$ is in amperes per square centimeter, $\mathbf{\mathcal{E}}$ in volts per centimeter, and σ in mho-cm^{-1}. The resistivity ρ is the reciprocal of σ.

In *anisotropic media*, σ and ρ are second-order tensors. Since all metals have an axis of symmetry, there are at most two independent resistivities, corresponding to directions parallel and perpendicular to this axis. The resistivity for flow in a direction making an angle θ with the axis is

$$\rho = \rho_{\parallel} \cos^2 \theta + \rho_{\perp} \sin^2 \theta \tag{6.2}$$

In cubic metals, ρ and σ are scalars. When the individual crystals are oriented at random in a polycrystalline noncubic metal, the mean resistivity is isotropic and equal to

$$\rho = \tfrac{1}{3}(\rho_{\parallel} + 2\rho_{\perp}) \tag{6.3}$$

Similar relations apply to thermal conduction. The heat current w is proportional to the temperature gradient:

$$w = -\kappa(dT/dx) \tag{6.4}$$

The constant of proportionality κ is the thermal conductivity. In pure metals, most of the heat is transported by electrons. In alloys of high electrical resistivity and in nonmetals, heat is carried mainly by phonons, the quanta of the lattice vibrations.

Drude's Theory. Soon after the discovery of the electron by J. J. Thomson in 1897, Drude [1]‡ developed a theory of conductivity based on the assumption that there is a gas of free electrons in a metal. The theory was developed more completely by Lorentz [2], who based his calculations on the Maxwell-Boltzmann equation of kinetic theory. The greatest success of the theory was the derivation of the Wiedemann-Franz law, which states that the ratio of thermal conductivity κ to electrical conductivity σ is the same for all metals and, as had been pointed out by Lorenz, is proportional to the absolute temperature T. Drude found the following expression for the Lorenz number L:

$$L = \kappa/\sigma T = 3(k_0/e)^2 \tag{6.5}$$

where k_0 is Boltzmann's constant and e the magnitude of the electronic charge. Modern theory gives $\pi^2/3$ in place of 3 for the numerical factor, in close although not exact agreement with observation.

Modern theories of conductivity, due mainly to Sommerfeld, Houston, and Bloch, differ from those of Drude and Lorentz in the use of Fermi-Dirac rather than Boltzmann statistics for the electron gas and in the use of an effective electron mass, or its equivalent, as derived from the energy-band structure. Much of the early theory can, in fact, be taken over *in toto* and applied to semiconductors, in which the electron density is generally small enough so that Boltzmann statistics can be used.

Drude's derivation of the expression for the electrical conductivity is quite simple. Let τ be the average time between collisions of an electron (relaxation time). The probability that an electron makes a collision in dt is then dt/τ. If v_0 is the velocity of an electron after its last collision, at time $t = -t_1$, the velocity at $t = 0$ is

$$\mathbf{v} = \mathbf{v}_0 - t_1 e \mathbf{\mathcal{E}}/m \tag{6.6}$$

The current density is obtained by multiplying by $-e$ and summing over all electrons in unit volume:

$$\mathbf{i} = -e\Sigma(v_0 - t_1 e \mathcal{E}/m) \qquad \text{esu} \tag{6.7}$$

We assume, for simplicity, that the average value of v_0 is zero. Since the average value of t_1 is

$$\langle t_1 \rangle_{\text{av}} = \tau^{-1} \int_0^\infty t_1 e^{-t_1/\tau}\, dt_1 = \tau \tag{6.8}$$

we find

$$\mathbf{i} = (ne^2\tau/m)\mathbf{\mathcal{E}} \tag{6.9}$$

The conductivity is therefore

$$\sigma = ne^2\tau/m \tag{6.10}$$

† The author is indebted to Dr. F. J. Blatt for help in preparation of the corresponding chapter in the first edition and to Dr. H. J. Stocker for help with revisions for this edition.

‡ Numbers in brackets refer to References at end of chapter.

TABLE 6.1. SOME CONSTANTS OF METALS

Metal	Crystal Structure	Number of atoms per cm³, $N_a \times 10^{-22}$	Conductivity at 0°C, (ohm-cm)$^{-1}$ $\times 10^4$	Hall coefficient at room temperature, R, cm³/coulomb $\times 10^4$	Debye temperature, Θ, °K	Absolute thermoelectric power at 0°C, ε, μ volt/°C
Li	b.c.	4.6	11.8	−1.7	363	10.3
Na	b.c.	2.5	23	−2.5	202	− 5.4
K	b.c.	1.3	15.9	−4.2	163	−12.7
Rb	b.c.	1.07	8.6	....	85	− 9.2
Cs	b.c.	0.85	5.6	−7.8	54	− 0.8
Cu	f.c.	8.5	64.5	−0.55	333	+ 1.6
Ag	f.c.	5.8	66.7	−0.84	223	+ 1.3
Au	f.c.	5.9	49	−0.72	175	+ 1.7
Be	hex.	12.3	18	+2.44	1,000	
Mg	hex.	4.3	25	−0.94	357	+ 1.3
Zn	hex.	6.6	18.1	+0.33	213	+ 1.9
Cd	hex.	4.6	15	+0.60	172	1.8
Hg	rhomb.	4.2	4.4	....	80	− 4.4
Ca	f.c.	2.3	23.5	....	230	− 9.2
Sr	f.c.	1.8	3.3	....	171	
Ba	f.c.	1.6	1.7	....	131	
Al	f.c.	6.0	40	−0.30	395	− 1.6
Ga	rhomb.	5.1	2.45	....	125	
In	tetr.	3.8	12	−0.07	198	1.2
Tl	hex.	3.5	7.1	+0.24	140	0.8
Sn	tetr.	3.7	10	−0.04	260	− 1.0
Pb	f.c.	3.3	5.2	+0.09	86	− 1.2
Ti	hex.	5.6	1.2	....	342	
Cr	b.c.	8.2	6.5	....	495	
Mn	b.c., tetr.	8.0	0.14, 1.1	−0.93	368	
Fe	b.c., f.c.	8.5	11.2	+0.245	420	15.0
Co	f.c., hex.	9.0	16	−1.33	401	−18.5
Ni	f.c.	9.0	16	−0.611	472	−18.8
Zr	hex., b.c.	4.3	2.4	....	288	
Nb	b.c.	5.5	4.4	....	184	
Mo	b.c.	6.4	23	+1.26	380	4.9
Ru	hex.	7.3	8.5	....	426	
Rh	f.c.	7.2	22	370	370	1.5
Pd	f.c.	6.5	10	−0.68	270	− 6.7
Hf	hex.	4.3	3.4	....	213	
Ta	b.c.	5.4	7.2	+1.01	228	− 2.0
W	b.c.	6.3	20	+1.18	333	0.5
Re	hex.	6.9	5.3	....	310	
Os	hex.	7.1	11	....	256	
Ir	f.c.	7.0	20	....	316	1.2
Pt	f.c.	6.6	10.2	−0.24	240	− 4.3
Rare earths:						
La	hex., f.c.	2.7	1.7	−0.8	132	
Ce	f.c., hex.	3.0	1.3	+0.181	.....	+ 6.0
Pr	hex.	2.8	1.1	+0.71		
Nd	hex.	2.9	1.3	+0.97		
Il						
Sm	....	3.2				
Eu	b.c.	2.1				
Gd	hex.	3.1	0.73	−0.95 (R_0)	152	
Tb	hex.	3.2				
Dy	hex.	3.2	0.73	−1.3 (R_d)		
Ho	hex.					
Er	hex.	3.3	0.57	−0.34		
Tm	hex.	3.3				
Yb	f.c.	2.4				
Lu	hex.	3.3				

SOURCES FOR TABLE 6.1

Column	*Source*
Structure, N_a	"Metals Handbook," p. 20, 1948.
Conductivity	Mott and Jones, "Theory of Metals and Alloys," p. 246, Oxford, 1936.
Conductivity, rare earths	"Metals Handbook," p. 20; also Legvold et al., *Revs. Mod. Phys.*, **25**: 129 (1953).
Hall coefficient	Meissner Handbuch der Experimentalphysik, vol. 11, p. 336; Seitz, "The Modern Theory of Solids," p. 183; Pugh and Rostoker, *Revs. Mod. Phys.*, **25**: 151 (1953).
Hall coefficient, rare earths	C. J. Kevane et al., *Phys. Rev.*, **91**: 1372 (1953).
Debye temperature	Mott and Jones, *op. cit.*, p. 246.
Debye temperature, rare earths	M. Griffel et al., *Phys. Rev.*, **93**: 657 (1954).
Thermoelectric power	Meissner, *op. cit.*, p. 421; value for Ag is from Wilson, "Theory of Metals," 2d ed., p. 207, Cambridge, 1953.

This expression does not depend on the use of Boltzmann statistics and can be applied to metals with a degenerate electron gas, provided that the simplifying assumptions made in the derivation are valid. In a typical metal, n is of the order of 10^{22} per cm³ and τ is of the order of 10^{-13} sec, giving $\sigma \sim 2 \times 10^{17}$ esu or $\sim 2 \times 10^5$ mho-cm^{-1}.

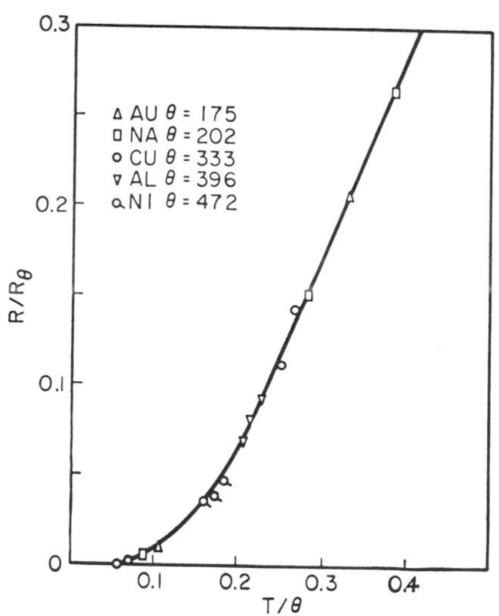

FIG. 6.1. Temperature variation of resistivity of various metals compared with the Bloch-Grüneisen function. [*After Bardeen, J. Appl. Phys.*, **11**: 88 (1940).]

To a first approximation, scattering by impurities and by thermal motion is additive. According to *Matthiessen's rule*, the resistivity of a relatively pure metal can be expressed as the sum of a temperature-independent term ρ_i resulting from impurity scattering and a temperature-dependent term ρ_T characteristic of the pure metal:

$$\rho = \rho_i + \rho_T \tag{6.11}$$

Grüneisen [3] has shown that ρ_T for most metals can be given to a close approximation by a universal function of the absolute temperature T, originally derived by Bloch [4]:

$$\rho_T = C \left(\frac{T}{\Theta}\right)^5 \int_0^{\Theta/T} \frac{x^5 e^x\, dx}{(e^x - 1)^2} \tag{6.12}$$

The characteristic temperature Θ which gives the best fit to the electrical conductivity is generally close to, although not exactly equal to, the Debye temperature for specific heats. Data for various metals are given in Table 6.1.

At high temperatures, $T \gg \Theta$, ρ_T is proportional to T, in accordance with the fact that the mean-square amplitude of the lattice vibrations increases linearly with T. At very low temperatures, ρ_T varies as T^5.

TABLE 6.2. PRESSURE COEFFICIENT OF RESISTIVITY OF VARIOUS METALS*

Metal	$-(d \log R/dp)10^{12}$ (cgs)	$(2\beta V_0/C_v)10^{12}$ (cgs)
Li	−4.0	21.
Na	73.	40.
Mg	5.9	9.
Al	4.8	6.
K	190.	91.
Ca	−8.9	
Fe	2.7	2.0
Co	1.1	2.0
Ni	2.1	2.0
Cu	2.3	3.0
Rb	200.	120.
Sr	−47.	
Mo	1.5	1.1
Ag	4.0	4.8
Cs	220.	157.
Ta	1.7	1.8
Pt	2.1	1.9
Au	3.4	3.3
Pb	15.4	12.5

* Reproduced from J. Bardeen, *J. Appl. Phys.*, **11**: 88 (1940).

Bridgman [5] has studied the *pressure variation* of resistance of a large number of metals. One of the most important factors is the change in amplitude of thermal motion with compression. According to the Debye theory, the mean-square amplitude varies as T/Θ^2, so that the relative change in resistivity due to the change in Θ is given by

$$d(\log \rho)/dp = -2d(\log \Theta)/dp \tag{6.13}$$

The change in Θ with pressure can be estimated from a formula due to Grüneisen:

$$d(\log \Theta)/dp = \beta V_0/C_v \tag{6.14}$$

In Eq. (6.14), β is the thermal coefficient of expansion, V_0 is the volume occupied by 1 g, and C_v is the specific heat.

Table 6.2 gives a comparison of $-d(\log \Theta)/dp$ from Bridgman's data with values of $2\beta V_0/C_v$. It can be seen that, although the agreement is close for a number of metals, there are exceptions. Some metals, for example, Li, Ca, Sr, have a negative coefficient of resistance. These anomalies undoubtedly result from changes in the energy-band structure with compression, which more than compensate for changes in Θ.

More recently, measurements of the electrical resistivity of metals have been made at pressures in the range from 100 to 600 kilobars. Quite frequently, large changes (either an increase or a decrease) are observed in the neighborhood of certain critical pressures. These may reflect significant changes in the band structure as well as phase changes involving atomic rearrangements. A review has been given by Drickamer [6].

TABLE 6.3. THE COEFFICIENT α IN EQ. (6.18) FOR THE THERMAL RESISTIVITY FOR SOME METALS

Metal	$\alpha \times 10^4$	Metal	$\alpha \times 10^4$
Cu	0.23	Sn	3.9
Ag	0.9	Pb	22
Au	1.8	Fe	1.8
Zn	2.1	Ni	2.2
Cd	14	Mo	0.75
Hg	200	Rh	2.2
		Pd	6.4
Al	0.22	W	0.88
		Ir	0.36
In	19	Pt	4.3

Resistivity changes also result from shearing strains. In general, both the resistivity change $\Delta\rho_{ij}$ and the strain ϵ_{kl} are tensors of the second rank, and the matrix of coefficients connecting them is one of the fourth rank. In a cubic crystal, there are three independent constants. If ρ is the isotropic resistivity of the cubic crystal and if the coordinate axes are those of the crystal, the equations are of the form

$$\begin{aligned} \Delta\rho_{xx}/\rho &= k_1\epsilon_{xx} + k_2(\epsilon_{yy} + \epsilon_{zz}) \\ \Delta\rho_{yz}/\rho &= 2k_3\epsilon_{yz} \end{aligned} \tag{6.15}$$

together with similar relations obtained by permutation of x, y, and z.

The strain gauge makes use of the change in resistance of a wire with tension. Part of the observed change results from change of shape and part from change in resistivity. Designating Poisson's ratio by μ and values of k_1 and k_2 for an isotropic polycrystalline material by K_1 and K_2, we have for the change in resistance R of a wire of length l

$$d \log R/d \log l = K_1 - 2\mu K_2 + 1 + 2\mu \tag{6.16}$$

Values of K_1 and K_2 are appropriate averages of k_1 and k_2 over directions.

The *thermal resistance* $1/\kappa$ may also be expressed approximately as the sum of contributions from impurity and thermal scattering:

$$\frac{1}{\kappa} = \frac{1}{\kappa_i} + \frac{1}{\kappa_T} \tag{6.17}$$

While both κ_i and κ_T obey the Lorenz relation (6.5) at high temperatures, only κ_i does so at low temperatures. Theory and experiment both indicate that κ_T varies as T^{-2} rather than as T^{-4}, as one might have expected from (6.5):

$$\kappa^{-1} = \alpha T^2 + \beta T^{-1} \tag{6.18}$$

where

$$\beta = \rho_i/L \tag{6.19}$$

Values of α as derived empirically by Hulm and by Mendelssohn and Rosenberg [7] for a number of metals are listed in Table 6.3. An outstanding discrepancy is that theoretical values of α are very much too large. A more complete discussion of conductivity at low temperatures is given in Sec. 6.

The reason that impurity scattering in metals is independent of temperature is that the velocity distribution of electrons in a degenerate gas does not vary much with temperature. In semiconductors in which the concentration is sufficiently small so that Boltzmann's statistics are applicable, the mean-square velocity is proportional to T. As shown in Sec. 5, impurity scattering may then vary rapidly with temperature.

Conduction in Alloys. Some alloys consist of a *heterogeneous mixture* of two or more distinct phases. An empirical rule for estimating the resistivity of such a mixture has been given by Lichtenecker [8]. Let ρ_1 and ρ_2 be the resistivities of two components whose volume concentrations are c_1 and c_2 (such that $c_1 + c_2 = 1$). The resistivity ρ of the mixture is then

$$\rho = \rho_1^{c_1} \cdot \rho_2^{c_2(1-ac_1)} \tag{6.20}$$

where a is a small number determined empirically for each case.

The resistivity of an alloy in which the components are mutually soluble, so as to form a single homogeneous phase, is usually much larger than that of either of the constituents. There is then a large temperature-independent contribution ρ_0 which comes from scattering by the random distribution of atoms. The total resistivity ρ is the sum of ρ_0 and the part due to thermal scattering ρ_T:

$$\rho = \rho_0 + \rho_T \tag{6.21}$$

The latter is similar to (6.11) and vanishes at $T = 0$.

In dilute solutions, the solute atoms may be considered to be impurities which scatter the conduction electrons of the solvent metal. Some examples are discussed in Sec. 5. Let x_A, x_B, etc., be the fractional atomic concentrations of solute atoms, A, B, etc. To the first order in the concentrations,

$$\rho_0 = x_A\rho_A + x_B\rho_B + \cdots \tag{6.22}$$

Values of ρ_A, ρ_B, etc., are much larger than ρ_T. Addition of 1 atomic per cent impurity may more than double the resistivity of a pure metal at room temperature. Exceptional purity is therefore required to get reliable values of the resistivities of pure metals. Scattering results from the *difference* in field of the solute and solvent atoms. Scattering tends to be large when the difference in valency is large.

Nordheim [9] has derived theoretically an approximate expression for the resistance of a binary solid solution. If x_A and $x_B = 1 - x_A$ are the fractional atomic concentrations of constituents A and B, he finds that, to a first approximation,

$$\rho_0 = Cx_A(1 - x_A) \tag{6.23}$$

Some examples (K-Rb, Pt-Pd, Ag-Au, In-Pb) in which the component metals are mutually soluble in all proportions are illustrated in Fig. 6.2.

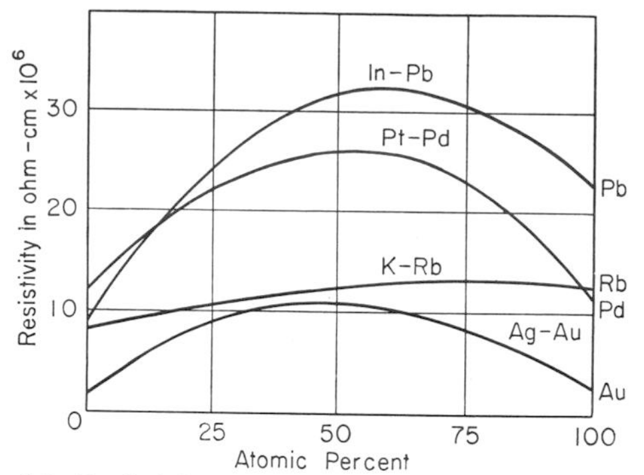

FIG. 6.2. Resistivity of K-Rb, Pt-Pd, Ag-Au, and In-Pb alloys at $T = 25°C$ as a function of atomic concentration.

Pronounced decreases in resistance are observed when an alloy becomes ordered. An example, which has been the subject of considerable study, is the copper-gold system. Ordered phases occur near the compositions CuAu and Cu_3Au if the alloy is annealed at a temperature below about 400°C. Figure 6.3 gives a plot of the resistivity of quenched and annealed alloys as a function of composition.

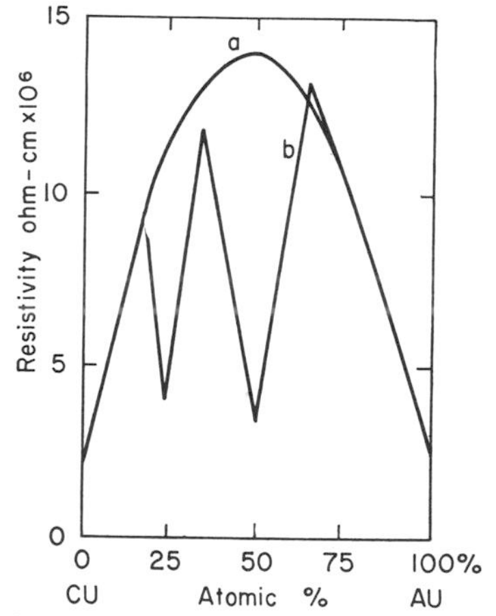

FIG. 6.3. Resistivity of Cu-Au alloys. (*a*) Quenched from 650°C. (*b*) Annealed at 200°C. [*Data from Johansson and Linde, after Bardeen, J. Appl. Phys.*, **11**: 88 (1940).]

The transition at the composition CuAu is not simply one of order-disorder. Copper and gold atoms tend to segregate on alternate planes. The spacing between these planes changes on ordering so that the structure changes from cubic to tetragonal. At the composition Cu_3Au, gold atoms tend to segregate on one of the four simple cubic lattices which form the face-centered cubic structure of the alloy.

The degree of long-distance order S in an order-disorder alloy has been defined by Bragg and Williams [10]: Designate sites for A atoms in the perfectly ordered alloy by A and sites for B atoms by B. Let n_A be the relative number of A sites and let p_A be the probability that an A site be occupied by an A atom. Then S is defined by

$$S = \frac{p_A - n_A}{1 - n_A} \tag{6.24}$$

Thus $S = 0$ in a random alloy, for which $p_A = n_A$, and $S = 1$ in a perfectly ordered alloy, for which $p_A = 1$. According to a theory of Muto [11], the resistance due to disorder should be proportional to $1 - S^2$.

Figure 6.4 is a plot of resistivity of annealed specimens of β brass (composition close to CuZn) as a function of temperature, together with plots of $1 - S^2$ according to theories of Bragg and Williams and of Bethe.

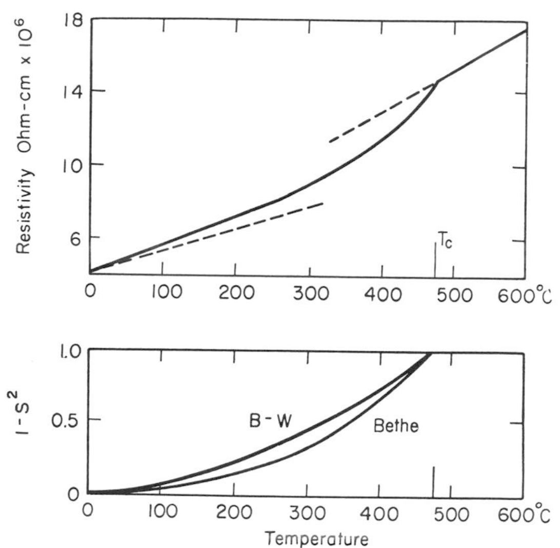

FIG. 6.4. Top: Resistance of β brass (51.25 atomic per cent Cu, 48.75 per cent Zn) as a function of temperature. Ordering sets in at about 475°C. Bottom: Plot of $1 - S^2$, where S is the long-distance order, according to theories of Bragg and Williams and of Bethe. [*After Bardeen, J. Appl. Phys.*, **11**: 88 (1940).]

Resistance measurements have proved to be a valuable tool for the study of order-disorder alloys. Such measurements are also useful for the study of precipitation-hardened alloys. One might expect a decrease in resistance as one of the constituents precipitates from a solid solution. However, one often observes an initial increase with precipitation, followed by a decrease. Mott has suggested that small clusters which occur in early stages of precipitation may be more effective in scattering than a random distribution.

2. Semiconductors

In a semiconductor, there is an energy gap between the normally filled levels of the valence band and the higher normally occupied levels of the conduction band. Current can be carried either by electrons in the conduction band or by holes (missing electrons) in the valence band. The concentrations of these carriers depend markedly on temperature and on impurities. At the absolute zero the ideal structure is an insulator. At elevated temperatures, electrons can be thermally excited from the valence band to the conduction band, giving *intrinsic* conductivity from equal numbers of conduction electrons and holes. More commonly, carriers are introduced by charged impurities, giving *extrinsic* conductivity, called *n-type* if conduction is primarily by conduction electrons (negative carriers) and *p*-type if conduction is primarily by holes (positive carriers).

Conductivities may be changed many orders of magnitude by changes in composition and temperature. It is therefore most convenient to express conductivity in terms of the concentrations n and p and the mobilities μ_n and μ_p of the conduction electrons and holes, respectively:

$$\sigma = \sigma_n + \sigma_p = ne\mu_n + pe\mu_p \tag{6.25}$$

The mobility is the magnitude of the drift velocity in unit electric field. Units are generally expressed as square centimeters per volt-second. If the concentration is in number per cubic centimeter and e in coulombs ($e = 1.6 \times 10^{-19}$ coulomb), σ is in $\text{ohm}^{-1}\ \text{cm}^{-1}$. The letters n and p refer to the sign of the carriers: $-e$ for conduction electrons, $+e$ for holes. Both types of carriers behave in most respects like particles with effective masses which may differ from but are generally of the same order of magnitude as the ordinary electron mass. There are exceptions, however. For example, the effective mass of conduction electrons in InSb is of the order of 10^{-2} m, and the mobility is abnormally large.

Impurities that are important in determining the conductivity are called donors and acceptors. Donors are impurities that may become positively ionized when introduced into the lattice; acceptors are those that may become negatively ionized. Both types have localized levels for electrons within the energy gap. As illustrated in Fig. 6.5, donors often have

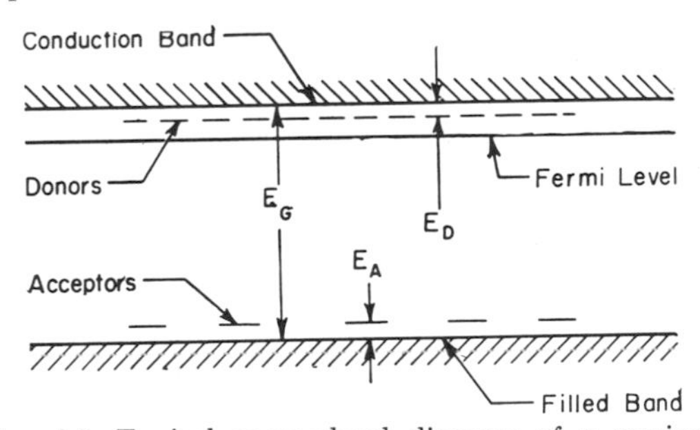

FIG. 6.5. Typical energy-level diagram of a semiconductor.

hydrogenlike levels with energies a little below the conduction band; the donor is neutral when the level is occupied and is positively ionized when empty. Correspondingly, acceptors often have levels with energies a little above the valence band and are negative ions when the level is occupied and neutral when unoccupied. The neutral site may be regarded as a negative ion with a bound hole. Double donors are those that may have a charge $+2e$; double acceptors may have a charge $-2e$. Some types of impurities have levels at intermediate positions in the gap. They are called traps if the probability of thermal release of an electron to the conduction band or of capture of an electron from the valence band, forming a hole, is small. There is no sharp distinction between traps and donors and acceptors. The term donor applies to any impurity that becomes positively ionized and acceptor to a negative impurity ion, regardless of the position of the level relative to the bands. Some impurities can exist in several different states of ionization.

In thermal equilibrium, the occupancy of the levels in the energy bands and impurities is determined by the Fermi level E_F. According to Fermi-Dirac statistics, the probability f_i that a level of energy E_i be occupied is

$$f_i = \frac{1}{1 + \exp\left[(E_i - E_F)/k_0T\right]} \tag{6.26}$$

The value of E_F is determined from the total number of electrons N:

$$N = \sum_i f_i \tag{6.27}$$

If impurity levels are degenerate or if excited states of these levels are important, the energy E_i should be an appropriate free energy. Usually N is such as to make the semiconductor electrically neutral.

The concentration of conduction electrons is obtained by summing (6.26) over the levels of the conduction band. When E_c, the energy of the lowest state of the conduction band, is above E_F, Boltzmann statistics may be used, and

$$n = N_c \exp\left[-(E_c - E_F)/k_0T\right] \tag{6.28}$$

where, for a single spherical band,

$$N_c = 2(2\pi m_n k_0 T/h^2)^{3/2}$$

Similarly, when the energy of the highest state of the valence band E_v is below E_F, Boltzmann statistics may be used for holes, and the concentration is

$$p = N_v \exp\left[-(E_F - E_v)/k_0T\right] \tag{6.29}$$

where $N_v = 2(2\pi m_p k_0 T/h^2)^{3/2}$. If there are degenerate bands, the total carrier concentrations are obtained by summing over the contributions from the different bands. Note that the product

$$\begin{aligned} np &= N_c N_v \exp\left[-(E_c - E_v)/k_0T\right] \\ &= N_c N_v \exp\left(-E_G/k_0T\right) \end{aligned} \tag{6.30}$$

is independent of the position of the Fermi level and thus of impurities. The only requirement for (6.30) is that Boltzmann statistics may be used for both n and p. Figure 6.6 is a plot of np as derived by Morin and Maita [12] from Hall and conductivity measurements for silicon.

In a semiconductor with $N^+{}_D$ donor ions/cm³ and $N^-{}_A$ acceptor ions/cm³, electrical neutrality requires that for singly ionized impurities

$$p - n + N^+{}_D - N^-{}_A = 0 \tag{6.31}$$

In an N-type semiconductor, $n \gg p$, and

$$n = N^+{}_D - N^-{}_A \tag{6.32}$$

In a P-type semiconductor, $p \gg n$, and

$$p = N^-_A - N^+_D \tag{6.33}$$

The semiconductor is N or P type, depending on whether donor or acceptor ions are in greatest abundance.

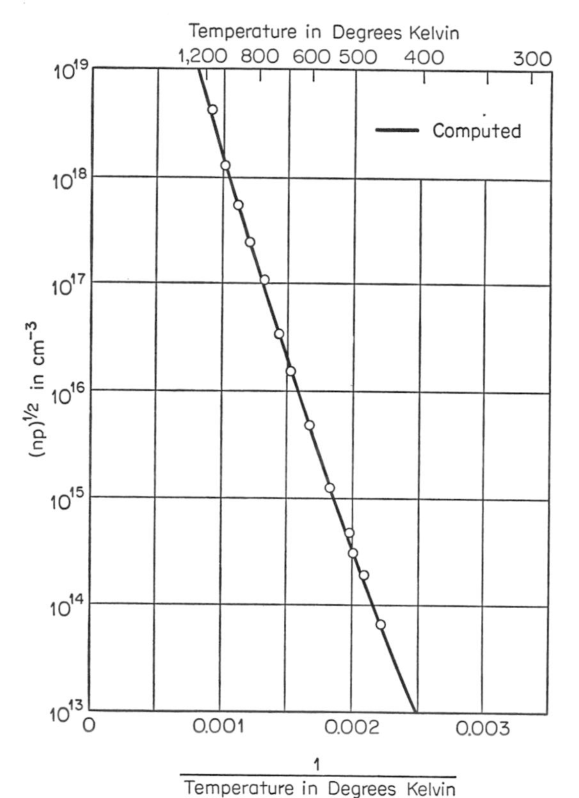

FIG. 6.6. Intrinsic carrier concentration $n_i = (np)^{1/2}$ in the intrinsic range of silicon as a function of the reciprocal absolute temperature compared with the theoretical curve. (*After Morin and Maita* [12].)

In an *intrinsic* semiconductor, the concentration of carriers introduced by thermal agitation is greater than the net concentration of impurity ions, so that

$$n, p \gg |N^+_D - N^-_A| \tag{6.34}$$

Electrical neutrality then requires that

$$n = p = n_i \tag{6.35}$$

where n_i, the intrinsic concentration, is equal to

$$n_i = (N_c N_v)^{1/2} \exp(-E_G/2k_0T) \tag{6.36}$$

The conductivity σ_n of an N-type semiconductor is equal to $ne\mu_n$. An additional measurement is required to determine n and μ_n separately. Estimates of n can be obtained from the Hall effect and, less accurately, from the thermoelectric power. The value of n may also be obtained from (6.32) if the concentrations of donor and acceptor ions are known. These concentrations are usually below the levels required for direct chemical analysis but have been estimated in suitably prepared specimens by radioactive tracer methods. Finally, the mobility may be determined directly by the Haynes-Shockley drift method [13].

Mobilities of carriers in several semiconductors and insulating crystals are listed in Table 6.4. In all cases, the samples are sufficiently pure so that the mobility is determined by lattice scattering.

Types of Semiconductors. There are two main classes of semiconductors: (1) valence crystals with bonding orbitals formed from atomic s and p electrons and (2) crystals in which the d shell of the metal ion is incomplete; these are mostly oxides of the transition metals.

We first discuss valence crystals, examples of which are as follows:

N type: ZnO, PbO_2, SnO_2, CdS, CdSe, Ag_2S, etc.

P type: Cu_2O, Cu_2S, Se, CuI, SnO_2, etc.

Either N or P: Si, Ge, Te, Pbs, PbTe, SiC, grey Sn, InSb, AlSb, etc.

Although we have called these valence crystals, the bonding in many is partially covalent and partially ionic. At high temperatures, the conductivity of most of these becomes intrinsic. Some (for example, Ge, Te, grey Sn) have been made sufficiently pure to be intrinsic at room temperature.

The nature of the imperfections or impurities responsible for the conductivity in some typical cases follows:

1. In ZnO, there is excess zinc in interstitial positions which can be easily ionized by thermal agitation to give conduction electrons.
2. Conduction in PbO_2 arises from excess lead; in fact, the composition can vary continuously from PbO_2 to PbO, giving material of very high conductivity at intermediate compositions.
3. Addition of 0.02 per cent Ga changes CdS from a good insulator to an N-type semiconductor with a resistivity of the order of 0.02 ohm-cm. The extra electron of trivalent gallium is freed for conduction. Replacement of cadmium by other trivalent ions, or even replacement of $S^=$ by Cl^-, also gives N-type conduction.
4. In Cu_2O there is excess oxygen, or, more accurately, a deficiency of copper, so that some copper-ion sites are vacant. Removel of a copper atom from the lattice requires removal of a copper ion and an electron from the valence band, leaving a hole. At low temperatures the hole is bound to the vicinity of the vacant lattice position, but at high temperatures it can be freed by thermal agitation to give P-type conductivity. The vacant copper-ion site gives an acceptor level about 0.3 ev above the top of the valence band.
5. Halogens act as acceptors in selenium.
6. Silicon and germanium, each with four valence electrons, have crystal structures similar to diamond, in which each atom is surrounded by four others with which it forms covalent bonds. Atoms from the fifth column of the periodic table (P, Sb, As) in substitutional positions have one more electron than is required for the valence bonds and act as donors. The fifth electron is attracted by the excess charge on the ion and forms a bound state with a hydrogen-like orbit at low temperatures. The ionization energy, decreased by the square of the dielectric constant, is so small that practically all such donors

TABLE 6.4. ENERGY GAPS AND MOBILITY OF CARRIERS FOR VARIOUS MATERIALS

Material	Energy gap ΔE, ev at T	Temperature T°K	Mobility at T μ (cm²/volt-sec)		Temperature dependence of mobility	
			Electron	Hole	Electron	Hole
Diamond[a]	7.2	300	1,800	1,200		
Si[b]	1.12	300	1,350	480	$2.1 \times 10^9 T^{-2.5}$	$2.3 \times 10^9 T^{-2.7}$
Ge[c]	0.67	300	3,900	1,900	$3.5 \times 10^7 T^{-1.6}$	$9.1 \times 10^8 T^{-2.3}$
AlSb[d]	1.6	300	>60	400	$\sim T^{-1.5}$	
GaAs[d]	1.43	300	8,500	435		$\sim T^{-2}$
GaP[d]	2.24	300	>130	150	$\sim T^{-1.5}$	$\sim T^{-1.5}$
InP[d]	1.26	300	4,500	150	$\sim T^{-2}$	$\sim T^{-2.4}$
GaSb[d]	0.7	300	2,500	1,420		
InAs[d]	0.36	300	27,000	450	$\sim T^{-1.5}$	$\sim T^{-2}$
InSb[d]	0.18	300	77,000	700	$\sim T^{-1.66}$	$\sim T^{-1.8}$
Te[e]	0.38	300	1,170	~560	$\sim T^{-1.5}$	$\sim T^{-1.5}$
Cu_2O[f]	1.8	300		70		10 exp (665/T)
ZnO[g]	3.2	300	160			
CdS[h]	2.4	300	210			
CdTe[i]	1.5	300	1,050	50		
PbS[e]	0.37	300	600	~250	$T^{-2.5}$	$T^{-2.5}$
PbTe[e]	0.32	300	2,100	~1,000	$T^{-2.5}$	$T^{-2.5}$
Mg_2Ge[j]	0.74	300	530	110		
Mg_2Sn[j]	0.36	300	210	150	$\sim T^{-2.3}$	
NaCl[k]	9.0	84	250		5.5 exp (370/T)	
KCl[l]	8.5	77	100		3.6 exp (300/T)	
KBr[m]	7.5	50	750		8.7 exp (244/T)	
KI[m]	6.5	77	200		10 exp (222/T)	
AgCl[n]	3.2	50	2,600		~exp (225/T)	
AgBr[o]	2.7	50	1,700		~exp (195/T)	
AgBr[o]		300	60	1.7		

[a] A. G. Redfield, *Phys. Rev.*, **94**: 526 (1954).
[b] G. W. Ludwig and R. L. Waters, *Phys. Rev.*, **101**: 1699 (1956).
[c] M. B. Prince, *Phys. Rev.*, **92**: 681 (1953).
[d] O. Madelung, "Physics of III-V Compounds," Wiley, New York, 1964, and references therein.
[e] R. A. Smith, "Semiconductors," Cambridge University Press, New York, 1959, and references therein.
[f] W. H. Brattain, *Revs. Mod. Phys.*, **23**: 203 (1951).
[g] A. R. Hutson, *Phys. Rev.*, **108**: 222 (1957).
[h] F. A. Kröger, H. J. Vink, and J. Volger, *Physica*, **20**: 1095 (1954).
[i] B. Segall, M. R. Lorenz, and R. E. Halstead, *Phys. Rev.*, **129**: 2471 (1963).
[j] J. M. Whelan, in N. B. Hannay (ed.), "Semiconductors," Rheinhold, New York, 1959, and references therein.
[k] A. G. Redfield, *Phys. Rev.*, **94**: 537 (1954).
[l] F. C. Brown and N. Inchauspé, *Phys. Rev.*, **121**: 1303 (1961).
[m] R. K. Ahrenkiel and F. C. Brown, *Phys. Rev.*, **136**: A223 (1964).
[n] T. Masumi, R. K. Ahrenkiel, and F. C. Brown, *Phys. Stat. Sol.*, **11**: 188 (1965).
[o] R. C. Hanson, *J. Phys. Chem.*, **66**: 2376 (1962).

are ionized at room temperature. In a similar way, atoms from the third column (B, Al, Ga) act as acceptors. Many other impurities and imperfections act as donors or acceptors with larger ionization energies. Figure 6.7 is a plot of the resistivity of several germanium specimens with varying concentrations of Sb.

7. The lead compounds, PbS, PbSe, and PbTe, have been studied extensively because of their utility as infrared detectors. While the conductivity is sensitive to deviations from stoichiometric composition (excess Pb gives N-type and excess S, P-type conductivity), impurities are also important. Monovalent ions, Ag^+ and Cu^+ substituted for Pb^{++}, give P-type conductivity, as does substitution of $S^=$ by As^{3-}. Trivalent ions, such as Bi^{3+}, Ga^{3+}, or In^{3+} substituted for Pb^{++}, or monovalent ions, such as Cl^- for $S^=$, give N-type conductivity.

8. The intermetallic compounds formed from third- and fifth-column elements, such as InSb and AlSb,

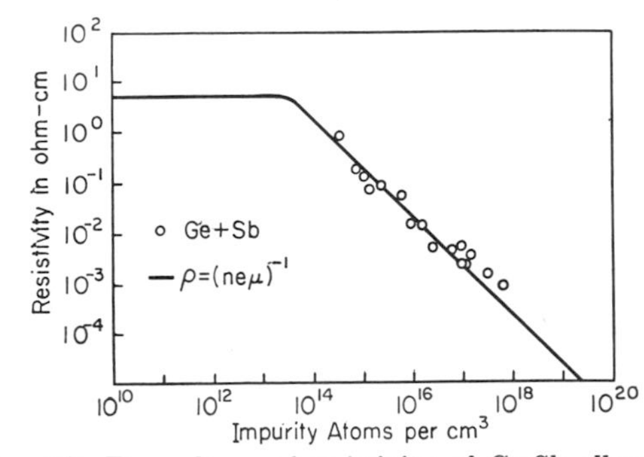

FIG. 6.7. Dependence of resistivity of Ge-Sb alloys on antimony concentration. (*Data from G. L. Pearson.*)

have a structure similar to diamond, in which each atom is surrounded tetrahedrally by four of the opposite type. Second-, fourth-, and sixth-column elements in substitutional positions act as donors or acceptors [14].

The impurity concentration required to affect the conductivity of valence crystals is very small. The mobility of the carriers is, in general, of the order of 10^2 to 10^3 cm^2/volt-sec, so that a concentration of only 10^{16} per cm^3, or 1 in 10^6, gives a conductivity of the order of 0.1 to 1 ohm^{-1} cm^{-1}. The energy-band picture gives a reasonably good account of conduction phenomena.

Conduction by d electrons is less well understood. Mobilities are so small, generally less than 1 cm^2/volt-sec, that the estimated mean free path is of the order of an interatomic distance or less. Impurity concentrations of the order of 1 part in 10^1 are required to produce substantial changes in conductivity. It is perhaps better to think of *d*-band conduction as a classical picture of exchange of electrons between neighboring atoms. Conduction occurs when ions of different valency (for example, Fe^{++}, Fe^{3+}) occupy equivalent lattice positions.

Examples of *d*-band semiconductors are NiO, Fe_2O_3, Fe_3O_4, MnO, TiO_2, $BaTiO_3$, WO_3, many spinels, perovskites, etc. Some are N type, some P type, and some can be either N or P type, depending on composition.

Let us first consider NiO, which has a cubic structure similar to NaCl. According to the band picture, one might expect NiO to have metallic conductivity, because the *d* band of the Ni ion is incompletely filled with 8 out of 10 possible *d* electrons per ion. Actually, NiO is a semiconductor with a very high resistivity ($\sim 10^{12}$ ohm-cm at room temperature for pure material with stoichiometric composition). Conductivity can be increased by heating in oxygen or by adding Li as an impurity [15]. With the first treatment, there is excess oxygen, and some Ni sites become vacant. Charge balance is maintained by changing two Ni^{++} ions to Ni^{3+}. At low temperatures, the Ni^{3+} ions will be located near the vacant site, but at high temperatures they will be distributed at random in the lattice. A Ni^{3+} ion can be considered to be a hole in a normal Ni^{++} configuration; conductivity resulting from motion of the hole (by transfer of an electron from a neighboring ion) is P-type. When Li^+ replaces Ni^{++}, another Ni^{++} is changed to Ni^{3+} to maintain charge balance. This also results in P-type conductivity, with one hole per added Li ion at high temperatures. An intrinsic conductivity with an activation energy of about 0.96 ev is also observed. This presumably results from thermal excitation removing electrons from Ni^{++} ions to create Ni^+ and Ni^{3+} ions, with resulting excess electron and hole conduction. Figure 6.8 gives a plot of conductivity vs. reciprocal temperature for several NiO samples as observed by Morin [16].

Another interesting case is Fe_2O_3, which also has been studied by Morin [17]. When Fe_2O_3 is heated in oxygen, to produce vacant Fe^{3+} sites, P-type conduction is observed. Some Fe^{3+} ions are then converted to Fe^{4+} to maintain charge balance. Titanium is an impurity which produces N-type conductivity. When a Ti^{4+} ion replaces Fe^{3+}, another Fe^{3+} is charged to Fe^{++}. The Fe^{++} ion acts as an excess electron in the normal Fe^{3+} configuration.

Spinels with the chemical formula AB_2O_3 form an interesting class of substances which have been investigated particularly by Verwey and associates [15]. The oxygen atoms form a close-packed cubic structure with 32 oxygen atoms in the unit cell. Metal ions occupy interstitial sites of two different types in the oxygen lattice. Eight *A* sites have a tetrahedral arrangement of oxygens as near neighbors, and 16 *B* sites are each surrounded by an octahedral arrangement of six oxygens. Ferrites are a special class of spinels in which two-thirds of the metal ions are iron. The formula for a normal ferrite is

$$M^{++}(Fe^{3+})_2O_4^{=}$$

The formula for an inverted ferrite is

$$Fe^{3+}(M^{++}Fe^{3+})O_4^{=}$$

Other spinels may also become inverted, with trivalent ions occupying tetrahedral positions. The trivalent ions which prefer tetrahedral positions and so tend to form inverted spinels are Fe^{3+}, In^{3+}, and Ga^{3+}. However, Zn^{++} and Cd^{++} have an even greater preference for the tetrahedral positions and so form normal spinels even when combined with Fe^{3+}, In^{3+}, and Ga^{3+}.

From their study, Verwey [15] and associates have found the following empirical rule for obtaining conduction: There must be an irregular distribution of ions of the same element but with charge differing by one unit on equivalent lattice positions. Some examples which illustrate applications of this rule follow:

1. Magnetite, Fe_3O_4, with an inverted spinel structure $Fe^{3+}(Fe^{++}, Fe^{3+})O_4$ is a fairly good conductor ($\sim$200 ohm^{-1} cm^{1}). The Fe^{++} and Fe^{3+} ions are irregularly distributed over the octahedral sites. There is a phase change at 150°K from a cubic to a rhombohedral structure, which is associated with a large increase in resistivity. It has been suggested that the Fe^{++} and Fe^{3+} ions become ordered at this temperature. The picture cannot be taken too literally, however, because the ordering energy is much less than would be computed for a purely ionic structure.

2. The resistance increases markedly when a mixed crystal is formed from Fe_3O_4, so that Fe^{3+} ions are replaced by Al^{3+}. At the 50–50 composition, the structure is $Fe^{3+}(Fe^{++}Al^{3+})O_4$, and the resistivity is of the order of 10^5 larger than Fe_3O_4. The Fe^{3+} and Fe^{++} ions are now on different sites, and the possibility of electron exchange is greatly reduced.

3. Magnesium ferrite, $Fe^{3+}(Mg^{++}, Fe^{3+})$, and zinc ferrite, $Zn^2(Fe^{3+}, Fe^{3+})O_4$, both have high resistivity. In the former the divalent and trivalent ions on octahedral sites come from different elements; in the latter the ions have the same charge.

4. Manganese oxide, with the structure

$$Mn^{++}(Mn^{4+}Mn^{++})O_4$$

has a high resistivity ($\sim 10^7$ ohm-cm). The charge of the manganese ions in the octahedral positions differs by two units, and this makes exchange of electrons difficult.

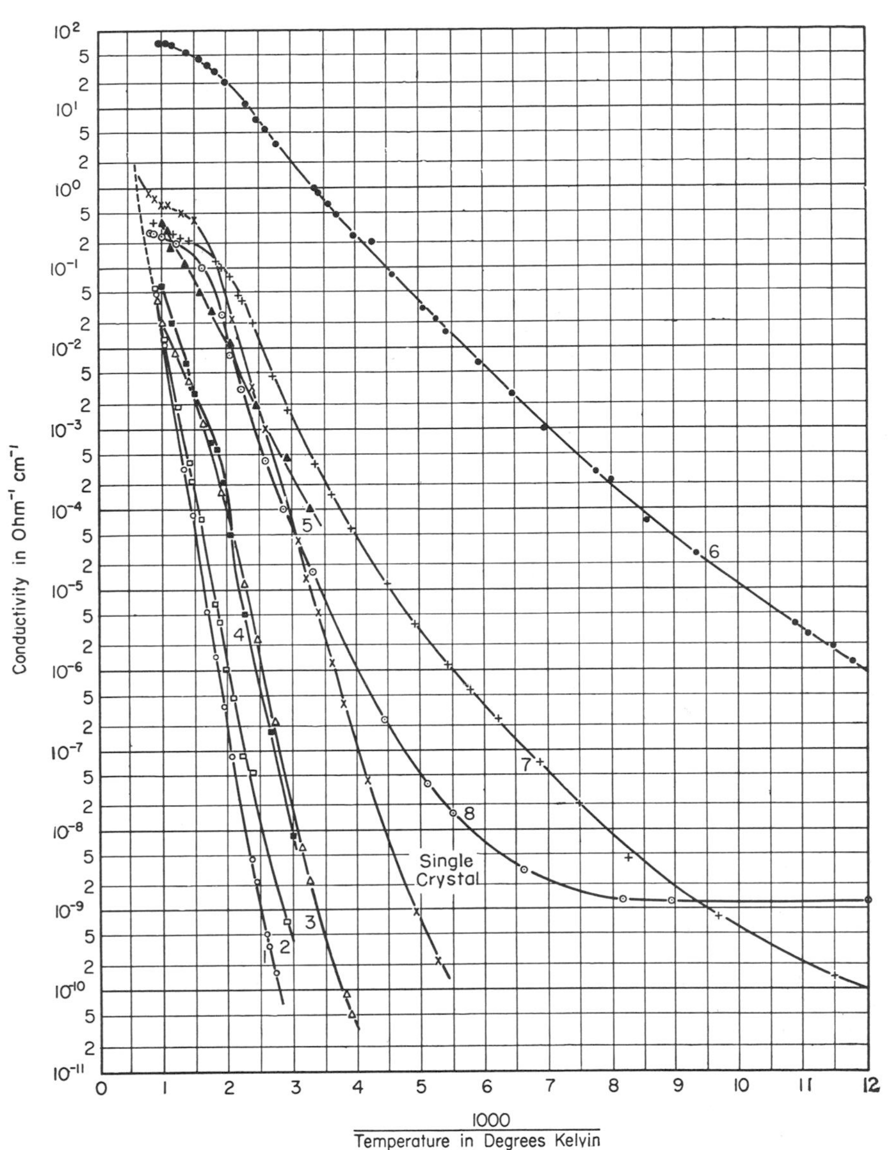

FIG. 6.8. Conductivity of nickel oxide samples as a function of reciprocal absolute temperature. Samples 4, 5, and 6 have increasing amounts of added Li; samples 7 and 8 were heated in an oxidizing atmosphere and have vacant Ni^{++} sites. (*After Morin* [16].)

5. The conductivity of $ZnFe_2O_4$ can be increased by adding titanium as an impurity. The titanium ion goes in as Ti^{4+} to maintain electrical neutrality, and an Fe^{3+} ion is changed to Fe^{++}. The excess electron on Fe^{++} may be transferred, giving N-type conductivity.

6. Jonker and Van Santen [18] have investigated conduction in compounds with composition

$$(La_{1-x}A_x)MnO_3$$

where A represents Ca, Sr, or Ba. These form the perovskite structure, which is cubic with Mn at the corners, A in the body centers, and 0 at the centers of the cube edges. Good electrical conductivity is found for $0.2 < x < 0.4$, and in the same range the compounds are ferromagnetic. For $x = 0$, the ionic formula is $La^{3+}Mn^{3+}O_3^{=}$. If some La^{3+} ions are replaced by a divalent ion, an equal number of Mn^{3+} are converted to Mn^{4+}. Conduction then results from exchange of electrons among the d shells of the Mn^{3+} and Mn^{4+} ions. Zener [19] has suggested that this exchange takes place through the oxygen ions by a process called *superexchange.* This can take place only if the spins of the d shells are parallel, and thus the conductivity is highest when there is ferromagnetism.

It is evident from the discussion of both valence

and d-band semiconductors that charge balance can be maintained in several ways when an ion is removed to form a vacancy or is replaced by an ion of different valency. If the valency of the additive is one greater than that of the normal anion, there is an extra positive charge which can be compensated by one of the following:

1. Creation of vacant cation sites.
2. Electrons in the conduction band or excess electrons on the normal constituent. In both cases N-type conductivity results. The latter has been called *controlled* valency, since the additive controls the valency of the normal ion.
3. Excess electrons on the additive, to reduce its valency to that of the normal constituent. This process, discovered by Selwood, is called *induced* valency.

In a similar way, if the valency of the additive is one less than that of the normal anion, charge can be compensated by creation of vacant anion sites, by holes, or by an increase by one of its normal valence. Similar rules also apply to additives which substitute for cations.

Oxides of the transition elements have found wide application in thermistors (thermally variable resistors). Ferrites are used for magnetic material of high resistivity.

Effect of Stress. Both the mobility and concentration of carriers may change with pressure. Concentration changes in an intrinsic semiconductor result from a shift in the energy gap with pressure. Since the resistance in the intrinsic region may be written

$$R = R_\infty \exp (E_G/2k_0T) \tag{6.37}$$

where R_∞ varies slowly with temperature, we may write

$$\frac{d \log R}{dp} = \frac{d \log R_\infty}{dp} + \frac{1}{2k_0T}\frac{dE_G}{dp} \tag{6.38}$$

The first term includes effects of mobility and effective mass variations and the second, the change in concentration from the shift in energy gap.

An interesting example is tellurium, which has been studied by Bridgman [20]. The resistivity decreases by a factor of about 600 at a pressure of 30 kilobars. From measurements made at two different temperatures, the energy gap at different pressures may be estimated from (6.37), with results shown in Fig. 6.9.

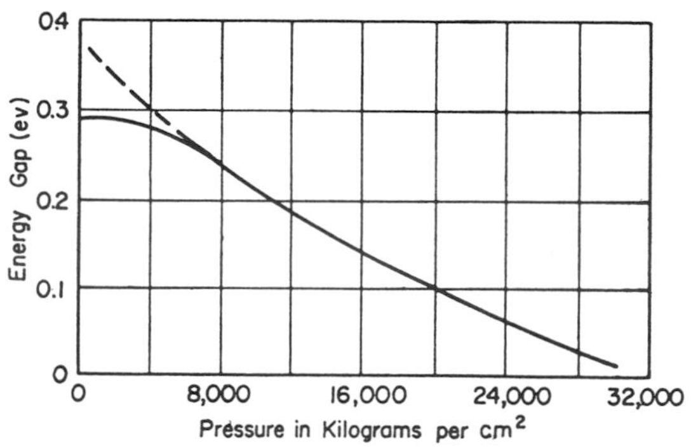

Fig. 6.9. Energy gap of tellurium as a function of pressure. [*After Bardeen, Phys. Rev.*, **75**: 1777 (1949).]

Bridgman's sample was apparently intrinsic at pressures above about 8 kilobars. The shift in energy gap is sufficient to account almost completely for the change in resistance with pressure. At somewhat higher pressures (about 40 kilobars), a phase transition to a close-packed metallic structure is observed.

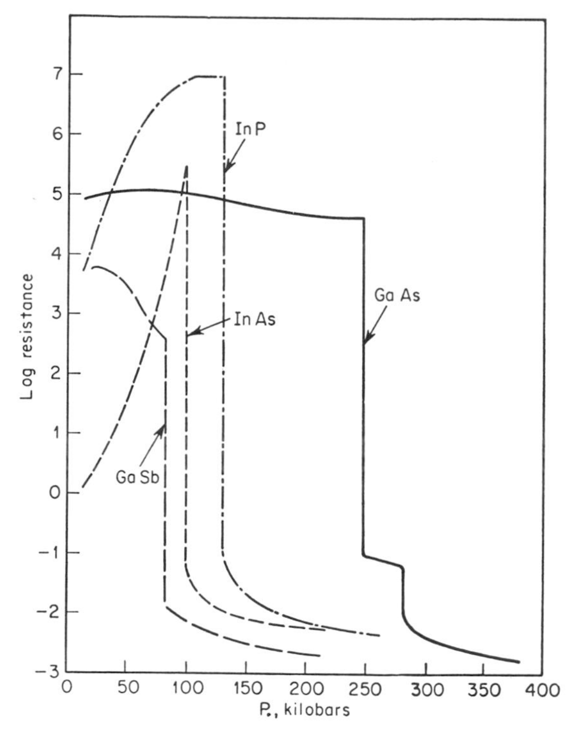

Fig. 6.10. Resistance versus pressure for several III-V compounds. After Drickamer [6].

It is found rather generally that with increasing pressure metallic structures are preferred over more open valence-bond structures. For example, there is an abrupt drop in the resistivity of germanium of several orders of magnitude at a pressure of about 120 kilobars and of silicon at about 190 kilobars.

Table 6.5. Values of the Piezoresistance Coefficients in Units of 10^{-12}cm²/dyne
Resistivities are in ohm-centimeters

Type	Material	Resistivity	Π_{44}	$\Pi_{11} - \Pi_{22}$	$-\frac{1}{P}\frac{\delta\rho}{\rho} = \Pi_{11} + 2\Pi_{12}$	
					Adiabatic	Isothermal
N	Ge	1.5	−138.1	+1.0	−8.7	−7.0
N	Ge	5.7	−136.8	+1.2	−10.5	−8.8
N	Ge	9.9	−137.9	+0.3	−14.7	−13.0
N	Ge	16.6	−138.7	+0.3	−16.2	−17.9
P	Ge	1.1	+96.7	−6.9	+2.7	+5.2
P	Ge	15.0	+98.6	−15.6	−0.6	+1.9
N	Si	11.7	−13.6	−155.6	+4.6	+5.7
P	Si	7.8	+138.1	+7.7	+4.4	+6.0

Similar changes to metallic structures have been observed in the related III–V compounds such as GaSb, InP, and GaAs and in II–VI compounds such as ZnS, ZnSe, and ZbTe. A plot of resistivity vs. pressure for several III–V compounds is given in Fig. 6.10, based on the work of Drickamer [6].

The complete piezoresistance coefficients for germanium and silicon have been determined by Smith [24] by measuring the longitudinal and transverse resistance of single crystal rods subject to tension. Just as for metals, there are three independent coefficients for a cubic crystal: a pressure coefficient, $\Pi_{11} + 2\Pi_{12}$, and two shear coefficients, Π_{44} and $\Pi_{11} - \Pi_{12}$. Their values as obtained by Smith are given in Table 6.5. For each sample studied, The large shear coefficients is small and one large. The large shear coefficients are at least an order of magnitude larger than expected from the effect of strain on thermal waves. A mechanism to explain this large anisotropy of the shear coefficients in terms of nonspherical energy surfaces has been advanced by C. Herring.

3. Thermoelectric and Transverse Effects

Thermoelectric effects occur when both electric and thermal currents are present. The so-called transverse effects occur when there is, in addition, a magnetic field normal to these currents. For simplicity, we assume in the following that the material is isotropic. We first discuss the thermoelectric effects and then show how the results may be generalized to include a transverse magnetic field.

There are three thermoelectric effects. The *Seebeck effect* (discovered in 1822) relates to the emf developed in a circuit made up of different conducting elements, not all of whose contacts are at the same temperature. The *Peltier effect* (discovered in 1834) refers to the reversible heating or cooling which occurs at a contact when current flows from one conductor to another. The *Thomson effect* (discovered by Lord Kelvin in 1856) refers to the reversible heat absorption which occurs when an electric current flows in a homogeneous conductor in which there is a temperature gradient. All three effects are related by thermodynamics. If one is known, the other two can be derived.

Seebeck Effect. Consider a circuit consisting of two different metals A and B; the temperatures of the two junctions will be designated T_1 and T_2, and the thermal electromotive force by $E_{AB}(T_1,T_2)$. The sense of E is conventionally defined to be positive if the sense of flow of current (direction of flow of positive charge) is from A to B at the junction of temperature T_2. Experiment shows that the thermal emf between all possible pairs of metals satisfies the following rule of combination:

$$E_{AB}(T_1,T_2) = E_{AX}(T_1,T_2) - E_{BX}(T_1,T_2) \tag{6.39}$$

Therefore, if the thermal emf of several materials A, B, . . . with respect to some standard reference material X is known, the thermal emf of any pair AB is equal to the emf for the pair AX minus that for the pair BX. The thermal emf between three temperatures T_1, T_2, T_3 also satisfies the relation

$$E_{AB}(T_1,T_3) = E_{AB}(T_1,T_2) - E_{AB}(T_3,T_2) \tag{6.40}$$

These two properties of $E_{AB}(T_1,T_2)$ show that the thermal emf $E_{AB}(T_1,T_2)$ can be written as the difference with regard to temperature and with regard to material of a single function of temperature for each material. Thus it is possible to represent the experimentally observed emfs for all pairs of materials and for all pairs of junction temperatures:

$$\begin{aligned} E_{AB}(T_1,T_2) &= [E_A(T_2) - E_A(T_1)] \\ &\qquad - [E_B(T_2) - E_B(T_1)] \\ &= [E_A(T_2) - E_B(T_2)] \\ &\qquad - [E_A(T_1) - E_B(T_1)] \end{aligned} \tag{6.41}$$

The functions $E_A(T)$ and $E_B(T)$ are functions of the temperature that are characteristic of each of the materials A and B. This gives the simplest possible formal representation of the thermal emf of a composite circuit. The function $E_A(T)$ is called the *absolute thermal electromotive* force of material A.

If the only data available are the thermal emfs of all pairs of metals as functions of the junction temperatures, it is not possible to derive, from such data alone, definite values of the functions $E_A(T)$ for each metal. For if $E_A(T)$ for various A and various T represent the given circuit emfs in the form (6.41), the same data could equally well be represented by adding the same arbitrary function $\phi(T)$ to the $E_A(T)$ for each material.

In particular, one could choose arbitrarily to make $E_X(T) = 0$ for some particular reference material X. This is the way thermoelectric data are usually presented in tables. Although this provides a convenient way to present data on thermoelectric emfs, it appears from the thermodynamic theory that the function $E_A(T)$ is a fundamental property of each material, which can be uniquely defined in terms of other physical measurements. If this function $E_X(T)$ is carefully measured for one material X, it then becomes possible to find the absolute functions $E_A(T)$ for any other materials from data on the thermal emfs of circuits in which this material is paired with X.

The reversible heat quantities are rather small and hence somewhat difficult to measure. Peltier found calorimetrically that the quantity of heat absorbed or given out at a junction is proportional to the total charge passing the junction. The proportionality factor is known as the *Peltier coefficient.* If Q is the heat absorbed when a charge q passes the junction, the Peltier coefficient Π_{AB} is measured in volts if Q is measured in joules and q in coulombs:

$$Q = \Pi_{AB}(T)q \tag{6.42}$$

The notation indicates that the Peltier coefficient depends on the temperature. The arbitrary sign convention is made that Π_{AB} is called positive if heat is absorbed when a positive charge passes from A to B. It follows that $\Pi_{BA}(T) = -\Pi_{AB}(T)$. It is also found experimentally that

$$\Pi_{AB}(T) = \Pi_{AX}(T) - \Pi_{BX}(T) \tag{6.43}$$

analogous to (6.39), and, therefore, it follows that the Peltier coefficient for a junction AB is representable as the difference of two intrinsic Peltier coefficients characteristic of each material separately:

$$\Pi_{AB}(T) = \Pi_A(T) - \Pi_B(T) \tag{6.44}$$

In contrast to the Peltier effect which occurs at a junction, the *Thomson effect* occurs in a homogeneous conductor. Experimentally, the rate at which heat is absorbed Q (joules) is equal to a *Thomson coefficient* $\sigma_{TA}(T)$, which depends on the material A and on the absolute temperature T multiplied by the temperature gradient dT/dx and by the charge q (coulombs) which passes:

$$Q = \sigma_{TA}(T)q(dT/dx) \tag{6.45}$$

The Thomson coefficient is reckoned positive if heat is absorbed when the flow of positive current is toward the warmer end of the wire (flow up the temperature gradient). The unit of the Thomson coefficient in (6.45) is volt per degree. Its actual magnitude is always of the order of a few microvolts per degree.

Application of the laws of thermodynamics to the circuit consisting of A, B at junction temperatures T_1 and T_2 permitted Kelvin to derive some important relations between $E_A(T)$, $\Pi_A(T)$, and $\sigma_{TA}(T)$.

When unit charge passes around the circuit so as to pass from A to B at the junction whose temperature is T_2, the electrical work delivered as output is $E_{AB}(T_1,T_2)$. This must, by the first law of thermodynamics, equal the net heat absorption, so that

$$E_{AB}(T_1,T_2) = \Pi_{AB}(T_1) - \Pi_{AB}(T_2) + \int_{T_1}^{T_2} [\sigma_{TA}(T) - \sigma_{TB}(T)]\, dT \tag{6.46}$$

Since this is valid for any pair of metals, it follows that

$$E_A(T) = \Pi_A(T) + \int^T \sigma_{TA}(T)\, dT \tag{6.47}$$

An arbitrary lower limit can be used for the integral because of the arbitrariness in $E_A(T)$ and $\Pi_A(T)$.

By an application of the second law of thermodynamics to the reversible heat, Kelvin found that

$$\frac{\Pi_A(T)}{T} + \int_0^T \frac{\sigma_{TA}(T)}{T}\, dT = 0 \tag{6.48}$$

Strictly speaking, from the second law it follows only that the quantity on the left must be a constant, but the third law of thermodynamics gives the result that this constant is equal to zero. Boltzmann [25] has objected that the applicability of thermodynamics to the thermoelectric circuit is questionable, since the same charge carriers involved in these effects are involved simultaneously in the irreversible generation of Joule heat and in the flow of heat by conduction along the temperature gradient. As we shall see, the results can be derived from the theory of thermodynamics of irreversible processes.

Equations (6.47) and (6.48) thus give two relations between $E(T)$, $\Pi(T)$, and $\sigma_T(T)$ for any material by which the other two can be expressed in terms of one of them. For example,

$$E(T) = \int_0^T \sigma_T(T)\, dT - T\int_0^T \frac{\sigma_T(T)}{T}\, dT \tag{6.49}$$

In this equation, $\sigma_T(T)$ can be regarded as a specific heat of electricity, in which case the first integral is the internal energy $U(T)$ of the electricity involved in carrying the current, and the second integral $S(T)$ is its entropy. Then (6.49) indicates that the absolute thermoelectromotive force is equal to the free energy per unit charge passing around the circuit.

This discussion shows that $E(T)$ has a fundamental significance of some importance. It is difficult to get $E(T)$ directly from (6.49) because the measurement of $\sigma_T(T)$ is so difficult. However, if $E_X(T)$ is obtained carefully by measuring $\sigma_{TX}(T)$ for only one reference material X, then $E_A(T)$ for all other materials is obtained by the comparatively easy measurement of $E_{AX}(T_1,T_2)$ of the thermal emf of A paired with X in a circuit.

Such careful measurements of $\sigma_T(T)$ have been made by Borelius, Keesom, Johansson, and Linde [26] for the Ag-Au alloy which contains 0.37 atomic per cent Au. From their data on $\sigma_{TX}(T)$ one can calculate

$$S_X(T) = \int_0^T [\sigma_{TX}(T)/T]\, dT$$

and

$$U_X(T) = \int_0^T \sigma_{TX}(T)\, dT$$

and finally $E_X(T)$ for this reference material.

If now the thermal emf of any other material A with respect to X is known for the junction temperatures T_0 and T, then one can find $E_A(T) - E_A(T_0)$ by the relation

$$E_A(T) - E_A(T_0) = E_{AX}(T_0,T) + E_X(T) - E_X(T_0)$$

which gives $E_A(T)$ except for an undetermined constant, the determination of which requires a knowledge of $E_{AX}(T_0,T)$ down to absolute zero, since $E(T) = 0$ at $T = 0$ for all materials.

In many discussions the *thermoelectric power* is made to play a central role in the presentation of the subject. The thermoelectric power $\epsilon_A(T)$ of material A at temperature T is defined as

$$\epsilon_A(T) = \frac{dE_A(T)}{dT} \tag{6.50}$$

Similarly, the thermoelectric power of a pair of metals is defined as

$$\epsilon_{AB}(T) = \frac{dE_A(T)}{dT} - \frac{dE_B(T)}{dT}$$

so that $\epsilon_{AB}(T)\, dT$ is the emf of the circuit with junction temperatures $T_1 = T$ and $T_2 = T + dT$.

The thermodynamic relations show that the Peltier coefficient is related to the thermoelectric power as

$$\Pi_A(T) = \epsilon_A(T)T \tag{6.51}$$

and

$$\frac{d\epsilon_A(T)}{dT} = -\frac{\sigma_{TA}(T)}{T} \tag{6.52}$$

We shall now show that (6.51) and (6.52) can be derived from the theory of irreversible thermodynamics. To present the irreversible thermodynamical theory of the thermoelectric and transverse effect from a unified point of view [27], it is convenient to introduce the electrochemical potential μ of the electrons. The thermodynamic definition of μ is

$$\mu = \partial G/\partial N \tag{6.53}$$

where

$$G = U - TS = \mu N \tag{6.54}$$

is the free energy of the system and N is the total number of electrons. The PV term has been omitted, since it is usually negligible in solids. Thus we need not distinguish between Gibbs and Helmholtz free energies. The energy μ includes the electrostatic potential energy of the electrons. There is some arbitrariness in the definition of μ, partly from the arbitrary origin for electrostatic energy and partly from the fact that only entropy differences are significant. The latter implies that one could add a specific entropy s_0 and thus a term s_0T to the values of μ for all metals without affecting any results. The value of s_0 can be specified by the third law, which requires that $S \to 0$ as $T \to 0$.

In the band picture, μ is just the Fermi energy of the electrons, E_F, which occurs in the Fermi-Dirac distribution law and which is presumed to include the electrostatic potential energy.

The value of μ is the same everywhere in a conductor in thermal equilibrium with no current flow. If two such conductors are in electrical contact, the values of μ are the same in both. It is often convenient to express μ as a potential rather than an energy, so that we define ϕ by the equation

$$\mu = -e\phi \tag{6.55}$$

Differences in ϕ between two conductors or between two positions on the same conductor correspond to differences in voltage as read on a voltmeter. Particularly for semiconductors, in which there can be large differences in concentration, differences in ϕ are *not* the same as differences in electrostatic potentials. However, in a homogeneous material with no temperature or concentration gradients, the electric field $\boldsymbol{\mathcal{E}}$ is given by the gradient of ϕ:

$$\boldsymbol{\mathcal{E}} = -\operatorname{grad}\phi \tag{6.56}$$

In general, we may regard grad ϕ as a voltage gradient.

Ohm's law may be written

$$\mathbf{j} = -e\mathbf{J} = \sigma\boldsymbol{\mathcal{E}} = -\sigma\operatorname{grad}\phi \tag{6.57}$$

Here $\mathbf{J}$ is the particle-current density.

The heat-current density $\mathbf{w}$ may be defined by

$$\mathbf{w} = T\mathbf{S} \tag{6.58}$$

where $\mathbf{S}$ is the entropy flux. Since

$$T\,\delta S = \delta U - \mu\,\delta N \tag{6.59}$$

it follows that

$$\mathbf{w} = T\mathbf{S} = \mathbf{W} - \mu\mathbf{J} = \mathbf{W} - \phi\mathbf{j} \tag{6.60}$$

where $\mathbf{W}$ is the total-energy flux.

To discuss thermal conductivity and the thermoelectric effects, we suppose that both potential and thermal gradients are present. There are linear relations between currents and gradients. In general, one might expect four independent coefficients in a matrix which connects the two currents with the two gradients. The number can, however, be reduced to three by an application of Onsager's relations. These three coefficients correspond to the electric and thermal conductivities and to the absolute thermoelectric power.

Onsager [28] has shown that, in the absence of a magnetic field, the matrix of coefficients is symmetric if the gradients are chosen appropriately. The gradients can be considered to be generalized forces F_i, which give rise to currents C_j. The linear relation between them may be written

$$C_j = \Sigma L_{ji}F_i \tag{6.61}$$

To apply Onsager's relations, the forces should be chosen in such a way that the rate of increase in entropy density as a result of the irreversible processes can be expressed in the form

$$\partial S/\partial t = \operatorname{div}\mathbf{S} = \Sigma F_iC_i \tag{6.62}$$

Onsager's theorem states that the matrix is then symmetric:

$$L_{ij} = L_{ji} \tag{6.63}$$

When a magnetic field H is present, the relation becomes

$$L_{ij}(H) = L_{ji}(-H) \tag{6.64}$$

To apply the theory to the present problem, we use (6.60) to derive

$$\operatorname{div}\mathbf{S} = \mathbf{W}\cdot\operatorname{grad}(1/T) - j\cdot\operatorname{grad}(\phi/T) \tag{6.65}$$

We have made use of the fact that

$$\operatorname{div}\mathbf{W} = \operatorname{div}\mathbf{j} = 0$$

under steady-state conditions. If we take the currents to be $\mathbf{w}$ and $\mathbf{j}$, we may replace $\mathbf{W}$ by $\mathbf{w} + \phi\mathbf{j}$ and obtain

$$\operatorname{div}\mathbf{S} = \mathbf{w}\cdot\operatorname{grad}(1/T) - (\mathbf{j}/T)\cdot\operatorname{grad}\phi \tag{6.66}$$

The appropriate generalized forces are therefore grad $(1/T)$ and $-(1/T)$ grad ϕ.

If the gradients are in the x direction,

$$j_x = L_{11}\left(-\frac{1}{T}\frac{\partial\phi}{\partial x}\right) + L_{12}\frac{\partial}{\partial x}\left(\frac{1}{T}\right) \tag{6.67}$$

$$w_x = L_{21}\left(-\frac{1}{T}\frac{\partial\phi}{\partial x}\right) + L_{22}\frac{\partial}{\partial x}\left(\frac{1}{T}\right) \tag{6.68}$$

By absorbing a factor of $1/T$ in the definitions of the coefficients, we obtain

$$j_x = K_1\left(-\frac{\partial\phi}{\partial x}\right) + K_2\left(-\frac{1}{T}\frac{\partial T}{\partial x}\right) \tag{6.69}$$

$$w_x = K_2\left(-\frac{\partial\phi}{\partial x}\right) + K_3\left(-\frac{1}{T}\frac{\partial T}{\partial x}\right) \tag{6.70}$$

This form is most convenient to compare with the theory.

In most experiments, electric currents and thermal gradients are under the control of the experimenter, and, as pointed out particularly by Mazur and Prigogine [29], it is often desirable to take these as the independent variables. Solving (6.67) for the voltage gradient, we have

$$-\frac{\partial\phi}{\partial x} = \mathfrak{L}_{11}j_x + \mathfrak{L}_{12}\frac{\partial T}{\partial x} \tag{6.71}$$

$$w_x = -T\mathfrak{L}_{21}j_x + \mathfrak{L}_{22}\left(-\frac{\partial T}{\partial x}\right) \tag{6.72}$$

where

$$\mathfrak{L}_{11} = 1/K_1 = \rho \tag{6.73}$$
$$\mathfrak{L}_{21} = \mathfrak{L}_{12} = K_2/TK_1 = \epsilon = -(d\phi/dT)_{j_x=0} \tag{6.74}$$
$$\mathfrak{L}_{22} = (K_1K_3 - K_2^2)/TK_1 = \kappa \tag{6.75}$$

The coefficients $\mathfrak{L}_{11}$, $\mathfrak{L}_{12}$, and $\mathfrak{L}_{22}$ are equal to the resistivity ρ, the thermoelectric power ϵ, and the thermal conductivity κ, respectively.

To find the Thomson coefficient, we calculate the rate at which heat is absorbed. The net heat produced, and which must be absorbed by the conductor, is the difference between the rate at which electrical energy is supplied, $j_x\mathcal{E}_x$, and the rate at which heat flows out, $\partial w_x/\partial x$. This gives

$$Q = j_x(-\partial\phi/\partial x) - \partial w_x/\partial x \tag{6.76}$$

$$Q = \rho j_x^2 + Tj_x\frac{\partial\epsilon}{\partial T}\frac{\partial T}{\partial x} + \frac{\partial}{\partial x}\left(\kappa\frac{\partial T}{\partial x}\right) \tag{6.77}$$

The second term represents the Thomson heat, and (6.52) follows as an immediate consequence.

The various transverse effects can be treated in exactly the same way. We now suppose that electric and thermal currents flow in the x and y directions in a slab and that there is a magnetic field in the z direction, normal to the faces of the slab. The equations analogous to (6.67) and (6.68) for the electric- and thermal-current densities are then

$$j_x = L_{11}\left(-\frac{1}{T}\frac{\partial\phi}{\partial x}\right) + L_{12}\frac{\partial}{\partial x}\left(\frac{1}{T}\right) + L_{13}\left(-\frac{1}{T}\frac{\partial\phi}{\partial y}\right) + L_{14}\frac{\partial}{\partial y}\left(\frac{1}{T}\right)$$

$$w_x = L_{21}\left(-\frac{1}{T}\frac{\partial\phi}{\partial x}\right) + L_{22}\frac{\partial}{\partial x}\left(\frac{1}{T}\right) + L_{23}\left(-\frac{1}{T}\frac{\partial\phi}{\partial y}\right) + L_{24}\frac{\partial}{\partial y}\left(\frac{1}{T}\right)$$

$$j_y = L_{31}\left(-\frac{1}{T}\frac{\partial\phi}{\partial x}\right) + L_{32}\frac{\partial}{\partial x}\left(\frac{1}{T}\right) + L_{33}\left(-\frac{1}{T}\frac{\partial\phi}{\partial y}\right) + L_{34}\frac{\partial}{\partial y}\left(\frac{1}{T}\right)$$

$$w_y = L_{41}\left(-\frac{1}{T}\frac{\partial\phi}{\partial x}\right) + L_{42}\frac{\partial}{\partial x}\left(\frac{1}{T}\right) + L_{43}\left(-\frac{1}{T}\frac{\partial\phi}{\partial y}\right) + L_{44}\frac{\partial}{\partial y}\left(\frac{1}{T}\right) \tag{6.78}$$

The Onsager relations are

$$L_{ij}(H) = L_{ji}(-H) \tag{6.79}$$

In an isotropic material, x and y directions are equivalent. There are six independent coefficients which may be chosen to be L_{11}, L_{12}, L_{13}, L_{14}, L_{22}, and L_{24}. In analogy with (6.69) and (6.70), we shall denote L_{11}, L_{12}, and L_{22}, which are even in H, by TK_1, TK_2, and TK_3 and denote L_{13}, L_{14}, and L_{24}, which are odd in H, by TL_1, TL_2, and TL_3, respectively. The equations are then

$$j_x = K_1\left(-\frac{\partial\phi}{\partial x}\right) + K_2\left(-\frac{1}{T}\frac{\partial T}{\partial x}\right) + L_1\left(-\frac{\partial\phi}{\partial y}\right) + L_2\left(-\frac{1}{T}\frac{\partial T}{\partial y}\right)$$

$$w_x = K_2\left(-\frac{\partial\phi}{\partial x}\right) + K_3\left(-\frac{1}{T}\frac{\partial T}{\partial x}\right) + L_2\left(-\frac{\partial\phi}{\partial y}\right) + L_3\left(-\frac{1}{T}\frac{\partial T}{\partial y}\right)$$

$$j_y = -L_1\left(-\frac{\partial\phi}{\partial x}\right) - L_2\left(-\frac{1}{T}\frac{\partial T}{\partial x}\right) + K_1\left(-\frac{\partial\phi}{\partial y}\right) + K_2\left(-\frac{1}{T}\frac{\partial T}{\partial y}\right)$$

$$w_y = -L_2\left(-\frac{\partial\phi}{\partial x}\right) - L_3\left(-\frac{1}{T}\frac{\partial T}{\partial x}\right) + K_2\left(-\frac{\partial\phi}{\partial y}\right) + K_3\left(-\frac{1}{T}\frac{\partial T}{\partial y}\right) \tag{6.80}$$

It is this form which is most useful for comparison with theory.

These equations can be written in a compact manner by using a complex notation. If we define

$$\begin{aligned} \mathcal{J} &= j_x - ij_y \qquad \mathcal{W} = w_x - iw_y \\ \mathcal{E} &= \mathcal{E}_x - i\mathcal{E}_y = -\frac{\partial\phi}{\partial x} + i\frac{\partial\phi}{\partial y} \\ \mathcal{G} &= \mathcal{G}_x - i\mathcal{G}_y = -\frac{1}{T}\frac{\partial T}{\partial x} + i\frac{1}{T}\frac{\partial T}{\partial y} \\ \mathcal{K}_n &= K_n + iL_n \end{aligned} \tag{6.81}$$

we may express Eqs. (6.80) in the form

$$\begin{aligned} \mathcal{J} &= \mathcal{K}_1\mathcal{E} + \mathcal{K}_2\mathcal{G} \\ \mathcal{W} &= \mathcal{K}_2\mathcal{E} + \mathcal{K}_3\mathcal{G} \end{aligned} \tag{6.82}$$

If $\mathcal{J}$ and $\mathcal{G}$ are taken as independent variables, these become

$$\begin{aligned} \mathcal{E} &= \mathcal{K}_1^{-1}\mathcal{J} - \mathcal{K}_2\mathcal{K}_1^{-1}\mathcal{G} \\ \mathcal{W} &= \mathcal{K}_2\mathcal{K}_1^{-1}\mathcal{J} + (\mathcal{K}_3 - \mathcal{K}_2^2\mathcal{K}_1^{-1})\mathcal{G} \end{aligned} \tag{6.83}$$

The term $\mathcal{K}_2^2\mathcal{K}_1^{-1}$ is generally negligible in comparison with $\mathcal{K}_3$ and will be omitted in the following equations.

The definitions of the various thermomagnetic effects are listed below. In the isothermal effects, the transverse temperature gradient is zero. In the adiabatic effects, the transverse heat flow is zero. These differ slightly. The isothermal effects can be expressed most simply in terms of the $\mathcal{K}$'s as defined above, but it is often the adiabatic coefficients which are actually measured. In the following, Re represents the real and Im the imaginary parts.

1. *Isothermal electrical conductivity*

$$\sigma_i = j_x/\mathcal{E}_x \qquad \text{with } \mathcal{G} = j_y = 0$$
$$\sigma_i = 1/\text{Re}\,(\mathcal{K}_1^{-1}) = (K_1^2 + L_1^2)/K_1$$

2. *Isothermal thermal conductivity*

$$\kappa_i = w_x/T\mathcal{G}_x \qquad \text{with } \mathcal{J} = \mathcal{G}_y = 0$$
$$\kappa_i = \text{Re}\,(\mathcal{K}_3)/T = K_3/T$$

3. *Absolute thermoelectric power*

$$\epsilon = -\frac{\mathcal{E}_x}{T\mathcal{G}_x} \qquad \mathcal{J} = \mathcal{G}_y = 0$$

$$\epsilon = \text{Re}\left(\frac{K_2}{TK_1}\right)$$

4. *Isothermal Hall effect*

$$R_i = \mathcal{E}_y/H\mathcal{J}_x \qquad \text{with } \mathcal{G} = \mathcal{J}_y = 0$$
$$R_i = -\text{ Im }(\mathcal{K}_1^{-1})/H = L_1/H(K_1^2 + L_1^2)$$

5. *Isothermal Nernst effect*

$$A_{Ni} = -\mathcal{E}_y/HT\mathcal{G}_x \qquad \text{with } \mathcal{J} = \mathcal{G}_y = 0$$
$$A_{Ni} = -\text{ Im }(\mathcal{K}_2/HT\mathcal{K}_1) = (L_1K_2 - K_1L_2)/HT(K_1^2 + L_1^2)$$

6. *Ettingshausen effect*

$$A_E = -T\mathcal{G}_y/H\mathcal{J}_x \qquad \text{with } j_y = w_y = \mathcal{G}_x = 0$$
$$A_E = -T\text{ Im }(\mathcal{K}_2/\mathcal{K}_1)/H\text{ Re }(\mathcal{K}_3) = T(L_1K_2 - K_1L_2)/HK_3(K_1^2 + L_1^2)$$

7. *Leduc-Righi effect*

$$A_L = \mathcal{G}_y/H\mathcal{G}_x \qquad \text{with } \mathcal{J} = w_y = 0$$
$$A_L = \text{Im }(\mathcal{K}_3)/H\text{ Re }(\mathcal{K}_3) = L_3/HK_3$$

The real and imaginary parts of $\mathcal{K}_1$ may be derived from definitions 1 and 4, of $\mathcal{K}_2$ from 3 and 5, and of $\mathcal{K}_3$ from 2 and 7. There is an additional relation between the coefficients, originally given by Bridgman and Lorentz:

$$\kappa_i A_E = TA_{Ni} \tag{6.84}$$

Callen, whose treatment we have followed, also gives relations between the adiabatic and isothermal coefficients:

$$\begin{aligned} \kappa_a - \kappa_i &= H^2\kappa_i A_L^2 \\ \rho_i - \rho_a &= H^2 A_{Ni} A_E \\ R_a - R_i &= \epsilon A_E \\ A_{Ni} - A_{Na} &= \epsilon A_L \end{aligned} \tag{6.85}$$

Experimental data are discussed after the theory is presented in Sec. 4. An extension of the equations to anisotropic media has been discussed by Domenicali [30].

4. Solutions of the Boltzmann Equation

This section formulates the Boltzmann equation for thermoelectric and transverse effects and solves it for electron scattering described by a relaxation time.

In the Bloch scheme, each electron is assumed to move independently in the periodic field of the crystal lattice. The wave functions and energies are designated by the wave vector $\mathbf{k}$ of magnitude $2\pi/\lambda$. The number of independent orbitals per unit volume with wave vectors in the interval

$$d\mathbf{k} = dk_x\, dk_y\, dk_z \qquad \text{is} \qquad d\mathbf{k}/8\pi^3$$

Each of these can be occupied by two electrons, of opposite spin.

The distribution of electrons in $\mathbf{k}$ space may be defined in terms of a function $f(\mathbf{k},\mathbf{r})$ such that the number of electrons, dN, per unit volume with wave vectors in $d\mathbf{k}$ is given by

$$dN = f(\mathbf{k},\mathbf{r})\, d\mathbf{k}/4\pi^3 \tag{6.86}$$

It is presumed that the space variation of $f(\mathbf{k},\mathbf{r})$ is sufficiently slow so that no difficulties arise from the uncertainty principle.

In thermal equilibrium, $f = f_0$ is given by the Fermi-Dirac distribution function (6.27):

$$f_0(\mathbf{k},\mathbf{r}) = \frac{1}{1 + \exp[(E - \mu)/k_0T]} \tag{6.87}$$

where μ is the Fermi level. We assume that both E and μ include the electrostatic potential energy, although the difference $E - \mu$ is independent of this assumption.

When both an electric field and a temperature gradient are present, a steady state is reached in which changes in f produced by applied electric and magnetic fields and by drift in space are balanced by changes in f as a result of collisions.

$$\left(\frac{df}{dt}\right)_{\text{drift}} + \left(\frac{df}{dt}\right)_{coll} = 0 \tag{6.88}$$

To express the drift terms, we use some of the basic equations of the Bloch theory. The change in $\mathbf{k}$ due to an applied force $\mathbf{F}$ is

$$\frac{d\mathbf{k}}{dt} = \frac{\mathbf{F}}{\hbar} \tag{6.89}$$

The velocity $\mathbf{v}$ is

$$\mathbf{v} = \hbar^{-1}\,\text{grad}_k\, E \tag{6.90}$$

An effective mass may be defined by

$$d\mathbf{v}/dt = \mathbf{F}/m^* \tag{6.91}$$

In general, $1/m^*$ is a tensor. For an isotropic energy band in which E depends only on $|\mathbf{k}| = k$ and not on the direction of $\mathbf{k}$, $1/m^*$ is a scalar:

$$\frac{1}{m^*} = \frac{1}{\hbar^2}\left(\frac{d^2E}{dk^2}\right) \tag{6.92}$$

If the energy band is filled up to a region of k-space, where $\partial^2E/\partial k^2$ is *negative*, it is more convenient to represent the distribution by giving the states which are unoccupied. These are called holes. The probability f_p that a hole is present is, for thermal equilibrium,

$$f_p = 1 - f_0 = \frac{1}{1 + \exp[(\mu - E)/k_0T]} \tag{6.93}$$

Holes behave in most respects like particles with positive effective mass and positive charge. A hole wave-packet moves in electric and magnetic fields as a positive charge would move.

The change in f due to a force $\mathbf{F}$ and to drift in space is

$$-\left(\frac{df}{dt}\right)_{\text{drift}} = \frac{F}{\hbar}\cdot\text{grad}_k\, f + \mathbf{v}\cdot\text{grad}_r\, f \tag{6.94}$$

The collision terms may be expressed in terms of the probability $P(\mathbf{k},\mathbf{k}')$ that an electron makes a transition from $\mathbf{k}$ to $\mathbf{k}'$ in unit time. The interesting problem of determining the scattering probability for impurities and for thermal motion will be discussed in Sec. 5. Because of the exclusion principle, the transition can take place only if the final state is unoccupied. Thus

$$\left(\frac{df}{dt}\right)_{coll} = (2\pi)^{-3} \int \{P(\mathbf{k}',\mathbf{k})f(\mathbf{k}')[1 - f(\mathbf{k})] - P(\mathbf{k},\mathbf{k}')f(\mathbf{k})[1 - f(\mathbf{k}')]\}\, d\mathbf{k}' \quad (6.95)$$

In general, one may assume that $P(\mathbf{k},\mathbf{k}')$ is independent of the presence or absence of a field.

For the equilibrium configuration $f = f_0$, there can be no net change due to collisions. Putting $f = f_0$ and setting the integrand equal to zero, we find

$$\exp\,(E_{k'}/k_0T)P(\mathbf{k},\mathbf{k}') = \exp\,(E_k/k_0T)P(\mathbf{k}',\mathbf{k}) \quad (6.96)$$

If the scattering is elastic ($E_k = E_{k'}$), $P(\mathbf{k},\mathbf{k}')$ is symmetric in $\mathbf{k}$ and $\mathbf{k}'$, and the collision term may be written

$$\left(\frac{df}{dt}\right)_{coll} = (2\pi)^{-3} \int P(\mathbf{k},\mathbf{k}')[f(\mathbf{k}\,) - f(\mathbf{k})]\, d\mathbf{k}' \quad (6.97)$$

This is the same as the expression which would have been obtained if the exclusion principle had been disregarded. Impurity scattering is elastic at all temperatures. Thermal scattering may be regarded as elastic at high but not at low temperatures.

If the further assumptions are made that $P(\mathbf{k},\mathbf{k}')$ is independent of the direction of $\mathbf{k}$ (although it may depend on the energy and on the angle θ between $\mathbf{k}$ and $\mathbf{k}'$) and that the energy is a function of $|\mathbf{k}|$ only, one may express the collision integral in terms of a relaxation time τ, which may be a function of the energy:

$$\left(\frac{df}{dt}\right)_{coll} = \frac{f_0 - f}{\tau} \quad (6.98)$$

where

$$1/\tau = 2\pi \int P(\theta)(1 - \cos\theta)\sin\theta\, d\theta \quad (6.99)$$

in which $P(\theta)\, d\Omega$ is the probability per unit time that an electron will be scattered through an angle θ into the solid angle $d\Omega$. For calculations of explicit expressions for τ, see Sec. 5.

It is often assumed that the collision integral may be expressed in terms of a relaxation time even when the above assumptions are not valid. The form (6.88) is used with τ as a general function of $\mathbf{k}$. The Boltzmann equation then becomes a differential rather than an integrodifferential equation and is much easier to solve. There is, however, no good justification for this procedure.

Consider an isotropic metal with an isotropic energy band, and suppose that there is an electric field and a temperature gradient in the x direction. To terms of the first order in the field and gradient, we may replace f by the equilibrium value f_0 in the drift terms. Since E includes the electrostatic potential energy, the force F due to the applied electric field is

$$F_x = -\partial E/\partial x \quad (6.100)$$

The drift terms then become

$$-\left(\frac{df_0}{dt}\right)_{\text{drift}} = -\frac{1}{\hbar}\frac{\partial E}{\partial x}\frac{\partial f_0}{\partial E}\frac{\partial E}{\partial k_x} + \frac{T}{\hbar}\frac{\partial E}{\partial k_x}\frac{\partial f_0}{\partial E}\frac{\partial}{\partial x}\left(\frac{E-\mu}{T}\right) = -\frac{1}{\hbar}\frac{\partial E}{\partial k_x}\frac{\partial f_0}{\partial E}\left[\frac{\partial \mu}{\partial x} + \left(\frac{E-\mu}{T}\right)\frac{\partial T}{\partial x}\right] \quad (6.101)$$

The solution of the Boltzmann equation,

$$-\left(\frac{df_0}{dt}\right)_{\text{drift}} = \left(\frac{df}{dt}\right)_{coll} = \frac{f_0 - f}{\tau} \quad (6.102)$$

is then

$$f = f_0 + \frac{\tau}{\hbar}\frac{\partial E}{\partial k_x}\frac{\partial f_0}{\partial E}\left(\frac{\partial \mu}{\partial x} + \frac{E-\mu}{T}\frac{\partial T}{\partial x}\right) \quad (6.103)$$

The electric- and thermal-current densities are obtained by substituting (6.103) into the general expressions

$$\mathbf{j} = -\frac{e}{4\pi^3}\int \mathbf{v} f\, d\mathbf{k} \quad (6.104)$$

$$\mathbf{w} = \frac{1}{4\pi^3}\int \mathbf{v}(E-\mu) f\, d\mathbf{k} \quad (6.105)$$

In an almost filled band, for which the hole description is appropriate,

$$\mathbf{j} = \frac{e}{4\pi^3}\int \mathbf{v} f_p\, d\mathbf{k} \quad (6.104a)$$

Equations of the form (6.69) are obtained with

$$K_n = -\frac{e^2}{4\pi^3}\int \left(\frac{1}{\hbar}\frac{\partial E}{\partial k_x}\right)^2 \left(\frac{\mu - E}{e}\right)^{n-1} \frac{\partial f_0}{\partial E}\,\tau\, d\mathbf{k} \quad (6.106)$$

Since μ enters only in the form $\mu - E$, the integrals are independent of the electrostatic potential, as they should be. The integration can be carried out by first integrating over a surface of constant energy E and then integrating with respect to E. Let the integral over the surface be denoted by

$$\sigma(E) = \frac{e^2}{4\pi^3}\int_E \left(\frac{1}{\hbar}\frac{\partial E}{\partial k_x}\right)^2 \frac{\tau\, dS}{|\text{grad}_k E|} \quad (6.107)$$

The expression for K_n is then

$$K_n = -\int \sigma(E)\left(\frac{\mu - E}{e}\right)^{n-1}\frac{\partial f_0}{\partial E}\, dE \quad (6.108)$$

For a degenerate electron gas, the integrals may be evaluated by keeping only the first nonvanishing terms in the expansion:

$$-\int g(E)\frac{\partial f_0}{\partial E}\, dE = g(\mu) + \frac{\pi^2}{6}(k_0T)^2\frac{d^2g}{dE^2} + \cdots \quad (6.109)$$

It is then found that

$$K_1 = \sigma(\mu) \quad (6.110)$$

$$K_2 = -\frac{\pi^2}{3}\frac{(k_0T)^2}{e}\left(\frac{d\sigma}{dE}\right)_{E=\mu} \quad (6.111)$$

$$K_3 = \frac{\pi^2}{3}\left(\frac{k_0T}{e}\right)^2 \sigma(\mu) \quad (6.112)$$

From Eqs. (6.73) to (6.75), which express σ, ϵ, and κ in terms of the K's, it is found that, to terms of lowest order in T,

$$\sigma = \sigma(\mu) \tag{6.113}$$

$$\epsilon = -\frac{\pi^2}{3}\frac{k_0{}^2T}{e}\left(\frac{d \log \sigma}{dE}\right)_{E=\mu} \tag{6.114}$$

$$\kappa \cong K_3/T = \pi^2 k_0{}^2 T\sigma(\mu)/3e^2 \tag{6.115}$$

Equation (6.113) justifies the notation we have used for the integral (6.107). The Wiedemann-Franz law is an immediate consequence of (6.113) and (6.115). The thermoelectric power is of opposite sign when conduction is by holes.

A simple explicit expression for (6.107) can be obtained if the energy varies quadratically with k:

$$E = \hbar^2 k^2/2m^* \tag{6.116}$$

Here m^* is the effective mass. The value of σ is then

$$\sigma(E) = ne^2\tau/m^* \tag{6.117}$$

where

$$n = \frac{1}{3\pi^2}\left(\frac{2m^*E}{\hbar^2}\right)^{3/2} \tag{6.118}$$

is the number of electrons per unit volume when the energy of the Fermi surface, measured from the lowest state in the band, is E. Equation (6.117) is exactly the same as that derived by elementary considerations in Sec. 1.

The thermoelectric power can also be expressed very simply if (6.116) is valid. The relaxation time τ is then proportional to $E^{3/2}$ [see Eqs. (6.170) and (6.171)], and making use of (6.117) and (6.118), one obtains

$$\epsilon = -\frac{\pi^2 k_0{}^2 T}{e\mu} \tag{6.119}$$

A brief account will now be given of the theory of the transverse effects. There are electric and thermal currents in the x and y directions and a magnetic field $\mathbf{H}$ in the z direction. We shall keep only first-order terms in the electric-field and temperature gradients, but suppose that H is unrestricted, since this generalization does not greatly complicate the equations.

The magnetic field adds a drift term:

$$(e/\hbar c)(\mathbf{v} \times \mathbf{H}) \cdot \text{grad}_k f \tag{6.120}$$

so that Boltzmann's equation becomes

$$\frac{f_0 - f}{\tau} + \frac{e}{\hbar c}(\mathbf{v} \times \mathbf{H}) \cdot \text{grad}_k f = \frac{F}{\hbar} \cdot \text{grad}_k f_0 + \mathbf{v} \cdot \text{grad}_r f_0 \tag{6.121}$$

Here $F = -\partial E/\partial x$ is the force due to the electric field, so that the right-hand side is similar to (6.101).

The solution is of the form

$$f = f_0 - \tau[v_x c_x(E) + v_y c_y(E)](\partial f_0/\partial E) \tag{6.122}$$

The magnetic terms are then

$$(e/\hbar c)(\mathbf{v} \times \mathbf{H}) \cdot \text{grad}_k f = \beta(v_y c_x - v_x c_y)(\partial f_0/\partial E) \tag{6.123}$$

where

$$\beta = eH\tau/m^*c \tag{6.124}$$

Actually, the theory is valid only for $\beta \ll 1$, since otherwise the quantization of the electron levels in the magnetic field would have to be taken into account [31].

The equations are expressed most simply by using the complex notation of Eqs. (6.81). In addition to the terminology used there, we define

$$C = c_x - ic_y \tag{6.125}$$

Boltzmann's equation (6.121) is satisfied if C is a solution of

$$C(1 + i\beta) = -e\mathcal{E} + (E - \mu)\mathcal{G} \tag{6.126}$$

Electric- and thermal-current densities are obtained by substituting the expression for f so obtained into Eqs. (6.104) and (6.105). The results can be expressed, using the complex notation, by equations of the form (6.82), with

$$\mathcal{K}_n = -\int \sigma(E)(1 - i\beta)\left(\frac{\mu - E}{e}\right)^{n-1}\frac{\partial f_0}{\partial E}\,dE \tag{6.127}$$

where

$$\sigma(E) = \frac{e^2}{4\pi^3}\int_E \frac{\tau v_x{}^2\,dS}{(1 + \beta^2)|\text{grad}_k E|} \tag{6.128}$$

Equation (6.128) differs from (6.107) only in the term β^2 in the denominator.

When the expansion (6.109) is used, the first nonvanishing terms are

$$\mathcal{K}_1 = \sigma(\mu)(1 - i\beta) \tag{6.129}$$

$$\mathcal{K}_2 = -\frac{\pi^2}{3}\frac{(k_0T)^2}{e}\left\{\frac{d[\sigma(1 - i\beta)]}{dE}\right\}_{E=\mu} \tag{6.130}$$

$$\mathcal{K}_3 = \frac{\pi^2 k_0 T}{3e}\sigma(1 - i\beta) \tag{6.131}$$

The various transverse coefficients are obtained from the expressions 1 to 7 toward the end of Sec. 3. To the lowest order in the magnetic field, they are

$$\begin{aligned}
&\sigma_i = \sigma \qquad \kappa_i = \pi^2 k_0{}^2 T\sigma/3e^2 \\
&\epsilon = -\frac{\pi^2 k_0 T}{3e}\left(\frac{d \log \sigma}{dE}\right)_{E=\mu} \\
&R_i = -\beta/H\sigma = -(nec)^{-1} \\
&A_N = -\frac{\pi^2 k_0{}^2 T}{3m^*c}\left(\frac{d\tau}{dE}\right)_{E=\mu} \qquad A_L = -\sigma R_i = e\tau/m^*c \\
&A_E = -\frac{Te^2}{m^*c\sigma}\left(\frac{d\tau}{dE}\right)_{E=\mu}
\end{aligned} \tag{6.132}$$

Effects which are linear in e (ϵ, R, A_L) are of opposite sign for hole conductors.

The measured Hall coefficients of most metals (see Table 6.1) are in reasonably good agreement with Eqs. (6.132) or with (6.147) in the case of two-band conduction. Since (6.132) is based on a free-electron model, one would expect to find best agreement for the monovalent metals, and such is the case. Even here there are exceptions; for example, the thermoelectric power of Li and Cu is positive rather than negative, as predicted by the free-electron theory.

As regards the other transverse effects, the experimental data are generally not as reliable as Hall effect data, the results of different workers showing considerable discrepancies. From Eqs. (6.132) and

(6.84) there are two relations among the transverse effects: $-A_L/\sigma R_i = 1$ and $A_{EKi}/A_{Ni}T = 1$. Tables 6.6 and 6.7 give numerical values of these two ratios, taken from two different sources. In view of the experimental error, the fluctuation of the results about the theoretical value of unity is not unreasonable.

TABLE 6.6. VALUES OF $-A_L/\sigma R_i$

	Borelius*	Meissner†
Cu	0.78	0.65
Ag	0.87	0.46
Au	1.03	0.89
Zn	0.85	0.70
Cd	0.74	1.24
Ir	0.69	
Pd	0.59	0.62
Pt	1.9	
Al	0.69	0.49
Bi	2.4	(0.035)
As	....	0.26
Sb	0.39	0.36
W	....	0.69
Mo	....	0.74

* G. Borelius, "Handbuch der Metallphysik," vol. 1, p. 431.
† W. Meissner, "Handbuch der Experimentalphysik," vol. 11, p. 383.

TABLE 6.7. VALUES OF $A_{EKi}/A_{Ni}T$

	Borelius*	Meissner†
Cu	1.05	0.76
Ag	1.22	1.34
Au	0.67	0.67
Zn	1.37	4.25
Cd	0.75	7.6
Pd	1.23	0.82
Al	1.43	1.73
Bi	0.94	2.88
As	0.98	
Sb	1.25	0.66

* G. Borelius, "Handbuch der Metallphysik," vol. 1, p. 431.
† W. Meissner, "Handbuch der Experimentalphysik," vol. 11, p. 383.

O. Madelung [32] has calculated coefficients for transverse effects in semiconductors, taking into account both thermal and ionic scattering.

Thin Films and the "Anomalous" Skin Effect. It has been heretofore assumed that the electrical properties of a sample were independent of the size or shape of the sample on which measurements are being made. This assumption is valid as long as the effective linear dimensions of the sample are large compared with the mean free path of the electrons or holes in the bulk. If, however, one of the dimensions of the sample is comparable with the mean free path, the boundary conditions will influence the electrical properties of the sample.

A simplified theory of conduction in *thin films* was first given by J. J. Thomson [33], and a more accurate treatment has been presented by Fuchs [34]. The results of Fuchs predict that the conductivity of a thin film is reduced below that in the bulk and is a function of the film thickness and of the scattering of electrons at the surface of the film. In the limit where the film thickness a is much less than the mean free path l one finds

$$\frac{\sigma_{\text{film}}}{\sigma_{\text{bulk}}} \cong \frac{3}{4}\frac{1+p}{1-p}\frac{a}{l}\log\frac{l}{a} \tag{6.133}$$

where p is the fraction of electrons suffering specular reflection at the surfaces of the film, the scattering of the remaining electrons being perfectly random.

The work of Fuchs has been extended by Sondheimer [35] and by Blatt [36] to calculations of the transverse effects and by Justi, Kohler, and Lautz [37] in an evaluation of the thermoelectric effects in thin films. The related problem of the conductivity of thin wires has been treated by MacDonald and Sarginson [38] and by Dingle [39].

There is another well-known case to which the solution of the Boltzmann equation in the bulk is not applicable. According to the classical theory of the skin effect, the surface conductivity of a metal is proportional to the square root of the d-c conductivity. Experimentally, however, it is observed that at high frequencies (microwaves) and low temperatures the surface conductivity approaches a limiting value independent of the d-c conductivity.

This *"anomalous" skin effect* appears at such high frequencies that the classical skin depth becomes smaller than the electronic mean free path, and Ohm's law is then no longer valid. The formal analysis of the anomalous skin effect is based on arguments similar to those used in the discussion of conductivity in thin films. From measurements of the optical reflectivity of metals, it appears that the assumption of diffuse scattering of electrons at the metal boundary is in better agreement with experiment than that of specular reflection.

For a detailed discussion of the anomalous skin effect, reference should be made to a series of articles by Dingle [40].

Transverse Effects in Semiconductors. In *semiconductors* for which Boltzmann statistics can be used, we have

$$f_0 \cong \exp{(\mu - E)/k_0T} \tag{6.134}$$

where, for an effective mass m^*,

$$E = E_c + \hbar^2k^2/2m^* = E_c + \tfrac{1}{2}m^*v^2 \tag{6.135}$$

It is most convenient to express the various quantities in terms of averages over the distribution, as follows:

$$\langle G \rangle = \frac{n}{4\pi^3}\int G(k)f_0\,d\mathbf{k} \tag{6.136}$$

where n is the total concentration of conduction electrons as given by (6.28). To evaluate (6.127), we set

$$-(\partial f_0/\partial E) = f_0/k_0T$$

and

$$v_x^2 = v^2/3$$

We then have

$$\mathcal{K}_n = \frac{e^{3-n}n<\tau v^2(1 - i\beta)(\mu - E_c - \frac{1}{2}m^*v^2)^{n-1}/(1 + \beta^2)>}{m^*<v^2>} \quad (6.137)$$

where we have replaced $3k_0T$ by its equivalent from the equipartition law:

$$\tfrac{1}{2}<m^*v^2> = \tfrac{3}{2}k_0T \quad (6.138)$$

We shall give explicit expressions only for σ, ϵ, and R in the limit of vanishing magnetic field ($\beta \to 0$). The expression for σ is

$$\sigma = \text{Re}\,(\mathcal{K}_1) = \frac{ne^2<\tau v^2>}{m^*<v^2>} \quad (6.139)$$

If τ is a constant, this expression for σ reduces to (6.117). The thermoelectric power is

$$\epsilon = \text{Re}\left(\frac{\mathcal{K}_2}{T\mathcal{K}_1}\right) = -\frac{<\tau v^2(E_c + \frac{1}{2}m^*v^2 - \mu)>}{eT<\tau v^2>} \quad (6.140)$$

If $E_c - \mu \gg k_0T$, this expression reduces to

$$\epsilon = -(E_c - \mu)/eT \quad (6.141)$$

The Peltier coefficient, which then becomes

$$\Pi = \epsilon T = -(E_c - \mu)/e \quad (6.142)$$

has a simple physical interpretation. Electrons which carry current in the semiconductor have an average energy roughly $E_c - \mu$ larger than those which carry current in the metal. When electrons flow from metal to semiconductor, the increase in electron energy comes from a cooling of the junction by $E_c - \mu$ per electron. Thermoelectric measurements are sometimes used to determine the position of the Fermi level relative to the conduction band and hence the concentration of electrons. The sign of the effect is opposite in P-type semiconductors; electrons *lose* an energy $\mu - E_v$ when flowing across the junction from the metal, and the junction is heated. It should be noted that thermoelectric effects are much larger in semiconductors than in metals, where the effect is of second order. The Hall coefficient is

$$R = -\text{Im}\,(\mathcal{K}_1^{-1})/H = -\frac{<v^2><\tau^2v^2>}{nec<\tau v^2>^2} \quad (6.143)$$

For a relaxation time independent of velocity, Eq. (6.143) reduces to $-1/nec$. If, as is the case for thermal scattering, the mean free path τv is independent of velocity, R becomes

$$R = -\frac{3\pi}{8}\frac{1}{nec} \quad (6.144)$$

Both ϵ and R are of opposite sign for hole conductors. Hall measurements have been widely used to determine the concentrations of carriers in semiconductors.

Rather complete analyses of data on σ, ϵ, and R for Ge and Si have been given by K. Lark-Horovitz, V. A. Johnson, and coworkers at Purdue University and by C. Herring, F. Morin, and others at the Bell Telephone Laboratories in papers published in the *Physical Review* and elsewhere.

If electrons are in *more than one band*, the current densities for each band may be added vectorially to get the total-current density. This implies, according to (6.82), that the total $\mathcal{K}_n$ is the sum of the $\mathcal{K}_n$'s for each band:

$$(\mathcal{K}_n)_{tot} = \sum_{\text{bands}} (\mathcal{K}_n)_{\text{band}} \quad (6.145)$$

Let us, for example, determine the Hall coefficient for a metal or semiconductor with two bands, denoted by the subscripts a and b. To the first order in H, we have for the separate bands

$$K_a = \sigma_a \qquad K_b = \sigma_b$$
$$L_a = H\sigma_a^2R_a \qquad L_b = H\sigma_b^2R_b \quad (6.146)$$

The Hall coefficient for the conductor, to first order in H, is therefore

$$R_{tot} = \frac{L_a + L_b}{H(K_a + K_b)^2} = \frac{\sigma_a^2R_a + \sigma_b^2R_b}{(\sigma_a + \sigma_b)^2} \quad (6.147)$$

In a semiconductor with conduction electrons and holes,

$$R_{tot} = -\frac{3\pi}{8ec}\frac{n\mu_n^2 - p\mu_p^2}{(n\mu_n + p\mu_p)^2} \quad (6.148)$$

A good review of galvanomagnetic properties of semiconductors is given by A. C. Beer (see Bibliography, *General References: Books* at chapter end).

Magnetoresistance. The resistance of both metals and semiconductors changes when they are placed in a magnetic field either transverse or longitudinal to the direction of current flow. Notable early measurements on metals are those of Kapitza [41], who used magnetic fields as high as 300,000 gauss, and those of Meissner and Scheffers [42], who used fields of the order of 10,000 gauss. These measurements were made at low temperatures in order to enhance the effect. The relative change in resistance, $\Delta\rho/\rho_0$, is proportional to H^2 for small fields but may increase less rapidly when the field is large. Observed values of $\Delta\rho/\rho_0$ for simple monovalent metal, such as Au, are small, only of the order of 10^{-3} at 10,000 gauss and 20°K, but may be large for more complex metals. Resistance changes of several orders of magnitude are found in Bi, Sb, and As at low temperature and high fields. For most metals, measurements made at different temperatures can be reduced to the same curve if $\Delta\rho/\rho_0$ is plotted as a function of H/ρ_0.

The theory for a single isotropic energy band gives no change in resistance for metals. According to (6.128), if τ, and thus β, is a constant over the energy surface,

$$\sigma(\mu) = \sigma_0(\mu)/(1 + \beta^2) \quad (6.149)$$

The resistivity, from (6.128), is independent of β and thus of H:

$$\rho = \text{Re}\left[\frac{1}{\sigma(\mu)(1 - i\beta)}\right] = \frac{1}{\sigma_0(\mu)} \quad (6.150)$$

The magnetic field creates an angle between the electric vector and the current, but the component of field in the direction of current flow is unchanged.

A nonvanishing effect is obtained if electrons occupy two or more isotropic energy bands. The theory, as worked out by Sondheimer and Wilson [43], gives results of a reasonable order of magnitude. Jones [44] has made detailed calculations for bismuth, assuming equal numbers of electrons and holes, and finds that, to obtain agreement with experiment, there must be only about 2.5×10^{-4} conduction electrons/atom.

Anisotropy of the energy surface or of the scattering probability also leads to a magnetoresistance and is required to account for a longitudinal effect. Davis [45] has given a formal theory obtained by assuming that E and τ are general functions of the wave vector $\mathbf{k}$.

Interesting magnetoresistance effects are also found in semiconductors. A nonvanishing result for the transverse effect is obtained for electrons in a single band if τ varies with energy. From (6.137), we find

$$\rho = \text{Re}\,(\mathcal{K}_1^{-1}) = \text{Re}\left[\frac{m^* \langle v^2 \rangle}{e^2 n \langle \tau v^2 (1 - i\beta)/(1 + \beta^2) \rangle}\right]$$

$$= \frac{\sigma_1}{\sigma_1^2 + \sigma_2^2} \tag{6.151}$$

where

$$\sigma_1 = \frac{ne^2 \langle \tau v^2/(1 + \beta^2) \rangle}{m^* \langle v^2 \rangle} \qquad \sigma_2 = \frac{ne^2 \langle \beta \tau v^2/(1 + \beta^2) \rangle}{m^* \langle v^2 \rangle} \tag{6.152}$$

Many semiconductors have a complex band structure and require a more complicated theory to account for their electrical properties. Experiments which give the best indication of band structure are magnetoresistance and piezoresistance measurements on single crystals and, most directly, cyclotron resonance, which gives effective masses for different orientations of the magnetic field. Rather complete studies have been made of the band structure of germanium and silicon. We shall give here only a brief summary of the important conclusions of this work.

Even though their crystal structures are cubic, the conduction and valence bands of both germanium and silicon are complex. The minimum energy of the conduction band does not occur at $\mathbf{k} = 0$ but at a number of equivalent points $\mathbf{k}_i$, at or near the surface of the first Brillouin zone. In Ge, these points lie along [111] directions, in Si along [100] directions. The effective mass in each "valley" is highly anisotropic. If it were not for spin, the top of the valence bands at $\mathbf{k} = 0$ would be triply degenerate. Spin-orbit effects split the levels into a double level at the top of the band and a single level of lower energy, analogous to the splitting of a P-type atomic level into $P_{3/2}$ and $P_{1/2}$. The energy is not given simply as a function of $\mathbf{k}$ but must be obtained by solution of secular equation.

The first evidence that the conduction band of Ge is complex came from magnetoresistance experiments [46], which gave a marked longitudinal effect, whereas none is to be expected for simple bands in a cubic crystal. That an adequate theory must take into account the complex band structure was emphasized by Shockley [47]. Theories based on the "many-valley" model of spheroidal energy bands with axes along [111] directions were developed independently by Abeles and Meiboom, Shibuya, and Herring [48]. Later measurements on N-type Si showed that the axes of the spheroidal bands lie along [100] directions in this material [49].

Band structure and effective masses of electrons and holes in Ge and Si are given most directly by cyclotron resonance experiments of Dresselhaus, Kip, and Kittel and of Lax, Zeiger, and Dexter [50]. Effective masses as given by the former for the conduction bands are, for Ge,

$$m_l = (1.58 \pm 0.04) \text{ and } m_t = (0.082 \pm 0.001)$$

and for Si,

$$m_l = (0.97 \pm 0.02) \text{ and } m_t = (0.19 \pm 0.01)$$

in electron mass units. Here m_l refers to the effective mass for motion along the axis of a spheroid and m_t to motion transverse to the axis.

Energies of electrons near the top of the valence bands can be represented by an expression of the form

$$E(\mathbf{k}) = Ak^2 \pm [B^2k^4 + C^2(k_x^2k_y^2 + k_y^2k_z^2 + k_x^2k_z^2)]^{1/2} \tag{6.152a}$$

The $\pm$ signs refer to the two bands which come together and are degenerate at $k = 0$. Values of A, B, and C for Ge are

$$A = -(13.0 \pm 0.2)(\hbar^2/2m)$$
$$|B| = (8.9 \pm 0.1)(\hbar^2/2m)$$
$$|C| = (10.3 \pm 0.2)(\hbar^2/2m)$$

Corresponding values for silicon are

$$A = -(4.1 \pm 0.2)(\hbar^2/2m)$$
$$|B| = (1.6 \pm 0.2)(\hbar^2/2m)$$
$$|C| = (3.3 \pm 0.5)(\hbar^2/2m)$$

The values obtained by Lax et al. are in good agreement with those quoted above. If C were equal to zero, (6.152a) would represent two spherical bands with different effective masses which come together at $k = 0$. The properties of the actual complex bands can be described approximately by assuming that there are two types of holes, one of small mass, present in small concentration, and one of larger mass, which are more abundant and give the main contribution to the conductance. The small-mass holes are of particular importance for the Hall effect and magnetoresistance. It is necessary to make measurements at quite small fields in order to determine the limiting value of the Hall coefficient as $H \to 0$. Willardson, Harmon, and Beer [51] have found that they could account for their measurements of the temperature dependence and magnetic-field dependence of the Hall effect and magnetoresistance of high-purity P-type Ge by use of a model with two types of holes with widely different masses.

5. Scattering Mechanisms

In this section we discuss scattering of electrons in metals and semiconductors and give some representative data to illustrate different aspects of the theory. Scattering may be caused by anything which gives a deviation from perfect periodicity of the lattice potential: (1) thermal motion of atoms; (2) impurity atoms; (3) lattice defects, such as vacant sites, interstitial atoms, dislocations, and grain boundaries; and (4) random distribution of atoms in an alloy. Topic 4 has been discussed briefly in Sec. 1; topics 1 to 3 are considered below.

The general expression, based on the Born approximation, for probability per unit time $P(\mathbf{k},\mathbf{k}')$ that an electron will be scattered from $\mathbf{k}$ to $\mathbf{k}'$ is [52]

$$P(\mathbf{k},\mathbf{k}') = \frac{2\pi}{\hbar} |M(\mathbf{k},\mathbf{k}')|^2 \delta[W(\mathbf{k}) - W(\mathbf{k}')] \quad (6.153)$$

where $W(\mathbf{k})$ and $W(\mathbf{k}')$ are the energies of the initial and final states, including vibrational energy of the lattice. Here $M(\mathbf{k},\mathbf{k}')$ is the matrix element for the transition

$$M(\mathbf{k},\mathbf{k}') = \int \psi(\mathbf{k}')^* U \psi(\mathbf{k})\, d\tau \quad (6.154)$$

where $\psi(\mathbf{k})$ and $\psi(\mathbf{k}')$ represent the wave function of the initial and final states, including vibrational coordinates. One of the main problems of the theory is to determine the interaction U which causes the transition.

For isotropic elastic scattering, in which E is a function of $|\mathbf{k}| = k$ only and $|M(\mathbf{k},\mathbf{k}')|^2$ depends only on $|\mathbf{k}|$ and the angle θ between $\mathbf{k}$ and $\mathbf{k}'$, it is most convenient to express the scattering in terms of $P(\theta)\, d\Omega$, the probability per unit time for scattering into the solid angle $d\Omega$. In terms of (6.153),

$$P(\theta) = \frac{1}{8\pi^3} \int P(\mathbf{k},\mathbf{k}') k^2 \frac{dk}{dE}\, dE = \frac{k^2}{4\pi^2\hbar} \frac{dk}{dE} |M(\mathbf{k},\mathbf{k}')|^2 \quad (6.155)$$

The Boltzmann equation may then be solved with use of the relaxation time defined by (6.99).

Thermal Motion. The normal modes of vibration of a crystal lattice are waves. They may be designated by a wave vector $\mathbf{q}$ which takes on values in the first Brillouin zone. The number of independent waves for a given $\mathbf{q}$ is three times the number of atoms in a unit cell. In cubic metals, with body-centered (b.c.) or face-centered (f.c.) structure, there is only one atom in a properly chosen unit cell and thus only one branch to the vibrational spectrum. The three directions of polarization correspond at long wavelengths to the longitudinal and two transverse acoustic waves. In alkali-halide crystals, in which there are two ions per unit cell, there are two branches: the acoustic in which the ions vibrate in phase and the polar in which neighboring ions are out of phase. Some monatomic crystals, such as diamond, have two atoms per unit cell and thus also have two branches to the spectrum. The higher-frequency branch is often called the polar branch, in analogy with ionic crystals, even though there is little or no electrical polarization associated with the motion.

The displacement δR_n of an atom at the lattice position $\mathbf{R}_n$ can be expressed in the form

$$\delta R_n = N^{-1/2} \Sigma \mathbf{n}_{qj} (a_{qj} e^{i\mathbf{q}\cdot\mathbf{R}_n} + a_{qj}^* e^{-i\mathbf{q}\cdot\mathbf{R}_n}) \quad (6.156)$$

where a_{qj} is the amplitude and $\mathbf{n}_{qj}$ is a unit vector in the direction of a wave designated by the wave vector $\mathbf{q}$. The integer j represents the branch and direction of polarization of the wave. The lattice waves are quantized; for an acoustic wave of velocity c, the angular frequency ω_q is $c|q|$, and the quantum of energy, called a phonon, is

$$\hbar\omega_q = hc|q| \quad (6.157)$$

When an electron is scattered, a phonon is emitted or absorbed. The selection rule is

$$\mathbf{k}' = \mathbf{k} \pm \mathbf{q} + \mathbf{K}_n \quad (6.158)$$

where the upper sign corresponds to absorption and the lower to emission of a phonon. The vector $\mathbf{K}_n$ is a lattice vector of the reciprocal lattice space, such that $\mathbf{K}_n \cdot \mathbf{R}_m$ is a multiple of 2π. Transitions for which $\mathbf{K}_n \neq 0$ are called, after Peierls, *Umklapp processes*. These occur when the vector difference between $\mathbf{k}$ and $\mathbf{k}'$ extends outside the first Brillouin zone. Conservation of energy requires that

$$W(\mathbf{k}) - W(\mathbf{k}') = E(\mathbf{k}) - E(\mathbf{k}') \pm \hbar\omega_q = 0 \quad (6.159)$$

At high temperature, $k_0T > \hbar\omega_q$, one may neglect the energy of the phonon and use an average amplitude appropriate to the temperature. From the equipartition law, the mean potential energy is

$$\tfrac{1}{2} M\omega_q^2 a_q^2 = \tfrac{1}{2} k_0 T$$

so that

$$a_q^2 = k_0T/M\omega_q^2 \quad (6.160)$$

At low temperatures, the quantization of the vibrations must be taken into account. Matrix elements of a_q are

$$(a_q)_{N_q-1,N_q} = \left(\frac{\hbar N_q}{2M\omega_q}\right)^{1/2} \quad \text{(absorption)} \quad (6.161a)$$

$$(a_q)_{N_q+1,N_q} = \left(\frac{\hbar (N_q+1)}{2M\omega_q}\right)^{1/2} \quad \text{(emission)} \quad (6.161b)$$

The high temperature value corresponds to taking $N_q = k_0T/\hbar\omega_q$ and summing the contributions from emission and absorption.

According to the Debye theory, the frequency is given by (6.157). The maximum value of q, the approximate radius of the first Brillouin zone, is

$$q_m = (6\pi^2 N)^{1/3} \quad (6.162)$$

The Debye characteristic temperature Θ is given by

$$k_0\Theta = \hbar c q_m \quad (6.163)$$

Actual vibrational spectra of crystals as measured by inelastic scattering of slow neutrons are far from those given by the Debye model. Rather complete data are now available for a number of metals, semiconductors, and other crystals. For accurate calculations of thermal scattering it is essential to use the true spectra rather than the rather poor Debye approximation.

Interaction Potential. Various methods have been used to estimate the interaction potential U for thermal motion. Bethe and Bloch [53] have used the hypothesis of a *deformable potential.* If $V(r)$ represents the potential at the point r in the undistorted crystal, the potential at the same point in a crystal subject to the distortion δR [as obtained from (6.156)] is assumed to be $V(r - \delta R)$. The interaction U is then the difference between the potentials in the distorted and undistorted crystals:

$$\begin{aligned} U &= V(r - \delta R) - V(r) \\ &= -\delta R \operatorname{grad} V \end{aligned} \tag{6.164}$$

The matrix element is expressed in terms of an interaction constant C which can be calculated if the Bloch wave functions are known. Houston [54] and Nordheim [55] have assumed that the ions move rigidly under a lattice distortion, so that if $v(r)$ is the potential of a single ion, the interaction potential is assumed to be

$$\begin{aligned} U &= \sum_n [v(r - R_n - \delta R_n) - v(r - R_n)] \\ &\simeq \sum_n [-\delta R_n \operatorname{grad} v(r - R_n)] \end{aligned} \tag{6.165}$$

Since the ions in a metal are shielded by the conduction electrons, one should take for $v(r)$ a shielded interaction, which for large r is of the form

$$v(r) = -\frac{Ze^2}{r} e^{-\alpha r} \tag{6.166}$$

Here Ze is the charge of the ion and α a screening constant, the magnitude of which can be estimated from the Thomas-Fermi model:

$$\alpha^2 = \frac{4m^*e^2}{\hbar^2} \left(\frac{3n}{\pi}\right)^{1/3} \tag{6.167}$$

In an improved calculation of the author [56], the screening of the ions has been determined by a Hartree self-consistent field method. The resulting expression for the electron-phonon interaction has been used in many of the subsequent theories of metallic conductivity.

Bohm and Pines [57] have shown that the long-range part of the coulomb interaction between electrons leads to coherent plasma oscillations of the electron gas. The frequencies of these oscillations are so high that normally they are not excited. When this long-range part of the interaction is separated out by a canonical transformation, one is left with a short-range screened interaction between electrons. This analysis shows why electron-electron collisions are not important for either electrical or thermal conduction. Estimates of E. Abrahams [58] indicate that the mfp for electron-electron scattering in Na is of the order of 5×10^{-4} cm at room temperature and about 2 cm at helium temperatures. While the cross sections are of a reasonable order of magnitude ($\sim 10^{-15}$ cm^2), the scatterings possible are so greatly restricted by the exclusion principle that collisions are infrequent. Initial and final energies of both electrons must have energies within k_0T of the Fermi surface. This reduces the effective scattering cross section by a factor of the order of $(k_0T/E_F)^2$.

For the calculation of the transport properties of the simpler metals (those with no close-lying d bands), much use has been made of the pseudo-potential method, in which the Bloch functions are replaced by slowly varying pseudo-wave functions and the interaction potential $v(r)$ by a pseudo potential. The pseudo-wave functions are often nearly plane waves. The pseudo potential is such that it gives the same matrix elements for scattering of the pseudo waves that the true potential does for the corresponding Bloch waves.

Ideal Resistivity of Metals. Calculation of the *absolute value of the conductivity* of metals is difficult because it requires a knowledge not only of the Bloch functions and the electron-phonon interaction but also the vibrational spectrum. In early calculations mostly for monovalent metals a number of simplifying assumptions were made: (1) the Debye theory for the vibrational spectrum, (2) vibrations isotropic and in thermal equilibrium, and (3) $E(k)$ a function of $|k|$ only. The Debye approximation does not give a good representation of the spectrum and leads to large errors in the calculation.

Another difficulty with the early calculations is the neglect of *Umklapp* scattering [see (6.158) and following], which contributes a large amount to the resistivity even at temperatures as low as a few degrees absolute. For a monovalent metal, the radius of the Fermi surface is $k_m = (3\pi^2N)^{1/3}$, and the approximate radius of the first Brillouin zone is $2^{1/3}k_m$; scattering through angles greater than

$$\theta_{\max} = 2 \sin^{-1} 2^{-2/3} \sim 79^\circ \tag{6.168}$$

takes place by *Umklapp* processes.

When (6.165) and (6.166) are inserted into (6.154), and it is further assumed that the electrons can be treated as plane waves, it is found that only the longitudinal part of the waves contributes to the scattering. For scattering through small angles ($q = |\mathbf{k} - \mathbf{k}'|$ small),

$$|M(\mathbf{k},\mathbf{k}')|^2 = \frac{4}{9N} |qa_q|^2 C^2 \tag{6.169}$$

where the interaction constant

$$C = \frac{6\pi e^2 N}{\alpha^2} = \frac{\hbar^2 k_m{}^2}{2m^*} = E_F \tag{6.170}$$

is just equal to the Fermi energy E_F. If we neglect *Umklapp* processes and assume, as Bethe and Bloch did, that C and thus $M(k,k')$ is a constant for all normal transitions, the relaxation time is

$$\begin{aligned} \frac{1}{\tau} &= \frac{k_m}{\pi\hbar} \left(\frac{dk}{dE}\right)_m |M(\mathbf{k},\mathbf{k}')|^2 \int_0^{\theta_{\max}} (1 - \cos\theta) \sin\theta \, d\theta \\ &= 2^{-5/3} m^* k_m |M(\mathbf{k},\mathbf{k}')|^2 / \pi\hbar^3 \end{aligned} \tag{6.171}$$

where $|M(\mathbf{k},\mathbf{k}')|^2$ is the matrix element for either emission or absorption and each includes the factor [from (6.161)]

$$|qa_q|^2 = k_0T/2Mc^2 - 2^{2/3}T(\hbar k_m)^2/2Mk_0\Theta^2$$

Using (6.169), we find, after some reduction,

$$\frac{\hbar}{\tau} = \frac{\pi m^* T C^2}{3k_0 M\Theta^2} \tag{6.172}$$

The conductivity is then

$$\sigma = \frac{Ne^2\tau}{m^*} = \frac{3k_0\hbar e^2}{\pi m^{*2}} \frac{M\Theta^2}{T} \frac{N}{C^2} = \frac{4k_0 e^2}{\pi^3\hbar^3 k_m} \frac{M\Theta^2}{T} \left(\frac{E_F}{C}\right)^2 \tag{6.173}$$

More generally, one may regard C in (6.172) and (6.173) as an appropriate average over scattering angles, including those requiring *Umklapp* processes. This average is approximately equal to E_F, the limiting value for small q. From (6.173), one can calculate the value of C/E_F required to give the observed conductivity; these are tabulated in Table 6.8 for the monovalent metals. It is seen that in all cases C/E_F is not far from unity.

TABLE 6.8. VALUES OF C/E_F DERIVED FROM CONDUCTIVITY

	σ (in esu) $\times 10^{-16}$, 0°C	M_A (mass units)	N/cm^3 $\times 10^{-21}$	h Θ k	C/E_F
Li	10.5	7	48	430	1.34
Na	20.8	23	26	160	0.82
K	13.3	39	14	100	0.87
Rb	7.7	85	11	60	1.09
Cs	5.0	133	9	45	1.27
Cu	57	64	85	310	1.30
Ag	60	108	59	220	1.23
Au	50	197	59	185	1.55

The self-consistent field procedure [56] and more recent calculations give the same scattering for small angles, but the scattering decreases with increasing angle. This decrease is, in large part, compensated by *Umklapp* processes which join smoothly with normal scattering at an angle of 79°.

In the *low-temperature* region, $T < \Theta$, it is not possible to express the collision integral in the Boltzmann equation in terms of a relaxation time, so that other approximate methods must be used. Bloch and Grüneisen first derived an approximate expression (6.12) for ρ_T based on use of the Debye model, taking C = const and neglecting *Umklapp* processes. As we have seen, this is in reasonably good agreement with experiment for a number of metals. Largest departures occur at very low temperatures, below 20°K. According to the Grüneisen formula, the ideal resistivity should vary as T^5 in this range. Measurements [59] give a variation as T^n, with n varying between 3 and 6 for various metals.

The type of solution used by Bloch and Grüneisen is not sufficiently accurate to determine thermal conductivity and thermoelectric power, and more powerful methods must be used to solve the Boltzmann equation. Wilson [60] derived an approximate solution for thermal conductivity κ, and Sondheimer [61] has extended the calculations to give an interpolation formula for the thermoelectric power ϵ. Köhler [62] has shown how the Boltzmann equation can be solved by a variational method which has been used by Sondheimer [63] and others in subsequent calculations. Sondheimer's calculations for σ, κ, and ϵ agreed with earlier derivations in first approximation, but he found that in higher approximations convergence is slower for κ and ϵ than for σ. An improved numerical solution by Klemens [64] resulted in a value for κ at low temperatures about 11 per cent higher than the value obtained by Sondheimer. Klemens' result is

$$\kappa(T) = \kappa(\infty)\left(\frac{\Theta}{T}\right)^2 (64N^{2/3})^{-1} \tag{6.174}$$

All calculated values of κ tend to be too small at low temperatures. Both theory and experiment indicate that the Lorentz number, $L = \kappa/\sigma T$, should decrease to a minimum with decreasing temperature and then increase to the high-temperature value when the residual resistance predominates.

More recently, attempts have been made to make more realistic calculations of the ideal resistivity of metals and its temperature dependence by use of (1) the phonon spectrum determined from neutron scattering, (2) calculated or semiempirical energy-band structure, (3) electron-phonon interaction from the pseudo-potential method or from the self-consistent field approach, (4) adequate treatment of the *Umklapp* process, and (5) approximate solution of the Boltzmann integral equation rather than use of a relaxation time. Some recent calculations which include some or all of these features are those of Bailyn [65], Ziman [66], Hasegawa [67], Greene and Kohn [68], and Wiser [69]. Values have been calculated for the liquid as well as the solid phase in some of these cases. While there are some exceptions, results are in general in as good accord with experiment as can be expected, considering the sensitivity of the result to the parameters involved.

Impurity Scattering in Metals. Scattering by impurities or other lattice imperfections gives the temperature-independent residual resistance in metals. This problem was first discussed on the basis of quantum mechanics by Nordheim and later by Mott and others. We shall follow Mott's development [70] as applied to impurity in metals, then give expressions for scattering by vacancies and dislocations, and finally discuss impurity scattering in semiconductors.

Mott expresses the scattering probability $P(\theta)$ in terms of the cross section $I(\theta)\,d\Omega$, for scattering into $d\Omega$. If $I(\theta)$ applies to a single impurity, and there are N_i impurities per unit volume, then

$$P(\theta) = N_i v I(\theta)$$

where v is the velocity of the electron. The total scattering cross section A of the impurity is the integral of $I(\theta)$ over the solid angle, with the weighting factor $(1 - \cos\theta)$:

$$A = 2\pi \int_0^\pi I(\theta)(1 - \cos\theta)\sin\theta\,d\theta \tag{6.175}$$

The added resistance due to the impurities is

$$\Delta\rho_i = mvN_iA/ne^2 \tag{6.176}$$

For a monovalent metal, for which the electron concentration n is equal to the atomic concentration N,

$$\Delta\rho_i = mvxA/e^2 \tag{6.177}$$

where $x = N_i/N$ is the relative concentration of impurities.

If the impurity can be represented as an effective potential-energy difference ΔE, extending over a volume of radius a, Mott finds that the cross section is

$$A = 0.81\pi a^2(\Delta E/E)^2 \tag{6.178}$$

He estimates values of ΔE for Cu, Ag, and Au dissolved in each other from the Hartree self-consistent fields, making use of the method of Wigner and Seitz. The values so obtained for ΔE, of the order of 1 to 3 ev, yield values of ρ_i in rough agreement with measurements of Linde [71].

Mott also estimated the scattering by a screened ionic field of the form (6.166), with $I(\theta)$ calculated by the Born approximation:

$$A = 2\pi(Ze^2/mv^2)^2f(y) \tag{6.179}$$

where $$f(y) = \log\left(1 + \frac{1}{y}\right) - \frac{1}{1+y}$$

and $$y = \hbar^2\alpha^2/4m^2v^2$$

Mott used this result to account for experimental results of Linde [71] on the resistance of dilute solutions of various metals in Cu, Ag, and Au. Linde's results, shown in Fig. 6.11, indicate that the added resistance is proportional to the square of the difference in valency between solute and solvent atoms. This result follows from (6.179) if Ze represents the difference in charge between the ion cores (the closed d shells). For example, $Z = 1$ for Zn in Cu, $Z = 2$ for Ga in Cu, etc. The value of the screening constant which gives the best fit with experiment is somewhat less than the Thomas-Fermi value (6.167). Friedel [72] has shown that there may be one or more bound orbits for the extra valence electrons, so that the Thomas-Fermi model may give a poor approximation to the actual charge distribution.

Jongenburger [73] has estimated the scattering of a *vacant site* in copper by assuming that the scattering field is the negative of the Hartree field of Cu^+, shielded by a spherical distribution of charge. This scattering field was approximated by a screened coulomb field, and the cross section was determined without use of the Born approximation. He found that

$$A = 2.2 \times 10^{-16}\ \text{cm}^2$$

giving a resistance of 1.25×10^{-6} ohm-cm for 1 per cent of vacancies. Values of A for other monovalent metals vary roughly in proportion to the two-thirds power of the atomic volume.

Several estimates of scattering by dislocations have been made; the best are those of Dexter [74] and of Hunter and Nabarro [75]. Scattering by dislocations is anisotropic. *For an edge dislocation* the added resistance is a maximum when current flows perpendicular to the dislocation axis and to the slip plane. Neither edge nor screw dislocations give scattering for current flow parallel to the axis. Hunter and Nabarro give the following values for $\Delta\rho$ for 1 cm of dislocation line, *randomly* oriented:

Copper, edge: $\Delta\rho_E = 0.59 \times 10^{-20}$ ohm-cm
screw: $\Delta\rho_S = 0.18 \times 10^{-20}$ ohm-cm
Sodium, edge: $\Delta\rho_E = 2.10 \times 10^{-20}$ ohm-cm
screw: $\Delta\rho_S = 0.18 \times 10^{-20}$ ohm-cm

If equal densities of edge and screw dislocations are assumed, it is found that about 2×10^{12} dislocation lines/cm² are required for $\Delta\rho = 2 \times 10^{-8}$ ohm-cm, in fair agreement with estimates of Dexter.

These estimates of scattering by lattice defects are of importance because resistivity measurements have been used widely in studies of cold work and of radiation damage in metals and also in experiments on the quenching in of defects, particularly vacancies, which exist in thermal equilibrium at high temperatures. By using pure metals and by making measurements of residual resistance at very low temperatures, it is possible to measure very small values of $\Delta\rho$.

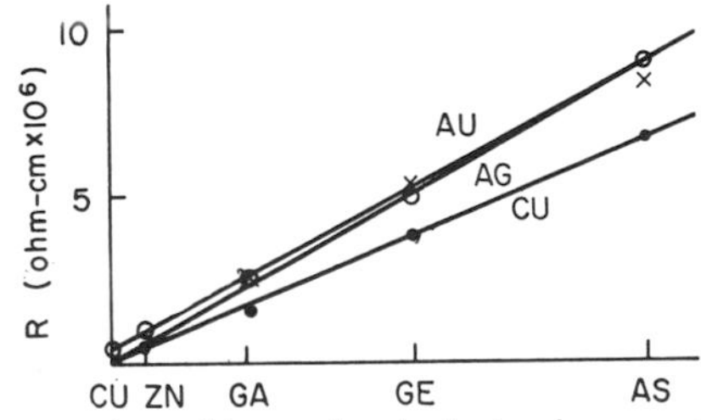

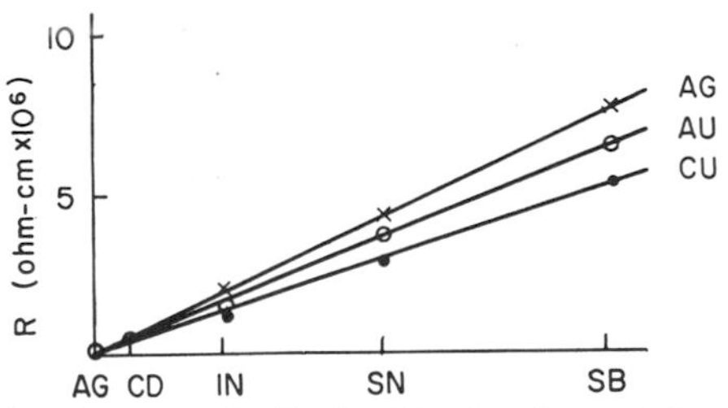

FIG. 6.11. Increase in resistance due to 1 atomic per cent of various metals dissolved in Cu, Ag, and Au. Abscissas are proportional to the square of the difference in valency between solvent and solute metals. (*After Bardeen, J. Appl. Phys.*, **11**: 88 (1940); *data from Linde* [60].)

A particularly striking phenomenon, long unexplained, is the minimum in resistance observed at low temperatures in a number of metals with magnetic impurities. The temperature at which the minimum is observed increases with concentration of magnetic impurity. The effect was first observed in a gold specimen by de Haas, de Boer, and van den Berg [76]. A tentative explanation involving resonant scattering of electrons with energies near the Fermi level has been advanced by Korringa and Gerritsen. By carrying the interaction between conduction electrons and magnetic ions to third order in perturbation theory, Kondo [77] has shown how such excess scattering arises. Kondo's theory accounts for the observed dependence of the temperature of the resistivity minimum on impurity concentration.

Thermal Scattering in Semiconductors. Wavelengths of electrons and phonons involved in thermal scattering in semiconductors are generally large compared with the interatomic distance. Wavelengths in Boltzmann distributions of electrons at room temperature are of the order of

$$\lambda = h/mv = h/\sqrt{3mk_0T} \qquad \sim 5 \times 10^{-7} \text{ cm} \qquad (6.180)$$

and phonons which can scatter these electrons have wavelengths of the same order. The energy of these phonons is small compared with the energies of the electrons, and this is true even at very low temperatures, as can be seen from the following calculation:

$$h\omega_q = \hbar cq \sim \hbar ck \sim cmv = (2c/v)(\tfrac{1}{2}mv^2) \qquad (6.181)$$

Here $\hbar k = mv$ is the magnitude of the electron momentum. At room temperature, $v \sim 10^7$ cm/sec and the sound velocity, $c \sim 10^5$ cm/sec, so that $(2c/v)$, the ratio of electron to phonon energy, is a small number. The ratio is less than unity even at helium temperatures, when $v \sim 10^6$ cm/sec.

Since the phonon energy is small, scattering of electrons by thermal motion is approximately elastic over nearly the entire temperature range. Provided that the scattering is isotropic, the Boltzmann equation can be solved with the use of a relaxation time.

The relatively long wavelengths of the lattice waves suggests that a continuum approximation might be valid. Scattering by acoustic waves has been treated by the deformation potential method [78]. It is assumed that the energy $E(\mathbf{k})$ of an electron with wave vector $\mathbf{k}$ depends only on the local strain. This implies that there is no polarization associated with the lattice wave which would give a potential variation depending on the strain pattern. The general expression for a strain ϵ_i is

$$E(\mathbf{k}) = E_0(\mathbf{k}) + \sum_{i,j=1,2,3} E_{ij}\epsilon_{ij} + \sum_{ijkl=1,2,3} E_{ijlm}\epsilon_{ij}k_lk_m + \cdots \qquad (6.182)$$

The tensor E_{ij} gives the shift in the energy of the lowest state and E_{ijlm} the change in effective mass. Both terms may contribute to the scattering. The number of independent coefficients is greatly reduced when the band has cubic symmetry. As is the case for Ge and Si, the bands may be degenerate and have a lower symmetry than the crystal itself.

The method has been applied to Ge, Si, and Te with the simplifying assumption that changes in effective mass may be neglected. For cubic symmetry (which is not really valid for Te), the energy then depends on the dilation:

$$E(\mathbf{k}) = E_0(\mathbf{k}) + E_1(\epsilon_{11} + \epsilon_{22} + \epsilon_{33}) \qquad (6.183)$$

Only longitudinal waves contribute to the scattering, since there is no dilation associated with transverse waves. The mobility is

$$\mu = \frac{(8\pi)^{1/2}\hbar^4 c_{ii}}{3E_1{}^2 m^{*5/2} k_0{}^{3/2}} T^{-3/2} \qquad (6.184)$$

where c_{ii} is the elastic constant for the longitudinal waves and m^* is the effective mass. Similar expressions apply to both conduction electrons and holes.

There is no independent method for determining $|E_{1c}|$ and $|E_{1v}|$ for the conduction and valence bands, respectively. They can be inferred from the measured mobility if values are assumed for the effective mass. Values for Ge are

$\lvert E_{1c}\rvert = 1.7$ ev	$\lvert E_{1v}\rvert = 2.4$ ev	for $m^* = m$
$\lvert E_{1c}\rvert = 9.6$ ev	$\lvert E_{1v}\rvert = 13.5$ ev	for $m^* = m/4$

The smaller effective mass is perhaps a reasonable average for the highly anisotropic energy bands of Ge. The difference between $|E_{1c}|$ and $|E_{1v}|$ is approximately equal to the change in energy gap with dilation.

The shift in the conduction bands of a multivalley semiconductor, such as Ge or Si, from shearing strains can be estimated from the measured elastoresistance.

In addition to the scattering of long-wavelength acoustic waves described above, there may also be scattering by the optical branch of the phonon spectrum. In multivalley semiconductors, scattering from one valley to another by acoustic phonons of relatively large wave vector q can also occur. Since phonons of relatively high energy are involved in both cases, the scattering probability depends strongly on temperature.

Impurity Scattering in Semiconductors. Since the average velocity of electrons in a Boltzmann gas varies with temperature, impurity scattering which depends on velocity will also vary with temperature. Given the scattering cross section $A(v)$ as a function of velocity, the mobility may be determined as follows: The contribution of impurity scattering to the relaxation time is

$$\frac{1}{\tau_i} = vN_iA \qquad (6.185)$$

The total relaxation time is the sum of lattice and impurity contributions:

$$\frac{1}{\tau} = \frac{1}{\tau_L} + \frac{1}{\tau_i} \qquad (6.186)$$

It follows from (6.139) that the appropriate average of τ for use in the expression for mobility μ is

$$\mu = \frac{e<\tau v^2>}{m^*<v^2>} \qquad (6.187)$$

It is often assumed as an *approximation* that

$$\frac{1}{\mu} = \frac{1}{\mu_L} + \frac{1}{\mu_i} \qquad (6.188)$$

but the value so derived may be in error by as much as 30 per cent.

Charged ions, such as ionized donors and acceptors, may scatter carriers in semiconductors. Because of large small-angle scattering, an unscreened coulomb field gives an infinite cross section. Conwell and Weisskopf [79] have derived an expression by arbitrarily cutting off the scattering at a distance equal to half the distance between neighboring ions. The cutoff distance is not very critical because it enters only in a logarithmic term. In an improved calculation, Brooks [80] calculated scattering by a random distribution of screened ions and obtained almost equivalent results. The Conwell-Weisskopf formula for the cross section of an ion of charge Ze is

$$A = \frac{4\pi Z^2 e^4}{\epsilon^2 m^{*2} v^4} \log\left(1 + \frac{\epsilon^2 m^{*2} v^4 d^2}{Z^2 e^4}\right) \tag{6.189}$$

where Ze is the charge of the ion, ϵ is the dielectric constant, m^* the effective mass of the electron (or hole), v is the velocity, and d is the cutoff distance. Brooks's expression differs only in the logarithm. The mobility derived from (6.187) is

$$\mu_i = \frac{2^{7/2}\epsilon^2 (k_0 T)^{3/2}}{\pi^{3/2} m^{*1/2} Z^2 e^3 N_i} \frac{1}{\log\,[1 + (6\epsilon k_0 T d/Ze^2)^2]} \tag{6.190}$$

Erginsoy [81] has modified the expression for scattering of electrons by hydrogen atoms to determine the scattering cross section for neutral donors and acceptors. When the kinetic energy of the electron is small compared with the ionization energy, the cross section A is approximately inversely proportional to v:

$$A \simeq 20\epsilon\hbar^3/m^{*2}e^2 v \tag{6.191}$$

The mobility μ_N for neutral scattering is then

$$\mu_N = \frac{1}{20} \frac{m^* e^3}{N_N \epsilon \hbar^3} \tag{6.192}$$

These expressions have been used to account for the variation of mobilities of carriers in germanium and silicon with temperature and impurity concentration. Figures 6.12 and 6.13 show the variation of mobility of electrons and holes at room temperature with concentration of donors or acceptors. These mobilities were obtained by the drift method. The solid line, calculated by combining lattice scattering with ionized impurity scattering according to the Conwell-Weisskopf formula, gives a good fit to the experimental data.

Figure 6.14 gives a plot of the variation of Hall mobility with temperature for silicon. Thermal scattering dominates at high temperatures, impurities at low temperatures. While a reasonably good fit to the data can be obtained with the use of the theory discussed in this section, the interpretation is complicated by the known complex band structure of these elements.

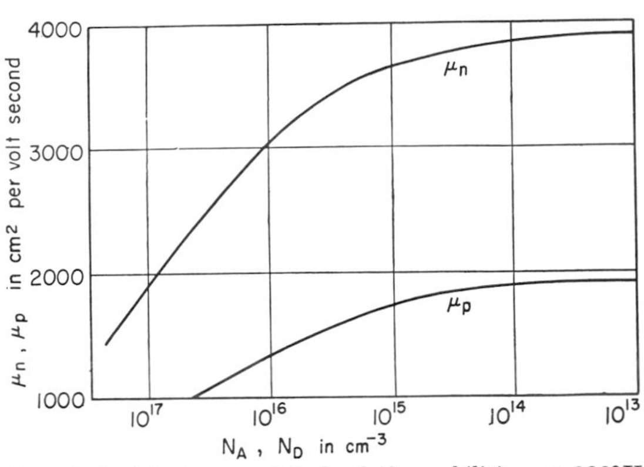

FIG. 6.12. Electron and hole drift mobilities at 300°K vs. concentration of acceptors in P-type and of donors in N-type germanium, respectively. [*After M. B. Prince, Phys. Rev.*, **92**: 681 (1953).]

Motion of Slow Electrons in Polar Crystals. Electrons in polar crystals interact strongly with the polar modes of vibration. This interaction not only leads to scattering but also makes a contribution to the energy and effective mass of the electron. As the electron moves through the crystal, its coulomb field will displace the positive and negative ions in the vicinity, and the resultant ionic polarization will modify the motion of the electron. The electron with the induced polarization which moves along with it is called a *polaron*.

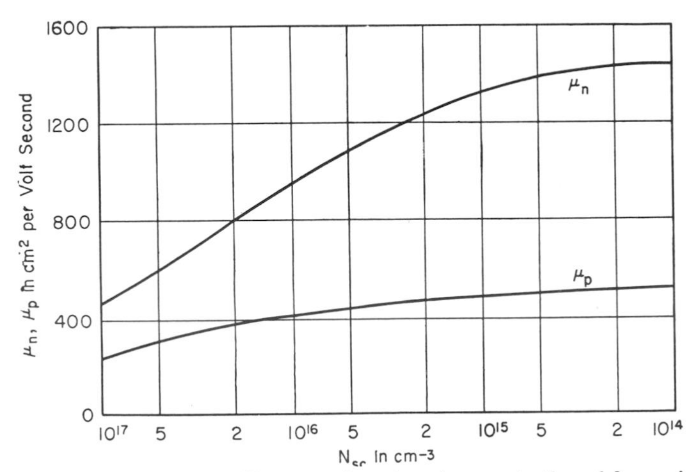

FIG. 6.13. Electron and hole drift mobilities in silicon as a function of concentration of donors in N-type and of acceptors in P-type silicon, respectively. [*After M. B. Prince, Phys. Rev.*, **93**: 1204 (1954).]

The strength of the interaction between electrons and polar vibrations can be characterized by a dimensionless *coupling constant:*

$$\alpha = \frac{e^2}{2\hbar c}\left(\frac{2mc^2}{\hbar\omega}\right)^{1/2}\left(\frac{1}{n^2} - \frac{1}{\epsilon}\right) \qquad (6.193)$$

where m is the effective mass for a stationary lattice, ω is the angular frequency of the polar modes, n is the index of refraction, and ϵ is the dielectric constant.

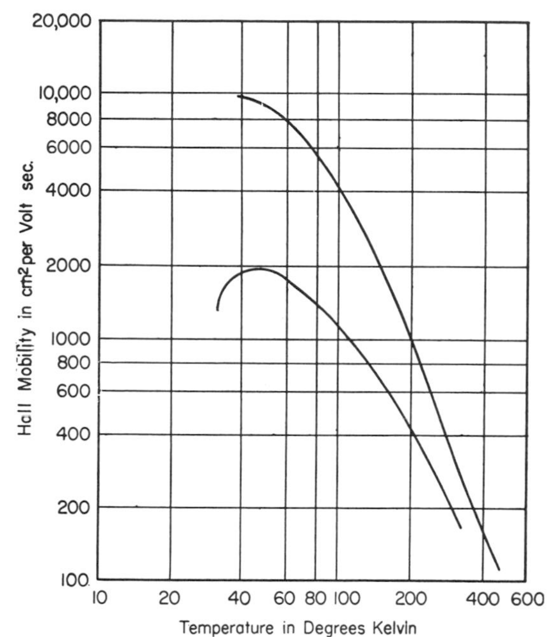

FIG. 6.14. Electron and hole mobilities in silicon samples containing boron as a function of reciprocal absolute temperature. Boron concentrations, upper curve about $8 \times 10^{14}/\text{cm}^3$, lower curve about $3 \times 10^{17}/\text{cm}^3$. *(After Morin and Maita* [12].)

The interaction can be taken into account by perturbation theoretic methods when α is small compared with unity. The appropriate theory has been given by Fröhlich and Mott [82] for the mobility and by Fröhlich, Pelzer, and Zienau [83] for the energy correction.

It can be shown that the strong coupling limit ($\alpha > 10$) is equivalent to the adiabatic approximation, which has been treated by Pekar [84] and by Markham and Seitz [85].

The actual range of coupling constants is of the order of magnitude of 3 to 6 for typical polar crystals, so that neither the weak-coupling nor the strong-coupling approximations are valid. A variational method of Tomonaga has been applied by Lee, Low, and Pines and independently by Gurari [86] to the case of intermediate coupling. They find for the shift in energy of a slow electron,

$$\Delta E = -\alpha\hbar\omega \qquad (6.194)$$

and for the effective mass,

$$m^* = m(1 + \alpha/6) \qquad (6.195)$$

The value of ΔE is equivalent to that obtained by perturbation theory. An expression for the mobility derived by Low and Pines [87], as modified by Langreth [88], is

$$\mu = (e/2\omega m^*)f(\alpha) \exp(h\omega/k_0T) \qquad (6.196)$$

where $f(\alpha)$ varies slowly with α and is approximately 1.2 for $3 < \alpha < 6$. The perturbation-theory limit corresponds to taking $f(\alpha) = 1$ and $m^* = m$. Thus the intermediate-coupling theory gives values quite close to those of perturbation theory.

Feynman [89] has given an approximate theory for the polaron energy valid over the entire range of coupling strengths from weak to strong, reducing to known results in the appropriate limits. A first-order approximation for the mobility using the Feynman method has been obtained by Osaka [90], giving results in the intermediate-coupling range not far from (6.196). Langreth and Kadanoff [91] have shown that the higher-order corrections to Osaka's theory tend to cancel and are probably not large.

References

1. Drude, P.: *Ann. Physik,* **1:** 566; **3:** 370, 869 (1900).
2. Lorentz, H. A.: "Electronentheorie der Metalle," Leipzig, 1909.
3. Grüneisen, E.: *Ann. Physik,* **16:** 530 (1933).
4. Bloch, F.: *Z. Physik,* **52:** 555 (1930).
5. Bridgman, P. W.: *Proc. Am. Acad. Arts Sci.,* **72:** 157 (1938).
6. Drickamer, H. G.: in F. Seitz and D. Turnbull (eds.), "Solid State Physics," vol. 17, pp. 1–33, Academic, New York, 1965.
7. Hulm, J. K.: *Proc. Roy. Soc. (London),* **A204:** 98 (1950). Mendelssohn, K., and H. M. Rosenberg: *Proc. Phys. Soc. (London),* **A65:** 385, 562 (1952).
8. Lichtenecker, K.: *Physik. Z.,* **25:** 169, 193, 225 (1924).
9. Nordheim, L.: *Ann. Physik,* **9:** 641 (1931).
10. Bragg, W. L., and E. V. Williams: *Proc. Roy. Soc. (London),* **A145:** 699; **A151:** 540 (1935).
11. Muto, T.: *Sci. Papers Inst. Phys. Chem. Research,* **30:** 99 (1936); 153 (1937).
12. Morin, F. J., and J. P. Maita: *Phys. Rev.,* **96:** 28 (1954).
13. Haynes, J. R., and W. Shockley: *Phys. Rev.,* **81:** 835–844 (1951).
14. Welker, H.: *Z. Naturforsch.,* **7a:** 744–749 (1952); *ibid.,* **8a:** 248 (1953); *Physica,* **20:** 893 (1954); R. G. Breckenridge: *Phys. Rev.,* **90:** 488; E. Justi and G. Lautz: *Z. Naturforsch.,* **7a:** 191–200, 602–613 (1952).
15. Verwey, E. J. W.: "Semi-conducting Materials," pp. 151–161, Butterworth, London, 1951.
16. Morin, F. J.: *Phys. Rev.,* **93:** 1199 (1954).
17. Morin, F. J.: *Phys. Rev.,* **93:** 1195 (1954).
18. Jonker, G. H., and J. H. Van Santen: *Physica,* **16:** 337, 599 (1950).
19. Zener, C.: *Phys. Rev.,* **82:** 403 (1951). Also see R. R. Heikes: *Phys. Rev.,* **99:** 1232 (1955).
20. Bridgman, P. W.: *Proc. Am. Acad. Arts Sci.,* **72:** 159 (1938).
21. Miller, P., and W. Taylor: *Phys. Rev.,* **76:** 179 (1949).
22. Bridgman, P. W.: *Proc. Am. Acad. Arts Sci.,* **79:** 139 (1951); **82:** 71 (1953).
23. Paul, W., and H. Brooks: *Phys. Rev.,* **94:** 1128 (1954).
24. Smith, C. S.: *Phys. Rev.,* **94:** 42 (1954).
25. Boltzmann, L.: *Ber. Wiener Akad.,* **96:** 1258 (1887).
26. Borelius, G., W. H. Keesom, C. H. Johansson, and J. O. Linde: *Commun. Kamerlingh Onnes Lab. Univ. Leiden Suppl.,* **70a** (1932).

27. Callen, H. B.: *Phys. Rev.*, **73**: 1349 (1949); **85**: 16 (1952).
28. Onsager, L.: *Phys. Rev.*, **37**: 405; **38**: 1265 (1932).
29. Mazur, P., and I. Prigogine: *J. phys. radium*, **12**: 616 (1951).
30. Domenicali, C. A.: *Phys. Rev.*, **92**: 877 (1953).
31. Teteica, S.: *Ann. Physik*, **22**: 129 (1935).
32. Madelung, O.: *Z. Naturforsch.*, **9a**: 667 (1954).
33. Thomson, J. J.: *Proc. Cambridge Phil. Soc.*, **11**: 119 (1901).
34. Fuchs, K.: *Proc. Cambridge Phil. Soc.*, **34**: 100 (1938).
35. Sondheimer, E. H.: *Phys. Rev.*, **80**: 401 (1950).
36. Blatt, F. J.: *Phys. Rev.*, **95**: 13 (1954).
37. Justi, E., M. Kohler, and G. Lautz: *Z. Naturforsch.*, **6a**: 456, 544 (1951).
38. MacDonald, D. K. C., and K. Sarginson: *Proc. Roy. Soc. (London)*, **A203**: 223 (1950).
39. Dingle, R. B.: *Proc. Roy. Soc. (London)*, **A201**: 545 (1950).
40. Dingle, R. B.: *Physica*, **19**: 311–347, 348–364, 729–736, 1187–1199 (1953).
41. Kapitza, P.: *Proc. Roy. Soc. (London)*, **A123**: 292 (1929).
42. Meissner, W., and E. Scheffers: *Physik. Z.*, **30**: 827 (1929).
43. Sondheimer, E. H., and A. H. Wilson: *Proc. Roy. Soc. (London)*, **A190**: 435 (1947).
44. Jones, H.: *Proc. Roy. Soc. (London)*, **A155**: 653 (1936).
45. Davis, L.: *Phys. Rev.*, **56**: 93 (1939).
46. Pearson, G. L., and H. Suhl: *Phys. Rev.*, **83**: 768 (1951).
47. Shockley, W.: *Phys. Rev.*, **78**: 173 (1950); **79**: 191 (1950).
48. Abeles, B., and S. Meiboom: *Phys. Rev.*, **95**: 31 (1954); M. Shibuya: *J. Phys. Soc. Japan*, **9**: 134 (1954) and *Phys. Rev.*, **95**: 1385 (1954); C. Herring: *Bell System Tech. J.*, **34**: 237 (1955). The latter paper gives a rather complete account of transport properties for the many-valley model, including mobility, thermoelectric power, piezoresistance, Hall effect, high-frequency dielectric constant, and magnetoresistance.
49. Pearson, G. L., and C. Herring: *Physica*, **20**: 975 (1954).
50. Dresselhaus, G., A. F. Kip, and C. Kittel: *Phys. Rev.*, **98**: 368 (1955); B. Lax, H. J. Zeiger, and R. N. Dexter: *Physica*, **20**: 818 (1954). References to earlier publications may be found in these two references.
51. Willardson, R. K., T. C. Harmon, and A. C. Beer: *Phys. Rev.*, **96**: 1512 (1954).
52. Seitz, F.: "The Modern Theory of Solids," p. 521, McGraw-Hill, New York, 1940.
53. Bloch, F.: *Z. Physik*, **52** (1928); A. Sommerfeld and H. Bethe: "Handbuch der Physik," vol. 24, part 2, Springer, Berlin, 1933.
54. Houston, W. V.: *Phys. Rev.*, **34**: 279 (1929); **88**: 1321 (1952).
55. Nordheim, L. W.: *Ann. Physik*, **9**: 607 (1931).
56. Bardeen, J.: *Phys. Rev.*, **52**: 688 (1937); also see J. Bardeen and D. Pines: *Phys. Rev.*, **99**: 1140 (1955).
57. Bohm, D., and D. Pines: *Phys. Rev.*, **82**: 625 (1951); **85**: 338 (1952); **92**: 609 (1953); **92**: 626 (1953).
58. Abrahams, E.: *Phys. Rev.*, **95**: 839 (1954).
59. De Haas, W. J., and G. J. Van den Berg: *Commun. Phys. Lab. Univ. Leiden Suppl.*, **82a** (1936); D. K. C. MacDonald and K. Mendelssohn: *Proc. Roy. Soc. (London)*, **A202**: 103 (1950).
60. Wilson, A. H.: *Proc. Cambridge Phil. Soc.*, **33**: 371 (1937).
61. Sondheimer, E. H.: *Proc. Cambridge Phil. Soc.*, **43**: 571 (1947).
62. Köhler, M.: *Z. Physik*, **124**: 772 (1948).
63. Sondheimer, E. H.: *Proc. Roy. Soc. (London)*, **A203**: 75 (1950).
64. Klemens, P. G.: *Australian J. Phys.*, **7**: 64 (1954).
65. Bailyn, M.: *Phys. Rev.*, **120**: 381 (1960).
66. Ziman, J. M.: *Phil. Mag.*, **6**: 1013 (1961); *Advan. Phys.*, **13**: 89 (1964).
67. Hasegawa, A.: *J. Phys. Soc. Japan*, **19**: 504 (1964).
68. Greene, M. P., and W. Kohn: *Phys. Rev.*, **137**: A513 (1965).
69. Wiser, N.: *Phys. Rev.*, **143**: 393 (1966).
70. Mott, N. F.: *Proc. Cambridge Phil. Soc.*, **32**: 281 (1936).
71. Linde, J. O.: *Ann. Physik*, **15**: 219 (1932).
72. Friedel, J.: *Phil. Mag.*, **43**: 153 (1952).
73. Jongenburger, P.: *Appl. Sci. Research*, **B, 3**: 237 (1953). An improved calculation has been made by F. J. Blatt: *Phys. Rev.*, **99**: 600 (1955).
74. Dexter, D. L.: *Phys. Rev.*, **86**: 770 (1952).
75. Hunter, S. C., and F. R. N. Nabarro: *Proc. Roy. Soc. (London)*, **A220**: 542 (1953).
76. De Haas, W. J., J. H. de Boer, and G. J. van den Berg: *Physica*, **1**: 1115 (1934).
77. Kondo, J.: *Progr. Theoret. Phys. (Kyoto)*, **32**: 37 (1964).
78. Bardeen, J., and W. Shockley: *Phys. Rev.*, **80**: 72 (1950).
79. Conwell, E., and V. F. Weisskopf: *Phys. Rev.*, **77**: 388 (1950).
80. Brooks, H.: *Phys. Rev.*, **83**: 879 (1951).
81. Erginsoy, C.: *Phys. Rev.*, **79**: 1013 (1950).
82. Fröhlich, H., and N. F. Mott: *Proc. Roy Soc. (London)*, **A171**: 496 (1939). Corrections to the original theory have been given in ref. 64. The correct expression is given following Eq. (6.196).
83. Fröhlich, Pelzer, and Zienau: *Phil. Mag.*, **41**: 221 (1950).
84. Pekar, S.: *J. Phys. Chem. (U.S.S.R.)*, **10**: 341 (1946).
85. Markham, J., and F. Seitz: *Phys. Rev.*, **74**: 1014 (1948).
86. Lee, Low, and Pines: *Phys. Rev.*, **90**: 297 (1953); H. Gurari: *Phil. Mag.*, **44**: 329 (1953). A review including a discussion of later work has been written by H. Fröhlich: *Phil. Mag. Suppl.*, **3**: 325 (1954). Also see R. Feynman: *Phys. Rev.*, **97**: 660 (1955).
87. Low, F. E., and D. Pines: *Phys. Rev.*, **98**: 414 (1955).
88. Langreth, D. C.: *Phys. Rev.*, **137**: A760 (1965).
89. Feynman, R. P.: *Phys. Rev.*, **97**: 660 (1955).
90. Osaka, Y.: *Progr. Theoret. Phys. (Kyoto)*, **25**: 517 (1961).
91. Langreth, D. C., and L. P. Kadanoff: *Phys. Rev.*, **133**: A1070 (1964).

Bibliography

General References: Review Articles

Bardeen, J.: Electrical Conductivity of Metals, *J. Appl. Phys.*, **11**: 88 (1940).
Brooks, H.: Theory of the Electrical Properties of Ge and Si, *Advan., Electron. Electron Phys.*, **7** (1955) (Academic, New York).
Broom, T.: Lattice Defects and Electrical Resistivity of Metals, *Advan. Phys.*, **3**: 26 (1954).
Burstein, E., and P. H. Egli: The Physics of Semiconductor Materials, *Advan., Electron. Electron Phys.*, **7** (1955) (Academic, New York).
Busch, G.: Electronenleitung in Nichtmetallen, *Z. Angew. Math. Phys.*, **1**: 3, 81 (1950).
Fröhlich, H.: Electrons in Lattice Fields, *Phil. Mag., Suppl.*, **3**: 325 (1954).
Gudden, B.: Electronic Phenomena in Non-metals, *Ergeb. Exakt. Naturw.*, **3**: 143 (1924); **13**: 223 (1934).

MacDonald, D. K. C.: Properties of Metals at Low Temperatures, *Progr. Metal Phys.*, **3**: 42 (1952).
Olsen, J. L., and H. M. Rosenberg: Thermal Conductivity of Metals at Low Temperatures, *Advan. Phys.*, **2**: 28 (1953).
Pippard, A. B.: Metallic Conduction at High Frequencies and Low Temperatures, *Advan., Electron. Electron Phys.*, **6** (1954) (Academic, New York).
Sommerfield, A., and H. Bethe: "Handbuch der Physik," vol. 24, pt. 2, Springer, Berlin, 1933.
Sondheimer, E. H.: The Mean Free Path of Electrons in Metals, *Advan. Phys.*, **1** (1952).
Proceedings of the International Conference on Electron Transport in Metals and Solids, *Can. J. Phys.*, **34**: 1171–1423 (1956).

General References: Books

Beer, A. C.: "Galvanomagnetic Effects in Semiconductors," Academic, New York, 1963.
Dekker, A. J.: "Solid State Physics," Prentice-Hall, Englewood Cliffs, N.J., 1957.
Ehrenberg, W.: "Electrical Conduction in Metals and Semiconductors," Oxford University Press, Fair Lawn, N.J., 1958.
Flugge, S. (ed.): "Encyclopedia of Physics," vols. 19 and 20, Springer, Berlin, 1956, 1957. Of particular interest are the following articles: A. N. Gerritsen, Metallic Conductivity, Experimental Part; H. Jones, Theory of Electrical and Thermal Conductivity of Metals; O. Madelung, Semiconductors.
Fröhlich, H.: "Elektronentheorie der Metalle," Springer, Berlin, 1936.
Gibson, A. F., and R. E. Burgess (eds.): "Progress in Semiconductors," vols. 1–8, Heywood, London, 1956–1964. (A series of annual volumes of review articles.)
Hannay, N. B. (ed.): "Semiconductors," Reinhold, New York, 1959.
Henisch, H. K., et al.: "Semiconducting Materials," Butterworth, London, 1951.
Kittel, C.: "Introduction to Solid State Physics," 2d ed., Wiley, New York, 1956.
Meissner, W.: "Handbuch der Experimentalphysik," vol. 11, pt. 2, Akad. Verlags., Leipzig, 1935.
Mott, N. F., and R. W. Gurney: "Electronic Processes in Ionic Crystals," Oxford University Press, London, 1940.
Mott, N. F., and H. Jones: "Theory of Metals and Alloys," Oxford University Press, London, 1936.
Schottky, W. (ed.): "Halbleiterprobleme," vols. I, II, III, Friedr. Vieweg & Sohn, Brunswick, Germany, 1955, 1956.
Seitz, F.: "The Modern Theory of Solids," McGraw-Hill, New York, 1940.
Seitz, F., and D. Turnbull (eds.): "Solid State Physics," vols. 1–17, Academic, New York, 1955–1965. Of particular interest are the following articles: H. Y. Fan, Valence Semiconductors, Germanium and Silicon, vol. 1; David Pines, Electron Interaction in Metals, vol. 2; H. Welker and H. Weiss, Group III-Group V Compounds, vol. 3; J. P. Jan, Galvanomagnetic and Thermomagnetic Effect in Metals, vol. 5; W. W. Scalon, Polar Semiconductors, vol. 9; R. W. Keyes, The Effects of Elastic Deformation on the Electrical Conductivity of Semiconductors, vol. 11; K. Mendelssohn and H. M. Rosenberg, The Thermal Conductivity of Metals at Low Temperature, vol. 12; L. J. Sham and J. M. Ziman, "The Electron-Phonon Interaction," vol. 15; H. G. Drickamer, "High Pressure and Electronic Structure," vol. 17.
Shockley, W.: "Electrons and Holes in Semiconductors," Van Nostrand, Princeton, N.J., 1950.
Smith, R. A.: "Semiconductors," Cambridge University Press, London, 1959.
Spenke, E.: "Electronische Halbleiter," Springer, Berlin, 1955.
Wilson, A. H.: "Theory of Metals," 2d ed., Cambridge University Press, London, 1953.
Ziman, J. M.: "Electrons and Phonons," Oxford University Press, London, 1960.

Chapter 7

Dielectrics

By A. VON HIPPEL, Massachusetts Institute of Technology

1. Introduction

In recent years, the designation *dielectric* [1]† has been applied to the broad expanse of *nonmetals* when considering their interaction with electric, magnetic, or electromagnetic fields. The storage and dissipation of electric and magnetic energy are the points at issue; polarization, magnetization, and conduction are the dielectric properties under consideration; and the discussion ranges from gases to liquids and solids, from insulators to semiconductors and up to metals.

A generally useful phenomenological description of dielectrics in sinusoidal fields is based on two complex parameters, describing the storage and dissipation of electric and of magnetic energy. The concepts *complex permittivity* (ϵ^*) and *complex permeability* (μ^*) are best suited for this purpose (Sec. 2). Incorporated into the field theory through the vectors *polarization* and *magnetization* (Sec. 3), they may be transformed into other sets of parameters, such as the complex propagation factor and impedance or the complex index of refraction (Sec. 4).

Polarization and magnetization measure the electric- and the magnetic-dipole moment per unit volume. These macroscopic moments are composed of elementary molecular moments. Induced electric-dipole moments result from the displacement of electrons or nuclei (electronic and atomic polarization); permanent dipole moments of molecules may be oriented (orientation polarization); apparent dipole moments, furthermore, result from field distortions when the travel of charge carriers is impeded (space charge or interfacial polarization) (Sec. 5). Electronic, atomic, and orientation polarization in gases lead to the resonance spectra of spectroscopy (Sec. 6); the orientation polarization in liquids and solids causes relaxation spectra (Sec. 7). Feedback coupling between dipole moments in special crystal structures produces piezoelectricity and ferroelectricity and macroscopic electroacoustic resonances (Sec. 8). The piling up of migrating charge carriers yields a relaxation polarization which may be indistinguishable from orientation polarization without special study (Sec. 9). Current transfer can destroy dielectrics by a variety of processes, of which the most interesting and debated one is the intrinsic electric breakdown by electron impact ionization (Sec. 10).

† Numbers in brackets refer to References at end of chapter.

2. Complex Permittivity and Permeability

A capacitor, connected to a sinusoidal voltage source

$$V = V_0 e^{j\omega t} \tag{7.1}$$

of *angular frequency* $\omega = 2\pi\nu$, stores, when vacuum is its dielectric, a charge

$$Q = C_0 V \tag{7.2}$$

and draws a *charging current*

$$I_c = \frac{dQ}{dt} = j\omega C_0 V \tag{7.3}$$

leading the voltage by a temporal-phase angle of 90° (Fig. 7.1). C_0 is the *vacuum* (or geometrical) *capacitance* of the capacitor.

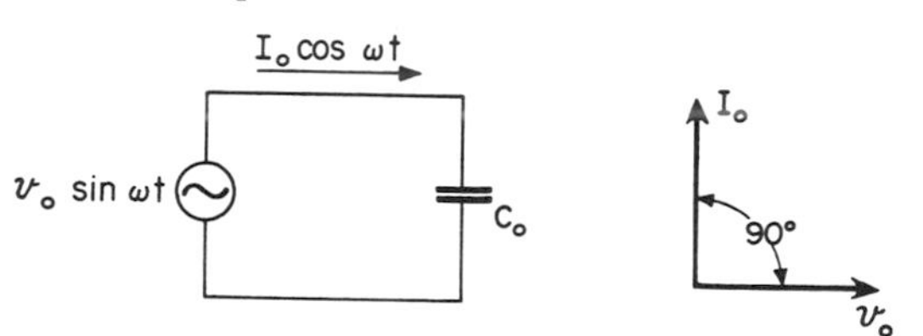

FIG. 7.1. Current-voltage relation in ideal capacitor.

When filled with some substance, the capacitor increases its capacitance to

$$C = C_0 \frac{\epsilon'}{\epsilon_0} = C_0 \kappa' \tag{7.4}$$

where ϵ' and ϵ_0 designate the real *permittivities*, or *dielectric constants*, of the dielectric and of vacuum, respectively, and their ratio κ' the *relative permittivity* (*relative dielectric constant*) of the material. Simultaneously, there may appear, in addition to the charging-current component I_c, a *loss-current* component

$$I_l = GV \tag{7.5}$$

in phase with the voltage; G represents the conductance of the dielectric. The total current traversing the capacitor,

$$I = I_c + I_l = (j\omega C + G)V \tag{7.6}$$

is inclined by a *power-factor angle* $\theta < 90°$ against the applied voltage V, that is, by a *loss angle* δ against the $+j$ axis (Fig. 7.2).

It is incorrect to suppose that the dielectric material corresponds in its electrical behavior to a capacitor paralleled by a resistor (*R-C* circuit) (Fig. 7.3). The

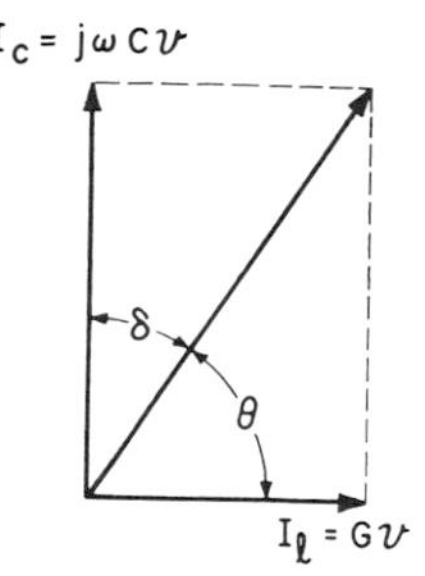

FIG. 7.2. Capacitor containing dielectric with loss.

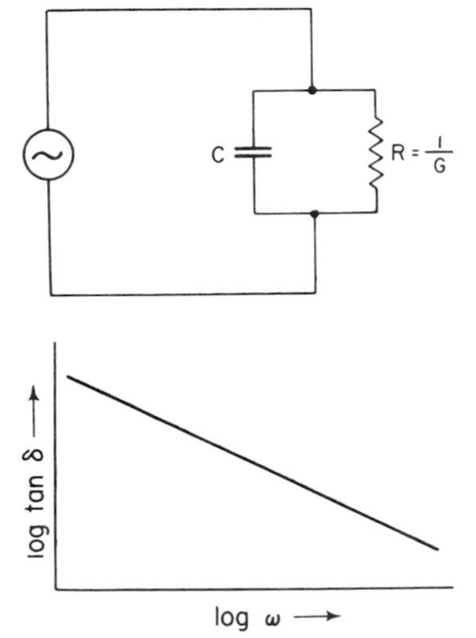

FIG. 7.3. *R-C* circuit and its frequency response.

frequency response of this circuit, which can be expressed by the ratio of loss current to charging current, i.e., the *dissipation factor D* or *loss tangent* tan δ, as

$$D = \tan \delta = \frac{I_l}{I_c} = \frac{1}{\omega RC} \tag{7.7}$$

may not at all agree with that actually observed, because the conductance term need not stem from a migration of charge carriers but can represent any other energy-consuming process. It has therefore become customary to refer to the existence of a loss current in addition to a charging current noncommittally by the introduction of a *complex permittivity*

$$\epsilon^* = \epsilon' - j\epsilon'' \tag{7.8}$$

The total current I of Eq. (7.6) may thus be rewritten

$$I = (j\omega\epsilon' + \omega\epsilon'') \frac{C_0}{\epsilon_0} V = j\omega C_0 \kappa^* V \tag{7.9}$$

where

$$\kappa^* \equiv \frac{\epsilon^*}{\epsilon_0} = \kappa' - j\kappa'' \tag{7.10}$$

is the *complex relative permittivity* of the material and ϵ'' and κ'' the *loss factor* and *relative loss factor*, respectively. The loss tangent becomes

$$\tan \delta = \frac{\epsilon''}{\epsilon'} = \frac{\kappa''}{\kappa'} \tag{7.11}$$

Since a parallel-plate capacitor of the area A and the plate separation d, fringing effects neglected, has the vacuum capacitance

$$C_0 = \frac{A}{d} \epsilon_0 \tag{7.12}$$

the current density J traversing a capacitor under the applied field strength $E = V/d$ becomes, according to Eq. (7.9),

$$J = (j\omega\epsilon' + \omega\epsilon'')E = \epsilon^* \frac{dE}{dt} \tag{7.13}$$

(Fig. 7.4). The product of angular frequency and loss factor is equivalent to a *dielectric conductivity*

$$\sigma = \omega\epsilon'' \tag{7.14}$$

which may be an actual conductivity caused by migrating charge carriers or may refer to some other source of friction such as the orientation of dipoles.

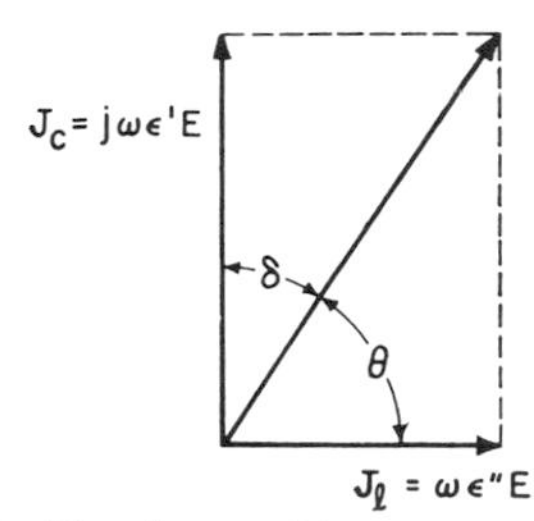

FIG. 7.4. Charging- and loss-current density.

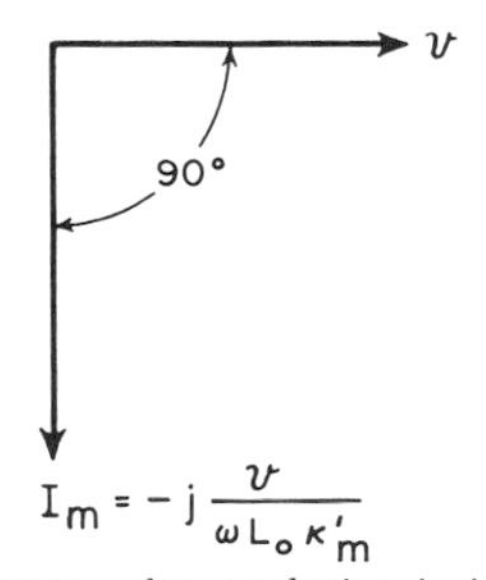

FIG. 7.5. Current-voltage relation in ideal inductor.

If the dielectric material is transferred from the electric field of the capacitor to the magnetic field of a coil, the voltage V drives through the coil a magnetization current I_m, according to Faraday's inductance law $[V = L(dI/dt)]$, as

$$I_m = \frac{V}{j\omega L_0(\mu'/\mu_0)} = -j \frac{V}{\omega L_0 \kappa'_m} \tag{7.15}$$

L represents the *inductance* and L_0 the *vacuum* (or geometrical) *inductance* of the coil. This magnetization current lags behind the applied voltage by 90° (Fig. 7.5). The *permeabilities* μ' and μ_0 designate the magnetization of the material and of vacuum, respectively, and their ratio

$$\kappa'_m \equiv \frac{\mu'}{\mu_0} \tag{7.16}$$

the *relative permeability* of the material in which the magnetic field of the coil resides.

Because of the resistance R of the coil windings, an ohmic-current component V/R exists. In addition, there may appear, in phase with V, a magnetic-loss current I_l caused by energy dissipation during the magnetization cycle. We shall allow for this magnetic loss by introducing a *complex permeability*

$$\mu^* = \mu' - j\mu'' \tag{7.17}$$

and a *complex relative permeability*

$$\kappa_m{}^* = \frac{\mu^*}{\mu_0} = \kappa'_m - j\kappa''_m \tag{7.18}$$

in complete analogy to the electric case. Thus we obtain the total magnetization current

$$I = I_m + I_l = \frac{V}{j\omega L_0 \kappa^*_m} = -\frac{jV(\mu' + j\mu'')}{\omega(L_0/\mu_0)(\mu'^2 + \mu''^2)} \tag{7.19}$$

Accordingly the macroscopic electric and magnetic behavior in sinusoidal fields is determined by the two complex parameters ϵ^* and μ^*. The real and imaginary parts (ϵ' and ϵ'' or μ' and μ'', respectively) of these complex parameters are not independent of each other. When one of them is given over the whole frequency spectrum, the other one is prescribed, because dispersion and absorption are two different aspects of the same phenomenon.

3. Polarization and Magnetization

A dielectric material increases the storage capacity of a capacitor by neutralizing charges at the electrode surfaces which otherwise would contribute to the external field. Faraday was the first to recognize this phenomenon of *dielectric polarization.* We may visualize it as the action of dipole chains which form under the influence of the applied field and bind countercharges with their free ends on the metal surfaces (Fig. 7.6).

By writing the voltage of the capacitor according to Eqs. (7.2) and (7.4) as

$$V = \frac{Q}{\kappa'}\frac{1}{C_0} \tag{7.20}$$

we may interpret this equation as stating that only a fraction of the *total charge* Q, the *free charge* Q/κ', contributes to the voltage, while the remainder, the *bound charge* $Q(1 - 1/\kappa')$, is neutralized by the polarization of the dielectric.

To obtain a clearer conception of the charge distribution and its effect in space, we correlate charge densities to field vectors. The *total* (or *true*) *charge* Q concentrated in the capacitor is distributed over the surface area A of the metal electrodes with a density s as

$$Q = \int_A s\,dA \tag{7.21}$$

We represent this true-charge density s by a vector $\mathbf{D}$, the *electric-flux density* (or *dielectric displacement*), such that the surface-charge density will be equal to the normal component of $\mathbf{D}$, or

$$s\,dA \equiv D\cos\alpha\,dA \equiv \mathbf{D}\cdot\mathbf{n}\,dA = D_n\,dA \tag{7.22}$$

A positive value of the *scalar* or *dot product* of the vector $\mathbf{D}$ and the unit normal vector $\mathbf{n}$ indicates a positive charge. (By postulating that the surface-charge densities are equal to the normal components of the field vectors instead of 4π times their magnitude, we have chosen a rationalized system of units.)

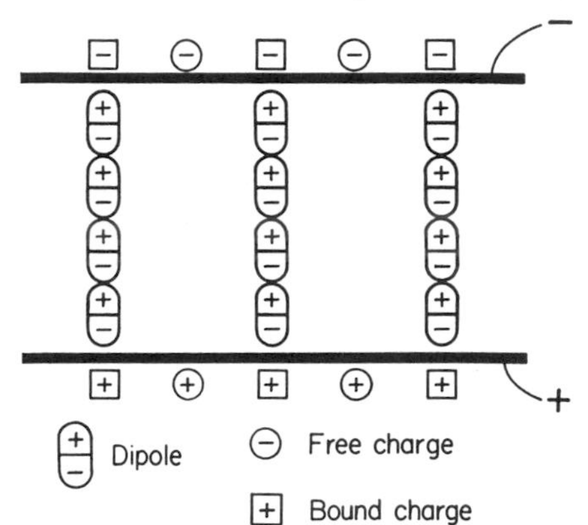

FIG. 7.6. Schematic representation of dielectric polarization.

Similarly, we allocate to the free-charge density s/κ' a vector $\mathbf{E}$, the *electric-field strength,* or *field intensity,* by defining

$$\frac{s}{\kappa'}\,da \equiv \epsilon_0\mathbf{E}\cdot\mathbf{n}\,dA = \epsilon_0 E_n\,dA \tag{7.23}$$

and to the bound-charge density a vector $\mathbf{P}$, called the *polarization,* as

$$s\left(1 - \frac{1}{\kappa'}\right) da \equiv \mathbf{P}\cdot\mathbf{n}\,dA = P_n\,dA \tag{7.24}$$

From Eqs. (7.22) and (7.23) follows the relation between dielectric-flux density and field strength

$$\mathbf{D} = \epsilon'\mathbf{E} \tag{7.25}$$

and from Eqs. (7.22) to (7.24) the interrelation between the three field vectors

$$\mathbf{P} = \mathbf{D} - \epsilon_0\mathbf{E} = (\epsilon' - \epsilon_0)\mathbf{E} \equiv \chi\epsilon_0\mathbf{E} \tag{7.26}$$

The factor

$$\chi = \frac{\mathbf{P}}{\epsilon_0\mathbf{E}} = \kappa' - 1 = \frac{\text{bound-charge density}}{\text{free-charge density}} \tag{7.27}$$

is known as the *electric susceptibility* of the dielectric material.

Electric-flux density $\mathbf{D}$ and polarization $\mathbf{P}$ have, according to their defining equations, the dimension *charge per unit area,* while the electric-field strength **E** can have a different physical meaning because the dimensions of the dielectric constant may yet be chosen. We take advantage of this possibility and extend the concept of the electric field into space by postulating that an electric-probe charge Q', placed in an electrostatic field of the intensity **E**, will be subjected to a force

$$\mathbf{F} \equiv Q'\mathbf{E} \tag{7.28}$$

Thus the electric-field strength **E** becomes equivalent in magnitude and direction to the force per unit charge acting on a detector charge, and the dielectric constant obtains the dimensions

$$[\epsilon] = \frac{\text{charge per unit area}}{\text{force per unit charge}} \tag{7.29}$$

Two electric charges of opposite polarity, $\pm Q$, separated by a distance d, represent a dipole of the moment

$$\boldsymbol{\mu} = Q\mathbf{d} \tag{7.30}$$

This *electric-dipole moment* is symbolized by a vector of the magnitude $|\boldsymbol{\mu}|$ pointing from the negative to the positive pole (Fig. 7.7). The polarization vector

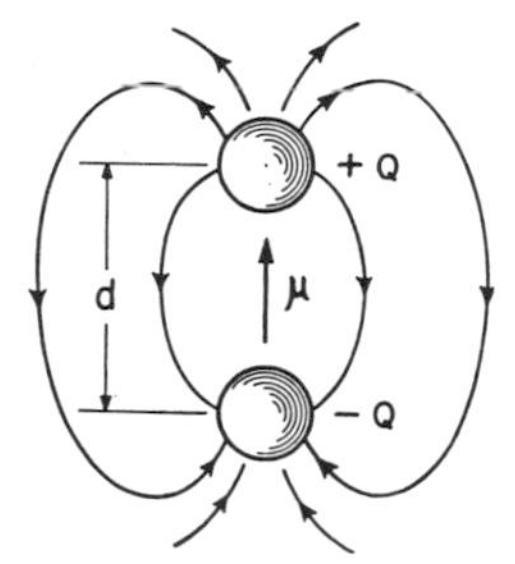

FIG. 7.7. Electric dipole of the moment $\boldsymbol{\mu} = Q\mathbf{d}$.

P corresponds in magnitude to the surface-charge density bound at the electrodes by the polarized dielectric, and it points in the direction of the applied field. The polarization **P** is therefore identical with the *electric-dipole moment per unit volume* of the dielectric material.

The electrostatic field in space obtains physical meaning because the field strength at any point can be measured by the force acting on a detector charge [Eq. (7.28)]. Alternatively, it could be measured by the torque **T** exercised by the electric field **E** on an electric dipole $\boldsymbol{\mu}$ as

$$\mathbf{T} = |\boldsymbol{\mu}|\,|\mathbf{E}|\sin\theta = \boldsymbol{\mu} \times \mathbf{E} \tag{7.31}$$

which tends to align this dipole in the field direction (Fig. 7.8).

The concepts developed for the electrostatic field apply for the magnetostatic field with the restriction that individual magnetic point charges of north and south polarity are not known to exist in nature. Hence the magnetic-dipole moment

$$\mathbf{m} = p\mathbf{d} \qquad p = \text{pole strength} \tag{7.32}$$

(Fig. 7.9) is the starting point. Visualizing that, under the influence of a magnetic field **H**, magnetic-dipole chains form in a dielectric analogy to the electric polarization of Fig. 7.6, we can introduce a *magnetization* vector **M**, which represents the *magnetic-dipole moment per unit volume* of the material. This magnetization **M**, together with the magnetic field **H** in the dielectric, determines the total *magnetic-flux density* (*magnetic induction*) **B**.

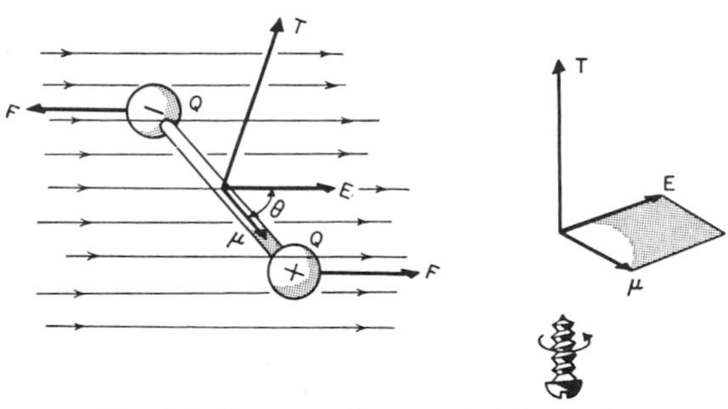

FIG. 7.8. Torque acting on electric dipole.

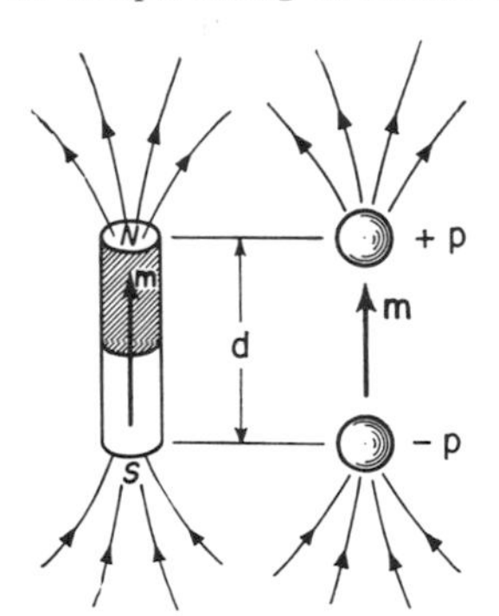

FIG. 7.9. Magnetic dipole.

At this point the magnetic-field theory deviates in its mathematical formulation from the electric theory, causing a great deal of confusion. Electric-flux density **D** and polarization **P** have identical dimensions (*surface-charge density* or *electric moment per unit volume*), and the interrelating equation

$$\mathbf{D} = \epsilon_0\mathbf{E} + \mathbf{P} = \epsilon'\mathbf{E} \tag{7.33}$$

allows us to define the electric-field strength **E** as *force per unit charge* or *torque per unit dipole moment* by the proper choice of the dimensions of the permittivity [see Eq. (7.29)]. The magnetic-flux density **B**, however, is defined by the equation

$$\mathbf{B} \equiv \mu_0\mathbf{H} + \mu_0\mathbf{M} = \mu'\mathbf{H} \tag{7.34}$$

where the factors μ' and μ_0 represent the *permeability* (*induced capacity*) of the material and of vacuum, respectively. Thus the magnetic-field strength **H** obtains the same dimension (*magnetic moment per unit volume*) as the magnetization **M**, and only the magnetic induction **B** can acquire, by a proper choice of the dimensions of the permeability, the meaning of torque per unit dipole moment. Thus **B** appears in the force and torque equations of the magnetic field, but not **H**, and the magnetic analogue to Eq. (7.31) is

$$\mathbf{T} = |\mathbf{m}|\,|\mathbf{B}|\sin\theta = \mathbf{m} \times \mathbf{B} \tag{7.35}$$

Rewriting Eq. (7.34) for the magnetization, we obtain

$$\mathbf{M} = \frac{1}{\mu_0}\mathbf{B} - \mathbf{H} = \chi_m \mathbf{H} \tag{7.36}$$

and define, in analogy to the electric susceptibility of Eq. (7.7), a *magnetic susceptibility*

$$\chi_m \equiv \frac{\mathbf{M}}{\mathbf{H}} = \frac{\mu'}{\mu_0} - 1 \equiv \kappa'_m - 1 \tag{7.37}$$

κ'_m is the *relative permeability*.

Thus far, electricity and magnetism appear as two new phenomena independent of each other and consequently requiring the introduction of two new fundamental quantities, for example, *electric charge* and *magnetic-dipole moment*, for their description. Actually, they are interlinked, and only one new, independent quantity may be introduced. One interrelation is given by the fact that an electric current creates a magnetic field, according to *Ampère's circuital law*

$$\oint \mathbf{H} \cdot d\mathbf{l} = I \tag{7.38}$$

(Fig. 7.10). The magnetic field encircling an electric current is a whirlpool field of closed lines, in contrast

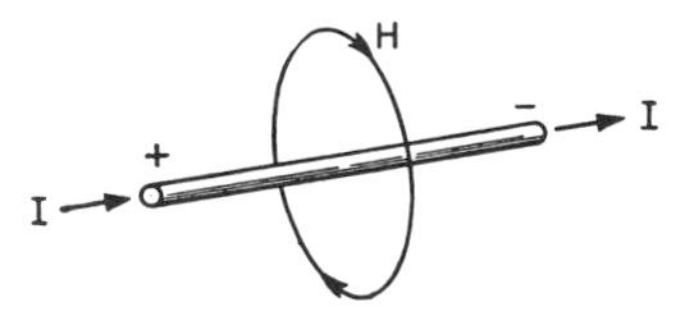

Fig. 7.10. Magnetic field encircling electric current.

to the magnetostatic field that originates in the free ends of dipole chains of a magnetic material. However, a ring current of the magnitude I encircling an area A produces at a distance, large in comparison with its radius, a magnetic field which is identical to that of a magnetic dipole of the moment

$$\mathbf{m} = IA\mathbf{n} \tag{7.39}$$

(Fig. 7.11). This equivalence between the fields of magnetic dipoles and of circular currents allows us to

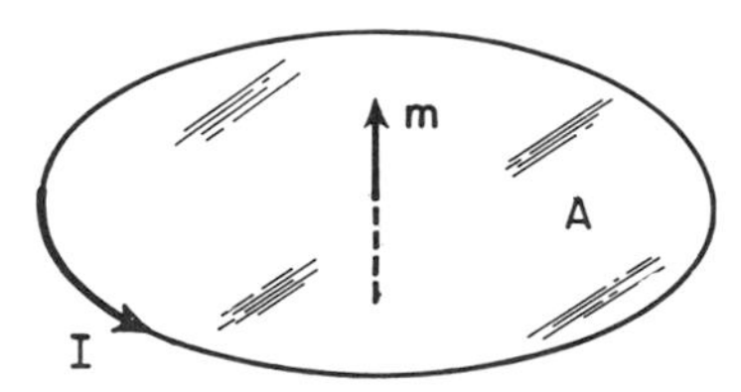

Fig. 7.11. Magnetic moment of circular current.

explain the phenomenon *magnetism* and the nonexistence of magnetic monopoles on the basis that the sources of magnetism may be *molecular-ring*, or *Ampère, currents*.

A second interrelation between electric and magnetic fields is given by *Faraday's induction law*. It states that when the magnetic flux

$$\Phi = \int_A \mathbf{B} \cdot \mathbf{n}\, dA \tag{7.40}$$

traversing a loop of wire, changes, whether by a change of the magnetic induction $\mathbf{B}$ or by a change in the position or shape of the loop, an emf is created along the wire, causing an induced voltage V_i to appear between its ends,

$$V_i = \int_a^b \mathbf{E} \cdot d\mathbf{l} = -\frac{d\Phi}{dt} \tag{7.41}$$

proportional to the speed of this change. If the loop is closed, a current I will flow through the loop resistance R,

$$I = \frac{V_i}{R} = -\frac{1}{R}\frac{d\Phi}{dt} \tag{7.42}$$

that causes a magnetic field opposing the change in flux.

Ampère's circuital law and Faraday's induction law were discovered and formulated for currents and voltages in wire loops. Maxwell postulated that these laws are valid in space quite independent of the presence of detector loops in which currents and voltages might develop, and thus he arrived at the electromagnetic-field equations.

A discussion of the current drawn by a capacitor will make evident the formulation of Maxwell's first field equation. The current of density J streaming into the electrode system is equal to the change of the true charge stored in the capacitor; that is,

$$\int_A \mathbf{J} \cdot \mathbf{n}\, dA = \int_A \frac{ds}{dt}\, dA \tag{7.43}$$

The surface-charge density s originates the electric-flux density $\mathbf{D}$ [see Eq. (7.22)]; hence,

$$\frac{ds}{dt}\, dA = \frac{d\mathbf{D}}{dt}\, \mathbf{n}\, dA \tag{7.44}$$

The current density may be measured as the time derivative of the electric-flux density at the electrode surface,

$$\mathbf{J} = \frac{d\mathbf{D}}{dt} \tag{7.45}$$

By postulating that this equivalence between the temporal change of the dielectric-flux density and an electric current holds also for the interior of a dielectric, i.e., that this change in flux produces a magnetic field just like a conduction current, Maxwell arrived at the concept that the conduction current charging a capacitor finds its continuation in a field current traversing the dielectric,

$$\int_A \mathbf{J} \cdot \mathbf{n}\, dA = \int_{A'} \frac{d\mathbf{D}}{dt}\, \mathbf{n}\, dA \tag{7.46}$$

This field current which extends through the cross section A' of the dielectric as far as the electric field of the capacitor reaches was named by Maxwell the *displacement current*. By including this displacement current in Ampère's circuit law, ***Maxwell's first field equation*** results in the integral formulation

$$\oint \mathbf{H} \cdot dl = \int_A \mathbf{J} \cdot \mathbf{n}\, dA + \int_{A'} \frac{\partial \mathbf{D}}{\partial t}\, n\, dA \tag{7.47}$$

With the help of Stokes's theorem, the line integral on the left can be transformed into a surface integral, and the differential formulation

$$\nabla \times \mathbf{H} = \mathbf{J} + \frac{\partial \mathbf{D}}{\partial t} \tag{7.48}$$

is obtained.

The current of the density $\mathbf{J}$ may be true conduction current obeying Ohm's law

$$J = \sigma \mathbf{E} \tag{7.49}$$

However, from a more general standpoint, the conductivity σ may be interpreted as the dielectric conductivity of Eq. (7.14) representing any energy-consuming process. Thus, by introducing the complex permittivity, the first field equation may be rewritten for sinusoidal fields and isotropic, linear dielectrics as

$$\nabla \times \mathbf{H} = \epsilon^* \frac{\partial \mathbf{E}}{\partial t} \tag{7.50}$$

The *second field equation* is a generalization of Faraday's induction law, claiming that the change of a magnetic-flux density creates an emf in space quite independently of the presence of a loop for its detection,

$$\oint \mathbf{E} \cdot dl = - \int_A \frac{\partial B}{\partial t} \mathbf{n}\, dA \tag{7.51}$$

In differential form,

$$\nabla \times \mathbf{E} = - \frac{\partial \mathbf{B}}{\partial t} \tag{7.52}$$

or, because magnetization may lead to energy dissipation, we introduce the complex permeability and arrive at a formulation completely symmetrical to that of the first field equation, except for the negative sign,

$$\nabla \times \mathbf{E} = -\mu^* \frac{\partial \mathbf{H}}{\partial t} \tag{7.53}$$

Maxwell's field equations thus describe the coupling between the electric- and magnetic-field vectors and their interaction with matter in space and time.

4. Macroscopic Description of Dielectrics by Various Sets of Parameters

An interpretation of Maxwell's field equations requires, as a first step, the separation of the field vectors $\mathbf{E}$ and $\mathbf{H}$. This can be done by differentiating the equations with respect to time and substituting from one equation into the other. One obtains in this way the *wave equations of the electromagnetic field,* which for our purpose may be simplified by assuming that $\mathbf{E}$ and $\mathbf{H}$ are a function of x and t only:

$$\begin{aligned} \frac{\partial^2 \mathbf{E}}{\partial x^2} &= \epsilon^* \mu^* \frac{\partial^2 \mathbf{E}}{\partial t^2} \\ \frac{\partial^2 \mathbf{H}}{\partial x^2} &= \epsilon^* \mu^* \frac{\partial^2 \mathbf{H}}{\partial t^2} \end{aligned} \tag{7.54}$$

The solution is a plane wave,

$$\begin{aligned} \mathbf{E} &= \mathbf{E}_0 e^{j\omega t - \gamma x} \\ \mathbf{H} &= \mathbf{H}_0 e^{j\omega t - \gamma x} \end{aligned} \tag{7.55}$$

varying periodically in time with the frequency $\nu = \omega/2\pi$ and advancing in the $+x$ direction through space with a *complex propagation factor*

$$\gamma = j\omega \sqrt{\epsilon^* \mu^*} = \alpha + j\beta \tag{7.56}$$

α is the *attenuation factor,* and β is the *phase factor* of the wave. Introducing these factors, we may rewrite Eqs. (7.55):

$$\begin{aligned} \mathbf{E} &= \mathbf{E}_0 e^{-\alpha x} \exp\,[j2\pi(\nu t - \beta x/2\pi)] \\ \mathbf{H} &= \mathbf{H}_0 e^{-\alpha x} \exp\,[j2\pi(\nu t - \beta x/2\pi)] \end{aligned} \tag{7.57}$$

The wave has a time period $T = 1/\nu$ and a space period $\lambda = 2\pi/\beta$. *Surfaces of constant phase* are given by

$$\nu t - \frac{x}{\lambda} = \text{const} \tag{7.58}$$

and hence propagate with the phase velocity

$$\frac{dx}{dt} = v = \nu\lambda = \frac{\omega}{\beta} \tag{7.59}$$

For a dielectric without loss ($\epsilon^* = \epsilon'$, $\mu^* = \mu'$), one obtains from Eq. (7.56) the phase factor $\beta = \omega(\epsilon'\mu')^{1/2}$, so that the *phase velocity* in a *loss-free unbounded medium* is $v = 1/(\epsilon'\mu')^{1/2}$.

To learn about the coupling between the $\mathbf{E}$ and $\mathbf{H}$ vectors, we have to return to the field equations and write out the field components. These component equations contain three statements:

1. The x components of the field vectors, the longitudinal field components of the electromagnetic wave, are independent of space and time and hence may be assumed to be zero. The plane wave is a *transverse electromagnetic,* or TEM, wave.

2. The coupled transversal components of the $\mathbf{E}$ and $\mathbf{H}$ waves are *perpendicular* to each other and form, together with the propagation direction, a right-hand coordinate system of the sequence $+x \rightarrow E_y \rightarrow H_z$ (Fig. 7.12).

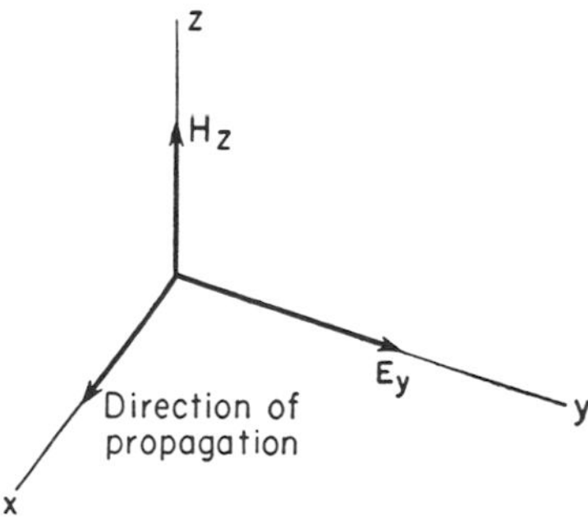

FIG. 7.12. Right-hand coordinate system for traveling TEM wave.

3. The ratio of the coupled electric- and magnetic-field vectors follows as

$$\frac{\mathbf{E}}{\mathbf{H}} = \frac{\gamma}{j\omega\epsilon^*} \equiv Z \tag{7.60}$$

This ratio Z, the *intrinsic impedance of the dielectric*, may be rewritten, using Eq. (7.56), in any one of three versions:

$$Z = \frac{\gamma}{j\omega\epsilon^*} = \sqrt{\frac{\mu^*}{\epsilon^*}} = \frac{j\omega\mu^*}{\gamma} \tag{7.61}$$

In Secs. 1 and 2 the response of a dielectric material to sinusoidal electric and magnetic fields was expressed by the two complex parameters ϵ^* and μ^*. These parameters were derived from the amplitude and temporal-phase relations between voltage and current in capacitors and coils, but one is not restricted to using ϵ^* and μ^* and may refer to alternative parameters.

The power engineer replaces the dielectric constant ϵ' and the loss factor ϵ'' by the combination of ϵ' and *power factor* $\cos\theta$, and the radio engineer may choose ϵ' and the *loss tangent* $\tan\delta$, where

$$\tan\delta = \frac{\epsilon''}{\epsilon'} = \frac{\text{loss current}}{\text{charging current}} \tag{7.62}$$

(Since $\cos\theta = \sin\delta$, the power factor and loss tangent may be considered equal only for sufficiently small *loss angles* δ.) Frequently the inverse of the loss tangent, the quality factor Q of the dielectric,

$$Q = \frac{1}{\tan\delta} = \frac{\omega\epsilon' E_0^2}{\omega\epsilon'' E_0^2} = 2\pi\nu\frac{\frac{1}{2}\epsilon' E_0^2}{\frac{1}{2}\sigma E_0^2}$$

$$= 2\pi\frac{\begin{matrix}\text{energy stored}\\ \text{per half cycle}\end{matrix}}{\begin{matrix}\text{energy dissipated}\\ \text{per half cycle}\end{matrix}} = \frac{\text{volt-amperes}}{\text{watts dissipated}} \tag{7.63}$$

serves as the *figure of merit*, especially in waveguide problems. An engineer interested in dielectric heating will probably refer to ϵ' and the *dielectric conductivity* $\sigma = \omega\epsilon''$ (ohm-meter)$^{-1}$ because the power absorbed per unit volume is $P = \sigma(E_0^2/2)$ watt/m^3.

If, instead of the time relation between current and voltage, the electromagnetic field in space is considered, new substitutes for ϵ^* and μ^* offer themselves. To derive them conveniently, we visualize the spatial electric-wave train at some moment t_1:

$$E_y = E_1 e^{-\gamma x} = E_1 e^{-\alpha x}\exp\left(-j2\pi\frac{x}{\lambda}\right) \tag{7.64}$$

[see Eqs. (7.57)]. The wave amplitude oscillates in space with a periodicity λ; it is enclosed between exponential envelopes determined by the attenuation constant α (Fig. 7.13*a*). Alternatively, in polar coordinates, the wave amplitude may be depicted as a radius vector which, rotating clockwise as the distance increases, describes a logarithmic spiral (Fig. 7.13*b*). The parameter x is replaced in the latter representation by the phase angle ϕ according to the relation $x/\lambda = \phi/2\pi$, and the electric-field strength is rewritten as

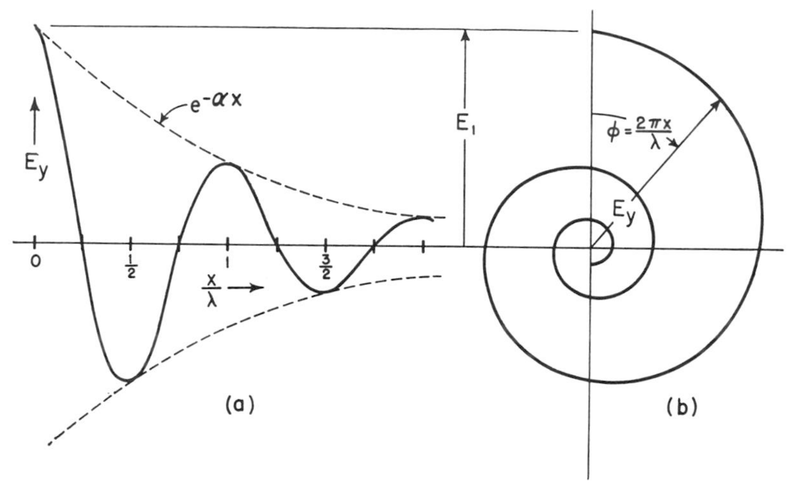

FIG. 7.13. Electric-wave train in space.

$$E_y = E_1\exp\left[-\phi\left(\frac{\alpha\lambda}{2\pi} + j\right)\right] \tag{7.65}$$

In vacuum the wavelength is λ_0, and the wave travels with the velocity of light: $c = \lambda_0\nu = 1/(\epsilon_0\mu_0)^{1/2}$. In other media the wavelength normally shortens, and the phase velocity slows down. The ratio of the wavelength or phase velocity in vacuum to that in the dielectric designates the *index of refraction* of the dielectric medium

$$n \equiv \frac{\lambda_0}{\lambda} = \frac{c}{v} = \frac{\lambda_0}{2\pi}\beta \tag{7.66}$$

For a loss-free medium this equation simplifies to

$$n = \sqrt{\frac{\epsilon'\mu'}{\epsilon_0\mu_0}} \equiv \sqrt{\kappa'\kappa'_m} \tag{7.67}$$

If, in addition, the magnetization can be neglected ($\mu' = \mu_0$), the well-known *Maxwell relation* results:

$$n^2 = \frac{\epsilon'}{\epsilon_0} = \kappa' \tag{7.68}$$

(This relation has been abused frequently in predicting static dielectric constants from optical-refraction data. Actually, it states only that the square of the index of refraction of a nonabsorbing, nonmagnetic material is equal to the relative permittivity at that frequency.)

The physicist normally uses the index of refraction as one of his parameters and pairs with it, by making use of the polar representation of the wave, the attenuation per radian called *index of absorption*

$$k = \frac{\alpha\lambda}{2\pi} = \frac{\alpha}{\beta} \tag{7.69}$$

By substituting these indices of refraction and absorption for the attenuation factor α and the phase factor β of the propagation factor in Eq. (7.56) we obtain

$$\gamma = j\frac{2\pi}{\lambda_0}n(1 - jk) = j\frac{2\pi}{\lambda_0}n^* \tag{7.70}$$

The propagation factor γ used by the communications engineer may thus be replaced by the *complex index of refraction* $n^* = n(1 - jk)$, employed in the calculations of physical optics.

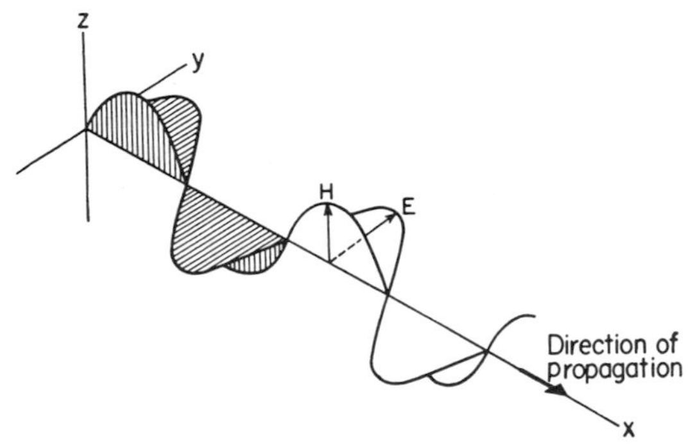

FIG. 7.14. Traveling TEM wave in loss-free dielectric.

The propagation factor γ is proportional to the product $(\epsilon^*\mu^*)^{1/2}$, while the intrinsic impedance Z is equal to the ratio $(\mu^*/\epsilon^*)^{1/2}$. Both complex quantities have to be determined to obtain ϵ^* and μ^* individually. From the intrinsic impedance

$$Z = \frac{E}{H} = \sqrt{\frac{\mu^*}{\epsilon^*}} \tag{7.71}$$

in polar form,

$$Z = |Z|e^{j\zeta} = \left[\frac{(\epsilon'\mu' + \epsilon''\mu'')^2 + (\epsilon''\mu' - \epsilon'\mu'')^2}{(\epsilon'^2 + \epsilon''^2)^2}\right]^{1/4} e^{j\zeta} \tag{7.72}$$

with

$$\tan 2\zeta = \frac{\epsilon''\mu' - \epsilon'\mu''}{\epsilon'\mu' + \epsilon''\mu''} \tag{7.73}$$

we can derive the phase relation between the electric and magnetic wave. The electric-field vector is advanced or retarded with respect to the magnetic vector in temporal phase, depending on the preponderance of the term pertaining to the electric or the magnetic loss. For negligible magnetic loss ($\mu'' = 0$),

$$\tan 2\zeta = \tan \delta = \frac{2k}{1 - k^2}$$

or

$$\tan \zeta = k$$

The phase advance of the electric wave is equal to the arc tangent of the index of absorption. In a loss-free medium in unbounded space the electric- and magnetic-field vectors of an electromagnetic wave are exactly in phase.

The general characterization of a dielectric as the carrier of an electromagnetic field requires two independent complex parameters which have to be determined by four independent measurements; however, the situation fortunately simplifies in practice. Ferromagnetics exempted, the magnetic polarization is, in general, so weak that μ^* may be replaced by the permeability μ_0 of free space for all practical purposes. Thus, two measurements normally suffice to determine the dielectric response of homogeneous isotropic materials at a given frequency. Consequently, in most cases, dielectric characteristics show only the specific permittivity $\kappa' = \epsilon'/\epsilon_0$ and the loss tangent $\tan \delta$ or loss factor κ''.

Equating the real and imaginary parts of Eq. (7.56), the attenuation factor of a transversal electromagnetic wave (TEM wave) is

$$\alpha = \frac{\lambda\omega^2}{4\pi}(\epsilon'\mu'' + \epsilon''\mu') \tag{7.74}$$

and for the phase factor

$$\beta = \frac{2\pi}{\lambda} = \omega\left\{\frac{\epsilon'\mu' - \epsilon''\mu''}{2}\left[1 + \sqrt{1 + \left(\frac{\epsilon'\mu'' + \epsilon''\mu'}{\epsilon'\mu' + \epsilon''\mu''}\right)^2}\right]\right\}^{1/2} \tag{7.75}$$

Thus for materials with negligible magnetic loss ($\mu'' = 0$) one obtains from Eq. (7.75) for the wavelength the simplified expression

$$\lambda = \frac{1}{\nu}\frac{1}{[\frac{1}{2}\epsilon'\mu'(1 + \sqrt{1 + \tan^2\delta})]^{1/2}} \tag{7.76}$$

If, in addition, the permeability is that of vacuum ($\mu' = \mu_0$), the index of refraction is

$$n = \frac{\lambda_0}{\lambda} = [\tfrac{1}{2}\kappa'(\sqrt{1 + \tan^2\delta} + 1)]^{1/2} \tag{7.77}$$

Similarly, the attenuation factor becomes

$$\alpha = \frac{2\pi}{\lambda_0}[\tfrac{1}{2}\kappa'(\sqrt{1 + \tan^2\delta} - 1)]^{1/2} \tag{7.78}$$

and the index of absorption

$$k = \frac{\alpha}{\beta} = \left(\frac{\sqrt{1 + \tan^2\delta} - 1}{\sqrt{1 + \tan^2\delta} + 1}\right)^{1/2} \tag{7.79}$$

These equations show that it is convenient to discuss the effect of the dielectric loss on other parameters by studying the three boundary cases: $\tan^2 \delta \ll 1$, $\tan^2 \delta \simeq 1$, and $\tan^2 \delta \gg 1$.

The attenuation produced by a dielectric is frequently expressed as the *attenuation distance* $1/\alpha$ through which the field strength decays to $1/e = 0.368$ of its original value,

$$\frac{1}{\alpha} = \frac{\lambda_0}{2\pi}\left[\frac{2}{\kappa'(\sqrt{1+\tan^2\delta}-1)}\right]^{1/2} \quad \text{m} \qquad (7.80)$$

or as the attenuation in *decibels per meter* produced by the material. If the field strength falls from $E(0)$ to $E(x)$ or the power from $P(0)$ to $P(x)$, over a length x of the dielectric, this decibel loss is defined as

$$20 \log \frac{E(0)}{E(x)} = 10 \log \frac{P(0)}{P(x)} = 8.686\alpha x \quad \text{db} \qquad (7.81)$$

that is, the decibel loss per meter is given as

$$8.686\alpha = 8.686 \frac{2\pi}{\lambda_0} [\tfrac{1}{2}\kappa'(\sqrt{1+\tan^2\delta} - 1)]^{1/2} \quad \text{db/m} \qquad (7.82)$$

For low-loss materials ($\tan \delta \ll 1$) this loss becomes simply

$$8.686 \frac{\pi}{\lambda_0} \sqrt{\kappa'} \tan \delta = 1637 \frac{\sigma}{\sqrt{\kappa'}} \quad \text{db/m} \qquad (7.83)$$

To arrive at numerical values requires two steps: a measurement of the velocity of light $c = 1/(\epsilon_0\mu_0)^{1/2}$, which establishes the product $\epsilon_0\mu_0$, and an agreement as to the value of ϵ_0 or μ_0. By international consent, the value of μ_0 has been fixed for the rationalized mks system as $\mu_0 = 4\pi \times 10^{-7} = 1.257 \times 10^{-6}$ henry/m. The velocity of light has been measured as

$$c = 2.9979 \times 10^8 \simeq 3 \times 10^8 \quad \text{m/sec}$$

Hence the dielectric constant of free space becomes

$$\epsilon_0 = \frac{1}{36\pi} \times 10^{-9} = 8.854 \times 10^{-12} \quad \text{farad/m} \qquad (7.84)$$

The intrinsic impedance Z_0 of free space, determined by the ratio of permeability to permittivity [see Eq. (7.71)], follows as

$$Z_0 = \sqrt{\frac{\mu_0}{\epsilon_0}} = 120\pi \simeq 376.6 \quad \text{ohms} \qquad (7.85)$$

5. Molecular Mechanisms of Polarization

A dielectric material can react to an electric field because it contains charge carriers that can be displaced. In Fig. 7.6 this polarization phenomenon was pictured schematically by the formation of dipole chains which line up parallel to the field and bind countercharges at the electrodes. The density of the neutralized surface charge is represented by the polarization vector

$$\mathbf{P} = (\epsilon' - \epsilon_0)\mathbf{E} = (\kappa' - 1)\epsilon_0 E \quad \text{coulomb/m}^2 \qquad (7.86)$$

Alternatively, the polarization $\mathbf{P}$ is equivalent to the dipole moment per unit volume of the material.

The dipole moment per unit volume results from the additive action of N elementary dipole moments $\boldsymbol{\mu}_{av}$: $\mathbf{P} = N\boldsymbol{\mu}_{av}$. The average dipole moment $\boldsymbol{\mu}_{av}$ of the elementary particle, furthermore, is proportional to the *local electric-field strength* E' that acts on the particle: $\boldsymbol{\mu}_{av} = \alpha\mathbf{E}'$. The proportionality factor α, called *polarizability*, measures the average dipole moment per unit field strength; its dimensions are sec² coulomb²/kg $= \epsilon\text{m}^3$ in the mks or cm³ in the esu system, respectively.

The equations above give two expressions for polarization:

$$\mathbf{P} = (\kappa' - 1)\epsilon_0\mathbf{E} = N\alpha\mathbf{E}' \qquad (7.87)$$

linking the macroscopically measured dielectric constant to three molecular parameters: the number N of contributing elementary particles per unit volume, their polarizability α, and the locally acting electric field $\mathbf{E}'$. This field will normally differ from the applied field $\mathbf{E}$ because of the polarization of the surrounding dielectric medium. It is the goal of the molecular theories to evaluate these parameters and thus to arrive at an understanding of the phenomenon *polarization* and its dependence on frequency, temperature, and applied field strength.

Matter consists of positive atomic nuclei surrounded by negative electron clouds. Upon application of an external electric field, electrons are displaced slightly with respect to the nuclei; *induced* dipole moments result and cause the so-called *electronic polarization* of materials. When atoms of different types form molecules, they will normally not share their electrons symmetrically; the electron clouds will be displaced eccentrically toward the stronger binding atoms. Thus atoms acquire charges of opposite polarity, and an external field acting on these net charges will tend to change the equilibrium positions of the atoms themselves. By this displacement of charged atoms or groups of atoms with respect to each other, a second type of induced dipole moment is created; it represents the *atomic polarization* of the dielectric. The asymmetric charge distribution between the unlike partners of a molecule gives rise, in addition, to *permanent* dipole moments which exist also in the absence of an external field. Such moments experience a torque in an applied field that tends to orient them in the field direction [see Eq. (7.31)]. Consequently, an *orientation* (or *dipole*) *polarization* can arise.

These three mechanisms of polarization (characterized by an *electronic polarizability* α_e, an *atomic polarizability* α_a, and an *orientation*, or *dipole*, polarizability α_d), are due to charges locally bound in atoms, in molecules or in the structures of solids and liquids. In addition, there usually exist charge carriers that can migrate for some distance through the dielectric. When such carriers are impeded in their motion, either because they become trapped in the material or on interfaces or because they cannot be freely discharged or replaced at the electrodes, space charges and a macroscopic field distortion result. Such a distortion appears as an increase in the capacitance of the sample and may be indistinguishable from a real rise of the dielectric permittivity. Thus we have to add to our polarization mechanisms a fourth one, a *space-charge* (or *interfacial*) polarization, characterized by a *space-charge* (or *interfacial*) *polarizability* α_s.

Assuming at present that the four polarization mechanisms indicated schematically in Fig. 7.15 act

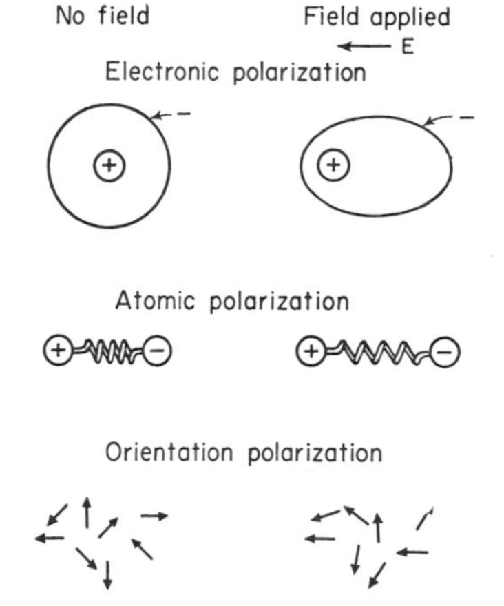

FIG. 7.15. Mechanisms of polarization.

independently of each other, we may write the total polarizability α of a dielectric material as the sum of the four terms:

$$\alpha = \alpha_e + \alpha_a + \alpha_d + \alpha_s \tag{7.88}$$

In Eq. (7.87) the polarizability α is regarded as a real quantity. Actually, in alternating fields, a temporal-phase shift may occur between the driving field and the resulting polarization, and a loss-current component appears (see Sec. 2). Thus α becomes complex, and Eq. (7.87) has to be replaced by the more general formulation

$$\mathbf{P} = (\kappa^* - 1)\mathbf{E} = N\alpha\mathbf{E}' \tag{7.89}$$

The parameter α contains the primary information on the electric-charge carriers and their polarizing action.

The locally acting field $\mathbf{E}'$ will be identical with the externally applied field $\mathbf{E}$ for gases at low pressure where interaction between molecules can be neglected. At high pressure, however, and especially in the condensed phases of solids and liquids, the field acting on a reference molecule A may be modified decisively by polarization of the surroundings (Fig. 7.16).

Assume molecule A surrounded by an imaginary sphere of such an extent that beyond it the dielectric can be treated as a continuum. If the molecules inside this sphere were removed while the polarization outside remains frozen, the field acting on A would stem from two sources: the free charges at the electrodes of the plate capacitor ($\mathbf{E}_1$), and the free ends of the dipole chains that line the cavity walls ($\mathbf{E}_2$). Actually, there are molecules inside the sphere, and they are so near A that their individual positions and shapes have to be considered. This adds an additional contribution $\mathbf{E}_3$ to the local field $\mathbf{E}'$, hence

$$\mathbf{E}' = \mathbf{E}_1 + \mathbf{E}_2 + \mathbf{E}_3 \tag{7.90}$$

The contribution from the free charges at the electrodes is, by definition, equal to the applied-field intensity: $\mathbf{E}_1 = \mathbf{E}$. To calculate $\mathbf{E}_2$, we recall that

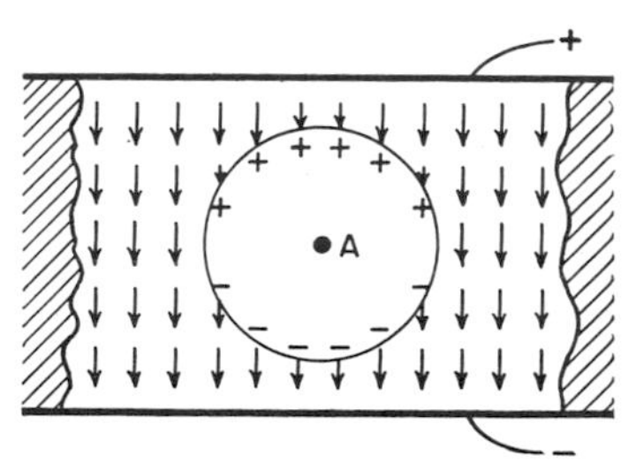

FIG. 7.16. Model for calculation of internal field.

the charge density lining the cavity walls stems from bound charges and is correspondingly determined by the normal component of the polarization vector $\mathbf{P}$ [see Eq. (7.24)] as

$$\mathbf{P} \cdot \mathbf{n}\, dA = P \cos\theta\, dA \tag{7.91}$$

(Fig. 7.17). Each surface element dA of the sphere contributes at A, according to Coulomb's law, a radial-field intensity

$$d\mathbf{E}_2 = \frac{P \cos\theta}{\epsilon_0 4\pi r^2}\, dA \tag{7.92}$$

For each surface element dA at a latitude position between θ and $\theta + d\theta$, there exists its counterpart which produces the same vertical but an equal and

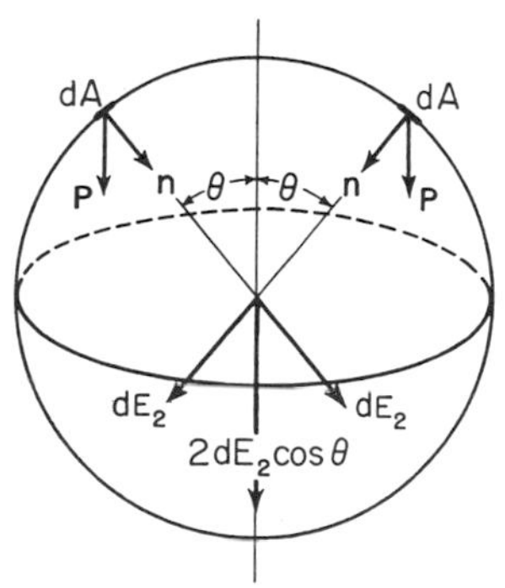

FIG. 7.17. Geometry for calculation of internal field.

opposite horizontal field component. Hence only the vertical components $d\mathbf{E}_2 \cos\theta$ count and create a field intensity

$$\mathbf{E}_2 = \oint_{\text{sphere}} \frac{P \cos^2\theta}{\epsilon_0 4\pi r^2}\, dA \tag{7.93}$$

oriented parallel to the applied field and strengthening it. By dividing the cavity walls into ring elements $dA = 2\pi \sin\theta\, r\, d\theta$ and integrating over θ, we obtain as the field contribution of the cavity-wall charge

$$\mathbf{E}_2 = \int_0^\pi \frac{P \cos^2 \theta}{\epsilon_0 4\pi r^2} 2\pi r^2 \sin \theta \, d\theta = \frac{1}{3}\frac{\mathbf{P}}{\epsilon_0} = \frac{\mathbf{E}}{3}(\kappa' - 1) \tag{7.94}$$

Evaluation of the field $\mathbf{E}_3$ which arises from the individual action of the molecules inside the sphere requires accurate information on the geometrical arrangement and polarizability of the contributing particles. The mathematical treatment is difficult. In a general way, we recognized the existence of these neighboring molecules in the calculation of $\mathbf{E}_2$ by assuming that the cavity is scooped out without disturbing the state of polarization of the remaining dielectric. Hence we postulate for the present that the additional individual field effects of the surrounding molecules on the particle at A will mutually cancel; that is, $\mathbf{E}_3 = 0$.

This assumption, first made by Mosotti in 1850, is a reasonable approximation when the elementary particles are neutral and without permanent dipole moment or when they are arranged either in complete disorder or in cubic or similar highly symmetrical arrays. It allows us to substitute for the unknown molecular parameter E'

$$\mathbf{E}' = \mathbf{E}_1 + \mathbf{E}_2 = \mathbf{E} + \frac{\mathbf{P}}{3\epsilon_0} = \frac{\mathbf{E}}{3}(\kappa' + 2) \tag{7.95}$$

that is, known macroscopic parameters.

By inserting this local *Mosotti field* into Eq. (7.87), we obtain the relation between the *polarizability per unit volume* $N\alpha$ and the relative permittivity of the dielectric κ':

$$\frac{N\alpha}{3\epsilon_0} = \frac{\kappa' - 1}{\kappa' + 2} \tag{7.96}$$

For gases at low pressure, $\kappa' - 1 \ll 1$; hence $\kappa' + 2$ may be replaced by the digit 3. This is the same as replacing the local field $\mathbf{E}'$ by the applied field $\mathbf{E}$, and Eq. (7.96) simplifies to

$$\frac{N\alpha}{\epsilon_0} = \kappa' - 1 = \chi \tag{7.97}$$

where χ is the *electric susceptibility of the gas* [see Eq. (7.27)].

Frequently one refers to an ideal gas under standard conditions (0°C, 760 mm Hg). The number of molecules per unit volume, N, is in this case identical with the *Loschmidt number*

$$N_L = 2.687 \times 10^{25} \quad \text{meter}^{-3} \tag{7.98}$$

In other cases, as long as the molecules themselves are the dipole carriers, it is convenient to eliminate the dependence of the polarization on the density of the material by referring to the *polarization per mole*. The number of molecules per mole is *Avogadro's number*

$$N_0 = \frac{NM}{\rho} = 6.023 \times 10^{23} \tag{7.99}$$

where M designates the molecular weight in kilograms, and ρ the density in kilograms per meter^3 if mks units are used.

Substituting N_0 for N in Eq. (7.96), the *polarizability per mole* (*molar polarization*) becomes

$$\Pi = \frac{N_0\alpha}{3\epsilon_0} = \frac{\kappa' - 1}{\kappa' + 2}\frac{M}{\rho} \tag{7.100}$$

This is the *Clausius-Mosotti equation.*

The same equation was formulated independently for the optical range by Lorentz in Holland and Lorenz in Denmark. In this case, the relative dielectric constant κ' of Eq. (7.100) may be replaced by the square of the index of refraction n^2, according to the Maxwell relation [Eq. (7.68)]. Thus we arrive at the *Lorentz-Lorenz equation*

$$\Pi = \frac{N_0\alpha}{3\epsilon_0} = \frac{n^2 - 1}{n^2 + 2}\frac{M}{\rho} \tag{7.101}$$

with Π called the *molar refraction.*

The equations are identical but not quite general enough, because they assume that α is real. By replacing the real permittivity or index of refraction with their complex counterparts [Eq. (7.89)], we arrive at the more general formulation of the *Clausius-Mosotti-Lorentz-Lorenz equation*

$$\Pi = \frac{N_0\alpha}{3\epsilon_0} = \frac{\kappa^* - 1}{\kappa^* + 2}\frac{M}{\rho} = \frac{n^{*2} - 1}{n^{*2} + 2}\frac{M}{\rho} \tag{7.102}$$

In using this equation, valuable information on the polarizability α can be obtained. However, erroneous conclusions may be drawn in case the "near field" E_3 cannot be neglected.

6. Resonance Polarization

Electronic polarization can be observed undisturbed by other effects in *monatomic* gases. If we assume that the electrons form a cloud of constant charge around the nucleus, inside a sphere of radius r_0, the electronic polarization of the atom is

$$\alpha_e = \epsilon_0 4\pi r_0^3 \tag{7.103}$$

The molar polarization defined in Eq. (7.100) becomes

$$\Pi = \frac{N_0\alpha_0}{3\epsilon_0} = N_0\frac{4\pi}{3}r_0^3 \tag{7.104}$$

It equals the volume actually filled by the spherical atoms.

The true value will be larger: Quantum mechanics leads to the picture of a more extended electron cloud, and since the distant parts of the electron atmosphere are more weakly bonded to the nucleus, they contribute appreciably to the polarization in spite of their rapidly decreasing density (Fig. 7.18).

The simple model of electrons quasi-elastically bound to equilibrium positions and reacting to field changes like linear harmonic oscillators leads to the equation of motion:

$$\frac{d^2z}{dt^2} + 2a\frac{dz}{dt} + \omega_0^2 z = \frac{e}{m}\mathbf{E}' \tag{7.105}$$

where $\mathbf{E}'$ represents the locally acting electric field displacing the electrons in the z direction.

The complex relative permittivity of the medium in molecular terms is

$$\kappa^* = 1 + \frac{\mathbf{P}}{\epsilon_0 \mathbf{E}} = 1 + \frac{Ne^2/\epsilon_0 m}{\omega_0{}^2 - \omega^2 + j\omega 2\alpha} \qquad (7.106)$$

In the more general case of s-oscillator types which contribute without mutual coupling,

$$\kappa^* = \kappa' - j\kappa'' = 1 + \sum_s \frac{N_s e^2/\epsilon_0 m_s}{\omega_s{}^2 - \omega^2 + j\omega 2\alpha_s} \qquad (7.107)$$

For nonmagnetic media,

$$\kappa^* = n^{*2} \qquad (7.108)$$

where n^* is the complex index of refraction of the medium [see text following Eq. (7.70)]. Equation (7.107) represents the *dispersion formula of classical*

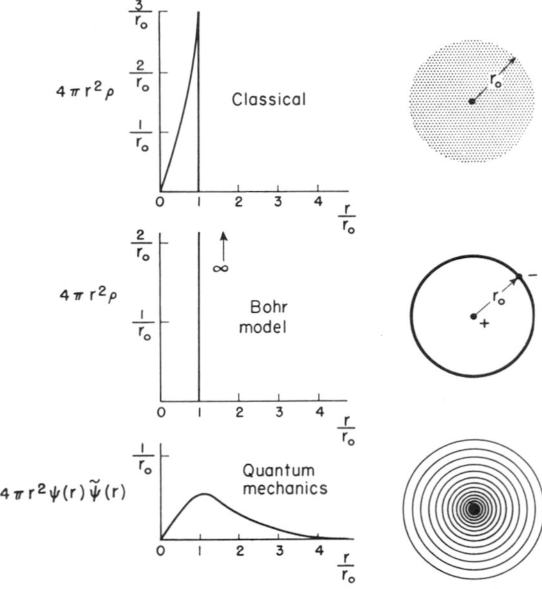

FIG. 7.18. Various models for electron charge distribution in the hydrogen atom: (*a*) electron cloud of uniform density, (*b*) Bohr model, (*c*) quantum mechanics (normalized to unit electronic charge).

physics; N_s designates the *number of dispersion electrons* per unit volume of the oscillator type s.

The frequency dependence of the real part of the relative permittivity of the dielectric constant κ' describes the *dispersion characteristic* of the dielectric medium (Fig. 7.19). It rises hyperbolically from the static contribution to a maximum at $\Delta\omega = +\alpha_s$, falls with a linear slope through the resonance frequency $\omega_s(\Delta\omega = 0)$, reaches a minimum at $\Delta\omega = -\alpha_s$, and then rises again asymptotically to the static contribution of the remaining dipoles.

The absorption characteristic of the dielectric, identified by the relative loss factor κ'', starts from zero at low frequencies, traverses its maximum at resonance, and falls again symmetrically to zero at high frequencies. The half-value points of this bell-shaped *absorption characteristic* (spectral line) are reached at the deviation from resonance $\Delta\omega = \pm\alpha_s$.

Since the real dielectric constant and index of refraction rise with increasing frequency over the major part of the dispersion characteristic, this behavior is called *normal dispersion,* in contrast to the anomalous dispersion in the half-width region of the spectral line, where the characteristic falls toward shorter wavelengths. This combination of normal and anomalous dispersion is typical for resonance phenomena.

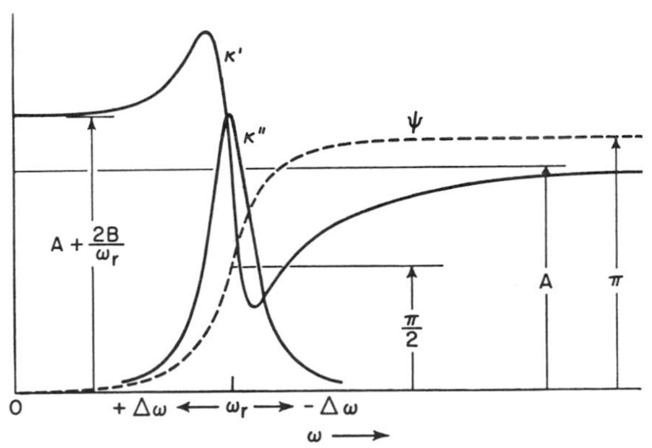

FIG. 7.19. Anomalous dispersion and resonance absorption.

Quantum physics leads to a reinterpretation of the classical model. Each resonance frequency, according to Bohr's frequency condition, corresponds to a transition between two stationary energy states $\mathcal{E}_i$ and $\mathcal{E}_j$,

$$\omega_s \rightarrow \omega_{ij} = \frac{\mathcal{E}_i - \mathcal{E}_j}{\hbar} \qquad (7.109)$$

The dipole moment of this *transition oscillator,* which determines the coupling strength to the electromagnetic field, is found as the average dipole moment of the mixed-wave functions of the two states

$$\mathbf{u}_{ij} = \int_{\text{all space}} \psi_i(e\mathbf{r})\tilde{\psi}_j \, d\tau \qquad (7.110)$$

To calculate the intensity of the spectral line, substitute in Eq. (7.107):

$$\frac{e^2}{m} \rightarrow \frac{2\omega_{ij}}{\hbar} |\mathbf{u}_{ij}|^2 \qquad (7.111)$$

Finally, the difference in occupation density enters in quantum physics: The atoms in the lower state (N_i) may absorb, while the atoms in the higher state (N_j) may be forced to return to the lower state by emission; hence

$$N_s \rightarrow N_i - N_j \qquad (7.112)$$

By these three substitutions the dispersion formula of classical physics [Eq. (7.107)] changes into the dispersion formula of quantum mechanics [2]:

$$\kappa^* = 1 + \frac{2}{\epsilon_0 \hbar} \sum_{i<j} \frac{\omega_{ij}|\mathbf{u}_{ij}|^2}{\omega_{ij}{}^2 - \omega^2 + j\omega 2\alpha} (N_i - N_j) \qquad \mathbf{(7.113)}$$

When unlike atoms form molecules, the attraction of the two partners for electrons, their *electronegativity*, is different, and *permanent electric-dipole moments* result. Since Debye was the first to realize the significance of such permanent moments [3], a convenient unit has been named in his honor:

$$1 \text{ debye} = 1 \times 10^{-18} \text{ esu} = 3.33 \times 10^{-30} \text{ coulomb-m}$$

The dipole moments of molecules, like the bond strength, the intermolecular distance, and other parameters, cannot be calculated with accuracy at present but must be measured [4].

Dipole moments are frequently determined by a quasi-static measurement of the polarization of the dielectric as a function of temperature. This method is based on a statistical theory of orientation, first developed for the permanent magnetic moments of paramagnetic substances by Langevin and applied to permanent electric moments by Debye [3].

Molecules carrying a permanent dipole moment experience a torque in an electric field that tends to align the dipole axis in the field direction [see Eq. (7.31)]. Thermal agitation, on the other hand, tends to maintain a random distribution.

The permanent dipole moment $\mathbf{\mu}$ of the molecules is nearly unaffected by the field; the density of the gas is so low that the dipolar interaction energy is small in comparison with the thermal equilibrium energy kT; finally, the dipoles can assume any direction with respect to the field axis. The statistical mean moment is

$$\frac{\mathbf{\mu}_{av}}{\mathbf{\mu}} = \coth x - \frac{1}{x} \equiv L(x) \qquad (7.114)$$

where $x = \mu E/kT$.

The function $L(x)$ is known as the *Langevin function*. Figure 7.20 shows the characteristic $L(x)$ as

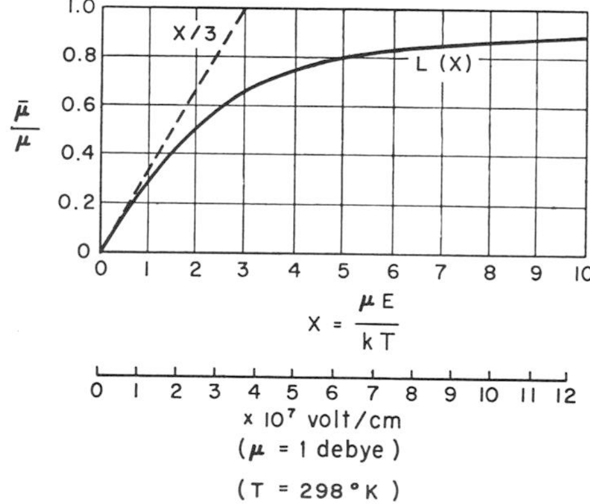

FIG. 7.20. Langevin function of dipole orientation.

applied by Debye for the electric case; the field scale refers to the orientation of polar molecules of unit moment, $\mathbf{\mu} = 1$ debye, against the thermal agitation at room temperature.

Very high field strengths are required to produce a deviation from linearity and an approach toward saturation. For the normal case of relatively small fields ($x \ll 1$), the Langevin function can be approximated by $x/3$, and so

$$\mathbf{\mu}_{av} \simeq \frac{|\mathbf{\mu}|^2}{3kT}\mathbf{E} \qquad (7.115)$$

The applied field will, in addition, induce a moment in the gas molecule by deforming its electron cloud (α_e) and by changing the spacing of the nuclei (α_a). Thus deformation and orientation polarization together produce the total moment per dipole molecule

$$\mathbf{\mu}_t = \left(\alpha_e + \alpha_a + \frac{\mathbf{\mu}^2}{3kT}\right)\mathbf{E} \qquad (7.116)$$

and the relative dielectric constant for static fields, κ'_s, becomes

$$\frac{N\mathbf{\mu}_t}{\epsilon_0 \mathbf{E}} = \kappa'_s - 1 = \frac{N}{\epsilon_0}\left(\alpha_e + \alpha_a + \frac{|\mathbf{\mu}|^2}{3kT}\right) \qquad (7.117)$$

If κ'_∞ designates the relative dielectric constant due to the induced moments only,

$$\kappa'_\infty \equiv 1 + \frac{N}{\epsilon_0}(\alpha_e + \alpha_a) \qquad (7.118)$$

the contribution of the permanent dipoles to the static dielectric constant is

$$\kappa'_s - \kappa'_\infty = \frac{N|\mathbf{\mu}|^2}{\epsilon_0 3kT} \qquad (7.119)$$

The fact that the orientation of the permanent dipole moments is strongly temperature-dependent in contrast to the practically temperature-independent contributions of the induced moments gives a convenient method of determining $\kappa'_s - \kappa'_\infty$ and with it the permanent dipole moment $\mathbf{\mu}$ of the molecules. To eliminate the effect of varying gas density, one refers to the polarizability per mole Π in the formulation for low gas pressure [see Eq. (7.100)]:

$$\Pi = \frac{N_0 \alpha}{3\epsilon_0} = \frac{\kappa'_s - 1}{3}\frac{M}{\rho} \qquad (7.120)$$

If Π is plotted as a function of $1/T$, one can distinguish easily between polar and nonpolar molecules (Fig. 7.21).

If we write the molar polarization of the gas in the form

$$\Pi = A + \frac{B}{T} \qquad \text{meter}^3 \qquad (7.121)$$

$$\text{where } A = \frac{N_0}{3\epsilon_0}(\alpha_e + \alpha_a) = 2.27 \times 10^{34}(\alpha_e + \alpha_a) \qquad (7.122)$$

$$B = \frac{N_0}{3\epsilon_0}\frac{|\mathbf{\mu}|^2}{3k} = 5.48 \times 10^{56}|\mathbf{\mu}|^2 \qquad (7.123)$$

the dipole moment is

$$\begin{aligned}\mathbf{\mu} &= 4.27 \times 10^{-29}\sqrt{B} \quad \text{coulomb-m} \\ &= 12.7\sqrt{B} \quad \text{debyes}\end{aligned} \qquad (7.124)$$

Figure 7.21 shows the results of measurements on some polar and nonpolar gases.

Electronic polarization can be determined from

measurements of the refractive index in the visible region, and so the molar polarization Π can be separated into its electronic, atomic, and dipole contributions. As a rule of thumb, the atomic polarization Π_a can be estimated as amounting to about 10 per cent of the electronic polarization Π_e for dipole moments greater than 1 debye [5].

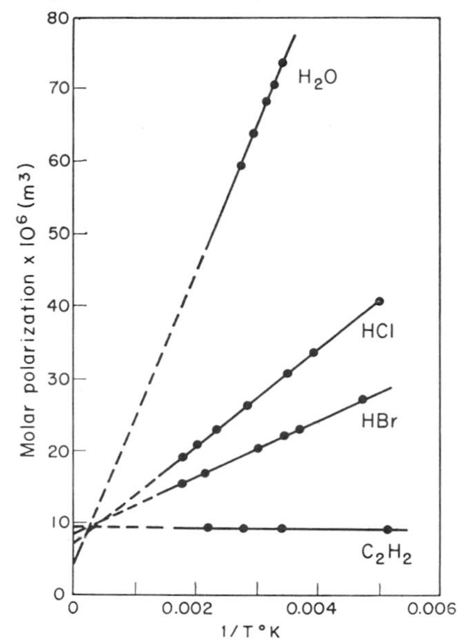

FIG. 7.21. Polarization of polar and nonpolar gases.

Frequently, a good first approximation of the dipole moment of polyatomic molecules can be obtained by regarding such molecules as composed of diatomic groups, for which the dipole moments are known. The over-all dipole moment results by a *vector addition* of the individual group moments. Dichlorobenzene in its ortho, meta, and para forms (Fig. 7.22)

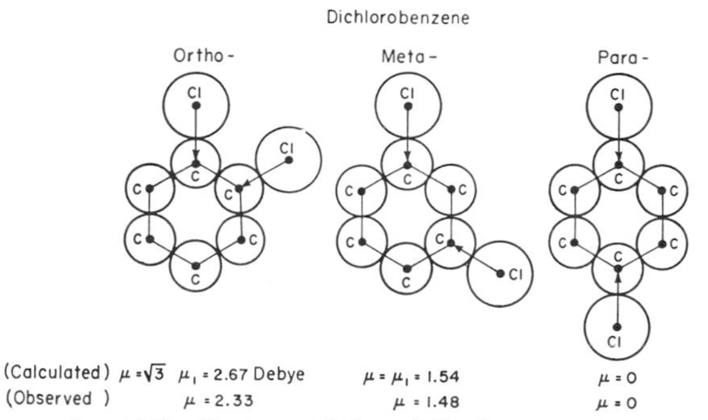

FIG. 7.22. Vector addition of dipole moments.

is the classical example by which Debye first illustrated this procedure.

The frequency response of molecular gases is characterized by resonance spectra. The vibration of nuclei relative to each other (*atomic polarization*) produces resonance states in the near infrared; the rotation of molecules as a whole or of molecular groups around internuclear axes (*orientation*, or *dipole, polarization*) gives rise to resonances in the far infrared down into the microwave region.

In liquids and solids there are also resonance spectra of electronic excitation and vibration. Many of them do not belong any more to individual particles but characterize electronic and vibrational states of the condensed phase. The characteristic colors of materials, for example, arise from their electronic absorption spectra in the visible spectral range, while the infrared absorption of ionic crystals is caused by vibrations of the positive against the negative ions of the crystal lattice. In comparison with the spectral lines of gases, these absorptions are greatly broadened, and, in consequence, much detailed information is lost that would be of help in their interpretation.

7. Relaxation Polarization

While the electronic excitation and vibration states keep a modified resonance character, the rotation states are altered completely in condensed materials. To turn molecules or molecular groups requires space. Moreover, the permanent dipole moments of molecular groups are characteristic construction elements in the formation of the condensed phases. They are built into their surroundings; free rotation as in gases becomes impossible, and the quantized rotation spectra disappear.

The dispersion formula of classical physics pictures resonating atoms or molecules in the gaseous state as L, R, C circuits shunted by a capacitance. The classical approach to the treatment of dipoles in the condensed phases of liquids and solids is to consider the polar molecules as rotating in a medium of dominating friction (Debye [3]). This corresponds to a neglect of the acceleration term, i.e., to the reduction of the L, R, C equivalent circuit to the R, C circuit of Fig. 7.23.

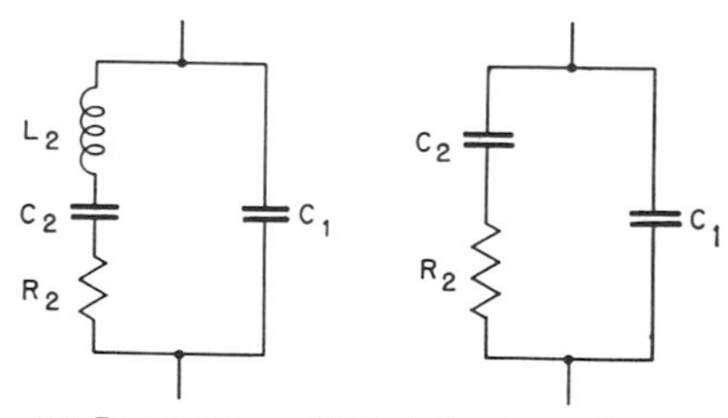

FIG. 7.23. Equivalent circuits of free resonator molecule and of rotator molecule in medium of dominating friction.

The complex dielectric constant assumes the general formulation

$$\kappa^* = \kappa'_\infty + \frac{\kappa'_s - \kappa'_\infty}{1 + j\omega\tau} \tag{7.125}$$

without reference to a specific equivalent picture. The resonance spectrum of rotation has thus been replaced by a relaxation spectrum. It is the task of molecular physics to reinterpret the static and optical permittivities κ'_s and κ'_∞ and the relaxation time τ by molecular quantities. The optical dielectric constant κ'_∞ corresponds to the electronic and atomic resonance polarization of the dielectric. In the low-frequency range of the relaxation polarization, this deformation polarization contributes constant induced moments [see Eq. (7.116)]:

$$\mathbf{u}_i = (\alpha_e + \alpha_a)\mathbf{E}' \tag{7.126}$$

The static dielectric constant κ'_s contains, in addition, the contribution resulting from the orientation of the permanent moments:

$$\mathbf{u}_{av} = \frac{\mathbf{u}^2}{3kT}\mathbf{E}' \tag{7.127}$$

If the local field $\mathbf{E}'$ represents the applied field $\mathbf{E}$, as in gases at low pressure, the complex permittivity becomes [Eq. (7.117)]

$$\kappa^* = 1 + \frac{N}{\epsilon_0}\left[(\alpha_e + \alpha_a) + \frac{\mathbf{u}^2}{3kT}\frac{1}{1 + j\omega\tau}\right] \tag{7.128}$$

Actually, the local field will differ from the applied field, and its evaluation will require a detailed structure analysis of the material in question. To escape this necessity, while making some reference to the influence of the polarized surroundings, one frequently introduces at this point the Mosotti field [see Eq. (7.95)]:

$$E' = E + \frac{\mathbf{P}}{3\epsilon_0} = \frac{\mathbf{E}}{3}(\kappa^* + 2) \tag{7.129}$$

In this case,

$$\kappa^* = \kappa'_\infty + \frac{\kappa'_s - \kappa'_\infty}{1 + j\omega\tau_e} \tag{7.130}$$

with the new time constant

$$\tau_e = \tau\frac{\kappa'_s + 2}{\kappa'_\infty + 2} \tag{7.131}$$

By replacing the applied field $\mathbf{E}$ with the Mosotti field $\mathbf{E}'$, the interpretation, but not the shape, of the relaxation spectrum has changed. The relaxation time has lengthened from τ to τ_e, and the molecular meaning of the static and optical dielectric constants has been altered.

The remaining problem is molecular interpretation of the relaxation time τ. The polar molecules are regarded as rotating under the torque $\mathbf{T}$ of the electric field with an angular velocity $d\theta/dt$ proportional to this torque, or

$$\mathbf{T} = \zeta\frac{d\theta}{dt} \tag{7.132}$$

The friction factor ζ will depend on the shape of the molecule and on the type of interaction it encounters. If one visualizes the molecule as a sphere of radius a rotating in a liquid of viscosity η according to Stokes' law, classical hydrodynamics leads to the value

$$\zeta = 8\pi\eta a^3 \tag{7.133}$$

In a static field, the spherical dipole carriers will have a slight preferential orientation parallel to this field and thus contribute the average moment of Eq. (7.127). A sudden removal of the external field will cause an exponential decay of this ordered state because of the randomizing agitation of the Brownian movement. The relaxation time τ (or τ_e) measures the time required to reduce the order to $1/e$ of its original value. Debye was able to calculate this time statistically by deriving the space orientation under the counteracting influences of the Brownian motion and of a time-dependent electric field and found

$$\tau = \frac{\zeta}{2kT} \tag{7.134}$$

Combining Eqs. (7.133) and (7.134), Debye obtained for the spherical molecule, if it behaves like a ball rotating in oil, the relaxation time

$$\tau = \frac{4\pi a^3\eta}{kT} = V\frac{3\eta}{kT} \tag{7.135}$$

The time constant is proportional to the volume of the sphere and to the macroscopic viscosity of the solution. Water at room temperature has a viscosity $\eta = 0.01$ poise; with a radius of ca. 2 A for the water molecule, a time constant of $\tau \simeq 0.25 \times 10^{-10}$ sec results.

Figure 7.24 shows that, indeed, the relaxation time of water is located near the wavelength of 1 cm.

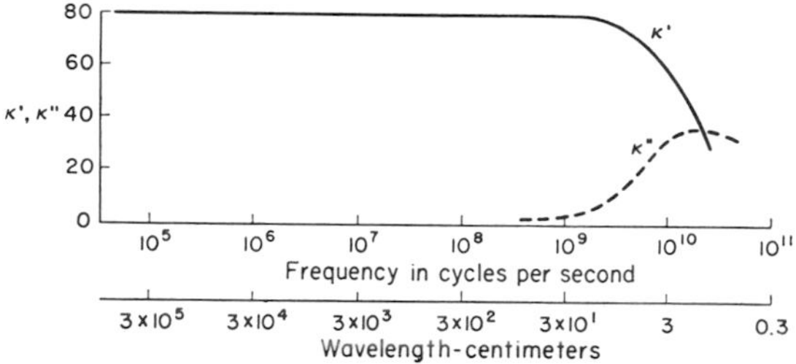

FIG. 7.24. Relaxation spectrum of water at room temperature.

The agreement, however, is somewhat marred by the realization that experimentally we should have determined $\tau_e \simeq 20\tau$ instead of τ itself. Obviously, the sphere model is only a rough approximation; the essence of Debye's approach is to postulate that the orientation of polar molecules in liquids and solids leads spectroscopically to a simple relaxation spectrum.

The *Debye equation* [Eq. (7.130)] may be written in various forms. Separating it into its real and imaginary parts, one obtains the standard version

$$\begin{aligned}\kappa' - \kappa'_\infty &= \frac{\kappa'_s - \kappa'_\infty}{1 + \omega^2\tau_e^2} \\ \kappa'' &= \frac{(\kappa'_s - \kappa'_\infty)\omega\tau_e}{1 + \omega^2\tau_e^2} \\ \tan\delta &= \frac{\kappa''}{\kappa'} = \frac{(\kappa'_s - \kappa'_\infty)\omega\tau_e}{\kappa'_s + \kappa'_\infty\omega^2\tau_e^2}\end{aligned} \tag{7.136}$$

If one introduces as a new variable

$$z = \ln\omega\tau_e \tag{7.137}$$

Eqs. (7.136) may be rewritten in a normalized form [6]:

$$\frac{\kappa' - \kappa'_\infty}{\kappa'_s - \kappa'_\infty} = \frac{1}{1 + e^{2z}} = \frac{e^{-z}}{e^z + e^{-z}}$$
$$\frac{\kappa''}{\kappa'_s - \kappa'_\infty} = \frac{1}{e^z + e^{-z}} \qquad (7.138)$$
$$\frac{\tan \delta}{\kappa'_s - \kappa'_\infty} = \frac{1}{\kappa'_\infty e^z + \kappa'_s e^{-z}}$$

Figure 7.25 shows this logarithmic plot of the dispersion and absorption characteristic, and, added to them, as a third curve, the relative dielectric conductivity [Eq. (7.14)]

$$\sigma = \omega\kappa'' \qquad (7.139)$$

in the normalized form

$$\frac{\sigma\tau_e}{\kappa'_s - \kappa'_\infty} = \frac{e^z}{e^z + e^{-z}} \qquad (7.140)$$

The conductivity curve is the mirror image of the κ' characteristic, that is, the orientation polarization leads to a constant maximum conductivity contribution beyond the range of the dispersion region.

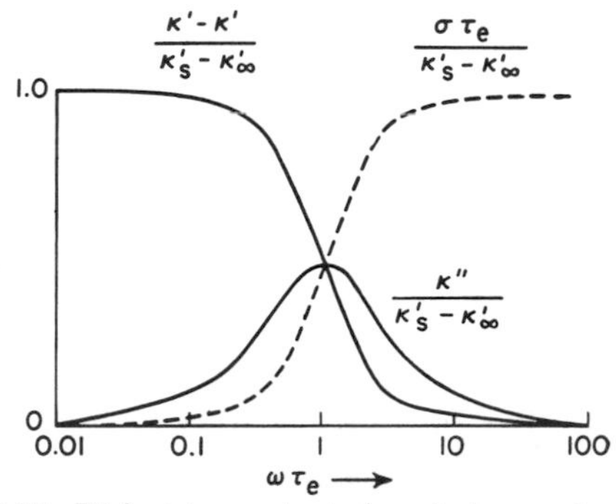

Fig. 7.25. Dielectric constant, loss factor, and conductivity of simple relaxation spectrum in normalized form.

After the polarizing action of the permanent dipoles has disappeared, their existence is still noted with full force as a conduction effect. The explanation is simple: As the frequencies range so high that the molecules have no time to turn, one does not notice that the two opposite dipole charges are coupled together; their effect on the conduction is therefore the full contribution of two ions of opposite polarity moving in the electric field according to Ohm's law.

Figure 7.25 shows clearly the frequency spread of the dispersion phenomenon. According to the decade scale, it is practically limited to one decade for κ' and to two decades for κ'' above and below the center frequency. One further graphical representation of the Debye equation proves of value in analyzing and extrapolating experimental data. If one plots κ'' against κ' in the complex plane, points obeying the Debye equation fall on a semicircle with its center at $(\kappa'_s + \kappa'_\infty)/2$ (Fig. 7.26), as Cole and Cole [7] first pointed out. This becomes evident when one rewrites Eq. (7.132) in the form

$$(\kappa^* - \kappa'_\infty) + j(\kappa^* - \kappa'_\infty)\omega\tau_e = \kappa'_s - \kappa'_\infty \qquad (7.141)$$

The first member on the left side corresponds to a vector **u**; the second member, as the factor j indicates, adds perpendicular to it and represents a vector **v**; the sum is the diagonal of the circle.

The loss factor κ'' reaches its maximum at the critical frequency

$$\omega_m = \frac{1}{\tau_e} \qquad (7.142)$$

that is, at the critical wavelength

$$\lambda_m = 2\pi c\tau_e \qquad (7.143)$$

at which the dipole polarization has fallen to its half value. Furthermore,

$$\kappa''_{\max} = \frac{\kappa'_s - \kappa'_\infty}{2} \equiv \frac{S}{2} \qquad (7.144)$$

The relaxation time and the contribution S of the orientation polarization to the permittivity can be determined by these relations from the absorption characteristic of a dielectric as long as the Debye equation is valid.

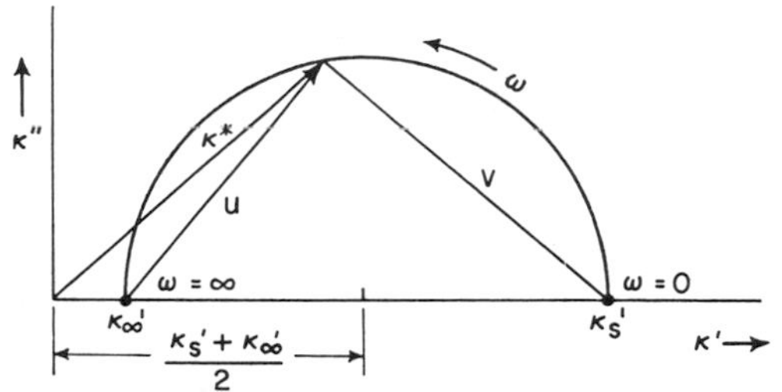

Fig. 7.26. Cole-Cole circle diagram of κ^* in complex plane.

The assumption of a simple relaxation spectrum fits satisfactorily the frequency response of a number of dielectrics, especially of dilute solutions of polar materials in nonpolar solvents. This fact, however, should not be construed as a confirmation of the special Debye equation [Eq. (7.130)], based on the Mosotti field [Eq. (7.129)]. By introducing this local field in the equation for the polarization

$$\mathbf{P} = (\kappa' - 1)\mathbf{E} = N\alpha\mathbf{E}$$

the polarization and susceptibility become

$$\mathbf{P} = \frac{N\alpha\mathbf{E}}{1 - N\alpha/3\epsilon_0} \qquad \chi = \frac{N\alpha/\epsilon_0}{1 - N\alpha/3\epsilon_0} \qquad (7.145)$$

and so they must approach infinity when the polarizability term of the denominator $N\alpha/3\epsilon_0$ approaches 1. Obviously, this is bound to happen at a critical or *Curie temperature* T_c, in case permanent moments contribute an orientation polarizability

$$\alpha_d = \frac{\mathbf{\mu}^2}{3kT} \qquad (7.146)$$

Forgetting about the deformation polarization, we

may replace α by α_d and obtain

$$\chi = \frac{3T_c}{T - T_c} \tag{7.147}$$

where
$$T_c = \frac{N\mu^2}{9\epsilon_0 k} \tag{7.148}$$

Equation (7.147) is the famous *Curie-Weiss* law of ferromagnetism, here derived for permanent electric- instead of magnetic-dipole moments. It predicts the spontaneous polarization (or magnetization) of dielectrics containing such moments.

Actually, a Mosotti-type "catastrophe" happens under very special conditions only and not at all as foreseen in the preceding derivation. The reason is that permanent electric-dipole moments are anchored in molecular groups that tend to lose their freedom of orientation in condensed phases through association and steric hindrance. Even if these groups could rotate like spheres in a medium of high friction, the Clausius-Mosotti formula would not apply, since the reference molecule A in the cavity of Fig. 7.16 is not a mathematical point but itself is a dipole carrier. As soon as the cavity is visualized as a molecular sphere in which a mathematical dipole is centered, calculations carried through by Onsager [8] show that spontaneous polarization does not occur.

In Onsager's treatment the surroundings of the dipole are still considered as a continuum. An improved model of Kirkwood [9] visualizes the dipole molecule with its first layer of neighbors as a structural unit, known in its statistical arrangement from X-ray patterns. The behavior of such a molecular island, floating in a dielectric continuum, is examined for static fields; the permittivities thus found are in fair agreement with experiment. To obtain still better results, the molecular island is extended to include the second nearest neighbors, but the mathematical problem becomes increasingly formidable [10].

8. Piezoelectricity and Ferroelectricity

Dipoles in liquids and polymers may be hindered in their orientation, but they will rotate sufficiently, at least at higher temperatures, to make their existence felt in relaxation spectra, spectra frequently characterized not by one time constant only but by a broad distribution of relaxation times (Fig. 7.27) [11]. Permanent electric dipoles in crystals, on the other hand, are, in general, completely immobilized as far as their individual rotation is concerned. In consequence, such dipoles are not detected in alternating fields by relaxation spectra, but, since they can couple to an electric field, they may, when properly arranged, give rise to resonance vibrations of the crystal as a whole (Fig. 7.28).

An electric field polarizes any material by *inducing* dipole moments. This displacement of charges from their equilibrium positions alters the mechanical dimensions of a solid; it causes *electrostriction*. However, mechanical stress applied to a neutral material cannot induce dipole moments; i.e., *electrostriction has no inverse*. If a mechanical distortion creates a voltage, the effect must be caused by permanent dipole moments anchored in the structure without a center of symmetry. This *piezoelectric effect* was discovered in 1880 by the brothers Curie [12] on certain asymmetrical crystals like quartz, tourmaline,

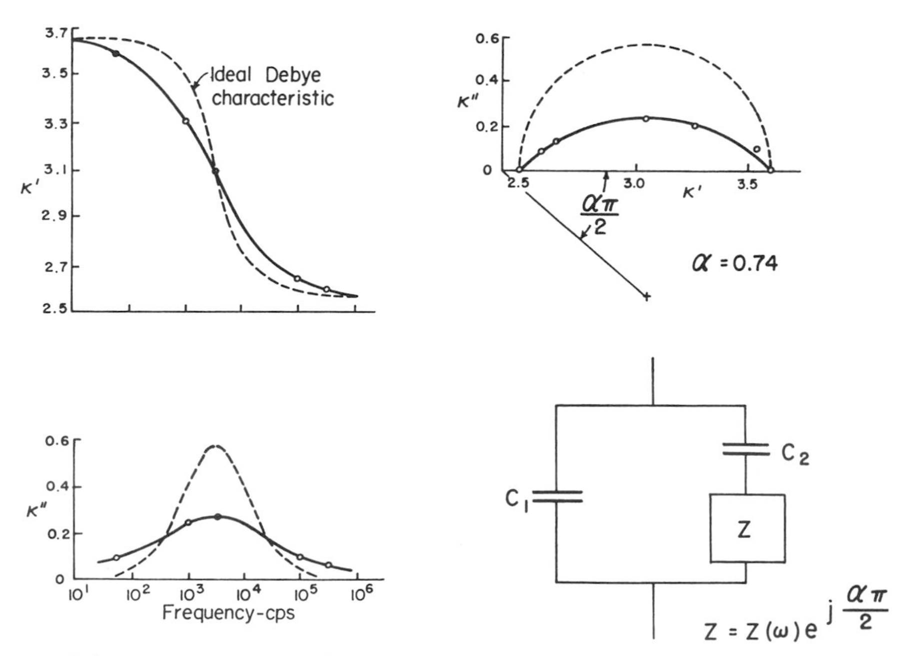

Fig. 7.27. Relaxation spectrum and Cole-Cole diagram of vulcanized rubber. (*After Kauzmann* [11].)

and rochelle salt. Compressed in specific directions, the materials develop a potential difference, and, vice versa, the application of an electric voltage creates a mechanical distortion. *Piezoelectricity is characterized by a one-to-one correspondence of direct and inverse effect;* it causes the electromechanical-resonance spectrum shown in Fig. 7.28.

Ferroelectricity, the *spontaneous* alignment of electric dipoles by mutual interaction, was not observed until

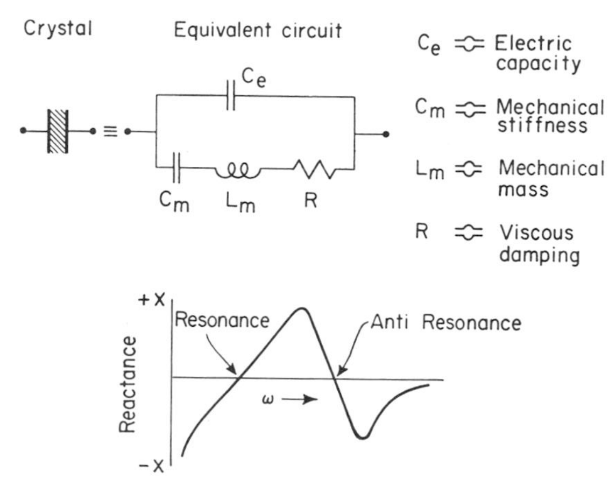

FIG. 7.28. Equivalent circuit and reactance of piezoelectric resonator.

recently, and few materials are, as yet, known to be true ferroelectrics. Obviously, very special conditions must prevail if polar groups can act as "free" dipoles in solids. Characteristic representatives are rochelle salt, the tetrahydrate of potassium sodium tartrate, recognized by Valasek [13] as a ferroelectric in 1921; potassium dihydrogen phosphate and arsenate by Busch and Scherrer [14] in 1935; and barium titanate, noticed for its unusual dielectric properties by Wainer and Salomon [15] in 1942–1943 and established as a new ferroelectric at the Massachusetts Institute of Technology in 1943–1944 [16]. Additional ferroelectrics related to the last type have been found by Matthias [17]. The systematic theory of ferroelectrics has been presented by Forsbergh [39].

The ferroelectric range of rochelle salt is very narrow and that of the phosphates and arsenates is limited to low temperatures. Both crystal types, furthermore, have a relatively complicated crystal structure, are piezoelectric above the Curie point, and develop ferroelectricity in one axis direction only. Barium titanate, in contrast, crystallizes in the simple perovskite structure (Fig. 7.29) and is cubic with a center of symmetry, hence not piezoelectric above its Curie point near 120°C. It may be employed as a single crystal or as rugged ceramic material which can be formed into any shape desired. Thus this substance lends itself better to fundamental investigations and applications [18].

When a multicrystalline sample of $BaTiO_3$ cools down through the temperature region near 120°C, a number of properties undergo rapid changes. The dielectric constant and loss traverse a sharp maximum and minimum, respectively; the slope of the thermal-expansion characteristic alters; and ferroelectric hysteresis loops appear (Fig. 7.30). The X-ray diagram of the cubic structure becomes transformed simultaneously and progressively into that of a tetragonal structure. The field-strength dependence of the dielectric constant makes the ceramics useful as nonlinear dielectrics. A biasing field transforms a barium titanate disk into a piezoelectric resonator. This piezoelectric response persists afterwards without an external field because of the remanence of the polarization [19].

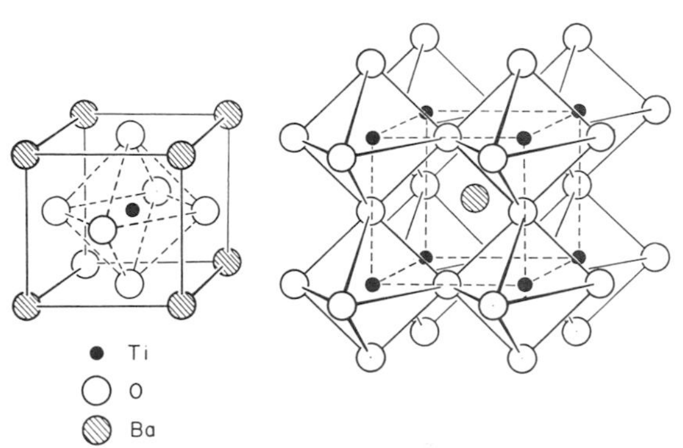

FIG. 7.29. Ideal perovskite structure.

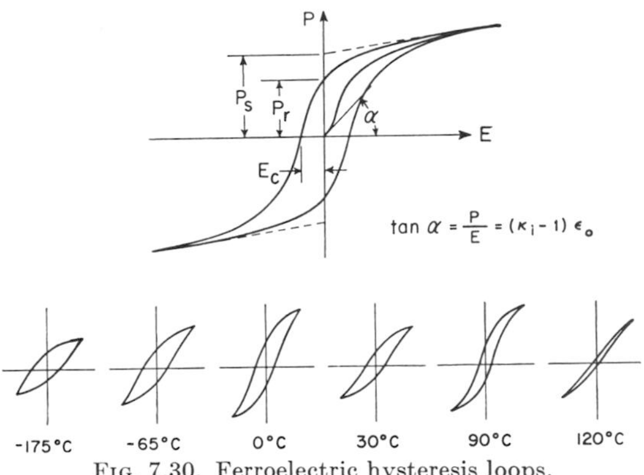

FIG. 7.30. Ferroelectric hysteresis loops.

A single crystal of $BaTiO_3$ is transparent above its Curie temperature but has a yellowish tinge. As it cools, a variety of shaded areas appears, a domain pattern which can be altered by temperature, pressure, or the application of an electric field [20]. The polar axis of the crystal develops at the Curie point in any one of the cube-edge directions. At two lower transitions it changes into the face diagonal (near 0°C) and space diagonal direction (near −75°C), respectively (Fig. 7.31) [21].

To understand how the spontaneous polarization arises, we return to the structure of Fig. 7.29 [22]. The titanium ions of $BaTiO_3$ are surrounded by six oxygen ions in an octahedral configuration. This coordination is to be expected from the radius ratio of the partners when visualized as ionic spheres: alternatively, the TiO_6 groups may be explained as resulting from covalent binding by octahedral *s, p, d* hybrid bonds. We find ourselves in a transition region between polar and nonpolar binding where slight changes in internuclear separation produce large changes in the electric-dipole moments. In the perovskite structure, all octahedra are placed in identical orientation, joined only at their corners and fastened in position by barium ions. Any displacement of one Ti ion toward a specific oxygen ion creates, by a kind of feedback coupling through the oxygen lattice, a tendency for the other titanium ions to move in the same direction. At a critical temperature the thermal agitation can be overcome, and the Mosotti catastrophe occurs [see Eqs. (7.145)] [23].

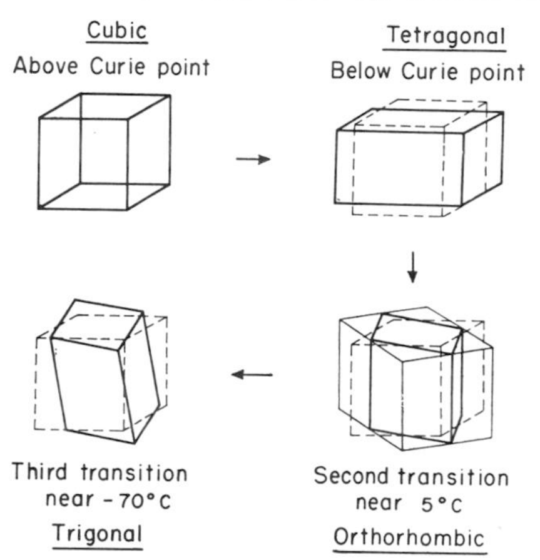

FIG. 7.31. Phase transitions of barium titanate.

This interpretation of the onset of spontaneous polarization and its dependence on structure parameters clarifies some of the prerequisites for the formation of a ferroelectric state. The old idea that the rotation of permanent moments leads to a Mosotti catastrophe has to be discarded; such moments are built in and not available for free rotation. Ferroelectricity arises, not from rotation, but from vibration states; the displacement of certain ions from their equilibrium positions strongly unbalances the equilibrium of the permanent moments. By a proper structural arrangement, this upset induces a motion of the neighboring ions in a supporting sense which increases the original displacement by feedback. The tendency to bring the vibrations of neighboring ions into ordered phase relations prevails at the Curie point against the random agitation, and the equilibrium position of the critical ions shifts to one side since the whole effect was made possible only by the displacement of these ions. The old balance of the permanent moments is destroyed and a polar axis created by the transformation of induced moments into permanent moments.

The preceding discussion of the formation of the ferroelectric state has clarified, by implication, the relation between ferroelectricity and piezoelectricity. A piezoelectric crystal, when pyroelectric, has a polar axis like a ferroelectric crystal, but the arrow direction of its axis is prescribed by the arrangement of the ions and cannot be reversed. For the ferroelectric crystal, in contrast, the possibility of reversal is inherent because the axis evolves at the Curie point from a state of higher symmetry. Only in ferroelectric crystals, therefore, can domain structures appear and the moment be inverted by a sufficiently strong opposing field. The response to pressure and voltage of a ferroelectric crystal is a truly piezoelectric one, involving the change of permanent moments, but complications arise through changes in the domain pattern.

A distortion of the oxygen lattice in the perovskite structure may destroy the feedback coupling (for example, $CaTiO_3$) or even reverse it, leading to *antiferroelectricity*, as in $PbZrO_3$ [24].

9. Polarization by Migrating Charge Carriers

The previously discussed effects of polarization have in common that they are produced by the displacement or orientation of bound-charge carriers. The remaining process mentioned in Sec. 5, the space-charge or interfacial polarization, is produced by traveling charges and causes large-scale field distortions.

The classical example of interfacial polarization is the Maxwell-Wagner two-layer capacitor [25] (Fig. 7.32). Here the dielectric consists of two

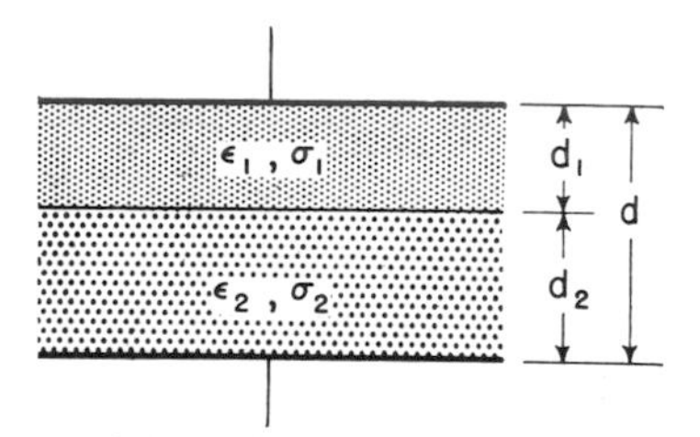

FIG. 7.32. Maxwell-Wagner two-layer capacitor.

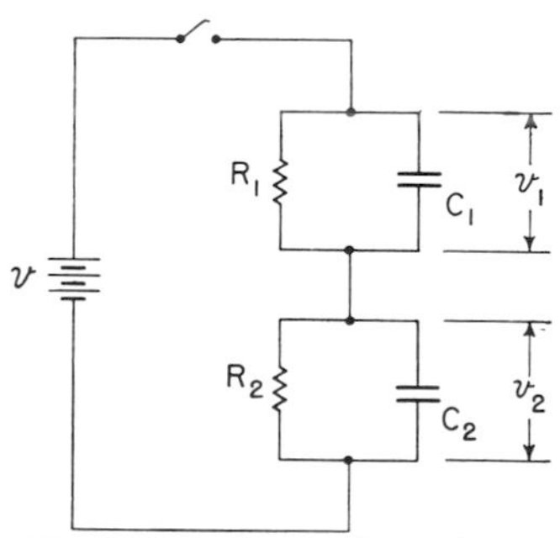

FIG. 7.33. Equivalent circuit of two-layer capacitor.

parallel sheets of different materials (1 and 2), characterized by their dielectric constant, conductivity, and thickness ($\epsilon'_1, \sigma_1, d_1$) and ($\epsilon'_2, \sigma_2, d_2$), respectively. When a d-c field is suddenly applied, the initial field distribution corresponds to the electrostatic requirement of constant flux density

$$D_1 = D_2 \quad \text{or} \quad E_1/E_2 = \epsilon'_2/\epsilon'_1 \tag{7.149a}$$

while the final distribution follows from the condition of current continuity

$$J_1 = J_2 \quad \text{or} \quad E_1/E_2 = \sigma_2/\sigma_1 \tag{7.149b}$$

According to the equivalent circuit of Fig. 7.33, the change-over from the one to the other condition takes place exponentially with a time constant

$$\tau = \frac{R_1\tau_2 + R_2\tau_1}{R_1 + R_2} \qquad (7.150)$$

where τ_1 and τ_2 are the time constants of the individual R, C circuits.

On a-c voltage as a function of frequency, the two-layer capacitor gives a relaxation spectrum indistinguishable from the simple orientation polarization of Eq. (7.125) as far as the dielectric constant κ' is concerned. The κ'' characteristic contains, in addition, the ohmic-conductivity term caused by the series resistor $R_1 + R_2$. The complex permittivity may thus be written

$$\kappa^* = \kappa'_\infty + \frac{\kappa'_s - \kappa'_\infty}{1 + j\omega\tau} - j\frac{\sigma}{\omega\epsilon_0} \qquad (7.151)$$

where

$$\kappa'_s = \frac{\tau_1 + \tau_2 - \tau}{C_0(R_1 + R_2)}$$

$$\kappa'_\infty = \frac{\tau_1\tau_2}{C_0(R_1 + R_2)\tau}$$

$$\sigma = \frac{\epsilon_0}{C_0(R_1 + R_2)}$$

From the molecular point of view, the optical dielectric constant κ'_∞ of the two-layer capacitor is determined by its spacings and the electronic, atomic, and orientation polarization as

$$\kappa'_\infty = \frac{d/\epsilon_0}{d_1/\epsilon'_1 + d_2/\epsilon'_2} \qquad (7.152)$$

The static dielectric constant

$$\kappa'_s = \kappa'_\infty \left\{1 + d_1 d_2 \left[\frac{(1/\sigma_1)\sqrt{\epsilon'_1/\epsilon'_2} - (1/\sigma_2)\sqrt{\epsilon'_2/\epsilon'_1}}{d_1/\sigma_1 + d_2/\sigma_2}\right]^2\right\} \qquad (7.153)$$

is larger because media 1 and 2 contain mobile-charge carriers of the densities N_1 and N_2, transporting charges e_1 and e_2 with the mobilities b_1 and b_2. Because of conductivities $\sigma_1 = N_1e_1b_1$ and $\sigma_2 = N_2e_2b_2$, charges pile up at the interface between 1 and 2 and change the field distribution, until the static conductivity

$$\sigma = \frac{d}{d_1/\sigma_1 + d_2/\sigma_2} \qquad (7.154)$$

is established.

Dividing media 1 and 2 into sublayers and stacking them in various ways does not affect the complex permittivity of the composite dielectric. However, if we change the geometry, for example, by dispersing medium 2 in medium 1 as spheres, ellipsoids, or rods (Fig. 7.34) [26], very different response characteristics result.

When the concentration of material 2 in base medium 1 is increased until the distorted field areas around each particle begin to overlap and affect each other, the static dielectric constant is determined by the detailed geometrical situation. Frequently a *logarithmic mixing rule* [27] is used to predict the dielectric constant of a mixture κ'_m, from the dielectric constants κ'_1 and κ'_2 and the volume ratios θ_1 and θ_2 of the components, as

$$\log \kappa'_m = \theta_1 \log \kappa'_1 + \theta_2 \log \kappa'_2 \qquad (7.155)$$

This is a rule of thumb which can hold accurately only in very specific types of distribution.

The dielectric aftereffect, well known in the discharging of capacitors, is, in general, the sign of space-charge build-up in the dielectric material. The charge carriers in a homogeneous material are required to migrate through the volume unimpeded, to discharge freely at the one electrode, and to be replaced spontaneously at the other electrode, if

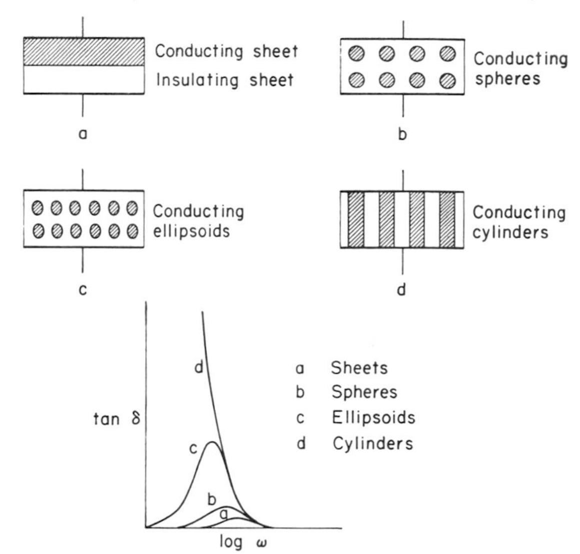

FIG. 7.34. Dependence of dielectric response on shape and orientation of particles of medium 2. (*After Sillars* [26].)

field distortion is to be avoided. This is a most unlikely combination of circumstances in nonmetals [28]. In consequence, space-charge polarization plays a most important and frequently not clearly recognized role in electric polarization phenomena and is, for example, the working principle of most *electrets*.

10. Electric Breakdown

The current-voltage characteristic of a gas at low pressure indicates that the mechanism of conduction traverses two stages (Fig. 7.35). At low field strength the charge carriers created by external ionization are gathered and drawn to the electrodes until a saturation current I_s is reached. At higher voltages the current rises rapidly by electron impact ionization until, at a critical voltage $V_{\max}$, breakdown occurs.

To explain this great increase of charge carriers, Townsend [29] introduced the avalanche concept. An electron, falling in field direction a distance dx, liberates a new electron with a probability $\alpha\, dx$; hence n electrons at x increase to $n + dn$, where $dn = n\alpha\, dx$. Thus n_0 electrons, starting at the cathode ($x = 0$), have become augmented to

$$N = n_0 e^{\alpha d} \qquad (7.156)$$

on arrival at the anode ($x = d$); that is, each starting electron produces an electron avalanche and, in consequence, leaves behind a positive ion avalanche of the height $H = e^{\alpha d} - 1$. Each positive ion, migrating in the opposite direction, may liberate an electron with the probability γ at the cathode. Hence, when the ion avalanche grows so large that regeneration of a starting electron becomes a certainty,

$$\gamma(e^{\alpha d} - 1) = 1 \qquad (7.157)$$

the discharge becomes self-supporting, and breakdown must occur.

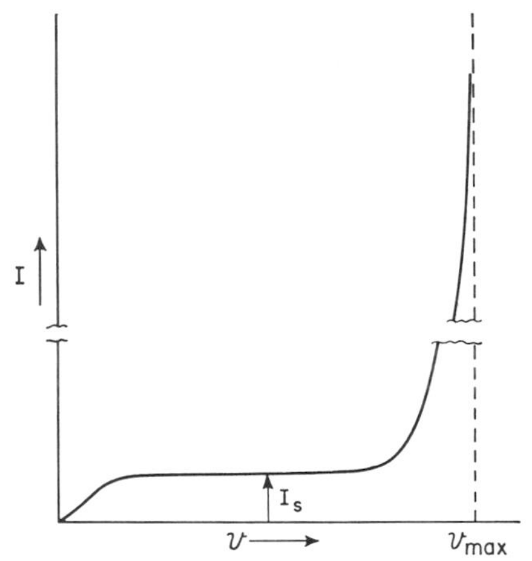

FIG. 7.35. Current-voltage characteristic at low gas pressure (schematic).

This breakdown condition of Townsend has, as its molecular parameters, an ionization probability α and a regeneration probability γ. Furthermore, the layer thickness d of the gas dielectric enters, and the origin of the starting electrons n_0 has to be established. Townsend's over-all picture has been essentially confirmed for gases up to atmospheric pressures. The detailed interpretation of the parameters α and γ and the various development stages in the formation of the final discharge after V_{max} is reached have been the subject of much careful research but are still being debated [30].

Some of these problems come into focus when one measures the breakdown voltage as a function of the gap distance. A slight increase in d should cause a very large increase in H and allow a drastic lowering of the breakdown voltage. This is observed for low pressures or short distances, whereas for high pressures or large distances the trend reverses (Fig. 7.36). As Paschen [31] first found, the breakdown voltage of a given gas is a function of the product of gas pressure p (more accurately, of the gas density ρ) and of the gap distance d. This statement is a typical *similarity law:* Reducing the thickness of the dielectric by some factor, while increasing the gas density by the same factor, leaves the decisive parameter, the voltage drop per free path, unaltered.

The minimum of the Paschen curve indicates that here the number of collisions is so adjusted that the kinetic energy of the electrons accumulates most efficiently for ionization. To the left of the minimum, the number of impacts is too small to reach the required height H of the avalanche without raising the voltage, while at the right too many collisions take place and an excessive amount of energy is squandered in electronic-excitation processes. The Paschen curve itself may be considered as a boundary line separating a lower region, in which the avalanche remains too small for regenerating one electron per starting electron ($1 \to <1$), from an upper region in which more than one electron per starting electron is reliberated at the cathode ($1 \to >1$).

Townsend's breakdown criterion corresponds to voltages on the Paschen curve and does not imply

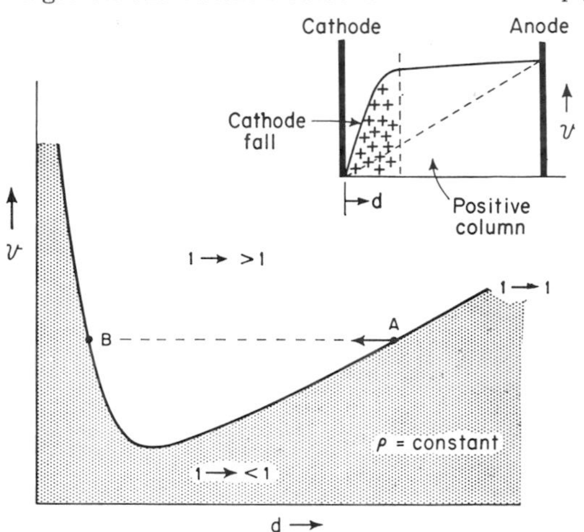

FIG. 7.36. Paschen curve and contraction into a glow discharge after breakdown voltage is reached at A.

instability. To produce breakdown requires an action pushing the operating point into the instability region $1 \to >1$. One internal mechanism might accomplish this: The positive space charge left behind by the emigrating-electron avalanches in front of the anode acts as a gap shortener [32]. The field toward the cathode increases, the virtual operating point moves from A toward the left, and the field contraction increases at an accelerated pace, until a steep cathode fall has developed and a new stabilization becomes possible at B near the left branch of the characteristic. The breakdown in this case terminates in a glow discharge, where the effective gap distance is represented by the dark space; the remaining length of the gap is bridged by the positive column, a well-conducting mixture of electrons and positive ions, that requires only a small voltage gradient for its maintenance.

While field contraction by space charge plays a decisive role, recent evidence seems to indicate that this effect may not come into play immediately but only after the applied voltage has risen into the instability region [30]. Furthermore, as photographs taken by Raether [33] with the Wilson cloud chamber testify, the regeneration of electrons by photoeffect in the gas may become of critical importance. At high gas densities, finally, and in the condensed phases, the regeneration of starting electrons ceases to be the decisive issue, because electrons can be provided by field emission. The Townsend criterion is also not applicable to high-frequency discharges when the electrons oscillate in the gap space; the loss of electrons by diffusion proves here to be the most important parameter [34].

In liquids and solids, various destructive effects may accompany the flow of current and produce breakdown of the dielectric. A simple mechanism is thermal breakdown, where the heat generated by electric conduction surpasses the heat loss by thermal conduction [35]. In liquids, gas bubbles may form, and thus a gas breakdown may trigger the liquid

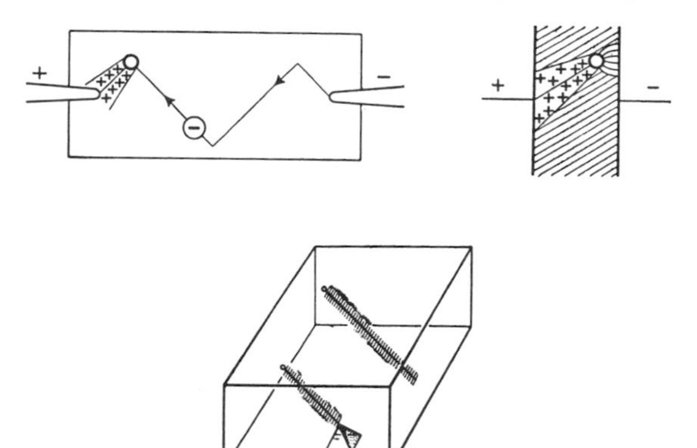

FIG. 7.37. Direction breakdown in NaCl at room temperature. (Breakdown direction is the face diagonal [110]; near the cathode the direction may change to [111] and characteristic pyramids form by cracking of the crystal between the [111] edges.)

breakdown; similarly, discharges in voids are frequently the cause of breakdown of commercial insulation. Current transfer, furthermore, may produce highly conducting decomposition products, for example, metal dendrites or carbonized paths bridging the electrodes; or radicals may break down the insulation by chemical action.

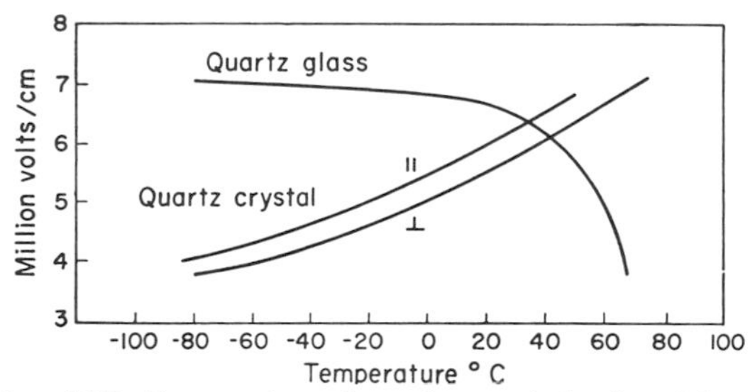

FIG. 7.38. Temperature dependence of the breakdown strength of quartz glass vs. quartz crystal.

If such destructive effects are avoided, one fundamental breakdown process for liquids and solids remains: the intrinsic breakdown by electron-impact ionization [36]. In gases, the electronic excitation of atoms and molecules is the barrier preventing early breakdown by this process; in liquids and solids, the excitation of vibration seems to be the decisive friction barrier. In periodic lattices, special quantum-mechanical features enter, as shown by the direction breakdown of crystals [36] (Fig. 7.37) or the difference in the temperature dependence of crystals vs. glasses (Fig. 7.38) [37]. Many interesting observations have been made and ingenious theoretical explanations proposed, but on this active frontier of research many issues are still in doubt [38].

11. Supplement

The connotation "dielectric" continues to expand with deepening understanding of how to tailor the properties of materials to order. Old boundaries between insulators, semiconductors, and metals disappear as one learns to create, adjust, and destroy electronic and ionic conductivity, whether by order and disorder effects, by vacancy formation and cation substitution, by oxidation or reduction, thermal treatment and high pressure, or by electron and hole injection. One also learned how to manipulate electric and magnetic moments and to influence theircoupling in lattice arrays. Long-forgotten electro- and magneto-optical effects were suddenly resurrected for device design, and new ones have been discovered. The subject "dielectrics" is now integrated into the wide expanse of modern materials research; only in that broad context can its full potentialities unfold.

The designing of materials starts with the atoms and their wave functions: the standing-wave modes of electrons, identified in their average radial distance from the nucleus by the principal quantum number n and in their eccentricity by the azimuthal quantum number l. The integer sequence $n = 1, 2, 3, 4, \ldots$ characterizes the shell structure of the atoms in the periodic system (Fig. 7.39); the $l = n - 1$ eccentric electron-cloud modes in their sequence $l = 0, 1, 2, 3, \ldots$ correspond to the $s, p, d, f, \ldots$ electron states. Each hydrogenlike orbital type (n,l) consists of $2l + 1$ individual orbitals; each orbital can accept two electrons with antiparallel spins.

The column number N of the atoms in the periodic system ranges from I to VIII, because $8 - N$ signifies the number of bonds needed to form the facsimile of an inert rare-gas shell (s^2p^6) around a reference atom. From $N = 8$ to $N = 4$ the individual atom can still supply at least one electron per orbital and thus establish a framework linking its electrons in fixed stereoconstellations to the surroundings. From $N = 3$ down, the atoms can provide only fractional electron charges per orbital from their own supply. If unable to steal electrons outright from electropositive neighbors, they must "resonate" electrons between various bond positions or pool them, as in metals. Hence, $N = 4$ designates a high-water mark of cohesion, as the boiling-point characteristics of elements completing s^2p^6 shells certify (Fig. 7.40), and a primary divide between metals (resonating bonds) and nonmetals (fixed bonds) (cf. Fig. 7.39). At high atomic numbers the metallic elements encroach toward the right, because the individual bonds begin to be submerged by multielectron attraction.

The designing of structures from identical atoms appears unambiguous at first sight: Molecules and crystals are created by forming joint wave functions between the partners through electronic overlap. However, the road forks almost immediately because of the possibility of choices. Sulfur and selenium, for example, can associate as diatomic, ring, or chain molecules (Fig. 7.41). And in condensing such molecules one can produce materials ranging from insulating, brittle ring structures to rubbery amor-

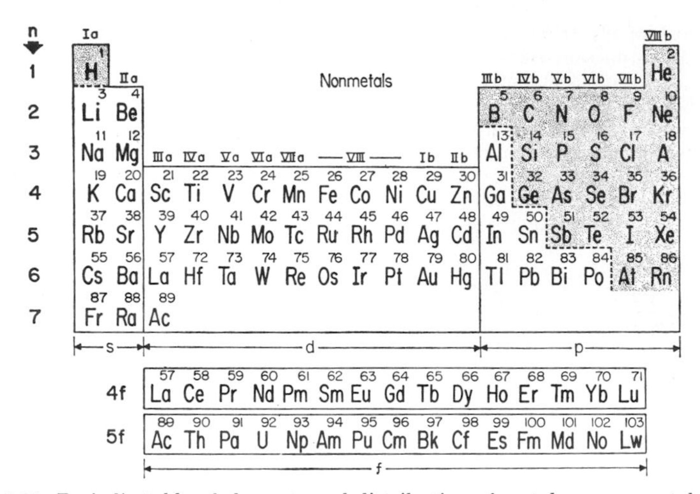

FIG. 7.39. Periodic table of elements and distribution of metals vs. nonmetals [41].

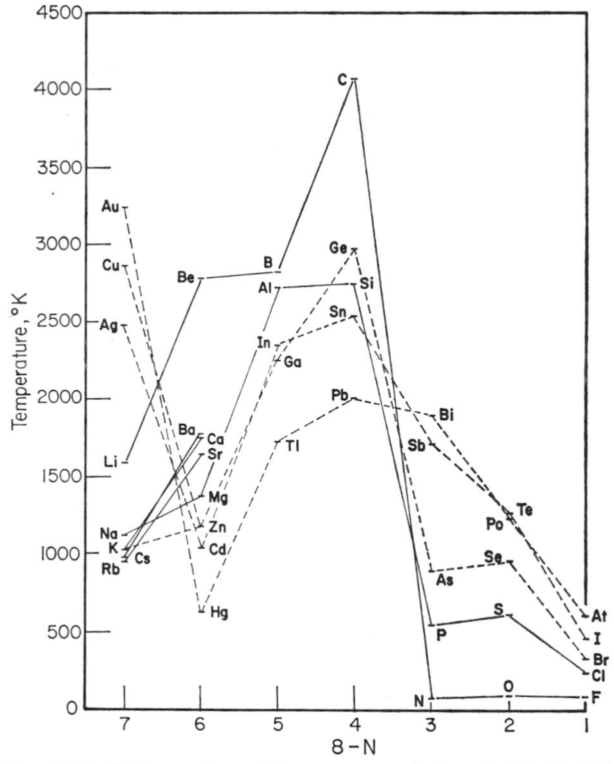

FIG. 7.40. Boiling points of elements completing s^2p^6 shells [41].

phous or crystalline semiconducting chain structures and finally—under pressure—to metals and even superconductors approaching the simple cubic structure of polonium (Fig. 7.42). This polymorphism emphasizes the haziness of boundaries between insulators and conductors, between polymers and crystals, between plastic and brittle.

When unlike atoms are admitted, a new design element enters through the polarity of the partners. Two types of rare-gas electron constellations can now be formed: the electronegative atoms that accept electrons and build up toward saturated configurations, and the electropositive ones that donate electrons and regress to subshells previously completed. Striving thus toward electric balance by oxidation and reduction, the anions and cations organize their electronic atmospheres so that their spins are in accord with the formal ionic charges Cl^-, $O^=$, Na^+, Ca^{++}, etc.

It is meaningful to assign average (Goldschmidt) radii to ions for packing considerations (Fig. 7.43)

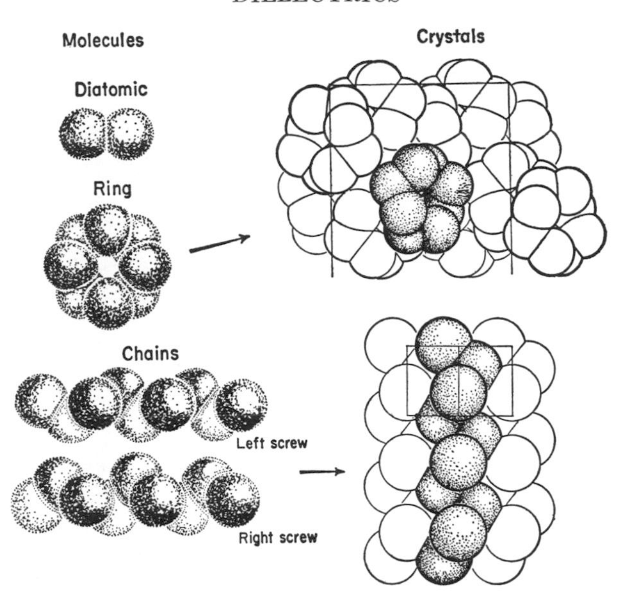

FIG. 7.41. Structure choices by polymorphism (S, Se).

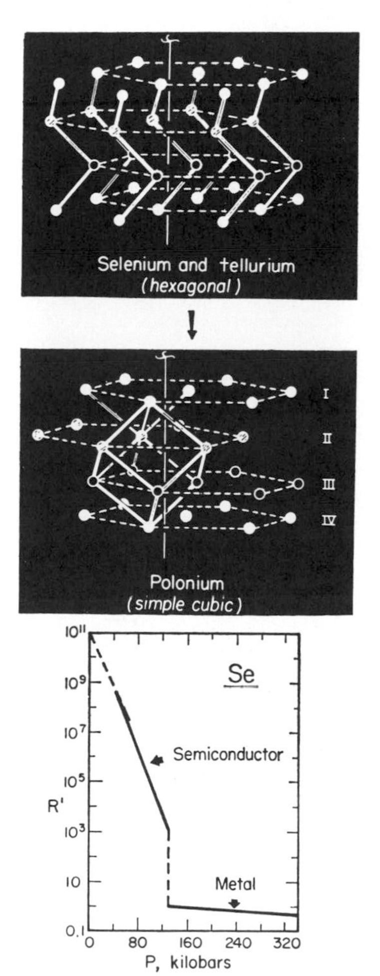

FIG. 7.42. Selenium-to-polonium-type transition [41].

but misleading to expect these ions to behave electrostatically like spheres thus charged. Large potential differences cannot be maintained in the presence of bonding electrons. The electrostatic effect of Fe^{3+}, for example, may be smaller than that of Fe^{++}, because the compensating electrons of the anion counteratmosphere penetrate closer to the nucleus.

A regular array of polar interpenetrating electron atmospheres, compared with a similar nonpolar array, will strengthen cohesion, i.e., raise the boiling points. Also the melting points will increase, because crystallization allows lowering of the free energy: Long-range order can maximize the polar attraction and minimize the polar repulsion terms. Since in well-compacted, strongly ionic crystals neither the ions have space to move nor the electrons can be activated at normal temperatures, such materials are insulators. However, as is well known, this deadlock can be upset in a variety of ways, e.g., by producing vacancies through irradiation, substitution, thermal treatment, or dislocations, thus creating a "structure-sensitive" ionic conductivity. Such disorder also favors electronic conduction by inserting a stepladder of localized energy states between valence and conduction band.

Electronic conduction can frequently be triggered by injecting either electrons from the cathode into the conduction band or holes from the anode into the valence band; however, such carriers must be able to traverse the crystal and leave at the opposite electrode. This may not be the case a priori, and the effect may depend on crystal orientation, as the behavior of rutile (TiO_2) demonstrates (Fig. 7.44). Here electron injection initially proceeds, but the electrons are trapped at various imperfections; the field gradient near the cathode flattens, and the current decreases. However, an anode fall builds up and, when it reaches a critical height, hole injection becomes possible. After a latency time the current suddenly rises: The holes compensate for the electron

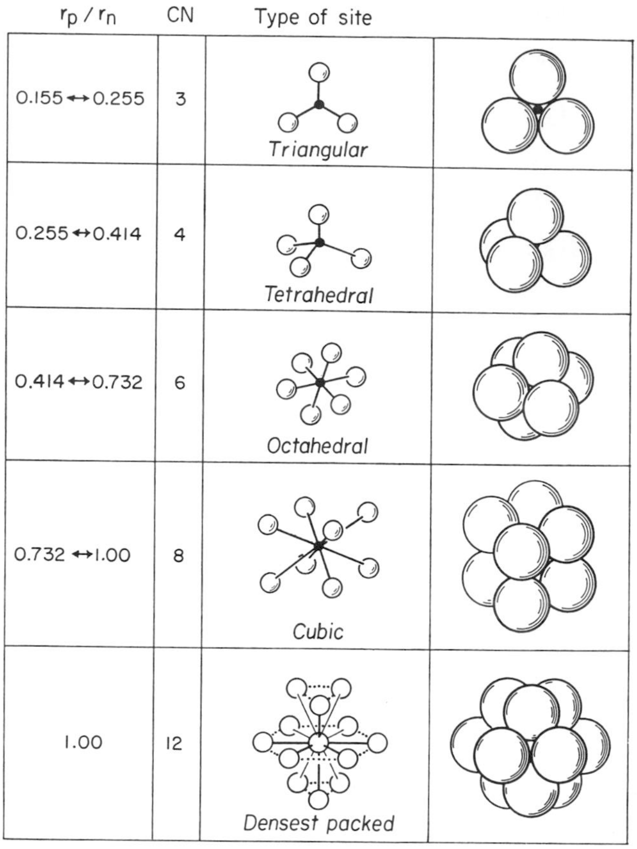

r_p/r_n	CN	Type of site	
0.155 ↔ 0.255	3	Triangular	
0.255 ↔ 0.414	4	Tetrahedral	
0.414 ↔ 0.732	6	Octahedral	
0.732 ↔ 1.00	8	Cubic	
1.00	12	Densest packed	

Fig. 7.43. Radius ratio, coordination number, and cation sites [40].

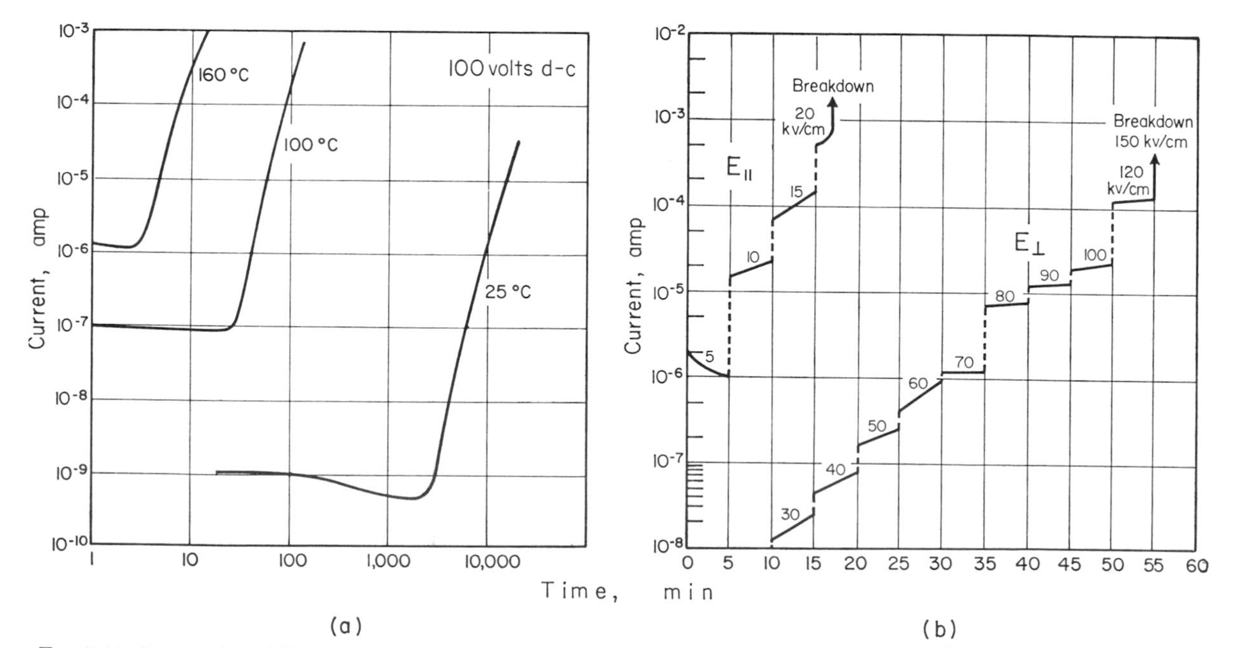

Fig. 7.44. Current instability and breakdown strength of rutile single crystals (d-c breakdown ‖ and ⊥ to c axis; impulse strength ‖ c reaches 690 kv/cm at 23°C and 10^{-6} sec rise time) [41].

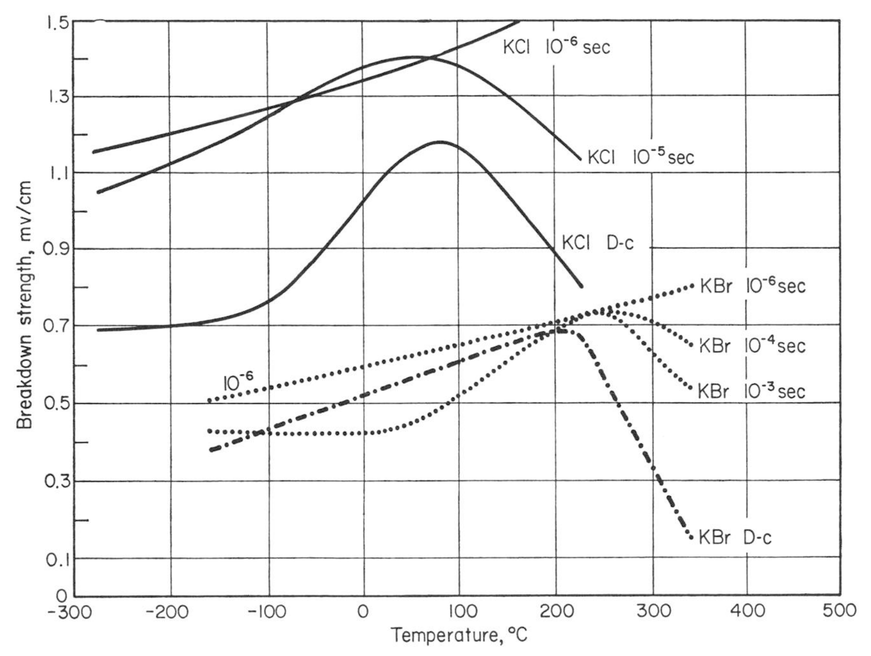

FIG. 7.45. Effect of rise time on the breakdown strength of alkali halide crystals.

space charge, the cathode fall increases, the electron injection rises, etc., until the crystal becomes incandescent and is destroyed in thermal breakdown.

Thermal breakdown is a relatively slow process; hence, the impulse strength of a crystal may greatly exceed its d-c strength and again depend on crystal orientation (cf. Fig. 7.44). However, this criterion in itself is not unambiguous. High-voltage application over sufficient time will practically always develop field distortion in dielectrics and force a lowering of $\mathcal{E}_{max}$, for example, because the onset of field emission or of impact ionization requires a critical gradient. The decrease of $\mathcal{E}_{max}$ for the alkali halide crystals at higher temperatures is due to such effects, as a comparison of d-c and impulse strength testifies (Fig. 7.45). The rising characteristic at low temperatures is caused by the scattering of electrons through lattice vibrations, as recently confirmed in our laboratory by direct experiment (Fig. 7.46): Increasing dislocation density with its disorder scattering systematically converts the shape of the breakdown curve from that of a crystal to that of a glass (cf. also Fig. 7.38).

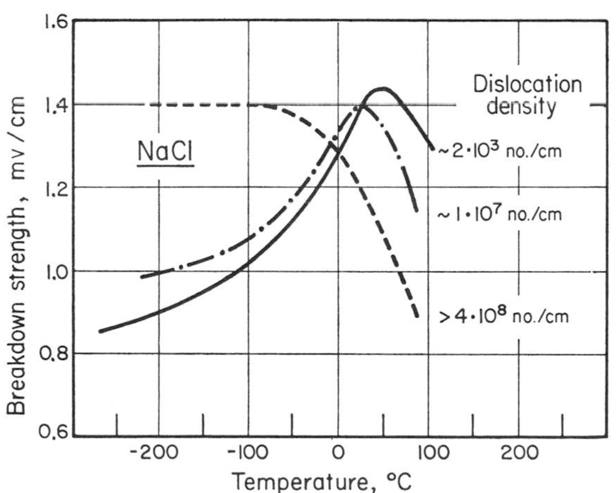

FIG. 7.46. Effect of dislocation density on the breakdown strength of alkali halide crystals [41].

References

1. Von Hippel, A.: "Dielectrics and Waves," Wiley, New York, 1954; "Dielectric Materials and Applications," The Technology Press, Massachusetts Institute of Technology, and Wiley, 1954.
2. Bethe, H.: "Handbuch der Physik," vol. 24, part 1, pp. 429ff., Springer, Berlin, 1933.
3. Debye, P.: *Physik. Z.*, **13**: 97 (1912); "Polar Molecules," Chemical Catalog, New York, 1929.
4. "Tables of Electric Dipole Moments," compiled by L. G. Wesson, The Technology Press, Massachusetts Institute of Technology, summarize the values published to 1948.
5. For recent data on polarizabilities and dipole moments, see, for example, Stuart, H. A.: "Die Struktur des freien Moleküls," Springer, Berlin, 1952.
6. Fröhlich, H.: "Theory of Dielectrics," Clarendon Press, Oxford, 1949.
7. Cole, K. S., and R. N. Cole: *J. Chem. Phys.*, **9**: 341 (1941).
8. Onsager, L.: *J. Am. Chem. Soc.*, **58**: 1486 (1936).
9. Kirkwood, J. G.: *J. Chem. Phys.*, **7**: 911 (1939).
10. An extensive treatment of the local field equations and their applications for dipole-moment determinations may be found in C. J. F. Böttcher: "Theory of Electric Polarization," Elsevier, Amsterdam and New York, 1952, and a concise summary of polarization phenomena in the low-field-strength region is given by W. F. Brown, Jr., "Encyclopedia of Physics," vol. 17, pp. 1–255, Springer, Berlin, 1956.

11. Kauzmann, W.: *Revs. Mod. Phys.*, **14:** 12 (1942).
12. Curie, P. and J.: *Compt. rend.*, **91:** 294 (1880).
13. Valasek, J.: *Phys. Rev.*, **17:** 475 (1921).
14. Busch, G., and P. Scherrer: *Naturwiss.*, **23:** 737 (1935).
15. Wainer, E., and A. N. Salomon: *Titanium Alloy Mfg. Co. Elec. Rept.* 8, 1942; 9 and 10, 1943.
16. Von Hippel, A., and coworkers: *NDRC Rept.* 300, August, 1944; A. von Hippel, R. C. Breckenridge, F. G. Chesley, and L. Tisza: *Ind. Eng. Chem.*, **38:** 1097 (1946).
17. Matthias, B. T.: *Phys. Rev.*, **75:** 1771 (1949); **76:** 175, 430, 1886 (1949); J. K. Hulm, B. T. Matthias, and A. Long: *Phys. Rev.*, **79:** 885 (1950).
18. Von Hippel, A.: *Revs. Mod. Phys.*, **22:** 221 (1950).
19. Roberts, S.: *Phys. Rev.*, **71:** 890 (1947).
20. Matthias, B., and A. von Hippel: *Phys. Rev.*, **73:** 1378 (1948).
21. Forsbergh, P. W., Jr.: *Phys. Rev.*, **76:** 1187 (1949).
22. Von Hippel, A.: *Z. Physik*, **133:** 171 (1952).
23. A thermodynamic theory for $BaTiO_3$ was formulated by A. F. Devonshire: *Phil. Mag.*, **40:** 1040 (1949); **42:** 1065 (1951); and by J. C. Slater: *Phys. Rev.*, **78:** 748 (1950). For additional references, see also E. T. Jaynes: "Ferroelectricity," Princeton University Press, Princeton, N.J., 1953.
24. Shirane, G.: *Phys. Rev.*, **86:** 219 (1952).
25. Maxwell, J. C.: "Electricity and Magnetism," vol. 1, p. 452, Clarendon Press, Oxford, 1892; K. W. Wagner in H. Schering (ed.): "Die Isolierstoffe der Elektrotechnik," pp. 1ff., Springer, Berlin, 1924.
26. Sillars, R. W.: *J. Inst. Elec. Engrs. (London)*, **80:** 378 (1937).
27. Lichtenecker, K.: *Physik. Z.*, **10:** 1005 (1909); D. A. G. Bruggeman: *Physik. Z.*, **37:** 906 (1936).
28. See, for example, A. von Hippel, E. P. Gross, J. G. Jelatis, and M. Geller: *Phys. Rev.*, **91:** 568 (1953).
29. Townsend, J. S.: "Electricity in Gases," Oxford University Press, 1914.
30. See, for example, F. Llewellyn Jones: Electrical Discharge, *Repts. Progr. in Phys.*, **16:** 216 (1953).
31. Paschen, F.: *Ann. Physik*, **37:** 69 (1889).
32. Von Hippel, A., and J. Franck: *Z. Physik*, **57:** 696 (1929).
33. Raether, H.: *Z. Physik*, **107:** 91 (1937); *Ergeb. exakt. Naturw.*, **22:** 73 (1948).
34. Allis, W. P., and S. C. Brown: *Phys. Rev.*, **87:** 419 (1952).
35. Wagner, K.: *Elec. Eng.*, **41:** 1034 (1922).
36. Von Hippel, A.: *Z. Physik*, **68:** 309 (1931); **75:** 145 (1932).
37. Von Hippel, A., and R. J. Maurer: *Phys. Rev.*, **59:** 820 (1941).
38. See S. Whitehead: "Dielectric Breakdown of Solids," Clarendon Press, Oxford, 1951; H. Fröhlich and J. H. Simpson: Intrinsic Breakdown in Solids, in "Advances in Electronics," vol. 2, Academic Press, Inc., New York, 1950; W. Franz: Theorie des reinen elektrischen Durchschlags fester Isolatoren, *Ergeb. exakt. Naturw.*, **27:** 1 (1953); "Encyclopedia of Physics," vol. 17, pp. 155–263, Springer, Berlin, 1956.
39. Forsbergh, P. W., Jr.: "Encyclopedia of Physics," vol. 17, pp. 264–392, Springer, Berlin, 1956.

Some Recent Books:

40. Von Hippel, A., and colleagues: "Molecular Science and Molecular Engineering," The Technology Press, Cambridge, Mass., and Wiley, New York, 1959.
41. Von Hippel, A. (ed.): "The Molecular Designing of Materials and Devices," The M.I.T. Press, Cambridge, Mass., 1965.
42. Birks, J. B. (gen. ed.): "Progress in Dielectrics," vol. 1, Wiley, New York, 1959, and subsequent volumes.
43. Schulman, J. H., and W. D. Compton: "Color Centers in Solids," Macmillan, New York, 1962.
44. Raether, H.: "Electron Avalanches and Breakdown in Gases," Butterworth's Advanced Physics Series, Butterworth, Washington, D.C., 1964.
45. O'Dwyer, J. J.: "The Theory of Dielectric Breakdown of Solids," Clarendon Press, Oxford, 1964.
46. Pauling, L.: "The Nature of the Chemical Bond," 3d ed., Cornell University Press, Ithaca, N.Y., 1960.
47. Jona, F., and G. Shirane: "Ferroelectric Crystals," Macmillan, New York, 1962.

Chapter 8

Magnetic Materials

By WILLIAM FULLER BROWN, JR., University of Minnesota

1. Basic Concepts [2, 8]†

In a physical theory of magnetic materials, the most important basic concept is *magnetic moment.*

Consider first a filamentary circuit in which a current I traverses a closed curve C in the positive direction of the line elements $d\mathbf{r}$. By Eq. (2.40) of Chap. 2, Part 4, the distant magnetic-field intensity $\mathbf{B}_0$ of the circuit is the curl of a vector potential

$$\mathbf{A}_0 = \frac{\gamma}{4\pi} \frac{\mathbf{m} \times \mathbf{R}_0}{R^2} \tag{8.1}$$

Here γ is 4π in Gaussian units and 1 in Lorentz-Heaviside and

$$\mathbf{m} = \frac{I}{2c} \oint_C \mathbf{r} \times d\mathbf{r} = \frac{I}{c} \int_S d\mathbf{S} \tag{8.2}$$

In the second integral in (8.2), S is any surface bounded by C; the component of $\mathbf{m}c/I$ along any direction is the projected area of S normal to this direction. Equation (8.2) defines the *magnetic moment* (= magnetic-dipole moment) of the circuit. From (8.1),

$$\mathbf{B}_0 = \nabla \times \mathbf{A}_0 = \frac{\gamma}{4\pi} \frac{-\mathbf{m} + 3\mathbf{m} \cdot \mathbf{R}_0\mathbf{R}_0}{R^3} \tag{8.3}$$

The forces exerted on the circuit by another circuit, whose field intensity is $\mathbf{B}'_0 = \nabla \times \mathbf{A}'_0$, can be calculated by differentiating (at constant $\mathbf{m}$) a potential energy

$$W = -\mathbf{m} \cdot \mathbf{B}'_0 \tag{8.4}$$

Thus the force $\mathbf{F}$ and torque $\mathbf{T}$ (found by differentiation with respect to positional and angular coordinates, respectively) are (see Part 4, Chap. 2, Sec. 7)

$$\mathbf{F} = \mathbf{m} \cdot \nabla \mathbf{B}'_0 \qquad \mathbf{T} = \mathbf{m} \times \mathbf{B}'_0 \tag{8.5}$$

Equations (8.3) and (8.4) can be used, instead of (8.2), to *define* magnetic moment. In this form, the definition can be extended to any objects that are found to behave in accordance with the defining equations, whether or not they are known to carry currents. Certain material bodies are observed to behave thus in their mechanical interactions with each other and with current circuits; they are then said to be *magnetized.* Their behavior at large distances of separation can be described by assigning to each body an appropriate magnetic moment. At distances of separation comparable with the dimensions of the bodies, it is necessary to assign a magnetic moment $\mathbf{M}\, d\tau$ to each volume element $d\tau$ and to integrate over the volume of each body. This defines the magnetic moment per unit volume, or *magnetization,* $\mathbf{M}$.

Equations (8.3) and (8.4) are identical in form with the equations that describe the interactions of dipoles in the sense *pairs of poles* (see Part 4, Chap. 2, Sec. 6). The field intensity of such a dipole of moment $\mathbf{m}$ is the negative gradient of a scalar potential

$$\psi = \frac{\gamma}{4\pi} \frac{\mathbf{m} \cdot \mathbf{R}_0}{R^2} \tag{8.6}$$

Small circuits and small magnetized bodies interact as if they were such dipoles, and finite circuits interact as if they were double layers. This equivalence of currents and poles persists throughout magnetostatics and permits many of the equations to be written in two forms; but it breaks down between the poles of a dipole, between the two sides of a double layer, and inside a magnetized body.

2. Megascopic Theory [2, 8]

Calculation of the mutual forces between magnetized bodies requires evaluation of the magnetic-field intensity of each body throughout the region where another body may be, i.e., outside the first body. In this region the field intensity can be calculated by any of three methods. Method 1 is to regard the body as a continuum of dipoles of density $\mathbf{M}$; that is, to replace $\mathbf{m}$ in (8.3) by $\mathbf{M}\, d\tau$ and to integrate the result over the body. Method 2 is to replace the magnetization by a distribution of fictitious currents inside the body and on its surface S and to calculate their field intensity. Method 3 is to replace the magnetization by a distribution of fictitious poles inside the body and on its surface S and to calculate their field intensity. The vector potential $\mathbf{A}$ for method 2 and the scalar potential ψ for method 3 are

$$\mathbf{A} = \frac{\gamma}{4\pi c} \int \frac{d\mathbf{I}_m}{R} \qquad \psi = \frac{\gamma}{4\pi} \int \frac{dp_m}{R} \tag{8.7}$$

where $d\mathbf{I}_m = c\nabla \times \mathbf{M}\, d\tau$ in the body and $-c\, d\mathbf{S} \times \mathbf{M}$ on its surface, and where $dp_m = -\nabla \cdot \mathbf{M}\, d\tau$ in the body

† Numbers in brackets refer to References at end of chapter.

and $d\mathbf{S} \cdot \mathbf{M}$ on its surface. The equivalence of the three methods can be shown by transforming the surface integrals in (8.7) to volume integrals; the results are the vector and scalar potentials corresponding to method 1.

At points inside the body whose field is being computed, all the integrals are improper. The integral corresponding to method 1 is only semi-convergent, and the others must be used with care. Two *distinct* vector functions $\mathbf{B}$ (the *magnetic-flux density*) and $\mathbf{H}$ (the *magnetizing force*) may be defined at all points by

$$\mathbf{B} = \mathbf{B}_0 + \nabla \times \mathbf{A} \qquad \mathbf{H} = \mathbf{B}_0 - \nabla\psi \tag{8.8}$$

where $\mathbf{B}_0$ is the field intensity of current circuits, and the integrals in (8.7) are now extended over all magnetic matter. The vectors $\mathbf{B}$ and $\mathbf{H}$ are equal at points outside magnetic matter but not, in general, at internal points, because the transformations that prove them equal fail in a region where $1/R$ has a singularity. In fact, their equality where $\mathbf{M} = 0$ is only a special case of the relation, valid everywhere,

$$\mathbf{B} - \mathbf{H} = \gamma\mathbf{M} \tag{8.9}$$

An interpretation of $\mathbf{B}$ and $\mathbf{H}$ in simple cases can be based on the solenoidal character of $\mathbf{B}$ and the irrotational character (in a region free of conduction currents) of $\mathbf{H}$, together with the formulas

$$\mathcal{E} = -\frac{(1/c)\,d\Phi}{dt} = -\frac{d}{c\,dt}\int_S \mathbf{B} \cdot d\mathbf{S} \qquad \oint_C \mathbf{H} \cdot d\mathbf{r} = \gamma I \tag{8.10}$$

The first relates the induced electromotive force $\mathcal{E}$ in a wire loop to the flux of $\mathbf{B}$ through it; the second relates the line integral of $\mathbf{H}$ around a path to the current linked by the path. When the surface S and line C are in free space, either $\mathbf{B}$ or $\mathbf{H}$ may be written in either integral; but when S and C are allowed to pass through magnetic matter, the choice indicated is necessary if the integrals are to remain constant as S and C are deformed. For a toroidally wound isotropic ring specimen with a closely wound test coil, a $\mathbf{B}$ vs. $\mathbf{H}$ curve is essentially a curve of fluxmeter readings against ammeter readings, with specimen dimensions and irrelevant constants divided out.

With less simple geometry, a specimen in an applied field usually becomes magnetized nonuniformly, in a manner that defies calculation. An important exception is an ellipsoidal specimen, which if uniformly magnetized produces a uniform internal $\mathbf{H} - \mathbf{B}_0$ and $\mathbf{B} - \mathbf{B}_0$. A uniform internal $\mathbf{H}$ and $\mathbf{B}$ can therefore be produced by use of a uniform applied field $\mathbf{B}_0$. Uniform magnetization along a principal axis is equivalent externally to the poles and currents shown in Fig. 8.1. The internal magnetizing force and flux density produced by them are

$$\mathbf{H}' = \mathbf{H} - \mathbf{B}_0 = -\gamma D\mathbf{M} \tag{8.11}$$

$$\mathbf{B}' = \mathbf{B} - \mathbf{B}_0 = +\gamma(1 - D)\mathbf{M} \tag{8.12}$$

The *demagnetizing factor* D (or γD) is positive and depends only on the axial ratios of the ellipsoid and on which axis $\mathbf{M}$ is along. The sum of the D's for the three axes is 1. For a long needle-ellipsoid, magnetized longitudinally, $D \ll 1$, and $\mathbf{H} \doteq \mathbf{B}_0$; for a thin disk magnetized along the short axis, $D \doteq 1$, and $\mathbf{B} \doteq \mathbf{B}_0$ [1, 3, 4, 7].

The induced emf (8.10) can be expressed, when the circuit is stationary, as $\mathcal{E} = -(1/c)\oint \dot{\mathbf{A}} \cdot d\mathbf{r}$; the rate of work against $\mathcal{E}$ is $-\mathcal{E}I$. If a small, fixed, rigid specimen is magnetized by current through the circuit, then by (8.1) its contribution to $\dot{\mathbf{A}}$ is $(\gamma/4\pi)\dot{\mathbf{m}} \times \mathbf{R}_0/R^2$. The expression for $-\mathcal{E}I$ in this case is equivalent to $\mathbf{B}_0 \cdot \dot{\mathbf{m}}$, where $\mathbf{B}_0$ is the field intensity of the circuit at the position of the specimen. Therefore the work done in changing the moment of the specimen by $d\mathbf{m}$ is $\mathbf{B}_0 \cdot d\mathbf{m}$. For an ellipsoid of volume τ magnetized along a principal axis,

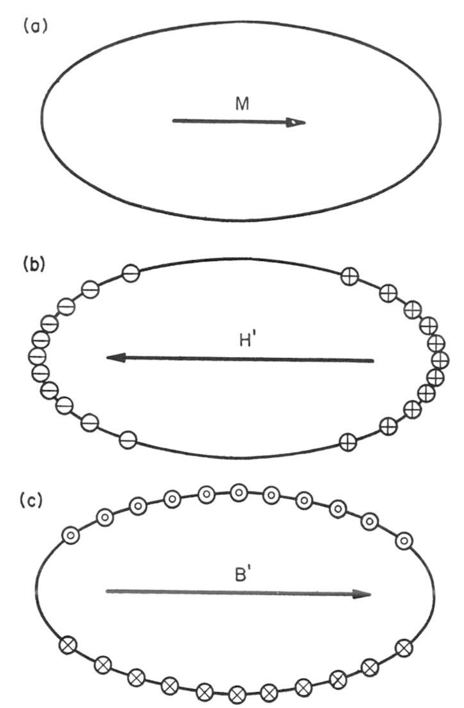

FIG. 8.1. A uniformly magnetized ellipsoid (*a*) and the equivalent poles (*b*) and currents (*c*).

$$\begin{aligned} dW &= \tau\mathbf{H} \cdot d\mathbf{M} + d(\tfrac{1}{2}\gamma D M^2 \tau) \\ &= \mathbf{B} \cdot d\mathbf{M} + d[-\tfrac{1}{2}\gamma(1 - D)M^2\tau] \end{aligned} \tag{8.13}$$

In a complete magnetic cycle, the energy dissipated per unit volume is

$$w = \oint \mathbf{H} \cdot d\mathbf{M} = \oint \mathbf{B} \cdot d\mathbf{M} = (1/\gamma)\oint \mathbf{H} \cdot d\mathbf{B} \tag{8.14}$$

In a reversible change, $\mathbf{H} \cdot d\mathbf{M} + T\,ds$ and $\mathbf{B} \cdot d\mathbf{M} + T\,ds$ (where s is entropy per unit volume) must be perfect differentials. This fact can be used to derive thermodynamic relations for a fluid at constant volume and for a solid at zero strain [2, 10].

The magnetic properties of a material can be described by presenting a relation between two of the three vectors $\mathbf{M}$, $\mathbf{B}$, and $\mathbf{H}$; the third is then determined by (8.9). Usually $\mathbf{M}$, $\mathbf{B}$, or $\mathbf{B} - \mathbf{H}$ is given as a function of $\mathbf{H}$. If the material is isotropic and if the $\mathbf{M}$ vs. $\mathbf{H}$ relation under specified conditions (e.g., constant temperature and density) is linear,

$$\mathbf{M} = \chi\mathbf{H} \qquad \mathbf{B} = \mu\mathbf{H} \qquad \mu = 1 + \gamma\chi \tag{8.15}$$

the magnetic properties can be described by giving the *susceptibility* χ or the *permeability* μ as a function of the parameters (e.g., temperature and density) on which it depends. For a crystal, χ and μ are tensors.

Magnetic materials fall experimentally into two general classes. The first class consists of materials for which (8.15) holds at all ordinary field strengths; these materials are further classified as *diamagnetic* ($\chi < 0$, $\mu < 1$) and *paramagnetic* ($\chi > 0$, $\mu > 1$). In either case, $\gamma\chi \ll 1$, so that μ is close to 1 and **B** is nearly equal to **H**. The second class consists of materials for which (8.15) does not hold; these are called *ferromagnetic* and include the practically useful magnetic elements iron, nickel, and cobalt. Ferromagnetic materials usually have $|B| \gg |H|$. In ferromagnetism the terms *susceptibility* and *permeability* refer to ratios M/H and B/H or to slopes dM/dH and dB/dH measured under specified conditions; usually the terms are preceded by qualifiers such as *initial* and *maximum* (see Fig. 8.2) [4].

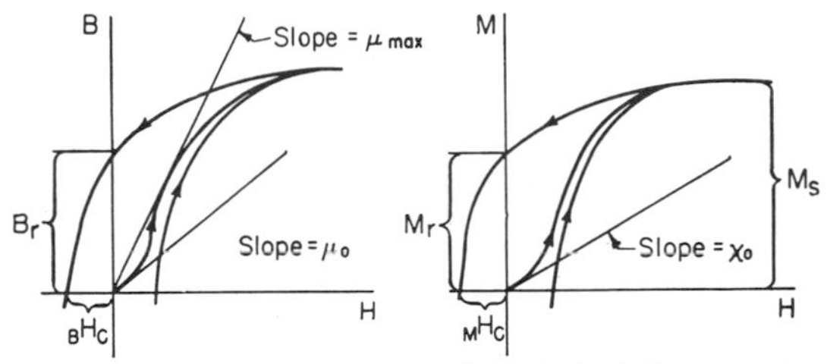

FIG. 8.2. Important constants of the technical magnetization curve: μ_0 and χ_0, initial permeability and susceptibility; μ_{max}, maximum permeability; B_r and M_r, remanent flux density (= induction) and magnetization; H_c, coercive force ($_BH_c$ = induction coercive force, $_MH_c$ = magnetization coercive force).

When $\mathbf{M} = \chi\mathbf{H}$ and $\gamma\chi \ll 1$, $\mathbf{H} \doteq \mathbf{B} \doteq \mathbf{B}_0$, and an ellipsoidal specimen acquires a moment $\tau\chi\mathbf{B}_0$. The force $\mathbf{m} \cdot \nabla\mathbf{B}_0$ [see Eqs. (8.5)] then becomes $\frac{1}{2}\tau\chi\nabla(B_0^2)$. Therefore a paramagnetic specimen is pulled into the field, and a diamagnetic specimen is pushed out of it. Properties of diamagnetic and paramagnetic specimens are usually determined by force or torque measurements. For ferromagnetic materials, the ammeter-fluxmeter technique and its variants are standard in practical work and useful in research, but research on ferromagnetic crystals and on special properties requires many other techniques as well [1, 4, 24, 25].

A *uniform* field aligns a needle, either paramagnetic or diamagnetic, along the field direction [1, p. 113].

3. Classical Microscopic Theory [8, 10, 11]

The connecting link between megascopic and microscopic theory is the interpretation of the magnetic moment $\mathbf{M}\,\Delta\tau$ of a physically small volume $\Delta\tau$ as the sum of moments $\boldsymbol{\mu}$ of atomic particles in $\Delta\tau$. (A *physically small* volume is one that is small on a megascopic scale but still contains many atoms.) The equality of the megascopic and microscopic quantities is subject to the approximation that the small-scale discontinuities and rapid fluctuations (spatial or temporal) in the latter are smoothed out in the former.

The microscopic magnetic moments that are important in magnetic materials are those associated with the orbital motion and with the spin of electrons. The former can be treated partly by classical methods.

A particle of charge e, with instantaneous position vector $\mathbf{r}$, produces an instantaneous vector potential $\gamma e\dot{\mathbf{r}}/4\pi cR$, where $\mathbf{R}$ is the vector from the particle to the field point; here retardation terms, of relative order $|\dot{\mathbf{r}}|^2/c^2$, have been neglected. If the particle moves in a closed orbit with period t_0, the time-average vector potential is $(\gamma e/4\pi ct_0)\oint d\mathbf{r}/R$, as if the orbit were carrying current e/t_0. The time-average distant field is that of a small circuit of moment [see Eq. (8.2)] $(e/2ct_0)\oint \mathbf{r} \times d\mathbf{r}$, which is the time average of

$$\boldsymbol{\mu} = \frac{e}{2c}\mathbf{r} \times \dot{\mathbf{r}} \tag{8.16}$$

The angular momentum of the particle, of mass m, with respect to the origin O of $\mathbf{r}$ is

$$\mathbf{J} = m\mathbf{r} \times \dot{\mathbf{r}} \tag{8.17}$$

provided the particle is in zero applied field $\mathbf{B}_0$ [see Eq. (8.23); in general, $\mathbf{J}$ is defined as $\mathbf{r} \times \mathbf{p}$, which is the same as $m\mathbf{r} \times \dot{\mathbf{r}}$ only if $\mathbf{A}_0 = 0$]. Therefore,

$$\boldsymbol{\mu} = (e/2mc)\mathbf{J} \tag{8.18}$$

If the particle moves under a central force directed toward O, $\mathbf{J}$ is a constant of the motion, so $\boldsymbol{\mu}$ is also constant. Thus (8.16) is usually taken as the definition of the instantaneous magnetic moment, which then depends on the point O chosen as origin.

In the distant field of many such particles, the time variations cancel if there is no correlation between the phases of different particles in a physically small volume $\Delta\tau$. It is then legitimate to replace $\Sigma_{\Delta\tau}\boldsymbol{\mu}$ by $\mathbf{M}\,\Delta\tau$ and to calculate the field by the formulas of Sec. 2 from a smoothed-out function $\mathbf{M}$. The field of conduction electrons in a circuit is similarly calculated from a smoothed-out current density.

If the particle is acted upon by the constant field $\mathbf{B}_0$ of distant sources, it is subject to an instantaneous force $e\dot{\mathbf{r}} \times \mathbf{B}_0/c$ and torque (about O) $e\mathbf{r} \times (\dot{\mathbf{r}} \times \mathbf{B}_0)/c$. The time-average force and torque are the same as for a circuit with current e/t_0, and they can therefore be computed by use of (8.5) with $\mathbf{m}$ replaced by $\boldsymbol{\mu}$.

If a particle is subject to a nonmagnetic force $\mathbf{F}_0 = -\nabla W_0$, derivable from a potential energy W_0, and also to a uniform and constant magnetic field of intensity $\mathbf{B}_0$, the equation of motion is

$$m\ddot{\mathbf{r}} = \mathbf{F}_0 + e\dot{\mathbf{r}} \times \mathbf{B}_0/c \tag{8.19}$$

It is assumed that W_0 is symmetric about an axis through the origin and along the field direction. Then if a transformation is made (see Part 2, Chap. 4, Sec. 2) to axes rotating about the origin with angular velocity

$$\boldsymbol{\omega} = -(e/2mc)\mathbf{B}_0 \tag{8.20}$$

the Coriolis term on the left of (8.19) cancels the term $e(D\mathbf{r}/dt) \times \mathbf{B}_0/c$ on the right. (Here D/dt means differentiation in rotating axes.) The terms in ω^2 or B_0^2 are negligible at ordinary fields; then the equation of motion in rotating axes is $mD^2\mathbf{r}/dt^2 = \mathbf{F}_0$. Thus the motion in rotating axes with the field present is identical with a possible motion in fixed axes without the field; the system precesses about the field direction with angular velocity ω. This is *Larmor's theorem.*

In a magnetic field $\mathbf{B}_0 = \nabla \times \mathbf{A}_0$, with associated electric field $\mathbf{E}_0 = -(1/c)\partial \mathbf{A}_0/\partial t$, the equation of motion of a charged particle can be derived from a Lagrangian function

$$\mathcal{L} = \tfrac{1}{2}m\dot{\mathbf{r}}^2 + \frac{e}{c}\dot{\mathbf{r}} \cdot \mathbf{A}_0 - W_0 \tag{8.21}$$

With the abbreviations $\partial/\partial\mathbf{r} = \mathbf{i}\partial/\partial x + \cdots$, $\partial/\partial\dot{\mathbf{r}} = \mathbf{i}\partial/\partial\dot{x} + \cdots$, the Lagrangian equations of motion of the particle can be written

$$\frac{d}{dt}\frac{\partial\mathcal{L}}{\partial\dot{\mathbf{r}}} - \frac{\partial\mathcal{L}}{\partial\mathbf{r}} = 0 \tag{8.22}$$

and are equivalent to (8.19) with an additional term $e\mathbf{E}_0$. The moments p_x, p_y, p_z conjugate to the components x, y, z of $\mathbf{r}$ form a vector

$$\mathbf{p} = \partial\mathcal{L}/\partial\dot{\mathbf{r}} = m\dot{\mathbf{r}} + e\mathbf{A}_0/c \tag{8.23}$$

The equations of motion can be put into Hamiltonian form, $\partial\mathcal{H}/\partial\mathbf{p} = \dot{\mathbf{r}}$, $\partial\mathcal{H}/\partial\mathbf{r} = -\dot{\mathbf{p}}$, with the Hamiltonian

$$\mathcal{H} = \mathbf{p} \cdot \dot{\mathbf{r}} - \mathcal{L} = \frac{1}{2m}\left(\mathbf{p} - \frac{e}{c}\mathbf{A}_0\right)^2 + W_0 \tag{8.24}$$

By (8.23), the quantity in parentheses in (8.24) is $m\dot{\mathbf{r}}$, and therefore $\mathcal{H}$ (when expressed in terms of $\dot{\mathbf{r}}$ rather than of $\mathbf{p}$) is the sum of the kinetic energy $\tfrac{1}{2}m\dot{\mathbf{r}}^2$ and the potential energy W_0; the effect of the field on the energy is contained in the Larmor contribution to $\dot{\mathbf{r}}^2$. The Hamiltonian (8.24) is useful for quantum-mechanical reformulation of the theory.

4. Quantum-mechanical Concepts [9, 10, 11]

In quantum mechanics the *instantaneous* values of such quantities as $\mathbf{r} \times \dot{\mathbf{r}}$ are not regarded as meaningful; instead, the theory deals with the *mean* value associated with each stationary state of the atomic system. The mean values of the *observable* μ_x in the various stationary states form the diagonal elements of a *matrix* μ_x; the mean values of the observable μ_x^2 are the corresponding diagonal elements of the matrix μ_x^2, where the operation of squaring is a matrix multiplication (depending on nondiagonal as well as diagonal elements of μ_x). The theoretical calculation of the magnetization $\mathbf{M}$ therefore requires evaluating the matrices μ_x, μ_y, and μ_z, or more concisely the vector matrix $\boldsymbol{\mu}$, for a typical atomic system. The diagonal elements of $\boldsymbol{\mu}$ give the mean atomic moments in the various stationary states. These states themselves must then be averaged over by use of statistical mechanics; here the classical Boltzmann formula is adequate except for conduction electrons in metals.

The magnetic-moment matrix $\boldsymbol{\mu}_L$ associated with the orbital motion of the electrons (of charge $-|e|$) in an atom is, in the absence of a field $\mathbf{B}_0$, $-(|e|/2mc)$ times the orbital-angular-momentum matrix. In simple cases, e.g. hydrogen-like atoms, the elements of $\boldsymbol{\mu}_L$ in zero applied magnetic field and their first-order coefficients with respect to the field $\mathbf{B}_0$ can be evaluated rigorously. This requires solution of the Schroedinger wave equation to the first order in $\mathbf{B}_0$, by perturbation methods; the matrix element $(n|\boldsymbol{\mu}_L|n')$ corresponding to the stationary states whose wave functions are ψ_n and ψ'_n (here n and n' represent sets of several quantum numbers) is $\int\psi^*_n\boldsymbol{\mu}_L\psi_{n'}\,d\tau$ ($d\tau$ = element of configuration-space volume). In this integral and in the wave equation $\mathcal{H}\psi = E\psi$, $\boldsymbol{\mu}$ of (8.16) and $\mathcal{H}$ of (8.24) must first be expressed in terms of coordinates q_i and their canonically conjugate momenta p_i, and then the momenta must be replaced by the corresponding operators $-i\hbar\partial/\partial q_i$. In a hydrogen-like atom, the component of angular momentum in the field direction is quantized to integral multiples of $\hbar$, and therefore the corresponding component of $\boldsymbol{\mu}_L$ in zero field is quantized to integral multiples of the *Bohr magneton* $|e|\hbar 2/mc$.

In addition, each electron has a spin and associated magnetic moment, for which no classical model (such as a spinning sphere) is very useful. For an individually quantized electron, the components of spin angular momentum and magnetic moment along the axis of quantization are, respectively, $\hbar/2$ and $|e|\hbar/2mc$ (1 Bohr magneton), so that the ratio of moment to angular momentum is twice that for orbital motion. In a many-electron atom, the spins of the various electrons interact strongly because of the Heisenberg "exchange" phenomenon. As an approximation adequate for magnetic purposes, Russell-Saunders coupling is usually assumed; that is, it is supposed that the electron spins interact to form a resultant spin angular momentum $\mathbf{S}\hbar$, that the electron orbits interact to form a resultant angular momentum $\mathbf{L}\hbar$, and that the orbital and spin angular momenta interact less strongly, approximately in accordance with a Hamiltonian term (const $\times$ $\mathbf{L} \cdot \mathbf{S}$). The total angular momentum is

$$\mathbf{J}\hbar = (\mathbf{L} + \mathbf{S})\hbar \tag{8.25}$$

and the total magnetic moment in zero field is

$$\boldsymbol{\mu} = \boldsymbol{\mu}_L + \boldsymbol{\mu}_S = -(|e|\hbar/2mc)(\mathbf{L} + 2\mathbf{S}) \tag{8.26}$$

The introduction of electron spin requires the addition to the Hamiltonian of a term

$$-\boldsymbol{\mu}_S \cdot \mathbf{B}_0 = +(|e|\hbar/mc)\mathbf{S} \cdot \mathbf{B}_0$$

The square of a diagonal matrix is itself diagonal, and its elements are the squares of the original elements; the corresponding observable has a definite value in each stationary state, since the mean of its square is equal to the square of its mean. The vector matrices $\mathbf{L}$, $\mathbf{S}$, and $\mathbf{J}$ of an atom are not diagonal, but their squares are, with elements of the form $L(L+1)$, $S(S+1)$, and $J(J+1)$, respectively. The possible values are $L = 0, 1, 2, \ldots$; $S = \tfrac{1}{2}n_e, \tfrac{1}{2}n_e - 1, \ldots, 0$ or $\tfrac{1}{2}$ (n_e = number of electrons); and $J = |L - S|, |L - S| + 1, \ldots, L + S$.

A stationary state corresponds to a set of values of L, S, J, and additional quantum numbers. In the presence of a small magnetic field $\mathbf{B}_0 = B_{0z}\mathbf{k}$, the component J_z of $\mathbf{J}$ is a diagonal matrix with elements $M_J = -J, -J+1, \ldots, J-1, J$. The $(2J+1)$ states distinguished by different values of M_J differ in energy when $B_{0z} \neq 0$, but for $\mathbf{B}_0 = 0$ they become a "degenerate" set of states of equal energy. The direction of the axis of quantization is then arbitrary. Under these conditions, the theorem of *spectroscopic stability* asserts that certain sums over the $(2J + 1)$ states of the degenerate set are independent of the choice of this axis. In particular, if x, y, z are cartesian coordinates of an electron in a spherically symmetric atom, the average over the $2J + 1$ states of the mean of the observable x^2, that is of the appropriate diagonal elements of the matrix x^2, is independent of the choice of the axis of quantization and therefore equal to the average for all orientations of the axis. By symmetry, this is the same as the average for y^2 or for z^2 and therefore equal to the average for $\frac{1}{3}r^2$.

The nonvanishing nondiagonal elements of **S** and **L** correspond to pairs of states (L, S, J, M_J) and $(L, S, J + 1, M_J)$.

5. Diamagnetism [5, 6, 10, 11]

For the calculation of diamagnetic and paramagnetic susceptibilities, the approximation $\mathbf{H} = \mathbf{B} = \mathbf{B}_0$ is legitimate; the susceptibility is then M/B_0. The theory is simplest for gases. A gas molecule may be considered isolated, except for the weak interactions necessary to keep it in statistical equilibrium with its environment at temperature T. If the mean moment of a molecule under these conditions is $<\boldsymbol{\mu}>_{av}$, the moment of a gram-molecule is $N<\boldsymbol{\mu}>_{av}$ (N = Avogadro's number), and the gram-molecular susceptibility (susceptibility times gram-molecular volume) is $N|<\boldsymbol{\mu}>_{av}|/B_0$. When the molecule has no moment in zero field ("permanent" moment), an applied field induces a negative moment, and the gas is diamagnetic. When the molecule has a permanent moment, the diamagnetic effect is still present, but there is also a much larger "orientation" effect; that is, in a field the directions of the moments are no longer distributed isotropically (as they must be, by symmetry, in zero field) but, on the average, have components in the field direction. The gas is then paramagnetic.

A *monatomic* gas molecule has zero magnetic moment in its normal state if this is a state with $S = L = 0$ (singlet S state). If the energy of excitation to the lowest excited state is large in comparison with kT (over the temperature range of interest), then the Boltzmann factor may be taken as 1 for the normal state and 0 for all others. Then only one element of the magnetic-moment matrix—the mean moment in the normal state—needs to be evaluated, and the gram-molecular susceptibility is independent of temperature.

For the electrons of a monatomic molecule in field $\mathbf{B}_0$, Larmor's theorem holds, with the origin O at the nucleus. In the field each electron (of charge $-|e|$) acquires an angular velocity $\boldsymbol{\omega} = +(|e|/2mc)\mathbf{B}_0$ [see Eq. (8.20)] about O. The resulting linear velocity is $\boldsymbol{\omega} \times \mathbf{r}$, and the resulting magnetic moment is $\boldsymbol{\mu} = -(|e|/2c)\mathbf{r} \times (\boldsymbol{\omega} \times \mathbf{r})$ [see Eq. (8.16)]. The component of $\boldsymbol{\mu}$ in the field direction is $-e^2(x^2 + y^2)B_0/4mc^2$, if the field is along the z axis. Since the moment is needed only to the first order in B_0, the coefficient of B_0 may be evaluated in zero field. This classical formula still holds in quantum mechanics if $\boldsymbol{\mu}$, x^2, and y^2 are interpreted as matrices. The mean moment in the normal state is $-e^2(<x^2>_{av} + <y^2>_{av})B_0/4mc^2$, where $<x^2>_{av}$ is the mean of those diagonal elements of the matrix x^2 that correspond to the $(2J + 1)$-fold degenerate normal state. By the theorem of spectroscopic stability, $<x^2>_{av} = <y^2>_{av} = \frac{1}{3}<r^2>_{av}$. Finally, the gram-molecular susceptibility χ_M is found by summing over the electrons of an atom and multiplying by N/B_0:

$$\chi_M = -\frac{Ne^2}{6mc^2} \sum <r^2>_{av} \tag{8.27}$$

The theoretical calculation of χ_M for any monatomic gas requires evaluation of the integral

$$<r^2>_{av} = \int r^2 |\psi|^2 \, d\tau$$

for each electron; here ψ is the wave function for the normal state in zero field. Except for atomic hydrogen (which is not diamagnetic), the wave functions are known only approximately. Methods that have been used for calculating χ_M include (1) the use of hydrogen wave functions modified by the introduction of screening constants; (2) the use of analytic approximations to the wave functions; and (3) the use of the wave functions calculated by the Hartree self-consistent field method.

The same methods may be used to calculate the diamagnetic susceptibility of ions; the experimental values are deduced from data on solutions and on solid salts, by assuming that the component ions contribute additively to the susceptibility. For polyatomic molecules the theory is complicated by two factors. There is no longer a single center of force, so that Larmor's theorem cannot be used in its usual simple form; and besides the diamagnetism, there is a temperature-independent paramagnetism, to be discussed in Sec. 6.

Table 8.1 gives some values of calculated and measured diamagnetic susceptibilities.

TABLE 8.1. DIAMAGNETIC SUSCEPTIBILITIES (Gram-molecular)

Substance	$-\chi_M \times 10^6$ (Gaussian units)				
	Calculated*				Observed*
	(1)	(2)	(3)	(4)	
He	1.54	1.64	1.86	1.85	1.88– 1.91
Ne	5.7	5.6	.....	....	6.6 – 7.7
A	13.6	18.5	25.	....	18.1 –25.3
Na^+	.....	.....	5.5	....	5.2 –10.4
Cl^-	.....	.....	40.4	....	20.4 –24.1
K^+	.....	.....	17.5	....	14.5 –16.9
Rb^+	.....	.....	30.1	....	23.2 –31.3
H_2	.....	.....	.....	4.2	3.9 – 4.0

* Basis of calculation: (1) hydrogenic wave functions with screening constants (Pauling-Van Vleck); (2) analytical approximations to wave functions (Slater); (3) wave functions by Hartree method (Stoner); (4) other approximations. Compiled from J. H. Van Vleck, "The Theory of Electric and Magnetic Susceptibilities," Oxford University Press, 1932, and Edmund C. Stoner, "Magnetism and Matter," Methuen, London, 1934.

Another type of diamagnetism occurs in metals, where the conduction electrons contribute a diamagnetic susceptibility of the same order as, but smaller than, the paramagnetic susceptibility to be discussed in Sec. 6. For electrons confined within a box of perfectly reflecting walls, but otherwise free, the projections of the classical paths on a plane perpendicular to $\mathbf{B}_0$ are circular arcs when $\mathbf{B}_0 \neq 0$. A circular path that never reaches the boundary contributes a negative mean moment, but a path composed of circular arcs with boundary reflections contributes a positive one, and according to classical statistical mechanics the net susceptibility is zero. In quantum statistics this cancellation is not complete [9, 10, 11].

6. Paramagnetism [5, 6, 10, 11]

In Langevin's classical theory, a paramagnetic molecule in zero field is assumed to have a magnetic moment μ along some axis fixed in the molecule. If this axis makes an angle θ with Oz (fixed in space), then a field $\mathbf{B}_0$ along Oz produces, to the first order in B_0, a change of energy from E^0 (independent of θ) to $E^0 - \mu B_0 \cos\theta$. The moment, to the first order, is $\mu \cos\theta + \alpha(\theta)B_0$; the second term in simple cases is the diamagnetic term computed in Sec. 5. The Boltzmann factor is changed by the field; therefore a new distribution of orientations will be established by collisions. The resulting mean moment along Oz will be

$$<\mu_z>_{\text{av}} = \frac{\displaystyle\int_0^\pi [\mu\cos\theta + \alpha(\theta)B_0] \exp[-(E^0 - \mu B_0\cos\theta)/kT]\sin\theta\, d\theta}{\displaystyle\int_0^\pi \exp[-(E^0 - \mu B_0\cos\theta)/kT]\sin\theta\, d\theta} \tag{8.28}$$

The Boltzmann factor must be expressed to the first order in B_0 in the coefficient of $\mu\cos\theta$ in the numerator; elsewhere the zeroth order is sufficient. Thus

$$\chi_M = \frac{N<\mu_z>_{\text{av}}}{B_0} = \frac{N\mu^2}{3kT} + N\alpha \tag{8.29}$$

where α is the isotropic average of $\alpha(\theta)$. Ordinarily the term $N\alpha$ is negligible. Langevin's theory explains Curie's law that the paramagnetic susceptibility varies inversely with the absolute temperature. (Debye later applied the same formula successfully to polar dielectrics.)

In the quantum-mechanical theory of Van Vleck and his collaborators, the integrals in (8.28) are replaced by sums over stationary states. For a uniform field $\mathbf{B}_0 = B_0\mathbf{k}$, $\mathbf{A}_0$ in (8.23) and (8.24) may be taken as $\frac{1}{2}\,\mathbf{B}_0 \times \mathbf{r}$. It follows that the first-order term in the Hamiltonian (8.24) is

$$-(e/2mc)\mathbf{p}\cdot\mathbf{B}_0 \times \mathbf{r} = -\boldsymbol{\mu}^0\cdot\mathbf{B}_0$$

where $\boldsymbol{\mu}^0$ is the moment in zero field [Eqs. (8.16) and (8.23) with $\mathbf{A}_0 = 0$]. This result remains true when the effects of all the electrons in an atom are summed and the spin term added [Eq. (8.26) and following sentence]. The quantum analogue of Eq. (8.28) is therefore

$$<\mu_z>_{\text{av}} = \frac{\displaystyle\sum_n [(n|\mu_z^0|n) + \alpha_n B_0]\exp\{-[E_n^0 - (n|\mu_z^0|n)B_0]/kT\}}{\displaystyle\sum_n \exp\{-[E_n^0 - (n|\mu_z^0|n)B_0]/kT\}} \tag{8.30}$$

and of Eq. (8.29)

$$\chi_M = \frac{N<\mu_z>_{\text{av}}}{B_0} = \frac{N}{kT}\frac{\displaystyle\sum_n (n|\mu_z^0|n)^2 e^{-E_n^0/kT}}{\displaystyle\sum_n e^{-E_n^0/kT}} + N\frac{\displaystyle\sum_n \alpha_n e^{-E_n^0/kT}}{\displaystyle\sum_n e^{-E_n^0/kT}} \tag{8.31}$$

Here E_n^0 is the energy in stationary state n in zero field; $(n|\mu_z^0|n)$ is the corresponding diagonal element of the matrix $\mu_z = \boldsymbol{\mu}\cdot\mathbf{k}$ in zero field; and $\alpha_n B_0$ is the first-order change of $(n|\mu_z|n)$ in the field. The symbol n represents a set of quantum numbers such as L, S, J, M_J.

The quantity α_n can be evaluated by calculating the matrix element $(n|\mu_z|n) = \int\psi^*{}_n\mu_z\psi_n\, d\tau$ to the first order in B_0. The effect of the field on ψ_n is found by first-order perturbation theory from the perturbing term $-\mu_z^0 B_0$ in the Hamiltonian; the effect of the field on μ_z (regarded as a function of canonical variables) is found from (8.16) and (8.23) to be a diamagnetic term $-\Sigma_e e^2(x^2 + y^2)B_0/4mc^2$, where Σ_e indicates a sum over electrons. The result is

$$\alpha_n = 2\sum_{n' \neq n} \frac{|(n|\mu_z^0|n')|^2}{E_{n'}^0 - E_n^0} - \sum_e \frac{e^2}{4mc^2}(n|x^2 + y^2|n) \tag{8.32}$$

For atoms and ions, the sums in (8.31) can be limited to the particular L and S that correspond to the states of lowest energy; all other terms have either negligible Boltzmann factors or vanishing matrix elements $(n|\mu_z^0|n')$. Within this group of states, the order of magnitude of the energy differences ΔE_J between adjacent states J and $J + 1$ determines to which of three forms (8.31) reduces.

1. If $\Delta E_J \gg kT$, the second term in (8.31) reduces to a temperature-independent quantity $N\alpha$; the first sum may be taken over components $M_J = -J, -J + 1, \ldots, J$ of the $(2J + 1)$-fold degenerate state corresponding to the lowest-energy J, since states of other J have negligible Boltzmann factors. The result is

$$\chi_M = \frac{N}{3kT}g^2\beta^2 J(J + 1) + N\alpha \qquad \Delta E_J \gg kT \tag{8.33}$$

where β is the Bohr magneton and g the Landé factor. The diamagnetic term has been neglected.

2. If $\Delta E_J \ll kT$, n and n' are to be interpreted as J and J', and the Boltzmann factors may be considered all equal; the terms of the second sum, for $\Delta E_J \to 0$, can be combined with those of the first

and the result simplified, with the final result

$$\chi_M = \frac{N}{3kT}\beta^2[4S(S+1) + L(L+1)] \qquad \Delta E_J \ll kT \tag{8.34}$$

3. If $\Delta E_J \cong kT$, Eq. (8.33) must be averaged over the various J's with appropriate Boltzmann weights.

Equations (8.33) and (8.34) are of the classical form (8.29), but the result in case 3 cannot be reduced to this form except by the formal artifice of defining μ^2 and α as functions of temperature.

The theory just outlined has been most successful for rare-earth salts. The moment is due to the electrons of the incomplete $4f$ group in the trivalent rare-earth ion. This group, well inside the atom, interacts only weakly with other atoms; thus the simple theory for isolated atoms is a fair approximation. At room temperature, formula (8.33), with α neglected, holds to within a few per cent except for samarium and europium, for which the energy levels are closer together and the more complicated formula of case 3 must be used. Table 8.2, upper part, compares some of the values calculated by the two methods with experimental values. The quantity tabulated is $(3kT\chi_M/N\beta^2)^{1/2}$ $(\beta = e\hbar/2mc)$; if (8.29) held with $\alpha = 0$, this quantity would be the permanent moment measured in Bohr magnetons. A striking success of the theory is its calculation of the details of the variation of susceptibility with temperature in Sm and Eu.

Table 8.2. Paramagnetic Susceptibilities

Substance	$(3kT\chi_M/N\beta^2)^{1/2}$ at room temperature			
	Calculated*			Observed*
	(1)	(2)	(3)	
Pr	3.58	3.62	....	3.4– 3.6
Sm	0.84	1.55–1.65	....	1.3– 1.6
Eu	0.00	3.40–3.51	....	3.1– 3.6
Ho	10.6	10.6	....	10.3–10.5
Cr^{3+}	0.77	2.97	3.87	3.7– 4.1
Fe^{++}	6.70	6.54	4.90	5.0– 5.6

* Basis of calculation: (1) formula (8.33), $\alpha = 0$; (2) formula for $\Delta E_J \cong kT$; (3) formula for $L = 0$. Compiled from J. H. Van Vleck, "The Theory of Electric and Magnetic Susceptibilities," Oxford University Press, 1932, and Edmund C. Stoner, "Magnetism and Matter," Methuen, London, 1934.

In salts of elements of the first transition series, the moment is due to the electrons of the incomplete $3d$ group, which is not well shielded from external interactions. The observed values of $(3kT\chi_M/N\beta^2)^{1/2}$ are in poor agreement with the theory for isolated atoms. The agreement is improved by supposing that only the spin contributes to the moment, i.e., by setting $L = 0$ in (8.34) despite spectroscopic indications that $L \neq 0$ (see Table 8.2). The accepted explanation is that in the salt the interatomic forces prevent orientation of the orbital moment by the field; theoretical calculations of this "quenching" have been made by replacing the interatomic forces by an asymmetric electrostatic field.

For diatomic molecules, different quantum numbers enter, and the formulas are slightly different in form. For polyatomic molecules, only the spin contributes to the first term of (8.29), and if there is no spin there remains only the term $N\alpha$; this consists of a small temperature-independent paramagnetic term and a diamagnetic term, either of which may be larger.

A small temperature-independent paramagnetism occurs in metals for a different reason. The conduction electrons obey Fermi-Dirac statistics. In zero magnetic field, the electrons occupy the cells of phase space in pairs, with opposing spins. In a small field, an electron with spin along the field has a slightly lower energy than one with spin opposite to the field, and therefore there will be a transfer of electrons from negative orientations in previously occupied cells to positive orientations in previously unoccupied ones. The theory shows that, in accordance with experiment, the variation of susceptibility with temperature is negligible. The diamagnetic term mentioned in Sec. 5 is one-third the paramagnetic [9, 10, 11].

7. Saturation in Paramagnetics and Spontaneous Magnetization in Ferromagnetics [4*a*, 5, 6, 7*a*, 7*b*, 11, 25*a*]

At large fields and low temperatures, the first-order approximation of Sec. 6 fails; but if α_n can be neglected, formula (8.30) holds for all values of B_0/kT. Under the conditions that led to (8.33), rigorous summation gives

$$M = M_\infty\left[\frac{2J+1}{2J}\coth\frac{(2J+1)x}{2J} - \frac{1}{2J}\coth\frac{x}{2J}\right] \equiv M_\infty B_J(x) \tag{8.35}$$

with

$$M_\infty = N_1 Jg\beta \qquad x = Jg\beta B_0/kT = M_\infty B_0/N_1kT \tag{8.36}$$

N_1 is the number of atoms in unit volume. For small x, $B_J(x)$ reduces to $(J+1)x/3J$, and the gram-atomic susceptibility reduces to the value (8.33) with $\alpha = 0$. For $x \to \infty$, that is for $B_0 \to \infty$ or $T \to 0$, M approaches the saturation value M_∞; this corresponds to alignment of all the moments as nearly as possible along the field. The *Brillouin function* $B_J(x)$ reduces for $J = \infty$ to the *Langevin function* $\coth x - 1/x$ obtained by using (8.28) instead of (8.30); it reduces for $J = 1/2$ (that is, $L = 0$, $S = 1/2$) to $\tanh x$.

Figure 8.3 shows $B_J(x)$ for several values of J. Measurements on gadolinium sulfate agree well with the theoretical curve corresponding to the known state of the gadolinium ion, viz., $^8S_{7/2}$ $(L = 0, S = J = 7/2, g = 2)$ [10, 11].

These paramagnetic saturation effects can be observed only at very high fields and low temperatures. In ferromagnetic materials, similar effects are observed at ordinary fields and temperatures.

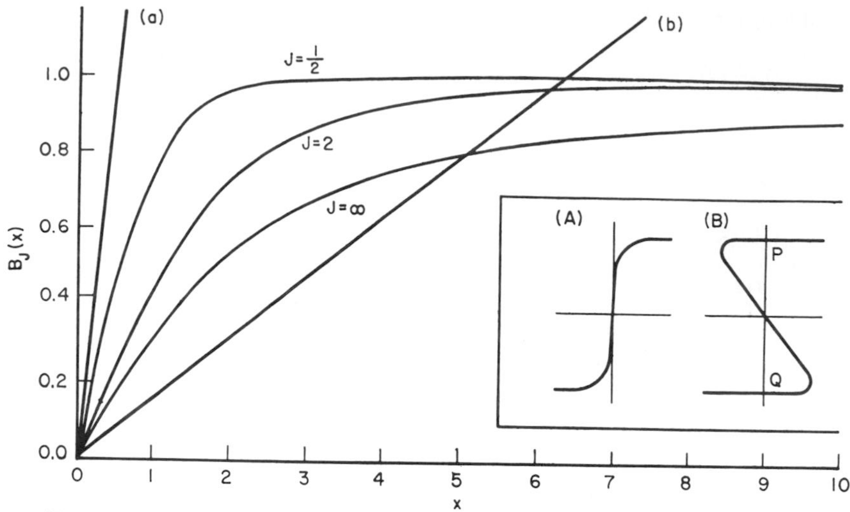

FIG. 8.3. The Brillouin function and the theory of spontaneous magnetization.

To explain this, Weiss postulated that the atomic moments of a ferromagnetic are subject to an intense *molecular field* $q\mathbf{M}$, where q is a positive constant. Then one of the curves in Fig. 8.3 will apply, provided that x is now interpreted as $M_\infty(B_0 + qM)/N_1kT$. To obtain M/M_∞ vs. $M_\infty B_0/N_1kT$, the abscissas must be measured from a slanting line of slope $N_1kT/qM_\infty{}^2$. According as this line is of type (a) or (b) in the main diagram, the theoretical magnetization curve will be of form (A) or (B) in the insert. For sufficiently low temperatures it will be of form (B); then for $B_0 = 0$ the specimen retains a magnetization corresponding to point P or Q (the intermediate points on the curve are unstable). The material is then ferromagnetic. As T increases, the line (b) increases in slope; its point of intersection with the $B_J(x)$ curve moves down the curve, so that the *spontaneous magnetization* decreases; and ultimately, at a critical temperature T_c, the situation changes from (b) to (a), and ferromagnetism disappears. This occurs when the slope $N_1kT_c/qM_\infty{}^2$ of the straight line becomes equal to the initial slope $(J + 1)/3J$ of the $B_J(x)$ curve. Therefore

$$T_c = \frac{J+1}{3J}\frac{qM_\infty{}^2}{N_1k} \tag{8.37}$$

Above the *Curie temperature* T_c, the material is paramagnetic but obeys the *Curie-Weiss* law,

$$\chi_M = C/(T - T_c)$$

rather than the Curie law $\chi_M = C/T$.

This theory predicts a definite spontaneous magnetization $M_s(<M_\infty)$ in zero external field. Ordinary ferromagnetic specimens show, instead, a remanent magnetization that depends on magnetic history and that can be removed by demagnetization. To circumvent this difficulty, Weiss further postulated that ordinary specimens are indeed spontaneously magnetized throughout any physically small volume but that over larger distances the spontaneous magnetization, while constant in magnitude, can vary in direction. This postulate has since been fully verified by experiment. It implies that the experimental quantity to be compared with the M_s of the Weiss theory is not the observed magnetization in zero field but an internal magnetization that must be evaluated indirectly. This can be done from measurements at high fields, which force the spontaneous magnetization to assume the same direction in all parts of the specimen. With this interpretation, the Weiss theory explains approximately the observed temperature variation of the *saturation* magnetization below the Curie point and of the susceptibility above it.

Weiss's original theory was based on the classical Langevin curve, $J = \infty$ in Fig. 8.3. According to gyromagnetic experiments (see Sec. 11), ferromagnetism is due mainly to electron spins and should therefore correspond more closely to the quantum curve $J = \frac{1}{2}$, $g = 2$. Use of this curve improves the agreement.

A more basic contribution of quantum mechanics has been the interpretation of Weiss's molecular field. Classical magnetic interactions give a q, in Gaussian units, of 4π or less, whereas the values required in Weiss's formula are of order 10^4. It was shown by Heisenberg that coupling energies of the required order of magnitude could result from exchange interaction, illustrated in the hydrogen molecule (Part 7, Chap. 1). This is formally equivalent to an interaction energy $A\mathbf{S}_i \cdot \mathbf{S}_j$ between the spins $\mathbf{S}_i$, $\mathbf{S}_j$ of neighboring atoms. For ferromagnetism, A must be negative, and this requires that an exchange integral be positive which in most cases is negative. This situation is consistent with the rarity of ferromagnetism among the elements.

From the Heisenberg coupling energy, the Weiss molecular-field formula follows as an approximation if the interaction of a typical atom i with its z nearest neighbors j, $-|A|\mathbf{S}_i \cdot \Sigma\mathbf{S}_j$, is replaced by $-z|A|\mathbf{S}_i \cdot \langle\mathbf{S}\rangle_{\text{av}}$, where $\langle\mathbf{S}\rangle_{\text{av}}$ is the mean spin for the whole lattice. For some purposes this modified molecular-field theory suffices, but it is neither theoretically rigorous nor experimentally precise. Attempts to improve the theory are concerned with two quite separate problems.

First, if the spins are assumed to be localized at lattice sites and to be coupled according to the Heisenberg formula, there is the problem of carrying out a statistical-mechanical calculation of the partition function, and hence of the magnetic moment and other properties, of the crystal. Because of the strong coupling, the system to be treated is the crystal as a whole and not, as in paramagnetism, a single atom. The molecular-field approximation is a device for reducing this calculation for a whole crystal to a calculation for a single atom, thus making the mathematics tractable. Improvement of the theory requires, if not a rigorous calculation, at least a less drastic approximation.

Second, although the localized-spin concept is legitimate for ferromagnetic materials that are electrical insulators, it is unrealistic for metals and alloys. In the ferromagnetic elements Fe, Ni, Co, the spins responsible for ferromagnetism are those of the $3d$ electrons, whose status is intermediate between that of the conduction electrons and that of the tightly bound inner shells. A satisfactory theory of ferromagnetism in metals must take account of this intermediate character of the $3d$ electrons—neither completely bound nor completely free.

The first of these problems, the statistical-mechanical one, has been attacked by several methods and with considerable success. Here the theoretical model is quite definite: a lattice of spins with nearest (and perhaps more distant) neighbors coupled according to a known formula. The problem is therefore purely mathematical. Besides the Heisenberg coupling, the simpler Ising coupling has often been used to simplify the mathematics; in it, $\mathbf{S}_i \cdot \mathbf{S}_j$ is replaced by S_iS_j, with $S_i = \pm 1$ [22, 24*a*].

An early improvement over the molecular-field approximation was the Bethe-Peierls-Weiss method, in which a group of spins, rather than a single spin, was treated as being in a molecular field [30]. More recent improvements begin with rigorous formulations of the problem and use a variety of techniques for obtaining approximate expressions for the quantities of interest. These include cluster and diagram methods, series expansions of various kinds, and other techniques [21, 24*a*, 25*b*]. Particular attention has been given to the behavior of the magnetization, susceptibility, and specific heat on approach to the Curie point from above and from below. A noteworthy achievement was Onsager's rigorous solution for the two-dimensional Ising lattice.

A somewhat different approach has been based on the "spin wave" concept [12, 21]. This applies to a lattice of almost perfectly aligned spins and was introduced by Bloch for the case of low temperatures; with it he obtained the rigorous result that, for $T \to 0$, $M_\infty - M_s \propto T^{3/2}$. The spin wave was originally presented as a quantum-mechanical concept; in a classical picture, the two small transverse components of the magnetization are propagated in a spin wave somewhat as the three components of displacement are propagated in an acoustic wave. An important difference is that the two transverse components of magnetization do not behave independently; the general behavior is a precession of the magnetization vector about its mean direction, with the phase of the precession different at different points. Bloch's original quantum-mechanical calculation has been extended to evaluate higher-order terms in $1/T$, but the difficulties are great. A number of "semiclassical" calculations have been made; these begin with a classical continuum treatment (analogous to a treatment of Debye thermal waves by continuous elasticity theory) and later introduce quantization or reinterpret classical quantities as quantum-mechanical operators. Spin-wave theory has thus been extended to take account of effects, such as magnetic anisotropy, that have not yet received a completely rigorous quantum-mechanical treatment. It has also been extended to higher temperatures by "renormalization" of the energies. A concept often used in spin-wave theory is the "magnon"; it has the same relation to a spin wave that a phonon has to a Debye wave.

More recently, much work has been based on the "Green function" method [21, 32]. This leads to a hierarchy of rigorous equations that involve successively larger numbers of spins; the equation that determines the one-spin function contains the two-spin function, and so on. The hierarchy can be reduced to a small finite number of equations, and thus made solvable, by introducing, in the n-spin equation, an approximate expression for the $(n + 1)$-spin function in terms of lower-order functions. Instead of introducing such a "decoupling" approximation, it is also possible to expand in powers of some parameter and truncate after a certain power.

These several methods have led to a number of illuminating results but not to a complete quantitative theory of the variation of spontaneous magnetization, susceptibility, and specific heat with temperature. The problem has much in common with the theories of the condensation of vapors and of order-disorder transitions; in fact, some of the models in these different fields have identical mathematical properties.

The second problem, realistic treatment of the $3d$ electrons in metals, has proved less tractable. The Heisenberg model, which localizes these electrons on particular atoms, is at one extreme. At the other is the *collective-electron* model of Slater, Stoner, and Wohlfarth [18, 25*a*, 29]. In it, the $3d$ electrons are supposed to travel freely from atom to atom and are therefore assigned not to atoms but to energy bands characteristic of the whole lattice. This method has led to useful calculations of the properties of alloys and of the interrelations of various properties, such as magnetization and specific heat. But an accurate theory of the coupling between $3d$ spins must take account of their intermediate character—neither completely bound nor completely free. This is a difficult problem, and no satisfactory method of attacking it has been found, although it has been discussed extensively and from a number of different points of view. Quite divergent opinions exist on the precise mechanism of the coupling and on the role played in it by the conduction electrons [18].

A number of alloys of manganese with other nonferromagnetic elements are ferromagnetic. The alignment of spins is attributed to exchange interactions, direct or indirect, which depend rather critically on the arrangement of the various types of atom in the lattice [4].

8. More Complicated Ordered Magnetic Structures [4*a*; 5; 6; 7*a*; 25, chaps. 1–5; 31]

In the ferromagnetic crystals so far discussed, the spins form a single lattice in which nearest neighbors (and to some extent next-nearest neighbors) are so coupled that they tend to be oriented parallel; this may be called *positive* or *ferromagnetic* coupling. In certain crystalline compounds, the spins form two sublattices, in which a spin on one sublattice and a neighboring spin on another sublattice are so coupled that they tend to be oriented antiparallel; this may be called *negative* or *antiferromagnetic* coupling. With the latter type of coupling, at sufficiently low temperatures each sublattice has its spins approximately aligned and therefore possesses spontaneous magnetization, but the spontaneous magnetizations of the two sublattices are oppositely oriented. When the two sublattices are identical, as in MnO, their spontaneous magnetizations cancel, and the material as a whole has zero spontaneous magnetization. The individual sublattice magnetizations, however, behave like the spontaneous magnetization of a ferromagnet: They decrease with increasing temperature and vanish at a critical temperature T_c, the *Néel temperature.* Their presence is revealed by specific-heat anomalies, by the behavior of the susceptibility as a function of direction and of temperature, and—most directly—by neutron-diffraction measurements. Such materials are known as *antiferromagnetic.*

In some materials with antiferromagnetic coupling of two sublattices, the sublattice magnetizations, though opposite in direction, are not equal in magnitude. There is then a spontaneous magnetization; the behavior is qualitatively like that of the ferromagnetic materials already discussed but is different in detail (for example, the two sublattice magnetizations may vary differently with temperature in such a way that one is larger below, and the other above, a certain "compensation temperature" at which the net spontaneous magnetization is zero). Typical of these materials are the ferrites Fe_3O_4, NiF_2O_4, and $Li_{0.5}Fe_{2.5}O_4$. Although one of these, magnetite (Fe_3O_4), is the oldest known ferromagnetic material, our understanding of their behavior, as well as of that of antiferromagnetic materials, is very recent and is largely due to Néel. This type of ferromagnetism might appropriately be called "Néel ferromagnetism" to distinguish it from the single-lattice "Weiss ferromagnetism" of iron, cobalt, and nickel. A common practice, however, is to call the Weiss type simply "ferromagnetism" and the Néel type "ferr*i*magnetism." (This terminology, like the use of *macroscopic* as an antonym to *m*i*croscopic,* is open to the objection that it entrusts an important distinction to a single not very distinguishable vowel.)

Ferrites, because of their high electrical resistivity, are free from the eddy-current effects that prevent the use of ferromagnetic metals at high frequencies. They are therefore useful in microwave devices and in devices that require microsecond response times. A very important device of the latter type is the magnetic digital storage element for electronic computers; ferrite toroids for this purpose have been produced in great quantities [28].

Like the ferrites but more complicated—with more than two sublattices—are the ferromagnetic garnets (e.g., yttrium iron garnet, $3Y_2O_3{\cdot}5Fe_2O_3$); their optical transparency has made possible the observation of internal domain structure (see Sec. 9) by magneto-optic methods, and their narrow line width has made them useful in ferromagnetic resonance experiments (see Sec. 11).

Certain crystals, such as α-Fe_2O_3 (hematite), exhibit in a certain temperature range a "weak ferromagnetism," with a spontaneous magnetization much smaller than that to be expected with perfect spin alignment. This phenomenon has now been explained by Dzyaloshinskiĭ as due to sublattice magnetizations that are equal in magnitude and almost, but not quite, opposite in direction [24].

The antiferromagnetic type of exchange interaction is generally interpreted as "superexchange": an indirect coupling, for example, of two Mn^{++} spins via the O^- ion that is their common neighbor. The quantitative theory, however, is incomplete [24]. The temperature dependence of the properties of antiferromagnets and ferrites can be treated approximately by a molecular-field method [26]; spin-wave, Green-function, and other methods discussed in Sec. 7 have also been applied to this problem [25, 25*b*].

More complicated spin arrays, such as spirals, occur in some materials (for example, $MnCr_2O_4$). Transitions between different structures can occur as the temperature is changed. Thus hematite changes from "weakly ferromagnetic" to antiferromagnetic on cooling below a certain temperature (in pure synthetic crystals, about −9°C).

In Dzyaloshinskiĭ's theory of weak ferromagnetism and in the theory of spin structures in general, symmetry considerations play a major role [14]. These, together with thermodynamics, provide a complete phenomenological theory that has been extremely

Table 8.3. Curie Points and Spontaneous Magnetization*

Material	Curie point, °C T_c	Magnetic moments, emu: At room temperature, per cm³ M_s	At 0°K, per cm³ M_∞	At 0°K, per g	Number of Bohr magnetons per atom or molecule
Fe	770	1,714 (20°C)	1,735	221.9	2.2
Co	1130	1,422 (20°C)	1,445	162.5	1.7
Ni	358	484 (15°C)	509	57.5	0.6
Gd	16	....	2,021	253.5	7.1
$NiFe_2O_4$	590	239	.....	.....	2.3
MnBi	360	621	.....	74.8	3.5
MnO	116	†			

* Compiled from Richard M. Bozorth, "Ferromagnetism," Van Nostrand, Princeton, N.J., 1951.

† Antiferromagnetic.

successful. The phenomenological theory, however, cannot predict absolute numerical values for any of the parameters involved, though it can relate the values of parameters that have not yet been measured to the values of ones that have. To predict absolute values is the problem of microscopic theory; that theory is very far from being complete.

Table 8.3 gives the Curie points, the spontaneous magnetization at $T = 0$, and the spontaneous magnetization at room temperature for a few ferromagnetic materials.

9. Ferromagnetic Domains and the Magnetization Curve; Ferromagnetic Films and Fine Particles [4, 7, 7b, 16, 17, 29a]

It remains to account for the spatial variation of magnetization in a ferromagnet. In this section the lattice will be supposed rigid. Then three types of interaction contribute to the energy W. (At temperatures far below the Curie point, the free energy $W - TS$ need not be distinguished from W.)

The *exchange forces* contribute $-\frac{1}{2}C'\Sigma_i\Sigma''_j\boldsymbol{\mu}_i \cdot \boldsymbol{\mu}_j$, where $\boldsymbol{\mu}_i$ is the mean moment of atom i at temperature T and Σ'' means a sum over the nearest neighbors. These forces dominate at short distances and prevent $\boldsymbol{\mu}_i$ from varying abruptly from atom to atom. Therefore $\boldsymbol{\mu}_i$ may be replaced by a continuous variable $\boldsymbol{\mu}$, and $\boldsymbol{\mu}_j$ by a Taylor's expansion truncated after the second-degree term; after summation over j, the sum over i may be replaced by an integral. For a cubic crystal, the result can be transformed to

$$W_e = \text{const} + \tfrac{1}{2}C\int[(\nabla\alpha_1)^2 + (\nabla\alpha_2)^2 + (\nabla\alpha_3)^2]\,d\tau \tag{8.38}$$

where $(\alpha_1, \alpha_2, \alpha_3)$ are the direction cosines of $\boldsymbol{\mu}$.

The *magnetic interactions*, neglected since Sec. 5, are not negligible in ferromagnetics. They fail by a factor 10^{-3} to account for the Weiss field, and therefore they perturb only slightly the parallelism of nearest-neighbor spins; but they remain effective at large distances, and therefore they can cause a variation of the direction of spontaneous magnetization over a distance of many lattice spacings. The mutual magnetic energy of all the atoms in the specimen is $-\frac{1}{2}\Sigma_i\boldsymbol{\mu}_i \cdot \Sigma'_j\mathbf{b}_{ij}$, where $\mathbf{b}_{ij}$ is the microscopic field intensity of atom j at atom i; it is given by formula (8.3) with obvious changes of notation. The sum Σ' is over $j \neq i$. It can be separated into a sum over atoms outside a physically small sphere of radius R about atom i, and a sum over atoms inside this sphere. The first sum can be replaced by $\mathbf{H}'_i - \mathbf{H}_{Ri}$, where $\mathbf{H}'_i$ is the megascopic quantity $\mathbf{H} - \mathbf{B}_0$ of (8.8) evaluated at atom i and where $\mathbf{H}_{Ri}$ is the contribution to $\mathbf{H}'_i$ from elements dp_m [see Eqs. (8.7)] inside sphere R. The magnetic energy is thus $-\frac{1}{2}\Sigma_i\boldsymbol{\mu}_i \cdot \mathbf{H}'_i - \frac{1}{2}\Sigma_i\boldsymbol{\mu}_i \cdot (\Sigma'_{r<R}\mathbf{b}_{ij} - \mathbf{H}_{Ri})$. The sum $\Sigma_i\boldsymbol{\mu}_i \cdot \mathbf{H}'_i$ can be replaced by an integral; this gives a quasi-magnetic energy

$$W = -\tfrac{1}{2}\int \mathbf{M} \cdot \mathbf{H}'\,d\tau = +\frac{1}{2\gamma}\int_{\text{space}} \mathbf{H}'^2\,d\tau \tag{8.39}$$

plus a "local" term $-\frac{1}{2}\Sigma_i\boldsymbol{\mu}_i \cdot (\Sigma'_{r<R}\mathbf{b}_{ij} - \mathbf{H}_{Ri})$.

The *short-range* interactions (other than exchange) result from spin-orbit coupling, quadrupole moments, etc. They are of the form $\Sigma_i f_i$, where f_i depends only on the orientation of the moments of atom i and a few of its neighbors. The "local" term in the magnetic energy also has this property and may therefore be combined with f_i to form a new function g_i; then the sum may be replaced by an integral

$$W_s = \int g(\alpha_1, \alpha_2, \alpha_3, \partial\alpha_1/\partial x, \ldots)\,d\tau \tag{8.40}$$

(If $\mathbf{B}'$ is used rather than $\mathbf{H}'$, a constant term

$$\tfrac{1}{2}\gamma\int M^2\,d\tau$$

will be transferred from W_m to W_s.)

In an external field $\mathbf{B}_0$, there is an additional energy $-\Sigma_i\boldsymbol{\mu}_i \cdot \mathbf{B}_0$, or

$$W_0 = -\int \mathbf{M} \cdot \mathbf{B}_0\,d\tau \tag{8.41}$$

The energy W_e attains its minimum when

$$\nabla\alpha_1 = \cdots = 0$$

that is, when the spontaneous magnetization is uniform in direction. The energy W_m attains its minimum when $\mathbf{H}' = 0$; that is, when the spontaneous magnetization is everywhere so oriented that there are no poles dp_m. [The equivalent formula $W_m = -(\frac{1}{2}\gamma)\int\mathbf{B}'^2\,d\tau$ + const leads to no simple conclusion about the distribution of Amperian currents $d\mathbf{I}_m$.] In a finite specimen these two requirements are incompatible. The expected compromise is a slow spatial variation of the direction of $\mathbf{M}$, such as to avoid both rapid changes of direction and large pole strengths. This distribution is further conditioned by the short-range term (8.40). For cubic crystals, the leading term in g is of the form $K(\alpha_1^2\alpha_2^2 + \alpha_2^2\alpha_3^2 + \alpha_3^2\alpha_1^2)$. The directions that minimize this are the six [100] directions in iron ($K > 0$) and the eight [111] directions in nickel at room temperature ($K < 0$). If $|K|$ is large, the specimen will consist mostly of regions magnetized nearly along one or another of these *directions of easy magnetization;* the transition layers between these *domains* will be thinner, the larger the *anisotropy constant* K. (With small anisotropy, the concept of distinct domains ceases to be useful.)

The problem of minimizing the total energy and thus determining the magnetization distribution is a very difficult one. Calculations of domain structure have usually been based on an approximate procedure devised by Landau and Lifshitz and elaborated by Néel. First a solution is found for a magnetization that is independent of x and y but changes from one direction of easy magnetization at $z = -\infty$ to another at $z = +\infty$. Then this one-dimensional solution is assumed to hold for each interdomain transition layer (*Bloch wall*) in a three-dimensional crystal; domain patterns are guessed at, their energies minimized by adjustment of parameters, and the patterns of lowest energy selected. The theory accounts for the complex patterns observed on the surface of single-crystal specimens by the magnetic powder or colloid technique and by magneto-optic and electron-microscopic methods [16; 17; 25, chap. 9].

When an external field $\mathbf{B}_0$ is applied, the magnetization can increase, according to domain theory, by three distinct processes.

1. At small B_0, domain walls are displaced in such directions that domains magnetized toward the field direction expand, while ones magnetized away from the field direction contract. This is a reversible process, conditioned by local variations of the short-range energy W_s (see Sec. 10).

2. At higher B_0, domain walls reach, locally, positions of unstable equilibrium, from which sections of them advance by discontinuous irreversible jumps. These are the Barkhausen jumps, originally detected as test-coil noise but now directly observable by the powder technique. They are the mechanism of hysteresis.

3. At very high B_0, only domains oriented close to the field direction remain; these now undergo a reversible rotation of the magnetization into the field direction.

The details of process 3 can be calculated accurately for single crystals; in polycrystalline materials, calculation is difficult because of magnetic and other interactions between the crystal grains. Process 1 permits only approximate calculation and process 2 only order-of-magnitude estimates. These processes are dependent on internal stresses, lattice imperfections, inclusions, etc.; the magnetic properties determined by them are therefore structure-sensitive. Such properties are initial susceptibility, coercive force, and, to a lesser degree, remanence (Fig. 8.2) [7, 7*b*].

The domain theory just outlined is based on a number of approximations of unknown validity. A rigorous theory would begin with free-energy expressions of the forms (8.38) to (8.41) and would find the functions $\alpha_1(x,y,z)$, $\alpha_2(x,y,z)$, and $\alpha_3(x,y,z)$ $\{= \pm[1 - (\alpha_1^2 + \alpha_2^2)]^{1/2}\}$ that minimize the total free energy. Such a theory, *micromagnetics*, leads in general to nonlinear partial differential equations that have been solved only in a few special cases, and only by laborious numerical methods, by use of electronic computers. However, certain exactly or approximately linear cases have been solved and have given illuminating results [15; 25, chap. 8; 30*a*]. A problem of particular interest is the following: Suppose that we start with an ideal crystalline ellipsoid, whose long axis coincides with an axis of easy magnetization, and apply along this axis a field large enough to insure uniform magnetization along the field. We now gradually decrease the applied field through zero to negative values. At what value of the reversed applied field will the original state of uniform magnetization become unstable, and what will be the form of the first deviation from this state? The calculated values of these *nucleation fields* are numerically much larger than those observed in bulk specimens; the discrepancy is attributed to imperfections not taken into account in the theory. This interpretation is supported by measurements on carefully prepared and selected iron whiskers, of transverse dimensions of order 10 μ; here the results agree with the theory. The probability of a flawless specimen decreases with increasing specimen size and is negligible in bulk specimens. The nucleation-field calculations have led to useful models of magnetization reversal; as particle size increases, the *rotation in unison* of a very small particle is replaced by a divergenceless reversal mode, *magnetization curling*. One can get an idea of the directions of magnetization, in a cylinder undergoing magnetization curling, by twisting a whiskbroom and observing the direction of the bristles at each point.

Thin ferromagnetic films have been the subject of extensive experimental study and of considerable theoretical work [28*a*]. The practical interest in films arises from the possibility of using them, instead of ferrite cores, as computer storage elements; the switching time can thereby be decreased from microseconds to nanoseconds. The magnetic properties of films are extremely sensitive to stresses set up in their manufacture or produced by their mechanical interaction with the substrate on which they are deposited. For this reason, most of the experimental research has been done on permalloy films of a composition (about 81 per cent Ni and 19 per cent Fe) at which the magnetostriction is nearly zero. Many properties remain very structure-sensitive, but certain properties are generally characteristic of ferromagnetic films as such.

First, theory indicates that a film only a few atoms thick should have a smaller spontaneous magnetization than the same material in bulk. The precise magnitude of this effect, however, has been somewhat controversial among theorists, and the experimental facts have also been controversial [25, chap. 6].

Second, the structure of the interdomain wall must change as the film thickness decreases. Let us take the x axis along the normal to the wall, and suppose that the direction of the magnetization changes from $-Oz$ at large negative x to $+Oz$ at large positive x. Then the change may occur by rotation about the x axis or about the y axis; the magnetization at the middle of the wall will point along Oy (or $-Oy$) in the former case and along Ox (or $-Ox$) in the latter. In bulk material, the former structure is energetically favored, because it makes $-\partial M_x/\partial x$ zero and thereby avoids the magnetostatic energy associated with a volume pole distribution. This is the structure of the Bloch wall. In a thin film with faces normal to Oy, the rotation about the y axis is favored because it makes the surface values of M_y zero and thereby avoids the magnetostatic energy associated with surface pole distributions. At small thickness this is the larger of the two energies, as was made plausible by Néel and has been confirmed through more accurate calculations. The existence of these *Néel walls*, in which the magnetization remains in the plane of the film, has been verified experimentally. However, the transition from Bloch walls at large thickness to Néel walls at small is more complicated than the simplified theory (which assumes variation only with x) predicts; in the transition range of thickness, more complicated two-dimensional structures ("cross-tie" walls) are observed [25, chap. 10].

The structure-sensitive properties of films include details of the domain structure and of the magnetization-reversal (switching) process; the theory of these is in a state comparable with that of the theory of domains in bulk material.

A sufficiently fine particle is incapable of a domain structure. This is obvious for a particle whose dimensions are comparable with the thickness of a domain wall (of order 10^2 A); there is then no room for the magnetization to vary its direction enough to avoid poles, without the creation of excessive exchange energy. The "critical size," at which a particle of given shape can no longer support a domain structure, has usually been estimated by comparing the energy of the particle in a state of uniform magnetization with its energy in some state of nonuniform magnetization; the former energy is minimized with respect to direction, the latter with respect to parameters included in the specification of the state. It is found that the former energy is smaller for particles of less than a certain size, the latter for particles of more than this size; the critical size is of order 10^2 A. Particles of less than this critical size are assumed to be uniformly magnetized [17]. By the methods of micromagnetics, it is possible to calculate rigorously a critical size of a somewhat different kind, that at which the mechanism of magnetization reversal changes from rotation in unison (in smaller particles) to curling (in larger). For spheres and cylinders of iron and nickel, these micromagnetic critical sizes range from about 200 to about 400 A [15; 25, chap. 8].

Given that a particle is uniformly magnetized (a "single-domain" particle), the calculation of its behavior in a magnetic field requires only minimization of its total energy with respect to the orientation angles θ and ϕ of its magnetization. For mathematical simplicity, the particle is often assumed to have uniaxial symmetry of material properties and of shape (the Stoner-Wohlfarth model). The particle then traverses a hysteresis loop of finite area, if the applied field is in any direction except perpendicular to the particle axis; the loop consists of two reversible curves connected by irreversible jumps at applied fields of the order of K/M_s or M_s, according as anisotropy or magnetostatic energy dominates. The jump field may be taken as a rough estimate of the coercive force of a powder composed of such fine particles; this calculation, however, neglects the magnetic interactions between the particles and, for this or other reasons, usually overestimates the coercive force. It nevertheless gives the right order of magnitude and predicts, correctly, that such fine-particle aggregates should have a much higher coercive force than the same material in bulk form [25, chap. 7; 30*a*].

Permanent magnets manufactured in this way are successful as magnets but, economically, have not displaced such materials as Alnico. The high coercive force of the latter materials is itself attributed to the fact that the microscopic structure is quite similar to that of a fine-particle magnet: it is very inhomogeneous, with particles of ferromagnetic material surrounded by a less magnetic matrix.

The special properties of fine particles are also exploited in the γ-Fe_2O_3 powders used in magnetic sound- and video-recording tapes [20].

If a single-domain particle is so small that the variations of its energy with magnetization direction are comparable with kT, it behaves like a paramagnetic atom, changing its orientation continuously in response to thermal agitation; and a powder of such particles has a reversible magnetization curve described by a Langevin function (somewhat modified, in general, by crystalline anisotropy or anisotropy of shape). This phenomenon is known as *superparamagnetism* [25, chap. 6]. At any temperature, particles below a certain range of sizes behave superparamagnetically; particles above this size range behave as stable ferromagnetic single-domain or many-domain particles; and particles within this range (which covers less than one order of magnitude) reverse their magnetization after a field reversal with a time constant comparable with experimental observation times. These last contribute to magnetic aftereffect (Sec. 11).

A particle can be superparamagnetic at a high temperature and stably ferromagnetic at a low temperature. This fact has been used by Néel in attempts to deduce, from the present magnetic state of rocks, information about the direction of the earth's field at the time the rocks were cooling through a critical range of temperature.

Other phenomena that warrant mention here are *induced anisotropy* and *exchange anisotropy* [24, chap. 5; 25, chap. 6]. Induced anisotropy is a uniaxial anisotropy that results from annealing in the presence of a magnetic field. Various detailed theories have been evolved to explain it, but in any case the essential process is this: In the presence of the field and at a high temperature, changes can occur that decrease the energy of the actual state of magnetization, so that afterward this state and ones energetically equivalent to it will be favored. Exchange anisotropy is a *unidirectional* (rather than *uniaxial*) anisotropy; that is, the hysteresis loop is displaced to the right or to the left of the usual symmetrical position. This is observed, for example, in cobalt particles with a cobaltous oxide shell, in the temperature range where the oxide is antiferromagnetic, and after cooling to this temperature in a large magnetic field. It is attributed to exchange coupling across the ferromagnet-antiferromagnet interface.

Table 8.4 shows the wide variation of structure-sensitive properties in ferromagnetic materials.

10. Magnetomechanical Phenomena in Ferromagnetics; Other Interaction Phenomena [3, 7, 14, 15, 19]

When the lattice is strained, additional terms are present in the free energy. The sphere R of Sec. 9 becomes an ellipsoid, if the strains are small and are uniform over a physically small volume. The change in the short-range terms W_e and W_s can be expressed as the volume integral of a function w'_s, the additional free energy per unit of unstrained volume; w'_s depends on the strains and on the direction cosines (α_1, α_2, α_3). It is sufficient to express this function to the second order in the strains and to assume that only the coefficients of the linear terms depend on α_1, α_2, α_3. For cubic crystals the leading terms are

$$\begin{aligned} w'_s = k_1[e_{xx}(\alpha_1^2 - \tfrac{1}{3}) + \cdots] + 2k_2(e_{yz}\alpha_2\alpha_3 + \cdots) \\ + \tfrac{1}{2}c_{11}(e_{xx}^2 + \cdots) + c_{12}(e_{yy}e_{zz} + \cdots) \\ + \tfrac{1}{2}c_{44}(e_{yz}^2 + \cdots) \end{aligned} \quad (8.42)$$

where the dots imply cyclic permutation. The quasi-magnetic energy (8.39) can be left in the same form as before if it is understood that $d\tau$ is now an element of *distorted* volume and that $\mathbf{H}'$ is now calculated from the pole distribution in the strained state. For the uniformly magnetized ellipsoid of Eq. (8.11), uniformly strained, $W_m = \frac{1}{2}\gamma D\tau M^2$, where D varies

TABLE 8.4. MAGNETIC PROPERTIES OF TYPICAL MATERIALS*

Material	Treatment	Initial permeability, $\mu_0 = (B/H)_{H\to 0}$	Maximum permeability, $\mu_{max} = (B/H)_{max}$	Coercive force, H_c (for $M = 0$)	Remanent flux density, B_r
Iron, 99.8% pure.....	Annealed	150	5,000	1.0	13,000
Iron, 99.95% pure....	Annealed in hydrogen	10,000	200,000	0.05	13,000
78 Permalloy.........	Annealed, quenched	8,000	100,000	0.05	7,000
Supermalloy.........	Annealed in hydrogen, with controlled cooling	100,000	1,000,000	0.002	7,000
Co, 99% pure........	Annealed	70	250	10	5,000
Ni, 99% pure........	Annealed	110	600	0.7	4,000
Steel, 0.9% C........	Quenched	50	≅100	70	10,300
Steel, 36% Co........	Quenched	....	....	240	9,500
Alnico 5.............	Cooled in magnetic field	≅4	....	575	12,500
Silmanal.............	Baked	....	....	6,000	550
Iron, fine powder.....	Pressed	....	....	470	6,000

* Compiled from Richard M. Bozorth, "Ferromagnetism," Van Nostrand, Princeton, N.J., 1951.

with the strain. The magnetostriction (strain due to magnetization) is found by minimizing the strain-dependent free energy with respect to the strains. Because of W_m, the magnetostriction in general contains a shape-dependent term, which has been calculated and observed in ellipsoids but which is often incorrectly omitted. It may be ignored if the specimen is a long thin needle magnetized longitudinally. Under these conditions, the magnetostriction of a *saturated* specimen is found by minimizing w'_s with respect to the strains. The theory explains satisfactorily the variation of saturation magnetostriction of single crystals with the orientation of the magnetization in crystal axes.

If tension σ is applied to a specimen, an additional term $-\lambda\sigma$ must be added to the free-energy density; here λ is the magnetostrictive elongation. The tension therefore tends to orient the magnetization in such a direction as to elongate the specimen as much as possible. This magnetostrictive elongation is superposed on the ordinary elastic elongation and causes the measured Young's modulus E to be smaller than it would be otherwise. A large magnetic field prevents orientation of the magnetization by the stress and so raises Young's modulus to its purely elastic value. This is the *ΔE effect*. A similar effect is observed in torsion. In many respects, the magnetic effect of an externally applied stress is similar to that of a field. The difference is that *opposite* orientations of the spontaneous magnetization are equivalent in the case of stress, so that a stress without a field, for instance, produces no observable large-scale magnetization. With this difference, the concepts of reversible and irreversible wall displacements and of rotation apply to stresses as well as to fields.

Because of the magnetomechanical interactions described by (8.42), *internal* stresses due to cold-working, impurities, precipitates, etc., contribute a spatially varying component of the short-range energy W_s; they therefore have a profound effect on magnetic properties below the knee of the magnetization curve. High internal stresses usually decrease the initial susceptibility and increase the coercive force. In calculations of Becker and Kersten, the internal-stress concept proved very useful for correlating magnetic properties. However, the theory did not include a detailed description of the "internal stress" and was never related to theories of slip, work-hardening, etc. Later work has shown that the coercive force, at least, depends also on other factors, in particular on inhomogeneities such as cavities, inclusions, and precipitates; and that local magnetic fields, stress fields, lattice distortions, and single-domain particle behavior must all be taken into account.

When internal stresses are important, the rather vague Becker-Kersten description of them can, in principle, be replaced by a detailed calculation based on *dislocations*. Such a calculation is extremely complicated; it has, however, been carried out in certain cases and especially in the theory of the approach to saturation, where the equations of micromagnetics can be linearized [15, 18*a*].

A long specimen with positive longitudinal magnetostriction becomes a single domain if subjected to sufficient longitudinal tension; this was observed by Sixtus and Tonks. The principle involved is the same as in the "fine-particle" experiments.

Table 8.5 lists some representative magnetomechanical data.

The theory outlined above carries over to a magnetized body certain concepts and formulas of small-displacement elasticity theory, which was developed for an unmagnetized body. This procedure involves certain errors; they can be avoided by revising the elastic derivation to suit the peculiarities of a body subject to magnetic forces. It is then found that the separation of an observable force into a "magnetic force" and a part due to "stresses" is nonunique and somewhat arbitrary and that the stress tensor is not necessarily symmetric. The basic analysis is best done by finite-strain theory [14*a*]. The errors of the usual theory, though important in principle, seem unlikely to be significantly large in ferromagnetic materials.

In antiferromagnetic crystals of sufficiently low symmetry, such as CoF_2 and MnF_2, the magnetic

TABLE 8.5. MAGNETOSTRICTIVE PROPERTIES AT ROOM TEMPERATURE*

Material	Crystal axis	10^{-6} × Saturation magnetostriction †	$\frac{\Delta E}{E}$‡
Fe	100	+(11–20)	
	111	−(13–20)	
	Polycrystal	− 8	0.002–0.003
Ni	100	−(50–52)	
	111	− 27	
	Polycrystal	−(25–47)	0.07
Co	Polycrystal	−(50–60)	

* Compiled from various sources.

† Fractional elongation at saturation, measured from the demagnetized state.

‡ Fractional change of Young's modulus from demagnetization to saturation.

analogue of piezoelectricity, *piezomagnetism*, can occur. Here the interaction-energy density consists of product terms in magnetization and strain components, so that they are linearly related to the field and stress components. *Magnetoelectric* interactions have also been observed in Cr_2O_3. These instances of piezomagnetism and magnetoelectricity were observed experimentally after Dzyaloshinskiĭ had predicted them theoretically, by arguments similar to those used to explain weak ferromagnetism (Sec. 8).

11. Dynamic Phenomena [23, 24, 25, 28, 28b]

An atom with a magnetic moment has also electronic angular momentum. Therefore the magnetization of a specimen has angular momentum associated with it. If the magnetization is suddenly changed, the associated angular momentum, which is that of the carriers of the moment (electron spins or orbits), is thereby changed; under conditions of constant total angular momentum, a compensating change must occur in the angular momentum of the massive part of the body. Observation of the change provides a method of determining the ratio of moment to angular momentum of the carriers. Conversely, rotation of the specimen produces magnetization. Either of these *gyromagnetic* experiments requires extremely precise measurements, even on ferromagnetics. The results show that ferromagnetism, except in pyrrhotite, is due mainly to electron spins; and they confirm in paramagnetics the predictions based on known spectroscopic states [4, 10, 11, 13].

The development of microwave techniques provided a new method of observing the gyromagnetic properties of atoms. Typically, the method consists in measuring the response of the specimen to a radio-frequency magnetic field applied at right angles to a static magnetic field $\mathbf{B}_0$. From a classical point of view, resonance is to be expected in isolated atoms when the radio frequency is equal to the Larmor precession frequency. In quantum theory, resonance occurs at a frequency $\nu = \Delta E/2\pi\hbar$, where ΔE is the difference in energy between two states for which the moments μ_z differ by $g\beta$, that is $\Delta E = g\beta B_0$; therefore $\nu = g\beta B_0/2\pi\hbar = g|e|B_0/4\pi mc$. In paramagnetic solids, the quantum states are more complicated, and the resonance experiments have yielded new information about the energy levels. The details of the resonance curve can be related to the relaxation times associated with spin-spin, spin-lattice, and other interactions. Information about relaxation times has also been obtained by absorption measurements at lower radio frequencies [5, 6, 23].

In ferromagnetics the resonance condition is more complicated because the torque on an atomic moment is not simply $\boldsymbol{\mu} \times \mathbf{B}_0$ but depends also on interatomic forces. The various forces considered in Secs. 9 and 10 can be combined into an equivalent field that must replace $\mathbf{B}_0$ in the Larmor formula; because of the term in $\mathbf{H}'$, the resonance frequency depends on the shape of the specimen [4, 5, 6].

In extended specimens, the precessions of the magnetization vectors at different points need not be in phase, and spatially nonuniform modes of oscillation are possible. The situation is simplest when a small exciting microwave field is superposed on a large constant field, so that the magnetization consists of small transverse alternating components superposed on a constant longitudinal component. The classical equations can then be linearized and are in fact the linearized equations of micromagnetics plus dynamic (gyromagnetic) terms. At sufficiently long wavelengths, terms from Eq. (8.39) are more important than terms from Eq. (8.38); then we have "magnetostatic modes" of natural oscillation and propagation of plane electromagnetic waves with rotation of the plane of polarization; the latter phenomenon has been applied in microwave ferrite devices such as *gyrators*. At shorter wavelengths, the terms from Eq. (8.38) are the more important; then we have "spin waves" that differ from those of Sec. 7 by being excited by an external field rather than thermally [24, 28].

Resonance of spin waves has been observed in thin films, but the interpretation of the results, though at first thought simple, has proved difficult because of uncertainty about the roles of inherent surface anisotropy, surface layers of chemically different material, and variation of properties through the thickness of the film.

The line width in ferromagnetic resonance and the transient behavior during a reversal of magnetization are determined by irreversible processes. These can be taken into account, for many purposes, by introducing a phenomenological damping term into the equation of motion of the magnetization vector. One form of the equation, suggested by Gilbert, is then

$$\frac{d\mathbf{M}}{dt} = \gamma_0 \mathbf{M} \times (\mathbf{H}_{\text{eff}} - \eta\dot{\mathbf{M}}) \qquad (8.43)$$

where γ_0 is the ratio of magnetic moment to angular momentum, $\mathbf{H}_{\text{eff}}$ is an "effective field" due to all the energies considered in Sec. 7, and η is a damping constant. For a more basic understanding of line width and transient behavior, it is necessary to consider relaxation processes on a microscopic scale. For this purpose, the semiclassical spin-wave theory mentioned in Sec. 7 has been much used; the general picture is that magnons are scattered by magnons, phonons, or conduction electrons.

Relaxation times and gyromagnetic inertial reactions must be taken into account in the description of domain wall displacements and rotations in ferrites at radio frequencies. In ferromagnetic metals, microscopic eddy currents are important even in slow changes, for they dissipate the energy released in a Barkhausen jump. This and other mechanisms lead to a magnetization-dependent energy dissipation in elastic vibrations [3, 4, 7].

Long-time lag (seconds or days) of the magnetization change behind the change of applied field is observed in some materials. When this is not due simply to extraneous disturbances, such as random mechanical vibrations, it must be attributed to the gradual surmounting of energy barriers under the influence of thermal agitation. It is then called *magnetic aftereffect*. The time constant for such a process is of the form $\tau = \tau_0 e^{W/kT}$, where W is the height of the energy barrier and where τ_0 varies only slowly with T. In observations covering a period $t_1 < t < t_2$, after a change of field at time $t = 0$, a particular process will contribute to observable aftereffect if its τ is neither much smaller than t_1 (in which case practically all the possible jumps will have occurred before the observations begin) nor much larger than t_2 (in which case practically none of the possible jumps will have occurred before the observations end). In fine-particle materials (Sec. 9), the process involved is a magnetization reversal of a single particle, and W is proportional to the particle volume; the particles that contribute to aftereffect are those neither small enough to be superparamagnetic nor large enough to be stably ferromagnetic. In bulk material, one process that can contribute is thermal fluctuations of domain-wall positions and of the magnetic fields dependent on them. Another much-studied process is diffusion of impurity atoms (especially hydrogen atoms) to new positions that have become energetically advantageous because of displacement of a domain wall. Still other processes have been studied in connection with induced anisotropy.

Although aftereffect gets its name from experiments on transient response, the same processes cause a lag of an alternating or rotating magnetization behind an alternating or rotating field; they therefore contribute to the energy loss a term distinct from the hysteresis and eddy-current losses. Magnetic aftereffect is also called *magnetic viscosity*. Some of the processes that contribute to it, for example, diffusion of atoms, cause also a change of initial permeability with time [3, 4, 7].

References

Basic and General References:

1. Bates, Leslie F.: "Modern Magnetism," 3d ed., Cambridge University Press, New York, 1951.
2. Becker, Richard: "Electromagnetic Fields and Interactions," vol. 1, edited by F. Sauter (trans.), Blaisdell Publishing Co., New York, 1964.
3. Becker, R., and W. Döring: "Ferromagnetismus," Springer, Berlin, 1939; Edwards, Ann Arbor, Mich., 1943.
4. Bozorth, Richard M.: "Ferromagnetism," Van Nostrand, Princeton, N.J., 1951.

4a. Chikazumi, Soshin: "Physics of Magnetism," Wiley, New York, 1964.

5. Dekker, Adrianus J.: "Solid State Physics," Prentice-Hall, Englewood Cliffs, N.J., 1957.
6. Kittel, Charles: "Introduction to Solid State Physics," 2d ed., Wiley, New York, 1956.
7. Kneller, Eckart: "Ferromagnetismus," Springer, Berlin, 1962.

7a. Mattis, Daniel C.: "The Theory of Magnetism: An Introduction to the Study of Cooperative Phenomena," Harper & Row, New York, 1965.

7b. Morrish, A. H.: "The Physical Principles of Magnetism," Wiley, New York, 1965.

8. Panofsky, W. K. H., and M. Philips: "Classical Electricity and Magnetism," 2d ed., Addison-Wesley, Reading, Mass., 1962.
9. Seitz, Frederick: "The Modern Theory of Solids," McGraw-Hill, New York, 1940.
10. Stoner, Edmund C.: "Magnetism and Matter," Methuen, London, 1934.
11. Van Vleck, J. H.: "The Theory of Electric and Magnetic Susceptibilities," Oxford University Press, London, 1932.

Note: Several books published before 1950 have been listed above because they contain some material that is useful and still not easily found elsewhere. They also contain some material that is obsolete, such as the theory of domains and coercive force in ref. 3. They should therefore be supplemented by more recent sources, especially on topics for which such sources are cited in the text of this article.

References on Specialized Topics:

12. Akhiezer, A. I., V. G. Bar'yakhtar, and M. I. Kaganov: Spin Waves in Ferromagnets and Antiferromagnets, *Soviet Phys.-Usp.*, **3**: 567–592 and 661–676 (1961).
13. Barnett, S. J.: Gyromagnetic and Electron Inertia Effects, *Revs. Mod. Phys.*, **7**: 129–166 (1935).
14. Birss, R. R.: "Symmetry and Magnetism," North-Holland Publishing Company, Amsterdam, 1964.

14a. Brown, William Fuller, Jr.: "Magnetoelastic Interactions," Springer, Berlin, 1966.

15. Brown, William Fuller, Jr.: "Micromagnetics," Interscience, New York, 1963.
16. Craik, D. J., and R. S. Tebble: "Ferromagnetism and Ferromagnetic Domains," North-Holland Publishing Company, Amsterdam, 1965.
17. Kittel, C., and J. K. Galt: Ferromagnetic Domain Theory, in "Solid State Physics," vol. 3, pp. 437–564, Academic, New York, 1956.
18. Kittel, C., C. Zener, R. R. Heikes, J. C. Slater, E. P. Wohlfarth, J. H. Van Vleck, and R. Smoluchowski: Symposium on Exchange, *Revs. Mod. Phys.*, **25**: 191–228 (1953).

18a. Kronmüller, H.: Magnetisierungskurve der Ferromagnetika. I. Mikromagnetische Grundlagen, Einmündung in die ferromagnetische Sättigung und Nachwirkungseffekte, in "Moderne Probleme der Metallphysik," A. Seeger (ed.), vol. 2, pp. 24–156, Springer, Berlin, 1966.

19. Lee, E. W.: Magnetostriction and Magnetomechanical Effects, *Rept. Progr. Phys.*, **18**: 184–229 (1955).
20. Mee, C. D.: "The Physics of Magnetic Recording," North-Holland Publishing Company, Amsterdam, 1964.
21. Morrish, A. H., and R. J. Prosen (eds.): "Magnetic Materials Digest: The Literature of 1963," M. W. Lads Publishing Co., Philadelphia, 1964.
22. Newell, G. F., and E. W. Montroll: On the Theory of the Ising Model of Ferromagnetism, *Revs. Mod. Phys.*, **25**: 353–389 (1953).
23. Pake, George E.: "Paramagnetic Resonance," Benjamin, New York, 1962.

24. Rado, G. T., and H. Suhl (eds.): "Magnetism: A Treatise on Modern Theory and Materials," vol. 1, "Magnetic Ions in Insulators: Their Interactions, Resonances, and Optical Properties," Academic, New York, 1963.
24*a*. Rado, G. T., and H. Suhl (eds.): "Magnetism: A Treatise on Modern Theory and Materials," vol. 2, part A, "Statistical Models, Magnetic Symmetry, Hyperfine Interactions, and Metals," Academic, New York, 1965.
25. Rado, G. T., and H. Suhl (eds.): "Magnetism: A Treatise on Modern Theory and Materials," vol. 3, "Spin Arrangements and Crystal Structure, Domains, and Micromagnetics," Academic, New York, 1963.
25*a*. Rado, G. T., and H. Suhl (eds.): "Magnetism: A Treatise on Modern Theory and Materials," vol. 4, "Exchange Interactions among Itinerant Electrons," Academic, New York, 1966.
25*b*. Smart, J. Samuel: "Effective Field Theories of Magnetism," W. B. Saunders Company, Philadelphia, 1966.
26. Smart, J. Samuel: The Néel Theory of Ferrimagnetism, *Am. J. Phys.*, **23**: 356–370 (1955).
27. Smit, J., and H. P. J. Wijn: "Ferrites," Wiley, New York, 1959.
28. Soohoo, Ronald F.: "Theory and Application of Ferrites," Prentice-Hall, Englewood Cliffs, N.J., 1960.
28*a*. Soohoo, Ronald F.: "Magnetic Thin Films," Harper & Row, New York, 1965.
28*b*. Sparks, Marshall: "Ferromagnetic-Relaxation Theory," McGraw-Hill, New York, 1964.
29. Stoner, Edmund C.: Ferromagnetism, *Rept. Progr. Phys.*, **11**: 43–112 (1948).
29*a*. Träuble, H.: Magnetisierungskurve und magnetische Hysterese ferromagnetischer Einkristalle, in "Moderne Probleme der Metallphysik," A. Seeger (ed.), vol. 2, pp. 157–475, Springer, Berlin, 1966.
30. Van Vleck, J. H.: A Survey of the Theory of Ferromagnetism, *Revs. Mod. Phys.*, **17**: 27–47 (1945).
30*a*. Wijn, H. P. J. (ed.): "Encyclopedia of Physics," vol. 18, part 2, "Ferromagnetism," Springer, Berlin, 1966.
31. Wolf, W. P.: Ferrimagnetism, *Rept. Progr. Phys.*, **24**: 212–303 (1961).
32. Zubarev, D. N.: Double-time Green Functions in Statistical Physics, *Soviet Phys.-Usp.* (*English Transl.*), **3**: 320–345 (1960).

See also the Proceedings of the Annual Conference on Magnetism and Magnetic Materials, published as a supplement to the March or April *J. Appl. Phys.* (1958–), and the Proceedings of the International Conferences on Magnetism, published in *J. Phys. Radium*, **12**: (March, 1951) and **20**: (February, 1959), in *J. Phys. Soc. Japan*, **17**: *Suppl.* B-I (1962), and by The Institute of Physics and The Physical Society, London (1965).

Chapter 9

Electrolytic Conductivity and Electrode Processes

By WALTER J. HAMER, National Bureau of Standards
and REUBEN E. WOOD, The George Washington University

1. Electrolytic and Electronic Conduction

Electrolytic conductors differ from *electronic*, or *metallic, conductors* in two ways: In electrolytic conductors the carriers of electrical energy are charged particles of atomic or molecular size, and a transfer of matter takes place; in metallic conductors no matter is transferred, and current flow involves electrons only. When electricity flows through a circuit composed of both types of conductors, a chemical reaction always occurs at each electronic-electrolytic boundary. These reactions are called *electrochemical* and involve oxidation and reduction. Where the flow of electrons is toward the electrolytic conductor, *reduction* occurs at the interface; where the flow of electrons is away from the electrolytic conductor, *oxidation* occurs at the interface. There will always be an equal number of these two types of boundaries in any circuit, so that oxidation and reduction both occur in any electrochemical cell; in fact, the extent of these two processes is always quantitatively equivalent. For example, in the electrolysis of copper sulfate between copper electrodes, copper ions are reduced to metallic copper at the interface where the electron flow is toward the electrolyte, and an equivalent amount of copper is oxidized to copper ions at the interfaces where the electron flow is away from the electrolytic conductor.

2. Electrolytic Conductors

Electrolytic conductors, for the most part, are liquid solutions composed of a solute and a solvent, although fused salts (ionic fluids), some pure liquids, and some solids (especially when near their melting points) show more or less electrolytic conductivity. Gases conduct electricity by an ionic process but are not called electrolytes or electrolytic conductors. Frequently only those substances showing marked conductance are called electrolytic conductors. For example, pure liquid water has such low conductivity that it is often classed as a nonelectrolyte.

In electrolytic solutions the carriers of electricity are charged particles of atomic or molecular size called *ions*. Each solution contains an equivalent number of positively charged and negatively charged ions whereby electroneutrality prevails. Under a potential gradient these ions move in opposite directions with their own characteristic velocity superimposed on their *Brownian movement*. Each type of ion moves with a different velocity, and hence each carries a different fraction of the total current through any one solution. This fraction for each type of ion is called its *transference* or *transport number* for the solution in which the ion exists.

As the temperature of electrolytic solutions is raised, the velocities of the different kinds of ions in a solution containing only two kinds tend to approach a common value under the same potential gradient, and hence the transference numbers of all types of ions tend toward a value of one-half. Furthermore, the mobility of ions increases with temperature, causing a corresponding increase in the conductivity of electrolytic solutions. This temperature effect on conductivity is opposite to that observed in most electronic conductors.

Electrolytes‡ are of two general classes: strong and weak. Strong electrolytes are considered to be totally or highly dissociated into ions at all concentrations. Weak electrolytes are solutes which are not totally dissociated into ions, except at infinite dilution, and contain both ions and neutral molecules. *Polyelectrolytes*, such as phosphoric acid, ionize in steps and belong to the class of weak electrolytes, although sulfuric acid is sometimes termed a *half electrolyte* in that the first ionization step is complete and the second incomplete. Most inorganic acids, bases, and salts give strong electrolytes when dissolved in water. On the other hand, organic compounds usually give weak electrolytes, if they are ionized at all.

3. Ionization

The process of *ionization* involves the dissociation of molecules or ionic crystals into ions when the substances are melted or dissolved in a suitable solvent. Although Arrhenius stated that molecules dissociated into *free ions* having no influence on each other, it is now known that ions are subject to coulombic forces; only at infinite dilution do ions obey the ideal solution laws wherein their behavior is independent

‡ The term *electrolyte* is used in two senses in the literature. It is used to refer to the conducting solution or to the solute making up a conducting solution. In this chapter it is used in both senses, as a convenience. The way in which it is being used will be evident from the context.

of the other ions present in the solution. The process of ionization is influenced by the nature of solute, nature of solvent, size of ion, and solute-solvent interaction. The dielectric constant of the solvent plays a predominant role in ionization; the higher the dielectric constant, the less are the electrostatic forces between ions, and the greater is the electrolytic conductivity. Moreover, electrostatically bound ion pairs which play no part in electrolytic conductivity and ion triplets can coexist in equilibrium with simple ions; these aggregates form mainly in media of relatively low dielectric constant where electrostatic forces are large.

Most inorganic salts are ionized in the crystalline state; i.e., they exist as *ionic lattices*. That the electrostatic forces between charged ions are responsible for the stability of a crystal has been established on considerations of the magnitude of the space-lattice energy. The comparative lack of conductivity in ionic crystals is due to the fact that the ions are fixed in the crystal lattice and are not able to migrate under an applied field. However, when the crystal is melted or dissolved, the ions can migrate, and the liquid or solution will conduct. Most molten salts can be correctly designated as ionic fluids. However, X-ray diffraction experiments show that molten salts, especially near their melting point, retain a short-range orderly arrangement of the ions comparable to the long-range order that exists in crystals. These forces are doubtless responsible for molten salts having less conductance than would otherwise be the case. Metals often show appreciable solubility in their own molten salts. The conductance of such solutions is partly ionic and partly electronic. That most salts are appreciably soluble only in solvents of high dielectric constants can be attributed to electrostatic attraction between oppositely charged ions, which is inversely proportional to the dielectric constant of the intervening medium. Thus, much less work is required to separate sodium and chloride ions from their crystal-lattice spacing in a dilute sodium chloride solution if the solvent is water, with a dielectric constant of about 80, than if it were benzene (in which sodium chloride is practically insoluble), with a dielectric constant of 2.3.

As distinct from most salts which are ionic in the solid state, most acids are not ionic in either the solid or the pure liquid state. Yet with the addition of water they give solutions that show strong electrical conductivity. Pure sulfuric acid is almost a nonconductor, but aqueous solutions of this acid, even when most concentrated, show strong electrolytic conductivity. Reduction of coulombic interionic forces by solvents of high dielectric constant doubtless plays a part in this process of ionization. However, these forces are not sufficient to explain entirely observed behavior, and the tendency of the solvent to form a bond with one of the ions, and thereby to stabilize it, accounts for a large part of the acid ionization process. In the case of aqueous solutions of acids, for example, the hydrogen ion is often written as H_3O^+ instead of as the unhydrated proton H^+. [Actually, the hydrated proton should be represented as $(H \cdot nH_2O)^+$ as there is no known method by which the actual degree of solvation of an ion can be determined.] Thus, the process of ionization involves the solvent in two ways: (1) by its altering the electrostatic forces and (2) by its reacting with the solute.

4. Degree of Ionization [1]‡

In a particular system the extent of ionization may vary with changes in concentration. For strong electrolytes, little, if any, change occurs; for weak electrolytes, the change in degree of ionization is marked. Arrhenius believed that all solutions showed changes in the degree of ionization as the concentration was altered and that complete ionization was realized only at infinite dilution. His concept received much support from the work of Van't Hoff, Ostwald, and others on the colligative properties of electrolytes (vapor-pressure lowering, boiling-point elevation, freezing-point depression, osmotic pressure). These experimenters found that solutions which conducted an electric current possessed colligative properties, the magnitudes of which were approximately some small integer times the magnitudes of the corresponding properties of an equivalent amount of a chemical which did not give a conducting solution. Thus, the vapor-pressure lowering of a dilute solution of sodium chloride was about twice that of a solution containing the same formal concentration of sugar. The sodium and chloride ions existed as separate entities, each producing a lowering in vapor pressure equivalent to a molecule of the nonelectrolyte sugar. The difference in the effectiveness of the electrolyte and nondissociated solutes in colligative phenomena became less as the solutions became more concentrated, leading to the conclusion that the degree of ionization of the solute became less as the concentration of the solution was increased.

Ostwald further postulated that the concentrations of ionized and un-ionized portions of the solution bore a simple relationship to one another, independent of the concentration of solution. Thus, for an electrolyte HA which dissociates to give ions H^+ and A^-, Ostwald stated:

$$\frac{(H^+)(A^-)}{(HA)} = K \tag{9.1}$$

(parentheses represent concentrations). This expression was found to hold fairly closely for certain solutions, now known to be weak electrolytes, but failed badly for others, now known to be strong electrolytes. In this method, Ostwald used either Λ/Λ^0 (see Sec. 9) or Van't Hoff's $i = 1 + \alpha(\nu - 1)$ to calculate the degree of ionization of the solute α, where ν represents the number of ions obtainable from one molecule of solute. It is now known that the behavior of strong electrolytes may be explained largely on the basis of interionic forces (see Sec. 12).

5. Ionic Charge and the Faraday

The charge on each ion is equal to the electronic charge or some integral multiple of it. Thus, any univalent negative ion has a charge equal in magni-

‡ Numbers in brackets refer to References at end of chapter.

tude to and of the same sign as a single electron. A bivalent ion has +2 or −2 electronic charges, depending on whether it is a positive or negative bivalent ion. The quantity of an element or molecular aggregate oxidized or reduced by one Avogadro's number of charges is called an *electrochemical equivalent* of the element or aggregate. For an element such as hydrogen, which forms univalent ions, an electrochemical equivalent is a gram atomic weight while for an element such as zinc, which forms bivalent ions, an electrochemical equivalent is one-half of a gram atomic weight, if the unit of mass is taken as the gram in each case.

In 1833 Faraday stated: "The chemical power of a current of electricity is in direct proportion to the absolute quantity of electricity which passes"; i.e., at all interfaces between electronic and electrolytic conductors in a series circuit, the amount of chemical change, or reaction, in chemical equivalents is the *same* and depends on the quantity of electricity passed through the interfaces. Since in electrolysis there is a definite quantity of electricity that will bring about 1 gram equivalent of chemical reaction, regardless of the system, it is important to determine the magnitude of this quantity of electricity. This quantity is the *faraday*, $\mathfrak{F}$, and represents an *Avogadro's number of charges:*

$$\mathfrak{F} = N\epsilon \qquad (9.2)$$

N being Avogadro's number and ϵ the magnitude of the charge of an electron. The units in which $\mathfrak{F}$ is most commonly expressed are coulombs per gram equivalent; it is also given in electromagnetic units per gram equivalent.

Theoretically, any electrolytic cell could be used to determine the faraday. However, very few electrolytic cells have been found to be sufficiently free of side reactions or other disturbing factors to permit a quantitative evaluation. Prior to 1951 only silver deposition and iodide oxidation appeared to be suitable for quantitative evaluations of the faraday. In 1951 two new methods were described, the oxalate coulometer and the omegatron method, the latter a strictly physical one [2]. Then in 1960 a new method involving silver dissolution [3] rather than silver deposition was used to obtain a value for the faraday. This method was free of the uncertainties inherent in the silver-deposition method.

The values of the faraday as found by these five methods in terms of the present unified scale of atomic weights based on $C^{12} = 12$ [4] and the legal units of electromotive force and resistance as disseminated by the National Bureau of Standards are:

Silver deposition:

$$\mathfrak{F} = 96{,}473 \text{ coulombs (gram equivalent)}^{-1} \qquad (9.3)$$

Iodide oxidation:

$$\mathfrak{F} = 96{,}486 \text{ coulombs (gram equivalent)}^{-1} \qquad (9.4)$$

Oxalate oxidation:

$$\mathfrak{F} = 96{,}488 \text{ coulombs (gram equivalent)}^{-1} \qquad (9.5)$$

Omegatron:

$$\mathfrak{F} = 96{,}492 \text{ coulombs (gram equivalent)}^{-1} \qquad (9.6)$$

Silver dissolution:

$$\mathfrak{F} = 96{,}487 \text{ coulombs (gram equivalent)}^{-1} \qquad (9.7)$$

The value obtained by the silver-deposition method is low, doubtless because of occlusions in the deposit which are difficult to eliminate. Of these values the National Academy of Sciences–National Research Council Committee on Fundamental Constants recommended the last value with an estimated error of 1.6, corresponding to a [5] limit based on the standard deviation for the experimental value, or

$$\mathfrak{F} = 96{,}487.0 \pm 1.6 \text{ coulombs (gram equivalent)}^{-1} \qquad (9.8)$$

The measurements of the faraday by electrochemical methods involve the measurement of the absolute current, the time, and the mass of material reacted. The omegatron method involves the measurement of ϵ/M, the charge-mass ratio of a substance of mass M and isotopic weight A. This quantity, theoretically, may be determined by any mass spectroscope, but the accuracy with which the voltages, dimensions, and magnetic-field intensities may be measured is insufficient for accurate determinations of $\mathfrak{F}$ except by the omegatron, a specialized form of mass spectroscope. The manner in which ϵ/M is used to evaluate the faraday is as follows: First we start with the defining Eq. (9.2), $\mathfrak{F} = N\epsilon$. Since Avogadro's number N is the number of atoms in a gram atomic weight of an element and if A is the atomic weight (in the omegatron procedure the weight of a particular isotope is involved so that A is a particular isotopic weight) and M the mass of one atom, then $A/M = N$, and

$$\mathfrak{F} = N\epsilon = A(\epsilon/M) \qquad (9.9)$$

Equation (9.9) may be further modified to put it in the form of quantities measurable in the omegatron. The cyclotron equation relating the angular velocity ω_c of an ion of specific charge-mass ratio ϵ/M in a magnetic field B is

$$\omega_c = \frac{\epsilon B}{M} \qquad (9.10)$$

The frequency of nuclear-resonance absorption for protons is proportional to the magnetic field:

$$\omega_n = \gamma_p B \qquad (9.11)$$

Combination of Eqs. (9.9) to (9.11) gives

$$\mathfrak{F} = A\gamma_p(\omega_c/\omega_n) \qquad (9.12)$$

Since A and γ_p have been determined with high precision, evaluation of the faraday involves the measurement of ω_c/ω_n. Sommer and Hipple [2] obtained for the proton $\omega_n/\omega_c = 2.792685 \pm 0.000025$. The method is not direct, in the sense that the gyromagnetic ratio γ_p must be obtained by other means. This method does, nevertheless, offer an approach to an evaluation of $\mathfrak{F}$ that is not electrochemical and thus provides an independent check on the purity of all electrochemical reactions. Deviations from Faraday's law, as applied to electrochemistry, may be caused by simultaneous electrode reactions, electrolytic reversal of electrode processes, and interaction of the products of one electrode with the products of the other electrode in an electrolytic cell.

6. Electrolytic Conductivity

Conductance is the reciprocal of resistance and is directly proportional to the cross-sectional area of the conductor and inversely proportional to its length:

$$\mathcal{C} = 1/R = \kappa \frac{A}{l} \tag{9.13}$$

where κ is a proportionality constant, termed *specific conductance* or *conductivity,* and has the units of $ohm^{-1}\ cm^{-1}$. Representative values are given in Table 9.1. Electrolytic solutions have much lower conductivity than do metals and much higher conductivity than the so-called *nonelectrolytes.* Molten salts, in general, have conductances of the same order of magnitude as aqueous solutions (see Table 9.1).

TABLE 9.1. REPRESENTATIVE VALUES OF ELECTROLYTIC CONDUCTIVITY

Substance	Type of conductor	Specific conductance	
		$ohm^{-1}\ cm^{-1}$	t°C
Copper	Metallic	6.4×10^5	0
Copper	Metallic	5.8×10^5	20
Lead	Metallic	4.9×10^5	0
Iron	Metallic	1.1×10^5	0
Sodium fluoride, molten	Electrolytic	5.09	1000
Sodium chloride, molten	Electrolytic	4.13	1000
4 molar H_2SO_4	Electrolytic	0.75	18
Zinc chloride, molten	Electrolytic	0.25	600
0.1 molar KCl	Electrolytic	0.0128	25
Beryllium (II) chloride, molten	Electrolytic	0.003	450
0.01 molar KCl	Electrolytic	0.00141	25
1 molar acetic acid	Electrolytic	0.0013	18
0.001 molar acetic acid	Electrolytic	4×10^{-5}	18
Water	Nonelectrolyte*	4×10^{-8}	18
Xylene	Nonelectrolyte*	1×10^{-19}	25

* The conductance these show is electrolytic in nature but is so low that the substances may be called *nonelectrolytes* or perhaps *nonconductors* (see Section 2).

In terms of Ohm's law, the defining equation (9.13) for specific conductance reduces to

$$\kappa = \frac{I}{E} \tag{9.14}$$

when I is the current and E the potential applied to a centimeter-cube sample of the conductor, the conductance being measured between a pair of opposite faces of the cube.

7. Equivalent and Molar Conductance

Although the specific conductance is useful in comparing metals, it has little direct importance in dealing with solutions. Since the concentration of solutions may be varied at will, comparisons of the conductance of solutions containing equivalents or fractions thereof are more significant. *Equivalent conductance* is defined:

$$\Lambda = \kappa/C \tag{9.15}$$

where C is in equivalents per cubic centimeter, or, more commonly,

$$\Lambda = 1{,}000\kappa/c \tag{9.16}$$

where c has units of equivalents per 1,000 cm^3. $1/C$ or $1{,}000/c$ replaces A/l in Eq. (9.13), and, therefore, the equivalent conductance is the conductance of that amount of an electrolytic solution which contains 1 equiv of electrolyte when placed between parallel planes 1 cm apart and of sufficient area to retain the volume of electrolyte; the conductance is measured normal to the planes.

The equivalent conductance may also be expressed as

$$\Lambda = \frac{I(1{,}000)}{Ec} \tag{9.17}$$

by substitution of Eq. (9.14) in (9.16). The equivalent conductance is numerically equal to the number of amperes that would pass through such a solution if a potential gradient of 1 volt were applied across the electrodes, all disturbing effects being absent.

In these definitions, how the electrolytic solution ionizes must be known, i.e., whether the solute ionizes simply or complexly; otherwise, the equivalent weight of the solute cannot be calculated. If the manner of ionization is not known, the *molar conductance* is used instead and is defined by

$$\Lambda_m = 1{,}000\kappa/m \tag{9.18}$$

where m is the gram moles of the solute dissolved in 1,000 cm^3 of solvent.

The relation between equivalent and molar conductance is given by

$$\Lambda_m = \nu_+ z_+ \Lambda = \nu_- z_- \Lambda = \nu z \Lambda \tag{9.19}$$

where ν_+ is the number of positive ions of charge z_+ formed in the dissociation of one molecule of solute and ν_- is the number of negative ions of charge z_- formed similarly. Since $\nu_+ z_+ = \nu_- z_-$, we can designate either by νz, where the product of quantities with like signs is implied. Since Λ is about 60 $ohm^{-1}\ cm^{-1}$ (see Table 9.6) for most ions, molar conductances give some information about the mode of ionization of electrolytes; e.g., a solution of uni-univalent salts would give a molar conductance of about 120 $ohm^{-1}\ cm^{-1}$ and a uni-bivalent salt, about 240 $ohm^{-1}\ cm^{-1}$.

8. Measurements of Electrolytic Conductivity

Conductivity of electrolytes usually is determined by a-c methods; d-c methods result in net electrolytic effects which change the equivalent concentration and produce concentration gradients at the electrode surface which give rise to polarization or back emf effects. It is common to use a frequency of 1,000 cps for precise measurements, although on occasions the equivalent conductance is determined as a function of frequency. Also, there is some evidence that the fractional (percentage) variation of equivalent conductance with frequency is the same for most electrolytes over a small range of changes in frequency. In practice, some reference solution is chosen for calibration of conductivity cells; the same frequency is used in this calibration and when the calibrated cells are used to measure the conductivity of another solution.

By Eq. (9.13) the ratio of $\mathcal{C}$ and κ is A/l; it is implicit in this definition that the current flows in parallel lines perpendicular to the electrode of area A. It is difficult to make a cell of suitable geometry to permit calculation of κ directly from $\mathcal{C}$. To circumvent this, specific conductances of a number of KCl solutions have been accurately determined so that they may be used to determine the cell constants of various conductivity cells of various sizes and shapes. The *cell constant* is defined as $\kappa/\mathcal{C}$ or as κR, where R is the resistance of the KCl solution in the conductivity cell under study. Specific conductances of standard KCl solutions for use in the calibration of conductivity cells are given in Table 9.2 [6].

TABLE 9.2. SPECIFIC CONDUCTANCES OF STANDARD KCl SOLUTIONS [6]*

Concentration demal†	Grams KCl per 1,000 g of solution in vacuum‡	Specific conductances		
		0°C	18°C	25°C
1.0	71.1352	0.065144	0.097790	0.111287
0.1	7.41913	0.0071344	0.0111612	0.0128496
0.01	0.745263	0.00077326	0.00121992	0.00140807

* Original data were expressed in international ohms and have been converted to absolute units here.

† A solution containing a gram mole of salt dissolved in a cubic decimeter of solution at 0° Celsius.

‡ Based on atomic weights of 1933.

9. Significance of Equivalent Conductance

Equivalent conductance varies in a significant way with concentration. On dilution, the equivalent conductance increases and approaches a limiting value Λ^0 at infinite dilution. Strong and weak electrolytes show widely different behavior (Figs. 9.1 and 9.2). In Fig. 9.1 the equivalent conductances for KCl, NaCl, and $CaCl_2$ at 25°C and for HF, CH_3COOH, and NH_4OH at 18°C are shown as a function of log c, where c is the equivalent concentration. The first three tend to approach a constant or limiting value on dilution, whereas the last three do not. In Fig. 9.2 the same data plotted against $(c)^{1/2}$ show the same phenomena. The first three electrolytes are *strong;* the last three *weak.* The variation of the equivalent conductance of the first three may be explained on the basis of interionic attraction (see Sec. 12) between ions of opposite sign and the last three on the changes in the degree of ionization of the solute with changes in solute concentration. Thus, measurements of equivalent conductance offer a most piquant method for distinguishing between *strong* and *weak* electrolytes. A method by which the limiting conductance Λ^0 of weak electrolytes may be obtained is discussed in Sec. 10. Figures 9.1 and 9.2 indicate that extrapolation to infinite dilution to obtain Λ^0 is feasible for *strong* electrolytes but not for *weak.*

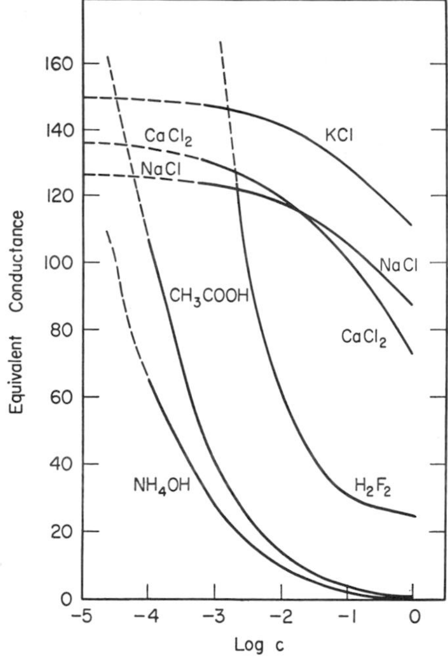

FIG. 9.1. The equivalent conductance of typical strong and weak electrolytes as a function of the logarithms of the concentration in equivalents per liter.

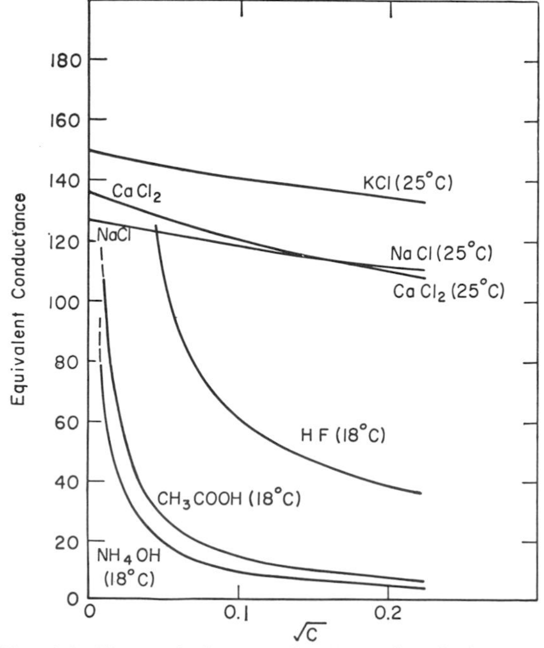

FIG. 9.2. The equivalent conductance of typical strong and weak electrolytes as a function of the square root of the concentration in equivalents per liter.

Measurements of equivalent conductances also serve to give information on the state of ions in nonaqueous and mixed solvents where electrostatic effects may vary markedly. Walden [7] measured the conductivity of tetraethylammonium iodide in different solvents with dielectric constants ranging from 8 to 80; qualitatively, the electrical conductivity increased with the dielectric constant of the solvent. He did not find a direct proportionality, for the tetraethylammonium and iodide ions are doubtless solvated to different extents in the various solvents. Nevertheless, Walden did show the role that the dielectric constant plays in the ionization of a dissolved solute.

Kraus and Fuoss [8], in an analogous study, limited their variables by using water-dioxane mixtures for solvents and, by varying the proportions, found that dielectric constants of the solvent mixtures varied from 2.2 to 78.6 at 25°C (Fig. 9.3). (They chose

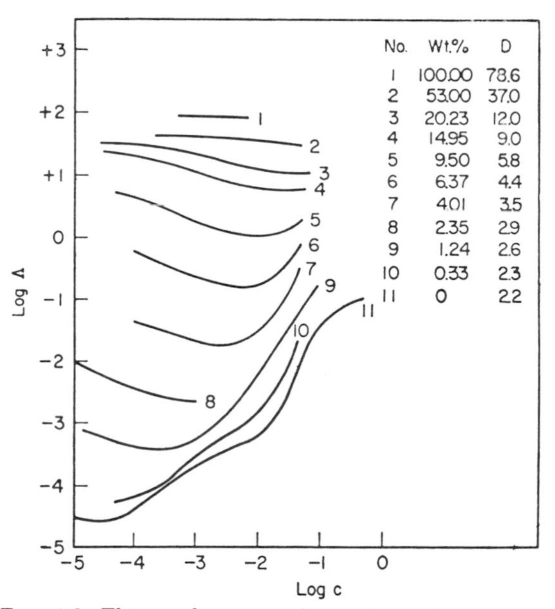

Fig. 9.3. The conductance of tetraisoamylammonium nitrate in dioxane-water mixtures at 25°C showing the influence of dielectric constant on the variation of conductance with solute concentration.

tetraisoamylammonium nitrate as the solute because it was readily soluble in all mixtures of dioxane and water.) The behavior of such systems varies markedly with a change in the dielectric constant. For high and intermediate values of D, the solute is completely or highly dissociated; for values of D from about 3 to 8, a minimum in the conductance curve is observed and may be accounted for by the formation of electrostatic ion pairs; for still lower values of D, minima and sigmoid flexures are observed in the conductance curve, and these may be explained by the interaction of nonconducting ion pairs with other ions to form conducting ion triplets. Since ion pairs do not contribute to electrolytic conductivity, they remove possible electrolytic carriers from the solution. They are different from neutral molecules whose formation would involve an electron shift, but they lead to the same result, namely, they do not contribute to electrical conductivity. On the other hand, ion triplets which bear a net charge do contribute to electrical conductivity.

Measurements of equivalent conductances in various solvents also give other information on the nature of ions in solution. For example, Stokes' law

$$v = f/6\pi\eta r \tag{9.20}$$

gives the velocity of a sphere of radius r in a fluid of viscosity η, moving under a force f. If ions behave ideally at infinite dilution (i.e., have no influence on each other), then their motion should depend only on their nature, the electric field, and the solvent. Thus, the product of the limiting equivalent conductance and the viscosity of the solvent for a particular solute should be a constant at any one temperature. In Table 9.3 some data on this product are presented. This rule, now known as *Walden's rule*, is only an approximation. However, here again strict conformity would not be expected, first, because the ions are probably not spherical in shape and, second, the radii of the ions (degree of solvation) undoubtedly vary from solvent to solvent. Even so, Walden's rule does approximate the situation. In Part 2 of Table 9.3 the value of the product $\Lambda^0\eta$ for water varies markedly from the values found for the same solute in other solvents. That $\Lambda^0\eta$ should approximate a constant value follows from Eqs. (9.20) and (9.31) wherein the force on an ion is given by $z\epsilon\mathbf{F}$, so that

Table 9.3. Tests of Walden's Rule

The product $\Lambda^0\eta$ for potassium iodide in various solvents at 25°C [7]

Solvent	Λ^0	η	$\Lambda^0\eta$
Sulfur dioxide*	265	0.00394	1.044
Acetonitrile	198.2	0.00345	0.684
Acetone	185.5	0.003158	0.586
Nitromethane	124	0.00611	0.758
Methyl alcohol	114.8	0.00546	0.627
Pyridine†	71.3	0.00958	0.682
Ethyl alcohol	50.9	0.01096	0.560
Furfurol	43.1	0.0149	0.642
Acetophenone	39.8	0.0162	0.644

The product $\Lambda^0\eta$ for sodium chloride in various solvents at 25°C [9]

Solvent	Λ^0	η	$\Lambda^0\eta$
Water	126.39	0.008949	1.131
Methyl alcohol	96.9	0.00546	0.529
Ethyl alcohol	42.5	0.01096	0.466

* 0°C.
† 20°C.

$$\Lambda^0\eta = \frac{z\epsilon\mathbf{F}}{6\pi r} = \text{const } 1/r \tag{9.21}$$

Since the forces restraining the movements of an ion under the influence of an electric field are the same as those which oppose thermal diffusion of the ion, there is a simple relationship between the diffusion coefficient D^*_i of an ion and its equivalent conductance λ_i:

$$D^*_i = \frac{\lambda_i RT}{\mathfrak{F}^2} \qquad (9.22)$$

Measurements of equivalent conductances of differing strong electrolytes in any one solvent also provide insight into the nature of ions and the role played by the solvent. Table 9.4 gives the equivalent conductances of a series of strong electrolytes in water as a function of concentration.

TABLE 9.4. EQUIVALENT CONDUCTANCES OF STRONG ELECTROLYTES IN DILUTE AQUEOUS SOLUTIONS AT 25°C

Solute	Infinite dilution	Concentrations, equiv/l					
		0.001	0.005	0.01	0.02	0.05	0.10
HCl	426.0	421.2	415.6	411.8	407.0	398.9	391.1
NaOH	247.9	244.9	239.9	236.9	232.9	226.9	220.9
$Ca(OH)_2$	257.9	.....	232.9	225.9	213.9		
NaCl	126.4	123.6	120.6	118.4	115.7	111.0	106.6
$CaCl_2$	135.7	130.3	124.2	120.3	115.6	108.4	102.4
KCl	149.8	146.9	143.5	141.2	138.2	133.3	128.9
NH_4Cl	149.6	146.8	143.5	141.2	138.3	133.2	128.7
$KHCO_3$	117.9	115.2	112.1	110.0	107.1		
$K_4Fe(CN)_6$	183.9	167.1	146.0	134.7	122.7	107.6	97.9
NaO_2CCH_3	91.0	88.5	85.7	83.8	81.2	76.9	72.8
$NaO_2CCH_2CH_3$	85.9	83.5	80.9	79.1	76.6		

The equivalent conductances of acids and bases are much higher in aqueous solutions than those of other strong electrolytes. This suggests that H^+ and OH^- play unique roles in a solvent containing both these ions. Bernal and Fowler [10] explained the abnormally high conductances of the hydrogen and hydroxyl ions in aqueous solutions by proton transfers between neighboring water dipoles. Unlike ions, which conduct an electric current by simple migration superimposed on their Brownian movement, electrons undergo displacement transfers from atom to atom in metallic conduction. A similar process is involved for H^+ and OH^- ions in water (HOH) except that such transfers are superimposed on the normal mobility of these ions.

10. Ionic Conductances and Transference Numbers

In 1875 Kohlrausch pointed out from the limiting equivalent conductances then available to him that ions behave independently at infinite dilution so that the equivalent conductance of an electrolyte is the sum of the ionic conductance of the ions composing the electrolyte, and, therefore,

$$\Lambda^0 = \lambda_+^0 + \lambda_-^0 \qquad (9.23)$$

His conclusion is supported by the fact that

$$\begin{aligned} \Lambda_{KCl}^0 - \Lambda_{NaCl}^0 &= 149.8 - 126.4 \\ &= 23.4 \text{ ohm}^{-1}\text{ cm}^2/\text{equiv} \\ \Lambda_{KI}^0 - \Lambda_{NaI}^0 &= 150.3 - 126.9 \\ &= 23.4 \text{ ohm}^{-1}\text{ cm}^2/\text{equiv} \end{aligned}$$

The same difference is obtained in the two cases, the logical conclusion being that the difference is $\lambda_K^0 - \lambda_{Na}^0$ in both.

By conductivity measurements alone, only combinations such as $\lambda_{Na^+} + \lambda_{Cl^-}$ or $\lambda_{K^+} - \lambda_{Na^+}$ can be determined. However, the transference number of Na^+ in a sodium chloride solution represents the fraction of current carried by the sodium ion, and this number t_+ times the equivalent conductance of the solution will equal the equivalent conductance of the sodium ion:

$$\lambda_+ = t_+\Lambda \qquad (9.24)$$

$$\lambda_- = t_-\Lambda \qquad (9.25)$$

Moreover, transference numbers in dilute solutions vary only slightly with concentration.

Once a single ionic conductance value, for example λ_{Na^+}, has been evaluated by determination of a transference number, all the other ionic conductances can be obtained from conductance data only by using relations such as

$$\lambda_{NH_4^+} = \Lambda_{NH_4Cl} - \lambda_{Cl^-} \qquad (9.26)$$

$$\lambda_{CH_3COO^-} = \Lambda_{CH_3COONa} - \lambda_{Na^+} \qquad (9.27)$$

$$\lambda_{OH^-} = \Lambda_{NaOH} - \lambda_{Na^+} \qquad (9.28)$$

$$\lambda_{H^+} = \Lambda_{HCl} - \lambda_{Cl^-} \qquad (9.29)$$

Transference numbers can be determined by measuring the changes in concentration in either or both of the electrode compartments of a divided electrolytic cell when a measured quantity of electricity is passed through the cell. They may also be determined by observing during electrolysis the rate of movement of a boundary between a pair of electrolytes such as KCl and LiCl. In this case the difference in refractive index of the two solutions is used to follow the movement of the boundary. It is much more convenient to follow a moving boundary if one of the ions is colored (visual spectrum). They can also be determined from measurements of the electromotive forces of special types of galvanic cells (see Sec. 16).

In Table 9.5 values of the transference number of the positive ion in various aqueous electrolytes are listed as a function of electrolyte concentration at 25°C. The transference numbers of the negative ions are given by the relation $t_- = 1 - t_+$, since the sum of the transference numbers in any electrolytic solution is unity.

Three important facts are shown in Table 9.5. First, the transference numbers vary only slightly with manifold concentration changes. Second, if the values of the transference numbers exceed 0.5, they increase in magnitude with increases in electrolyte concentration; if they are below 0.5, they decrease (silver nitrate excepted). This behavior is in accord with the theory of interionic attraction (see Sec. 12). The anomalous behavior exhibited by $AgNO_3$ suggests that it ionizes complexly. Third, the transference numbers of the hydrogen and hydroxyl ions (negative ion in NaOH) are of much greater magnitude than those of all other ions. The reason for this has been discussed in Sec. 9.

In Table 9.6 some values of the limiting equivalent conductances for various ions are listed for 25°C.

TABLE 9.5. TRANSFERENCE NUMBER OF THE POSITIVE ION IN VARIOUS AQUEOUS ELECTROLYTE SOLUTIONS AT 25°C AS A FUNCTION OF CONCENTRATION

Solute	Concentration, equiv/l					
	0.01	0.02	0.05	0.10	0.20	0.50
HCl	0.8251	0.8266	0.8292	0.8314	0.8337	
H_2SO_4	0.813*	......	0.819	0.819	0.819	0.815
LiCl	0.3289	0.3261	0.3211	0.3168	0.3112	
NaCl	0.3918	0.3902	0.3876	0.3854	0.3821	
KCl	0.4902	0.4901	0.4899	0.4898	0.4894	0.4888
$BaCl_2$	0.4400	0.4375	0.4317	0.4253	0.4162	0.3986
$LaCl_3$	0.4625	0.4576	0.4482	0.4375	0.4233	
NH_4Cl	0.4907	0.4906	0.4905	0.4907	0.4911	
Na_2SO_4	0.3848	0.3836	0.3829	0.3828	0.3823	
K_2SO_4	0.4829	0.4848	0.4870	0.4890	0.4910	0.4909
KNO_3	0.5084	0.5087	0.5093	0.5103	0.5120	
$AgNO_3$	0.4648	0.4652	0.4664	0.4682		
NaOH	0.202*					

* Infinite dilution.

TABLE 9.6. LIMITING EQUIVALENT CONDUCTANCES OF VARIOUS IONS IN AQUEOUS SOLUTIONS AT 25°C [11]*

(+) Ions	λ_+°	a	(−) Ions	λ_-°	a
H^+	349.66	0.0135	OH^-	198.2	0.0197
NH_4^+	73.51	0.0206	$Fe(CN)_6^{4-}$	110.4	
K^+	73.46	0.0200	$Fe(CN)_6^{3-}$	100.9	
La^{3+}	69.72	0.0210	$SO_4^=$	79.98	0.0207
Ba^{++}	63.60	0.0203	Br^-	78.10	0.0203
Ag^+	61.87	0.0194	I^-	76.80	0.0202
Ca^{++}	59.47	0.0230	Cl^-	76.31	0.0208
Sr^{++}	59.42	0.0205	NO_3^-	71.42	0.0196
Mg^{++}	53.02	0.0219	HCO_3^-	44.48	
Na^+	50.08	0.0228	CH_3COO^-	40.88	0.0206
Li^+	38.66	0.0194	$C_2H_5COO^-$	35.8	

* Original data were based on international ohms; converted to absolute units here.

Their variation with temperature may be expressed by

$$\lambda_t^0 = \lambda_{25°}[1 + a(t - 25)] \qquad (9.30)$$

in the region near 25°C. Values of a in Eq. (9.30) are also included in Table 9.6 [11]. These values make possible the calculation of the limiting equivalent conductance of a large number of electrolytes and are especially valuable in computing the limiting conductance of weak electrolytes, those whose limiting equivalent conductances cannot be obtained by extrapolation to infinite dilution (see Figs. 9.1 and 9.2). Thus, the limiting equivalent conductance of NH_4OH would be the sum of the ionic conductances of the NH_4^+ and OH^- ions [Eq. (9.26) + Eq. (9.28)], or 73.5 + 198.2 = 271.7, and that of acetic acid, CH_3COOH, the sum of the ionic conductances of the CH_3COO^- and H^+ ions [Eq. (9.27) + Eq. (9.29)], or 40.88 + 349.66 = 390.54.

The temperature coefficient of the ionic conductances is nearly the same for all ions: about 2 per cent per degree, which is about the same but of opposite sign as the temperature coefficient for the viscosity of water. This is consistent with the relation between conductance and viscosity (Sec. 9). Highly hydrated ions have larger temperature coefficients for their conductances than less hydrated ones because they lose relatively more water on heating.

11. Ionic Mobilities

Equivalent conductances of ions approach a limiting value as the solution approaches infinite dilution, because at great dilution interionic-attraction effects approach zero; also, the solute is considered to be completely dissociated. Differences in the equivalent conductances among different ions, therefore, imply a difference in the ease with which different ions can move through the solution. Such differences certainly would be expected because of differences in shapes, sizes, and degree of solvation of ions, as well as the possibility of special transfer mechanisms in certain cases (for example, H^+ and OH^- in aqueous solutions).

Ionic mobility of a particular kind of ion is defined as the rate in centimeters per second at which such ions will move under a potential gradient of 1 volt/cm. For the usual case, in which the electrolyte contains only two kinds of ions, there are some simple, easily derived relationships between the positive and negative ion mobilities U_+ and U_-:

$$\lambda_+ = \mathfrak{F}U_+ \qquad \lambda_- = \mathfrak{F}U_-$$
$$\text{and} \qquad \Lambda = \mathfrak{F}(U_+ + U_-) \qquad (9.31)$$

With $\mathfrak{F}$, the faraday constant, in coulombs per equivalent and conductances in ohm^{-1} cm^2, mobilities are in units of cm^2 sec^{-1} volt^{-1}. These relations are valid at any concentration. However, U_+ and U_- vary in magnitude with the concentration; thus so does the equivalent conductance. The ionic mobilities at 25°C are of the order of 4×10^{-4} to 36×10^{-4} cm/sec, with the hydrogen ion having the highest mobility.

12. Interionic Attraction and Electrolytic Conductivity [9, 12]

Ions in an electrolytic solution are subject to coulombic forces which influence the electrical conductivity. In treating these forces, Debye and Hückel introduced the concept of the *ion atmosphere:* At a distance r from an ion there is, on a time average, an *ionic cloud* of opposite charge which sets up a field of potential whose magnitude depends on the distance of the cloud from the central ion. Thermal vibration of the ions tends to disturb this arrangement but not entirely so, the Boltzmann principle expressing the ionic distribution as a function of the ratio of electrical and thermal energies. In an element of volume dV around a point P in space at a distance r from a central ion there will then be, over a period of time, an excess of ions of opposite sign which sets up a field of potential ψ, whose average value will depend on the distance of the cloud from the central ion. The work required to bring a positive ion from infinity to P will be $z_+\epsilon\psi$, and for a negative ion $z_-\epsilon\psi$, where z_+ and z_- are the valences of the positive and negative ions, respectively.

The ionic distribution in a total volume V as a function of the ratio of electrical and thermal energies in the volume element is given by the Boltzmann expression

$$dn_+ = n_+ \exp \frac{-z_+\epsilon\psi}{kT} \, dV \tag{9.32}$$

$$dn_- = n_- \exp \frac{-z_-\epsilon\psi}{kT} \, dV \tag{9.33}$$

where n_+ and n_- are the total number of positive and negative ions in volume V, k is the Boltzmann constant, T is absolute temperature, and ϵ is electronic charge. For a uni-univalent electrolyte, z_+ and z_- are both unity, and, since for electroneutrality $n_+ = n_- = n$, the net charge per unit volume, or the electrical density, is

$$\rho = n\epsilon(e^{-\epsilon\psi/kT} - e^{\epsilon\psi/kT}) \tag{9.34}$$

Upon expansion, neglecting higher terms,

$$\rho = -\frac{2n\epsilon^2\psi}{kT} \tag{9.35}$$

Poisson's equation gives another relation between average electrical density and the potential:

$$\frac{1}{r^2}\frac{d}{dr}\left(r^2 \frac{d\psi}{dr}\right) = -\frac{4\pi\rho}{D} \tag{9.36}$$

D is the dielectric constant of the *medium*, and r has the meaning given above. Eliminating ρ between Eqs. (9.35) and (9.36),

$$\frac{1}{r^2}\frac{d}{dr}\left(r^2 \frac{d\psi}{dr}\right) = \kappa^2\psi \tag{9.37}$$

where κ represents

$$\kappa = \left(\frac{8\pi n\epsilon^2}{DkT}\right)^{1/2} \tag{9.38}$$

κ has the dimensions of reciprocal length so that its reciprocal represents the thickness of the *ion atmosphere* around a chosen central ion. Integration of Eq. (9.37) gives

$$\psi = A\frac{e^{-\kappa r}}{r} + A'\frac{e^{\kappa r}}{r} \tag{9.39}$$

where A and A' are constants of integration. Since the potential approaches zero as r becomes infinitely large, $e^{\kappa r}/r$ becomes infinite in the limit, and constant A' must be zero. As a result, the potential at a point of distance r from a central ion is

$$\psi = A\frac{e^{-\kappa r}}{r} \tag{9.40}$$

where we still have an unevaluated constant. However, at infinite dilution the thickness of the ion atmosphere must be zero [see Eq. (9.38)], and thus ψ would equal A/r since $e^{-\kappa r}$ would be unity for $\kappa = 0$. Furthermore, for such a dilute solution the potential in the vicinity of the central ion would be that of the ion alone, as all other ions would be so far distant as to have no influence. If the ion is taken to be a point charge,

$$\frac{A}{r} = \frac{z_i\epsilon}{Dr} \tag{9.41}$$

and

$$A = \frac{z_i\epsilon}{D} \tag{9.42}$$

from which it follows that the potential ψ is

$$\psi = \frac{z_i\epsilon e^{-\kappa r}}{Dr} \tag{9.43}$$

and

$$\psi = \frac{z_i\epsilon}{Dr} - \frac{z_i\epsilon}{Dr}(1 - e^{-\kappa r}) \tag{9.44}$$

if the solution is very dilute so that κ is small and $(1 - e^{-\kappa r}) \approx \kappa r$,

$$\psi = \frac{z_i\epsilon}{Dr} - \frac{z_i\epsilon\kappa}{D} \tag{9.45}$$

The second term is due to the ion atmosphere, and the first term is the potential due to the ion itself.

If the ion is not taken as a point charge, we integrate from a "distance of closest approach" a_i (the sum of the radii of the oppositely charged ions in contact) to a distance r taken in any direction. The potential becomes

$$\psi = \frac{z_i\epsilon}{Dr}\frac{e^{\kappa(a_i - r)}}{1 + \kappa a_i} \tag{9.46}$$

and, since at the closest distance of approach $r = a_i$,

$$\psi = \frac{z_i\epsilon}{Da_i} - \frac{z_i\epsilon\kappa}{D(1 + \kappa a_i)} \tag{9.47}$$

where the first term is the potential at the surface of the ion due to the charge on the ion and the second term is that of the ion atmosphere.

Using Poisson's equation, the distribution of space charge or electrical density in the ion atmosphere as a function of r is obtained by substitution of the value of ψ given by Eqs. (9.47) and (9.46) and subsequent differentiation, or by

$$\rho = -\frac{z_i\epsilon\kappa^2 e^{\kappa a_i} e^{-\kappa r}}{4\pi(1 + \kappa a_i)r} \tag{9.48}$$

If an electrolytic solution is subjected to a potential gradient of intensity J, a chosen ion will tend to move with its own characteristic velocity v_{di}^0, which is independent of other ions in the solution. However, the ion atmosphere being of opposite sign, the ion will tend to move in the opposite direction. This introduces a retarding effect known as the *electrophoretic effect*, since in its action it appears to involve the movement of the solvent. Each element in the ion atmosphere is acted upon by a force per unit volume which is the product of the space charge or electrical density and the potential gradient of intensity J. Thus, using Eq. (9.48) for the space charge, the electrophoretic force acting on the ion atmosphere

$$f_e = -J\kappa^2 \frac{z_i\epsilon e^{\kappa a_i} e^{-\kappa r}}{4\pi r(1 + \kappa a_i)} \tag{9.49}$$

Furthermore, the ion atmosphere may be considered as shells of thickness dr and area $4\pi r^2$. The total force f acting on this ion atmosphere is

$$df = 4\pi r^2 f\, dr \qquad (9.50)$$

The ion would move with a velocity v_{re} in a direction reverse to the normal motion of the ion. If it is assumed that the ions are spheres, this velocity is also given by Stokes' law [see Eq. (9.20)]. Combining the above three equations gives

$$dv_{re} = -1.5J\kappa^2 \frac{z_i \epsilon e^{\kappa a_i} e^{-\kappa r}}{4\pi\eta(1 + \kappa a_i)}\, dr \qquad (9.51)$$

which, when integrated from the closest distance of approach to infinity, gives

$$v_{re} = -\frac{J\epsilon z_i}{6\pi\eta} \frac{\kappa}{1 + \kappa a_i} \qquad (9.52)$$

For very dilute solutions, κa_i negligible,

$$v_{re} = \frac{J\epsilon z_i}{6\pi\eta} \kappa \qquad (9.53)$$

the velocity of the ion in a direction reverse to its normal direction.

Superimposed on this electrophoretic effect is the *relaxation-time effect*. When a chosen ion tends to move in its normal direction under the influence of an applied potential it tends to move its ion atmosphere with it. This ion atmosphere will adjust to its new location in time, but the delay results in a dissymmetry in the potential field around the ion; the field is greater behind the ion than in its immediate vicinity. This effect, independent of the viscosity of the medium, depends on the limiting conductance of the ion. When a central ion is removed, the ion atmosphere falls to zero potential in a time dependent on the velocity of the ions. For very dilute solutions, the velocity in a direction reverse to the normal as a result of the dissymmetry effects is given by

$$v'_{re} = \frac{-J\epsilon^3 z_i \kappa}{6DkT} \omega \qquad (9.54)$$

where $\omega = z_+ z_- \dfrac{2q}{1 + \sqrt{q}}$

and $q = \dfrac{z_+ z_-}{z_+ + z_-} \dfrac{\lambda_+ + \lambda_-}{z_+\lambda_+ + z_-\lambda_-}$

Thus, the velocity of an ion as it moves through a solution under a potential gradient is

$$v_{di} = v_{di}^0 - \frac{J\epsilon z_i \kappa}{6\pi\eta} - \frac{J\epsilon^3 z_i \kappa\omega}{6DkT} \qquad (9.55)$$

or, for a potential gradient of 1 volt/cm,

$$U_{di} = U_{di}^0 - \frac{\epsilon z_i \kappa}{1{,}800\pi\eta} - \frac{\epsilon^3 z_i \kappa\omega}{1{,}800DkT} \qquad (9.56)$$

where the first term on the right is the limiting value dependent on the nature of the ion only, the second is the relaxation-time effect, and the third is the electrophoretic effect. Since $\mathfrak{F}U_i^0 = \lambda_i^0$ [see Eqs. (9.31)],

$$\lambda_i = \lambda_i^0 - \frac{\epsilon\kappa}{300}\left(\frac{z_i}{6\pi\eta}\mathfrak{F} + \frac{300\epsilon}{6DkT}\lambda_i^0\omega\right) \qquad (9.57)$$

because J as 1 volt/cm is equivalent to $\frac{1}{300}$ esu. When the known values of the various constants and the value of κ are introduced into Eq. (9.57),

$$\lambda_i = \lambda_i^0 - \left[\frac{29.16 z_i}{(DT)^{1/2}\eta} + \frac{9.905 \times 10^5}{(DT)^{3/2}}\lambda_i^0\omega\right] \sqrt{c_+ z_+^2 + c_- z_-^2} \qquad (9.58)$$

for the equivalent conductance of an ion. Since the equivalent conductance of an electrolyte is equal to the sum of the conductances of the constituent ions,

$$\Lambda = \Lambda^0 - \left[\frac{29.16(z_+ + z_-)}{(DT)^{1/2}\eta} + \frac{9.905 \times 10^5}{(DT)^{3/2}}\Lambda^0\omega\right] \sqrt{c(z_+ + z_-)} \qquad (9.59)$$

for the equivalent conductance of an electrolytic solution. For a uni-univalent electrolyte,

$$\Lambda = \Lambda^0 - \left[\frac{82.48}{(DT)^{1/2}\eta} + \frac{8.20 \times 10^5}{(DT)^{3/2}}\Lambda^0\right]\sqrt{c} \qquad (9.60)$$

The equivalent conductance decreases linearly with the square root of the equivalent concentration. This equation may be written

$$\Lambda = \Lambda^0 - (a + b\Lambda^0)\sqrt{c} \qquad (9.61)$$

and is now known as the *Onsager equation* [13]. These equations were derived on the assumption that the solute of the electrolyte was completely dissociated into ions at all concentrations; when this is not so, αc is substituted for c in the above equations, where α is the degree of ionization.

In Table 9.7 a comparison of experimental results on the equivalent conductance of aqueous solutions of KCl and those calculated by the Onsager equation is given for a temperature of 25°C. The theory agrees with experimental results only at extremely low concentrations, where agreement is excellent. At higher concentrations, assumptions in the theory are no longer valid, and empirical equations which include addition terms to the Onsager equation have been proposed that give excellent agreement with experimental results [9]. The data of Table 9.7 are typical; in general, similar agreements between experimental and theoretical data are obtained.

The Onsager equations may also be used to give the theoretical equation for the variation of transference numbers with concentration;

$$t^+ = \frac{\lambda_+^0 - (0.5a + b\lambda_+^0)\sqrt{c}}{\Lambda^0 - (a + b\Lambda^0)\sqrt{c}} \qquad (9.62)$$

for the transference number of a positive ion, since the transference number is given by λ_+^0/Λ. Upon differentiation of Eq. (9.62) with respect to $(c)^{1/2}$,

$$\left(\frac{dt_+}{d\sqrt{c}}\right)_{c\to 0} = \frac{2t^0 - 1}{\Lambda^0} a \qquad (9.63)$$

TABLE 9.7. EQUIVALENT CONDUCTANCE OF AQUEOUS SOLUTIONS OF POTASSIUM CHLORIDE AT 25°C (OBSERVED AND CALCULATED)

Concentration, moles/liter	Equivalent conductance (observed)	Equivalent conductance (calculated)*	Difference (o − c)
Infinite dilution	149.84	149.84	0
0.000032576	149.30	149.30	0
0.00010445	148.88	148.87	0.01
0.00026570	148.35	148.30	0.05
0.00033277	148.16	148.12	0.04
0.00035217	148.09	148.06	0.03
0.00046948	147.86	147.79	0.07
0.00060895	147.49	147.51	−0.02
0.00084200	147.20	147.09	0.11
0.00092856	147.04	146.96	0.08
0.0011321	146.73	146.66	0.07
0.0014080	146.43	146.29	0.14
0.0015959	146.23	146.06	0.17
0.0020291	145.69	145.58	0.11
0.0020568	145.68	145.55	0.13
0.0023379	145.45	145.27	0.18
0.0027848	144.97	144.85	0.12
0.0028777	144.96	144.77	0.19
0.0032827	144.61	144.42	0.19
0.005	143.48	143.15	0.33
0.01	141.20	140.39	0.81
0.02	138.27	136.47	1.80
0.05	133.30	128.70	4.60
0.10	128.90	119.94	8.96
0.12	127.63	117.09	10.54

* Onsager equation. See L. Onsager, *Physik. Z.*, **27**: 388 (1926): **28**: 277 (1927).

for the variation of the transference numbers with concentration in dilute solutions. If t_+^0 is less than 0.5, variation with concentration is negative; if t_+^0 is greater than 0.5, variation of transference numbers is positive (Sec. 10).

13. High-field Effects in Conductance

In the preceding section it was pointed out that ions under a low potential gradient of the order of 1 volt/cm are subject to two effects: the electrophoretic effect and the relaxation-time effect. If time is insufficient for these two effects to be operative, the expected equivalent conductance will be higher and approach that value expected if interionic-attraction effects were absent. Wien [14], using potential gradients of the order of 20,000 volts/cm, found that the equivalent conductances were higher than those measured at low potential gradients and that the effects were more marked for electrolytes composed of high-valence ions in which the interionic forces are more intense. In Fig. 9.4 his data are given for four different electrolytes for concentrations where the specific conductance was 4.5×10^{-5} ohm^{-1} cm^{-1} for all solutions in the absence of the high electrical field; $\Delta\Lambda$ is the increase in equivalent con-

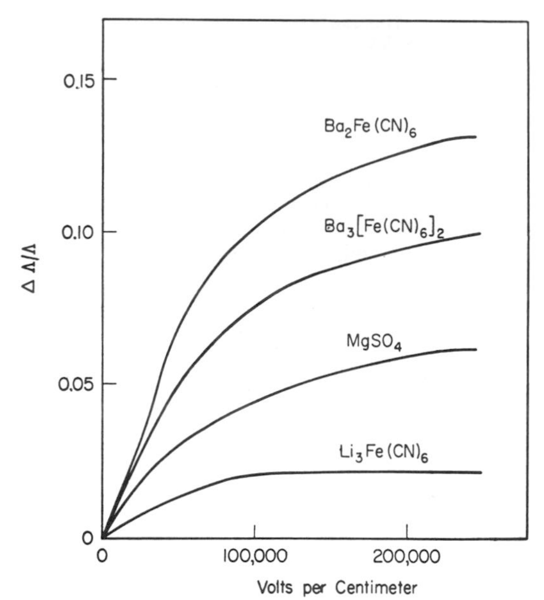

FIG. 9.4. The Wien effect for salts of different valence types having the same low field conductance.

ductance resulting from the applied potential gradient. These results give excellent confirmation of the Debye-Hückel-Onsager theory of interionic attraction: At high ion velocities the ion atmosphere has not sufficient time to form so that ion velocities and equivalent conductances approach values obtained at high dilution where ion-attraction effects are absent.

For weak electrolytes, high potential gradients lead not only to a decrease in any interionic-attraction effects but also to an increase in the ionization of the weak electrolyte. Thus, high potential gradients cannot be used to determine the degree of ionization of weak electrolytes.

14. Conductance at High Frequencies

When an ion moves under a potential gradient, it tends to pull its own ion atmosphere with it, producing a dissymmetry in the field around the ion; the time required for decay of this dissymmetry has a retarding effect on the normal migration of the ion. Debye and Falkenhagen [15] predicted that, if measurements of equivalent conductances were made with alternating currents of very high frequencies of the order of 10 Mc instead of the usual 1,000 to 8,000 Hz, the effects due to dissymmetry should largely disappear. This prediction was checked by Sack and confirmed for a number of electrolytes. Figure 9.5 presents data obtained by several investigators, collected by Geest [16]. The conductance increases by several per cent at the higher frequencies.

15. Electrochemical Thermodynamics

Electrochemical systems must, of course, conform to the laws of thermodynamics. The application of these laws to systems which can undergo reversible or nearly reversible changes in state has been most

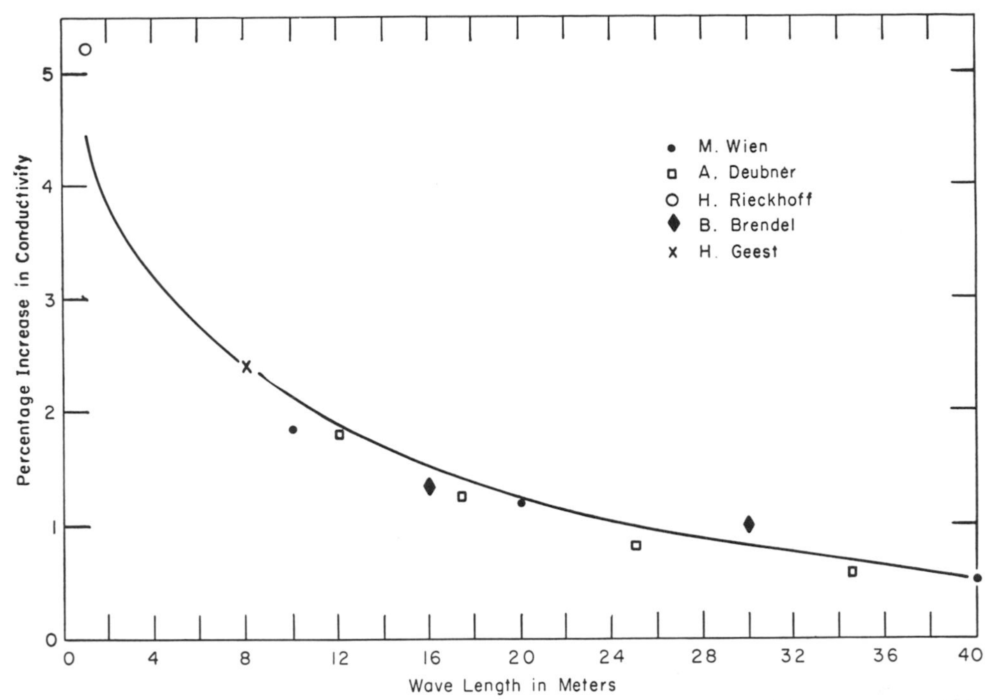

Fig. 9.5. Variation in conductivity with frequency for 0.001M magnesium sulfate solutions at 18°C.

useful. Such applications will be considered in this section and in Sec. 16. In Sec. 17 irreversible phenomena such as overvoltage will be discussed from the somewhat thermodynamic viewpoint which underlies the concept of Gibbs energies (free energies) of activation in absolute reaction-rate theory.

For a closed system,

$$dE = T\,dS - p\,dV \qquad (9.64)$$

For an open system, the boundary of which is so chosen that matter passes into or out of the system during a change in state,

$$dE = T\,dS - p\,dV + \mu_A\,dn_A + \mu_B\,dn_B + \cdots + \mu_Z\,dn_Z \qquad (9.65)$$

E is energy; T, absolute temperature; S, entropy; p, pressure; V, volume; n, the number of moles of components designated by the various subscripts; and μ's are *Gibbs's chemical potentials* of the various components. Definitions of other thermodynamic variables are $H = E + pV$, $A = E - TS$, and

$$G = H - TS$$

where H is heat content, A is work content, and G is Gibbs energy. In view of these and Eq. (9.65), one can obtain expressions for the chemical potential of component I, for example,

$$\mu_I = \left(\frac{dE}{dn_I}\right)_{SVn'} = \left(\frac{dH}{dn_I}\right)_{Spn'} = \left(\frac{dA}{dn_I}\right)_{TVn'} = \left(\frac{dG}{dn_I}\right)_{Tpn'} \qquad (9.66)$$

where n' indicates the constancy of all the components except I. As experiments are most commonly done at constant temperature and pressure, the usual expression for μ is in terms of the partial molal Gibbs energy $\bar{G}$,

$$\bar{G}_I = \left(\frac{dF}{dn_I}\right)_{Tpn'} = \mu_I \qquad (9.67)$$

The absolute value of the chemical potential is indeterminate. However, changes in $\bar{G}_I$ when component I undergoes a change in state can be evaluated. This change is related to the activity a_I of component I by the defining equation

$$\bar{G}_I - \bar{G}_I{}^0 = RT \ln a_I \qquad (9.68)$$

where $\bar{G}_I{}^0$ is the partial molal Gibbs energy of component I in some arbitrarily chosen standard state. The activity is considered to be an effective concentration. (For solutions, a scale of molality is frequently chosen. *Molality* is the gram-molecular weight of a solute in 1,000 g of solvent; this definition is chosen since the molality of a solution would be the same at all temperatures; concentrations on a volume basis vary with changes in temperature.) The standard state is chosen so that the ratio of the activity and the molality is equal to unity at infinite dilution. The ratio of activity to molality at any concentration is called the *activity coefficient*, and its magnitude gives a measure of the deviation of solutions from ideal behavior. An ideal solution obeys *Raoult's law:*

$$p_I = p_I{}^0 N_I \qquad (9.69)$$

where p_I is the partial pressure of component I from a solution of mole fraction N_I and p_I^0 is the vapor pressure of pure component I. When the solution is not ideal, as is usually true, Eq. (9.69) becomes

$$p_I = p_I^0 N_I f_I \tag{9.70}$$

where f_I is the activity coefficient. Thus Eq. (9.68) may be written

$$\bar{G}_I = \bar{G}_I^0 = RT \ln N_I + RT \ln f_I \tag{9.71}$$

where the first term on the right is for an ideal solution and the second term gives the magnitude of the deviations from ideality. Thus, if the solution were ideal, its changes in partial molal Gibbs energy would be directly proportional to its mole fraction. Similar relations may be given for solid solutions, partially miscible liquids, gaseous mixtures, etc. The activity of solvents in very dilute solutions and of pure solids taking part in chemical reactions is taken conventionally as unity since their concentrations undergo no change.

For nonideal solutions, the activity coefficient has to be evaluated from the defining equation

$$f = a/N \tag{9.72}$$

That is, $\bar{G}_I - \bar{G}_I^0$ may be evaluated by direct or indirect experimental means. Thus the activity is no more simply determinable than the partial molal Gibbs energy; often it is, however, somewhat more convenient to deal with, especially where it may be treated as equal or nearly equal to a concentration or partial pressure. Concentration units other than mole fraction may also be used. In very dilute solutions the concentrations or molalities are proportional to mole fraction so that the activity may also be defined in terms of concentrations or molalities, the standard state then being different in each case.

It is impossible to measure the activity a_+ or a_-, or the activity coefficient f_+ or f_-, of a single kind of ions in solution. They can be approximated in some cases [see (9.87)]. The relation of a_+ and a_- to the measurable activity a of the electrolyte is given by the first part of Eq. (9.73), and the definition of the mean ionic activity $a_\pm$ is implied in the second part:

$$a = a_+^{\nu_+} a_-^{\nu_-} = a_\pm^{(\nu_+ + \nu_-)} \tag{9.73}$$

Here ν_+ and ν_- have the significance stated in Sec. 7. Entirely analogous equations apply to the activity coefficients f, f_+, f_-, and $f_\pm$. Combining Eqs. (9.68) and (9.73) gives for electrolytic solutions

$$\bar{G}_I - \bar{G}_I^0 = \nu RT \ln a_\pm \tag{9.74}$$

where $\nu = \nu_+ + \nu_-$.

The Gibbs-Duhem equation gives the following relation between the activity of one component and that of the other (solvent and solute) for a two-component solution:

$$\int d \ln a_2 = - \int \frac{N_1}{N_2} d \ln a_1 \tag{9.75}$$

where a_2 is the activity of the solute, a_1 is the activity of the solvent, and N_1 and N_2 are the respective mole fractions. Thus activities and activity coefficients of solutes can be measured either directly by determinations of their partial molal Gibbs energies or indirectly by determination of the partial molal Gibbs energy of the solvent.

For a chemical reaction represented by

$$aA + bB \rightarrow cC + dD \tag{9.76}$$

the Gibbs energies of each reactant and product are represented by the expression in Eq. (9.68) or (9.74) if they are electrolytes. On substitution of (9.74) for A, B, C, and D in (9.76), using proper subscripts, the change in Gibbs energy for the over-all reaction is

$$(\bar{G}_C + \bar{G}_D - \bar{G}_A - \bar{G}_B) - (\bar{G}_C^0 + \bar{G}_D^0 - \bar{G}_A^0 - \bar{G}_B^0) = RT \ln \frac{(a_C)^c (a_D)^d}{(a_A)^a (a_B)^b} \tag{9.77}$$

which may be written

$$\Delta G = \Delta G^0 + RT \ln \frac{(a_C)^c (a_D)^d}{(a_A)^a (a_B)^b} \tag{9.78}$$

and $\Delta G = \Delta G^0$ when the activities of all the constituents in the reaction are unity. At equilibrium, $\Delta G = 0$, so that the left member of Eq. (9.78) becomes 0, giving

$$\Delta G^0 = -RT \ln K \tag{9.79}$$

where K is the equilibrium constant of the reaction.

Equation (9.79) applies to all electrochemical reactions. For a reversible constant-temperature-and-pressure process, the work done by the system, not counting expansion work, is equal to $-\Delta G$ for the change in state. In a galvanic cell operating reversibly at constant temperature and pressure, the change in Gibbs energy ΔG is, therefore,

$$\Delta G = -nE\mathfrak{F} \tag{9.80}$$

where n is the number of faradays involved, $\mathfrak{F}$ is the faraday constant, and E is the electromotive force. Substitution of (9.80) in (9.78) gives

$$E = E^0 - \frac{RT}{n\mathfrak{F}} \ln \frac{(a_C)^c (a_D)^d}{(a_A)^a (a_B)^b} \tag{9.81}$$

which expresses the variation of the emf of an electrochemical cell with the activity of the reactants and products of the cell where E^0 represents the standard potential of the cell. If concentrations are substituted for the activities, the Nernst equation results.

Activity coefficients may be determined experimentally by means of (1) freezing-point depression, (2) boiling-point elevation, (3) vapor-pressure lowering, (4) isotonic or vapor-pressure equilibration, (5) emf of galvanic cells without liquid junction, (6) transference numbers and emf of galvanic cells with liquid junction, and (7) solubility. The first four involve measurements of the escaping tendency of the solvent; the last three give a direct measure of solute activity. Activity coefficients of electrolytes in dilute solutions are less than unity. As the concentration of electrolytic solutions increases, the activity coefficients, in general, pass through a minimum, then increase, and for some electrolytes exceed a value of unity. They therefore do not represent a degree of dissociation or ionization but give a measure of the deviation of ions from ideal behavior, regardless of the cause of these deviations.

Interionic attraction also plays a dominant role in the colligative properties of electrolytic solutions. Equation (9.78) may be rearranged to give

$$\Delta G = \Delta G^0 + RT \ln \frac{(c_C)^c(c_D)^d}{(c_A)^a(c_B)^b} + RT \ln \frac{(f_C)^c(f_D)^d}{(f_A)^a(f_B)^b} \qquad (9.82)$$

where the first two terms on the right are the change in Gibbs energy if the components obey the ideal solution laws and the third term represents a correction for deviations from ideality, or

$$\Delta G \text{ (nonideal, electrical)} = RT \ln \frac{(f_C)^c(f_D)^d}{(f_A)^a(f_B)^b}$$

To illustrate how interionic attraction affects this Gibbs-energy change, consider the hypothetical cell (see Sec. 16)

$$\text{Pt, } H_2(g, \text{ 1 atm})|HCl(c_0)|AgCl, Ag, AgCl|HCl(c_1)|H_2(g, \text{ 1 atm}) \text{ Pt} \qquad (9.83)$$

where the vertical lines indicate the interface between electrodes and electrolytes. The emf of this hypothetical cell gives the Gibbs-energy change for the transfer of 1 mole of HCl from the more concentrated solution to one which approaches infinite dilution. The potential associated with the ion in the more concentrated solution is greater than that associated with the ion in the solution which approaches infinite dilution. These potentials as given by Eq. (9.47) may be written as

$$\psi = \frac{\epsilon}{Da_i} - \frac{\epsilon\kappa}{D(1 + \kappa a_i)} \qquad (9.84)$$

where the first term on the right is the portion of the potential due to the ion itself and is, therefore, the same in both the concentrated and the infinitely dilute solution: it may thus be neglected in determining the Gibbs-energy change associated with the charge-discharge process. The second term on the right may be integrated from zero to infinity to give the Gibbs-energy change associated with the charging of the ion, for example, the H^+ ion, which already has the excess potential ψ, thus:

$$\Delta G \text{ (nonideal)} = RT \ln f_{H^+} = \int_0^\infty \frac{\epsilon\kappa}{D(1 + \kappa a_i)} d\epsilon \qquad (9.85)$$

or

$$\Delta G \text{ (nonideal)} = -\frac{\epsilon^2\kappa}{2D(1 + \kappa a_i)} \qquad (9.86)$$

Using the expression for κ given in Eq. (9.38),

$$-\ln f = \frac{\epsilon^3 \sqrt{2\pi n}}{(DkT)^{3/2}(1 + \sqrt{8\pi n\epsilon a_i^2/DkT})} \qquad (9.87)$$

which may also be written as

$$-\log f = \frac{(1.8123 \times 10^6) \sqrt{c}}{(DT)^{3/2}[1 + 50.288 \times 10^8 a_i \sqrt{c}/(DT)^{1/2}]} \qquad (9.88)$$

since n, the number of ions per cubic centimeter, is equal to $cN/1{,}000$, where N is the Avogadro number. Equation (9.88) may be put in the form

$$-\log f = \frac{A\sqrt{c}}{1 + \beta a_i \sqrt{c}} \qquad (9.89)$$

or, for very dilute solutions,

$$-\log f = A\sqrt{c} \qquad (9.90)$$

According to this equation, the values of activity coefficients decrease with increases in concentration. Equation (9.89) holds quite well for solutions up to about 0.1 or 0.2 molar but fails for more concentrated solutions where the assumptions used in the derivation are no longer valid. For more concentrated solutions an additional term $-B'c$, to the right of Eq. (9.89), suffices in most cases to reproduce experimental values. B' is an empirical constant introduced by Hückel [17] to represent possible changes in the dielectric constant of the solution as a function of solute concentration [in the derivation of Eq. (9.88) the dielectric constant of the pure solvent was used]. For aqueous solutions at 25°C the constant A of Eqs. (9.89) and (9.90) is approximately 0.50.

Values found for a_i, the "closest distance of approach" of the ions, are listed in Table 9.8 for a series of simple salts and are compared with the sum of the crystallographic radii s of the salts [18]. For most salts, values of a_i exceed the sum of the crystallographic radii. This is explained on the basis that the ions are hydrated in aqueous solution. Cesium salts represent an exception and probably contain ion pairs. Bjerrum considered that ions of opposite sign could associate in certain solutions to form ion pairs. According to the Boltzmann distribution law, the number dn_i ions in a spherical shell of radius r and thickness dr surrounding a central ion of opposite sign is given by

$$dn_i = n_i 4\pi r^2 e^{-\epsilon\psi/kT} dV = n_i 4\pi r^2 e^{-W/kT} dV \qquad (9.91)$$

where W represents the work required to separate one of the i ions from the central ion. If the ions are considered to be point charges and Coulomb's

TABLE 9.8. VALUES* OF THE CLOSEST DISTANCE OF APPROACH OF IONS IN A SERIES OF SIMPLE SALTS IN AQUEOUS SOLUTION AS COMPARED WITH THE SUM OF THE CRYSTALLOGRAPHIC RADII OF THE SALTS IN THE CRYSTALLINE STATE [18]

Salt	a_i	s	Salt	a_i	s
LiI	5.05	2.77	KCl	3.8	3.14
LiBr	4.3	2.56	RbCl	3.6	3.29
LiCl	4.25	2.41	RbBr	3.55	3.43
NaI	4.2	3.13	RbI	3.5	3.35
NaBr	4.1	2.91	CsCl	3.0	3.46
NaCl	4.0	2.76	CsBr	2.93	3.61
KI	3.94	3.50	CsI	2.87	3.82
KBr	3.84	3.28			

* In angstrom units.

law is assumed to apply for small interionic distances, the work required to separate the ions from the distance r to infinity is

$$W = \frac{|z_+ z_-|\epsilon^2}{Dr} \tag{9.92}$$

where z_+ and z_- are the valences of the positive and negative ions, respectively. Combining Eqs. (9.91) and (9.92) gives

$$\frac{dn_i}{dr} = n_i 4\pi r^2 \exp\left(\frac{-|z_+ z_-|\epsilon^2}{DrkT}\right) = p(r) \tag{9.93}$$

where dn_i/dr gives a measure of the probability p of finding an i ion at a distance r from a central ion of opposite sign. By differentiating (9.93) with respect to r and equating to zero, the minimum value for r for finding two ions of opposite sign not associated is obtained:

$$r_{\min} = \frac{|z_+ z_-|\epsilon^2}{2DkT} \tag{9.94}$$

At 25°C for aqueous solutions, $r_{\min}$ for a uni-univalent electrolyte is 3.57 A. Bjerrum made the assumption then that all ions within a sphere of radius 3.57 A were associated into ion pairs and all those lying beyond this sphere would be "free" in the sense of Arrhenius. Thus, all salts listed in Table 9.8 would be completely dissociated into ions except the cesium salts, and possibly RbBr and RbI.

Equation (9.94) shows that ion association would be more marked, the higher the valence of the ions or the lower the dielectric constant of the solvent. Therefore, formation of ion pairs would be more likely for highly charged ions, bivalent for example, than for uni-univalent electrolytes. By integration of Eq. (9.93) between a_i and $r_{\min}$, the fraction of ions θ associated into ion pairs is

$$\theta = n_i 4\pi \int_{a_i}^{r_{\min}} r^2 \exp\left(\frac{-|z_+ z_-|\epsilon^2}{DrkT}\right) \tag{9.95}$$

for use in the equilibrium expression $K = (1 - \theta)^2 c/\theta$, where K is the association constant for the reaction $A^+ + B^- \rightarrow {}^+AB^-$; ${}^+AB^-$ represents an ion pair formed from the ions A^+ and B^-.

16. Galvanic Cells and Reversible Electrode Reactions

When a cell operates in a thermodynamically reversible manner, the measured electromotive force of the cell corresponds to that represented by Eq. (9.81). Some cells cannot be made to approximate reversible behavior under any practical conditions. For others, the difficulty of achieving a desired approach to reversibility varies greatly among different systems. It is primarily a function of the reversibility of the process at the two electrodes of the cell. (A *reversible* electrode is composed of an electronic conductor and the constituents of the surrounding electrolytic solution that take part in the electrode reaction; all reversible electrodes contain an element in two states of oxidation.) Most reversible electrodes can be classified as of the "first," "second," "third," or "fourth" kind. (Those of the fourth kind are sometimes called *oxidation-reduction* electrodes; this is a misnomer since all types of reversible electrodes involve oxidation-reduction.) Electrodes of the first kind consist of an element in a solution of its ions, for example, copper in copper sulfate. The second kind consists of a metal, a sparingly soluble salt of the metal, and a solution of a soluble electrolyte with the same negative ion as the insoluble salt, for example, silver–silver chloride in a solution of hydrochloric acid. The third kind consists of a metal, one of its sparingly soluble salts, a somewhat less insoluble salt with the same negative ion, and a solution of a soluble salt with the same positive ion as the latter salt, for example, lead–lead oxalate–calcium oxalate in a solution of calcium chloride. Electrodes of the fourth kind consist of two compounds of an element in two different states of oxidation and, if neither is an electronic conductor, an inert metal in a solution containing the compounds or in a solution whose constituents involve the compounds in electrochemical reaction, for example, platinum in an aqueous solution of ferrous and ferric chlorides.

Since metallic-electrolytic interfaces always occur in pairs in any cell, the absolute potential of single electrodes cannot be measured [1, 2]. However, whole-cell electromotive forces may be considered as composed of the sum of two potentials, each characteristic of one of the electrodes. Because these electrodes' (half-cell) potentials cannot be measured individually, convention ascribes zero potential to the reversible *hydrogen electrode* when the hydrogen gas and the hydrogen ions are at unit activity and for which the reaction is H_2 (gas) $= 2H^+$ (solution) $+ 2\epsilon$. This convention is adopted for all temperatures.

A *hydrogen electrode* consists of a noble metal, usually platinum, of large surface area covered with hydrogen gas in a solution of hydrogen ion saturated with hydrogen gas. A large surface area is obtained by plating the surface of the metal with platinum sponge, using a solution of chloroplatinic acid. As a rule, metal foil is used and is welded to a wire of the same metal sealed in the bottom of a glass tube partially filled with mercury and through which electrical contact is made to an external circuit usually by a copper wire. The potential of the hydrogen electrode varies logarithmically with hydrogen-ion activity and hydrogen-gas pressure according to

$$E_{\mathrm{H}} = -\frac{RT}{\mathfrak{F}} \ln a_{\mathrm{H}^+} + \frac{RT}{2\mathfrak{F}} \ln p_{\mathrm{H}_2} \tag{9.96}$$

where R, T, and $\mathfrak{F}$ have the significance given above. At 25°C this relation is

$$E_{\mathrm{H}} = (-0.059157 \log a_{\mathrm{H}^+} + 0.029579 \log p_{\mathrm{H}_2}) \text{ volts} \tag{9.97}$$

and thus for a 10-fold change in a_{H^+} at constant p_{H_2} the potential of the hydrogen electrode changes by 0.059157 volt. At constant a_{H^+} a 10-fold change in p_{H_2} produces a change of 0.029579 volt in E_{H}. In determining the pressure of the hydrogen gas, the vapor pressure p of the solution should be subtracted from the observed barometric pressure P. The observed potential of the electrode is corrected to

1 atm hydrogen pressure by the relation

$$E_{1\,\text{atm}} = E_{\text{obs}} + \frac{RT}{2\mathfrak{F}} \ln \frac{760}{P - p} \qquad (9.98)$$

As the hydrogen electrode can be directly or indirectly combined with any other reversible electrode, a series of standard electrode or half-cell potentials can be set up on the basis of the hydrogen-electrode convention. Table 9.9 lists a number of electrode potentials for aqueous solutions at 25°C; in each case the potential is for the metal ion at unit activity. These ionic activity other than unit activity may be obtained from the standard potentials given in Table 9.9 through Eq. (9.81), the activity quotient being written without reference to the ϵ that appears in the half-cell reaction equations. These potentials cannot be used indiscriminately for other pure solvents or for molten salts. For nonaqueous solvents, ethyl alcohol, for example, the potentials would differ somewhat from those given in Table 9.9, owing to a difference in standard state. For molten salts, an electromotive series of the elements is based on the free energy of formation of the electrolytic phase, these phases being of like type, such as chlorides, bromides, oxides, etc. [21]. Table 9.10 gives some values for molten chlorides at 800°C. These values correspond to the decomposition voltages of the respective chlorides. If the chlorine electrode, in each case, were assumed to have a potential of zero the potential for each metal, by convention, would be equal to the emf of the whole cell.

TABLE 9.9. STANDARD ELECTRODE (OXIDATION) POTENTIALS OF SOME ELECTRODES IN AQUEOUS SOLUTION AT 25°C [19]

Electrode reaction	Standard potential, volts	Electrode reaction	Standard potential, volts
$Li = Li^+ + \epsilon$	3.045	$Ag + I^- = AgI + \epsilon$	0.151
$K = K^+ + \epsilon$	2.925	$Sn = Sn^{++} + 2\epsilon$	0.136
$Na = Na^+ + \epsilon$	2.714	$Pb = Pb^{++} + 2\epsilon$	0.036
$Mg = Mg^{++} + 2\epsilon$	2.37	$H_2 = 2H^+ + 2\epsilon$	0.000
$Al = Al^{3+} + 3\epsilon$	1.66	$Cu = Cu^{++} + 2\epsilon$	−0.337
$Zn = Zn^{++} + 2\epsilon$	0.763	$2OH^- = H_2O + \frac{1}{2}O_2 + \epsilon$	−0.401
$Fe = Fe^{++} + 2\epsilon$	0.440	$2Hg = Hg_2^{++} + 2\epsilon$	−0.789
$Cd = Cd^{++} + 2\epsilon$	0.403	$Ag = Ag^+ + \epsilon$	−0.799
$Ni = Ni^{++} + 2\epsilon$	0.250	$2Cl^- = Cl_2 + 2\epsilon$	−1.360

TABLE 9.10. STANDARD ELECTRODE POTENTIALS OF SOME ELECTRODES IN MOLTEN CHLORIDES AT 800°C [21]

Cell reaction	Standard emf, volts	Cell reaction	Standard emf, volts
$K + \frac{1}{2}Cl_2 = KCl$	3.441	$Fe + Cl_2 = FeCl_2$	1.118
$Cs + \frac{1}{2}Cl_2 = CsCl$	3.362	$Pb + Cl_2 = PbCl_2$	1.112
$Mg + Cl_2 = MgCl_2$	2.460	$Co + Cl_2 = CoCl_2$	0.977
$Th + 2Cl_2 = ThCl_4$	2.264	$Cu + \frac{1}{2}Cl_2 = CuCl$	0.970
$Mn + Cl_2 = MnCl_2$	1.807	$Ag + \frac{1}{2}Cl_2 = AgCl$	0.826
$Tl + \frac{1}{2}Cl_2 = TlCl$	1.473	$Pd + Cl_2 = PdCl_2$	0.331
$Cd + Cl_2 = CdCl_2$	1.193	$Rh + Cl_2 = RhCl_2$	0.142

potentials also correspond to oxidation potentials since the reaction as written, if given a positive sign in potential, will occur spontaneously with a decrease in Gibbs energy; the magnitude of the Gibbs energy decreases and changes sign at hydrogen as one reads down the table. In 1953 the International Union of Pure and Applied Chemistry, in what is now known as the "Stockholm convention" [20], issued a statement in support of the opposite sign to that given in Table 9.9; by this convention the electrode potentials are listed in a decreasing order of reducing power. The positive-sign convention is retained here to be consistent with the reference [19] cited and with the first edition of this Handbook.

Some of the electrodes referred to in Table 9.9 and similar tables are not readily reversible (for example, the Ni, Ni^{++} electrode). In such cases, the values listed have been calculated from Gibbs energies through Eq. (9.80). An electrode potential at an

The data of Table 9.10 are for theoretical cells, i.e., for ones in which there is no interaction between the metals, the chlorine, or the chlorides; no solubilities of the metals in their chlorides or vice versa; or any other factors that would lead to deviations of the metals, the chlorides, or the chlorine from their standard states. Delimarskii and Markov [22] have reviewed the reactions that may occur between metals and molten (or fused) salts.

Galvanic cells made with reversible electrodes are of two general types: without and with liquid junction.

In the first, the two electrodes are dipped into the same solution. An example would be a cell composed of hydrogen and chlorine electrodes in a solution of hydrochloric acid; the hydrogen electrode is reversible to the hydrogen ion, the chlorine electrode to the chloride ion. The most common examples of cells with liquid junction are those used in pH determination (pH represents the negative of the common logarithm of the hydrogen-ion activity).‡ These cells may be represented by

$$\text{Pt, H}_2\ (g,\ 1\ \text{atm})|\text{solution}||\text{KCl}\cdot(\text{sat.})|\text{Hg}_2\text{Cl}_2,\ \text{Hg} \qquad (9.99)$$

where the electrode compartment [KCl (sat.)|Hg_2Cl_2, Hg] is commonly known as the *saturated calomel half cell.* Normal and decinormal calomel half cells made, respectively, with normal and 0.1 normal KCl are also used. The Hg_2Cl_2 and Hg are isolated from the solution under study by the solution of KCl. This introduces an unknown but small error in potential owing to the liquid-junction potential produced at the interface of the two solutions represented by the double vertical lines. The pH of the solution under study is given by

$$\text{pH} = \frac{E - E'_{\text{cal}} - E_j}{RT/\mathfrak{F}} = \frac{E - E_{\text{cal}}}{RT/\mathfrak{F}} \qquad (9.100)$$

where E_{cal} includes average values of E_j. Potentials of the saturated, normal, and decinormal calomel electrodes not including E_j are, respectively, 0.2415, 0.2800, and 0.3337 volt at 25°C on the hydrogen scale [24].

Inclusion of E_j raises these potentials by 2 to 5 mV, depending on whether the calomel half cell is used with a buffer or strong acid. The pH of a solution may also be determined by cells without liquid junction of the type

$$\text{Pt, H}_2\ (g,\ 1\ \text{atm})|\text{solution, KCl (m)}|\text{AgCl, Ag} \qquad (9.101)$$

and thus be free of the uncertainties arising from a liquid-junction potential. Values for solutions containing no KCl are obtained by extrapolating values of pH for solutions of various m's to $m = 0$.

A hydrogen electrode cannot be used in oxidizing or reducing solutions nor in ones that might poison its surface. In practice, therefore, it is usually replaced by a *glass electrode* which functions as a hydrogen electrode. A glass electrode consists of a glass tube having a special thin glass membrane (flat surface, bulb, spiral, or pinpoint) at one end and containing a sealed-in inner electrode immersed in a suitable solution contained within the glass tube. Connection to the external circuit is made through the inner electrode. Soda-lime glasses give good performance, and Corning 015 glass having a composition of 72 per cent SiO_2, 22 per cent Na_2O, and 6 per cent CaO has been frequently used. Today, glasses of special composition are also made for optimum behavior over various temperature ranges. Various kinds of inner electrodes are used, including platinum or gold wire dipped in a buffer solution containing quinhydrone, or a silver chloride or calomel electrode in a buffered chloride solution. The potential of the inner and outer surfaces of the glass membrane may differ, giving rise to an asymmetry potential, the magnitude of which may be determined by placing the electrode in a solution of identical composition to the inner solution. In practice, however, measurements of the asymmetry potential are not made, and the glass electrode (or pH meter) is calibrated with a solution of known pH. For this purpose a 0.05-molal solution of acid potassium phthalate is generally used. Such a solution has the pH values [25] listed in Table 9.11, obtained from measurements of cells without liquid junction of the type

‡ Since the activity of an individual ion is thermodynamically inaccessible and approximate values rest upon extra thermodynamic considerations, pH values are sometimes regarded merely as numbers expressing a convenient scale of acidities and alkalinities, although attempts are made to place them as closely as feasible to thermodynamic requirements.

TABLE 9.11. pH VALUES OF STANDARD BUFFER SOLUTIONS*

Temperature, °C	KH phthalate (0.05 m)	KH tartrate (sat. 25°C)	KH_2PO_4 (0.025 m), Na_2HPO_4 (0.025 m)	KH_2PO_4 (0.008695 m), Na_2HPO_4 (0.03043 m)	Borax (0.01 m)
0	4.006	.....	6.984	7.534	9.464
5	3.999	.....	6.951	7.500	9.395
10	3.997	.....	6.923	7.472	9.332
15	3.997	.....	6.900	7.448	9.276
20	4.000	.....	6.881	7.429	9.225
25	4.008	3.557	6.865	7.413	9.180
30	4.016	3.552	6.853	7.400	9.139
35	4.024	3.549	6.844	7.389	9.102
38	4.030	3.548	6.840	7.384	9.081
40	4.035	3.547	6.838	7.380	9.068
45	4.047	3.547	6.834	7.373	9.038
50	4.063	3.549	5.833	7.367	9.011
55	4.079	3.554	6.834	.....	8.985
60	4.094	3.560	6.836	.....	8.962

* Original data corrected to conform with new values for the potential of the silver–silver chloride electrode and with the presently accepted values for the physical constants.

(9.101). When used to calibrate a glass-calomel assembly the entire assembly should be at the same temperature. Values of the pH of four additional solutions, recommended as pH standards by the National Bureau of Standards, are also given in Table 9.11 [26].

The glass electrode functions as a hydrogen electrode because its surface potential changes logarithmically with changes in hydrogen-ion activity a_{H^+}, except for very low and very high values of hydrogen-ion activity. For very high pH values (above 8) the glass electrode is subject to positive errors; i.e., it gives too low a value for pH, arising from response to Na^+ ions in the glass as well as to H^+ ions in the solution. For use in solutions containing large amounts of sodium ion, special sodium-free glass electrodes are frequently used. For very low pH values (below 2) the glass electrode is subject to negative errors; i.e., it gives too high a value for pH, arising from anion participation or appreciable transport of hydrated protons across the glass membrane.

For cells of the type

$$\text{Ag, AgCl}|\text{NaCl}(c_1)\|\text{KCl}(c_2)|\text{AgCl, Ag} \qquad (9.102)$$

the potential at the junction of the two solutions may be approximated by the *Henderson equation:*

$$E_j = \frac{RT}{\mathfrak{F}} \frac{c_1(U_1^+ - U_1^-) - c_2(U_2^+ - U_2^-)}{c_1(U_1^+ + U_1^-) - c_2(U_2^+ + U_2^-)} \ln \frac{U_1^+ + U_1^-}{U_2^+ + U_2^-} \qquad (9.103)$$

where the c's are in equivalents per unit volume and the U's are the mobilities of the positive and negative ions in the solutions of concentrations c_1 and c_2, respectively. If $c_1 = c_2$,

$$E_j = \frac{RT}{\mathfrak{F}} \ln \frac{U_1^+ + U_1^-}{U_2^+ + U_2^-} = \frac{RT}{\mathfrak{F}} \ln \frac{\Lambda_1}{\Lambda_2} \qquad (9.104)$$

where the Λ's are the equivalent conductances of the solutions.

Measurements of galvanic cells with and without liquid junction may also be used to measure the transference number of an ion (Sec. 10). Thus, if the emf of cells

$$\text{Hg, Hg}_2\text{SO}_4|\text{H}_2\text{SO}_4(c_1)\|\text{H}_2\text{SO}_4(c_2)|\text{Hg}_2\text{SO}_4, \text{[illegible]}; E_t$$
$$\text{Hg, Hg}_2\text{SO}_4|\text{H}_2\text{SO}_4(c_1)|\text{H}_2(g, 1\text{ atm}), \text{[illegible]}$$
$$\text{Hg, Hg}_2\text{SO}_4|\text{H}_2\text{SO}_4(c_2)|\text{H}_2(g, 1\text{ atm}), \text{Pt}; E_2$$

were measured, the transference number of the hydrogen ion is given by

$$t_{H^+} = \frac{E_t}{E_1 - E_2} \qquad (9.105)$$

if t_{H^+} is constant in value within the range of c_1 to c_2. If not, the transference number is then given by

$$t_{H^+} = \int_{c_1}^{c_2} \frac{dE_t}{d(E_1 - E_2)} \qquad (9.106)$$

Equations (9.105) and (9.106) follow from the basic thermodynamic relationships expressed in Eq. (9.80), by comparing the changes in state per faraday in each of the above three cells.

The most reversible cells are *standard cells* [27]. They are highly reproducible, have low emf-temperature coefficients, and a constant emf over long periods of time. They are of two types, the saturated and unsaturated. The saturated type may be represented by

$$\text{Cd, Hg (2-phase)} \; |3\text{CdSO}_4\cdot 8\text{H}_2\text{O}(s)|\text{CdSO}_4 \text{ (sat. aq. sol.)} \; |3\text{CdSO}_4\cdot 8\text{H}_2\text{O}(s)|\text{Hg}_2\text{SO}_4(s), \text{Hg}(l)$$

The unsaturated cell is free of the crystals of cadmium sulfate and is usually prepared with a solution of cadmium sulfate that is saturated at about 4°C; the solution is then unsaturated at higher temperatures. The unsaturated cell has the lower emf-temperature coefficient (± 4 to ± 10 μV/°C, depending on the age of the cell) and is made portable by placing septa over both electrodes. Its life span is about 15 years; its emf decreases, on an average, at a rate of 20 to 40 μV per year [27]. The saturated type has a higher emf-temperature coefficient of about 40 μV/°C at room temperature, but because it represents a nonvariant system of phases, it can maintain a precisely constant potential for a virtually unlimited time under the small current drains in its normal use. The standard cell is, therefore, used in the maintenance of standards of electromotive force.

17. Galvanic Cells on Charge and Discharge

When a galvanic cell is subjected to a charging or a discharging current of electricity it no longer functions in a reversible fashion and its emf is higher or lower than the reversible, or standard, value. This deviation from the reversible value may be divided into three parts. One part is the IR drop through the main body of the electrolyte. This part is readily dealt with, when desirable, in terms of cell geometry and ionic conductance values. Another part is the potential drop caused by the fact that, as ions are discharged and removed from the solution layer (or as ions are formed and accumulate in the solution layer) at the electrode, they must be replaced (or removed) in part by thermal diffusion which will occur only after concentration gradients have been set up. The third part is *overvoltage* consisting of *overpotential* at each electrode and includes the effect of all parts of the mechanism of the electrode reactions that require activation energy to occur. The magnitude of overpotential depends on the current density, the electrode material, temperature, impurities, and other factors. Overpotential is low in value for the discharge of most metallic ions but is appreciable for the discharge of hydrogen, hydroxyl, halogen, and other nonmetallic ions. It is believed to arise from the slowness of some electrode process such as ionic discharge, ionic diffusion, or the establishment of equilibrium, for example, between hydrogen atoms and hydrogen molecules on an electrode surface. These slow reactions set up a potential barrier over which the ions must pass to be discharged, and the electrodes are said to be polarized. It is considered that they must receive a Gibbs energy of activation in order to pass over this energy barrier or go through an "activated complex" state before they can be discharged.

The rate for the discharge of a univalent metallic ion is given by

$$\overrightarrow{V} = k_1 a_+ e^{\alpha \mathfrak{F} E/RT} \qquad (9.107)$$

where k_1 is the specific rate constant for the ionic discharge, a_+ is the activity of the ion being discharged, E is the reversible electrode potential, and α is a fraction of the potential operative for the ionic discharge. The rate of the reverse process is

$$\overleftarrow{V} = k_2 e^{-(1-\alpha)\mathfrak{F} E/RT} \qquad (9.108)$$

where k_2 is the specific rate constant for the metallic dissolution. When the electrode functions reversibly, i.e., there is no net flow of electricity across the electrode-electrolyte interface, $\overrightarrow{V} = \overleftarrow{V}$ and

$$k_1 a_+ e^{\alpha \mathfrak{F} E/RT} = k_2 e^{-(1-\alpha)\mathfrak{F} E/RT} \qquad (9.109)$$

or

$$e^{-\mathfrak{F} E/RT} = a_+ \frac{k_2}{k_1} \qquad (9.110)$$

or

$$E = \frac{RT}{\mathfrak{F}} \ln \frac{k_1}{k_2} - \frac{RT}{\mathfrak{F}} \ln a_+ \qquad (9.111)$$

or

$$E = E^0 - \frac{RT}{\mathfrak{F}} \ln a_+ \qquad (9.112)$$

where $(RT/\mathfrak{F}) \ln (k_1/k_2) = \text{const}$ (or E^0), which is the Nernst equation [see Eq. (9.81)] for reversible conditions. The cathodic and anodic currents for unit area, or the current densities, are, respectively,

$$\overrightarrow{I} = \mathfrak{F} k_1 a_+ e^{\alpha \mathfrak{F} E^*/RT} = \mathfrak{F} k_1 a_+ e^{\alpha \mathfrak{F}(E+\eta)/RT} \qquad (9.113)$$

$$\overleftarrow{I} = \mathfrak{F} k_2 e^{-(1-\alpha)\mathfrak{F} E^*/RT} = \mathfrak{F} k_2 e^{-(1-\alpha)\mathfrak{F}(E+\eta)/RT} \qquad (9.114)$$

where E^* is now used to indicate that it differs from E, the reversible value, by η, the overpotential. When there is excess current in the forward direction, $\overrightarrow{V} > \overleftarrow{V}$, and the net forward current is

$$I_n = \overrightarrow{I} - \overleftarrow{I} = \mathfrak{F}(k_1 a_+ e^{\alpha \mathfrak{F}(E+\eta)/RT} k_2 e^{-(1-\alpha)\mathfrak{F}(E+\eta)/RT}) \qquad (9.115)$$

At the reversible potential,

$$k_1 a_+ e^{\alpha \mathfrak{F} E/RT} = k_2 e^{-(1-\alpha)\mathfrak{F} E/RT}$$

then

$$I_n = \mathfrak{F} k_1 a_+ e^{\alpha \mathfrak{F} E/RT}(e^{\alpha \mathfrak{F} \eta/RT} - e^{-(1-\alpha)\mathfrak{F}\eta/RT}) \qquad (9.116)$$

Under reversible conditions $\mathfrak{F} k_1 a_+ e^{\alpha \mathfrak{F} E/RT}$ gives the magnitude of the cathodic and hence the anodic current at the reversible potential; this current is known as the *exchange current* and may be designated by the symbol I_0, or

$$I_n = I_0(e^{\alpha \mathfrak{F}\eta/RT} - e^{-(1-\alpha)\mathfrak{F}\eta/RT}) \qquad (9.117)$$

The exponentials can be expanded; if η is small, all terms beyond the first may be neglected and

$$I_n = I_0(\mathfrak{F}\eta/RT) \qquad (9.118)$$

or

$$\eta = \frac{RT}{\mathfrak{F}}\left(\frac{I}{I_0}\right) = aI \qquad (9.119)$$

where a is a constant at a specified termperature and equal to $RT/\mathfrak{F} I_0$. If η is large, at relatively high currents, the second exponential becomes negligible relative to the first and

$$I_n = I_0 e^{\alpha \eta \mathfrak{F}/RT} \qquad (9.120)$$

$$\eta = \frac{RT}{\alpha \mathfrak{F}} \ln \left(\frac{I}{I_0}\right) \qquad (9.121)$$

which is sometimes written in the form

$$\eta = a' + b \log I \qquad (9.122)$$

where $a' = -(2.303RT/\alpha\mathfrak{F}) \log I_0$, $b = 2.303RT/\alpha\mathfrak{F}$, and b is known as the *Tafel slope*. The value of η generally lies between 0.4 and 1 although values as high as 2 have been reported [2, 28, 29]. Equations (9.119) and (9.121) fit some of the experimental data on overpotential. Values of the hydrogen overpotential of various metals are listed in Table 9.12.

TABLE 9.12. HYDROGEN OVERPOTENTIAL OF VARIOUS METALS AT 25°C [2, 29, 30]

Metal	Solution	I_0 amp cm^{-2}	α	Current density, amp cm^{-2}	
				0.01 volts	0.10 volts
Platinized platinum	2 N H_2SO_4	1.8×10^{-3}	2.96	0.035	0.055
Smooth platinum	1 N HCl	1.0×10^{-3}	1.5–2.0	0.03–0.04	0.08–0.09
Gold	1 N HCl	1.0×10^{-5}	0.7	0.34	0.42
Silver	0.1 N HCl	5.0×10^{-7}	0.66	0.39	0.48
Nickel	1 N HCl	4.0×10^{-6}	0.53	0.38	0.49
Copper	0.1 N HCl	2.0×10^{-7}	0.50	0.56	0.67
Iron	1 N HCl	1.0×10^{-6}	0.4	0.59	0.74
Palladium	0.6 N HCl	2.0×10^{-4}	2.0	0.50	0.80
Zinc	2 N H_2SO_4	3.8×10^{-5}	0.19	0.75	1.06
Lead	0.01–8 N HCl	2.0×10^{-13}	0.48	0.94	1.06
Cadmium	1 N HCl	1.0×10^{-7}	0.3	0.99	1.18
Tin	1 N HCl	1.0×10^{-10}	0.45	1.05	1.18
Mercury	1 N HCl	2.0×10^{-12}	0.49	1.17	1.29

Since mercury has such a high hydrogen overpotential, it is frequently used to amalgamate other metals to increase their hydrogen overpotential. For example, the zinc in the common dry cell is amalgamated to increase the storage life of the cell even though the difference in the hydrogen overpotential of zinc and mercury is not large.

The relation between current, cell potentials, and the thermal diffusion of ions can be derived from principles previously discussed. Suppose that ions of one particular kind (i) of charge z are discharged and removed from solution in the electrode reaction. Per faraday, t_i equivalents of these ions will move by transference into the layer of electrolyte on the electrode. One equivalent of these ions will be removed from that layer by the electrode reaction. Therefore, when the steady state has been established, $1 - t_i$ equivalents must come into the layer by thermal diffusion. Let the diffusion layer have a thickness δ; that is, assume that the solution changes from its bulk activity a to the lesser activity a' which exists on the electrode through a distance δ. Then *Fick's law* gives for the rate of diffusion in moles per unit time

$$\frac{dn}{dt} = \frac{AD^*}{\delta}(a - a') \qquad (9.123)$$

where A is the electrode area and D^* the diffusion coefficient.

As the rate (in moles) at which these ions must diffuse is $(1 - t_i)/z$ per faraday, it will be $IA(1 - t_i)/z\mathfrak{F}$ per second, where I is the current density in amperes per square centimeter. Equating this expression to Eq. (9.123) gives, for current density,

$$I = \frac{z\mathfrak{F}D^*}{(1 - t_i)\delta}(a - a') = \frac{z\mathfrak{F}D^*}{t\delta}(a - a') \qquad (9.124)$$

where t is the sum of all transference numbers except t_i.

The potential at the electrode with no electricity flowing will be

$$E = E^0 - \frac{RT}{z\mathfrak{F}}\ln a$$

With current I, it will be

$$E' = E^0 - \frac{RT}{z\mathfrak{F}}\ln a'$$

so that the concentration polarization $\Delta E(=E - E')$ will be

$$\Delta E = \frac{RT}{z\mathfrak{F}}\ln\frac{a'}{a} \qquad (9.125)$$

which, from Eq. (9.124), becomes

$$\Delta E = \frac{RT}{z\mathfrak{F}}\ln\left[1 - \left(\frac{t\delta}{aD^*z\mathfrak{F}}I\right)\right] \qquad (9.126)$$

This equation indicates that, when the current density becomes equal to $D^*z\mathfrak{F}a/t\delta$, ΔE would become infinite. Actually, before this happens, some electrode reaction other than the reduction of i ions would take place. However, as the current density approaches this value, a rapid rise in the ΔE occurs, and the *limiting current density* I_d, for the deposition of an ion, is defined by

$$I_d = \frac{D^*z\mathfrak{F}}{t\delta}a \qquad (9.127)$$

If the ion i is of negligible concentration compared with the total ionic concentration, $t = 1$, and the current density under these conditions, I'_d, is sometimes called the *diffusion current:*

$$I'_d = \frac{D^*z\mathfrak{F}}{\delta}a \qquad (9.128)$$

which, by comparison with Eq. (9.22), becomes

$$I'_d = \frac{zRT\lambda_i}{\delta\mathfrak{F}}a \qquad (9.129)$$

For a solution containing a number of dischargeable ions, each one then would have its own characteristic limiting diffusion current. This fact is utilized in polarographic analysis of trace constituents. Equation (9.128) shows I'_d is proportional to the ion activity or concentration; therefore, the magnitude of the limiting current density gives a measure of the amount of ions present in solution. For most ions, λ_i is about 60 ohm^{-1} cm^2, and it has been found that the thickness of the diffusion layer is of the order of 0.05 cm. Therefore, Eq. (9.129) becomes

$$I'_d \approx 0.03cz \qquad (9.130)$$

The foregoing has dealt with some aspects of irreversible discharge of ions. There are also irreversible processes that occur when metals or other substances go into solution as ions during electrolysis. One of these is the phenomenon of passivity. An example of this is the resistance of aluminum toward electrolytic oxidation in many electrolytes. A potential difference of many volts may be applied between an aluminum conductor and a phosphate solution without causing any appreciable dissolving of the aluminum. The electrolytic rectifier and electrolytic capacitor operate on the basis of this fact. The effect is attributed to the formation of an adherent insoluble, nonconducting film on the metal, but the phenomenon of passivity is not completely understood.

In almost any cell there is more than one conceivably possible electrode process. When polarization effects are absent, the electrode process that will occur is the one requiring the least energy. Thus, in an electrolysis of a solution containing zinc and copper chlorides both at unit activity, copper will deposit before zinc because less energy is required for the discharge of the copper ions (Table 9.9). When the concentration of the copper ions is reduced, as a result of deposition, until the potentials of the copper and zinc electrodes are equal, then zinc and copper will be codeposited. In most cases where polarizations come into play, their effects will be explained by the considerations discussed above. For example, less energy is required for the discharge of hydrogen ions from neutral solutions than for zinc ions at any possible concentrations. However, on electrolyzing an aqueous solution of a zinc salt between platinum electrodes, zinc and not hydrogen gas is the chief product at the negative electrode.

This results because the hydrogen overpotential on zinc is large.

18. Batteries

Cells which are constructed and used for the purpose of providing electrical energy occupy an important place in electrochemistry. There are many considerations that apply to power cells and batteries of them, which have not been dealt with in the foregoing sections or which have been discussed only implicitly.

The ideal battery would embody the following desirable characteristics, most of which are interrelated in such a way that, to maximize one, others must be compromised:

1. It requires inexpensive materials and construction.
2. It should be compact, have high energy on a volume, weight, and cell basis.
3. It should have low internal resistance—the capability of high discharge rates.
4. It should be physically rugged, portable, and nonspilling.
5. It should have low self-discharge rate.
6. It should have a large range of effective operating temperatures.
7. It should not include hazards such as the production of explosive or noxious gas or of being itself especially combustible or explosive.
8. It should have constant potential and resistance during discharge.

For rechargeable (secondary) batteries:

9. It should be efficient in terms of the ratio of discharge energy to charging energy.
10. It should be capable of many cycles of operation.
11. It should be unharmed by high rates of charge or discharge, or by overcharge, or by overdischarge, or by long periods of standing in a charged or in an uncharged or discharged condition.
12. It should require the minimum of attention, such as electrolyte replacement.

Only a few of the implications of these requirements can be discussed here, and these only briefly, but they are very restrictive, a fact that is borne out by the paucity of types of power cells that have been developed commercially. For low self-discharge rates, diffusion of oxidizing and reducing substances between the electrodes must be slight. This condition cannot be achieved by wide separation of electrodes without making batteries bulky. It cannot be achieved by increasing the viscosity of the electrolyte without increasing the internal resistance of the cells. Barriers or diaphragms between electrodes also increase internal resistance. If it is achieved by using nearly insoluble electrode materials, these must be electronic conductors themselves or must be finely divided, porous, and in intimate contact with the electronic conductors to avoid serious polarization effects. Metals as reducing agents meet these restrictions well, but there are relatively very few known oxidizing agents, for example, PbO_2, which have conductivity at all comparable with that of metals. To make a grid to provide an electronic conductor in intimate contact with a nonconducting insoluble oxidizing material, one would have to choose from the list of noble metals such as gold and platinum or from a small list of materials such as nickel, silver, or steel in alkali or lead antimony or other lead alloys in sulfuric acid, which because of polarization or passivity are not rapidly oxidized.

There are only a few good ionizing solvents which are liquid over usual operating ranges of temperature and pressure. Except for a few very special applications, no nonaqueous electrolytes have been used in practical power cells. But this fact itself constitutes a serious problem in cells for low-temperature operation.

As hydrogen and hydroxyl ions have far greater conductivity in aqueous solutions than other ions, the electrolyte conductance criterion is somewhat compromised if electrolytes which are not strongly basic or strongly acidic are used.

If the cell reaction produces a metallic conductor, the conductor must not "tree" between electrodes and cause a short circuit. This is an important problem in batteries involving metallic silver because of its tendency to be reduced in the form of long needles or filaments. However, this same tendency is responsible for the very low internal resistance of such batteries and their capability of extremely high discharge rates.

If the electrodes of a secondary battery comprise insoluble powders held in grids, there is the problem of keeping this material in place as it undergoes chemical changes and changes in density during cycling and also as it is acted upon by the mechanical forces accompanying any gas formation during operation.

To get high potentials per cell, one would wish to use electrodes with reducing potentials greater than that of hydrogen. If this is to be possible, the electrode must have a sufficiently high hydrogen overvoltage so that it will not be self-discharged by replacement of hydrogen from the water in the electrolyte. The satisfactory performances of zinc in the Leclanché cell and of lead in the lead-acid storage battery depend on this overpotential.

Only a few types of batteries are in any considerable use. There are now five types of secondary cells. They are commonly known as the lead-acid, the nickel-iron (Edison), the nickel-cadmium (Jungner), the zinc–silver oxide, and the cadmium–silver oxide cell. These five secondary cells may be represented as follows:

Lead acid:

$$Pb(s),\ PbSO_4(s)|H_2SO_4(35\%\ aq.)|PbSO_4(s),\ PbO_2(s)$$

Nickel-iron:

$$Fe(s),\ FeO(s)|KOH(21\%\ aq.)|Ni(OH)_4(s),\ Ni(OH)_2(s),\ Ni(s)$$

Nickel-cadmium:

$$Cd(s),\ CdO(s)|KOH(20\%\ aq.)|Ni(OH)_4(s),\ Ni(OH)_2(s),\ Ni(s)$$

Zinc–silver oxide:

$$Zn(s)|ZnO(sat.\ aq.),\ KOH(40\%\ aq.)|AgO(s),\ Ag(s)$$

Cadmium–silver oxide:

$$Cd(s)|KOH(40\%\ aq.)|AgO(s),\ Ag(s)$$

where s denotes solid; aq., aqueous solution; and sat., saturated solution; and the vertical lines represent the interface between electrodes (or plates) and electrolyte. As written, the electrode on the left has a negative potential with respect to the one on the right in each of these cells.

The over-all chemical reactions of these cells are considered to be as follows:

Lead-acid:

$$Pb(s) + PbO_2(s) + 2H_2SO_4(35\% \text{ aq.}) \underset{\text{charge}}{\overset{\text{discharge}}{\rightleftharpoons}} 2PbSO_4(s) + 2H_2O(l) \quad (9.131)$$

Nickel-iron:

$$Fe(s) + Ni(OH)_4 \underset{\text{charge}}{\overset{\text{discharge}}{\rightleftharpoons}} FeO(s) + Ni(OH)_2(s) + H_2O(l) \quad (9.132)$$

Nickel-cadmium:

$$Cd(s) + Ni(OH)_4 \underset{\text{charge}}{\overset{\text{discharge}}{\rightleftharpoons}} CdO(s) + Ni(OH)_2(s) + H_2O(l) \quad (9.133)$$

Zinc–silver oxide:

$$Zn(s) + AgO(s) + 2KOH(40\% \text{ aq.}) \underset{\text{charge}}{\overset{\text{discharge}}{\rightleftharpoons}} Ag(s) + K_2ZnO_2(\text{aq.}) + H_2O(l) \quad (9.134)$$

Cadmium–silver oxide:

$$Cd(s) + AgO(s) + H_2O(l) \underset{\text{charge}}{\overset{\text{discharge}}{\rightleftharpoons}} Ag(s) + Cd(OH)_2(s) \quad (9.135)$$

where l represents liquid.

The dependence of the thermodynamic emfs or open-circuit voltages (OCV) of those power cells on their activities is as follows:

Lead-acid:

$$E = E^0 - \frac{RT}{\mathfrak{F}} \ln a_{H_2O} + \frac{RT}{\mathfrak{F}} \ln a_{H^+}^2 a_{SO_4^-} \quad (9.136)$$

Nickel-iron:

$$E = E^0 - \frac{RT}{2\mathfrak{F}} \ln a_{H_2O} \quad (9.137)$$

Nickel-cadmium:

$$E = E^0 - \frac{RT}{2\mathfrak{F}} \ln a_{H_2O} \quad (9.138)$$

Zinc–silver oxide:

$$E = E^0 - \frac{RT}{2\mathfrak{F}} \ln \frac{a_{ZnO_2^-}}{a_{HO^-}^2} - \frac{RT}{2\mathfrak{F}} \ln a_{H_2O} \quad (9.139)$$

Cadmium–silver oxide:

$$E = E^0 + \frac{RT}{2\mathfrak{F}} \ln a_{H_2O} \quad (9.140)$$

The activities of the solid phases are taken conventionally as unity. Reference to these equations shows that the emf of the lead-acid cell is a function of the concentration of the solution, whereas the other four are relatively independent of the concentration of the solution. (Of course, the solution cannot get too dilute, for then other than the primary electrode reactions would ensue.) For the lead-acid storage battery the specific gravity of the electrolyte increases during charging and decreases during discharging. On the other hand, the other four show very little change in specific gravity during charging or discharging; the slight changes that are observed are due to the fact that some of the metallic oxides are hydrated and water molecules are involved in the cell reactions. (For the zinc–silver oxide cell the slight increase in specific gravity on discharge is due to the formation of potassium zincate.)

Of primary cells, the Leclanché is the best known. It is built in various forms and may be represented by

$$Zn(s)|ZnCl_2(1\text{–}4\%), NH_4Cl(\text{sat. aq.})|MnO_2(s) + C(s), C(s)$$

where, as above, the electrode on the left is the negative electrode. MnO_2 is a poor conductor and is, therefore, mixed with carbon in the form of acetylene black. The MnO_2-C mix extends nearly to the zinc surface; electrical contact is prevented by a porous separator. It is called a *dry cell* because the electrolyte is absorbed in the porous materials in the cell. A modified form, in which the chlorides are replaced by sodium hydroxide, and known as the alkaline MnO_2 cell, will withstand higher discharge rates than the conventional form. The magnesium dry cell is constructed similarly to Leclanché cells except that magnesium replaces the zinc, an electrolyte of magnesium bromide replaces the saturated solution of ammonium chloride, and no zinc chloride is used. The mercury dry cell may be represented by

$$Zn(s)|ZnO(\text{sat. aq.}), KOH(7.7 \text{ m aq.})|HgO(s), Hg(l)$$

and the over-all reaction on discharge is given by

$$Zn(s) + HgO(s) \rightarrow ZnO(s) + Hg(l) \quad (9.141)$$

The electrolyte is absorbed in an absorbent pad.

The silver chloride dry cell may be represented by

$$Zn(s)|NaCl, ZnCl_2, \text{ or } NH_4Cl(\text{sat. aq.})|AgCl(s), Ag(s)$$

there still being some uncertainty as to the best electrolyte to use. This cell has a long life but is not suitable for very heavy current drains. Its internal resistance is high but decreases on use. The air dry cell is used mostly for hearing aids; it is similar to the Leclanché dry cell except that MnO_2 is not used as the oxidizing agent; instead the cell "breathes" air which acts as the oxidizing agent.

A large number of wet primary cells (those in which the electrolyte is not largely immobilized in absorbent materials) have been proposed throughout the years. Only three are used extensively today: the Lalande copper oxide alkaline primary cell, the *wet* air cell, and the Drumm cell which for a while was considered to show promise as a secondary cell. These three wet primary cells may be represented by

Lalande:

$$Zn(s)|NaOH(18\% \text{ aq.})|CuO(s), Cu(s)$$

Air cell:

$$Zn(s)|NaOH(20\%\ aq.)|O_2(\text{gas from air}),\ C$$

Drumm cell:

$$Zn(s)|KOH(20\%\ aq.) + ZnO|,\ Ni(OH)_4(s),\ Ni(OH)_2(s),\ Ni(s)$$

In some air cells lime is added to the electrolyte to react with the sodium zincate formed in the cell reaction and regenerate sodium hydroxide which constitutes the electrolyte.

TABLE 9.13. NOMINAL VOLTAGES AND THEORETICAL AND AVERAGE PRACTICAL OUTPUTS OF BATTERIES ON A WEIGHT BASIS

Type	Nominal voltage	Ampere-hours per pound	
		Theoretical	Practical
Secondary batteries:			
Lead-lead dioxide acid	2.00	25.6	10
Nickel-iron alkaline	1.35	133.2	10
Nickel-cadmium alkaline	1.34	101.7	11
Zinc-silver oxide alkaline	1.86	51.8	35
Cadmium-silver oxide alkaline	1.41	52.9	30
Primary batteries:			
Leclanché dry cell	1.54	101.7	18
Alkaline MnO_2 dry cell	1.52	101.7	26
Magnesium dry cell	2.04	122.7	43
Mercury dry cell	1.35	86.3	52
Silver chloride dry cell	1.06	69.1	9
Air cell (20 % NaOH)	1.46	50.5	40
Lalande alkaline wet cell	0.98	41.3	14
Drumm alkaline wet cell	1.80	32.3	20
Reserve batteries:			
Perchloric acid	1.92	21.8	9
Fluoboric acid	1.86	19.0	6
Chlorine	2.05	178.5	14
CuCl water-activated	1.8	109.4	30
AgCl water-activated	1.6	78.2	60
Zinc-silver oxide alkaline	1.86	51.8	35
$Zn:H_2SO_4:PbO_2$	2.3	29.0	26
$Cd:H_2SO_4:PbO_2$	2.1	27.5	14

Reserve batteries are those activated at or shortly before the time of use. Activation may be carried out by one of several means. Some presently used are listed in the last part of Table 9.13; these batteries are used for special purposes and find more use in military applications than in civilian use. The perchloric acid and fluoboric acid batteries make use of lead and lead dioxide electrodes, as do the lead-sulfuric acid batteries. The chlorine battery is similar to air cells except that chlorine is used instead of air and the electrolyte is a solution of zinc chloride. The water-activated batteries are made to act by immersion in water, preferably sea water. The last two reserve batteries listed in Table 9.13 are similar to the familiar lead-acid storage battery except that either zinc or cadmium is used instead of lead; they are not rechargeable.

The nominal open-circuit voltage (OCV) and the maximum theoretical and average practical outputs in ampere-hours per unit weight of a number of primary and secondary cells are given in Table 9.13. The theoretical ampere-hour-per-pound data were calculated on the basis of the assumed cell reaction, considering only the stoichiometric quantities of reactants appearing in the reaction equation and taking no account of practical requirements in cell design. For example, the electrolyte does not appear in the nickel-iron or nickel-cadmium cell-reaction equations and so was not considered in calculating the weight of these cells even though such cells could not operate without the electrolyte. The theoretical values do, nevertheless, represent values that cannot be exceeded under any circumstances. The last column of Table 9.13 lists average outputs that may be obtained from the batteries under moderate current drains; somewhat greater outputs may be obtained at lower drains and lesser outputs at higher drains.

The unavailability of the materials in a battery for electrochemical action may result from electrode polarization; the nature of the electrode surface, whether it is porous or nonporous; current distribution; heat losses; chemical side reactions; and other factors. The magnitude of the effects resulting from these phenomena will depend on such things as the rate of discharge, the temperature of the discharge, and the age of the battery. In general, less output is obtained at higher rates, at lower temperatures, and after the batteries have aged. Because batteries give less than their theoretical output, they are given rated capacities based on some normal discharge rate. Discharges exceeding this rate will give lesser output; discharges at rates less than the normal rate will give more than the rated capacity unless the rate is so low that effects of self-discharge come into play. Thus, electric cells are designed and assembled for the service for which they are intended; they may be altered to meet special requirements.

Another type of cell, known as a *fuel cell,* has received much attention in recent years. A fuel cell is one in which the electrode reactants are added continuously during cell operation (also electrode products are sometimes removed periodically or continuously during cell operation). Potentially, fuel cells have advantages over conventional types in that electrode polarizations are kept at a minimum, periods of operation are relatively long owing to the continuous replenishing of electrode materials, and materials may be used that could not otherwise be used in cells. The materials that are continuously added can reasonably be called fuels, hence the name fuel cell; they are consumed at one electrode. Oxygen or air is used as the oxidant at the other electrode. The major advantage of the electrochemical production of electrical energy is that chemical energy is converted directly into electricity without the preliminary conversion into heat. As a consequence, this conversion is not subject to the limitations of the Carnot cycle. The fuel cell that has received the most attention and research is the one utilizing hydrogen and oxygen. The reactions of these cells are:

Anodic:

$$H_2(g) + 2OH^-(aq.) \rightarrow 2H_2O(l) + 2\epsilon \quad (9.142)$$

Cathodic:

$$(1)\ O_2(g) + H_2O(l) + 2\epsilon \rightarrow HO_2^-(aq.) + OH^-(aq.) \quad (9.143)$$

$$(2)\ HO_2^-(aq.) \xrightarrow[\text{catalytic}]{\text{chemicals}} OH^-(aq.) + \tfrac{1}{2}O_2(g) \quad (9.144)$$

Cell reaction:

$$H_2(g) + \tfrac{1}{2}O_2(g) \rightarrow H_2O(l) \quad (9.145)$$

where the peroxide ion, HO_2^-, formed in the primary reaction at the cathode is chemically decomposed as fast as it is formed by catalysts embedded in the cathode to give oxygen which is utilized at the cathode. Solutions of potassium hydroxide and electrodes of either porous nickel [31] or semiwaterproofed activated carbon containing platinum or other catalyst [32] are generally used. Some cells are constructed to operate at 50 to 60°C [32] and others at 240°C under a pressure of about 800 lb/in.^{2} [31]. Studies have also been made of hydrocarbons, alcohols, and carbon monoxide as fuels in these systems. To date, the limiting factor in fuel cells is the development of efficient, inexpensive, and durable catalysts to promote the electrode reactions.

References

1. Hamer, W. J.: Fifty Years of Electrochemical Theory, *J. Electrochem. Soc.*, **99**(12): 331C (1952).
2. Electrochemical Constants, *Natl. Bur. Std. (U.S.), Circ.* 524, Aug. 14, 1953.
3. Craig, D. N., J. I. Hoffman, C. A. Law, and W. J. Hamer: *J. Res. Natl. Bur. Std., A*, Physics and Chemistry, **64**: 381 (1960).
4. Cameron, A. E., and E. Wichers: *J. Am. Chem. Soc.*, **84**: 4175 (1962).
5. New Values for the Physical Constants—Recommended by NAS—NRC, *Natl. Bur. Std. (U.S.), Tech. News Bull.*, **47** (10): 175 (1963). Consistent Set of Physical Constants Proposed, *Chem. Eng. News*, **41**(46): 43 (1963).
6. Jones, G., and B. C. Bradshaw: *J. Am. Chem. Soc.*, **55**: 1780 (1933).
7. Walden, P.: "Salts, Acids and Bases," McGraw-Hill, New York, 1929.
8. Kraus, C. A., and R. M. Fuoss: *J. Am. Chem. Soc.*, **55**: 21 (1933).
9. MacInnes, D. A.: "The Principles of Electrochemistry," Reinhold, New York, 1939.
10. Bernal, J. D., and R. H. Fowler: *J. Chem. Phys.*, **1**: 515 (1933).
11. Robinson, R. A., and R. H. Stokes: "Electrolyte Solutions," 2d ed., Butterworth, London, 1959.
12. Hamer, W. J. (ed.): "The Structure of Electrolytic Solutions," Wiley, New York, 1959.
13. Onsager, L.: *Physik. Z.*, **27**: 388 (1926); **28**: 277 (1927).
14. Wien, M.: *Ann. Physik*, **1**: S393, 400 (1929).
15. Debye, P., and H. Falkenhagen: *Physik. Z.*, **29**: 401 (1928).
16. Geest, Von H.: *Physik. Z.*, **34**: 660 (1933).
17. Hückel, E.: *Physik. Z.*, **26**: 93 (1925).
18. Harned, H. S., and B. B. Owen: "The Physical Chemistry of Electrolytic Solutions," Reinhold, New York, 1943.
19. Latimer, W. M.: "The Oxidation States of the Elements and Their Potentials in Aqueous Solutions," 2d ed., Prentice-Hall, Englewood Cliffs, N.J., 1952.
20. Christiansen, J. A., and M. Pourbaix: *Compt. Rend. Conf. Union Intern. Chim. Pure Appl., 17th, Stockholm*, p. 83, 1953.
21. Hamer, W. J., M. S. Malmberg, and B. Rubin: *J. Electrochem. Soc.*, **103**: 8 (1956).
22. Delimarskii, IU. K., and B. F. Markov: "Electrochemistry of Fused Salts," transl. by A. Peiperl and R. E. Wood, Sigma Press, Washington, D.C., 1961.
23. American Standard Definition of Electrical Terms, Group 60—Electrochemistry & Electrometallurgy, ASA C42.60, New York, 1957.
24. Hamer, W. J.: *Trans. Electrochem. Soc.*, **72**: 45 (1937).
25. Hamer, W. J., and S. F. Acree: *J. Res. Natl. Bur. Std.*, **32**: 215 (1944).
26. Bates, R. G.: "Determination of pH, Theory and Practice," p. 76, Wiley, New York, 1964.
27. Hamer, W. J.: Standard Cells, Their Construction, Maintenance, and Characteristics, *Natl. Bur. Std. (U.S.), Monograph* 84, 1965.
28. Bockris, J. O'M.: "Modern Aspects of Electrochemistry," Academic, New York, 1954.
29. Conway, B. E.: "Electrochemical Data," American Elsevier Publishing Company, New York, 1952.
30. International Critical Tables, **6**: 339 (1933).
31. Bacon, F.: *British Patents* 667,298; 725,661; *Beama J.*, **61**: 6 (1954); *ibid.*, **63**: 2 (1956).
32. Kordesch, K.: in G. J. Young (ed.), "Fuel Cells," chap. 2, Reinhold, New York, 1960.

Chapter 10

Conduction of Electricity in Gases

By SANBORN C. BROWN *and* JOHN C. INGRAHAM, Massachusetts Institute of Technology

The physics of the conduction of electricity in ionized gases involves the interactions of the electrons and ions in the gas among themselves, with the ground-state and excited gas atoms, with any surfaces that may be present, and with any electric or magnetic fields that may exist in the gas.

Section 1 is developed in a manner that exhibits the effect of the electron-velocity distribution on various measurements involving a swarm or distribution of electron velocities. This is essential in order properly to interpret measurements. In this section lowly ionized-gas phenomena are primarily considered; in this case electron-ion collisions are negligible relative to electron-atom collisions in so far as they affect electron mobility, diffusion, or energy loss due to elastic-recoil collisions. Since the electron-atom interactions are, in many cases of physical interest, of a range short compared with the average gas-atom separation, they may be interpreted on the basis of a model that assumes the interactions to be binary, neglecting the effects of neighboring gas atoms on a given electron-atom encounter. In this case, measurements or calculations regarding such interactions in an ionized gas, where the electrons are distributed both in velocities and positions, can be directly related to measurements or calculations regarding beam experiments where the collisions are binary and the electrons have approximately one velocity, provided the distribution of electron velocities is known in the case of the ionized-gas interactions.

Section 2 discusses ion mobility, ion-ion and electron-ion recombination, electron attachment, ionization by electron collision, and other processes that affect the conduction of electricity in gases. Section 3 discusses the problem of production of an ionized gas, using a-c and d-c fields.

1. Electron-Atom Elastic Collisions and Electron-transport Phenomena

Electron-beam Experiments. In 1903 Lenard [1]† determined the attenuation of a beam of monoenergetic electrons by some of the common gases. He measured the fraction of an electron beam that was transmitted, without scattering, through a field-free region containing the gas target. An electron beam of density n_e electrons/cm^3, traveling with velocity v, passes through the gas a distance dx. Let N be the number of atoms per cubic centimeter. The number of electrons colliding with gas atoms per unit time is then

$$dn_e/dt = -NvQn_e \tag{10.1}$$

where Q is the effective collision cross-section area for electrons of velocity v. Writing $v\,dt = dx$,

$$n_e = n_{e0}e^{-NQx} \tag{10.2}$$

In many cases it is convenient to define a "probability" of collision, P_c, as the average number of collisions made by an electron in traveling 1 cm in a gas at 1 torr pressure and 0°C temperature. In a gas of temperature T_g° Kelvin and p torr,

$$NQ = \frac{273}{T_g} P_c p \tag{10.3}$$

A mean collision frequency may also be defined:

$$\nu_c = NQv = \frac{273}{T_g} P_c pv \tag{10.4}$$

so that

$$n_e = n_{e0}e^{-\nu_c t}$$

The average distance an electron travels before colliding with a gas atom is the mean free path, given by

$$l = \frac{T_g/273}{pP_c}$$

Experimental determinations of the probability of elastic collision were summarized by Brode [2] and are shown in Fig. 10.1. The abscissa is given as the square root of the voltage used to accelerate the electrons and so is proportional to the electron velocity. Also shown is the elastic-collision cross section. Above energies of a few volts, the collision probability decreases as the electron velocity increases. At the higher velocities in monatomic elements, P_c is inversely proportional to the ionization potentials and directly proportional to the gas polarizability.

The fact that P_c takes on extremely low values for slow electrons in certain gases is known as the *Ramsauer effect*. Great similarity exists between curves for atoms and molecules with similar external electron arrangements. A quantum-mechanical explanation of the Ramsauer effect was derived by

† Numbers in brackets refer to References at end of chapter.

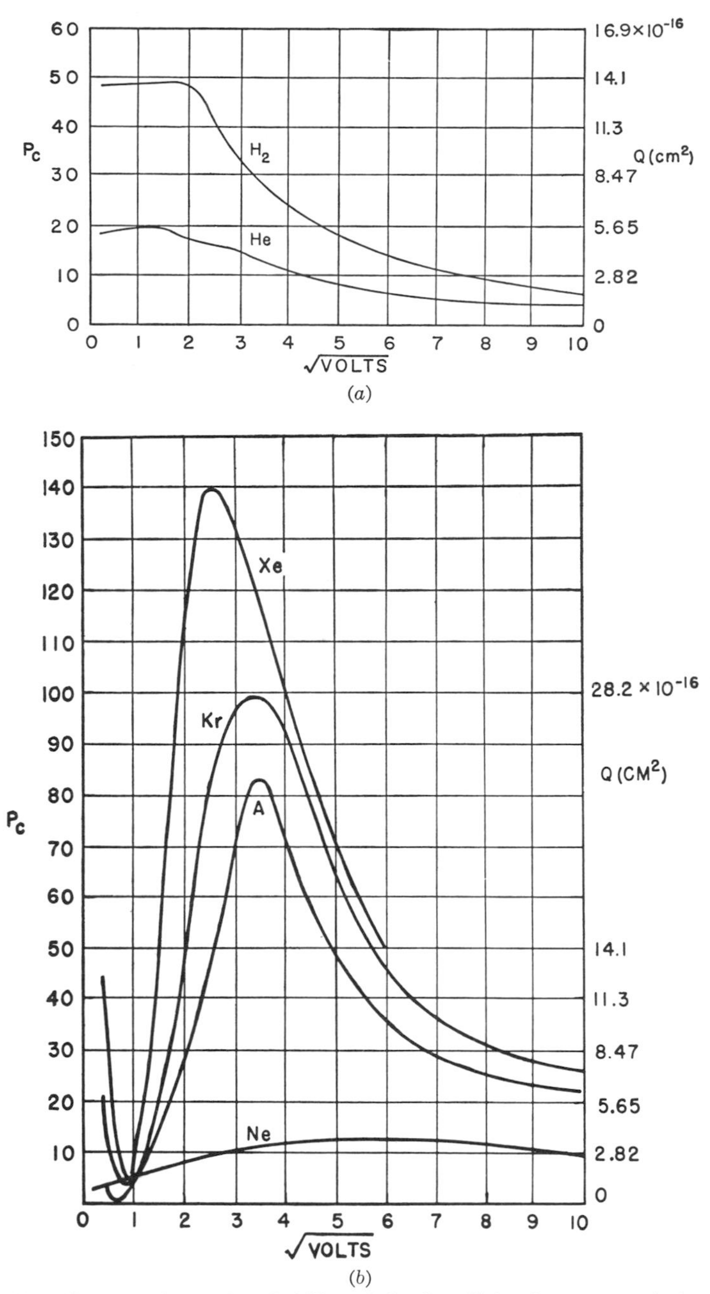

Fig. 10.1. Cross section and probability of elastic collision for some typical gases.

Allis and Morse [3] by considering the diffraction of the electron wave by the potential field of the atom.

Electron-transport Phenomena. Electron-transport phenomena in an ionized gas, such as diffusion under the influence of a density gradient and drift under the influence of an electric field, are directly related to the electron particle current $\mathbf{\Gamma}_e$ which is caused by these perturbing influences. In order to calculate $\mathbf{\Gamma}_e$ it is necessary to know the electron spatial and velocity density distribution $f(\mathbf{v},\mathbf{r},t)$. The electron density n_e and current $\mathbf{\Gamma}_e$ are then given by

$$n_e(\mathbf{r},t) = \int f(\mathbf{v},\mathbf{r},t)\, d^3v \tag{10.5}$$

and

$$\mathbf{\Gamma}_e(\mathbf{r},t) = \int \mathbf{v} f(\mathbf{v},\mathbf{r},t)\, d^3v \tag{10.6}$$

The electron drift velocity $\mathbf{v}_{de}$ is related to $\mathbf{\Gamma}_e$ and n_e by

$$\mathbf{v}_{de} = \mathbf{\Gamma}_e / n_e \tag{10.7}$$

This drift velocity is generally small compared with

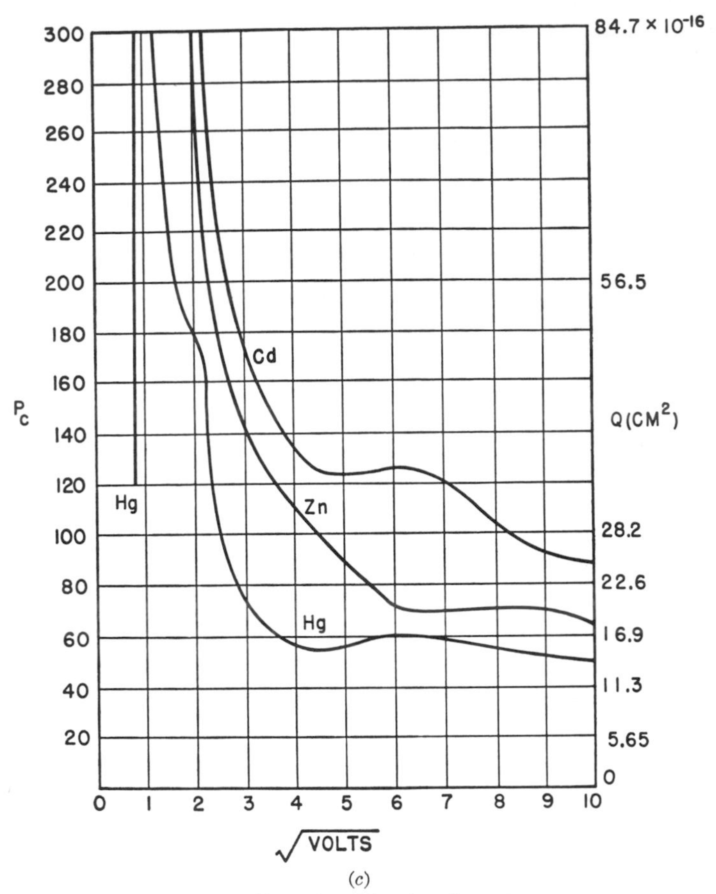

(*c*)

FIG. 10.1. (*Continued*)

the electron random velocities. The other limiting case is discussed in connection with *ionic mobilities* in Sec. 2.

(*a*). *The Boltzmann Equation.* The density distribution $f(\mathbf{v},\mathbf{r},t)$ of a given kind of particle in phase space (configuration and velocity space) is determined by the combined effect of all the interactions to which the kind of particle is subject. If there are no sources of loss or production of the particle, the number of particles in a volume element of this six-dimensional space which moves *with the particles* should not change with time. As viewed from a coordinate system, which does not move with the particles, this becomes a continuity equation for f. In configuration space alone this would have the familiar form $\partial f/\partial t + \nabla_r \cdot \mathbf{v}f = 0$. In phase space the continuity equation has the form

$$\partial f/\partial t + \nabla_r \cdot \mathbf{v}f + \nabla_v \cdot \mathbf{a}f = 0 \qquad (10.8)$$

where the subscripts on the divergence operators denote the independent variable, $\mathbf{v}$ is the particle flow velocity in configuration space, and $\mathbf{a}$ is the particle acceleration ("flow velocity" in velocity space) which may be written as $q/m(\mathbf{E} + \mathbf{v} \times \mathbf{B})$ for a particle of charge q ($q = -e = -1.6 \times 10^{-19}$ coulomb for the electron) and mass m, in the presence of electric and magnetic fields $\mathbf{E}$ and $\mathbf{B}$. Equation (10.8) is the Boltzmann equation for the case where there are no source terms.

In principle, the effect of elastic and inelastic collisions may be included in the expressions for $\mathbf{E}$ and $\mathbf{B}$, but what is generally done is to treat the collisions as source terms which have the property of transporting the colliding particles *instantaneously* from one volume element of velocity space to another. This is valid to a good approximation, for example, in the case of electron-atom collisions, but is not valid for the longer-range electron-ion and electron-electron interactions in the presence of high-frequency electric fields. Other collisions, such as ionizing collisions and collisions where electrons are lost, act to change the total electron density rather than simply to transport electrons from one element of velocity space to another. Their effect may also be included in the source term.

The electron-electron collisions are essential in order to produce the Maxwellian velocity distribution in the absence of perturbing mechanisms to the distribution function. Hence, their effect should always be included in the Boltzmann equation. There are two regimes of electron density where this may be done with ease: When the electron-electron interaction is very strong, the distribution function may be assumed to be very nearly Maxwellian and the electron-electron collision term then

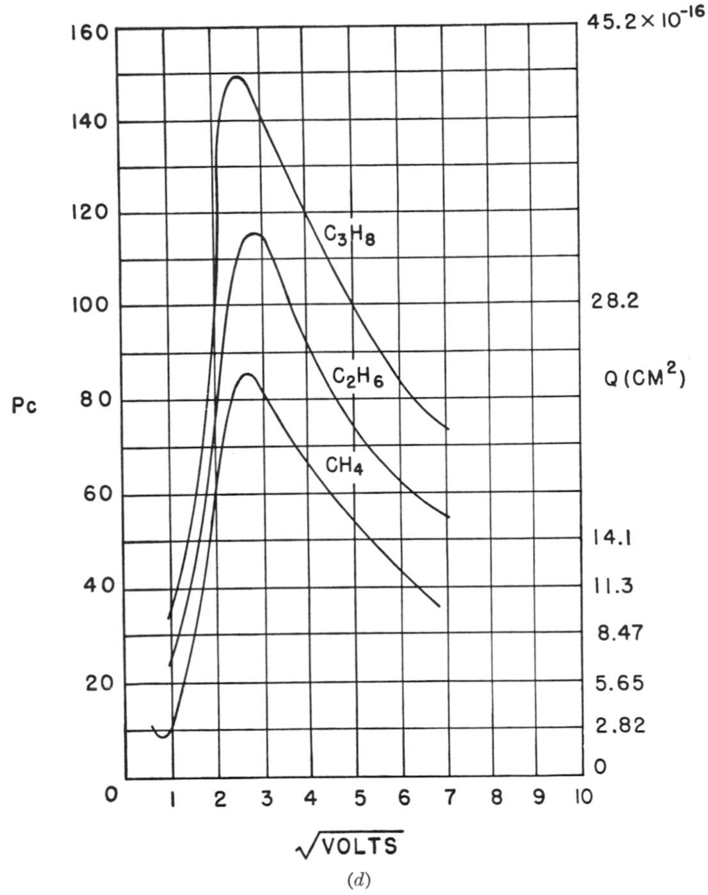

(d)

Fig. 10.1. (*Continued*)

assumed to be zero; and when the electron-electron interaction is wholly negligible so that the distribution function is assumed to be determined by the perturbing mechanisms alone. Setting the electron-electron collision term to zero in the first case does not introduce a serious error in cases where the perturbations are due to charge-dependent forces, since the electrons drift in the same direction under the influence of the perturbations and electron-electron collisions would have only a second-order effect in diverting this drift, as opposed to the effect of electron-atom or electron-ion collisions.

Spitzer [4], Dreicer [5], and Butler [6] treat cases where the electron-electron term is included explicitly.

Including collisions as a source term and utilizing the fact that $\nabla_r \cdot \mathbf{v} = 0$ and $\nabla_v \cdot \mathbf{a} = 0$, Eq. (10.8) may be written as

$$\partial f/\partial t + \mathbf{v} \cdot \nabla_r f + \mathbf{a} \cdot \nabla_v f = (\partial f/\partial t)_{\text{collisions}} \qquad (10.9)$$

where $(\partial f/\partial t)_{\text{collisions}}$ represents the source term due to all types of collisions.

b. Electron-Atom Elastic Collisions. For electrons in a weakly ionized gas, where only electron-atom elastic collisions need be explicitly considered, the collision term in Eq. (10.9) is given by the Boltzmann collision integral [7]. This integral describes the rate at which electrons are brought to and removed from an element of volume in velocity space in terms of the differential elastic-scattering cross section $Q(v,\theta)$ [Eq. (10.15)].

Allis [8] has given a perturbation method for solving the Boltzmann equation in this form where the electric and magnetic fields and density gradients are the perturbing quantities. The distribution function is expanded in spherical harmonics of the angles in velocity space, the first term in the expansion representing the unperturbed isotropic distribution and the rest of the terms representing the anisotropy introduced by the perturbing fields. Keeping only the unperturbed term and the term proportional to the first-order spherical harmonics, f may be written as

$$f(\mathbf{v},\mathbf{r},t) \cong f^0(v,\mathbf{r},t) + \frac{\mathbf{v} \cdot \mathbf{f}^1(v,\mathbf{r},t)}{v} \qquad (10.10)$$

where the three components of $\mathbf{f}^1$ correspond to the three first-order spherical harmonics. Referring to Eqs. (10.5) and (10.6), we see, after taking account of the independence of $f^0(v,\mathbf{r},t)$ and $\mathbf{f}^1(v,\mathbf{r},t)$ of the direction of $\mathbf{v}$, that

$$n_e = \int f^0\, d^3v = \int f^0 4\pi v^2\, dv \qquad (10.11)$$

$$\mathbf{\Gamma}_e = \int \mathbf{v}\left(\frac{\mathbf{v} \cdot \mathbf{f}^1}{v}\right) d^3v = \int \frac{v}{3}\, \mathbf{f}^1 4\pi v^2\, dv \qquad (10.12)$$

The expression for the distribution function, which neglects higher-order terms, is then substituted into Eq. (10.8), and four coupled equations are formed by equating the terms with no angular dependence, and by equating the terms proportional to the three first-order spherical harmonics. In addition, the approximation that the gas atoms are much more massive than the electrons is used to simplify the Boltzmann collision integral. The equations thus obtained are

$$\frac{\partial f^0}{\partial t} + \frac{v}{3} \nabla_r \cdot \mathbf{f}^1 - \frac{e}{3mv^2} \frac{\partial}{\partial v} (v^2 \mathbf{E} \cdot \mathbf{f}^1) = g \frac{1}{2v^2} \frac{\partial}{\partial v} \left[v^3 \nu_m \left(f^0 + \frac{kT_g}{mv} \frac{\partial f^0}{\partial v} \right) \right] \quad (10.13)$$

$$\frac{\partial \mathbf{f}^1}{\partial t} + v \, \nabla_r f^0 - \frac{e\mathbf{E}}{m} \frac{\partial f^0}{\partial v} + \boldsymbol{\omega}_B \times \mathbf{f}^1 = -\nu_m \mathbf{f}^1 \quad (10.14)$$

where ν_m in both equations arises from the Boltzmann collision integral and is called the momentum-loss collision frequency. It is given by

$$\nu_m(v) = Nv \int Q(v,\theta)(1 - \cos\theta)\, d^2\Omega \quad (10.15)$$

integrated over all solid angles, where N is the gas-atom density, θ the angle at which the electron is scattered, and $Q(v,\theta)$ is the differential cross section for scattering at the angle θ per unit solid angle. Here $\nu_m(v)$ is not exactly equal to the collision frequency $\nu_c(v)$ as given by Eq. (10.4); in (10.4)

$$Q(v) = \int Q(v,\theta)\, d^2\Omega$$

which weights all scattering angles equally, whereas the integral of Eq. (10.15) gives zero weight to small scattering angles and double weight to large scattering angles.

The right-hand term of Eq. (10.13) is the first-order correction for the finite mass M of the atom and gives rise to an exchange of kinetic energy between the electrons and atoms due to elastic recoil. The quantity g is equal to $2m/(M + m)$, where m is the electron mass, and represents the fraction of energy lost per collision by an energetic light (or heavy) particle in colliding with an unenergetic heavy (or light) particle.

The term containing $\mathbf{E}$ on the left of Eq. (10.13) represents work done on the electrons by the electric field and contributes to a change of the average electron energy and to a change of shape of the distribution f^0. The middle term on the left of Eq. (10.13) contributes to particle loss due to drift of the particles under the influence of electric fields or density gradients.

In Eq. (10.14) $\boldsymbol{\omega}_B = e\mathbf{B}/m$ is the vector angular frequency of rotation of a particle of charge $-e$ in a magnetic field. The magnitude of $\boldsymbol{\omega}_B$ is referred to as the cyclotron frequency. For the case of the electron, for example, this equation predicts that the electron rotates counterclockwise in the xy plane if $\mathbf{B}$ is in the positive z direction. The term proportional to $\mathbf{E}$ represents the change in the number of electrons in a volume element of velocity space due to the acceleration of the electric field, and the term $v\nabla_r f^0$ represents an analogous change in the number of electrons in a spatial volume element due to electron drift across a density gradient. The term from the Boltzmann collision integral, $-\nu_m \mathbf{f}^1$, exhibits the tendency of the electron-atom collisions to randomize the direction of the electron velocity, since the $\mathbf{f}^1$ corresponds to a preferred direction for the velocity.

It is valid to neglect the higher-order terms in order to obtain the closed set of equations (10.13) and (10.14), provided the electrons gain an amount of energy between collisions that is small relative to their random energy and provided, also, that the term f^0 changes by a small fraction of itself in the distance equal to the mean electron speed times the time between collisions. These are the most stringent conditions and may be relaxed to a certain extent in some cases. For instance, for a particle orbiting in a d-c magnetic field in the presence of a d-c electric field that is transverse to the magnetic field, the time required for one complete cyclotron orbit, if shorter than the time between collisions, replaces the time between collisions in the above criteria. If the transverse electric field is a-c at a frequency equal to the cyclotron frequency, then the criteria revert to the original criteria. For electric-field frequencies off of resonance the criteria are again less stringent. Allis [8] and Ginzburg [9] discuss these problems of convergence.

The Electron Drift Current Due to Density Gradients and D-C Electric Field. The first step in calculating electron-transport phenomena is to calculate the electron drift current $\boldsymbol{\Gamma}_e$ given in Eq. (10.12). To find $\boldsymbol{\Gamma}_e$ for a given experimental situation we must solve Eq. (10.14) for $\mathbf{f}^1$. This can be done for a rather general type of situation. Let us assume that $\omega_B = 0$ and that the electric field $\mathbf{E}$ *is constant in time.* Then if we assume that f^0 changes at a rate slow compared with ν_m, we may neglect $\partial\mathbf{f}^1/\partial t$ to a first approximation. This assumption is valid if the convergence criteria at the end of the preceding subsection are satisfied. Under this assumption

$$\mathbf{f}^1 = -\frac{1}{\nu_m} \left(v \nabla_r n_e f_v{}^0 - \frac{e\mathbf{E}}{m} n_e \frac{\partial f_v{}^0}{\partial v} \right) \quad (10.16)$$

where we have defined $f^0(v,r,t) = n_e(r,t) f_v{}^0(v,r,t)$.

Using Eq. (10.12), we calculate

$$\boldsymbol{\Gamma}_e = -\nabla_r D_e n_e - n_e \mu_e \mathbf{E} \quad (10.17)$$

which defines the electron-free-diffusion coefficient

$$D_e = \int \frac{v^2}{3\nu_m} f_v{}^0 4\pi v^2 \, dv = \frac{\langle v^2 \rangle_{\mathrm{av}}}{3\nu_m} \quad (10.18)$$

and the d-c electron mobility

$$\mu_e = -\frac{e}{m} \int \frac{v}{3\nu_m} \frac{\partial f_v{}^0}{\partial v} 4\pi v^2 \, dv \quad (10.19)$$

Allis points out that including D_e inside the gradient operator is somewhat deceptive since the only part of D_e that should be operated on by ∇_r of Eq. (10.17) is that part whose spatial dependence enters through $f_v{}^0$, such as a variation of electron temperature with position. However, if the gas pressure varies with position (which would affect ν_m) this should not be operated on by ∇_r.

a. Electron Free Diffusion. When the ionization density is low enough, the electrons diffuse freely to the walls, being unaffected by the space-charge field caused by the ions, which diffuse more slowly than the electrons. Setting $\mathbf{E} = 0$ in Eq. (10.17), we have for the electron current due to diffusion

$$\mathbf{\Gamma}_e = -\nabla_r D_e n_e \qquad (10.20)$$

Integrating Eq. (10.13) over all electron velocities and noting that the electric-heating term and the elastic-recoil term do not contribute to a net gain or loss of electrons gives the free-diffusion equation for a decaying electron density

$$\frac{\partial n_e}{\partial t} + \nabla_r \cdot \mathbf{\Gamma}_e = \frac{\partial n_e}{\partial t} - \nabla_r{}^2 D_e n_e = 0 \qquad (10.21)$$

When D_e may be considered independent of position, solutions of Eq. (10.21) are readily obtained. For instance, for a long cylindrical container of radius R in the case where the electron density depends only on the distance r from the cylinder axis,

$$\frac{\partial n_e}{\partial t} = D_e \frac{1}{r}\frac{\partial}{\partial r}\left(r\frac{\partial n_e}{\partial r}\right) \qquad (10.22)$$

The general solution for n_e satisfying $n_e(R,t) = 0$ is $n_e = \sum_{m=0} b_m e^{-t/\tau_m} J_0(a_m r/R)$, where J_0 is the zeroth-order Bessel function, and $J_0(a_m) = 0$, $a_{m+1} > a_m$, and $1/\tau_m = D_e(a_m/R)^2$. The b_m are determined by the radial dependence of n_e at $t = 0$. The time constants τ_m decrease rapidly with increasing m so that for sufficiently long times after a density distribution begins to relax

$$n_e \approx b_0 e^{-t/\tau_0} J_0\left(\frac{a_0 r}{R}\right) = b_0 e^{-t/\tau_0} J_0\left(\frac{2 \cdot 4r}{R}\right) \qquad (10.23)$$

This is called the *fundamental diffusion mode.* The above treatment gives a good approximation to n_e provided the electron mean free path is much less than R.

For the case where ν_m is velocity-independent and f^0 is a Maxwellian of temperature T, Eq. (10.18) becomes

$$D_e = \frac{kT_e}{m\nu_m} \qquad (10.24)$$

For the case where a magnetic field is applied to the ionized gas parallel to the cylinder axis so as to impede electron diffusion to the walls, the free-diffusion coefficient in Eq. (10.22) is replaced by

$$D_{eB} = \int \frac{\nu_m v^2 f_v{}^0}{3(\nu_m{}^2 + \omega_B{}^2)} 4\pi v^2\, dv = \frac{\langle \nu_m v^2 \rangle_{\text{av}}}{3(\nu_m{}^2 + \omega_B{}^2)} \qquad (10.25)$$

In this case the electrons cannot diffuse to the walls *unless* they make collisions that interrupt their spiraling motion about a certain line of magnetic force.

b. Electron D-C Mobility. The electron current induced by an electric field $\mathbf{E}$ from Eq. (10.17) is $\mathbf{\Gamma}_e = -n_e\mu_e\mathbf{E}$, and the electron drift velocity is

$$\mathbf{v}_d = -\mu_e \mathbf{E} \qquad (10.26)$$

Measurement of the electron drift velocity by a time-of-drift measurement can, in principle, through the use of Eq. (10.19) provide some information about ν_m *provided $f_v{}^0$ is known.* Since a d-c electric field may perturb considerably the distribution function and since the electron density is likely to be low in such measurements, $f_v{}^0$ should be determined using Eq. (10.13) after having eliminated $\mathbf{f}^1$ using Eq. (10.14). In general, the electron drift velocity has a complex dependence on the electric field. In the case where ν_m is independent of electron velocity, however, Eq. (10.19) after one integration by parts reduces to

$$\mu_e = \frac{e}{m\nu_m} \int f_v{}^0 4\pi v^2\, dv = \frac{e}{m\nu_m} \qquad (10.27)$$

which is independent of the distribution function. Two simple cases exist in which ν_m is nearly constant: helium and hydrogen for electron energies above 3 or 4 ev. Referring to Brode's curves in Fig. 10.1*a*, one may deduce for helium $\nu_m \simeq 2.3 \times 10^9 p$, where p is in torrs, and for hydrogen $\nu_m \simeq 4.8 \times 10^9 p$. These expressions are fairly good approximations for ν_m even though it is ν_c that is obtained for the P_c curves. Using Eqs. (10.26) and (10.27), we find for helium that

$$v_d \text{ (cm/sec)} = -7.6 \times 10^5\ (E/p) \qquad \text{(volts/cm-torr)}$$

and for hydrogen

$$v_d \text{ (cm/sec)} = -3.7 \times 10^5\ (E/p) \qquad \text{(volts/cm-torr)}$$

Electron drift velocities for a number of gases as a function of E/p were measured by Bradbury [10] and Nielsen [11]. The gases studied included H_2 and He as well as N_2, Ne, and Ar.

c. The Ratio of D_e/μ_e for Electrons. The ratio of the free-diffusion coefficient to the d-c mobility is important since it is a measure of the average electron energy in some cases. From Eqs. (10.18) and (10.19) we have

$$\frac{D_e}{\mu_e} = -\frac{m}{e} \frac{\int (v^4/\nu_m) f_v{}^0\, dv}{\int (v^3/\nu_m)(\partial f_v{}^0/\partial v)\, dv} \qquad (10.28)$$

When $f_v{}^0$ is Maxwellian, $\partial f_v{}^0/\partial v = -(mv/kT_e) f_v{}^0$ so that

$$\frac{D_e}{\mu_e} = \frac{kT_e}{e} \qquad (10.29)$$

becomes a measure of the electron energy. Also, when ν_m is velocity-independent the ratio again depends only on the average energy.

For an experiment in which the electrons are drifting in the steady state because of the combined action of diffusion and an electric field in the z direction, the equation for the electron density is

$$\nabla^2 n_e = -\frac{\mu_e}{D_e} E_z \frac{\partial n_e}{\partial z} \qquad (10.30)$$

Townsend [12] designed an experiment to determine μ_e/D_e for which he was able to solve this differential equation for the geometrical conditions of his apparatus. The results of these types of measurements are shown in Fig. 10.2, after Healey and Reed [13]. Ordinates give the ratio of electron to gas temperature T_e/T_g for several gases.

Another method of determining the average electron energy is by measuring a microwave gas-discharge breakdown as modified by an applied d-c electric field. A gas will break down when the

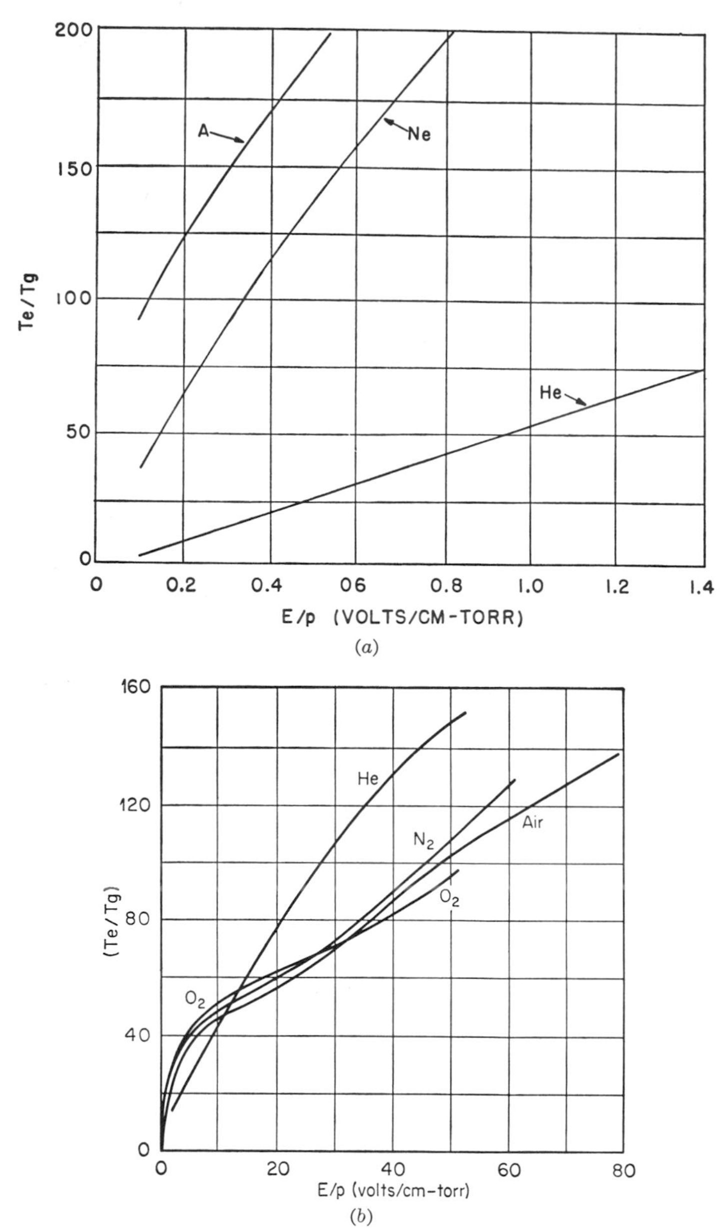

FIG. 10.2. Ratio of electron to gas temperatures for a number of common gases.

losses of electrons to the walls of a container are replaced by ionization in the body of the gas. When an a-c field alone is applied, electrons are lost by diffusion. When a small d-c sweeping field is also applied, electrons are lost by both diffusion and mobility, thus changing the breakdown condition. The experimental determination of the average electron energy u_{av} by Townsend and Bailey by the Townsend method and by Varnerin and Brown [14] by the microwave method are given in Fig. 10.3.

More recent electron-drift studies have been made by Englehardt, Phelps, Frost, and Risk [15, 16, 17] and have been used to study both elastic and inelastic processes as a function of electron energy.

d. Ambipolar Diffusion. In electron free diffusion, discussed in Sec. 1*b*, diffusion of the electrons (and other charge species) is unaffected by space-charge fields caused by an imbalance of positive and negative charges. When the density is high, this is not the case. Figure 10.4*a* shows an arbitrary density distribution in space at $t = 0$. One might expect that, since the electrons, when diffusing freely, move faster than the positive ions, the distribution in Fig. 10.4*b* would result. Although this tends to occur, the resulting electric field retards the electrons and enhances the positive-ion motion. In this limit the flow of positive ions and of electrons is equal, that is, $\mathbf{\Gamma}_e = \mathbf{\Gamma}_+$.

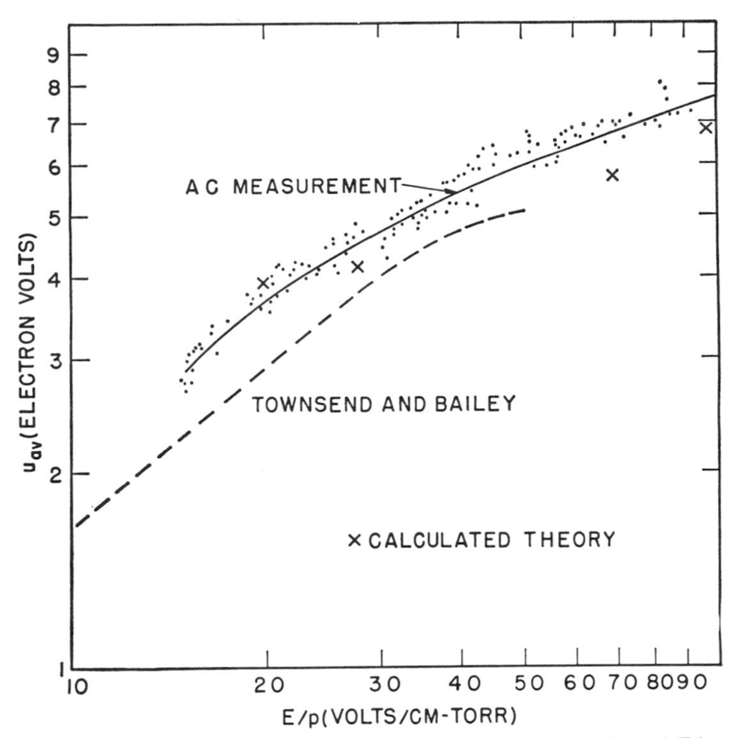

FIG. 10.3. Average energy of electrons in hydrogen as a function of E/p.

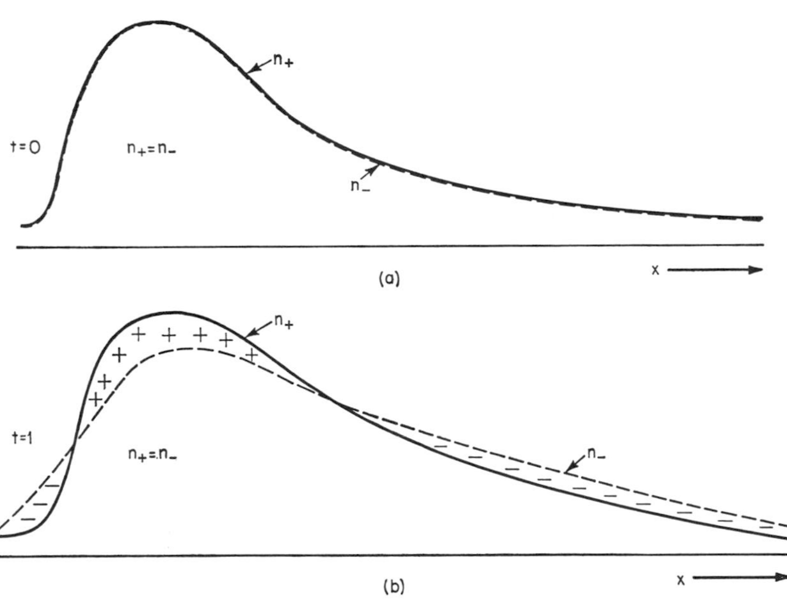

FIG. 10.4. Sketch of the ion-density distribution: (a) $t = 0$; (b) some time later.

For electrons in the absence of temperature gradients

$$\mathbf{\Gamma}_e = -D_e \nabla_e n - n_e \mu_e \mathbf{E}$$

where D_e and μ_e are given by Eqs. (10.18) and (10.19), respectively.

The analogous equation for the positive ions is

$$\mathbf{\Gamma}_+ = -D_+ \nabla n_+ + n_+ \mu_+ \mathbf{E}$$

Eliminating $\mathbf{E}$ between these two equations and setting $n_e \approx n_+ = n$ and $\mathbf{\Gamma}_e \approx \mathbf{\Gamma}_+ = \mathbf{\Gamma}$,

$$\mathbf{\Gamma} = -\left(\frac{D_+\mu_e + D_e\mu_+}{\mu_e + \mu_+}\right) \nabla n \qquad (10.31)$$

The quantity in parentheses is a diffusion coefficient that is applicable both to the electrons and to the ions, since they are interacting so that they diffuse together. This quantity is called the *ambipolar-diffusion coefficient,*

$$D_a = \frac{D_+\mu_e + D_e\mu_+}{\mu_+ + \mu_e} \tag{10.32}$$

In the case where the electrons have a temperature T_e and the ions a temperature T_+,

$$\frac{D_+}{\mu_+} = \frac{kT_+}{e} \quad \text{and} \quad \frac{D_e}{\mu_e} = \frac{kT_e}{e} \tag{10.33}$$

and since $\mu_+ \ll \mu_e$

$$D_a \cong D_+\left(1 + \frac{T_e}{T_+}\right) = \mu_+ \frac{kT_+}{e}\left(1 + \frac{T_e}{T_+}\right) \tag{10.34}$$

Allis and Rose [18] have studied theoretically the transition from free to ambipolar diffusion as the electron density is increased.

For the case in which the electrons and ions are diffusing together in the radial direction across a d-c magnetic field in the z direction and the electron-atom and ion-atom collision rates may be assumed independent of temperature, the magneto–ambipolar-diffusion coefficient $D_a(B)$ is given by

$$D_a(B) = \frac{D_a}{1 + \mu_+\mu_e B^2} \tag{10.35}$$

A review of the theoretical and experimental work on diffusion in a magnetic field has been given by Hoh [19].

Electron A-C Mobility and Electromagnetic Waves. The study of wave propagation in an ionized gas requires a knowledge of the electric current induced in the gas by the wave. This current is directly connected with the electron a-c mobility which is calculated below. Using Eq. (10.14) with $\omega_B = 0$, $\nabla_r f^0 = 0$ and $\mathbf{E} = \mathbf{E}_{0y}e^{i\omega(t-nx/c)}$, where n is the index of refraction, the equation

$$\frac{\partial \mathbf{f}^1}{\partial t} - \frac{e\mathbf{E}_{0y}}{m} e^{i\omega(t-nx/c)} \frac{\partial f^0}{\partial v} = -\nu_m \mathbf{f}^1 \tag{10.36}$$

is obtained for $\mathbf{f}^1$. In many cases of interest, such as the microwave measurement of electron densities, $\mathbf{E}_{0y}$ may be assumed to perturb f^0 negligibly, in which case the linearized solution

$$\mathbf{f}^1 = \frac{e\mathbf{E}_{0y}e^{i\omega(t-nx/c)}}{m(\nu_m + i\omega)} \frac{\partial f^0}{\partial v} \tag{10.37}$$

is obtained. Combining Eqs. (10.12) and (10.37), we obtain

$$\mathbf{\Gamma}_e = -n_e\mathbf{E}\left(-\frac{e}{3m}\int \frac{v}{\nu_m + i\omega}\frac{\partial f_v{}^0}{\partial v} 4\pi v^2\,dv\right) \equiv -n_e\mathbf{E}\mu_e \tag{10.38}$$

For a Maxwellian velocity distribution,

$$\frac{\partial f_v{}^0}{\partial v} = -(mv/kT_e)f_v{}^0$$

and

$$\mu_e = \left(\frac{e}{3kT_e}\int \frac{v^2 f_v{}^0}{\nu_m + i\omega} 4\pi v^2\,dv\right) \equiv \mu_{er} - i\mu_{ei} \tag{10.39}$$

where μ_{er} and $-\mu_{ei}$ are defined as the real and imaginary parts of μ_e.

The electric-current density carried by the electrons is $\mathbf{J} = -e\mathbf{\Gamma}_e = n_e e\mathbf{E}\mu_e$, and the quantity $n_e e\mu_e$ is the conductivity of the electrons. To determine how the electrons affect the propagation of the wave, $\mathbf{J}$ is substituted into Maxwell's equations. With the assumption that $\mathbf{E} = \mathbf{E}_{0y}e^{i\omega(t-nx/c)}$, the equation for the index of refraction is

$$n^2 = \left(1 - \frac{n_e e\mu_{ei}}{\omega\epsilon_0}\right) - i\left(\frac{n_e e}{\omega\epsilon_0}\mu_{er}\right)$$

where ϵ_0 is the permittivity of free space. For low electron densities

$$n \approx \left(1 - \frac{n_e e\mu_{ei}}{2\omega\epsilon_0}\right) - i\left(\frac{n_e e}{2\omega\epsilon_0}\mu_{er}\right)$$

The plasma affects the phase velocity of the wave through the real part of n and contributes to the wave's absorption through the imaginary part of n. Study of an ionized gas through observation of electromagnetic wave propagation through the gas can, in principle, yield the electron density n_e and some information about ν_m. A review of the experiments using this method is given by Brown [20] and McDaniel [21]. The theory for wave propagation in plasma in the presence of a d-c magnetic field is treated by Allis, Buchsbaum, and Bers [22].

a. Plasma Cutoff. Consider Eq. (10.40) for the case where $\nu_m = 0$. Then

$$n = \left(1 - \frac{n_e e}{\omega\epsilon_0}\mu_{ei}\right)^{1/2} = \left(1 - \frac{\omega_p{}^2}{\omega^2}\right)^{1/2}$$

where

$$\omega_p = \left(\frac{n_e e^2}{m\epsilon_0}\right)^{1/2}$$

is called the plasma frequency. For an electron density large enough that $\omega_p = \omega$, the index of refraction is zero. All waves for which $\omega \leq \omega_p$ cannot propagate in the plasma and will be reflected at the plasma boundary. This phenomenon, called plasma cutoff, is discussed at the end of Sec. 3 in connection with steady-state high-frequency discharges.

Plasma Oscillations. The frequency for plasma cutoff is also equal to the natural frequency of oscillation of a slab of plasma because of relative displacement of the electrons with respect to the ions in the direction perpendicular to the slab.

In 1925 Langmuir investigated the velocity-scattering properties of a low-pressure d-c mercury-arc plasma by means of a probe. The electron-velocity distribution of a discharge plasma is roughly Maxwellian. At very low pressures a beam of high-energy electrons should also be present in the plasma region adjacent to the cathode fall. This high-energy beam is formed from the electrons emitted by the cathode which are accelerated in the cathode fall of the arc but which do not undergo collisions with the gas atoms. The experimental data indicated that the beam electrons were scattered in velocity over a wider range than expected and that some electrons possessed velocities corresponding to energy gains higher than any d-c potential existing in the discharge tube. Tonks and Langmuir [23] proposed a mechanism of charged-particle oscillation to account for the anomalous scattering and detected the presence of the oscillations experimentally.

The plasma region of an arc contains equal numbers of positive and negative charges uniformly distributed so that the resultant space charge is zero. The electrons, which are much lighter than the positive ions, oscillate about their equilibrium positions when the neutrality of charge is disturbed in any manner. Thus an elementary theory of plasma electron oscillations can be derived by considering the positive ions to have fixed positions and the electrons to be displaced an amount ζ from their initial position. Consider the parallel planes X_1 and X_2 and the electrons displaced into the volume bounded by $X_1 + \zeta_1$ and $X_2 + \zeta_2$. Let A be the area of each plane and n_e the density of electrons contained in the undisplaced volume. The resulting electron concentration n_e' is

$$n'_e = \frac{n_e A(X_2 - X_1)}{A(X_2 + \zeta_2 - X_1 - \zeta_1)}$$

and the change in concentration Δn_e is

$$\Delta n_e = n'_e - n_e = -\frac{n_e(\zeta_2 - \zeta_1)}{X_2 + \zeta_2 - X_1 - \zeta_1}$$

$$\cong -n_e \frac{\zeta_2 - \zeta_1}{X_2 - X_1}$$

since the displacement ζ is small compared with the distance between the two boundary planes. Considered as a limiting process,

$$\Delta n_e = -n_e \frac{\partial \zeta}{\partial X}$$

The resultant space charge creates an electric field which can be calculated from Poisson's equation:

$$\nabla \cdot \mathbf{E} = -\frac{e\Delta n_e}{\epsilon_0} \rightarrow \frac{\partial E}{\partial X} = \frac{n_e e}{\epsilon_0}\frac{\partial \zeta}{\partial X}$$

Since the space-charge field is zero when the displacement is zero,

$$E = \frac{n_e e}{\epsilon_0}\zeta$$

This field, which acts on the electrons, arises solely from their displacement and is a longitudinal field in the direction of the displacement. Applying the force law,

$$m\frac{d^2\zeta}{dt^2} = -Ee = -\frac{n_e e^2}{\epsilon_0}\zeta$$

$$\frac{d^2\zeta}{dt^2} + \frac{n_e e^2}{m\epsilon_0}\zeta = 0$$

This equation of motion of the displaced electrons shows that the electrons execute simple harmonic motion about their equilibrium position with a frequency given by

$$\omega_p = \left(\frac{n_e e^2}{m\epsilon_0}\right)^{1/2} \qquad (10.40)$$

2. Atomic and Molecular Processes

Ionic Mobility. Ion-atom collisions impede the drift motion of an ion that is being acted upon by an electric field. The drift velocity is related to the electric field by the equation

$$\mathbf{v}_{d_i} = \mu_i \mathbf{E}$$

where the subscript i refers to a particular kind of ion and μ_i is its mobility.

The measurement of μ_i involves the measurement of the length of time an ion takes to drift a known distance under the influence of a known uniform electric field. Townsend and Tizard [24] and Tyndall and Powell [25] introduced early methods of ion-mobility measurement. Recent refinements of these techniques have been used by other experimenters [26–29].

Another measurement of μ_i can be obtained from a measurement of electron-density decay in a regime where ambipolar diffusion is the dominant loss mechanism [30–32] as shown by Eq. (10.34). This method is difficult to apply if the electron temperature is not known, if two or more kinds of ion are present, or if other loss mechanisms for the electron density are comparable with the diffusion loss.

The importance of identifying the ion whose mobility is being measured has been emphatically demonstrated by recent measurements of ionic mobilities where even in noble gases such as helium and argon [29] as many as three different positive ions are suspected to be present. The separation of kinds of ion has involved the use of time-resolved drift-velocity measurements, in which different-mobility ions arrive at the detector at different times, and also the use of mass spectrometry [30, 33, 34].

Table 10.1 gives typical measurements of mobilities of ions in their parent gases. In the case of the noble gases especially, it is essential to use very pure gas samples because of the formation of ion clusters which give rise to ion mobilities that are too small. The noble-gas data here refer to well-purified gas samples. The formation of several different ions in the same gas is shown by the data.

The mobility of ions through gas mixtures (provided there is no tendency for cluster formation) depends in a straightforward manner on the relative concentration of the gases. David and Munson [35] showed that the mobility of alkali ions in He-Xe mixtures gave reciprocal mobility values which were linearly proportional to the relative concentrations in the binary gas mixture. Some of their results are shown in Fig. 10.5. Lithium ions formed in a mixture of noble gases and water vapor do not follow this simple rule. These results are illustrated in Fig. 10.6.

The mobility of ions passing through a gas is related to the collision frequency. The distance an ion moves in time t between collisions is the sum of $v_r t$, the average random velocity times the time, and the motion due to the applied electric field S_d, which is $S_d = at^2/2 = eEt^2/2m$. Provided the random velocity v_r is much greater than the velocity induced by the field, the average time between collisions is $<t> = l/v_r$, where l is the mean free path, and the average drift velocity is

$$v_d = \frac{S_d}{<t>} = \frac{el}{2M_i v_{r_i}}E$$

The proportionality constant is the mobility

$$\mu_i = \frac{el}{2M_i v_{r_i}}$$

The condition that $v_{d_i} \ll v_{r_i}$ is satisfied provided

TABLE 10.1. MOBILITY OF POSITIVE AND NEGATIVE IONS IN THEIR PARENT GASES FOR A TEMPERATURE OF 273°K AND 760-TORR (2.69×10^{19} CM^{-3} GAS ATOMS) GAS PRESSURE

Gas	$\mu_-\left(\frac{\text{cm/sec}}{\text{volt/cm}}\right)$	$\mu_+\left(\frac{\text{cm/sec}}{\text{volt/cm}}\right)$
He	....	10.4(He^+); 16.2(He_2^+?); 20(He_2^+?)
Ne	....	4.2(Ne^+); 6.2(Ne_2^+)
Ar	....	1.6(Ar^+); 1.9(Ar_2^+)
Kr	....	0.9(Kr^+); 1.1–1.2(Kr_2^+)
Xe	....	0.6(Xe^+); 0.7(Xe_2^+)
Air (dry)	2.1	1.36
Air (very dry)	2.5	1.8
N_2	....	3.3(N^+); 1.8(N_2^+); 3.1(N_3^+); 2.4(N_4^+)
O_2	2.6 (probably O_2^- or O_3^-)	2.2 (probably O_2^+)
H_2	....	12.3 (H_3^+)
Cl_2	0.74	0.74
CCl_4	0.31	0.30
C_2H_2	0.83	0.78
C_2H_5Cl	0.38	0.36
C_2H_5OH	0.37	0.36
CO	1.14	1.10
CO_2 (dry)	0.98	0.84
HCl	0.62	0.53
H_2O (at 100°C)	0.95	1.1
H_2S	0.56	0.62
NH_3	0.66	0.56
N_2O	0.90	0.82
SO_2	0.41	0.41
SF_6	0.57 (SF_6^-)	

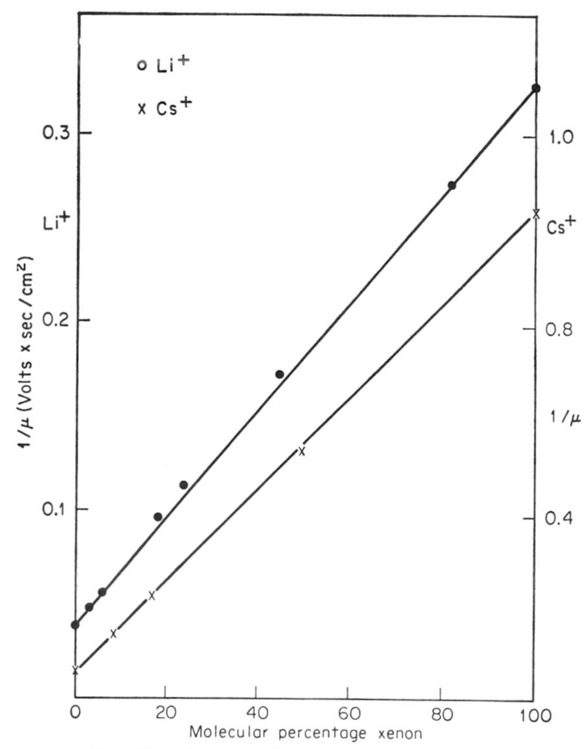

FIG. 10.5. Ion mobility in He-Xe mixtures.

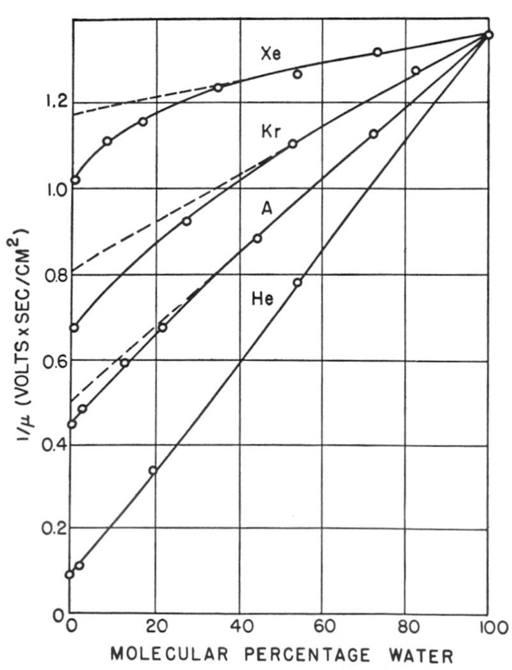

FIG. 10.6. Ion mobility showing the effect of cluster formation.

$El \ll kT$, the energy gained per mean free path, is much less than the random energy. Furthermore, to insure that the ion temperature is not raised significantly above the gas temperature because of heating by the electric field, the condition

$$El\left(\frac{M_i}{M_g} + \frac{M_g}{M_i}\right) \ll kT \tag{10.41}$$

must be satisfied. Here the mass ratios take into account the inefficiency of ion-atom energy transfer through elastic recoil if the ion mass M_i and the atom mass M_g are not equal.

Langevin [36] introduced a theory for ion mobility which assumed the ion and atom to interact through a hard-sphere collision (constant mean free path l) and which accounted for the recoil of the target atom. He obtained

$$\mu_i = \frac{0.75el}{M_i v_{r_i}}\left(1 + \frac{M_i}{M_g}\right)^{1/2}$$

Since the kinetic energies of the gas atoms and ions are the same, this may also be written as

$$\mu_i = \frac{0.75el}{M_g v_{r_g}}\left(1 + \frac{M_g}{M_i}\right)^{1/2}$$

The numerical factor 0.75 arises from the particular averaging process used by Langevin in the calculation. The equation predicts the variation of mobility coefficient as $(1 + M_g/M_i)^{1/2}$. This is shown in Fig. 10.7 due to Mitchell and Ridler [37]. The experimental data were obtained for various ions traveling through H_2. However, the magnitude of the mobility as predicted by Langevin's theory is not in agreement with experiment.

Langevin developed a second mobility theory [38] which assumed the ions and atoms to interact through an attractive force due to the polarization of the atom in the ion's field, in addition to the hard-sphere repulsion for close distances of approach. He obtained

$$\mu_i = \frac{A}{(\alpha M_i/\epsilon_0)^{1/2} N_g}\left(1 + \frac{M_i}{M_g}\right)^{1/2} \tag{10.42}$$

where A is approximately ½ for gases in which the polarization interaction dominates the hard-sphere interaction. The excellent agreement of this expression with experimental data is shown in Table 10.2 [39].

TABLE 10.2. MOBILITY OF VARIOUS POSITIVE IONS IN INERT GASES*

	He		Ne		A		Kr		Xe	
	μ_{th}	μ_{exp}	μ_{th}	μ_{exp}	μ_{th}	μ_{exp}	μ_{th}	μ_{exp}	μ_{th}	μ_{exp}
Li^+	22.9	25.8	12.0	11.85	5.17	4.99	4.17	3.97	3.19	3.04
Na^+	20.0	24.2	8.20	8.70	3.44	3.23	2.50	2.34	1.88	1.80
K^+	19.4	22.9	7.49	8.00	2.97	2.81	2.09	1.98	1.53	1.44
Rb^+	18.5	21.4	6.73	7.18	2.57	2.39	1.66	1.57	1.17	1.10
Cs^+	17.3	19.6	6.48	6.50	2.40	2.24	1.52	1.42	1.04	0.97

*From Tyndall [39].

This theory is not expected to predict correct values of an ion in its parent gas (for example, He^+ in He) since the quantum-mechanical effects involving the resonance exchange of the outer electron can dominate over polarization or hard-sphere forces.

From Eq. (10.42), the mobility is proportional to $1/p$. Thus the drift velocity should be proportional to E/p, the experimental parameter that measures the energy gained per mean free path. Measurements of Mitchell and Ridler [37] are very extensive. In the low E/p region, their results show that the positive-ion mobilities times the pressure are constant up to a fixed point and thereafter increase linearly with E/p. This critical value of E/p is smaller, the larger the mass of the ion, and the slope of the $\mu_i p$ versus E/p line decreases with increasing ionic mass. These observations reflect the regime of E/p where the conditions of Eq. (10.41) are not satisfied so that one would not expect $\mu_i p$ to be independent of E/p.

At high values of E/p, where the drift velocity is larger than the random velocity, the energy gained between collisions is proportional to the kinetic energy of the ion, and so the drift velocity is proportional to the square root of E/p. The fact that the drift velocity is proportional to E/p at low E/p, where polarization and in some cases charge-transfer collisions are important, and proportional to $(E/p)^{1/2}$ at

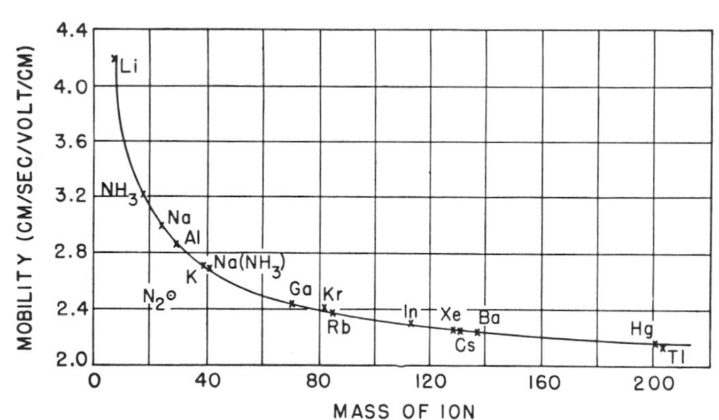

FIG. 10.7. Mobility of ions in hydrogen as a function of ionic mass.

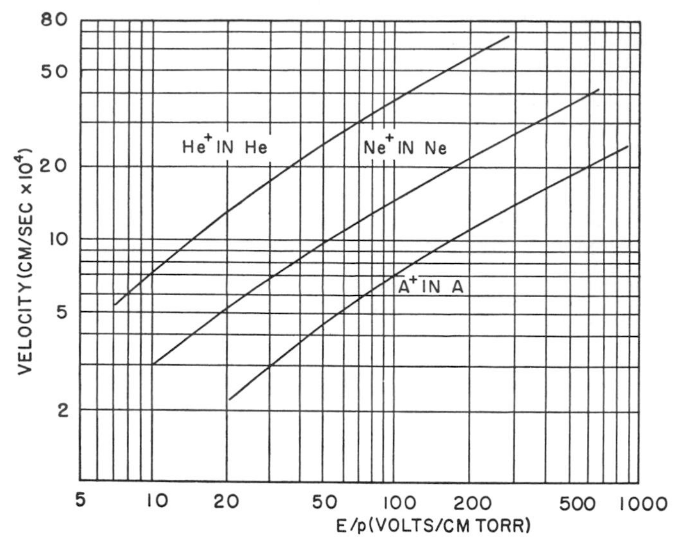

FIG. 10.8. Drift velocity of ions as a function of E/p.

high E/p, where the hard-sphere type of collision dominates, is well illustrated in the results of Hornbeck and Wannier [40] shown in Fig. 10.8.

Ion-Ion Recombination. One of the most common loss mechanisms for ions is the recombination of negative ions and electrons with positive ions. The loss of ions due to recombinations is proportional to the product of the ion concentrations,

$$\frac{dn_+}{dt} = \frac{dn_-}{dt} = -\alpha n_- n_+$$

Here α is called the *recombination coefficient* and n_- is the negative-ion concentration. In almost all discharges the positive- and negative-ion concentrations are equal, and so

$$\frac{dn}{dt} = -\alpha n^2$$

which, on integration, gives

$$\frac{1}{n} = \frac{1}{n_0} + \alpha t$$

where n_0 is the initial ion concentration at $t = 0$. Recombination phenomena thus exhibit a linear relation between $1/n$ and t.

Two theories for recombination between negative and positive ions have been known for a long time, one postulated by Langevin [41] and applicable to high pressures, above 1 atm, and the other due to Thomson [42], useful in the low-pressure region.

Langevin's theory of recombination postulates that ions of opposite sign move toward one another under the influence of their mutual attraction and that their relative drift velocity is mobility-controlled. This theory neglects the effects of thermal motion. This type of theory is valid only when the ion mean free path is much shorter than the range of relative separations over which two interacting ions move under their mutual influences before they actually combine and hence is applicable only at high pressures.

Measurements in the high-pressure region have been made to test the Langevin recombination theory, which predicts that the recombination coefficient should be inversely proportional to the pressure. Above a pressure of 1 or 2 atm this is found to be true.

Thomson's theory of recombination assumes that two ions of opposite sign do not combine unless they are closer than a critical distance r. If the ions approach each other within this critical distance, they will recombine only if there is present a gas molecule to carry off the energy released in the recombination process. Therefore this is a three-body collision process. The critical distance r is such that the recombining ion has a potential energy equal to the mean energy of thermal agitation.

The Thomson theory predicts a linear dependence of α at very low pressure and a recombination coefficient independent of pressure at higher pressure.

Natason [43] has developed a theory for a wider pressure range which agrees with the Langevin theory at high pressures and is close to the Thomson theory at low pressures. Figure 10.9 [44] displays the agreement between theory and experiment in dry air. The experimental data at low pressures are due to

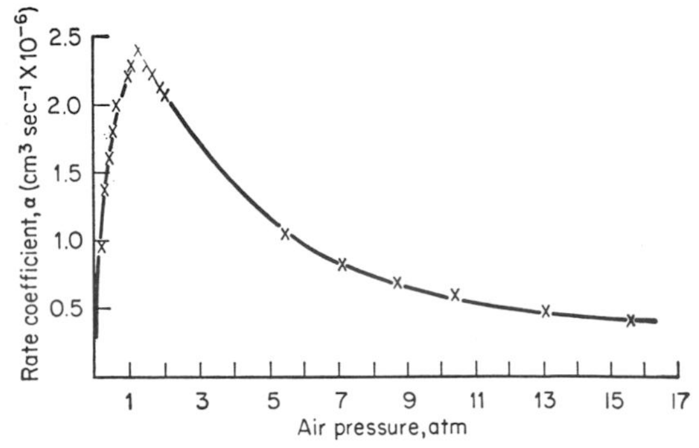

FIG. 10.9. Relation between rate coefficient for ionic recombination and pressure for dry air at normal temperature.

Sayers [45] and at high pressures to Machler [46]. To calculate absolute values of α from theory the kind of ion present must be assumed, as well as the ion mean free path and ion thermal velocity. Hence, experimental determination of these additional quantities would allow a more meaningful comparison between theory and experiment.

In addition to the recombination processes described above which require the presence of at least a third body, two-body recombination may also occur. In this case the ions do not combine but neutralize each other through the transfer of an electron from the negative to the positive ion. The energy liberated in this process results in electron excitation and the kinetic energy of the resulting two atoms. Theoretical calculations for this process have been made by Bates and Lewis [47] and Bates and Boyd [48]. For atomic ions the theories predict $\alpha \approx 10^{-7}$ cm^3 sec^{-1}. Yeung [49] and Greaves [50] have made experimental determinations of the two-body recombination coefficient for iodine and bromine. They found α of the order of 10^{-7} cm^3 sec^{-1} and pressure-independent, as theory predicts.

Electron-Ion Recombination. The mechanisms for electron-ion recombination differ from those for ion-ion recombination. Radiative recombination in which the capture of a free electron by an ion is accompanied by the emission of a photon is one mechanism that can contribute to electron loss. This is recognized spectroscopically by the afterglow emission of a continuum lying above the ionization limit of the atom or molecule. Stueckelberg and Morse [51] calculated quantum-mechanically the transition probability of electrons with positive initial energy into excited levels of the resulting atom and predicted a recombination coefficient of the order of 10^{-12} cm^3 sec^{-1}. All similar calculations of the radiative-recombination coefficient arrive at about the same value but experimental measurements of this quantity do not agree with this value.

Dielectronic recombination, in which the liberated energy is taken up in the excitation of two electrons of the resulting atom, may in some cases exceed the theoretical radiative-recombination coefficient. Bates and Dalgarno [52] have pointed out that under favorable conditions the dielectronic coefficient may be as large as 10^{-10} but no case has as yet been found for which such conditions exist.

The three-body recombination process in which the third body is a gas atom, which is so important in ion recombination, is usually not important for electrons since the gas atom is relatively ineffective in carrying away the excess kinetic energy of the recombining electron in an elastic collision, because of the large mass ratio. At pressures where such a process is likely to be important, ion-ion recombination is likely to dominate.

Dissociative recombination, in which an electron combines with a positive molecular ion and the excess energy is carried away by the atoms resulting from the subsequent dissociation of the molecule, appears to play an important role in determining electron loss. Theoretical calculations of this process show reasonable agreement with experiment [53] although the latest experimental data have made the agreement somewhat poorer.

In plasmas of high electron density, recombination processes involving two or more electrons and the ion become important. Bates, Kingston, and McWhirter [54] have calculated a collisional-radiation recombination coefficient in which both collisions and radiation play important roles in the complex process of recombination for the medium- and high-density cases. They assume that the plasma is optically thin to the radiation emitted but that the electron density is sufficiently high so that the populations of the excited atomic states are determined by electron collisions and hence may be determined from the Saha equation. They also assume that the excited-state populations are much less than the ground-state populations and the electron density. With these assumptions, they can determine the net rate at which the ground state is populated, and from this they can determine an effective recombination coefficient.

In most measurements of recombination coefficients for low-density plasmas microwave techniques have been used to measure electron-density decay in the afterglow. The recombination coefficients appear to be due to dissociative recombination, as first suggested by Bates [53]. Measurements under different experimental conditions yield rather widely scattered results, probably because of varying impurity concentrations, different ionic species of the parent gas being present in different proportions, varying concentrations of metastable atoms which through a metastable-metastable ionizing collision can produce an electron [55], and only partially accounted-for effects of ambipolar diffusion. Gray and Kerr [56] have discussed the latter point in detail. Hasted [57] summarizes the different measured values of dissociative-recombination coefficients, showing the wide range of disagreement. Typical values are (in units of cm^3 sec^{-1}) helium, 2×10^{-9}; neon, 2×10^{-7}; and argon, 7×10^{-7}. Collisional-radiative recombination has been observed by Kuckes et al. [58] and reasonable agreement obtained with the theory of Bates, Kingston, and McWhirter.

Electron Attachment. The phenomenon of electron attachment to a neutral atom or molecule to form a negative ion is a common occurrence for gases whose outer electronic shells are nearly filled. The energy of formation of a negative ion is called the electron affinity. This varies from about 3.5 volts (for the halogens) to nearly zero among the gases that exhibit electron attachment. Atoms having closed electronic shells do not form negative ions. These atoms, which have $^1S_0$ ground states, include the noble gases. Molecules in a $^1\Sigma$ ground state are characterized by no resulting spin or angular momentum. Their electrons form closed groups, and the molecules do not form negative ions (except in the case of H_2).

A general experimental technique for measurement of attachment coefficients involves passing electrons through a gas target and measuring the attenuation of the electrons due to attachment and/or the negative ions produced in the process. This type of experiment can be made with an electron beam or electron swarm. Lozier [59] developed a beam technique, which has been utilized recently in a refined form by Buchel'nikova [60]. Schulz [61] has also used a beam technique for study of attachment.

Recent swarm experiments have been carried out by Herreng [62], Doehring [63], McAfee [64], Chanin [65], Kuffel [66], Huxley [67], and Harrison [68]. The measurement of the decay of electron density with time, using microwave techniques, was suggested by Biondi [69] as a means of measuring attachment coefficients. Muschlitz [70], Fox [71], and Curran [72] have used mass-spectrometric techniques in identifying the negative ions produced in the attachment process.

Table 10.3 lists the processes that contribute to electron attachment.

Figures 10.10 and 10.11 show Chanin's [65] results for the recombination coefficient in oxygen as a function of electron energy. The K plotted in Fig. 10.10 is the three-body attachment coefficient for a process of the type corresponding to the second entry in Table 10.3. The β plotted in Fig. 10.11 is the rate coefficient for dissociative attachment, the fourth entry in Table 10.3. Also shown in Fig. 10.10 is the value of β obtained from the data of Craggs, Thorburn, and Tozer [73].

Neutral Atoms and Molecules. An electron of kinetic ϵ may collide with a neutral gas molecule XY, thereby supplying energy to produce a positive ion X^+ and a negative ion Y^- with a relative kinetic energy ϵ_{XY}. The energy balance for this reaction is

$$\epsilon_{XY} = \epsilon + A_Y - u_X - D_{XY}$$

where A_Y is the electron affinity of Y, u_X is the ionization energy of atom X, and D_{XY} the energy required to dissociate XY into the neutral atoms X and Y. The electron's energy ϵ must be greater than a certain threshold energy for the reaction to occur. This phenomenon is observed, for example, in oxygen for electron energies of 21 volts.

Excited gas atoms may produce ionization when they collide with another gas atom. Particularly, those atoms with long effective lifetimes in the ionized gas, such as metastable atoms or atoms excited by resonance radiation [24], contribute most strongly to such ionization phenomena.

The collision between a metastable atom and another atom can produce ionization in the reaction $X^m + Y \rightarrow X + Y^+ + e$ provided the metastable energy of X exceeds the ionization energy of Y. The cross section for this process is so great that very small admixtures of the ionizable gas produce large effects on the characteristics of the main gas. Because Penning [75] and his colleagues were responsible for a great deal of the study of this phenomenon, it is often called the Penning effect.

The collision between a pair of similar metastable atoms can produce ionization according to the relation $X^m + X^m \rightarrow X^+ + e + X$. This has been extensively studied by Biondi [76].

Ionization by Electron Collision. For many plasmas and ionized gases the ionization density is produced by electron collisions with atoms and molecules.

Figure 10.12 shows some typical measurements of the cross section for ionization, σ_i, given by Hasted [77] from the data of Tate and Smith for a number of different gases; an electron-beam technique was used.

TABLE 10.3. PROCESSES OF ELECTRON ATTACHMENT

Process	Remarks	Examples
Capture of free electrons by atom accompanied by radiation	Long wave limit of radiation = hc/A	$H + e \rightarrow H^-$
Capture of free electrons by atom in presence of a third body	Not at very low pressures	$NO + NO + e \rightarrow NO^- + NO$
Capture of free electrons by molecules with vibration but no dissociation	Low-energy electrons	$SO_2 + e \rightarrow SO_2-$ $O_2 + e \rightarrow O_2-$
Capture of free electrons with dissociation of molecules	Very common	NH_3, H_2S, SO_2, O_2, and N_2O

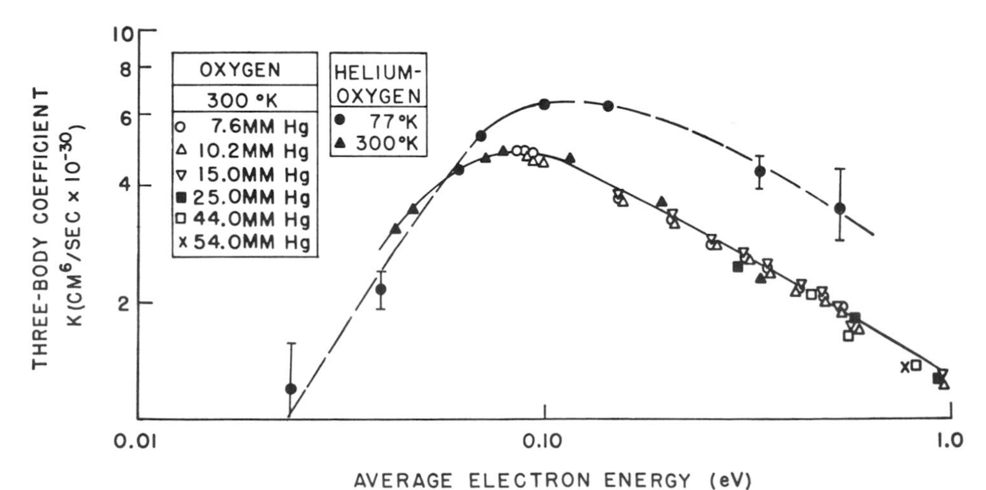

FIG. 10.10. Energy dependence of the three-body attachment coefficient for oxygen at 300 and 77°K. The solid points were obtained from measurements of oxygen-helium mixtures containing 1 to 5 per cent of oxygen.

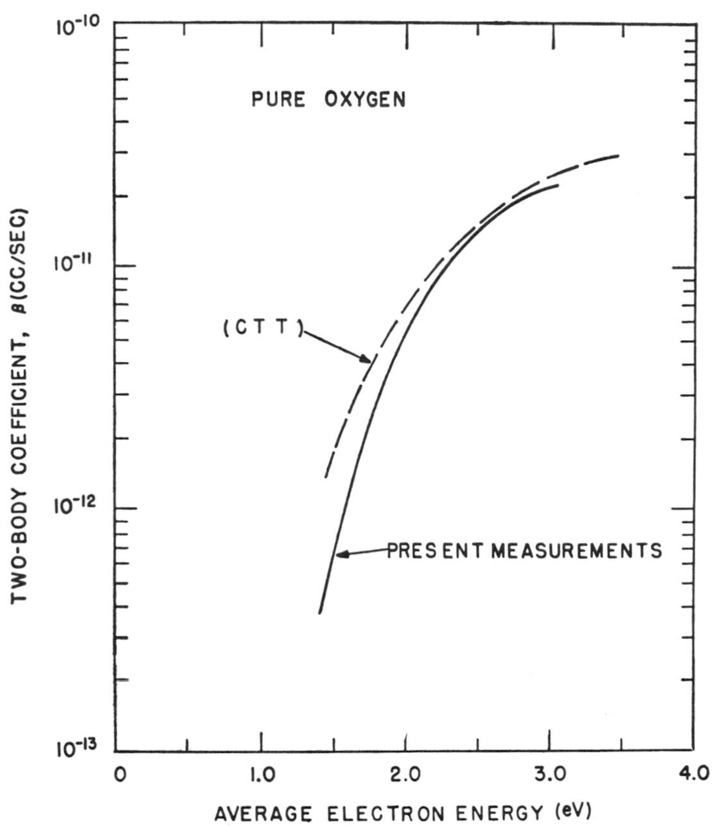

FIG. 10.11. Energy dependence of the two-body attachment coefficient. The dashed curve was obtained by averaging the cross sections measured by Craggs, Thorburn, and Tozer (C.T.T.) over a Druyvesteyn electron-energy distribution for various average electron energies.

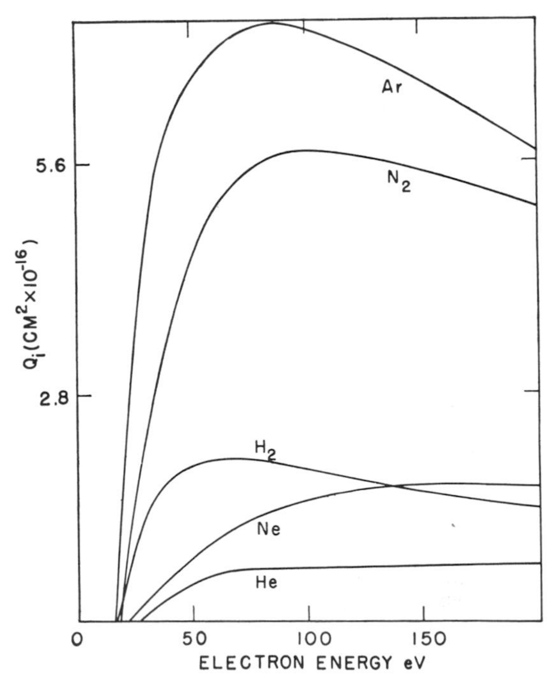

FIG. 10.12. Total ionization cross sections. (*From Hasted* [77], *after Tate and Smith.*)

In a plasma where the electrons have a distribution of velocities $f_v(v,r,t)$ a rate of ionization per electron may be defined as

$$\nu_i = N \int_{v_i}^{\infty} Q_i v f_v(v,r,t) \, d^3v \qquad (10.43)$$

where N is the gas-atom density, and v_i the minimum electron velocity that can cause ionization. Hence, if $f_v(v,r,t)$ and σ_i are known for a plasma, it is possible to calculate ν_i. Emeleus, Lunt, and Meek [78] have studied this integral for various distribution functions and forms of Q_i. For a Maxwellian distribution, ν_i is an increasing function of temperature.

3. Electrical Discharges in Gases

High-frequency Breakdown. In the theory for high-frequency breakdown, the electron distribution function is calculated as a function of the applied a-c electric field, using Eqs. (10.13) and (10.14); from this, ν_i of Eq. (10.43) is calculated, so that the rate of ion production can be expressed in terms of the applied electric field. This ν_i is then equated to the rate at which electrons are being lost because of diffusion, attachment, recombination, or a combination of these. This equation then determines the electric field necessary to produce breakdown.

Writing Eq. (10.38) for a constant ν_m and a Maxwellian velocity distribution, we have for the electric-current density $\mathbf{J}$

$$\mathbf{J} = -e\mathbf{\Gamma}_e = \frac{n_e{}^2\mathbf{E}_0 e^{i\omega t}}{m(\nu_m + i\omega)}$$

where the time dependence of $\mathbf{E}$ is written explicitly. The energy gained per unit volume and time, P, is the time average of $\mathbf{J} \cdot \mathbf{E}$:

$$P = \overline{\mathbf{J} \cdot \mathbf{E}} = \frac{n_e{}^2\nu_m}{m(\nu_m{}^2 + \omega^2)} \frac{E_0{}^2}{2}$$

For a given electric field the power absorbed is seen to have a maximum value at $\nu_m = \omega$. If the electron energy loss is due to elastic-recoil collisions with gas atoms of temperature T_g this may be determined directly by integrating the product of the electron energy $\frac{mv^2}{2}$ and the term on the right side of Eq. (10.13). For a constant ν_m and a Maxwellian velocity distribution, this energy-loss term is

$$L = +n_e\nu_m(kT_e - kT_g)3\%_2g$$

Equating gain and loss of energy gives an equation for temperature as a function of the electric field which in turn can be used to determine ν_i as a function of the electric field.

MacDonald and Brown [79] have carried out a theoretical and experimental analysis of breakdown in hydrogen and in helium-mercury mixtures when the electron particle loss was due to diffusion.

Low-pressure D-C Breakdown. Analysis of d-c breakdown, in principle, is the same as a-c breakdown; that is, breakdown occurs when the electron production rate equals the rate of loss of electrons. However, the problem is now more complicated because the loss mechanism, in addition to being due to diffusion, etc., is also contributed to by the sweeping action of the d-c electric field which causes electrons to be removed at the anode. Also, the production mechanism may be complicated by secondary electrons ejected from the cathode into the gas when the cathode is struck by the positive ions. The breakdown potential of a gas then depends on whether the cathode is heated (and emitting electrons) or cold, in which case the cathode material affects the breakdown potential through the secondary-electron-emission coefficient. Von Hipple, in Chap. 7, Sec. 4, discusses some aspects of d-c breakdown in a dielectric medium. Loeb [80] discusses d-c breakdown in gases in further detail.

Steady-state D-C Discharge—The Positive Column. The positive column [81] of a low-pressure d-c discharge is frequently used as a source of ionized gas of low-percentage ionization. It is possible to obtain a wide range of electron densities since the electron density varies nearly linearly with the discharge current. The electron temperatures encountered are typically between 10,000 and 100,000° in the 0.1- to 10.0-torr pressure range in gases such as helium, neon, argon, hydrogen, etc.

The electron temperature corresponds to the value of ν_i that just equals the loss rate of electrons. In other words, it is expected that the temperature will increase as the loss rate is increased. Furthermore, the electron temperatures are higher in the gases with high ionization potentials. Cobine [81] discusses the theories of gas discharges in more detail.

Steady-state A-C High-frequency Discharges. The electrodeless gas discharge is most suited to experiments demanding a very high degree of purity. However, there is a limit on the maximum attainable density for a given electric-field frequency, because of the inability for an electromagnetic wave to penetrate a plasma whose density exceeds a critical value n_{ec}. From Eq. (10.41) this critical density is given approximately by

$$n_{ec} \cong \left(\frac{f}{8{,}980}\right)^2 \times 10^{12} \text{ cm}^{-3}$$

where f is in megacycles.

Physically this reflection of the wave at the plasma boundary occurs because the electric current induced in the electrons by the wave becomes of sufficient magnitude to cancel the contribution of the wave's displacement current to Maxwell's equation, at which point the wave can no longer propagate into the plasma. However, if the plasma is small compared with the radiation wavelength some radiation can still be transmitted through the plasma because of the finite skin depth.

References

1. Lenard, P.: *Ann. Physik*, **12**: 714 (1903).
2. Brode, R. B.: *Revs. Mod. Phys.*, **5**: 257 (1933).
3. Allis, W. P., and P. M. Morse: *Z. Physik*, **70**: 567 (1931).
4. Spitzer, L.: "Physics of Fully Ionized Gases," p. 80, Interscience, New York, 1956.
5. Dreicer, H.: *Phys. Rev.*, **117**: 343 (1960).
6. Butler, S. T., and M. J. Buckingham: *Phys. Rev.*, **126**: 1 (1962).
7. Kennard, E. H.: "Kinetic Theory of Gases," pp. 34–40, McGraw-Hill, New York, 1938.
8. Allis, W. P.: Motions of Ions and Electrons, "Encyclopedia of Physics," vol. 21, pp. 383–444, Springer, Berlin, 1956.
9. Ginzburg, V. L.: "The Propagation of Electromagnetic Waves in Plasmas," pp. 20–25, Addison-Wesley, Reading, Mass., 1964.
10. Bradbury, N. E., and R. A. Nielsen: *Phys. Rev.*, **49**: 388 (1936).
11. Nielsen, R. A.: *Phys. Rev.*, **50**: 950 (1936).
12. Townsend, J. S.: "Electricity in Gases," p. 166, Clarendon Press, Oxford, 1915.
13. Healey, R. H., and J. W. Reed: "The Behavior of Slow Electrons in Gases," Amalgamated Wireless Ltd., Sydney, 1941.
14. Varnerin, L. J., and S. C. Brown: *Phys. Rev.*, **79**: 946 (1950).
15. Englehardt, A. G., and A. V. Phelps: *Phys. Rev.*, **133**: A375 (1964).
16. Englehardt, A. G., A. V. Phelps, and C. G. Risk: *Phys. Rev.*, **135**: A1566 (1964).
17. Frost, L. S., and A. V. Phelps: *Phys. Rev.*, **13**:. A1538 (1964).
18. Allis, W. P., and D. J. Rose: *Phys. Rev.*, **93**: 84 (1954)**6**
19. Hoh, F. C.: *Revs. Mod. Phys.*, **34**: 267 (1962).
20. Brown, S. C.: "Basic Data of Plasma Physics," Wiley, New York, 1959.
21. McDaniel, E. W.: "Collision Phenomena in Ionized Gases," Wiley, New York, 1964.
22. Allis, W. P., S. J. Buchsbaum, and A. Bers: "Waves in Anisotropic Plasmas," M.I.T., Cambridge, Mass., 1963.

23. Tonks, L., and I. Langmuir: *Phys. Rev.*, **33**: 195 (1929).
24. Townsend, J. S., and H. T. Tizard: *Proc. Roy. Soc. (London)*, **A88**: 366 (1913).
25. Tyndall, A. M., and C. F. Powell: *Proc. Roy. Soc. (London)*, **A136**: 145 (1932).
26. Hornbeck, J. A.: *Phys. Rev.*, **84**: 615 (1951).
27. Varney, R. N.: *Phys. Rev.*, **88**: 362 (1952).
28. Biondi, M. A., and L. M. Chanin: *Phys. Rev.*, **94**: 910 (1954); **106**: 473 (1957).
29. Beaty, E. C., and P. Patterson: Sixth International Conference on Ionization Phenomena in Gases, Paris, 1963.
30. Phelps, A. V., and S. C. Brown: *Phys. Rev.*, **84**: 102 (1952).
31. Oskam, H. J., and V. R. Mittelstadt: *Phys. Rev.*, **132**: 1435 (1963).
32. Brown, S. C.: "Basic Data of Plasma Physics," p. 88, Wiley, New York, 1959.
33. McDaniel, E. W., et al.: *IEEE Trans. Nuclear Sci.*, **NS-10**: 111 (1963).
34. McAfee, K. E., and D. Edelson: Sixth International Conference on Ionization Phenomena in Gases, Paris, 1963.
35. David, H. G., and R. J. Munson: *Proc. Roy. Soc. (London)*, **A177**: 192 (1941).
36. Langevin, P.: *Ann. Chim. Phys.*, **28**: 289 (1903).
37. Mitchell, J. H., and K. E. W. Ridler: *Proc. Roy. Soc. (London)*, **A146**: 911 (1934).
38. Langevin, P.: *Ann. Chim. Phys.*, **5**: 245 (1905).
39. Tyndall, E. P. T.: "The Mobility of Positive Ions in Gases," Cambridge University Press, London, 1938.
40. Hornbeck, J. A., and G. H. Wannier: *Phys. Rev.*, **82**: 458 (1951).
41. Langevin, P.: *Ann. Chim. Phys.*, **28**: 433 (1903).
42. Thomson, J. J.: *Phil. Mag.*, **47**: 337 (1924).
43. Natason, G. L.: *Zh. Tekh. Fiz.*, **29**: 1373 (1959).
44. Bates, D. R. (ed.): "Atomic and Molecular Processes," p. 274, Academic, New York, (1962).
45. Sayers, J.: *Proc. Roy. Soc. (London)*, **88**: 488 (1938).
46. Machler, W.: *Z. Phys.*, **164**: 1 (1936).
47. Bates, D. R., and J. T. Lewis: *Proc. Phys. Soc. (London)*, **A68**: 173 (1955).
48. Bates, D. R., and T. J. Boyd: *Proc. Phys. Soc. (London)*, **A69**: 910 (1956).
49. Yeung, H. Y.: *Proc. Phys. Soc. (London)*, **71**: 341 (1958).
50. Greaves, C.: Thesis, University of Birmingham, England, 1959.
51. Stueckelberg, E. C. G., and P. M. Morse: *Phys. Rev.*, **36**: 16 (1930).
52. Bates, D. R., and A. Dalgarno: in D. R. Bates (ed.), "Atomic and Molecular Processes," Academic, New York, 1962.
53. Bates, D. R.: *Phys. Rev.*, **77**: 781; **78**: 492 (1950).
54. Bates, D. R., A. E. Kingston, and R. W. P. McWhirter: *Proc. Roy. Soc. (London)*, **A267**: 297 (1962).
55. Biondi, M. A.: *Phys. Rev.*, **82**: 453 (1951); **83**: 653 (1951).
56. Gray, E. P., and D. E. Kerr: *Ann. Phys.*, **17**: 276 (1962).
57. Hasted, J. B.: "Physics of Atomic Collisions," p. 267, Butterworth, Washington, D.C., 1964.
58. Kuckes, A. F., R. W. Motley, E. Hinnov, and J. G. Hirechberg: *Phys. Rev. Letters*, **6**: 337 (1961).
59. Lozier, W. W.: *Phys. Rev.*, **46**: 268 (1934).
60. Buchel'nikova, N. S.: *Zh. Eksperim. Teor. Fiz.*, **35**: 1119 (1958).
61. Schulz, G. J.: *Phys. Rev.*, **128**: 178 (1962).
62. Herreng, P.: *Cahiers Phys.*, **38**: 7 (1952).
63. Doehring, A.: *Z. Naturforsch.*, **79**: 253 (1952).
64. McAfee, K. B.: *J. Chem. Phys.*, **23**: 1435 (1955).
65. Chanin, L. M., A. V. Phelps, and M. A. Biondi: *Phys. Rev. Letters*, **2**: 344 (1959).
66. Kuffel, E.: *Proc. Phys. Soc. (London)*, **74**: 297 (1959).
67. Huxley, L. G. H.: *Australian J. Phys.*, **12**: 171 (1959).
68. Harrison, M. A., and R. Geballe: *Phys. Rev.*, **91**: 1 (1953).
69. Biondi, M. A.: *Phys. Rev.*, **84**: 1072 (1951).
70. Muschlitz, E. E.: *J. Appl. Phys.*, **28**: 1414 (1957).
71. Fox, R. E.: *Phys. Rev.*, **109**: 2008 (1958).
72. Curran, R. K.: *J. Chem. Phys.*, **35**: 1849 (1961).
73. Craggs, J. D., R. Thorburn, and B. A. Tozer: *Proc. Roy. Soc. (London)*, **A240**: 473 (1957).
74. Mitchell, A. C. G., and M. W. Zemansky: "Resonance Radiation and Excited Atoms," Cambridge University Press, New York, 1961.
75. Druyvesteyn, M. J., and F. M. Penning: *Revs. Mod. Phys.*, **12**: 87 (1940).
76. Biondi, M. A.: *Phys. Rev.*, **82**: 453 (1951).
77. Hasted, J. B.: "Physics of Atomic Collisions," p. 230, Butterworth, Washington, D.C., 1964.
78. Emeleus, K. G., R. W. Lunt, and C. A. Meek: *Proc. Roy. Soc. (London)*, **156**: 394 (1936).
79. MacDonald, A. D., and S. C. Brown: *Phys. Rev.*, **75**: 411 (1949); **76**: 1634 (1949).
80. Loeb, L. B.: "Basic Processes of Gaseous Electronics," University of California Press, Berkeley, Calif., 1955.
81. Cobine, J. D.: "Gaseous Conductors," Dover, New York, 1958.

General References

Bates, D. R.: "Atomic and Molecular Processes," Academic, New York, 1962.

Brown, S. C.: "Basic Data of Plasma Physics," Wiley, New York, 1959.

Francis, Gordon: "Ionization Phenomena in Gases," Academic, New York, 1960.

Hasted, J. B.: "Physics of Atomic Collisions," Butterworth, Washington, D.C., 1964.

Loeb, L. B.: "Basic Processes of Gaseous Electronics," University of California Press, Berkeley, Calif., 1955.

Massey, H. S. W., and E. H. S. Burhop: "Electronic and Ionic Impact Phenomena," Oxford University Press, Fair Lawn, N.J., 1952.

McDaniel, E. W.: "Collision Phenomena in Ionized Gases," Wiley, New York, 1964.

Reed, R. I.: "Ion Production by Electron Impact," Academic, New York, 1962.

Chapter 11

Plasma Physics

By JOHN C. INGRAHAM, Massachusetts Institute of Technology

1. General Plasma Behavior

1.1. The Debye Length and Plasma Sheaths. A plasma, being an ionized medium, is subject to the phenomena described in the preceding chapter. The phenomena that occur in a plasma and distinguish it from any arbitrary collection of charged particles are the near equality of positive and negative charges throughout the plasma volume and the ability of the charges to participate in plasma oscillations (Chap. 10, Sec. 1). In order to qualify as a plasma, a body of ionized gas must be larger than the Debye length l_D, and furthermore l_D must be less than any other length characterizing the plasma.

Small-scale fluctuations of charge density are caused by the random thermal motions of the charged particles. An estimate of the spatial size of these fluctuations may be obtained by equating the electric potential energy V of a sphere of uniform positive charge of radius R to the thermal energy of the electrons that would have to leave this sphere to produce this net positive charge:

$$V = \frac{\frac{3}{5}[(4\pi R^3/3)n + Ze]^2}{4\pi\epsilon_0 R} = \left(\frac{4\pi}{3}R^3\right) n_e \left(\frac{3}{2}kT_e\right)$$

where Ze is the average ionic charge and

$$\epsilon_0 = 8.854 \times 10^{-12} \text{ farad/m}$$

is the permittivity of free space. The rationalized mks system of units is used throughout this chapter. When the units of a quantity are not of the mks system, they are given explicitly. Taking $Zn_+ = n_e$, we have

$$R = \left(\frac{15}{2}\right)^{1/2} \left(\frac{kTe\epsilon_0}{n_e e^2}\right)^{1/2} \equiv \left(\frac{15}{2}\right)^{1/2} l_D$$

where
$$l_D = \left(\frac{kT_e\epsilon_0}{n_e e^2}\right)^{1/2}$$

is called the *Debye length.* Thus the average size of the charge-density fluctuations increases with electron temperature and decreases with increasing electron density. From Chap. 10, Sec. 1 the plasma frequency $\omega_p = (n_e e^2/m_e\epsilon_0)^{1/2}$, and the Debye length may be written as

$$l_D = \left(\frac{kTe\epsilon_0}{n_e e^2}\right)^{1/2} = \frac{1}{\omega_p}\left(\frac{\langle v_e^2\rangle_{av}}{3}\right)^{1/2} \tag{11.1}$$

where $\langle v_e e^2\rangle_{av} = 3kT_e/m_e$ is the root-mean-square electron velocity.

The Debye length also appears in the calculation of the average radius of the spherical volume surrounding a positive ion, in which the ion's electric field is approximately unshielded by electrons. Outside this Debye sphere the ion potential falls rapidly to zero. Solving Poisson's equation for the electric potential ϕ around an ion under the assumption that the electrons are distributed according to $n_e = n_{e\infty}e^{+\phi/kT_e}$ and that $e\phi/kT_e \ll 1$, we find the potential to be given by

$$\phi = \frac{e}{4\pi\epsilon_0 r}\exp(-r/l_D) \tag{11.2}$$

which is a shielded Coulomb potential. This discussion neglects the ion thermal motion. This results in only a small error as long as the ion temperature does not greatly exceed the electron temperature.

The concept of a Debye length is valid only so long as the number of electrons, n_{lD}, in a sphere of radius l_D is appreciably greater than 1:

$$n_{lD} = \frac{4\pi}{3} l_D{}^3 n_e = \frac{4\pi}{3n_e{}^{1/2}}\left(\frac{kT\epsilon_0}{e^2}\right)^{3/2} \gg 1 \tag{11.3}$$

For this condition to be satisfied, the Debye length must be greater than the interparticle spacing.

Figure 11.1 [1]† shows how l_D and ω_p vary for the various plasmas encountered. Also indicated are the region in which $n_{lD} > 1$ and the region in which $n_{lD} < 1$.

Unless the ionized medium has dimensions much larger than l_D, it will be subject to large charge-density fluctuations and cannot qualify as a plasma. Furthermore, from the derivation of Chap. 10, Sec. 1 for plasma oscillations, it can be seen that if the plasma volume is only of the order of l_D in thickness the random thermal velocity $\bar{v}_e \sim (kT_e/m_e)^{1/2}$ is able to damp the oscillation. This is because the average electron transit time across the volume is, in this case, $t_T \sim (\bar{v}_e/l_D)^{-1} \approx 1/\omega_p$, that is, of the same order as the oscillation frequency. Thus, the plasma must have dimensions much greater than l_D in order to support oscillations. If electron collision or loss rates are of the order of ω_p, plasma oscillations are damped out. All the above criteria for the existence of plasma oscillations may be summarized by stating the l_D must

† Numbers in brackets refer to References at end of chapter.

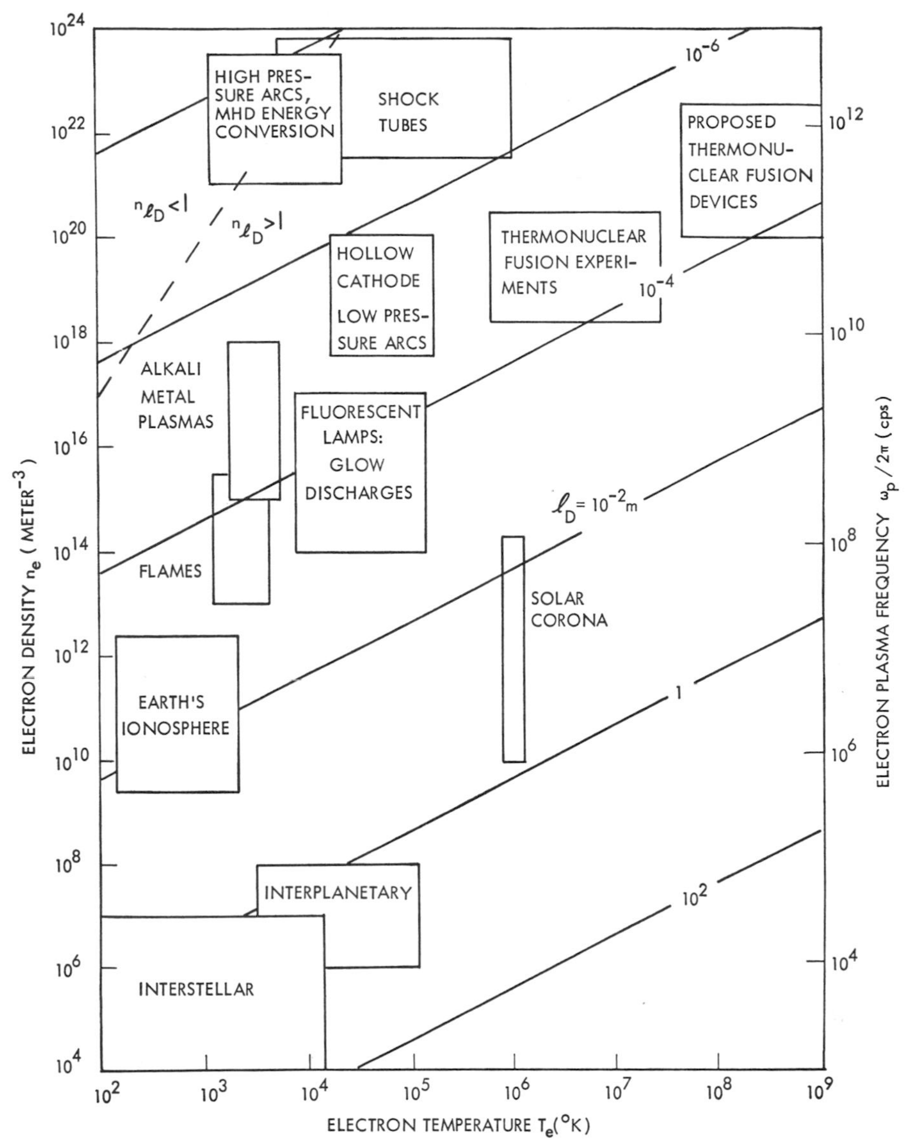

FIG. 11.1. Electron density, temperature, plasma frequency, and Debye length for typical plasmas.

be less than any of the other characteristic plasma lengths (except the average interparticle spacing).

From these arguments on charge neutrality it appears that any charged layer or sheath that exists at a plasma boundary will be approximately l_D in thickness. These sheaths generally do exist since the electrons and ions tend to diffuse and drift from the volume at different rates, and it is only through the presence of a charged sheath that retards the more mobile charged constituents that the plasma retains its neutrality.

1.2. Longitudinal Electron Plasma Waves and the Vlasov Equation. The derivation of the plasma oscillation frequency in Chap. 10, Sec. 1 neglects electron-temperature effects. Just as is the case for sound waves in a gas, it was found that this longitudinal oscillation does not propagate in the zero-temperature medium. The inclusion of the electron temperature (or, equivalently, pressure) gives rise to a propagating *longitudinal* plasma wave, so-called because the associated electric field is in the direction of propagation.

To solve this problem it is convenient to use the *Vlasov equation* or *collisionless Boltzmann equation* in which the charged particles interact only through the mutually induced space-charge field and collisions are assumed to be negligible. In this approximation Eq. (10.9) becomes, for the electron distribution function f,

$$\frac{\partial f}{\partial t} + \mathbf{v} \cdot \nabla_r f - \frac{e}{m_e} \mathbf{E} \cdot \nabla_v f = 0 \qquad (11.4)$$

Poisson's equation is used to determine **E**,

$$\nabla \cdot \mathbf{E} = \frac{e}{\epsilon_0}\left(n_+ - \int f\, d^3v\right) \tag{11.5}$$

The longitudinal electron wave is solved for by linearizing this equation, assuming the perturbation $\mathbf{E} = \mathbf{E}_z(z,t)$ to be small and the distribution function $f = f^0(\mathbf{v}) + f^1(\mathbf{v},z,t)$. This is a somewhat different approximation than that of Eq. (10.10). The resulting equations are

$$\frac{\partial f^1}{\partial t} + v_z \frac{\partial f^1}{\partial z} - \frac{eE_z}{m_e}\frac{\partial f^0}{\partial v_z} = 0 \tag{11.6}$$

$$\frac{\partial E_z}{\partial z} = -\frac{e}{\epsilon_0}\int f^1\, d^3v \tag{11.7}$$

where use has been made of the fact that

$$\int f^0\, d^3v = n_{e0} = n_{+0} \tag{11.8}$$

and the ion motion has been neglected so that $n_+ = n_{+0}$. Now E_z and f^1 are assumed to vary as $e^{i(\omega t - kz)}$, and the solution of f_1 is obtained from Eq. (11.6):

$$f^1 = \frac{(eE_z/m_e)(\partial f^0/\partial v_z)}{i(\omega - kv_z)} \tag{11.9}$$

If this expression is substituted into Eq. (11.7) an equation relating ω and k (the dispersion relation) is obtained:

$$-1 = \frac{n_{e0}e^2}{km\epsilon_0}\int \frac{\partial f_v{}^0/\partial v_z}{\omega - kv_z}\, d^3v = \frac{\omega_p{}^2}{k}\int \frac{\partial f_v{}^0/\partial v_z}{\omega - kv_z}\, d^3v \tag{11.10}$$

where $f^0 \equiv n_{e0} f_v{}^0$ and $\omega_p = \left(\frac{n_{e0}e^2}{m_e\epsilon_0}\right)^{1/2}$

is the electron plasma frequency.

The singularity of Eq. (11.10) arises at $v_z = \omega/k$. Physically this is due to the electrons of the distribution which move with the wave and hence are acted on by a d-c field rather than an oscillating electric field. Depending on the nature of $f_v{}^0$, this can give rise either to amplification or attenuation of the wave. The latter case is referred to as *Landau damping* [2]. In any case, this does not give a stationary oscillatory solution as assumed. If k or ω is allowed to be complex, the singularity is removed and the solution of Eq. (11.10) yields a growing or damped wave, depending upon $f_v{}^0$, as stated above. In many cases the electron thermal velocity is much less than the wave velocity ω/k, and a good approximation to the solution is obtained by neglecting the singularity and expanding:

$$\frac{1}{\omega - kv_z} \approx \frac{1}{\omega}\left[1 + \frac{kv_z}{\omega} + \left(\frac{kv_z}{\omega}\right)^2 + \cdots\right]$$

Assuming $f_v{}^0$ to be Maxwellian, we find, keeping only terms up to $(kv_z/\omega)^2$,

$$1 = \frac{\omega_p{}^2}{\omega^2}\left[1 + \left(\frac{k}{\omega}\right)^2 \frac{3kT_e}{m_e}\right] = \frac{\omega_p{}^2}{\omega^2}\left[1 + \left(\frac{k}{\omega}\right)^2 \langle v_e{}^2\rangle_{\text{av}}\right] \tag{11.11}$$

which is valid only for $(k/\omega)^2 \langle v_e{}^2\rangle_{\text{av}} \ll 1$. This equation may be solved for the wave's phase velocity ω/k,

$$\left(\frac{\omega}{k}\right)^2 = \frac{\langle v_e{}^2\rangle_{\text{av}}}{\omega^2/\omega_p{}^2 - 1} \tag{11.12}$$

This equation reveals that the electron plasma wave propagates at frequencies $\omega > \omega_p$; at $\omega = \omega_p$ the wavelength becomes infinite, and the wave becomes a nonpropagating oscillation. This expression is valid only for ω slightly greater than ω_p since for $\omega \geq \sqrt{2}\,\omega_p$ the wave phase velocity is equal to or less than $(\langle v_e{}^2\rangle_{\text{av}})^{1/2}$ and Landau damping would be expected to become important. In Fig. 11.2 is plotted the positive part of $(\omega/k)^2$ of Eq. (11.12), showing qualitatively at what frequencies the wave can propagate.

1.3. Longitudinal Ion Plasma Waves. In the previous section motion of the ions was neglected For *sufficiently high electron temperatures* and low fre

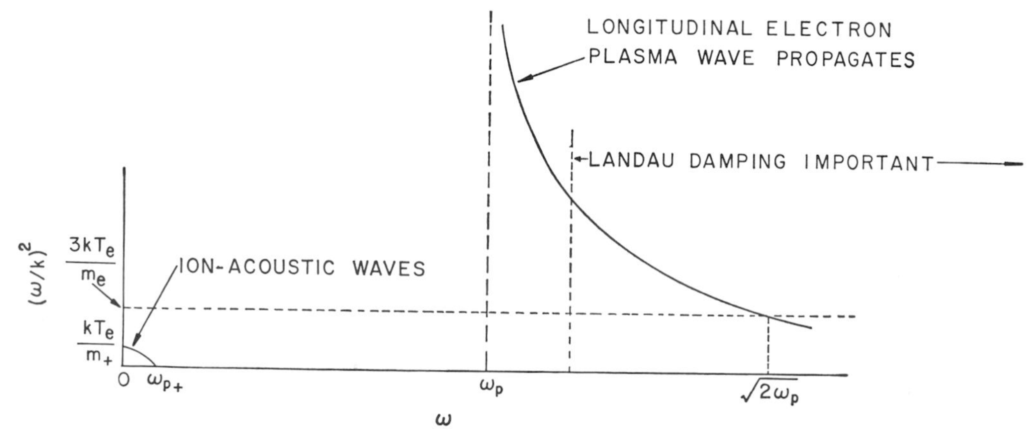

FIG. 11.2. Square of the wave phase velocity for longitudinal electron and ion plasma waves as a function of ω (not to scale).

quencies, however, the ions oscillate, giving rise to a propagating compression wave in the ion density. Recalling the derivation for electron plasma oscillations of Chap. 10, Sec. 1, we express the ion density n_+ as

$$n_+ = n_{+0}\left(1 - \frac{\partial \zeta}{\partial x}\right) \tag{11.13}$$

where ζ now represents the position of a volume element of ion density with respect to its equilibrium position and is governed by Newton's equation

$$m_+ \frac{\partial^2 \zeta}{\partial t^2} = -e\frac{\partial \phi}{\partial x} \tag{11.14}$$

where ϕ is the electric potential. Poisson's equation for ϕ has the form

$$\frac{\partial^2 \phi}{\partial x^2} = -\frac{e}{\epsilon_0}\left[n_{+0}\left(1 - \frac{\partial \zeta}{\partial x}\right) - n_{e0}e^{e\phi/kT_e}\right] \tag{11.15}$$

Here we assume that the ion oscillation frequencies are so low that electrons remain distributed spatially according to the Boltzmann factor $e^{e\phi/KT_e}$. Setting $e^{e/\phi KT_e} \approx 1 + e\phi/kT_e$ and $n_{+0} = n_{e0}$, we obtain

$$\frac{\partial^2 \phi}{\partial x^2} = \frac{n_{e0}e}{\epsilon_0}\left(\frac{\partial \zeta}{\partial x} + \frac{e\phi}{kT_e}\right) \tag{11.16}$$

Equations (11.14) and (11.16) together can be solved by assuming ϕ and ζ to be described by a longitudinal disturbance $e^{i(\omega t - kx)}$. The resulting dispersion relation relating ω and k is

$$\omega^2 = \frac{(kl_D)^2 \omega_{p+}{}^2}{1 + (kl_D)^2} \tag{11.17}$$

where $$\omega_{p+} = \left(\frac{n_{e0}e^2}{m_+\epsilon_0}\right)^{1/2}$$

is the ion "plasma frequency" and l_D is the Debye length. For low frequencies the wave's phase velocity ω/k is given approximately by

$$\frac{\omega}{k} \approx l_D\omega_{p+} = \left(\frac{kT_e}{m_+}\right)^{1/2} \tag{11.18}$$

The wave is dispersionless in this regime and is referred to as the ion acoustic wave. It may be pictured as arising from a combination of the ion inertia (m_+) and the electron pressure ($n_{e0}kT_e$). The maximum frequency is $\omega = \omega_{p+}$ at which point $\omega/k = 0$. The positive part of the square of the phase velocity of this wave is shown in Fig. 11.2. If the electron temperature is not considerably greater than the ion temperature, the wave moves with a velocity comparable with the ion thermal velocity and suffers Landau damping.

Alexeff and Neidigh [3] and Wong, D'Angelo, and Motley [4] have performed experimental studies of the damping and propagation of ion acoustic waves.

1.4. Beam-Plasma Interaction: An Example of an Instability. Equation (11.9) relates the perturbed part of the distribution function f^1 to the unperturbed part f^0. In order to study the propagation of the longitudinal wave through a plasma that is simultaneously being traversed (in the wave-propagation direction) by an electron beam, f^0 may be assumed to be composed of a stationary distribution due to the plasma electrons plus a drifting distribution due to the beam electrons. Thermal motions are neglected in both groups of electrons. Substituting into Eq. (11.10) and using ω_{p_1} and ω_{p_2} to represent the plasma-electron and beam-electron plasma frequencies, respectively, we find for the dispersion relation

$$1 = \frac{\omega_{p_1}{}^2}{\omega^2} + \frac{\omega_{p_2}{}^2}{(\omega - kv_B)^2} \tag{11.19}$$

where v_B is the beam drift velocity.

If $\omega_{p_2} = 0$ we recover the relationship between ω_p and ω for plasma oscillations of a zero-temperature plasma as derived in Chap. 10, Sec. 1. The term proportional to $\omega_{p_2}{}^2$ is analogous to the plasma-electron term, but since the beam electrons are drifting they are subject to a Doppler-shifted oscillation frequency $\omega' = \omega - kv_B$.

Equation (11.19) may be solved for the real and imaginary parts of k, k_r, and k_i, assuming ω to be real. The solution thus found reveals that for frequencies $-\omega_{p_1} < \omega < \omega_{p_1}$ two waves are propagating, each with a phase velocity $\omega/k_r = v_B$. As it propagates, one of these waves grows at a rate

$$k_i = k_r \frac{\omega_{p_2}}{(\omega_{p_1}{}^2 - \omega^2)^{1/2}}$$

and the other attenuates at a rate

$$k_i = -k_r \frac{\omega_{p_2}}{(\omega_{p_1}{}^2 - \omega^2)^{1/2}}$$

The presence of growing or decaying waves when $\omega/k = v_B$ is reminiscent of the Landau damping mentioned in Sec. 1.2, which arises when the electron distribution contains some electrons that move with the wave.

The growing wave has a maximum growth rate in the vicinity of $\omega \approx \omega_{p_1}$. This type of instability contributes to the noise generated in d-c gas discharges. The original oscillations noted by Tonks and Langmuir were caused by the plasma interaction with the electron beam that was produced as the electrons leaving the cathode were accelerated through the cathode fall. Ion oscillations and waves may also be excited by drifting electrons when the electron drift or beam velocity becomes of the order of the ion acoustic-wave velocity. They have been studied theoretically by Pines and Schrieffer and by Ichimaru [5] and experimentally by V. Arunasalam [6]. Several experimenters [7] have studied the interaction between an electron beam and a plasma. Bers and Briggs [8] have developed a rather general stability criterion which may be applied to plasma-wave systems. Stix [9] gives a treatment of beam-plasma interaction.

1.5. Plasma Collisional Relaxation Rates. The collision between two charged particles in a plasma is not a simple binary collision. Because of the long-range nature of the Coulomb interaction, the fields of many particles act during a scattering event, and the scattering angle due to each event, on an average, is small. Spitzer [10] has treated this problem by considering the effective collision frequency to be the average reciprocal time required for the path of a charged particle to be turned through an angle of 90° owing to the cumulative effect of many small-angle collisions. The electron-ion collision frequency may be estimated another way by using Eq. (10.15), in which the differential scattering cross section corresponding to the shielded Coulomb potential (Sec. 1.1) of the ion is used. From such calculation, rates for energy distribution among the particles (thermalization times) may be determined.

Three basic energy relaxation times are given by Spitzer; we shall designate them as τ_{ee}, τ_{++}, and τ_{e+}. They represent, respectively, the rates at which electrons come to energy equilibrium among themselves, ions come to energy equilibrium among themselves, and electrons come to energy equilibrium with the ions. They have the values

$$\tau_{ee} = \frac{11.4 T_e^{3/2}}{n \ln \Lambda} A_e^{1/2} \quad \text{sec} \tag{11.20a}$$

$$\tau_{++} = \frac{11.4 T_+^{3/2}}{n \ln \Lambda} \frac{A_+^{1/2}}{Z_+^4} \quad \text{sec} \tag{11.20b}$$

$$\tau_{e+} = \frac{5.87}{n \ln \Lambda} \frac{A_e A_+}{Z_+^2} \left(\frac{T_e}{A_e} + \frac{T_+}{A_+}\right)^{3/2} \quad \text{sec} \tag{11.20c}$$

where n is the ion or electron density in cm^{-3}, $A_e = 1/1{,}823$ for electrons, A_+ is the ion atomic weight, Z_+ the ion charge number, T_e and T_+ the temperatures in degrees Kelvin, and Λ is the ratio of the electron Debye length l_D to the impact parameter P_0, corresponding to a 90° scattering of an electron in a two-particle (binary) Coulomb scattering event. The $\ln \Lambda$ term represents the effect of the shielding of the Coulomb fields. Generally l_D/P_0 is much greater than 1, and $\ln \Lambda$ is between 10 and 20 for most plasmas of interest. Taking $T_+ = T_e$, $Z_+ = 1$, $A_+ = 1$, we can compare the relaxation times for protons and electrons:

$$\begin{aligned} \tau_{ee} &= \frac{11.4 T_e^{3/2}}{n \ln \Lambda} \left(\frac{1}{1{,}823}\right)^{1/2} = \frac{11.4 T_e^{3/2}}{n \ln \Lambda} \frac{1}{43} \\ \tau_{++} &= \frac{11.4 T_e^{3/2}}{n \ln \Lambda} = 43 \tau_{ee} \\ \tau_{e+} &\cong \frac{11.4 T_e^{3/2}}{n \ln \Lambda} \frac{5.87}{11.4} (43) \cong \frac{1{,}823}{2} \tau_{ee} \end{aligned} \tag{11.21}$$

The time scales for the thermalization between the different particles are quite different because of the electron-proton mass ratio. Thus, if a plasma is produced with nonthermal distributions of energies, one may expect that first the energy of the electrons will have a Maxwellian distribution, followed at a much slower rate by the thermalization of the ions among themselves, and finally, if $T_+ \neq T_e$, the two temperatures will equalize on an even slower scale of time. For an electron temperature of 10,000°K, an electron density of $10^{12}\ \text{cm}^{-3}$, $\ln \Lambda = 9.43$ and $\tau_{ee} = 2.8 \times 10^{-8}$ sec.

1.6. Plasma Electrical Conductivity, Viscosity, and Thermal Conductivity. Another quantity of interest in connection with the electron-electron and electron-ion collisions is the plasma d-c conductivity. The electron-ion collision frequency ν_e is equal to $1/\tau_{ee}$ of Eq. (11.20*a*). This is equivalent to an electron-moment-loss frequency as defined in Eq. (10.15) and can be used to define a d-c mobility as in Eq. (10.27), $\mu = e/m\nu_e$, from which the electrical conductivity $\sigma = ne\mu = ne^2/m\nu_e$ can be calculated. Spitzer gives such a value for the reciprocal electrical conductivity but which is corrected for the effect of electron-electron collisions:

$$\frac{1}{\sigma} = 6.53 \times 10^3 \frac{\ln \Lambda}{T^{3/2}} \quad \text{ohm-cm} \tag{11.22}$$

where $T_e = T_+ = T$, $Z_+ = 1$, and electron-atom collisions are negligible. The conductivity is independent of electron density because it is the ions that impede the electron motion. This equation also describes a-c plasma conductivity provided ω is much less than ω_p and ν_e. Heald and Wharton [11] review the various plasma-conductivity calculations.

The viscosity and thermal-conduction coefficients of a fully ionized plasma in the absence of a magnetic field may be estimated by using the elementary expressions given in Part 5, Chap. 2, Sec. 3 for the viscosity and thermal-conduction coefficients for particles having a mean free path l. The viscosity coefficient as given there is $\eta = \frac{1}{3} nm\bar{v}l$ and the thermal-conduction coefficient is $\kappa = \frac{1}{3} nc_v\bar{v}l$, where c_v is the heat capacity per particle. The electrons contribute predominantly to the plasma-transport equations because their velocities are generally much greater than the ion velocities. The electron mean free path in the plasma can be approximated by $l \approx \bar{v}/\nu_e$. Using the expression for the electrical conductivity $\sigma \approx ne^2/m\nu_e$ and setting $\bar{v}^2 \approx 3kT/m$, we may write, for η and κ, $\eta = kTm\sigma/e^2$ and

$$\kappa = (3/2) kTk\sigma/e^2$$

Using Eq. (11.22), we obtain

$$\eta = 7.5 \times 10^{-18} \frac{T^{5/2}}{\ln \Lambda} \quad \text{kg/m-sec}$$

and $$\kappa = 4.1 \times 10^{-13} \frac{T^{5/2}}{\ln \Lambda} \quad \text{cal/(°K-cm)(sec)}$$

The thermal conductivity given here is about two-fifths of the more exact value of Spitzer [10] for the case of a plasma of electrons and singly ionized ions.

1.7. Plasma Particle Correlations and the Scattering of Electromagnetic Radiation. The Debye length may be regarded as representing a correlation distance between two charges; that is, if the particles are separated by distances greater than l_D their fields are shielded from one another and their motions are uncorrelated. The correlated motion of particles separated by less than l_D plays an important role in the scattering of electromagnetic radiation from a plasma.

The experiments of Bowles [12] in which the scattering of radar signals from the ionosphere was measured were the first experimental verification of the correlation effects. The spectrum of the scattered radiation was not broadened by a Doppler effect due to the electron thermal velocities but by a Doppler effect due to the ion thermal motion even though the radiation is scattered from the electrons. This result is in agreement with the theory for scattering from the correlated charge fluctuations of a plasma. The most recent treatment of this type of scattering is by Rosenbluth and Rostoker [13], in which references to earlier papers by other authors can be found.

2. Single-particle Motions in Electric and Magnetic Fields

Although in most plasmas collective effects among the charged particles or collisions with gas atoms are not negligible, an understanding of the motion of a single particle in d-c electric and magnetic fields is often helpful in analyzing the behavior of the plasma as a whole.

2.1. Particle Motion in a Constant Homogeneous Magnetic Field and the Particle Magnetic Moment. A charged particle moving in a time-independent and homogeneous magnetic field moves on a helical path, with the helix axis parallel to the magnetic-field lines. The circulating component of the particle motion has a vector angular frequency $\boldsymbol{\omega}_B = -q\mathbf{B}/m$, where q and m are the charge and mass of the particle and $\mathbf{B}$ the magnetic field. The charge q is $+e$ and $-e$ for ions and electrons, respectively. This frequency, called the *cyclotron frequency,* is independent of particle velocity so long as relativistic effects may be ignored. The vector velocity $\mathbf{v}_B$ of the particle prependicular to the magnetic field is $\mathbf{v}_B = \boldsymbol{\omega}_B \times \mathbf{r}_B$, where $\mathbf{r}_B$ is the vector radius of the circular motion. The center of the particle orbit drifts with a constant velocity $v_{d\|}$ parallel to the B lines, and the pitch angle for the helical path is defined in the equation $\tan \theta_p = v_{d\|}/v_B$.

The magnetic moment of the particle due to its orbiting motion is [14]

$$\boldsymbol{\mu}_B = \frac{1}{2}\int \mathbf{r} \times \mathbf{J}\, dV = \frac{q\omega_B}{2\pi}\pi r_B{}^2 = \frac{-mv_B{}^2}{2}\frac{\mathbf{B}}{B^2} \tag{11.23}$$

and is seen to be directed opposite to the magnetic field, proportional to the transverse kinetic energy, and independent of charge.

2.2. Adiabatic Invariance of the Magnetic Moment. The magnetic moment of a charged particle subject to a changing magnetic field remains invariant so long as the magnetic field changes by a small fraction of itself during one orbit of the particle. Note that this change can be due either to a time-dependent magnetic field or to the particle's motion along the lines of a spatially varying magnetic field. From Eq. (11.23) we see that $mv_B{}^2/2$ must be proportional to B in order that $\boldsymbol{\mu}_B$ be invariant.

If B changes by only a small fraction of itself in one cyclotron orbit, the particle motion will be very nearly circular, and the magnetic force $q\mathbf{v}_B \times \mathbf{B}$ may be equated to the centripetal force $+m\boldsymbol{\omega} \times (\boldsymbol{\omega} \times \mathbf{r})$. Since $\mathbf{v}_B = \boldsymbol{\omega} \times \mathbf{r}$, this gives

$$q\mathbf{v}_B \times \mathbf{B} = +m\boldsymbol{\omega} \times \mathbf{v}_B = -m\mathbf{v}_B \times \boldsymbol{\omega}$$

or $\boldsymbol{\omega} = -q\mathbf{B}/m = \boldsymbol{\omega}_B$ as before, only now $\mathbf{B}$ may be varying. The induced emf around one nearly circular orbit is $\oint \mathbf{E} \cdot d\mathbf{s} = 2\pi r_B \bar{E} = -\dot{B}\pi r_B{}^2$, where $\bar{E}$ is the average electric field tangent to the orbit. This gives $\bar{E} = -\dot{B}r_B/2$. From the work-energy theorem

$$\frac{d}{dt}\frac{mv_B{}^2}{2} = \overline{q\mathbf{E} \cdot \mathbf{v}_B} = -\frac{qv_B\dot{B}r_B}{2} \tag{11.24}$$

The magnetic force is always at right angles to the velocity so that it does no work. The velocity v_B on the right-hand side of Eq. (11.24), although written as a scalar, still contains a sign dependence to account for the sign of the work done by the induced electric field. To obtain the correct sign for the work, $v_B = -(r_B qB/m)$. Setting $r_B = -(v_B m/qB)$, we obtain

$$\frac{d}{dt}\frac{mv_B{}^2}{2} = -\frac{qv_B}{2}\dot{B}\left(-\frac{v_B m}{qB}\right) = \frac{mv_B{}^2}{2}\frac{\dot{B}}{B} \tag{11.25}$$

or

$$\frac{d}{dt}\ln\frac{mv_B{}^2}{2} = \frac{d}{dt}\ln B$$

giving

$$\frac{mv_B{}^2}{2} = B \times \text{const} \tag{11.26}$$

which demonstrates that $\boldsymbol{\mu}_B$ is an adiabatic invariant.

2.3. The Guiding-center Approximation and a Second Adiabatic Invariant. We have shown that $\boldsymbol{\mu}_B$ remains constant in a slowly varying magnetic field. It is convenient to consider this magnetic moment as an intrinsic part of the particle and to discuss the motion of the *center* of the particle's orbital motion. The motion of this *guiding center* is determined by the usual Lorentz forces on a charged particle plus the forces exerted on a magnetic dipole in an inhomogeneous magnetic field. This procedure is called the guiding-center approximation. Figure 11.3 shows the type of description for the particle position that is used: the position of the guiding center of the particle, $\mathbf{r}_d$, with respect to the center of sym-

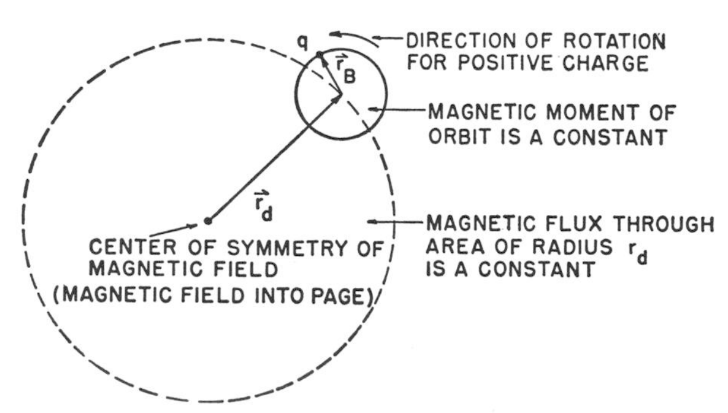

FIG. 11.3. Guiding-center description of particle motion in a plane perpendicular to **B**.

metry of the magnetic field, and the position of the orbiting particle with respect to $\mathbf{r}_d$, as given by $\mathbf{r}_B$.

We begin by assuming the particle velocity to be the sum of the orbiting velocity $\mathbf{v}_B = \mathbf{r}_B \times \boldsymbol{\omega}_B$ described in the previous sections plus the drift velocity $\mathbf{v}_d = (d/dt)\mathbf{r}_d$ of the guiding center, $\mathbf{v} = \mathbf{v}_d + \mathbf{v}_B$. Substituting into the force equation and writing $\mathbf{v}_d$ in terms of components parallel to and perpendicular to $\mathbf{B}$ gives

$$\frac{d}{dt}(\mathbf{v}_B + \mathbf{v}_{d\perp} + \mathbf{v}_{d\|}) = \frac{\mathbf{F}}{m} + \frac{q}{m}\mathbf{v}_{d\perp} \times \mathbf{B} + \frac{q}{m}\mathbf{v}_B \times \mathbf{B} \tag{11.27}$$

The force $\mathbf{F}$ includes electrical forces as well as any inertial forces arising from viewing the motion from an accelerated frame. We neglect $(d/dt)\mathbf{v}_{d\perp}$ in (11.27), assume that $(d/dt)\mathbf{v}_B = (q/m)(\mathbf{v}_B \times \mathbf{B}_0)$, where $\mathbf{B}_0$ is the magnetic field at the position of the guiding center, and set $(q/m)(\mathbf{v}_{d\perp} \times \mathbf{B}) \approx (q/m)(\mathbf{v}_{d\perp} \times \mathbf{B}_0)$. Three separate equations may then be written:

$$\frac{d\mathbf{v}_{d\|}}{dt} = \frac{\mathbf{F}_\|'}{m} \tag{11.28a}$$

$$\frac{d\mathbf{v}_B}{dt} = \frac{q}{m}\mathbf{v}_B \times \mathbf{B}_0 \tag{11.28b}$$

$$0 = \frac{\mathbf{F}_\perp'}{m} + \frac{q}{m}\mathbf{v}_d \times \mathbf{B}_0 \tag{11.28c}$$

The force $\mathbf{F}'$ now includes effects due to the variation of $\mathbf{B}$ across the particle orbit which are averaged over one orbit of the particle. These average forces are calculated from the force on a magnetic dipole in an inhomogeneous magnetic field in the following sections. Equation (11.28*b*) is used in Secs. 2.1 and 2.2 to calculate $\boldsymbol{\mu}_B$. The solution of Eq. (11.28*c*) is

$$\mathbf{v}_{d\perp} = \frac{\mathbf{F}_\perp' \times \mathbf{B}_0}{qB_0^2} \tag{11.29}$$

which is a drift perpendicular to both $\mathbf{F}_\perp'$ and $\mathbf{B}_0$. If $\mathbf{F}_\perp' = q\mathbf{E}_\perp$, then $\mathbf{v}_d = (\mathbf{E}_\perp \times \mathbf{B}_0)/B_0^2$, which is the familiar drift motion encountered in crossed $\mathbf{E}$ and $\mathbf{B}$ fields. For an electric field $E = 1{,}000$ volts/m and $B_0 = 0.1$ weber/m^2 (1,000 gauss), $|v_d| = 10^5$ m/sec independent of the particle charge or mass. For comparison, the thermal velocity of room-temperature electrons is about 10^5 m/sec.

A second adiabatic invariant of great importance is the magnetic flux through the area πr_d^2, $\phi_d = \pi r_d^2 B$. The invariance of this quantity demonstrates that the guiding center always stays with a given B line so that as the magnetic field increases the guiding centers move toward the origin (see Fig. 11.3). Thus by increasing the magnetic field the plasma is compressed. The guiding centers are not actually restricted to a single B line but may (for a cylindrically symmetric $\mathbf{B}$ field, for instance) move in a circle about the origin moving between *equivalent B lines*.

To demonstrate the invariance of ϕ_d we consider Eq. (11.29) for the case where $\mathbf{F}_\perp'$ is due only to an induced electric field due to a changing $\mathbf{B}$. If we assume $\mathbf{B}$ to be cylindrically symmetric about the origin of Fig. 11.3, then

$$\mathbf{E} = \frac{\mathbf{r}_d \times \dot{\mathbf{B}}}{2} \tag{11.30}$$

and assuming that $\mathbf{B}$ is parallel or antiparallel to $\mathbf{B}$, Eq. (11.29) becomes

$$\mathbf{v}_d = \dot{\mathbf{r}}_d = q\left(\frac{\mathbf{r}_d \times \dot{\mathbf{B}}}{2}\right) \times \frac{\mathbf{B}}{qB^2} = -\mathbf{r}_d\frac{\dot{B}}{2B} \tag{11.31}$$

where $\dot{B}$ is a scalar whose sign is positive if $\dot{\mathbf{B}}$ is parallel to $\mathbf{B}$. Equation (11.31) states that $\dot{\mathbf{r}}_d$ is parallel or antiparallel to $\mathbf{r}_d$ so that it may be written as a scalar equation $\dot{r}_d = -(r_d\dot{B}/2B)$, from which follows

$$2B\dot{r}_d + r_d\dot{B} = \frac{1}{r_d}\frac{d}{dt}(r_d^2B) = 0 \tag{11.32}$$

thus demonstrating that $\phi_d = \pi r_d^2 B$ is an adiabatic invariant.

In the adiabatic approximation, guiding centers always remain attached to the same, or *equivalent*, B lines, so that if B increases, the plasma is compressed. From this it follows that

$$B/n = \text{const} \tag{11.33}$$

where n is the volume density of the charged particle being considered. This equation is not true if plasma compression *along the lines* of $\mathbf{B}$ occurs.

2.4. Motion in an Inhomogeneous Magnetic Field. A spatial variation of $\mathbf{B}$ across the plane of the cyclotron orbit causes the particle orbit to be more sharply curved where $\mathbf{B}$ is greatest, thus giving rise to a drift of the particle perpendicular to $\mathbf{B}$ and perpendicular to the transverse component of ∇B^2. Figure 11.4 illustrates this drift for an extreme case where the particle moves along the interface between two different values of B (B_1 and B_2), where $B_2 = \frac{3}{2}B_1$. If in addition the $\mathbf{B}$ lines linking the

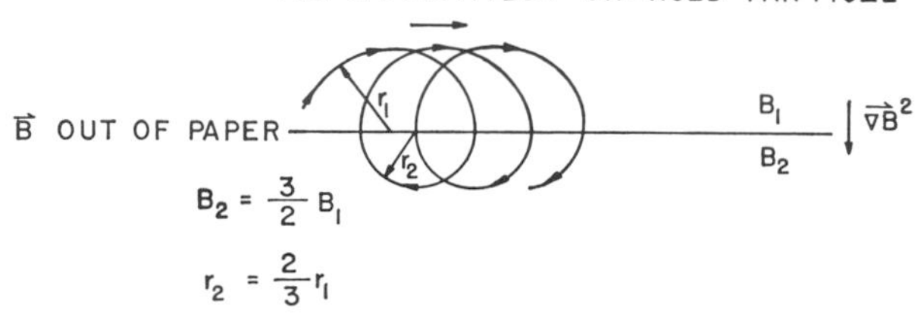

FIG. 11.4. Guiding-center drift due to a gradient of $\mathbf{B}$ perpendicular to $\mathbf{B}$. The case shown is for a positively charged particle. A negative particle drifts in the opposite direction.

cyclotron orbit are diverging or converging this will give rise to a net $q\mathbf{v} \times \mathbf{B}$ force directed parallel to the lines of **B**. Both these forces may be accounted for by considering the force exerted on a magnetic dipole by an inhomogeneous magnetic field [14] $\mathbf{F}_{\mu B} = \nabla(\boldsymbol{\mu}_B \cdot \mathbf{B})$, where the gradient operator operates only on **B**. This expression is valid provided $\nabla \times \mathbf{B} = 0$; hence the current density **J** must be small. Using $\boldsymbol{\mu}_B = -mv_B^2/2B^2\mathbf{B}$ and accounting for the fact that ∇ does not operate on $\boldsymbol{\mu}_B$, we have

$$\mathbf{F}_{\mu B} = -\frac{mv_B^2}{2B}\nabla B = -\frac{mv_B^2}{4B^2}\nabla B^2 \tag{11.34}$$

This force has components parallel and perpendicular to **B**. The parallel component is substituted for $F_\parallel'$ in Eq. (11.28*a*) to obtain the particle acceleration parallel to **B**. It gives an acceleration toward the region of weakest B field and is the basis of plasma containment by use of a magnetic mirror (Sec. 2.6). The force transverse to **B** gives rise to a drift velocity when substituted into Eq. (11.29):

$$\mathbf{v}_{d\perp} = -\frac{mv_B^2}{4B^2}\frac{\nabla B^2 \times \mathbf{B}}{qB^2} = \frac{mv_B^2\mathbf{B} \times \nabla B^2}{4qB^4} \tag{11.35}$$

which is dependent on the particle charge and perpendicular to **B** and ∇B^2 as predicted above.

2.5. Drift Motion Due to Curvature of the Magnetic-field Lines. If the lines of **B** are curved, the guiding-center motion describes a curved path as it follows along the lines of **B**. If this motion is viewed from a coordinate system moving with the guiding center, there is an inertial force (the centrifugal force) acting on the particle and directed *away* from the center of curvature of the **B** lines. This force is equal to $(mv_{d\parallel}^2/R)(\mathbf{R}/R)$, where $\mathbf{v}_{d\parallel}$ is the particle drift velocity along **B**, and **R** is the radius of curvature of the **B** lines as shown in Fig. 11.5. This transverse inertial force when substituted into Eq. (11.29) gives rise to a drift

$$\mathbf{v}_{d\perp} = \frac{mv_{d\parallel}^2}{R}\frac{\mathbf{R} \times \mathbf{B}}{RqB^2} = -\frac{mv_{d\parallel}^2}{qB^2}\mathbf{B} \times \frac{\mathbf{R}}{R^2} \tag{11.36}$$

Setting [15] $\mathbf{R}/R^2 = -\left(\frac{(\mathbf{B}\cdot\nabla)}{B^2}\right)\mathbf{B}$, we obtain

$$\mathbf{v}_{d\perp} = \frac{mv_{d\parallel}^2}{qB^4}\mathbf{B} \times (\mathbf{B}\cdot\nabla)\mathbf{B} \tag{11.37}$$

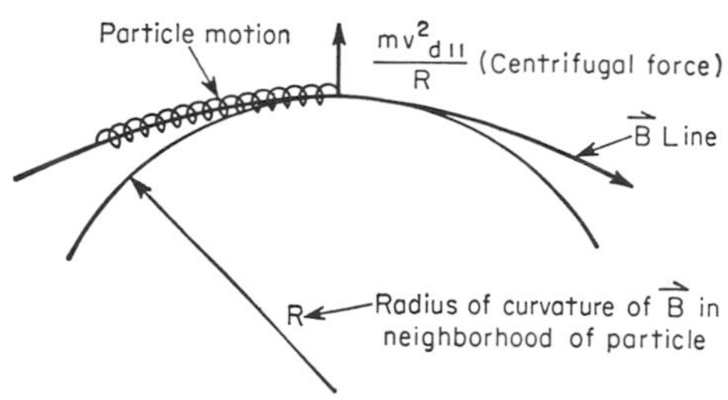

FIG. 11.5. Particle motion along the curved **B** line showing the centrifugal force $mv_{d\parallel}^2/R$ that arises when motion is viewed from a frame moving with the guiding center.

This drift, like the drift of Sec. 2.4, is charge-dependent. Its direction is perpendicular to **B** and **R**.

Such charge-dependent drifts can give rise to electric fields owing to charge separation in certain geometries. This electric field in turn gives rise to an $\mathbf{E} \times \mathbf{B}$ drift of the charged particles.

A word of caution should be given about $\mathbf{v}_d$ in connection with calculating the electric-current density **J**. Here $\mathbf{v}_d$ does not represent the *actual motion* of a charge and is not therefore directly connected to **J**. The great usefulness of the guiding-center method is in determining individual-particle trajectories. Rose and Clarke [15] further discuss the connection between $\mathbf{v}_d$ and **J**.

2.6. Particle Confinement in a Magnetic-mirror System. The magnetic mirror is one means of confining charged particles in a magnetic field. The field-and-coil configuration for a typical magnetic mirror is shown in Fig. 11.6. The charged particles that are trapped in the mirror system travel back and forth between the ends of the system, always moving along equivalent **B** lines. The curvature of their paths and the inhomogeneity of the **B** field induce azimuthal drifts, causing the particles to move on a cylindrically symmetric surface containing equivalent B lines. Such a surface is generated by rotating any B line shown in Fig. 11.6 about the axis of symmetry of the system. The reflection of the particles occurs because of the component of $\mathbf{F}_{\mu B}$ in Eq. (11.34) that is parallel to **B**. Substitution of the parallel component into Eq. (11.28*a*) would give the point, if it exists, at which the particle is reflected.

Another method of solution employs conservation of kinetic energy of the particle and the adiabatic

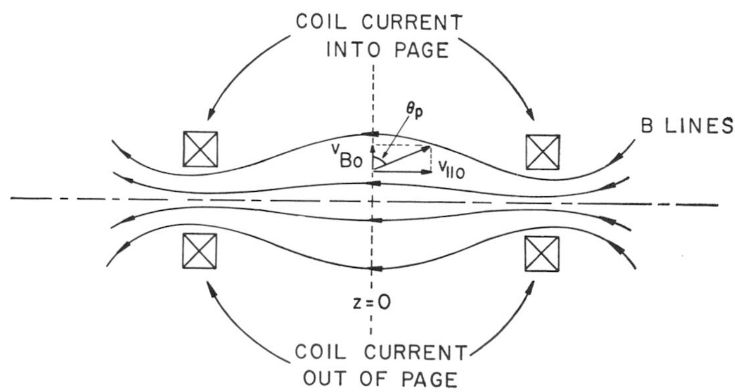

FIG. 11.6. Magnetic-mirror arrangement showing magnet coils and field lines.

invariance of the magnitude of $\boldsymbol{\mu}_B$. The particle kinetic energy $E_k = \frac{1}{2}mv_B^2 + \frac{1}{2}mv_{d\|}^2$ must remain constant (note that we neglect $\frac{1}{2}mv_{d\perp}^2$) as the particle moves in the mirror system since the magnetic force can do no work on the particle. The magnitude of $\boldsymbol{\mu}_B$, $\mu_B = mv_B^2/2B$, is constant and so we may write

$$E_k = \mu_B B + \tfrac{1}{2}mv_{d\|}^2 = \text{const} \qquad (11.38a)$$

or

$$\tfrac{1}{2}mv_{d\|}^2 = E_k - \mu_B B \qquad (11.38b)$$

Equation (11.38*b*) shows that as the magnitude of **B** increases $v_{d\|}$ decreases and will actually go to zero if B becomes large enough. This is the point at which the guiding center reverses direction and the particle is reflected. If, at $z = 0$ in Fig. 11.6, $B = B_0$, $\mathbf{v}_B = \mathbf{v}_{B0}$, $v_{d\|} = v_{d\|0}$, then at any other point

$$\tfrac{1}{2}mv_{d\|}^2 = \tfrac{1}{2}mv_{d\|0}^2 + \tfrac{1}{2}mv_{B0}^2\left(\frac{1 - B}{B_0}\right)$$

Setting B equal to its maximum possible value B_m along the particular trajectory and $v_{d\|} = 0$, we determine the maximum value of $v_{d\|0}$ that can be contained in the system: $v_{d\|0,\max} = v_{B0}[(B_m - B_0)/B_0]^{1/2}$. Since $v_{d\|0}/v_{B0} = \tan\theta_p$ is the tangent of the helical pitch at $z = 0$, $v_{d\|0,\max}/v_{B0} = [(B_m - B_0)/B_0]^{1/2}$ defines a maximum pitch angle, $\theta_{p,\max}$, such that all particles having $\theta_p > \theta_{p,\max}$ at $z = 0$ will be lost out the ends of the mirror system. This defines a *loss cone* in velocity space for the particles.

3. Waves in Plasmas

In free space two independent polarizations of transverse electromagnetic radiation can propagate. In the presence of a plasma these waves propagate with an altered phase velocity (Chap. 10, Sec. 1) and for frequencies less than the plasma frequency cannot propagate (neglecting all collisions). With plasma present, waves in which the electric field is parallel to the direction of propagation may also propagate when the electron-temperature effects are included. These latter types of waves have been discussed in Sec. 1.2 and 1.3. Stix [9] and Allis, Bers, and Buchsbaum [16] give detailed treatments of wave propagation in plasmas.

When a d-c magnetic field is applied to the plasma the electromagnetic waves have propagation characteristics depending on the direction of propagation with respect to the magnetic field. The following discussion outlines the method generally used to solve such wave-propagation problems and discusses the particular cases of propagation parallel to and perpendicular to the direction of the d-c magnetic field. The plasma is assumed uniform in space, and collisional effects are neglected. Only two species of particles are assumed present, electrons and ions. The waves are assumed to be of small amplitude so that linear theory applies. The effects of electron and ion temperature (pressure) are not included in the final discussion since many of the important properties of the transverse waves are affected only slightly by temperature.

The basic equations for this problem are a Boltzmann equation for each kind of particle in the plasma [Eq. (10.9)], describing the behavior of the respective distribution functions (f_e for electrons, f_+ for ions), and Maxwell's equations. The general procedure is to assume a traveling wave $\mathbf{E} = \mathbf{E}_0 e^{i\omega(t - \mathbf{n}\cdot\mathbf{r}/c)}$ (Chap. 10, Sec. 1) to be propagating through the plasma, and, using the Boltzmann equations and Maxwell's equations, to obtain an expression for n^2 in terms of the direction of propagation, the d-c magnetic field, the plasma density, and the frequency. This relation is called the *dispersion relation*. The vector **n** has a magnitude equal to the plasma index of refraction (for a lossless medium), and its direction is the direction of propagation of the wave. It is equal to $\mathbf{k}/k_0$, where **k** is the propagation vector of the wave in the plasma (magnitude $2\pi/\lambda$) and $k_0 = \omega/c = 2\pi/\lambda_0$ is the propagation constant in free space.

Rather than use the Boltzmann equations for the particles, the *transport equations* will be used. These equations are formed by integrating the Boltzmann equation over all particle velocities after having multiplied by some combination of velocity components (for example, $v_x^0 = 1$, v_x, v_x^2, v_xv_z, etc.). This method eliminates the possibility of studying effects due to wave interaction with some portion of the velocity distribution, as in the case of Landau damping (Sec. 1.2), but is used in this section and in the sections to follow where the hydromagnetic plasma equations are discussed.

3.1. The Particle- and Momentum-transport Equations. Integrating the Boltzmann equation over all electron (or ion) velocities and assuming no volume production or loss of charge yield particle-transport (continuity) equations for the electron and ion densities

$$\frac{\partial}{\partial t} n_{\substack{e\\+}} + \nabla\cdot n_{\substack{e\\+}}\bar{\mathbf{v}}_{\substack{e\\+}} = 0 \qquad (11.39)$$

where $n\bar{\mathbf{v}} = \int \mathbf{v} f\, d^3v$. Multiplying the Boltzmann equation by **v** and integrating over all electron (or ion) velocities give the momentum-transport equations (divided by m)

$$\frac{\partial}{\partial t} n_{\substack{e\\+}}\bar{\mathbf{v}}_{\substack{e\\+}} + \frac{\nabla\cdot(nm\overline{\mathbf{v}\mathbf{v}})_{\substack{e\\+}}}{m_{\substack{e\\+}}} \pm \frac{en_{\substack{e\\+}}}{m_{\substack{e\\+}}}(\mathbf{E} + \bar{\mathbf{v}}_{\substack{e\\+}} \times \mathbf{B}) = \int \mathbf{v}\left(\frac{\partial f_{\substack{e\\+}}}{\partial t}\right)_{\text{coll}} d^3v \qquad (11.40)$$

where the upper sign in the third term is for electrons and the lower for ions. The notation $(nm\overline{\mathbf{v}\mathbf{v}})$ is dyadic notation representing a second-order tensor with components, for example,

$$(nm\overline{\mathbf{v}\mathbf{v}})_{xy} = m\int v_xv_yf\,d^3v$$

It is convenient to set $\mathbf{v}_{\substack{e\\+}} = \bar{\mathbf{v}}_{\substack{e\\+}} + \mathbf{v}_{r\substack{e\\+}}$, where $\mathbf{v}_r$ represents the random component of the particle motion and, by definition, $\bar{\mathbf{v}}_r = 0$. Then

$$(nm\overline{\mathbf{v}\mathbf{v}})_{\substack{e\\+}} = (nm\bar{\mathbf{v}}\bar{\mathbf{v}})_{\substack{e\\+}} + (nm\overline{\mathbf{v}_r\mathbf{v}_r})_{\substack{e\\+}} = (nm\bar{\mathbf{v}}\bar{\mathbf{v}})_{\substack{e\\+}} + \mathbf{p}_{\substack{e\\+}} \qquad (11.41)$$

where **p** is called the kinetic stress tensor or pressure tensor. The diagonal components of **p** are related to the kinetic energy and hence to the scalar pressure $p = nm\langle v_r^2\rangle_{\text{av}}/3$. The off-diagonal components of p give rise to momentum transfer transverse to the direction of the momentum and hence to viscosity effects. The tensor $(nm\bar{\mathbf{v}}\bar{\mathbf{v}})_{\underset{+}{e}}$ is the rate at which momentum density $(nm\bar{\mathbf{v}})_{\underset{+}{e}}$ is transported across a unit area because of the drift motion of the particles. To obtain an equation for the components of $\mathbf{p}$, higher moments of the Boltzmann equation must be calculated. With each higher moment another variable is introduced, however, so that to obtain a closed set of equations some approximation must be made.

3.2. Solution of the Transport Equations for a Plane-wave Perturbation. In studying the wave-propagation problem we shall neglect $(nm\overline{\mathbf{v}\mathbf{v}})_{\underset{+}{e}}$ altogether, which means that the drift velocities must be small and that temperature and viscosity effects are neglected. The collision term on the right-hand side of Eq. (11.40) will also be neglected. With these assumptions Eq. (11.40) becomes

$$\frac{\partial}{\partial t}\underset{+}{n_e}\underset{+}{\bar{\mathbf{v}}_e} \pm \frac{e\underset{+}{n_e}}{\underset{+}{m_e}}(\mathbf{E} + \underset{+}{\bar{\mathbf{v}}_e} \times \mathbf{B}) = 0 \qquad (11.42)$$

If the particle densities are constant, these equations reduce to the equations of motion for electrons or ions.

Equation (11.42) is linearized for a plane-wave perturbation by setting

$$\underset{+}{n_e} = \underset{+}{n_{0e}} + \underset{+}{n_{1e}}e^{i\omega[t-(\mathbf{n}\cdot\mathbf{r})/c]} \qquad \underset{+}{\bar{\mathbf{v}}_e} = \underset{+}{\bar{\mathbf{v}}_{1e}}e^{i\omega[t-(\mathbf{n}\cdot\mathbf{r})/c]}$$
$$\mathbf{E} = \mathbf{E}_0e^{i\omega[t-(\mathbf{n}\cdot\mathbf{r})/c}$$

and keeping only linear terms in the small quantities. Also we set $\mathbf{B} = \mathbf{B}_0$, the d-c magnetic field produced by currents outside the plasma. Note that when the above complex notation is used actually

$$\underset{+}{\bar{\mathbf{v}}_e} = \operatorname{Re} \underset{+}{\mathbf{v}_{1e}}e^{i\omega[t-(\mathbf{n}\cdot\mathbf{r})/c]}$$

and likewise for the other quantities. From Eq. (11.39) we obtain, neglecting the effect of the electron motion on the phase of the wave,

$$i\omega\left(\underset{+}{n_{1e}} - \underset{+}{n_{0e}}\frac{\mathbf{n}\cdot\underset{+}{\mathbf{v}_{1e}}}{c}\right)e^{i\omega[t-(\mathbf{n}\cdot\mathbf{r})/c]} = 0 \qquad (11.43)$$

This has the solution

$$\underset{+}{n_{1e}} = \underset{+}{n_{0e}}\frac{\mathbf{n}\cdot\underset{+}{\mathbf{v}_{1e}}}{c} \qquad (11.44)$$

so that space-charge oscillations occur only when the induced electron velocity has a component parallel to $\mathbf{n}$, such as in the longitudinal waves of Secs. 1.2 and 1.3. From Eq. (11.40) we obtain

$$\left[i\omega \underset{+}{n_{0e}}\underset{+}{\mathbf{v}_{1e}} \pm \frac{e\underset{+}{n_{0e}}}{\underset{+}{m_e}}(\mathbf{E}_0 + \underset{+}{\mathbf{v}_{1e}} \times \mathbf{B}_0)\right]e^{i\omega[t-(\mathbf{n}\cdot\mathbf{r}/c)]} = 0 \qquad (11.45)$$

To solve this equation it is convenient to consider separately the components of this vector equation parallel to and perpendicular to $\mathbf{B}_0$. Thus,

$$(\underset{+}{\mathbf{v}_{1e}})_\parallel = \mp\frac{e}{\underset{+}{m_e}i\omega}\mathbf{E}_{0\parallel} \qquad (11.46a)$$

and

$$(\underset{+}{\mathbf{v}_{1e}})_\perp = \mp\frac{e}{\underset{+}{m_e}i\omega}[\mathbf{E}_{0\perp} + (\underset{+}{\mathbf{v}_{1e}})_\perp \times \mathbf{B}_0] \qquad (11.46b)$$

The solution to Eq. (11.46b) is obtained by taking the cross product of $\mathbf{B}_0$ into the equation and using this newly formed equation to eliminate $\underset{+}{\mathbf{v}_{1e}} \times \mathbf{B}_0$ in Eq. (11.46b). The result is

$$(\underset{+}{\mathbf{v}_{1e}}) = \mp\frac{(e/\underset{+}{m_e}i\omega)[\mathbf{E}_{0\perp} \mp (e/\underset{+}{m_e}i\omega)\mathbf{E}_{0\perp} \times \mathbf{B}_0]}{1 - e^2B_0{}^2/\underset{+}{m_e}{}^2\omega^2} \qquad (11.47)$$

This equation shows the cyclotron resonance at

$$e^2B_0{}^2/\underset{+}{m_e}{}^2\omega^2 = \underset{+}{\omega_{Be}}{}^2/\omega^2 = 1$$

It also shows that the induced particle velocity is formed of two orthogonal components which are 90° out of phase in time because of the factor i. This describes an elliptical orbit for the particle in the plane perpendicular to $\mathbf{B}_0$ for the case where $\mathbf{E}_{0\perp}$ represents a plane-polarized wave (that is, $\mathbf{E}_{0\perp}$ having no imaginary part).

For example, we take $\mathbf{E}_{0\perp} = \boldsymbol{\varepsilon}_xE_0$ and $\mathbf{B}_0 = \boldsymbol{\varepsilon}_zB_0$, where $\boldsymbol{\varepsilon}_x$ and $\boldsymbol{\varepsilon}_z$ are unit vectors in the x and z directions and E_0 is real. Then $\mathbf{E}_{0\perp} \times \mathbf{B}_0 = -\boldsymbol{\varepsilon}_yE_0B_0$ and for electrons Eq. (11.47) becomes

$$\mathbf{v}_{1e} = -\frac{(eE_0/m_ei\omega)[\boldsymbol{\varepsilon}_x + \boldsymbol{\varepsilon}_y(e/m_ei\omega)B_0]}{1 - e^2B_0{}^2/m_e{}^2\omega^2}$$

Since $\bar{\mathbf{v}}_e = \operatorname{Re}\mathbf{v}_{1e}e^{i\omega[t-(\mathbf{n}\cdot\mathbf{r}/c)]}$ we have

$$\bar{\mathbf{v}}_e = +\frac{eE_0/m_e\omega}{1 - e^2B_0{}^2/m_e{}^2\omega^2}\left[-\boldsymbol{\varepsilon}_x \sin\omega\left(t - \frac{\mathbf{n}\cdot\mathbf{r}}{c}\right) + \frac{eB_0}{m_e\omega}\boldsymbol{\varepsilon}_y\cos\omega\left(t - \frac{\mathbf{n}\cdot\mathbf{r}}{c}\right)\right]$$

In discussing the motion of a single electron we may assume that $\mathbf{r}$ is constant in the oscillatory terms of the equation, provided that the amplitude of the electron's motion in the direction of propagation of the wave is much smaller than the wavelength. In this case $\bar{\mathbf{v}}_e$ shows the electron to move in an elliptical orbit in the *right-hand* sense. For the same electric field the ion moves in the *left-hand* sense.

It is convenient to relate the induced electron and ion velocities to the electric field by using a mobility tensor $\boldsymbol{\mu}_e$. For the case where $\mathbf{B}_0$ points in the positive z direction and the equation is written in terms of the cartesian coordinates, we have

$$\underset{+}{\mathbf{v}_{1e}} = \mp\underset{+}{\boldsymbol{\mu}_e}\cdot\mathbf{E}_0$$
$$= \mp\frac{e}{\underset{+}{m_e}i\omega}\begin{pmatrix} \underset{+}{\gamma_e} & \pm i\underset{+}{\gamma_e}\dfrac{\underset{+}{\omega_{Be}}}{\omega} & 0 \\ \mp i\underset{+}{\gamma_e}\dfrac{\underset{+}{\omega_{Be}}}{\omega} & \underset{+}{\gamma_e} & 0 \\ 0 & 0 & 1 \end{pmatrix}\begin{pmatrix} E_{0x} \\ E_{0y} \\ E_{0z} \end{pmatrix} \qquad (11.48)$$

where $\underset{+}{\gamma_e} = \dfrac{1}{1 - \underset{+}{\omega_{Be}}{}^2/\omega^2}$ and $\underset{+}{\omega_{Be}} = \dfrac{eB_0}{\underset{+}{m_e}}$

Note that ω_{Be} and ω_{B+} are both positive, the sign differences being written explicitly in Eq. (11.48).

3.3. The Plasma Dispersion Relation. The electric current **J** is given by

$$\mathbf{J} = e(n_{0+}\mathbf{v}_{1+} - n_{0e}\mathbf{v}_{1e}) \equiv \boldsymbol{\delta} \cdot \mathbf{E} \qquad (11.49)$$

where $\boldsymbol{\delta}$ is the tensor conductivity of the plasma. This may now be substituted into Maxwell's equations:

$$\nabla \times \mathbf{E} = -\mu_0 \frac{\partial \mathbf{H}}{\partial t} \qquad (11.50a)$$

$$\nabla \times \mathbf{H} = \mathbf{J} + \epsilon_0 \frac{\partial \mathbf{E}}{\partial t} = \boldsymbol{\delta} \cdot \mathbf{E} + \epsilon_0 \frac{\partial \mathbf{E}}{\partial t} \qquad (11.50b)$$

where $\mu_0 = 4\pi \times 10^{-7}$ weber/amp-m is the permeability of free space. Assuming that all quantities vary as $e^{i\omega[t-(\mathbf{n}\cdot\mathbf{r}/c)]}$ and eliminating **H** from the equation yield a single equation for $\mathbf{E}_0$:

$$\mathbf{n} \times (\mathbf{n} \times \mathbf{E}_0) + \mathbf{K} \cdot \mathbf{E}_0 = 0 \qquad (11.51)$$

where **K** may be thought of as the plasma dielectric tensor and is given by

$$\mathbf{K} = \mathbf{I} + \frac{\boldsymbol{\delta}}{i\omega\epsilon_0} \qquad (11.52)$$

where **I** is the diagonal unity matrix.

Assuming that **n** lies in the xz plane and that it makes an angle θ with B_0, Eq. (11.51) becomes

$$\begin{pmatrix} S - n^2\cos^2\theta & iD & n^2\cos\theta\sin\theta \\ -iD & S - n^2 & 0 \\ n^2\cos\theta\sin\theta & 0 & P - n^2\sin^2\theta \end{pmatrix} \begin{pmatrix} E_{0x} \\ E_{0y} \\ E_{0z} \end{pmatrix} = 0 \qquad (11.53)$$

where S, D, and P are determined from Eqs. (11.48), (11.49), (11.51), and (11.52). The notation of Eq. (11.53) is similar to that of Stix [9] except that the signs of D are reversed because he assumes $e^{-i[\omega t-(\mathbf{n}\cdot\mathbf{r}/c)]}$ for the plane-wave phase. The quantities S, D, and P are

$$S = 1 - \frac{\omega_{p+}^2}{\omega^2 - \omega_{B+}^2} - \frac{\omega_{pe}^2}{\omega^2 - \omega_{Be}^2} \qquad (11.54a)$$

$$D = \frac{\omega_{B+}\omega_{p+}^2}{\omega(\omega^2 - \omega_{B+}^2)} - \frac{\omega_{Be}\omega_{pe}^2}{\omega(\omega^2 - \omega_{Be}^2)} \qquad (11.54b)$$

$$P = 1 - \frac{\omega_{p+}^2}{\omega^2} - \frac{\omega_{pe}^2}{\omega^2} = 1 - \frac{\omega_p^2}{\omega^2} \qquad (11.54c)$$

where $\omega_{p+}^2 = \dfrac{n_{0+}e^2}{m_+\epsilon_0}$ and $\omega_{pe}^2 = \dfrac{n_{0e}e^2}{m_e\epsilon_0}$

are the squares of the ion and electron plasma frequencies (Secs. 1.2 and 1.3) and $\omega_p^2 = \omega_{pe}^2 + \omega_{p+}^2$.

Solutions to Eq. (11.53) exist only if the determinant of the matrix multiplying $\mathbf{E}_0$ is zero. This determinantal equation determines the eigenvalues of n for a given angle θ of propagation with respect to $\mathbf{B}_0$. These eigenvalues when substituted into Eq. (11.53) determine the relations between the components of $\mathbf{E}_0$, or the polarization of the waves. If, for a given frequency ω, an eigenvalue of n is imaginary then the wave cannot propagate in the plasma. We wish to determine the frequency regions of allowed and unallowed propagation. At the boundaries between these regions of allowed and unallowed propagation n^2 either approaches infinity or zero. If n^2 is infinite a *resonance* is said to occur, and n^2 equal to zero corresponds to a *plasma cutoff* [16].

We shall investigate the two principal directions of propagation only, parallel to $\mathbf{B}_0$ ($\theta = 0$) and perpendicular to $\mathbf{B}_0$ ($\theta = 90°$). For the coordinate system being used, the z direction corresponds to propagation parallel to $\mathbf{B}_0$ and the x direction to propagation perpendicular to $\mathbf{B}_0$.

We define

$$R = S + D = 1 - \frac{\omega_p^2}{(\omega + \omega_{B+})(\omega - \omega_{Be})} \qquad (11.55a)$$

$$L = S - D = 1 - \frac{\omega_p^2}{(\omega - \omega_{B+})(\omega + \omega_{Be})} \qquad (11.55b)$$

where we have assumed $n_{0e} = n_{0+}$. The determinantal equation can now be conveniently written in a form [16] that relates $\tan^2\theta$ to n^2 and the plasma parameters:

$$\tan^2\theta = -\frac{P(n^2 - R)(n^2 - L)}{(Sn^2 - RL)(n^2 - P)} \qquad (11.56)$$

For $\theta = 0$ or $\theta = 90°$ the roots of this equation are readily found. For other values of θ the waves of the principal propagation directions are altered owing to an intermixing arising because $\tan^2\theta$ is neither zero nor infinite.

3.4. Wave Propagation Parallel to $\mathbf{B}_0$. To satisfy Eq. (11.56) for $\theta = 0$ we have either $P = 0$, $n_R^2 = R$, or $n_L^2 = L$. The equation $P = 0$ corresponds to $\omega = \omega_p$ and is simply the nonpropagating (for zero-temperature electrons) plasma oscillation with $E_{0x} = E_{0y} = 0$, $E_{0z} \neq 0$. The two values of n^2 (n_R^2 and n_L^2) correspond to right-hand and left-hand circularly polarized waves propagating parallel to $\mathbf{B}_0$ and having $E_{0z} = 0$. Consider, for instance, $n_R^2 = R$. When this is substituted into Eq. (11.53) we find $(S - R)E_{0x} + iDE_{0y} = 0$ and $E_{0z} = 0$ for the wave. From Eq. (11.55a) $S - R = -D$, and we obtain $E_{0x} = +iE_{0y}$. To calculate the sense of rotation we take

$$\mathbf{E} = \operatorname{Re} \mathbf{E}_0 e^{i\omega[t-(\mathbf{n}\cdot\mathbf{r})/c]} = E_{0y} \operatorname{Re}(i\boldsymbol{\varepsilon}_x + \boldsymbol{\varepsilon}_y)e^{i\omega[t-(\mathbf{n}\cdot\mathbf{r}/c)]}$$
$$= E_{0y}\left[-\boldsymbol{\varepsilon}_x \sin\omega\left(t - \frac{\mathbf{n}\cdot\mathbf{r}}{c}\right) + \boldsymbol{\varepsilon}_y \cos\omega\left(t - \frac{\mathbf{n}\cdot\mathbf{r}}{c}\right)\right]$$

This is a circularly polarized wave rotating in the right-hand sense with respect to the magnetic-field vector. The resonance [Eq. (11.55a)] in R at $\omega = \omega_{Be}$ occurs when the electric field rotates synchronously with the orbiting electron so that the electron is subject to a constant electric force and can exchange energy with the wave. The left-hand wave exhibits a resonance at $\omega = \omega_{B^+}$.

If a plane-polarized wave is incident normally on a plasma that has a d-c magnetic field applied normal to its surface, the wave is split into its right- and left-hand rotating components which travel at different speeds as determined by n_R^2 and n_L^2. When the wave, after traveling a distance d, emerges on the other side of the plasma its plane of polarization will have been rotated by an angle $\alpha = (n_L - n_R)(\pi d/\lambda_0)$, where λ_0 is the wavelength in free space of the wave. This is the phenomenon of Faraday rotation in a

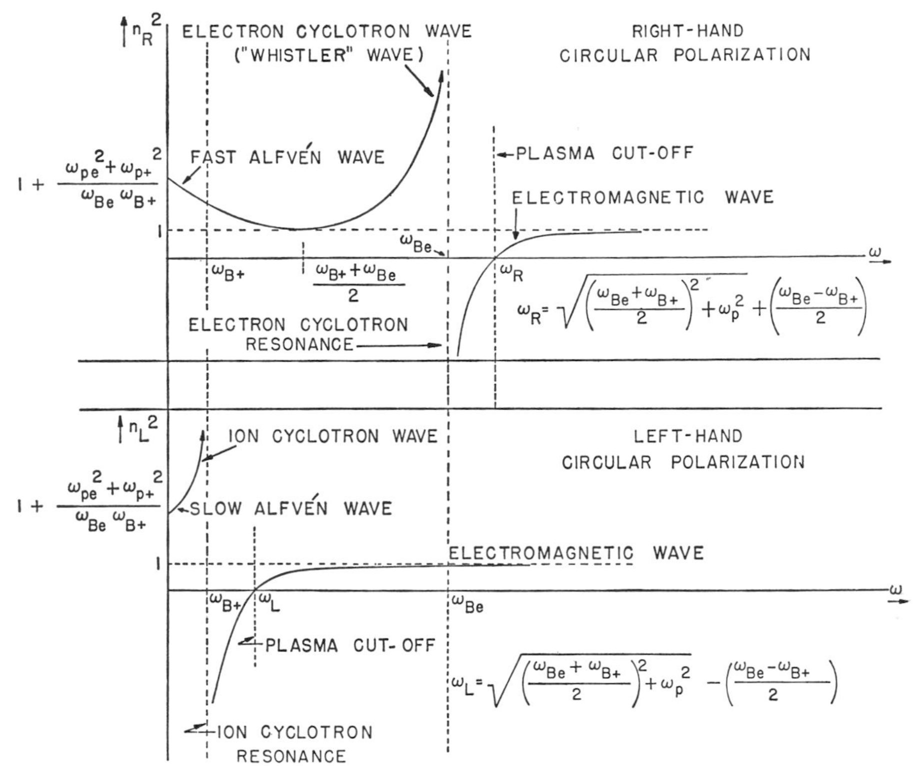

FIG. 11.7. The square of the indices of refraction of the right-hand and left-hand waves versus ω. Propagation is parallel to $\mathbf{B}_0$.

plasma. The equation is valid provided both waves can propagate; that is, n_R and n_L must be real.

The squares of the indices of refraction of the right- and left-hand waves have been plotted in Fig. 11.7 for a case where $\omega_{pe}^2 \ll \omega_{Be}^2$ and are only approximately to scale. At frequencies much less than ω_{B^+} both waves may propagate; the right-hand wave with smaller index of refraction propagates faster. The plasma participates strongly in the wave propagation in this region. These waves were first treated by Alfvén [17]. The plasma velocity of these waves for ω near zero is

$$v_{\text{ph}} = \frac{c}{n_{\substack{R\\L}}} = \frac{c}{(1 + \omega_p^2/\omega_{Be}\omega_{B^+})^{1/2}}$$

$$= \frac{c}{[1 + (n_{0e}m_e + n_{0+}m_+)/\epsilon_0 B_0^2]^{1/2}} \qquad (11.57)$$

and is called the Alfvén velocity. For an H_2^+ density in a hydrogen plasma of 10^{18} m^{-3} and $B_0 = 0.1$ weber/m^2, $v_{\text{ph}} \cong 5 \times 10^4$ m/sec.

As the ion cyclotron frequency is approached, the left-hand wave experiences cyclotron resonance. In the vicinity of the resonance the wave is described as the ion cyclotron wave. Thereafter the left-hand wave cannot propagate until its frequency exceeds

$$\omega_L = \sqrt{\left(\frac{\omega_{Be} + \omega_{B^+}}{2}\right)^2 + \omega_p^2} - \frac{\omega_{Be} - \omega_{B^+}}{2}$$

as shown in Fig. 11.7. The cutoff frequency $\omega = \omega_L$ reduces to the normal plasma cutoff $\omega = \omega_p$ for $B_0 = 0$ and to $\omega = \omega_{B^+}$ for $\omega_p = 0$. For higher frequencies the left-hand wave's index of refraction approaches that of an electromagnetic wave in vacuum.

The right-hand wave experiences effects analogous to the left-hand wave, but they occur in the vicinity of the electron cyclotron resonance. The electron cyclotron wave has been called the "whistler wave" in connection with the propagation of waves along the earth's magnetic-field lines in the ionosphere [18].

3.5. Wave Propagation Perpendicular to $\mathbf{B}_0$. For propagation perpendicular to $\mathbf{B}_0$ the eigenvalues of n^2 from Eq. (11.56) are

$$n_O^2 = P = 1 - \frac{\omega_p^2}{\omega^2} \qquad (11.58a)$$

and

$$n_X^2 = \frac{RL}{S} = 1 - \omega_p^2 \frac{\omega^2 - \omega_p^2 - \omega_{B+}\omega_{Be}}{(\omega^2 - \omega_{Be}^2)(\omega^2 - \omega_{B^+}^2) - \omega_p^2(\omega^2 - \omega_{B^+}\omega_{Be})} \qquad (11.58b)$$

The wave of Eq. (11.58a) is linearly polarized with its electric vector parallel to $\mathbf{B}_0$. Thus the propagation of this wave is unaffected by $\mathbf{B}_0$. This wave is called [16] the "ordinary" wave since it is unaffected by the magnetic field. The value of n_O^2 versus ω is plotted in Fig. 11.8. The plasma cutoff at $\omega = \omega_p$ has been described in Chap. 10, Sec. 1.

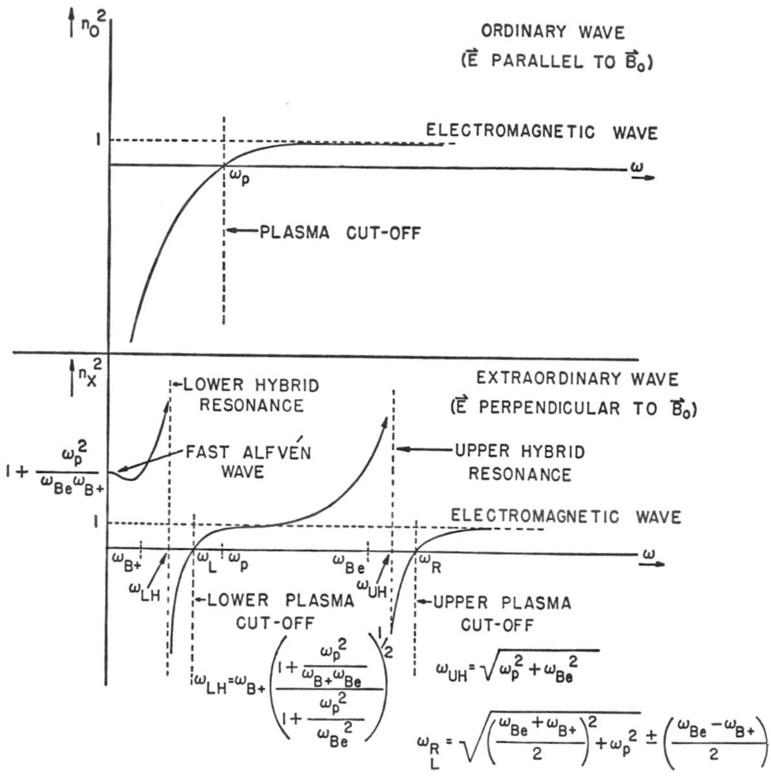

FIG. 11.8. The square of the indices of refraction of the ordinary and extraordinary waves versus ω. Propagation is perpendicular to $\mathbf{B}_0$.

The wave corresponding to Eq. (11.58*b*) is called [16] the "extraordinary" wave. Its polarization is found by substituting Eq. (11.58*b*) into (11.53) and setting $\theta = 90°$, which gives $E_{0z} = 0$ and $SE_{0x} + iDE_{0y} = 0$ or

$$E_{0x} = -\frac{iD}{S}E_{0y} = -i\frac{R-L}{R+L}E_{0y} \quad (11.59)$$

The extraordinary wave has a nonzero electric vector component E_{0x} in the direction of propagation which is 90° out of phase in time with the transverse component E_{0y}. This wave is then elliptically polarized with the electric vector rotating in a plane perpendicular to $\mathbf{B}_0$. From Eq. (11.58*b*) it is found that the extraordinary wave experiences cutoffs ($n_X^2 = 0$) at the same frequencies (ω_R and ω_L), as do the right- and left-hand waves. The general behavior of n_X^2 with frequency is shown in Fig. 11.8. The graph is drawn for the case where $\omega_p^2 \ll \omega_{Be}^2$ and is not to scale. Since $R = L$ at $\omega = 0$ [Eqs. (11.55*a*) and (11.55*b*)], the wave is predominantly of transverse polarization for low frequencies, according to Eq. (11.59). The two resonances experienced by the extraordinary wave occur at frequencies different from ω_{B^+} and ω_{Be} because this wave has space charge associated with it [see Eqs. (11.44), (11.48), and (11.59)] which causes the resonances to be shifted. The resonant frequencies ω_{LH} and ω_{UH} are referred to as the lower and upper hybrid frequencies, respectively. The upper hybrid resonance is shifted to a frequency higher than ω_{Be}, $\omega_{UH} = \sqrt{\omega_{Be}^2 + \omega_p^2}$. The lower hybrid frequency approaches ω_{B^+} for $\omega_p \approx 0$, and for $\omega_p^2 \gg \omega_{Be}^2$ it approaches the geometric mean of ω_{Be} and ω_{B^+}, $\omega_{LH} \approx \sqrt{\omega_{Be}\omega_{B^+}}$.

At points near resonances the wave velocity is small. When it becomes comparable with the phase velocity of the longitudinal wave propagating across $\mathbf{B}_0$ (which we have ignored), the two waves interact strongly and the conditions very near resonance are not as described above. Allis, Bers, and Buchsbaum discuss this effect [16].

4. The Plasma Hydromagnetic Equations

The plasma hydromagnetic equations describe the behavior of certain properties characteristic of the plasma as a whole rather than characteristic of the ions or electrons separately. A continuity equation for the plasma mass density $\rho_m = n_+m_+ + n_em_e$ is obtained by adding the continuity equations for electrons and ions [see Eq. (11.39)] after having multiplied by the respective masses. Similarly the equation for the plasma charge density $\rho = e(n_+ - n_e)$ is found by taking the difference of the continuity equations. The conservation of charge and mass equations are

$$\frac{\partial \rho}{\partial t} + \nabla \cdot \mathbf{J} = 0 \qquad (11.60a)$$

$$\frac{\partial \rho_m}{\partial t} + \nabla \cdot \rho_m \mathbf{V} = 0 \qquad (11.60b)$$

where $\mathbf{J} = e(n_+\bar{\mathbf{v}}_+ - n_e\bar{\mathbf{v}}_e)$ is the plasma electric current, and $\mathbf{V} = (n_+m_+\bar{\mathbf{v}}_+ + n_em_e\bar{\mathbf{v}}_e)/\rho_m$ is the plasma mass-flow velocity. These equations state that the rate of change of charge or mass density in some plasma volume element arises only from the net flow of charge or mass from the volume element.

To obtain equations for **V** and **J** we first combine the results of Eqs. (11.40) and (11.41) to obtain

$$\frac{\partial}{\partial t} n_{\underset{+}{e}}\bar{\mathbf{v}}_{\underset{+}{e}} + \nabla \cdot n_{\underset{+}{e}}\bar{\mathbf{v}}_{\underset{+}{e}}\bar{\mathbf{v}}_{\underset{+}{e}} + \frac{\nabla \cdot \mathbf{p}_{\underset{+}{e}}}{m_{\underset{+}{e}}} \pm \frac{en_{\underset{+}{e}}}{m_{\underset{+}{e}}}(\mathbf{E} + \bar{\mathbf{v}}_{\underset{+}{e}} \times \mathbf{B})$$

$$= \int \mathbf{v}_{\underset{+}{e}} \left(\frac{\partial f}{\partial t}\right)_{\text{coll}} d^3v = \mp n_{\underset{+}{e}}\nu_{\underset{+}{e}}(\bar{\mathbf{v}}_e - \bar{\mathbf{v}}_+) \qquad (11.61)$$

where the effect of the collisions between electrons and ions is accounted for by using the collision frequency ν_e for electron-ion collisions (Sec. 1.6) and ν_+ for ion-electron collisions. The frequency ν_e is approximately equal to $1/\tau_{ee}$, where τ_{ee} is the electron-electron relaxation time given by Eq. (11.21). This is a valid approximation provided that the frequencies of the phenomena being studied are low compared with ν_e and ω_p. Since momentum must be conserved between electrons and ions, we have

$$m_en_e\nu_e(\bar{\mathbf{v}}_e - \bar{\mathbf{v}}_+) = m_+n_+\nu_+(\bar{\mathbf{v}}_e - \bar{\mathbf{v}}_+)$$

or, assuming $n_e = n_+$, $\nu_+ = (m_e/m_+)\nu_e \ll \nu_e$.

The pressure tensor $\mathbf{p}_{\underset{+}{e}}$ is now assumed to have equal diagonal elements and zero nondiagonal elements so that it behaves as a scalar pressure. This assumption greatly simplifies the hydromagnetic equations but eliminates any effects due to viscosity or thermal conduction. The scalar pressure is

$$p_{\underset{+}{e}} = \left(\frac{nm\langle v_r^2\rangle_{\text{av}}}{3}\right)_{\underset{+}{e}}$$

where the factor of ⅓ arises because

$$\langle v_{rx}^2\rangle_{\text{av}} = \langle v_{ry}^2\rangle_{\text{av}} = \langle v_{rz}^2\rangle_{\text{av}} = v_r^2\rangle_{\text{av}}/3$$

for a velocity distribution that is isotropic in the three dimensions.

We can now form the equation for **V** and **J** by taking the sum and difference of the ion and electron momentum-transport equations [Eq. (11.61)]. The conditions $m_e \ll m_+$ and $|n_e - n_+| \ll n_e$ or n_+ imply that $\rho_m/m_+ \approx n_e \approx n_+$. Using these approximations and writing $p = p_e + p_+$ we obtain

$$\frac{\partial}{\partial t}(\rho_m\mathbf{V}) = -\nabla \cdot \left[\rho_m\mathbf{VV} + \left(\frac{m_e}{n_ee^2}\right)\mathbf{JJ}\right] - \nabla p + \rho\mathbf{E} + \mathbf{J} \times \mathbf{B} \qquad (11.62a)$$

$$\frac{\partial}{\partial t}\mathbf{J} = -\nabla \cdot \left(\mathbf{VJ} + \mathbf{JV} - \frac{1}{n_ee}\mathbf{JJ}\right) + \frac{e}{m_e}\nabla p_e + \frac{n_ee^2}{m_e}(\mathbf{E} + \mathbf{V} \times \mathbf{B}) - \frac{e}{m_e}\mathbf{J} \times \mathbf{B} - \nu_e\mathbf{J} \qquad (11.62b)$$

where $(VJ)_{xy} = V_xJ_y$, and so forth. Equation (11.62a) may be compared directly with the fluid equation preceding Eq. (1.15) in Part 3, Chap. 1. The external forces correspond to $\rho\mathbf{E}$ and $\mathbf{J} \times \mathbf{B}$ for the plasma fluid. The only terms that cause the equations to differ are the $(m_e/n_ee^2)\mathbf{JJ}$ term which arises because the plasma is composed of two fluids (ions and electrons) and the term $\sum_{i\neq j} G_{ij}$ in the equation of Part 3, which represents the short-range forces between molecules. These short-range forces (van der Waals' and repulsive "hard-core" forces) are negligible in plasmas because of the greater average particle separations. The ∇p term in Eq. (11.62a) corresponds to $\nabla \cdot (\rho\boldsymbol{\beta})$ in the equation of Part 3 except that in the plasma case we have assumed the pressure tensor to be diagonal.

For most situations of interest the electron thermal velocity is much greater than the relative drift velocity of the ions and electrons, in which case terms involving **JJ** and **VJ** may be neglected. Furthermore, $\rho \approx 0$ within the plasma volume. The equations for **V** and **J** then become

$$\frac{\partial}{\partial t}(\rho_m\mathbf{V}) = -\nabla \cdot (\rho_m\mathbf{VV}) - \nabla p + \mathbf{J} \times \mathbf{B} \qquad (11.63a)$$

$$\frac{m_e}{n_ee^2}\frac{\partial \mathbf{J}}{\partial t} = \frac{1}{n_ee}[\nabla p_e + n_ee(\mathbf{E} + \mathbf{V} \times \mathbf{B}) - \mathbf{J} \times \mathbf{B}] - \frac{1}{\sigma}\mathbf{J} \qquad (11.63b)$$

where $\sigma = n_ee^2/m_e\nu_e$ is the plasma conductivity for low-frequency disturbances as given by Eq. (11.22) (which is corrected for electron-electron collisions). For regions within a Debye length of the plasma boundary an unbalance of charge generally exists so that the space-charge term may not be neglected near plasma boundaries.

Since $\rho_m \approx n_+m_+$, Eq. (11.63a) resembles the equation of motion of the ion; thus we expect the time response of **V** to be governed by the ion inertia. Equation (11.63b) for **J** suggests that the time response of **J** is governed by the electron inertia. Furthermore, the term $\mathbf{E} + \mathbf{V} \times \mathbf{B}$ in Eq. (11.63b) represents the electric field $\mathbf{E}'$ which is observed from a coordinate system moving with the mass velocity **V** so that if $\nabla p_e - \mathbf{J} \times \mathbf{B}$ is neglected in Eq. (11.63b) the equation becomes equivalent to the equation of motion for a single electron whose motion is viewed from a frame moving with the velocity **V**.

4.1. Energy-transport Equations. We have not yet determined an equation for the scalar pressure. Multiplying Eq. (10.9) by $(m\mathbf{v} \cdot \mathbf{v})/2$ and integrating over all particle velocities for electrons or ions, and neglecting energy exchange between them due to elastic recoil, give

$$\left\{\frac{\partial}{\partial t}\left(\frac{nm\bar{\mathbf{v}} \cdot \bar{\mathbf{v}}}{2} + \frac{3}{2}p\right) + \nabla \cdot \left[\bar{\mathbf{v}}\left(\frac{nm\bar{\mathbf{v}} \cdot \bar{\mathbf{v}}}{2} + \frac{5}{2}p\right)\right] = \mp ne\mathbf{E} \cdot \bar{\mathbf{v}}\right\}_{\underset{+}{e}} \qquad (11.64)$$

where thermal-conduction effects are absent because $\mathbf{p}_{\underset{+}{e}}$ was assumed diagonal. This equation states that the time rate of change of the energy density in a volume element of the plasma is due to work done by the electric field and to the net flow of energy density from the volume element due to convection. A factor $5/2$ appears in front of p inside the second term on the left because the particles with random velocity components parallel to $\bar{\mathbf{v}}$ contribute more to the energy flow than those particles with random velocity components antiparallel to $\bar{\mathbf{v}}$ subtract from the energy flow.

The addition of the ion and electron energy equations gives an equation for the plasma energy transport. Assuming $|\bar{\mathbf{v}}_e - \bar{\mathbf{v}}_+| \ll |\bar{\mathbf{v}}_e + \bar{\mathbf{v}}_+|$ and $m_e \ll m_+$, we obtain

$$\frac{\partial}{\partial t}\left(\frac{\rho_m V^2}{2} + \frac{3}{2}p\right) + \nabla \cdot \left[\mathbf{V}\left(\frac{\rho_m V^2}{2} + \frac{5}{2}p\right)\right] = \mathbf{J} \cdot \mathbf{E} \tag{11.65}$$

Since thermal-conduction effects have been assumed negligible it is possible to derive the adiabatic- (constant entropy) compression law by using Eq. (11.64) for electrons and ions separately. Note that momentum transfer between the ions and electrons must be neglected since a finite plasma conductivity leads to dissipation which increases the entropy. The result for electrons and ions is $p_{e,+}/n_{e,+}^{5/3} = \text{const}$, where $5/3$ is the specific-heat ratio for either ions or electrons. An adiabatic-compression law relating the plasma mass density $\rho_m = n_e m_e + n_+ m_+$ and the plasma pressure $p = p_e + p_+$ can now be derived. Assuming n_e approximately equal to n_+ and $\rho_m \approx n_+ m_+$, we obtain

$$\frac{p}{\rho_m^{5/3}} \cong \frac{p_e + p_+}{(n_+ m_+)^{5/3}} \cong \frac{1}{m_+^{5/3}}\left(\frac{p_e}{n_e^{5/3}} + \frac{p_+}{n_+^{5/3}}\right) = \text{const} \tag{11.66}$$

4.2. The Magnetohydrodynamic Approximation. The magnetohydrodynamic (MHD) approximation is valid for very low-frequency plasma behavior, steady state, or equilibrium configurations. For this reason it is useful in studying the *stability* of plasma configurations. In keeping with this approximation, the displacement current is neglected in Maxwell's equations. Maxwell's equations and the hydromagnetic equations in the MHD approximation become

$$\frac{\partial}{\partial t}(\rho_m \mathbf{V}) = -\nabla \cdot (\rho_m \mathbf{V}\mathbf{V}) - \nabla p + \mathbf{J} \times \mathbf{B} \tag{11.67a}$$

$$\mathbf{J} = \sigma(\mathbf{E} + \mathbf{V} \times \mathbf{B}) \tag{11.67b}$$

$$\nabla \times \mathbf{E} = -\frac{\partial \mathbf{B}}{\partial t} \tag{11.67c}$$

$$\nabla \times \mathbf{B} = \mu_0 \mathbf{J} \tag{11.67d}$$

$$\nabla \cdot \mathbf{B} = 0 \tag{11.67e}$$

$$\nabla \cdot \mathbf{E} = \rho/\epsilon_0 \tag{11.67f}$$

plus the continuity equations (11.60*a*) and (11.60*b*). Equation (11.67*b*) is the modified form of Ohm's law and is valid provided that the frequencies of the disturbances are much less than ω_{Be} and ω_p.

4.3. Magnetic Pressure, Magnetic Stress, and Plasma Confinement. The expression $\mathbf{J} \times \mathbf{B}$, which is the magnetic force influencing the mass motion in Eq. (11.67*a*), can be written in terms of the magnetic field, using Maxwell's equations. From (11.67*d*),

$$\mathbf{J} \times \mathbf{B} = \frac{\nabla \times \mathbf{B}}{\mu_0} \times \mathbf{B} = \frac{1}{\mu_0}(\mathbf{B} \cdot \nabla)\mathbf{B} - \frac{1}{2\mu_0}\nabla(\mathbf{B} \cdot \mathbf{B}) \tag{11.68}$$

The second term in Eq. (11.68) is called the magnetic pressure. Consider the case where the lines of **B** are not curved and the plasma is in equilibrium. Then $\mathbf{J} \times \mathbf{B} = -(1/2\mu_0)\nabla(\mathbf{B} \cdot \mathbf{B})$ and Eq. (11.67*a*) becomes at equilibrium

$$\nabla\left(p + \frac{\mathbf{B} \cdot \mathbf{B}}{2\mu_0}\right) = 0$$

or

$$p + \frac{\mathbf{B} \cdot \mathbf{B}}{2\mu_0} = \text{const} \tag{11.69}$$

Thus, the presence of a magnetic field in some region tends to exclude the plasma from the region, leading to the possibility of confining a plasma by surrounding it with a suitable arrangement of magnetic fields.

Consider, for example, two semi-infinite regions as shown in Fig. 11.9 with their mutual boundary region (transition region) containing the z axis. One region contains primarily plasma, and the other is immersed in a magnetic field that is parallel to the z axis. If the plasma pressure in the region of zero magnetic field is p_0 and the magnetic field in the region of zero plasma density is B_0, then Eq. (11.69) gives

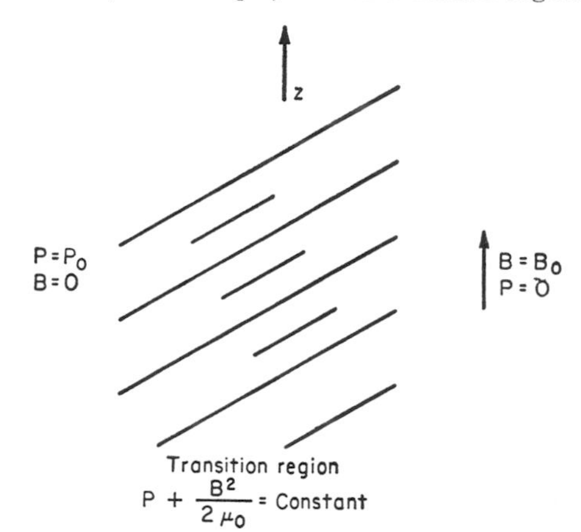

FIG. 11.9. Magnetic-field and plasma configuration to demonstrate plasma confinement and magnetic-field exclusion.

$$p + \frac{B^2}{2\mu_0} = \frac{B_0^2}{2\mu_0} = p_0 \tag{11.70}$$

In order for the magnetic field to contain the plasma, in this case, the external magnetic pressure must equal the internal plasma pressure. Conversely, for a plasma to be able completely to exclude a magnetic field from its interior the above condition must be satisfied.

In Eq. (11.68) the first term $1/\mu_0(\mathbf{B} \cdot \nabla)\mathbf{B}$ on the right gives rise to a force on the plasma if the B-field lines are curved or stretched analogous to an elastic medium. Since $\nabla \cdot \mathbf{B} = 0$, this term may be written

$$\frac{1}{\mu_0}(\mathbf{B} \cdot \nabla)\mathbf{B} = \frac{1}{\mu_0}\nabla \cdot (\mathbf{B}\mathbf{B}) \tag{11.71}$$

In analogy with the mechanical-stress tensor $\mathfrak{T}$ of Eq. (1.6) in Part 3, Chap. 1, whose divergence gives rise to forces on a volume element of an elastic or deformable medium, the dyadic $\mathbf{BB}/\mu_0$ may be regarded as a magnetic-stress tensor. The magnetic-field lines in the presence of plasma exhibit elastic properties that can give rise to plasma waves. The Alfvén waves discussed in Secs. 3.4 and 3.5 are examples of waves where the elastic properties of the magnetic field play an important role.

4.4. The Infinite Conductivity Approximation and Frozen Flux Lines. For a high-conductivity plasma Eq. (11.67*b*) is approximately given by

$$(\mathbf{E} + \mathbf{V} \times \mathbf{B}) = J/\sigma \approx 0 \qquad (11.72)$$

From Eq. (11.72) it follows that the number of magnetic-field lines linking a bounded surface moving with the plasma (at velocity $\mathbf{V}$) is a constant. Faraday's induction law for such a surface is

$$\frac{d}{dt}\int_S \mathbf{B} \cdot \mathbf{dS} = -\oint \mathbf{E}' \cdot \mathbf{ds} = -\oint (\mathbf{E} + \mathbf{V} \times \mathbf{B}) \cdot \mathbf{ds} = 0 \qquad (11.73)$$

which proves the statement. Use has been made of the fact that the electric field as seen from a coordinate system moving with velocity $\mathbf{V}$ is $\mathbf{E}' = \mathbf{E} + \mathbf{V} \times \mathbf{B}$.

If the number of magnetic-field lines linking a bounded surface remains constant, it follows that B/n = const, where n is the plasma density. This expression is valid unless there is plasma compression along the lines of B. Equation (11.33) is an equivalent statement derived in the adiabatic approximation for very low-density plasmas.

The approximation of Eq. (11.72) gives rise to the solution for $\mathbf{V}$

$$\mathbf{V} \approx \frac{\mathbf{E} \times \mathbf{B}}{B^2} \qquad (11.74)$$

which is the same as the motion of a single-particle guiding center given in Eq. (11.29) for the case where an electric field $\mathbf{E}$ acts on the particle transverse to $\mathbf{B}$. In the present case, however, the plasma may have strong perturbing effects on the electric and magnetic fields.

4.5. The Diffusion of a Fully Ionized Plasma across a Magnetic Field. Electron-ion collisions introduce diffusion of the plasma across the magnetic-field lines. Assuming the plasma to be nearly stationary and neglecting $\nabla \cdot (\rho_m \mathbf{VV})$, Eq. (11.65*a*) gives

$$\nabla p = \mathbf{J} \times \mathbf{B} \qquad (11.75)$$

Using Eq. (11.67*b*), we obtain

$$\nabla p = \sigma(\mathbf{E} \times \mathbf{B} + (\mathbf{V} \times \mathbf{B}) \times \mathbf{B}) = \sigma(\mathbf{E} \times \mathbf{B} - \mathbf{V}_\perp B^2) \qquad (11.76)$$

where $\mathbf{V}_\perp$ is the component of $\mathbf{V}$ that is perpendicular to $\mathbf{B}$. Setting $p = n_e kT_e + n_+ kT_+ \approx n_e(kT_e + kT_+)$ and assuming that temperature is independent of position, we obtain for $\mathbf{V}_\perp$

$$\mathbf{V}_\perp = -\frac{(kT_e + kT_+)\nabla n_e}{\sigma B^2} + \frac{\mathbf{E} \times \mathbf{B}}{\sigma B^2} \qquad (11.77)$$

The plasma drift parallel to ∇n_e is a diffusional drift. Substituting into Eq. (11.60*b*) and neglecting the $\mathbf{E} \times \mathbf{B}$ term, we have

$$\frac{\partial \rho_m}{\partial t} = \nabla \cdot \rho_m \frac{kT_e + kT_+}{\sigma B^2} \nabla n_e$$

or, since $\rho_m \approx n_e m_+$, we may write

$$\frac{\partial \rho_m}{\partial t} = \nabla \cdot n_e \frac{kT_e + kT_+}{\sigma B^2} \nabla \rho_m \qquad (11.78)$$

The diffusion coefficient is $D_B = n_e[(kT_e + kT_+)/\sigma B^2]$. The ambipolar-diffusion coefficient in a magnetic field as derived in the preceding chapter is, using Eqs. (10.34) and (10.35) in the limit $\mu_+\mu_e B^2 \gg 1$,

$$D_{aB} = \frac{\mu_+}{e}\frac{kT_+ + kT_e}{1 + \mu_+\mu_e B^2} \approx \frac{kT_+ + kT_e}{e\mu_e B^2}$$

Since $\sigma = n_e e \mu_e$, we have

$$D_{aB} \approx \frac{n_e(kT_+ + kT_e)}{\sigma B^2} = D_B$$

If we replace the right-hand side of Eq. (11.78) by $-D_B\rho_m/L^2$, where L^2 is the square of a typical plasma dimension, we see that an estimate for the diffusion-loss time may be obtained: $\tau_D \approx L^2/D_B$. A thermonuclear plasma with $L \approx 1$ m, $kT_e/e \approx kT_+/e = 10{,}000$ ev, $B = 10{,}000$ gauss, and $n_e = n_+ = 10^{20}$ m^{-3} would have a diffusion-loss time [using (Eq. 11.22)] of about 1,000 sec. In practice this diffusion-loss time for plasma density is not observed, because of the presence of enhanced diffusion which varies as $1/B$ rather than $1/B^2$. Such enhanced diffusion is observed in stellarators (see Sec. 6) and is probably due to turbulence in the plasma which gives rise to fluctuating electric fields that in turn produce a fluctuating drift velocity $(\mathbf{E} \times \mathbf{B})/B^2 \approx E/B$. Because of the random nature of $\mathbf{E}$ this would give rise to a drift in the direction of $-\nabla n_e$ and hence would contribute to particle loss.

4.6. The Linear Cylindrical Pinch. The pinch effect occurs when the external magnetic field produced by the plasma electric current is sufficiently large so that the plasma is compressed. The linear cylindrical pinch is produced by passing a current I through a plasma that is located between two electrodes. From Ampere's law the magnetic field outside the current channel at a distance r from its center is

$$B_\theta = \frac{\mu_0 I}{2\pi r} \qquad (11.79)$$

where θ is the azimuthal angle measured about the channel axis. Assuming the plasma density to be zero outside the current channel and neglecting forces on the plasma due to the curvature of the magnetic-field lines about the plasma, we may use Eq. (11.70) to obtain

$$p_0{}^2 = \frac{B_\theta{}^2}{2\mu_0} = \frac{\mu_0 I^2}{(2\pi R)^2} \qquad (11.80)$$

where p_0 is the plasma pressure at the channel axis and R is the radius of the current channel. It is valid to neglect the force on the plasma due to the curvature of the magnetic-field lines provided that the radial thickness of the region of transition of the plasma to zero pressure is small relative to R (sharp plasma boundary).

Equation (11.80) represents the equilibrium balance of plasma and magnetic-field pressure. The compression of the pinch is a dynamic phenomenon, however, so that in general the complete MHD equations (11.67) including the plasma acceleration must be employed.

4.7. Shock Waves in Plasmas. In most fluid media a compression wave travels fastest where the particle temperature is greatest. Thus an initially sinusoidal disturbance has a tendency to distort in shape because the peaks of the disturbance (high-temperature regions) travel faster than the troughs (low-temperature regions). If the amplitude of the wave is sufficient, a shock wave, which is a sharp discontinuity in the parameters of the medium, forms. The thickness of the shock is determined by dissipative effects such as viscosity and thermal conduction, which act to decrease the gradients of the fluid parameters.

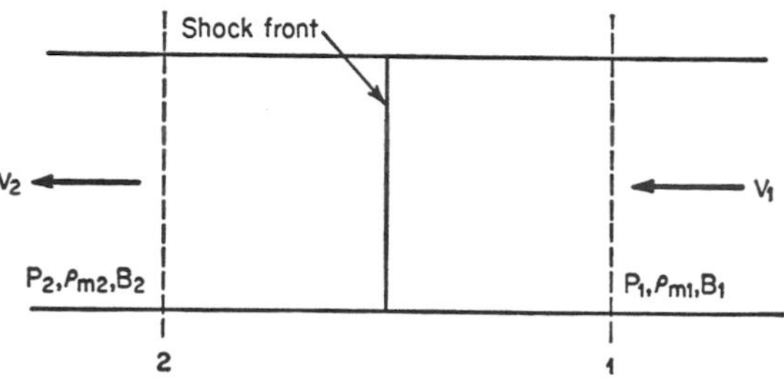

FIG. 11.10. The shock-wave problem as viewed from a coordinate system moving with the shock front.

We shall consider a one-dimensional shock-wave problem where the shock front propagates perpendicular to the direction of the magnetic field (Fig. 11.10). The shock-wave problem is solved from a coordinate system moving with the shock front, and a steady-state flow of plasma past the shock front is assumed. By choosing boundaries sufficiently far from the shock front (labeled 1 and 2 in Fig. 11.10) it can be assumed that thermal equilibrium has been established and that any gradients due to the presence of the shock front are negligible. Since in the steady state no particles, momentum, or energy may accumulate in the region between 1 and 2, the conservation laws may be applied across the shock region.

For a steady flow $\partial\rho_m/\partial t = 0$ at all points, and Eq. (11.60*b*) gives $\nabla \cdot (\rho_m \mathbf{V}) = 0$. Integrating this quantity over the volume between 1 and 2 and converting the volume integral to a surface integral over the boundary of the shock region give for the one-dimensional case

$$\rho_{m1}V_1 = \rho_{m2}V_2 \tag{11.81}$$

where the subscript 1 labels a quantity to the right of boundary 1 and the subscript 2 labels a quantity to the left of boundary 2. Equation (11.81) is a statement that the rate at which mass enters the shock region equals the rate at which it leaves.

A similar equation for the momentum flow is obtained by integrating Eq. (11.67*a*) over the shock region, replacing $\mathbf{J} \times \mathbf{B}$ by the gradient of the magnetic pressure [Eq. (11.68)] and assuming $\mathbf{B}$ always points perpendicular to the velocity flow so that $(\mathbf{B} \cdot \nabla)\mathbf{B}/\mu_0 = 0$. The conservation-of-momentum equation is

$$p_1 + \rho_{m1}V_1^2 + \frac{B_1^2}{2\mu_0} = p_2 + \rho_{m2}V_2^2 + \frac{B_2^2}{2\mu_0} \tag{11.82}$$

The energy equation is found by integrating Eq. (11.65) across the shock region. The equation thus obtained is

$$\tfrac{5}{2}p_1V_1 + \tfrac{1}{2}\rho_{m1}V_1^3 = \tfrac{5}{2}p_2V_2^2 + \tfrac{1}{2}\rho_{m2}V_2^3 + \int_1^2 \mathbf{J}\cdot\mathbf{E}\,dx$$

where the integral represents work done on the plasma by the electromagnetic forces. This integral must be equal to the difference in the electromagnetic energy flow across boundaries 1 and 2

$$\int_1^2 \mathbf{J}\cdot\mathbf{E}\,dx = \frac{\mathbf{E}_2 \times \mathbf{B}_2}{\mu_0} - \frac{\mathbf{E}_1 \times \mathbf{B}_1}{\mu_0}$$

Since thermodynamic equilibrium exists at boundaries 1 and 2, $\mathbf{J}_1 = \mathbf{J}_2 = 0$ and, from Eq. (11.67*b*),

$$\mathbf{E}_1 = -\mathbf{V}_1 \times \mathbf{B}_1$$

and $\mathbf{E}_2 = -\mathbf{V}_2 \times \mathbf{B}_2$ so that $\mathbf{E}$ is parallel to boundaries 1 and 2. It then follows that

$$\int_1^2 \mathbf{J}\cdot\mathbf{E}\,dx = \frac{V_2B_2^2}{\mu_0} - \frac{V_1B_1^2}{\mu_0}$$

The energy equation then becomes

$$\frac{5}{2}p_1V_1 + \frac{1}{2}\rho_{m1}V_1^3 + \frac{V_1B_1^2}{\mu_0} = \frac{5}{2}p_2V_2 + \frac{1}{2}\rho_{m2}V_2^3 + \frac{V_2B_2^2}{\mu_0} \tag{11.83}$$

The adiabatic-compression law does not apply across the shock region in general, since the entropy increases because of the dissipative losses at the shock front.

The one remaining equation is obtained by noting for a steady state that $\partial B/\partial t = \nabla \times \mathbf{E} = 0$ so that, from Stokes' theorem, $\mathbf{E}_1 = \mathbf{E}_2$ from which follows

$$\mathbf{V}_1 \times \mathbf{B}_1 = \mathbf{V}_2 \times \mathbf{B}_2$$

or, since $\mathbf{V}$ is perpendicular to $\mathbf{B}$,

$$V_1B_1 = V_2B_2 \tag{11.84}$$

Equations (11.81) to (11.84) relate the plasma parameters on either side of the shock region. If all the quantities in region 1 are known, for instance, those in region 2 may be determined. The velocity of the shock is the magnitude of V_1 for the case where the plasma ahead of the shock is at rest as viewed from laboratory coordinates. This shock velocity depends on the relative magnitudes of the plasma parameters.

In the limit of a small-amplitude shock, all the quantities in region 2 will be close to the values in region 1. Writing $V_2 = V_1 + \delta V_1$, $p_2 = p_1 + \delta p_1$, $\rho_{m2} = \rho_{m1} + \delta_{m1}$, and $B_2 = B_1 + \delta B_1$, substituting into the shock equations, and saving only terms linear in the perturbed quantities give for the shock speed

$$V_1 = \left(\frac{5}{3}\frac{p_1}{\rho_{m1}} + \frac{B_1^2}{\mu_0\rho_{m1}}\right)^{1/2}$$

which is the square root of the sum of the square of the sound velocity in the absence of a magnetic field plus the square of a velocity that is approximately the Alfvén velocity [see Eq. (11.57)]. The quantity

$$\frac{5}{3}\frac{p_1}{\rho_{m1}} \approx \frac{5}{3}\frac{kTe + kT_+}{m_+}$$

differs from the acoustic-wave velocity of Eq. (11.18) because there the ion temperature was neglected and the electron temperature was assumed uniform (isothermal compression) whereas the above calculation contains the sound speed for adiabatic compressions with a specific heat ratio of 5/3. The adiabatic-compression law is valid in the limit of small amplitude shock waves.

5. Plasma Instabilities

An important facet of plasma physics involves the containment of the hot plasma through the application of a suitable configuration of d-c magnetic fields. For any of these field configurations the question of the stability of the contained-plasma volume arises. That is, when the equilibrium plasma is disturbed in some manner, will the plasma be unstable and the disturbance grow with time, thus destroying the containment? Plasma instabilities can be classified as "microscopic" or "macroscopic." The microscopic instabilities arise from some characteristic of the velocity distributions of the ions and electrons so that it is necessary to use Boltzmann's equation in studying them. The macroscopic instabilities (or hydromagnetic instabilities) do not depend on the particle-velocity distributions so that the hydromagnetic equations may be used in the analysis.

5.1. Microscopic Instabilities. The beam-plasma instability of Sec. 1.4 is an example of a microscopic instability although some aspects of this particular case may be treated by using the fluid equations and replacing the average unperturbed electron velocity by the beam velocity. This instability occurs because the electron beam energy is transferred to the longitudinal plasma wave, causing the wave amplitude to grow exponentially.

The microscopic instabilities arising because of the gradients of density, temperature, and magnetic field [19] that always exist at the plasma boundary are referred to as universal instabilities. As in the case of the beam-plasma instability, these instabilities occur because the gradients produce a distorted distribution function which has an excess number of particles traveling at velocities near the phase velocity of one of the longitudinal plasma waves that can propagate in the plasma.

An instability in which a *transverse* electromagnetic wave takes energy from the electron-velocity distribution [20] can also occur under the proper conditions. In Sec. 1 of Chap. 10, Eq. (10.38) gives an expression for the a-c mobility of electrons moving under the influence of a transverse electromagnetic field and having a momentum-loss frequency ν_m due to electron-atom collisions. The equations for the plasma index of refraction that follow Eq. (10.38) show that the electromagnetic wave is unstable and grows in amplitude as it propagates through the plasma if the real part of the electron mobility μ_{er} is *negative*. From Eq. (10.38)

$$\mu_{er} = -\frac{e}{3m}\int_0^\infty \frac{v\nu_m}{\nu_m^2 + \omega^2}\frac{\partial f_v^0}{\partial v} 4\pi v^2\, dv \qquad (11.85)$$

After integrating once by parts and using the fact that f_v^0 goes to zero as v goes to infinity, we obtain

$$\mu_{er} = \frac{4\pi e}{3m}\int_0^\infty f_v^0 \frac{\partial}{\partial v}\frac{v^3\nu_m}{\nu_m^2 + \omega^2}\, dv \qquad (11.86)$$

It is clear from the form of the two integral expressions in Eqs. (11.85) and (11.86) that μ_{er} can be negative *only* in cases where $\partial f_v^0/\partial v > 0$ over a sufficient range of velocities, and the quantity $v^3\nu_m/(\nu_m^2 + \omega^2)$ is a decreasing function of velocity over a sufficient range of velocities. The first condition is a requirement that the distribution function have more electrons at higher than at lower energies over part of its range (for instance, a beam of electrons).

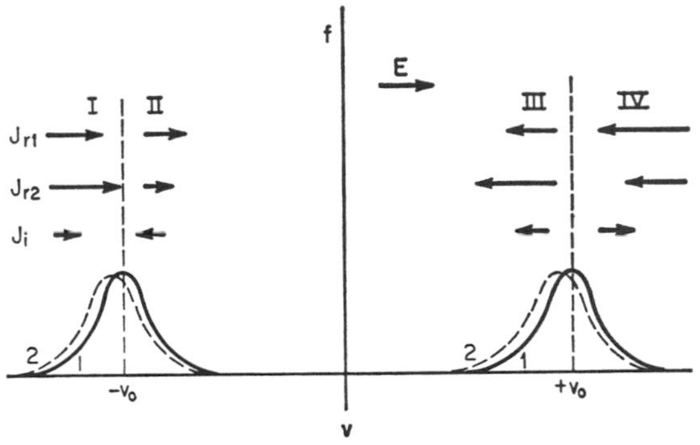

FIG. 11.11. A spherical-shell velocity distribution plotted along an axis parallel to an applied electric field. The dotted line represents the shift of the distribution function due to the effect of the electric field. The induced current J_i in a particular velocity interval is the difference in the electron random currents J_{r1} and J_{r2} before and after the field is applied.

Musha and Yoshida [21] have given a physical picture of this instability or "negative absorption." For purposes of discussion, we consider a spherical-shell velocity distribution for which the electrons travel in random directions but have speeds distributed in a narrow region about the speed v_0. If this distribution function is plotted along an axis parallel to the direction of the applied electric field, it has the form shown in Fig. 11.11. For an applied electric field E of frequency low compared with the frequency ν_m, the electrons move in phase with $-eE$, and when the electric field is in the direction shown in the figure the distribution function is shifted as shown. The induced current for a given interval of electron velocities is calculated from the difference of the random currents J_{r1} and J_{r2}, before and after the field is applied. The induced current is indicated qualitatively for four regions in Fig. 11.11, which shows that for electrons of speed *less* than v_0 the induced current flows in a direction *opposite* to E. Since it is the momentum-loss frequency ν_m that inhibits the flow of induced current, the counterflowing current of the low-velocity electrons ($|v| < v_0$) can actually *exceed* the induced current parallel to E of the high-velocity electrons ($|v| > v_0$) if ν_m is a strongly *increasing* function of electron speed. This would correspond to a *negative* conductivity and hence the mobility would be negative, thus giving rise to the possibility of negative absorption. Quantitatively this requirement on ν_m from Eq. (11.86) would be $[\partial/\partial v(v^3/\nu_m)]_{v \approx v_0} < 0$ in the limit $\omega \ll \nu_m$. This requires that ν_m increase with velocity more strongly than v^3 in order that negative absorption occur. This effect of negative conductivities has been observed by Forman [22] and Ohara [23].

At electric-field frequencies large compared with ν_m the electrons perform many oscillations between collisions at a phase of 90° with respect to the electric field. Only when a collision occurs is there a component of the current that is in phase with the electric field (and hence can exchange energy). Thus, greater in-phase currents occur for the larger values of ν_m in this frequency limit, and ν_m must now be a strongly *decreasing* function of electron speed in order that negative absorption occur. The negative-absorption condition on ν_m from Eq. (11.86) in this frequency limit is $[\partial/\partial v(v^3\nu_m/\omega^2)]_{v \approx v_0} < 0$ so that ν_m must decrease more strongly than v^{-3} in order for negative absorption to be possible.

5.2. Macroscopic Instabilities. The investigation of the stability of a plasma configuration may take two forms. One is an energy analysis, and the other is referred to as a normal-mode analysis. The latter method involves assuming a *small* perturbation of the equilibrium plasma parameters $\{\rho_m = \rho_{m0} + \rho_{m1}e^{i[\omega t - (\mathbf{n} \cdot \mathbf{r})/c]}$, and so forth$\}$, substituting into the hydromagnetic equations, and linearizing in the perturbed quantities. The final step is to determine whether the perturbation grows with time and is unstable or whether it is bounded and stable. This is done by determining the relation between ω and $\mathbf{n}$. This is similar to the method of Sec. 3, where the relation between ω and $\mathbf{n}$ was determined for plane waves propagating in an infinite medium. The plasma boundaries and the magnetic-field nonuniformities in the present problem make the normal-mode stability analysis considerably more complicated. R. J. Taylor [24] has applied this technique to a linear cylindrical pinch. A variational method [25] has been developed to determine the normal modes of the system without needing to know the exact spatial form of the perturbation. This variational technique is equivalent to the energy-analysis method described below.

The total energy content of a plasma-and-field configuration is

$$H = \int \left(\frac{1}{2}\rho_m V^2 + \frac{3}{2}p + \frac{1}{2}\frac{B^2}{\mu_0} + \frac{K\epsilon_0 E^2}{2}\right) dV$$

where K is the plasma dielectric constant. This can be written as $H = \int(\frac{1}{2}\rho_m V^2)\,dV + Q$, where Q represents the energy content other than the mass-flow energy. The term Q may be regarded as a *potential energy* of the plasma-and-field configuration. At equilibrium the mass-flow energy is zero. The energy analysis proceeds by perturbing the shape of the equilibrium configuration and calculating the resultant change in Q. If Q *decreases*, the system is unstable since, for a system of constant *total* energy H, the mass-flow energy will have increased. Likewise if Q increases, the system is stable since there is no available energy for mass flow if such a perturbation occurs.

a. The Flute or Ripple Instability. This type of instability occurs at the interface of the plasma and magnetic field and is present in both high-density and low-density plasmas. The existence of this instability depends upon the presence of a *charge-independent force* acting on the plasma at right angles to the magnetic-field lines and in the direction away from the plasma. This is analogous to the regular Taylor instability in which in the presence of a gravitational field a heavy fluid (plasma) is supported by a light fluid (magnetic field).

The following discussion treats a case where the plasma pressure is so low that the magnetic field is unchanged by the presence of the plasma. This argument also applies to high-pressure plasmas because the boundary of a high-pressure plasma (where the instability occurs) involves a transition to low-pressure plasma at which point the following arguments apply.

The instability is brought about by the drifts of the ions and electrons in the crossed magnetic field and charge-independent force. We consider the plasma boundary to be initially flat and perpendicular to a gravitational field. The plasma is supported by the magnetic field as shown in Fig. 11.12*a*. In Fig. 11.12*b*, the interface is perturbed by a sinusoidal ripple which has troughs running *parallel* to $\mathbf{B}$. From Eq. (11.29) the drift velocity for electrons is

$$\mathbf{v}_{de} = -\frac{m_e\mathbf{g} \times \mathbf{B}}{eB^2}$$

and for ions

$$\mathbf{v}_{d+} = +\frac{m_+\mathbf{g} \times \mathbf{B}}{eB^2}$$

where $\mathbf{g}$ is the vector gravitational field. The oppositely directed drift velocities give rise to an electric field $\mathbf{E}$ within the ripple disturbance as shown in Fig. 11.11*b*. This electric field is in such a direction that the $\mathbf{E} \times \mathbf{B}$ drift of the ions and electrons *within* each ripple is in the direction that increases the ripple amplitude. If $\mathbf{g}$ were in the opposite direction the electric field would be reversed and the ripple would decrease with time. A calculation of the time behavior of the ripple for a sharp boundary between plasma and magnetic field is given by Rose and Clarke [15]; they find the ripple to vary with time as $e^{\gamma t}$, where $\gamma = \sqrt{g2\pi/\lambda}$ and λ is the wavelength of the ripple. The quantity g is positive if $\mathbf{g}$ is directed away from the plasma; hence the solution is oscillatory (stable) if $\mathbf{g}$ is in the other direction, since γ is then imaginary.

From this discussion it follows that magnetic-field lines that are concave toward the plasma so that their center of curvature lies within the plasma region (Fig. 11.13*a*) are unstable to a ripple perturbation that has troughs parallel to the magnetic-field lines. The particles, as they follow along the curved magnetic-field lines, are subject to an outward-directed inertial force (see Fig. 11.5, for example) which is charge-independent. Since this force is directed away from the plasma it leads to instability. Also, since $\nabla \times \mathbf{B} = 0$, it follows that ∇B^2 points toward the

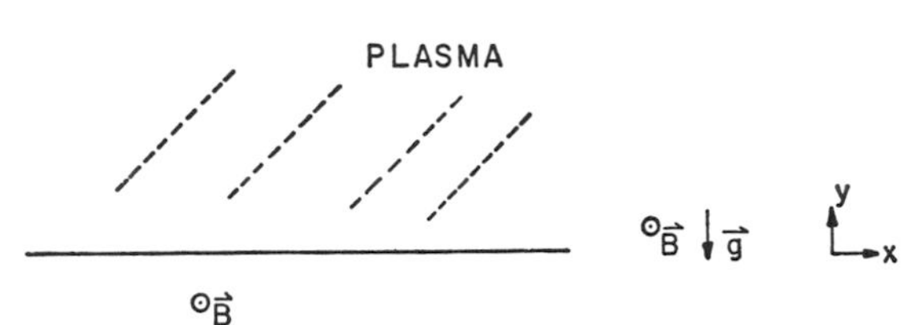

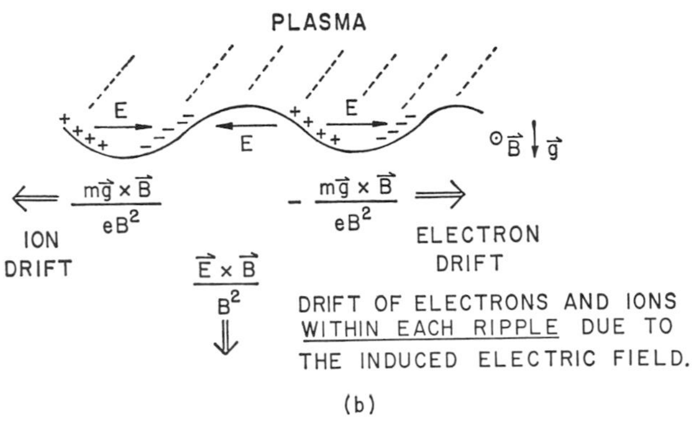

FIG. 11.12. The plasma is supported by magnetic pressure in the presence of a charge-independent force (gravity, in this case). A ripple perturbation gives rise to a space-charge electric field, which in turn causes the perturbation to grow.

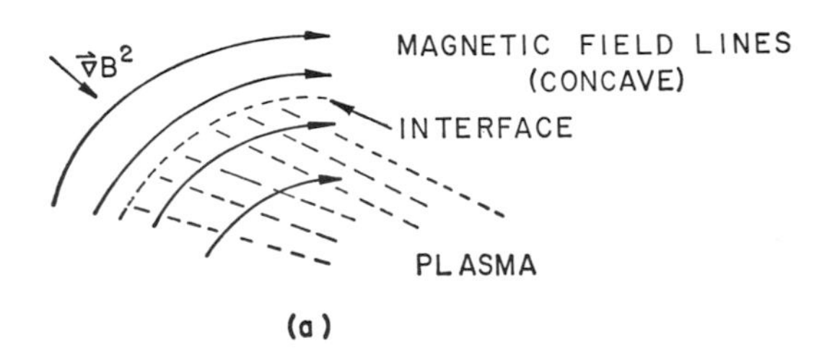

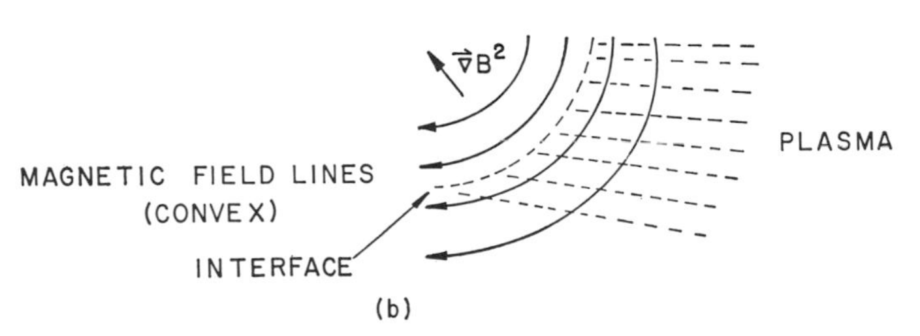

FIG. 11.13. Magnetic-field lines that curve toward the plasma (concave) are unstable to a ripple perturbation. Magnetic-field lines that curve away from the plasma (convex) are stable.

plasma. This gives rise to a charge-independent force on the charged-particle magnetic moments [from Eq. (11.34), $\mathbf{F}_{\mu_B} = -(mv_B{}^2/4B^2)\nabla B^2$] which is directed away from the plasma and therefore also contributes to the instability. Estimates of the growth rate of the ripple can be obtained by substituting for $\mathbf{g}$ in the previous treatment the above forces divided by the particle mass.

In a converse manner, field lines that are convex toward the plasma (Fig. 11.13*b*) are inherently stable to the ripple perturbation.

b. Experimental Examples. Figure 11.6 shows that the magnetic mirror has field lines concave toward the plasma near the mid-plane and convex toward the plasma near the ends. The stability in this case depends on the difference of the induced charge in a ripple perturbation in the two regions, since particles can flow freely along the magnetic-field lines to neutralize charge differences.

In the case of the linear pinch described in Sec. 4.6, the magnetic-field lines are always concave toward the plasma so that the system is inherently unstable. By trapping a longitudinal magnetic field within the pinch current channel the tendency for the ripple to form ("sausage" instability) can be reduced (see Fig. 11.14*a*). At points along the pinch where the ripple reduces the cross-sectional area of the pinch, the internal longitudinal-field lines are forced closer together, thus increasing the magnetic pressure which tends to expand the indented portions of the pinch. A sidewise displacement of a portion of the pinch channel with no change in channel cross section is not stabilized by the trapped longitudinal field (see Fig. 11.14*b*). This "kink" or "firehose" instability can

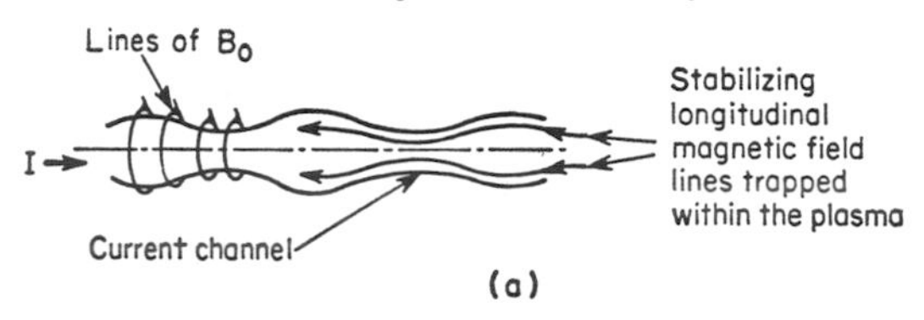

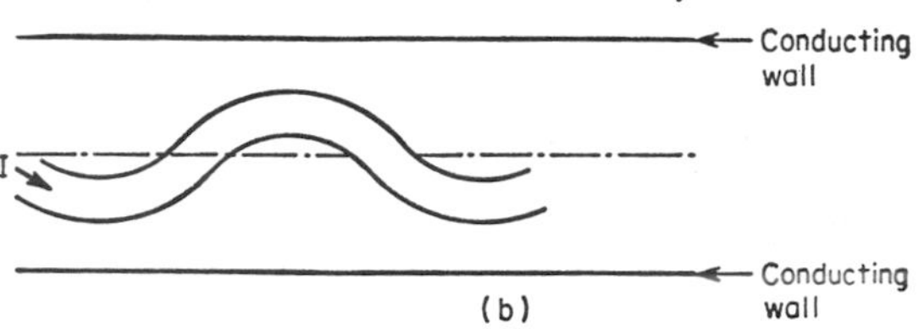

FIG. 11.14. The "sausage" and "kink" instabilities of the linear cylindrical pinch and the means by which they are stabilized.

be stabilized by surrounding the current channel with a conducting wall. The currents induced in the wall as the kink approaches create magnetic fields that repel the kink.

To obtain stable convex curvature of magnetic-field lines the "cusp" geometry has been used. Such a configuration of field lines is obtained, for instance, if one of the coil currents in a mirror field geometry is reversed. As shown in Fig. 11.6, the plasma is then contained near the mid-plane between the two coils in the region where the magnetic field is near zero. The plasma particle motion is no longer adiabatic because of the zero-field region; consequently all particles may eventually be lost by traveling outward along the mid-plane.

The line cusp is formed by arranging equally spaced parallel conducting wires as shown in Fig. 11.15. Gott, Ioffe, and Tel'kovskii [26] have combined the stability of the line-cusp geometry of Fig. 11.15 with the containment properties of the magnetic-mirror system of Fig. 11.6. The wire conductors ("Ioffe bars") are oriented parallel to the mirror axis and located on a cylindrically symmetric surface outside the mirror volume which runs the length of the mirror system. These experimenters observe that the life-

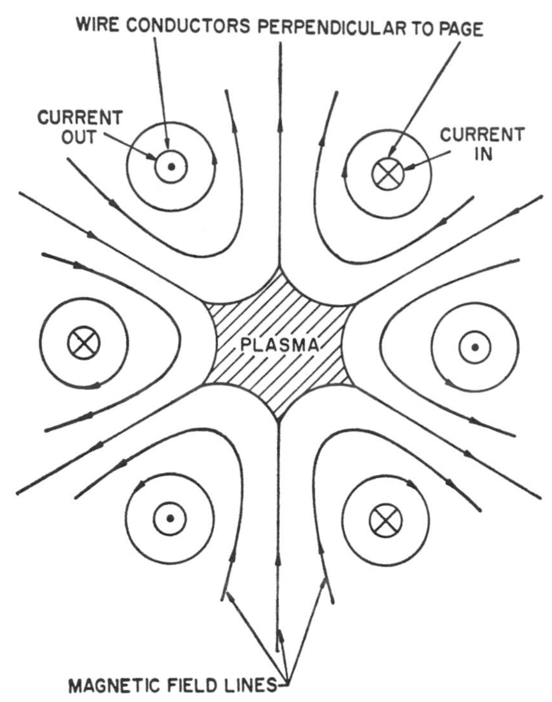

FIG. 11.15. Line-cusp configuration showing current directions, magnetic field, and plasma.

time of the plasma in the system is increased over its lifetime in the usual mirror system to the point where the limiting loss factor appears to be charge exchange with the background of neutral gas atoms.

6. Plasma Applications

The fundamentals of plasma physics apply to a widely divergent range of problems. Figure 11.1 summarizes a number of applications where characteristic plasma phenomena are present. In this section we touch lightly on a few of these applications.

6.1. Controlled Fusion. The controlled-fusion reaction, if the great physical and engineering problems associated with it can be overcome, will be a source of power having a fuel supply of almost unlimited extent. The deuteron (proton plus neutron) is the heart of the fusion reaction and has an abundance relative to hydrogen atoms of 1/7,000 in water. The fusion of two deuterons produces with equal probability a tritium nucleus (two neutrons plus proton) and a proton (4 Mev released) or a He^3 nucleus and a neutron (3.3 Mev released). The total cross section for the process is 0.01×10^{-24} cm^2 at 50 kev energy.

The fusion of a deuteron with a tritium nucleus has a cross section a hundred times larger at the same energy and therefore seems a more practical reaction to use. The products of this fusion are He^4 (3.5 Mev) and a neutron (14.1 Mev). The neutron energy must be converted into heat. This could be done by surrounding the fusion reactor with a neutron moderator ("blanket") which slows the neutron through elastic-recoil collisions and captures it through a nuclear reaction. Because tritium is not available in sufficient abundance this type of reactor would have to rely on the regeneration of tritium (T) through capture of the neutrons by lithium, for example, in the reactions

$$Li^7 + n \text{ (fast)} \rightarrow T + He^4 + n$$
and
$$Li^6 + n \text{ (slow)} \rightarrow T + He^4 + 4.8 \text{ Mev}$$

A sketch of a hypothetical controlled-fusion reactor by Brown et al. [1] is reproduced in Fig. 11.16.

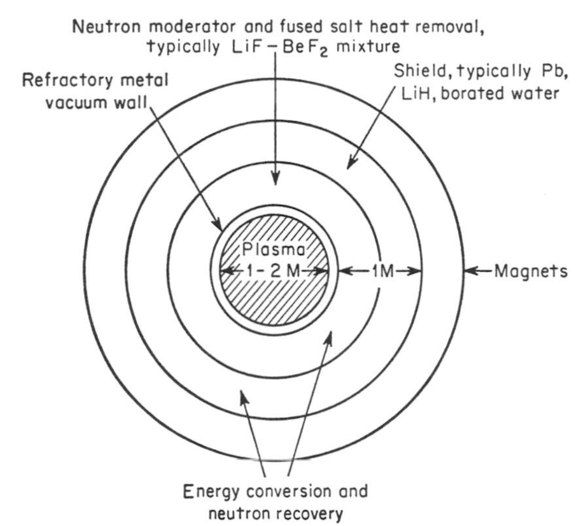

FIG. 11.16. Cylindrical configuration for a hypothetical controlled-fusion reactor.

An estimate of the size of the reacting volume and the particle temperatures needed to produce a reasonable amount of power may be made. The magnetic field necessary to confine such a plasma can also be estimated. For deuterium and tritium densities of 10^{14} cm^{-3} and a temperature corresponding to 50 kev, a power density of 10^7 watts/m^3 is generated in the reacting volume if the fusion cross section is 10^{-24} cm^2. In order to "burn" an appreciable amount of the fuel in the reacting volume during one reaction cycle, the plasma must be confined for a sufficient length of time. For a cross section of 10^{-24} cm^2 the mean lifetime for a deuteron before undergoing fusion is about 40 sec so that confinement times of the order of seconds are required.

For a plasma temperature corresponding to 50 kev, the total pressure, of a plasma of electrons and equal densities (10^{14} cm^3) of deuterons and tritons, is approximately 30 atm. To contain such a pressure a magnetic field [see Eq. (11.70)] greater than 30,000 gauss is required.

An "ideal" fusion reactor might then have an electron density of 10^{14} cm^{-3}, a particle temperature 50 kev, a volume of 10^6 cm^3, and a confinement time of 1 sec. Table 11.1 [1] gives a comparison between the ideal machine and some typical fusion machines in terms of the departure of the actual machine from the ideal in powers of ten of the electron density, plasma temperature, reacting volume, and confinement time.

a. Plasma Confinement Using a Magnetic Mirror. The confinement of the plasma can be achieved in steady-state mirror systems such as the DCX of the United States and OGRA of the Soviet Union. The DCX is about 300 cm in length, not including the mirror end coils. The central magnetic field is about 12,000 gauss and in the mirror regions is about 40,000 gauss.

One of the main problems of steady-state mirror confinement is the creation of a hot plasma inside the containing region. If a particle beam can be fired into the system and trapped, a hot plasma can be generated. Because of the adiabatic nature of the particle motion in the mirror, however, the particles escape the mirror system in a short time, leaving in the same manner as they entered. However, if their motion can be disturbed in some way while they are in the mirror system (by a collision, for example) the probability is high that they will be trapped. In the case of the DCX and OGRA plasma production is accomplished by directing a high-energy beam of deuterium molecule-ions into the mirror from outside in a direction transverse to the magnetic-field lines in such a manner that they dissociate by collision inside the mirror system. The deuterium ion is thus trapped since the radius of curvature of its cyclotron orbit is one-half the molecule-ion's radius of curvature.

Another method of injection involves a neutral beam which is ionized by collision once it is inside the mirror system. It is also possible to ionize an atom by passing it through a magnetic field, provided that it is in a high-lying state of excitation. The atom, moving with a velocity **V**, is subject to a **V × B** force which tends to separate the electron from the positive nucleus so that there is a certain probability that the atom will be ionized. This latter scheme is quite desirable since the injection can be done so that only plasma particles (ions and electrons) are introduced by the injection. The other injection methods introduce neutral atoms ("background gas") as well. Since one of the main sources of ion loss is through charge transfer to a neutral atom, background-gas pressure must be kept to a minimum.

The plasma can also be produced by rapidly increasing the mirror fields after first pre-ionizing the plasma by some other means. As the mirror field builds up in strength the plasma is compressed and heated by the increasing magnetic pressure.

b. The Stellarator. The stellarator is a cylindrical channel that closes on itself so that it has the shape of a torus. Solenoidal windings on the outside of the channel produce a magnetic field within the channel parallel to the channel axis. However, the gradients of the longitudinal magnetic field and the inertial force due to the curvature of the field lines cause the particles to drift to the outer part of the torus wall. This drift occurs because these charge-independent forces displace the ions toward the top of the channel (for instance) and the electrons toward the bottom of the channel so that the resulting electric field causes an **E × B** drift that is always toward the outer wall.

The feature that distinguishes a stellarator from a torus with solenoidal windings alone is the presence of helical windings around the channel which cause the magnetic-field lines to be twisted in such a way that as the particles drift they are constantly turned back toward the channel center. The helical windings are similar to the solenoidal windings but have a much larger pitch angle of winding.

The plasma is produced by pre-ionizing the gas in the channel with radio-frequency fields when the confining fields, which vary slowly with time, have reached a maximum. The plasma density is further increased through ohmic heating by an induced longitudinal electric field that is produced by operating the plasma torus as the secondary winding of a pulsed transformer. The heating of the plasma through application of radio-frequency fields at the ion cyclotron frequency has also been studied.

Turbulent diffusion (see Sec. 4.5) is apparently the mechanism that presently limits the confinement time of plasmas in the stellator. Table 11.1 compares the stellarator plasma characteristics with the "ideal" reactor plasma characteristics.

6.2. The MHD Generator. The MHD (magnetohydrodynamic) generator [27] offers a means of converting the thermal energy of combustion, fission, or fusion directly to electric power without the use of mechanical generators. The principle of operation involves allowing a hot plasma to expand through a channel that has a magnetic field applied transverse to the direction of the plasma flow velocity. The $\mathbf{V} \times \mathbf{B}$ force on the flowing ions and electrons drives them to opposite sides of the channel, causing an electric field to be set up across the channel in a direction antiparallel to $\mathbf{V} \times \mathbf{B}$. This potential difference across the plasma is then used to drive current through an external circuit.

To lower the conductivity of the plasma, alkali metals of low ionization potentials are added to the plasma to obtain high ionization densities at the plasma temperatures produced by the thermal-energy source. Since no moving parts are in contact with the high-temperature surfaces, high reaction temperatures may be employed, leading to high Carnot efficiencies of the generator. For large power-generation plants (large-area channels) the efficiency of this generator is expected to exceed steam-generator efficiencies, since the relative importance of heat losses to the channel walls of the MHD generator decreases as the ratio of the wall area to the channel area is decreased.

6.3. Extraterrestrial Plasmas. The interactions of electric and magnetic fields with plasmas play an important role in the physics of stars and of galactic and intergalactic plasma [28]. An understanding of the features of the solar-system plasmas will lead to better understanding of the farther-removed plasmas. In Table 11.2 are compared the electron density, temperature, and magnetic fields (in gauss) of the plasmas existing in the solar system [1].

The exterior of the sun is divided into three regions. The outer layer of the luminous solar disk is called the photosphere. The emission from the photosphere has a continuous spectrum broken by absorption lines due to the presence of unexcited ions and atoms in the outer reaches of the photosphere. The stormy environment of the solar surface gives rise to fluctuations in its appearance. Although the average magnetic fields at the photosphere are only a few gauss, magnetic fields in the vicinity of a sunspot as large as 4,000 gauss have been measured by the Zeeman effect.

Outside the photosphere lies the chromosphere from which an emission-line spectrum is observed. Outside the chromosphere and extending to a distance of about ten solar radii is the corona.

The high temperature of the corona, as determined by optical and radio observations, is thought to be caused by heating of the plasma by magnetohydrodynamic waves. These waves are generated as small disturbances at the solar surface. As they travel outward into regions of low density the energy content per particle of these waves increases (assuming the wave energy to be constant) so that when they are absorbed in the low-density regions they have a relatively strong heating effect. Radio-frequency studies of the apparent positions of radio stars as the sun passes between the earth and the star give information about the electron-density profile of the corona, since the path of the radiation is bent (refracted) if it propagates at an angle with respect to the gradient of the density (plasma index of refraction). Studies of the thermal emission of power from the corona at different radio frequencies given information about the temperature profile (see Sec. 7.3*a*) since the emission comes predominantly from the region in which $\omega \approx \omega_p$.

Bursts of plasma emitted from the solar surfaces are called solar flares. They are detected optically by observing the emission of H_α radiation from the chromosphere. As these jets of plasma traverse the corona they induce plasma oscillations which in turn radiate electromagnetic energy [29]. The center frequency of this radiation decreases with time during the period of approximately 10 min over which the radiation was measured. This suggests that the flare excites plasma oscillations as it moves outward through the corona in the direction of decreasing density. Also observed are the harmonics of the plasma frequency which could be due to nonlinear effects caused by the large amplitude of the oscillations. Large flares also cause sudden changes (2-min rise time) in the earth's magnetic field 1 to 2 days following the flare. This is thought to be due to shock waves induced by the flare which propagate through the interplanetary plasma compressing the earth's magnetic field on arrival.

Charged particles in the solar atmosphere with sufficiently great thermal velocities [30] can escape the sun's gravitational field. Thus the sun is constantly emitting electrons and protons into space in addition to the sporadic bursts of particles associated with nonquiescent solar activity. The presence of this "solar wind" is indicated by the fact that comet tails are observed to point away from the sun [31]. These electrons and protons are prevented from reaching the earth by the earth's dipole magnetic field. This region from which interplanetary plasma is excluded is called the magnetosphere.

The magnetosphere contains high-energy charged particles, the source of which is not well understood, which are trapped in the earth's magnetic field. They travel along the magnetic-field lines, undergoing reflection in the polar regions where the magnetic-field intensity is greatest.

The earth's ionosphere is produced by ionizing radiation from the sun incident on the upper atmosphere. Reflection of radio waves from the surfaces of the ionosphere and propagation of waves within it are understood through application of wave-propagation theory for plasma in the presence of a d-c magnetic field.

6.4. Plasma Applications in Solids. The conduction electrons (and "holes") of solid conductors and semiconductors are free to move through the crystal lattice under the influence of externally applied electric and magnetic fields. (A "hole" behaves like a positive electron in its motion.) Thus these particles can support plasma wave propagation in the solid just as in a gaseous plasma. The electrons and holes behave as free particles but with effective masses that differ from the actual electron mass because of the presence of the periodic potential of the lattice. Since this potential is anisotropic, the effective mass of the particles depends on their direction of motion within the lattice. Thus the study of plasma wave propagation in solids may be used to study the nature of the potential of the crystal lattice. A plasma in a solid has a number of features that make comparison of experiment with theory simpler than in the case of a gaseous plasma: The solid-state plasma has a very uniform density with sharply defined boundaries, there is no problem of instabilities generated in the production of the plasma since the plasma is always present, and the effect of impurities may be carefully controlled.

To observe plasma-wave phenomena in solids the relaxation time of the charge carriers due to scattering from thermal lattice vibrations, impurities, and faults must be made sufficiently low so that the wave phenomena are not damped out. This requires using samples of high purity at liquid-helium temperatures.

The cutoff frequency corresponding to the electron densities in metals is in the optical region. Most experimenters have therefore studied plasma-wave phenomena in materials that are immersed in a d-c magnetic field so that wave propagation in the material at low frequencies is possible (see Figs. 11.7 and 11.8). For frequencies of propagation much less than the electron cyclotron frequency one may expect wave propagation in the form of Alfvén waves or electron cyclotron waves, as discussed in Sec. 3. Equations (11.54) and (11.55) may be applied directly to wave propagation in a conducting solid provided the following modifications are made: ω_{p+} and ω_{B+} are now understood to apply to holes rather than positive ions; the effective masses are used for the electron and hole masses; and the dielectric constant of the lattice K in the absence of plasma effects is used so that $\omega_{pe}^2 = n_e e^2/m_e K\epsilon_0$, for example. Furthermore, it is no longer valid to assume that the electron and hole concentrations are equal, since in a metal, for example, the hole concentration is zero.

As an example we consider propagation along the magnetic field of a wave for which $\omega \ll \omega_{Be}$ and $\omega \ll \omega_{B+}$. Equation (11.55*a*) when written for *unequal* electron and hole concentrations [refer to Eqs. (11.54*a*) and (11.54*b*)] is

$$R = 1 - \frac{\omega_{p+}^2}{\omega(\omega + \omega_{B+})} - \frac{\omega_{pe}^2}{\omega(\omega - \omega_{Be})}$$

where the "+" subscript now refers to the holes. We apply this to two cases that have been treated experimentally and theoretically. The first is the case where the electron and hole concentrations are equal, as in bismuth [32]. This case is analogous to the gaseous-plasma result, and we obtain Alfvén wave propagation at a phase velocity [see Eq. (11.57)]

$$v_{\text{ph}} = \frac{c}{[1 + n_0(m_e + m_+)/K\epsilon_0 B_0^2]^{1/2}}$$

where n_0 represents electron or hole concentration. This velocity is independent of frequency and for electron densities found in metals and effective masses equal to the electron mass

$$v_{\text{ph}} = \frac{6.6 \times 10^3 B_0' K^{1/2}}{n_0'^{1/2}} \qquad \text{cm/sec}$$

where B'_0 is in kilogauss and n'_0 in units of 10^{24} cm^{-3}.

The other type of wave to be discussed is the helicon wave [33] which occurs in cases where the hole density is small compared with the electron density. The name of the wave is derived from the fact that the wave is circularly polarized. Experiments [34] find good agreement with theory. Setting $\omega_{p+}^2 = 0$ in the above expression for R and assuming $\omega \ll \omega_{Be}$, we have $R \approx 1 + \omega_{pe}^2/\omega\omega_{Be}$. The phase velocity for the right-handed circularly polarized wave is determined by its index of refraction $n_R^2 = R$ (see Sec. 3.4) in the equation

$$v_{\text{ph}} = \frac{c}{n_R} \approx \frac{c}{(1 + \omega_{pe}^2/\omega\omega_{Be})^{1/2}}$$

which is independent of the effective electron mass. For high electron densities such as in metals $v_{\text{ph}} \approx 170(fB'_0K/n'_0)^{1/2}$ cm/sec, where f is in megacycles, B'_0 in kilogauss, and n'_0 in units of 10^{24} cm^{-3}. Thus, low-frequency helicon waves travel at very low velocities.

Damping effects due to electron scattering can be much stronger in the case of Alfvén waves than in the case of helicon waves [35]. For damping to be unimportant in the former case, $\omega\tau$ must be much greater than 1, where τ is the electron relaxation time for scattering. In the case of helicon waves $\omega_{Be}\tau$ must be much greater than 1 and since $\omega \ll \omega_{Be}$, the condition is more difficult to satisfy for Alfvén waves.

7. Plasma Diagnostics

The complete study of a plasma involves the measurement of the densities of the particle constituents and their velocity distributions. If the plasma is changing with time because of wave propagation in it or because of its natural decay, time-resolved measurements of the above quantities must be made. Several recent books listed in the General References at the end of this chapter give quite complete accounts of plasma diagnostic techniques. Griem's book "Plasma Spectroscopy" deals with the measurement of optical radiation emitted from plasmas as a means of plasma study. Heald and Wharton's book "Plasma Diagnostics with Microwaves" covers microwave techniques, and Anderson, Springer, and Warder have edited a book that covers a rather wide range of diagnostic techniques: "Physico-Chemical Diagnostics of Plasma." Huddlestone and Leonard have edited "Plasma Diagnostic Techniques,'' which also covers a wide range of diagnostic methods.

7.1. General Tabulation of Diagnostic Techniques. Heald and Wharton [36] have given a useful tabulation of diagnostic techniques and the ranges of parameters for which they are applicable. This tabulation is given below.

Electron Density and Distribution

1. Microwave interferometer; $10^{10} < n_e < 10^{14}/\text{cm}^3$
2. Microwave cavity perturbation; $10^8 < n_e < 10^{12}/\text{cm}^3$
3. R-F conductivity probes; $10^8 < n_e < 10^{15}/\text{cm}^3$; for high collision rates
4. Microwave scattering; $10^{12} < n_e < 10^{14}/\text{cm}^3$; sensitive to instabilities
5. Optical interferometer; $10^{14} < n_e < 10^{19}/\text{cm}^3$
6. Optical Faraday rotation; $n_e > 10^{16}/\text{cm}^3$ for 10,000 gauss
7. Optical spectroscopic intensities; $n_e > 10^{13}$; equilibrium plasmas
8. Optical scattering; $10^{14} < n_e < 10^{19}/\text{cm}^e$
9. Optical Balmer-series limit; $10^{13} < n_e < 10^{15}/\text{cm}^3$
10. Particle collectors; $10^{-6} < j_e < 1$ amp/cm^2; yields product qn_ev
11. Electron or ion beam scattering; sensitive to potential fluctuations

Electron Temperature

1. Microwave radiation intensities; $T_e \gtrsim 0.1$ ev; stable plasmas
2. Doppler broadening of cyclotron radiation line; $T_e \gtrsim 50$ ev
3. Infrared and optical intensities; $T_e \gtrsim 10$ ev; equilibrium plasmas
4. X-ray intensities; $T_e \gtrsim 6$ kev; wall problems
5. Relative intensities of spectral lines; $1 < T_e < 50$ ev
6. Relative intensities of bremsstrahlung and recombination radiation
7. Doppler broadening of optical (Thomson) scattering; $T_e \gtrsim 5$ ev
8. Langmuir probes; $0.1 < T_e < 1{,}000$ ev; moderate densities

Ion Density and Distribution

1. Stark broadening of spectral lines; $n_i \gtrsim 10^{15}/\text{cm}^3$
2. Langmuir probes, single and double
3. Electron, ion, neutral-atom, or neutron-beam probes; $n_i \gtrsim 10^{14}/\text{cm}^3$
4. Diamagnetic effect (requires knowledge of temperature)
5. Alfvén and sound-wave propagation; dense plasmas
6. Calorimetry (requires knowledge of temperature)
7. Radioactive gas tracers and collimated detectors
8. Charge-exchange neutral detectors

Ion Temperature and Energy

1. Calorimetry; total energy and momentum
2. Doppler broadening of spectral lines; $T_i > 5$ ev
3. External energy-momentum analyzer; samples escaping ions
4. Time of flight; gives particle or shock-front velocity
5. Diamagnetic effect, using magnetic probes inside and outside plasma

Neutral Density, Distribution, and Identity

1. Shielded ionization gauge
2. Ion or neutral-atom beam scattering
3. Rayleigh scattering and resonance absorption of infrared and light photons by bound electrons
4. Schlieren and Mach-Zender photography
5. Charge-exchange detectors; fast neutrals
6. Molecular resonance spectroscopy; r-f and infrared
7. Re-ionization by delayed ionizing pulses

Drift Velocity, Shock Velocity, Rotation, and Thrust

1. Doppler frequency of reflected microwaves
2. Doppler shift of synchrotron radiation
3. Doppler shift of spectral lines
4. Ballistics and calorimetry
5. Time flight; probes, light, and microwave sampling
6. Nonreciprocity of phase shift for space charge and EM wave propagation

Instabilities and Turbulence

1. Electron and ion energy-momentum analysis; external measurement
2. Microwave radiation (nonthermal effects)
3. Microwave scattering from turbulence
4. Electron and ion beam scattering
5. Fast photography; time-resolved spectroscopy and total light; high densities
6. Magnetic probes and Rogowsky loops
7. R-F and Langmuir probes
8. External voltage-current measurements
9. X rays (if high-energy electrons are generated)
10. Neutron energy analysis (if high-energy ions are present)

Sheath Regions

1. Langmuir probes
2. R-F probes (sheath oscillations and space-charge waves)
3. Electron- and ion-beam probes
4. Microwave scattering from sheath oscillations

Constituent Identity (Purity)

1. Optical and atomic-resonance spectroscopy
2. Ion cyclotron-resonance absorption; e/m ratio
3. Magnetic analyzer; escaping ions
4. Mass spectrometer; neutral gas

7.2. Electromagnetic-wave Diagnostics. A large family of plasma diagnostic techniques center on the measurement of the real and imaginary parts of the plasma index of refraction in the absence of a magnetic field by studying the propagation of an *electromagnetic wave* through the plasma (see Chap. 10, Sec. 1). The absorption of the wave (real part of mobility) measures an average over the velocity distribution of the electron-collision frequency; when combined with a measurement of the phase velocity of the wave in the plasma (imaginary part of mobility) it yields the electron density. The phase velocity of the wave in the plasma is measured by using an interferometer technique where part of the incident wave is sent through the plasma and is then recombined with the part that bypassed the plasma so that a change in the interference of the two waves is observed as the plasma density changes. If the plasma can be contained in a resonant cavity, it is also possible to measure the real and imaginary parts of the mobility by observing the broadening and shift of the cavity resonance due to the plasma.

Such electromagnetic techniques measure only electron parameters. For the case where a magnetic field is present and the incident wave is polarized so as to experience cyclotron resonance, ion parameters may be measured.

These techniques may be employed at radio, microwave, infrared [37], and optical frequencies. At optical frequencies the laser has proved a useful source of highly monochromatic radiation which can be used in interferometers [38]. The Mach-Zender interferometer [39] yields an interference pattern of the whole plasma at once and can be used to determine density contours. At optical frequencies the background gas affects the index of refraction and must be corrected for.

The necessary frequency for study of a plasma is determined by the electron density. The general criterion is that $\omega \gtrless \omega_p$ (see Sec. 3 in Chap. 10). Thus the maximum electron density that can be studied at $f = 10^6$ cps (radio) is about $n_{\max} \approx 10^4$ cm^{-3}, at $f = 10^{10}$ cps (microwave) $n_{\max} \approx 10^{12}$, cm^{-3}, and at $f = 10^{15}$ cps (optical) $n_{\max} \approx 10^{22}$ cm^{-3}. The range of densities measurable with a given frequency depends on the sensitivity of the experiment but for good conditions is about three orders of magnitude.

In the presence of a magnetic field the Faraday effect described in Sec. 3.4 may be used to measure electron density.

7.3. Plasma Radiation Emission. The two prime sources of radiation emission from plasmas are the free electrons as they undergo transitions to other free states or to bound states of atoms and ions and the bound electrons as they undergo transitions to other bound states. The former type of emission gives rise to emission of radiation over continuous intervals of frequency and is referred to as continuum emission. The latter type gives rise to line spectra and is called line emission.

a. Continuum Emission. This radiation arises from bremsstrahlung radiation emitted by electrons undergoing collisions and from cyclotron radiation emission of such electrons between collisions if a magnetic field is applied to the plasma. Measurements of the intensity of this emitted radiation can be related to the electron temperature if the electrons have a Maxwellian distribution of velocities; if the velocity distribution is non-Maxwellian, it is possible in some cases to determine its nature from such emission measurements.

At microwave frequencies, for example, a plasma placed in a waveguide in such a manner that its boundaries do not reflect microwave radiation incident upon it emits an amount of power per interval of frequency [40]

$$P_f = AkT_R$$

where k is the Boltzmann constant and T_R the radiation temperature of the emitting particles in the plasma (electrons). The quantity A is the fraction of power of a wave that is absorbed as the wave travels the length of the plasma in the *direction of emission*. The radiation temperature equals the electron temperature for the case where the electrons have a Maxwellian distribution of velocities; thus, by making a simultaneous measurement [40] of A, an AkT_R, the electron temperature, is determined.

An expression has been derived [41] that relates T_R to the electron-velocity distribution and the radiation-emission mechanisms (bremsstrahlung and cyclotron emission) so that the deviation from Maxwellian may be studied [42].

At optical frequencies, in addition to the continuum emission due to bremsstrahlung, there is also continuum emission arising from the capture of free electrons into various excited states of the atoms and ions. The recombination continuum corresponding to a given state into which the electron is captured is distinguished by the presence of a low-frequency cutoff edge corresponding to the capture of a zero-energy free electron. McWhirter [43] has used the frequency dependence of the recombination continuum near these cutoff edges to determine electron temperatures.

b. Line Emission. Optical measurements of the line spectra emitted from plasmas can be used for constituent identification. Also, lines emitted by plasma particles are broadened because of the thermal motion of the emitting particle (Doppler effect) and the random electric-field fluctuations (Stark effect) due to the fluctuating charge density around the emitting particle. The Doppler broadening can be used to determine the temperature of the emitting particle; Stark broadening depends mainly on the electron density and only weakly on the electron temperature.

The remaining optical techniques involve the measurement of the absolute intensity of spectral lines and the relative intensities of spectral lines from different excited states of the same ion or atom or from excited states of ions of different degrees of ionization. Relative line-to-continuum measurements can also be made. These types of measurements generally require that local thermodynamic equilibrium exist; that is, that all particle constituents have the same temperature and that the excited states are populated in a thermal-equilibrium distribution. For the latter requirement to be satisfied, the electron density must be sufficiently high so that the collisional depopulation rate of the states by electrons greatly exceeds the radiational depopulation of the states. These types of measurements are discussed in detail in Griem's book.

7.4. Plasma Probes. *a. The Langmuir Probe and the Double Floating Probe.* The Langmuir probe [44] is in direct electrical contact with the plasma. By observing the current flowing to the probe from the plasma as the probe potential is varied, the electron temperature and density may be determined. When the probe potential is biased positive with respect to the plasma potential, only electrons from the plasma are collected and the current to the probe is negative. Further increases in the probe potential do not change the current greatly because it is limited to a value

TABLE 11.1. COMPARISON OF CONTROLLED-FUSION EXPERIMENTAL DEVICES*

Machine \ Ideal	Electron density n	Temperature T	Volume V	Confinement time τ, sec	Departure from ideal
	10^{14}/cm³	10–50 kev	10^6 cm³	1 sec	0
Zeta II Unstabilized toroidal pinch	10^{13}/cm³ (−1)	100 ev (−2)	10^5 cm³ (−1)	10^{-3} (−3)	−7
C-Stell Stabilized steady-state torus	10^{13} (−1)	100 ev (−2)	10^5 cm³ (−1)	10^{-2} (−2)	−6
DCX-1 Magnetic mirror, steady-state	10^8 (−6)	300 kev (0)	10^4 cm³ (−2)	40 (+1)	−7
Multiple compression pulsed magnetic mirror	10^{14} (0)	3 kev (−1)	10^3 cm³ (−3)	10^{-4} (−4)	−8
Pharos High-field pulsed one-turn mirror	10^{17} (+3)	1 kev (−1)	10^2 cm³ (−4)	10^{-5} (−5)	−7
OGRA Steady-state magnetic mirror	10^8 (−6)	100 kev (0)	10^6 cm³ (0)	10^{-2} (−2)	−8

* From Brown et al. [1].

determined by the random electron current in the plasma. As the potential is decreased, not all the electrons are able to reach the probe, because of the potential barrier; the number having sufficient energy to reach the probe is proportional to $e^{eV_p kT_e}$, where V_p is the probe potential and T_e the electron temperature. For probe potentials sufficiently negative, the probe current has reversed sign because only the ion current is now collected. Probe measurements are valid provided the current drawn from the plasma represents only a small perturbation on the plasma. Surrounding the probe is a sheath, and the criterion that the plasma not be disturbed by the probe is satisfied if the dimensions of the probe and the sheath together are small compared with the particle mean free paths.

In the presence of a magnetic field, the probe data are considerably altered if the Larmor radius of the particle being collected is not large compared with the probe-and-sheath dimensions. Thus the region of the probe curve where predominantly electrons are collected is most strongly affected by the presence of a magnetic field. By the same token, the positive-ion saturation current drawn by the probe may be correctly interpreted even in the presence of fairly large magnetic fields. Bohm, Burhop, and Massey [45] discuss probe theory and experiment with and without a magnetic field present.

TABLE 11.2. SOLAR-SYSTEM PLASMAS AND THEIR DENSITIES, TEMPERATURES, AND MAGNETIC FIELDS*

Plasma		$\log_{10} N$ cm⁻³	$\log_{10} T°K$	$\log_{10} BG$
Center	of sun	25	7	
Photosphere	of sun	14	4	0
Corona	of sun	7	6	−1
Interplanetary	of earth	1	6	−4
Magnetosphere	of earth	1 to 2	?	−3 to 0
Ionosphere	of earth	>6	3	0

* From Brown et al. [1].

The double floating probe [46] is a modification of the Langmuir probe whereby two identical probes, separated by about 1 cm, are introduced into the plasma; the current flowing between them, as a function of their potential difference, is measured. The maximum current drawn is limited by the ion saturation current so that considerable success has been obtained with double-probe measurements in a magnetic field [47].

Electronic methods of rapidly sweeping the probe potential (single or double probe) through a complete cycle have been developed to allow the study of transient plasma phenomena. Harp [48] has given such a circuit for a single (Langmuir) probe, and Olsen and Skarsgard [47] have studied a transient helium plasma, using a double probe.

b. R-F Plasma Probes. Various types of probes may be introduced into a plasma to measure its r-f conductivity. A metallic probe [49] immersed in a plasma exhibits a resonance in the d-c component of current drawn when an r-f voltage of frequency equal to the plasma frequency is simultaneously applied to the probe. The a-c conductivity of the plasma may be measured by measuring the a-c impedance of a coil [50] that is immersed in the plasma.

References

1. Brown, S. C., et al.: *Am. J. Phys.*, **31**: 637–691 (1963).
2. Landau, L. D.: *J. Phys.* (*U.S.S.R.*), **10**: 25 (1946).

3. Alexeff, I., and R. V. Neidigh: *Phys. Rev.*, **129**: 516 (1963).
4. Wong, A. Y., N. D'Angelo, and R. W. Motley: *Phys. Rev. Letters*, **9**: 415 (1962).
5. Pines, D., and J. R. Schrieffer: *Phys. Rev.*, **124**: 1387 (1961). S. Ichimaru, *Phys. Fluids*, **5**: 1264 (1962).
6. Arunasalam, V.: Ph.D. Thesis, Massachusetts Institute of Technology, Department of Physics, 1964 (unpublished); *Phys. Rev.*, **140**: A471 (1965).
7. Getty, W. D., and L. D. Smullin: *J. Appl. Phys.*, **34**: 3421 (1963). I. Alexeff, R. V. Neidigh, W. F. Reed, E. D. Shipley, and E. G. Harris: *Phys. Rev. Letters*, **10**: 273 (1963). M. A. Allen and G. S. Kino: *Phys. Rev. Letters*, **6**: 163 (1961). A. K. Berezin, G. P. Berezina, L. I. Boloton, and Ya. B. Fainberg: *J. Nucl. Energy, Pt. C*, Plasma Physics, **6**: 173 (1964).
8. Bers, A., and R. J. Briggs: *Bull. Am. Phys. Soc.*, II, **9**: 304–305 (1964); *Mass. Inst. Technol., Res. Lab. Electron. Quart. Progr. Rept.* 71, pp. 122–131, Oct. 15, 1963.
9. Stix, T. H.: "The Theory of Plasma Waves," McGraw-Hill, New York, 1962.
10. Spitzer, L.: "Physics of Fully Ionized Gases," Interscience, New York, 1956.
11. Heald, M. A., and C. B. Wharton: "Plasma Diagnostics with Microwaves," p. 80, Wiley, New York, 1965.
12. Bowles, K. L.: *Phys. Rev. Letters*, **1**: 454 (1958).
13. Rosenbluth, M. N., and N. Rostoker: *Phys. Fluids*, **5**: 776 (1962).
14. Jackson, J. D.: "Classical Electrodynamics," pp. 146–149, Wiley, New York, 1962.
15. Rose, D. J., and M. Clarke: "Plasmas and Controlled Fusion," Wiley, New York, 1961.
16. Allis, W. P., A. Bers, and S. J. Buchsbaum: "Waves in Anisotropic Plasmas," M.I.T., Cambridge, Mass., 1963.
17. Alfvén, H.: *Nature*, **3805,** p. 405, Oct. 3, 1942.
18. Barkhausen, H.: *Proc. I.R.E.*, **18**: 1155 (1930). T. L. Eckersley, *Nature*, **135**: 104 (1935).
19. Krall, N. A., and M. N. Rosenbluth: *Phys. Fluids*, **5**: 1435 (1962). B. B. Kadomtsev: *Soviet Phys. JETP* (*English Transl.*).
20. Twiss, R. Q.: *Australian J. Phys.*, **11**: 564 (1958). G. Bekefi, J. L. Hirshfield, and S. C. Brown: *Phys. Fluids*, **4**: 173 (1961).
21. Musha, T., and F. Yoshida: *Phys. Rev.*, **133**: A1303 (1964).
22. Forman, R.: *Phys. Rev.*, **128**: 1487 (1962). R. Forman, J. A. Ghormley, and J. R. Reiss: *Phys. Rev.*, **128**: 1493 (1962).
23. Ohara, S.: *Phys. Fluids*, **5**: 1483 (1962); *J. Phys. Soc. Japan*, **18**: 852 (1963).
24. Tayler, R. J.: *Proc. U.N. Intern. Conf. Peaceful Uses At. Energy, 2nd*, **31**: 160 (1958).
25. Bernstein, I. B., E. A. Frieman, M. D. Kruskal, and R. M. Kulsrud: *Proc. Roy. Soc.* (*London*), **A244**: 17 (1958).
26. Gott, Yu. V., M. S. Ioffe, and V. G. Tel'kovskii: *Nucl. Fusion*, 1962 *Suppl., Pt.* 3, 1045 (1962).
27. Sporn, P., and A. Kantrowitz: *Power*, **103**: 62 (1959).
28. See, for example, I. S. Shklovsky: "Cosmic Radio Waves," Harvard, Cambridge, Mass., 1960.
29. Wild, J. P., J. D. Murray, and W. C. Rose: *Nature*, **172**: 533 (1953).
30. Kiepenheuer, K. O.: *Naturforsch.*, **6a**: 627 (1951).
31. Biermann, L.: *Z. Astrophys.*, **29**: 274 (1951).
32. Buchsbaum, S. J., and J. K. Galt: *Phys. Fluids*, **4**: 1514 (1961). G. A. Williams and G. E. Smith: *IBM J. Res. Develop.*, **8**: 276 (1964).
33. Aigrain, P.: Proceedings International Conference on Semi-conductor Physics, Prague, 1960, p. 224, Academic, New York, 1961.
34. Bowers, R., Legendy, and F. Rose: *Phys. Rev. Letters*, **7**: 339 (1961). C. C. Grimes: *Bull. Am. Phys. Soc.*, **9**: 239 (1964).
35. Buchsbaum, S. J.: in Proceedings of 7th International Conference on the Physics of Semiconductors, vol. 2, Plasma Effects in Solids, Dunod, Paris, 1965.
36. Heald, M. A., and C. B. Wharton: "Plasma Diagnostics with Microwaves," p. 366, Wiley, New York, 1965. See also C. B. Wharton: in T. P. Anderson, R. W. Springer, and R. C. Warder, Jr. (eds.), "Physico-Chemical Diagnostics of Plasmas," p. 4, Northwestern University Press, Evanston, Ill., 1963.
37. Brown, S. C., G. Bekefi, and R. E. Whitney: *J. Opt. Soc. Am.*, **53**: 448 (1963). D. T. Llewellyn-Jones, S. C. Brown, and G. Bekefi: *Proc. Intern. Conf. Ionization Phenomena Gases, 6th*, 1963.
38. Ashby, D. E. T. F., and D. F. Jephcott: *Appl. Phys. Letters*, **3**: 13 (1963). J. B. Gerardo and J. T. Verdeyen: *Proc. I.E.E.E.*, **52**: 690 (1964).
39. See, for example, W. C. Elmore, E. M. Little, and W. E. Quinn: *Proc. U.N. Intern. Conf. Peaceful Uses At. Energy* (*Geneva*), *2nd*, **32**: 337 (1958).
40. Bekefi, G., and S. C. Brown: *J. Appl. Phys.*, **32**: 25 (1961).
41. Bekefi, G., J. L. Hirshfield, and S. C. Brown: *Phys. Rev.*, **122**: 1037 (1961).
42. Fields, H., G. Bekefi, and S. C. Brown: *Phys. Rev.*, **129**: 506 (1963).
43. McWhirter, R. W. P.: *Bull. Am. Phys. Soc.*, **8**: 164 (1963).
44. Tonks, L., and I. Langmuir: *Phys. Rev.*, **34**: 876 (1929).
45. Bohm, D., E. H. S. Burhop, and H. S. W. Massey: in A. Guthrie and R. K. Wakerling (eds.), "Characteristics of Electrical Discharges in Magnetic Fields," chap. 2, McGraw-Hill, New York, 1949.
46. Johnson, E. O., and L. Malter: *Phys. Rev.*, **80**: 58 (1950).
47. Olsen, O. D., and H. M. Skarsgard: *Can. J. Phys.*, **41**: 391 (1963).
48. Harp, R. S.: *Rev. Sci. Instr.*, **34**: 416 (1963).
49. Takayama, K. H., H. Ikegami, and S. Miyasaki: *Phys. Rev. Letters*, **5**: 238 (1960).
50. Wharton, C. B., and R. Hawke: *Univ. Calif.* (*Livermore*), *Lawrence Radiation Lab, Electron. Eng. Rept.* LEL.

Some General References

51. Alfvén, H.: "Cosmical Electrodynamics," Oxford University Press, London, 1953.
52. Allis, W. P., S. J. Buchsbaum, and A. Bers: "Waves in Anisotropic Plasmas," M.I.T., Cambridge, Mass., 1963.
53. Allis, W. P.: Motions of Ions and Electrons, in S. Flügge (ed.), "Encyclopedia of Physics," vol. 21, pp. 383–444, Springer, Berlin, 1956.
54. Anderson, T. P., R. W. Springer, and R. C. Warder, Jr. (eds.): "Physico-Chemical Diagnostics of Plasmas," Northwestern University Press, Evanston, Ill., 1963.
55. Artsimovich, L. A.: "Controlled Thermonuclear Reactions," Gordon and Breach, Science Publishers, New York, 1964.
56. Bishop, A. S.: "Project Sherwood," Addison-Wesley, Reading, Mass., 1958.
57. Bok, J. (ed.): Proceedings of the 7th International Conference on Semi-Conductor Physics, vol. 2, Plasma Effects in Solids, Dunod, Paris, 1965.
58. Cambel, A. B.: "Plasma Physics and Magnetofluidmechanics," McGraw-Hill, New York, 1963.
59. Ferraro, V. C. A., and C. Plumpton: "An Introduction to Magnetofluidmechanics," Oxford University Press, London, 1961.

60. Glasstone, S., and R. H. Lovberg: "Controlled Thermonuclear Reactions," Van Nostrand, Princeton, N.J., 1960.
61. Griem, H. R.: "Plasma Spectroscopy," McGraw-Hill, New York, 1964.
62. Heald, M. A., and C. B. Wharton: "Plasma Diagnostics with Microwaves," Wiley, New York, 1965.
63. Huddlestone, R., and S. Leonard (eds.): "Plasma Diagnostic Techniques," Academic, New York, 1965.
64. Kuiper, G. P. (ed.): "The Sun," University of Chicago Press, Chicago, 1953.
65. Montgomery, D. C., and D. A. Tidman: "Plasma Kinetic Theory," McGraw-Hill, New York, 1964.
66. Rose, D. J., and M. Clarke: "Plasmas and Controlled Fusion," Wiley, New York, 1961.
67. Rosenbluth, M. N. (ed.): "Advanced Plasma Theory," Academic, New York, 1964.
68. Simon, A.: "An Introduction to Thermonuclear Research," Pergamon Press, New York, 1959.
69. Shklovsky, I. S.: "Cosmic Radio Waves," Harvard, Cambridge, Mass., 1960.
70. Spitzer, L.: "Physics of Fully Ionized Gases," Interscience, New York, 1956.
71. Stix, T. H.: "The Theory of Plasma Waves," McGraw-Hill, New York, 1962.
72. Thompson, W. B.: "An Introduction to Plasma Physics," Pergamon Press, New York, 1962.

Particularly Applicable References

Section	*Reference Numbers*
1	66, 70
2	53
3	52, 71
4	58, 59, 66, 68, 70, 72
5	55, 56, 60, 66, 67, 72
6	51, 55, 56, 57, 60, 64, 66, 69
7	54, 61, 62, 63

Part 5 · Heat and Thermodynamics

Chapter 1

Principles of Thermodynamics

By E. U. CONDON, University of Colorado

1. The Nature of Heat

Heat is a form of energy associated with random and chaotic motions of the molecules of which matter is composed. It is therefore fundamentally measured in mechanical energy units, such as ergs or joules. In modern calorimetry heat input is measured electrically as the joule heat generated by an electric current flowing in a resistance coil in the calorimeter.

Traditionally heat was measured in terms of quantities defined in terms of the amount of heat needed to produce a specified change in temperature of a specified quantity of water. The *calorie* is the amount of heat needed to increase the temperature of water by 1°C, and the *Calorie* or *kilogram-calorie* (kg-cal) is 1,000 times larger, being the amount of heat needed to increase the temperature of 1 kg of water by 1°C. The *British Thermal Unit* (Btu) is the amount of heat needed to increase the temperature of 1 lb of water by 1°F.

More precisely it is necessary to state through which change in temperature the 1°C or 1°F increase takes place, for the heat capacity of water depends slightly on the temperature. The value of a calorie in absolute joules is known as the *mechanical equivalent of heat*. An idea of the variability of the heat capacity of water is given in Table 1.1. In view of

TABLE 1.1. HEAT CAPACITY OF WATER*

Temperature, °C	*Joules/gram degree*
0	4.2176
15	4.1857
17	4.1839
18	4.1831
25	4.1795

* Osborne, Stimson, and Ginnings, *J. Research, Natl. Bur. Standards*, **23**(2): 197–260 (August, 1939).

the fact that heat energy is measured electrically in terms of joules in modern calorimetry, the calorie is no longer needed logically. Its use is, however, firmly entrenched in the literature. To avoid confusion arising from changing absolute determinations of the heat capacity of water, most modern experimental work is expressed in terms of the *thermochemical calorie*, arbitrarily defined as equal to 4.1840 absolute joules. (The word "absolute" serves to distinguish the joule of the cgs system of mechanical units from the international joule, based on the international electrical units formerly in use.) Unless the contrary is stated, the calorie will be assumed to be defined as 4.1840 joules.

Thermodynamics deals with the laws governing changes in the thermal state of materials in states not departing greatly from thermal equilibrium. The study of heat divides into four closely related parts: (1) general principles of thermodynamics and correlation of thermal properties of matter in terms of them, (2) principles of statistical mechanics which interpret the principles of thermodynamics in terms of the statistical behavior of ensembles of molecules, (3) application of statistical mechanics to detailed molecular interpretation of thermal properties of matter, and (4) thermal behavior of matter when there is departure from thermal equilibrium.

In thermodynamics the material under study is referred to as a system. It may consist of a number of different *components* or distinct chemical substances. The amounts of the components may be fixed or may change if chemical reactions occur in the changes which the system undergoes. It may also include several *phases*, as gaseous, liquid, and solid. There is never more than one gaseous phase present. Solids and liquids, however, show limited mutual solubility, and a system may include any number of solid and liquid phases in addition to the single gaseous phase. Systems also differ in the ways in which they may exchange energy with their surroundings. In cases most often studied, this exchange is by the flow of heat across the boundaries and by changes of volume in which the system exerts a simple isotropic pressure on its boundaries. Other modes of exchange of energy are possible, as when the surface area changes while exerting a surface tension or when an electric current passes into the system and out again while an internal electromotive force is exerted. Likewise, in the case of solids showing rigidity, more general stress systems than isotropic pressure may act across the boundaries, and it is necessary to take into account changes in the shape of the body as well as changes in its volume. The generalization in such cases is fairly straightforward so that, in the following development, it will be supposed that the system interacts with its surroundings only through volume changes and isotropic pressure, in addition to the flow of heat across its boundaries (except where other cases are explicitly under consideration).

2. First Law of Thermodynamics

The first law consists simply in the recognition that heat is a form of energy and, therefore, has to be included in reckoning changes in the internal energy content of the system. The law asserts that the system possesses an internal energy content U which is a function of the variables necessary to specify the state. For a simple one-component system the state is fully specified by giving the volume V and temperature T. For such a system there are *two equations of state*. The first gives the pressure $p(V,T)$, and the second gives the internal energy $U(V,T)$ as a function of the independent variables.

If the system undergoes a small change in which it remains close to thermal equilibrium and the quantity of heat energy entering through the walls is δQ while the volume increases by dV against the pressure p, then the change in internal energy dU is

$$\delta Q = dU + p\,dV \tag{1.1}$$

which expresses the fact that the system only exchanges energy with its surroundings by taking in heat, δQ, or by doing external work, $p\,dV$. Here δQ is written instead of dQ to emphasize that it is not the exact differential of some function of the variables defining the state. This comes about because $p\,dV$ is not an exact differential, since

$$\int_A^B p\,dV$$

depends on the path of intermediate states followed in going from state A to state B.

If the change occurs at *constant volume*, no external work is done. All the heat added goes to increase the internal energy.

The *heat capacity* at constant volume, c_v, is defined as $c_v = (\delta Q/\delta T)_v$ and is given by

$$c_v = \left(\frac{\partial U}{\partial T}\right)_v \tag{1.2}$$

The term heat capacity may be used for the extensive property defined by (1.2), where U is the internal energy of the whole system, or the intensive quantity resulting from (1.2) if U is the internal energy of unit mass or unit molal quantity of material.

Because changes at constant pressure are so often studied experimentally, it is convenient to write (1.1) in another form,

$$\delta Q = dH - V\,dp \tag{1.3}$$

in which H is a new function of the state of the system called its heat content or *enthalpy*, defined as

$$H = U + pV \tag{1.4}$$

From (1.3) it follows that in a change occurring at *constant* pressure (also called *isopiestic*) all the heat added to the system goes to increase the enthalpy.

The heat capacity at constant pressure, c_p, is defined as $c_p = (\delta Q/\delta T)_p$ and is therefore given by

$$c_p = \left(\frac{\partial H}{\partial T}\right)_p \tag{1.5}$$

which may be an extensive or intensive quantity according as H is taken to be the enthalpy of the whole system or the enthalpy of unit quantity of material.

From the definitions it follows that

$$c_p - c_v = \left[p + \left(\frac{\partial U}{\partial V}\right)_T\right]\left(\frac{\partial V}{\partial T}\right)_p \tag{1.6}$$

Only differences in U and H are of physical significance; thus the states of the system from which U and H are reckoned as zero may be arbitrarily chosen. In what follows it will be supposed, unless otherwise stated, that U is expressed in terms of the independent variables V and T, while H is expressed as a function of p and T.

An *adiabatic* change is one in which no heat is taken in or given out; thus $\delta Q = 0$ in (1.3).

If a one-component system consists of but a single phase, its properties are fully described by giving $p(V,T)$ and either $U(V,T)$ or $H(p,T)$. But if two phases are present, another variable is needed, say x, to describe what fraction of the total material is in the first phase, the fraction $(1 - x)$ being in the second phase. In general, the specific volume (volume of unit quantity, either mass or molal) of the two phases in equilibrium and the internal energy content and enthalpy of the two phases in equilibrium will be different.

3. Second Law of Thermodynamics

It will be supposed in this section that the changes which a system undergoes are reversible; that is, the departure from equilibrium is so small at all stages that a minute change in the circumstances would serve to reverse the direction of the change.

Because δQ is not an exact differential of the variables of a state, it is possible to carry a system through a closed cycle of changes by which it is restored to its initial state, but in which the net amount of heat absorbed (total amount absorbed q_1 minus that given off q_2) by the system is finite. Since the internal energy will have returned to its original value, this finite net amount of heat taken in must be equal to the finite net amount of external work done $\int p\,dV$ in the cycle. Such a cycle provides a means of transforming heat energy into mechanical energy.

The *efficiency*, η, of a heat engine is defined as the ratio of the work output $(q_1 - q_2)$ to the heat taken in q_1,

$$\eta = (q_1 - q_2)/q_1 \tag{1.7}$$

A particularly basic cycle of operations is that known as the *Carnot cycle*. This consists of four parts: (1) the system undergoes isothermal expansion, taking in heat and doing external work; (2) it undergoes further adiabatic expansion, not taking in heat but doing more external work, with consequent drop in temperature; (3) it undergoes isothermal compression, giving out heat and having work done on it; (4) it undergoes adiabatic compression, with no exchange of heat with the outside, but having work done on it with a consequent rise in temperature. It is supposed that the magnitudes of these four stages are so proportioned that the system is returned to its initial state at the end of the fourth operation.

The importance of the Carnot cycle lies in the fact that the heat taken in, q_1, is all taken in at a single temperature τ_1 and that given out is given out at a single lower temperature τ_2. Any more general cycle may be regarded as made up of an approximate combination of infinitesimal Carnot cycles. Here τ is the temperature on any arbitrary temperature scale.

The *second law of thermodynamics* says that all

self-acting cyclic heat engines require a difference of temperature over which to operate. It has been stated in various ways: one way, due to Planck, is the proposition that *it is impossible to construct a system operating in a complete cycle which produces no effect except the exchange of heat with a reservoir at one temperature and the output of mechanical work.* Like the principle of conservation of energy the second law is a generalization from experience. A machine which violates the principle of conservation of energy would produce *perpetual motion of the first kind,* and one which violates the second law of thermodynamics would produce *perpetual motion of the second kind.*

Important consequences can be derived from the simple negation which is the second law:

1. Of all engines working between two heat reservoirs of temperature τ_1 and $\tau_2 (\tau_1 > \tau_2)$, the reversible engine is the one of maximum efficiency. Proof follows from the fact that if another engine A had greater efficiency, it could be used to drive the reversible engine backward, returning to the temperature τ_1 the heat given out by A at temperature τ_2 so that the net effect would be that only heat would be taken in at temperature τ_1 and mechanical work given out, in violation of the second law.

2. All reversible engines working between the same two heat reservoirs must have the same efficiency. The proof is as in (1), for the more efficient one could be used to drive the less efficient one in reverse, producing an over-all result in violation of the second law.

3. The heat taken in at the higher temperature is to the heat given out at the lower temperature as the ratio of $T(\tau_1)$ to $T(\tau_2)$, where $T(\tau)$ is a function of the arbitrary empirical temperature τ.

The efficiency of a reversible engine operating between temperatures τ_1 and τ_2 is $\eta = 1 - q_2/q_1$. This must be a function of τ_1 and τ_2; therefore write $q_2/q_1 = \varphi(\tau_1,\tau_2)$. Another reversible engine operating to take in heat at τ_1 and giving out heat at temperature τ will have $q/q_1 = \varphi(\tau_1,\tau)$. If this is coupled with another reversible engine which takes in heat q at temperature τ and gives out q_2 at temperature T_2, then $q_2/q = \varphi(\tau,\tau_2)$. The function of the two temperatures must be such that $\varphi(\tau_1,\tau_2) = \varphi(\tau_1,\tau)\varphi(\tau,\tau_2)$ for any value of τ. This requires that $\varphi(\tau_1,\tau_2)$ be of the form $\varphi(\tau_1,\tau_2) = T(\tau_2)/T(\tau_1)$, where T is a universal function of the temperature, independent of the working material in the reversible engine.

Hence there exists a function $T(\tau)$ such that the heat q taken in at temperature τ and the q_2 given out at temperature τ_2 by any reversible engine operating between these temperatures is proportional to $T(\tau)$, that is,

$$\frac{q}{T(\tau)} = \frac{q_2}{T(\tau_2)} \tag{1.8}$$

Since the q's as defined are essentially positive, $T(\tau)$ must be a function that is positive and increases with increasing temperature.

4. In an arbitrary cyclic process

$$\oint \frac{dQ}{T(\tau)} = 0$$

if the process is reversible. This follows on breaking up the general process into a series of cyclic reversible processes operating between τ and a fixed temperature τ_2 and using the relation (1.8). If the cyclic process is not wholly reversible, then, since the efficiency of the irreversible cycle elements is lower than the reversible case,

$$\oint \frac{dQ}{T(\tau)} \leq 0 \tag{1.9}$$

where the inequality applies to an irreversible cycle and the equality applies to a reversible cycle. This result is known as the *Clausius inequality.*

5. It follows that

$$\int_A^B \frac{dQ}{T(\tau)}$$

when evaluated along a series of reversible changes is independent of the path followed from A to B and is therefore equal to the difference of the value of a function of state at B minus that at A. This function of state is defined as the *entropy* S of the system, and

$$S_B - S_A = \int_A^B \frac{dQ}{T(\tau)} \tag{1.10}$$

6. From the Clausius inequality it follows that if the process by which the system passes from A to B is irreversible, the change in entropy is greater than

$$\int_A^B \frac{dQ}{T(\tau)}$$

so that

$$S_B - S_A \geq \int_A^B \frac{dQ}{T(\tau)} \tag{1.11}$$

where the equality applies if the process in going from A to B is reversible; otherwise the inequality applies.

In particular, for an isolated system in which there is no exchange of heat with the outside, the integral vanishes. It follows that the *entropy always increases* in any spontaneously occurring irreversible process occurring within an isolated system.

4. Absolute Temperature Scale

In Sec. 3 it was supposed that temperature was measured according to any scale such that larger numbers are associated with higher temperatures, but otherwise arbitrary. In empirical work various scales have been set up based on the differential expansion of a liquid and a solid as with the mercury-in-glass thermometer, or on some other conveniently observed property. Lord Kelvin recognized that the theory of the reversible heat engine makes possible the definition of an absolute temperature scale which is universal, i.e., independent of the properties of a particular physical substance.

If τ is an arbitrary empirical temperature scale, then $T(\tau)$ as defined by (1.8) has the property of being positive for all actual temperatures and of increasing with increase in temperatures, i.e., heat always flows spontaneously from a place of greater $T(\tau)$ to one of lower $T(\tau)$. This suggests that T itself can be taken as the measure of temperature. This is known as the *absolute temperature scale.*

The function $T(\tau)$ as introduced in (1.8) is not fully determined since it can be multiplied by an arbitrary constant and still satisfy that equation. This corresponds to a freedom of choice of the size of the degree. The choice adopted in 1954 by the Tenth General Conference on Weights and Measures assigns the arbitrary value

$$T_0 = 273.16°\text{K} \tag{1.12}$$

to the temperature of the *triple point* of water, that is, the temperature at which ice, water, and water vapor coexist in equilibrium (Sec. 8). Temperatures expressed on such a scale are called Kelvin temperatures and are written °K.

The melting point of ice under standard atmospheric pressure (1,013,246 dynes/cm²) is experimentally found to be 0.010°K lower than T_0, and so the ice point is 273.15°K. Similarly, the boiling point of water under standard atmospheric pressure is found experimentally to be 373.15°K within present attainable accuracy.

In the original definition of the Celsius scale, the size of the degree was so defined to produce 100° as the difference between the steam point and the ice point. In fact, the arbitrary choice contained in (1.12) was made to preserve this result. But the present intention is to adhere to (1.12) as the basic definition of the size of the degree, with the possible result that future experimental determinations of the ice point and the steam point may have slightly different values from those given above and hence possibly also a slightly different difference between the steam point and the ice point.

It is clear that temperature could also be specified by using any monotonic function of T, rather than T itself. For example, if a reversible heat engine works between temperatures T and $T - dT$, its efficiency is dq/q, which from (1.8) is equal to dT/T or $d(\log T)$. Thus if temperature is measured by $\theta = \log T$, the efficiency is given by $d\theta$. Kelvin made such suggestions but they have not found general acceptance.

The experimental establishment of the absolute temperature scale (Part 5, Chap. 3) is by means of a gas thermometer. By *perfect gas* is meant the limiting behavior of any gas at low densities and at temperatures well above its critical point. For such a gas the $p(V,T)$ relation takes the form

$$pV = \frac{m}{M} RT \tag{1.13}$$

where m is the total mass of gas involved, and M is the molecular weight. Thus the quantity m/M is expressed in moles. For such a gas it is also found that the internal energy per mole, $U(V,T)$, is actually a function of T alone. It follows that

$$dS = Q/T = dU/T + Rd(\log V)$$

is an exact differential, and so the temperature in (1.13) can be identified with the absolute thermodynamic temperature.

From (1.10) we see that the extensive unit of entropy is any energy unit per degree, as calorie per degree or joule per degree. The corresponding intensive unit is usually expressed as energy per degree per mole and thus, from (1.13), has the same unit as the molal gas constant R. It is convenient to express molal entropies as values of S/R, which is therefore a pure number. The experimental value of R is

$$R = 8.31696 \pm 0.00034 \text{ joules/mole deg}$$

Another important attribute of the perfect gas is that its internal energy is independent of its volume, so that $U(V,T)$ is actually a function of T alone. It follows that $dS = \delta Q/T = dU/T + Rd(\log V)$ is an exact differential when the perfect gas temperature scale is used and therefore the perfect gas temperature scale may be identified with the absolute thermodynamic scale of temperature.

5. Third Law of Thermodynamics

The definition of entropy, given in (1.10), serves only to define difference of entropy between two states of a system. In statistical mechanics entropy is interpreted as related to the state of disorder of the molecular motions (Chap. 2). It is natural to suppose that the state of least disorder for a pure crystalline substance is the state which it approaches at the absolute zero of temperature and that therefore the most natural zero from which to reckon entropy would be to suppose $S = 0$ at $T = 0$ for such a substance.

This property of having vanishing entropy at absolute zero of temperature is known as the *third law of thermodynamics*.

Since the dQ is equal to $c\,dT$, where c is the specific heat, this implies that the specific heat tends to zero as $T \to 0$; otherwise there would not be a finite entropy difference between the state at absolute zero and the state at higher temperatures. This is in agreement with experimental facts.

Entropies referred to the level $S = 0$ at $T = 0$ will be called *absolute entropy* whenever necessary, to distinguish from the values defined by (1.10) together with some arbitrary choice as to the state of zero entropy.

6. Equilibrium Conditions

In Sec. 3 it was shown that the change occurring in any spontaneously irreversible change in which no heat is taken in from the outside is one in which the entropy increases. Therefore, the equilibrium state of such an isolated system must be one in which the entropy is a maximum; otherwise spontaneous irreversible processes would take place in which the entropy would increase still further.

Hence under conditions of heat isolation the spontaneous changes are such that $\delta S > 0$ and the equilibrium state will be such that $\delta S = 0$ (S maximum).

In an actual change, which may not be reversible, from Sec. 3,

$$\delta S \geq \frac{\delta U + p\delta V}{T}$$

which can be written

$$\delta U + p\,\delta V - T\,\delta S \leq 0 \tag{1.14}$$

If now the system is held at constant volume ($\delta V = 0$) and at constant temperature, then the spontaneously occurring irreversible changes occurring within it will be such that $\delta(U - TS) \leq 0$, the equality sign holding only for reversible changes. The combination

$(U - TS)$ is known as the *free energy* (sometimes the *Helmholtz free energy*) and will be denoted by A:

$$A = U - TS \qquad (1.15)$$

Therefore a system held at constant volume and constant temperature will spontaneously move toward lower values of A. At equilibrium the Helmholtz free energy is a *minimum*.

The equilibrium situation at constant temperature and constant volume is therefore not that of minimum internal energy. If by going to a state of somewhat higher internal energy the entropy is so much increased that the decrease in A due to increase in the TS term outweighs the increase in A due to increase in U, then the equilibrium state will be the state of higher internal energy.

If the system is held at constant temperature and constant pressure, then it is necessary to make allowance for the $p\,\delta V$ in (1.14) which can be written as $\delta(U + pV - TS) \leq 0$ when $\delta p = 0$. This combination $(U + pV - TS)$ is also called free energy, but is called the *Gibbs free energy* when it is necessary to distinguish it from the Helmholtz free energy. It will be denoted by

$$G = H - TS = A + pV \qquad (1.16)$$

Because of the relation of G to A, it has also been called *free enthalpy*. In a system held at constant temperature and pressure the spontaneously occurring irreversible changes will be such as to diminish the Gibbs free energy. Therefore the equilibrium state will be such that the *Gibbs free energy is a minimum*.

These relations play an important role in determining equilibria in chemical reactions and in equilibria between phases (Part 5, Chap. 11, and Part 8, Chap. 6). For example, above the melting point the Gibbs free energy of the solid is higher than that of the liquid and therefore a crystal would spontaneously melt in an irreversible process leading to a diminution of the Gibbs free energy per unit quantity. If G is the same for solid as for liquid, both phases can coexist in equilibrium. The melting point itself is the temperature at which the Gibbs free energy of the solid is equal to that of the liquid.

7. Relations between Thermodynamic Functions

If volume and temperature are used as independent variables, there are two equations of state, $p = p(V,T)$ and $U = U(V,T)$, for the pressure and internal energy.

However, these functions are not independent. They must be related to avoid violation of the second law. Since

$$dU = \left(\frac{\partial U}{\partial T}\right)_V dT + \left(\frac{\partial U}{\partial V}\right)_T dV$$

it follows that

$$dS = \frac{1}{T}\left(\frac{\partial U}{\partial T}\right) dT + \frac{1}{T}\left(p + \frac{\partial U}{\partial V}\right) dV$$

Since this is an exact differential,

$$\frac{\partial}{\partial T}\left(\frac{p}{T} + \frac{1}{T}\frac{\partial U}{\partial V}\right) = \frac{\partial}{\partial V}\left(\frac{1}{T}\frac{\partial U}{\partial T}\right)$$

which gives

$$\frac{\partial}{\partial T}\left(\frac{p}{T}\right) = \frac{\partial}{\partial V}\left(\frac{U}{T^2}\right) \qquad (1.17)$$

This relation between the functions $p(V,T)$ and $U(V,T)$ must be satisfied for every real substance.

Equation (1.17) can be satisfied by introducing a function G so that

$$\frac{p}{T} = \frac{\partial G}{\partial V} \qquad \text{and} \qquad \frac{U}{T^2} = \frac{\partial G}{\partial T}$$

On substituting this in the expression for dS, it is found that dS is the differential of $G + U/T$. By proper choice of integration constant, G can be identified with $-A/T$, where $A(V,T)$ is the Helmholtz free energy. It thus appears that the Helmholtz free energy is the basic function in terms of which pressure, entropy, and internal energy may be expressed. The relations are

$$p = -\left(\frac{\partial A}{\partial V}\right)_T \qquad S = -\left(\frac{\partial A}{\partial T}\right)_V \qquad (1.18)$$

$$U = A - T\left(\frac{\partial A}{\partial T}\right)_V = \left[\frac{\partial(A/T)}{\partial(1/T)}\right]_V \qquad (1.19)$$

If *pressure and temperature* are the independent variables, then the functions $V(p,T)$ and $H(p,T)$ are related by the second law. It now turns out that the Gibbs free energy $G(p,T)$ is the simple function in terms of which the other properties can be expressed. The relations are

$$V = \left(\frac{\partial G}{\partial P}\right)_T \qquad S = -\left(\frac{\partial G}{\partial T}\right)_p \qquad (1.20)$$

$$H = G - T\frac{\partial G}{\partial T} = \left[\frac{\partial(G/T)}{\partial(1/T)}\right]_p \qquad (1.21)$$

If *entropy and volume* are taken as the independent variables, internal energy is the basic function in terms of which the others are expressed:

$$T = \left(\frac{\partial U}{\partial S}\right)_V \qquad p = -\left(\frac{\partial U}{\partial V}\right)_S \qquad (1.22)$$

$$A = U - S\left(\frac{\partial U}{\partial S}\right)_V \qquad (1.23)$$

Finally, if *entropy and pressure* are taken as the independent variables, enthalpy is the basic function in terms of which the others are simply expressed:

$$T = \left(\frac{\partial H}{\partial S}\right)_p \qquad V = \left(\frac{\partial H}{\partial p}\right)_S \qquad (1.24)$$

$$G = H - S\left(\frac{\partial H}{\partial S}\right)_p \qquad (1.25)$$

In differential form the preceding results imply the following:

$$dU = T\,dS - p\,dV \qquad (1.26)$$
$$dH = T\,dS + V\,dp \qquad (1.27)$$
$$dA = -S\,dT - p\,dV \qquad (1.28)$$
$$dG = -S\,dT - V\,dp \qquad (1.29)$$

There are four relations, known as *Maxwell's relations*, which follow from the fact that these are all perfect differentials:

$$\left(\frac{\partial T}{\partial V}\right)_S = -\left(\frac{\partial p}{\partial S}\right)_V \qquad (1.30)$$

$$\left(\frac{\partial T}{\partial p}\right)_S = \left(\frac{\partial V}{\partial S}\right)_p \qquad (1.31)$$

$$\left(\frac{\partial S}{\partial V}\right)_T = \left(\frac{\partial p}{\partial T}\right)_V \qquad (1.32)$$

$$\left(\frac{\partial S}{\partial p}\right)_T = -\left(\frac{\partial V}{\partial T}\right)_p \qquad (1.33)$$

In statistical mechanics direct calculations of $A(V,T)$ are made from a molecular theory for the system in question. From this the more directly observable thermodynamic properties like pressure and internal energy are calculable by differentiation. The problem of the experimental determination of $A(V,T)$ or $G(p,T)$ is that of working back from the observed values of $p(V,T)$ and $U(V,T)$ or of $V(p,T)$ and $H(p,T)$ to a construction of the free energy functions by a process of integration.

Equation (1.17) can be written

$$p + \left(\frac{\partial U}{\partial V}\right)_T = T\left(\frac{\partial p}{\partial T}\right)_V$$

This permits elimination of the U function from (1.6), giving

$$c_p - c_V = T\left(\frac{\partial p}{\partial T}\right)_V \left(\frac{\partial V}{\partial T}\right)_P \qquad (1.34)$$

8. Phase Equilibria of Single-component Systems

If a system consists of a single component that can exist in two phases, each phase will have its own free energy function $G_1(p,T)$ and $G_2(p,T)$; if x is the fraction of the material that is in phase 2, the free energy of unit quantity in the system is

$$G(p,T) = (1 - x)G_1(p,T) + xG_2(p,T) \qquad (1.35)$$

Since spontaneous changes tend to make F a minimum, the equilibrium value of x will be zero if $G_2 > G_1$ and unity if $G_2 < G_1$. Only if $G_2 = G_1$ can x assume intermediate values; that is, both phases coexist in equilibrium.

Thus the condition of equilibrium will be fulfilled for a series of functionally related values of p and T. This is called the *vapor-pressure curve* if one phase is the vapor and the other is solid or liquid. It is the *melting curve* if one phase is solid and the other liquid. The fundamental condition is that it is the curve along which

$$G_1(p,T) = G_2(p,T) \qquad (1.36)$$

for the two phases in question.

Likewise, if there is a third phase, then phases 1 and 3 can coexist along the curve

$$G_1(p,T) = G_3(p,T) \qquad (1.37)$$

while phases 2 and 3 can coexist along the curve $G_2(p,T) = G_3(p,T)$. For all three phases to coexist in equilibrium, both (1.36) and (1.37) must be satisfied simultaneously. In general, this will happen only at a single pair of values (p_t, T_t) where the three equilibrium curves between pairs of phases intersect, i.e., the *triple point*.

If there is a fourth phase with the free energy function $G_4(p,T)$, then for all four phases to coexist in equilibrium would require the additional condition $G_1(p,T) = G_4(p,T)$ to be satisfied, which will not in general be possible, although it could happen if the several free energy functions were related in just this way. There is no known case of four phases of a single component being in equilibrium.

The *Clausius-Clapeyron equation* is an important differential relation connecting points on the equilibrium curve between two phases. On the equilibrium curve $G_1(p,T) = G_2(p,T)$ and also

$$G_1(p + dp,\ T + dT) = G_2(p + dp,\ T + dT)$$

so that $d(G_2 - G_1) = 0$. From (1.29) this gives $-(S_2 - S_1)\,dT + (V_2 - V_1)\,dp = 0$ and since

$$(S_2 - S_1)T = L_{21}$$

where L_{21} is the latent heat absorbed by the system in going from phase 1 to phase 2,

$$\frac{dp}{dT} = \frac{L_{21}}{T(V_2 - V_1)} \qquad (1.38)$$

which is the Clausius-Clapeyron equation. If phase 2 is a gas of low density so that the perfect gas law applies and $V_1 \gg V_2$ and L_{21} is regarded as independent of temperature, (1.38) can be integrated to give

$$p = P \exp\left(-\frac{L_{21}}{RT}\right) \qquad (1.39)$$

for the approximate temperature dependence of the vapor pressure of a condensed phase, solid or liquid.

In the case of equilibrium between two solid phases or a solid and a liquid phase, V_2 and V_1 are nearly of the same small magnitude so that dp/dT is very great; that is, large changes in pressure are required to produce an appreciable change in the equilibrium temperature. Assuming the phases are so numbered that $L_{21} > 0$, then the sign of dp/dT is that of $V_2 - V_1$. Most solids expand on melting and therefore an increase in pressure raises the melting point; but a few, like ice, contract on melting and, for these, increase of pressure lowers the melting point.

In the case of the liquid-gas phase change, the change in specific volume on vaporization tends to zero as the temperature is increased. At a particular state called the *critical point* (p_c, V_c, T_c) all the properties of the liquid phase become the same as those of the gas phase, and at higher temperatures there is no distinction between liquid and gas so that the *vapor-pressure curve ends at the critical point.*

Qualitatively, therefore, the typical phase diagram for a pure substance is like Fig. 1.1 where T is the triple point and C is the critical point. The fact that the vapor-pressure curve ends at C gives rise to a *continuity of the liquid and vapor states* in the sense that by following a suitable path of change up around the end of the vapor-pressure curve a sub-

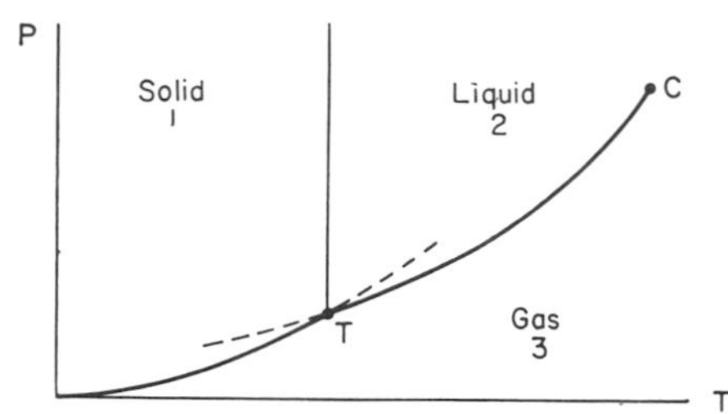

Fig. 1.1. Qualitative phase diagram for single component.

stance can be taken continuously from a liquid state to a gaseous state without passing through a condition in which two distinct phases are present simultaneously. At the triple point the vapor-pressure curve has a small discontinuity in slope as indicated by the dotted extensions of the two parts. These show that at temperatures below the triple point the liquid has a higher vapor pressure than the solid and therefore cannot exist in equilibrium with the solid.

Given unit quantity of material at a temperature and pressure such that two phases are in equilibrium, the actual amounts, x and $1 - x$, of each phase present will determine the volume and the enthalpy and other properties,

$$V = V_1 + x(V_2 - V_1)$$
$$H = H_1 + x(H_2 - H_1)$$

and so on, so that a determination of any one of them such as V serves to determine x. At the triple point, where three phases are in equilibrium, let x be the fraction in the liquid and y that in the gaseous phase; thus $1 - x - y$ is the fraction in the solid phase. In this case, the volume and enthalpy are

$$V = V_1(1 - x - y) + V_2x + V_3y$$
$$H = H_1(1 - x - y) + H_2x + H_3y$$

so that a determination of two of the properties independently is needed to fix x and y.

9. Systems of Several Components

The most general kind of system is one consisting of several phases and several components. If interfacial energies at the boundaries between phases are neglected, then the volume or energy and other extensive properties of the system will be the sum of the contributions due to each phase.

All such properties for a phase depend not only on the general variables such as temperature and pressure but also on the composition of the phase. Let m_{ab} denote the quantity (in mass or molal units) of component a in phase b. Then the various thermodynamic properties for phase b depend on variables such as T and p, but also are homogeneous functions of the first degree in the composition variables m_{ab} of that phase.

The equilibrium conditions of Sec. 6 can then be generalized by considering variations corresponding to the physical process in which a quantity ϵ of component a passes from phase b to phase c. If the whole system is held *under conditions of heat isolation,* the spontaneously occurring changes will be in a direction tending to increase the entropy, and the equilibrium condition will be that in which the entropy is a maximum.

For the change of ϵ of component a from phase b to phase c,

$$\delta S = \left(\frac{\partial S_c}{\partial m_{ac}} - \frac{\partial S_b}{\partial m_{ab}}\right)\epsilon \tag{1.40}$$

where S_c is the entropy of phase c and S_t is the entropy of phase b. This change will occur spontaneously if

$$\frac{\partial S_c}{\partial m_{ac}} > \frac{\partial S_b}{\partial m_{ab}}$$

while under equilibrium conditions these two partial entropies must be equal.

Similarly, if the *whole system is maintained at constant* temperature and pressure, the spontaneously occurring changes are those which diminish the Gibbs free energy G. For the change of ϵ of component a from phase b to phase c,

$$\delta G = \left(\frac{\partial G_c}{\partial m_{ac}} - \frac{\partial G_b}{\partial m_{ab}}\right)\epsilon \tag{1.41}$$

and therefore this change can occur spontaneously if

$$\frac{\partial G_c}{\partial m_{ac}} < \frac{\partial G_b}{\partial m_{ab}} \tag{1.42}$$

and at equilibrium these two partials must be equal. Likewise if the system is maintained under constant *temperature and volume,* the criteria are of the same form as (1.42) except that the Helmholtz free energy A appears in place of G.

Because systems are so often studied under conditions of constant temperature and pressure, the partials of the Gibbs free energy with regard to the quantity of each component are particularly important. Gibbs called

$$\mu_{ab} = \frac{\partial G_b}{\partial m_{ab}} \tag{1.43}$$

the *potential* or *chemical potential* of component a in phase b. Accordingly the natural or spontaneous direction of exchange of material between phases is from a phase in which it is at high potential toward one where it is at low potential, and equilibrium occurs when the potential of each component is the same in each of the phases. Later writers, particularly G. N. Lewis, referred to quantities like $\partial G_b/\partial m_{ab}$ as the *partial molal free energy* of component a in phase b, the designation serving to emphasize that the m's are expressed in molal rather than mass quantities.

Dropping for the moment the index b which designates a particular phase, the free energy G of any phase is a function of T, p and m_1, m_2, m_3—the quantities of components it contains. The generalization of (1.29) is

$$dG = -S\,dT + V\,dp + \sum_a \mu_a\,dm_a \tag{1.44}$$

Since one can pass from G to the other functions U, H, and A by changes of variable not affecting the

m's, the generalizations of (1.26), (1.27), and (1.28) are

$$\begin{aligned} dU &= T\,dS - p\,dV + \Sigma\mu_a\,dm_a \\ dH &= T\,dS + V\,dp + \Sigma\mu_a\,dm_a \\ dA &= -S\,dt - p\,dU + \Sigma\mu_a\,dm_a \end{aligned} \tag{1.45}$$

where μ_a is the same quantity in all these equations. Thus μ_a may equally well be defined as $\partial H/\partial m_a$ taken at constant entropy and pressure—equivalent to $\partial G/\partial m_a$ taken at constant temperature and pressure.

Since G is homogeneous of the first degree in the m_a, *Euler's theorem* for homogeneous functions gives

$$G = \Sigma\mu_a m_a \tag{1.46}$$

and the partial free energies μ_a are homogeneous of the zeroth degree. If there are C components in the phase, the *composition* of the phase is determined by the $C - 1$ independent ratios of the m_a, and therefore the intrinsic properties (properties other than the total amount) of a phase of c components require $C + 1$ variables for their specification; thus such a phase is said to possess $C + 1$ *degrees of freedom.*

If P phases each having C components are considered independently, they require $P(C + 1)$ variables for their complete specification and this would be the number of degrees of freedom for the system. But if the P phases coexist in equilibrium, then for each component there are $P - 1$ restrictions on the degrees of freedom because the value of the chemical potential of any component must, for equilibrium, be the same in all phases. Thus there are $(P - 1)C$ such equilibrium conditions and in addition $2(P - 1)$ conditions which require the temperature and pressure of all phases to be equal, reducing the number of degrees of freedom to

$$P(C + 1) - (P - 1)(C + 2) = C + 2 - P \tag{1.47}$$

This result for the number of degrees of freedom of a system of P phases and C components is known as *Gibbs' phase rule.*

For some purposes it is convenient to regard the μ_a as the independent variables for describing the composition instead of the m_a. Just as the $C - 1$ ratios of the m_a are all that are needed to describe the composition, so also the C different μ_a are not independent. There is a relation between the changes in T, p and the μ_a (the *Gibbs-Duhem relation*) which must also be satisfied; from (1.46),

$$dG = \Sigma\mu_a\,dm_a + \Sigma m_a\,d\mu_a$$

but, using (1.44), this becomes

$$S\,dT - V\,dp + \Sigma m_a\,d\mu_a = 0 \tag{1.48}$$

as the relation which restricts the variation of the μ's when these are treated as independent variables.

10. Chemical Equilibrium

In Sec. 9 it was tacitly assumed that no chemical reactions were taking place and therefore the total amount, $\Sigma_b m_{ab}$, of the component A in the whole system remains constant. Variations in the m_{ab} resulted only from the passage of some of the component A from one phase to another.

The discussion of Sec. 9 is readily generalized to provide the theory of chemical equilibria, that is, of the behavior of multicomponent systems in which the amounts of some or all of the components present may also change in virtue of the occurrence of chemical reactions. A general chemical reaction may be symbolized as follows:

$$\Sigma n_A A \rightarrow \Sigma n_B B \tag{1.49}$$

in which the substances A are called the *reactants* and the substances B are called the *products.* In special examples A and B will be replaced by the chemical formula for the components in question and the n's will be determined (except for a common factor) by the requirement that the total number of atoms of each chemical element entering into the reactants must be the same as appear in the products. (In dealing with nuclear reactions, this requirement is replaced by the one that the numbers of protons and neutrons must balance.) In this notation one may regard A as standing for one formula weight M_A of the component A in which case n_A is the number of moles of A entering into the reaction.

Whether the reaction tends to go spontaneously in either direction is now determined, according to the other conditions, as in Sec. 9. The most frequent case in practice is that of constant temperature and pressure, for which spontaneous change is in the direction of diminishing the total Gibbs free energy G, and equilibrium occurs for conditions in which G is a minimum.

If now one regards the quantities m_a of Sec. 9 as expressed in moles, then for the reaction to go forward by an amount ϵ will be understood as meaning that the amounts of reactants A present are diminished by $n_A\epsilon$ moles, while the amounts of products B are increased by $n_B\epsilon$ moles. This results in a change in the free energy given by

$$dG = (-\Sigma n_A\mu_A + \Sigma n_B\mu_B)\epsilon$$

and therefore the reaction will go spontaneously forward if

$$\Sigma n_B\mu_B < \Sigma n_A\mu_A \tag{1.50}$$

that is, if there is a decrease in the total chemical potential of the components involved. Equilibrium occurs if (1.50) holds with an equality sign.

In practical applications to reactions occurring in the gas phase it is usual to express the partial molal free energy in terms of its value at a standard pressure (usually 1 atm) and a modifying term showing the pressure dependence of each partial molal free energy. These pressure-dependent terms that enter into the usual definition of the equilibrium constant of the gas reaction arise as factors in the partial molal free energies of the materials taking part in the reaction. A similar treatment is also given for reactions taking place in dilute aqueous solution in which the concentration dependence of the partial molal free energies enters into the equilibrium constant of the reaction through the use of certain auxiliary measures of the effective concentration known as activities of the several dissolved substances. These matters are dealt with more fully in Chap. 8.

Bibliography

Allis, W. P., and M. A. Herlin: "Thermodynamics and Statistical Mechanics," McGraw-Hill, New York, 1952.

Callen, H. B.: "Thermodynamics, an Introduction to the Physical Theories of Equilibrium Thermostatics and Irreversible Thermodynamics," Wiley, New York, 1960.

Guggenheim, E. A.: "Thermodynamics," Interscience, New York, 1950.

Pitzer, K. S., and L. Brewer: "Thermodynamics," 2d ed. (revision of Lewis and Randall), McGraw-Hill, New York, 1961.

Pippard, A. B.: "Elements of Classical Thermodynamics," Cambridge University Press, New York, 1957.

Tribus, Myron: "Thermostatics and Thermodynamics," Van Nostrand, Princeton, N.J., 1961.

Zemansky, M. W.: "Heat and Thermodynamics," 4th ed., McGraw-Hill, New York, 1957.

Chapter 2

Principles of Statistical Mechanics and Kinetic Theory of Gases

By E. W. MONTROLL, University of Rochester

1. Scope of Statistical Mechanics

Statistical mechanics develops a formalism which relates macroscopic behavior of physical systems composed of a large number of molecules to the properties of the individual molecular species in the system, the laws of force which govern intermolecular interactions, and the external forces acting on the system.

The subject divides naturally into equilibrium and nonequilibrium statistical mechanics. The equilibrium theory is concerned with systems which have been aged long enough under fixed external conditions so that their macroscopic properties have become stationary. It forms a molecular basis for the laws of thermodynamics. Some typical equilibrium problems are: to deduce the equation of state of imperfect gases and the magnetic equation of state of ferromagnetic materials; to develop a theory of various kinds of phase transitions from molecular models; to discuss the variation of the concentration of the products of various chemical reactions with temperature and pressure; and to compute heat capacities of gases composed of complex molecules.

Nonequilibrium statistical mechanics is concerned with rate processes. It attempts to derive phenomenological equations, such as the equation of heat conduction in solids and the diffusion equation of large molecules in liquids from molecular models, and to compute the parameters of the equations (diffusion constant, viscosity coefficient, etc.). It also tries to predict chemical reaction rates and the mechanism of irreversible processes such as the decay of turbulence.

In the present state of development general equations exist from which one can in principle compute equilibrium thermodynamic properties from molecular models. These equations have been solved for many systems of independent or almost independent particles, such as perfect gases, imperfect gases at low density, gases of complex molecules, paramagnetic systems, and electrons in metals. The difficulties of the equilibrium theory lie in the mathematical problems encountered in almost every investigation of systems of interacting particles—liquids, ferromagnets, superconductors, etc.

While there has been considerable progress in the past 10 years in the development of the foundations of nonequilibrium statistical mechanics, the mathematical problems encountered in the solution of specific problems are at least as difficult as those found in the equilibrium theory. This chapter is a brief survey of the ideas and formulas of equilibrium statistical mechanics and of the kinetic theory of gases. A few remarks will be made concerning the general theory of nonequilibrium statistical mechanics. More complete discussions of the foundations are given in the books by Gibbs [1],† Tolman [2], Fowler [3], Mayer and Mayer [4], and Khinchin [5], and in the article of Ehrenfest [6]. Details of kinetic theory are available in the works of Kennard [7], Chapman and Cowling [8], Grad [9], Herzfeld [10], Ter Haar [12], and Huang [18].

2. Identification of Temperature with Molecular Motion and the Maxwell Velocity Distribution

Consider N perfectly elastic spherically symmetrical noninteracting molecules of mass m in a cube of volume V. Also let $n_v\,d\mathbf{v}$ be the number of particles per unit volume with velocity between $\mathbf{v}$ and $\mathbf{v} + d\mathbf{v}$. Then the number in this velocity range which collide in time t with an area A of a face of the wall parallel to the (y,z) plane is $n_v v_x t A\ d\mathbf{v}$ (we repre-

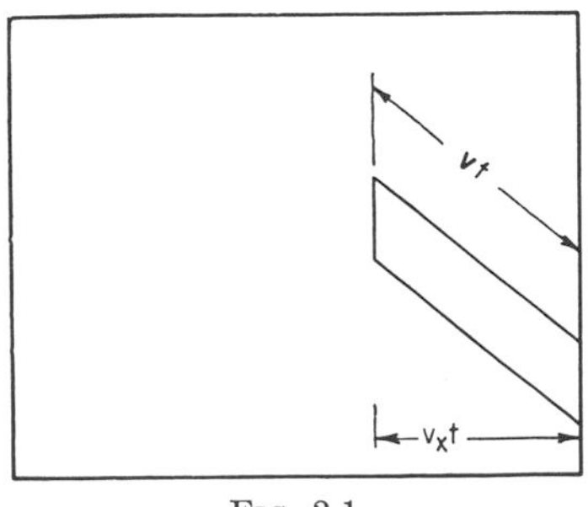

Fig. 2.1

† Numbers in brackets refer to References at end of chapter.

sent $dv_x\, dv_y\, dv_z$ by $d\mathbf{v}$). The momentum transferred to the wall as a result of one collision of this type is $2mv_x$, and the total momentum transferred to an area A in time t is

$$2mv_x{}^2 n_v tA\, d\mathbf{v}$$

The rate of change of the integral of this quantity over all molecular velocities is the force acting per unit area on the wall, or the pressure of the gas:

$$p = \int_{v_x>0} 2mv_x{}^2 n_v\, d\mathbf{v}$$

The restriction to positive v_x is eliminated by dividing the unrestricted integral on the right by 2, on the hypothesis that at a given time as many molecules are moving to the left as to the right. Hence

$$p = \tfrac{1}{2}2mn <v_x{}^2>_{\text{av}}$$

where n is the number of molecules per unit volume and

$$n <v_x{}^2>_{\text{av}} = \int n_v v_x{}^2\, d\mathbf{v}$$

The molecular chaos hypothesis

$$<v_x{}^2>_{\text{av}} = <v_y{}^2>_{\text{av}} = <v_z{}^2>_{\text{av}} = \tfrac{1}{3}<v^2>_{\text{av}}$$

yields

$$p = \tfrac{2}{3}n <\tfrac{1}{2}mv^2>_{\text{av}} \tag{2.1}$$

The total kinetic energy of our system is

$$U = \tfrac{1}{2}Nm <v^2>_{\text{av}} \tag{2.2}$$

so that

$$pV = \tfrac{2}{3}U \tag{2.3}$$

If the temperature scale is defined by the gas thermometer with

$$pV = NkT$$

then

$$\tfrac{2}{3}U = NkT$$

and

$$<\tfrac{1}{2}mv^2>_{\text{av}} = \tfrac{3}{2}kT \tag{2.4}$$

This is the *classical equipartition law;* the average kinetic energy of a system of noninteracting particles is $\tfrac{1}{2}kT$ per degree of freedom.

The actual velocity distribution function n_v can be deduced by the first Maxwell argument on the assumption (1) of the independence and equivalence of distributions of velocity components in x, y, and z directions of a given molecule and (2) that the distribution function is an even function of v_x, v_y, and v_z.

Assumption 2 implies that the velocity distribution function depends on v^2 so that velocities in forward and backward directions occur with equal probability.

Assumption 1 implies that the velocity distribution function $n_v = f(v^2)$ is factorable into a product of distribution functions of the three velocity components:

$$f(v^2) = f(v_x{}^2 + v_y{}^2 + v_z{}^2) = \varphi(v_x{}^2)\varphi(v_y{}^2)\varphi(v_z{}^2)$$

This functional equation for the unknown functions f and φ can be solved in the same manner as the simpler equation $f(x + y) = \varphi(x)\varphi(y)$. We differentiate with respect to x to obtain $f'(x + y) = \varphi'(x)\varphi(y)$. Then

$$f'(x + y)/f(x + y) = \varphi'(x)/\varphi(x)$$

so the fact that $\varphi'(x)/\varphi(x)$ is independent of y implies that $f'(x + y)/f(x + y)$ is also independent of y. Similarly, differentiation with respect to y indicates that

$$\frac{f'(x + y)}{f(x + y)} = \frac{\varphi'(x)}{\varphi(x)} = \frac{\varphi'(y)}{\varphi(y)} = -\lambda$$

is also independent of x and that λ is a constant. Hence

$$\varphi(x) = Ae^{-\lambda x} \qquad \varphi(y) = Ae^{-\lambda y}$$

and

$$f(x + y) = A^2 e^{-\lambda(x+y)}$$

where A is a constant. The generalization to three variables and identification of x with $v_x{}^2$, y with $v_y{}^2$, and z with $v_z{}^2$ yields

$$f(v^2) = A^3 \exp(-\lambda v^2)$$
$$\varphi(v_x{}^2) = A \exp(-\lambda v_x{}^2) \qquad \text{etc.}$$

The integration over all velocities must give the number of particles per unit volume,

$$n = \int_0^\infty 4\pi v^2 A^3 \exp(-\lambda v^2)\, dv = \left(\frac{\pi}{\lambda}\right)^{3/2} A^3$$

so that $A^3 = n(\lambda/\pi)^{3/2}$ and

$$n_v = f(v^2) = n(\lambda/\pi)^{3/2} \exp(-\lambda v^2)$$

The constant λ is related to the temperature by (2.4):

$$\tfrac{3}{2}kT = <\tfrac{1}{2}mv^2>_{\text{av}}$$
$$= \frac{1}{2}\left(\frac{\lambda}{\pi}\right)^{3/2} m \int_0^\infty 4\pi v^4 \exp(-\lambda v^2)\, dv = \frac{3m}{4\lambda}$$

so that

$$n_v = \left(\frac{m}{2kT\pi}\right)^{3/2} \exp\left(-\frac{mv^2}{2kT}\right) \tag{2.5}$$

the famous Maxwell velocity distribution function.

The average magnitude of the velocity is

$$\bar{v} = \int_0^\infty v \left(\frac{m}{2\pi kT}\right)^{3/2} 4\pi v^2 \exp\left(-\frac{mv^2}{2kT}\right) dv$$
$$= 2\left(\frac{2kT}{\pi m}\right)^{1/2} \tag{2.6}$$

3. Mean Free Path and Elementary Theory of Transport Processes

The transport properties of rarefied gases can be described in terms of the mean free path or average distance traversed by a molecule before it collides with another one. This distance depends on the physical process under investigation, the gas density, and the nature of intermolecular forces.

The simplest case is that of a low-density beam of rigid spherical molecules of radius r_1 shot into a rarefied gas composed of randomly distributed rigid molecules of radius r_2. Consider a 1 cm² cross section of the beam. The number of molecules deflected from the beam as it traverses a region of thickness dx between x and $x + dx$ is proportional to the number per unit volume which enter the region. If $n(x)$ is this number density, and $n(0) = n_0$, then

$$-dn = l^{-1}n(x)\, dx \tag{2.7}$$

and

$$n(x) = n_0 \exp\left(-\frac{x}{l}\right) \tag{2.8}$$

The probability of a particle traveling a distance x before collision is then $n(x)/n_0$, and the mean free path is

$$\frac{\int_0^\infty x n_0^{-1} n(x)\,dx}{\int_0^\infty n_0^{-1} n(x)\,dx} = \frac{\int_0^\infty x e^{-(x/l)}\,dx}{\int_0^\infty e^{-x/l}\,dx} = l$$

The mean free path l in (2.7) depends on the radii r_1 and r_2. Any time two particles appear within each other's sphere of influence, a sphere of radius $(r_1 + r_2)$, a collision occurs. If n_2 is the number of particles of radius r_2 per unit volume, the average number of collisions in a region of 1 cm^2 cross section and thickness dx at x is, for very small gas densities,

$$\pi(r_1 + r_2)^2 n_2 n(x)\,dx$$

Each collision leads to a deflection of a particle from the beam; thus this number has the same meaning as the right-hand side of (2.7). Hence the mean free path is given by

$$l = [n\pi(r_1 + r_2)^2]^{-1} \tag{2.9}$$

The mean free path of a system of identical molecules of diameter $d = 2r$ moving at random with a Maxwellian velocity distribution is

$$l = (n\pi d^2 \sqrt{2})^{-1} \tag{2.10}$$

Certain physical quantities are transported by molecular collisions through an arbitrary plane in a nonuniform gas in order for the gas to relax into a more uniform state. For example, if the temperature on one side of the plane is higher than that on the other, heat flows through the plane.

We examine the mechanism of transport through rarefied gases close to equilibrium by considering a system whose transport variable of interest is constant on any z-constant plane. Let n_+ be the number of particles per unit volume moving in the positive z direction, v_+ their average velocity, and n_- and v_- the corresponding properties of particles moving in the $-z$ direction. Let $Q(z)$ be the amount of the transport variable Q associated with a particle at z. On the average, a particle going in the negative direction through a plane at z has had its last collision at $z + l$. Hence the average amount of Q associated with such a particle is $Q(z + l)$. The amount associated with a particle moving in the positive direction at z is $Q(z - l)$. Hence the net rate of transport of Q in the positive z direction per unit area is

$$n_+v_+Q(z - l) - n_-v_-Q(z + l) \simeq (n_+v_+ - n_-v_-)Q(z) - l(n_+v_+ + n_-v_-)\frac{dQ}{dz}$$

The net flow of particles through a given plane is very small in a system close to equilibrium. Hence

$$n_+v_+ = n_-v_- = \tfrac{1}{6}n\bar{v}$$

where n is the average particle density and $\bar{v}$ the mean velocity. The upward flux of Q per unit area per unit time is then

$$-\tfrac{1}{3}\,n\bar{v}l\,\frac{dQ}{dz} \tag{2.11}$$

An important special example is obtained by letting Q be the momentum of the gas molecules, $m\bar{v}$. Then the rate of transport of momentum is

$$-\tfrac{1}{3}n\bar{v}lm\,\frac{d\bar{v}}{dz} \tag{2.12}$$

However, the phenomenological law for the viscous force between a unit area of two planes in a fluid when a velocity gradient exists is

$$F = \eta\,\frac{dv}{dz} \tag{2.13}$$

η being the viscosity. When dv/dz is positive, momentum is transferred at a rate $\eta\,dv/dz$ downward through the plane at z. The corresponding transfer on a molecular basis is the negative of (2.12); thus the viscosity coefficient is given by

$$\eta = \tfrac{1}{3}n\bar{v}lm \tag{2.14}$$

If we substitute (2.10) and (2.6) into this equation,

$$\eta = \frac{2}{3\pi d^2}\left(\frac{mkT}{\pi}\right)^{1/2} \tag{2.15}$$

The viscosity depends only on the temperature and not on the pressure of a rarefied gas.

The heat conductivity of a gas is obtained by choosing $Q(z)$ to be the average energy per molecule, ϵ. Then the heat flux is given by

$$-\tfrac{1}{3}n\bar{v}l\,\frac{d\epsilon}{dz} = -\tfrac{1}{3}\,n\bar{v}lc_v\,\frac{dT}{dz}$$

since the heat capacity per molecule, c_v, is $d\epsilon/dT$. According to the phenomenological Fourier law of heat conduction the heat flux is

$$-\kappa\,\frac{dT}{dz}$$

where κ is the heat conductivity. Hence

$$\kappa = \tfrac{1}{3}n\bar{v}lc_v \tag{2.16}$$

An interesting relation exists between viscosity and heat conductivity

$$\frac{\eta c_v}{\kappa m} = 1 \tag{2.17}$$

Although experimental data does not show this ratio to be 1, it is fairly constant over the monatomic gases, as is predicted by more elaborate theories.

Fick's law of diffusion can be derived by generalizing the above argument to the transport of particles in a two-component system.

4. Boltzmann Equation and the Systematic Kinetic Theory of Gases

The systematic analysis of the transport properties of rarefied gases is based on the Boltzmann equation for the distribution function $f(\mathbf{x},\mathbf{v},t)$. This function is defined so that $f(\mathbf{x},\mathbf{v},t)\,\delta\mathbf{x}\,\delta\mathbf{v}$ is the number of molecules in the volume element $\delta\mathbf{x}$ and in the velocity range $\delta\mathbf{v}$ at $(\mathbf{x},\mathbf{v})$ in position and velocity space at time t. Here $\mathbf{x}$ represents the position and $\mathbf{v}$ the velocity associated with the volume element.

The number of particles per unit volume $n(\mathbf{x},t)$ is given by

$$n(\mathbf{x},t) = \int f(\mathbf{x},\mathbf{v},t)\, d\mathbf{v} \tag{2.18}$$

where the integration extends over all possible particle velocities. Hence the density is

$$\rho(\mathbf{x},t) = mn(\mathbf{x},t)$$

Certain moments of f are of central importance in gas dynamics. The mean velocity u is represented by

$$\mathbf{u}(\mathbf{x},t) = [n(\mathbf{x},t)]^{-1}\int \mathbf{v}f(\mathbf{x},\mathbf{v},t)\, d\mathbf{v} \tag{2.19}$$

thus the intrinsic velocity as measured by an observer moving with the mean velocity is

$$\delta(\mathbf{x},t) = \mathbf{v} - \mathbf{u}(\mathbf{x},t) \tag{2.20}$$

Clearly

$$\int \mathbf{c}f\, d\mathbf{v} = 0 \tag{2.21}$$

We write c_1, c_2, c_3 as the components of the vector $\mathbf{c}$ (and indeed use an analogous notation for all vectors) and define a set of second and third moments by the symmetric tensors with components

$$P_{ij}(x,t) = m\int c_ic_jf\, dv \qquad S_{ijk}(x,t) = m\int c_ic_jc_kf\, dv \tag{2.22}$$

The contractions†

$$\tfrac{1}{3}P_{ii} = p \qquad \text{and} \qquad \tfrac{1}{2}S_i = \tfrac{1}{2}S_{ijj} \tag{2.23}$$

correspond to the pressure and heat flux vector while the nonvanishing elements of

$$p_{ij} = P_{ij} - p\delta_{ij} \tag{2.24}$$

represent the stress tensor. The δ_{ij} vanishes unless $i = j$, in which case

$$\delta_{ii} = \delta_{11} + \delta_{22} + \delta_{33} = 3$$

The Boltzmann equation‡ for $f(\mathbf{x},\mathbf{v},t)$ is a consequence of the hypotheses (1) the probability of triple or higher-order collisions in rarefied gases is negligibly small and (2) the joint probability of finding two particles at two preassigned volume elements in phase space is the product of their individual probabilities of being found in the same regions. Hypothesis (2) is frequently called the molecular chaos hypothesis.

A detailed analysis of molecular streaming and collisions based on the above hypotheses leads to the Boltzmann equation for f in a system of identical, spherically symmetrical molecules without internal degrees of freedom

$$\frac{\partial f}{\partial t} + X_i\frac{\partial f}{\partial v_i} + v_i\frac{\partial f}{\partial x_i} = J(f) \tag{2.25a}$$

where $\quad J(f) = \int d\mathbf{v}_0 \int (f'_0f' - ff_0)gI(g,\theta)\, d\Omega \quad$ (2.25b)

is the rate per unit phase space volume at which particles accumulate at $(\mathbf{x},\mathbf{v})$ due to collisions. The second term on the left of (2.25a) is the time rate of change of f which results from streaming of particles in velocity space, while the third term is that due to streaming in position space. Here X_i is the ith component of the acceleration caused by an external force.

† Repeated indices imply summation:

$$P_{ii} = P_{11} + P_{22} + P_{33}$$

‡ Critical discussions of the derivation of the Boltzmann equation have recently been made by N. N. Bogoliubov, M. S. Green, and G. Uhlenbeck and J. W. Ford; these are reviewed in [19].

Also,

$$g = |\mathbf{v} - \mathbf{v}_0| = |\mathbf{v}' - \mathbf{v}'_0| \tag{2.26}$$

where the subscript 0 refers to particles which collide with those that are either knocked into or out of a neighborhood of $(\mathbf{x},\mathbf{v})$ of interest. $I(g,\theta)\, d\Omega$ is the differential cross section for collision, while the integral (2.25b) extends over all velocities $\mathbf{v}_0$ and over all angles of approach for collision. The primed quantities are values of functions after collision.

The macroscopic equations of motion of a low-density gas (which were first derived by Maxwell) follow immediately from the Boltzmann equation. Let φ be a function of $\mathbf{v}$ and t which is conserved in the collision process (for example momentum, energy, etc.). If Eq. (2.25a) is multiplied by φ and integrated with respect to $\mathbf{v}$, the integral over the collision term can be shown to vanish so that

$$\frac{\partial(n\bar{\varphi})}{\partial t} + \frac{\partial}{\partial x_i}n\overline{\varphi v_i} = n\left(\frac{\partial\bar{\varphi}}{\partial t} + \overline{v_i\frac{\partial\varphi}{\partial x_i}} + X_i\overline{\frac{\partial\varphi}{\partial v_i}}\right) \tag{2.27}$$

where $n\bar{\varphi} = \int \varphi f\, d\mathbf{v}$. In the derivation of (2.27) several integrations by parts are made and it is noted that f vanishes for very large velocities.

The three special choices $\varphi = 1$, $\varphi = v_i$, and $\varphi = v_iv_i = v^2$ lead

$$\frac{\partial\rho}{\partial t} + \frac{\partial}{\partial x_i}(\rho u_i) = 0 \tag{2.28a}$$

$$\frac{\partial u_j}{\partial t} + u_i\frac{\partial u_j}{\partial x_i} + \frac{1}{\rho}\frac{\partial P_{ij}}{\partial x_i} = X_j \tag{2.28b}$$

$$\frac{\partial p}{\partial t} + \frac{\partial}{\partial x_i}(u_ip) + \tfrac{2}{3}P_{ij}\frac{\partial u_j}{\partial x_i} + \tfrac{1}{3}\frac{\partial S_i}{\partial x_i} = 0 \tag{2.28c}$$

As an example of the manner in which these equations are derived, let $\varphi \equiv 1$. Then $\bar{\varphi} = 1$, the right-hand side of (2.27) vanishes, and (2.27) becomes

$$\frac{\partial n}{\partial t} + \frac{\partial}{\partial x_i}(n\bar{v}_i) = 0$$

If we multiply through by m and take cognizance of (2.19), we have (2.28a). If we identify the variables ρ, u, p, p_{ij}, and S_j with the quantities mentioned below (2.18) and (2.23), Eqs. (2.28) are in the familiar form for the continuity equation and for the equations of conservation of momentum and energy.

In principle, a discussion of any of the dynamical properties of monatomic gases can be based on (2.28) and the equation of state of the gas. However, the exact forms of ρ, p, p_{ij}, and S_j are unknown until one solves the Boltzmann equation and employs the moment formulas (2.20) to (2.22). Since this equation is nonlinear, one usually makes the assumption that the deviation of f from its equilibrium value is so small that terms quadratic in this deviation can be neglected. In this case the Boltzmann equation becomes a linear integrodifferential equation in the deviation. The equilibrium f in the absence of external forces is, of course, Maxwellian:

$$f_0 = n\left(\frac{m}{2kT\pi}\right)^{3/2}\exp\left(-\frac{mc^2}{2kT}\right)$$

The investigation of transport properties is simplified in either the very-low-density so-called Knudsen range where it is assumed that the ratio (the Knudsen number $m = d/l$) of the dimensions of the container d to the mean free path l is small or from the relatively high-density or Clausius range which corresponds to a large Knudsen number. Collisions can, to a first approximation, be neglected in the Knudsen range, while streaming can be treated as a small perturbation in the Clausius range.

One of the first systematic procedures to solve the Boltzmann equation was developed by Lorentz in connection with his theory of electrical conductivity of metals. Lorentz examined the case of a mixture of very massive and very small particles with so few of the smaller species that the number of their collisions with each other was small compared with the number with the larger particles. The equipartition equation (2.4) implies that the random motion of the small particles at a given temperature occurs with so much greater velocity than that of the large ones that for all practical purposes the large particles might be considered to be at rest. The collision problem then reduces to the scattering of the small particles from a matrix of fixed larger scatterers. The small particles in the Lorentz theory correspond to electrons. When the number of free electrons is small compared with the number of neutral atoms, this model is quite reasonable. It is known that the quantum-mechanical generalization of the Boltzmann equation must be used to obtain the right order of magnitude for the electrical conductivity of metals.

Hilbert extended the work of Lorentz. The modern theory of the solution of the Boltzmann equation for particles with an arbitrary force law and for mixtures of particles of arbitrary mass was finally developed by Enskog and Chapman. This theory is clearly developed and presented by Chapman and Cowling. Since the solution comes out in powers of the mean free path, it is appropriate for the Clausius region.

The mathematics of the Chapman-Enskog-Hilbert approach is quite lengthy. It is readily available in the book by Chapman and Cowling [8]. Only a brief discussion of its qualitative features is given here. The distribution function is essentially an expansion in powers of the mean free path. The zeroth order distribution function is the Maxwellian distribution. When the various moments of f are computed and substituted into (2.28), one obtains the hydrodynamic equation of an incompressible fluid. When terms up to the first power in the mean free path are included in the analysis of a nonuniform gas, (2.28) reduce to the classical Stokes-Navier phenomenological equations. There the heat flux vector S_i is proportional to the negative of the temperature gradient. Formulas for the proportionality constant, the heat conductivity of the gas, are obtained in term of the force law. The stress tensor p_{ij} depends linearly on the deformation tensor

$$D_{ij} = \frac{1}{2}\left(\frac{\partial u_i}{\partial x_j} + \frac{\partial u_j}{\partial x_i}\right)$$

the coefficient being the viscosity coefficient.

The second approximation to the distribution function (qualitatively this corresponds to terms proportional to the square of the mean free path) leads to much more complicated expressions for the heat flux vector and stress tensors. The p_{ij}'s were derived by Burnett and the heat flux vector by Chapman and Cowling (a slight mistake, which was corrected by Wang Chang and Uhlenbeck [11], exists in their expression). The corresponding approximations to the hydrodynamic equations are called the Burnett equations.

Certain terms have been obtained by Wang Chang and Uhlenbeck for the third approximation. With each added approximation one must include more boundary conditions to solve the Boltzmann equation. This is because each further approximation gives more information about the molecular quantities while in the Navier-Stokes approach all results are given in terms of macroscopic quantities. A detailed discussion of boundary conditions is given by Wang Chang and Uhlenbeck and by Grad.

In a new approach to the solution of the Boltzmann equation Grad [9] represents the distribution function f as a product of the Maxwell distribution function and an infinite series of "generalized" Hermite vector polynomials in the dimensionless velocity $\mathbf{v}' = \mathbf{c}(kT)^{-1/2}$. Then

$$f = f_0\left(a^{(0)}\mathcal{H}^{(0)} + a_i^{(1)}\mathcal{H}_i^{(1)} + \frac{1}{2!}a_{ij}^{(2)}\mathcal{H}_{ij}^{(2)} + \cdots\right) \tag{2.29}$$

The first few polynomials are $\mathcal{H}^{(0)} = 1$, $\mathcal{H}_i^{(1)} = v'_i$, $\mathcal{H}_{ij}^{(2)} = v'_iv'_j - \delta_{ij}$, etc. At equilibrium all the a's are zero except $a^{(0)}$, which represents the equilibrium spatial distribution of particles. The Hermite polynomials form an orthonormal set with the weight function f_0. Hence the coefficients $a_{ij\cdots}^{(m)}$ are related to f through

$$a_{ij\cdots}^{(m)} = \frac{1}{n(\mathbf{x},t)}\int f\mathcal{H}_{ij\cdots}^{(m)}\,d\mathbf{v}$$

The first few a's are connected to the moments (2.22) of f by

$$a^{(0)} = 1 \qquad a_i^{(1)} = 0 \qquad a_{ij}^{(2)} = \frac{p_{ij}}{p}$$

$$a_{ijk}^{(3)} = \frac{S_{ijk}}{p(kT)^{1/2}} \qquad \text{etc.}$$

Hence in the course of solving Boltzmann's equation one obtains expressions for the tensors of interest in terms of the intermolecular forces.

Grad breaks off the series for f by retaining terms only up to the third Hermite polynomial. When (2.29) is substituted into the Boltzmann equation with terms retained up to $a^{(3)}$, and with terms quadratic in the a's neglected, a set of differential equations is obtained for the a's or for the thermodynamic quantities, ρ, u, T, p_{ij}, S_{ijk}. When he approximates S_{ijk} by a linear combination of the contractions, he can discuss all quantities in terms of the thirteen moments of f, ρ, u, T, p_{ij}, and S_i (the equation of state relates ρ and p_{ii}). Grad believes the complete thirteen-moment approximation yields hydrodynamic equations which are more reliable than those of Burnett.

Unfortunately neither the Chapman-Enskog nor Grad solutions apply when one goes very far from the extreme Clausius range. Convergence of the various approximations is slow (and indeed has never been proved); thus it is doubtful that one can proceed very far into the Knudsen range by these methods. Wang Chang and Uhlenbeck have tried to close the gap between the Clausius and Knudsen ranges.

5. The Boltzmann *H* Theorem

The classical statistical discussion of irreversibility of behavior of a system of a large number of particles was made through the Boltzmann *H* theorem. This theorem is the analogue of the thermodynamic statement that the entropy of a system tends to a maximum.

Consider a rarefied gas in the absence of external fields. The Boltzmann *H* function is defined as

$$H = \int\int f \log f \, d\mathbf{v} \, d\mathbf{r} \tag{2.30}$$

the integration ranging over all position and momentum space. In a system at equilibrium the *H* function is, to within an additive constant, the negative of the entropy divided by Boltzmann's constant [see Eq. (2.64*a*)]. The statement of the *H* theorem is that *H* is a decreasing function of time unless f is Maxwellian, in which case *H* is constant. Hence in the relaxation of a gas from a nonequilibrium state *H* decreases until f becomes Maxwellian and equilibrium is achieved.

Properties of the average

$$\begin{aligned} J\{\varphi\} &= \int \varphi J(f) \, d\mathbf{v} \\ &= \int\int\int \varphi(\mathbf{v}) \, d\mathbf{v}_0 \, d\mathbf{v}(f'_0 f' - f f_0) g I(g,\theta) \, d\Omega \end{aligned} \tag{2.31}$$

where φ is an arbitrary function of the velocity $\mathbf{v}$ and $J(f)$ is the collision integral (2.25*b*), are used in derivation of (2.27) as well as the *H* theorem. Here $\mathbf{v}$ and $\mathbf{v}_0$ are the velocities of a pair of particles before their collision and $\mathbf{v}'$ and $\mathbf{v}'_0$ those after collision, while f' is to be interpreted as the distribution function after collision. The collision is equivalent to a mathematical operation which transforms the pair of vectors $(\mathbf{v},\mathbf{v}_0)$ into a new pair $(\mathbf{v}',\mathbf{v}'_0)$. It can be shown that the Jacobian of this transformation is 1 and, therefore, $d\mathbf{v} \, d\mathbf{v}_0 = d\mathbf{v}' \, d\mathbf{v}'_0$. It can also be shown that the collision transformation which takes $(\mathbf{v},\mathbf{v}_0)$ into $(\mathbf{v}',\mathbf{v}'_0)$ also takes $(\mathbf{v}',\mathbf{v}'_0)$ into $(\mathbf{v},\mathbf{v}_0)$. This is a result of the dynamical reversibility of collisions. Since $g = |\mathbf{v} - \mathbf{v}_0| = |\mathbf{v}' - \mathbf{v}'_0|$ is unchanged by the collision, if we formally interchange $(\mathbf{v},\mathbf{v}_0)$ with $(\mathbf{v}',\mathbf{v}'_0)$ in (2.31) and note that $f'_0 f' - f f_0 \rightarrow f f_0 - f' f'_0$ in this process while

$$d\mathbf{v}_0 \, d\mathbf{v} \rightarrow d\mathbf{v}'_0 \, d\mathbf{v}' = d\mathbf{v}_0 \, d\mathbf{v}$$

we have

$$J\{\varphi'\} = -J\{\varphi\} \tag{2.32a}$$

where $\varphi(\mathbf{v}') \equiv \varphi'$. All factors of the integrand of (2.31) with the exception of $\varphi(\mathbf{v})$ are symmetrical with respect to $\mathbf{v}_0$ and $\mathbf{v}$; hence

$$J\{\varphi\} = J\{\varphi_0\} = -J\{\varphi'_0\} \tag{2.32b}$$

where $\varphi_0 \equiv \varphi(\mathbf{v}_0)$ and $\varphi'_0 \equiv \varphi(\mathbf{v}'_0)$.

To derive the *H* theorem, note that

$$\begin{aligned} \frac{\partial H}{\partial t} &= \iint (1 + \log f) \frac{\partial f}{\partial t} \, d\mathbf{v} \, d\mathbf{r} \\ &= \iint (1 + \log f) \left[J(f) - v_i \frac{\partial f}{\partial x_i} \right] d\mathbf{v} \, d\mathbf{r} \end{aligned}$$

But, on application of (2.32) to the special case $\varphi = 1 + \log f$

$$\begin{aligned} \int (1 + \log f) J(f) \, d\mathbf{v} &= \tfrac{1}{4}(J\{1 + \log f\} + J\{1 + \log f_0\} - J\{1 + \log f'\} - J\{1 + \log f'_0\}) \\ &= \int\int d\mathbf{v} \, d\mathbf{v}_0 [f' f'_0 - f f_0] \log (f f_0 / f' f'_0) \int g I(g,\theta) \, d\Omega \end{aligned}$$

Since $gI(g,\theta)$ is non-negative, and

$$(x - y) \log \frac{y}{x} \begin{cases} <0 & \text{if } x \neq y \\ =0 & \text{if } x = y \end{cases}$$

$$\int\int J(f)(1 + \log f) \, dr \, dv \leq 0$$

Now

$$\iint (1 + \log f) v_i \frac{\partial f}{\partial x_i} \, d\mathbf{r} \, d\mathbf{v} = \int d\mathbf{r} \frac{\partial}{\partial x_i} \int v_i f \log f \, d\mathbf{v}$$

This integral is transformed, through the divergence theorem, to the surface integral

$$\int_S d\mathbf{A} \int \mathbf{v} f \log f \, d\mathbf{v}$$

which extends over the boundary of the container. Physically it represents the rate at which the quantity $f \log f$ streams through the boundary. Since there is no streaming of particles across the boundary, this term vanishes, and therefore

$$\frac{\partial H}{\partial t} \leq 0 \tag{2.33}$$

The equality exists when $f' f'_0 = f f_0$, the condition which is satisfied only when f is Maxwellian. In this case *H* remains constant. Hence *H* decreases until equilibrium is established, after which it does not change.

Many interesting controversies exist in the literature on the reconciliation of this theorem with the reversibility of the equations of motion. These have been clearly analyzed by P. and T. Ehrenfest [6]; see also Tolman [2].

6. Averages in Equilibrium Statistical Mechanics and the Liouville Equation

Equilibrium statistical mechanics generally starts from an ensemble of identical systems which contains all those systems whose microscopic states are compatible with a set of preassigned macroscopic constraints. The so-called microcanonical ensemble is composed of all classical systems of a fixed number of particles whose positions and momenta are compatible with a preassigned total energy. Another example of an ensemble is one whose systems have a fixed angular momentum as well as energy and numbers of particles.

Consider an ensemble of systems under identical external constraints and with identical molecular populations. If each system contains N particles, it is characterized by $3N$ position and $3N$ momentum coordinates, and hence by a point in a $6N$-dimensional phase space. Denote such a point by $(\mathbf{r}^N,\mathbf{p}^N)$. The entire ensemble can be identified by the set of representative points of its systems in phase space.

We define a probability density of systems in phase space, $P_N(\mathbf{r}^N,\mathbf{p}^N)$, so that the probability of a randomly chosen system of our ensemble being at $(\mathbf{r}^N,\mathbf{p}^N)$ in the $6N$-dimensional volume element $\delta\mathbf{p}^N\,\delta\mathbf{r}^N$ is $P_N(\mathbf{r}^N,\mathbf{p}^N)\,\delta\mathbf{p}^N\,\delta\mathbf{r}^N$. This density function is normalized to unity when integrated over all of phase space,

$$\int P_N(\mathbf{r}^N,\mathbf{p}^N)\,d\mathbf{r}^N\,d\mathbf{p}^N = 1 \tag{2.34}$$

One of the fundamental postulates of statistical mechanics is, given $P_N(\mathbf{r}^N,\mathbf{p}^N)$, the average observed values of any function $G(\mathbf{r}^N,\mathbf{p}^N)$ of $(\mathbf{r}^N,\mathbf{p}^N)$ in the systems of our ensemble is

$$\langle G\rangle_{\text{av}} = \int G(\mathbf{r}^N,\mathbf{p}^N)P_N(\mathbf{r}^N,\mathbf{p}^N)\,d\mathbf{p}^N\,d\mathbf{r}^N \tag{2.35}$$

The usefulness of this postulate depends on the fact that as $N \to \infty$ certain choices of P_N and G lead to vanishingly small fluctuations of G from $\langle G\rangle_{\text{av}}$. In this case one feels that the value of G observed in most typical systems of the ensemble should be very close to the ensemble average.

As time goes on, each dynamical system of the ensemble generates a trajectory in phase space. If no systems are created or destroyed, we expect P_N to satisfy a continuity equation analogous to that associated with the motion of a fluid. If ρ is the local density of a fluid and $\mathbf{v}$ the velocity vector of a small volume located at the point $\mathbf{r}$, the continuity equation of fluid dynamics is

$$\frac{\partial\rho}{\partial t} + \operatorname{div}(\rho\mathbf{v}) = 0 \tag{2.36}$$

The generalization of this equation to a $6N$-dimensional fluid with coordinates $x_1,y_1,\ \ldots\ ,z_N,p_{x_1},p_{y_1},\ \ldots\ ,p_{z_N}$ and density function P_N is the Liouville equation

$$\frac{\partial P_N}{\partial t} + \sum_{j=1}^{N}\left(\nabla_j\frac{P_N\mathbf{p}_j}{m} + \delta_j P_N\dot{\mathbf{p}}_j\right) = 0 \tag{2.37}$$

Here

$$\nabla_s = i\frac{\partial}{\partial x_s} + j\frac{\partial}{\partial y_s} + k\frac{\partial}{\partial z_s}$$

$$\delta_s = i\frac{\partial}{\partial p_{x_s}} + j\frac{\partial}{\partial p_{y_s}} + k\frac{\partial}{\partial p_{z_s}}$$

the phase space velocity associated with $\mathbf{r}_j$ is $\mathbf{v}_j = \mathbf{p}_j/m$ and that associated with $\mathbf{p}_j$ is $\dot{\mathbf{p}}_j$ or the acceleration of the particle. Newton's equations of motion for a conservative system imply that

$$\dot{\mathbf{p}}_j = -\nabla_j\Phi$$

where Φ is the total potential energy of particles in a representative system of our ensemble (including both the contribution of external fields and interparticle interactions). Liouville's equation is equivalent to

$$\frac{\partial P_N}{\partial t} + \sum_{j=1}^{N}\left(\frac{\mathbf{p}_j}{m}\nabla_j P_N - \nabla_j\Phi\,\delta_j P_N\right) = 0 \tag{2.38}$$

with $\partial P_N/\partial t = 0$ at equilibrium. At equilibrium any function $P_N = F(E)$ of the total energy

$$E = (1/2m)\Sigma p_j^2 + \Phi$$

satisfies Liouville's equation.

Liouville's equation is equivalent to the statement that the total derivative dP_N/dt is identically zero. Another consequence of this statement is that if a certain bounded region R of phase space has a constant phase space density at a given time, those regions that it is transformed into at later times have the same density even though the shape of R may change with time.

7. The Microcanonical and Canonical Ensembles

A simple distribution function P_N which is a function of only the energy and which satisfies the Liouville equation is that of the "microcanonical ensemble" of Gibbs. This ensemble is the class of all isolated systems (say of N particles) of energy E. The distribution function is

$$P_N(\mathbf{r}^N,\mathbf{p}^N) = A\,\delta[H(\mathbf{p}^N,\mathbf{r}^N) - E]$$

where $H(\mathbf{r}^N,\mathbf{p}^N)$ is the Hamiltonian of a representative system and $\delta(x)$ is the Dirac delta function

$$\delta(x) = 0 \qquad \text{unless } x = 0$$
$$\int\delta(x)\,dx = 1$$

All regions of equal "area" on the constant energy surface in phase space are given the same statistical weight. The justification for the use of this ensemble would be a consequence of an ergodic theorem; that is, a theorem which states that almost all systems in the ensemble spend the same amount of time in any regions of the same (nonvanishing) area on the constant energy surface. Time averages on a single system are equal to ensemble averages in ergodic systems. A discussion of the present status of this theorem for dynamical systems is given in the book by Khinchin [5].

The Fourier integral representation of the δ function

$$\delta(x) = \frac{1}{2\pi}\int_{-\infty}^{\infty} e^{-ixy}\,dy$$

is convenient for the calculation of averages over the microcanonical ensemble. The average value of the function $G(\mathbf{r}^N,\mathbf{p}^N)$ is

$$\langle G(\mathbf{r}^N,\mathbf{p}^N)\rangle_{\text{av}} = Z_M^{-1}\int_{-\infty}^{\infty}\exp(iEy)\,dy\int\cdots\int G(\mathbf{r}^N,\mathbf{p}^N)\exp[-iyH(\mathbf{p}^N,\mathbf{r}^N)]\,d\mathbf{p}^N\,d\mathbf{r}^N \tag{2.39a}$$

where
$$Z_M = \int_{-\infty}^{\infty} e^{-iEy}e^{-NyA(y)i}\,dy \tag{2.39b}$$

and

$$\exp[-NyA(y)i] = \int\cdots\int \exp[-iyH(\mathbf{p}^N,\mathbf{r}^N)]\,d\mathbf{p}^N\,d\mathbf{r}^N \tag{2.39c}$$

The multiple integrals extend over all momentum space and over the volume V of the container of the particles. If one writes $E = N\epsilon$, with ϵ being the average energy per particle, (2.39*b*) can be evaluated in terms of $A(y)$ by the method of steepest descents. The calculation of the averages (2.39*a*) for choices of G which correspond to the usual thermodynamic variables proceeds in essentially the same manner as discussed in Sec. 8 for averages over the canonical ensemble.

Suppose our system to be composed of two parts such that (1) one part has many less degrees of freedom than the other, (2) no direct mechanical interaction exists between the two parts, but (3) heat can be transferred between them. Then it can be shown that the energy distribution function of the small system is the Maxwell-Boltzmann distribution, proportional to $\exp(-E/kT)$. The larger of the two systems can be interpreted as a thermostat which regulates the equilibrium temperature of the smaller system.

The Maxwell-Boltzmann distribution function may be derived from the canonical ensemble rather than from a division of the microcanonical ensemble into two parts. Our results will be valid for quantum-mechanical as well as classical systems. Consider an ensemble of a large number of identical systems which are in thermal contact with each other but which do not interact mechanically. Also, let the total energy of all these systems be E. This ensemble is called the canonical ensemble.

Suppose that our systems are quantum-mechanical ones with the possible energy states $E_1, E_2, E_1, \ldots$ and that the number of systems in the ensemble with these energies are, respectively, $n_1, n_2, n_3, \ldots$. Then if the total number of systems is N,

$$\Sigma n_j = N \qquad \Sigma n_j E_j = E \qquad (2.40)$$

We also assume that, if the jth energy state has a degeneracy ω_j, the a priori probability of a system selected at random [without conditions (2.40) being applied] being in the jth state is proportional to ω_j.

When our systems are classical, we suppose the space available to each system to be broken into a large number of cells, with the average energy of the jth cell being E_j, and the phase space volume of the cell being ω_j. We now determine the most probable number of systems in each energy state as the total number of systems becomes large and E/N remains fixed.

The hypothesis that the a priori probability of finding a system in the jth energy region is ω_j implies that the probability, $P(n_1, n_2, \ldots)$, of finding n_j particles in the jth region ($j = 1,2,3, \ldots$) is proportional to

$$\frac{N!}{n_1! n_2! \cdots} \omega_1^{n_1} \omega_2^{n_2} \cdots \qquad (2.41)$$

Our required distribution is that set of n_j's which maximize (2.41) [and hence $\log P(n_1, n_2, \ldots)$] under the auxiliary conditions (2.40). Using the method of Lagrange multipliers, introduce multipliers β and λ and find the set of n_j's which make

$$\delta[\log P - \lambda(N - \Sigma n_j) + \beta(E - \Sigma n_j E_j)] = 0$$

If n_j is changed by an amount δn_j in this variation (using Sterling's approximation $\log n! = n \log n - n$)

$$\sum_i \delta n_i(\log \omega_i - \log n_i - \lambda - \beta E_i) = 0$$

For arbitrary variations in n_i the sum vanishes only if

$$n_i = \omega_i \exp(-\lambda - \beta E_i) \qquad \text{for all } i \qquad (2.42)$$

The first Lagrange multiplier λ is chosen so that the n_i's satisfy (2.40). It is given by

$$\exp(\lambda) = N^{-1} \Sigma_i \omega_i \exp(-\beta E_i)$$

Hence

$$n_i = N \frac{\omega_i \exp(-\beta E_i)}{\Sigma \omega_i \exp(-\beta E_i)} \qquad i = 1,2,3, \ldots \qquad (2.43)$$

The second multiplier β is related to the reciprocal of the temperature by proving that, if ensembles of two different kinds of systems are at equilibrium with each other, they have the same value of β. If one system is a perfect gas, in the absence of any external fields the value of β would have to be chosen as $1/kT$ to be consistent with the discussion of Sec. 2. Consider an ensemble of two types of connected systems with N_1 systems of the first type, N_2 of the second, and with combined energy E. Furthermore, let the energy regions of phase space be $E_1^{(1)}, E_2^{(1)}, E_3^{(1)}, \ldots$ for systems of the first type with statistical weights $\omega_1^{(1)}, \omega_2^{(1)}, \ldots$ and $E_1^{(2)}, E_2^{(2)}, E_3^{(2)}, \ldots$ for systems of the second type with statistical weights $\omega_1^{(2)}, \omega_2^{(2)}, \omega_3^{(2)}, \ldots$.

If $n_j^{(1)}$ is the number of systems of the first type in the jth energy region $E_j^{(1)}$ and $n_k^{(2)}$, the number in the kth region of the second type of system then is

$$\sum n_j^{(1)} = N_1 \qquad \sum n_j^{(2)} = N_2 \qquad (2.44a)$$

$$\sum_j (n_j^{(1)} E_j^{(1)} + n_j^{(2)} E_j^{(2)}) = E \qquad (2.44b)$$

The probability of having $n_j^{(1)}$ systems of the first kind in the jth energy region available to them and $n_k^{(2)}$ of the second in kth region available to them ($j,k = 1,2,3, \ldots$) is proportional to

$$N_1! N_2! \left[\prod_j \frac{(\omega_j^{(1)})^{n_j^{(1)}}}{n_j^{(1)}!}\right] \left[\prod_k \frac{(\omega_k^{(2)})^{n_k^{(2)}}}{n_k^{(2)}!}\right]$$

To find the set of $n^{(1)}$'s and $n^{(2)}$'s which maximize this probability under the auxiliary conditions (2.44) three Lagrange multipliers must be introduced. Proceeding as in the case of one type of system, we find [after eliminating two of the multipliers through (2.44)]

$$n_i^{(j)} = N_j \frac{\omega_i^{(j)} \exp(-\beta E_i^{(j)})}{\sum_i \omega_i^{(j)} \exp(-\beta E_i^{(j)})} \qquad j = 1,2, \ldots$$

where the constant β is common to both distributions. Since the temperature of two systems at equilibrium in contact with each other is the same in both systems, this result suggests that β is a function of the temperature. Finally, if one of the systems is a perfect gas, without external forces, Sec. 2 implies that $\beta = 1/kT$. Since one can hardly expect the state of a system to depend on the mechanical details of a system with which it is in thermal equilibrium, we can universally choose $\beta = 1/kT$.

Remaining sections of this chapter are based on the energy distribution function of the canonical ensemble (2.43). The fluctuations in thermodynamic quantities from the values given by averaging over (2.43) can be shown to be small in systems with a large number of degrees of freedom. The use of this distribution function is generally based on the fact that it is the appropriate one for a small system in contact with an enormous heat reservoir and that the probability of fluctuations from ensemble averages is small.

8. The Partition Function and the Statistical Basis of Thermodynamics

The denominator of the function (2.42) which gives the fraction of systems $f_i = n_i/N$ in the ith energy state,

$$Z = \sum_i \omega_i \exp(-\beta E_i) \tag{2.45}$$

is called the *partition function* or sum over states. We shall now show how various thermodynamic quantities can be expressed in terms of the partition function.

When the ith energy state is interpreted to be a region in classical phase space, the sum in (2.45) is replaced by an integral over all of phase space. It is advantageous to make this conversion in a pseudo-quantum-mechanical manner so that the thermodynamic quantities (in particular the entropy) calculated from the integral are consistent with their quantum-mechanical equivalents.

According to the uncertainty principle the product of the dispersion in a momentum measurement by that of a simultaneous position measurement satisfies

$$\Delta p \, \Delta q \geq h \tag{2.46}$$

where h is Planck's constant. Hence it is unnecessary to enumerate energies in phase space cells which have a volume smaller than h^{3N}. If the value of the Hamiltonian at a point $(\mathbf{p}_1, \ldots, \mathbf{r}_N)$ in a cell of volume $\delta\mathbf{p}_1 \cdots \delta\mathbf{r}_N$ in phase space is $H(\mathbf{r}^N,\mathbf{p}^N)$, the a priori statistical weight of that cell can be taken as the number of smallest cells of volume h^{3N} which are contained in $\delta\mathbf{p}_1, \ldots, \delta\mathbf{r}_N$; $h^{-3N}\delta\mathbf{p}_1 \cdots \delta\mathbf{r}_N$ and the contribution of this cell to the partition function is

$$h^{-3N} \exp[-\beta H(\mathbf{r}^N,\mathbf{p}^N)] \, \delta\mathbf{p}_1 \cdots \delta\mathbf{r}_N \tag{2.47}$$

To be consistant with the quantum theory we consider as a single state all configurations which differ only by a permutation of particles. On this basis a factor $1/N!$ is included when (2.47) is summed over all cells in the calculation of the classical partition function:

$$Z = \frac{1}{h^{3N}N!} \int \cdots \int \exp[-\beta H(\mathbf{r}^N,\mathbf{p}^N)] \, d\mathbf{p}_1 \cdots d\mathbf{r}_N \tag{2.48}$$

The integration extends from $-\infty$ to ∞ on the momentum variables and over the volume of the container of the system in configuration space.

The connection between statistical mechanical and thermodynamic variables is made by noticing the similarity of certain statistical mechanical formulas to some consequences of the first and second laws of thermodynamics.

Consider a system in which certain extensive parameters $\xi_1, \xi_2, \ldots, \xi_n$ have been varied reversibly by the infinitesimal amounts $\delta\xi_1, \ldots, \delta\xi_n$ as the result of the application of generalized external forces

$$X_i = -\frac{\partial H}{\partial \xi_i} \qquad i = 1, 2, \ldots, n \tag{2.49}$$

where H is the Hamiltonian of the system of interest. Two special examples of generalized external forces are the pressure p and an external magnetic field H. The corresponding extensive parameters are the volume V and the magnetization M. According to the first law of thermodynamics the heat δQ generated by our infinitesimal process is (here E is the internal energy of the system)

$$\delta Q = \delta E - \delta W = \delta E + \Sigma X_i \, \delta\xi_i \tag{2.50}$$

since the work δW done on the system by varying the external conditions is

$$\delta W = -\Sigma X_i \, \delta\xi_i$$

The second law gives the relation between the heat generated by the system and the entropy change δS in the system

$$\delta Q = T \, \delta S \tag{2.51}$$

Since the Helmholtz free energy A is defined as

$$A = E - TS \tag{2.52}$$

the second law can be written in terms of A instead of S:

$$\begin{aligned} \delta A &= \delta E - S\,\delta T - \delta Q \\ &= \delta E + (A - E)\frac{\delta T}{T} - \delta Q \\ &= (A - E)\frac{\delta T}{T} - \sum X_i \, \delta\xi_i \end{aligned} \tag{2.53}$$

The fraction of systems f_i in the ith state are

$$f_i = \omega_i \exp \beta(\psi - E_i) \tag{2.54}$$

where $\qquad \exp(-\beta\psi) = Z$

and $\qquad \Sigma_i f_i = 1$

As external conditions are varied, the sum over the f_i must remain normalized to unity; hence

$$\delta \sum f_i = 0$$

or

$$\begin{aligned} \delta \sum f_i &= \sum_i \delta[\omega_i \exp(\psi - E_i)\beta] \\ &= \sum_i f_i \left[T\,\delta\psi - (\psi - E_i)\,\delta T - T \sum_j \frac{\partial E_i}{\partial \xi_j} \delta\xi_j \right] \end{aligned} \tag{2.55}$$

The weight factors ω_i do not change in the variation of the external parameters.

The average energy over the ensemble is

$$<E>_{\text{av}} = \Sigma E_i f_i \tag{2.56}$$

and the average generalized force of the jth kind acting on the system is

$$<X_j>_{\text{av}} = -\left\langle \frac{\partial E}{\partial \xi_j} \right\rangle_{\text{av}} = -\sum_i f_i \frac{\partial E_i}{\partial \xi_j} \tag{2.57}$$

Hence (2.55) becomes

$$\delta\psi = (\psi - <E>_{\text{av}})\, \delta T/T - \Sigma <X>_{\text{av}}\, \delta\xi_j \tag{2.58}$$

This equation is equivalent to the thermodynamic relation (2.53) if ψ is identified with the Helmholtz free energy A, $<E>_{\text{av}}$, with the internal energy, and $<X_j>_{\text{av}}$ with the jth generalized force.

These correspondences lead to

$$A = -kT \log Z \tag{2.59}$$

and

$$E = kT^2\, \partial \log Z/\partial T \tag{2.60}$$

When the only generalized force is the pressure,

$$\delta A = (A - E)\, \delta T/T - p\, \delta V$$

so that

$$p = -\left(\frac{\partial A}{\partial V}\right)_T \tag{2.61}$$

Hence the statistical mechanical formula for the pressure is

$$p = kT \frac{\partial \log Z}{\partial V} \tag{2.62}$$

The heat capacity at constant volume is

$$C_v = \left(\frac{\partial E}{\partial T}\right)_v = -\frac{\partial}{\partial T}\left(kT^2 \frac{\partial \log Z}{\partial T}\right) \tag{2.63}$$

An expression for the entropy results from a combination of (2.52), (2.59), and (2.60). However, an alternative form follows from the consideration of

$$\sum_i f_i \log f_i = \sum \omega_i \exp\left(\frac{\psi - E_i}{kT}\right)\left(\frac{\psi - E_i}{kT} + \log \omega_i\right)$$

$$= (A - E)/kT + \sum f_i \log \omega_i$$

$$= -S/k + \sum f_i \log \omega_i$$

Then

$$S = -k \sum f_i \log f_i/\omega_i \tag{2.64}$$

When all ω_i's are equal this reduces to the Boltzmann expression

$$S = -k\Sigma f_i \log f_i + \text{const} \tag{2.64a}$$

The chemical potential μ is defined thermodynamically as

$$\mu = \frac{\partial A}{\partial N}$$

Hence if we let A_N be the Helmholtz free energy of a system of N particles, we have as $N \to \infty$

$$\mu = A_N - A_{N-1} = -kT \log \frac{Z_N}{Z_{N-1}} \tag{2.65a}$$

or

$$\frac{Z_N}{Z_{N-1}} = \exp\left(\frac{-\mu}{kT}\right) \tag{2.65b}$$

where Z_N is the partition function of a system of N particles.

The partition function of a system composed of several independent parts factors into a product of the individual partition functions of each of the separate parts. Let the energy states of the individual parts be $E_{i_1}{}^{(1)}$, $E_{i_2}{}^{(2)}$, . . . with degeneracies $\omega_{i_1}{}^{(1)}$, $\omega_{i_2}{}^{(2)}$, ... Then the energy of a typical state of the total system is

$$E_{i_1, i_2, \ldots} = E_{i_1}{}^{(1)} + E_{i_2}{}^{(2)} + \cdots$$

and the partition function is

$$Z = \sum_{i_1, i_2, \ldots} \omega_{i_1}{}^{(1)} \omega_{i_2}{}^{(2)} \cdots \exp\left[-\beta(E_{i_1}{}^{(1)} + E_{i_2}{}^{(2)} + \cdots)\right]$$

$$= \left[\sum_{i_1} \omega_{i_1}{}^{(1)} \exp(-\beta E_{i_1}{}^{(1)})\right]\left[\sum_{i_2} \omega_{i_2}{}^{(2)} \exp(-\beta E_{i_2}{}^{(2)})\right] \cdots$$

$$= Z_1 Z_2 Z_3 \cdots \tag{2.66}$$

To a first approximation the energy levels of electrons in a metal are independent of the lattice vibrations. Hence for a metal

$$Z \simeq Z_{\text{elec}} Z_{\text{lat vib}}$$

In a rarefied gas composed of diatomic molecules the rotational energy levels are to a first approximation independent of the vibrational levels. Both of these are independent of the translational motion of the gas molecules. Hence

$$Z \simeq Z_{\text{trans}} Z_{\text{rot}} Z_{\text{vib}}$$

Many other examples of the division of the energy of a complicated system into a sum of independent terms exist.

Since the total energy of a classical system of spherically symmetrical particles with no internal degrees of freedom is the sum of the kinetic energy (which depends only on the momentum) and the potential energy (which depends only on the positions of the particles), (2.48) becomes

$$Z = \frac{Q_N}{h^{3N}} \int_{-\infty}^{\infty} \cdots \int_{-\infty}^{\infty} \exp\left[\frac{-(p_1{}^2 + p_2{}^2 + \cdots + p_N{}^2)}{2mkT}\right] d\mathbf{p}_1\, d\mathbf{p}_2 \cdots d\mathbf{p}_N$$

$$= \frac{Q_N}{h^{3N}}\left[\int_{-\infty}^{\infty} \exp\left(\frac{-p^2}{2mkT}\right) dp\right]^{3N}$$

$$= \frac{Q_N}{\lambda^{3N}} \tag{2.67}$$

where we shall refer to

$$Q_N = \frac{1}{N!} \int_V \cdots \int \exp\left(\frac{-\Phi}{kT}\right) d\mathbf{r}^N \qquad (2.68a)$$

as the configurational partition function and define

$$\lambda^2 = \frac{h^2}{2\pi mkT} \qquad (2.68b)$$

9. Some Simple Examples

The various formulas derived in the last section are now applied to some simple but important systems.

Perfect Gas. The classical partition function of the perfect monatomic gas is derived from (2.67) and (2.68) by setting the potential energy $\Phi \equiv 0$. Then $Q_N = V^N/N!$, and

$$Z = \frac{V^N}{\lambda^{3N} N!} \qquad (2.69)$$

so that (2.52), (2.59), (2.60), and (2.62) yield

$$A = -kT\left(N \log V + \frac{3N}{2} \log T - \frac{3N}{2} \log \frac{h^2}{2\pi mk} - \log N!\right) \qquad (2.70a)$$

$$E = \tfrac{3}{2}NkT \qquad (2.70b)$$

$$PV = NkT \qquad (2.70c)$$

and after employing Sterling's formula for $\log N!$, the entropy is given by the Sackur-Tetrode equation

$$S = k\left(\frac{5}{2}N + \frac{3}{2}N \log \frac{2\pi mkTv^{2/3}}{h^2}\right) \qquad (2.70d)$$

where $v = V/N$, the specific volume of the gas.

The same results are a consequence of the substitution of the quantum-mechanical energy levels of a free particle in a cube of volume V into (2.45):

$$E_{l,m,n} = \frac{(l^2 + m^2 + n^2)h^2}{8mV^{2/3}} \qquad l,m,n = 0,1,2, \ldots \qquad (2.71)$$

Then the partition function per particle is

$$Z_p = \sum_{l,m,n=0}^{\infty} \exp\left[\frac{-h^2(l^2 + m^2 + n^2)}{8mkTV^{2/3}}\right] = \left[\sum_{l=0}^{\infty} \exp\left(-\frac{l^2h^2}{8mkTV^{2/3}}\right)\right]^3 \qquad (2.72)$$

If α is a very small number,

$$\sum_{l=0}^{\infty} \exp(-\alpha^2 l^2) = \frac{1}{\alpha} \sum_{l=0}^{\infty} \alpha \exp(-\alpha^2 l^2) \sim \frac{1}{\alpha} \int_0^{\infty} \exp(-x^2)\, dx = \frac{\pi^{1/2}}{2\alpha}$$

Hence, as long as $1 \ll 8mkTV^{2/3}/h^2 = 4V^{3/2}/\pi\lambda^2 = \alpha^{-2}$ (high temperatures and large specific volumes, the range in which Einstein-Bose and Fermi-Dirac statistics reduce to the Maxwell-Boltzmann statistics), the partition function per particle is

$$Z_p = \left(\frac{1}{2}\pi^{1/2} \frac{2V^{1/3}}{\pi^{1/2}\lambda}\right)^3 = \frac{V}{\lambda^3} \qquad (2.73)$$

The complete quantum-mechanical partition function of a system of N indistinguishable noninteracting particles under the above-stated conditions is then exactly the same as the classical result (2.69):

$$Z = \frac{V^N}{\lambda^{3N} N!} \qquad (2.74)$$

The factor $1/N!$ appears as a result of the indistinguishability of the particles. Each distribution of energy levels between the particles is counted only once because any interchange of the particles cannot, according to quantum theory, be detected experimentally.

Harmonic Oscillator of Frequency ν. The energy levels of a harmonic oscillator of frequency ν are

$$E_j = (j + \tfrac{1}{2})h\nu \qquad j = 0, 1, 2, \ldots \qquad (2.75)$$

Hence the partition function per oscillator of an ensemble of identical oscillators is

$$Z = \sum_{j=0}^{\infty} e^{-(j+1/2)\Theta} = \frac{e^{1/2\Theta}}{1 - e^{-\Theta}} = \tfrac{1}{2} \operatorname{csch} \tfrac{1}{2}\Theta \qquad \Theta = \frac{h\nu}{kT} \qquad (2.76)$$

The internal energy per oscillator is

$$E = \frac{-h\nu\partial \log Z}{\partial\Theta} = h\nu\left(\frac{1}{2} + \frac{1}{e^{\Theta} - 1}\right) \qquad (2.77)$$

and the heat capacity is

$$C = k\frac{(\frac{1}{2}\Theta)^2}{\sinh^2(\frac{1}{2}\Theta)} \qquad (2.78)$$

These results of Planck are especially important in the quantum theory of radiation and were first applied by Einstein to the theory of vibrations of crystal lattices.

At high temperatures $E \rightarrow kT$ and $C \rightarrow k$, the classical results.

Electric Dipole in a Uniform Electric Field. The potential energy of a dipole of dipole moment μ_0 oriented at an angle θ with respect to a constant electric field is

$$V(\theta) = -\mu_0 E \cos\theta \qquad (2.79)$$

Hence the classical partition function per dipole of a gas of independent dipoles is (here we ignore the various kinetic energy factors and consider only the orientational factor in the partition function)

$$Z = 2\pi \int_0^{\pi} \exp\left(\frac{\mu_0 E \cos\theta}{kT}\right) \sin\theta\, d\theta = \frac{4\pi kT}{\mu_0 E} \sinh \frac{\mu_0 E}{kT} \qquad (2.80)$$

Hence the average dipole moment per gas particle is

$$\mu_{av} = \mu_0 <\cos\theta>_{av} = \mu_0 \frac{\partial \log Z}{\partial(\mu_0 E/kT)} = \mu_0 \left(\coth \frac{\mu_0 E}{kT} - \frac{kT}{\mu_0 E}\right) \quad (2.81)$$

This is the well-known Langevin equation. In the limit of high temperatures or low fields it reduces to

$$\frac{\mu_{av}}{\mu_0} \sim \frac{1}{3}\left(\frac{\mu_0 E}{kT}\right) \qquad \text{as } \frac{\mu_0 E}{kT} \to 0 \quad (2.82)$$

The complete curve of μ_{av}/μ_0 is plotted in Fig. 2.2.

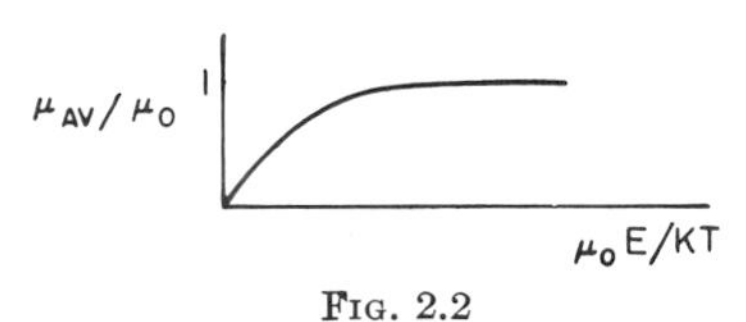

FIG. 2.2

Einstein-Bose and Fermi-Dirac Statistics. In the quantum theory identical particles are indistinguishable (see Part 2, Chap. 6, Sec. 10).

Fermi-Dirac particles are characterized by the antisymmetry of their wave functions. They obey the Pauli exclusion principle; no two particles can be in the same quantum state. Let n_s correspond to the occupation number and E_s the energy of the sth quantum state. By counting every state separately (whether its energy is the same as that of another state or not), the degeneracy factor ω_s in (2.45) can be omitted. A typical energy state of a system of N Fermi-Dirac particles is

$$E = \Sigma n_s E_s \qquad \text{with } N = \Sigma n_s \quad (2.83)$$

the n_s's being restricted to the values 0 and 1.

Einstein-Bose particles have symmetrical wave functions and no exclusion principle. The possible occupation numbers for the various states are the non-negative integers $n_s = 0,1,2, \ldots$.

The average number of particles in a given state of a system of N Einstein-Bose particles can be found as follows: The Gibbs formula for statistical averages gives as the expected number of particles in the first state

$$<n_N(1)>_{av} = Z_N^{-1} \sum_{\substack{n_1,n_2,=0 \\ n_1+n_2+\ldots=N}}^{\infty} n_1 \exp[-\beta(n_1\epsilon_1 + n_2\epsilon_2 + \cdots)] \quad (2.84)$$

Z_N being the partition function and the n_s's ranging over all non-negative integers subject to the restriction $\Sigma n_s = N$. Now $n'_1 = n_1 - 1$ has the range $-1,0,1,2, \ldots$. However, if we use it as a summation variable in place of n_1 in (2.84), it can be chosen to range only over $0,1,2, \ldots$ because the term which corresponds to $n_1 = 0$ does not contribute to the sum. Hence

$$<n_N^{(1)}>_{av} = Z_N^{-1} \exp(-\beta\epsilon_1) \sum_{\substack{n'_1,n_2,\ldots=0 \\ n'_1+n_2+\ldots=N-1}}^{\infty} (1 + n'_1) \exp[-\beta(n'_1\epsilon_1 + n_2\epsilon_2 + \cdots)] = Z_N^{-1}[1 + <n_{N-1}^{(1)}>_{av}]Z_{N-1} \exp(-\beta\epsilon_1)$$

As $N \to \infty$, $\lim <n_{N-1}^{(1)}>_{av} = \lim <n_N^{(1)}>_{av}$ which we shall call $\bar{n}_1$. Also $Z_N/Z_{N-1} \to \exp(-\mu\beta)$, μ being the chemical potential. We then have

$$\bar{n}_1 = \frac{1}{\exp[\beta(\epsilon_1 - \mu)] - 1}$$

Since this argument could be applied to any state, the average number of Einstein-Bose particles in the ith state becomes

$$\bar{n}_i = \{\exp[\beta(\epsilon_i - \mu)] - 1\}^{-1} \quad (2.85)$$

The chemical potential μ must be chosen so that the total number of particles is N, i.e., so that

$$N = \sum_{i=1}^{\infty} \bar{n}_i = \sum_{i=1}^{\infty} \{\exp[\beta(\epsilon_i - \mu)] - 1\}^{-1}$$

The corresponding formulas for Fermi-Dirac particles follow from consideration of the average

$$<n_N(1)>_{av} = Z_N^{-1} \sum_{\substack{n_1 n_2 \cdots = 0,1 \\ \Sigma n_s = N}} n_1 \exp[-\beta(n_1\epsilon_1 + n_2\epsilon_2 + \cdots)] \quad (2.86)$$

The variable $n'_1 = n_1 - 1$ now has the range $0,-1$. Since the term $n_1 = 0$ or $n'_1 = -1$ does not contribute to (2.86) we can omit it. When $n'_1 = 0$, $1 + n'_1 = 1 - n'_1 = 1$. The quantity $1 - n'_1$ vanishes when $n'_1 = 1$. Hence if we introduce the new variable n'_1 into (2.86) we can replace $n_1 = 1 + n'_1$ by $1 - n'_1$ in the coefficient of the exponential and sum over $n'_1 = 0,1$. Then (2.86) becomes

$$<n_N(1)>_{av} = \exp(-\beta\epsilon_1)Z_N^{-1} \sum_{\substack{n'_1,n_2,\cdots=0,1 \\ n'_1+n_2+\cdots=N-1}} (1 - n'_1) \exp[-\beta(n'_1\epsilon_1 + n_2\epsilon_2 + \cdots)] = \exp(-\beta\epsilon_1)(Z_{N-1}/Z_N)[1 - <n_{N-1}(1)>_{av}]$$

A generalization of the same procedure to the ith state and a repetition of the limit argument used in the Einstein-Bose case yields

$$\bar{n}_i = \{\exp[\beta(\epsilon_i - \mu)] + 1\}^{-1} \quad (2.87)$$

with the chemical potential μ being determined by:

$$N = \sum \bar{n}_i = \sum_{i=1}^{\infty} \{\exp[\beta(\epsilon_i - \mu)] + 1\}^{-1}$$

The internal energy of an Einstein-Bose system is

$$E_{\text{E-B}} = \sum_{i=1}^{\infty} \frac{\epsilon_i}{\exp[\beta(\epsilon_i - \mu)] - 1}$$

while that of a Fermi-Dirac system is

$$E_{\text{F-D}} = \sum_{i=1}^{\infty} \frac{\epsilon_i}{\exp[\beta(\epsilon_i - \mu)] + 1}$$

The derivations of this section on E-B and F-D gases are based on the work of H. Schmidt.

The Ising Model of Ferromagnetism. The Ising model is probably the simplest example of a system of interacting particles. A scalar variable σ whose possible values are ± 1 is associated with each lattice point of a crystal. These two values represent two possible electron spin directions in a ferromagnet.

The interaction energy between two particles located at the jth and kth lattice points and in the states σ_j and σ_k is postulated to be

$$E_{jk} = \begin{cases} -J\sigma_j\sigma_k & \text{if } j \text{ and } k \text{ are nearest neighbors} \\ 0 & \text{otherwise} \end{cases} \tag{2.88}$$

The constant J is a measure of strength of the coupling between spins. It is positive in a ferromagnetic system and negative in an antiferromagnetic one.

The Ising model is important not only as a model of ferromagnetism but also as a model of order-disorder effects in binary alloys and of dense gases and liquids. The two values of σ represent the two possible types of atomic occupants of a given lattice point in a binary alloy. In a model of a liquid or gas a container is divided into many small cells, each large enough to hold a single molecule. The two values of σ associated with a cell represent its two possible states, occupied or empty.

The partition function of a linear Ising model chain with N lattice points (numbered successively $1, 2, \ldots, N$) is

$$Z_N = \sum_{\sigma_1 = \pm 1} \cdots \sum_{\sigma_N = \pm 1} \exp\left[\beta J(\sigma_1\sigma_2 + \sigma_2\sigma_3 + \cdots + \sigma_{N-1}\sigma_N)\right]$$

Since $\sigma_{N-1}\sigma_N$ has the two values ± 1, $(\sigma_{N-1}\sigma_N)^2 = 1$. Hence

$$\exp(\beta J\sigma_{N-1}\sigma_N) = \cosh\beta J + \sigma_{N-1}\sigma_N \sinh\beta J$$

and

$$\sum_{\sigma_N = \pm 1} \exp(\beta J\sigma_{N-1}\sigma_N) = 2\cosh\beta J$$

Hence

$$Z_N = (2\cosh\beta J)Z_{N-1} = (2\cosh\beta J)^2 Z_{N-2} = \cdots = (2\cosh\beta J)^{N-2}Z_2$$

but

$$Z_2 = (2\cosh\beta J)\sum_{\sigma_1 = \pm 1} 1 = 4\cosh\beta J$$

so that

$$Z_N = 2(2\cosh\beta J)^{N-1}$$

The internal energy is given by

$$E = -\frac{\partial \log Z_N}{\partial\beta} = -J(N-1)\tanh\beta J \tag{2.89}$$

and the heat capacity by

$$C = \frac{\partial E}{\partial T} = (N-1)k\left(\frac{J}{kT}\right)^2 \operatorname{sech}^2\frac{J}{kT} \tag{2.90}$$

When plotted as a function of temperature, the heat capacity has a maximum.

A general review of the Ising model and its applications has been made by Newell and Montroll [14].

Black-body Radiation. Quantum statistics originated with Planck's theory of the energy density of black-body radiation in a given frequency range. Planck's formula is a consequence of the equation (2.77)

$$E_\nu - E_\nu^{(0)} = \frac{h\nu}{-1 + \exp(-h\nu/kT)}$$

($E_\nu^{(0)} = \frac{1}{2}h\nu$, and we shall calculate the energy density above that of the zero point energy.) (See Part 6, Chap. 1, Sec. 9.)

When many oscillators of different frequencies are at equilibrium at a temperature T, the average energy density in the range between ν and $\nu + d\nu$ is

$$\rho_\nu\, d\nu = \frac{V^{-1}g(\nu)h\nu\, d\nu}{-1 + \exp(-h\nu/kT)} \tag{2.91}$$

where V is the volume of the oscillating system and $g(\nu)\, d\nu$ is the number of oscillators with frequencies between ν and $\nu + d\bar{\nu}$.

Since black-body radiation is composed of electromagnetic waves, the number of oscillators with frequencies in a given frequency range is found by examining the solutions of the scalar wave equation

$$c^2\nabla^2\varphi = \frac{\partial^2\varphi}{\partial t^2}$$

which satisfy the conditions $\varphi = 0$ at the boundary of the region. Weyl showed that the frequency spectrum is independent of the shape of the container for those wavelengths which are short compared with the dimensions of the container (long wavelengths are important only at low temperatures, an uninteresting range in the theory of black-body radiation). Indeed, the frequency spectrum is independent of the detailed nature of the boundary conditions in the short wavelength range as long as no net energy transfer occurs at the boundaries.

Consider a rectangular black body with sides of length l_1, l_2, l_3. Solutions of the wave equation are

$$\varphi = Ae^{2\pi i\nu t}\sin\frac{\pi n_1 x}{l_1}\sin\frac{\pi n_2 y}{l_2}\sin\frac{\pi n_3 z}{l_3}$$

with n_1, n_2, and n_3 ranging over the non-negative integers $0, 1, 2, \ldots$. The corresponding frequencies are

$$\nu^2 = \frac{1}{4}c^2\left[\left(\frac{n_1}{l_1}\right)^2 + \left(\frac{n_2}{l_2}\right)^2 + \left(\frac{n_3}{l_3}\right)^2\right] \qquad n_1,n_2,n_3 = 0,1,2 \ldots \tag{2.92}$$

The total number of frequencies less than ν, $N(\nu)$ is proportional to the volume of the ellipsoid defined by (2.92), $\frac{32}{3}\pi\nu^3 l_1l_2l_3/c^3 = 32\pi V\nu^3/3c^3$. Since all n_j's are positive, only one-eighth of this volume contributes to $N(\nu)$. There is exactly one frequency per unit volume in (n_1,n_2,n_3) space. Hence

$$N(\nu) = \frac{\frac{4}{3}\pi V\nu^3}{c^3}$$

Since two states of polarization exist, this number is to be doubled.

The number of frequencies between ν and $\nu + d\nu$ is

$$g(\nu)\, d\nu = \frac{\partial N}{\partial\nu}\, d\nu = \frac{8\pi V\nu^2}{c^3}$$

therefore the radiation density at frequency ν is

$$\rho_\nu = \frac{8\pi\nu^3 h/c^3}{-1 + \exp(-h\nu/kT)}$$

the Planck formula.

10. Molecular Distribution Functions

Thermodynamic quantities can be determined from certain molecular distribution functions as well as from the partition function. This approach to statistical mechanics has been especially emphasized by Kirkwood, Yvon, Mayer, Born and Green, and their collaborators. A review has recently been written by De Boer [15].

The probability density in phase space of a system of N particles $P(\mathbf{r}^N,\mathbf{p}^N)$ was defined at the beginning of Sec. 6. The corresponding momentum and position probability densities are

$$P_N(\mathbf{p}^N) = \int P_N(\mathbf{r}^N,\mathbf{p}^N)\, d\mathbf{r}^N \tag{2.93a}$$
$$P_N(\mathbf{r}^N) = \int P_N(\mathbf{r}^N,\mathbf{p}^N)\, d\mathbf{p}^N \tag{2.93b}$$

The probability density $P_h(r_i{}^h,p_i{}^h)$ of a specified subset $i_1, i_2, \ldots, i_h$ of h particles at the point $(r_{i_1},r_{i_2}, \ldots ,r_{i_h}; p_{i_1}, \ldots , p_{i_h})$ in the $6h$-dimensional space of the particles is related to $P_N(\mathbf{r}^N,\mathbf{p}^N)$ by

$$P_h(r_i{}^h,p_i{}^h) \tag{2.94}$$

(i represents the set of $N - h$ particles which is the complement of the set h). The position and momentum probability densities of specified subsets of h particles are given by

$$P_h(r_i{}^h) = \int P_h(r_i{}^h,p_i{}^h)\, dp_i{}^h \tag{2.94a}$$
$$P_h(p_i{}^h) = \int P_h(r_i{}^h,p_i{}^h)\, dr_i{}^h \tag{2.94b}$$

Thermodynamic quantities are better discussed in terms of probability densities of sets of particles which are fixed in numbers but unspecified in detail. The quantity $f_h(r^h,p^h)\, dr^h\, dp^h$ is defined as the probability of a sample system having *any* h particles in volume elements $dr_1,dp_1, \ldots ,dr_h,dp_h$ (here the subscripts do not correspond to specified particles but merely identify volume elements). Since the subsets of h particles can be selected from the N of the original system in $N!/(N - h)!$ ways,

$$f_h(r^h,p^h) = \frac{N!}{(N-h)!} P_h(r^h,p^h) \tag{2.95a}$$

and

$$f_h(r^h,p^h) = \frac{1}{(N-h)!} \int \cdots \int f_N(r^N,p^N)\, dr^{N-h}\, dp^{N-h} \tag{2.95b}$$

If we define

$$n_h(r^h) = \int f_h(r^h,p^h)\, dp^h \tag{2.96}$$

we find

$$n_h(r^h) = \frac{1}{(N-h)!} \int \cdots \int n_N(r^N)\, dr^{N-h} \tag{2.97}$$

The quantity n_1 is exactly the density of particles and

$$\int_V n_1\, dr^1 = N \tag{2.97a}$$

Expressions for n_h can be obtained directly from Liouville's equation (2.37) if one postulates the equilibrium distribution function $f_N(\mathbf{p}_1 \cdots \mathbf{r}_N)$ of systems in phase space to be factorable into a function of the particles positions multiplied by the Maxwell momentum distribution function (A is the normalization constant):

$$f_N = A n_N(\mathbf{r}_1 \cdots \mathbf{r}_N) \exp\left(-\frac{\beta\Sigma p_j{}^2}{2m}\right) = N!P_N \tag{2.98}$$

The Liouville equation reduces to

$$\sum_k \mathbf{p}_k(\nabla_k n_N + n_N\beta\nabla_k\Phi) = 0$$

Since the momenta p_k are all independent variables, the only way this equation can be satisfied is to set

$$\nabla_k n_N + \beta n_N \nabla_k \Phi = 0 \qquad \text{for all } k \tag{2.99}$$

An appropriately normalized solution of this equation is

$$n_N = Q_N{}^{-1} e^{-\beta\Phi}$$
$$Q_N = \frac{1}{N!}\int \cdots \int e^{-\beta\Phi}\, d\mathbf{r}_1 - d\mathbf{r}_N \tag{2.100}$$

Here Q_N is just the configurational factor of the partition function Z.

An expression for n_h is then

$$n_h(\mathbf{r}_1, \ldots ,\mathbf{r}_h) = \frac{1}{(N-h)!Q_N} \int_V \cdots \int \exp[-\beta\Phi(r_1, \ldots ,r_N)]\, d\mathbf{r}_{h+1} \cdots d\mathbf{r}_N \tag{2.101}$$

11. Calculation of Thermodynamic Quantities from Molecular Distribution Functions

The internal energy E and the pressure p can be deduced from the pair distribution function n_2 when the potential energy has the form

$$\Phi = \sum_{i>j} v(|\mathbf{r}_i - \mathbf{r}_j|) = \sum_{i>j} v_{ij} \tag{2.102}$$

and all particles are identical. A typical plot of $v(r)$ is given in Fig. 2.3. Notice the strong short-

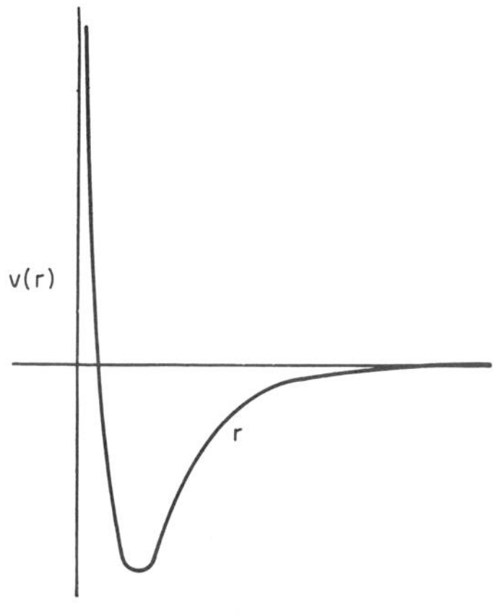

Fig. 2.3

range repulsion and the rapidity with which $v(r) \to 0$ as r increases. A typical analytic expression frequently used for $v(r)$ is the Lennard-Jones potential $Ar^{-12} - Br^{-6}$. A detailed review of the theory of intermolecular forces is given in a treatise on gases by Hirschfelder, Curtiss, and Bird (see ref. 7a of Chapter 4).

Equations (2.60) and (2.68) lead to

$$E = -\frac{\partial \log Z}{\partial \beta} = \tfrac{3}{2}NkT - \frac{\partial \log Q_N}{\partial \beta}$$
$$= \tfrac{3}{2}NkT + \frac{1}{Q_N N!}\int \cdots \int \left(\sum_{i>j} v_{ij}\right) e^{-\beta\Phi}\, d\mathbf{r}_1 \cdots d\mathbf{r}_N \quad (2.103)$$

Since the interactions between all pairs of particles are equivalent

$$\int \cdots \int v_{ij} e^{-\beta\Phi}\, d\mathbf{r}_1 \cdots d\mathbf{r}_N = \int \cdots \int v_{12} e^{-\beta\Phi}\, d\mathbf{r}_1 \cdots d\mathbf{r}_N$$

Hence

$$E = \tfrac{3}{2}NkT + \frac{N(N-1)}{2Q_N N!}\int \cdots \int v_{12} e^{-\beta\Phi}\, d\mathbf{r}_1 \cdots d\mathbf{r}_N$$

This equation combines with (2.101) when $n = 2$ to yield

$$E = \tfrac{3}{2}NkT + \tfrac{1}{2}\int\int v(\mathbf{r}_1,\mathbf{r}_2) n_2(\mathbf{r}_1,\mathbf{r}_2)\, d\mathbf{r}_1\, d\mathbf{r}_2 \quad (2.104)$$

If $v(\mathbf{r}_1,\mathbf{r}_2)$ depends only on $|\mathbf{r}_1 - \mathbf{r}_2|$, so also does $n(\mathbf{r}_1,\mathbf{r}_2)$. We let $r = |r_1 - r_2|$, θ, and ϕ be the spherical coordinates of particle 2 relative to particle 1. Then

$$E = \tfrac{3}{2}NkT + \tfrac{1}{2}\int d\mathbf{r}_1 \int v(r) n_2(r) r^2 \sin\theta\, d\theta\, d\phi\, dr$$

The limits of integration are such that both particles 1 and 2 must be inside the volume V; $v(r)$, however, is usually negligible when $r \gg r_0$, r_0 being the effective "range of force" (small compared with the linear dimensions of V). Hence, the integral over r,θ,ϕ is nearly independent of $\mathbf{r}_1$ except when $\mathbf{r}_1$ is within a distance r_0 of the boundary of V. The integral over r,θ,ϕ can be approximated by an integral over the entire space, independent of $\mathbf{r}_1$. The integral over $\mathbf{r}_1$ then gives V and the integrals over θ,ϕ give 4π so that

$$E = \tfrac{3}{2}NkT + 2\pi V \int_0^\infty v(r) n_2(r) r^2\, dr \quad (2.105)$$

The pressure is related to Q_N by $p = kT\partial \log Q_N/\partial V$. Now Q_N does not involve the volume V explicitly, but only implicitly in terms of $\mathbf{r}_i$. To introduce V explicitly, one may use the following trick due to H. S. Green. Write $\mathbf{r}_i = l\mathbf{r}'_i$, where l = length of a face of the container (which we assume to be a cube), $V = l^3$. Then

$$p = -\frac{kT}{3l^2 Q_N}\frac{\partial Q_N}{\partial l}$$

but

$$Q_N = \frac{l^{3N}}{N!}\int_0^1 \cdots \int_0^1 d\mathbf{r}'_1 \cdots d\mathbf{r}'_N \exp\left[-\beta\Sigma_{i>j} v(l|\mathbf{r}'_i - r'_j|)\right]$$

$$\frac{\partial Q_N}{\partial l} = \frac{3NQ_N}{l} - \frac{N(N-1)\beta}{2N!l}\int \cdots \int d\mathbf{r}_1 \cdots d\mathbf{r}_N |\mathbf{r}_1 - \mathbf{r}_2| \frac{\partial v(|\mathbf{r}_1 - \mathbf{r}_2|)}{\partial(|\mathbf{r}_1 - \mathbf{r}_2|)} e^{-\beta\Phi}$$

and

$$\frac{1}{l^2 Q_N}\frac{\partial Q_N}{\partial l} = \frac{3N}{V} - \frac{\beta}{2V}\iint |\mathbf{r}_1 - \mathbf{r}_2| \frac{\partial v_{12}}{\partial r_{12}} n_2(r_{12})\, d\mathbf{r}_1\, d\mathbf{r}_2$$

Hence

$$p = \frac{NkT}{V} - \frac{2\pi}{3}\int_0^\infty \frac{\partial v}{\partial r} n_2(r) r^3\, dr \quad (2.106)$$

This equation was first derived by Debye from the virial theorem. The general statement of the virial theorem is that $-2T = -3pV + \Xi$, where T is the average kinetic energy of a system and Ξ is the average

$$\Xi = \langle \Sigma \mathbf{r}_i \cdot \mathbf{F}_i \rangle_{av}$$

$\mathbf{r}_i$ being the position vector of the ith particle and $\mathbf{F}_i$ the force acting on it as a result of its interaction with the remaining $N - 1$ particles.

The scattering of light, electrons, or X rays by gases and liquids can also be discussed in terms of the distribution function $n_2(r)$.

The scattered wave function of an isolated scatterer at a large distance, R, from the scatterer at an angle θ with the direction of the incident beam is of the form $\psi = R^{-1}f(\theta)\exp(ik_0R)$, where $k_0 = 2\pi/\lambda$ (λ being the wavelength of the radiation) and the function $f(\theta)$ depends on the nature of the scattering atoms and molecules. The observed beam scattered from a system of many scatterers is the result of the interference of waves scattered by every pair of particles in the system (here we shall neglect multiple scattering). Normally, intensity observations are made at distances many wavelengths from the scatterers (and indeed at distances large compared with the container of the system of interest).

Consider a fixed distribution of atoms located at points $\{\mathbf{r}_j\}$ in a gas or liquid. At a point $\mathbf{r}$ the wave scattered by the jth particle is described by the wave function

$$\psi_j = r^{-1}\exp(ik_0r) f(\theta) \exp[ik_0(\mathbf{s}_0 - \mathbf{s}_1)\cdot \mathbf{r}_j]$$

where $\mathbf{s}_0$ and $\mathbf{s}_j$ are, respectively, unit vectors in the direction of the incident and scattered waves

$$(\cos\theta = \mathbf{s}_0 \cdot \mathbf{s}_1)$$

The sum of the individual scattered wave functions

$$\Psi = \Sigma\psi_j = r^{-1}\exp(ik_0r)\Sigma_j f(\theta)\exp[ik_0(\mathbf{s}_0 - \mathbf{s}_1)\cdot\mathbf{r}_j]$$

gives the total wave function at the point of observation. The observed intensity is then

$$I(\theta) = r^2\langle|\Psi|^2\rangle_{av} = |f(\theta)|^2 \cdot \left\langle \sum_{j>m} \exp[ik_0(\mathbf{s}_0 - \mathbf{s}_1)\cdot(\mathbf{r}_j - \mathbf{r}_m)] \right\rangle_{av}$$

where the average is taken over all possible atomic positions. That is, the function inside the brackets is to be weighted with

$$\frac{1}{N!Q_N}\exp[-\beta\Phi(\mathbf{r}_1 \cdots \mathbf{r}_N)]$$

Hence

$$I(\theta) = \frac{|f(\theta)|^2}{Q_N N!} \int \cdots \int \left\{ N + \frac{N(N-1)}{2} \exp [ik_0(\mathbf{s}_0 - \mathbf{s}_1) \cdot \mathbf{r}_{12}] \right\} \exp(-\beta\Phi)\, d\mathbf{r}_1 \cdots d\mathbf{r}_N$$

$$= |f(\theta)|^2 \{ N + \textstyle\iint n_2(r_{12}) \exp [ik_0(\mathbf{s}_0 - \mathbf{s}_1) \cdot \mathbf{r}_{12}]\, d\mathbf{r}_1\, d\mathbf{r}_2 \}$$

$$= N|f(\theta)|^2(1 + I_1 + I_2)$$

where

$$I_1 = \frac{N}{V^2} \iint \exp [ik_0(\mathbf{s}_0 - \mathbf{s}_1) \cdot \mathbf{r}_{12}]\, d\mathbf{r}_1\, d\mathbf{r}_2 \qquad (2.107)$$

depends only on the geometry of the container of the fluid and the incident radiation, and

$$I_2 = v \int_0^\infty 4\pi[n_2(r) - n_1{}^2] \frac{\sin(2k_0 r \sin \frac{1}{2}\theta)}{2k_0 r \sin \frac{1}{2}\theta} r^2\, dr$$

depends on the distribution of the scatterers.

12. The Integrodifferential Equations for the Distribution Functions

An important chain of integrodifferential equations follows from an integration of (2.99) when the total potential energy of interaction is the sum of pair interactions (2.102). Then

$$\nabla_k n_N + \beta n_N \sum_{j=1}^{N} \nabla_k v_{jk} = 0$$

To exclude interaction of a particle with itself we set $v_{jj} = 0$.

To obtain equations for n_h, we integrate the above equation over the coordinates of the $(h + 1)$st to the Nth particles, recalling that

$$\int n_N\, dr_{h+1} \cdots dr_N = n_h(N - h)!$$

Then

$$(N - h)! \nabla_k n_h + \beta \sum_{j=1}^{N} \int n_N \nabla_k v_{jk}\, dr_{h+1} \cdots dr_N = 0$$

$$k = 1,2, \ldots ,h$$

If $j \leq h$ and $k \leq h$, then $\nabla_k v_{jk}$ does not involve coordinates $h + 1$ to N; thus the first h terms of the sum are

$$\beta \sum_{j=1}^{h} n_h(N - h)! \nabla_k v_{jk}$$

The remaining terms involve integrals of v_{jk} with $k > h$. Since n_N depends symmetrically on $\mathbf{r}_{h+1}$, $\mathbf{r}_{h+2}, \ldots , \mathbf{r}_N$ each term in the set with $j = h + 1$ to N is the same and is equal to the term for $j = h + 1$. Hence the remaining terms contribute

$$\beta(N - h) \int n_N \nabla_k v_{h+1,k}\, d\mathbf{r}_{h+1} \cdots d\mathbf{r}_N = \beta(N - h)! \int n_{h+1} \nabla_k v_{h+1,k}\, d\mathbf{r}_{h+1}$$

The equation for n_h becomes

$$\nabla_k n_h + \beta n_h \nabla_k \sum_{j=1}^{h} v_{jk} + \beta \int n_{h+1} \nabla_k v_{h+1,k}\, d\mathbf{r}_{h+1} = 0 \qquad k \leq h \qquad (2.108)$$

When $h = 2$, we have

$$\nabla_1 n_2(\mathbf{r}_1\mathbf{r}_2) + \beta n_2(\mathbf{r}_1,\mathbf{r}_2) \nabla_1 v_{12} + \beta \int n_3(\mathbf{r}_1\mathbf{r}_2\mathbf{r}_3) \nabla_1 v_{13}\, d\mathbf{r}_3 = 0 \qquad (2.109)$$

which relates n_2 to n_3. In general each equation relates n_h to n_{h+1}.

The problem of solving these equations for all n_h is tremendously difficult. In order to reduce the set of equations to a single equation in n_2, Kirkwood introduced a simplification. The total potential energy of interaction between atoms can to a good approximation be represented as a sum of functions which depend only on the positions of pairs of atoms (2.102). Kirkwood assumed that the potential of average force acting on three particles in a complex system can also be expressed as a sum of pair functions. On this basis a set of equations equivalent to (2.108) which have been derived by Kirkwood reduce to a single nonlinear equation for n_2. This approximation is called the "superposition principle." Born and Green expressed this principle in a slightly different form (which is equivalent to Kirkwood's to within the order of terms neglected in Kirkwood's theory) by assuming that

$$n_3(r_1,r_2,r_3) = \frac{n_2(r_1r_2)n_2(r_2r_3)n_2(r_3r_1)}{n_1(r_1)n_1(r_2)n_1(r_3)} \qquad (2.110)$$

If this expression is substituted into (2.109), one arrives at a single nonlinear integrodifferential equation for $n_2(r)$.

The superposition equation (2.110) is quite reasonable for a rarefied gas because if two particles out of a set of three are close together it is very likely that the third is far enough away from the other two to be independent of them. Kirkwood has shown that (2.110) is also valid in a solid when Hooke's law forces only exist. On this basis he assumed that (2.110) is also valid in the intermediate liquid region. There has been considerable criticism of the superposition principle.

13. Theory of Fluctuations and the Grand Canonical Ensemble

Although the thermodynamic properties of systems of large numbers of particles can on the average be computed from the various equations derived in this chapter, fluctuations do exist and their probabilities can be calculated. We shall in particular be concerned with fluctuations from the mean number of particles in a given subvolume of a large system. Our analysis is similar to that of Pauli [16].

Let a system of N particles in a volume V be divided into two parts, one of N_1 particles in volume V_1 and the remaining $N - N_1$ in a volume $V_2 = V - V_1$. Then the probability of finding N_1 specified particles in V_1 and the remaining ones in V_2 is

$$W(N_1,V_1,T) = \frac{\int_{V_1}\int_{V_2} \exp\left[-\beta H(p_1,\ldots,r_{N_1};p_{N_1+1},\ldots,r_N)\right]d^{6N_1}\Omega_1 d^{6N_2}\Omega_2}{\int_V \exp(-\beta H)d^{6N}\Omega} \tag{2.111}$$

where the range of integration of the positions of the first N_1 particles extends over the volume V_1 and that of the remaining N_2 extends over V_2.

If the two volumes V_1 and V_2 are sufficiently great to contain a large number of particles, the interactions of particles across the regional boundaries can be neglected when compared with those between particles which exist in the same region (we assume all forces to be short-ranged). Then we write

$$H \simeq H_1(p_1, \ldots, r_{N_1}) + H_2(p_{N_1+1}, \ldots, r_{N_2})$$

and $\exp -\beta F(V_j,N_j) = \int_{V_j} \exp(-\beta H_j)d^{6N_j}\Omega_j$

we have as N, N_1, and $N_2 \to \infty$

$$\log W(N_1,V_1,T) = \beta[F(V,N) - F(V_1,N_1) - F(V_2,N_2)]$$

The probability of finding an arbitrary set of N_1 particles in V_1 and the remaining N_2 in V_2 is

$$P(N_1,V_1) = \frac{N!}{N_1!N_2!} W(N_1,V_1,T)$$

Since we define the partition function as

$$Z(N,V) = \frac{1}{N!h^{3N}} \exp[-\beta F(V,N)]$$

and the Helmholtz free energy as

$$-kT \log Z(N,V) = A(N,V)$$

we have

$$\log P(N_1,V_1) = \beta[A(V,N) - A_1(V_1,N_1) - A_2(V_2,N_2)] \tag{2.112}$$

The most probable partition of N into N_1 and N_2 is that which maximizes $P(N_1,V_1)$. This partition is determined by the conditions

$$\frac{\partial \log P}{\partial N_j} = 0 \quad \text{and} \quad \frac{\partial^2 \log P}{\partial N_j^2} < 0 \quad \text{for } j = 1,2 \ldots \tag{2.113}$$

This implies $\partial A_1/\partial N_1 = \partial A_2/\partial N_2 = 0$, but $\partial A/\partial N$ is the chemical potential u of a system. Hence the condition for maximum probability is

$$\mu_1(N_1,V_1) = \mu_2(N_2,V_2) \tag{2.114}$$

μ_1 refers to the subsystem of volume V_1 and μ_2 to that of volume V_2, and

$$-\frac{\partial^2 A_j}{\partial N_j^2} = -\frac{\partial \mu_j}{\partial N_j} < 0 \qquad \text{for } j = 1,2 \ldots$$

If we let P_0 represent the probability of the most probable partition, we can write the probability of an arbitrary partition as

$$\log P = \log P_0 - \frac{1}{2}\beta\left[\frac{\partial^2 A_1}{\partial N_1^2}(\Delta N_1)^2 + \frac{\partial^2 A_2}{\partial N_2^2}(\Delta N_2)^2\right] + O(\Delta N)^3$$

ΔN_j being the deviation of N_j from its most probable value. Since $\Delta N_1 = -\Delta N_2$, we have in the limit of small fluctuations

$$P = P_0 \exp\left[-\frac{(\Delta N_1)^2}{2kT}\left(\frac{\partial^2 A_1}{\partial N_1^2} + \frac{\partial^2 A_2}{\partial N_2^2}\right)\right] \tag{2.115a}$$

The mean value of $(\Delta N_1)^2$ is easily found to be

$$<(\Delta N_1)^2>_{\text{av}} = \frac{kT}{\partial^2 A_1/\partial N_1^2 + \partial^2 A_2/\partial N_2^2} = \Delta^2$$

Equation (2.115*a*) is valid as long as $(\Delta N_1)^2$ is small compared with N_1^2. The normalized P is

$$P = \frac{1}{\Delta\sqrt{2\pi}} \exp\left[-\frac{(\Delta N_1)^2}{2\Delta^2}\right] \tag{2.115b}$$

Equation (2.70*a*) for the Helmholtz free energy of a perfect gas implies (after an application of Sterling's theorem) that

$$\mu = \frac{\partial A}{\partial N} = -kT\left[\log\frac{V}{N} + f(T)\right]$$

$f(T)$ being a function of temperature. Hence the most probable partition is $V_1/N_1 = V_2/N_2$ as would have been expected. Since $\partial^2 A/\partial N^2 = kT/N$

$$\left\langle\frac{(\Delta N_1)^2}{N_1^2}\right\rangle_{\text{av}} = N_1^{-2}(N_1^{-1} + N_2^{-2})^{-1} = \frac{N_2}{NN_1} \tag{2.115c}$$

This quantity is small as long as N_1 is large. When V_1 is chosen to be so small that on the average only one or two particles are expected to be included in V_1, $<(\Delta N_1)^2/N_1^2>_{\text{av}}$, is not small and the expression (2.115*a*) for P may need higher-order terms. It is to be noted that $-\partial^2 A_i/\partial N_i^2 = -kT/N_i < 0$ so that a most probable distribution does exist.

To proceed further with a general system of interacting particles we shall use Van Hove's [17] theorem which states that if attractive forces between particles are short-ranged and if a hard-core repulsion exists between particles, then (1) the Helmholtz free energy can be written in the form $A(V,N) = Nf(V/N)$, and (2) the pressure is a nonincreasing function of the volume, $\partial p/\partial V \leq 0$; experimentally the equality exists in a two-phase region. For further analysis, see [20].

The Van Hove theorem then implies that $A(V,N)$ is a homogeneous function of order 1, for

$$A(\lambda V,\lambda N) = \lambda Nf(V/N) = \lambda A(V,N)$$

However, Euler's theorem on homogeneous functions states that if $F(x_1,x_2, \ldots, x_n)$ is a homogeneous function of order α with

$$F(\lambda x_1,\lambda x_2, \ldots) = \lambda^\alpha F(x_1,x_2, \ldots)$$

then

$$x_1\frac{\partial F}{\partial x_1} + x_2\frac{\partial F}{\partial x_2} + \cdots = \alpha F$$

Hence $$A = N\frac{\partial A}{\partial N} + V\frac{\partial A}{\partial V} = N\mu - pV \tag{2.116}$$

since $\mu = \partial A/\partial N$ and the pressure p is equal to $-\partial A/\partial V$. Also, since $\mu = (A + pV)/N$ is a homogeneous function of order zero,

$$N\frac{\partial \mu}{\partial N} + V\frac{\partial \mu}{\partial V} = 0$$

or $$\frac{\partial^2 A}{\partial N^2} = \frac{V}{N}\frac{\partial^2 A}{\partial V\,\partial N} = \frac{V}{N}\frac{\partial p}{\partial N}$$

On the other hand p is also a homogeneous function of order zero, so that

$$N\frac{\partial p}{\partial N} + V\frac{\partial p}{\partial V} = 0$$

and $$\frac{\partial^2 A}{\partial N^2} = -\left(\frac{V}{N}\right)^2\frac{\partial p}{\partial V} = -\frac{v^2}{N}\frac{\partial p}{\partial v} \tag{2.117}$$

where $v = V/N$.

The condition that a sharp most probable partition of N into $N_1 + N_2$ exists, $\partial^2 A/\partial N^2 > 0$, is then satisfied as long as $\partial p/\partial v < 0$, which is valid in a single-phase region. In a two-phase region in which $\partial p/\partial V = 0$ fluctuations become enormous. In a gas-liquid equilibrium, a given region in space may sometimes be occupied by a large number of particles in the condensed phase or a relatively small number in the gas phase.

When $N \simeq N_2 \gg N_1 \gg 1$, the substitution of (2.117) into the equation below (2.115*a*) yields

$$\left\langle\left(\frac{\Delta N_1}{N_1}\right)^2\right\rangle_{\text{av}} = -\frac{kT}{N_1{}^3 v_1{}^2}\frac{\partial p}{\partial v_1} = -\frac{kT}{V_1{}^2}\frac{\partial p}{\partial V_1}$$

This reduces to (2.115*c*) when $pV_1 = N_1kT$ and $N_2 \simeq N_1$.

We shall now restrict ourselves to sufficiently large subregions of a volume V so that (2.115) is valid. A grand canonical ensemble of systems whose number of particles is not fixed can be discussed in terms of (2.115).

The probability of there being n particles in a region of volume V and that these particles are in the jth energy state available to n particles is

$$\begin{aligned} p(n, j, V) &= \exp\{\beta[A(V', N) - A(V, n) \\ &\quad - A(V' - V, N - n)]\}\frac{\exp[-\beta E_j(n)]}{\exp[-\beta A(V, n)]} \\ &= \exp\{\beta[A(V', N) - A(V' - V, N - n)]\} \\ &\qquad \exp[-\beta E_j(n)] \end{aligned}$$

Here we postulate the total number of particles to be N and the total volume V'. When $N \gg n$ and $V' \gg V$,

$$\begin{aligned} A(V', N) - A(V' - V, N - n) &\simeq A(V', N) \\ &\quad - A(V', N) + V\frac{\partial A}{\partial V'} + n\frac{\partial A}{\partial N} + \cdots \end{aligned}$$

Since $p = -\partial A/\partial V'$ and $\mu = \partial A/\partial N$, we have

$$p(n,j,V) = \exp \beta[-pV + \mu n - E_j(n)]$$

In order for the distribution to be normalized, we must have

$$\exp\left(\frac{pV}{kT}\right) = \sum_{n,j} \exp(\beta\mu n)\exp[-\beta E_j(n)] \tag{2.118}$$

We shall call this quantity the **grand partition function**. Then the pressure of a system can be determined from Z_G through

$$\frac{pV}{kT} = \log Z_G \tag{2.119}$$

The average number of particles in the subregion V is

$$\bar{n} = kT\frac{\partial \log Z_G}{\partial \mu} \tag{2.120}$$

Hence the average density of particles in the volume of interest is

$$\rho = \frac{\bar{n}}{V} = \frac{kT}{V}\frac{\partial \log Z_G}{\partial \mu} \tag{2.121}$$

Equation (2.118) can be generalized to

$$\begin{aligned} Z_G &= \exp\left(\frac{pV}{kT}\right) && (2.122a) \\ &= \sum_{n_1,n_2\ldots,j} \exp\left[\beta\sum n_i\mu_i - \beta E_j(n_1,n_2,\cdots)\right] \end{aligned}$$

for systems which contain several molecular species. The sum extends over all molecular concentrations. The variable μ_s represents the chemical potential of the sth species and the average number of molecules of the sth species in volume V is

$$\bar{n}_s = \frac{kT\partial \log Z_G}{\partial \mu_s} \tag{2.122b}$$

It is sometimes easier to calculate the grand partition function with an unrestricted number of particles in a given volume than it is to calculate the ordinary partition function for a fixed molecular composition. It can be shown that with the choice of the chemical potentials (2.120) and (2.122*b*), the relative fluctuations $<[(n_s - \bar{n}_s)/\bar{n}_s]^2>_{\text{av}}$ approach zero as $\bar{n}_s \to \infty$. Hence the thermodynamic properties of systems with large but fixed numbers of particles can be calculated from the grand partition function as well as from the ordinary one.

As an example of the application of the grand partition function let us find the equation of state of a Fermi-Dirac and Einstein-Bose gas. As was discussed in Sec. 9 in the paragraph on Einstein-Bose and Fermi-Dirac statistics, the possible energy states of a Fermi-Dirac gas are $E = n_1\epsilon_1 + n_2\epsilon_2 + \cdots$, where the n_j's are 0 or 1. When the total number of particles is $N = \Sigma n_j$, the grand partition function is given by the Fermi-Dirac equations

$$\begin{aligned} Z_G &= \sum_{n_j=0,1} \exp[-\beta(n_1\epsilon_1 + n_2\epsilon_2 + \cdots)] \\ &\qquad \exp[\mu\beta(n_1 + n_2 + \cdots)] \\ &= \sum_{n_j=0}^{1}\prod_{j=1}^{\infty} \exp[-\beta(\epsilon_j - \mu)n_j] \\ &= \prod_{j=1}^{\infty}\{1 + \exp[-\beta(\epsilon_j - \mu)]\} \end{aligned}$$

or $$\frac{pV}{kT} = \sum_{j=1}^{\infty} \log\{1 + \exp[\beta(\mu - \epsilon_j)]\}$$

The occupation numbers of Einstein-Bose states range over all the non-negative integers; hence the Einstein-Bose equations

$$Z_G = \sum_{n_j=0}^{\infty} \prod_{j=1}^{\infty} \exp[-\beta(\epsilon_j - \mu)n_j] = \prod_{j=1}^{\infty} \frac{\epsilon_j}{1 - \exp[\beta(\mu - \epsilon_j)]}$$

and

$$\frac{pV}{kT} = -\sum_j \log\{1 - \exp[\beta(\mu - \epsilon_j)]\}$$

The chemical potential μ is to be chosen so that the average number of particles is the preassigned number N; see Eq. (2.120).

The theory of the fluctuation in the number of particles in a given subvolume of a large system can easily be generalized to that of small fluctuations in other macroscopic variables. Let $a^{(1)}, a^{(2)}, \ldots, a^{(n)}$ be a set of macroscopic variables which satisfy

$$a_1^{(j)} + a_2^{(j)} = a^{(j)} \qquad j = 1,2, \ldots, n \tag{2.123}$$

where $a_1^{(j)}$ represents the value of $a^{(j)}$ in subvolume V_1 of our system and $a_2^{(j)}$ that in V_2 with $V_1 + V_2 = V$, the total volume. In particular, we might choose $N = a^{(1)}$.

The Helmholtz free energy of a system whose values of the macroscopic variables are fixed at $a^{(1)}, a^{(2)}, \ldots$ is

$$\exp[-\beta A(V, a^{(1)}, \ldots, a^{(n)})] = \frac{1}{N!h^{3N}} \int \cdots \int_{V,a^{(1)},a^{(2)},\ldots} \exp(-\beta H) d^{6N}\Omega \tag{2.124}$$

where the integration proceeds over that region of phase space in which V, $a^{(1)} \equiv N$, $a^{(2)}, \ldots$ have their postulated values. The logarithm probability of the subvolume V_1 having its $a^{(j)}$ values as $\{a_1^{(j)}\}$ is

$$\log P(V_1, \{a_1^{(j)}\}) = \beta[A(V, \{a^{(j)}\}) - A_1(V_1, \{a_1^{(j)}\}) - A_2(V_2, \{a_2^{(j)}\})]$$

Hence, if $\delta_i^{(j)}$ is the deviation of $a_i^{(j)}$ from its equilibrium value, the logarithm of the probability of a preassigned set of small deviations, $\delta_1^{(j)}$, is (since $\delta_1^{(j)} = -\delta_2^{(j)}$)

$$\log P = \log P_0 - \frac{1}{2}\beta \sum_{i,j=1}^{n} \delta_1^{(j)}\delta_1^{(i)} \left\{\frac{\partial^2 A}{\partial a_1^{(i)}\,\partial a_1^{(j)}} + \frac{\partial^2 A}{\partial a_2^{(i)}\,\partial a_2^{(j)}}\right\} \tag{2.125}$$

We abbreviate

$$\frac{\partial^2 A_k}{\partial a_k^{(i)}\,\partial a_k^{j}} = g_{ij}{}^k \tag{2.126}$$

Then the distribution can be normalized only if the quadratic form

$$G = \Sigma g_{ij}\delta^{(i)}\delta^{(j)} \tag{2.127}$$

is positive definite. It is only in this case that the system is stable with respect to small fluctuations. When $V_2 \gg V_1$, the Helmholtz free energy A_2 varies much more slowly than A_1 as fluctuations occur in V_1. Under this condition the normalized distribution function for $\{\delta_1^{(j)}\}$ is

$$P(V_1, \{\delta_1^{(j)}\}) = (2\pi kT)^{-1/2n}(\det G)^{1/2} \exp\left(-\frac{1}{2kT}\sum_{ij} g_{ij}{}^1\delta_1^{(j)}\delta_1^{(i)}\right) \tag{2.128a}$$

The average value of $<\delta_1^{(i)}\delta_1^{(j)}>$ is easily found to be

$$<\delta_1^{(i)}\delta_1^{(j)}>_{\text{av}} = \frac{kT\partial \log(\det G)}{\partial g_{ij}} = kTg_{ji}{}^{(-1)} \tag{2.128b}$$

$g_{ij}{}^{(-1)}$ being an element of the inverse of the matrix of G.

The formulas (2.128) are also valid for fluctuations in a system in contact with an enormous reservoir which can exchange the quantities $a^{(1)}, a^{(2)}, \ldots$ with the system of interest even though the reservoir may be a different kind of system. Similar formulas exist when the large volume V is divided into many smaller volumes (with each containing a large number of particles) and the a's are associated with different subregions of V.

14. The General Theory of Transport Coefficients

When the systems of an ensemble are driven by an external time-dependent force the phase space distribution function of the ensemble changes with time. The average response of some function $F(p,q)$ is given by

$$<F> = \int F(p,q)\rho(p,q;t)\,dp\,dq \tag{2.129}$$

where $\rho(p,q,t)$ is the appropriate time-dependent solution.

Although no phase space distribution function exists for a quantum-mechanical assembly, the response of a function F can be followed through the effect of the external driving force on the density matrix, which is defined below.

The wave function of an assembly can be expanded as a linear combination of a set of orthonormal functions $\{\psi_j\}$:

$$\psi(r,t) = \sum_n c_n(t)\psi_n(r) \tag{2.130}$$

A set of functions $\{c_n(t)\}$ exists for each assembly of an ensemble, and an ensemble average of $\{c^*_n(t)c_m(t)\}$ also exists. A matrix whose elements are given by the averages

$$\rho_{mn}(t) = <c^*_n(t)c_m(t)> \tag{2.131}$$

is called the density matrix of the ensemble. The quantum-mechanical generalization of (2.36) can be derived for the ensemble average of a dynamical variable F in terms of the matrix elements of F and the density matrix. The matrix elements of F are

$$F_{nm} = \int \psi^*_n(r)F\psi_m(r)\,dr \tag{2.132}$$

while the ensemble average of F at time t is

$$\begin{aligned} <F> &= <\int \psi^*(t,r) F \psi(t,r)\, dr> \\ &= \left\langle \int \sum_{nm} c^*_n(t) \psi^*_n(r) F c_m(t) \psi_m(r)\, dr \right\rangle \\ &= \sum_{nm} <c^*_n(t) c_m(t)> F_{nm} = \sum_{n_1 m} \rho_{mn} F_{nm} \\ &= \text{tr}\, \rho F \end{aligned} \tag{2.133}$$

The quantum analogue of the Liouville equation is a differential equation for the density matrix. It is obtained by differentiating (2.130) with respect to time, performing some elementary manipulations, and incorporating the Schroedinger equation. One finds

$$\frac{\partial \rho}{\partial t} = \frac{1}{i\hbar}[H,\rho] \tag{2.134}$$

H being the Hamiltonian of the system, and the bracket term represents the commutator of the H and ρ operators.

Now suppose that the Hamiltonian H can be divided into two parts

$$H = H_0 + H'(t) \tag{2.135}$$

where $H'(t)$ is due to an external driving force while H_0 is the Hamiltonian of our assembly of interacting particles in the absence of the external force. The density matrix of a canonical ensemble of such assemblies is

$$\rho_0 = Z^{-1} \exp(-\beta H_0) \tag{2.136}$$

It is easily verified that the differential equation (2.134) is equivalent to the integral equation

$$\rho(t) = \rho_0 - \frac{i}{\hbar} \int_{-\infty}^{t} e^{-i(t-t')H_0/\hbar} [H'(t'), \rho(t')] e^{i(t-t')H_0/\hbar}\, dt' \tag{2.137}$$

Now suppose

$$H'(t) = -AF(t) \tag{2.138}$$

where A is an operator and $F(t)$ is a scalar which characterizes the magnitude and temporal variation of an external driving force. For example, if a system is driven by an external electric field $E(t)$,

$$H'(t) = -\Sigma r_i e_i E(t) \tag{2.139a}$$

so that $A = \Sigma e_i r_i$ and $F(t) = E(t)$ (2.139b)

while (2.138) becomes

$$\rho(t) = \rho_0 + \hbar^{-1} \int_0^\infty \exp\left(-\frac{i\tau H_0}{\hbar}\right) [A, \rho(t-\tau)] \exp\left(\frac{i\tau H_0}{\hbar}\right) F(t-\tau)\, d\tau \tag{2.140}$$

This equation can be used to derive formulas for transport coefficients in very general systems.

The ensemble average of any dynamical variable B at time t is given by

$$<B> = \text{Tr}\, \rho B \tag{2.141}$$

The response of a variable B to the driving potential (2.138) is obtained by combining (2.140) and (2.141). If $<B>$ vanishes in the absence of $H'(t)$, one finds

$$<B> = \text{Tr}\, \frac{i}{\hbar} \int_0^\infty \exp\left(-\frac{i\tau H_0}{\hbar}\right) [A, \rho_0] \exp\left(\frac{i\tau H_0}{\hbar}\right) F(t-\tau) B\, d\tau \tag{2.142}$$

The time variation of the quantum-mechanical operator associated with a dynamical variable B is generally given by

$$B(\tau) = \exp\left(\frac{i\tau H}{\hbar}\right) B \exp\left(-\frac{i\tau H}{\hbar}\right)$$

if one defines $B(0) \equiv B$. To first order in F [see (2.135) and (2.138)]

$$B(\tau) = \exp\left(\frac{i\tau H_0}{\hbar}\right) B \exp\left(-\frac{i\tau H_0}{\hbar}\right) \tag{2.143}$$

so that to the same order in F

$$<B> = \text{Tr}\, \frac{i}{\hbar} \int_0^\infty [A, \rho_0] B(\tau) F(t-\tau)\, d\tau \tag{2.144}$$

In an initially canonical ensemble with ρ_0 given by (2.136), this equation can be transformed into a form symmetrical in A and B by employing the following identity of Kubo [21]:

$$\begin{aligned} [A, e^{-\beta H_0}] &= \frac{\hbar}{i} e^{-\beta H_0} \int_0^\beta e^{\beta H_0} \dot{A} e^{-\beta H_0}\, d\lambda \\ &= e^{-\beta H_0} \int_0^\beta e^{\beta H_0} [A, H_0] e^{-\beta H_0}\, d\lambda \end{aligned} \tag{2.145}$$

This can be proved by showing that the quantities on the left- and right-hand sides of (2.145) satisfy the same differential equation. They are identical at $\beta = 0$, both vanishing at that point.

Then

$$<B> = Z^{-1}\, \text{Tr} \int_0^\infty F(t-\tau)\, d\tau \int_0^\beta \exp[-(\beta-\lambda)H_0] \dot{A} \exp(\lambda H_0) B(\tau)\, d\lambda \tag{2.146}$$

so that, if $G \equiv \dot{A}$,

$$\begin{aligned} <B> &= Z^{-1}\, \text{Tr} \int_0^\infty F(t-\tau)\, d\tau \int_0^\beta e^{-\beta H_0} G(-i\hbar\lambda) B(\tau)\, d\lambda \\ &= Z^{-1}\, \text{Tr} \int_0^\infty F(t-\tau)\, d\tau \int_0^\beta e^{-iH_0\tau/\hbar} G(\tau) e^{iH_0\tau/\hbar} e^{-\lambda H_0} B(\tau) \exp[-(\beta-\lambda)H_0]\, d\lambda \end{aligned} \tag{2.147}$$

The reader can verify that the τ in $G(\tau)$ can be changed to $G(\tau+\tau')$ for any τ' provided that $B(\tau)$ is also changed to $B(\tau+\tau')$.

Equation (2.147) can be used to derive Ohm's law for an assembly of charges in an external electric field. Let

$$F(t) \equiv E(t) \equiv E e^{(i\omega+\alpha)t} \tag{2.148}$$

E being the magnitude and ω the circular frequency of the field in the limit $\alpha \to 0$. We keep α small but finite so that $F(t) \to 0$ as $t \to -\infty$. The potential energy of interaction between the external field and a set of particles with charges $\{e_j\}$ is

$$H'(t) = -\Sigma e_j r_j E(t) \tag{2.149}$$

so that in Eq. (2.138) $F(t)$ is given by (2.148) and

$$A = \Sigma e_j r_j \tag{2.150}$$

The electric current whose response we wish to find is

$$J = \Sigma e_j v_j = \Sigma e_j p_j / m_j \tag{2.151}$$

Also

$$G \equiv \dot{A} = \Sigma e_j \dot{r}_j = J \tag{2.152}$$

Since (2.148) implies that

$$F(t - \tau) = F(t) e^{-(i\omega+\alpha)\tau} \tag{2.153}$$

we find from (2.147) that the ensemble average of the current J is

$$\langle J \rangle = E(t)\sigma(\omega) \tag{2.154}$$

where as $\alpha \to 0$

$$\sigma(\omega) = \lim_{\alpha \to 0} Z^{-1} \operatorname{Tr} \int_0^\infty e^{-(i\omega+\alpha)\tau} \, d\tau \int_0^\beta e^{-\beta H_0} J(-i\hbar\lambda) J(\tau) \, d\lambda \tag{2.155}$$

is the electrical conductivity of a medium at circular frequency ω, Eq. (2.154) being Ohm's law. Equation (2.155) states that the conductivity is an average over the autocorrelation function of a current at a real time with itself at an imaginary time. By changing the contours of integration one can reexpress σ as an integral over a more standard autocorrelation function. Indeed one finds

$$\sigma(\omega) = \lim_{\alpha \to 0} \int_0^\infty e^{-(\alpha+i\omega)\tau} \, d\tau \int_{-\infty}^\infty \Gamma(\tau - t) \psi_{JJ}(t) \, dt \tag{2.156}$$

$$\text{where } \Gamma(t) = \frac{2}{\pi\hbar} \log \coth \left(\frac{\pi|t|}{2\beta\hbar} \right) \tag{2.157a}$$

$$\begin{aligned} \text{and} \quad \psi_{JJ}(t) &= \tfrac{1}{2} \operatorname{Tr} Z^{-1} e^{-\beta H_0} [J(0)J(t) + J(t)J(0)] \\ &= \tfrac{1}{2} \langle J(0)J(t) + J(t)J(0) \rangle \end{aligned} \tag{2.157b}$$

the average being taken over a canonical ensemble at time $\tau = 0$. In view of (2.151), (2.157b) is a superposition of auto and joint momentum correlation functions.

In the classical limit as $\hbar \to 0$

$$\Gamma(t) = \beta\delta(t) \tag{2.158}$$

so that

$$\begin{aligned} \sigma(\omega) &= \lim_{\alpha \to 0} \beta \int_0^\infty e^{-(\alpha+i\omega)\tau} \psi_{JJ}(\tau) \, d\tau \\ &= \lim_{\alpha \to 0} \beta \int_0^\infty e^{-(\alpha+i\omega)\tau} \langle J(0)J(\tau) \rangle \, d\tau \end{aligned} \tag{2.159}$$

Similar expressions can be obtained for other transport coefficients such as the diffusion constant viscosity, thermal conductivity, thermal diffusion, etc. The appropriate interpretations for the current $J(t)$ in the first three cases are the particle number current, the velocity current, and the thermal current, respectively, while in the case of thermal diffusion Eq. (2.159) becomes a joint correlation function between the particle number current and the thermal current. The relation between transport coefficients and current correlation functions was first noted by M. S. Green. The analysis given here follows the work of Kubo [21]. Reviews of this subject have been given in [22] and [23].

References

1. Gibbs, J. W.: "Elementary Principles of Statistical Mechanics," Yale University Press, New Haven, 1902.
2. Tolman, R. C.: "Principles of Statistical Mechanics," Oxford University Press, New York and London, 1938.
3. Fowler, R. H.: "Statistical Mechanics," 2d ed., Cambridge University Press, New York and London, 1936.
4. Mayer, J. E., and M. G. Mayer: "Statistical Mechanics," Wiley, New York, 1940.
5. Khinchin, A. I.: "Mathematical Foundations of Statistical Mechanics," Dover, New York, 1949.
6. Ehrenfest, P., and T. Ehrenfest: *Encykl. math. Wiss. IV*, 2, ii. Heft 6, 1911.
7. Kennard, E. H.: "Kinetic Theory of Gases," McGraw-Hill, New York, 1938.
8. Chapman, S., and T. G. Cowling: "The Mathematical Theory of Non-uniform Gases," 2d ed., Cambridge University Press, New York and London, 1952.
9. Grad, H.: "Kinetic Theory and Statistical Mechanics," Lecture Notes, New York University, 1950; and *Commun. Pure and Appl. Math.*, vol. 2, p. 331, 1949.
10. Herzfeld, K.: *Handbuch und Jahrbuch der Chem. Phys.*, vol. 3, part 2, p. 96, 1939.
11. Wang Chang, C., and G. Uhlenbeck: "Reports on Transport Phenomenon in Very Dilute Gases," University of Michigan, 1948–1953.
12. Ter Haar, D.: "Elements of Statistical Mechanics," Rinehart, New York, 1954.
13. Schmidt, H.: *Z. Physik*, **134**: 430 (1953).
14. Newell, G., and E. Montroll: *Revs. Mod. Phys.*, **25**: 353 (1953).
15. De Boer, J.: *Repts. Prog. Phys.*, **12**: 305 (1948).
16. Pauli, W.: "Statistische Mechanik," lecture notes, Zurich, 1951.
17. Van Hove, L.: *Physica*, **15**: 951 (1949).
18. Huang, K.: "Statistical Mechanics," Wiley, New York, 1963.
19. Uhlenbeck, G., and J. W. Ford: "Lectures in Statistical Mechanics," American Mathematical Society, New York, 1963.
20. Ruelle, D.: "Lectures in Theoretical Physics," vol. 5, University of Colorado, Boulder, Colo., 1963.
21. Kubo, R.: *J. Phys. Soc. Japan*, **12**: 570 (1957), and "Lectures in Theoretical Physics," vol. 1, University of Colorado, Boulder, Colo., 1958.
22. Montroll, E.: "Lectures in Theoretical Physics," vol. 3, University of Colorado, Boulder, Colo., 1961.
23. Zwanzig, R.: *Ann. Rev. Phys. Chem.*, **16**: 67 (1965).

Chapter 3

Thermometry and Pyrometry

By JAMES F. SWINDELLS, National Bureau of Standards

A meaningful measurement of temperature requires some instrument for indicating temperatures and a scale so defined that stated values of temperature have universal meaning. In Part 5, Chap. 1 the bases for several scales that do not depend on the physical properties of a particular thermometric substance are given. These scales are to be contrasted with those which depend upon the change of a specific property of a substance with temperature, e.g., the expansion of mercury in glass. Historically, the latter scales appeared long before those which were independent of the thermometric substance. Apparently beginning with the conception of a thermometer by Galileo (ca. 1592) many early scientists have contributed to the thermometer's development [1]†.

Most of these thermometers were based upon fluid expansion. Credit for each contribution to the art cannot be authoritatively assigned because many of the records were unpublished, and in some cases correspondence between scientists is our only source of information. It is certain, however, that among the most important contributions were those of Fahrenheit, Réaumur, and Celsius. Fahrenheit's work in the early eighteenth century culminated (1717) in a mercury-in-glass thermometer whose scale was fixed by assigning a temperature of 32° to the ice point and 96° to the normal temperature of the human body [2]. The zero of this scale came close to the temperature of an ice–ammonium chloride–water mixture, and the boiling point of water was at about 212°.

In 1730 Réaumur [3] reported on his alcohol-filled thermometers which were calibrated by assigning 0 and 80° to the freezing and boiling points of water. Later (1742) Anders Celsius published a paper [4] on a mercury-filled thermometer whose scale was also based upon the freezing and boiling points of water but with the interval between the two being divided into 100°. The Fahrenheit, Réaumur, and Celsius scales have survived to the present time. The Fahrenheit scale is used in English-speaking countries; Réaumur is still used in some parts of northern Europe; Celsius is used in non-English-speaking countries and by scientists in all countries.

It should be emphasized that temperatures on these early scales depended not only on the values assigned to the fixed points but also on the behavior of the thermometric substance between and beyond the fixed points. A similar situation exists today with the International Practical Temperature Scale of 1948, discussed in Sec. 2.

† Numbers in brackets refer to References at end of chapter.

1. Thermodynamic Temperature Scale

Ideally, only one temperature scale would be used for all measurements, and that would be an *absolute thermodynamic scale*.

The word "absolute" is used to indicate that all finite temperatures on the scale define thermal states hotter than that of the zero of the scale (absolute zero); and the word "thermodynamic" indicates that the temperature scale conforms to the laws of thermodynamics and is independent of the properties of any thermometric substance. The present internationally accepted thermodynamic scale is called the Kelvin scale. In a resolution adopted in 1954, the Tenth General Conference on Weights and Measures redefined the size of the Kelvin degree by selecting the triple point of water as the sole defining fixed point, assigning to it the temperature 273.16°K [5]. Previous to this time the size of the degree on the Kelvin scale had been defined (1854) by assigning 100° to the temperature interval between the ice point and the steam point. The new Kelvin degree, however, does not differ from the earlier degree by an amount that has been definitely determined experimentally up to the present time.

Although the Kelvin scale has thus been exactly defined with ideal simplicity, its accurate realization is a difficult experimental undertaking. In principle, this temperature scale may be deduced from any thermodynamic formula based on the first and second laws of thermodynamics. The choice of a method in a particular instance is usually governed by the temperature range of interest, the accuracy required, and any specific restrictions inherent in the situation of the measurement. This has led to a wide variety of approaches to the temperature measurement, but many of them are so restricted in their application that only a few will be discussed here. By far the most widely used method has been gas thermometry. By painstaking work, fairly accurate values have been determined for thermodynamic temperatures ranging from those of liquid helium through the freezing point of gold (1336°K). For the Kelvin scale above the gold point, techniques based on the laws of radiation are commonly used. This part of the scale is usually based upon a value for the temperature of the gold point as determined by gas thermometry.

Realization of the Thermodynamic Temperature Scale by Gas Thermometry. In Part 5, Chap. 1 it is shown that the perfect gas temperature scale is the same as the corresponding thermodynamic scale. From *Boyle's law,* which states that the pressure-volume product pV of a definite mass of a perfect gas is constant at a definite temperature T, and *Joule's law,* which states that the energy of a definite mass of a perfect gas is independent of the volume when the temperature is constant, the relation

$$pV = nRT \tag{3.1}$$

is derived. In this equation n is the number of moles of perfect gas, R is the universal gas constant, and T is the Kelvin temperature.

In principle, each gas-thermometer measurement is based upon observations of p, V, and T in two thermodynamic states. If p_0V_0 and pV are the pressure-volume products of a definite mass of perfect gas at temperatures T_0 and T, Eq. (3.1) gives

$$\frac{pV}{p_0V_0} = \frac{T}{T_0} \qquad n = \text{const} \tag{3.2}$$

Equation (3.2) is the basis for gas thermometry.

There are no perfect gases, but for a few real gases, at pressures up to 2 atm or so, sufficiently high accuracy is attained by replacing Eq. (3.1) with the simple equation of state

$$pV = nRT(1 + Bp) \tag{3.3}$$

where B, usually called the second virial coefficient, is a function of temperature. Using this equation, with n constant in the two states, Eq. (3.2) becomes

$$\frac{T}{T_0} = \frac{pV}{p_0V_0}\frac{1 + B_0p_0}{1 + Bp} \qquad n = \text{const} \tag{3.4}$$

where B_0 and B are values of the virial coefficient at T_0 and T. If T_0 is taken as the temperature assigned to the triple point of water on the Kelvin scale (the defining fixed point), Eq. (3.4) permits one to determine experimentally higher and lower temperatures on the scale within the range of the gas thermometer.

Four different methods of gas thermometry may be used [6, 7]: the constant-volume method, the constant-pressure method, the constant-bulb-temperature method, and the pV-isotherm method. In all four methods the accuracy with which an unknown temperature can be determined is limited by the accuracy with which the pressure-volume products can be measured. The discrimination of small differences in temperature is exacting and becomes more so at higher temperatures; for instance, at the steam point a change of a thousandth of a degree changes the pressure-volume product of the gas by only 1 part in about 373,000.

In the idealized constant-volume method (constant-density method), n moles of a perfect gas is contained in a bulb of invariant volume V. A pressure indicator of negligible volume is connected by a tube of negligible volume to the thermometer bulb. When the bulb is first at the unknown temperature T and then at the reference temperature $T_0 = 273.16°\text{K}$, the corresponding pressures, p and p_0, are measured. The ratio of the Kelvin temperatures of the gas in the bulb is given by the equation

$$\frac{T}{T_0} = \frac{p}{p_0} \tag{3.5}$$

In the idealized constant-pressure method, a bulb of invariant volume V is connected by a tube of negligible volume to a pressure indicator of negligible volume and a pipette of variable volume which is thermostated at the reference temperature, $T_0 = 273.16°\text{K}$. The gas pressure is the same in the bulb and the pipette. The bulb contains all n moles of the gas at the lower temperature. When the temperature is changed to a higher temperature, part of the gas is transferred into a volume ΔV in the pipette in order to keep the pressure constant in the system. The accounting for mass in the system reduces to the equations

$$\frac{V}{T} = \frac{V - \Delta V}{T_0} \qquad \text{when } T > T_0 \tag{3.6}$$

and

$$\frac{V}{T} = \frac{V + \Delta V}{T_0} \qquad \text{when } T < T_0 \tag{3.7}$$

In the idealized constant-bulb-temperature method the bulb is connected to the pressure indicator and pipette, as in the previous method, but the bulb temperature is not changed during a set of measurements. The pressure p_0 is first measured when all the gas is in the bulb. Some of the gas is then transferred into the pipette until the pressure p is reduced to one-half, $p/2$. At this pressure the accounting for mass in the system reduces to the equation

$$\frac{V}{T} = \frac{\Delta V}{T_0} \tag{3.8}$$

In the idealized pV-isotherm method the gas constant R is assumed to be known exactly. Measuring the pressure p, the volume V, and the number of moles, n, in the bulb gives the Kelvin temperature by means of the equation

$$T = \frac{pV}{nR} \tag{3.9}$$

The value of R, however, must have been determined by similar measurements for $T_0 = 273.16°\text{K}$; hence the ratio of the unknown to the defined temperature has been determined indirectly. When a real gas is used, the pV isotherms are measured for different values of n and extrapolated to zero pressure or zero density.

The experimental gas thermometer necessarily differs from the idealized instrument in many details. For high accuracy, a number of corrections must be made for these departures from ideality. Fortunately, most of the corrections are small, but some are difficult to evaluate with certainty. Some corrections can be minimized by the adoption of certain refinements in the design of the apparatus, but few can be entirely eliminated.

Temperature Scales Based on Black-body Radiation. Certain characteristics of black-body radiation may be deduced from thermodynamic considerations alone. One is that the total intensity of the radiation is proportional to the fourth power of the Kelvin temperature of the body. Another is that $\lambda_m T$ is a universal constant where λ_m is the wavelength at which the maximum power per unit range of wavelength is radiated from a body at Kelvin temperature T.

The radiation law is based upon the quantum mechanisms of emission and absorption of radiation and upon the assumption of a Maxwell-Boltzmann distribution among energy levels in a system in thermal equilibrium. This is *Planck's law.*

$$J = 2hc^2\lambda^{-5}(e^{hc/\lambda kT} - 1)^{-1} \quad \text{ergs/sec-cm}^2 \tag{3.10}$$

where J is the power radiated, per unit projected area, per unit solid angle, per unit wavelength interval, at wavelength λ, from a black body at Kelvin temperature T. The presently accepted value of the constant hc/k (Planck's second radiation constant c_2) is 1.43879 ± 0.00019 cm-deg [8]. The following temperature scales are based upon the characteristics of black-body radiation using various temperature parameters, or measured quantities.

Monochromatic. The temperature parameter is the ratio of the power radiated from a black body at temperature T to that radiated from a black body at a fixed temperature T_0 at a single wavelength.

$$R_m = \frac{J(T,\lambda)}{J(T_0,\lambda)} \tag{3.11}$$

Two-color. The parameter is the ratio of the power radiated from a black body at two wavelengths.

$$R_c = \frac{J(T,\lambda_1)}{J(T,\lambda_2)} \tag{3.12}$$

Wien Displacement. This scale is based upon the constance of $\lambda_m T$. The parameter is λ_m.

$$\lambda_m T = \lambda_{m_0} T_0 \tag{3.13}$$

where T_0 is a fixed temperature. From Planck's law the value of the universal constant may be determined. From this results the scale

$$\lambda_m T = \frac{c_2}{4.965} = 0.28978 \pm 0.00004 \text{ cm-deg [8]} \tag{3.14}$$

Total Radiation. Proportionality of radiated power to the fourth power of the temperature leads to a scale having the parameter $\int_0^\infty J\,d\lambda$,

$$T^4 \int_0^\infty J(T_0,\lambda)\,d\lambda = T_0^4 \int_0^\infty J(T,\lambda)\,d\lambda \tag{3.15}$$

Again Planck's law gives an absolute value for the constant of proportionality,

$$\int_0^\infty J(T,\lambda)\,d\lambda = \frac{\sigma}{\pi} T^4 \tag{3.16}$$

where [8]

$$\sigma = (5.6697 \pm 0.0029) \times 10^{-5} \text{ erg/(cm}^2\text{)(sec)(deg}^4\text{)}$$

Scales Based on Black-body Radiation. A general expression for the radiation emitted by a real body and detected by a real instrument is

$$D_{\lambda_i} = \int^\Omega \int^A \int_0^\infty J_\lambda e_\lambda S_{\lambda_i}\,d\lambda_i\,dA\,d\Omega \tag{3.17}$$

where D_{λ_i} = response of instrument adjusted to select a range of frequencies in neighborhood of λ_i

e_λ = monochromatic emissivity of emitting object

S_{λ_i} = sensitivity function of instrument when set at λ_i; width of band of frequencies in which $S\lambda_i$ is appreciable is an inverse measure of resolving power of instrument

dA = an elementary area on source projected normal to a pencil of rays of solid angle $d\Omega$

If e_λ and S_{λ_i} are independent of position (dA) and ray direction ($d\Omega$) within the integral limits set by the geometry of the instrument optics, the integrations with respect to A and Ω are separate and

$$D_{\lambda_i} = \int_0^\infty J_\lambda e_\lambda S_{\lambda_i}\,d\lambda \int^A dA \int^\Omega d\Omega \tag{3.18}$$

In all the scales only ratios of D's are involved and hence $\int dA \int d\Omega$ cancel out. These factors are therefore dropped from now on. The advantages of small aperture and small area are apparent. In what follows the limits of the integration with respect to λ will be zero to infinity.

As the first practical scale consider the *optical pyrometer,* which uses the *monochromatic scale.* Equation (3.11) will apply but in the form

$$R = \frac{\int J(T,\lambda)S_{\lambda_i}\,d\lambda}{\int J(T_0,\lambda)S_{\lambda_i}\,d\lambda} = \frac{J(T,\lambda_e)}{J(T_0,\lambda_e)} \tag{3.19}$$

A determination of T from the parameter R requires knowledge of the effective wavelength λ_e. If S_{λ_i} were a very narrow function of λ, then λ_e would be equal to λ_i. For an S_{λ_i} wide enough so that J is not essentially constant within its range, the effective wavelength can be calculated from (3.19), assuming a value of T and carrying out the integrations. Thus λ_e will be a function of T and this must be taken into account in calibrating the instrument.

In measuring temperatures with the optical pyrometer, a null, rather than a ratio, method is used. If the object whose temperature is to be measured has an emissivity e, less than unity, then

$$\int e_\lambda J(T,\lambda)S_{\lambda_i}\,d\lambda = \int J(T',\lambda)S_{\lambda_i}\,d\lambda \tag{3.20}$$

where T' is the temperature indicated by the instrument, which has been calibrated against a black body. T' is the brightness temperature of the object.

In the *two-color* method, Eq. (3.12) takes the form

$$R = \frac{\int J(T,\lambda)e_\lambda S_{\lambda_a}\,d\lambda}{\int J(T,\lambda)e_\lambda S_{\lambda_b}\,d\lambda} \tag{3.21}$$

A *radiation pyrometer* uses the parameter

$$D = \int J S_\lambda\,d\lambda \tag{3.22}$$

and has no defined scale but is calibrated against a black body or a calibrated source. In this instrument S_λ is made very wide to enhance its sensitivity.

Measurement of Gas Temperatures Using Black-body Radiation Laws and Kirchhoff's Law. The emissivity of a gas is a rapidly varying function of wavelength. It depends also upon the path length through which the observed radiation passes and upon the temperature, pressure, and composition of the gas. In general it is necessary to determine the emissivity for each measurement of temperature. This is accomplished by taking advantage of Kirchhoff's law, which states that the monochromatic absorptivity of an object in thermal equilibrium is equal to its monochromatic emissivity, at all wavelengths. The validity of Kirchhoff's law rests upon the second law of thermodynamics. The observed radiation traversing a body of gas from a source of intensity I_0 (ergs/sec-cm²) will produce an instrumental response

$$D_1 = D_0 - \int I_0 e_\lambda S_{\lambda_i}\, d\lambda \tag{3.23}$$

where $D_0 = \int I_0 S_{\lambda_i}\, d\lambda$ is the response due to the unabsorbed radiation from the source. The radiation emitted by the gas will produce a response

$$D_2 = \int J e_\lambda S_{\lambda_i}\, d\lambda \tag{3.24}$$

If I_0 and J are essentially constant in the wavelength range in which S_λ is appreciable, then

$$\frac{D_2 D_0}{D_0 - D_1} = \frac{J I_0 \int e_\lambda S_{\lambda_i}\, d\lambda \int S_{\lambda_i}\, d\lambda}{I_0 \int e_\lambda S_{\lambda_i}\, d\lambda} = J(\lambda_i) \int S_{\lambda_i}\, d\lambda \tag{3.25}$$

The quantity $\int S_{\lambda i}\, d\lambda$ may be determined by calibrating the instrument against a black body of known temperature. Separation of the quantities D_1 and D_2 may be physically accomplished by modulation of I_0.

For a similar technique involving a null method, D_1 and D_2 are not separated and I_0 is adjusted until $D_1 + D_2 = D_0$ or

$$[J(\lambda_i) - I_0(\lambda_i)]\int e_\lambda S_{\lambda_i}\, d\lambda = 0 \tag{3.26}$$

whence J may be determined by calibration of I_0.

A variation of the null method is the line reversal method. Here, a continuum of intensity I_0 is observed through the body of gas at two wavelengths at which the emissivities are different. The difference between the observed intensities at the two wavelengths will be

$$\Delta D = \int S_{\lambda_a}[I_0 + (J - I_0)e_\lambda]\, d\lambda - \int S_{\lambda_b}[I_0 + (J - I_0)e_\lambda]\, d\lambda \tag{3.27}$$

If I_0 and J are essentially constant in the range covered by S_{λ_a} and S_{λ_b} and if $\int S_{\lambda_a}\, d\lambda = \int S_{\lambda_b}\, d\lambda$, then

$$\Delta D = (J - I_0)\left(\int_0^\infty S_{\lambda_a} e_\lambda\, d\lambda - \int_0^\infty S_{\lambda_b} e_\lambda\, d\lambda\right) \tag{3.28}$$

Adjustment of I_0 to make ΔD vanish will enable J to be determined.

Measurement of Gas Temperature Based on Quantum Mechanics. Under certain conditions it is possible to determine the temperature of a gas in equilibrium by observation of individual spectral lines and by making use of quantum-mechanical theory of emission and absorption of radiation by gas molecules. The instrumental response due to the observed power radiated in a given line (index i) with the instrument set at $\lambda_{i'}$ is

$$D_{i,i'} = \int J e_{\lambda_i} S_{\lambda_{i'}}\, d\lambda = J(\lambda_i) \int e_{\lambda_i} S_{\lambda_{i'}}\, d\lambda$$

$$e_{\lambda_i} = 1 - \exp\left\{-Q^{-1} g_{ui} \frac{A_i \lambda_i^4}{8\pi c} \left[\exp\left(-\frac{E_{l_i}}{kT}\right) - \exp\left(\frac{-E_{u_i}}{kT}\right)\right] \int_0^l N_0 f_{\lambda i}\, dl\right\} \tag{3.29}$$

where Q = partition function of molecule
g_{u_i} = spatial multiplicity of upper state involved in transition giving rise to line i
A_i = Einstein probability of transition per sec
E_{u_i}, E_{l_i} = energies of upper and lower states, respectively
N_0 = density of molecules of species giving rise to observed line
f_{λ_i} = shape function of the line, normalized so that

$$\int_0^\infty f_{\lambda_i}\, d\lambda = 1$$

l = path length followed by observed ray

No Self-absorption. If e_λ is sufficiently small over the entire line so that the exponential in e_λ may be approximated by unity minus the exponent, then D simplifies to

$$D_{i,i'} = \int S_{\lambda_{i'}} Q^{-1} g_{u_i} \frac{A_i ch}{4\pi\lambda_i} \exp\left(\frac{-E_{u_i}}{kT}\right) \int_0^l N_0 f_{\lambda_i}\, dl\, d\lambda \tag{3.30}$$

If we further assume that $f_{\lambda i}$ is independent of l, as it will be if temperature and pressure are constant along l, then

$$D_{i_i} = \int S_{\lambda_{i'}} f_{\lambda_i} B_i\, d\lambda = B_i(\lambda_i) \int S_{\lambda_{i'}} f_{\lambda_i}\, d\lambda \tag{3.31}$$

with

$$B_i = Q^{-1} g_{u_i} \frac{A_i ch}{4\pi\lambda_i} \exp\left(\frac{-E_{u_i}}{kT}\right) \int_0^l N_0\, dl \tag{3.32}$$

From the ratio of the values of B for two different lines the temperature may be determined.

$$\frac{B_a}{B_b} = \frac{g_{u_a} A_a \lambda_b}{g_{u^b} A_b \lambda_a} \exp \frac{E_{u_b} - E_{u_a}}{kT} \tag{3.33}$$

Certain of the quantities in Eq. (3.32) are the same for both lines and the remaining ones, appearing in Eq. (3.33), must be independently known, either from theory or experiment.

The method of determination of B depends upon the resolving power of the instrument. For high resolution, that is, $S_{\lambda_{i'}}$ much narrower than f_{λ_i}, for the line peaks ($\lambda_i = \lambda_{i'}$)

$$\frac{D_{aa'}}{D_{bb'}} = \frac{B_a f_{\lambda_a}(\lambda_a) \int S_{\lambda_a}\, d\lambda}{B_b f_{\lambda_b}(\lambda_b) \int S_{\lambda_{b'}}\, d\lambda} \tag{3.34}$$

The ratio of the instrumental integrals $\int S\, d\lambda$ may be determined experimentally, but determination of the ratio of the f's depends upon a knowledge of the nature of the line broadening, that is, the shape of the f's. However, if the integration over the instrument wavelengths $\lambda_{a'}$

$$\int D_{aa'}\,d\lambda_{a'} = B_a \int f_{\lambda_a}(\lambda_{a'}) \int S_{\lambda_{a'}}\,d\lambda\,d\lambda_{a'} = B_a \int S_{\lambda_{a'}}\,d\lambda \tag{3.35}$$

is carried out, then f is eliminated by the property $\int f_\lambda\,d\lambda = 1$, and the fact that $\int S_{\lambda_{a'}}\,d\lambda = \int S_{\lambda_{a'}}\,d\lambda_{a'}$ is a constant over the limited region of $\lambda_{a'}$ involved.

For an intermediate resolution, that is, $S_{\lambda_{a'}}$ about the same width as $f_{\lambda_{a'}}$, we have

$$\int D_{aa'}\,d\lambda_{a'} = B_a \int\int f_{\lambda_a} S_{\lambda_{a'}}\,d\lambda\,d\lambda_{a'} = B_a \int S_{\lambda_{a'}}\,d\lambda_{a'} \tag{3.36}$$

For low resolution, $S_{\lambda_{a'}}$ much wider than f_{λ_a} but still narrow enough so that B_a is constant within the range of $S_{\lambda_{a'}}$, the correct expression at the line peak is

$$D_{aa'} = S_{\lambda_{a'}}(\lambda_a) B_a \int f_{\lambda_a}\,d\lambda = S_{\lambda_{a'}}(\lambda_a) B_a \qquad \lambda_a = \lambda_{a'} \tag{3.37}$$

Low resolution would appear desirable since no integration is required to eliminate f. However, a practical limitation is imposed by the necessity of completely resolving individual lines. In all of these cases if $\lambda_a \approx \lambda_b$ then, for practical purposes,

$$\int f_{\lambda_a} S_{\lambda_{a'}}\,d\lambda = \int f_{\lambda_b} S_{\lambda_{b'}}\,d\lambda$$

and the ratio of B's is equal to the ratio of D's at the line peaks ($\lambda_{a'} = \lambda_a$, $\lambda_{b'} = \lambda_b$) without recourse to integration with respect to $\lambda_{a'}$.

It is important to ascertain that the self-absorption is indeed small enough for the approximation for small e_λ to be valid. This fact may be determined by high resolution absorption measurements or in some cases by theoretical calculation.

Self-absorption. In cases where e_λ is not sufficiently small, but it is possible to find a pair of lines of equal intensity and approximately equal wavelength, so that $J(\lambda_a) \approx J(\lambda_b)$, then, from Eqs. (3.29),

$$\int e_{\lambda_a} S_{\lambda_{a'}}\,d\lambda = \int e_{\lambda_b} S_{\lambda_{b'}}\,d\lambda \quad (\lambda_{a'} = \lambda_a,\ \lambda_{b'} = \lambda_b,\ \lambda_a \approx \lambda_b) \tag{3.38}$$

Since the line shapes are the same, the peak emissivities must be the same and, again from Eqs. (3.29),

$$\frac{g_{u_a} A_a}{g_{u_b} A_b} = \exp \frac{E_{l_a} - E_{l_b}}{kT} = \exp \frac{E_{u_a} - E_{u_b}}{kT} \tag{3.39}$$

In the absence of such isointensity, isofrequency pairs of lines, it is possible to apply Eqs. (3.29) to a series of lines, choosing T by successive approximation until a fit is obtained. Since f_λ may not be removed by integration, it is necessary to use high resolution and to know the manner in which the peak value of f_λ varies from line to line.

Miscellaneous Methods of Measuring Thermodynamic Temperatures. Through the years, a number of thermodynamic relations have been suggested for the measurement of temperature. Examples of some of those which have met with some success are mentioned below.

Noise Thermometer. In principle the thermodynamic temperature may be deduced from any thermodynamic formula based on the first and second laws of thermodynamics. One such method uses the discovery by J. B. Johnson [9] of a nonperiodic alternating electromotive force in conductors, related in a simple manner to the temperature of the conductor and due to the thermal agitation of the electrons in the conductor. Nyquist [10], using thermodynamic reasoning, related the mean-square voltage developed across a conductor to its Kelvin temperature by means of the following expression:

$$\bar{V}^2 = \int_{f_1}^{f_2} 4kT\,\mathrm{Re}\,[Z(f)]\,df \tag{3.40}$$

where $\bar{V}^2$ is the mean-square thermal voltage developed in a frequency interval $(f_2 - f_1)$, Re $[Z(f)]$ is the real part of the frequency-dependent impedance of the two-terminal passive network, and k is Boltzmann's constant.

This voltage fluctuation, called Johnson noise, which places a limit on the ultimate sensitivity of amplifiers, provides a method for measuring thermodynamic temperature. If two networks having impedances Z_1 and Z_2, respectively, are maintained at different temperatures T_1 and T_2, the mean-square voltages $\bar{V}_1^2$ and $\bar{V}_2^2$ generated by each over the same bandwidth $f_2 - f_1$ may be made equal by adjusting either Re $[Z_1]$ or Re $[Z_2]$. When this is done, we have

$$\int_{f_1}^{f_2} 4kT_1\,\mathrm{Re}\,[Z_1]\,df = \int_{f_1}^{f_2} 4kT_2\,\mathrm{Re}\,[Z_2]\,df \tag{3.41}$$

or

$$\frac{T_2}{T_1} = \frac{\int_{f_1}^{f_2} \mathrm{Re}\,[Z_1(f)]\,df}{\int_{f_1}^{f_2} \mathrm{Re}\,[Z_2(f)]\,df} \tag{3.42}$$

If Re $[Z]$ can be expressed, for each network, as the product of R, a frequency-independent measurable resistance in the network times $P(f)$, where $P(f)$ gives the frequency dependence of the network, then

$$\frac{T_2}{T_1} = \frac{R_1}{R_2} \frac{\int_{f_1}^{f_2} P_1(f)\,df}{\int_{f_1}^{f_2} P_2(f)\,df} \tag{3.43}$$

Furthermore, if $P_1(f)$ and $P_2(f)$ can be arranged to be equal over the band $f_2 - f_1$, then

$$\frac{T_2}{T_1} = \frac{R_1}{R_2} \tag{3.44}$$

from which the ratio of the absolute thermodynamic temperatures of the two networks can be obtained by a simple resistance ratio measurement.

Several thermometers have been developed based on this principle [11–15], but, in general, an accuracy better than 1 per cent has been found very difficult to achieve.

Acoustical Thermometer. The velocity of sound, W, in a real gas may be related to the thermodynamic temperature by the pressure expansion,

$$W^2 = \left(\frac{C_p}{C_v}\right)_{p=0} \frac{R}{M} T[1 + \alpha p + \beta p^2 + \cdots] \tag{3.45}$$

in which $(C_p/C_v)_{p=0}$ is the ratio of the specific heats at zero pressure, R is the universal gas constant, M is the molecular weight of the gas, p is the gas pressure, and α and β are constants into which the virial coefficients enter [16]. This well-known relation has served as the basis for thermometry at both high and low temperatures.

Barrett and Suomi [17] determined atmospheric temperatures from measurements of sound velocity in the air. The travel times of sound pulses between two transducers at a measured separation distance (ca. 40 cm) provided the basic information.

In most acoustical thermometers the design requires both the generating and receiving transducers to be at the temperature being measured. Apfel [18], however, developed a system for use in nuclear reactors in which the transducers are remote from the region where the temperature is measured. The apparatus consists of an acoustically resonant cavity coupled to a transceiver by a long small-diameter tube. The transceiver is used to generate and detect acoustic waves traveling through the coupling tube to the cavity which is at the temperature to be measured. The temperature is determined from a measurement of the resonant frequency of the gas in the cavity. Temperature errors of less than 1 per cent are claimed for measurements up to about 1300°K.

Recently an accurate acoustical thermometer was developed by Plumb [19, 20] for use primarily in the temperature range 2 to 20°K. In this apparatus the wavelength is measured, and, with a known generating frequency, the velocity is calculated. Helium gas is contained in a small cylindrical cell of variable length maintained at the temperature to be measured. A single quartz crystal is used as both the emitter and receiver of the ultrasonic waves. The waves traverse the gas and are reflected back to the quartz crystal from the face of a piston whose position is controlled by a micrometer. When the reflecting surface is moved toward or away from the crystal, the gas column is set in resonance each time the separation becomes an integral number of half wavelengths. The resonant load on the quartz crystal results in an impedance change which may be detected by voltage measurements across the crystal. A precision of 0.1 per cent or better has been achieved at temperatures near the helium and hydrogen boiling points. An analysis of possible systematic errors has not yet been completed (1965), but it is expected that the accuracy of absolute measurements of thermodynamic temperatures will be of the same order as the precision of the instrument.

Nuclear Resonance Thermometer. Following the suggestion of Dean and Pound in 1952 [21], several investigations have been made of the possibility of realizing a temperature scale based upon the temperature dependence of the resonant frequency of pure nuclear quadrupole splitting of the energy levels in a solid. A basic parameter characterizing this interaction is the quadrupole coupling constant

$$\eta_k = eQq \tag{3.46}$$

where e = electron charge
Q = quadrupole moment
q = gradient of electric field

The frequency corresponding to pure quadrupole resonance absorption depends upon the temperature and certain second-order effects. A change in temperature produces a change in the gradient of the electric field due to thermal excitation of the crystal lattice of the substance. Since, through use of modern methods for frequency measurement, η_k can be determined with high accuracy, it becomes feasible to calculate the temperature with an accuracy depending primarily on a knowledge of the other parameters in Eq. (3.46).

Benedek and Kushida [22] were the first to investigate the possibilities of the method, using the resonance of Cl^{35} in $KClO_3$ at temperatures from 77 to 300°K. Using the same salt, Vanier [23] extended the investigation down to 15°K. Precisions found by these investigators were ±0.05° at 20°K, ±0.005° at 60°K, and ±0.002° at 273°K. The development of this method is continuing, but on the basis of work reported up to 1965, it is concluded that the method will not independently yield thermodynamic temperatures, but, with calibration, such a thermometer can provide sensitive temperature measurements.

2. Practical Temperature Scales

The accurate realization of the Kelvin Thermodynamic Temperature Scale requires apparatus and skill such as are found only in a few laboratories specializing in thermometric researches. It long ago became evident, therefore, that, for uniformity in the precise measurement of temperature, a practical scale was needed which could be used for stating temperatures on the same basis in laboratories all over the world. This need led to discussions, extending over nearly two decades, between the national laboratories of Germany, Great Britain, and the United States. As a result, an "International Temperature Scale" was adopted by the Seventh General Conference on Weights and Measures in 1927 [24]. It was understood that it would not replace the thermodynamic scale but would represent it in a practical manner with sufficient accuracy to serve the everyday needs of scientific and industrial laboratories. It was recommended, however, that studies on temperature scales be continued. As a result, the scale was revised in 1948 [25]. Changes in the basis on which the scale is defined were adopted in 1960 [26]. In 1960, the name of the scale was changed to the International Practical Temperature Scale of 1948. The changes in the definition of the scale adopted in 1960 do not change the value of any temperature on the 1948 scale by as much as the experimental error of measurement.

It should be emphasized that in 1948 the General Conference on Weights and Measures decided to designate the degree as degree Celsius in place of degree centigrade [27]. While more and more editors are coming to insist on the use of the name Celsius for the scale, many scientists are still unaware of this change in name or, for some reason, are reluctant to accept it. To make the nomenclature of temperature uniform in all countries this reluctance should be discouraged.

The International Practical Temperature Scale. Temperatures on the International Practical Temperature Scale of 1948 are expressed in degrees Celsius, designated by °C or °C (Int. 1948), and are denoted here by the symbol t or t_{int}

The International Practical Temperature Scale [28] is based on six reproducible temperatures (defining fixed points), to which numerical values are assigned, and on formulas establishing the relation between temperature and the indications of instruments using values assigned to the six defining fixed points for the calibration. These fixed points are defined by specified equilibrium states, each of which, except for the triple point of water, is under a pressure of 101,325 newtons/m^2 (1 standard atmosphere).

The defining points of the scale and the exact numerical values assigned to them are given in Table 3.1.

TABLE 3.1. DEFINING FIXED POINTS

	Temperature °C (*Int.* 1948)
Oxygen point	−182.97
Triple point of water	+ 0.01
Steam point	100.
Sulfur point†	444.6
Silver point	960.8
Gold point	1063.

† In place of the sulfur point, it is recommended to use the temperature of equilibrium between solid zinc and liquid zinc (zinc point) with the value 419.505°C (Int. 1948). The zinc point is more reproducible than the sulfur point, and the value assigned to it has been so chosen that its use leads to the same values of temperature on the International Practical Temperature Scale as does the use of the sulfur point.

The procedures for interpolation lead to a division of the scale into four parts.

From 0 to 630.5°C (antimony point) the temperature t is defined by a standard platinum resistance thermometer using the formula

$$R_t = R_0(1 + At + Bt^2) \tag{3.47}$$

where R_t is the resistance at temperature t of the platinum wire resistor, and R_0 is its resistance at 0°C. The constants R_0, A, and B are to be determined from the values of R_t at the triple point of water, at the steam point, and at the sulfur point (or the zinc point). The platinum wire of a standard resistance thermometer should be annealed and its purity be such that R_{100}/R_0 is not less than 1.3920.

From the oxygen point to 0°C, the temperature t is defined by the formula

$$R_t = R_0[1 + At + Bt^2 + C(t - t_{100})t^3] \tag{3.48}$$

where R_0, A, and B are determined in the same manner as above; the constant C is to be determined from the value of R_t at the oxygen point, and $t_{100} = 100°C$.

From 630.5°C to the gold point the temperature t is defined by the formula

$$E = a + bt + ct^2 \tag{3.49}$$

where E is the electromotive force of a standard thermocouple of platinum and platinum-rhodium alloy, when one of the junctions is at 0°C and the other at the temperature t. The constants a, b, and c are to be determined from the values of E at 630.5°C, at the silver point, and at the gold point. The value of the electromotive force at 630.5°C is to be determined by measuring this temperature with a standard resistance thermometer.

The wires of the standard thermocouple must be annealed and the purity of the platinum wire be such that the ratio R_{100}/R_0 is not less than 1.3920. The platinum-rhodium wire must consist nominally of 90 per cent platinum and 10 per cent rhodium by weight. When one junction of the thermocouple is at 0°C and the other is successively at 630.5°C, the silver point, and the gold point, the completed thermocouple should have electromotive forces such that

$$\begin{aligned} E_{Au} &= 10{,}300\ \mu v \pm 50\ \mu v \\ E_{Au} - E_{Ag} &= 1{,}183\ \mu v \\ &\quad + 0.158(E_{Au} - 10{,}300\ \mu v) \pm 4\ \mu v \\ E_{Au} - E_{630.5} &= 4{,}766\ \mu v \\ &\quad + 0.631(E_{Au} - 10{,}300\ \mu v) \pm 8\ \mu v \end{aligned}$$

Above the gold point the temperature t is defined by the formula

$$\frac{J_t}{J_{Au}} = \frac{\exp\,[C_2/\lambda(t_{Au} + T_0)] - 1}{\exp\,[C_2/\lambda(t + T_0)] - 1} \tag{3.50}$$

where J_t and J_{Au} are the radiant energies per unit wavelength interval at wavelength λ, emitted per unit time per unit solid angle per unit area of a black body at the temperature t and the gold point, respectively; C_2 is the second radiation constant with the value $C_2 = 0.014\ 38$ m-deg; λ is in meters; and $T_0 = 273.15°$.

Relation between the International Practical Temperature Scale and the Thermodynamic Scale. When it is desired to use a value of temperature on the thermodynamic scale, the usual procedure is to obtain the value on the International Practical Temperature Scale and then to convert it to the thermodynamic scale by adding the appropriate difference between the scales. These differences, however, have to be determined by experiment. They are difficult to determine accurately because they are small compared with their Kelvin temperature. Some of these differences obtained in various parts of the scale are given below in order to show the present state of information about the agreement of the two scales.

In the range from 0°C to the sulfur point, intercomparisons of two nitrogen-gas thermometers with standard resistance thermometers were reported in 1955 by Beattie [6]. The differences found between the thermodynamic Celsius temperature t_{th} (definition of 1954) and the temperature t_{int} (1948 scale) are now formulated [29] as follows:

$$t_{th} - t_{int} = \frac{t}{t_{100}} - 0.0060 + \left(\frac{t}{t_{100}} - 1\right)(0.041\ 06 - 7.363 \times 10^{-5}\ \text{deg}^{-1}t)\ \text{deg} \tag{3.51}$$

This relation gives 99.994°C (therm.) for the steam point and 444.70°C (therm.) for the sulfur point. The results obtained with the two gas thermometers differed by 0.005° at the steam point and 0.05° at the sulfur point. In 1958 the Physikalisch-Technische Bundesanstalt reported the value 444.66°C (therm.) for the sulfur point.

In the range from the oxygen point to 0°C, investigations at the Physikalisch-Technische Reichsanstalt, reported in 1932, and at the University of Leiden, reported in 1935, give a set of values indicating that the differences $t_{th} - t_{int}$ have a maximum of about $+0.04°$ at about $-80°C$. Below $-100°C$ some of the reported differences have opposite signs. These differences are of the order of magnitude of the possible uncertainties in the gas-thermometer measurements. For the oxygen point, the results published since 1927 by five laboratories [30–34] have been recalculated on the basis of the value $T_0 = 273.15°K$ (adopted in 1954). These values are 90.191°K from the Physikalisch-Technische Reichsanstalt (1932), 90.17°K from the Tohoku University, Sendai, Japan (1935), 90.160°K from the University of Leiden (1940), 90.150°K from the Pennsylvania State University (1953), and 90.177°K from the National Physical Laboratory, United Kingdom (1960). The average of these five results is 90.170°K, or $-182.980°C$ (therm.).

The International Practical Temperature Scale is not defined below the oxygen point, but interim scales have been devised for use until international agreement can be reached (see following subsection).

In the neighborhood of 1000°C new determinations of the thermodynamic temperature of the silver point and gold point have been made in recent years in Germany, Japan, and the Soviet Union. At the Physikalisch-Technische Bundesanstalt (1963) 961.93°C (therm.) was obtained for the silver point and 1064.48°C (therm.) for the gold point [35]. At the Tokyo Institute of Technology (1956), the values of 961.28°C (therm.) and 1063.69°C (therm.) were obtained for these points [36]. The All-Union Institute of Metrology, Leningrad (1960), reported 1064.54°C (therm.) for the value of the gold point [37]. All these determinations at the gold point are under critical review in the various laboratories (1965), however, and the values given here are not necessarily final.

In the range above the gold point the Planck radiation formula is used to define the scale. The Planck formula is consistent with the laws of thermodynamics and hence it would give true values of Kelvin temperature if the correct values were used for the Kelvin temperature of the gold point and for the second radiation constant C_2 (see Sec. 3).

The international practical Kelvin temperatures are obtained by adding the value of $T_0 = 273.15°$ to the international practical Celsius temperatures, defined above. Values of the thermodynamic Celsius temperatures are obtained by subtracting T_0 from the thermodynamic Kelvin temperatures. Table 3.2 gives the recommended designations; the arrows point from the defined temperatures to the temperatures derived by changing the origin.

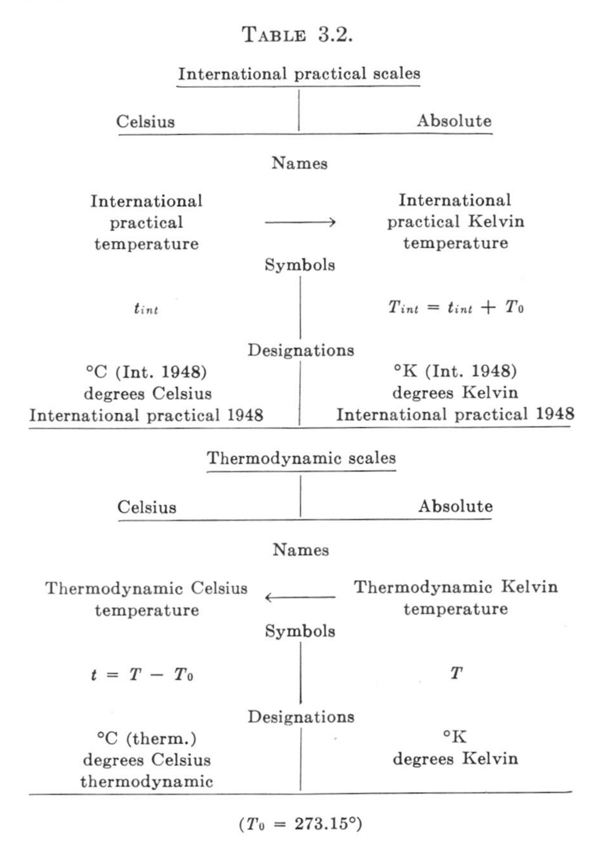

TABLE 3.2.

International practical scales	Celsius		Absolute
Names	International practical temperature	→	International practical Kelvin temperature
Symbols	t_{int}		$T_{int} = t_{int} + T_0$
Designations	°C (Int. 1948) degrees Celsius International practical 1948		°K (Int. 1948) degrees Kelvin International practical 1948

Thermodynamic scales	Celsius		Absolute
Names	Thermodynamic Celsius temperature	←	Thermodynamic Kelvin temperature
Symbols	$t = T - T_0$		T
Designations	°C (therm.) degrees Celsius thermodynamic		°K degrees Kelvin

($T_0 = 273.15°$)

NOTE. For the international practical temperature, the subscript "*int*" after t may be omitted if there is no possibility of confusion.

Temperature Scales below the Oxygen Point. The International Practical Temperature Scale is not defined below the oxygen point, but several scales have been devised for use at low temperatures. In the United States the National Bureau of Standards Provisional Scale of 1955 is most commonly used to define temperatures between 11 and 90°K. This scale is defined by the resistance-temperature relationship of a group of platinum resistance thermometers which were originally calibrated by means of a gas thermometer. Similar scales are maintained in several other national laboratories. Through a cooperative circulation of a group of platinum resistance thermometers, it has been established that these scales are in good agreement (ca. $\pm 0.02°K$) above 20°K. Disagreement as large as 0.06°K has been found between certain laboratories at 12°K.

The vapor pressure of liquid He^4 has been used for a number of years as a standard for thermometry between 1 and 5.2°K. In 1958 the International Committee on Weights and Measures approved the "1958 He^4 Vapor Pressure Scale of Temperatures" and recommended its general use. The scale is defined by a table giving the vapor pressure of He^4 every millidegree from 0.5 to 5.22°K [38]. At a temperature just below 2.2°K, however, liquid He^4 undergoes a transformation to a superfluid liquid state having a number of remarkable properties. With the liquid in this superfluid state, accurate vapor measurements are subject to experimental difficulties which result in uncertainties in the realization of the scale below 2.2°K. In contrast, the light isotope of helium, He^3, does not become a superfluid and, in addition, has a much higher vapor pressure at low temperatures. For this reason, as early as 1950 a He^3 vapor-pressure scale was proposed for use between 1 and 3.3°K. At the 1962 meeting of the Advisory Committee on Thermometry of the International Committee on Weights and Measures a He^3 vapor-pressure scale was approved and was recommended for the determination of temperatures between 0.25° and the critical temperature, 3.324°K. The defining equation for this scale and a table of computed vapor pressures with entries every millidegree between 0.20 and 3.325°K have been published by Sherman et al. [39]. Temperatures on the He^3 and He^4 scales are reported [40] to be in agreement within 0.4 millidegree at temperatures below 2.0°K.

At temperatures below 1°K paramagnetic salts have been used extensively in the dual role of refrigerator and thermometer [41]. The thermometric property used is the magnetic susceptibility χ. For an ideal paramagnetic salt the relation is given by the Curie law,

$$\chi = \frac{C}{T} \tag{3.52}$$

where T is the thermodynamic temperature and C is the Curie constant. Real salts all depart from this relation, however, and the Curie-Weiss law

$$\chi = \frac{C}{T + \Delta} \tag{3.53}$$

more closely represents the behavior of real materials. The constant Δ is the Weiss constant. Usually the constants C and Δ are evaluated at temperatures defined by the helium-vapor-pressure scales. For each paramagnetic material, however, there is some low temperature at which the Curie-Weiss law becomes inaccurate. Below that, values of temperature derived directly by means of the law are called magnetic temperatures and are denoted by T^*. Data on the differences between T^* and T have been published by de Klerk [42].

Up until recently, no practical scale was available for use in the range 5.2 to 11°K. In 1965 a new scale was established [43] based upon the resistance-temperature relation of a group of germanium thermometers which were calibrated against an acoustical thermometer [19] over the range 2 to 20°K. These germanium thermometers are selected four-lead, single-crystal thermometers which have been demonstrated to be extremely stable.

3. Practical Temperature Measurements

At low temperatures the magnetic thermometer or the vapor-pressure thermometer is usually built into the experimental apparatus, and the appropriate scale can be used directly to obtain the temperature. At higher temperatures nearly all temperature measurements in the United States are ultimately based upon the accepted temperature scales as reproduced at the National Bureau of Standards. Values of tempera-

TABLE 3.3. ESTIMATED UNCERTAINTIES ASSOCIATED WITH THE REALIZATION OF TEMPERATURE SCALES

Temperature		Scale	Thermometer used to realize scale	Estimated uncertainties	
				In realizing scales	In thermodynamic temperature
°K	°C			Deg	Deg
0.01		Magnetic	Magnetic	0.002	0.002
1.		Magnetic	Magnetic	0.002	0.002
1.		He3	Vapor pressure	0.001	0.003
5.		He4	Vapor pressure	0.002	0.010
2.		Ultrasonic	Ultrasonic	0.002	
20.		Ultrasonic	Ultrasonic	0.007	
11.	−262.15	NBS 55	Pt. Res.	0.004	0.03
20.	−253.15	NBS 55	Pt. Res.	0.001	0.01
98.18	−182.97	NBS 55	Pt. Res.	0.001	0.01
90.18	−182.97	IPTS	Pt. Res.	0.005	0.01
273.16	0.01	IPTS	Pt. Res.	0.0002	0.0002
373.15	100.	IPTS	Pt. Res.	0.0005	0.003
717.75	444.6	IPTS	Pt. Res.	0.002	0.04
903.65	630.5	IPTS	Pt. Res.	0.01	0.06
303.65	630.5	IPTS	Thermocouple	0.2	
1233.95	960.8	IPTS	Thermocouple	0.2	
1336.15	1063.	IPTS	Thermocouple	0.2	0.3
1336.15	1063.	IPTS	Photo. Pyr.	0.07	0.4
2273.	2000.	IPTS	Photo. Pyr.	0.4	1.4
3073.	2800.	IPTS	Photo. Pyr.	1.3	3.
3800.	3527.	IPTS	Photo. Pyr.	1.9	5.

ture as defined by these scales are made available through calibration services provided for selected types of temperature-measuring instruments. Scientific and industrial laboratories may thus have standards by means of which uniform values of temperature may be attained within the limitations of the instrumentation involved.

Table 3.3 lists scales used to define temperatures up to that of the positive crater of a carbon arc (3800°K). The estimates given for the uncertainties associated with the realization of the scales are based upon estimated magnitudes of known sources of error, but the estimates are largely based upon judgment. In nearly all cases, there are insufficient data of a kind to permit statistical analysis of the contribution of a particular potential source of error. The uncertainties assigned to the realization of thermodynamic temperatures were obtained by combining the uncertainties in the realization of the practical scales with best current (1965) estimates of the differences between the practical scales and the thermodynamic Kelvin scale. No estimate is given for the accuracy of thermodynamic temperatures as determined by the ultrasonic thermometer, since this work has not been completely evaluated as yet.

The uncertainties listed for temperatures of 1336°K and higher relate to the realization of the International Practical Temperature Scale at the National Bureau of Standards by means of a specially designed photoelectric pyrometer. In addition, knowledge of thermodynamic temperatures depend upon uncertainties in the values of the temperature of the gold point and the second radiation constant C_2. The uncertainties given in Table 3.3 are based upon a value of 1337.6°K as determined by Moser [35] but using an uncertainty of ±0.3, which is somewhat higher than Moser's estimate. The value of C_2 is taken as 1.43879 ± 0.00019 [8].

Table 3.4 gives some indication of the uncertainties involved in putting the available temperature scales to practical use. Uncertainties are listed for commonly used types of temperature-measuring instruments when used under favorable conditions and when good laboratory practices are followed. In some instances better accuracy is possible with special techniques, but usually the listed uncertainties are difficult to attain.

Since the standard platinum resistance thermometer in the range −183 to 630.5°C and the standard platinum vs. platinum–10 per cent rhodium thermocouple between 630.5 and 1063°C are the prescribed instruments for interpolation between the defining points on the International Practical Temperature Scale, the uncertainties listed for these instruments in Table 3.4 are the same as the uncertainties associated with the reproduction of the scale, as shown in Table 3.3. Below −183°C, low-temperature resistance thermometers conforming to the requirements of a standard on the International Practical Temperature Scale can be calibrated in terms of the National Bureau of Standards 1955 scale with an accuracy comparable to that with which the scale itself is reproduced. This is not the case with other types of instruments calibrated in the range −262 to 1063°C. Such instruments as liquid-in-glass thermometers and base-metal thermocouples are inherently less accurate than the standard resistance thermometer and standard thermocouple; consequently the uncertainties associated with the calibration and use of such instruments are appropriately higher. For optical and photoelectric pyrometers black-body conditions are assumed. Under other conditions the spectral emittance of the incandescent surface being viewed must be accurately known for precise measurements. In the vicinity of 1063°C and a wavelength of 6,500 A, an error of 10 per cent in spectral emittance results in an error of about 9°.

It should be pointed out that commercially available photoelectric pyrometers are new products on the market at this time (1965) and therefore are still under extensive development. It is expected that the uncertainties associated with the calibration and use of these instruments will be materially reduced in the near future.

TABLE 3.4. UNCERTAINTIES ASSOCIATED WITH THE CALIBRATION AND USE OF TEMPERATURE-MEASURING INSTRUMENTS

Type	Temperature, °C	Estimated uncertainty, °C
Resistance thermometers		
Capsule, 4-lead	−262	0.004
Capsule, 4-lead	−253 to −183	0.001
IPTS standard	−183.	0.005
IPTS standard	+ 0.01	0.0002
IPTS standard	100.	0.0005
IPTS standard	445.	0.002
IPTS standard	630.	0.01
Liquid-in-glass thermometers		
Organic liquid filled	−200 to 0	0.5
Mercury-thallium filled	− 56 to 0	0.05
Mercury filled	− 35 to +100	0.02
Mercury filled	100 to 200	0.1
Mercury filled	200 to 300	0.2
Mercury filled	300 to 400	0.5
Thermocouples		
Base metal	−183 to +300	0.1
Base metal	0 to 1100	0.5
Pt vs. Pt-Rh	0 to 1100	0.3
Pt vs. Pt-Rh	1450	2.
Pyrometers		
Optical	1100	3
Optical	2000	6
Optical	2800	8.
Optical	3500	20.
Photoelectric	1100	1.
Photoelectric	2000	3.
Photoelectric	2800	5.
Photoelectric	3500	8.

References

1. Landsberg, H. E.: A Note on the History of Thermometer Scales, *Weather*, **19**(1): 2 (January, 1964).
2. Dorsey, E. N.: *J. Wash. Acad. Sci.*, **36**: 361 (1946).
3. Réaumur, R. A. de: *Hist. Mém. Acad. R. Sci.*, Paris, (1967), p. 452.
4. Celsius, A.: *Kgl. Svenska Vetenskaps akad. Handl.*, **4**: 197 (1742).
5. Comptes Rendus de la Dixième Conférence Générale des Poids et Mesures, p. 79, 1954.
6. Beattie, J. A.: "Temperature, Its Measurement and Control in Science and Industry," vol. 2., p. 63, Reinhold, New York, 1955.
7. Moser, H.: "Temperature, Its Measurement and Control in Science and Industry," vol. 2, p. 103, Reinhold, New York, 1955.
8. New Values for the Physical Constants, Recommended by National Academy of Sciences–National Research Council, *Natl. Bur. Std.* (*U.S.*), *Tech. News Bull.* 47, p. 175, October, 1963.
9. Johnson, J. B.: *Phys. Rev.*, **32**: 97 (1928).
10. Nyquist, H.: *Phys. Rev.*, **32**: 110 (1928).
11. Garrison, J. B., and A. W. Lawson: *Rev. Sci. Instr.*, **20**: 785 (1949).
12. Hogue, E. W.: Factors Affecting the Precision and Accuracy of an Absolute Noise Thermometer, *ASTIA AD*-46864.
13. Fink, H. J.: *Can. J. Phys.*, **37**: 1397 (1959).
14. Patronis, E. T., Jr., H. Marshak, C. A. Reynolds, V. L. Sailor, and F. J. Shore: *Rev. Sci. Instr.*, **30**: 578 (1959).
15. Savateev, A. V.: *Meas. Tech.* (*USSR*) (*English Transl.*), January–June, 1962, p. 114. (From *Izmeritel'naya Tekhnika*, no. 2, p. 19, February, 1962.)
16. Van Itterbeek, A.: "Progress in Low Temperature Physics," vol. 1, p. 362, Interscience, New York, 1955.
17 Barrett, E. W., and V. E. Suomi: *J. Meteorol.*, **6**: 273 (1949).
18 Apfel, J. H.: *Rev. Sci. Instr.*, **33**: 428 (1962).
19 Cataland, G., M. Edlow, and H. H. Plumb: "Temperature, Its Measurement and Control in Science and Industry," vol. 3, pt. 1, p. 129, Reinhold, New York, 1962.
20. Cataland, G., and H. H. Plumb: *J. Acoust. Soc. Am.*, **34**: 1145 (1962).
21. Dean, C., and R. V. Pound: *J. Chem. Phys.*, **20**: 195 (1952).
22. Benedek, G. B., and T. Kushida: *Rev. Sci. Instr.*, **28**: 92 (1957).
23. Vanier, J.: *Can. J. Phys.*, **38**: 1397 (1960).
24. Comptes Rendus de la Septième Conférence Générale des Poids et Mesures, p. 94, 1927.
25. Comptes Rendus de la Neuvième Conférence Générale des Poids et Mesures, p. 89, 1948.
26. Comptes Rendus de la Onzième Conférence Générale des Poids et Mesures, p. 64, 1960.
27. Comptes Rendus de la Neuvième Conférence Générale des Poids et Mesures, p. 64, 1948.
28. Stimson, H. F.: International Practical Temperature Scale of 1948, Text Revision of 1960, *J. Res. Natl. Bur. Std.*, **65A**: 1939 (1961), and *Natl. Bur. Std.* (*U.S.*), *Monograph* 37, 1961.
29. Beattie, J. A., M. Benedict, B. E. Blaisdell, and J. Kaye: *J. Chem. Phys.* (April, 1965).
30. Heuse, W., and J. Otto: *Ann. Physik*, **9**: 486 (1931); **14**: 185 (1932).
31. Aoyama, S., and E. Kanda: *Bull. Chem. Soc. Japan*, **10**: 472 (1935).
32. Van der Horst, H.: Thesis, University of Leiden, 1940, p. 79. Van Dijk, H.: *Proces-Verbaua Seances, Comite Inter-Natl. Poids Mesures* (2e sèr.), **23B**: T44 (1952).
33. Aston, J. G., and G. W. Moessen: *J. Chem. Phys.*, **21**: 948 (1953).
34. Barber, C. R.: "Temperature, Its Measurement and Control in Science and Industry," vol. 3, pt. 1, p. 345, Reinhold, New York, 1962.
35. Moser, H., J. Otto, and W. Thomas: *Z. Physik.*, **175**: 327 (1963).
36. Oishi, J., M. Awano, and T. Mochizuki: *J. Phys. Soc. Japan*, **11**: 311 (1956).
37. Arefjewa, H. W., U. W. Dijkow, K. S. Israilow, J. J. Kirenkow, and N. W. Schemetillo: *All-Union Sci. Res. Inst. Metrology*, Leningrad, ed. 49 (109): 13 (1960).
38. Brickwedde, F. G.: *J. Res. Natl. Bur. Std.*, **64A**: 1 (1960).
39. Sherman, R. H., S. G. Sydoriak, and T. R. Roberts: *J. Res. Natl. Bur. Sta.*, **68A**: 579 (1964).
40. Sydoriak, S. G., T. R. Roberts, and R. H. Sherman: *J. Res. Natl. Bur. Std.*, **68A**: 559 (1964).
41. Hudson, R. P.: "Temperature, Its Measurement and Control in Science and Industry," vol. 3, pt. 1, p. 51, Reinhold, New York, 1962.
42. Klerk, D. de: in S. Flügge (ed.), "Encyclopedia of Physics," vol. 15, p. 38, Springer, Berlin, 1956.
43. Cataland, G., and H. H. Plumb: *J. Res. Natl. Bur. Std.*, **70A**: 243 (1966).

Chapter 4

The Equation of State and Transport Properties of Gases and Liquids

By R. B. BIRD, J. O. HIRSCHFELDER, AND C. F. CURTISS, University of Wisconsin

In this chapter we present a brief survey of the present status of the statistical mechanical theory of the properties of gases and liquids and give formulas and tables for the calculation of the properties of nonpolar and polar substances. We also indicate the additional tabulations that are available in the literature.

In the first section a brief discussion is given of the nature of the intermolecular forces in nonpolar and in polar substances. The next two sections deal with the results of *equilibrium statistical mechanics*—the calculation of the equation of state at low densities and at high densities. The following two sections contain the results of *nonequilibrium statistical mechanics* (more commonly referred to as *kinetic theory*)—the calculation of the transport properties at low and high densities. Finally, the last section indicates that many equilibrium and nonequilibrium properties of gases and liquids may be estimated with the help of the principle of corresponding states in the absence of a satisfactory molecular theory.

The properties discussed in this chapter are those which one needs to know for the solution of the "equations of change" for multicomponent, nonisothermal fluid mixtures; for a binary mixture, with gravity as the only external body force, these equations are as follows:

1. *The equation of continuity for species A:*

$$\frac{\partial}{\partial t}\rho_A = -(\nabla\cdot\rho_A\mathbf{v}) - (\nabla\cdot\mathbf{j}_A) + r_A$$

Here ρ_A is the mass concentration of species A, $\mathbf{v}$ is the mass-average velocity, r_A is the mass rate of production of species A by chemical reaction, and $\mathbf{j}_A$ is the mass flux (g cm^{-2} sec^{-1}) of A with respect to the mass-average velocity $\mathbf{v}$. A similar equation can be written for species B.

2. *The equation of motion:*

$$\frac{\partial}{\partial t}\rho\mathbf{v} = -[\nabla\cdot\rho\mathbf{v}\mathbf{v}] - \nabla p - [\nabla\cdot\boldsymbol{\tau}] + \rho\mathbf{g}$$

Here ρ is the mass density of the mixture, p is the isotropic pressure, $\mathbf{g}$ is the acceleration of gravity, and $\boldsymbol{\tau}$ is the momentum flux tensor.

3. *The equation of energy:*

$$\frac{\partial}{\partial t}(\rho\hat{U} + \tfrac{1}{2}\rho v^2) = -(\nabla\cdot(\rho\hat{U} + \tfrac{1}{2}\rho v^2)\mathbf{v}) - (\nabla\cdot\mathbf{q}) - (\nabla\cdot p\mathbf{v}) - (\nabla\cdot[\boldsymbol{\tau}\cdot\mathbf{v}]) + (\mathbf{v}\cdot\rho\mathbf{g})$$

Here $\hat{U}$ is the internal energy (per unit mass) and $\mathbf{q}$ is the energy flux (cal cm^{-2} sec^{-1}).

The fluxes appearing in these equations are:

1. The mass flux of species A (in the absence of pressure and forced diffusion) is

$$\mathbf{j}_A = -\rho\mathfrak{D}_{AB}\left[\nabla\omega_A + \frac{c^2}{\rho^2}M_AM_B\frac{k_T}{(\partial\ln a_A/\partial\ln x_A)_{T,p}}\nabla\ln T\right]$$

Here $\mathfrak{D}_{AB}$ is the *diffusivity* for the pair A-B; ρ and c are the mass and molar densities of the fluid mixture, ω_A is the mass fraction ρ_A/ρ, x_A is the mole fraction of A, and a_A is the activity of A. The M_i are molecular weights. The quantity k_T is the *thermal-diffusion ratio,* which is a measure of the tendency for diffusion to occur in the presence of a temperature gradient. (Note that the coefficient of $\nabla\ln T$ is used by some authors as the definition of k_T[7*d*]).‡ The expression for $\mathbf{j}_B$ is similar to that for $\mathbf{j}_A$, except that the coefficient of the $\nabla\ln T$ term is negative. It is conventional to label the heavier of the two species as "A" and the lighter as "B". With this convention, k_T is usually positive, and species A moves to the colder region.

2. The momentum flux is

$$\boldsymbol{\tau} = -\eta[\nabla\mathbf{v} + (\nabla\mathbf{v})^\dagger] + (\tfrac{2}{3}\eta - \kappa)(\nabla\cdot\mathbf{v})\boldsymbol{\delta}$$

Here η is the *viscosity* and κ is the *bulk viscosity.* The quantity $\boldsymbol{\delta}$ is the unit tensor, and $(\nabla\mathbf{v})^\dagger$ is the transpose of $\nabla\mathbf{v}$.

3. The energy flux (in the absence of the "Dufour effect") is

$$\mathbf{q} = -\lambda\nabla T + \left(\frac{\bar{H}_A}{M_A}\mathbf{j}_A + \frac{\bar{H}_B}{M_B}\mathbf{j}_B\right)$$

Here λ is the *thermal conductivity.* The $\bar{H}_i$ are partial molar enthalpies. The Dufour effect would contribute an extra term proportional to the concentration gradient, which contains k_T; it is normally unimportant.

In this chapter we are concerned with the evaluation of the *equation of state* $p = p(\rho,T)$ and the *transport properties* $\mathfrak{D}_{AB}$, k_T, η, κ, and λ. For more details the reader may wish to consult "Molecular Theory of

‡ Numbers in brackets refer to References at end of chapter.

Gases and Liquids," by J. O. Hirschfelder, C. F. Curtiss, and R. B. Bird [7a].

1. The Potential Energy of Interaction between Two Molecules

By means of statistical mechanical theory it is possible to express many of the bulk properties of matter in terms of the potential energy of interaction φ between a pair of molecules in the substance. (If φ depends only upon the intermolecular distance r, then $-d\varphi/dr$ is the force acting between the molecules.) The dependence of this potential function on the separation between the molecules and their mutual orientation can in principle be obtained by quantum mechanical calculation. Attempts along this line have yielded results for only the very simplest atoms and molecules [see ref. 7a, sec. 1.3 (elementary discussion) and chaps. 12 to 14 (detailed discussion)]. Accordingly it is customary to use empirical potential functions with several adjustable parameters and to determine these parameters from experimental measurements of bulk properties in conjunction with the statistical mechanical calculations of the same properties.

Empirical Potential Energy Functions for Nonpolar and Polar Molecules. At the present time the most used potential function for nonpolar molecules is the Lennard-Jones potential and for polar molecules the Stockmayer potential. These functions are

Lennard-Jones (nonpolar):

$$\varphi(r) = 4\epsilon\left[\left(\frac{\sigma}{r}\right)^{12} - \left(\frac{\sigma}{r}\right)^{6}\right] \tag{4.1}$$

Stockmayer (polar):

$$\varphi(r, \theta_1, \theta_2, \phi_2 - \phi_1) = 4\epsilon\left[\left(\frac{\sigma}{r}\right)^{12} - \left(\frac{\sigma}{r}\right)^{6}\right] - \frac{\mu^2}{r^3} f(\theta_1, \theta_2, \phi_2 - \phi_1) \tag{4.2}$$

in which σ and ϵ are the adjustable parameters. In the Lennard-Jones potential the r^{-6} term represents quite accurately the long-range attractive forces (sometimes called London dispersion forces) and the r^{-12} is an approximation to the short-range repulsive forces (sometimes called *valence* or *chemical* forces). The parameter ϵ is the maximum energy of attraction and σ is the value of r for which $\varphi(r) = 0$ (see Fig. 4.1). In the Stockmayer potential the r^{-6} and r^{-12} terms have the same significance as in the Lennard-Jones potential, but the parameters σ and ϵ have to be interpreted somewhat differently. The r^{-3} term represents the interaction between two ideal dipoles, the mutual orientation of which is described by the angles θ_1, θ_2, and $\phi_2 - \phi_1$. The angles θ_i and ϕ_i are the usual spherical polar angles of the dipoles in a coordinate system in which the z axis is the intermolecular axis. The function

$$f(\theta_1, \theta_2, \phi_2 - \phi_1) = 2\cos\theta_1\cos\theta_2 - \sin\theta_1\sin\theta_2\cos(\phi_2 - \phi_1)$$

and μ is the dipole moment of a single molecule. The Lennard-Jones potential is reasonably good for

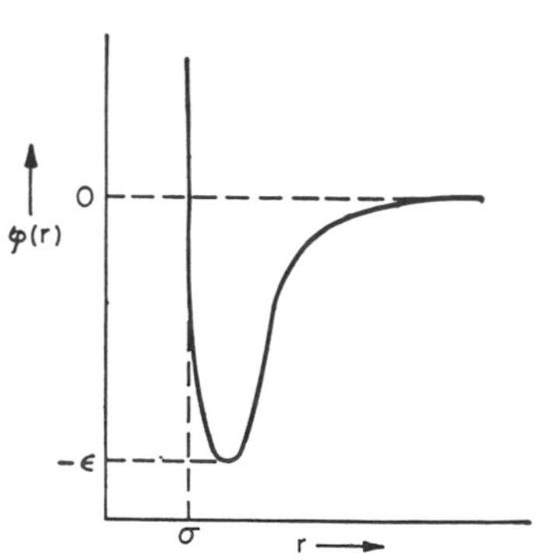

FIG. 4.1. Lennard-Jones intermolecular potential; $|\varphi|$ drops off to 0.01ϵ at about $r/\sigma = 3$.

spherical (and almost spherical) nonpolar molecules. The Stockmayer potential is a useful approximation for simple polar molecules. For complex polar molecules it is less appropriate inasmuch as the angle dependence of the short-range repulsive forces and the dipole-quadrupole interaction are not taken into account.

Determination of Parameters in Empirical Potential-Energy Functions. The parameters σ and ϵ in the Lennard-Jones potential are most satisfactorily determined from experimental viscosities and second virial coefficients in conjunction with the statistical mechanical formulas for these properties which are given in Secs. 2 and 4. In Table 4.1 is given a tabulation of these parameters.

TABLE 4.1. CONSTANTS FOR THE LENNARD-JONES (6–12) POTENTIAL‡

Gas	Constants from viscosity		Constants from second virial coefficients	
	ϵ/k (°K)	σ (A)	ϵ/k (°K)	σ (A)
He	10.22	2.576	10.8	2.57
Ne	35.7	2.789	35.8	2.75
Ar	124.	3.418	119.7	3.406
Kr	190.	3.61	173.	3.59
Xe	229.	4.055	225.3	4.070
H_2	33.3	2.968	36.7	2.959
N_2	91.5	3.681	95.05	3.698
O_2	113.	3.433	117.5	3.58
F_2	112.	3.653	121.	3.61
CO	110.	3.590	100.2	3.763
CO_2	190.	4.000	187.5	4.47
CH_4	140.	3.808	148.1	3.809
CCl_4	327.	5.881		
SF_6	155.	5.46	188.7	5.91

‡ For complete tables and references, see [7a, pp. 1110–1113, 1212–1215] and [1m.].

Frequently viscosity and second virial coefficient data are not available, so that the potential parameters cannot be obtained in this manner. In such cases there are several rough empirical relationships which may be used to estimate the σ and ϵ of the Lennard-Jones potential from the properties of the substance at its critical point (c), its melting point (m), its boiling point (b), its Boyle point (B), or the absolute zero (z). These relations are given in terms of ϵ/k and b_0:

$$\begin{aligned} \frac{\epsilon}{k} &= 0.77T_c & b_0 &= 0.75\tilde{V}_c = \frac{18.4T_c}{p_c} \\ \frac{\epsilon}{k} &= 1.15T_b & b_0 &= 2.0\tilde{V}_b^{(\text{liq})} \\ \frac{\epsilon}{k} &= 1.92T_m & b_0 &= 2.3\tilde{V}_m^{(\text{sol})} \\ \frac{\epsilon}{k} &= 0.292T_B & & \cdots\cdots\cdots \\ \cdots\cdots & & b_0 &= 2.293\tilde{V}_z^{(\text{sol})} \end{aligned} \tag{4.3}$$

These formulas may be used to estimate potential parameters for the Lennard-Jones potential and the parameters in turn may be used to calculate the equation of state and the transport coefficients with the aid of the formulas and tables given in subsequent sections.

In Table 4.2 are given the parameters σ and ϵ for the Stockmayer potential for several polar molecules

TABLE 4.2. CONSTANTS FOR THE STOCKMAYER POTENTIAL‡

Gas	μ (debyes)	Constants from viscosity			Constants from second virial coefficients		
		ϵ/k (°K)	σ (A)	$\sqrt{2}\,t^*$	ϵ/k (°K)	σ (A)	t^*
H_2O	1.85	775	2.52	1.0	380	2.65	1.2
NH_3	1.47	358	3.15	0.7	320	2.60	1.0
CH_3Cl	1.88	414	3.94	0.5	380	3.43	0.6
CH_3OH	1.68	417	3.67	0.5	630	2.40	0.8
n-C_3H_7OH	1.69	495	4.71	0.2	866	2.61	0.5
HCl	1.08	328	3.36	0.34			
HI	0.42	313	4.13	0.029			

‡ For complete tables and references, see [7a, pp. 214, 1190, 1200].

as determined from viscosities and second virial coefficients. In addition the parameter

$$t^* = \mu^{*2}/\sqrt{8} = \mu^2/\epsilon\sigma^3\sqrt{8}$$

is given. This quantity is a measure of the deviation from nonpolar behavior.

Empirical Combining Laws for Interactions between Two Dissimilar Molecules. The potential parameters given in Tables 4.1 and 4.2 are for interactions between molecules of the same chemical species. In the calculation of the bulk properties of mixtures it is necessary to know the potential energy function describing the interaction between pairs of molecules of different species. The best way to obtain this information is from the temperature-dependence of gaseous binary diffusion coefficients. Since only meager diffusion data exist, it is often necessary to employ empirical *combining laws*, which have been demonstrated to be reasonably adequate [7a, chap. 8]. These laws are given in the following paragraphs:

Nonpolar-Nonpolar Interaction. The interaction of two nonpolar molecules of species 1 and 2 is described by the Lennard-Jones potential, where the parameters σ and ϵ are given by

$$\sigma_{12} = \tfrac{1}{2}(\sigma_1 + \sigma_2) \qquad \epsilon_{12} = \sqrt{\epsilon_1\epsilon_2} \tag{4.4}$$

and σ_i and ϵ_i are the parameters appropriate for molecules of the ith species as given in Table 4.1.

Polar-Polar Interaction. The interaction of two polar molecules of species 1 and 2 is described by the Stockmayer potential, in which the parameters σ and ϵ are given by Eqs. (4.4), where σ_i and ϵ_i are now the parameters given in Table 4.2. Also the parameter t^* characteristic of the unlike interaction is given by

$$t^*_{12} = \frac{\mu_1\mu_2}{\sqrt{8}\,\epsilon_{12}\sigma_{12}^3} \doteq \sqrt{t^*_1 t^*_2} \tag{4.5}$$

where μ_i is the dipole moment of the ith type of molecule and is the quantity previously defined and tabulated in Table 4.2.

Nonpolar-Polar Interaction. The interaction between a polar molecule p and a nonpolar molecule n is described approximately by a Lennard-Jones potential, in which the parameters σ and ϵ are given by

$$\sigma_{pn} = \tfrac{1}{2}(\sigma_p + \sigma_n)\xi^{-1/6} \qquad \epsilon_{pn} = \sqrt{\epsilon_p\epsilon_n}\,\xi^2 \tag{4.6}$$

and the factor ξ is given by

$$\xi = \left(1 + \frac{\alpha^*_n t^*_p}{\sqrt{2}}\sqrt{\frac{\epsilon_p}{\epsilon_n}}\right) \tag{4.7}$$

The parameters σ_n and ϵ_n are obtained from Table 4.1, and σ_p, ϵ_p, t^*_p are obtained from Table 4.2. The quantity α^*_n is the polarizability of the nonpolar molecule α_n divided by σ^3.

Present Status of Theory of Intermolecular Forces. Quantum-mechanical calculations have given us a considerable amount of information about the theory of the forces between spherical molecules at distances from one another which are large with respect to molecular dimensions [1a, 1b, 1c]. The theory of these long-range forces has also been worked out for asymmetric molecules by London [1d] and for long conjugated double-bond molecules by Coulson and Davies [1e]. The a priori calculation of the potential energy of interaction for all values of the separation has progressed very little beyond the interaction of two noble gas atoms [1f] and of two hydrogen molecules [1g].

Another approach to the study of intermolecular forces is to determine the adjustable parameters in empirical potential-energy functions which are more realistic and hence more elaborate than the Lennard-Jones and Stockmayer potentials. For spherical nonpolar molecules Buckingham and Corner [1h] have proposed a potential-energy function which contains an additional attractive term proportional to r^{-8} to account for the induced dipole-induced quadrupole interaction, and the repulsive part of which is of an exponential form. For nonspherical molecules several models (which are extensions of the Lennard-Jones potential) have been proposed—one by Corner [1*i*] and a somewhat simpler one by Kihara [1*j*].

For polar molecules Rowlinson [1*k*] has suggested that the Stockmayer potential be modified by the inclusion of a term proportional to r^{-4}, which represents the dipole-quadrupole interaction. For polar, nonspherical molecules a model consisting of rigid ellipsoids with imbedded point dipoles has been used [1*l*]. The success of this semiempirical sort of an approach is definitely limited by the lack of very accurate experimental measurements of the bulk properties which are needed for the unique determination of the adjustable parameters in the potential functions. Furthermore, a considerable amount of information needs yet to be compiled about the quadrupole and higher multipole moments of molecules.

2. The Equation of State of Dilute and Moderately Dense Gases

At very low pressures and high temperatures the pressure of a gas is given by the ideal gas law

$$pV = NkT$$

where V is the volume occupied by N molecules. The thermodynamic properties of a gas under such conditions (that is, the so-called "zero-pressure" properties) can be calculated by means of statistical mechanics for substances consisting of polyatomic molecules. The complete details of the theory are given in the textbooks of Mayer and Mayer [2*a*], and Münster [2*b*].

A formal development of the equation of state may be made in two ways. One method leads to an expression for the pressure in terms of the *configurational integral* Q_N, which (for an angle-independent potential function) is

$$\iint \cdots \int \exp\left[-\tfrac{1}{2}\sum_{i=1}^{N}\sum_{j=1}^{N}\varphi(r_{ij})/kT\right] \times dr_1\, dr_2 \cdots dr_N$$

The quantity $\frac{1}{2}\Sigma_i\Sigma_j\varphi(r_{ij})$ is just the total potential energy of the molecules in the gas in a given configuration, in the approximation that the potential energies are pairwise additive. The second method leads to an expression for the pressure in terms of the *radial distribution function* $g(r)$. The latter is so defined that $(N^2/2V)g(r)4\pi r^2\,dr$ is the number of molecules for which the separation lies between r and $r + dr$. In terms of these quantities the equation of state is written as

$$pV = kT\left(\frac{\partial \ln Q_N}{\partial \ln V}\right)_T \tag{4.8}$$

$$pV = NkT - \frac{N^2}{6V}\int g(r)r\frac{d\varphi}{dr}4\pi r^2\,dr \tag{4.9}$$

It has been established that these two expressions for the equation of state are equivalent [7*b*]. These expressions form the starting point for the theoretical and computational results discussed in this and the following section.

Virial Form of the Equation of State. It is possible to develop both $\ln Q_N$ and $g(r)$ as power series in the reciprocal of the molar volume, $\tilde{V} = V/\tilde{N}$, where $\tilde{N}$ is Avogadro's number. This allows the equation of state to be written as the "virial expansion"

$$\frac{p\tilde{V}}{RT} = 1 + \frac{B(T)}{\tilde{V}} + \frac{C(T)}{\tilde{V}^2} + \frac{D(T)}{\tilde{V}^3} + \cdots \tag{4.10}$$

in which $B, C, D, \ldots$ are the second, third, fourth, $\ldots$ virial coefficients. For some purposes it is more convenient to write this expansion in terms of powers of the pressure:

$$\frac{p\tilde{V}}{RT} = 1 + B'(T)p + C'(T)p^2 + D'(T)p^3 + \cdots \tag{4.11}$$

in which $B' = B/RT$ and $C' = (C - B^2)/(RT)^2$. The virial coefficients are functions of the temperature only and are given as integrals which contain the intermolecular potential function φ. For angle-independent potentials (such as the Lennard-Jones potential) the second and third virial coefficients are given by

$$B(T) = -2\pi\tilde{N}\int_0^\infty f_{12}r_{12}^2\,dr_{12} \tag{4.12}$$

$$C(T) = -\frac{8\pi\tilde{N}^2}{3}\iiint f_{12}f_{13}f_{23}r_{12}r_{13}r_{23}\,dr_{12}\,dr_{13}\,dr_{23}$$

(integral over all r_{12}, r_{13}, r_{23} which form a triangle) (4.13)

where $f_{12} = \{\exp[-\varphi(r_{12})/kT] - 1\}$. For angle-dependent potentials (such as the Stockmayer potential) $B(T)$ and $C(T)$ involve additional integrations over angular variables [7*a*, sec. 3.4]. The second virial coefficient basically describes the deviations from the ideal gas law $p\tilde{V} = RT$ due to binary collisions. The third virial coefficient describes the deviations due to ternary collisions. Thus as the density of a gas is increased more virial coefficients are needed to explain the pVT behavior.

Virial Coefficients for Nonpolar Gases. The second and third virial coefficients have been evaluated for nonpolar molecules (Lennard-Jones potential) and the results may be expressed in the very simple form:

$$B(T) = b_0B^\star(T^*) \tag{4.14}$$

$$C(T) = b_0^2C^\star(T^*) \tag{4.15}$$

The functions $B^\star(T^*)$ and $C^\star(T^*)$ are given in Table 4.3. The parameters σ and ϵ in the Lennard-Jones potential enter into the expression through the unit of volume $b_0 = \frac{2}{3}\pi\tilde{N}\sigma^3$ and the reduced temperature $T^* = kT/\epsilon$. Thus the potential parameters in Table 4.1 can be used to calculate $B(T)$ and $C(T)$ for nonpolar molecules from Table 4.3. It has been found that the agreement between the experimental and calculated $B(T)$ is excellent, and that for $C(T)$ it is moderately good. Tables are also available [7*a*,

TABLE 4.3. FUNCTIONS USED IN VIRIAL-COEFFICIENT CALCULATIONS‡
(Entries are tabulated for the Stockmayer potential; values listed under $t^* = 0$ are for the Lennard-Jones potential.)

$B^\star(T^*,t^*)$

T^* \ t^*	0	0.2	0.4	0.6	0.8	1.0
0.30	−27.88	−42.97				
0.50	−8.720	−10.40	−17.03	−36.36		
0.75	−4.176	−4.630	−6.163	−9.413	−16.05	−30.4
1.00	−2.538	−2.744	−3.402	−4.657	−6.820	−10.54
1.25	−1.704	−1.821	−2.187	−2.852	−3.915	−5.559
1.50	−1.201	−1.277	−1.511	−1.925	−2.561	−3.490
2.00	−0.6276	−0.6671	−0.7875	−0.9953	−1.302	−1.727
2.50	−0.3126	−0.3370	−0.4108	−0.5368	−0.7194	−0.9658
3.00	−0.1152	−0.1318	−0.1820	−0.2671	−0.3892	−0.5517
4.00	−0.1154	0.1062	0.0784	0.0316	−0.0349	−0.1221
5.00	0.2433	0.2374	0.2197	0.1898	0.1476	0.0926
10.00	0.4609	0.4593	0.4547	0.4469	0.4359	0.4218
50.00	0.5084	0.5083	0.5080	0.5076	0.5071	0.5064
100.00	0.4641	0.4641	0.4640	0.4639	0.4637	0.4635
400.00	0.3584	0.3584	0.3583	0.3583	0.3583	0.3583
			$C^\star(T^*,t^*)$			
1.0	0.4297	0.5304				
2.0	0.4371	0.4826	0.6496	0.995	1.595	2.46
2.5	0.3811	0.4076	0.5195	0.6871	0.999	1.482
3.0	0.3523	0.3692	0.4275	0.5403	0.7248	1.002
4.0	0.3266	0.3350	0.3630	0.4156	0.4986	0.6194
10.0	0.2861	0.2871	0.2902	0.2957	0.3039	0.3151

‡ For complete tables and references, see [7a, pp. 1114–1119, 1147–1154, 1215].

Table IB] for the function $T^* \, dB^\star/dT^* - B^\star$ which is simply related to the zero-pressure Joule-Thomson coefficient μ^0:

$$\mu^0 \tilde{C}_p{}^0 = b_0 \left(T^* \frac{dB^\star}{dT^*} - B^\star \right) \tag{4.16}$$

in which $\tilde{C}_p{}^0$ is the zero-pressure heat capacity per mole at constant pressure. Agreement between experimental and calculated values of $\mu^0 \tilde{C}_p{}^0$ has been found to be extremely good.

Virial Coefficients for Polar Gases. The second and third virial coefficients have also been evaluated for polar molecules (Stockmayer potential) and the results are given simply as:

$$B(T) = b_0 B^\star(T^*,t^*) \tag{4.17}$$

$$C(T) = b_0^2 C^\star(T^*,t^*) \tag{4.18}$$

The functions $B^\star(T^*,t^*)$ and $C^\star(T^*,t^*)$ are given in Table 4.3. Here the σ and ϵ of the Stockmayer potential enter into the expressions through $b_0 = \frac{2}{3}\pi \tilde{N} \sigma^3$ and $T^* = kT/\epsilon$ and, in addition, in the quantity t^*, which is a measure of the deviation from nonpolar behavior. Hence the potential parameters in Table 4.2 may be used to calculate $B(T)$ and $C(T)$ from Table 4.3 for polar molecules. What few good experimental measurements are available indicate that the use of these tables gives reasonably good agreement with experiment. Tables have also been prepared for calculating Joule-Thomson coefficients of polar gases (see [7a, Table IIB]).

Virial Coefficients for Mixtures of Gases. Thus far the discussion has been restricted to the calculation of the equation of state of pure substances. The virial equation of state [Eq. (4.10)] may also be used for multicomponent mixtures. For a mixture containing ν components the second and third virial coefficients are given by

$$B_{\text{mix}} = \sum_{i=1}^{\nu} \sum_{j=1}^{\nu} x_i x_j B_{ij} \tag{4.19}$$

$$C_{\text{mix}} = \sum_{i=1}^{\nu} \sum_{j=1}^{\nu} \sum_{k=1}^{\nu} x_i x_j x_k C_{ijk} \tag{4.20}$$

The quantity B_{jj} is the second virial coefficient for the pure jth component calculated according to Eq. (4.14) or (4.17). The quantity B_{ij} is the second virial coefficient for a hypothetical substance, calculated with the parameters σ_{ij} and ϵ_{ij} characteristic of the interactions between pairs of dissimilar molecules in the gas. [For an illustrative example on the use of Eq. (4.19) see ref. 7a, p. 224.]

Present Status of Theory. The status of the theoretical development of the equation of state at moderate densities is quite satisfactory. In fact, the virial expansion is one of the cleanest-cut developments in the subject of statistical mechanics. The quantum-mechanical theory has been developed as well as the classical theory [7b]. The assumptions which go into calculations based upon the theory are the pair-potential-energy function and the pairwise additivity of the molecular interaction.

Few calculations of virial coefficients higher than the third have been made. The higher coefficients are exceedingly sensitive to the exact shape of the potential-energy function, and hence little would seem to be gained by their calculation until more information is obtained about the nature of the intermolecular interaction. Some calculations of the second virial coefficient have been made for potentials more complex than the Lennard-Jones and the Stockmayer potentials [2c].

For spherical nonpolar molecules Rice and Hirschfelder [2d] have evaluated $B(T)$ for a three-parameter Buckingham potential, which includes an r^{-6} attraction term and an exponential repulsion term. For elongated molecules calculations of $B(T)$ have been made by both Corner [1i] and Kihara [1j]. Kihara also has shown how the second virial coefficient can be calculated for flat, disk-shaped molecules, such as benzene and cyclohexane, and has shown how his work can be extended to still other shapes [2e, 2f].

3. The Equation of State of Dense Gases and Liquids

It would be possible to describe the equation of state of dense gases and liquids by means of the virial equation of state [Eq. (4.10)] but the numerical evaluation of many virial coefficients would be necessary. Furthermore, there is some question as to the convergence of the virial expansion in the neighborhood of the critical point. This approach has been used in the theoretical study of condensation, but no quantitative results are obtained [2a]. The most significant theoretical and computational advances have been made by going back either to Eq. (4.8) or to Eq. (4.9). The developments from either of these two equations necessitate at the present time the introduction of certain assumptions. For this reason the results of the two methods are not in agreement, and the discrepancy is a measure of the reasonableness of the assumptions involved. The derivation based on Eq. (4.8) leads to the so-called "lattice theories," in which a dense gas or liquid is pictured as a crystal lattice in which the molecules are free to roam to a certain extent from their lattice points. The derivation based on Eq. (4.9) involves the substitution of an approximate expression for $g(r)$ into the integral. Let us now summarize the results obtained by these two approaches.

Calculations Based on an Approximate Expression for the Configurational Integral. In the development of lattice theories one first supposes that the molecules of the dense gas or liquid are frozen at the lattice points of some sort of regular network just as though the substance were a perfect crystal. Then *one* molecule (called the *wanderer*) is allowed to stray from its lattice point, all other molecules being held at their lattice points. The wanderer can move in the "cage" formed by its nearest neighbors. This volume is called the *free volume* and hence lattice theories are sometimes also referred to as *free-volume theories*. The configurational integral for the wanderer may be calculated by taking into account the intermolecular forces between the wanderer and its nearest neighbors. The calculations may be refined by considering also the interactions with the next-nearest neighbors, and so forth. Now each molecule in the dense gas or liquid may be assumed to behave in very nearly the same manner as the wanderer. Hence it is assumed that the total configurational integral is just the configurational integral for the one wanderer raised to the Nth power. This type of approach is due originally to Lennard-Jones and Devonshire [3a].

Extensive calculations have been made [3b] for such a model in which the Lennard-Jones potential energy of interaction is used, and in which interactions between the wanderer and the first three shells of neighbors are included. The results can be presented in a simple tabular form as shown in Table 4.4A. The compressibility factor $p\tilde{V}/RT$ is there given in terms of the reduced volume $v^* = v/\sigma^3$ and the reduced temperature $T^* = kT/\epsilon$. Hence, given a value of the specific volume v and the temperature T, one may compute the reduced variables v^* and T^* by means of the parameters σ and ϵ given in Table 4.1; and then obtain the compressibility factor from Table 4.4A. Similar tables for the various thermodynamic functions have also been prepared [3b]. These tables have been shown to give good results at very high densities, but the agreement with experiment becomes less satisfactory as the density is decreased. This is easily understood, for at high densities the molecules are in fact restrained to move in a sort of free volume in a manner similar to that described by the model used in the calculations. At lower density, however, the molecules are free to roam out of their cages, and hence the model no longer describes physical reality.

A number of attempts have been made to improve

TABLE 4.4. COMPRESSIBILITY FACTOR $p\tilde{V}/RT$‡

A. Lennard-Jones-Devonshire 3-Shell Model

T^* \ v^*	1.131	1.414	2.121	3.536	4.243
0.80	−1.442	−2.413	−0.885	−0.0661	0.0990
1.00	0.1881	−0.8515	−0.0547	0.3985	0.4897
1.30	1.6128	0.5151	0.6933	0.8230	0.8340
1.60	2.4417	1.3168	1.1453	1.084	1.030
2.50	3.570	2.460	1.806	1.471	1.266
5.00	4.253	3.185	2.291	1.700	1.311
400.00	2.543	2.082	1.296	1.026	1.007

B. Radial Distribution Function Method [3m]

T^* \ v^*	1.222	1.483	2.260	3.632	13.82
0.833	−2.829	−2.433	−1.445	−0.594	
1.000	−1.382	−1.268	−0.734	−0.156	0.629
1.250	0.052	−0.115	−0.038	0.264	0.768
1.677	1.467	1.018	0.649	0.670	0.883
2.500	2.856	2.139	1.326	1.064	
5.000	4.223	3.242	1.998	1.456	
∞	5.567	4.333	2.667	1.833	1.167

‡ For complete tables and references, see [7a, pp. 1122–1125].

the lattice type of calculations [3c, 3d, 3w]. One type of modification is to consider the possibility that a cell may be doubly occupied [3e, 3f]. Another approach [3g] is to assume that there are vacant lattice sites or "holes," and that the number of holes increases with decreasing density. Unfortunately these "hole theories" have not been very successful.

De Boer and collaborators [3h, 3i, 3j] have developed a generalization of the cell methods in which the collective motion of sets of molecules in "cell clusters" is taken into account in a systematic manner. This treatment has been extended to mixtures by Cohen [3k]; the latter has also discussed the introduction of holes into the cell-cluster theory as a special case of the binary-mixture theory. Other generalizations of the cell theories are the "tunnel model" of Barker [3d] and the "worm model" or "supercell model" of Chung and Dahler [3l]; these models appear to be promising in their agreement with limited experimental data.

Calculations Based on an Approximate Expression for the Radial Distribution Function. The lattice theories and their various modifications may be regarded as satisfactory at very high densities and unsatisfactory at intermediate densities. An alternative approach is based on the use of the radial distribution function. It is possible to obtain an integral equation for the radial distribution function $g(r)$ (or the distribution function for "pairs," which is closely related to it) in terms of the next higher distribution function, that is, the distribution function for "triples." Since the latter is not known, it is impossible to solve the equation for $g(r)$. It is customary to obviate this difficulty by writing the distribution function for triples as the product of three distribution functions for pairs. This is known as the *superposition approximation*, the validity of which has not yet been fully assessed. When this approximation has been introduced, the integral equation may be solved to give an approximate curve for $g(r)$. This in turn can be used in Eq. (4.9) to obtain the equation of state. The radial distribution function has been evaluated for a (slightly modified) Lennard-Jones potential [3m]. The curves of $g(r)$ for various temperatures were used to obtain the compressibility factors given in Table 4.4B.

An alternative to the use of the superposition approximation has been developed by Percus and Yevick [3n, 3o]. The resulting equation for the radial distribution function has been solved exactly for the special case of rigid spheres of diameter σ [3p]; this leads to the equation of state

$$p\tilde{V}/\tilde{N}kT = (1 + \eta + \eta^2)/(1 - \eta)^3$$

where $\eta = b_0/4\tilde{V}$, and $b_0 = \frac{2}{3}\pi\tilde{N}\sigma^3$.

Calculations Based on Monte Carlo Methods and Molecular-dynamics Methods. Some direct calculations of the equation of state have been made using Monte Carlo methods and molecular-dynamics methods. The Monte Carlo approach provides a direct numerical evaluation of the configurational integral Q_N in Eq. (4.8) for a small number of particles. These techniques have been applied extensively by Wood and his collaborators for calculating the equation of state for the rigid-sphere and Lennard-Jones models [3q, 3r, 3s, 3t].

The molecular-dynamics method is a numerical method in which the detailed molecular trajectories of a few hundred molecules are computed; these results lead to an approximate evaluation of the radial distribution function $g(r)$. Alder and his collaborators have applied this method to the calculation of the equation of state of rigid spheres and mixtures of rigid spheres of different sizes [3u, 3v].

4. The Transport Properties of Dilute Gases

Diffusivity, viscosity, and thermal conductivity are referred to as *transport* properties, inasmuch as these properties describe the transport of mass, momentum, and energy by means of molecular motion and molecular collisions. Rigorous expressions for the transport properties of dilute gases were first obtained by Chapman and by Enskog through the solution of the Boltzmann equation for the distribution function. Their rigorous developments also predicted the transport of mass by thermal gradients (called *thermal diffusion* or the *Soret effect*) and the transport of energy resulting from concentration gradients (the *Dufour effect*). A full account of their theory and the results for pure gases and binary mixtures may be found in the treatise of Chapman and Cowling [7c]. The theory has also been extended to include multicomponent mixtures [4a]. The discussion in this section is restricted to the application of the latter theories and is valid for a classical, low-density gas composed of structureless particles. The extension to moderate densities is discussed in the following section.

Attempts to remove the various restrictions on the Chapman-Enskog theory have been made:

1. The Boltzmann equation for a low-density quantum gas of structureless particles was first proposed by Uehling and Uhlenbeck [4b]. The solution of this modified Boltzmann equation leads to expressions for the quantum-mechanical transport properties; the deviations from classical behavior (i.e., quantum corrections) are important only for the low-molecular-weight gases at low temperatures. The calculations for the quantum corrections, based on the Lennard-Jones potential, have been performed, and tabulated functions are available [4c, 4d].

2. Several studies of transport properties of dilute classical gases of molecules with rotational degrees of freedom have been made; these theories have been restricted to the rough-sphere model [4e], the loaded-sphere model [4f], and rigid ovaloids [4g].

3. A Boltzmann equation for a low-density gas of particles with structure was proposed by Wang Chang, Uhlenbeck, and de Boer [4h]. They have solved this modified Boltzmann equation to obtain expressions for the transport properties. A statistical mechanical derivation of the Boltzmann equation for this case has been given by Waldmann [4i] and by Snider [4j]. The resulting equation differs in detail from that proposed by Wang Chang, Uhlenbeck, and de Boer; however, in one special case (in fact, the only physically important case), the resulting expressions for the transport properties are identical, except for some additional terms which are presumed to be small, provided that the cross sections of Wang Chang, Uhlenbeck, and de Boer are interpreted as "degeneracy-averaged" cross sections [4k]. The transport-property expressions have been evaluated only for the

loaded-sphere model [4l]; they have also been used for the development of an approximate theory for transport properties of polar gases and gas mixtures [4m, 4n].

Summary of Theoretical Development. The Chapman-Enskog development differs from the earlier mean-free-path theories in that it attacks the fundamental problem of solving the Boltzmann integrodifferential equation for the molecular distribution function. The solution of the Boltzmann equation enables one to express the transport coefficients in terms of a set of double integrals, $\Omega^{(l,s)}$, which involve explicitly the dynamics of a binary collision and hence the intermolecular force law. The theoretical development is intricate, and the mathematical calculations lengthy. The results apply at low densities (where termolecular collisions can be considered unimportant) but not at such low densities that the dimensions of the vessel are large compared with the mean free path of the molecules (i.e., "Knudsen gases"). Strictly speaking the results apply to monatomic gases only. Practically, however, it is found that the internal motions of the molecules are relatively unimportant in mass and momentum transfer, and hence the results of the Chapman-Enskog theory may be applied to the calculation of the diffusivity and viscosity of polyatomic molecules. The same is not true for thermal conductivity where the internal modes of motion contribute substantially to the energy transport in the gas. This effect can be corrected for approximately, however, by the "Eucken correction" which is discussed subsequently. The Chapman-Enskog theory can be used only for angle-independent potential functions (such as the Lennard-Jones potential) and hence cannot be used to calculate rigorously the transport coefficients for polar molecules.

The evaluation of the $\Omega^{(l,s)}$ for the Lennard-Jones potential has been performed independently by several groups of investigators [4o, 4p, 4q, 4r, 4m]. The results are tabulated in terms of the functions $\Omega^{(l,s)\star}(T^*)$, which are just the values of the $\Omega^{(l,s)}$ for the Lennard-Jones potential divided by the corresponding values for rigid spheres. These results are given in Table 4.5. The function $\Omega^{(1,1)\star}$ is used for calculating the diffusivity; the function $\Omega^{(2,2)\star}$ is used for calculating the viscosity and thermal conductivity of pure gases. In the same table we also give the functions $A^\star$, $B^\star$, and $C^\star$ which are used in calculating the transport coefficients of mixtures. Similar tables of the $\Omega^{(l,s)\star}$ have been prepared for other nonpolar potential energy functions; in fact, a general program is now available for computing the $\Omega^{(l,s)\star}$ for an arbitrary potential energy function [4s].

For polar molecules "orientation-averaged" $\Omega^{(l,s)\star}$ have been evaluated and tabulated by Monchick and Mason [4m]. For polar gases the transport properties can be expressed approximately [4m, 4n] in terms of these orientation-averaged $\Omega^{(l,s)\star}$.

We now summarize the transport-property formulas to be used with the tabulations of $\Omega^{(l,s)\star}$ for polar and nonpolar molecules. These formulas are written in the forms most convenient for practical calculations. The following units are used: η(g/cm-sec), $\mathfrak{D}$(cm²/sec), λ(cal/cm − sec − deg), T(°K), p(atm), ϵ/k(°K), σ(A). When no subscripts are attached to the quantities $\Omega^{(l,s)\star}$, $A^\star$, $B^\star$, $C^\star$, they are to be calculated as functions of $T^* = kT/\epsilon$, where ϵ/k is the parameter given in Table 4.1 or 4.2 for interactions between pairs of similar molecules. When the subscripts ij are attached, these quantities are to be computed as functions of $T_{ij}^* = kT/\epsilon_{ij}$, where the ϵ_{ij}/k are the parameters characteristic of interactions between pairs of dissimilar molecules as given by the combining laws in Eq. (4.4) or (4.6). No comparison is given here between the experimental results and the values calculated according to these tables and formulas, since extensive tables have been presented elsewhere for this purpose [4r, 4m, 4n]. Generally the viscosity

TABLE 4.5. FUNCTIONS USED IN TRANSPORT-PROPERTY CALCULATIONS (LENNARD-JONES POTENTIAL)‡

T^*	$\Omega^{(1,1)\star}$	$\Omega^{(2,2)\star}$	$A^\star$	$B^\star$	$C^\star$
0.30	2.662	2.785	1.046	1.289	0.848
0.50	2.066	2.257	1.093	1.284	0.825
0.75	1.667	1.841	1.105	1.233	0.825
1.00	1.439	1.587	1.103	1.192	0.837
1.25	1.296	1.424	1.099	1.165	0.851
1.50	1.198	1.314	1.097	1.143	0.863
2.00	1.075	1.175	1.094	1.119	0.884
2.50	0.9996	1.093	1.094	1.106	0.899
3.00	0.9490	1.039	1.095	1.101	0.911
4.00	0.8836	0.9700	1.098	1.095	0.924
5.00	0.8422	0.9269	1.101	1.092	0.932
10.00	0.7424	0.8242	1.110	1.094	0.945
50.00	0.5756	0.6504	1.130	1.095	0.948
100.00	0.5130	0.5882	1.138	1.095	0.948
400.00	0.4170	0.4811	1.154	1.095	0.948

‡ For complete tables and references see [7a, pp. 1126–1131, 1215].

and diffusivity can be calculated to within 2 to 5 per cent over a range of temperature of about 400°K. The agreement for thermal conductivity is not quite so good since the theory for polyatomic molecules (the Eucken correction) is rough. The thermal diffusion ratio may be in error by as much as 10 per cent, for this property is highly sensitive to the exact form of the potential function which is used.

Viscosity. For a pure gas the viscosity is given by

$$\eta = (2.6693 \times 10^{-5}) \frac{\sqrt{MT}}{\sigma^2 \Omega^{(2,2)\star}} \tag{4.21}$$

where $\Omega^{(2,2)\star}$ is a tabulated function of the dimensionless temperature $T^* = kT/\epsilon$, depending on the choice of intermolecular potential.

In order to discuss the viscosity of mixtures it is convenient to define a quantity η_{12}, thus:

$$\eta_{12} = \frac{2.6693 \times 10^{-5} \sqrt{2M_1M_2T/(M_1 + M_2)}}{\sigma_{12}{}^2 \Omega_{12}{}^{(2,2)\star}} \tag{4.22}$$

This quantity may be regarded as the coefficient of viscosity of a hypothetical pure substance, the molecules of which have a molecular weight $2M_1M_2/(M_1 + M_2)$ and interact according to a potential curve specified by the interaction parameters σ_{12} and ϵ_{12}. [This is similar to the quantity B_{12} defined in Eq.

(4.19).] If in Eq. (4.22) the subscript 2 is replaced by 1, then the resulting expression is identical with the formula given in Eq. (4.21) for the pure substance 1.

It should be pointed out that the quantity η_{12} is closely related to the diffusivity:

$$\eta_{12} = \frac{5}{3} \frac{M_1 M_2}{(M_1 + M_2)} \frac{p\mathfrak{D}_{12}}{A^\star_{12} RT} \tag{4.23}$$

If diffusion data were available, it would be better to make use of them to obtain the quantity η_{12} than to use Eq. (4.22).

For multicomponent gas mixtures the coefficient of viscosity is (see [7*a*, p. 572] for illustrative example)

$$\eta_{\text{mix}} = -\frac{\begin{vmatrix} H_{11} & H_{12} & H_{13} & \cdots & H_{1\nu} & x_1 \\ H_{12} & H_{22} & H_{23} & \cdots & H_{2\nu} & x_2 \\ H_{13} & H_{23} & H_{33} & \cdots & H_{3\nu} & x_3 \\ \vdots & \vdots & \vdots & & \vdots & \vdots \\ H_{1\nu} & H_{2\nu} & H_{3\nu} & \cdots & H_{\nu\nu} & x_\nu \\ x_1 & x_2 & x_3 & \cdots & x_\nu & 0 \end{vmatrix}}{\begin{vmatrix} H_{11} & H_{12} & H_{13} & \cdots & H_{1\nu} \\ H_{12} & H_{22} & H_{23} & \cdots & H_{2\nu} \\ H_{13} & H_{23} & H_{33} & \cdots & H_{3\nu} \\ \vdots & \vdots & \vdots & & \vdots \\ H_{1\nu} & H_{2\nu} & H_{3\nu} & \cdots & H_{\nu\nu} \end{vmatrix}} \tag{4.24}$$

in which the matrix elements are

$$H_{ii} = \frac{x_1^2}{\eta_i} + \sum_{k \neq i}^{\nu} \frac{2x_i x_k}{\eta_{ik}} \frac{M_i M_k}{(M_i + M_k)^2} \left[\frac{5}{3A^\star_{ik}} + \frac{M_k}{M_i}\right]$$

$$\underset{i \neq j}{H_{ij}} = -\frac{2x_i x_j}{\eta_{ij}} \frac{M_i M_j}{(M_i + M_j)^2} \left[\frac{5}{3A^\star_{ij}} - 1\right] \tag{4.25}$$

Here, the $A_{ik}^\star$ are tabulated functions of the dimensionless temperature $T^*_{ik} = kT/\epsilon_{ik}$.

Because of the complexity of this formula, empirical expressions are often found to be useful. A particularly simple formula is that of Wilke [4*t*]:

$$\eta_{\text{mix}} = \sum_{i=1}^{\nu} \frac{x_i \eta_i}{\sum_{j=1}^{\nu} x_j \Phi_{ij}} \tag{4.26}$$

in which

$$\Phi_{ij} = \frac{1}{\sqrt{8}} \left(1 + \frac{M_i}{M_j}\right)^{-1/2} \left[1 + \left(\frac{\eta_i}{\eta_j}\right)^{1/2} \left(\frac{M_j}{M_i}\right)^{1/4}\right]^2 \tag{4.27}$$

where the x_i are mole fractions. Other approximate formulas for the viscosities of mixtures have been suggested [4*u*].

Thermal Conductivity. For a pure *monatomic gas* the thermal conductivity is given by

$$\lambda = (1.9891 \times 10^{-4}) \frac{\sqrt{T/M}}{\sigma^2 \Omega^{(2,2)\star}} = \frac{15}{4} \frac{R}{M} \eta \tag{4.28}$$

For a pure *polyatomic gas* the following approximate relation may be used

$$\lambda^{\text{Eucken}} = \frac{15}{4} \frac{R}{M} \eta \left(\frac{4}{15} \frac{\tilde{C}_v}{R} + \frac{3}{5}\right) \tag{4.29}$$

The factor in parentheses, which is unity for monatomic gases, is called the "Eucken correction." It takes into account approximately the transfer of energy between the translational and internal degrees of freedom when polyatomic molecules collide. Although the derivation of this Eucken correction involves several approximations, it agrees surprisingly well with experimental data. More rigorous expressions and detailed discussion of the theory have been given by Hirschfelder [4*v*] and by Mason and Monchick [4*n*].

To describe the thermal conductivity of mixtures, it is convenient to introduce the quantity λ_{12} (analogous to η_{12}) defined as

$$\lambda_{12} = (1.9891 \times 10^{-4}) \frac{\sqrt{T(M_1 + M_2)/2M_1 M_2}}{\sigma_{12}^2 \Omega_{12}^{(2,2)\star}} \tag{4.30}$$

The quantity λ_{12} is closely related to the diffusivity:

$$\lambda_{12} = \frac{25}{8} \frac{p\mathfrak{D}_{12}}{A^\star_{12} T} \tag{4.31}$$

When diffusivity data are available, they should be used to replace the calculated quantities λ_{12}.

The rigorous kinetic theory for multicomponent gas mixtures gives the following formula for the thermal conductivity of mixtures (see [7*a*, p. 1197, Note to p. 537]):

$$\lambda_{\text{mix}} = 4 \frac{\begin{vmatrix} L_{11} & \cdots & L_{1\nu} & x_1 \\ \vdots & & \vdots & \vdots \\ L_{\nu 1} & \cdots & L_{\nu\nu} & x_\nu \\ x_1 & \cdots & x_\nu & 0 \end{vmatrix}}{\begin{vmatrix} L_{11} & \cdots & L_{1\nu} \\ \vdots & & \vdots \\ L_{\nu 1} & \cdots & L_{\nu\nu} \end{vmatrix}} \tag{4.32}$$

in which

$$L_{ii} = -\frac{4x_i^2}{[\lambda_i]} - \sum_{k \neq i} \frac{2x_i x_k [\frac{15}{2} M_i^2 + \frac{25}{4} M_k^2 - 3M_k^2 B^\star_{ik} + 4M_i M_k A^\star_{ik}]}{(M_i + M_k)^2 A^\star_{ik} [\lambda_{ik}]} \tag{4.33}$$

$$L_{ij} = \frac{2x_i x_j M_i M_j [\frac{55}{4} - 3B^\star_{ij} - 4A^\star_{ij}]}{(M_i + M_j)^2 A^\star_{ij} [\lambda_{ij}]}, \quad i \neq j \tag{4.34}$$

This expression has been compared numerically with that predicted by several semiempirical expressions and experimental values for several ternary mixtures of noble gases by Srivastava and Saxena [4*w*].

For polyatomic gas mixtures it is possible to modify Eq. (4.32) to account approximately for the effect of the internal degrees of freedom by a generalization of the Eucken correction discussed above. The result is:

$$\lambda_{\text{mix}}^{\text{Eucken}} = \lambda_{\text{mix}} + \sum_{i=1}^{\nu} \frac{x_i(\lambda_i^{\text{Eucken}} - \lambda_i)}{\sum_{j=1}^{\nu} x_j \mathfrak{D}_{ii}/\mathfrak{D}_{ij}} \tag{4.35}$$

in which λ_{mix}, $\lambda_i^{\text{Eucken}}$, and λ_i are the quantities defined in Eqs. (4.32), (4.29), and (4.28), respectively. The Eucken correction can be further generalized for chemically reacting mixtures (see [7*a*, p. 1196, Note to p. 536]).

Because of the complexity of these formulas, empirical expressions are often useful. The following formula is particularly simple [4*x*, 7*e*]:

$$\lambda_{\text{mix}} = \sum_{i=1}^{\nu} \frac{x_i \lambda_i}{\sum_{j=1}^{\nu} x_j \Phi_{ij}} \tag{4.36}$$

in which the Φ_{ij} are given by Eq. (4.27). Other approximations for thermal conductivity of mixtures have been suggested [4*u*].

Diffusivity. The diffusivity of a binary gas mixture is given by

$$\mathfrak{D}_{12} = 0.0026280 \frac{\sqrt{T^3(M_1 + M_2)/2M_1M_2}}{p\sigma_{12}^2\Omega_{12}^{(1,1)\star}} \tag{4.37}$$

The fact that $\mathfrak{D}_{12}$ depends solely on the forces between dissimilar molecules means that the temperature dependence of the diffusivity provides an excellent method for evaluating the parameters σ_{12} and ϵ_{12} between unlike pairs of molecules.

When Eq. (4.37) is written for a single component, we obtain the formula for self-diffusivity:

$$\mathfrak{D} = 0.0026280 \frac{\sqrt{T^3/M}}{p\sigma^2\Omega^{(1,1)\star}} \tag{4.38}$$

Let us now inquire as to the meaning of self-diffusivity. Clearly, if the molecules are all physically identical, it is impossible to measure their interdiffusion. It is, however, possible to measure quantities experimentally which are very nearly self-diffusivities:

1. *Interdiffusion of heavy isotopes* for which

$$\sigma_{12} = \sigma_1 = \sigma_2$$

$\epsilon_{12} = \epsilon_1 = \epsilon_2$, and $2M_1M_2/(M_1 + M_2) \cong M_1$ or M_2.

2. *Interdiffusion of ortho and para forms* for which $2M_1M_2/(M_1 + M_2) = M_1 = M_2$ and $\sigma_{12} \cong \sigma_1$ or σ_2, $\epsilon_{12} \cong \epsilon_1$ or ϵ_2, slight differences in the interaction potential resulting from the effects of the different rotational states.

It is apparent that the self-diffusivity must be regarded as somewhat artificial. It is more correct simply to consider it as a limiting form of the binary diffusivity. It is interesting to note that in this limit of equal masses and equal intermolecular forces Eq. (4.23) simplifies to

$$\frac{\rho\mathfrak{D}}{\eta} = \frac{6}{5}A^\star \tag{4.39}$$

The function $A^\star$ is very slowly varying in T^* approximately equal to 1.1 (see Table 4.5), and hence $\rho\mathfrak{D}/\eta$ has very nearly the constant value 4/3.

The rigorous theory of multicomponent mixtures [4*a*] also gives expressions for the coefficients of diffusion in gases containing more than two components. (See [7*a*, pp. 1204, 1205, Notes to pp. 714 and 718].)

Thermal-diffusion Ratio. The thermal-diffusion ratio for a binary mixture of heavy isotopes is given by

$$k_T = \frac{15(2A^\star + 5)(6C^\star - 5)}{2A^\star(16A^\star - 12B^\star + 55)} \frac{(M_1 - M_2)}{(M_1 + M_2)} x_1 x_2 \tag{4.40}$$

in which $A^\star$, $B^\star$, and $C^\star$ are the functions of the reduced temperature T^* given in Table 4.5. Formulas are given in ref. 7*a* (pp. 541 et seq.) for computing thermal diffusion in mixtures of light isotopes, nonisotopic mixtures, and multicomponent mixtures.

5. The Transport Properties of Dense Gases and Liquids

The theory of transport phenomena in dense gases and liquids has been developed by means of nonequilibrium statistical mechanics [7*f*, 5*a*]. The development involves the nonequilibrium radial distribution function just as the equation of state of dense gases and liquids involves the equilibrium radial distribution function. The time evolution of the distribution functions is described by the Liouville equation [7*a*, Eq. 4.9–13] or by the set of contracted equations [7*a*, Eq. 4.9–15] which are known as the Bogoliubov-Born-Green-Kirkwood-Yvon (BBGKY) equations. Since the equation for each distribution function involves the distribution function of one higher order, it is necessary to truncate the series in some manner. Several methods of truncation have been suggested [7*g*, 5*b*, 5*c*]. The solution of these equations justifies the results obtained by Enskog, which are discussed below, and also leads to generalizations of these results. The theoretical development is not yet complete.

A theory of transport phenomena in liquids has also been developed by Eyring as a special application of the theory of absolute reaction rates [7*h*, 7*i*]. This theory has been moderately successful in predicting the transport coefficients in the liquid phase and explaining them on a simple pictorial basis. It is not possible by means of this approach to obtain expressions for the transport coefficients in terms of the intermolecular forces. Rather one obtains relationships among various macroscopic quantities.

We summarize briefly the results of the Enskog and Eyring theories.

The Enskog Theory (Dense Gases). The rigorous kinetic theory of dilute gases which was described in Sec. 4 is based on the Boltzmann equation. In its derivation it is assumed that there are two-body collisions only and that the molecules have no finite extension in space (or, more correctly, that the molecular diameter is negligibly small in comparison with the average distance between the molecules). Both of these assumptions are certainly valid in dilute gases. In dense gases, however, these two assumptions have to be reconsidered.

Enskog [5d, 7c] was the first to make an advance in this direction by developing a kinetic theory of dense gases made up of rigid spherical molecules of diameter σ. For this special molecular model there are no three-body and higher-order collisions. By thus considering only two-body collisions and by taking into account the finite size of the molecules he succeeded in grafting a theory of dense gases onto his earlier theory of dilute gases. As a gas is compressed, there are two effects which become important because molecules have volume: (1) *the "collisional transfer" of momentum and energy*—when two rigid-sphere molecules undergo a collision, there is an instantaneous transport of energy and momentum from the center of one molecule to the center of the other—and (2) *the change in the rate of collisions*—the frequency of collisions tends to become greater because σ is not negligibly small compared with the average distance between the molecules, and on the other hand the frequency of collisions tends to be smaller because the molecules are close enough to shield one another from oncoming molecules. The net increase in frequency of collisions is related to the radial distribution function. Enskog introduced the equilibrium radial distribution function, which, for rigid spheres, can be expressed explicitly in terms of the equation of state.

The results of the Enskog theory for the transport properties may be written in the form

$$\frac{\eta}{\eta^0}\left(\frac{\tilde{V}}{b_0}\right) = \frac{1}{y} + 0.8 + 0.7614y \qquad (4.41)$$

$$\frac{\kappa}{\eta^0}\left(\frac{\tilde{V}}{b_0}\right) = 1.002y \qquad (4.42)$$

$$\frac{\lambda}{\lambda^0}\left(\frac{\tilde{V}}{b_0}\right) = \frac{1}{y} + 1.2 + 0.7574y \qquad (4.43)$$

$$\frac{p\mathfrak{D}}{(p\mathfrak{D})^0}\left(\frac{\tilde{V}}{b_0}\right) = \frac{1}{y} + 1 \qquad (4.44)$$

In these relations $b_0 = \frac{2}{3}\pi\tilde{N}\sigma^3$ and $y = (p\tilde{V}/RT) - 1$; the properties labeled with a superscript "0" are those calculated from the low-pressure formulas [η^0, Eq. (4.21); λ^0, Eq. (4.28); and $(p\mathfrak{D})^0$, Eq. (4.38)]. At low to moderate densities, y is given by the virial equation for rigid spheres:

$$y = \left(\frac{b_0}{\tilde{V}}\right) + 0.6250\left(\frac{b_0}{\tilde{V}}\right)^2 + 0.2869\left(\frac{b_0}{\tilde{V}}\right)^3 + \cdots \qquad (4.45)$$

Substitution of Eq. (4.45) into Eqs. (4.41) to (4.44) leads to expressions for the transport properties as power series in the density.

Although the formulas for the transport properties given above were obtained for gases composed of rigid spherical molecules, Enskog showed that these results can be applied to real gases with reasonable success. In order to use the above formulas, it is necessary to specify b_0 and y. Enskog suggested that the pressure p in the expression for y be replaced by the "thermal pressure" $T(\partial p/\partial T)_V$ so that y may be determined from the experimental pVT data from the relation

$$y = \frac{\tilde{V}}{RT}\left[T\left(\frac{\partial p}{\partial T}\right)_{\tilde{V}}\right] - 1 \qquad (4.46)$$

Enskog also suggested that b_0 be evaluated by fitting the minimum in the curve of $(\eta/\eta^0)\tilde{V}$ as a function of y. The usefulness of this empirical method of Enskog has been demonstrated by numerical comparisons for N_2 and CO_2 for which high-density viscosity and equation-of-state data are available. The pressure dependence of the viscosity of N_2 is described quite accurately over a pressure range of 1,000 atm, and that of CO_2 over a 100-atm range. It should be pointed out that the Enskog theory does not include the effect of the internal degrees of freedom and hence does not apply to polyatomic gases. These effects are probably not important in shear viscosity and diffusivity; the effects of internal degrees of freedom on thermal conductivity and bulk viscosity are not known.

The Enskog theory has been generalized to binary mixtures of rigid spheres [7c, p. 292; 7a, pp. 646, 1202]. The extension of the theory to nonrigid spherical molecules has been explored, and calculations of the first density correction have been made [5e]. The errors introduced by Enskog's use of the equilibrium distribution function have also been examined [5f].

The effect of stable and metastable bound pairs of molecules on the first density corrections to the transport properties has been investigated by Stogryn and Hirschfelder [5g] and by Kim and Ross [5h]. These effects are important at low temperatures.

The Eyring Theory (Liquids). In the Eyring theory one considers as the fundamental process the movement of a molecule between two adjacent lattice sites; in order to obtain the viscosity of a fluid, it is necessary to know how the activation energy of the fundamental process is altered by a velocity gradient. This approach ultimately leads to the following formula:

$$\eta = nh \exp\left(\frac{0.408\Delta\tilde{U}_{\text{vap}}}{RT}\right) \qquad (4.47)$$

The general form of this relation is found from the theory, and the numerical constant 0.408 is obtained from experimental data. According to Trouton's rule $\Delta\tilde{U}_{\text{vap}} = 9.4RT_b$, where T_b is the boiling temperature at 1 atm pressure. Substitution of this into Eq. (4.47) gives the relation

$$\eta = nh \exp\left(\frac{3.8T_b}{T}\right) \qquad (4.48)$$

which is useful for rough estimates. The theory also predicts the following relation between viscosity and self-diffusivity:

$$\eta\mathfrak{D} = n^{1/3}kT \qquad (4.49)$$

Similar relationships, differing only by a numerical factor, have been derived by means of the "hydrodynamical theory" of diffusion—the Stokes-Einstein equation for diffusion of large molecules in a solvent, and the Sutherland-Einstein equation for self-diffusivity in liquids [5i, 5j]. A critical comparison of the hydrodynamic and activated-state theories has been given by Li and Chang [5k, 7i].

For a polyatomic liquid the thermal conductivity may be related to the speed of sound, c, by [5l]

$$\lambda = 2.80kn^{2/3}\gamma^{-1/2}c \qquad (4.50)$$

Here $\gamma = C_p/C_v$ should be taken to be those values which are found in speed of sound measurements. This formula has been found satisfactory for a large number of liquids (a mean deviation of around 10 per cent). Bridgman [5m] has pointed out that Eq. (4.50) gives the correct temperature dependence of the thermal conductivity of liquids at 1 atm. However, for most liquids the thermal conductivity increases by a factor of 2 when the pressure is raised to 12,000 atm, whereas Eq. (4.50) indicates that the thermal conductivity should increase by a factor of 4.

The Eyring theory has been used to study the viscosities of the liquids composed of molecules which are nonspherical, such as long-chain hydrocarbons and long-chain polymers. The influence of pressure on the viscosity of liquids has also been examined [7i].

6. The Principle of Corresponding States

As we have seen in the preceding sections, there are numerous gaps in the theory, the calculations, and the experimental data. When other methods cannot be used, dimensional analysis coupled with a limited amount of experimental data often provides a useful tool. Hence we summarize here the basic ideas connected with this method which leads to the principle of corresponding states.

Reduction in Terms of Critical Parameters. In the early studies of the behavior of matter it was recognized that all substances behave in a similar manner. For example, the pV isotherms of all substances were found to be of hyperbolic form at high temperatures, to exhibit a critical point, and to show condensation phenomena at low temperatures. The critical point occupies a unique place on the pV isotherms and represents a certain specific state of aggregation of the substance. It was suggested by van der Waals that the critical point is therefore a "point of corresponding states" and that the properties of various substances should be compared under conditions where the variables are the same multiples of the variables at the critical point. Accordingly the compressibility factor may be written as a universal function of the reduced pressure $p_r = p/p_c$ and the reduced temperature $T_r = T/T_c$:

$$\frac{p\tilde{V}}{RT} = Z(p_r, T_r) \tag{4.51}$$

in which $Z(p_r, T_r)$ is the same function for all substances (see Fig. 4.2). This is the basis for the widely used "generalized compressibility charts" and the "generalized charts of thermodynamic functions" of Hougen and Watson [6a]. These charts provide an excellent means for calculations because they are quick and easy to use.

In a similar fashion the principle of corresponding states may be used in the correlation of transport-property data. These quantities may also be expressed as universal functions of p_r and T_r:

$$\frac{\mathfrak{D}}{\mathfrak{D}_c} = \mathfrak{D}_r(p_r, T_r) \tag{4.52}$$

$$\frac{\eta}{\eta_c} = \eta_r(p_r, T_r) \tag{4.53}$$

$$\frac{\lambda}{\lambda_c} = \lambda_r(p_r, T_r) \tag{4.54}$$

Equation (4.53) provides the basis for Uyehara and Watson's "generalized viscosity chart" [6b] as well as more recent correlations [6u, 6v]. Equation (4.54)

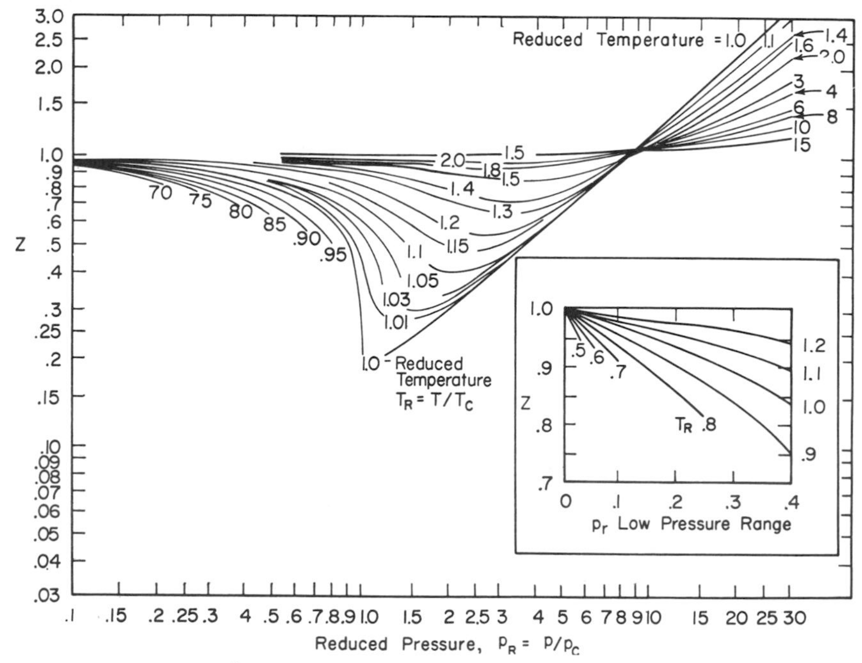

FIG. 4.2. Compressibility factor, $Z = p\tilde{V}/RT$, for gases and liquids. *(From: Hougen, O. A., and K. M. Watson: "Chemical Process Principles," chap.* XII, *John Wiley,* 1947.)

was used by Gamson [6c] and later workers [6w] to construct a similar chart for thermal conductivity of dense gases.

Other charts have been prepared by plotting the transport properties divided by their limiting values at zero pressure, thus:

$$\frac{p\mathfrak{D}}{(p\mathfrak{D})^0} = \mathfrak{D}\#\,(p_r,T_r) \tag{4.55}$$

$$\frac{\eta}{\eta^0} = \eta\#\,(p_r,T_r) \tag{4.56}$$

$$\frac{\lambda}{\lambda^0} = \lambda\#\,(p_r,T_r) \tag{4.57}$$

Equation (4.56) has been used by numerous investigators to correlate existing dense gas viscosity data, the most refined chart being that of Carr, Parent, and Peck [6d]. From viscosity data and the Enskog relations given in Eqs. (4.41) and (4.43), Comings and Nathan [6e] prepared a chart of the form of Eq. (4.57) for predicting dense gas thermal conductivities. Similar charts of the form of Eq. (4.55) have been prepared for self-diffusivity [6f, 6x]. High-density transport-property data are summarized in the monographs of Comings [6g] and Reid and Sherwood [6h].

Reduction in Terms of Critical Parameters, with Additional Correlating Parameters. In the early development it was recognized that all substances did not truly obey the principle of corresponding states. However, those substances containing molecules which are similar in structure tend to exhibit closely similar behavior. Thus the halogens obey one principle of corresponding states, whereas the saturated hydrocarbons obey another. This correlation of the bulk behavior of substances with the molecular structure was first suggested by Kamerlingh Onnes and was called by him the "principle of mechanical equivalence." Its usefulness lies in the fact that if data are available for the physical properties of several chemically related substances, then the properties of other substances of the same general structure may be estimated by the application of the principle of corresponding states.

The principle of mechanical equivalence has been applied in a semiempirical fashion with considerable success. Several approaches have been used; all of which involve the introduction of a "third parameter" into relations such as Eq. (4.51):

(*a*) Meissner and Seferian [6i] have suggested that all substances with the same value of the parameter $Z_c = p_c\tilde{V}_c/RT_c$ obey the same principle of corresponding states. That is, Eq. (4.51) should then be replaced by

$$\frac{p\tilde{V}}{RT} = Z(p_r,T_r;\, Z_c) \tag{4.58}$$

indicating explicit parametric dependence on Z_c (which varies from about 0.2 to 0.3). Equation (4.58) has been used by Lydersen, Greenkorn, and Hougen [6j, 6k] to prepare an extensive correlation of pVT data and thermodynamic properties of gases and liquids.

(*b*) Riedel [6l] proposed that all substances with the same value of $Y_c = (d \ln p_{\text{vap}}/d \ln T)_c$ follow the same principle of corresponding states. That is, Eq. (4.51) should be replaced by

$$\frac{p\tilde{V}}{RT} = Z(p_r,T_r;\, Y_c) \tag{4.59}$$

Riedel has had excellent success in correlating many physical properties by means of this principle of corresponding states [6m]. An almost identical suggestion was made subsequently by Pitzer [6n] and used by him and his colleagues for correlating compressibility data [6o].

(*c*) Hirschfelder, Buehler, McGee, and Sutton have developed an analytical equation of state [6p] and analytical relations for the thermodynamical excess functions which cover the full range of gases and liquids. The standard form of these equations makes use of both the parameter Z_c and the parameter Y_c. However, the various constants in the equations can be adjusted to give a very close fit either to the properties for any particular substance or to any of the tabular forms of the generalized corresponding states equation of state. For example, at densities less than the critical density a useful equation of state is

$$\left(p_r + \frac{\beta}{V_r^2}\right)[\beta(3\beta - 1)V_r - (3\beta^2 - 6\beta - 1) + \beta(\beta - 3)V_r^{-1}] = (\beta + 1)^3 T_r \tag{4.60}$$

in which $p_r = p/p_c$, $V_r = V/V_c$, $T_r = T/T_c$, and β is a parameter closely related to $Z_c = p_c\tilde{V}_c/RT_c$:

$$Z_c = \frac{\beta(3\beta - 1)}{(\beta + 1)^3} \tag{4.61}$$

For noble gases $\beta \cong 6$, for light hydrocarbons $\beta \cong 7$, and for water $\beta \cong 9$. It should be noted that Eq. (4.60) simplifies to the van der Waals equation when β is assigned the very unrealistic value of $\beta = 3$.

The above discussion concerns correlation of pVT data; other properties may be correlated as well. For example, the Z_c and Y_c schemes described have been applied to surface tension to yield the following results [6q]:

$$\gamma'_r = \left(-0.951 + \frac{0.432}{Z_c}\right)(1 - T_r)^{11/9} \tag{4.62}$$

$$\gamma'_r = (-0.281 + 0.133Y_c)(1 - T_r)^{11/9} \tag{4.63}$$

in which $\gamma'_r = \gamma'/p_c^{2/3}T_c^{1/3}$ is a reduced surface tension.

Reduction in Terms of Molecular Parameters. The principle of corresponding states has been studied extensively from a molecular point of view [6r]. If the molecules in a substance obey a two-constant potential function of the form

$$\varphi(r) = \epsilon f\left(\frac{r}{\sigma}\right) \tag{4.64}$$

(the Lennard-Jones potential is of this form), then the various variables describing the properties of the substance and the state of the system may be expressed in reduced units by means of the appropriate combinations of the parameters ϵ and σ: $p^* = p\sigma^3/\epsilon$, $T^* = kT/\epsilon$, and $v^* = v/\sigma^3$. Hence the equation of state assumes the reduced form

$$p^* = p^*(v^*,T^*) \tag{4.65}$$

That is, the reduced pressure is a function of the reduced volume and the reduced temperature and is the same for all substances. The parameters σ and ϵ may be expressed in terms of critical properties for many nonpolar substances [see Eq. (4.3)]. If these relations are used in Eq. (4.65), we can obtain the fact that $p_r = p_r(V_r,T_r)$, which brings out the connection with the older statement of the principle of corresponding state based on critical properties.

Similarly the transport properties may be expressed in terms of corresponding states relationships, based on reduction of quantities with the σ and ϵ of the potential function:

$$\mathfrak{D}^* = \frac{\mathfrak{D}\sigma^2}{\sqrt{m\epsilon}}\frac{m}{\sigma^3} = \mathfrak{D}^*(v^*,T^*) \tag{4.66}$$

$$\eta^* = \frac{\eta\sigma^2}{\sqrt{m\epsilon}} = \eta^*(v^*,T^*) \tag{4.67}$$

$$\lambda^* = \frac{\lambda\sigma^2}{\sqrt{m\epsilon}}\frac{m}{k} = \lambda^*(v^*,T^*) \tag{4.68}$$

Thus the reduced transport properties are universal functions of the reduced volume and the reduced temperature. Other properties, such as vapor pressure and surface tension, may also be expressed in this fashion.

For molecules which do not have spherically symmetric potential functions, the use of reduced variables leads to a generalization of this result. As an example, we consider those polar molecules which obey the Stockmayer potential. Then the equation of state in dimensionless form for a substance containing molecules with dipole moment μ assumes the form

$$p^* = p^*(v^*,T^*;\mu^*) \tag{4.69}$$

where $\mu^* = \mu/\sqrt{\epsilon\sigma^3}$ is the reduced dipole moment of the molecule. Thus, molecules that have the same value of μ^* obey the same principle of corresponding states. Similar results are found for elongated molecules of length l and "width" σ:

$$p^* = p^*(v^*,T^*;l^*) \tag{4.70}$$

where $l^* = l/\sigma$ is the length-to-width ratio of the molecules. Equations (4.69) and (4.70) are examples of the principle of mechanical equivalence of Kamerlingh Onnes.

The quantum effects in the noble gas series have been studied extensively by means of the quantum-mechanical principle of corresponding states [6s, 6t]. According to this principle, the equation of state is given by the reduced equation

$$p^* = p^*(v^*,T^*;\Lambda^*) \tag{4.71}$$

in which Λ^* is the de Boer quantum-mechanical parameter $h/\sigma\sqrt{m\epsilon}$. The value of Λ^* is a measure of the importance of quantum effects for various substances. The larger the value of Λ^*, the more a substance is expected to deviate from classical behavior at a given reduced temperature. For the noble gases, the values of Λ^* are: He^3, 3.08; He^4, 2.67; Ne, 0.59; A, 0.19; Kr, 0.10; and Xe, 0.06. $\Lambda^* = 0$ corresponds to classical behavior. For most substances Λ^* is small compared with unity, and quantum effects are negligible except at extremely low temperatures. For the isotopes of helium and hydrogen, however, quantum effects are observable (though small) at room temperature and become quite large at very low temperatures. The quantum-mechanical principle of corresponding states was used to predict the properties of the light helium isotope [6t; 7a, p. 430] and the hydrogen isotopes [6s].

List of Symbols

R = gas constant
k = Boltzmann constant = $R/\tilde{N}$
$\tilde{N}$ = Avogadro number
h = Planck's constant
x_i, M_i = mole fraction and molecular weight of the ith chemical species in a mixture
T = temperature
p = pressure
p_{vap} = vapor pressure
$\tilde{V}$ = volume per mole
v = volume per molecule
ρ = density (mass per unit volume)
n = number density (number per unit volume)
φ = intermolecular potential energy function
σ, ϵ = parameters in intermolecular potential energy function
μ = dipole moment of a molecule
α = polarizability of a molecule
B, C, D = second, third, and fourth virial coefficients
$\mathfrak{D}$, η, λ = diffusivity, viscosity, thermal conductivity
k_T = thermal-diffusion ratio
μ^0 = zero pressure Joule-Thomson coefficient
γ = ratio of specific heats
γ' = surface tension
$\Omega^{(l,s)\star}$ = integrals in terms of which the transport properties are expressed
$A^\star$, $B^\star$, $C^\star$ = special combinations of the $\Omega^{(l,s)\star}$ which occur in the formulas for the transport properties of mixtures
0 (superscript) = limiting value at zero pressure
~ (above symbol) = quantity per mole
c (subscript) = quantity evaluated at the critical point
r (subscript) = quantity reduced by dividing by the value of the quantity at the critical point
* (superscript) = quantity reduced by dividing by the proper combination of σ and ϵ
★ (superscript) = quantity reduced by dividing by combinations of σ and ϵ in such a way as to make use of rigid-sphere quantities

$T^* = kT/\epsilon$ $\quad t^* = \mu^{*2}/\sqrt{8}$
$v^* = v/\sigma^3$ $\quad \alpha^* = \alpha/\sigma^3$
$p^* = p\sigma^3/\epsilon$ $\quad B^\star = B/b_0$
$\mu^* = \mu/\sqrt{\epsilon\sigma^3}$ $\quad C^\star = C/b_0^2$

References

Intermolecular Forces:

1*a*. London, F.: *Trans. Faraday Soc.*, **33**: pt. 1, 8–26 (1937).
1*b*. Margenau, H.: *Revs. Mod. Phys.*, **11**: 1–35 (1939).
1*c*. De Boer, J.: *Ned. Tijdschr. Natuurk.*, **19**: 231–250 (1953).
1*d*. London, F.: *J. Phys. Chem.*, **46**: 305–316 (1942).
1*e*. Coulson, C. A., and P. L. Davies: *Trans. Faraday Soc.*, **48**: 777–788 (1952).
1*f*. Bleick, W. E., and J. E. Mayer: *J. Chem. Phys.*, **2**: 252–259 (1934).
1*g*. De Boer, J.: *Physica*, **9**: 363–382 (1942).
1*h*. Buckingham, R. A., and J. Corner: *Proc. Roy. Soc. (London)*, **A189**: 118–129 (1947).
1*i*. Corner, J.: *Proc. Roy. Soc. (London)*, **A192**: 275–291 (1948).
1*j*. Kihara, T.: *J. Phys. Soc. Japan*, **6**: 289–296 (1941).
1*k*. Rowlinson, J. S.: *Trans. Faraday Soc.*, **47**: 120–129 (1951).
1*l*. Muckenfuss, C., C. F. Curtiss, and R. B. Bird: *J. Chem. Phys.*, **23**: 1542–1543 (1955).
1*m*. Tee, L. S., S. Gotoh, and W. E. Stewart, *Ind. Eng. Chem. Fund.*, **5**: 56–363 (1966).

Virial Coefficients:

2*a*. Mayer, J. E., and M. G. Mayer: "Statistical Mechanics," Wiley, New York, 1940.
2*b*. Münster, A.: "Statistische Thermodynamik," Springer, Berlin, 1956.
2*c*. Bird, R. B., and J. R. Brock: *Am. Inst. Chem. Engrs. J.*, **5**: 436 (1959).
2*d*. Rice, W. E., and J. O. Hirschfelder: *J. Chem. Phys.*, **22**: 187–192 (1954).
2*e*. Kihara, T.: *Revs. Mod. Phys.*, **25**: 831–843 (1953).
2*f*. Kihara, T.: *Revs. Mod. Phys.*, **27**: 412–423 (1955).

Liquid State and Cell Theories:

3*a*. Lennard-Jones, J. E., and A. F. Devonshire: *Proc. Roy. Soc. (London)*, **A163**: 53 (1937).
3*b*. Wentorf, R. H., Jr., R. J. Buehler, J. O. Hirschfelder, and C. F. Curtiss: *J. Chem. Phys.*, **18**: 1484 (1950).
3*c*. Rowlinson, J. S., and C. F. Curtiss: *J. Chem. Phys.*, **19**: 1519 (1951).
3*d*. Barker, J. A.: "Lattice Theories of the Liquid State," Pergamon Press, New York, 1963.
3*e*. Janssens, P., and I. Prigogine: *Physica*, **16**: 851 (1950).
3*f*. Pople, J.: *Phil. Mag.*, **42**: 459 (1951).
3*g*. Ono, S.: *Mem. Fac. Eng., Kyushu Univ.*, **10**(4): 190 (1947).
3*h*. De Boer, J.: *Physica*, **20**: 655 (1954).
3*i*. Cohen, E. G. D., J. de Boer, and Z. W. Salzburg: *Physica*, **21**: 137 (1955); **23**: 389 (1957).
3*j*. Salzburg, Z. W., E. G. D. Cohen, B. C. Reithmeir, and J. de Boer: *Physica*, **23**: 407 (1957).
3*k*. Cohen, E. G. D.: On the Theory of the Liquid State, Ph.D. Thesis, University of Amsterdam, 1957.
3*l*. Chung, H. S., and J. S. Dahler: *J. Chem. Phys.*, **40**: 2868 (1964); **42**: 2374 (1965).
3*m*. Kirkwood, J. G., V. A. Lewinson, and B. J. Alder: *J. Chem. Phys.*, **20**: 929–938 (1952).
3*n*. Percus, J. K., and G. J. Yevick: *Phys. Rev.*, **110**: 1 (1957).
3*o*. Percus, J. K.: *Phys. Rev. Letters*, **8**: 462 (1962).
3*p*. Wertheim, M. S.: *Phys. Rev. Letters*, **10**: 321 (1963).
3*q*. Wood, W. W., and F. R. Parker: *J. Chem. Phys.*, **27**: 720 (1957).
3*r*. Wood, W. W., and J. D. Jacobson: *J. Chem. Phys.*, **27**: 1207 (1957).
3*s*. Wood, W. W., F. R. Parker, and J. D. Jacobson: *Nuovo Cimento, Suppl.* **9** (ser. X): 133 (1958).
3*t*. Salsburg, Z. W.: *J. Chem. Phys.*, **37**: 798 (1962).
3*u*. Alder, B. J., and T. E. Wainwright: *J. Chem. Phys.*, **33**: 1439 (1960).
3*v*. Alder, B. J.: *J. Chem. Phys.*, **40**: 2724 (1964).
3*w*. Levelt, J. M. H., and E. D. G. Cohen: "Studies in Statistical Mechanics," II, Part B, Wiley, New York, 1964.

Gas Transport Properties:

4*a*. Curtiss, C. F., and J. O. Hirschfelder: *J. Chem. Phys.*, **17**: 550 (1949).
4*b*. Uehling, E. A., and G. E. Uhlenbeck: *Phys. Rev.*, **43**: 552 (1933).
4*c*. Imam-Rahajoe, S., C. F. Curtiss, and R. B. Bernstein: *J. Chem. Phys.*, **42**: 530 (1965).
4*d*. Munn, R. J., F. J. Smith, and E. A. Mason: *J. Chem. Phys.*, **42**: 537 (1965).
4*e*. Condiff, D. W., Wer-Koa Lu, and J. S. Dahler: *J. Chem. Phys.*, **42**: 3445 (1965).
4*f*. Dahler, J. S., and N. F. Sather: *J. Chem. Phys.*, **38**: 2363 (1963).
4*g*. Curtiss, C. F., and C. Muckenfuss: *J. Chem. Phys.*, **26**: 1619 (1957).
4*h*. Wang Chang, C. S., G. E. Uhlenbeck, and J. de Boer: "Studies in Statistical Mechanics," II, Part C, Wiley, New York, 1964.
4*i*. Waldmann, L.: *Z. Naturforsch.*, **12a**: 660 (1957).
4*j*. Snider, R. F.: *J. Chem. Phys.*, **32**: 1051 (1960).
4*k*. McCourt, F. R., and R. F. Snider: *J. Chem. Phys.*, **41**: 3185 (1965).
4*l*. Mueller, J. J., and C. F. Curtiss: *J. Chem. Phys.*, **45** (1966).
4*m*. Monchick, L., and E. A. Mason: *J. Chem. Phys.*, **35**: 1676 (1961).
4*n*. Mason, E. A., and L. Monchick: *J. Chem. Phys.*, **36**: 1622, 2746 (1962).
4*o*. Kihara, T., and M. Kotani: *Proc. Phys. Math. Soc. Japan*, **24**: 76 (1942).
4*p*. De Boer, J., and J. van Kranendonk: *Physica*, **14**: 442 (1948).
4*q*. Rowlinson, J. S.: *J. Chem. Phys.*, **17**: 101 (1949).
4*r*. Hirschfelder, J. O., R. B. Bird, and E. L. Spotz: *J. Chem. Phys.*, **16**: 968 (1948); **17**: 149 (1950); *Chem. Revs.*, **44**: 205 (1949).
4*s*. Smith, F. J., and R. J. Munn: *J. Chem. Phys.*, **41**: 3560 (1964).
4*t*. Wilke, C. R.: *J. Chem. Phys.*, **18**: 517 (1950).
4*u*. Brokaw, R. S.: *J. Chem. Phys.*, **42**: 1140 (1965).
4*v*. Hirschfelder, J. O.: *J. Chem. Phys.*, **26**: 282 (1957).
4*w*. Srivastava, B. N., and S. C. Saxena: *J. Chem. Phys.*, **27**: 583 (1957).
4*x*. Mason, E. A., and S. C. Saxena: *Phys. Fluids*, **1**: 361 (1958).

Liquid Transport Properties:

5*a*. Kirkwood, J. G., and colleagues: *J. Chem. Phys.* **14**: 180–201, 346 (1946); **15**: 72–76, 155 (1947); **17**: 988–994 (1949); **18**: 901–902 (1950); **18**: 817–829 (1950); **19**: 1173–1180 (1951); **21**: 2050–2055 (1953).
5*b*. Hollinger, H. B., and C. F. Curtiss: *J. Chem. Phys.*, **33**: 1386 (1960).
5*c*. Hoffman, D. K., and C. F. Curtiss: *Phys. Fluids*, **7**: 1887 (1964).
5*d*. Enskog, D.: *Kgl. Svenska Vetenskapsakad. Handl.*, **63**(4): (1922) (in German).
5*e*. Hoffman, D. K., and C. F. Curtiss: *Phys. Fluids*, **8**: 667, 890 (1965).
5*f*. Livingston, P. M., and C. F. Curtiss: *Phys. Fluids*, **4**: 816 (1961).

5g. Stogryn, D. E., and J. O. Hirschfelder: *J. Chem. Phys.*, **31**: 1531 (1960).
5h. Kim, S. K., and J. Ross: *J. Chem. Phys.*, **42**: 263 (1965).
5i. Sutherland, W.: *Phil. Mag.*, **9**: 781 (1905).
5j. Fürth, R.: in "Handbuch der Physikalischen und Technischen Mechanik," vol. 7, pp. 635 et seq., Barth, Leipzig, 1931.
5k. Li, J. C. M., and P. Chang: *J. Chem. Phys.*, **23**: 518 (1955).
5l. Powell, R. E., W. E. Roseveare, and H. Eyring; *Ind. Eng. Chem.*, **33**: 430 (1941).
5m. Bridgman, P. W.: *Proc. Am. Acad. Arts Sci.*, **59**: 141 (1923).

Corresponding States:

6a. Hougen, O. A., and K. M. Watson: "Chemical Processes Principles," chap. 12, Wiley, New York, 1947.
6b. Uyehara, O. A., and K. M. Watson: *Natl. Petroleum News, Tech. Sec.*, **36**: R764 (Oct. 4, 1944).
6c. Gamson, B. W.: *Chem. Eng. Progr.*, **45**: 154 (1949).
6d. Carr, N. L., J. D. Parent, and R. E. Peck: *Chem. Eng. Progr., Symp. Ser.*, **51**(16): 91 (1955).
6e. Comings, E. W., and M. F. Nathan: *Ind. Eng. Chem.*, **39**: 964 (1947).
6f. Slattery, J. C.: M.S. Thesis, University of Wisconsin, 1955.
6g. Comings, E. W.: "High Pressure Technology," McGraw-Hill, New York, 1956.
6h. Reid, R. C., and T. K. Sherwood: "The Properties of Gases and Liquids," McGraw-Hill, New York, 1958.
6i. Meissner, H. P., and R. Seferian: *Chem. Eng. Progr.*, **47**: 579 (1951).
6j. Lydersen, A. L., R. A. Greenkorn, and O. A. Hougen: Thermodynamic Properties of Pure Fluids from the Theorem of Corresponding States, *Univ. Wisconsin Eng. Expt. Sta. Rept.* 4, 1955.
6k. Hougen, O. A., K. M. Watson, and R. A. Ragatz: "Chemical Process Principles," Wiley, New York, 1959.
6l. Riedel, L.: *Chem.-Ing.-Tech.*, **26**: 83 (1954).
6m. Riedel, L.: *Chem.-Ing.-Tech.*, **26**: 259, 679 (1955); **27**: 209, 475 (1956).
6n. Pitzer, K. S.: *J. Am. Chem. Soc.*, **77**: 3427–3433 (1955).
6o. Pitzer, K. S., D. Z. Lipmann, R. F. Curl, Jr., C. M. Huggins, and D. E. Petersen: *J. Am. Chem. Soc.*, **77**: 3433–3440 (1955).
6p. Hirschfelder, J. O., R. J. Buehler, H. A. McGee, and J. R. Sutton: *Ind. Eng. Chem.*, **50**: 375, 386 (1958); [see *errata, Ind. Eng. Chem., Fund.*, **1**: 224 (1962)].
6q. Brock, J. R., and R. B. Bird: *Am. Inst. Chem. Eng. J.*, **1**: 174–177 (1955).
6r. De Boer, J.: Doctoral Dissertation, University of Amsterdam, 1940; *Physica*, **14**: 139 (1948).
6s. De Boer, J., and B. S. Blaisse: *Physica*, **14**: 149 (1948).
6t. De Boer, J., and R. J. Lunbeck: *Physica*, **14**: 318, 510, 520 (1948).
6u. Carr, N. L., R. Kobayashi, and D. B. Burroughs: *Am. Inst. Mining and Met. Engrs., Petroleum Technology*, **6**: 47 (1954).
6v. Jossi, J. A., L. I. Stiel, and G. Thodos: *Am. Inst. Chem. Eng. J.*, **8**: 59–63 (1963).
6w. Owens, E. J., and G. Thodos: *Am. Inst. Chem. Eng. J.*, **3**: 454–461 (1957).
6x. Tee, L. S., R. C. Robinson, G. F. Kuether, and W. E. Stewart: *Trans. American Petroleum Institute* (in press).

General:

7a. Hirschfelder, J. O., C. F. Curtiss, and R. B. Bird: "Molecular Theory of Gases and Liquids," 2d printing with notes added, Wiley, New York, 1964.
7b. De Boer, J.: *Repts. Progr. Phys.*, **12**: 305 (1949).
7c. Chapman, S., and T. G. Cowling: "Mathematical Theory of Non-uniform Gases," 2d printing with notes added, Cambridge University Press, New York, 1951.
7d. Landau, L. D., and E. M. Lifshitz: "Fluid Mechanics," p. 224, Addison-Wesley, Reading, Mass., 1959.
7e. Bird, R. B., W. E. Stewart, and E. N. Lightfoot: "Transport Phenomena," 7th corrected printing, Wiley, New York, 1966.
7f. Born, M., and H. S. Green: "A General Kinetic Theory of Liquids," Cambridge University Press, New York, 1949.
7g. Bogoliubov, N. N.: "Studies in Statistical Mechanics," vol. 1, North Holland Publishing Company, Amsterdam, 1962.
7h. Glasstone, S., K. J. Laidler, and H. Eyring: "The Theory of Rate Processes," McGraw-Hill, New York, 1941.
7i. Eyring, H., D. Henderson, B. J. Stover, and E. M. Eyring: "Statistical Mechanics and Dynamics," Wiley, New York, 1964.

Chapter 5

Heat Transfer

By E. U. CONDON, University of Colorado

1. Heat Conductivity

When a temperature inequality exists in a body there is a net transfer of thermal energy from warmer to colder regions, tending to produce a state of temperature uniformity. The net transfer of energy in unit time across the vector element of area $\mathbf{dS}$ may be characterized by a vector field $\mathbf{I}$ (watts/cm²) or (cal/cm²-sec) such that the net transfer of heat across $\mathbf{dS}$ is $\mathbf{I} \cdot \mathbf{dS}$ (watts) or (cal/sec).

Experiment shows that in isotropic materials $\mathbf{I}$ is proportional to the temperature gradient grad T. The proportionality factor k is called the heat or thermal conductivity of the material. The basic law of heat conduction is thus

$$\mathbf{I} = -k \text{ grad } T \tag{5.1}$$

where k is to be replaced by a symmetric second-order tensor in the case of anisotropic materials.

Conversion factors for various units of k are

$$1 \frac{\text{cal/cm}^2\text{-sec}}{°\text{C/cm}} = 4.180 \frac{\text{watt/cm}^2}{°\text{C/cm}}$$

$$1 \frac{\text{Btu/ft}^2\text{-hr}}{°\text{F/in.}} = \tfrac{1}{12} \frac{\text{Btu/ft}^2\text{-hr}}{°\text{F/ft}}$$

$$= 5.191 \frac{\text{watts/cm}^2}{°\text{C/cm}}$$

$$= 1.24 \frac{\text{cal/cm}^2\text{-sec}}{°\text{C/cm}}$$

$$1 \frac{\text{cal/cm}^2\text{-sec}}{°\text{C/cm}} = 242 \frac{\text{Btu/ft}^2\text{-hr}}{°\text{F/ft}}$$

$$= 360 \frac{\text{kcal/m}^2\text{-hr}}{°\text{C/m}}$$

Some typical values of k for various materials are given in Table 5.1.

2. Equation of Heat Conduction

The net rate at which heat flows out of any volume V bounded by the closed surface S is

$$\int \mathbf{I} \cdot \mathbf{dS} = \int \text{div } \mathbf{I} \, dV \tag{5.2}$$

If Q is the rate at which heat is being generated in unit volume, and if ρ and c are the density and specific heat, respectively, then

$$\int Q \, dV = \int \left(\rho c \frac{\partial T}{\partial t} + \text{div } \mathbf{I} \right) dV \tag{5.3}$$

Using (5.1), this gives the dynamical differential equation for heat conduction:

$$\rho c \frac{\partial T}{\partial t} = \text{div } (k \text{ grad } T) + Q \tag{5.4}$$

In case k, ρ and c are constant, it is convenient to write $Q/\rho c = H$ and $\alpha = k/\rho c$ so that (5.4) becomes

$$\frac{\partial T}{\partial t} = \alpha \, \Delta T + H \tag{5.5}$$

The combination of constants α (cm²/sec) is known as the *thermal diffusivity*.

TABLE 5.1A. THERMAL CONDUCTIVITY OF GASES
Unit is 10^{-5} (cal/cm²-sec)/(°C/cm)

Gas	Temperature, °C						
	−180	−100	−50	0	20	100	200
Hydrogen	....	21.8	35.0	41.9	44.5	54.7	63.4
Deuterium	....	....	....	30.6	....	37.7	
He	16.3	24.6	29.6	34.3	36.1	40.8	
Ne	4.9	8.2	10.9	.....	....	13.3	
A	1.4	2.6	3.2	3.9	4.2	5.2	
Kr	....	....	....	2.1			
Xe	....	....	....	1.2	....	1.6	
Air (dry)	2.1	3.9	4.9	5.76	6.1	7.4	8.8
O_2	2.0	3.9	4.9	5.8	6.2	7.6	
N_2	2.1	3.9	4.9	5.7	6.1	7.3	8.5
H_2S	....	....	....	3.0			
SO_2	....	....	....	2.0			
CS_2	....	....	....	1.6			
NH_3	....	....	....	5.2			
N_2O	....	....	....	3.6			
NO	....	....	....	4.6			
CO	....	....	....	5.3			
CO_2	....	....	....	3.4			
CH_4	....	....	....	7.3			
C_2H_6	....	....	....	4.3			
C_2H_4	....	....	....	4.0			
C_2H_2	....	....	....	4.4			
C_3H_8	....	....	....	3.6			
n-C_4H_{10}	....	....	....	3.2			
Benzene C_6H_6	....	....	....	2.1			

TABLE 5.1B. THERMAL CONDUCTIVITY OF LIQUIDS
Unit is 10^{-5} (cal/cm²-sec)/(°C/cm)

Substance	Temp, °C	Conductivity
Water	0	139
	4	138
	20	143
	75	160
Acetone	0	44
Carbon tetrachloride	. .	28
Pentane	0	33
Glycerin	0	68
Methyl alcohol	0	51
Ethyl alcohol	0	45
Propyl alcohol	0	41
Butyl alcohol	0	41
Formic acid	12	65
Acetic acid	12	47
Propionic acid	12	39
Diethyl ether	. .	34

TABLE 5.1C. THERMAL CONDUCTIVITY OF SOLIDS (NONMETALS)
Unit is 10^{-5} (cal/cm²-sec)/(°C/cm)

Fibrous materials loose, such as kapok, cotton, silk, and wool, approximately fall in the range 8 to 14, depending on density of packing.

Paper: the approximate range for various kinds is 33 to 83.

Material	Conductivity	Material	Conductivity
Minerals and rocks:		Minerals and rocks:	
Fire brick	28	Basalt	400
Portland cement	71	Feldspar	580
Red brick	150	Granite	700
Chalk	200	Limestone	530
Diatom. earth	13	Marble	670
Graphite	1,200	Soda-lime glass	170
Magnesia	16	Pyrex glass	253
		Fused silica	320
Woods (dry across grain):		Woods (dry across grain):	
Cypress	23	Oak	35
White pine	27	Maple	38
Mahogany	31	Cork	72

In steady-state problems, $\partial T/\partial t = 0$, so that (5.4) takes on the form of the Poisson equation of electrostatics. If $k(T)$ depends on the temperature, one may introduce an effective temperature scale,

$$U = \int^T k(T)\,dT \qquad \text{watts/cm} \tag{5.6}$$

in which case the equation for the effective temperature U takes the form

$$\Delta U + Q = 0$$

so that the heat source Q is a source of the field scalar U just as electrostatic charge is a source for the field of electrostatic potential. Therefore a steady-state heat-flow problem corresponds to every electrostatic field problem, and all the mathematical devices used for electrostatic field problems are applicable also in steady-state heat-flow problems. In particular, in an infinite medium, the temperature distribution $U(\mathbf{r})$, resulting from a given distribution of heat sources, is

$$U(\mathbf{r}) = \frac{1}{4\pi}\int \frac{Q(\mathbf{r}')\,dV'}{|\mathbf{r}-\mathbf{r}'|} \tag{5.7}$$

As another example, if there is a point source of heat near one plane wall of a semi-infinite medium which is maintained at $U = 0$, by good thermal contact with a thermostat at this temperature, then the resulting temperature distribution will be the same as that in an infinite medium in which there is a point heat sink (negative Q) at the point which is the mirror image of the point source in the wall. Likewise, other calculations based on image methods or on conformal mapping are applicable.

Suppose that heat flows from a body that is a good conductor at effective temperature U_1 to another at U_2 through an intervening medium. The distribution of temperature $U(\mathbf{r})$ will be the same as that in free space of the electrostatic potential, if the potential of the first body were U_1 and

TABLE 5.1D. THERMAL CONDUCTIVITY OF SOLIDS (METALS)
Unit is 1 (cal/cm²-sec)/(°C/cm)

Material	Temp, °C	Conductivity
Lithium	−100	0.19
	0	0.16
	100	0.17
Sodium	−100	0.37
	0	0.33
	100	0.20
Potassium	0	0.23
	100	0.36
Beryllium	−100	0.30
	0	0.38
	100	0.45
Magnesium	−100	0.43
	0	0.41
	100	0.40
Aluminum	−100	0.50
	0	0.50
	100	0.52
Silver (very pure)	Constant at 1.0 from −100 to 100°C	
Copper (electrolytic)	−100	1.1
	0	0.94
	100	0.94
Iron (Armco)	0	0.21
	100	0.16
	500	0.10
Platinum	0	0.17
	300	0.18
	1000	0.21
Bismuth	0	0.02
Antimony	0	0.55
Zinc	0	0.30
Gold	0	0.75

of the second U_2. The total heat flowing from the first body in unit time is

$$P = \int \text{grad } U \cdot d\mathbf{S}$$

the integral being extended over the surface of the first body. In the analogous electrostatic problem, $P/4\pi$ is the total charge on the first body, which is equal to $C(U_1 - U_2)$, where C is the electrostatic capacity in centimeters of the capacitor formed by the two conductors. Therefore the power flow due to drop in effective temperature $U_1 - U_2$ is

$$P = \frac{C}{4\pi}(U_1 - U_2) \tag{5.8}$$

or, in the case of constant heat conductivity,

$$P = \frac{kC}{4\pi}(T_1 - T_2) \tag{5.9}$$

which is analogous also to the problem of the electric resistance between two electrodes in an extended conducting medium.

A special example of frequent occurrence in applications is that of steady radial heat flow in a long, circular cylinder of internal radius r_1 at temperature T_1, and external radius r_2 at T_2, for which

$$T(r) = T_1 + (T_2 - T_1)\frac{\ln (r/r_1)}{\ln (r_2/r_1)}$$

giving for the power flow from length A cm axially,

$$P = \frac{2\pi kA}{\ln (r_2/r_1)}(T_1 - T_2) \tag{5.10}$$

Let $r_2/r_1 = x$. If the value of a heat unit lost through the insulation on a pipe is C_2 and the cost of insulating material is C_1 in dollars per unit volume, and the interest rate is i per sec including other charges such as maintenance and depreciation, then the most economical thickness ratio x is given by

$$x \ln x = \sqrt{\frac{k(T_1 - T_2)}{r_1^2}\frac{C_2}{iC_1}} \tag{5.11}$$

on balancing the value of the heat lost against the cost of insulation.

3. Simple Boundary-value Problems

Conditions at the boundary of a finite body in which heat is flowing are usually described in terms of one of three idealizations, of which the last includes the first two as special cases:

1. The boundary is assumed to be maintained at a fixed preassigned temperature T, by good thermal contact with a well-stirred reservoir.

2. The boundary is assumed to be impervious to heat flow, which means that the normal component of the heat current, $\mathbf{I}$, and hence $\mathbf{n} \cdot \text{grad } T$ vanishes at the boundary.

3. Transfer of heat across the boundary is assumed to be proportional to the temperature of the boundary surface (this means the temperature relative to the surrounding cooling medium). This is known as *Newton's law of cooling.* The transfer of heat across the boundary is characterized by a heat-transfer coefficient h cal/(cm²)(sec)(°C) or Btu/(ft²)(hr)(°F) such that at the surface the normal component of $\mathbf{I}$ is hT and therefore if $\mathbf{n}$ is the outward drawn unit normal vector to the surface, the boundary condition at the surface is

$$\mathbf{n} \cdot \text{grad } T = \frac{h}{k} T \tag{5.12}$$

The nature of the thermal contact is thus characterized by a parameter k/h which has the dimensions of length. Case 1 corresponds to $k/h = 0$ and case 2 corresponds to k/h becoming infinite. Mixed cases may arise in which h has different values on different parts of the bounding surface, as when some surfaces are in good thermal contact with a reservoir, or reservoirs, and other surfaces are insulated. The value of h itself often involves analysis of convection conditions in a surrounding fluid medium (Secs. 7 and 8).

The general problem is that in which the heat sources are given, the initial temperature distribution is given, the boundary conditions are given, and it is desired to find the temperature distribution at later times. Here only the case of no heat sources is considered. The problem is to solve

$$\frac{\partial u}{\partial t} = \alpha \, \Delta u$$

being given $u(\mathbf{r},0)$ and boundary conditions of the form (5.12). The solution is in terms of an expansion in terms of a set of orthogonal functions, $v_s(r)$ with associated proper values, λ_s, where

$$\Delta v_s + \lambda_s v_s = 0 \tag{5.13}$$

and each v_s satisfies the boundary conditions. Then

$$u(\mathbf{r},t) = \sum_s c_s v_s(r) \exp(-\alpha\lambda_s t) \tag{5.14}$$

This completes the formal solution of the problem.

The characteristic values λ_s have dimensions per square centimeter or per square foot, so that when multiplied by α they give characteristic time constants for decay of each term in the expansion (5.14). The values of λ_s are determined not only by the scale of linear dimensions of the body, but also by the magnitude of the thermal contact length k/h which enters into the boundary conditions, except in the limiting cases (1) and (2). Physically the limiting cases will never be exactly realized: the important thing will be the magnitude of $(k/h)^2/\lambda_s$ for a given special mode of decay of the transient temperature distribution.

To illustrate, consider one-dimensional heat flow in a bar of length L, in the range $0 < x < L$, supposed perfectly insulated at $x = 0$ and subject to (5.12) at the end $x = L$. The condition at $x = 0$ determines that v_s is of the form $\cos k_s x$, and the condition at $x = L$ requires that k_s be determined by

$$\tan k_s L = -\frac{h}{kk_s}$$

Hence for $h/k \ll k_s$ the values of k_s will be nearly the same as if $h = 0$, while for $h/k \gg k_s$ the values will be nearly the same as if $h = \infty$. Thus the modes of low k_s are in good thermal contact with the boundaries, whereas those of high k_s decay at a rate that is mostly determined by internal equalization of temperature inequalities, and therefore behave nearly the same as if the boundary were thermally insulated.

Detailed solution of special problems can lead to long and intricate calculations, although the formal solution is quite elegant. Often the practical needs are satisfied by a study of only the leading term in the series, since this indicates the behavior of the terms which approach the equilibrium condition most slowly.

4. Cooling of Simple Bodies

Suppose that a body is initially at uniform temperature T_0 and at $t = 0$ is plunged into a perfect quenching medium which from that time on is able to hold its surface at the temperature $T = 0$. At first, the temperature will be nearly uniform throughout, except for a thin surface layer where the temperature drops abruptly to zero.

Approximately, if the layer thickness is δ, the mean temperature gradient in it is T_0/δ; thus kT_0/δ is the rate at which heat leaves unit area of the surface. This heat is supplied bv the cooling of the layer material through an average temperature drop of $T_0/2$, which supplies heat at the rate $c\rho\,\dot{\delta}T_0/2$, where $\dot{\delta}$ is the rate of increase of thickness of the layer. Therefore after time t the thickness of the layer is

$$\delta^2 = 4\alpha t \tag{5.15}$$

Thus δ increases very rapidly at first when the layer's temperature gradient is very high and its heat content is small. Later it increases more slowly because the temperature gradient is smaller and the heat content of the thicker layer is greater.

As long as δ is small compared to the radius of curvature of any convex part of the surface this qualitative behavior continues. As the thickness δ becomes greater, the cooled surface layer rounds off the sharp corners, and special features of the heat flow that are determined by the particular shape become evident.

Four special cases will be presented. In all of these a is a characteristic linear dimension. It is convenient to introduce a reduced or dimensionless time variable u, defined as

$$u = \frac{\alpha t}{a^2} = \frac{\delta^2}{4a^2} \tag{5.16}$$

so u of the order unity corresponds to the case in which the skin layer has become comparable with the principal linear dimensions:

1. *Infinite plane slab*, thickness $2a$, bounded by $-a < x < +a$. The temperature $T(x,u)$ is given by

$$\frac{T(x,u)}{T_0} = \frac{4}{\pi}\sum_{n=1}^{\infty}\frac{(-1)^n}{2n+1}\exp\left(-\frac{n^2\pi^2 u}{4}\right) \cdot \cos\,(2n+1)\frac{\pi x}{2a} \tag{5.17}$$

2. *Infinite cylinder of square cross section.*

$$-a < x < +a \quad \text{and} \quad -a < y < +a$$

$$\frac{T(x,y,u)}{T_0} = \left(\frac{4}{\pi}\right)^2 \sum_{m,n=1}^{\infty}\frac{(-1)^{m+n}}{(2m+1)(2n+1)} \cdot \exp\left[\frac{-(m^2+n^2)\pi^2 u}{4}\right]\cos\frac{(2m+1)\pi y}{2a} \cdot \cos\frac{(2n+1)\pi x}{2a} \tag{5.18}$$

3. *Infinite circular cylinder* of radius a.

$$\frac{T(r,u)}{T_0} = 2\sum_{n=1}^{\infty}\exp\,(-\beta_n^2 u)\frac{J_0(\beta_n r/a)}{\beta_n J_1(\beta_n)} \tag{5.19}$$

in which β_n is the nth root of the equation $J_0(\beta_n) = 0$, and J_0 and J_1 are the Bessel functions usually denoted in this way.

4. *Sphere of radius a.*

$$\frac{T(r,u)}{T_0} = 2\sum_{n=1}^{\infty}(-1)^n\exp\,(-n^2\pi^2 u)\frac{(\sin n\pi r/a)}{n\pi r/a} \tag{5.20}$$

Williamson and Adams* calculated Table 5.2, which gives $T(r,u)/T_0$ for the sphere for various values of u and of r/a. They also calculated Table 5.3 which shows the time dependence of the *central* temperature as a function of y, for bodies of various shapes.

Table 5.2. Cooling of a Sphere

r/a	u							
	0.004	0.016	0.036	0.064	0.100	0.196	0.256	0.400
0.00	1.000	1.000	0.994	0.910	0.707	0.288	0.160	0.0386
0.05	1.000	1.000	0.994	0.908	0.705	0.287	0.159	0.0385
0.25	1.000	1.000	0.979	0.858	0.647	0.260	0.144	0.0347
0.333	1.000	0.999	0.961	0.813	0.600	0.239	0.132	0.0319
0.50	1.000	0.990	0.875	0.676	0.474	0.184	0.102	0.0246
0.667	1.000	0.906	0.679	0.473	0.316	0.119	0.0661	0.0160
0.75	0.993	0.792	0.531	0.354	0.232	0.0869	0.0480	0.0116
0.95	0.394	0.179	0.103	0.0644	0.0411	0.0152	0.0084	0.0020

Effects of different degrees of thermal contact [the h of (5.12)] on bodies of several shapes have been calculated by Groeber† and by Heisler.‡ Rather full sets of curves for (1) an infinite plane slab, (2) an infinitely long circular cylinder, and (3) a sphere of

* Williamson and Adams, *Phys. Rev.*, **14**: 99 (1919).
† Groeber, *Z. Ver. deut. Ing.*, **69**: 705 (1925).
‡ Heisler, *Trans. ASME*, **69**: 227 (1947).

TABLE 5.3. CENTRAL COOLING OF DIFFERENT BODIES

u	(1)	(2)	(3)	(4)	(5)	(6)
0.032	1.000	1.000	1.000	0.999	0.999	0.998
0.080	0.975	0.951	0.927	0.918	0.895	0.828
0.100	0.949	0.901	0.856	0.848	0.805	0.707
0.160	0.846	0.715	0.605	0.627	0.530	0.409
0.240	0.702	0.493	0.346	0.399	0.280	0.187
0.320	0.578	0.334	0.193	0.252	0.145	0.085
0.800	0.177	0.031	0.006	0.016	0.003	0.001
1.600	0.025	0.001	0.000	0.000	0.000	0.000
3.200	0.000	0.000	0.000	0.000	0.000	0.000

(1) Infinite plane slab; (2) infinite square cylinder; (3) cube; (4) infinite circular cylinder; (5) circular cylinder of length equal to diameter; (6) sphere.

radius a, for different degrees of thermal contact are reproduced in ref. 7, pp. 274–291.

5. Point-source Solutions

If a source of heat generating Q heat units is released instantaneously at the origin at $t = 0$, it will give rise to an infinite temperature there, and because of the infinite temperature gradient the heat will very quickly diffuse out, causing a temperature rise in the region around the origin. As a result of the spreading of the heat over a finite region, the temperature becomes finite at the origin for all but the first instant. Similar situations exist for instantaneous line or plane sources of heat, and useful solutions for more general problems may be built up by superpositions of such solutions.

The strength of the source, Q heat units, is conveniently expressed as $H = Q/\rho c$, where H is expressed in degrees per cubic centimeter. The temperature after time t at distance r from such a point source is

$$T = \frac{H}{\pi^{3/2}\delta^3} \exp\left(-\frac{r^2}{\delta^2}\right) \tag{5.21}$$

where $\delta^2 = 4\alpha t$, the skin thickness introduced in the approximate argument leading to (5.15). This has the property that the volume integral of T over all space is equal to H at all times. The temperature distribution is that of a spherical Gauss-error function whose effective radius grows as the square root of the time.

This solution (5.21) also finds application as a way of describing the flow of heat in an infinite medium, if the initial temperature distribution is given. Let the initial temperature at (x,y,z) be $U(x,y,z)$. If actually the temperature distribution were $U(x',y',z')$ in a volume element dV' and zero elsewhere, then the temperature at (x,y,z) at time t would be given by (5.21) on using $U(x',y',z')\,dV'$ for H and interpreting r as the distance between (x,y,z) and (x',y',z'). Therefore, by linear superposition,

$$T(x,y,z,t) = \pi^{-3/2}\delta^{-3} \iiint U(x',y',z') \exp\left(-\frac{r^2}{\delta^2}\right) dV' \tag{5.22}$$

where $r^2 = |\mathbf{r} - \mathbf{r}'|^2$. From this point of view the temperature at time t at any point P is a weighted average of the initial temperature distribution in the surrounding region for distances of the order of δ.

Alternatively, if the heat sources in the past had been variable in amount, so that the strength of the source at time t' at (x',y',z') is $H(x',y',z',t')\,dt'$, then (5.22) can be further generalized to include an integration over past times, $-\infty < t' < t$, to include the contribution of all heat sources which have been active up to the time under consideration. In such an integration, $\delta^2 = 4\alpha(t - t')$, for the effective size of the region influenced by a particular source is determined by the total elapsed time since its heat was released. For example, if a constant heat source of strength Q has been operating since the infinite past at the origin, it will give rise to a steady temperature distribution $T(r) = Q/4\pi kr$, and if the heat source, though constant, has only been operating since a time t'_0 in the finite past, this method may be used to determine the way in which the temperature distribution approaches the steady-state condition.

For one-dimensional problems, and problems with axial symmetry, analogous forms exist. Suppose that at $t = t'$ there is an instantaneous plane source of heat, Q heat units per unit area, all over the $x = x'$ plane. By symmetry the temperature distribution will be an even function of $(x - x')$ and be independent of y and z. The solution corresponding to (5.21) is

$$T(x,t) = \frac{H}{\pi^{1/2}\delta} \exp\left[-\frac{(x - x')^2}{\delta^2}\right] \tag{5.23}$$

and so, analogous to (5.2), the temperature at time t, if the initial temperature distribution is $U(x)$, is

$$T(x,t) = \pi^{-1/2}\delta^{-1} \int U(x') \exp\left[-\frac{(x - x')^2}{\delta^2}\right] dx' \tag{5.24}$$

Similarly, if there is an instantaneous line source along the line $x = x'$ and $y = y'$, which releases Q heat units per unit length, independently of z, the temperature distribution is

$$T(x,y,t) = \frac{H}{\pi\delta^2} \exp\left(-\frac{r^2}{\delta^2}\right) \tag{5.25}$$

where $r^2 = (x - x')^2 + (y - y')^2$, and the expression analogous to (5.4) is

$$T(x,y,t) = \pi^{-1}\,\delta^{-2} \iint U(x',y') \exp\left(-\frac{r^2}{\delta^2}\right) dx'\,dy' \tag{5.26}$$

An important application of (5.24) is to the penetration of a temperature change into a medium which extends over the entire half-space in which $x > 0$, initially at the uniform temperature U, and whose surface is maintained at the temperature $T = 0$. The surface condition at $x = 0$ will be fulfilled by imagining the medium to fill the whole space, with the condition that the initial temperature is $-U$ throughout the region $x < 0$. In this case, the temperature distribution becomes, for $x > 0$,

$$T(x,t) = U\left(\frac{2}{\pi}\right)^{1/2} \int_0^{x/\delta} \exp(-u^2)\,du \quad (5.27)$$

The temperature gradient at the surface is $2U/\pi^{1/2}\delta$. The total amount of heat which passes across unit area from $t = 0$ to $t = t$ is

$$Q = \frac{U\rho c\delta}{\pi^{1/2}} = U\left(\frac{4k\rho ct}{\pi}\right)^{1/2} \quad (5.28)$$

so that the total heat transfer increases with the square root of the time, a result which finds application in analysis of metallurgical quenching operations.

A variant is this: the material in the region $x > 0$ has a diffusivity α_1 and that in the region $x < 0$ has a diffusivity α_2. Initially the first material is at temperature U, while the second is at temperature 0. At $t = 0$ they are brought into perfect thermal contact. The interface, $x = 0$, assumes a temperature T_0 between 0 and U such that at each instant the flow of heat out of the $x > 0$ region just equals the rate at which heat flows into the $x < 0$ region. Using (5.28), this determines T_0 by the relation

$$\frac{U - T_0}{T_0} = \left(\frac{k_2\rho_2c_2}{k_1\rho_1c_1}\right)^{1/2} \quad (5.29)$$

The total heat which flows across the boundary between $t = 0$ and $t = t$ is then given by (5.28), using T_0 in place of U and $k_2\rho_2c_2$ in place of $k\rho c$.

6. Periodic Temperature Change

Consider a semi-infinite medium in the space $x > 0$, whose surface temperature, the plane $x = 0$, is made to undergo a cyclic time variation

$$T(0,t) = U\cos\omega t$$

After this surface variation has gone on for a long enough time a steady state is established in which each point of the medium undergoes a cyclic temperature variation of the same frequency, but of reduced amplitude at points deep within the material, and also with a phase lag characteristic of wave propagation at finite velocity.

The skin depth associated with the time $(2\omega)^{-1}$ in which $(4\pi)^{-1}$ of a cycle occurs is

$$\delta^2 = \frac{2\alpha}{\omega}$$

In terms of this the solution for the temperature wave is

$$T(x,t) = Ue^{-x/\delta}\cos\left(\omega t - \frac{x}{\delta}\right) \quad (5.30)$$

At a depth equal to $\pi\delta$ the temperature variations will occur in exactly opposite phase to those occurring at the surface, although the amplitude at this depth is only $e^{-\pi} = 0.043$ as great as on the surface.

These results are useful in deciding the thickness of insulating material needed to shield against penetration of periodic temperature variations. For example, for loose sawdust, $k = 12 \times 10^{-5}$, $\rho = 0.20$ and $c = 0.42$; thus $\alpha = 143 \times 10^{-5}$ cm²/sec. For an annual seasonal variation,

$$(2\omega)^{-1} = 29 \text{ days} = 25 \times 10^5 \text{ sec}$$

giving $\delta = 120$ cm, so that a depth of 376 cm would be needed to bring about complete phase reversal and reduction of the seasonal temperature amplitude to 0.043 of its surface value.

Imperfect thermal contact between the medium in the region $x < 0$, represented by a Newton's law of cooling, may also be incorporated. The boundary condition is

$$\frac{\partial T}{\partial x} + \frac{h}{k}[T - U(t)] = 0 \quad (5.31)$$

at $x = 0$, where $U(t)$ is time-dependent temperature in the $x < 0$ medium. Then, from (6.1) and (6.2),

$$U(t) = \left(\frac{kT_0}{h\delta}\right)\left[\left(\frac{h\delta}{k} - 1\right)\cos\omega t + \sin\omega t\right]$$

Therefore the effect of imperfect thermal contact is to introduce a phase shift between the cyclic temperature variations in the two media and to make the amplitude T_0 be less than the amplitude of $U(t)$, the variation of temperature of the other medium.

7. Natural Heat Convection

Convection refers to transfer of heat through the action of a moving fluid. Free, or natural, heat convection refers to the situation in which the motion is principally the result of gravity acting on density differences due to fluid expansion.

For a moving fluid (5.1) becomes

$$\mathbf{I} = \rho cT\mathbf{v} - k\,\text{grad}\,T \quad (5.32)$$

the first term allowing for the convective heat transfer. The modified heat-conduction equation is

$$\rho c\left(\frac{\partial T}{\partial t} + \mathbf{v}\cdot\text{grad}\,T\right) + \rho cT\,\text{div}\,\mathbf{v} = \text{div}\,(k\,\text{grad}\,T) + Q \quad (5.33)$$

The corresponding dynamical equation for fluid flow is Eq. (1.41) of Chap. 1, Part 3.

$$\frac{d\mathbf{v}}{dt} + \text{grad}\,(P + U - \tfrac{4}{3}\nu\,\text{div}\,\mathbf{v}) + \nu\,\text{curl}\,\text{curl}\,\mathbf{v} = 0 \quad (5.34)$$

Some remarks on combined effects of fluid flow and heat transfer are made in Part 3, Chap. 2, Sec. 9. Problems of interest always take the form of calculation of the heat exchange between a solid boundary and the fluid flowing relative to it. Let $T = 0$ be the temperature of the main body of the fluid and $T = U$ that of the solid boundary. The heat crossing unit area in unit time is

$$H = k\frac{\partial T}{\partial n} \quad (5.35)$$

where $\partial T/\partial n$ is the normal temperature gradient evaluated at the solid boundary. The problems are quite complicated when attacked fundamentally from (5.33) and (5.34) with appropriate boundary conditions. However, a good deal of progress can be made by correlating results of experimental tests in terms of significant dimensionless parameters.

The mean value of H over a body, H_{av}, will be related as in (5.35) to $(\partial T/\partial n)_{av}$, which in turn can be written as U/δ, where δ is an effective film thickness of the fluid over which the full temperature drop U takes place. If L is a characteristic linear dimension of the body, it is customary to define the dimensionless *Nusselt number* as

$$\mathrm{Nu} = \frac{H_{av}L}{kU} \tag{5.36}$$

In terms of the effective film thickness δ,

$$\mathrm{Nu} = \frac{L}{\delta}$$

In most common circumstances the film thickness is small compared with an over-all characteristic length, such as the vertical height of a vertical plate, or the diameter of a sphere, or the like; therefore in most situations the Nusselt number is large compared to unity. However, in the case of fine wires the film thickness may exceed the diameter of the wire, giving $\mathrm{Nu} < 1$, as Langmuir* observed in experiments on convective loss of heat from filaments. Because of his early recognition of the importance of the more or less stagnant layer of fluid around a body this layer is often called the *Langmuir film*.

The driving force in natural convection is the buoyancy due to density differences caused by natural expansion, and the velocity pattern that is set up results from a balance between the effects of this driving force and the sluggishness arising from viscosity. This balance may be described by the dimensionless *Grashof number* defined as

$$\mathrm{Gr} = \frac{g\beta UL^3}{\nu^2} \tag{5.37}$$

where β is the volume coefficient of thermal expansion. A very small β or U, or correspondingly a large viscosity, will give rise to a small Gr and obviously to slow convection currents.

The fluid itself is characterized by another dimensionless parameter known as the *Prandtl number*, defined as the ratio of the kinematic viscosity to the thermal diffusivity,

$$\mathrm{Pr} = \frac{\nu}{\alpha} \tag{5.38}$$

By the principles of dimensional analysis for a given shape of boundary,

$$\mathrm{Nu} = f(\mathrm{Gr},\mathrm{Pr})$$

but from such reasoning alone the form of the functional relation cannot be determined. Experimentally it is found that Nu is principally determined as a function of the product (Gr · Pr), and is only secondarily also dependent on Pr. This main factor can be written

$$(\mathrm{Gr}\cdot\mathrm{Pr}) = \frac{g\beta}{\nu\alpha}UL^3 \tag{5.39}$$

where the first factor is an intrinsic property of the fluid, sometimes called its convection modulus. A large convection modulus is favorable to a high rate of heat transfer by the fluid.

For very low values of (Gr · Pr) the convection velocities tend to zero and the loss of heat by the solid body tends to that given by pure conduction. At larger values of (Gr · Pr) convection currents begin that are slow enough that the flow remains laminar everywhere (small $\mathrm{Re} = Lv/\nu$), but at larger values the flow becomes more vigorous, as indicated by a large Reynolds number which is accompanied by instability of the laminar flow. The laminar flow breaks up into a turbulent boundary layer, with a very thin laminar sublayer in the immediate neighborhood of the solid boundary.

The case of natural convection along a heated vertical plate of constant width in laminar flow has been given an analytical solution by Pohlhausen (see ref. 7, Chap. 22). For the particular case, $\mathrm{Pr} = 0.73$, approximately the value for air, numerical integration of the differential equations shows that

$$H_{av} = \frac{kU}{L}\mathrm{Nu} \qquad \text{with } \mathrm{Nu} = 0.48(\mathrm{Gr})^{1/4} \tag{5.40}$$

where L is the vertical height of the plate, supposed small enough that the flow in the boundary layer remains laminar. The flow becomes turbulent* when $\mathrm{Gr}\cdot\mathrm{Pr} > 2\cdot 10^9$. The thickness of laminar boundary layer increases as $x^{1/4}$, where x is the distance measured vertically upward from the bottom edge of the heated plate and is of the order of a characteristic length a, where

$$a^3 = \frac{4\nu^2}{g\beta U} = \frac{4L^3}{\mathrm{Gr}} \tag{5.41}$$

which affords an interpretation of the Grashof number as relating the boundary layer thickness to the plate height, L. The thickness y at height x is of the order $y = a(x/a)^{1/4}$. The maximum vertical velocity v_x occurs at this distance out from the wall, and its magnitude is approximately equal to $(\nu/a)(x/a)^{1/2}$. Thus for air at 20°C, a temperature drop of 10°C gives $(\nu/a) = 1$ cm/sec. This theoretical result agrees quite well with observed values for the upward natural convective motion.

The heat transfer locally at distance x up from the bottom edge may be expressed in terms of Nu_x, Gr_x, and Pr, which are defined by using x for L in (7.5) and (7.6). E. R. G. Eckert [3, p. 162] has given an approximate analysis leading to

$$\mathrm{Nu}_x = 0.508\,\mathrm{Pr}^{1/2}(0.952 + \mathrm{Pr})^{-1/4}(\mathrm{Gr}_x)^{1/4}$$

* I. Langmuir, *Phys. Rev.*, **34**: 405 (1912).

* O. A. Saunders, *Proc. Roy. Soc.* (*London*), **A172**: 55 (1939).

which is in good agreement with the experiments of Saunders. Eckert [3, p. 168] gives a beautiful picture of the interference fringes due to density changes in the natural convective flow past a heated vertical plate.

Natural convective flow past a horizontal circular cylinder of diameter L represents a case of considerable practical importance. The experimental results are quite well represented by an empirical formula (see ref. 1, p. 101):

$$\mathrm{Nu} = [0.63 + 0.35(\mathrm{Gr} \cdot \mathrm{Pr})^{1/6}]^2 \tag{5.42}$$

A theoretical solution has been given for this case by R. Hermann.* In the case of fine wires, for which the boundary layer is comparable with or larger than the diameter L, a different expression must be used:

$$\mathrm{Nu} = \frac{2}{\ln\left[1 + \dfrac{2}{0.4(\mathrm{Gr})^{1/4}}\right]}$$

as shown by C. W. Rice.†

Experimental results for heat loss by natural convection past vertical cylinders were obtained by Y. S. Touloukian, G. A. Hawkins, and M. Jacob‡ and found to be governed by the following empirical equations:

Laminar range:

$$\mathrm{Nu} = 0.726(\mathrm{Gr} \cdot \mathrm{Pr})^{1/4}$$

Turbulent range:

$$\mathrm{Nu} = 0.0674(\mathrm{Gr} \cdot \mathrm{Pr})^{1/3}(\mathrm{Pr})^{1/10}$$

A full discussion of the available data on free convection, and their correlation in terms of dimensionless parameters may be found in ref. 7, Chap. 25.

8. Forced Heat Convection

Heat convection is said to be forced if the motion of the fluid is principally due to other driving forces than those arising from buoyancy differences due to thermal expansion. The general remarks in Sec. 7 leading to definition of the dimensionless Nusselt number for describing the heat-transfer coefficient are also applicable here.

In forced heat convection the effect of the buoyant force is neglected, which means passing to the limit $\mathrm{Gr} = 0$. In place of the Grashof number, the Reynolds number

$$\mathrm{Re} = \frac{v_0 L}{\nu} \tag{5.43}$$

appears as a dimensionless parameter characterizing the fluid-flow pattern. One is usually interested in heat transfer from a solid object immersed in a stream such that v_0 is the uniform fluid velocity in the undisturbed regions far from the solid body and L is a characteristic linear dimension of the solid.

* R. Hermann, *Z. angew. Math. u. Mech.*, **13**: 433 (1933).

† C. W. Rice, *Trans. AIEE*, **42**: 653 (1923).

‡ Y. S. Touloukian, G. A. Hawkins, and M. Jacob, *Trans. ASME*, **70**: 13 (1948).

The following discussion is based on Eckert [3, Chap. 3]. In case of flow past a flat plate, it will be convenient to write $\mathrm{Re}_x = v_0 x/\nu$ for the Reynolds number calculated for a place at a distance x along the plate from the leading edge. Similarly Nu_x may be written for the local heat-transfer rate per unit area at distance x in terms of kU/x. Experiment shows that for small values of x or Re_x the flow in the boundary layer near the leading edge is laminar. For values of x such that Re_x is of the order 10^5, the exact value being smaller for smooth surfaces and undisturbed initial flow, a transition range develops which passes over into a region at larger values of x in which the boundary layer is fully turbulent.

In the laminar-flow region, the thickness of the boundary layer increases as $x^{1/2}$, being given approximately by

$$\delta = 4.64\left(\frac{\nu x}{v_0}\right)^{1/2} = \frac{4.64x}{(\mathrm{Re}_x)^{1/2}} \tag{5.44}$$

where δ is the parameter in the following approximate representation of the velocity $v(y)$ at distance y from the plate,

$$v(y) = \frac{v_0}{2}\left[3\left(\frac{y}{\delta}\right) - \left(\frac{y}{\delta}\right)^3\right] \tag{5.45}$$

for $y < \delta$ and $v(y) = v_0$ for $y > \delta$. The shearing stress at the wall is then

$$\mu\left(\frac{\partial v}{\partial y}\right)_0 = \frac{(\rho v_0^2/2)}{4.64(\mathrm{Re}_x)^{1/2}} \tag{5.46}$$

which is simply related to the kinetic energy in unit volume of the freely moving fluid.

In the region of turbulence the velocity profile is reasonably well represented by an expression due to Prandtl,

$$v(y) = v_0\left(\frac{y}{\delta}\right)^{1/7} \tag{5.47}$$

with
$$\delta = 0.376x(\mathrm{Re}_x)^{-1/5}$$

In the immediate neighborhood of the plate there is a laminar sublayer of thickness

$$\delta' = 191\delta(\mathrm{Re}_x)^{-0.7}$$

and the velocity at the edge of the laminar sublayer is

$$v_b = 2.11 v_0(\mathrm{Re}_x)^{-0.1}$$

Thus the velocity $v(y)$ increases linearly with y in the range $y < \delta'$, and then joins on to the $v(y)$ given by (8.5) in the range $\delta' < y < \delta$, after which $v(y) = v_0$ for $y > \delta$.

In case the plate has the temperature $T = 0$ in the range $0 < x < x_0$, and has the temperature $T = U$ for $x > x_0$, the laminar boundary layer starts to develop at $x = 0$, but a corresponding temperature drop only occurs for $x > x_0$. The heat transfer across unit area in unit time at x can be written

$$Q = \frac{kU}{x}\mathrm{Nu}_x$$

with

$$\mathrm{Nu}_x = 0.331(\mathrm{Pr})^{1/3}(\mathrm{Re}_x)^{1/2}\left[1 - \left(\frac{x_0}{x}\right)^{3/4}\right]^{-1/3} \tag{5.48}$$

in the region in which the boundary layer is laminar. If the entire plate is heated, $x_0 = 0$ and the bracket factor equals unity. For $x_0 = 0$, and for x in the turbulent boundary-layer region, the Nusselt number is

$$\mathrm{Nu}_x = \frac{0.0296\ \mathrm{Pr}(\mathrm{Re}_x)^{4/5}}{1 + 1.30(\mathrm{Pr})^{-1/6}(\mathrm{Re}_x)^{-1/10}(\mathrm{Pr} - 1)} \tag{5.49}$$

9. Condensation and Evaporation

Vapor condenses on a wall whose temperature is less than the saturation temperature at the pressure in question, liberating the latent heat of vaporization. If the liquid wets the surface it forms a continuous film on the wall which drains down under gravity at a rate limited by the liquid viscosity. The rate of condensation is limited by the thermal flow resistance of the liquid film. At low rates the fluid flow is laminar, but if the draining fluid film exceeds a critical thickness, the flow becomes turbulent, resulting in an increased rate of heat transfer through it, and therefore an increased rate of condensation.

Under some conditions the condensing liquid forms droplets, which increase in size until they run down the wall, sweeping down with them the small droplets in their path. Heat transfer is much greater in this case than when a continuous liquid film is formed. Clean surfaces and pure vapor favor continuous film formation, while rough surfaces and impure vapor favor formation of droplets. If the condensing vapor is diluted by a noncondensing gas, then there is a tendency for the concentration of the noncondensing gas to build up at the condensing surface, which further acts to reduce the rate of condensation.

Nusselt* analyzed the heat transfer through a film draining from a vertical plate in laminar flow. Measuring x down from the upper edge of the vertical plate, let $\delta(x)$ be the liquid film thickness at x. The downward velocity $v(y)$ of liquid flow in the film at distance y from the plate is

$$v(y) = V\left(\frac{2y}{\delta} - \frac{y^2}{\delta^2}\right) \tag{5.50}$$

Under steady conditions, neglecting acceleration in the fluid flow, the viscous shear on the wall, $2\eta V/\delta$, is balanced by the weight of the fluid on unit area of the plate, $\rho g\delta$; therefore $V = \rho g\delta^2/2\eta$.

The mass of fluid, $M(x)$, crossing unit horizontal length of film at level x in unit time is

$$M(x) = \frac{2\rho\ \delta V}{3} = \frac{\rho^2 g\delta^3}{3\eta}$$

If L is the latent heat of vaporization of unit mass, the rate at which heat is supplied to the free surface per unit area by condensation is $L(dM/dx)$, and this heat has to be conducted through the film, with a temperature drop of U, where U is the difference in temperature between the vapor at the film and the vertical plate, so that $L(dM/dx) = kU/\delta$. These permit an integration for the film thickness giving

$$\left(\frac{\delta}{a}\right)^4 = \frac{x}{a}$$

where a is a characteristic length defined by

$$a^3 = \frac{4kU(\eta/\rho)}{\rho gL} \tag{5.51}$$

Hence the rate of heat transfer per unit area at the level x is

$$Q = \frac{kU}{a}\left(\frac{a}{x}\right)^{1/4} \tag{5.52}$$

and the average heat-transfer rate across unit area in the range from 0 to x is four-thirds of the value given by (5.52).

Forming the Reynolds number, with the mean velocity $2V/3$, and the film thickness δ, this becomes

$$\mathrm{Re} = \frac{4kU}{3\eta L}\left(\frac{\delta}{a}\right)^3 \tag{5.53}$$

Experimental results show that the rate of heat transfer is in good agreement with the foregoing calculations, provided this Reynolds number is less than about 300. U. Grigull* has made corresponding calculations for the rate of heat transfer in a condensing film on a vertical plate in which there is turbulent flow.

Nusselt also made calculations for the rate of condensation with a laminar liquid film on a horizontal circular cylindrical tube, and found that the average film heat-transfer rate is the same as the average on a vertical plate whose height is 2.5 times the diameter of the tube. Drew, Hottel, and McAdams† give results of experimental work on conditions affecting drop formation rather than film formation in condensation. With drop formation the heat-transfer rate may easily become ten times what it is in film condensation. A. P. Colburn and also C. E. Kirkbride‡ have given a careful analysis of the experimental results in terms of dimensionless parameters related to the foregoing analysis.

The case in which the condensed material is a solid and therefore adheres to the wall involves a condensed film of growing thickness and therefore the heat conduction must necessarily be in a transient state. It has been studied experimentally by Brun and Feniger and by Feniger.¶

In case the main body of the vapor is superheated

* Nusselt, *Z. Ver. deut. Ing.*, **60**: 541, 1916.

* U. Grigull, *Forsch. Gebiete Ingenieurw.*, **13**: 49 (1942).

† Drew, Hottel, and McAdams, *Trans. AICE*, **32**: 271 (1936).

‡ A. P. Colburn, *Ind. Eng. Chem.*, **22**: 967 (1930), and **26**: 432 (1934); *Ind. Eng. Chem.*, **26**: 425 (1934).

¶ Brun and Feniger, *Compt. rend.*, **226**: 1966 (1948); Feniger, *Bull. Ind. Inst. Refrig.*, **29**: A39 (1948).

relative to the temperature of the condensed film, then there is a temperature drop in the vapor near the wall which has a relatively high thermal resistance, being governed by the ordinary laws of natural convention. In consequence the over-all flow of heat for considerable degree of superheat of the vapor is not very much greater than if the vapor is not superheated.

Heat Transfer to Boiling Liquids. If a solid body is immersed in a liquid and its temperature is raised by an internal heat source, it will transfer heat to the liquid at a rate governed by the general discussion in preceding sections for free or forced convection, according to circumstances. Here it will be supposed that the case of free or natural convection only is under consideration. As the temperature is further increased, the surface of the solid will finally begin to exceed the boiling temperature of the liquid, that is, the temperature at which its vapor pressure is equal to the prevailing hydrostatic pressure in the liquid. It will be supposed that natural convection has brought the bulk of the liquid to a substantially uniform temperature near the boiling point.

When the temperature of the solid is substantially above the boiling point, bubbles of vapor form at certain sensitive spots on the surface of the solid. Usually this takes place at only a few well-defined points so that streams of vapor bubbles arise from these places and from nowhere else on the surface of the hot solid. This is the beginning of the process known as *nucleate boiling*. Under these conditions the rate of heat transfer from solid to liquid is essentially still governed by the processes of free convection. To be sure, the rising streams of vapor bubbles act as an additional agency to set up convection currents in the liquid and therefore the rate of heat transfer is somewhat greater than it would be with all other conditions the same but without the formation of vapor bubbles. The size of the bubbles at the instant at which they break free from the solid surface is determined by counterbalancing of surface tension σ and the angle of contact β, defined as the angle, measured in the liquid, between the solid-liquid interface and the gas-liquid interface and the buoyant forces due to gravity. In all such problems a characteristic length known as the Laplace constant, $a = (2\sigma/\rho g)^{1/2}$, plays a role. Approximately, the bubble radius is given by

$$r = \tfrac{2}{5}\beta a$$

For water, using $\sigma = 75$ dynes/cm, one finds $a = 0.39$ cm. The sensitive spots are thus those where surface conditions are such that β is small. M. Jakob [7, pp. 624–635] found experimentally that the bubble sizes in nucleate boiling are in accord with the values given by capillarity theory. He also found that there was a pause of about $\frac{1}{40}$ sec after each bubble broke off before the next one started to form and that the bubble required approximately an equal time in which to grow up to its maximum size; therefore the frequency of bubble generation from such a sensitive spot was about 20 per second for bubbles about 5 mm in diameter, and that this frequency varies approximately inversely as the diameter, so that the product, $w = 10$ cm/sec, is a velocity characteristic of the over-all bubble growth process.

As the internal heat supply to the solid is increased, the heat-transfer rate increases with an increasing temperature drop between the solid and the liquid outside the boundary layer. When the heat flux has reached values around 0.05 to 0.10 cal/cm²-sec for vertical surfaces and somewhat more than twice that for horizontal heating surfaces, there is a radical change in the character of the process, characterized by the fact that now the streams of vapor bubbles have become so numerous that the convection currents which they set up in the liquid begin to dominate the flow pattern, producing much more vigorous circulation than would occur in the absence of boiling.

As the heat-transfer rate is still further increased, a larger and larger fraction of the surface begins to act as a source of bubbles until a value in the neighborhood of 25 to 30 cal/cm²-sec is reached (for water) which occurs for a temperature drop between solid and bulk of the liquid of about 30°C. Then the entire surface of the solid becomes covered with a continuous film of vapor which acts as a relatively poor conductor of heat, so that the heat transfer across unit area rather quickly falls to about one-tenth of its maximum value as the temperature drop is increased to 50°C and more. This condition is known as *film boiling*. Naturally it is a condition to be avoided in efficient heat-transfer apparatus; thus there is an intrinsic limit to the rate of heat transfer from a solid to a boiling liquid.

Film boiling plays an important role in metallurgical quenching operations and can be demonstrated vividly in a way described by R. C. L. Bosworth.* A small 1-oz block of copper is heated to dull redness and plunged into a beaker of alcohol. Immediately it is surrounded by a continuous blanket of vapor which reduces the heat-transfer rate to such a value that the alcohol boils quietly until the temperature drops to a point where the film is no longer continuous. Then there is suddenly a very great increase in the rate of boiling which continues until the temperature of the copper is below the boiling point of alcohol when boiling subsides and the block continues to cool by natural convection.

Details of the behavior are sensitive to surface conditions, both mechanical and chemical.† F. H. Rhodes and C. H. Bridges‡ found that the rate of boiling of water on steel was considerably reduced by a trace of mineral oil. They also found that when the temperature drop was so great that they were in the film boiling regime, a little sodium carbonate would restore the condition of nucleate boiling and increase the heat-transfer rate by a factor of roughly ten. All these effects, with regard to organic liquids, as well as steam are of obvious importance in chemical engineering and additional specific data are to be found in papers in *Chemical Engineering Progress*.

* R. C. L. Bosworth, *J. Roy. Soc. N.S.W.*, **80:** 20 (1946).

† E. K. Spring, P. T. Lansdale, and L. W. Alexander, *Trans. A.S.M.*, **33:** 42 (1944).

‡ F. H. Rhodes and C. H. Bridges, *Ind. Eng. Chem.*, **30:** 1401 (1938).

10. Radiative Heat Transfer

Thermal radiation is here considered as a mode of heat transfer: its equilibrium properties are presented in Part 6, Chap. 1, Sec. 10.

A small area element of a black body dS_1 at temperature T_1 radiates at the rate $(\sigma/\pi)T_1^4 \cos\theta_1\, dS_1\, d\omega$ into an element of solid angle in a direction making an angle θ_1 with the normal to dS_1. The solid angle subtended by another element dS_2 whose normal makes an angle θ_2 with the line joining dS_1 and dS_2 is $d\omega = dS_2 \cos\theta_2/r^2$; thus the rate at which radiation passes from dS_1 to dS_2 is

$$\frac{\sigma}{\pi} T_1^4 \frac{\cos\theta_1 \cos\theta_2}{r^2}\, dS_1\, dS_2$$

and the amount passing in the opposite direction is the same expression with T_2 in place of T_1. Therefore the net rate of transfer from dS_1 to dS_2 is

$$\frac{\sigma}{\pi}(T_1^4 - T_2^4)\frac{\cos\theta_1 \cos\theta_2}{r^2}\, dS_1\, dS_2 \tag{5.54}$$

The total rate of transfer between two extended bodies is obtained by integrating over dS_1 and dS_2 where the integration over dS_2 is over that part of the area of the second body that is directly visible from dS_1, followed by integration over all of the first body. Writing S_1F_{12} for the resulting integral, where S_1 is the total area of the first body, the quantity F_{12} is called the *geometrical factor* applicable to radiative heat transfer between the two bodies. Calculation of the geometrical factor is quite involved.*

If a small body of emissivity e_1 at temperature T_1 is immersed in a region traversed by black-body radiation at temperature T_2, then the net rate of loss of energy from unit area of the small body is

$$Q = e_1\sigma(T_1^4 - T_2^4) \tag{5.55}$$

where $\sigma = 5.672 \times 10^{-5}$ erg/cm^2-deg^4 is the Stefan-Boltzmann constant.

A convenient approximation is that of the *gray body*, which is defined as having constant emissivity at all temperatures and frequencies. The net heat transfer per unit area between two infinite parallel planes of emissivity e_1 at temperature T_1 and emissivity e_2 at temperature T_2 is

$$Q = \frac{e_1e_2}{e_2 + e_1(1 - e_2)}\sigma(T_1^4 - T_2^4) \tag{5.56}$$

For concentric cylinders of radii r_1 and r_2 with r_1 less than r_2

$$Q = \frac{e_1e_2}{e_2 + e_1(1 - e_2)(r_1/r_2)}\sigma(T_1^4 - T_2^4) \tag{5.57}$$

The same formula holds for concentric spheres on replacing r_1/r_2 by its square.

A standard procedure for experimental realization of a black body from materials whose emissivity is less than unity is to make a hollow enclosure with only a small hole in one wall through which the radiation can be observed. To get an idea of the extent to which this corrects for the imperfect emissivity of the enclosure walls the following result is useful. Let the enclosure shape be spherical and let the fraction of its total internal area that is cut away by the hole be f; then the effective emissivity e' of the hole is

$$e' = \frac{e}{e + f(1 - e)} \tag{5.58}$$

Therefore even if $e = 0.5$, if $f = 0.01$, the value of e' will be 0.99; thus in this way a black body can be achieved in practice with materials of low emissivity.

In quasi-transparent media the thermal radiation will be absorbed and reemitted throughout the volume. If A (per centimeter) is the absorption coefficient and the medium is large in extent relative to A^{-1}, then the transport of radiant energy will be diffusive in character with A^{-1} playing a role analogous to that of the mean free path in the molecular interpretation of heat conductivity of gases (Part 5, Chap. 4, Sec. 4).

In such a medium let g be the radiation emitted in all directions by unit volume in unit time, so that the outward flux in solid angle $d\omega$ from a volume element dV is $g\,dV\,d\omega/4\pi$. A volume element dV' at distance r sends an intensity $g'\,dV'\,e^{-Ar}/4\pi r^2$ to dV and the rate at which this energy is absorbed in dV is this expression times $A\,dV$. Here g' is the rate of radiation appropriate to the temperature conditions at dV'; thus if the temperature is not uniform, g' is a function of position. Let r, θ, ϕ be the coordinates of dV' relative to dV; then the total rate of absorption of energy in dV is

$$\frac{dV}{4\pi}\int_0^\infty \int_0^\pi \int_0^{2\pi} g'e^{-Ar}d(Ar)\sin\theta\, d\theta\, d\phi \tag{5.59}$$

If g' is constant and equal to g, this becomes $g\,dV$ so that the total rate of absorption equals the rate of emission which expresses Kirchhoff's law for quasi-transparent media.

If $A(\nu)$ is the absorption coefficient for frequency, then the rate of isotropic emission from dV is

$$c\,dV \int_0^\infty A(\nu)\rho(\nu,T)n^2\,d\nu \tag{5.60}$$

where n is the refractive index and $\rho(\nu,T)$ is the vacuum black-body radiation density. The density in a medium of refractive index n is $n^3\rho$, but one factor of n is canceled in calculating the rate of radiation since the velocity in the medium is c/n. The rate of absorption in dV is

$$c\,dV \int_0^\infty A(\nu)\left[\int_0^\infty \int_0^\pi \int_0^{2\pi} n^2\rho(\nu,T') \frac{e^{-Ar}}{4\pi} d(Ar)\sin\theta\, d\theta\, d\zeta\right] d\nu \tag{5.61}$$

Thus this expression gives a transport to dV from the region around for a distance of the order of A^{-1}. This has as a consequence that the equation for the time dependence of the temperature distribution

* W. H. McAdams, "Heat Transmission," p. 55, McGraw-Hill, New York, 1942.

has an integrodifferential form because of the action at a distance character of (5.61). If A^{-1} is small compared with the distances over which T changes appreciably, then $\rho(\nu,T')$ can be developed in powers of $(x' - x)$, etc., and the leading term representing the combined effects of (5.60) and (5.61) becomes

$$\frac{c}{6}\int_0^\infty A(\nu)n^2\left(\int_0^\infty r^2e^{-Ar}d(Ar)\right)(\Delta\rho)d\nu$$

which can be written

$$\frac{c}{3}n^2\Delta[L(T)\rho(T)]$$

where Δ is the Laplace operator and $L(T)$ is an effective free path for the radiation

$$L(T) = \frac{\int_0^\infty A^{-1}\rho(\nu,T)\,d\nu}{\int_0^\infty \rho(\nu,T)\,d\nu} \tag{5.62}$$

For small temperature gradients therefore the quasi-transparent medium behaves as if it had an extra thermal conductivity k' which is given by

$$k' = \tfrac{16}{3}n^2L(T)\sigma T^3 \tag{5.63}$$

For hot glass $L(T)$ may be of the order of some centimeters, so that k' becomes much greater than the k due to molecular motions at temperatures of the order of 1000°C. Thus transfer of heat by diffusive radiation flow becomes an important mechanism of heat transfer in glass-melting tanks. However, because $L(T)$ is large compared with usual thicknesses of formed glass objects, the procedure of replacing the expressions (5.60) and (5.61) by the approximation (5.63) is not applicable. Transfer of heat by radiation is of great importance in these thinner objects but obeys different detailed laws than those which obtain in molecular heat conduction.

References

1. Bosworth, R. C. L.: "Heat Transfer Phenomena," Wiley, New York, 1952.
2. Carslaw, H. S., and J. C. Jaeger: "Conduction of Heat in Solids," Oxford University Press, London, 1947.
3. Eckert, E. R. G.: "Introduction to the Transfer of Heat and Mass," McGraw-Hill, New York, 1950.
4. Fishenden, M., and O. A. Saunders: "An Introduction to Heat Transfer," Oxford University Press, London, 1950.
5. *Heat Transfer*, a Symposium held at the University of Michigan, summer, 1952; Engineering Research Institute, University of Michigan, Ann Arbor, 1953.
6. Ingersoll, L. R., O. J. Zobel, and A. C. Ingersoll: "Heat Conduction with Engineering and Geological Applications," McGraw-Hill, New York, 1948.
7. Jakob, M.: "Heat Transfer," vol. 1, Wiley, New York, 1949.
8. Kern, D. Q.: "Process Heat Transfer," McGraw-Hill, New York, 1950.
9. Knudsen, J. G., and D. L. Katz: Fluid Dynamics and Heat Transfer, *Univ. Mich. Eng. Research Bull.* **37**, **1954.**

Chapter 6

Vacuum Technique

By ANDREW GUTHRIE, California State College at Hayward

1. The Vacuum Circuit—Conductance

In general, a high vacuum system consists of equipment (pumps) for the removal of gases from the enclosure of interest, connecting lines, and pressure-measuring equipment. The pressure in the system is normally measured in terms of atmospheric pressure, namely, that required to support a column of mercury 760 mm high at a temperature of 0°C and with $g = 980.665$ cm/sec^2. Commonly used units of pressure are the micron, μ (10^{-3} mm Hg), and the microbar (1 dyne/cm^2); 1 micron equals 1.333 microbars. In recent years the unit torr, which is very nearly equal to 1 mm Hg, has been adopted as a standard by the American Vacuum Society. This unit will be used throughout the rest of this chapter.

The *quantity* of a gas is given by the product of pressure and volume at a given temperature. The *gas flow* Q is the net quantity of gas per unit time passing through an isothermal plane. In most cases it is given by $Q = PV'$, where the pressure P is measured in the plane and V' is the time derivative of the volume. Q is proportional to the net number of molecules crossing the plane per unit time. Typical units are micron liters (μl) per second or microbar cubic centimeters per second.

Assume a small leak from the atmosphere into the vacuum enclosure. Air passing through this leak passes through the system into the pump and, hence, back into the atmosphere. This closed circuit is analogous to an electric circuit where the pressure P plays the role of electric potential and the gas flow Q plays the role of electric current. The rate of flow of gas per unit difference of pressure for any segment of the circuit is defined as the conductance of this segment and can be expressed as

$$C = \frac{Q}{P_1 - P_2} \tag{6.1}$$

where P_1 and P_2 are the pressures upstream and downstream, respectively. The same pressure units must be used for Q, P_1, and P_2. Typical units for C are cubic centimeters per second or liters per second. The conductance C corresponds to the same term as used in the conduction of electricity. The impedance (or resistance) of a segment of a vacuum circuit is given by

$$Z = \frac{1}{C} = \frac{P_1 - P_2}{Q} \tag{6.2}$$

with the quantities defined as for Eq. (6.1). For series and parallel connections of impedances

$$Z_s = \sum_i Z_i \quad \text{and} \quad \frac{1}{Z_p} = \sum_i \frac{1}{Z_i} \tag{6.3}$$

or

$$\frac{1}{C_s} = \sum_i \frac{1}{C_i} \quad \text{and} \quad C_p = \sum_i C_i \tag{6.4}$$

2. Flow of Gases through Tubes

In the design and operation of a vacuum system, the conductances of various component parts must be known. These will depend on the geometrical properties of the parts, on the operating pressures, on the nature of the gas, and on the gas temperature. Emphasis is placed on cylindrical tubes which are most commonly used as connecting lines in vacuum work. In the pressure region where intermolecular collisions predominate over collisions of molecules with the tube walls and there is no turbulence, the term *viscous flow* is applied. At low pressures when the mean free path λ of the molecules is greater than the cross-sectional dimensions of the tube, there is *molecular* or *free-molecule* flow. A *transition flow region* exists in the intermediate pressure range.

Viscous, laminar flow of gases through cylindrical tubes is governed by Poiseuille's law [1].* Application of this law gives the conductance of a cylindrical tube of diameter D and length L as

$$C = \frac{\pi}{128} \frac{D^4 \bar{P}}{\eta L} \quad \text{cgs units} \tag{6.5}$$

where η is the viscosity of the gas and $\bar{P}$ is the average pressure in the tube. For air at 20°C, Eq. (6.5) reduces to

$$C = 0.182 \frac{D^4 \bar{P}}{L} \quad \text{liters/sec} \tag{6.6}$$

where D and L are in centimeters and $\bar{P}$ is in microns. For a long tube, where the entrance-aperture impedance can be neglected, Eq. (6.6) is accurate to within about 10 per cent as long as $D\bar{P} \geq 500$ μ-cm. Relatively little information is available concerning the viscous conductance for tubes of noncircular cross sections. The viscous conductance of a long rectangular duct for air at 20°C is given by

$$C = 0.26 \frac{a^2 b^2}{L} \bar{P} Y \quad \text{liters/sec} \tag{6.7}$$

* Numbers in brackets refer to References at end of chapter.

where a and b are the cross-sectional dimensions in centimeters, L is the duct length in centimeters, $\bar{P}$ is the average pressure in microns of Hg, and Y is a correction factor given below:

a/b	1.0	0.9	0.8	0.7	0.6	0.5	0.4	0.3	0.2	0.1
Y	1.00	0.99	0.98	0.95	0.90	0.82	0.71	0.58	0.42	0.23

Where the entrance-aperture impedance is of consequence, it can be combined in series with that of the pipe. Prandtl [2] has considered the tube entrance-impedance, and applications of some of the results may be found in the references at the end of this chapter.

Consider a tube of length L, with varying cross section A and perimeter H. In the molecular flow region and following Knudsen's treatment [3], it can be shown that the conductance is given by

$$C = \frac{4}{3} K \frac{V_a}{\int_0^L \frac{H}{A^2}\,dL} \quad \text{cgs units} \tag{6.8}$$

where V_a is the average molecular velocity and it is assumed that the tube conductance is much less that that of the ends. V_a is given by $\sqrt{8kT/\pi M}$, where k is Boltzmann's constant, T is the absolute temperature, K is a dimensionless constant, and M is the mass of a molecule in grams. In Knudsen's formulation K is 1. However, a more rigorous treatment introduces certain integrals which depend on the geometrical form of the cross section. Although Eq. (6.8) is again obtained, for noncircular cross sections K is generally not equal to 1. For a tube of circular, uniform cross section and again neglecting end effects, Eq. (6.8) reduces to

$$C = \frac{1}{6}\sqrt{\frac{2\pi kT}{M}}\frac{D^3}{L} \quad \text{cgs units} \tag{6.9}$$

where D and L are the tube diameter and length, respectively. For air at 20°C and with D and L in centimeters, Eq. (6.9) becomes

$$C = 12.1\frac{D^3}{L} \quad \text{liters/sec} \tag{6.10}$$

When necessary to include the tube end conductance, it is possible to consider the tube end and the tube as two parallel conductances and combine them according to Eq. (6.4). The tube entrance conductance is given by

$$C_E = \sqrt{\frac{kT}{2\pi M}}\frac{A_0 A}{A_0 - A} \quad \text{cgs units} \tag{6.11}$$

where A_0 and A are the cross-sectional areas of the region from which gas flows into the tube and of the tube entrance, respectively. The total conductance of a tube of varying cross section and perimeter, including end conductance, is then given by combining Eq. (6.11) with Eq. (6.8) as parallel conductances. For $A_0 \gg A$ (a common case), Eq. (6.11) reduces to

$$C_E = \sqrt{\frac{kT}{2\pi M}}A$$

which becomes

$$C_E = \sqrt{\frac{\pi kT}{32M}}D^2$$

for cylindrical tubes of diameter D. In this latter case for air at 20°C the tube end conductance is given by $C_E = 9.11D^2$ l/sec where D is in centimeters. Combining this expression for the tube end conductance in parallel with the tube conductance as given by Eq. (6.10) leads to the following expression for the total conductance of a cylindrical tube for air at 20°C:

$$C_T = 12.1\frac{D^3}{L}\alpha \quad \text{liters/sec} \tag{6.12}$$

Here, α is given by $1/(1 + \frac{4}{3}D/L)$. Clausing [4] has treated short cylindrical tubes exhaustively and has shown that the simple parallel conductance picture is not strictly correct. His treatment leads to Eq. (6.12) but with a different expression for α. Clausing's α is given by

$$\alpha = \frac{15L/D + 12(L/D)^2}{20 + 38L/D + 12(L/D)^2} \tag{6.13}$$

Figure 6.1 gives the conductances of cylindrical tubes as a function of L for air at 20°C and for several values of D, using Eqs. (6.12) and (6.13). (Molecular flow occurs when $D\bar{P} \leq 10$ μ-cm.)

The molecular conductances for several shapes of ducts have been given by Barrett and Bosanquet [5]. Expressions for the conductances of several shapes of ducts are given below. These expressions are for long ducts, when the impedance of the duct is large compared with that of the ends. When this is not the case, the end impedance, as given by Eq. (6.11), is introduced in series with the duct impedance. Unless otherwise noted, the values of K listed are those derived by Barrett and Bosanquet.

For a *circular annulus* the conductance is given by

$$C = \frac{\pi}{3}\sqrt{\frac{kT}{2M}}\frac{(D_1 - D_2)^2(D_1 + D_2)}{L}K \quad \text{cgs units} \tag{6.14}$$

Here D_1 and D_2 are the diameters of the two concentric pipes and L is the length of each pipe. k, M, and T have been defined previously, and recur in the other cases following. The K values are as follows:

D_2/D_1	0	0.259	0.500	0.707	0.866	0.966
K	1	1.072	1.154	1.254	1.430	1.675

The conductance of a *rectangular duct* is given by

$$C = \frac{8}{3}\sqrt{\frac{kT}{2\pi M}}\frac{U^2V^2}{(U + V)L}K \quad \text{cgs units} \tag{6.15}$$

where U and V are the cross-sectional dimensions of the duct and L is the length of the duct. Values of K are as follows:

V/U	1	0.667	0.500	0.333	0.200	0.125	0.100
K	1.108	1.126	1.151	1.198	1.297	1.400	1.444

For an *equilaterally triangular duct*, K can be taken as 1.24 and the conductance is then given by

$$C = 0.413 \sqrt{\frac{kT}{2\pi M}} \frac{U^3}{L} \qquad \text{cgs units} \qquad (6.16)$$

where U is the side of the triangular cross section. In the case of a *thin slitlike tube*, with cross-sectional dimensions U and V $(U \gg V)$ and length L, the conductance is given by

$$C = \frac{8}{3} \sqrt{\frac{kT}{2\pi M}} \frac{UV^2}{L} K \qquad \text{cgs units} \qquad (6.17)$$

Values for K have been given by Clausing as follows:

L/V	0.1	0.2	0.4	0.8	1	2	3	4	5	10	>10
K	0.036	0.068	0.13	0.22	0.26	0.40	0.52	0.60	0.67	0.94	$\ln \frac{3}{8} (L/V)$

The transition flow region has been treated by introducing slippage at the tube wall into the viscous flow treatment and also in a semiempirical manner by Knudsen [3].

Knudsen arrived at the expression for the conductance of a tube in the transition flow region:

$$C = \frac{\pi}{128} \frac{D^4 \bar{P}}{\eta L} + \frac{1}{6} \sqrt{\frac{2\pi kT}{M}} \frac{D^3}{L} \cdot \left(\frac{1 + \sqrt{M/kT}\, D\bar{P}/\eta}{1 + 1.24 \sqrt{M/kT}\, D\bar{P}/\eta} \right) \qquad \text{cgs units} \qquad (6.18)$$

The symbols here have been previously defined. Ebert [6] has experimentally verified this expression. An examination of Eq. (6.18) reveals that for $\bar{P}$ negligible it reduces to Eq. (6.9) and for large values of $\bar{P}$ the third term is constant and Eq. (6.5) is obtained. For air at 20°C, Eq. (6.18) reduces to

$$C = \left(0.182 \frac{D^4 \bar{P}}{L} \right) + 12.1 \frac{D^3}{L} \left(\frac{1 + 0.256 D\bar{P}}{1 + 0.316 D\bar{P}} \right) \qquad \text{liters/sec} \qquad (6.19)$$

for D and L in centimeters and $\bar{P}$ in microns. Where end effects must be considered in the transition flow region, it is usually sufficient to replace $12.1D^3/L$ in Eq. (6.19) by $12.1D^3\alpha/L$, as discussed above. Correction factors to be applied to the conductances of round tubes at very low pressures (Fig. 6.1) in order to obtain the conductances for the transition region are given by Millen [7].

3. Pumping Speed and Evacuation Rate

The pressure against which a pump may be operated is referred to as the *exhaust pressure.* The *degree of vacuum attainable* is the lower limit of pressure (limiting pressure) attained in a closed vessel connected to a pump. This pressure is determined by the characteristics of the pumps and particularly by the exhaust pressures used. With vapor pumps the limiting pressure theoretically is the pressure of the pump vapor itself. With leaks or evolution of gas and vapor in the vacuum vessel the lower limit of pressure cannot be attained. When static conditions are reached, the vacuum vessel and the pump itself then have certain base or operating pressures.

In general, the *pumping speed* for any part of a vacuum system is given by $S = Q/P$, where S and P must be measured in the same region and Q represents the *total* gas flow in the system. When Q represents only the flow over and above that due to leaks, then the effective pumping speed or *speed of exhaust* is obtained. It is readily shown that the above definition of pumping speed reduces to $S = Q/\Delta P$ if ΔP is the pressure difference corresponding to a gas flow Q. This form of the definition finds the most general application. The *intrinsic speed, Sp,* of a pump is determined by the characteristics of the pump itself. Pumping speed has the same units as conductance, e.g., liters per second.

Suppose that S is the pumping speed at a vessel of volume V. It is readily shown by making use of $S = Q/\Delta P$ and the gas law that

$$-\frac{dP}{dt} = \frac{S}{V}(P - P_s) \qquad (6.20)$$

where P is the instantaneous pressure in the vessel and P_s is the base pressure of the vessel. Integration of Eq. (6.20), assuming S independent of pressure, leads to

$$P_t - P_s = (P_0 - P_s)e^{-(S/V)t} \qquad (6.21)$$

where P_t and P_0 are the pressures at times t and 0, respectively, and e is the base of the natural system of logarithms. Normally $P_s \ll P_0$ so that the pressure approaches its base value exponentially. When S depends on P, Eq. (6.21) can still be applied by taking an average value of S over the pressure region of interest. Equation (6.20) may be expressed as $-dP/dt = S_E P/V$ where

$$S_E = S\left(\frac{1 - P_s}{P}\right) \qquad (6.22)$$

Here S_E is the *effective pumping speed* or speed of exhaust at the vessel. For $P \gg P_s$, $S_E = S$. Since the effective pumping speed is determined by the ratio of base pressure to instantaneous pressure, when these are equal, $S_E = 0$. However, with leaks in the system there is still a flow of gas. Let C be the conductance of connecting lines and portions of the pump between the pumping region and the vacuum vessel. Then it is readily shown that

$$\frac{1}{S} = \frac{1}{C} + \frac{1}{S_p} \qquad (6.23)$$

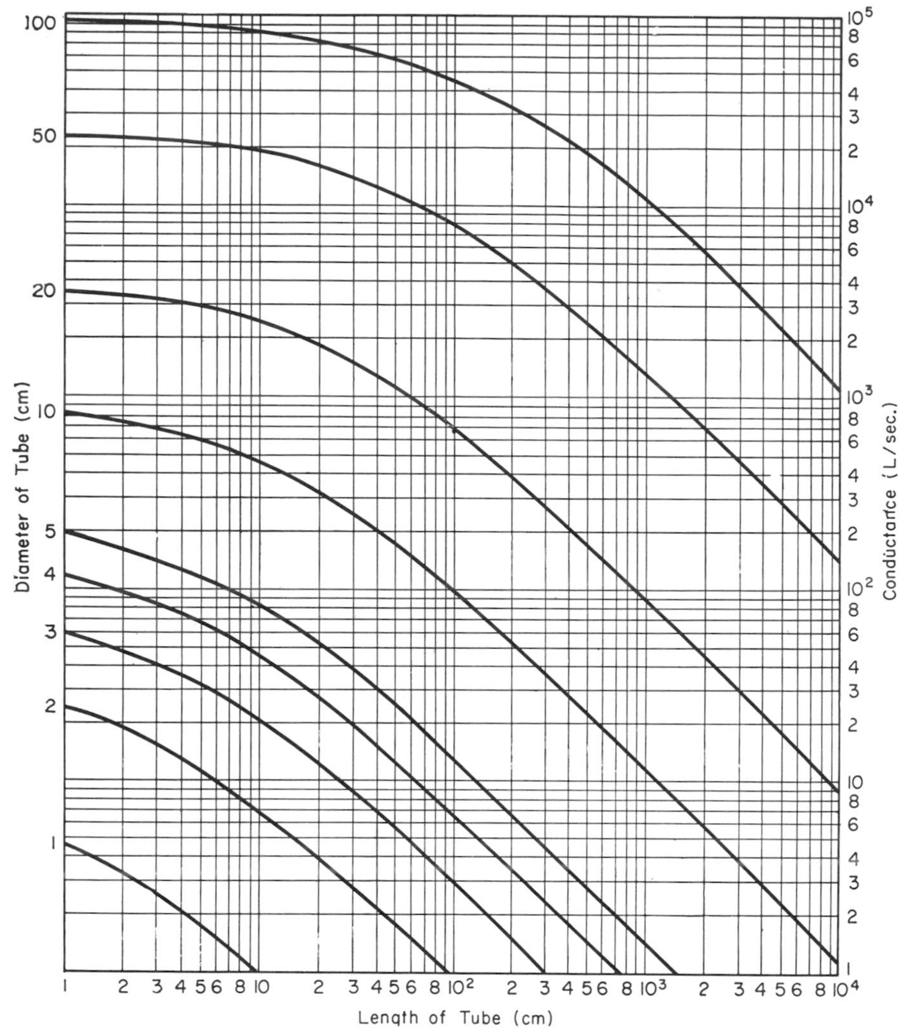

FIG 6.1. Molecular conductance of cylindrical tubes for air at 20°C.

This is a fundamental equation of vacuum design. Equation (6.22) can be applied to give the effective pumping speed at the pump by using S_p instead of S and with P_s representing the base pressure of the pump. In order to utilize to the greatest extent the speed of a given pump, the connections between the pump and the rest of the system should be as large as practicable.

A cold trap in the system acts as a pump for condensible vapors. The pumping speed of a trap for molecular flow is given by

$$S = 11.6 \sqrt{\frac{29}{M}} \left(1 - \frac{P_2}{P_1}\right) A \qquad \text{liters/sec} \qquad (6.24)$$

where M is the molecular weight of the condensible vapor, P_1 is the partial pressure of the vapor in the vacuum system near the trap, and P_2 is the vapor pressure of the vapor at the cold-trap temperature. A is an effective area for the trap which acts as an aperture and for many designs of traps is simply the cooled area. The condensible vapor of most common interest is water vapor and it is generally removed by a mechanical refrigerating system, a dry-ice trap, or a liquid-nitrogen trap. The working temperatures for the latter substances as well as the water vapor pressures for these temperatures are to be found in such handbooks as are listed at the end of this chapter.

Methods of measuring pumping speeds actually measure intrinsic speed S_p rather than effective speed S_E. Two common methods are used, the constant-pressure or metered-leak method and the constant-volume or rate-of-rise method. In the first method, a known leak at atmospheric pressure is introduced into the system. The pumping speed is then given by $P_a V_1 / \Delta P$, where P_a is atmospheric pressure, V_1 is the volumetric rate of flow measured at atmospheric pressure, and ΔP is the change of pressure introduced by the leak. In the second method, the rates of rise of pressure in the system produced by closing off the system from the pumps, with and without artificial leaks, are observed. Only the rates of rise of pressure and the volume of the system need be known.

4. Mechanical and Diffusion Pumps

Two distinct classes of pumps are used in high vacuum systems, namely, mechanical and vapor pumps [8]. Mechanical pumps are either of the rotary or molecular drag type. Reciprocating pumps are very rarely used. Vapor pumps use a condensible vapor as the driving force and are of the ejector or vapor-stream type, the latter being referred to as diffusion-condensation or simply diffusion pumps. In general, mechanical and ejector pumps have their highest efficiency in the viscous-flow region, while diffusion pumps have highest efficiency in the molecular-flow region. To attain pressures below 1 μ Hg, a combination of these pumps is generally used.

The modern rotary pumps are essentially of the same basic design as that used by Gaede [9]. They consist of a rotor rotating in a housing of steel of larger diameter than itself so that a point on the circumference is always in contact with the housing to provide a seal. A sliding or "scraping" vane is commonly used to provide the seal. This vane sweeps out a crescent-shaped volume, compressing the gas and forcing it out of a nonreturn exhaust valve to the atmosphere. The pump is either immersed in or continuously fed with oil for sealing, lubrication, and cooling. Clearances between moving parts must be kept to within a few mils to get good sealing. Some common designs of rotary pumps are shown in Fig. 6.2.

For each of these pumps there is a constant volumetric displacement per cycle. The speed is given by the product of volume of the exhaust chamber times the rps. The ultimate pressure and efficiency are limited by (1) limits of accuracy in machining which determine leakage of gas back past vanes and rotors and (2) vapor pressures of oil and contamination by condensible vapors. Since the outlet of the pump is in direct contact with the atmosphere, some air is absorbed by the oil. This is partially expelled until finally equilibrium is reached. This problem can be minimized by using a two-stage pump or by connecting two pumps in series. The capacity of a pumping system can be increased by using two or more pumps in parallel.

The ultimate pressure produced by a rotary pump is usually less than the vapor pressure of the oil because the vapor is swept out faster than it is produced. Normally, a good-quality light petroleum oil is used, with the high vapor-pressure fractions removed. Clean oil should be used and it should be replaced periodically. The characteristics of several commonly used mechanical pump oils, together with suppliers, are tabulated in several of the references in the Bibliography at the end of this chapter; for example, Kohl or Guthrie and Wakerling. The most serious reduction in the efficiency and ultimate pressure is due to contaminating condensible vapors which are trapped in the pump oil. These can be eliminated to a large degree by the use of oil separators, centrifuges, hot oil (>100°C), or traps (mechanical or cold). Probably cold traps are most commonly used for this purpose. Commercial oil-sealed rotary pumps have speeds ranging from a fraction of a cubic foot per minute to over 700 cfm, with ultimate pressures between 10^{-1} and 10^{-5} torr.

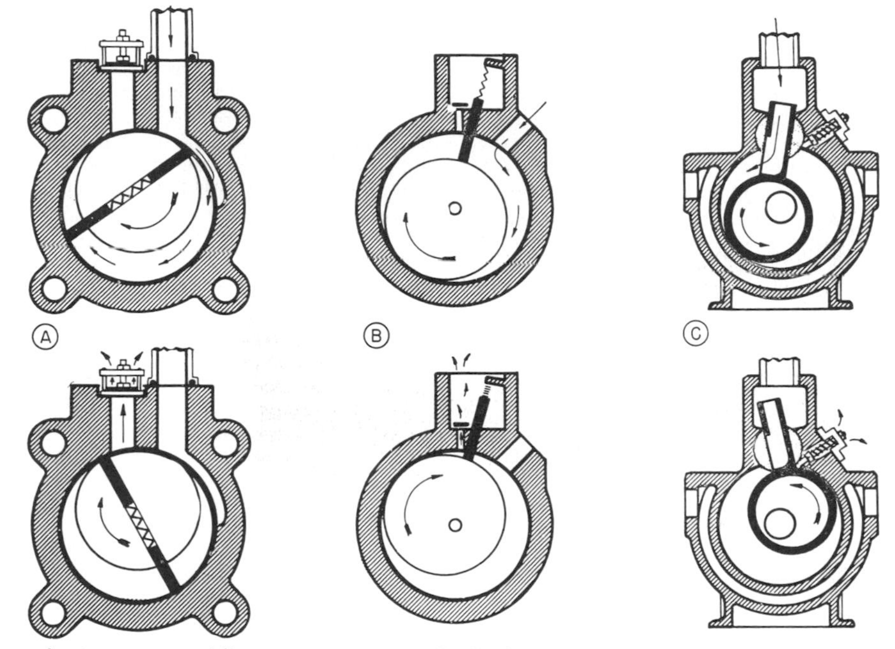

FIG. 6.2. Types of rotary pumps: (*A*) rotary vane type; (*B*) fixed spring-loaded vane type; (*C*) Kinney type. (*Reproduced from "Vacuum Physics," J. Sci. Instr., suppl.* 1, *October*, 1951.)

The choice of pump is determined by (1) working

pressure range required, (2) capacity required (chamber size, pump-down time, and size of diffusion pumps), (3) presence of vapors, and (4) mechanical and general details (size, noise, and vibration). The Siegbahn [10] molecular drag pump consists of a moving, smooth surface next to a stationary one. The rotor is rotated rapidly at around 10,000 rpm or more, and pumping speeds up to 80 l/sec have been obtained. Ultimate pressures down to 10^{-6} torr are possible. These pumps will pump gases and vapors without traps and will pump heavier gases faster than light ones. Precision construction is required. A form of molecular drag pump using slotted disks has been reported by Becker [11]. Pressures below 10^{-8} torr have been claimed for this type of pump.

For all vapor pumps a stream of high-velocity vapor is the driving force. Steam and oil ejectors consist essentially of a nozzle and exhaust tube attached to the vacuum chamber, with the vapor boiler separate for the first type. Diffusion pumps are the same plus condensing surfaces and with the boiler generally part of the pump [12]. The jet nozzle is used to transform the pressure energy of the vapor into velocity energy in a directed stream.

In the viscous-flow region and for vapor velocities less than that of sound, the vapor stream is a confined jet with eddies at the gas-vapor boundary which entrain gas in the stream. The ultimate pressure is limited by back-scattering of vapor when the gas pressure becomes less than the vapor pressure. Actual ejectors have supersonic vapor velocities as the pressure is reduced and very little is scattered back, which gives lower pressures.

In the molecular-flow region the vapor comes out of the nozzle at supersonic speeds, with a small fraction flowing backward. The vapor stream imparts a forward component of velocity to the gas molecules. As the vapor stream expands, the gas loses its streaming velocity and the density increases to a maximum. The advantage of supersonic flow is lost unless the vapor is condensed. This is true for oil and steam ejectors as well as diffusion pumps. This essential feature was first recognized by Langmuir [13].

At low pressures, the velocity must be high enough to prevent back diffusion of compressed gas. Thus, density and viscosity of vapor stream, nozzle design, width of diffusion slot, and disposition of condenser affect the general pump design. The nozzle design determines the proportion of vapor going in the forward direction. The vapor molecules should move in nearly parallel directions with few collisions between molecules, so as to reduce back streaming. The density and viscosity of the vapor stream determine the pressure range over which the pump will function and the back pressure required. The ultimate pressure is limited by the vapor pressure of the pump fluid and, to some extent, by back diffusion of gas molecules as well as vapor molecules. The speed factor or Ho coefficient is the ratio of observed pump speed measured as the volume of gas pumped per unit time to the theoretical speed of diffusion through an orifice with the same areas as the pump throat (11.67 l/sec-cm^2 for air). The speed factor is usually less than 0.5 due to impedance of pump head and back streaming of both gas and vapor. The most important factor affecting performance of a vapor pump is the heat input, which determines the density and viscosity of the vapor stream and consequently the critical backing pressure and pumping speed. Some types of vapor pumps are illustrated in Fig. 6.3. Steam ejectors are most efficient above 0.5 torr although by connecting several nozzles in series with intermediate condensers it is possible to get down to 50 μ with high pumping speeds. These ejectors handle condensible vapors directly.

Glass and steel are commonly used for the pump body. The condenser surface is usually conical and water-cooled, although air cooling can be used. Two or more stages of jets are often used so that a higher backing pressure can be used. In this case the stages on the atmospheric side of the pump act more as ejector than diffusion pumps. Mercury pumps have been built with pumping speeds from several cubic centimeters per second up to several hundred liters per second at around 10^{-5} torr and are usually of the conventional Langmuir form with umbrella-type annular jets. The disadvantage of a mercury pump is in the high vapor pressure of mercury, which is about 10^{-3} torr at 20°C. To reach lower ultimate pressures than this, it is necessary to use cold traps. However, these pumps offer one important advantage over oil-diffusion pumps since mercury cannot dissociate. The advent of oils with low vapor pressures led to oil-diffusion pumps largely through the work of Burch [14] and Hickman [14].

In general, even with very low vapor-pressure oils, these pumps are operated with traps to eliminate back diffusion of oil vapor. In some cases, baffles or a combination of baffles and traps are used. The main disadvantage with baffles is in a serious reduction in pumping speed. Large boilers and vapor columns should be used to reduce the heat input so as not to decompose the oil. The amount of oil in the boiler should be kept to a minimum so as to avoid excessive quantities of high vapor-pressure decomposition products accumulating. Good conducting materials such as aluminum or copper should be used for the jet system. Most oil-diffusion pumps are made of metal, usually steel, with two or more stages of jets in series. Pumps with barrels from as small as 2 in. up to around 48 in. have been designed and used. The pumping speeds range from as little as 10 l/sec up to around 30,000 l/sec at around 10^{-5} torr.

The oils used are usually vacuum distilled mineral oils, such as the Apiezon group, or synthetic oils, particularly the higher-order esters such as the phthalates and sebacates. Newer oils which have come into use since the Second World War are the silicones and the chlorinated hydrocarbon oils which are extremely stable.

Oil-diffusion pumps suffer from the disadvantage of oil decomposition. The problem can be minimized by using a fractionation design (Fig. 6.3*c*). In this type of pump the lighter components of the oil are boiled off in the last jet stages and are pumped off by the forepump. Only the lowest vapor-pressure fraction reaches the jets nearest the vacuum chamber resulting in a low ultimate pressure. The choice of a vapor pump is determined by the required speed, ultimate pressure, and the backing pump characteristics.

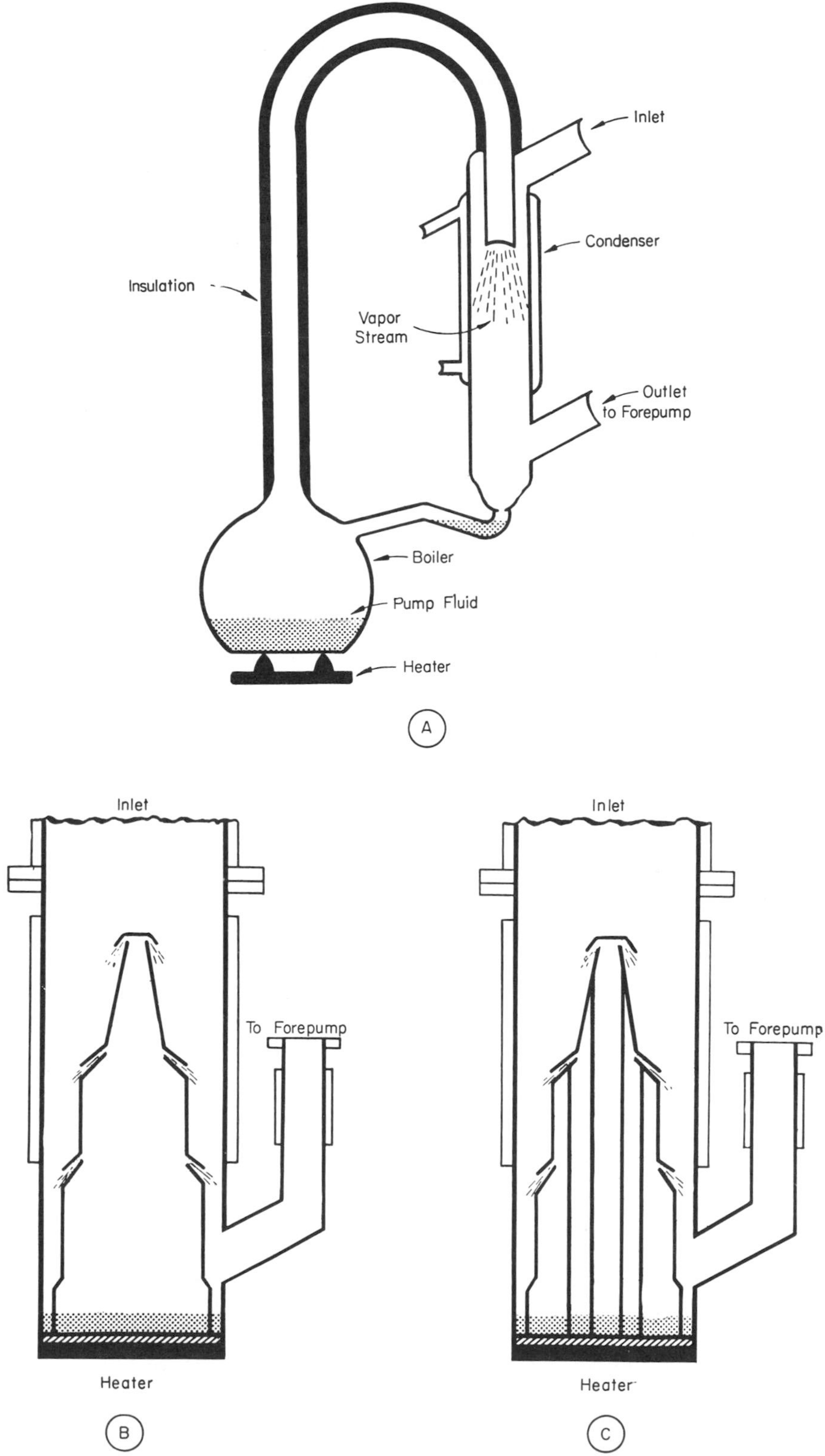

FIG. 6.3. Vapor pumps: (*A*) Langmuir-type condensation pump; (*B*) nonfractionating metal oil-vapor pump; (*C*) fractionating metal oil-vapor pump.

TABLE 6.1. VAPOR-PUMP OILS

Trade name	Chemical composition	Ultimate vacuum, torr	Resistance to oxidation	Ref.†
Amoil	*i*-Diamyl phthalate	1.3×10^{-5}	....	D
Amoil-S	*i*-Diamyl sebacate	1×10^{-6}	Poor	D
Apiezon A	Mixture of hydrocarbon; refined from petroleum sources	2×10^{-6}	Fair	D
Apiezon B	Mixture of hydrocarbons; refined from petroleum sources	4×10^{-7}	Fair	D
Apiezon oil C	Mixture of hydrocarbons	10^{-8}	Fair	
Apiezon oil G	Mixture of hydrocarbons	10^{-6}	Fair	
Arochlor 1254	Chlorinated hydrocarbon	8×10^{-6}	Fair	D
Butyl phthalate	*n*-Dibutyl phthalate	2.4×10^{-5}	Fair	D
Butyl sebacate	*n*-Dibutyl sebacate	2×10^{-5}	Fair	
Eimac type A	Mixture of hydrocarbons; refined from petroleum	4×10^{-7}	Fair	M
Litton molecular C	Mixture of hydrocarbons; refined from petroleum	2×10^{-6}	Fair	M
Myvane-20	Mixture of hydrocarbons; refined from petroleum	10^{-6}	Fair	S
Narcoil-10	Mixture of polychlorinated biphenyls similar to pentachlor biphenyl	2×10^{-5}	Fair to good	M
Narcoil-20	2-Ethyl hexyl sebacate	5×10^{-8}	Poor	M
Narcoil-30	2-Ethyl hexyl phthalate	2×10^{-7}	Poor	M
Octoil	2-Ethyl hexyl phthalate	3×10^{-7}	Poor	D
Octoil-S	Di-2-ethyl hexyl sebacate	2×10^{-8}	Poor	D
m-Cr	Tri-*m*-cresyl phosphate	6×10^{-8}	....	D
p-Cr	Tri-*p*-cresyl phosphate	2×10^{-8}	....	D
b-S	Dibenzyl sebacate	4×10^{-9}	....	D
Silicone DC-702	Mixture of organic silicone molecules	2×10^{-7}	Good	S
Silicone DC-703	Mixture of organic silicone molecules	5×10^{-8}	Good	S

* *References:*
D: S. Dushman, "Scientific Foundations of Vacuum Technique," 2d ed., Wiley, New York, 1962.
S: H. M. Sullivan, Vacuum Pumping Equipment and Systems, *Rev. Sci. Inst.*, **19**: 1 (1948).
M: Manufacturer's data.

The ultimate pressure depends on the pump oil and the design of the pump. The choice of backing pump is important and is determined by the critical backing pressure and the speed of the vapor pump. The speed of the vapor pump is usually independent of backing pressure below the critical value. Many modern vapor pumps will work against backing pressures between 200 μ and 500 μ. The characteristics of some commonly used oils are given in Table 6.1. The values listed in column 3 correspond to the pressures attainable with vapor pumps using these oils. These values cover measurements made by static methods using different forms of hypsometers as well as measurements made by evaporation methods. Details of the methods used for several of the oils can be found in the references at the end of the table. In general, it has been found that the experimental vapor-pressure data for these liquids can be represented by $\log P = A - B/T$, where A and B are constants for each liquid and T is the absolute temperature. This relationship has been used to obtain the pressure values at 20°C. In general, values quoted for ultimate vacuums must be looked on with considerable caution. The conditioning, as well as possible maltreatment of the oil, are important factors. Also, measurements made with an ionization gauge are always difficult to interpret because the response depends on the types of ion present.

For the names of suppliers of the oils listed in Table 6.2, see the Bibliography at the end of this chapter, e.g., Guthrie: "Vacuum Technology," p. 507.

5. Getter-Ion Pumps and Cryopumps

Sorption is the process of taking up gas (or vapor) by absorption, adsorption, chemisorption, or any combination of them. Absorption and adsorption are physical processes, the first being the binding of gas in the interior of a solid (or liquid) and the second the trapping of gas on the surface of a solid. Chemisorption is the binding of gas (or vapor) on the surface or in the interior of a solid (or liquid) by chemical action.

Getters, which have been used widely in the tube industry for many years, are examples of sorption materials. Getters are normally thin films or metal ribbons and powders (solid getters). Evaporated films have large surface areas because of their granular structure, and they clean up gases rapidly at normal temperatures because of the freshness of their surfaces. Solid getters are generally used at high temperatures to aid diffusion of the adsorbed gas into the solid. They are usually selected from elements with high melting points, such as titanium, zirconium, tantalum, molybdenum, tungsten, and thorium. Thin-film getters are usually chosen from the alkaline earth metals, such as barium, strontium, calcium, and magnesium, because of their chemical activity and the ease with which they can be volatilized in vacuum. The properties of a number of getters have been summarized by Holland [15] and by Espe, Knoll, and Wilder [16].

In recent years, there have been rapid developments in the use of sorption materials (getters) as pumps to achieve pressures much lower than are generally possible with other types of pumps [17, 18]. Although attempts have been made to use getters alone, the general approach has been to combine gettering with ionization. Getters will not pump the inert gases that occur in the atmosphere, and so ionization of the residual gases and vapors is used. It is clear that before the ionization process can be used it is necessary to reduce the pressure so that electrons can be accelerated to a high enough energy. Attempts have been made to pump purely by ionization but electric-power requirements are excessive. The gettering action can be provided by evaporation or sputtering. The trend has been to sputtering, which is used in most commercial pumps. Such pumps are called sputter-ion pumps or simply ion pumps. The pumping action of various devices involving ionization and gettering has been observed for many years, e.g., ionization and cold-cathode (Philips or Penning) vacuum gauges. Such gauges can be used as effective pumps for small volumes.

The principle of operation of a sputter-ion pump is shown in Fig. 6.4. Two cathodes and one anode are

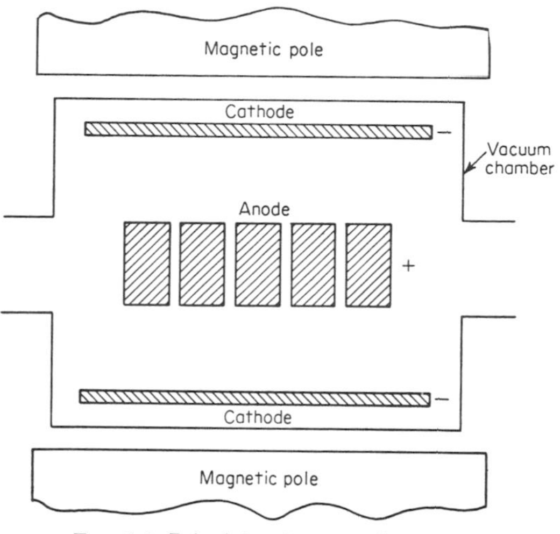

FIG. 6.4. Principle of sputter-ion pump.

mounted inside an enclosure which can be connected to some type of roughing pump and to a vacuum chamber. The anode shown is a hollow, cell structure, and the cathodes are flat plates. Many designs of anodes and cathodes have been reported. The cathodes made of titanium are good gettering materials, while the anode is often made of copper. The electrode assembly is mounted, as shown, between the poles of a permanent magnet which usually provides field strengths between 2,000 and 4,000 gauss.

When several thousand volts (usually 2 to 10 kv or more) are applied between the anode and the cathodes, electrons are speeded up toward the anode but are constrained to move in helical paths by the magnetic field. These electrons oscillate between the cathodes, ionizing the gases and vapors present and migrating to the anode. The positive ions strike the cathodes, knocking out small particles (sputtering) and being adsorbed or absorbed in the process. The sputtered material is deposited on the anode structure (and surroundings) and combines with the chemically active gases, thus pumping them. The inert gases are also pumped, probably by being covered with sputtered materials at the cathodes and by entrapment at the anode. The total number of ions produced is a function of the pressure, so that measuring the current from the power supply which is connected to anode and cathodes gives a measure of the pressure.

To reduce the pressure to the point where voltage can be applied to these pumps, mechanical pumps are commonly used. To obtain a clean system (one free of contaminating vapors), various types of sorption pumps are used. Materials such as activated charcoal, zeolite, and activated alumina are effective as sorption agents, particularly when cooled to liquid-nitrogen temperatures. With large systems, in order to reduce the amounts of sorption materials and refrigerant, it is possible to do a preliminary (and clean) rough pumping with a pump such as a water aspirator or a steam-jet aspirator. In small systems a primer, which is a getter material such as titanium, is sometimes used in conjunction with a roughing pump to reduce the pressure to a low enough value so that the ion pump can be turned on.

It is possible to achieve pressures of 10^{-8} torr or somewhat better without heating the vacuum system (baking) to drive off adsorbed and absorbed gases and vapors. With baking it is possible to achieve pressures of the order of 10^{-12} torr or better. Claims have been made of pressures as low as 10^{-14} torr. However, there is always the question of the accuracy of the pressure measurement. If the magnets are not removed during the baking process, it is not possible to use temperatures much above 250°C. In view of the fact that the rate of removal of gases and vapors increases rapidly with temperature, it is advisable to remove the magnets and to make a careful choice of construction materials. A temperature of around 450°C is in fairly common use. Both the pump and the vacuum system should be baked. The lifetime of a sputter-ion pump is a function of the pressure. Typically, at a pressure of 10^{-7} torr, the lifetime is of the order of 200,000 hr. In principle, there is no limit to the pumping speed of this type of pump. This is basically a matter of engineering design.

The pumping speeds are considerably higher for chemically active gases such as hydrogen, oxygen, and nitrogen than for the inert gases. There appears to be some evidence that when hydrogen is bled into a system (as in the case of nuclear accelerators) this type of pump will not pump this gas after a while. With a diode-type pump (Fig. 6.4) using a continuous air leak, periodic pressure fluctuations may occur; they are associated with the argon normally present in the atmosphere. This is often referred to as "argon instability." Various changes in pump design have been made to overcome this difficulty, including auxiliary electrode collectors and different shapes of cathodes [19].

Cryopumps are used to reduce the pressure by low-temperature condensation of gases in the vacuum system. The pressure in a system is that of the vapor pressure of the condensate on low-temperature surfaces. When the temperature is low enough so that the condensate freezes, a slight decrease in the temperature of the solid usually causes a sharp reduction in the vapor pressure. Cold traps are forms of cryopumps although they are usually used in conjunction with other types of pumps. To reduce the pressure effectively, temperatures low enough to freeze the bulk of gases and vapors present are required. The pumping speed for a cold surface in the molecular-flow region is given by Eq. (6.24).

Table 6.2 gives the vapor pressures of some substances at the temperatures of liquid air and liquid

TABLE 6.2. VAPOR PRESSURES OF SOME SUBSTANCES AT LIQUID-AIR AND LIQUID-HELIUM TEMPERATURES

Substance	Approx. vapor pressure, torr	
	At −190°C	At −268.8°C
Argon (A)	500	10^{-90}
Carbon dioxide (CO_2)	10^{-7}	
Carbon monoxide (CO)	760	10^{-94}
Helium (He)	......	760
Hydrogen (H_2)	......	10^{-6}
Mercury (Hg)	10^{-32}	
Oxygen (O_2)	350	10^{-104}
Neon (Ne)	......	10^{-26}
Nitrogen (N_2)	760	10^{-81}
Pump oils	10^{-35}	
Water (H_2O)	10^{-22}	

helium. It is clear that liquid air is completely ineffective for argon, carbon monoxide, oxygen, and nitrogen. Carbon dioxide can also be troublesome if pressures in the ultrahigh-vacuum region are desired. Liquid helium is effective for all gases except helium, although hydrogen may be a factor in the ultrahigh-vacuum region. It should be kept in mind that the ultimate pressure achieved with cryopumping depends on the quantities of various gases present at the start.

With liquid hydrogen (b.p. = −252.7°C), all gases except neon, hydrogen, and helium are frozen effectively. Under standard atmospheric conditions, the pressures of these gases are, in the order listed above, 1.14×10^{-2} torr, 7.60×10^{-2} torr, and 0.38×10^{-2} torr. The total pressure under these conditions is more than 0.1 torr; this is the lowest pressure obtainable under these conditions. The corresponding pressure value for liquid helium is 0.38×10^{-2} torr. The obvious way to achieve lower pressures is to reduce the pressure by other means before starting the cryopumping. Pumping to 1 torr first gives final pressures of about 10^{-4} and 4×10^{-6} torr for liquid hydrogen and liquid helium, respectively. It is possible to achieve pressures in the ultrahigh-vacuum region (10^{-9} torr or lower) by first pumping to considerably lower pressures than the 1 torr mentioned above. Even better results are obtained by first flushing with a gas that can be readily frozen, e.g., carbon dioxide, and then carrying out the above pumping procedure.

It is evident that, with cryopumping, gases and vapors are held within the vacuum system, as is the case with various sorption materials. However, in cryopumping the gases and vapors are simply frozen whereas in the cases of sorption materials various complex physical and chemical processes are involved. In using cryogenic materials, it is necessary to design the cooling surfaces so as to minimize heat flow. Often traps or panels used for liquid hydrogen or helium are placed adjacent to surfaces which are cooled by liquid nitrogen. This substantially reduces the rate of consumption of hydrogen or helium. Recommended procedures for handling liquid hydrogen must be followed. Various manufacturers will provide panels of assorted shapes and sizes to give efficient pumping in small or large systems.

6. Vacuum Gauges

Pressures from a fraction of a torr to atmospheric are readily measured by making use of the force exerted by the gas, usually by means of manometers or Bourdon gauges. In the pressure region from a few millimeters down to around 10^{-8} torr [20], use is made of some pressure-dependent property of the gas such as Boyle's law, heat conductivity, viscosity, thermal molecular pressure, or ionization.

A *McLeod gauge* [21] can still be considered to be the only absolute measuring device for pressures above about 10^{-6} torr and is generally used to calibrate other types of gauges. It operates on the basis of Boyle's law and can be calibrated from its physical dimensions. A diagrammatic sketch of such a gauge is shown in Fig. 6.5. A volume V of rarefied gas at

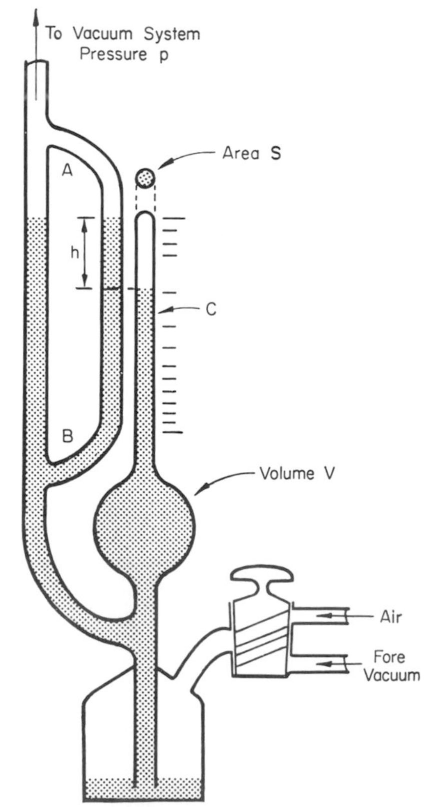

FIG. 6.5. Diagrammatic sketch of a McLeod gauge.

the pressure P, which must be determined, is compressed into the small capillary C by raising the mercury beyond the cutoff. The tube AB, of the same cross-sectional area as C, is added to reduce errors due to capillarity. Usually the mercury is raised to a level in AB even with the top of C. Then, by Boyle's law, $P = kh^2/(V - kh)$, where k is the capillary constant in cubic centimeters per millimeter of length and h is the difference in levels of the mercury columns in AB and C (in millimeters). For small values of P, say 1 torr or less, $P = kh^2/V$ and a quadratic scale results. For small values of h, the gas in C is often compressed to a definite volume v and the height h_1 (in millimeters) of the mercury in AB above this level is observed. In this case $P = vh_1/V$ and a linear scale results. V is usually between 100 and 200 cc, although values as great as 1,400 cc have been used [22].

Capillary diameters ranging between a fraction of 1 mm and 10 mm have been used. Values much less than 1 mm make it necessary to use extreme care in the construction and operation of the gauge because of the sticking of the mercury in the capillary. The mercury in a McLeod gauge is normally raised pneumatically or by mechanically lifting the reservoir.

The most serious disadvantages of this type of gauge are in the fact that condensible vapors are not measured directly and the operation is discontinuous. In addition, the mercury usually must be trapped from the rest of the system with cold traps, and imperfect operation of these traps leads to errors. Discrepancies may arise due to the surface conditions of the glass and the mercury. In the tilting McLeod gauge as originally described [23] the reservoir is always in communication with the vacuum being measured and is run into the bulb by tilting. Because of its compactness, this type of gauge is very convenient, particularly for fore-vacuum measurements. McLeod gauges using low vapor-pressure oils as the operating fluid, such as butyl phthalate, butyl sebacate, and silicone oils, have been proposed but they suffer from the types of difficulties associated with oil manometer work.

Thermal conductivity gauges consist of heated wire filaments mounted in envelopes connected to the vacuum system. The thermal conductivity of the gas surrounding the filament is pressure-dependent below about 10 torr. At low pressures where the mfp is of the same order as, or larger than, the dimensions of the gauge, the energy loss by conduction in the gas is proportional to pressure, and also depends on the mechanism of energy interchange between the solid surface of the wire and walls and the gas molecules, i.e., the accommodation coefficient of the surface, the structure, and the molecular weight of the gas. The heat lost by conduction through a gas of molecular weight M at pressure P between two surfaces at temperatures T_2 and T_1 is $KC_aP(T_2 - T_1)/(MT_1)^{1/2}$ cal cm^2-sec, where K depends on the molecular structure of the gas and C_a is the accommodation coefficient [24].

At very low pressures, the energy loss by conduction is much less than the radiation losses so that a natural limit for pressure measurements is about 10^{-4} torr. With the heated wires in these gauges operated under nearly constant energy input, the temperature will decrease with increased pressure. Different gauges employ different methods for measuring this change in temperature. This may be a thermocouple fixed to the wire (thermocouple gauge) or the wire may be made of high-temperature coefficient material and its resistivity measured by connecting it in one arm of a Wheatstone bridge (Pirani gauge). In both cases the indicating meter may be directly calibrated in terms of pressure. Thermal conductivity gauges have to be referred to other gauges such as the McLeod gauge for their calibration. The accuracy of the gauges is highly dependent on the surface condition of the filament. They are commonly used in the medium vacuum range for control purposes and for the recording of pressures in vacuum process work as well as for leak testing. These gauges will directly measure the pressure of condensible vapors.

The gas surrounding a moving mechanical system causes damping depending on the viscosity of the gas. As in the case of thermal conductivity, the effect becomes pressure dependent at low pressures. Several methods of applying this effect to the measurement of low pressures have been reported [25].

Knudsen [3] has shown that at low pressures when the mfp of the gas is long relative to the distance between two plates at temperatures T_1 and T_2 ($T_1 > T_2$) in an enclosure at T_2 there is a repulsion between the plates which is proportional to the gas pressure P. On the average, molecules at T_2 have more momentum than molecules at T_1 and this difference in momentum is used to rotate a suspended vane. The rotation of the vane is usually measured by a mirror system. Various forms of the Knudsen gauge have been discussed by Klumb and Schwartz [26] and by Steckelmacher [26].

By using different vane suspensions, this type of gauge can readily be used to measure pressures from about 10^{-7} to 10^{-2} torr. The calibration over this range is substantially linear, and it is independent of type of gas or vapor. At pressures above about 10^{-2} torr the scale becomes nonlinear. These gauges can be constructed so that direct calibration can be made from physical characteristics. However, in practice they are constructed for ruggedness and must be calibrated against some other form of gauge, usually the McLeod.

All types of ionization gauges depend on ionization of gas, including glow-discharge tubes, Philips gauges, thermionic ionization gauges, and certain special types. The glow-discharge device consists of a self-maintained electrical discharge between two electrodes and through the gas of interest. A measurement of the pressure is obtained from the color of the glow or by measurement of the discharge current. The voltage must be higher as the pressure is reduced and finally a point is reached where the discharge cannot be maintained. The range from about 10^{-2} to 10 torr can be covered. The Penning or Philips gauge extends the maintained glow discharge to lower pressures by increasing the average electron path and therefore the ionization of the gas. This gauge uses magnetic and electric fields disposed so as to produce electron oscillations in helical paths [27].

The electrodes generally consist of two parallel-plate cathodes with a ring anode between. The magnetic field is perpendicular to the planes of these

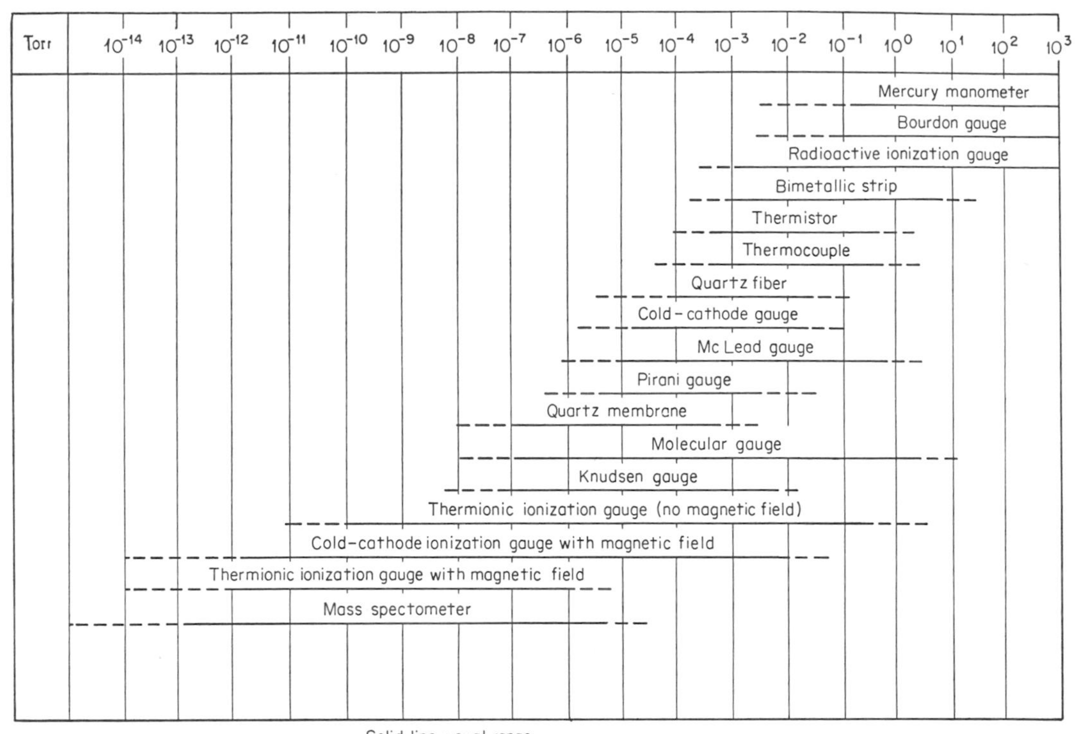

FIG. 6.6. Pressure ranges covered by various types of gauges. (*a*) Solid line, usual range; (*b*) dashed line, range with special gauge design.

electrodes. Higher sensitivity is obtained with d-c than with a-c potential. This gauge readily covers the range 10^{-6} to 10^{-3} torr, although the calibration depends on the gas. Magnetic fields upwards of 500 gauss and electric potentials around 2,000 volts are commonly used.

The term ionization gauge is often specifically applied to the thermionic form which depends on electron emission from a hot filament. Positive ions are formed by electron bombardment of the gas in the gauge. Usually, a triode tube is used with the positive ions collected on the grid and the negative ions collected on the positively charged anode, although the roles of grid and anode are often interchanged. The pressure in millimeters of mercury is given by kci_+/i_- where i_+ is the positive ion current (microamperes), i_- is the electron current (milliamperes), k is a function of the potentials and geometry of the gauge, and c is a factor depending on the relative probability of ionization of different gases, referred to air. The calibration is linear with fixed potentials and constant emission for a given gas. The positive ion currents are quite small and can be measured by galvanometers or d-c amplifiers, the latter being most commonly used. For continuous pressure readings, constant electron emission is necessary.

This type of gauge is normally used to measure pressures from about 10^{-2} to 10^{-7} torr. A lower limit appears to be set by sensitivity of the current measuring method, electron current from collector to anode and the production of X rays at the anode. Special arrangements of electrodes together with low atomic number anodes and extreme care in elimination of leakage currents and photoelectric effects have resulted in gauges that will read pressures down to about 10^{-11} torr. The upper pressure limit is set by filament life rather than departure from linearity. The gauge suffers from poisoning of the cathode due to dissociation of gases and vapors. Considerable care must be exercised in interpreting the readings made with such a gauge, particularly at low pressures [28]. The nature of the gas or vapor is particularly pertinent.

In recent years considerable effort has been devoted to the development of instruments for measuring pressures well below 10^{-11} torr. A number of approaches have been taken. Hobson and Redhead [29] have described a cold-cathode inverted magnetron gauge which is claimed to measure pressures somewhat below 10^{-12} torr. Houston [30] has designed a special ionization gauge which is similar to a Penning gauge with a hot cathode added. It has been estimated that this gauge may measure down to almost 10^{-15} torr. Lafferty [31] has described a hot-cathode magnetron gauge which measures pressures down to about 10^{-14} torr. Several types of mass spectrometers have been developed to analyze residual gases but which also give some measure of the pressure. The Omegatron [32], which operates on the principle of a cyclotron, has come into fairly general use. Several types of mass spectrometers using magnetic deflection have been reported [33].

For the pressure range from 10^{-3} to 10 torrs, a form of ionization gauge called the Alphatron has been used [34]. This gauge makes use of a radium capsule, the alpha particles from which ionize a small percentage of the gas in the gauge. The pressure ranges covered by the various types of gauges are shown in Fig. 6.6.

7. Vacuum Systems

We shall refer to systems of such design that they cannot be baked as conventional systems can. Without baking, gases and vapors are gradually released from the surfaces; this limits the lowest pressure that can be achieved. Oil or mercury vapor from diffusion pumps also contributes to the ultimate pressure. To achieve the lowest possible pressures, it is necessary to use cold traps or some cooled sorption material such as zeolite, activated alumina, or activated charcoal. With elastomeric gaskets, pressures of the order of 10^{-8} torr can be obtained. The lowest pressures are achieved by using special gasket materials such as fluorinated elastomers. These are marketed under various trade names such as Viton and Teflon (E. I. du Pont de Nemours & Company). Cooling the gaskets results in lower pressures.

Conventional vacuum systems can be designed for a wide range of pressures (several torr to around 10^{-8} torr) and pumping speeds (a few liters per second to many thousands of liters per second). The most common pumps used are rotary mechanical pumps and diffusion pumps (oil or mercury). The choice of pumps depends on the vacuum process involved and on the size of the system. Where large quantities of water vapor are involved, rotary ballast pumps can be used. Large industrial systems used in processes requiring pressures in the torr or micron range are often designed around mechanical and steam ejector pumps.

During recent years there has been a mounting interest in achieving extremely low pressures. In certain research areas such as low-energy collision phenomena, surface physics, and the study of thin films, it is necessary to reduce the residual gases to an absolute minimum. This same requirement holds in the production of various materials which must be of controlled composition. To achieve pressures much lower than those obtainable with conventional vacuum systems, the general approach has been to use different types of pumps and construction materials and to bake the system.

The most common type of pump used is a sputter-ion or getter-ion pump. To achieve the lowest possible pressures, it is necessary to bake the pumps and vacuum system. The rate of evolution of adsorbed and absorbed gases and vapors is rapidly increased as the baking temperature is raised. Construction materials, including the pump magnets, set an upper limit on the temperature that can be used. By removing the magnets, a temperature of around 450°C can be used, which eliminates the bulk of the gases and vapors in a reasonable time. Various baking techniques include the use of ovens with resistance heating (wire or tape). Metal rather than elastomeric gaskets are used, and greased seals or mercury cutoffs are avoided. Also, special bakeable valves must be used. The design must be such that the baking process does not introduce leaks through differential expansion of various joints. By proper constructional techniques and adequate cleaning and baking techniques, it is possible to achieve pressures below 10^{-12} torr, although there is some question about the accuracy of pressure measurements in this range.

In recent years, there has been considerable interest in very large systems operating in the ultrahigh-vacuum region, say less than 10^{-9} torr, largely because of the space program. All the problems encountered with small systems are present but to a greater degree. The methods to be described are, of course, applicable to smaller systems. Also, various combinations of these methods have been used. Sputter-ion pumps can be made with very large pumping speeds; such pumps, with appropriate roughing pumps, are commonly used with large systems.

Many attempts have been made to use oil diffusion pumps (and mercury). The systems differ from conventional systems in that the pressure rise across the diffusion pump is kept to a minimum (often backing with a booster pump). Also, bakeable traps and baffles (or cooled sorption traps) are used, as well as low-vapor-pressure elastomers such as Viton. Often the gaskets are cooled. There are many reports in the literature of pressures below 10^{-9} torr being achieved by the above methods, using vacuum systems of volume 100 ft^3 or larger.

Differential pumping systems have also been used to reach very low pressures. This type of system has two vacuum chambers, one inside the other. The inside chamber is pumped down to the desired low pressure, often with a diffusion-pump system of the type discussed above. The space between the two chambers is pumped down separately, usually with a conventional diffusion-pump system. The inner chamber can be of light-weight construction because of the small pressure differential. Also, small leaks in the inner chamber are not as critical as with single chambers, and the outer chamber (which is exposed to the atmosphere) can be made of an economical material such as mild steel.

It is also possible to obtain ultrahigh vacuum by using cryopumps. As has been noted, the lowest pressures attainable with liquid hydrogen or liquid helium are well above the ultrahigh-vacuum region if pumping is started at atmospheric pressure. To reach very low pressures it is necessary to use first other types of pumps to obtain pressures of the order of 10^{-6} torr (or better), perhaps first flushing with a gas that can readily be frozen, before any cryopumping. Diffusion pumps or sputter-ion pumps are commonly used for this purpose.

The actual bakeout procedure for large systems is determined by the nature of the system. The power requirements rise rapidly as the size of the system increases. It is impractical completely to surround the chamber (and pumps) with an oven, which is commonly done with small systems. Some methods that have been used are (1) passing electric current through the chamber walls, (2) induction heating, (3) use of internally mounted radiation sources, and (4) resistance heating (strip or tape). Most systems do not lend themselves to method 1, and method 2 is expensive and awkward. Methods 3 and 4 are commonly used, with method 4 gaining favor. Of course, vapors and gases can be held on the inside surfaces of the system by cooling these surfaces with liquid hydrogen or liquid helium. As systems designed to operate in the ultrahigh-vacuum region become larger, it is likely that combinations of the above pumping methods will be used, e.g., sputter-ion and cryogenic pumping.

8. Components and Materials

The choice of materials and components for a particular vacuum system is dictated largely by the process for which the system is to be used. One of the main characteristics of the materials that must be considered is the vapor pressure. Good vacuum practice requires proper attention to cleanliness and elimination of extraneous materials.

The trend has been to use steel in the larger systems for the tank and connecting pipe lines because of its low cost and suitability for welding. Brass and copper are also commonly used, although their cost becomes prohibitive for the larger systems. The smaller systems can be constructed of glass or metal. In the case of glass systems, occluded gases can be removed by heating of the glass and ultimate pressures of 10^{-8} torr or lower can be achieved with continuous pumping. For static systems, considerably lower pressures, say down to 10^{-12} torr, can be achieved by the use of getters. The welding, soldering, or brazing of metal vacuum systems must be carried out with great care. Wherever possible, subassemblies should be checked for leaks and corrected prior to making the final assembly. Particular care should be exercised in checking any cast parts that are used.

In most vacuum systems, it is necessary to provide various types of seals. The most common type of sealing material, particularly for metal systems, is rubber, either natural or synthetic. Certain plastic materials such as Teflon are also used. The choice of sealing material is dictated by the ultimate vacuum needed, and by the required mechanical, electrical, and chemical properties of the sealant. In general, the amount of sealant exposed to the vacuum should be kept to a minimum. It must be kept in mind that rubber is essentially an incompressible material and gasket grooves should be designed accordingly. All rubbers change in hardness with aging and assume some permanent set which increases with temperature.

Commercial O rings made in several types of rubbers are readily available. In making up gaskets, the ends, in general, should be vulcanized rather than cemented. Rubber gaskets are commonly used in packing glands, pipe-flange gaskets and for general sealing of parts to a vacuum system. Glass-to-metal seals are commonly used in both glass and metal vacuum systems. Such seals are particularly useful in metal systems when electrical insulation is required. Small wires, up to, say, 10 mils can be sealed directly to glass by proper choice of glass and metal. Platinum and tungsten are often sealed in this manner. A rather wide range of sizes and shapes of metals can be sealed to glass by making use of special alloys of iron, nickel, and cobalt. In particular, Kovar seals find wide application. Large insulated seals using glass, quartz, or some ceramic are often installed by use of rubber gaskets.

For high-temperature operation, metal gaskets made of such soft metals as lead, copper, or aluminum are used. Several methods have been used for introducing motion through the vacuum wall without breaking the vacuum. Magnetic forces have been used by controlling the motion of a magnetic material inside with a permanent magnet or electromagnet outside the system. This method requires nonmagnetic vacuum walls. Stopcocks are also used, usually being made of glass. They can also be constructed of a soft metal such as brass or copper, but in this case great care in construction is needed. A bellows or sylphon-type seal is in some cases useful, although it suffers from limited motion and from ultimate cracking of the metal with resulting leaks. Seals with rubber sealant, such as the Wilson seal or O rings, are commonly used.

Valves for use in vacuum systems can be broken down into two general classes, those which withstand atmospheric pressure from either side without leakage, and those that hold only when excess pressure is exerted in one direction. The first class is generally made in sizes up to around 6-in. opening and is primarily used in roughing lines. The second class ranges in sizes from several inches to 3 ft or more in opening, and is primarily for use on the high-vacuum side of the system.

Commercial valves in several sizes designed specifically for vacuum practice are available. However, in many cases commercial fluid valves have been modified for use in vacuum systems. The smaller valves can be either gate or globe valves, but the larger valves are, in general, gate or disk valves so as to reduce impedance. The control of the valve is usually made through either a bellows or a Wilson seal. A mercury cutoff can be used as a valve in some processes. The pressure in a system can be held within certain limits by throttling the pump line or by introducing gas into the system. The first method generally involves the use of valves or a pressure regulator such as the Hershberg-Huntress type. A pressure gauge such as a Pirani gauge can be used to control the mechanical pumps or valves. To regulate the flow of the gas into the system, needle valves are often used. Considerable care must be exercised in the design of such valves. Specialized methods have been used to introduce certain gases into the system, including unglazed porcelain, silver (oxygen), and palladium (hydrogen, deuterium). Collapsed metal or rubber tubing has also been used. In the operating of vacuum systems, provision must be made for reducing loss of operation time or damage to parts of the system through water or power failures, leaks, and loss of refrigerant. Pressure-indicating devices are generally used to give an alarm or to operate the proper mechanism to eliminate trouble. Several types of such devices are described in the literature [35].

In order to achieve and maintain ultrahigh vacuums, say, 10^{-9} torr or less, it is necessary to dispense with conventional methods using greased seals and valves, mercury cutoffs, rubber seals, etc. The pumping action of ionization gauges is often used in achieving such vacuums. The choice of components for ultrahigh-vacuum systems is dictated by the need for baking to drive off gases and vapors. To minimize the area available for the adsorption of gases, metals are generally polished. Stainless steel is commonly used in the construction of vacuum chambers. Metal gaskets (copper, aluminum, etc.) are available commercially in various designs that avoid leakage through differential expansion while baking and cooling. Various types of metal valves that can be baked are also available, usually involving a seal between a soft metal and a very hard metal. Glass-to-metal seals (Kovar, etc.) are available commercially.

9. Leak-detection Instruments and Techniques

The size of leak that must be detected cover a very wide range, depending on the base pressure required and the characteristics of the pumping system. For a pump with a speed of 5 l/sec and a required base pressure of 10^{-8} torr, a leak of 5×10^{-5} μl/sec could just be tolerated; whereas, for a pumping speed of 30,000 l/sec and a working pressure of 10^{-5} torr, the leak could be as great as 300 μl/sec.

A quantitative formulation of the flow through small holes is difficult because of variation in hole shape. However, in many cases the hole approximates a cylinder which can be treated quantitatively by application of Knudsen's equation. Ochert and Steckelmacher [36] have applied this equation to the flow of gas at 20°C through a cylinder of length L (centimeters) and diameter D (micron). Let η and η_a be the viscosities of the gas flowing through the cylinder and of air, respectively, with M and M_a the corresponding molecular weights. Then the treatment shows that for large values of D, say, $D > 100\ \mu$, the ratio of flow of a given gas to that of air approaches η_a/η. For very small values of D ($D < 0.01\ \mu$) this ratio becomes M_a/M so that in this case a low molecular weight gas flows through the leak most readily. This is important in reducing loss of time in detecting leaks and in increasing sensitivity.

Two general methods for the detection of leaks are in common use. The first of these involves the use of a fluid, usually a gas, under pressure inside the system with detection outside. The second method involves the use of a probe material (gas or liquid) on the outside of the system with detection inside. The first method has been covered in some detail by Jacobs and Zuhr [37]. The most common application of the method is to put air or nitrogen into the apparatus under test and look for leaks by applying soap solution (indication, bubbles), by using a flame on the suspected surface (indication, flame wavering), or by listening for escaping gas (large leaks). This procedure is most applicable to parts of an apparatus prior to assembly. An organic halide can be used with detection by the change in flame color of a torch. Gases or vapors which interact chemically to give a visual indication, one inside the apparatus and the other outside, can be used.

In the second method, the general procedure is to apply the probe material to suspected areas and look for a change in pressure as indicated by a vacuum gauge. The probe material is chosen on the basis of its physical and chemical properties (boiling point, viscosity, molecular weight, etc.) and its effect on the reading of the particular gauge used. A substance which produces a temporary or "permanent" seal can be used, with a drop in pressure as the indication. The results achieved by this procedure are uncertain. Soap solution is included here. Various methods are used to apply the probe material to isolated areas, such as the use of hoods, dams, etc. The location of the indicating gauge in the system is dictated by its useful pressure range. Pirani and thermocouple gauges generally are inserted between the fore-pump and diffusion pump, while ionization gauges are placed on the high-vacuum side of the diffusion pump.

Pirani and thermocouple gauges are used for leak detection by choosing a probe gas with a thermal conductivity different from that of air. Hydrogen has a high thermal conductivity and, due to its low molecular weight, gives rapid response. Other materials commonly used are carbon dioxide, butane, and acetone. Ionization gauges, including both Philips and thermionic types, detect probe gas molecules through the fact that they have a different probability of ionization than air molecules. With helium passing through a leak, the gauge reading will drop, while with organic vapors such as ether or acetone it will increase (vapor not trapped). Ionization gauges are sometimes used with a palladium interface between vacuum system and gauge. The palladium is heated to around 800°C, which allows hydrogen to diffuse freely but not other gases. Hydrogen passing through a leak then gives an increased gauge reading.

The change in work function of an electron-emitting surface due to the presence of the probe gas is also used to detect the presence of a leak. Diode tubes with tungsten filaments and hydrogen or oxygen as probe gas are often used in this connection. These gases reduce the electron emission and thus the pressure reading. Higher sensitivity can be achieved by using a triode tube and choosing a probe gas that decreases the filament work function and also has a high probability of ionization. Some metals, such as platinum, when heated emit positive ions. This emission is increased markedly when vapors of compounds containing a halogen strike the electrode surface. "Thermionic halogen vapor detectors" are based on this principle and probe materials such as Freon, carbon tetrachloride, and chloroform are used.

The most sensitive method of leak detection presently known is that involving the use of a mass spectrometer. In this method, the mass spectrometer is used as a selective ionization gauge, being set so as to be responsive only to a particular gas. Helium is usually used as the probe gas so that the spectrometer is then set for mass 4. This device will detect the presence of helium in the system to a concentration of about 1 part in 200,000, so that leaks as low as 10^{-6} μl/sec are detectable. A natural lower limit of sensitivity is set by the natural helium content of air which is about 1 part in 250,000. Details of design and application techniques are described in the literature [37, 38].

Even without application of probe gas, pressure fluctuations will occur due to instability in operation of the diffusion pumps. These are observed on both the high-vacuum and fore-vacuum sides of the system. In addition, the gauges generally used have inherent fluctuations, even when no gas is being pumped. These fluctuations can be reduced by constricting the gas flow to the pump or by constricting the gauge tubulation until the gauge time constant is several times the period of the fluctuations. A more elaborate method is to use a differential gauge system. Here, identical gauges are mounted adjacent to each other on the vacuum system and are connected in a bridge circuit to balance out the mean pressure and pressure fluctuations. The probe gas is allowed to enter only one gauge.

The gauge deflection of a given probe gas is governed by the partial pressure of this gas in the atmosphere over the leak, by change in flow rate through the leak when the probe gas is substituted for air, by change in flow rate through the diffusion pumps under these circumstances, by the relative sensitivity of the gauge for air and probe gas, and in the case of the differential gauge by the effectiveness of condensation or absorption. In general, liquids are not as desirable as gases due to temporary blocking of the leak.

Blears and Leck [40] have discussed the steady-state pressure and transient pressure-time equations. This consideration shows that for the more common leaks, i.e., those between 10^{-4} and 10 μl/sec, the flow through the leak is viscous, and it follows that, for maximum change in gauge reading following probe application, there must be complete coverage with probe material, high gauge sensitivity for probe material, low viscosity and high molecular weight for probe material, and small value of conductance between gauge and pump. The response time for the detection system is determined by the time required for the probe gas to pass through the leak and to displace the air from the vacuum system, plus the time lag of the detecting system itself.

No discussion of leak detection would be complete without mention of the time-honored method of the Tesla coil (or other type of spark coil). In this method, a leak at any point is indicated by a characteristic pink discharge, due to the presence of nitrogen in the air. If ether, carbon tetrachloride, or acetone is poured on the suspected spot, a leak is indicated by the characteristic color of the glow due to the presence in the system of the vapors of these compounds. Of course, the spark-coil test is useless for leaks in metal parts. Also, the test is not very sensitive. Hermetically sealed components can be leak-tested by use of a radioactive gas [39], with a claimed sensitivity of 10^{-12} std cc/sec. Usually krypton 85 is allowed to diffuse into the leaky component by "soaking" it in a dilute atmosphere of the radioactive gas (about 2 torr). The krypton 85 can be pumped out for reuse, and the component is flushed with air and then counted for gamma rays.

For detecting leaks in ultrahigh-vacuum systems, it might be thought that a helium leak detector would be the obvious choice. However, this is not the case [28]. The pressure in the system being tested is usually considerably lower than that in the leak detector. The standard method for leak-testing such a system is to use the ultrahigh-vacuum gauge, which is part of the system, with a probe material (acetone, ether, etc.). If vapor from the probe material enters the system, the pressure reading goes up; if the leak is clogged, the reading drops. Another technique is the rate-of-rise method where the system is isolated from the pumps and the rate of pressure rise is measured. It is usually necessary to operate the gauge intermittently because of its pumping action.

Before attempting to detect a leak in an existing system, care must be taken to establish that the leak exists. The time variation in the reading of a pressure gauge together with the value of the reading itself can be used in this connection. The pressure, pumping speed, and pressure fluctuation in the gauge should be calculated so as to determine whether the available gauge is suitable for detecting the possible change in pressure due to probing. Finally, the time constant of the system should be determined in order that the probe gas will be applied sufficiently long to a suspected area. In designing new systems, care should be taken to provide for leak-detection gauges on both high-vacuum and fore-vacuum sides of the system.

References

1. Herzfeld, K. F., and H. M. Smallwood: "Taylor's Treatise on Physical Chemistry," 2d ed., vol. I, p. 175, Van Nostrand, Princeton, N.J., 1931.
2. Prandtl, L.: "The Physics of Solids and Fluids," part II, Blackie & Son, Ltd., Glasgow, 1936.
3. Knudsen, M.: *Ann. Physik*, **28**: 75, 999 (1909).
4. Clausing, P.: *Ann. Physik*, **12**: 961 (1932).
5. Barrett, A. S. D., and C. H. Bosanquet: "Resistance of Ducts to Molecular Flow," Imperial Industries, Ltd. (Billingham Division), *Rept. BR*-296, November, 1944.
6. Ebert, H.: *Physik. Z.*, **33**(4): 145 (1932).
7. Millen, A.: Vacuum Physics, *J. Sci. Instr.*, *suppl.* 1, October, 1951.
8. Sullivan, H. M.: *Rev. Sci. Instr.*, **19**: 1 (1948).
   R. Newmann: High Vacuum Pumps, Their History and Development, *Elec. Eng.*, **20**: January–May, 1948.
9. Encyclopaedia Brittanica, 14th ed., vol. 22, p. 929.
10. Siegbahn, M.: *Archives for Mathematics, Astronomy and Physics* (*Royal Swedish Academy*), vol. 30B, no. 2, 1943.
11. Becker, W.: *Vakuum-Tech.*, October, 1958.
12. Witty, R.: *Brit. J. Appl. Phys.*, **9**: 232 (1950).
13. Langmuir, I.: *Gen. Elec. Rev.*, **19**: 1060 (1916); *J. Franklin Inst.*, **182**: 719 (1916).
14. Burch, C. R.: *Nature*, **122**: 729 (1928); *Proc. Roy. Soc.* (*London*), **123**: 271 (1929).
15. Holland, L.: *J. Sci. Instr.*, **36**: 105 (1959).
16. Espe, W., M. Knoll, and M. P. Wilder: *Electronics*, October, 1950.
17. Holland, L.: *J. Sci. Instr.*, **36**: 105 (1959).
18. Holland, L., L. L. Laurenson, and J. M. Holden: *Nature*, **182**: 851 (1958).
19. Jepson, R. L., S. L. Mercer, and M. J. Callaghan: *Rev. Sci. Instr.*, **30**: 377 (1959).
20. Pirani, M., and R. Neumann: *Electron. Eng.*, December, 1944, January, February, March, 1945.
21. Barr, W. E., and V. J. Anhorn: *Instruments*, **19**: 666 (1949).

22. Rosenberg, P.: *Rev. Sci. Instr.*, **10**: 131 (1939).
23. Reiff, H. J.: *Z. Instrumentenk.*, **34**: 97 (1914).
24. Knudsen, M.: "Kinetic Theory of Gases," p. 49, Methuen, London, 1933.
25. Langmuir, I.: *J. Am. Chem. Soc.*, **35**: 107 (1913). Coolidge, A. S.: *J. Am. Chem. Soc.*, **45**: 1637 (1937).
26. Klumb, H., and H. Schwartz: *Z. Physik.*, **122**: 418 (1944). Steckelmacher, W.: *Vacuum*, **1**: 266, 1951.
27. Penning, F. M.: *Physica*, **4**: 71 (1937). Makinson, R. E. B., and P. B. Treacy: *J. Sci. Instr.*, **25**: 298 (1948). Valle, G.: *Nuovo Cimento*, **7**: 174 (1950); **9**: 145 (1952).
28. Bayard, R. T., and D. Alpert: *Rev. Sci. Instr.*, **21**: 571 (1950). Alpert, D.: *J. Appl. Phys.*, **24**: 860 (1953).
29. Hobson, J. P., and P. A. Redhead: *Can. J. Phys.*, **36**: 271, 255, (1958).
30. Houston, J. M.: *Bull. Am. Phys. Soc.*, II, **1**: 301 (1956).
31. Lafferty, J. M.: *J. Appl. Phys.*, **32**: 424 (1961).
32. Sommer, H., H. A. Thomas, and J. A. Hipple: *Phys. Rev.*, **82**: 697 (1951). Alpert, D., and R. S. Buritz: *J. Appl. Phys.*, **32**: 202, 265 (1954).
33. Reynolds, J. H.: *Rev Sci. Instr.*, **27**: 928 (1956).
34. Downing, J. R., and G. Mellen: *Rev. Sci. Instr.*, **17**: 218 (1946).
35. Oliver, G. D., and W. G. Bickford: *Rev. Sci. Instr.*, **16**: 130 (1945). Picard, R. G., P. C. Smith, and S. M. Zollus: *Rev. Sci. Instr.*, **17**: 125 (1946).
36. Ochert, N., W. Steckelmacher, A. Guthrie, and R. K. Wakerling: *Brit. J. Appl. Phys.*, **2**: 332 (1951).
37. Jacobs, R. B., and H. F. Zuhr: *J. Appl. Phys.*, **18**: 34 (1947).
38. Nier, A. O., C. M. Stevens, A. Hustulid, and T. A. Abbott: *J. Appl. Phys.*, **18**: 30 (1947).
39. Cassen, B., and D. Burnham: *Intern. J. Appl. Radiation Isotopes*, **9**: 54 (1960).
40. Blears, J., and J. H. Leck: *J. Sci. Instr., suppl. 1*, **28**: 20 (1951). See also Leck's book below.

Bibliography

Diels, K., and R. Jaeckel: "Leybold Vakuum-Taschenbuch," Springer, Berlin, 1958.

Dushman, S.: "Scientific Foundations of Vacuum Technique," 2d ed., Wiley, New York, 1962.

Guthrie, A.: "Vacuum Technology," Wiley, New York, 1963.

Guthrie, A., and R. K. Wakerling: "Vacuum Equipment and Techniques," McGraw-Hill, New York, 1949.

"Handbook of Chemistry and Physics," Chemical Rubber Publishing Company, Cleveland, Ohio, 1964.

Holland, L.: "The Vacuum Deposition of Thin Films," Wiley, New York, 1956.

Jaeckel, R.: "Kleinste Drucke," Springer, Berlin, 1950.

Jnanananda, S.: "High Vacua," Van Nostrand, Princeton, N.J., 1947.

Leck, J. H.: "Pressure Measurement in Vacuum Systems," Chapman & Hall, London, 1964.

Pirani, M., and J. Yarwood: "High Vacuum Technique," Chapman & Hall, London, 1961.

Roberts, Richard W., and Thomas A. Vanderslice: "Ultrahigh Vacuum and Its Applications," Prentice-Hall, Englewood Cliffs, N.J., 1963.

Spinks, W. S.: "Vacuum Technology," Chapman & Hall, London, 1963.

Steinherz, H. A.: "Handbook of High Vacuum Engineering," Chapman & Hall, London 1964.

Yarwood, J.: "High Vacuum Technique," 3d ed., Chapman & Hall, London, 1955.

Chapter 7

Surface Tension, Adsorption

By STEPHEN BRUNAUER Clarkson College of Technology
and L. E. COPELAND, Portland Cement Association

SURFACE TENSION

1. The Thermodynamic Theory of Capillarity

It is usually assumed in thermodynamic considerations that a system is composed of one or more homogeneous phases bounded by sharply defined geometric surfaces. This is an oversimplification. The boundary between two phases is a thin layer across which the properties change from those of one homogeneous phase to those of the other. Ordinarily, the energy or entropy of these inhomogeneous regions is small in comparison with the total energy or entropy of the system, and the contributions of the interfaces are therefore neglected. However, when we consider materials in a fine state of subdivision, or porous materials with fine capillary systems, the surface properties can no longer be neglected. It is the properties of these inhomogeneous, interfacial regions that constitute the subject matter of this chapter.

The thermodynamic theory of capillarity presented here is essentially that of Gibbs [9].*

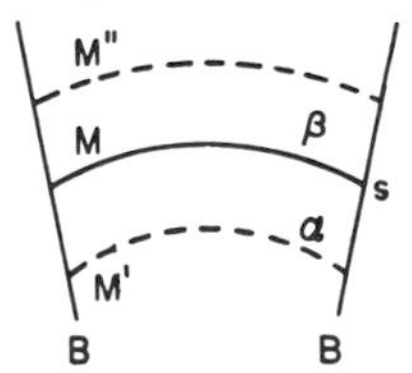

FIG. 7.1. Gibbs' concept of the interfacial region.

Fluid-fluid Interfaces. Consider in Fig. 7.1 the interfacial region between two fluids in contact, confined by an envelope BB. M' designates a region within one of the homogeneous phases, and M'' a region within the other homogeneous phase. The intervening part, M, includes some of each homogeneous phase and the nonhomogeneous, interfacial region. s represents a surface passing through an arbitrarily chosen point in the interfacial region and through all other points similarly situated with respect to the homogeneous phases M' and M''. The envelope BB is normal to the surface s at all points of intersection. Whatever variation the system undergoes, the boundaries of M are to be considered fixed.

* Numbers in brackets refer to References at end of chapter.

It can be shown that the conditions for internal equilibrium of the system are the same as for homogeneous phases.

$$T = T' = T'' \qquad \mu_i = \mu'_i = \mu''_i \tag{7.1}$$

where T, T', and T'' are the temperatures and μ_i, μ'_i, and μ''_i are the chemical potentials of component i within the regions M, M', and M'', respectively.

For all *reversible** variations occurring within the region M

$$\delta E = T\,\delta S + \sum_i \mu_i\,\delta m_i \tag{7.2}$$

where E is the total energy, S is the entropy, and m_i is the quantity of component i in M. The dividing surface s divides M into two parts: α, adjacent to M', and β, adjacent to M''. Let us imagine these regions as being filled with the same substances, at the same densities of mass, energy, and entropy, and at the same chemical potentials as the contiguous, homogeneous phases. Then for reversible variations at constant volume

$$\begin{aligned} \delta E^\alpha &= T\,\delta S^\alpha + \sum_i \mu_i\,\delta m_i^\alpha \\ \delta E^\beta &= T\,\delta S^\beta + \sum_i \mu_i\,\delta m_i^\beta \end{aligned} \tag{7.3}$$

Combining Eqs. (7.2) and (7.3) gives

$$\delta E - \delta E^\alpha - \delta E^\beta = T(\delta S - \delta S^\alpha - \delta S^\beta) + \sum_i \mu_i(\delta m_i - \delta m_i^\alpha - \delta m_i^\beta)$$

or

$$\delta E^s = T\,\delta S^s + \sum_i \mu_i\,\delta m_i^s \tag{7.4}$$

where E^s is the excess of the energy of the actual system over that which it would possess if the density of energy were uniform up to the dividing surface. S^s and m_i^s have analogous significance.

Equation (7.4) is true for reversible variations in which the surface is fixed. E^s is affected only slightly by the assumed position of the dividing surface, as long as it is within the inhomogeneous

* See Gibbs, ref. 9, footnote on p. 222, for his definition of "reversible."

region. Gibbs showed that the dividing surface can be so placed that only a single term, accounting for possible variations in the area of the system, need be added to (7.4). Then

$$\delta E^s = T\,\delta S^s + \gamma\,\delta A + \sum_i \mu_i\,\delta m_i{}^s \qquad (7.5)$$

where γ is the surface tension and A the area of the surface. The product $\gamma\,\delta A$ is the two-dimensional analogue of $p\,\delta v$, γ being analogous to pressure and A to volume.

Gibbs called the dividing surface the "surface of tension." He showed that the dividing surface, which makes Eq. (7.5) valid for surface radii of curvature large in comparison with the thickness of the inhomogeneous region, is essentially the physical surface of discontinuity.

Tolman [226] deduced the relationship between the surface tension and the radius of curvature of the surface for the case when the thickness, t, of the inhomogeneous region is not small in comparison with the radius of curvature r. For $t/r = 0.02$, his equation predicts a 4 per cent decrease in surface tension. This value of t/r corresponds to a sphere radius of 10^{-6} cm.

Kelvin's Equation. Consider the system (Fig. 7.1) divided into two parts by the dividing surface, with the energy, entropy, and quantities of components calculated on the assumption that each part is homogeneous up to the dividing surface. Then the general condition of equilibrium is

$$\delta E = \delta E^s + \delta E^\alpha + \delta E^\beta > 0 \qquad (7.6)$$

If the total volume, the total entropy, and the quantities of the components are constant, this reduces to

$$\gamma\,\delta A - p^\alpha\,\delta v^\alpha - p^\beta\,\delta v^\beta = 0 \qquad (7.7)$$

If all parts of the dividing surface move a uniform normal distance, δN, then we have three conditions:

$$\begin{aligned} \delta A &= \left(\frac{1}{r_1} + \frac{1}{r_2}\right) A\,\delta N \\ \delta v^\alpha &= A\,\delta N \\ \delta v^\beta &= -A\,\delta N \end{aligned} \qquad (7.8)$$

where $1/r_1$ and $1/r_2$ are the principal curvatures. It follows then that

$$\gamma\left(\frac{1}{r_1} + \frac{1}{r_2}\right) = p^\alpha - p^\beta \qquad (7.9)$$

Positive curvature is in the direction of the region α. This equation was first derived by Lord Kelvin from mechanical considerations [225]. La Mer and Gruen [161] recently verified the equation experimentally by measuring the sizes of droplets of a monodisperse aerosol in equilibrium with vapor over a plane, liquid surface.

Kelvin's equation is most frequently applied to liquid-vapor interfaces in cylindrical capillaries and to adsorption of vapors by porous solids. In these applications it is usually assumed that the vapor is a perfect gas, and that $r_1 = r_2$; so the equation is rewritten in the form

$$\frac{2\gamma}{(\rho_l - \rho_v)r} = \frac{RT}{M}\ln\frac{p}{p_s} \qquad (7.10)$$

where ρ_l and ρ_v are the densities of the liquid and vapor, respectively, M is the molecular weight, and p_s the saturation pressure of the vapor.

The Gibbs Adsorption Equation. For those variations in which the system remains in equilibrium, we may write d for δ. A Legendre transformation of Eq. (7.5) gives

$$S^s\,dT + A\,d\gamma + \sum_i m_i{}^s\,d\mu_i = 0 \qquad (7.11)$$

Dividing by A and writing

$$\Gamma_i = \frac{m_i{}^s}{A} \qquad S_s = \frac{S^s}{A}$$

we obtain

$$d\gamma = -S_s\,dT - \sum_i \Gamma_i\,d\mu_i \qquad (7.12)$$

For plane surfaces Eq. (7.5) is valid, regardless of the assumed position of the dividing plane in the interfacial region. Moving the dividing surface does not affect γ, but it affects Γ. We may choose the position of the surface so as to make $\Gamma = 0$ for one of the components. This corresponds to uniform densities of that particular component up to the dividing surface on both sides. Let us designate this component by the subscript 1, then $\Gamma_1 = 0$, and Eq. (7.12) can be written

$$d\gamma = -S'_s\,dT - \Gamma'_2\,d\mu_2 - \Gamma'_3\,d\mu_3 \cdots \qquad (7.13)$$

The adsorption coefficients Γ'_i are surface excess quantities and can be either positive or negative. Equations (7.12) and (7.13) show that positive adsorption decreases the surface tension.

In a two-component system, in which neither of the components is confined to the interface alone, one can choose the individual phases as the components. If the temperature and pressure are then chosen as the independent variables, Eq. (7.12) may be put in a simple form by introducing the relationships between the chemical potentials, the temperature, and the pressure:

$$d\gamma = -\left(S_s - \frac{\Gamma_1}{\rho_1}S_1 - \frac{\Gamma_2}{\rho_2}S_2\right)dT - \left(\frac{\Gamma_1}{\rho_1} + \frac{\Gamma_2}{\rho_2}\right)dp \qquad (7.14)$$

where S_1 and S_2 are the *entropies per unit volume* of the two bulk phases and ρ_1 and ρ_2 are their densities. This procedure is always possible unless the two phases have identical compositions.

One confirmation of the Gibbs adsorption equation was obtained by McBain and his coworkers [180, 181]. They succeeded in cutting off rapidly, by means of a microtome, a thin layer from the surface of a solution contained in a long trough. Γ, calculated from surface-tension measurements of the solutions, was found to agree with the experimentally measured quantity.

Solid-fluid Interfaces. The interfaces between solids and fluids can be considered in terms of the same concepts as were introduced for the fluid-fluid interface. The volume of the system is assumed to be constant. Component 1 will be the solid component. The quantity γ is defined by

$$\gamma = E_{s(1)} - TS_{s(1)} - \mu_2\Gamma_{2(1)} - \mu_3\Gamma_{3(1)} - \cdots \quad (7.15)$$

where the subscript (1) refers to the situation in which the dividing surface was chosen to make $\Gamma_1 = 0$. γ, thus defined, is the free surface energy, or the work necessary to form an additional unit surface. In fluid-fluid systems the free surface energy is always numerically equal to the surface tension, but this equality does not hold for solid-fluid systems because of the rigidity of the solid. Experimental methods so far applied give γ, the free surface energy, though, by analogy with the liquid, it is frequently *called* the surface tension.

As with fluid-fluid systems, equilibrium requires that $T_1 = T_2 = T_3$, but the condition $\mu_1 = \mu_2 = \mu_3$ will be satisfied only for isotropic amorphous solids. In crystalline solids the chemical potentials and the free surface energies of the different crystal faces will be different.

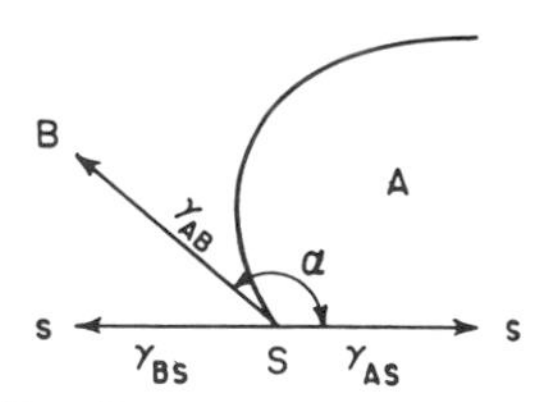

Fig. 7.2. The contact angle.

The equilibrium at the junction of two fluid phases and one solid phase can be described by the Neumann triangle of forces. If the surface of the solid is continuous at the line formed at the junction of the three phases (Fig. 7.2) equilibrium will be attained when

$$\gamma_{AB} \cos \alpha = \gamma_{BS} - \gamma_{AS} \quad (7.16)$$

If A is a liquid phase and B a vapor phase, the angle α is called the contact angle.

For any constant state of strain in the solid and for those variations which allow the system to remain at equilibrium, we may write

$$dE_{s(1)} = T\, dS_{s(1)} + \mu_2\, d\Gamma_{2(1)} + \mu_3\, d\Gamma_{3(1)} + \cdots \quad (7.17)$$

From (7.15) and (7.17) we obtain

$$d\gamma = -S_{s(1)}\, dT - \Gamma_{2(1)}\, d\mu_2 - \Gamma_{3(1)}\, d\mu_3 \quad (7.18)$$

This equation is strictly analogous to Eq. (7.13) derived for fluid-fluid systems.

If the area of the interface is increased by unity, and at the same time the chemical potentials of the components are kept constant, the heat absorbed, Q, is equal to $TS_{s(1)}$, and it follows from Eq. (7.18) that

$$Q = -T\left(\frac{\partial\gamma}{\partial T}\right)_{\mu_i} \quad (7.19)$$

The work expended directly in extending the interface is γ. The total increase in surface energy, $E_{s(1)}$, is given by

$$E_{s(1)} = \gamma + TS_{s(1)} = \gamma - T\left(\frac{\partial\gamma}{\partial T}\right)_{\mu_i} \quad (7.20)$$

When a solid is immersed in a liquid, heat is evolved. The heat of immersion for a unit surface is given by

$$h_I = E_s - E_{SL} \quad (7.21)$$

where E_s and E_{SL} are the total surface energies of the solid prior to and after immersion. From Eq. (7.20) it follows that

$$h_I = \gamma_S - \gamma_{SL} - T\left(\frac{\partial\gamma_S}{\partial T}\right)_{\mu_i} + T\left(\frac{\partial\gamma_{SL}}{\partial T}\right)_{\mu_i} \quad (7.22)$$

Statistical Theories of Capillarity. Kirkwood and Buff [155] developed a statistical mechanical theory for plane interfaces. Their equations were derived rigorously, but approximations were made in solving the equations for the surface tension and total surface energy of liquid argon. The calculated values were within 25 to 30 per cent of the experimental values. The authors also calculated the distance between the Gibbs surface of tension and the dividing surface placed to make $\Gamma_1 = 0$, and obtained the value +3.6 A. The positive sign is taken in the same direction in which the positive radius is defined. The fact that the sign is positive confirms Tolman's conclusion that surface tension decreases with decreasing radius of curvature. Buff [50] extended this theoretical treatment to curved interfaces.

Hill [136] developed an approximate theory for the purpose of making a complete set of calculations of interfacial phenomena. His calculated values for the surface tension and total surface energy of liquid argon agreed with the experimental values within a factor of 2.

Jura [145] calculated the total surface energies of argon and mercury using a modification of a procedure suggested by Fowler and Guggenheim [6]. He obtained agreement with experiment within 5 to 10 per cent.

2. The Surface Tension and Total Surface Energy of Liquids and Solids

The Surface Tension of Liquids. Both static and dynamic methods have been used for measuring the surface tension of liquids [1, 13, 70, 123]. The static methods capable of greatest accuracy are those in which the pressure difference across a curved interface is measured. The surface tension is calculated from the Kelvin equation (7.10). The most rapid methods of measuring the surface tension are based upon measuring the force required to extend a film of liquid, as, for example, the ring method and the Wilhelmy slide method.

The dynamic methods depend upon measuring periodic extensions and contractions of the surface accompanying certain vibrations. In one such method the wavelength of the oscillations of a stream of liquid, issuing from an elliptical orifice, is measured.

Among the liquids, mercury has the highest surface tension, about 475 dynes/cm, at room temperature.

The surface tension of water at 20°C is 72.75 dynes/cm, while that of most organic liquids lies between 25 and 30 dynes/cm. Most fused salts have surface tensions between 300 and 475 dynes/cm.

Since the surface tension is due to asymmetric intermolecular forces acting upon a molecule in the interfacial region, any variation in the state of the system which increases the kinetic energy of the molecules causes a decrease in the surface tension. Thus, the surface tension decreases with increasing temperature and vanishes at the critical temperature. Eötvös [84] derived a relationship between the surface tension γ and the temperature T from theoretical considerations of corresponding states. Ramsay and Shields [195] proposed a modification of the Eötvös equation which made it fit experimental data more accurately:

$$\gamma(M_v)^{1/3} = k(T - T_c - 6) \tag{7.23}$$

where M_v is the molar volume of the liquid and T_c is the critical temperature. The Eötvös constant k was found to average 2.12 for many normal liquids. The value of k was found to be low in many liquids, and this was attributed to association of the molecules. However, some liquids gave high values of k, and in most cases this could not be attributed to the reverse effect. The theory of the motions of the molecules of liquids, developed by Born and Courant [38], suggests that k should be influenced by the number of degrees of freedom for molecular motion. Indeed, it seems that k should be influenced not only by the association of molecules and the degrees of freedom for thermal motion but also by such factors as the orientation of the molecules in the surface, molecular shape and packing, mutual interactions between parts of molecules, and the thickness of the surface layer. At any rate, k is not a universal constant.

Several empirical relations between the surface tension and the temperature have been proposed. Two of the simplest forms are

$$\gamma = \gamma_0(1 - bT) \tag{7.24}$$

$$\gamma = \gamma_0\left(1 - b'\frac{T}{T_c}\right) \tag{7.25}$$

where b and b' are constants. Equation (7.24) does not fit experimental data as well as Eq. (7.25). Numerous modifications of these empirical equations have been made to obtain a better fit with experimental data. One such modification is that of Katayama [149]:

$$\gamma\left(\frac{M}{\rho_l - \rho_v}\right)^{2/3} = kT_c\left(1 - \frac{T}{T_c}\right) \tag{7.26}$$

where M is the molecular weight, and ρ_l and ρ_v are the densities of the liquid and vapor, respectively. Equation (7.26) was found to fit data well even close to the critical temperature.

Van der Waals [231] proposed the equation

$$\gamma = K_2T_c^{1/3}p_c^{2/3}\left(1 - \frac{T}{T_c}\right)^n \tag{7.27}$$

with K_2 a universal constant for liquids. Furgusen [100] found n to average 1.21 for many liquids.

Macleod [176] proposed another equation, which is obtained by eliminating the factor $(1 - T/T_c)$ between van der Waals' and Katayama's equation

$$\frac{\gamma}{(\rho_l - \rho_v)^4} = C \tag{7.28}$$

Fowler [97] derived this equation from statistical considerations. Sugden [216] suggested that Macleod's equation offered a method of comparing liquids under conditions of equivalent surface tension. The fourth root of the constant in Macleod's equation multiplied by the molecular weight of the liquid is a molecular volume independent of the temperature. Sugden called this molecular volume the *parachor*. It can be divided into contributions characteristic of atoms or radicals present in the molecule. This seems to be the best method for comparing molecular volumes.

The Total Surface Energy of Liquids. The total surface energy is obtained by measuring the surface tension at different temperatures and applying Eq. (7.20). The total surface energy of a liquid varies very little with temperature until the critical temperature is reached, where it falls to zero. Within a homologous series of organic compounds the chain length has little effect on the total surface energy. For polar compounds, like the alcohols, E_s is only slightly greater than for the hydrocarbons. This small difference in the surface energy between the two series of compounds has been interpreted as evidence in favor of an orientation of unsymmetrical molecules in the surface in such a manner that the surface is predominantly composed of the nonpolar parts of the molecule.

The total surface energy is not completely independent of molecular structure. Langmuir [162] suggested that, as a first approximation, the field of force around particular groups in a molecule was characteristic of the group and independent of the rest of the molecule. This led to the conclusion that each group in a molecule made its specific contribution to the total surface energy. Langmuir used this principle of "independent surface action" to calculate the relative probabilities of the possible orientations of unsymmetrical molecules in the surface. The limitations of this principle have been discussed by Smyth and Engel [214].

The vaporization of a liquid can be thought of as a two-step process: in the first step the molecule moves from the interior of the liquid into the surface region; in the second, it moves from the surface region into the vapor phase. Stefan [215] suggested that half of the energy of vaporization would be expanded in the first step of this process. It was found that with symmetrical molecules like nitrogen or carbon dioxide the total surface energy is approximately half of the heat of vaporization. With other molecules, however, the total surface energy is a smaller fraction of the energy of vaporization, and the more unsymmetrical the molecule, the smaller is the fraction. For methyl alcohol, for example, the total surface energy is about one-eighth of the heat of vaporization.

The Surface Tension of Solids. The free surface energy of solids is much more difficult to measure than the free surface energy of liquids because of the ability of solids to support shear stresses. Since the molecules of a liquid are mobile and rearrange themselves to relieve anisotropic stresses, the surface tension of a liquid is constant over the entire surface. It follows from this, from the condition of equilibrium that $\int \gamma \, \delta A$ be a minimum, and from the Kelvin equation, that $1/r_1 + 1/r_2$ must be constant. A sphere is the only closed surface that satisfies these conditions; consequently, liquid drops are spherical. In crystalline solids γ is constant over a particular face of a crystal, but it is not the same for all faces. We therefore find that that combination of plane faces which is consistent with the minimum energy requirement bounds the solid. The modification of crystal growth habits by adsorbable impurities in the system may be due to differences in the relative free energy lowering of the different crystal faces by the same absorbed impurity [64].

The measurement of the free surface energy of solids is complicated by the fact that the conditions for equilibrium cannot be made explicit without a knowledge of the state of stress. The usual procedure in experimental work is to try to achieve a condition of zero strain energy.

The free surface energy of most solids is much higher than that of liquids, and the lowering of the free surface energy by adsorbed material is correspondingly much greater than in liquids. Orowan [190] found, by a method proposed by Obreimoff [189], that the free surface energy of mica when cleaved in a vacuum was 4,500 ergs/cm^2, but when mica was cleaved in air the free surface energy was only 375 ergs/cm^2. The free surface energy of copper [229] under its own vapor is 1,430 ergs/cm^2, but in an atmosphere of lead vapor [19, 210] it is 760 ergs/cm^2.

It has long been known that thin metal foils shrink when heated [89]. Chapman and Porter [56] ascribed this "creep" to a free surface energy effect; they and others [205, 207, 218] calculated free surface energies from creep measurements. Fischer and Dunn [92] presented equations relating the force required to prevent creep under various experimental conditions to the free surface energy.

Rymer and Butler [204] calculated that a tension of 500 dynes/cm would account for the anomalous lattice spacings obtained in electron-diffraction studies of thin gold foil. The surface tension of gold calculated from creep measurements is 1510 dynes/cm.

Much of the information concerning free surface energies in solids has been obtained by applying the Dupré conditions to the observations of the angles formed by three interfaces meeting at a line. These conditions are given by

$$\frac{\gamma_{12}}{\sin \alpha_{12}} = \frac{\gamma_{13}}{\sin \alpha_{13}} = \frac{\gamma_{23}}{\sin \alpha_{23}} \tag{7.29}$$

where γ_{ij} is the interfacial free energy of the ij interface, and α_{ij} is the angle opposite the interface. This procedure was applied to copper in contact with liquid lead, and also to grain boundaries in metals [75, 99]. It was found that the interfacial energy at grain boundaries of copper was 550 ergs/cm^2, and that changes of 15 to 20 per cent could be produced by slight differences in the angles of orientation of grains. The interfacial energy of coherent twin grain boundaries was very low, about 20 ergs/cm^2, but that of the noncoherent twin grain boundaries was much higher, 440 ergs/cm^2. Similar results were found for other metals. Further information on the free surface energies of solids, and especially of metals, may be found in recent books [10, 15].

The Total Surface Energy of Solids. The total surface energies of solids differ only slightly from the free surface energies, because the temperature dependence of the free surface energy is ordinarily very slight. For example, the free surface energies of copper, gold, and silver are almost independent of the temperature.

The total surface energy of a solid is usually measured by determining the heats of solution of samples of the solid having different specific surface areas and measuring the surface areas. The difference between the heats of solution divided by the difference between the specific surface areas gives the total surface energy per unit surface.

The first experimental determination of the total surface energy of a solid was made by Lipsett, Johnson, and Maas [170], who obtained 400 ergs/cm^2 for the surface energy of sodium chloride. They determined the area by microscopic examination, which method frequently gives too low surface area values. Benson and Benson [34] obtained 305 ergs/cm^2 for the surface energy of sodium chloride, using the BET method for surface area measurement. (This method will be discussed later.) Jura and Garland [147] obtained 1,090 ergs/cm^2 for the surface energy of magnesium oxide at 298°K; Brunauer, Kantro, and Weise [48] obtained for calcium oxide, calcium hydroxide, amorphous silica, and hydrous amorphous silica the values 1,310, 1,180, 260, and 130 ergs/cm^2, respectively, at 296°K.

Jura and Garland determined also the free surface energy of magnesium oxide. At 0°K the free surface energy and the total surface energy are the same, 1,040 ergs/cm^2. In the temperature range from 0 to 298°K, the former decreases by 40 ergs/cm^2; the latter increases by 50 ergs/cm^2. The surface entropy at 298°K is 0.28 erg/cm^2-deg.

Theoretical calculations of the surface energies of solids crystallizing in the cubic lattice were made by Born and Stern [39], and later by Lennard-Jones and his coworkers [169]. The values obtained for the alkali halides were far too low, but those for calcium oxide and magnesium oxide were about right. Recently, van der Hoff and Benson [230] obtained a better approximation for lithium fluoride by using a quantum-mechanical model.

The Heat of Immersion. The experimental results on heats of immersion, obtained by various workers, are discordant. Nevertheless, certain carefully executed experiments show that the heats of immersion in water of hydrophyllic solids, like $BaSO_4$ or TiO_2, are high, but those of hydrophobic solids, like graphite, are low [115]. On the other hand, the heat of immersion of $BaSO_4$ in dry benzene is low, whereas that of graphite is high. If $BaSO_4$ is immersed in benzene containing dissolved water, the heat of immersion of $BaSO_4$ is practically the same as in water. It is probable that the variations in the reported results are due largely to traces of impurities present in the chosen liquids.

The Work of Cohesion and Adhesion. The work necessary to pull apart a bar of liquid 1 cm^2 in cross section is equal to twice the surface tension, since 2 cm^2 of surface is formed in the process. Harkins [117] called this work the work of cohesion, or free energy of cohesion.

$$W_c = 2\gamma \tag{7.30}$$

Dupré [4] defined the work of adhesion as the work necessary to separate a unit cross-sectional area of the interface between two liquids, A and B.

$$W_{AB} = \gamma_A + \gamma_B - \gamma_{AB} \tag{7.31}$$

where γ_{AB} is the interfacial tension between A and B. Hardy [112] found that the work of adhesion between a series of organic liquids and water was determined almost entirely by the water-attracting groups in the molecule. Harkins and his coworkers [118, 120, 122] found that organic compounds containing oxygen gave higher work of adhesion to water than organic compounds containing halogens, whereas for the adhesion to mercury the reverse result was obtained.

The concept of the work of cohesion has been extended to solids, and that of the work of adhesion to solid-solid and solid-liquid systems. Harkins [13] showed that the work of adhesion between a liquid and a solid is given by

$$W_{SL} = \pi_e + \gamma_L(1 + \cos\alpha) \tag{7.32}$$

where γ_L is the surface tension of the liquid, α the contact angle, and π_e—the equilibrium spreading pressure of the liquid on the solid—is the decrease in the free surface energy of the solid due to an adsorbed film of vapor in equilibrium with the liquid.

Harkins and his coworkers found for the paraffins that the work of cohesion and the work of adhesion to water were almost equal. The work of cohesion of the aliphatic molecules containing polar groups (e.g., alcohols, acids) is almost exactly equal to that of the paraffins, but the work of adhesion to water is increased greatly by the presence of the polar group. These results support the theory advanced by Hardy [113], Langmuir [163], and Harkins [116, 119] that molecules are oriented at interfaces. At a liquid-vapor interface, for example, alcohols would be oriented with the hydroxyl groups directed toward the interior of the liquid. At an alcohol-water interface the hydroxyl groups of the alcohol would be directed toward the water phase.

The work of cohesion of solids cannot be measured directly because an uncertain amount of work is expended in producing deformations, such as plastic flow, before rupture. The same difficulty is encountered in the determination of the work of adhesion between two solids. However, the effects of cohesion and adhesion between solid surfaces have been demonstrated in many experiments, especially by Bowden and his collaborators [41].

Few solid surfaces are smooth on a molecular scale; consequently, the area of contact between two surfaces is limited to more or less isolated surface asperities. Very small forces are sufficient to produce plastic deformation at these minute spots of contact, resulting in a "cold welding" at the junctions. Thus, large adhesive forces are produced by pressing a steel ball against a freshly scraped indium surface; the coefficient of adhesion (defined by analogy with the coefficient of friction as the ratio of the adhesive force to the total load) is about unity. In experiments with a steel block on an indium surface the coefficient of adhesion was found to have the same order of magnitude as the coefficient of friction, indicating that the tensile strengths of the junctions were about the same as their shear strengths.

If iron surfaces are carefully cleaned by outgassing in a vacuum at an elevated temperature, and, after cooling, are placed in contact, the coefficient of friction is very high, and so is the coefficient of adhesion. The presence of an oxide film, or of small quantities of adsorbed gases or vapors, produces large decreases in the coefficients of adhesion and friction, unless sufficiently large loads are used to break through the films and allow metal-to-metal contacts.

By using thin mica sheets, Bowden and his coworkers produced and measured areas of contact between molecularly smooth regions. The shear strength of such a contact was about 400 g/mm^2, and the tensile strength was approximately 500 g/mm^2, or 5,000 times greater than the work of cohesion of mica (9,000 ergs/cm^2). Monolayers deposited on the mica sheets were found to reduce the area of contact as well as the friction and adhesion between the surfaces.

Film Formation and the Spreading Coefficient. A drop of liquid placed on the surface of a second liquid will either form a lens or spread in a thin layer over the surface. A condition for spreading is that the total surface energy be decreased by the spreading process, which condition implies that the attraction between unlike molecules is greater than that between like molecules. Dupré [4] first formulated this condition, and Harkins [121] defined the spreading coefficient

$$\phi = W_{AB} - W_C = \gamma_A - \gamma_B - \gamma_{AB} \tag{7.33}$$

Films are formed if ϕ is positive. The equation shows that if liquid A spreads on liquid B, liquid B will not necessarily spread on liquid A. As an example, most organic liquids will spread on a water surface, but water will not spread on the surface of any organic liquid when the two are mutually insoluble.

Unimolecular films are formed if the spreading coefficient is high. If the spreading coefficient is low, a relatively thick film forms initially. However, thick films are unstable and form monolayers, the excess liquid forming small lenses over the surface. Mixtures of oils may form thick films which persist for a long time.

Since a small quantity of solute influences the surface tension of a liquid, the value of the spreading coefficient depends upon whether the two liquids are pure or mutually saturated. The instability of thick films can be attributed to the change in the value of the spreading coefficient as the liquids become mutually saturated.

The nature and the properties of unimolecular and multimolecular films on liquids and solids are discussed under "Adsorption."

ADSORPTION

3. Adsorption on Liquid Surfaces

Unimolecular Films of Insoluble Substances. The spreading of a monolayer of adsorbed substance is always accompanied by a decrease in the free surface energy of the system. This decrease is given by

$$\pi = \gamma_L - \gamma_{LF} \tag{7.34}$$

where γ_L is the surface tension of the pure liquid and γ_{LF} is the surface tension of the liquid with the adsorbed film on it. π is called the *film pressure* or *surface pressure*. It can be measured by any of the static methods for measuring surface tension, or by measuring the force exerted on a two-dimensional piston lying in the surface and separating the film from the pure liquid surface.

If a small quantity of an insoluble surface-active material is placed upon a clean liquid surface, the substance will spread to form a monolayer. If the surface area is so small that some of the bulk material remains, equilibrium between the monolayer and the bulk material will be reached at a film pressure which is characteristic of the adsorbed substance, of the liquid upon which it spreads, and of the temperature. This equilibrium pressure, π_e, is called the *spreading pressure*.

The properties of adsorbed monolayers justify their classification as two-dimensional gas, liquid, or solid [114]. At very large areas per molecule, σ, the films are gases. At intermediate areas phases exist which have no analogy in three dimensions. At small molecular areas liquids exist, and at still smaller areas these liquids change to solids. There are two or more phases which exhibit properties characteristic of liquids.

The equation of state for a perfect gas in two dimensions is

$$\pi\sigma = kT \tag{7.35}$$

At sufficiently large areas this equation is obeyed, but as the film is compressed, it becomes an imperfect gas. As the gas film is further compressed, it may change continuously to a state called *vapor expanded*, or it may undergo a first-order phase change to form a *liquid expanded* film [1]. (The concept of the different orders of phase changes was developed by Ehrenfest [76].) The isotherms for these two types of films appear very much alike, the chief difference being that the liquid expanded film exists at smaller areas per molecule. The liquid expanded film is a condensed film and is classed as one of the liquid modifications, whereas the vapor expanded film is gaseous. Either one of these states, but never both, may be found in any given system. Langmuir [166] proposed an equation of state for liquid expanded films

$$(\pi + \pi_0)(\sigma - \sigma_0) = kT \tag{7.36}$$

π_0 and σ_0 are constants in some systems, but prove not to be constant for all systems.

Liquid condensed films are formed at relatively small molecular areas, and they are characterized by low compressibility. In the region between the liquid expanded and condensed films the *intermediate* film exists, which is a coherent film of high compressibility. Whether or not this is a homogeneous film or a heterogeneous film in a transition region is a moot question. A fourth type of film possessing both liquid and solid properties was found in some systems [62]. This film, called the *LS* film, has anomalous viscosity properties and very low compressibility.

Solid films possess very low compressibility, and, on the whole, much higher viscosity than any of the four liquid films. The area per molecule is very close to that occupied by the molecules in the three-dimensional crystalline phase. The film usually has a rigidity that is not possessed by any other type of film.

The transition between the gaseous state and the condensed or liquid expanded state is accompanied by a first-order phase change. When a vapor expanded state exists, it goes into the liquid state by a second-order phase change. Second-order transitions are the most frequent in two-dimensional systems. Practically all phase transitions between condensed films and between expanded and condensed films are second-order changes. However, first-order transitions between two condensed states have also been found in a few instances.

The viscosity of monolayers may be either Newtonian—independent of the rate of shear—or non-Newtonian. In general it decreases with increasing temperature, and increases with increasing pressure, but some films have been found in which the viscosity is practically independent of the pressure and increases with the temperature. Solid films are either rigid or highly viscous. Liquid films may have high or low viscosities, depending upon their structures. Gas films usually have viscosities too low to measure. In the region of transition between two phases in a monolayer, the viscosity is generally found to change very rapidly; thus the viscosity curve may be used to locate the positions of phase transitions.

The Thermodynamics of Monolayers. It can be shown from the definitions of the quantities involved that

$$\gamma_f = \left(\frac{\partial E}{\partial a_f}\right)_{s,v,\Gamma_2} = \left(\frac{\partial G}{\partial a_f}\right)_{T,p,\mu_2} \tag{7.37}$$

where γ_f is the surface tension of a film on a liquid, a_f is the area of the film, E is the total energy, G the Gibbs free energy of the system, and Γ_2 and μ_2 are the surface excess and chemical potential of the adsorbed film.

The area of a film may be increased in either of two ways: by *extension* or by *spreading* [126].

In the process of extension, the area of the film together with that of the subphase is increased, while the film-free area of the subphase is held constant. An example of this process is the raising of a ring from the surface, in determining the surface tension. The free energy of extension is

$$dF = \left(\frac{\partial G}{\partial a_f}\right)_{T,p,a_w} da_f = \gamma_f\, da_f \tag{7.38}$$

where a_w is the area of the film-free subphase. For unit area increase of surface

$$\Delta F(\Delta a_f = 1 \text{ cm}^2) \equiv f_e = \left(\frac{\partial G}{\partial a_f}\right)_{T,p,a_w} \quad (7.39)$$

The entropy of extension is

$$s_e \equiv \left(\frac{\partial S}{\partial a_f}\right)_{T,a_w} = -\left(\frac{\partial \gamma_f}{\partial T}\right)_{a_f} \quad (7.40)$$

The heat absorbed is

$$q_e = Ts_e \quad (7.41)$$

The increment of heat content in extension is

$$h_e \equiv \left(\frac{\partial H}{\partial a_f}\right)_{T,a_w} = \left(\frac{\partial \frac{\gamma_f}{T}}{\partial \frac{1}{T}}\right)_{a_f} \quad (7.42)$$

These equations are applicable to pure liquid surfaces also.

In the process of spreading, the area of the film is increased, while the area of the pure liquid is decreased by an equal amount. This may be done by separating the film covered surface from the clean surface by a barrier. If the barrier is moved to increase the area of the film by unity, then the area of the clean surface is decreased by unity. The free energy of spreading is given by

$$dF_s = \left(\frac{\partial G}{\partial a_f}\right)_{T,\Sigma} da_f + \left(\frac{\partial G}{\partial a_w}\right)_{T,\Sigma} da_w$$
$$= -(\gamma_L - \gamma_{LF})\, da_f = -\pi\, da_f \quad (7.43)$$

where $\Sigma = a_w + a_f$; and

$$f_s \equiv \left(\frac{\partial G}{\partial a_f}\right)_{T,\Sigma} = -\pi \quad (7.44)$$

The entropy of spreading is

$$s_s \equiv \left(\frac{\partial S}{\partial a_f}\right)_{T,\Sigma} = +\left(\frac{\partial \pi}{\partial T}\right)_{a_f,\Sigma} \quad (7.45)$$

The heat absorbed is

$$q_s = Ts_s \quad (7.46)$$

The increase in heat content is

$$h_s \equiv \left(\frac{\partial H}{\partial a_f}\right)_{T,\Sigma} = \left(\frac{\partial \frac{\pi}{T}}{\partial \frac{1}{T}}\right)_{a_f,\Sigma} \quad (7.47)$$

From these equations it can be seen that

$$s_e = s_s + s_w$$
and
$$h_e = h_s + h_w \quad (7.48)$$

where s_w and h_w are the entropy and heat content per unit area of the film-free subphase.

The energy changes associated with the isothermal expansion or compression of a film may be calculated with the help of these equations. For example, the free energy change in the spreading of a film from an area a_1 to an area a_2 is given by

$$\Delta G = \int_{a_1}^{a_2} -\pi\, da_f \quad (7.49)$$

The heat absorbed in this process is

$$q_s = T \int_{a_1}^{a_2} \left(\frac{\partial \pi}{\partial T}\right)_{a_f,\Sigma} da_f \quad (7.50)$$

It is possible to calculate the latent heat of spreading of a solid, $\Delta H_{S\rightarrow F}$, or a liquid, $\Delta H_{L\rightarrow F}$, to form a monolayer by an application of the Clapeyron equation in the form

$$\Delta H_{S\rightarrow F} = T\left(\frac{d\pi_e}{dT}\right)(a_f - a_s) \quad (7.51)$$

where a_s is the area occupied on the surface of the liquid by the solid remaining in the bulk state. a_s can be made very small in comparison with a_f. It is found experimentally that $\Delta H_{S\rightarrow F}$ is positive, but at the melting point the slope $\partial\pi_e/\partial T$ becomes negative, so that $\Delta H_{L\rightarrow F}$ is negative. The difference between the two heats of spreading at the melting point is equal to the heat of fusion of the solid.

Deposition of Monolayers to Form Multilayers. Blodgett and Langmuir [36] found that monolayers of calcium stearate and other substances can be transferred from a liquid subphase onto a solid by simply dipping a slide through the monolayer into the subphase. At high pH values the deposition of calcium stearate occurs as the solid passes down through the monolayer. They called this type of film the *x* film. At lower values of pH a monolayer is deposited on both the down and up trips of the slide. These films were called *y* films. In other instances the deposition occurred only on the up trip, and these were called *z* films. The thicknesses of these films can be measured by the interference colors with white polarized light or by the interference intensities with monochromatic light. Blodgett and Langmuir [37] found the thickness of barium stearate films to be 24.4 Å per layer.

Adsorption of Soluble Substances. An application of the Gibbs adsorption equation shows that if the concentration of a solute is greater in the surface region than it is in the bulk solution, the surface tension of the solution decreases with increasing concentration of the solute. This class of solutes is called surface active, or capillary active. On the other hand, if the concentration of the solute in the interfacial region is less than in the solution, the surface tension increases with increase in concentration. These solutes are capillary inactive. The capillary-active materials can have a very large effect upon the surface tension of the liquid, but the capillary-inactive materials do not.

Films of organic solutes adsorbed from dilute solutions are predominantly gaseous in nature. Traube [228] discovered that the reduction in the surface tension produced by members of a homologous series increased by a constant amount for each additional methylene group. Langmuir [163] showed that this phenomenon corresponded to a constant decrease in the amount of work necessary to bring the organic compound into the interface for each additional CH_2 group in the chain. He further deduced that these adsorbed films were two-dimensional gases. Szyszkowski's [217] work on organic acids indicates that the adsorption approaches a limiting value as the concentration increases. These adsorbed films are probably unimolecular, although there is some evidence that the surface region may be somewhat more than a molecule thick in concentrated solutions [52].

4. Adsorption on Solid Surfaces. Physical Adsorption of Gases and Vapors [3, 5, 14]

General Remarks. Historically, the investigation of adsorption on solid surfaces antedates the study of adsorption on liquid surfaces. The phenomenon of adsorption was discovered by C. W. Scheele in 1773 and by the Abbé F. Fontana in 1777, who both noted the removal of gases by charcoal. Adsorption from solutions was discovered in 1785 by T. Lowitz, who found that charcoal took the coloring matter out of solutions.

The terminology used in the field of adsorption on solids is different from that used in adsorption on liquids. The subphase is called the *adsorbent*, and the adsorbed film is called the *adsorbate*. In addition, two types of adsorption are distinguished. If the interaction between adsorbent and adsorbate is weak, the phenomenon is called *physical adsorption;* if it is strong, it is called *chemical adsorption* or *chemisorption*.

Adsorption on solids, as on liquids, is spontaneous; consequently, the free energy of the system decreases in the process. The adsorbed particles are either rigidly held to the surface or they can move over the surface in two dimensions. Since prior to adsorption the gas molecules moved freely in three dimensions, the adsorption of a gas or vapor by a solid is accompanied by a *decrease in entropy*. The change in the heat content of the system is given by

$$\Delta H = \Delta F + T\,\Delta S$$

and since both ΔF and ΔS are negative, ΔH must also be negative. This means that all such *adsorption processes are exothermic*, a result found invariably, whether the measurements are made by direct or indirect methods. The decrease in the heat content of the system is called the *heat of adsorption*. In physical adsorption it is of the same order of magnitude as the heats of condensation of gases; in chemisorption, as the heats of chemical reactions. Thus, for example, the heat of the physical adsorption of nitrogen on an iron catalyst is about 3000 cal/mole, but the heat of chemisorption is an order of magnitude larger: about 35,000 cal/mole. Evidence is strong that in the first case nitrogen is adsorbed in the molecular form; in the second case it dissociates into atoms. The surface iron atoms and the nitrogen atoms mutually saturate each other's free valence forces; thus, only one chemisorbed layer forms on the surface of the adsorbent (unimolecular adsorption). In physical adsorption, however, the residual surface forces and the attraction of the already adsorbed molecules enable the iron surface to adsorb several layers of nitrogen molecules below the critical temperature (multimolecular adsorption).

When a gas or vapor is admitted to a thoroughly evacuated adsorbent, its molecules distribute themselves between the gas phase and the adsorbed phase. The disappearance of the molecules from the gas phase occurs with great rapidity in some cases; in others, at a measurable rate. After a while the process stops and a state of equilibrium is reached. The amount of gas adsorbed per gram of adsorbent at equilibrium is a function of the temperature, the pressure, and the nature of the adsorbent and the adsorbate.

The kinetics of physical adsorption has been investigated much less adequately than its thermodynamics. We shall discuss the latter first.

Adsorption Equilibrium. Although there is hardly any type of experimental technique known to the physicist and chemist that has not been used in the study of adsorption, the vast majority of measurements have been of two types: determination of the amount adsorbed, and determination of the heat liberated in the adsorption process.

For a given gas or vapor and unit weight of a given adsorbent, the amount of gas adsorbed at equilibrium is a function of the final pressure and temperature only

$$a = f(p,T) \tag{7.52}$$

where a is the amount of gas adsorbed per gram of adsorbent, p is the equilibrium pressure, and T is the absolute temperature. The generality of Eq. (7.52) is restricted by two factors:

1. It is assumed that a given weight of adsorbent implies a definite surface area. Adsorbents can be prepared in different states of subdivision; however, Eq. (7.52) holds only if, for example, two silica gel samples having different specific surface areas are considered two different adsorbents.

2. In physical adsorption the amount adsorbed in certain pressure ranges is frequently not the same if the equilibrium is approached from the low-pressure side (*adsorption*) and from the high-pressure side (*desorption*). This phenomenon, called *hysteresis*, will be discussed later.

In experimental work usually either the pressure or the temperature alone is varied, while the other is kept constant. When the pressure of the gas is varied and the temperature is kept constant, the plot of the amount adsorbed against the pressure is called the *adsorption isotherm*.

$$a = f(p) \qquad T = \text{const} \tag{7.53}$$

The explicit forms of Eq. (7.53) will be discussed shortly. When the temperature is varied and the pressure is kept constant, one obtains the adsorption *isobar*

$$a = f(T) \qquad p = \text{const} \tag{7.54}$$

A third method of expressing the results of adsorption is by means of the *adsorption isostere*, i.e., the variation of the equilibrium pressure with respect to the temperature for a definite amount of gas adsorbed

$$p = f(T) \qquad a = \text{const} \tag{7.55}$$

The adsorption isotherm is by far the most frequently determined experimental relation in the field of adsorption. It should be noted that a theory that gives an account of the adsorption isotherm accounts for the isobar and isostere as well, since a series of adsorption isotherms can always be replotted in the form of a series of isobars or isosteres.

The amount of gas or vapor adsorbed is usually expressed as the volume of gas at 0°C and 760 mm pressure (STP) taken up per gram of adsorbent, or as the weight of gas adsorbed per gram of adsorbent. The number of moles adsorbed per mole of adsorbent is also often used.

The heat liberated in the adsorption process is measured by means of isothermal or adiabatic calorimeters, or it can be calculated from the adsorption isotherms. Other thermodynamic functions of the adsorbent-adsorbate system can also be calculated from the isotherms, as will be seen later.

The Adsorption Isotherm. The experimental vapor adsorption isotherms may be grouped into the five types shown in Fig. 7.3 [44]. An example

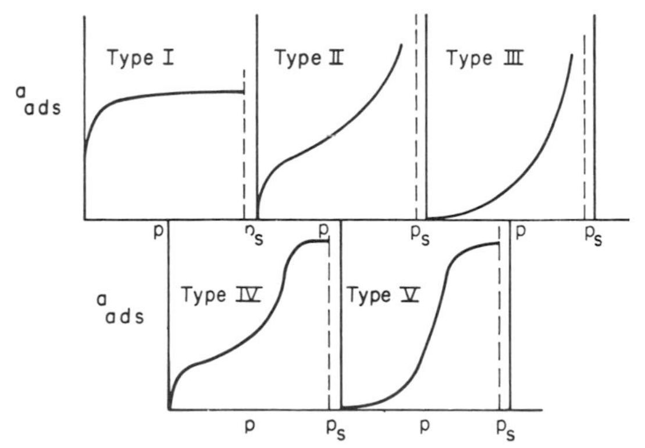

Fig. 7.3. The five types of vapor adsorption isotherms.

of each type may be given here: ethyl chloride on charcoal [102] (Type I), water on hydrated portland cement [194] (Type II), bromine on silica gel [197] (Type III), benzene on ferric oxide gel [160] (Type IV), and water on coconut charcoal [60] (Type V). The idealized isotherms of the figure represent the adsorption from zero pressure to saturation pressure (p_s).

The gas adsorption isotherms either look like the Type I isotherm in Fig. 7.3, i.e., the amount adsorbed tends toward a constant value, beyond which the adsorption does not increase with increasing pressure; or the isotherms look like the lower part of the Type II isotherm, i.e., the amount adsorbed does not tend toward a saturation value with increasing pressure. In either case the adsorption is believed to be unimolecular; i.e., the adsorbed gas covers only a part of the surface of the adsorbent, or at most the entire surface, with a single layer of adsorbed molecules. The Type II to Type V isotherms appear only in vapor adsorption, and they are believed to represent multimolecular adsorption.

The Langmuir Equation. The first theoretical treatments of the adsorption isotherm were advanced in 1915 independently by Langmuir [164] and Polanyi [193]. Their approach was quite different: Langmuir believed that adsorption was similar to a chemical process and that the adsorbed layer was unimolecular; Polanyi believed that adsorption was like a condensation process and that the adsorbed phase was many layers thick. Both treatments have been successful in fitting experimental data in many instances, and both have their limitations. The Polanyi theory applies to physical adsorption only; the Langmuir theory applies, within limits, to both physical and chemical adsorption. Polanyi's theory will not be further discussed here.

Langmuir reasoned that at equilibrium the amount adsorbed on a given adsorbent was constant at a given temperature and pressure; consequently, the rate of adsorption must be equal to the rate of desorption. He assumed that (1) adsorption took place only on the bare surface of the adsorbent, i.e., the adsorption was unimolecular; and (2) the rate of desorption was directly proportional to the amount of gas or vapor adsorbed on the surface. The second assumption is equivalent to using an average heat of adsorption throughout the entire range of adsorption.

On the basis of these two assumptions Langmuir derived the equation

$$\Theta = \frac{v}{v_m} = \frac{bp}{1 + bp} \tag{7.56}$$

where Θ is the fraction of the surface covered with adsorbed molecules, v is the volume of gas adsorbed, v_m is the volume of gas necessary to cover the entire surface with a unimolecular layer of adsorbed gas, p is the equilibrium pressure, and b is a constant at constant temperature. Fowler [96] derived an equation, formally identical with the Langmuir equation, on the basis of statistical mechanics, and obtained for the adsorption coefficient b the expression

$$b = \frac{h^3}{(2\pi m)^{3/2}(kT)^{5/2}} \frac{f_a(T)}{f_g(T)} e^{q/kT} \tag{7.57}$$

where h is Planck's constant, k is Boltzmann's constant, m is the mass of a gas molecule, $f_a(T)$ is the partition function of the adsorbed states attached to a specified surface atom, $f_g(T)$ is the partition function for the rotations and vibrations of the free gas molecule, and q is the energy required to transfer the molecule from the lowest gaseous state to the lowest adsorbed state.

At low pressure ($bp \ll 1$), Eq. (7.56) reduces to

$$f = v_m bp \tag{7.58}$$

and the isotherm obeys Henry's law. At high pressures ($bp \gg 1$), it reduces to

$$v = v_m \tag{7.59}$$

i.e., the adsorption reaches a saturation value. In the middle pressure range, the complete equation (7.56) must be used. The Langmuir equation has been successful in interpreting many experimental data [3]. Obviously, it must break down when one or the other of its fundamental assumptions does not hold. If the adsorption is multimolecular, the simple Langmuir equation is replaced by a generalized form of it, called the BET equation. This will be discussed later. If the heat of adsorption in a range of surface covering strongly deviates from the assumed average, the Langmuir equation does not hold even for unimolecular adsorption.

For most adsorbent-adsorbate systems the adsorption coefficient b varies with the amount adsorbed. As Eq. (7.57) shows, b is dependent on an energy term ($e^{q/RT}$) and on an entropy term (the rest of the expression). A good deal is known about the variation of the heat of adsorption with the amount adsorbed [3], but the variation of the entropy of adsorption with the amount adsorbed is known only for very few adsorbent-adsorbate systems [138].

The variation of the heat of adsorption with the amount adsorbed is largely due to two factors. First of all, the surface of the adsorbent is heterogeneous in nature. At the lowest pressures adsorption takes place on the most active centers of the surface, i.e., on spots where the free surface energy of the solid is the largest; at higher pressures the less active parts of the surface become covered with adsorbed molecules. The effect of this is that the heat of adsorption decreases with increasing adsorption. In the second place, the adsorbed molecules attract each other, the energy of attraction increasing as the amount adsorbed increases. The two effects are in the opposite direction and partly compensate for each other. In addition, the variation of the entropy factor in Eq. (7.57) frequently compensates partly for the variation of the energy factor. Thus it happens at times in unimolecular adsorption that b remains approximately constant over a wide range of pressures, and Langmuir's equation is obeyed [45]. This, however, is rather the exception than the rule.

Various theoretical treatments have been advanced which take into consideration the heterogeneity of the surface ("configurational" effects) or the interaction of adsorbed molecules ("cooperative" effects) or both. These are discussed by Hill [137].

Graham [103] suggested that an equilibrium function could be used to characterize adsorption systems. The function would be an equilibrium constant for localized, noninteracting adsorption on a uniform surface. The change in value of the equilibrium function with coverage shows departure from the ideal system. (1) The equilibrium function decreases with adsorption on a nonuniform surface; this is noticeable at low coverages. (2) The equilibrium function increases with adsorption in the presence of lateral interaction. This is noticeable only after an appreciable fraction of the surface is covered. (3) The equilibrium function first decreases, then increases; this indicates nonuniform surface and lateral interactions.

The Freundlich Equation. The Freundlich equation [8] is the oldest isotherm equation (often referred to as the "classical" equation), and it is still widely used by investigators, particularly in industrial practice. The equation is

$$v = kp^n \tag{7.60}$$

or

$$\log v = \log k + n \log p$$

where k and n are empirical constants and $1 > n > 0$. As the pressure increases, the adsorption increases, and it does not approach a saturation value.

For decades this equation was regarded as purely empirical; however, recently Zeldowitsch [237], Baly [21], Halsey and Taylor [107], and Sips [213] advanced theoretical derivations, based on the Langmuir equation. The surface of the adsorbent is assumed to be heterogeneous, and a certain type of distribution of the heat of adsorption over the surface is postulated.

Halsey and Taylor assumed an exponential distribution of the heat of adsorption, i.e.,

$$N_q = c \exp - \frac{q}{q_m} \tag{7.61}$$

where N_q is the number of adsorbing centers whose energy of adsorption is q and c and q_m are constants. Such a distribution of adsorption energies does lead to Eq. (7.60) and gives the empirical constants k and n as functions of c and q_m. A good example of this type of adsorption was found in the chemisorption of hydrogen on tungsten, which will be discussed later.

Sips stated on a theoretical basis that the exponential distribution of adsorption energies is the only type that can lead to the Freundlich equation. This point has not yet been investigated experimentally, but the fact is that exponential distribution of adsorption energies has seldom been found, whereas the Freundlich equation is frequently obeyed in gas adsorption [3].

Sips also advanced a three-parameter isotherm equation, which is a combination of the Langmuir and Freundlich equations:

$$\frac{v}{v_m} = \frac{kp^n}{1 + kp^n} \tag{7.62}$$

This equation was used successfully by Honig and Reyerson [139] and by Deitz and Loebenstein [67] to fit their adsorption data above as well as below the critical temperature. The covering of the surface in these experiments was always less than a monolayer. The latter investigators found that the v_m values obtained from the Sips equation agreed well with the BET values for v_m.

The BET Equation. In 1938 Brunauer, Emmett, and Teller [47] derived an equation that accounted for the Type II, or S-shaped, isotherm—the type most frequently found in vapor adsorption. They discarded the idea that vapor adsorption was unimolecular, and used, instead, the following assumptions: (1) adsorption was multimolecular, and each separate layer obeyed a Langmuir equation; (2) the average heat of adsorption in the second adsorbed layer was the same as in the third and higher layers, and it was equal to the heat of condensation of the vapor; and (3) the average heat of adsorption in the first adsorbed layer was different from that of the second and higher layers. The equation is

$$\frac{v}{v_m} = \frac{cx}{(1 - x)(1 - x + cx)} \tag{7.63}$$

where $x = p/p_s$, and p_s is the saturation pressure of the vapor (or the vapor pressure); v_m has the same meaning as in Eq. (7.56), and c is a constant which includes the net heat of adsorption:

$$c = c_1 \exp\left(\frac{E_1 - E_L}{RT}\right) \tag{7.64}$$

where c_1 is another constant, E_1 is the average heat of adsorption in the first layer, and E_L is the heat of liquefaction of the vapor. The difference $E_1 - E_L$ is called the net heat of adsorption.

Cassie [55] and Hill [129] derived an equation, formally identical with the BET equation, by statistical mechanics and obtained for c the expression

$$c = \frac{j_a}{j_e} \exp\left(\frac{E_1 - E_L}{RT}\right) \tag{7.65}$$

where j_a and j_e are the pressure independent parts of the partition functions of the adsorbate and of the liquid, respectively.

The BET equation can be put into the linear form:

$$\frac{p}{v(p_s - p)} = \frac{1}{v_m c} + \frac{c-1}{v_m c}\frac{p}{p_s} \tag{7.66}$$

Most vapor-adsorption isotherms plotted according to Eq. (7.66) give straight lines in the range $p/p_s = 0.05$ to 0.35. The constants v_m and c can be calculated from the slope and intercept of the straight line. Figure 7.4 [47] gives an example of such straight-line plots.

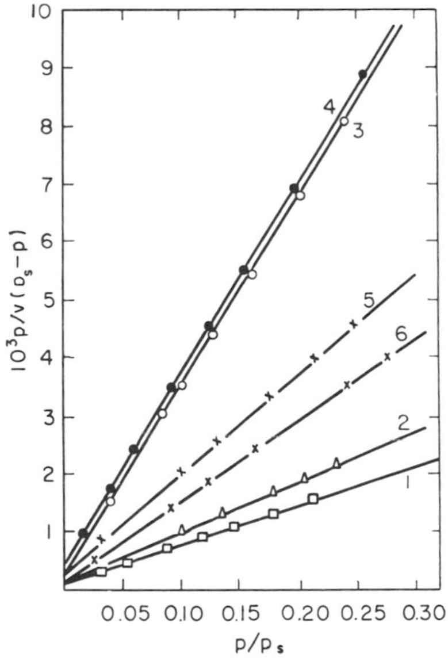

FIG. 7.4. Straight line BET plots for the adsorption of nitrogen on various adsorbents.

The surface area of the adsorbent can be evaluated from v_m. The number of molecules that cover the adsorbent with a complete monolayer is given by v_m, and the area covered by a molecule is obtained (1) by calculating a spherical volume for the molecule from the density of the liquid, and (2) by assuming hexagonal close-packing of spheres on the surface [82]. An alternative method of obtaining the area of a molecule is to use a two-dimensional analogue of the van der Waals equation [53, 131, 171]. The two methods give not too different molecular surface area values. Occasionally, however, the assumption of liquidlike packing of the adsorbed molecules on the surface may lead to a considerable error. The worst case of this sort is found in the adsorption of helium. Long and Meyer [174], as well as others, reported that the spacing of the helium atoms in the adsorbed phase was 2 A, whereas the spacing in the liquid phase was 4 A. The former value corresponds roughly to the gas kinetic diameter of 2.1 A.

The surface area values obtained for a variety of adsorbents by means of the BET equation agreed in all cases within a few per cent with surface areas determined by direct visual means (e.g., by electron microscope), where such determinations were possible. For porous adsorbents a number of indirect surface measurements confirmed the BET values. Thus, at present, the method is regarded as the most reliable one for the determination of the surface areas of finely divided substances, and largely because of this the BET equation is the most widely used isotherm equation in the field of adsorption.

From the constant c one can obtain an approximate estimate of the average heat of adsorption in the first layer, and the values so obtained are in semiquantitative agreement with the experimental results. Equation (7.63) also describes correctly the temperature dependence of the adsorption isotherms.

The equation gives too low adsorption values below $p/p_s = 0.05$, because of the heterogeneity of the surface. McMillan [183] and Walker and Zettlemoyer [233] took this factor into account. In the range $p/p_s > 0.35$, the BET equation gives too high adsorption values. There are several reasons for this, one of which was considered by Brunauer, Emmett, and Teller [47].

The BET equation was derived for adsorption on a free surface, and it was assumed that at p_s an infinite number of layers could build up on the adsorbent. In a porous adsorbent even at saturation pressure only a finite number of adsorbed layers can build up. For this case the isotherm equation is

$$\frac{v}{v_m} = \frac{cx}{1-x}\,\frac{1-(n+1)x^n + nx^{n+1}}{1+(c-1)x - cx^{n+1}} \tag{7.67}$$

where n is the maximum number of layers that can be adsorbed and the other terms have the same meaning as in Eq. (7.63). For $n = \infty$, Eq. (7.67) reduces to Eq. (7.63), and for $n = 1$ it reduces to the Langmuir equation.

The three-constant BET equation gives a better fit at higher relative pressures (p/p_s) than the two-constant equation, but it does not fit the data up to saturation pressure. Pickett [192], Anderson [18], Cook [59], Dole [69], Hüttig [140], and others introduced semiempirical modifications of the BET equation, and obtained better fits at higher relative pressures for certain isotherms. There exists, however, no isotherm equation that can give a good agreement with experiment throughout the entire pressure range, from $p = 0$ to $p = p_s$.

The Hüttig equation was analyzed and tested to a

greater extent than the other modifications of the BET equation. The equation is

$$\frac{v}{v_m} = \frac{cx(1 + x)}{1 + cx} \tag{7.68}$$

At saturation pressure $v = 2v_m$ or smaller. This definitely contradicts experimental facts and, in addition, one of Hüttig's assumptions contradicts the principle of microscopic reversibility, as was pointed out by Hill [137]. The equation gives too small adsorption values at higher relative pressures. Lopez-Gonzalez and Deitz [175] showed that an empirical linear combination of the BET and the Hüttig equations gives a good fit of certain experimental data up to $x = 0.8$.

The theoretical interpretation of the Type III, IV, and V isotherms of Fig. 7.3 was given by Brunauer, Deming, Deming, and Teller [44]. They showed that the BET equation (7.63) represented a Type III isotherm if $c = 1$ or smaller. The data of Reyerson and Cameron [197] for the adsorption of bromine and iodine on silica gel were well fitted by the equation. Also, the surface area of the adsorbent, the average heat of adsorption, and the temperature dependence of the isotherms were accurately given by the equation.

In the Type IV and V isotherms the attainment of saturation adsorption below saturation pressure was attributed to capillary condensation. A four-constant equation was developed in which the fourth constant was dependent on the heat of capillary condensation. The equations fitted certain experimental data fairly well; the surface area and the temperature dependence were accurately obtained, and reasonable values were found for the heat of adsorption and the heat of capillary condensation. Since Type IV and V isotherms are rather rare, and because the four-constant equation is complex, it has found little use so far. Recently Joyner and Emmett [143] applied it for the adsorption of nitrogen or porous glass.

The Frenkel-Halsey-Hill Equation. In 1948 Halsey [105] deduced by semiempirical reasoning the isotherm equation

$$\ln x = \frac{k}{v^s} \tag{7.69}$$

where k and s are constants. The same equation was deduced independently somewhat earlier by Frenkel [7] and later by Hill [133].

The assumptions underlying this equation are (1) the entropy of adsorption in the second and higher adsorbed layers is the same as the entropy of the liquid, but the heat of adsorption is not the same; (2) the net heat of adsorption diminishes according to the power law

$$E_r - E_L = a'r^{-s} \tag{7.70}$$

where a' is a constant and E_r is the energy of adsorption at the distance r from the surface of the adsorbent; and (3) r is proportional to v/v_m. This last assumption is meaningful only for large values of v/v_m. For smaller values a stepwise isotherm should be obtained. However, Halsey showed that the heterogeneity of the surface may smooth out the isotherms [106].

Hill [137] deduced Eq. (7.69) on theoretical grounds and found that $s = 3$ and that k can be expressed in terms of certain physical constants of the adsorbent and of the liquid. McMillan and Teller [184] obtained the value of k by an approximate theoretical argument and found it to have the order of magnitude of the experimental value.

Equation (7.69) has been tested so far only in a few instances. One adsorption isotherm of nitrogen and one of water on anatase, obtained by Harkins and Jura [125], were fairly well fitted up to the neighborhood of the saturation pressure, with $s = 2.67$ and 2.5, respectively. Drain and Morrison [74] found a good fit for argon on rutile, with $s = 2$. Harkins and Jura earlier derived a semiempirical equation of the form of Eq. (7.69), with $s = 2$. Bowers [42] measured the adsorption of nitrogen, oxygen and argon on aluminum foil and found that the equation was obeyed, with $s = 3$.

The temperature dependence of the Frenkel-Halsey-Hill equation has not been tested. Only considerable further investigation can establish the extent of the validity and usefulness of the equation. Its usefulness is limited by the fact that it furnishes no information about the surface area and the pore structure of the adsorbent.

Capillary Condensation and Hysteresis. The oldest theory of adsorption was advanced by Zsigmondy [238] in 1911. While examining the pore structure of silica gel under the ultramicroscope, he came to the conclusion that in such fine pores condensation of liquid should occur at much lower pressures than the saturation pressure. The vapor-pressure lowering in a cylindrical capillary is given by the Kelvin equation (7.10). If it is assumed that γ and ρ are the same for the adsorbate as for the bulk liquid, r can be calculated for any value of p/p_s. The radii so obtained indicate that it is not unreasonable to expect capillary condensation for $p/p_s > 0.3$ or 0.4. The theory predicts discontinuous adsorption, but the continuity of the adsorption isotherms is explained by assuming a continuous distribution of the capillary radii in the adsorbent.

In porous adsorbents it is almost invariably found that the desorption isotherm does not coincide with the adsorption isotherm in the range of higher pressures. The amounts held at a given p/p_s are greater for desorption than for adsorption. This phenomenon is called *hysteresis*. The hysteresis loop usually closes around $p/p_s = 0.3$; at lower pressures the isotherm is reversible. Figure 7.5 [160] shows isotherms with hysteresis loops.

Hysteresis was first explained in terms of the capillary condensation theory. Kraemer [156] and McBain [179] showed that the shape of the capillary may account for greater amounts of liquid condensed during desorption than adsorption (the "ink-bottle" theory). Hill [135] showed that hysteresis may be produced in capillary condensation, regardless of the shape of the pore. However, occasionally, hysteresis persists to such low pressures that capillary condensation cannot occur. Hill pointed out that phase changes in the adsorbate may also be responsible for hysteresis.

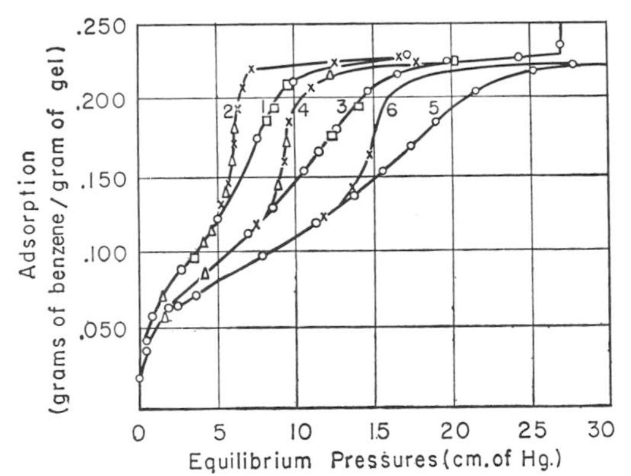

FIG. 7.5. Hysteresis in the adsorption of benzene on ferric oxide gel.

Everett and his coworkers [86 to 88] developed a treatment of the independent domain model of hysteresis, which they applied successfully to adsorption hysteresis. The theory postulates that a system is made up of domains, each of which exists in one of two (or more) possible states. The transition from one state to another, I → II, may be brought about by increasing an external variable x to x_u. The reverse transition, II → I, is brought about by decreasing x; but it occurs at x_e, a value lower than x_u. Thus, at least one of the transitions must be irreversible. Each of the domains is independent of the others. The transitions I → II occur at different values of x_u, and the transitions II → I occur at different values of x_e. The state of a system made up of such domains cannot be specified by fixing the usual external variables. An internal variable, the particular distribution of domains among their possible states, must also be specified. Everett showed that such a model could explain the reproducible and time-independent hysteresis loops that have been observed in many systems.

Patrick and his coworkers [182], Kubelka [157], Schuchowitzky [208], and Foster [93] among others investigated the capillary condensation theory of adsorption. Foster showed convincingly that the desorption branch of the hysteresis loop is caused by capillary condensation. He suggested that hysteresis was due to a delay in the formation of the meniscus in open pores during the adsorption process. Rao [196] and Cohan [58] made further contributions to the study of hysteresis.

The capillary condensation theory proved to be of great value in the investigation of the pore structure of the adsorbent. Anderson [17] was the first in this field, and further significant advances were made by Kubelka [158], Foster [94], Wheeler [235], and Barrett and Joyner [24, 141]. Barrett, Joyner, and their collaborators, carrying further Wheeler's ideas, calculated the pore volume distribution in certain adsorbents by using a combination of the Kelvin equation and the BET three-parameter equation. They assumed (1) that the desorption isotherms represented capillary condensation and (2) that the capillaries were cylindrical. The distribution curves so obtained were supported by three different lines of evidence: (1) different adsorbates gave approximately the same pore-volume distribution for the same adsorbents [144]; (2) the surface areas calculated from the pore distribution curves agreed well with the BET surface areas; and (3) pore-volume distribution curves, obtained by means of a high-pressure mercury porosimeter [202], confirmed the curves obtained from the adsorption data.

Adsorption Thermodynamics. No single treatment of adsorption thermodynamics has been adopted universally as yet. The introduction of a function of the surface as an additional independent variable into thermodynamics doubles the number of thermodynamic functions one can define, and it is a matter of opinion which set of functions is the most convenient to use generally [12, 85]. Gibbs's treatment, although abstract, gives simple and exact equations. In considering the equilibrium of fluids at the surface of unreactive solids, Gibbs suggested a treatment, the principles of which have been used by many workers in the field of adsorption. The film pressure, π, is chosen as the free surface energy function; the Gibbs adsorption equation then takes the form

$$d\pi = S_s\, dT + \Gamma\, d\mu \tag{7.71}$$

for a two-component system. Here S_s is the difference between the surface excess entropy of the solid in equilibrium with its own vapor and of the solid in equilibrium with the fluid, and is analogous to the entropy of spreading of a film [Eq. (7.45)]. Γ is the surface excess of adsorbate, expressed in moles per unit area. Bangham [22, 23] was the first to use the Gibbs adsorption equation to calculate the reduction in free surface energy from an adsorption isotherm by integrating Eq. (7.71):

$$\pi = RT \int_0^p \frac{\Gamma}{p}\, dp \tag{7.72}$$

A large error may be introduced unless the adsorption isotherm is determined to low enough pressures to give a gas film. The extrapolation to zero pressure can then be made using the equation of state of a two-dimensional gas.

The thermodynamic function most frequently determined experimentally is the heat of adsorption, i.e., the change in the heat content of the system occurring in the adsorption process. This quantity can be measured directly by calorimetric means [28, 73, 153, 185] or it can be calculated from experimental isotherms [138, 142, 198]. The calorimeter measures the *integral heats of adsorption;* the slopes of plots of the integral heats of adsorption against the amount adsorbed give the *differential heats of adsorption.* The differential heats of adsorption can also be calculated directly from adsorption isotherms with the help of the Clapeyron-Clausius equation

$$q_{st} = RT^2 \left(\frac{\partial \ln p}{\partial T}\right)_\Gamma \tag{7.73}$$

q_{st} is called the isosteric heat of adsorption. Few comparisons between calorimetric and isosteric heats of adsorption, performed on the same adsorbent-adsorbate systems, are available. Joyner and Emmett [142] compared their isosteric heats with the calorimetric heats of Beebe and his coworkers [28] and found good agreement.

Recently Hill [132, 134], among others, discussed the use of Eq. (7.73). Kington and Aston [153] showed that excellent agreement is obtained between their calorimetric and isosteric heats of adsorption, if the proper correction terms are applied.

Harkins and Jura [124] showed that the integral heat of adsorption can be determined from two immersion experiments. First, the heat of immersion of the clean adsorbent in the liquid adsorbate is determined; then the heat of immersion of the adsorbent, with an adsorbed film on it, is measured. If Γ moles of adsorbate per unit area of surface are adsorbed, the heat of adsorption per unit area of surface, h_a, is given by

$$h_a = h_{I_{(S \to L)}} - h_{I_{(S_f \to L)}} - \Gamma\lambda \qquad (7.74)$$

where λ is the molar heat of vaporization of the liquid.

The surface excess entropy can be determined from adsorption isotherms by applying Eq. (7.71). By a transformation of variables this equation becomes

$$d\pi = (S_s - \Gamma S_g)\, dT + \Gamma v_g\, dp \qquad (7.75)$$

where S_g and v_g are the molar entropy and volume of the gas in equilibrium with the adsorbed film. From Eq. (7.75) it follows that

$$\left(\frac{\partial \pi}{\partial T}\right)_p = S_s - \Gamma S_g \qquad (7.76)$$

and

$$\left(\frac{\partial \ln p}{\partial T}\right)_\pi = \frac{(S_s/\Gamma) - S_g}{RT} \qquad (7.77)$$

Hill [137] derived Eq. (7.77) by a slightly different approach, and he called S_s/Γ the *integral entropy*.

Hill also derived an equation for the *differential entropy*, $(\partial S/\partial N)_{A,T}$

$$\left(\frac{\partial \ln p}{\partial T}\right)_\Gamma = \frac{S_g - (\partial S/\partial N)_{A,T}}{RT} \qquad (7.78)$$

where N is the number of moles of gas adsorbed on area A ($\Gamma = N/A$) and S is the total entropy of the adsorbent-adsorbate system. Hill relates the integral entropy to the number of quantum states, Ω, by the Boltzmann equation $S = R \ln \Omega$. The entropies of the adsorption of nitrogen on two carbon blacks were evaluated by means of Eqs. (7.77) and (7.78) [138]. The entropy of adsorption was recently discussed by Kemball [150].

The differential entropy in Eq. (7.78) is the partial molal surface entropy of the adsorbate. There is also a partial molal surface entropy of the adsorbent, even though few workers in the field seem to be interested in studying it. The relationship between the partial molal surface entropy of the adsorbent and the entropy functions discussed above can be shown to be

$$\bar{S}_1 - S_1^\circ = \Gamma \Sigma \left(\frac{S_s}{\Gamma} - \bar{S}_2\right) \qquad (7.79)$$

where $\bar{S}_1$ and $\bar{S}_2$ are the partial molal surface entropies of the adsorbent and adsorbate, respectively; S_1° is the molal surface entropy of the adsorbent in vacuum, and Σ is the area per mole of adsorbent. Published results [74, 138] for $(S_s/\Gamma) - S_L$ and $\bar{S}_2 - S_L$ (where S_L is the entropy of the adsorbate in the liquid state) show that these quantities vary over approximately the same range of values as $(S_s/\Gamma) - \bar{S}_2$. Thus, $\bar{S}_1 - S_1^\circ$ is a significant thermodynamic quantity. Similar considerations hold for the partial molal surface free energy and heat content of the adsorbent [61].

The Adsorbent and the Adsorbate. The most important property of the adsorbent is its specific surface. In porous adsorbents it is also important to know the distribution of pore sizes, and occasionally it is of value to know the surface area distribution in the pores of different diameters.

The BET equation gives the number of molecules necessary to cover the adsorbent with a monolayer, but it does not give the area covered by a molecule. Harkins and Jura [125] developed an "absolute" method for determining the surface areas of nonporous adsorbents, which does not involve the assumption of a given area for a molecule. The surface area obtained for TiO_2 by this method checked the BET surface area, obtained by the adsorption of nitrogen at $-195.8°C$, within less than 1 per cent. This may be considered a confirmation of the molecular area of nitrogen used in the BET surface area determinations, as well as a mutual confirmation of the assumptions involved in both types of determinations. In addition, Harkins and Jura showed that the areas assumed by Emmett and Brunauer for some other molecules, especially for polar and for elongated nonpolar molecules, were incorrect, and they gave the correct molecular areas to be used in the BET method. Livingston [171] made a detailed analysis of the molecular areas of a number of adsorbates.

The Harkins-Jura absolute method is very laborious and it is not applicable to porous adsorbents. The BET method is the simplest and most reliable method of surface area determination for practically all adsorbents. A variety of other methods have been proposed, many of which involve adsorption techniques [3, 79]. Among these the "relative method" of Harkins and Jura [125] is especially useful.

Frequently, it is important to know the sizes and shapes of the adsorbent particles. Adsorption experiments supply no information about these. However, the order of magnitude of the particle size can be obtained by assuming that all the particles are spheres and uniform in size. The radius of a particle is then given by

$$r = \frac{3}{A} \qquad (7.80)$$

where A is the surface area per cubic centimeter of adsorbent.

The best method known at present for the determination of the pore-volume distribution of the adsorbent is that developed by Barrett and Joyner [24], discussed before. The method also gives the surface areas assignable to pores of different sizes.

The forces of interaction between adsorbent and adsorbate in physical adsorption are called van der

Waals forces. The energy of interaction (the van der Waals energy) between the surface of a heteropolar adsorbent and a polar adsorbate consists of three parts: the nonpolar interaction term E_D, which is the "dispersion energy" of London [173]; the interaction between the electrostatic field of the surface and the dipole moment of the adsorbate molecules E_0, analogous to the "orientation energy" of Keesom [152]; and the term arising from the polarization of the adsorbate by the surface, E_I, analogous to the "induction energy" of Debye [66]. The heat of adsorption contains a fourth term, the interaction energy between the adsorbate molecules. If it is assumed that the interaction energy between the adsorbate molecules in the adsorbed state is the same as in the liquid state, the net heat of adsorption, $E_1 - E_L$, is equal to the sum of the three terms, E_D, E_0, and E_I.

Zettlemoyer and his coworkers [57, 128] investigated the adsorbent-adsorbate interactions in detail. They evaluated the net heats of adsorption from the heat of immersion of the solid in the liquid and the total surface energy of the liquid. The net heat of adsorption of polar adsorbates on rutile was found to be a linear function of the dipole moment of the adsorbate. From the slope of the line they calculated the electrostatic field strength of the rutile surface at the distance of the dipole to be 2.75×10^{15} esu. With the help of this value they were able to estimate E_0 and E_I, and E_D was obtained as the difference between the net heat of adsorption and the sum of E_0 and E_I. They found that in the adsorption of alcohols on rutile about 68 per cent of the net heat was due to E_0, about 6 per cent to E_I, and about 26 per cent to E_D; and in the adsorption of hydrocarbons on rutile about two-thirds of the net heat was due to E_D and one-third to E_I. Graphon has no electrostatic surface field, which was shown by the fact that the net heats were independent of the dipole moments of the adsorbates. Thus, the net heat of the adsorption of both hydrocarbons and alcohols on graphon is due entirely to E_D.

Calorimetric and adsorption isotherm measurements supply much useful information about the properties of the adsorbent-adsorbate system, but little has been done so far to ascertain the properties of the adsorbate by itself. The heat contents, free energies, and entropies evaluated by the methods of adsorption thermodynamics, discussed before, refer to the adsorbent-adsorbate system and not to the adsorbate alone. It is customary [137] to call these thermodynamic functions properties of the adsorbate, but this may introduce large errors.

The adsorbate behaves in many respects like a three-dimensional substance. The essential difference is that its molecules cannot move freely in the direction perpendicular to the surface of the adsorbent; thus, one translational degree of freedom is replaced by a vibration. The mobility of the adsorbate over the surface was first demonstrated by Volmer [232].

Adsorbed films on solids exhibit phase changes similar to those exhibited on liquid subphases, as was demonstrated by Jura and his coworkers [146, 148]. They observed mostly second-order phase changes, but in the transition from the gas to the liquid they found also first-order changes. An example of this is shown in Fig. 7.6. The gas-to-liquid transition usually occurs at very low pressures, $p/p_s = 0.02$, or less. Thus, throughout practically the whole range of the isotherm, the adsorbate below the critical temperature behaves like a liquid.

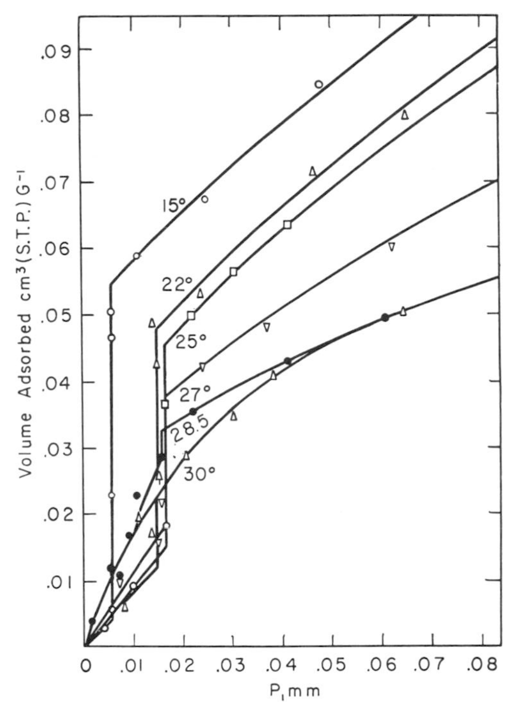

FIG. 7.6. First order phase change in the adsorption of *n*-heptane on ferric oxide.

A number of equations of state have been proposed for the adsorbate, the most important being the two-dimensional analogue of the van der Waals equation [53, 131]:

$$\left(\pi + \frac{a'N^2}{A^2}\right)(A - Nb') = NkT \qquad (7.81)$$

where π is the surface pressure, A is the surface area, N is the number of molecules adsorbed, and a' and b' are the two-dimensional van der Waals constants. These constants have definite relations to the three-dimensional van der Waals constants a and b [137]. The relationships between the two- and three-dimensional critical constants were evaluated from the van der Waals equation. A more detailed theoretical attack on this subject was made by Devonshire [68], but only his calculated critical temperatures checked the experimental results. The two-dimensional critical temperature was found to be about one-half of the three-dimensional critical temperature.

Mixed Adsorption. Three types of mixed adsorption may be distinguished: (1) a single adsorbent and a mixture of gases, (2) a mixture of adsorbents and a single gas, and (3) a mixture of adsorbents and a mixture of gases. Mixed adsorption plays an important role in a number of industrial processes.

The adsorption of a single gas on a one-component

adsorbent is a complicated enough phenomenon, but the complexity greatly increases in mixed adsorption. Nevertheless, because of its practical importance, some investigations were conducted in the field. A part of this work was discussed by McBain [14] and by Brunauer [3].

Markham and Benton [177] derived equations similar to the Langmuir equation for the adsorption of gas mixtures. The theory showed fair agreement with experiment. Hill [130] derived a theory of multimolecular adsorption for mixtures of vapors, based on the BET model.

Adsorption Kinetics. The rate of physical adsorption on nonporous adsorbents is very high. In porous adsorbents, however, especially if the pores are very narrow, equilibration may take minutes, hours, or even longer periods. The rate-determining step in this case is the process of diffusion of gas molecules into the pores. McBain [178] was the first to apply Fick's diffusion law to the adsorption process, and later Damköhler [63], Wicke [236], and others derived and tested somewhat different diffusion equations. The diffusion of gases in and through solids was discussed in detail by Barrer [2].

The kinetics of physical adsorption has not been studied as thoroughly as the field of adsorption equilibrium. A part of the theoretical and experimental investigations was discussed by Brunauer [3].

5. Chemical Adsorption of Gases on Solids [3, 11, 16]

Physical vs. Chemical Adsorption. The main differences between physical adsorption and chemisorption may be summarized as follows:

1. The forces of interaction between adsorbent and adsorbate are predominantly van der Waals forces in physical adsorption and valence forces in chemical adsorption.
2. The energy of binding between adsorbent and adsorbate is several times as great in chemisorption as in physical adsorption.
3. Chemical adsorption is specific; physical adsorption is nonspecific. The latter takes place at sufficiently low temperatures between any surface and any gas, but the former depends on chemical affinity between the particular adsorbent and adsorbate.
4. The adsorption isobars of gases that can be adsorbed both chemically and physically on the same adsorbent show two regions in which the adsorption decreases with increasing temperature, one corresponding to physical, the other to chemical adsorption.
5. The adsorption isotherm in chemisorption always indicates unimolecular adsorption; physical adsorption may be unimolecular or multimolecular.
6. Physical adsorption is rapid; the gas molecules are adsorbed as rapidly as they can reach the surface. In chemisorption an energy of activation must be supplied to the system before the adsorbent-adsorbate complex can form. It is true, however, that this energy of activation may be so small that chemisorption proceeds at a very high rate even at low temperatures.

Chemisorption and Surface Reaction. Chemical adsorption may be regarded as a surface reaction; it is a chemical reaction that takes place between the ions, atoms, or molecules of the surface of the adsorbent and the molecules of the adsorbate.

Emmett and Brunauer [80] obtained chemisorption isotherms of nitrogen on iron catalysts used in the synthesis of ammonia at temperatures of 400 and 450°C and at pressures between 25 and 760 mm. Under these conditions iron and nitrogen do not react to form a solid-phase iron nitride. The dissociation pressure of Fe_4N at 450°C is 4,600 atm, and Fe_3N and Fe_2N have even higher dissociation pressures. Yet nitrogen was readily taken up by the iron catalyst at pressures of 25 mm and lower. The solubility of nitrogen in iron at 450°C is much smaller than the amounts taken up in these experiments, and it is a function of the volume of the iron, whereas here three iron catalysts having equal volumes but surface areas in the proportion 1:5:10, took up nitrogen in the proportion 1:5:10. It is clear, therefore, that nitrogen reacted only with the surface of the iron catalyst, to form a surface nitride. Evidence is strong [46] that nitrogen is chemisorbed in the form of atoms and not molecules. These adsorbed atoms, or *adatoms*, can be removed from the surface by passing a stream of hydrogen over it, and they come off as ammonia.

Activated carbon adsorbents adsorb oxygen at temperatures so high that the physical adsorption of the gas is negligibly small. Loebenstein and Deitz [172] investigated the chemisorption of oxygen on charcoal, bone char, and carbon black at 200°C, and they found that the gas desorbed at 400°C only partly as molecular oxygen, partly as CO, CO_2, and H_2O. Here again the adsorption of oxygen is probably atomic, and the binding between the adatoms and the surface carbon and hydrogen atoms is so strong that some of them come off the surface together with the oxygen atoms.

Langmuir [167] found that oxygen was adsorbed on incandescent tungsten at a pressure of 10^{-5} mm; and the binding was so strong that the adsorbed layer did not come off below 1700°K, whereas WO_3 evaporates at 1200°K. In these and all other chemisorption experiments the maximum amount adsorbed never exceeded a single adsorbed layer.

The specificity of chemisorption can be well illustrated with the chemisorptions of CO, CO_2, H_2, and N_2 on unpromoted and promoted iron catalysts [46, 81]. The activity of an iron catalyst toward the synthesis of ammonia is greatly increased if it contains a small quantity of Al_2O_3 as promoter (singly promoted catalyst), and it is further increased if it contains two promoters, Al_2O_3 and K_2O (doubly promoted catalyst). At liquid air temperatures an unpromoted iron catalyst adsorbs chemically a complete monolayer of carbon monoxide. This chemisorption is molecular. The singly promoted catalyst chemisorbs less than a monolayer of CO; only the iron atoms on the surface chemisorb CO, and Al_2O_3 molecules do not. From the amount of CO chemisorbed, one can calculate the concentration of Al_2O_3 molecules on the surface of the catalyst.

Iron does not adsorb CO_2 chemically, nor does the Al_2O_3 promoter, but the alkali promoter in doubly promoted catalysts chemisorbs CO_2 in molecular form at dry ice temperature. The chemisorption of CO_2 thus gives a measure of the K_2O (or $K_2Al_2O_4$) molecules on the surface of the catalyst, and the CO chemisorption a measure of the iron atoms on the surface. It was found that the sum of the two chemisorptions corresponded to a single monolayer, as determined by the BET method.

The chemisorption of hydrogen on iron catalysts was investigated by Emmett and Harkness [83], who found two different types of chemisorption in different temperature ranges. It was shown [46] that both types were atomic, and both covered the surface at maximum absorption with a single layer of atoms. Recently, even a third type of hydrogen chemisorption was found at −195°C [159].

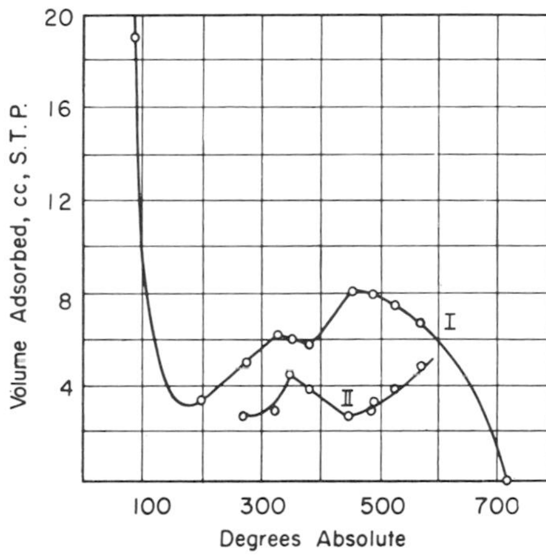

FIG. 7.7. Adsorption isobar of hydrogen on zinc oxide.

In physical adsorption, carbon monoxide and nitrogen behave in a very similar manner on iron catalysts or on any other adsorbent, but the chemisorptions of the two gases show no similarity whatever. Carbon monoxide is chemisorbed at liquid air temperature in the molecular form, analogous to iron carbonyl formation. The adsorption decreases with temperature, and it is negligible at room temperature. Nitrogen is chemisorbed at temperatures of 200°C and higher in the atomic form, analogous to iron nitride formation.

The Adsorption Isobar. Benton and White [35] found that for a gas that can be attached to the surface of an adsorbent in two different forms, the adsorption isobar shows two descending portions with an ascending portion between them. Taylor and Strother [224] found three descending portions on an isobar of hydrogen on zinc oxide at 760 mm pressure, as shown in Fig. 7.7. The first descending portion corresponds to physical adsorption; the other two, at higher temperatures, to two different types of chemisorption. Emmett and Harkness [83] likewise found three descending portions on their isobars of hydrogen on iron catalysts at 760 mm pressure.

Taylor [221] demonstrated that the points on the ascending portions of the isobar do not correspond to equilibrium conditions. If one approaches these points from the desorption side, one obtains higher values. Adsorption, both physical and chemical, is exothermic; consequently, according to the principle of Le Châtelier at true equilibrium the amount adsorbed must always decrease with increasing temperature. Taylor interpreted the increase in adsorption with increasing temperature as being due to an increase in the *rate of adsorption* with temperature. This concept of *activated adsorption* will be discussed later.

The Adsorption Isotherm. Since chemisorption is always unimolecular, the isotherms should tend toward saturation at sufficiently high pressures. This *sufficiently* high pressure may be a very low pressure, as we have seen in the adsorption of carbon monoxide on iron or oxygen on tungsten, and then again it may be a pressure quite beyond the experimental range. For example, Frankenburg [98] made a very thorough study of the chemisorption of hydrogen on tungsten powder between −195 and 600°C and between 10^{-5} and 30 mm pressure. He estimated by extrapolation of his data that saturation of the surface would be reached only at a pressure a hundred times as great as the highest pressure used in his experiments. His chemisorption isotherms are shown in Fig. 7.8.

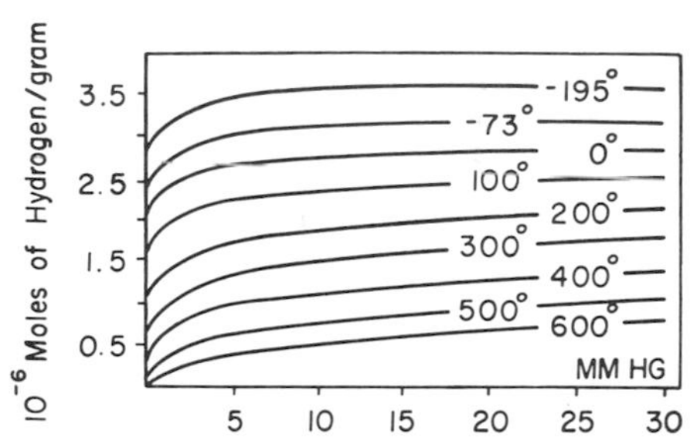

FIG. 7.8. Adsorption isotherms of hydrogen on tungsten.

Chemisorption isotherms are rather scarce in the literature, and not much attempt has been made to interpret them. If the heterogeneity of the surface is not too great, and the heat of adsorption can be assumed to be roughly constant over the surface, the Langmuir equation (7.56) should be obeyed. This is seldom, if ever, true in chemisorption over a wide range of surface covering.

For some systems the heat of adsorption varies linearly over the surface, as was found, for example, by Roberts [203] for the adsorption of hydrogen on tungsten wire. The isotherm equation for this case is [49]

$$\ln p = A + Bv \tag{7.82}$$

where A and B are constants. This equation, called the Temkin equation, fits the chemisorption isotherms of nitrogen on iron catalysts [80] very well.

If the heat of adsorption varies exponentially over the surface, the Freundlich equation (7.60) should be obeyed [107]. Frankenburg's data illustrate this case: the heat of adsorption does vary exponentially over most of the surface, and the isotherms obey the Freundlich equation over most of the pressure range.

The Kinetics of Chemisorption. Taylor [221] was the first to point out that chemisorption is not always instantaneous but that, on the contrary, it frequently proceeds at a measurable rate. Figure 7.9 represents such rate curves for the adsorption of

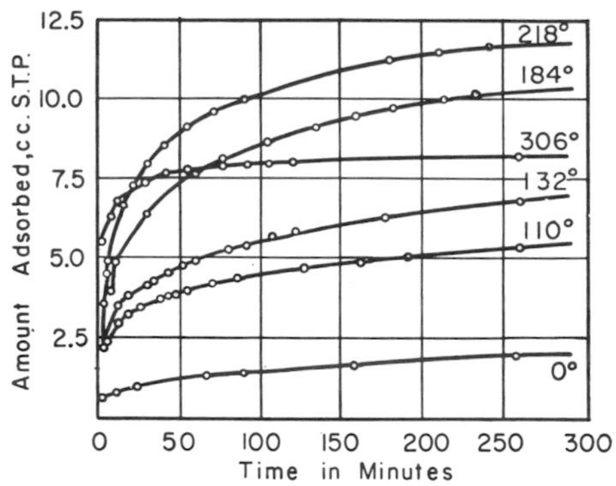

FIG. 7.9. Rate of adsorption of hydrogen on zinc oxide.

hydrogen on a zinc oxide catalyst, obtained by Taylor and Sickman [223]. The 306°C curve crosses the 218 and 184°C curves because the initial rate is higher, but the final amount adsorbed is less than that at the lower temperatures.

Taylor showed that the rate of chemisorption increases exponentially with temperature, just like the rates of chemical reactions. He calculated the activation energies from the Arrhenius equation

$$\ln \frac{t_1}{t_2} = \frac{E_a}{R}\left(\frac{1}{T_1} - \frac{1}{T_2}\right) \tag{7.83}$$

where t_1 and t_2 are the times required for the adsorption of a given volume of gas at absolute temperatures T_1 and T_2, and E_a is the energy of activation of adsorption. Because this type of chemisorption requires measurable activation energies, Taylor named it activated adsorption.

Since the surface of an adsorbent has unsaturated valence forces, the chemisorption process has a certain degree of similarity to a reaction between an atom, or a free radical, and a molecule in the gas phase. That such reactions often require energies of activation was established by Hartel and Polanyi [127]. The activation energies measured for adsorption processes are of the same order of magnitude (about 10,000 cal/mole) as the activation energies between free radicals and saturated molecules.

Some chemisorption processes require much smaller energy of activation than 10,000 cal/mole. Trapnell [16] calculated the energy of activation of oxygen on tungsten at room temperature from the results of Morrison and Roberts [186], and obtained a value of 600 cal/mole for Θ (the fraction of the surface covered with adsorbed molecules) = 0.2. A similar calculation of E_a for the chemisorption of nitrogen on tungsten at 300°K, based on the results of Becker and Hartman [27], gave a value of 400 cal/mole at $\Theta = 0$. Chemisorptions occurring at still lower temperatures frequently have immeasurably small activation energies.

Quantum-mechanical calculations of activation energies of a few simple adsorbent-adsorbate systems were made by Sherman and Eyring [211] and others. The results are in semiquantitative agreement with the experimental values.

The energy of activation of adsorption ordinarily increases with the extent of surface covering. The first portions of gas adsorbed give the lowest E_a values and, as we have seen before, they also give the highest heats of adsorption. This is what one would expect, since adsorption takes place first on the most unsaturated, most active, part of the surface—the initial adsorption therefore needs the least energy of activation.

The energy of activation of desorption is given by

$$E_d \doteq E_a + q \tag{7.84}$$

where q is the heat of adsorption. Since the heat of chemisorption is large, E_d is large. This means that ordinarily desorption is a very slow process. In catalytic reactions it is not uncommon that the desorption of the reaction products from the catalyst becomes the slowest and, consequently, the rate determining step.

The velocity of chemisorption is represented by the equation

$$u_a = K_a f_a(\Theta) \exp\left(-\frac{E_a}{RT}\right) \tag{7.85}$$

and the velocity of desorption by the equation

$$u_d = K_d f_d(\Theta) \exp\left(-\frac{E_d}{RT}\right) \tag{7.86}$$

where K_a and K_d are the velocity constants for adsorption and desorption and $f_a(\Theta)$ and $f_d(\Theta)$ are functions of the number of sites available for adsorption and desorption, respectively. Theoretical evaluations of the velocity constants and the dependence of the velocity upon coverage, as well as the available experimental data were discussed by Trapnell [16].

Recently, H. A. Taylor and Thon [219] investigated the applicability of the Elovich [77] rate equation

$$\frac{dv}{dt} = ae^{-\alpha v} \tag{7.87}$$

where v is the volume of gas adsorbed in time t and a and α are constants. They found that this equation describes a large body of rates of chemisorption data with considerable success.

The Adsorbent and the Adsorbate. The great practical importance of chemisorption lies in the role it plays in catalytic processes. In heterogeneous catalysis the chemical reaction takes place on the surface of the catalyst; consequently, at least one of the reactants must be adsorbed before the reaction can take place, and at least one of the reaction products must desorb from the catalyst surface. The rates of these adsorption or desorption processes may determine the rate of the catalytic reaction.

Chemisorption experiments furnish important information about the adsorbent-adsorbate complex. Because of the strong interaction between the adsorbent (almost always a catalyst) and the adsorbate, it is even more difficult to separate the properties of the adsorbent from those of the adsorbate than it is in physical desorption.

From physical adsorption the extent of the total surface area of an adsorbent can be determined; from chemisorption, information about the "active" surface of an adsorbent is obtained. Taylor [220] was the first to demonstrate the heterogeneous nature of catalyst surfaces by arguments drawn from experiments in the fields of adsorption and catalysis. His theory of "active centers," though questioned by some investigators, received ever firmer foundations with the passing of years. One of the most convincing proofs came from Taylor [222] himself. He showed experimentally that at one temperature a part of the catalyst surface may be covered with chemisorbed gas, and another part of the surface may be bare, whereas at a different temperature the two parts of the surface may change roles: the one previously covered with adsorbed gas may be bare, and the one previously empty may be covered.

The energy of interaction between the bare surface and the adsorbate in chemisorption is usually large. For example, Beebe and Taylor [29] obtained a value of 20,600 cal/mole for the chemisorption of hydrogen on nickel powder, and Beeck [32], 31,000 cal/mole for hydrogen on an evaporated film of nickel. The heat of physical adsorption of hydrogen is an order of magnitude smaller than these values. The heat of chemisorption, itself, may show wide differences, if the mechanism of the interaction changes. Nitrogen on evaporated iron films is rapidly chemisorbed at liquid air temperatures with an initial heat of 10,000 cal/mole; at room temperature the chemisorption is slow and the initial heat is 40,000 cal/mole [31]. In the first case, the adsorption is molecular, in the second, atomic.

The heat of chemical adsorption usually decreases with increasing surface covering, Θ, as we have seen before. This decrease may be due to three factors: one primarily dependent on the adsorbate, one on the adsorbent, and one on the interaction of adsorbent and adsorbate.

If the interaction of adsorbed atoms or molecules with each other is repulsive, the heat of adsorption decreases with Θ. However, at low values of Θ the adsorbed particles are probably too far from each other to interact, and this is the region where the heat of adsorption usually diminishes most strongly with increasing Θ. At high values of Θ repulsive interactions may be appreciable, but even here they may not account fully for the observed decrease. Rideal and Trapnell [200] found that in the adsorption of hydrogen atoms on tungsten repulsive dipole-dipole forces can account only for a small fraction of the total decrease in the heat of adsorption.

The heterogeneous nature of the adsorbent surface is probably the most important factor in accounting for the variation of the heat of adsorption with Θ. The heterogeneity may be due to different exposed crystal planes, to edges and corners of crystals, to amorphous patches or extra lattice atoms, to cracks and faults in the crystals, to grain boundaries, to impurities, and possibly to other causes. Among these factors the effect of different crystal planes was investigated most thoroughly. Burk [51] and Balandin [20] were the first to point out the importance of the geometry of the surface in catalysis. Especially valuable contributions to the understanding of the geometric factor in adsorption and catalysis were made by Beeck and his coworkers [31]. Recently, several investigators studied either chemisorption or catalysis or both on individual crystal planes [27, 168, 188, 198]. For discussions of the geometric factor, see also Griffith [104] and Trapnell [227].

A third important factor in chemisorption and catalysis is the so-called electronic factor. The interaction of adsorbent and adsorbate produces changes in the electronic structures of both. Dowden [71] pointed out that the chemisorption of gases by metals may be explained by assuming the passage of electrons from the gas into vacant atomic d orbitals of the metal, thus forming covalent bonds. This is the reason why the transition and the near-transition metals are the most active in chemisorption and catalysis. Oxygen, which is chemisorbed by practically all metals, forms probably largely ionic bonds with the nontransition metals, with electron donation from the s and p bands of the metal [16]. Numerous experimental investigations support the electronic theory; important contributions were made, among others, by Schwab [209] and Boudart [40].

It is not always easy to decide whether the geometric or the electronic factor is predominant in a given chemisorption or catalytic reaction [16]. Beeck [30] showed that the catalytic activity of metals in ethylene hydrogenation correlates well with their lattice spacings, and later he showed [32] that it correlates also well with their d character. The reason for this is that the d character controls the lattice spacing, which was shown by Pauling [191]. In a number of cases, however, a clear-cut distinction can be made. When an oriented and a nonoriented evaporated metal film show different catalytic activities, it is clear that the geometric factor is operative; when a nontransition metal that has the optimum lattice spacing is inactive, it is clear that the electronic factor is responsible for this.

Chemisorption changes the work function of the adsorbent. The influence of the work function on chemisorption and catalysis was studied by several investigators [72, 151, 234]. Because of the strong binding between adsorbent and adsorbate, the mobility of adsorbed particles is far more restricted than in physical adsorption. In chemisorption, migration over the surface always requires activation energy. In most cases the activation energy of migration is smaller than the activation energy of desorption. For example, Brattain and Becker [43] obtained a value of 110,000 cal/g atom for the activation energy of migration of thorium atoms over a tungsten surface; the energy of activation of desorption is 191,000 cal/g atom. If a gas dissociates on adsorption, the energy of activation of migration of the atoms over the surface is smaller than the energy of activation of desorption *as atoms*, but it may be larger than the energy of activation of desorption *as molecules*. This may lead to an immobile adsorbed layer [16].

E. W. Müller's field emission microscope proved itself to be a powerful tool in studying migration in chemisorption, as well as in the investigation of other aspects of chemisorption [26, 33, 101, 187]. Electron diffraction, first used in chemisorption by Davisson and Germer [65], was recently revived as a tool by Farnsworth and his coworkers [206].

The "active" surface which plays a role in chemisorption is not necessarily identical with the catalytically active surface. Occasionally the entire surface of a catalyst can chemically adsorb a given gas, and all the chemisorbed gas can be active in a given catalytic reaction, but this is the exception rather than the rule. Most frequently, only a part of the surface can chemisorb a given gas and only part of the chemisorbed gas is active catalytically.

The catalytic reaction in the chemisorbed layer may take place by one of two mechanisms: (1) two particles chemisorbed on adjacent sites react with each other, or (2) a chemisorbed particle reacts with a molecule striking it from the gas phase or physically adsorbed on an adjacent site. The first is called the Langmuir mechanism [165], the second, the Rideal mechanism [199]. They may be illustrated by the exchange reaction between hydrogen and deuterium over a metal catalyst [16]. Hydrogen and deuterium are chemisorbed as atoms. If the Langmuir mechanism operates, the reaction is

```
H   D
|   |     |   |
M—M→M—M + H—D
```

If the Rideal mechanism operates, the reaction is

```
H   D2     HD D
|   :      :  |
M—M    →   M—M
```

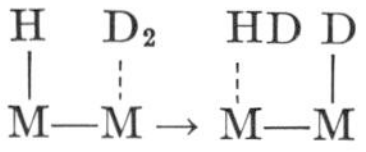

The full lines represent covalent bonds, the dotted lines, physical adsorption. In the first mechanism two covalent bonds are broken and only one is formed; consequently, it requires more activation energy than the second.

The three simplest catalytic reactions, the conversion of para to ortho hydrogen, the conversion of ortho to para deuterium and the hydrogen-deuterium exchange reaction, were most extensively investigated by Farkas and Farkas [90, 91]. The conversion reactions do not require dissociation into atoms; reversal of the nuclear spin may occur in an adsorbed molecule because of the inhomogeneous magnetic field existing near an adsorbent surface. The exchange reaction proceeds through the Langmuir mechanism [16]. Outside of these three reactions, the mechanisms of only a few of the simplest catalytic reactions are partially understood.

Not only chemical adsorption but physical adsorption also can furnish valuable information about catalysts. The importance of the surface area has long been recognized, but the realization of the importance of the pore structure is a relatively new development, largely due to Wheeler [235]. The usefulness of physical adsorption studies in furnishing information about the structure and certain properties of catalysts is illustrated by the work of Ries and his coworkers [201] on cracking catalysts.

6. Adsorption on Solids from Solutions [8, 11]

In applying the Gibbs adsorption equation to three-component systems, it is possible to fix the interfacial plane so that the surface excess of the solid Γ_s is zero, but then Γ for neither of the other two components is equal to zero. In the three-component system, solution in contact with a solid, both the solvent and the solute are adsorbed at the interface. The amount of adsorption is usually measured by measuring the concentration of the solute in the solution before and after the adsorption process. From this change in concentration it is possible to calculate the *apparent adsorption* of the solute. For capillary-active solutes the apparent adsorption first increases to a maximum value as the concentration is increased; then it decreases, and may become negative and show a minimum value. It is not possible to calculate the true adsorption of the solute without adopting some assumptions as to the nature of the adsorption process.

Bartell and Sloan [25] gave a relationship between the individual adsorption isotherms and the apparent adsorption of the solute in a two-component system.

$$\frac{n_0\,\Delta x}{m} = n_1^s(1 - x) - n_2^s x \tag{7.88}$$

where n_0 is the total number of moles in the original solution, Δx is the change in the mole fraction of component 1 (the solute) produced by adsorption, x is the mole fraction of component 1 after adsorption, m is the mass of the adsorbent, and n_1^s and n_2^s are the number of moles of component 1 and component 2 (the solvent) adsorbed per gram of adsorbent. They assumed that the adsorption isotherms of the individual components followed the Freundlich equation (7.60), and they obtained a four-parameter equation relating the apparent adsorption to the final mole fraction of the solute. By successive approximations, they evaluated the equation that gave the best fit to their experimental data and from the parameters calculated the individual adsorption isotherms.

Kipling and Tester [154] observed that the adsorption of pure adsorbates on porous adsorbents usually obeyed the Langmuir equation (7.56). They derived a four-parameter equation for apparent adsorption, based on the Langmuir equation, and they were able to represent the data for the benzene-ethyl alcohol-charcoal system as well as Bartell and Sloan did. However, the individual isotherms calculated from their equation were entirely different from those of Bartell and Sloan, so Kipling and Tester concluded that probably neither treatment was correct. They deduced, therefore, by two different methods, the relationship between apparent adsorption and the adsorption isotherms of the individual components.

1. They observed, as Elton [78] did earlier, that in adsorption from solution the adsorbent surface is always covered completely with adsorbate. They assumed that a monolayer was adsorbed and deduced that

$$\frac{n_1^s}{(n_1^s)^\circ} + \frac{n_2^s}{(n_2^s)^\circ} = 1 \tag{7.89}$$

where n_1^s and n_2^s have the same meaning as in Eq. (7.88), and $(n_1^s)^\circ$ and $(n_2^s)^\circ$ are the values obtained for the adsorption of the saturated vapors of the pure components. From Eqs. (7.88) and (7.89) they were able to calculate the individual adsorption isotherms.

2. From kinetic considerations they deduced the equation

$$\Theta_1 = \frac{a_1}{a_1 + ka_2} \tag{7.90}$$

where Θ_1 is the fraction of the surface covered by component 1, a_1 and a_2 are the activities of the components in the solution after adsorption, and k is a constant. From Eqs. (7.88) and (7.90) they then calculated the individual adsorption isotherms. The results obtained by the two methods were in good agreement. Kipling and Tester measured the adsorption from the vapor phase in equilibrium with the solutions. They calculated the adsorption of the individual components from the total weight adsorbed and from the change in the concentration of the solution.

Hansen and Hansen [111] showed that Kipling and Tester did not need the assumption of unimolecular adsorption in porous adsorbents. The pore volume limits the amount adsorbed, and Eq. (7.89) directly follows from this fact.

Hansen and Fackler [109] applied a modified Polanyi potential theory [193] to the adsorption of propanol-1 and butanol-1 from aqueous solutions on a nonporous adsorbent Spheron-6 (a carbon black). They measured the vapor adsorption isotherms of the two alcohols and water individually, calculated the Polanyi potentials, and deduced the isotherms for the (apparent) adsorption of each alcohol from its aqueous solutions. The calculated and experimental isotherms were in semiquantitative agreement.

In earlier investigations the adsorption of the solvent was generally ignored. Then it was found that the Freundlich adsorption equation (7.60) frequently held with good precision. Organic solutes usually form monolayers which are generally imperfect gases [95]. Harkins and Jura [124] reported that titanium dioxide adsorbed about 70 per cent as much stearic acid from dry benzene as would be required to make a condensed monolayer.

It is possible to show a correlation between adsorption and one property or another common to a group of related adsorbates, but such correlations are rather limited because of the complexity of three-component systems. An example of this is found in the adsorption of fatty acids from ligroin solutions [54] on various adsorbents. On charcoal Traube's rule, discussed in Sec. 3, applies, and the order of adsorption is lauric $<$ myristic $<$ palmitic $<$ stearic acid. But on silica gel the reverse is true: stearic $<$ lauric acid; and on alumina, the adsorptions of lauric, myristic, and palmitic acid are equal.

Hansen and Craig [108] studied the adsorption by nonporous carbons of a series of aliphatic acids and a series of primary alcohols from aqueous solutions. The adsorption isotherms of different members of a homologous series were remarkably similar when the amount adsorbed was plotted as a function of the activity of the solute. In fact, at activities below 0.1 the number of moles of solute adsorbed was practically independent of the chain length.

For a group of related solutes in a given solvent, the adsorption, in general, decreases with increasing solubility of the solute. For any one solute in various solvents, the adsorption is generally greatest from the solution containing the poorest solvent.

Hansen, Fu, and Bartell [110] studied the adsorption of various solutes, having limited solubilities, upon three different forms of carbon. They were able to fit their data with the three-parameter BET equation (7.67) [47], but the value of n could not be interpreted as the number of layers of molecules formed at the interface, nor did the v_m values give entirely consistent surface areas for the different adsorbents. Nevertheless, the authors were able to conclude that films adsorbed from nearly saturated solutions may be multimolecular. Hansen and Hansen [111] believe that adsorption from solutions is almost always multimolecular.

References

1. Adam, N. K.: "The Physics and Chemistry of Surfaces," Oxford University Press, London, 1941.
2. Barrer, R. M.: "Diffusion in and through Solids," Macmillan, New York, 1941.
3. Brunauer, Stephen: "The Adsorption of Gases and Vapors," vol. I, "Physical Adsorption," Princeton University Press, Princeton, N.J., 1943.
4. Dupré, A.: "Théorie mécanique de la chaleur," Paris, 1869.
5. DeBoer, J. H.: "The Dynamical Character of Adsorption," Oxford University Press, New York and London, 1953.
6. Fowler, R. H., and E. A. Guggenheim: "Statistical Thermodynamics," Cambridge University Press, New York and London, 1939.
7. Frenkel, J.: "Kinetic Theory of Liquids," Oxford University Press, New York and London, 1946.
8. Freundlich, H.: "Colloid and Capillary Chemistry," Dutton, New York, 1926.
9. Gibbs, J. W.: "Collected Works," vol. I, Longmans, New York.
10. Gomer, R., and C. S. Smith: "Structure and Properties of Solid Surfaces," University of Chicago Press, Chicago, 1953.
11. Gregg, S. J.: "The Surface Chemistry of Solids," Reinhold, New York, 1951.
12. Guggenheim, A. E.: "Modern Thermodynamics," Methuen, London, 1933.
13. Harkins, W. D.: "The Physical Chemistry of Surface Films," Reinhold, New York, 1952.
14. McBain, J. W.: "The Sorption of Gases and Vapours by Solids," Van Nostrand, Princeton, N.J., 1932.
15. Shockley, W., J. H. Hollomon, R. Maurer, and F. Seitz: "Imperfections in Nearly Perfect Crystals," Wiley, New York, 1952.
16. Trapnell, B. M. W.: "Chemisorption," Academic Press, New York, 1955.
17. Anderson, J. S.: *Z. physik. Chem.*, **88**: 191 (1914).
18. Anderson, R. B.: *J. Am. Chem. Soc.*, **68**: 686 (1946).
19. Bailey, C. L. J., and H. C. Watkins: *Proc. Phys. Soc. (London)*, **63B**: 350 (1950).
20. Balandin, A. A.: *Z. physik. Chem.*, **B2**: 289 (1929).
21. Baly, E. C. C.: *Proc. Roy. Soc. (London)*, **A160**: 465 (1937).
22. Bangham, P. H.: *Trans. Faraday Soc.*, **33**: 805 (1937).
23. Bangham, P. H., and R. I. Razouk: *Trans. Faraday Soc.*, **33**: 1463 (1937).
24. Barrett, E. P., L. G. Joyner, and P. P. Halenda: *J. Am. Chem. Soc.*, **73**: 373 (1951).
25. Bartell, F. E., and C. K. Sloan: *J. Am. Chem. Soc.*, **51**: 1637, 1643 (1920).
26. Becker, J. A.: *Bell System Tech. J.*, **30**: 907 (1951).

27. Becker, J. A., and C. D. Hartman: *J. Phys. Chem.*, **57**: 153 (1953).
28. Beebe, R. A., J. Biscoe, W. R. Smith, and C. B. Wendell: *J. Am. Chem. Soc.*, **69**: 95 (1947).
29. Beebe, R. A., and H. S. Taylor: *J. Am. Chem. Soc.*, **46**: 43 (1924).
30. Beeck, O.: *Revs. Mod. Phys.*, **17**: 61 (1945).
31. Beeck, O.: "Advances in Catalysis," vol. II, p. 151, Academic Press, Inc., New York, 1950.
32. Beeck, O.: *Discussions Faraday Soc.*, **8**: 118 (1950).
33. Benjamin, M., and R. O. Jenkins: *Proc. Roy. Soc. (London)*, **A180**: 225 (1942).
34. Benson, G. C., and G. W. Benson: *Can. J. Chem.*, **33**: 232 (1955).
35. Benton, A. F., and T. A. White: *J. Am. Chem. Soc.*, **52**: 2325 (1930).
36. Blodgett, K. B., and I. Langmuir: *Phys. Rev.*, **51**: 964 (1937).
37. Blodgett, K. B., and I. Langmuir: *Phys. Rev.*, **55**: 391 (1939).
38. Born, M., and R. Courant: *Physik. Z.*, **14**: 731 (1913).
39. Born, M., and O. Stern: *Sitzber. preuss. Akad. Wiss. Physik. math. Kl.*, 901 (1919).
40. Boudart, M.: *J. Am. Chem. Soc.*, **74**: 3556 (1952).
41. Bowden, F. P., and D. Tabor: reference 15, p. 203.
42. Bowers, R.: *Phil. Mag.* (7), **44**: 467 (1953).
43. Brattain, W. H., and J. A. Becker: *Phys. Rev.*, **43**: 428 (1933).
44. Brunauer, S., L. S. Deming, W. E. Deming, and E. Teller: *J. Am. Chem. Soc.*, **62**: 1723 (1940).
45. Brunauer, S., and P. H. Emmett: *J. Am. Chem. Soc.*, **59**, 2682 (1937).
46. Brunauer, S., and P. H. Emmett: *J. Am. Chem. Soc.*, **62**: 1732 (1940).
47. Brunauer, S., P. H. Emmett, and E. Teller: *J. Am. Chem. Soc.*, **60**: 309 (1938).
48. Brunauer, S., D. L. Kantro, and C. H. Weise: *Can. J. Chem.*, **34**: 729, 1483 (1956).
49. Brunauer, S., K. S. Love, and R. G. Keenan: *J. Am. Chem. Soc.*, **64**: 751 (1942).
50. Buff, F. P.: *Phys. Rev.*, **82**: 773(T) (1951).
51. Burk, R. E.: *J. Phys. Chem.*, **30**: 1134 (1926).
52. Butler, J. A. V., and A. Wightman: *J. Chem. Soc.*, 2094 (1932).
53. Cassel, H. M.: *J. Phys. Chem.*, **48**: 195 (1944).
54. Cassidy, H. G.: "Technique of Organic Chem. V," Adsorption and Chromatography.
55. Cassie, A. B. D.: *Trans. Faraday Soc.*, **41**: 450 (1945).
56. Chapman, J. C., and H. L. Porter: *Proc. Roy. Soc. (London)*, **A83**: 65 (1910).
57. Chessick, J. J., A. C. Zettlemoyer, F. H. Healey, and G. J. Young: *Can. J. Chem.*, **33**: 251 (1955).
58. Cohan, L. H.: *J. Am. Chem. Soc.*, **60**: 433 (1938).
59. Cook, M. A.: *J. Am. Chem. Soc.*, **70**: 2925 (1948).
60. Coolidge, A. S.: *J. Am. Chem. Soc.*, **49**: 708 (1927).
61. Copeland, L. E.: *Gordon Conference, AAAS*, July, 1955.
62. Copeland, L. E., W. D. Harkins, and G. E. Boyd: *J. Chem. Phys.*, **10**: 357 (1942).
63. Damköhler, G.: *Z. physik. Chem.*, **A174**: 222 (1935).
64. Davis, P. B., and W. G. France: *J. Phys. Chem.*, **40**: 811 (1936).
65. Davisson, C. J., and L. H. Germer: *Phys. Rev.*, **30**: 705 (1927).
66. Debye, P.: *Physik. Z.*, **21**: 178 (1920).
67. Deitz, V. R., and W. V. Loebenstein: to be published in *J. Phys. Chem.*
68. Devonshire, A. F.: *Proc. Roy. Soc. (London)*, **A163**: 132 (1937).
69. Dole, M.: *J. Chem. Phys.*, **16**: 25 (1948).
70. Dorsey, N. E.: *Natl. Bur. Standards Sci. Paper* 540, 1926.
71. Dowden, D. A.: *Research*, **1**: 239 (1948).
72. Dowden, D. A., and P. Reynolds: *Discussions Faraday Soc.*, **8**: 184 (1950).
73. Drain, L. E., and J. A. Morrison: *Trans. Faraday Soc.*, **48**: 316 (1952).
74. Drain, L. E., and J. A. Morrison: *Trans. Faraday Soc.*, **48**: 840 (1952).
75. Dunn, C. G., F. W. Daniels, and M. J. Bolton: *Trans. AIME*, **188**: 368 (1950).
76. Ehrenfest, P.: *Proc. Acad. Sci. Amsterdam*, **36**: 153 (1933).
77. Elovich, S. Y., and G. M. Zhabrova: *Zhur. Fiz. Khim.*, **13**: 1761, 1775 (1939).
78. Elton, G. A. H.: *J. Chem. Soc.*, 2958 (1951).
79. Emmett, P. H.: "Advances in Colloid Science," vol. I, p. 1, New York, 1942.
80. Emmett, P. H., and S. Brunauer: *J. Am. Chem. Soc.*, **56**: 35 (1934).
81. Emmett, P. H., and S. Brunauer: *J. Am. Chem. Soc.*, **59**: 310 (1937).
82. Emmett, P. H., and S. Brunauer: *J. Am. Chem. Soc.*, **59**: 1553 (1937).
83. Emmett, P. H., and R. W. Harkness: *J. Am. Chem. Soc.*, **57**: 1631 (1935).
84. Eötvös, L.: *Ann. Physik*, **27**: 448 (1886).
85. Everett, D. H.: *Trans. Faraday Soc.*, **46**: 453 (1950).
86. Everett, D. H.: *Trans. Faraday Soc.*, **50**: 1077 (1954).
87. Everett, D. H., and F. W. Smith: *Trans. Faraday Soc.*, **50**: 187 (1954).
88. Everett, D. H., and W. I. Whitton: *Trans. Faraday Soc.*, **48**: 749 (1952).
89. Faraday, M.: *Trans. Roy. Soc. (London)*, **147**: 145 (1849).
90. Farkas, A.: *Trans. Faraday Soc.*, **35**: 943 (1939).
91. Farkas, A., and L. Farkas: *J. Am. Chem. Soc.*, **60**: 22 (1938).
92. Fischer, J. C., and C. G. Dunn: ref. 15, p. 317.
93. Foster, A. G.: *Trans. Faraday Soc.*, **28**: 645 (1932).
94. Foster, A. G.: *Proc. Roy. Soc. (London)*, **A147**: 128 (1934); **A150**: 77 (1935).
95. Fowkes, F. M., and W. D. Harkins: *J. Am. Chem. Soc.*, **62**: 3377 (1940).
96. Fowler, R. H.: *Proc. Cambridge Phil. Soc.*, **31**: 260 (1935).
97. Fowler, R. H.: *Proc. Roy. Soc. (London)*, **A159**: 229 (1937).
98. Frankenburg, W.: *J. Am. Chem. Soc.*, **66**: 1827, 1838 (1944).
99. Fullman, R. L.: *J. Appl. Phys.*, **22**: 448 (1951).
100. Furgusen, A.: *Phil. Mag.*, **31**: 37 (1916); *Trans. Faraday Soc.*, **19**: 403 (1923).
101. Goldman, F., and M. Polanyi: *Z. physik. Chem.*, **132**: 321 (1928).
102. Gomer, R.: *J. Chem. Phys.*, **21**: 1869 (1953).
103. Graham, D.: *J. Phys. Chem.*, **58**: 869 (1954).
104. Griffith, R. H.: "Advances in Catalysis," vol. I, p. 91, Academic Press, Inc., New York, 1948.
105. Halsey, G.: *J. Chem. Phys.*, **16**: 931 (1948).
106. Halsey, G.: *J. Am. Chem. Soc.*, **73**: 2693 (1951).
107. Halsey, G., and H. S. Taylor: *J. Chem. Phys.*, **15**: 624 (1947).
108. Hansen, R. S., and R. P. Craig: *J. Phys. Chem.*, **58**: 211 (1954).
109. Hansen, R. S., and W. V. Fackler, Jr.: *J. Phys. Chem.*, **57**: 643 (1953).
110. Hansen, R. S., Y. Fu, and F. E. Bartell: *J. Phys. & Colloid Chem.*, **53**: 769 (1949).
111. Hansen, R. D., and R. S. Hansen: *J. Colloid Sci.*, **9**: 6 (1954).
112. Hardy, W. B.: *Proc. Roy. Soc. (London)*, **A86**: 634 (1911); **A88**: 303 (1913).
113. Hardy, W. B.: *Proc. Roy. Soc. (London)*, **A88**: 311 (1913).

114. Harkins, W. D., and G. E. Boyd: *J. Phys. Chem.*, **45**: 20 (1941).
115. Harkins, W. D., and G. E. Boyd: *J. Am. Chem. Soc.*, **64**: 1195 (1942).
116. Harkins, W. D., F. E. Brown, and E. C. H. Davies: *J. Am. Chem. Soc.*, **39**: 354 (1917).
117. Harkins, W. D., and V. C. Cheng: *J. Am. Chem. Soc.*, **43**: 35 (1921).
118. Harkins, W. D., G. L. Clark, and L. E. Roberts: *J. Am. Chem. Soc.*, **42**: 700 (1920).
119. Harkins, W. D., E. C. H. Davies, and G. L. Clark: *J. Am. Chem. Soc.*, **39**: 541 (1917).
120. Harkins, W. D., and W. W. Ewing: *J. Am. Chem. Soc.*, **42**: 2539 (1920).
121. Harkins, W. D., and A. Feldman: *J. Am. Chem. Soc.*, **44**: 2665 (1922).
122. Harkins, W. D., and E. H. Grafton: *J. Am. Chem. Soc.*, **42**: 2534 (1920).
123. Harkins, W. D., and H. F. Jordan: *J. Am. Chem. Soc.*, **52**: 1751 (1930).
124. Harkins, W. D., and G. Jura: *J. Am. Chem. Soc.*, **66**: 919 (1944).
125. Harkins, W. D., and G. Jura: *J. Am. Chem. Soc.*, **66**: 1362, 1366 (1944).
126. Harkins, W. D., T. F. Young, and E. Boyd: *J. Chem. Phys.*, **8**: 954 (1940).
127. Hartel, H. v., and M. Polanyi: *Z. physik. Chem.*, **B11**: 97 (1930).
128. Healey, F. H., J. J. Chessick, A. C. Zettlemoyer, and G. J. Young: *J. Phys. Chem.*, **58**: 887 (1954).
129. Hill, T. L.: *J. Chem. Phys.*, **14**: 263 (1946).
130. Hill, T. L.: *J. Chem. Phys.*, **14**: 268 (1946).
131. Hill, T. L.: *J. Chem. Phys.*, **14**: 441 (1946).
132. Hill, T. L.: *J. Chem. Phys.*, **17**: 520 (1949).
133. Hill, T. L.: *J. Chem. Phys.*, **17**: 590, 668 (1949).
134. Hill, T. L.: *J. Chem. Phys.*, **18**: 246 (1950).
135. Hill, T. L.: *J. Phys. & Colloid Chem.*, **54**: 1186 (1950).
136. Hill, T. L.: *J. Chem. Phys.*, **20**: 141 (1952).
137. Hill, T. L.: "Advances in Catalysis," vol. IV, p. 211, Academic Press, Inc., New York, 1952.
138. Hill, T. L., P. H. Emmett, and L. G. Joyner: *J. Am. Chem. Soc.*, **73**: 5102, 5933 (1951).
139. Honig, J. M., and L. H. Reyerson: *J. Phys. Chem.*, **56**: 140 (1952).
140. Hüttig, G. F.: *Monatsh.*, **78**: 177 (1948).
141. Joyner, L. G., E. P. Barrett, and R. Skold: *J. Am. Chem. Soc.*, **73**: 3155 (1951).
142. Joyner, L. G., and P. H. Emmett: *J. Am. Chem. Soc.*, **70**: 2353 (1948).
143. Joyner, L. G., and P. H. Emmett: *J. Am. Chem. Soc.*, **70**: 2359 (1948).
144. Juhola, A. J., A. J. Palumbo, and S. B. Smith: *J. Am. Chem. Soc.*, **74**: 61 (1952).
145. Jura, G.: *J. Phys. & Colloid Chem.*, **52**: 40 (1948).
146. Jura, G., and D. Criddle; *J. Phys. & Colloid Chem.*, **55**: 163 (1951).
147. Jura, G., and C. W. Garland: *J. Am. Chem. Soc.*, **74**: 6033 (1952).
148. Jura, G., E. H. Loeser, P. R. Basford, and W. D. Harkins: *J. Chem. Phys.*, **14**: 117 (1946).
149. Katayama, M.: *Sci. Repts. Tôhoku Imp. Univ.*, ser. 1, **4**: 373 (1916).
150. Kemball, C.: "Advances in Catalysis," vol. II, p. 233, Academic Press, Inc., New York, 1950.
151. Kemball, C.: *Proc. Roy. Soc.* (*London*), **A214**: 413 (1952).
152. Keesom, W. H.: *Physik. Z.*, **22**: 129, 643 (1921).
153. Kington, G. L., and J. G. Aston: *J. Am. Chem. Soc.*, **73**: 1929 (1951).
154. Kipling, J. J., and D. A. Tester: *J. Chem. Soc.*, 4123 (1952).
155. Kirkwood, J. G., and F. P. Buff: *J. Chem. Phys.*, **17**: 338 (1949).
156. Kraemer, E. O., in H. S. Taylor: "A Treatise on Physical Chemistry," chap. XX, p. 1661, New York, 1931.
157. Kubelka, P.: *Z. Elektrochem.*, **37**: 637 (1931).
158. Kubelka, P.: *Kolloid Z.*, **55**: 129 (1931).
159. Kummer, J. T., and P. H. Emmett: *J. Phys. Chem.*, **56**: 258 (1952).
160. Lambert, B., and A. M. Clark: *Proc. Roy. Soc.* (*London*), **A122**: 497 (1929).
161. La Mer, V. K., and R. Gruen: *Trans. Faraday Soc.*, **48**: 410 (1952).
162. Langmuir, I.: *J. Am. Chem. Soc.*, **38**: 2222 (1916).
163. Langmuir, I.: *J. Am. Chem. Soc.*, **39**: 1883 (1917).
164. Langmuir, I.: *J. Am. Chem. Soc.*, **40**: 1361 (1918).
165. Langmuir, I.: *Trans. Faraday Soc.*, **17**: 621 (1922).
166. Langmuir, I.: "Colloid Chemistry," vol. I, p. 543, Alexander, 1926.
167. Langmuir, I.: *J. Am. Chem. Soc.*, **53**: 486 (1931).
168. Leidheiser, H., and A. T. Gwathmay: *J. Am. Chem. Soc.*, **70**: 1200 (1948).
169. Lennard-Jones, J. E., and P. A. Taylor: *Proc. Roy. Soc.* (*London*), **109**: 476 (1954).
170. Lipsett, S. G., F. M. G. Johnson, and O. Maas: *J. Am. Chem. Soc.*, **49**: 925 (1927); **50**: 2701 (1928).
171. Livingston, H. K.: *J. Colloid Sci.*, **4**: 447 (1949).
172. Loebenstein, W. V., and V. R. Deitz: *J. Phys. Chem.*, **59**: 481 (1955).
173. London, F.: *Z. phys. Chem.*, **B11**: 222 (1930).
174. Long, E., and L. Meyer: *Phil. Mag. Suppl.* **2**: 1 (1953).
175. Lopez-Gonzalez, J. D., and V. R. Deitz: *J. Research Natl. Bur. Standards*, **48**: 325 (1952).
176. Macleod, D. B.: *Trans. Faraday Soc.*, **19**: 38 (1923).
177. Markham, E. C., and A. F. Benton: *J. Am. Chem. Soc.*, **53**: 497 (1931).
178. McBain, J. W.: *Z. phys. Chem.*, **68**: 471 (1909).
179. McBain, J. W.: *J. Am. Chem. Soc.*, **57**: 699 (1935).
180. McBain, J. W., and C. W. Humphreys: *J. Phys. Chem.*, **36**: 300 (1932).
181. McBain, J. W., and R. C. Swain: *Proc. Roy. Soc.* (*London*), **A154**: 608 (1936).
182. McGavack, J., Jr., and W. A. Patrick: *J. Am. Chem. Soc.*, **42**: 946 (1920).
183. McMillan, W. G.: *J. Chem. Phys.*, **15**: 390 (1947).
184. McMillan, W. G., and E. Teller: *J. Chem. Phys.*, **19**: 25 (1951).
185. Morrison, J. A., J. M. Los, and L. E. Drain: *Trans. Faraday Soc.*, **47**: 1023 (1951).
186. Morrison, J. L., and J. K. Roberts: *Proc. Roy. Soc.* (*London*), **A173**: 1 (1939).
187. Müller, E. W.: *Z. Naturforsch.*, **5a**: 473 (1950).
188. Müller, E. W.: ref. 10, p. 72.
189. Obreimoff, J. W.: *Proc. Roy. Soc.* (*London*), **A127**: 290 (1930).
190. Orowan, E.: *Z. Physik*, **82**: 239, 259 (1933).
191. Pauling, L.: *Proc. Roy. Soc.* (*London*), **A196**: 343 (1949).
192. Pickett, G.: *J. Am. Chem. Soc.*, **67**: 1958 (1945).
193. Polanyi, M.: *Verh. deut. physik. Ges.*, **15**: 55 (1916).
194. Powers, T. C., and T. L. Brownyard· *J. Am. Concrete Inst., Proceedings*, **43**: 249 (1947).
195. Ramsay, W., and J. Shields: *Phil. Trans.*, **A184**: 647 (1893).
196. Rao, K. S.: *J. Phys. Chem.*, **45**: 506, 513, 517 (1941).
197. Reyerson, L. H., and A. E. Cameron: *J. Phys. Chem.*, **39**: 181 (1935).
198. Rhodin, T. N., Jr.: *J. Am. Chem. Soc.*, **72**: 5692 (1950).
199. Rideal, E. K.: *Proc. Cambridge Phil. Soc.*, **35**: 130 (1938).
200. Rideal, E. K., and B. M. W. Trapnell: *J. chim. phys.*, **47**: 126 (1950).
201. Ries, H. E., Jr.: "Advances in Catalysis," vol. IV. p. 87, New York, 1952.

202. Ritter, H. L., and L. C. Drake: *Ind. Eng. Chem., Anal. Ed.*, **17**: 782 (1945).
203. Roberts, J. K.: *Proc. Roy. Soc. (London)*, **A152**: 445 (1935).
204. Rymer, T. B., and C. C. Butler: *Proc. Phys. Soc. (London)*, **59**: 541 (1947).
205. Sawai, I., and M. Nishida: *Z. anorg. u. allgem. Chem.*, **190**: 375 (1950).
206. Schlier, R. E., and H. E. Farnsworth: *Phys. Rev.*, **90**: 351 (1953).
207. Schottky, H.: *Nachr. Ges. Wiss. Göttingen, Math.-phys. Klasse*, **4**: 480 (1912).
208. Schuchowitzky, A. A.: *Kolloid Z.*, **66**: 139 (1934).
209. Schwab, G. M.: *Trans. Faraday Soc.*, **42**: 689 (1946).
210. Sears, G. W.: *J. Appl. Phys.*, **21**: 721 (1950).
211. Sherman, A., and H. Eyring: *J. Am. Chem. Soc.*, **54**: 2661 (1932).
212. Shuttleworth, R.: ref. 15, p. 343.
213. Sips, R.: *J. Chem. Phys.*, **16**: 490 (1948).
214. Smyth, C. P., and E. W. Engel: *J. Am. Chem. Soc.*, **51**: 2646, 2660 (1929).
215. Stefan, J.: *Wied. Ann.*, **29**: 655 (1896).
216. Sugden, S. J.: *J. Chem. Soc.*, **125**: 1177 (1924).
217. Szyszkowski, B.: *Z. physik. Chem.*, **64**: 385 (1908).
218. Tamman, G., and W. Boehme: *Ann. Phys.*, **12**: 820 (1932).
219. Taylor, H. Austin, and N. Thon: *J. Am. Chem. Soc.*, **74**: 4169 (1952).
220. Taylor, H. S.: *Proc. Roy. Soc. (London)*, **A103**: 105 (1925).
221. Taylor, H. S.: *J. Am. Chem. Soc.*, **53**: 578 (1931).
222. Taylor, H. S.: "Advances in Catalysis," vol. I, p. 1, Academic Press Inc., New York, 1948.
223. Taylor, H. S., and D. V. Sickman: *J. Am. Chem. Soc.*, **54**: 602 (1932).
224. Taylor, H. S., and C. O. Strother: *J. Am. Chem. Soc.*, **56**: 586 (1934).
225. Thomson, W.: *Phil. Mag.* (4), **42**: 448 (1871).
226. Tolman, R. C.: *J. Chem. Phys.*, **17**: 333 (1949).
227. Trapnell, B. M. W.: "Advances in Catalysis," vol. III, p. 1, Academic Press Inc., New York, 1951.
228. Traube, J.: *Annalen*, **265**: 27 (1891).
229. Udin, H., A. J. Shaler, and J. Wulff: *J. Metals*, **1**: 186 (1949).
230. van der Hoff, B. M. E., and G. C. Benson: *J. Chem. Phys.*, **22**: 475 (1954).
231. van der Waals, J. D.: *Z. physik. Chem.*, **13**: 716 (1894).
232. Volmer, M., and G. Adhikari: *Z. physik. Chem.*, **119**: 46 (1926).
233. Walker, W. C., and A. C. Zettlemoyer: *J. Phys. & Colloid Chem.*, **52**: 47, 58 (1948).
234. Wansbrough-Jones, O. H., and E. K. Rideal: *Proc. Roy. Soc. (London)*, **A123**: 202 (1929).
235. Wheeler, A.: "Advances in Catalysis," vol. III, p. 249, New York, 1951.
236. Wicke, E.: *Kolloid Z.*, **86**: 167 (1939).
237. Zeldowitsch, J.: *Acta Physicochim. USSR*, **1**: 961 (1934).
238. Zsigmondy, R.: *Z. anorg. Chem.*, **71**: 356 (1911).

More Recent References

239. Adamson, A. W.: "Physical Chemistry of Surfaces," Interscience, New York, 1960.
240. Derjaguin, B. V.: "Research in Surface Forces," Consultants Bureau, New York, 1963. (Trans. from Russian.)
241. Garner, W. E.: "Chemisorption," Academic, New York, 1957.
242. Likhtman, V. I., E. D. Shehukin, and P. A. Rehbinder: "Physico-chemical Mechanics of Metals," U.S.S.R. Academy of Sciences, Moscow, 1962. (In Russian.)
243. Kuznetsov, V. D.: "Surface Energy of Solids," H. M. Stationary Office, London, 1957. (Trans. from Russian.)
244. Ross, Sydney, and J. P. Olivier: "On Physical Adsorption," Interscience, New York, 1964.
245. Young, D. M., and A. D. Crowell: "Physical Adsorption of Gases," Butterworth, London, 1962.
246. Aston, J. G., E. S. Tomezsko, and Hakze Chon: Effect of Lateral Interaction on Monolayer Adsorption, *Advan. Chem. Ser.*, **33**: 325 (1961).
247. Bascom, W. D., R. L. Cottington, and C. R. Singleterry: Dynamic Surface Phenomena in the Spontaneous Spreading of Oils on Solids, *Advan. Chem. Ser.*, **43**: 355 (1964).
248. Benson, G. C., P. I. Freeman, and E. Dempsey: Calculation of the Distortion in the Surface Region of an Alkali Halide Crystal Bounded by a (100) Face, *Advan. Chem. Ser.*, **33**: 26 (1961).
249. Benson, G. C., and K. S. Yun: Surface Tension of the (100) Face of Alkali Halide Crystals, *J. Chem. Phys.*, **42**: 3085 (1965).
250. Bewig, K. W., and W. A. Zisman: Surface Potentials and Induced Polarization in Nonpolar Liquids Adsorbed on Metals, *U.S. Naval Res. Lab., NRL Rept.* 6068, June 19, 1964.
251. Brunauer, S.: Surface Energies of Solids, "Proceedings 2d International Congress of Surface Activity," vol. 2, p. 17, Butterworth, London, 1957.
252. Brunauer, S.: Surfaces of Solids, plenary lecture at XXth Congress on Pure and Applied Chemistry, Moscow, July, 1965.
253. Brunauer, S., and L. E. Copeland: Physical Adsorption of Gases and Vapors on Solids, ASTM Materials Science Series, 4, Symposium on Properties of Surfaces, p. 59, American Society for Testing and Materials, Philadelphia, 1963.
254. Buff, F. P.: Statistical Mechanical Verification of the Gibbs Adsorption Equation, *Advan. Chem. Ser.*, **33**: 340 (1961).
255. Cannon, P.: Some Surface Properties of an Artificial Halozeolite; Examination of the Contributions of Dispersion and Electrostatic Terms to the Heat of Adsorption of Aluminosilicates, *Advan. Chem. Ser.*, **33**: 122 (1961).
256. Clampitt, B. H., and D. E. German: Heats of Vaporization of Molecules at Liquid-Vapor Interfaces, *J. Phys. Chem.*, **62**: 438 (1958).
257. Cook, M. A.: Surface Condensation Forces, *Advan. Chem. Ser.*, **33**: 220 (1961).
258. Copeland, L. E., and T. F. Young: Thermodynamic Theory of Adsorption, *Advan. Chem. Ser.*, **33**: 348 American Chemical Society, Washington, D.C. (1961).
259. Cranston, R. W., and F. A. Inkley: Determination of Pore Structures from Nitrogen Adsorption Isotherms, "Advances in Catalysis," vol. 9, 143, Academic, New York, 1957.
260. De Boer, J. H.: The Shapes of Capillaries, in Everett and Stone (eds.), "The Structure and Properties of Porous Materials," p. 68, Academic, New York, 1958.
261. Deeds, C. T., and H. van Olphen: Density Studies in Clay-Liquid Systems: I. The Density of Water Adsorbed by Expanding Clays, *Advan. Chem. Ser.*, **33**: 332 (1961).
262. Deitz, V. R., and F. G. Carpenter: The Rate of Physical Adsorption at Low Surface Coverage, *Advan. Chem. Ser.*, **33**: 146 (1961).
263. Dowden, D. A., and D. Wells: A Crystal Field Interpretation of Some Activity Patterns, "Proceedings of the Second International Congress of Catalysis," vol. 2, p. 1479, Technip, Paris, 1960.
264. Dubinin, M. M.: The Potential Theory of Adsorption of Gases and Vapors for Adsorbents with Energetically Non-uniform Surfaces, *Chem. Rev.*, **60**: 235 (1960).
265. Dunning, W. J.: Structure of Surfaces, "Physics and Chemistry of the Organic Solid State," vol. 1, p. 411, Interscience, New York, 1963.

266. Ehrlich, G.: Molecular Dissociation and Reconstitution on Solids, *J. Chem. Phys.*, **31**: 1111 (1959).
267. Eischens, R. P., and W. A. Pliskin: The Infrared Spectra of Adsorbed Molecules, "Advances in Catalysis," vol. 10, p. 1. Academic, New York, 1958.
268. Emmett, P. H.: Chemisorption and Catalysis, ASTM Materials Science Series, 4, Symposium on Properties of Surfaces, p. 42, American Society for Testing and Materials, Philadelphia, 1963.
269. Farnsworth, H. E., and H. H. Madden: The Mechanism of Oxygen Chemisorption on Nickel, *Advan. Chem. Ser.*, **33**: 114 (1961).
270. Flood, E. A.: Adsorption Potentials, Adsorbent Self-potentials, and Thermodynamic Equilibria, *Advan. Chem. Ser.*, **33**: 248 (1961).
271. Fowkes, F. M.: Dispersion Force Contributions to Surface and Interfacial Tensions, Contact Angles, and Heats of Immersion, *Advan. Chem. Ser.*, **43**: 99 (1964).
272. Gaines, G. L., Jr.: Adsorption of Gases on Ion Exchanged Mica, *Advan. Chem. Ser.*, **33**: 264 (1961).
273. Galwey, A. K., and C. Kemball: Dissociation Adsorption of Hydrocarbons on a Supported Nickel Adsorbent, *Trans. Faraday Soc.*, **55**: 1959 (1959).
274. Germer, L. H., A. U. MacRae, and C. D. Hartman: (110) Nickel Surface, *J. Appl. Phys.*, **32**: 2432 (1961).
275. Goates, J. R., and C. V. Hatch: Standard Adsorption Potentials of Water Vapor on Soil Colloids, *Soil Sci.*, **75**: 275 (1953).
276. Gomer, R.: Field Emission Microscopy and Some Applications to Catalysis and Chemisorption, "Advances in Catalysis," vol. 7, p. 93, Academic, New York, 1957.
277. Good, R. J.: Theory for the Estimation of Surface and Interfacial Energies, *Advan. Chem. Ser.*, **43**: 74 (1964).
278. Graham, D.: Physical Adsorption on Low Energy Solids: (I) Adsorption of *e*, Ar, and N on Poly-(tetra-fluoro) ethylene, *J. Phys. Chem.*, **66**: 1815 (1962).
279. Hall, W. K., and P. H. Emmett: Studies of the Hydrogenation of Ethylene over Copper-Nickel Alloys, *J. Phys. Chem.*, **63**: 1102 (1959).
280. Halsey, G. D.: The Role of Surface Heterogeneity in Adsorption, "Advances in Catalysis," vol. 4, p. 259, Academic, New York, 1952.
281. Holmes, J. M., and R. A. Beebe: Adsorption Studies on Bone Mineral, *Advan. Chem. Ser.*, **33**: 291 (1961).
282. Honig, J. M.: Utilization of Order-Disorder Theory in Physical Adsorption, *Advan. Chem. Ser.*, **33**: 239 (1961).
283. Johnson, R. E., and R. H. Dettre: Contact Angle Hysteresis: I. Study of an Idealized Rough Surface, *Advan. Chem. Ser.*, **43**: 112 (1964).
284. Jura, G., and K. S. Pitzer: The Specific Heat of Small Particles at Low Temperatures, *J. Am. Chem. Soc.*, **74**: 6030 (1952).
285. Kantro, D. L., S. Brunauer, and C. H. Weise: Development of Surface in the Hydration of Calcium Silicates, *Advan. Chem. Ser.*, **33**: 199 (1961).
286. Kiselev, A. V.: Adsorbate-Adsorbate Interactions in the Adsorption of Vapor on Graphitized Carbon Black, *Kolloidn. Zh.*, **20**: 338, 444 (1958).
287. Kiselev, A. V., and D. P. Poshkus: Statistical Calculation of Thermodynamic Functions of an Adsorbed Substance on Graphite, *Trans. Faraday Soc.*, **59**: 176, 428 (1963).
288. Michaels, A. S.: Fundamentals of Surface Chemistry and Surface Physics, ASTM Materials Science Series, 4, Symposium on Properties of Surfaces, p. 3, American Society for Testing and Materials, Philadelphia, 1963.
289. Mikhail, R. Sh., L. E. Copeland, and S. Brunauer: Pore Structures and Surface Areas of Hardened Portland Cement Pastes by Nitrogen Adsorption, *Can. J. Chem.*, **42**: 426 (1964).
290. Morrison, J. A., and D. Patterson: The Heat Capacity of Small Particles of Sodium Chloride, *Trans. Faraday Soc.*, **52**: 764 (1956).
291. Müller, E. W.: Observation of Almost Perfect Metal Crystals and of Point Defects in the Field Ion Microscope, *Z. Physik.*, **156**: 399 (1959).
292. Nicolson, M. M.: Surface Tension in Ionic Crystals, *Proc. Roy. Soc. (London)*, **A228**: 490 (1955).
293. Parravano, G., and M. Boudart: Chemisorption and Catalysis on Oxide Semiconductors, "Advances in Catalysis," vol. 7, p. 47, Academic, New York, 1955.
294. Razouk, R. I., R. Sh. Mikhail, and B. S. Girgis: Adsorption of Cyclohexane and Methanol on Iron Oxide, *Advan. Chem. Ser.*, **33**: 42 (1961).
295. Reyerson, L. H., and A. Solbakken: Effect of Hydrogen Adsorption on the Magnetic Susceptibility of Palladium Dispersed on Silica Gel, *Advan. Chem. Ser.*, **33**: 86 (1961).
296. Roberts, B. F.: A Procedure for Estimating Pore Volume and Area Distributions from Sorption Isotherms, presented at the 145th National Meeting of the American Chemical Society, New York, September, 1963. To be published.
297. Schay, G., P. Fejes, I. Halasz, and J. Kiraly: Determination of Adsorption Isotherms by Frontal Gas Chromatography, *Acta Chim. Acad. Sci. Hung.*, **11**: 381 (1957).
298. Schay, G., P. Fejes, and J. Szathmary: Studies on the Adsorption of Gas Mixtures: I. Statistical Theory of Physical Adsorption of the Langmuir Type in Multicomponent Systems, *Acta Chim. Acad. Sci. Hung.*, **12**: 299 (1957).
299. Schwab, G. M., and J. Block: Oxidation of CO and Decomposition of N_2O on Defined Semiconducting Oxides, *Z. physik. Chem. (N.F.)*, **1**: 42 (1954).
300. Selwood, P. W.: The Mechanism of Chemisorption: Ethylene and Ethane on Nickel, *J. Am. Chem. Soc.*, **79**: 3346 (1957).
301. Steele, W. A.: General Theory of Monolayer Physical Adsorption, *Advan. Chem. Ser.*, **33**: 269 (1961).
302. Tchenrekdjian, N., A. C. Zettlemoyer, and J. J. Chessick: The Adsorption of Water Vapor onto Silver Iodide, *J. Phys. Chem.*, **68**: 773 (1964).
303. Ter-Minassian-Saraga, L.: Chemisorption and Dewetting of Glass and Silica, *Advan. Chem. Ser.*, **43**: 232 (1964).
304. Tompkins, F. C.: Adsorption Isotherms for Nonuniform Surfaces, *Trans. Faraday Soc.*, **46**: 580 (1950).
305. Volkenstein, T.: The Electron Theory of Catalysis on Semiconductors, "Advances in Catalysis," vol. 12, p. 189, Academic, New York, 1960.
306. Wade, W. H., and Norman Hackerman: Thermodynamics of Wetting of Solid Oxides, *Advan. Chem. Ser.*, **43**: 222 (1964).
307. Wanlass, F. M., and H. Eyring: Sticking Coefficients, *Advan. Chem. Ser.*, **33**: 140 (1961).
308. Weyl, W. A.: A New Approach to Surface Chemistry and to Heterogeneous Catalysis, *State Univ., Penn., Mineral Ind. Expt. Sta. Bull.* 57, 1951.
309. Weyl, W. A.: Effect of the Environment upon the Properties of Solids, *Advan. Chem. Ser.*, **33**: 72 (1961).
310. Wu, Y. C., and L. E. Copeland: Thermodynamics of Adsorption; Barium Sulfate–Water System, *Advan. Chem. Ser.*, **33**: 357 (1961).
311. Zettlemoyer, A. C., and J. J. Chessick: Wettability by Heats of Immersion, *Advan. Chem. Ser.*, **43**: 88 (1964).
312. Zisman, W. A.: Relation of Equilibrium Contact Angle to Liquid and Solid Constitution, *Advan. Chem. Ser.*, **43**: 1 (1964).

Chapter 8

Chemical Thermodynamics†

By FREDERICK D. ROSSINI, University of Notre Dame

1. Introduction

The fundamental thermodynamic properties are taken as the pressure P, the volume V, the temperature T, the energy E, and the entropy S, and the laws of thermodynamics are stated in the following manner:

When a given system participates in any process, the first law states

$$dE = \delta q + \delta w + \delta u \tag{8.1}$$

where dE is the differential increase in energy of the system, δq is the infinitesimal quantity of heat energy absorbed by the system, δw is the infinitesimal quantity of PV work energy absorbed by the system, and δu is all other energy absorbed by the system. The energy E is a property of the system the values of which depend entirely upon the states of the system and not upon the path followed during the process. The energies represented by δq, δw, and δu depend upon the path followed during the process and may have any values so long as their algebraic sum is equal to dE. If the given system is in pressure equilibrium with its confining boundaries during the process, then

$$\delta w = -P\,dV \tag{8.2}$$

where P is the pressure of the system and V its volume.

According to the second law, when a given system participates in any process, its increase in entropy may be measured in terms of the absolute temperature and the heat energy it would absorb if the process were carried on in a reversible manner, such that all pressures, forces, etc., operating in the process differ only by infinitesimal amounts. For a reversible process,

$$dS = \frac{\delta q}{T} \tag{8.3}$$

where dS is the differential increase in entropy of the given system, δq the infinitesimal quantity of heat energy absorbed by the system, and T its absolute temperature.

The third law may be stated formally as follows: The entropy of any substance of which all component parts are in complete internal equilibrium becomes zero at the absolute zero of temperature. By internal equilibrium in a given substance is meant that each of the atoms or molecules comprising the given substance has free access to all the permitted states in accordance with the governing distribution law.

† Based on "Chemical Thermodynamics" by Frederick D. Rossini [2].

2. Useful Energy; Free Energy; Criteria of Equilibrium

One of the most important relations in thermodynamics is obtained by applying the first and second laws of thermodynamics to a reversible process which involves the transfer, between the system and its surroundings, of energy in addition to heat energy and PV work energy. For such a general process,

$$dE = T\,dS - P\,dV + \delta u \tag{8.4}$$

where δu represents all energy other than the heat energy and the PV work energy absorbed by the system from the surroundings. The quantity δu is identified as useful energy absorbed by the system from the surroundings. Thus the useful energy made available by the system to the surroundings is

$$-\delta u = -(dE + P\,dV - T\,dS) \tag{8.5}$$

This gives, for any process performed reversibly, the useful energy obtainable from the system in terms of the five fundamental thermodynamic properties, P, V, T, E, and S.

For special cases, this assumes simpler forms:

At constant volume,

$$-\delta u = -(dE - T\,dS) \tag{8.6}$$

At constant pressure,

$$-\delta u = -[d(E + PV) - T\,dS] = -(dH - T\,dS) \tag{8.7}$$

At constant temperature,

$$-\delta u = -[d(E - TS) + P\,dV] = -(dA + P\,dV) \tag{8.8}$$

At constant entropy

$$-\delta u = -(dE + P\,dV) \tag{8.9}$$

At constant volume and entropy,

$$-\delta u = -dE \tag{8.10}$$

At constant pressure and entropy,

$$-\delta u = -d(E + PV) = -dH \tag{8.11}$$

At constant volume and temperature,

$$-\delta u = -d(E - TS) = -dA \tag{8.12}$$

At constant pressure and temperature,

$$-\delta u = -d(E + PV - TS) = -d(H - TS) = -dG \quad (8.13)$$

In the foregoing, abbreviating definitions have been used for the heat content or enthalpy H, for the Helmholtz free energy A, and for the Gibbs free energy G:

$$H = E + PV \quad (8.14)$$
$$A = E - TS \quad (8.15)$$
$$G = E + PV - TS = H - TS \quad (8.16)$$

Many processes occur at constant pressure and temperature, and the decrement in the particular combination of properties given in Eq. (8.13) represents the energy which is free to be put to some useful purpose under these conditions of constant temperature and pressure.

Every system if left to itself tends to change toward a final state of rest or equilibrium. If a given system has some capacity for spontaneous change, an appropriate mechanism can, in principle, always be introduced to harness the system to obtain some useful energy as the system passes reversibly from its initial state toward the state of equilibrium. For every process that occurs spontaneously, therefore, some useful energy may be obtained from the system. To make the given system move away from the state of equilibrium, that is, in a nonspontaneous or unnatural direction, it is necessary to supply some useful energy to the system. The algebraic sign of the useful energy obtainable from the system, therefore, serves to tell whether the given process is one in which the system is moving toward or away from equilibrium. If δu is negative, useful energy is obtainable from the system and the change is a naturally occurring one in the direction toward equilibrium. If δu is positive, useful energy is required to be supplied to bring about the desired change and the change is an unnatural one in the direction away from equilibrium. If δu is zero, the system is already at equilibrium with respect to the prescribed change.

The condition that δu is zero for a reversible process may be used as a general criterion of equilibrium. That is, equilibrium exists with respect to any prescribed change occurring reversibly whenever

$$\delta u = dE + P\,dV - T\,dS = 0 \quad (8.17)$$

The criterion of equilibrium reduces to the following for the indicated special cases:

Variables Held Constant	*Criterion of Equilibrium*	
Volume:	$dE - T\,dS = 0$	(8.18)
Pressure:	$dH - T\,dS = 0$	(8.19)
Temperature:	$d(E - TS) + P\,dV = 0$	(8.20)
Entropy:	$dE + P\,dV = 0$	(8.21)
Energy:	$P\,dV - T\,dS = 0$	(8.22)
Volume, entropy:	$dE = 0$	(8.23)
Pressure, entropy:	$dH = 0$	(8.24)
Energy, entropy:	$dV = 0$	(8.25)
Volume, temperature:	$d(E - TS) = 0$	(8.26)
Pressure, temperature:	$dG = 0$	(8.27)

3. Equilibrium Constant and Change in Free Energy for Reactions of Ideal Gases

When a given system participates in a reversible process in which only heat energy and PV work energy occur, at constant temperature,

$$dG = V\,dP \quad (8.28)$$

For an ideal gas this becomes

$$dG = \frac{RT}{P}\,dP = RT\,d\ln P \quad (8.29)$$

giving the variation of molal free energy of an ideal gas with pressure at constant temperature. On integrating between states, A and B, of different pressure, at constant temperature,

$$G_B - G_A = RT \ln \frac{P_B}{P_A} \quad (8.30)$$

The entropy of mixing different ideal gases, for the case of bringing together gases initially at identical pressures to form a mixture at the same pressure, calculated *per mole* of the mixture, is

$$\Delta S\ (\text{mixing}) = -R\Sigma N_i \ln N_i \quad (8.31)$$

where N_i is the *mole fraction* of component i in the mixture and the summation is carried over all the components. If number of moles in the entire mixture is n, the entropy of mixing for the entire mixture is the value given by Eq. (8.31) multiplied by n.

From the definition of the ideal gas, for constant temperature,

$$\Delta E\ (\text{mixing}) = \Delta H\ (\text{mixing}) = 0 \quad (8.32)$$

From the definition of free energy for a given substance, for a process or reaction at constant temperature,

$$(G_B - G_A) = (H_B - H_A) - T(S_B - S_A) \quad (8.33)$$

or

$$\Delta G = \Delta H - T\,\Delta S \quad (8.34)$$

Applying this to the process of mixing ideal gases, and substituting from Eqs. (8.31) and (8.32) at constant temperature,

$$\Delta G\ (\text{mixing}) = RT\Sigma N_i \ln N_i \quad (8.35)$$

This gives the change in free energy, *per mole* of mixture, on mixing different ideal gases brought together in different amounts from initially identical pressures to form a mixture at the same pressure. For the entire mixture, the free energy is given by Eq. (8.35) multiplied by n, the number of moles in the entire mixture.

The change in free energy on mixing ideal gases originally at different pressures is calculated by first taking account of the change in free energy on bringing each gas separately to the common pressure before mixing. The change in free energy with pressure at constant temperature for the mixture of ideal gases is given by Eq. (8.30), assuming the mixture to remain fixed in composition.

By proceeding in this way, it may be shown that,

as in the case of entropy, the free energy of an ideal gas at constant temperature is fixed by the volume it occupies, and is independent of the pressure of other gases. Thus, if two ideal gases are mixed, such that the volume of the mixture is the same as the volume of each gas before mixing, then the change in free energy (as also the change in entropy) is zero.

It is convenient to define the "proper quotient of pressures," as well as the equilibrium constant, for reactions of ideal gases. Consider the following reaction in which each reactant and product is an ideal gas:

$$bB(g) + cC(g) = mM(g) + nN(g) \qquad (8.36)$$

This states that b moles of the gas B react with c moles of the gas C to form m moles of the gas M and n moles of the gas N. For any such reaction, the "proper quotient of pressures" is defined as

$$Q_P = \frac{(P_M)^m(P_N)^n}{(P_B)^b(P_C)^c} \qquad (8.37)$$

The numerator is the product of the pressure of each of the product gases with each pressure raised to a power equal to the stoichiometrical number of moles of the given gas, and the denominator is the product of the pressure of each of the reactant gases, with each pressure raised to a power equal to the stoichiometrical number of moles of the given gas.

The "proper quotient of pressures," Q_P, may be written for any reaction of gases in any states. When the pressure of each reactant and each product gas is that which it has at thermodynamic equilibrium for the given reaction, the "proper quotient of pressures" is then specified for the equilibrium states, as Q_P (equilibrium) or Q_{P^e}. When each of the gases involved is in the ideal state, we may define Q_{P^e} as the equilibrium constant K. For the reaction of ideal gases given by Eq. (8.36), the equilibrium constant is

$$K = Q_{P^e} = \frac{(P_{M^e})^m(P_{N^e})^n}{(P_{B^e})^b(P_{C^e})^c} \qquad (8.38)$$

Consider the reaction given by Eq. (8.36) as occurring at some constant temperature T. The change in free energy for the given reaction, for any specified set of conditions, is

$$\Delta G = mG_M + nG_N - bG_B - cG_C \qquad (8.39)$$

where G_M, G_N, G_B, and G_C are the free energies, per mole, for the gases M, N, B, and C, respectively, and m, n, b, and c, respectively, are the number of moles of the given gases participating in the reaction. If this reaction takes place under two different sets of conditions, one in which the participating gases are in certain states s, and the other in which the participating gases are in certain states e, the difference in the change in free energy for the reaction under the two sets of conditions may be calculated as follows. For the states s,

$$\Delta G^s = m(G_{M^s}) + n(G_{N^s}) - b(G_{B^s}) - c(G_{C^s}) \qquad (8.40)$$

And for the states e,

$$\Delta G^e = m(G_{M^e}) + n(G_{N^e}) - b(G_{B^e}) - c(G_{C^e}) \qquad (8.41)$$

Subtraction gives

$$\Delta G^s - \Delta G^e = m(G_{M^s} - G_{M^e}) + n(G_{N^s} - G_{N^e}) - b(G_{B^s} - G_{B^e}) - c(G_{C^s} - G_{C^e}) \qquad (8.42)$$

But, from Eq. (8.30) for the gas M,

$$G_{M^s} - G_{M^e} = RT \ln \frac{P_{M^s}}{P_{N^e}} \qquad (8.43)$$

and

$$m(G_{M^s} - G_{M^e}) = RT \ln \left(\frac{P_{M^s}}{P_{M^e}}\right)^m \qquad (8.44)$$

Similar relations hold for the other gases. Therefore Eq. (8.42) gives

$$\Delta G^s - \Delta G^e = RT \ln \left(\frac{P_{M^s}}{P_{M^e}}\right)^m + RT \ln \left(\frac{P_{N^s}}{P_{N^e}}\right)^n - RT \ln \left(\frac{P_{B^s}}{P_{B^e}}\right)^b - RT \ln \left(\frac{P_{C^s}}{P_{C^e}}\right)^c \qquad (8.45)$$

which may be rearranged to bring together the factors involving the same states rather than the same molecules, to give

$$\Delta G^s - \Delta G^e = RT \ln \frac{(P_{M^s})^m(P_{N^s})^n}{(p_{B^s})^b(P_{C^s})^c} - RT \ln \frac{(p_{M^e})^m(P_{N^e})^n}{(p_{B^e})^b(P_{C^e})^c} \qquad (8.46)$$

or

$$\Delta G^s - \Delta G^e = RT \ln Q_{P^s} - RT \ln Q_{P^e} \qquad (8.47)$$

Identifying the states s as the reference, or standard, states in which each of the ideal gases has a pressure of unity, so that

$$(P_{M^s}) = (P_{N^s}) = (P_{B^s}) = (P_{C^s}) = 1 \qquad (8.48)$$

Then

$$Q_{P^s} = 1 \quad \text{and} \quad \ln Q_{P^s} = 0 \qquad (8.49)$$

The reference or standard state is usually indicated by the superscript °; so the change in free energy for the s states becomes the standard change in free energy:

$$\Delta G^s = \Delta G^\circ \qquad (8.50)$$

Identifying the e states as those of thermodynamic equilibrium for the given reaction at the given temperature, the proper quotient of pressures is then the equilibrium constant,

$$Q_{P^e} = K \qquad (8.51)$$

At constant temperature and constant pressure, the change in free energy at thermodynamic equilibrium is equal to zero; thus

$$\Delta G^e = 0 \qquad (8.52)$$

and therefore

$$\Delta G^\circ = -RT \ln K \qquad (8.53)$$

Of the two pairs of corresponding terms in Eq. (8.47), one term applying to the standard states and one term applying to the equilibrium states have been eliminated, so that in Eq. (8.53) there remain two noncorresponding terms, one relating to the standard states, ΔG°, and one, $RT \ln K$, to the equilibrium states.

Equation (8.53), derived for any reaction of ideal gases at a given temperature, is one of the most powerful relations in the practical application of chemical thermodynamics. With it, the equilibrium constant may be evaluated from a knowledge of the standard change in free energy, and vice versa. The powerful simplicity of Eq. (8.53) can also be retained for use in reactions of any substances in any states, solid, liquid, or gas.

From the definition of free energy,

$$\Delta G = \Delta H - \Delta(TS) \tag{8.54}$$

At constant temperature,

$$\Delta G = \Delta H - T\,\Delta S \tag{8.55}$$

If each of the reactants and products of the given reaction are in their respective standard reference states,

$$\Delta G^\circ = \Delta H^\circ - T\,\Delta S^\circ \tag{8.56}$$

where each of the thermodynamic properties applies to each reactant and product in its standard reference state. Combination of Eqs. (8.53) and (8.56) gives

$$\Delta H^\circ - T\,\Delta S^\circ = -RT \ln K \tag{8.57}$$

or

$$\ln K = \frac{\Delta S^\circ}{R} - \frac{\Delta H^\circ}{RT} \tag{8.58}$$

From Eq. (8.58), one may calculate the equilibrium constant for a given reaction at a given temperature, if the value of the standard heat of the reaction, ΔH°, and the value of the standard change in entropy for the reaction, ΔS°, are known. The value of ΔH° may normally be determined from appropriate thermochemical measurements. The value of ΔS° is obtained as the sum of the entropies of the products less the sum of the entropies of the reactants, each in its standard reference state:

$$\Delta S^\circ = \Sigma S^\circ \text{ (products)} - \Sigma S^\circ \text{ (reactants)} \tag{8.59}$$

The entropy of a given substance may be determined from statistical calculations utilizing spectroscopic and other molecular data, or from application of the third law to calorimetric measurements of heats of vaporization, fusion, and transition, and heat capacities, down to low temperatures.

When two free neutral gaseous hydrogen atoms combine to form a gaseous hydrogen molecule,

$$H(g) + H(g) = H_2(g) \tag{8.60}$$

there is a decrease in the energy of the system of more than 100 kcal/mole, since the hydrogen molecule, with one H—H bond, has a lower energy than two separated hydrogen atoms not bonded together. Whenever the atoms in a given system rearrange themselves so that they are more securely bound, the energy of the system is lowered. In general, the change in energy of a reaction, ΔE [and for a simple approximation this is not significantly different from the value of ΔH since the two differ only by the term $\Delta(PV)$], is a measure of the security of binding of the atoms in the molecules; the lower the energy, the greater the security of binding. The decrease in energy (or approximately the heat content) of a system is thus a measure of the security of existence possessed by the given system.

The entropy of a system is a measure of the total number of quantum states of existence available to it. If a given assembly of atoms has a greater number of states of existence in one molecular configuration than in a second, its entropy is greater in the first configuration than in the second. Usually, the entropy of a given number of atoms will be greater, the greater the number of separate molecules formed by the given atoms. For example, in the combination of two hydrogen atoms to form the hydrogen molecule, the two atoms enjoy a greater freedom of existence as separate atoms than they do combined in the molecule. At room temperature, the entropy of the hydrogen molecule is less than the entropy of two separate hydrogen atoms by about 24 cal/deg mole. In general, the change in entropy accompanying a given reaction is a measure of the increase or decrease in the total number of states of existence available to the system in the form of the products over the number of states of existence available in the form of the reactants. The total number of states of existence available to a given system or assembly of atoms is a measure of the freedom enjoyed by the given system.

Equation (8.58) may be written explicitly in terms of the equilibrium constant

$$K = \left[\exp\left(\frac{\Delta S^\circ}{R}\right)\right]\left[\exp\left(-\frac{\Delta H^\circ}{RT}\right)\right] \tag{8.61}$$

Given any reaction, as for example,

$$CO(g) + H_2O(g) = CO_2(g) + H_2(g) \tag{8.62}$$

when the value of the equilibrium constant is large, greater than unity, the component atoms, at equilibrium, tend to go preferentially into forming the products, and when the equilibrium constant is small, less than unity, the component atoms tend to go preferentially into forming the reactants. The equilibrium constant will be greatest for the reaction written with products having a positive value of the change in entropy, ΔS°, and a negative value of the change in heat content, ΔH°. Therefore, the atoms constituting the molecules will tend to go into those molecular configurations in which the entropy is greatest and in which the heat content (approximately the energy) is lowest (algebraically). The greatest entropy is in general associated with molecular configurations having the largest number of states of existence. On the other hand, the lowest heat content (approximately the energy) is in general associated with those molecular configurations in which the atoms are bound most securely one to another. The final state of equilibrium is a compromise between these two opposing tendencies, toward maximum freedom on the one hand and maximum security on the other.

Equation (8.61) shows also that, in general, the equilibrium constant at high temperatures is determined largely by the value of ΔS°, the "freedom" factor, whereas at low temperatures the equilibrium constant is determined largely by the value of ΔH°, the "security" factor. At thermodynamic equilibrium in a given process, the atoms involved prefer a secure molecular configuration at low temperatures and a loose or free molecular configuration at high temperatures.

The foregoing statements regarding behavior of atoms in chemical processes or reactions apply only to those states between which the path is sufficiently open to permit the establishment of thermodynamic equilibrium.

4. Fugacity; Standard States

The molal free energy is a measure of the *escaping tendency* of a given molecular species from a given phase. In the equilibrium involving one pure substance in several different phases, the given substance has a tendency to leave each phase in which it exists and to pass into every other phase which is open to it. At equilibrium, the escaping tendency has a constant value throughout every phase of the system, and similarly for the molal free energy. When the phases under consideration are condensed ones, as liquid and solid, the molal free energy serves as a satisfactory quantitative measure of the escaping tendency. However, when the gaseous phase is involved, the molal free energy is a rather inconvenient measure of the escaping tendency because the value of the free energy of an ideal gas approaches minus infinity as the pressure approaches zero. Since any real gas approaches the ideal gas as the pressure is indefinitely reduced, the free energy of any real gas approaches minus infinity as the pressure approaches zero. A more convenient measure of the escaping tendency is therefore needed in the case of gases.

For the ideal gas the pressure itself is a satisfactory measure of escaping tendency, because for the ideal gas, at constant temperature, the pressure is simply related to the molal free energy,

$$G_P - G_{(P=1)} = RT \ln P \tag{8.63}$$

Real gases depart significantly from Eq. (8.63). Nevertheless, it is possible to retain the form of this equation in the case of real gases by using the thermodynamic function invented by G. N. Lewis, called the *fugacity* and labeled f. The fugacity f is evaluated in such a way that the substitution of values of the fugacity for pressure will make Eq. (8.63) valid for real gases as well as for the ideal gas.

For any gas at constant temperature

$$dG = V\,dP \tag{8.64}$$

For any gas in its real state,

$$dG_{\text{real}} = V_{\text{real}}\,dP \tag{8.65}$$

For the same gas in its hypothetical ideal condition,

$$dG_{\text{ideal}} = V_{\text{ideal}}\,dP \tag{8.66}$$

so

$$d(G_{\text{real}} - G_{\text{ideal}}) = (V_{\text{real}} - V_{\text{ideal}})\,dP \tag{8.67}$$

The experimentally observable quantity representing, at some given pressure, the difference in the molal volumes of the gas in the ideal and the real conditions will be written

$$\alpha = V_{\text{ideal}} - V_{\text{real}} = \frac{RT}{P} - V_{\text{real}} \tag{8.68}$$

In terms of α,

$$\int_0^P d(G_{\text{real}} - G_{\text{ideal}}) = -\int_0^P \alpha\,dP \tag{8.69}$$

so that

$$(G_{\text{real}} - G_{\text{ideal}})^P - (G_{\text{real}} - G_{\text{ideal}})^{P=0} = -\int_0^P \alpha\,dP \tag{8.70}$$

The difference in free energy between real and ideal conditions approaches zero at zero pressure, so that

$$(G_{\text{real}} - G_{\text{ideal}})^{P=0} = 0 \tag{8.71}$$

giving

$$G_{\text{real}} - G_{\text{ideal}} = -\int_0^P \alpha\,dP \tag{8.72}$$

or

$$G_{\text{real}} = G_{\text{ideal}} - \int_0^P \alpha\,dP \tag{8.73}$$

The relation between free energy and pressure for an ideal gas at constant temperature and the pressure P is

$$G_{\text{ideal}} = G_{\text{ideal}}{}^{P=1} + RT \ln P \tag{8.74}$$

Whereas for a real gas

$$G_{\text{real}} = G_{\text{ideal}}{}^{P=1} + RT \ln P - \int_0^P \alpha\,dP \tag{8.75}$$

The fugacity, f, is defined to make

$$G_{\text{real}} = G_{\text{ideal}}{}^{P=1} + RT \ln f \tag{8.76}$$

so the fugacity must be defined by the relation

$$RT \ln f = RT \ln P - \int_0^P \alpha\,dP \tag{8.77}$$

giving

$$\ln \frac{f}{P} = -\frac{1}{RT}\int_0^P \alpha\,dP \tag{8.78}$$

or

$$\frac{f}{P} = \exp\left[-\left(\frac{1}{RT}\right)\int_0^P \alpha\,dP\right] \tag{8.79}$$

For any gas at constant temperature the fugacity may be evaluated as a function of pressure from values of the molal volume of the actual gas at various pressures. If $\alpha = [(RT/P) - V]$ is plotted on the scale of ordinates for various pressures, the area under the curve from zero pressure to the pressure P, multiplied by the negative of the reciprocal of RT, gives the natural logarithm of the ratio of the fugacity to the pressure P at the given pressure.

The fugacity of the ideal gas is equal to its pressure. The fugacity is a corrected pressure that permits retention of the simplicity of the relation with free energy in Eq. (8.76). This is important in the relations between the equilibrium constant and the free energy for reactions involving real gases.

It is convenient to use a reference state of unit pressure from which to measure changes in free energy. In practical application of chemical thermodynamics, such reference states are used quite frequently. The thermodynamic standard reference state that is used for gases is taken to the *ideal* gaseous state at unit pressure (in atmospheres unless otherwise specified) at each temperature. This standard state is designated by the usual superscript on the appropriate thermodynamic symbols, as $G°$ or $f°$.

Any real gas may be taken from its real state at the

pressure P to the ideal state at unit pressure by proceeding along the path of the real gas from the pressure P to zero pressure and then passing from zero pressure to the pressure P along the path of the ideal gas. In Fig. 8.1 is plotted schematically the ratio

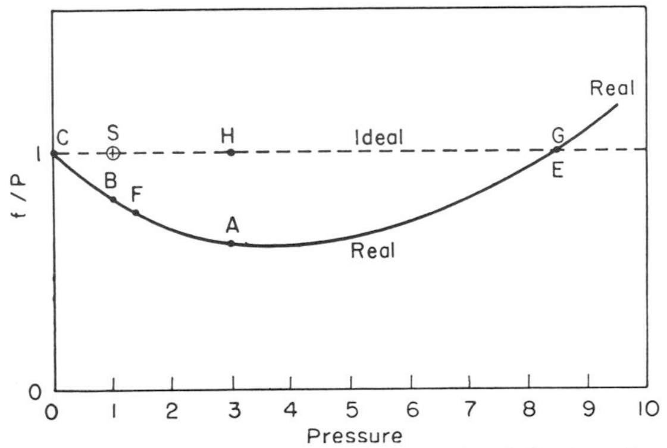

FIG. 8.1. Schematic diagram of the ratio of the fugacity to pressure for a gas as a function of pressure.

of fugacity to pressure, f/p, as a function of pressure at constant temperature. The process consists in taking the real gas from the state A along the path of the real gas through B to zero pressure at C and then along the path of the ideal gas to S, the ideal state at unit pressure, which is the thermodynamic standard reference state. Point B represents the real gas at unit pressure, point E represents the real gas at some particular higher pressure at which its fugacity is equal to the pressure, and point F represents the real gas at a pressure where its fugacity is unity. At constant temperature, the energy and heat content of a gas in the hypothetical ideal condition are constant. As shown in Fig. 8.2, for a

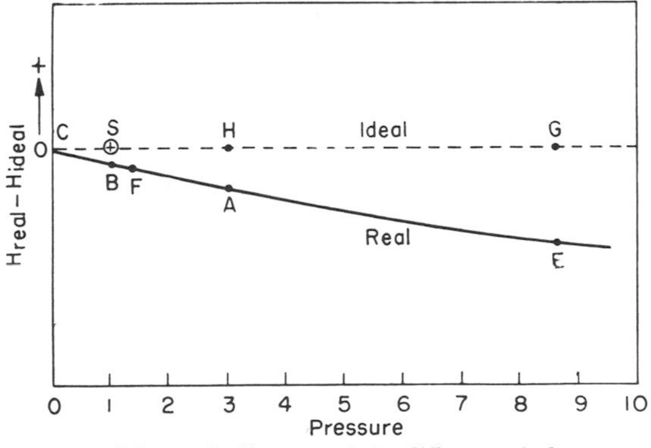

FIG. 8.2. Schematic diagram of the difference in heat content of a gas between the real and ideal states as a function of pressure.

gas in the ideal condition, the heat content at zero pressure, point C, is the same as the heat content at the standard state of unit pressure, point S, and as the heat content at any other pressure in the ideal condition, point G.

At constant temperature, the energy and heat content of a gas in its real condition vary with the pressure. The heat content of the real gas at a given pressure, as at point A (Fig. 8.2), relative to its heat content at zero pressure, measures the difference in heat content of the gas at the given pressure, point A, relative to its heat content in the standard state, point S, or relative to its heat content in the ideal condition at the same pressure, point H.

Although at the pressure given by point E the fugacity and pressure of the given gas have the same value, the real gas at the pressure P_E is not in the same thermodynamic state as for the ideal gas at the pressure P_E because the heat content of the real gas at the pressure P_E has a value different from that for the ideal gas at the same pressure. Although at the pressure P_E the fugacity is the same in the two states, real and ideal, and the free energy is the same in the two states, the heat content in the two states is different. Figure 8.2 gives a schematic diagram of the relation between the heat content of the gas in the real condition less that at zero pressure, as a function of the pressure.

With reference to the thermodynamic standard state, for any real gas at a given pressure at constant temperature,

$$G - G^\circ = RT \ln \frac{f}{f^\circ} \tag{8.80}$$

or, since for the standard state f° is unity,

$$G - G^\circ = RT \ln f \tag{8.81}$$

From Chap. 1, we have

$$\left(\frac{\partial S}{\partial P}\right)_T = -\left(\frac{\partial V}{\partial T}\right)_P \tag{8.82}$$

or at constant temperature,

$$dS = -\left(\frac{\partial V}{\partial T}\right)_P dP \tag{8.83}$$

Then

$$dS_{\text{real}} = -\left[\frac{R}{P} - \left(\frac{\partial \alpha}{\partial T}\right)_P\right] dP \tag{8.84}$$

and

$$dS_{\text{ideal}} = -\frac{R}{P}\, dP \tag{8.85}$$

From these relations

$$(S_{\text{real}} - S_{\text{ideal}})^P - (S_{\text{real}} - S_{\text{ideal}})^{P=0} = \int_0^P \frac{\partial \alpha}{\partial T}\, dP \tag{8.86}$$

Since the difference in entropy between the real and ideal states approaches zero at zero pressure,

$$(S_{\text{real}} - S_{\text{ideal}})^P = \int_0^P \frac{\partial \alpha}{\partial T}\, dP \tag{8.87}$$

From the definition of the free energy, for constant temperature,

$$(G_{\text{real}} - G_{\text{ideal}})^P = (H_{\text{real}} - H_{\text{ideal}})^P - T(S_{\text{real}} - S_{\text{ideal}})^P \tag{8.88}$$

therefore

$$(H_{\text{real}} - H_{\text{ideal}}) = -\int_0^P \alpha\, dP + T \int_0^T \frac{\partial \alpha}{\partial T}\, dP \tag{8.89}$$

The foregoing equations permit evaluation of the difference in the properties of free energy, entropy, and heat content of a gas between the real and ideal states at any pressure P at a given temperature.

If the volume of a gas can be expressed explicitly as a function of pressure and temperature, it becomes a simple matter to evaluate the difference in the properties of a gas between the real and ideal states. For example, from the Berthelot equation of state,

$$V = \frac{RT}{P} + \frac{9}{128}\frac{RT_c}{P_c}\left(1 - \frac{6T_c^2}{T^2}\right) \quad (8.90)$$

where T_c and P_c represent the critical temperature and critical pressure.

$$\alpha = \frac{RT}{P} - V = -\frac{9}{128}\frac{RT_c}{P_c}\left(1 - \frac{6T_c^2}{T^2}\right) \quad (8.91)$$

This value of α gives

$$G_{\text{real}} - G_{\text{ideal}} = \frac{9}{128}\frac{RT_c}{P_c}\left(1 - \frac{6T_c^2}{T^2}\right)P \quad (8.92)$$

$$S_{\text{real}} - S_{\text{ideal}} = \frac{27}{32}\frac{R}{P_c}\left(\frac{T_c}{T}\right)^3 P \quad (8.93)$$

and $$H_{\text{real}} - H_{\text{ideal}} = \frac{9}{128}\frac{RT_c}{P_c}\left(1 - \frac{18T_c^2}{T^2}\right)P \quad (8.94)$$

These are valid only over the range of applicability of the equation of state and within its limits of accuracy.

It is sometimes desired to evaluate the fugacity of a substance in a condensed phase. This may be done by evaluating the fugacity in the gaseous phase that is in equilibrium with the liquid or solid phase. Given, under certain conditions, the equilibrium,

$$MX(l) = MX(g) \quad (8.95)$$

then, for the given equilibrium

$$f(l) = f(g) \quad (8.96)$$

Similarly, given the equilibrium

$$MX(c) = MX(g) \quad (8.97)$$

then $$f(c) = f(g) \quad (8.98)$$

The vapor pressure of liquids and solids under ordinary conditions is usually low, under which circumstances the fugacity of the gaseous phase will not differ greatly from its pressure, so the fugacity of ordinary pure liquid and solid substances may be taken as approximately equal to the vapor pressure.

For pure liquids and solids, it is also convenient to select a standard state to use for reference. Because the fugacity of a substance in the liquid and solid phase is usually very low, it is desirable to use a different reference state than that of the gaseous phase. A satisfactory thermodynamic standard reference state for pure liquids and solids at a given temperature is the real state of the liquid or solid substance at a pressure of 1 atm. As in the case of gases,

$$G - G^\circ = RT \ln \frac{f}{f^\circ} \quad (8.99)$$

Whereas for the gaseous phase the fugacity of the standard state, f°, is unity, for the liquid and solid phases the fugacity of the standard state differs from unity, except by coincidence, and will normally be less than unity.

From the definition of fugacity

$$\ln f - \ln f^\circ = \frac{1}{RT}(G - G^\circ) \quad (8.100)$$

Differentiation with pressure at constant temperature gives

$$\left(\frac{\partial \ln f}{\partial P}\right)_T = \frac{1}{RT}\left(\frac{\partial G}{\partial P}\right)_T \quad (8.101)$$

since f° and G° are constant at constant temperature. But

$$\left(\frac{\partial G}{\partial P}\right)_T = V \quad (8.102)$$

and, therefore,

$$\left(\frac{\partial \ln f}{\partial P}\right)_T = \frac{V}{RT} \quad (8.103)$$

Thus

$$d \ln f = \frac{V}{RT}\,dP \quad (8.104)$$

or, between two specified pressures, P_A and P_B, at the same temperature, as

$$\ln \frac{f_B}{f_A} = \int_{P_A}^{P_B} \frac{V}{RT}\,dP \quad (8.105)$$

From the definition of fugacity

$$\ln f - \ln f^\circ = \frac{G/T - G^\circ/T}{R} \quad (8.106)$$

Differentiation with temperature at constant pressure gives

$$\left(\frac{\partial \ln f}{\partial T}\right)_P - \left(\frac{\partial \ln f^\circ}{\partial T}\right)_P = \frac{[\partial(G/T)/\partial T]_P - [\partial(G^\circ/T)/\partial T]_P}{R} \quad (8.107)$$

But, at constant pressure,

$$\frac{d(G/T)}{dT} = -\frac{H}{T^2} \quad (8.108)$$

and, therefore,

$$\left(\frac{\partial \ln f}{\partial T}\right)_P - \left(\frac{\partial \ln f^\circ}{\partial T}\right)_P = -\frac{(H/T^2 - H^\circ/T^2)}{R} = -\frac{H - H^\circ}{RT^2} \quad (8.109)$$

For gases, f° is unity at all temperatures,

$$\left(\frac{\partial \ln f^\circ}{\partial T}\right)_P = 0 \quad (8.110)$$

Therefore, for gases,

$$\left(\frac{\partial \ln f}{\partial T}\right)_P = -\frac{H - H^\circ}{RT^2} \quad (8.111)$$

For constant pressure,

$$d \ln f = -\left(\frac{H - H^\circ}{RT^2}\right) dT \qquad (8.112)$$

or, between two specified temperatures, T_A and T_B, at constant pressure, as

$$\ln \frac{f_A}{f_B} = -\frac{1}{R}\int_{T_A}^{T_B}\left(\frac{H - H^\circ}{T^2}\right) dT = \frac{1}{R}\int_{T_A}^{T_B}(H - H^\circ) d\left(\frac{1}{T}\right) \qquad (8.113)$$

From the manner of definition of fugacity, at any given temperature, a knowledge of α over the range from zero pressure to the pressure P provides a completed evaluation of the fugacity. Since the operation is one at constant temperature,

$$\ln \frac{f}{P} = -\int_0^P \left(\frac{\alpha}{RT}\right) dP \qquad (8.114)$$

But

$$\frac{\alpha}{RT} = \frac{1}{P} - \frac{V}{RT} = \frac{1}{P}\left(1 - \frac{PV}{RT}\right) = \frac{1 - z}{P} \qquad (8.115)$$

where z is the compressibility factor PV/RT. If values of the compressibility factor are known as a function of the pressure, at constant temperature, then

$$\ln \frac{f}{P} = -\int_0^P \left(\frac{1 - z}{P}\right) dP = -\int_0^P (1 - z) d \ln P \qquad (8.116)$$

Whenever PVT data on a given gas are not available or not adequate for evaluating the fugacity, a reasonably good approximation can be made by reason of the fact that, for the same value of the reduced pressure P/P_c and the reduced temperature T/T_c, the ratio of fugacity to pressure, f/P, has about the same value for nearly all gases.† Accordingly, if available data on different gases are all placed on one plot of f/P against P/P_c, generalized curves are formed, one for each value of the reduced temperature, from which values of f/P can be read for given values of P/P_c at given values of the reduced temperature T/T_c, for any gas for which data do not exist.

5. Solutions: Apparent and Partial Molal Properties

For systems consisting of one component, that is, pure substances, the thermodynamic state of the system under ordinary conditions can be adequately specified in terms of two variables, such as the pressure and temperature. With a system of two or more components, comprising a solution, the thermodynamic state of the system is not adequately specified until the composition is given. For solutions, the thermodynamic state is, under ordinary conditions, adequately specified in terms of three or more variables, one or more of which is the composition. The composition of a solution may be specified by giving the number of moles, n, of each of the components, 1, 2, 3, etc., as n_1, n_2, n_3, etc.,† using the subscript 1 to denote the component present in largest amount, subscript 2 for the component present in next largest amount, etc.

Instead of specifying the actual number of moles of each component, it is usually more convenient to specify the composition as an intensive property by means of the mole fraction,† N, which, for any given component i, is the number of moles of component i, N_i, divided by the total number of moles of all components,

$$N_1 = \frac{n_1}{\Sigma n_i} \qquad N_2 = \frac{n_2}{\Sigma n_i} \qquad N_3 = \frac{n_3}{\Sigma n_i} \qquad \text{etc.} \qquad (8.117)$$

The sum of the mole fractions for all components equals unity.

Specification of the composition of a solution by means of mole fractions is simple and exact. In the case of aqueous solutions, there has grown up a system of expressing the composition in terms of the *concentration*, defined as the number of moles of solute per liter of the solution, or the *molality*, which is defined as the number of moles of solute per 1,000 grams of the solvent, water. Expression of composition in terms of the number of moles per liter of solution is disadvantageous because, for a given solution, the volume changes with temperature, and, as the temperature changes, the concentration changes, without addition or removal of any components. Further, the specification of composition is not complete unless there is given, along with the concentration, the density of the solution at the given temperature. For this reason, the composition of aqueous solutions is usually more advantageously expressed in terms of the molality, or number of moles of solute per 1,000 grams of water, rather than in concentration. The molality is, for any given solution, independent of the temperature, and provides a complete specification of the composition. For any solute i the molality‡ is given by

$$m_i = \frac{1{,}000}{M_i}\,\frac{\text{mass of component } i}{\text{mass of water}} \qquad (8.118)$$

where M_i represents the molecular weight of the component i.

For a system of two or more components, it is important to know what part of the total value of a given thermodynamic property of the solution is due to each component. It is convenient to define (G. N. Lewis) the partial molal quantity, for a component i, as

$$B_i = \left(\frac{\partial B}{\partial n_i}\right)_{P,T,n_1,n_2,n_3,\ldots,n_j} \qquad (8.119)$$

† The ratio of the given absolute temperature to the critical temperature is called the *reduced* temperature, T/T_c. Similarly, the ratio of the pressure to the critical pressure is called the reduced pressure, P/P_c.

† Lower case n is used as the symbol for the number of moles.

‡ Italic lower case m is used for molality. Italic lower case c may be used for concentration in moles per liter of solution.

For any property B of the solution, the partial molal quantity for component i is denoted by $\bar{B}_i$ and is defined as the rate of change of the property B of the solution with change in the number of moles of component i, with the number of moles of all other components being held constant, all at constant pressure and temperature. If the property involved is the volume, then the partial molal volume of component i is the change in the volume of the solution, ΔV, which takes place on the addition of Δn_i moles of component i, divided by the number of moles of component i added, the ratio being taken in the limit as Δn_i approaches zero, all at constant pressure and temperature:

$$\bar{V}_i = \text{limit, as } \Delta n_i \to 0, \text{ of } \left(\frac{\Delta V}{\Delta n_i}\right)_{P,T} \tag{8.120}$$

The partial molal volume of component i is the change in the volume occurring on addition of one mole of component i to an exceedingly large quantity of solution, the quantity of solution being so large that addition of one mole of component i does not significantly change the composition, all at constant pressure and temperature.

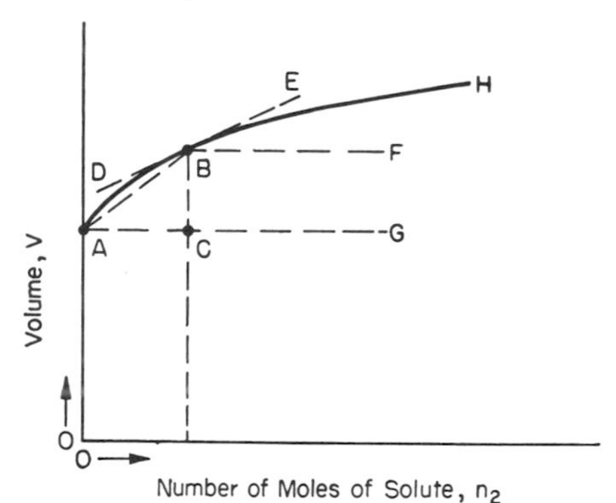

Fig. 8.3. Schematic diagram showing the relation between the partial molal property (volume) and the apparent molal property (volume).

This is illustrated in Fig. 8.3, in which the volume V of a system composed of two components is plotted as a function of the number of moles of the solute n_2. At A is shown the volume of the pure solvent, or, what is the same thing, of a solution of components 1 and 2 infinitely dilute in component 2. At C is shown the volume of a given solution containing the original amount of the solvent, component 1, plus a finite amount of the solute, component 2. The rate of change of the volume of the solution with change in the number of moles of component is given by the slope of the line DE, drawn tangent to the curve at C. That is, $\bar{V}_2$ is the tangent of the angle EBF.

The partial molal property may not always be conveniently measurable experimentally with the required precision. A simpler means of evaluating the partial molal quantity uses the apparent molal property of a solute. The apparent molal property of a given solute component j in a solution of two or more components is defined as the value of the property B of the given solution less the value of the property B for the same solution with all the given solute component j removed, divided by the number of moles of solute component j in the original solution. For example, the given solution contains n_1, n_2, n_3, n_j, $+ \cdots + n_i$ moles of components, and its volume is V_B. With n_j moles of component j removed, the solution contains $n_1, n_2, n_3, + \cdots + n_i$ moles of components, and the volume is now V_A. The apparent molal volume of solute component j is then

$$\phi V_j = \frac{V_B - V_A}{n_j} \tag{8.121}$$

For any property B the apparent molal property for component ϕB_j is

$$\phi B_j = \frac{(B)_{\text{given solution}} - (B)_{n_j=0}}{n_j} \tag{8.122}$$

In another form

$$\phi B_j = \Delta B / n_j \tag{8.123}$$

Usually the apparent molal property is used in connection with binary solutions in which water is the main component and the apparent molal property is applied to the solute. Then the apparent molal property, ϕB_2, of the solute is equal to the value of the property B for the solution containing n_1 moles of water and n_2 moles of the solute less the value of the property B for n_1 moles of water, divided by the number of moles, n_2, of the solute,

$$\phi B_2 = \frac{B_{(n_1+n_2)} - B_{(n_1)}}{n_2} = \frac{\Delta B}{n_2} \tag{8.124}$$

In Fig. 8.3 is plotted the volume of a solution of two components, 1 and 2, as a function of the number of moles of the solute, component 2. The apparent molal volume of the solute in the solution B is given by the slope of the chord drawn from the origin, A, to the point B. That is, ϕV_2 is the slope of AB or the tangent of the angle BAC. The line BC represents ΔV and the line AC represents n_2 in the relation

$$\phi V_2 = \frac{V_{(n_1+n_2)} - V_{(n_1)}}{n_2} = \frac{\Delta V}{n_2} \tag{8.125}$$

For the solution denoted by B, the partial molal volume of the solute differs from the value of the apparent molal volume of the solute by the difference in the slopes of the lines DE and AB. In the solution denoted by A, which is pure solvent, or a solution containing a small finite number of moles of solute in an infinite amount of solvent, the tangent at A is identical with the chord at A. Hence in the infinitely dilute solution the apparent molal property of a solute is identical with the partial molal property.

The change in any property B of any solution is, at constant pressure and temperature, a function of the components only, and may be expressed as

$$dB = \sum \left(\frac{\partial B}{\partial n_i}\right)_{n_j} dn_i \tag{8.126}$$

where j refers to all components except component i. This is equivalent to

$$\begin{aligned} dB &= \bar{B}_1\,dn_1 + \bar{B}_2\,dn_2 + \bar{B}_3\,dn_3 + \cdots + \bar{B}_i\,dn_i \\ &= \Sigma \bar{B}_i\,dn_i \end{aligned} \tag{8.127}$$

This equation may be integrated at constant relative composition to give, with pressure and temperature constant, the first basic partial molal equation

$$B = \bar{B}_1 n_1 + \bar{B}_2 n_2 + \bar{B}_3 n_3 + \cdots + \bar{B}_i n_i = \Sigma \bar{B}_i n_i \quad (8.128)$$

This equation may be differentiated at constant temperature and pressure, with all the components variable, to give

$$\begin{aligned} dB &= n_1\, d\bar{B}_1 + \bar{B}_1\, dn_1 + n_2\, d\bar{B}_2 + \bar{B}_2\, dn_2 + n_3\, d\bar{B}_3 \\ &\quad + \bar{B}_3\, dn_3 + \cdots + n_i\, d\bar{B}_i + \bar{B}_i\, dn_i \\ &= \Sigma n_i\, d\bar{B}_i + \Sigma \bar{B}_i\, dn_i \end{aligned} \quad (8.129)$$

therefore

$$n_1\, d\bar{B}_1 + n_2\, d\bar{B}_2 + n_3\, d\bar{B}_3 + \cdots + n_i\, d\bar{B}_i = \Sigma n_i\, d\bar{B}_i = 0 \quad (8.130)$$

If each term is divided by the total number of moles in the solution, corresponding equations with the composition in terms of mole fraction are obtained:

$$B = \bar{B}_1 N_1 + \bar{B}_2 N_2 + \bar{B}_3 N_3 + \cdots + \bar{B}_i N_i = \Sigma \bar{B}_i N_i \quad (8.131)$$

$$N_1\, d\bar{B}_1 + N_2\, d\bar{B}_2 + N_3\, d\bar{B}_3 + \cdots + N_i\, d\bar{B}_i = \Sigma N_i\, d\bar{B}_i = 0 \quad (8.132)$$

For two components, the basic partial molal equations are

$$B = \bar{B}_1 n_1 + \bar{B}_2 n_2 \quad (8.133)$$

$$n_1\, d\bar{B}_1 + n_2\, d\bar{B}_2 = 0 \quad (8.134)$$

$$B = \bar{B}_1 N_1 + \bar{B}_2 N_2 \quad (8.135)$$

and

$$N_1\, d\bar{B}_1 + N_2\, d\bar{B}_2 = 0 \quad (8.136)$$

Dividing by dN_2,

$$\frac{\dfrac{d\bar{B}_1}{dN_2}}{\dfrac{d\bar{B}_2}{dN_2}} = -\frac{N_2}{N_1} \quad (8.137)$$

This gives several important relations involving the slopes of curves of $\bar{B}_1$ and $\bar{B}_2$ plotted as a function of the mole fraction, N_1 or N_2.

For a binary solution consisting of n_1 moles of solvent and n_2 moles of solute, the value of a given property is

$$B = n_1 \bar{B}_1 + n_2 \bar{B}_2 \quad (8.138)$$

When n_2 is zero, as in the infinitely dilute solution or pure solvent, the value of the given property of the solution is the same as that of the pure solvent. Using a superscript asterisk to denote pure solvent for zero moles of solute,

$$B^* = n_1 \bar{B}^*_1 \quad (8.139)$$

The value of the apparent molal property of the solute in the given solution, containing n_1 moles of solvent and n_2 moles of solute, is

$$\phi B_2 = \frac{B - B^*}{n_2} \quad (8.140)$$

therefore

$$B = n_1 \bar{B}^*_1 + n_2 \phi B_2 \quad (8.141)$$

Solving for the apparent molal property of the solute,

$$\phi B_2 = \bar{B}_2 + \frac{n_1}{n_2}(\bar{B}_1 - \bar{B}^*_1) \quad (8.142)$$

This serves to evaluate either $\bar{B}_1$ or $\bar{B}_2$ when $\phi \bar{B}_2$ and either $\bar{B}_2$ or $\bar{B}_1$ are known, as

$$\bar{B}_1 - \bar{B}^*_1 = \frac{n_2}{n_1}(\phi B_2 - \bar{B}_2) \quad (8.143)$$

or

$$\bar{B}_2 = \phi B_2 - \frac{n_1}{n_2}(\bar{B}_1 - \bar{B}^*_1) \quad (8.144)$$

also

$$\bar{B}_2 = \phi B_2 + \frac{n_2 d(\phi B_2)}{dn_2} \quad (8.145)$$

This shows how the partial molal property of the solute $\bar{B}_2$ may be calculated if the apparent molal property of the solute ϕB_2 is known as a function of the number of moles of component 2. On a plot of ϕB_2 against n_2, B_2 is, for a given value of n_2, equal to the ordinate plus the product of the slope and the abscissa.

Combination of Eqs. (8.142) and (8.145) gives

$$\bar{B}_1 - \bar{B}^*_1 = -\frac{n_2^2}{n_1}\frac{d(\phi B_2)}{dn_2} \quad (8.146)$$

This shows how the partial molal property of the solvent, referred to pure solvent, may be evaluated if the apparent molal property is known as a function of the numbers of moles of solute. If ϕB_2 is plotted against n_2, $\bar{B}_1 - \bar{B}^*_1$ is, for any given value of n_2, equal to the slope of the curve multiplied by n_2^2/n_1.

If the given binary solution is an aqueous one, consisting of 1,000 g or 55.506 moles of water and m moles of solute, then $n_1 = 55.506$ and $n_2 = m$, so that

$$B = 55.506\bar{B}_1 + m\bar{B}_2 \quad (8.147)$$

and

$$B = 55.506\bar{B}^*_1 + m\phi B_2 \quad (8.148)$$

Then

$$\phi B_2 = \bar{B}_2 + \frac{55.506}{m}(\bar{B}_1 - \bar{B}^*_1) \quad (8.149)$$

and

$$\frac{dB}{dm} = m\frac{d(\phi B_2)}{dm} + \phi B_2 \quad (8.150)$$

also

$$\bar{B}_2 = \phi B_2 + m\frac{d(\phi B_2)}{dm} \quad (8.151)$$

and

$$\bar{B}_1 - \bar{B}^*_1 = -\frac{m^2}{55.506}\frac{d(\phi B_2)}{dm} \quad (8.152)$$

All the thermodynamic relations given previously, involving properties such as heat capacity, heat content, free energy, entropy, and volume, for one mole of substance hold for the corresponding partial molal properties. Thus

$$\bar{G}_1 = \bar{H}_1 - T\bar{S}_1 \quad (8.153)$$

$$\left(\frac{\partial \bar{G}_1}{\partial T}\right)_P = -\bar{S}_1 \quad (8.154)$$

$$\left(\frac{\partial \bar{G}_1}{\partial P}\right)_T = \bar{V}_1 \quad (8.155)$$

$$\bar{C}_{P_1} = \left(\frac{\partial \bar{H}_1}{\partial T}\right)_P \quad (8.156)$$

and

$$\left[\frac{\partial(\bar{G}_1/T)}{\partial T}\right]_P = -\frac{\bar{H}_1}{T^2} \quad (8.157)$$

In these, the partial molal property has been labeled for component 1 in the solution; the same relations hold for each of the other components.

6. The Ideal Solution

To systematize understanding of properties of solutions and to focus attention upon the important variables involved, it is desirable to deal with an ideal solution, which is a hypothetical solution having properties not possessed by any real solution. Properties of a given real solution may approach those of the ideal solution, according as the components of the solution become more nearly alike in their intermolecular properties. The ideal solution, which may be gaseous, liquid, or solid, is defined as one in which the fugacity of each component is proportional to its mole fraction, over the entire range of composition, at all temperatures and pressures. That is,

$$f_i = k_i N_i \tag{8.158}$$

The proportionality factor k_i is a constant for a given temperature and pressure, and holds for all concentrations at that temperature and pressure. The constant of proportionality k_i is equal to f^*_i, the fugacity of pure component i in the same state as the solution, gas, liquid, or solid, at the given temperature and pressure,

$$f_i = N_i f^*_i \tag{8.159}$$

Consider component i in a pure gas phase in equilibrium with component i in an ideal solution, gas, liquid, or solid. From the definition of fugacity, for component i in the pure gas phase,

$$G_i(g) = G_i^\circ(g) RT \ln f_i(g) \tag{8.160}$$

If component i in the pure gas phase is in equilibrium with component i in the solution,

$$G_i(g) = \bar{G}_i \text{ (sol)} \qquad f_i(g) = f_i \text{ (sol)} \tag{8.161}$$

Therefore $G_i = G_i^\circ(g) + RT \ln f_i \text{ (sol)}$ (8.162)

This applies to any mole fraction in the solution: Thus for $N_i = 1$,

$$\bar{G}^*_i = G_i^\circ(g) + RT \ln f^*_i \tag{8.163}$$

Thus for an ideal solution

$$\bar{G}_i = \bar{G}^*_i = RT \ln N_i \tag{8.164}$$

So the partial molal free energy, referred to its molal free energy in the pure state, gas, liquid, or solid, at the given temperature and pressure, is equal to RT multiplied by the natural logarithm of the mole fraction. Since this applies to constant temperature and pressure,

$$\left(\frac{\partial \bar{G}_i}{\partial N_i}\right)_{P,T} = \frac{RT}{N_i} \tag{8.165}$$

Differentiating Eq. (8.164) with respect to pressure at constant temperature,

$$\left(\frac{\partial \bar{G}_i}{\partial P}\right)_T - \left(\frac{\partial \bar{G}^*_i}{\partial P}\right)_T = 0 \tag{8.166}$$

or

$$\bar{V}_i = \bar{V}^*_i \tag{8.167}$$

For the ideal solution at a given temperature and pressure, the partial molal volume of each component at every mole fraction is equal to its volume in the pure state at the same temperature and pressure. Thus there is zero change in volume on mixing components to form an ideal solution. Differentiating Eq. (8.164) with respect to temperature at constant pressure,

$$\left[\frac{\partial(\bar{G}_i/T)}{\partial T}\right]_P - \left[\frac{\partial(\bar{G}^*_i/T)}{\partial T}\right]_P = 0 \tag{8.168}$$

or

$$\bar{H}_i = \bar{H}^*_i \tag{8.169}$$

For the ideal solution at a given temperature and pressure, the partial molal heat content of each component at every mole fraction is equal to its molal heat content in the pure state at the same temperature and pressure. Thus the heat of mixing components to form an ideal solution is zero. Since

$$\bar{C}_{P_i} = \left(\frac{\partial \bar{H}_i}{\partial T}\right)_P \qquad \bar{C}^*_{P_i} = \left(\frac{\partial \bar{H}^*_i}{\partial T}\right)_P \tag{8.170}$$

it follows that

$$\bar{C}_{P_i} = \bar{C}^*_{P_i} \tag{8.171}$$

For the ideal solution at a given temperature and pressure, the partial molal heat capacity of each component at every mole fraction is equal to its partial molal heat capacity in the pure state at the same temperature and pressure.

From the definition of the free energy,

$$\bar{G}_i = \bar{H}_i - T\bar{S}_i \qquad \bar{G}^*_i = \bar{H}^*_i - T\bar{S}^*_i \tag{8.172}$$

So from Eq. (8.169),

$$\bar{S}_i - \bar{S}^*_i = -R \ln N_i \tag{8.173}$$

For component i in an ideal solution, the partial molal entropy, referred to its molal entropy in the pure state, gas, liquid, or solid, at the given temperature and pressure, is equal to the negative of the gas constant R multiplied by the natural logarithm of the mole fraction.

These relations of the partial molal properties apply to any solution, whether gaseous, liquid, or solid, that conforms to the definition of the ideal solution given by Eq. (8.159).

Equation (8.173) may be used to calculate the entropy of mixing components to form an ideal solution, as N_1 moles of pure component 1, N_2 moles of pure component 2, N_3 moles of pure component 3, and N_i moles of pure component i, to form one mole of an ideal solution. The change in entropy for introducing each component into the solution is

$$N_i(\bar{S}_i - \bar{S}^*_i) = -N_i R \ln N_i \tag{8.174}$$

The sum for all components is the entropy of mixing all the components to form the ideal solution,

$$\Delta S \text{ (mixing)} = -R\Sigma N_i \ln N_i \tag{8.175}$$

The thermodynamic equilibrium between a pure substance A in the solid state and an ideal liquid solution in which it is one of the components, involves the reaction,

(Component 1) (c) = (component 1) (in ideal liq sol with component 2) **(8.176)**

Where component 2 represents one or more other components, which appear only in the liquid phase, not in the solid phase. At equilibrium, the molal free energy of component A in the solid phase must be equal to that in the liquid solution. For equilibrium, the change in the molal free energy of component 1 in the solid state must equal the change in the partial molal free energy of component 1 in the solution,

$$dG_1(c) = d\bar{G}_1 \tag{8.177}$$

The free energy of component 1 in the solid state may be expressed as a function of pressure and temperature,

$$dG_1(c) = V_1(c)\, dP - S_1(c)\, dT \tag{8.178}$$

and the free energy of component 1 in the solution may be expressed as

$$d\bar{G}_1 = \bar{V}_1\, dP - \bar{S}_1\, dT + RT\, d \ln N_1 \tag{8.179}$$

Therefore

$$RT\, d \ln N_1 = -[\bar{V}_1 - V_1(c)]\, dP + [\bar{S}_1 - S_1(c)]\, dT \tag{8.180}$$

At constant temperature, with $\bar{V}_1 = V^*_1$,

$$\frac{d \ln N_1}{dP} = -\frac{\bar{V}^*_1 - V_1(c)}{RT} \tag{8.181}$$

This gives, for the equilibrium between a pure solid and an ideal solution in which it is one of the components, the change in mole fraction of the solution with pressure at constant temperature in terms of the change in volume on melting. If there is an increase in volume on melting, then $\bar{V}^*_1 - V_1(c)$ is positive and the mole fraction of component 1 in the solution (which measures the solubility of the given solid component 1 in the solution) decreases with increase in pressure. With increasing pressure, the equilibrium is shifted in the direction of the state of lesser volume. If the solubility is known as a function of pressure, the foregoing would serve to evaluate the change in volume on melting of the pure solid, component 1.

At constant pressure,

$$\frac{d \ln N_1}{dT} = \frac{\bar{S}_1 - S_1(c)}{RT} \tag{8.182}$$

But at equilibrium,

$$\bar{G}_1 = G_1(c) \qquad \bar{H}_1 - T\bar{S}_1 = H_1(c) - TS_1(c) \tag{8.183}$$

so that

$$\bar{S}_1 - S_1(c) = \frac{\bar{H}_1 - H_1(c)}{T} \tag{8.184}$$

and therefore

$$\frac{d \ln N_1}{dT} = \frac{\bar{H}^*_1 - H_1(c)}{RT^2} \tag{8.185}$$

This gives, for equilibrium between a pure solid and an ideal solution in which it is one of the components, the change in mole fraction of the solution with temperature at constant pressure in terms of the change in heat content on melting or the heat of fusion or melting. The heat of melting is always positive; thus with increasing temperature the mole fraction of component 1 in the solution (which measures the solubility of the given solid component 1 in the solution) increases with increase in temperature. The equilibrium is thus shifted toward that phase which has a greater heat content.

Equation (8.185) may be integrated from the composition $N_1 = 1$, at which the solution is pure component A and the temperature is T^*_1, to any other mole fraction N_A and corresponding temperature T as

$$\ln N_1 = \frac{1}{R}\int_{T^*_1}^{T} \frac{[\bar{H}^*_1 - H_1(c)]}{T^2}\, dT \tag{8.186}$$

giving

$$\ln N_1 = -\frac{A(\Delta T)}{1 - \Delta T/T^*_1} + J\left(\ln \frac{1 - \Delta T}{T^*_1} + \frac{\Delta T}{T^*_1 - \Delta T}\right) \tag{8.187}$$

where

$$\Delta T = T^*_1 - T \tag{8.188}$$

$$A = \Delta H^*\, m_1/RT^{*2}_1 \tag{8.189}$$

$$J = \frac{\Delta C_{P^*_1}}{R} \tag{8.190}$$

Subscript 1 refers to the major component, the superscript asterisk refers to the pure component, when $N_1 = 1$, $\Delta H\, m$ is the heat of fusion or melting, T is the equilibrium temperature, and ΔC_P is the heat capacity of the given component in the liquid state less that in the solid state. Equation (8.187) covers the entire range of composition since it has no mathematical approximations involving N_1 or ΔT, but it does involve the assumption that the heat of fusion of pure component 1 is a linear function of the temperature, that is, that $\Delta C_{P^*_1}$ is constant.

The form taken by Eq. (8.187) is shown in Fig. 8.4 by curves 1 and 2. Curve 1 is based entirely on the values of the properties of pure component 1 and curve 2 is based entirely on the values of the proper-

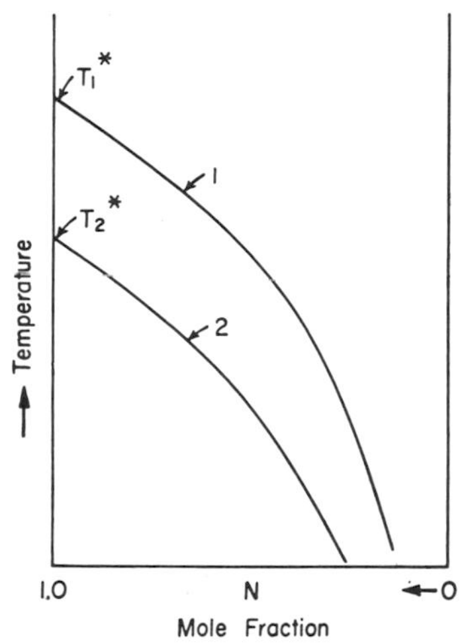

FIG. 8.4. Schematic diagram showing relation between temperature and composition of the solution for the equilibrium between a pure solid substance, component 1, and an ideal liquid solution in which it is one of the components, and similarly for the equilibrium between a pure solid substance, component 2, and an ideal liquid solution in which it is one of the components.

ties of pure component 2. Curve 1 gives the relation between the mole fraction of the pure component 1 in an ideal liquid solution and the temperature at which this solution, at any given mole fraction, is in equilibrium with pure component 1 in the solid phase. Similarly, curve 2 gives the relation between the mole fraction of the pure component 2 in an ideal liquid solution and the temperature at which this solution, at any given mole fraction, is in equilibrium with pure component 2 in the solid phase. In each case, the given curve depends only on the properties of the component that is in both phases and not at all upon the properties of any other components.

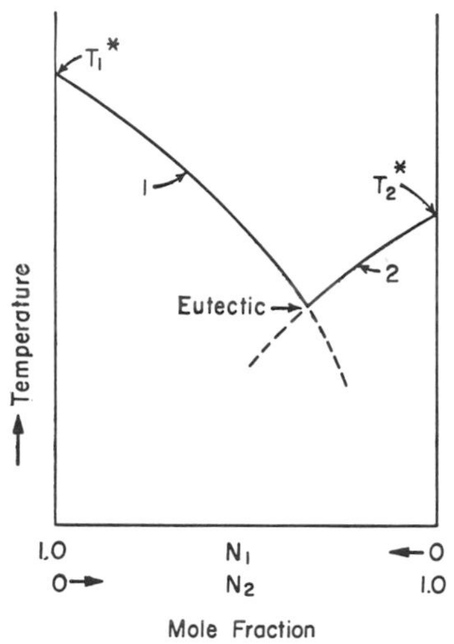

FIG. 8.5. Schematic diagram showing the relation between temperature and composition of the solution for the liquid–solid equilibrium in a binary system of two components, 1 and 2, which form an ideal liquid solution, but are not soluble in one another in the solid phase. From $N_1 = 1$ to the eutectic composition at the intersection of the two curves, the solid phase in equilibrium with the solution is pure component 1. From the eutectic composition to $N_2 = 1$, the solid phase is pure component 2. The components 1 and 2 are the same ones represented in Fig. 8.4.

Figure 8.5 shows the two curves of Fig. 8.4 combined to form the temperature-composition diagram for the liquid–solid equilibrium involving an ideal binary system. From $N_1 = 1$ to the intersection of the two curves, the solid phase is pure component 1. From $N_2 = 1$ to the intersection of the two curves, the solid phase is pure component 2. At the intersection of the two curves, which gives the eutectic temperature and composition, both pure solid phases, 1 and 2, are in equilibrium with the solution of the eutectic composition. For any two substances that form a substantially ideal liquid solution, and that do not form a solid solution, it is possible to calculate the complete temperature-composition diagram, including the eutectic composition and temperature, entirely from the properties of the two pure components, in accordance with Eq. (8.187) applied to each of the substances in turn.

Equation (8.186) may be integrated in such a way as to provide a simpler relation to cover the range of composition in the neighborhood of $N_1 = 1$:

$$\ln N_1 = -A(\Delta T)[1 + B(\Delta T) + C(\Delta T)^2 + \cdots] \tag{8.191}$$

Here ΔT and A have the same significance as before, and

$$B = \frac{(1 - \Delta C_{P^*_1} T^*_1 / 2\Delta Hm^*_1)}{T^*_1} \tag{8.192}$$

$$C = \frac{(1 - 2\Delta C_{P^*_1} I^*_1 / 3\Delta Hm^*_1)}{T^{*2}_1} \tag{8.193}$$

As ΔT approaches zero, Eq. (8.191) reduces to

$$N_2 = A(\Delta T) \tag{8.194}$$

Equation (8.187), or any of its equivalent or modified forms, properly applied, is one of the most important relations in chemistry and physics, because it permits one to evaluate, under appropriate conditions, the purity of a given substance *without knowing the identity of the components constituting the impurity* in the given substance. Equation (8.187) is the basic equation for the lowering of the freezing point on the addition of a solute that is liquid-soluble and solid-insoluble.

The corresponding equations covering equilibrium between a pure gaseous substance and an ideal liquid solution in which all other components are nonvolatile, in which the gaseous state is substituted in each case for the solid state, are:

At constant temperature,

$$\frac{d \ln N_1}{dP} = \frac{V_1(g) - \bar{V}^*_1}{RT} \tag{8.195}$$

At constant pressure,

$$\frac{d \ln N_1}{dT} = -\frac{[H_1(g) - \bar{H}^*_1]}{RT^2} \tag{8.196}$$

The latter gives the change in mole fraction with temperature in terms of the heat of vaporization of pure component 1, for the equilibrium between component 1 in its ideal liquid solution with other nonvolatile components and pure component 1 in the gaseous phase. It is the basic equation for elevation of the boiling point on addition of a nonvolatile solute.

7. The Dilute Real Solution

The dilute real solution is one for which, from $N_2 = 0$ to some small value of N_2, the fugacity of the solute is proportional to its mole fraction, as

$$f_2 = k_2 N_2 \tag{8.197}$$

The range of composition covered by different real solutions satisfying the foregoing definition may be quite different. For some the range of applicability of Eq. (8.197) may be large and for others it may be exceedingly small. The value of the constant of proportionality in Eq. (8.197) is different for each solute and must be determined experimentally for each case. Experimental determination of k will also usually determine the range of composition over which the ratio of f_2/N_2 is constant. From the definition of the dilute real solution the fugacity of the solute is given directly in terms of the mole fraction of the solute and the value of the constant k, over the range of composition from $N_2 = 0$ to that value of N_2 for which the relation is still applicable. In the very dilute range, the molality is proportional to the mole fraction; thus in this range, for aqueous solution,

$$f_2 = \frac{k_2 m}{55.506} \tag{8.198}$$

A similar linear relation holds in the dilute range between the fugacity of the solute and the concentration expressed in moles per liter of solution.

If the gaseous phase of component 2 in equilibrium with the solution may be taken as ideal, then the partial pressure of component 2 may be substituted for the fugacity

$$P_2 = k_2 N_2 \tag{8.199}$$

and

$$P_2 = \frac{k_2 m}{55.506} \tag{8.200}$$

These are expressions of *Henry's law* for the vapor pressure of the solute in a dilute solution.

To the dilute real solution of two components, at constant temperature and pressure, may be applied one of the basic partial molal equations, together with the appropriate equations relating partial molal free energy and fugacity, to obtain

$$f_1 = N_1 f^*_1 \tag{8.201}$$

Thus in a dilute real solution the fugacity of the solvent is proportional to its mole fraction. Over the range of composition for which they are applicable, Eqs. (8.197) and (8.201) for the dilute real solution are identical with corresponding equations for the ideal solution. In fact, over the range of concentration for which the fugacity of the solute is proportioned to its mole fraction, the dilute real solution has all the properties of an ideal solution in which the solute is a hypothetical one whose fugacity up to $N_2 = 1$ is a continuation of the linear relationship between f_2 and N_2 in the dilute range. It follows that, within the foregoing limitations, in the dilute real solution, the partial molal volumes, heat contents, and heat capacities of the components are, as in the ideal solution, independent of the composition over the given range of composition near $N_1 = 1$.

If the gaseous phase of component 1 in equilibrium with the dilute solution may be taken as ideal, then, as an approximation,

$$P_1 = N_1 P^*_1 \tag{8.202}$$

This equation is a statement of *Raoult's law* of the lowering of the vapor pressure of the solvent.

In considering equilibrium between a pure solid and a dilute real solution, with the solid phase being pure component 1, the following relation is applicable from $N_1 = 1$ to a value of N_1 not far from unity:

$$RT\,d\ln N_1 = -[\bar{V}_1 - V_1(c)]\,dP + \frac{\bar{H}_1 - H_1(c)}{T}\,dT \tag{8.203}$$

This and all the relations derivable from it are applicable over the range of composition for which Eq. (8.197) is valid.

A constant temperature,

$$\frac{d\ln N_1}{dP} = -\frac{\bar{V}^*_1 - V_1(c)}{RT} \tag{8.204}$$

which gives the change, with pressure at constant temperature, of the mole fraction of the solvent in the dilute real solution.

At constant pressure,

$$\frac{d\ln N_1}{dT} = \frac{\bar{H}^*_1 - H_1(c)}{RT^2} \tag{8.205}$$

which gives the change, with temperature at constant pressure, of the mole fraction of the solvent in the dilute real solution. This equation also evaluates the lowering of the freezing point of the solvent on addition of a given amount of solute. It may be integrated over the desired range of composition, to give

$$\ln N_1 = -A(\Delta T)[1 + B(\Delta T) + C(\Delta T)^2 + \cdots] \tag{8.206}$$

If the temperature of a given substance in the liquid state is gradually lowered, the temperature at which an infinitesimal amount of crystals of the major component exist in thermodynamic equilibrium with the liquid phase may be defined as the freezing point. Similarly, if the temperature of the given substance in the solid state is gradually increased, the temperature at which the last infinitesimal amount of crystals in equilibrium with the liquid phase disappears is also the freezing point, identically the same temperature. Since in each case the quantity of the given sample in the crystal phase is infinitesimal, the composition of the liquid phase is identical with the composition of the original substance, and the freezing point may be used as a measure of the purity of the given substance in terms of the mole fraction of the main component, in accordance with Eq. (8.206).

As in the preceding section, the gaseous phase may be substituted for the solid phase to obtain the following relation for equilibrium between the dilute real solution and a pure gaseous phase consisting of the major component only:

$$RT\,d\ln N_1 = [V_1(g) - \bar{V}_1]\,dP = \frac{H_1(g) - \bar{H}_1}{T}\,dT \tag{8.207}$$

This equation is applicable over the entire range of composition for which Eq. (8.197) is valid. The equilibrium being considered is one in which the solute exists only in the liquid phase, with no measurable amount of it being in the gaseous phase. At constant pressure, this reduces to

$$\frac{d \ln N_1}{dT} = -\frac{H_1(g) - \bar{H}^*_1}{RT^2} \qquad (8.208)$$

What was said about Eq. (8.205) applies equally well to Eq. (8.208).

The equilibrium distribution of a given solute, C, between two solvents, A and B, which are in contact but not completely miscible with one another, involves the reaction

$$C \text{ (sol in } A, N_c) = C \text{ (sol in } B, N_c) \qquad (8.209)$$

At equilibrium,

$$(\bar{G}_c)^A = (\bar{G}_c)^B \qquad (f_c)^A = (f_c)^B \qquad (8.210)$$

But for the dilute real solution,

$$(f_c)^A = (k_c)^A(N_c)^A \qquad (f_c)^B = (k_c)^B(N_c)^B \qquad (8.211)$$

Hence,

$$\frac{(N_c)^A}{(M_c)^B} = \frac{(k_c)^B}{(k_c)^A} = k' \qquad (8.212)$$

(Here A and B are superscripts, not exponents.) This is the *Nernst law* of the distribution of a solute between two solvents, and shows that, over the range of composition where Eq. (8.197) is applicable, the ratio of the mole fractions of the solute in the two solvents remains constant.

At constant temperature, the equilibrium between a pure liquid solvent, 1, under a pressure, P^*, may be in equilibrium, through a membrane permeable only to this solvent, with a solution of the solvent and a solute, 2, the solution being under a pressure P and qualifying as a dilute real solution. As the concentration of solute in the solution increases, the pressure P is required to be increased to maintain equilibrium. The value of $P - P^*$ for any given composition of the solution is defined as the *osmotic pressure* for that composition.

At equilibrium between the two phases,

$$(\bar{G}^*_1)^{P^*} = (\bar{G}_1)^P \qquad (f^*_1)^{P^*} = (f_1)^P \qquad (8.213)$$

But, at constant temperature,

$$d\bar{G}_1 = \bar{V}_1\, dP \qquad (8.214)$$

This may be integrated between the pressure P^* and the pressure P, assuming $\bar{V}_1$ to be constant,

$$(\bar{G}_1)^P - (\bar{G}_1)^{P^*} = \bar{V}_1(P - P^*) \qquad (8.215)$$

Substituting,

$$(\bar{G}_1 - \bar{G}^*_1)^{P^*} = -\bar{V}_1(P - P^*) \qquad (8.216)$$

hence

$$(P - P^*) = -\frac{RT}{\bar{V}_1} \ln N_1 \qquad (8.217)$$

In the limit, this becomes

$$P - P^* = \frac{N_2 RT}{\bar{V}_1} \qquad (8.218)$$

These are statements of the *van't Hoff law of osmotic pressure,* with the latter equation resembling in form the equation of state of the ideal gas.

8. Equilibrium Constant and the Standard Change in Free Energy

In studying the change in free energy accompanying any reaction or process, it is convenient to evaluate such changes with each reactant and product in its thermodynamic standard reference state. For any reaction at any given temperature and pressure,

$$bB + cC = mM + nN \qquad (8.219)$$

the change in free energy for any given set of states of the reactants and products is

$$\Delta G = mG_M + nG_N - bG_B - cG_C \qquad (8.220)$$

and the standard change in free energy, with each reactant and product in its thermodynamic standard reference state, is

$$\Delta G^\circ = mG^\circ_M + nG^\circ_N - bG^\circ_B - cG^\circ_C \qquad (8.221)$$

The difference in the change in free energy for the given set of states and for the standard states is

$$\Delta G - \Delta G^\circ = m(G - G^\circ)_M + n(G - G^\circ)_N - b(G - G^\circ)_B - c(G - G^\circ)_C \qquad (8.222)$$

For any substance in any state, with the *activity* a_i equal to f_i/f°_i,

$$G_i - G^\circ_i = \frac{RT \ln f_i}{f^\circ_i} = RT \ln a_i \qquad (8.223)$$

where f_i, G_i, and a_i refer to the same given state for substance i and f°_i and G°_i refer to the standard state,

$$\Delta G - \Delta G^\circ = RT \ln \left[\frac{(a_M)^m(a_N)^n}{(a_B)^b(a_C)^c}\right] \qquad (8.224)$$

The *proper quotient of activities* for the reaction given by Eq. (8.219) is defined as

$$Q_a = \frac{(a_M)^m(a_N)^n}{(a_B)^b(a_C)^c} \qquad (8.225)$$

The numerator is the product of the activity of each of the products of the reaction, each raised to a power equal to the number of moles of the given product, and the denominator is the corresponding product for the reactants of the reaction. The proper quotient of activities is thus identical with the terms inside the bracket on the right side of Eq. (8.224).

The *equilibrium constant* is defined as the proper quotient of activities when each reactant and product is in its equilibrium state for the given reaction, as

$$Q_a \text{ (equilibrium)} = K = \frac{(a_M{}^e)^m(a_N{}^e)^n}{(a_B{}^e)^b(a_C{}^e)^c} \qquad (8.226)$$

When the given reaction takes place with each reactant and product in its standard state, the change in free energy is the standard change in free energy, ΔG°, and is equal to the value by Eq. (8.221). When the given reaction takes place with each reactant and product in its equilibrium concentration or pressure for the given conditions, then

$$\Delta G \text{ (equilibrium)} = \Delta G^e = mG_M{}^e + nG_N{}^e - bG_B{}^e - cG_C{}^e \qquad (8.227)$$

The difference in the change in free energy for the two sets of conditions, standard states and equilibrium states, is

$$\Delta G^e - \Delta G^\circ = m(G^e - G^\circ)_M + n(G^e - G^\circ)_N - b(G^e - G^\circ)_B - c(G^e - G^\circ)_C \quad (8.228)$$

For any substance in its equilibrium state,

$$G_i^e - G_i^\circ = RT \ln a_i^e \quad (8.229)$$

so

$$\Delta G^e - \Delta G^\circ = RT \ln \frac{(a_M{}^e)^m (a_N{}^e)^n}{(a_C{}^e)^b (a_B{}^e)^c} \quad (8.230)$$

At constant temperature and pressure, the change in free energy at equilibrium is zero,

$$\Delta G^e = 0 \quad (8.231)$$

Substituting K for the terms in the fraction,

$$\Delta G^\circ = -RT \ln K \quad (8.232)$$

an important equation relating the standard change in free energy and the equilibrium constant.

Differentiation of G/T at constant pressure gives

$$\frac{d(G/T)}{dT} = \frac{-TS + G}{T^2} = \frac{-H}{T^2} \quad (8.233)$$

which holds for each reactant and product of the given reaction, so that for the reaction, at constant pressure,

$$d\left(\frac{\Delta G}{T}\right) = -\frac{\Delta H}{T^2}\, dT \quad (8.234)$$

Applied to the reaction with each substance in its standard state, this is written

$$d\left(\frac{\Delta G^\circ}{T}\right) = -\frac{\Delta H^\circ}{T^2}\, dT \quad (8.235)$$

For ΔH as a function of temperature:

$$\Delta H^\circ = \Delta H_*{}^\circ + \frac{(\Delta a)T + (\Delta b)T}{2} + \frac{(\Delta c)T^3}{3} \quad (8.236)$$

which is valid over the range of temperature for which the original expressions for heat capacity as a function of temperature were valid. $\Delta H_*{}^\circ$ is not the value of ΔH° at 0°K. When this is substituted into Eq. (8.235) and the latter integrated,

$$\frac{\Delta G^\circ}{T} = \frac{\Delta H_*{}^\circ}{T} - (\Delta a) \ln T - \frac{(\Delta b)T}{2} - \frac{(\Delta c)T^2}{6} + \cdots + I \quad (8.237)$$

In this equation, I is the constant of integration. This may also be written

$$\Delta G^\circ = \Delta H_*{}^\circ - (\Delta a)T \ln T - \frac{(\Delta b)T^2}{2} - \frac{(\Delta c)T^3}{6} + \cdots + IT \quad (8.238)$$

These are valid only over the range of temperature for which the original expressions for the heat capacities of the reactants and products are valid. The constant of integration I must be evaluated from one value of ΔG° within the given range of temperature. If the change in heat content were constant with temperature, the factors (Δa), (Δb), and (Δc) would be zero, and

$$\Delta G^\circ = \Delta H^\circ + IT \quad (8.239)$$

By comparison with the relation

$$\Delta G^\circ = \Delta H^\circ - T\,\Delta S^\circ \quad (8.240)$$

the constant of integration I is related to the negative of the standard change in entropy for the reaction.

Evaluation of the standard change in free energy by Eqs. (8.237) and (8.238) is subject to the disadvantage that, for a given set of constants, the range is limited and further, the form of the equations may not be such as to reproduce experimental observations at different temperatures within their limits of accuracy. A more direct and more accurate way of evaluating ΔG° at different temperatures is by using the values of the free energy function of the reactants and products if these have been tabulated at different temperatures. Values of $\Delta G^\circ/T$ for a series of temperatures in the range of interest are calculated and the value for any particular temperature may be obtained by suitable interpolation.

For the given reaction, the value of $\Delta H_0{}^\circ$ is obtained from the reference value of $\Delta H^\circ{}_{298.16}$ and values of the heat content at 298.15°K, relative to 0°K, of each of the reactants and products:

$$\Delta H_0{}^\circ = H^\circ{}_{298.15} - \Delta(H^\circ{}_{298.15} - H_0{}^\circ) \quad (8.241)$$

Since

$$\frac{\Delta G^\circ}{T} = \Delta H_0{}^\circ + \frac{\Delta(G^\circ - H_0{}^\circ)}{T} \quad (8.242)$$

the value of $\Delta G^\circ/T$ for the given reaction at the temperature T is equal to the change in heat content at 0°K, $\Delta H_0{}^\circ$ plus the increment, for the given reaction, of the free energy function,

$$\frac{\Delta(G^\circ - H_0{}^\circ)}{T} = \frac{\Sigma(G^\circ - H_0{}^\circ)}{T} \ \text{(products)} - \frac{\Sigma(G^\circ - H_0{}^\circ)}{T} \ \text{(reactants)} \quad (8.243)$$

The equilibrium constant is then

$$\ln K = -\frac{\Delta H_0{}^\circ}{R} - \frac{\Delta(G^\circ - H_0{}^\circ)}{RT} \quad (8.244)$$

Substitution of $-R \ln K$ in Eq. (8.234) gives

$$\frac{d \ln K}{d(1/T)} = -\frac{\Delta H^\circ}{R} \quad (8.245)$$

For any given temperature the equilibrium constant is directly related to the standard change in free energy. Since, at any given temperature, the free energy in the standard state for each reactant and product, G_i°, is independent of the pressure, the standard change in free energy for the reaction ΔG° is independent of the pressure. At constant temperature, the equilibrium constant K, as defined by Eq. (8.226) and related to $\Delta G°$ by Eq. (8.232), is also independent of the pressure,

$$\left(\frac{\partial K}{\partial P}\right)_T = 0 \tag{8.246}$$

The fact that the equilibrium constant does not change in value with pressure does not necessarily mean that concentrations of the components of a chemical reaction at equilibrium remain constant as the pressure is increased. In the following reaction at equilibrium at some temperature T, with each component in the gaseous phase:

$$bB(g) + cC(g) = mM(g) + nN(g) \tag{8.247}$$

the equilibrium constant for this reaction is

$$K = \frac{(a_M{}^e)^m(a_N{}^e)^n}{(a_B{}^e)^b(a_C{}^e)^c} \tag{8.248}$$

For gases, the activity is equal to the fugacity, since $f° = 1$,

$$K = \frac{(f_M{}^e)^m(f_N{}^e)^n}{(f_B{}^e)^b(f_C{}^e)^c} \tag{8.249}$$

and the activity is proportional to the partial pressure $\gamma_i = f_i/P_i$ and the partial pressure is $P_i = N_iP$, where N_i is the mole fraction of the given component and P is the total pressure. Then $f_i = \gamma_iN_iP$, so that

$$K = \frac{(\gamma_M{}^e)^m(\gamma_N{}^e)^n}{(\gamma_B{}^e)^b(\gamma_C{}^e)^c}\left[\frac{(N_M{}^e)^m(N_N{}^e)^n}{(N_B{}^e)^b(N_C{}^e)^c}\right]P^{(m+n-b-c)} \tag{8.250}$$

Thus the equilibrium constant varies with the absolute pressure to the $(m + n - B - C)$ power. Introduction of an inert gas into the system has no effect on these relations, except to change the activity coefficients, a little in the case of nonideal gases, provided the partial pressure of the inert gas is not included in the value of the total pressure P. That is, the total pressure P given above must be counted as the sum of the partial pressures of the gaseous components participating in the given reaction. If the number of moles of gaseous products is less than the number of moles of gaseous reactants, then the mole fractions of the products will increase at the expense of the mole fractions of the reactants. That is, with increase in pressure, the equilibrium is shifted in favor of the products, and conversely. For real gases, the effect of pressure on concentrations of the components of a chemical reaction at equilibrium also includes the change of the activity coefficient with pressure. These effects may be very large.

It is occasionally necessary to make calculations of equilibrium concentrations of the components of a chemical reaction at equilibrium in a vessel at constant volume. In lieu of the pressures, there will be known the volume of the space available to the gaseous components and the number of moles of the substances at the beginning. If the temperature is high enough and the volume is large enough, in relation to the number of moles of gaseous reacting substances, the gases may be assumed to be ideal. Usually, in such problems, the concentration is given in moles per liter, as c_i. For the reaction given by Eq. (8.247), since the gases are assumed ideal, the equilibrium constant is the proper quotient of pressures at equilibrium,

$$K = \frac{(P_M{}^e)^m(P_N{}^e)^n}{(p_B{}^e)^b(P_C{}^e)^c} = Q_P{}^e \tag{8.251}$$

For ideal gases, the concentration in moles per liter is related to the pressure by

$$c_i = \frac{n_i}{V} \qquad P_i = \frac{n_iRT}{V} = c_iRT \tag{8.252}$$

where V is the whole volume of the space available to the gaseous components. Thus

$$K = \frac{(c_M{}^e)^m(c_N{}^e)^n}{(c_B{}^e)^b(c_C{}^e)^c}(RT)^{(m+n-b-c)} \tag{8.253}$$

or

$$Q_c{}^e = K(RT)^{-(m+n-b-c)} \tag{8.254}$$

This equation permits evaluation, for a reaction involving gases assumed ideal, of the proper quotient of concentrations at equilibrium from the value of the equilibrium constant and the stoichiometrical change in the number of moles of gaseous components between reactants and products. If the stoichiometrical number of moles of gaseous products is the same for the gaseous reactants, $m + n = b - c$, then

$$Q_c{}^e = K$$

With concentration in liters, and the equilibrium constant K in terms of atmospheres, the value of R here must be expressed in liter-atmospheres.

9. Thermodynamic Calculations

In modern compilations of selected values of thermodynamic properties, as in the tables of "Selected Values of Physical and Thermodynamic Properties of Hydrocarbons and Related Compounds" [3],† values of the following thermodynamic properties are given for the thermodynamic standard state from the absolute zero to high temperatures, for a large number of hydrocarbons and related compounds:

$(G° - H_0°)/T$ = free energy function
$(H° - H_0°)/T$ = heat content function
$S°$ = entropy
$H° - H_0°$ = heat content (referred to 0°K)
$C_P°$ = heat capacity
$\Delta Hf°$ = heat of formation (from the elements)
$\Delta Gf°$ = free energy of formation (from the elements)
$\log_{10} Kf$ = logarithm of the equilibrium constant of formation (from the elements)

In making such a compilation, the basic values used are:

† Numbers in brackets refer to References at end of chapter.

$\Delta Hf^\circ_{298.16}$, the standard heat of formation (from the elements) at 25°C

$(H^\circ - H_0^\circ)/T$, for each temperature for which the thermodynamic values are to be tabulated

$(G^\circ - H_0^\circ)/T$, for each temperature for which the thermodynamic values are to be tabulated

C_P°, for each temperature for which the thermodynamic values are to be tabulated

The entire table is constructed from these as follows:

From the heat content function, $H^\circ - H_0^\circ$, the heat content at the given temperature referred to the absolute zero, is obtained by multiplication of the heat content function by the temperature,

$$H^\circ - H_0^\circ = (T)\left(\frac{H^\circ - H_0^\circ}{T}\right) \tag{8.255}$$

Standard entropy, S°, is obtained as the heat content function less the free energy function,

$$S^\circ = \left(\frac{H^\circ - H_0^\circ}{T}\right) - \left(\frac{G^\circ - H_0^\circ}{T}\right) \tag{8.256}$$

The value of ΔHf_0° is obtained from the relation

$$\Delta Hf_0^\circ = \Delta Hf^\circ_{298.15} - \Delta(H^\circ_{298.15} - H_0^\circ) \tag{8.257}$$

where

$$\Delta(H^\circ_{298.15} - H_0^\circ) = (H^\circ_{298.15} - H_0^\circ) \text{ (compound)} - \Sigma(H^\circ_{298.15} - H_0^\circ) \text{ (elements)} \tag{8.258}$$

When the value of ΔHf_0° is obtained, the values of ΔHf° at all other temperatures are obtained from the relation,

$$\Delta Hf^\circ = \Delta Hf_0^\circ + \Delta(H^\circ - H_0^\circ) \tag{8.259}$$

where

$$\Delta(H^\circ - H_0^\circ) = (H^\circ - H_0^\circ) \text{ (compound)} - \Sigma(H^\circ - H_0^\circ) \text{ (elements)} \tag{8.260}$$

Values of ΔGf° at each temperature are obtained from the relation

$$\Delta Gf^\circ = \Delta Hf^\circ - T(\Delta Sf^\circ) \tag{8.261}$$

where

$$\Delta Sf^\circ = S^\circ \text{ (compound)} - \Sigma S^\circ \text{ (elements)} \tag{8.262}$$

Values of log Kf are obtained at each temperature from the values of ΔGf° from the relation

$$\Delta Gf^\circ = -RT \ln Kf \tag{8.263}$$

or

$$\log Kf = -\frac{\Delta Gf^\circ}{2.302585RT} \tag{8.264}$$

The tables of "Selected Values of Physical and Thermodynamic Properties of Hydrocarbons and Related Compounds" [3] also give values for the following properties:

Heat of formation, free energy of formation, and entropy, for the liquid state, at 25°C

Heat and entropy of fusion

Heat and entropy of vaporization

Standard heat, entropy, and free energy of vaporization, at 25°C

Heat of combustion, for the gaseous and liquid states, at 25°C

Vapor pressures, from 10 to 1,500 mm Hg

The tables of "Selected Values of Chemical Thermodynamic Properties" [4] include the following properties: Series I, heat of formation, free energy of formation, logarithm of the equilibrium constant of formation, entropy, and heat capacity, each at 25°C, for solid, liquid, and gaseous states, as appropriate and known; Series II, heat, temperature, and entropy of transition, fusion, and vaporization, including values of pressure as appropriate and known.

The "NBS-NACA Tables of Thermal Properties" [5] contain values of the physical thermodynamic properties for many of the simpler gases.

Whenever values of log Kf are available for the temperature or temperatures of interest for each of the reactants and products in a given chemical reaction, the calculation of log K or the equilibrium constant for the given reaction may be made very simply. For the reaction $bB + cC = mM + nN$

$$\log K = \Sigma \log Kf \text{ (products)} - \Sigma \log Kf \text{ (reactants)} = m \log Kf_M + n \log Kf_N - b \log Kf_B - c \log Kf_C \tag{8.265}$$

For each reactant or product that is an element in the standard state, the value of log Kf is given in the table as zero since ΔGf° is zero. In this manner, values of the logarithm of the equilibrium constant as a function of temperature for a number of reactions involving O_2, H_2, H_2O, C(graphite), CO, CO_2, and CH_4 were calculated and plotted by Wagman, Kilpatrick, Taylor, Pitzer, and Rossini [6], as shown in Fig. 8.6.

In most of the foregoing calculations, the various properties dealt with have usually applied to the thermodynamic reference state, which for gases is the ideal state of unit fugacity (1 atm, unless otherwise specified) and for pure liquids and solids is the real state at actual unit pressure (1 atm, unless otherwise specified). A simple framework can be assembled to give values for all the important thermodynamic properties as a function of temperature for the thermodynamic standard state. In the solution of a given problem involving real states at some high pressure P, at some elevated temperature T, the procedure is to calculate the results for the standard states at the given temperature, and then to compute, at that constant temperature, the change in the values of the given properties in going from the standard states to the given real states at the pressure P_1. This means that, for example, a complete description and evaluation of the thermodynamic properties of a substance can be provided by means of (1) a tabulation giving values of the several thermodynamic properties for the standard state at each temperature of interest and (2) a tabulation giving, for the same temperature, values of the difference in the given thermodynamic property between

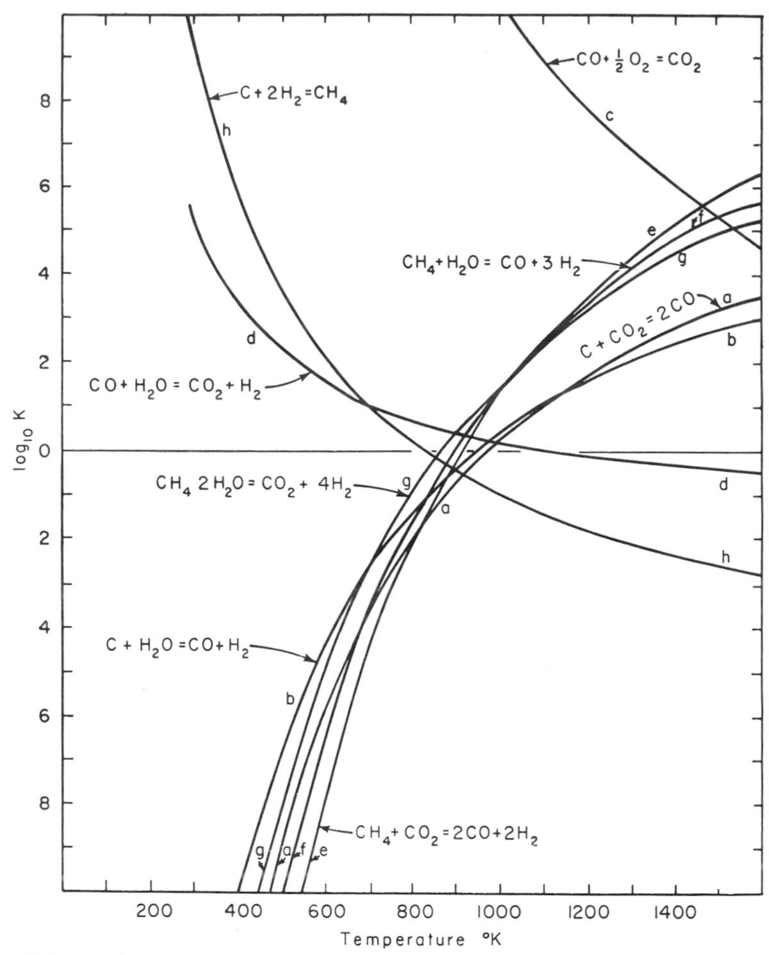

FIG. 8.6. Logarithm of the equilibrium constant for eight reactions involving O_2, H_2, H_2O, C (graphite), CO, CO_2, and CH_4. Ordinates give $\log_{10} K$ for the given reaction. Abscissas give the temperature in degrees Kelvin. The curves apply to the following reactions:

(*a*) C(solid, graphite) + CO_2(gas) = 2 CO(gas)
(*b*) C(solid, graphite) + H_2O(gas) = CO(gas) + H_2(gas)
(*c*) CO(gas) + $\frac{1}{2}$ O_2(gas) = CO_2(gas)
(*d*) CO(gas) + H_2O(gas) = CO_2(gas) + H_2(gas)
(*e*) CH_4(gas) + CO_2(gas) = 2 CO(gas) + $2H_2$(gas)
(*f*) CH_4(gas) + H_2O(gas) = CO(gas) + $3H_2$(gas)
(*g*) CH_4(gas) + $2H_2$(gas) = CO_2(gas) + $4H_2$(gas)
(*h*) C(solid, graphite) + $2H_2$(gas) = CH_4(gas)

(*From D. D. Wagman, J. E. Kilpatrick, W. J. Taylor, K. S. Pitzer, and F. D. Rossini* [6].)

various pressures and the standard state, covering each pressure of interest. Values of the several thermodynamic properties may be assembled for the standard state at various temperatures. The next step is to evaluate the change of the given thermodynamic property with pressure at the given temperature. If, at the given temperature, the change of energy, entropy, and volume with pressure is known, the picture is complete. That is, if $(\partial E/\partial P)_T$, $(\partial S/\partial P)_T$, and $[\partial(PV)/\partial P]_T$ are known as functions of pressure, all other defined thermodynamic properties become known as a function of pressure at the given temperature. From the various relations involved a knowledge of the PVT relations for the given substance, over the range of temperature and pressure to be covered, serves to give a complete evaluation of the difference in the given thermodynamic property between the standard state and the real state at each pressure and temperature. Similarly, it suffices to have at each temperature values of $PV = f(P)$ together with values at the given temperature of $[\partial(PV)\ \partial T]_P$ at the various pressures involved.

The procedure for calculating mole fractions of the components present at equilibrium in a reaction proceeding at constant volume in the gaseous phase may be illustrated in a simple way for the case where the concentrations are small enough and the temperatures high enough for the gases to be assumed ideal. For the reaction

$$CO(g) + \tfrac{1}{2}O_2(g) \rightarrow CO_2(g) \quad (8.266)$$

the value of the standard change in free energy at 3000°K is

$$\Delta G^\circ_{3000} = -6.44 \text{ kcal/mole} \quad (8.267)$$

giving

$$\log K_{3000} = 0.4691 \qquad K_{3000} = 2.945 \quad (8.268)$$

One mole of carbon monoxide and one-half mole of oxygen are placed in a vessel having a volume of 50 l. Assume the gases ideal and that equilibrium is established at 3000°K. It is desired to calculate the mole fractions of the three gases at equilibrium at 3000°K. Now

$$K = (a_{CO_2})/(a_{CO})(a_{O_2})^{1/2} \quad (8.269)$$

If the gases are ideal, the activity may be replaced by the pressure. Accordingly,

$$K = P_{CO_2}/P_{CO}P_{O_2}{}^{1/2} \quad (8.270)$$

If x is the number of moles of CO_2 present at equilibrium, then the number of moles of CO and O_2 at equilibrium are $1 - x$ and $(1 - x)/2$, as summarized in the following table:

TABLE 8.1

Gas	Number of moles, n			Mole fraction at equilibrium
	Initial	Change	At equilibrium	
CO	1	$-x$	$1 - x$	$\frac{2(1-x)}{3-x}$
O_2	$\frac{1}{2}$	$-\frac{x}{2}$	$\frac{1-x}{2}$	$\frac{1-x}{3-x}$
CO_2	0	x	x	$\frac{2x}{3-x}$
Total.......	$1\frac{1}{2}$	$-\frac{x}{2}$	$\frac{3-x}{2}$	1

The pressure of each gas at equilibrium is equal to the number of moles of that gas at equilibrium multiplied by RT/V. So thus, at equilibrium, for a volume of 50 l,

$$P_{CO} = (1 - x)\frac{RT}{V} \quad (8.271)$$

where RT/V is constant,

$$\frac{RT}{V} = \frac{(0.0820544)(3000)}{(50)} = 4.923 \quad (8.272)$$

since the gas constant must be given in liter-atmospheres. Similarly,

$$P_{O_2} = \frac{(1 - x)}{2}\frac{RT}{V} \quad (8.273)$$

and

$$P_{CO_2} = \frac{xRT}{V} \quad (8.274)$$

Then

$$K = \frac{x}{(1 - x)^{3/2}}\left(\frac{2V}{RT}\right)^{1/2} \quad (8.275)$$

So

$$\frac{x}{(1 - x)^{3/2}} = (2.945)(2.462)^{3/2} \quad (8.276)$$

The value of x may be determined by squaring both sides and solving the resulting cubic equation for the root between 0 and 1.

For many calculations in thermodynamics it is convenient to have analytical expressions for the vapor pressure of condensed substances, liquid or solid, as a function of the temperature. Of the many different equations proposed the most generally useful equation appears to be the Antoine equation of three constants [2, 7, 8]:

$$\log P = A - \frac{B}{C + T} \quad (8.277)$$

The Antoine equation may also be expressed explicity in terms of temperature,

$$T = \frac{B}{(A - \log P)} - C \quad (8.278)$$

Over moderately large ranges of temperature, this reproduces the experimental observations with great accuracy; therefore it can be extrapolated reasonable distances beyond the range of the experimental observations. Some other equations that contain five or six constants, values of which have been juggled to fit the experimental observations accurately over a small range of temperature, may be quite useless beyond the range of measurement.

Calculation of the heat of vaporization of a given substance under its own vapor pressure at a given temperature resolves itself into obtaining an accurate value for the change of vapor pressure with temperature at the given temperature and a value for the difference in the molal volumes in the gaseous and liquid phases. The former is obtained from experimental observations of vapor pressure at various temperatures, or, by calculation, from an equation, such as the Antoine equation, set up to represent experimental observations. The difference in the molal volumes of the substance in the gaseous and liquid phases is calculated from values of the compressibility factor for the gas at the given temperature and pressure and the density of the liquid at the given temperature and pressure. Accurate calculation of the heat of vaporization in this way may be illustrated as follows: Given, for a given substance, the Antoine equation representing the vapor pressure of the liquid over the required range of temperature, one may derive, from the fundamental thermodynamic equations, the following expression for the heat of vaporization at saturation pressure:

$$\Delta H_v = (2.30259)BRT^2z\frac{[1 - PV(\text{liq})/zRT]}{(C + T)^2} \quad (8.279)$$

This shows that the heat of vaporization at a temperature T may be calculated if, for the given temperature and pressure, the compressibility factor z, the molal volume of the liquid, and the constants B and C of the Antoine equation are known.

The process of the thermal ionization of gases can be subjected to thermodynamic scrutiny [9]. Consider the reaction of ionization of a neutral atom to form a positive ion of the same element and a free electron, all in the gaseous state at some appropriate high temperature:

$$M(\text{gas}) = M^+(\text{gas}) + \text{electron (gas)} \qquad (8.280)$$

From the appropriate equations (see Part 5, Chaps. 1 and 2), one may evaluate the free energy function, $(G^\circ - H_0^\circ)/T$, for each of the species in the foregoing equation at the given high temperature. From appropriate other data, one may obtain the value for the change in heat content (enthalpy) of the foregoing process at 0°K, ΔH_0°, which is simply the heat of ionization.

From Eq. (8.244), we have the relation

$$\ln K = -\frac{1}{R}\left[\frac{\Delta H_0^\circ}{T} + \Delta\left(\frac{G^\circ - H_0^\circ}{T}\right)\right] \qquad (8.281)$$

When this equation is applied to the reaction given by Eq. (8.280), ΔH_0° is given by the heat of ionization at 0°K, and the last term inside the brackets on the right side is given by the relation

$$\Delta\left(\frac{G^\circ - H_0^\circ}{T}\right) = \left(\frac{G^\circ - H_0^\circ}{T}\right)_{\text{electron}} + \left(\frac{G^\circ - H_0^\circ}{T}\right)_{M^+} - \left(\frac{G^\circ - H_0^\circ}{T}\right)_M \qquad (8.282)$$

For any monatomic gaseous species, having only translational and electronic excitation energy,

$$\left(\frac{G^\circ - H_0^\circ}{T}\right) = \left(\frac{G^\circ - H_0^\circ}{T}\right)_{\text{translation}} + \left(\frac{G^\circ - H_0^\circ}{T}\right)_{\text{electronic excitation}} \qquad (8.283)$$

$$\left(\frac{G^\circ - H_0^\circ}{T}\right)_{\text{translation}} = R - R\ln\left[\left(\frac{2\pi mkT}{h^2}\right)^{3/2}\frac{V}{N}\right] \qquad (8.284)$$

$$\left(\frac{G^\circ - H_0^\circ}{T}\right)_{\text{electronic excitation}} = -R\ln\Sigma A_i \qquad (8.285)$$

In the foregoing equations, m is the mass of an individual particle of the species, k is the Boltzmann constant, h is the Planck constant, V is the volume of one mole of the species at the given pressure, N is the Avogadro number, and

$$\Sigma A_i = g_0 + g_1 e^{-(\epsilon_1-\epsilon_0)/kT} + g_2 e^{-(\epsilon_2-\epsilon_0)/kT} + \cdots + g_i e^{-(\epsilon_i-\epsilon_0)/kT} \qquad (8.286)$$

In this equation, g_0 is the multiplicity of the ground state of the given species, and g_i is that for the ith state. The subscripts refer to energy levels 0, 1, 2, etc., to the ith state. The quantity $\epsilon_i - \epsilon_0$ is simply the energy of the ith state referred to the energy of the ground state.

With the appropriate numerical values substituted into Eq. (8.281), for the process given by Eq. (8.280), one obtains the value of the equilibrium constant K. For the reaction given by Eq. (8.280),

$$K = \frac{f[\text{electron (gas)}]f[M^+(\text{gas})]}{f[M(\text{gas})]} \qquad (8.287)$$

In the temperatures likely to be encountered in such processes, the fugacity f can be replaced, without much error, by the partial pressure for each species. The extent of the ionization of the gas, M, can thus be readily calculated.

References

1. Lewis, G. N., and M. Randall: "Thermodynamics and the Free Energy of Chemical Substances," McGraw-Hill, New York, 1923. Revised by Pitzer, K. S., and L. Brewer: McGraw-Hill, New York, 1961.
2. Rossini, F. D.: "Chemical Thermodynamics," Wiley, New York, 1950.
3. Rossini, F. D., K. S. Pitzer, R. L. Arnett, R. M. Braun, and G. C. Pimentel: "Selected Values of Physical and Thermodynamic Properties of Hydrocarbons and Related Compounds," American Petroleum Institute Research Project 44, Carnegie Institute of Technology Press, Pittsburgh, Pa., 1953.
4. Rossini, F. D., D. D. Wagman, W. H. Evans, S. Levine, and I. Jaffe: "Selected Values of Chemical Thermodynamic Properties," *Natl. Bur. Standards Circ.* 500, 1952.
5. "NBS-NACA Tables of Thermal Properties," National Bureau of Standards, Washington, D.C.
6. Wagman, D. D., J. E. Kilpatrick, W. J. Taylor, K. S. Pitzer, and F. D. Rossini: *J. Research Natl. Bur. Standards* **34**: 143 (1945).
7. Antoine, C.: *Compt. rend.*, **107**: 681, 836, 1143 (1888).
8. Thomson, G. W.: *Chem. Revs.*, **38**: 1 (1946).
9. Saha, M. H., and B. N. Srivastava: "A Treatise on Heat," The Indian Press, Calcutta, India, 1950.

Chapter 9

Chemical Kinetics

By RICHARD M. NOYES, University of Oregon

RESULTS OF KINETIC OBSERVATIONS

1. Introduction

Chemical kinetics is concerned with the rate of chemical change and with the use of rate measurements to elucidate the mechanisms by which chemical systems approach thermodynamic equilibrium. Although the *direction* of chemical change is determined solely by thermodynamic considerations, the *rate* of the change does not necessarily bear any simple relationship to the stoichiometry of the reaction or to the changes in any thermodynamic quantities.

2. Experimental Techniques [1]†

Chemical reactions can be classified roughly as slow or fast. *Slow* reactions require at least a few seconds to proceed half way to equilibrium. Reaction systems can be prepared by conventional mixing of reagents, and the extent of chemical change can be followed by chemical analysis or by measurement of any appropriate physical property at suitable intervals. Because many measurements are taken and because the differences between them are of prime importance, the procedure must give accurate results with a minimum of labor. In the study of slow reactions, the measurement of time is usually much more accurate than that of extent of chemical change.

Fast reactions having reaction times of a few milliseconds or more can be studied by techniques that mix solutions rapidly and follow their subsequent behavior by physical methods. Shorter reaction times down to the order of microseconds can be studied by *relaxation* techniques. An existing equilibrium may be suddenly perturbed by altering temperature, pressure, or electric field, and the subsequent behavior can be followed with appropriate circuitry. The perturbation may also be periodic, as with high-frequency sound. Eigen and de Maeyer [2] have prepared an excellent review of techniques for the study of fast reactions.

Temperature has a profound effect on the rates of most chemical reactions, and it must be carefully controlled in a kinetic study.

3. Dependence of Rate on Concentration [3]

The rate of a chemical reaction can frequently (but by no means invariably) be described over a wide range of concentrations by an equation of the form

$$\frac{d[D]}{dt} = k[A]^m[B]^n[C]^p \tag{9.1}$$

where the terms in brackets are concentrations of chemical species. The species A, B, and C may be reactants or products in the net chemical reaction or may not even appear in the equation for that reaction. The exponents m, n, and p may be positive, negative, or zero and may be integral or nonintegral. The quantity k is called the *specific reaction-rate constant* or *rate constant;* it is independent of changes in concentrations of reactant species but is frequently strongly dependent on the temperature and on the solvent (if any). Such a reaction is said to be mth order in A and nth order in B; the over-all reaction is $(m + n + p)$th order.

Fit of experimental data to a simple reaction order is often accomplished with an *integrated* form of (9.1). This is done most easily when the rate depends only upon concentrations of reactant species and when these are present in stoichiometric amounts so that the ratio of concentrations does not change as reaction proceeds. If a is the initial concentration and x is the extent of reaction at time t, the integrated form is given in Table 9.1. Thus, if the reaction is first order, a plot of ln $(a - x)$ against t will give a straight line of slope $-k$.

TABLE 9.1

Order	Rate equation	Integrated equation
0	$\frac{dx}{dt} = k$	$kt = x$
½	$\frac{dx}{dt} = k(a - x)^{1/2}$	$kt = 2[a^{1/2} - (a - x)^{1/2}]$
1	$\frac{dx}{dt} = k(a - x)$	$kt = \ln \frac{a}{a - x}$
2	$\frac{dx}{dt} = k(a - x)^2$	$kt = \frac{1}{a - x} - \frac{1}{a} = \frac{x}{a(a - x)}$
3	$\frac{dx}{dt} = k(a - x)^3$	$kt = \frac{2ax - x^2}{2a^2(a - x)^2}$

A *differential* method of treating the data is to plot $a - x$ against t and to estimate dx/dt with a tangent meter or otherwise. If m is the order of the reaction, a plot of log (dx/dt) against log $(a - x)$ gives a straight line of slope m. This method suffers from

† Numbers in brackets refer to References at end of chapter.

the fact that dx/dt is not known as accurately as x at a given point. However, the differential method is more apt to reveal deviations from the exact form of Eq. (9.1).

It must be emphasized that the *order* of a chemical reaction is an empirical quantity used to describe experimental data over a range of concentrations. No reaction will continue to obey Eq. (9.1) as the system approaches equilibrium, and for many reactions an equation of this form does not fit the data over any significant range of concentrations. However, this equation describes the rates of many reactions over a sufficiently wide range of conditions to make the concept of order empirically useful.

4. Reversible Reactions

The discussion in Sec. 3 applies to systems so far removed from thermodynamic equilibrium that net rate of change can be equated to the rate of reaction in one direction only. Eventually the forward and reverse rates become comparable and are of course equal when equilibrium is attained. If the kinetics of the forward and reverse reactions are known, it is usually possible to integrate the combined rate equation to determine the net chemical change as a function of time.

If the kinetics of a reaction are known in one direction only, the form of the rate equation for the reverse reaction must be such that the two rates are equal at all conditions of thermodynamic equilibrium. This requirement greatly limits the kinetic equations which may be proposed for the reverse reaction, but it does not uniquely determine the proper equation [4].

5. Effect of Temperature

The rates of most chemical reactions increase rapidly with increasing temperature. The data can usually be fitted empirically by the Arrhenius equation of the form

$$k = Ae^{-E/RT} \tag{9.2}$$

A is called the *frequency factor* (although the dimensions of A include concentration terms for all except first-order reactions), E is called the *energy of activation* and is almost invariably reported in calories per mole, R is the gas constant, and T is the absolute temperature. The energy of activation is calculated from the slope of the straight line obtained by plotting $\log k$ against $1/T$. The frequency factor can be evaluated from the equation of the best straight line in this plot; alternatively, $\log A$ is the slope of a plot of $T \log k$ against T.

6. Other Factors Affecting Reaction Rate

Although kinetic studies of most chemical reactions have been limited to examining effects of concentration and of temperature, changes in other controllable factors can influence rates. Such factors include pressure, solvent (often with relation to properties such as dielectric constant and viscosity), concentration of inert electrolyte, isotopic substitution, or changing structure in the nonreacting portion of a molecule. Effects of changing such factors often provide very useful information about the mechanism of a reaction.

THEORETICAL INTERPRETATION OF CHEMICAL KINETICS

7. Introduction

Kinetic data may be examined from two objectives which are not mutually exclusive. One objective is the development of a theoretical interpretation of the general process of chemical change. The other objective is the elucidation of the molecular species that actually undergo reaction in a specific example.

It is generally assumed that any reaction, no matter how complicated, can be explained in terms of *elementary processes.* Such a process takes place in a single step and involves one, two, or possibly three molecules. Such steps are called unimolecular, bimolecular, or termolecular, respectively. The rate of a virtually irreversible unimolecular process is first order in the reacting species; the rate of such a bimolecular process is second order, etc. However, a first-order reaction is not necessarily a unimolecular elementary process. The *order* of a reaction is experimentally determinable; the *molecularity* is a hypothetical interpretation of the observations and can never be demonstrated unequivocally.

Theories of chemical kinetics attempt first to understand the rates at which elementary processes proceed. Elucidation of mechanism requires the combination of elementary processes in a manner consistent with experimental observations.

The theories of chemical kinetics can be further subdivided. The so-called *collision theory* attempts a detailed mechanistic picture in terms of the equations for molecular collisions developed in Chap. 2. The so-called *absolute reaction rate theory* is a quasi-thermodynamic approach which assumes that molecules at the instant of reaction are still in statistical equilibrium with the other species in the system. The two approaches lead to equations which differ considerably in general form. However, the conclusions are in agreement for those systems to which both approaches are easily applicable.

8. Collision Theory of Bimolecular Gas Reactions [5]

The collision theory for bimolecular reactions assumes that reaction occurs between two molecules when they collide in the proper relative orientation with sufficient energy. The formulas are developed from the simple kinetic theory of gases. This involves assuming that the total volume of the individual molecules is small compared with the total volume of the gas and that the distribution of translational energies is that predicted for particles exerting no forces at a distance and undergoing elastic collisions. For such a gas, it was shown in Chap. 2 that the number of collisions between A and B molecules per cubic centimeter per second is

$$z = \left[\frac{8\pi RT(M_A + M_B)}{M_A M_B}\right]^{1/2} \sigma_{AB}{}^2 n_A n_B = Z n_A n_B \tag{9.3}$$

where subscripts A and B refer to different molecular species, M is molecular weight, n is number of molecules per cubic centimeter, and σ_{AB} is the minimum separation of the centers of A and B (assumed to be spherical) during what is regarded as a collision. If the two colliding molecules are identical,

$$z = \left(\frac{4\pi RT}{M}\right)^{1/2} \sigma^2 n^2 = Zn^2 \tag{9.4}$$

In its simplest form, the collision theory assumes that the only energy available for producing reaction is the energy of relative motion along the line of centers at the instant of collision. The number of collisions per cubic centimeter per second for which this relative energy exceeds E' cal/mole is $ze^{-E'/RT}$. Then the rate constant for the second-order reaction is given by

$$k = PZe^{-E'/RT} \qquad \text{cm}^3\text{/molecule-sec} \tag{9.5}$$

where P is a probability factor introduced because steric or other reasons may prevent reaction during some collisions which satisfy the energy requirements.

Equation (9.5) is not quite identical with the Arrhenius equation (9.2) because Z varies with T. However, the temperature variation of the exponential term is so great that experimental errors completely mask the slight curvature which (9.5) predicts for a plot of ln k against $1/T$. The energies of activation in the two equations are related by

$$E = E' + \frac{RT}{2} \tag{9.6}$$

This theory has been tested for a number of second-order gas reactions by comparing observed rate constants with those calculated using experimentally determined activation energies. If reasonable collision diameters are assumed, values of P often lie between 0.01 and 1. A few values of 10 or more have been reported, but it is not clear whether the deviation from unity is beyond the uncertainty introduced by the possible error in E'. The values of P calculated for some reactions are very much less than 10^{-2}. Since the probability or steric factor cannot be predicted reliably, the theory has had only limited success but has shown that rates are not too fast to be seriously inconsistent with a collision mechanism in which the activation energy comes from relative translational motion.

9. Collision Theory of Unimolecular Gas Reactions [6]

Some substances at pressures more than a few millimeters decompose with first-order kinetics at rates independent of the presence of other substances. The data seem to require that the molecules are "activated" by a collisional process but are not influenced by other molecules at the instant of decomposition.

The first-order rate constant will be

$$k = \frac{1}{n} \sum_j k_j n_j \tag{9.7}$$

where n is the total number of molecules per cubic centimeter and the summation is over the activated states. The energy of an activated molecule is very much greater than the average energy per molecule, and it appears that deactivation almost invariably occurs at the first collision. Then

$$Znn_j^\circ = Znn_j + k_j n_j \tag{9.8}$$

where n_j°, the number of molecules that would be in the jth state if there were no decomposition, is calculable by customary methods from the energy and degeneracy of the state. The left side of (9.8) is the rate of activation (and of deactivation in the absence of decomposition); the right is the sum of the rates of deactivation and decomposition. From (9.8)

$$n_j = \frac{n_j^\circ}{1 + k_j/Zn} \tag{9.9}$$

As long as n is sufficiently large, n_j/n is almost independent of n and the first-order rate constant is independent of pressure. As the pressure is reduced, n_j/n becomes proportional to n and the first-order rate constant becomes proportional to pressure. The equations for a bimolecular reaction are approached as a limit when most of the activated molecules decompose before they can be deactivated.

Any extension of the theory must assign values to the k_j's. Available data definitely rule out the simple assumption that k_j is negligibly small for states having less than a certain critical energy and is constant for states having more than this energy. They also seem to be inconsistent with a theory by Slater [7] that treats the internal motion of the molecule as the sum of noninteracting orthogonal vibrations of different frequency and that assumes reaction to occur when superposition of random phases contributes enough extension to a particular bond.

The best theoretical treatment still seems to be that developed by Kassel [6]. Energy of vibrational excitation is distributed at random among m quantized oscillators and is allowed to migrate from one oscillator to another; k_j is then proportional to the probability that at least a certain critical energy will be associated with a specific mode of vibration. Although this theory can hardly be a unique description, reasonable values of the parameters permit data to be fitted over a very wide range of pressures [8]. More refined treatments [9] have been applied to specific reactions.

To be satisfactory, the collision theory must provide a sufficiently high rate of activation. The observed rate is much more than the rate of collisions having relative translational energy in excess of E. However, if the rotational and vibrational energy of the colliding molecules is also available for transfer during collision, it is possible to account for the rate of production of activated molecules with the required internal energy. Therefore, the collisional theory of unimolecular gas reactions appears to be in satisfactory agreement with experimental observations.

10. Statistical-Thermodynamic Theory of Reaction Kinetics [10]

During the reaction of a molecule, the positions of the individual atoms may be regarded as changing by a continuous process and at a rate which is slow compared to the motion of the electrons. During this process, one configuration is critical in the sense that the probability of completion of the reaction is great once this configuration has been attained. Thus, if the process can be regarded simply as passage over an energy barrier, the critical configuration will be the one of maximum potential energy. This critical configuration for reaction is called the *activated complex* or *transition state*.

The absolute reaction-rate theory assumes that the transition state may be regarded as a distinct molecular species whose concentration can be calculated by statistical methods. It also assumes that reaction can be described as passage along an extension of one of the normal vibrational modes of the reactant molecule while the energies of the other modes retain a normal statistical distribution. The first assumption requires that the subsequent reaction of molecules that attain the transition state does not seriously disturb the equilibrium distribution of molecules capable of attaining that state easily. This assumption fails badly for unimolecular gas reactions at low pressures (Sec. 9) and for very fast reactions in solution (Sec. 12), but it is probably valid for most other situations of interest. The assumption that the other modes can be treated independently requires frequent interaction with other molecular systems during passage along the reaction coordinate. This assumption is probably quite good for reactions in solution but poor for reactions in dilute gases [11].

The theory also assumes that passage through the transition state may be regarded as a translation through a distance δ which is made as small as is permissible by the uncertainty principle. If a system is constrained between limits δ apart, the classical average time per passage in a specified direction is given by

$$\tau = \delta \left(\frac{2\pi m^*}{\kappa T}\right)^{1/2} \tag{9.10}$$

where m^* is the effective mass for the system undergoing translation and κ is the Boltzmann constant. If $n\ddagger$ is the concentration of transition state species in thermodynamic equilibrium with reactants and products, the rate of reaction in one direction is $n\ddagger/\tau$. If reaction in the direction of interest involves bringing together molecules of A and B, and if passage through the transition state almost always leads to the formation of stable product, then

$$k = \frac{n\ddagger}{n_A n_B \tau} = \frac{K\ddagger}{\tau} \tag{9.11}$$

where $K\ddagger$, the equilibrium constant for the formation of the transition state from A and B, may be regarded as a ratio of partition functions. In principle, $K\ddagger$ can be calculated from the masses, moments of inertia, and fundamental vibrational frequencies of the reactant molecules and the transition state. In practice, many of these quantities are often unknown for the molecules in their ground states and are never known for the transition state. However, it is often possible to estimate values of partition functions with reasonable confidence. One of the fundamental modes of motion of the transition state is the translation between limits separated by width δ; the partition function for this motion is

$$f\ddagger_{tr} = \frac{(2\pi m^* \kappa T)^{1/2}\delta}{h} \tag{9.12}$$

where h is Planck's constant. Combining these equations and using the definitions of thermodynamic functions, we obtain the usual form of the rate constant

$$k = \varkappa \frac{\kappa T}{h} \exp\left(\frac{\Delta S\ddagger}{R}\right) \exp\left(\frac{-\Delta H\ddagger}{RT}\right) \tag{9.13}$$

Here $\varkappa$, the transmission coefficient, is introduced to correct for those systems which enter the transition state but do not continue to reaction. In general, $\varkappa$ is assumed to be unity.

One mode of motion in the transition state has been approximated by a translation, and the partition function for this mode has been separated from the others and incorporated in the $\kappa T/h$ term. The remaining partition functions for the transition state of s atoms include contributions from translation and rotation but from only $3s - 7$ (if transition state is nonlinear) instead of the usual $3s - 6$ vibrational modes. The partition functions for the ground state include contributions from all modes of motion. Hence $\Delta S\ddagger$ and $\Delta H\ddagger$ indicate differences in entropy and enthalpy between a transition state in which one mode of motion has been neglected and a ground state in which all modes are included.

The quantity $\kappa T/h$ is of the order of 10^{13} sec^{-1} at moderate temperatures. Since translation and rotation make much bigger contributions than vibration to the entropies of many simple molecules, and since these contributions can also be estimated well for transition states, the theory permits rough estimates of the A term in the Arrhenius equation (9.2). For unimolecular reactions, this term is of the order of 10^{13} sec^{-1} unless the vibrational entropy of the transition state is considerably different from that of the ground state. For bimolecular gas reactions, the loss of translational entropy in forming a transition state causes A to be of the order of 10^9 to 10^{11} l/mole-sec. Applications to spherical molecules that neglect vibrational contributions give A values identical with those computed by the simple collision theory (Sec. 8). Some reactions having abnormally small values of A appear to involve changes in electronic spin, and the transmission coefficient $\varkappa$ is small for these.

11. Theoretical Estimation of Energies of Activation

Theories of elementary gas reactions discussed in Secs. 8 to 10 permit the A term in the Arrhenius equation (9.2) to be estimated within about an order of magnitude and sometimes a bit better. Even this modest success does not permit reliable estimation of reaction rates unless the energy of activation, E, can also be predicted.

Calculation of an activation energy from first principles would require complete solution of the wave equation for the nuclei and electrons in both ground and transition states; it is out of the question with present computational techniques. Approximate treatments have been attempted for exchange reactions of the type $A + BC \rightarrow AB + C$, where the letters designate atoms or simple groups. These groups are assumed to lie on a line, and a *potential energy surface* is constructed by calculating energies for different AB and BC distances. The calculations are so approximate that barrier heights computed in this way are often in unsatisfactory agreement with experiment. However, this kind of approach sometimes provides valid predictions of activation energy changes resulting from changing one of the species A or C [12].

Activation energies clearly are raised as the dissociation energies of the reactant bonds are increased, are lowered as the reactions become more exothermic, and are also lowered as the reactant atoms become more polarizable. These qualitative generalizations have not yet been expressed in quantitative terms.

Noyes [13] has examined a number of reactions of diatomic molecules of the type $AB + CD \rightarrow AC + BD$ and has developed a method for predicting activation energies empirically if bond distances, dissociation energies, and force constants are known for reactant and product molecules. Benson and co-workers [14] have used a theory based on dipole-dipole interactions to predict activation energies for some of the same reactions and also for the addition of diatomic species to olefinic bonds. Westheimer [15] has successfully predicted activation energies for racemization of biphenyls by considering interatomic repulsions and bending and stretching force constants for the atoms and bonds most affected by the formation of the transition state. These treatments are all too specialized to provide generally valid procedures for estimating activation energies.

12. Reactions in Solution

Collisional theories do not lend themselves readily to a description of reactions in liquid phase because of difficulties associated with defining what is meant by a collision. The absolute reaction-rate theory obviates the need for a detailed picture of the liquid state, and it can appropriately be applied to that great majority of reactions in which molecular species are distributed at random with respect to each other. If the reactant molecules and the transition states are solvated about equally, rates in solution and gas should differ little. If anything, a bimolecular reaction might be slightly faster in solution because not quite so much entropy of translation is lost in forming the transition state. These predictions are satisfied for a number of reactions that seem to go by the same mechanism in gas phase and in nonpolar solvents [16].

Neither absolute reaction rate nor collisional theories can be applied to some reactions, such as the quenching of fluorescence or the recombination of atoms, which take place at every encounter between potential reactants. Unlike the situation in a gas, two inert molecules that encounter each other in a liquid may make several subsequent encounters before they diffuse apart. If a highly reactive reference molecule is suddenly created in a solution of potential reactants distributed at random among inert solvent molecules, the probability of reaction per unit of time is initially determined by the frequency of encounters anticipated for species distributed at random. However, if the particular reference molecule has existed for a finite time without reacting, some of its encounters will be with molecules it has encountered before, and these molecules will necessarily be inert solvent molecules. Hence, the probability of reaction per unit of time is less for an old reference molecule than its initial a priori reactivity was, and the probability of finding a potential reactant near it is less than it would be near a randomly located point in the solution. After a period of about 10^{-9} second in most liquids, the probability of reaction per unit of time for a still unreacted reference molecule will be the probability calculable from Fick's first law for steady-state diffusion of potential reactants into a sink the size of the reference molecule.

Because reactive molecules in solution are not distributed at random with respect to each other, no theory based on such a distribution will apply. In fact, the change of reactivity with time makes it impossible to talk of a rate "constant" for molecules that have not lived long enough for the limiting situation to apply. General kinetic equations have been developed to treat this type of situation and to handle the intermediate case where the probability of reaction per encounter is less than, but significant in comparison with, unity [17].

All the previous discussion has assumed ideal behavior of gases and solutions. Ionic solutions in particular will deviate greatly from ideal behavior. In general, $K\ddagger$ [Eq. (9.11)] is defined as

$$K\ddagger = \frac{[X]}{[A][B]} \frac{\gamma_X}{\gamma_A \gamma_B} \tag{9.14}$$

where X denotes the transition state, brackets denote concentration, and the γ's are thermodynamic activity coefficients referred to ideality at infinite dilution. Since the rate is $[X]/\tau$,

$$k = k_0 \frac{\gamma_A \gamma_B}{\gamma_X} \tag{9.15}$$

where k_0 is the rate constant in infinitely dilute solution. In its simplest form, the Debye-Hückel theory of electrolytes predicts that for an ion

$$\ln \gamma = -\frac{N^2 \epsilon^3 (2\pi)^{1/2} z^2 \mu^{1/2}}{(DRT)^{3/2}(1{,}000)^{1/2}} = -Gz^2\mu^{1/2} \tag{9.16}$$

where N is Avogadro's number, ϵ is the electronic charge in esu, D is the dielectric constant, z is the algebraic charge on the ion in electronic units, and μ is the ionic strength defined by

$$\mu = \tfrac{1}{2} \sum_i c_i z_i^2 \tag{9.17}$$

where c is concentration in moles per liter and the summation is taken over all ionic species present in the solution. Since the charge on X is the algebraic sum of the charges on A and B,

$$\ln k = \ln k_0 + 2Gz_A z_B \mu^{1/2} \tag{9.18}$$

The theory predicts that reactions between ions of like charge are accelerated by addition of electrolyte containing no common ion, and reactions between ions of opposite charge are decelerated. A reaction between an ion and a neutral molecule will be affected very much less. Reactions between ions of opposite charge appear to satisfy these predictions quantitatively as long as the ionic strength is low enough for Eq. (9.16) to be applicable. For reactions of ions with the same charge, ionic strength is a less satisfactory criterion, and the rate often shows specific dependence on the nature of ions of charge opposite to the reactants [18].

In liquid phase, solvent molecules undoubtedly take part in many reactions including those involving ionic species. Kinetic measurements cannot provide information on the degree of solvent participation because the concentration of solvent cannot be varied without affecting the nature of the medium.

ELUCIDATION OF CHEMICAL MECHANISM

13. Criteria for a Satisfactory Mechanism [19]

The *mechanism* of a chemical reaction is the detailed combination of reversible and irreversible elementary processes whose net consequence is the reaction of interest. Kinetic data furnish the most direct clues for discovering the mechanism in any specific example. However, in principle an infinite number of microscopic mechanisms can be proposed to explain any set of macroscopic observations. A molecular mechanism can never be proved in a rigorous sense, but kinetic measurements can often disprove a mechanism and show it is not consistent with observations. Confidence increases with the volume of data consistent with a simple interpretation. Mechanisms of a small but significant number of reactions now seem to be established beyond question.

The rate is proportional to the concentration of transition states, and this concentration is determined by an equilibrium with reactants. Therefore, the order with respect to reactants is a direct measure of the number of reactant molecules in the transition state. Although the kinetic order tells the empirical formula of the transition state, it does not furnish any information as to the detailed structure or the mode of formation. As shown in Sec. 15, a first-order reaction does not necessarily proceed by the unimolecular decomposition of an isolated molecule.

No mechanism is acceptable if the formation of any intermediates from the reactants is endothermic by more than the energy of activation. Thus, the decomposition of gaseous nitrogen pentoxide cannot involve the step $N_2O_5 \rightarrow N_2O_4 + O$, for this would require a minimum of 61 kcal/mole, and the energy of activation is only about 25 kcal/mole.

The principle of microscopic reversibility or detailed balancing requires for a system in equilibrium that forward and reverse rates must be identical for every individual reaction path. If any intermediate is postulated prior to the rate-determining step, the break-up of the intermediate must go by a path that is the exact reverse of its formation. This requirement is easy to overlook in the zeal to develop a mechanism consistent with the data.

Mechanisms frequently lead to predictions that are capable of experimental test. Such tests might include rate of isotopic exchange, formation or failure to form isomers, rate of loss of optical activity, etc. Many mechanisms for organic reactions have been discarded because it was shown that proposed intermediates were stable under the conditions employed for the over-all reaction. Any satisfactory mechanism must meet all those challenges which the most ingenious investigator can devise.

14. Consecutive Reactions

An over-all chemical reaction may take place by a sequence of elementary processes, many or all of which are potentially reversible. A detailed discussion of this situation has been published elsewhere [20], but a few of the principles can be illustrated here.

Such a sequence can be represented schematically as in Fig. 9.1. Here A is the initial state of the reactant molecules, D is the final state of the product molecules, and B and C are metastable states formed during the reaction. Although single letters have been

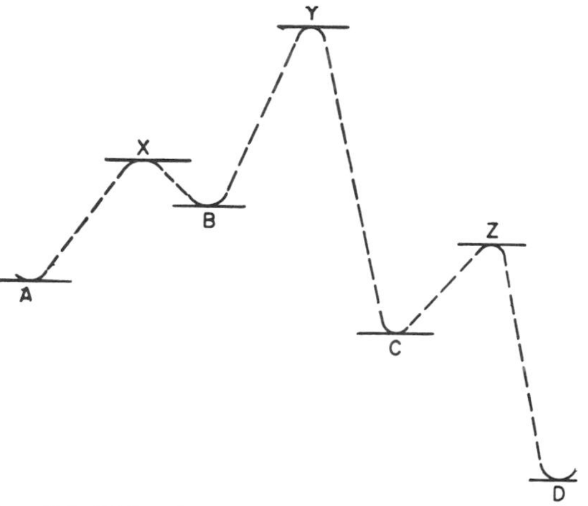

FIG. 9.1. Molar free energy as a function of reaction coordinate.

used in the figure, each state represented there must contain enough different molecules so that its chemical stoichiometry is the same as that of any other state represented. Species present only in B and C are called *intermediates,* a term that should never be applied to transition states like X, Y, and Z.

The abscissa or reaction coordinate in Fig. 9.1 can be given a precise definition only if the extent of reaction can be described in terms of a single parameter such as in the stretching dissociation of a single bond. Often no precise definition is intended, and the plot merely shows the sequence in which reaction steps take place.

Sometimes the ordinate in Fig. 9.1 is taken as the potential energy of the reacting molecules. Since the rate of reaction is determined by $\Delta G\ddagger$ (which is $\Delta H\ddagger - T\,\Delta S\ddagger$), many authors designate the ordinate as molar free energy. In this case, it is difficult to construct a continuous curve because the free energies of the transition states X, Y, and Z are calculated with the use of one less mode of motion than with the other states. The difficulty can be devoutly exorcised by including all modes of motion in the transition state but by choosing a value of δ such that the H and TS terms exactly cancel for the pseudo-translational motion along the reaction coordinate.

The kinetic behavior of the system can be described by differential equations like

$$\frac{d[A]}{dt} = -k_1[A] + k_{-1}[B] \tag{9.19}$$

$$\frac{d[B]}{dt} = k_1[A] - k_{-1}[B] - k_2[B] + k_{-2}[C] \tag{9.20}$$

where $K_1 = k_1/k_{-1}$ is the equilibrium constant for the reaction $A \rightarrow B$ and $K_2 = k_2/k_{-2}$ is that for $B \rightarrow C$. Although exact solution of these equations is often impossible for reactions of higher than first order, certain approximations are frequently useful.

One such approximation is that of a *rate-determining step*. If one transition state (such as Y in the figure) has a much greater free energy than any other, and if (as in Fig. 9.1) a subsequent intermediate like C reacts to form final products with a rate constant greater than that for attainment of transition state Y, the rate of the over-all reaction $A \rightarrow D$ is

$$\frac{d[D]}{dt} = K_1k_2[A] = k_2[B] \tag{9.21}$$

The approximation assumes that k_{-1} is so much greater than k_2 that almost every B reverts to A rather than going on to C. Hence A and B can be assumed to be in a thermodynamic equilibrium not significantly perturbed by the reaction of B to form C.

If B is much less stable than A (as in the figure) and if the free energies of X and Y are approximately equal, it is permissible to apply the *uniform flux sequence* approximation that the net rate of $A \rightarrow B$ is the same as that of $B \rightarrow C$. This rate is $k_\alpha[A]$, where

$$\frac{1}{k_\alpha} = \frac{1}{k_1} + \frac{1}{K_1k_2} \tag{9.22}$$

This equation reduces to the rate-determining step approximation if $k_{-1} \gg k_2$.

The above discussion has been confined to differential equations describing rate as a function of concentrations of reactants. Experimental measurements often need to be compared with integrated equations for total chemical change. For consecutive first-order reactions (such as the decompositions of a series of radioactive nuclei), the mathematical solutions are well known [21]. Integration is much more difficult for consecutive reactions that are not first-order, but various numerical and approximation methods have been proposed for specific examples [22].

15. Decomposition of Nitrogen Pentoxide

The decomposition of nitrogen pentoxide, N_2O_5, to nitrogen dioxide and oxygen provides an excellent example of the establishment of a mechanism involving consecutive steps. The rate of decomposition of the pure material is clearly first-order in N_2O_5 over a very wide range of pressures. However, the reaction is not an elementary unimolecular process.

The reaction was long thought to follow the mechanism

$$N_2O_5 \rightarrow N_2O_3 + O_2 \tag{9.23}$$

$$N_2O_3 \rightleftharpoons NO_2 + NO \tag{9.24}$$

$$NO + N_2O_5 \rightarrow 3NO_2 \tag{9.25}$$

It was postulated that (9.23) was rate-determining. Some difficulties were always recognized. Since O_2 is in a triplet state, (9.23) involves a change in multiplicity, which is inconsistent with the "normal" frequency factor [Eq. (9.2)] of 10^{13} sec^{-1}. Since the first-order rate constant for the gas reaction does not fall off until pressures well below 1 mm, this mechanism requires unreasonably long lifetimes for the activated molecules. Finally, (9.25) cannot be a simple bimolecular process, for separate studies have shown that the rate is first order in N_2O_5 but is zero order in NO.

A more satisfactory mechanism is.

$$N_2O_5 \rightarrow NO_2 + NO_3 \tag{9.26}$$

$$NO_2 + NO_3 \rightarrow N_2O_5 \tag{9.27}$$

$$NO_2 + NO_3 \rightarrow NO + O_2 + NO_2 \tag{9.28}$$

$$NO + NO_3 \rightarrow 2NO_2 \tag{9.29}$$

If (9.28) is rate-determining, the rapid reactions (9.26) and (9.27) serve merely to establish an equilibrium among the species involved. In this mechanism, the transition state is formed in a bimolecular reaction between NO_2 and NO_3, but it has the same empirical formula (and will show the same kinetics) as the transition state for a unimolecular dissociation of N_2O_5. If enough NO is added to the system, it competes so efficiently with NO_2 for the available NO_3 that (9.29) can be made much faster than (9.27) and (9.28). Under these conditions, (9.26) becomes the rate-determining step for the decomposition of N_2O_5. The reaction with added NO has been studied over a 10^5-fold range of total pressures (0.07 to 7,000 mm), and the observed rates for (9.26) are in good agreement [23] with the theory of unimolecular gas reactions developed in Sec. 9.

16. Parallel Reactions

Sometimes the products of a chemical reaction can be formed by independent parallel paths that proceed at comparable rates under certain conditions. The total rate of reaction is then the sum of the rates by each of these paths. If the kinetics of the paths are different, the observed rate of reaction bears little resemblance to the form of Eq. (9.1). However, each of the parallel paths that can be isolated can be treated as a sequence of elementary processes.

An example is provided by the oxidation of oxalate by manganese (III). The complexes $MnC_2O_4^+$, $Mn(C_2O_4)_2^-$, and $Mn(C_2O_4)_3^{3-}$ have all been identified, and each can decompose in a rate-determining step to initiate formation of manganese (II) and carbon dioxide. The total rate of reaction is the sum of the independent contributions from the different complexes [24].

17. Chain Reactions [25]

Most of the stable molecular species encountered in nature contain an even number of electrons arranged in pairs with opposite spins. Species having an unpaired electron are called radicals and are usually very reactive. The reaction of a radical with a normal molecule will produce one (or a larger odd number) of radicals. If a radical is formed in a system, all the subsequent reactions will continue to produce radical products until there is a reaction with another radical, and the production of a few radicals in a system that is not at thermodynamic equilibrium may produce a considerable amount of chemical change.

A reaction of this sort, called a *chain reaction,* can be illustrated by the behavior of hydrogen and bromine. The rate can be described empirically by the equation

$$\frac{d[\mathrm{HBr}]}{dt} = \frac{k[\mathrm{H_2}][\mathrm{Br_2}]^{1/2}}{1 + k'[\mathrm{HBr}]/[\mathrm{Br_2}]} \tag{9.30}$$

All the data are consistent with the mechanism

$$\mathrm{Br_2} \rightarrow 2\mathrm{Br} \quad (k_1) \tag{9.31}$$
$$\mathrm{Br} + \mathrm{H_2} \rightarrow \mathrm{HBr} + \mathrm{H} \quad (k_2) \tag{9.32}$$
$$\mathrm{H} + \mathrm{Br_2} \rightarrow \mathrm{HBr} + \mathrm{Br} \quad (k_3) \tag{9.33}$$
$$\mathrm{H} + \mathrm{HBr} \rightarrow \mathrm{H_2} + \mathrm{Br} \quad (k_4) \tag{9.34}$$
$$\mathrm{Br} + \mathrm{Br} \rightarrow \mathrm{Br_2} \quad (k_5) \tag{9.35}$$

In a reaction of this sort, (9.31) is called *initiation,* (9.32) and (9.33) are called *propagation,* (9.34) is called *retardation* or *inhibition,* and (9.35) is called *termination*

Mathematical treatment of chain reactions involves the assumption that the radical species are present in very low concentrations. Usually the rates of change in the absolute concentrations of these species are negligible compared to the rate of net chemical change. Then $d[\mathrm{Br}]/dt = 0$ is a satisfactory approximation. Since the rate of initiation of chains must equal the rate of termination at any time, we can equate the rates of (9.31) and (9.35). Then

$$[\mathrm{Br}] = \left(\frac{k_1}{k_5}\right)^{1/2} [\mathrm{Br_2}]^{1/2} \tag{9.36}$$

The rate of (9.32) is $k_2[\mathrm{Br}][\mathrm{H_2}]$. If (9.32) is followed by (9.33), a bromine atom is regenerated and a link of the chain is completed with the net production of two molecules of HBr. If (9.32) is followed by (9.34), no net chemical change is produced by the two steps. The total rate of production of HBr is twice the rate of (9.32) multiplied by the probability that a hydrogen atom will react by (9.33) rather than by (9.34). If this probability is defined as the ratio of the rate of (9.33) to the sum of the rates of (9.33) and (9.34), it is $k_3[\mathrm{Br_2}]/(k_3[\mathrm{Br_2}] + k_4[\mathrm{HBr}])$. The combination of these results gives

$$\frac{d[\mathrm{HBr}]}{dt} = \frac{2k_2(k_1/k_5)^{1/2}[\mathrm{H_2}][\mathrm{Br_2}]^{1/2}}{1 + k_4[\mathrm{HBr}]/k_3[\mathrm{Br_2}]} \tag{9.37}$$

Although the proposed mechanism gives the proper kinetic behavior, it is necessary to justify the omission of certain steps which may seem as probable as those included. The bond energies of H_2, Br_2, and HBr are 103.6, 45.5, and 86.9 kcal/mole, respectively. Thus (9.31) is the least endothermic initiation reaction we could expect from these molecules. We have omitted the reaction $\mathrm{Br} + \mathrm{HBr} \rightarrow \mathrm{Br_2} + \mathrm{H}$. This reaction is endothermic by 41.4 kcal/mole, and the activation energy must be at least this great. On the other hand, (9.32) is endothermic by only 16.7 kcal/mole, and (9.33) and (9.34) are exothermic. These three reactions should all be much faster than the one that was omitted. Finally, we have omitted termination steps involving H + H and H + Br. Since we can predict from energy considerations that k_3 and k_4 will be much greater than k_2, the time required for a hydrogen atom to generate a bromine atom will be much less than the time required for a bromine atom to generate a hydrogen atom. Almost all the radical species in the system at any instant will be bromine atoms, and (9.35) is by far the most probable termination step.

The above discussion illustrates the type of argument necessary to establish confidence in a mechanism. Unfortunately, few complex mechanisms can be presented with as much certainty.

Although most chain reactions seem to involve radical species, some nonradical chains have been studied. Also, some radicals are so stable that they are unable to propagate reaction chains in certain systems.

18. Branching Chains [26]

For some gaseous oxidation reactions, very little reaction is observed at certain conditions of temperature and pressure, while an explosion occurs upon a slight change in conditions. It is postulated that a chain-carrying intermediate can occasionally react to produce two or more species capable of propagating chains. If this *branching* reaction becomes even infinitesimally more probable than termination, the number of chains increases rapidly and explosion results.

At a given temperature, the explosive region frequently shows both lower and upper pressure limits. The low-pressure limit depends upon the dimensions of the vessel and results because chain carriers diffuse to the walls where heterogeneous termination keeps the average chain length finite. The upper limit is independent of the dimensions and results because some termination reaction involves a higher order than the branching reaction. These reactions are not understood in detail. Apparently their unusual behavior is related to the fact that an oxygen atom is a very reactive nonradical species which can cause production of two radicals.

Chain branching also takes place in a uranium pile or nuclear explosion.

19. Photochemistry [27]

Electromagnetic radiation may cause profound chemical changes in a system. Although photons may interact with molecules in many ways, chemically active species are almost never produced except by absorptions that involve electronic transitions. Therefore, visible and ultraviolet regions are of most interest to the photochemist. Four general processes are available to the molecule after absorption of the photon:

1. The absorbed energy may be dissipated as excess thermal energy. This is the common fate of quanta having wavelengths longer than in the visible.

2. The molecule may fluoresce.

3. The absorbed energy may be used to excite chemical change in another species. Thus, mercury atoms absorb radiation at 2537 Å, and the excited atoms can dissociate hydrogen molecules in subsequent collisions even though hydrogen itself is transparent at this wavelength. Irradiation of hydrogen containing mercury vapor can initiate a number of chain reactions involving hydrogen atoms.

4. The absorbing molecule may undergo chemical change itself. It is customary to treat the initial act of absorption and any subsequent electronic transitions by the Franck-Condon principle [28], which assumes the positions and momenta of the nuclei are not affected during the rapid transition. Frequently the chemical effect is dissociation of the absorbing molecule. If the upper state is repulsive or if its vibrational excitation is greater than that necessary for dissociation, the fragments will separate in a time of the order of that necessary for one vibration. If the upper level is metastable, the molecule may make some vibrations and then undergo a radiationless transition (*predissociation*) to a repulsive state. The inhomogeneous fields produced by neighboring molecules seem to facilitate predissociative transitions. The fragments from the dissociation are usually radicals and may instigate subsequent chemical reactions.

Interpretations of photochemistry use the *postulate of Grotthus* that only the absorbed light is effective, and the *postulate of Einstein* that only one quantum is involved in the primary process. It is customary to measure the *quantum yield* or the number of molecules undergoing chemical change per quantum absorbed. Quantum yields greatly in excess of unity are not uncommon and are interpreted as chain reactions initiated by the species produced from the original absorption of a quantum.

If chains are initiated by continuous homogeneous illumination of the whole system and if they are terminated by recombination of pairs of radicals, the concentration of radicals and the rate of chemical change produced by them will both be proportional to the square root of the rate of absorption of radiation. If the light is regularly interrupted for periods n times as long as the periods of illumination, the total chemical change for a slow cycle will be $1/n$ of that for continuous illumination. If the cycle is very rapid so that the concentration of radicals is not significantly depleted during the dark periods, the chemical change is $(1/n)^{1/2}$ of that for continuous illumination. The dependence of reaction rate on duration of cycle permits calculation of the average lifetime of a radical chain. If this information is combined with measurements of radical concentration or of rate of initiation of chains, it is possible to evaluate absolute rate constants for individual processes in chain reactions [29].

A similar situation exists if the initiating light is intermittent in space rather than in time. If the light is continuous but illuminates only certain regions of the cell, the space-average concentration of radicals in light and dark areas depends upon the ability of radicals to diffuse from illuminated areas and hence upon the size of those areas. Dependence of reaction rate upon the size of the individual illuminated areas makes it possible to calculate diffusion coefficients of these very reactive radicals [30].

20. Heterogeneous Reactions [31]

Many solids react with or catalyze reactions of species in gas or liquid phases. These reactions presumably involve five consecutive steps: (1) diffusion of the reacting molecules to the surface, (2) adsorption on the surface, (3) reaction of the adsorbed molecules, (4) desorption from the surface, and (5) diffusion of the desorbed products away from the surface. The diffusion steps are almost never rate-determining for reactions involving gases but may be so for fast reactions of liquids in contact with solids.

The adsorption steps may involve rapidly reversible *physical adsorption* resulting from van der Waals' (polarization) forces between the adsorbed molecule and the surface. Most chemical reactions involve *chemisorption* which requires an energy of activation and results in binding forces of the order of those in chemical bonds. For a reaction involving chemisorption, step 2, 3, or 4 may be rate-determining.

The kinetics of heterogeneous reactions depend upon the extent to which the surface is covered by various species. Most treatments make use of an *adsorption isotherm* relating the number of adsorbed molecules to the partial pressure of the adsorbed substance. Interpretations are complicated because surfaces are not uniform, and most of the reaction may take place on a very small fraction of the total surface.

References

1. Laidler, K. J.: "Chemical Kinetics," chap. 2, McGraw-Hill, New York, 1950.
2. Eigen, M., and L. de Maeyer: Investigation of Rates and Mechanisms of Reactions, Very Fast Reactions in Solution, p. 895ff., in S. L. Friess, E. S. Lewis, and A. Weissberger (ed.), "Technique of Organic Chemistry," 2d ed., vol. 8, pt. II, Interscience, New York, 1963.
3. Laidler, K. J.: "Chemical Kinetics," chap. 1, McGraw-Hill, New York, 1950. Frost, A. A., and R. G. Pearson: "Kinetics and Mechanism," 2d ed., chaps. 2, 3, Wiley, New York, 1961.
4. Hollingsworth, C. A.: *J. Chem. Phys.*, **20**: 921 (1952).
5. Moelwyn-Hughes, E. A.: "The Kinetics of Reactions in Solution," 2d ed., chap. 1, Oxford University Press, London, 1947.
6. Kassel, L. S.: "Kinetics of Homogeneous Gas Reactions," chap. 5, Reinhold, New York, 1932.
7. Slater, N. B.: "Theory of Unimolecular Reactions," Cornell University Press, Ithaca, N.Y., 1959.
8. Vreeland, R. W., and D. F. Swinehart: *J. Am. Chem. Soc.*, **85**: 3349 (1963).
9. Marcus, R. A.: *J. Chem. Phys.*, **20**: 352, 355, 359, 364 (1952). Johnston, H. S., and J. R. White, *J. Chem. Phys.*, **22**: 1969 (1954).
10. Glasstone, S., K. J. Laidler, and H. Eyring: "The Theory of Rate Processes," chap. 4, McGraw-Hill, New York, 1941.
11. Noyes, R. M.: *J. Phys. Chem.*, **66**: 1058 (1962).
12. Glasstone, S., K. J. Laidler, and H. Eyring: "The Theory of Rate Processes," chap. 3, McGraw-Hill, New York, 1941.

13. Noyes, R. M.: *J. Am. Chem. Soc.*, **88**: 4311, 4318, 4324 (1966).
14. Benson, S. W., and A. N. Bose: *J. Chem. Phys.*, **39**: 3463 (1963). Benson, S. W., and G. R. Haugen, *J. Am. Chem. Soc.*, **87**: 4036 (1965).
15. Westheimer, F. H., and J. E. Mayer: *J. Chem. Phys.*, **14**: 733 (1946). Westheimer, F. H.: *J. Chem. Phys.*, **15**: 252 (1947). Rieger, M., and F. H. Westheimer: *J. Am. Chem. Soc.*, **72**: 19 (1950).
16. Laidler, K. J.: "Chemical Kinetics," chap. 5, McGraw-Hill, New York, 1950.
17. Noyes, R. M.: *Progr. Reaction Kinetics*, **1**: 129 (1961) and previous papers referred to there.
18. Olson, A. R., and T. R. Simonson: *J. Chem. Phys.*, **17**: 1167 (1949). Indelli, A., G. Nolan, Jr., and E. S. Amis: *J. Am. Chem. Soc.*, **82**: 332, 3233, 3237 (1960).
19. Frost, A. A., and R. G. Pearson: "Kinetics and Mechanism," 2d ed., chap. 12, Wiley, New York, 1961.
20. Noyes, R. M.: *Progr. Reaction Kinetics*, **2**: 337 (1964).
21. Frost, A. A., and R. G. Pearson: "Kinetics and Mechanism," 2d ed., pp. 173ff., Wiley, New York, 1961.
22. Frost, A. A., and R. G. Pearson: "Kinetics and Mechanism," 2d ed., chap. 8, Wiley, New York, 1961. Widequist, S.: *Arkiv Kemi*, **8**: 325 (1955).
23. Mills, R. L., and H. S. Johnston: *J. Am. Chem. Soc.*, **73**: 938 (1951). Johnston, H. S., and R. L. Perrine: *J. Am. Chem. Soc.*, **73**: 4782 (1951).
24. Adler, S. J., and R. M. Noyes: *J. Am. Chem. Soc.*, **77**: 2036 (1955).
25. Dainton, F. S.: "Chain Reactions, an Introduction," Methuen, London, and Wiley, New York, 1956.
26. Semenoff, N.: "Chemical Kinetics and Chain Reactions," Oxford University Press, London, 1935.
27. Noyes, W. A., Jr., and P. A. Leighton: "The Photochemistry of Gases," Reinhold, New York, 1941.
28. Franck, J.: *Trans. Faraday Soc.*, **21**: 536 (1926). Condon, E. U.: *Phys. Rev.*, **28**: 1182 (1926).
29. Burnett, G. M., and H. W. Melville: Investigation of Rates and Mechanisms of Reactions, in S. L. Friess and A. Weissberger (eds.), "Technique of Organic Chemistry," vol. 8, pp. 133–168, Interscience, New York, 1953.
30. Noyes, R. M.: *J. Am. Chem. Soc.*, **81**: 566 (1959); **86**: 4529 (1964); *J. Phys. Chem.*, **69**: 3182 (1965). Levison, S. A., and R. M. Noyes: *J. Am. Chem. Soc.*, **81**: 4525 (1964).
31. Laidler, K. J.: "Chemical Kinetics," chap. 6, McGraw-Hill, New York, 1950.

Bibliography

Benson, S. W.: "Foundations of Chemical Kinetics," McGraw-Hill, New York, 1960.

Friess, S. L., E. S. Lewis, and A. Weissberger: Investigation of Rates and Mechanisms of Reactions, "Technique of Organic Chemistry," 2d ed., vol. 8, Interscience, New York, 1962–1963. (Experimental techniques including fast reactions.)

Frost, A. A., and R. G. Pearson: "Kinetics and Mechanism," 2d ed., Wiley, New York, 1961.

Glasstone, S., K. J. Laidler, and H. Eyring: "The Theory of Rate Processes," McGraw-Hill, New York, 1941. (Absolute reaction-rate theory.)

Kassel, L. S.: "The Kinetics of Homogeneous Gas Reactions," Reinhold, New York, 1932. (Collisional theories.)

King, E. L.: "How Chemical Reactions Occur," Benjamin, New York, 1963. (Introduction to field.)

Laidler, K. J.: "Chemical Kinetics," McGraw-Hill, New York, 1950.

National Bureau of Standards: Tables of Chemical Kinetics: Homogeneous Reactions, *Natl. Bur. Stds. (U.S.) Circ.* 510, 1951, and supplements, 1956.

Noyes, W. A., Jr., and P. A. Leighton: "The Photochemistry of Gases," Reinhold, New York, 1941.

Semenoff, N.: "Chemical Kinetics and Chain Reactions," Oxford University Press, London, 1935.

Chapter 10

Vibrations of Crystal Lattices and Thermodynamic Properties of Solids

By E. W. MONTROLL, University of Rochester

1. Introduction

The heat capacity, C_x, of a system of N interacting particles under an external constraint x is defined by $C_x = (\partial Q/\partial T)_x$, the ratio of the added heat to the corresponding temperature rise of the system. The heat capacities of solids are generally measured at constant pressure, while statistical mechanics leads more naturally to formulas for the constant-volume quantity. The thermodynamic formula

$$C_p - C_v = \frac{9\alpha^2 VT}{\kappa}$$

relates C_p to C_v. The quantities α, V, and κ represent, respectively, the coefficient of thermal expansion, the volume, and the compressibility. The heat capacity at constant magnetic field strength H is of considerable importance in low-temperature physics.

Various types of microscopic degrees of freedom in solids make their characteristic contributions to the heat capacity. The constants associated with these degrees of freedom determine their energy levels and hence the temperature at which their effect on the heat capacity becomes important. Measurements of the temperature variation of the heat capacity have been a rich source of information on the electronic, atomic, and molecular dynamics of crystals.

The constituent atoms and molecules of all solids undergo (either through thermal agitation or quantum-mechanical zero-point energy) small oscillations about their equilibrium positions. At temperatures not too close to the melting point a crystalline solid is essentially a set of a large number of coupled oscillators. Einstein pointed out many years ago that the vibrational contribution to thermodynamic properties of solids can be calculated through the application of the quantum theory of the harmonic oscillator. The free electrons in metals are responsible for most of the heat capacity at very low temperatures. Since they obey Fermi-Dirac statistics, their heat capacity is proportional to T as $T \to 0$, while lattice vibrations have $C_v \alpha T^3$. For discussion of the electron theory of metals, see Part 8. Experimental low-temperature C_v data of metals are frequently fitted to

$$C_v = 464.4 \left(\frac{T}{\theta}\right)^3 + aT$$

The appropriate units are calories per mole per degree Kelvin. The θ and a values of a variety of metals are given in Table 10.1. A more complete list can be found in Appendix C of "Phenomena at the Temperature of Liquid Helium" by Burton, Smith, and Wilhelm [1].*

TABLE 10.1. CONSTANTS θ_D AND a FOR LOW-TEMPERATURE HEAT CAPACITIES
[Units are chosen so that $C_v = 464.4(T/\theta_D)^3 + aT$ is in calories per mole per degree Kelvin]

Metal	θ_D	$a \times 10^4$
Al	419	3.48
Ag	229	1.54
Cu	335	1.78
Pt	233	16.07
Pb	90	7.15
Mg	410	42.1
Sn	185	4.0

Anomalies in the form of "λ points" or discontinuities in C_v occur in binary substitution alloys, ferromagnets, ferroelectrics, hydrogen halides, superconductors, various molecular crystals (especially organic ones), etc. These anomalies are generally associated with the disappearance of some kind of long-range order. They are discussed in the chapter on phase transitions. We shall merely point out here that if one wishes to compare the results of a quantitative theory of an anomalous effect with experimental measurements, it is necessary to have an accurate theory of the normal contribution of lattice vibrations to the heat capacities so that the anomalous effect can be obtained by subtraction.

This chapter is a brief survey of the theory of lattice vibrations and their influence on heat capacities of crystals. The equation of state of solids will also be discussed.

The coupled oscillator model of a crystal can be decomposed into its independent normal modes.

* Numbers in brackets refer to References at end of chapter.

If the normal mode frequencies of a crystal with $3N$ degrees of freedom are $\nu_1, \nu_2, \ldots, \nu_{3N}$, then Eq. (2.77), Chap. 2, implies that the internal energy of the crystal is

$$E = \sum_{j=1}^{3N} h\nu_j \left(\frac{1}{2} + \frac{1}{-1 + \exp \theta_j}\right) \qquad \theta_j = \frac{h\nu_j}{kT} \tag{10.1}$$

while its heat capacity is

$$C_v = k \sum_{j=1}^{3N} \frac{(\frac{1}{2}\theta_j)^2}{\sinh^2 (\frac{1}{2}\theta_j)} \tag{10.2}$$

Generally as the number of degrees of freedom becomes infinite, the normal mode frequencies become so dense that a frequency distribution function $g(\nu)$ exists with the property that

$$\int_0^\nu g(\nu)\, d\nu$$

is the number of frequencies less than ν. The heat capacity can be expressed as an integral with respect to ν:

$$C_v = k \int_0^{\nu_L} g(\nu) \frac{(\frac{1}{2}\, h\nu/kT)^2}{\sinh^2 (\frac{1}{2}\, h\nu/kT)}\, d\nu \tag{10.3}$$

where ν_L is the largest normal mode frequency.

The power series expansion of the exponential and hyperbolic functions of (10.1) and (10.2) leads naturally to the following high-temperature formulas [2, 3]:

$$E = 3NkT \left[1 - \sum_{n=0}^{\infty} \frac{(-1)^n B_n}{(2n)!} \left(\frac{h\nu_L}{kT}\right)^{2n} \frac{\mu_{2n}}{\nu_L{}^{2n}}\right] \tag{10.4}$$

$$C_v = 3NkT \left[1 + \sum_{n=1}^{\infty} \frac{(2n-1)(-1)^n B_{2n}}{(2n)!} \left(\frac{h\nu_L}{kT}\right)^{2n} \frac{\mu_{2n}}{\nu_L{}^{2n}}\right] \tag{10.5}$$

The series converge when $h\nu_L/kT < 2\pi$; the B_n's are Bernoulli numbers

$$B_1 = \tfrac{1}{6} \qquad B_2 = \tfrac{1}{30} \qquad B_3 = \tfrac{1}{42} \qquad B_4 = \tfrac{1}{30}$$
$$B_5 = \tfrac{5}{66} \qquad B_6 = \tfrac{691}{2730} \quad \cdots$$

and the μ_n's are the moments of the frequency distribution $g(\nu)$:

$$\mu_n = \sum \nu_j{}^n = \frac{1}{\nu_L} \int_0^{\nu_L} \nu^n g(\nu)\, d\nu \tag{10.6}$$

The calculation of these moments will be discussed in the next sections. It is clear that as $T \to \infty$, (10.5) yields the Dulong-Petit result $C_v \sim 3Nk$.

The low-temperature behavior of C_v depends on the low-frequency form of $g(\nu)$. For example, if there are two constants A and α such that as $\nu \to 0$

$$g(\nu) \sim 3NA\nu^\alpha \tag{10.7}$$

Equation (10.3), when written in terms of the variable $x = h\nu/kT$, has the low-temperature form

$$C_v = \frac{3}{4} NAk \left(\frac{kT}{h}\right)^{\alpha+1} \int_0^\infty \frac{x^{2+\alpha}}{\sinh^2 \frac{1}{2}x}\, dx$$
$$= 3ANk \left(\frac{kT}{h}\right)^{\alpha+1} (2+\alpha)!\,\zeta(2+\alpha) \tag{10.8}$$

where $\zeta(y)$ is the Riemann zeta function,

$$\zeta(y) = 1 + 2^{-y} + 3^{-y} + \cdots$$

[$\zeta(2) = \pi^2/6$ and $\zeta(4) = \pi^4/90$]. Since low-frequency vibrations in a crystal have wavelengths which are very long compared with lattice spacings, the low-frequency behavior of $g(\nu)$ should have the same form as that of an elastic continuum. Hence in a two-dimensional crystal the appropriate value of α should be 1, while it should be 2 in a three-dimensional crystal. The low-temperature heat capacity of a two-dimensional structure should vary as T^2, while that of a three-dimensional material should vary as T^3.

The asymptotic T^3 law was first observed experimentally by Nernst and his collaborators many years ago and has been verified for most materials. The results of Pitzer [4] and De Sorbo [5] on diamond are typical of a nonmetal.

Several important exceptions to the T^3 law [6–9] exist in graphite, gallium, and BN. Experimental C_v behavior of these materials is better fitted by the form $C_v \alpha T^2$; which is, according to (10.8), more appropriate for two-dimensional systems. This is not surprising because the flakiness of pure graphite and BN indicates that the binding forces between crystal layers are very weak.

Thermodynamic properties are then derived by considering these materials as formed of almost independent two-dimensional hexagonal arrays of atoms. Transverse vibrations give the main contribution to thermodynamic quantities. G. F. Newell [10] has made a detailed investigation of weakly interacting layer structures and has reviewed the work of earlier authors. Various other layer structures can be expected to have a T^2 law for the low-temperature heat capacities.

There is also some evidence of a T law in materials such as Se and S which are composed of weakly connected atomic chains [11].

The systematic theory of the variation of thermal properties of polymers and glasses is still in its formative stage. These materials have a network structure of chains with varying length. The work of Dole and his collaborators [12] on the heat capacity of polystyrene is representative of that on polymers. Reversible temperature changes seem to be hard to achieve in polymer systems.

According to Winkelmann [13] the specific heat of glasses can be expressed very well as a linear function of the composition

$$C = \Sigma C_r x_r$$

the x_r's being the weight fraction of oxides occurring in the glasses. The empirical ratio of the thermal factor C_r (so-called Winkelmann factors) to the classical high-temperature specific heat of the particular components has the correct qualitative behavior in that strongly bound atoms in the network contribute low C_r's (reflecting quantum effects) while those of weakly bound atoms are close to the classical value (see Condon [14]).

Dyson [15] has made a first step toward the theory of the heat capacity of networks of particles with varying force constant and masses by studying the normal mode distribution function $g(\nu)$ of a linear chain of coupled springs and mass with random force constants and masses.

At high temperatures anharmonic vibrations yield a contribution above the Dulong-Petit value which is proportional to the temperature (see Born and Brody [16]).

An excellent bibliography of experimental data on heat capacities of solids (as well as other thermal properties) can be found in Partington's treatise on physical chemistry [17].

The complete frequency distribution function $g(\nu)$ is necessary for the description of the behavior of C_v at temperatures out of the range of the two asymptotic formulas (10.5) and (10.8).

2. Debye Theory of Heat Capacities

The first calculation of the distribution of frequencies of normal modes was made by Debye [18]. He postulated a solid to be an elastic continuum. Since a continuum has an infinite number of normal modes, Debye cut the frequency spectrum off at a frequency such that the total number of normal modes was equal to the number of degrees of freedom of the solid. Debye's work was one of the great successes of the early quantum theory. The theoretical heat capacities based on his frequency spectrum are in good agreement with experimental results. The simplicity of the Debye theory combined with this fact has given it a long and fruitful life as the dominant theory of the heat capacity of solids. The classical exposition of the Debye theory was written by Schroedinger [19] (see also Mayer and Mayer [20] and Fowler [21]). Schroedinger collected an enormous amount of heat-capacity data and compared it with the Debye theory.

The equations for wave propagation in an elastic continuum are

$$\frac{1}{c^2}\frac{\partial^2\varphi}{\partial t^2} = \nabla^2\varphi \tag{10.9}$$

where c is the velocity of propagation and φ is the displacement of a point which at equilibrium is located at (x,y,z). The normal modes of vibration which correspond to stationary crystal boundaries are found as follows. Let our solid have the shape of a rectangular box with sides of length l_1, l_2, and l_3. Solutions of (10.9) of the form

$$\varphi = Ae^{2\pi i\nu t}\sin\frac{x\pi n_1}{l_1}\sin\frac{y\pi n_2}{l_2}\sin\frac{z\pi n_3}{l_3} \tag{10.10}$$

exist when (n_1,n_2,n_3) range through the integers 0, 1, 2, 3, ... By substituting (10.10) into (10.9), we find that the frequency ν is related to (n_1,n_2,n_3) by

$$\nu^2 = \frac{c^2}{4}\left[\left(\frac{n_1}{l_1}\right)^2 + \left(\frac{n_2}{l_2}\right)^2 + \left(\frac{n_3}{l_3}\right)^2\right] \qquad n_1,\ n_2,\ n_3 = 0,\ 1,\ 2,\ \ldots \tag{10.11}$$

The total number of frequencies less than ν, $N(\nu)$, is proportional to the volume of the ellipsoid defined by (10.11) in (n_x,n_y,n_z) space. The volume of the ellipsoid is $(\frac{3}{3}{}^2)\pi\nu^3 l_1 l_2 l_3/c^3 = (\frac{3}{3}{}^2)\pi V\nu^3/c^3$, V being the volume of our box. Since all n_j's are positive, only one-eighth of this volume contributes to $N(\nu)$. There is exactly one frequency per unit volume in (n_1,n_2,n_3) space. Hence

$$N(\nu) = \frac{\frac{4}{3}\pi V\nu^3}{c^3} \tag{10.12}$$

There are two kinds of waves propagated in a continuum model of a solid, transverse and longitudinal, each with its characteristic velocity (which we represent, respectively, by c_t and c_l); indeed there are two transverse waves for every longitudinal one. Hence the total number of normal modes of both types with frequency less than ν are $\frac{4}{3}\pi V\nu^3(2c_t^{-3} + c_l^{-3})$ and the number of normal modes between ν and $\nu + d\nu$ is

$$g(\nu) = 4\pi V\nu^2(2c_t^{-3} + c_l^{-3}) \tag{10.13}$$

We mentioned above that the frequency spectrum must be cut off at a frequency ν_L such that $N(\nu_L) = 3N$, the number of degrees of freedom of a lattice of N particles. Hence

$$\frac{4\pi}{3}V\nu_L{}^3(2c_t^{-3} + c_l^{-3}) = 3N$$

so that

$$\nu_L = \left\{\frac{9N}{[4\pi V(2c_t^{-3} + c_l^{-3})]}\right\}^{1/2}$$

and

$$\nu_L g(\nu) = \begin{cases} 9N\left(\dfrac{\nu}{\nu_L}\right)^2 & \text{if } \nu < \nu_L \\ 0 & \text{if } \nu > \nu_L \end{cases} \tag{10.14}$$

The internal energy of the Debye model is

$$E = \frac{9}{8}Nh\nu_L + 3NkTD\left(\frac{\theta_D}{T}\right) \tag{10.15a}$$

where θ_D is the so-called Debye temperature

$$\theta_D = \frac{h\nu_L}{k} \tag{10.15b}$$

and $D(u)$ is the Debye function

$$D(u) = \frac{3}{u^3}\int_0^u \frac{x^3\,dx}{e^x - 1} \tag{10.15c}$$

The heat capacity at constant volume is

$$\begin{aligned} C &= 3NkD\frac{\theta_D}{T} + 3NkT\frac{\partial}{\partial T}D\frac{\theta_D}{T} \\ &= 3Nk\left[4D\frac{\theta_D}{T} - \frac{3\theta_D/T}{\exp(\theta_D/T) - 1}\right] \end{aligned} \tag{10.16}$$

and the entropy is

$$S = 3Nk\left\{\frac{4}{3}D\left(\frac{\theta_D}{T}\right) - \log\left[1 - \exp\left(-\frac{\theta_D}{T}\right)\right]\right\}$$

The functions required for calculation of the various thermodynamic quantities have been tabulated by Beattie [22]. An important feature of the Debye theory is that only a single parameter is required to characterize a material. By choosing the temperature scale properly all heat-capacity data can be fitted to a single universal curve. At low temperatures we can apply Eqs. (10.7) and (10.8). The parameters of those equations are $A = 9/\nu_L^3$, $\alpha = 2$, and $\zeta(4) = \pi^4/90$. Hence as $T \to 0$ we have

$$C_v = 3Nk\left[\frac{4}{5}\pi^4\left(\frac{T}{\theta_D}\right)^3 + \cdots\right] \qquad (10.17)$$

The even moments of $g(\nu)$ are

$$u_{2n} = \sum \nu_i^{2n} = \int_0^{\nu_L} \nu^{2n}g(\nu)\,d\nu = \frac{9N}{\nu_L^3}\int_0^{\nu_L} \nu^{2n+2}\,d\nu = \frac{9N\nu_L^{2n}}{2n+3}$$

Hence the first few terms in the high-temperature expansion of C_v are [Eq. (10.5)]:

$$C_v = 3Nk\left[1 - \frac{1}{20}\left(\frac{\theta_D}{T}\right)^2 + \frac{1}{560}\left(\frac{\theta_D}{T}\right)^4 + \cdots\right]$$

while the entropy goes as

$$S = 3Nk\left[\frac{4}{3} - \log\frac{\theta_D}{T} + \frac{1}{40}\left(\frac{\theta_D}{T}\right)^2 - \cdots\right]$$

The highest frequency ν should correspond to the vibration of shortest wavelength in the lattice. This wavelength would be of the order of the distance between a pair of nearest neighbors and the associated normal mode would involve nearest neighbors oscillating 180° out of phase. This vibration could be excited by an external electric field of the proper frequency in an ionic lattice such as NaCl or KCl in which components of pairs of nearest neighbors have unlike charges. The frequency ν_L therefore corresponds to the highest *reststrahlen* frequency of the crystal.

The largest frequency ν_L, being related to the velocity of elastic waves in the crystal, can also be expressed in terms of its elastic constants. An internal check of the Debye theory has been made by comparing the values of ν_L determined from elastic constants with those which give the best fit of heat-capacity data with the Debye theory (see Blackman [23]). In metals it is assumed that at low temperatures $C_v = aT + bT^3$, the linear term being the electronic contribution to C_v. Some examples of the type of agreement obtained are given in (here we discuss $\theta_D = h\nu_L/k$ rather than ν_L) Table 10.2. The temperatures in the third and fifth columns are those at which the elastic constants have been determined.

If one plots C_v data over a wide temperature range the apparent agreement between experiment and theory is almost unbelievable when he considers the enormous physical differences between various kinds

TABLE 10.2. COMPARISON OF θ_D AS DETERMINED BY THERMAL AND ELASTIC DATA [23]

Material	θ_D (thermal)	T, °K	θ_D (elastic)	T, °K
Ag	237	4	216	~290
Zn	308	4	305	~290
NaCl	308	10	320	0
KCl	230	3	246	0

of crystals and the coarseness of the continuum model.

A better appreciation of the lack of agreement is obtained by relating each experiment point to that value of θ_D which is required in Eq. (10.16) to give the measured C_v at the appropriate temperature. If complete agreement were to exist, the set of θ_D's computed would be temperature independent. A typical variation of θ_D with temperature T is plotted in Fig. 10.1. Plots of this type were first proposed by Blackman [23, 24]. The extreme values of θ_D differ by about 15 per cent. In very anisotropic substances such as Li, Zn, and Cd, the deviations are of the order of 30 to 50 per cent, while in gold and tungsten (face-centered crystals which are almost isotropic) the deviations are only 10 per cent.

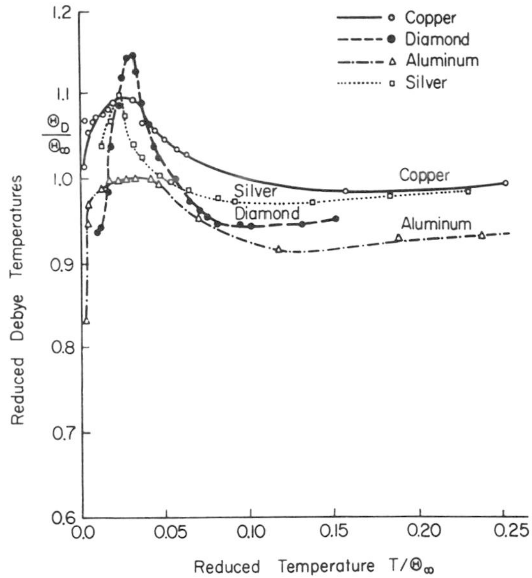

FIG. 10.1. Variation of θ_D/θ_∞ with temperature. Her θ_∞ is the asymptotic Debye temperature at high ten peratures [5]. The face-centered metals Cu, Al, and A and diamond are shown as examples.

It is to be expected that some improvement shoul result from the use of a discrete lattice model rath than a continuum one. Unfortunately the theo becomes much more complicated. No simple un versal formulas for thermodynamic properties see to exist, each example being a special case. T discrete model was first analyzed by Born and v Kármán [25].

3. Theory of Born and von Kármán

In the Born–von Kármán theory a crystalline solid is postulated to be a set of coupled springs and masses. The force constant of a spring which is assumed to connect a given pair of atoms is determined from the interatomic force law. In nonionic crystals the short-ranged nature of the interatomic forces (van der Waals, etc.) allows one to restrict the interactions to those between close neighbors (generally nearest and next nearest). The nearest and next nearest neighbor force constants in cubic lattice can be expressed in terms of the macroscopic force constants. Since the Coulomb forces in ionic lattices are long-ranged, more force constants must be accounted for and certain lattice sums must be calculated; thus complicating the theory. Born and Huang [26] have recently published a comprehensive account of all aspects of the dynamics of lattice vibrations (optical, elastic, and electronic properties are discussed as well as thermal).

An important feature of the theory of coupled harmonic oscillators is that the normal mode frequencies resulting from classical calculations are identical with those deduced from quantum theory. Hence, for simplicity, we shall proceed with a classical determination of $g(\nu)$. We shall sketch the procedure for finding the frequency spectrum, $g(\nu)$, of a two-dimensional lattice [27]; the generalization to three dimensions will be obvious to the reader.

Consider a square lattice which contains N rows and N columns of identical particles of mass M. The lattice points are identified by (l,m) (where both l and m range through the integers $1, 2, \ldots, N$). The components of the displacement of the particle (l,m) are designated by $u_{l,m}$ and $v_{l,m}$.

We shall assume that our lattice is formed on a torus so that we can apply the simplifying Born–von Kármán boundary conditions:

$$u_{l,m} = u_{l+N,m} \qquad v_{l,m} = v_{l+N,m} \tag{10.18a}$$

$$u_{l,m} = u_{l,m+N} \qquad v_{l,m} = v_{l,m+N} \tag{10.18b}$$

Physically one expects thermodynamic functions, which are proportional to the number of particles in a system, to be independent of surface effects in three dimensions and of boundary effects in two dimensions. Hence our final $g(\nu)$ should be independent of the boundary conditions (see Ledermann [28]). The boundary conditions (10.18) are generally used in solid-state problems.

If Φ is the harmonic approximation to the total potential energy of interaction between particles, the forces which act in the horizontal and vertical directions respectively on the (l,m)th particle are

$$-\frac{\partial\Phi}{\partial u_{l,m}} = -\sum_{\lambda,u} a_{\lambda,u}^{(1,1)} u_{l+\lambda,m+u} - \sum_{\lambda,u} a_{\lambda,u}^{(1,2)} v_{l+\lambda,m+u} \tag{10.19a}$$

$$-\frac{\partial\Phi}{\partial v_{l,m}} = -\sum_{\lambda,u} a_{\lambda,u}^{(2,1)} u_{l+\lambda,m+u} - \sum_{\lambda,u} a_{\lambda,u}^{(2,2)} v_{l+\lambda,m+u} \tag{10.19b}$$

where $-a_{\lambda,u}^{(1,1)} u_{l+\lambda,m+u}$ is the contribution to the horizontal component at (l,m) due to a horizontal displacement of a particle λ lattice distances on the right and μ above (l,m), etc.

The equations of motion of our system of coupled particles are

$$\frac{M\,d^2u_{l,m}}{dt^2} = -\frac{\partial\Phi}{\partial u_{l,m}} \tag{10.20a}$$

$$\frac{M\,d^2v_{l,m}}{dt^2} = -\frac{\partial\Phi}{\partial v_{l,m}} \tag{10.20b}$$

These equations are linear in the displacements and contain only constant coefficients to the $u_{l,m}$'s and $v_{l,m}$'s. Hence we can seek solutions of the form

$$u_{l,m} = u \exp(-2\pi i\nu t) \exp[i(\varphi_1 l + \varphi_2 m)] \tag{10.21a}$$

$$v_{l,m} = v \exp(-2\pi i\nu t) \exp[i(\varphi_1 l + \varphi_2 m)] \tag{10.21b}$$

where u, v, ν, φ_1, and φ_2 are constants which must be chosen so that (10.21) satisfies (10.20), the boundary conditions (10.18), and whatever initial conditions that might be imposed.

The boundary conditions (10.18) are satisfied if φ_1 and φ_2 are solutions of

$$\exp(iN\varphi_1) = 1 \qquad \text{and} \qquad \exp(iN\varphi_2) = 1$$

that is, if

$$\varphi_1 = \frac{2\pi k}{N} \qquad \text{and} \qquad \varphi_2 = \frac{2\pi j}{N} \tag{10.22}$$

with $k, j = 1, 2, \ldots, N$ or (if N is even) k, j may run through $-\frac{1}{2}N + 1, -\frac{1}{2}N + 2, \ldots, 0, 1, 2, \ldots, \frac{1}{2}N$. This gives N^2 possible choices of the pair (φ_1,φ_2). The frequency ν is related to φ_1 and φ_2 by noting that if (10.21) is substituted into (10.19) and (10.20) the constants u and v are solutions of the homogeneous linear equations

$$u[F_{11}(\varphi_1,\varphi_2) - 4\pi^2\nu^2 M] + vF_{12}(\varphi_1,\varphi_2) = 0 \tag{10.23a}$$

$$uF_{21}(\varphi_1,\varphi_2) + v[F_{22}(\varphi_1,\varphi_2) - 4\pi^2\nu^2 M] = 0 \tag{10.23b}$$

where

$$F_{jk}(\varphi_1,\varphi_2) = \sum_{\lambda,u} a_{\lambda,\mu}^{(j,k)} \exp i(\varphi_1\lambda + \varphi_2\mu) \tag{10.23c}$$

In order for solutions of (10.23) to exist, the determinant of the coefficients of u and v must vanish:

$$\begin{vmatrix} F_{11}(\varphi_1,\varphi_2) - 4\pi^2\nu^2 M & F_{12}(\varphi_1,\varphi_2) \\ F_{21}(\varphi_1,\varphi_2) & F_{22}(\varphi_1,\varphi_2) - 4\pi^2\nu^2 M \end{vmatrix} = 0 \tag{10.24}$$

This equation relates the frequency ν to the constants (φ_1,φ_2). There are two values of ν^2 for each (φ_1,φ_2) pair. Hence there is a total of $2N^2$ normal modes, or solutions (10.21) of our equations of motion (10.20). The most general solution is a linear combination of these $2N^2$ normal modes.

The formulas appropriate for a three-dimensional simple cubic lattice are essentially the same as these. The displacements would have three components so that (10.20) and (10.21), (10.22) and (10.23) would become sets of three equations and a φ_3 is to be added to the set (φ_1,φ_2). The characteristic equation analogous to (10.24) would be a 3×3 determinant with $3N^3$ normal modes in an $N \times N \times N$ lattice.

In a general periodic lattice the φ_1, φ_2, and φ_3 of the generalization of (10.21) range over the lattice points in the appropriate Brillouin zone of the reciprocal lattice of the system of interest. The displacement components $u_{l,m,n}$, etc., are taken along the axes of the fundamental unit cell in the lattice. The detailed formulation for the general case is discussed in the books of Seitz [29] and Born and Huang [26]. The determinants which correspond to (10.24) in the case of the monatomic cubic lattices are 3×3 while those of more complicated lattices are larger.

Equation (10.20) is a quadratic equation in the square of the circular frequency $\omega = 2\pi\nu$. Since each F_{ij} is a doubly periodic function of φ_1 and φ_2 with periods 2π, $\omega(\varphi_1,\varphi_2)$ also has this character. We shall refer to all those frequencies which are generated as (φ_1,φ_2) runs through the N^2 values defined by (10.22) as a branch of the frequency spectrum. Our lattice has two branches. We shall refer to the (φ_1,φ_2) space over which our doubly periodic function $\omega(\varphi_1,\varphi_2)$ is defined as a doubly periodic reciprocal lattice. A simple cubic lattice has three branches corresponding to the three roots of cubic equations.

In the thermodynamic limit, $N \to \infty$, the uniformly distributed points in (φ_1,φ_2) space [those defined in (10.22)] which correspond to a branch of frequencies of normal modes become dense. The number of normal modes associated with a closed region in (φ_1,φ_2) space is proportional to the area of that region. The proportionality constant is $N^2/4\pi^2$ since the area of a single period of our doubly periodic (φ_1,φ_2) space is associated with N^2 normal modes.

It is clear that a doubly periodic function defined on a two-dimensional space is equivalent to a function defined on a torus. Since the F_{ij}'s are continuous functions of (φ_1,φ_2), two neighboring points in (φ_1,φ_2) space yield frequencies which differ only slightly from each other. Curves of constant frequency can be constructed in (φ_1,φ_2) space. They are closed curves on the (φ_1,φ_2) torus. The curve $\omega = \omega_1 =$ constant separates the torus into two regions, one which corresponds to frequencies which are less than $\nu_1 = \omega_1/2\pi$ and the other to frequencies greater than ν_1 (there may actually be several closed curves of a single frequency of interest; in this case the regions of frequencies $<\nu_1$ may not be connected).

If we let $N_\alpha(\nu)$ be the number of frequencies less than ν in the αth branch, we have in the limit as $N \to \infty$

$$N_\alpha(\nu) = \frac{N^2}{4\pi^2} \iint_R d\varphi_1\, d\varphi_2 \qquad (10.25)$$

where the integration extends over the set R of all values of (φ_1,φ_2) for which $\nu_\alpha^2(\varphi_1,\varphi_2)$ is less than ν^2. The integral gives the area enclosed by the curve $\nu_\alpha(\varphi_1,\varphi_2) = \nu$. Since the number of frequencies between ν and $\nu + d\nu$ in the αth branch is

$$g_\alpha(\nu)\, d\nu = N_\alpha(\nu + d\nu) - N_\alpha(\nu) = \frac{\partial N_\alpha}{\partial \nu}\, d\nu$$

we have

$$g_\alpha(\nu) = \frac{\partial N_\alpha}{\partial \nu} = \frac{N^2}{4\pi^2}\frac{\partial}{\partial \nu} \iint_R d\varphi_1\, d\varphi_2 \qquad (10.26)$$

The three-dimensional generalization of this formula is obvious.

Van Hove [30] has converted the three-dimensional volume integral to the surface integral

$$g_\alpha(\nu) = 2\nu K_\alpha \iint \frac{dS}{\sqrt{(\partial\nu/\partial\varphi_1)^2 + (\partial\nu/\partial\varphi_2)^2 + (\partial\nu/\partial\varphi_3)^2}} \qquad (10.27)$$

over the entire surface of constant frequency ν. Here dS is an infinitesimal surface element and k_α a constant depending on the lattice type.

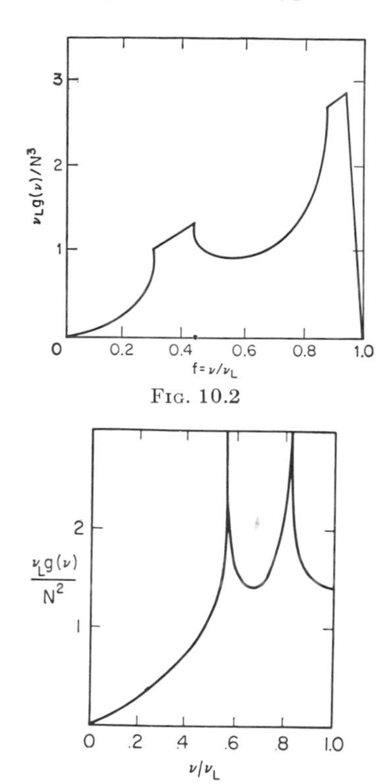

Fig. 10.2

Fig. 10.3. Frequency spectrum of 2-dimensional square lattice with ratio of central to noncentral force constants equal to $\frac{1}{2}$.

These formulas have been applied to the calculation of the frequency spectrum of various types of two-dimensional lattices [31–35] and of simple cubic lattices of the following types: (*a*) lattices with interactions between nearest neighbors only [32, 36, 37] (central and noncentral forces are included) and (*b*) lattices with interactions between both nearest and next nearest neighbors [38] (with central forces only). We have plotted $g(\nu)$ for case (*a*) in Fig. 10.2. The ratio of noncentral to central force constants is chosen to be $\frac{1}{12}$. For comparison the frequency spectrum of a two-dimensional square lattice with nearest neighbor interactions only is plotted in Fig. 10.3.

The $g(\nu)$ in the two-dimensional case has logarithmic singularities, while in the simple cubic lattice $dg/d\nu$ varies as $\pm(|\nu - \nu_c|)^{-1/2}$ on one side of a singularity and remains rather constant on the other. It was pointed out by Smollett [33] that the singularities occur at those frequencies at which $\nu(\varphi_1,\varphi_2)$ has a saddle point in the two-dimensional case. Van Hove [30] generalized this observation to three dimensions and also showed that the existence of singularities is a consequence of certain topological theorems derived by M. Morse [38] and is not an accident in the models presented above. A special case of one of Morse's theorems states that functions [of an appropriate class which includes the frequencies as functions of (φ_1,φ_2)] which are defined on a torus (that is, on a doubly periodic space) have at least two saddle points.

In a symmetrical square lattice both saddle points occur at the same frequency. Van Hove showed that in general the frequency spectrum of a two-dimensional crystal has at least one logarithmic singularity per branch and at least a finite discontinuity which occurs at the upper end of the spectrum. The Morse theorems which are relevant for three-dimensional lattices led Van Hove to the conclusion that $g(\nu)$ is continuous and $dg/d\nu$ has a least two infinite discontinuities and takes the value $-\infty$ at the upper end of the spectrum. The author [27] has given an elementary discussion of the Van Hove theory and discussed various examples to show how the removal of symmetries leads to the splitting of singularities and the increase of their number.

Unfortunately no real monatomic crystals form simple cubic lattices. Hence although the examples mentioned above have led to a deeper insight into the general nature of the frequency spectrum for all crystals, they cannot be used to discuss thermodynamic properties of real solids. The integrals (10.26) and (10.27) which correspond to lattices other than simple cubic and to diatomic simple cubic (NaCl, etc.) have not been evaluated analytically. However, the singularities in the $g(\nu)$ have been located by Rosenstock [39] and Golovin [40] for all monatomic cubic lattices with nearest and next nearest neighbor central-force interactions and for the model discussed by Fuchs [41] and de Launay [42] for the electron gas contribution to the noncentral forces. The tables of Rosenstock and Golovin should be a useful adjunct in the synthesis of real $g(\nu)$ curves by various approximation methods.

Two approximation methods have been applied to the determination of $g(\nu)$. In the first one samples a large number of points in the reciprocal lattice or chooses the crystal of interest to be sufficiently small so that it is feasible to calculate all normal mode frequencies. The machine time on a high-speed computer which is required to find the spectrum of a crystal of 10^6 lattice points is about 7 to 10 hr. In trial calculations Golovin has found that no signs of singularities show up in this size lattice. However, if the total frequency range is divided into 16 parts, the histograms associated with lattices of $(48)^3$ and $(96)^3$ lattice points are essentially the same. The sampling method was used by Blackman [24] in his first investigation of the Born–von Kármán model. The approximate $g(\nu)$ of many solids (for example Li [43], Ag [44], W [45], diamond [46], KCl [47], and NaCl [48]) has been found by this method. A review of these results has been given by Leibfried and Brenig [49]. The frequency spectrum of Ag as calculated by Leighton is given in Fig. 10.4.

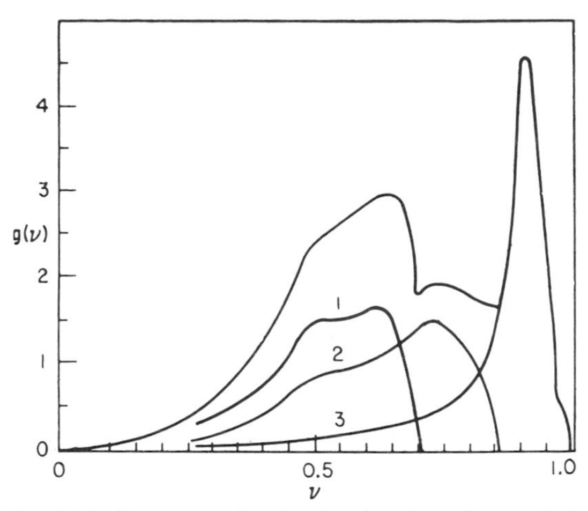

FIG. 10.4. Frequency distribution for Ag. Curves 1, 2, and 3 represent various branches of the spectrum. Top curve, the complete spectrum, is the sum of the other three curves [44].

The second scheme, the method of moments [50], depends on the fact that a distribution function can be well approximated if a sufficiently large number of moments are known. The squares of the normal mode frequencies are the characteristic values of a matrix M. Hence [see (10.6)]

$$\sum_j \nu_j^2 = \text{Trace } M = \mu_2$$

while

$$\sum_j \nu_j^{2m} = \text{Trace } M^m = \mu_{2m}$$

These moments can be substituted into (10.4) and (10.5) for the determination of high-temperature thermodynamic quantities. Various formulas exist for the expression of distribution functions in terms of their moments. For example, the series expansion of $g(\nu)$ in Legendre polynomials $P_n(\nu/\nu_L)$ is

$$g(\nu) = \sum_{n=0}^{\infty} a_{2n} P_{2n}(\nu/\nu_L)$$

where

$$a_{2n} = \frac{3N(4n+1)}{(2n)!2^{2n}\nu_L} \left[\frac{d^{2n}}{dy^{2n}} (y^2 - 1)^{2n} \right]_{y^{2n} = \mu_{2n}/3N\nu_L^{2n}}$$

the variable y^{2n} being replaced by $\mu_{2n}/3N\nu_L^{2n}$ after differentiation.

The frequency distributions of various types of lattices have been calculated by these methods [51, 52]. They are in general agreement with those obtained by the sampling procedure but do not have singularities. The singularities can be put into the analysis by taking advantage of the tables of singularities located by Golovin and Rosenstock. One can choose expressions for $g(\nu)$ which have singularities at the correct points and which include a certain number of parameters. These parameters could then be evaluated so that the first few moments of $g(\nu)$ could be given correctly. Such calculations have been made by Lax and Leibowitz [53] for square lattices and agree very well with the exact $g(\nu)$ curves.

The frequency spectrum of polyatomic crystals generally consists of several branches. For example, in the case of a diatomic simple cubic lattice with interactions between nearest neighbors only, the normal mode frequencies are

$$\omega^2 = \frac{(\gamma_1 + 2\gamma_2)(M+m)}{Mm} \pm (mM)^{-1}\left[(\gamma_1 + 2\gamma_2)^2 (M-m)^2 + 4mM\left(\sum_1^3 \gamma_j \cos\varphi_j\right)^2\right]^{1/2}$$

where γ_1 is the central-force constant, $\gamma_2 = \gamma_3$ the noncentral one, and M and m (with $M > m$) are the masses of the two atomic species. Also $\varphi_j = 2\pi n_j/N$ ranges over the values which correspond to $nj = 1, 2, \ldots, N$. As N^3, the number of atoms in the lattice approach ∞, the frequencies given above fall into two dense bands. The low-frequency band which corresponds to the $-$ sign has its maximum frequency at $\omega_1^2 = 2(\gamma_1 + 2\gamma_2)/M$, while the upper band ($+$ sign) ranges from $\omega_2^2 = 2(\gamma_1 + 2\gamma_2)/m$ to $\omega_L^2 = 2(\gamma_1 + 2\gamma_2)(M+m)/Mm$. Generally those bands whose lowest frequency does not approach 0 as $N \to \infty$ contain "optical modes." The modes whose frequencies are in a band with a limit frequency $\omega = 0$ are called acoustical modes. It can be shown that in an ionic lattice with the particles M and m of opposite charge the optical modes are optically active and can be excited by an electromagnetic field. The relation of lattice vibrations to optical properties of crystals is discussed in detail by Born and Huang [26].

Generally the available information on interatomic forces is not sufficient to deduce the force constants used in our equations from first principles. However, consistent results can be obtained by relating the atomic force constants to the elastic constants of the crystal. In the analyses of nonionic lattices one frequently takes into account only central-force interactions between nearest and next nearest neighbors in the lattice. In cubic lattices two force constants, α for nearest neighbor interactions and γ for those of second neighbors, are related to the elastic constants c_{11}, c_{12}, and c_{44}. The relations are deduced by comparing the strain energy for various stresses when expressed in terms of continuum elasticity theory with that in terms of molecular theory. The central-force hypothesis leads to

$$c_{11} = \frac{2(\alpha + 3\gamma)}{a} \qquad c_{12} = \frac{2\alpha}{3a} = c_{44} \qquad \text{for body-centered cubic lattices}$$

and

$$c_{11} = \frac{2(\alpha + 2\gamma)}{a} \qquad c_{12} = \frac{\alpha}{a} = c_{44} \qquad \text{for face-centered cubic lattices}$$

a being the length of a side of a unit cell and the Cauchy relations $c_{12} = c_{44}$ being satisfied in both cases. Experimental data generally yield deviations from this equality. The deviations have been discussed in metals by Fuchs [41] and de Launay [42], who consider the effect of the electron gas in a metal on the macroscopic elastic constants. When this contribution is subtracted from compressibility of the metal the purely "atomic" part of the elastic constants yield the Cauchy equation to a good approximation. The theory and data on elastic constants are given in Born and Huang and are concisely discussed by Kittel [54].

4. Equation of State of Crystals

The statistical mechanical formula for the equation of state of a solid is

$$p = -\left(\frac{\partial A}{\partial V}\right)_T \tag{10.28}$$

A being the Helmholtz free energy. Now, in the harmonic approximation used so far in this chapter the atomic equilibrium positions are postulated to remain fixed at all temperatures. Any lattice expansion with increasing temperature would be associated with anharmonicities in the interatomic force law. A useful theory of the equation of state of solids has been obtained by Grueneisen [55] without explicitly discussing anharmonicities. He assumes that normal mode frequencies are volume (or lattice spacing) dependent. He then evaluates both the equation of state and expansion coefficients in terms of the parameter which characterizes the volume dependence, the so-called Grueneisen constant γ. Then γ is determined experimentally from measurements of expansion coefficients and is used in the equation of state.

The total free energy of a crystal is

$$A = E_0(V) + A_D(T,V)$$

where $E_0(V)$ represents the total potential energy of all atoms in the crystal when they are located at their equilibrium positions. A_D is the vibrational contribution to the free energy. If we restrict our discussion to the Debye theory, we note from (10.13) and (10.14) that the frequency distribution depends on the volume only through the θ_D. This parameter is explicitly proportional to $V^{-1/2}$ and implicitly related to V through the dependence of the wave propagation velocities c_t and c_l on the density of the lattice.

Now the Helmholtz free energy of a Debye solid is of the form

$$A_D(T,V) = Tf\left(\frac{\theta_D}{T}\right) \tag{10.29}$$

so that

$$\frac{\partial A_D}{\partial \theta_D} = f'\left(\frac{\theta_D}{T}\right) = -\frac{T^2}{\theta_D}\frac{\partial f}{\partial T} = \theta_D^{-1}\frac{\partial(A_D/T)}{\partial(1/T)} \tag{10.30}$$

However, the thermodynamic relations $A = E - TS$ and $-S = (\partial A/\partial T)_V$ imply

$$E = A - \left(\frac{\partial A}{\partial T}\right)_V T = \frac{\partial(A/T)}{\partial(1/T)} \tag{10.31a}$$

so that

$$\frac{\partial A_D}{\partial \theta_D} = \frac{E_D}{\theta_D} \tag{10.31b}$$

If we define the Grueneisen constant γ by

$$\gamma = -\frac{\partial \log \theta_D}{\partial \log V} \tag{10.32}$$

we find the equation of state of Mie and Grueneisen:

$$\begin{aligned} p &= -\frac{\partial E_0}{\partial V} + \frac{\partial A_D}{\partial \theta_D}\frac{\partial \theta_D}{\partial V} \\ &= -\frac{\partial E_0}{\partial V} + \frac{\gamma E_D}{V} \end{aligned} \tag{10.33}$$

where E_D is the internal energy of our Debye crystal. The quantity E_0 depends on the interatomic force law.

The Grueneisen constant is related to the linear expansion coefficient by recalling the relation

$$\beta = \frac{1}{3}\kappa\left(\frac{\partial p}{\partial T}\right)_V$$

κ being the compressibility. Since $(\partial E_D/\partial T)_V = C_V$, our equation of state yields $(\partial p/\partial T)_V = \gamma C_V/V$ so that

$$\beta = \frac{\frac{1}{3}\kappa\gamma C_V}{V} \tag{10.34}$$

which relates γ to the compressibility and expansion coefficient. Grueneisen has shown by examining experimental data that γ remains constant in many metals over a wide range of temperature and density.

TABLE 10.3. VALUES OF THE GRUENEISEN CONSTANT γ

Metal	γ
Fe	1.60
Co	1.87
Ni	1.88
Cu	1.96
Pd	2.23
Ag	2.40
W	1.62
Pt	2.54

Some values of γ are given in Table 10.3. Mayer and Helmholz [56] have tabulated values of the γ's for the alkali halides.

5. Scattering of Thermal Neutrons by Lattice Vibrations

The most direct experimental procedure for the determination of the dispersion curves $\omega = \omega(\mathbf{q})$ as well as the complete frequency spectrum is through neutron spectroscopy. As with X rays, a crystal acts as a diffraction grating for neutrons. However, since the energy of thermal neutrons is of the same order as that of a lattice-vibration phonon, a careful measurement of the energy and momentum change of a neutron as it passes through a small crystal gives a measure of the frequency and wave vector of the phonon with which it interacted, provided that it interacted with only one phonon. The change of energy of the neutron is given by

$$E' - E = \frac{\hbar^2}{2m}(\mathbf{k}'^2 - \mathbf{k}^2) = \hbar\omega(\mathbf{q})$$

$\mathbf{k}'$ and $\mathbf{k}$ being the final and initial wave vectors of the neutron. The momentum transfer is determined by the Bragg condition

$$\mathbf{k}' - \mathbf{k} = 2\pi\boldsymbol{\tau} + \mathbf{q}$$

where $\boldsymbol{\tau}$ is an arbitrary translation vector of the reciprocal lattice. Since $\omega(\mathbf{q} + 2\pi\boldsymbol{\tau}) = \omega(\mathbf{q})$ (the generalization of the periodicity of the φ space discussed in Sec. 3), the energy transfer equation can be written as

$$E' - E = \frac{\hbar^2}{2m}(\mathbf{k}^2 - \mathbf{k}^2) = \hbar\omega(\mathbf{k}' - \mathbf{k}) \equiv \hbar\omega(\mathbf{q})$$

The difference $\mathbf{k}' - \mathbf{k}$ is a measure of the scattering angle of an observed neutron. A determination of its energy, say by time-of-flight measurement, yields the dispersion curve $\omega = \omega(\mathbf{q})$. The complete spectrum is obtained by application of the 3D generalization of (10.26) or by (10.27) to the experimental data.

The basic theory of neutron scattering by crystals was given by Weinstock [57], Cassels [58], and Waller and Froman [59]. It is well presented and further developed in refs. 60 and 61. The early experiments were done at Brookhaven [62] and Chalk River [63] in 1955. An excellent review of experimental techniques and results is given by Brockhouse in [64]. The 1963 Aarhus Lecture proceedings on phonons and phonon interactions (edited by Thor A. Bak [65]) and ref. 66 give a broad survey of almost all aspects of lattice vibrations.

References

1. Burton, E. F., H. Grayson Smith, and J. O. Wilhelm: "Phenomena at the Temperature of Liquid Helium," Reinhold, New York, 1940.
2. Thirring, H.: *Physik. Z.*, **14:** 867 (1913); **15:** 127 (1914); **15:** 180 (1914).
3. Montroll, E.: *J. Chem. Phys.*, **11:** 481 (1943).
4. Pitzer, K.: *J. Chem. Phys.*, **6:** 68 (1938).
5. De Sorbo, W.: *J. Chem. Phys.*, **21:** 876 (1953).
6. Estermann, I., and G. Kirkland: private communication.
7. De Sorbo, W., and W. Tyler: *J. Chem. Phys.*, **21:** 1660 (1953).
8. De Sorbo, W.: *J. Chem. Phys.*, **21:** 168 (1953).
9. Dworkin, A. S., D. J. Sasmor, and E. R. van Artsdalen: *J. Chem. Phys.*, **21:** 954 (1953).
10. Newell, G. F.: *J. Chem. Phys.*, **23:** 2341 (1955).
11. De Sorbo, W.: *J. Chem. Phys.*, **21:** 1144 (1953).
12. Dole, M., W. P. Hettinger, Jr., N. R. Larson, and J. A. Werthington, Jr.: *J. Chem. Phys.*, **20:** 781 (1952).
13. Winkelmann, A.: *Ann. Physik*, **49:** 401 (1893).
14. Condon, E. U.: *Am. J. Phys.*, **22:** 43 (1954).
15. Dyson, F.: *Phys. Rev.*, **92:** 1331 (1953).
16. Born, M., and E. Brody: *Z. Physik*, **6:** 132 (1921).
17. Partington, J. R.: "An Advanced Treatise on Physical Chemistry," vol. III, Longmans, 1952.
18. Debye, P.: *Ann. phys.*, **39:** 789 (1912).
19. Schroedinger, E.: "Handbuch der Physik," vol. 10, Springer, Berlin (1926).
20. Mayer, J. E., and M. G. Mayer: "Statistical Mechanics," Wiley, New York, 1941.
21. Fowler, R. H.: "Statistical Mechanics," 2d ed., Cambridge University Press, New York and London, 1936.
22. Beattie, J. A.: *J. Math. and Phys.*, **6:** 1 (1926).
23. Blackman, M.: *Repts. Progr. Phys.*, **VIII:** 11 (1942).
24. Blackman, M.: *Proc. Roy. Soc. (London)*, **A148:** 365 (1935); **148:** 384 (1935); **149:** 117 (1935); **149:** 126 (1935); **159:** 416 (1937).

25. Born, M., and T. von Kármán: *Physik. Z.*, **13**: 297 (1912); **14**: 15 (1913).
26. Born, M., and K. Huang: "Dynamical Theory of Crystal Lattices," Oxford University Press, 1954.
27. Montroll, E.: *Am. Math. Monthly*, **61** (pt. II): 46 (1954).
28. Ledermann, W.: *Proc. Roy. Soc. (London)*, **A182**: 362 (1944).
29. Seitz, F.: "The Modern Theory of Solids," McGraw-Hill, New York, 1940.
30. Van Hove, L.: *Phys. Rev.*, **89**: 1189 (1953).
31. Montroll, E.: *J. Chem. Phys.*, **15**: 575 (1947).
32. Bowers, W., and H. Rosenstock: *J. Chem. Phys.*, **18**: 1056 (1950).
33. Smollett, M.: *Proc. Phys. Soc. (London)*, **A65**: 109 (1952).
34. Holsen, J., and W. Nirenberg: *Phys. Rev.*, **89**: 662 (1953).
35. Rosenstock, H. B.: *J. Chem. Phys.*, **21**: 2064 (1953).
36. Rosenstock, H. B., and G. F. Newell: *J. Chem. Phys.*, **26**: 1607 (1953).
37. Montroll, E.: *Proc. 3d Berkeley Symposium on Mathematical Statistics and Probability*, Berkeley, Calif., 1955.
38. Morse, M.: *Trans. Am. Math. Soc.*, **27**: 345 (1925); "Calculus of Variations in the Large," Colloquium Lectures, American Mathematical Societies, 1934.
39. Rosenstock, H. B.: *Phys. Rev.*, **97**: 290 (1955).
40. Golovin, N.: Ph. D. dissertation, George Washington University.
41. Fuchs, K.: *Proc. Roy. Soc. (London)*, **A157**: 144 (1936).
42. De Launay, J.: *J. Chem. Phys.*, **21**: 1975 (1953).
43. Leibfried, G., and W. Brenig: *Z. Physik*, **134**: 451 (1953).
44. Leighton, R. B.: *Revs. Mod. Phys.*, **20**: 165 (1948).
45. Fine, P. C.: *Phys. Rev.*, **56**: 355 (1939).
46. Smith, H.: *Trans. Roy. Soc. (London)*, **A241**: 105 (1948).
47. Iona, M.: *Phys. Rev.*, **60**: 822 (1941).
48. Kellermann, E. W.: *Trans. Roy. Soc. (London)*, **A238**: 513 (1940).
49. Leibfried, G., and W. Brenig: *Fortschr. Physik*, **1**: 187 (1953).
50. Montroll, E.: *J. Chem. Phys.*, **10**: 218 (1942); **11**: 481 (1943).
51. Montroll, E., and D. Peaslee: *J. Chem. Phys.*, **12**: 98 (1944).
52. Garland, C. W., and G. Jura: *J. Chem. Phys.*, **22**: 1108 (1954).
53. Lax, M., and J. L. Leibowitz: *Phys. Rev.*, **95**: 594 (1954).
54. Kittel, C.: "Introduction to Solid State Physics," Wiley, New York, 1953.
55. Grueneisen, E.: "Handbuch der Physik," **10**: 22, Springer, Berlin, 1926.
56. Mayer, J. E., and L. Helmholz: *Z. Physik*, **75**: 19 (1932).
57. Weinstock, R.: *Phys. Rev.*, **65**: 1 (1944).
58. Cassels, J. M.: *Progr. Nucl. Phys.*, **1**: 185 (1950).
59. Waller, I., and P. O. Froman: *Arkiv Fysik*, **4**: 183 (1952).
60. Kothari, L. S., and K. S. Singivi: "Solid State Physics," vol. 8, p. 109, Academic, New York, 1959.
61. Van Hove, L.: *Phys. Rev.*, **95**: 249 (1954).
62. Carter, R. S., D. J. Hughes, and H. Paleosky: *Phys. Rev.*, **106**: 1168 (1957).
63. Brockhouse, B. N., and A. T. Stewart: *Revs. Mod. Phys.*, **30**: 236 (1958); *Phys. Rev.*, **100**: 756 (1955).
64. Brockhouse, B. N., "Phonons and Neutron Scattering," p. 233 in "*Phonon and Phonon Interactions*," T. Bak (ed.), Benjamin, New York, 1964.
65. Bak, T. (ed.), "Phonons and Phonon Interactions," Benjamin, New York, 1964.
66. Maradudin, A. A., E. W. Montroll, and G. H. Weiss, "Theory of Lattice Dynamics in the Harmonic Approximation," supplement 3 of "Solid State Physics," Academic, New York, 1963.

Chapter 11

Superfluids

By K. R. ATKINS, University of Pennsylvania

1. Introduction

The He^4 atom, with two electrons completely filling the K shell, is a stable, spherically symmetrical structure with no electric or magnetic dipole moment. It is not easily polarized, and the mutual force between two atoms is of the van der Waals type and very weak. It seems, therefore, that the unusual properties of liquid He^4, described below, must depend upon the fact that this liquid exists only at very low temperatures—the critical temperature is 5.2°K and the boiling point 4.2°K—and at these temperatures certain quantum effects become important.

The Phase Diagram (Fig. 11.1). The liquid exists right down to 0°K and the only way to produce the

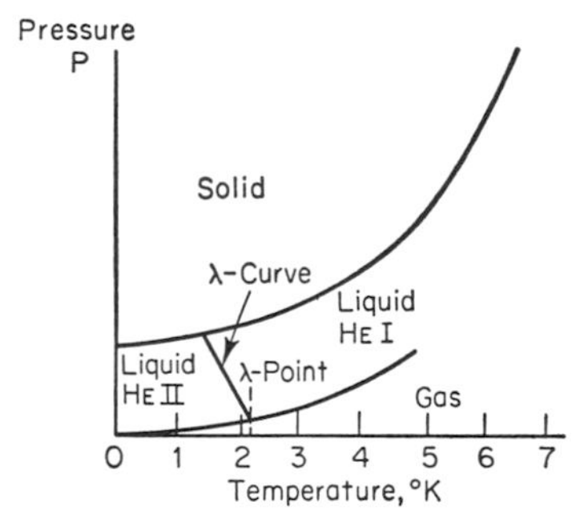

Fig. 11.1. The phase diagram of liquid helium (schematic).

solid is to apply a pressure of at least 25 atm. The explanation of this [1]* is that a helium atom is confined by its neighbors within a space of dimensions $\Delta x \sim 4 \times 10^{-8}$ cm and therefore has an uncertainty in its momentum of $\Delta p \sim \hbar/\Delta x$ and an average zero-point energy of the order of $\hbar^2/2m(\Delta x)^2$, where m is the mass of the atom. To keep this energy term small, the atoms move further apart than they would if only their mutual potential energy were involved, and there results the loose structure of a liquid in which any atom can move fairly easily through the gap between its neighbors.

The λ Transition. At 2.17°K the liquid undergoes a transition from the comparatively normal high-temperature modification, He I, to the very abnormal low-temperature modification, He II. This transition is not accompanied by a latent heat or a discontinuous change in volume, but there are anomalies in the specific heat [2] (Fig. 11.2) and coefficient of expansion [3] (Fig. 11.3). X-ray [4] and neutron diffraction [5] experiments reveal that the λ transition involves no appreciable change in spatial structure.

* Numbers in brackets refer to References at end of chapter.

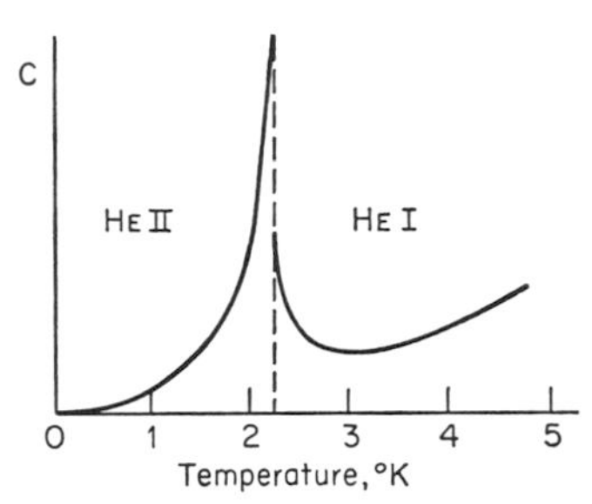

Fig. 11.2. The specific heat of liquid helium (schematic).

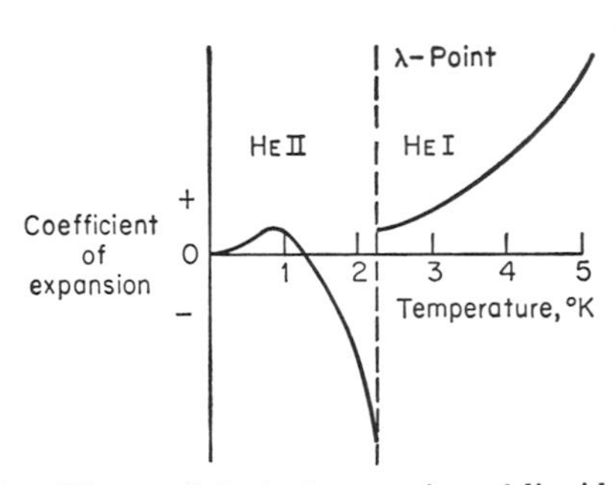

Fig. 11.3. The coefficient of expansion of liquid helium (schematic).

2. Superfluidity and the Two-fluid Theory

One of the most unusual of the properties of liquid helium II is its ability to flow readily through very narrow channels [6, 7]. In Fig. 11.4*a* the exit from the vessel is made by polishing together two glass or quartz plates until the gap between them is about 10^{-5} cm. The flow of a normal liquid, such as He I, through this narrow gap is imperceptible, but He II flows through with a velocity of some tens of centimeters per second. If this phenomenon could be described in terms of a viscosity, its value would be less than

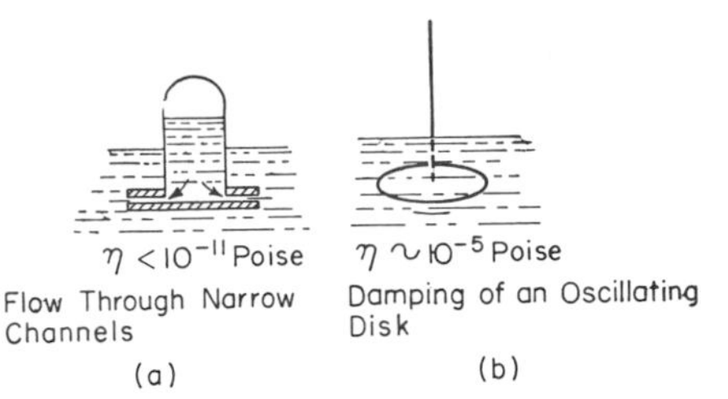

FIG. 11.4. The two different methods of measuring the viscosity of liquid helium II. [*From Atkins, K. R.: "Advances in Physics," Philosophical Magazine Supplement*, **I**: (2) 169 (1952).]

10^{-11} cgs units, or 10^6 times smaller than for any other known substance, but in fact the flow varies in a complicated way with the pressure head and the length and width of the channel, so that the simple concept of viscous flow is inapplicable. However, if one attempts to measure the viscosity of He II by observing the damping of the motion of an oscillating disk [8] (Fig. 11.4*b*), the liquid then behaves normally and a well-defined value of the viscosity of the order of 10^{-5} cgs units is obtained.

The discrepancy between these two methods of measuring the viscosity is resolved by the two-fluid theory of Tisza [9], which assumes that the total density of the liquid may be divided into two parts

$$\rho = \rho_s + \rho_n \qquad (11.1)$$

The superfluid component ρ_s has negligible viscosity and is the component which flows so readily through very narrow channels, whereas the normal component ρ_n is associated with a normal type of viscosity and thus cannot pass through the narrow channels but can damp down the motion of the oscillating disk. This hypothesis was verified by an experiment due to Andronikashvili [10]. A pile of disks performed torsional oscillations in the liquid (Fig. 11.5). The gap between the disks was 0.21 mm, which was sufficiently small so that, above the λ point, all the liquid between the disks was dragged round by them and contributed to the effective moment of inertia of the system. Below the λ point, however, only the normal component moved with the disks, the superfluid component being subject to no frictional force which would have brought it into motion. Therefore, as the temperature was lowered below the λ point, the period of oscillation steadily decreased and it was possible to deduce the fraction, ρ_n/ρ, of liquid contributing to the moment of inertia of the system. The results are shown in Fig. 11.5, from which it will be seen that the fraction, ρ_s/ρ, of superfluid increases from zero at the λ point to unity at 0°K.

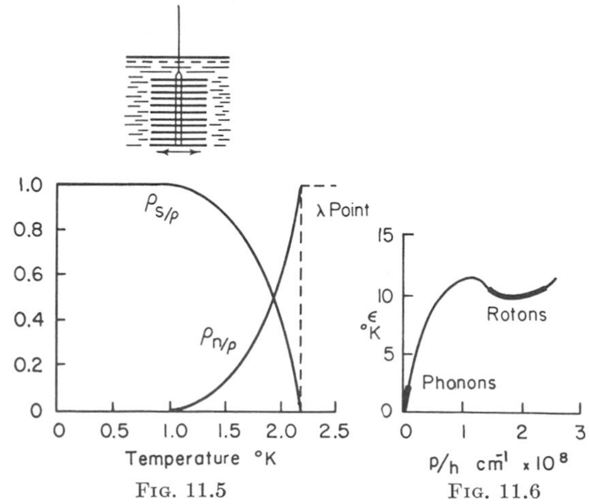

FIG. 11.5 FIG. 11.6

FIG. 11.5. Andronikashvili's experiment. (*Source: same as Fig.* 11.4.)

FIG. 11.6. The elementary excitations of the Landau theory. (*Source: same as Fig.* 11.4.)

Thermal Effects. The superfluid component has zero entropy associated with it, and all the heat content is carried by the normal component. This was first proved conclusively by Kapitza [11], using an apparatus similar to that of Fig. 11.4*a* except that the inside of the vessel was thermally isolated by a vacuum, as in a Dewar vessel, and there were arrangements for supplying heat to the inside and measuring the temperatures inside and outside. When the superfluid component flowed into the vessel, the inside was cooled (the *mechanocaloric effect*). Heat was then supplied to the inside to keep its temperature the same as that outside, and it was found that the entropy which had to be supplied per gram of inflowing superfluid was equal to the total entropy per gram of the bulk liquid. The superfluid was bringing in zero entropy to within the accuracy of the experiment. When the inside was maintained at a steady temperature greater than the outside temperature by ΔT, the inner liquid level came to rest above the outer level, so that in equilibrium the pressure inside the vessel was greater than the outside pressure by Δp. This is the *thermomechanical effect*, and H. London [12] has proved thermodynamically that, if the entropy of the superfluid component is zero, then

$$\frac{\Delta p}{\Delta T} = \frac{S}{V} \qquad (11.2)$$

where S and V are the entropy and volume of the bulk liquid. Equation (11.2) is in agreement with the experimental results to within the accuracy that S is known.

The thermal conductivity in narrow channels and capillaries is unusually high, in some circumstances more than 1,000 times greater than that of copper, but again the heat current varies in a complicated way with the temperature difference and the geometry of the liquid column, so that a unique coefficient of thermal conductivity cannot be defined [13]. The process responsible for the flow of heat is a counterflow of the two components, the superfluid component moving toward the source of heat and the normal component moving away from it. The forces which oppose this counterflow are hydrodynamical in character and in the bulk liquid are very small, or even zero, so that no detectable temperature differences can be set up in the bulk liquid even when there is a large flow of heat. For this reason liquid helium II can never be made to boil, since the formation of a bubble at a point below the liquid surface requires a local increase in temperature to produce the increased vapor pressure needed to overcome the hydrostatic pressure head.

3. The Theory of Bose-Einstein Condensation

The He^4 atom contains an even number of fundamental particles and therefore obeys Bose-Einstein statistics. London and Tisza [14] have therefore suggested that the properties of liquid helium II may be related to a peculiar condensation phenomenon which occurs in an ideal Bose-Einstein gas at low temperatures. Because of a mathematical peculiarity in the distribution function for Bose-Einstein statistics, a transition occurs at a certain temperature which would be in the region of 3°K for an ideal gas having the same density as liquid helium. Above this temperature the atoms distribute themselves amongst the various energy levels in the usual way and there are very few atoms in the ground state. Below the transition temperature, a finite fraction of atoms falls into the ground state, this fraction increasing from zero at the transition temperature to unity at 0°K. As the atoms in the ground state have no thermal energy, it is an obvious step to equate them to the superfluid component. As liquid helium is not an ideal gas, but a condensed phase in which the interatomic forces are important, the theory is not able to make good quantitative predictions of the various properties of the liquid.

However, Penrose and Onsager [15] have shown that, even when the strong interatomic forces are not ignored, in the ground state about 8 per cent of the helium atoms have identical momenta (perhaps zero, but not necessarily so). Because of the nature of Bose-Einstein statistics, all the atoms have a strong tendency to move into the same frame of reference as the 8 per cent "condensate" and so, at 0°K, the whole liquid is superfluid.

4. Landau's Theory

Landau [16] treats liquid helium as more analogous to a solid than a gas and uses an approach similar to the Debye theory of solids. In the Debye theory the elementary excitations, or normal modes, are taken to be longitudinal and transverse sound waves. A liquid cannot support transverse waves; therefore Landau retains only the longitudinal waves and introduces a new type of excitation called rotons corresponding to rotational motions of the liquid. The experimental results can be best explained if the elementary excitations have the form shown in Fig. 11.6, which shows how the energy ϵ of an excitation varies with its momentum p. The elementary excitations of small energy are the longitudinal sound waves, or phonons, and it is well known that their energy is linearly proportional to their momentum.

Phonons: $$\epsilon = cp \qquad (11.3)$$

c being the velocity of ordinary sound. At higher values of p there is a minimum in the curve near which

Rotons: $$\epsilon = \Delta + \frac{(p - p_0)^2}{2\mu} \qquad (11.4)$$

These are the rotons.

This form for the spectrum of elementary excitations has been verified experimentally [17] by studying the inelastic scattering of slow neutrons off the liquid. The incident neutron creates an elementary excitation in the liquid and is scattered sideways with diminished energy. The decrease in energy of the neutron gives the energy ϵ of the excitation. From the laws of conservation of energy and momentum it can be shown that the neutrons scattered through a fixed angle have all created excitations with the same p, which can be calculated. Repeating this procedure at various angles of scattering, the relationship between ϵ and p can be determined and is found to be as shown in Fig. 11.6.

The specific heat can be expressed as the sum of two parts, due to phonons and rotons separately. The phonon and roton contributions are, respectively,

$$C_{\text{ph}} = \frac{2\pi^2}{15} \frac{k^4 T^3}{\rho \hbar^3 c^3} \qquad (11.5)$$

$$C_{\text{r}} = \frac{2\mu^{1/2} p_0^2 \Delta^2}{(2\pi)^{3/2} \rho k^{1/2} T^{3/2} \hbar^3} \left[1 + \frac{kT}{\Delta} + \frac{3}{4} \left(\frac{kT}{\Delta} \right)^2 \right] e^{-\Delta/kT} \qquad (11.6)$$

Because of the factor $e^{-\Delta/kT}$ the roton contribution is dominant above 1°K, but can be neglected below about 0.6°K. Below 0.6°K the observed specific heat does vary as T^3 and is in good numerical agreement with Eq. (11.5). At higher temperatures there is an extra contribution to the specific heat which can be explained quite well by Eq. (11.6) using the values of Δ, p_0, and μ obtained from the neutron-scattering experiments (in particular $\Delta = 8.6$°K). Above 1.6°K the simple theory breaks down because there is then a high density of interacting rotons and phonons, and one gets the intricate complications which are always associated with a high-order transition in a cooperative assembly. Under these circumstances, the neutron-scattering experiments reveal that the effective value of Δ decreases as the λ point is approached.

The coefficient of expansion α may be derived from the above expressions for the specific heat [3], using the equations

$$\alpha = \frac{1}{V} \left(\frac{\partial V}{\partial T} \right)_p = -\frac{1}{V} \left(\frac{\partial S}{\partial p} \right)_T \qquad (11.7)$$

Referring to Fig. 11.3, α is positive below 1.15°K because the phonons predominate and the liquid is similar to a solid (mathematically $\partial c/\partial p$ is positive). The negative coefficient above 1.15°K is presumably connected with the rotons and can be explained if $\partial \Delta/\partial p$ is negative.

The normal component is to be considered as a gas of elementary excitations and the superfluid component is the background in which these excitations are embedded. Thus, in the experiment of Fig. 11.4*a*, the superfluid background flows through the slit, but the rotons and phonons collide with the walls and are held back. It is easy to see why the superfluid component carries no entropy. In the oscillating disk experiment of Fig. 11.4*b* the rotons and phonons collide with the disk and produce the damping of its motion which is associated with the normal component. If the elementary excitations are given a uniform drift velocity, it is a straightforward calculation in statistical mechanics to deduce the linear momentum associated with this velocity and hence calculate the effective mass of the elementary excitations per unit volume, which is the same thing as the density of the normal component ρ_n. The contributions from the phonons and rotons are

$$\rho_{\mathrm{ph}} = \frac{2\pi^2}{45}\frac{k^4T^4}{\hbar^3c^5} \tag{11.8}$$

$$\rho_{\mathrm{r}} = \frac{2\mu^{1/2}p_0{}^4e^{-\Delta/kT}}{3(2\pi)^{3/2}(kT)^{1/2}\hbar^3} \tag{11.9}$$

Feynman's Theory. Feynman [18, 30] has advanced an atomistic interpretation of the spectrum of elementary excitations assumed by Landau. If ϕ is the wave function of the ground state, he assumes that, in a first approximation, the wave function of an excited state may be written

$$\psi = \phi \sum_i f(\mathbf{R}_i) \tag{11.10}$$

where $\mathbf{R}_i$ is the position vector of the ith atom. Applying the variational principle, it is found that the energy is a minimum if $f(\mathbf{R}_i)$ is $\exp(i\mathbf{k}\cdot\mathbf{R}_i)$ so that

$$\psi = \phi \sum_i \exp(i\mathbf{k}\cdot\mathbf{R}_i) \tag{11.11}$$

This is well known to be the wave function for the phonons with wavelengths embracing many atoms, but it now applies also to wavelengths comparable with the interatomic spacing. The energy associated with the wave function is

$$\epsilon = \frac{\hbar^2k^2}{2m\int p(\mathbf{R}_{ij})\exp(i\mathbf{k}\cdot\mathbf{R}_{ij})d^3\mathbf{R}_{ij}} \tag{11.12}$$

where $p(\mathbf{R}_{ij})$ is the radial distribution function giving the probability that the ith atom is at a distance $\mathbf{R}_{ij}$ from the jth atom. $p(\mathbf{R}_{ij})$ is known from X-ray [4] or neutron diffraction [5] experiments and so the energy ϵ of the excitation can be deduced as a function of its momentum $p = \hbar k$. The resulting curve is similar to Fig. 11.6, the minimum which gives rise to the rotons corresponding to the maximum in $p(\mathbf{R}_{ij})$ produced by the first nearest neighbors.

The Viscosity of the Normal Component. The viscosity of the normal component η_n may be measured [19] by the oscillating disk experiment of Fig. 11.4*b* if one remembers that this type of experiment measures the product $\eta\rho$, which has to be interpreted as $\eta_n\rho_n$. As the temperature is lowered, η_n decreases from 20 μp at the λ point to a minimum value of about 10 μp near 1.8°K and then increases rapidly as the temperature is lowered still further. If the normal component can be considered as an ideal gas of rotons and phonons, the viscosity of this gas may be deduced [20] from simple kinetic theory considerations and has two terms corresponding to sideways transfer of drift momentum by rotons and phonons separately.

$$\eta_n = \frac{\pi}{10}\rho_{\mathrm{r}}\bar{v}_{\mathrm{r}}l_{\mathrm{r}} + \alpha\rho_{\mathrm{ph}}cl_{\mathrm{ph}} \tag{11.13}$$

Here $\bar{v}_{\mathrm{r}}$ is the mean thermal velocity of a roton, α is a numerical constant, and l_{r} and l_{ph} are the mean free paths of rotons and phonons, respectively. A roton free path is terminated by collision with another roton and the roton viscosity is independent of temperature, since the main effect of temperature is to change the density of rotons and the viscosity of an ideal gas is independent of its density. A phonon free path is also terminated by collision with a roton above 1°K and, as the temperature is lowered, the number of rotons decreases rapidly and l_{ph} therefore increases rapidly. This explains the rapid rise in η_n as the temperature is lowered toward 1°K. Below 1°K the mean free paths become very large and may become comparable with the dimensions of the experimental apparatus.

5. First and Second Sound

Two types of wave propagation with different velocities are possible in liquid helium II [14, 16]. First sound, or ordinary sound, results when the superfluid and normal components oscillate in phase with one another to produce periodic density variations. Second sound is a temperature wave which results when the two components oscillate out of phase, so that the cold, superfluid component collects at a point of low temperature while the normal component collects at a point of high temperature half a wavelength away. To generate second sound [21], a plane resistive element is immersed in the liquid and an alternating current (or square pulse of current) is fed into it. At these low temperatures the thermal capacity of the element is so small that its temperature is able to vary in sympathy with the periodic variation of the joule heat. A temperature variation is therefore produced in the liquid near the transmitter and this propagates as a second sound wave through the liquid. The receiver is a plane resistance thermometer carrying a steady current and the voltage across this varies with the incident temperature fluctuation and can be amplified and displayed on a cathode ray oscilloscope.

The velocity of *first* sound at 0°K is 237 m/sec. The attenuation of first sound as a function of temperature is shown schematically in Fig. 11.7 [22].

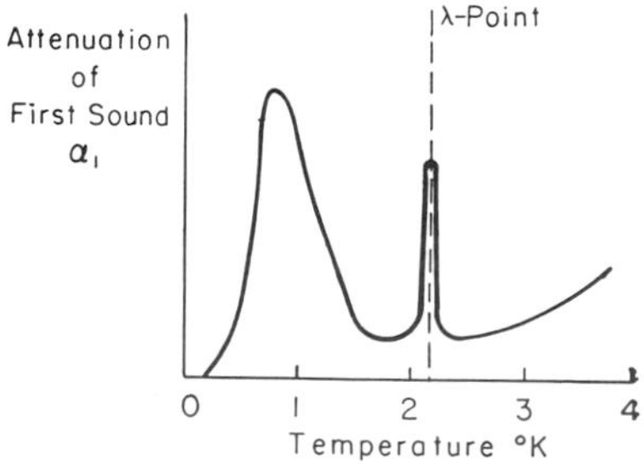

Fig. 11.7. The attenuation of first sound in liquid helium (schematic).

The part of the curve above 3°K is in agreement with the classical theory which attributes the attenuation to the viscosity and thermal conductivity of the liquid. The sharp peak at the λ point is connected with the cooperative phenomena which occur there. Below 2.0°K the attenuation can be explained [23] in terms of an interesting relaxation mechanism similar to the relaxation effect which gives rise to the absorption of sound in a diatomic gas, when the rotational and vibrational energy levels of the molecules are not able to adjust themselves immediately to the changes in temperature accompanying the adiabatic compressions and rarefactions of the sound wave. When liquid helium II is compressed adiabatically, the number of rotons and phonons present has to adjust itself to the new conditions of density and temperature, but can only do so after a characteristic relaxation time determined by those collision processes which create new phonons and rotons. Actually, there are two relaxation times corresponding to two creative processes, a collision of two phonons from which three phonons emerge and a collision of two rotons from which a single roton and a high energy phonon emerge. The two relaxation times τ_1 and τ_2 increase rapidly as the temperature is lowered.

The velocity of *second* sound, u_2, is given by

$$u_2{}^2 = \frac{\rho_s}{\rho_n}\frac{TS^2}{C} \tag{11.14}$$

as can be shown using the hydrodynamical equations which will be discussed later. The variation of this velocity with temperature is shown in Fig. 11.8

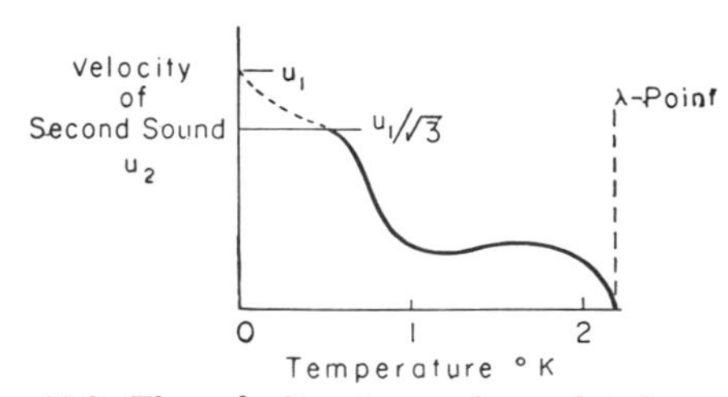

FIG. 11.8. The velocity of second sound (schematic).

[21, 24]. Above 1°K, u_2 is consistent with the known values of ρ_s, ρ_n, S and C. Below 0.5°K only phonons need be considered and it is readily shown that Eq. (11.14) then reduces to $u_2 = u_1/\sqrt{3}$, where u_1 is the velocity of first sound which is assumed to be the same as the velocity of the phonons. In Fig. 11.8 it will be seen that u_2 shows a tendency to level off at $u_1/\sqrt{3}$ near 0.5°K, but at still lower temperatures appears to rise toward a value u_1. However, below 0.5°K the mean free path of a phonon is much greater than either the wavelength of the second sound or the dimensions of the apparatus. Under these conditions, true wave propagation is not possible and the phenomena observed are more correctly described as the diffusion of phonons, the first phonon arriving with a velocity $c = u_1$. If a square temperature pulse is fed into the second sound transmitter at a temperature below 0.5°K, phonons are created in the immediate vicinity of the transmitter and then fly off in all directions, some going directly to the receiver and others colliding many times with the walls before they reach the receiver. Experimentally the received pulse is found to be much broader than the transmitted pulse.

The attenuation of second sound [25] is very large near the λ point, falls to a minimum near 1.8°K, and then rises rapidly as the temperature is lowered further. At temperatures below the minimum the attenuation can be explained in terms of the viscosity of the normal component, the relaxation effects which occur in first sound, and a type of thermal conductivity in the phonon-roton gas [23].

6. Third and Fourth Sound

When boundary conditions are important, other types of wave propagation become possible. Third sound [26, 27] is a surface wave on a liquid-helium film. Fourth sound [26, 28] is a pressure and temperature wave in a very narrow channel. In both cases the normal component remains stationary and the superfluid component alone oscillates.

7. The New Hydrodynamics

The hydrodynamics of liquid helium II is complicated by the necessity to consider the motion of the two components separately and to assign independent velocities $\mathbf{v}_s$ and $\mathbf{v}_n$ to them. The thermal effects must also be taken into account. Ignoring second-order terms, the equations, as written by Khalatnikov [23], take the following form. The total mass current $\mathbf{j}$ is

$$\mathbf{j} = \rho_s\mathbf{v}_s + \rho_n\mathbf{v}_n \tag{11.15}$$

The two equations of motion are

$$\frac{\partial \mathbf{j}}{\partial t} + \text{grad } p = \eta_n(\nabla^2\mathbf{v}_n + \tfrac{1}{3}\text{ grad div } \mathbf{v}_n) + \xi_1 \text{ grad div } (\mathbf{j} - \rho\mathbf{v}_n) + \xi_2 \text{ grad div } \mathbf{v}_n \tag{11.16}$$

$$\frac{\partial \mathbf{v}_s}{\partial t} + \frac{1}{\rho}\text{grad } p - S \text{ grad } T = \xi_3 \text{ grad div } (\mathbf{j} - \rho\mathbf{v}_n) + \xi_4 \text{ grad div } \mathbf{v}_n \tag{11.17}$$

The term S grad T in (11.17) arises because the two components can be accelerated by temperature gradients as well as pressure gradients. The above equations must be supplemented by an equation expressing conservation of mass.

$$\frac{\partial \rho}{\partial t} + \text{div } \mathbf{j} = 0 \tag{11.18}$$

and an equation describing the flow of entropy

$$\frac{\partial}{\partial t}(\rho S) + \text{div } (\rho S\mathbf{v}_n) = \frac{\chi}{T}\nabla^2 T \tag{11.19}$$

The coefficients ξ_1, ξ_2, ξ_3, and ξ_4 arise from the relaxation processes which were discussed in connection with first sound. The attenuation of first sound of frequency ν is

$$\alpha_1 = \frac{2\pi^2\nu^2}{\rho u_1{}^3}\left(\frac{4}{3}\eta_n + \xi_2\right) \tag{11.20}$$

The coefficient χ corresponds to a type of thermal conduction and appears in the expression for the attenuation of second sound

$$\alpha_2 = \frac{2\pi^2\nu^2}{\rho u_2^3}\left[\frac{\rho_s}{\rho_n}\left(\frac{4}{3}\eta_n + \xi_2 - \rho\xi_1 - \rho\xi_4 + \rho^2\xi_3\right) + \frac{\chi}{C}\right] \quad (11.21)$$

The Critical Velocity. One very important feature of the flow of the superfluid component cannot be included in the hydrodynamical equations. In the narrowest channels ($d \sim 10^{-5}$ cm) the velocity of the superfluid is found to be a constant, varying only slightly with the pressure head or the length of the channel. It is as though the dissipative forces are zero until a critical velocity v_c is reached, but become very large as soon as v_c is exceeded. In order of magnitude the critical velocity is given by

$$mv_cd \sim \hbar \quad (11.22)$$

where m is the mass of the helium atom and d is the width of the channel. v_c therefore increases rapidly as the channel is made narrower. Experimentally, v_cd is not quite constant but increases slowly with d.

8. Quantized Vortex Lines

Onsager [29] and Feynman [30] have suggested that the only way to give the liquid a rotational motion is to embed in it quantized vortex lines. These are cylindrical rotational motions about an axis, with the velocity at a distance r from the axis given by

$$v_s = \frac{\hbar}{mr} \quad (11.23)$$

This ensures that the wave function is single-valued, because during a complete rotation around the axis its phase changes by 2π. Moreover, curl $v_s = 0$ everywhere, except near the axis, where a few atoms rotate with an angular momentum quantized in a manner already familiar in connection with polyatomic molecules. A rotational flow can be simulated by embedding in the liquid a large number of vortex lines. Imagine a plane surface of area A which is small but contains many atoms, and let there be N^+_v anticlockwise and N^-_v clockwise vortex lines inside it. Then the hydrodynamical situation is almost equivalent to having within the area an average curl

$$[\text{curl } \mathbf{v}_s]_{\text{av}} = \frac{N^+_v - N^-_v}{A}\frac{2\pi\hbar}{m} \quad (11.24)$$

When the superfluid flows through a channel the only way in which the walls can bring it to rest is by the creation of rotational motion, that is, by the creation of vortex lines. It can be argued very plausibly [16, 30] that the processes which create vortex lines are not possible until the velocity of flow exceeds the value given by Eq. (11.22). Exact details are still obscure.

The ideas are supported by the fact that, when second sound is propagated in the liquid inside a rotating vessel, there is an extra attenuation proportional to the angular velocity ω [31]. The suggested explanation is that the rotation can arise only from a density of vortex lines proportional to ω. The second sound wave consists of a relative oscillatory motion of the superfluid component and the elementary excitations, phonons and rotons. The phonons and rotons collide with the vortex lines and therefore experience a resistive force which attenuates the second sound.

Further evidence comes from an ingenious experiment performed by Vinen [32] on a vibrating wire stretched along the axis of a vessel of rotating liquid. A net circulation around the wire caused the vibration to precess at a rate proportional to the line integral of the velocity around a closed curve surrounding the wire. The results were consistent with Eq. (11.23). Hall [33] has repeated Andronikashvili's experiment (Fig. 11.5) in rotating liquid and has shown that, under these circumstances, circularly polarized transverse waves travel along the cores of the vortex lines.

9. Ions in Liquid Helium

The nature of ions in liquid helium is not yet properly understood, but is has been suggested that the positive ion is surrounded by a region of increased liquid density [34], whereas the negative ion may be an electron inside a small bubble [35]. As the ion drifts through the liquid, its progress is hindered by repeated collisions with phonons and rotons. The ionic mobility therefore increases rapidly as the temperature is lowered and the number of rotons and phonons present decreases [36]. At temperatures near 0.5°K it is possible to achieve a drift velocity of 4×10^3 cm/sec, and the ion is then capable of creating rotons, so that it becomes increasingly more difficult to raise its velocity further. However, if a very high electric field is applied, the ion suddenly changes its nature and is able to travel long distances through the liquid before its energy is appreciably reduced by collisions [37]. Further investigation reveals that the drift velocity has fallen to about 10^2 cm/sec and that it decreases as the energy is increased. This extraordinary behavior can be explained by the assumption that the ion has in fact created and attached itself to a "quantized vortex ring." This is a length of quantized vortex line with its axis, or "core," bent into a circle, so that the velocity field in its neighborhood is similar to the magnetic field produced by a circular current loop. The total kinetic energy E of the flow around this ring can be shown to be approximately proportional to its radius R:

$$E = \frac{\rho h^2 R}{2m^2}\left[\ln\left(\frac{8R}{a}\right) - \frac{7}{4}\right] \quad (11.25)$$

Here a ($\simeq 1.3 \times 10^{-8}$ cm) is the radius of the hollow core. The ring travels through the liquid with a velocity

$$v = \frac{h}{4\pi mR}\left[\ln\left(\frac{8R}{a}\right) - \frac{1}{4}\right] \quad (11.26)$$

which is inversely proportional to R. The velocity v therefore varies approximately as $1/E$. From their measurements of v as a function of E, Rayfield and Reif [37] were able to check Eqs. (11.25) and (11.26), providing impressive evidence that circular flow in liquid helium II is quantized.

Nonlinear Frictional Forces. In channels of width greater than 10^{-5} cm the velocity of flow is not constant but varies slowly in a nonlinear manner with the pressure head [38]. The variation becomes more marked the wider the channel. Similar effects are observed if the flow is produced by a temperature gradient rather than by a pressure gradient [39]. A critical velocity probably still exists, but it is now possible to exceed the critical velocity at the expense of introducing frictional forces which vary nonlinearly with the excess velocity. These forces also manifest themselves in the oscillating disk experiment of Fig. 11.4*b* at high amplitudes of oscillation [40]. The damping of the oscillation is due to the viscosity of the normal component alone only if the peripheral velocity of the disk does not exceed v_c, but at higher velocities the logarithmic decrement increases rapidly and nonlinearly with amplitude. Similarly, in the Andronikashvili experiment of Fig. 11.5, the effective density of the liquid carried round with the disks is ρ_n at small amplitudes but increases toward ρ when the amplitude becomes large. These nonlinear forces are very complicated and not very well understood, but they may be caused by turbulence.

10. Helium Films (Fig. 11.9)

If an empty beaker is partially immersed in liquid helium II then a thick liquid film forms on its wall and the liquid flows through the film over the rim and fills

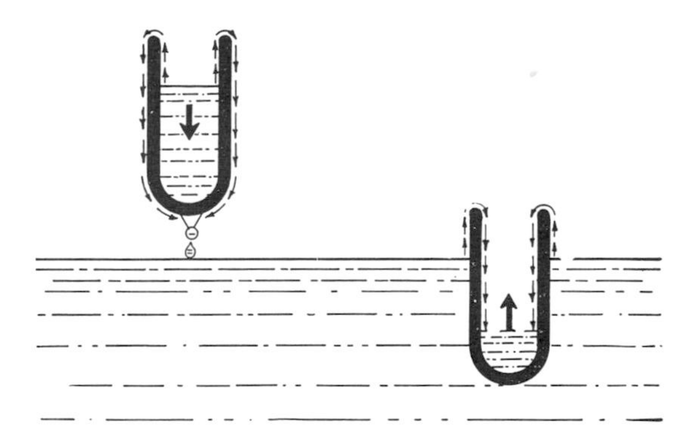

FIG. 11.9. Flow of the liquid helium film.

the beaker until the levels are equal inside and outside [41]. If the beaker is then raised, the liquid flows out through the film until the levels are again equal. The thickness of the film is about 3×10^{-6} cm at a height of 1 cm above the surface of the bulk liquid and decreases with height [42]. It is difficult to be more precise, as the experiments are strongly influenced by the cleanliness and smoothness of the surface on which the film is formed. The flow of the film is similar to flow through narrow channels. There is a critical velocity and the rate of flow is almost independent of the pressure head or length of the film and is limited at the point where the perimeter of the surface is least.

One of the factors in the formation of the film is almost certainly the van der Waals forces between the helium atoms and the atoms of the wall [43]. Another important consideration [44] is that the zero-point energy may be different from that for the bulk liquid and may vary with film thickness. The equilibrium values of film thickness and film density would then correspond to a minimum free energy with all these effects taken into account.

Thin adsorbed films in equilibrium with vapor at a pressure less than the saturated vapor pressure have also been studied [45]. The first one or two layers near the wall are probably solid, but superfluid flow has been observed in only a single liquid layer on top of the solid layer. For these very thin films, however, the superfluid flow ceases at a temperature below the bulk λ point, the temperature becoming lower the thinner the film.

11. Liquid He^3

He^3 has a critical temperature of 3.35°K and a boiling point of 3.2°K [46]. Like He^4 it is liquid down to 0°K and can be solidified only by the application of pressure [47]. However, it shows no λ transition and no superfluidity at all temperatures down to at least 0.01°K. This is presumably because the atom has a nuclear spin of $\frac{1}{2}\hbar$ and obeys Fermi-Dirac statistics. The nuclear spin can have two possible orientations associated with an entropy of $R \ln 2$ per mole at high temperatures. The entropy of the liquid can be deduced from measurements of its specific heat and from the vapor pressure curve [48]. At 0.5°K the total entropy is already less than $R \ln 2$, so there must be some correlated alignment of the nuclear spins antiparallel to one another at this temperature. Nuclear resonance experiments in a magnetic field give the relative numbers of spins parallel and antiparallel to the field [49]. Above 1°K these relative numbers are what might be expected for an ideal paramagnetic gas, but at 0.5°K the alignment produced by the field is already appreciably less than for the ideal gas, revealing a cooperative antiferromagnetic alignment of the spins. These phenomena cannot be readily explained by applying the formulae relevant to Fermi-Dirac degeneracy in an ideal gas.

12. The Fermi Liquid

At temperatures near 0.1°K, considerable success is achieved by Landau's theory of the Fermi *liquid* [50]. In this theory the simple "bare" He^3 atom is replaced by a "quasi particle," which is a "dressed" He^3 atom accompanied by its cloud of virtual excitations. For this quasi particle the relationship between energy ϵ and momentum p is not unique but depends upon the distribution function of all the other quasi particles. Nevertheless, the specific heat is still proportional to T at low temperatures, with an effective mass about three times the actual mass of the He^3 atom. Because of the Pauli exclusion principle and the scarcity of available states into which the quasi particle can be scattered, its mean free path increases rapidly as the temperature is lowered, varying as $1/T^2$. The viscosity therefore varies as $1/T^2$, the diffusion coefficient as $1/T^2$, and the thermal conductivity as $1/T$ [51].

Zero Sound. According to Landau [50], ordinary first sound can no longer be propagated in liquid He^3 when the mean free path of the quasi particles exceeds the wavelength. It is then replaced by "zero sound," which is best visualized as an unusual oscillatory distortion of the Fermi surface. There is some indirect experimental evidence that zero sound actually does replace first sound at low temperatures [52].

A Possible λ Transition in Liquid He³. It has been suggested [53] that the He^3 atoms can form bound pairs at very low temperatures and that this results in a transition very similar to the superconducting transition (Part 5, Chap. 12). At the time of this writing, Peshkov [54] has obtained an experimental indication of a specific heat anomaly at 0.0055K°, but Wheatley et al. [55] have found no evidence for a transition down to 0.0035°K.

Liquid He³-He⁴ Mixtures. The phenomena encountered when He^3 is dissolved in He^4 can be satisfactorily explained on the assumption that the dissolved He^3 atoms form part of the normal component. For example, if we perform the Andronikashvili experiment (Fig. 11.5) for the mixture, the values of ρ_n are greater than for the pure liquid and tend at 0°K to a value about twice the partial density of He^3 [56]. At 0°K there are no phonons or rotons and ρ_n is due entirely to the He^3 in solution. The effective mass of an He^3 atom in solution must therefore be about twice its actual mass. The velocity of second sound in the mixtures can be explained in a similar way, using Eq. (11.14). The principal difference from Fig. 11.8 is that there is a maximum below 1°K and u_2 tends to zero at 0°K [57].

References

1. London, F.: *Proc. Roy. Soc. (London)*, **A153**: 576 (1936).
2. Buckingham, M. J., and W. M. Fairbank: article in "Progress in Low Temperature Physics," vol. 3, North Holland Publishing Company, Amsterdam, 1961.
3. Atkins, K. R., and M. H. Edwards: *Phys. Rev.*, **97**: 1429 (1955).
4. Reekie, J., and T. S. Hutchinson: *Phys. Rev.*, **92**: 827, (1953).
5. Henshaw, D. G., and D. G. Hurst: *Phys. Rev.*, **91**: 1222 (1953).
6. Kapitza, P.: *Nature*, **141**: 74 (1938).
7. Allen, J. F., and A. D. Misener: *Nature*, **141**: 75 (1938).
8. Keesom, W. H., and G. E. MacWood: *Physica*, **5**: 737 (1938).
9. Tisza, L.: *Nature*, **141**: 913 (1938).
10. Andronikashvili, E. L.: *J. Phys. (U.S.S.R.)*, **10**: 201 (1946).
11. Kapitza, P. L.: *J. Phys. (U.S.S.R.)*, **5**: 59 (1941).
12. London, H.: *Proc. Roy. Soc. (London)*, **A171**: 484 (1939).
13. Keesom, W. H., B. F. Saris, and L. Meyer: *Physica*, **7**: 817 (1940).
14. London, F.: *Phys. Rev.*, **54**: 947 (1938). Tisza, L.: *Phys. Rev.*, **72**: 838 (1947).
15. Penrose, O., and L. Onsager: *Phys. Rev.*, **104**: 576 (1956).
16. Landau, L.: *J. Phys. (U.S.S.R.)*, **5**: 71 (1941); **11**: 91 (1947).
17. Palevsky, Otnes, Larsson, Pauli, and Stedman: *Phys. Rev.*, **108**: 1346 (1957). Yarnell, Arnold, Bendt, and Kerr: *Phys. Rev.*, **113**: 1379 (1959).
18. Feynman, R. P.: *Phys. Rev.*, **94**: 262 (1954).
19. Andronikashvili, E. L.: *J. Expl. Theoret. Phys. (U.S.S.R.)*, **18**: 429 (1948).
20. Landau, L., and I. M. Khalatnikov: *J. Expl. Theoret. Phys. (U.S.S.R.)*, **19**: 637, 709 (1949).
21. Peshkov, V.: *J. Phys. (U.S.S.R.)*, **10**: 389 (1946).
22. Pellam, J. R., and C. F. Squire: *Phys. Rev.*, **72**: 1245 (1947). Atkins, K. R., and C. E. Chase: *Proc. Phys. Soc. (London)*, **A64**: 826 (1951). Herlin, M. A., and C. E. Chase: *Phys. Rev.*, **97**: 1447 (1955).
23. Khalatnikov, I. M.: *J. Expl. Theoret. Phys. (U.S.S.R.)*, **23**: 8 (1952).
24. De Klerk, D., R. P. Hudson, and J. R. Pellam: *Phys. Rev.*, **93**: 28 (1954).
25. Atkins, K. R., and K. H. Hart: *Can. J. Phys.*, **32**: 381 (1954). Pellam, J. R., and W. B. Hanson: *Phys. Rev.*, 321 (1954).
26. Atkins, K. R.: *Phys. Rev.*, **113**: 962 (1959).
27. Everitt, C. W. F., K. R. Atkins, and A. Denenstein: *Phys. Rev.*, **136**: A1494 (1964).
28. Pellam, J. R.: *Phys. Rev.*, **73**: 608 (1948). Rudnick, I., and K. A. Shapiro: *Phys. Rev. Letters*, **9**: 191 (1962).
29. Onsager, L.: *Nuovo Cimento*, vol. 6, supp. 2, 249 (1949).
30. Feynman, R. P.: Article in "Progress in Low Temperature Physics," vol. 1, North Holland Publishing Company, Amsterdam, 1955.
31. Hall, H. E., and W. F. Vinen: *Phil. Mag.*, **46**: 546 (1955).
32. Vinen, W. F.: *Proc. Roy. Soc. (London)*, **A260**: 218 (1961); article in "Progress in Low Temperature Physics," vol. 3, North Holland Publishing Company, Amsterdam, 1961.
33. Hall, H. E.: *Phil. Mag. Suppl.* **9**: 89 (1960).
34. Atkins, K. R.: *Phys. Rev.*, **116**: 1339 (1959).
35. Kuper, C. G.: *Phys. Rev.*, **122**: 1007 (1961).
36. Careri, G.: article in "Progress in Low Temperature Physics," vol. 3, North Holland Publishing Company, Amsterdam, 1961. Reif, F., and L. Meyer: *Phys. Rev.*, **119**: 1164 (1960); **123**: 727 (1961).
37. Rayfield, G. W., and F. Reif: *Phys. Rev.*, **A136**: 1194 (1964).
38. Allen, J. F., and A. D. Misener: *Proc. Roy. Soc. (London)*, **A172**: 467 (1939).
39. Winkel, P., A. M. G. Delsing, and C. J. Gorter: *Physica*, **21**: 312 (1955).
40. A. C. Hollis-Hallett: *Proc. Phys. Soc. (London)*, **A63**: 1367 (1950).
41. Daunt, J. G., and K. Mendelssohn: *Proc. Roy. Soc. (London)*, **A170**: 423, 439 (1939).
42. Burge, E. J., and L. C. Jackson: *Proc. Roy. Soc. (London)*, **A205**: 270 (1951). Bowers, R.: *Phys. Rev.*, **91**: 1016 (1953). Hemming, D.: *Bull. Am. Phys. Soc.*, **8**: 91 (1963).
43. Frenkel, J.: *J. Phys. (U.S.S.R.)*, **2**: 345 (1940). Schiff, L. I.: *Phys. Rev.*, **59**: 838 (1941). Dsyaloshinskii, I. E., E. M. Lifshitz, and L. P. Pitaevskii: *Advan. Phys.*, **10**: 165 (1961).
44. Byl, A., A. de Boer, and J. Michels: *Physica*, **8**: 655 (1941). Atkins, K. R.: *Can. J. Phys.*, **32**: 347 (1954).
45. Long, E., and L. Meyer: Advances in Physics, *Phil. Mag.*, supp. 2, p. 1, 1953. Brewer, D. F., A. J. Symonds, and A. L. Thomson: *Phys. Rev. Letters*, **15**: 182 (1965).
46. Abraham, B. M., D. W. Osborne, and B. Weinstock: *Phys. Rev.*, **80**: 366 (1950).
47. Osborne, D. W., B. M. Abraham, and B. Weinstock: *Phys. Rev.*, **85**: 158 (1952).
48. De Vries, G., and J. G. Daunt: *Phys. Rev.*, **93**: 631 (1954). Roberts, T. R., and S. G. Sydoriak: *Phys. Rev.*, **93**: 1418 (1954). Osborne, D. W., B. M. Abraham, and B. Weinstock: *Phys. Rev.*, **94**: 202 (1954).

49. Fairbank, W. M., W. B. Ard, and G. K. Walters: *Phys. Rev.*, **95**: 566 (1954).
50. Landau, L. D.: *Soviet Phys., JETP (English Transl.)*, **3**: 920 (1956).
51. Wheatley et al.: *Phys. Rev. Letters*, **6**: 443 (1961); **7**: 220, 299 (1961).
52. Keen, B. E., P. W. Matthews, and J. Wilks: *Proc. Roy. Soc. (London)*, **A284**: 125 (1965).
53. Brueckner, K. A., T. Soda, P. W. Anderson, and P. Morel: *Phys. Rev.*, **118**: 1442 (1960). Emery, V. J., and A. M. Sessler: *Phys. Rev.*, **119**: 43 (1960). Pitaevskii, L. P., *Soviet Phys., JETP (English Transl.)*, **10**: 1267 (1960).
54. Peshkov, V. P.: *Soviet Phys. JETP (English Trans.)*, **19**: 1023 (1964); **21**: 663 (1965).
55. Abel, W. R., A. C. Anderson, W. C. Black, and J. C. Wheatley: *Phys. Rev. Letters*, **14**: 129 (1965).
56. Pellam, J. R.: *Phys. Rev.*, **99**: 1327 (1955).
57. King, J. C., and H. A. Fairbank: *Phys. Rev.*, **93**: 21 (1954).

Bibliography

Atkins, K. R.: "Liquid Helium," Cambridge University Press, London, 1959.

Jackson, L. C.: "Low Temperature Physics," Methuen, London, 1962.

Khalatnikov, I. M.: "Introduction to the Theory of Superfluidity," W. A. Benjamin, Inc., New York, 1965.

Keesom, W. H.: "Helium," Elsevier, Amsterdam, 1942.

Lane, C. T.: "Superfluid Physics," McGraw-Hill, New York, 1962.

Lifshitz, E. M., and E. L. Andronikashvili: "A Supplement to Helium," Consultants Bureau, Inc., New York, 1959.

London, F.: "Superfluids," vol. 2, Wiley, New York, 1954.

Simon, F. E., N. Kurti, J. F. Allen, and K. Mendelssohn: "Low Temperature Physics," Pergamon Press, New York, 1952.

Squire, C. F.: "Low Temperature Physics," McGraw-Hill, New York, 1953.

Chapter 12

Superconductivity

By WILLIAM M. FAIRBANK *and* ALEXANDER L. FETTER, Stanford University

1. Introduction

The most striking feature of the onset of superconductivity is the disappearance of electrical resistance below a critical temperature T_c. This effect, which was found by Onnes [1],† was the first experimental indication of the existence of the superconducting transition. The sharpness of the transition depends on the state and purity of the sample, but in favorable situations it can occur within a temperature interval of less than 0.001°K. Present estimates of the resistivity of superconducting metals are derived from experiments on the lifetime of persistent currents; these studies yield only an upper bound so that the resistivity of a superconductor may in fact be zero.

2. Flux Expulsion and Meissner Effect

A second dramatic effect is the destruction of superconductivity by a sufficiently large magnetic field

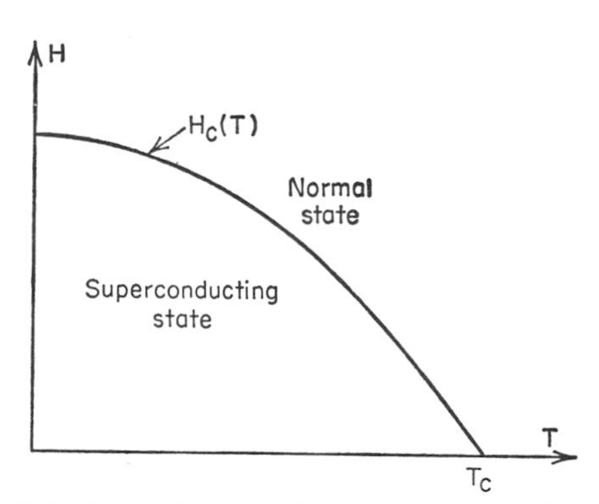

Fig. 12.1. Phase diagram of a metal showing the critical curve $H_c(T)$ separating the normal and superconducting states.

($\approx$1,000 gauss). The critical field $H_c(T)$ is largest at zero temperature and decreases to zero at T_c. The critical curve in the H-T plane is well approximated by a parabola of the form

$$H_c(T) = H_c(0)[1 - (T/T_c)^2] \qquad (12.1)$$

as shown in Fig. 12.1. If a material becomes superconducting in a region free of magnetic fields, the property of infinite conductivity ($\sigma = \infty$) implies that the magnetic flux in the sample will remain zero as long as the material is actually superconducting. This follows from Maxwell's equation

$$c^{-1}\, \partial \mathbf{B}/\partial t = -\text{curl } \mathbf{E} \qquad (12.2)$$

since $\mathbf{E} = 0$ in the limit $\sigma \to \infty$. As the magnetic field is increased through the critical value, the material reverts to its normal state and the flux suddenly penetrates the sample.

The inverse effect also occurs, as was demonstrated by Meissner and Ochsenfeld [2]. They placed a normal sample in a magnetic field (Fig. 12.2*a*) and lowered the temperature. When the metal became

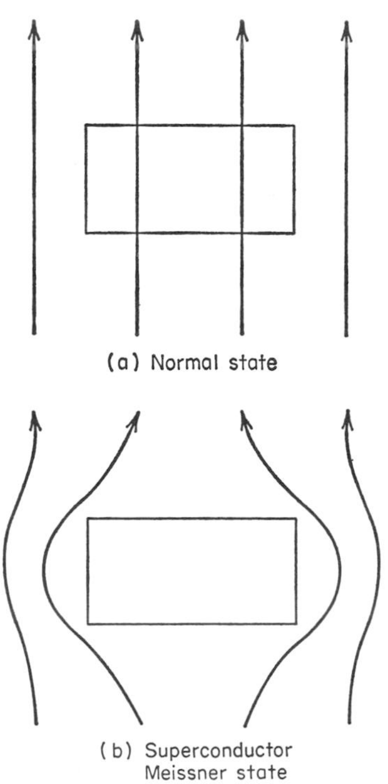

Fig. 12.2. Magnetic field penetration of a metal in (*a*) the normal state (*b*) the superconducting (Meissner) state.

† Numbers in brackets refer to References at end of chapter.

superconducting, the magnetic flux inside the material was expelled, producing a distortion of the nearby field lines (Fig. 12.2*b*); this is known as the Meissner effect. Thus a simply connected superconductor in a magnetic field contains zero flux ($\mathbf{B} = 0$), independent of whether the transition occurred in zero field or in finite field. The existence of a unique final state is of great importance, for it means that the state of a superconductor does not depend on its history and allows the application of reversible thermodynamics [3]. The physical origin of the vanishing magnetic flux is the presence of superconducting surface currents which create a magnetization **M** in the specimen. The general relation

$$\mathbf{B} = \mathbf{H} + 4\pi\mathbf{M} \tag{12.3}$$

may be used to find the magnetization, which is given by

$$\mathbf{M} = -(4\pi)^{-1}\mathbf{H} \tag{12.4}$$

since $\mathbf{B} = 0$. Equation (12.4) is frequently expressed by the statement that a superconductor is a perfect diamagnet. The metal becomes normal if H exceeds the critical field; the surface supercurrents and associated magnetization then vanish, as is shown in Fig. 12.3.

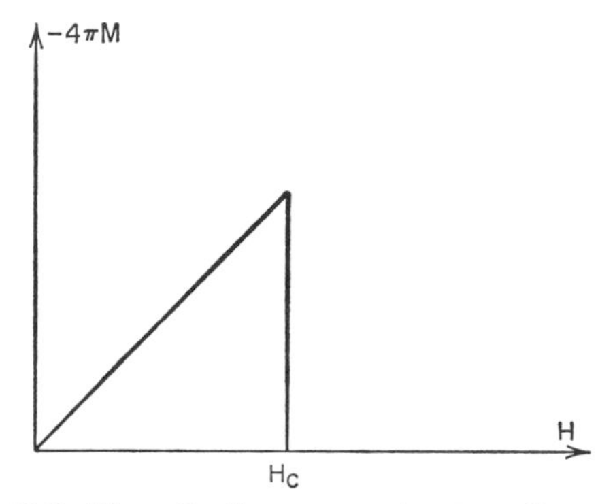

Fig. 12.3. Magnetization curve of a type I superconductor.

3. Thermodynamics

The thermodynamics of a superconductor can be derived from the equation relating the change in the internal energy U to the changes in the entropy S and magnetic moment MV:

$$dU = T\,dS + Hd(MV) \tag{12.5}$$

where V is the volume of the sample and the (small) effects associated with the changes of volume are neglected. It is convenient to introduce a generalized Gibbs free energy

$$G = U - TS - HMV \tag{12.6}$$

with the corresponding differential relation

$$dG = -S\,dT - MV\,dH \tag{12.7}$$

The appropriate Maxwell's relation yields

$$\left(\frac{\partial S}{\partial H}\right)_T = \left(\frac{\partial(MV)}{\partial T}\right)_H \tag{12.8}$$

The right side of Eq. (12.8) vanishes since the magnetization in a superconductor depends only on H [Eq. (12.4)] hence the entropy is independent of magnetic field and depends only on the temperature. At a fixed value of the temperature, Eqs. (12.4) and (12.7) imply that

$$\left(\frac{\partial G}{\partial H}\right)_T = -MV = \frac{VH}{4\pi} \tag{12.9}$$

which can be integrated to obtain the Gibbs free energy of a superconductor in a magnetic field H,

$$G_s(H) = G_s(0) + \frac{VH^2}{8\pi} \tag{12.10}$$

At the critical field, the Gibbs free energy of the superconducting and normal states must be equal, so that

$$G_n = G_s(H_c) = G_s(0) + \frac{VH_c^2}{8\pi} \tag{12.11}$$

Thus the superconducting state in zero field has a lower free energy than the normal state; this energy difference arises from the ordered (condensed) nature of the superelectrons, and the quantity $-H_c^2/8\pi$ may be considered as the condensation energy per unit volume associated with the phase transition in zero field.

Thermodynamics can also be used to verify directly that the superconductor represents a more ordered state than the normal metal. It follows from the definition of equilibrium that the Gibbs free energy of the normal and of the superconducting states are equal all along the critical curve $H_c(T)$ (Fig. 12.1). If we vary H and T by small amounts, remaining on the critical curve, the corresponding changes in G_s and G_n must also be equal. Since the magnetization vanishes in the normal metal, this condition leads to the equation

$$-S_s\,dT - MV\,dH_c = -S_n\,dT \tag{12.12}$$

which may be rewritten as

$$S_n - S_s = -\frac{VH_c}{4\pi}\frac{dH_c}{dT} \tag{12.13}$$

Figure 12.1 shows that $dH_c/dT \leq 0$; hence the entropy of the normal state is never less than that of the superconducting state. The critical field vanishes at T_c, and dH_c/dT vanishes at $T = 0$, so that the entropy difference between the two states is zero both at $T = T_c$ and $T = 0$. The normal-superconducting transition in zero magnetic field is a second-order phase transition because there is no change of entropy or latent heat; in finite magnetic field, however, the transition is first order. From Eq. (12.13), it is easy to calculate the difference between specific heat in the normal and in the superconducting states:

$$\begin{aligned} C_n - C_s &= T\frac{\partial(S_n - S_s)}{\partial T} \\ &= \frac{-VT}{4\pi}\left(\frac{dH_c}{dT}\right)^2 - \frac{VH_c}{4\pi}\frac{d^2H_c}{dT^2} \end{aligned} \tag{12.14}$$

which is negative at T_c but becomes positive at lower temperatures [4].

The specific heat of a metal can be separated into contributions from the electrons and the lattice. Experimental studies confirm that the lattice specific heat agrees with the Debye T^3 law and is unchanged when the metal becomes superconducting. The electronic specific heat of the normal state is proportional to T, so that the total specific heat of the normal metal is of the form

$$C_n = \alpha T + \beta T^3 \tag{12.15a}$$

where α and β are constants. The electronic specific heat of the superconducting phase C_{es} may be determined experimentally from the relation

$$C_s = C_{es} + \beta T^3 \tag{12.15b}$$

At T_c in zero field, the specific heat has a finite discontinuity [5] as shown in Fig. 12.4. Well below the critical temperature ($T < 0.1\ T_c$), C_{es} exhibits an exponential temperature dependence

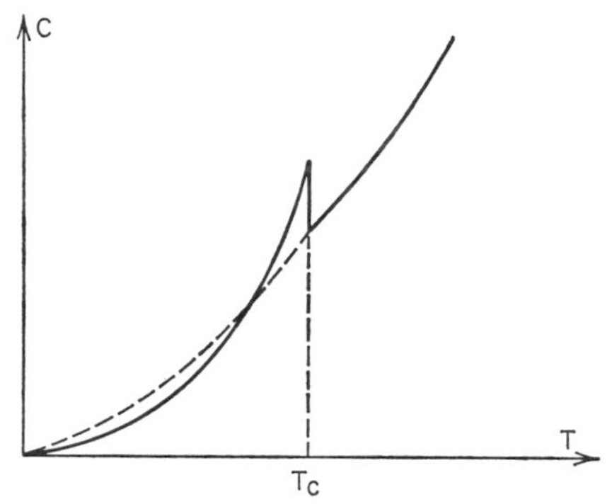

FIG. 12.4. Comparison of the specific heat of a metal in the superconducting phase (solid line) and in the normal phase (dotted line).

$$C_{es} \propto \exp(-\Delta/kT) \tag{12.16}$$

where k is Boltzmann's constant [6]. The quantity Δ is called the energy gap and is interpreted as the minimum energy that the system can absorb; experimental studies indicate that the ratio Δ/T_c is the same for most superconductors. The energy gap plays a fundamental role in the microscopic theory of superconductivity, which is discussed below.

4. Phenomenological Theories

London Theory. We shall now consider two phenomenological theories that describe the electrodynamics of superconducting metals: the London theory [7, 8] and the Ginzburg-Landau theory [9]. The London theory is based on the analogy between the condensed superelectrons and a classical inviscid fluid; the supercurrent $\mathbf{j}_s$ is written as

$$\mathbf{j}_s = n_s^* e^* \mathbf{v}_s \tag{12.17}$$

where n_s^* and e^* are the effective number density and charge of the superelectrons and $\mathbf{v}_s$ is their velocity field. Here the asterisk has been used to denote the effective quantities since the actual superconducting entities are bound pairs of electrons. A perfectly conducting fluid will not in general display a Meissner effect, however, and it is necessary to postulate an additional equation, which is known as the London equation:

$$\text{curl}\ \mathbf{j}_s = -\left(\frac{c}{4\pi\lambda^2}\right)\mathbf{H} \tag{12.18}$$

where λ is a constant with the dimensions of a length. Equation (12.18) may be combined with the static Maxwell's equation

$$\text{curl}\ \mathbf{H} = \left(\frac{4\pi}{c}\right)\mathbf{j}_s \tag{12.19}$$

to yield

$$\nabla^2\mathbf{H} = \lambda^{-2}\mathbf{H} \tag{12.20}$$

In a semi-infinite superconducting medium ($x > 0$) the appropriate solution of Eq. (12.20) is

$$H = H_0 e^{-x/\lambda} \tag{12.21}$$

This result shows that the magnetic field and other electromagnetic effects are confined to a thin surface layer of thickness λ, which is just the Meissner effect. The quantity λ is known as the London penetration depth, and in pure superconductors

$$\lambda = \left[\frac{m^* c^2}{4\pi\, n_s^* (e^*)^2}\right]^{1/2} \approx 500\text{Å} \tag{12.22}$$

If the magnetic field is described by a vector potential $\mathbf{A}$, the London equation may be rewritten as

$$\text{curl}\left[m^*\mathbf{v}_s + \frac{e^*\mathbf{A}}{c}\right] = \text{curl}\ \mathbf{p}_s = 0 \tag{12.23}$$

where the canonical momentum is given by

$$\mathbf{p}_s = m^*\mathbf{v}_s + \frac{e^*\mathbf{A}}{c}$$

In classical hydrodynamics, irrotational flow is defined by the condition

$$\text{curl}\ \mathbf{v} = 0 \tag{12.24}$$

and the London equation may be considered a gauge-invariant generalization of Eq. (12.24) to charged systems. The physical assumption inherent in Eq. (12.18) is that the superfluid condensate cannot assume an arbitrary flow pattern; quantum mechanical restrictions require that any physical motion satisfy Eq. (12.23) [or Eq. (12.24) in the case of neutral liquid He II]. The condition of "irrotational flow" arises from the single-valuedness of the wave function and is very closely related to the quantization of circulation in liquid He II and of magnetic flux in superconductors.

In addition to the penetration depth λ, a superconductor has another distinct characteristic length ξ, which is called the coherence length. The necessity for such a concept was suggested by Pippard [10], who had studied the electrodynamics of superconducting alloys. His experiments implied that the mean supercurrent at a given point depended on a spatial average of the electromagnetic field within a sphere of radius $\approx \xi$; thus the simple London relation between

$\mathbf{j}_s$ and $\mathbf{H}$ [Eq. (12.18)] must be replaced by an integral relation, in which ξ determines the extent of nonlocality,

$$\mathbf{j}_s(\mathbf{r}) = -\frac{c}{4\pi\lambda^2}\frac{3}{4\pi\xi}\int \frac{\mathbf{R}[\mathbf{R}\cdot\mathbf{A}(\mathbf{r}')]}{R^4}\,e^{-R/\xi}\,d^3r' \quad (12.25)$$

where $\mathbf{R} = \mathbf{r} - \mathbf{r}'$ and $R = |\mathbf{R}|$. The coherence length is therefore a measure of the distance through which the properties at one point of a superconductor can affect nearby points.

Ginzburg-Landau Theory. An alternative approach to the coherence length originates in the Ginzburg-Landau theory, which is based on Landau's theory of phase transitions. They describe the ordered state of a superconductor with a complex order parameter ψ, which is interpreted as the wave function of the condensed superelectrons. Since the ordering vanishes at the critical temperature, the order parameter is small in the temperature range $(T_c - T) \ll T_c$, and the free energy of the superconductor may be expanded in ascending powers of ψ. A variational calculation leads to the following set of nonlinear coupled equations for the order parameter and the vector potential:

$$\frac{1}{2m^*}\left(-i\hbar\nabla - \frac{e^*\mathbf{A}}{c}\right)^2\psi + \alpha\psi + \beta\psi|\psi|^2 = 0$$

$$\nabla^2\mathbf{A} = \frac{-4\pi\mathbf{j}}{c} \quad (12.26)$$

$$\mathbf{j} = -\frac{ie^*\hbar}{2m^*}(\psi^*\nabla\psi - \psi\nabla\psi^*) - \frac{(e^*)^2}{m^*c}|\psi|^2\mathbf{A}$$

which are the Ginzburg-Landau equations. Here α and β are phenomenological constants related to the critical field curve in Fig. 12.1. Although Eqs. (12.26) resemble typical quantum mechanical wave equations, the quantity ψ must be considered as describing the *whole* condensate rather than a single charged particle; ψ is therefore normalized to the total number of superelectrons in the sample. Since a rapid spatial variation in ψ implies a corresponding large kinetic energy, changes in the order parameter occur with a characteristic length, which turns out to be the coherence length ξ. Thus if ψ vanishes at some point, such as a normal-superconducting boundary, the order parameter requires a length ξ to rise to its value in the bulk sample. For this reason, ξ is sometimes called the healing length. In pure superconductors, ξ is the order of 10^4 Å. The Ginzburg-Landau equations have also been derived from the microscopic theory. This derivation shows that the effective charge and mass are twice the values for single electrons and leads to a precise definition of the order parameter as the wave function for the condensed bound electron pairs.

5. Quantized Flux Lines

The properties of a superconductor depend sensitively on the ratio of the penetration depth λ to the coherence length ξ, and it is convenient to define a dimensionless parameter $\kappa = \lambda/\xi$. The importance of κ may be understood by considering a plane surface between a superconductor and a normal metal. It is clear from the previous thermodynamic discussion that the transition between a normal metal and a superconductor involves two distinct energy changes, each with a different characteristic length: The penetration depth λ is associated with the positive energy of flux expulsion, while the coherence length ξ is associated with the negative condensation energy. If $\kappa \ll 1$ ($\lambda \ll \xi$), the superconductor is conventionally called type I (Fig. 12.5*a*). The magnetic field in the normal metal must equal the critical field H_c, and the field penetrates only a small distance λ into the superconductor. The order parameter is zero in the normal metal and rises to its full value in a relatively large distance ξ. The system is in equilibrium deep in the superconductor ($x \to +\infty$) where the negative condensation energy just cancels the positive energy of

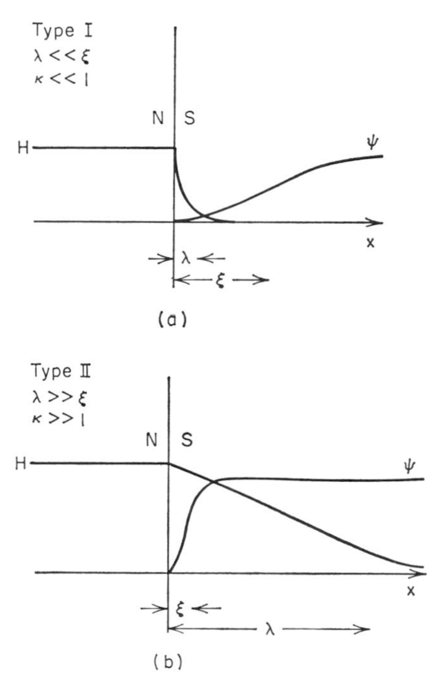

FIG. 12.5. A plane boundary separating normal and superconducting phases for (*a*) type I sample (*b*) type II sample.

flux expulsion. In contrast, the cancellation is incomplete in the surface region $\xi \gtrsim x \gtrsim \lambda$, where the condensation energy falls to zero but the magnetic field is still wholly shielded. The magnetic energy therefore dominates near the surface and leads to a positive energy associated with a normal-superconducting surface. Hence it is energetically favorable for a type I superconductor to remain in a uniform state since the formation of a surface region requires positive energy.

In the opposite limit of large $\kappa(\lambda \gg \xi)$, the metal is called type II (Fig. 12.5*b*). The magnetic field now penetrates a relatively great distance into the superconductor, while the order parameter rises rapidly to its value far in the interior of the sample. In the surface region $\lambda \gtrsim x \gtrsim \xi$, the positive energy of flux expulsion vanishes, while the condensation energy continues to make a full negative contribution. The net effect is a *negative* energy associated with the formation of a normal-superconducting surface, and a type II superconductor can lower its energy by increasing the surface area between normal and superconducting regions. This subdivision is eventually limited by the quantization of magnetic flux, because any normal region surrounded by superconducting metal must contain an integral number of flux quanta. It is not entirely obvious what spatial geometry of normal and superconducting metal represents the lowest energy state of a type II superconductor in a magnetic field, but detailed calculations [11] have shown that the magnetic flux penetrates the sample in the form of quantized flux lines (Fig. 12.6). This configuration is known as the mixed state and represents an intermediate situation between the normal state (Fig. 12.2*a*) and the Meissner state (Fig. 12.2*b*).

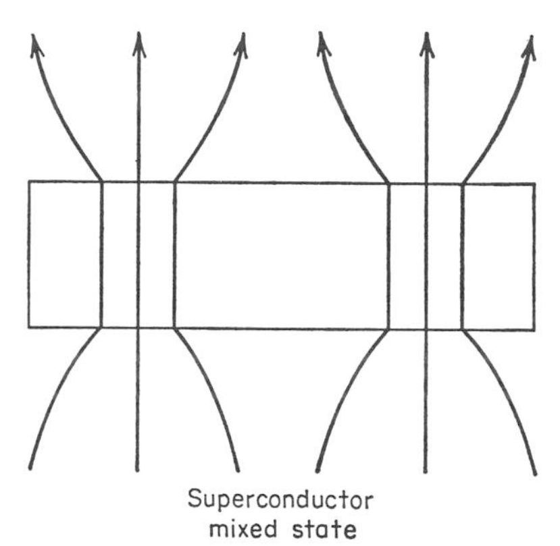

FIG. 12.6. Magnetic field penetration of a superconducting metal in the mixed state.

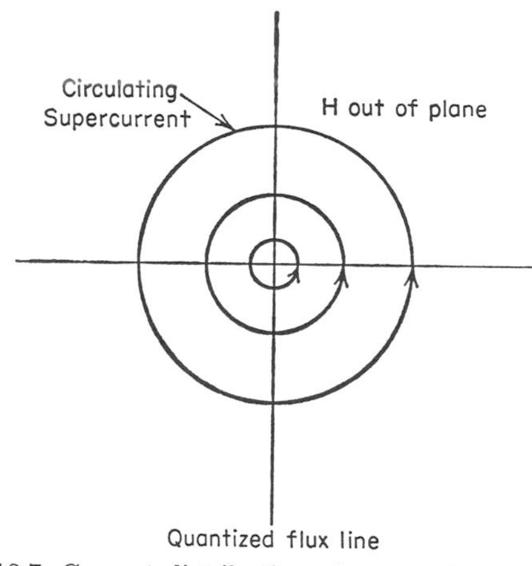

FIG. 12.7. Current distribution of a quantized flux line.

The mixed state has been a subject of great interest in recent years; the bulk of the metal remains superconducting so that no energy dissipation is associated with the flow of electric current. This situation can persist up to very high fields ($\approx 10^5$ gauss) and is clearly important in the design of high-field superconducting magnets. The detailed structure of an individual flux line is very similar to that of a classical vortex in an inviscid incompressible fluid (Fig. 12.7). At the center of the line is a core of normal metal of radius approximately equal to ξ, surrounded by circulating supercurrents. In the region $\xi \lesssim r \lesssim \lambda$, the velocity field is inversely proportional to the distance from the axis of the flux line; at large distances from the core ($r \gg \lambda$), the Meissner effect predominates and the currents vanish. Thus the metal is superconducting everywhere except inside the small core ($r \lesssim \xi$); the single quantum of magnetic flux is spread out over a much larger region ($r \lesssim \lambda$).

Most pure superconductors are type I, since typical values of $\lambda(\approx 500$ Å) and $\xi(\approx 10^4$ Å) lead to small $\kappa \ll 1$. In contrast, most alloys and thin films are type II and exhibit the unusual properties associated with the presence of quantized flux lines. In both cases, the origin of the type II behavior is the reduced electronic mean free path l. In alloys, l is the order of the distance between impurities, while in thin films, l is the order of the film thickness. The significance of the electronic mean free path may be seen as follows: On the one hand, the coherence length ξ characterizes the range of nonlocal influence between two nearby points and decreases as l decreases; on the other hand, the penetration depth λ characterizes the distance within which superelectrons are unable to screen the electromagnetic field and increases as l decreases. Thus sufficiently impure alloys and sufficiently thin films will always exhibit type II behavior.

A precise description of a flux line in a bulk type II superconductor is easily found from the London equation, which must be modified to account for the singularity at the center of the line. In the limit $\kappa \gg 1$, the extended London equation is

$$\text{curl}\,\mathbf{j}_s + \left(\frac{c}{4\pi\lambda^2}\right)\mathbf{H} = \left(\frac{\varphi_0 c}{4\pi\lambda^2}\right)\hat{z}\delta(\mathbf{r}) \qquad (12.27)$$

where φ_0 ($=hc/2e$) is the quantum of magnetic flux. Here $\hat{z}$ is a unit vector along the axis of the vortex, which is placed at the origin, and $\mathbf{r}$ is a two-dimensional vector. A combination of Eqs. (12.19) and (12.27) shows that the magnetic field due to a single flux line is

$$\mathbf{H}(\mathbf{r}) = \left(\frac{\varphi_0}{2\pi\lambda^2}\right)\hat{z}K_0\left(\frac{r}{\lambda}\right) \qquad (12.28)$$

while the corresponding supercurrent is

$$\mathbf{j}_s(\mathbf{r}) = \left(\frac{c}{4\pi}\right)\text{curl}\,\mathbf{H} = \left(\frac{\varphi_0 c}{8\pi^2\lambda^2 r}\right)\hat{z}\times\mathbf{r}K_1\left(\frac{r}{\lambda}\right) \qquad (12.29)$$

The K functions are the modified Bessel functions that vanish exponentially for $r \gg \lambda$; for $r \ll \lambda$, Eqs. (12.28) and (12.29) behave like

$$\begin{aligned}\mathbf{H}(\mathbf{r}) &\approx \hat{z}\left(\frac{\varphi_0}{2\pi\lambda^2}\right)\ln\left(\frac{\lambda}{r}\right)\\ \mathbf{j}_s(\mathbf{r}) &\approx \left(\frac{\varphi_0 c}{8\pi^2\lambda^2}\right)(\hat{z}\times\mathbf{r})r^{-2}\end{aligned} \qquad (12.30)$$

The interaction energy V_{12} per unit length between two flux lines at $\mathbf{r}_1$ and $\mathbf{r}_2$ is

$$V_{12} = \left(\frac{\varphi_0^2}{8\pi^2\lambda^2}\right)K_0(r_{12}/\lambda) \qquad (12.31)$$

where $r_{12} = |\mathbf{r}_1 - \mathbf{r}_2|$. Equation (12.31) represents a finite-range interaction which vanishes for $r \gg \lambda$.

The properties of a flux line in a thin film of thickness d are similar to those in bulk superconductors, but the derivation is rather more complicated [12]. The most important difference is the long-range behavior, which obeys a power law instead of an exponential:

$$\begin{aligned} j_s &\sim \left(\frac{\varphi_0 c}{4\pi^2 d}\right)(\hat{z} \times \hat{r})r^{-2} \\ V_{12} &\sim \left(\frac{\varphi_0^2}{4\pi^2 d r_{12}}\right) \end{aligned} \tag{12.32}$$

for $r \gtrsim \lambda_{\text{eff}} = 2\lambda^2/d$. The physical difference between the bulk material and thin film arises from the fringing fields as the flux lines leave the film (Fig. 12.6). In a bulk sample, the flux lines interact by means of the magnetic field in the metal, which is shielded by the Meissner effect at distances large compared to λ. In a thin film, the field lines expand on entering the vacuum and most of the interaction occurs in the space surrounding the thin film. Thus the interaction energy between two flux lines in a thin film is long range, and it is necessary to consider the effect of the boundaries of the sample.

The magnetic properties of the mixed state are particularly interesting, for there are two critical magnetic fields: H_{c1} where the flux first penetrates the sample and H_{c2} where the sample reverts to the normal state. The theoretical magnetization curve is shown in Fig. 12.8, both for bulk samples and for thin films.

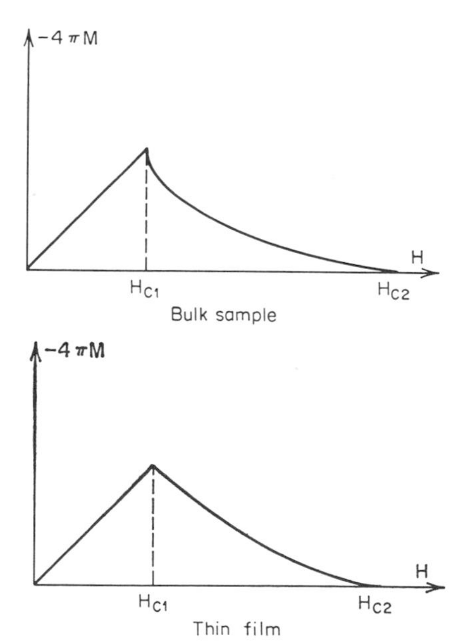

FIG. 12.8. Magnetization curve of a type II superconductor, showing the difference between a bulk sample and a thin film.

Below H_{c1}, the system behaves like a type I superconductor, and no magnetic flux penetrates the sample. At H_{c1} it becomes energetically favorable for the flux to enter the material and M decreases; the slope of the magnetization curve just above H_{c1} is infinite for a bulk sample but is finite for a thin film. As H increases, the density of flux lines increases until H_{c2} is reached, when the cores of adjacent flux lines start to overlap and the system becomes normal. Experiments confirm the general shape shown in Fig. 12.8, but the singularity near H_{c1} is generally rounded off. In bulk samples, theory predicts that the lines form a triangular lattice for all values of H ($H_{c1} \leq H \leq H_{c2}$) [11, 13]. The corresponding calculation has not been carried out in detail for thin films; the long-range interaction introduces a dependence on the sample geometry, and it is not clear whether a regular lattice is expected. Neutron diffraction experiments on bulk type II superconductors confirm the expected triangular structure [14].

6. Flux Quantization

The discovery of flux quantization in 1961 [15, 16] has had a major effect on both the theoretical and experimental development of superconductivity. It seems worth while at this point to recall the ideas which led F. London to suggest in 1946, long before any of the modern microscopic theories had been developed, that flux trapped in a superconducting ring might be quantized, or more exactly that the fluxoid

$$\varphi_0 = \frac{c}{e} \oint \mathbf{p}_s \cdot d\mathbf{s} \tag{12.33}$$

might be quantized. We have seen in Eq. (12.23) that the London equation for the supercurrent due to a magnetic field can be written

$$\text{curl } (m^*\mathbf{v}_s + e^*\mathbf{A}/c) = \text{curl } \mathbf{p}_s = 0$$

where $\mathbf{p}_s$ is the canonical momentum of the superconducting electrons and determines the de Broglie wavelength of the electron in the presence of a vector potential $\mathbf{A}$. London suggested that superconductivity is characterized by long-range order in the canonical momentum $\mathbf{p}_s$ and must be the result of a macroscopic quantum effect.

In the last chapter of his book on superfluids [17], London discusses a program for the development of a proper microscopic theory of superconductivity. He notes that, in the presence of a magnetic field, the electrons in the superconducting state differ drastically from free electrons. Instead of curling up in small circles, the Meissner effect shows that the electrons necessarily behave cooperatively. The electrons near the surface within the penetration layer flow in such a way as to exclude completely the magnetic field from the interior of the superconductor; hence the electrons in the bulk of the superconductor behave as if the magnetic field were not present. London pointed out that "superconductivity would result if the eigenfunction of a fraction of the electrons were not disturbed at all when the system is brought into a magnetic field less than H_c." He suggested that "it would be sufficient if the eigenfunction were to stay essentially as it was without the magnetic field, as if, so to speak, it were rigid." This would represent a quantum structure on a macroscopic scale. London suggested that such a rigid wave function would represent a sufficient, but perhaps not necessary explanation of superconductivity.

To explore the consequences of such a rigid wave function, London considered a superconducting ring with a fluxoid φ trapped in the center. He suggested that if the wave function of the superconducting system were indeed rigid, then for the wave function ψ to be single-valued, it is necessary that

$$\oint \bar{\mathbf{p}}_s \cdot d\mathbf{s} = nh \tag{12.34}$$

where n is an integer, h is Planck's constant, and $\bar{\mathbf{p}}_s$ is the average value of $\mathbf{p}_s$. London called the quantity $(c/e)\oint \bar{\mathbf{p}}_s \cdot d\mathbf{s}$ the fluxoid and noted that if the above quantum condition were true, there exists a universal unit for the fluxoid

$$\varphi = \frac{hc}{e} = 4 \times 10^{-7} \text{ gauss cm}^2 \tag{12.35}$$

For thick rings the quantity φ is the magnetic flux since there is no current flowing in the bulk of the superconducting ring and the contribution to $\oint \bar{\mathbf{p}}_s \cdot d\mathbf{s}$ from the velocity of the electrons is zero. For thin rings the fluxoid includes a term

$$(c/e)\oint m\bar{\mathbf{v}}_s \cdot d\mathbf{s} = \oint c\Lambda \mathbf{j}_s \cdot d\mathbf{s}$$

where $\Lambda = m/n_s e^2$

This prediction of quantized flux was checked experimentally by Deaver and Fairbank [15] and independently by Doll and Näbauer [16]. Deaver and Fairbank found that when a superconducting cylinder is cooled below its transition temperature in the presence of a magnetic field, the magnetic flux contained in the hole in the cylinder changes to the nearest quantized flux value. Both Deaver and Fairbank and Doll and Näbauer demonstrated the quantization of the magnetic flux trapped in a hollow superconducting cylinder after the magnetic field is turned off. The unit of quantization was found to be one-half of the unit predicted by London or $\varphi_0 = hc/2e = 2 \times 10^{-7}$ gauss cm^2. This factor of 2 is explained as being due to electron pairing and constitutes a direct proof of pairing in superconductors. Byers and Yang [18] considered the quantum mechanical problem of an ideal electron gas in a thin cylinder and showed that no flux quantization resulted. However, if the BCS theory of pairing is introduced, flux quantization in integral units of $hc/2e$ results. Similarly Onsager [19], Blatt [20], and Bloch and Rorschach [21] showed that a condensed Bose gas of electron pairs exhibits flux quantization in units of $hc/2e$. Many theoretical papers have been written on flux quantization, and many experiments have been performed which give further proof that the fluxoid in a superconductor is quantized. Indeed, as stated earlier, when flux is trapped in a singly connected superconductor, it is trapped in quantized fluxoid vortices. Each vortex contains a normal region at its core and generally contains one quantum of flux. In type II superconductors containing strains and impurity centers, these strains and impurity centers serve as the center for quantized vortices.

In the experiments of Deaver and Fairbank, it was demonstrated that the quantum number of magnetic flux trapped in a hollow superconducting cylinder cooled below T_c always takes the value nearest to the flux originally passing through the cylinder in the normal state. Thus it becomes possible to obtain a spatial region of truly zero magnetic field. A hollow superconducting shield should trap no flux at all if it is cooled down in a magnetic field sufficiently small so that less than one-half a quantum of flux passes through the shield before cooling. We will return to this point in the section on applications of superconductivity.

7. Tunneling and Josephson Effect

Superconductivity exists both in pure metals and in very impure alloys. An interesting question arises as to the size of a potential barrier which is sufficient to interrupt the long-range order of superconductivity. Giaever [22] experimented with electrons tunneling between superconductors through insulating layers of 100 to 200 Å. He demonstrated that the tunneling current increases sharply when the voltage across the insulator exceeds Δ/e. Measurement of this voltage has given a very accurate measure of the energy gap in superconductors. Single electrons tunnel through such barriers, and the tunneling current increases when the voltage is sufficient to raise the energy of the electrons above the energy gap.

Josephson [23] asked the question, what would happen if the insulator were made progressively thinner and thinner. He predicted, from theoretical calculations, that for sufficiently small insulating junctions of the order of 10 Å electron pairs would tunnel through the junction. Under these conditions the tunneling current would flow without any voltage across the junction between two superconductors. When a critical current was exceeded, a d-c voltage would develop across the junction, and the voltage would be given by $2eV = h\nu$, where e is the charge on an electron, h is Planck's constant, and ν is the frequency of the photon radiated by the electron pair when tunneling across the junction. The maximum zero-voltage current across the junction is given by the following equation

$$J = J_0 \sin\left(\delta_0 + \frac{2e}{c\hbar}\int \mathbf{A} \cdot d\mathbf{s}\right) \tag{12.36}$$

where δ_0 is a constant. Thus the current is a periodic function of the flux passing through the junction at right angles to the current, the period being the quantum of flux $hc/2e$.

The Josephson predictions were demonstrated by Anderson and Rowell [24], who showed that a zero-voltage current will flow through a thin insulating junction between two superconductors. The maximum value of this current oscillates with the external magnetic field, giving exactly the single-slit diffraction pattern predicted by Josephson (Fig. 12.9) [25]. Shapiro [26] has shown that the zero-voltage current can be changed by the application of an r-f voltage, proving that the a-c Josephson effect exists. Langenberg, Scalapino, Taylor, and Eck [27] have observed microwave radiation from a Josephson junction and verified the relationship $2eV = h\nu$ to one part in 10^4.

The Josephson current through a double junction has been measured by Lambe, Silver, Mercereau, and Jaklevic [28] as a function of the magnetic field in the area between the junctions. The maximum zero-resistance current was a periodic function of the external magnetic field with the period of the flux quantum and a shape similar to the diffraction pattern for a double slit, as predicted by the Josephson equations (Fig. 12.10). This is found both when the junctions are

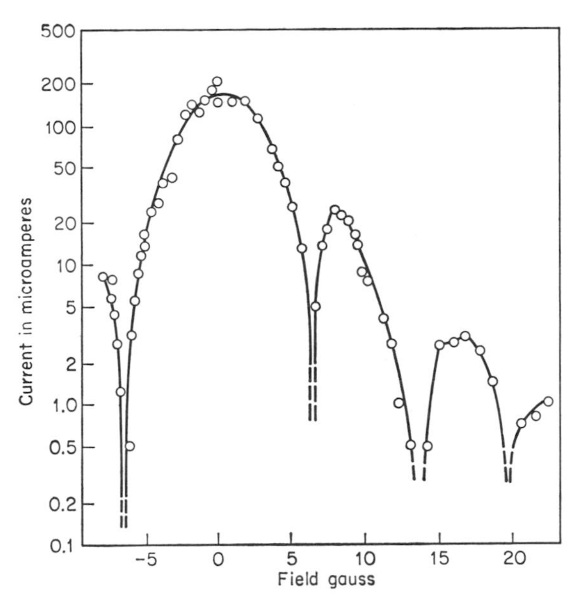

FIG. 12.9. Josephson tunneling current through a single junction, as a function of the applied magnetic field. (*From Rowell* [25].)

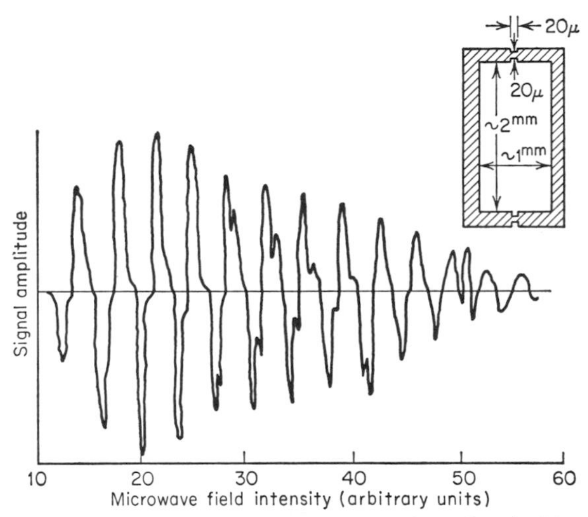

FIG. 12.10. Josephson tunneling current through a double junction as a function of the applied magnetic field. (*From Lambe and Silver* [28].) A schematic diagram of the apparatus is shown at right.

insulating junctions and when the junctions are thin superconducting sections in which the critical current is exceeded in the junctions when the voltage appears.

In all these weak-junction experiments, the current flows without resistance in certain quantized configurations. When a voltage appears, the current distribution changes suddenly to a different allowed configuration and a quantized unit of flux passes across the weak junction. When the junction or junctions form a resonant circuit, the quantized flux units pass across the junction at the resonant frequency ν, and the currents oscillate between two possible quantized configurations. This oscillation produces a d-c voltage V given by $2eV = h\nu$. The quantity $h\nu$ represents the energy of a microwave photon emitted by an electron pair of charge $2e$ crossing the junction with the voltage V. The total number N of pairs crossing the junction is given by $E/h\nu$, where E is the energy difference between the two microscopic current states. This represents another dramatic demonstration of quantized fluxoid or phase coherence in a superconductor.

Anderson [29] has explained the Josephson effect in terms of a macroscopic wave equation in the spirit of the London prediction of a rigid wave function. The important point is that the existence of a wave function with a definite phase precludes precise knowledge of the number of particles involved. Anderson's approach is in contrast to the more usual description of a many-body system as an assembly of a large but definite number of particles. In this case, the existence of superconductivity is generally discussed in

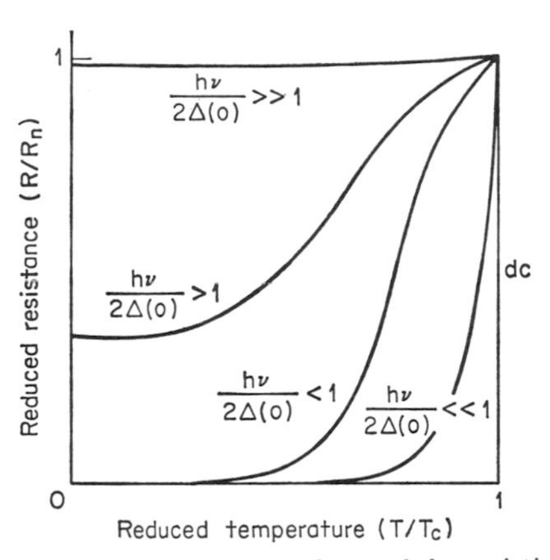

FIG. 12.11. Temperature dependence of the resistivity of a superconductor at various frequencies.

terms of the presence or absence of "off-diagonal long-range order" in the reduced density matrix [30].

Goldman, Kreisman, and Scalapino [31] have demonstrated that persistent currents can flow around a ring containing two insulating Josephson junctions. They have observed persistent currents with one, two, and three quanta of flux. The currents were observed for more than one hour. Persistent currents in a larger ring containing Josephson junctions have been observed by Smith [32].

When an alternating current is applied to a superconductor, a resistance is observed due to the presence of normal electrons in the penetration layer where the a-c magnetic field penetrates [33]. Figure 12.11 shows a family of such resistance curves as a function of frequency. When the frequency exceeds twice the energy gap additional resistance appears. This has been used to make accurate measurements on the energy gap [34, 35].

8. Microscopic Theories

Current microscopic theories of superconductivity are based almost exclusively on the fundamental work of Bardeen, Cooper, and Schrieffer [36]. The BCS theory has been remarkably successful in computing the observed properties of superconductors in terms of a very few constants that characterize the electronic properties of the particular metal. A complete discussion of this theory is impossible here, and we shall merely attempt a qualitative description. A much more complete analysis may be found in various excellent review articles and monographs which treat the BCS theory in detail [37–41].

The theory arose from Cooper's observation [42] that the usual ground state of an electron gas (filled Fermi sphere) is unstable with respect to an arbitrarily weak attractive interaction between electrons near the Fermi surface. A simple calculation shows that the formation of a bound electron pair lowers the energy of the system by a finite amount, so that the normal-state Fermi distribution no longer represents the configuration of lowest energy. This new state containing bound pairs is identified with the superconducting ground state. The energy difference between the normal state and the superconducting state is proportional to

$$\exp(-\text{const}/V) \tag{12.37}$$

where V represents the strength of the interaction. Equation (12.37) has no series expansion in ascending powers of V, and a perturbation analysis of the transition to a superconducting state is therefore impossible. Fröhlich had previously noted the difficulty with a perturbation calculation [43].

The original BCS theory assumes that the superconducting system can be described as an assembly of spin one-half fermions interacting through an attractive potential; this represents a generalization of Cooper's work. No detailed attempt is made to compute the actual interparticle potential; instead, the theory studies an artificial model of a constant attraction of strength V acting between particles of opposite spin and momentum lying very near the Fermi surface [44]. It is energetically favorable to form bound pairs with a binding energy $2\Delta(T)$, which depends on the temperature T. The quantity $\Delta(T)$ vanishes above a transition temperature T_c and rises rapidly to a constant value $\Delta(0)$ for $T \ll T_c$ (Fig. 12.12).

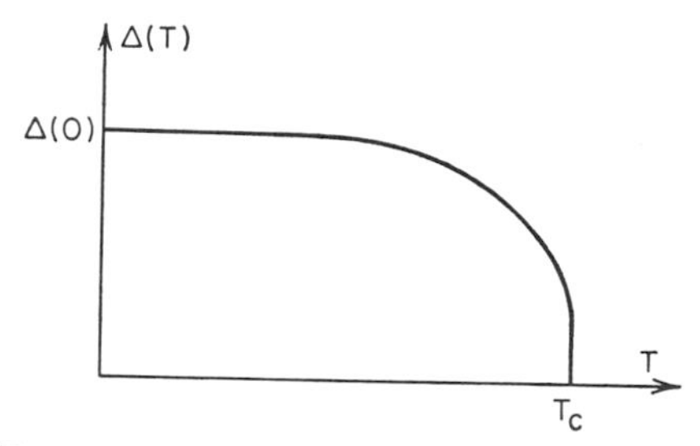

FIG. 12.12. Variation of the energy gap $\Delta(T)$.

The theory predicts the numerical values T_c and $\Delta(0)$ in terms of a cutoff frequency ω_0 and the density of states $N(0)$ at the Fermi surface for particles of one spin orientation:

$$\Delta(0) = 2\hbar\omega_0 \exp[-1/N(0)V] \tag{12.38}$$
$$kT_c = 1.13\,\hbar\omega_0 \exp[-1/N(0)V] \tag{12.39}$$

Although both expressions depend sensitively on the parameters of the theory, their ratio is a universal constant

$$2\Delta(0)/kT_c = 3.53 \tag{12.40}$$

which should be valid for all superconductors. $\Delta(0)$ can be measured directly because it determines the microwave surface impedance of a superconducting metal [34, 35], the infrared transmission [45] and absorption [46] of a thin superconducting film, and the low-temperature specific heat [6]; Eq. (12.40) is well satisfied for a wide class of superconductors.

In most metals, the attraction between the electrons is due to the coupling with the phonons. The relevance of the lattice vibrations is shown by the isotope effect: different isotopes of a given element have different transition temperatures, and the variation of T_c with the isotopic mass M is given by

$$T_c \propto M^{-1/2} \tag{12.41}$$

This effect was predicted theoretically by Fröhlich [47] and found simultaneously in experiments on mercury [48, 49]. Since lattice vibrations in a crystal are equivalent to an assembly of coupled harmonic oscillators, the lattice frequencies are expected to vary as $M^{-1/2}$. In the BCS model the cutoff frequency ω_0 is taken to be the Debye frequency of the metal. Hence $\omega_0 \propto M^{-1/2}$, and Eq. (12.39) automatically incorporates the isotope effect. The electron-phonon coupling modifies the interaction energy between two electrons; a calculation using second-order perturbation theory predicts an attraction between electrons near the Fermi surface with opposite momenta. This attraction tends to cancel the Coulomb repulsion, and a superconducting transition can occur only if the net *effective* interaction between electrons is attractive.

The superconducting ground state is determined variationally and consists of a mixture of bound electron pairs and normal electrons. Since the binding energy 2Δ is small, each pair spreads out over a large size, which turns out to be the coherence length ξ_0; detailed calculations show that

$$\xi_0 = \frac{\hbar v_F}{\pi\Delta(0)} \tag{12.42}$$

where v_F is the mean velocity of an electron at the Fermi surface. The higher states of a superconducting system can be described as the creation of one or more quasiparticles characterized by a temperature-dependent dispersion relation

$$E(p) = \{[\Delta(T)]^2 + [(p^2 - p_F^2)/2m]^2\}^{1/2} \tag{12.43}$$

where $p_F = mv_F$ is the Fermi momentum. Equation (12.43) shows that $E(p)$ is never less than $\Delta(T)$, so that the excited states are separated from the ground state by a finite energy $\Delta(T)$, often called the energy gap. Thus a superconductor cannot absorb an arbitrarily small amount of energy. $\Delta(0)$ is approximately 10^{-4} eV, corresponding to the energy of an infrared photon, which allows a direct measurement of the energy gap in infrared experiments [45, 46]. In the BCS theory, the thermodynamics of a superconductor follows immediately from Eq. (12.43). At low temperatures, the specific heat exhibits an exponential temperature dependence in agreement with Eq. (12.16); the observed discontinuity in the specific heat at T_c (Fig. 12.4) is also given correctly by the theory. Finally, for most superconductors, the observed critical magnetic-field curve deviates slightly from the parabolic approximation [Eq. (12.1)]; detailed calculations explain both the sign and the magnitude of this effect [50].

One of the most characteristic features of the BCS theory is the description of an excited state (quasiparticle) as a coherent superposition of a particle and a hole. This aspect appears to be a general property of superfluids and applies both to liquid He II [45] and to superconductors [52–54]. Thus the wave function of the quasiparticle is written as a linear combination of wave functions for a particle and a hole; any process involving the excitation of a quasiparticle necessarily leads to quantum mechanical interference between the two terms. The sign of the interference can be either positive or negative and depends on the particular mechanism considered. Calculations have been carried out for ultrasonic attenuation, absorption of electromagnetic radiation, and nuclear-spin relaxation. Each of these physical processes exhibits a distinct temperature dependence, and all agree with the experimental observations. Such complete correspondence between theory and experiment provides striking confirmation both of the general BCS model and of the concept of coherence.

The BCS theory also predicts the electrodynamic properties of a superconductor. At $T = 0$, supercurrents obey the London equation (12.18) if the spatial variations in $\mathbf{j}_s$ have sufficiently long wavelengths. Thus the model predicts a Meissner effect, and magnetic fields are expelled from the bulk of the sample. The supercurrent density depends nonlocally on the applied field, exactly as in Pippard's equation (12.25); a detailed evaluation shows that the kernel of the BCS expression is essentially identical with Pippard's exponential kernel $\exp(-|\mathbf{r} - \mathbf{r}'|/\xi_0)$. It is this comparison that yields the value of ξ_0 given in Eq. (12.42). The penetration depth can also be calculated; apart from small corrections, the temperature dependence of λ is very closely approximated by

$$[\lambda(0)/\lambda(T)]^2 = 1 - (T/T_c)^4 \qquad (12.44)$$

which was predicted earlier by a two-fluid model [55].

Probably the most remarkable aspect of the BCS theory is that it reproduces the successful features of all previous theories; thus it predicts the thermodynamics and the various temperature dependences of the two-fluid model, as well as the electrodynamics and long-range macroscopic quantum effects of the Londons' theory and of Pippard's nonlocal equations. In addition, Gorkov [56] has derived the phenomenological Ginzburg-Landau equations from the BCS theory. His proof shows that the quantities e^* and m^* [Eq. (12.26)] must be taken as $2e$ and $2m$, because of the electron pairing. It follows that the trapped magnetic flux in a superconductor is quantized in units of $hc/2e$, although this connection was noticed only after the experimental measurements of the flux quantum [15, 16].

In recent years, the BCS theory has been extended in several ways, of which we shall discuss only two: superconducting alloys and strong-coupling superconductors. The most surprising fact about superconducting alloys is their mere existence, since the presence of random impurities might be expected to reduce the electronic conductivity, exactly as in a normal metal. Experiments show that rather large concentrations of nonmagnetic impurities scarcely affect the transition temperature [57]. This observation was explained by Anderson [58], who emphasized the necessity for pairing between electrons in time-reversed states. The addition of nonmagnetic impurities alters the structure of the single-particle states, while preserving the time-reversal invariance of the theory. In particular, momentum is no longer a good quantum number because of the electron scattering. Superconductivity occurs in such alloys, but a different choice of pairing states is required from the original BCS choice of opposite momentum and spin. In contrast, spin-dependent interactions (or other mechanisms such as external currents or magnetic fields) affect each member of the pair differently and lead to breakup of the pair. A small concentration of magnetic impurities therefore produces a drastic reduction of the transition temperature [59]. The theory of a superconducting alloy containing paramagnetic impurities [60] is of particular interest, for it demonstrates the crucial role of the superfluid condensate. Abrikosov and Gorkov [60] showed that increased paramagnetic impurity concentration affects both the energy gap Δ, which appears in the excitation spectrum, and the transition temperature. These two quantities decrease at a different rate, however, and a finite range of concentration exists where Δ vanishes but both the wave function of the condensed superelectrons and the transition temperature remain finite. This surprising prediction of *gapless* superconductivity was fully confirmed by experiments of Reif and Woolf [59]. The study of superconducting alloys has now become a fully developed field, which has recently been reviewed in detail by de Gennes [61].

A second development from the BCS theory has been the careful study of the phonon coupling in superconductors. The original theory was developed by Bogoliubov, Tolmachev, and Shirkov [62] and in a different form by Eliashberg [63]. These calculations have been successfully applied to strong-coupling superconductors such as Pb and Hg, which do not fit the simple BCS predictions [64]. The phonons and electrons must both be included in the theory, and rather extensive numerical computations are necessary. The original calculations assumed simple models for the spectrum of the lattice vibrations and then predicted the properties of the superconductor [65, 66]. Recently, the opposite approach has also been developed, and the phonon spectrum of Pb has been computed from its superconducting density of states, which can be measured directly in tunneling experiments [67].

The essential feature of the BCS theory is the existence of an attractive interaction between electrons. Although the observed superconductivity in metals is due primarily to the phonon coupling, other mechanisms can occur in principle, and the exciting possibility of a room temperature superconductor has recently been suggested [68]. In this case, the phonon coupling to the ionic lattice would be replaced by a coupling to other electrons in a one-dimensional system (such as a long organic molecule). The isotope effect predicts that the reduction of the relevant mass would lead to a dramatic increase in T_c. At present, no superconducting materials are known to exist above 20°K, and the search for nonmetallic superconductors is being actively pursued [69].

9. Applications of Superconductivity

Since the original discovery of zero resistance in a superconductor, there has been much discussion of the practical applications of superconductivity. However, until recently, no actual practical applications have been apparent. In 1956 Buck [70] suggested a circuit called a cryotron in which a flip-flop circuit of two different superconductors could be made and used as a computer element. Since that time considerable effort has gone into trying to develop superconducting computers including persistent-current memories. Such computers, in principle, could be very reliable, very fast, and very small. However, the entire technology of superconducting circuit elements must be developed and made sufficiently better than room-temperature circuitry to warrant the complexity of maintaining the circuits at helium temperature.

In 1961 a breakthrough was obtained by Kunzler [71] in superconducting magnets. He discovered that type II superconducting alloys could be used as magnets to produce critical fields much higher than the critical fields of pure superconductors. Magnetic fields in excess of 100 kilogauss have been obtained with niobium tin magnets.

Superconducting magnetometers [72] with sensitivities capable of measuring magnetic fields less than 10^{-9} gauss have been made using a double superconducting loop in which the inductance of one of the loops is modulated at high frequency. A magnetometer using multiply connected Josephson or weak link junctions has also been built. Coupled with the possibility of zero magnetic field inside a superconducting shield, these magnetometers increase the available sensitivity for experiments in small magnetic fields.

A recent application using large-scale low-temperature refrigeration below the lambda point is the superconducting linear electron accelerator [73–75] being built at Stanford. Figure 12.13 is a diagram of an electron linear accelerator. The electrons are accelerated by a large electric field maintained along the axis of the accelerator cavities. The electrons move through the successive accelerator cavities in phase with the accelerating voltage. In order to obtain several million volts per foot in a room-temperature cavity, several megawatts of r-f klystron power must be supplied to the cavities. This is very expensive and can be done only for a few microseconds after which the klystrons must be turned off for about 1,000 to 10,000 times longer than the time the r-f voltage is on. This low duty cycle means that the electrons in the accelerator beam are bunched into very high current peaks. When coincidence experiments are attempted on very weak events, the high background peak current makes high resolution difficult. For such experiments there is a linear gain in sensitivity as the duty cycle is increased to unity. The losses in superconducting lead cavities operating at 1.85°K have been reduced by more than 100,000 below the losses in copper cavities operating at room temperature. This low loss has been observed for r-f fields up to the critical d-c magnetic field of lead and for a-c electric fields of several million volts per foot. Thus it appears possible to build a superconducting linear electron accelerator with unit duty cycle and several million volts per foot accelerating voltage. Such an accelerator to produce 400-Mev electrons is being built at Stanford. It will include a 300-watt refrigerator to remove this heat at 1.85°K. It is necessary to operate below the lambda point in order to make use of the superfluid helium for heat transfer. We have described this particular application because it requires large-scale refrigeration below the lambda point where heat can be transferred over large distances with very little temperature drop with the aid of superfluid helium.

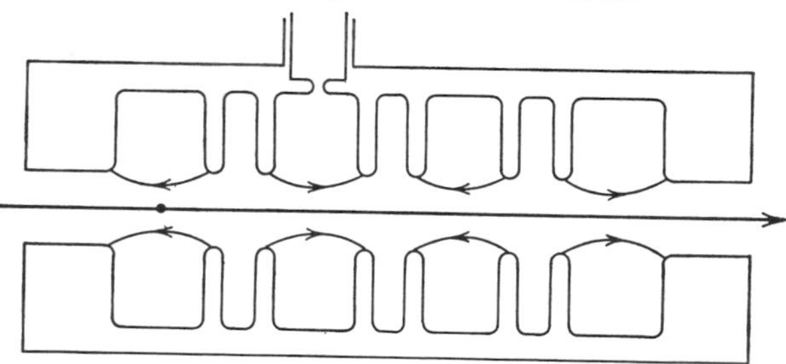

Fig. 12.13. Schematic diagram of one section of a superconducting linear electron accelerator.

Further applications including a superconducting gyroscope and other experiments in space are being developed. The day of large-scale applications of superconductivity seems to be almost here.

References

1. Onnes, H. K.: *Commun. Phys. Lab. Univ. Leiden, Suppl.*, **34** (1913).
2. Meissner, W., and R. Ochsenfeld: *Naturwissenschaften*, **21**: 787 (1933).
3. Gorter, C. J.: *Arch. Mus. Teyler*, **7**: 378 (1933).
4. Rutgers, A. J.: *Physica*, **1**: 1055 (1934).
5. Keesom, W. H., and J. A. Kok: *Commun. Phys. Lab. Univ. Leiden*, no. 221e (1932).
6. Corak, W. S., B. B. Goodman, C. B. Satterthwaite, and A. Wexler: *Phys. Rev.*, **96**: 1442 (1954).
7. London, F., and H. London: *Proc. Roy. Soc. (London)*, **A149**: 71 (1935); *Physica*, **2**: 341 (1935).
8. London, F.: "Une conception nouvelle de la supraconductibilité," Hermann & Cie., Paris, 1937.
9. Ginzburg, V. L., and L. D. Landau: *Zh. Eksperim. i Teor. Fiz.*, **20**: 1064 (1950).

10. Pippard, A. B.: *Proc. Roy. Soc. (London)*, **A216**: 547 (1953).
11. Abrikosov, A. A.: *Zh. Eksperim. i Teor. Fiz.*, **32**: 1442 (1957) [English transl.: *Soviet Phys.—JETP*, **5**: 1174 (1957)].
12. Pearl, J.: "Vortex Theory of Superconductive Memories," Ph.D. thesis, Polytechnic Institute of Brooklyn, unpublished; "Low Temperature Physics, LT 9," Plenum Press, New York, 1965, p. 566.
13. Kleiner, W. H., L. M. Roth, and S. H. Autler: *Phys. Rev.*, **133**: A 1226 (1964); J. Matricon, *Phys. Let.*, **9**: 289 (1964); A. L. Fetter, P. C. Hohenberg, and P. Pincus, *Phys. Rev.*, **147**: 140 (1966).
14. Cribier, D., B. Jacrot, L. M. Rao, and B. Farnoux, *Phys. Let.*, **9**: 106 (1964); "Quantum Fluids," D. F. Brewer (ed.) North-Holland Publishing Co., Amsterdam, 1966, p. 121.
15. Deaver, B. S., Jr., and W. M. Fairbank: *Phys. Rev. Let.*, **7**: 43 (1961).
16. Doll, R., and M. Näbauer: *Phys. Rev. Let.*, **7**: 51 (1961).
17. London, F.: "Superfluids," Dover Publications, Inc., New York, 1961.
18. Byers, N., and C. N. Yang: *Phys. Rev. Let.*, **7**: 46 (1961).
19. Onsager, L.: *Phys. Rev. Let.*, **7**: 50 (1961).
20. Blatt, J. M.: *Phys. Rev. Let.*, **7**: 82 (1961).
21. Bloch, F., and H. E. Rorschach: *Phys. Rev.*, **128**: 1697 (1962).
22. Giaever, I.: *Phys. Rev. Let.*, **5**: 147 (1960).
23. Josephson, B. D.: *Phys. Let.*, **1**: 251 (1962).
24. Anderson, P. W., and J. M. Rowell: *Phys. Rev. Let.*, **10**: 230 (1963).
25. Rowell, J. M.: *Phys. Rev. Let.*, **11**: 200 (1963).
26. Shapiro, S.: *Phys. Rev. Let.*, **11**: 80 (1963).
27. Langenberg, D. N., D. J. Scalapino, B. N. Taylor, and R. E. Eck: *Phys. Rev. Let.*, **15**: 294 (1965).
28. Lambe, J., A. H. Silver, J. E. Mercereau, and R. C. Jaklevic: *Phys. Rev. Let.*, **11**: 16 (1964).
29. Anderson, P. W.: "Lectures on The Many-Body Problem," E. R. Caianiello (ed.), Academic Press, Inc., New York, 1964, vol. 2, p. 113; *Rev. Mod. Phys.*, **38**: 298 (1966).
30. Yang, C. N.: *Rev. Mod. Phys.*, **34**: 694 (1962).
31. Goldman, A. M., P. J. Kreisman, and D. J. Scalapino: *Phys. Rev. Let.*, **15**: 495 (1965).
32. Smith, T. I.: *Phys. Rev. Let.*, **15**: 460 (1965).
33. London, H.: *Nature*, **133**: 497 (1934); *Proc. Roy. Soc. (London)*, **A176**: 522 (1940).
34. Biondi, M. A., M. P. Garfunkel, and A. O. McCoubrey: *Phys. Rev.*, **108**: 495 (1957).
35. Biondi, M. A., A. T. Forrester, and M. P. Garfunkel: *Phys. Rev.*, **108**: 497 (1957).
36. Bardeen, J., L. N. Cooper, and J. R. Schrieffer: *Phys. Rev.*, **108**: 1175 (1957).
37. Bardeen, J., and J. R. Schrieffer: article in "Progress in Low Temperature Physics," vol. III, C. J. Gorter (ed.), North-Holland Publishing Company, Amsterdam, 1961, p. 170.
38. Tinkham, M.: article in "Low Temperature Physics," C. de Witt, (ed.), Gordon and Breach, Publishers, New York, 1962, p. 149.
39. Schrieffer, J. R.: "Theory of Superconductivity," W. A. Benjamin, Inc., New York, 1964.
40. Blatt, J. M.: "Theory of Superconductivity," Academic Press, Inc., New York, 1964.
41. Rickayzen, G.: "Theory of Superconductivity," Interscience Publishers, New York, 1965.
42. Cooper, L. N.: *Phys. Rev.*, **104**: 1189 (1956).
43. Fröhlich, H., *Proc. Roy. Soc. (London)*, **A223**: 296 (1954).
44. See also: Gorkov, L. P.: *Zh. Eksperim. i Teor. Fiz.*, **34**: 735 (1958) [English transl., *Soviet Phys.—JETP*, **7**: 505 (1958)].
45. Glover, R. E., and M. Tinkham: *Phys. Rev.*, **108**: 1094 (1957).
46. Richards, P. L., and M. Tinkham: *Phys. Rev.*, **119**: 575 (1960).
47. Fröhlich, H.: *Phys. Rev.*, **79**: 845 (1950).
48. Maxwell, E.: *Phys. Rev.*, **78**: 477 (1950).
49. Reynolds, C. A., B. Serin, W. H. Wright, and L. B. Nesbitt: *Phys. Rev.*, **78**: 487 (1950).
50. Mapother, D. E., *IBM J. Research Develop.*, **6**: 77 (1962).
51. Bogoliubov, N. N., *J. Phys. (USSR)*, **11**: 23 (1947).
52. Bogoliubov, N. N.: *Nuovo Cimento*, **7**: 794 (1958).
53. Valatin, J. G.: *Nuovo Cimento*, **7**: 843 (1958).
54. A general discussion of these questions may be found in Nozières, P.: article in "Quantum Fluids," D. F. Brewer, (ed.) North-Holland Publishing Company, Amsterdam, 1966.
55. Gorter, C. J., and H. B. G. Casimir: *Phys. Z.*, **35**: 963 (1934).
56. Gorkov, L. P., *Zh. Eksperim. i Teor. Fiz.*, **36**: 1918 (1959) [English transl., *Soviet Phys.—JETP*, **9**: 1364 (1959)].
57. Chanin, G., E. A. Lynton, and B. Serin, *Phys. Rev.*, **114**: 719 (1959).
58. Anderson, P. W., *J. Phys. Chem. Solids*, **11**: 26 (1959).
59. Reif, F., and M. A. Woolf: *Phys. Rev. Let.*, **9**: 315 (1962).
60. Abrikosov, A. A., and L. P. Gorkov, *Zh. Eksperim. i Teor. Fiz.*, **39**: 1781 (1960) [English transl., *Soviet Phys.—JETP*, **12**: 1243 (1961).
61. de Gennes, P. G.: "Superconductivity of Metals and Alloys," W. A. Benjamin, Inc., New York, 1966.
62. Bogoliubov, N. N., V. V. Tolmachev, and D. V. Shirkov: "A New Method in the Theory of Superconductivity," *Acad. Sci. USSR*, 1958 [English transl., Consultants Bureau, New York, 1959].
63. Eliashberg, G. M.: *Zh. Eksperim. i Teor. Fiz.*, **38**: 966 (1960) [English transl., *Soviet Phys.—JETP*, **11**: 696 (1960)].
64. Schrieffer, J. R., D. J. Scalapino, and J. W. Wilkins: *Phys. Rev. Let.*, **10**: 336 (1963).
65. Scalapino, D. J., Y. Wada, and J. C. Swihart: *Phys. Rev. Let.*, **14**: 102 (1965).
66. Swihart, J. C., D. J. Scalapino, and Y. Wada: *Phys. Rev. Let.*, **14**: 106 (1965).
67. McMillan, W. L., and J. M. Rowell: *Phys. Rev. Let.*, **14**: 108 (1965).
68. Little, W. A.: *Phys. Rev.*, **134**: A 1416 (1964); *Sci. Amer.*, **212**: no. 2, 21 (1965).
69. Matthias, B. T., T. H. Geballe, and V. B. Compton: *Rev. Mod. Phys.*, **35**: 1 (1963).
70. Buck, D. A.: *Proc. IRE*, **44**: 482 (1956).
71. Kunzler, J. E., E. Buehler, F. S. L. Hsu, and J. H. Wernick: *Phys. Rev. Let.*, **6**: 89 (1961).
72. Deaver, B. S., Jr., and W. M. Fairbank: in "Proceedings of the Eighth International Conference on Low Temperature Physics," Butterworths Scientific Publications, London, 1963, p. 116.
73. Schwettman, H. A., P. B. Wilson, J. M. Pierce, and W. M. Fairbank: in "International Advances in Cryogenic Engineering," K. D. Timmerhaus (ed.), Plenum Press, New York, 1965, vol. 10, p. 88.
74. Schwettman, H. A., T. I. Smith, and W. M. Fairbank: in "Proceedings of the Tenth International Conference on Low Temperature Physics," (to be published).
75. Smith, T. I., H. A. Schwettman, W. M. Fairbank, and P. B. Wilson: in "Proceedings of the Tenth International Conference on Low Temperature Physics," (to be published).

Part 6 · Optics

Chapter 1 Electromagnetic Waves *by* E. U. Condon *6-3*

Chapter 2 Geometrical Optics *by* Max Herzberger *6-20*

Chapter 1

Electromagnetic Waves

By E. U. CONDON, University of Colorado

1. Nature of Light

In a narrow sense light refers to a form of radiant energy which gives rise to visual sensations on entering the eye. The visual sensation is known to be caused by electromagnetic waves described by wave solutions of Maxwell's equations for the electromagnetic field [1].† Such waves are propagated through space with the characteristic velocity $c \sim 3 \times 10^{10}$ cm/sec. Their properties in regard to interaction with matter are strongly dependent on the wavelength λ (centimeters per cycle) or the frequency ν (cycles per second), which are related by $\lambda\nu = c$.

The eye (Part 6, Chap. 4) responds only to waves extending roughly from $\lambda = 3.8 \times 10^{-5}$ cm (violet) to 7.8×10^{-5} cm (red). It is convenient to use the term *light* in a more generalized sense to refer to the entire subject of the physics of such electromagnetic radiations of all frequencies.

Radiations for which $\lambda < 3.8 \times 10^{-5}$ cm are called *ultraviolet* down to about $\lambda \sim 100 \times 10^{-8}$ cm, below which they are called *X rays*. In the older literature the term *gamma ray* referred to X rays of very high energy or extremely short wavelength. Nowadays the tendency is to distinguish between X rays and gamma rays on the basis of their mode of generation, calling them X rays if generated by processes occurring outside the atomic nucleus and gamma rays if generated within an atomic nucleus. This distinction is not really important and is not always made.

Radiations for which $\lambda > 7.8 \times 10^{-5}$ cm belong to the *infrared*. For $\lambda < 10\ \mu$ or 10^{-3} cm the radiations are said to be in the *near infrared*, and for $\lambda > 10\ \mu$ they are said to be in the *far infrared*. The boundary at 10 μ is arbitrary and not always used in exactly the same way. Radiations whose wavelengths are of the order of millimeters and centimeters are called *microwaves* (Part 7, Chap. 6), and those of still longer wavelength are called *radio waves*, or *Hertzian waves*.

Wave solutions of Maxwell's equations exist in many forms as plane waves, expanding and converging spherical waves, cylindrical waves, and the like. For most purposes analysis is made in terms of harmonic plane waves and solutions derived from them by superposition.

† Numbers in brackets refer to References at end of chapter.

A harmonic plane wave is, strictly speaking, infinite in extent, although solutions filling only finite regions of space that are, however, large compared with the wavelength are obtainable by superposition of plane waves, which vary only slightly in regard to their directions of propagation and the magnitude of the wavelength. Phenomena associated with the edges of such superposed wave systems are called *diffraction phenomena* (Part 6, Chap. 5).

The plane wave is specified as to direction of propagation and wavelength by giving the wave-number vector $\boldsymbol{\sigma}$ or $\mathbf{k} = 2\pi\boldsymbol{\sigma}$. This has the direction of propagation of the wave and a magnitude equal to the number of waves in unit length or in 2π units of length, respectively. In terms of the frequency ν or $\omega = 2\pi\nu$ the *phase* is given by $2\pi(\nu t - \boldsymbol{\sigma} \cdot \mathbf{r})$ or $\omega t - \mathbf{k} \cdot \mathbf{r}$.

The *amplitude* of the wave in empty space is specified by two mutually orthogonal vectors, the electric and magnetic vectors **E** and **H**. These are equal in magnitude when **E** is measured in stat-volts per centimeter and **H** in oersteds. Both **E** and **H** are orthogonal to $\boldsymbol{\sigma}$ in such a way that **E**, **H**, and $\boldsymbol{\sigma}$ form a right-handed basis in that order. Thus the electric and magnetic fields for a monochromatic plane wave are given by the real parts of

$$\begin{aligned} \mathbf{E} &= \mathbf{A}e^{i(\omega t - \mathbf{k}\cdot\mathbf{r})} \\ \mathbf{H} &= \mathbf{n} \times \mathbf{A}e^{i(\omega t - \mathbf{k}\cdot\mathbf{r})} \end{aligned} \tag{1.1}$$

Here **A** is a constant vector normal to **k**, and **n** is a *unit* vector in the direction of **k**. The fact that **A** can be itself real or complex anywhere in the plane normal to **k** gives rise to different possibilities with regard to polarization of light (Sec. 2).

The energy transported in unit time across unit area normal to **k** is given by the Poynting vector **S**,

$$\mathbf{S} = \frac{c}{4\pi}\mathbf{E} \times \mathbf{H} \tag{1.2}$$

where **S** is in ergs per square centimeter per second if **E** and **H** are in stat-volts per centimeter and oersteds, respectively. In full sunlight at normal incidence at the distance of the earth the power flux is 1.34×10^6 ergs/cm$^2 \cdot$ sec; so the root-mean-square (rms) value of **E** is 7.1 volts/cm.

Such a plane wave also transports electromagnetic momentum. The volume density of electromagnetic momentum is given by $\mathbf{S}/c^2$, which is moving in the direction of **k** with velocity c; so $\mathbf{S}/c$ is the rate at which electromagnetic momentum crosses unit area normal to $\mathbf{k}$. If the radiation falls on a body and is absorbed, this electromagnetic momentum acts to exert a pressure on the body known as *radiation pressure*. Thus for full sunlight the radiation pressure is 4.46×10^{-5} dyne/cm^2 at the distance of the earth and 2.06 dynes/cm^2 at the surface of the sun.

Light also has quantum aspects in interaction with matter. Radiation of frequency ν, traveling in the direction of $\mathbf{k}$, behaves in some respects like a stream of *light quanta* or *photons*, each possessing energy $h\nu = \hbar\omega$ and momentum $(h\nu/c)\mathbf{n} = h\boldsymbol{\sigma} = \hbar\mathbf{k}$. Here h is Planck's constant, and $\hbar$ is Planck's constant divided by 2π. For many purposes we can think of the light as a stream of such light quanta whose motions are statistically determined by the motion of the associated classical electromagnetic wave, in the same way as the quantum-mechanical wave function gives the probability of finding a particle in a given region of space (Part 7, Chap. 1). More accurately, one must work with a form of solutions of the electromagnetic-field equations in which the amplitudes of plane waves are regarded as quantum-mechanical variables (Sec. 11).

The sizes of the light quanta need to be compared with energies characteristic of atomic and molecular structures, which are usually expressed in electron volts (ev). The wavelength of radiation whose quantum energy is one absolute electron volt is

$$\lambda_0 = 10^{-8}\frac{h}{e'}c^2 = (12{,}398.1 \pm 0.4) \times 10^{-8}\text{ cm}$$

and likewise the wave number associated with one absolute electron volt is

$$\lambda_0^{-1} = 8{,}065.73 \pm 0.23\text{ cycles/cm}$$

(Part 7, Chap. 1).

2. States of Polarization

By *polarization* of light is meant the properties which depend on the vectorial character of the amplitude $\mathbf{A}$ occurring in (1.1). The simplest instance is that in which $\mathbf{A}$ is a real vector (or one whose imaginary part is parallel to its real part). In this case the light is said to be *plane-polarized*, and the plane containing the vectors $\mathbf{E}$ (or $\mathbf{A}$) and $\boldsymbol{\sigma}$ is called the *plane of polarization* [2].†

Light from most light sources is unpolarized. This means that the direction of $\mathbf{A}$ shifts around in an irregular way in the plane normal to $\boldsymbol{\sigma}$ so that the average value of its mean-square projection on any plane through $\boldsymbol{\sigma}$ is the same. Although polarized light waves are emitted in individual quantum transitions of atoms and molecules, the light observed from ordinary light sources represents the superposition of such polarized waves, polarized in all possible ways, resulting from many uncoordinated quantum transitions in the source.

Polarized light is obtained only from light sources in which the emitting atoms are in some way aligned with respect to a preferred axis, as in emission of light from gas excited by an electric discharge while in a magnetic field. Polarized light may also be obtained from unpolarized light by various means affecting its transmission. For example, when light is partially reflected at an oblique angle from a non-conducting material such as glass, the fraction reflected depends on whether the plane of polarization is in the plane of incidence or perpendicular to it. Other materials, such as Polaroid, have the property of strongly absorbing the part of the light polarized in one plane while transmitting the other with relatively little absorption. Thus an unpolarized beam becomes polarized on transmission through it.

Some crystals are *doubly refracting*, which means that the velocity of propagation of a wave through the crystal depends on the relation of the plane of polarization to the axes of the crystal. In consequence, when an unpolarized beam is transmitted into the crystal, the beam will be separated into two parts having different directions of travel associated with the different states of polarization (Part 6, Chap. 6).

The most general state of polarization of a plane wave of light is that in which the real and imaginary parts of $\mathbf{A}$, namely, $\mathbf{A}_r$ and $\mathbf{A}_i$, are not parallel. In this case the end of the $\mathbf{E}$ vector describes an ellipse; so this is called *elliptical polarization*. Linear polarization is a limiting case in which the minor axis of the ellipse vanishes.

A state of elliptic polarization requires for its description a statement of the orientation of the principal axes of the ellipse described by $\mathbf{E}$ and the *sense* in which the ellipse is described. Conventionally the sense is called *right* if the description of the ellipse appears *clockwise* to a person faced oppositely to the direction of propagation, that is, faced as an observer into whose eyes the light enters. The opposite sense is called *left*.

Suppose $\mathbf{i}$ and $\mathbf{j}$ are two unit vectors along the axes of the ellipse in such a sense that $\mathbf{i}$, $\mathbf{j}$, $\mathbf{n}$ form a right-handed basis. The state of polarization may be described by a complex unit vector

$$\mathbf{l}(\delta) = \mathbf{i}\cos\frac{\delta}{2} + i\mathbf{j}\sin\frac{\delta}{2} \tag{1.3}$$

which is a unit vector in the sense $\mathbf{l}\cdot\mathbf{l}^* = 1$. All possible states of polarization, having $\mathbf{i}$ and $\mathbf{j}$ as principal axes, are described by assigning to δ various values from 0 to 2π according to the scheme

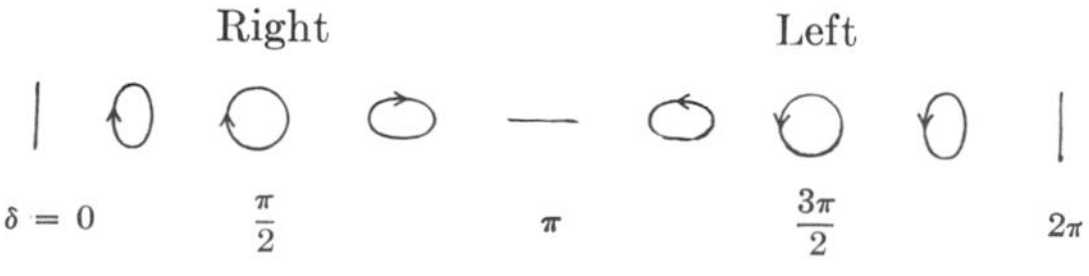

Right Left

$\delta = 0$ $\quad \frac{\pi}{2} \quad \pi \quad \frac{3\pi}{2} \quad 2\pi$

The unit vector defined as

$$\mathbf{m}(\delta) = \mathbf{l}(\delta + \pi) = \left(i\mathbf{i}\sin\frac{\delta}{2} - \mathbf{j}\cos\frac{\delta}{2}\right) \tag{1.4}$$

is orthogonal to $\mathbf{l}(\delta)$ in the sense $\mathbf{l}\cdot\mathbf{m}^* = 0$. Accordingly any state of polarization, represented by an arbitrary complex $\mathbf{A}$ in the plane normal to $\mathbf{n}$, can be regarded as arising from the superposition of

† In the older literature the plane through $\boldsymbol{\sigma}$ and normal to $\mathbf{E}$ is called the *plane of polarization*, but it is preferable to give the plane of $\mathbf{E}$ and $\boldsymbol{\sigma}$ this designation because the interaction of light and matter is principally through the action of $\mathbf{E}$ on the electrons in matter.

two elliptically polarized waves having their axes chosen in an arbitrary relation to the axes of the ellipse described by **A**. The formula for representing **A** in this way, in terms of **l** and **m**, is

$$\mathbf{A} = (\mathbf{A} \cdot \mathbf{l}^*)\mathbf{l} + (\mathbf{A} \cdot \mathbf{m}^*)\mathbf{m} \tag{1.5}$$

This embraces a large number of special cases usually given separate treatment in books on optics:

a. If **A** represents linear polarization in a plane making the angle θ with the (**i**,**n**) plane and if $\delta = 0$ so **l** and **m** correspond to linear states of polarization along the **i** and **j** axes, $\mathbf{l} = \mathbf{i}$ and $\mathbf{m} = -\mathbf{j}$, then the components have the amplitudes $A \cos \theta$ and $-A \sin \theta$, respectively, as with simple vector projection.

b. If **A** represents the same linear polarization and one chooses $\delta = \pi/2$, then the linearly polarized wave is represented as the superposition of a right circularly polarized wave and a left circularly polarized wave.

$$\mathbf{l}(\pi/2) = 2^{-1/2}(\mathbf{i} + i\mathbf{j}) \qquad \text{right circular}$$
$$\mathbf{m}(\pi/2) = 2^{-1/2}(i\mathbf{i} - \mathbf{j}) \qquad \text{left circular}$$

The linearly polarized wave at angle θ is

$$\mathbf{A} = \frac{A}{\sqrt{2}}(\mathbf{l}e^{-i\theta} - i\mathbf{m}e^{i\theta}) \tag{1.6}$$

This representation is used in relating the rotation of the plane of polarization in optically active substances to a difference in refractive index for right and left circularly polarized waves (Part 6, Chap. 6).

3. Maxwell Field Equations

The Maxwell field equations applicable inside matter have the form

$$\operatorname{div} \mathbf{D} = 4\pi\rho \qquad \operatorname{div} \mathbf{B} = 0$$
$$\operatorname{curl} \mathbf{E} + \frac{1}{c}\dot{\mathbf{B}} = 0 \qquad \operatorname{curl} \mathbf{H} - \frac{1}{c}\dot{\mathbf{D}} = 4\pi\mathbf{i} \tag{1.7}$$

This way of writing the equations is based on our supposed ability to distinguish between two kinds of charge and current densities making up the total charge and current density present. The total amounts will be written $\rho + \rho'$ and $\mathbf{i} + \mathbf{i}'$, respectively.

The part written ρ and **i** is regarded as that which is relatively free to move over distances large compared with the size of individual atoms and which gives rise to the macroscopic charge distributions and currents. In addition there are other charges, and currents arising from their motion, which are confined to motions within atoms or molecules which are represented by the additional charge and current density ρ' and i' (Part 4, Chap. 1).

Conservation of charge requires that these be connected by equations of continuity

$$\operatorname{div} \mathbf{i} + \frac{1}{c}\dot{\rho} = 0 \qquad \operatorname{div} \mathbf{i}' + \frac{1}{c}\dot{\rho}' = 0 \tag{1.8}$$

in which c appears because ρ is in electrostatic units (esu) and **i** in electromagnetic units (emu).

The effects of ρ' and $\mathbf{i}'$ are usually described in terms of the electric dipole moment in unit volume **P** and the magnetic moment in unit volume **I**. In terms of these

$$\rho' = -\operatorname{div} \mathbf{P}$$
$$\mathbf{i}' = \operatorname{curl} \mathbf{I} + \frac{1}{c}\dot{\mathbf{P}} \tag{1.9}$$

Also the field vectors **D** and **B** are related to **E** and **H** as

$$\mathbf{D} = \mathbf{E} + 4\pi\mathbf{P}$$
$$\mathbf{B} = \mathbf{H} + 4\pi\mathbf{I} \tag{1.10}$$

Substituting (1.10) in (1.7) to eliminate **D** and **H**, one can also use (1.9) to eliminate **P** and **I** and find the field equations for **E** and **B**,

$$\operatorname{div} \mathbf{E} = 4\pi(\rho + \rho') \qquad \operatorname{div} \mathbf{B} = 0$$
$$\operatorname{curl} \mathbf{E} + \frac{1}{c}\dot{\mathbf{B}} = 0 \qquad \operatorname{curl} \mathbf{B} - \frac{1}{c}\dot{\mathbf{E}} = 4\pi(\mathbf{i} + \mathbf{i}') \tag{1.11}$$

A complete account of the electromagnetic properties of materials requires that these equations be supplemented by telling how the atomic charges and currents, or the electric and magnetic moments in unit volume, depend on the field vectors. In the simplest case this is done by introducing dielectric constant ϵ and magnetic permeability μ, defined by

$$\mathbf{D} = \epsilon\mathbf{E} \qquad \mathbf{B} = \mu\mathbf{H} \tag{1.12}$$

Alternatively these same assumed connections can be described in terms of the susceptibilities K and K_m,

$$\mathbf{P} = K\mathbf{E} \qquad \mathbf{I} = K_m\mathbf{H} \tag{1.13}$$

where, in virtue of (1.10),

$$\epsilon = 1 + 4\pi K \qquad \mu = 1 + 4\pi K_m \tag{1.14}$$

In dynamic situations these connections are not applicable in such a simple form. When the fields **E** and **H** are rapidly time-varying, the electric and magnetic moments do not follow in exact time-phase relation with **E** and **H**. This complication can be handled formally by writing $\epsilon' - i\epsilon''$ for ϵ where the time dependence of all the field quantities is through a factor $e^{i\omega t}$. This amounts to assuming that **P** is made up of a part that is proportional to **E** plus another part proportional to $\dot{\mathbf{E}}$. Similar assumptions are also sometimes made for μ, although these are usually not needed to describe observed phenomena in the field of optics (Part 4, Chap. 7).

In crystalline materials further complications arise in that ϵ and μ become second-order tensors. In optics it is usually the tensor character of ϵ that plays an important role in interpretation of the optical properties (Part 6, Chap. 6).

At surfaces of discontinuity between two kinds of matter further conditions have to be introduced, which are the surface divergence and curl analogues of the field equations. Let **n** be the normal to the bounding surface, pointing from medium 1 to medium 2, and let $\mathbf{P}_1$ and $\mathbf{P}_2$ be the values of **P** at the boundary in the two media. Then on the bounding surface there will be a *surface charge density* η' of charge that is bound to the atomic structure like the volume density ρ', given by

$$\eta' = -\mathbf{n} \cdot (\mathbf{P}_2 - \mathbf{P}_1) \tag{1.15}$$

Likewise a discontinuity in $\mathbf{I}$ across the boundary is accompanied by a bound *surface distribution of current* $\mathbf{K}'$ given by

$$\mathbf{K}' = \mathbf{n} \times (\mathbf{I}_2 - \mathbf{I}_1) \tag{1.16}$$

The surface boundary conditions, in the absence of free surface charge and current densities on the bounding surface, are that there must be continuity of

Normal components of **D** and **B**
Tangential components of **E** and **H** (1.17)

4. Poynting Theorem

The theorem relating to the flow of electromagnetic energy in the field is called the *Poynting theorem.* In (1.11) multiply the curl **E** equation by $\mathbf{B}\cdot$ and the curl **B** equation by $\mathbf{E}\cdot$, subtract, and use

$$\operatorname{div}(\mathbf{E}\times\mathbf{B}) = \mathbf{B}\cdot\operatorname{curl}\mathbf{E} - \mathbf{E}\cdot\operatorname{curl}\mathbf{B}$$

The result is

$$\operatorname{div}\mathbf{S} + \frac{1}{8\pi}\frac{\partial}{\partial t}(\mathbf{B}^2 + \mathbf{E}^2) + c\mathbf{E}\cdot(\mathbf{i} + \mathbf{i}') = 0 \tag{1.18}$$

in which

$$\mathbf{S} = \frac{c}{4\pi}(\mathbf{E}\times\mathbf{B})$$

is defined as the Poynting vector.

The last term is the rate at which the electric field does work on all the moving charges, in unit volume, a rate of doing work which is accomplished at the expense of the amount of energy stored in the electromagnetic field. The second term is recognized as the amount of energy stored in unit volume of the electric and magnetic fields at the location in question. The first term is therefore to be interpreted as the rate at which the field energy is diminishing because of a net outward flow of energy; so **S** is interpreted as the amount of electromagnetic energy crossing unit area normal to **S** in unit time. This argument does not determine **S** uniquely since its curl is left undetermined. Therefore the flow may be equally well represented by $\mathbf{S} + \mathbf{S}'$, where $\mathbf{S}'$ is any vector field whose divergence vanishes.

For some purposes it is desirable not to have $\mathbf{i}'$ appearing explicitly in (1.18). Using (1.9), this becomes

$$\operatorname{div}\mathbf{S} + \frac{1}{4\pi}\left(\mathbf{E}\cdot\frac{\partial\mathbf{D}}{\partial t} + \mathbf{H}\cdot\frac{\partial\mathbf{B}}{\partial t}\right) + c\mathbf{E}\cdot\mathbf{i} = 0 \tag{1.19}$$

in which **S** is now defined as

$$\mathbf{S} = \frac{c}{4\pi}\mathbf{E}\times\mathbf{H}$$

In the case of nonmagnetic materials for which $\mathbf{I} = 0$, and therefore $\mathbf{B} = \mathbf{H}$, the two definitions of the Poynting vector become the same, but for magnetic materials the distinction is important.

In (1.19) the second term represents not only the stored electromagnetic-field energy but also some energy stored in the material in virtue of the work necessary to change its condition of electric and magnetic polarization and in addition the rate of dissipation of field energy due to atomic currents being in phase with **E**. Thus, if ϵ and μ have complex forms, this implies

$$\mathbf{D} = (\epsilon'\cos\omega t + \epsilon''\sin\omega t)\mathbf{E}$$
$$\mathbf{B} = (\mu'\cos\omega t + \mu''\sin\omega t)\mathbf{H}$$

where **E** and **H** stand for the amplitudes of fields varying as $\cos\omega t$. The second terms of (1.19) take the form

$$\frac{\partial}{\partial t}\left(\frac{\epsilon'\mathbf{E}^2 + \mu'\mathbf{H}^2}{8\pi}\right) + \omega\,\frac{\epsilon''\mathbf{E}^2 + \mu''\mathbf{H}^2}{8\pi}\,2\cos^2\omega t \tag{1.20}$$

The first terms represent the time rate of change of the energy stored purely in the field and in the polarizations of the medium. The second terms, which are always positive, correspond to dissipation of electromagnetic energy as a result of irreversible processes occurring in the material, the energy loss per cycle and per unit volume being

$$\tfrac{1}{4}(\epsilon''\mathbf{E}^2 + \mu''\mathbf{H}^2) \tag{1.21}$$

If the fields are represented as the real parts of complex vectors multiplying into $e^{i\omega t}$ and $e^{-i\omega t}$, then the Poynting vector takes the form

$$\mathbf{S} = \frac{c}{16\pi}(\mathbf{E}\times\mathbf{H}e^{2i\omega t} + \mathbf{E}^*\times\mathbf{H}^*e^{-2i\omega t})$$
$$+ \frac{c}{16\pi}(\mathbf{E}\times\mathbf{H}^* + \mathbf{E}^*\times\mathbf{H}) \tag{1.22}$$

It thus consists of a constant part (second line) which gives the time average flow in the field and a double frequency part which corresponds to energy flows which average to zero in each half cycle. In many calculations only the average energy flow is of interest: this is given by the real part of the complex Poynting vector

$$\frac{c}{8\pi}(\mathbf{E}\times\mathbf{H}^*)$$

Note the 8π here as contrasted with 4π in (1.18) and (1.19).

The flow of electromagnetic momentum in the field is handled similarly, except that, whereas a vector is required to describe the transport of the scalar energy density, a tensor is needed to describe the transport of the vector momentum density.

The force on unit volume of material, acting on the free charge and current density, is

$$\mathbf{F} = \rho\mathbf{E} + \mathbf{i}\times\mathbf{B}$$

From (1.18)

$$\mathbf{F} = \left(\frac{1}{4\pi}\operatorname{div}\mathbf{D} - \operatorname{div}\mathbf{P}\right)\mathbf{E}$$
$$+ \left(\frac{1}{4\pi}\operatorname{curl}\mathbf{H} - \frac{1}{4\pi c}\dot{\mathbf{D}} + \frac{1}{c}\dot{\mathbf{P}} + \operatorname{curl}\mathbf{I}\right)\times\mathbf{B}$$

In terms of the second-order symmetrical tensor

$$\mathfrak{T} = \frac{1}{8\pi}(\mathbf{E}^2 + \mathbf{B}^2)\mathfrak{J} - \frac{1}{4\pi}(\mathbf{EE} + \mathbf{BB}) \tag{1.23}$$

this becomes

$$\frac{\partial}{\partial t}\left(\frac{1}{4\pi c}\mathbf{E}\times\mathbf{B}\right)+\operatorname{div}\mathfrak{T}=-\mathbf{F} \qquad (1.24)$$

the reduction being effected by using the tensor formula

$$\operatorname{div}\mathfrak{T}=\frac{1}{4\pi}[\mathbf{E}\operatorname{div}\mathbf{E}+(\operatorname{curl}\mathbf{E})\times\mathbf{E}+(\operatorname{curl}\mathbf{B})\times\mathbf{B}]$$

in which a term $\mathbf{B}\operatorname{div}\mathbf{B}$ vanishes since $\operatorname{div}\mathbf{B}=0$.

Since $\mathbf{F}$ is the rate of increase of momentum of the moving charges in unit volume, conservation of momentum will be satisfied if $-\mathbf{F}$ is regarded as the time rate of increase of the electromagnetic momentum in unit volume. Then in (1.24) one interprets

$$\mathbf{G}=\frac{1}{4\pi c}(\mathbf{E}\times\mathbf{B}) \qquad (1.25)$$

as the volume density of electromagnetic momentum in the field. The term $\operatorname{div}\mathfrak{T}$ gives the rate of decrease of the electromagnetic momentum due to the transport of this quantity to other parts of the field by purely electrodynamic action. The tensor $\mathfrak{T}$ is interpreted in the sense that $\mathfrak{T}\cdot d\mathbf{A}$ gives the amount of electromagnetic momentum as a vector which crosses the area $d\mathbf{A}$ in unit time toward the positive direction of $d\mathbf{A}$.

5. Plane Waves in Isotropic Media

In the field equations (1.7) suppose ρ and $\mathbf{i}$ are zero, and that all field quantities vary as $e^{i\omega t}$, the letter $\mathbf{E}$ standing for the complex vector amplitude, and suppose that the material's properties are described by complex ϵ and μ as described in Sec. 3.

The equations for the complex amplitudes become

$$\begin{aligned}\operatorname{div}\mathbf{E}=0 \qquad \operatorname{div}\mathbf{H}&=0\\ \operatorname{curl}\mathbf{E}+ik_0\mu\mathbf{H}&=0\\ \operatorname{curl}\mathbf{H}-ik_0\epsilon\mathbf{E}&=0\end{aligned} \qquad (1.26)$$

in which $k_0=\omega/c$. Both $\mathbf{E}$ and $\mathbf{H}$ satisfy the wave equation

$$\nabla^2\mathbf{A}+k_0^2\epsilon\mu\mathbf{A}=0 \qquad (1.27)$$

It is sufficient to discuss the solutions for $\mathbf{E}$ since the accompanying $\mathbf{H}$ field can be found by calculation of curl $\mathbf{E}$.

The *complex index of refraction* is usually written $n(1-i\kappa)$, the coefficient κ being known as the *absorption index* (not to be confused with the absorption coefficient). It is related to the product $\epsilon\mu$ as

$$n^2(1-i\kappa)^2=\epsilon\mu \qquad (1.28)$$

Plane-wave solutions have the form $\mathbf{E}=\mathbf{A}e^{-i\mathbf{k}\cdot\mathbf{r}}$, in which $\mathbf{A}$ describes the polarization as in Sec. 2 and $\mathbf{k}=\mathbf{k}_r-i\mathbf{k}_i$ is the complex propagation vector [3]; so $\mathbf{E}=\mathbf{A}\exp[i(\omega t-\mathbf{k}_r\cdot\mathbf{r})]\exp(-\mathbf{k}_i\cdot\mathbf{r})$. The planes of constant phase of the oscillatory part are those given by $\mathbf{k}_r\cdot\mathbf{r}=$ const and are called *wavefronts*. The *phase velocity* v is the speed of motion in the direction of $\mathbf{k}_r$ needed to stay at a place of constant phase, $v=\omega/k_r$. The planes of constant value of the real exponential are called *damping fronts* and are given by $\mathbf{k}_i\cdot\mathbf{r}=$ const. Motion in the direction of $\mathbf{k}_i$ gives the maximum rate of decrease of amplitude of the wave. From (1.27)

$$\mathbf{k}^2=k_0^2n^2(1-i\kappa)^2$$

giving

$$\begin{aligned}\mathbf{k}_r^2-\mathbf{k}_i^2&=k_0^2n^2(1-\kappa^2)\\ \mathbf{k}_r\cdot\mathbf{k}_i&=k_0^2n^2\kappa\end{aligned} \qquad (1.29)$$

The second of these indicates that, in the case of an absorbing medium ($\kappa>0$), the damping vector $\mathbf{k}_i$ must make an acute angle with the wave vector $\mathbf{k}_r$, whereas, in the case of a nonabsorbing medium ($\kappa=0$), they must be orthogonal.

In a nonabsorbing medium $\mathbf{k}_i$ must be orthogonal to $\mathbf{k}_r$ and $\mathbf{k}_r$ is increased in virtue of the damping vector $\mathbf{k}_i$. The simplest case is that of an undamped plane wave, $\mathbf{k}_i=0$, for which $k_r=k_0n$, giving

$$v=\frac{c}{n} \qquad (1.30)$$

for the phase velocity, which is the relation in terms of which the index of refraction is often defined.

For an absorbing medium the least value of the damping occurs when $\mathbf{k}_i$ is parallel to $\mathbf{k}_r$. Values of k_r and k_i are given by solving (1.29). In the case of small damping, in which $\kappa\ll 1$, the approximate solution is

$$k_r=k_0n \qquad k_i=k_0n\kappa$$

So, if x is a coordinate measured in the direction of propagation, the damping factor on the amplitude may be $\exp(-k_0n\kappa x)$ and therefore the energy in the beam diminishes according to $\exp(-2k_0n\kappa x)$. The *absorption coefficient* α of the medium is defined as the quantity α in the relation $I=I_0e^{-\alpha x}$, where I_0 is the initial intensity and I the intensity after traversing a distance x. The relation between absorption coefficient α and absorption index κ is

$$\alpha=\frac{4\pi\kappa}{\lambda}=\frac{4\pi n\kappa}{\lambda_0} \qquad (1.31)$$

where λ is the wavelength in the medium and λ_0 is the vacuum wavelength. Therefore the intensity falls by a factor e^{-1} in a distance equal to $(4\pi\kappa)^{-1}$ wavelengths in the medium.

6. Reflection and Refraction at a Plane Boundary

When a beam of light falls on the plane interface between two media, part of the light is *reflected* and part is transmitted into the other medium with a change in direction of propagation, which is called *refraction*. The relations involved are shown in Fig. 1.1. The medium occupying the space $z<0$ is supposed to have index of refraction n_a, and n_b in the space $z>0$. The direction of propagation of the incident beam will be supposed to make an angle θ, called the *angle of incidence*, with the normal to the interface and to lie in the (x,z) plane. Thus the direction of the incident beam is given by the unit vector

$$\mathbf{n}_1=\mathbf{i}\sin\theta+\mathbf{k}\cos\theta$$

and therefore the phase of the wave at time t at a

place whose coordinates are $(x,y,0)$ is

$$\omega t - k_0 n_a x \sin \theta$$

In order that the boundary conditions (1.17) at $z = 0$ can be satisfied at all points on the plane, the reflected and refracted waves must show the same dependence of phase on t and x, from which it follows that they must have the same frequency as the incident wave, that their directions, $\mathbf{n}_2$ and $\mathbf{n}_3$, respectively, must also lie in the xz plane, that the angle of reflection must equal the angle of incidence so that

$$\mathbf{n}_2 = \mathbf{i} \sin \theta - \mathbf{k} \cos \theta$$

and that the angle of the refracted ray must be such that

$$n_a \sin \theta = n_b \sin \varphi \tag{1.32}$$

where φ is the *angle of refraction* so that

$$\mathbf{n}_3 = \mathbf{i} \sin \varphi + \mathbf{k} \cos \varphi$$

The relation (1.32) is known as *Snell's law of refraction.* The ratio $n_b/n_a = n$ is known as the *relative index of refraction* of the two media. The ray is bent toward the normal in going into a medium of higher index and away from the normal in going into a medium of lower index.

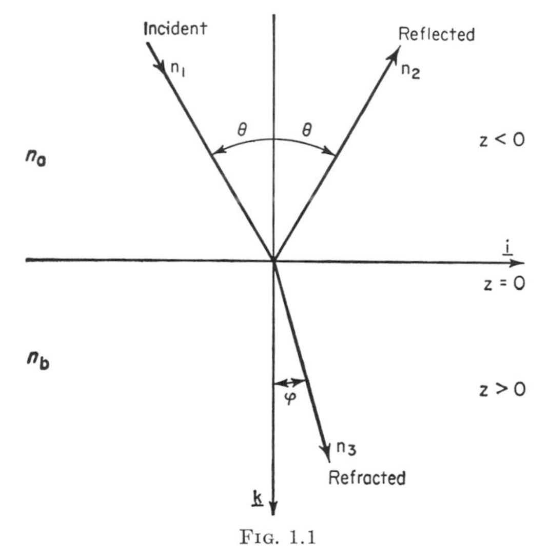

FIG. 1.1

Considering the case in which the light enters a medium of lower index ($n_b < n_a$), so that $\varphi > \theta$, as the angle of incidence is increased φ will equal $\pi/2$ when θ is such that $\sin \theta = n$. For angles of incidence greater than this limiting angle the light undergoes total reflection at the interface. The incident and reflected waves are then of equal amplitude. In the high-index medium there is propagated a wave that is exponentially damped on fronts parallel to the interface and whose wavefronts are in the plane interface entirely in the x direction. This wave is of negligibly small intensity except for distances of the order of a few wavelengths from the interface.

Conditions are more complicated if the light enters an absorbing medium so that n_b is complex. Then the wave in the absorbing medium has a complex propagation vector $\mathbf{k}_r - i\mathbf{k}_i$. The boundary conditions require that $\mathbf{k}_i$, the normal to the damping fronts, be normal to the interface and that the direction of $\mathbf{k}_r$ be determined by Snell's law. Thus the light that enters an absorbing medium at oblique incidence is a wave in which $\mathbf{k}_r$ and $\mathbf{k}_i$ are not parallel.

Application of the boundary conditions (1.17) gives relations between the amplitudes of the reflected and refracted waves and the incident waves which are known as the *Fresnel formulas.* The relations are different for light linearly polarized with its electric vector in the plane of incidence than for light linearly polarized at right angles to the plane of incidence.

Let the incident amplitude of the electric vector be written

$$\mathbf{A}_1 = A_{p1}(\mathbf{i} \cos \theta - \mathbf{k} \sin \theta) + A_{s1}\mathbf{j}$$

so that A_{p1} represents the amplitude of the part in the plane of incidence and A_{s1} represents the part at right angles to the plane of incidence. Similarly, for the reflected and refracted rays,

$$\mathbf{A}_2 = A_{p2}(\mathbf{i} \cos \theta + \mathbf{k} \sin \theta) + A_{s2}\mathbf{j}$$
$$\mathbf{A}_3 = A_{p3}(\mathbf{i} \cos \varphi - \mathbf{k} \sin \varphi) + A_{s3}\mathbf{j}$$

Then the boundary conditions (1.17) require, for the amplitudes of the reflected waves,

$$A_{p2} = -A_{p1} \frac{\tan (\theta - \varphi)}{\tan (\theta + \varphi)} \tag{1.33}$$

$$A_{s2} = -A_{s1} \frac{\sin (\theta - \varphi)}{\sin (\theta + \varphi)} \tag{1.34}$$

and, for the refracted waves,

$$A_{p3} = A_{p1} \frac{2 \sin \varphi \cos \theta}{\sin (\theta + \varphi) \cos (\theta - \varphi)} \tag{1.35}$$

$$A_{s3} = A_{s1} \frac{2 \sin \varphi \cos \theta}{\sin (\theta + \varphi)} \tag{1.36}$$

If the reflection occurs where the beam strikes an optically denser medium so that $\theta > \varphi$, then the negative signs in (1.33) and (1.34) show that there is a reversal in sign of the reflected wave relative to the incident which is the equivalent to the phase change that would be introduced by an extra path length of half a wavelength. This phase change is important in certain interference phenomena (Part 6, Chap. 5).

In (1.33), if $\theta + \varphi = \pi/2$, that is, if the reflected ray's and the refracted ray's directions of propagation are at right angles, then the parallel component's reflected amplitude vanishes and therefore the reflected light is *completely linearly* polarized perpendicular to the plane of incidence. The angle θ_B at which this occurs is known as *Brewster's angle* and is given by

$$\tan \theta_B = \frac{n_b}{n_a} \tag{1.37}$$

It follows from (1.33) and (1.34) that at all angles

$$\frac{A_{s2}}{A_{p2}} > \frac{A_{s1}}{A_{p1}}$$

which means that, if the incident light is linearly polarized, the reflected light is linearly polarized in a plane more nearly perpendicular to the plane of incidence than the incident light, which goes exactly into the perpendicular condition at the Brewster angle. Likewise the polarization of the refracted ray is always turned toward the plane of incidence, but there is no angle at which it is completely polarized.

The formulas (1.33) to (1.36) as written assume indeterminate forms in the limit of normal incidence when θ and φ each tend to zero with $\theta \to n\varphi$. The limiting forms are applicable at normal incidence, where, of course, there is no distinction between parallel and perpendicular polarization and

$$\begin{aligned} &\text{Reflected:} \qquad A_2 = -A_1 \frac{n-1}{n+1} \\ &\text{Transmitted:} \qquad A_3 = A_1 \frac{2}{n+1} \end{aligned} \tag{1.38}$$

The reflecting power R at normal incidence is thus given by $(A_2/A_1)^2$ so that

$$R = \left(\frac{n-1}{n+1}\right)^2 \tag{1.39}$$

For example, for $n = 1.5$, a typical value for common transparent materials, the reflecting power is 4 per cent.

In the Fresnel formulas (1.33) and (1.36) and also (1.38) and (1.39) all quantities are real, indicating that the reflected and transmitted waves are in phase (or exactly out of phase) with the incident waves. In case the light is incident on an absorbing medium [4] whose refractive index relative to the medium in which the incident light is transmitted is now complex, $n(1 - i\kappa)$, one can use the Fresnel formulas but with complex values for $\sin \varphi$ and $\cos \varphi$ given by

$$\sin \varphi = \frac{\sin \theta}{n(1 - i\kappa)}$$

$$\text{and} \quad \cos \varphi = \frac{1}{n(1 - i\kappa)} \sqrt{n^2(1 - i\kappa)^2 - \sin^2 \theta} \tag{1.40}$$

This gives a complex ratio for A_{s2}/A_{p2} which implies that linear polarized incident light is reflected as elliptically polarized light. Measurements of the ellipticity of the reflected light, at various angles of incidence, permit experimental determination of the complex index of refraction.

At normal incidence the amplitude of the reflected wave in terms of the incident wave is

$$A_2 = A_1 \frac{n(1 - i\kappa) - 1}{n(1 - i\kappa) + 1}$$

So the reflecting power becomes

$$R = \frac{(n-1)^2 + n^2\kappa^2}{(n+1)^2 + n^2\kappa^2} \tag{1.41}$$

In the case of a weakly absorbing medium ($n\kappa \ll 1$) the absorption makes only a small correction to (1.39). If the medium absorbs very strongly at certain frequencies so that for these $n\kappa \gg 1$, then the reflecting power tends to unity; thus those frequencies are strongly reflected which in the transmitted ray are strongly absorbed. This result is important in the interpretation of the colors of metals and strongly absorbing materials.

7. Plane Waves in Anisotropic Media

Propagation of light in ordinary doubly refracting crystals is described by assuming ϵ to be a second-order tensor. Formally similar effects could be obtained by a similar assumption for the magnetic permeability, but in actuality all crystals behave as if $\mu = 1$, and so this will be assumed in this section.

For nonabsorbing crystals ϵ is a real symmetric tensor, and undamped plane waves can be propagated in them. Damped plane waves in crystals also play a role in a complete analysis, such as in total reflection where the less dense medium is a doubly refracting crystal. In this section all field vectors are assumed to depend on space and time through the factor $e^{i(\omega t - \mathbf{k}\cdot\mathbf{r})}$ with $\mathbf{k}$ real.

The equations div $\mathbf{D} = 0$ and div $\mathbf{B} = 0$ give the result that $\mathbf{D}$ and $\mathbf{B}$ are orthogonal to $\mathbf{k}$ (and also $\mathbf{H}$ since $\mu = 1$). Thus the waves are transverse in $\mathbf{D}$, but as $\mathbf{D}$ and $\mathbf{E}$ are in general not parallel, it also turns out that $\mathbf{E}$ in general does not lie in the plane wavefronts.

The two curl equations in (1.7) give

$$\begin{aligned} \mathbf{H} &= \frac{c}{\omega} \mathbf{k} \times \mathbf{E} \\ \mathbf{D} &= -\frac{c}{\omega} \mathbf{k} \times \mathbf{H} \end{aligned} \tag{1.42}$$

Eliminating $\mathbf{H}$, writing $\mathbf{k} = k\mathbf{n}$, with $\mathbf{n}$ a unit vector, and $\omega/kc = n^{-1}$, where n is the index of refraction,

$$(\mathfrak{I} - \mathbf{nn}) \cdot \mathbf{E} = n^{-2}\mathbf{D} \tag{1.43}$$

Instead of ε it is convenient to introduce the tensor reciprocal to ε, called $\mathfrak{n}$, so that $\mathbf{E} = \mathfrak{n} \cdot \mathbf{D}$,

$$(\mathfrak{I} - \mathbf{nn}) \cdot \mathfrak{n} \cdot \mathbf{D} = n^{-2}\mathbf{D} \tag{1.44}$$

Suppose $\mathbf{l}$, $\mathbf{m}$, $\mathbf{n}$ are three mutually orthogonal unit vectors. Writing

$$\begin{aligned} \mathfrak{n} = \ & \eta_{11}\mathbf{ll} + \eta_{12}\mathbf{lm} + \eta_{13}\mathbf{ln} \\ & + \eta_{21}\mathbf{ml} + \cdots \end{aligned}$$

where $\eta_{21} = \eta_{12}$, etc., then (1.44) takes the form

$$\begin{bmatrix} \eta_{11} & \eta_{12} \\ \eta_{21} & \eta_{22} \end{bmatrix} \begin{bmatrix} D_1 \\ D_2 \end{bmatrix} = n^{-2} \begin{bmatrix} D_1 \\ D_2 \end{bmatrix} \tag{1.45}$$

Thus the allowed values of this two-dimensional matrix of the part of $\mathfrak{n}$ projected on the plane normal to the direction of propagation are the allowed values of n^{-2}, and the associated proper vectors give the corresponding states of polarization of $\mathbf{D}$.

If $\mathbf{l}$ and $\mathbf{m}$ are chosen along the two allowed, or normal, states of polarization then $\eta_{12} = 0$; so this equation can be put down as a condition for the determination of the proper orientation of $\mathbf{l}$ and $\mathbf{m}$ in the plane normal to $\mathbf{n}$,

$$\eta_{12} = \mathbf{l} \cdot \mathfrak{n} \cdot \mathbf{m} = 0$$

If **i**, **j**, and **k** are the principal axes of $\boldsymbol{\eta}$ so that

$$\boldsymbol{\eta} = \eta_1 \mathbf{ii} + \eta_2 \mathbf{jj} + \eta_3 \mathbf{kk}$$

and if the direction of **n** is specified by θ and φ in the usual way,

$$\mathbf{n} = \mathbf{k} \cos \theta + (\mathbf{i} \cos \varphi + \mathbf{j} \sin \varphi) \sin \theta$$

then one may take

$$\mathbf{l}' = \mathbf{i} \sin \varphi - \mathbf{j} \cos \varphi$$
$$\mathbf{m}' = -\mathbf{k} \sin \theta + (\mathbf{i} \cos \varphi + \mathbf{j} \sin \varphi) \cos \theta$$

and, for **l** and **m**,

$$\mathbf{l} = \mathbf{l}' \cos \psi - \mathbf{m}' \sin \psi$$
$$\mathbf{m} = \mathbf{l}' \sin \psi + \mathbf{m}' \cos \psi$$

and therefore

$$\mathbf{l} \cdot \boldsymbol{\eta} \cdot \mathbf{m} = \tfrac{1}{2}(\mathbf{l}' \cdot \boldsymbol{\eta} \cdot \mathbf{l}' - \mathbf{m}' \cdot \boldsymbol{\eta} \cdot \mathbf{m}') \sin 2\psi + (\mathbf{l}' \cdot \boldsymbol{\eta} \cdot \mathbf{m}') \cos 2\psi$$

So $\eta_{12} = 0$ if

$$\tan 2\psi = \frac{-2(\mathbf{l}' \cdot \boldsymbol{\eta} \cdot \mathbf{m}')}{(\mathbf{l}' \cdot \boldsymbol{\eta} \cdot \mathbf{l}') - (\mathbf{m}' \cdot \boldsymbol{\eta} \cdot \mathbf{m}')} \tag{1.46}$$

This determines ψ as an explicit function of θ and φ, if the expressions for $\mathbf{l}'$ and $\mathbf{m}'$ in terms of θ and φ are used. The two values of ψ so determined differ by $\pi/2$, giving orthogonal normal modes of plane polarization, and the two allowed values of the index of refraction become

$$n_a^{-2} = \mathbf{l} \cdot \boldsymbol{\eta} \cdot \mathbf{l} \qquad \text{and} \qquad n_b^{-2} = \mathbf{m} \cdot \boldsymbol{\eta} \cdot \mathbf{m} \tag{1.47}$$

For propagation in directions which lie in a principal plane ($\varphi = 0$ or $\pi/2$), $\psi = 0$ and $\pi/2$, and so the normal modes of polarization are along **j** and $\mathbf{m}'$ and the two indices are

$$n_a^{-2} = \eta_2 \qquad \text{for polarization along } \mathbf{j}$$

and

$$n_b^{-2} = \eta_3 \sin^2 \theta + \eta_1 \cos^2 \theta \qquad \text{for polarization along } \mathbf{m}'$$

If the axes have been so chosen that η_2 is intermediate in value between η_1 and η_3, then there will be two directions in this plane ($\varphi = 0$ and π) for which these two indices are equal, $n_a = n_b$.

Such directions, in which the double refraction vanishes, are called *optical axes*. Thus the general crystal in which the three principal values of η are different will have two optical axes. Such crystals are called *biaxial*.

If two of the principal values of $\boldsymbol{\varepsilon}$ are equal, then the two optical axes coincide with the axis of the odd principal value of $\boldsymbol{\varepsilon}$ and the crystal is called *uniaxial*. If the odd value is greater than the two equal values, the crystal is called *positive* uniaxial; otherwise it is *negative* uniaxial. Zircon is positive uniaxial, and calcite is negative uniaxial. Data for some common crystals are given in Table 1.1.

In a uniaxial crystal the plane containing the optical axis and the direction of the normal to the wavefront, **k**, is called the *principal plane*. The wave which is polarized in such a way that **D** is normal to the principal plane is called the *ordinary wave*, this name being due to the fact that its phase velocity is the same for all directions of **k**, and therefore it obeys Snell's law for refraction at a boundary as in the case of isotropic materials. The wave which is polarized with **D** in the principal plane is called the *extraordinary wave*. Its phase velocity v for a direction of **k** which makes an angle θ with the optical axis is

$$v^2 = v_o^2 \cos^2 \theta + v_e^2 \sin^2 \theta$$

and therefore the phase velocity of the extraordinary wave is v_e for propagation in directions normal to the optical axis. The two principal phase velocities v_o and v_e are related to the principal values of ϵ by

$$\left(\frac{c}{v_o}\right)^2 = n_o^2 = \epsilon_x = \epsilon_y$$
$$\left(\frac{c}{v_e}\right)^2 = n_e^2 = \epsilon_z$$

By the Poynting vector the flow of energy is given by $S = (c/4\pi)(\mathbf{E} \times \mathbf{H})$. Since in a doubly refracting medium **E** is not in general normal to **n**, it follows that the direction of energy flow is not that of the normal to the wavefronts, **n**. The direction of **S** is called the direction of the *ray*, and the speed with which a point of constant wave phase moves along the ray is called the *ray velocity*. Equation (1.43) shows that **E** lies in the plane of **D** and **n**, and therefore **E** and **H** are perpendicular. If $\mathbf{n}'$ is a unit vector in the direction of the ray and $\mu = c/v$, where v is the ray velocity, then **E** is determined by an equation similar to (1.45),

$$(\mathfrak{J} - \mathbf{n}'\mathbf{n}') \cdot \epsilon \cdot \mathbf{E} = \mu^2 \mathbf{E} \tag{1.48}$$

Thus the theory of the proper values of **E** and μ in relation to the ray direction $\mathbf{n}'$ closely parallels that which relates **D** to the wave propagation direction **n** and the phase velocity νc. A graphical construction for the allowed planes of polarization may be given

TABLE 1.1. REFRACTIVE INDICES FOR Na D LIGHT OF SOME UNIAXIAL CRYSTALS

Name	Formula	n_o	n_e
Positive uniaxial ($n_o < n_e$):			
Ice	H_2O	1.309	1.313
Quartz	SiO_2	1.544	1.553
Willemite	$2ZnO{\cdot}SiO_2$	1.691	1.719
Zircon	$ZrO_2{\cdot}SiO_2$	1.923	1.968
Iodyrite	AgI	2.210	2.220
Wurtzite	ZnS	2.356	2.378
Greenockite	CdS	2.506	2.529
Rutile	TiO_2	2.616	2.903
Cinnabar	HgS	2.854	3.201
Negative uniaxial ($n_o > n_e$):			
Calcite	$CaO{\cdot}CO_2$	1.658	1.486
Dolomite	$CaOMgO{\cdot}2CO_2$	1.681	1.500
Magnesite	$MgO{\cdot}CO_2$	1.700	1.509
Corundum	Al_2O_3	1.768	1.760
Siderite	$FeO{\cdot}CO_2$	1.875	1.635
Octahedrite	TiO_2	2.554	2.493
Hematite	Fe_2O_3	3.220	2.940 (Li line)

in relation to Fig. 1.2. Let AA' and BB' be the intersections of the two optical axes with the unit sphere and N be the intersection of the unit vector **n** with the unit sphere. On the sphere the two allowed planes of polarization are indicated by the vectors **l**

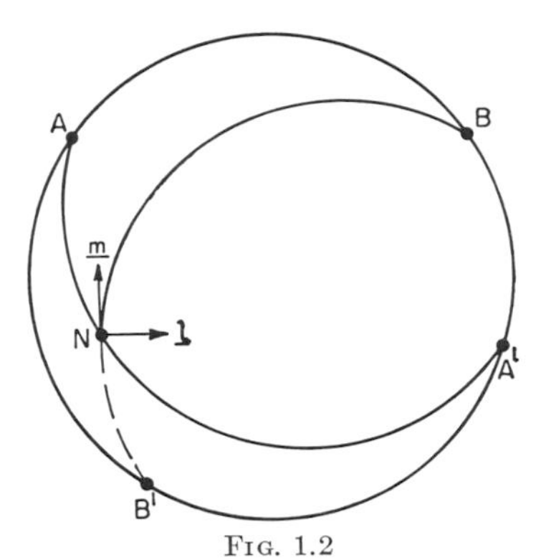

Fig. 1.2

and **m**, which bisect the angles ANB and $A'NB$. If the crystal is uniaxial, AA' and BB' coincide; so one allowed plane of polarization is that which contains **n** and the single optical axis, while the other is perpendicular thereto.

A number of surfaces have been introduced to show geometrically the dependence of the allowed velocities of propagation on **n**. Let

$$u_1 = \frac{c}{\sqrt{\epsilon_1}} \qquad u_2 = \frac{c}{\sqrt{\epsilon_2}} \qquad u_3 = \frac{c}{\sqrt{\epsilon_3}}$$

be the three principal values of the wave velocity. Then one can introduce (*a*) the *Fresnel ellipsoid,*

$$\frac{x^2}{u_1^2} + \frac{y^2}{u_2^2} + \frac{z^2}{u_3^2} = 1 \tag{1.49}$$

and (*b*) the *index ellipsoid,*

$$u_1^2x^2 + u_2^2y^2 + u_3^2z^2 = 1 \tag{1.50}$$

The Fresnel ellipsoid has this property: For a given direction of *ray,* **n**$'$, the plane central section normal to **n**$'$ is an ellipse whose principal axes determine the planes of polarization of **E**, the lengths of the semiaxes giving the associated *ray velocities.* Suppose that $u_1 > u_2 > u_3$. Then for the direction in the xz plane which makes an angle Θ with the x axis where

$$\tan^2 \Theta = \frac{u_1^2 - u_3^2}{u_2^2 - u_3^2}$$

the normal central section of the Fresnel ellipsoid becomes a circle, and so a *ray* is propagated along these directions with velocity independent of the state of polarization. These directions are called the *biradials* and are not to be confused with the optical axes.

The index ellipsoid has this property: For a given direction **n** of *wave* propagation the plane central section normal to **n** is an ellipse, the principal axes of this ellipse determine the planes of polarization of **D**, and the *reciprocals* of its semiaxes give the associated wave velocities.

Suppose that $u_1^{-1} > u_2^{-1} > u_3^{-1}$. Then for the directions in the xz plane having an angle Θ' with the x axis where

$$\tan^2 \Theta = \frac{u_1^{-2} - u_3^{-2}}{u_2^{-2} - u_3^{-2}}$$

the elliptical section becomes a circle, and so a *wave* propagated in this direction travels with a velocity independent of the state of polarization. These directions are sometimes called *binormals* although the more common usage is to call them optical axes.

Two ovaloid surfaces are also used, (*a*) the *ray ovaloid*

$$\frac{x^2}{u_1^2} + \frac{y^2}{u_2^2} + \frac{z^2}{u_3^2} = (x^2 + y^2 + z^2)^2 \tag{1.51}$$

and (*b*) the *Fresnel ovaloid*

$$u_1^2x^2 + u_2^2y^2 + u_3^2z^2 = (x^2 + y^2 + z^2)^2 \tag{1.52}$$

The ray ovaloid's central section normal to the ray direction **n**$'$ gives the axes of polarization of **E** and the associated ray velocities. Similarly the Fresnel ovaloid's central section normal to the wave direction **n** gives the axes of polarization of **D** and the associated wave velocities.

Two different two-sheeted surfaces have also been used:

(*a*) *Ray surface.* Let $\mathbf{r} = x\mathbf{i} + y\mathbf{j} + z\mathbf{k}$ be the radius vector to a point on the surface. Then the equation for the two-sheeted ray surface is

$$\frac{x^2}{u_1^{-2} - r^2} + \frac{y^2}{u_2^{-2} - r^2} + \frac{z^2}{u_3^{-2} - r^2} = 0 \tag{1.53}$$

(This is called the wave surface in Walker's "Analytical Theory of Light," Cambridge University Press, p. 206.) This surface has the simple physical significance of being the locus of all points reached in unit time by an electromagnetic disturbance starting from the origin. Tangent planes to the ray surface give the wavefronts of plane waves associated with the ray direction of the point of tangency.

The ray surface intersects each of the coordinate planes (y,z), (z,x), (x,y) in a circle and an ellipse. In the (z,x) plane, associating the greatest and least principal values of ε, the circle and ellipse intersect, and the lines drawn from the origin to these points of intersection are the biradials. *Outer conical refraction* is an optical property associated with propagation of light in directions close to the biradials.

(*b*) *Normal surface.* Its equation is

$$\frac{x^2}{u_1^2 - r^2} + \frac{y^2}{u_2^2 - r^2} + \frac{z^2}{u_3^2 - r^2} = 0 \tag{1.54}$$

Walker calls this the *surface of wave quickness.* The two values of the radius in the direction **n** are the two values of the wave velocity associated with wavefronts normal to **n**.

The normal surface is the pedal of the ray surface. Suppose A is a point on the ray surface. The tangent plane to the ray surface at A is the wavefront for a wave whose *ray* direction is the vector OA, and its ray velocity is the magnitude of OA. From O drop a perpendicular to this tangent plane, and let B be the place where the perpendicular intersects the tangent plane. Then B is a point on the normal surface: as A moves over both sheets of the ray surface, B will generate the normal surface.

The intersections of the normal surface with the principal coordinate planes are circles and ovals. In the (z,x) plane the circle and oval intersect in points which determine the directions of the optical axes. *Inner conical* refraction is an optical property associated with properties of the normal surface near the optical axes.

The *surface of wave slowness*,

$$\frac{x^2}{u_1^2r^2 - 1} + \frac{y^2}{u_2^2r^2 - 1} + \frac{z^2}{u_3^2r^2 - 1} = 0 \quad (1.55)$$

is obtained from the ray surface by inversion with respect to the unit sphere about the origin. This is the surface which has to be used in applying Huygens' principle to discussions of wave propagation in crystals.

There is a good deal of confusion in the literature arising from the fact that the terminology of these various surfaces is not fully standardized.

The theory for *absorbing crystals* involves the introduction of a complex dielectric tensor $\boldsymbol{\varepsilon}' - i\boldsymbol{\varepsilon}''$, but the resulting formulas are too complicated to be of general interest.

8. Optical Activity

By *optical activity* is meant the property of a medium to rotate the plane of polarization of a wave traveling through it [5]. Fresnel recognized in 1825 that this phenomenon arises from circular double refraction, in which the normal modes of wave propagation are right and left circularly polarized waves.

Assigning indices n_r and n_l to such waves, and considering them to move in the $+z$ direction, the vector amplitudes for the two parts are

$$D_r(\mathbf{i} + i\mathbf{j}) \exp\left[i\omega\left(t - \frac{zn_r}{c}\right)\right]$$

and

$$D_l(\mathbf{i} - i\mathbf{j}) \exp\left[i\omega\left(t - \frac{zn_l}{c}\right)\right]$$

If a plane-polarized wave, of amplitude D_0, that is polarized along $\mathbf{i}$ enters the medium at $z = 0$, it will be represented by the sum of two such waves with $D_r = D_l = D_0/2$. Then at distance z in the medium the resultant will be

$$\frac{D_0}{2}\left\{(\mathbf{i} + i\mathbf{j}) \exp\left[-\frac{i\omega z(n_r - n_l)}{2c}\right] + (\mathbf{i} - i\mathbf{j}) \exp\left[+\frac{i\omega z(n_r - n_l)}{2c}\right]\right\} \cdot \exp\left\{i\omega\left[t - \frac{z(n_r + n_l)}{2c}\right]\right\} \quad (1.56)$$

The factor in the outside braces is real, therefore representing a plane-polarized wave whose plane of polarization has been turned in the counterclockwise sense from $\mathbf{i}$ through an angle $\omega z(n_r - n_l)2c$. Thus the plane of polarization is turned in unit distance of travel through the medium through an angle

$$\Phi = \left(\frac{\pi}{\lambda}\right)(n_r - n_l) \quad (1.57)$$

By this formula Φ is positive if $n_r > n_l$. Thus the plane of polarization turns in the same sense as the circularly polarized wave which travels with the greater phase velocity.

The medium is said to have *dextrorotation* if the rotation is in the clockwise sense and *levulorotation* if in the counterclockwise sense; so

Dextro- if $n_r < n_l$
Levulo- if $n_r > n_l$

Natural quartz crystals occur having both signs of rotatory power for light traveling along the optical axis. The magnitude of the rotation for the Na D lines is 216°/cm, giving

$$|n_r - n_l| = 7.06 \times 10^{-5}$$

from which it is evident that rotation of the plane of polarization is an extremely sensitive way of measuring very small amounts of circular double refraction.

Optical activity was first discovered in quartz crystals. Later it was discovered that regular crystals of $NaClO_3$ and $NaBrO_3$ show optical activity for waves propagated in all directions in the crystal. Still later it was discovered that sucrose and sodium potassium tartrate (rochelle salt), both of which have biaxial crystals, show optical rotation for light propagated along their optical axes. Optical activity is also shown by many liquids, where it is a property of the structural arrangement of atoms in the randomly oriented molecules.

The *Faraday effect* is the property of transparent substances by which the plane of polarization is rotated when the material is placed in a magnetic field, for light propagated along the magnetic field. More accurately, the rotation is proportional to the component of the magnetic field along the direction of propagation of the light.

Circular double refraction occurs if the relation between $\mathbf{E}$ and $\mathbf{D}$ includes a term dependent on the time derivative $\dot{\mathbf{D}}$ as follows:

$$\mathbf{E} = \boldsymbol{\eta} \cdot \mathbf{D} - \mathbf{A} \times \dot{\mathbf{D}} \quad (1.58)$$

Here $\mathbf{A}$ is a characteristic of the medium known as the *gyration vector* and $\boldsymbol{\eta}$ is the reciprocal of the dielectric constant tensor introduced in Sec. 7. Carrying out the time differentiation, this becomes

$$\mathbf{E} = \boldsymbol{\eta} \cdot \mathbf{D} - i\omega\mathbf{A} \times \mathbf{D} \quad (1.58a)$$

and with this more general connection the equation for the determination of $\mathbf{D}$ takes the form

$$\begin{bmatrix} \eta_{11} - n^{-2} & \eta_{12} - i\omega A_n \\ \eta_{12} + i\omega A_n & \eta_{22} - n^{-2} \end{bmatrix} \begin{bmatrix} D_1 \\ D_2 \end{bmatrix} = 0 \quad (1.59)$$

The vanishing of the determinant of this matrix leads to two roots for the refractive index,

$$\eta^{-2} = \tfrac{1}{2}(\eta_{11} + \eta_{22}) \pm \sqrt{\eta_{12}^2 + \omega^2 A_n^2 + \tfrac{1}{4}(\eta_{11} - \eta_{22})^2}$$

and the determination of the normal modes of polarization shows that these are two orthogonal states of elliptic polarization. Here A_n is the component of $\mathbf{A}$ in the direction of propagation, or $\mathbf{A} \cdot \mathbf{n}$.

This result gives the general interplay between ordinary double refraction and rotatory power. The simpler case is that in which $\mathbf{n}$ is a scalar, so that $\eta = \eta_{11} = \eta_{22}$ and $\eta_{12} = 0$. In this case the two values of the refractive index are given by

$$n^{-2} = \eta \pm \omega A_n$$

The normal mode of propagation associated with the upper sign is a right circularly polarized wave, while that associated with the lower sign is the orthogonal left circularly polarized wave.

Different kinds of optically active media can be classified by the properties of the gyration vector $\mathbf{A}$. In the case of optically active liquids, $\mathbf{A}$ is a scalar multiple of the vector $\mathbf{n}$, or, in other words, A_n has the same value for all directions of propagation.

In the Faraday effect the gyration vector is proportional to the external magnetic field $\mathbf{H}$,

$$\mathbf{A} = \beta \mathbf{H} \tag{1.60}$$

in which β is the significant magnetooptic parameter of the medium. In an optically active liquid the direction of rotation bears a fixed relation to the direction of propagation so that if a beam of light is reflected back on itself the net rotation is zero, whereas in the Faraday effect rotation bears a fixed relation to the direction of $\mathbf{H}$ so that reflection back on itself doubles the over-all rotation.

Optically active crystals have an associated *gyration tensor* such that

$$\mathbf{A} = \boldsymbol{\alpha} \cdot \mathbf{n} \tag{1.60a}$$

so that the rotation is different in amount for different directions of propagation.

9. Waveguides and Transmission Lines

Use is often made of the propagation of electromagnetic waves along metallic conductors and in hollow metallic tubes. In a first approximation the resistivity may be neglected so that tangential $\mathbf{E} = 0$ at conducting surfaces. The conductors are cylindrical, extending indefinitely in the z direction.

Solutions of Maxwell's equations can be found in which all vectors depend on t and z through the factor $\exp i(\omega t - k_3 z)$. Writing any vector in the form $\mathbf{A} = \mathbf{A}_s + A_z\mathbf{k}$, the curl equations can be put in a form which emphasizes the separation of longitudinal and transverse components.

$$ik_3\mathbf{k} \times \mathbf{E}_s - i\frac{\mu\omega}{c}\mathbf{H}_s = \left(\mathbf{i}\frac{\partial}{\partial y} - \mathbf{j}\frac{\partial}{\partial x}\right)E_z \tag{1.61a}$$

$$-i\frac{\mu\omega}{c}H_z = \frac{\partial E_y}{\partial x} - \frac{\partial E_x}{\partial y} \tag{1.61b}$$

$$ik_3\mathbf{k} \times \mathbf{H}_s + i\frac{\epsilon\omega}{c}\mathbf{E}_s = \left(\mathbf{i}\frac{\partial}{\partial y} - \mathbf{j}\frac{\partial}{\partial x}\right)H_z \tag{1.61c}$$

$$+i\frac{\epsilon\omega}{c}E_z = \frac{\partial H_y}{\partial x} - \frac{\partial H_x}{\partial y} \tag{1.61d}$$

Three kinds of solutions for these exist:

1. TEH (transverse electric and magnetic), for which both $E_z = 0$ and $H_z = 0$.
2. TH (transverse magnetic), with $E_z \neq 0$ and $H_z = 0$.
3. TE (transverse electric), with $E_z = 0$ and $H_z \neq 0$.

Considering TEH waves, (1.61*b*) and (1.61*d*) show that $\mathbf{E}_s$ and $\mathbf{H}_s$ can each separately be represented as the gradient of a scalar function. Also (1.61*a*) and (1.61*c*) show that these are mutually orthogonal in such a way that k_3 must have the value

$$\frac{\omega}{k_3} = \frac{c}{\sqrt{\epsilon\mu}} \tag{1.62a}$$

and

$$\mathbf{E}_s + i\sqrt{\frac{\mu}{\epsilon}}\mathbf{H}_s = -\operatorname{grad} w \tag{1.62b}$$

where w is an analytic function of the complex variable $x + iy$ satisfying the Cauchy-Riemann conditions,

$$u_x = v_y \quad \text{and} \quad u_y = -v_x$$
$$\text{and} \quad w = u + iv$$

The value assigned to k_3 shows that TEH waves travel with the same phase velocity as does an infinite plane wave in the same medium.

The boundary condition on $\mathbf{E}$ is satisfied if the metallic conductors coincide with curves u = const. Thus the problem of fitting the boundary conditions is reduced to that of finding the appropriate function $w(x + iy)$ for which this is true. Conversely every analytic function defines the solution to a variety of transmission line problems.

No TEH wave can be propagated down the inside of a hollow conductor, for it is known that the only analytic function of a complex variable that is constant on a closed curve is also constant everywhere inside that curve and therefore its gradient vanishes, giving no solution. The functions which can be used for description of possible TEH waves are ones which have poles outside the region to which the solution applies, which usually means inside the metallic conductors.

As a simple example,

$$w = A \log(x + iy) = A(\log r + i\theta)$$

has its real part constant along circles centered on the origin. It is therefore appropriate for discussing waves traveling in coaxial cable, that is, in the medium between an inner conductor of radius a and an outer conductor of radius b. The total current flowing in the inner conductor is calculated from the line integral of $\mathbf{H}_s$ around a path surrounding the inner conductor and is

$$I = \frac{(\epsilon/\mu)^{1/2}A}{2} \quad \text{abamp}$$

while the potential difference between the two conductors (line integral of $\mathbf{E}_s$ from one to the other) is $A \log (b/a)$ statvolts. Therefore the *characteristic impedance* Z of the line, which is the ratio of potential difference to current, is

$$Z = 60\left(\frac{\mu}{\epsilon}\right)^{1/2} \ln\frac{b}{a} \quad \text{ohms}$$

Similarly for a parallel-wire transmission line the appropriate function is

$$x + iy = c \tanh (u + iv)$$

for which the curves of constant u are the bipolar family of circles

$$(x - c \coth 2u)^2 + y^2 = c^2 \operatorname{csch}^2 2u$$

and the curves of constant v are the bipolar family of circles orthogonal to these,

$$x^2 + (y + c \cot 2v)^2 = c^2 \csc^2 2v$$

In particular, therefore, if the center-to-center distance of the two wires is d and each wire is of radius a, one wire is represented by

$$\cosh \frac{d}{2a} = 2u$$

and the other wire is represented by the negative value of u of equal magnitude.

Calculation of current and voltage as before shows that the characteristic impedance of this line is

$$Z = 120 \left(\frac{\mu}{\epsilon}\right)^{1/2} \cosh^{-1} \frac{d}{2a} \qquad \text{ohms}$$

Next, considering TH waves, multiplying (1.61a) by $\mathbf{k} \times$ and combining with (1.61c) gives

$$\left(k_3^2 - \frac{\omega^2}{c^2}\right) \mathbf{E}_s = ik_3 \operatorname{grad} E_z \qquad (1.63a)$$

which, by expressing $\mathbf{E}_s$ as a gradient, also satisfies (1.61b). From (1.61c), on multiplying by $\mathbf{k} \times$,

$$\mathbf{H}_s = \frac{\epsilon\omega}{k_3 c} \mathbf{k} \times \mathbf{E}_s \qquad (1.63b)$$

and so the transverse magnetic field is also expressible in terms of the single function $E_z(x,y)$. Using these results in (1.61d), one finds that E_z must satisfy a wave equation,

$$\left(\frac{\partial^2}{\partial x^2} + \frac{\partial^2}{\partial y^2}\right) E_z + k^2 E_z = 0$$

in which

$$k^2 = \frac{\epsilon\mu\omega^2}{c^2} - k_3^2$$

To satisfy the boundary conditions, one must have $E_z = 0$ on the walls of the conductor. Therefore every solution of the two-dimensional wave equation for E_z which satisfies this boundary condition leads to the wave pattern for a possible kind of TH wave.

The boundary-value problem defines a sequence of allowed values of k, say, k_a, k_b, k_c, and so on. The relation between k, k_3, and ω shows that ω must be greater than a lower limit characteristic of each mode of propagation for this equation to be satisfied. The lower limiting frequency for propagation of a particular kind of wave is called its *cutoff frequency*. If an attempt is made to transmit a wave at lower than its cutoff frequency, k_3 becomes pure imaginary and the wave is exponentially attenuated rather than having real propagation along the waveguide.

Similarly for TE waves, all the vectors can be expressed in terms of H_z,

$$(k_3^2 - \epsilon\mu\omega^2/c^2)\mathbf{H}_s = ik_3 \operatorname{grad} H_z \qquad (1.64a)$$

and

$$\mathbf{E}_s = -\left(\frac{\mu\omega}{k_3 c}\right) \mathbf{k} \times \mathbf{H}_s \qquad (1.64b)$$

Here H_z satisfies the same two-dimensional wave equation as is satisfied by E_z in the case of the TH waves. However, in this case the solutions which satisfy the boundary conditions on $\mathbf{E}$ are different: for H_z one must take solutions of the two-dimensional wave equation whose gradient normal to the surface of the conductor vanishes. Therefore in general these modes of propagation are characterized by different cutoff frequencies and different distributions of the electromagnetic field within the guide.

10. Black-body Radiation

The electromagnetic radiation present in any region of empty space at thermodynamic equilibrium at temperature T is known as *black-body radiation*, or thermal radiation. It is of practical importance as being the maximum amount of radiation that can be emitted by hot solid bodies. It is of theoretical importance in the history of physics because through study of its properties Planck was led in 1900 to the initial ideas of the quantum theory [6].

Black-body radiation is isotropic and unpolarized and has a continuous distribution of frequencies. The density of radiant energy in unit volume in the frequency range ν to $\nu + d\nu$ will be written $\rho(\nu)\, d\nu$. It is given by the experimentally verified *Planck distribution law*,

$$\rho(\nu)\, d\nu = \frac{h\nu}{e^{h\nu/kT} - 1} \frac{8\pi\nu^2\, d\nu}{c^3} \qquad (1.65)$$

Here h is Planck's constant, c is the velocity of light, and k is the Boltzmann constant. In the experimental literature this is often written as a distribution with respect to wavelengths,

$$\rho(\lambda)\, d\lambda = \frac{c_1}{\lambda^5} \frac{d\lambda}{\exp (c_2/\lambda T) - 1} \qquad (1.66)$$

in which c_1 and c_2 are called the *first* and *second Planck radiation constants*,

$$c_1 = 8\pi hc \qquad c_2 = \frac{hc}{k} \qquad (1.67)$$

The first factor in (1.65) is the mean energy at temperature T of a quantized harmonic oscillator having as allowed levels the integral multiples of $h\nu$ (Part 5, Chap. 2). The second factor gives the number of degrees of freedom of the radiation field in unit volume whose frequency lies between ν and $\nu + d\nu$.

The *total density* ρ of radiant energy in unit volume is the integral of (1.65) over all frequencies, giving

$$\rho = \frac{8\pi(kT)^4}{h^3c^3} \int_0^\infty \frac{u^3\, du}{e^u - 1} = \frac{8\pi^5}{15} \frac{(kT)^4}{h^3c^3} = aT^4 \qquad (1.68)$$

The fact that the total density of radiant energy increases as the fourth power of the temperature is known as the *Stefan-Boltzmann law*.

Ordinarily radiant energy is observed in terms of the flux of energy radiated from a hot body to colder surroundings. The density of radiant energy at ordinary temperatures is extremely small compared with the volume density of energy in the molecular motions of a gas at extremely high vacuum, so it is not observed as a contribution to the heat capacity of the contents of a calorimeter even though in principle it should be counted. Average molecular energies go up as T, but radiant energy density goes up as T^4; so it becomes relatively important at the extremely high temperatures prevailing in the interior of stars. At much lower temperatures, in moderately transparent materials, the radiant energy flux makes an important contribution to heat transfer because of its high mobility in spite of its small total amount. In a material of refractive index n one writes c/n for c in (1.65) and (1.68) so that the density is n^3 times as great as in empty space.

The factor $c/4$ between energy density and total flux across unit area holds for each frequency. Therefore the spectral distribution of flux between ν and $\nu + d\nu$ is

$$F(\nu)\, d\nu = \frac{c}{4}\rho(\nu)\, d\nu = \frac{2\pi\nu^2\, d\nu}{c^2} \frac{h\nu}{e^{h\nu/kT} - 1} \tag{1.70}$$

which can be written

$$F(\nu)\, d\nu = \sigma T^4 \frac{15\pi^{-4}u^3}{e^u - 1}\, du \tag{1.71}$$

where $u = h\nu/kT$.

Figure 1.3 is a graph of the second factor of (1.71). The whole curve has unit area under it, and so the area between any two ordinates gives the fraction of the radiant energy in the frequency range between the ordinates. The figure shows the locations of the ordinates which divide the total radiation into tenths,

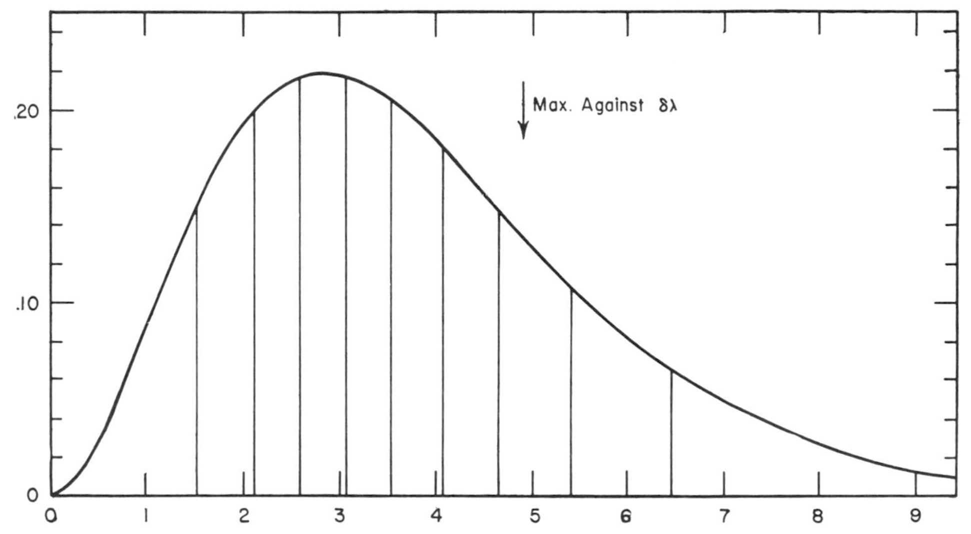

FIG. 1.3. Plot of $15\pi^{-4}u^3/(e^u - 1)$ against u, with ordinates marking each ten per cent of the total area.

The *flux of energy* in unit time across unit area of a plane in the medium, crossing it in a particular sense, may be derived as follows: As the energy flows with velocity c and is isotropically distributed, the energy flow across unit area normal to the axis of a small cone of solid angle opening $d\omega$ is $c\rho\, d\omega/4\pi$. If the axis of the cone is inclined at an angle θ to the unit area, the projection factor $\cos\theta$ must be included; so the total flux from one side to the other becomes

$$2\pi \int_0^{\pi/2} \cos\theta \sin\theta\, d\theta \frac{c\rho}{4\pi} = \frac{c\rho}{4}$$

The total radiation crossing unit area in unit time in all directions in one hemisphere is usually written σT^4, where σ is called the *Stefan-Boltzmann constant*. Accordingly

$$\sigma = \frac{2\pi^5}{15} \frac{k^4}{h^3c^2} = 5.669 \times 10^{-5}\, \text{erg/cm}^2 \cdot \text{deg}^4 \cdot \text{sec} \tag{1.69}$$

For practical calculations it is convenient to note that $(T/645)^4$ gives the total radiant flux in watts per square centimeter.

which is an adequate characterization of the distribution for many practical applications [7].

The locations of these dividing ordinates are:

Fraction	u	*Fraction*	u
0.0	0.0		
0.1	1.55	0.6	4.05
0.2	2.15	0.7	4.64
0.3	2.60	0.8	5.40
0.4	3.05	0.9	6.45
0.5	3.55	1.0	∞

The maximum of the curve occurs at $u = 2.80$. When the distribution is expressed in terms of $d\lambda$, the function is $u^5/(e^u - 1)$ and its maximum is at $u = 4.96$.

The fact that (1.71) can be written as a function of $u = h\nu/kT$ shows that the relative shape of the distribution of energy over frequencies remains the same, the range on the frequency scale being proportional to the absolute temperature. This property is known as *Wien's displacement law*.

For $u \ll 1$ the mean energy of a quantized oscillator approaches kT, and so (1.65) assumes a limiting form

$$\rho(\nu)\,d\nu \rightarrow kT\,\frac{8\pi\nu^2 d\nu}{c^3} \tag{1.72}$$

which is the form which would be expected to be valid at all frequencies on the basis of classical statistical mechanics. This form is known as the *Rayleigh-Jeans law*. For $u \gg 1$ the fractional difference between $e^u - 1$ and e^u becomes negligible, and the distribution law approaches another form, known as the *Wien law*, obtained by writing e^u in place of $e^u - 1$ in the preceding equations.

When an atom or molecule is in thermodynamic equilibrium with the thermal radiation, there must be a statistical balance between the rate at which atoms go from a lower level 1 to an upper level 2 by absorbing radiation of frequency ν, such that $h\nu = E_2 - E_1$, and the rate at which they return from level 2 to level 1 by emitting radiation of this frequency. Einstein used this fact to derive important consequences concerning the nature of the radiative process [8].

The number of upward transitions in unit volume in unit time is proportional to the number of atoms in unit volume in the lower level, n_1, and to the density of the radiation being absorbed, $\rho(\nu)$, and to a coefficient B_{12} characteristic of the coupling of the atom to the radiation field. Similarly the downward rate is the sum of a spontaneous rate A_{21} and an induced emission rate $B_{21}\rho(\nu)$, each multiplied by n_2, so the rate balance for equilibrium is

$$n_1 B_{12}\rho = n_2(A_{21} + B_{21}\rho)$$

giving

$$\rho = \frac{A_{21}/B_{21}}{(n_1/n_2)(B_{12}/B_{21}) - 1}$$

In equilibrium $n_1/n_2 = (g_1/g_2)e^u$, by the Boltzmann law, where g_1 and g_2 are the statistical weights of the levels. Comparing with (1.65), this indicates that the laws of interaction of radiation and matter must be such that

$$g_1 B_{12} = g_2 B_{21} = C_{12}$$

where C_{12} is a characteristic of the transition that is symmetric in the initial and final levels, and

$$\frac{A_{21}}{B_{21}} = \frac{8\pi h\nu^3}{c^3} \tag{1.73}$$

Since $(e^u - 1)^{-1}$ is the mean value of the quantum number $\bar{n}$ of a harmonic oscillator of frequency ν at temperature T, the rates of absorptive and emissive transitions are

$$\begin{aligned} &\frac{n_1}{g_1} C_{12} \frac{8\pi h\nu^3}{c^3}\,\bar{n}(\nu) \\ &\frac{n_2}{g_2} C_{12} \frac{8\pi h\nu^3}{c^3}\,[\bar{n}(\nu) + 1] \end{aligned} \tag{1.74}$$

respectively.

The same argument can be narrowed down to a discussion of equilibrium between an atom and the energy in any one degree of freedom of the radiation field to indicate that the absorption rate is proportional to n and the emission rate to $n + 1$, where n is the quantum number of the particular degree of freedom of the radiation field in question.

It is customary to characterize materials by an effective total emissivity e, defined such that $e\sigma T^4$ is the actual rate of radiation of unit area in unit time at temperature T. If the body at temperature T is surrounded by materials at a temperature T_0 which have emissivity e_0, then the net rate of loss of energy is $eT^4 - e_0T_0^4$. This fact that the net loss by radiant energy is a difference between a rate of emission and a rate of absorption is known as *Prévost's law of exchanges*.

TABLE 1.2. TYPICAL TOTAL EMISSIVITIES OF VARIOUS SOLIDS†

Material	Temperature, °C		
	200	400	600
Silver	0.020	0.030	0.038
Platinum	0.060	0.086	0.110
Oxidized aluminum	0.113	0.153	0.192
Oxidized monel	0.411	0.439	0.463
Oxidized copper	0.568	0.568	0.568
Oxidized steel	0.790	0.788	0.787

† Randolph and Overholzer, *Phys. Rev.*, **2:** 144 (1913).

TABLE 1.3. SPECTRAL EMISSIVITIES OF VARIOUS SOLIDS FOR $\lambda = 0.55\ \mu$ AND $0.65\ \mu$†

	Emissivity e	
	$\lambda = 0.55\mu$	$\lambda = 0.65\mu$
Metals:		
Rh	0.29	0.29
Ag	0.35	0.04
Cu	0.38	0.10
Pt	0.38	0.33
Ni	0.44	0.36
Cr	0.53	0.39
Fe	....	0.37
Oxides:		
NiO	....	0.89
Fe_3O_4	....	0.63
TiO_2	....	0.52
ThO_2	....	0.57
U_3O_8	....	0.30

† Burgess and Waltenberg, *Bur. Standards Bull.*, **11:** 591 (1914).

Values are nearly constant from room temperature to melting point.

Real materials may be characterized by an emissivity $e(\nu)$ at each frequency defined as the fraction of the black-body flux (1.70) radiated at that frequency where this is integrated over the full hemisphere of directions away from the material. If θ is the angle with the normal, then a further refinement is to introduce a dependence of the emissivity on θ. The black-body radiation from unit area into the solid angle element $d\omega$ at angle θ is $(c\rho/4\pi)\cos\theta\, d\omega$, and so if $e(\nu,\theta)$ is the emissivity at angle θ with the normal at frequency ν, the radiation from unit area into solid angle $d\omega$ in this direction is $e(\nu,\theta)(c\rho/4\pi)\cos\theta\, d\omega$. A material which radiates the same angular distribution as in black-body radiation, so that $e(\nu,\theta)$ is independent of θ, is said to obey *Lambert's law*. Since a $\cos\theta$ factor appears in converting the actual area of the radiator to the projected area normal to the line of sight to an observer, a material radiating according to Lambert's law will appear equally bright when viewed from any direction.

Some general indications of the magnitude of total emissivities and of emissivities at wavelengths of $\lambda = 0.55\ \mu$ and $0.65\ \mu$ are given in Tables 1.2 and 1.3.

Very complicated problems arise when a transparent material is in a nonequilibrium condition owing to losing heat by radiation such as occurs in the transport of radiation through the outer layers of a star or into or out of a transparent material [9].

11. Radiation from Oscillating Charge Distribution

Suppose that there exists a charge and current distribution described by charge density ρ (esu) and current density $\mathbf{i}$ (emu) in a finite region around the origin of coordinates, where ρ and $\mathbf{i}$ are time-dependent and related by the equation of continuity

$$\dot{\rho} + c \operatorname{div} \mathbf{i} = 0 \tag{1.75}$$

The associated electromagnetic field is derivable from a scalar φ and a vector potential $\mathbf{A}$, which it is convenient to relate by the Lorentz condition

$$\dot{\varphi} + c \operatorname{div} \mathbf{A} = 0 \tag{1.76}$$

the formulas for $\mathbf{E}$ and $\mathbf{H}$ being

$$\mathbf{E} = -\frac{1}{c}\dot{\mathbf{A}} - \operatorname{grad}\varphi \qquad \mathbf{H} = \operatorname{curl}\mathbf{A} \tag{1.77}$$

Substitution in (1.7), using $\mathbf{D} = \mathbf{E}$ and $\mathbf{B} = \mathbf{H}$ for free space, gives

$$\begin{aligned} -\Delta^2\varphi + \frac{1}{c^2}\frac{\partial^2\varphi}{\partial t^2} &= 4\pi\rho \\ -\Delta^2\mathbf{A} + \frac{1}{c^2}\frac{\partial^2\mathbf{A}}{\partial t^2} &= 4\pi\mathbf{i} \end{aligned} \tag{1.78}$$

These equations possess a *retarded potential* solution which expresses the fact that electromagnetic-field changes are propagated with velocity c. Let $\mathbf{r}$, t be the time and place for which the potentials are calculated and $\mathbf{r}'$, t' the position at time t' of a charge element, and write R for $|\mathbf{r} - \mathbf{r}'|$. Then the charge or current distribution which contributes to the value of the potential at $\mathbf{r}$ at time t has to be evaluated at the earlier time $t' = t - R/c$; so

$$\begin{aligned} \varphi(\mathbf{r},t) &= \iiint \frac{\rho(\mathbf{r}', t - R/c)}{R}\, dx'\, dy'\, dz' \\ \mathbf{A}(\mathbf{r},t) &= \iiint \frac{\mathbf{i}(\mathbf{r}', t - R/c)}{R}\, dx'\, dy'\, dz' \end{aligned} \tag{1.79}$$

If the charge and current vary sufficiently slowly, then the fact that in the integral $t - R/c$ occurs in place of t will have a negligible effect and so the fields are practically the same as if the finite speed of propagation were regarded as infinite.

In many applications to radiation one is interested only in the fields set up at large distances from a relatively localized distribution of charges and currents. Moreover it is usually sufficient to suppose harmonic time dependence through a factor $e^{i\omega t}$ for ρ and $\mathbf{i}$. By using the expansion

$$\frac{e^{-ikR}}{R} = \sum_\lambda (2\lambda + 1)\frac{\zeta_\lambda(kr)}{ir}\frac{\psi_\lambda(kr')}{kr'} P_\lambda(\cos\omega)$$

where ω is the angle between $\mathbf{r}$ and $\mathbf{r}'$ and the ζ_λ and ψ_λ are spherical Bessel functions,

$$\begin{aligned} \psi_\lambda(kr') &= \sqrt{\tfrac{1}{2}\pi kr'}\, J_{\lambda+1/2}(kr') \\ \zeta_\lambda(kr) &= \sqrt{\tfrac{1}{2}\pi kr}\, [J_{\lambda+1/2}(kr) + (-1)^\lambda i J_{-\lambda-1/2}(kr)] \end{aligned} \tag{1.80}$$

The potentials in (1.79) are represented as a set of outgoing spherical waves because, asymptotically for large kr',

$$\frac{\zeta_\lambda(kr)}{ir} = i^\lambda \frac{\exp(-ikr)}{r}\left[1 - i\frac{\lambda(\lambda+1)}{2kr} + \cdots\right]$$

It may be further supposed that ρ and $\mathbf{i}$ are localized to such a region of space that $kr \ll 1$. In this case it is sufficient to use the first terms of

$$\frac{\psi_\lambda(kr')}{kr'} = \frac{(kr')^\lambda}{1\cdot 3\cdot 5\cdots(2\lambda+1)}\left[1 - \frac{k^2r'^2}{2(2\lambda+3)} + \cdots\right] \tag{1.81}$$

in calculating the integrals in (1.79). Thus the terms of successive order in λ involve increasingly higher moments of the oscillating charge and current distributions. The moments of the current distribution are related to those of the charge distribution because the two distributions are connected through the equation of continuity. If g is any scalar, vector, tensor field finite everywhere that ρ and $\mathbf{i}$ are not zero,

$$ik\int \rho g\, dv = \int \mathbf{i}\cdot\nabla g\, dv \tag{1.82}$$

The dipole and quadrupole moments of the charge distribution are, respectively,

$$\mathbf{P} = \int \rho\mathbf{r}\, dv \qquad \mathbf{N} = \int \rho\mathbf{rr}\, dv \tag{1.83}$$

and, using $g = \mathbf{r}$ and $\mathbf{rr}$, respectively,

$$\int \mathbf{i}\, dv = ik\mathbf{P} \qquad \int(\mathbf{ir} + \mathbf{ri})\, dv = ik\mathbf{N} \tag{1.84}$$

The complete term $\int \mathbf{ir}\, dv$ is the sum of a symmetric tensor related to the quadrupole charge moment and an antisymmetric tensor related to the magnetic moment $\mathbf{M}$ of the current distribution,

$$\mathbf{M} = \frac{1}{2}\int \mathbf{r}\times\mathbf{i}\, dv \tag{1.85}$$

Thus the leading terms in the potentials are represented as the radiated fields due to electric-dipole, electric-quadrupole, and magnetic-dipole moments as follows:

Electric dipole:

$$\varphi = \frac{e^{i(\omega t-kr)}}{r} ik\left(1 - \frac{i}{kr}\right) \mathbf{r}_0 \cdot \mathbf{P}$$
$$\mathbf{A} = \frac{e^{i(\omega t-kr)}}{r} ik\mathbf{P} \tag{1.86}$$

Electric quadrupole:

$$\varphi = -\frac{e^{i(\omega t-kr)}}{r}\frac{1}{2}k^2 \left[\mathbf{r}_0 \cdot \mathbf{N} \cdot \mathbf{r}_0 - \frac{i}{kr}(3\mathbf{r}_0 \cdot \mathbf{N} \cdot \mathbf{r}_0 - N_s)\right] \tag{1.87}$$
$$\mathbf{A} = -\frac{e^{i(\omega t-kr)}}{r}\frac{1}{2}k^2\left(1 - \frac{i}{kr}\right) \mathbf{r}_0 \cdot \mathbf{N}$$

Magnetic dipole:

$$\varphi = 0$$
$$\mathbf{A} = -\frac{e^{i(\omega t-kr)}}{r} ik\left(1 - \frac{i}{kr}\right) \mathbf{r}_0 \times \mathbf{M} \tag{1.88}$$

Here $\mathbf{r}_0$ is the unit vector in the direction of $\mathbf{r}$, and $N_s = \int \rho r^2\, dv$ is the sum of the diagonal elements of $\mathbf{N}$.

Calculation of the electric-field vector for the electric-dipole case gives

$$\mathbf{E} = \frac{e^{i(\omega t-kr)}}{r}\left[\left(k^2 - \frac{ik}{r}\right)(\mathfrak{J} - \mathbf{r}_0\mathbf{r}_0) \cdot \mathbf{P} + \frac{2ik}{r}\mathbf{r}_0\mathbf{r}_0 \cdot \mathbf{P}\right] \tag{1.89}$$

and so the state of polarization of the radiation is that of the transverse component of the electric-dipole moment. Also, the average value of the Poynting vector, giving the radiation flow at large distances, is

$$\mathbf{S}_{\text{av}} = \frac{c}{8\pi}\frac{k^4}{r^2}|(\mathfrak{J} - \mathbf{r}_0\mathbf{r}_0) \cdot \mathbf{P}|^2\mathbf{r}_0 \tag{1.90}$$

In particular, if $\mathbf{P}$ is real, then $(\mathfrak{J} - \mathbf{rr}) \cdot \mathbf{P}$ is real and so the radiation is linearly polarized in all directions, falling to zero intensity in directions parallel to $\pm\mathbf{P}$. The intensity varies as $\sin^2\theta$, where θ is the angle between $\mathbf{r}$ and $\mathbf{P}$. The entire rate of radiation through a large sphere in all directions is

$$\frac{(2\pi\sigma)^4}{3} cP^2 \tag{1.91}$$

Correspondingly the parts of the electric and magnetic field that are responsible for radiation flow in the electric-quadrupole field are

$$\mathbf{E} = \tfrac{1}{2}ik^3\frac{e^{i(\omega t-kr)}}{r} \cdot (\mathfrak{J} - \mathbf{r}_0\mathbf{r}_0) \cdot (\mathbf{N} \cdot \mathbf{r}_0)$$
$$\mathbf{H} = \tfrac{1}{2}ik^3\frac{e^{i(\omega t-kr)}}{r} r_0 \times (\mathbf{N} \cdot \mathbf{r}_0) \tag{1.92}$$

and the average energy flow is

$$\mathbf{S}_{\text{av}} = \frac{c}{32\pi}\frac{k^6}{r^2}|(\mathfrak{J} - \mathbf{r}_0\mathbf{r}_0) \cdot (\mathbf{N} \cdot \mathbf{r}_0)|^2\mathbf{r}_0 \tag{1.93}$$

12. Quantization of the Radiation Field

Description of the interaction of radiation and matter requires a quantum modification of the theory of electromagnetic waves [10]. This is usually done by representing the field in terms of a vector potential $\mathbf{A}$,

$$\mathbf{E} = -\frac{1}{c}\dot{\mathbf{A}} \qquad \mathbf{H} = \text{curl } \mathbf{A} \tag{1.94}$$

where div $\mathbf{A} = 0$, and $\mathbf{A}$ satisfies the wave equation

$$\nabla^2\mathbf{A} - c^{-2}\ddot{\mathbf{A}} = 0 \tag{1.95}$$

If boundary conditions are assigned for a particular region of space this will define proper values k_r^2 and proper vector wave functions $\mathbf{A}_r$ satisfying

$$\nabla^2\mathbf{A} + k_r^2\mathbf{A}_r = 0 \tag{1.96}$$

which are orthogonal functions over the region of space in question. The general field satisfying the boundary conditions is then an expansion in terms of these orthogonal proper functions,

$$\mathbf{A} = \Sigma q_r\mathbf{A}_r \tag{1.97}$$

where the field equations require that each q_r vary harmonically with the time with radian frequency $\omega_r = ck_r$.

Quantization of the field consists basically in regarding these q as quantum-mechanical coordinates of a set of equivalent harmonic oscillators (Part 2, Chap. 6). This is sometimes called *second quantization* in the sense that the boundary conditions have determined one allowed-value problem in determining the k_r and the $\mathbf{A}_r$, and there is another quantizing procedure in treating the amplitude q_r as a quantum-mechanical coordinate. The allowed energy levels become

$$W = \sum_r \left(n_r + \frac{1}{2}\right)\hbar\omega_r \tag{1.98}$$

in which n_r is the harmonic-oscillator quantum number associated with the rth field mode, and the associated wave function Ψ is an infinite product of harmonic-oscillator wave functions $u(n_r,q_r)$ for each coordinate,

$$\Psi(n_r \cdots q_r \cdots) = \prod_r u(n_r,q_r) \tag{1.99}$$

The wave function Ψ representing a general state of the radiation field is

$$\Psi = \sum_{n_r} c(n_r \cdots)\psi(n_r \cdots)e^{-iW(n_r)/\hbar} \tag{1.100}$$

where the summation is over all values of n_r for each coordinate.

In (1.98) the lowest energy state (all $n_r = 0$) is itself infinite, but this causes no difficulty because only differences in energy occur in actual physical calculations. For this reason it is customary to omit the $\frac{1}{2}h\omega_r$ terms from (1.98).

In nearly all cases the first quantization is merely a mathematical artifice, the purpose of which is to replace the essentially continuous spectrum of radiation-frequency modes by a denumerably infinite set. This is analogous to replacing a Fourier integral by a

Fourier series whose period is large compared with the range over which the function has appreciable values. A convenient choice for the $\mathbf{A}_r$ is thus a set of functions representing plane-polarized standing plane waves with period L in x, y, and z, where L is large.

In this case the allowed values of the wave vector $\mathbf{k}$ are

$$\mathbf{k} = \frac{2\pi}{L}(r_x\mathbf{i} + r_y\mathbf{j} + r_z\mathbf{k}) \tag{1.101}$$

in which the r are positive or negative integers. For each k there will be standing-wave functions of type $\cos \mathbf{k}\cdot\mathbf{r}$ and $\sin \mathbf{k}\cdot\mathbf{r}$, but as no new wave functions are obtained by considering $-\mathbf{k}$ as well as $\mathbf{k}$, it is essential in (1.101) to restrict one of the r, say, r_z, to positive values. Associated with each $\mathbf{k}$, one may choose arbitrarily two unit vectors such that $\mathbf{e}_1$, $\mathbf{e}_2$, and $\mathbf{k}$ form axes of a right-handed orthogonal coordinate system. Thus with each $\mathbf{k}$ there are four different wave functions, and so four q, corresponding to the cos or sin forms and to plane polarization along $\mathbf{e}_1$ and $\mathbf{e}_2$, respectively. Normalizing to unity on the basic cube, $0 < x < L$, $0 < y < L$, $0 < z < L$, it is convenient to use the normalized functions

$$\varphi_{11\mathbf{k}} = \sqrt{\frac{8\pi}{L^3}}\,\mathbf{e}_1 \cos \mathbf{k}\cdot\mathbf{r} \qquad \varphi_{21\mathbf{k}} = \sqrt{\frac{8\pi}{L^3}}\,\mathbf{e}_2 \cos \mathbf{k}\cdot\mathbf{r}$$

$$\varphi_{12\mathbf{k}} = \sqrt{\frac{8\pi}{L^3}}\,\mathbf{e}_1 \sin \mathbf{k}\cdot\mathbf{r} \qquad \varphi_{22\mathbf{k}} = \sqrt{\frac{8\pi}{L^3}}\,\mathbf{e}_2 \sin \mathbf{k}\cdot\mathbf{r} \tag{1.102}$$

So the first two-valued index refers to the state of plane polarization and the second to the choice of cos or sin forms. Analogously one may write $q_{11\mathbf{k}}$ for the amplitude associated with the function $\varphi_{11\mathbf{k}}$ in the expansion of $\mathbf{A}$.

Thus

$$\mathbf{A} = -c\Sigma q_{ab\mathbf{k}}\varphi_{ab\mathbf{k}} \tag{1.103}$$

where the summation is over the two values 1 and 2 of a and b and over the values of $\mathbf{k}$ given by (1.101). The series for $\mathbf{E}$ is

$$\mathbf{E} = \Sigma \dot{q}_{ab\mathbf{k}}\varphi_{ab\mathbf{k}} \tag{1.104}$$

For $\mathbf{H}$ and for application of the dynamical field equations (1.7) one needs the result

$$\operatorname{curl}\begin{bmatrix}\varphi_{11}\\ \varphi_{21}\\ \varphi_{12}\\ \varphi_{22}\end{bmatrix} = \mathbf{k}\begin{bmatrix}0 & 0 & 0 & -1\\ 0 & 0 & 1 & 0\\ 0 & 1 & 0 & 0\\ 1 & 0 & 0 & 0\end{bmatrix}\begin{bmatrix}\varphi_{11}\\ \varphi_{21}\\ \varphi_{12}\\ \varphi_{22}\end{bmatrix} \tag{1.105}$$

or

$$\operatorname{curl} \varphi_{ab\mathbf{k}} = k(-1)^{a+b+1}\varphi_{a+1,b+1,\mathbf{k}} \tag{1.105a}$$

The dynamical field equations

$$\dot{\mathbf{E}} = c \operatorname{curl} \mathbf{H} \qquad \text{and} \qquad \dot{\mathbf{H}} = -c \operatorname{curl} \mathbf{E}$$

are thus, in Hamiltonian form,

$$\dot{p}_{ab\mathbf{k}} + (ck)^2 q_{ab\mathbf{k}} = 0 \qquad \dot{q}_{ab\mathbf{k}} = p_{ab\mathbf{k}} \tag{1.106}$$

the Hamiltonian function being

$$H = \frac{1}{2}\sum [p^2{}_{ab\mathbf{k}} + (ck)^2 q^2{}_{ab\mathbf{k}}] \tag{1.107}$$

Treated quantum-mechanically, this will have the allowed values in (1.98). The electromagnetic-field energy in the basic cube is the volume integral of $(\mathbf{E}^2 + \mathbf{H}^2)/8\pi$, which is just equal to the Hamiltonian with the choice of normalizing factor already made in (1.102).

References

1. Stratton, J. A.: "Electromagnetic Theory," McGraw-Hill, New York, 1941. Born, M., and E. Wolf: "Principles of Optics," Pergamon Press, New York, 1959. Jenkins, F. A., and H. E. White: "Fundamentals of Optics," 2d ed., McGraw-Hill, New York, 1950. Rossi, Bruno: "Optics," Addison-Wesley, Reading, Mass., 1957. Strong, John: "Concepts of Classcal Optics," Freeman, San Francisco, 1958. Stone, John M.: "Radiation and Optics," McGraw-Hill, New York, 1963.
2. Shurcliff, William A.: "Polarized Light, Production and Use," Harvard University Press, Cambridge, Mass., 1962. Shurcliff, William A. and Stanley Ballard: "Polarized Light," Van Nostrand, Princeton, N.J., 1964. Fano, U.: *J. Opt. Soc. Am.*, **39**: 859 (1949). McMaster, W. H.: *Am. J. Phys.*, **22**: 351 (1954). Walker, M. J.: *Am. J. Phys.*, **22**: 170 (1954). Jerrard, H. G.: *J. Opt. Soc. Am.*, **44**: 634 (1954).
3. Ketteler, E.: "Theoretische Optik," p. 122, Vieweg-Verlag, Brunswick, Germany, 1885. Lorentz, H. A.: *Ann. Physik*, **46**: 244 (1892). Fry, T. C.: *J. Opt. Soc. Am.*, **15**: 137 (1927); **16**: 1 (1928).
4. Simon, I.: *J. Opt. Soc. Am.*, **41**: 336 (1951).
5. Lowry, T. M.: "Optical Rotatory Power," Longmans, New York, 1935. Condon, E. U.: *Revs. Mod. Phys.*, **9**: 432 (1937).
6. Planck, M.: *Ann. Physik*, **4**: 553 (1901); see also, "The Theory of Heat Radiation," Blakiston, Philadelphia, 1914; Westphal, W.: "Handbuch der Astrophysik," vol. 3, Wärmestrahlung, Springer, Berlin, 1930.
7. Lowan, A. N., and G. Blanch: *J. Opt. Soc. Am.*, **30**: 70 (1940).
8. Einstein, A.: *Physik, Z.*, **18**: 121 (1917). Dirac, P. A. M.: *Proc. Roy. Soc. (London)*, **A106**: 581 (1924). Eddington, A. S.: "The Internal Constitution of the Stars," chap. 3, Cambridge University Press, New York, 1930, Van Vleck, J. H.: *Phys. Rev.*, **24**: 330, 347 (1924).
9. Chandrasekhar, S.: "Radiative Transfer," Dover, New York, 1960. Aller, L. H.: "Astrophysics: the Atmospheres of the Sun and Stars," 2d ed., chap. 7, Ronald, New York, 1963. Kourganoff, V.: "Basic Methods in Transfer Problems," Dover, New York, 1963.
10. Dirac, P. A. M.: *Proc. Roy. Soc. (London)*, **A114**: 237, 710 (1927). Fermi, E.: *Revs. Mod. Phys.*, **4**: 131 (1931). Heitler, W.: "Quantum Theory of Radiation," 3d ed., chap. 2, Oxford University Press, Fair Lawn, N.J., 1954.

Chapter 2

Geometrical Optics

By MAX HERZBERGER, Research Laboratories, Eastman Kodak Company, and Eidg. Technische Hochschule

1. Introduction

Geometrical optics can be built as an exact science, starting from Fermat's principle and deducing the laws of image formation in optical systems. Or it can be considered as an applied science, which studies the ailments of optical systems and searches for ways of curing them. As such it has some resemblance to the field of medicine, trying hard to find a scientific basis for its intuitions and "hunches." In presenting this side of optics we divide the field parallel to the fields of medicine. First we treat the *anatomy* of optical systems: ray tracing and the basic laws of image formation (Secs. 5 to 16). Next, we deal with problems of *diagnosis:* the problem of analyzing a given optical system (Secs. 17 to 28). Then attempts to combat these ailments are considered, by *therapy* (Sec. 29) and by *prophylaxis* (Secs. 30 to 32). General theory is taken up initially (Secs. 2 to 4).

I. GENERAL THEORY

2. Optical Form of the General Variation Problem

The general variation problem considers an integral

$$E = \int L(X_i, X_i', t)\, dt \tag{2.1}$$

where X_i is a point in k-dimensional space, t is an arbitrary parameter, usually the time, and X'_i is the differential quotient of X_i with respect to t, the integral being taken on a curve C connecting two points P_1 and P_2. The problem is to find the curves C such that E is an extremum.

We may alternately consider the problem in $(k + 1)$-dimensional space, taking t as one of the variables, and considering X_i and t as functions of a parameter s which for the moment is left arbitrary. Then

$$\begin{aligned} x_i &= X_i & \frac{dx_i}{ds} &= \dot{x}_i = X'_i \dot{x}_{k+1} \\ x_{k+1} &= t & \frac{dt}{ds} &= \dot{x}_{k+1} \end{aligned} \tag{2.2}$$

and $$E = \int L\left(x_i, \frac{\dot{x}_i}{\dot{x}_{k+1}}, x_{k+1}\right) \dot{x}_{k+1}\, ds = \int n\, ds \tag{2.3}$$

where $$n = L\dot{x}_{k+1} \tag{2.4}$$

As a function of the $\dot{x}_i$, n is homogeneous of first order. So E is independent of the chosen parameter, and we can take as parameter s the arc of length along the extremal.

Thus the variation problem is transformed into one in which the quantity n in (2.4) is a function of two "$(k + 1)$-dimensional vectors," the point vector $\mathbf{a}$ with the coordinates x_i and the unit vector $\dot{\mathbf{a}}$ tangential to the extremal with coordinates $\dot{x}_i$. The function n, the *refractive index*, as a function of $\dot{\mathbf{a}}$, is homogeneous of first order, which means that the *momentum vector* $\mathbf{s}$ with coordinates $\partial n/\partial \dot{x}_i$ obeys Euler's equation,

$$\mathbf{s}\dot{\mathbf{a}} = \frac{\partial n}{\partial \dot{x}_i} \dot{x}_i = n \tag{2.5}$$

using the summation convention with regard to the repeated index. So instead of (2.3) the vector integral is

$$E = \int n\, ds = \int \mathbf{s} \frac{d\mathbf{a}}{ds}\, ds = \int \mathbf{s}\, d\mathbf{a} \tag{2.6}$$

Thus the general variation problem is reduced to the study of an optical problem in $k + 1$ dimensions.

3. General Problem of Geometrical Optics

The general problem of geometrical optics is that discussed in the preceding paragraph if the space is three-dimensional. We assume the space to be divided into a finite number of media with any two of these media separated by a smooth surface, i.e., a surface which is continuous and whose tangential plane and curvature change continuously.

In the inside of each of these media a function n, the refractive index, is given, where n shall be a function (continuous and differentiable as often as necessary) of the position vector $\mathbf{a}$ (from an arbitrary but fixed origin) and the direction vector $\dot{\mathbf{a}}$. The function n shall be homogeneous of first order with respect to the direction cosines dx_i/ds.

For each position and direction inside the media there is given a momentum vector $\mathbf{s}$, which in optics we usually call the *normal vector*, such that

$$\mathbf{s}\dot{\mathbf{a}} = n \tag{2.7}$$

We then find the equations of the extremals, or light rays, from the Euler differential equations, which can be written in vector form, introducing the gradient vector

$$\mathbf{g} = \frac{\partial n}{\partial x_i} \tag{2.8}$$

as

$$\dot{\mathbf{s}} = \mathbf{g} \tag{2.9}$$

or, in coordinates,

$$\frac{d}{ds}\left(\frac{\partial n}{\partial \dot{x}_i}\right) = \frac{\partial n}{\partial x_i} \tag{2.10}$$

Since n is a function of x_i and $\dot{x}_i$, these equations are equivalent with the set of second-order differential equations,

$$\frac{\partial^2 n}{\partial \dot{x}_i\, \partial \dot{x}_k}\ddot{x}_k + \frac{\partial^2 n}{\partial \dot{x}_i\, \partial x_k}\dot{x}_k = \frac{\partial n}{\partial x_i} \tag{2.11}$$

These equations can be shown to give in general at each point and in each direction a well-defined light ray.

In the special case that, within one medium, n is *homogeneous*, that is, $\mathbf{g} = \partial n/\partial x_i = 0$, we find from (2.11) $\ddot{x}_k = 0$, and so the rays form straight lines. Equation (2.9) shows that along these straight lines the normal vector remains constant. The normal vector, however, does not necessarily have the direction of the ray, i.e., it is not necessary to have $\dot{\mathbf{a}}$ proportional to $\mathbf{s}$.

That is, however, the case if n is the same for all directions. Then because of the homogeneity condition

$$n = n_0(x_i)\sqrt{\dot{x}_i^2} \tag{2.12}$$

or

$$\mathbf{s} = \frac{\partial n}{\partial \dot{x}_i} = n_0\dot{x}_i = n_0\dot{\mathbf{a}} \tag{2.13}$$

So in this case we can say that the normal has the direction of the (in general curved) light ray. Such a medium is called an *isotropic* medium.

In general we shall assume n to be homogeneous and isotropic. The light transport through crystals forms an example of an anisotropic medium which is homogeneous; the light path through the atmosphere gives an example of light going through an inhomogeneous medium.

In a homogeneous isotropic medium light goes in straight lines, and the normal vector has the direction of the light ray.

Some of the basic laws of optics hold in anisotropic, inhomogeneous media, and we therefore shall describe them before restricting ourselves to the simplest and most important practical case.

4. Characteristic Function of Hamilton. Laws of Fermat and of Malus-Dupin. Descartes' Law of Refraction. Lagrange Bracket

W. R. Hamilton [14]* discovered the following important theorem: If a one-, two-, or higher-dimensional manifold of light rays is given and we assume on each ray an initial point (vector $\mathbf{a}$) and a terminal point (vector $\mathbf{a}'$) such that the initial and terminal points form continuous and continuously differentiable manifolds (according to the surface parameters), then we always have the equation

$$dE = \mathbf{s}'\, d\mathbf{a}' - \mathbf{s}\, d\mathbf{a} \tag{2.14}$$

That is, the expression on the right side forms a total differential, and it is the differential of the light path $E = \int n\, ds$ taken from initial point $\mathbf{a}$ along the extremal to the final point $\mathbf{a}'$. This expression can be said to contain all the basic laws of optics. It is a direct consequence of (2.6). Since it contains only the differentials $d\mathbf{a}$ and $d\mathbf{a}'$, we could even choose different origins for the initial and the terminal manifolds, a procedure which is convenient in practical applications. Let us apply this formula to some special cases.

Consider a continuous system of light rays starting at an initial point $\mathbf{a}$, and assume that they all go through the same terminal point $\mathbf{a}'$. This means,

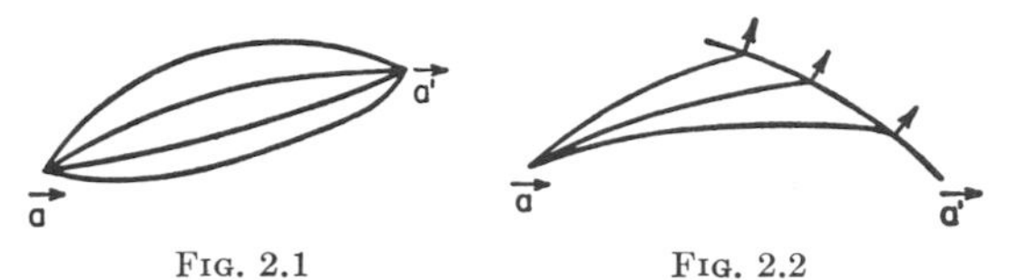

FIG. 2.1 FIG. 2.2

FIG. 2.1. Theorem of Fermat.
FIG. 2.2. Normal surface.

since $d\mathbf{a} = d\mathbf{a}' = 0$, that $dE = 0$, or the *theorem of Fermat.* If a continuous manifold of rays starting at an object point is united at an image point, the optical path between the object and image points ($\int n\, ds$) must be the same on all rays.

On the other hand, consider the rays coming from a point $d\mathbf{a} = 0$ and a manifold of terminal points $\mathbf{a}'$ which have the same optical distance $\int n\, ds$ from $\mathbf{a}$, that is, $dE = 0$. Equation (2.14) gives

$$\mathbf{s}'\, d\mathbf{a}' = 0 \tag{2.15}$$

Thus the manifold of terminal points at each of its points is normal to the normal vector $\mathbf{s}$. Such a surface is called a *wave surface*, and such a system of rays, which has wave surfaces, is called a *normal* system. The wave surfaces in anisotropic media are normal not to the rays (vector $\dot{\mathbf{a}}'$) but to the normal vectors $\mathbf{s}$, and only in isotropic systems do the two everywhere have the same directions.

Equation (2.14) can be derived from (2.6) only for the interior of each medium. If we assume them (or prove them from the extremum principle) to hold generally, Eq. (2.14) permits us to coordinate the rays at points of the surfaces of discontinuity separating two media. Let $d\mathbf{a}$ and $d\mathbf{a}'$ represent two surface elements approaching each other at the separating surface from the two respective media. Then dE converges toward zero, and one finds

$$(\mathbf{s}' - \mathbf{s})\, d\mathbf{a} = 0 \tag{2.16}$$

which is equivalent to saying that $\mathbf{s}' - \mathbf{s}$ has the direction of the vector $\mathbf{o}$, normal to the surface of discontinuity at the point of intersection. Thus

$$\mathbf{s}' - \mathbf{s} = \Gamma\mathbf{o} \tag{2.17}$$

This equation or the vector equation

$$\mathbf{s}' \times \mathbf{o} = \mathbf{s} \times \mathbf{o} \tag{2.18}$$

* Numbers in brackets refer to References at end of chapter.

is equivalent to *Descartes' law of refraction* for a general medium. It permits one to compute $\mathbf{s}'$, if $\mathbf{s}$ and $\mathbf{o}$ are

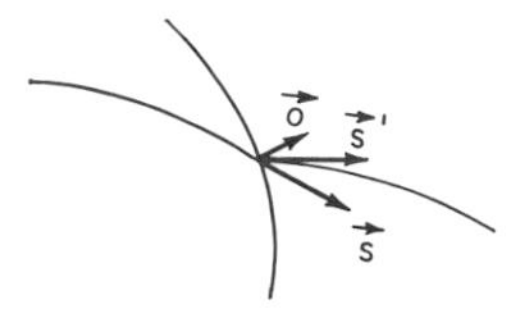

FIG. 2.3. General refraction law.

given, and thus to connect the extremals at the surface of discontinuity.

For two isotropic media $\mathbf{s}' = n'\mathring{\mathbf{a}}'$, and $\mathbf{s} = n\mathring{\mathbf{a}}$, and Eq. (2.18) gives

$$n \sin i = n' \sin i' \tag{2.19}$$

where i, i' are the angles between the incident ray and surface normal, and the emerging ray and surface normal, respectively.

The theorem of Lagrange. Given a two-dimensional manifold of rays; i.e., let the initial manifold $\mathbf{a}$, terminal manifold $\mathbf{a}'$, and the light path E be functions of two parameters u and v.

Equation (2.14) then gives

$$\begin{aligned} \frac{\partial E}{\partial u} &= \mathbf{s}'\frac{\partial \mathbf{a}'}{\partial u} - \mathbf{s}\frac{\partial \mathbf{a}}{\partial u} \\ \frac{\partial E}{\partial v} &= \mathbf{s}'\frac{\partial \mathbf{a}'}{\partial v} - \mathbf{s}\frac{\partial \mathbf{a}}{\partial v} \end{aligned} \tag{2.20}$$

or, representing differentiation with respect to a variable by the subscript,

$$\begin{aligned} E_u &= \mathbf{s}'\mathbf{a}'_u - \mathbf{s}\mathbf{a}_u \\ E_v &= \mathbf{s}'\mathbf{a}_v - \mathbf{s}\mathbf{a}_v \end{aligned} \tag{2.21}$$

We can eliminate E by differentiating the first equation with respect to v and the second with respect to u and equating. This gives

$$\mathbf{s}'_u\mathbf{a}'_v - \mathbf{s}'_v\mathbf{a}'_u = \mathbf{s}_u\mathbf{a}_v - \mathbf{s}_v\mathbf{a}_u \tag{2.22}$$

an equation used by Lagrange in many fields of mathematics. The quantity on either side is known in the theory of partial differential equations as the *Lagrange bracket.*

The differential expression on the right (and therefore on the left) is zero if and only if $\mathbf{s}\,d\mathbf{a}$ is a total differential in u and v. This is the case if and only if the manifold of rays has one (and therefore infinitely many) wave surfaces, and this gives the generalized theorem of *Malus-Dupin-Gergonne: A normal system remains a normal system after refraction.*

In the following we shall restrict ourselves to optical systems consisting of homogeneous isotropic media.

II. ANATOMY

RAY TRACING

5. The Refraction Law

Here we show how to follow the path of a ray through a number of optical surfaces separating optical media, in each of which the refractive index has a constant value for light of a given color (wavelength). The problem of ray tracing can be defined as follows: The initial ray is given by the coordinates (x,y,z) of the initial point $\mathbf{a}$ and by the coordinates (ξ,η,ζ) of the direction vector $\mathbf{s}$, whose length equals n. The problem is to find the intersection point with the refracting surface and to find the direction of the refracted ray $\mathbf{s}'$.

To avoid ambiguities, make the following assumptions: Light is assumed to be going in the positive direction of the z axis, which for systems with symmetry of rotation shall coincide with the optical axis. The only exception to this rule is that light reflected an odd number of times shall have the negative direction of the z axis; that is, ζ is normally positive, but negative for an odd number of reflections.

Furthermore, a surface in general has two sides; we determine the vector $\mathbf{o}$ normal to the surface as the one whose z component is positive.

The refraction law

$$n \sin i = n' \sin i' \tag{2.23}$$

has in general two solutions since with i' the angle $\pi - i'$ forms a solution. But only one of these rays goes into the second medium, namely, the one for

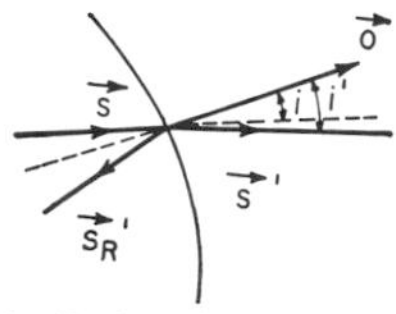

FIG. 2.4. Refraction and reflection.

which $\cos i$ and $\cos i'$ have the same sign. The other ray, for which $\cos i$ and $\cos i'$ have the opposite sign, lies in the same medium: i.e., it is the reflected ray. The reflected ray thus is given by $i = \pi - i'$, or

$$\begin{aligned} \sin i &= \sin i' \\ \cos i &= -\cos i' \end{aligned} \tag{2.24}$$

The vectorial form of the law of refraction and reflection is

$$\mathbf{s}' - \mathbf{s} = \Gamma\mathbf{o}$$

with

$$\begin{aligned} \Gamma &= n' \cos i' - n \cos i && \text{for refraction} \\ \Gamma &= -2n \cos i && \text{for reflection} \end{aligned} \tag{2.25}$$

The optical systems used in industry consist mostly of sets of surfaces with a common axis of rotation, and these surfaces are for practical reasons usually spheres or planes, and very seldom second-order surfaces or higher.

6. Tracing a Ray through a Surface of Rotation

The point where the surface of rotation intersects the system axis is called the *vertex* of the surface. If one chooses the origin at the vertex and the z axis in the direction of the optical axis, the entering ray may be given by the vector $\mathbf{a}(x,y,0)$ of the intersection point of the ray with the vertex plane $z = 0$ and by the direction vector $\mathbf{s}$ of length n (n = refractive index) and coordinates ξ, η, ζ with $\xi^2 + \eta^2 + \zeta^2 = n^2$. The intersection point $\mathbf{b}(X,Y,Z)$ with the refracting surface is to be found, i.e., the scalar quantity λ such that

$$\mathbf{a} + \lambda\mathbf{s} = \mathbf{b} \tag{2.26}$$

lies on the surface. The second problem is to find the normal vector at the point of intersection. This normal vector lies in the plane determined by the axis and the point of intersection. If $\mathbf{n}$ denotes the

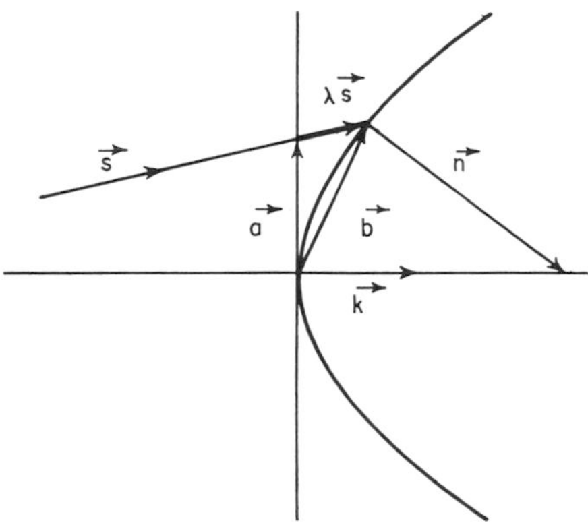

FIG. 2.5. Finding intersection point and normal vector.

vector from the surface point to its intersection point with the axis, the above fact is equivalent to the existence of two constants α, β such that*

$$\mathbf{n} = \alpha\mathbf{b} + \beta\mathbf{k} \tag{2.27}$$

The refraction law finally requires finding a scalar ϕ such that

$$\mathbf{s}' = \mathbf{s} - \phi\mathbf{n} \tag{2.28}$$

Knowing α, β, ϕ, and λ, one determines the refracted ray by one of its points (vector $\mathbf{b}$) and its direction vector $\mathbf{s}'$ of length n'.

Combining the three equations,

$$\begin{aligned} \mathbf{b} &= \mathbf{a} + \lambda\mathbf{s} \\ \mathbf{s}' &= -\phi\alpha\mathbf{a} + (1 - \lambda\phi\alpha)\mathbf{s} - \phi\beta\mathbf{k} \\ \mathbf{k} &= \mathbf{k} \end{aligned} \tag{2.29}$$

Equations (2.29) have a determinant equal to unity, which is equivalent to saying that the determinant products (see Part 1, Chap. 9) are equal,

$$[\mathbf{b}\mathbf{s}'\mathbf{k}] = [\mathbf{a}\mathbf{s}\mathbf{k}] \tag{2.30}$$

Methods of calculating the quantities λ, α, β, ϕ have been studied a great deal in the literature. As long as logarithms were used, great ingenuity was applied to transform the formulas to a pattern fitted for logarithmic calculation. The users of electrical machines showed preference for algebraic formulas. High-speed computers suggest the use of iteration formulas. Much effort has also been applied to improve the accuracy of the computation. For this and similar topics consult the papers by T. Smith [37] on ray tracing.

* In case of the surface normal being parallel to the axis, $\mathbf{n}$ shall designate the unit vector in its direction, that is, $\mathbf{n} = \mathbf{k}$.

The rotation surface is generally given by its vertex equation. Abbreviate with

$$\begin{aligned} \tfrac{1}{2}\mathbf{b}^2 &= \tfrac{1}{2}(X^2 + Y^2 + Z^2) = U \\ \mathbf{bk} &= Z \end{aligned} \tag{2.31}$$

and let

$$f(U,Z) = 0 \tag{2.32}$$

be the equation of the surface. The value λ in (2.29) is found by substituting

$$\begin{aligned} \mathbf{b}^2 &= (\mathbf{a} + \lambda\mathbf{s})^2 = \mathbf{a}^2 + 2\lambda\mathbf{as} + \lambda^2 n^2 \\ \text{and} \quad \mathbf{bk} &= (\mathbf{a} + \lambda\mathbf{s})\mathbf{k} = \lambda\zeta \end{aligned} \tag{2.33}$$

into (2.32) and finding the smallest value of λ which satisfies this equation. Having found λ and thus vector $\mathbf{b}$, we find $\mathbf{n}$ by

$$\mathbf{n} = \frac{\partial f}{\partial Z}\mathbf{k} + \frac{\partial f}{\partial U}\mathbf{b} \tag{2.34}$$

Finally ϕ can be shown to be given (for the case of refraction) by

$$\phi = \frac{1}{\mathbf{n}^2}\left[\sqrt{n^2\mathbf{n}^2 - (\mathbf{s} \times \mathbf{n})^2} - \sqrt{n'^2\mathbf{n}^2 - (\mathbf{s} \times \mathbf{n})^2}\right] \tag{2.35}$$

which solves the problem in question.

7. Special Surfaces

For a *plane* one obtains

$$\begin{aligned} Z &= 0 \\ \mathbf{n} &= \mathbf{k} \\ \phi &= \zeta - \sqrt{n'^2 - n^2 + \zeta^2} \end{aligned} \tag{2.36}$$

The vertex equation of a *sphere* is

$$Z - \frac{1}{r}U = 0 \tag{2.37}$$

(The radius of a sphere in optics is counted from vertex to center, i.e., positive, if the system is convex toward the entering ray.) Thus the unit normal vector is given by

$$\mathbf{k} - \frac{1}{r}\mathbf{b} \tag{2.38}$$

The plane therefore can be considered as a sphere with zero curvature ($\rho = 1/r \to 0$).

A *general second-order surface* is given by

$$Z + (B - A)Z^2 = 2BU \tag{2.39}$$

and is an ellipsoid if $BA > 0$, a hyperboloid for $BA < 0$, a paraboloid for $A = 0$, a sphere for

$$B = A = \tfrac{1}{2}\rho$$

The normal vector $\mathbf{n}$ is given by

$$\mathbf{n} = [1 + 2Z(B - A)]\mathbf{k} - 2B\mathbf{b} \tag{2.40}$$

and ϕ again by (2.35).

8. Transfer Formulas

Having found the point of intersection with the surface (vector $\mathbf{b}$) and the direction vector $\mathbf{s}'$, it is necessary to shift the origin to the next vertex and compute the intersection point with the next vertex plane (the plane through the next vertex perpendicular to the axis). Let d be the distance between the vertex planes; the transformation formulas then are

$$\begin{aligned} \mathbf{a}' &= \mathbf{b} + \lambda'\mathbf{s}' - d\mathbf{k} \\ \mathbf{s}' &= \mathbf{s}' \\ \mathbf{k} &= \mathbf{k} \end{aligned} \tag{2.41}$$

where, since the z component of $\mathbf{a}'$ vanishes, one obtains

$$Z + \lambda'\zeta' - d = 0$$

$$\lambda' = \frac{d - Z}{\zeta'} \tag{2.42}$$

For this transformation, too,

$$[\mathbf{a}'\mathbf{s}'\mathbf{k}] = [\mathbf{b}\mathbf{s}'\mathbf{k}] \tag{2.43}$$

and so the optical *moment* of a ray around the axis, the triple product of a point vector from an arbitrary origin on the axis, the direction vector $\mathbf{s}$ (of length n), and vector $\mathbf{k}$ along the axis, is invariant.

A ray for which this *moment* vanishes is called a *meridional ray;* it lies in a plane through the axis (*meridian plane*) and will not leave this plane after multiple refraction and reflection. There are short cuts for the tracing formulas in case of a meridional ray.

A nonmeridional ray is called a *skew* ray. Many designers incorrectly assume that the knowledge of the path of meridional rays through an optical system is sufficient for the study of the image formation. The value in (2.43) may be called the *skewness* of the ray and designated by the letter σ.

9. General Formulas. Diapoint Computation

Given an optical system, choose arbitrary origins on the image and object sides such that both origins lie on the optical axis, for instance, at the first and last vertex, respectively. If $\mathbf{a}$ is a vector from the object origin to an arbitrary *initial* point on the object ray, and analogously $\mathbf{a}'$ a vector from the image origin to an arbitrary terminal point on the image side, and $\mathbf{s}$, $\mathbf{s}'$ the direction vectors of length n, n' along the rays, we then have equations of the form

$$\begin{aligned} \mathbf{a}' &= \alpha\mathbf{a} + \beta\mathbf{s} + C_1\mathbf{k} \\ \mathbf{s}' &= \gamma\mathbf{a} + \delta\mathbf{s} + C_2\mathbf{k} \\ \mathbf{k} &= \mathbf{k} \end{aligned} \tag{2.44}$$

where the coefficients can be obtained by multiplying systems of transformation matrices corresponding to the refraction at each surface (2.28) and to the transition from surface to surface (2.41).

The existence of the scalar invariant

$$[\mathbf{a}'\mathbf{s}'\mathbf{k}] = [\mathbf{a}\mathbf{s}\mathbf{k}] \tag{2.45}$$

requires that the determinant of (2.44) equals unity, i.e.,

$$\alpha\delta - \beta\gamma = 1 \tag{2.46}$$

This important fact suggests a special one-to-one correspondence between the points of the object and image rays. One coordinates to an arbitrary initial point $\mathbf{a} + \lambda\mathbf{s}$ that point $\mathbf{a}' + \lambda'\mathbf{s}'$ on the image ray which lies in the same plane through the axis, i.e., for which

$$[\mathbf{a} + \lambda\mathbf{s}, \mathbf{a}' + \lambda'\mathbf{s}', \mathbf{k}] = 0 \tag{2.47}$$

This point is called the *diapoint* of the initial point. It is found that the determinant

$$\begin{vmatrix} \alpha + \lambda'\gamma & 1 \\ \beta + \lambda'\delta & \lambda \end{vmatrix} = 0 \tag{2.48}$$

or that

$$\lambda\lambda'\gamma + \lambda\alpha - \lambda'\delta - \beta = 0$$

$$\lambda' = -\frac{\lambda\alpha - \beta}{\lambda\gamma - \delta} \qquad \lambda = \frac{\lambda'\delta + \beta}{\lambda'\gamma + \alpha} \tag{2.49}$$

One then obtains

$$\begin{aligned} \mathbf{a}'_D &= \mathbf{a}' + \lambda'\mathbf{s}' = m(\mathbf{a} + \lambda\mathbf{s}) + C'\mathbf{k} \\ \text{where} \quad m &= -\frac{1}{\lambda\gamma - \delta} = \alpha + \lambda'\gamma \\ C' &= C_1 + \lambda'C_2 \end{aligned} \tag{2.50}$$

The quantity m is called the *diamagnification.*

To every point of the object ray belongs a diapoint on the image ray. This statement is correct only if we coordinate to the point $\mathbf{a} + \lambda\mathbf{s}$ given by

$$\lambda = \frac{\delta}{\gamma} \tag{2.51}$$

the infinite point of the image ray, since λ' and m converge toward infinity for this point. The point where $\lambda = \delta/\gamma$ is called the *diafocal point* on the object ray. Analogously (2.50) shows that to the point given on the image ray by

$$\lambda' = -\frac{\alpha}{\gamma} \tag{2.52}$$

belongs no finite object point. This point is called the *image-side diafocal point* ($m \to 0$).

Equations (2.44) and the relationship discussed assume a very simple form if the initial and terminal points are chosen suitably.

Initial and terminal points ($\lambda = \lambda' = 0$) *are diafocal points.* Here (2.51) and (2.52) give $\alpha = \delta = 0$; and because of (2.46)

$$\beta\gamma = -1 \tag{2.53}$$

Thus (2.44) become

$$\begin{aligned} \mathbf{a}' &= -\frac{1}{\gamma}\mathbf{s} + C_1\mathbf{k} \\ \mathbf{s}' &= \gamma\mathbf{a} + C_2\mathbf{k} \\ \mathbf{k} &= \mathbf{k} \end{aligned} \tag{2.54}$$

and (2.49) and (2.50) transform to

$$\begin{aligned} \lambda\lambda'\gamma^2 &= -1 \\ m &= -\frac{1}{\lambda\gamma} = \lambda'\gamma \end{aligned} \tag{2.55}$$

The quantity $(-\gamma)$ in (2.54) is called in optics the *power* ϕ of the system for the ray in question. The distance of the object point and its diapoint from the respective diafocal points and the diamagnification m are related by (2.55), which are generalizations of well-known formulas of *Newton.*

Initial and terminal points ($\lambda = \lambda' = 0$) *are conjugated points of magnification* 1. Then one has by (2.50)

$$\delta = \alpha = 1 \qquad \beta = 0 \tag{2.56}$$

or instead of (2.44)

$$\begin{aligned} \mathbf{a}' &= \mathbf{a} + \qquad C_1\mathbf{k} \\ \mathbf{s}' &= \gamma\mathbf{a} + \mathbf{s} + C_2\mathbf{k} \\ \mathbf{k} &= \qquad\qquad \mathbf{k} \end{aligned} \tag{2.57}$$

and the distances λ, λ' of point, diapoint and their diamagnification are related by

$$\lambda\lambda'\gamma + \lambda - \lambda' = 0$$

or

$$\frac{1}{\lambda'} - \frac{1}{\lambda} = -\gamma \tag{2.58}$$

$$m = 1 + \lambda'\gamma = \frac{1}{1 - \lambda\gamma}$$

equations which are generalizations of some formulas of Euler.

In the case that the object and image ray gradually converge against the axis, the point-diapoint theory reduces to the theory of first-order image formation known under the name of *Gaussian optics* (Sec. 18).

BASIC TOOLS OF OPTICS

10. The Characteristic Functions

The preceding section showed how to trace a single ray through an optical system with symmetry of rotation. Other methods will now be discussed for giving a mathematical representation of the object-image ray coordination for all rays at once. Emphasis is on systems with an axis of symmetry and on the application of the laws of Hamilton and Lagrange to these systems.

The object and image rays are given as in the preceding chapters, by an initial point (vector $\mathbf{a}$ from the object origin on the axis) and the direction vector $\mathbf{s}$ (of length n) and by a terminal point $\mathbf{a}'$ and the direction vector $\mathbf{s}'$ of length n'.

Assume that, for all rays under consideration, $\mathbf{a}$, $\mathbf{a}'$, $\mathbf{s}$, and $\mathbf{s}'$ are continuous and continuously differentiable vector functions of the variables.

Let E be the light path from the initial to the terminal point; Hamilton's formula (2.14) then gives

$$dE = \mathbf{s}'\,d\mathbf{a}' - \mathbf{s}\,d\mathbf{a} \tag{2.59}$$

If the starting and terminal points are chosen so that through each initial and terminal point goes only one ray, for instance, if the initial and terminal points lie on a suitably chosen plane normal to the axis, (2.59) is equivalent to four equations. For instance, if $(x,y,0)$ and $(x',y',0)$ are the independent coordinates of $\mathbf{a}$ and $\mathbf{a}'$ and analogously $(\xi,\eta,\zeta)(\xi',\eta',\zeta')$ of $\mathbf{s}$ and $\mathbf{s}'$, (2.59) can be split up into

$$\begin{aligned} E_x &= -\xi \qquad & E_{x'} &= \xi' \\ E_y &= -\eta \qquad & E_{y'} &= \eta' \end{aligned} \tag{2.60}$$

as Bruns [9] has shown.

In a system with rotation symmetry E is a function of three symmetric functions of $\mathbf{a}$ and $\mathbf{a}'$, for instance, of $\frac{1}{2}\mathbf{a}^2 = e_1$, $\mathbf{a}\mathbf{a}' = e_2$, $\frac{1}{2}\mathbf{a}'^2 = e_3$, and Eqs. (2.60) become

$$\begin{aligned} \mathbf{s} &= -(E_1\mathbf{a} + E_2\mathbf{a}') + C_1\mathbf{k} \\ \mathbf{s}' &= E_2\mathbf{a} + E_3\mathbf{a}' + C_2\mathbf{k} \end{aligned} \tag{2.61}$$

where E_ν is the derivative of E with respect to e_ν and the functions C_1 and C_2 can be found by the condition that $\mathbf{s}$ and $\mathbf{s}'$ must have the lengths n and n', respectively.

With (2.59) the functions

$$\begin{aligned} dV &= \mathbf{s}\,d\mathbf{a} + \mathbf{a}'\,d\mathbf{s}' \\ dW &= \mathbf{a}\,d\mathbf{s} - \mathbf{a}'\,d\mathbf{s}' \end{aligned} \tag{2.62}$$

are total differentials of their variables, and each can be split up into four equations, if, for V, no two rays from the same object point have parallel image rays or, for W, no two rays entering the system parallel to each other (same $\mathbf{s}$) leave the system parallel to each other (same $\mathbf{s}'$).

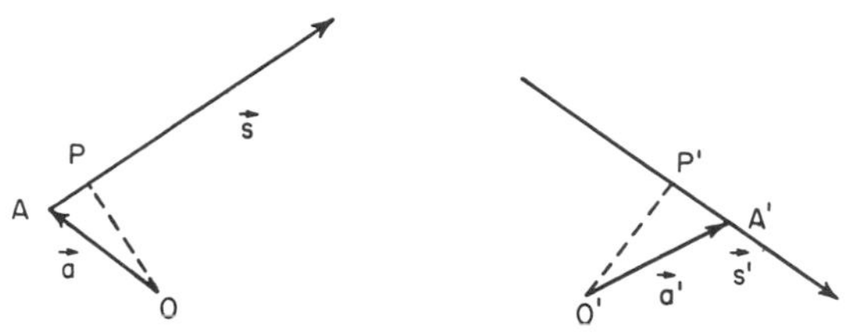

FIG. 2.6. Characteristic functions: E = optical path A to A' (point characteristic); $-V$ = optical path A to P' (mixed characteristic); W = optical path P to P' (angle characteristic).

The function E is called the *point characteristic* (*point eikonal*), V the *mixed characteristic* (*eikonal*), and W the *angle characteristic* (*eikonal*). The geometrical significance can be ascertained from

$$\begin{aligned} V + E &= (\mathbf{a}'\mathbf{s}') \\ E - W &= (\mathbf{a}'\mathbf{s}') - (\mathbf{a}\mathbf{s}) \end{aligned} \tag{2.63}$$

where $(\mathbf{as})$ is the length on the ray from the initial point to the foot point of the perpendicular dropped from object origin to object ray, with a corresponding meaning for $(\mathbf{a}'\mathbf{s}')$ on the image side.

Again, for systems with symmetry of rotation, we can consider V and W each as a function of three variables. Assuming the initial points as points of the plane $z = 0$, it is most convenient (for V) to take for the terminal points the diapoints using as coordinates

$$v_1 = \tfrac{1}{2}\mathbf{a}^2 = \tfrac{1}{2}(x^2 + y^2) \qquad v_2 = \mathbf{a}\mathbf{s}' = x\xi' + y\eta' \qquad v_3 = \zeta' \tag{2.64}$$

Then instead of (2.63)

$$\begin{aligned} \mathbf{a}' &= V_2\mathbf{a} + V_3\mathbf{k} \\ \mathbf{s} &= V_1\mathbf{a} + V_2\mathbf{s}' + C\mathbf{k} \end{aligned} \tag{2.65}$$

where again C is to be determined by the fact that $\mathbf{s}^2 = n^2$.

Comparing (2.65) with (2.50),

$$V_2 = m \tag{2.66}$$

where m is the diamagnification and

$$V_3 = z' \tag{2.67}$$

the longitudinal distance of the diapoint from the origin.

For W we again choose the initial point on the plane $z = 0$ and the terminal point on the plane $z' = 0$. As coordinates a good choice is

$$w_1 = \zeta \qquad w_2 = \mathbf{ss}' = \xi\xi' + \eta\eta' + \zeta\zeta' \qquad w_3 = \zeta' \tag{2.68}$$

with the result that $\mathbf{a}$ and $\mathbf{a}'$ can be calculated from

$$\begin{aligned} \mathbf{a} &= C\mathbf{s} + W_2\mathbf{s}' + W_1\mathbf{k} \\ \mathbf{a}' &= -(W_2\mathbf{s} + C'\mathbf{s} + W_3\mathbf{k}) \end{aligned} \tag{2.69}$$

where C and C' are functions determined by the fact that z and z', the axial components of $\mathbf{a}$ and $\mathbf{a}'$, are zero according to definition.

In these characteristic functions the ray is given for E by the vectors $\mathbf{a}$ and $\mathbf{a}'$, for V by $\mathbf{a}$ and $\mathbf{s}'$, and for W by $\mathbf{s}$ and $\mathbf{s}'$.

The attempt to compute the data of the image ray if those of the object ray are given leads in each of the above cases to an elimination problem which cannot be generally solved, i.e., the image theory for finite aperture and field cannot easily be coordinated to the results of ray tracing.

It is therefore natural to look for an analytic method of expressing the image data *directly* in terms of the object data.

11. The Direct Method

Given $\mathbf{a}$ and $\mathbf{s}$, to find $\mathbf{a}'$ and $\mathbf{s}'$. Let $\mathbf{a}$, the initial point, lie on a plane $z = 0$, analogously $\mathbf{a}'$ on an arbitrary image plane $z' = 0$. We then can find [see (2.44)] functions α, β, γ, δ with

$$\alpha\delta - \beta\gamma = 1 \tag{2.70}$$

such that

$$\begin{aligned} \mathbf{a}' &= \alpha\mathbf{a} + \beta\mathbf{s} + C_1\mathbf{k} \\ \mathbf{s}' &= \gamma\mathbf{a} + \delta\mathbf{s} + C_2\mathbf{k} \end{aligned} \tag{2.71}$$

the (unessential) functions C_1 and C_2 being determined by the fact that the z' component of $\mathbf{a}'$ is zero, and that the vector $\mathbf{s}'$ has the length n'.

Unfortunately α, β, γ, δ cannot be expressed in terms of the derivatives of a single characteristic function. However, Lagrange's formulas give all the differential relations between these quantities. Eliminating α from (2.70) and considering as variables

$$\begin{aligned} u_1 &= \tfrac{1}{2}(x^2 + y^2) \\ u_2 &= x\xi + y\eta \\ u_3 &= \tfrac{1}{2}(\xi^2 + \eta^2) \end{aligned} \tag{2.72}$$

the differential equations between α, β, γ, δ then can be written in the following way:

$$\begin{vmatrix} \beta_2 & 2u_1\gamma\gamma_2 + u_2(\gamma\delta_2 + \gamma_2\delta) + 2u_3\,\delta\delta_2 + \gamma\delta \\ \beta_3 & 2u_1\gamma\gamma_3 + u_2(\gamma\delta_3 + \gamma_3\delta) + 2u_3\,\delta\delta_3 + \delta^2 \end{vmatrix} = \begin{vmatrix} \delta_2 & 2u_1\gamma_2\alpha + \beta\gamma_2u_2 + \alpha\delta \\ \delta_3 & 2u_1\gamma_3\alpha + \beta\gamma_2u_2 + \beta\delta \end{vmatrix}$$

$$\begin{aligned} \beta_1 &= \frac{\begin{vmatrix} \alpha & \gamma_2 \\ \beta & \gamma_3 \end{vmatrix} + \begin{vmatrix} \beta_2 & u_2\gamma_1 + 2u_3\gamma_2 \\ \beta_3 & -2u_1\gamma_1 + 2u_3\gamma_3 - \gamma \end{vmatrix}}{-(2u_1\gamma_2 + u_2\gamma_3 + \delta)} \\ \delta_1 &= \frac{\begin{vmatrix} \delta_2 & u_2\gamma_1 + 2u_3\gamma_2 \\ \delta_3 & -2u_1\gamma_1 + 2u_3\gamma_3 - \gamma \end{vmatrix} + \begin{vmatrix} \gamma & \gamma_2 \\ \delta & \gamma_3 \end{vmatrix}}{-(2u_1\gamma_2 + u_2\gamma_3 + \delta)} \end{aligned} \tag{2.73}$$

This means that we can eliminate β_1 and δ_1 and obtain a symmetric relationship between β_2, γ_2, δ_2 and β_3, γ_3, δ_3.

LAWS OF IMAGE FORMATION

12. Image of a Point. Caustic Surface

In a system with symmetry of rotation the following possibilities for imaging a point exist:

a. The object point is sharply imaged; i.e., all its rays go through the same point, the *image point.* This point for reasons of symmetry must lie in the meridian plane (the plane through the object point and the axis); i.e., in this case the diapoints of the object point must coincide for all rays. (If the object wave is a plane wave, i.e., if the object point is at infinity, the meridian plane is the plane through the axis parallel to the entering rays.)

b. The object point (finite or infinite) is symmetrically imaged. In this case all the rays from the object point intersect a curve in the image space. This curve must, for reasons of symmetry, lie in the meridian plane; i.e., it is the locus of the diapoints.

Hamilton's formula (2.14) applied to such a manifold of rays proves that the rays through a single point of the diapoint curve form a cone symmetrical about the tangent to the diapoint curve; i.e., the image rays can be considered as consisting of a manifold of cones bisected by the diapoint curve which is the curve of their vertices. The image rays are, however, in general tangential to a second surface, the *caustic surface.* In this case the diapoint curve may be considered as a degenerated branch of the caustic surface.

c. The rays from the object point form a system of rays normal to its wave surface, i.e., a normal system of rays. The rays are tangential to a well-defined single surface, the caustic surface, and each ray is tangential in two points. The points of the caustic surface form a single surface and not two surfaces, as frequently stated in the literature.

These facts can be described analytically with the help of the mixed characteristic V [see (2.64) to (2.66)].

A fixed object point is given by a constant value of $v_1 = \frac{1}{2}(x^2 + y^2)$. For the rays from this point V becomes a function of v_2 and v_3 alone.

The coordinates of the diapoint are obtained from (2.65) as

$$\begin{aligned} x'_D &= V_2x \\ y'_D &= V_2y \\ z'_D &= V_3 \end{aligned} \tag{2.74}$$

The image is sharp if V_2 and V_3 are constant as functions of v_2 and v_3 or if

$$V_{22} = V_{23} = V_{33} = 0 \qquad \textit{sharpness condition} \tag{2.75}$$

The image is symmetric if z'_D is a function of $y'^2_D + x'^2_D$ or if V_3 is a function of V_2, that is, if

$$\begin{vmatrix} V_{22} & V_{23} \\ V_{23} & V_{33} \end{vmatrix} = 0 \qquad \textit{symmetry condition} \tag{2.76}$$

If (2.76) is not fulfilled, the image is of a general character.

The points $\mathbf{a}' + k\mathbf{s}'$ of the caustic are given by the following quadratic equation (k being the distance of the caustic from the diapoints divided by n'):

$$[\mathbf{a}'_2 + k\mathbf{s}'_2, \mathbf{a}'_3 + k\mathbf{s}'_3, \mathbf{s}'] = k^2[\mathbf{s}'_2\mathbf{s}'_3\mathbf{s}'] + k([\mathbf{a}'_2\mathbf{s}'_3\mathbf{s}'] + [\mathbf{s}'_2\mathbf{a}'_3\mathbf{s}']) + [\mathbf{a}'_2\mathbf{a}'_3\mathbf{s}'] = 0 \quad (2.77)$$

or, inserting the derivatives of V,

$$n'^2k^2 + k[V_{22}(2v_1n'^2 - v_2^2) - 2V_{23}v_2v_3 + V_{33}(n'^2 - v_3^2)] + [2v_1(n'^2 - v_3^2) - v_2^2](V_{23}V_{33} - V_{23}^2) = 0 \quad (2.78)$$

Equation (2.78) verifies the statement that

$$k_1 = k_2 = 0$$

for a *sharp* image ($V_{22} = V_{23} = V_{33} = 0$).

Analogously, for a *symmetric* image

$$(V_{22}V_{33} - V_{23}^2 = 0)$$

the diapoints ($k = 0$) form part of the caustic, the other part being given by

$$k = -\frac{1}{n'^2}[V_{22}(2v_1n'^2 - v_2^2) - 2V_{23}v_2v_3 + V_{33}(n'^2 - v_3^2)] \quad (2.79)$$

In the general case the caustic cannot be decomposed into two separate surfaces, the discriminant given by (2.78) not being a total square.

For an axis point ($x = y = v_1 = v_2 = 0$) the diapoints always lie on the axis; i.e., an axis point always has a symmetric image, with the system axis as line of symmetry.

13. Image of the Points of a Plane

The preceding section showed that, for an object point for which $v_1 = \frac{1}{2}(x^2 + y^2)$ has a fixed value (and therefore for the points x, y of the circle of points given by $x^2 + y^2 = 2v_1$), two functions, V_2 and V_3, of two variables give information about the image formation, and because of (2.74) we can state that V_2 gives the *lateral* aberration, whereas V_3 gives the *longitudinal* aberration. Even in the ideal case where every point of the plane is sharply imaged, i.e., where V_2 and V_3 for every value of v_1 are constant as functions of v_2 and v_3, we might not have a desirable image. If V_3 varies as a function of v_1, we find that the image points do not lie on a plane; i.e., we have *curvature of field.*

But even if V_3 vanishes, i.e., if V_3 is constant as a function of v_1, the diamagnification V_2 may still be dependent on it. In this case, we have different magnifications for different points in the object plane, which means that we have a distorted image of an object in this plane.

Thus one function V of three variables gives complete information about the image of the points of a plane. Especially its second-order derivatives V_{12}, V_{13}, V_{22}, V_{23}, V_{33}, regarded as functions of the three variables, are of importance. To give names to these error functions, we recall well-known concepts, which are here generalized. For the rays from the axial point (that is, $\mathbf{a} = 0$, $x = y = v_1 = v_2 = 0$), V is a function of v_3 alone. If $V_{33} \neq 0$, the rays from the axial point form cones in the image space with different vertices; i.e., the system has *spherical aberration.* If $V_{23} \neq 0$, the *magnification* changes from cone to cone. This is called *error against the sine condition.*

Thus we shall call the function V_{33} the *spherical-aberration function,* V_{23} the *sine error function,* and the function V_{22} the *error of astigmatism.* The error function

$$V_{22}V_{33} - V_{23}^2 = S \quad (2.80)$$

is called the *error against symmetry,* or the *coma error.* The function V_{13} is designated as the *curvature error* and the function V_{12} as the *error of distortion.*

All these names are generalizations of an approximation theory in which V is developed only to its quadratic terms, i.e., in which the V_{ik} are numbers and not functions. This theory has been developed by L. Seidel and will be sketched in Sec. 26.

14. The Image of the Points of Space

The preceding paragraphs have shown that the problem of analyzing the image of the points of a plane is excellently solved theoretically with the help of the function V. However, the use of this function has its disadvantages. It is, for instance, impossible to give V in closed form for even as simple a case as a single sphere. Moreover one cannot in closed form solve the elimination problem necessary to express the function V for a *shifted* object as a function of the V and its variables for the plane, $z = 0$.

Though this problem cannot be solved for V, it can be solved for its derivatives V_i and for the error functions V_{ik}. The method consists in expressing the V_i and V_{ik} with the help of the W_i and W_{ik}, and vice versa.

The geometric definition of W [see (2.63)] as the light path between the foot points of the perpendiculars dropped from the respective object and image origins to the object and image rays brings with it the fact that a change of origins by z, z', respectively, changes W linearly,

$$\bar{W} = W + zw_1 - z'w_3 \quad (2.81)$$

which shows that

$$\bar{W}_1 = W_1 + z \qquad \bar{W}_2 = W_2$$
$$\bar{W}_3 = W_3 - z' \qquad \bar{W}_{ik} = W_{ik} \quad (2.82)$$

showing that the W_{ik} are *invariant* against shift of origin.

We give some of the results without derivation. All the points of space can be sharply imaged if, and only if,

$$n'\mathbf{a}' = \pm n\mathbf{a} \quad (2.83)$$

In this case any plane is imaged sharply with the magnification $\pm(n/n')$ upon another plane.

If two object points (with finite difference in z value) are sharply imaged (in this case because of symmetry the points of two circles must be sharply imaged), it can be proved that the diamagnifications follow the equation

$$n'^2m_1m_2 = n^2 \quad (2.84)$$

Two surfaces can be sharply imaged only if they are of second order and the first object and second image surface and the second object and first image surface are similar in the ratio $\pm(n/n')$ (see ref. 7).

All concentric systems (systems of spheres with a common center) give a symmetric image for every point where the "pseudoaxis" (the line uniting the object point and the common center) is the axis of symmetry (see ref. 8).

Besides concentric systems there is only one type of freak system, the system with the angle characteristic

$$W = C \times \sqrt{kw_1w_3 + w_2 + (n'w_1 + nw_3)\sqrt{k(k+2)} + nn'(k+1)} \tag{2.85}$$

(where C and k are two constants), where every point has a symmetric image. In this case the diapoints of a point lie on a hyperboloid through the origin (Boegehold, Herzberger [5, 7, 20]).

15. The Characteristic Function W for a Single Surface

T. Smith [36] has given a general method for finding the characteristic function W for refraction at a single surface. For refraction at a sphere (object and image origins at the center) one obtains by his method

$$\begin{aligned} W &= r\sqrt{(\xi' - \xi)^2 + (\eta' - \eta)^2 + (\zeta' - \zeta)^2} \\ &= r\sqrt{n'^2 + n^2 - 2w_2} \end{aligned} \tag{2.86}$$

The general rule given by T. Smith is as follows: Let the surface equation be given in homogeneous form, replacing x, y, z by x/a, y/a, z/a as

$$f(x,y,z,a) = 0 \tag{2.87}$$

Differentiate with respect to x, y, z, a, and eliminate these variables, obtaining a function

$$\phi(fa,fx,fy,fz) = 0 \tag{2.88}$$

The characteristic function then is given by

$$\Phi(W,L,M,N) = 0 \tag{2.89}$$

where

$$L = \xi' - \xi \qquad M = \eta' - \eta \qquad N = \zeta' - \zeta \tag{2.90}$$

For rotation symmetric systems W becomes a function of

$$\begin{aligned} \tfrac{1}{2}(L^2 + M^2) &= \tfrac{1}{2}[(\xi' - \xi)^2 + (\eta' - \eta)^2] \\ &= \tfrac{1}{2}[n^2 + n'^2 - 2w_2 + (w_1 - w_3)^2] \end{aligned} \tag{2.91}$$

and $N = \zeta' - \zeta = w_3 - w_1$.

16. The Direct Method and the Addition of Systems

Useful as the characteristic functions are in analyzing an optical system, it is difficult to use them to investigate a system in terms of its components, i.e., to solve the problem: Given two optical systems and their characteristic functions, to find the characteristic function of the combined system. This problem is clumsily solvable for an approximate theory and impossible to solve for finite aperture and field.

The problem, for the case of the angle characteristic, is stated thus: Given origins O, O', O'' and two characteristic functions $\bar{W}$ and $\bar{\bar{W}}$, to find the characteristic for the combined system. It is geometrically easy to see that

$$W = \bar{W} + \bar{\bar{W}} \tag{2.92}$$

But $\bar{W}$ is a function of $\bar{w}_1 = \zeta$, $\bar{w}_2 = \xi\xi' + \eta\eta' + \zeta\zeta'$, $\bar{w}_3 = \zeta'$; $\bar{\bar{W}}$ of $\bar{\bar{w}}_1 = \bar{w}_3 = \zeta'$, $\bar{\bar{w}}_2 = \xi'\xi'' + \eta'\eta'' + \zeta'\zeta''$, $\bar{\bar{w}}_3 = \zeta''$, and we want to find W as a function of $w_1 = \bar{w}_1$, $w_3 = \bar{\bar{w}}_3$, and $w_2 = \xi\xi'' + \eta\eta'' + \zeta\zeta''$.

We therefore have to eliminate the intermediate quantities. This can be done theoretically from (2.69) since

$$\begin{aligned} \mathbf{a} &= \bar{C}\mathbf{s} + \bar{W}_2\mathbf{s}' + \bar{W}_1\mathbf{k} = C\mathbf{s} + W_2\mathbf{s}'' + W_1\mathbf{k} \\ \mathbf{a}' &= -(\bar{W}_2\mathbf{s} + C'\mathbf{s}' + \bar{W}_3k) \\ &= \bar{\bar{C}}\mathbf{s}' + \bar{\bar{W}}_2\mathbf{s}'' + \bar{\bar{W}}_1\mathbf{k} \\ \mathbf{a}'' &= -(\bar{\bar{W}}_2\mathbf{s}' + \bar{\bar{C}}\mathbf{s}'' + \bar{\bar{W}}_3\mathbf{k}) \\ &= -(W_2\mathbf{s} + C'\mathbf{s}'' + W_3\mathbf{k}) \end{aligned} \tag{2.93}$$

But it cannot be done explicitly in the general case, nor even for the case of two spheres. This is the reason why the direct method is suggested.

If x', y' are given as functions of x, y and x'', y'' as functions of x', y' as in (2.71), a simple substitution gives x'', y'' as functions of x, y and therefore permits one to solve the problem of addition of systems in closed form.

In using the direct method

$$2a_1 = x^2 + y^2 \qquad a_2 = x\xi + y\eta \qquad 2a_3 = \xi^2 + \eta^2 \tag{2.94}$$

were used as coordinates. It is frequently useful to employ instead of the direction cosines the coordinates of the intersection point with a second plane, e.g., the entrance pupil, i.e., choosing b_1, b_2, b_3 as

$$2b_1 = x^2 + y^2 \qquad b_2 = xx_p + yy_p \qquad 2b_3 = x_p^2 + y_p^2 \tag{2.95}$$

An interesting suggestion is to use polar coordinates

$$\begin{aligned} x &= r\cos\phi \qquad & x_p &= r_p \cos\psi \\ y &= r\sin\phi \qquad & y_p &= r_p \sin\psi \end{aligned} \tag{2.96}$$

or $2b_1 = r^2 \qquad b_2 = rr_p\cos(\psi - \phi) \qquad 2b_3 = r_p^2$ and to replace in the series development the powers of $\cos(\psi - \phi)$ by the multiples of $\cos k(\psi - \phi)$, $k = 1, 2, \ldots$ (see Nijboer [28]). Since $\psi - \phi = 0$ describes meridional rays, the introduction of $\sin(\psi - \phi)$ as a variable might permit one to separate the meridional and skew errors.

Many other variables have been suggested especially to simplify the computation of diffraction problems. Nijboer [28] and Zernike [41] use special normed polynomials [circle polynomials of the variables in (2.95)], and v. Heel suggested the use of Chebyshev polynomials.

III. DIAGNOSIS

GAUSSIAN OPTICS

17. Introduction

In dealing with an abstract system and developing the general laws we could rightly assume that the functional relationships which we investigated were known exactly and therefrom derive the general laws of image formation. In applying this knowledge to a practical optical system given by its numerical data an abstract analysis is neither desirable nor sufficient. The designer makes a mathematical model of the system, a model which is at the same time elaborate enough to give the desired approximation and simple enough to be understandable. In the language of mathematics we must find a sufficient approximation for the functions which characterize optical-image formation.

We shall consider the following approximation schemes: first, Gaussian optics and Seidel optics. Second, we shall see what information can be obtained from the traditional trace of meridional rays. Third, we shall give interpolation formulas which, from the trace of a few meridional and skew rays, permit one to obtain all the desired information and to analyze the result.

One must keep in mind that the refracting surfaces are not of an abstract mathematical nature. They have boundaries, and thus not all the light from the object point will be transmitted. Moreover it is necessary to consider the intensity of the light distribution in the plane where the image is fixed.

18. General Laws

Let us restrict ourselves to considering only those rays which lie so near the axis that the angle u with the axis (in radians), $\sin u$, and $\tan u$, have approximately the same value; i.e., in the development of $\sin u$ and $\tan u$, etc., we can without appreciable error neglect powers of u higher than the first.

$$\begin{aligned} \sin u &= u - \left[\frac{u^3}{3!} - \frac{u^5}{5!} + \cdots\right] \sim u \\ \tan u &= u + \left[\frac{2}{3!}u^3 + \frac{6}{5!}u^5 + \cdots\right] \sim u \end{aligned} \tag{2.97}$$

This means that, in equations containing the characteristic function V, we can assume V_1, V_2, V_3 to be constant. Or, in the direct method, we can assume α, β, γ, δ to be constant with the side condition

$$\alpha\delta - \beta\gamma = 1 \tag{2.98}$$

Let $\mathbf{A}$, $\mathbf{A}'$, $\mathbf{S}$, $\mathbf{S}'$ be the projection of the vectors $\mathbf{a}$, $\mathbf{a}'$, $\mathbf{s}$, $\mathbf{s}'$ on a plane perpendicular to the axis, i.e.,

$$\begin{aligned} \mathbf{a} &= \mathbf{A} & \mathbf{a}' &= \mathbf{A}' \\ \mathbf{s} &= \mathbf{S} + \zeta\mathbf{k} & \mathbf{s}' &= \mathbf{S}' + \zeta'\mathbf{k} \end{aligned} \tag{2.99}$$

($\mathbf{k}$ being the unit vector in the direction of the axis), where for the rays in the neighborhood of the axis (the paraxial rays) ζ and ζ' can be set equal to n, n', respectively. We have

$$\begin{aligned} \mathbf{A}' &= \alpha\mathbf{A} + \beta\mathbf{S} \\ \mathbf{S}' &= \gamma\mathbf{A} + \delta\mathbf{S} \end{aligned} \tag{2.100}$$

This crude approximation is of enormous importance for optical problems since it gives a first-hand survey of the positions of object and image and of the size of its magnification, as well as of the positions of the focal points.

For the imagery of the axis points (but not the off-axis points) the Gaussian image points can be interpreted in the following way: The rays from a general point are tangential to a surface, the caustic surface. For an axial point the caustic surface splits into two parts, the diapoints on the axis forming one part of it and the other part being formed by a surface symmetric around the axis, and having a cusp. This cusp, where there is an especially good image formation since it is the point common to both parts of the caustic, is the *Gaussian* image point, i.e., the focus of the paraxial rays.

The laws of Gaussian imagery are the laws to which the diapoint laws described in Sec. 9 converge for points near the axis. Shifting the origins by the amounts z and z' gives the following changes:

$$\begin{aligned} \tilde{\mathbf{A}} &= \mathbf{A} + \frac{z}{n}\mathbf{S} & \tilde{\mathbf{S}} &= \mathbf{S} \\ \tilde{\mathbf{A}}' &= \mathbf{A}' + \frac{z'}{n'}\mathbf{S}' & \tilde{\mathbf{S}}' &= \mathbf{S}' \end{aligned} \tag{2.101}$$

or

$$\begin{aligned} \tilde{\alpha} &= \alpha + \frac{z'}{n'}\gamma & \tilde{\beta} &= \beta - \frac{z}{n}\alpha + \frac{z'}{n'}\delta - \frac{zz'}{nn'}\gamma \\ \tilde{\gamma} &= \gamma & \tilde{\delta} &= \delta - \frac{z}{n}\gamma \end{aligned} \tag{2.102}$$

The invariant γ or, better, its negative value is called the *power* of the system.

19. Focal Points and Nodal Points

Each of the coefficients α, β, γ, δ connected by $\alpha\delta - \beta\gamma = 1$ could be equal to zero in special cases.

Focal Point (Image Side) ($\alpha = 0$). The case $\alpha = 0$ [which, according to (2.102), can be achieved by choosing a special point as image origin] leads to

$$\begin{aligned} \mathbf{A}' &= -\frac{1}{\gamma}\mathbf{S} \\ \mathbf{S}' &= \gamma\mathbf{A} + \delta\mathbf{S} \end{aligned} \tag{2.103}$$

Equations (2.103) show that the rays with $\mathbf{S} = 0$ ($\mathbf{s} = n\mathbf{k}$), that is, the rays parallel to the axis, come to a focus at the image origin. This point, the image of the infinite point, is called the *focal point*. Equations (2.103) show that any set of parallel rays, not necessarily parallel to the axis, comes to a focus at a fixed point

$$\mathbf{A}' = -\left(\frac{1}{\gamma}\right)\mathbf{S} = \left(\frac{1}{\phi}\right)\mathbf{S} = \left(\frac{f}{n}\right)\mathbf{S}$$

of the coordinate plane $z' = 0$, which is customarily called the *focal plane*.

The constant f defined above is called the *focal length*.

Conjugate Points ($\beta = 0$). The specification $\beta = 0$, which according to (2.102) can be achieved for a fixed object origin (for instance, $z = 0$) by choosing a well-defined image origin or for a fixed image origin by a well-chosen object origin, leads to

$$\begin{aligned} \mathbf{A}' &= \alpha\mathbf{A} \\ \mathbf{S}' &= \gamma\mathbf{A} + \frac{1}{\alpha}\mathbf{S} \end{aligned} \tag{2.104}$$

Here the rays from the object origin come to a focus at the image origin; the object and image are *conjugated* in the sense of Gaussian optics. Equations (2.104) show that all the rays from a point of the object plane come to a focus (within the validity of Gaussian optics) at a point of the image plane. The magnification $m = \alpha$ is constant for all points of the object plane.

Under these circumstances the rays from the object origin ($\mathbf{A} = 0$) meet at the image origin ($\mathbf{A}' = 0$) so that

$$\mathbf{S}' = \frac{1}{\alpha}\mathbf{S} \tag{2.105}$$

i.e., the so-called "angular magnification" of the rays from an axis point is reciprocal to the magnification. In the special case $\alpha = m = 1$ the conjugated points are called *principal points*. In the case $m = n/n'$, the object and image origins are called *nodal points*. Equations (2.105) and the conditions $\mathbf{s}^2 = n^2$, $\mathbf{s}'^2 = n'^2$ show that corresponding rays through the nodal points in the object and image spaces are parallel.

Object Origin at Focal Point ($\delta = 0$). This can be achieved by choosing a special *object* origin, which is called the *object focal point*. We have

$$\begin{aligned} \mathbf{A}' &= \alpha\mathbf{A} - \frac{1}{\gamma}\mathbf{S} \\ \mathbf{S}' &= \gamma\mathbf{A} \end{aligned} \tag{2.106}$$

The rays through the object origin (object focal point) emerge parallel to the axis ($\mathbf{A} = 0$, $\mathbf{s}' = n'\mathbf{k}$). The rays through an arbitrary point $\mathbf{A}$ of the object focal plane emerge parallel to each other at an inclination $\mathbf{S}' = \gamma\mathbf{A}$.

Afocal Systems ($\gamma = 0$). γ is an invariant of the optical system, and so the condition $\gamma = 0$, $\alpha\delta = 1$ determines a special type of optical system. Equations (2.102) give for the case $\gamma = 0$

$$\begin{aligned} \tilde{\alpha} &= \alpha & \tilde{\beta} &= \beta - \frac{z}{n}\alpha + \frac{z'}{n'}\frac{1}{\alpha} \\ \tilde{\gamma} &= \gamma & \tilde{\delta} &= \delta = \frac{1}{\alpha} \end{aligned} \tag{2.107}$$

which shows that for afocal systems there exist no object-side focal points or image-side focal points (hence the name *afocal*) because neither α nor δ can vanish, since $\alpha\delta = 1$. However, we can assume by choice of the object and (or) image origins that the origins are conjugated points, that is, $\beta = 0$. Then

$$\begin{aligned} \mathbf{A}' &= \alpha\mathbf{A} \\ \mathbf{S}' &= \frac{1}{\alpha}\mathbf{S} \end{aligned} \tag{2.108}$$

which says that a bundle of rays parallel to the axis ($\mathbf{S} = 0$) emerges parallel to the axis ($\mathbf{S}' = 0$). Any bundle of parallel rays emerges as a bundle of parallel rays with an angular magnification equal to $1/\alpha$, which is the same for *all* such bundles. The object plane $z' = 0$ is imaged sharply on the image plane $z = 0$ with the magnification α. The image magnification and angular magnification thus again prove to be reciprocal, though this time they are constant for all rays. Notice that the formulas of the last four examples are typical for the four basic classes of optical instruments:

$\alpha = 0$, object point at infinity, image finite; characteristic of *photographic systems*, where the image of a distant object is formed on a plane, that of the photographic plate or film.

$\beta = 0$, object and image points finite and conjugated; may be exemplified by a reproduction objective or an enlarger lens.

$\gamma = 0$, exemplified by a *telescope*, which gives an apparent magnification of a distant object. The image offered to the unaccommodated eye, like the object, is far away.

$\delta = 0$, object near the focal point; describes instruments which offer an apparent magnification of a near object to the accommodated eye. Such instruments are called *loupes*, or *magnifying glasses*, if consisting of a single (simple or cemented) lens and having a small apparent magnification, and *microscopes* if they are compound lenses with a high apparent magnification.

20. Viewing through an Instrument

If the object and image origins are not conjugated, we can assume that we place our (unaccommodated) eye at the object origin and look at the points of the image plane. The following definitions are self-evident. If

$$\begin{aligned} \mathbf{A}' &= \alpha\mathbf{A} + \beta\mathbf{S} \\ \mathbf{S}' &= \gamma\mathbf{A} + \delta\mathbf{S} \end{aligned} \tag{2.109}$$

we call the *apparent distance* z'_A from the object origin the distance at which the object would be if seen with the naked eye under the same angle as through the instrument. The apparent distance of the image origin from the object origin is then given by

$$\frac{z'_A}{n} = \beta \tag{2.110}$$

Conversely we find for the apparent distance of the object origin from the image origin $z'_A/n' = -\beta$.

We thus see that conjugated points can be defined also as points of apparent distance zero. The apparent distance z_A of an image at the distance z' (from the image origin) from an object at the distance z (from the object origin) is given by (2.102) as

$$\frac{z_A}{n} = \beta - \frac{z}{n}\alpha + \frac{z'}{n'}\delta - \frac{zz'}{nn'}\gamma = -\frac{z'_A}{n'} \tag{2.111}$$

If the object origin ($z = 0$) is at the focal point ($\delta = 0$, $\beta = 1/\gamma$), we have

$$\frac{z_A}{n} = \frac{1}{\gamma} \tag{2.112}$$

That is, all the image points have the same apparent distance from the focal point. This distance is called the *focal length*.

21. Distance of Conjugated Points from the Origins and Their Magnification

Conjugated points are given by $\tilde{\beta} = 0$. In this case α equals the magnification, and we have $\tilde{\delta} = 1/\alpha$.

Inserting into (2.102) and applying the identity $\alpha\delta - \beta\gamma = 1$, we find for the distances of conjugated points from *arbitrary* origins and their magnification the equations (setting $\gamma = -\phi$)

$$\tilde{m} = \alpha - \frac{z'}{n'}\phi \qquad \frac{1}{\tilde{m}} = \delta + \frac{z}{n}\phi \tag{2.113}$$

and

$$\left(\alpha - \frac{z'}{n'}\phi\right)\left(\delta + \frac{z}{n}\phi\right) = 1$$

These formulas are more familiar if the object and image origins are specialized as in the following cases.

***a.* Object and Image Origins Are at Focal Points.** This means $\alpha = \delta = 0$, $\beta = -1/\gamma = 1/\phi$, and

$$zz' = -\frac{nn'}{\phi^2} \qquad m = -\frac{z'}{n'}\phi = \frac{n}{z\phi} \tag{2.114}$$

The distances of the principal points ($m = 1$) from the focal points are given by

$$z = \frac{n}{\phi} = -f \qquad z' = -\frac{n'}{\phi} = -f' \tag{2.115}$$

The distances of the *nodal points* ($m = n/n'$) from the *focal points* are given by

$$z = \frac{n'}{\phi} = f' \qquad z' = -\frac{n}{\phi} = f \tag{2.116}$$

The quantities f and f' thus defined are frequently called the *object-* and *image-side focal lengths*. They play a great role in elementary optical treatises.

***b.* Object and Image Origin in Conjugated Points of Magnification *m*.** We have $\beta = 0$, $\alpha = 1/\delta = m_0$. Equations (2.113) give

$$\frac{n'm_0}{z'} - \frac{n}{zm_0} = \phi = \frac{n'}{f'} = -\frac{n}{f}$$
$$m = m_0 - \frac{z'}{n'}\phi = m_0 - \frac{z'}{f'} \tag{2.117}$$
$$\frac{1}{m} = \frac{1}{m_0} + \frac{z}{n}\phi = \frac{1}{m_0} - \frac{z}{f}$$

In particular for $m_0 = 1$ (origins at the *principal points*) we have

$$\frac{n'}{z'} - \frac{n}{z} = \phi = \frac{n'}{f'} = -\frac{n}{f}$$
$$m - 1 = -\frac{z'}{f'} \qquad \frac{1}{m} - 1 = -\frac{z}{f} \tag{2.118}$$

and for $m_0 = n/n'$ (origins at the nodal points)

$$\frac{1}{n'z'} - \frac{1}{nz} = \frac{\phi}{nn'} = \frac{1}{nf'} = -\frac{1}{n'f}$$
$$\frac{n'}{n}m = 1 + \frac{z'}{f} \qquad \frac{n}{n'm} = 1 + \frac{z}{f'} \tag{2.119}$$

22. Gaussian Brackets

A Gaussian bracket of n symbols is a polynomial which is linear in each of the symbols and fulfills the following equations: The brackets fulfill the relationship

$$[a_1 \cdots a_n] = [a_1 \cdots a_{n-1}]a_n + [a_1 \cdots a_{n-2}]$$
$$[a_1] = a_1 \qquad [\ ] = 1 \tag{2.120}$$

From this recursion formula we can derive $[a_1 \cdots a_n]$. We find in particular

$$[a_1a_2] = a_1a_2 + 1$$
$$[a_1a_2a_3] = a_1a_2a_3 + a_1 + a_3 \tag{2.121}$$
$$[a_1a_2a_3a_4] = a_1a_2a_3a_4 + a_1a_2 + a_1a_4 + a_3a_4 + 1$$

In general $[a_1 \cdots a_n]$ can be formed as follows: Take the product of all n elements. Then take all products of $n - 2$ elements which can be written with increasing indices, starting with an odd index and alternating odd and even indices. Repeat with $n - 4$, $n - 6$, etc., until one ends finally with $a_1 + a_3 + \cdots + a_n$ in case of odd n, or the number 1 in the case of even n. The bracket is the sum of all these terms. To exemplify further,

$$\begin{aligned}[a_1 \cdots a_7] = {} & a_1a_2a_3a_4a_5a_6a_7 + a_1a_2a_3a_4a_5 \\ & + a_1a_2a_3a_4a_7 + a_1a_2a_3a_6a_7 + a_1a_2a_5a_6a_7 \\ & + a_1a_4a_5a_6a_7 + a_3a_4a_5a_6a_7 + a_1a_2a_3 \\ & + a_1a_2a_5 + a_1a_2a_7 + a_1a_4a_5 + a_1a_4a_7 \\ & + a_1a_6a_7 + a_3a_4a_5 + a_3a_4a_7 + a_3a_6a_7 \\ & + a_5a_6a_7 + a_1 + a_3 + a_5 + a_7\end{aligned} \tag{2.122}$$

A Gaussian bracket is linear in each of its elements. The following development formulas can be shown to be true:

$$\begin{aligned}[a_1 \cdots a_n] &= a_1[a_2 \cdots a_n] + [a_3 \cdots a_n] \\ &= a_k[a_1 \cdots a_{k-1}][a_{k+1} \cdots a_n] \\ &\quad + [a_1 \cdots a_{k-1} + a_{k+1} \cdots a_n] \\ &= [a_1 \cdots a_{n-1}]a_n + [a_1 \cdots a_{n-2}]\end{aligned} \tag{2.123}$$

The following determinant relation holds between four brackets:

$$\begin{vmatrix} [a_1 \cdots a_n] & [a_2 \cdots a_n] \\ [a_1 \cdots a_{n-1}] & [a_2 \cdots a_{n-1}] \end{vmatrix} = (-1)^n \tag{2.124}$$

Equation (2.123) permits one to differentiate a Gaussian bracket with respect to each of its members:

$$\frac{\partial}{\partial\lambda}[a_1 \cdots a_n] = \sum_{k=1}^{n} \frac{\partial a_k}{\partial\lambda}[a_1 \cdots a_{k-1}][a_{k+1} \cdots a_n] \tag{2.125}$$

A finite continued fraction can be expressed with the help of Gaussian brackets. The following formula holds:

$$\cfrac{1}{a_1 + \cfrac{1}{a_2 + \cfrac{1}{a_3 + \cfrac{}{\ddots + \cfrac{1}{a_n}}}}} = \frac{[a_2 \cdots a_n]}{[a_1a_2 \cdots a_n]} \tag{2.126}$$

23. Expression of Basic Data of Gaussian Optics with the Help of Gaussian Brackets

A single refracting surface with rotation symmetry is, with respect to Gaussian optics, and only with respect to Gaussian optics, indistinguishable from a sphere with the same vertex curvature, $\rho = 1/r$. If both the origins are placed at the vertex, which is the principal point for both the image and the object sides, the Gaussian equation may be written

$$\begin{aligned} \mathrm{A}' &= \alpha\mathrm{A} + \beta\mathrm{S} \\ \mathrm{S}' &= \gamma\mathrm{A} + \delta\mathrm{S} \end{aligned} \tag{2.127}$$

with

$$\alpha = 1 \qquad \beta = 0$$
$$\gamma = -\phi = \frac{n - n'}{r} \qquad \delta = 1 \tag{2.128}$$

For a plane, $\gamma = -\phi = 0$.

Transforming to the next surface, if d is the distance of the second vertex from the first, one obtains

$$\begin{aligned} \mathrm{A}_2 &= \mathrm{A}' - \frac{d}{n}\mathrm{S} \\ \mathrm{S}_2 &= \mathrm{S}' \end{aligned} \tag{2.129}$$

where n = refractive index.

Successive applications of these formulas lead, in the case where the object origin is at the first vertex and the image origin at the last vertex, to

$$\begin{aligned} \mathrm{A}' &= \alpha\mathrm{A} + \beta\mathrm{S} \\ \mathrm{S}' &= \gamma\mathrm{A} + \delta\mathrm{S} \end{aligned} \tag{2.130}$$

with
$$\alpha = \left[-\phi, \frac{d_1}{n_1}, -\phi_2, \ldots, \frac{d_{k-1}}{n_{k-1}}\right]$$
$$\beta = \left[\frac{d_1}{n_1}, -\phi_2, \frac{d_2}{n_2}, \ldots, \frac{d_{k-1}}{n_{k-1}}\right]$$
$$\gamma = \left[-\phi_1, \frac{d_1}{n_1}, -\phi_2, \ldots, -\phi_k\right] = -\Phi$$
$$\delta = \left[\frac{d_1}{n_1}, -\phi_2, \frac{d_2}{n_2}, \ldots, -\phi_k\right]$$

Having the object and image origins at the distances z and z' from the first and last vertices, respectively, we find

$$\begin{aligned} \tilde{\mathrm{A}}' &= \tilde{\alpha}\tilde{\mathrm{A}} + \tilde{\beta}\mathrm{S} \\ \mathrm{S}' &= \tilde{\gamma}\tilde{\mathrm{A}} + \tilde{\delta}\mathrm{S} \end{aligned} \tag{2.131}$$

with [see (2.102)]

$$\begin{aligned} \tilde{\alpha} &= \alpha + \frac{z'}{n'}\gamma = \left[-\phi_1, \frac{d_{k-1}}{n_{k-1}}, \ldots, -\phi_k, \frac{z'}{n'}\right] \\ \tilde{\beta} &= \left[-\frac{z}{n}, -\phi_1, \frac{d_1}{n_1}, -\phi_k, \ldots, \frac{z'}{n'}\right] \\ \tilde{\gamma} &= \gamma = [-\phi_1 - \phi_k] \\ \tilde{\delta} &= \delta = \left[-\frac{z}{n}, -\phi_1, \frac{d_1}{n}, \ldots, -\phi_k\right] \end{aligned} \tag{2.132}$$

If z and z' denote conjugate distances $\tilde{\beta} = 0$, we find the magnification from

$$\begin{aligned} \tilde{\alpha} &= m = -\Phi\frac{z'}{n'} + \left[-\phi_1, \ldots, \frac{d_{k-1}}{n_{k-1}}\right] \\ \gamma &= -\Phi = [-\phi_1, \ldots, -\phi_k] \\ \delta &= \frac{1}{m} = -\frac{z}{n}\Phi + \left[\frac{d_1}{n_1}, \ldots, -\phi_k\right] \end{aligned} \tag{2.133}$$

The focal distance (front focus) is given by

$$\frac{z_F}{n} = -\frac{\delta}{\Phi} \qquad \text{or} \qquad z_F = -\delta f \tag{2.134}$$

and the focal length by

$$f = \frac{n}{\Phi}$$

The image focal distance (back focus) is given by

$$\frac{z'_F}{n'} = \frac{\alpha}{\Phi} \qquad z'_F = -\alpha f' \tag{2.135}$$

Especially simple formulas result if we neglect the thicknesses of the individual lenses, but not their distances. We find, if ϕ_i is the power of the ith lens,

$$\phi_i = (n_i - 1)(\rho_{2i} - \rho_{1i})$$

ρ_{2i} and ρ_{1i} being the first and second curvatures of the ith lens; and if e_i is the distance between the ith and $(i + 1)$th lenses, if we put the origins at the first and last lenses, then

$$\begin{aligned} \mathrm{A}' &= \alpha\mathrm{A} + \beta\mathrm{S} \\ \mathrm{S}' &= \gamma\mathrm{A} + \delta\mathrm{S} \end{aligned} \tag{2.136}$$

with
$$\begin{aligned} \alpha &= [-\phi_1, e_1, -\phi_2, e_2, \cdots, e_k] \\ \beta &= [e_1, -\phi_2, \ldots, e_k] \\ \gamma &= [-\phi_1, e_1, -\phi_2, \ldots, -\phi_k] = -\Phi \\ \delta &= [e_1, -\phi_2, \ldots, -\phi_k] \end{aligned} \tag{2.137}$$

For example, for a single thin lens, we have the formulas

$$\alpha = 1 \qquad \beta = 0$$
$$\gamma = -\phi_1 \qquad \delta = 1$$

For a doublet, two thin lenses with finite distance, we have

$$\begin{aligned} &\alpha = 1 + \phi_1 e_1 \\ &\beta = e_1 \\ &-\gamma = \Phi = \phi_1 + \phi_2 - e_1\phi_1\phi_2 \qquad \delta = 1 + e_1\phi_2 \end{aligned} \tag{2.138}$$

24. Vignetting

L. Schleiermacher [32], and later Ernst Abbe [1] and his school, drew the attention of optical designers to the fact that ray tracing alone or even the mathematical coordination of object and image rays alone does not determine the image quality of an optical instrument. An optical lens does not consist of two mathematical surfaces extending to infinity. The two surfaces intersect, and the diameter of this intersection gives a *natural* limitation for the opening (aperture) of the lens. In general, a system of lenses mounted in a barrel has a still smaller aperture than the natural one, and frequently the mounting and especially the diaphragm, a changeable aperture stop, restrict the rays still more.

It is absolutely necessary for two reasons for the optical designer to know approximately which rays from every object point succeed in getting into the image space. First, especially in photographic systems, it is necessary to know the light intensity at the points of the image plane, in order to give the picture the right exposure. Second, the image quality may suffer if the light rays from some points are cut off unsymmetrically. To study this, take an uncorrected lens, and place a point source somewhere on the axis. The image viewed from a suitable image point will be a disk which is fairly evenly illuminated. Now put an opaque screen between the light source and the system so that the light is cut off unsymmetrically. The image will show all the qualities of an unsymmetrical image, one which, in the language of opticians, has coma errors.

The problem of finding out exactly which rays go through a given system is quite complex, but fortunately an approximate estimate is all that is needed for practical purposes. Optical systems studied by lens designers are fairly well corrected, and, for the purpose of computing intensity and vignetting, we can assume the image rays to come to a point in the image plane conjugated to the object according to Gaussian optics. If we project all the surfaces, according to Gaussian optics, into the image space and consider the cone of rays from each projected aperture to the Gaussian image point, the cone of rays common to these cones will be a fairly good approximation to the manifold of rays emerging from the system. As a matter of fact, it will in general suffice to project the first lens surface and the changeable diaphragm aperture into the image space. These two apertures, together with the aperture of the last lens surface, give in general the necessary information. Figure 2.7 shows how in this case the axis point

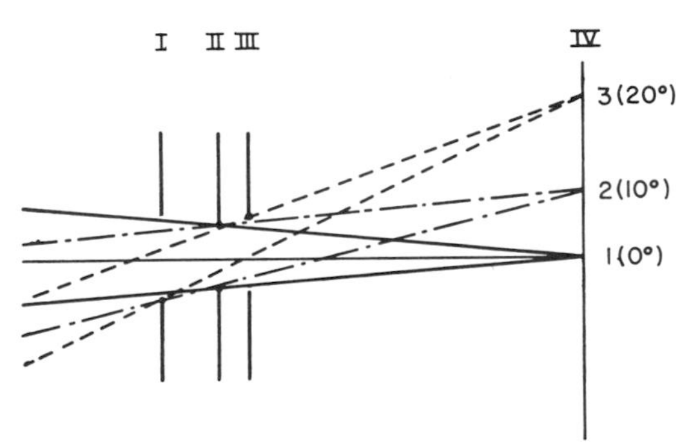

FIG. 2.7. Vignetting: I, image of first surface; II, exit pupil; III, last surface; IV, image plane; 1, axis point (0°); 2, point in the field (10°); 3, limit of field (20°).

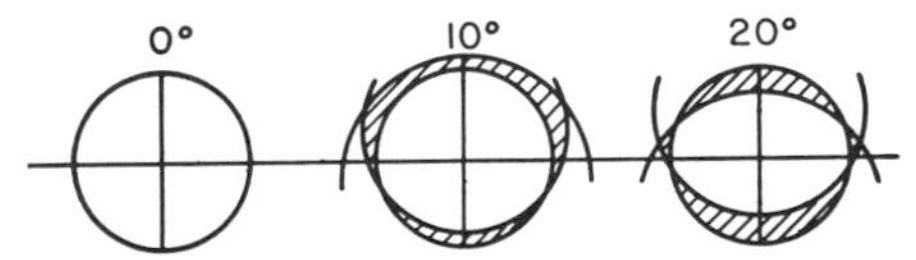

FIG. 2.8. Vignetting. Apertures in Fig. 2.7 projected into plane II, the exit pupil. The areas are proportional to the amount of light going to the different image points.

receives a circular cone of light, whereas for an off-axis point in general some of the light is cut off at the first and the last lens.

By projecting the three circular apertures from the image point into the exit pupil (Fig. 2.8), a region is obtained bordered by three concentric circles for an axis point and by three eccentric circles for an off-axis point. The cone from the image point to the area common to these three circles gives approximately the cone of light rays emerging from the system.

The ratio of the areas for the light rays from an axis point and from an off-axis point forms the basis for calculation of the relative light intensity in the image.

The same construction just performed for the image space can be performed in the object space. We can project the first lens diaphragm and the last lens into the object space and thus construct for each object point the cone of object rays which traverses the optical system.

ANALYSIS OF A GIVEN OPTICAL SYSTEM

25. Introduction

The direct method of optical calculation permits one in principle to compute the coefficients α, β, γ, δ for a system with a finite number of lenses as functions of the curvatures ρ_i, thicknesses d_i, and air spaces e_i. However, the method would give α, β, γ, δ in a form which is not well adapted to mathematical analysis. They are found to be polynomials of roots, or roots of roots, etc., of algebraic functions of the object data. Many attempts have been made to obtain the expressions in such a form that the image can be easily analyzed. Two methods suggest themselves.

***a.* The Approximation Method.** This can be applied either to the characteristic function or to the direct method. We saw in Sec. 10 that the knowledge of the characteristic function permits one to analyze the optical image. Let us develop V into a series with respect to its coordinates,

$$V = \alpha_0 + \alpha_1 v_1 + \alpha_2 v_2 + \alpha_3 v_3 + \tfrac{1}{2}(\alpha_{11} v_1^2 + 2\alpha_{12} v_1 v_2 + 2\alpha_{13} v_1 v_3 + \alpha_{22} v_2^2 + 2\alpha_{23} v_2 v_3 + \alpha_{33} v_3^2) + \cdots \quad (2.139)$$

and assume that an approximate expression is sufficient to describe the object-image coordination for all the rays going through the optical system.

Theoretically the coefficients α_i and α_{ik} should be chosen as the differential quotients of V and calculated by differential analysis of the rays near the axis.

The first-order members alone will not suffice because in this case we have in (2.65) V_1, V_2, and V_3 equal to α_1, α_2, α_3; that is, they would be constant. That would lead only to the Gaussian approximation, and with respect to the Gaussian approximation all optical systems are identical and indistinguishable. Thus it was suggested that one compute the *six* coefficients α_{ik}. This leads to Seidel's approximation theory [34]. If the cubic terms in the development of V are investigated, we find the *ten* coefficients of the Schwarzschild-Kohlschütter aberration theory [25, 33], lately corrected by M. Herzberger [21] and F. Wachendorf [38].

However, all these approximations, complex as they are, do not enable one to compute the light intensity in the image with even a fair degree of accuracy.

A much better approximation method, among others, is that suggested by H. Hopkins [22].

Consider, not the image of all object points, but of only a few, for instance, three: $v_1 = 0$ (the axis point) and two other values of v_1, one corresponding to a point at the outer corner of the object to be imaged (diagonal of the picture) and one at a point about two-thirds out in the field, thus choosing in all cases v_1 constant so that V is a function of only two variables.

Moreover, V is thought of as developed, not in the neighborhood of the ray $v_2 = v_3 = 0$, but in the neighborhood of the principal ray, i.e., the ray through the center of the entrance pupil, which, if k is the distance between the object and the entrance pupil, leads to

$$v_2^0 = -\frac{2nkv_1}{\sqrt{k^2 + 2v_1}} \qquad v_3^0 = \sqrt{\frac{(n'^2 - n^2k^2)2v_1 + n'^2k^2}{k^2 + 2v_1}} \tag{2.140}$$

and V is developed as a function of $(v_2 - v_2^0)$ and $(v_3 - v_3^0)$. Again analytical methods can be applied to determine the coefficients.

***b.* The Interpolation Method.** The author contends that the following method of giving a picture of the light distribution in the optical image is more satisfying. As above, one tries (for a given object point, $v_1 = v_1^0$) to fit a characteristic function to the ray data. From (2.66) and (2.68) one obtains

$$\begin{aligned} m = V_2 &= \alpha_2 + \alpha_{22}v_2 + \alpha_{23}v_3 \\ &\quad + \tfrac{1}{2}(\alpha_{222}v_2^2 + 2\alpha_{223}v_2v_3 + \alpha_{233}v_3^2) + \cdots \\ z' = V_3 &= \alpha_3 + \alpha_{23}v_2 + \alpha_{33}v_3 \\ &\quad + \tfrac{1}{2}(\alpha_{223}v_2^2 + 2\alpha_{233}v_2v_3 + \alpha_{333}v_3^2) + \cdots \end{aligned} \tag{2.141}$$

but this time we shall not determine the coefficients in (2.141) analytically. We shall trace a sufficient number of rays from the object point through the system and shall determine the coefficients of V_2 and V_3 by least-square methods to fit the computed data. Having computed a sufficient number of the rays distributed evenly over the vignetted exit pupil, the remainder of (2.141) for the computed rays gives a measure for the accuracy of the approximation.

After V_2 and V_3 are found, Eqs. (2.141) give an easy method for computing the ray data for a large number of intermediate data.

The two methods thus described can be applied also to find the α, β, γ, δ of the direct method, either by approximation or by interpolation.

In the following paragraphs we shall sketch Seidel's theory and the interpolation method and give suggestions on how to evaluate ray-tracing data.

26. The Seidel Aberrations

Choose the coordinate origin on the object side at the entrance pupil and on the image side at the plane where the image is to be inspected. Let x, y, k be the coordinates of the intersection point of the object ray with the plane $z = k$ (the object plane), ξ, η, ζ the optical direction cosines, and x_p, y_p, 0 the intersection of the object ray with the plane of the entrance pupil. We then can find functions α, β, γ, δ such that

$$\begin{aligned} x' &= \alpha x + \beta x_p & \frac{k}{\zeta}\xi' &= \gamma x + \delta x_p \\ y' &= \alpha y + \beta y_p & \frac{k\eta'}{\zeta'} &= \gamma y + \delta y_p \end{aligned} \tag{2.142}$$

where α, β, γ, δ are functions of

$$s_1 = \tfrac{1}{2}(x^2 + y^2) \qquad s_2 = xx_p + yy_p \qquad s_3 = \tfrac{1}{2}(x_p^2 + y_p^2) \tag{2.143}$$

and k is the distance between the object and entrance pupils.

The functions α, β, γ, δ can be shown to be connected with the diamagnification m of the object point, the diapoint distance z, and the corresponding quantities for the entrance pupil ($\bar{m}$, $\bar{z}'$) by

$$\begin{aligned} m &= \frac{1}{\delta} & \frac{z'}{m} &= -\beta k\frac{\zeta'}{\zeta} \\ \bar{m} &= -\frac{1}{\gamma} & \frac{\bar{z}'}{\bar{m}} &= \alpha k\frac{\zeta'}{\zeta} \end{aligned} \tag{2.144}$$

Since for rays near the axis ζ and ζ' have approximately the values n and n', and never vanish, δ and β can be considered as designating the error in magnification and the longitudinal aberration, respectively.

The functions α, β, γ, δ are connected by differential equations obtained by applying Lagrange's theorem.

Seidel's aberration theory now consists in considering α, β, γ, δ as functions which can be linearly approximated so that we can write

$$\begin{aligned} \alpha &= \alpha_0 + \alpha_1 s_1 + \alpha_2 s_2 + \alpha_3 s_3 \\ \beta &= \beta_0 + \beta_1 s_1 + \beta_2 s_2 + \beta_3 s_3 \end{aligned} \tag{2.145}$$

and so on. The Gaussian image plane is given by $\beta_0 = 0$.

We find for this approximation, since

$$\alpha_0 = m_0 = \frac{1}{\delta_0}$$

is the Gaussian magnification,

$$\begin{aligned} x' - m_0x &= (\alpha_1s_1 + \alpha_2s_2 + \alpha_3s_3)x \\ &\quad + (\beta_1s_1 + \beta_2s_2 + \beta_3s_3)x_p \\ y' - m_0y &= (\alpha_1s_1 + \alpha_2s_2 + \alpha_3s_3)y \\ &\quad + (\beta_1s_1 + \beta_2s_2 + \beta_3s_3)y_p \end{aligned} \tag{2.146}$$

The Lagrange formulas lead to

$$\alpha_3 = \beta_2 \tag{2.147}$$

Assuming the object point in the yz plane (no loss of generality), we have

$$\begin{aligned} &s_1 = \tfrac{1}{2}y^2 \qquad s_2 = yy_p \qquad s_3 = \tfrac{1}{2}(x_p^2 + y_p^2) \\ &x' = (\beta_1s_1 + \alpha_3s_2 + \beta_3s_3)x_p \\ &y' - m_0y = (\alpha_1s_1 + \alpha_2s_2 + \alpha_3s_3)y \\ &\qquad + (\beta_1s_1 + \alpha_3s_2 + \beta_3s_3)y_p \end{aligned} \tag{2.148}$$

The quantity x' gives the sagittal deviation and $y' - m_0y$ the meridional deviation of a ray from the object point. The geometrical meaning of the five coefficients is easily ascertained

If $\alpha_1 = \alpha_2 = \alpha_3 = \beta_2 = \beta_3 = 0$, all rays go through the Gaussian image point,

$$\begin{aligned} x' &= 0 \\ y' &= m_0 y \end{aligned} \tag{2.149}$$

That is, the system is free of any error.

If α_1 is the only nonzero coefficient, we have

$$\begin{aligned} x' &= 0 \\ y' &= (m_0 + \alpha_1 y^2) y \end{aligned} \tag{2.150}$$

All points have a sharp image, but the magnification changes with the object distance. Then a square is not imaged as a square, i.e., the image is *distorted.* This distortion is called *pincushion distortion* if $\alpha_1 > 0$ and *barrel distortion* if $\alpha_1 < 0$.

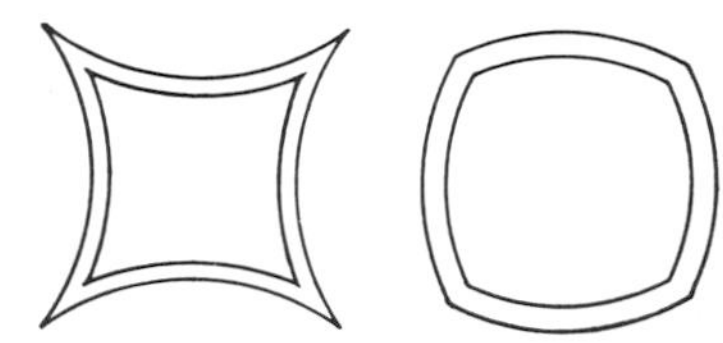

Fig. 2.9. Pincushion and barrel distortion.

The quantities α_2 and β_1 are usually treated together. Consider the case where all the other coefficients vanish. From (2.147) we then have

$$\begin{aligned} x' &= \tfrac{1}{2}\beta_1 y^2 x_p \\ y' - m_0 y &= (\alpha_2 + \tfrac{1}{2}\beta_1) y^2 y_p \end{aligned} \tag{2.151}$$

Inserting $x_p = \sqrt{2s_3} \cos \phi$ and $y_p = \sqrt{2s_3} \sin \phi$ and eliminating ϕ in (2.151) we obtain

$$\frac{x'^2}{\beta_1^2} + \frac{(y' - m_0 y)^2}{(2\alpha_2 + \beta_1)^2} = \tfrac{1}{2} s_3 y^4 \tag{2.152}$$

which proves that the rays from an object point to a set of concentric circles (s_3 constant) in the entrance pupil intersect the image plane in a set of concentric ellipses. The ratio of the axes of these ellipses equals $\beta_1/2\alpha_2 + \beta_1$. This error is called *astigmatism.* Closer study shows that this is also the ratio of the two distances of the image plane from the caustic.

The quantity β_1 is proportional to the distance of the image plane from the diapoint. It is called the *coefficient of sagittal curvature,* whereas $2\alpha_2 + \beta_1$

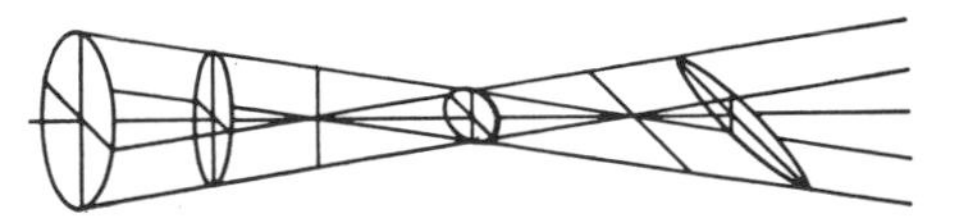

Fig. 2.10. Astigmatism in different planes.

is the *coefficient of meridional curvature.* If α_2 vanishes, the ellipses become circles and the rays come to a focus; i.e., the object is imaged without astigmatism (anastigmatically). The quantity α_2 is therefore called the *coefficient of astigmatism.*

If α_3 is the only nonzero coefficient, we have

$$\begin{aligned} x' &= \alpha_3 y x_p y_p \\ y' - m_0 y &= \tfrac{1}{2}\alpha_3 (3y_p^2 + x_p^2) y \end{aligned} \tag{2.153}$$

which is equivalent to

$$(y' - m_0 y - \alpha_3 y s_3)^2 + x'^2 = 2(\alpha_3 y y_p)^2 s_3 \tag{2.154}$$

showing that the rays from the object point to a series of concentric circles in the entrance pupil go to a series of eccentric circles in the image plane. These circles are tangential to two lines originating at the Gaussian image point and forming an angle of 60°

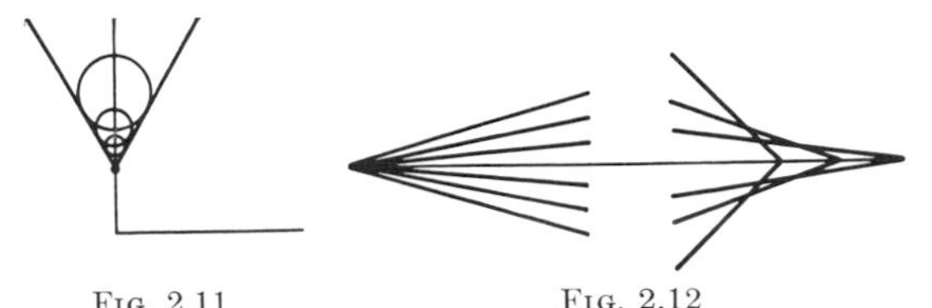

Fig. 2.11 Fig. 2.12

Fig. 2.11. Coma.
Fig. 2.12. Spherical aberration.

with each other. Then the light distribution takes an unsymmetric cometlike form, which has led to the name *coma* for the error designated by α_3.

Finally, if β_3 is the only nonzero coefficient, we have

$$\begin{aligned} x' &= \beta_3 x_p (x_p^2 + y_p^2) \\ y' - m_0 y &= \beta_3 y_p (x_p^2 + y_p^2) \end{aligned} \tag{2.155}$$

which leads to

$$x'^2 + (y' - m_0 y)^2 = 8\beta_3^2 s_3^2 \tag{2.156}$$

This equation coordinates to each object point a set of concentric circles, i.e., concentrates the light in a little disk the radius of which is proportional to β_3. This is called the *error of spherical aberration.*

In a general system, of course, these errors will not be isolated but will overlap each other, and the combination of errors has to be studied.

For the rays from the axis point

$$(x = y = s_1 = s_2 = 0)$$

the so-called "aperture" rays (2.148) give

$$\begin{aligned} x' &= \beta_3 s_3 x_p \\ y' &= \beta_3 s_3 y_p \end{aligned} \tag{2.157}$$

That is, spherical aberration is the only error. For the rays through the entrance pupil, often called *principal* rays because optical systems are usually so constructed that each of these rays forms the central ray of the bundle going through the system, we have

$$\begin{aligned} x' &= 0 \\ y' - m_0 y &= \alpha_1 s_1 y \end{aligned} \tag{2.158}$$

That is, distortion is the only error.

27. Extension of Seidel Theory to Finite Aperture and Field

The coefficients in (2.146) can be developed with respect to s_1, s_2, s_3 into series containing quadratic and higher-order members. However, the number of error coefficients increases rapidly, and it is laborious to study the influence of the higher-order coefficients on image formation. Moreover, the computation of the correct coefficients by developing them in Taylor series is frequently misleading since the series do not converge very rapidly.

Thus, for investigation of the image of an object

point for finite aperture and field, the lens designer must trace a number of distinct rays through the optical system and evaluate the results of ray tracing.

The proposals to be made by the author deviate from those generally used, which will therefore be discussed first. The popularity of Seidel's theory led to an attempt to generalize his conceptions for finite aperture and field. This can be done, and the corresponding "errors" can be computed and measured. The only difficulty is that for finite aperture and field they do not give sufficient information about the image quality. To amplify, let us investigate spherical aberration.

Tracing a number of rays (usually two but sometimes three) from an axis point, we can calculate their intersection points with the axis (the diapoints) or their intersection points with one or several image planes perpendicular to the axis. Owing to the symmetry of the problem, it suffices to trace rays in the yz plane, the meridian plane.

The result of this computation gives sufficient information about the image of the axis point. Figures 2.13 to 2.15 show for a given system the

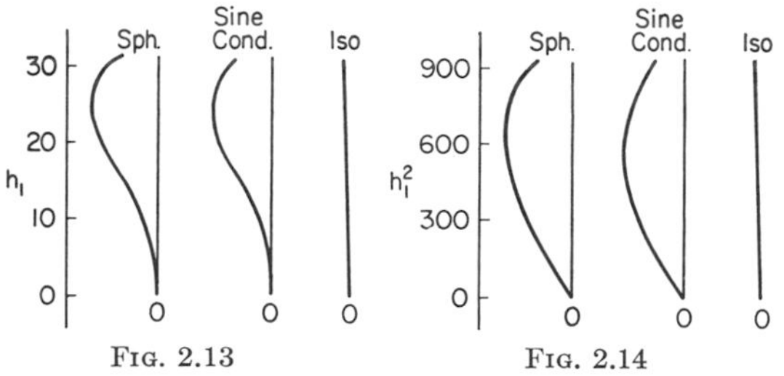

FIG. 2.13 FIG. 2.14

FIG. 2.13. Spherical aberration, sine condition, and isoplanatic condition plotted against height.

FIG. 2.14. Spherical aberration, sine condition, and isoplanatic condition plotted against square of height. The slope of the tangent at the origin corresponds to the Seidel coefficients for the above errors.

results of such a computation. In the first two plots the longitudinal aberration z' is plotted against the entrance height and the square of the height, respectively. The latter plot seems to be preferable, since z' is a function of the square of the height and in most practical cases the curve is sufficiently approximated by a parabola. If Seidel aberrations were solely responsible, this curve would be a straight line. Thus the slope of this curve at the axial point is a measure for the Seidel aberration.

Figure 2.15, in which the height in the image plane is plotted against the tangent of the exit angle, has the advantage that we can easily evaluate the best plane for spherical aberration, since the slope of the best straight line through the points is proportional to the distance of the best focal plane.

Correspondingly, a number of principal rays (two or three) is computed, one corresponding to an object point of medium distance and another to a point near the limit of the field to be investigated. For these points, we calculate first the intersection point with the image plane and then plot the distance from the Gaussian image point (the distortion) as a function of the object height (usually in percentages).

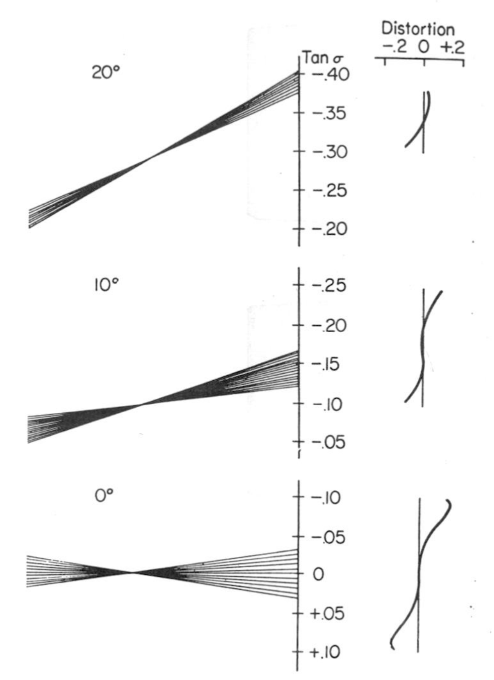

FIG. 2.15. Right: Bertele's curves for the axial point and rim rays. Left: Schade drawings of the rim-ray correction.

It is more difficult to transfer the conceptions of astigmatism and coma to finite aperture and field. The rays from an off-axis point are tangential in two points to their caustic surface. On a single meridian ray, it is easy to trace these points from surface to surface (Coddington's formulas). It is customary, in general, to trace these points for the principal

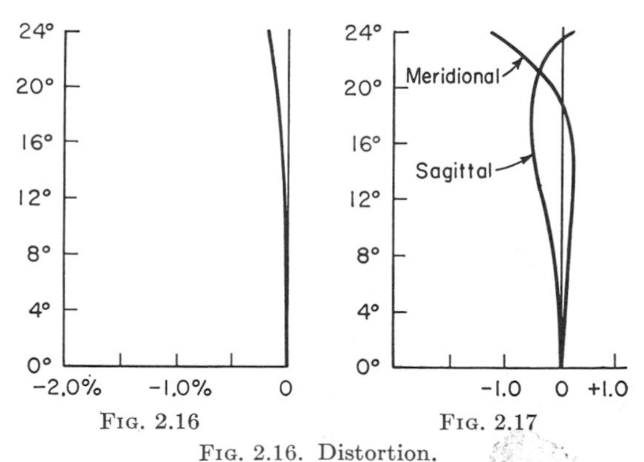

FIG. 2.16 FIG. 2.17

FIG. 2.16. Distortion.

FIG. 2.17. Astigmatism (meridional and sagittal).

rays. In this case one of them coincides with the diapoint (the sagittal focus), whereas the other, the intersection of two approximate meridian rays, is called the *meridional focus*. One usually plots the lengths of the perpendiculars from the sagittal and meridional image points to the Gaussian focal plane as functions of the tangent of the exit angle. Obviously the knowledge of astigmatism, i.e., the knowledge of two points of the caustic for each object point, is no substitute for a method which allows one to compute the entire caustic.

The Coddington equations for tracing astigmatism through an optical system are, for a single surface,

$$\frac{m'}{s'} - \frac{m}{s} = \frac{n' \cos i' - n \cos i}{r}$$
$$\frac{n' \cos^2 i'}{t'} - \frac{n \cos^2 i}{t} = \frac{n' \cos i' - n \cos i}{r} \quad (2.159)$$

where i, i' are the angles of incidence and refraction at the surface and s and t, s' and t', respectively, are the distances of the sagittal and meridional (tangential) points from the intersection point of the ray with the surface before and after refraction.

The same objection holds with respect to the attempt to predict the asymmetry of the image of an off-axis point by generalizing the coma error.

Consider the image of an off-axis point. Using the characteristic function V as a function of

$$v_1 = \tfrac{1}{2}(x^2 + y^2) \qquad v_2 = x\xi' + y\eta' \qquad v_3 = \tfrac{1}{2}(\xi'^2 + \eta'^2) \quad (2.160)$$

one can derive

$$-x' = V_2 x + V_3 \xi' \qquad -\xi = V_1 x + V_2 \xi'$$
$$-y' = V_2 y + V_3 \eta' \qquad -\eta = V_1 y + V_2 \eta' \quad (2.161)$$

Consider an object point so near to the axis that x^2 and y^2 can be neglected, whereas ξ, η shall be permitted to have finite values, which is equivalent to assuming v_1 to be zero, v_2 to be small, v_3 to be finite. Developing V as a function of v_2 with v_3 as a parameter,

$$-x' = V_2{}^0 x + V_3 \xi' \quad (2.162)$$

where V_2 and V_3 are functions of v_3. If the axis point is sharply imaged, we have $V_3 = 0$; that is, the system has no spherical aberration. In this case a neighboring point is imaged sharply if and only if V_2 is constant as a function of v_3. But the direction cosines of the rays from the axis point $x = 0$ obey [because of (2.161)] the equations

$$-\xi = V_2 \xi' \qquad -\eta = V_2 \eta' \quad (2.163)$$

which is equivalent to saying that the ratio of the sines of the angles with the axis is constant if and only if the point near the axis is sharply imaged. This is the reason why the ratios of these sines, the Abbe sine condition, is plotted by most designers (see Figs. 2.13 and 2.14). The sine condition has meaning even if the spherical aberration is not corrected. In this case a point near the axis is *symmetrically* imaged if V_2 and V_3 are linearly dependent (see Figs. 2.13 and 2.14).

This leads to the Staeble-Lihotzky condition for symmetry, often called *isoplanatic condition* (see Figs. 2.13 and 2.14). In the case of spherical aberration it is advisable to construct the system so that the exit pupil coincides with the point where the symmetry axis of the bundle from the near-axis point intersects the system axis. The knowledge of the sine condition is of great importance to the optical designer, and it must be very well corrected in a good optical system, but it is not a sufficient condition that a point far off the axis shall be well corrected.

The following other methods are applied in some factories to give information about the image of far-off-axis points, at least with respect to the meridional rays.

A set of meridional rays, usually five, two on each side of the principal ray, are traced from the off-axis point, more or less evenly distributed over the (vignetted) entrance pupil. The intersection points of these rays with the Gaussian image plane are computed, and the intersection height minus its Gaussian equivalent plotted (as suggested by L. Bertele) as a function of the tangent of the exit angle.

This plot (right side of Fig. 2.15), which is made for an axis point and at least two object points in the field, shows at a glance some of the defects of the image.

Let us assume that all the rays go through a point which is given by its height H and its distance Z from the Gaussian image plane. The plot of Bertele then gives a straight line, namely,

$$h - h_0 = H - h_0 + Z \tan u' \quad (2.164)$$

In general the Bertele curve can be interpreted as follows: It shows for each ray the distortion. The slope of the best straight line through the points gives the distance Z of the best focal plane. Moreover, the slope of the plotted curve at each of its points is a measure of the meridional astigmatism for the corresponding ray.

If the object point had a symmetric image, the rays could be gathered in cones with vertices on its axis of symmetry; i.e., two rays on opposite sides of the principal ray would go through the same point. This would make the aberration curve have the form of an S, i.e., quasisymmetric. If the aberration curve is convex or concave, we can be sure that no symmetric image exists.

Another suggestion for utilizing the same data follows: We compute the intersection heights with two planes and make a graphical interpolation which helps to predict the trace of a large number of rays. Drawing these rays (Fig. 2.15) gives a very good graphical picture which shows the meridional trace of the caustic, the plane where the meridional image has the best contrast, and where the meridional image has the best resolution (in general these do not fall together).

28. The Spot-diagram Analysis and the Diapoint Plot

Not content with these methods, we have developed methods to ascertain the contribution of skew rays to the image formation in an effort to obtain more comprehensive information about the optical image, especially since, for systems of high aperture, in general the vignetting takes place only in the meridional plane and the skew-ray contributions to the image errors carry more weight than the meridional corrections.

The ideal would be to trace a large number of rays evenly distributed over the vignetted entrance

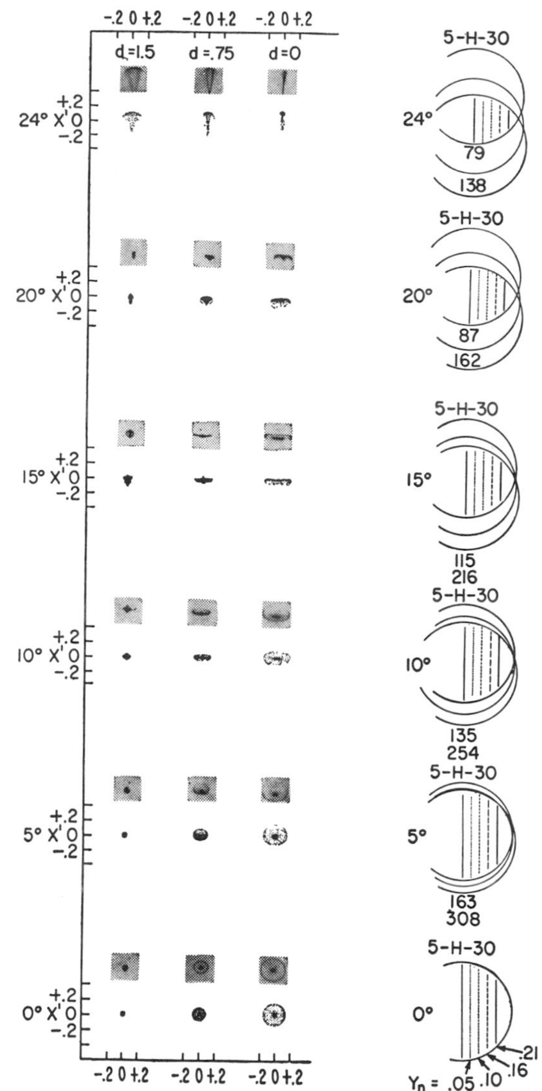

FIG. 2.18. Right: distribution of rays (traced or interpolated) at vignetted entrance pupil. Left: intersection of rays with planes perpendicular to the axis at distance 0, 0.75, 1.5 from the Gaussian focus. Upper figures are microphotographs of luminous points for the actual manufactured system.

pupil (so that every ray stands for the same amount of light) through the optical system, and plot (*a*) the intersection points with a set of image planes perpendicular to the axis, (*b*) the intersection points of the rays with the meridian plane (the diapoint). This has been done with the help of punched-card machines for a series of optical systems, including a wide-angle lens and a lens of extreme aperture ($f/0.7$).

However, the trace of a large number of skew rays is too laborious a task and cannot be undertaken in the general analysis of an optical system, where variations of all the correction elements are studied. The author has tried, therefore, to reduce the problem to the trace of a few rays, using interpolation formulas to compute the trace of a large number of rays.

This is the exact generalization of the method described in Sec. 27, where we traced a few meridional rays and plotted the results, combining the plotted points by curves, with the exception that the interpolation now becomes a two-dimensional interpolation, which has to be done analytically.

It has been found that the trace of four skew rays (aside from the traditional five meridional rays) from a given object point is sufficient to compute the coefficients of an interpolation formula. The vignetting diagram discloses the position of the chosen rays by their coordinates in the exit pupil. We trace the 12 rays through the optical system and determine the diapoints relating to the object point and the entrance pupil, i.e., the functions m, $\tilde{m}$, $Z' = z'/m$, $\tilde{Z}' = \tilde{z}'/\tilde{m}$.

These four functions are functions of

$$s_2 = xx_p + yy_p \qquad s_3 = \tfrac{1}{2}(x_p{}^2 + y_p{}^2) \qquad (2.165)$$

where x, y are the coordinates of the object point x_p and y_p of the point of intersection with the entrance pupil (it is no lack of generality to assume the object point in the yz plane, i.e., to assume $x = 0$, $s_2 = yy_p$, $2s_3 = x_p{}^2 + y_p{}^2$), and we determine

$$m = m_0 + m_2 s_2 + m_3 s_3 + \tfrac{1}{2}(m_{22}s_2{}^2 + 2m_{23}s_2 s_3 + m_{33}s_3{}^2) \qquad (2.166)$$

and analogously for $\tilde{m}$, Z', $\tilde{Z}'$, and for ζ'. The determination of the coefficients is done with the help of least-squares methods. Once these results have been obtained, (2.144) and (2.142) can be used to compute the intersection points with a set of planes perpendicular to the axis.

After thus obtaining interpolation formulas, it is an easy task to use these formulas to interpolate a large number of rays evenly distributed over the entrance pupil, i.e., the rays going through the points of the vignetted entrance pupil marked on the right of Fig. 2.18. We compute and plot first the intersection points with a series of planes perpendicular to the axis. A comparison of the figures on the left side of Fig. 2.18 signifies that such a plot (*a*) is in surprisingly good agreement with the plot obtained by tracing all the rays through the optical system and (*b*) is in good agreement with enlarged photographs taken of the image of a point light source through the manufactured optical system.

This plot, which we call a *spot diagram*, gives a good picture of the light distribution in the optical image; the intersection of the caustic with different planes is plainly evident by the maximum concentration of the spots. It is not difficult to find for each object point the position of maximum sharpness, where most of the stray light is captured in the large blob, and the position of maximum resolving power, where most of the light is concentrated at the expense of a generally fairly large amount of flare.

The analysis of the spot diagram can be used to draw curves showing the intensity distribution in the image, if the intensity distribution in the object is given.

Thus for the diagnosis. The spot diagrams seem to give for most optical systems a fairly good picture of their imaging qualities, and it is the experience of the author that in most optical systems the geometrical optical errors are so large that diffraction does not have to be considered. However, in those systems in which diffraction plays a role, the method here developed permits, in principle, a computation of diffraction effects.

It is assumed, though only recently proved by *G. Toraldo di Francia,* that diffraction through an optical instrument can be approximately calculated if one considers the geometric optical wave coming from the object point and calculates the diffraction pattern originating from the emerging wave.

Since the characteristic function can be used to compute the optical path from an object point to any point of the exit pupil, we obtain the distribution of phases along the exit pupil. Assuming the points of the vignetted exit pupil as coherent light sources with known amplitudes and phase differences, we can use Fresnel's method to determine the light vector at every point in space (excepting points near the plane of the exit pupil).

The spot diagrams, though given in our theory by a small number of parameters, are not too easy to analyze geometrically because in the planes in which we investigate them we have in general a heaping of singularities which defy simple analytical discussion. This is why the *approximation* methods

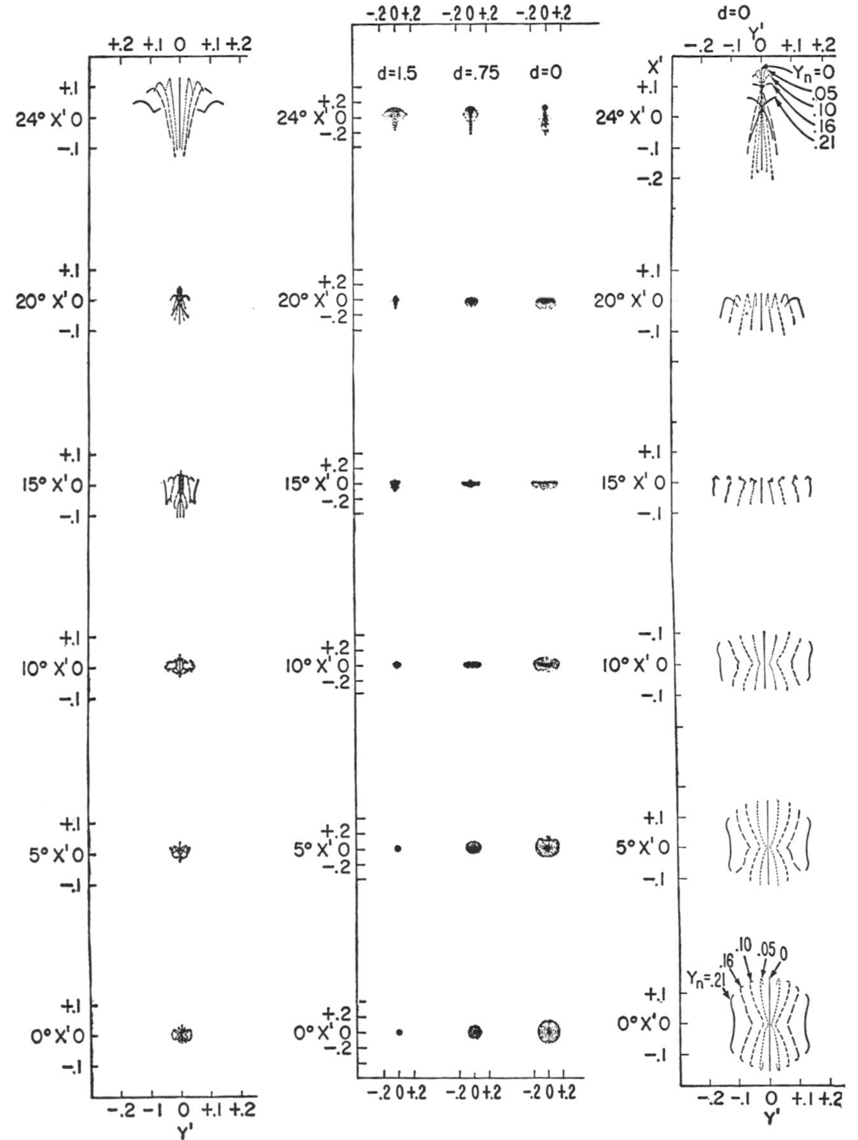

FIG. 2.19. Scanning of the image. The figures on the right and left are drawings of the curves corresponding to the rays through vertical lines of Fig. 2.18. The figures on the right correspond to $d = 0$, those on the left to $d = 1.5$. The reader will notice the heaping of singularities.

discussed in the literature have proved so unsatisfactory. To see this, we have in the next figure scanned the rays going through a set of parallel lines in the entrance pupil and given magnified pictures of the intersection with two planes. Viewing these alongside the photographed images of the point gives an idea of the complex process of image forming. The lines in the image plane are not conducive to simple analytical approximation.

Fortunately the analysis of the intersection points with the meridian plane is sufficient to give a complete analysis of the image. In Fig. 2.20 are plotted the intersection lines of the rays studied above with the meridian plane, or, more accurately, we have plotted m against z'. The diapoints belonging to equidistant points are marked on these lines with crosses.

These lines are fairly simple curves, and it can be shown that the coefficients of the development can

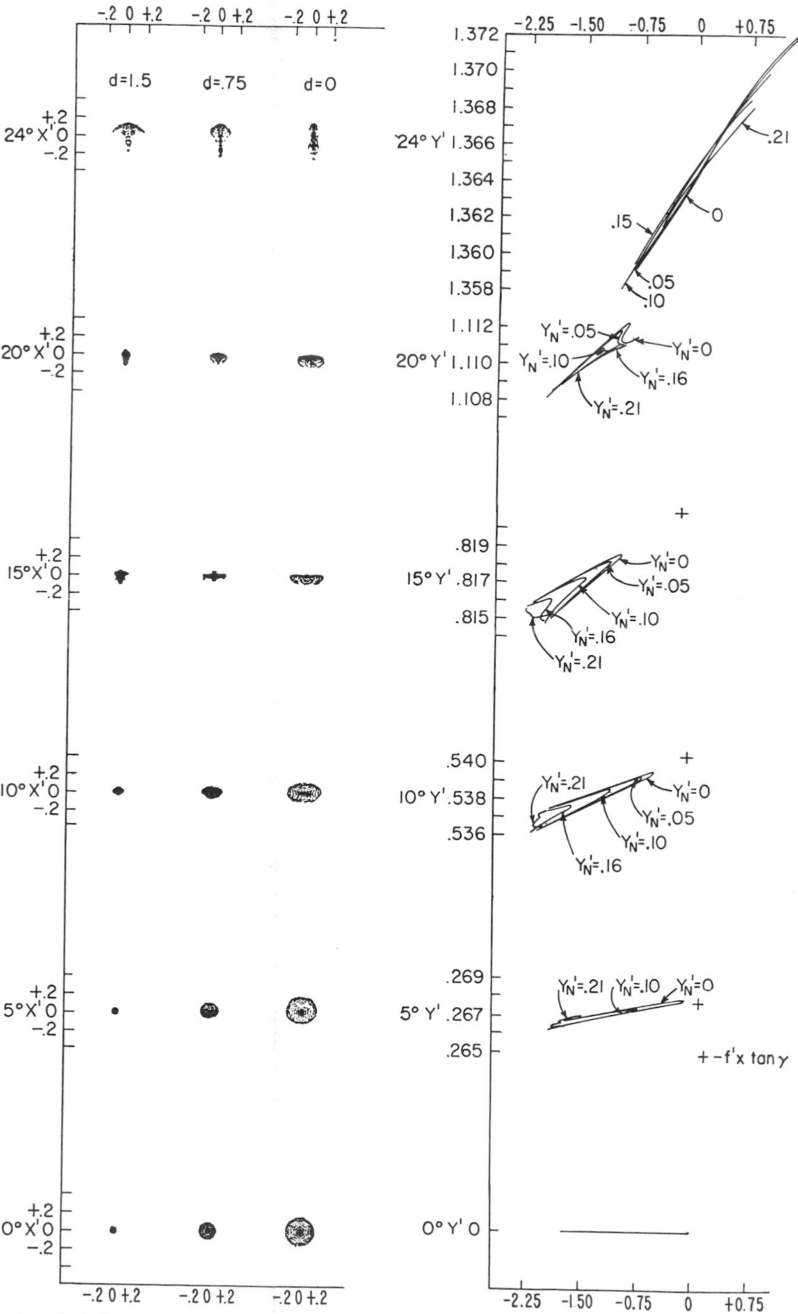

FIG. 2.20. Scanning the diapoint image for the same points as in Fig. 2.18 and Fig. 2.19. The reader will note the lack of singular points in these diagrams.

be easily interpreted. We recognize at a glance the best image plane. Since all the points lie on a line if and only if the image is symmetrical, we recognize the deviations from symmetry.

IV. THERAPY

29. Correction of an Optical System

The methods of correcting an optical system have not yet been developed in a satisfactory scientific manner. We shall try to give only some hints which may lead to a scientific method in the future.

The method applied most commonly is that of trial and error; i.e., one of the data of the system, a radius, a glass thickness, an air space, or a glass, is varied. All the rays are traced through the changed system. A change table is made out for some of the errors, and an attempt is made to balance small changes, until an optimum is achieved. Unfortunately this method leads in general only to a gross type of correction.

A small change in a correction element usually has practically the same effect on all rays. Combinations of small changes can generally not correct so-called "zonal" errors. *The art of the designer is to combine two or more large changes in opposite directions to obtain the desired results.*

The designer is here helped by some practical considerations. Rays going through the center of a sphere are undeviated, and therefore a small variation of such a surface will have much less influence on those rays than on rays which have large angles of incidence.

The rays through the vertex of a sphere and the rays through its aplanatic points are sharply imaged. All these facts are of great value to the lens designer. The following method of analyzing the diapoint diagram ray by ray and surface by surface is an attempt to give a scientific background to the trial-and-error methods.

Given an optical system consisting of a number of spheres of radii r_ν (positive if convex, negative if concave toward the light) and center distances c_ν. We have

$$c_\nu = r_{\nu+1} - r_\nu + d_\nu \tag{2.167}$$

where d_ν is the lens thickness or the distance between two consecutive vertices. The diapoint calculation surface by surface can now be given for a system of spheres if we assume the object point and therefore all its diapoints to lie in the yz plane. Assuming z_ν to be the distance from the νth center, we have

$$\begin{aligned} y_2 &= m_1 y_1 & y_3 &= m_2 y_2 \\ z_2 &= m_1 z_1 + c_1 & z_3 &= m_2 z_2 + c_2 \end{aligned} \tag{2.168a}$$

etc., which gives combined

$$\begin{aligned} y' &= m_1 m_2 \cdots m_{k-1} y \\ z' &= m_1 m_2 \cdots m_{k-1} z + m_2 \cdots m_{k-1} c_1 \\ &\quad + m_3 \cdots m_{k-1} c_2 + \cdots \\ &\quad + m_{k-1} c_{k-2} + c_{k-1} \end{aligned} \tag{2.168b}$$

Inserting for the c_i the more tangible lens thicknesses, we obtain, taking instead of the center distance z the vertex distance s,

$$\begin{aligned} y' &= My \\ s' &= Ms + C \end{aligned} \tag{2.169}$$

with $M = m_1 m_2 \cdots m_{k-1}$

$$\begin{aligned} C = {} & d_1 m_2 \cdots m_{k-1} + d_2 m_3 \cdots m_{k-1} \\ & + \cdots + d_{k-2} m_{k-1} + d_{k-1} \\ & + r_1(m_1 - 1) m_2 m_3 \cdots m_{k-1} \\ & + r_2(m_2 - 1) m_3 \cdots m_{k-1} + \cdots \end{aligned} \tag{2.170}$$

Plotting log M and considering it as the sum of the logarithms of the partial magnifications, we can see ray by ray the lateral contribution of the different surfaces. The factors of d_ν and r_ν in (2.170) will give a hint about the contributions of radii and thicknesses on the longitudinal contributions. This analysis takes account only of the diapoint coordinates. The following system of formulas, which are interesting in themselves, take account of the directions of the rays:

The tracing formulas for the direction cosines are

$$\begin{aligned} \eta_{\nu+1} &= -\phi_\nu y_\nu + \frac{1}{m_\nu} \eta_\nu \\ \zeta_{\nu+1} &= -\phi_\nu z_\nu + \frac{1}{m_\nu} \zeta_\nu \end{aligned} \tag{2.171}$$

Combining these with (2.167), i.e., eliminating m_ν, we find for the y coordinates

$$\frac{1}{y_{\nu+1}\eta_{\nu+1}} = \frac{\phi_\nu}{\eta_\nu \eta_{\nu+1}} + \frac{1}{y_\nu \eta_\nu} \tag{2.172}$$

Applied to a set of surfaces, we find

$$\frac{1}{y'\eta'} = \frac{1}{y\eta} + \sum \frac{\phi_\nu}{\eta_\nu \eta_{\nu+1}} \tag{2.173}$$

The formulas for z' and ϕ' are not quite so simple; we have

$$\frac{1}{z'_\nu \zeta_{\nu+1}} = \frac{\phi_\nu}{\zeta_\nu \zeta_{\nu+1}} + \frac{1}{z_\nu \zeta_\nu} \tag{2.174}$$

where $z_{\nu+1}$ is not equal to z'_ν but

$$z_{\nu+1} = z'_\nu - c_\nu \tag{2.175}$$

where c_ν is the center distance of two consecutive spheres. Hence we can express z' with the help of Gaussian brackets,

$$z'\zeta' = \frac{\left[z\zeta, \dfrac{\phi_1}{\zeta_1\zeta_2}, -c_1\zeta_2, \dfrac{\phi_2}{\zeta_2\zeta_3}, -c_2\zeta_3, \ldots, -c_{n-1}\zeta_n\right]}{\left[z\zeta, \dfrac{\phi_1}{\zeta_1\zeta_2}, -c_1\zeta_2, \dfrac{\phi_2}{\zeta_2\zeta_3}, -c_2\zeta_3, \ldots, \dfrac{\phi_n}{\zeta_n\zeta_{n+1}}\right]} \tag{2.176}$$

a formula which again permits an estimate of the influence of a correction element. In (2.176) we can for approximate calculation set

$$\frac{\phi_\nu}{\zeta_\nu \zeta_{\nu+1}} \sim \frac{n_{\nu+1} - n_\nu}{r_\nu n_\nu n_{\nu+1}} = \frac{1}{r_\nu n_\nu} - \frac{1}{r_\nu n_{\nu+1}} \tag{2.177}$$

which is frequently called the *Petzval contribution* of the surface.

Since the diapoint coincides with the sagittal focus for meridional rays, (2.173) can be used to compute sagittal focus on the meridional rays. In this form it was found and applied by Max Lange.

Similarly we can derive from (2.171) and (2.167) the formula

$$\eta_{\nu+1}y_{\nu+1} = -\phi_\nu y_\nu y_{\nu+1} + \eta_\nu y_\nu$$

or

$$\eta' y' = \eta y - \Sigma\phi_\nu y_\nu y_{\nu+1} \qquad (2.178)$$

and correspondingly

$$z'\zeta' = z\zeta - \Sigma\phi_\nu m_\nu z_\nu^2 + \Sigma c_\nu \zeta_{\nu+1} \qquad (2.179)$$

V. PROPHYLAXIS

30. Introduction

The precalculation of optical systems is an art which can be performed in many different ways. The main object is to compute a system which has only a few errors to start with.

The concept of a thin lens or of a system of thin lenses is very helpful. We neglect the lens thicknesses but not the distances between the lenses. Since lens thicknesses in optical systems are usually small compared with the focal length of the lenses, the introduction of thicknesses into the final lens system will be a minor correction problem, which can generally be solved by differential changes.

The Gaussian optics of a thin lens in air is very simple; the power is given by

$$\phi = (n - 1)(\rho_1 - \rho_2) \qquad (2.180)$$

where n is the refractive index of the lens and ρ_1 and ρ_2 the curvatures of the front and back surfaces.

The nodal points and cardinal points coincide at the "vertex" of the thin lens. This shows that changing the two radii in such a way that $\rho_1 - \rho_2$ remains constant does not influence the first-order qualities of a lens. This operation is called *bending*, and it is one of the tools of the designer, who by bending can change the imaging qualities of a system without influencing the Gaussian correction.

Attempts have been made to compute the Seidel errors and the fifth-order errors of a thin lens or of a system of thin lenses with finite distances. But the formulas are, on the whole, difficult to use, and the approximation is not good enough to warrant the amount of work involved.

An attempt of M. Berek [3] to use finite calculation and Seidel aberration coefficients together is ingenious and, in spite of some inaccuracies, beneficial (see Boegehold [6]).

One result of Seidel's theory is that an object can have a sharp image on a plane if and only if the so-called "Petzval sum" is corrected.

The Petzval condition can be expressed as

$$\sum \frac{\phi_k}{n_k} = 0 \qquad (2.181)$$

It is the problem of the optical designer to watch the value of the Petzval sum in his designs.

It has been found that the condition that an optical system is corrected within the limits of Gaussian optics for color and color magnification is such a strong restriction that, taken together with the fulfillment of the Petzval condition, it constitutes a basis for precalculation.

Having found a system of thin lenses with the desired geometrical qualities (back focus, focal length, and usable field angle), thicknesses have to be introduced to obtain the necessary aperture. Bending of the individual lenses is made first to correct spherical aberration and sine condition, and then combinations of bendings are introduced which do not change the correction of the axis point but help to correct the field.

We shall first study the properties of glass and finally give formulas to correct the color aberrations of thin lenses with finite distances.

31. Dispersion of Glass

Throughout this chapter we have considered the refractive index of an optical material as constant for a homogeneous isotropic medium. This is exactly true only if we consider monochromatic light, i.e., light of only one wavelength. In general, n depends on the color of the light; i.e., the refractive index n is a function of the wavelength λ. The dependence of n on the wavelength λ for a transparent substance is a complicated phenomenon, and many hypothetical "dispersion" formulas have been proposed to describe it.

The most commonly used formula is the Sellmeier-Helmholtz formula, which can be interpreted as follows: Let $\lambda_1, \lambda_2, \ldots, \lambda_k$ be the absorption bands of the optical material. Then the square of the refractive index can be given by a formula

$$n^2 = n_1^2 + \sum \frac{A_i}{\lambda^2 - \lambda_i^2} \qquad (2.182)$$

Quantum theory suggests that the λ_i do not correspond exactly to the points of discontinuity.

The author [16, 17], following some slightly deviating ideas, has suggested

$$n = n_1 + \sum \frac{A_i}{\lambda^2 - \lambda_i^2} \qquad (2.183)$$

as a dispersion formula and has found that such a formula fits the available data just as well as or even better than the Sellmeier-Helmholtz formula. In particular, for the most important optical region, the visual spectrum from $\lambda = 4000$ to $\lambda = 7680$ A, the dispersion is primarily determined by one absorption band which is in the near ultraviolet and a second band in the far infrared.

For optical purposes it would be very convenient to have a dispersion formula which is linear in its constants. The following four-constant formula has been shown to represent the refractive indices of optical materials with an excellent degree of accuracy (± 1 in the fifth decimal) over the visible spectrum and with sufficient accuracy over the whole optically important spectrum from 3650 A in the ultraviolet to 10,000 A in the infrared. The formula is

$$n = n_0 + n_1\lambda^2 + \frac{n_2}{\lambda^2 - 28 \times 10^5} + \frac{n_3}{(\lambda^2 - 28 \times 10^5)^2} \qquad (2.184)$$

The formula consists of the first terms of a Taylor series corresponding to the influence of the far absorption band and the first terms of a Laurent series corresponding to an absorption band near $\lambda = 1680$ A.

The author considers (2.183) as a theoretically satisfactory formula and (2.184) as an excellent practical approximation for values within the limits mentioned above. It has been verified that all the glasses commercially available, as well as crystals like fluorite, lithium fluoride, quartz, etc., can be represented by formula (2.184). This means that within the spectral region from 3650 to 10,000, optical materials form a four-dimensional manifold and any four data determine the dispersion of the glass, since one can compute n_0, n_1, n_3 in (2.184). Most of the regular glasses are, however, already determined by two data, which in the optical literature usually are n_D, the refractive index for the D line in the solar spectrum ($\lambda = 5893$), and the so-called "Abbe number,"

$$\nu = \frac{n_D - 1}{n_C - n_F} \tag{2.185}$$

where C and F are lines with wavelengths of $\lambda_C = 6563$ and $\lambda_F = 4861$.

However, two data are not sufficient to characterize optical materials. Equation (2.184) suggests that four data might suffice. Let us define $\lambda_* = 10{,}140$ A and $\lambda_{**} = 3650$ A. Taking, in addition, $\lambda_C = 6563$ A and $\lambda_F = 4861$ A, we have four such data.

We now claim that there exist four universal functions of λ, namely, a_*, a_C, a_F, and a_{**}, such that, for all substances, n_λ is determined by

$$n_\lambda - n_F = a_*(n_* - n_F) + a_C(n_C - n_F) + a_{**}(n_{**} - n_F)$$

if n_*, n_C, n_F, and n_{**} are given. This is equivalent to

$$n_\lambda = a_*n_* + a_Cn_C + a_Fn_F + a_{**}n_{**} \tag{2.186}$$

with

$$a_* + a_C + a_F + a_{**} \equiv 1$$

We can show that a_*, a_C, a_F, and a_{**} can be taken as the universal functions of the form (2.184) which assume the value unity for their respective wavelengths and the value zero for all other wavelengths. They are therefore determined theoretically, without use of empirical data of the materials. The value of the four functions is given in Table 2.1.

TABLE 2.1. VALUES OF UNIVERSAL FUNCTIONS APPEARING IN EQ. (2.186) FOR TWELVE SELECTED WAVELENGTHS†

λ(angstroms)	Line	a_*	a_C	a_F	a_{**}
10,140	*	+1.000000	0.000000	0.000000	0.000000
7682	A′	+0.192687	+1.051955	−0.276197	+0.031555
6563	C	0.000000	+1.000000	0.000000	0.000000
6438	C′	−0.012813	+0.966511	+0.051075	−0.004774
5893	D	−0.045919	+0.744598	+0.326272	−0.024952
5876	d	−0.046326	+0.735480	+0.336338	−0.025492
5461	e	−0.043764	+0.479049	+0.597795	−0.033080
4861	F	0.000000	0.000000	1.000000	0.000000
4800	F′	+0.006659	−0.052699	+1.036699	+0.009340
4358	g	+0.057036	−0.394569	+1.193553	+0.143980
4047	h	+0.075791	+0.487634	+1.045064	+0.366779
3650	**	0.000000	0.000000	0.000000	+1.000000

† From *Applied Optics*, June, 1963, p. 555, a copyrighted publication. Permission has been granted to reprint this table in this edition of "Handbook of Physics."

The treatment of color problems is simplified if we replace Abbe's ν value (2.185) by a similar concept. We define

$$\nu_C = \frac{n_F - 1}{n_C - n_F} \tag{2.187}$$

and P_λ, the partial dispersion, by

$$P_\lambda = \frac{n_\lambda - n_F}{n_C - n_F} \tag{2.188}$$

Introducing the abbreviation

$$\nu_\lambda = \frac{n_F - 1}{n_\lambda - n_F} \tag{2.189}$$

we see that

$$P_\lambda = \frac{\nu_C}{\nu_\lambda} \tag{2.190}$$

For a thin lens, the color aberration for an arbitrary wavelength is then given by

$$\phi_\lambda - \phi_F = \frac{\phi}{\nu_\lambda} = \frac{\phi}{\nu_C} P_\lambda \tag{2.191}$$

and for a system of thin lenses we obtain

$$\Phi_\lambda - \Phi_F = \sum \frac{\phi}{\nu_C} P_\lambda \tag{2.192}$$

With two lenses we can, in general, correct a lens for only two colors [see (2.202)]. The correction for a third color would demand that

$$\begin{aligned} &\frac{\phi_1}{\nu_{1C}} + \frac{\phi_2}{\nu_{2C}} = 0 \\ &\frac{\phi_1}{\nu_{1\lambda}} + \frac{\phi_2}{\nu_{2\lambda}} = \frac{\phi_1}{\nu_{1C}} P_{1\lambda} + \frac{\phi_2}{\nu_{2C}} P_{2\lambda} = 0 \end{aligned} \tag{2.193}$$

This requires

$$P_{1\lambda} = P_{2\lambda} \tag{2.194}$$

which is rare, and it is impossible to correct two lenses for more than three colors.

However, with three lenses correction can be achieved for four colors, for instance, λ_C, λ_F, λ_*, and λ_{**}. If ϕ_1, ϕ_2, and ϕ_3 are the powers of the three lenses for λ_F, the conditions are

$$\begin{aligned} &\frac{\phi_1}{\nu_1} + \frac{\phi_2}{\nu_2} + \frac{\phi_3}{\nu_3} = 0 \\ &\frac{\phi_1}{\nu_1} P_{1*} + \frac{\phi_2}{\nu_2} P_{2*} + \frac{\phi_3}{\nu_3} P_{3*} = 0 \\ &\frac{\phi_1}{\nu_1} P_{1**} + \frac{\phi_2}{\nu_2} P_{2**} + \frac{\phi_3}{\nu_3} P_{3**} = 0 \end{aligned} \tag{2.195}$$

These three equations can be fulfilled only if the plot of P_* versus P_{**} for all glasses shows three glasses lying on a straight line. This was formerly considered impossible, but Fig. 2.21 shows that there are many triads of glasses obeying this condition. Lens systems utilizing such triads and corrected for four wavelengths are called *superachromats* [19]. Reference 18 shows that they are then corrected not only for four wavelengths but for all intermediate wavelengths, that is, for the whole visible spectrum. Examples of such systems are given in [19].

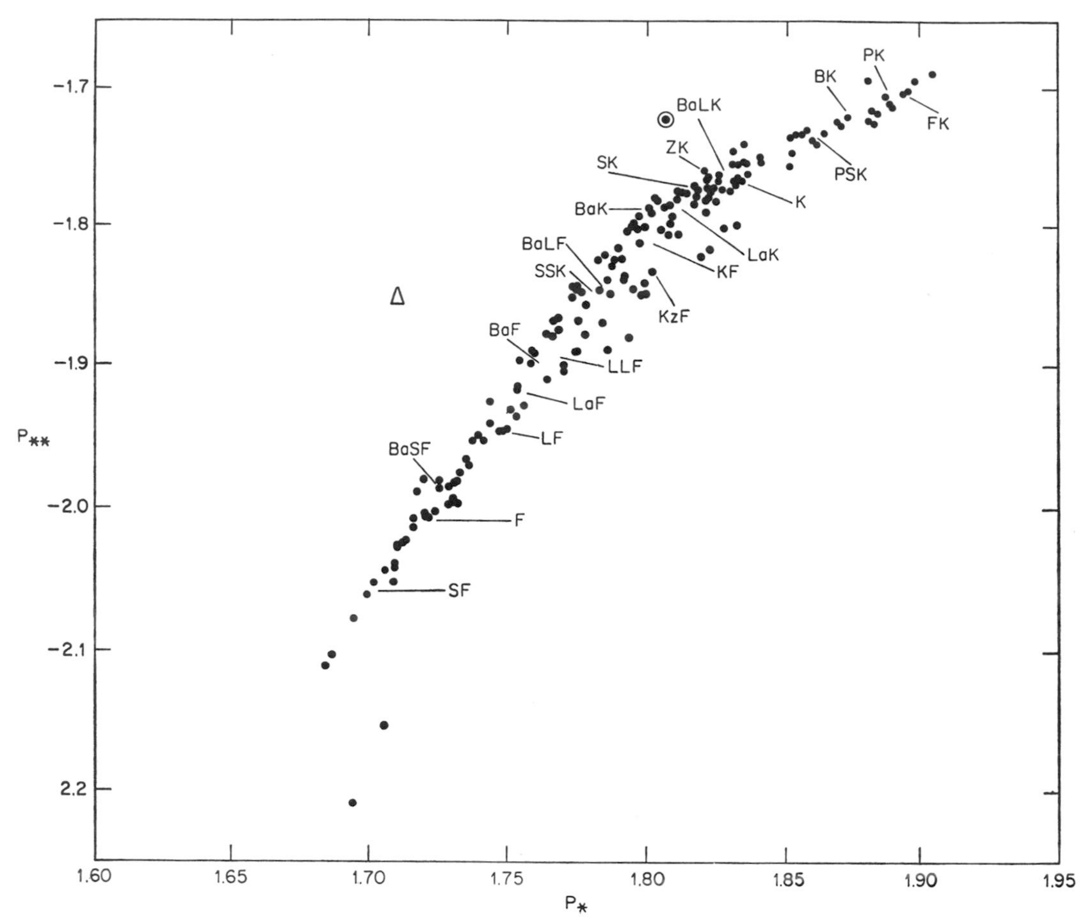

FIG. 2.21. Plot of $P_{**} = (n_{**} - n_F)/(n_C - n_F)$ versus $P_* = (n_* - n_F)/(n_C - n_F)$ for selected glasses, largely from Schott catalogue 350-E, and certain other materials (O—fluorite; Δ—methyl metacrylate). The types of glass are designated as in the Schott catalogue. The materials used for a superachromat must lie on a straight line in this plot. *(From Applied Optics, June,* 1963, *p.* 555, *a copyrighted publication. Permission has been granted for its use.)*

32. Color-corrected System of Thin Lenses

We can now compute the color correction of an optical system of thin lenses with finite distances and write down the condition that a system be color-corrected. For an object at a distance s from the first lens, assuming Gaussian optics, we find by the method of Sec. 23 that the image distance s from the last lens and the magnification are given by Gaussian brackets as

$$s' = \frac{[s, \phi_1, -e_1, \ldots, -e_{k-1}]}{[s, \phi_1, -e_{k-1}, \ldots, \phi_k]} \qquad m = \frac{1}{[s, \phi_1, \ldots, \phi_k]} \tag{2.196}$$

The ϕ_i are the powers of the single lenses. The power of a lens is given by

$$\phi = (n - 1)(\rho_1 - \rho_2)$$

where ρ_1 and ρ_2 are the front and back curvatures of the lens. Because of the linearity of the equation, logarithmic differentiation gives

$$\frac{1}{\phi}\frac{d\phi}{d\lambda} = \frac{1}{n-1}\frac{dn}{d\lambda}$$

which is approximately equal to $1/\nu_\lambda$:

$$\frac{1}{\nu_\lambda} = \frac{\phi_\lambda - \phi_F}{\phi_F} = \frac{1}{\nu_C} P_\lambda$$

Assuming s to be the same for all wavelengths, i.e., considering a fixed object, the image is free from longitudinal and lateral color if and only if

$$\frac{ds'}{d\lambda} = \frac{dm}{d\lambda} = 0 \tag{2.197}$$

or if

$$\frac{d}{d\lambda}(s[\phi_1 \cdots \phi_k] + [-e_1 \cdots \phi_k]) = 0$$
$$\frac{d}{d\lambda}(s[\phi_1 \cdots -e_{k-1}] + [-e_1 \cdots -e_{k-1}]) = 0 \tag{2.198}$$

Introducing

$$\begin{aligned} A &= \frac{d}{d\lambda}[\phi_1, -e_1 \ldots -e_{k-1}] \\ B &= -\frac{d}{d\lambda}[-e_1,\phi_2 \ldots -e_{k-1}] \\ C &= -\frac{d}{d\lambda}[\phi_1 \cdots \phi_k] \\ D &= \frac{d}{d\lambda}[-e_1 \cdots \phi_k] \end{aligned} \tag{2.199}$$

or

$$\begin{aligned} A &= \frac{\phi_1}{\nu_1}[-e_1 \cdots -e_{k-1}] \\ &\quad + \frac{\phi_2}{\nu_2}[-e_1][-e_2 \cdots -e_{k-2}] + \cdots \\ -B &= \frac{\phi_2}{\nu_2}[-e_1][-e_2 \cdots -e_{k-1}] \\ &\quad + \frac{\phi_3}{\nu_3}[-e_1, \phi_2, -e_2][-e_3 \cdots -e_{k-1}] + \cdots \\ -C &= \frac{\phi_1}{\nu_1}[-e_1 \cdots \phi_k] \\ &\quad + \frac{\phi_2}{\nu_2}[\phi_1, -e_1][-e_2 \cdots \phi_k] + \cdots \\ D &= \frac{\phi_2}{\nu_2}[-e_1][-e_2 \cdots \phi_k] \\ &\quad + \frac{\phi_3}{\nu_3}[-e_1, \phi_2, -e_2][-e_3 \cdots \phi_k] + \cdots \end{aligned}$$

(where the ν value should be taken for the colors for which we want to correct the system), we find that an object is color-corrected if and only if

$$AD = BC \tag{2.200}$$

and we have in this case for the object and image distances

$$\begin{aligned} s &= \frac{B}{A} = \frac{D}{C} \\ s' &= \frac{B}{D} = \frac{A}{C} \end{aligned} \tag{2.201}$$

If we want to investigate the color correction for a third color, we have to replace in the above formulas ϕ/ν by $(\phi/\nu)P_\lambda$.

For a system of thin lenses without separations (a *monoplet*) we have

$$\begin{aligned} A &= 1 \qquad -C = \sum \frac{\phi_k}{\nu_k} \\ B &= 0 \qquad D = 0 \end{aligned} \tag{2.202}$$

That is, the monoplet is corrected for all object distances and all magnifications if $\Sigma\phi_k/\nu_k$ is corrected, i.e., if it is an achromat.

For a *duplet*, a system of lenses with a single distance e_1, we have

$$\begin{aligned} A &= \frac{\phi_1}{\nu_1}e_1 \qquad -C = \frac{\phi_1}{\nu_1}(1 - e_1\phi_2) \\ &\qquad\qquad\qquad + \frac{\phi_2}{\nu_2}(1 - e_1\phi_1) \\ B &= 0 \qquad -D = e_1\frac{\phi_2}{\nu_2} \end{aligned} \tag{2.203}$$

The condition for color correction $(AD - BC = 0)$ shows that, in general, a duplet cannot be corrected for longitudinal and lateral color, even for two wavelengths.

For a *triplet*, however, we find

$$\begin{aligned} A &= \frac{\phi_1}{\nu_1}[-e_1,\phi_2, -e_2] + \frac{\phi_2}{\nu_2}[\phi_1, -e_1][-e_2] \\ -B &= \frac{\phi_2}{\nu_2}[-e_1][-e_2] \\ -C &= \frac{\phi_1}{\nu_1}[-e_1,\phi_2, -e_2,\phi_3] \\ &\quad + \frac{\phi_2}{\nu_2}[\phi_1, -e_1][-e_2,\phi_2] + \frac{\phi_3}{\nu_3}[\phi_1, -e_1,\phi_2, -e_2] \\ D &= \frac{\phi_2}{\nu_2}[-e_1][-e_2,\phi_3] + \frac{\phi_3}{\nu_3}[-e_1,\phi_2, -e_2] \end{aligned} \tag{2.204}$$

Color correction is achieved if

$$\begin{aligned} AD &= BC = 0 = \frac{\phi_1\phi_2}{\nu_1\nu_2}[-e_1]^2 \\ &\quad + \frac{\phi_1}{\nu_1}\frac{\phi_3}{\nu_3}[-e_1,\phi_2, -e_2]^2 + \frac{\phi_2\phi_3}{\nu_2\nu_3}[-e_2]^2 = 0 \end{aligned} \tag{2.205}$$

Correcting for three or four wavelengths would demand these equations to hold for all the respective ν_λ's.

APPENDIX

33. Intensity Considerations

An important problem is the study of the intensity of light transmitted through an optical system. The Lagrange invariant leads to the conclusion that the energy flux through a surface element is independent of the position of the surface element on which we catch it, a statement which is equivalent to saying that the total brightness in an optical instrument cannot be increased. However, brightness can be diminished:

By reflection. One finds about 4 per cent reflection loss per untreated surface, a fact which has led manufacturers of optical systems in recent years to treat their lenses, i.e., to put on the surfaces a thin interference layer which has the property that the multiply reflected light suffers destructive interference, so that more light is transmitted.

By absorption. Absorption increases exponentially with glass thickness and is one reason why thick lenses are not so frequently used in optics as the purely geometrical properties would demand. Absorption is also a function of wavelength and increases toward the violet end of the spectrum.

Every luminous object (every light source) has a radiation diagram; i.e., it radiates with different intensities in different directions. For most objects one obtains a good approximation by assuming that the radiated intensity is proportional to the square of the cosine which the principal ray of the bundle forms with the normal to the radiating object (and analogously with the normal to the receiving image plane).

If the aperture of the objective is small, *if* the

exit pupil is at the nodal point, and *if* the principal rays have no distortion, the illumination at a plane perpendicular to the axis would be proportional to the fourth power of the cosine of the angle which the principal rays forms with the axis on object and image side.

This rule of thumb has been frequently misused and applied to cases where one or all of the mentioned "ifs" are not fulfilled. This problem has found extensive attention in the papers by G. Slussareff [35], M. Reiss [31], and I. C. Gardner [12] and in the book by E. Wandersleb [40].

34. Some Historical Remarks

There are allegedly two books by *Euclid* [11], one on perspective and one on reflection. *Hero* of Alexandria [15] derived the *reflection laws* from a minimum principle and discussed the images in cylindrical and conical mirrors.

Ptolemy [30] and Alhazen [2] gave correct data, Vitellius false ones, for the connection between incident and refracted ray for air and glass, air and water, etc.

Kepler [24], in his commentaries on Vitellius, would have found the law of refraction if he had not trusted the data of his predecessor too much. He developed the theory of Gaussian optics for systems of thin lenses. *Descartes* [10] found the laws of refraction and discussed the cartesian surfaces, surfaces that image an object point sharply.

Newton [27] and *Huygens* [23] made an important science of geometrical optics. Huygens discussed the wave surfaces of light and showed that in refraction and reflection the light rays are normal to a system of wave surfaces. He discussed the caustic surface. Newton treated the dispersion of light in one of the most readable original works of science. He found the error of spherical aberration and discussed the color aberrations of lenses and prisms.

W. R. Hamilton [14], probably the most profound thinker in this field, developed optics and mechanics parallel to each other, starting from the variation principle. He introduced the different types of characteristic functions and thus laid the foundations for attacking optical problems in the large.

L. Schleiermacher [32] developed the theory of vignetting. He alone, up to the present, suggested using the method of least squares to investigate improvement in image quality, for a variation of system data.

L. Seidel [34] and *J. Petzval* [29] developed systematically the approximation theory of image errors. Seidel developed the so-called "Seidel aberrations," whereas a report on Petzval's lost investigations shows that he considered aberrations of higher order.

The greatest contributions to optical knowledge in our time have been made by *A. Gullstrand* [13], the only scientist in the field of geometrical optics to win the Nobel prize, and by *T. Smith* and *H. Boegehold*. (F. Zernike was awarded the Nobel prize in 1953 for his work in phase contrast.)

Gullstrand investigated especially the approximation theory for the rays near a principal ray, trying to find geometrical interpretations for the terms in the Taylor development. He has fought many of the erroneous assumptions which have entered optical textbooks, for instance, the assumption that the rays in the neighborhood of a principal ray can justly be represented by a *Sturm* conoid, i.e., by the ray going through two line elements perpendicular to itself and to the principal ray. His studies of the human eye have given a deeper insight into this most important of all optical instruments.

T. Smith's [36–38] analysis of the problem of ray tracing in optical systems, his versatility in using the angle characteristic and computing its form for optical systems which have special image-forming qualities, his analysis of higher-order image errors in dependence on the position of object and stop, and especially the variety of his ideas and techniques in naming and solving problems of optics have made him the teacher of the current generation of optical scientists.

H. Boegehold has published many papers on historical aspects of geometrical optics. Moreover, he has used the different characteristic functions of Hamilton-Bruns in solving many of the problems in the large and showing the limitations of optical image formation. His paper [4] gives in concise form a survey of all the results in the first thirty years of this century.

References and Bibliography

1. Abbe, Ernst: "Gesammelte Abhandlungen," Fischer, Jena, 1904–1906.
2. Alhazen: "Opticae thesaurus Alhazeni, etc.," Risner (ed.), Basileae, 1572.
3. Berek, M.: "Grundlagen der praktischen Optik," de Gruyter, Berlin, 1930.
4. Boegehold, H.: Ueber die Entwicklung der Theorie der optischen Instrumente seit Abbe, *Ergeb. exakt. Naturwiss.*, vol. 8, 1929.
5. ———: Raumsymmetrische Abbildung, *Z. Instrumentenk.*, **56**: 98–109 (1936).
6. ———: Review of "Grundlagen der praktischen Optik" by Berek, *Naturwiss.*, **19**: 425–427 (1931).
7. ——— and M. Herzberger: Kann man zwei verschieedene Flächen durch dieselbe Folge von Umdrehungsflächen scharf abbilden?, *Compositio Mathematica*, **1**: 1–29 (1935).
8. ——— and M. Herzberger: Kugelsymmetrische Systeme, *Z. angew. Math. und Mech.*, **15**: 157–178 (1935).
9. Bruns, H.: Das Eikonal, *Leipzig. Sitzber.*, **21**: 321–436 (1895); also as a book, S. Hirzel, Leipzig, 1895.
10. Descartes, René: La dioptrique in "Oeuvres de Descartes," vol. VI, C. Adam and P. Tannery (eds.), Paris, 1902.
11. Euclid: Catoptrica in "Collected Works of Euclid," L. I. Heiberg (ed.), Teubner, Leipzig, 1895.
12. Gardner, I. C.: Validity of the Cosine Fourth-power Law of Illumination, *J. Research Natl. Bur. Standards*, **39**: 213–219 (1947).
13. Gullstrand, A.: Allgemeine Theorie der monochromatischens Aberrationen, etc., *Acta Regial Soc. Sci. Upsala*, V, **3**: (1900); Die Reelle optische Abbildung, *Svenska Vetensk. Handl.*, **41**: 1–119 (1906); Tatsachen und Fiktionen in der Lehre der optischen Abbildung, *Arch. Optik*, **1**: 1–4, 81–97 (1907); Das allgemeine optischen Abbildungssystem, *Svenska Vetensk. Handl.*, **55**: 1–139 (1915).
14. Hamilton, Sir William Rowan: "The Mathematical Papers of Sir William Rowan Hamilton," vol. I, Geometrical Optics, Cambridge University Press, New York, 1931; also German edition by George Prange, "Abhandlung zur Strahlenoptik," Akademische Verlagsgesellschaft, Leipzig, 1933.

15. Hero: See E. Wilde, "Geschichte der Optik," Teil I, Rücker and Püchler, Berlin, 1833.
16. Herzberger, M.: The Dispersion of Optical Glass, *J. Opt. Soc. Am.*, **32**: 70–77 (1942).
17. ——— and H. Jenkins: Color Correction in Optical Systems and Types of Glass, *J. Opt. Soc. Am.*, **39**: 984–989 (1949).
18. ———: Colour Correction in Optical Systems and a New Dispersion Formula, *Opt. Acta*, **6**: 197–215 (1959).
19. ——— and N. R. McClure: The Design of Superachromatic Lenses, *Appl. Opt.*, **2**: 553–560 (1963).
20. ———: The Limitations of Optical Image Formation, *Ann. N.Y. Acad. Sci.*, **48**: 1–30 (1946). (Cressy Morrison prize.)
21. ———: Theory of Image Errors of the Fifth Order in Rotationally Symmetrical Systems, I, *J. Opt. Soc. Am.*, **29**: 395–406 (1939).
22. Hopkins, H. H.: "Wave Theory of Aberrations," Oxford University Press, Fair Lawn, N.J., 1950.
23. Huygens, C.: "Traité de la lumière," Van der Aa, Leiden, 1690; reprinted by Gressner and Schramm, Leipzig, 1885. (English translation by S. P. Thompson, Macmillan, London, 1912.)
24. Kepler, J.: "Ad Vitellionem Paralipomena, etc.," Marnium and Aubrii (eds.), Frankfurt, 1604.
25. Kohlschütter, A.: Die Bildfehler fünfter Ordnung optischer Systeme, inaugural dissertation, Kaestner, Göttingen, 1908.
26. Maxwell, J. C.: On the Application of Hamilton's Characteristic Function, etc., *Proc. London Math. Soc.*, **6**, 117–122, 1874–1875.
27. Newton, I.: A New Theory about Light and Colours, *Phil. Trans.*, **6** (1672); see also "Opticks," Smith and Walford (eds.), London, 1704; reprinted by Dover, New York, 1952.
28. Nijboer, R. A.: The Diffraction Theory of Optical Aberrations, *Physica*, **13**: 605–620 (1947).
29. Petzval, J.: Bericht über dioptrische Untersuchungen, *Wien. Sitzber.*, **26**: 33–90 (1857).
30. Ptolemy: "L'ottica di Claudio Tolomeo," G. Gooi (ed.), G. B. Paravin, Turin, 1885.
31. Reiss, M.: The $\cos^4$ Law of Illumination, *J. Opt. Soc. Am.*, **35**: 283–288 (1945); see also Notes on the $\cos^4$ Law of Illumination, *J. Opt. Soc. Am.*, **38**: 980–986 (1948).
32. Schleiermacher, L.: Ueber den Gebrauch der analytischen Optik, etc., *Pogg. Ann.*, **14** (1828); Analytische Optik, *Z. Physik Math.*, **9**: 1–35, 161–178, 454–474 (1831); **10**: 171–200, 329–357 (1832); "Analytische Optik," Darmstadt, 1842.
33. Schwarzschild, K.: Untersuchungen zur geometrischen Optik, *Gött. Abh. N. F.*, **4**, Nos. 1–3 (1905).
34. Seidel, L.: Zur Dioptrik, Ueber die Entwicklung der Gliedern 3ter Ordnung, etc., *Astron. Nachr.*, **43**: 289–332 (1856).
35. Slussareff, G.: L'Eclairement de l'image formée par les objectifs photographiques grands angulaires, *J. Phys. (U.S.S.R.)*, **4**: 537–545 (1941).
36. Smith, T.: Presidential Address: Some Uncultivated Optical Fields, *Trans. Opt. Soc.*, **28**: (5) 225–284 (1926–1927).
37. ———: On Tracing Rays through an Optical System, First Paper, *Proc. Phys. Soc.*, **28**: 502 (1915); Second Paper, *Proc. Phys. Soc.*, **30**: 221 (1918); Third Paper, *Proc. Phys. Soc.*, **32**: 252 (1920); Fourth Paper, *Proc. Phys. Soc.*, **33**: 174 (1921); Fifth Paper, *Proc. Phys. Soc.*, **57**: 286 (1945).
38. ———: The Changes in Aberrations When the Object and Stop Are Moved, *Trans. Opt. Soc.*, **23**: (5) 311–322 (1921–1922).
39. Wachendorf, F.: Bestimmung der Bildfehler 5 Ordnung in zentierten optischen Systemen, *Optik*, **5**: 80–122 (1949).
40. Wandersleb, E.: Tücken der Cosinuspotenzen, etc., *Z. wiss. Phot.*, **46**: 16–60 (1951); "Die Lichtverteilung im Grossen im der Brennebene des photographischen Objectivs," Akademie Verlag, Berlin, 1952.
41. Zernike, F.: Diffraction Theory of the Knife-edge and Phase-contrast Tests for Mirrors, *Physica*, **1**: 689–704 (1934).

Textbooks

42. Buchdahl, H. A.: "Optical Aberration Coefficients," Oxford University Press, London, 1954.
43. Chrétien, H.: "Cours de calcul des combinaisons optiques," *Revue d'optique*, Paris, 1938. (Probably the most explicit book in discussing formulas for precalculation of optical systems.)
44. Conrady, A. E.: "Applied Optics and Optical Design," Oxford University Press, Fair Lawn, N.J., 1929. (Thorough analysis of lens tracing formulas, containing valuable hints on chromatic correction of lenses.)
45. Czapski-Eppenstein: "Der Theorie der Optischen Instrumente," Barth, Leipzig, 1924. (Contains, beyond geometrical optics, a theory of optical instruments up to the date of publication and the most extensive bibliography in the field.)
46. Herzberger, M.: "Strahlenoptik," Springer, Berlin, 1931. (Attempts to derive the results of geometrical optical research from the basic ideas of W. R. Hamilton.)
47. Herzberger, M.: "Modern Geometrical Optics," Interscience, New York, 1958. (Develops a simple mathematical model of an optical system.)
48. Linfoot, E. H.: "Recent Advances in Optics," Clarendon Press, Oxford, 1955.
49. Slussareff, G.: "Metodi di calcolo dei sistemi ottici," Italian translation by G. Toraldo, G. Filipini, Florence, 1943. (Thorough analysis of Seidel aberrations and of their use for optical computations; well written; occasional mistakes in the formulas. A second book by Slussareff, "Geometrical Optics," Acad. Sci. U.S.S.R., Moscow, 1946, has not been available to the author.)
50. Strong, J.: "Concepts of Classical Optics," Freeman, San Francisco, 1958.

Chapter 3

Photometry and Illumination

By E. S. STEEB, JR., *and* W. E. FORSYTHE, Lamp Division, General Electric Company

1. Visual Photometry

Photometry is that part of radiometry which deals with the measurement of light [1].* Light is radiant energy evaluated according to its capacity to produce visual sensation. The radiant energy which strikes a surface per unit time, or is emitted by a source per unit time, is called *radiant flux* and is preferably measured in watts or in ergs per second. Radiant flux evaluated according to its capacity to evoke the sensation of brightness is called *luminous flux*. The unit in which this is commonly measured is the *lumen*. The ratio of the luminous flux to the corresponding radiant flux is known as the *luminosity* (visibility) *efficiency* and is expressed in lumens per watt. The *luminous intensity* of a source of light in any given direction is the solid angular luminous flux density in the direction in question. The unit of luminous intensity is one lumen per steradian, or one candle. The *brightness* in a given direction of a surface emitting light is the quotient of the luminous intensity measured in that direction divided by the area of this surface projected on a plane perpendicular to the direction considered. Brightness is expressed in candles per unit area, or in terms of the lumens per unit area emitted by a perfectly diffusing surface of equal brightness.

The field of photometry involves *candlepower photometry*, or the measurement of the light output of sources in candles or lumens, and *illumination photometry*, which deals with the illumination produced by a light source. In candlepower photometry many selective types of receivers have been used, including the eye.

Visual Photometers. The eye is incapable of comparing two luminous sources with accuracy, but it can compare the brightness of adjacent and similar surfaces with remarkable precision. Many devices have been used which bring into the field of view two surfaces which are illuminated by two light sources for comparison. Two very simple devices for making such a comparison are Bouquer's photometer and Rumford's photometer, on which the distance of the sources was varied until the brightness on adjoining screens was equal to the eye [2]. The first really accurate photometer head was developed by Bunsen [3]. Only the two photometers almost universally used in visual direct-comparison photometry will be considered.

* Numbers in brackets refer to References at end of chapter.

Lummer-Brodhun Photometer Cubes. In 1888 Lummer and Brodhun developed their photometer cubes; one an equality-of-brightness photometer cube, the other a contrast photometer cube [4]. The field in the equality-of-brightness photometer (Fig. 3.1) is divided into two parts. One part is illuminated

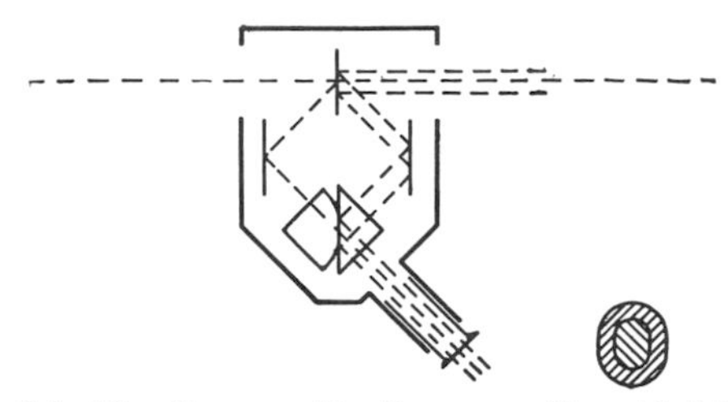

FIG. 3.1. The Lummer-Brodhun equality-of-brightness photometer head.

by light from the standard source and the other part by the light from the comparison source. The two sides of a disk made of magnesium oxide or some other white diffusing substance are illuminated by the two sources being compared. Two silvered glass mirrors reflect the light to the prisms of the cube. One of the prisms' principal surfaces is spherical instead of flat but has a small area in the center which is flat and which makes optical contact with the flat principal surface of the other prism. Thus light will pass through the prism system undeviated through the portion in contact, while it will be reflected at all other points. The eye then sees the light from one source directly through the center of the prism system as a central spot and the light from the other source as a ring around the spot with a sharp boundary line between them. A photometric balance is obtained by visually setting a brightness match of the two parts of the field, which causes the boundary to disappear. Such a match can be made visually with greater accuracy if the sources are identical in color.

The contrast cube (Fig. 3.2) developed by Lummer and Brodhun divides the field into two halves, but with a trapezoid etched or sandblasted into the flat surface of both prisms. When the prisms are pressed together, the arrangement of the trapezoids results in a trapezoidal figure which is illuminated by light from the opposite source from the main part of this side. The light intensity of the trapezoids is reduced 8 per cent by the addition of thin glass plates, and thus two additional glass surface reflections. At the position of balance the contrast between a trapezoid and its background will be 8 per cent on both sides of the field of view, and thus contrast will increase on one side and decrease on the other as the photometer is moved from the position of balance.

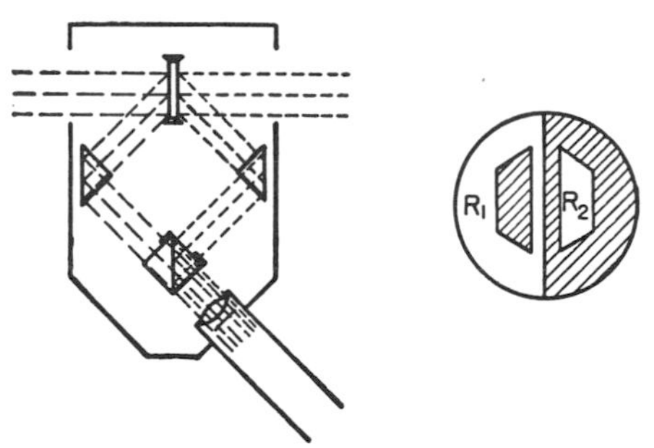

FIG. 3.2. The Lummer-Brodhun contrast photometer head.

The fields of both photometer cubes are viewed with an eyepiece having a visual angle of 2°. The two sources whose horizontal intensities are to be compared are generally mounted on opposite ends of an optical bench with the photometer cube or head between them. The photometer head or either source may be moved until a balance is reached; then the candlepowers of the two sources are to each other as the squares of the distances.

Several precautions should be observed in using these methods. The light source should be at least ten to fifteen times its largest dimension from the photometer head in order to fulfill the requirements of the inverse-square law. And a potentiometer should be used whenever possible to measure the electrical characteristics of the lamp while it is being measured, as the light output varies rapidly with a small change in electrical values.

Flicker Photometer. In this instrument [5] the two halves of the photometric field are presented to the observer in rapid succession. The two fields are balanced in brightness when the flicker produced by presenting them alternately to the eye disappears at a minimum frequency. The critical frequency [6] is connected to the logarithm of brightness by

$$\text{Frequency} = A \log \text{brightness} + B$$

where A and B are constants depending on the color of the light, the ratio of the lengths of the light and dark periods, and the steady brightness, if any, on which the flicker brightness is superimposed. With a large color difference the speed of alternation, which is normally in the order of 10 per second, becomes higher. With the instrument reasonably precise, photometric results can be obtained for color differences which are practically impossible in the steady equality-of-brightness method.

In the methods mentioned the eye is used to judge equality of brightness of the two surfaces which are presented to it. Vision, which involves physical and physiological processes, is liable to considerable variation; so physical instruments have been developed which may be used instead of the eye for measuring light. While the eye can compare brightness of two adjacent surfaces, physical photometry can be used for direct measurement of any illumination without simultaneously comparing with a standard illuminant. Any physical photometer must fulfill several requirements. It must give an indication which is proportional to the illumination, and the calibration curve must remain constant. If it is to be used to measure sources of different spectral distributions, it must respond as the eye does to the various frequencies.

Some time around 1914 it was decided to give the output of lamps in lumens rather than in mean horizontal candles, and the best solution for the difficulties involved seemed to be to measure the mean spherical output of the sources. The first device of this kind was the lumen meter [7], but today the spherical integrator is used whenever the mean spherical candlepower is to be measured.

2. Physical Photometry: The Spherical Integrator

The Ulbricht Sphere. The luminous flux emitted from a light source can be determined by observing the spatial distribution of the luminous intensity and then integrating the results mathematically. A more practical method is to integrate the luminous flux before measurement. This can be done by use of the Ulbricht sphere [8].

The principle of spherical integration was first described by Sumpner in 1892, but Ulbricht in 1900 first used the sphere as a photometer. The modern integrating sphere is a hollow enclosure, the inside surface of which should have three reflection characteristics:

1. Reflection as nearly as possible according to Lambert's cosine law (as perfectly diffusing as possible).
2. Spectral reflection, over the wavelength range to be used, as uniform as possible (as white as can be attained).
3. High reflection factor.

These requirements would be best met by a spherical envelope carved from a magnesium-oxide block. However, there are more practical methods of obtaining a satisfactory spherical integrator.

The brightness, due to reflected light only, of any part of the inner surface of a perfect spherical integrator is a measure of the total luminous flux of a source of any spatial distribution and at any location in the sphere. The theoretical expression for this relation in terms of the sphere and the reflection factor of its inner surface, assumed uniform over the whole surface, is

$$B_r = \frac{\rho}{1 - \rho} \frac{L}{S} \tag{3.1}$$

where B_r is the brightness due to reflected light only; L is the luminous flux from the lamp in lumens; S is the area of the sphere's surface; and $\rho/(1 - \rho)$ is the reflectance factor, where ρ is the single reflectance value of the surface. Therefore, the brightness at any point on the sphere wall due to reflected light is directly proportional to the luminous output of the lamp since the reflectance of the sphere wall and the sphere itself are constant at any one time. The spherical enclosure can also be considered as a black-body radiator whose inner surface has a high reflection factor [9]. The walls of a cavity composed of some solid and opaque material and having a small opening for viewing will have a brightness equal to that of an ideal black body. In the case of a spherical photometer the walls will not cover the entire 4π steradians because of the measurement opening. In consequence the observed brightness is

$$B = \frac{B_0 e}{1 - Sr} \tag{3.2}$$

where B_0 is the brightness of an ideal surface at the same temperature, e is the emissivity of the surface, S the fraction of completeness of the sphere, and r the fraction the radiation is reduced at each reflection from the sphere walls.

If the emissivity is high, the measurement opening can be a large fraction of the sphere wall area before the difference between B and B_0 exceeds 1 per cent; but if the emissivity is low, as with photometric spheres, the opening must be relatively small. A measurement opening not greater than 6 in.^{2} in a sphere 60 in. in diameter is indicated.

Actual accomplishment is difficult. It is not possible to obtain for practical use a perfectly diffusing surface, or one in which ρ is constant over the entire surface of the sphere, or even a perfectly white one. Even if it were, introduction of the source to be measured, its supporting apparatus, and the necessary baffles to allow only reflected light to fall on the measurement point are inherent departures from the theoretically perfect integrating instrument.

Practicable Structures. The spherical enclosure can be made of any material or structure that will retain physical shape and that can be satisfactorily painted on the inside surface. Spheres made of sheet metal are satisfactory, as they are sturdy and can be modified easily with doors, windows, or other openings and supporting mechanisms can be conveniently attached.

Small spheres, 30 in. or less in diameter, can best be made of hemispheres spun from aluminum or copper. This process provides a very light and well-formed sphere that can easily be modified.

It is difficult to state a general rule in regard to the relation of sphere size to source size. If the lamp used as a standard of luminous flux to calibrate the sphere is of the same type and general dimensions as the test source to be measured, the relation of source to sphere size is less critical. Practice has proved that in work with incandescent lamps, clear or inside-frosted, sources up to 3 in. in diameter can be satisfactorily measured in 60-in. spheres. This will be true only of relatively new and clean lamps. Special care must be taken in the case of lamps which have burned most of their operating life and have darkened.

If extreme precision is desired or if standard and test source involved are unlike in general size or self-absorption, it is desirable to determine a correction for differential self-absorption. This can be done by providing a very stable light source at some point in the sphere remote from the center, where the standard or test source will be located. This separate source should be baffled so that no radiation from it can fall directly on the measurement window or either the test source or the luminous-flux standard when in position for measurement or calibration. With the separate source lighted, the measurement-window brightness is observed with only the luminous-flux standard in position, but unlighted, and again with only the test source in position, but unlighted. Equality in the two observations indicates that the absorption of the two sources involved is equal, and no correction is needed. If they are not equal, the calibration by the standard or the observation of the test source must be corrected appropriately by the ratio of the two observations of absorption.

In fluorescent-lamp photometry the size of the sphere also depends on the size of the lamps to be tested, but as a general rule the length of the lamp under test should be not greater than 85 per cent of the inside diameter of the sphere. The sphere should be equipped with a center socket mounted on an arm that can be raised or lowered to the center of the sphere to accommodate incandescent standard lamps. It is necessary to have two sliding arms, one from each side of the sphere, which will support the fluorescent lamps. A support, which can be a tight wire across the stationary hemisphere, should also be included to hold the baffle.

If discharge lamps are stabilized in the sphere, care should be exercised to avoid a rise in temperature within the sphere [10], due to heat generated by the lamp, as temperature changes affect emitted flux. This can be assured by leaving the sphere open until the measurements are made. If the lamps are stabilized outside of the sphere, care must be taken not to rotate the lamp, but to place it in exactly the same position as during the stabilization period so that its temperature equilibrium is not altered.

Surface Requirements. The surface should be such that all light striking it is completely diffused according to Lambert's cosine law. To the degree that this surface condition exists it can be said that a unit of flux falling on a unit of surface area on any part of the sphere results in a measurement-window illumination the same as that due to a unit of flux falling on a unit of surface on any other part of the sphere. For a perfect integrator, source location could be at any point; however, the optimum location at sphere center results in fewer errors due to deviations from a perfect integrator.

A high reflection factor is desirable to increase the sphere-wall brightness and provide a more measurable quantity of flux at the measurement window from a given light source. For example, a reflectance of 90 per cent ($\rho = 0.9$) results in a factor of 9, compared with a factor of only 4 if the reflectance is 80 per cent ($\rho = 0.8$). This becomes increasingly important for measurement of sources of small amounts of luminous flux. A high reflection factor also reduces the error due to baffles and lamp supports in the sphere.

One of the problems connected with spherical photometry concerns the paint used in the spheres. As yet no paint has been found which is spectrally nonselective, and since the integrating action of the sphere amplifies any nonuniformity in the spectral reflectance, the problem is one of importance. If a perfectly nonselective paint were available (along with other components having equally sensitive responses to all parts of the spectrum), the addition of a standard luminosity filter would complete the requirements. Since such a paint has not been developed, the characteristics of the painted sphere must be determined.

Various paints and painting methods can be compared by means of spectrophotometric analysis of a sample of the painted surface. A baffle used in the

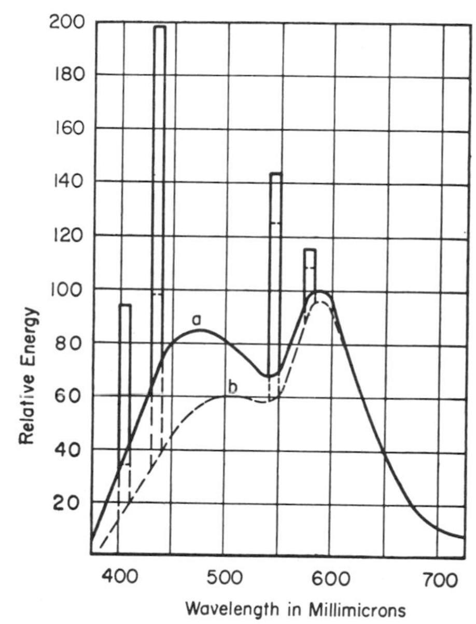

FIG. 3.3. Spectral distribution of a daylight fluorescent lamp (*a*). The same lamp, after being modified by multiple reflections within the sphere, is represented by curve *b*.

sphere, for example, can be used to test the paint's characteristics. This method has the advantage of ease in obtaining reflectance values of the paint at any wavelength in the visible spectrum, but it has the disadvantage that the sample used gives a measurement of single reflectance, whereas in the sphere multiple reflection exists. Thus the single-reflectance measurement does not represent the total effect of a sphere coated with this paint. It also fails to include the effect of supporting apparatus and baffles in the sphere. A preferable method is to obtain spectral data on light emitted from any source within the closed sphere and to compare it with data from the same source outside the photometer and away from reflective material.

Any type of lamp could conceivably be used for this purpose; however, greater accuracy can be attained if two spectrally different sources are used. An incandescent lamp and a daylight fluorescent lamp, or an incandescent lamp with a blue bulb, are typical. Either of the latter two sources is satisfactory, their purpose being only to increase the relative amount of blue light emitted from the source and thus increase the accuracy of the paint-absorption data in the blue region.

The tungsten lamp has a spectral-emission curve which produces sufficient energy in the longer wavelengths. In utilizing this method on spheres used for fluorescent-lamp photometry a tungsten lamp of 150 or 200 watts is used to give sufficient brightness along with a 40-watt daylight fluorescent lamp. A spectrometer is placed in such a manner that, with the receiver and filter housing removed from the sphere, the diffused light from the measurement window fills the extreme slit. A spectral curve of both sources, successively, is obtained with the sphere closed. Both lamps are then removed from the sphere, and the spectral curves are again obtained, with the spectrometer in such a position as to view the entire source.

A comparison of the two spectral curves of each source, at each wavelength, results in the spectral selectivity of the paint used. Figure 3.3 indicates the effect sphere paint has on the spectral emission of a fluorescent lamp. It is necessary to obtain the paint-selectivity curve for each sphere used, for, while they do not differ markedly in their selectivity, they are not identical.

Surface of the Sphere. Since proper integration depends chiefly upon the quality and condition of the paint, the interior of the sphere should be cleaned periodically. Low-pressure air or a vacuum cleaner, used as a blower, can be used for this purpose, since cleaning with a cloth may mar the surface.

The integrating qualities of a sphere may be checked from time to time with a vacuum lamp having a C9 ring-type filament in a single horizontal plane and a bulb coated with white enamel from the base to a point opposite the filament. It is necessary that it be a vacuum lamp so that there will be no gas currents to affect the filament when the lamp is inverted. This lamp should be placed in the sphere, properly centered in a base-up position, and measured for luminous flux. It should then be inverted (base down) and again measured. When differences between base-up and base-down readings exceed 3 per cent, the sphere should be repainted. This 3 per cent criterion applies only in the case of direct substitution measurements, that is, when standard and test source have similar flux distributions. In a newly painted sphere this difference should be less than 1 per cent.

Measurement Window. The use of a measurement window provides a convenient means of observing a close equivalent of sphere-wall brightness at the measurement point. The window should be no larger than is necessary to provide a good measurable quantity of flux for the receiver. A 3-in.-diameter circular window is quite satisfactory. It should be so located as smoothly to complete the inner surface of the sphere; otherwise part of the incident flux will fail to reach the window.

Choice of material for the window presents some problems. Since the window must provide a means of measuring sphere-wall brightness, the ratio of outer-face brightness to inner-face brightness must be constant. Since the direction at which the outer face is viewed is generally constant, the inner-face brightness should be independent of the direction of incident flux. Opal glass is often used, but because of its spectral selectivity visually clear glass is more generally utilized. Although such glass is used, the ground or etched surfaces, which must be made to obtain diffusion, will cause the window to be somewhat selective spectrally, because the diffusing surface causes greater scattering of the energy at shorter wavelengths, resulting in a reduced over-all transmission of the window in the blue end of the spectrum. For this reason the measurement window is included in determination of total spectral reflectivity of the sphere. In all cases it is desirable that the window should be removable so that it can be cleaned, and so that the interior of the sphere can be repainted without damaging the window.

Baffles. In order that the sphere-wall brightness at the measurement window be due only to reflected light, a baffle must be interposed between the source and the point of measurement. The baffle should be coated with the same surface material that is applied to the sphere wall. The size and position of the baffle do affect the distribution of flux within a sphere. Two wall areas will be shielded from direct radiation. One of these wall areas is that which includes the measurement window and will receive no direct radiation from the source. The other shielded area is located in the opposite hemisphere and cannot be viewed from the measurement window and thus can contribute no flux directly to the measurement. The influence on the illumination at the window is zero when the substitution method is used and the sources are the same, and thus the screened areas are equal for both lamps. One of the requirements of a good spherical integrator is a paint having a high reflection factor, and it should be noted that, as it increases, any error due to different screened areas diminishes.

To reduce both screened areas as much as possible, the baffle should be only of sufficient size to shield all points of the source from all parts of the measurement window. When the light source is in the center of the sphere, the minimum value of screened areas is obtained if the baffle is placed one-third of the distance from the source to the measurement window. The exact baffle position is not critical.

Apparatus is necessary to support the source being measured. This should be coated white and be as small as practicable. Any parts to be handled frequently are preferably chrome-plated with satin finish, so that the disturbing effect of their presence remains constant even though not so minute as if painted white.

Physical Receivers. There are now available many physical receivers which offer advantages in practicability, precision, and accuracy over methods employing the eye as a measuring device. While some photometric work is still being done by the visual comparison process, as Parry Moon [50] says, "We must regard the visual methods as giving mere approximation to the true values obtained by physical measurements." Phototubes and photomultiplier tubes in envelopes that may be either evacuated or gas-filled are often used. Although gas-filled tubes have greater sensitivities, vacuum phototubes are widely used because of their greater stability and linearity in response to light and thus they are better suited for photometric measurements. Since the response time of the vacuum phototube is limited only by the usual vacuum-electron-tube parameters, it is often used in the measurement of light transients, such as photoflash lamps and flashtubes. Phototubes require associated current-amplifying apparatus and high-resistance circuits, in which leakage and dark-cell currents may become a problem.

Photoconductive cells, commonly called *selenium cells,* show a change of resistance due to change of light on the cathode. These have not been used widely for routine photometry.

The thermopile can be used as a receiver and has the advantage over all others on the point of spectral response, being uniform throughout the visible spectrum. This can then be modified by a standard luminosity filter such as has been developed at the National Bureau of Standards [11]. However, since speed of response is low and there is no absolute zero, a difference of two readings must be observed. Since ambient temperature is a critical condition and the low currents are difficult to measure, the thermopile is not practical for routine photometric work, although it is a good precise receiver when used under carefully controlled and observed laboratory conditions.

The photovoltaic-cell type of receiver, such as commonly used in light meters and photographic-exposure meters, requires no auxiliary power supply. The photovoltaic cell, sometimes called the *blocking-layer dry type,* is made of a flat steel plate with a layer of selenium on one surface. Over this layer are deposited a number of extremely thin layers of conducting metal. When light strikes the sensitive surface, voltage is generated between the front and back surfaces of the cell.

The spectral response is sufficient throughout the visible spectrum to allow satisfactory modification by color filters to meet the standard luminosity curve. When used with a measuring instrument of reasonably low resistance (500 ohms or less) or with a current-balance circuit (resulting in zero resistance apparent to the cell), the linearity of response is very good and effects due to temperature variations are minimized. The response time, although too long for measurement of fast light transients, such as photoflash lamps, is quite satisfactory for routine measurements of static light sources.

Compensation must be made for effects of temporary fatigue by exposing the cell to illumination of approximately the same level as will prevail during actual measurement. A period of 30 min will usually suffice; however, no harm is done and complete stabilization is assured by illuminating the cell continuously between measurements.

The spectral-response curve of any physical receiver is dependent upon its photoemissive material. For spherical photometry other conditions influence the choice of receiver as much as the cathode response does. The receiver used must be stable and linear to changes in radiant flux, and its response curve should peak near 550 mμ, the region of maximum eye sensitivity.

Since the spectral response of any receiver varies with wavelength, a complete examination of the spectrum is necessary. A tungsten incandescent lamp which has a continuous spectrum is used as a source and is viewed by a monochromator. The receiver to be examined is placed at the exit slit, and the monochromatic light allowed to fall on the sensitive surface. The receiver *r* is connected through the appropriate circuit to either a galvanometer or an electrometer and readings taken on successive wavelengths. A thermopile *t* is then used to measure the same wavelengths, and by use of the relationship r/t the data will form a composite spectral-response curve.

It should be recognized that none of the receivers considered is ideal, the principal deviations from the theoretical being due to differences between the spectral-response curve of the detector and the standard visibility curve, deviations from the cosine law of illumination, and measurement errors due to temperature and fatigue effects. It is essential that the seriousness of these errors be recognized, especially during the photometry of discharge sources.

Blackened targets, such as thermopiles and bolometers, follow the cosine law rather closely. Yet the relative simplicity of the circuitry required and the large response of the barrier-layer cell make this device invaluable for ordinary commercial measurements.

The receiver current can be measured by several different means. Most common are the current balance and the microammeter. The former means is preferred because greatest linearity of the cell can be achieved through the low impedance inherent in a current-balance circuit. The condition of maximum linearity would be a short circuit across the cell. The current balance consists of a series of resistance slide wires in the form of a potentiometer. The externally connected batteries produce a current that balances the current produced by the receiver. This balancing current is regulated by means of rheostats to produce an opposing current equivalent to that produced by the receiver. Under this condition the galvanometer is in balance, and the readings on the dials become a measure of the luminous flux produced by the test lamp.

Because of the high series-resistance current supplied by the current balance [12], part of the network is linear with dial settings, regardless of galvanometer resistance, provided it is not too high. A self-contained galvanometer having a resistance of 325 to 375 ohms, a sensitivity of 0.012 μa/mm scale division, and a period of 2 or 3 sec is typical.

A microammeter connected directly to the receiver can be used instead of the current balance for ordinary accuracy. Current from the receiver should be kept low, preferably below 50 μa, because the resistance of the meter will affect the linearity of the cell. Direct luminous-flux values cannot be measured on the microammeter, but a constant can easily be determined from the standard lamp calibration and actual lamp values calculated.

3. Photometry Spectral Response vs. Luminosity Curve

Luminosity Curve—Standard Observer. Many determinations have been made, by several methods, of the spectral response of the eye (Part 6, Chap. 4). At the National Bureau of Standards the spectral-luminosity curve of 52 experienced observers was obtained. Their findings indicate that the eye responds to light between 3800 and 7600 A, reaching the maximum at 5550 A in the yellow-green region. These results were in close agreement with other data and were adopted by the International Commission on Illumination as the standard luminosity curve. The luminosity curve is often referred to as the eye response of the standard observer. The luminosity values vs. wavelength are given in Part 6, Chap. 4, Table 4.2, as the $\bar{y}$ functions.

Choice of Filter. An important goal of photometry is to rate lamps of any type as the standard observer does. This can be accomplished only if the response of the photometer is almost identical to that of the average eye. Both the sphere paint and the receiver itself have selectivity curves that differ enough from the luminosity curve to require suitable filtering to correct these differences. The transmission of the filter must be such that the combined spectral qualities of sphere paint, measurement window, and receiver will match those of the luminosity curve as closely as possible [13].

Before a sphere is put into operation, characteristics of the receivers and filters should be examined to find a combination which, with the sphere paint, will produce a curve that duplicates the luminosity curve as closely as possible (Fig. 3.4).

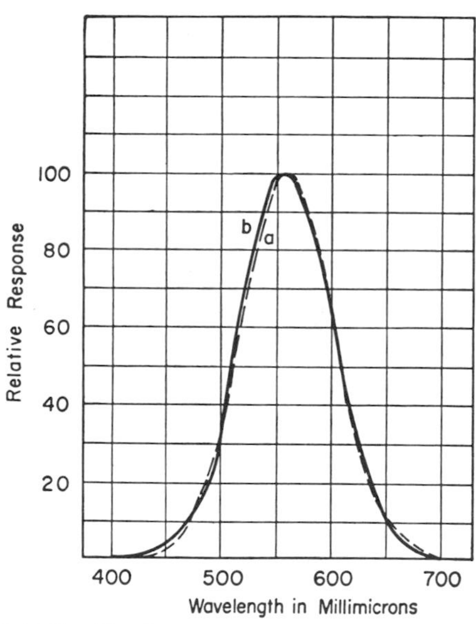

FIG. 3.4. Combined-component response curve, *a*, compared with luminosity curve, *b*.

Determination of Physical Photometers' Spectral Corrections. Determination of the spectral selectivity of each spherical photometer makes it possible to compensate for the variance of the components by applying correction factors for each type of lamp used in the sphere. Through judicious choice of equipment, photometric results with an error of less than 2 per cent, except for saturated colors, can be obtained without a thorough study of the photometer components, but these variations must be considered if good photometric data are to result [14].

Larger corrections have to be applied in the case of fluorescent-lamp photometry than in incandescent photometry because there is a higher percentage of blue energy emitted from fluorescent lamps in comparison with an incandescent standard. A sphere used only for incandescent photometry will have negligible corrections (if the standard lamp and the test lamp are the same) because of similarity of the incandescent test lamp's spectral distribution to that of the incandescent standard lamp. Since both lamps are the same spectrally, the amounts of sphere-paint absorption will be similar, as will the spectral response of the components. In fluorescent-lamp photometry where there is a large difference between the spectral distribution of the incandescent standard and the variety of fluorescent lamps measured a more detailed study must be made.

To obtain the effective coefficient at each wavelength for the entire instrument, it is necessary to know the spectral characteristics of all the parts of the photometer. The effective coefficient S_λ is made up of the sphere paint's spectral selectivity multiplied by the spectral response of the receiver and the spectral transmission of the correcting filter. The transmission of the measurement window is automatically a part of the sphere-paint data. Multiplying the spectral data of the three components at each wavelength gives the effective coefficient at that wavelength for the one particular photometer made up of these components. Since the effective coefficient is a function of the wavelength, the spectral characteristics of all the sources to be measured in that sphere, as well as the incandescent standard, must be known for each wavelength. Since the response of the spherical photometer is compared with the luminosity curve, a ratio can be obtained for a fluorescent lamp,

$$\frac{\Sigma E_{F\lambda} \cdot S_\lambda}{\Sigma E_{F\lambda} \cdot \bar{y}} = R_F \tag{3.3}$$

where S_λ is the effective coefficient at any wavelength $\bar{y}$ is the luminosity coefficient at that wavelength, and $E_{F\lambda}$ is the flux of a fluorescent lamp at the same wavelength. In using the incandescent standard lamp,

$$\frac{\Sigma E_{\omega\lambda} \cdot S_\lambda}{\Sigma E_{\omega\lambda} \cdot \bar{y}} = R_\omega \tag{3.4}$$

where $E_{\omega\lambda}$ is the flux of an incandescent standard lamp at a specific wavelength. The correction factor is then

$$\frac{R_\omega}{R_F}$$

The photometric reading of the fluorescent lamp can be multiplied by the resultant correction factor to obtain correct photometric value. Different combinations of incandescent standard lamps and fluorescent test lamps will, by virtue of their changing spectral ratios, result in different correction factors. A typical set of corrections for use in a fluorescent spherical photometer is shown in Table 3.1.

Table 3.1. Corrections for Fluorescent Spherical Photometer Using 100-watt Incandescent Lamp as Standard

Fluorescent-lamp type	*Correction*
Standard warm-white	0.995
Standard cool-white	0.992
Daylight	0.989

If for some reason it is impractical to determine the effect that the sphere paint, detector, and filter play in approaching the desired luminosity curve and so deriving correction factors, a more direct, though less reliable, method may be used. Lumen standards available in several incandescent- and fluorescent-lamp types can be used as a measurement basis. The particular standard that most closely approximates the lamps to be measured in luminous output and color is put into the spherical photometer, and the photometer is calibrated against that standard. The calibration factor used to correct the photometer is then applied, either as a meter attenuation or through calculation, to the readings obtained on the test lamps. Because differences exist between the response of the photometer and the desired luminosity curve, failure to use a standard similar to the lamp to be measured can result in misleading values arising from these spectral differences and their effect on dissimilar spectral emissions of the sources.

4. Production of Light

Light is electromagnetic radiation, evaluated according to its capacity to produce visual sensations. Thus, to produce light, something must be caused to give out radiant energy of sufficient intensity and within proper wavelength limits (0.38 μ to 0.78 μ) to affect the eye. The sensitivity of the eye is such that, unless radiant energy with an intensity of at least 2 to 6 $\times 10^{-10}$ erg per 10 minutes cone, per 0.001 sec, falls upon the cornea, no visual sensation results.

Simple flame sources led to the candle (about 0.1 lumen/watt) and to oil-burning lamps. Use of flames as light sources was extended by the introduction of various gas-burning lamps about 1800. The gas lamp was improved by the use of the Welsbach mantle about a century later. The first of the modern oil-burning light sources, the kerosene lamp (about 0.3 lumen/watt) was introduced in the latter part of the nineteenth century. These early sources produced light by heating something in the flame, generally the partly burned carbon.

In 1877 Brush introduced a practical carbon arc. This was a big step from producing light by burning the material, and although the arc material, carbon, was destroyed, this light source could be controlled. The next step was taken in 1878, when Edison showed how to produce a practical light source by heating a filament of carbon, in a vacuum, by means of an electric current, so that this light source was not destroyed in giving out light. Both these sources operated from a source of electrical power. Thus was introduced the electrical age. Light sources since that time have mostly been electrically operated.

5. Radiant Energy

Radiant energy is the result of acceleration of an electric charge, or charges, which, except for some special cases in the infrared and the gamma-ray regions, are always electrons. To cause any material to radiate, energy must be applied to it to cause some of its electrons to be accelerated.

Materials radiate as they do as a consequence of their atomic and molecular structure, and of the conditions of excitation of these atoms. The electron or electrons responsible for emission of radiation drop back to a state of lower energy from an excited quantum state and radiate the difference in energy at a definite wavelength.

Any radiator, when in a definite physical condition and excited in a definite manner, emits the same type of radiation; that is, it gives radiation having a definite distribution with wavelengths. This holds true for the radiation from the atoms of a gas, such as hydrogen at a very low pressure, in a tube of definite size and with electrodes of definite shape, size, and material excited by passage of an electric current, and also for the radiation from a solid radiator, such as tungsten filament in a definite physical condition caused to radiate by being heated by an electric current.

All materials when excited emit their characteristic radiation, because the electrons of the various atoms or molecules that do the radiating make transitions between quantum levels determined by the structure of the material.

A good example of the effect of neighboring atoms on the emitted radiation is given by the radiation from mercury vapor at various pressures. At low pressure a small fraction of a millimeter of mercury, the radiation consists of very sharply defined and extremely narrow lines. As the pressure in the discharge tube is increased and the molecules are brought closer together, the lines broaden until at a pressure of about 300 atm almost all evidence of the lines has vanished (Fig. 3.3). At the high pressures the electrons of any one atom are so much influenced by neighboring atoms that they move under constraints approaching those found in a solid.

A good illustration of the extreme conditions—a solid and a vapor—is given by tungsten and mercury, the metals most used for converting electrical energy into light and into the neighboring infrared and ultraviolet radiation. These two metals have, respectively, the highest and the lowest melting and boiling points of any of the metallic elements. The reason for this contrast is that there are two ways of securing satisfactory lamp life: the use of a filament that will remain solid, neither melting nor evaporating at an unduly rapid rate, and the use of a conductor that will not become solid under ordinary conditions and will flow back to its original location when displaced.

The atoms of a heated tungsten filament interact so continuously that no part of the radiation, distributed continuously over the region of the spectrum permitted by thermodynamic considerations, can be regarded as the product of any particular atom; in mercury vapor at low pressure the atoms are far enough apart so that interactions are negligible.

In both cases the number of atoms involved is large. The quantum of energy for the well-known yellow light of sodium vapor is 3.4×10^{-12} erg. Thus it requires 2×10^6 quanta/cm²/sec to give just perceptible vision.

In the germicidal lamp, which has the lowest pressure (7 to 10 μ Hg) of any type of commercial mercury-vapor lamp, there are approximately 2.5×10^{14} mercury atoms/cm³. At a pressure of 100 atm or more, in the ultrahigh-pressure mercury-vapor lamps, where the distinction between the two types of radiation tends to become blurred, there are approximately 10^{21} mercury atoms/cm³.

The type of radiation depends upon the material studied, its physical state, and the way in which the exciting energy is supplied. The energy may be supplied by application of heat, by chemical means, by electrical or electromagnetic means, or by irradiation. An example of exciting by chemical means is the burning of gas, where the light is given by the unburned carbon that is raised to a high temperature by the flame. If a gas mantle is used, the light is given by the heated mantle, the character of the radiation depending upon the materials in the mantle and its final temperature. Products of combustion may give a great part of the light, as in certain photoflash lamps, where light is partly given by the hot aluminum oxide.

Heating by electrical means is illustrated by the carbon- or tungsten-filament incandescent lamps. The material in the form of a gas or vapor may be excited by the passage of an electric current through it. Good examples are the mercury- or sodium-arc lamps. Radiation stimulated by irradiation is the basis of the fluorescent lamp.

The principal methods of excitation are incandescence and luminescence. *Incandescence* refers to radiation that is due to the temperature of the source; its intensity increases rapidly with the temperature of the source, and the wavelength of the maximum of the intensity shifts toward shorter wavelengths as the temperature is increased. *Luminescence* is the name applied to all types of radiation due to causes other than temperature. Other kinds of excitation are: thermoluminescence, electroluminescence, chemiluminescence, triboluminescence, and photoluminescence.

Photoluminescence may be further divided into phosphorescence and fluorescence. Phosphorescent radiation continues after the exciting cause has been removed, and fluorescent radiation stops when the exciting cause is removed. Luminescence is caused by irradiating certain materials called *phosphors* with selected types of radiation that may consist of other electromagnetic radiations, by electrons or other charged particles, by the products of radiative decomposition, by chemical means, and by certain physical means.

There is no sharp line of separation between fluorescence and phosphorescence (Part 6, Chap. 7). Some materials cease giving off radiation in a very short time—measured in small fractions of a second—after the exciting radiation has been removed; other materials may continue to give this secondary radiation for several hours after the exciting radiation has been removed. Even the most nearly instantaneous fluorescence must last for a time long enough to complete the radiation process, of the order of 10^{-9} to 10^{-10} sec.

Some materials, such as certain sulfides, when excited by radiation from the sun, continue to give off light 10 to 15 hr after the excitation is removed. Some other materials remain in an excited state for a long time unless they are reexcited with certain types of radiations.

Some crystals may be excited by electron bombardment and, if immediately cooled by liquid air, will remain excited and give off phosphorescence when warmed up. Some chemical reactions, besides burning, produce radiations.

Some animals have developed methods of producing light, some of which is quite bright, as, for instance, that of some fireflies. This seems to be caused by the animal developing two types of solutions that when mixed produce the light [15]. The animal has some control over the mixing of these solutions.

Thermal radiation is more generally associated with heated solid bodies than with liquids, gases, or vapors. However, some vapors, gases, and liquids may be excited to radiation by application of heat. Temperature radiators may be divided into two classes: black bodies and nonblack bodies. Nonblack bodies may be further divided into gray bodies and selective radiators. A black body is one that absorbs all the radiation that falls upon it, neither reflecting nor transmitting any of the incident radiation. The black body at any temperature radiates more energy, both in the total spectrum and also for each wavelength interval, per unit time per unit area, than any other temperature radiator, and more than any nonblack body at the same temperature.

No known substance has the radiating characteristics of the black body, although some, such as lampblack and some finely divided metals, approach it in certain parts of the spectrum. An enclosure with opaque walls at uniform temperature and with finite emissivity for all wavelengths contains black-body radiation of the same temperature as that of the walls. If a small hole, small as compared with the area of the walls, is made in the enclosing walls, the radiation that escapes will approximate that of the black body at this temperature. The blackness of the cavity black body depends upon the emissivity of the material of the walls and the size of the observing hole relative to the area of the walls [16]. For an emissivity of 50 per cent and an opening of about 1 per cent of the surface of the cavity the radiation will be about 99 per cent black; for an opening of one-half of 1 per cent the blackness will be about 99.5 per cent for this same emissivity. Higher emissivity gives higher blackness.

Many different types of the cavity black body have been used. Two are shown diagrammatically in Fig. 3.5. The basic importance of this radiator

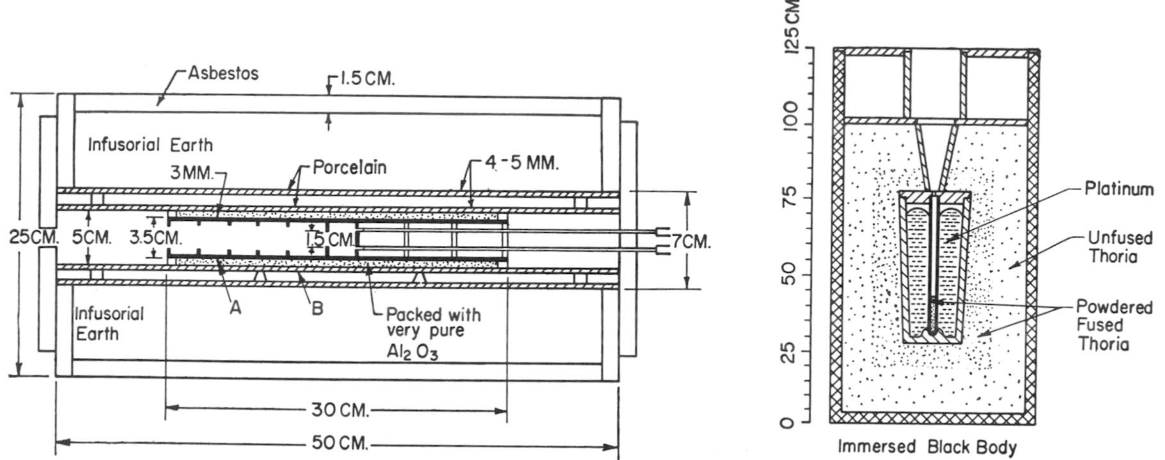

FIG. 3.5. Diagram of two forms of the cavity black body.

is due to the fact that the intensity of the total radiation for any temperature and the spectral radiation for any wavelength interval are a function of the temperature alone. Total intensity is given by the Stefan-Boltzmann law

$$W = \sigma T^4 \tag{3.5}$$

and the spectral intensity is given by the Planck radiation law

$$J_\lambda = \frac{Ac_1\lambda^{-5}}{\exp (c_2/\lambda T) - 1} \tag{3.6}$$

Equations [17] have been developed giving the relation between the temperature of the black body and the emission of photons. Efficiency of production of radiant energy by a black body depends upon the temperature [18].

By expressing the Planck radiation law as a function of T an expression for the fraction of the spectra radiation given by a black body between $\lambda = 0$ and any value of λ can be obtained [19].

Thus the black body is of value as a standard of radiation that can be operated to give a definite known intensity of radiation for any part of the spectrum. The black body is also used as a standard of color of the light of incandescent sources, by giving their color temperatures. The color temperature of a source is defined as the temperature at which it is necessary to operate the black body so that the emitted light of the black body will match in color that of the source studies (Part 6, Chap. 4).

The radiation given by a nonblack body may be represented in terms of the radiation laws given above, by the use of factors that give the relative intensity of radiation of the nonblack body and of the black body at the same temperature. Such factors—less than unity—are called *emissivities* and may be either for the total energy radiated—total emissivities—or for the radiation of any spectral interval—spectral emissivities. Such factors must be experimentally determined. Emissivities for a number of materials are given in Tables 3.2 and 3.3.

TABLE 3.2. SPECTRAL EMISSIVITIES OF SOME MATERIALS*

Material	Temperature, °K	Wavelength	Emissivity	Wavelength	Emissivity
Carbon	1600	0.66	0.89		
	2500	.....	0.84		
Copper	1275	.....	0.105		
	1375	.....	0.150		
	1500	.....	0.13		
Iron	1000	.....	0.27		
	1480–1500	.....	0.29		
Molybdenum	300	0.665	0.420	0.467	0.425
	1300	.....	0.378	.....	0.395
	2000	.....	0.353	.....	0.380
	2750	.....	0.332	.....	0.365
	1300–2100†	.....	0.382		
Nickel	1200–1260	0.665	0.375	.....	0.450
	1200–1400†	.....	0.350		
Tantalum	300	.....	0.493	.....	0.565
	1400	.....	0.442	.....	0.505
	2800	.....	0.390		
Tungsten	1200	.....	0.452	.....	0.482
	2000	.....	0.435	.....	0.469
	2800	.....	0.419	.....	0.458
	3400	.....	0.407	.....	0.450
	1200–2200†	0.66	0.46		

* For a more complete list see "Smithsonian Physical Tables," 9th ed., Table 78.
† Well outgassed.

TABLE 3.3. TOTAL EMISSIVITIES OF A NUMBER OF MATERIALS*

Material	Condition	Temperature, °K	Emissivity
Aluminum	Polished	375	0.095
Brass	Polished	375	0.059
Carbon	Rough plate	500	0.72
Copper	Polished	375	0.052
Iron	Roughly polished	375	0.27
Molybdenum	Polished	375	0.071
Radiator paint	White	375	0.79
Nickel	Polished	375	0.072
Silver	Polished	375	0.052
Tungsten	Polished	375	0.066

* For a more complete list see "Smithsonian Physical Tables," 9th ed., Table 80.

Drude developed a theoretical relation between the resistivity of a metal and its emissivity for a definite wavelength,

$$e_\lambda = 0.365 \sqrt{\frac{\rho}{\lambda}} \tag{3.7}$$

This holds for long wavelengths, but not for short wavelengths.

A gray body is defined as a radiator that has the same spectral emissivity for all wavelengths. A selective radiator is one that has different emissivities for different wavelength intervals.

6. Light Sources

Four methods are now commonly used to produce light: (1) heating a filament by an electric current; (2) some type of arc; (3) passage of an electric current through some gas or vapor; (4) photoluminescence. Some work has also been done on production of light by applying an alternating field to selected powders in a solid dielectric [20].

Getting the energy into the light-producing part of any type of lamp, whether an incandescent lamp, an arc lamp, or gas- or vapor-discharge lamp, requires expenditure of energy, which results in a loss in efficiency. Conduction of energy away from the filaments of an incandescent lamp by the leads and filament supports results in such a loss. For the 120-volt tungsten-filament lamp these losses in efficiency amount to 3 to 4 per cent but may amount to 30 to 50 per cent for some of the miniature lamps with a very short filament. Attempts have been made to reduce this end loss for short filaments by tapering the ends of the filament, but this was not found practical. The loss has, in part, been corrected for some of the high-wattage projection lamps, with a filament consisting of four or five coils in one plane, by winding the outer coils with smaller pitch than the inner coils so that the mutual heating causes these outer coils to operate at about the same temperature as the inner coils. The ends of the inner coils may also be wound with a smaller pitch, thus keeping the end turns at about the same temperature as the others.

All gaseous- (or vapor-) discharge lamps show a variation of the light intensity near the electrodes, where part of the energy input is used in ionization, in accelerating electrons, and in maintaining the space charges set up on the walls. These end losses are different for the two electrodes and are about constant for one tube size, independent of the length of the particular tube. Thus their percentage effect on the radiation output is less for the longer tube. These losses also vary with the gas used and the type of electrodes. The voltage drop for a 40-watt fluorescent lamp in a tube 48 in. long and $1\frac{1}{2}$ in. in diameter, with heated oxide-coated electrodes, is about 108 volts; 12 to 16 per cent of this represents end losses. If the same lamp has flat uncoated iron electrodes, the voltage drop will be about 500; and 82 per cent of this will be end losses. Similar relations hold for other gas or vapor discharges.

While the end losses for the production of the 2537-A mercury radiation in this 40-watt low-pressure tube, 48 in. long and $1\frac{1}{2}$ in. in diameter, amount to about 14 per cent, for the 30-watt 300-volt lamp in a tube 96 in. long and 1 in. in diameter the loss is only about 4 per cent. For the fluorescent lamp there are additional end losses due to variation of the brightness of the phosphor near the ends. For the 40-watt fluorescent lamp the end losses amount to about 17 per cent, and for the fluorescent lamp in the longer tube, the 30-watt lamp, in a tube 96 in. long, such losses are only about 5.4 per cent.

Incandescent Lamps. The problem that confronted Edison in developing a practical incandescent lamp was to find a filament and a method of mounting it. The resulting lamp would then produce light at an efficiency of such value that with regard to the convenience of its operation it could compete with existing flame sources.

Before Edison, many attempts had been made to make an electric incandescent lamp using some of the well-known materials such as platinum or carbon for the filament, but the resulting lamps were unsatisfactory because they were too low in intensity, had too short a life, or were of too low voltage. Edison tried many materials for the filament and decided that carbon was best. His big problem then was to make a filament of carbon small enough and at the same time strong enough to make a successful lamp for the high voltage desired. He decided that a multiple-operated lamp was necessary and that a voltage of at least 100 must be used to avoid excessive line losses. His contribution was a filament of carbon of such diameter and length that, when mounted in a vacuum, it consumed only about 100 watts from a 110-volt line. It gave sufficient light for a life of 600 hr and competed with existing flame sources. The first filaments were made by carbonizing pieces of sewing thread.

The search was continued for a better filament material. A process of squirting the carbon filament from a paste was developed [21] which gave more uniform filaments than those obtained from carbonizing threads or from strips of bamboo. W. R. Whitney [21] developed a method of improving the carbon filament by heat-treatment. These improvements gave an increase in efficiency from about 5.5 watts per mean horizontal candle (1.8 lumens/watt) for Edison's early lamp to about 2.5 watts per mean horizontal candle (4.0 lumens/watt) for Whitney's treated carbon filament, the gem lamp. The incandescent lamp with a carbon filament or a treated carbon filament held the field of general lighting for about 27 years (1878–1905).

The Nernst glower, which consisted of a short rod made of a mixture of some rare earth oxides, was the first real competitor of the carbon lamp. Shortly after it was introduced, other filament materials were developed which gave a better lamp than the Nernst glower. Tantalum and osmium were found to be about twice as efficient as the old carbon lamp, but only about 20 per cent more efficient than the treated carbon filament. The metal tungsten was introduced as an incandescent-lamp filament about 1905. Data on some of these early filaments are given in Table 3.4.

To make a good lamp filament, the substance must so radiate, with respect to wavelengths, that a large fraction of its radiated energy is in the visible part of the spectrum. In addition it must be of a material that can be made into a filament uniform enough to be heated to a uniform temperature along its length; the filament must be strong enough to permit mounting and to stand up under the use of the lamp.

The data in Table 3.5 show that, for tungsten filament the size of that for a 120-volt 500-watt lamp, about 15 per cent of the radiated energy is in the visible spectrum. Up to temperatures well beyond those attainable with any known substance the proportion of the energy radiated in the visible spectrum increases with an increase in the temperature; so a high operating temperature is desirable. Tungsten satisfies these conditions the best of any known substance because it has a very low vapor pressure, which permits operating at a very high temperature, and because it radiates selectively in favor of the visible spectrum. Despite many statements to the contrary, no substance is known that can be used as a filament that satisfies the radiating conditions for a life beyond a few hours as well as tungsten, even disregarding some of its other valuable properties.

When tungsten lamps were first made, it was necessary to develop means of drawing tungsten wire. Coolidge [22] developed a method of drawing tungsten wire of very small diameter. This new wire replaced the old fragile pressed filaments and enabled the lamp engineer to make lamps of very exact electrical characteristics. The difficulty of making the early carbon filaments of exact physical dimensions was one of the reasons for different voltages of electric services at various places. With the ability to make accurate tungsten wire, and thus

Table 3.4. Efficiencies of Some Early Incandescent Lamps of About 60-watt Size

	Lumens/watt	Life, hr
Edison's early carbon lamp	1.8	600
Treated carbon lamp	3.2	600
Gem lamp	4.0	600
Nernst glower	5.0	600
Tantalum lamp	4.9	900
Osmium lamp	4.9	
Tungsten lamp (1907)	7.8	1,000
Tungsten lamp (1950)	13.9	1,000

Table 3.5. Some Operating Characteristics of Gas-filled Tungsten Lamps

Watts	Lumens/watt	Filament temp, °K	Per cent input radiated within visible spectrum	End-loss per cent input	Gas-loss per cent input	Per cent input radiated beyond bulb by filament	Per cent efficiency loss due to end losses
40	11.8	2680	7.1	1.6	24.4	66.8	3.8
60	14.3	2780	7.6	1.6	22.2	69.1	3.9
150	18.2	2910	9.7	1.7	16.1	75.1	4.1
200	19.7	2960	10.2	1.7	13.7	77.4	4.2
300	20.8	2990	10.7	1.8	11.6	79.8	4.3
500	21.0	3000	11.4	1.8	9.2	82.3	4.3

lamps with definite electrical characteristics, there is a move toward a standard voltage of 120.

Many improvements have been made in tungsten lamps. Among them are better wire, better supports for the filament, better lead-in wires, inside frosting of the bulb, tipless exhaust, reflector bulbs, and lamps for special purposes. The outstanding improvement, however, was the gas-filled lamp. Gas was introduced into the bulb to reduce the rate of evaporation from the filament so that the filament could be operated at a higher temperature and thus at a higher efficiency for the same life. Molecules of the heated filament cannot escape so rapidly when the filament is surrounded by an atmosphere of a neutral gas. For a carbon lamp it was found that, while the filament could be operated at a higher temperature for the same life, there was no gain in efficiency since losses due to conduction away from the filament by the surrounding gas more than made up for the gain due to the higher temperature.

Langmuir [23] found that, when a filament is operated in an atmosphere of an inert gas, it is surrounded by a more or less stationary sheath of the gas and that the fraction of the input energy that is carried away by the gas is much less for filaments of larger diameter. These losses were markedly reduced by coiling the filament into a closely spaced helix, since the gas sheath surrounded the entire coil and thus the gas losses depended upon the diameter of the coil and not that of the wire. By the use of the coiled filament engineers made a 120-volt 500-watt gas-filled lamp that was about 40 per cent more efficient than the corresponding vacuum lamp. The filaments for lamps smaller than about the 60-watt lamp would not stand up when coiled into a helix of large enough diameter to make a gas-filled lamp that would be as efficient as the vacuum lamp. It was later found that this helix could be coiled a second time, which resulted in a stable coiled-coil filament of such diameter that it is possible to make a 40-watt 120-volt gas-filled lamp that is more efficient than the vacuum lamp. Evaporation loss from a coiled filament is less than that from a straight filament at the same temperature owing to interference of parts of the coil with escaping molecules. To reduce the lamp size and manufacturing costs, the filaments of the vacuum lamps were also coiled into a helix.

The first gas used was nitrogen. Heavier gases, such as argon or krypton, are now known to be better, for two reasons: their heat conductivity is less, and the larger molecules offer more interference to the escaping molecules of tungsten. There was some discharge between different parts of the coil if pure argon was used. A mixture of about 80 per cent argon with the remainder nitrogen overcame this discharge and gave about 20 per cent gain in efficiency over the nitrogen-filled lamp. If krypton could be obtained at low enough cost, another gain of about 30 to 40 per cent would result.

Many special types of lamps are now commercially available. Among these are lamps for general lighting in homes and other buildings, street lighting, automobile lighting, and the large list of lamps for special purposes. For some services the bulb is made the reflector to control the light, as in the sealed-beam automobile headlight lamps. Some characteristics of a number of tungsten lamps are given in Table 3.6.

TABLE 3.6. CHARACTERISTICS OF SOME 120-VOLT TUNGSTEN LAMPS

Watts	Amp	Approx. initial lumens	Rated initial lumens/watt	Rated average life, hr	Un-coiled-filament length, cm	Filament diameter, cm	Filament temp, °K
6*	0.050	44	7.4	1,500	36.9	0.0012	2400
25*	0.21	265	10.6	1,100	56.9	0.0031	2570
40	0.34	470	11.8	1,000	43.3	0.0035	2680
75†	0.63	1,180	15.7	750	53.7	0.0053	2840
100†	0.83	1,750	17.5	750	56.4	0.0063	2900
100‡	0.83	1,920	17.2	50	48.2	0.0063	2965
200	1.67	3,940	19.7	750	72.7	0.0101	2960
500	4.17	10,500	21.0	1,000	92.3	0.0181	3000
1,000	8.3	23,300	23.3	1,000	103.0	0.0283	3100
1,000‡	8.3	28,000	22.5	50	82.8	0.0274	3350
3,000§	93.8	88,500	29.5	100	34.4	0.122	3250
10,000	83.4	330,000	33.0	75	138.5	0.114	3350

* Vacuum lamp.
† Coiled-coil filament.
‡ Projection lamp.
§ A 32-volt lamp.

Carbon Arc. Light from the arc comes from the hot electrodes and from the discharge through the heated vapors between these electrodes. Such arcs may be operated in the open or may be partially enclosed, which makes them easier to control. To increase their light output and to produce different colors, the carbons are cored with selected materials.

The carbon arc was at first used for street lighting and the lighting of large areas. This field now has been taken over by the incandescent lamp and some other arcs or discharge lamps. Carbon arcs are now used for the most part in searchlights, where the small, intense source is of great advantage. They also find use in the motion-picture industry as sources for both taking the pictures and projecting them. Some characteristics of carbon arcs are given in Table 3.7.

The carbon arc is started by momentarily bringing the ends of the carbon together in an inductive circuit and then suddenly pulling them apart, thereby breaking the circuit. The inductive impulse, together with the ionization due to the resulting spark, starts the arc. By use of a selected transformer or inductance the current is maintained at the proper value.

Discharges through Gases or Vapors. Many gases or vapors will produce light if excited by passage of an electric current through the gas, or vapor, enclosed in a tube with electrodes at each end. These discharge tubes have negative current-voltage relations (Part 4, Chap. 10); hence the current will increase without limit for a constant applied voltage and the tube destroyed unless some means are taken to limit the current. The voltage needed for starting

TABLE 3.7. CHARACTERISTICS OF SOME CARBON ARCS

Arc	Current	Volts	Color temp, °K	Brightness, candles/mm²	Total lumens
8-mm Nat. M. P Studio carbons	40	37.5	4650		200,000
12-mm low-intensity	30	55.	3550	180	25,900
6-mm Suprex carbons	40	32	5850	600	43,400
7-mm Suprex carbons	50	36	5965	625	60,500
8-mm Suprex carbons	65	38	6400	650	82,600
11-mm high-intensity carbons	90	57	6400	640	106,000
11-mm superhigh-intensity carbons	135	75		1,200	303,000
13.6-mm high-intensity carbons	125	65	5650	700	250,000
13.6-mm superhigh intensity carbons	180	75	5480	900	383,000
16-mm high-intensity carbons	150	70	6000	730	330,000

is much higher than that at which it is safe to operate the tube.

Several methods are available to start the current in an alternating-current (a-c) circuit, and the same means will help control the current. An inductance may be included in the circuit and the lamp short-circuited; the short circuit will be removed suddenly so that the resulting high-voltage impulse across the tube will start the discharge. This same inductance will keep the current at the proper value for the tube. A transformer giving a high enough voltage to start the discharge is also used, the transformer voltage being dependent on the current so that as the current increases the voltage drops, keeping the current at the proper value. The principle of the electrical finger, which is a source of momentary high voltage, may be used by applying it to one end of the tube; thus the high voltage starts a small arc at that end, and the resulting ionization enables the applied voltage to start the arc through the tube. Sometimes a conducting strip is put on the outside of the long tube, thus producing a shorter gap—through the glass. A shorter arc is started, and again the resulting ionization helps start the main discharge.

Many different types of special manually or automatically operated switches are used to open the circuit that short-circuits the lamp and applies the inductive "kick" to start the lamp. The added equipment necessary to start and operate the arcs or discharge lamps consumes energy and thus reduces the over-all efficiency. Makers of the lamps generally give the lamp efficiency rather than the over-all efficiency, which depends upon the starting and operating devices used.

A mercury arc, if liquid mercury is present in the tube, may be started by tipping the tube so as to produce and break a contact through the liquid mercury. In many cases a few millimeters pressure of one of the inert gases is put in the tube with the mercury. The discharge starts through the gas and heats the mercury so that enough vapor is formed to carry the current.

In the excitation of a gas to emit radiation the passing electrons must make a collision with the molecules of the gas to excite them. The number of the collisions, which determines the intensity of the emitted radiation, will depend upon the number of electrons, the electron current, and the number of atoms or molecules of the gas or vapor present. An atom may be excited to emit radiation of different wavelength intervals. Thus one has to consider the probability of production of the particular type of radiation desired. Much work has been done in this field, and so for some gases the proper arrangement may be made to increase the probability of producing the radiation desired [24].

Gaseous-discharge Lamps. Gaseous-discharge lamps have been used since the 1890s for special purposes where their color and low intensity have special values, such as in the sign-lighting field. They are also used as indicator lamps to show when a circuit is in operation. Discharge lamps used in the sign-lighting field are generally tubular in form, of small diameter, very long, and with cold metal electrodes. These tubes may be bent into many shapes. They were operated from special transformers that give high enough voltage to start the lamps. They are so constructed that the voltage depends upon the current drawn, so that the lamps continue to operate satisfactorily. The gas or vapor used depends upon the color of light desired. Neon is used when red light is desired, mercury for green, carbon dioxide for white, xenon for bluish white, helium for purplish white. For some signs fluorescent tubes are used.

Mercury-vapor Lamp. The first successful mercury-vapor arc lamp was brought out by Cooper-Hewitt [25] in 1902. This lamp consists of a tube 125 cm long, 2.5 cm in diameter, operates on 72 volts, consumes 266 watts in the lamp, and gives light at a lamp efficiency of 24.5 lumens/watt. Special starting and operating apparatus is required, which is true for all arc and vapor lamps. Mercury-vapor lamps in quartz tubes were introduced as sources of ultraviolet radiation. The mercury-arc lamp of lowest mercury-vapor pressure made is used for two purposes: as the germicidal lamp, sometimes called the *bactericidal lamp* or "*Sterilamp*," and as the source of the exciting radiation for the fluorescent lamp. A large part of the input electrical energy is radiated by the line 2537 A. The glass used for the two lamps is different; the one for the germicidal lamp has a very high transmission for 2537 A, while that for fluorescent lamps is practically opaque for this part of the spectrum.

Higher-pressure mercury-vapor lamps were developed later [26]. First a 400-watt lamp operating at a pressure of 1.2 atm was produced, followed by lamps of higher pressure and higher wattage. The first lamps had double-walled bulbs to keep the mercury vapor at the temperature necessary to reach this high pressure. Some of the superhigh-pressure lamps had to be air- or water-cooled and made of quartz to withstand the high temperature and pressure. Some laboratory lamps have been operated at pressures of about 1,000 atm. Some data on high-pressure mercury lamps are given in Table 3.8.

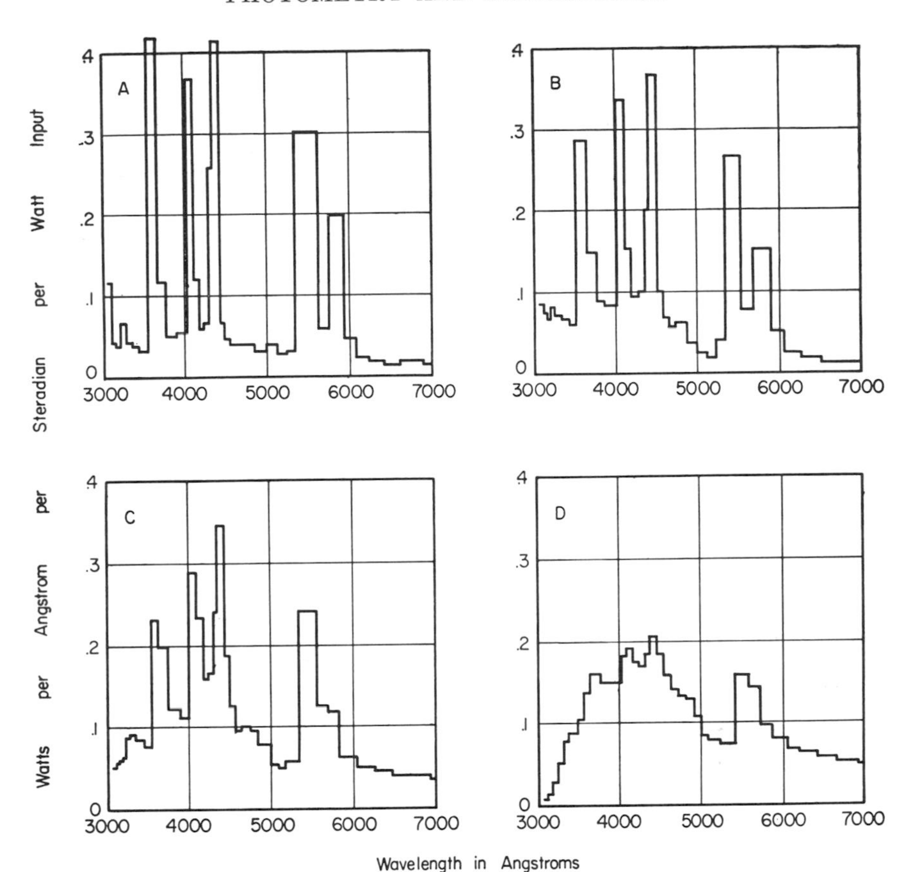

FIG. 3.6. Spectral distribution of energy radiated by a mercury arc in a tube 2 × 25 mm for various pressures. Curve *A*, for mercury vapor pressure of 54 atmospheres; curve *B*, for 102 atmospheres; curve *C*, for a pressure of 197 atmospheres; and curve *D*, for a pressure of 319 atmospheres.

TABLE 3.8. CHARACTERISTICS OF SOME MERCURY ARCS

Lamp	Arc volts	Arc watts	Lumens	LPW	Operating pressure, atm
H38-4HT	130	100	3,650	36.5	9.1
H37-22KB	130	175	7,300	41.5	3.5
H37-5KB	130	250	11,000	44.0	4.8
H33-1CD	135	400	20,500	51.3	3.9
A-H6	840	1,000	65,000	65.0	110
H9FJ	535	3,000	132,000	44.0	0.7

The spectrum of a low-pressure mercury-vapor lamp consists of a great number of lines in the ultraviolet and visible spectrum. As the operating pressure is increased, the lines broaden and finally at the higher pressure, about 300 atm, all appearance of lines disappears (Fig. 3.6). Mercury-vapor lamps must have special apparatus for starting and operating them.

Table 3.9 gives the dependence of intensity of some of the lines on temperature of the tube, and therefore on pressure of the mercury vapor.

The mercury arc gives a greenish-colored light. This has been somewhat color-corrected by the addition of some red light by use of a phosphor on the inside of the tube that fluoresces owing to the irradiation by the mercury arc. This phosphor absorbs some of the light given by the arc, but the color is corrected as a compensation.

Sodium-arc Lamp. Two problems were encountered in the development of the sodium lamp. Sodium is a solid at ordinary temperatures and has to be heated to give the vapor necessary to carry the current. The heating was accomplished by putting some neon in the bulb to first carry the discharge and furnish the necessary heat to vaporize the sodium. Then the sodium vapor carries the discharge, since it has a lower ionization potential, and furnishes the necessary heat to keep the sodium in the vapor state. The second difficulty to be overcome is that hot sodium vapor attacks the glass. This was stopped by coating the inside of the bulb with a transparent material that is not attacked by sodium vapor.

The sodium-arc lamp gives light, of the familiar yellow color, at an efficiency of about 50 lumens/watt for a 180-watt 10,000-lumen lamp. These lamps are used for outdoor lighting where the color of the light is not objectionable.

Table 3.9. Efficiency of Production of Mercury Line Radiation

Current amp	Bulb-wall temp, °C	Per cent input* converted into these lines							Sum	Proportion in 2,537 A, %	Lumens/watt
		2537 A	3129 A	3654 A	4047 A	4538 A	5461 A	5780 A			
0.25	48	62	0.53	0.45	0.59	1.30	0.86	0.18	65.9	94	5.5
0.50	57	55	0.68	0.62	0.86	1.45	1.26	0.27	60.1	92	7.3
1.00	69	40	0.89	0.82	1.02	1.79	1.49	0.39	46.4	86	10.0
360-volt Uviarc†		1.8	2.2	2.9	1.1	1.7	2.0	2.2			

* The first three lines are for the positive column of a low-pressure tube 1 in. in diameter and operated in still air at 25°C.
† This arc operates at a pressure of about 1 atm.

Fluorescent Lamp. A new light source, using fluorescence, was offered to the public early in 1938. This problem had been studied for a long time, but some of the underlying principles were not well known. Several experimenters worked with fluorescence to produce light many years ago, including Edison and Andrews (1910). They put their phosphors on the inside of the bulb; Edison used X rays or electrons as the exciting radiation, Andrews mercury-arc radiation. Neither succeeded in producing a satisfactory light source. The basis of the fluorescent lamp, as developed today, is the low-pressure mercury arc, which converts a large proportion of the input energy into the 2537-A radiation.

Modern study of phosphors has been successful in finding materials that can be excited by this 2537-A radiation to give light of various colors. By suitable combinations various desired colors are produced.

Fluorescent lamps are generally operated on 60-cycle a-c with a choke coil in the circuit to regulate the current. The choke coil causes the current to lag behind the voltage, resulting in low power factor, which can be brought back to a value of about 90 per cent by the use of a selected capacitor across the line.

Another disadvantage results from operation on alternating current. Radiation from the low-pressure mercury arc follows very closely the variation of the current, which passes through zero 120 times per second. This results in a flicker when moving objects are observed. Many of the phosphors used have enough phosphorescence radiation so that the excited radiation decays slowly enough to help reduce the flicker.

Two methods are used further to reduce this flicker. If a number of fluorescent lamps are used together, flicker can be greatly reduced by operating a part of these lamps on each of the phases of a three-phase power supply. On an ordinary supply they may be operated in pairs, with a selected capacitor in one circuit, which will cause the current in that circuit to lead the voltage by about the same amount that the current lags in the inductive circuit; this arrangement makes the maximum light output of one lamp overlap the minimum light output of the other. This method of operation also results in a high power factor.

The fluorescent lamp has a very high efficiency in light production. Lamps giving light of almost any color can be produced at a much higher efficiency than colored light produced with tungsten lamps. Some problems of illumination can be better solved by the use of long, tubular light sources.

Disadvantages are that it requires special equipment to operate the lamps and, unless care is taken, there is a bad flicker and a low power factor. These last two disadvantages can be corrected. Equipment is available also for operating these lamps on direct current.

The fluorescent lamp is manufactured in many colors, including blue, daylight, warm white, cool white, green, gold, and red.

Several factors enter into their efficiency. If all the input electrical energy could be converted to a yellow-green light (5550 A), to which the eye is most sensitive, the efficiency would be about 680 lumens/watt. In the 40-watt warm-white fluorescent lamp the phosphors produce light over a range of wavelengths with an average luminous efficiency of 57 per cent. This reduces the efficiency to 388 lumens/watt. Conversion of the 2537-A radiation to the longer visible wavelengths is accomplished by the phosphor at an efficiency of 45 per cent. About 60 per cent of the energy delivered to the lamp is converted to 2537-A radiation. About 15 per cent of this is lost owing to bulb and phosphor absorption and nonutilization of the exciting radiation by the phosphor. These losses result in an efficiency of about 89 lumens/watt. Phosphor imperfections and depreciation in the first 100 hr of life reduce this to about 60 to 65 lumens/watt. But even after allowing for the 20 to 25 per cent in ballast losses, this lamp is much more efficient than the tungsten lamp of the same light output.

For the same light output the fluorescent lamp radiates into the room much less heat energy than the corresponding incandescent lamp.

The fluorescent lamp is more expensive but has a much longer life. The net result is that the new lamp delivers light to the user at a much lower cost per lumen-hour than older light sources, considering the cost of the lamp and the cost of the electricity to operate it.

Table 3.10 shows some of the sizes and some of the operating characteristics of a number of fluorescent lamps.

Photographic Lamps. Photoflood Lamps. Photography requires directly and indirectly a number of special lamps. Photoflood lamps, which are tungsten-filament lamps of very short life, operate

TABLE 3.10. CHARACTERISTICS OF SOME FLUORESCENT LAMPS

	Min. bipin				Medium bipin					Mogul bipin		Monopin			Recessed double contact
	T-5 bulb		T-8 bulb		T-12 bulb					T-17 bulb		T-12 bulb			PG17 bulb
Nominal wattage	4	8	15	30	14	15	20	30	40	40	90				
Nominal length (in.)	6	12	18	36	15	18	24	36	48	60	60	48	72	96	96
Lamp															
Current (ma)	135	170	308	360	385	324	380	430	425	425	1,550	425	425	425	1,500
Volts	32.0	56.5	54.5	97.5	37.5	46.8	57.5	81	106	104	65	100	147	200	175
Watts	3.8	8.2	15.0	30.6	13.6	14.4	20.4	33.0	41.8	41.0	90	39.5	57.0	76.0	215
Lumens—CW	113	365	780	2,000	620	740	1,170	2,250	3,150	2,800	6,000	2,900	4,250	6,100	15,500

at a very high temperature to give about twice the lumens and three times the photographic effectiveness —for the same wattage—as the regular lamps of like wattage. These lamps in clear glass bulbs are furnished at different color temperatures ranging from about 3200°K to about 3400°K, and with blue bulbs the color temperature is extended to about 6000°K for use with daylight color film. These photoflood lamps are available in reflector bulbs.

Photoflash Lamps. Sometimes there is a need for a flash of light of short duration. Such short-time flashes used to be produced by burning a selected flash powder. Anderson produced such a source by discharging a capacitor charged to about 20,000 volts through a small iron wire. For wires about 5 cm long and weighing about 2 mg, the time of the discharge was about 10^{-5} sec and the brightness about 16 million candles/cm^2.

A very intense flash of light of relatively short duration is given by the photoflash lamps, some of which consist of a bulb filled with very fine shredded aluminum foil in an atmosphere of oxygen. A small filament covered with a primer serves to ignite the shredded foil when the filament is heated by passage of a small current. Other photoflash lamps contain no aluminum foil but have an extra amount of a selected primer on the filament that furnishes the light when ignited.

Data on some of these lamps are given in Table 3.11.

TABLE 3.11. CHARACTERISTICS OF SOME PHOTOFLASH LAMPS

Lamp designation	Approx. time to full peak, msec	Approx. lumen-sec	Approx. peak lumens	Mean color temperature, °K
Speed midget SM	6	4,700	900,000	3300
Synchro-press No. 5	21	16,000	1,200,000	3800
Synchro-press No. 5B	21	7,000	530,000	6000
Focal plane No. 6	...	15,500	620,000	3800
Synchro-press No. 11	21	29,000	1,800,000	3800
Synchro-press No. 22	21	63,000	4,000,000	6000
Synchro-press No. 22B	21	27,000	1,800,000	3800
Focal plane No. 31	...	77,000	1,500,000	3800
Photoflash No. 50	30	95,000	5,200,000	3800
Synchro-press No. 50B	30	43,000	2,500,000	6000

Flash Tubes. For taking pictures where intense flashes of light of shorter duration than a few thousandths of a second are needed, flash tubes were developed. The first flash tube contained mercury in a tube about 30 cm long which was flashed by discharging through it a capacitor charged to about 2,000 volts. To start the discharge, the electrical finger was applied to one end of the tube. This was actuated and the lamp flashed by the discharge of a capacitor charged through a step-up transformer, thus applying a high voltage to the end of the tube which started the flash through the tube. By choosing proper constants of the circuit, flashes are obtained of high intensity and very short duration, measured in microseconds. Flash tubes were developed using other gases to produce the light. Many of the flash tubes for photographic purposes contain xenon because this gas gives, as excited, a good approximation of daylight.

A compact flash tube is made by coiling the tube into a helix about 3.5 cm in diameter and about 3 cm long. When this flash tube is operated from a capacitor of 112 μf charged to 2,000 volts, a light output of about 10,000 lumen-sec, having a maximum intensity of about 30 million lumens, is produced. The time from start of the flash to maximum is about 80 μsec, and the intensity drops to about one-half value in about 300 μsec. With this type of flash tube a flash picture of a shaft of about 3 in. in diameter rotating at a speed of about 21,000 rpm shows almost no trace of blur.

Projection Lamps. A number of lamps are designed for projecting pictures on the screen. These have special filaments designed to get as much light as possible in the beam. Single coils or coiled-coil filaments are used. The larger lamps have four or five coils in a plane, and some have four more coils behind these, so placed that they send light through the space between the front coils. These projection lamps range in size from 75 watts for projecting 8-mm film pictures to 2,100-watt lamps for use in small theaters.

Reflector-type Lamps. There are a number of lamps of this type, including the silver bowl, used in indirect lighting, and several for spot- and flood-lighting. The sealed beam lamps are used mostly for automobile and other headlights, where the bulb, coated with aluminum, is the reflector and the cover, the lens, helps to direct the light onto the road.

References

1. American Standard for Illuminating Engineering Nomenclature and Photometric Standards, ASA Z7.1—1942.
2. Kunerth, W.: "Text Book of Illumination," p. 48, Wiley, New York, 1929.
3. Walsh, J. W. T.: "Photometry," p. 152, Van Nostrand, Princeton, N.J., 1926.
4. ———: "Photometry," p. 155, Van Nostrand, Princeton, N.J., 1926.
5. Forsythe, W. E.: "Measurement of Radiant Energy," p. 395, McGraw-Hill, New York, 1937.
6. Ives, H. E.: *Phil. Mag.*, **24**: 149, 352, 744, 845, 853 (1912).
7. Blondel, A.: *J. phys.*, **5**: 222 (1896).
8. Walsh, J. W. T.: "Photometry," p. 205, Van Nostrand, Princeton, N.J., 1926.
9. Benford, F.: The Blackbody, Part I, *Gen. Elec. Rev.*, **46**: 377 (July, 1943).
10. Marden, J. W., and N. C. Beese: Effect of Temperature on Fluorescent Lamps, *Trans. Illum. Eng. Soc. (N.Y.)*, **34**: 55 (January, 1939).
11. Teele, R. P., and K. S. Gibson: *J. Opt. Soc. Am.*, **38**: 1096 (1948).
12. Barbrow, I. E.: Photometry with Barrier Layer Cells, *J. Research Natl. Bur. Standards*, **25**: 703 (1946).
13. Preston, J. S.: *J. Sci. Instr.*, **27**: 479 (Mar. 25, 1949).
14. Gabriel, M. H.: Photometry, *Gen. Elec. Rev.*, **54**: 23 (October, 1951).
15. Harvey, E. Newton: "The Nature of Animal Light," Lippincott, Philadelphia, 1920.
16. Benford, Frank: *Gen. Elec. Rev.*, **43**: 377 (1940).
17. Worthing, A. G.: *J. Opt. Soc. Am.*, **29**: 97 (1938). Haas and Guth: *Phys. Rev.*, **53**: 324 (1938).
18. Benford, Frank: *J. Opt. Soc. Am.*, **29**: 92 (1938).
19. Holladay, L. L.: *J. Opt. Soc. Am.*, **17**: 329 (1928).
20. Destrian, G.: *Phil. Mag.*, **7**, Ser. 38, pp. 700–739, 774–793, 880–888 (1947).
21. Powell: "The History of the Incandescent Lamp," Howell and Schroder, 1927.
22. Coolidge, W. D.: *AIEE J.*, **29**: 961 (1910).
23. Langmuir, I.: *Trans. Faraday Soc.*, **17**: 621 (1922).
24. Dushman, S.: *Gen. Elec. Rev.*, **40**: 260 (1934); *J. Opt. Soc. Am.*, **27**: 1 (1937); *Elec. Eng.*, 1934, p. 1204.
25. Buttolph, L. J.: *Gen. Elec. Rev.*, **23**: 741 (1920).
26. Ryde, J. W.: *Elec. Rev.* (London), **113**: 538 (1933).
27. Babcock, H. W.: Integrating Photometer for Low Light Levels, *J. Opt. Soc. Am.*, **40**: 409–411 (July, 1950).
28. Barrows, W. E.: "Light, Photometry, and Illumination," 1st ed., chaps. 6–8, McGraw-Hill, New York, 1912.
29. Baumgartner, G. R.: Practical Photometry of Fluorescent Lamps and Reflectors, *Illum. Eng.*, **36**: 1340–1356 (October, 1941).
30. Cady, F. E., and H. B. Dates: "Illuminating Engineering," chap. 3, McGraw-Hill, New York, 1928.
31. Forsythe, W. E.: "Measurement of Radiant Energy," chap. 13, McGraw-Hill, New York, 1937.
32. Einhorn, H. D., and J. D. Sauermann: Fluorescent Lamp Photometry, *J. Inst. Elec. Engrs. (London)*, **95** (pt. 2): 319–324 (June, 1948).
33. Horton, G. A.: Modern Photometry of Fluorescent Luminaires, *Illum. Eng.*, **45**: 458–465 (July, 1950).
34. Kunerth, W.: "Textbook of Illumination," chap. 3, Wiley, New York, 1929.
35. Morton, C. A.: Cosine Response of Photocells and the Photometry of Linear Light Sources, *Light & Lighting*, **38**: 157–160 (November, 1943).
36. Plymak, W. S., Jr., and G. T. Hicks: Physical Photometry in the Purkinje Range, *J. Opt. Soc. Am.*, **42**: 344–348 (May, 1952).
37. Sears, F. W.: "Principles of Physics," Part III, Optics, chap. 12, Addison-Wesley, Reading, Mass., 1945.
38. Teele, Ray P.: Photometer for Luminescent Materials, *J. Research Natl. Bur. Standards*, RP-1646, **34**: 325–332 (April, 1945).
39. ———: A Physical Photometer, *J. Research Natl. Bur. Standards*, RP-1415, **27**: 217–228 (September, 1941).
40. Waldram, J. M.: Photometry of Projected Light, *Illum. Eng.*, **47**: 397–408 (July, 1952).
41. Walsh, E. G.: Internal Standard Flame Photometry, *J. Sci. Instr.*, **29**: 23–25 (January, 1952).
42. Walsh, J. W. T.: "Photometry," Van Nostrand, Princeton, N.J., 1926.
43. Wilson, R.: Fundamental Limit of Sensitivity of Photometers, *Rev. Sci. Instr.*, **23**: 217–223 (May, 1952).
44. Winch, G. T.: Photometry and Colorimetry of Electric Discharge Lamps, *Light & Lighting*, **39**: 41–42 (1946).
45. Wrobel, H. T., and H. H. Chamberlain: Photometric Equipment for Blocking-layer Light-sensitive Cells, *Gen. Elec. Rev.*, **49**: 25–29 (April, 1946).
46. I. E. S. Guide for Photometric Testing of Fluorescent Lamps 1948, *Illum. Eng.*, **44**: 423–424 (July, 1949).
47. Integrating Photometer, *Elec. Rev.* (London), **144**: 479 (Mar. 25, 1949).
48. A Large Photometric Sphere, *Natl. Bur. Standards (U.S.), Tech. News Bull.*, **32**: (6): 65–66 (June, 1948).
49. Large Photometric Sphere, *J. Franklin Inst.*, **246**: 61–3, (July, 1948).
50. Moon, P.: "The Scientific Basis of Illuminating Engineering," p. 212, Dover, New York, 1961.

Chapter 4

Color Vision and Colorimetry

By DEANE B. JUDD, National Bureau of Standards

1. Definition of Color

Color is the property of radiant energy that permits a living organism to distinguish by eye between two uniform, structure-free patches of identical size and shape. If two such patches cannot be distinguished by eye from each other, they are said to be a *color match* for that organism.

2. Types of Color Vision

Some organisms (houseflies, cats, bulls, totally color-blind human beings) require for a color match simply that the two patches appear equally bright. For such organisms the condition for a color match between two patches of spectral radiance (areal density of spectral radiant intensity), N_1 and N_3, is

$$\int_0^\infty N_1 S\, d\lambda = \int_0^\infty N_3 S\, d\lambda \qquad (4.1)$$

where S is the relative spectral sensitivity of the organism to radiant energy. Table 4.1 gives the relative spectral sensitivity of the most common type of totally color-blind human organism. This wavelength function also applies well to all normal, dark-adapted human eyes and is known as the *scotopic luminous-efficiency function* (Commission Internationale de l'Éclairage, 1951).*

Most human observers, however, can still distinguish two patches even though the radiant energy flowing from them toward the observer is so adjusted as to make them appear equally bright; that is, the patches may also differ in the red-green sense and in the yellow-blue sense. Thus for a light-adapted observer of normal color vision the two patches must simultaneously satisfy three conditions if one is to be indistinguishable from the other,

$$\begin{aligned} \int_0^\infty N_1 \bar{x}\, d\lambda &= \int_0^\infty N_2 \bar{x}\, d\lambda \\ \int_0^\infty N_1 \bar{y}\, d\lambda &= \int_0^\infty N_2 \bar{y}\, d\lambda \\ \int_0^\infty N_1 \bar{z}\, d\lambda &= \int_0^\infty N_2 \bar{z}\, d\lambda \end{aligned} \qquad (4.2)$$

where for that observer $\bar{x}$, $\bar{y}$, $\bar{z}$ are the tristimulus values of a spectrum of unit radiance per unit wavelength. Table 4.2 gives the tristimulus values of such a spectrum for the 1931 CIE standard observer (Commission Internationale de l'Éclairage, 1931). Figure 4.1*a* shows a plot of them against wavelength. The $\bar{y}$ function is the luminous-efficiency function.

Table 4.1. 1951 CIE Scotopic Luminous-efficiency Function S

Wavelength, mμ	Scotopic luminous efficiency S	Wavelength, mμ	Scotopic luminous efficiency S
380	0.000589		
390	0.002209		
400	0.00929	600	0.03315
410	0.03484	610	0.01593
420	0.0966	620	0.00737
430	0.1998	630	0.003335
440	0.3281	640	0.001497
450	0.455	650	0.000677
460	0.567	660	0.0003129
470	0.676	670	0.0001480
480	0.793	680	0.0000715
490	0.904	690	0.00003533
500	0.982	700	0.00001780
510	0.997	710	0.00000914
520	0.935	720	0.00000478
530	0.811	730	0.000002546
540	0.650	740	0.000001379
550	0.481	750	0.000000760
560	0.3288	760	0.000000425
570	0.2076	770	0.0000002413
580	0.1212	780	0.0000001390
590	0.0655		

Other organisms (partially color-blind human beings) require but two wavelength functions for the statement of the conditions of color match. For protanopic vision the conditions are

$$\begin{aligned} \int_0^\infty N_1 \bar{w}_p\, d\lambda &= \int_0^\infty N_2 \bar{w}_p\, d\lambda \\ \int_0^\infty N_1 \bar{z}\, d\lambda &= \int_0^\infty N_2 \bar{z}\, d\lambda \end{aligned} \qquad (4.3)$$

* Names and years in parentheses refer to Bibliography at end of chapter.

Table 4.2. 1931 CIE Standard Observer for Colorimetry

Wave-length, mμ	Tristimulus specifications of equal-energy spectrum			Wave-length, mμ	Tristimulus specifications of equal-energy spectrum		
	$\bar{x}(\lambda)$	$\bar{y}(\lambda)$	$\bar{z}(\lambda)$		$\bar{x}(\lambda)$	$\bar{y}(\lambda)$	$\bar{z}(\lambda)$
380	0.0014	0.0000	0.0065	585	0.9786	0.8163	0.0014
385	0.0022	0.0001	0.0105	590	1.0263	0.7570	0.0011
390	0.0042	0.0001	0.0201	595	1.0567	0.6949	0.0010
395	0.0076	0.0002	0.0362	600	1.0622	0.6310	0.0008
400	0.0143	0.0004	0.0679	605	1.0456	0.5668	0.0006
405	0.0232	0.0006	0.1102	610	1.0026	0.5030	0.0003
410	0.0435	0.0012	0.2074	615	0.9384	0.4412	0.0002
415	0.0776	0.0022	0.3713	620	0.8544	0.3810	0.0002
420	0.1344	0.0040	0.6456	625	0.7514	0.3210	0.0001
425	0.2148	0.0073	1.0391	630	0.6424	0.2650	0.0000
430	0.2839	0.0116	1.3856	635	0.5419	0.2170	0.0000
435	0.3285	0.0168	1.6230	640	0.4479	0.1750	0.0000
440	0.3483	0.0230	1.7471	645	0.3608	0.1382	0.0000
445	0.3481	0.0298	1.7826	650	0.2835	0.1070	0.0000
450	0.3362	0.0380	1.7721	655	0.2187	0.0816	0.0000
455	0.3187	0.0480	1.7441	660	0.1649	0.0610	0.0000
460	0.2908	0.0600	1.6692	665	0.1212	0.0446	0.0000
465	0.2511	0.0739	1.5281	670	0.0874	0.0320	0.0000
470	0.1954	0.0910	1.2876	675	0.0636	0.0232	0.0000
475	0.1421	0.1126	1.0419	680	0.0468	0.0170	0.0000
480	0.0956	0.1390	0.8130	685	0.0329	0.0119	0.0000
485	0.0580	0.1693	0.6162	690	0.0227	0.0082	0.0000
490	0.0320	0.2080	0.4652	695	0.0158	0.0057	0.0000
495	0.0147	0.2586	0.3533	700	0.0114	0.0041	0.0000
500	0.0049	0.3230	0.2720	705	0.0081	0.0029	0.0000
505	0.0024	0.4073	0.2123	710	0.0058	0.0021	0.0000
510	0.0093	0.5030	0.1582	715	0.0041	0.0015	0.0000
515	0.0291	0.6082	0.1117	720	0.0029	0.0010	0.0000
520	0.0633	0.7100	0.0782	725	0.0020	0.0007	0.0000
525	0.1096	0.7932	0.0573	730	0.0014	0.0005	0.0000
530	0.1655	0.8620	0.0422	735	0.0010	0.0004	0.0000
535	0.2257	0.9149	0.0298	740	0.0007	0.0003	0.0000
540	0.2904	0.9540	0.0203	745	0.0005	0.0002	0.0000
545	0.3597	0.9803	0.0134	750	0.0003	0.0001	0.0000
550	0.4334	0.9950	0.0087	755	0.0002	0.0001	0.0000
555	0.5121	1.0002	0.0057	760	0.0002	0.0001	0.0000
560	0.5945	0.9950	0.0039	765	0.0001	0.0000	0.0000
565	0.6784	0.9786	0.0027	770	0.0001	0.0000	0.0000
570	0.7621	0.9520	0.0021	775	0.0000	0.0000	0.0000
575	0.8425	0.9154	0.0018	780	0.0000	0.0000	0.0000
580	0.9163	0.8700	0.0017		21.3713	21.3714	21.3715

where $\bar{w} = -0.460\bar{x} + 1.359\bar{y} + 0.101\bar{z}$ (see Fig. 4.1*d*). For deuteranopic vision of type I (Willmer, 1949) the conditions are

$$\int_0^\infty N_1\bar{y}\,d\lambda = \int_0^\infty N_2\bar{y}\,d\lambda \qquad \int_0^\infty N_1\bar{z}\,d\lambda = \int_0^\infty N_2\bar{z}\,d\lambda \tag{4.4}$$

For deuteranopic vision of type II (Nuberg and Justova, 1955), substitute for $\bar{y}$ in (4.4) the function $0.312\bar{x} + 0.757\bar{y} - 0.069\bar{z}$. And for tritanopic vision the conditions are

$$\int_0^\infty N_1\bar{y}\,d\lambda = \int_0^\infty N_2\bar{y}\,d\lambda \qquad \int_0^\infty N_1\bar{w}_p\,d\lambda = \int_0^\infty N_2\bar{w}_p\,d\lambda \tag{4.5}$$

If two spectrally dissimilar patches satisfy (4.2), they must also satisfy (4.3), (4.4), and (4.5). Thus

Table 4.3. Spectral Transmittance of Ocular Media*

Wave-length, mμ	Spectral transmittance of cornea, lens, and aqueous and vitreous humors (Ludvigh-McCarthy)	Spectral internal transmittance of macula lutea (Wald)	Spectral transmittance of ocular media, including the macula lutea
360	0.052*	0.859	0.045
370	0.056*	0.826	0.046
380	0.062*	0.762	0.047
390	0.069*	0.695	0.048
400	0.086	0.577	0.050
410	0.106	0.506	0.054
420	0.160	0.396	0.063
430	0.248	0.316	0.078
440	0.318	0.305	0.097
450	0.388	0.212	0.082
460	0.426	0.206	0.088
470	0.438	0.299	0.131
480	0.458	0.250	0.115
490	0.481	0.263	0.126
500	0.495	0.516	0.256
510	0.510	0.798	0.407
520	0.525	0.935	0.491
530	0.543	0.968	0.526
540	0.559	0.977	0.546
550	0.566	0.985	0.557
560	0.572	0.989	0.566
570	0.583	0.989	0.577
580	0.594	0.989	0.587
590	0.602	0.989	0.595
600	0.610	1.000*	0.610
610	0.619	1.000*	0.619
620	0.631	1.000*	0.631
630	0.641	1.000*	0.641
640	0.649	1.000*	0.649
650	0.657	1.000*	0.657
660	0.664	1.000*	0.664
670	0.676	1.000*	0.676
680	0.690	1.000*	0.690
690	0.698	1.000*	0.698
700	0.705	1.000*	0.705
710	0.707	1.000*	0.707
720	0.708	1.000*	0.708
730	0.710	1.000*	0.710
740	0.711	1.000*	0.711
750	0.713	1.000*	0.713

* Extrapolated.

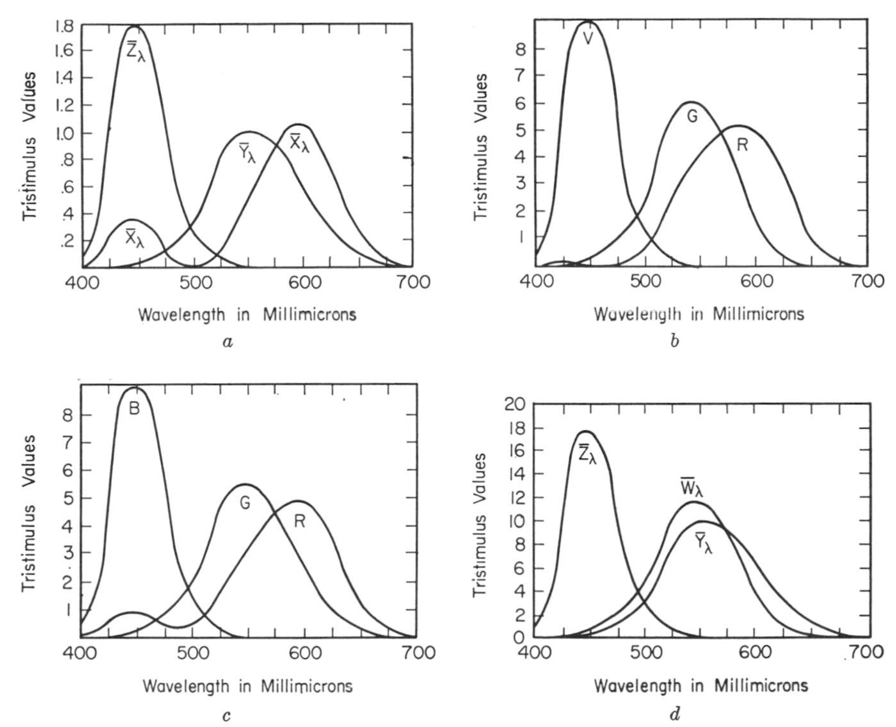

FIG. 4.1*a*. CIE standard observer and coordinate system (cone stage of Adams theory).
FIG. 4.1*b*. Young theory (cone-pigment stage of Adams and Müller theories).
FIG. 4.1*c*. Ladd-Franklin theory (cone stage of v. Kries–Schrödinger theory).
FIG. 4.1*d*. Dominator-modulator theory, late König.

a color match set up for normal human vision also holds for typical protanopic, deuteranopic, and tritanopic vision, which are, on this account, called *reduction forms* of normal vision.

Finally there are two groups of human observers requiring three wavelength functions for stating the conditions of color match different from those defined in Table 4.2. The largest group, embracing the great majority of human beings, have a pigmentation of the eye media (lens, macula) significantly heavier or lighter than standard. The required wavelength functions may be written $T\bar{x}$, $T\bar{y}$, $T\bar{z}$, where T is the ratio of the spectral transmittance of the eye media of the particular observer to that embodied in the standard observer defined by Table 4.2. Table 4.3 gives the internal spectral transmittance of the macula according to Wald (1945) and the spectral internal transmittance of the rest of the ocular media (cornea, lens, humors) according to Ludvigh and McCarthy (1938). The color-matching functions defining the 1931 CIE standard observer for colorimetry refer to observation of fields subtending from 1 to 4°. The color-matching functions defining the 1964 CIE supplementary observer (Judd and Wyszecki, 1963) refer to fields subtending more than 4°. The difference between these two sets of color-matching functions corresponds chiefly to less macular pigment for the latter; see column 3, Table 4.3.

The second group have vision intermediate between either normal and protanopic or normal and deuteranopic vision and are known as *anomalous trichromats*. For protanomalous vision write an equation like (4.2) in which N_4 replaces N_2, $\bar{w}_p$ replaces $\bar{x}$, and $(0.32\bar{x} + 0.25\bar{y} - 0.07\bar{z})^n$ replaces $\bar{y}$, where n is a number, greater than 1, characterizing each individual protanomalous trichromat (Pitt, 1949). For deuteranomalous vision write an equation like (4.2) in which N_5 replaces N_2 and the same new function, $(0.32\bar{x} + 0.25\bar{y} - 0.07\bar{z})^n$, used for protanomalous vision, replaces $\bar{x}$ (De Vries, 1948). Protanopic vision is a reduction form both of normal and of protanomalous vision; deuteranopic, both of normal and deuteranomalous.

3. Tristimulus Values

Equations (4.2) express the fundamental basis for colorimetry. The values of the integrals on the left are designated as tristimulus values X_1, Y_1, Z_1 and specify the color of the first patch. Those on the right are designated X_2, Y_2, Z_2 and specify that of the second. The amounts R, G, B of any three samples of radiant flux required to produce by additive mixture any color X, Y, Z may be computed from (4.2) provided no two of the samples may be added to produce a color match for the third. Let unit amount of the first sample (say, red) have tristimulus values X_r, Y_r, Z_r; the second (green), X_g, Y_g, Z_g; and the third (blue), X_b, Y_b, Z_b. It follows from (4.2) that the tristimulus values X, Y, Z of the color produced by adding R units of the red sample to G units of the green and to B units of the blue are

$$\begin{aligned} X &= X_rR + X_gG + X_bB \\ Y &= Y_rR + Y_gG + Y_bB \\ Z &= Z_rR + Z_gG + Z_bB \end{aligned} \tag{4.6}$$

The amounts R, G, B of the three samples of radiant flux required to match color X, Y, Z are found by solving (4.6).

$$\begin{aligned} DR &= (Y_gZ_b - Y_bZ_g)X + (X_bZ_g - X_gZ_b)Y + (X_gY_b - X_bY_g)Z \\ DG &= (Y_bZ_r - Y_rZ_b)X + (X_rZ_b - X_bZ_r)Y + (X_bY_r - X_rY_b)Z \\ DB &= (Y_rZ_g - Y_gZ_r)X + (X_gZ_r - X_rZ_g)Y + (X_rY_g - X_gY_r)Z \end{aligned} \tag{4.7}$$

provided the determinant D of the system (equal to $X_rY_gZ_b + X_bY_rZ_g + X_gY_bZ_r - X_bY_gZ_r - X_rY_bZ_g - X_gY_rZ_b$) is different from zero, as it will be if no two of the samples can be combined to color-match the third. The amounts R, G, B are called *tristimulus values* of the color in the RGB system, just as the amounts X, Y, Z are its tristimulus values in the XYZ system. The samples of radiant flux, which may be arbitrarily chosen, are known as *primaries* of the RGB system.

In a visual tristimulus colorimeter there are two juxtaposed patches, one for the sample of radiant flux to be measured, the other for the three-part sum whose amounts have to be adjusted by the observer until the two fields are indistinguishable. The amounts R, G, B, so found, are the tristimulus values of the unknown color relative to the working primaries; the tristimulus values X, Y, Z of this color in the standard coordinate system may be found from Eqs. (4.6).

If the unknown color has tristimulus values X, Y, Z, such that one of R, G, B computed from (4.7) is less than zero, it will be found impossible to set the three-part mixture field of the colorimeter to produce a match for this color. If, however, the indicated amount of the corresponding primary be added to the test patch, this two-part mixture will be found to be indistinguishable from the mixture of the two remaining primaries. A tristimulus colorimeter, if it is to be capable of measuring any color, must be provided with means for adding known amounts of the primaries to either patch.

Although tristimulus values of unknown colors are still occasionally found by direct use of a tristimulus colorimeter, it is more common practice to evaluate them either by spectroradiometry in accordance with (4.2) or by visual or photoelectric comparison with color standards so evaluated. Detailed instructions for evaluating, by summation, approximations to the integrals of (4.2) are available elsewhere (Hardy, 1936; OSA Committee on Colorimetry, 1944, 1953; Judd, 1950a,b, 1952).

4. Theories of Color Vision

A theory of color vision accounts for the facts of color matching by the normal-eyed observer expressed by (4.1) and (4.2) in terms of processes assumed to take place within the retinal rods and cones. All the theories of current interest assume that these processes are photochemical decomposition of pigments contained in the receptors (rods and cones). The photosensitive pigment contained by the retinal rods is known to be rhodopsin (Wald, 1937–1938), and the scotopic luminous-efficiency function (Table 4.1) is closely proportional to the product of the spectral absorptance (ratio of the flux absorbed by a layer to that incident on it) of rhodopsin by the spectral transmittance of the ocular media (Table 4.3). The photosensitive pigments contained by the retinal cones are not yet positively identified, but the wavelength dependence of their absorptances has been approximately determined directly by microspectrophotometry of single receptors (Brown and Wald, 1964) and indirectly by threshold measurements (Stiles, 1959; Wald, 1964) before and after exposure to suitable chromatic lights. This dependence is found to be closely such as to predict color matches in accordance with (4.2).

Equations (4.2) identify various groups of samples of radiant energy as indistinguishable from each other by the CIE standard observer by reference to the tristimulus values of the spectrum expressed relative to the primaries of the CIE coordinate system. These same groups may be identified by tristimulus values of the spectrum expressed relative to any other set of primaries, real or imaginary, in accordance with (4.7). [An imaginary primary is a color requiring negative radiance for some parts of the spectrum to produce it in accordance with (4.2). One or two of the tristimulus values of an imaginary color expressed in the CIE coordinate system may be less than zero.] By defining for each theory the primaries of the corresponding coordinate system in terms of the CIE system, it is possible to give in condensed form precisely the wavelength dependence of absorptance of the pigments, or pigment mixtures, assumed to be present in the cones. This is done in Table 4.4. The reverse-transformation constants are given in matrix form from (4.6),

$$\begin{bmatrix} X_r, & X_g, & X_b \\ Y_r, & Y_g, & Y_b \\ Z_r, & Z_g, & Z_b \end{bmatrix}$$

and at least one primary in each coordinate system is seen from the negative values in Table 4.4 to be imaginary. The constants of transformation from the CIE system, given in Table 4.4, similarly correspond to (4.7). Figure 4.1*b* shows the wavelength dependence of absorptance of the photosensitive pigments assumed by the Young theory (Judd, 1950*b*) multiplied by the spectral transmittance of the ocular media (Table 4.3). These products were computed from (4.7) by inserting the coefficients of X, Y, Z from Table 4.4 and substituting for X, Y, Z the tristimulus values of the spectrum from Table 4.2. The Young theory has been used for the cone-pigment stage of the Adams (1923) theory and the Müller (1930) theory as quantified by Judd (1949). Figure 4.1*c* gives similar information for the Ladd-Franklin (1892, 1929) theory, or early König (1892, 1903) theory. The v. Kries–Schrödinger (Schrödinger, 1925) theory uses the same formulation but takes each wavelength function to refer to whatever mixture of the three pigments may be present in that type of cone without assuming perfect segregation as in the Young theory.

Each theory of color vision also seeks to give a

simple explanation of the various types of defective color vision. The reduction systems [(4.3) to (4.5)] may be explained by faults in any of the stages of the visual mechanism, that is, by failure of one cone pigment (Young), by two kinds of cone having identical segregation of pigment (Fick, 1879; Adams, 1923), by failure of one kind of cone (late König, 1897, 1903), by failure of one kind of secondary chemical process in the cones (Müller, 1930), or by failure of one kind of process in the optic nerve (Müller, 1930). The Helmholtz, Müller, and late König theories give valid accounts of dichromatic vision; but the late König theory gives the simplest account, and the functions generated by its coordinate system (see Table 4.4) were used to state the facts of dichromatic vision in (4.3) to (4.5). The first column of reverse-transformation constants in Table 4.4 for this theory defines the missing primary for protanopic vision. It is the same primary (except for a scale constant) used for the *yR-bG* process of the cone stage according to the Müller theory. The second column defines the missing primary for deuteranopic vision. It is the same as the primary used for the *R-G* process of the optic-nerve stage of the Müller theory. The third column defines the missing primary for tritanopic vision; it is closely the same primary used for the *gY-rB* process of the cone stage of the Müller theory. The Müller theory has the unique advantage of describing correctly both the color confusions and the color perceptions of color-blind observers (Judd, 1949). The Young primaries are extreme spectrum red, extreme spectrum violet, and an extraspectral green (see columns of reverse-transformation constants in Table 4.4). The Young theory accounts well for tritanopia but only approximately for protanopia and deuteranopia.

Figure 4.1*a*, *c*, and *d* shows the distribution of the cone responses throughout a spectrum of unit radiance according to the Adams, the v. Kries–Schrödinger, and the late König theories, respectively. Figure 4.2*b* refers to the chromatic responses of the cones according to the Müller theory.

Several theories (Hering, Ladd-Franklin, v. Kries–Schrödinger, Adams, Müller) lay stress on the stimuli required by a neutrally adapted subject to yield the unitary hues, red, yellow, green, and blue. Unitary red is one perceived as neither yellowish nor bluish; unitary blue, neither greenish nor reddish; and so on. The *Y* primary of the CIE system refers to a stimulus for yellowish green carrying all the luminous aspect of color, the *Z* primary to a stimulus of zero luminosity for unitary blue, and the *X* primary to a stimulus of zero luminosity for unitary red. Table 4.4 shows that this system, intended to be nontheoretical, is nevertheless adapted for use unchanged in the cone stage of the Adams theory. The Hering theory was the first to take explicit account of these stimuli for unitary hues (sometimes called the *psychological primaries*). It is an opponent-colors theory, postulating in the optic nerve a white process, *W*, whose negative is black, a red process, *R-G*, whose negative is green, and a yellow process, *Y-B*, whose negative is blue. From the first column of reverse-transformation constants it is seen that the Hering *R-G* primary is close to the CIE *X* primary; they are both unitary reds. Similarly the second column shows the Hering *Y-B* primary to be close to the negative of the CIE *Z* primary; that is, the Hering blue and the CIE *Z* primary are both unitary blues. The third column shows that the Hering *W* primary corresponds to equal amounts of the CIE primaries, that is, to the color of a source having an equal-energy spectrum. This identity of the three numbers in a column of the reverse-transformation constants marks the applicability of the coordinate system to an opponent-colors theory. Table 4.4 shows that the three other formulations (v. Kries–Schrödinger, Adams, Müller) for events at the optic-nerve stage have the opponent-colors form and that in the latter two of these formulations the chromatic primaries are identical to the *X* and *Z* primaries of the CIE system, which correspond to unitary red and unitary blue, respectively.

Theories of color vision also seek to account for the fact that the luminous aspect of color can be perceived more or less separate from the chromatic aspect as in equality-of-brightness heterochromatic photometry. Since *Y* evaluates the luminous aspect of the color *X*, *Y*, *Z*, the second row of reverse-transformation constants indicates the relative amounts of luminosity associated with each fundamental process of each theory.. In the Young theory the luminosity is distributed about equally between the red and green processes, with the blue process carrying a negligible amount. In the Helmholtz theory, probably highest in current favor, much more of the luminosity is placed in the red process; and in the late König theory, all of it is placed there. Theories, like the late König theory, postulating a separate kind of cone for the luminous aspect of color, the other color processes being unassociated with any luminous aspect, are called *dominator-modulator theories*, following Granit's (1947) pioneer microelectrode work on responses of single fibers of the optic nerves of various animals. The coordinate systems adapted to such theories may be identified by the first and third constants of transformation in some one row being zero, and this requires that two of the second row of reverse-transformation constants be zero. Table 4.4 shows that the cone stage of the Adams theory is a dominator-modulator theory. Other theories (v. Kries–Schrödinger, Müller) assume that separation of the luminous from the chromatic aspect of a color perception does not take place until the optic-nerve stage. The Hering theory assumes that this separation is never complete. Although most of the luminosity is associated with the white process, the red process *R-G* carries a small fraction of it and the yellow process *Y-B* a still smaller fraction. The cone stage of the Müller theory also carries this idea. Technical color reproduction (screen-plate printing, color television) also pays tribute to this idea whenever the common practice is followed of having a separate black-white process (black impression in process printing, principle of mixed highs in color television), and the corresponding coordinate systems show similarities to those of the Hering theory and the cone stage of the Müller theory.

Theories of color vision also seek to explain various accessory phenomena not involving color matching: perceptibility of color differences, hue change with

TABLE 4.4. CONSTANTS DEFINING THE RELATIONS BETWEEN THE 1931 CIE STANDARD COORDINATE SYSTEM FOR COLORIMETRY AND THE COORDINATE SYSTEMS CORRESPONDING TO SOME OF THE BETTER-KNOWN THEORIES OF COLOR VISION*

Name of theory	Constants of transformation from CIE system				Constants of reverse transformation				Anatomical location of assumed processes
		X	Y	Z		R	G	V	
Young, three-component	R	3.1956	2.4478	−0.6434	X	0.24513	−0.08512	0.03999	Cone pigments
	G	−2.5455	7.0492	0.4963	Y	0.08852	0.11112	0.00036	
	V	0.0000	0.0000	5.0000	Z	0.00000	0.00000	0.20000	
Helmholtz, three-component	R	0.355	4.725	−0.080	X	0.5117	−0.3558	0.0441	Cones
	G	−2.300	6.795	0.505	Y	0.1732	0.0267	0.0001	
	V	0.000	0.000	5.000	Z	0.0000	0.0000	0.2000	
Dominator-modulator, late König	R	0.000	1.000	0.000	X	2.954	−2.174	0.220	Cones
	G	−0.460	1.359	0.101	Y	1.000	0.000	0.000	
	V	0.000	0.000	1.000	Z	0.000	0.000	1.000	
						R	G	B	
Ladd-Franklin, early König	R	3.7656	1.4635	−0.2291	X	0.24395	−0.05825	0.01430	Cone pigments
	G	−1.3973	6.1289	0.2683	Y	0.05562	0.14988	−0.00549	
	B	0.0000	0.0000	5.0000	Z	0.00000	0.00000	0.20000	
						R-G	Y-B	W	
Hering, opponent-color	R-G	1.000	−1.000	0.000	X	1.040	0.010	1.000	Optic nerve
	Y-B	0.000	0.400	−0.400	Y	0.040	0.010	1.000	
	W	−0.040	1.036	0.004	Z	0.040	−2.490	1.000	
						R	G	B	
v. Kries–Schrödinger, zone or stage	R	3.7656	1.4635	−0.2291	X	0.24395	−0.05825	0.01430	Cones
	G	−1.3973	6.1289	0.2683	Y	0.05562	0.14988	−0.00549	
	B	0.0000	0.0000	5.0000	Z	0.00000	0.00000	0.20000	
						G-R	B-Y	W	
	G-R	−3.537	3.196	0.341	X	−0.2749	0.0206	1.0000	Optic nerve
	B-Y	1.341	−5.884	4.542	Y	0.0000	0.0000	1.0000	
	W	0.000	1.000	0.000	Z	0.0812	0.2141	1.0000	
						R	G	V	
Adams, zone or stage	R	3.1956	2.4478	−0.6434	X	0.24513	−0.08512	0.03999	Cone pigments
	G	−2.5455	7.0492	0.4963	Y	0.08852	0.11112	0.00036	
	V	0.0000	0.0000	5.0000	Z	0.00000	0.00000	0.20000	
						R	G	B	
	R	1.000	0.000	0.000	X	1.000	0.000	0.000	Cones
	G	0.000	1.000	0.000	Y	0.000	1.000	0.000	
	B	0.000	0.000	1.000	Z	0.000	0.000	1.000	
						R-G	B-Y	W	
	R-G	1.000	−1.000	0.000	X	1.000	0.000	1.000	Optic nerve
	B-Y	0.000	−0.400	0.400	Y	0.000	0.000	1.000	
	W	0.000	1.000	0.000	Z	0.000	2.500	1.000	
						R	G	V	
Müller, zone or stage	R	3.1956	2.4478	−0.6434	X	0.24513	−0.08512	0.03999	Cone pigments
	G	−2.5455	7.0492	0.4963	Y	0.08852	0.11112	0.00036	
	V	0.0000	0.0000	5.0000	Z	0.00000	0.00000	0.20000	
						yR-bG	gY-rB	W_p	
	yR-bG	5.741	−4.601	−1.140	X	0.2391	−0.0967	1.0000	Cones
	gY-rB	−0.932	2.751	−1.819	Y	0.0810	0.0024	1.0000	
	W_p	−0.463	1.366	0.097	Z	0.0000	−0.4966	1.0000	
						R-G	Y-B	W_d	
	R-G	6.325	−6.325	0.000	X	0.1581	0.0000	1.0000	Optic nerve
	Y-B	0.000	2.004	−2.004	Y	0.0000	0.0000	1.0000	
	W_d	0.000	1.000	0.000	Z	0.0000	−0.4990	1.0000	

* See Eqs. (4.6) and (4.7).

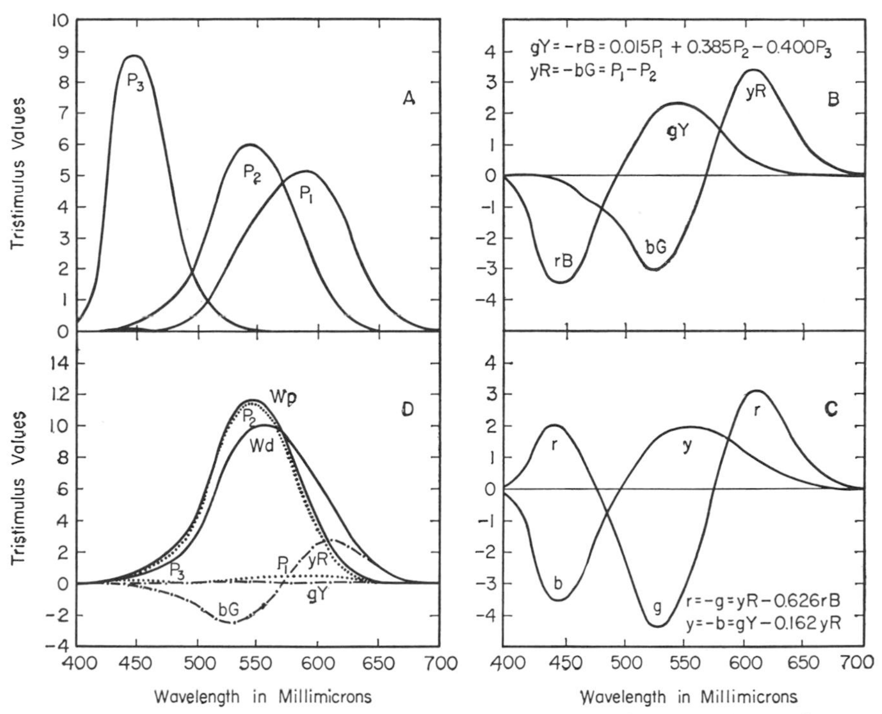

FIG. 4.2. Tristimulus values for the three stages of Müller theory. (*A*) Initial photosensitive-substance stage (same as Young theory); (*B*) stage of chromatic retinal sensory processes; (*C*) chromatic processes in optic-nerve fiber stage (same as Hering theory); (*D*) components in luminosity function both for normal, W_d, and protanopic, W_p, vision.

luminance (Bezold, 1873; Brücke, 1879), hue change by dilution with achromatic light (Abney, 1912, 1913), influence of neighboring areas on color perception (Jaensch, 1921; Land, 1959, 1964; Judd, 1960; Walls, 1960), color perception with chromatic adaptation (Wright, 1934; Helson, Judd, Warren, 1952), and the like. For dealing with such phenomena the optic-nerve stage must be used (Hurvich and Jameson, 1957), and the wavelength functions linearly connected (Table 4.4) to the tristimulus values of the spectrum in the CIE system (Table 4.2) usually do not apply. The response of a stimulated cone takes the form of a burst of nerve discharges of constant intensity (all-or-none principle) signaling the onset of the stimulus, the frequency of discharges not being linearly connected to the radiance of a stimulus of constant spectral quality. Instead, because of adaptation, the connection is one of diminishing returns, and the magnitude of response is variously assumed to correspond to the logarithm (Fechner), a power less than 1 (Plateau), a hyperbolic function (Adams, 1922), or the Munsell value function V (Newhall, Nickerson, and Judd, 1943) of luminance Y. The Munsell value function V is defined implicitly in terms of luminance Y.

$$Y = 1.2219V - 0.23111V^2 + 0.23951V^3 - 0.021009V^4 + 0.0008404V^5 \quad (4.8)$$

Tables of this function are available elsewhere (Newhall, Nickerson, and Judd, 1943; Nickerson, 1950; Judd, 1952). The hyperbolic form adjusted to the same end points (Judd, 1952) would be

$$V = \frac{10Y(Y_b + 1)}{Y + Y_b} \quad (4.9)$$

where Y_b is the luminance to which the observer is adapted.

Figure 4.2 summarizes the Müller theory, which is the most completely developed of all, though anatomically implausible (Talbot, 1951). No completely satisfactory theory of color vision has yet been developed. The chief limitations of various theories have been briefly summarized elsewhere (Judd, 1951).

5. Chromaticity Diagrams

The complete graphical representation of color data requires a tridimensional plot. Each set of tristimulus values (such as X, Y, Z) determines a vector in 3-space, starting from the origin of an orthogonal coordinate system, whose length is interpreted as the amount of the color and whose direction indicates the quality, or chromaticity, of the color. These directions are sometimes defined by the intersections of the vectors with the plane passing through unit amounts on each coordinate axis. The location of the point on such a plane corresponding to tristimulus values X, Y, Z is given by the ratios $X/(X + Y + Z)$, $Y/(X + Y + Z)$, $Z/(X + Y + Z)$, called *chromaticity coordinates* x, y, z. The location may be found by plotting any one of the chromaticity coordinates x, y, z against any other in 60° triangular coordinates. It has been customary, however, to plot x against y in rectangular coordinates, and this chromaticity diagram is now accepted as the American Standard (American Standards Association, 1951). Table 4.5 shows the chromaticity coordinates x, y, z of the spectrum for the 1931 CIE standard observer computed from seven-place tristimulus values of which the first four are given in Table 4.2 (Smith and Guild, 1931–1932), and Fig. 4.3 shows them plotted on the American Standard Chromaticity Diagram. This diagram is equivalent to the parallel projection of the unit plane onto the XY plane.

On a chromaticity diagram the points representing the additive combination of two component colors are found on the straight line connecting the points representing the two components. The straight line connecting the extremes of the spectrum locus in Fig. 4.3 thus represents the chromaticities producible by mixing radiant energy from the long-wave extreme (700 to 780 mμ) with that from the short-wave extreme (380 mμ), and the area inclosed by the spectrum locus and this straight line represents the locus of chromaticities producible by mixtures of the spectrum colors, that is, the locus of all real chromaticities. From the convex curvature of the spectrum locus it follows that any coordinate system based on real chromaticities cannot specify all colors with positive tristimulus values; for some real colors negative amounts of the real primaries will be required.

Each chromaticity diagram is characterized by a straight line corresponding to zero luminosity and known as the *alychne*. Chromaticities represented by points on the same side of the alychne as the spectrum locus correspond to positive luminosity, and the interpretation of such points is straightforward even though they represent unreal chromaticities. Those on the opposite side of the alychne refer to negative luminosity, and negative amounts of the colors so designated refer to a positive amount of the complementary color (Wyszecki, 1954). For example, the Helmholtz G primary defined by the middle column of the reverse-transformation constants in Table 4.4 has chromaticity coordinates in the American Standard Chromaticity Diagram:

$$x = \frac{-0.3558}{-0.3291} = 1.081$$

$$y = \frac{0.0267}{-0.3291} = -0.081$$

$$z = \frac{0.0000}{-0.3291} = 0.000$$

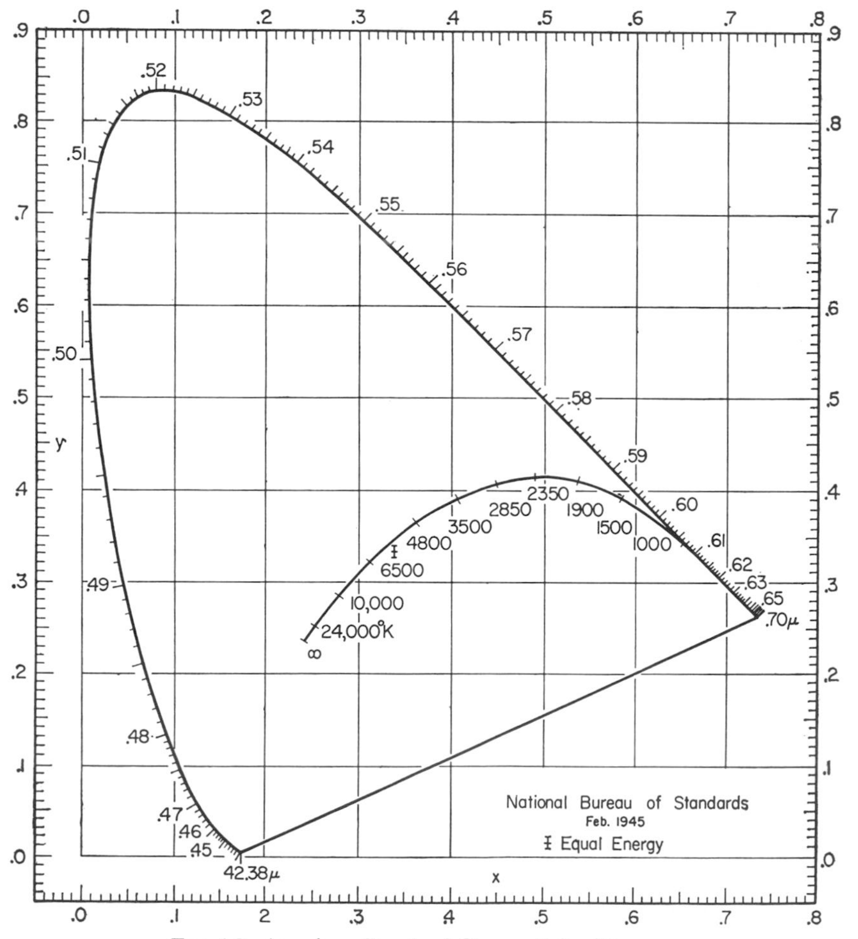

Fig. 4.3. American Standard Chromaticity Diagram.

TABLE 4.5. CHROMATICITY COORDINATES OF THE SPECTRUM IN THE 1931 CIE COORDINATE SYSTEM

Wave-length, mμ	Chromaticity coordinates			Wave-length, mμ	Chromaticity coordinates		
	x	y	z		x	y	z
380	0.1741	0.0050	0.8209	580	0.5125	0.4866	0.0009
385	0.1740	0.0050	0.8210	585	0.5448	0.4544	0.0008
390	0.1738	0.0049	0.8213	590	0.5752	0.4242	0.0006
395	0.1736	0.0049	0.8215	595	0.6029	0.3965	0.0006
400	0.1733	0.0048	0.8219	600	0.6270	0.3725	0.0005
405	0.1730	0.0048	0.8222	605	0.6482	0.3514	0.0004
410	0.1726	0.0048	0.8226	610	0.6658	0.3340	0.0002
415	0.1721	0.0048	0.8231	615	0.6801	0.3197	0.0002
420	0.1714	0.0051	0.8235	620	0.6915	0.3083	0.0002
425	0.1703	0.0058	0.8239	625	0.7006	0.2993	0.0001
430	0.1689	0.0069	0.8242	630	0.7079	0.2920	0.0001
435	0.1669	0.0086	0.8245	635	0.7140	0.2859	0.0001
440	0.1644	0.0109	0.8247	640	0.7190	0.2809	0.0001
445	0.1611	0.0138	0.8251	645	0.7230	0.2770	0.0000
450	0.1566	0.0177	0.8257	650	0.7260	0.2740	0.0000
455	0.1510	0.0227	0.8263	655	0.7283	0.2717	0.0000
460	0.1440	0.0297	0.8263	660	0.7300	0.2700	0.0000
465	0.1355	0.0399	0.8246	665	0.7311	0.2689	0.0000
470	0.1241	0.0578	0.8181	670	0.7320	0.2680	0.0000
475	0.1096	0.0868	0.8036	675	0.7327	0.2673	0.0000
480	0.0913	0.1327	0.7760	680	0.7334	0.2666	0.0000
485	0.0687	0.2007	0.7306	685	0.7340	0.2660	0.0000
490	0.0454	0.2950	0.6596	690	0.7344	0.2656	0.0000
495	0.0235	0.4127	0.5638	695	0.7346	0.2654	0.0000
500	0.0082	0.5384	0.5434	700	0.7347	0.2653	0.0000
505	0.0039	0.6548	0.3413	705	0.7347	0.2653	0.0000
510	0.0139	0.7502	0.2359	710	0.7347	0.2653	0.0000
515	0.0389	0.8120	0.1491	715	0.7347	0.2653	0.0000
520	0.0743	0.8338	0.0919	720	0.7347	0.2653	0.0000
525	0.1142	0.8262	0.0596	725	0.7347	0.2653	0.0000
530	0.1547	0.8059	0.0394	730	0.7347	0.2653	0.0000
535	0.1929	0.7816	0.0255	735	0.7347	0.2653	0.0000
540	0.2296	0.7543	0.0161	740	0.7347	0.2653	0.0000
545	0.2658	0.7243	0.0099	745	0.7347	0.2653	0.0000
550	0.3016	0.6923	0.0061	750	0.7347	0.2653	0.0000
555	0.3373	0.6589	0.0038	755	0.7347	0.2653	0.0000
560	0.3731	0.6245	0.0024	760	0.7347	0.2653	0.0000
565	0.4087	0.5896	0.0017	765	0.7347	0.2653	0.0000
570	0.4441	0.5547	0.0012	770	0.7347	0.2653	0.0000
575	0.4788	0.5202	0.0010	775	0.7347	0.2653	0.0000
580	0.5125	0.4866	0.0009	780	0.7347	0.2653	0.0000

In this diagram the alychne corresponds to $y = 0$; so this primary of the Helmholtz theory corresponding to a negative amount of X on the opposite side of the alychne from the spectrum locus is correctly designated as a positive amount of green. On the other hand, the third primary of the early König theory, which also plots on the negative side of the alychne, corresponds to a positive amount of Z and is therefore a positive amount of blue associated with a negative luminosity.

The American Standard Chromaticity Diagram is widely used as a map showing the relation between the chromaticity of the sample of radiant flux under consideration to those of such colorimetric landmarks as the spectrum chromaticities, the chromaticities of the complete radiator at various temperatures (see Fig. 4.3), and the chromaticity of various well-established systems of color standards such as the Munsell Book of Color (Glenn and Killian, 1940; Kelly, Gibson, and Nickerson, 1943; Granville, Nickerson, and Foss, 1943), the Color Harmony Manual (Granville and Jacobson, 1944), and the Lovibond glasses (Schofield, 1939; Judd, Chamberlin, and Haupt, 1962). As a map this diagram suffers from scale distortion, and there have been frequent attempts to correct this, both by linear and nonlinear transformations. Table 4.6 gives the constants of transformation according to Eq. (4.7), and the reverse-transformation constants according to (4.6) for some of the most used transformations of the 1931 CIE coordinate system.

The first three coordinate systems defined in Table 4.6 refer to physically realizable primaries that can be used in tristimulus colorimeters. These primaries are defined in the CIE system by the columns of reverse-transformation constants which give the tristimulus values X, Y, Z for the corresponding primary, and these values have proportions that correspond to points within the area representing real colors in Fig. 4.3. The remaining seven coordinate systems are based on primaries, at least one of which is imaginary, chosen to yield a chromaticity diagram of a particular spacing. The MacAdam (1937) diagram has been recommended by the CIE and is now known as the 1960 CIE-UCS diagram. Several of these chromaticity diagrams have been combined with value scales, such as given by (4.8), to define variables O, P, Q that, when plotted on mutually perpendicular axes, define a color space in which distance between two points correlates more or less well with visual estimates ΔE, of the size of the corresponding color difference (Judd, 1952). That is, the visual estimate ΔE is found to be predicted with good approximation by the relation

$$\overline{\Delta E^2} = \overline{\Delta O^2} + \overline{\Delta P^2} + \overline{\Delta Q^2} \tag{4.10}$$

The spacing of the Munsell color scales can be described with considerable fidelity in this way (Adams, 1942; Saunderson and Milner, 1946) by writing O, P, Q in terms of the Munsell value functions V_x, V_y, V_z [see Eq. (4.8)] of the tristimulus values X, Y, Z of the color, as suggested by the optic-nerve stage of the Adams theory (see Table 4.4).

The definitions of O, P, Q recommended by the CIE in 1964 are (Wyszecki, 1963)

$$Q = 25Y^{1/3} - 17 \qquad 1 \leq Y \leq 100$$
$$O = 13Q(u - u_0)$$
$$P = 13Q(v - v_0)$$

where
$$u = 4X/(X + 15Y + 3Z)$$
$$v = 6Y/(X + 15Y + 3Z)$$

as in the MacAdam (1937) diagram, and u_0, v_0 are the chromaticity coordinates of the color to which the observer is adapted.

TABLE 4.6. CONSTANTS DEFINING THE RELATIONS BETWEEN THE 1931 CIE STANDARD COORDINATE SYSTEM FOR COLORIMETRY AND SOME OF THE MOST USED NONSTANDARD COORDINATE SYSTEMS*

Identification		Constants of transformation from CIE system†				Constants of reverse transformation‡			Intended use
		X	Y	Z		R	G	B	
WDW system (Wright, 1928–1929; Judd, 1944)	R	1.0559	−0.1953	−0.1769	X	1.018	0.121	0.144	Visual research to give results independent of ocular pigmentation of observer, transformation to CIE system valid only for standard observer, primaries 460, 530, 650 mμ
	G	−0.6478	1.7175	0.0507	Y	0.384	0.627	0.030	
	B	0.0235	−0.0623	1.2087	Z	0.000	0.030	0.826	
						R	G	B	
NPL system (Smith and Guild, 1931–1932)	R	1.85221	−0.70226	−0.36664	X	0.62555	0.37963	0.20000	Color specification by National Physical Laboratories, England, prior to 1931, primaries 436.1, 546.1, 700 mμ
	G	−0.42067	1.16478	0.07247	Y	0.22592	0.99487	0.01063	
	B	0.00520	−0.01441	1.00921	Z	0.00000	0.01225	0.99000	
						R	G	B	
Color television (Judd, 1952)......	R	1.735	−0.527	−0.232	X	0.675	0.220	0.130	Color-television receivers, example of primaries realizable by suitable filter-phosphor combinations
	G	−0.842	1.749	0.038	Y	0.325	0.680	0.080	
	B	0.106	−0.221	1.271	Z	0.000	0.100	0.790	
						R	G	B	
Uniform chromaticity scale, UCS (Judd, 1935)	R	3.1956	2.4478	−0.1434	X	0.24513	−0.08512	0.11996	Chromaticity-difference estimation among self-luminous colors subtending 2°, triangular plot (60°)
	G	−2.5455	7.0492	0.9963	Y	0.08852	0.11112	−0.09802	
	B	0.0000	0.0000	1.0000	Z	0.00000	0.00000	1.00000	
						X''	Y''	Z''	
Rectangular uniform chromaticity scale, RUCS (Breckenridge and Schaub, 1939)	X''	0.0960	1.3040	−1.4000	X	0.01245	−0.13283	0.04556	Chromaticity-difference estimation among self-luminous colors subtending 2°, opponent-colors type, rectangular plot
	Y''	−5.2963	4.1986	1.0977	Y	0.16262	0.05870	0.04556	
	Z''	6.4805	11.8846	3.5828	Z	−0.56195	0.04556	0.04556	
						$\bar{\alpha}$	$\bar{\beta}$	$\bar{\gamma}$	
(α,β) system (Hunter, 1942)......	$\bar{\alpha}$	2.1052	−1.6845	−0.3214	X	0.4750	0.2418	0.1455	Chromaticity-difference estimation approximately valid for all colors subtending between 2 and 20°, rectangular plot, opponent-colors type based on CIE source C (daylight)
	$\bar{\beta}$	0.0000	0.6739	−0.5708	Y	0.0000	0.5194	0.1484	
	$\bar{\gamma}$	0.0000	4.3793	1.9976	Z	0.0000	−1.1387	0.1753	
						U	V	W	
Uniform chromaticity scale, simple UVW (MacAdam, 1937) now known as the 1960 CIE-UCS diagram	U	0.6667	0.0000	0.0000	X	1.5000	0.0000	0.0000	Chromaticity-difference estimation, 2° self-luminous colors, dominator-modulator type, rectangular plot
	V	0.0000	1.0000	0.0000	Y	0.0000	1.0000	0.0000	
	W	−0.5000	1.5000	0.5000	Z	1.5000	−3.0000	2.0000	
						U	V	W	
Uniform chromaticity scale, UVW (OSA Colorimetry Committee, 1944, 1953)	U	0.4661	0.1593	0.0000	X	2.1455	−0.5193	0.0000	Chromaticity-difference estimation, same spacing as UCS system, dominator-modulator type, rectangular plot
	V	0.0000	0.6581	0.0000	Y	0.0000	1.5195	0.0000	
	W	−0.38105	0.4250	0.2424	Z	3.3727	−3.4806	4.1255	
						X'	Y'	Z'	
Constant-luminance system (Adams, 1942)	X'	3.0000	0.0000	0.0000	X	0.3333	0.0000	0.0000	Chromaticity-difference estimation, 2° self-luminous colors except saturated blue and violet, equiluminous primaries, rectangular plot
	Y'	−3.0000	1.0000	−1.0000	Y	1.0000	1.0000	1.0000	
	Z'	0.0000	0.0000	1.0000	Z	0.0000	0.0000	1.0000	
						R-G	B-Y	W	
Chromatic-valence system (Adams, 1942)	R-G	1.0000	−1.0000	0.0000	X	1.0000	0.0000	1.0000	Chromaticity-difference estimation, 5 to 20° object colors, dominator-modulator, opponent-colors type, rectangular plot
	B-Y	0.0000	−0.4000	0.4000	Y	0.0000	0.0000	1.0000	
	W	0.0000	1.0000	0.0000	Z	0.0000	2.5000	1.0000	

* See Eqs. (4.6) and (4.7). † See. Eq. (4.7). ‡ See Eq. (4.8).

6. Photoelectric Colorimeters

Essentially a photoelectric tristimulus colorimeter consists of three filter-photocell combinations whose spectral sensitivities S_r, S_g, and S_b satisfy approximately throughout the whole spectrum the condition

$$\begin{aligned} S_r &= k_{11}\bar{x} + k_{12}\bar{y} + k_{13}\bar{z} \\ S_g &= k_{21}\bar{x} + k_{22}\bar{y} + k_{23}\bar{z} \\ S_b &= k_{31}\bar{x} + k_{32}\bar{y} + k_{33}\bar{z} \end{aligned} \tag{4.11}$$

and whose three outputs of photocurrent, I_r, I_g, I_b, when their fields of "view" are filled with an area of spectral radiance, N, are

$$\begin{aligned} I_r &= k_r \int_0^\infty S_r N \, d\lambda \\ I_g &= k_g \int_0^\infty S_g N \, d\lambda \\ I_b &= k_b \int_0^\infty S_b N \, d\lambda \end{aligned} \tag{4.12}$$

where $\bar{x}$, $\bar{y}$, $\bar{z}$ are the tristimulus values of a spectrum of unit radiance, N, per unit wavelength for the CIE standard observer (Table 4.2), k_{11}, k_{12}, and so on, are any arbitrary constants whose determinant

$$\begin{vmatrix} k_{11} & k_{12} & k_{13} \\ k_{21} & k_{22} & k_{23} \\ k_{31} & k_{32} & k_{33} \end{vmatrix}$$

is far from being zero, and k_r, k_g, k_b are scale constants determined by calibration.

If the filter-photocell combinations satisfy (4.11) precisely, then it follows that any two patches of spectral radiance, N_1 and N_2, that satisfy the condition for color match given in (4.2) will yield by (4.12) identical triads of current output from the photoelectric colorimeter, thus justifying the name "colorimeter." [Devices for measuring the absorptance of solutions for radiant flux of more or less restricted wavelength range, frequently used to determine the concentration of an energy-absorbing constituent in solution, are often called colorimeters, but they do not measure color in the above sense and might more aptly be called "absorptimeters," as suggested by Mellon (1950).]

The photocell may be of the barrier-layer type for greatest portability and convenience, the vacuum phototube for greatest stability, the photomultiplier tube for greatest sensitivity, or the image-orthicon tube for responding to pattern and movement as in a camera for color television. A photoelectric colorimeter may also include a built-in light source and specimen holder for measuring the colors of material specimens by transmitted or reflected light, or both. The reliability and convenience of the current-measuring device and the geometric conditions of irradiation of the specimen and collection of the transmitted or reflected flux for measurement are important properties of a photoelectric colorimeter, determining its usefulness for specific applications, but the most important feature of the instrument is the design of the filters, determining how well the instrument readings correlate with visual judgments carried out under the light source of interest, usually daylight. If S be the spectral sensitivity of the photocell and T_r, T_g, and T_b the spectral transmittances of the filters combined with it so that $S_r = ST_r$, $S_g = ST_g$, and $S_b = ST_b$, if H_A be the spectral irradiance of the light source (usually an incandescent lamp) used in the colorimeter, and if H_C be the spectral irradiance of the light source of interest (usually CIE source C, representative of average daylight), the conditions to be satisfied throughout the whole spectrum (Hunter, 1942; Gibson, 1936) are an extension of (4.11),

$$\begin{aligned} H_A T_r S &= H_C(k_{11}\bar{x} + k_{12}\bar{y} + k_{13}\bar{z}) \\ H_A T_g S &= H_C(k_{21}\bar{x} + k_{22}\bar{y} + k_{23}\bar{z}) \\ H_A T_b S &= H_C(k_{31}\bar{x} + k_{32}\bar{y} + k_{33}\bar{z}) \end{aligned} \tag{4.13}$$

which reverts to (4.11) if the source of interest is built directly into the colorimeter, that is, if $H_A = H_C$. If the instrument is to read tristimulus values X, Y, Z of the unknown specimen directly ($I_r = X$, $I_g = Y$, $I_b = Z$), then k_{11}, k_{22}, k_{33} must differ from zero and the other constants (off the diagonal of the matrix) must equal zero. Because of the difficulty of constructing a permanent filter of spectral transmittance $T_r = k_{11}H_C\bar{x}/H_AS$, which must have a major long-wave maximum and a minor short-wave maximum, to fit $\bar{x}$ (see Table 4.2 and Fig. 4.1a), a common choice of constants is $k_{11} = 1.18$, $k_{12} = 0.00$, $k_{13} = -0.18$, which gives a curve of transmittance against wavelength that can be approximately duplicated by available glass filters having a unimodal curve of spectral transmittance. Alternatively the choice of $k_{11} = 1$, $k_{12} = k_{13} = 0$ may be made and the resulting bimodal curve fitted by two separate filter-photocell combinations whose currents are added electrically (Barnes, 1939; Hunter, 1948) to evaluate I_r or by two separate filters each covering a portion of the sensitive surface of the same photocell (Nimeroff and Wilson, 1954).

The relation between the readings I_r, I_g, I_b of the photoelectric colorimeter whose filters satisfy (4.13) and the tristimulus values X, Y, Z of an opaque specimen of spectral reflectance R, irradiated by flux of spectral distribution H_C, may be derived from an extension of (4.12),

$$\begin{aligned} I_r &= k_r \int_0^\infty H_A R T_r S \, d\lambda \\ I_g &= k_g \int_0^\infty H_A R T_g S \, d\lambda \\ I_b &= k_b \int_0^\infty H_A R T_b S \, d\lambda \end{aligned} \tag{4.14}$$

From the definition of tristimulus values in (4.2) and (4.6) ($X = \int_0^\infty H_C R \bar{x} \, d\lambda$, $Y = \int_0^\infty H_C R \bar{y} \, d\lambda$, $Z = \int_0^\infty H_C R \bar{z} \, d\lambda$, where $H_C R = N$), and from the expressions for T_r, T_g, T_b taken from (4.13), (4.14) may be written

$$\begin{aligned} I_r &= k_r(k_{11}X + k_{12}Y + k_{13}Z) \\ I_g &= k_g(k_{21}X + k_{22}Y + k_{23}Z) \\ I_b &= k_b(k_{31}X + k_{32}Y + k_{33}Z) \end{aligned} \tag{4.15}$$

To calibrate such a colorimeter requires a single reflectance standard of known tristimulus values for the source of spectral irradiance, H_C, such as magnesium oxide (ASTM Method D 986-50). The triad of current outputs for this standard is measured, and the scale constants k_r, k_g, k_b evaluated from (4.15).

Comparison of (4.15) and (4.7) shows immediately that the instrument readings I_r, I_g, I_b of a photoelectric colorimeter satisfying (4.13) perfectly are themselves tristimulus values in accord with the CIE standard observer expressed relative to primaries defined by the constants k_{ij} of (4.13) chosen for the filter design. These readings may be converted to tristimulus values X, Y, Z in the standard coordinate system by solving explicitly for X, Y, Z, as in (4.6) and (4.7).

Photoelectric tristimulus colorimeters finding considerable application in American industry are the Lumetron transmissometer and photoelectric reflection meter (Photovolt Corporation), the Hunter multipurpose reflectometer and color-difference meter (Hunter 1940, 1942, 1948), the PPG-IDL Color Eye (Bentley, 1951), and the Colormaster (Glasser and Troy, 1952).

7. Colorimetry by Difference

Although it is possible in theory to design a set of permanent filters satisfying (4.13) exactly by making each filter consist of a dispersing device (combination of prism or grating with lenses and slits) with a template of the desired shape placed in the plane of the spectrum, so formed, of the source to be measured (Winch and Machin, 1940), the photoelectric tristimulus colorimeters presently used by American industry satisfy (4.13) only approximately. The filters consist of one or more layers of glass through which the radiant energy to be evaluated passes in succession, the glass for the layers and their thicknesses being chosen to satisfy (4.13) to whatever degree of approximation is believed to be sufficient for the intended purposes. Such a photoelectric colorimeter has the properties of an anomalous trichromat (see Sec. 2), not necessarily of either the protanomalous or the deuteranomalous type. Two patches of spectral radiance, N_1 and N_2, which are a color match for the CIE standard observer because they satisfy (4.2), will not necessarily be found to be equivalent by the colorimeter. If these patches satisfy (4.2) in spite of large differences in some part of the spectrum between N_1 and N_2 (strongly metameric match), the colorimeter may find [in accordance with (4.14)] that the two patches yield widely different readings; if they satisfy (4.2) with only slight spectral differences (slightly metameric match), the colorimeter will, in general, yield only slightly different readings; and, of course, if they satisfy (4.2) by virtue of the fact that $N_1 = N_2$ throughout the whole spectrum (nonmetameric match), the colorimeter will necessarily yield identical readings for each, regardless of how badly the filters fail to satisfy (4.13). The usefulness of photoelectric colorimeters with simple glass filters arises from the fact that the great majority of color-measurement problems in industry involve comparisons of spectrally similar specimens (evaluation of nonmetameric or slightly metameric differences). Examples of problems involving spectrally similar specimens are (*a*) checking the correctness of a colorant formulation made of the same colorants as the standard, (*b*) measurement of the fading of a specimen exposed to various deteriorative agents (weathering, laundering, sunlight, corrosive fumes), and (*c*) measuring the correct end point of a commercial operation by color (cooking, titration).

To obtain the tristimulus values of such specimens it is necessary to determine the scale constants k_r, k_g, k_b from (4.15) by means of a color standard that is spectrally similar to the specimen itself. There must be available, therefore, a color standard for each group of spectrally similar specimens of interest. The method is called *colorimetry by difference*.

For calibrating photoelectric tristimulus colorimeters the National Bureau of Standards has issued two sets of standards having spectral reflectances of general interest, but this is a small part of the total interest. In many cases a special standard of permanent material having the desired spectral characteristics must be procured and calibrated spectrophotometrically. If the design of the filters satisfies (4.13) to a close approximation, only a limited number of special standards will be required, but if the design is poor, so many special standards would be required that evaluation of the colors in the fundamental CIE terms becomes economically unfeasible.

Bibliography

Abney, W. de W.: On the Change of Hue of Spectrum Colours by Dilution with White Light, *Proc. Roy. Soc. (London)*, ser. A, **87** (1912).

———: "Researches in Colour Vision and the Trichromatic Theory," p. 255, Longmans, London, New York, 1913.

Adams, E. Q.: A Comparison of the Fechner and Munsell Scales of Luminous Sensation Value, *J. Opt. Soc. Am.* and *Rev. Sci. Instr.*, **6**: 932 (1922).

———: A Theory of Color Vision, *Psychol. Rev.*, **30**: 56 (1923).

———: X-Z Planes in the 1931 I.C.I. System of Colorimetry, *J. Opt. Soc. Am.*, **32**: 168 (1942).

American Standard Method for Determination of Color Specifications, Z58.7.2–1951, American Standards Association, New York.

Barnes, B. T.: A Four-filter Photoelectric Colorimeter, *J. Opt. Soc. Am.*, **29**: 448 (1939).

Bentley, G. P.: Industrial Tristimulus Color Matcher, *Electronics*, **24**: 102 (1951).

Bezold, W. v.: Ueber das Gesetz der Farbenmischung und die physiologischen Grundfarben, *Pogg. Ann.*, **150**: 221 (1873).

Breckenridge, F. C., and W. R. Schaub: Rectangular Uniform-chromaticity-scale Coordinates, *J. Opt. Soc. Am.*, **29**: 370 (1939).

Brown, P. K., and G. Wald: *Science*, **144**: 45 (1964).

Brücke, E.: Ueber einige Empfindungen im Gebiete der Sehnerven, *Sitzber. Akad. Wiss. Wien*, ser. III, **77**: 39 (1879).

Commission Internationale de l'Éclairage: *Proc. 8th Session*, pp. 19–29, Cambridge, England, September, 1931.

———: *Proc. 12th Session*, vol. I, pt. 4, p. 11, Central Bureau CIE, New York, 1951.

———: *Proc. 14th Session*, vol. A, p. 34, Brussels, Belgium, June, 1959.

———: *Proc. 15th Session*, vol. A, p. 113, Vienna, Austria, June, 1963.

Fick, A.: Die Lehre von der Lichtempfindung, *Hermann's Handb. Physiol.* (Leipzig), **3**: 139 (1879).

Gibson, K. S.: Photoelectric Photometers and Colorimeters, *Instruments*, **9**: 309, 335 (1936).

Glasser, L. G., and D. J. Troy: A New High Sensitivity Differential Colorimeter, *J. Opt. Soc. Am.*, **42**: 652 (1952).

Glenn, J. J., and J. T. Killian: Trichromatic Analysis of the Munsell Book of Color, *J. Opt. Soc. Am.*, **30**: 609 (1940).

Granit, R.: "Sensory Mechanisms of the Retina," Oxford University Press, Fair Lawn, N.J., 1947.

Granville, W. C., and E. Jacobson: Colorimetric Specification of the Color Harmony Manual from Spectrophotometric Measurements, *J. Opt. Soc. Am.*, **34**: 382 (1944).

———, D. Nickerson, and C. E. Foss: Trichromatic Specifications for Intermediate and Special Colors of the Munsell System, *J. Opt. Soc. Am.*, **33**: 376 (1943).

Hardy, A. C.: "Handbook of Colorimetry," Technology Press, Cambridge, Mass., 1936.

Helson, H., D. B. Judd, and M. H. Warren: Object-color Changes from Daylight to Incandescent Filament Illumination, *Illum. Eng.*, **47**: 221 (1952).

Hunter, R. S.: Photoelectric Tristimulus Colorimetry with Three Filters, *Natl. Bur. Standards Circ.* C429, July 30, 1942.

———: A Multipurpose Photoelectric Reflectometer, *J. Research Natl. Bur. Standards*, **25**: 581 (RP1345) (1940); also *J. Opt. Soc. Am.*, **30**: 536 (1940).

———: Photoelectric Color-difference Meter, *J. Opt. Soc. Am.*, **38**: 661 (1948); Accuracy, Precision, and Stability of a New Photoelectric Color-difference Meter, *J. Opt. Soc. Am.*, **38**: 1094 (1948).

Hurvich, L. M., and D. Jameson: *Psychol. Rev.*, **64**: 384 (1957).

Jaensch, E. R.: Ueber den Farbencontrast und die sog. Berücksichtigung der farbigen Beleuchtung, *Z. Sinnesphysiol.*, **52**: 165 (1921).

Judd, D. B.: A Maxwell Triangle Yielding Uniform Chromaticity Scales, *J. Research Natl. Bur. Standards*, **14**: 41 (1935); also *J. Opt. Soc. Am.*, **25**: 24 (1935).

———: Standard Response Functions for Protanopic and Deuteranopic Vision, *J. Research Natl. Bur. Standards*, **33**: 407 (1944); also *J. Opt. Soc. Am.*, **35**: 199 (1945).

———: Response Functions for Types of Vision According to the Müller Theory, *J. Research Natl. Bur. Standards*, **42**: 1 (1949).

———: Colorimetry, *Natl. Bur. Standards Circ.* 478, Mar. 1, 1950a.

———: Vision: Color, "Medical Physics," vol. II, p. 1152, Year Book Publishers, Chicago, 1950b.

———: Measurement and Specification of Color, "Mellon's Analytical Absorption Spectroscopy," Wiley, New York, 1950.

———: Basic Correlates of the Visual Stimulus, "Stevens' Handbook of Experimental Psychology," pp. 830–836, Wiley, New York, 1951.

———: "Color in Business, Science, and Industry," Wiley, New York, 1952.

———: *J. Opt. Soc. Am.*, **50**: 254 (1960).

———, G. J. Chamberlin, and G. W. Haupt: *J. Research Natl. Bur. Standards*, **66C**: 121 (1962).

——— and G. Wyszecki: "Color in Business, Science, and Industry," 2d ed., Wiley, New York, 1963.

Kelly, K. L., K. S. Gibson, and D. Nickerson: Tristimulus Specification of the Munsell Book of Color from Spectrophotometric Measurements, *J. Research Natl. Bur. Standards*, **31**: 55 (1943); also *J. Opt. Soc. Am.*, **33**: 355 (1943).

König, A., and C. Dieterici: Die Grundempfindungen in normalen und anomalen Farbensystemen und ihre Intensitätsverteilung im Spectrum, *Z. Psychol.*, **4**: 241 (1892).

———: Ueber "Blaublindheit," *Sitzber. Akad. Wiss.* (Berlin), July 8, 1897, 718.

———: "Gesammelte Abhandlungen," pp. 214, 396, Barth, Leipzig, 1903.

Ladd-Franklin, C.: Eine neue Theorie der Lichtempfindungen, *Z. Psychol. Physiol. Sinnesorg.*, **4**: 211 (1892).

———: "Colour and Colour Theories," Harcourt, Brace, New York, 1929.

Land, E. H.: *Proc. Natl. Acad. Sci.* (*U.S.*), **45**: 115, 636 (1959).

———: *Am. Scientist*, **52**: 247 (1964).

Ludvigh, E., and E. F. McCarthy: Absorption of Visible Light by the Refractive Media of the Human Eye, *Arch. Ophthalmol.* (*Chicago*), **20**: 37 (1938).

MacAdam, D. L.: Projective Transformations of I.C.I. Color Specifications, *J. Opt. Soc. Am.*, **27**: 294 (1937).

Mellon, M. G.: "Analytical Absorption Spectroscopy," p. 117, Wiley, New York, 1950.

Müller, G. E.: Ueber die Farbenempfindungen, *Z. Psychol. Ergänzungsb.*, **17**: 18 (1930).

Newhall, S. M., D. Nickerson, and D. B. Judd: Final Report of the O.S.A. Subcommittee on the Spacing of the Munsell Colors, *J. Opt. Soc. Am.*, **33**: 385 (1943).

Nickerson, D.: Tables for Use in Computing Small Color Differences, *Am. Dyestuff Reptr.*, **39**: 541 (Aug. 21, 1950).

Nimeroff, I., and S. W. Wilson: A Colorimeter for Pyrotechnic Smokes, *J. Research Natl. Bur. Standards*, **52**: 195 (1954); RP2488.

Nuberg, N. D., and E. N. Yustova: Lenin Optical Institute, **24**: 33 (Moscow, 1955).

OSA Committee on Colorimetry: Quantitative Data and Methods for Colorimetry, *J. Opt. Soc. Am.*, **34**: 647 (1944).

———: "The Science of Color," Crowell, New York, 1953.

Pitt, F. H. G.: Some Aspects of Anomalous Vision, *Doc. Ophthalmol.*, **3**: 307 (1949).

Saunderson, J. A., and B. I. Milner: Modified Chromatic Value Color Space, *J. Opt. Soc. Am.*, **36**: 36 (1946).

Schofield, R. K.: The Lovibond Tintometer Adapted by Means of the Rothamsted Device to Measure Colours on the CIE System, *J. Sci. Instr.*, **16**: 74 (1939).

Schrödinger, E.: Ueber das Verhältnis der Vierfarben zur Dreifarbentheorie, *Sitzber. Akad. Wiss.* (*Wien*), ser. IIa, **134**: 471 (1925).

Smith, T., and J. Guild: The C.I.E. Colorimetric Standards and Their Use, *Trans. Opt. Soc.* (*London*), **33**: 73 (1931–1932).

Stiles, W. S.: *Proc. Natl. Acad. Sci.* (*U.S.*), **45**: 100 (1959).

Talbot, S. A.: Recent Concepts of Retinal Color Mechanism, *J. Opt. Soc. Am.*, **41**: 895, 918 (1951).

Vries, H. De: The Fundamental Response Curves of Normal and Abnormal Dichromatic and Trichromatic Eyes, *Physics*, **14**: 367 (1948).

Wald, G.: On Rhodopsin in Solution, *J. Gen. Physiol.*, **21**: 795 (1937–1938).

———: Human Vision and the Spectrum, *Science*, **101**: 653 (June 29, 1945).

———: *Science*, **145**: 1007 (1964).

Walls, G. L.: *Psychol. Bull.*, **57**: 1 (1960).

Willmer, E. N.: *Doc. Ophthalmologica*, **3**: 194 ('S,Gravenhage, W. Junk, 1949).

Winch, G. T., and C. F. Machin: The Physical Realization of the CIE Average Eye, *Trans. I.E.S.* (*London*), **5**: 93 (1940).

Wright, W. D.: A Re-determination of the Trichromatic Coefficients of the Spectral Colors, *Trans. Opt. Soc.* (*London*), **30**: 141 (1928–1929).

———: The Measurement and Analysis of Colour Adaptation Phenomena, *Proc. Roy. Soc.* (*London*), ser. B, **115**: 49 (1934).

Wyszecki, G.: Invariance of Insidedness in Projective Transformations of the Maxwell Triangle, *J. Opt. Soc. Am.*, **44**: 524 (1954).

———: *J. Opt. Soc. Am.*, **53**: 1318 (1963).

Chapter 5

Diffraction and Interference

By J. G. HIRSCHBERG, University of Miami, J. E. MACK, University of Wisconsin,
and F. L. ROESLER, University of Wisconsin

1. Geometrical Optics as an Approximation

In geometrical optics (Part 6, Chap. 2) light is considered as traveling in rays, which in a homogeneous medium are rectilinear and which, upon striking the boundary between two such media, are reflected back into the original medium or refracted into another according to certain simple rules.

Optical phenomena may be treated quantitatively as wave phenomena satisfying the equations of Maxwell [Part 6, Eq. (1.7)]. The rules of geometrical optics are valid only in the approximation to which path differences of the order of a wavelength can be neglected. The wavelength of visible light occupies a range in the neighborhood of 5×10^{-7} m.

The absolute refractive index n of a substance satisfies [Part 6, Eq. (1.30)]

$$n = \frac{c}{v} \tag{5.1}$$

where c (2.997925×10^8 m/sec) is the velocity of light in vacuum and v is the phase velocity of the light wave in the substance.

Fermat's and Huygens' Principles. The optical path $\int_A^B n\, ds$ between two points A and B, where ds is the element of length along the path, is from (5.1) just c times the time required for the light to travel from A to B. In the sense of the calculus of variations one may imagine, running alongside and infinitesimally separated from the actual path, a continuum of neighboring paths between A and B. *Fermat's principle*, as somewhat extended and expressed in modern form, states that any path from A to B whose length is stationary with respect to a variation of the path, and only such a path, is an actual path from A to B. Usually the actual path length is a minimum, but it may be a maximum, and in the case of image formation it is invariant. *Huygens' principle* states that every point on any wavefront may be considered as a new source of waves, and all the points on any wavefront may be used alternatively with the source to predict any later wavefront. Although sufficient for predicting the position, Huygens' principle is not adequate to predict the amplitude of the later front. Kirchhoff (Sec. 2) developed it from general principles into a complete system for predicting the amplitude and phase at any point, arising from a wave segment, to a very close approximation.

The range of validity and the limitations of Fermat's and Huygens' principles may be understood from the more general treatment of optics by Fresnel and his successors. Fermat's principle, narrowly interpreted as though light were propagated in mathematical rays, yields the geometrical optical rules of rectilinear propagation, reflection, and refraction. In the actual (i.e., the wave) process, wavelets arriving by slightly different paths reinforce one another almost as much as by strictly equal paths, if the path difference is small compared with a wavelength. As the path difference increases, there is a gradual, rather than a sudden, decrease in the resultant amplitude. Thus the true picture of the passage of light from A to B is not of a ray, but rather of an unsharply bounded filament of light with an appreciable cross section. To the extent that the ray picture and the concomitant strict requirement of a stationary path in Fermat's formulation are inexact, the rules of geometrical optics are invalid.

Geometrical Optics Analogous to Classical Mechanics. There is a close analogy between the inexactness of geometrical optics and that of classical mechanics: classical mechanics is an approximation to quantum mechanics (Part 2, Chap. 6), valid only so long as Planck's constant h may be neglected; and geometrical optics, so elegantly formulated by Hamilton (Part 6, Chap. 2), is an approximation to the true, or physical, optics, valid only so long as the wavelength λ may be neglected. (Where the phase of a light wave is important, it is often convenient to express distances in terms of "lambda bar,"

$$\lambdabar \equiv \frac{\lambda}{2\pi} \tag{5.2}$$

which is the path in which the phase changes by one radian.)

2. General Aspects of Diffraction and Interference

Consider light in a homogeneous medium, emanating from a source Q of negligible extent, passing near or through a limited aperture at O_1 in an opaque diaphragm B, and consider more particularly the light that may fall upon a screen S that intercepts the beam from Q in the general direction of QO_1 extended (Fig. 5.1).

The area of S may be divided into two categories (omitting certain physically irrelevant mathematical refinements): those area elements such as dS_1 in the neighborhood of P_1, for which the light QP_1 intersects the diaphragm, and which are therefore said to be in the *geometric shadow* of the diaphragm, and those such as dS_2 in the neighborhood of P_2, for which QP_2 may be drawn without encountering any part of the diaphragm. Geometrical optics states that all elements dS_1 remain completely without illuminance and all elements dS_2 are (if one neglects minor corrections due to inequality of distance from Q) equally illuminated; and in the case where Q is an extended uniform source, the illuminance on any element dS is proportional to the area of Q from which light can reach dS rectilinearly.

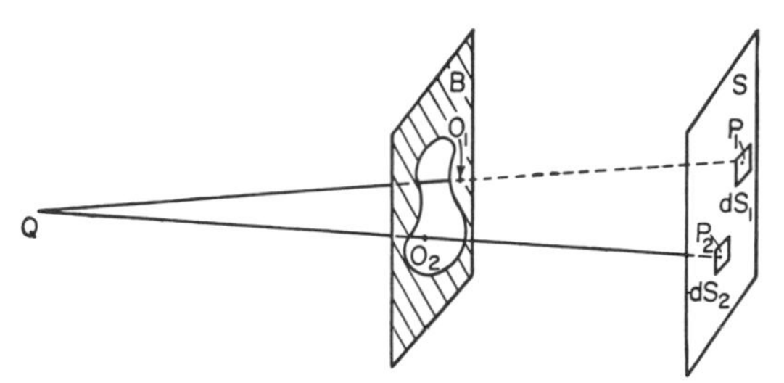

FIG. 5.1. Geometric shadow. Geometrically, the ray QP_1 is interrupted at O_1; thus, according to geometrical optics P_1 would not be illuminated, although P_2 would be.

In fine detail, however, the real situation is different in two respects. First, the light does not travel only rectilinearly. Any departure of the actual light path from that prescribed by geometrical optics is called *diffraction*. Diffraction is exhibited most obviously by the occurrence of light in the geometric shadow, as shown in some of the examples below. Second the illuminance at a point is not always the simple sum of the elements of illuminance that reach the point by different paths. This nonadditivity of illuminance (or of intensity; cf. third paragraph below) is called *interference*. There is no categorical distinction between interference and diffraction. The term interference is usually employed to refer to the vector addition of the displacements in a number of beams, and diffraction to the addition of the infinitesimal elements of a wavefront in a single beam.

Coherence. Two or more wave sources are said to be *coherent sources* if the phase difference between a pair of points, one in each source, remains constant. Sources that are not coherent are called *incoherent sources*. Waves from coherent sources are called *coherent waves,* and waves from incoherent sources, *incoherent waves*. The phase of a light wave [Part 6, Eq. (1.1)] depends upon the emitter. In a real physical light source there are many emitters, all sending out finite wave trains enduring not longer than about 10^{-8} sec. Since observations usually require time intervals long compared with the duration of a wave train, usually separate light sources, or different parts of the same source, are for practical purposes, incoherent, and only waves derived from a common luminous origin are coherent; a unique exception occurs in the case of a *laser,* in which the wave train passing repeatedly through the active medium between partially transmitting mirrors is built up by the process of stimulated emission from a highly populated metastable state. This process adds to the original wave train in exactly the same phase and gives rise to coherence over a region of space as large as the laser surface. Also, two separate laser sources can be coherent.

Following rather widespread practice, in this chapter the term "displacement" or "amplitude" of an electromagnetic wave, unless otherwise specified, means the electric displacement or amplitude; and "intensity," defined as scalar amplitude squared and differing from illuminance by a constant coefficient, is sometimes used in place of illuminance where the coefficient is not important. Thus, if the vector amplitude is $\mathbf{u}$, the intensity I may be written $\mathbf{uu}^*$ or $|\mathbf{u}|^2$.

Superposition of Waves; the Interference Term in the Intensity. Consider two coherent

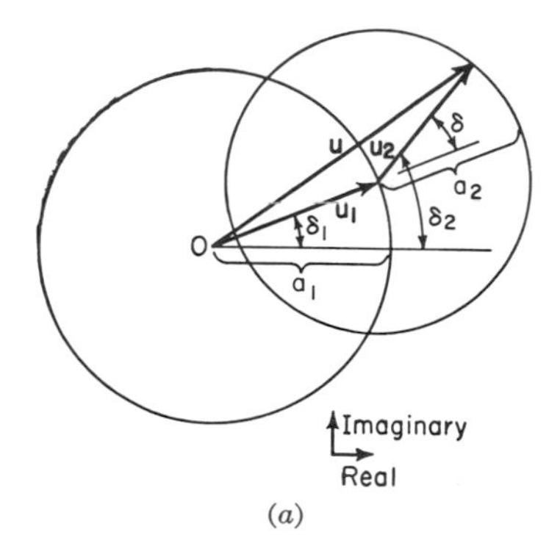

(a)

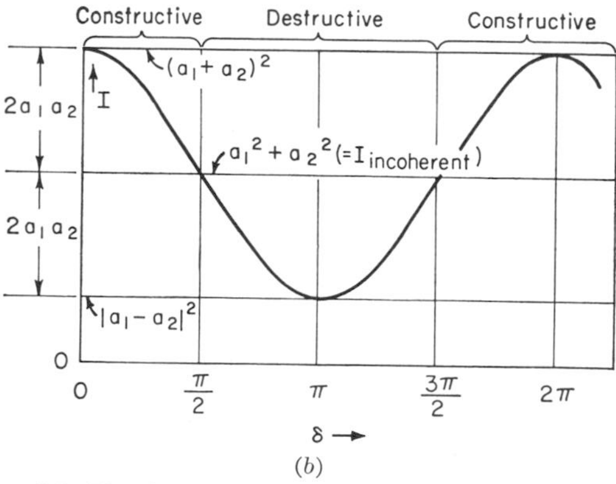

(b)

FIG. 5.2. Two-beam interference. (a) Vector amplitude diagram. (b) Dependence of intensity upon the phase difference, δ. The ordinate value $a_1^2 + a_2^2$, labeled $I_{\text{incoherent}}$, would have been the intensity if the beams had been incoherent.

waves or beams of the same frequency traveling in the z direction. If the components of the displacement in a certain transverse direction, say, the x direction (Fig. 5.2a), are the real parts, respectively, of

$$\mathbf{u}_1 = a_1 \exp [i(-z/\lambda\!\!\!^{-} + \delta_1)]$$

and

$$\mathbf{u}_2 = a_2 \exp [i(-z/\lambda + \delta_2)]$$

where the complex quantities $\mathbf{u}_1$ and $\mathbf{u}_2$ are called the *vector amplitudes* and a_1 and a_2 are real coefficients called the *scalar amplitudes*, then the intensities of the beams considered separately would be, respectively, $I_1 = \mathbf{u}_1\mathbf{u}^*_1 = a_1^2$ and $I_2 = \mathbf{u}_2\mathbf{u}^*_2 = a_2^2$, and the intensity I of the resultant is

$$\begin{aligned} I_{\text{two coherent waves}} &= (\mathbf{u}_1 + \mathbf{u}_2)(\mathbf{u}^*_1 + \mathbf{u}^*_2) \\ &= I_1 + I_2 + 2I_1^{1/2}I_2^{1/2} \cos \delta \end{aligned} \quad (5.3)$$

where δ stands for the *phase difference* $\delta_2 - \delta_1$. The quantity $2I_1^{1/2}I_2^{1/2} \cos \delta$ may be called the *interference term* in the intensity. When $\cos \delta > 0$, the interference is called *constructive*, and when $\cos \delta < 0$, *destructive*. Figure 5.2*b* shows I as a function of δ.

A special case of (5.3), two-wave interference, occurs when $a_1 = a_2$. Then, and only then, the amplitude of the interference term (not the wave amplitude) is as large as the sum $I_1 + I_2$:

$$I_{\text{two equal coherent waves}} = 2I_1(1 + \cos \delta) \quad (5.4)$$

and I ranges from $4I_1$ down to 0, depending upon δ.

It will be noticed from (5.3) that the average value of I over all phases is $I_1 + I_2$; in incoherent sources the interference term averages out, and there is no interference.

$$I_{\text{two incoherent waves}} = I_1 + I_2 \quad (5.5)$$

More generally, for any number of coherent waves,

$$I_{\text{coherent}} = \left(\sum_k \mathbf{u}_k\right)\left(\sum_k \mathbf{u}^*_k\right) = \left|\sum_k \mathbf{u}_k\right|^2 \quad (5.6)$$

and, for any number of incoherent waves,

$$I_{\text{incoherent}} = \sum_k (\mathbf{u}_k\mathbf{u}^*_k) = \sum_k |\mathbf{u}_k|^2 \quad (5.7)$$

Equations (5.6) and (5.7) state that interference occurs for coherent, but not for incoherent, waves.

The Huygens-Fresnel-Kirchhoff Treatment. Kirchhoff developed Huygens' principle into a tool for the quantitative determination of the amplitude at a point, using the wave equation and Green's theorem, applied to a region of space including the illuminated point P but no luminous source, and having the diffracting aperture as a part of its boundary. The displacement of a spherical wave at a distance r_0 from the source is proportional to $r_0^{-1} \exp (ir_0/\lambda)$. The displacement $\mathbf{u}_P$ transmitted from a point source of light at Q to P through an opening in an opaque barrier is found to be, to a close approximation, proportional to

$$u_P = -\frac{i}{2\lambda} \iint \frac{\exp [i(r + r_0)/\lambda]}{|r||r_0|} [\cos (\boldsymbol{\nu},\mathbf{r}) - \cos (\boldsymbol{\nu},\mathbf{r}_0)]\, dB \quad (5.8)$$

where $\mathbf{r}_0$, $\mathbf{r}$ are radius vectors drawn to the surface element of the opening, dB, from Q and P, respectively, and $\boldsymbol{\nu}$ is a vector normal to dB drawn away from P. The most serious geometric limitation here is the requirement $|\mathbf{r}|, |\mathbf{r}_0| \gg \lambda$; although this requirement is the same as that of geometrical optics, it does not offer such a severe limitation because the exponential factor determines the phase. It is convenient, in most numerical calculations, to limit the opening to plane surfaces.

In the examples below consider a rectangular cartesian or a cylindrical coordinate system (Fig. 5.3), with Q on or near the $-z$ axis, P on or near the $+z$ axis, and the diffracting opening near the origin in the plane $z = 0$. The x, y, z coordinates of Q, P, and a point in the opening may be designated by (x_0,y_0,z_0), (x,y,z), and $(\xi,\eta,0)$, respectively.

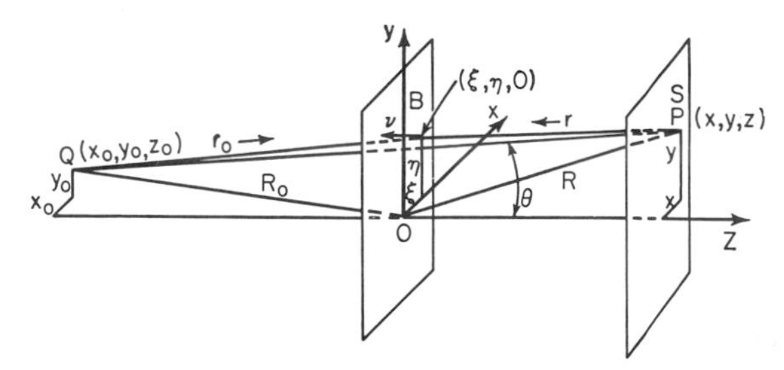

FIG. 5.3. Symbols used in Kirchhoff formulation.

The phase factor in (5.8) varies so much faster than the rest of the integrand that for small openings the rest may be removed from under the integration. In the physically interesting case where P is near the edge of the geometric shadow, the factor $\cos (\boldsymbol{\nu},\mathbf{r}) - \cos (\boldsymbol{\nu},\mathbf{r}_0)$ may be replaced by $2 \cos \theta$, where θ is the angle between QP and the z axis. Thus, with $R = (x^2 + y^2 + z^2)^{1/2}$ and $R_0 = (x_0^2 + y_0^2 + z_0^2)^{1/2}$, and the direction cosines $-x_0/R_0$, x/R, $-y_0/R_0$, y/R, $-z_0/R_0$, and z/R designated by α_0, α, β_0, β, γ_0, and γ, respectively, (5.8) may be rewritten,

$$\mathbf{u}_P = -i \cos \theta \exp \left[\frac{i(R + R_0)}{\lambda}\right] \iint e^{i\delta(\xi,\eta)}(RR_0)^{-1}\, d\xi\, d\eta \quad (5.9)$$

where the phase factor δ, expanded in inverse powers of R_0 and R, satisfies the relation

$$\begin{aligned} \lambda\delta(\xi,\eta) &= (\alpha_0 - \alpha)\xi + (\beta_0 - \beta)\eta \\ &\quad + \tfrac{1}{2}[(R_0^{-1} + R^{-1})(\xi^2 + \eta^2) - R_0^{-1}(\alpha_0\xi + \beta_0\eta)^2 \\ &\quad - R^{-1}(\alpha\xi + \beta\eta)^2 + \cdots] \end{aligned} \quad (5.10)$$

where the dots stand for terms of higher power in R_0^{-1} and R^{-1}.

Babinet's Principle. In any optical system consider an area S that has no illuminance, or only a negligible amount. Now suppose that an obstacle, or stop, B_1 is placed in any cross section of the system so that S is illuminated by diffraction (not by geometrical transmission or reflection). The obstacle, or stop, B_2, *complementary* to B_1, is defined as the body which allows the passage of that light, and only that light, which B_1 intercepts. Thus B_1 and B_2 together would intercept all the light passing through the system. According to *Babinet's principle* the diffraction pattern on S arising from the presence of B_1 (without B_2) in the system is identical with that arising from the presence of B_2 (without B_1). The principle follows from the necessity that, since

$\mathbf{u}_1 + \mathbf{u}_2 = 0$ for each point of S, $\mathbf{u}_1 = -\mathbf{u}_2$ and $|\mathbf{u}_1|^2 = |\mathbf{u}_2|^2$, where $\mathbf{u}_k$ is the amplitude arising from the presence of B_k.

Reciprocity Law for Diffraction. Equation (5.8) remains unchanged if r_0 and r are interchanged and at the same time the direction of the normal, $\mathbf{v}$, is reversed, or, in other words, if the source and the illuminated point are interchanged. This is a general rule, known as the *reciprocity law for diffraction:* If any point P is illuminated at a certain amplitude from a source point Q, then Q would be illuminated with exactly the same amplitude from the same source at P, the geometry of the system being otherwise unchanged.

Obliquity Factor. The factor $[\cos(\mathbf{v},\mathbf{r}) - \cos(\mathbf{v},\mathbf{r}_0)]$ in (5.8) takes an especially simple form in the important case where the light is coming from a single spherical or point source and the surface element coincides with a wavefront. Then $(\mathbf{v},\mathbf{r})$ equals the angle φ between the direction of propagation and the direction from dB to P, and $\cos(\mathbf{v},\mathbf{r}_0) = -1$. Thus the factor becomes the obliquity factor $(1 + \cos\varphi)$, which ranges from 2 to 0, and eliminates the backward wavelets, the prediction of which, contrary to experience, was a defect in the original form of Huygens' principle.

Post-Kirchhoff Developments in Diffraction-Interference Theory. Progress has been made beyond Kirchhoff's formulation in only a few special cases. Sommerfeld developed an exact treatment from Maxwell's equations, applicable to the diffraction from a sharp, straight edge of a perfect reflector. The recent burgeoning importance of microwaves has redirected attention to some unsolved problems in the field. Bouwkamp has found an exact solution for a plane wave incident normally on a circular aperture. Levine and Schwinger have developed variational methods for a plane wave incident in any direction upon an aperture of arbitrary shape in a plane reflecting screen. For normal incidence on a circular aperture their approximation converges rapidly to the exact solution.

Classifications of Fringes. Interference and diffraction phenomena generally are characterized by patterns of maximum and minimum intensity, or *fringe systems,* as exemplified in (5.3) for two beams with varying δ. A region of maximum or minimum intensity in such a system is called a bright or dark *interference* (or *diffraction*) *fringe.* It is convenient, for different purposes, to classify fringes in several ways according to different criteria. The list given here is not exhaustive.

According to the number of separate beams involved, they may be classified as single-beam (diffraction), two-beam interference, or more-than-two-beam (i.e., multibeam) interference fringes. Multibeam fringes and the interference phenomena giving rise to them may be classified according as the amplitudes of the several beams are equal in magnitude (*equal-amplitude case*) or not. Among unequal-amplitude multibeam interference fringes the most important case is that in which the scalar amplitude for each succeeding beam is less than that for its predecessor in a constant ratio; this case may be called the *geometrically degraded amplitude case.*

According to the means of separation of the original beam into interfering beams, interference fringes may be classified as *fringes from division of wavefront,* e.g., in a diffraction grating, or *fringes from division of amplitude,* e.g., in a semireflecting mirror.

Fringes from interference in plane-parallel plates, as in a well-adjusted Fabry-Perot system, an example of which is described below (Sec. 7), are called *fringes of equal inclination* (*Haidinger fringes*), while monochromatic-light fringes from interference in other geometrical situations are called *fringes of equal thickness* (*Fizeau fringes*). Haidinger fringes from plane surfaces are always infinitely distant (parallel-light) fringes, while all other fringes are localized in finite space.

Fringes arising from path differences of the order of one wavelength can be seen even in white light, owing to the circumstance that the visible spectrum covers only about one octave; they are called *white-light fringes* or, in the case of long, separate paths, *Brewster fringes.* When the path differences are great, fringes are ordinarily detected only for nearly monochromatic light; they are loosely called *monochromatic-light fringes.* The greater the path difference between extremes of the beam in the case of diffraction, or among the several beams in the case of interference, the more stringent the requirement that the range of wavelength be small.

In the case of multibeam fringes with divided and geometrically degraded amplitude, which follow the law given below in (5.26*b*), Tolansky distinguishes between two classes among the fringes in which the wavelength varies: constant angle θ_n, and variable thickness t, called *fringes of equal chromatic order,* and constant t and variable θ_n, which may be called *generalized Edser-Butler fringes,* or *white-light Fabry-Perot fringes.*

3. Diffraction

When light approaches and leaves an object in plane waves, the resulting diffraction falls into an especially simple class called *Fraunhofer diffraction.* Diffraction that does not satisfy the conditions for Fraunhofer diffraction is called *Fresnel diffraction.* The simplicity of Fraunhofer diffraction lies in the circumstance that the pattern is a pattern of only the direction, not the spatial distribution, of the emergent light. The screen, like the source, may be effectively at an infinite distance or may be in the principal focal plane of a lens. In practice the source is usually a specially designed arc lamp, in which the actual source is less than a millimeter in extent, or a brightly illuminated small aperture such as a pinhole, which may be at the principal focus of a collimating lens. An ideal laser beam is tantamount to a beam collimated from a point source; in practice a laser may emit in several *modes* (not further discussed here) departing somewhat from the ideal. (For diffraction by a slit or a straight edge it can be shown that the arc or pinhole may be replaced by a parallel incandescent wire or a narrow slit.) The essential step in the calculation of the intensity pattern in Fraunhofer diffraction consists in the determination of the resultant amplitude of the diffracted wave as a function of the angle of diffraction. In the calculation of Fresnel diffraction, which is always more complicated, the distance of the source or the illuminated region or both must be considered. The resultant amplitude in either case is the vector resultant amplitude of the wavelets, in which the phase retardation is the pertinent path difference divided by $\lambda\!\!\!^{-}$.

Fraunhofer diffraction problems may be treated by solving (5.9), taking into account only that part of (5.10) which is independent of R_0 and R, that is, which is linear in ξ and η. Fresnel diffraction requires taking into account the dependence of the phase upon R and R_0, at least as far as the terms explicitly set down in (5.10), i.e., at least as far as the terms in R_0^{-1} and R^{-1} (except that, if $R_0 \gg R$ or $R \gg R_0$, the reciprocal of the larger may sometimes be dropped).

Fraunhofer Diffraction at a Rectangular Aperture. If parallel light of wavelength λ, propagated in the $+z$ direction through a rectangular aperture in the $z = 0$ plane extending from $-A/2$ to $+A/2$ in the x direction and from $-B/2$ to $+B/2$ in the y direction, falls upon a screen at z = const, the intensity in the direction defined by (a,b) is

$$I(a,b)_{\text{rectangle}} = I_0 \left[\frac{\sin (aA/2\lambda\!\!\!^{-})}{aA/2\lambda\!\!\!^{-}}\right]^2 \left[\frac{\sin (bB/2\lambda\!\!\!^{-})}{bB/2\lambda\!\!\!^{-}}\right]^2 \tag{5.11}$$

where I_0 is the intensity in the $+z$ direction and a, b are, respectively, the sines of the projections on the xz and yz planes of the angle between the z direction and the direction of the light emerging from the aperture. The function $\left(\frac{\sin f}{f}\right)^2$ is shown in Fig. 5.4. It has zeros at all integral multiples of π except 0 and maxima at all values for which $\tan f = f$. The principal maximum is at $f = 0$, and the others at $|f|$ slightly less than $3\pi/2$, $5\pi/2$, $7\pi/2$, ... The pattern exhibits a property common to the diffraction patterns of all apertures: the width of any feature of the pattern, e.g., the distance $\Delta a = \lambda/A$ between

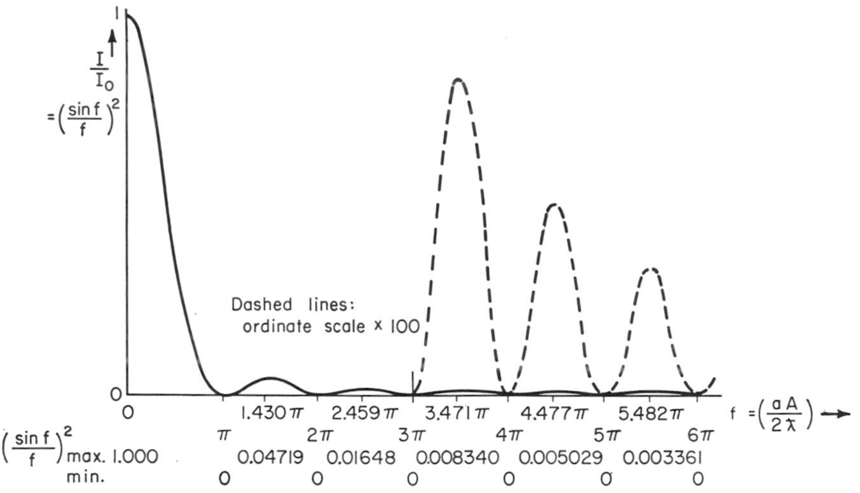

FIG. 5.4. Fraunhofer diffraction at a slit, or one of the two symmetric factors in Fraunhofer diffraction at a rectangular aperture.

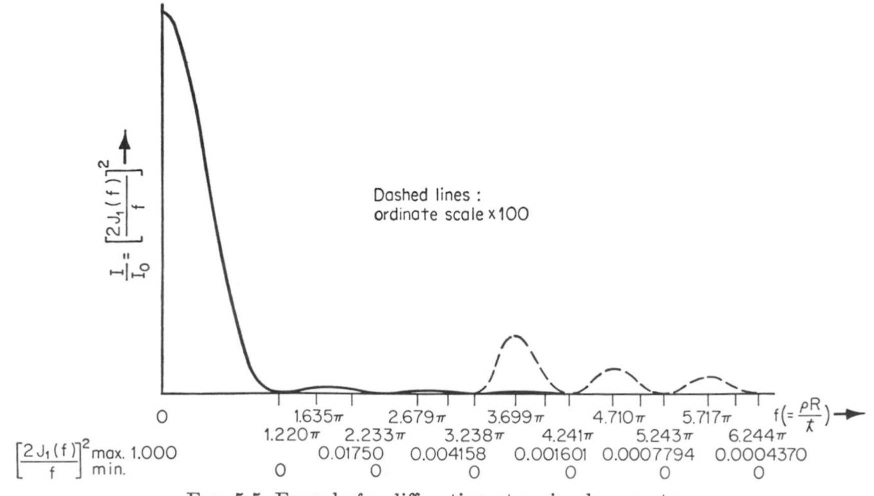

FIG. 5.5. Fraunhofer diffraction at a circular aperture.

zeros on each side of the axis, varies not directly but inversely as the width of the aperture.

Fraunhofer Diffraction at a Slit. An important case of (5.11) is $B \gg A$, with the source not a point, but a long, narrow, incoherent line extending parallel to the opening. Since $\int_{-\infty}^{+\infty} f^{-2} \sin^2 f = \pi$,

$$I(a)_{\text{slit}} = I_0 \left[\frac{\sin (aA/2\lambda)}{aA/2\lambda}\right]^2 \tag{5.12}$$

Fraunhofer Diffraction at a Circular Aperture. If the rectangular aperture of the examples above is replaced by a circular aperture of radius R, centered at the origin, in a plane normal to the z axis, the intensity is

$$I(\rho)_{\text{circle}} = I_0 \left[\frac{2J_1(f)}{f}\right]^2 \tag{5.13}$$

where $f = \rho R/\lambda$, J_1 is the first-order Bessel function, and ρ is $\sin \theta$, the sine of the angle of deviation. The function $[2J_1(f)/f]^2$ is shown in Fig. 5.5.

The rectangular and the circular patterns have in common that the intensity minima are zeros. However, while the zeros of the former make a rectangular network, in each direction equally spaced at intervals λ/A except for the middle line, where the central maximum occurs instead of a zero, the zeros of the latter are a single set of concentric circles with differences in the radial parameter, ρ, a little greater than $\lambda/2R$. The deviation angle $\theta_{\text{first min}}$ at the first zero is $0.61\lambda/R$, and the relative height of the first subordinate maximum in the circular case is 0.01750.

Fresnel Zones. Fresnel developed the qualitatively useful idea of diffraction *half-period zones*, usually called *Fresnel zones*, for determining the effect at a point P of a spherical wave of monochromatic light emitted from a point Q. In Fig. 5.6

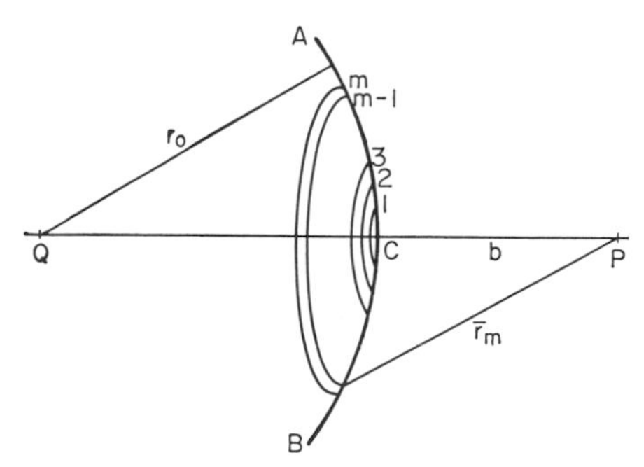

FIG. 5.6. Fresnel zones.

the arc ACB represents a section of such a wave of radius r_0, where QCP is a straight line. Representing the distance CP by b where $b \gg \lambda$), (imagine the front ACB cut by a series of spheres centered on P, with successive radii $b + m\lambda/2$ ($m = 1, 2, 3, \ldots$), and consider the contributions of the zones thus produced to the amplitude at P. The area of the mth zone, i.e., the area between the spheres of radii $b + m\lambda/2$ and $b + (m - 1)\lambda/2$, is

$$A_m = \frac{\pi r_0 b\lambda}{r_0 + b}\left[1 + \frac{(m - \frac{1}{2})\lambda}{2b}\right] \tag{5.14}$$

and the average distance of the zone from P is

$$\bar{r}_m = b + \frac{(m - \frac{1}{2})\lambda}{2} = b\left[1 + \frac{(m - \frac{1}{2})\lambda}{2b}\right] \tag{5.15}$$

Since the successive optical paths differ by $\lambda/2$, the successive phases differ by π, that is, they are directly opposed. The amplitude contributed by each zone is thus proportional to $(-1)^m(1 + \cos \varphi)A_m/\bar{r}_m$, or, because of the equality of the bracketed factors in (5.14) and (5.15), simply to

$$\mathbf{u}_{P,m} = (-1)^m(1 + \cos \varphi)\frac{\pi r_0\lambda}{r_0 + b} \tag{5.16}$$

Thus the contributions from successive zones decrease very slowly with increasing m. The sum of the contributions for any two successive zones is almost zero; where the aperture encompasses a certain integral number of zones, say, m, the amplitude is the average of that from its first and last zones,

$$\mathbf{u}_P = \frac{\mathbf{u}_{P,1} + \mathbf{u}_{P,m}}{2} \tag{5.17}$$

and in the case where there is no limited aperture the amplitude from the outermost zones becomes negligible and the total amplitude is half that of the first zone (Fig. 5.7).

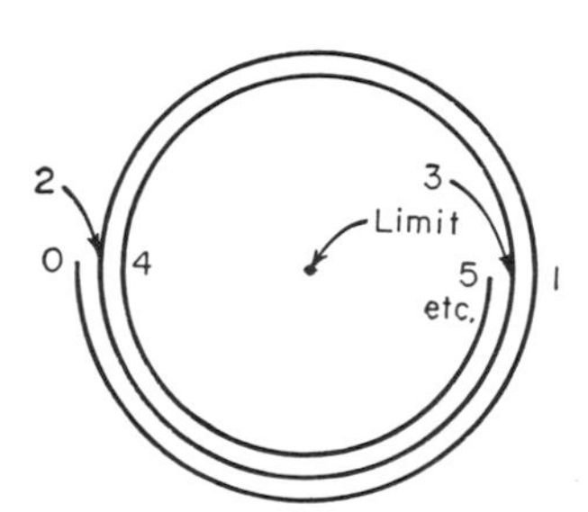

FIG. 5.7. Fresnel-zone vector amplitude diagram. Each semicircular segment between successive integers represents the integrated amplitude contributed by one zone.

Bright Center in a Circular Shadow. The shadow of an axially symmetric obstacle illuminated by a concentrated source on the axis exhibits a bright spot and a few concentric fainter rings in the center of its shadow. Since for a spherical obstacle there is no uniquely defined axis, the details of a source consisting of a small aperture or a silhouette pattern may be faithfully reproduced in the center of the shadow of a ball; the ball intercepts a few zones, but the amplitude on the axis is still comparable with that in the absence of any obstruction.

Fresnel Diffraction at a Slit and at a Straightedge; Cornu Spiral. Fresnel diffraction at an obstacle or aperture of a particular shape, such as one of those for which the Fraunhofer diffraction has been solved above, is in general more complicated than its special case, Fraunhofer diffraction, on account of the terms in R_0^{-1} and R^{-1} in (5.10), and no such simple expressions as (5.11), (5.12), (5.13) can be presented to show its general behavior. For instance, in the case of a circular aperture the center of the pattern may exhibit either a maximum or a minimum, several intensity maxima may occur within a nearly uniformly illuminated central region, and even outside that region the minima in general show nonzero intensity. Nevertheless, certain problems in Fresnel diffraction can be solved completely with some degree of generality.

Suppose that there exist a concentrated source at Q and an aperture in a plane B between Q and a surface S whose illumination is to be studied. Let P be a typical point of the surface S. A special choice of rectangular cartesian coordinate system within the previously detailed conventions (Fig. 5.3) is that (Fig. 5.8) in which the $z = 0$ plane is the plane of the

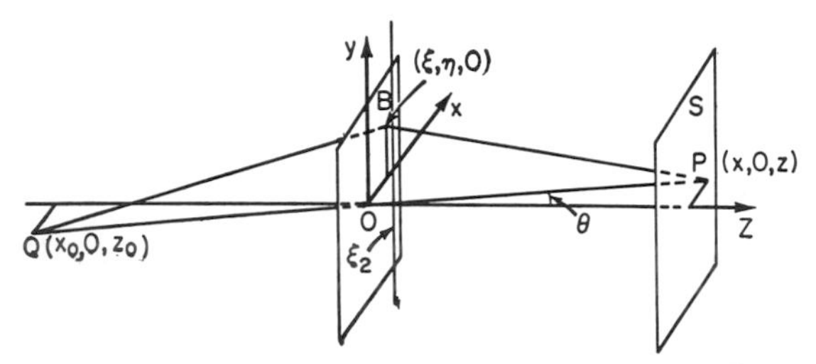

FIG. 5.8. Coordinates for Fresnel diffraction at a slit or at a straightedge: special case of fig. 5.3.

aperture, the origin is the intersection of the straight line QP with the $z = 0$ plane, and the x axis is the projection of QP on that plane. (Note that there is, in general, a different coordinate system for each P.)

Referring to the general Kirchhoff formulation, the direction cosines γ_0, γ designated just before (5.9) become $\cos\theta$, and in the approximation specifically expressed in (5.10) the phase difference between a path from Q to P through $(\xi,\eta,0)$ and the straight path QOP is

$$\delta = \tfrac{1}{2}\lambda^{-1}(R_0^{-1} + R^{-1})(\xi^2\cos^2\theta + \eta^2) \qquad (5.18)$$

With the notational simplifications

$$u^2 = 2\lambda^{-1}(R_0^{-1} + R^{-1})\xi^2\cos^2\theta \qquad (5.19a)$$

$$v^2 = 2\lambda^{-1}(R_0^{-1} + R^{-1})\eta^2 \qquad (5.19b)$$

$$A = -i\cos\theta\exp\left[\frac{i(R_0 + R)}{\lambda}\right](\lambda R_0 R)^{-1} \qquad (5.19c)$$

$$a = \lambda[2(R_0^{-1} + R^{-1})\cos\theta]^{-1} = d\xi\,d\eta(du\,dv)^{-1} \qquad (5.19d)$$

$$C = \iint \cos\delta\,d\xi\,d\eta = a\iint\cos[\tfrac{1}{2}\pi(u^2 + v^2)]\,du\,dv \qquad (5.19e)$$

$$S = \iint \sin\delta\,d\xi\,d\eta = a\iint\sin[\tfrac{1}{2}\pi(u^2 + v^2)]\,du\,dv \qquad (5.19f)$$

where the integrations in the uv plane are carried out over the region corresponding to the opening in the $\xi\eta$ plane, the amplitude (5.9) and intensity at P may be written

$$\mathbf{u}_P = A(C + iS) \qquad (5.20a)$$

$$I_P = |\mathbf{u}_P|^2 = |A|^2(C^2 + S^2) \qquad (5.20b)$$

In certain cases, especially when the opening is a rectangle with sides ξ_1, ξ_2, η_1, η_2 in the coordinate directions, $\mathbf{u}_P$ and I_P may be readily found, analytically with the aid of the Fresnel integrals or graphically with the Cornu spiral, which were devised essentially for the solution of (5.19e) and (5.19f).

The *Fresnel integrals* are defined as the real integrals

$$U(w) = \int_0^w \cos\frac{\pi w^2}{2}\,dw \qquad V(w) = \int_0^w \sin\frac{\pi w^2}{2}\,dw \qquad (5.21)$$

It will be seen that $U(-w) = -U(w)$ and

$$V(-w) = -V(w)$$

The U, V plot of the parameter w, shown in Fig. 5.9, is called the *Cornu spiral*. It is a plot in the complex plane of the function

$$U(w) + iV(w) = \int_0^w \exp(i\pi w^2/2)\,dw \qquad (5.22)$$

The element of length along the Cornu spiral is simply dw, and the angle of inclination to the U axis is $\pi w^2/2$. When w is the square root of an integer, the curve is parallel to one of the axes (even integers, horizontal; odd, vertical) and successive unit lengths starting from the origin sweep out angles $(m - \frac{1}{2})\pi$. The curve approaches the asymptotic points

$$U(\pm\infty) = \pm\tfrac{1}{2},\ V(\pm\infty) = \pm\tfrac{1}{2}$$

The scale factors between $d\xi$ and du and between $d\eta$ and dv are given by (5.19a) and (5.19b), and the limits u_1, v_1, u_2, v_2 depend upon the coordinate system (cf. Fig. 5.8). With the application of (5.21) to (5.19e), (5.19f), and (5.20b),

$$C = a\{[U(u_2) - U(u_1)][U(v_2) - U(v_1)] - [V(u_2) - V(u_1)][V(v_2) - V(v_1)]\} \qquad (5.23a)$$

$$S = a\{[U(u_2) - U(u_1)][V(v_2) - V(v_1)] + [V(u_2) - V(u_1)][U(v_2) - U(v_1)]\} \qquad (5.23b)$$

$$I_P = \frac{[U(u_2) - U(u_1)]^2[U(v_2) - U(v_1)]^2 + [V(u_2) - V(u_1)]^2[V(v_2) - V(v_1)]^2}{2(R_0 + R)^2} \qquad (5.23c)$$

Analytically u and v are special cases of the argument w of the Fresnel integral, and geometrically the bracketed expressions in (5.23) are chords of the Cornu spiral (Fig. 5.9).

For an infinitely long slit (5.23) is simplified:

$$\eta_2 \to +\infty, \qquad \eta_1 \to -\infty,$$
$$U(v_2) - U(v_1) = V(v_2) - V(v_1) = 1$$

For example, suppose Na D light ($\lambda = 5.892 \times 10^{-7}$ m) passes from a small source Q through a long slit 1.358 mm wide in a plane 1 m away to a screen 1 m farther away

$$(z_0 = -1\text{ m},\ z = +1\text{ m},\ \xi_2 - \xi_1 = 1.358 \times 10^{-3}\text{ m})$$

Then, from (5.19a), the slit width corresponds to an interval of approximately $w_2 - w_1 = 3.356$ on the spiral, and the amplitude at any P is proportional to the chord between w_2 and w_1 (Fig. 5.10a).

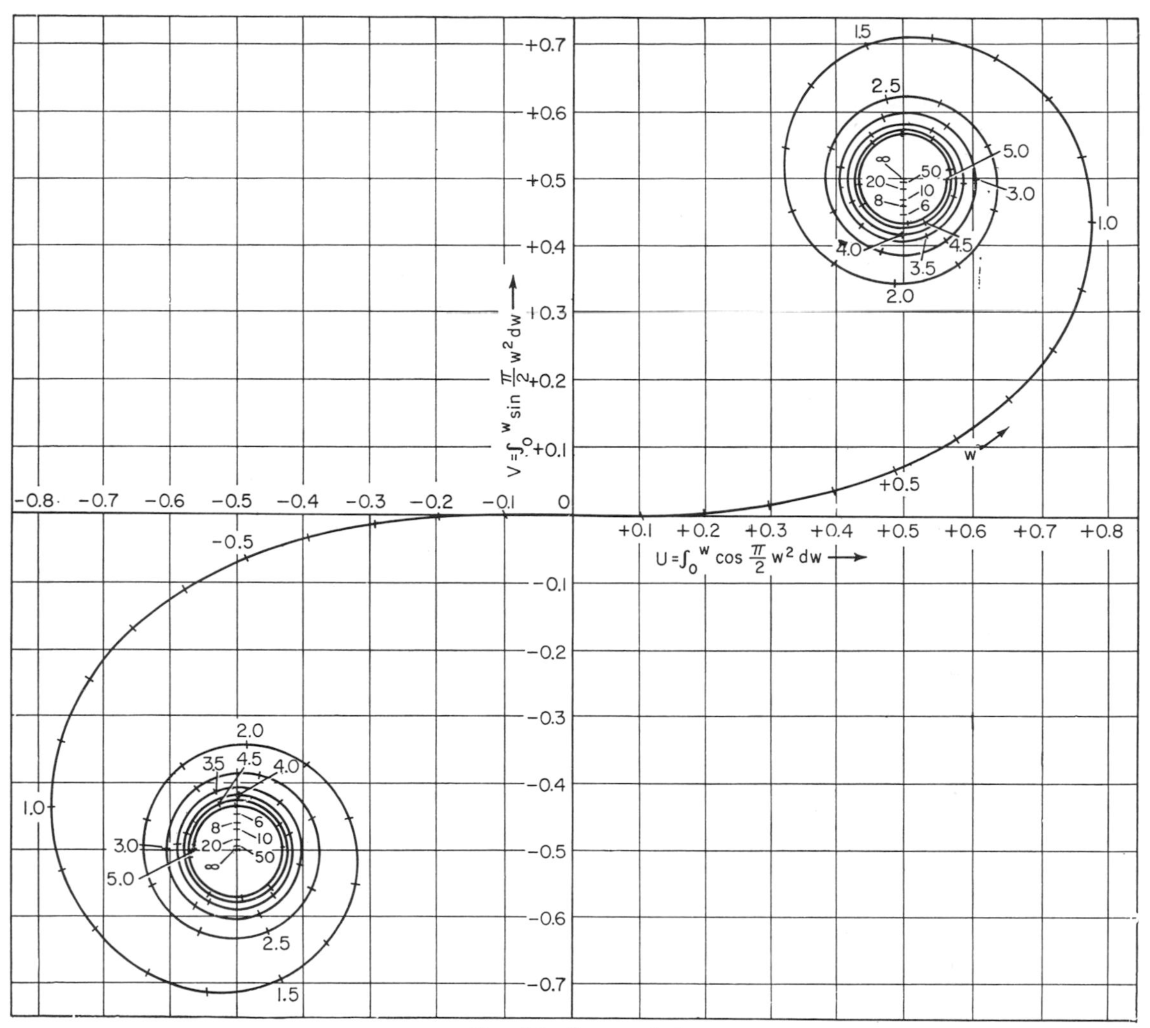

FIG. 5.9. Cornu spiral.

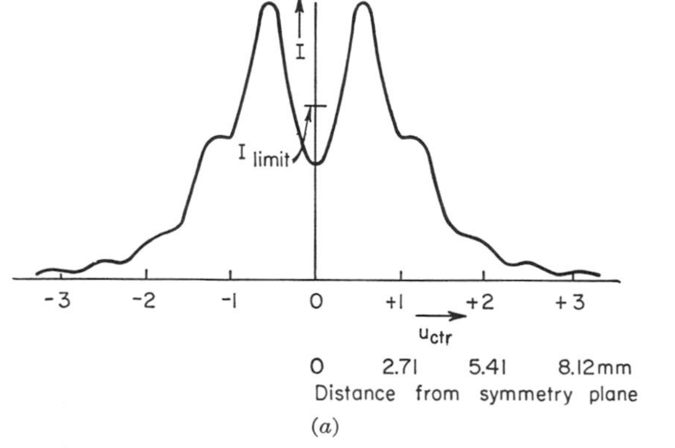

(a)

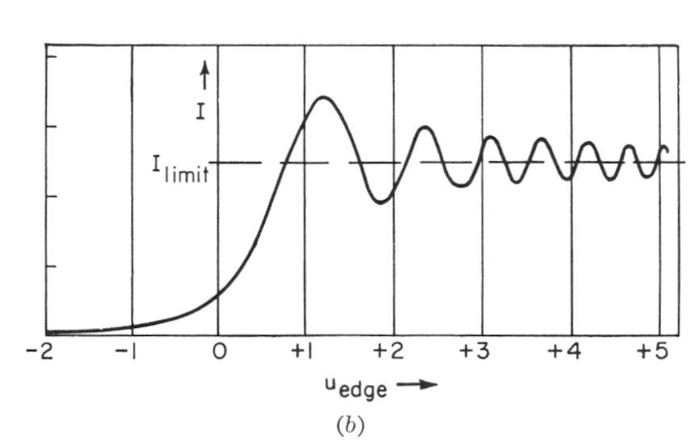

(b)

FIG. 5.10. Fresnel diffraction. (a) At a slit 1.358 mm wide, 1 m from source and 1 m from screen, $\lambda = 5.892 . 10^{-7}$ m. (b) At a straightedge also 1 m from source and 1 m from screen.

For a single straightedge $\xi_1 \to -\infty$,

$$U(w_1) = V(w_1) = -\tfrac{1}{2}$$

and (5.23) is further simplified to the universal expressions

$$C = a(U - V) \qquad (5.24a)$$
$$S = a(U + V + 1) \qquad (5.24b)$$
$$I = \tfrac{1}{2}[(U + \tfrac{1}{2})^2 + (V + \tfrac{1}{2})^2](R_0 + R)^{-2} \qquad (5.24c)$$

where U means $U(u_{\text{edge}})$ and V means $V(u_{\text{edge}})$. Figure 5.10*b* can be applied to any straightedge by adjustment of the abscissa scale according to (5.19*a*). It is derived from Fig. 5.9 by plotting as ordinate the square of the chord from the asymptote $(-\tfrac{1}{2}, -\tfrac{1}{2})$ to u_{edge}.

4. Resolution and Fringe Shape

Resolving Power of an Optical Instrument. The ability of an instrument to produce separately distinguishable images of closely neighboring object points may be expressed by one of several measures, depending partly upon the function of the instrument. Commonly encountered measures are the *limit of resolution in object space*, which is the distance $\Delta_r x_0$, usually normal to the line of sight, between two points whose images are just distinguishable (*resolved*); the *limit of resolution in image space*, the corresponding distance $\Delta_r x_1$; the corresponding *angular limit of resolution* $\Delta_r\theta$; and the *resolving power*, or "resolution," of the instrument, which is often expressed as the reciprocal of $\Delta_r x_0$, $\Delta_r x_1$, or $\Delta_r\theta$, but which in the case of a spectroscopic instrument is usually the *spectroscopic resolving power* $\lambda/\Delta_r\lambda$ or $\sigma/\Delta_r\sigma$, where $\Delta_r\lambda$ and $\Delta_r\sigma$ are the just-resolved difference in wavelength and in wave number, respectively.

Image-width Criteria. According to the *Rayleigh criterion* it is customary in the calculation of Fraunhofer diffraction by narrow slits illuminated by monochromatic light (Fig. 5.4) to consider two images resolved when the central maximum of each falls on the first minimum (zero) of the other. For two equally illuminated images, in the ideal case, this situation yields a minimum with $8\pi^{-2}$, or 0.8106 as much illuminance as the two neighboring maxima. Under the wide range of conditions found in practice the actual resolving power of an instrument may be somewhat greater or, especially when the objects differ greatly in brightness or the instrument is in poor condition, much less than that indicated by the Rayleigh criterion. There are several other measures. One modern measure of the width of a line or other image feature, often used in resolution studies instead of the Rayleigh criterion, is the *half-intensity* width, $\Delta_{1/2}x$ (sometimes inexcusably called the "half width," which obviously means something quite different), which is the distance between two points, one on each side of the point of maximum illuminance, at each of which the illuminance is one-half that at the maximum. Other fractional-intensity widths $\Delta_{1/z}x$ with analogous meaning, such as the one-tenth-intensity width, $\Delta_{1/10}x$, are sometimes used. Here x stands for any of the quantities x_0, x_1, θ, λ, . . . , referred to in the previous paragraph.

Equation (5.11) and those following show that the Fraunhofer diffraction width of the image of a narrow object varies inversely as the width of the beam admitted through the instrument. Thus the resolving power is proportional to the width of the beam unless it is further limited by other factors.

Fringe Shape. The resolving power of an instrument depends also upon the shape of its fringe patterns. The mathematical problem of determining the shape of a fringe pattern, i.e., of the curve, intensity vs. a function of the phase difference between a given pair of wave elements, is essentially that of finding a chord length in vector amplitude space and squaring. This problem has already been treated for two-beam interference as shown in (5.3) and Fig. 5.2, for Fraunhofer diffraction at rectangular and circular openings as shown in (5.11) to (5.13) and Figs. 5.4 and 5.5, and for Fresnel diffraction at a straight edge as shown in Fig. 5.10, the chord in the last case being a chord of the Cornu spiral (Fig. 5.9). It will be considered below for the two principal cases of importance in multibeam interference: divided wavefront with equal amplitudes, and divided, geometrically degraded amplitude.

But first consider the phase difference δ, introduced when a plane wavefront is divided laterally, as by a periodic ruling pattern in the case of a grating, and when the amplitude is divided, as in the case of the Fabry-Perot interferometer.

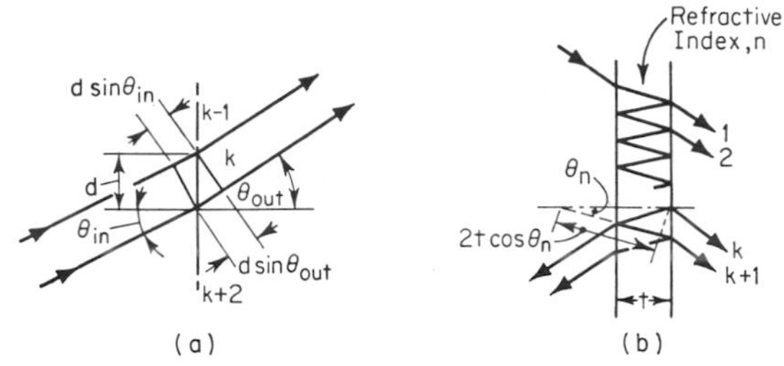

FIG. 5.11. Geometry of multibeam interference among plane waves. (*a*) Laterally divided wavefront. (*b*) Divided amplitude.

In the former case, δ between adjacent wavefront segments is (Fig. 5.11*a*)

$$\delta_{\text{divided wavefront}} = \delta_k - \delta_{k-1} = \lambda^{-1}nd(\sin\theta_{\text{in}} \pm \sin\theta_{\text{out}}) \qquad (5.25a)$$

and, for the special condition of maximum constructive interference across the whole front (bright fringes), $\delta = 2\pi m$ and

$$m\lambda_{0,\text{ bright fringe, divided wavefront}} = nd(\sin\theta_{\text{in}} \pm \sin\theta_{\text{out}}) \qquad (5.25b)$$

where d is the so-called *grating constant*, the distance between corresponding parts of successive rulings, m is an integer called the *order* of the spectrum, designating the number of wavelengths in the path difference between successive beams, n is the index of the medium outside the grating, and θ_{in} and θ_{out} are the angles of incidence and emergence, respectively; the sign is minus for transmitted and plus for reflected light.

In the latter case, δ between successive beams is (Fig. 5.11b)

$$\delta_{\text{divided amplitude}} = \delta_k - \delta_{k-1} = 2\lambda_0^{-1}nt \cos \theta_n \quad (5.26a)$$

and for bright fringes

$$m\lambda_{0,\text{ bright fringe, divided amplitude}} = 2nt \cos \theta_n \quad (5.26b)$$

where t is the distance or gap, and n the refractive index of the medium, between the surfaces where the amplitude is divided, and θ_n is the magnitude of the angle of incidence and emergence with respect to the surface, within the medium.

In equal-amplitude multibeam interference there is a finite number N of amplitude elements (N being in practice sometimes as large as $\sim 10^5$) identical in magnitude a_k, and in the example of the ideal plane diffraction grating (Sec. 6) the phase of each succeeding element is shifted from that of its predecessor by the same amount, δ [but for the concave grating see (5.49)]. As a quantity yielded by an integration over a range of phase, the scalar amplitude a_k is itself a slowly varying function of δ (Fig. 5.12a). The intensity from a plane grating, with grooves (or slits) of width qd and center-to-center spacing d, illuminated by a plane wavefront, is

$$I_{\text{equal amplitude, uniform phase difference}} = I_0 \left[\frac{\sin (q\delta/2)}{q\delta/2}\right]^2 \left[\frac{\sin (N\delta/2)}{\sin (\delta/2)}\right]^2 \quad (5.27)$$

Here the quantities shown in brackets are, respectively, the diffraction factor from (5.12) and the interference factor. The quantity $q\delta/2$ is one-half the phase difference across one line, and $\delta/2$ is one-half the phase difference between successive line centers. The diffraction factor is shown in Fig. 5.12b for a system in which s is one-half of d; the maximum comes at the image point predicted by geometrical optics. The numerator and denominator of the interference factor are shown in Figs. 5.12c and d, respectively, and their ratio in Fig. 5.12e, for the case $N = 10$; Fig. 5.12f shows the whole expression (5.27) for a particular relationship of Fig. 5.12b to Fig. 5.12e. Figure 5.12g shows the pattern for a case where N is so large that neither the amplitudes nor the separations of the minor peaks can be shown to scale. Equation (5.36), below, comes directly from (5.27).

In the geometrically degraded amplitude case, exemplified by the Haidinger fringes in the beam transmitted by the Fabry-Perot interferometer (Sec. 7), there is an infinite series of amplitude elements, with a constant phase shift δ between successive elements as in the example above, but with the magnitude a_k of each succeeding element less than that of its predecessor by a factor R, that is, $a_k = a_{k-1}R = a_0R^k$, and the total length of the broken curve in vector amplitude space is the series sum $a_0(1 - R)^{-1}$. In the example cited R is the *reflectance*, i.e., the relative intensity reflected or the square of the relative scalar amplitude ratio for reflection at one of the surfaces alone. Figure 5.12h shows the first 10 transmitted amplitude elements for $R = 0.50$ and several values of δ, and Fig. 5.12i for $R = 0.90$. The calculated relative intensity transmitted as a function of δ, known as *Airy's equation* and shown in Fig. 5.12j (and on log scales in Fig. 5.12k) is

$$I_{\text{geometrically degraded amplitude, uniform phase difference}} = \frac{I_0T^2}{(1 - R)^2 \left[1 + \dfrac{4R \sin^2 (\delta/2)}{(1 - R)^2}\right]} \quad (5.28)$$

where I_0 is the incident intensity and T is the *transmittance* of, i.e., the relative intensity transmitted by, one of the surfaces alone.

The *contrast range* is the intensity ratio

$$\frac{I_{\max}}{I_{\min}} = \frac{(1 + R)^2}{(1 - R)^2} \quad (5.29)$$

of the brightest ($\delta = 2m\pi$, where m is an integer) to the darkest [$\delta = (2m + 1)\pi$] part of the fringe system described in (5.28).

A convenient index of relative fringe width in the case of degraded amplitude is N_{eff}, the *finesse* (but see the next paragraph) or resolving power per order, or the "effective number of reflections," analogous to the number of lines in the plane-grating case (5.36). The concept of N_{eff} is fully useful only when R is high enough so that the contrast factor is of the order of 10^2 or more, i.e., when $R \gtrsim 0.8$. For sufficiently flat parallel plates and uniform coatings, where N_{eff} is not limited geometrically, N_{eff} depends only upon the reflectance; for sufficiently high R, N_{eff} and R are related approximately by

$$N_{\text{eff}} \doteq \frac{\pi R^{1/2}}{1 - R} \quad (5.30)$$

Some British authors (following Fabry) refer to the coefficient $4R(1 - R)^{-2}$, which appears in (5.28) and is proportional to the square of N_{eff}, as the "finesse" or "coefficient of finesse" of the system, and some French authors use the same word for its square root.

Table 5.1 shows the values of R required for several values of N_{eff}, with the corresponding contrast ranges and some fractional-intensity widths expressed in terms of the free spectral range discussed below. Figure 5.12l shows the relation between R and N_{eff}.

TABLE 5.1. REFLECTANCE R, CONTRAST RANGE $I_{\max}/I_{\min}$, AND CERTAIN RELATIVE FRACTIONAL-INTENSITY WIDTHS FOR SOME FRACTIONS $1/Z$,

$$\Delta_{1/Z}\sigma/\Delta_F\sigma = 2\pi^{-1} \arcsin [3(Z - 1)^{1/2}/2N_{\text{eff}}]$$

FOR SOME VALUES OF THE FINESSE N_{eff}, WITH GEOMETRICALLY DEGRADED AMPLITUDE MULTIBEAM FRINGES

R	$\frac{I_{\max}}{I_{\min}}$	N_{eff}	$\frac{\Delta_{1/2}\sigma}{\Delta_F\sigma}$	$\frac{\Delta_{1/10}\sigma}{\Delta_F\sigma}$	$\frac{\Delta_{1/50}\sigma}{\Delta_F\sigma}$	$\frac{\Delta_{1/100}\sigma}{\Delta_F\sigma}$
0.6	11	5	0.194	0.713		
0.74	45	10	0.096	0.297		
0.856	1.8×10^2	20	0.048	0.145	0.352	
0.942	1.1×10^3	50	0.019	0.057	0.135	0.193
0.970	4.5×10^3	100	0.0095	0.029	0.067	0.095
0.985	1.8×10^4	200	0.0048	0.014	0.033	0.048
0.994	1.1×10^5	500	0.0019	0.0057	0.0134	0.0191
0.997	4.5×10^5	1,000	0.0010	0.0029	0.0067	0.0096
1.	∞	∞	0	0	0	0

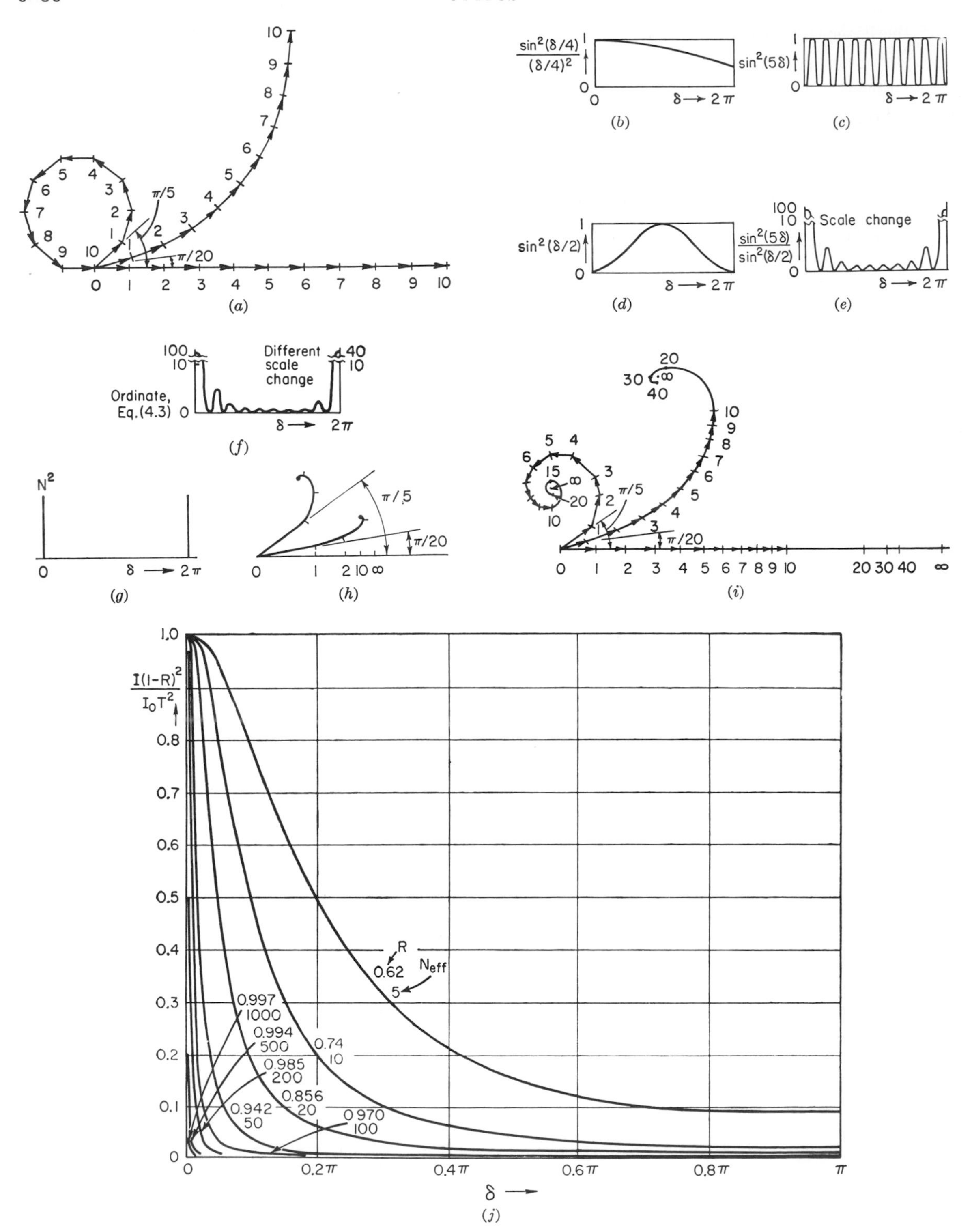

FIG. 5.12. Multibeam interference: (*a*) to (*f*), equal amplitude, uniform phase difference, ten elements; (*h*) to (*k*) geometrically degraded amplitude, uniform phase difference; (*g*) applicable to both classes. All the vector-amplitude diagrams are drawn to the same scale. (*a*) Vector-amplitude diagram for $\delta = 0$, $\pi/20$, and $\pi/5$. (*b*) Diffraction factor for slit width of half the slit center-to-center spacing; see Fig. 5.4. (*c*) Interference factor numerator; see Fig. 5.2*b*. (*d*) Interference factor denominator; see Fig. 5.2*b*. (*e*) Interference factor: note the change of scale; the ends rise to

Free Spectral Range. Allied to the problem of resolution is that of the overlap of successive orders. The *free spectral range* of a system is traditionally defined as the wavelength interval $|\Delta_F\lambda|$ equal to the difference between a wavelength in order m and the wavelength in the next higher order that coincides with it geometrically; alternatively, and usually more simply, the free spectral range, may be expressed as $\Delta_F\sigma$, in terms of the vacuum wave number σ, where $\sigma = \lambda_0^{-1}$. In general (neglecting a dispersion term with the relative value $m\,\Delta n/n$), $|\Delta_F\lambda| = \lambda/m$, and $\Delta_F\sigma = \sigma/m$. In the cases treated in (5.25) and (5.26),

$$|\Delta_F\lambda_0|_{\text{divided wavefront}} = m^{-2}d(\sin\theta_{\text{in}} \pm \sin\theta_{\text{out}}) \qquad (5.31a)$$

$$|\Delta_F\lambda_0|_{\text{divided amplitude}} = 2m^{-2}nt\cos\theta_n \qquad (5.32a)$$

or, expressed in terms of wave number,

$$(\Delta_F\sigma)_{\text{divided wavefront}} = [d(\sin\theta_{\text{in}} \pm \sin\theta_{\text{out}})]^{-1} \qquad (5.31b)$$

$$(\Delta_F\sigma)_{\text{divided amplitude}} = (2nt\cos\theta_n)^{-1} \qquad (5.32b)$$

Resolving Power of Typical Instruments. The angular limit of resolution, $\Delta_r\theta$, of a photographic or telescope lens is, following Fig. 5.5,

$$\Delta_r\theta_{\text{telescope}} = 1.220\lambda a^{-1} \qquad (5.33)$$

where a is the lens diameter. The limit of resolution in object space, $\Delta_r x_0$, for a microscope, is

$$\Delta_r x_0 = \frac{\lambda}{2\alpha} \qquad (5.34)$$

where α is the so-called *numerical aperture* of the lens, defined as the product of the refractive index of the medium between the object and the lens (air, for a dry lens, and oil, for an oil-immersion lens) and the sine of half the angle subtended by the lens diameter at the object. The proportionality of the resolving power to the beam width is implicit, but somewhat obscured, in the usual expressions for the spectroscopic resolving power of a prism at minimum deviation,

$$\left(\frac{\lambda}{\Delta_r\lambda}\right)_{\text{prism}} = B\frac{dn}{d\lambda} \qquad (5.35)$$

where B is the length of the prism base (or, more strictly, the difference between the distances traversed through the prism on the side near the base and on the side near the vertex) and $dn/d\lambda$ the wavelength dispersion (Part 6, Chap. 2) of the prism substance; and of an instrument (e.g., a grating) in which the wavefront is divided laterally into equally retarded interfering beams,

$$\left(\frac{\lambda}{\Delta_r\lambda}\right)_{\text{laterally divided front}} = mN \qquad (5.36)$$

[cf. (5.27)], where N is the number of lines in the grating or, more generally, the number of interfering beams.

In the case of geometrically degraded amplitude fringes from plane-parallel plates the beam width does not enter at all into the characteristic expression [cf. (5.30)] for the resolving power.

$$\left(\frac{\lambda}{\Delta_r\lambda}\right)_{\text{geometrically degraded amplitude}} = mN_{\text{eff}} \doteq \frac{\pi m R^{1/2}}{1-R} \qquad (5.37)$$

However, if, in an unusual case, such as an exceedingly small lens aperture, (5.37) indicates a value higher than (5.33), with the angular dispersion [(5.52) below] taken into account, then (5.33) limits the resolving power.

Optimum Slit Width for Line Image Formation. Now consider the problem of finding the optimum value s_{opt} of the slit width s of a spectro-

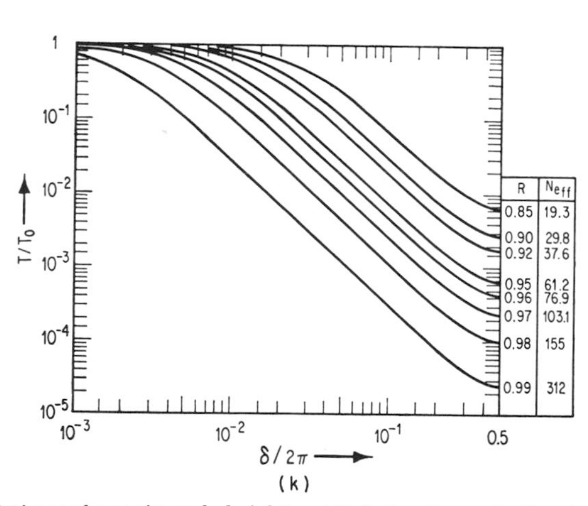

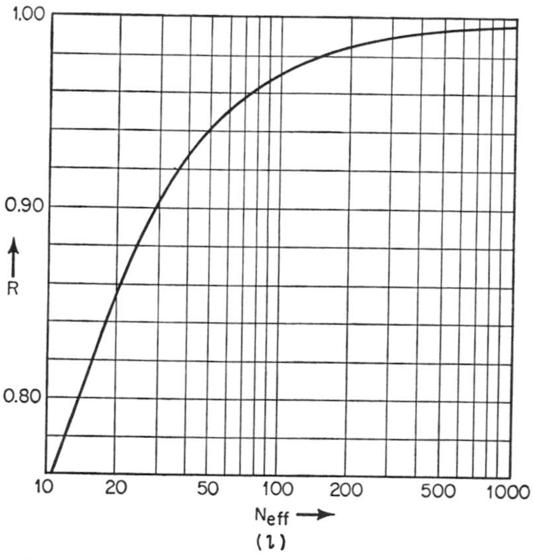

10 times the main scale height. (*f*) Intensity: note the changes of scale; the ends rise by different amounts. (*g*) Intensity, limiting case: equal amplitude, number of elements $N \to \infty$, or geometrically degraded amplitude, reflectance $R \to 1$. (*h*) Vector-amplitude diagram, $R = 0.50$, $\delta = 0$, $\pi/20$, and $\pi/5$; the numbers show the element number, k. (*i*) Vector-amplitude diagram, $R = 0.90$, $\delta = 0$, $\pi/20$, $\pi/5$. (*j*) Intensity (Airy diagram). (*k*) Airy diagram, log log scale. (*l*) Finesse N_{eff}, as a function of reflectance, R.

graph or other optical instrument (Fig. 5.13*a*) in which an image of the slit, of width s', is formed by an image-forming element such as a lens, mirror, or grating, of width W. For simplicity suppose that the image-forming element is distant r from s and r' from s' and that its longitudinal extent may be neglected. The criterion for choosing the boundaries of s' need not, for the present purpose, be specified. The optimum is a compromise between image brightness and image sharpness. Depending upon the

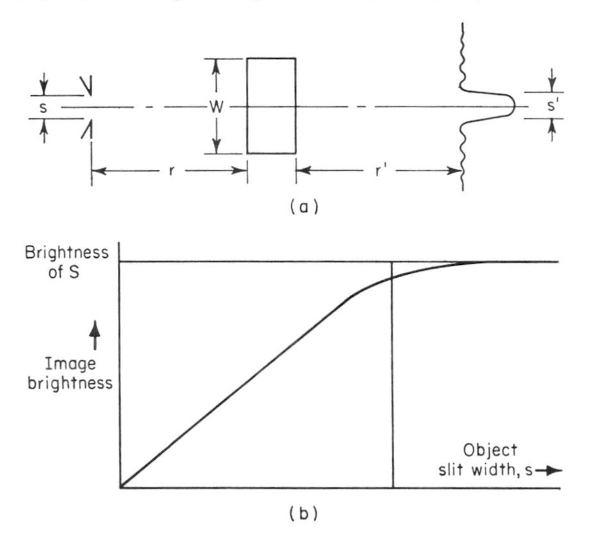

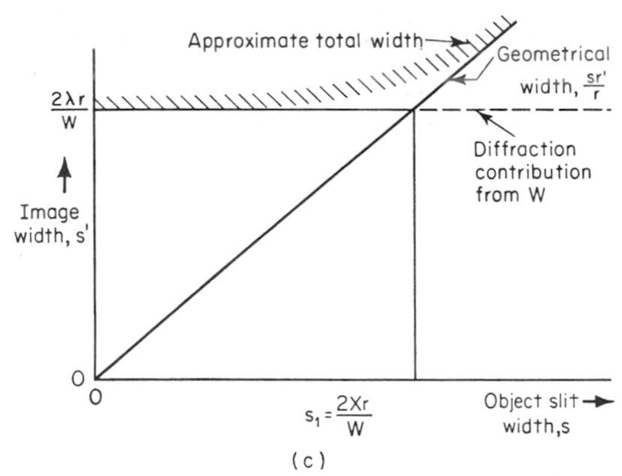

FIG. 5.13. Line image width (schematic). (*a*) Geometry. (*b*) Brightness as a function of slit width. (*c*) Image width as a function of slit width.

purpose of the instrumental use, one of the desiderata may have precedence over the other; so s_{opt} in general is not precisely defined. Moreover, the physical situation depends upon the coherence of the illuminance at s. The incoherent and partly coherent cases will be treated no further here; they yield values of s_{opt} of the same order as the coherent case. It will now be shown qualitatively for coherent illumination with wavelength λ that the curves for the brightness and the width as functions of s behave approximately according to geometrical optics for values of s down to a certain s value, $s_1 = 2\lambda r W^{-1}$, and less favorably for successively lower s values (greater than proportional width, less brightness). This value, then, is approximately the optimum,

$$s_{opt} \cong s_1 = \frac{2\lambda r}{W} \tag{5.38}$$

The slit width for which the first minima of (5.11) fall at the edges of the image-forming element is just s_1. For larger s the diffraction pattern from s is smaller than W, and for smaller s much of the light is lost by being diverted outside of W. According to geometrical optics the image of an object, produced by transmission through perfectly transparent media, within the pertinent directional range is as bright as the object. In Fig. 5.13*b* the brightness is less only on account of the excess of s' over its geometrical value and the diversion described above, which rapidly becomes very important as s decreases below s_1.

Although the beam diffracted by a narrow opening spreads to a width inversely as the opening, here there are two openings in series and consequently the image width s' is directly proportional to s, except that it cannot go below a certain lower boundary value determined by W. Half the angular width between first minima of the beam diffracted at s is, in accordance with (5.11),

$$\frac{\lambda}{s} = \frac{w_{dif}}{2r} \tag{5.39}$$

where w_{dif} measures the distance between first minima in the diffraction pattern in the plane of the image-forming element. That element has the function of directing the light to the geometrical image at s'. Similarly, if the amplitude variation across w is neglected, half the angular width at the slit image, on account of the limited beam at W, is

$$\frac{\lambda}{w} = \frac{s'}{2r'} \tag{5.40}$$

where w is the smaller of the quantities, W and w_{dif}.

Figure 5.13*c* qualitatively shows the limitation imposed upon the image sharpness by the narrowness of the image-forming element. From (5.39) and (5.40) s' is approximately the width of the geometrical image $sr'r^{-1}$, unless $w_{dif} > W$, in which case W limits the beam, taking over the role of w_{dif} in (5.39). It is just at s_1 that $w_{dif} = W$; so s' has the lower bounds

$$s' = \frac{2\lambda r'}{W} \qquad s < s_1 \tag{5.41a}$$

$$s' = \frac{sr'}{r} \qquad s > s_1 \tag{5.41b}$$

List of Spectroscopic Properties. The most important spectroscopic properties of a diffraction-interference system are all closely related and are:

Angular condition for maximum intensity [(5.25), (5.26)]
Fringe shape [(5.27),(5.28)]
Free spectral range [(5.31),(5.32)]
Resolving power [(5.36),(5.37)]
Dispersion [(5.48),(5.52)]

Apodization. If any cross section of a beam, e.g., the surface of a lens, has deposited upon it a nonscattering, partly transmitting layer, such as a thin metallic layer, whose transmittance varies across the section, the simple conditions leading to (5.11), (5.12), (5.13), (5.27), or (5.28) are inapplicable and the fringe shape may be quite different from those discussed. For instance, the reduction of the relative amplitude transmitted near the periphery of a beam can eliminate the subordinate maxima and transform a pattern like that of Figs. 5.4, 5.5, or 5.12*f* into one more like Fig. 5.12*j*. Such transmittance control is called *apodization*. It was suggested by Straubel and developed into a high state of advancement by several workers, especially Jacquinot.

In fact, Fraunhofer diffraction may be thought of as yielding at an image a Fourier transform of the transmission at an aperture stop. For example, the expression on the right side of (5.11) is proportional to the Fourier transform of a rectangular function (cf. Part 1, Chap. 3). The property alluded to above, the width of the pattern varying inversely as the width of the aperture, now appears very naturally as a well-known property of Fourier transforms. A Gaussian absorbing screen placed at an aperture will, for example, result in a smooth Gaussian image of a point without the wings, or feet (or *podia;* hence "apodization") of Fig. 5.4, since the Fourier transform of a Gaussian is also a Gaussian (cf. Part 1, Chap. 3, Table 3.6).

Apodization is often used where the elimination of wings is desirable, for example, in spectroscopy, for the detection of a weak line in the neighborhood of a strong one. In Fourier-transform spectroscopy, slow variations in amplitude result in improved line shapes; in the construction of radio antennas, proper attenuation of parts of the structure removes unwanted sensitivity lobes, which correspond to the wings of a diffraction pattern.

5. Two-beam Interference

Young's Experiment, Fresnel's Bimirror and Biprism, Billet's Split Lens, Lloyd's Mirror. The crucial experiment for the establishment of the wave nature of light was the double-slit experiment performed by Thomas Young in 1801. Two narrow parallel slits Q_1, Q_2 (Fig. 5.14*a*) in an opaque screen *B*, a small distance *d* apart, designed to serve as effectively coherent sources, are illuminated by a linear source *Q* or a slit *Q* parallel to Q_1, Q_2, lying far from *B* in the axial plane (which is the perpendicular bisector of a line running normally between Q_1 and Q_2). A screen *S* parallel to *B* and distant *a* from it on the shadow side is found to be illuminated in the neighborhood of the axis, where the diffraction patterns of QQ_1 and QQ_2 overlap, by a series of narrow fringes parallel to *Q*. If the paths from Q_1 and Q_2 to a point *P* of *S*, distant *p* from the axis, are called r_1 and r_2, respectively, then

$$|r_2^2 - r_1^2| = 2pd = |r_2 - r_1|(r_2 + r_1)$$

If $p \ll a$ and $d \ll a$, then $r_2 + r_1 = 2a$ and the path difference *D* is

$$D = \frac{pd}{a}$$

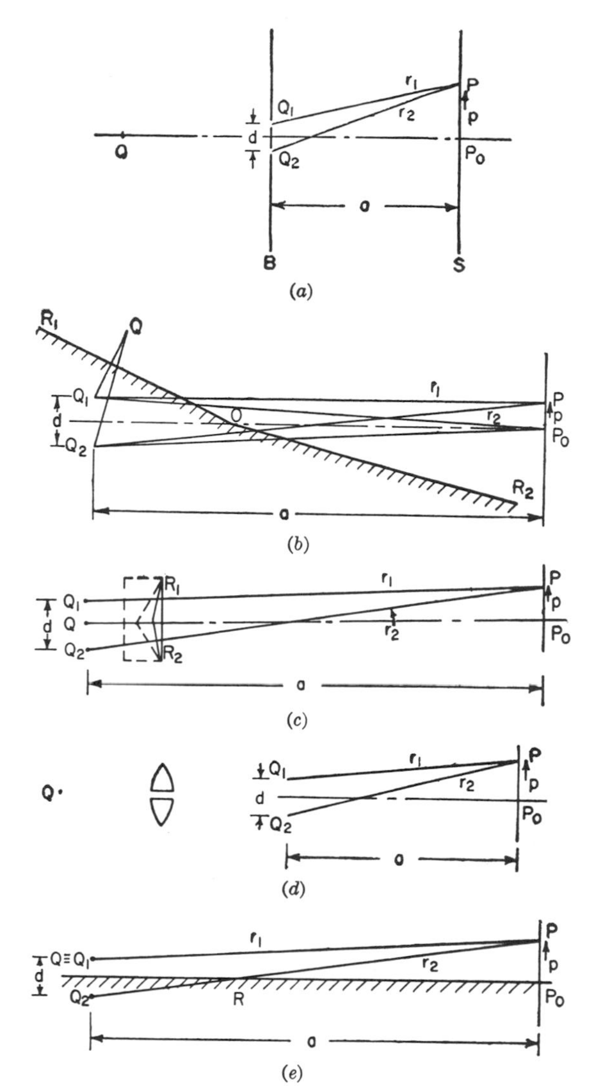

FIG. 5.14. Two-beam interference, historical geometries: all lateral distances P_0P are greatly exaggerated compared with longitudinal distances QP_0. (*a*) Young's crucial experiment. (*b*) Fresnel's bimirror. (*c*) Fresnel's biprism. (*d*) Billet's split lens. (*e*) Lloyd's (asymmetric) mirror.

yielding a phase difference, from (5.2),

$$\delta = \frac{D}{\lambda} = \frac{pd}{\lambda a}$$

There are bright (constructive) fringes centered at points

$$p_{\text{bright}} = \frac{m\lambda a}{d} \qquad m = 0, \pm 1, \pm 2, \ldots \qquad (5.42a)$$

including the axial point P_0, and complete darkness at

$$p_{\text{dark}} = \frac{(m + \frac{1}{2})\lambda a}{d} \qquad m = 0, \pm 1, \pm 2, \ldots \qquad (5.42b)$$

Although the discussion has been in terms of the plane section in Fig. 5.14*a*, it is valid for the three-dimensional case, because integration over the slit length, normal to the plane of the figure, does not alter the phase differences.

Instead of being attained with slits Q_1, Q_2 as described above, the essential coherence of Q_1 and Q_2 may be attained by a device that produces two images of the slit Q, very close together. Fresnel's bimirror (Fig. 5.14*b*) uses two almost coplanar mirrors OR_1 and OR_2 to yield the nearly equal paths QR_1P and QR_2P; his biprism (Fig. 5.14*c*) uses a pair of very small-angle prisms or, in usual practice, larger-angle prisms incorporated in one wall of a cell (dashed lines, Fig. 5.14*c*) containing a liquid with a refractive index near enough to that of the prism glass to make d very small; Billet's split lens (Fig. 5.14*d*) consists of two halves that form real or virtual images Q_1 and Q_2 of Q. Lloyd's mirror (Fig. 5.14*e*), one mirror RP_0 in a plane close to Q, sending light to P directly and by reflection at almost grazing incidence, is somewhat different in principle from all those described above, and less symmetric, in that Q_1 is Q itself, the path Q_1P is always the shorter, and only in the longer path $Q_2P = QRP$ is there a reflection. Regardless of the nature of the reflecting substance there is always a phase change of π in light reflected at grazing incidence; so all the fringes (5.42) are shifted by $\lambda a/2d$ and the center of the system, P_0, is dark instead of bright. Whereas in all the other above-mentioned two-beam examples one part of the original beam is laterally displaced by its whole width with respect to the other, in Lloyd's mirror the beam is folded back on itself. In practice, the interference of originally close-neighboring parts of the front, being relatively free of outside disturbances, follows the theory especially well.

Newton's Rings; Colors of Thin Films, Low-reflectance Case. When the smooth surfaces of two transparent bodies, convex to one another, are actually or nearly in contact, on being viewed in an approximately normal direction the neighborhood of contact exhibits colored fringes of equal thickness, which are circular (*Newton's rings*) if the surfaces are spherical. In this thin-film case, but not in the interferometers about to be discussed, multiple reflection occurs. The discussion in this subsection is based on the presumption that the reflectances of the surfaces in question are so low that no beams reflected more than once need be considered. Multibeam interference in thin films, arising when more than one reflection occurs with appreciable amplitude, is discussed below (Sec. 7). At a spot where the distance between the surfaces is t (Fig. 5.15*a*), for light viewed from a direction approximately normal to the surface the optical path difference between the waves reflected at the two surfaces is $2nt$, where n is the refractive index of the intervening medium (e.g., air). However, on account of the change of phase of π upon reflection against a body of higher index occurring at the second surface, the effective path difference may be written $2nt - \lambda_0/2$, yielding

Phase change of π at only one interface:

$$t_{\text{bright}} = \frac{(m + \frac{1}{2})\lambda_0}{2n} \qquad (5.43a)$$

$$t_{\text{dark}} = \frac{m\lambda_0}{2n} \qquad (5.43b)$$

for the film thickness of the medium at the bright and dark fringes. Here λ_0 is the vacuum wavelength of the radiation, and m is a nonnegative integer.

For a film of a transparent liquid or solid in air, such as a soap bubble or a thin glass film (Fig. 5.15*b*), the situation is the same except that the phase change occurs at the first surface and Eqs. (5.43) are still valid.

For a film between two media of different index Eqs. (5.43) are valid if the indices of media on the two sides are both higher or both lower than that of the film, but if one is higher and the other lower (1 and 3, Fig. 5.15*c*), as in the case of a low-index ($1 < n < n_{\text{glass}}$) coating on glass, the bright and the dark fringes are interchanged.

Phase change of 0 or π at both interfaces:

$$t_{\text{bright}} = \frac{m\lambda_0}{2n} \qquad (5.43c)$$

$$t_{\text{dark}} = \frac{(m + \frac{1}{2})\lambda_0}{2n} \qquad (5.43d)$$

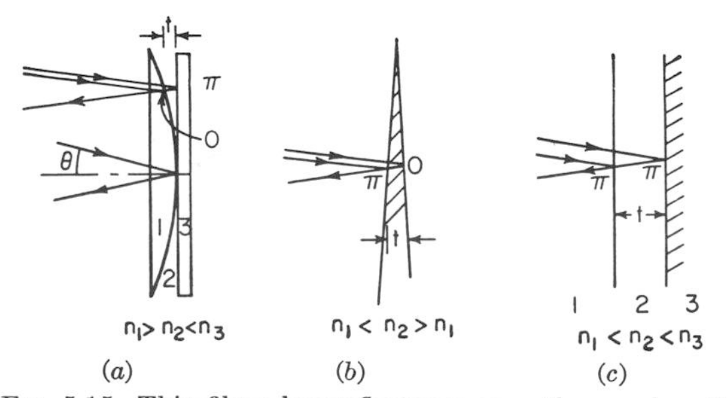

Fig. 5.15. Thin films, low-reflectance case; the numbers 0 and π are the phase changes on reflection. (*a*) Newton's rings. (*b*) Film of optically dense material in air. (*c*) Low-index film on higher-index substrate.

Michelson, Twyman-Green Interferometers. The *Michelson interferometer* (Fig. 5.16*a*) employs two-beam interference in the practical comparison of two optical paths with an accuracy on the order of one-tenth wavelength. Its introduction in 1882 marked a major step in the science of metrology. A light beam from an extended source S is split by a rear-semisilvered flat mirror G at an angle of $\pi/4$ into two beams normal to one another. The direct beam passes through a compensating plate G' parallel to G. The beams are reflected back to G by flat mirrors M_1, M_2, one of which is mounted for translation in the direction GM, and are viewed from a direction normal to the original direction of incidence. If the one mirror and the image of the other in G are nearly parallel and the optical paths GM_1G and GM_2G are nearly equal, white-light fringes may be seen. The occurrence of these white-light fringes, including an obvious central dark one (on account of different phase shift at G), is a convenient indicator of path equality. The greater the difference of the paths, the more nearly monochromatic must be the light in order that any fringes be observable. If the fringes shift on account of a displacement of M_1, since the path difference between successive bright fringes is $\lambda/2$ the passage of one bright fringe to the position previously occupied by the adjacent one implies the translation of M_1 by a distance $\lambda/4$ in the direction GM_1. Distances up to the order of decimeters, using conventional monochromatic light sources, can be measured directly in a Michelson interferometer, and longer distances may be laid out with the aid of spacer blocks measured with the interferometer; with laser sources fringes can be observed for distances up to many meters.

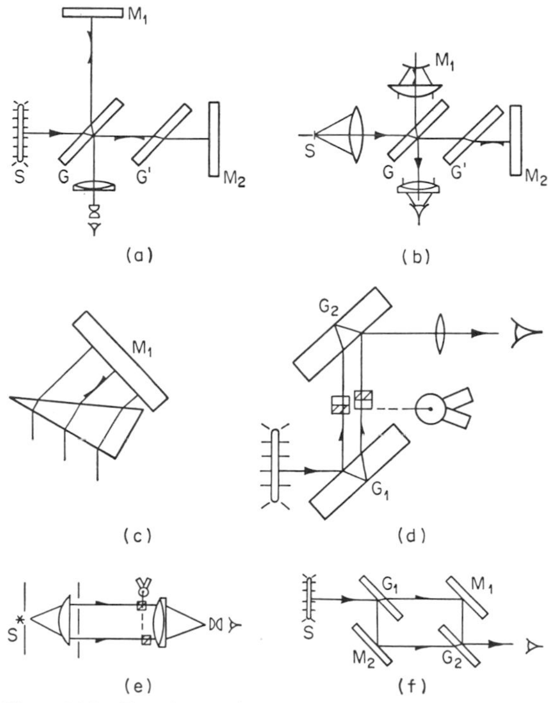

FIG. 5.16. Two-beam interferometers, schematic. (*a*) Michelson. (*b*) Twyman-Green, for lens testing. (*c*) Detail, Twyman-Green, for prism testing. (*d*) Jamin. (*e*) Rayleigh. (*f*) Mach.

The *Twyman-Green interferometer* (Fig. 5.16*b,c*) is a modification of the Michelson interferometer especially designed for stringent lens and prism tests. The beam is collimated from a pinhole. One of the mirrors is shaped, for a lens, or oriented, for a prism, so that in the ideal case the light returns to G as a parallel beam. Not only surface geometrical imperfections, but the combined effects of such imperfections and internal inhomogeneities, appear in the recombined beam.

Jamin, Rayleigh, Mach Refractometer-Interferometers. Any interferometer can be used for refractive-index measurement; the interferometers described in this subsection are especially designed for it. In the *Jamin refractometer* (Fig. 5.16*d*), which preceded the Michelson interferometer by almost half a century, light from an extended source S is split by a thick plane-parallel glass plate G_1 at $\pi/4$ to the incident beam, into two parallel beams as shown, and recombined by an identical plate G_2. Cells may be placed in the two beams: one with the substance under study, and the other a blank or with a standard substance. The orientations of a pair of compensating plates, one in each path, and rotating about a common axis normal to the beams, can be controlled, not only to adjust the path difference to zero, but also to adjust the sensitivity of the instrument. In the *Rayleigh refractometer* (Fig. 5.16*e*) coherent light collimated by a lens L_1 passes through two wide slits to form the two beams, which are recombined by a telescope lens L_2 large enough to cover both beams.

In the Jamin and Rayleigh instruments the geometrical separation of the beam is limited, although the latter can be modified by mirror systems, as in the *Michelson stellar interferometer*, to accept beams separated by a considerably greater distance than the diameter of L_2.

The *Mach refractometer* (Fig. 5.16*f*) allows the beams to be separated by a beam splitter G_1 at $\pi/4$ and to be recombined, after another reflection in a mirror M_1 or M_2 at $\pi/4$, in a second, reversed beam splitter G_2. Because of the geometrical latitude allowed by its design, the Mach instrument is frequently used for the study of transient phenomena, such as those occurring in turbulent flow and especially in shock waves; by suitable tilting of the plates the fringes may be localized anywhere along the path, e.g., at the disturbed region.

The null point, or point of zero path difference, in a Michelson, Jamin, Rayleigh, or Mach interferometer can be identified by white-light fringes.

Location and Distinctness of Fringes from Two Plane Surfaces. If the two surfaces giving rise to two-beam interference are strictly plane, the following analysis, due essentially to Michelson, is applicable: Suppose the two surfaces, of which the one farther from the observer at P is called M', are inclined to one another at a small angle $\varphi/2$ (Fig. 5.17).

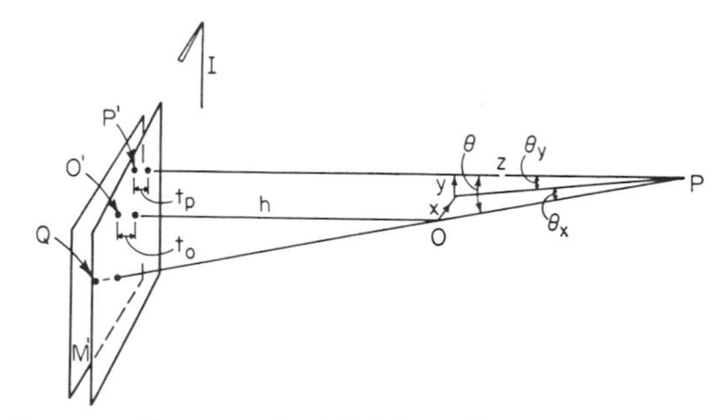

FIG. 5.17. Geometry for Michelson discussion of location and distinctness of two-beam interference fringes from flat plates.

Let the z axis be normal to M' and the y axis parallel to the intersection I of the surfaces. The light to be investigated comes approximately normally from the surfaces, through a fixed point which will be taken as the origin O, to a limited opening, or pupil, in the neighborhood of $P(x,y,z;\ z > 0)$. Drop normals OO' and PP' to M', and call the segments of these normals between the surfaces t_O and t_P, respectively. Let $OO' = h$. Call the intersection of PO extended, with M', Q. Call the small angle OPP', θ, and the projections of OPP' on the $y = 0$ and the $x = 0$ planes, θ_x and θ_y, respectively.

Now the interference-path difference D, arising in the neighborhood of Q, between the two beams through P as seen through a pupil at O is, in the small-angle approximation,

$$D = 2(t_O + h\varphi\theta_x)(1 + \theta_x^2 + \theta_y^2)^{-1/2} \quad (5.44)$$

The range of D over the pupil is least, and the fringes consequently most distinct, when $\theta_y = 0$ and h has the value

$$h_{\text{opt}} = \frac{t_O\theta_x}{\varphi} \quad (5.45)$$

The distance f from O at which the fringes are localized falls in the following ranges:

$$\begin{array}{lll} f \text{ is indeterminate} & \begin{cases} \text{if } \varphi = 0 \text{ and } t_O = 0 \\ \text{or } \varphi = 0 \text{ and } \theta_x = 0 \end{cases} & \\ f = \infty & \text{if } \varphi = 0 & \\ h < f < \infty & \text{if } QI < O'I & (5.46) \\ f = h & \begin{cases} \text{if } t_O = 0 \\ \text{or } \theta = 0 \end{cases} & \\ f < h & \text{if } QI > O'I & \end{array}$$

The shape of a fringe (i.e., of a locus of constant D) satisfies the conic-section equation

$$D^2y^2 = [4(z + h)^2\varphi^2 - D^2]x^2 + 8t_Pz(z + h)\varphi x + (4t_P^2 - D^2)z^2 \quad (5.47)$$

and consequently:

If	*the curve is*
$D = 0$	a straight line
$0 < D < 2(z + h)\varphi$	a hyperbola
$D = 2(z + h)\varphi$	a parabola
$D > 2(z + h)\varphi$	an ellipse
$\varphi = 0$	a circle (Haidinger fringe)

Holography. A unique case of two-beam interference is involved in the process of wavefront reconstruction, or holography, which has taken on considerable importance since the advent of laser sources. Light scattered or reflected from an object interferes with a plane wave of exactly the same frequency to produce a standing-wave pattern which can be recorded directly on a photographic plate. When the developed photographic plate is illuminated by a plane monochromatic wave, the light leaving the plate contains a wave that reproduces the characteristics of the wave scattered from the object and therefore forms an image of the object. Although this process was discovered before laser sources were available, the high coherence and intensity of laser sources permit better recording of the standing-wave pattern and better reconstruction of the wavefront.

Phase-contrast Microscope; Interference Microscope. Microscopic methods utilizing phase contrast, interference, and polarized light have been extensively reviewed by Barer. In a typical form of the Zernike *phase-contrast microscope* (Fig. 5.18*a*) light emerging from an annular opening Q in a source housing passes through a high-quality condensing lens C to the object O and thence through the microscope objective L and a phase plate P to the image O' formed by L. At P that annular region Q' which coincides with the image of Q is made, as, for instance, by the deposition of a ring of an absorbing medium, to give all the light arriving at O' through Q' a path differing by $\lambda/4$ from that of the light that misses Q'.

Suppose O occupies a thin, transparent, but inhomogeneous layer consisting, for instance, of an optically denser body O_A embedded in a rarer surrounding medium O_B: $n_A > n_B$, with a sharp boundary between the two. Let $\mathbf{u}$ be the vector amplitude of the light forming an element of the image O near the boundary, and suppose for the moment that P is absent.

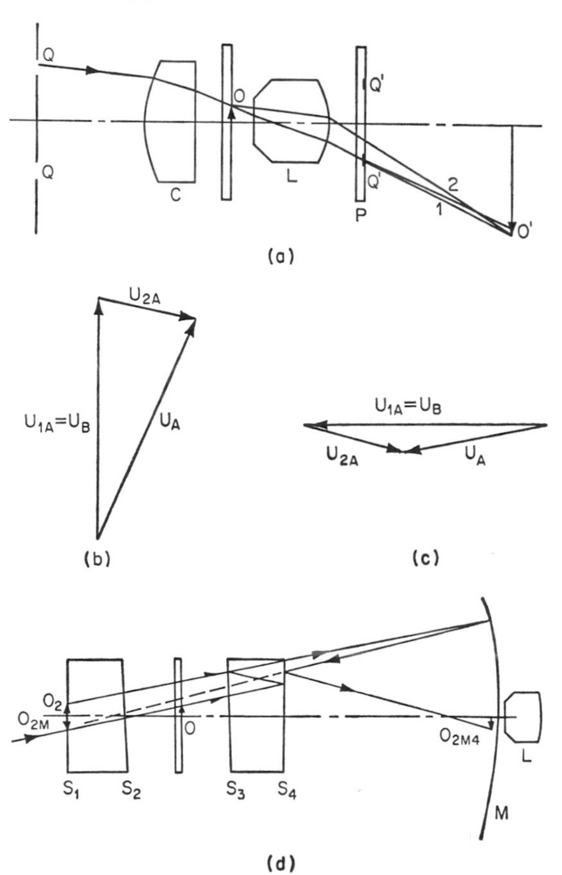

FIG. 5.18. Microscopes. (*a*) Phase microscope. (*b*) Vector amplitude diagram for the light from two thin, transparent objects, O_A and O_B, with different refractive index: ordinary microscope. (*c*) Like (*b*), but for phase microscope. (*d*) Interference microscope.

The equal transmittance of A and B implies $|\mathbf{u}_A| = |\mathbf{u}_B|$, although there is a phase difference δ (here assumed small) between the two vectors arising from the difference in index. Notice (Fig. 5.18*b*, *c*) that u_A can be resolved as $\mathbf{u}_{1A} + \mathbf{u}_{2A}$, where $\mathbf{u}_{1A}$ is identical with $\mathbf{u}_B$ and (neglecting any incoherent scattering) the scattered part, $\mathbf{u}_{2A}$, is approximately equal to $\mathbf{u}_Be^{-i\pi/2}\delta$. In an ordinary microscope, the mere equality in magnitude,

$$|\mathbf{u}_A| = |\mathbf{u}_B|$$

ensures equality of intensity, and the image O'_A is indistinguishable from the adjacent image O'_B, unless one depends upon minute amplitude differences (not treated here) in the diffraction pattern at the boundary to make the outline of O_A visible. In the interference microscope with P present, the image O'_A of O_A is either brighter or fainter than that of the surroundings, depending on the phase difference introduced at the phase plate, i.e., upon whether $\mathbf{u}_1$ and $\mathbf{u}_2$ interfere destructively or constructively in the image (cf. Fig. 5.2). Diffraction at Q' produces halos around O'_A and O'_B, and with large objects the brightness at the center approaches that of the surroundings. It appears that a slight turbidity of the phase ring Q' would spread out the light not scattered at O and make the halo effects less disturbing.

Among the simplest forms of *interference microscope* is that used by Twyman and by Linnik, respectively, for the testing of microscope objectives and the study of the surfaces of opaque bodies. It is constructed by inserting matched objectives before the mirrors M_1, M_2 of a Michelson interferometer (Fig. 5.16*a*), with parallel light incident from a concentrated source. The light is focused by the objectives upon the mirrors and returns through the objectives. Among the several more versatile modern forms that of Dyson (Fig. 5.18*d*) employs a pair of semireflecting surfaces S_1, S_2, below the object O and another pair S_3, S_4 above it, as a Jamin interferometer (generalized as to incidence angle), and uses a spherical mirror M to image O nearly free from aberration close to the objective L. S_4 has a small central opaque patch to shield L from direct light. In practice (omitting obvious refraction corrections), if S_2 is midway between O and S_1 and if the mirror surface is centered on S_1, then if O is near the axis, it has a virtual image O_2 at S_1, by reflection in S_2, and O_2 is reflected by M almost on itself at O_{2M}, which in turn by reflection at S_4 has a real image O_{2M4}, which lies in position before L if S_1S_4 is slightly less than half the radius of M. Suppose, now, that O is transparent but inhomogeneous, with two parts A and B, where $n_A > n_B$. The essential interference feature is that, whereas with no object or a homogeneous object the field would consist of a simple two-beam interference pattern, the phase difference δ_{AB} introduced by the inhomogeneity shifts the phase (abscissa, Fig. 5.2*b*) by an amount δ_{AB} and makes A visible against B. Quantitative measurement of tissue masses is among the potentialities of the instrument. A serious disadvantage of this instrument is its loss of light at the several semireflecting surfaces.

Two-beam Interference Spectroscopy. Following the early tradition of Fizeau and of Michelson, two-beam interferometry has increasingly been used for spectroscopy, especially in the infrared. The great luminosity of the Fabry-Perot interferometer pointed out by Jacquinot is combined with the advantage of a photographic spectrograph in that the whole spectrum is under observation at once. A disadvantage of the two-beam method is that the recorded result is the Fourier transform of the desired spectrum, and, in cases where the resolution desired is high, the reverse transformation may be tedious.

The method commonly used is to employ a modified Michelson interferometer (see Fig. 5.16*a*), employing corner-cube reflectors for M_1 and M_2. M_2 is slowly moved along the optical axis at a constant rate, and an aperture is used with the source, as in Fig. 5.16*b*. A photodetector placed at the eye position records the Fourier transform of the spectrum as a function of time. The theoretical resolving power equals twice the number of wavelengths transversed by the moving mirror.

Wide-angle Interference. Coherent light can interfere for any angle of beam divergence, but the greater the divergence angle, the narrower must be the source in order to keep the phase differences among beam elements from different parts of the source small enough to maintain an appreciable net interference effect. For the greatest divergence angles ($\sim\pi$) the source width ought to be less than $\sim\lambda/8$. Selenyi used as a source a very thin layer of fluorescent liquid between a glass block and a thin mica cover sheet with a reflecting upper surface, to verify the occurrence of interference at very wide angles.

The multipolarity of the elementary oscillations comprising any radiation determines the angular dependence of the coherence of the radiation, which in turn can be determined through the study of wide-angle interference.

6. Equal-amplitude Multibeam Interference

Plane Diffraction Grating, Echelette, Echelle, Echelon. Advantages of the *diffraction grating* over the prism for spectroscopy are its high, geometrically calculable, and nearly wavelength-independent dispersion and resolving power. Jacquinot has shown, in spite of tradition to the contrary, that for a given spectroscopic resolving power a grating that is well blazed (cf. second paragraph below) typically has many times the light-gathering power of a prism of comparable size. A grating spectrum follows the law (5.25*b*). The angular dispersion, in terms of the vacuum wave number σ, is

$$\frac{d\theta}{d\sigma} \doteq \pm \frac{m}{\sigma^2 d \cos \theta_{\text{out}}} \tag{5.48}$$

In a plane grating illuminated with parallel light the phase difference δ is the same for successive elements over the whole area, and (5.25*b*) and (5.48) express the condition $\delta = 2\pi m$.

Crude wire transmission gratings were made by Rittenhouse in 1785 and independently by Fraunhofer about 1821, but it was only when Rowland developed the art of diamond-ruling reflection gratings and invented the concave grating about 1882 that serious grating spectroscopy began. The "ghosts," or false lines from periodic errors, that marred the perfection of Rowland's gratings have been practically eliminated through the efforts of Gale, Wood, Siegbahn, Harrison, and others. Many modern gratings are replicas (second-generation gratings), inexpensively made and practically perfect copies, of a few excellent originals, or "masters." In a typical modern replica process the original receives two evaporated coatings, first a release agent and second a metal coating; then there is added a thin liquid adhesive plastic coating, and finally a glass optical flat is pressed on. Upon separation the adhesive and the metal stay with the glass. The "negative" groove shape attainable in a replica is said to be superior to the shape that can be ruled on an original. Replicas of replicas (third-generation gratings) are sometimes made to avoid wear on the originals.

Almost all modern gratings are *blazed,* or ruled with such line shape as to throw a large fraction of the incident radiation into one general direction or, for a given wavelength, into one order. Blazing was suggested by Rayleigh. Blazed gratings designed and developed by Wood and by Randall for infrared spectroscopy are called *echelettes.*

Harrison has developed, by iterative grinding and polishing, a coarse, blazed grating, known as the *echelle,* with a spacing on the order of hundreds of wavelengths, that can attain very high resolution ($\sim 10^6$) through working in an order intermediate between those ordinarily used in a grating (~ 1) and those used in an echelon ($\sim 10^5$). The echelle usually requires an auxiliary dispersion to avoid overlapping of orders. The echelette and the echelle are intermediate in their properties between the ordinary grating and the echelon, now to be described.

Noting that the expression (5.36) for the resolving power of a grating is symmetric in the order number and the number of lines, Michelson produced a transmission model of an instrument called the *echelon,* consisting of about 20 plane-parallel plates, about 1 cm thick, from the same original glass sheet, each extending beyond its next neighbor on one side by a millimeter or so; the resolving power attainable is of the order of 10^6. A few reflection echelons have been constructed. Echelons are expensive and not very adaptable and are rarely used.

Concave Grating. The plane grating alone cannot produce a real image spectrum from a real slit. The *concave grating* is ruled on a concave spherical mirror. Let the radius of the sphere (Fig. 5.19*a*) be r, and consider an arrangement in the plane of the center of the sphere and the normal bisector of a groove. The *Rowland circle* is a circle of radius $r/2$ in the plane described, internally tangent to the grating at the center of the grooves. Its important property is this: If the slit s is oriented parallel to the rulings with its center anywhere on the Rowland circle, its sharp-line (tangential) astigmatic image and all its colored images (i.e., the spectrum lines) also lie on the Rowland circle in the directions given by (5.25*b*). Many, but not all, of the conventional concave-grating mounts make use of the Rowland circle.

In a concave grating it is the projections of the rulings on a tangential plane, and not the rulings themselves, that are equidistant. The phase difference between successive elements varies over the face of the grating, and there arises a fourth-order aberration that gives the amplitude plot the shape of Fig. 5.19*b* rather than that of (5.25) and Fig. 5.12*a*, limiting the useful width of the grating as shown by Mack, Stehn, and Edlén, to approximately

$$W_{\text{opt}} = \left[\frac{40\lambda r^3}{\sin\theta_{\text{in}}\tan\theta_{\text{in}} + (m\lambda/d - \sin\theta_{\text{in}})\tan\arcsin(m\lambda/d - \sin\theta_{\text{in}})}\right]^{1/4} \tag{5.49}$$

The limitation is important only at near-grazing incidence or with exceptionally wide gratings. The fringe shape is different from that for a plane grating, as Fig. 5.19*c* shows. For mountings not using the Rowland circle there is a third-order aberration.

Grating Mounts. Of the many grating *mounts* (*mountings,* geometries) that have been proposed and used, only a representative few will be described here. Recent improvements in blazing, the increasing appreciation of *multipass* schemes (where the radiation strikes the grating more than once), and emphasis on nonphotographic recording have strongly influenced mounting practice.

For the plane grating G (Fig. 5.20) the two mounts mentioned below have much in common: the collimator acts also as the image-forming element, with the slit S at its principal focus, and the advantages in freedom from aberration inherent in a system in which the light remains near the axis are partly offset by the necessity of departing from symmetry about the plane normally bisecting the rulings in order to keep S and the plateholder P spatially separated. In the *Littrow mount* (Fig. 5.20*a*) the collimator is a lens L (in a common modification shown in dashed lines, S is replaced by its image S' in a small, or totally reflecting prism), and P may be tilted or curved to agree with the focal properties of L for any one wavelength region. In the modified *Ebert-Fastie mount* (Fig. 5.20*b*) the collimator is a spherical mirror M.

For the concave grating, blazing has overthrown the previously predominant *Paschen mount* (Fig. 5.19*a*), which covered the whole spectrum without moving parts, an advantage offset by the disadvantages of astigmatism and the preemption of a large laboratory area. Two adaptations of the Paschen mount have continuing currency. The *Eagle mount* (Fig. 5.20*c*), simplest of all, takes advantage of the blaze but is highly astigmatic. Below about 500 A, consideration of the great effective reflectance at high θ_{in} dictates the use of the *grazing-incidence mount* (Fig. 5.20*d*), in which θ_{in} is within a few degrees of 90°. The grazing-incidence mount was first used at Chicago, after A. H. Compton and Doan's pioneer work on grating diffraction of X rays, by Osgood and by Hoag, and was developed by Siegbahn and his students, notably Edlén, into the medium through which have come almost all the fine details known about radiation in the vacuum region.

In one form of *Wadsworth stigmatic mount* (Fig. 5.20*e*) S is at the principal focus of a concave mirror M, whence parallel light falls upon G, which can rotate keeping its normal at a point P_0 of P, while the distance GP remains at $GP_0 = r/(1 + \cos\theta_{\text{in}})$, where r is the radius of curvature of the grating and θ_{in} is calculated from (5.48) with $\theta_{\text{out}} = 0$. The curvature radius ρ of P follows the *Shenstone relation*

$$\rho = \frac{r\cos^2\theta_{\text{out}}}{\cos\theta_{\text{in}} + \cos\theta_{\text{out}}} \tag{5.50}$$

In the *Hulthén-Lind double-pass stigmatic mount* with parallel light striking a plane mirror (Fig. 5.20*f*), if $SG = r/(1 + \cos\theta_{\text{out}})$, where r is the radius of curvature of the grating and θ_{out} is a solution of (5.25*b*) with $\theta_{\text{in}} = 0$, then a spectrum is in focus about

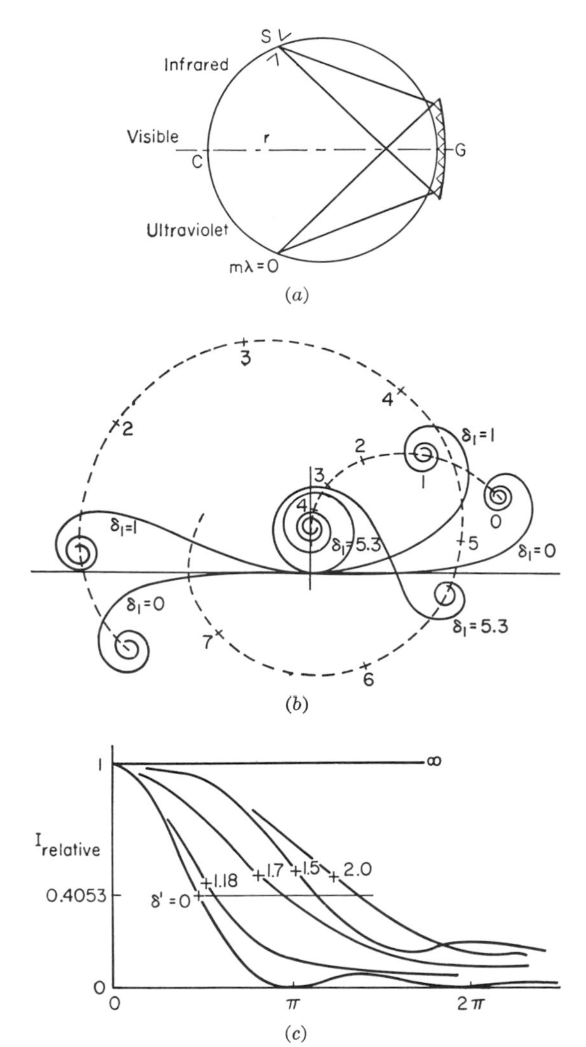

FIG. 5.19. Concave grating. (*a*) Geometry. (*b*) Vector amplitude diagram, analogous to Fig. 4.2*a*, but for many elements (grooves) on a spherical surface; δ_1 is a generalized coordinate proportional to displacement along the grating. (*c*) Intensity, analogous to left-hand part of Fig. 4.2*f* and more closely to Fig. 3.1; δ' is proportional to the grating width and to the magnitude of the fourth-order aberration at unit width; the $\delta' = 0$ curve is identical with Fig. 3.1, and the + marks indicate the resolution half width for several values of δ'.

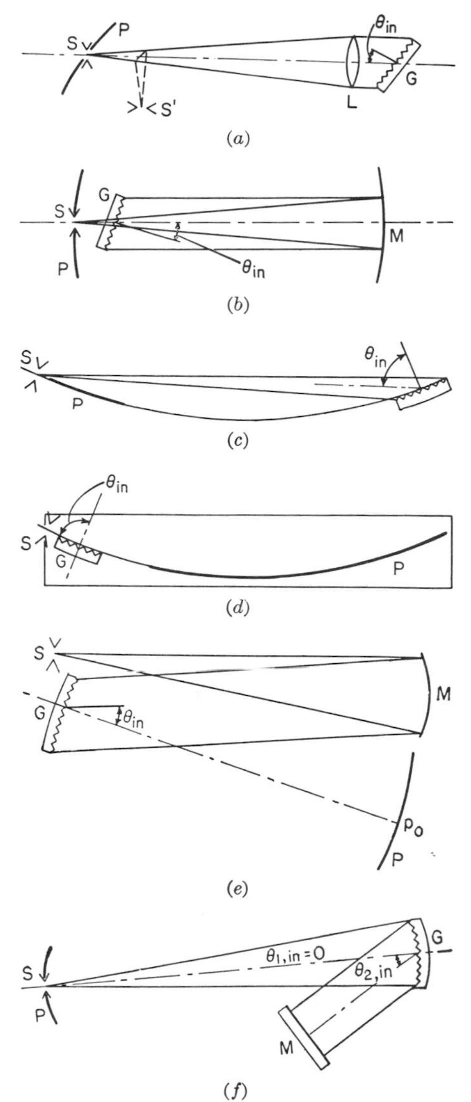

FIG. 5.20. Grating mounts: (*a*) and (*b*) Plane grating; (*c*) to (*f*) concave grating. (*a*) Littrow. (*b*) Modified Ebert-Fastie. (*c*) Eagle. (*d*) Grazing incidence. (*e*) Typical Wadsworth. (*f*) Typical Hulthén-Lind double pass.

S with twice the dispersion and twice the theoretical resolving power of the corresponding Wadsworth mount.

In all the stigmatic mounts the overlapping of orders may be avoided by a weak cross dispersion, such as that of a thin-wedge prism, usually placed before the slit with a field lens at the slit, but sometimes inserted near the grating when the beam there is collimated.

Grid Spectrometer. A large increase in flux-handling capacity without loss in resolution can sometimes be realized by replacing the slits of a grating monochromator by grids consisting of opaque and transparent areas carefully patterned to have large transmittance only when the entrance grid image is coincident with the exit grid. The increased solid angle subtended by the open areas of the grid, as compared with a slit, accounts for the flux gain. Grids may also be employed in spectrographs, in which case the developed photographic plate must be scanned with the grid to retrieve the information.

Zone Plate. The radii of the boundaries of a Fresnel zone (Fig. 5.6) are approximately $[(m - 1)b\lambda]^{1/2}$ and $[mb\lambda]^{1/2}$. An ordinary *zone plate* is a plate with alternately opaque and transparent regions bounded by concentric circles with radii proportional to the square roots of successive integers. A zone plate with N regions, employed in such a geometry that for certain monochromatic light its regions are successive zones, yields an amplitude N times, and an intensity N^2 times, as great as if the plate were not present (Fig. 5.21*a*). A *phase-reversal zone plate,* suggested by Rayleigh and developed by Wood, is transparent throughout but in alternate zones has an extra optical thickness of an odd half-integral number of wavelengths, thus reversing instead of suppressing the displacement from alternate zones, and ideally producing an intensity $4N^2$ times as great as if the plate were not present (Fig. 5.21*b*). Zone plates practical for laboratory use can be made with some tens, but hardly with hundreds, of zones. The peculiar properties of a zone plate extend a limited distance away from the axis. The plate has focal properties obeying the lens equation (Part 6, Chap. 2) and multiorder chromatic properties somewhat like those of a diffraction grating. A zone plate is not strictly an equal-amplitude instrument, but (5.16) shows it to be very nearly one.

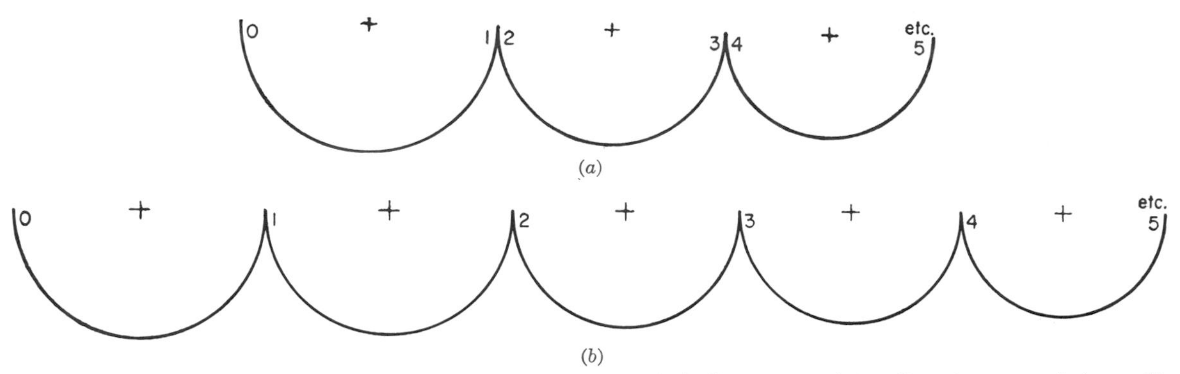

FIG. 5.21. Zone-plate vector-amplitude diagrams; see Fig. 5.7. (*a*) Ordinary zone plate, alternate zones missing. (*b*) Phase-reversal zone plate, alternate zones reversed.

Grating Grooves as Resonators. Smith and Purcell have noticed that the passage of a high-velocity electron beam across the surface of a grating, very close to the grooves, with velocity v directed toward $\theta = +\pi/2$, produces radiation, polarized with the electric vector normal to the grooves, satisfying the relation

$$m\lambda = d(cv^{-1} - \sin \theta_{\text{out}}) \tag{5.51}$$

where c is the velocity of light.

At least superficially similar is an old discovery of Wood, who found that in certain gratings, ruled in soft metal and probably possessing very narrow, fragile ridges pressed above the general level, the spectrum of a continuum of light polarized normal to the grooves showed an intensity anomaly. The anomaly occurs at wavelengths that, in higher orders, just graze the grating ($\theta_{\text{out}} = \pi/2$); for light just able to emerge in the higher order the spectrum is anomalously dark, changing to anomalously bright for slightly greater wavelengths, and the opposite anomaly occurs in the central image. According to Wood, Rayleigh attributed the phenomenon to resonance.

7. Geometrically Degraded Amplitude Multibeam Interference

Fabry-Perot, Lummer-Gehrcke Interferometers. The *Fabry-Perot interferometer* (Fig. 5.22*a*) consists of two plates with facing, partly reflecting, extremely flat surfaces M, adjusted to precise parallelism for producing Haidinger fringes. The adjustment is usually made with the aid of three pairs of accurately ground contact platforms, spaced at $2\pi/3$ in azimuth on a replaceable tubular *etalon* ("standard," spacer) made of fused silica or invar (36 per cent nickel steel) to minimize temperature effects. The range of usual spacer, or gap, values t extends approximately from 10^{-1} to 10 cm. The plates, which may be of glass, crystal quartz sliced normal to its optic axis, or fused silica, are slightly wedge-shaped ($\sim 10^{-2}$ radian) to displace unwanted reflections. The outside faces may be of ordinary optical quality. In the traditional form the etalon is held in a tubular case by three strong springs, used for achieving and maintaining parallelism, but in the improved form of Williams and of Meissner the etalon is in optical contact with the plates and no strong spring is required. In some recent mountings parallelism is maintained by a servo system controlled by Brewster fringes produced by a light beam double-passed through the edges of the interferometer. The very first Fabry-Perot had a movable plate.

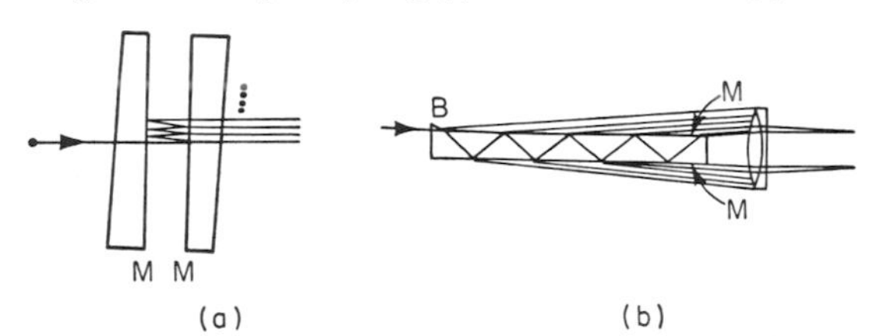

FIG. 5.22. Geometrically degraded amplitude multibeam interferometers. (*a*) Fabry-Perot; faces M are coated for high reflectance, slight transmittance. (*b*) Lummer-Gehrcke.

While the Fabry-Perot has had important uses in metrology, for instance, in the comparison of the former metallic meter with a standard spectrum line, now that the meter is defined in terms of an atomic spectrum line, it is as an interference spectrograph that it enjoys its widest use.

Suppose that the surfaces have equal reflectance R and equal transmittance T. Of the light impinging upon the plates at approximately normal incidence a fraction T^2 passes directly through, but for high R a large portion makes several to-and-fro trips within the gap, and some of it leaks out in successive interfering beams to join the first, as already illustrated in Fig. 5.11*b* and vectorially in Fig. 5.12*h*, *i*.

A strong factor in the outstanding importance of the Fabry-Perot as an interference instrument is the possibility of attaining very sharp fringes, and consequently great resolving power, through the use of surfaces with high reflectance [cf. (5.37)]. The practical importance of this R-dependence was appreciated by Bouloch in 1893, a few years before the work of Fabry and Perot.

Connes has described a *spherical-surface Fabry-Perot interferometer* with two concave surfaces of equal radius r, separated along their mutual axis by one radius. An important advantage is that, since the only alignment consideration is that the center of each sphere lie in the opposite sphere, the instrument is much easier to adjust for large spacers than the conventional Fabry-Perot. In a first-order approximation, the path difference is the same for all rays, so that the interference condition does not depend on the angle, as with the ordinary Fabry-Perot interferometer; for the spherical-surface instrument, $n\lambda = 4r$. Actually, because of third-order aberrations, there is an angular dependence, so that rings are formed, but the central one, which is the only one useful for spectroscopy, subtends a much larger angle for a given $\Delta\lambda$ than the ordinary Fabry-Perot. The absence of useful rings means that the spherical Fabry-Perot cannot be used photographically in the same way as a plane Fabry-Perot; however, it can be usefully employed in photoelectric scanning spectrometers (see below), in which only the central ring is used. Aberrations also limit the useful diameter of the spherical interferometer so that, in spite of the large solid angle subtended by the central fringe, the spherical Fabry-Perot has a higher flux-handling capacity than a plane Fabry-Perot only when the separation of the spherical plates becomes greater than about one-third the useful diameter of the plane plates. For practical cases this corresponds to a spectroscopic resolving power of about 4×10^6. Because of its advantages at large plate spacings, the spherical Fabry-Perot has found great currency as a laser resonator.

The *Lummer-Gehrcke plate* (Fig. 5.22*b*), now rarely used, is simpler than the Fabry-Perot system in that the Haidinger fringes are produced from the two faces of a single piece of glass or quartz, so that there is no occasion for gap adjustment. The serious limitation imposed by its fixed spacing is partly overcome by accepting the light in a system whose axis is parallel, instead of normal, to the plane surfaces. Thus the parallel light, which is introduced through an auxiliary prism B, has an almost grazing emergence angle. The limited length of the plate usually curtails the effective number of reflections, N_{eff}, in the Lummer-Gehrcke, in contrast with the Fabry-Perot, where for fringes near the center N_{eff} is limited only by (5.30) (cf. Table 5.1).

Fabry-Perot Emission Spectroscopy. The spectrum lines arising from a Fabry-Perot interferometer are ring-shaped Haidinger fringes satisfying (5.26*b*); with a camera lens of focal length f the Fabry-Perot dispersion and fringe location near the center are

$$\frac{dx}{d\sigma} \doteq \frac{f^2}{x\sigma_{\text{center}}} \tag{5.52a}$$

$$x \doteq 2^{1/2} f \left(1 - \frac{\sigma_{\text{center}}}{\sigma}\right)^{1/2} \tag{5.52b}$$

where x is the radial distance along the fringe pattern, measured from the center as zero, σ is $1/\lambda_0$, and σ_{center} is $m/2nt$. The free spectral range $\Delta_F\sigma$ is

$$\Delta_F\sigma = \frac{1}{2nt\cos\theta} \doteq \frac{1}{2t} \tag{5.53a}$$

where n is the index of the gap medium, usually air, and the resolving power [cf. (5.37)] is

$$\left|\frac{\lambda}{\Delta_r\lambda}\right| = \left|\frac{\sigma}{\Delta_r\sigma}\right| = 2t\sigma N_{\text{eff}} \doteq \frac{2\pi t\sigma R^{1/2}}{1 - R} \tag{5.53b}$$

In (5.52) and (5.53) the slight distinction between θ_{FP}, the inclination of the ray in the gap, and θ_H, the inclination with respect to the center of the Haidinger fringe system after the passage of the light through the wedge-shaped plate, is neglected.

In order to avoid the overlapping of rings throughout the spectrum, for spectroscopic use the etalon is usually placed in series ("crossed," i.e., having mutually normal dispersions) with a stigmatic auxiliary prism or grating spectrograph, the etalon being either before the slit, with the fringes focused on the slit, or in a beam of parallel light within the auxiliary spectrograph. When the fringe system is centered, fringe diameters may be measured along the lines. For the analysis of a very small region the etalon may be placed after the auxiliary dispersion, which then need not be stigmatic; only one line then has centered fringes.

With the improved quality of narrow-pass filters it has been possible to replace the auxiliary spectrograph by filters, in cases where the spectral region under study is not too line-rich.

When spectroscopy is carried out nonphotographically, as, for instance, by directly recording with a photomultiplier or photoconductor, usually the photographic plate is replaced by a slit and the spectrum is scanned by sweeping the spectrum with respect to the slit; the slit may be moved or the auxiliary disperser rotated. In high-resolution Fabry-Perot analysis of a narrow-line complex the scanning problem is especially simple because of the short spectral range of the scanning path. The technical problem is usually that of sweeping the two spectrographs simultaneously (i.e., keeping the wave number of maximum intensity the same) across the slit in a known way (e.g., linearly) with respect to the time, as suggested by Jaffé and carried out by Jacquinot, by Rank, and by Hirschberg and Mack. Often the sweeping is done geometrically. Sometimes the Fabry-Perot is placed in a sealed chamber and swept by control of the pressure of air or another gas, e.g., carbon dioxide, while the gap geometry stays fixed. If the index of the gas at 1 atm pressure is n_{gas}, the wave-number change $\Delta_P\sigma$ occurring for a given θ when the chamber is evacuated from P atm is $-\sigma P(n_{gas} - 1)$ and the number of orders swept out, $\Delta_P m$, is $mP(n_{gas} - 1)$. If two interference instruments, such as two Fabry-Perots or a Fabry-Perot and a grating, are placed in communicating pressure chambers and put into adjustment, since the geometrical conditions for both instruments, (5.25*b*), (5.26*b*), depend only on $m\lambda_0/n$, the instruments stay in synchronism as the pressure is varied. In scanning applications the Fabry-Perot is best employed by restricting the incident light so that only the central fringe is filled, i.e., to a bundle centered about normal incidence of solid angle $8\pi/R_0$, where R_0 is the resolving power given by (5.53*b*).

Hirschberg has suggested a multichannel Fabry-Perot which eliminates the need for scanning when evaluating line breadths. If a system of concentric zones with equal area is placed in the focus where the Fabry-Perot fringes are formed, the zones will intercept equal increments of the spectrum. By inclining each zone mirror in a different azimuth so that the light collected falls on separate detectors, an analysis of the spectrum is obtained. This method has essentially unlimited time resolution and has proved very useful in the analysis of rapidly changing sources.

The Fabry-Perot comparison of wavelengths has been developed into one of the most precise kinds of physical measurement. Relative values have been published implying as many as 10 significant digits. The principal potential source of systematic error in the best work lies in the dispersion of the phase shift accompanying reflection. In metallic coatings such a shift is well understood in principle but difficult to measure quantitatively. Dielectrics, themselves are subject to no such characteristic shifts, but they are dispersive media and even if they were not, the phase of the reflected wave would depend upon the exact thicknesses of all the layers. Their indices in thin evaporated layers are probably variable and certainly not well known. Even if the thicknesses were ideal to produce no shift at one wavelength, there would be shifts at all others (cf. Fig. 5.24*b* and the accompanying example below). Studies of these shifts have hardly begun. However, by using a slightly modified data-reduction method it has been possible to make accurate wavelength measurements without exact knowledge of phase shifts.

Flatness has attained new importance for Fabry-Perot plates since the advent of high reflectance. Crude considerations show that the departure of a plate from flatness harms the fringe quality if it is greater than about $\lambda/4N_{eff}$. Now with $N_{eff} \sim 10^2$ to 10^3 possible in multilayers, departures from flatness on the order of a few times the interatomic instance may affect plate performance. (Table 5.1.)

Compound Fabry-Perot Interference Spectroscope. The resolving power (5.37) of a geometrically ideal Fabry-Perot is proportional to the order number, i.e., to the gap value; the free spectral range (5.53), however, is inversely proportional to the gap value. Consequently the necessity of avoiding the overlapping of orders often limits the usable resolution. A means of attaining the resolving power of a long gap with the free spectral range of a short gap is the use of the *compound Fabry-Perot interference spectroscope*, which consists of two Fabry-Perot interferometers in series (Fig. 5.23*a*), the gaps of which may be in a simple ratio, such as 1:3 or 2:7. Originated by Houston, it has been used by Meissner, Jackson, Jacquinot, and Kuhn, and others. The gap ratio may be adjusted by putting one of the instruments in a controlled pressure chamber. When one gap is precisely an integral multiple, say, a factor K, of the other, Brewster fringes may be observed between the light reflected within one gap and that reflected K times at a surface of the other. Interference effects between the surfaces of the first and of the second instrument may be eliminated or made negligible by slight tilting or by putting the two $\frac{1}{2}$ m or so apart or by placing one internal and the other external to the auxiliary spectrograph. If the properties of the two instruments separately are identified by ′ and ″, and if $t'' = Kt'$, where K is exactly an integer, then the phases are related, $\delta'' = K\delta$, and, if reflections between the instruments can be neglected, the relative intensity transmitted by the compound instrument is, following (5.28),

$$\frac{I}{I_0} = \frac{(T')^2(T'')^2}{\{[(1 - R')^2 + 4R' \sin^2 (\delta'/2)] [(1 - R'')^2 + 4R'' \sin^2 (K\delta'/2)]\}} \tag{5.54}$$

Figure 5.23*b* shows the fringe shapes from t', t'', and the compound instrument, for $K = 3$. Principal maxima occur only when both $\sin^2$ terms in the denominator are zero; this circumstance gives the compound instrument its advantage in range. A serious disadvantage of the compound instrument is that minor maxima or "ghosts" occur when only one of the $\sin^2$ terms is zero, corresponding to a fringe from the wider gap. The presence of two factors in the denominator makes the fringes sharper than the fringes from a simple Fabry-Perot with the wider gap. Even two identical instruments in series ($K = 1$) have considerably better resolving power than one. Meissner gives for the relative advantage of the compound instrument in this respect over that of a simple instrument with the wider gap, if $T' = T''$ and $R' = R''$, 1.61, 1.25, 1.12, and 1.07, for $K = 1$, 2, 3, and 4, respectively. Until recently only very bright lines have been susceptible to analysis with the compound Fabry-Perot, but the advent of multilayers with high R and with T equal to almost $1 - R$ (see below) has permitted high-resolution studies of very weak lines, using compound instruments with a transmission near 0.5 and ghost rejection of about 3×10^{-3}.

The gap ratios in a compound instrument may also be adjusted in a vernier ratio, i.e., nearly equal, and with the high values of finesse now available peak suppression may be achieved sufficiently far out so that an interference filter may be used to realize suppression of all unwanted peaks. The PEPSIOS spectrometer employs three Fabry-Perots in series with an interference filter to achieve isolation of a single peak and suppression of all others to below 10^{-3} times the main peak transmission. Figure 5.23*c*, which gives results of an experimental test, illustrates the basis of the PEPSIOS spectrometer.

Fabry-Perot Absorption Spectroscopy. If a Fabry-Perot and an auxiliary prism or grating spec-

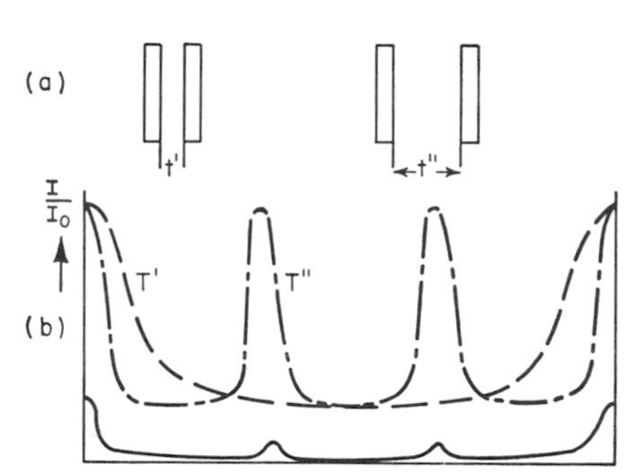

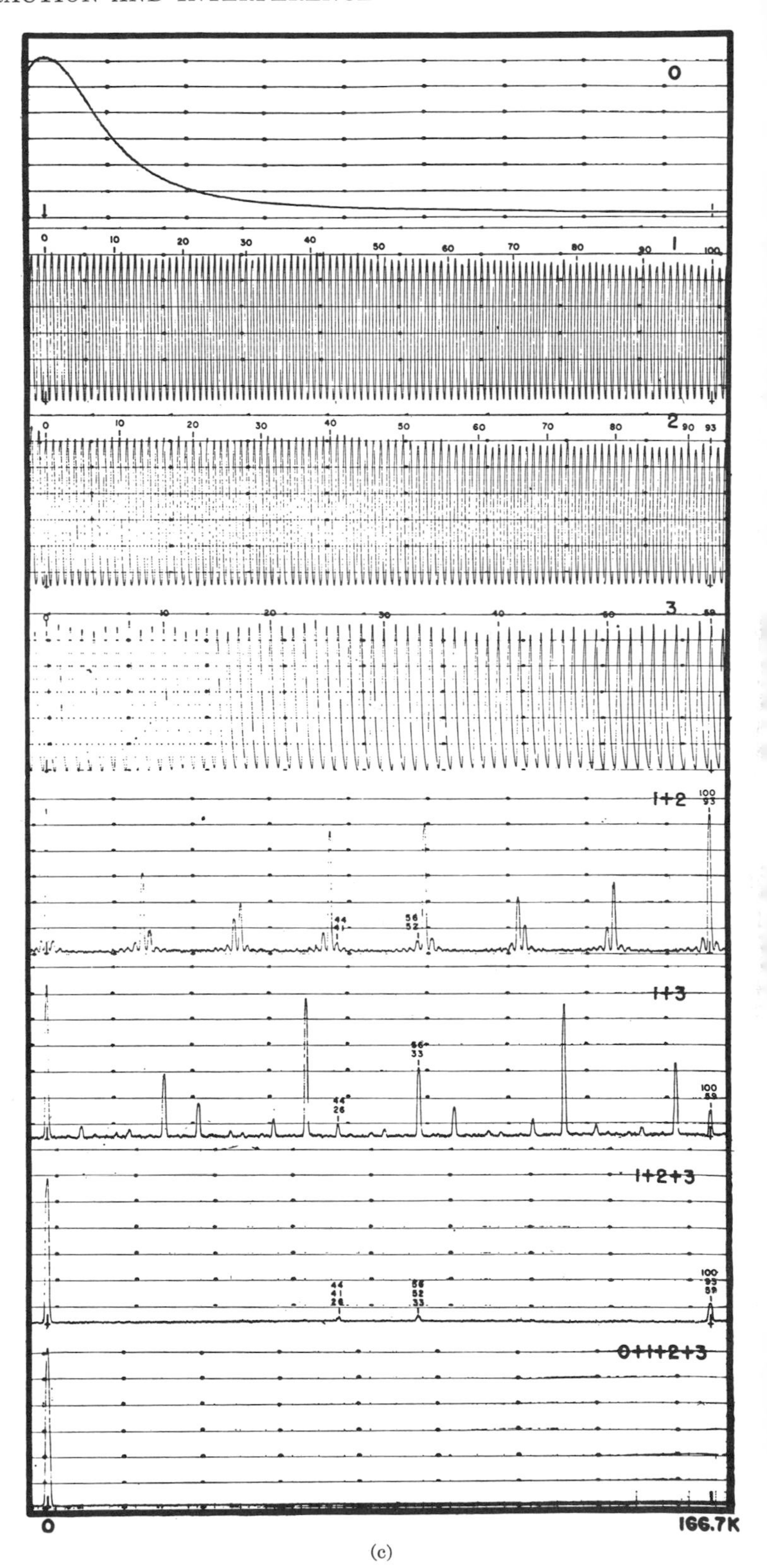

(c)

FIG. 5.23. Compound Fabry-Perot interference spectroscope, case $K = 3$. (*a*) Geometry. (*b*) Intensity, qualitative. (*c*) Monochromatization in the PEPSIOS spectrometer by suppression of parasitic light. The source is a continuum and the light arbitrarily selected for "σ_0", that is, for transmission, is yellow light in the neighborhood of the sodium resonance doublet. The light from the source was sent through a filter to make the amplitudes of the peaks approximately uniform. The large digits stand for elements in the train: 0 for the interference filter, with $\delta_{1/2}\sigma = 30$ K = 11 A; 1 for an etalon with $t_1 = 3.000$ mm; 2 for an etalon with $t_2 = 0.930\ t_1 = 2.790$ mm; 3 for an etalon with $t_3 = 0.5897\ t_1 = 1.769$ mm. The small digits stand for interference order numbers, measured from σ_0. For each trace, after an intial tuning to maximize the transmittance of σ_0, the pressure in each chamber was kept static and the transmitted light analyzed by means of a grating monochromator; since the grating has an inferior resolution, the traces show the positions, but not the shapes, of the peaks; they are not PEPSIOS spectrometer traces. In trace "1 + 2," the near equality in amplitude between orders 0, 0 and 100, 93 and the symmetry of the intervening ghost pattern show that the design spacing ratio of 100:93 was effectively attained. The asymmetry of trace "1 + 3," on the contrary, shows a departure from the ratio 100:59, and a study of the asymmetry of the ghost pattern affords a basis for the quantitative determination of the departure. Traces "1 + 2 + 3" and "0 + 1 + 2 + 3" indicate that effective monochromatism has been attained. The improvement in the ghost system arising out of the departure of t_3/t_1 from exactly 0.5900 came about in the first instance by accident. The reproduction of Fig. 1 is defective, especially in that it loses certain peaks near order 0 and order 14 of the fifth line ("1 + 2") and incompletely reproduces certain other material. Line "1 + 2" is practically symmetric, left to right, and every line has a peak of approximately the same size in order 0.

trograph with a narrow slit are crossed and exposed to a continuous spectrum, a system of continuous fringes is swept out, each fringe being heterochromatic, with a shape (supposing the image slit is straight) dependent upon the relative dispersion of the crossed elements. The fringes are generalized Edser-Butler fringes; in the original fringes of Edser and Butler each fringe was the central ($\cos\theta \cong 1$), approximately monochromatic part of a fringe of a similar system, produced by placing a silvered air gap just before a prism spectrograph slit.

The fringes are parabolas (modified by any wave-number nonlinearity in the auxiliary dispersion), convex to the red of the auxiliary dispersion. Consider the case where the auxiliary dispersion is that of a grating, and let the fringe symmetry axis, parallel to the grating dispersion, be the y axis (Fig. 5.24a).

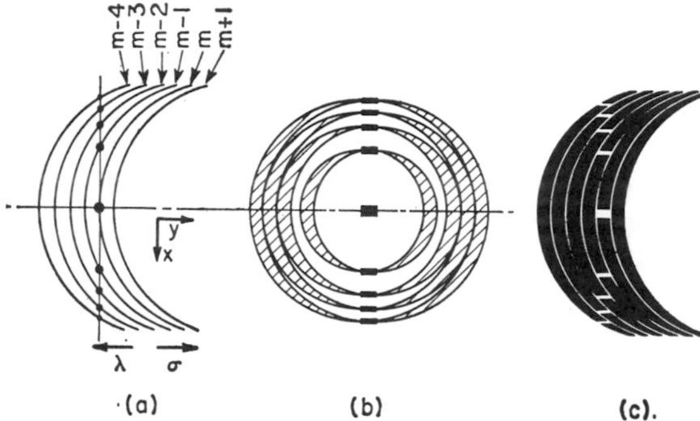

Fig. 5.24. Fabry-Perot absorption spectroscopy. (a) Intersection of white-light Fabry-Perot fringes with the locus of constant σ from the auxiliary dispersion. (b) Effect of slit width upon a monochromatic-light fringe system; the cross hatched area shows the spread-out rings (complete Haidinger fringes) of which only the diametral region, with sharpness almost unaffected by the spreading, occurs within the image slit. (c) Monochromatic absorption (white) in region of continuous emission (black). The pattern of one absorption line in each generalized Edser-Butler fringe is geometrically identical with that of an emission line, but with the reverse intensity pattern.

The slope is the ratio of the grating image distance, v_g, times the angular dispersion of the grating (5.48), to the linear dispersion of the Fabry-Perot (5.52a), modified by any magnification introduced between the two dispersions. If the dispersions are in series, with the Fabry-Perot first, then so long as $x^2 \ll f^2$,

$$\frac{dy}{dx} = \frac{|m_g|v_g x}{\sigma M f_{FP}{}^2 d_g \cos\theta_{\text{out},g}} = \frac{2|m_g|v_g n_{FP} t_{FP} x}{d_g M_g \cos\theta_{\text{out},g} m_{FP} f_{FP}{}^2} \qquad (5.55a)$$

$$y - y_{\text{vertex}} = \frac{|m_g|v_g x^2}{2\sigma M f_{FP}{}^2 d_g \cos\theta_{\text{out},g}} \qquad (5.55b)$$

where M is the magnification of the auxiliary spectrograph and the meaning of the subscripts is evident. If the arrangement is internal, $v_g = f_{PF}$, and

$$y - y_{\text{vertex}} = \frac{|m_g|x^2}{2\sigma f d_g \cos\theta_{\text{out},g}} \qquad (5.55c)$$

Fringes of successive order are spaced in the y direction at intervals corresponding in grating dispersion to $(2nt)^{-1}$, the free spectral range of the Fabry-Perot (5.53a). In the x direction (σ = const) the spacing of successive fringes is identical with that of the monochromatic-light fringes in the case of an emission-line spectrum (Fig. 5.24b).

If the slit were to be widened until its width corresponded, in grating dispersion, to $(2nt)^{-1}$, the free spectral range of the Fabry-Perot, the whole focal plane would be covered, not with an ordinary continuous spectrum, but with parabolic segments in each of which σ would be a function of x alone. In practice the slit is made somewhat less wide than this so that the separate orders remain distinct; the slit width is not critical, but an upper limit may be set by scattering. The absence of radiation of a certain wave number produces a straight pattern of blanks (Fig. 5.24c) in successive segments that is geometrically identical with the bright-fringe pattern (5.26b) of an emission line. In any practical system, while the resolution is proportional to the gap t, the t value is limited by the requirement that the prism or grating be able to resolve the Fabry-Perot free spectral range $(2nt)^{-1}$.

Spectrograph Calibration with Fringes. Rubens in 1892 used (unsilvered) air-gap fringes to calibrate a prism in the infrared. Edser and Butler used a silvered gap in the visible and ultraviolet in 1898. The scheme is still useful in regions where standards are scarce. Recently, closely spaced Michelson or Fabry-Perot fringes have been used in conjunction with scanned Fabry-Perot spectrometers to obtain high precision in optical measurements of fine and hyperfine structures at spectral resolutions in the neighborhood of 10^6.

Enhanced and Reduced Reflection: Introduction. The normal reflectance R of a boundary between two dielectrics of refractive index n_1 and n_2 is given by Fresnel's equation [Part 6, (1.39)],

$$R = \left(\frac{n_1 - n_2}{n_1 + n_2}\right)^2$$

The reflectance is 4 per cent for an air-glass boundary when the glass has an index of 1.5; common optical glasses have indices a little greater than this. In optical systems where many interfaces exist the loss of light in the transmitted beam may be serious; for instance, in a system of eight separated glass components the loss may be more than half.

Toward the end of the nineteenth century several observers, notably Taylor, noticed that lenses such as telescope objectives, upon being exposed to the weather and becoming tarnished with age, exhibited an improved performance due to a reduction of reflections. During the First World War, Kollmorgen was able to produce this tarnish by chemical means. In the 1930's Bauer measured the thickness of thin halide films by their reflection maxima and minima; Pfund produced a high-index dielectric film on glass that enhanced the reflectance of the surface by constructive interference; Strong used destructive interference from a film to reduce the reflectance; and Cartwright and Turner developed the *dielectric multilayer* to enhance the effect of a single layer. Since then, such coatings have been widely produced in quantity, both for reducing and for increasing the reflectance. Extremely low absorption is one of the useful properties of partly transmitting mirrors produced from dielectric films.

For an approximate picture of the process that takes place at such surfaces, multiple reflections may be neglected. Consider a thin film of dielectric of index n_2 and thickness t_2 overlying a glass substrate of index n_3 (Fig. 5.25*a* with $M = 1$; cf. Fig. 5.15*c*).

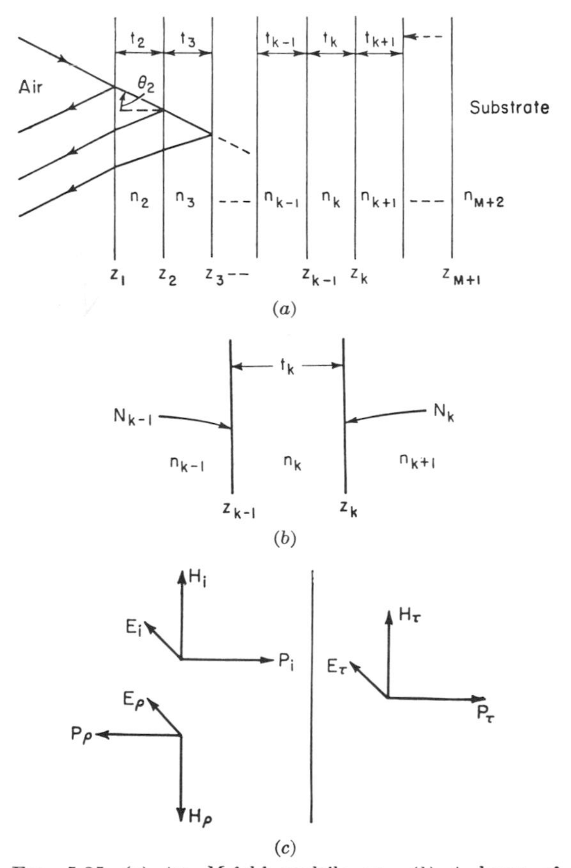

FIG. 5.25. (*a*) An *M*-fold multilayer. (*b*) A layer of index n_k, showing input and output admittances N_{k-1} and N_k, respectively. (*c*) A boundary, showing the preservation of the right-handedness of the system *E*, *H*, *P* (electric, magnetic, and Poynting vectors) on reflection and transmission. The subscripts i, ρ, and τ indicate incident, reflected, and transmitted waves.

If light of amplitude a_0 falls on the surfaces so that the ray reflected in the medium is at an angle θ, there are two reflected rays, of scalar amplitude a_1 and a_2.

If $n_2 < n_3$, there is a phase change of π at each interface, so that the condition for destructive interference in the reflected beam is, following (5.43*d*),

$$2n_2t_2 \cos \theta = (m + \tfrac{1}{2})\lambda_0 \tag{5.56}$$

where *m* is an integer.

Treatment of nonnormal incidence throughout would complicate the equations without adding greatly to their usefulness. From this point on, normal incidence is assumed. When *m* equals zero, n_2t_2 is seen to be equal to $\lambda_0/4$. Single-layer antireflecting films have a thickness of $\lambda_0/4n$, where n is the index of refraction of the film. For complete cancellation of the reflection, a_1 should equal a_2 (cf. Fig. 5.2). Taking multiple reflections into account leads to the condition that the reflectance at the interfaces, considered separately, should be equal. Fresnel's formula then yields $n_2 = n_3^{1/2}$. Since most common glasses have n about 1.5, the antireflecting film should have an index of $1.5^{1/2}$ or 1.22 to satisfy this condition exactly. No suitable material has been found with so low an index; in practice the reflectance can be reduced only to about 1 per cent with a single film. Also, since the film can be a quarter wave thick at only one wavelength, greater reflection losses occur at other wavelengths.

If, in Fig. 5.25*a*, $n_2 > n_3$, there is a phase change of π at interface z_1, as before, but no change at interface z_2. Here, the situation leads to constructive interference for the reflected light and hence to enhanced reflection. The beam amplitudes a_1 and a_2 add, and an increase in the value of n_2/n_3 will give a correspondingly increased reflectance. Titanium dioxide with $n = 2.4$ and zinc sulfide with $n = 2.30$ are commonly used to provide enhanced reflection, the former being applied by a chemical process and the latter by evaporation.

In Fig. 5.25*a*, if the n_{even} are high indices of refraction and the n_{odd} lower indices, it will be seen that, if $n_2t_2 = n_3t_3 = n_4t_4 = \cdot \cdot \cdot = (m + \tfrac{1}{2})\lambda_0/2$, the amplitudes will add and the reflectance will be enhanced. As the number of such layers, *M*, is increased, the reflectance is enhanced, and to obtain full reinforcement, *M* should be odd, with one more high- than low-index layer. Multilayers have been used to attain reflectances of the order of 98 per cent.

For odd *M*, (5.63), below, may be used to calculate the maximum reflectance for nonabsorbing dielectrics. Values are shown in Fig. 5.26. In practice, the choice of coating substances depends upon the circumstances, and especially upon the pertinent spectrum region. Penselin and Steudel, and especially Baumeister and Stone, have shown that very good reflectance (0.95 per cent) can be attained over a wide spectrum region by the use of somewhat different thicknesses for the different layers. Table 5.2 shows some properties of several media used or proposed for coatings.

Enhanced and Reduced Reflection: Theoretical Details. For a closer examination of the behavior of multilayers, taking multiple reflection into account, consider Maxwell's equations [Part 6, (1.7)] for homogeneous charge-free dielectric regions. For plane-polarized waves, traveling in the *z* direction,

$$\frac{\partial E_x}{\partial z} = -\frac{\mu}{c}\frac{\partial H_y}{\partial t} \qquad \frac{\partial H_y}{\partial z} = -\frac{\epsilon}{c}\frac{\partial E_x}{\partial t} \tag{5.57}$$

For simplicity the subscripts on *E* and *H* are henceforth omitted, and absorption is neglected.

A convenient concept, borrowed from transmission-line theory by Muchmore and others, is the *wave admittance N*, taken here as the ratio of *H* to *E*. In a given layer, for example the *k*th layer of Fig.

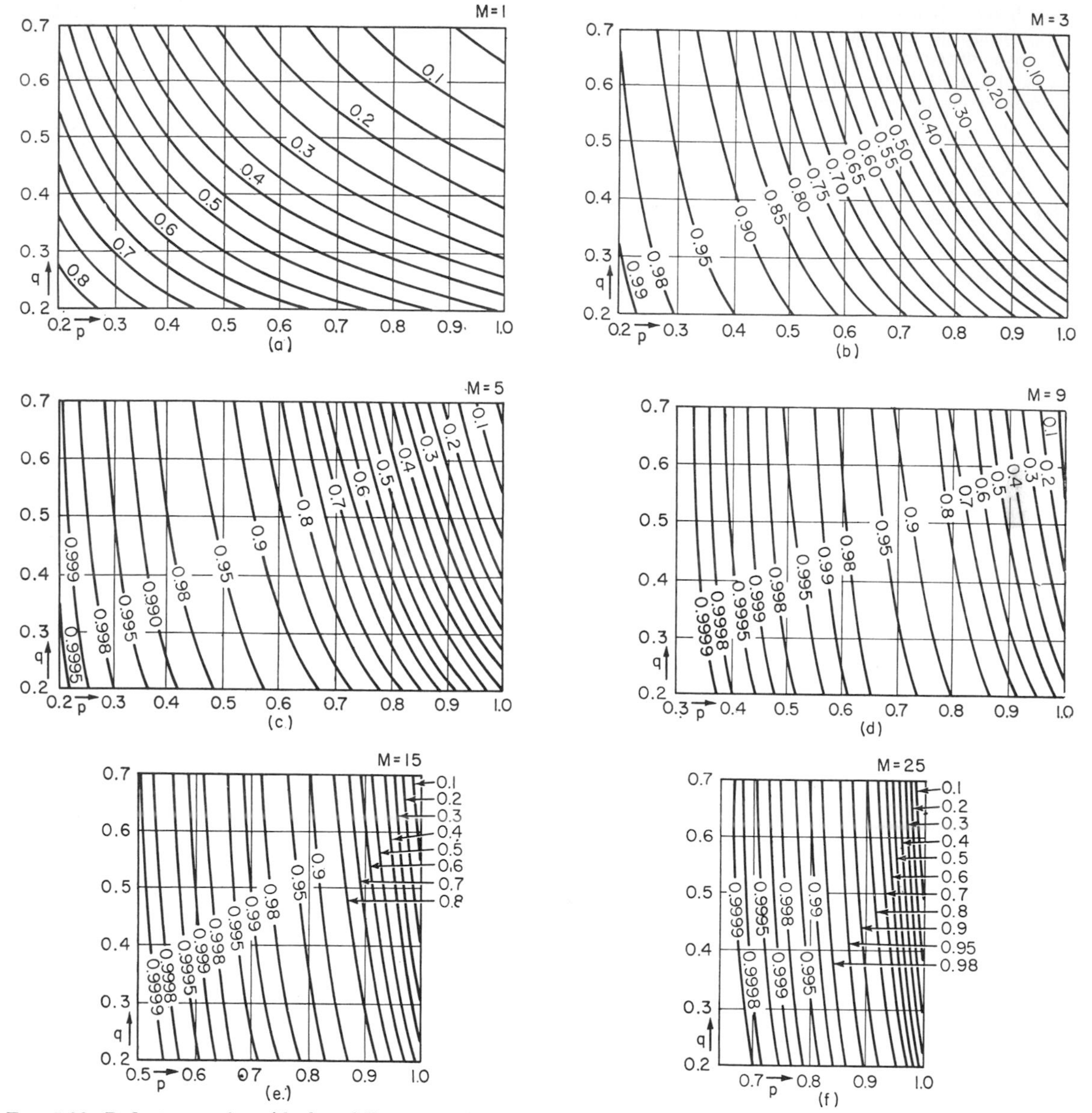

FIG. 5.26. Reflectance of an ideal multilayer consisting of an odd number, M, of perfect, nonabsorbing quarter-wave layers with alternately high index, n_H, and low index, n_L, there being one more high- than low-index layer, laid upon a substrate with index n_g, in a surrounding medium with index n_0. The coordinates are $p = n_0n_g/n_Ln_H$ and $q = n_L/n_H$. (a) $M = 1$. (b) $M = 3$. (c) $M = 5$. (d) $M = 9$. (e) $M = 15$. (f) $M = 25$.

5.26b, the input admittance N_{k-1}, the index of refraction of the dielectric, n_k, and the output, or load, admittance N_k are related by the following expression, which is derived from the above equations by assuming a harmonic time dependence of E and H:

$$N_{k-1} = n_k \frac{Q_k + i \tan \delta_k}{1 + iQ_k \tan \delta_k} \tag{5.58a}$$

where N_{k-1} is the input admittance of the kth layer, Q_k is N_k/n_k, the ratio of the output admittance of the kth layer to the index of refraction of this layer, and δ_k is the phase angle, $n_k\lambda_0^{-1}t_k$. In form, (5.58) is identical with the admittance equation of transmission-line theory.

Since $N_k = H_k/E_k$, (5.58a) may be written in matrix form

$$\begin{pmatrix} E_{k-1} \\ H_{k-1} \end{pmatrix} = \begin{pmatrix} \cos \delta_k & in_k^{-1} \sin \delta_k \\ in_k \sin \delta_k & \cos \delta_k \end{pmatrix} \begin{pmatrix} E_k \\ H_k \end{pmatrix} \tag{5.58b}$$

Equation (5.58a) gives the input admittance N_{k-1} for any layer if n_k, t_k, and in particular N_k, the output admittance of the layer, are known. It will be

TABLE 5.2. SOME PROPERTIES OF COMMONLY USED DIELECTRICS

	n (approx.)	Method of deposition	Advantages	Disadvantages	Source
Low-index materials:					
$3NaF \cdot AlF_3$ (cryolite)	1.35	Evaporate	Easy to evaporate	Not so hard as MgF_2, and not so stable dimensionally	Dufour, Hermansen
CaF_2 (fluorite)	1.22	Evaporate	Very low index	Very fragile	Dufour
LiF	1.29	Evaporate	Very low index	Hygroscopic	Dufour
MgF_2	1.38	Evaporate	Very hard, tenacious	Substrate must be hot (~300°C) for hard film	Dufour
NaCl	~1.5	Evaporate	Easy to obtain thick film, necessary for infrared	Water soluble	Greenler
High-index materials:					
ZnS	2.3	Evaporate	Easy to evaporate, tough and adherent	High absorption for $\lambda \lesssim 4000$ A	Abelès, Dufour, Hermansen
TiO_2	2.4–2.6	Evaporate Ti, heat in air to TiO_2; or burn a $TiCl_4$ product in air	Very tough, high index, low absorption	Complicated method makes thickness control difficult	Bunning, Haas
Sb_2S_3 (stibnite)	2.8–3.0	Evaporate	Very high index; easy to evaporate	High absorption for $\lambda \lesssim 5500$ A	Dufour
CdS	2.26–2.52	Evaporate	High index	Relatively high absorption	Hall and Ferguson
Sb_2O_3	2.1	Evaporate	Transparent in ultraviolet to $\lambda < 3000$ A	Relatively low index; soft surface	Barr and Jenkins
CsI	1.9	Evaporate	Transparent to $\lambda < 2500$ A	Fragile; relatively low index	Steudel
Ge	3.	Evaporate	Very high index in infrared	Opaque for $\lambda < 1.6\ \mu$	Heavens
Te	5.	Evaporate	Very high index in infrared; useful from 5 to 20 μ	Opaque for $\lambda < 2.5\ \mu$	Greenler
CeO_2	~2.4	Evaporate	Robust	n depends on substrate temp.; absorbs if $\lambda < 3800$ A	Rouard and Bousquet
$PbCl_2$	2.3	Evaporate	Useful to $\lambda = 3000$ in near ultraviolet		Penselin and Steudel
CsBr	~1.9	Evaporate	Useful to $\lambda = 2300$	Very fragile; must be protected	Steudel and Stotz

noticed that N_k is the input admittance of layer $k + 1$, so that an iterative process can be used to obtain the properties of a number of layers, as will be explained below. R and T, the reflectance and transmittance, can be calculated directly from Q_1.

In the iterative method it is necessary to start at the substrate and work backward to the first surface of the multilayer. This necessity arises from a fundamental asymmetry about reflection, the simplest example of which is the fact that, when radiation is reflected, an incident and a reflected ray are both on one side of the boundary, while a single transmitted ray appears on the other.

For a semi-infinite slab of dielectric separated by a plane boundary $z = z_1$ from a semi-infinite slab of another dielectric (Fig. 5.25*c*) Maxwell's equations lead to the conditions that the tangential components of **E** and **H** are continuous across the boundary. In a dielectric, the ratio of the displacements, H/E, is just n, the index of refraction. Letting H_i and E_i be the magnetic and electric displacements of the incident radiation in the body of the first dielectric, $H_i/E_i = n_k$, while just on the other side of the boundary $H_\tau/E_\tau = N_k$, where H_τ and E_τ are the transmitted displacements. Unless $n_k = N_k$, there must be a reflected wave, which may be described by H_ρ and E_ρ. From the boundary conditions,

$$H_\rho + H_i = H_\tau \tag{5.59a}$$

$$E_\rho + E_i = E_\tau \tag{5.59b}$$

$$\frac{H_\rho}{E_\rho} = -n_k \tag{5.59c}$$

The negative sign in (5.59*c*) results from the fact that only one of the two vectors **E** and **H** changes sign on reflection; if **P** is the Poynting vector, this can be seen from the requirement that the right-handedness of the system **E**, **H**, **P** is invariant upon reflection. With the *amplitude reflection coefficient* defined as $\rho = (E_\rho/E_i)_{\text{boundary}}$ consideration of (5.59) yields the result

$$\rho_k = \frac{1 - Q_k}{1 + Q_k} \tag{5.60}$$

where, as before, $Q_k = N_k/n_k$. Similarly, the *amplitude transmission coefficient* τ is defined as $\tau = (E_\tau/E_i)_{\text{boundary}}$, and

$$\tau_k = \frac{2Q_k^{1/2}}{1 + Q_k} \tag{5.61}$$

The relative intensities reflected and transmitted are the normal reflectance $R = |\rho|^2$ and the normal transmittance $T = |\tau|^2$. Closed solutions for arrays of low multiplicity have been obtained by Mooney and others.

Accurate numerical methods, borrowed from transmission-line theory, have been used by Abelès, Herpin, Osterberg, Stone, and others. The passage of the wave through each layer is represented by a 2×2 matrix operator (5.58*b*) and the Q_k for each layer, expressed as a column matrix, is transformed by the operator into Q_{k-1}. Here the iterative process can be performed as before. For purposes of computation the matrix form is often to be preferred over the repeated application of (5.58) because a symmetry of many multilayer arrays reduces the length of the computations, as pointed out by Stone. These calculations can be adapated to off-axis illumination and absorbing media.

Cotton pioneered in graphical methods of calculating the properties of multilayers. Leurgans has pointed out the usefulness of the Smith chart, already familiar to electrical engineers, an example of the use of which is given below.

Polack has proposed a transmission-line analogue computer for multilayers, using the fact that the admittance of a transmission line is analogous to the refractive index of an optical layer. MacNeille and Dixon have made an analogue computer with each section representing a layer and containing an adjustable phase shifter and attenuators, so that various conditions can be simulated by simple adjustments.

Enhanced and Reduced Reflection: Example. Consider a threefold multilayer of zinc sulfide and cryolite on a substrate of glass layers numbered air 1, sulfide 2, cryolite 3, sulfide 4, and glass 5. Call the refractive indices n_s and n_c, respectively, for the films, and n_g for the glass. Consider $n_k t_k = \lambda_0/4 = 1/4\sigma$ for a given wave number, say, $\sigma = 15{,}000$ cm^{-1}. Then $\delta_k = \pi/2$, and (5.58) assumes a particularly simple form,

$$N_{k-1} = n_k/Q_k \tag{5.62a}$$

or

$$N_k = \frac{n_{k+1}^2}{N_{k+1}} \tag{5.62b}$$

The glass substrate is analogous to an infinite transmission line; i.e., coherent reflection from its far face is neglected so that the output admittance N_k for layer 4, or N_4, is just the index of refraction of the glass, n_g. Substituting in (5.62*b*) for layer 3,

$$N_k = N_3 = \frac{n_s^2}{n_g} \tag{5.62c}$$

Similarly $N_2 = n_3^2/N_3 = n_c^2 n_g/n_s^2$; and finally $N_1 = n_s^4/n_c^2 n_g$. For the air the output admittance is N_1, and since $n_1 = 1$, Q for the first face, i.e., for

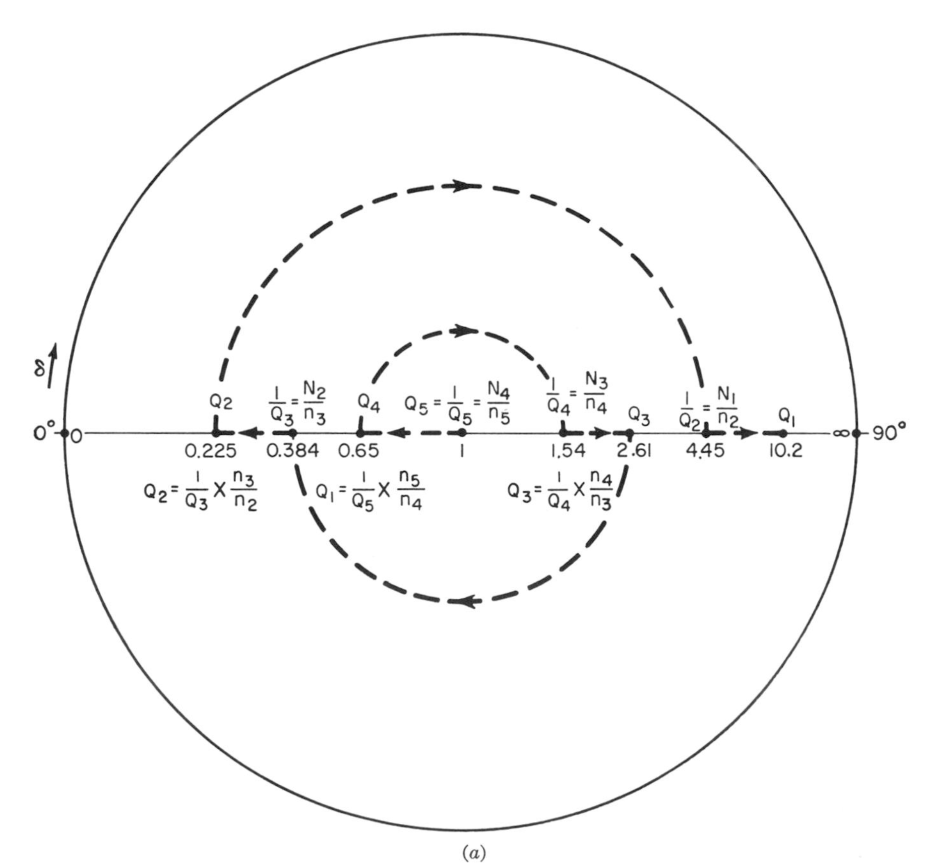

(*a*)

Fig. 5.27. Smith chart. (*a*) In-phase example.

the multilayer as a whole, equals N_1/n_1 or $n_s^4/n_c^2n_g$. The reflectance is then $(1 - n_s^4/n_c^2n_g)^2/(1 + n_s^4/n_c^2n_g)^2$, or about 68 per cent if $n_g = 1.50$, $n_c = 1.35$, and $n_s = 2.30$. Extension of this calculation to a larger number of layers is obvious. If we write n_0 for the index of refraction of the surrounding medium, n_H, n_L, and n_g for the indices of the high-index, low-index, and substrate materials respectively (1, n_s, n_0, and n_g above), and generalize the previous result, we obtain for any odd number M of layers the reflectance R, given by

$$R = (1 - qp^M)^2(1 + qp^M)^{-2} \tag{5.63}$$

where $q = n_Hn_L/n_0n_g$ and $p = n_H/n_L$

Since all the reflected rays are in phase, this is a maximum reflectance of the multilayer: other maxima will occur for $\lambda_0 = 2n_kt_k/(m + \frac{1}{2})$ with integral $m > 0$.

To determine the behavior of the multilayer at other wavelengths than that of maximum reflectance, Leurgans has pointed out that the Smith "transmission-line calculator," familiar to electrical engineers, can be employed as an amplitude-reflection coefficient chart to shorten the calculations. The Smith chart (Fig. 5.27) is a graphic solution of (5.58), the orthogonal coordinates representing the real and imaginary parts of Q_k and N_{k-1}/n_k. The axis of reals is the horizontal diameter and that of imaginaries the outside circumference. Rotation about the center represents the phase angle δ_k; notice that in the conformal representation an arc of double magnitude is drawn to represent any δ. To use the chart, a circular arc is drawn with the center point, $1 + 0i$, as center, through the value of Q_k for each layer, and followed through the angle represented as δ_k clockwise, to N_{k-1}/n_k. One proceeds from the substrate back by steps to Q_1, which is Q for the whole multilayer; the first step is trivial since Q for the finally transmitted beam is unity and the corresponding circle has zero radius. After any N_{k-1}/n_k is known, Q_{k-1} (that is, Q_k for the preceding layer) is then found by multiplying N_{k-1}/n_k by the ratio of the refractive indices n_k/n_{k-1} and the process is repeated until all the layers have been traversed and Q_1 determined.

For the in-phase case already discussed, Figure 5.27*a* shows the calculation graphically, each arc being a semicircle since δ_k is $\pi/2$. For radiation at 5000 A in the same example, for which $\delta = 2\pi/3$, $Q_5 = 1$ as before (the center of the chart),

$$1/Q_5 = 1 = N_4/N_5$$

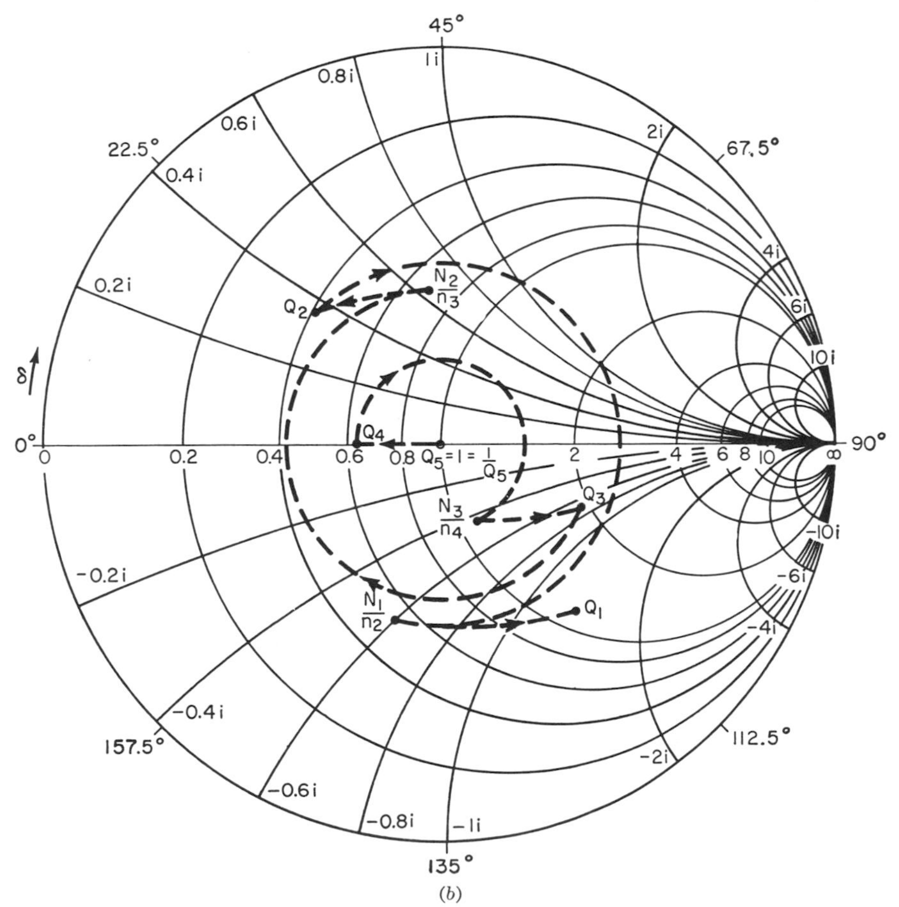

(b)

plotted on the real axis. (b) Out-of-phase example.

as in the in-phase case, and multiplying by n_5/n_4 one obtains

$$Q_4 = N_4/n_4 = n_g/n_s = 0.65$$

Starting at $Q = 0.65$, with the center at 1 describe a circle through the phase angle of $2\pi/3$ on the scale clockwise to point N_3/n_4, which is $1.1 - 0.45i$. Multiplying by n_4/n_3 or 1.7, one reaches $Q_3 = 1.9 - 0.76i$. Thence $2\pi/3$ clockwise to point N_2/n_3, $0.69 + 0.65i$. Q_2 is n_3/n_2, or 0.59 times N_2/n_3, or $0.41 + 0.38i$. N_1/n_2 is found in the same way to be $0.50 - 0.60i$. Finally Q_1 is $1.15 - 1.38i$.

The reflectance R can be found from (5.60) or directly from the chart: it is the square of the distance from the center of the chart to point Q_1 if the radius of the chart is taken as 1. Thus $R \doteq 0.29$.

Interference Filters. Several varieties of light filter are known that employ interference phenomena:

a. Broad-band dielectric multilayer
b. Fabry-Perot-type transmission filter
  *b*1. Metallic
  *b*2. Multilayer
  *b*3. Frustrated total reflection
c. Reflection filter
  *c*1. Metallic
  *c*2. Multilayer
d. Echelon-type filter
e. Lyot filter

(Although neither the Lyot nor the echelon-type filter is a geometrically degraded amplitude device, both are listed in this section for completeness.)

a. The *broad-band multilayer* is the enhanced reflection multilayer described just above. It is often used to separate large segments of the spectrum with negligible absorption, e.g., in color photography and color television.

*b*1. In (5.32*b*) it is shown that, for divided amplitude interference, the free spectral range is inversely proportional to the gap t and transmission peaks occur when $\lambda_0 = 2nm^{-1}t \cos \theta$, where m is an integer. In 1942 Geffcken made an interferometer with a gap of the order of only one wavelength of light (Fig. 5.28*a*); the free spectral range is thousands of angstroms, and only one or two transmission maxima appear in, say, the visible region. Unwanted maxima are eliminated by means of ordinary dye filters. In this way very sharp filters are obtained which pass certain wavelengths, depending on the gap. These filters are usually made by deposition in a vacuum and may consist of two partly transmitting metallic films separated by a dielectric layer forming the gap. Absorption in the metal limits the maximum transmittance T_{max}, which equals $T^2/(1 - R)^2$, where T is the transmittance and R the reflectance of the individual reflecting layers (5.28). The reflectance of the layers is usually limited to about 90 per cent, giving filters transmitting bands of half-intensity width of the order of 100 A, with T_{max} usually less than 30 per cent.

*b*2. With dielectric multilayers as reflectors (Fig. 5.28*b*) the absorption (except near an absorption band of a constituent) is generally negligible. Moreover, extremely high reflectances, of the order of 97 per cent or more, are obtainable, and since with high R the half-intensity width is low (Table 5.1) very narrow-band filters can be produced. Stone has reported that with 11-layer zinc sulfide–cryolite multilayers half-intensity widths of 15 A and transmittances of as high as 60 per cent were obtained. A serious disadvantage of dielectric filters is that the $\lambda/4$ condition in the multilayer reflecting layers is approximately satisfied only for a limited spectral region, and so the reflectance R of the multilayers

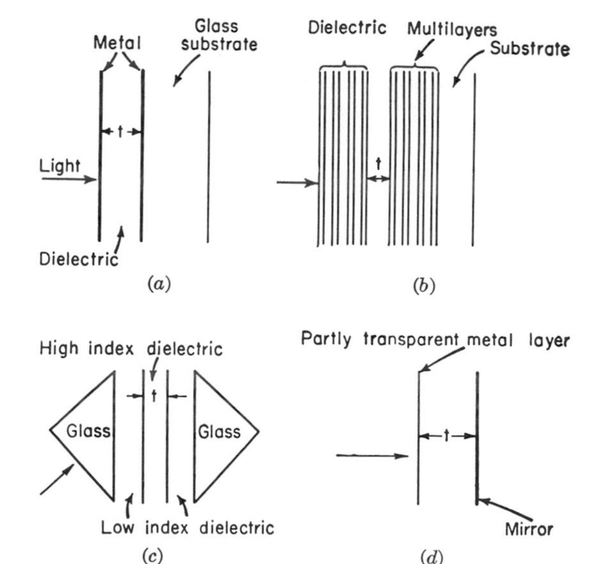

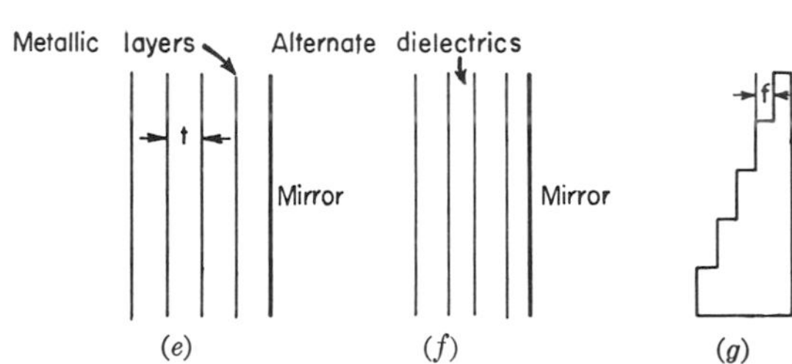

FIG. 5.28. Interference filters. (*a*) Fabry-Perot, with metal reflectors and glass substrate. (*b*) Fabry-Perot, with dielectric multilayer reflectors. (*c*) Frustrated total reflection. (*d*) Single-layer reflection. (*e*) Multilayer reflection, employing metallic partly reflecting surfaces. (*f*) Multilayer reflection, employing dielectrics. (*g*) Stepwise.

decreases on either side of the optimum wavelength. The contrast range (5.29), or ratio of maximum to minimum intensity, $(1 + R)^2/(1 - R)^2$, drops off (see Table 5.1), so that unwanted radiation is admitted. The addition of dye filters or, in severe cases, other interference filters is usually necessary. Multilayer filters of the highest quality now employ layer configurations which are analogus to compound, rather than simple, Fabry-Perot systems.

*b*3. The *frustrated total-reflection filter*, described by Leurgans and Turner, is constructed as shown in Fig. 5.29*c*. The low-to-high index boundaries show frustrated total reflection if light is incident at the proper angle (in the case of zinc sulfide and cryolite, 60°), and hence show very high reflection with negligible absorption. Very narrow filters can be obtained, but unfortunately the non-normal incidence results in two transmitted peaks polarized at right angles to each other. Whereas most interference filters are analogous to the Fabry-Perot, this one also resembles the Lummer-Gehrcke.

c. The broad-band multilayers, besides being transmission filters, reflect the light not transmitted (neglecting absorption). As the number of layers is increased, the sides become increasingly steep, and for a large number of layers, say 100, the bands reflected would be quite square. Multilayer arrays have not been reported with so many layers, but as early as 1810, Seebeck, as reported by Goethe, noticed that a photosensitive substance exposed to colored light took on the color of the light, an effect later exploited by Lippman as a means of color photography and explained by Rayleigh and others as being due a multilayer array formed in the photosensitive material by bleaching at the electric-vector antinodes in a standing-wave interference pattern. The intense colors of opalescent crystals, pearls, feathers, and the surfaces of insects are explained as resulting from natural multilayers.

*c*1. In the infrared region a totally reflecting metallic surface combined with a dielectric gap and a partly reflecting metallic surface (Fig. 5.28*d*) has been studied by Hadley and Dennison, who have given the reflectance of such a filter as

$$R = \frac{(1 - f)^2 + n^2 \cot^2 (2\pi nt/\lambda_0)}{(1 + f)^2 + n^2 \cot^2 (2\pi nt/\lambda_0)} \qquad (5.64)$$

where f is defined as $377/r$, n is the index of refraction of the material in the gap, and r is the numerical value of the resistance in ohms between the opposite sides of a square of the film.

*c*2. Multiple metallic and dielectric layers giving sharper reflection bands have also been used in a similar way (Fig. 5.28*e* and *f*). The multiple, partly reflecting surfaces may be either boundaries between dielectrics of different index or thin metallic layers, but in each case the spacing must be such that the condition $\delta = 2\pi m$ is satisfied for maxima.

d. Hirschberg suggests a filter analogous to the Michelson echelon. Equally thick layers of dielectric would be evaporated on a substrate, while moving a shutter so that each succeeding layer covers a smaller portion of the plate, forming a series of steps (Fig. 5.28*g*). For normal incidence the thickness of each layer, t, must be $m\lambda_0/2n_{\text{air}}$ for a reflection filter and $m\lambda_0/(n - n_{\text{air}})$ for a transmission filter, where m is an integer and n is the refractive index of the dielectric. In the reflection stepwise filter a reflecting film is deposited on the stepwise layer, while, in the transmission type, the substrate and evaporated film must be transparent to the desired radiation.

The reflection stepwise filter, especially if deposited on a concave blank, might be used in the far infrared and at grazing incidence in the far ultraviolet regions, where filters are scarcely, if at all, available.

e. The Lyot filter is listed among the interference filters for completeness, as noted above; it employs interference of polarized light and is not treated.

Topographical Studies: Multibeam Fizeau and Equal-chromatic-order Fringes. On account of the unsharpness of the fringes it yields, two-beam interference is not well suited to the detailed study of the shapes of surfaces. Multibeam interference, as recently developed, can be used under the most favorable circumstances to determine contours with an accuracy of 10^{-7} cm or better, corresponding, in the case of a wide-spaced crystal lattice, to less than a single atomic layer. All the geometrically degraded amplitude multibeam fringes discussed above are Haidinger fringes, which are by definition inappropriate for topographical studies. Of the several topographical-study methods one due to Buckley, using Fizeau fringes, is among the most successful. The surface under study and a flat, both coated with uniform high-reflectance low-absorbance films, placed within a few wavelengths of one another, are illuminated with normally incident monochromatic light. With the surfaces as nearly parallel and the fringes thus as wide as possible, great sensitivity in contour measurements is possible. Tolansky has extended the technique, superimposing on such a photograph another, or two others at right angles, of fringes spaced relatively closely by adjusting the wedge angle between the two surfaces, to give rough guides to the contours in order to simplify the interpretation of the wide-fringe picture.

With surfaces very nearly flat the problem is somewhat simpler. Rasmussen has studied the flatness of coated Fabry-Perot plates, moderately close together, with a slight wedge angle (cf. Fig. 5.17) between them. Another method of checking the flatness of plates is to mount them parallel in a windowed pressure-tight chamber and back-illuminate them with an extended monochromatic source. By means of a large lens free from spherical aberration, fringes are produced in the principal focal plane upon an opaque screen pierced with a circular hole so that only a small portion of the central fringe passes through. A camera at this aperture, focused on the large lens, would show uniform illumination of the lens by the light of the central Haidinger fringe, if the spacing t were an integral multiple of $\lambda_0/2n$ and the plates were strictly flat; but, with not quite flat plates, a contour map of the plates is produced when a succession of exposures is made with the gas pressure between the plates varied by known small equal amounts. If, instead of being photographed, the camera field is scanned with a photomultiplier, sensitivity to a few angstroms spacing change can be achieved by virtue of the rapid change in transmittance with spacing at the half-intensity point of the fringe.

Successive transmitted beams contributing to a Fizeau fringe show an angular deviation and a lateral displacement that increase with the wedge angle φ. Thus the fringe width is an increasing function of φ, and the topographically useful magnification of a Fizeau-fringe photograph is limited. Fringes of equal chromatic order are subject to no such limitation in φ and have certain other technical advantages over Fizeau fringes, especially for the topographical study of opaque objects. The arrangement for such fringes is the same as with Fizeau fringes except that white light is used, and, instead of the whole focal plane of the fringes being viewed or photographed, one strip or line of that plane is thrown onto the slit of a stigmatic spectrograph and the variation of wavelength along the line is studied.

In topological studies of large flats of the highest quality by multibeam fringes, nonuniformity in the required coatings may cause an appreciable error. Methods of high sensitivity which avoid this problem by making use of interference in light reflected from the uncoated surfaces have been described by Dew and by Roesler.

Bibliography

Sec. 1

Born, Max, and E. Wolf: "Principles of Optics," 2d ed., chap. 3, Macmillan, New York, 1964.
Schrödinger, E.: Quantisierung als Eigenwertproblem, *Ann. Physik,* **79**: 489 (1926).

Sec. 2

Born, Max, and E. Wolf: "Principles of Optics," 2d ed., chap. 8, Macmillan, New York, 1964. (Most of the treatment in Secs. 2 and 3 uses approximately Born's notation and method of approach.)
Bouwkamp, C. J.: Thesis, Groningen, 1941.
Jenkins, F. A., and H. E. White: "Fundamentals of Optics," 3d ed., chaps. 11, 12, McGraw-Hill, New York, 1957.
Kirchhoff, G.: *Ges. Abhandl. Nachtr,* p. 23; *Ann. Physik Chem.,* **18**: 663 (1883).
Levine, H., and J. Schwinger: *Phys. Rev.,* **74**: 958 (1948). **75**: 1423, 1608 (1949); *Communs. Pure Appl. Math.,* **3**: 355 (1950).
Sommerfeld, A.: *Math. Ann.,* **47**: 317 (1896); *Z. Math. Physik,* **46**: 11 (1901).
Tolansky, S.: *Phil. Mag.,* **34**: 555 (1943), **35**: 120, 179, 229 (1944), **36**: 225, 236 (1945).
Wood, R. W.: "Physical Optics," 3d ed., Macmillan, New York, 1934.

Sec. 3

Born, Max, and E. Wolf: "Principles of Optics," 2d ed., Macmillan, New York, 1964.
Cornu, A.: *J. phys.,* **1**: 44 (1874).
Fresnel, A.: "Oeuvres," vol. 1, p. 321.
Jenkins, F. A., and H. E. White: "Fundamentals of Optics," 3d ed., chaps. 15, 18, McGraw-Hill, New York, 1957.
Lommel, E.: *Abhandl. Math.-Wiss. Kl. bayer. Akad.,* **15**: 229, 529 (1886).
Meyer, C. F.: "The Diffraction of Light, X-rays, and Material Particles," University of Chicago Press, Chicago, 1934.

Sec. 4

Airy, G. B.: *Phil. Mag.,* **2**: 20 (1833); *Pogg. Ann.,* **41**: 512 (1837).
Born, Max, and E. Wolf: "Principles of Optics," 2d ed., Macmillan, New York, 1964.
Childs, W. H. J.: *J. Sci. Instr.,* **3**: 97 (1926), **3**: 129 (1926).
Dossier, B., P. Boughon, and P. Jacquinot: *J. Rech. Centre Natl. Rech. Sci., Lab. Bellevue (Paris),* no. 11, p. 1, 1950.
Haidinger, W.: *Pogg. Ann.,* **77**: 217 (1849).
Jacquinot, P., and B. Roizen-Dossier: Apodization, *Progr. Optics,* **3**: 31 (1963).
Jenkins, F., and H. E. White: "Fundamentals of Optics," 3d ed., chaps. 14–17, McGraw-Hill, New York, 1957.
Rayleigh: *Phil. Mag.,* **8**: 261 (1879), **9**: 40 (1880), **10**: 116 (1880); "Encyclopaedia Britannica," 9th ed., **24**: 421 (1888).
Straubel, C. R.: Dissertation, Jena, 1888; "Theorie der Beugungserscheinungen Kreisförmig begrenzter, symmetrischer, nicht sphärischer Wellen," Akademische, Munich, 1893.

Sec. 5

Barer, R.: Chap. 3, in R. C. Mellors, "Analytical Cytology," McGraw-Hill, New York, 1955.
Bennett, A. H., H. Jupnik, H. Osterberg, and O. W. Richards: "Phase Microscopy," Wiley, New York, 1951.
Born, Max, and E. Wolf: "Principles of Optics," 2d ed., Macmillan, New York, 1964.
Colloque international sur les progrès recent in spectroscopic interferantielle, Sept. 9–13, 1957. Fifty-one papers in *J. Phys. Radium,* **19**: 185–436 (1958).
Dyson, J.: *Proc. Roy. Soc. (London),* **A204**: 170 (1950).
Connes, Janine and Pierre: *J. Opt. Soc. Am.,* **56**: 896 (1966).
Gabor, D.: *Nature,* **161**: 777 (1948); *Proc. Roy. Soc. (London),* **A197**: 454 (1949); *Proc. Phys. Soc. (London),* **B64**: 449 (1951).
Jamin, J.: *Pogg. Ann.,* **98**: 345 (1856).
Michelson, A. A.: *Phil. Mag.,* **13**: 236 (1882), **31**: 338 (1891), **34**: 280 (1892).
Newton, I.: "Opticks," 4th ed., book 2, pt. 1, Innys, London, 1730; reprinted McGraw-Hill, New York, 1931
Selenyi, P.: *Ann. Physik,* **35**: 444 (1911); *Math. u. Naturw. Ber. Ungarn,* **27**: 76 (1912).
Young, T.: *Phil. Trans. Roy. Soc. (London),* **91**: 12, 387 (1802)
Williams, W. E.: "Applications of Interferometry," Methuen, London, 1930.
Zernike, F.: *Med. Tijdschr. Natuurk.,* **4**: 357 (1942); *Physica,* **9**: 686, 974 (1942).

Sec. 6

Edlén, B.: *Nova Acta Reg. Soc. Sci. Upsaliensis,* **9** (6), (1933).
Fastie, W. G.: *J. Opt. Soc. Am.,* **42**: 641 (1952).
Harrison, G. R.: *J. Opt. Soc. Am.,* **39**: 413, 522 (1949).
Heavens, O. S.: "Optical Properties of Thin Films," Academic Press, New York, 1953.
Hulthén, E., and E. Lind: *Arkiv Fysik,* **2**: 253 (1950).
Kayser, H.: "Handbuch der Spektroscopie," S. Hirzel Verlag, Leipzig, 1900. (Especially vol. 1, chap. 4.)
Mack, J. E., J. R. Stehn, and B. Edlén: *J. Opt. Soc. Am.,* **22**: 245, (1932), **23**: 184 (1933).
Meissner, K. W.: *J. Opt. Soc. Am.,* **32**: 185 (1942). (Especially §§ 21–27.)
Michelson, A. A.: *Astrophys. J.,* **8**: 36 (1898).
Randall, H. M.: *J. Opt. Soc. Am.,* **44**: 97 (1954).
Sawyer, R. A.: "Experimental Spectroscopy," 2d ed., Prentice-Hall, Englewood Cliffs, N.J., 1951. (Especially chaps. 6, 7, 12.)
Smith, S. J., and E. M. Purcell: *Phys. Rev.,* **92**: 1069L (1953)
Wadsworth, F. L. O.: *Astrophys. J.,* **2**: 370 (1895), **3**: 47 (1896).
Williams, W. E.: private communication.
Wood, R. W.: *Phys. Rev.,* **48**: 928 (1935).

Sec. 7

Abelès, F.: *Rev. opt.,* **28**: 11 (1949).
Barr, W. L., and F. A. Jenkins: *J. Opt. Soc. Am.,* **46**: 141 (1956).

Bauer, G.: *Ann. Physik*, **19**: 434 (1934).
Baumeister, P. W., and J. M. Stone: *J. Opt. Soc. Am.*, **46**: 228 (1956).
Bouloch, R.: *J. phys.*, **2**: 316 (1893).
Bruin, F.: *Koninkl. Ned. Akad. Wetenschap.* (Amsterdam), **56B**: 517, 526 (1953), **57**: 125 (1954).
Cartwright, C. H., and A. F. Turner: *Phys. Rev.*, **55**: 595 (1935).
Chabbal, R.: *J. Rech. Centre Natl. Rech. Sci., Lab. Bellevue (Paris)*, 1953, p. 138. (English translation by R. B. Jacobi, UKAEA Research Group, Harwell 1958/JMR HX 4128.)
Connes, P.: *Rev. opt.*, **35**: 37 (1956).
Cotton, P.: *Ann. phys.*, **2**: 209 (1947).
Dew, G. D.: *J. Sci. Instr.*, **41**: 160 (1964).
Dufour, C.: Theses, Paris, 1951.
Edser, E., and C. P. Butler: *Phil. Mag.*, **46**: 207 (1898).
Fabry, Ch., and A. Perot: *Ann. chim. et phys.*, **12**: 459 (1897), **16**: 115 (1899), **22**: 564 (1901).
Geffcken, W.: D.R. Patent 716153, 1942.
Gehrcke, E.: "Handbuch der Physikalischen Optik," Barth, Leipzig, 1927.
Goethe, J. W. v.: "Zur Farbenlehre" (1810).
Greenler, R. G.: *J. Opt. Soc. Am.*, **45**: 788 (1955).
Hall, J. F., and W. F. C. Ferguson: *J. Opt. Soc. Am.*, **45**: 714 (1955).
Heavens, O. S.: *Rept. Progr. Phys.*, **23**: 1 (1960).
Hermansen, A.: *Danske Math.-fys. Medd.*, **29** (13): (1955).
Herpin, A.: *Compt. rend.*, **225**: 182 (1947).
Hirschberg, J. G., and R. R. Kadesch: *J. Opt. Soc. Am.*, **48**: 1772 (1958).
———: *J. Opt. Soc. Am.*, **50**: 514 (1960).
——— and P. Platz: *Appl. Opt.*, **4**: 1375 (1965).
Holland, L.: "Vacuum Deposition of Thin Films," Wiley, New York, 1956.
Houston, W. V.: *Phys. Rev.*, **29**: 478 (1927).
Jacquinot, P.: *J. Opt. Soc. Am.*, **44**: 761 (1954).
——— and C. Dufour: *J. Phys. Radium*, **11**: 427 (1950).
Jaffe, J. H.: *Nature*, **168**: 381 (1951).
King, P., and L. B. Lockhart: *J. Opt. Soc. Am.*, **36**: 513 (1946).
Kollmorgen, F.: *Trans. Soc. Illum. Eng.*, **11**: 220 (1916).
Koppelmann, G. and K. Krebs: *Optik.*, **18**: 349, 358 (1961).
Leurgans, P. J.: *J. Opt. Soc. Am.*, **41**: 574 (1951).
——— and A. F. Turner: *Phys. Rev.*, **55**: 595 (1935).
Lummer, O., and E. Gehrcke: *Berlin, Ber.*, 11, (1902).
Mack, J. E., D. P. McNutt, F. L. Roesler, and R. Chabbal: *Appl. Opt.*, **2**: 873 (1963).
MacNeille, S. M., and E. O. Dixon: *J. Opt. Soc. Am.*, **44**: 805 (1954).
Madden, R. P.: "Physics of Thin Films," vol. 1, p. 123, Academic, New York, 1963.
McNutt, D. P.: *J. Opt. Soc. Am.*, **55**: 288 (1965).
Meissner, K. W.: *J. Opt. Soc. Am.*, **31**: 45 (1941), **32**: 185 (1942), errata, **32**: 211 (1942).
Mooney, R. L.: *J. Opt. Soc. Am.*, **35**: 574 (1945).
Muchmore, R. B.: *J. Opt. Soc. Am.*, **38**: 20 (1948).
Osterberg, H., and N. E. Page: *J. Opt. Soc. Am.*, **43**: 728 (1953).
Penselin, S., and A. Steudel: *Z. Physik*, **142**: 21 (1955).
Pfund, A. H.: *J. Opt. Soc. Am.*, **24**: 99 (1934).
Pohlack, H.: *Feingerätetechnik*, **2**: 499 (1953).
Rasmussen, E.: *Danske Math.-fys. Medd.*, **30** (6) (1955).
Rayleigh: *Phil. Mag.*, **26**: 256 (1888).
Roesler, F. L. and W. Traub: *Appl. Opt.*, **5**: 463 (1966).
Rouard, P., and P. Bousquet.: *Progr. Optics*, **4**: 147 (1965).
Rubens, H.: *Ann. Physik Chemie*, **45**: 238 (1842).
Smith, P. H.: *Electronics*, **12**: 29 (1939), **17**: 131 (1944).
Stanley, R. W., and K. L. Andrew: *J. Opt. Soc. Am.*, **54**: 625 (1964).
Steudel, A., and S. Stotz: *Z. Physik.*, **151**: 233 (1958).
Stone, J. M.: *J. Opt. Soc. Am.*, **43**: 927 (1953); Ph.D. thesis, California, 1954.
Strong, J.: *J. Opt. Soc. Am.*, **26**: 73 (1936).
Taylor, H. D.: "The Adjustment and Testing of Telescope Objectives," 2d ed., Cooke, York, England, 1896.
Tolansky, S.: "Multiple-beam Interferometry of Surfaces and Films," Oxford University Press, Fair Lawn, N.J., 1948.
Vasicek, A.: *J. Opt. Soc. Am.*, **37**: 623 (1947).

Chapter 6

Molecular Optics

By E. U. CONDON, University of Colorado

1. Molecular Refractivity

From (1.28) the square of the refractive index for a nonabsorbing medium is equal to the dielectric constant,

$$n^2 = \epsilon \tag{6.1}$$

This is known as *Maxwell's relation* and was one of the earliest deductions from the electromagnetic theory of light. Although it is valid only for n and ϵ measured at the same frequency, there are many nonpolar materials for which the relation is approximately true when the refractive index n_D for the yellow D lines of sodium is compared with the dielectric constant measured at frequencies less than 10^6 cps.

The significant attribute of individual atoms or molecules here is the frequency-dependent polarizability α, which gives the relation between the induced dipole moment $\mathbf{p}$ and the locally effective electric field $\mathbf{E}'$.

$$\mathbf{p} = \alpha \mathbf{E}' \tag{6.2}$$

If $\mathbf{p}$ and $\mathbf{E}'$ are in cgs electrostatic units, then α has the dimensions of cubic centimeters and is of the order of 10^{-24} cm^3. According to (6.2) the dipole moment is in the same direction as the field $\mathbf{E}'$ and in time phase with it. Absorption is taken into account by recognizing the existence of a component of $\mathbf{p}$ in time quadrature with $\mathbf{E}'$. Actually individual molecules are anisotropic, and α is a tensor, but when the molecules of a medium are oriented at random, the component of $\mathbf{p}$ transverse to $\mathbf{E}'$ averages out in physically small volume elements and the effective value of α in (6.2) becomes the mean of the three principal values of the tensor, that is, its scalar invariant.

As in the theory of dielectrics (Part 4, Chap. 7) the effective field $\mathbf{E}'$ differs from $\mathbf{E}$ by a contribution that is proportional to $\mathbf{P}$, the polarization or electric moment of unit volume,

$$\mathbf{E}' = \mathbf{E} + 4\pi f \mathbf{P} \tag{6.3}$$

where $f = \frac{1}{3}$ according to the results of Lorentz and Lorenz for amorphous materials. If the medium consists of a mixture of different molecular species distinguished by the subscript a, having polarizabilities α_a, of which N_a is the number of species a in unit volume, the polarization is

$$\mathbf{P} = \beta \mathbf{E}' \qquad \text{where} \qquad \beta = \sum_a N_a \alpha_a \tag{6.4}$$

giving

$$\mathbf{E}' = (1 - 4\pi f \beta)^{-1} \mathbf{E} \tag{6.5}$$

and so the effective field $\mathbf{E}'$ is greater than $\mathbf{E}$. Accordingly, since $\mathbf{D} = \mathbf{E} + 4\pi\mathbf{P}$, the refractive index is given by

$$n^2 = 1 + \frac{4\pi\beta}{1 - 4\pi f \beta} \tag{6.6}$$

Using the value $f = \frac{1}{3}$, this can be written,

$$\frac{n^2 - 1}{n^2 + 2} = \frac{4\pi}{3} \beta$$

Letting ρ_a be the mass concentration of species a, whose molecular weight is M_a, it is convenient to define the *molecular refractivity* P_a of each constituent as

$$P_a = \frac{4\pi N}{3} \alpha_a \tag{6.7}$$

in terms of which the formula for the index of refraction is

$$\frac{n^2 - 1}{n^2 + 2} = \sum_a \frac{\rho_a}{M_a} P_a \tag{6.8}$$

in which the factors ρ_a/M_a are the concentrations of the several constituents expressed in moles per cubic centimeter. Here N is the Avogadro number.

In considering the application of (6.8) it must be realized that what is here called "molecular species" means any structural unit which preserves its character with regard to polarizability in the different physical conditions which are being compared. The mildest test of (6.8) is to apply it to the variation of refractive index of a gas with density, resulting from variation of pressure at constant temperature, because throughout this change of conditions the molecules remain well separated.

For gases at ordinary temperatures and pressures $n - 1$ is observed to be proportional to the density, because $n - 1$ is so small that one can write $\frac{2}{3}(n - 1)$ in place of $(n^2 - 1)/(n^2 + 2)$. Van Vleck [1]* reproduces experimental data obtained for air up to nearly 200 atm, which indicates that $(n^2 - 1)/(n^2 + 2)$

* Numbers in brackets refer to References at end of chapter.

TABLE 6.1. MOLECULAR REFRACTIVITIES OF VARIOUS SUBSTANCES CALCULATED FROM GASEOUS- AND LIQUID-STATE DATA*

Substance	M	$(n_g - 1) \times 10^4$	n_l	ρ_l	P_g	P_l
Hydrogen, H_2	2	1.32	1.10	0.071	1.98	1.85
Water, H_2O	18	2.49	1.334	1.000	3.74	3.71
Ammonia, NH_3	17	3.73	1.325	0.616	5.60	5.54
Nitrogen, N_2	28	2.96	1.205	0.808	4.34	4.52
Oxygen, O_2	32	2.71	1.221	1.124	4.06	4.04
Nitric oxide, NO	30	2.97	1.330	1.269	4.45	4.83
Nitrous oxide, N_2O	44	5.16	1.193	0.870	7.74	6.24
Chlorine, Cl_2	71	7.73	1.367	1.33	11.58	11.92
Hydrochloric, HCl	36.5	4.47	1.245	0.95	6.71	5.95
Bromine, Br_2	160	11.32	1.659	3.12	17.00	15.30
Hydrobromic, HBr	81	5.73	1.352	1.630	8.57	10.72
Sulfur:						
S_2	64	11.11	1.929	2.04	16.6	14.9
H_2S	34	6.23	1.384	0.91	9.35	8.72
SO_2	64	6.90	1.410	1.359	10.32	11.61
CS_2	76	14.7	1.628	1.264	22.0	21.4
CO_2	44	4.49	1.192	0.796	6.74	6.78
Methanol, CH_3OH	32	5.49	1.331	0.794	8.24	8.23
Ethanol, C_2H_5OH	46	8.71	1.3623	0.800	13.05	12.72
Acetaldehyde, CH_3CHO	44	8.11	1.3316	0.800	12.16	11.40
Acetone, CH_3COCH_3	58	10.8	1.3589	0.791	16.20	16.05
Phosphorus, P_2	62	12.12	2.144	1.83	18.20	18.15

* The data refer to the Na D lines.

is more accurately proportional to the density than $n^2 - 1$, which is the form obtained if the distinction between $\mathbf{E}'$ and $\mathbf{E}$ is ignored, that is, if f is set equal to zero. At a pressure of 176 atm $n_D = 1.05213$; so the correction for the effective field amounts to only about 3.5 per cent and, although given correctly by the theory, is small.

A more severe test is that given by comparing the index of refraction of the same material in the gaseous and the liquid or solid states. In such condensed states the density is several thousand times the density of the gas at standard conditions, and so it could well be that the molecules affect each other's electronic structure when packed densely, making the α for the same molecule in a condensed phase different from the value in the gaseous phase. Therefore a comparison of this kind has the meaning of seeking to determine the extent to which the α and hence the P remain unaffected by such a change of state. For all gases at standard conditions the molal volume is 22,414 cm³, and therefore the molecular refractivity is just equal to $1.500(n_g - 1) \times 10^4$, where n_g is the refractive index for the gas under standard conditions. Thus, for water vapor, $n_g = 1.000254$, and therefore the molecular refractivity inferred from this is $P_g = 3.80$. For liquid water $n_l = 1.334$, and the molar volume is 18 cm³, giving $P_l = 3.71$ in good agreement with the value calculated from observations on the vapor. Thus one may conclude that the average polarizability of water molecules for oscillating fields of the D-line frequency is not much affected by the packing together of the water molecules in the liquid state.

In Table 6.1 are given some additional examples for comparing the molecular refractivity at D-line frequency as calculated from index of refraction and density in the gaseous and liquid states, indicating that agreement like that calculated for water is a fairly general phenomenon.

Another test of Eq. (6.8) involves the refractive index of mixtures of miscible liquids. If the weight fraction of the ath constituent is x_a and the over-all density of the mixture is ρ, then $\rho_a = x_a\rho$, and (6.8) may be conveniently put in the form

$$\frac{1}{\rho}\frac{n^2 - 1}{n^2 + 2} = \sum_a x_a \frac{P_a}{M_a}$$

indicating that the quantity on the left ought to be a linear function of the weight fractions which express the composition. As an example, take the observed data on water-ethanol mixtures, measured at 15°C for the F line. The water-ethanol mixture is chosen because ρ^{-1} itself is a nonlinear function of the weight fraction x of ethanol, $1 - x$ being the weight fraction of water. The data are given in Table 6.2.

In Fig. 6.1, $\rho^{-1}(n^2 - 1)/(n^2 + 2)$ is graphed against x. The points at $x = 0$ and $x = 1$ that are marked with arrows indicate the values of P/M for

TABLE 6.2. REFRACTIVE DATA FOR WATER-ETHANOL MIXTURES FOR F LINE AT 15°C

x	ρ	n	$\rho^{-1}\frac{n^2 - 1}{n^2 + 2}$
0.0000	0.998	1.337	0.2080
0.2075	0.970	1.3508	0.2220
0.4089	0.939	1.3616	0.2357
0.5998	0.899	1.3670	0.2500
0.7999	0.854	1.3693	0.2645
1.0000	0.805	1.3676	0.2800

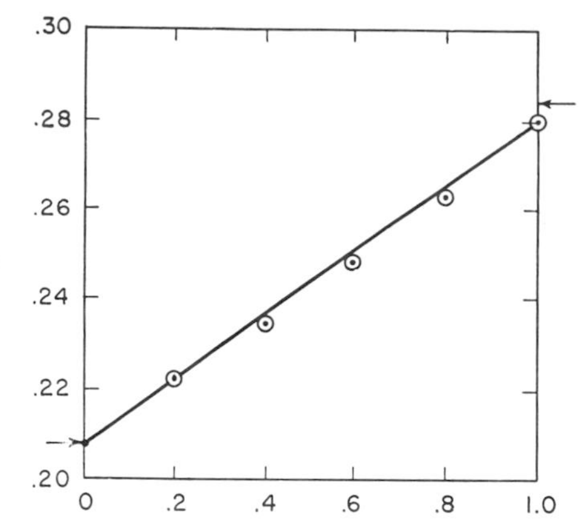

Fig. 6.1. Linear dependence of $\rho^{-1}(n^2 - 1)/(n^2 + 2)$ on weight fraction of ethanol in water-ethanol mixtures.

H_2O and C_2H_5OH as calculated from the observed values of the refractive index of the vapor, as given in Table 6.1. The linearity is quite accurate, indicating that the strong liquid-state interactions that are responsible for the nonlinearity of the specific volume ρ^{-1} do not make appreciable changes in the molecular values of the polarizability α, for water and alcohol molecules.

Many attempts have been made to go further, representing the molecular refractivity for chemical compounds additively in terms of contributions from individual atoms [2]. Such attempts are fairly successful in the case of organic compounds, and considerably less so in the case of alkali halide crystals.

Table 6.3. Molecular Refractivity (D Lines) of Alkali Halide Crystals

	F		Cl		Br		I
Li	2.64	*4.69*	7.38	*2.80*	10.13	*4.71*	14.84
	0.34		*0.93*		*0.96*		*1.25*
Na	2.98	*5.28*	8.26	*2.83*	11.09	*5.00*	16.09
	2.12		*2.24*		*2.31*		*2.69*
K	5.10	*5.40*	10.50	*2.90*	13.40	*5.38*	18.78
	3.58		*1.84*		*2.80*		*1.90*
Rb	8.68	*3.66*	12.34	*2.86*	15.20	*5.48*	20.68
	3.20		*2.24*		*2.71*		*2.52*
Cs	11.88	*2.70*	14.58	*3.33*	17.91	*5.29*	23.20

Table 6.3 gives the molecular refractivity for the D lines of the alkali halides. If these were accurately representable as a sum of a constant atomic or ionic contribution from each of the two constituents, then the first differences in rows or columns ought to be constant, and this is only very roughly true. This may well [3] be due to inaccuracy of the expression $(4\pi/3)\mathbf{P}$ for the difference between $\mathbf{E}'$ and $\mathbf{E}$.

Table 6.4 gives a set of atomic refractivity values for additive calculation from the structural formula of the molecular refractivity of organic compounds which will be found to fit the facts about as well as any such set of additive factors. The values are applicable for the D-line refractivity.

Table 6.4. Atomic-refractivity Values for D Lines*

Structural unit	*Contribution to P*
Hydrogen, —H	1.100
Carbon, C	2.418
Increment for double bond, C=C	1.733
Increment for triple bond, C≡C	2.336
Methylene group, $—CH_2—$	4.711
Oxygen:	
In hydroxyl, —OH	1.525
In ether, —O—	1.643
In carbonyl, =O	2.211
In peroxide, $—O_2—$	4.035
Nitrogen:	
In O—N=C	3.901
In C—N=C	4.10
In N—N=C	3.46
Cyanide group, —CN	5.415
Isonitrile group, —NC	6.136
Chlorine:	
On alkyl group	5.967
On carbonyl group	6.336
Bromine	8.865
Iodine	13.90
Sulfur:	
Divalent	7.80
Quadrivalent	6.98
Hexavalent	5.34

* Eisenlohr, *Jahrb. Radioakt. u. Elektronik*, **9**: 315 (1912).

The values are applicable for organic compounds to give the molecular refractivity P from the structural formula.

In all such studies it has been found necessary to take into account the nature of the chemical bonding involved in the molecule as is implied in the values given in Table 6.4. This has led to an alternative description [5] of molecular refractivities as a sum of contributions arising from different chemical bonds, rather than as a sum of contributions over the constituent atoms, which seems to give a better representation of the observed molecular refractivities. Table 6.5 gives a table of *bond refractivities* for D-line frequency as derived from this point of view.

2. Dispersion

The dependence of the refractive index n on the light frequency is known as *dispersion* because it is the property that makes it possible for a prism to disperse the several colors of mixed light into emergent beams traveling in different directions. Normally the refractive index increases with increasing frequency. Under special circumstances in a narrow range of frequencies near absorption bands the refractive index decreases with increasing frequency. This is known as *anomalous dispersion*.

Dispersion results from the frequency dependence of the molecular polarizability α. In classical electron theory it is assumed that α is due to the forced motions of electronic oscillators as set up by the effective field $\mathbf{E}'$. These are treated as damped harmonic oscillators having an equation of motion,

$$\ddot{\mathbf{r}} + \gamma\dot{\mathbf{r}} + \omega_0^2\mathbf{r} = \frac{e}{m}\mathbf{E}'e^{i\omega t}$$

The contribution of any one electron to the electric-dipole moment is then $e\mathbf{r}$, and the whole polarizability

TABLE 6.5. BOND REFRACTIVITIES FOR D-LINE FREQUENCY

C—C	1.296	C—S	4.58	N—N	1.97
C—H	1.676	C=S	11.82	N—O	2.37
C—O ethers	1.51	S—S	8.11	N—O	1.74
C—O acetals	1.45	S—H	4.78	N=O	4.03
O—H alcohols	1.66	S—O	4.92	C≡C terminal	5.87
O—H acids	1.86	S—O	−0.20	C≡C nonterminal	6.24
C=O	3.26	S—Cl	10.63	C=C	4.15
C=O methyl ketones	3.42	C—N	1.55	C—C 3-ring	1.49
C—F	1.41	C—N tertiary aliphatic	1.53	C—C 4-ring	1.38
C—Cl	6.47	N—H	1.76	C—C 5-ring	1.26
C—Br	9.36	C=N	3.63	C—C 6-ring	1.27
C—I	11.59	C=N	4.80		

of the molecule is obtained by summing over all the electrons in the molecule. The result is that the polarizability α is expressed by a formula of the type,

$$\alpha = \frac{e^2}{m} \sum_s \frac{f_s}{(\omega_s^2 - \omega^2) + i\gamma_s\omega} \tag{6.9}$$

Here the sum extends over each resonant frequency of the molecule, and f_s is regarded as being the number of electrons in the molecule which have this particular resonant frequency, and therefore the sum of all the f_s ought to be equal to the total number of electrons in the molecule.

This result is capable of various formal generalizations suitable for application to crystal optics, by supposing that the restoring force and the damping force acting on each electron have a tensor character with different proper values along different crystalline axes. The damping terms γ_s give rise to a complex α, and hence through (1.28) to a complex index of refraction describing an absorbing medium.

The model leading to (6.9) has more than historical interest today because the quantum-mechanical treatment of dispersion [6] also leads to a result of the form of (6.9) with an altered interpretation of the symbols. In the quantum-mechanical theory there is a term of the form of those in (6.9) associated with each possible quantum transition in the molecule. The quantities f_s are replaced by quantities known as *oscillator strengths* (Part 7, Chap. 3) f_{mn}. associated with each spectral line by the formula

$$f_{mn} = \frac{2}{3}\frac{m}{\hbar}\,\omega_{mn}|\mathbf{r}_{mn}|^2 \tag{6.10}$$

in which ω_{mn} is the (radian) frequency associated with the quantum jump between the mth and the nth levels and $\mathbf{r}_{mn}$ is the matrix component of the vector coordinates of the electron connecting these two states. The theory also associates a damping term, γ_{mn}, with each quantum transition.

The quantum theoretic formula has a feature that has no classical analogue in the existence [7] of *negative dispersion*. In (6.10) it is tacitly supposed that the quantum transition $m \to n$ is one corresponding to a jump from a lower to a higher level, that is, an absorption. However, if the initial state m is an excited state, then there will be lower states, n, for which the associated quantum jump corresponds to an emission, and for these the frequency ω_{mn}, and hence the oscillator strength f_{mn}, must count as negative in (6.10) as it is counted as positive for absorption transitions. Thus the contribution to dispersion due to molecules in excited states, in regard to critical frequencies associated with emission transitions, is negative and subtracts from the total of α instead of adding to it.

Accordingly, if N is the total number of molecules in unit volume and N_m is the number in unit volume in the mth quantum state, a particular jump frequency will occur twice in the formula for α, once positively as an absorption frequency and once negatively as an emission frequency. The total resulting formula for α is thus given by

$$\alpha = \frac{1}{N}\frac{e^2}{m} \sum_m {\sum_n}' \frac{(N_m - N_n) f_{mn}}{(\omega_{mn}^2 - \omega^2) + i\gamma_{mn}\omega} \tag{6.11}$$

where the prime on the summation over n indicates that this sum is to be taken only for levels which are higher in energy than the mth level.

The experimental proof of the existence of negative dispersion was an important confirmation of the quantum-mechanical theory. It is, however, unimportant in discussing optical properties of materials under ordinary circumstances because N_n is usually negligibly small compared with N_m, especially for the ultraviolet frequencies, which contribute mainly to the refractivity in the visible part of the spectrum.

The damping terms γ_{mn} are always small compared with the ω_{mn}; so absorption is negligible except very close to the quantum jump frequency. For ordinary materials that are transparent in the visible, the effective values of ω_{mn} are those corresponding to electronic absorption bands in the ultraviolet. These bands are quite complicated, involving a large number of vibrational and rotational initial levels which are appreciably populated at ordinary temperatures, from which transitions can be made to a large number of upper levels. There is no case in which the bands are known in sufficient detail to make a close comparison between the predictions of the quantum-mechanical dispersion formula and the observed refraction.

At ordinary temperatures the number of molecules in the upper state for ultraviolet frequencies is entirely negligible, and so negative dispersion plays no role with reference to these terms. As temperature is increased, there will be a shift in the values of the N_m which will be the origin of a variation with temperature of α. Thus the temperature dependence of refractive index stems partly from the temperature variation of the density ρ and partly from a temperature variation of α.

Increasing temperature puts more molecules in higher vibrational states. If the predominant transitions from these are to the same excited states, this will have the effect of bringing smaller values of ω_{mn} into play, resulting in a moving of the ultraviolet absorption bands toward the visible, causing an increase of α which may more than offset the decrease of ρ, thus giving rise to a case in which the refractive index increases with temperature. It may further happen that the higher vibrational levels of the lower electronic state favor transitions to lower vibrational levels of the electronic state, a circumstance which would contribute even more to causing α to increase with increasing temperature. But the opposite effect may also happen, resulting in a decrease of α with temperature. Therefore to predict which way α will change with temperature requires much more detailed knowledge of the vibrational transitional probabilities than is at present available in any specific case.

As already mentioned, the polarizability α has the dimensions of volume and has numerical values of the order of the actual spatial volumes of atoms, that is, 10^{-24} cm³. How this comes about may be seen by making a few notational changes in (6.11). Write $\omega_{mn} = 2\pi c R \sigma'_{mn}$ where R is the Rydberg constant and σ'_{mn} is the wave number of the absorption line measured in Rydberg units. An absorption line at $\lambda = 1800$ A corresponds to $\sigma' \sim \frac{1}{2}$; so this is a natural unit in which to express the optical frequencies. Neglecting damping, the polarizability α becomes, from (2.3),

$$\alpha = a^3 \frac{1}{N} \sum_m \sum_n{}' \frac{4(N_m - N_n) f_{mn}}{\sigma'^2_{mn} - \sigma'^2} \tag{6.12}$$

in which $a = \hbar^2/me^2$ is the radius of the first Bohr orbit, and so a^3 is a natural unit of atomic volume, and the terms in the sum are of the order of unity. Atoms of larger Z give rise to larger sums of the oscillator strengths, and for corresponding electronic structures (same column of the periodic table) their strong absorption lines lie less deep in the ultraviolet than is the case for atoms of smaller Z, which is the basic quantum-theoretic reason why the larger atoms have larger polarizabilities.

The quantum-mechanical theory of dispersion [3] yields a result that is more complicated than (6.11) in two essential respects. In the preceding discussion attention was focused on the contribution due to forced vibrations of electrons, but in the quantum mechanics all transitions contribute to the sum in (6.11), including pure rotation and rotation-vibration transitions in the infrared spectrum of molecules in which there is no transition of the electronic quantum numbers. These terms make a negligibly small contribution to α in the visible but make important contributions in the far infrared and radio frequencies; they are thus the main source of the difference between n^2 for visible light and ϵ for radio frequencies (Part 8, Chap. 3). The other complication is that in reality m and n in (6.11) stand for the totality of all quantum numbers necessary to specify the individual quantum states. Included among these, therefore, is the particular quantum number which specified the spatial orientation of the total angular momentum of the molecule. The more detailed calculations show that α is a tensor for any pair of such spatially oriented quantum states. This tensor reduces to a scalar in the summation over all quantum states if the population of such states is isotropic; therefore, the complete derivation of (6.11) implies that such an isotropic averaging over individually anisotropic contributions has been carried out. It is this feature which also gives the possibility of double refraction being induced by external electric and magnetic fields.

3. Absorption and Selective Reflection

Absorbing media are described by means of a complex index of refraction, $n(1 - i\kappa)$. The theory of reflection and refraction at a plane boundary (Part 6, Chap. 1, Sec. 6) involves the same formulas as for a nonabsorbing medium if the complex index is substituted for the real index. The effect of κ is to introduce an exponential damping of the light wave as it travels in the absorbing medium. It also introduces a phase shift between the parallel and perpendicular polarized components of both the transmitted and the reflected waves; so these show elliptic polarization even if the incident light is unpolarized.

If $\kappa \ll 1$, the medium is said to be weakly absorbing. In this case the absorption can be measured directly by measurement of the attenuation of light intensity on traversing a measured path length in the medium. But for $\kappa \sim 1$ or $\kappa \gg 1$ the penetration of waves is inappreciable beyond a few wavelengths, and so direct measurement of κ becomes difficult. Likewise it is difficult in this case to measure n by measuring the deviation of a ray through a prism, as the angle of the prism must be extremely narrow, for otherwise the light will be unable to penetrate its thicker portions. In spite of these difficulties some work has been done with narrow-angle prisms of highly absorbing materials which agrees with the theory within the rather poor limits of accuracy.

For strongly absorbing materials the optical properties are usually studied by measurements made on the reflected light, with regard to its intensity and state of polarization. The approximate reflecting powers R_s and R_p for light whose electric vector is perpendicular and parallel to the plane of incidence, respectively, for angle of incidence θ, are given [for $n^2(1 + \kappa^2) \gg 1$] by

$$R_s = \frac{n^2(1 + \kappa^2)\cos^2\theta - 2n\cos\theta + 1}{n^2(1 + \kappa^2)\cos^2\theta + 2n\cos\theta + 1}$$

$$R_p = \frac{n^2(1 + \kappa^2) - 2n\cos\theta + \cos^2\theta}{n^2(1 + \kappa^2) + 2n\cos\theta + \cos^2\theta}$$

For nonabsorbing media, for which $\kappa = 0$, we have $R_s = 0$ at Brewster's angle, but there is no angle at which $R_s = 0$ for reflection from an absorbing medium, although R_s shows a minimum at angles near the Brewster angle.

The accurate formulas for reflection are obtained by observing that (1.40) (Part 6, Chap. 1) defines a complex angle of refraction, $\alpha + i\beta$, if the refractive index is complex and the angle of incidence is real. In terms of it the accurate formulas are

$$R_s = \frac{\sin^2 (\theta - \alpha) + \sinh^2 \beta}{\sin^2 (\theta + \alpha) + \sinh^2 \beta}$$

$$R_p = R_s \frac{\cos^2 (\theta + \alpha) + \sinh^2 \beta}{\cos^2 (\theta - \alpha) + \sinh^2 \beta}$$

For $\kappa \gg 1$, $\sinh \beta \gg 1$, and both reflecting powers tend to unity.

In a colloquial sense this is a paradoxical result: a strongly absorbing medium is one which attenuates rapidly that part of the light wave that penetrates inside it, but since most of the incident light is reflected from its surface, a strongly absorbing medium actually is a poor absorber when measured by the fraction of the total incident light that is absorbed.

Designating by Δ the phase of the parallel component of the reflected beam relative to that of the perpendicular component, this may be expressed in terms of the angle of incidence and the complex angle of refraction,

$$\tan \Delta = \frac{\sinh 2\beta \cdot \sin 2\theta}{\cos 2\theta + \cos 2\phi \cosh 2\beta}$$

The phase difference vanishes for $\theta = 0$ and increases to π for $\theta = \pi/2$; so for normal and for grazing incidence the reflected light is plane-polarized if the incident light is plane-polarized. The angle of incidence Θ for which $\Delta = \pi/2$ is called the *principal angle of incidence*. When unpolarized light is reflected with this angle of incidence, the phase shift of $\pi/2$ appears in the reflected light. This results in elliptic polarization of the reflected light because in general the amplitudes of parallel and perpendicular reflection are unequal. If now a quarter-wave plate is introduced into the beam with its slow direction in the plane of incidence, the light will be plane-polarized after transmission through it. The angle between the electric vector of this plane-polarized wave and the normal to the plane of incidence is called the *principal azimuth*, Ψ.

The complex index of refraction can be calculated from measurements of Θ and Ψ by the relations

$$\kappa = \tan 2\Psi$$

$$n^2(1 + \kappa^2) = \sin^2 \Theta \tan^2 \Theta$$

One way of subjecting the foregoing results to experimental test is to determine n and κ for a particular wavelength and then to calculate R, the reflecting power at normal incidence, from the theoretical formula which relates R to n and κ, and then to compare this calculated R with an experimentally determined R. Table 6.6 gives a comparison of this kind, the values of n and κ being those due to Drude [9], with red light not precisely defined as to wavelength. The values of R were obtained [10] with different specimens. The values of R depend considerably on the manner of polishing, which probably accounts for the discrepancies.

Surface conditions affect the results in two different ways. The values of n and κ will depend on the mode of preparation of the surface. Second the properties may vary with depth in the surface layers, in which case the simple theory that regards the surface as a simple discontinuous boundary between two homogeneous media rather than as a transition layer of finite thickness will be inappropriate.

TABLE 6.6. REFLECTING POWER OF SOME METALS FOR RED LIGHT

Metal	n	κ	R (calc.)	R (obs.)
Cu	0.641	4.08	0.732	0.890
Au	0.366	7.70	0.851	0.882
Ni	1.79	1.86	0.620	0.659
Pt	2.06	2.06	0.701	0.663
Ag	0.181	20.2	0.953	0.935

Although it is true that for a fixed value of n the reflecting power increases with increasing κ, the situation with regard to selective reflection of colors by strongly absorbing materials is more complicated because both n and κ depend on the color of the light. Thus R. W. Wood [11] points out that for the organic dye, cyanine, the center of the absorption band is in the yellow and the color of the reflected light is purple. Examination of the spectrum of the reflected light shows that the reflecting power in the green is very low, corresponding to the fact that the refractive index drops down to about 1.1 in that part of the spectrum.

The strong reflection associated with strong absorbing power is the basis of the discovery by E. F. Nichols in 1897 of the rest-strahlen, or residual rays, in the far infrared part of the spectrum. In the near infrared the reflecting power of NaCl crystal is less than 0.05, but at wavelengths longer than 40 μ there is a rapid increase of reflecting power, reaching a maximum of 0.82 at $\lambda = 52\ \mu$. Analogous behavior is shown by other ionic crystals. These broad absorbing regions are interpreted as originating in the vibrations of the positive-ion lattice relative to the negative-ion lattice. Owing to the strong variation of n with wavelength, maximum reflecting power does not come at the center of the absorption band, which is at $\lambda = 61\ \mu$ for NaCl. The experimental observations are on the whole in agreement with theory except that the experimental data show a fine structure in these absorption bands that has not been interpreted [12].

4. Crystalline Double Refraction

Double refraction by crystalline media is related to the tensor character of the dielectric constant ϵ, as developed in Part 6, Chap. 1, Sec. 7. The problem of relating the tensor properties back to the fundamental polarizability properties of the atoms and molecules and their arrangement in the crystal is a difficult one on which not much progress has been made.

Individual monatomic atoms or ions when isolated from each other are optically isotropic. When molecule formation occurs, the electronic structure of the molecules is no longer isotropic and so the polarizability of individual molecules must be represented by a polarizability tensor. In a gas in which the molecules are statistically oriented at random the average behavior will be the same as if the tensor were replaced by a scalar equal to the mean of its three principal values.

In a crystal these individually anisotropic units are not oriented at random but have fixed relations to the basic lattice. Therefore for this reason alone the dielectric tensor will be anisotropic. In addition the anisotropic distribution of the individual dipoles which contribute to **P**, the dipole moment in unit volume, results in an anisotropy of the local field correction by which **E**′ differs from **E**. Thus in (6.3) one must assume that the local field factor which is there designated by f must be replaced by a tensor **f**. Hence the anisotropic behavior of ϵ has its origin in two distinct causes: (*a*) the anisotropy of individual structural units whereby the α_a are tensors, and (*b*) the anisotropy of the local field correction represented by making **f** be a tensor [13].

If the principal axes of the **f** tensor and the resultant α tensor are parallel, then the calculations leading to (6.6) go through as there indicated for **E** along each of the principal axes, with the distinction that for f and β there one must now use the principal values f_x, f_y, f_z, β_x, β_y, and β_z appropriate to the particular axis of polarization of **E**.

Some general remarks on the relation of double refraction to structure may be made [14]. Strong double refraction occurs for crystals in which the atoms or molecules are arranged in layers or in parallel chains. Where a compound ion is planar, there will be a strong tendency to anisotropy, with the axes of strong polarizability being in the plane of the ion and the axis of weaker polarizability being normal to the plane of the ion [15]. This is the origin

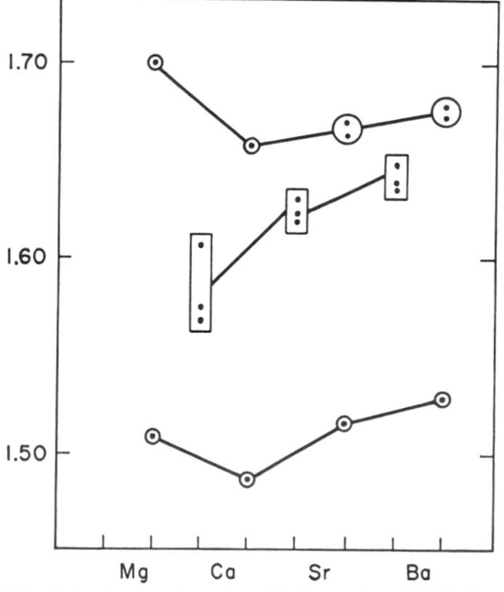

FIG. 6.2. Principal refractive indices (for NaD light) of carbonates, ○, and sulphates, □, of the alkaline earths, showing the large double refraction associated with the planar carbonate ion as compared with the small double refraction associated with the nearly isotropic sulphate ions.

of the strong double refraction shown by the alkaline-earth carbonates, since the CO_3^{--} ion is known from X-ray studies to have a planar structure. More symmetric ions like SO_4^{--} give rise to much smaller double refraction. This behavior is shown in Fig. 6.2, where comparison is made of the refractive indices of the crystalline sulfates and carbonates of the alkaline earths. The carbonates of strontium and barium (strontianite and witherite) are slightly biaxial, but two of the principal indices of refraction are nearly equal as compared with their large difference relative to the third.

5. Faraday Effect; Cotton-Mouton Effect

Both the Faraday effect and the Cotton-Mouton effect are kinds of double refraction produced in transparent media by the action of a magnetic field. The Faraday effect [16], discovered in 1845, consists of a rotation of the plane of polarization of linearly polarized light that is proportional to the path length in the medium and to the component of the magnetic field in the direction of propagation. Its discovery was of great importance historically as being the first experimental indication of an interaction between light and electromagnetic effects.

The Cotton-Mouton effect [17] consists of a double refraction for light propagated in directions at right angles to the magnetic field. Here the normal modes of polarization are, respectively, plane polarization with **D** of the light wave parallel to the external **H**, for which the index of refraction is n_p, and plane polarization with **D** normal to the external **H**, whose index will be written n_s.

In the Faraday effect the angle of rotation of the plane of polarization, α, is given by

$$\alpha = VHd \tag{6.13}$$

in which the coefficient V, known as the *Verdet constant*, depends on the frequency of the light and the material of the medium. The kinematic origin of the rotation of the plane of polarization, as shown in Part 6, Chap. 1, Sec. 8, is a kind of double refraction in which the normal modes of polarization are right and left circularly polarized waves. Accordingly the Verdet constant is related to the difference of the refractive indices for n_r and n_l for the right and left circular polarization by

$$n_r - n_l = \frac{\lambda_0}{\pi} V\mathbf{H}\cdot\mathbf{n} \tag{6.14}$$

in which **n** is the unit vector in the direction of propagation of the light waves.

Conventionally positive Verdet constant means that the plane of polarization is rotated, by passage through the medium, in the same sense as the direction of flow of positive electric current flowing in a solenoid which could produce the magnetic field. The rotation is in the same sense for either direction of propagation of the light, and so the total rotation can be multiplied by reflecting the light back and forth several times through the magnetic field. For a material of positive Verdet constant and for propagation in the direction of **H**, the rotation will therefore appear as left, because the observer is facing in the direction $-\mathbf{H}$.

This therefore corresponds to $n_r > n_l$, as in (6.14). For propagation in the opposite direction a minus sign is introduced by the $\mathbf{H} \cdot \mathbf{n}$ but as the meanings of right and left are now interchanged owing to change in orientation of the observer, this gives the same absolute sense of rotation in space as for propagation in the direction $+\mathbf{H}$.

The Verdet constant is commonly given in minutes per gauss-centimeter. For Na D light in liquid water at 0°C,

$$V = +1.31 \times 10^{-2} \text{ min/gauss-cm}$$

and therefore $n_r - n_l = 7.15 \times 10^{-11} H$

where H is in gauss. The Faraday effect for most solids and liquids (Table 6.7) is of this order of magnitude, indicating that the double refraction produced by the magnetic field is extremely small even for fields of the order of 10^4 gauss. The Verdet constant for gases under standard conditions is approximately 10^{-3} smaller and is consequently difficult to measure [18]. The Faraday effect falls rapidly to zero with increasing wavelength [19], as indicated in Table 6.8.

TABLE 6.7. VERDET CONSTANT FOR SOLIDS AND LIQUIDS (Na D LIGHT)*

Material	Temperature, °C	V (10^{-3} min/gauss-cm)
H_2O	0	13.11
C_2H_5OH	25	11.12
n-C_3H_7OH	17.3	11.81
Acetone	15.1	11.09
CH_3Cl	18	12.9
CCl_4	15	16.03
C_6H_6	20	29.7
CS_2	0	43.41
ZnS (sphalerite)	16	225
NaCl	16	35.85
KCl	16	28.58
$PbO \cdot SiO_2$ (glass)	16	77.9

* For more complete data see "International Critical Tables," vol. VI, p. 425, McGraw-Hill, New York, 1926–1930.

TABLE 6.8. DISPERSION OF VERDET CONSTANT*

Material	Wavelength λ, μ				
	0.6	0.8	1.0	1.5	2.0
H_2O	12.6	7.0	4.4	(2.9 at λ = 1.25)	
CCl_4	16.1	8.9	5.7	2.5	1.3
C_6H_6	28.1	15.3	9.5	3.9	2.2
CS_2	39.4	21.4	13.5	5.8	3.1

* V in 10^{-3} min/gauss-cm. Temperature 23°C.

The first theory [20] of the Faraday effect was based on the classical theory of the Zeeman effect and the related theorem of Larmor concerning the effect of a magnetic field on the classical motion of electrons. According to the Larmor theorem for electrons moving in an axially symmetric force field a magnetic field applied along the axis of symmetry causes all the motions to precess around the magnetic-field axis, so that they are the same when referred to a coordinate system rotating with angular velocity,

$$\omega_L = \frac{e}{2mc}\mathbf{H}$$

as they are relative to a fixed coordinate system in the absence of a magnetic field Therefore oscillatory motion parallel to the magnetic field is unaffected by it, while electrons moving in circular orbits at right angles to the magnetic field will have their frequency increased if their undisturbed motion is in the same sense as that of the Larmor precession and decreased if it is in the opposite sense.

In consequence a term of the form $A/(\omega_0^2 - \omega^2)$ in the dispersion formula becomes modified to

$$\frac{A}{(\omega_0 \pm \omega_L)^2 - \omega^2}$$

in the presence of a magnetic field, the plus sign appearing for motions in the same sense as the Larmor precession and the minus sign for those opposed to the Larmor precession. Thus for a fixed light frequency the contributions of the like-sense motions to the refractive index are reduced and those of the unlike sense enhanced, the amount being to the first order,

$$\frac{\pm \omega_L A}{(\omega_0^2 - \omega^2)^2}$$

Following out this argument in detail gives Becquerel's classical formula for the Verdet constant,

$$V = -\frac{e}{2mc^2}\lambda\frac{\partial n}{\partial \lambda} \tag{6.15}$$

in which n is the refractive index in the absence of a magnetic field and λ is the vacuum wavelength of the light, while e and m are the charge and mass of the electron. Since $\partial n/\partial\lambda$ is negative, the theory gives a positive Verdet constant in accordance with observation for most materials.

The quantum-mechanical theory of the Faraday effect [21] is based on the general theory of dispersion, taking into account the effect of the magnetic field on the energy levels of the atoms or molecules. The theory becomes more complicated than the simpler Becquerel picture, for several reasons. One is the need to consider the anomalous magnetic moment of electron spin, which complicates the Zeeman effect of atomic spectral lines. Another is the complications arising from the lack of axial symmetry in the fields acting on ions in crystals or in atoms that are combined in molecules. Still another is the complications due to temperature dependence of the relative population of Zeeman sublevels of low-lying states which are split magnetically [22], as in the theory of paramagnetic susceptibilities, giving rise to a corresponding temperature dependence of the Verdet constant over and above that which is due simply to variation of density with temperature.

The Cotton-Mouton effect may also be considered from the elementary point of view used by Becquerel.

For light polarized with **D** along the external **H** the electron oscillations which contribute to the dispersion are those which are unaffected by the external magnetic field. For **D** transverse to the external **H** the circular motions as modified by Larmor precession contribute equally, giving a term of the form

$$\frac{1}{2}\left[\frac{A}{(\omega_0+\omega_L)^2-\omega^2}+\frac{A}{(\omega_0-\omega_L)^2-\omega^2}\right] = \frac{A}{\omega_0^2-\omega^2} - \frac{\omega_L^2 A}{(\omega_0^2-\omega^2)^2} + \cdots$$

Therefore, according to this simple picture, the refractive index n_p ought to be greater than that for the perpendicular polarization n_s by an amount that is proportional to the square of the applied magnetic field.

These results are in accordance with experiment for most materials. The Cotton-Mouton constant C is defined as giving the phase difference Δ between the components for a path length d of light of vacuum wavelength λ by

$$\Delta = \frac{(n_p - n_s)d}{\lambda} = CdH^2 \qquad (6.16)$$

Some representative values of the Cotton-Mouton constant are given in Table 6.9.

TABLE 6.9. COTTON-MOUTON CONSTANT FOR LIQUIDS

Liquid	Temp, °C	λ, mμ	$C \times 10^{13}$
Acetone	20.2	578	37.6
Benzene	26.5	580	7.5
Chloroform	17.2	578	−65.8
CS_2	28.0	580	− 4.0

From the observed values it follows that for usual field strengths the Cotton-Mouton effect is small compared with the Faraday effect. Thus, for CS_2,

$$n_r - n_l = 2.36 \times 10^{-10}H$$
$$n_p - n_s = -2.32 \times 10^{-17}H^2$$

Therefore a field of 10^7 gauss would be needed to produce as great Cotton-Mouton double refraction transverse to the field as the Faraday double refraction along the magnetic field. With fields of the order 10^4 gauss the Faraday effect will be some 10^3 times greater than the Cotton-Mouton double refraction.

In the preceding discussion the two effects have for convenience been discussed separately. Actually both are present together. This means that the normal modes of polarization for light propagated at a direction making an angle θ with **H** are elliptically polarized orthogonal modes which become the opposite circular modes in the limit $\theta = 0$ and go over continuously into the parallel and transverse linear modes at $\theta = \pi/2$. If, as is usually the case, the Faraday birefringence is large compared with the Cotton-Mouton birefringence, then the transition from nearly circular modes over to linear modes takes place in a very small range of directions near $\theta = \pi/2$, which is the justification for the usual simplifying assumption which neglects the Cotton-Mouton effect at all directions of propagation except $\theta = \pi/2$, where the Faraday effect vanishes.

6. Kerr Effect

Isotropic transparent substances become doubly refracting when placed in an electric field, the behavior being like that of a uniaxial crystal with the optic axis in the direction of the applied electric field. This property was discovered in 1875 by Kerr [23] and is known as the *Kerr effect.* The effect is observed in gases and liquids as well as in solids, indicating that it is due to a direct action of the electric field on the optical properties of the medium, rather than being entirely due to the indirect action of the electric field in producing electrostriction, which then produces stress birefringence.

Writing n_p and n_s for the refractive indices for waves with **D** of the light wave parallel and perpendicular, respectively, to the applied **E** and λ for the vacuum wavelength of the light, the Kerr effect is characterized by the Kerr constant B, defined in

$$n_p - n_s = \lambda B E^2 \qquad (6.17)$$

In the terminology of crystal optics n_p refers to the extraordinary ray and n_s to the ordinary ray.

Some representative values of B are given in Table 6.10. For most substances B is positive, but negative values also occur.

TABLE 6.10. KERR CONSTANT FOR SEVERAL LIQUIDS (20°C AND Na D LIGHT)*

Liquid	Formula	$B \times 10^7$
Benzene	C_6H_6	0.60
Carbon disulfide	CS_2	3.21
Chloroform	$CHCl_3$	−3.46
Water	H_2O	4.7
Chlorobenzene	C_6H_5Cl	10.0
Nitrotoluene	$C_5H_7NO_2$	123.
Nitrobenzene	$C_6H_5NO_2$	220.

* E in Eq. (6.17) is in esu per centimeter or 300 volts/cm.

The molecular interpretation [24] of the Kerr effect is based on recognition that the individual molecules of the medium are anisotropic. Without an applied electric field the molecules are oriented at random, and so the anisotropic properties are lost on the average. The applied electric field acts to produce a partial statistical orientation so that now the anisotropic properties of individual molecules no longer average out completely, giving rise to the observed birefringence. In solids the situation is more complicated and less well understood, because orientation of molecules cannot be simply analyzed in this case, and there are also complications [25] because of the simultaneous presence of an electrostriction and a birefringence due to the elastic deformation thus produced. This acts to make a negative contribution to B.

According to the orientation theory for fluids the effective electric field tending to orient a molecule is $\mathbf{F} = \mathbf{E}(\epsilon + 2)/3$, where ϵ is the static dielectric constant rather than $\mathbf{E}$ itself. The theory gives, for the Kerr constant,

$$B = N \frac{n^2 + 2}{2n} \left(\frac{\epsilon + 2}{3}\right)^2 \frac{b}{\lambda} \tag{6.18}$$

where N is the number of molecules in unit volume and b is a statistical molecular parameter discussed below.

An individual molecule is characterized by three quantities: α, the polarizability tensor for optical frequencies, which determines its interaction with the electric field of the light wave; β, the static-polarizability tensor, which determines its interaction with the applied static electric field; and $\mathbf{p}$, the permanent electric-dipole moment. Both β and $\mathbf{p}$ interact with the effective external field $\mathbf{F}$ to produce orientation tendencies since the energy of a molecule in the field is given by

$$W = -\mathbf{p} \cdot \mathbf{F} - \tfrac{1}{2}\mathbf{F} \cdot \beta \cdot \mathbf{F} \tag{6.19}$$

This expression for W has to be used in a Boltzmann distribution formula to determine the distribution of angular orientations in the same way as is done in the theory of the dielectric constant of polar molecule-containing media. The effect of the first term is to tend to line up $\mathbf{p}$ parallel to $\mathbf{F}$, while the second term tends to line up the axis of largest static polarizability along $\mathbf{F}$. As $W \ll kT$ for actually applicable fields, the orientation produced is only a slight statistical tendency toward these preferred conditions.

In terms of these molecular properties the parameter b in (6.18) is

$$b = \frac{b_1}{kT} + \frac{b_2}{(kT)^2} \tag{6.20}$$

in which

$$b_1 = \tfrac{1}{45}[(\alpha_x - \alpha_y)(\beta_x - \beta_y) + (\alpha_y - \alpha_z)(\beta_y - \beta_z) + (\alpha_z - \alpha_x)(\beta_z - \beta_x)]$$
$$b_2 = \tfrac{1}{45}[(\alpha_x - \alpha_y)(p_x^2 - p_y^2) + (\alpha_y - \alpha_z)(p_y^2 - p_z^2) + (\alpha_z - \alpha_x)(p_z^2 - p_x^2)]$$

Therefore b_1 measures the optical anisotropy introduced by the orienting action of the anisotropy of the static polarizability, and b_2 measures that due to the orienting action of the permanent-dipole moment.

For a molecule having the z axis as an axis of symmetry $p_x = p_y = 0$, $\alpha_x = \alpha_y$, and $\beta_x = \beta_y$; so the foregoing expressions become

$$b_1 = \tfrac{2}{45}(\alpha_z - \alpha_x)(\beta_z - \beta_x)$$
$$b_2 = \tfrac{2}{45}(\alpha_z - \alpha_x)p_z^2$$

The difference between α and β is that due to dispersion; thus one expects (though it need not be true) that both factors in b_1 will be of the same sign so that b_1 will usually be positive. It is, however, conceivable that if $\alpha_z - \alpha_x$ is small it might have a different sign from $\beta_z - \beta_x$, giving a negative b_1. If b_1 is positive, then a negative Kerr constant must arise from b_2 being sufficiently negative to outweigh the positive b_1 term. This seems to be the fundamental reason why $CHCl_3$ has a negative Kerr constant in contrast to CH_3Cl, namely, because the triangle of the three Cl atoms has a larger polarizability in its own plane than for directions at right angles to this plane and hence directed along the axis of symmetry of the molecule. H. A. Stuart [26] has devoted attention to showing how theoretical interpretations of the Kerr effect of vapors can contribute to knowledge of molecular structure.

Of course, in the foregoing discussion, static polarizability β really means low-frequency polarizability if the electric fields used are not static. Some work has been done on the dependence of the Kerr effect on time of application of the electric field and the relaxation time for Kerr birefringence when the applied electric field is put on at radio frequencies.

7. Optical Rotatory Power

Optical activity, or optical rotatory power, is the property of some media of rotating the plane of polarization of linearly polarized light [27]. It is due to a natural circular double refraction of the medium in which the normal modes of polarization are right and left circularly polarized waves. If ϕ is the angle turned through by the plane of polarization in traversing unit path length through the medium, λ is the vacuum wavelength, and n_l and n_r the refractive indices for left and right circularly polarized light,

$$\phi = \frac{\pi}{\lambda}(n_l - n_r) \tag{6.21}$$

Positive rotation is that which takes place in the clockwise sense as seen by the observer and occurs when $n_l > n_r$. The sense of turning of the plane of polarization is the same as that of the circularly polarized wave which has the greater phase velocity.

Rotation in the clockwise sense is said to be *dextro-*, or *d-* , *rotation*, and that in the opposite sense is called *levulo-* , or *l-* , *rotation*.

Natural optical activity is distinguished from that induced by a magnetic field (Faraday effect) in that the sense of turning bears a fixed relation to the direction of propagation. Thus, when a beam of light is reflected back through the medium, the rotation, in keeping the same relation to the reversed direction of propagation, reverses its absolute sense in space and so the net rotation is zero in such a double traversal of the path.

Optical activity occurs in some crystals in combination with ordinary double refraction [28]. Thus quartz crystals occur in right- and left-handed forms which are structurally related as mirror images and which have equal and opposite rotation. The rotation for Na D light along the optical axis is 21.6°/mm and therefore $n_l - n_r = 7.06 \times 10^{-5}$. This circular double refraction is present simultaneously with ordinary double refraction. Along the optical axis the ordinary double refraction vanishes, and the normal modes of polarization are the two circular forms. At right angles to the optical axis the two indices for the two states of plane polarization differ by a much greater amount, $n_e - n_o = 900 \times 10^{-5}$. At intermediate directions the normal modes are orthogonal states of elliptic polarization, but since the ordinary double refraction is so much larger than the optical rotatory power, the elliptic states become practically plane-polarized states for angles inclined by a small amount to the optical axis.

Cinnabar, HgS, is another example of a uniaxial crystal which shows large double refraction and large optical activity. The Na D indices for this hexagonal crystal are $n_o = 2.854$ and $n_e = 3.201$, while the rotatory power along the optical axis is 540°/mm.

Some regular crystals, of which $NaClO_3$ and $NaBrO_3$ are examples, show optical activity without ordinary double refraction. The rotations of these for Na D light are 3.16 and 2.13 °/mm, respectively. Here the rotation is due to the arrangement of the atoms in the crystal because these materials show no rotation when in aqueous solution. Optical rotation is also shown by biaxial crystals, but here, too, the circular double refraction is so small compared with the ordinary double refraction that the rotatory power is observable only along the directions of the optical axes.

In the case of more complicated molecules the rotatory power is at least partly an attribute of the molecules themselves, as evidenced by the fact that aqueous solutions as well as their crystals show rotatory power. In this respect curious reversals may occur. Thus with rochelle salt, $NaKC_4H_4O_6 \cdot 4H_2O$, solutions that are *d*-rotatory give rise to crystals that are *d*-rotatory, but, with rubidium tartrate, solutions that are *d*-rotatory give rise to crystals that are *l*-rotatory.

A great many naturally occurring organic compounds show optical rotatory power in solution or in the liquid state. In these cases the rotatory power has to be a structural attribute of individual molecules, although the amount of the effect may be modified by loose associative combinations with molecules of the solvent. The study of optical rotatory power has been of great importance historically in the development of organic chemistry, as it provided the original basis for the ideas of the stereochemistry of carbon compounds and of the tetrahedral directions of the four valencies of carbon.

Optical rotatory power originates in properties of a molecule represented by a parameter β occurring in the equations

$$\begin{aligned} \mathbf{p} &= \alpha\mathbf{E}' - \frac{\beta}{c}\dot{\mathbf{H}} \\ \mathbf{m} &= +\frac{\beta}{c}\dot{\mathbf{E}}' \end{aligned} \qquad (6.22)$$

in which $\mathbf{p}$ and $\mathbf{m}$ are, respectively, the electric- and magnetic-dipole moments of the molecule. A nonvanishing β arises from some kind of helical structure in the molecule. The direct action of an electric field on the molecule is the displacement in opposite directions of positive and negative charges toward $\mathbf{E}'$ and $-\mathbf{E}'$, respectively, to generate the electric-dipole moment represented by $\alpha\mathbf{E}'$. Currents have to flow in the molecule in phase with $\dot{\mathbf{E}}'$ to produce the time variations of this electric dipole. Now, if the molecular structure is such that these currents do not move directly in the direction of $\dot{\mathbf{E}}'$ but are constrained to move in a somewhat helical way so that they have a circulatory component around $\dot{\mathbf{E}}'$, then this circulatory component will give rise to a dipole magnetic moment that is given by the second of Eqs. (6.22). Conversely, when such a molecule is in a changing magnetic field, the changing magnetic flux through the molecule sets up induced currents circulating around $\dot{\mathbf{H}}$ in the sense given by Lenz's law. The same helical constraints now will give rise to a separation of charges in the direction of $\dot{\mathbf{H}}$, thus setting up an electric-dipole moment which is represented by the term in β in the first of Eqs. (6.22).

If now there are N_s molecules of kind s in unit volume, then

$$\mathbf{P} = \sum_s N_s\mathbf{p}_s \qquad \text{and} \qquad \mathbf{I} = \sum_s N_s\mathbf{m}_s$$

which leads to the connections

$$\begin{aligned} \mathbf{D} &= \epsilon\mathbf{E} - g\dot{\mathbf{H}} \\ \mathbf{B} &= \mathbf{H} + g\dot{\mathbf{E}} \end{aligned}$$

in which ϵ and g are defined by

$$\begin{aligned} \epsilon &= 1 + \frac{4\pi\Sigma N_s\alpha_s}{1 - (4\pi/3)\Sigma N_s\alpha_s} \\ cg &= \frac{4\pi\Sigma N_s\beta_s}{1 - (4\pi/3)\Sigma N_s\alpha_s} \end{aligned} \qquad (6.23)$$

These connections lead, by the methods outlined in Part 6, Chap. 1, to the result that the normal modes of polarization are right and left circularly polarized waves in which the rotation ϕ per unit length is

$$\phi = \frac{4\pi^2}{\lambda^2}cg = \frac{16\pi^3}{\lambda^2}\frac{n^2+2}{3}\sum N_s\beta_s \qquad (6.24)$$

in which $n^2 = \epsilon$ is the ordinary index of refraction of the mixture. In this way the optical rotatory power of the medium is interpreted in terms of the parameters β_s of the molecules of which it is made.

The quantum mechanics [29] of optical rotatory power involves the calculation of β from a specific molecular model. In the usual development of dispersion theory the ratio of the molecular size to the wavelength of the light is treated as negligibly small. To get optical rotatory power, it is necessary to carry the calculations to a higher order, taking into account the first power of the ratio of molecular size to wavelength. This gives, for molecules in the quantum state a, the value

$$\beta_a = \frac{c}{3\pi h}\sum_b \frac{R_{ba}}{\nu_{ba}{}^2 - \nu^2} \qquad (6.25)$$

where R_{ba} is a quantity called the *rotational strength* of the absorption line associated with the quantum frequency ν_{ba} involved in the transition $a \to b$. The value of β to be used for a system of molecules at a given temperature has to be calculated in the usual way by making a weighted average of the values of β_a for each quantum state weighted according to the

Boltzmann distribution. Equation (6.25) is the analogue of the corresponding quantum-mechanical expression for the optical polarizability which is responsible for the ordinary refractive index.

The quantum-mechanical theory further gives

$$R_{ba} = \text{Im}\,[(a|\mathbf{p}|b) \cdot (b|\mathbf{m}|a)] \qquad (6.26)$$

which relates it to the matrix components of electric- and magnetic-dipole moments associated with the transition $a \rightarrow b$. Since **p** is a polar vector and **m** an axial vector, their scalar product is a pseudoscalar whose numerical value reverses sign on passing from left-handed to right-handed coordinates. Therefore the numerical values of R_{ba} reverse sign on passing from a particular molecular model to its own mirror image. Hence, if the molecule possesses a center of symmetry, all the R_{ba} vanish and so such a molecule can have no optical rotatory power.

Very little has been accomplished in the way of explicit calculation of the parameter β for realistic molecular models. In the older electron theory Drude based a calculation on a model which postulated that the electrons were elastically bound to equilibrium positions but were constrained to move in helical paths. However, in 1933, Kuhn showed that there was an error in Drude's calculations and that the optical activity of such a model really vanishes. In 1915 Born, Oseen, and Gray, working independently, showed that coupled electron oscillators could show optical rotatory power, and this model has been applied in some detail to chemical problems by Kuhn [30]. After he had discredited the Drude model, it was believed that no one-electron model would give optical rotatory power, and therefore it must necessarily arise from a dynamical coupling of electrons located at different parts of the molecule. However, this was shown not to be the case [31] by a calculation of the rotatory power of molecules in which a single electron moves in the potential field,

$$V = \tfrac{1}{2}(k_1x^2 + k_2y^2 + k_3z^2) + Axyz$$

which is the simplest form having the necessary dissymmetry.

The simplest kind of molecule capable of showing rotatory power is one containing a single asymmetric carbon atom, that is, a carbon atom to which four different groups are attached. For example, rotatory powers have been measured [32] for a large number of asymmetric monohydroxy alcohols,

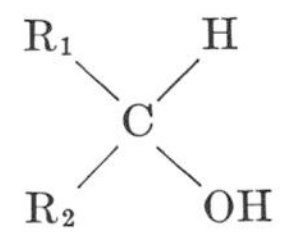

where R_1 and R_2 stand for different aliphatic radicals.

The *specific rotation* is defined as ϕ/ρ, where ϕ is usually in degrees per decimeter and ρ is the density in grams per cubic centimeter or, in the case of a solution, is the concentration in grams per cubic centimeter of active material per cubic centimeter of solution. The *molecular rotation* is the product of the specific rotation by the molecular weight M of the active material, and therefore it is the rotation divided by the concentration of active material expressed in gram moles per cubic centimeter.

In this unit, for Na D light the molecular rotation is about 1,100 for R_1 = methyl and R_2 = ethyl, increasing gradually to about 1,400 as R_2 moves along the homologous series up to *n*-decyl. In the case in which R_1 = *n*-propyl and R_2 = isopropyl the molecular rotation is 2,470, indicating that strong asymmetry is produced when two of the groups attached to the asymmetric carbon atom are isomers.

8. Photoelasticity

The double refraction of solid materials is changed when they are strained by being put in a state of stress. This phenomenon was discovered by David Brewster in 1816 and is variously known as *photoelasticity, mechanical birefringence,* and *stress birefringence.* The effect is used as a technique for studying stress distributions in transparent models (Part 3, Chap. 6).

The simplest case is that of a material which is optically isotropic when in an unstressed condition. When a uniaxial stress is applied along the z axis, the material becomes weakly doubly refracting like a uniaxial crystal whose optical axis is the axis of the applied stress. Two photoelastic constants are needed to characterize the behavior in the range of stress for which Hooke's law holds. Let n be the refractive index for the unstressed solid, n_z that for the stressed solid for the extraordinary wave which is polarized with **D** in the plane containing the direction of propagation and the direction of stress, and $n_x = n_y$ that for the ordinary waves that are polarized with **D** at right angles to the principal plane. Then, if the stress tensor is $-P_z\mathbf{kk}$, so that positive P_z means a compressive uniaxial stress along the z axis, the stress optical coefficients B_1 and B_2 are defined by the linear relations

$$n_x - n = B_1P_z \qquad n_z - n = B_2P_z \qquad (6.27)$$

In most experiments all that is measured is the difference of these, which is usually denoted as B,

$$B = B_1 - B_2 = \frac{(n_x - n_z)}{P_z} \qquad (6.28)$$

From these definitions it follows that positive B means that the refractive index of the ordinary wave is greater than that for the extraordinary wave; that is, the behavior is like that of a negative uniaxial crystal. This is the actual sign of B for most materials.

In the case of biaxial or triaxial stress the material becomes like a biaxial crystal, the behavior being described by the same relations as in (6.27) applied three times over for each of the principal uniaxial stress components. The case of pure hydrostatic pressure leads to vanishing double refraction, but to a change of refractive index with pressure given by

$$\Delta n = (2B_1 + B_2)P \qquad (6.29)$$

From the definitions it follows that the B coefficients have the dimensions of reciprocal pressure, or square centimeters per dyne in absolute cgs units. In the specialized literature the values are commonly expressed in Brewsters, to honor the discoverer of the effect, where

$$1 \text{ Brewster} = 10^{-13} \text{ cm}^2/\text{dyne} = 10^{-7} \text{ (bar)}^{-1}$$

and it is found empirically that the stress optical coefficients are of the order of 1 to 10 Brewsters.

From general theoretical considerations it is probably more significant to describe the effect in terms of strain optical coefficients rather than stress optical coefficients. This view is supported by the fact that experimental results show the effect to be directly correlated with strain rather than stress in various organic materials which show a large elastic hysteresis. The strain optical coefficients were first introduced by F. Neumann [33], who developed an early theory of the photoelastic effect. Thus for a uniaxial strain in the z direction (Part 3, Chap. 1, Sec. 6) whose only component is $e_{33} = \partial r_3/\partial x_3$, a stretching along the z axis, Neumann's strain optical coefficients P and Q are defined as

$$\begin{aligned} n_x - n = n_y - n &= -n^2 P e_{33} \\ n_z - n &= -n^2 Q e_{33} \end{aligned} \qquad (6.30)$$

In view of the relations between stress and strain for an isotropic solid (Part 3, Chap. 1, Sec. 7) the strain optical coefficients and the stress optical coefficients are related as follows,

$$\begin{aligned} (P - Q)n^2 &= \frac{EB}{1 + \sigma} \\ (2P + Q)n^2 &= \frac{E(2B_1 + B_2)}{1 - 2\sigma} \end{aligned} \qquad (6.31)$$

and the change of index of refraction with change in density produced by isotropic compression is

$$\frac{dn}{d\rho} = \frac{1}{3}(2P + Q)n^2 \qquad (6.32)$$

Present views on the molecular optics of photoelasticity are largely based on the work of Mueller [34], who pointed out that two distinct effects contribute to the double refraction produced by the deformation, (*a*) an anisotropic alteration of the relation between the local field $\mathbf{E}'$ and $\mathbf{E}$ ordinarily given for amorphous materials by the Lorentz-Lorenz correction, and (*b*) an alteration of the polarizability of the supposed initially isotropic ions themselves as a result of the anisotropy of their surroundings caused by the deformation. Mueller's analysis shows that the first of these effects tends to make B positive and the second has the opposite sign, so that the actual effect is the difference of the two parts. (*c*) Later work [35] recognizes another kind of contribution, which is believed to be more important than (*b*) in the behavior of organic polymers. The refractivity of molecules is the sum of contributions associated with chemical bonds. Evidence from light scattering indicates that these are individually anisotropic, and therefore the isotropy of an amorphous material must arise from uniformity of random orientation of such units. Deformation of the material upsets this uniformity of distribution of orientations, giving some resultant anisotropy. In (*b*) anisotropy results from distortion of originally isotropic units, whereas in (*c*) it results from distortion of an originally random distribution in orientation of individually anisotropic units.

9. Flow Birefringence: Maxwell Effect

Viscous liquids consisting of anisotropic molecules and flowing in such a way that there is a shearing velocity gradient show optical birefringence. This is often called the *Maxwell effect.* The simplest case is that in which the fluid is flowing parallel to a plane boundary (the plane $y = 0$) with velocity v_x, which is a function of y.

For moderate velocity gradients it is observed that the principal planes of polarization for light traveling along the z axis are the $x'y'$ axes, which are inclined at an angle of 45° to the xy axes. The difference in refractive indices along these two axes is found to be proportional to the shearing velocity gradient,

$$\Delta n = c\left(\frac{\partial v_x}{\partial y}\right) \qquad (6.33)$$

The coefficient c, which is the measure of flow birefringence, is known as *Maxwell's constant.*

These principal axes and this proportionality to the velocity gradient are verified experimentally up to velocity gradients of the order of 10^4 sec^{-1}. At about this value a change in the principal axes takes place whereby these turn from $x'y'$ toward xy.

The theory of the effect has been discussed by Frenkel [36], who shows that the orientation forces acting on rod-shaped molecules in a fluid which has a shearing velocity gradient tend to orient the molecules in such a way as to make xy be the principal axes. Brownian movement causes a diffusional disorientation which tends to destroy birefringence, but in such a way as to favor the $x'y'$ axes as principal axes. At low gradients the resultant effect is mostly determined by the Brownian movement establishing $x'y'$ as the principal axes.

The effect is an important tool for the study of high polymers and macromolecular substances of biochemical interest [37].

10. Pleochroism

Since in all cases optically absorbing media are characterized by a complex index of refraction (Sec. 3), crystalline absorbing media therefore show a dependence of the optical absorption coefficient on direction of propagation of the light through the crystal and on the state of polarization of the light traveling in a particular direction. Similar effects are shown by complex polycrystalline materials in which there is a partial statistical orientation of the anisotropic crystallites.

The name *pleochroism*, or many-coloredness, refers to the variety of effects arising from this dependence of absorption coefficient on direction and polarization. The word *dichroism* is often used for the same physical phenomenon, this name emphasizing the *two* different absorption coefficients associated with the two normal modes of propagation in a particular direction. Pleochroism is the word mostly used in mineralogical literature, while dichroism is more often found in the physical literature.

Optically active materials show circular dichroism, that is, a difference of absorption coefficient for right and left circularly polarized light. But this difference is always so small that it is observed only indirectly. If plane-polarized light is sent through such a medium, the initially equal right and left circularly polarized components, after traversing a finite path length, combine in different phase to produce rotation of the plane of polarization. But, being now also of different amplitude, they recombine to produce an emergent beam of elliptically polarized light. It is by the measurement of this ellipticity that circular dichroism is observed.

The classic example of a strongly dichroic crystal is the natural mineral *tourmaline.* These crystals are aluminoborosilicates of variable composition, usually having B_2O_3 about 10 per cent, Al_2O_3 about 20 to 40 per cent, and SiO_2 about 35 to 40 per cent, with small amounts of Fe and Cr replacing some of the Al. The crystals belong to the hexagonal system, without a center of symmetry and having a polar principal axis. Tourmalines absorb the ordinary ray strongly for all colors of the visible, and so a plate a few millimeters thick that is cut parallel to the principal axis and used as a filter in a beam of unpolarized light will give an emergent beam that is almost entirely the extraordinary ray.

The manufactured polarizing material called Polaroid consists of an oriented sheet of small organic crystals that are strongly dichroic [38].

An interesting example of pleochroism is afforded by cutting a cube of *diaspore,* $Al_2O_3{\cdot}H_2O$, with the faces normal to the principal axes of the index-of-refraction tensor. This crystal belongs to the orthorhombic system and has D-line refractive indices, 1.664, 1.671, and 1.694. The colors resulting from dichroism by absorption of white light are potentially six in number, for the two normal modes of linear polarization associated with each of the three mutually orthogonal directions of propagation. However, these are equal in pairs because the absorption depends only on the direction of polarization and not on the direction of propagation. The three principal tints of diaspore are sky blue, violet, and clear yellow.

In uniaxial crystals the principal tints are two in number, an ordinary tint that is the same for all directions of propagation, and an extraordinary tint that like n_e depends on the direction of propagation. The mineral *penninite* has green for its ordinary tint and orange red for its extraordinary tint observed in directions of propagation normal to the optical axis. In *smoky quartz* the ordinary tint is pale red, and the extraordinary tint is pale yellow.

Every kind of birefringence presumably has its counterpart in a corresponding kind of dichroism. For example, flow birefringence gives rise to flow dichroism, which has been investigated theoretically and experimentally in the case of fluorescin dye in flowing viscous solutions in glycerin [39].

11. Light Scattering

When a beam of light traverses a material medium, its intensity is reduced, partly by absorption (Sec. 3) and partly by scattering. In scattering, light energy is taken from the beam and reradiated as spherical waves from each scattering center and is thus lost to the beam because it now travels in an altered direction.

Scattering always arises from some kind of heterogeneity of the material. The most obvious kind of heterogeneity is that due to suspended particles as in smokes, dusts, or clouds of liquid particles in air. Because of his pioneer studies in this field [40], this light-scattering property is often called the *Tyndall effect.* Tyndall thought the blue of the sky resulted from scattering of sunlight by suspended dust particles in the atmosphere.

Lord Rayleigh recognized [41], however, that the deep blue color and brightness of the sky in daytime are due to scattering of sunlight by dust-free air. Additional scattering on occasion arises from dust, smoke, or water droplets, but the main effect arises from the air itself. In this case each molecule acts as a scattering center, taking energy from the incident beam and radiating a spherical wave. The resultant effect in any direction is obtained by summing the amplitudes of all such scattered waves, taking due regard for their relative phases.

If the molecules were all alike and were arranged exactly in a lattice, the resultant amplitude would vanish in every direction except that of the incident beam, giving no scattering. In the direction of the incident beam the scattered amplitudes add constructively to give a wave which combines with the incident wave to yield a resultant which propagates with a phase velocity c/n. In this way the index of refraction is intimately related to the coherent scattering by individual molecules.

But as the gas molecules are not arranged in such an exact way, such complete destructive interference of the light scattered at finite angles does not occur. The actually observed scattering is a residual effect arising from fluctuations in density of the gas due to thermal motions of its molecules. In addition, if the molecules are anisotropic, fluctuations in the orientational distribution of the molecules will occur, giving rise to additional scattering. If the gas is a mixture of several components, an additional source of fluctuations giving rise to scattering is that the composition of particular volume elements will depart from the average uniform composition of the whole gas.

In liquids and solids analogous fluctuations produce light scattering. Similarly the scattering of light by dilute liquid solutions is greater than that of the pure solvent because of the heterogeneity introduced by fluctuations in the concentration of the solute. Light scattering has become an important tool in the study of high polymers, colloids, and materials of biochemical interest [42].

Most experimental work in this field does not involve high resolution of the incident and scattered light. However, such study by C. V. Raman [43] led him in 1927 to the discovery of the *Raman effect.* In this effect some of the scattered light is modified in frequency from that of the incident light. Such modifications are always consequences of motions occurring in the scattering medium, which may take many forms, and which are governed in important ways by the quantum dynamics of the scattering molecules.

Classically, if light of frequency ν is scattered by a

molecule in which there are motions of frequencies ν_1 and ν_2, then the scattered waves will consist of several different combination frequencies,

$$\nu + m\nu_1 + n\nu_2$$

where m and n are positive or negative integers or zero. Likewise the translational motion of the scattering molecule modifies the frequency by a Doppler effect. This gives rise to structure of the scattered spectrum line from a monochromatic source because of the distribution of translational velocities.

Quantum mechanically a molecule in its lowest energy level may scatter an incident light quantum of energy $h\nu$ in a process which raises the molecule to a state higher in energy by amount $h\nu_1$, thereby leaving $h(\nu - \nu_1)$ for the energy of the scattered quantum. Molecules in the higher state may scatter light and undergo simultaneously a transition to the lower energy level, adding $h\nu_1$ to the energy of the incident quantum, so that the scattered quantum has the energy $h(\nu + \nu_1)$. Study of the spectrally resolved Raman scattered light has become an important tool for studying molecular energy levels, supplementing the study of infrared absorption spectroscopy. Nevertheless, in much of the experimental work on light scattering, because of the feeble intensity of the scattered light, such spectral resolution is not achieved, and so the intensity observed is the sum of the intensities of the unmodified and modified lines. In what follows, therefore, no further consideration is given to the Raman effect.

Observations are usually made of the intensity of the scattered light as a function of the angle θ of scattering relative to the incident beam [44]. If the scattering centers are small compared with the wavelength λ' in the medium ($\lambda' = \lambda/n_0$), then this dependence, for unpolarized incident light, is through a factor

$$\tfrac{1}{2}(1 + \cos^2\theta)$$

which is symmetrical about $\theta = \pi/2$. If the scattering centers are larger, the scattering shows dissymmetry, which is often characterized by stating the ratio $I(45)/I(135)$. For larger colloidal particles the angular dependence of the scattering becomes more complicated, showing a number of maxima and minima, and cannot be adequately characterized by any single parameter like the dissymmetry.

The terminology for describing polarization is based on the fact that usual experimental arrangements have both the incident and the observed scattered beams in the horizontal plane, and therefore θ is measured in a horizontal plane. Then $H(\theta)$ and $V(\theta)$ are used to represent the two components of scattered light at angle θ that are polarized with the electric vector horizontal and vertical, respectively. Likewise the incident intensity I_0 may be broken into components called H_0 and V_0. For isotropic scatterers that are small compared with λ', theory gives

$$H(\theta) = aH_0\cos^2\theta$$
$$V(\theta) = aV_0$$

If the incident light is unpolarized so that $H_0 = V_0$, this gives the factor for angular dependence already stated.

According to these formulas the light scattered at 90° from isotropic scatterers is perfectly polarized vertically even though the incident light is unpolarized. This is in accordance with observation for monatomic gases like argon.

However, for most gases experimental results show that the scattered light at 90° is not completely polarized, that is, it is partly depolarized. This is interpreted [45] as the result of the anisotropy of the polarizability of the molecules. Calling the principal values of the polarizability tensor α_1, α_2, α_3, Rayleigh showed that the depolarization of the 90° scattered light is

$$\rho_u = \frac{H_h + H_v}{V_h + V_v} = \frac{1}{2}\frac{(\alpha_1 - \alpha_2)^2 + (\alpha_2 - \alpha_3)^2 + (\alpha_3 - \alpha_1)^2}{(\alpha_1 + \alpha_2)^2 + (\alpha_2 + \alpha_3)^2 + (\alpha_3 + \alpha_1)^2}$$

in which H_v is written for the 90° scattered intensity of light with horizontal polarization due to incident light with vertical polarization, etc. If the molecule has an axis of symmetry so that $\alpha_1 = \alpha_2$, then observation of ρ_u allows one to infer the ratio α_3/α_1 and this together with the measured value of the index of refraction which gives the mean of the three principal values of α fully determines the polarizability tensor. Table 6.11 gives some representative values of the observed depolarization and the calculated values of the ratio-of-polarizability principal values for gases.

TABLE 6.11. DEPOLARIZATION OF LIGHT SCATTERING BY GASES*

Name	Formula	100 ρ	$\frac{\alpha_3}{\alpha_1}$
Argon	A	0	1.0
Methane	CH_4	0	1.0
Nitrogen	N_2	3.6	1.64
Oxygen	O_2	6.5	1.95
Chlorine	Cl_2	4.1	
Air	.....	4.2	
Nitric oxide	NO	2.7	1.54
Carbon monoxide	CO	1.3	1.35
Hydrogen chloride	HCl	0.7	1.25
Hydrogen bromide	HBr	0.8	1.27
Hydrogen iodide	.....	1.3	1.35
Carbon dioxide	CO_2	9.7	2.30
Carbon disulfide	CS_2	11.5	2.50
Carbon oxysulfide	COS	8.8	2.19
Water	H_2O	2.0	
Hydrogen sulfide	H_2S	0.3	
Sulfur dioxide	SO_2	3.1	
Ethane	C_2H_6	0.5	1.21
Ammonia	NH_3	1.0	0.75

* From S. Bhagavantam, "Light Scattering and the Raman Effect," Chemical Publishing, New York, 1942.

The formula for ρ_u is derived on the assumption that the scattering center is small compared with λ'. Experimentally an observed lack of dissymmetry gives assurance that this condition is fulfilled, after which the measured value of ρ_u gives information about the anisotropy of the polarizability.

For small isotropic scattering centers the intensity $I(\theta)$ observed at distance r at angle θ from a scattering volume V containing N scatterers per unit volume each of polarizability α is

$$\frac{I(\theta)}{I_0} = NV\left(\frac{2\pi}{\lambda'}\right)^4 \frac{\alpha^2}{r^2}\frac{1}{2}(1 + \cos^2\theta) \qquad (6.34)$$

The total light power scattered through all angles from unit volume and unit incident intensity across a sphere of radius r is

$$\tau = \frac{4\pi}{3}\left(\frac{2\pi}{\lambda'}\right)^4 N\alpha^2 \qquad (6.35)$$

This quantity is also the fractional reduction of intensity of the incident beam in going unit distance through the medium due to scattering. If β is the coefficient of absorption, then the attenuation of the beam intensity in going a distance x is

$$I = I_0 e^{-(\beta+\tau)x}$$

The quantity τ is variously called the *scattering coefficient* and the *turbidity coefficient.*

In the case of anisotropic scattering centers that are small compared with λ', the right side of (6.34) must include as another factor

$$\frac{1 + \rho_u}{1 - \frac{7}{6}\rho_u}$$

and α has to be interpreted as the mean of the three principal values of the tensor. Likewise in (6.35) the formula for τ has to be modified by inclusion on the right side of the factor

$$\frac{1 + \frac{1}{2}\rho_u}{1 - \frac{7}{6}\rho_u}$$

The maximum depolarization possible is that which corresponds to rodlike scatterers for which

$$\alpha_1 = \alpha_2 = 0$$

For these $\rho_u = 0.5$, and the intensity at 90° is 3.6 times, while the turbidity is 3.0 times, what it would be for isotropic scatterers having the same mean polarizability.

It is convenient to characterize the medium by the experimental value of its *Rayleigh's ratio,* which is defined as

$$R_u(90) = \frac{I(90)r^2}{I_0 V}$$

in which $I(90)$ is the scattering at 90° at distance r from a scattering volume V. According to the Rayleigh theory, therefore,

$$R_u(90) = \frac{1}{2}\left(\frac{2\pi}{\lambda'}\right)^4 N\alpha^2 \frac{1 + \rho_u}{1 - \frac{7}{6}\rho_u}$$

For gases this expression has been tested carefully as to the variation as inverse fourth power of the wavelength. It is this strong relative scattering of short wavelengths that makes the sky blue.

Another kind of test of the theory is to measure R_u for a known wavelength, measure ρ_u, and get an experimental value of α from the measured index of refraction by

$$\alpha = \frac{n^2 - 1}{4\pi N}$$

The theoretical equation thus leads to an experimental determination of N and hence, since $N_a = NM/\rho$, where M is the molecular weight and ρ the density, to an experimental value for the Avogadro number, N_a.

The difficulties of experimental measurement are such that this is not a precision method of determining the Avogadro number. The measurements do agree with other better determinations within their own experimental accuracy, thus giving support to the point of view that the scattering from a dust-free gas is due to the individual molecules.

Conversely the same theory can be usefully applied to determine [46] the molecular weight M of proteins and other macromolecules in aqueous solution. The Rayleigh ratio for the *increase* of light scattering of the solution over that of the pure solvent is measured for a known weight concentration ρ of the macromolecular substance, whose index of refraction, n, in the pure state, as well as n_0, that of the pure solvent, is supposed known. In this case

$$n^2 - n_0^2 = 4\pi N\alpha$$

and $N = N_a\rho/M$, and so the theoretical formula for the Rayleigh ratio becomes

$$R_u(90) = \frac{\pi^2}{2}(\lambda')^{-4}\frac{(n^2 - n_0^2)^2}{N_a\rho} M \frac{1 + \rho_u}{1 - \frac{7}{6}\rho_u}$$

From this M can be calculated, as all the other quantities are known. Light scattering is much used in this way for molecular-weight determinations.

Similarly the molecular weight may be obtained from an experimental measurement of the turbidity coefficient of a solution of known concentration, the corresponding formula being

$$\tau = \frac{4\pi^3}{3}(\lambda')^{-4}\frac{(n^2 - n_0^2)^2}{N_a\rho} M \frac{1 + \frac{1}{2}\rho_u}{1 - \frac{7}{6}\rho_u}$$

If the diameter of the scattering particles is greater than about one-tenth the wavelength in the medium, the simple theory no longer applies. The correct treatment becomes quite complicated. Only the case of spherical particles has been worked out in detail [47]. The Mie theory treats the scattering as an electromagnetic-wave boundary-value problem in which the scattered waves are determined by the condition that together with the incident plane wave the appropriate boundary conditions at the surface of the sphere must be satisfied. If $m = n/n_0$ is the index of refraction of the sphere relative to that of the medium, R is the radius of the sphere, $\lambda' = \lambda/n_0$ is the wavelength of the light in the medium, and $x = 2\pi R/\lambda'$, then the scattered wave amplitudes are determined as infinite series of Bessel functions of the radius multiplied by spherical harmonics in θ.

In the limit $mx \to 0$ the series reduces to its first term, which represents the Rayleigh scattering. This corresponds to the reradiation from the induced electric-dipole moment in the sphere. The next approximation involves the inclusion as well of terms representing the induced electric-quadripole moment and the induced magnetic moment of the sphere. This approximation suffices, for example, to give a good account of the scattering with water droplets in air, $m = 1.33$, up to a radius about one-sixth wavelength.

Writing $H_u(\theta)$ and $V_u(\theta)$ for the horizontal and vertical polarized scattered intensity at distance r from one such sphere for unit intensity of unpolarized incident light, the theory gives

$$V_u(\theta) = \frac{\lambda'^2}{8\pi r^2} |a_1 + (a_2 + p_1) \cos \theta|^2$$

$$H_u(\theta) = \frac{\lambda'^2}{8\pi r^2} |a_1 \cos \theta + a_2 \cos 2\theta + 2p_1|^2$$

where
$$a_1 = x^3 \frac{m^2 - 1}{m^2 + 2}$$

$$a_2 = -\tfrac{1}{12} x^5 \frac{m^2 - 1}{m^2 + \frac{3}{2}}$$

$$p_1 = -\tfrac{1}{30} x^5 (m^2 - 1)$$

The same formulas hold for metallic particles if the appropriate complex value of m is used. For larger particles more terms in the infinite series become appreciable and must be included. There is a considerable literature [48] of calculations of these complicated functions.

The main qualitative effects in the first correction to Rayleigh dipole scattering are: (*a*) dissymmetry appears in a way that makes forward scattering more intense than backward for dielectric spheres, and (*b*) there is depolarization of the light scattered through 90° even for these spherical particles. Integration of the Mie series over θ gives formulas for the turbidity [49] as a function of m, λ', and R. The contribution to τ from N spheres of radius R in unit volume is of the form

$$\tau = N\pi R^2 f\left(\frac{2\pi R}{\lambda'}\right)$$

Here $f(x)$ is a complicated function which has the following limiting forms:

Proportional to x^4	for $x \ll 1$
Proportional to x^2	for $x \sim 1$
Approaches the value 2	for $x \gg 1$

Thus large scattering particles scatter white light more nearly as white light than do molecules, which is why clouds are white though the sky is blue [50].

The approach of $f(x)$ to 2 for large spheres gives rise to the paradoxical result that such large spheres scatter twice as much light as they intercept geometrically from the incident beam. More detailed consideration shows that half this amount corresponds to the scattering through large angles, as common sense would suggest, and the other πR^2 corresponds to scattering through extremely small angles which in most experiments would not be counted as removing light from the beam [51].

This general character of the result for turbidity shows that, for a fixed amount of scattering material in unit volume, the total scattering increases with finer and finer subdivision of the material until the particles have radii of the order of the wavelength, after which further subdivision into still smaller particles diminishes the turbidity or scattering.

Metallic colloids [52] are also described by the Mie theory provided appropriate complex values of m are used in the formulas. These show pronounced color-selective effects because of the variation with wavelength of both n and κ in the complex index $n(1 - i\kappa)$. They usually show opposite dissymmetry, that is, more scattering in the backward than in the forward hemisphere.

Light scattering by liquids and liquid mixtures also finds its interpretation in thermal fluctuations in their structure [53]. Any thermodynamic quantity x will fluctuate about its mean value $\bar{x}$. The equilibrium state is that which makes the free energy $A = E - TS$ take a minimum value, and the probability of a state occurring which has the value $A + \Delta A$ is proportional to $e^{-\Delta A/kT}$.

At equilibrium the first derivative of A with respect to x must vanish, and so ΔA is proportional to $(\Delta x)^2$. Thus the probability distribution for Δx is Gaussian in the first approximation. This gives the following formula for the mean-square fluctuation of Δx,

$$<(\Delta x)^2>_{\text{av}} = \frac{kT}{(\partial^2 A/\partial x^2)}$$

in which the second derivative is to be evaluated at the equilibrium value $\bar{x}$.

The light scattering from a volume element δV which is small compared with the wavelength and in which for whatever reason the dielectric constant is $\epsilon + \Delta\epsilon$ is given by

$$\frac{I(\theta)}{I_0} = \frac{\pi^2}{\lambda^4 r^2} \cdot <(\Delta\epsilon)^2>_{\text{av}} \delta V \tfrac{1}{2}(1 + \cos^2 \theta)$$

and the total light scattered in all directions by such an element is

$$\frac{8\pi^3}{3\lambda^4} <(\Delta\epsilon)^2>_{\text{av}} \delta V$$

For a pure liquid the dielectric constant ϵ fluctuates because of spontaneous changes in its density ρ at local volume elements and also because of spontaneous local fluctuations in temperature. The latter contribution proves to be negligible compared with the former. Likewise dependence of the free energy on density leads to the following formula for the spontaneous fluctuations in density,

$$<(\Delta\rho)^2>_{\text{av}} = \rho^2 \beta \frac{kT}{\delta V}$$

in which β is the isothermal compressibility of the liquid. Therefore the scattering of such a volume element is given by substituting

$$<(\Delta\epsilon)^2>_{\text{av}} = \left[\frac{\partial\epsilon}{\partial\rho}\right]_T^2 \rho^2 \frac{\beta kT}{\delta V}$$

which makes the total light scattering independent of the size of the volume element, on the implicit supposition that there is no correlation in the fluctuations of adjacent volume elements.

Various suppositions can be made about the variation of ϵ with ρ to evaluate $\partial\epsilon/\partial\rho$, which appears in the formula for fluctuation of the dielectric constant. The traditional one is the Clausius-Mosotti relation, which makes $(\epsilon - 1)/(\epsilon + 2)$ proportional to the density. However, the experimental data on light scattering are more closely in accord with calculations based on the use of the Onsager relation [54] according to which $(\epsilon - 1)(2\epsilon + 1)/9\epsilon$ is proportional to the density.

Einstein pointed out that in liquid mixtures an additional source of fluctuations in the dielectric constant would be local fluctuations in the composition of the liquid. Thus, with a binary mixture, if c is the weight concentration of one component so that $\rho - c$ is that of the other, then one must add to the value of $<(\Delta\epsilon)^2>_{av}$ already calculated for the density fluctuations another part due to the concentration fluctuations. This additional part can be reduced to the form

$$\left(\frac{\partial\epsilon}{\partial c}\right)^2 \frac{c}{(\partial P/\partial c)} \frac{kT}{\delta V}$$

in which P is the osmotic pressure of the component c in the solution. Clearly the concentration fluctuations are of importance only when the dielectric constant depends strongly on the composition.

Light scattering by liquids becomes very great as the critical point is reached. This phenomenon is called *critical opalescence.* An analogous great scattering occurs under conditions in which a liquid solution is at the point of separating into two phases. At the critical point the preceding calculations break down because β tends to infinity. In this region strong correlations between the fluctuations of adjacent volume elements [55] occur so that the fluctuation formulas previously discussed have to be modified by replacing kT by

$$\frac{kT}{\beta^{-1} + \left(\frac{2\pi g}{\lambda'} \sin \frac{\theta}{2}\right)^2}$$

in which g is a linear dimension that is related to the range of intermolecular forces [56]. With this modification in the formulas, as β^{-1} tends to zero, the theory predicts the development of a strong dissymmetry of the scattered light and a change from λ^{-4} to λ^{-2} dependence of scattering on wavelength, both of which are experimentally confirmed.

References

1. Van Vleck, J. H.: "The Theory of Electric and Magnetic Susceptibilities," p. 15, Oxford University Press, New York, 1932.
2. Partington, J. R.: "Advanced Treatise on Physical Chemistry," vol. 4, pp. 42–78, Longmans, New York, 1953.
3. Mott, N. F., and R. W. Gurney: "Electronic Processes in Ionic Crystals," chap. 1, Oxford University Press, New York, 1948.
4. Eisenlohr: *Jahrb. Radioakt. u. Elektronik,* **9**: 315 (1912).
5. Denbigh: *Trans. Faraday Soc.,* **36**: 936 (1940). Warwick: *J. Am. Chem. Soc.,* **68**: 2455 (1946). Cresswell, Jeffery, Leicester, and Vogel: *Research (London),* **1**: 719 (1948), **2**: 46 (1949). Vogel, Cresswell, Jeffery, and Leicester: *J. Chem. Soc.* (London), pt. I, p. 514, 1952.
6. Kramers, H. A., and W. Heisenberg: *Z. Physik,* **31**: 681 (1925).
7. Ladenburg, R.: *Z. Physik,* **65**: 167, 189 (1930); *Revs. Mod. Phys.,* **5**: 243 (1933).
8. Born, M., and P. Jordan: "Elementare Quantenmechanik," p. 240, Springer, Berlin, 1930.
9. Drude, P.: *Ann. Physik,* **39**: 481 (1890).
10. Hagen, E., and H. Rubens: *Ann. Physik,* **11**: 873 (1903).
11. Wood, R. W.: "Physical Optics," 3d ed., p. 510, Macmillan, New York, 1934.
12. Born, M., and K. Huang: "Dynamical Theory of Crystal Lattices," p. 116, Oxford University Press, New York, 1954.
13. Ewald, P. P.: *Ann. Physik,* **49** (1916). M. Born: "Optik," p. 327, Springer, Berlin, 1933. E. Hylleraas: *Physik. Z.,* **26**: 811 (1925); *Z. Physik,* **36**: 859 (1926); *Z. Krist.,* **65**: 469 (1927).
14. Wooster, W. A.: "A Textbook on Crystal Physics," p. 177, Cambridge University Press, New York, 1938.
15. Bragg, W. L.: *Proc. Roy. Soc. (London),* **A105**: 370 (1924), **A106**: 346 (1924).
16. Faraday, M.: *Trans. Roy. Soc. (London),* **136**: 1 (1846).
17. Cotton, A., and H. Mouton: *Compt. rend.,* **145**: 229 (1907).
18. Gabiano, P.: *Ann. phys.,* **20**: 68 (1933). (Gives precise modern measurements for the Verdet constant for gases and comparison with the Verdet constant for the liquids.)
19. Ingersoll, L. R.: *Phys. Rev.,* **9**: 257 (1917).
20. Becquerel, H.: *Compt. rend.,* **125**: 679 (1897).
21. Born, M., and P. Jordan: "Elementare Quantenmechanik," p. 267, Springer, Berlin, 1930. J. H. Van Vleck: "The Theory of Electric and Magnetic Susceptibilities," p. 367, Oxford University Press, New York, 1932. R. Serber: *Phys. Rev.,* **41**: 489 (1932). R. de Malleman: *Congr. Intern. élec. Paris, Rapp.* 31, 1932.
22. Gorter, C. J.: *Physik. Z.,* **34**: 238 (1933).
23. Kerr, J.: *Phil. Mag.,* **50**: 337, 446 (1875), **8**: 85, 229 (1879). J. W. Beams: *Revs. Mod. Phys.,* **4**: 133 (1932).
24. Born, M.: "Optik," p. 365, Springer, Berlin, 1933. [Note that in his equations (1) to (3) on pp. 366, 367 the subscripts *e* and *o* are interchanged.]
25. Tauern, O. D.: *Ann. Physik,* **32**: 1064 (1910).
26. Stuart, H. A.: *Z. Physik,* **55**: 358 (1929), **59**: 13 (1930), **63**: 533 (1930).
27. Lowry, T. M.: "Optical Rotatory Power," Longmans, New York 1935. E. U. Condon: *Revs. Mod. Phys.,* **9**: 432 (1937).
28. "International Critical Tables," vol. 7, p. 353, McGraw-Hill, New York, 1926–1930.
29. Rosenfeld, L.: *Z. Physik,* **52**: 161 (1928). M. Born and P. Jordan: "Elementare Quantenmechanik," p. 250, Springer, Berlin, 1930.
30. Kuhn and Friedenberg: "Hand- und Jahrbuch der chemischen Physik," vol. 8, pt. 3, p. 47, 1932. Discussion on optical rotatory power, *Trans. Faraday Soc.,* 1930.
31. Condon, E. U., W. Altar, and H. Eyring: *J. Chem. Phys.,* **5**: 753 (1937).
32. "International Critical Tables," vol. 7, p. 360, McGraw-Hill, New York, 1926–1930.
33. Neumann, F.: *Ann. Physik,* **54**: 449 (1841).
34. Mueller, H.: *Physics,* **6**: 179 (1935).
35. Kolsky, H., and A. C. Shearman: *Proc. Phys. Soc. (London),* **55**: 383 (1943). L. R. G. Treloar: *Trans. Faraday Soc.,* **43**: 277 (1947), **50**: 119 (1951). J. E. H. Braybon: *Proc. Phys. Soc. (London),* **66B**: 617 (1953).
36. Frenkel, J.: "Kinetic Theory of Liquids," p. 288, Oxford University Press, New York, 1946.

37. Cerf, R., and H. Scheraga: *Chem. Revs.*, **51**: 185 (1952). H. A. Scheraga, J. T. Edsall, and J. O. Gadd, Jr.: *J. Chem. Phys.*, **19**: 1101 (1951). M. Wales: *J. Phys. & Colloid Chem.*, **52**: 976 (1948).
38. Graham, M.: *J. Opt. Soc. Am.*, **27**: 420 (1937).
39. Nikitine, S.: *Compt. rend.*, **205**: 124 (1937), **208**: 513 (1939).
40. Tyndall, John: *Phil. Mag.*, **17**: 92, 222, 317 (1868), **37**: 384 (1869).
41. Rayleigh, Lord: "Scientific Papers," vol. 4, p. 397, Cambridge University Press, New York, 1903.
42. Oster, Gerald: *Chem. Revs.*, **43**: 319 (1948).
43. Raman, C. V.: *Indian J. Phys.*, **2**: 387 (1928).
44. Cabannes, J., and Y. Rocard: "La diffusion moleculaire de la lumière," Paris, 1929. S. Bhagavantam: "Light Scattering and the Raman Effect," Chemical Publishing, New York, 1942.
45. Rayleigh, Lord: *Phil. Mag.*, **35**: 373 (1918).
46. Putzeys, P., and J. Brosteux: *Trans. Faraday Soc.*, **31**: 1314 (1935); *Mededeel. Koninkl. Vlaam- Ackad. Wetenschap. Belg. Kl. Wetenschap.*, **3**: 3 (1941).
47. Mie, G.: *Ann. Physik*, **25**: 377 (1908). P. Debye: *Ann. Physik*, **30**: 59 (1909). Lord Rayleigh: *Proc. Roy Soc. (London)*, **A85**: 25 (1911). J. A. Stratton: "Electromagnetic Theory," chap. 9, McGraw-Hill, New York, 1941.
48. Van de Hulst, H. C.: "Light Scattering by Small Particles," Wiley, New York, 1957.
49. Jobst, G.: *Ann. Physik*, **78**: 157 (1925). V. K. LaMer: *J. Phys. & Colloid Chem.*, **52**: 65 (1948).
50. Middleton, W. E. K.: "Visibility in Meteorology," 2d ed., University of Toronto Press, Toronto, 1941.
51. Sinclair, D.: *J. Opt. Soc. Am.*, **37**: 475 (1947).
52. Swedberg, T.: "Colloid Chemistry," 2d ed., Chemical Catalog, New York, 1928. H. Freundlich, "Capillarchemie," 4th ed., Akademische Verlagsgesellschaft m.b.H., Leipzig, 1930.
53. Smoluchowski, M.: *Ann. Physik*, **25**: 205 (1908); *Phil. Mag.*, **23**: 165 (1912). A. Einstein: *Ann. Physik*, **33**: 1275 (1910).
54. Onsager, L.: *J. Am. Chem. Soc.*, **58**: 1486 (1936). J. S. Rosen: *J. Opt. Soc. Am.*, **37**: 932 (1947).
55. Ornstein, L., and F. Zernike: *Proc. Roy. Soc. Amsterdam*, **17**: 793 (1914).
56. Placzek, G.: *Physik. Z.*, **31**: 1052 (1930). Y. Rocard: *J. phys.*, **4**: 165 (1933). H. Mueller: *Proc. Roy. Soc. (London)*, **A166**: 425 (1938).

Chapter 7

Fluorescence and Phosphorescence

By J. G. WINANS, State University of New York at Buffalo, *and* E. J. SELDIN, Union Carbide Corp.

1. Introduction

Fluorescence [1]† means a luminescence stimulated by radiation, not continuing more than about 10^{-8} sec after the stimulating radiation is cut off.

Phosphorescence [2–5] is a stimulated luminescence which persists for an appreciable time after the stimulating process has ceased. The time for decay to a certain fraction of the initial intensity depends upon the nature of the phosphorescing matter and on the temperature.

Fluorescence and phosphorescence are special cases of *photoluminescence,* which is luminescence stimulated by light, particularly ultraviolet light.

Some materials after exposure to a stimulating process show luminescence on later raising the temperature. This is known as *thermoluminescence.* In some cases the original stimulation may have come from radioactive matter or cosmic radiation many years before the temperature increase causing thermoluminescence.

Slow fluorescence [9] means a luminescence which persists for more than a fraction of a second after cessation of stimulation but one whose intensity is independent of the temperature.

Delayed fluorescence is like slow fluorescence but it is considered to result from a different mechanism.

Some investigators like to specify fluorescence as luminescence during excitation and phosphorescence as luminescence after excitation has ceased for more than 10^{-8} sec. This designation has the disadvantage that the luminescence specified as fluorescence is in reality a mixture of both fluorescent and phosphorescent radiation.

Resonance radiation [6] and *reemission* [7–8] are types of fluorescence in which a gas or vapor emits radiation of a certain wavelength when illuminated by radiation of that same wavelength. Resonance radiation is obtained with matter which shows a line absorption spectrum, while reemission is obtained with matter which shows a continuous or unresolved band absorption spectrum.

Selective reflection [10] is closely related to fluorescence. A vapor or a gas at high pressure acts as a reflector for radiation of wavelength nearly equal to that of its resonance radiation but does not reflect radiation of other wavelengths.

† Numbers in brackets refer to References at end of chapter.

Sensitized fluorescence [1, 11] is luminescence from one component in a mixture when the mixture is illuminated by radiation which only another component can absorb.

Other types of luminescence closely related to fluorescence and phosphorescence are the following:

Sonoluminescence is luminescence produced by high-frequency sound waves or phonons.

Cathodoluminescence is luminescence produced by the impact of electrons on crystals [106, p. 288; 150].

Electroluminescence is luminescence produced by the application of an electric field to matter. The matter generally constitutes the dielectric of a capacitor [106, p. 237].

Radioluminescence is luminescence produced by the impact of high-energy particles from accelerators, cosmic rays, radioactive matter, or other sources [110].

Triboluminescence is luminescence caused by friction or by mechanical deformation or rupture.

Chemiluminescence and *bioluminescence* are examples of luminescence resulting from chemical reactions that can occur in inorganic or organic matter.

Some special observations of luminescence have been given special names, such as the following:

Optical stimulation, the enhancement of luminescence by long-wavelength visible or infrared radiation.

Optical quenching, the reduction of luminescent intensity by long-wavelength visible or infrared radiation.

Fluorescence differs from phosphorescence and slow fluorescence primarily in the decay times. This is the time, with no stimulation, for the intensity of luminescence to decay to e^{-1} times the intensity at $t = 0$. The decay time for fluorescence is of the order of 10^{-8} sec, while that for phosphorescence and slow fluorescence may be from 10^{-3} sec to many days.

Fluorescence and phosphorescence are described in terms of energy changes for atoms and molecules (Fig. 7.1). G symbolizes the ground state of the molecule or atom, and D and M are metastable states.

Resonance radiation, reemission, and selective reflection are represented by transitions (1). Absorption of radiation of energy E_F produces an energy change such that the absorbing molecule is changed from state G to state F. The molecule then changes back to state G after about 10^{-8} sec, radiating energy at the same wavelength as that absorbed. A delayed reemission, or phosphorescence with wavelength equal to that absorbed, is represented by transitions (2). In this process the energy is stored for a time in a metastable state M. Atoms or molecules in state F are changed to the metastable state M by collisions of the second kind, with the energy difference between states F and M going into kinetic energy of the colliding molecules. At some later time, molecules in the metastable state M may be changed back to the nonmetastable state F by collisions of the first kind, the energy needed being taken from the kinetic energy of the colliding molecules. Atoms thus raised to state F return to state G with emission of luminescence.

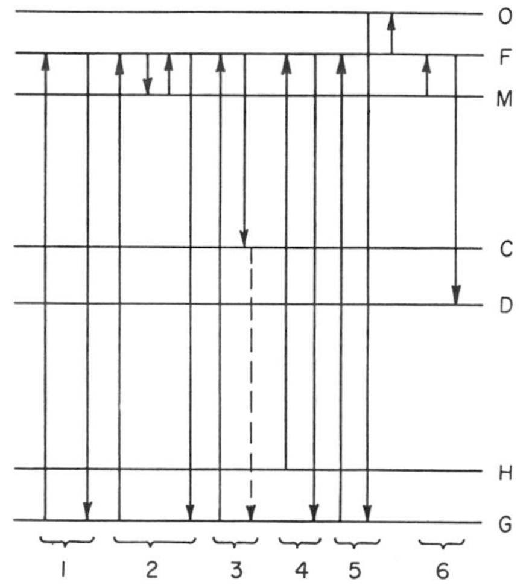

FIG. 7.1. Energy transitions for various types of luminescence.

Fluorescence in accordance with *Stokes' law*, which states that the wavelength for fluorescent radiation is greater than the wavelength of the stimulating radiation, is represented by transitions (3). In this case the molecule after absorption of radiation changes from state F to a lower state C by radiation at frequency such that $h\nu = E_F - E_C$.

Fluorescence in violation of Stokes' law (anti-Stokes radiation) is represented by transitions (4) and (5). Molecules in an initial excited state H may absorb radiation at one wavelength in changing from H to F and emit radiation at a shorter wavelength in changing from F to G. Also, excited molecules in state F may be raised to a higher energy state O, either by collisions of the first kind or by absorption of additional radiation, and then return to the ground state G by emission of radiation of wavelength less than that originally absorbed.

Thermoluminescence and phosphorescence are represented by transitions (6). Molecules left in a metastable state M by an earlier process may be returned to state F by thermal energy, and this may require an increase in temperature. From state F the molecule may change to a lower state through the emission of radiation.

Slow fluorescence is represented by transitions from M to lower levels through radiation.

Delayed fluorescence is considered to occur when two molecules in metastable states (D, C, or M) collide and produce one molecule in a nonmetastable state (F) which thereafter radiates.

The decay time for fluorescence is the average life of the molecule in the higher energy state F, while that for phosphorescence and slow fluorescence is the average life for molecules in the metastable state M.

Let τ_0 be the average life for the excited state associated with the persistence of fluorescence or phosphorescence. The intensity I decays with time according to the relation

$$I = -K\frac{dN}{dt} = \frac{KN}{\tau_0} \tag{7.1}$$

The death rate dN/dt may be replaced by N/τ_0, where N is the average population of the excited state. K is a constant. The integral of (7.1) shows that the population decreases with time according to the relations

$$N = N_0 \exp\left(-\frac{t}{\tau_0}\right)$$

$$\text{and} \qquad \frac{dN}{dt} = -\frac{N_0}{\tau_0}\exp\left(-\frac{t}{\tau_0}\right) \tag{7.2}$$

where N_0 is the population at a time specified as $t = 0$ following the cessation of stimulation.

For slow fluorescence the decay time is the average life of state M. For phosphorescence [transitions (2), Fig. 7.1] the decay of intensity is like that for the radioactive decay of a short-lived daughter (state F) of a long-lived parent (state M). The decay time corresponds to the average life of the parent.

The process for anti-Stokes fluorescence and for thermoluminescence is like that for a unimolecular reaction with an activation energy. The speed of such a reaction is described by

$$\text{Reaction rate } p = se^{-\frac{w}{kT}}$$

where w is the activation energy and $p = \dfrac{ds}{dt}$ and s is the quantity of reacting matter.

The intensity of anti-Stokes fluorescence or thermoluminescence is proportional to the reaction rate, and therefore $I = Ke^{-\frac{w}{kT}}$, where w is the energy difference between states F and O in Fig. 7.1 for anti-Stokes fluorescence and states M and F for thermoluminescence.

Intensities at two different temperatures are described by

$$\frac{I_1}{I_2} = \exp\left[\frac{-w}{k}\left(\frac{1}{T_1} - \frac{1}{T_2}\right)\right] \tag{7.3}$$

2. Fluorescence of Gases and Vapors

Resonance Radiation. Observation of resonance radiation requires a light source in which the resonance lines are not reversed. This is accomplished for mercury by using either a water- or an air-cooled mercury arc or a hot cathode arc discharge through an inert gas such as argon containing a small quantity of mercury vapor. An arrangement like that in Fig.

7.2 assures that some resonance radiation will escape reabsorption in the vapor.

Resonance radiation has been observed at wavelengths 1849 ($6\ ^1S - 6\ ^1P$) and 2537 ($6\ ^1S - 6\ ^3P_1$) for mercury [12–13] and 5890 and 5896 ($^2S - {}^2P_{\frac{3}{2},\frac{1}{2}}$) for sodium [14]. Because of interference by absorption bands due to O_2 in the atmosphere, it is not convenient to study the λ1849 resonance radiation of mercury. Other vapors for which resonance radiation has been observed are cadmium [15], lithium [16], zinc [17], calcium [18], potassium [19], and iodine [20].

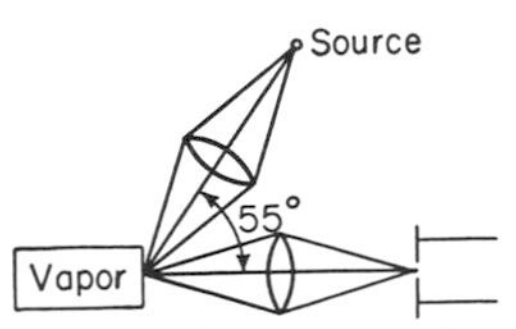

FIG. 7.2. Arrangement of apparatus for observing resonance radiation with no effect from depolarization.

Quenching of Resonance Radiation. When a foreign gas is admitted to a fluorescence cell arranged as in Fig. 7.2, the intensity of resonance radiation is diminished by an amount depending on the pressure and kind of admixed gas. The angle of incidence for the stimulating radiation should be 55° from the direction of observation of the resonance radiation to minimize any changes in intensity due to changes in polarization caused by the admixture of the foreign gas [6]. The angle 55° is the angle of incidence that will provide the same intensity of resonance radiation for 100 per cent polarization and complete depolarization. The intensity of resonance radiation changes with pressure as in Fig. 7.3 and Fig. 7.4. I_0 is the intensity of resonance radiation with no foreign gas present, and I_1 is the intensity with a foreign gas.

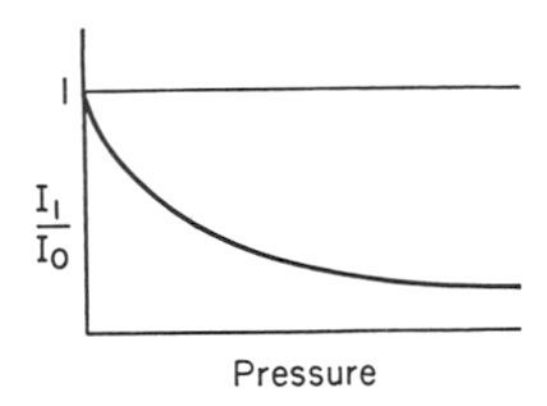

FIG. 7.3. Decrease in resonance radiation intensity from addition of a foreign gas.

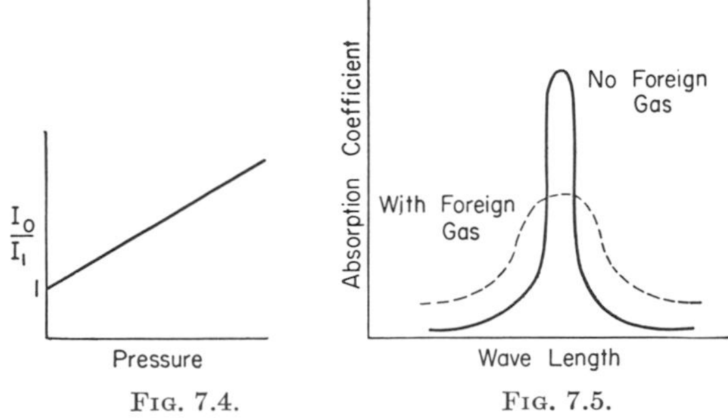

FIG. 7.4. FIG. 7.5.

FIG. 7.4. Plot of the Stern-Volmer relation.
FIG. 7.5. Effect of foreign-gas pressure on absorption coefficient.

It might be expected that the effect of the foreign gas on the contour of the absorption coefficient, as in Fig. 7.5, would cause a decrease in the power absorbed from the stimulating radiation and thus cause a decrease in intensity of resonance radiation even with no quenching from collisions.

With the arrangement of Fig. 7.2 the stimulating radiation penetrates the vapor to whatever depth is necessary for total absorption, either with or without a foreign gas present. A broadening of the absorption-coefficient contour because of the foreign-gas pressure therefore increases the depth of penetration of the stimulating radiation and also increases the vapor path through which the resonance radiation must pass to reach the observer, and at the same time decreases the absorption coefficient for this path. These two effects cancel so that the intensity of resonance radiation is proportional to the intensity of incident radiation, regardless of the effect of foreign-gas pressure on the wavelength contour of the absorption coefficient of the vapor. The foreign gas does, however, produce quenching through collisions of the second kind in which excitation energy is transferred to kinetic energy of the separating molecules.

For an atom with a single excited state participating in resonance radiation and no foreign gas there will result (Fig. 7.6) an equilibrium population n_0 for which the birth rate KI_e and the death rate n_0a_0 are equal.

$$KI_e = n_0a_0 \tag{7.4}$$

where $a_0 = 1/\tau_0$. τ_0 is the decay time for radiation, and I_e is the intensity of the stimulating radiation. K is a constant.

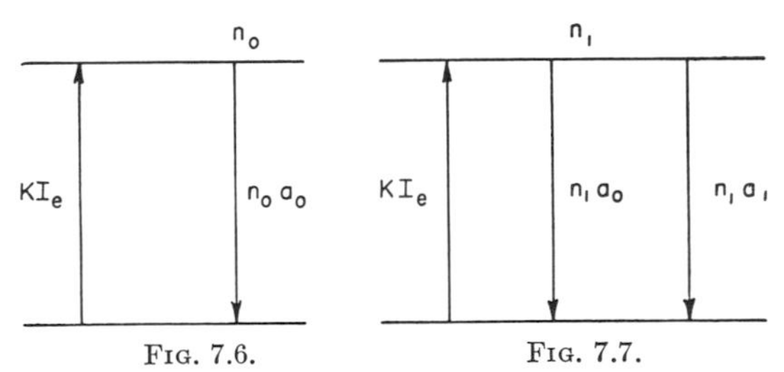

FIG. 7.6. FIG. 7.7.

FIG. 7.6. Energy transitions for resonance radiation with no foreign gas.
FIG. 7.7. Energy transitions for resonance radiation with a foreign gas.

When a foreign gas is added, the birth rate remains the same, independent of pressure broadening, and the death rate is increased. As shown in Fig. 7.7, for equilibrium with a foreign gas present

$$KI_e = n_1(a_0 + a_1) \tag{7.5}$$

where $a_1 = 1/\tau_1$. τ_1 is the average time for an excited atom to make a collision of the second kind. Combining Eqs. (7.4) and (7.5) gives

$$\frac{n_0}{n_1} = 1 + \frac{a_1}{a_0} \tag{7.6}$$

If I_0 is the intensity of resonance radiation with no foreign gas and I_1 the intensity with a foreign gas,

$$\frac{I_0}{I_1} = \frac{n_0a_0}{n_1a_1} = \frac{n_0}{n_1} \tag{7.7}$$

and therefore

$$\frac{I_0}{I_1} = 1 + \frac{a_1}{a_0} \tag{7.8}$$

If every collision between an excited atom and foreign-gas atom were a quenching collision, $1/a_1$ would be the average free time before a collision. This average free time is inversely proportional to the number N of foreign-gas molecules per cubic centimeter, and therefore a_1 is proportional to the pressure, provided the temperature remains constant.

Let $a_1 = Hp$, where H is a constant and p is the pressure. Then (7.8) becomes the Stern-Volmer equation:

$$\frac{I_0}{I_1} = 1 + H\tau_0 p \tag{7.9}$$

Equation (7.9) describes the curve shown in Fig. 7.4. From the slope of the line and τ_0 the constant H may be determined. From H the effective cross section for collisions of the second kind may be found, as shown below.

In the case of resonance radiation of sodium vapor there are two excited states, $^2P_{\frac{3}{2}}$ and $^2P_{\frac{1}{2}}$, with an energy separation of 0.02 volt. There will be transfers of sodium atoms from $^2P_{\frac{3}{2}}$ to $^2P_{\frac{1}{2}}$ through collisions of the second kind and transfers from $^2P_{\frac{1}{2}}$ to $^2P_{\frac{3}{2}}$ through collisions of the first kind with foreign gas and with unexcited sodium atoms. Since the average thermal energy for atoms in quenching experiments is only about 0.04 electron volt, there will tend to be an accumulation of sodium atoms in the $^2P_{\frac{1}{2}}$ state. If $^2P_{\frac{3}{2}}$ and $^2P_{\frac{1}{2}}$ have the same average life periods (or decay times), the effects due to transfers between $^2P_{\frac{3}{2}}$ and $^2P_{\frac{1}{2}}$ cancel out and Eqs. (7.4) and (7.5) apply exactly, with n_1 and n_0 representing the total population of sodium atoms in both $^2P_{\frac{3}{2}}$ plus $^2P_{\frac{1}{2}}$ states with and without a foreign gas, respectively.

The quenching constant $a_1 = Hp$ in Eq. (7.9) equals the number of collisions made per second between an excited atom and foreign-gas atoms multiplied by the probability that a collision shall be one of the second kind. No method of determining this probability by other measurements has been devised, and so it is usually assumed that this probability is unity. With this assumption, $a_1 = z$ = number of collisions per second per resonating atom. Statistical mechanics [105] gives

$$a_1 = SN\sqrt{\frac{8kT}{\pi m}} \tag{7.10}$$

where S is the effective cross section for quenching, N is the number of foreign-gas molecules per cubic centimeter, and m is the reduced mass,

$$1/m = 1/m_1 + 1/m_2$$

with m_1 the mass of the fluorescing molecules and m_2 the mass of the quenching molecules. T is the absolute temperature, and k is Boltzmann's constant.

The average velocity of molecules with respect to the walls of their container is given by (2.6) of Part 5, which is the same as the square-root part of (7.10), with m denoting the mass of one molecule. The average velocity of one molecule with respect to another molecule is given by

$$\bar{v}_{12}^2 = \bar{v}_1^2 + \bar{v}_2^2 \tag{7.11}$$

where $\bar{v}_1^2 = \frac{8kT}{\pi m_1}$ and $\bar{v}_2^2 = \frac{8kT}{\pi m_2}$

The average collision between two molecules is a right-angle collision:

$$v_{12} = \sqrt{\frac{8kT}{\pi m}} \quad \text{with} \quad \frac{1}{m} = \frac{1}{m_1} + \frac{1}{m_2}$$

The number of collisions per second per excited atom is therefore

$$a_1 = SNv_{12} = SN\sqrt{\frac{8kT}{\pi m}} \tag{7.12}$$

At the low pressures used in quenching experiments (0 to 20 mm Hg) $N = p/kT$, where p is the pressure of the gas. This gives

$$\frac{I_0}{I_1} = 1 + \frac{Sp}{a_0}\sqrt{\frac{8}{\pi mkT}} \tag{7.13}$$

If the experimental measurements fit a straight line like Fig. 7.4 with I_0/I_1 intercept 1, it can be concluded that there is no appreciable influence of line broadening or depolarization. The effective cross section for quenching may be determined from the slope of the line if the average life for radiation, $1/a_0$, is known from other experiments.

Effective cross sections for quenching for a number of gases are different from the kinetic-theory cross sections. For some gases the quenching cross sections are several times larger and for others several times smaller than the kinetic-theory cross sections.

In the study of quenching of sodium resonance radiation possible errors due to pressure broadening may be eliminated by using NaI, NaBr, or NaCl as the fluorescing vapor. When NaI is illuminated by radiation of wavelength less than 2400 A, the D lines of Na are obtained in fluorescence. The intensity of the Na fluorescence may be measured at different pressures of a foreign gas added to the NaI vapor. NaI is dissociated by the absorption of the radiation into excited sodium atoms (Na′) and normal iodine atoms. The energy difference between the energy of the photon absorbed and that needed to dissociate NaI and excite Na goes into kinetic energy of Na′ and I.

Since the absorption spectrum of NaI is continuous for wavelengths less than 2400 A, the effect of adding a foreign gas is to broaden each line in a continuous spectrum. This does not change the total absorption coefficient at any one wavelength. The total radiation energy absorbed by the NaI should therefore be independent of the pressure of the admixed gas. Measurements of quenching of NaI fluorescence have been carried out by Winans [21] and Terenin and Prileshajewa [22]. Effective cross sections determined from these experiments are given in Table 7.1.

Other measurements of quenching of NaI fluorescence using improved methods have been made by Hanson [23].

TABLE 7.1. SAMPLE CROSS SECTIONS FOR QUENCHING OF RESONANCE RADIATION

Fluorescing gas	Resonance-line wave-length	Quenching gas	Effective quenching cross section in 10^{-16} sq cm
Hg	2536	H_2O	1.43
		D_2O	0.46
		NH_3	4.2
		NO	35.3
		CO_2	3.54
		N_2	0.27
		H_2	8.6
		D_2	11.9
		O_2	19.9
		Ne	0.325
		A	0.223
		CH_4	0.085
		C_6H_6	59.5
Cd	3260	H_2	3.54
		D_2	1.80
		CO	0.14
		NH_3	0.052
		N_2	0.021
		C_2H_6	0.024
		C_6H_6	28.4
Na	5893	CH_4	0.11
		C_2H_6	0.17
		C_6H_6	75.

Curves of I_0/I_1 versus pressure of foreign gas obtained by Hanson are shown in Fig. 7.8. The quenching curve for CO_2 is a straight line with I_0/I_1 intercept 1 in agreement with the Stern-Volmer equation (7.9). The curve for H_2 is a straight line at the

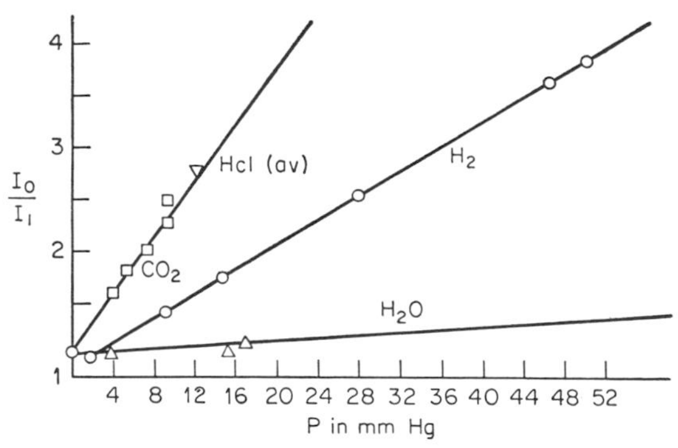

FIG. 7.8. Quenching curves for NaI fluorescence.

higher pressures, but this line does not have an I_0/I_1 intercept of 1. The shape of this curve for low pressures is shown in Fig. 7.9. A small quantity of H_2 causes an increase in the intensity of fluorescence, while a larger amount causes a decrease.

The increase of fluorescence intensity for small additions of H_2 may be due to a prevention of sublimation of NaI, thus causing an increase in temperature; or it may be due to the depolarizing effect of H_2, causing an increase in intensity which exceeds the quenching at low pressures. With the arrangement used, depolarization with no quenching should increase the fluorescence intensity by 33.3 per cent, and with quenching to somewhat less than this. The observed increase in intensity was 25 per cent, a value consistent with the effect of depolarization.

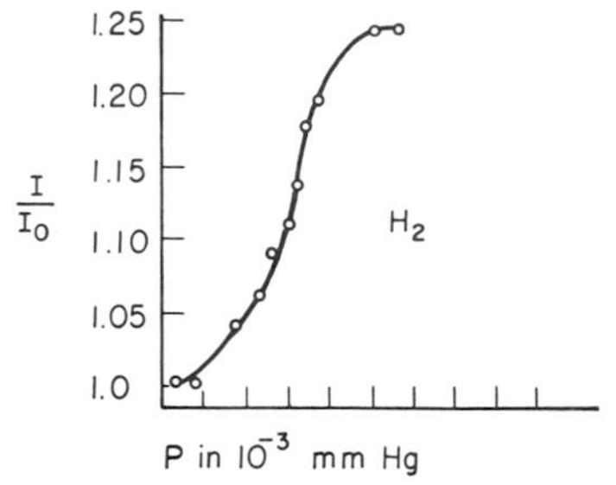

FIG. 7.9. The enhancement of NaI fluorescence by the addition of small pressures of a weakly quenching gas.

Water vapor showed very little if any quenching. This was surprising considering the chemical activity of liquid water and metallic sodium. E. Warhurst [24] reports that there is little or no chemical action between water vapor and sodium vapor at temperatures above 300°C.

An extension of the Stern-Volmer theory, taking into account the high velocity of Na′ produced in the photochemical dissociation of NaI, gives

$$\frac{I_0}{I} = 1 + z\tau_0 \tag{7.14}$$

where τ_0 is the decay time for radiation alone and z is the number of collisions per second made by an Na′ atom,

$$z = N \frac{S}{\sqrt{\pi}} V \frac{\psi(x)}{x^2} \tag{7.15}$$

where V is the average velocity of Na′ after photodissociation of NaI. $x = V/V_Q$, where V_Q is the average velocity of the quenching-gas molecules,

$$\psi(x) = x \exp(-x^2) + (2x^2 + 1) \int_0^x \exp(-u^2)\, du \tag{7.16}$$

and S is the effective cross section for quenching of Na′ by foreign-gas molecules. N is the number of quenching-gas molecules per cubic centimeter.

The number of collisions per second for Na′ may also be obtained by considering the Na′ to have an effective mass less than the true mass of Na [21].

$$\text{Effective mass} = \frac{\text{thermal energy} \times \text{true mass}}{\text{actual kinetic energy of Na}'} \tag{7.17}$$

The only quenching gases which showed any change in S with change in velocity of Na′ were HCl and CO_2. Values of S for various velocities are shown in Fig. 7.10.

The agreement between the data shown in Fig. 7.8 and Eq. (7.14) for all gases except H_2 shows that the simple form of the Stern-Volmer theory is applicable.

The fact that only HCl and CO_2 showed changes in S with change in velocity of Na′ probably means that only HCl and CO_2 have a quenching efficiency nearly 1. For other gases there are probably several ordinary collisions between Na′ and gas molecules before a quenching collision occurs. The excess velocity of Na′ would be lost in these preliminary collisions, and no change in quenching cross section with change in original velocity of Na′ would be observable.

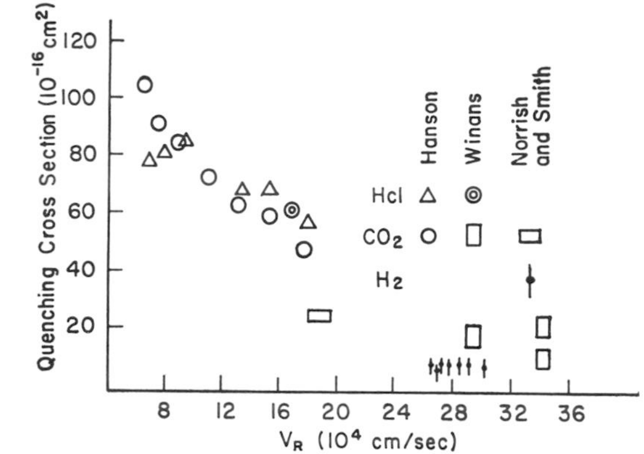

FIG. 7.10. Quenching cross sections for various relative velocities of Na′.

Measurements of the quenching of mercury resonance radiation were made by Stuart [25]. The experimental points fell on a straight line with I_0/I intercept 1. This showed that the simple Stern-Volmer equation applied for his apparatus, with no observable errors due to pressure broadening or depolarization.

The Stern-Volmer equation applies to the quenching of mercury resonance radiation 2536 (6 ^1S–6 $^3P_1$) even though there is a metastable state $^3P_0$ at energy 0.22 volt below $^3P_1$.

3. General Theory of Quenching of Fluorescence

Consider fluorescence of one kind of gas (A) in a mixture of gases to be stimulated by illuminating the mixture by radiation which only molecules A can absorb. The energy changes are illustrated in Fig. 7.11, where it is considered that there are $l + 1$

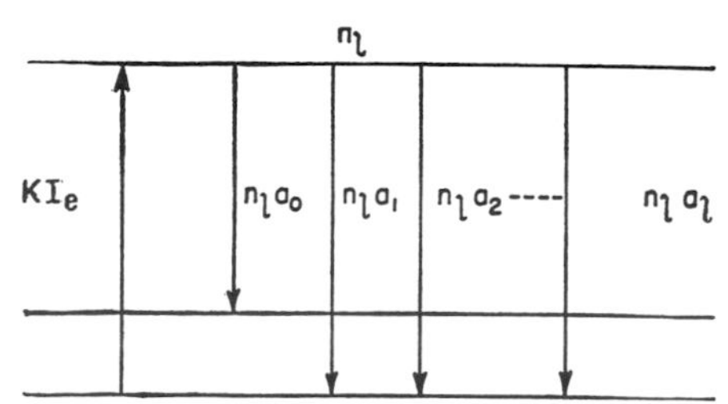

FIG. 7.11. Energy transitions in general case (no foreign gases).

different and independent ways for the excited atoms A' to die, or lose their energy, before any additional gases are added. If KI_e is the number of photons per second absorbed per cubic centimeter in the line of sight of the detecting instrument, then

$$KI_e = n_l(a_0 + a_1 + a_2 + \cdots a_l) \tag{7.18}$$

where n_l is the equilibrium population per cubic cm along the line of sight. It is assumed that there is no loss by diffusion.

If now an additional $m - l$ different gases or kinds of quenching processes are introduced, to make a total of $m + 1$ quenching processes, the population will be changed because of the increase in the death

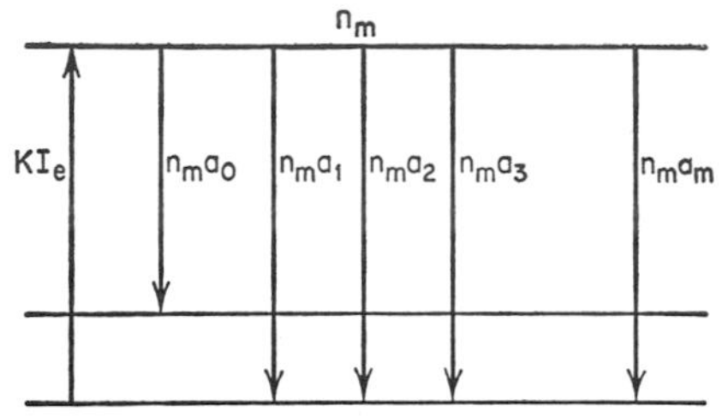

FIG. 7.12. Energy transitions in general case (with foreign gases).

rate. With $m + 1$ quenching processes as illustrated in Fig. 7.12

$$KI_e = n_m(a_0 + a_1 + \cdots + a_l) + n_m(a_{l+1} + a_{l+2} + \cdots + a_m) \tag{7.19}$$

Combining (7.18) and (7.19) gives

$$\frac{n_l}{n_m} = 1 + \frac{a_{l+1} + a_{l+2} + \cdots + a_m}{a_0 + a_1 + \cdots + a} \tag{7.20}$$

Let I_l be the intensity of fluorescence with $l + 1$ quenching processes and I_m the intensity with $m + 1$ processes. Then

$$\frac{I_l}{I_m} = \frac{n_l}{n_m} \tag{7.21}$$

Expressing the coefficients in terms of average life periods or decay times, $a_0 = 1/\tau_0$, where τ_0 is the average life for radiation alone and $a_l = 1/\tau_l$, where τ_l is the average life if process l were the only quenching process. Let $a_0 + a_1 + \cdots + a_l = a_{0-l}$ or

$$\frac{1}{\tau_0} + \frac{1}{\tau_1} + \cdots \frac{1}{\tau_l} = \frac{1}{\tau_{0-l}} \tag{7.22}$$

τ_{0-l} is the reduced life or life period when $l + 1$ different quenching processes act at once. Let $a_{l+1} + a_{l+2} + \cdots + a_m$, or

$$a_{0-m} - a_{0-l} = \frac{1}{\tau_{0-m}} - \frac{1}{\tau_{0-l}} \tag{7.23}$$

Thus on substitution in (7.20)

$$\frac{I_l}{I_m} = 1 + \frac{1/\tau_{0-m} - 1/\tau_{0-l}}{1/\tau_{0-l}} \tag{7.24}$$

Rearranging gives

$$\frac{I_l}{I_m} = \frac{\tau_{0-l}}{\tau_{0-m}} \tag{7.25}$$

or

$$\frac{I_l}{\tau_{0-l}} = \frac{I_m}{\tau_{0-m}} = \frac{I}{\tau} = \text{const} \tag{7.26}$$

Thus the addition of a gas to a mixture reduces the intensity I of fluorescence of any component in the mixture and also reduces the reduced life so that the ratio I/τ remains constant.

In case $l = 0$ and $m = 1$ as for a single quenching gas, Eq. (7.25) reduces to

$$\frac{I_0}{I_1} = \tau_0\left(\frac{1}{\tau_0} + \frac{1}{\tau_1}\right) = 1 + \frac{\tau_0}{\tau_1} \tag{7.27}$$

which is the Stern-Volmer equation.

Quenching of Resonance Radiation by Collisions of the Second Kind. Table 7.1 shows that the effective cross sections of excited atoms for quenching by a strongly quenching gas is several times the cross sections of the molecules as determined from van der Waals constant b or from the coefficient of viscosity.

The process of quenching by collisions of the second kind may be described by potential-energy curves as shown in Fig. 7.13 and by the Franck-Condon

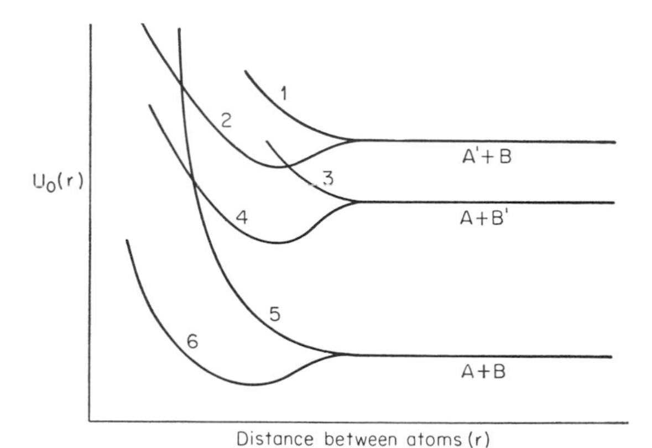

FIG. 7.13. Potential curves for head-on collisions.

principle. Two atoms in collision behave as a temporary molecule (quasi molecule). Energy changes between states for this quasi molecule can be expected to resemble the corresponding energy changes for diatomic molecules.

In general, when two atoms approach there result several possible potential-energy curves, some with and some without minima, depending on the spin orientation of the electrons in each of the two atoms.

If an excited atom A' collides with an unexcited atom B and the potential energy follows curve (2) of Fig. 7.13 on the approach, there is the possibility that the quasi molecule may change its potential energy and follow curve (5) during the separation. The change can occur with no change in position or momentum of the nuclei by occurring at the crossing point for curves (2) and (5). For atoms separating and following curve (5) all the excitation energy of A' is converted into thermal kinetic energy of the separating atoms.

In case atom B has an excited state of energy below that of atom A, the two atoms could approach following curve (2) and separate following curve (3). This would result in the production of excited atoms B' and the appearance of an amount of thermal kinetic energy equal to the energy difference between atoms A' and B'. Atoms B' could then either radiate energy as sensitized fluorescence or make other collisions following curves like (4) and (5) to dissipate the excitation energy into thermal energy.

When two atoms such as A' and B collide, the usual collision will be a noncentral collision rather than a head-on collision. In noncentral collisions the angular momentum of the system remains constant, and the potential-energy curve followed is the same as that for a rotating molecule,

$$U(r) = U_0(r) + \frac{P^2}{2\mu r^2} \tag{7.28}$$

where P is the angular momentum and $U_0(r)$ is the potential energy for $P = 0$. μ is the reduced mass.

The curves are like those shown in Fig. 7.14. With potential energies like these an excited atom A' col-

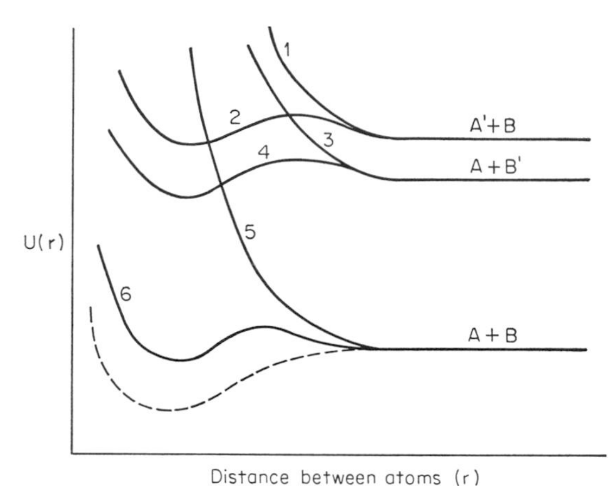

FIG. 7.14. Potential curves for noncentral collisions.

liding with an unexcited atom B can be quenched by transferring from curve (2) to curve (3) at a much larger distance than for a head-on collision. Thus, for a strongly quenching gas, the effective cross section of an excited atom is several times that given by kinetic theory.

The small effective cross sections for quenching by some gases may be associated with a small probability of a collision being a quenching collision rather than with a small distance between atoms during collision.

Two atoms colliding in noncentral collisions may, if they have the right energy and angular momentum, spend considerable time spiraling around their common center of mass at a separation equal to that near the flat maximum of the curves of Fig. 7.14.

Quenching resembles closely the process of predissociation of molecules. It is to be expected, therefore, that the selection rules which apply to predissociation will apply also to the quenching of fluorescence by foreign gases. These rules are $\Delta J = 0$, $\Delta\Lambda = 0, \pm 1$, $\Delta S = 0$, and the Franck-Condon principle. Here J is the total angular-momentum quantum number, Λ is the component of electron orbital angular momentum along the internuclear axis and S is the total electron-spin angular-momentum number [28].

Quenching processes may also involve chemical

reactions between excited atoms and the quenching-gas molecules as discussed by Pringsheim [26].

Imprisonment of Resonance Radiation. In experiments on quenching of resonance radiation, self-quenching or quenching of excited atoms A' by unexcited atoms of the same kind can be neglected at the pressures used. The effect of collisions between A and A' is simply to exchange function. The excited atom reverts to normal, and the normal atom becomes excited. This does not change the number of excited atoms or produce any effect on the average life of excited atoms for radiation. The probability for a change of the excitation energy into thermal energy through a collision between A and A' is negligible compared with the probability of an exchange of function.

It is possible for resonance radiation emitted by one atom to be absorbed by another. This provides an additional opportunity for the energy to be converted into thermal energy through collisions of the second kind. Resonance radiation emitted in the interior of a cell escapes from the cell by diffusion, i.e., by a succession of absorptions and emissions, until it reaches the wall of the container and escapes. Effects due to imprisonment of resonance radiation have been described by Zemansky [29] and Kenty [30]. Zemansky used imprisonment of radiation with a special arrangement of apparatus to provide measurements of quenching cross sections of foreign gases. Theory shows that these measurements are not influenced by pressure broadening. There may, however, be some uncorrected effects due to depolarization, since the angles between the stimulating radiation and the directions of observation were not those for no polarization effects. A shift of phase, from resonance-radiation transfer, has been observed [151].

The effect of isotopes on imprisonment of radiation has been investigated by Holstein, Alpert, and McCoubrey [31].

4. Polarization of Resonance Radiation

Effect of Foreign Gases. The degree of polarization is specified in terms of the intensity of radiation,

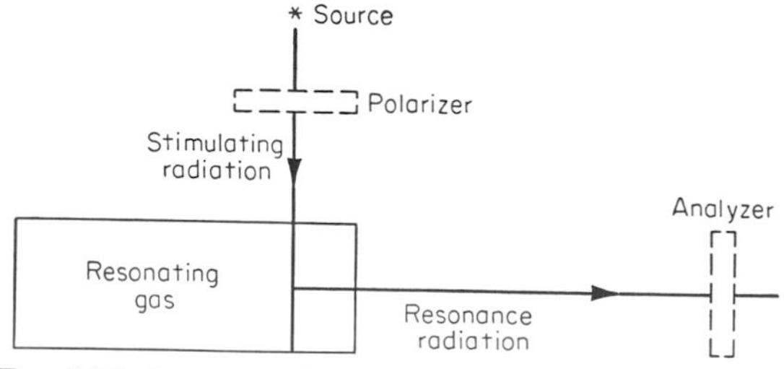

FIG. 7.15. Apparatus for measuring polarization of resonance radiation.

I, transmitted through an analyzer A (Fig. 7.15). The degree of polarization is

$$P = \frac{I_{\max} - I_{\min}}{I_{\max} + I_{\min}} \tag{7.29}$$

With polarized stimulating radiation having its electric vector perpendicular to the plane of Fig. 7.15, the resonance radiation of mercury is found to be partially polarized with its electric vector parallel to that of the stimulating radiation [32]. As the pressure of the mercury vapor is reduced, the degree of polarization increases to about 80 per cent at the lowest pressures for which resonance radiation can be observed.

The degree of polarization was shown by Keussler [33] to diminish with the addition of a foreign gas, as shown in Fig. 7.16.

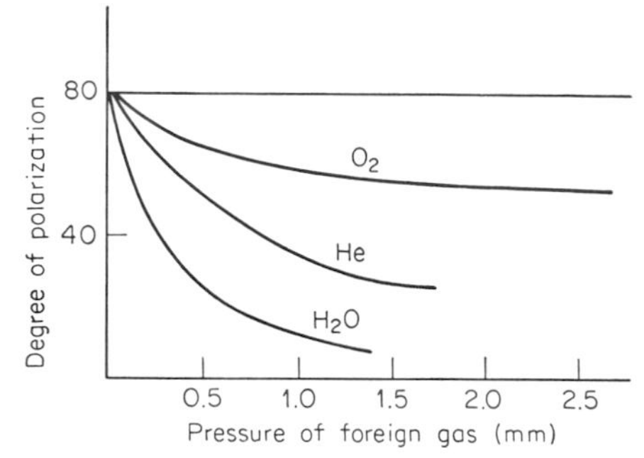

FIG. 7.16. Change in degree of polarization of resonance radiation with foreign-gas pressure.

The depolarizing effectiveness of the strong quenching gases O_2 and H_2 was less than that of the weak quenching gases A and He. The most effective depolarizing gas was water vapor.

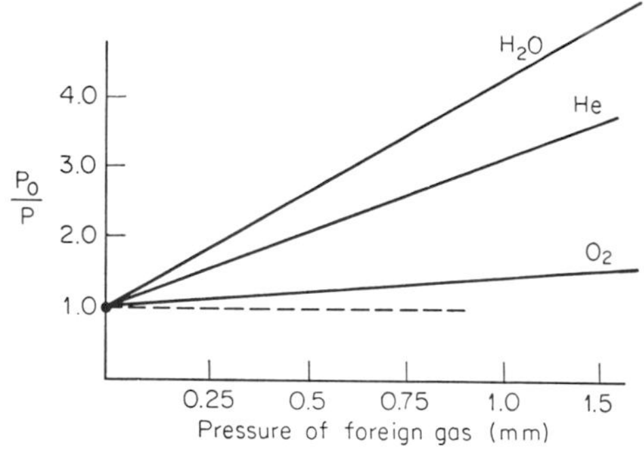

FIG. 7.17. Stern-Volmer plot for degree of polarization.

Keussler's data follow a Stern-Volmer formula (Fig. 7.17),

$$\frac{P_0}{P} = 1 + \frac{\tau_0}{t} \tag{7.30}$$

where τ_0 is the average life of an excited atom before radiation and t is the average time between excitation and a depolarizing collision. Effective cross sections for depolarization are given in Table 7.2. These are about the same as the kinetic-theory cross sections.

The effective cross section for depolarization by collisions with atoms of the same kind are much greater than for collisions with most foreign-gas molecules. A notable exception is the depolarization

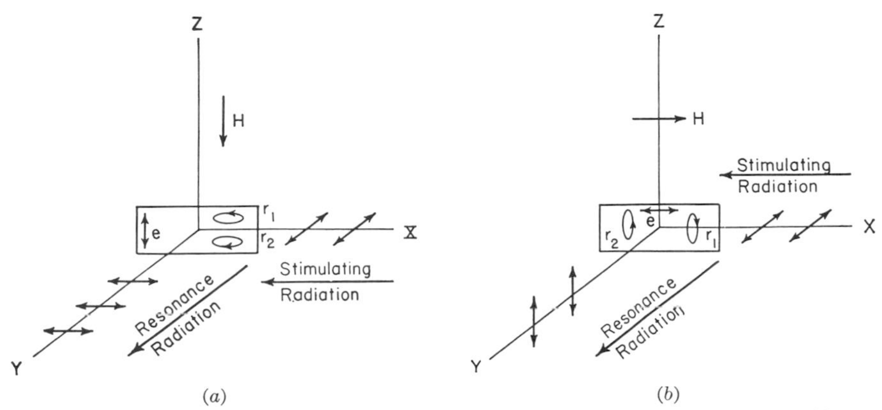

FIG. 7.18. Polarization of mercury resonance radiation in a strong magnetic field.

of sodium resonance radiation by potassium vapor, which shows an effective cross section over 1,000 times the kinetic-theory cross section [34].

TABLE 7.2. CROSS SECTION FOR DEPOLARIZATION OF Hg RESONANCE RADIATION

Gas	*Cross section in* 10^{-16} *sq cm*
O_2	5.28
H_2	3.29
CO_2	49.5
H_2O	47.7
N_2	33.8
He	9.92
Ar	17.6

The abnormally high cross section for depolarization by potassium vapor is attributed to the effect of the magnetic moment of potassium atoms which arises from the spin of the external *s* electron. A near collision between a potassium atom and an excited sodium atom can bring about a deorientation of the excited atom through precession in the magnetic field from the potassium atom. For inert gases and molecules with zero magnetic moment, a definite collision is needed to deorient an excited atom.

The low-depolarization cross section for strongly quenching gases is attributed to the quenching action of these gases so that the resonance radiation which is observed comes primarily from atoms which have not collided. Thus a high degree of polarization is obtained. The effective cross sections for depolarization (Table 7.2) are to be associated with a combination of quenching and depolarizing effects. The separation of these two effects was accomplished by Suppe [35] to obtain depolarization cross sections for gases independent of the quenching cross sections.

Effect of Magnetic Field. If a fluorescing cell containing mercury vapor is placed in a strong magnetic field (5,000 to 10,000 oersteds) and resonance radiation stimulated by polarized radiation as shown in Fig. 7.18*a*, the resonance radiation is nearly 100 per cent polarized, with electric vector **E** perpendicular to **H** and to the electric vector of the stimulating radiation [33, 36]. A 90° rotation of the magnetic field to the direction shown in Fig. 7.18*b* produces a 90° rotation of the electric vector of the resonance radiation.

If a fluorescence cell containing sodium vapor is placed in a strong magnetic field, the resonance radiation is less than 50 per cent polarized for any direction of the magnetic field [32].

The polarization observed with Hg resonance radiation can be described by the classical theory as applied in the description of the Zeeman effect. The fluorescing mercury atoms in a strong magnetic field are considered to carry three oscillators; one linear, *e*, with vibrations parallel to the magnetic field and two circular, r_1 clockwise and r_2 counterclockwise, about the direction of the field (Fig. 7.18). One circular oscillator has a frequency higher, and the other a frequency lower, than that of the linear oscillator. The stimulating radiation is considered to have a broad enough frequency range to be absorbed by all three of the resonating oscillators.

Polarized stimulating radiation excites the linear oscillator *e* when its electric vector is parallel to *H* and the circular oscillators r_1 and r_2 when its electric vector is perpendicular to *H*.

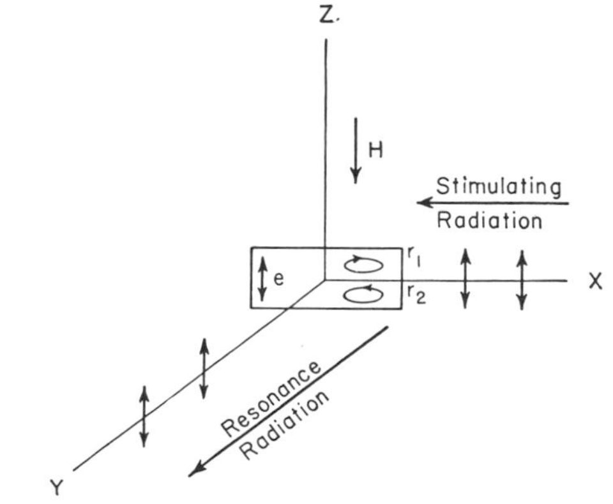

FIG. 7.19. Arrangement to stimulate only plane-polarized resonance radiation.

With the arrangement of Fig. 7.19 the linear oscillator *e* is alone stimulated and the resonance radiation is polarized with its electric vector parallel to *Z*.

With the arrangement of Fig. 7.18*a* the two circular

oscillators are stimulated, and the resonance radiation observed along Y consists of the X components of r_1 and r_2. It is thus plane-polarized with its electric vector parallel to X.

The polarization observed for mercury resonance radiation in a strong magnetic field may also be described in terms of quantum theory. Neglecting the effect of hyperfine structure, the energy states associated with 2536 radiation in a magnetic field are shown in Fig. 7.20. Transitions with $\Delta M_J = \pm 1$

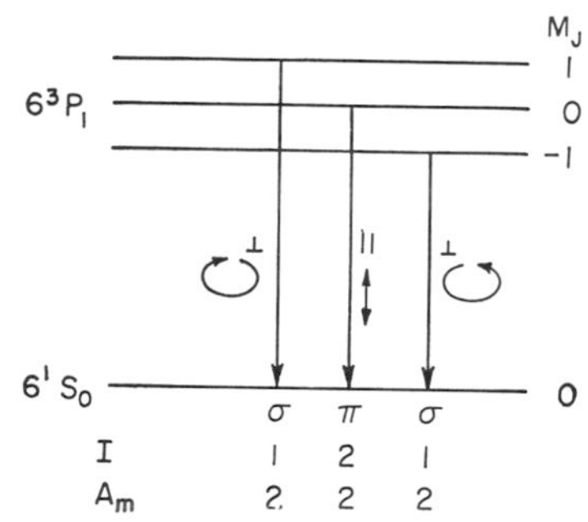

FIG. 7.20. Energy states for Hg 2536 in a weak magnetic field.

correspond to radiation circularly polarized about the direction of the field. These are the σ components and correspond to r_1 and r_2 in the classical description.

Transitions with $\Delta M = 0$ correspond to linearly polarized radiation with its electric vector parallel to the magnetic field. This is the π component and corresponds to e in the classical description.

If the arrangement is such that only the π component is stimulated (as in Fig. 7.19), the resonance radiation will be 100 per cent plane-polarized, with vibrations parallel to the field. If the arrangement is like that in Fig. 6.18a, only the σ components are stimulated and, since observation is along Y at right angles to the field, the resonance radiation is observed to be plane-polarized with its electric vector parallel to X. Resonance radiation in direction Z appears either unpolarized or, with high resolution, as two circularly polarized beams, one clockwise and one counterclockwise.

The polarization of sodium resonance radiation appears incapable of description by the classical model. The failure to obtain a degree of polarization of more than 50 per cent under any conditions does not agree with the classical model. Kastler [37] showed that the small degree of polarization cannot be attributed to depolarization through collisions with normal sodium atoms since it persists to low pressures where a change of pressure produces no change in the polarization. Gaviola and Pringsheim [38] and Datta [34] and Larrick [39] observed the polarization of the D lines separately and found D_1 always unpolarized with or without a magnetic field. D_2 was partially polarized, and the polarization changed with a change in magnetic-field strength. The maximum polarization observed for D_2 was in a strong magnetic field and was 58 per cent [34].

The energy states for Na in a magnetic field are shown in Fig. 7.21.

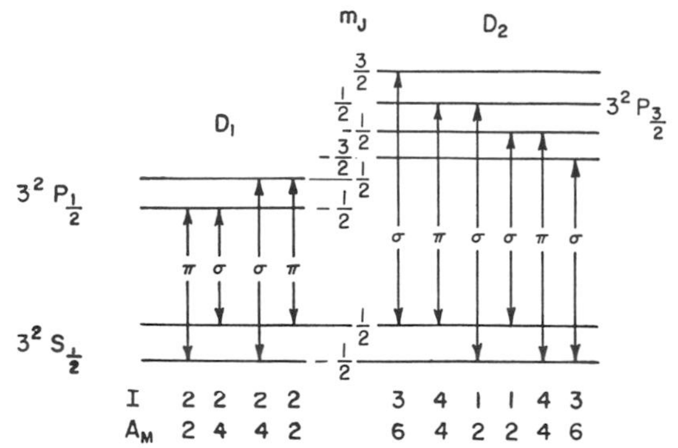

FIG. 7.21. Energy states for Na in a magnetic field.

With the arrangement of Fig. 7.22 only the π components are absorbed. From Fig. 7.21 the D_1 resonance radiation will be unpolarized since each state can radiate both a π and σ component, and since in the direction of observation they have equal intensities. The Zeeman-effect intensities as determined from the sum rule are shown in Fig. 7.21 as I.

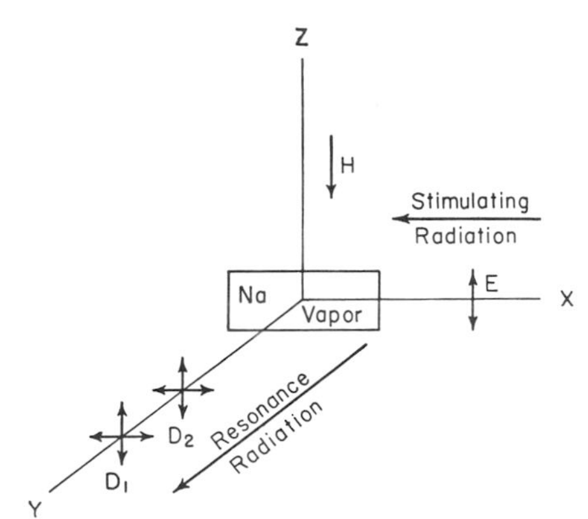

FIG. 7.22. Polarization for sodium resonance radiation.

For D_2, absorption of π components produces states which can radiate π components with total intensity 8 and σ components with total intensity 2. This gives a degree of polarization $P = (8 - 2)/(8 + 2) = 60$ per cent. Thus the maximum observed polarization of 58 per cent for D_2 is consistent with theoretical predictions.

If the two D lines are not resolved, the intensity of D_2 will be twice that of D_1 in the stimulating radiation and four times that of D_1 in the resonance radiation. Excitation solely by π components produces a total radiation intensity 36 for the π and 12 for the σ components. This gives a theoretical maximum of 50 per cent for the degree of polarization of the D lines, in agreement with the maximum observed polarization of 46 per cent [39].

Van Vleck has derived equations for the intensities of σ and π transitions observed in a direction perpendicular to the magnetic field. These have been combined by Mitchell and Zemansky [6 p. 274] into Eq. (7.31). This expresses the degree of polarization of any resonance radiation in a magnetic field in terms of the transition probabilities for the Zeeman components.

$$P = \frac{\left[3 \sum_{\mu} (A_{\pi}\mu)^2 - \frac{2J+1}{3} A^2\right] (3\cos^2\theta - 1)}{(3\cos^2\theta - 1) \sum_{\mu} (A_{\pi}^{\mu})^2 + \frac{2J+1}{3} A^2(3 - \cos^2\theta)}$$

Here J is the quantum number $L + S$ for the upper state, and μ represents one of the $2J + 1$ states into which the upper energy state splits in a magnetic field. Σ_{μ} means a sum over all the magnetic sublevels of the upper state. A_{π}^{μ} is the Zeeman transition probability for substate μ (shown as A_M in Figs. 7.20 and 7.21). $A = A_{\pi}^{\mu} + A_{\sigma}^{\mu}$, where A_{σ}^{μ} is the Zeeman transition probability for the σ transition. θ is the angle between the electric vector of the stimulating radiation and the magnetic field. For sodium resonance radiation excited as in Fig. 7.22 Eq. (7.31) reduces to

$$P = 0 \qquad \text{for } D_1$$

and

$$P = 60\% \qquad \text{for } D_2$$

Equation (7.31) shows that when

$$3\cos^2\theta = 1 \qquad \text{or} \qquad \theta = 54°45' \qquad P = 0$$

When the stimulating-radiation electric vector makes an angle 54°45′ to the magnetic field, any resonance radiation is unpolarized.

This proves to be the same as the angle between unpolarized stimulating radiation and resonance radiation for no polarization effect in the resonance radiation.

Effect of Hyperfine Structure. The maximum degree of polarization obtained with very weak or zero magnetic fields is about 80 per cent for mercury and about 16 per cent for sodium. This is considerably below the predicted maximum of 100 per cent for Hg and 50 per cent for Na. Von Keussler [41] suggested that these discrepancies are due to hyperfine structure.

The hyperfine structure of spectrum lines is associated with the nuclear spin (see Part 7, Chap. 4). Suppose that resonance radiation is observed for an atom consisting of a single isotope with nuclear-spin angular momentum I. The electronic angular momentum J combines with the nuclear-spin angular momentum to give a vector resultant angular momentum F. This gives $2J + 1$ or $2I + 1$ separate energy states, whichever is the smaller. In a weak magnetic field each separate level associated with a single value of F splits into $2F + 1$ states differing as to the component of angular momentum parallel to the field m_F ($F \geq m_F \geq -F$). The effects observed in a weak magnetic field are expected to be like those for the Zeeman effect in weak fields. The relative intensities and transition probabilities for the various transitions may be determined from the equations derived for the Zeeman effect by substituting F for J.

In general, resonance radiation will correspond to transitions from $2F_b + 1$ upper magnetic sublevels to $2F_a + 1$ lower magnetic sublevels. These transitions follow the selection rules $\Delta F = 0, \pm 1, \Delta m_F = 0, \pm 1$.

When the separate transitions give lines which are not resolved, the total intensity may be calculated with each line weighted properly by replacing each state by $2F + 1$ magnetic states, each of them states having the same total transition probability.

A general equation for the degree of polarization of an unresolved line taking into account the hyperfine structure is given by Mitchell and Zemansky [6]. The equation contains $3\cos^2\theta - 1$ as a factor, where θ is the angle between the electric vector of the stimulating radiation and the magnetic field. Thus the angle for zero polarization is $\cos^{-1}(1/\sqrt{3})$, or 55°, and is the same for all resonance lines with or without hyperfine structure.

The resultant polarization depends also on the intensity of lines in the stimulating radiation, and these are usually not known. If the source is a cooled vacuum arc deflected by a magnetic field to reduce reversal of lines, the stimulating radiation may be considered to cover a broad enough spectral range for all hyperfine lines to be exposed to the same intensity, but, with low-pressure vacuum arcs or certain other sources, the assumption of broad-line excitation does not result in values of P in agreement with experiment.

Ellett and Larrick [42] measured the degree of polarization of the cadmium resonance lines 2288 and 3261 at the lowest possible vapor pressures. The observed maximum polarization was 76.3 per cent for 2288 and 86 to 87 per cent for 3261. Assuming (*a*) that 2288 resonance radiation resulted from broad-line stimulation and (*b*) that the hyperfine lines of the even-atomic-weight isotopes coincided with the strongest hyperfine line of the odd isotopes, the degree of polarization was calculated to be $P = 80.5$ per cent, using 3.34 for the ratio of even- to odd-atomic-weight isotopes. The observed maximum degree of polarization was $P = 76.3$ per cent. The difference was considered greater than the experimental error and is not explained.

Calculations of the degree of polarization for Hg resonance radiation give $P = 73.5$ per cent for $\theta = \pi/2$ and $P = 84.7$ per cent for $\theta = 0$. The maximum observed value for $\theta = 0$ was 86 per cent, in good agreement with the theory. θ = angle between stimulating radiation and direction of observation.

For sodium there is only one isotope, but the resonance radiation consists of several overlapping hyperfine lines. Various values of nuclear spin for Na were assumed and P calculated for each value of I. The observed degree of polarization of 16.5 per cent indicated a nuclear spin 1, while independent determinations showed $I = \frac{3}{2}$. Breit [43] showed that, for small separations of the hyperfine levels, corrections to the theory would give $\frac{3}{2}$ for the spin determined from the observed degree of polarization.

Effect in Very Weak Magnetic Fields. If there is no external magnetic field but only local fields oriented at random near resonating atoms, the degree of polarization, as calculated by Heisenberg, should be 27 per cent for Hg 2537, 14 per cent for Na D_2, and 0 for Na D_1. The experiments showed a degree of polarization of 80 per cent for Hg 2537, 60 per cent for Na D_2, and 0 for Na D_1.

To account for this discrepancy, Bohr [44] introduced the principle of spectroscopic stability, which states that, for zero magnetic field, resonance radiation has the same polarization as that for a classical oscillator in a strong magnetic field parallel to the electric vector of the stimulating radiation. The rule was generalized by Heisenberg [45] to apply to atoms which show an anomalous Zeeman effect. According to the generalized rule, if the stimulation is by polarized radiation, the resonance radiation for zero magnetic field is like that for a magnetic field parallel to the electric vector of the stimulating radiation. If the stimulating radiation is unpolarized or circularly polarized, the resonance radiation will be like that for a magnetic field parallel to the direction of incidence of the stimulating radiation.

The polarizations with zero magnetic fields will resemble the polarization with magnetic fields of magnitude below that needed for a Paschen-Back effect on the hyperfine structure. Thus, for an atom such as Hg, the polarization observed with zero magnetic field would be 85 per cent, or the same as that calculated from the hyperfine structure in a magnetic field not strong enough for a Paschen-Back effect for the hyperfine structure.

A general equation for the polarization of resonance radiation in zero magnetic field when stimulated by partially polarized radiation is derived using the principle of spectroscopic stability in Appendix XIII of Mitchell and Zemansky's book [6].

5. Stepwise Excitation of Fluorescence in Gases

Excitation of Hg Fluorescence. When a fluorescence cell containing mercury vapor at low pressure is illuminated by the entire radiation from a low-pressure mercury arc, the vapor gives a fluorescence whose spectrum consists of the resonance lines and a number of other lines of the arc spectrum of mercury [1, p. 40]. Energy levels of mercury are shown in Fig. 7.23.

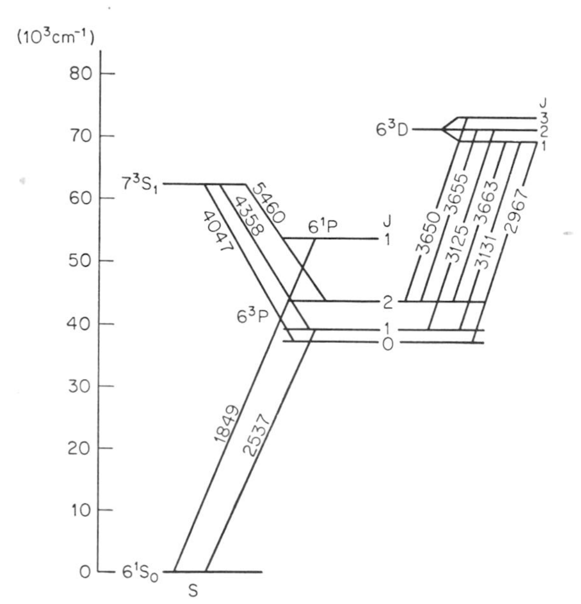

Fig. 7.23. Energy transitions for Hg.

Illumination of Hg vapor by the entire radiation of a single water-cooled mercury arc gave fluorescence whose spectrum showed lines corresponding to the lines in Fig. 7.23 (Fuchtbauer [47], Wood [48]). A screen which reduced the intensity of all wavelengths nearly equally could be inserted between the fluorescence cell and the spectrograph or between the stimulating light source and the fluorescence cell. Comparison of spectra for these two arrangements showed some fluorescence lines to have the same intensity in the two spectra but others to be much weaker for the spectra photographed with the absorbing screen between the stimulating light source and the fluorescence cell. These observations show that some of the fluorescence lines are being stimulated by double and some by triple absorption of radiation.

The primary absorption is of 2536, resulting in a transfer of mercury atoms from the 6 $^1S_0$ to the 6 $^3P_1$ state. The resulting population of 6 $^3P_1$ permits a secondary absorption of 4358, causing transitions from 6 $^3P_1$ to 7 $^3S_1$. From 7 $^3S_1$ Hg atoms can reach 6 $^3P_2$, 6 $^3P_1$, and 6 $^3P_0$ by radiation of 5460, 4358, and 4047, respectively. This creates populations of excited atoms in the metastable states $^3P_0$ and $^3P_2$, which in turn permits the tertiary absorption and emission of lines 6 $^3P_2$ − 6 $^3D_{3,2,1}$ at 3650, 3655, 3663 and 6 $^3P_0$ − 6 $^3D_1$ at 2967.

When the exciting light contains only radiation below 4000 A, secondary absorption of 6 $^3P_1$ − 6 $^3D_1$ (3131) and 6 $^3P_1$ − 6 $^3D_2$ (3125) gives 6 $^3D_1$ and 6 $^3D_2$ states but no 6 $^3D_3$ states. In the transitions 6 $^3P_2$ − 6 $^3D_{3,2,1}$ the resulting fluorescence shows 3663 and 3655 strong and 3650 very weak (Fig. 7.24). 3650 is obtained, but with very small intensity, by tertiary absorption from 6 $^3P_2$ atoms.

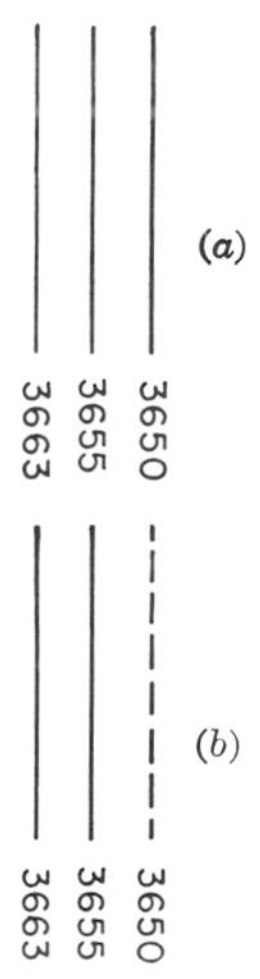

Fig. 7.24. Hg fluorescence. (a) Stimulated by all radiation from a cooled Hg arc. (b) Stimulated by all radiation of wavelength less than 4000 A from a cooled Hg arc.

When the stimulating radiation contains 6 $^3P_1$ − 7 $^3S_1$ (4358), 7 $^3S_1$ atoms change to 6 $^3P_2$ with the emission of 5460 and the extra population of 6 $^3P_2$ so produced enhances 3650 until it is the strongest member of the triplet, as shown in Fig. 7.24*a*.

The presence of a few millimeters of nitrogen in the fluorescence cell enhances those lines corresponding to absorption by $6\,^3P_0$ atoms. The effect of the N_2 is to transfer Hg atoms from $6\,^3P_1$ to $6\,^3P_0$ through collisions of the second kind. Wood [48] was able to demonstrate production of $6\,^3P_1$ and $6\,^3P_0$ Hg atoms by radiation from a cooled mercury arc and the further excitation of these atoms by light from a second hot mercury arc which alone would produce no direct excitation of mercury lines in fluorescence.

Optical excitation has been observed for Hg, Cd [49], Zn [50], Na [51], Li [52], Ag [53], Ca [54], Tl [55, 56], Pb [57, 58], Bi, Sb, As, P [57], Mn [59], Cs [52], He [60, 61], O [62].

Polarization of Stepwise Fluorescence. If the fluorescence of mercury vapor is stimulated by polarized radiation with the electric vector perpendicular to a magnetic field as shown in Fig. 7.25, the Hg line 4358 will be polarized with its electric vector parallel to that of the stimulating radiation, while the line 4047 will be polarized with its electric vector perpendicular to that of the stimulating radiation. This polarization of 4047 is known as *negative polarization.*

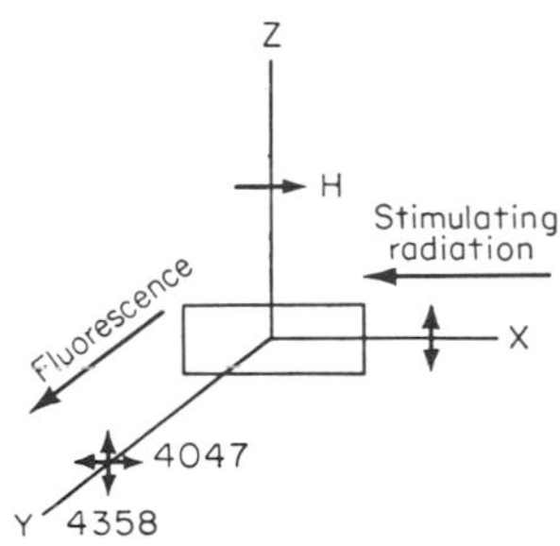

FIG. 7.25. Negative polarization of stepwise fluorescence.

If a small quantity of nitrogen is added to the fluorescence cell, the polarizations of these two lines are reversed. Thus, with nitrogen present, 4047 is polarized positively and 4358 negatively.

Neglecting the effect of hyperfine structure, positive and negative polarization of stepwise excited fluorescence can result as shown in Fig. 7.26. With the arrangement of Fig. 7.25, and no nitrogen in the fluorescence cell, only the two σ components of 2536 can be absorbed, thus populating the $^3P_1$ states with $M_J \pm 1$ and not the state with $M_J = 0$. The further absorption of 4358 populates only the state $M_J = 0$ of $7\,^3S_1$ because the stimulating radiation can be absorbed only as σ components. Hg atoms in state $M_J = 0$ of $7\,^3S_1$ can revert to $M_J = \pm 1$ of $^3P_1$ with the emission of the circularly polarized σ components of 4358. For the observation direction in Fig. 7.25, 4358 will appear 100 per cent plane-polarized positively, that is, with its electric vector parallel to that of the stimulating radiation.

For the emission of 4047 in fluorescence corresponding to a transition from $M_J = 0$ of $7\,^3S_1$ to $6\,^3P_0$, the only radiation possible is for $\Delta M_J = 0$ giving a π component. This gives plane-polarized light with its electric vector parallel to the magnetic field. Thus 4047 should show 100 per cent negative polarization, that is, with its electric vector perpendicular to that of the stimulating radiation.

If nitrogen is added to the fluorescence cell, many Hg $^3P_1$ atoms will be transferred to $^3P_0$ by collisions of the second kind with the destruction of any special orientation. With N_2 present the secondary absorption for stepwise excitation originates nearly completely in $^3P_0$. Since the stimulating radiation contains only σ components, the absorption corresponds to $\Delta M_J = \pm 1$, and only the states $M_J = \pm 1$ of $^3S_1$ are populated by the absorption of 4047. These atoms can then radiate the components of 4047, giving circularly polarized fluorescence which appears plane-polarized 100 per cent positively in the direction of observation. The states can also radiate 4358 with $\Delta M_J = \pm 1$, 0, giving two σ and two π components. The π components have twice the intensity of the σ components so that $P = (I_\pi - I_\sigma)/(I_\pi + I_\sigma) = -33$ per cent for 4358. Thus the addition of nitrogen reverses the polarization of 4047 and 4358 excited in stepwise fluorescence.

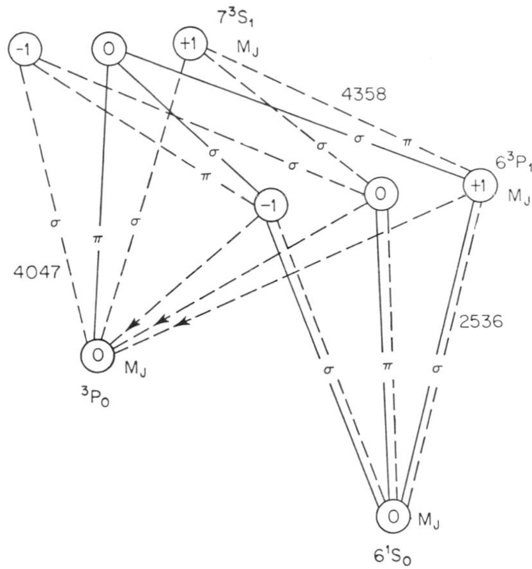

FIG. 7.26. Polarization of stepwise fluorescence in Hg vapor:
——— transitions for Hg alone
- - - - - transitions for Hg with N_2 added

Because of the effects of hyperfine structure the resulting polarizations of stepwise excited fluorescence are not so great as calculated.

Hanle and Richter [63] first observed that when fluorescence in Hg vapor containing N_2 is stimulated by 4047 and also by the ordinary and extraordinary beams obtained by passing 2536 radiation through a block of calcite, the fluorescence produced by these beams differs in color. Fluorescence from one is blue-green and from the other is yellow-green. Fluorescence from one shows lines polarized positively and from the other shows lines polarized negatively. Hanle and Richter showed that such observations were to be expected from the energy states of Hg.

6. Optical Orientation of Nuclei

It was first pointed out by Kastler [64] that resonance radiation provided a means of obtaining an inequality of population in the Zeeman sublevels in the ground state of atoms. Such orientation was first successfully accomplished by Brossel, Kastler, and Winter [65]. Orientation followed by deorientation by a radio-frequency field was accomplished by Brossel, Cagnac, and Kastler [66]. The procedure for doing this is illustrated in Fig. 7.27. A beam of

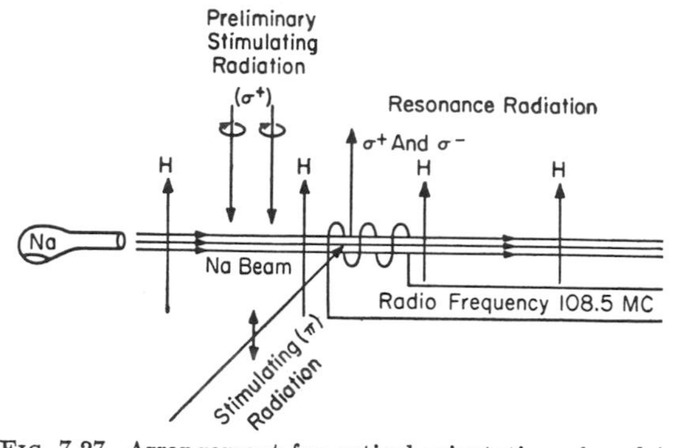

FIG. 7.27. Arrangement for optical orientation of nuclei.

sodium atoms is projected through a magnetic field H and stimulated to emit resonance radiation by a 25-cm-long beam of right circularly polarized light from a sodium lamp [(σ^+) in Fig. 7.27].

The transitions stimulated by the D_1 line are shown in Fig. 7.28. Absorption of σ^+ causes transitions from $m_J = -\frac{1}{2}$ of ${}^2S_{\frac{1}{2}}$ to $m_J = +\frac{1}{2}$ of ${}^2P_{\frac{1}{2}}$. There

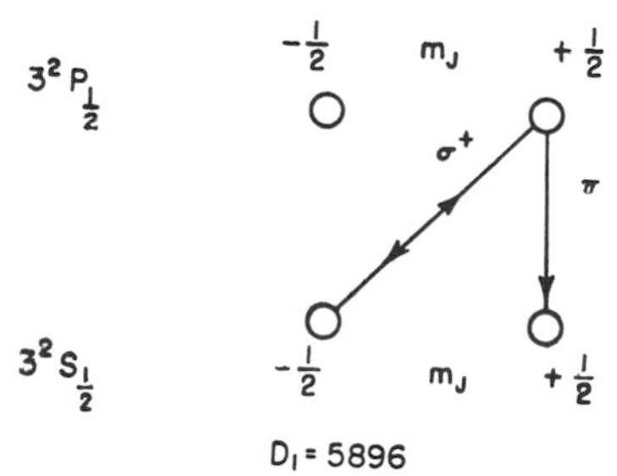

FIG. 7.28. Resonance transitions for D₁ in magnetic field which decouples J and I

is no absorption by $m_J = +\frac{1}{2}$ of ${}^2S_{\frac{1}{2}}$. The resulting resonance radiation consists in part of π transitions to $m_J = +\frac{1}{2}$ and σ^+ transitions to $m_J = -\frac{1}{2}$. These transitions give to $m_J = +\frac{1}{2}$, an excess of population over $m_J = -\frac{1}{2}$. There is set up an optical pumping action which populates $m_J = +\frac{1}{2}$ and depopulates $m_J = -\frac{1}{2}$ of the ground state.

This action of concentration occurs for magnetic fields H sufficiently strong to decouple electronic and the nuclear-spin angular momenta. If the field H is not sufficient to do this, the pumping action will populate some of the Zeeman sublevels of the resultant angular momentum $F = J + I$ and depopulate others. Figure 7.29 shows the Zeeman sublevels for states $F = 1$ of $3\ {}^2S_{\frac{1}{2}}$ and $F = 1$ of $3\ {}^2P_{\frac{1}{2}}$. The pumping action due to absorption of σ^+ and radiation of π, σ^+, and σ^- tends to populate the state $m_F = +1$ of ${}^2S_{\frac{1}{2}}$. This extra population of $m_F = +1$ may be detected by stimulating the Na beam with π radiation at a point downstream from the excitation region as shown in Fig. 7.27. Resonance radiation observed in the direction of H consists primarily of σ^+ with a small amount of σ^-. Resonance radiation σ^+ is received by one photomultiplier cell and σ^- by another connected in opposition. When σ^+ and σ^- resonance radiation

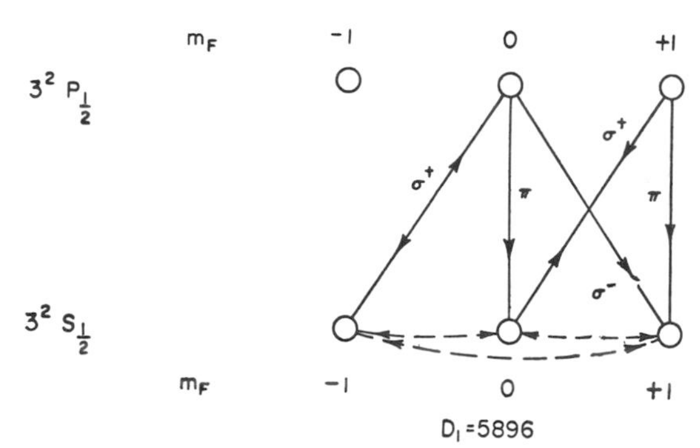

FIG. 7.29. Resonance transitions for D₁ in weak magnetic field.

differ in intensity, a signal is obtained. The ratio $R = (I_+ - I_-)/(I_+ + I_-)$ was 35 per cent with the preliminary stimulating radiation acting and zero with no preliminary stimulation. The actual experiment was conducted with a mixture of D_1 and D_2 stimulating radiation.

By changing the strength of the magnetic field H the wave-number difference between two of the Zeeman sublevels could be made to equal 108.5 Mc, and the excess population in states $m_F = +1$ destroyed by transitions to other states as shown by dotted lines in Fig. 7.29. Figure 7.30 shows

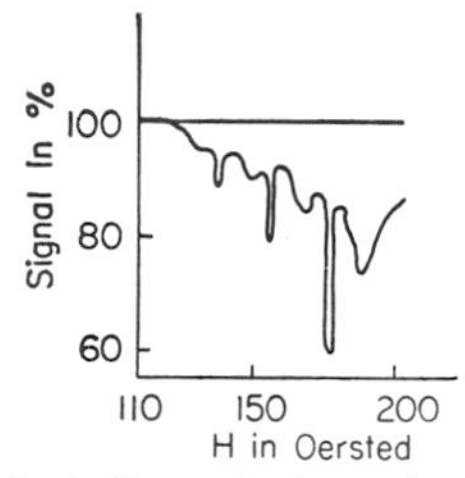

FIG. 7.30. Deorientation of atoms by radio-frequency field.

sample decreases in signal intensity at the depopulating resonance frequencies. With sufficient radio-frequency power, transitions with $\Delta m_F = \pm 1$, ± 2, and ± 3 were obtained.

The combination of optical and radio-frequency resonance has been used by Brossel and Bitter [67] and by Bitter, Davis, Richter, and Young [68] in the study of the energy levels of mercury, including radioactive mercury.

More recent developments in optical pumping have been described by Kastler [114]. Nuclear orientation through optical pumping has been obtained in a stationary volume of vapor. This has permitted the study of quenching or relaxation of the oriented nuclei by collisions with the walls of the resonance cell and by the action of light-stimulated relaxation.

In some experiments with mercury vapor a resonance cell containing only vapor of Hg^{199} (nuclear spin $\frac{1}{2}$) is irradiated by 2537 radiation from a light source containing only Hg^{204}. The single hyperfine component of 2537 from Hg^{204} coincides with the hyperfine component ($i = \frac{1}{2} \rightarrow F = \frac{1}{2}$) of Hg^{199} so that only this component is excited. The Hg^{199} cell is placed in a magnetic field whose direction is that of the incident light, and the incident light is made circularly polarized. The Zeeman sublevels are the same as those shown in Fig. 7.28 for sodium. The values of m for the lower state correspond to m_i for a nuclear spin $\frac{1}{2}$ while those for the upper state correspond to m_F for the hyperfine level $F = \frac{1}{2}$ of the $^3P_1$ state of Hg. Absorption of the exciting radiation (σ^+) produces only excited Hg^{199} atoms in state $m_F = +\frac{1}{2}$. These can radiate by σ^+ and π transitions to provide an increase in population ($N_{+\frac{1}{2}}$) of state $m_i = +\frac{1}{2}$. There results a decrease in population ($N_{-\frac{1}{2}}$) of state $m_i = -\frac{1}{2}$.

The intensity of π and σ radiation in all directions is proportional to the population of the excited state $m_F = +\frac{1}{2}$; this population is proportional to $N_{-\frac{1}{2}}$ and to the intensity of the pumping light (I). For constant pumping intensity (I) the σ^+ resonance-radiation intensity (L) provides a measure of the population in the ground state $N_{-\frac{1}{2}}$. The extent of nuclear orientation achieved is described by a polarization

$$P = \frac{N_{+\frac{1}{2}} - N_{-\frac{1}{2}}}{N_{+\frac{1}{2}} + N_{-\frac{1}{2}}} = 1 - 2\frac{N_{-\frac{1}{2}}}{N} \tag{7.32}$$

where $N = N_{+\frac{1}{2}} + N_{-\frac{1}{2}}$.

$$L = CN_{-\frac{1}{2}} = C\frac{N}{2}(1 - P) \tag{7.33}$$

A time variation of the light signal L corresponds to a time variation in P.

With the resonance cell in the dark, thermal distribution makes $N_{-\frac{1}{2}} = N_{+\frac{1}{2}} = N/2$ because of the very small energy difference between the levels. When the shutter is opened, the initial light signal corresponds to the equilibrium population of $N_{-\frac{1}{2}}$. As radiation continues and optical pumping starts, the light signal L and $N_{-\frac{1}{2}}$ decrease exponentially to a steady-state signal L_∞ in a manner described by $L_0 - L = (L_0 - L_\infty)e^{-t/\tau_1}$, where τ_1 is a decay constant. τ_1 is the average life of atoms in state $m_i = -\frac{1}{2}$ under the action of radiation. τ_1 depends on the intensity of the pumping light L. The curve of dark relaxation with time can be obtained as

$$L - L_\infty = (L_0 - L_\infty)e^{-t/\phi_1}$$

where ϕ_1 represents the average life of atoms in state $m_i = +\frac{1}{2}$ before being destroyed by a collision with a wall.

For a steady state ($t = \infty$), there is an equilibrium population ($N_{+\frac{1}{2}}$) for which the gain from radiation transfer from $m_i = -\frac{1}{2}$ through $m_F = +\frac{1}{2}$ equals the loss through wall collisions plus the loss through light absorption to $m_F = +\frac{1}{2}$ and back to $m_i = -\frac{1}{2}$ by stimulated emissions. The rate of gain for $m_i = +\frac{1}{2}$ equals the rate of loss for $m_i = -\frac{1}{2}$. Thus

$$(N_{-\frac{1}{2}})_\infty \frac{1}{\tau_1} = (N_{+\frac{1}{2}})_\infty \left(\frac{1}{\phi_1} + \frac{1}{T_1}\right) \tag{7.34}$$

where T_1 is the average life for $m_i = +\frac{1}{2}$ before being changed to state $m_i = -\frac{1}{2}$ by the action of the pumping light. Since for any time

$$N_{+\frac{1}{2}} = N - N_{-\frac{1}{2}} \quad \text{and} \quad N = \frac{2L_0}{C}$$

$$(N_{-\frac{1}{2}})_\infty = \frac{L_\infty}{C} \quad \text{and} \quad (N_{+\frac{1}{2}})_\infty = \frac{2L_0}{C} - \frac{L_\infty}{C}$$

Substitution into (7.34) gives

$$\frac{L_\infty}{C}\frac{1}{\tau_1} = \frac{2L_0 - L_\infty}{C}\left(\frac{1}{\phi_1} + \frac{1}{T_1}\right)$$

from which the average life for relaxation by light stimulation can be determined through

$$\frac{1}{T_1} = \frac{L_\infty}{2L_0 - L_\infty}\frac{1}{\tau_1} - \frac{1}{\phi_1} \tag{7.35}$$

Since the lifetime for atoms in state $m_i = +\frac{1}{2}$ depends on the intensity of the pumping light, the uncertainty principle requires that the width of the energy level also depend on the intensity of the pumping light. The effect has been observed through the use of nuclear magnetic resonance.

It was also found that Zeeman levels were displaced by the action of the pumping light, the effect being designated as the Lamp effect.

Optical pumping has been utilized by Colegrove and Franken to obtain alignment of He^4 atoms in the $^3S_1$ metastable state [120]. The technique has been extended to He^3 by Colegrove, Schearer, and Walters [121] and Greenhow [122]. He^3 with nuclear spin $\frac{1}{2}$ has Zeeman levels in the ground state $^1S_0$. Aligned atoms in the metastable state $^3S_1$ can make collisions with $^1S_0$ atoms with exchange of excitation state, thereby producing aligned He^3 atoms in the ground state. By this means a ground-state polarization of 40 per cent was achieved in He^3 gas at 1 mm pressure.

Alignment of molecules has been achieved by the method of photoselection as described by Albrecht [141] and by Lombardi, Raymonda, and Albrecht [142].

7. Sensitized Fluorescence

When a mixture of two gases is illuminated by radiation which only one of the gases can absorb, the fluorescent radiation may consist of radiation from both the gases in the mixture.

This was first observed by Cario and Franck [69] in a mixture of Hg and Tl vapors. When the mixture was illuminated by 2536 A from a cooled Hg arc, the spectrum of the fluorescence showed both Hg and Tl lines [70].

Sensitized fluorescence has since been observed for mixtures of mercury vapor with the following vapors:

Ag [71], Cd [71], Pb [72], Bi [73], K [74], Zn [75], Fe [74], Cr [76], Cu [77], Au [78], Na [79].

The mechanism of sensitized fluorescence consists in absorption of radiation by one component A in the mixture to produce excited atoms A^*, the collision between excited atoms A^* and unexcited atoms B producing excited atoms B^* and normal atoms A, and emission of radiation by atoms (B^*). The difference in energy between A^* and B^* (energy discrepancy) either goes into or is taken from the thermal kinetic energy of the colliding atoms. The process may also take place with an original absorption by molecules such as Hg_2.

The intensity of fluorescence from atoms B^* depends in large part on the energy discrepancy, being greater for the smaller discrepancy. This was well demonstrated for a mixture of Na and Hg vapors by Beutler and Josephy [79].

The intensities of lines in sensitized fluorescence also are described by the rule first stated by Wigner [80], that those transitions are most probable for which the change in electron spin for one atom is equal and opposite to that for the other.

The selection rules which apply to quenching of fluorescence by foreign gases and to predissociation of diatomic molecules also apply to sensitized fluorescence.

An additional selection rule which describes collisions of the second kind was pointed out by Winans [78]. This rule states that the resultant electronic angular momentum for a quasi molecule remains constant during a collision of the second kind. The total electron angular-momentum number $= J = J_1 + J_2$, where J_1 and J_2 are the numbers for the colliding atoms. For example, in the sensitized fluorescence of Sn in Hg vapor, at pressure to provide a large population of Hg $^3P_0$ atoms, the value of J before collision of the second kind is $J = J_{\text{Hg}}\ {}^3P_0 + J_{\text{Sn}}\ {}^3P_0 = 0$. After collision it is $J = J_{\text{Hg}}\ {}^1S_0 + J_{\text{Sn}}\ {}^3P_{2,1,0} = 2$, 1, or 0.

If J after collision is to equal that before, the state Sn $^3P_0$ will be the most abundantly excited. Experiments show that the intensity of line 3034 A from Sn $^3P_0$ is much stronger in sensitized fluorescence than the line 3009 from Sn $^3P_1$, while they have nearly equal intensities in the spectrum of a carbon arc containing Sn.

Measurements of intensities by Swanson and McFarland [81] and by Anderson and McFarland [83] provide further confirmation of the rule $\Delta J = 0$. They measured the intensity of sensitized fluorescence of Tl at various mercury temperatures, keeping the Tl pressure constant, and also at various pressures of Ar and He, keeping admixed Tl and Hg temperatures constant. The quasi molecules formed during collisions of the second kind resemble molecules with case (c) coupling [28].

Sensitized fluorescence with very small energy discrepancies has been observed by Buhl [82]. Mercury vapor was illuminated by resonance radiation from only one of the Hg isotopes. At low pressures of Hg vapor in the fluorescence cell the fluorescence spectrum showed only lines from the Hg isotope corresponding to the stimulating radiation. A slight increase in pressure of Hg vapor caused the appearance of fluorescence lines from the other isotopes. Also, the application of a magnetic field to the fluorescence tube caused the appearance of fluorescence lines from the other isotopes. The effect of the magnetic field was to lower the energy discrepancy and thus increase the intensity of fluorescence. The effective cross section for sensitized fluorescence was over 1,000 times the kinetic-theory collision cross section.

Sensitized fluorescence has also been obtained for molecules. The effects have been described by Pringsheim [1, p. 218].

8. Selective Reflection

When the pressure of a vapor showing resonance radiation is increased from a few millimeters of mercury to above atmospheric pressure, the region in the tube showing fluorescence emission shrinks to the surface of the tube at the position of entrance of the stimulating radiation. The vapor then behaves as a mirror for wavelengths near that of the resonance radiation. This was first observed for mercury vapor by Wood [84]. For cadmium vapor at a pressure of 20 atm Welsh, Kastner, and Lauriston [10] have shown that selective reflection occurs over a range of several thousand wave numbers. Lauriston and Welsh [85] measured the reflecting power of vapors of Na, K, Rb, and Cs at pressures between 10 mm and 1 atm. At the lower pressures the reflecting power changes with wavelength in the same manner as does the index of refraction for anomalous dispersion.

Selective reflection by Hg has been studied up to a pressure of 340 atm [46]. The results were described by the classical theory of reflection from an absorbing medium. The damping constant γ was found to vary directly with the density of Hg vapor. At high pressures selective reflection was also observed in the Hg_2 band at 2540. The influence of foreign gases was studied up to pressures of 1,500 atm. The damping constant varied linearly with foreign-gas pressure [119].

9. Reemission

When the vapor pressure in a cell containing a vapor such as Cd is increased beyond a few centimeters of Hg, it is observed that fluorescence can be stimulated by any radiation of wavelength within an absorption band of the vapor. The spectrum of the fluorescence will in general show bands extending only toward longer wavelengths from the wavelength of the stimulating radiation.

It was first observed by Kapuscinski [86] that the spectrum of fluorescein vapor showed not only bands but also lines at the wavelengths of the stimulating radiation. This effect is known as *reemission.* Cram [87] showed that reemission by cadmium vapor occurred for Cu spark lines of wavelength between 2288 and 2212 A, but not for lines of wavelength less than 2212.

Jablonski and Pringsheim [88] showed that, if the vapor pressure in a fluorescing cell containing Na is increased from a few millimeters of mercury to nearly 1 atm, the beam of resonance radiation in the cell first shrinks to the surface at which the stimulating radiation enters and may then expand from the surface to reestablish a beam of yellow light through the cell. The expansion is observed if the stimulating radiation consists of broadened or self-reversed sodium D lines.

The spectrum of the fluorescence obtained at these high pressures shows the two D lines. If the stimulating radiation is polarized, the fluorescence radiation for D_1 is unpolarized and for D_2 more than 30 per cent polarized, at a temperature of 250°C in the cell. A magnetic field of 80 Oersteds parallel to the direction of observation did not change the degree of polarization within the experimental accuracy. The polarization of ordinary sodium resonance radiation would have been destroyed under these circumstances. This effect in Na is another example of reemission of radiation of wavelength within the wavelength range for absorption by Na_2 molecules.

The mechanism of reemission is shown in Fig. 7.31. A loosely bound molecule absorbs radiation

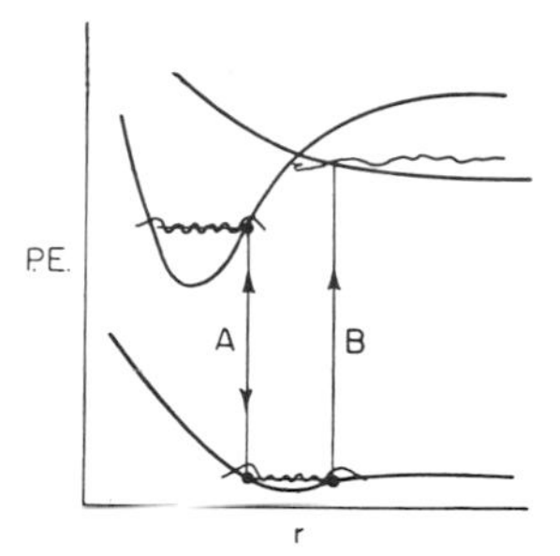

FIG. 7.31. Potential curves for loosely bound molecules.

corresponding to the transition A. This absorption places excited molecules on a vibrational level from which they may return to the ground state with reemission of radiation of wavelength equal to that absorbed. If the lifetime of the excited state is of the order of the time between collisions, some of the molecules will reemit radiation, while others will be knocked into lower vibrational levels and emit radiation corresponding to the band spectrum of the molecule. If the excited state reached by an absorption such as B (Fig. 7.31) is a repulsive state, the atoms will fly apart and no reemission will occur. The appearance of reemission can be used to establish the stability of an excited molecule and thus to locate a dissociation limit as done by Cram [87].

10. Fluorescence in Liquids

Pure liquids or solids usually show only faint fluorescence, but a given substance shows brighter fluorescence as a solid than as a liquid. For example, the fluorescence of pure benzene is faint, and the intensity of fluorescence decreases as the temperature is raised but increases appreciably when the benzene is frozen. Another example is naphthalene, which at room temperature is a solid and shows bright fluorescence. When the naphthalene is melted, the fluorescence almost completely disappears.

Small amounts of impurities in solutions and in crystals play a very important part in fluorescence. A pure crystal of anthracene shows violet fluorescence when stimulated by ultraviolet light or by X rays. Ordinary crystals of anthracene, however, show green fluorescence when stimulated in this manner. The green fluorescence has been traced to small amounts of naphthalene as an impurity in the anthracene crystals. The green fluorescence is due mainly to the naphthalene. Since naphthalene occurs in very small quantities in the anthracene crystals, it is believed that most of the incident energy is absorbed by anthracene molecules and then transported through the crystal to the naphthalene molecules. These then emit their characteristic fluorescent radiation.

A set of careful experiments on the effect of the addition of controlled amounts of fluorescent molecules as impurities in pure solvents has been carried out by Kallmann and Furst [89]. Using a gamma-ray source for excitation, they measured the intensity of fluorescence for a given solvent as a function of the concentration of fluorescent molecules as solute. Various combinations of more than a dozen different solvents and about two dozen different solutes were used. The solutes were, in all cases, molecules which could by themselves show fluorescence, though the pure solvents did not always show fluorescence. Concentrations were of the order of 1 part solute to 1,000 parts solvent.

The added solute caused either an increase or a decrease in the intensity of fluorescence. The actual effect observed depended upon the nature of both the solvent and the solute. The same solute, in fact, might cause an increase in fluorescence in one solvent and a decrease in fluorescence in another.

When the solute quenched the fluorescence, there was little to observe, because the fluorescence of a pure solvent is very faint. When the solute caused an increase in the intensity of fluorescence, it was found that (*a*) the increase in intensity is caused by the addition of the solute; (*b*) the fluorescence intensity increases with increase in concentration of solute at a rapid rate until an optimum concentration is reached, after which it decreases; (*c*) the increase in intensity of fluorescence upon adding a solute is greater than the increase expected from direct absorption of stimulating radiation by the solute molecules; (*d*) the spectrum of the fluorescence shifts from that of the solvent to that of the solute; (*e*) the shift in the spectrum of the fluorescence is a shift toward longer wavelengths.

In those cases where the addition of a solute caused a decrease in the intensity of fluorescence it was shown that the fluorescence spectrum of the solute lay at shorter wavelengths than the fluorescence spectrum of the solvent.

For liquids or for gases at high pressures the effect of collisions of excited molecules is either to change the frequency of the fluorescence or to quench it. In the liquid state molecules may be considered as in a constant state of collision whose effect is partially or sometimes completely to quench the fluorescence.

The experimental observations lead to the conclusion that in dilute solutions the stimulating radiation is directly absorbed primarily by molecules of the solvent. A solvent molecule after absorption is left in an excited state with excitation energy distributed between electronic, vibrational, and rotational energies. The vibrational and rotational energy is quickly lost through collisions and appears as heat. The electronic energy of an excited molecule may be given up through any of the following processes: (*a*) Transition to a lower energy state through radiation as fluorescence. (*b*) Radiationless transition into vibrational, rotational, or translational energy through a collision of the second kind. (*c*) Transfer to another molecule of the same type on collision. Since the two colliding molecules are in resonance, this transfer can be expected to occur with a large effective cross section. The energy is eventually emitted as fluorescence or changed to thermal energy through collisions of the second kind. (*d*) Transfer to a molecule of the solute on collision. If the solute molecule has an excited state of energy less than that of an excited solvent molecule, the solute molecule will trap the excitation energy and not transfer it back to the solvent molecules. The excited solute molecules can then radiate this energy as fluorescence characteristic of the solute molecules or can lose the energy to thermal energy through collisions of the second kind or internal conversion. The process is analogous to sensitized fluorescence for gases.

Kallmann and Furst [91], in another set of experiments, tried adding an additional solute to a solvent already containing an optimum concentration of fluorescent molecules of one type. They distinguish between three different types of additional solute, each of which gives different results:

1. If the original solute has an emission and absorption spectrum of shorter wavelength than that of the second solute, then a small amount of second solute strongly decreases the fluorescence. The second solute acts like an additional quencher.

2. If the original solute has an emission and absorption spectrum close to that of the second solute, then a large amount of second solute is needed to strongly decrease the intensity of fluorescence. The second solute in this case absorbs energy from the solvent in competition with the molecules of the original solute, and the intensity of fluorescence decreases only as self-quenching among the two solutes becomes pronounced.

3. If the original solute has an emission and absorption spectrum of longer wavelength than that of the second solute, the second solute can act like an additional solvent. The second solute can trap the excitation energy of the solvent molecules and transfer this energy to the original solute, the energy transfer taking place in two stages instead of one. The intensity of the fluorescence depends to a great degree on the internal quenching of the second solute and tends to be smaller than it was without the second solute.

The fluorescence and absorption spectra of a dilute solution of some solute-solvent combinations show vibrational structure and mirror symmetry. Energy absorbed by the solvent is lost by transfer to the solute molecules which in turn can lose their energy (1) by fluorescence, (2) by radiationless transfer to a metastable state and subsequent quenching by collisions of the second kind, and (3) by decomposing into free radicals.

The fluorescence for a number of aromatic compounds is quenched by the addition of CCl_4. It has been shown by Gisela Kallmann Oster [124] that the quenching of fluorescence of naphthalene in cyclohexane by addition of CCl_4 is directly related to the production of free radicals. The concentration of solute was great enough to insure that the stimulating radiation was absorbed directly by the solute molecules. The excited solute molecules transferred their energy to CCl_4 molecules on collision, and the CCl_4 molecules decomposed into free radicals. The rate of production of free radicals follows a curve complementary to the intensity of fluorescence.

The depolarization of fluorescence in liquid solutions has been studied by Jablonski [125]. Effects that cause depolarization in liquids are (1) rotation and torsional vibrations of luminescent molecules, (2) migration of energy between like molecules followed by secondary fluorescence, and (3) transfer of vibrational energy to the solution after light absorption through what was called initial shock.

If a molecule follows potential curves like those of Fig. 7.38, the molecule is left in a state of high vibrational energy immediately after absorption of light. The first collision after absorption is therefore a violent collision for the molecule and thus can produce disorientation and consequent depolarization of the fluorescent radiation. Because the amount of vibrational energy possessed after absorption depends on the frequency of the light absorbed, there should be a dependence between the depolarization and the absorbed frequency. Polarization was described by the emission anisotropy r determined by

$$r = \frac{I^{\parallel} - I^{\perp}}{I^{\parallel} + 2I^{\perp}} = \frac{I^{\parallel} - I^{\perp}}{I}$$

where $I^{\parallel}$ and $I^{\perp}$ are the intensities of fluorescence for vibrations parallel and perpendicular to that of the exciting light. I represents the total actual intensity of fluorescence emitted in all directions.

For some examples of the change in intensity of fluorescence as a function of concentration of the solute, see Stolz [126]. The intensity is described by the Kallmann-Furst equation [127]

$$I = \frac{Pc}{(Q + c)(R + c)}$$

where c is the concentration of the solute and P, Q, and R are constants. Q is determined by the solvent and P and R by the solute. The agreement between theory and experiment is shown by the fit of the points on the curves.

Additional characteristics of fluorescence in solutions are described by Garlick [107] and by Bowen and Wokes [128].

Jabloński [131] has derived an equation for the intensity of fluorescence as it depends on concentration of foreign additives, on the assumption of the existence of an active sphere surrounding the excited solute molecule. Good agreement with experiment was obtained. Comparison of experimental observations and theories of Jablonski, Forster, and Ore [135] have been made by Kawski [132]. Further comparisons with experiment were made by Szalay and Sarkany [133]. Another development of interest is the observation of emission from excited dimers called eximers; some of the observations are described by Birks [134]. A summary of quenching studies of energy transfer in organic systems has been given by Brown, Furst, and Kallmann [136]. Ewald [137] showed that the effect of pressure on quenching in liquids was through the effect of pressure on viscosity.

The ground electronic state of solute molecules is, in general, a singlet state due to the closed electron shell structure of neutral stable molecules. Excitation of one electron produces excited states resulting from combinations of the excited electron and a positive hole core due to removal of an electron from a closed shell. The electron and hole can thus form singlet and triplet excited states, and the triplet states are metastable. Molecules in a triplet state can diffuse through the liquid and react with other metastable molecules to produce higher-energy nonmetastable singlet-state molecules which can then radiate by delayed fluorescence. Some examples are described by Parker, Hatchard, and Joyce [137].

The excited electron-hole combination constitutes what is called an *exciton*. An exciton formed from a solvent molecule can migrate by transfer of excitation from one solvent molecule to another. The energy of an exciton may be transferred to a solute molecule, thereby exciting it to a state which may be metastable. A metastable solute molecule may migrate and combine with another solute molecule to form an excited dimer or eximer.

11. Thermoluminescence

When a solid material such as limestone or granite is heated in a dark place, it is usually observed to emit light during the period of rising temperature. If the material is held at constant temperature, the intensity of luminescence decreases about 90 per cent per minute, the rate depending on the temperature. The intensity of the radiation depends upon the sample, the temperature, and the rate of rise of temperature. A rock taken from the ground gives luminescence on first being heated but gives no luminescence if cooled and heated again. If, after having been once heated, the rock is exposed to high-intensity X rays or γ rays, the thermoluminescent property is restored. The first heating after X-ray or γ-ray irradiation produces luminescence, and a later heating with no intervening irradiation produces no luminescence.

Some minerals, when taken from the ground, have a color which disappears upon heating. These minerals have both their thermoluminescent property and their color restored by X-ray or γ-ray irradiation.

To obtain quantitative measurements of thermoluminescence, the intensity of radiation has been measured with photomultiplier cell and recording potentiometer, by Daniels and Saunders [92]. Measurements were made, under controlled conditions, of temperature and rate of rise of temperature, using various samples of the same size and shape. Specimens investigated were generally in the form of a slice of solid material 1 cm square and ½ mm thick.

For measurements from temperatures −180°C to 50°C a sample was mounted in a specially constructed container and irradiated with X rays while at the temperature of liquid air. The container with the sample was then placed in a light tight box with a photocell, the box evacuated, and the intensity of radiation measured, as the temperature was caused to rise to 50°C at a constant rate of about 0.75°/sec.

For measurements between room temperature and 500°C the specimen was mounted in a special furnace which consisted of a light tight box containing a photomultiplier tube and a heater. A sample was placed on a silver block which received heat from below at a varying rate so that the temperature rise was 0.8°/sec between room temperature and 450°C. Temperatures were measured by a thermocouple embedded in the silver block.

A graph of light intensity vs. time for a uniform rate of rise of temperature shows peaks and intensity changes characteristic of a particular mineral. Sample curves are shown in Figs. 7.32 to 7.35. These curves, known as *glow curves*, have been obtained for over 4,000 samples.

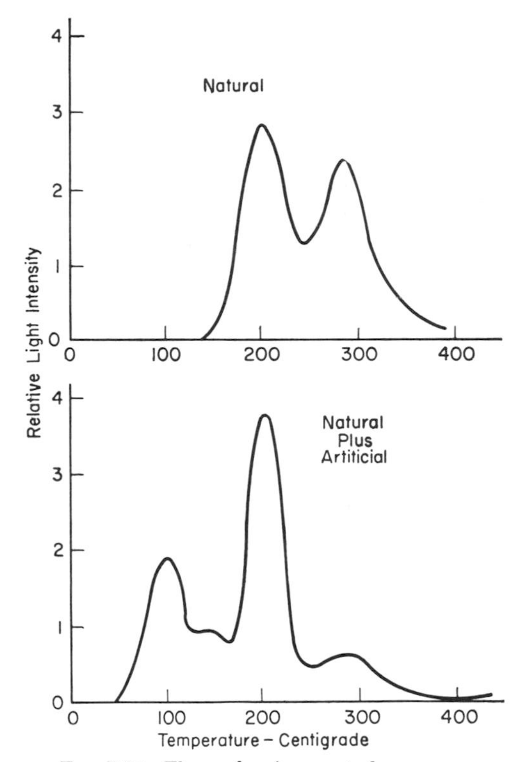

FIG. 7.32. Thermoluminescent glow curves.

The glow-curve peaks obtained upon first heating a fresh sample are duplicated and additional lower temperature peaks obtained if the sample is cooled, exposed to X rays or γ rays, and then reheated. An original glow curve and one obtained from a like sample after irradiation are shown in Fig. 7.32. With the radiation dosages used, the intensity for artificial thermoluminescence exceeded the intensity for natural thermoluminescence. The natural glow curves showed no peaks below about 200°C.

Artificial thermoluminescence has been produced in a number of carefully prepared crystals. The glow curves are reproducible for both the low-temperature (−180°C to 50°C) and high-temperature (20 to

500°C) ranges. An irradiated sample can have the low-temperature peaks annealed out by warming to room temperature without the high-temperature peaks being affected.

The shape of the glow curves depends upon the chemical nature of the crystal, the impurities in it, the number of imperfections in the crystal lattice, and the history of the crystal with respect to X-ray or γ-ray irradiation. Artificial thermoluminescence glow curves for LiF and LiCl are shown in Figs. 7.33 and 7.34.

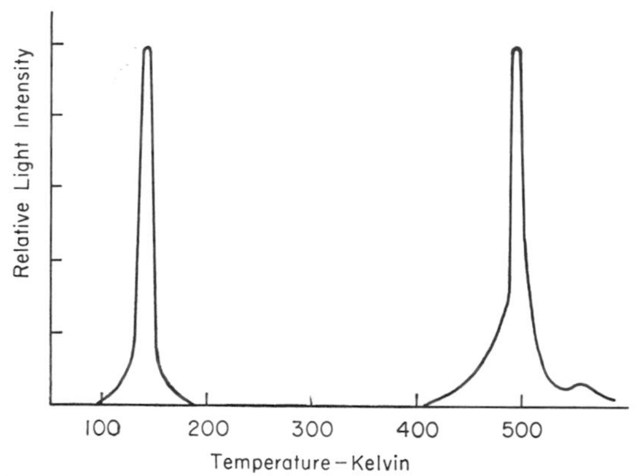

FIG. 7.33. Glow curve for LiF.

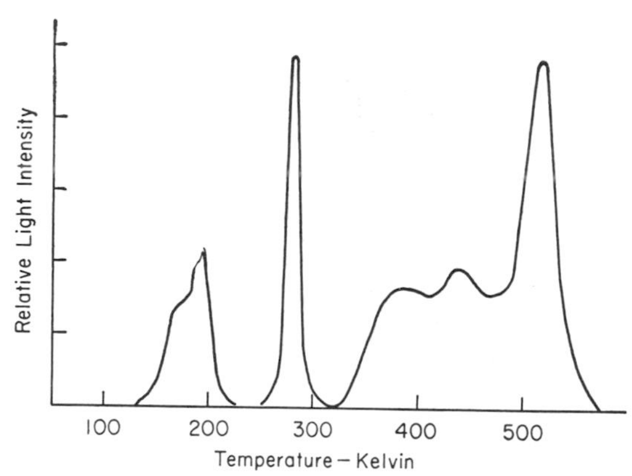

FIG. 7.34. Glow curve for LiCl.

Increasing the quantity of radiation causes a rise, then a fall, in the intensities of some peaks but may result in the stimulation of an additional higher-temperature peak as shown in Fig. 7.35.

Measurements have also been made of the intensity during decay when the sample was held at various constant temperatures.

The spectrum of thermoluminescence consists of a series of bands for most materials but shows both bands and lines in some fluorites.

The presently accepted mechanism of thermoluminescence was first proposed by Meyer and Przibram, who stated "certain groups of electrons are displaced by radiation from their normal positions and take up new metastable position among the atoms."

It was first suggested by Jablonski [97] that the

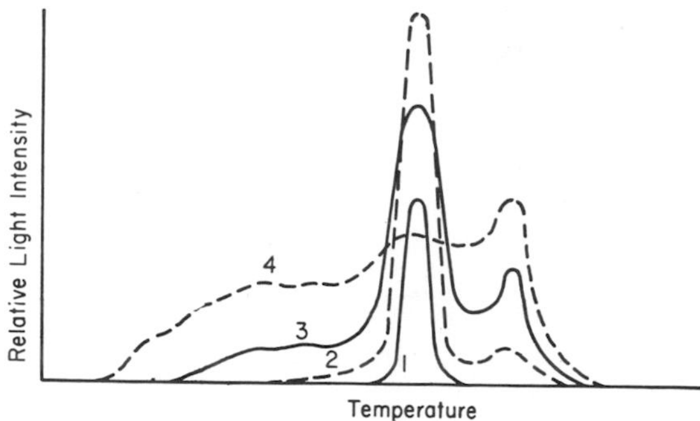

FIG. 7.35. Typical glow curves for varying amounts of radiation from 1 (minimum) to 4 (maximum).

electron traps could be considered as metastable states of various energies. The theory is described in more detail by Mott and Gurney [93] and is illustrated in Fig. 7.36.

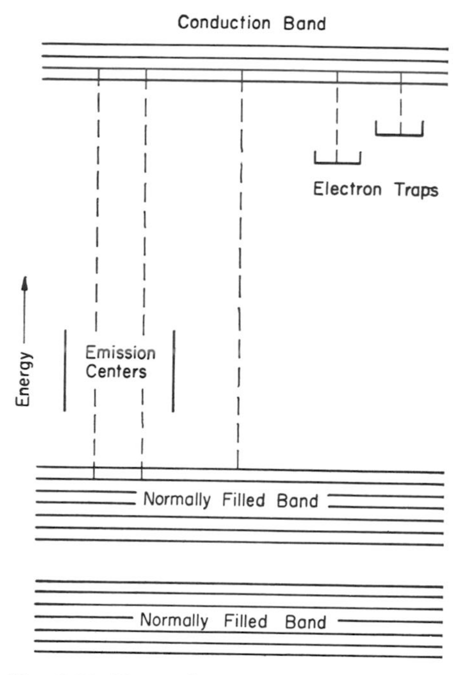

FIG. 7.36. Energy-band scheme for electrons.

In a freshly grown crystal, electrons occupy fixed positions with energies in the normally filled bands. Irradiation with X rays or γ rays changes the electrons to a higher-energy conduction band. Electrons with energy in the conduction band move through the crystal until they become bound into electron traps or F centers, by radiation or other means.

The electrons remain in the traps until they are transferred back to the conduction band through an elevation in temperature of the crystal. Once again in the conduction band, electrons move through the crystal until they reach an emission center. Then they can lose energy by radiation and become fixed in the crystal at the point of the emission center.

With the gradual increase in temperature the higher-energy electron traps are emptied first, giving rise to the low-temperature peaks in the glow curves. The lower-energy electron traps become emptied at the higher temperatures, giving rise to the high-temperature peaks.

Natural minerals have had their higher-energy (low-temperature) traps emptied because of ground temperatures, but their lower-energy (high-temperature) electron traps still contain electrons in proportion to the total quantity of X-ray or γ-ray irradiation in the history of the sample. The color lost through heating and restored by irradiation is associated with the absorption of light by the displaced electrons while in their metastable positions or *F* centers.

Boyd [94] has determined the binding energies of electrons in *F* centers of LiF from the intensity-time curves for various constant temperatures. Daniels [95] demonstrated that all thermoluminescence, natural and artificial, is a result of radiation by γ rays, cosmic rays, or X rays. Minerals which show natural thermoluminescence are found in close proximity to radioactive deposits. By correlating the intensities in glow curves with the amount of radioactive material in the rock Saunders [96] showed that it is possible to use thermoluminescence as a means for estimating the history and age of the rock. It is also possible to use thermoluminescence as a tool in prospecting for uranium.

Thermoluminescence is found to be a more sensitive detector of radioactive deposits than the Geiger counter. This results since the thermoluminescence shows the accumulated effect of γ-ray and cosmic-ray exposure over millions of years.

Since the intensity of thermoluminescence of a crystal such as LiCl is proportional to the quantity of irradiation, a crystal of LiCl can be used as a radiation dosimeter.

Thermoluminescence can also be stimulated in crystals by mechanical crushing. Crystals of NaCl which show no thermoluminescence can be ground up, after which they will show thermoluminescence. This indicates that a piezoelectric or triboelectric effect of sufficient intensity to raise electrons to the conduction level has taken place in the grinding process.

12. Phosphorescence

Observed Effects. When some solid substances are irradiated with ultraviolet light or X rays, they emit a strong luminescence during irradiation and a weak afterglow luminescence after the irradiation has ceased.

The spectrum of the luminescence during irradiation usually shows one or more regions or bands of high intensity and one or more bands of low intensity. After irradiation has ceased, there is observed a rapid drop in intensity in some bands and only a slight change in intensity in others. The relative intensity of bands in the afterglow and the rate of change of relative intensity depend on the temperature. Bands which were strongest at room temperature may be weakest at liquid-air temperature.

A good example of these effects is the luminescence of a solid solution of fluorescein in boric acid. During irradiation at room temperature the luminescence spectrum shows a strong band with peak at 4800 and two weak bands with peaks at 5300 and 5750. The afterglow spectrum at room temperature shows also a strong band at 4800 and weak bands at 5300 and 5750. The afterglow spectrum at $-180°C$, however, shows only the band at 5750. A change in temperature produces little change in the intensity of the 5750 band but causes a large change in the intensity of the 4800 and 5300 bands.

Some crystals of the type ZnS(Cu) show photoconduction accompanying phosphorescence, while others such as KCl(Tl) show phosphorescence with no appearance of electrical conductivity. (The preceding notation means a crystal of ZnS or KCl containing a small amount of Cu or Tl impurity.)

Some solid materials may be stimulated by radiation, immersed in liquid air, and stored for several weeks. When later removed and warmed to room temperature, they resume phosphorescence with an intensity only somewhat less than that just before they were immersed.

Some crystals after they have been stimulated with radiation and allowed to decay until the intensity of phosphorescence has become negligible can be stimulated to luminescence again by infrared radiation.

Theory of Phosphorescence. Phosphorescence may be considered as a special case of thermoluminescence. It corresponds to thermoluminescence which can be stimulated at room temperature.

The energy levels and transitions needed to describe phosphorescence are shown in Fig. 7.37. Absorption

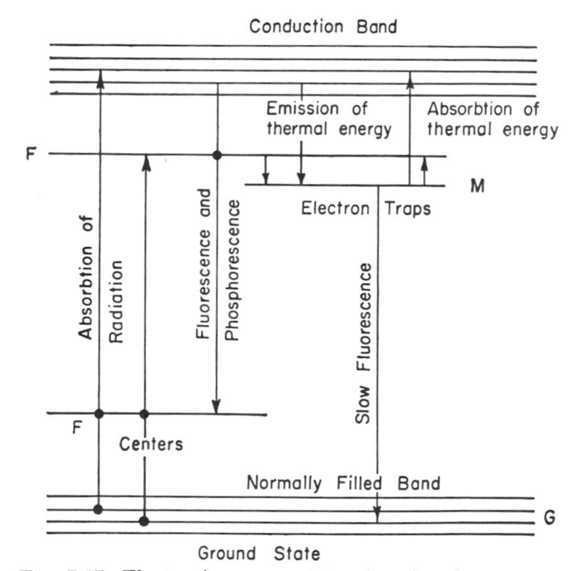

FIG. 7.37. Electronic energy states for phosphorescence.

of the stimulating radiation by crystals which are photoconductors is considered to change electrons from the valence energy band in the lattice to the conduction band, most of the changes occurring at luminous centers or *F* centers. Electrons in the conduction band either return to the valence band with the emission of radiation as fluorescence or fall into electron traps. Trapping requires the transfer of some energy to the thermal energy of the lattice through the equivalent of a collision of the second kind in a gas.

For crystals which show phosphorescence without photoconduction absorption of radiation is considered to produce excitation of an activator atom or F center to an energy state (F in Fig. 7.37) which lies below the conduction band. Electrons in state F may return to the ground state through emission of radiation (strong fluorescence) or may become bound into an electron trap. An electron in a trap corresponds to a metastable atom. Metastable atoms or electrons in traps may revert to the ground state through the emission of radiation by a forbidden transition (slow fluorescence). They may also be raised to a radiating state F or to the conduction band by absorption of thermal energy or absorption of infrared radiation, after which they can revert to the ground state with the emission of phosphorescence.

Phosphorescence and thermoluminescence are therefore the same type of phenomenon. The difference between them is that, for phosphorescence, molecules in state M may be changed to state F through thermal collisions at room temperature, while, for thermoluminescence, the transfer from M to F may require a temperature of 400 to 500°C.

The intensity of the phosphorescence bands relative to that of the slow-fluorescence forbidden bands thus depends on the temperature. At very low temperatures, when the thermal energy is not sufficient to cause a transfer from M to F or from M to the conduction band, the luminescence consists of only the slow-fluorescence bands.

The rapid drop of intensity when the stimulating radiation is cut off is associated with the cessation of the strong fluorescence, which drops off with a decay time of 10^{-5} to 10^{-8} sec, corresponding to the average lifetime of the electron in the conduction band or to the average life of an activator atom in a radiating state F. The decay time for phosphorescence and slow fluorescence is the lifetime of the metastable state M or the lifetime of electrons in traps.

Decay of Intensity in Phosphorescence. The intensity of phosphorescence decays with the population of the traps, and therefore from Eqs. (7.2) $I = I_0 e^{-t/\tau}$, where I_0 is the intensity at $t = 0$ and τ is the decay time.

In general, for a phosphorescent crystal which contains traps of different energies and different populations, the decrease of intensity with time will be

$$I = I_1 \exp\left(-\frac{t}{\tau_1}\right) + I_2 \exp\left(-\frac{t}{\tau_2}\right) + \cdots + I_n \exp\left(-\frac{t}{\tau_n}\right)$$

where I_n is the intensity of phosphorescence due to electrons in traps of energy depth E_n such that $1/\tau_n = A \exp(-E_n/kT)$. E_n is the energy difference between an electron in a trap and an electron in either a radiating state or the conduction band, depending on whether or not the crystal is a photoconductor.

By representing the intensity decay curve as a sum of exponentials the coefficients $I_1, I_2, \ldots, I_n$ may be determined. From decay curves at different temperatures the constants A and E_n may be determined. Since the relative populations in the traps are proportional to the intensities $I_1, I_2, \ldots, I_n$, the distribution of electrons in traps of various energies will thus be known.

The distribution of electrons in traps may also be obtained from the analysis of thermoluminescence glow curves. The two methods give the same results within experimental errors.

For ZnS(Cu) there have been obtained trap depths 0.30, 0.40, 0.50, 0.57, 0.65 to 0.70 electron volts.

Distribution of Intensity in Phosphorescence. The spectrum of phosphorescence generally consists of a wide band extending toward longer wavelengths from the wavelength of the stimulating radiation.

The relation between absorption and emission spectra may be described as first done by von Hippel [123]. In a crystal such as KCl(Tl) an ion Tl^+ replaces an ion K^+ and forms an emission center. The Tl^+ ion then occupies a position of equilibrium in the lattice such that a displacement from that position produces an increase in potential energy for the entire lattice system near the Tl^+ ion. The potential energy can be plotted as a function of a configuration coordinate representing the distance from some fixed point in the lattice to the position of a hypothetical particle, whose potential energy equals the total potential energy of the Tl^+ + neighboring lattice system. The position of this hypothetical particle for a minimum of potential energy depends on the state of excitation of the outer electrons of the Tl^+ ion. A plot of potential energy vs. distance for this particle, as shown in Fig. 7.38, resembles plots

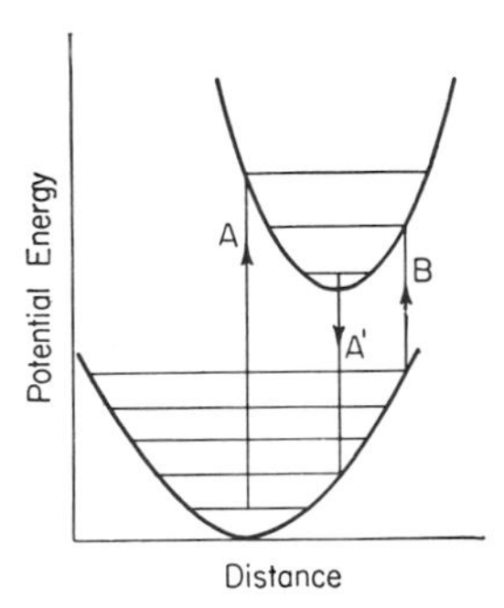

FIG. 7.38. Configuration curves for luminescence.

of the potential energy for a hypothetical particle of mass equal to the reduced mass for diatomic molecules. It can be expected that experimental observations can be described in terms of the Franck-Condon principle that the most probable electronic transitions are those for which the position and momentum are constant.

As shown in Fig. 7.38, absorption of radiation (transition A) changes the Tl^+ ion to an excited state for which the potential-energy curve has a minimum at a position different from that in the ground state. The energy absorbed corresponds to the vertical distance between the curves along line A, and absorption leaves the system in a state of high vibrational energy. Since the excited hypothetical particle is in a dense medium, it makes many effective collisions during its lifetime of 10^{-5} to 10^{-8} sec and thereby dissipates its surplus vibrational energy to the lattice. The excited particles acquire a Boltzmann distribution. The most intense emission therefore corresponds to transition A' at a wavelength different from that for maximum absorption intensity.

If the temperature is sufficient to give an appreciable population to the higher vibrational levels of the ground state, there is the possibility of absorption such as B in Fig. 7.38 followed by emission of radiation of wavelength shorter than that absorbed.

If the potential curve for the excited state is broad, with the position of the minimum considerably displaced from that in the ground state, the emission and absorption bands will be widely separated in the spectrum. If the upper curve minimum is not much displaced, the emission and absorption bands will overlap.

Some molecules show "mirror symmetry" of absorption and emission bands as shown in Fig. 7.39.

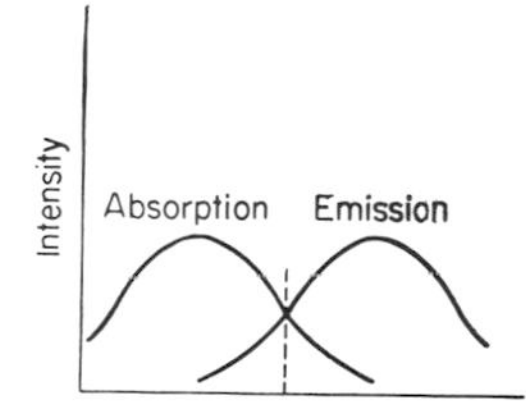

FIG. 7.39. Mirror symmetry of absorption and emission.

This results if the curves for the excited and ground states are parabolic, with the same width but with minima at different positions. For these molecules the electronic energy of the excited state equals the average of the energies for the absorption and emission peaks. A good example of mirror symmetry is the spectrum of rhodamine G in ethanol, as described by Levshin [98].

Impurities play an important role in phosphorescence. They can exist in the crystalline base material: (*a*) as ions replacing ions of base material; (*b*) as extra atoms, thus causing a deformation of the lattice; (*c*) as ions or atoms at positions of lattice defects.

An impurity may or may not activate phosphorescence, depending upon its concentration and upon the base material.

Increasing the concentration of an activator increases the intensity of fluorescence with respect to the intensity of phosphorescence. The optimum concentration for fluorescence or for phosphorescence depends on the nature of the activator and on the base material. For example, the optimum concentration of Ag or Cu in ZnS for phosphorescence is about 10^{-4} per cent but is about five to ten times larger in SrS or CaS.

The quenching of phosphorescence at concentrations of activators greater than the optimum can be ascribed to increase of fluorescence, with a consequent decrease in the number of electrons trapped in the long-lived states.

The quenching of phosphorescence also depends on the location of the quenching atoms in the lattice. ZnS(Cu) containing 10^{-3} per cent Fe introduced at 270°C is phosphorescent, but after it has been heated to 950°C, it shows no phosphorescence.

A good example of phosphorescence from high-energy stimulation has been described by Furst and Kallmann [99]. An extensive discussion of color centers in crystals has been given by Seitz [100]. The preparation of phosphorescence materials has been treated by Garlick [101] and Fonda and Seitz [102]. A discussion of concentration quenching in inorganic phosphors has been given by Dexter and Schulman [103]. A good summary of various aspects of phosphorescence has been given by Curie and Curie [104].

A good example of the description of luminescent phenomena in terms of the properties of the separate components of the phosphor is provided through investigations of KCl(Tl). When one part of TlCl powder is mixed with 100 parts of KCl powder and the mixture fused and cooled there result crystals of KCl containing Tl to 1 part in 10,000. The Tl exists in the lattice as Tl^+ replacing a K^+ ion. A comparison of the absorption spectrum of KCl(Tl) with that of pure KCl shows two strong absorption bands associated with the presence of Tl^+ in the KCl.

The maxima of intensity are at wavelengths 1960 and 2500 A, with the 1960 maximum considerably the stronger [106, p. 32]. There is also a weaker band at 2100 A. When KCl(Tl) is irradiated by 2537 from a Hg discharge, there result an intense fluorescence emission at 3050 and a blue emission between 4750 and 4900 A.

These observations may be correlated with the energy levels of a free Tl^+ ion as modified by the crystalline lattice. A free Tl^+ ion resembles a mercury atom in that the outer electron shell is a closed s shell. The ground state results from $6s^2$ and is therefore ${}^1S_0$. The excited states resulting from $6s6p$ are ${}^1P_1$ and ${}^3P_{2,1,0}$, as for Hg (Fig. 7.23). F. E. Williams [138] has calculated that the ground state lies 9.5 ev below the base of the conduction band, while the valence band lies 9.4 volts below the conduction band. The ground state of Tl^+ therefore lies 0.1 volt below the top of the valence band. This means simply that it takes 0.1 volt more to change an electron from Tl^+ than from Cl^- to the conduction band. The emission and absorption bands can be correlated with the energy levels of Tl^+ in KCl as follows:

Strongest absorption band at 1960	1^1S_0–2^1P_1
Strong absorption band at 2500	1^1S_0–2^3P_1
Weak absorption band at 2100	1^1S_0–2^3P_2
Emission band at 2470	1^1S_0–2^1P_1
Emission band at 3050	1^1S_0–2^3P_1
Visible blue emission 4750–4900 not explained	

The differences in wavelengths for absorption and emission can be correlated with a shift in position of the potential minimum, as shown in Fig. 7.38.

The metastable states of Tl^+, ${}^3P_2$ and ${}^3P_0$, constitute one form of electron trap. Tl^+ in a metastable state can absorb radiation to become Tl^{++} with the electron entering the conduction band or another electron trap. A simplified configurational model for KCl(Tl) is given by Johnson and Williams [148].

Another type of emission and absorption center is one in which an ion such as Cl^- in KCl is replaced by a single electron. The electron would be subjected to the same forces as the Cl^- ions but the electron could absorb and radiate energy at a shorter wavelength. The chlorine vacancy in a crystal such as KCl would act like a positive hole and bind an electron to the position that would normally be occupied by a Cl^- ion.

An electron bound in this way would constitute a color center or *F* center. That this is the nature of a color center is shown by experiments on electron spin resonance [106, p. 45]. A crystal is mounted in a coil through which a radio-frequency current is passed. The coil and specimen are in a static magnetic field (H_z). Free electrons are made to precess about the direction of H_z with a component of spin angular momentum parallel to the field equal to $\pm\frac{1}{2}$. When the magnetic-field strength H_z is made equal to the value for which the energy difference between the two orientations equals the energy of a photon of frequency equal to the frequency of the current in the coil, there can be an absorption of energy from the coil by electrons with subsequent dissipation of this energy as heat through collisions of the second kind. The circuit for measuring loss will show the condition of resonance as an increased loss. Observation of such a signal shows the existence of isolated bound electrons in the sample.

An alkali-halide crystal with many color centers can be formed by heating the crystal to a high temperature in the alkali vapor and then cooling quickly. The crystal will be colored, because of absorption of visible light by *F* centers. If the alkali-halide crystal is heated in an atmosphere of halide vapor at high pressure and quickly cooled, a crystal is formed with many positive-ion vacancies. These vacancies will be filled with positive holes to form *V* centers.

Color centers may also be produced by electric fields. A crystal of KCl with two embedded point electrodes can be heated until it becomes conducting. Colored regions will be seen to grow radially away from each electrode. The centers growing from the negative electrode are *F* centers with trapped electrons. Those growing from the positive electrode are *V*-type centers, with trapped holes. The two regions have different colors. When the colored regions have grown to an extent to cause overlapping, the region of overlap remains colorless. This corresponds to the mutual annihilation of *F* centers (electrons) and *V* centers (holes). The effect of the electric field is to pull ions out of the crystal, thus leaving vacancies which become filled with electrons or holes. In a region where the electric fields are pulling out positive and negative ions in equal numbers, no vacancies are left for trapping either electrons or holes and the crystal remains colorless.

A comparison of the absorption spectra of a crystal with and without *F*-center coloration shows that *F* centers give rise to a single absorption band with a bell-shaped contour and half width about 0.5 volt at 20°C. This is called the *F* band. The width decreases with decrease in temperature. The wavelength at the maximum for alkali-halides with NaCl-type structure depends on the lattice spacing *d*. Experimental results can be approximately described by $\lambda_m = 600d^2$, where λ_m and *d* are expressed in angstrom units. A good discussion of the experimental observations that led to the model of an *F* center as an electron in a negative-ion vacancy has been given by Schulman and Compton [139, chap. 3]. This model is illustrated in Fig. 7.40 (after Schulman

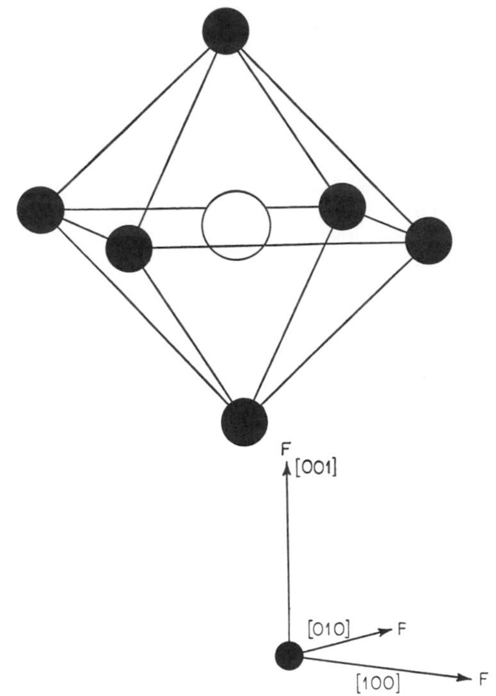

FIG. 7.40. With permission from J. H. Schulman and W. D. Compton, "Color Centers in Solids," Pergamon Press Ltd., 1963.

and Compton) which shows an electron in its relation to the six nearest positive ions. Absorption of light in the *F* band puts the trapped electron into an excited state, with the neighboring lattice ions in a high-energy vibrational state. A possible configuration coordinate diagram is shown in Fig. 7.41 (also after Schulman and Compton). After absorption, represented by $A \to B$, the system relaxes to a Boltzmann distribution in the excited state.

The maximum intensity for emission corresponds to the transition $C \to D$. The type of vibration for both electron states is considered to be the breathing mode for the system shown in part in Fig. 7.40. If the part of the excited-state curve near *B* can be approximated by a straight line, the shape of the *F* absorption band would be Gaussian as calculated according to the Franck-Condon principle. Calculation would be like that for hydrogen [140]. The emission spectrum should also be Gaussian with a slightly different half width in case the vibration wave number ω' for the excited state differs from that for the ground state ω''. The half width can be shown to depend on the temperature according to the relation:

$$\text{half width} \propto [\coth (\hbar\omega/kT)]^{1/2}$$

If the temperature is such that there are many *F* centers with vibrational energy *U* (Fig. 7.41), there will be radiationless transfers to the ground state with the dissipation of energy $U + (E_c - E_a)$ as heat in the lattice. Luminescence intensities at higher temperatures should be much lower than intensities at lower temperatures. This is in agreement with observations.

Curves like Fig. 7.41 can be constructed from the following experimental observations: (1) energy at the

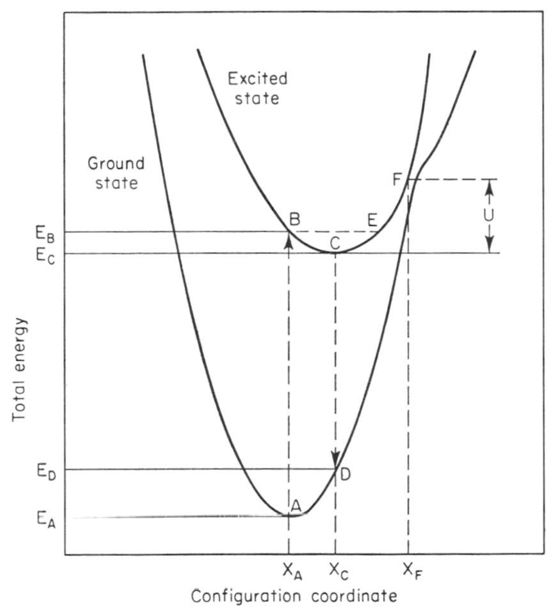

Fig. 7.41. With permission from J. H. Schulman and W. D. Compton, "Color Centers in Solids," Pergamon Press Ltd., 1963.

maximum for the absorption band, (2) energy at the maximum for the emission band, (3) the half width of the absorption band at 0°K, and (4) the temperature variation of the absorption-band half width. From these curves, the half width of the emission band can be calculated and compared with experimental results. A reasonable agreement is obtained.

The lifetime of excited *F* centers has been determined to be as long as 10^{-6} sec for some crystals [144]. Park [145] was able to excite *F* centers of KI by light from a ruby laser.

Improved experimental arrangements show additional absorption bands on the short-wave side of the *F* band. These have been designated as *K*, L_1, L_2, L_3 bands. They are attributed to higher-energy excited states for the *F* center.

Absorption of light in the *F* band produces luminescence from excited *F* centers but also produces photoconduction in the crystal even at low temperatures. The dependence of photoconduction current on temperature indicates that the photoconduction arises from thermal ejection of electrons from excited *F* centers.

Continued absorption in the *F* band causes a reduction in intensity of *F*-band absorption and the appearance of a new broad overlapping absorption band with a maximum at longer wavelength. This new band, called the *F′* band, has been attributed to a negative-ion vacancy that has trapped two electrons.

Prolonged excitations of alkali-halide crystals by X rays or by *F*-band light causes the appearance of additional absorption bands on the long-wave side of the *F* band. These have been attributed to different types of luminescence centers designated as *M*, R_1, R_2, *N*, and *O* centers.

Experiments with polarized light show that an *M* center is a combination of two neighboring *F* centers with two electrons shared between them, as illustrated in Fig. 7.42. Various aggregates of *F* centers have been proposed as models for the R_1, R_2, *N*, and *O* centers. Schneider and Rabin [149] have observed optical absorption bands which they attribute to transitions of ionized *F* aggregate centers.

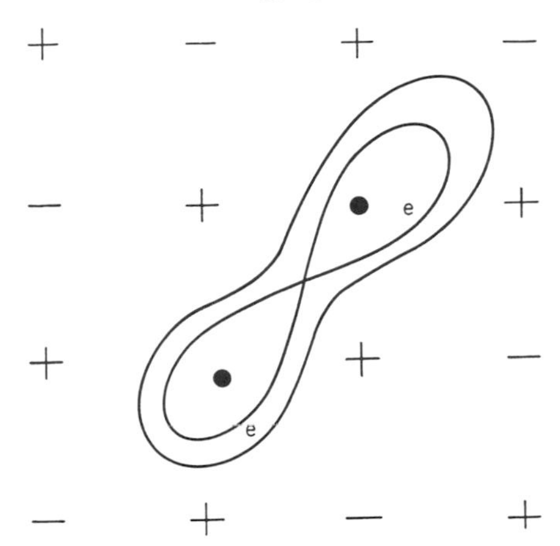

Fig. 7.42. F_2 model for *M* center.

Absorption spectra of crystals such as CdS, HgI_2, and others show an absorption edge in the ultraviolet region corresponding to the energy gap between the valence and conduction bands. At low temperatures, a series of absorption lines appear near the absorption edge at wavelengths that fit a hydrogenlike equation. These lines are attributed to the formation of electron-hole combinations or excitons.

An exciton in a crystal such as KCl is considered to result from the transfer of an electron from Cl^- to K^+, forming an excited and neutral KCl combination. The energy required is less than that needed to remove an electron from Cl^- and put it into the conduction band. Excitons can migrate through the crystal by exchange of condition with lattice ions. Excitons can radiate energy at a longer wavelength than that absorbed, with considerable energy going into heat in the lattice both before and after emission. The rapid relaxation of the lower state after emission makes possible an inversion of population. The use of exciton emission to provide laser action has been shown by Fink [143]. Other characteristics of exciton absorption and emission are described by Curie [106] and Schulman and Compton [139].

Irradiation of KCl, KCl(Tl), KCl(Ag), and KCl(Pb) by X rays while at 77°K produced other centers that caused absorption bands at 3650 and 7500. These were shown by Delbecq, Smaller, and Yuster to be due to $Cl^-{}_2$ molecule ions [146]. Absorption spectra associated with $F^-{}_2$, $Br^-{}_2$, and $I^-{}_2$ in alkali-halide crystals have also been obtained [147].

The extensive activity in the field of luminescence

has provided many interesting observations and models to describe them. A review of two conferences at which many of these developments were reported has been given by Spruch [129, 130].

References

1. Pringsheim, Peter: "Fluorescence and Phosphorescence," Interscience, New York, 1949.
2. Curie, Maurice: "Luminescence des corps solides," Presses Universitaires de France, Paris, 1934.
3. Faraday Society: "Luminescence—A General Discussion," Gurney, London, 1938.
4. Kroger, F. A.: "Some Aspects of the Luminescence of Solids," Elsevier Pub. Co., Amsterdam, 1948.
5. Leverenz, Humboldt W.: "An Introduction to Luminescence of Solids," Wiley, New York, 1950.
6. Mitchell, A. C. G., and M. W. Zemansky: "Resonance Radiation and Excited Atoms," Macmillan, New York, 1934.
7. Cram, S. W.: *Phys. Rev.*, **46**: 205 (1934).
8. Pringsheim, Peter: "Fluorescence and Phosphorescence," p. 238, Interscience, New York, 1949.
9. ———: "Fluorescence and Phosphorescence," p. 290, Interscience, New York, 1949.
10. Welsh, H. L., J. Kastner, and A. C. Lauriston: *Can. J. Research*, sec. **A28**: 93 (1950).
11. Willey, E. J. B.: "Collisions of the Second Kind," E. Arnold, London, 1937.
12. Wood, Robert W.: "Physical Optics," Macmillan, New York, 1921.
13. Rump, W.: *Z. Physik*, **31**: 901 (1925).
14. Wood, R. W., and L. Dunoyer: *Phil. Mag.*, **27**: 1018 (1914).
15. Ellett, A.: Polarization of Cadmium Resonance Radiation 1 $^1S_0 - 2\ ^3P_1$, *Phys. Rev.*, **33**: 124 (1929).
16. Bogros, A.: *Compt. rend.*, **187**: 124 (1926); *Ann. Physik*, **17**: 249 (1932).
17. Billeter, W.: *Helv. Phys. Acta*, **7**: 505, 841 (1934).
18. Steinhaeuser, A.: *Z. Physik*, **99**: 300 (1936).
19. Nielsen, J. R., and N. Wright: *J. Opt. Soc. Am.*, **20**: 27 (1930).
20. Dymond, E. G.: *Z. Physik*, **34**: 553 (1925).
21. Winans, J. G.: *Z. Physik*, **60**: 631 (1930).
22. Terenin, A., and N. Prileshajewa: *Z. physik. Chem.*, **B13**: 72 (1931).
23. Hanson, Howard: *J. Chem. Phys.*, **23**: 1391 (1955).
24. Warhurst, E.: Private communication to J. G. Winans.
25. Stuart, H. A.: *Z. Physik*, **32**: 262 (1925).
26. Pringsheim, Peter: "Fluorescence and Phosphorescence," p. 218, Interscience, New York, 1949.
27. Swanson, R., and R. McFarland: *Phys. Rev.*, **98**: 1063 (1955).
28. Herzberg, G.: "Spectra of Diatomic Molecules," Van Nostrand, Princeton, N.J., 1950.
29. Zemansky, M. W.: *Phys. Rev.*, **36**: 919 (1930).
30. Kenty, C.: *Phys. Rev.*, **42**: 823 (1932).
31. Holstein, T. D. Alpert, and A. O. McCoubrey: *Phys. Rev.*, **85**: 985 (1952).
32. Ellett, A., L. O. Olsen, and R. Peterson: *Phys. Rev.*, **60**: 107 (1941).
33. Keussler, V. von: *Ann. Physik*, **82**: 793 (1927).
34. Datta, G. L.: *Z. Physik*, **37**: 625 (1926).
35. Suppe, F.: *Z. Physik*, **113**: 141 (1939).
36. Wood, R. W.: *Phil. Mag.*, **44**: 1107 (1922).
37. Kastler, A.: *Physica*, **12**: 619 (1946).
38. Gaviola, E., and P. Pringsheim: *Z. Physik*, **25**: 367 (1924).
39. Larrick, L.: *Phys. Rev.*, **46**: 581 (1934).
40. Van Vleck, J. H.: *Proc. Natl. Acad. Sci.*, **11**: 612 (1925).
41. Keussler, V. von: *Physik. Z.*, **27**: 313 (1926).
42. Ellett, A., and L. Larrick: *Phys. Rev.*, **39**: 294 (1932).
43. Breit, G.: *Revs. Mod. Phys.*, **5**: 91 (1933).
44. Bohr, N.: *Naturwiss.*, **12**: 1115 (1924).
45. Heisenberg, W.: *Z. Physik*, **31**: 617 (1925).
46. Galt, J. A., and H. L. Welch: *Can. J. Phys.*, **35**: 98 (1957).
47. Fuchtbauer, C.: *Physik. Z.*, **21**: 635 (1920).
48. Wood, R. W.: *Proc. Roy. Soc.* (*London*), **A106**: 697 (1924); *Phil. Mag.*, **50**: 774 (1925).
49. Bender, P.: *Phys. Rev.*, **36**: 1535, 1543 (1930).
50. Winans, J. G.: *Proc. Natl. Acad. Sci.*, **11**: 738 (1925); *Phys. Rev.*, **30**: 1 (1927), **31**: 710 (1928), **32**: 427 (1928).
51. Strutt, R. J. (Lord Rayleigh): *Proc. Roy. Soc.* (*London*), **96**: 272 (1919).
52. Boeckner, C.: *Bur. Standards J. Research*, **13** (1930); *Phys. Rev.*, **35**: 664 (1930).
53. Donat, K.: *Z. Physik*, **29**: 345 (1924).
54. Steinhaeuser, A.: *Z. Physik*, **95**: 669 (1935), **99**: 300 (1936).
55. Cario, G., and J. Franck: *Z. Physik*, **17**: 202 (1923).
56. Mueller, F.: *Helv. Phys. Acta*, **8**: 55 (1935).
57. Terenin, A.: *Z. Physik*, **31**: 26 (1925), **37**: 98 (1926).
58. Winans, J. G.: *Phys. Rev.*, **55**: 242 (1939).
59. Fridrichson, J.: *Z. Physik*, **64**: 43 (1930), **68**: 550 (1931); *Compt. rend. soc. polon. phys.*, **5**: 337 (1931).
60. Lees, J. H., and H. W. B. Skinner: *Proc. Roy. Soc.* (*London*), **A137**: 186 (1932).
61. Maurer, W., and R. Wolf: *Z. Physik*, **92**: 100 (1934).
62. Bowen, I. S.: *Astrophys. J.*, **81**: 1 (1935).
63. Hanle, W., and E. F. Richter: *Z. Physik*, **54**: 811 (1929).
64. Kastler, A.: *J. phys. radium*, **11**: 255 (1950); *Physica*, **17**: 191 (1951).
65. Brossel, J., A. Kastler, and J. Winter: *J. phys. radium*, **13**: 668 (1952).
66. ———, B. Cagnac, and A. Kastler: *J. phys. radium*, **15**: 6 (1954); *Compt. rend.*, **237**: 984 (1953).
67. ——— and F. Bitter: *Phys. Rev.*, **86**: 308 (1952).
68. Bitter, F., S. P. Davis, B. Richter, and J. E. R. Young: *Phys. Rev.*, **96**: 1531 (1954).
69. Cario, G., and J. Franck: *Z. Physik*, **10**: 185 (1922), **37**: 619 (1926).
70. Loria, S.: *Phys. Rev.*, **26**: 573 (1925).
71. Donat, K.: *Z. Physik*, **29**: 345 (1924).
72. Kopferman, H.: *Z. Physik*, **21**: 316 (1924).
73. Krause, E. A.: *Phys. Rev.*, **55**: 164 (1939).
74. Winans, J. G.: *Phys. Rev.*, **30**: 1 (1927).
75. ——— and S. Breen: *Phys. Rev.*, **62**: 297 (1942).
76. ——— and W. J. Pearce: *Phys. Rev.*, **64**: 43 (1943).
77. McGowan, F. K., and J. G. Winans: *Phys. Rev.*, **65**: 349 (1944).
78. Winans, J. G.: *Revs. Mod. Phys.*, **16**: 175 (1944).
79. Beutler, H., and B. Josephy: *Z. Physik*, **53**: 747 (1929).
80. Wigner, E. P.: *Göttingen Nachr.* (1927), p. 375. O. Buhl: *Z. Physik*, **109**: 180 (1938), **110**: 395 (1938).
81. Swanson, R. E., and R. H. McFarland: *Phys. Rev.*, **98**: 1063 (1955).
82. Wigner, E. P.: *Göttingen Nachr.* (1927), p. 375. O. Buhl: *Z. Physik*, **109**: 180 (1938), **110**: 395 (1938).
83. Anderson, R. A., and R. H. McFarland: *Phys. Rev.*, **119**: 643 (1963).
84. Wood, R. W.: "Physical Optics," p. 430, Macmillan, New York, 1921.
85. Lauriston, A. C., and H. L. Welsh: *Can. J. Research*, **29**: 217 (1951).
86. Kapuscinski, W.: *Nature*, **116**: 170 (1925); *Z. Physik*, **41**: 214 (1927).
87. Cram, S. W.: *Phys. Rev.*, **46**: 205 (1934).
88. Jablonski, A., and P. Pringsheim: *Z. Physik*, **70**: 593 (1931).

89. Kallmann, H., and M. Furst: *Phys. Rev.*, **79**: 857 (1950), **85**: 816 (1952).
90. Franck, J., and R. Livingston: *Revs. Mod. Phys.*, **21**: 505 (1949).
91. Kallmann, H., and M. Furst: *Phys. Rev.*, **81**: 853 (1951).
92. Saunders, D., F. F. Morehead, and F. Daniels: *J. Am. Chem. Soc.*, **75**: 3096 (1953).
93. Mott, N. F., and R. W. Gurney: "Electronic Processes in Ionic Crystals," Oxford University Press, Fair Lawn, N.J. 1940.
94. Boyd, Charles A.: *J. Chem. Phys.*, **12**: 1221 (1949).
95. Daniels, F.: *Science*, **117**: 343 (1953).
96. Saunders, D.: *Bull. Am. Assoc. Petroleum Geolo.*, **37**: 114 (1953).
97. Jablonski, A.: *Z. Physik*, **73**: 460 (1931).
98. Levshin: *Z. Physik*, **72**: 368 (1931); *Acta Physico chim. U.R.R.S.*, **1**: 685 (1934), **2**: 221 (1935); *Compt. rend. Leningrad*, **I**: 474 (1935).
99. Furst, M., and H. Kallmann: *Phys. Rev.*, **91**: 1356 (1953).
100. Seitz, F.: *Revs. Mod. Phys.*, **18**: 384 (1946), **26**: 7 (1954).
101. Garlick, G. F.: "Luminescent Materials," Oxford University Press, Fair Lawn, N.J., (1949).
102. Fonda, G. R., and F. Seitz: "Preparation and Characteristics of Solid Luminescent Materials" (Cornell Symposium of the American Physical Society), Wiley, New York, 1948.
103. Dexter, D. L., and J. H. Schulman: *J. Chem. Phys.*, **22**: 1063 (1954).
104. Curie, M., and D. Curie: *Cahiers d. phys.*, **55**(1) (1955).
105. Tolman, R. C.: "Statistical Mechanics with Applications to Physics and Chemistry," p. 69, Chemical Catalog Co., 1927.
106. Curie, D.: "Luminescence in Crystals" (trans. by G. F. J. Garlick), Wiley, New York, 1963.
107. Garlick, G. F. J.: "Encyclopedia of Physics," vol. 26, pp. 1–128, Springer, Berlin, 1958.
108. Proceedings of the International Conference on Luminescence at Torun, Poland, September, 1963, A. Jablonski (ed.), *Acta Phys. Polon.*, **26**: 309–843 (1964).
109. International Conference of Organic and Inorganic Materials, H. P. Kallmann and G. M. Spruch (eds.), Wiley, New York, 1962.
110. Colloquium on Luminescence, Turtu (Esthony), 1957, *Bull. Acad. Sci. S.S.S.R.; Columbia Tech. Transl.*, **21**: 475–786.
111. Colloque international sur la luminescence les corps anorganiques, M. Curie and G. Destriau (eds.), *J. phys. radium*, **17**: 609–832 (1956).
112. Conference on Luminescence, Cavendish Laboratory, Cambridge, April, 1954, *Brit. J. Appl. Phys.*, **6** (Suppl. 4): 1–120 (1955).
113. Henisch, H. K.: "Electroluminescence," Pergamon Press, New York, 1962.
114. Kastler, A.: *Acta Phys. Polon.*, **26**: 311 (1964); *J. Opt. Soc. Am.*, **53**: 902 (1963).
115. Frenzel, H., and H. Schultes: *Z. Phys. Chem.*, **27B**: 421 (1934).
116. Klarner, B., and R. O. Prudhomme: *Compt. rend.*, **256**: 4891 (1963).
117. Acoustics Congress, Copenhagen, 21–28, Aug., 1962, *Phys. Abstr.*, **67**: 14117 (1964); **66**: 21494 (1963).
118. Heckelsburg, L. F., and F. Daniels: *J. Phys. Chem.*, **61**: 414 (1957).
119. Galt, J. A., and H. A. Welch: *Can. J. Phys.*, **35**: 114 (1957).
120. Colegrove, F. D., and P. E. Franken: *Phys. Rev.*, **119**: 680 (1960).
121. Colegrove, F. D., L. D. Schearer, and G. K. Walters: *Phys. Rev.*, **132**: 2561 (1963).
122. Greenhow, R. C.: *Phys. Rev.*, **136**: A660 (1964).
123. Hipple, A. v.: *Z. Physik*, **101**: 680 (1936).
124. Oster, Gisela Kallmann: *Acta Phys. Polon.*, **26**: 435 (1964).
125. Jablonski, A.: *Acta Phys. Polon.*, **26**: 427 (1964).
126. Stolz, W.: *Acta Phys. Polon.*, **26**: 501 (1964).
127. Kallmann, H., and M. Furst: *Phys. Rev.*, **79**: 857 (1950). Furst, M., and H. Kallmann: *Phys. Rev.*, **109**: 646 (1958).
128. Bowen, E. J., and F. Wokes: "Fluorescence of Solutions," Longmans, London, 1953.
129. Spruch, Grace Marmor: Report on the 1963 Luminescence Conference at Turin, Poland, *Phys. Today*, **17**: 40 (1964).
130. Spruch, Grace Marmor: Report on the 1961 Conference on Luminescence at New York University at Washington Square, *Phys. Today*, **15**: 24 (1962).
131. Jablonski, A.: *Bull Acad. Polon. Sci., Ser. Sci., Math., Astron.*, **6**: 663 (1958).
132. Kawski, A.: *Z. Naturforsch.*, **18a**: 961 (1963).
133. Szalay, L., and B. Sarkany: *Acta Phys. Chem. Szeged (Hungary)*, **8**: 25 (1962).
134. Birks, J. B.: *Acta. Phys. Polon.*, **26**: 367 (1964).
135. Ore, A.: *J. Chem. Phys.*, **31**: 442 (1959).
136. Brown, F. H., M. Furst, and H. P. Kallmann: "Luminescence of Organic and Inorganic Materials," p. 100, Wiley, New York, 1962.
137. Parker, C. A., G. S. Hatchard, and Thelma A. Joyce: *J. Mol. Spectry.*, **14**: 311 (1964).
138. Williams, F. E.: *Phys. Rev.*, **80**: 306 (1950); **82**, 281 (1957); *J. Chem. Phys.*, **19**: 457 (1951).
139. Schulman, James H., and W. Dale Compton: "Color Centers in Solids," Macmillan, New York, 1962.
140. Winans, J. G., and E. C. G. Stuckelberg: *Proc. Natl. Acad. Sci. (U.S)*., **14**: 867 (1928).
141. Albrecht, Andreas C.: *J. Mol. Spectry.*, **6**: 84 (1961).
142. Lombardi, J. R., J. W. Raymonda, and A. C. Albrecht: *J. Chem. Phys.*, **40**: 1148 (1964).
143. Fink, E. L.: *Appl. Phys. Letters*, **7**: 103 (1965).
144. Swank, Robert K., and Frederick C. Brown: *Phys. Rev.*, **140**: A1735 (1965).
145. Park, Kwangajai: *Phys. Rev.*, **140**: A1735 (1965).
146. Delbecq, C. J., B. Smaller, and P. H. Yuster: *Phys. Rev.*, **111**: 1235 (1958).
147. Delbecq, C. J., W. Hayes, and P. H. Yuster: *Phys. Rev.*, **121**: 1043 (1961).
148. Johnson, P. D., and F. E. Williams: *Phys. Rev.*, **117**: 964 (1960).
149. Schneider, Irwin, and Herbert Rabin: *Phys. Rev.*, **140**: A1983 (1965).
150. Suppel, Robert F., and Everett D. Glover: *Science*, **150**: 1283 (1965).
151. Mochizuki, H., and R. G. Fowler: *Phys. Rev.*, **137**: A17 (1965).

Chapter 8

Optics and Relativity Theory

By E. L. HILL, University of Minnesota

1. Introduction

The propagation of optical and electromagnetic fields through space has long occupied the attention of physicists [1].* The ultimate recognition that optical phenomena represent but one manifestation of the general electromagnetic field simplified the problem greatly. During the nineteenth century, when physical theory was much concerned with the mechanical conceptions associated with hydrodynamics and elastic solids,† the prevailing view was that there must exist a physical medium through which propagation takes place. The medium which was hypothesized to perform this function for the electromagnetic field has come to be known as the *aether*, or *ether*.‡

If we accept Maxwell's field equations as representing the empirical facts correctly for the propagation of electromagnetic phenomena in free space, we are faced with the difficulty that they contain no direct trace of quantities, other than the speed of light, which might be interpreted as elastic or dynamical characteristics of a physical medium. The simplest interpretation which can be made from this point of view is that Maxwell's equations refer to propagation with respect to the ether and are expressed in terms of a coordinate system which is itself at rest in the medium. It would not be surprising under such circumstances to find that the only dynamical property which entered into the equations of vibrations in the medium would be the speed of propagation.

The adoption of this point of view raises questions at once on the properties of the ether for which the interpretation of Maxwell's equations is not free from ambiguity. Can the ether be used to define a standard of absolute rest? When a material body is set in motion, does the ether permeate it freely and so remain at rest or does the ether partake of the motion, wholly or in part? Finally, what form should the field equations take when expressed in a system of coordinates in motion with respect to the ether? Such questions provided the battleground for physical theory and experiment during much of the nineteenth century, but did not receive a fully significant solution until the development of the special (or restricted) theory of relativity by A. Einstein (1879–1955) in 1905 [2].

Einstein's theory has shown that the whole complex of experimental results and theoretical ideas associated with the concept of the ether can be systematized in a self-consistent manner without reference to the properties of any such medium. The question of the "mechanism" by which an electromagnetic field is propagated in free space is relegated to the class of pseudo problems. It is accepted as a simple fact requiring no further explanation that the field is propagated in the manner indicated by Maxwell's equations. From this point of view the ether is no longer required for the explanation of the empirical facts and has become an unnecessary, if not a strictly disproved, appendage of physical theory.

We shall not be concerned further with the problem of the ether but adopt the special theory of relativity as the cornerstone of our treatment and formulate the subject from a logical and deductive viewpoint rather than following the line of historical development.

2. The Special Theory of Relativity

One of the most important features of the restricted theory of relativity is the critical revision of the concepts of space and time to which it leads. As a result of this analysis physicists at present are inclined to pose the essential questions concerning the propagation of the electromagnetic field in a more abstract manner than was followed in the prerelativistic theories based on the idea of the ether. When any of the laws of physics are formulated mathematically, as differential equations or otherwise, we must assume that there exists at least one reference system in which the equations are valid. One can then ask whether this system is unique, or whether there exist other coordinate systems which could have served equally well in the sense that the equations would have taken exactly the same form in the new reference systems that they had in the originally chosen one. In other words, do there exist transformations of coordinates under which the equations are form-invariant? Furthermore, if such transformations exist, what physical interpretation can

* Numbers in brackets refer to References at end of chapter.

† It is of interest that the development of the theory of elastic solids was encouraged by the earlier work on the theory of the ether. See Whittaker [1, vol. 1, chap. 5].

‡ The Americanized spelling is *ether* and is used throughout this book. It is not profitable to identify the mechanical ether of the eighteenth and nineteenth centuries with the "polarizable vacuum" of modern quantum field theories.

be attached to the relationships subsisting among these various reference systems?

The functional purpose of the special theory of relativity was originally, and in large measure still is, to show that there does exist such a family of coordinate transformations under which Maxwell's equations preserve a strict form invariance, and to provide a consistent kinematic interpretation of the associated coordinate systems as equivalent cartesian reference systems moving with respect to each other with constant velocities in space. Whether this formal property of the field equations corresponds to an actual physical equivalence is a matter which can be settled only by an appeal to the experimental evidence.

The group of symmetry transformations which appears as the mathematical basis of the special theory of relativity is known as the *Lorentz transformation group,* in honor of H. A. Lorentz (1853–1928). These transformations are very different in structure from the corresponding group of *Galilean transformations,* which are employed ordinarily in the development of the equations of Newtonian mechanics in moving reference systems, although they contain the latter as a limiting case. The novel implications of the Lorentz transformations had been guessed in part in the proposals made for the null result of the Michelson-Morley experiment prior to the formulation of the special theory of relativity.

The basic assumption of Einstein's theory, that this mathematical-equivalence property of Maxwell's field equations expresses also a real physical equivalence of the associated set of moving reference systems, has proved to be valid, not only for the electromagnetic field, but for *all* physical phenomena, so far as we are aware at the present time. The conflict between the interpretations of the Lorentz and the Galilean groups as the proper transformations to uniformly moving reference systems has resulted in the generalization of the Newtonian equations of mechanics for application to particles moving with speeds nearly equal to that of light, and the experimental results have provided powerful support for the theory outside the field of optical phenomena [3].*

This expansion of the basis on which the special theory of relativity rests naturally has diminished in considerable measure the importance of the experiments which first gave it observational support. The situation has altered to such a degree during the last half century and so many phenomena have been found which depend on the theory of relativity for their interpretation that it is now certain that any successful attempt to overthrow the theory would require much stronger support than could possibly be provided by mere criticism of the Michelson-Morley and other early experiments.

3. The Transformation Formulas of Special Relativity [4]

Our considerations will be limited to the relationships which exist between the results of experiments obtained in two cartesian reference systems which are moving with respect to each other with constant speeds in a fixed direction. We can make the motion appear to take place along the common (x-x') axis by a suitable orientation of the two sets of axes, as indicated in Fig. 8.1, and by a corresponding choice

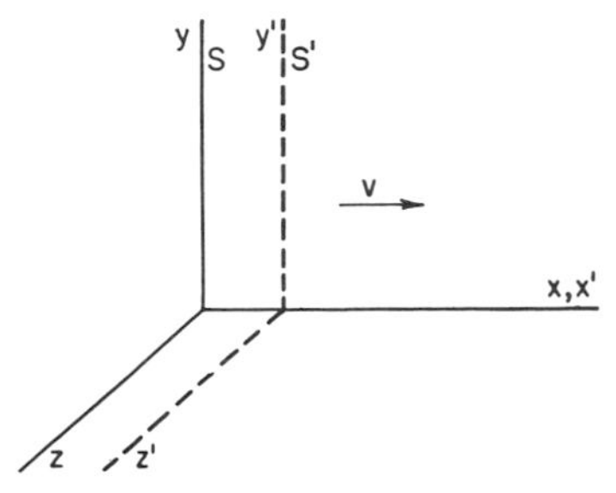

Fig. 8.1

of the origins of the time scales in the two systems one can make the instant at which the origins of the two systems are coincident correspond to the values $t = t' = 0$. The particular transformation equations between the sets of space and time coordinates (x,y,z,t) and (x',y',z',t') are

$$x' = \frac{x - \beta ct}{(1 - \beta^2)^{1/2}} \qquad y' = y \qquad z' = z$$

$$t' = \frac{t - \beta x/c}{(1 - \beta^2)^{1/2}} \qquad (8.1)$$

with $\beta = v/c$. The inverse form of this transformation is obtained by interchanging the primed and unprimed variables and changing the sign of the parameter β in the above equations.

The relative speed of each of the reference systems, as measured with respect to the other, is v, with system S' moving to the right with respect to S in Fig. 8.1. The two time coordinates t and t' are interpretable in the ordinary sense as covering times for each of the separate space reference systems, but the exact form of (8.1) must be used in the interpretation of events as observed in one system in terms of observations on the same physical events in the other system. The kinematical consequences of these equations have been discussed in Part 2, Chaps. 1 and 6.

Suppose now that an electromagnetic field is present. In each of the reference systems the corresponding observed fields will be describable by giving the electric- and magnetic-field vectors as functions of the appropriate space-time coordinates associated with that system, so that we have the two sets of field vectors

$$S\colon \quad \mathbf{E}(x,y,z,t) \qquad \mathbf{H}(x,y,z,t)$$
$$S'\colon \quad \mathbf{E}'(x',y',z',t') \qquad \mathbf{H}'(x',y',z',t')$$

It is a fundamental assumption of the special theory of relativity that these two sets of vectors will be solutions of Maxwell's equations and that the field equations will have the ordinary form in *both* reference systems. The mathematical self-consistency of this assumption is assured by the fact that a suitable transformation scheme can be established between

* One should emphasize here the modern experiments on the betatron, synchro-cyclotron, bevatron, and other accelerators of high-energy particles whose design depends on the use of relativistic formulas.

the two sets of field vectors such that Maxwell's equations will be completely form-invariant on making the transformation from the one to the other reference system. This scheme is

$$\begin{aligned} E'_x &= E_x & H'_x &= H_x \\ E'_y &= \frac{E_y - \beta H_z}{(1-\beta^2)^{1/2}} & H'_y &= \frac{H_y + \beta E_z}{(1-\beta^2)^{1/2}} \\ E'_z &= \frac{E_z + \beta H_y}{(1-\beta^2)^{1/2}} & H'_z &= \frac{H_z - \beta E_y}{(1-\beta^2)^{1/2}} \end{aligned} \quad (8.2)$$

The transformation of the field equations requires a knowledge also of the corresponding relations connecting the charge and current densities, which are

$$J'_x = \frac{J_x - \beta c\rho}{(1-\beta^2)^{1/2}} \qquad J'_y = J_y \qquad J'_z = J_z$$
$$\rho' = \frac{\rho - \beta J_x/c}{(1-\beta^2)^{1/2}} \quad (8.3)$$

The inverse transformations to (8.2) and (8.3) are obtained by interchanging the primed and unprimed quantities and changing the sign of β. In these two sets of equations the quantities on the left-hand sides are to be expressed directly in terms of the space and time variables of system S', while those on the right-hand sides are expressed in terms of the corresponding quantities for system S. The elimination of variables is then carried out using (8.1).

Equations (8.3) can be shown to be compatible with the requirement that the total electrical charge on a particle is independent of the reference system in which it is observed. The quantities

$$\Theta_1 = \mathbf{E} \cdot \mathbf{H} \qquad \Theta_2 = \mathbf{H}^2 - \mathbf{E}^2 \quad (8.4)$$

are invariants under transformation (8.2). They are, in fact, invariants under the complete group of Lorentz transformations.

4. The Transformation Equations for Plane Waves

The transformation equations for the field vectors associated with a plane-polarized plane wave can be written out as a special case of Eqs. (8.2). If the wave has an angular frequency $\omega = 2\pi\nu$, and a wave number $k = 2\pi/\lambda$, in the system S, then the solution of Maxwell's equations representing such a wave takes the form

$$\mathbf{E} = \mathbf{E}_0 f(\mathbf{k} \cdot \mathbf{r} - \omega t) \qquad \mathbf{H} = \mathbf{H}_0 f(\mathbf{k} \cdot \mathbf{r} - \omega t) \quad (8.5)$$

provided the following relations are satisfied,

$$\mathbf{k} \cdot \mathbf{E}_0 = 0 \qquad \mathbf{H}_0 = \mathbf{n} \times \mathbf{E}_0 \qquad k = \frac{\omega}{c} \quad (8.6)$$

where $\mathbf{n}$ is the unit vector $\mathbf{n} = \mathbf{k}/k$. These relations show that the magnitudes of the constant vectors, $\mathbf{E}_0$ and $\mathbf{H}_0$, are equal (in Gaussian units) and that the vectors $(\mathbf{E}_0, \mathbf{H}_0, \mathbf{k})$ form a right-handed orthogonal set. The functional form of the function f is immaterial so long as it depends only on the *phase* of the wave

$$\phi = \mathbf{k} \cdot \mathbf{r} - \omega t = k_1 x + k_2 y + k_3 z - \omega t \quad (8.7)$$

The surfaces of constant phase form the wavefronts at each instant, in this case being the family of planes having the common unit normal $\mathbf{n}$. On applying the transformations (8.2) one finds that the field, as observed in S', is also of the form of a plane-polarized wave

$$\mathbf{E}' = \mathbf{E}'_0 f(k' \cdot r' - \omega' t) \qquad \mathbf{H}' = \mathbf{H}'_0 f(k' \cdot r' - \omega' t) \quad (8.8)$$

Here

$$k'_1 = \gamma\left(k_1 - \frac{\beta\omega}{c}\right) \qquad k'_2 = k_2 \qquad k'_3 = k_3 \quad (8.9)$$

$$\begin{aligned} \omega' &= \gamma(\omega - k_1 \beta c) \\ \mathbf{E}'_0 &= (1-\gamma)(\mathbf{E}_0 \cdot \mathbf{i})\mathbf{i} + \gamma(\mathbf{E}_0 + \beta \mathbf{i} \times \mathbf{H}_0) \\ \mathbf{H}'_0 &= (1-\gamma)(\mathbf{H}_0 \cdot \mathbf{i})\mathbf{i} + \gamma(\mathbf{H}_0 - \beta \mathbf{i} \times \mathbf{E}_0) \end{aligned} \quad (8.10)$$

with $\gamma = \dfrac{1}{(1-\beta^2)^{1/2}}$

It can be verified that these relations satisfy all the conditions for a plane wave. In particular, the two invariants of (8.4) show that the field vectors in S' are orthogonal and have the same magnitude, since this is true in S.

The relation between the unit vectors $\mathbf{n}$ and $\mathbf{n}'$ giving the apparent directions of the wave normals as observed in S and S' is found to be

$$n'_1 = \frac{n_1 - \beta}{1 - \beta n_1} \qquad n'_2 = n_2 \frac{(1-\beta^2)^{1/2}}{1 - \beta n_1}$$
$$n'_3 = n_3 \frac{(1-\beta^2)^{1/2}}{1 - \beta n_1} \quad (8.11)$$

If θ and θ' are the angles which these unit vectors make with the common $(\mathbf{i}, \mathbf{i}')$ axis, in their respective systems, we have

$$\tan\theta = \frac{(n_2^2 + n_3^2)^{1/2}}{n_1} = \frac{(1 - n_1^2)^{1/2}}{n_1}$$
$$\tan\theta' = \frac{(n_2'^2 + n_3'^2)^{1/2}}{n'_1} = \frac{(1 - n_1'^2)^{1/2}}{n'_1} \quad (8.12)$$

$$\tan\theta' = \tan\theta \frac{(1-\beta^2)^{1/2}\cos\theta}{\cos\theta - \beta} \quad (8.13)$$

All these relations can be inverted simply by interchanging the primed and unprimed quantities and changing the sign of β.

5. The Dynamical Properties of Photons

According to the wave-particle duality which underlies the modern quantum theory of fields, an electromagnetic field can be considered to be composed of "light quanta," or *photons*. The relations of Sec. 4 give the dynamical properties of the photons associated with plane waves. The fundamental relation between the momentum of the photons and the wave-number vector is

$$\mathbf{p} = \mathbf{k}\hbar \quad (8.14)$$

where $h = 2\pi\hbar$ is Planck's constant. The corresponding relation between the energy of the photon, w, and the frequency of the wave is

$$w = h\nu = \hbar\omega \quad (8.15)$$

The invariant $k^2 - \omega^2/c^2$ for the wave thus leads to the invariant $p^2 - w^2/c^2$ for the photons, each of these quantities having the numerical value zero. The latter invariant is a special case of the more general relation

$$w^2 - p^2c^2 = (m_0c^2)^2 \tag{8.16}$$

which is the relativistic energy-momentum relation for a particle of rest mass m_0. The field vectors for the wave provide the interpretation of the polarization of the photons [5].

The behavior of the energy and momentum components of a photon under the transformation (8.1) is given by the equations

$$p'_1 = \frac{p_1 - \beta w/c}{(1-\beta^2)^{1/2}} \qquad p'_2 = p_2 \qquad p'_3 = p_3$$
$$w' = \frac{w - p_1\beta c}{(1-\beta^2)^{1/2}} \tag{8.17}$$

The dynamical properties of photons have been tested thoroughly in the work on the Compton effect, pair production, and other phenomena involving photons of high energy [6]. (See Part 7, Chap. 8.)

6. Aberration of Light

The aberration of light is defined as the apparent angular change of the line of sight to a distant body with the velocity of the observer. It was discovered in 1725 by Bradley as a small shift in the celestial latitude of a star with season of the year. Bradley explained the effect correctly as depending on the relative velocity of the earth and the star.

Let the system S be at rest with respect to the star, and let S' be at rest with respect to the earth. If $\bar{\theta}$ and $\bar{\theta}'$ are the angles made by the incoming light beam with respect to the earth's velocity vector, as shown in Fig. 8.2, then in terms of the notation

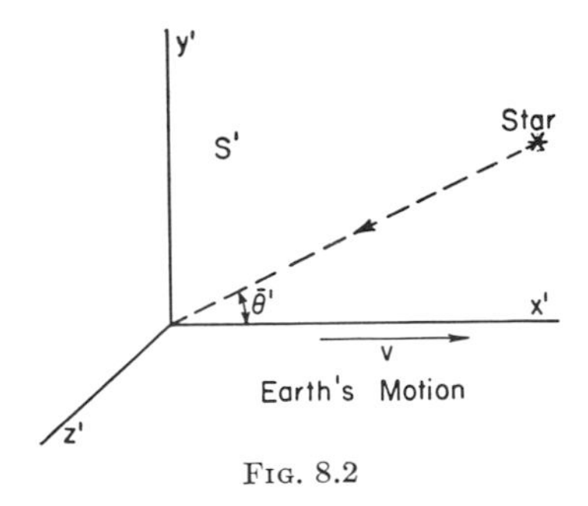

FIG. 8.2

of Sec. 4 we have $\bar{\theta} = \pi - \theta$ and $\bar{\theta}' = \pi - \theta'$. Formula (8.13) now takes the form

$$\tan \bar{\theta}' = \tan \bar{\theta} \frac{(1-\beta^2)^{1/2} \cos \bar{\theta}}{\beta + \cos \bar{\theta}} \tag{8.18}$$

The angular deviation of the line of the incoming light beam from the star, as compared in the two reference systems, is $\alpha = \bar{\theta}' - \bar{\theta}$. This is a small angle, and as only first-order terms in β can be measured experimentally, we calculate it from (8.18) with neglect of all higher powers. The final formula is

$$\tan \alpha = -\beta \sin \bar{\theta} \tag{8.19}$$

A star lying on one of the celestial poles appears to perform a small annual motion in a circle of angular diameter $2 \tan^{-1} \beta$, while one lying in the plane of the ecliptic performs a motion along a line having the same angular diameter. A star lying between these extremes on the celestial sphere performs a small elliptical motion with an eccentricity which depends on its latitude [7]. Astronomical observations give $\tan^{-1} \beta = 20.47''$. Calculating the speed of the earth from this, using the empirical value of the speed of light (c = 299,776 km/sec), one finds a value of 29.75 km/sec (18.49 mps), in quite satisfactory agreement with other evidence.

In elementary explanations of the aberration effect an analogy is often drawn with the falling of raindrops, and it is stated that the orientation of the telescope must be altered with the velocity of the observer in order to allow the light to travel from the objective to the eyepiece of the telescope, during which time the telescope is also in motion. This explanation would lead one to expect that the magnitude of the aberration would depend on the speed of light in the telescope. Attempts to detect such a phenomenon were made by Airy, using a telescope filled with water, with negative results. From the point of view of the ether theory this requires a special explanation involving the concept of the Fresnel ether drag coefficient (Sec. 12).

7. Doppler Effect

The dependence of the apparent frequency of a light source on its motion relative to the observer was first remarked in 1842 by C. Doppler. The theory of the Doppler effect follows directly from the relations of Sec. 4. Let S' be the co-moving system of the source, so that ν' and $\lambda' = c/\nu'$ are the frequency and wavelength of the radiation as emitted by the source, while ν and $\lambda = c/\nu$ are the corresponding quantities as observed in the laboratory system S. From the last of (8.9)

$$\nu = \nu' \frac{(1-\beta^2)^{1/2}}{1 - \beta \cos \theta} \tag{8.20}$$

To first-order terms in β this reduces to

$$\delta\nu = \nu - \nu' = \nu\beta \cos \theta \tag{8.21}$$

which agrees with the result obtained from elementary arguments.

When $\theta = 0$ or π, the light beam travels along the axis of the relative velocity of the source and the observer and the magnitude of the shift in frequency has a maximum value. This particular case is referred to as the *longitudinal Doppler effect.*

When $\theta = \pi/2$, the elementary formula (8.21) predicts no frequency shift, but according to the complete relativistic formula (8.20) there should be a *transverse Doppler shift* of amount

$$(\delta\nu)_{tr} = (\nu - \nu')_{\theta=\pi/2} = \nu[1 - (1-\beta^2)^{-\frac{1}{2}}] \tag{8.22}$$

It would be of particular interest to measure the transverse shift, since it is a purely relativistic (second-order) effect. The experiment is complicated seriously by the difficulty of obtaining an angle of relative motion of exactly $\pi/2$, since small deviations from this value would introduce appreciable errors from the first-order longitudinal effect. Ives and Stilwell have devised a different method of measuring the second-order effect, discussed in the next section.

8. The Experiment of Ives and Stilwell [8]

The principle of this experiment is that of measuring the mean values of the longitudinal Doppler shifts of spectral lines for the forward and backward directions of emission with respect to the velocity of the source. The first-order shift cancels out in this averaging, and one has left only a second-order effect. Equation (8.20) shows that the mean wavelength is

$$(\lambda)_{av} = \frac{\lambda_0}{(1 - \beta^2)^{1/2}} \tag{8.23}$$

if λ_0 is the wavelength of the line emitted by the source at rest. The mean position of the line is shifted to the red by the amount

$$\Delta_2\lambda = \lambda_0[(1 - \beta^2)^{-1/2} - 1] \cong \frac{\lambda_0\beta^2}{2} \tag{8.24}$$

The moving light sources used were molecular hydrogen ions, H_2^+ and H_3^+, accelerated in a specially designed canal ray tube by an electric field. The maximum field used corresponded to a value of β of about 0.0072 for H_2^+ ions. The experimental results are shown in Fig. 8.3. To avoid using the

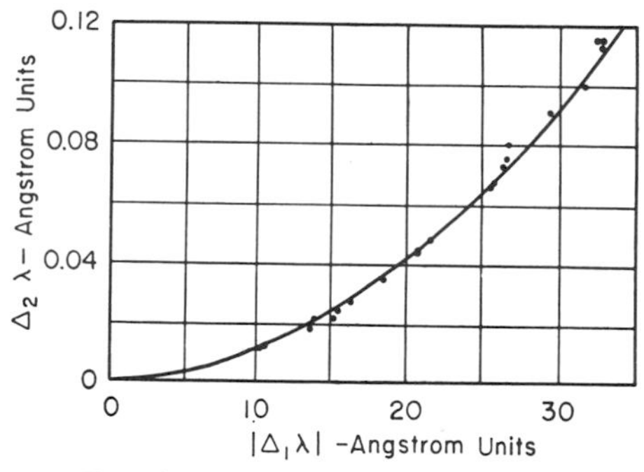

FIG. 8.3. Experimental measurements on the second-order longitudinal Doppler effect.

measured voltages on the tube, which were somewhat inaccurate, the magnitude of the longitudinal first-order Doppler shift $|\Delta_1\lambda|$ is used as a measure of the speed of the ions and is plotted as the abscissas in the figure. The solid curve gives the theoretical relation between the second- and first-order Doppler shifts ($\Delta_2\lambda = |\Delta_1\lambda|^2/2\lambda_0$) as given by (8.21) and (8.24), for small values of β. The dots indicate the observed measurements of the second-order shifts. The agreement between theory and experiment is excellent.

9. The Michelson-Morley Experiment [9]

This experiment, one of the most famous in the history of physics, was conceived as a direct test of the optical effect of motion in the ether. Its failure to detect any such influence was directly responsible for the initiation of the chain of reasoning which led to the special theory of relativity. It no longer occupies a central place in physical theory because more positive verifications of the concepts of the theory of relativity have been found in the intervening period.

The idea of the experiment is to detect the effect of the orbital motion of the earth on the speed of light emitted from a source in the laboratory. According to the theory of relativity no such effect exists, and the conclusion is drawn that the experiment can lead only to a negative result. According to the simpler form of the ether hypothesis, in which the ether forms a reference system at absolute rest, the velocities of the light and of the observer should be directly additive, following the usual vector law. Light from a source in the laboratory should therefore travel with a different speed along the direction of the earth's instantaneous orbital velocity from its speed at right angles to this direction. The Michelson-Morley experiment attempts to measure this difference to terms in β^2.

A schematic diagram of the apparatus is shown in Fig. 8.4. A source of light, s, sends a beam along

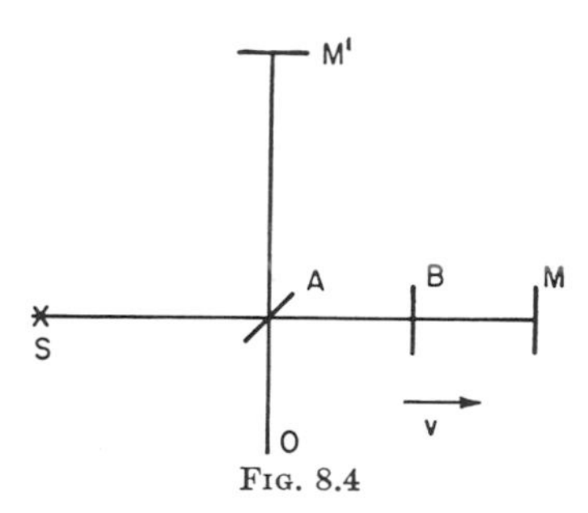

FIG. 8.4

the arm sAM of a Michelson interferometer, which we suppose to be oriented along the axis of the earth's instantaneous orbital motion. One surface of the transparent plate A is half-silvered, so that the beam is broken into two parts, one traveling along and the other perpendicular to the earth's motion. After reflections at the mirrors M and M' these beams return to the plate A and are brought together to form an interference pattern at O. The beams $AM'AO$ make three traversals of the plate A, while the beams $AMAO$ make but one, for which reason a compensating plate B is inserted in the arm AM. After the instrument is adjusted, it is rotated about a vertical axis and any shift in the pattern of interference fringes is noted. Much more complex designs of the interferometer were used in the actual experiments, involving multiple reflections from a set of mirrors giving a total optical path of about 11 m. The instrument was kept in slow rotation to avoid the strains incident to starting and stopping. The experiment has been repeated by several independent observers in recent years [9].

Consider the standard orientation of the interferometer as shown in Fig. 8.4. We can analyze the experiment from the point of view of the ether hypothesis by making use either of a reference system at rest with respect to the ether or of one at rest with respect to the apparatus, provided that in the latter case we take care to consider the relative velocity of light with respect to the apparatus, which then depends on its direction of motion in the laboratory reference system. We choose the latter method for simplicity [10].* It is now necessary only to compare the two beams from their initial separation at the surface of plate A to their return to this point, exclusive of the effects of the double traversal through plates A and B, since these compensate each other.

The beam of light which travels along the axis AM has a speed $c - v$ on the outward journey and a speed $c + v$ on the return, both measured relative to the apparatus. If the length of the geometrical path is L, the time required for the total trip is

$$t_{AMA} = \frac{L}{c - v} + \frac{L}{c + v} = \frac{2L}{c} \frac{1}{1 - \beta^2} \tag{8.25}$$

The other beam, which travels along the arm AM', must move in space (i.e., in the ether) along directions which are slightly in the forward sense in order that it will travel along AM' as viewed in the laboratory reference system. The angle of shift is $\sin^{-1} \beta$. The speed of this beam relative to the apparatus is $(c^2 - v^2)^{1/2}$, so that the time required for the total journey is

$$t_{AM'A} = \frac{2L}{(c^2 - v^2)^{1/2}} = \frac{2L}{c(1 - \beta^2)^{1/2}} \tag{8.26}$$

if the geometrical length of the arm AM' is also L.

Equations (8.25) and (8.26) show that the beam which travels the path AMA is retarded with respect to that which travels $AM'A$ by an amount

$$t_{AMA} - t_{AM'A} = \frac{2}{c(1 - \beta^2)} [L - L(1 - \beta^2)^{1/2}] \tag{8.27}$$

Retaining only the lowest ordered terms in β,

$$t_{AMA} - t_{AM'A} = \frac{L\beta^2}{c} \tag{8.28}$$

Terms of first order in β cancel owing to the necessity of having the beams complete a return journey before being reunited.

A phase retardation between the beams could not be observed with a single setting of the interferometer, because of the necessity of adjusting the instrument, but once this is done, a further rotation about the vertical axis should give rise to a steady shifting in the interference pattern according to this theory. In the experiment as performed by Michelson and Morley, the observed fringe displacements did not exceed 5 per cent of the expected value. Later observers concur in the conclusion that the experiment shows no effect of the ether, in agreement with Einstein's arguments [1].†

FitzGerald and Lorentz suggested that the negative result of the Michelson-Morley experiment could be understood on the assumption that the dimensions of the apparatus in the direction of the earth's motion are reduced by a factor $(1 - \beta^2)^{1/2}$, as compared with its dimensions in the transverse direction. This would amount to the assumption that the true length of the arm AM of the interferometer in the standard configuration of Fig. 8.4 is $L(1 - \beta^2)^{1/2}$ instead of L. The discussion leading to (8.27) shows that this would decrease the time t_{AMA} to make it just equal to $t_{AM'A}$, from which the null result would follow. This postulated contraction is known as the *FitzGerald-Lorentz contraction.* It was considered originally that it arose from an interaction between the material of which the apparatus is made and the ether and that this should give rise to elastic stresses in the apparatus. Numerous experiments have been made to detect such stresses, but uniformly without positive results. The explanation given by the theory of relativity is that the contraction effect is real but that it is a purely kinematical result of the new definitions used in the measurement of the dimensions of moving objects, being a direct kinematical consequence of the Lorentz transformations. As such it applies to all materials and requires no dynamical explanation.

10. The Kennedy-Thorndike Experiment [12]

The kinematical interpretation of the Lorentz transformations to moving reference systems leads to a second novel result, which is known as the *time-dilatation effect.* Let two events take place at a fixed space point in a moving reference system S', but at different instants of time in that system. Employing the notation of Sec. 3, by inversion of (8.1)

$$t = \frac{t' + \beta x'/c}{(1 - \beta^2)^{1/2}} \tag{8.29}$$

Since the two events take place at a point of fixed coordinates (x',y',z'), the time interval between them, Δt, *as measured in system* S, is related to the time interval between them, $\Delta t'$, as measured in system S', by the formula

$$\Delta t = \frac{\Delta t'}{(1 - \beta^2)^{1/2}} \tag{8.30}$$

which follows at once from (8.29). This shows that the time interval between the events appears to be *longer* in system S than it is in the co-moving system S'.

Kennedy and Thorndike attempted to measure this time dilatation directly by a modification of the Michelson-Morley experiment. In their arrangement the arms of the interferometer were made unequal in length and were not at right angles to each other. In this case the FitzGerald-Lorentz contraction would not by itself lead to a null result for the experiment. Kennedy and Thorndike showed

* A discussion using a reference system other than the laboratory system is given by E. Cunningham [10]. This is useful in verifying that the conditions for interference of the reunited beams are satisfied.

† In an experiment using ammonia maser beams [11] the accuracy of the measurement has been increased to about 0.03 km/sec.

that a detailed relativistic analysis, inclusive of the time dilatation, would predict such a null result, which was what they actually found.

A somewhat indirect, but very striking, verification of formula (8.30) is provided by the observed increase in the apparent lifetimes of mesons of high energies [13].

11. Generalizations of the Lorentz Transformation Group

Ives [14] has remarked that the explanation of the Michelson-Morley and the Kennedy-Thorndike experiments in terms of a contraction of the apparatus along the direction of motion and a time dilatation is not unique. The latter experiment can be accounted for equally well, for instance, by a contraction factor $1 - \beta^2$ along, and a contraction factor $(1 - \beta^2)^{1/2}$ transverse to, the direction of motion of the apparatus, without any consideration of the time dilatation. Even more general systems of contractions and associated time dilatations can be devised which will predict the null results of both experiments.

This lack of uniqueness has its origin in the circumstance that the simple Lorentz group of transformations of the special theory of relativity does not exhaust the group of symmetry transformations of the electromagnetic-field equations. It was discovered in 1910 by Cunningham and Bateman that there exists a larger group of continuous transformations under which Maxwell's equations preserve their form invariance, the complete group being a 15-parameter Lie group of conformal transformations in space time [15].

When all *optical* experiments are considered to be explicable by Maxwell's equations alone, it follows that the results predicted from any and all of the transformations of this complete symmetry group of the field equations cannot be proved to be in error *by optical experiments alone.* However, the kinematical interpretation of the conformal group involves the introduction of uniformly accelerated coordinate systems [16] so that a proper study of their implications must involve mechanical as well as optical considerations. The physical interpretation of this generalization of the restricted theory of relativity is still a matter of uncertainty.

12. Electromagnetic Phenomena in Moving Media

Prior to the development of an adequate molecular theory of matter the assumption prevailed that the equations of the electromagnetic field in material media were of the same generality as those for free space. More recent work has shown clearly that this is not the case, but that the fields in matter are to be obtained from those in free space by a process of averaging over the molecular fields [17]. This implies that the electromagnetic properties of matter in bulk no longer provide a test of the theory of relativity, and conversely that relativistic considerations alone will not be adequate to establish a definitive form for the averaged field equations in material bodies.

Two general methods have been developed for the formulation of relativistically invariant field equations in material bodies. In the first, one recognizes that the medium itself provides a standard of rest so that, starting from any reasonable set of field equations in the corresponding coordinate system, relativistic considerations can be used to formulate appropriate equations for the fields as defined in a moving reference system. Theories of this type have been formulated by Minkowski, Hertz, and Lorentz.* A quantum-mechanical version has been given by Jauch and Watson [19]. In the second method a formal tensor generalization of Maxwell's equations is employed, the material constants being interpreted as appropriate tensor quantities. A theory of this nature has been worked out by Tamm [20].

Much of the early work in this field has been robbed of its importance by the development of the theory of relativity, but there is one point which is of more than historical interest. In 1818 Fresnel suggested that a measurement of the apparent speed of light in a moving medium would provide a test of the hypothesis that the ether is dragged along with the body. Fresnel proposed the formula

$$u = \frac{c}{n} \pm v\left(1 - \frac{1}{n^2}\right) \tag{8.31}$$

for the apparent speed of light in the laboratory system, in which v is the velocity of the moving medium, n is the index of refraction, and the upper or lower sign is to be taken according as the light moves with or against the motion of the medium. The quantity

$$\mathcal{E}_F = 1 - \frac{1}{n^2} \tag{8.32}$$

is called the *Fresnel drag coefficient* since it was supposed to indicate the fractional part of the velocity of the body by which the ether was dragged along with it. Fresnel's formula was tested experimentally in 1851 by Fizeau and again in 1886–1887 by Michelson and Morley, with improved techniques. The results were in fair agreement with (8.31).

It is not difficult to see that Fresnel's formula follows from the kinematical relations of the special theory of relativity. The formula for the composition of velocities (see Part 2, Chap. 6, Sec. 10) is, for motion along the x axis,

$$u = \frac{v + v'}{1 + vv'/c^2} \tag{8.33}$$

We use the notation of Sec. 3, letting S' be the comoving system of the moving medium and limiting the formula to the case in which the light moves in the direction of motion of the medium. In this case $v' = c/n$, which is the speed of light measured with respect to the moving medium, so that from (8.33)

$$u = \frac{v + c/n}{1 + v/cn} \tag{8.34}$$

* A review of these theories is given in the book by Cunningham [18].

On expanding in powers of v/c, keeping only first-order terms, we obtain Fresnel's result for this case. The formula for the case in which the light and the medium move in opposite directions is found by changing the sign of v.

It was pointed out by Lorentz [21] that Fresnel's formula is subject to correction if the medium is dispersive since the frequency of the light as observed in the co-moving system S' differs from that in the laboratory system. He gave the corrected formula as

$$u = \frac{c}{n} \pm v\left(1 - \frac{1}{n^2} - \frac{\lambda}{n}\frac{dn}{d\lambda}\right) \tag{8.35}$$

Lorentz's formula has been made the subject of extensive measurements by Zeeman, using both liquid and solid media, with results favorable to it [22].

13. The Special Theory of Relativity and Quantum Mechanics

Relativistic considerations have always played an important role in quantum-mechanical theory. The wavelength-momentum formula for material particles as given by De Broglie is just our (8.14) written in the form [23] $p = \hbar k = h/\lambda$. Dirac's development in 1928 of his wave equation for the electron [24] not only showed that the spin of the electron is of relativistic origin but also provided a theoretical interpretation of a number of small correction terms which had been found to be necessary in the interpretation of fine structures in atomic spectra [25]. When applied to the hydrogen atom it leads to the same energy formula as was derived by Sommerfeld [26] on the basis of the old quantum theory, taking into consideration the relativistic variation of mass of the electron with its speed.

The foundations of a quantum theory of the electromagnetic field were laid in 1927 by Heisenberg and Pauli [27], and this has since developed into a general theory of quantized fields which attempts to explain the properties of all kinds of elementary particles [28]. This theory presents many severe mathematical ambiguities, but it has some notable successes to its credit. Perhaps the most definitive of these is the theory of the Lamb-Retherford shift. According to the Sommerfeld-Dirac energy-level formula for hydrogen the two levels ($2\,^2S_{1/2}$, $2\,^2P_{1/2}$) should have exactly the same energy and so form a degenerate pair. Lamb and Retherford [29] have shown experimentally that in fact the level $2\,^2S_{1/2}$ lies above the level $2\,^2P_{1/2}$, the absorption frequency being, $1{,}062 \pm 5$ Mc/sec. A similar shift in the energy levels of the ion He^+ is of amount $14{,}020 \pm 100$ Mc/sec [30]. Theoretical values calculated on the basis of quantum electrodynamics are in good agreement with the measurements.

References

1. Whittaker, E. T.: "A History of the Theories of Aether and Electricity," 2 vols., Nelson, London, 1951–1953.
2. Einstein, A.: *Ann. Physik*, **17**(4): 891 (1905).
3. McCrea, W. H.: "Relativity Physics," 4th ed., Wiley, New York, 1954.
4. Einstein, A.: "The Meaning of Relativity," 5th ed., Princeton University Press, Princeton, N.J., 1955. R. C. Tolman: "Relativity, Thermodynamics, and Cosmology," Oxford University Press, Fair Lawn, N.J., 1934. H. P. Robertson: *Revs. Mod. Phys.*, **21**: 374 (1949).
5. Heitler, W.: "The Quantum Theory of Radiation," 3d ed., Oxford University Press, New York, 1954; J. M. Jauch and F. Rohrlich: "Theory of Photons and Electrons," Sec. 2–8, Addison-Wesley, Reading, Mass., 1955.
6. Rossi, B.: "High-energy Particles," Prentice-Hall, Englewood Cliffs, N.J., 1952.
7. Russell, Dugan, and Stewart: "Astronomy," vol. 1, secs. 162–164, Ginn, Boston, 1945.
8. Ives, H. E., and G. R. Stilwell: *J. Opt. Soc. Am.*, **28**: 215 (1938), **31**: 369 (1941). H. P. Robertson: *Revs. Mod. Phys.*, **21**: 374 (1949).
9. Michelson, A. A., and E. W. Morley: *Am. J. Sci.*, **34**: 333 (1887). K. K. Illingworth: *Phys. Rev.*, **30**: 692 (1927). G. Joos: *Ann. Physik*, **7**(5): 385 (1930).
10. Cunningham, E.: "The Principle of Relativity," p. 17, Cambridge University Press, New York, 1921.
11. Cedarholm, J. P., G. F. Bland, B. L. Havens, and C. H. Townes: *Phys. Rev. Letters*, **1**: 342 (1958).
12. Kennedy, R. J., and E. W. Thorndike: *Phys. Rev.*, **42**: 400 (1932). H. P. Robertson: *Revs. Mod. Phys.*, **21**: 374 (1949).
13. Rossi, B.: "High-energy Particles," p. 157, Prentice-Hall, Englewood Cliffs, N.J., 1952. L. Jánossy: "Cosmic Rays," 2d ed., p. 22, Oxford University Press, Fair Lawn, N.J., 1950.
14. Ives, H. E.: *J. Opt. Soc. Am.*, **27**: 177 (1937).
15. Cunningham, E.: *Proc. London Math. Soc.*, **8**(2): 77(1910). H. Bateman: *Proc. London Math. Soc.*, **8**(2): 223 (1910).
16. Kotteler, F.: *Ann. Physik*, **44**(4): 701 (1914). J. Haantjes; *Proc. Koninklijke Akad. Wetenschap. Amsterdam*, **43**: 1288 (1940). E. L. Hill: *Phys. Rev.*, **72**: 143 (1947).
17. Frenkel, J.: "Lehrbuch der Elektrodynamik," vol. 2, chaps. 1, 2, Springer, Berlin, 1928. L. Rosenfeld:, "Theory of Electrons," North Holland Pub. Co., Amsterdam, 1951.
18. Cunningham, E.: "The Principle of Relativity," chap. 10, Cambridge University Press, New York, 1921.
19. Jauch, J. M., and K. M. Watson: *Phys. Rev.*, **74**: 950, 1485 (1948), **75**: 1249 (1949).
20. Tamm, I.: *J. Russ. Phys.-Chem. Soc.*, **56**: 248 (1924); **57**: 209 (1925). L. Mandelstam and I. Tamm: *Mathematische Annalen*, **95**: 154 (1926).
21. Lorentz, H. A.: "Lectures on Theoretical Physics," vol. 3, p. 303, Macmillan, New York, 1931. W. H. McCrae: "Relativity Physics," p. 39, Wiley, New York, 1950.
22. Zeeman, P., et al.: *Koninklijke Akad. Wetenschap. Amsterdam*, **29**: 1252 (1921).
23. De Broglie, L: *Ondes et mouvements*, chap. 1, Gauthier-Villars, Paris, 1926. E. C. Kemble: "The Fundamental Principles of Quantum Mechanics," p. 13, McGraw-Hill, New York, 1937.
24. Dirac, P. A. M.: *Proc. Roy. Soc. (London)*, **A117**: 610 (1928), **A118**: 351 (1928).
25. Condon, E. U., and G. H. Shortley: "The Theory of Atomic Spectra," p. 125, Cambridge University Press, New York, 1935. E. L. Hill and R. Landshoff: *Revs. Mod. Phys.*, **10**: 87 (1938).
26. Sommerfeld, A.: "Atomic Structure and Spectral Lines," chap. 5, Dutton, New York, 1935.
27. Heisenberg, W., and W. Pauli: *Z. Physik*, **56**: 1 (1927).
28. Wentzel, G.: "Quantum Theory of Fields," Interscience, New York, 1949.
29. Lamb, W. E., and R. C. Retherford: *Phys. Rev.*, **72**: 241 (1947).
30. Lamb, W. E., and M. Skinner: *Phys. Rev.*, **78**: 539 (1950). W. E. Lamb: Anomalous Fine Structure of Hydrogen and Singly Ionized Helium, in "Reports on Progress in Physics," vol. 14, p. 19, Physical Society, London, 1951.

Part 7 · Atomic Physics

Chapter 1

Quantum Mechanics and Atomic Structure

By E. U. CONDON, University of Colorado

1. Particle Waves

In 1924 there began a set of new developments in the dynamics of atomic phenomena known as quantum mechanics, quantum dynamics, or wave mechanics. The new ideas are largely due to L. de Broglie, E. Schroedinger, W. Heisenberg, P. A. M. Dirac, and many others [1].‡ Ever since Bohr's basic formulation of the quantum theory of atomic structure in 1913, it was recognized that the mechanics of systems of atomic dimensions must obey essentially different laws than the larger systems which are governed by the classical mechanics of Newton and his successors.

These laws of quantum dynamics must involve the universal Planck constant, $h = 6.62 \times 10^{-27}$ erg-sec, in an essential way; and the quantum laws must go over asymptotically into the classical laws, not involving h, as the scale of the phenomena is increased in an appropriate way. This guiding idea, variously formulated, was called *Bohr's correspondence principle.*

The most characteristic difference between quantum dynamics and classical mechanics lies in the fact that certain of the dynamical variables are *quantized,* i.e., they may only assume values belonging to a discrete set of *allowed values.* The system is never observed to be in a state in which a quantized variable has other than one of its allowed values. Important examples of quantized variables are the total energy of a closed system and its angular momentum.

A closed atomic system may pass more or less discontinuously from one quantized energy level to another by interaction with another system, in particular by interaction with the radiation field, involving emission or absorption or Raman scattering of light. When the system passes from a state of higher energy W_1 to one of lower energy W_2 the frequency of radiation emitted, ν, is such that one quantum, of energy $h\nu$, is emitted, equal to the energy difference,

$$h\nu = W_1 - W_2 = \hbar\omega \tag{1.1}$$

Likewise the same frequency can be absorbed by a system in the state of energy W_2 which is thereby raised to a state of energy W_1. This rule is known as *Bohr's frequency condition* and is basic to the interpretation of atomic and molecular spectra.

‡ Numbers in brackets refer to References at end of chapter.

The idea of a dual aspect to light goes back to Planck's derivation of the law of distribution of black-body radiation in 1900, and came more vividly into physics in 1905 when Einstein applied an equation like (1.1) to the photoelectric emission of electrons from solids. It was natural that, if light energy was regarded as behaving in some respects as little particles of energy, such particles would also be endowed with momentum. On relativistic grounds it would be expected that the number of waves per unit length in the direction of propagation, the wave-number vector, $\boldsymbol{\sigma}$, would be related to the momentum $\mathbf{p}$ in the same way as the frequency is to the energy. In other words

$$\begin{aligned} E &= h\nu \qquad \mathbf{p} = h\boldsymbol{\sigma} \\ \text{or} \qquad E &= \hbar\omega \qquad \mathbf{p} = \hbar\mathbf{k} \end{aligned} \tag{1.2}$$

together are the four equations which say that the relativistic energy-momentum 4-vector is equal to h times the wave-propagation 4-vector [2].

This momentum associated with a quantum of visible light is too small to give rise to easily observable effects. But it is much larger in the X-ray region. The *Compton effect,* discovered in 1924 by A. H. Compton, was a direct confirmation of this idea. A study of the scattering of X rays by materials of low atomic number revealed that there is a shift in frequency toward lower values exactly of the amount to be expected if the X-ray quantum made an elastic impact with the almost free electrons of the scattering material (Part 7, Chap. 8).

De Broglie's basic idea was that this same wave-particle duality also occurs in the basic mechanics of the electron and other atomic particles, whereby a wave motion of frequency ν and wave number $\boldsymbol{\sigma}$ is associated with the particle of energy E and momentum $\mathbf{p}$ by the same relations (1.2). This general suggestion was developed by Schroedinger into a general calculus of the dynamics of atomic systems.

Schroedinger formulated a wave equation for a scalar wave function, ψ, whose amplitude is that of the associated wave motion. This wave equation contains the total energy of the system as a parameter, and the allowed quantized values of the system are the values of this parameter for which the wave equation possesses physically admissible finite continuous and single-valued solutions for ψ.

Born gave the interpretation that $\psi^*\psi$, where ψ^* is the complex conjugate of ψ, is the relative probability of finding the system in unit volume of that part of configuration space. This has the implication that the detailed motion of the system is not completely, but only statistically, determinate. This interpretation has been extraordinarily fruitful in the interpretation of the mathematical formalism, although some physicists, especially Einstein, regard this as a sign of incompleteness of the present formulation rather than as a fundamental attribute of nature. Heisenberg and Bohr, however, have given basic analyses of the measurement process which lead them to maintain that the fundamental laws of quantum dynamics really must have an indeterminate character.

2. The Schroedinger Wave Equation

Let $H(p,q)$ be the Hamiltonian for a simple system in which the q's represent cartesian coordinates of individual particles. According to Schroedinger's rule the wave equation for such a system is obtained by replacing each p_r by the corresponding $\frac{\hbar}{i}\frac{\partial}{\partial q_r}$ so that a kinetic energy term of the form $(1/2m)(p_x^2 + p_y^2 + p_z^2)$ for a particle of mass m is converted into a Laplace operator

$$-\frac{\hbar^2}{2m}\left[\frac{\partial^2}{\partial x^2} + \frac{\partial^2}{\partial y^2} + \frac{\partial^2}{\partial z^2}\right]$$

Thus the Hamiltonian $H(p,q)$ becomes converted into a differential operator. The wave equation, which is basic to the nonrelativistic quantum mechanics of the system having this Hamiltonian is then

$$i\hbar\frac{\partial\psi}{\partial t} = H\left(\frac{\hbar}{i}\frac{\partial}{\partial q}, q\right)\psi \tag{1.3}$$

In particular, for a single particle of mass m moving in a field of potential energy $V(x,y,z)$ the equation is

$$i\hbar\frac{\partial\psi}{\partial t} = -\frac{\hbar^2}{2m}\left(\frac{\partial^2\psi}{\partial x^2} + \frac{\partial^2\psi}{\partial y^2} + \frac{\partial^2\psi}{\partial z^2}\right) + V(x,y,z)\psi$$

The stationary states are those whose time dependence is expressed through a factor $\exp(-iWt/\hbar)$ so that

$$\psi(x,y,z,t) = u(x,y,z)\exp(-iWt/\hbar) \tag{1.4}$$

and the equation satisfied by the spatial wave function, $u(x,y,z)$, is

$$-\frac{\hbar^2}{2m}\nabla^2 u + Vu = Wu \tag{1.5}$$

The allowed values of W are those for which this equation possesses solutions that are finite, continuous, and single-valued throughout the whole configuration space. Suppose the allowed values are $W_1, W_2, \ldots, W_n$ and the associated solutions for u are written $u_1, u_2, \ldots, u_n$. Then the quantity $|u_n|^2\,dx\,dy\,dz$ is interpreted as giving the relative probability that the particle will be found in the volume element $dx\,dy\,dz$ at (x,y,z) when the system is known to be in the state W_n. Examples of particular solvable problems are given in later sections.

Equation (1.5) can be written in the symbolic form

$$Hu = Wu \tag{1.6}$$

where H is the differential operator which represents the Hamiltonian function of the system. The form of this suggests a generalization: that other quantities than the total energy may also be represented by operators in quantum mechanics, that their observable, allowed values are also determined by an equation of the same form as (1.6), and that the associated solutions tell the probability of finding the system in a given volume element when the other dynamical quantity in question is known to have a particular one of its allowed values.

An example of this is the operator equation for the cartesian component of momentum, p, that is, conjugate to x, which is represented by the differential operator

$$\frac{\hbar}{i}\frac{\partial}{\partial x}$$

so that the analogue of (1.6) is

$$\frac{\hbar}{i}\frac{\partial u}{\partial x} = p'u \tag{1.7}$$

where p' is written for an allowed value of the momentum. This has the solution $u = C\exp(ip'x/\hbar)$ which is finite, continuous, and single-valued for all real values of p'. Hence momentum is not a quantized variable. Like the coordinate itself all values are allowed. Also, u^*u is constant, so when the momentum is known to have any precise value, all values of the coordinate x have equal probability. This is an extreme special case of the *Heisenberg uncertainty principle.*

The classical mechanical expression for the angular momentum of a particle's motion about the L_z axis is $L_z = xp_y - yp_x$. The quantum operator for L_z is assumed to be

$$L_z = \frac{\hbar}{i}\left(x\frac{\partial}{\partial y} - y\frac{\partial}{\partial x}\right) = \frac{\hbar}{i}\frac{\partial}{\partial\phi} \tag{1.8}$$

where ϕ is the usual polar angle in the x,y plane. Hence the wave equation to determine the allowed values L'_z of this component of angular momentum is of the form of (1.7) with obvious changes of notation so that u must contain a factor $\exp(iL'_z\phi/\hbar)$. In this case, however, the physical coordinate system is periodic, with period 2π in ϕ; thus to obtain a single-valued function, one must require that the coefficient of i in the exponent assume integral values, $m = 0, \pm1, \pm2 \ldots$. Hence the allowed values of any component of orbital angular momentum are

$$L_z = m\hbar \tag{1.9}$$

where m is a positive or negative integer or zero.

In the general formulation of the theory each physically significant variable is represented by a linear operator which acts on the ψ of the system. In the simplest case such a variable may be a function $f(x_1,x_2, \ldots, x_n)$ of the coordinates of the system. The operator for such a variable is taken to be the function itself; thus f acting on ψ is simply the function $f(x_1x_2 \cdots x_n)\psi(x_1x_2 \cdots x_n)$. In view of the probability interpretation of $\psi^*\psi$ it follows that

$$f_{\text{av}} = \int\psi^*f\psi\,d\tau \tag{1.10}$$

is the average value of the physical quantity f in the state represented by ψ. Here it is assumed that ψ is "normalized" so that $\int \psi^*\psi\, d\tau = 1$, the integrals being extended over the entire configuration space.

In the more general case the operator representing the variable may involve differential operators, or it may involve a permutation of coordinates in the argument of ψ. It is a basic postulate of the theory that if α represents such a more general operator, then α is a linear operator so that

$$\alpha(\psi_1 + \psi_2) = \alpha\psi_1 + \alpha\psi_2$$

and (1.10) gives the average value of α in the state represented by ψ; thus

$$\alpha_{\text{av}} = \int \psi^*(\alpha\psi)\, d\tau \tag{1.11}$$

The operators which represent real physical quantities must be such that α_{av} is a real number for all states, ψ. This restricts the operators representing real physical quantities to a class known as *Hermitian operators*. Associated with every operator α is another operator $\alpha^\dagger$ known as the *Hermitian conjugate* of α which has the property that

$$(\alpha^\dagger)_{\text{av}} = \alpha_{\text{av}} = \int \psi^*(\alpha^\dagger\psi)\, d\tau$$

for all states ψ.

More generally, if ψ_a and ψ_b are the wave functions for any two different states, then $\alpha^\dagger$ has the property that

$$\int \psi^*{}_a(\alpha^\dagger\psi_b)\, d\tau = \int (\alpha\psi_a)^*\psi_b\, d\tau \tag{1.12}$$

If α represents a real physical quantity, then $\alpha^*{}_{\text{av}} = \alpha_{\text{av}}$ for all ψ and hence the operator which represents it must satisfy the *Hermitian condition* that

$$\alpha^\dagger = \alpha \tag{1.13}$$

Likewise if α represents a quantity which is purely imaginary its operator must satisfy the anti-Hermitian condition, $\alpha^\dagger = -\alpha$.

If the particular state in question is one for which ψ_s satisfies the equation $\alpha\psi_s = \alpha_s\psi_s$, where α_s denotes one of the ordinary number allowed values of the operator α, then by (1.12) α_{av} for such a state is equal to α_s. Moreover $[(\alpha - \alpha_s)^2]_{\text{av}} = 0$ and therefore in this state α has the *precise value* α_s and not merely a statistical distribution of values whose average is α_s.

It will now be proved that ψ_r and ψ_s, the *wave functions* associated with two unequal precise values of a real operator α, are *orthogonal:*

$$\alpha\psi_r = \alpha_r\psi_r, \qquad (\alpha\psi_s)^* = \alpha_s\psi_s^*$$

Multiply the first of these by ψ_s^* and integrate, and the second by ψ_r and integrate. The left-hand sides of the two equations are equal by (1.12). Hence, subtracting,

$$(\alpha_r - \alpha_s)\int \psi_s^*\psi_r\, d\tau = 0 \tag{1.14}$$

so that the integral vanishes if $\alpha_r \neq \alpha_s$.

If α and β represent two different real physical variables, then it may happen that these operators do not commute, that is,

$$\alpha\beta - \beta\alpha = \gamma \neq 0$$

It follows that states ψ such that $\int \psi^*\gamma\psi\, d\tau \neq 0$ cannot represent states in which α and β each have precise values, α_s, β_s, for such a state would satisfy both the equations $\alpha\psi_s = \alpha_s\psi_s$ and $\beta\psi_s = \beta_s\psi_s$ leading to $\gamma_{\text{av}} = 0$, contradicting the hypothesis. Physically this means that in such states the physical quantities α and β do not simultaneously possess precise values. The existence of this situation is an essential feature of quantum dynamics that is known as the *Heisenberg indeterminacy principle.*

The commonest example is that of a cartesian coordinate x and its conjugate momentum p for which the operators obey the commutation rule,

$$px - xp = \frac{\hbar}{i} \tag{1.15}$$

This operator equation is satisfied by representing p by the operator

$$p = \frac{\hbar}{i}\frac{\partial}{\partial x} + f(x) \tag{1.16}$$

where $f(x)$ is an arbitrary function of x. As no useful extra generality is obtained by introducing $f(x)$, it is usual to work with a representation in which $f(x) \equiv 0$.

3. Matrix Representations

In the preceding section the state of the system is described by means of a wave function ψ which is a function of the configuration-space coordinates. The physical variables are then represented by linear Hermitian operators which operate on such ψ functions. This is only one of many forms of representation of the laws of quantum mechanics. For brevity all the coordinates are designated by x. Let $\psi_n(x)$ be any complete set of normal orthogonal functions, where n may actually stand for a multiple set of discrete indices.

Then an arbitrary ψ function may be expanded in terms of this set of normalized functions

$$\psi = \sum_n \psi_n(x)C_n \tag{1.17}$$

and in this representation the set of C_n may be regarded as completely specifying ψ. The C_n may be regarded as a one-column matrix representing ψ.

In such a representation any linear operator is represented by a matrix, for the result of operating with the operator α on any ψ is known if the result of operating with α on any $\psi_n(x)$ is known. Since $\alpha\psi_n(x)$ is another ψ and thus can be expanded in a series of $\psi_n(x)$,

$$\alpha\psi_n(x) = \sum_m \psi_m(x)\alpha_{mn} \tag{1.18}$$

The array of two-index coefficients α_{mn} is known as the matrix representation of the operator α based on the particular set of functions $\psi_n(x)$. In this notation the effect of α operating on a general ψ is

$$\alpha\psi = \sum_n \alpha\psi_n(x)C_n = \sum_{m,n} \psi_m(x)\alpha_{mn}C_n \tag{1.19}$$

In other words, the one-column matrix which represents $\alpha\psi$ is obtained by multiplying the matrix α_{mn} into the matrix C_n by the ordinary rules of matrix multiplication (Part 1, Chap. 2).

Similarly it is convenient to write

$$\psi^* = \sum_m C^*_m \psi^*_m(x) \tag{1.20}$$

the order of factors being chosen for purely formal reasons. The quantity $\int \psi^*(x)\psi(x)\,d\tau$ becomes, because of the normal, orthogonal property of the $\psi_n(x)$,

$$\int \psi^*\psi\,d\tau = \sum_n C^*_n C_n \tag{1.21}$$

To regard this as a matrix equation, it is formally necessary to regard $\psi^*(x)$ as represented by a one-row matrix whose elements are C^*_n while $\psi(x)$ is represented by a one-column matrix with the elements C_n.

The explicit formula for α_{mn},

$$\alpha_{mn} = \int \psi^*_m(x)\alpha\psi_n(x)\,d\tau \tag{1.22}$$

follows from (1.18) in view of the orthogonal property of the $\psi_n(x)$.

The quantity which is conjugate imaginary to $\alpha\psi$ is $(\alpha\psi)^*$ and like all ψ^*'s must be represented by a row matrix. From (1.19)

$$(\alpha\psi)^* = \sum_{m,n} \psi^*_m(x)\alpha^*_{mn}C^*_n$$

but in order to get the coefficients of $\psi^*_m(x)$ in the form of a single-row matrix, while writing C^*_n as a row matrix it is necessary to introduce the matrix $\alpha^\dagger$ whose matrix elements are obtained from those of α by interchanging rows and columns and taking conjugate complex elements; thus

$$\alpha_{mn}{}^\dagger = \alpha^*_{nm} \tag{1.23}$$

in which case $(\alpha\psi)^* = \sum_{m,n} C_n\alpha_{nm}{}^\dagger\psi^*_n(x)$ so that $(\alpha\psi)^*$ is represented by the row matrix whose elements are $\Sigma_n C^*_n\alpha_{nm}{}^\dagger$.

The explicit formula for $\alpha_{mn}{}^\dagger$, that is analogous to (1.22), is

$$\alpha_{mn}{}^\dagger = \int(\alpha\psi_m(x))^*\psi_n(x)\,d\tau \tag{1.24}$$

An operator representing a real physical quantity must be such that $\int \psi^*(x)\alpha\psi(x)\,d\tau$ is real for every state which in matrix notation requires that

$$[\alpha]_{\text{av}} = \sum_{m,n} C^*_m\alpha_{mn}C_n$$

be real for any set of C's. Taking, first, all C's zero except for a particular index m, this reduces to α_{mm}, and therefore all diagonal elements must be real. Taking, second, all C's equal to zero except for two particular indices m and n, and choosing these each equal to $1/\sqrt{2}$,

$$[\alpha]_{\text{av}} = \tfrac{1}{2}(\alpha_{mm} + \alpha_{nn}) + \tfrac{1}{2}(\alpha_{mn} + \alpha_{nm})$$

and therefore α_{nm} must be complex conjugate of α_{mn}. *Therefore α_{mn} as a matrix is equal to its own Hermitian conjugate matrix* if α represents a real variable.

Since the normal orthogonal set of functions $\psi_n(x)$ is arbitrary, it follows that if some other set is introduced, say $\phi_s(x)$, then a ψ function and an operator α will be represented by different matrices than when the matrices refer to the expansion with regard to the $\psi_n(x)$. The rules of transforming the C_n and α_{nm} when the basis is changed from the $\psi_n(x)$ to the $\phi_s(x)$ are analogous to the rules for transforming vector and tensor components with rotation of the coordinate frame (Part 1, Chap. 9).

Let the relation between the $\phi_s(x)$ and the $\psi_n(x)$ be given by the transformation matrix T_{ns}:

$$\phi_s(x) = \sum_n \psi_n(x)T_{ns} \tag{1.25}$$

Here, and in what follows, m and n are used as indices referring to the $\psi_n(x)$ basis, and s and t as indices referring to $\phi_s(x)$ basis. In actual physical problems the indices m and n will have physical significance as being equal to allowed values of one set of physical variables, while the s and t will be sets of allowed values of a different set of physical variables.

Correspondingly

$$\phi^*_t(x) = \sum_m T_{tm}{}^\dagger\psi^*_m(x)$$

where $T_{tm}{}^\dagger = T^*_{mt}$. Since ϕ's are also assumed to be normal and orthogonal

$$\int\phi^*_t(x)\phi_s(x)\,d\tau = \delta_{ts} = \sum_m T_{tm}{}^\dagger T_{ms} \tag{1.26}$$

The transformation matrix must therefore be such that *its Hermitian conjugate is its own reciprocal.* Hence the transformation matrices are of a different kind than the matrices representing real physical variables; they are called *unitary* matrices.

From (1.25) it follows that T_{ns} has the explicit expression

$$T_{ns} = \int \psi^*_n(x)\phi_s(x)\,d\tau$$

which is analogous to the relation of the transformation coefficients on a rotation of axes to the cosines of angles between old and new basis vectors. Similarly, if T^{-1} denotes the transformation reciprocal to T,

$$\psi_n(x) = \sum_s \phi_s(x)(T^{-1})_{sn}$$

so that $\qquad (T^{-1})_{sn} = \int\phi^*_s(x)\psi_n(x)\,d\tau$

and therefore $\qquad (T^{-1})_{sn} = T^*_{ns}$

which is another way of deriving the result (1.26) that the Hermitian conjugate of a transformation matrix is its own reciprocal.

On the ϕ basis, an arbitrary ψ is represented by

$$\psi = \sum_s \phi_s(x)B_s = \sum_{n,s}\psi_n(x)T_{ns}B_s$$

and therefore the law of transformation of the C's is

like that of the components of a vector:

$$C_n = \Sigma T_{ns} B_s \tag{1.27}$$

and accordingly

$$\psi^* = \sum_s B^*_s \phi^*_s(x) = \sum_{s,n} B^*_s T_{sn}{}^\dagger \psi^*_n(x)$$

$$C^*_n = \sum_s B^*_s T_{sn}{}^\dagger$$

The property (1.26) that $T^\dagger T = 1$ ensures the result that

$$\sum_n C^*_n C_n = \sum_s B^*_s B_s \qquad \text{for all } \psi\text{'s}$$

The law of transformation of the α_{mn} follows from (1.22):

$$\begin{aligned}\alpha_{st} &= \int \phi^*_s(x) \alpha \phi_t(x)\, d\tau \\ &= \sum_{mn} T_{sm}{}^\dagger \alpha_{mn} T_{nt}\end{aligned}$$

In terms of T itself, in view of (1.26), this is

$$\alpha_{st} = \sum_{m,n} (T^{-1})_{sm} \alpha_{mn} T_{nt} \tag{1.28}$$

which is the basic rule for changing the matrix of a physical variable from one representation to another.

This result affords a new insight into the mathematical meaning of the equation $\alpha\psi = a\psi$ used for determination of the allowed values and proper functions of an operator α. Suppose α is given as a matrix α_{mn} on the $\psi_n(x)$ basis. Let a particular solution of the allowed values equation be denoted by ψ_s with the proper value a_s, where $\psi_s(x) = \Sigma_n \psi_n(x) C_{ns}$ where the C_{ns} satisfy the equations $\Sigma_m \alpha_{nm} C_{ms} = a_s C_{ns}$. If the $\psi_s(x)$, determined in this way, are regarded as defining a new basis of representation, then it follows that the C_{ns} are components of the transformation matrix from the n basis to the s basis and that, in the coordinate system so defined, α_{st} will have no nondiagonal elements connecting states associated with unequal allowed values $a_s \neq a_t$. However, if $a_s = a_t$, then α_{st} need not vanish.

The notation of the preceding discussion is quite satisfactory when the general principles of the formalism are being developed. However, in actual work what is written as a single index m may stand for a large number of different quantum numbers. For this reason, instead of writing indices as subscripts, it is desirable to use a notation due to Dirac. According to this one would write

$$\psi = \Sigma \psi(n)(n|) \tag{1.29}$$

where the bracket symbol $(n|)$ replaces C_n, and for matrix components

$$(m|\alpha|n) = \int \psi(m)^* \alpha \psi(n)\, d\tau \tag{1.30}$$

The components of a transformation matrix are written $(n|s)$ as

$$\phi(s) = \sum_n \psi(n)(n|s) \tag{1.31}$$

and the equation $T\dagger T = 1$ takes the form

$$\sum_m (s|m)(m|t) = \delta(s,t)$$

while that for matrix transformation is

$$(s|\alpha|t) = \sum_{m,n} (s|m)(m|\alpha|n)(n|t) \tag{1.32}$$

This notation is preferable where many degrees of freedom are involved, in contrast to working with general theory or simple problems where the subscript notation is more convenient.

4. The Harmonic Oscillator

The quantum dynamics of the harmonic oscillator provides a particularly simple example whose results find wide applicability. The Hamiltonian function for a particle of mass μ acted on by a Hooke's law restoring force of force constant k is

$$H = \frac{p^2}{2\mu} + \frac{1}{2} kx^2 \tag{1.33}$$

so that the wave equation becomes

$$-\frac{\hbar^2}{2\mu}\frac{d^2\psi}{dx^2} + \frac{1}{2} kx^2\psi = W\psi \tag{1.34}$$

Writing $u = x/a$ and $W = \frac{1}{2}\lambda\, \hbar w\omega$, where $\omega = \sqrt{k/\mu}$, and $a = \sqrt{\hbar/\mu\omega}$, this takes the form

$$\frac{d^2\psi}{du^2} + (\lambda - u^2)\psi = 0$$

which has solutions (Part 1, Chap. 3)

$$\psi_n = a_n \exp\left(-\frac{u^2}{2}\right) H_n(u) \tag{1.35}$$

with $\lambda_n = 2n + 1$, where n is an integer. Here $H_n(u)$ is the nth Hermite polynomial. Therefore the allowed values of the energy are equally spaced with the interval between successive allowed values equal to $\hbar\omega$, where ω is the classical radian frequency of the oscillator,

$$W_n = (n + \tfrac{1}{2})\hbar\omega \tag{1.36}$$

The normalized wave functions are (for integration on du)

$$\psi_n(u) = (2^n n!\, \sqrt{\pi})^{-1/2} \exp\left(-\frac{u^2}{2}\right) H_n(u)$$

where

$$H_n(u) = (-1)^n \exp(u^2) \frac{d^n}{du^n}[\exp(-u^2)] \tag{1.37}$$

The characteristic length, a, which measures the spread of the wave function in the x axis is equal to the classical amplitude of motion in a state of energy $\frac{1}{2}\hbar w$, the lowest energy state.

In Fig. 1.1 are shown the wave functions for the states $n = 0$ to 5. The heavy horizontal line in each case indicates the region traversed by the classical harmonic oscillator of the same total energy.

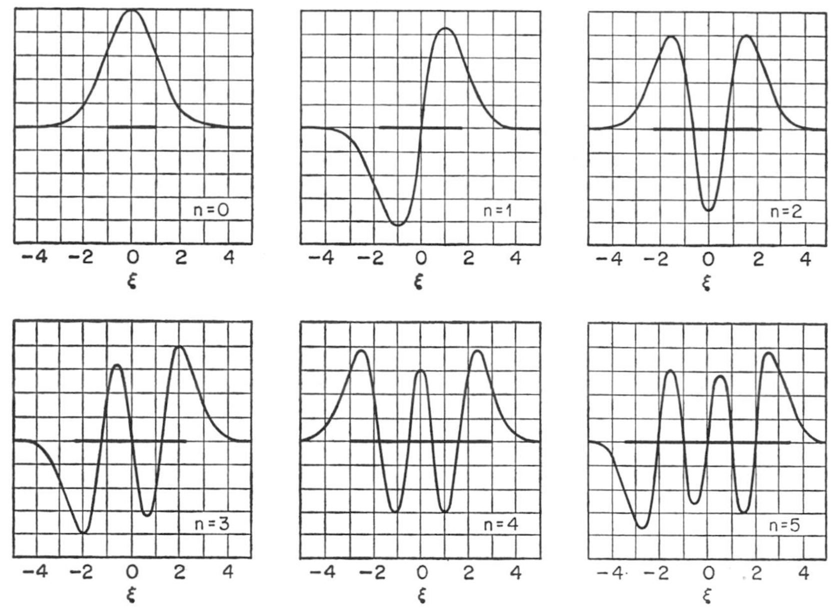

Fig. 1.1. Wave functions for the first six states of the harmonic oscillator. *(From Pauling and Wilson* [1*b*].)

The wave functions give quite appreciable probabilities of finding the particle outside the classical range of motion, where on classical ideas the system would have more potential energy than its total energy, and the amplitude of the waves is larger near the classical turning points of the motion where the classical particle moves more slowly. Moreover, a simple calculation shows that the quasi wavelength (twice the distance between nodes) is approximately given by the de Broglie formula $\lambda = h/p$, where p is the classical momentum in the region of motion in question. This correlation becomes more definite and precise for the larger values of n.

The harmonic oscillator problem can also be solved by a purely symbolic method based on the algebra of matrices.

On writing $\alpha = p/\sqrt{2\mu\hbar w}$ and $\beta = x\sqrt{k/2\hbar w}$, the commutation rule becomes $\alpha\beta - \beta\alpha = 1/2i$, and the Hamiltonian function is $H = \hbar w(\alpha^2 + \beta^2)$.

Writing $\gamma = \alpha + i\beta$ and $\gamma^* = \alpha - i\beta$, the relations

$$\gamma\gamma^* = (\alpha^2 + \beta^2 - \tfrac{1}{2})$$
and
$$\gamma^*\gamma = (\alpha^2 + \beta^2 + \tfrac{1}{2}) \tag{1.38}$$

follow from the commutation rule of α and β.

Let n be an allowed value of $\gamma\gamma^*$ and $\psi(n)$ its wave function, so that $\gamma\gamma^*(n) = n\psi(n)$. Applying γ to both sides and using the commutation rule, $\gamma\gamma^*[\gamma^*\psi(n)] = (n-1)[\gamma^*\psi(n)]$. Therefore if n is an allowed value, $n - 1$ is also an allowed value, unless $\gamma^*\psi(n)$ vanishes identically, in which case also $\gamma\gamma^*\psi(n) = 0$; therefore the least value of n is $n = 0$ and the allowed values of $\gamma\gamma^*$ are the positive integers and zero. Similarly

$$\gamma\gamma^*[\gamma\psi(n)] = (n+1)[\gamma\psi(n)]$$

and therefore $\gamma\psi(n)$ is proportional to $\psi(n+1)$.

Writing $\gamma\psi(n) = C(n)\psi(n+1)$ and the conjugate complex equation $\gamma^*\psi(n+1) = C^*(n)\psi(n)$, it follows that $|C(n)|^2 = (n+1)$; thus with a proper choice of normalizing phases

$$\begin{aligned} \gamma\psi(n) &= \sqrt{n+1}\,\psi(n+1) \\ \gamma^*\psi(n) &= \sqrt{n}\,\psi(n-1) \end{aligned} \tag{1.39}$$

which with the use of the definition of γ in terms of p and x give the matrix components of the position and momentum for the harmonic oscillator.

The operators γ and γ^* play an important role in the quantization of the electromagnetic field.

5. Angular Momentum [3]

The three components of angular momentum L_x, L_y, L_z do not commute with each other. Their commutation relations follow from their definitions:

$$L_x = yp_z - zp_y$$

and the commutation rules

$$xp_x - p_xx = i\hbar$$
and are
$$L_xL_y - L_yL_x = i\hbar L_z \tag{1.40}$$

Each of the three components commutes with the quantity

$$L^2 = L_x^2 + L_y^2 + L_z^2$$

Hence it is possible to find states in which L^2 and one component, say L_z, have precisely given values. The determination of the allowed values and the matrices can be done by a symbolic method like that used for the harmonic oscillator. Let $l(l+1)\hbar^2$ and $m\hbar$ stand for allowed values of L^2 and L_z and $\psi(l,m)$ for the associated wave function. Then

$$\begin{aligned} L^2\psi(l,m) &= l(l+1)\hbar^2\psi(l,m) \\ L_z\psi(l,m) &= m\hbar\psi(l,m) \end{aligned}$$

For convenience one may get rid of $\hbar$ by writing L_x for $L_x/\hbar$, etc. Then

$$(L_x^2 + L_y^2)\psi(l,m) = [l(l+1) - m^2]\psi(l,m)$$

Since $L_x^2 + L_y^2$ is positive, it follows that, for a given l, the quantity m is restricted to values such that $l(l+1) - m^2$ is positive. From the commutation rules

$$L_z(L_x \pm iL_y) = (L_x \pm iL_y)(L_z \pm 1)$$

Hence

$$L_z(L_x \pm iL_y)\psi(l,m) = (m \pm 1)(L_x \pm iL_y)\psi(l,m)$$

and if m is an allowed value, so also is $m \pm 1$ and the corresponding wave function is $(L_x \pm iL_y)\psi(l,m)$, unless $(L_x \pm iL_y)\psi(l,m) = 0$. Let m_1 be the largest m associated with l; then $(L_x + iL_y)\psi(l,m_1) = 0$; and operating with $(L_x - iL_y)$,

$$(L_x^2 + L_y^2 - L_z)\psi(l,m_1) = 0$$

giving $[l(l+1) - m_1(m_1+1)] = 0$ and therefore the largest value of m going with l is $m_1 = l$. In the same way, if m_2 is the least value of m going with l, it follows that $m_2 = -l$. Since the successive values of m differ by an integer, it follows that $m_1 - m_2$ is an integer; hence $2l$ is an integer and so the allowed values of l are $l = 0, \frac{1}{2}, 1, \frac{3}{2}, 2 \ldots$, which in turn gives the allowed values of L^2. The normalization of the ψ's can be so chosen that

$$(L_x \pm iL_y)\psi(l,m) = \sqrt{(l \mp m)(l \pm m + 1)}\,\psi(l,m \pm 1) \tag{1.41}$$

and therefore the nonvanishing matrix elements for L_x and L_y in this representation are

$$\begin{aligned}
(l\,m+1|L_x|l\,m) &= \tfrac{1}{2}\sqrt{(l-m)(l+m+1)} \\
(l\,m-1|L_x|l\,m) &= \tfrac{1}{2}\sqrt{(l+m)(l-m+1)} \\
(l\,m+1|L_y|l\,m) &= -\tfrac{1}{2}i\sqrt{(l-m)(l+m+1)} \\
(l\,m-1|L_y|l\,m) &= \tfrac{1}{2}i\sqrt{(l+m)(l+m+1)}
\end{aligned} \tag{1.42}$$

In the trivial case in which $l = 0$, m must also vanish and L_x, L_y, and L_z are simultaneously equal to zero.

In case $l = \frac{1}{2}$, the two possible values of m are $+\frac{1}{2}$ and $-\frac{1}{2}$ and the matrices are

$$2L_x = \begin{bmatrix} 0 & 1 \\ 1 & 0 \end{bmatrix} \quad 2L_y = \begin{bmatrix} 0 & -i \\ i & 0 \end{bmatrix} \quad 2L_z = \begin{bmatrix} 1 & 0 \\ 0 & -1 \end{bmatrix} \tag{1.43}$$

These matrices are often called the *Pauli spin matrices*, as they were first introduced by him to give a nonrelativistic description of electron spin.

In Sec. 2 it was shown that the requirement that $\psi(x,y,z)$ be a single-valued function of position requires that m assume integral values. Hence, for the *orbital angular momentum* only integral values of l and m are allowed. However, this does not preclude the use of the fractional values for representation of an internal intrinsic angular momentum of an elementary particle. These matrices of Pauli are so used in atomic physics.

In terms of spherical polar coordinates, (r,θ,ϕ) for position of a particle, the operators for components of the orbital angular momentum are (with $\hbar$ as unit)

$$\begin{aligned}
(L_x \pm iL_y) &= e^{\pm i\phi}\left(\pm\frac{\partial}{\partial\theta} + i\cot\theta\,\frac{\partial}{\partial\phi}\right) \\
L_z &= -i\frac{\partial}{\partial\phi}
\end{aligned} \tag{1.44}$$

The wave function's dependence on θ and ϕ is derivable at once from these forms: $\psi(lm)$ as a function of (θ,ϕ) is a product of a function $\Theta(l,m)$ into the function $\Phi(\phi) = e^{im\phi}/\sqrt{2\pi}$. Since, by (1.41),

$$(L_x + iL_y)\psi(l,l) = 0$$

it follows from (1.44) that

$$\left(\frac{\partial}{\partial\theta} - l\cot\theta\right)\Theta(l,l) = 0$$

The normalized $\Theta(l,l)$ is then

$$\Theta(l,l) = (-1)^l\sqrt{\frac{(2l+1)!}{2}}\,\frac{1}{2^l l!}\sin^l\theta \tag{1.45}$$

the $(-1)^l$ being so chosen to conform to the common notation for spherical harmonics.

For a function of the form $e^{im\phi}f(\theta)$,

$$(L_x \pm iL_y)^k e^{im\phi}f(\theta) = (\mp 1)^k e^{i(m+k)\phi}\sin^{k+m}\theta\,\frac{d^k}{d(\cos\theta)^k}[\sin^{\mp m}\theta f(\theta)]$$

Hence the general formula for the $\Theta(l,m)$ is

$$\Theta(l,m) = A\sqrt{\frac{2l+1}{2}\,\frac{(l-m)!}{(l+m)!}}\sin^{|m|}\theta\,\frac{d^{|m|}}{d(\cos\theta)^{|m|}}P_l(\cos\theta) \tag{1.46}$$

where A is to be replaced by $(-1)^m$ for $m > 0$ and by $+1$ for $m < 0$. Here $P_l(\cos\theta)$ is the lth Legendre polynomial (Part 1, Chap. 3).

By use of (1.44) it follows that the operator for L^2 is (with $\hbar$ as unit)

$$L^2 = -\frac{1}{\sin\theta}\frac{\partial}{\partial\theta}\left(\sin\theta\,\frac{\partial}{\partial\theta}\right) - \frac{1}{\sin^2\theta}\frac{\partial^2}{\partial\phi^2}$$

which is the angular part of the Laplace operator in spherical polar coordinates.

In addition to the position vector $\mathbf{r}$, the electron requires a two-valued, $s = \pm 1$, coordinate to describe the spin, in terms of which the z component of spin angular momentum is $s\hbar/2$. The wave function is now a function of s as well as of $\mathbf{r}$, that is, $\psi(\mathbf{r},s)$. The interpretation is that $|\psi(\mathbf{r},+1)|^2\,dv$ gives the probability of finding the electron in dv with its z component of spin equal to $\hbar/2$, and similarly for $|\psi(\mathbf{r},-1)|^2$ for the opposite spin.

The two components are written as a column matrix

$$\begin{matrix} \psi(\mathbf{r},+1) \\ \psi(\mathbf{r},-1) \end{matrix}$$

and the three components of $\mathbf{S}$, the spin vector, are represented by using (1.43) in the vector matrix form

$$\mathbf{S} = \tfrac{1}{2}\hbar \begin{bmatrix} \mathbf{k} & \mathbf{i} - i\mathbf{j} \\ \mathbf{i} + i\mathbf{j} & -\mathbf{k} \end{bmatrix}$$

A ψ which is an eigenstate of $2S_z/\hbar$ for the value ± 1 contains a factor $\delta(s, \pm 1)$. These are distinguished by the introduction of a quantum number m_s, and so the wave function for these eigenstates is a function of position times the factor $\delta(s,m_s)$.

In Sec. 6 we shall see that the ψ for an electron in a central field includes also a factor $Y(l,m_l)$ in which Y is the normalized spherical harmonic $\Theta(l,m_l)e^{im\phi}/\sqrt{2\pi}$, for which $l(l+1)\hbar^2$ and $m_l\hbar$ are the $\mathbf{L}^2$ and L_z of the orbital angular momentum, respectively.

For a state characterized by $\psi(l,m_l,m_s)$ we can find the eigenstates in which the resultant angular momentum, $\mathbf{J} = \mathbf{L} + \mathbf{S}$, has precise values of its magnitude $j(j+1)\hbar^2$ and $J_z = m_j\hbar$. A value of m_j can be realized in two ways: $m_l = m_j \mp \tfrac{1}{2}$, $m_s = \pm\tfrac{1}{2}$, except that for $m_j = l + \tfrac{1}{2}$ only the upper sign is applicable.

Writing $\phi(l,j,m_j)$ for the wave functions that are eigenstates of $\mathbf{J}^2$ and J_z, we therefore have

$$\phi(l, l + \tfrac{1}{2}, l + \tfrac{1}{2}) = \psi(l,l,\tfrac{1}{2})$$

and therefore a possible allowed value of $\mathbf{J}^2$ is $(l + \tfrac{1}{2})(l + \tfrac{3}{2})\hbar^2$, with which $2l + 2$ distinct values of m_j are possible. Altogether there are $2(2l+1)$ states in the m_l, m_s scheme, which shows that the other value of $\mathbf{J}^2$ must be $(l - \tfrac{1}{2})(l + \tfrac{1}{2})\hbar^2$, requiring the remaining $2l$ substates.

Thus the possible values of $\mathbf{J}^2$ and J_z are given by the quantum numbers

$$j = l - \tfrac{1}{2} \quad \text{with} \quad m_j = -(l - \tfrac{1}{2}) \cdots + (l - \tfrac{1}{2})$$

and

$$j = l + \tfrac{1}{2} \quad \text{with} \quad m_j = -(l + \tfrac{1}{2}) \cdots + (l + \tfrac{1}{2})$$

More generally, if the system contains two independent parts described by $\mathbf{L}$, L_{z1} and $\mathbf{L}_{z2}$, L_{z2}, the associated values of $\mathbf{J}^2$, where $\mathbf{J} = \mathbf{L}_1 + \mathbf{L}_2$, are given by $j = l_1 + l_2, l_1 + l_2 - 1, \ldots, |l_1 - l_2|$. Calculations involving vector addition of angular momenta play a large role in atomic and nuclear physics.

6. Central-force Problems

If two particles of masses μ_1 and μ_2 interact with central forces described by $V(r)$, where r is the distance between them, as in classical mechanics, it is convenient to introduce the coordinates of the center of mass, X, Y, Z, and of particle 2 relative to particle 1 as x, y, z:

$$X = \frac{\mu_1 x_1 + \mu_2 x_2}{\mu_1 + \mu_2} \cdots \qquad x = (x_2 - x_1) \cdots \tag{1.47}$$

in which case the Hamiltonian takes the form

$$H = \frac{1}{2M}(P_x^2 + \cdots + \cdots) + \frac{1}{2\mu}(p_x^2 + \cdots + \cdots) + V(r) \tag{1.48}$$

where $M = \mu_1 + \mu_2$ and $\mu = \mu_1\mu_2/(\mu_1 + \mu_2)$ is the reduced mass, while $P_x \ldots$ are the momenta conjugate to X, Y, Z, and $p_x \ldots$ are the momenta conjugate to x, y, z.

The problem is separable. The dependence of the wave function on X, Y, Z can be taken to be through a factor

$$\exp \frac{i(P_x X + \cdots + \cdots)}{\hbar}$$

corresponding to a linear translational motion of the center of mass with precisely given momentum components, $P_x \ldots$ of its center of mass and with translational kinetic energy,

$$W_0 = \frac{1}{2M}(P_x^2 + \cdots + \cdots)$$

The internal degrees of freedom for the relative coordinates give rise to the following wave equation for the factor of the complete wave function which depends on x, y, z:

$$\frac{-\hbar^2}{2\mu}\nabla\psi + V(r)\psi = W\psi$$

where W is the part of the total energy associated with the internal motion. Because the Hamiltonian operator on the left commutes with any orbital angular momentum component, as L_z and also L^2, it follows that states of motion can be characterized by constant precise values m and $l(l+1)$ for these quantities (with $\hbar$ as unit). The wave function is thus a product of a function of the radius and the functions $\Theta(l,m)\Phi(m)$ described in Sec. 5. Denoting the radial function by $R(r)/r$, the differential equation which determines it is

$$\frac{\hbar^2}{2\mu}R''(r) + \left[W - V(r) - \frac{\hbar^2}{2\mu}\frac{l(l+1)}{r^2}\right]R = 0 \tag{1.49}$$

where $R(0) = 0$ and $R(r)/r$ is finite as $r \to \infty$. This equation determines the allowed energy levels and wave functions for states of orbital angular momentum l. The fact that the equation does not depend on m shows that the allowed energy levels do not depend on the orientation of the angular-momentum vector, as expected for a central-force problem.

The simplest case is that in which the two particles do not interact; thus $V(r) = 0$, all positive values of W are allowed, and the corresponding radial wave functions is (Part 1, Chap. 3, Sec. 11)

$$R(r) = \psi_l(kr) = \sqrt{\tfrac{1}{2}\pi kr}\, J_{l+\frac{1}{2}}(kr) \tag{1.50}$$

where $k = \sqrt{2\mu W}/\hbar$, and ψ and J denote Bessel functions of half-integral order. These wave functions find application in certain scattering problems.

Another special case of great importance is that in which the particles interact according to the Coulomb law of attraction between a nucleus of charge Ze and an electron of charge $-e$, so that $V(r) = -Ze^2/r$. This gives the solutions appropriate to the *nonrelativistic hydrogen atom* and the related hydrogenlike ions. Making the substitutions

$$W = -Z^2 \frac{\mu e^4}{2\hbar^2 n^2} \qquad r = \frac{\rho n a}{2Z} \qquad a = \frac{\hbar^2}{\mu e^2} \tag{1.51}$$

the equation for $R(\rho)$ becomes

$$R''(\rho) + \left[-\frac{1}{4} + \frac{n}{\rho} - \frac{l(l+1)}{\rho^2} \right] R(\rho) = 0$$

which is the equation of the confluent hypergeometric function. Its solution is of the form

$$R = \rho^{l+1} e^{-\rho/2} f(\rho)$$

where $f(\rho)$ is a polynomial of degree $n - l - 1$, if n is an integer greater than l. Otherwise $f(\rho)$ is a transcendental function which becomes infinite like e^ρ as $\rho \to \infty$. Therefore, the allowed *negative* values of the energy are discrete, given by (1.51) with $n \geq l + 1$.

The polynomials $f(\rho)$ are known as *associated Laguerre polynomials*. The Laguerre polynomials are defined by (Part 1, Chap. 3, Sec. 11)

$$L_n(x) = e^x D^n (x^n e^{-x})$$

where D means differentiation with regard to x and the associated Laguerre polynomials are

$$L_n{}^m(x) = D^m L_n(x)$$

In terms of these

$$f(\rho) = L^{2l+1}{}_{n+l}(\rho)$$

From the definitions the explicit formula for the polynomials is

$$L^{2l+1}{}_{n+l}(\rho) = -[(n+l)!]^2 \sum_{\lambda=0}^{n-l-1} \frac{(-\rho)^\lambda}{(n-l-1-\lambda)!(2l+1+\lambda)!}$$

The radial functions $R(n,l)$, normalized in the sense

$$\int_0^\infty R^2(nl)\, dr = 1$$

are given by

$$R(nl) = \sqrt{\frac{Z(n-l+1)!}{n^2 a[(n+1)!]^3}}\, e^{-\rho/2} \rho^{l+1} L^{2l+1}{}_{n+l}(\rho) \tag{1.52}$$

Figure 1.2 shows several of the $R^2(nl)$ plotted against r/a.

For positive values of the energy there are finite and continuous solutions for $R(r)$ for all values of the energy. The $R(r)$ are expressible in terms of the hypergeometric function.

Another important special example is that of the *rigid rotator*. In its idealized form this supposes that $V(r)$ is in the form of a narrow deep well near some value, say $r = a$. In this case solutions exist such that the lowest value of W is nearly as low as the bottom of the well and $R(r)$ has values differing appreciably from zero only in a narrow range of values near $r = a$; the system behaves as if $r = a$, and the dependence of the energy levels on l is given

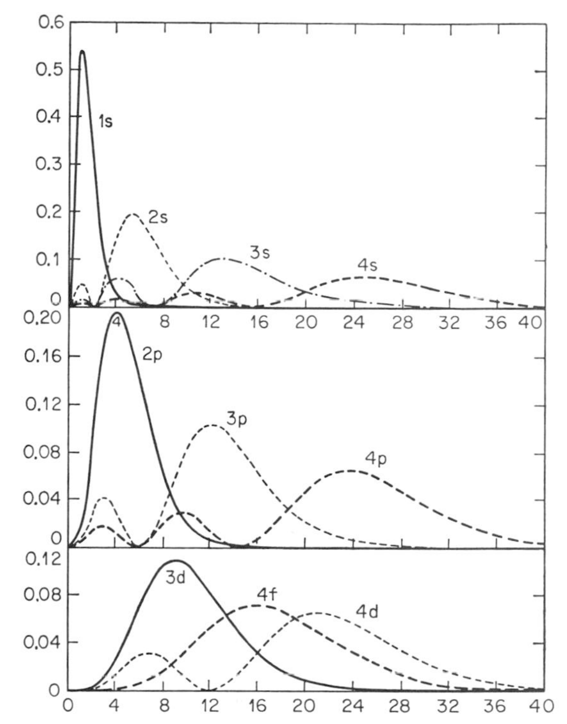

FIG. 1.2. Radial probability distribution $aR^2(nl)$ for several of the lowest levels in hydrogen. (Abscissa is the radius in atomic units.)

by an additive term

$$W = W_0 + \frac{l(l+1)}{2I} \hbar^2$$

where $I = \mu a^2$ is the moment of inertia. This model is used in approximations to the rotational behavior of diatomic molecules (Part 7, Chap. 5).

More accurate models for the vibrational-rotational motion of diatomic molecules are based on various analytic expressions for $V(r)$ which are qualitatively like Fig. 1.3, corresponding to very large repulsive forces when the two atoms are close together, an equilibrium position followed by a rapid falling off to zero of the forces of interaction as the distance of separation becomes several times that of the equilibrium position.

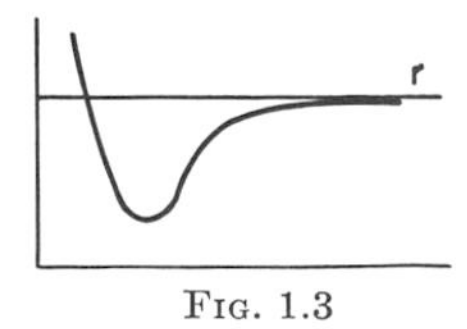

FIG. 1.3

Both with regard to molecular rotation-vibration problems and also in the theory of the deuteron, attention has been given to the inverse problem: given the empirically observed allowed energy levels, infer the interaction energy, $V(r)$. This inverse problem also arises in angular scattering problems: given the angular law of scattering, infer $V(r)$.

7. The Dynamical Equation

The equation giving the time dependence of ψ is

$$i\hbar \frac{\partial \psi}{\partial t} = H\psi \tag{1.53}$$

where H is the Hamiltonian operator for the system. If the allowed values of the energy were W_n and the associated wave functions are u_n, then a ψ which is a pure u_n has a time-dependent ψ given by

$$\psi = u_n \exp\left(-\frac{iW_n t}{\hbar}\right)$$

and likewise a general ψ is a sum of such terms

$$\psi = \sum_n u_n c_n \exp\left(-\frac{iW_n t}{\hbar}\right) \tag{1.54}$$

The time dependence of $[\alpha]_{\text{av}}$, where α is any physical quantity, is given by

$$\begin{aligned}[\alpha]_{\text{av}} &= \int \psi^* \alpha \psi \, d\tau \\ &= \sum_{m,n} c^*_m c_n (m|\alpha|n) \exp\left[\frac{i(W_m - W_n)t}{\hbar}\right] \end{aligned} \tag{1.55}$$

Hence the diagonal matrix elements give constant parts of the average and the nondiagonal matrix elements give parts which vary harmonically with the time with frequencies given by the *Bohr frequency condition* [Eq. (1.1)].

If α is a quantity which commutes with H, then $(m|\alpha|n) = 0$ for $W_m \neq W_n$ and thus $[\alpha]_{\text{av}}$ is constant for such a quantity, which is therefore appropriately called *a constant of the motion.*

It is convenient to define the time derivative of an operator as that operator whose average in any state is equal to the time derivative of the average. This can be done by defining $\dot{\alpha}$ so that each of its matrix components is equal to the time derivative of the corresponding matrix component of α. Since

$$(a|\alpha|b) = \int \psi^*_a \alpha \psi_b \, d\tau$$

$$\frac{\partial}{\partial t}(a|\alpha|b) = \int \frac{\partial \psi^*_a}{\partial t} \alpha \psi_b \, d\tau + \int \psi^*_a \alpha \frac{\partial \psi_b}{\partial t} \, d\tau$$

and therefore

$$i\hbar \frac{\partial}{\partial t}(a|\alpha|b) = \int \psi^*_a (\alpha H - H\alpha) \psi_b \, d\tau$$

so that the desired operator for $\dot{\alpha}$ is given by

$$i\hbar \dot{\alpha} = \alpha H - H\alpha \tag{1.56}$$

There is an important class of problems in which the solution for the allowed energies and corresponding wave functions is known, and in which the system is acted on by a *perturbation which depends on the time.* It is possible to consider in this way the effect on an atom of switching on and off an external magnetic field as a case in point, or the response of an atom to the various magnetic fields in a molecular beam apparatus as it goes through them.

Here the extra or perturbation term in the Hamiltonian contains the time explicitly. Even though the system is initially in a definite allowed energy level, it does not remain so, and the perturbation may be regarded as inducing or stimulating transitions from one state to another. Let H_0 denote the unperturbed Hamiltonian for which u_n and W_n are the wave functions and allowed values, respectively, and let $V(t)$ be the small time-dependent perturbation. The dynamical equation is

$$i\hbar \frac{\partial \psi}{\partial t} = (H_0 + V)\psi$$

A solution of the form $\psi = \Sigma_n u_n \exp(-iW_n t/\hbar) c_n$ is assumed, where now c_n is not constant but a function of time to be determined. Substituting in the equation of motion, and writing

$$V(t) u_m = \sum_n u_n (n|V(t)|m)$$

the equation for the time variation of the c_n's becomes

$$\begin{aligned} i\hbar \dot{c}_n &= \sum_m (n|V(t)m) \exp\left[\frac{i(W_n - W_m)t}{\hbar}\right] c_m \\ &= \sum_m [n|V(t)|m] c_m \end{aligned} \tag{1.56a}$$

using square brackets in the second line of Eq. (1.56*a*) to denote the matrix component inclusive of the time factor. These are exact equations for the $c_n(t)$. At any instant $|c_n(t)|^2$ is the probability of finding the system in the nth state. This way of regarding the matter is particularly appropriate if $V(t)$ vanishes after a time.

An approximate integration of (1.56) can be carried out if the rates of change of c_n induced by the perturbation are regarded as small. Write $c_n(t) = c_{n0} + c_{n1}(t) + c_{n2}(t) + \cdots$, and determine the successive approximations by the chain of equations

$$i\hbar \dot{c}_{n1}(t) = \sum_m [n|V|m] c_{m0}$$

$$i\hbar \dot{c}_{n2}(t) = \sum_m [n|V|m] c_{m1}$$

$$\cdots\cdots\cdots\cdots\cdots$$

where the c_{n0} are the initially given constant values. The solution of these is

$$c_{n1}(t) = -i\hbar^{-1} \sum_m c_{m0} \int_0^t [n|V|m] \, dt$$

$$c_{nk}(t) = i\hbar^{-1} \sum_m \int_0^t [n|V|m] c_{m(k-1)} \, dt$$

The probability $P_n(t)$ of the system's being in the nth state at time t is

$$\begin{aligned} P_n(t) = |c_{n0}|^2 &+ (c^*_{n1}(t) c_{n0} + c_{n1}(t) c^*_{n0}) + (c^*_{n2}(t) c_{n0} \\ &+ c^*_{n1}(t) c_{n1}(t) + c^*_{n0} c_{n2}(t)) + \cdots \end{aligned} \tag{1.57}$$

The results take a simpler form if the system is known to be in a definite state initially so that $c_{s0} = 1$ for the sth state and all other $c_{n0} = 0$. Then

$$c_{n1}(t) = \frac{i}{\hbar}\int_0^t [n|V|s]\,dt$$

and the first-order term vanishes. Likewise if n is not the same state as s, so that there really is a transition, the only nonvanishing term in the second approximation is $|c_{n1}(t)|^2$ which is

$$|c_{n1}(t)|^2 = \hbar^{-2}\left|\int_0^t [n|V|s]\,dt\right|^2 \tag{1.58}$$

This first-order transition probability is *symmetric in the initial* and final states: If initially in state s, the probability of a transition to state n is the same as the probability of transition to state s, if the system is initially in state n. The transition probability is simply related to a Fourier integral component of $(n|V|s)$. It is convenient to suppose the initial time to be $t = -\infty$ and that the perturbation extends to $t = +\infty$, supposing the integrals to be convergent. After the perturbation, writing c_{n1} for $c_{n1}(\infty)$,

$$c_{n1} = -\frac{i}{\hbar}\int_{-\infty}^{+\infty} (n|V|s)e^{2\pi i\nu t}\,dt$$

where ν is the frequency associated with the transition by the Bohr frequency condition $h\nu = W_n - W_s$. The integral has the dimensions of energy times time, or action, so that c_{n1} is a pure number, as it should be. The approximation is valid only when it produces a transition probability small compared to unity, and hence is valid only when the action in the Fourier integral component is small compared with $\hbar$.

Several examples of particular time dependence are of interest in applications: Suppose $V(t) = 0$ for $t < 0$ and that for $t > 0$

$$V(t) = V_0 e^{-\lambda t}\cos(2\pi\nu_0 t + \delta)$$

The transition probability is

$$P_n = \hbar^{-2}|(n|V_0|s)|^2\cdot\frac{1}{4}\left|\frac{e^{i\delta}}{\lambda + 2\pi i(\nu_0 - \nu)} + \frac{e^{-i\delta}}{\lambda - 2\pi i(\nu_0 - \nu)}\right|^2$$

In practice it would usually happen that δ is not known and that one would want the statistical average of P_n over all values of δ. This is

$$[P_n]_{\text{av}} = \hbar^{-2}|(n|V_0|s)|^2\cdot\frac{1}{4}\left[\frac{1}{\lambda^2 + 4\pi^2(\nu_0 - \nu)^2} + \frac{1}{\lambda^2 + 4\pi^2(\nu_0 + \nu)^2}\right] \tag{1.59}$$

In most cases the perturbation would be one of small decrement so that $\lambda \ll \nu_0$. In that case the first term, regarded as a function of ν_0, has a sharp maximum at $\nu_0 = \nu$ and has sunk to half its maximum value at $\nu_0 = \nu \pm \lambda/2\pi$. The width of such a curve is often specified by giving the *full width at half-maximum points,* which is λ/π. If $\lambda \ll \nu_0$, then the second term is negligible. This result finds application in the theory of the finite width of spectrum lines.

If the perturbation is due to undamped harmonic wave trains of length T, the transition probability averaged over all values of the phase δ is

$$P = \hbar^{-2}|(n|V_0|s)|^2\frac{\sin^2\pi(\nu_0 - \nu)T}{4\pi^2(\nu_0 - \nu)^2} \tag{1.60}$$

The average effect produced by a statistical distribution of durations of various finite wave trains, distributed so that $\lambda e^{-\lambda T}\,dT$ is the probability of occurrence of one of duration between T and $T + dT$, is given by $2(P_n)_{\text{av}}$ from (1.59) so that one cannot tell from the form of the frequency dependence whether the perturbation is a damped harmonic wave train or a statistical distribution of undamped wave trains of finite length.

Returning to the general series integration for the $c_n(t)$, suppose that the system is initially in the state s but suppose that $(n|V_0|s) = 0$ so that the first-order transition probability given by (1.58) vanishes. In such a case the next nonvanishing term in the general expression (1.57) is given by $|c_{n2}(t)|^2$.

As a special case suppose that V has an undamped harmonic time dependence of duration T, so that $V = V_0\cos(2\pi\nu_0 t + \delta)$ for $0 < t < T$. Assuming ν_0 nearly equal to the Bohr frequency ν_{ns}, the second-order probability for the transition from state s to state n is, averaged over δ,

$$P = \hbar^{-4}\left|\sum_m \frac{(n|V_0|m)(m|V_0|s)}{2\pi(\nu_{ms} - \nu_0)}\right|^2\frac{\sin^2\pi(\nu_{ns} - \nu_0)T}{4\pi^2(\nu_{ns} - \nu_0)} \tag{1.61}$$

This is of the same form as (1.60) as to line width but shows that it is possible as a second-order effect for a harmonic perturbation whose frequency is nearly that of the Bohr condition to induce transitions, $s \to n$, even though the direct matrix component of the perturbation $(n|V_0|s)$ between the initial and final states vanishes. What is necessary is that there be a class of *intermediate states, m,* such that the perturbation has a nonvanishing matrix component with n and with s. This result plays an important role in certain problems in atomic physics.

8. Perturbation Theory for Discrete States

Let H_0 be the Hamiltonian of a system, having a discrete set of allowed energy levels, whose exact solution is known. The allowed energy levels will be designated $W_0(n)$ and the associated wave functions will be designated $u_0(n)$. Here n symbolizes all the quantum numbers needed to specify a state, and it may happen that $W_0(n)$ is actually independent of several of these, in which case the system is said to be *degenerate.*

Perturbation theory provides an approximate means of finding the allowed energies and wave functions of a system whose Hamiltonian is $H_0 + V$, where V is "small" in a sense which is made more precise in what follows. The problem is that of finding an approximation to the transformation which transforms the basic representation from that in which the operator H_0 is referred to its principal axes to one in which $(H_0 + V)$ is referred to its principal axes.

The matrix elements of the perturbation operator V with regard to the states of the unperturbed system will be denoted by

$$(m|V|n) = \int u_0^*(m) V u_0(n)\, d\tau$$

It is convenient to discuss separately the effects of different classes of matrix elements of V. First consider the part of V which consists only of the diagonal matrix elements in this representation. In the equation $(H_0 + V)\psi = W\psi$ write

$$\psi = \Sigma\, u_0(n)c(n)$$

where the $c(n)$ are to be determined. The allowed-values equation becomes

$$\sum_n u_0(n)[W_0(n) + (n|V|n) - W]c(n) + \sum_{\substack{n,m \\ (m\neq n)}} u_0(n)(n|V|m)c(m) = 0$$

If the nondiagonal components of the V matrix are neglected, this becomes

$$\sum_n u_0(n)[W_0(n) + (n|V|n) - W]c(n) = 0$$

Hence $\qquad W_1(n) = W_0(n) + (n|V|n) \qquad (1.62)$

and $\qquad c(m) = \delta(n,m)$

Therefore the *diagonal elements of V* give an *additive correction to the unperturbed energy levels* and do not affect the wave functions. This result takes the diagonal elements into account rigorously. The result of equating to zero the coefficient of each $u_0(n)$ in the allowed-values equation is

$$[W_1(n) - W]c(n) + \sum_{(m\neq n)} (n|V|m)c(m) = 0 \qquad (1.63)$$

there being one such equation for each n.

In this equation the index m runs over all the states other than n. It often happens in the applications that large numbers of the matrix components vanish. If the quantum numbers n are written more explicitly as (n,r) and m are written (m,s) such that $(n,r|V|m,s) = (n,r|V|m,r)\delta(r,s)$, then the perturbed Hamiltonian is diagonal in the indices (r,s). In this case (1.63) becomes

$$[W_1(n) - W]c(n,r) + \sum_m (n,r|V|m,r)c(m,r) = 0$$

Therefore the complete set of equations breaks up into separate sets for each value of the index r. In consequence the expansion of the ψ function for a perturbed state of index r contains only $u_0(n,r)$ referring to the same index r. This is an exact result.

To take a simple special case suppose that the value $W_1(n)$ is fairly well isolated on the energy scale so that $|W_1(m) - W_1(n)|$ for all the other states is large compared to the matrix components of the perturbation. Then there will be one value of W such that $[W_1(n) - W]$ is small, and the $c(m)$ will all be small except $c(n)$ which will be nearly equal to unity. In the equation

$$[W_1(n) - W]c(m) + \sum_{p\neq m} (m|V|p)c(p) = 0 \qquad (1.64)$$

all the quantities in the sum will be small, of the second order except for $p = n$, and thus an approximate value of $c(n)$ will be obtained by neglecting all but this term,

$$c(m) = \frac{(m|V|n)}{W - W_1(m)}\, c(n) \qquad (1.65)$$

Hence, using this approximation in (1.64),

$$c(n)\left[W_1(n) + \sum_{m\neq n} \frac{(n|V|m)(m|V|n)}{W - W_1(m)} - W\right] = 0$$

and therefore, approximately,

$$W = W_1(n) + \sum_{m\neq n} \frac{|(n|V|m)|^2}{W - W_1(m)} \qquad (1.66)$$

Since, by hypothesis, the sum in (1.66) is small, W will be nearly equal to $W_1(n)$; therefore it will be a good approximation in (1.66) to write $W_1(n)$ for W in the denominators of the sum, giving as an explicit second approximation formula

$$W_2(n) = W_1(n) + \sum_{m\neq n} \frac{|(n|V|m)|^2}{[W_1(n) - W_1(m)]} \qquad (1.67)$$

Since the numerators are all essentially positive, the effect of the matrix component $(n|V|m)$ is to produce an *apparent repulsion* between these two energy levels, the lower one of the two being lowered, and the upper one of the two raised, as a result of the perturbation. For the same magnitude of $(n|V|m)$ the effect is greater, the nearer together on the energy scale are the two interacting states because of the *resonance denominator* $W_1(n) - W_1(m)$.

It remains to discuss the case of perturbation of energy levels which are not widely separated on the energy scale. Suppose that the index n be replaced by (n,r) as before and m by (m,s) but with the meaning that changes in the first index result in large changes in $W_1(n,r)$ but changes in the second index result in small changes in $W_1(n,r)$. In other words, all the quantities $[W_1(n,r) - W_1(n,s)]$ have to be regarded as small. Such a group of close-lying levels will be called a *cluster*. Just how close together the levels must be to be regarded as belonging to a cluster depends on the magnitude of the perturbation matrix components.

From the argument leading to (1.65) it is evident that if $c(m)$ refers to an unperturbed state in the same cluster as n, then $c(m)$ will no longer be small, which contradicts the hypothesis on which (1.65) was derived. Therefore it is necessary to deal with (1.64) more exactly when the $W_1(n)$ occur in clusters. In many presentations of perturbation theory a sharp distinction is made between nondegenerate and degenerate systems, i.e., according to whether $W_1(n)$ is a single state or whether there are several states whose energy is *exactly equal* to $W_1(n)$.

The essential thing is not whether the $W_1(n)$ for the states belonging to a cluster are exactly equal, but whether their differences are of the order of the nondiagonal matrix elements of the perturbation energy connecting them.

In the case of a perturbed cluster one must assume that all the $c(n,r)$ for states of the same cluster are of the order of unity in dealing with the perturbed energy for states in that cluster and regard the $c(p,t)$ for states in other clusters as small compared with unity. In this case (1.64) takes the form

$$[W_1(m,s) - W]c(m,s) + \sum_{t=s} (m,s|V|m,t)c(m,t) + \sum_{(n,r)\neq(m,s)} (m,s|V|n,r)c(n,r) = 0 \quad (1.68)$$

An approximation to the solution of this set of equations is obtained by neglecting the matrix elements in the second summation, i.e., those connecting different clusters. In this case (1.68) breaks up into separate finite systems of equations, one system for each cluster which can be solved by finite algebraic methods. The work is exactly like that involved in finding the allowed frequencies and normal modes of vibration for M coupled oscillators, where M is the number of states belonging to the mth cluster.

Writing $s = 1, 2, \ldots, M$, one has

$$\begin{aligned} &[W_1(m,1) - W]c(m,1) + (m,1|V|m,2)c(m,2) \\ &\qquad + \cdots + (m,1|V|m,M)c(m,M) = 0 \\ &(m,2|V|m,1)c(m,1) + [W_1(m,2) - W]c(m,2) \\ &\qquad + \cdots + (m,2)|V|m,M)c(m,M) = 0 \\ &(m,M|V|m,1)c(m,1) + \cdots \\ &\qquad + [W_1(m,M) - W]c(m,M) = 0 \end{aligned} \quad (1.69)$$

The allowed values of W are the roots of the secular equation obtained by setting the determinant of the coefficients of the c's equal to zero. Let $W_{11}(m,q)$ denote the qth root of the secular equation and $c(m,s|q)$ be the solution for normalized $c(m,s)$ associated with the qth root. Then the $c(m,s|q)$ are the coefficients of a transformation matrix which determines the transformation from the unperturbed states $u_0(m,s)$ to the perturbed states $u_1(m,q)$.

The states $u_1(m,q)$, determined in this way, represent an *exact* solution for the problem which takes into account all the diagonal matrix elements of V, and all the nondiagonal elements connecting states belonging to the same cluster.

Qualitatively, the levels repel each other, so that the spread of perturbed levels of a cluster on the energy scale is always greater than the spread of the unperturbed values. It may happen in this way that after the perturbation some members of the cluster are no longer well separated from members of the next cluster. One way to deal with this situation is to regard the two nearby clusters as forming a single cluster. However, it is generally not feasible to deal with secular equations of high order; therefore the practical utility of this formal method is severely limited. What usually will happen is that only a few levels of the two nearby clusters are brought close together by the first-order treatment. In this case these levels may be regarded as forming new clusters and their interaction dealt with in a higher approximation by a repetition of the secular equation method.

To get a higher approximation for the levels in the mth cluster it is convenient to transform the basis of the representation in the mth cluster from the states $u_0(m,s)$ to the states $u_1(m,q)$. The algebraic process by which this is done makes the nondiagonal elements of V within the mth cluster vanish. Therefore (1.68) takes the form

$$[W_{11}(m,q) - W]c(m,q) + \sum_{n\neq m} (m,q|V|n,r)c(n,r) = 0$$

and

$$[W_1(n,r) - W]c(n,r) + \sum_{m\neq n} (n,r|V|m,q)c(m,q) = 0$$

The perturbed state will be supposed to be such that W is near to one particular $W_{11}(m,q)$; thus for this one $[W_1(n,r) - W]$ is larger. In consequence $c(m,q)$ for that particular state will be nearly unity and the other c's will be small so that, approximately,

$$c(n,r) = \frac{(n,r|V|m,q)}{W - W_1(n,r)}\, c(m,q)$$

analogous to (1.65) and, therefore, analogous to (1.66), writing $W_{11}(m,q)$ in place of $W_1(n,r)$ one gets as an improved approximation, analogous to (1.67),

$$W_2(m,q) = W_{11}(m,q) + \sum_{\substack{n,r \\ (n\neq m)}} \frac{|(n,r|V|m,q)|^2}{W_{11}(m,q) - W_1(n,r)} \quad (1.70)$$

Alternatively, it may sometimes be convenient to incorporate the terms involving the squares of the matrix components of V into the problem before solving the secular equation. Write $W_a(n)$ for the mean value of the energies $W_1(n,r)$ of the states in cluster n. Then in

$$[W_1(n,r) - W]c(n,r) + \sum (n,r|V|m,s)c(n,s) + \sum_{m\neq n} (n,r|V|m,s)c(m,s) = 0$$

one may neglect the matrix components connecting states within the nth cluster, and assume that $W_1(n,r)$ can be replaced by $W_a(n)$ and that the only terms of importance in the sum over the other clusters is that referring to the mth cluster,

$$c(n,r) = \sum_t \frac{(n,r|V|m,t)c(m,t)}{W_a(m) - W_a(n)}$$

Using this approximation in (1.68), there results a system of linear equations for the $c(m,t)$ of the same form as (1.69) except that $(m,s|V|m,t)$ is replaced by $\{m,s|V|m,t\}$, where

$$\{m,s|V|m,t\} = (m,s|V|m,t) + \sum_{\substack{n,r \\ n \neq m}} \frac{(m,s|V|n,r)(n,r|V|m,t)}{W_1(m) - W_1(n)} \qquad (1.71)$$

This set of equations can now be solved to give an approximation in which the main effects of interaction with other clusters is taken into account.

Perturbation theory is often presented in a formal way in which great stress is placed on formal orders of approximation according to a particular process. The preceding discussion will have indicated that there is some flexibility possible in the way in which this is done and experience shows that a proper investigation of a problem requires study to show which arrangement of the procedure gives the most satisfactory result in a particular case. In formal presentations it is possible to write down expressions for third and higher approximations, but these are almost always too complicated to be of practical use.

Often the problem is given as one in which the Hamiltonian contains a parameter, say, $H_0 + \lambda V$ and it is desired to know the dependence of the allowed energies on λ. An example is that of the behavior of an atom or molecule in an external magnetic field. In this case the dependence on λ is given by writing λV for V in all the preceding work.

A particular case of importance will now be considered. Suppose $W(1)$ and $W(2)$ are two levels determined as functions of λ to an approximation which neglects $(1|V|2)$. If $|W(1) - W(2)|$ remains large for all λ of interest, then the neglect of $(1|V|2)$ is justified. The case in which $[W(1) - W(2)]$ vanishes in the interval for a particular value of λ, say $\lambda = \lambda_0$, requires special consideration, for in this vicinity the interaction matrix element $(1|V|2)$ produces large effects. Writing $u(1)$ and $u(2)$ for the first-order wave functions, then the second approximation is

$$\psi_a = u(1)a_1 + u(2)a_2$$
$$\psi_b = u(1)b_1 + u(2)b_2$$

The diagonal matrix elements of V are already supposed to be incorporated into $W_1(1)$ and $W_1(2)$ so that the equation for

$$\begin{bmatrix} a_1 \\ a_2 \end{bmatrix}$$

becomes

$$\begin{bmatrix} W_1(1) - W & V_{12} \\ V_{21} & W(2) - W \end{bmatrix} \begin{bmatrix} a_1 \\ a_2 \end{bmatrix} = 0$$

where $(1|V|2) = V_{12}$. Writing

$$\Delta = \tfrac{1}{2}[W_1(2) - W_1(1)]$$

and $z = 2\Delta/|V_{12}|$, the allowed values of W are

$$\left.\begin{matrix} W(a) \\ W(b) \end{matrix}\right\} = \frac{1}{2}[W_1(2) + W_1(1)] \pm |V_{12}| \sqrt{1 + \left(\frac{z}{2}\right)^2} \qquad (1.72)$$

Hence even at $\lambda = \lambda_0$, where $z = 0$, the two roots are unequal when V_{12} is taken into account. The

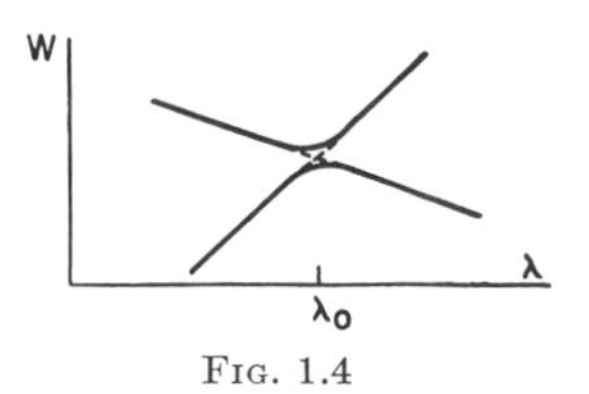

FIG. 1.4

qualitative behavior is as in Fig. 1.4. This is known as the *noncrossing rule*. Defining θ by

$$\tan\theta = [z + (z^2 + 1)^{1/2}]^{-1}$$

and δ by $V_{12} = |V_{12}|e^{i\delta}$, the wave functions become

$$\begin{bmatrix} a_1 \\ a_2 \end{bmatrix} = \begin{bmatrix} \cos\theta \\ -e^{-i\delta}\sin\theta \end{bmatrix} \qquad \begin{bmatrix} b_1 \\ b_2 \end{bmatrix} = \begin{bmatrix} e^{i\delta}\sin\theta \\ \cos\theta \end{bmatrix} \qquad (1.73)$$

As λ increases from $\lambda \ll \lambda_0$ to $\lambda \gg \lambda_0$, the auxiliary z varies from large positive through zero to large negative values, so that θ varies from 0 through $\pi/4$ to $\pi/2$. Hence

$$\begin{bmatrix} a_1 \\ a_2 \end{bmatrix} \text{ goes from } \begin{bmatrix} 1 \\ 0 \end{bmatrix} \text{ to } \frac{1}{\sqrt{2}}\begin{bmatrix} 1 \\ -e^{-i\delta} \end{bmatrix} \text{ to } \begin{bmatrix} 0 \\ -e^{-i\delta} \end{bmatrix}$$

Therefore the one state makes a *continuous* change-over from being represented by a wave function that is purely $u(1)$ to one that is purely $u(2)$. The other state

$$\begin{bmatrix} b_1 \\ b_2 \end{bmatrix}$$

shows a similar continuous change-over from pure $u(2)$ to pure $u(1)$.

The quantum number N which has the value 1 for $u(1)$ and 2 for $u(2)$ does not have an exact value for the states which are allowed energy levels in the perturbed motion. N is represented by the matrix

$$\begin{bmatrix} 1 & 0 \\ 0 & 2 \end{bmatrix}$$

and therefore for the state

$$\begin{bmatrix} a_1 \\ a_2 \end{bmatrix}$$

$$[N]_{\text{av}} = (a^*_1 \quad a^*_2) \begin{bmatrix} 1 & 0 \\ 0 & 2 \end{bmatrix} \begin{bmatrix} a_1 \\ a_2 \end{bmatrix} = 1 + \sin^2\theta$$

which makes a continuous change from 1 to 2 in this state as λ goes from $\lambda \ll \lambda_0$ to $\lambda \gg \lambda_0$. Likewise, a simple calculation with the dynamical equation shows that N is not a constant of the motion.

9. Variation Method [4]

Suppose the allowed values of an operator α are $\alpha_1 < \alpha_2 < \alpha_3 \ldots$ and the associated wave functions are $u(1)$, $u(2)$, $u(3)$ . . . , so that a general wave function is $\psi = \Sigma\, u(k)c(k)$. For such a state, assuming $\Sigma|c(k)|^2 = 1$, the average value of α is

$$[\alpha]_{\text{av}} = \sum_k \alpha_k |c(k)|^2 \qquad (1.74)$$

The minimum value of this is α_1 when $c(1) = 1$ and all other c's are zero. Hence in terms of the wave functions of the coordinates in configuration space where

$$[\alpha]_{\text{av}} = \int \psi^* \alpha \psi \, d\tau$$

the wave function $u(1)$ of the lowest state is the function ψ which makes this integral a minimum, subject to $\int \psi^* \psi \, d\tau = 1$, and the lowest allowed value is equal to this least value of the integral.

This is the basis of a method of approximate calculation of wave functions and allowed values for the lowest state known as the *Rayleigh-Ritz method* or the *Ritz method.* Where α is given as a differential operator, the standard methods of calculus of variations lead to $\alpha\psi = \alpha\psi$ as the equation to determine the ψ which minimizes $\int \psi^* \alpha \psi \, d\tau$. In the Rayleigh-Ritz method a functional form is chosen for ψ which contains several parameters, say $\lambda, \mu, \nu, \ldots$, and which from general study of the problem is believed to be a good approximation to the right form for ψ. On substituting this form for ψ and evaluating $\int \psi^* \alpha \psi \, d\tau$ one obtains a function, say $A(\lambda,\mu,\nu, \ldots)$ of the parameters. The best approximation in this family of trial functions to the unknown exact ψ is then obtained by choosing $\lambda, \mu, \nu, \ldots$ by methods of differential calculus so as to make A a minimum.

The value of A so obtained is then an approximation to the lowest value. The method has the advantage that the true value of the lowest value is always lower than the estimate, and thus an upper bound to the true value is obtained.

Going back to (1.74): If the class of functions admitted is such that $c(1) = 0$, that is, the trial function is orthogonal to the wave function of the lowest state, then the minimum value of $[\alpha]_{\text{av}}$ is α_2. Hence this provides in principle a way of using the Rayleigh-Ritz method to find the first excited state, and so on.

In practice, however, the wave function of the lowest state is usually not accurately known so that it is not possible to fulfill exactly the condition that the trial function for the first excited state be orthogonal to the wave function of the lowest state. As this condition needs to be rather accurately fulfilled, this proves in practice to be a severe limitation on the applicability of the method to excited states.

If the wave functions are partly known, then sometimes this helps to fulfill the orthogonality condition rigorously. For example, in central field problems the dependence on angle coordinates is known exactly, and states of different total angular momentum are orthogonal. Therefore, one can deal with the lowest state of energy for *each* value of angular momentum as easily as for the particular one which is the lowest energy level of all.

10. Identical Particles

An important characteristic of many atomic systems to which quantum dynamics is applied is that they consist of several identical particles, such as electrons, protons, and neutrons. Particles are said to be identical if the Hamiltonian function for the system is a symmetric function of the dynamical variables representing such particles.

Let P_{st} be an operator which interchanges the location of the variables representing the sth particle and the tth particle in the function on which it operates. Then P_{st}^2, which represents the making of such an interchange twice, gives in every case the original function. Hence $P_{st}^2 = 1$ and therefore the allowed values of P_{st} are ± 1. Since the Hamiltonian is a symmetric function of the variables labeled by the indices s and t, it follows that P_{st} commutes with the Hamiltonian $(P_{st}H - HP_{st})\psi = 0$ for all ψ and therefore P_{st} is a constant of the motion.

This argument permits a classification of all wave functions into two classes: (1) *symmetric wave functions* which are unaltered by interchange of the variables representing any two identical particles and thus correspond to the value $P_{st} = +1$ for every pair of indices s and t, and (2) *antisymmetric wave functions* which reverse sign on interchange of the variables representing any two identical particles, and thus correspond to the value $P_{st} = -1$ for every pair of indices s and t.

Since P_{st} is an absolute constant of the motion, if the particles are truly identical, there will be no transitions between states of the symmetric kind and those of the antisymmetric kind. Which class of wave functions really occurs in nature for each kind of fundamental particle remains an experimental question.

Experiment shows that the three basic kinds of fundamental particles of which matter is composed (electrons, protons, and neutrons) are alike in that the states occurring in nature are those corresponding to the *antisymmetric* wave functions. Such particles are said to obey the *Fermi-Dirac statistics.*

If the system is composed of several kinds of identical particles, then the wave function must be separately antisymmetric with regard to interchanges among each kind of identical particle. Thus the wave function for the H_2 molecule must be antisymmetric with regard to interchange of its two protons, and also antisymmetric with regard to interchange of its two electrons.

This antisymmetry property of the wave function is a characteristic feature of quantum mechanics that is without analogue in classical mechanics. It has a number of important applications in atomic, molecular, and nuclear physics.

If in a particular problem a composite particle remains stable and undissociated in the states of interest (for example, any atomic nucleus throughout the course of ordinary physical and chemical processes), then the composite particle may be treated as if it had no internal structure, *provided* that the correct symmetry property is used for the wave function of such composite particles. Thus in problems involving the physical-chemical behavior of heavy hydrogen, the deuteron may be treated as a single particle. However, the fact that it is really a proton and a neutron manifests itself in the fact that interchange of two deuterons in the wave function must produce no change in sign, for it is fundamentally an interchange of two protons and of two neutrons, each of which produces a change in sign. The wave function must therefore be a *symmetric function* of the coordinates of centers of mass of the deuterons. Identical particles which are represented by symmetric wave functions are said to obey the *Einstein-Bose statistics.*

This consideration furnishes a decisive test between the old and the new view of the constitution of atomic nuclei. Before neutrons were discovered (1932), it was supposed, for example, that the N^{14} nucleus consisted of 14 protons and 7 electrons, giving an odd total number of particles and therefore obeying the Fermi-Dirac statistics. But the new view is that it consists of 7 protons and 7 neutrons, giving an even total number of particles and therefore obeying the Einstein-Bose statistics. This latter view is supported by observations on the alternating intensity of rotational lines in the electronic band spectrum of N_2 molecules.

The requirement of over-all antisymmetry of the wave function of particles possessing spin coordinates gives rise to an apparent dependence of the energy levels of the magnitude of the resultant spin even when there are no spin-energy terms in the Hamiltonian function. This attribute of the theory plays an important role in interpreting atomic and molecular spectra and also the properties of ferromagnetic materials. For example, suppose a system contains two electrons and that the spin coordinates do not occur explicitly in the Hamiltonian. The coordinates of each electron will be denoted x_1s_1 and x_2s_2, where x represents the three continuous positional coordinates x, y, z and s the two-valued spin variable $s = \pm 1$, the physical interpretation of s being that it is the z component of spin angular momentum measured in units of $\frac{1}{2}\hbar$.

Since the spin coordinates are absent from the Hamiltonian it follows that the wave functions are of the form $\psi = U(x_1,x_2)V(s_1,s_2)$, that is, the spin variables are separable from the position variables and the energy associated with ψ is independent of the choice of the spin dependence. However, the interchange operator P_{12} acts separately on the positional and spin factors. Therefore, if the over-all wave function is to be antisymmetric, this can only be accomplished in two ways: (1) for $U(x_1,x_2)$ to be symmetric and $V(s_1,s_2)$ antisymmetric and (2) for $U(x_1,x_2)$ to be antisymmetric and $V(s_1,s_2)$ symmetric. The other two possibilities of associating symmetric U with symmetric V or antisymmetric U with antisymmetric V would give an over-all symmetric function, applicable in case the particles obey Einstein-Bose rather than Fermi-Dirac statistics.

The spin factor $V(s_1,s_2)$ can be written as the product of two factors $v_a(s_1)v_b(s_2)$ or their linear combinations. Suppose $v_a(s_1)$ corresponds to spin pointing in the $+z$ direction. Then $v_a(1) = 1$ and $v_a(-1) = 0$. Likewise if $v_b(s_1)$ corresponds to spin pointing in the $-z$ direction, $v_b(1) = 0$ and

$$v_b(-1) = 1$$

Since the energy is independent of the quantum number a, b, there will be four possibilities (a,a) $(a,b)(b,a)$ and (b,b) for the quantum numbers. Going with the first and last the spin factor of the wave function is symmetric, giving $v_a(s_1)v_a(s_s)$ and $v_b(s_1)$ $v_b(s_2)$ as the spin factors, respectively. With the other two there are a symmetric and an antisymmetric linear combination:

$$V(s_1,s_2) = 2^{-1/2}[v_a(s_1)v_b(s_2) + v_a(s_2)v_b(s_1)]$$
$$V(s_1,s_2) = 2^{-1/2}[v_a(s_1)v_b(s_2) - v_a(s_2)v_b(s_1)]$$

Thus three of the states have a symmetric spin factor, while but one has an antisymmetric spin factor. The three states associated with symmetric spin factors belong to a resultant spin of 1 for the squared sum of the vector resultant spin angular momentum $(\mathbf{S}_1 + \mathbf{S}_2)^2$ of the two electrons, while the antisymmetric spin factor corresponds to a resultant spin of 0 for the two electrons.

Thus the requirement of over-all antisymmetry has a consequence that symmetric spatial dependence occurs only with zero resultant spin and antisymmetric spatial dependence occurs only with unit resultant spin. In general the energy will be dependent on the spatial symmetry of the wave function and therefore the energy of the state will be correlated with the value of the resultant spin, even though the Hamiltonian of the system contains no energy terms depending explicitly on the spin variables.

The case of two composite particles of larger spin is handled in the same way. If the spin is s, then there will be $(2s + 1)$ possibilities so that instead of (a,b) this quantum number can assume $(2s + 1)$ values a, b, c . . . . There will then be $(2s + 1)$ symmetric wave functions of the form $v_a(s_1)v_a(s_2)$ in which both quantum numbers are the same. In addition there will be $s(2s + 1)$ symmetric wave functions and $s(2s + 1)$ antisymmetric wave functions in which different quantum numbers appear. Therefore the total number of states having symmetrical spin functions is $(s + 1)(2s + 1)$ and the number having antisymmetric spin functions is $s(2s + 1)$. The ratio is $s:(s + 1)$. The result is used in interpreting alternating intensities in rotational lines in band spectra of diatomic molecules.

11. Collision Problems [5]

The simplest kind of collision problem is that representing the relative motion of two particles whose reduced mass is μ and which interact according to the potential energy function $V(r)$, where $V(r)$ tends to zero for large r.

The wave equation takes the form

$$\nabla^2\psi + [k^2 - U(r)]\psi = 0 \tag{1.75}$$

where $k^2 = 2mW/\hbar^2$ and $U(r) = 2mV(r)/\hbar^2$, W being the total energy of relative motion. A solution has to be found which has the asymptotic form

$$\psi \to e^{ikz} + \frac{g(\theta)e^{ikr}}{r} \tag{1.76}$$

at large values of r. The first term corresponds to an incident plane wave of unit particle density and therefore to a flux of v particles per unit area per unit time, moving in the $+z$ direction, where $v = \hbar k/m$ is the classical velocity. The second term corresponds to the outgoing stream of scattered particles. It has an intensity of $|g(\theta)|^2/r^2$ so that the number crossing a sphere of radius r in the element of solid angle in unit time $d\Omega$ is $v|g(\theta)|^2\, d\Omega$.

The number scattered into the solid angle $d\Omega$ is thus equal to the number contained in an area equal to $|g(\theta)|^2\, d\Omega$ normal to the direction of motion of the incident beam. This area is known as the effective *collision cross section* for the scattering process. The total cross section, often denoted by σ, is the integral over all directions of scattering

$$\sigma = 2\pi \int_0^\pi |g(\theta)|^2 \sin\theta \, d\theta \tag{1.77}$$

The foregoing discussion applies in the simplest case in which the two particles are not identical and in which the interaction is independent of spin if there is any.

In the case of two identical particles the wave function must be either symmetric or antisymmetric with regard to interchange of the particles, i.e., substitution of $-z$ for z and substitution of $\pi - \theta$ for θ. Therefore the solution to be used instead of (1.76) must be

$$\psi \to [e^{ikz} \pm e^{-ikz}] + \frac{[g(\theta) \pm g(\pi - \theta)]e^{ikr}}{r} \tag{1.78}$$

The cross section for scattering of a particle into the solid angle element $d\Omega$ at θ is then given by

$$\{|g(\theta)|^2 + |g(\pi - \theta)|^2 \pm 2\mathrm{Re}[g(\theta)g^*(\pi - \theta)]\}\, d\Omega \tag{1.79}$$

The first two terms together correspond to the fact that either of the two particles may appear in the given solid angle. The third term corresponds to a characteristic additional part arising from interference of waves due to the symmetrization of the wave function.

In (1.79) the upper sign applies in the case of the scattering of two spinless particles obeying the Einstein-Bose statistics, as, for example, two helium atoms. The lower sign would apply for two spinless particles obeying Fermi-Dirac statistics, a case which does not occur in nature.

In case the identical particles have spin s, then since the even values of $2s$ always are associated with Einstein-Bose statistics and odd values of $2s$ are always associated with Fermi-Dirac statistics, it follows that the $(s + 1)(2s + 1)$ symmetric spin states will be associated with one sign in (1.79) and the $s(2s + 1)$ antisymmetric spin states with the other sign. The over-all result for the effective cross section for scattering into $d\Omega$ at θ is therefore

$$\left\{|g(\theta)|^2 + |g(\pi - \theta)|^2 + \frac{(-1)^{2s}}{2s + 1} 2\mathrm{Re}[g(\theta)g^*(\pi - \theta)]\right\} d\Omega \tag{1.80}$$

In these expressions for the scattering cross section it is implied, of course, that the reference system is one in which the center of mass of the two particles is at rest. The transformation between a center-of-mass system and a laboratory system in which one of the particles is initially at rest is made by the same relations as are applicable in classical theory, as discussed in Part 2, Chap. 2, Sec. 7.

The actual determination of the scattering amplitude $g(\theta)$ is often done by means of an expansion of ψ in spherical harmonics. The wave equation (1.75) will have, for positive W and real k, solutions which are of the form $F_l(r)P_l(\cos\theta)$, where $P_l(\cos\theta)$ is the lth Legendre polynomial and $F_l(r)$ is a solution of the radial wave equation which is finite at $r = 0$. This determines that $F_l(r)$ oscillates like a standing sine wave for large r so that its asymptotic form is $F_l(r) \to (kr)^{-1} \sin(kr - \frac{1}{2}l\pi + \eta_l)$, where the η_l are called *phase shifts*. The explicit occurrence of $\frac{1}{2}l\pi$ is written so that all the phase shifts η_l vanish when $V(r) \equiv 0$. The η_l have to be calculated from the radial wave equation as functions of k for each l. The requirement that ψ be of the form (1.76) then determines $g(\theta)$ in terms of the phase shifts,

$$g(\theta) = (2ik)^{-1} \sum_l (2l + 1)[\exp(2i\eta_l) - 1]P_l(\cos\theta) \tag{1.81}$$

In the case of short-range forces only the first few values of η_l will be appreciably different from zero. In the literature it is common to denote the values of l by the same code as in spectroscopy, namely, s,p,d,f for $l = 0, 1, 2, 3$.

In case only the s wave is appreciable, the series in (1.81) reduces to a single term giving

$$|g(\theta)|^2 = (2k)^{-2}2(1 + \cos 2\eta_0) \tag{1.82}$$

which is independent of θ and is therefore spherically symmetric in the center-of-mass system. An important feature of (1.82) is that large values of η_0, such that $2\eta_0$ is an odd multiple of π, can give small values of the scattering cross section. Thus in quantum mechanics the scattering may be small at certain energies even though the interaction energy is quite large (Ramsauer effect, Part 4, Chap. 10).

In the case of scattering under the Coulomb law, $V(r) = -ZZ'e^2/r$, of two particles of charges Ze and $Z'e$, the mathematical treatment is considerably more complicated since the Coulomb field extends so far that the asymptotic character of the incident beam is not that of a plane wave, e^{ikz}, but a distorted plane wave $\exp(ikz + i\alpha \log k(r - z))$, where $\alpha = ZZ'e^2/hv$, and the corresponding modification in the radial function of the scattered wave is $r^{-1}\exp(ikr - i\alpha \log r)$. The angular function $g(\theta)$ becomes

$$g(\theta) = \frac{ZZ'e^2}{2mv^2} \cdot \operatorname{cosec}^2 \frac{\theta}{2} \cdot \exp[i\alpha \log(1 - \cos\theta) + i\pi + 2i\eta_0] \tag{1.83}$$

where $\exp(2i\eta_0) = \Gamma(1 + i\alpha)/\Gamma(1 - i\alpha)$.

In spite of the extra complication in $g(\theta)$ in (1.83) the scattering cross section $|g(\theta)|^2$ for nonidentical particles becomes

$$|g(\theta)|^2 = \left(\frac{ZZ'e^2}{2mv^2}\right)^2 \operatorname{cosec}^4 \frac{1}{2}\theta \tag{1.84}$$

which is exactly the same as in the classical formula of Rutherford (Part 9, Chap. 3, Sec. 3).

In studying the law of force between protons by interpretation of experimentally measured angular distributions in proton-proton scattering, all these complications occur. The object of study is the departure of the actual energy of interaction from the exact Coulomb law so that the phase shifts are defined as the departure of the phase of the radial wave functions from what occur with Coulomb law interaction, rather than relative to the phases in no interaction. Then because the particles have spin $\frac{1}{2}$

and obey Fermi-Dirac statistics it is necessary to use a formula analogous to (1.80) to take into account the effects of over-all antisymmetry of the wave function.

12. Nuclear Atom Model

Modern ideas on atomic structure stem from the concept of the nuclear atom developed in 1912 by Sir Ernest Rutherford [6]. On this model the positive electricity in an atom is associated with most of the mass of the atom in a central particle called the *nucleus*. Surrounding the nucleus are enough negatively charged electrons to make the normal atom neutral.

In the hands of Niels Bohr [7] this model proved extraordinarily fruitful in the interpretation of atomic spectra. To accomplish this, he had to introduce ideas of quantization of the motion of the electrons around the nucleus. These ideas were later developed into quantum mechanics.

The period up to 1926 brought about a general understanding of many atomic phenomena in terms of this model. With the discovery of the principles of quantum mechanics, together with addition of important details such as the discovery of electron spin by G. E. Uhlenbeck and S. A. Goudsmit [8] in 1925, the theory rapidly assumed the form in which it is known today.

In all this development it is hardly necessary to say anything about the structure of the nucleus itself and even now this part of the subject can be handled quite separately (Part 9) from the part of physics which deals with the extranuclear structure of the atom. Quantum mechanics made possible also a rapid development of nuclear physics. In 1932 the heavy isotope of hydrogen called *deuterium* was discovered by Urey [9] and the *neutron* was discovered by Chadwick [10]. This gave the present view that the nucleus is a compound of protons and neutrons for all nuclei except the simplest one, that of the ordinary abundant isotope of hydrogen, which is just a single proton. The word *nucleon* is used as a generic term meaning proton or neutron.

It turns out that the spectroscopic and chemical properties of atoms are almost entirely determined by the *atomic number* Z, which is the integer giving the nuclear charge in electronic units (hence the number of protons in the nucleus) and is also equal to the number of extranuclear electrons in the neutral atom. This classification of atoms by the atomic number corresponds to the chemists' classification into chemical elements: there is a one-to-one correlation between the chemical elements and the integral values of Z. Most of the chemical elements that occur in nature are a mixture of several different *isotopes*, that is, several different species characterized by the same Z, but by different values of N, the numbers of neutrons in the nucleus. The total number of nucleons in the nucleus will be denoted by A, where

$$A = Z + N \tag{1.85}$$

A is called the *nucleon number* or sometimes *mass number*.

Elements are known to occur in nature for Z up to 92, uranium, those of higher Z being unstable and undergoing radioactive transformation. Small amounts of elements of Z beyond 92 have been made by artificial means. These too are unstable and have transient existence. For small values of Z, less than 20, the values of A for the stable isotopes are around $2Z$; then as Z increases, A becomes increasingly greater than $2Z$ for stable isotopes. Thus for uranium, $2Z = 184$, but the most abundant isotopes have $A = 235$ and $A = 238$.

Because of the great similarity of their chemical properties the relative abundance of the different isotopes occurring in nature is almost constant wherever they are found. Therefore the ordinary chemical symbol of the element may be interpreted as referring to one average atom of such a constant mixture of the natural isotopes. For convenience, the Z is sometimes attached as a prefixed subscript to the symbol, as ${}_{92}\mathrm{U}$, for uranium. To distinguish specific isotopes, it is customary to attach the value of A as a superscript; thus the common uranium isotopes become ${}^{235}_{92}\mathrm{U}$ and ${}^{238}_{92}\mathrm{U}$. Since U implies $Z = 92$ it is often more convenient to omit the explicit writing of the Z subscript.

In the case of hydrogen, the very large fractional difference in masses of the isotopes is so much greater that the chemical differences become more appreciable. For this reason these isotopes have been given special names and symbols: *deuterium* is the isotope for which $A = 2$ whose nucleus called the *deuteron* is therefore a binary compound of one proton and one neutron, and *tritium* [11] is the isotope of mass number 3 whose nucleus is called the *triton*. Thus ordinary hydrogen is a mixture of hydrogen, deuterium, and tritium for which the chemical symbols are H, D, and T, respectively.

13. Periodic Table of the Elements

In the first half of the nineteenth century a number of investigators had noted a more or less regular recurrence of elements with similar chemical properties in the list when the elements were arranged in order of increasing atomic weight. The concept of atomic number was not known at that time. This work culminated in that of D. Mendéleeff [12] in 1869, who proposed the most systematic classification of elements possible at that time in the form of a periodic table of the elements. This led to some correct predictions of the properties of several elements that had not been discovered at that time.

Because the "periods" are far from regular in length a large number of ways of presenting the periodic relationships of the elements in the form of a periodic table are to be found in the literature. It is now known that this periodic behavior has its origin in the following general rules: (1) although the electrons strongly interact with each other, the structure of the atom may be well described in a first approximation by assuming that the electrons move in an effective central field that is made up of their direct interaction with the nucleus (Coulomb attraction) and an effective reduction or average screening of this field by the repulsive action of all the other electrons on the one in question; (2) the possible electronic states may be classified by a quantum number, l, which is equal to the orbital angular momentum of the electron measured in quantum units, four quantum numbers (n, l, m_l, m_s) being required to specify a state completely, where n is called the total quantum number, and m_l and m_s have the interpretation of giving the projection on the z axis of the orbital and electron-spin angular momenta, respectively; (3) the filling of electron states is governed by the rule that successive electrons go into the lowest energy state possible consistent with the Pauli exclusion principle, which says that no two electrons can have the same set of four quantum numbers.

It is conventional to distinguish the various types of orbits by a code for the l values:

Code:	*s*	*p*	*d*	*f*	*g*
l:	0	1	2	3	4

The permissible values of n are the integer series, $n \geq (l + 1)$. An electronic state for which $n = 3$ and $l = 1$ is written $3p$. The permissible values of m_l are the $2l + 1$ ranging from $-l$ to $+l$. In a central field the energy does not depend on m_l so there is a $(2l + 1)$-fold degeneracy related to different possibilities of orientation of the orbital angular momentum. The permissible values of m_s are $\pm\frac{1}{2}$, corresponding to two possibilities of orientation of the spin angular momentum, giving a 2-fold spin degeneracy. Therefore in all a level specified by n and l is $2(2l + 1)$-fold degenerate.

The binding energy of the electron states depends mainly on n, and secondarily on l, being independent of l in hydrogen except for relativistic effects. In other atoms for a given n, the binding decreases with increasing l but in the lighter elements this is a smaller effect than the effect of increasing n. However, as Z increases, the dependence of the binding energy on l becomes greater. At first, then, the order of tightness of binding (magnitude of negative energy relative to the ion) goes with the quantum numbers as follows:

$$1s \quad 2s \quad 2p \quad 3s \quad 3p$$

It is customary to speak of electrons which have the same nl values as belonging to the same *shell*, and to speak of the 1*s* shell, 2*p* shell, etc.

These important properties of the energy levels of one electron in an effective central field are shown in Fig. 1.5. Here the levels (as observed spectroscopically) of hydrogen and the alkali metals are plotted against Z. At $Z = 1$, the energies depend on n alone. Moving to larger Z, the levels of s type become rapidly much more tightly bound. This corresponds to the fact that the s wave functions penetrate in deeply toward the nucleus where the effective central field is much stronger than the Coulombic, $-e^2/r$.

A similar trend is shown by the p levels, but in lesser degree, corresponding to the fact that their wave functions are less penetrating than those of the s levels. The tendency toward tighter binding with increasing Z is still smaller for the less penetrating d levels. Finally, for the $4f$ and higher nf levels, whose wave functions lie well outside the central core of closed shells, the energy values remain quite

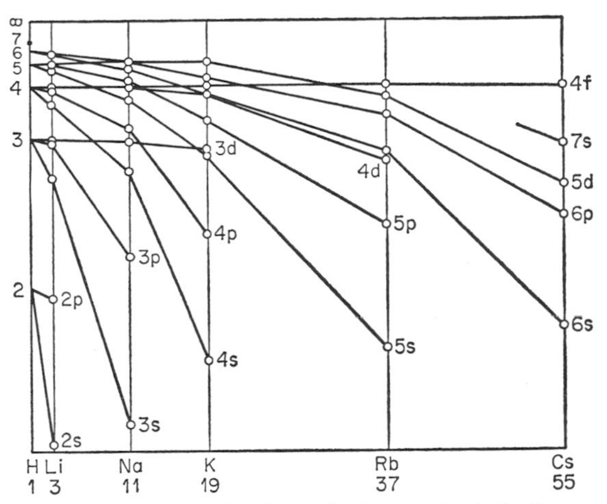

FIG. 1.5. Energy levels of one-electron states in hydrogen and the alkali metals plotted against atomic number Z. The hydrogenic values, indicated by n values on the left, are $109{,}737/n^2$ cm^{-1}.

accurately at the hydrogenic value all the way up to $Z = 55$.

Therefore the normal state of hydrogen is that in which its one electron is in the 1*s* shell, and the normal state of helium is that in which both of its electrons are in the 1*s* shell, written $(1s)^2$. A shell is said to be filled, or closed, when it has in it the maximum number of electrons allowed by the Pauli exclusion principle $2(2l + 1)$ or 2 for an s shell. Therefore helium has a closed 1*s* shell.

The normal state of the next element, lithium, is found by adding an electron in the next most tightly bound unfilled shell, in this case the 2*s* shell, giving $1s^22s$ for its *electron configuration*, which is the term used to list the nl values of the occupied orbitals. The fourth element adds another electron to the 2*s* shell, closing it, so that beryllium's normal electron configuration is $1s^22s^2$.

After that the next six elements successively add one 2*p* electron. The electron configurations therefore in addition to containing the underlying closed shells $1s^22s^2$ have the following configurations:

B	C	N	O	F	Ne
$2p$	$2p^2$	$2p^3$	$2p^4$	$2p^5$	$2p^6$

In traditional forms of the periodic table the first two elements are said to form the zeroth period, and the eight elements from Li to Ne inclusive form the first period.

The next eight elements have the same underlying closed shells as neon, and in addition fill the 3*s* shell, and follow by filling the 3*p* shell:

Na	Mg	Al	Si	P	S	Cl	Ar
$3s$	$3s^2$	$3p$	$3p^2$	$3p^3$	$3p^4$	$3p^5$	$3p^6$

Up to here the behavior is quite straightforward. As Z increases, the dependence of the binding energy on l increases. Also with increasing n the levels come closer together rapidly. Thus it comes about that after argon the next most tightly bound electron for the next element is not 3*d* but 4*s* so the 4*s* shell is filled by adding electrons for the next two elements.

These are K ($4s$) and Ca ($4s^2$). However, $3d$ electrons are more tightly bound than the $4p$ electrons, and a d shell requires 10 electrons to fill it. Therefore the next ten elements after calcium are those in which the $3d$ shell is filled. This group of elements are called the iron group or the *transition* elements. After them comes a group of six elements in which the $4p$ shell is filled, culminating in the rare gas, krypton.

Thus the eighteen elements from K (19) to Kr (36) inclusive are the ones in which the $4s$, $3d$, and $4p$ shells are filled. These elements are said to make up the first long period. A similar development occurs with the next eighteen elements. These are called the second long period and involve the successive filling of the $5s$, $4d$, and $5p$ shells, ending with Xe (54).

The third long period is extra long, because with the use of $n = 4$, a $4f$ shell becomes possible which can contain 14 electrons. Just where it comes in will naturally depend on finer details of the effective central field in which the electrons move. Actually what happens next is that the $6s$ shell is filled with Cs (55) and Ba (56). Next comes La (57) with a single $5d$ electron. At this point the $4f$ level becomes more tightly bound than the $5d$ so that the next fourteen elements are those which see the filling of this shell (lanthanide series). After the filling of the $4f$ shell the filling of the $5d$ shell continues, which is then followed in the normal way by the development of the $6p$ shell which ends at Rn (86).

The periodic table ends with the partial development of a fourth long period. It starts with the $7s$ shell with francium (87) and radium (88). Next comes a development of the filling of the $6d$ shell and the $5f$ shell.

Following this procedure, one sees that elements with similar chemical properties have electron configurations of like kind, regarding s^2p^3 as of like kind whether it be $2s^22p^3$ or $2s^22p^3$. Thus the inert gases all correspond to closed p shell configurations except helium which is in a class by itself with the very stable $1s^2$ configuration. The alkali metals all have a single s electron outside closed shells, the alkaline earths all have two s electrons outside closed shells, the halogens have p^5 configurations, that is, one less than a closed shell.

There is just one more point that needs to be mentioned in the general survey. As will be seen in Chap. 2 the complexity of the scheme of energy levels becomes greatest in the middle of a shell. Configurations like p^3, but especially d^4, d^5, d^6 and f^6, f^7, and f^8 give rise to many different energy levels because of different ways in which the electrons can interact with each other. Thus it may happen that such structure may slightly overshadow the general structure of filling the shells by successive addition of more electrons of a given type. This happens in all of the long periods but, except for the recognition of the possibility of its occurring, does not need to be taken into account in a general survey of the subject. Thus the normal state of vanadium is $4s^23d^3$ but that of the next element, chromium, is $4s3d^5$, which is followed by manganese whose normal state is $4s^23d^5$. Energy differences of this kind are usually smaller than those involved in the formation of chemical bonds; thus such effects are not important for description of the chemical properties.

Although many forms of periodic table are to be found in the chemical literature, probably the commonest is that shown in Table 1.1. Columns are called groups and rows are called periods. Three elements are lumped together in group VIII in each long period to squeeze the ten elements of a d shell into eight columns. Also to recognize that the chemical properties of elements with partly filled d shells are quite different from those with partly filled p shells, the groups, other than group VIII, are split into subgroups, usually called A and B, so that the elements of the A subgroups are those having open s or p shells, while those in the B subgroups correspond to filling d shells. In this arrangement the fourteen rare earth elements between lanthanum and hafnium are usually listed separately somewhat as a footnote to the main table.

The chemical similarity of the fourteen lanthanides is in marked contrast to the difference between the different elements involved in the filling of p or d shells. This is due to the fact that the chemical properties are mainly determined by the nature of the states of the electrons which are on the outermost part of the atom where they can more readily take part in molecular combinations, but the $4f$ electronic states tend to be buried inside the general electron structure of the atom so that the properties are not sensitive to how many of them there are.

Table 1.2 gives a form of periodic table which is coming to be preferred over the more traditional Table 1.1 in that it emphasizes more clearly the electron configurations in the normal states of the atoms, although it glosses over the interplay between s and d electrons in the filling of the d shells.

14. Atomic Units, or Hartree Units

It is often convenient to use units which assign simple numerical values to quantities of fundamental importance. As it is conventional to base a system of derived mechanical and electrical units on three primary units of mass, length, and time, and as there are more than three independent atomic quantities of fundamental importance, such systems can be constructed in many ways, and so an arbitrary choice must be made.

One selection was suggested by Hartree [13]. The system is called *atomic units*, or sometimes Hartree units. In this system the value unity is assigned to

$\hbar$, the quantum unit of angular momentum
m, the rest mass of the electron
e, the electrostatic measure of the magnitude of the proton or electron charge (1.86)

An alternative system, which assigns the value unity to $\hbar$, c, and m, where c is the speed of light, is also used in relativistic quantum mechanics.

Some authors, having adopted one system, write basic theoretical relations in a way that is applicable only to their particular system, suppressing the letters $\hbar$, e, and m or $\hbar$, c, and m because they have been assigned the value unity. The equations are then not invariant to the choice of units. It is more

TABLE 1.1. PERIODIC TABLE OF THE ELEMENTS IN TRADITIONAL FORM

Period	0	I_a	I_b	II_a	II_b	III_a	III_b	IV_a	IV_b	V_a	V_b	VI_a	VI_b	VII_a	VII_b	VIII		
0		$_{1}$H																
1	$_{2}$He	$_{3}$Li		$_{4}$Be		$_{5}$B		$_{6}$C		$_{7}$N		$_{8}$O		$_{9}$F				
2	$_{10}$Ne	$_{11}$Na		$_{12}$Mg		$_{13}$Al		$_{14}$Si		$_{15}$P		$_{16}$S		$_{17}$Cl				
3	$_{18}$Ar	$_{19}$K		$_{20}$Ca			$_{21}$Sc		$_{22}$Ti		$_{23}$V		$_{24}$Cr		$_{25}$Mn	$_{26}$Fe	$_{27}$Co	$_{28}$Ni
			$_{29}$Cu		$_{30}$Zn	$_{31}$Ga		$_{32}$Ge		$_{33}$As		$_{34}$Se		$_{35}$Br				
4	$_{36}$Kr	$_{37}$Rb		$_{38}$Sr			$_{39}$Y		$_{40}$Zr		$_{41}$Nb		$_{42}$Mo		$_{43}$Tc	$_{44}$Ru	$_{45}$Rh	$_{46}$Pd
			$_{47}$Ag		$_{48}$Cd	$_{49}$In		$_{50}$Sn		$_{57}$Sb		$_{52}$Te		$_{53}$I				
5	$_{54}$Xe	$_{55}$Cs		$_{56}$Ba			$_{57}$La	Rare earths	$_{72}$Hf		$_{73}$Ta		$_{74}$W		$_{75}$Re	$_{76}$Os	$_{77}$Ir	$_{78}$Pt
			$_{79}$Au		$_{80}$Hg	$_{81}$Tl		$_{82}$Pb		$_{83}$Bi		$_{84}$Po		$_{85}$At				
6	$_{86}$Rn	$_{87}$Fr		$_{88}$Ra			$_{89}$Ac											

4*f*, Lanthanide series	$_{58}$Ce	$_{59}$Pr	$_{60}$Nd	$_{61}$Pm	$_{62}$Sm	$_{63}$Eu	$_{64}$Gd	$_{65}$Tb	$_{66}$Dy	$_{67}$Ho	$_{68}$Er	$_{69}$Tm	$_{70}$Yb	$_{71}$Lu
5*f*, Actinide series	$_{90}$Th	$_{96}$Pa	$_{92}$U	$_{93}$Np	$_{94}$Pu	$_{95}$Am	$_{96}$Cm	$_{97}$Bk	$_{98}$Cf					

TABLE 1.2. PERIODIC TABLE OF THE ELEMENTS BY ELECTRON CONFIGURATIONS

n	s	s^2	p	p^2	p^3	p^4	p^5	p^6	d	d^2	d^3	d^4	d^5	d^6	d^7	d^8	d^9	d^{10}
1	$_{1}$H	$_{2}$He																
2	$_{3}$Li	$_{4}$Be	$_{5}$B	$_{6}$C	$_{7}$N	$_{8}$O	$_{9}$F	$_{10}$Ne										
3	$_{11}$Na	$_{12}$Mg	$_{13}$Al	$_{14}$Si	$_{15}$P	$_{16}$S	$_{17}$Cl	$_{18}$Ar										
$4s$, $3d$	$_{19}$K	$_{20}$Ca							$_{21}$Sc	$_{22}$Ti	$_{23}$V	$_{24}$Cr	$_{25}$Mn	$_{26}$Fe	$_{27}$Co	$_{28}$Ni	$_{29}$Cu	$_{30}$Zn
$4p$			$_{31}$Ga	$_{32}$Ge	$_{33}$As	$_{34}$Se	$_{35}$Br	$_{36}$Kr										
$5s$, $4d$	$_{37}$Rb	$_{38}$Sr							$_{39}$Y	$_{40}$Zr	$_{41}$Nb	$_{42}$Mo	$_{43}$Tc	$_{44}$Ru	$_{45}$Rh	$_{46}$Pd	$_{47}$Ag	$_{48}$Cd
$5p$			$_{49}$In	$_{50}$Sn	$_{57}$Sb	$_{52}$Te	$_{53}$I	$_{54}$Xe										
$6s$, $5d$, $4f$	$_{55}$Cs	$_{56}$Ba							$_{57}$La (lanthanide series below)	$_{72}$Hf	$_{73}$Ta	$_{74}$W	$_{75}$Re	$_{76}$Os	$_{77}$Ir	$_{78}$Pt	$_{79}$Au	$_{80}$Hg
$6p$			$_{81}$Tl	$_{82}$Pb	$_{83}$Bi	$_{84}$Po	$_{85}$At	$_{86}$Rn										
$7s$, $6d$, $5f$	$_{87}$Fr	$_{88}$Ra							(actinide series below)									

	df	$5d+f$	f^2	f^3	f^4	f^5	f^6	f^7	f^8	f^9	f^{10}	f^{11}	f^{12}	f^{13}	f^{14}
Lanthanides		$_{58}$Ce	$_{59}$Pr	$_{60}$Nd	$_{61}$Pm	$_{62}$Sm	$_{63}$Eu	$_{64}$Gd	$_{65}$Tb	$_{66}$Dy	$_{67}$Ho	$_{68}$Er	$_{69}$Tm	$_{70}$Yb	$_{71}$Lu
Actinides		$_{89}$Ac	$_{90}$Th	$_{91}$Pa	$_{92}$U	$_{93}$Np	$_{94}$Pu	$_{95}$Am	$_{96}$Cm	$_{97}$Bk	$_{98}$Cf				

TABLE 1.3. INDEX TO THE PERIODIC TABLES

Name		*Z*
Actinium	Ac	89
Aluminum	Al	13
Americium	Am	95
Antimony	Sb	51
Argon	Ar	18
Arsenic	As	33
Astatine	At	85
Barium	Ba	56
Berkelium	Bk	97
Beryllium	Be	4
Bismuth	Bi	83
Boron	B	5
Bromine	Br	35
Cadmium	Cd	48
Calcium	Ca	20
Californium	Cf	98
Carbon	C	6
Cerium	Ce	58
Cesium	Cs	55
Chlorine	Cl	17
Chromium	Cr	24
Cobalt	Co	27
Copper	Cu	29
Curium	Cm	96
Dysprosium	Dy	66
Einsteinium	Es	99
Erbium	Er	68
Europium	Eu	63
Fermium	Fm	100
Fluorine	F	9
Francium	Fr	87
Gadolinium	Gd	64
Gallium	Ga	31
Germanium	Ge	32
Gold	Au	79
Hafnium	Hf	72
Helium	He	2
Holmium	Ho	67
Hydrogen	H	1
Indium	In	49
Iodine	I	53
Iridium	Ir	77
Iron	Fe	26
Krypton	Kr	36
Lanthanum	La	57
Lead	Pb	82
Lithium	Li	3
Lutetium	Lu	71
Magnesium	Mg	12
Manganese	Mn	25
Mendelevium	Md	101
Mercury	Hg	80
Molybdenum	Mo	42
Neodymium	Nd	60
Neon	Ne	10
Neptunium	Np	93
Nickel	Ni	28
Niobium	Nb	41
Nitrogen	N	7
Nobelium	No	102
Osmium	Os	76
Oxygen	O	8
Palladium	Pd	46
Phosphorus	P	15
Platinum	Pt	78
Plutonium	Pu	94
Polonium	Po	84
Potassium	K	19
Praseodymium	Pr	59
Promethium	Pm	61
Protactinium	Pa	91
Radium	Ra	88
Radon	Rn	86
Rhenium	Re	75
Rhodium	Rh	45
Rubidium	Rb	37
Ruthenium	Ru	44
Samarium	Sm	62
Scandium	Sc	21
Selenium	Se	34
Silicon	Si	14
Silver	Ag	47
Sodium	Na	11
Strontium	Sr	38
Sulfur	S	16
Tantalum	Ta	73
Technetium	Tc	43
Tellurium	Te	52
Terbium	Tb	65
Thallium	Tl	81
Thorium	Th	90
Thulium	Tm	69
Tin	Sn	50
Titanium	Ti	22
Tungsten	W	74
Uranium	U	92
Vanadium	V	23
Xenon	Xe	54
Ytterbium	Yb	70
Yttrium	Y	39
Zinc	Zn	30
Zirconium	Zr	40

convenient to carry the basic constants in the equations until calculations are made by substitution of numerical values.

From the choice (1.86), units of length and time are built, and from these, together with that of mass included in (1.86), derived units of other mechanical and electromagnetic quantities are built in the usual way. The units of length and time, denoted by a and τ, respectively, are

$$a = \hbar^2/me^2 \qquad \text{and} \qquad \tau = \hbar^3/me^4 \tag{1.87}$$

The length unit a is called the *Bohr radius*. It equals the radius of the first Bohr orbit in hydrogen as given in the original 1913 Bohr theory, using nonrelativistic mechanics and neglecting m/M_p, in which M_p is the mass of the proton. This length also appears in the analogous quantum-mechanical problem (1.51).

The time unit in (1.87) has not appeared as often in the literature. It is the time required for an electron in the first Bohr orbit in hydrogen to traverse 1 rad of its path. The angular velocity of the electron in the first Bohr orbit is therefore

$$1 \text{ rad}/\tau = (2\pi)^{-1} \text{ cycle}/\tau$$

From these units we find a unit of velocity, which equals the electron speed in the first Bohr orbit,

$$a/\tau = e^2/\hbar \tag{1.88}$$

Of equally fundamental importance is c, the velocity of light in vacuum, which is also the ratio between the electrostatic measure of charge and the electromagnetic measure of the same charge. The ratio, $\alpha = e^2/\hbar c$, of these two velocities first came into prominence in Sommerfeld's theory of the relativistic fine structure of the hydrogen energy levels. He introduced the name *fine-structure constant* for the ratio.

The Hartree unit of acceleration, $a\tau^{-2}$, is the centripetal acceleration of the electron in the first Bohr orbit. Since this arises from the Coulomb interaction between the electron and the proton, it is also expressible as e^2/ma^2. Similarly, the unit of force is that needed to give this acceleration to one electron mass and so is e^2/a^2.

The Hartree unit of energy is the work done when unit force acts through unit distance. This is known at the atomic unit (au). It is equal to e^2/a and is therefore equal to the magnitude of the (negative) potential energy of interaction between proton and electron in the first Bohr orbit; it is therefore equal to twice the ionization energy of hydrogen, $e^2/2a$. Some writers adopt $e^2/2a$ as their atomic unit of energy, but we dislike this practice as introducing a confusing departure from the usual relation between units of force and energy. We have

$$1 \text{ au} = e^2/a = me^4/\hbar^2 = \hbar/\tau \tag{1.89}$$

A mixed unit of energy is called the electron volt (ev), defined as the amount of energy transferred when an electron moves through a potential difference of 1 volt. Since 1 volt is defined as $10^8/c$ statvolts (Part 4, Chap. 1, Sec. 2), with c expressed in centimeters per second, it follows that

$$1 \text{ ev} = 10^8(e/c) \qquad \text{ergs} \tag{1.90}$$

in which e/c is the charge of the electron expressed in cgs electromagnetic measure.

Another way of regarding the Hartree unit of energy is important in relation to the Bohr frequency condition (1.1). In $E = \hbar\omega$, when ω is expressed in radians τ or τ^{-1}, the energy is in the atomic unit e^2/a. Therefore $e^2/a = \hbar/\tau$.

Unit *momentum* is $me^2/\hbar = \hbar/a$, the momentum of the electron in the first Bohr orbit. The corresponding quantity of importance in relativistic electron dynamics is $mc = \alpha^{-1}me^2/\hbar$.

Unit *angular momentum* is $ma^2\tau^{-1}$, which on substitution from (1.87) reduces to $\hbar$, as it should. It is the orbital angular momentum of the electron in the first Bohr orbit.

We turn again to (1.1) in the form that gives the wave number σ of the radiation. The wave number σ is defined as the reciprocal of the wavelength λ; therefore $\sigma = \nu/c$, where ν is the frequency in cycles per unit time. A natural unit of frequency is that associated with 1 au of energy through (1.1), which is therefore $\nu_1 = (2\pi\hbar)^{-1}e^2/a$, giving the corresponding atomic units of wave number and wavelength:

$$\sigma_1 = (\alpha/2\pi)a^{-1} \qquad \text{and} \qquad \lambda_1 = (2\pi\alpha^{-1})a \tag{1.91}$$

Since the ionization energy of hydrogen is ½ au in energy, it follows that the limit of the Lyman series in this approximation (which neglects m/M_p) is $2\lambda_1$.

The wave number corresponding to $e^2/2a$, the ionization energy of hydrogen, is equal to $\sigma_1/2$ and is a quantity widely used in atomic physics. It is called the *Rydberg constant*, R or R_∞, when it is desired to emphasize that m/M_p has been neglected:

$$R = \sigma_1/2 = (\alpha/4\pi)a^{-1} \tag{1.92}$$

Here the fine-structure constant has entered to give λ_1, another length unit of basic importance, which is approximately $862a$. In theoretical work it is usually more convenient to regard $\lambdabar = \lambda/2\pi$ as basic; it is the distance over which the light wave undergoes a phase change of 1 rad. From (1.91) the natural unit is $\lambdabar_1 = \alpha^{-1}a$.

The system of length units may also be extended to units smaller than a, by means of powers of α. In the Compton effect (Part 7, Chap. 8) the quantity h/mc is the basic increase in wavelength occurring when an X-ray photon is scattered through $\pi/2$ by a free electron. Using $\lambdabar$ in place of λ, the quantity of interest is

$$\lambdabar_c = \hbar/mc = \alpha a \tag{1.93}$$

The electromagnetic radius of the electron is e^2/mc^2. This is

$$r_e = e^2/mc^2 = \alpha^2 a \tag{1.94}$$

Summarizing:

(*a*)	$\lambdabar_1$, for a quantum of energy e^2/a:	$\alpha^{-1}a$
(*b*)	Bohr radius:	a
(*c*)	$\lambdabar_c = \hbar/mc$, Compton shift for $\pi/2$:	αa
(*d*)	$r_e = e^2/mc^2$, electromagnetic radius:	$\alpha^2 a$

Similarly, powers of α provide natural extensions of energy units. The relativistic energy equivalent of the electron mass is

$$mc^2 = \alpha^{-2}e^2/a \qquad \text{or} \qquad e^2/a = \alpha^2 mc^2 \tag{1.95}$$

the latter form being preferred in the system that assigns the value unity to $\hbar$, m, and c.

Similarly the energy quantity

$$\alpha^2 e^2/a = \alpha^4 mc^2 \tag{1.96}$$

occurs as a natural unit for the relativistic fine structure and the magnetic spin-orbit interaction in atoms (Sec. 19).

Other natural energy units (Sec. 16) are based on an extension of the system to include atomic masses involving $\mu_p = M_p/m$, the ratio of the proton mass to that of the electron.

15. Atomic Electromagnetic Units

The electron charge in electrostatic measure, e, is assigned unit value in the Hartree system. It follows that the units for electron-dipole, quadrupole, . . . , 2^n-pole moments of a charge distribution are ea^n, where $n = 1, 2, \ldots, n$.

Electrostatic potential (Part 4, Chap. 1, Sec. 2) in electrostatic measure has the atomic unit e/a. Electric-field strength is measured in the unit e/a^2.

Electric surface-charge density is measured in the same unit as electric-field strength. However, as these units are not rationalized, a unit surface-charge density gives rise to a discontinuity of the normal electric-field component of $4\pi(e/a^2)$. Electric volume-charge density has the atomic unit e/a^3, and a region where the charge density is 1 in this unit gives rise to a numerical value of 4π of these units for div $\mathbf{E}$.

Electric current in electrostatic measure is the time rate at which charge crosses a chosen surface. The atomic unit therefore is e/τ. The current flowing in the first Bohr orbit is $(2\pi)^{-1}e/\tau$, since the electron passes a fixed point in its orbit once in the time interval $2\pi\tau$.

Electric current in electromagnetic measure is defined by Eq. (1.65) of Part 4, on setting $\mu_0/4\pi = 1$. This shows that unit current is the same as the square root of the force unit. The cgs unit, called the abampere, applies when the force is measured in dynes. The atomic unit, $(ma\tau^{-2})^{1/2}$, is applicable when the force is measured in the atomic unit e^2/a^2.

Electric charge in electromagnetic measure is the charge transferred in unit time by unit current. The cgs emu unit is called the abcoulomb and is equal to 10 coulombs. The atomic unit is that charge transferred during time τ by a unit current, $(ma)^{1/2}$. When this is compared with the electrostatic measure of charge, a charge Q_e atomic esu must be divided by a velocity c to get the electromagnetic measure. The factor occurring here is equal to the velocity of light (Part 4, Chap. 1, Sec. 8) and so, in atomic units, $c = \alpha^{-1}a\tau^{-1}$. Hence the value of the charge or current in atomic emu, Q_m or I_m, is related to its value in atomic electrostatic measure by

$$Q_m = \alpha Q_e \qquad I_m = \alpha I_e$$

The magnetic-dipole moment produced by current I_m flowing in a plane closed path of area A is given by I_mA (Part 4, Chap. 1, Sec. 8). The electrostatic current of the electron in the first Bohr orbit is $e/2\pi\tau$; in electromagnetic measure this is $(\alpha/2\pi)ea$. Therefore the magnetic-dipole moment due to this current is

$$\mu_B = (\alpha/2\pi)ea \cdot \pi a^2 = e\hbar/2mc \tag{1.97}$$

This magnetic moment is the natural unit associated with the orbital motion of electrons and is commonly called the *Bohr magneton*. It was long thought also to be equal to the intrinsic magnetic moment of the electron, μ_s, associated with its spin angular momentum $\hbar/2$, but this is now known [14] experimentally and theoretically to be somewhat larger,

$$\mu_s = \mu_B[1 + (\alpha/2\pi) + 1.312(\alpha/2\pi)^2] \tag{1.98}$$

Unit magnetic induction, $\mathbf{B}$, is defined so that $\mathbf{B}$ at the center of a one-turn circular coil of radius R, carrying a current I_m, is $2\pi I_m/R$. Therefore unit magnetic induction is defined as $(2\pi)^{-1}$ times the field at the center due to unit current in electromagnetic measure in a circle of unit radius. In the cgs system this is called 1 *gauss* and is $(2\pi)^{-1}$ times the field at the center of a coil of 1-cm radius carrying 1 abamp. The atomic unit is $(2\pi)^{-1}$ times the magnetic field at the nucleus due to the orbital motion of the electron in the first Bohr orbit. Since $I_m = (\alpha/2\pi)e/a$, the value is $(\alpha/2\pi)(e/a^2)$. The magnetic field at the proton due to the orbital motion of an electron in the first Bohr orbit is 2π atgauss = 124,000 gauss.

The magnetic induction $\mathbf{B}$ at a distance a from a Bohr magneton μ_B along its axis is $2\mu_B/a^3$; at the same distance in a direction orthogonal to its axis it is half as much, and so the two fields are approximately 124,000 and 62,000 gauss, respectively.

Intensity of magnetization, $\mathbf{M}$, is defined as magnetic moment per unit volume [Part 4, Eq. (1.49)] and so the atomic unit is $\mu_B/a^3 = (\alpha/2)e/a^2 = 62{,}000$ gauss.

16. Atomic Mass Units

We now consider extensions of the atomic-unit system to include M_p and M_n, the masses of the proton and neutron.

$M(Z,A)$ is written for the mass of the *neutral* atom having atomic number Z and nucleon number A. Alternatively, this may be written $M(^ASy)$, where Sy is the chemical symbol of the element of atomic number Z. By $M(Z)$ or $M(Sy)$ is to be understood the weighted mean of the atomic masses of a particular element, the weighting being according to the relative natural abundance of the stable isotopes of the element.

Formerly $M(A,Z)$ and $M(Z)$ were expressed on a scale in which the unit was chosen so that $M(8,16) = M(^{16}O) = 16$ exactly. In 1961 a decision was made by the International Union of Pure and Applied Chemistry to adopt a new scale defined by the choice that $M(6,12) = M(^{12}C) = 12$ exactly. The unit of mass on this scale is commonly called the atomic mass unit (amu), which is *defined* as $M(^{12}C)/12$.

An important pure number is μ_C, the value of the amu expressed in electronic mass units,

$$\mu_C = \frac{1}{12}\frac{M(^{12}C)}{m} \tag{1.99}$$

We write $\mu_p = M_p/m$, and $\mu_H = M_H/m$.

By atomic mass is usually meant the mass of an atom, $M(Z,A)$ or $M(Z)$, expressed in amu.

Another important ratio between mass units is commonly called the *Avogadro number* N. This is usually defined as the number of molecules in a gram-molecular weight, or gram mole, of any substance. As a gram mole is defined as being as many grams of the material as its molecular weight when expressed in amu, it follows that the Avogadro number is also definable as the mass of 1 g expressed in amu:

$$1 \text{ g} = N \quad \text{amu} \tag{1.100}$$

Thus the Avogadro number is the link between the amu mass scale and the cgs unit of mass.

The nonrelativistic energy levels of a hydrogenic atom of nuclear atomic number Z and nucleon number A are given by [Eqs. (1.47) and (1.51)]

$$W_n = -\frac{R(A)hcZ^2}{n^2}$$

in which $R(A)$ is the *Rydberg constant* for the nuclide of nucleon number A,

$$R(A) = \frac{R}{1 + m/M(A)} \tag{1.101}$$

Associated with the amu is the corresponding energy unit, $M_C c^2 = \mu_C mc^2$. The mass defect $\Delta(Z,A)$ of the (Z,A) nuclide is defined as

$$\Delta(Z,A) = [ZM(1,1) + (A - Z)M_n] - M(Z,A) \tag{1.102}$$

Therefore $\Delta(Z,A)c^2$ is the energy required to dissociate the (Z,A) nuclide into Z hydrogen atoms in their normal state and $(A - Z)$ free neutrons.

Natural units for the energy of vibration and rotation of molecules (Part 7, Chap. 5) are easily defined. A diatomic molecule has a vibration-energy quantum approximately equal to $\hbar\omega$, where $\omega = (k/M)^{1/2}$, in which k is the force constant for small displacements of the internuclear distance from that of the minimum of the potential-energy curve describing the bond between the two atoms, and M is the reduced mass of the two atoms. As the bond is a result of the dependence of the electronic energy of the molecule on the internuclear distance, the natural atomic unit for k is the atomic unit of force per atomic radius, e^2/a^3. The natural unit for M is the amu, and so for M we write $M\mu_C m$, where M is now expressed in amu; k is also assumed to be written as ke^2/a^3. The quantum unit of vibrational energy becomes

$$E_{\text{vib}} = (k/M)^{1/2}(e^2/a\sqrt{\mu_C}) \tag{1.103}$$

Thus the natural unit for molecular vibrational energy becomes the Hartree unit divided by $\sqrt{\mu_C}$.

The rotational-energy levels of a diatomic molecule are given approximately by $E_{\text{rot}} = J(J+1)\hbar^2/2I$, where $I = Mr_e^2$, M being the reduced mass and r_e the equilibrium internuclear distance. Replacing M by $M\mu_C m$ and r_e by $r_e a$, we are led to

$$E_{\text{rot}} = \frac{J(J+1)}{2Mr_e^2}\frac{e^2}{a\mu_C} \tag{1.104}$$

which shows that $e^2/a\mu_C$ is a natural atomic unit for the rotational energy of molecules.

Molecular-vibration amplitudes lead naturally to a derived unit of length. If the vibration is treated as a harmonic oscillator (Sec. 4), the basic amplitude, which we here call b, is

$$b = (h/M\omega)^{1/2} = a/\sqrt[4]{\mu_C} \tag{1.105}$$

An appropriate atomic unit of temperature (Part 5, Chap. 1, Sec. 4) is obtained by adopting various values for the Boltzmann constant k. The Celsius degree unit for temperature will continue to be used, as will the fundamental quantity of energy, kT, where k is the Boltzmann constant. Writing k for the value as usually expressed in ergs per degree we see that ka/e^2 is the factor by which T must be multiplied to give kT in atomic units. Alternatively, one may write T_a, T_v, T_r, respectively, for the temperatures at which kT is equal to e^2/a, $e^2/a\sqrt{\mu_C}$, and $e^2/a\mu_C$.

17. Numerical Values of Atomic Units

The values of the relation of atomic units to the metric system are inferred by critical analyses of the results of a wide variety of experiments. Critical studies of this kind were started by R. T. Birge in 1929 and continued by him for many years. Further studies were later made by J. W. M. DuMond and E. R. Cohen. Their most recent revision (Part 7, Chap. 10) was prepared in 1963 [15] and has been recommended for general use by the Commission on Nuclidic Masses and Related General Constants of the International Union of Pure and Applied Physics and also by the Committee on Fundamental Constants of the National Academy of Sciences–National Research Council.

Definitions

The *meter* is defined as 1,650,763.73 vacuum wavelengths of the unperturbed transition $2p_{10} - 5d_5$ in ^{86}Kr.

The *kilogram* is defined as the mass of the international kilogram at Sèvres, France.

The *second* is defined as $(31{,}556{,}925.9747)^{-1}$ of the tropical year at 12^h ET, 0 January 1900 (Part 2, Chap. 8) [17].

The degree Kelvin is defined in the thermodynamic scale (Part 5, Chap. 1) by assigning 273.16°K to the triple point of water. The freezing point 273.15°K is taken as 0°C, the zero of the Celsius scale (Part 5, Chap. 3).

The *mole* is defined as an amount of substance containing the same number of molecules as 12 g of pure ^{12}C.

Normal atmospheric pressure (atm) is defined as 1,013,250 dynes/cm².

Standard acceleration of free fall (g_n) is defined as 980.665 cm/sec² (Part 2, Chap. 7).

The thermochemical calorie (cal$_{\text{th}}$) is defined as 4.1840×10^7 ergs (Part 5, Chap. 1), and the International Steam Table calorie as 4.1968×10^7 ergs.

The liter is defined as 1,000.028 cm³.

Table 1.4 gives the adjusted values of constants, as presented to the Conference [15], with minor modi-

TABLE 1.4. ADJUSTED VALUES OF CONSTANTS (1963)

(Digits in parentheses give the standard-deviation error limit expressed in units of the last digit. The fractional error is the ratio of the standard deviation to the adjusted value expressed as parts per million.)

Name	Symbol	Value (cgs)	Fractional error
Speed of light in vacuum	c	$= 2.9917925(1) \times 10^{10}$ cm/sec	0.01
Electron charge	e	$= 4.80298(7) \times 10^{-10}$ statcoulomb	14.
	e/c	$= 1.60210(2.3) \times 10^{-20}$ abcoulomb	14.
Electron mass	m	$= 9.10910(1.3) \times 10^{-28}$ g	14.
Quantum constant	$\hbar$	$= 1.05450(2.3) \times 10^{-27}$ erg-sec	25.
	h	$= 6.6256(1.7) \times 10^{-27}$ erg-sec	25.
Avogadro number	N_A	$= 6.02252(9) \times 10^{23}$ amu/g	15.
Atomic mass unit $M(^{12}C)/12$	N_A^{-1}	$= 1.66043(2) \times 10^{-24}$ g	15.
Proton mass	M_P	$= 1.00727663(8)$ amu	0.08
		$= 1.67252(1.7) \times 10^{-24}$ g	18.
Hydrogen mass	$M(^1H)$	$= 1.00782549(4.3)$ amu	0.04
		$= 1.67343(2.7) \times 10^{-24}$ g	18.
Neutron mass	M_n	$= 1.0086654(4.3)$ amu	1.3
		$= 1.67482(27) \times 10^{-24}$ g	18.
Amu/electron-mass ratio	μ_c	$= 1822.83(3)$	17.
Fine-structure constant, $e^2/\hbar c$	α	$= 7.29720(3.3) \times 10^{-3}$	4.5
	α^{-1}	$= 137.0388(6.3)$	4.5
	$\alpha/2\pi$	$= 1.161385(5.3) \times 10^{-3}$	4.5
	α^2	$= 5.32692(4.7) \times 10^{-5}$	9.0
Bohr radius $\hbar^2/me^2$	a	$= 0.529167(2.3) \times 10^{-8}$ cm	4.4
Lyman limit $(4\pi/\alpha)(1 + m/M_p)a$	λ_L	$= 911.76(1) \times 10^{-8}$ cm	( )
	$\bar{\lambda}_L$	= ( )	( )
Compton shift h/mc	λ_c	$= 2.42621(2) \times 10^{-10}$ cm	8.3
	$\bar{\lambda}_c$	$= 3.86144(3) \times 10^{-11}$ cm	8.3
Vibrational amplitude	$a/\sqrt[4]{\mu_c}$	$= 8.09853(7) \times 10^{-10}$ cm	9.
Electron radius $r_e = \alpha^2 a = e^2/mc^2$	r_e	$= 2.81777(4) \times 10^{-13}$ cm	14.
Nuclear radius $R = r_n A^{1/3}$	r_n	$= 1.37 \times 10^{-13}$ cm	approximate relation
Atomic time $\tau = \hbar^3/me^4$	τ	$= 2.4189(3) \times 10^{-17}$ sec	125.
Atomic velocity $e^2/\hbar = \alpha c$	a/τ	$= 2.18764(1) \times 10^8$ cm/sec	5.
Atomic acceleration $me^6/\hbar^4$	a/τ^2	$= 9.0440(12) \times 10^{24}$ cm/sec²	130.
Atomic momentum $me^2/\hbar = \hbar/a$	ma/τ	$= 1.99275(5.3) \times 10^{-19}$ g-cm/sec	26.
	$mc = \alpha^{-1}\, ma/\tau$	$= 2.730840(5) \times 10^{-17}$	
Atomic force $e^2/a^2 = m^2e^6/\hbar^4$	ma/τ^2	$= 8.2383(3.7) \times 10^{-3}$ dyne	45.
Atomic force constant	e^2/a^3	$= 1.5568(1) \times 10^6$ dynes/cm	67.
Atomic energy $e^2/a = me^4/\hbar^2 = \hbar/\tau$	ma^2/τ^2	$= 4.3594(1.3) \times 10^{-11}$ erg	30.
Electron equivalent mc^2	$\alpha^{-2}ma^2/\tau^2$	$= 8.18685(2.7) \times 10^{-7}$ erg	3.3
Electron volt (ev)	$(e/c) \times 10^8$	$= 1.60210(2.3)$ ergs	18.
Amu equivalent	$N_A^{-1}C^2$	$= 1.49232(2.3) \times 10^{-3}$ erg	160.
Spin orbit	$\alpha^2e^2/2a$	$= 1.16067(4.5) \times 10^{-15}$ erg	39.
Vibrational	$(e^2/2a)/\sqrt{\mu_c}$	$= 5.1053(1) \times 10^{-13}$ erg	20.
Rotational	$(e^2/2a)/\mu_c$	$= 1.19578(3) \times 10^{-14}$ erg	25.
Nuclear spin orbit	$(e^2/2a)\alpha^2(m/M_P)$	$= 6.3214(3.7) \times 10^{-19}$ erg	60.
Wave number E/hc		kayser (1 cycle/cm)	
Atomic unit 2 Rydberg	e^2/ahc	$= (\alpha/2\pi)a^{-1}$	
		$= 219{,}474.62(2)$ cm⁻¹	0.1
Rydberg (R)	$(\alpha/4\pi)a^{-1}$	$= 109{,}737.31(1)$ cm⁻¹	0.1
	$R(^1H)$	$= 109{,}677.58(1)$	0.1
	$R(^2H)$	$= 109{,}707.46(1)$	0.1
	$R(^3H)$	$= 109{,}717.41(1)$	0.1
	$R(^4He)$	$= 109{,}722.39(1)$	0.1
	$R(^7Li)$	$= 109{,}728.78(1)$	0.1
	$R(^{23}Na)$	$= 109{,}734.71(1)$	0.1

TABLE 1.4. ADJUSTED VALUES OF CONSTANTS (1963) (*Continued*)

Name	Symbol	Value (cgs)	Fractional error
$2p_{10} - 5d_5$ in ^{86}Kr (definition of meter)		16,507.6373(0) cm^{-1}	0.00
Spin orbit	$\frac{1}{2}\alpha^2 R$	= 2.67171(0.5) cm^{-1}	2.
Vibrational	$R/\sqrt{\mu_e}$	= 2,570.3(1) cm^{-1}	40.
Rotational	R/μ_e	= 60.20(1) cm^{-1}	170.
Nuclear spin orbit	$\alpha^2 R(m/M_P)$	= 3.1825(1) $\times 10^{-3}$ cm^{-1}	31.
Electric dipole	ea	= 2.54158(14) $\times 10^{-18}$ statcoulomb-cm	
Quadrupole	ea^2	= 1.34492(9) $\times 10^{-26}$ statcoulomb-cm^2	67.
Potential	e/a	= 9.0756(2.7) $\times 10^{-2}$ statvolt	30.
	$(e/a)(c/108)$	= ( ) volt	( )
Electric field	e/a^2	= 1.71524(4) $\times 10^7$ statvolts/cm	23.
	$(e/a^2)c/108$	= ( ) volts/cm	( )
Current in first Bohr orbit	$e/2\pi\tau$	= ( ) statamp	( )
	$(e/2\pi\tau)(c/10)$	= ( ) amp	( )
Bohr magneton $e\hbar/2mc = \alpha ea/2$	μ_B	= 9.2732(2) $\times 10^{-21}$ erg/gauss	22.
Nuclear magneton $e\hbar/2M_Pc$	μ_N	= 5.0505(1.3) $\times 10^{-24}$ erg/gauss	26.
Proton moment	μ_P	= 2.79268(2.3)μ_N	8.
		= 1.41049(4.3) $\times 10^{-23}$ erg/gauss	30.
Neutron moment	μ_n	= ( )	( )
Magnetic induction	μ_B/a^3	= 62,582(4.3). gauss	70.
(At nucleus due to electron in 1_1 orbit)	$2\mu_B/a^3$	= 125,164(8.6). gauss	70.
Zeeman splitting factor	μ_B/hc	= 4.66858(1) $\times 10^{-5}$ kayser/gauss	2.2
Proton gyromagnetic ratio	γ	= 2.67519(1) $\times 10^4$ $\frac{\text{rad/sec}}{\text{gauss}}$	3.8

fications of notation and abbreviations; it is arranged in an order more suitable for discussion of the Hartree system.

18. Atomic Energy Levels [16]

The allowed energy levels are given by the proper values of the Hamiltonian, as in (1.6). A good approximation to the Hamiltonian uses $p_i^2/2m$ for the kinetic energy of each electron and includes only the Coulomb interaction of the electrons with the nucleus and with each other:

$$H = \sum_i \left(-\left(\frac{\hbar^2}{2m}\right) \Delta_i - \frac{Ze^2}{r_i} \right) + \sum_{i>j} \frac{e^2}{r_{ij}} \tag{1.106}$$

A more accurate Hamiltonian includes relativistic corrections to the kinetic energy, magnetic interactions between the electrons' orbital motion, interaction of the electrons with the magnetic moment of the nucleus and with its electric-quadrapole moment, and terms allowing for the finite size and mass of the nucleus. Usually these effects are small compared with those included in (1.106).

Of those neglected, the magnetic spin-orbit interaction (Sec. 19) is next in magnitude. When this is small compared with the electrostatic terms there is *Russell-Saunders coupling;* when not, there is *intermediate coupling.*

The whole terminology for naming energy levels is based on the *central-field approximation.* In this, each electron is regarded as moving in a central field

$$V(r) = -Z(r)e^2/r \tag{1.107}$$

in which $Z(r)$ varies from Z close to the nucleus to

$$Z_0 = Z - N + 1 \tag{1.108}$$

at distances well outside the charge cloud of the electrons. In labeling ions it is customary to use the chemical symbol for Z followed by a roman numeral for Z_0. Thus Fe IX refers to an iron ion $Z = 26$ which has lost eight electrons, and so $N = 18$. A sequence of ions of constant N is called *isoelectronic.*

The Hamiltonian can then be separated into

$$H = H_0 + H_1$$

with

$$H_0 = \sum_i \left[-\frac{\hbar^2}{2m} \Delta_i + V(r_i) \right]$$
$$H_1 = \sum_i - \left[V(r_i) + \frac{Ze^2}{r_i} \right] + \sum_{i>j} \frac{e^2}{r_{ij}} \tag{1.109}$$

To characterize states of H_0 it is necessary to have N individual sets of four quantum numbers $(n_a l_a m_{la} m_{sa})$. The set of N such individual sets $a_1 a_2 \cdots a_N$ is called the complete set A.

As in Secs. 5 and 6, the one-electron wave functions are of the form

$$u = [R(nl)/r]Y(l_1, m_l)\delta(s, m_s) \tag{1.110}$$

where $R(nl)$ is an appropriate solution of (1.49). The corresponding one-electron energies are $E_0(nl)$, and so the zeroth approximation to the energy is

$$E_0(A) = \sum_a E_0(n_a l_a) \tag{1.111}$$

which does not depend on the m_l's and m_s's in the set A.

All states having the same (nl)'s in the N individual states are said to belong to the same *configuration.* They are therefore degenerate in H_0.

To calculate the first-order energies it is necessary to calculate the matrix elements $(A|H|B)$, where A and B belong to the same configuration. In some cases $E_0(A)$ for two different configurations may lie fairly close together; then it is necessary to treat the two configurations together as a cluster in the sense of Sec. 8. In that case the wave functions obtained by solving the equation $H\psi = W\psi$ for the two configurations involve A's from both configurations, an effect that is called *configuration interaction* or configuration *mixing*.

A configuration is called *even* or *odd* according to the parity of the sum of the l values of the occupied orbitals. Closed shells always have an even number of occupied orbitals, and so the parity of the configuration is determined by that of the occupied orbitals of the open shells. It follows from the invariance of the Hamiltonian to inversion, that is, the substitution of $-\mathbf{r}$ for $\mathbf{r}$ and $-\mathbf{p}$ for $\mathbf{p}$ for each electron, that the matrix components of H which connect states of opposite parity vanish. For this reason, characterization of a level by its parity remains exact even where configuration interaction is strong.

For each closed shell in A there is only one possible set of (m_l,m_s) values (Sec. 13). For an open (partially filled shell) which could contain $2(2l+1)$ electrons but which actually contains k electrons, the number of distinct ways of assigning (m_l,m_s) values is $\binom{2(2l+1)}{k}$, the binomial coefficient. The total order of degeneracy of a configuration is the product of such factors for each open shell in it. This can become quite high for partially filled d and f shells, which accounts for the great complexity of the level schemes of the transition metals and of the lanthanides.

The Hamiltonian (1.106) does not contain the spin and therefore commutes with

$$\mathbf{S} = \Sigma \mathbf{S}_i$$

the vector sum of the electron spins. Hence the levels can be labeled by exact values of S, where $S(S+1)\hbar^2$ is an allowed value of $\mathbf{S}^2$. They are called singlets, doublets, triplets, etc., according to the values of $2S+1$. The states can also be labeled by $M_s = \sum_a m_{sa}$, where $S_z = M_s\hbar$ since the operator S_z also commutes with H.

Similarly the Hamiltonian commutes with

$$\mathbf{L} = \Sigma \mathbf{L}_i$$

the vector sum of the orbital angular momenta and so the states can be labeled by L and M_L, where $L(L+1)\hbar^2$ is an allowed value of $\mathbf{L}^2$ and $M_L = \Sigma m_l$, an allowed value of $L_z/\hbar$.

The L values are usually given in the S, P, D, F code with the multiplicity $(2S+1)$ written as a left superscript; thus ^3F means $S = 2$ and $L = 3$. A set of such $(2S+1)(2L+1)$ states all have the same energy (in this approximation) and are said to make up a *term*.

When a weak spin-orbit interaction (Sec. 19) is considered, it is found that $\mathbf{L}$ and $\mathbf{S}$ no longer separately commute with this part, but $\mathbf{J} = \mathbf{L} + \mathbf{S}$ does. In consequence the $(2S+1)(2L+1)$ states split into separate *levels*, characterized by J, where $\mathbf{J}^2 = J(J+1)\hbar^2$ and J ranges from $|L-S|$ by unit steps up to $L+S$.

Thus the terms of a configuration are characterized by $^{(2S+1)}L$ and the individual states by M_L and M_S. The method of finding the terms is given in ref. 16*a*, chap. 7. For closed shells only, the term is always ^1S. Results for other configurations of importance are summarized.

For one-electron (nlm_lm_s) outside closed shells there is one term ^2L. This is the situation in the alkali metals and their isoelectronic sequences.

For $(ns,n'l)$ outside closed shells, the terms are ^1L and ^3L, and for two equivalent s electrons, there is only ^1S, the triplet being ruled out by the Pauli exclusion principle. This situation obtains in the helium and the alkaline-earth isoelectronic sequences.

The diagonal matrix elements of the first sum in H_1 [Eq. (1.109)] are independent of (m_l,m_s) and so displace the configuration as a whole on the energy scale. Their contribution is

$$\sum_i I(a_i,a_i) = \sum_i \left(a_i \left| -V(r) - \frac{Ze^2}{r} \right| a_i\right) \qquad (1.112)$$

to be added to the zero-order energy from (1.111).

Calculation of the matrix elements of the Coulomb interaction $\sum_{i>j} e^2/r_{ij}$ in H_1 is a little more elaborate. The result for a diagonal matrix element is

$$\left(A \left| \sum e^2/r_{ij} \right| A\right) = \sum_{i>j} J(a_i,a_j) - K(a_i,a_j)\delta(m_{sa_i},m_{sa_j}) \qquad (1.113)$$

in which

$$\begin{aligned} J(a_i,a_j) &= \iint |u_1(a_i)|^2(e^2/r_{12})|u_2(a_j)|^2 \, d\tau_1 \, d\tau_2 \\ K(a_ia_j) &= \iint u^*_1(a_i)u_1(a_j)(e^2/r_{12})u_2(a_i)u^*_2(a_j) \, d\tau_1 \, d\tau_2 \end{aligned} \qquad (1.114)$$

J and K are called *direct* and *exchange* integrals, respectively. It is easily shown that they are positive in all cases.

Because of the form (1.110) of the one-electron functions it is possible to express the J's and K's in terms of double radial integrals and certain integrals over spherical harmonics which have been tabulated. This leads to

$$\begin{aligned} J(a_ia_j) &= \sum_k a^k(l_{a_i}m_{la_i},l_{a_j}m_{la_j})F^k(n_{a_i}l_{a_i},n_{a_j}l_{a_j}) \\ K(a_ia_j) &= \sum_k b^k(l_{a_i}m_{la_i},l_{a_j}m_{la_j})G^k(n_{a_i}l_{a_i},n_{a_j}l_{a_j}) \end{aligned} \qquad (1.115)$$

where

$$\begin{aligned} F^k(n_al_a,n_bl_b) &= \int_0^\infty \int_0^\infty \frac{e^2r_<^k}{r_>^{k+1}} R_1^2(a)R_2^2(b) \, dr_1 \, dr_2 \\ G^k(n_al_a,n_bl_b) &= \int_0^\infty \int_0^\infty \frac{e^2r_<^k}{r_>^{k+1}} R_1(a)R_1(b)R_2(a)R_2(b) \, dr_1 \, dr_2 \end{aligned} \qquad (1.116)$$

The sums on k are quite finite because the a's and b's vanish unless $k + l_{a_i} + l_{a_j}$ is even and $|l_{a_i} - l_{a_j}| \leq k \leq l_{a_i} + l_{a_j}$.

Since $a^0(l_a m_a, l_b m_b) = 1$ for all values of its arguments, this part of the J integrals also gives the same value for all states of the same configuration, serving only to displace the configuration as a whole. The same is true for $b^0(l_a m_a, l_b m_b)$ when $l_a + l_b$ is even, and so these G integrals also displace the configuration as a whole.

Thus the parts of the $(A|H|A)$ that give rise to separation of the terms are the parts of (1.115) for $k \neq 0$. They represent the electrostatic interaction of two electrons in orbitals of different relative orientations. Thus the interval between terms of the same configuration is reduced to the computation of the integrals in (1.116) for $k \neq 0$.

It is convenient to express $(A|H_1|A)$ as the mean of the quantity for all the A's of a configuration plus the deviation of this matrix element from that mean. We write $[\alpha|H_1|\alpha]$ for the mean for a configuration α, and $[A|H_1|A]$ for the deviation of $(A|H_1|A)$ from that mean,

$$(A|H_1|A) = [\alpha|H_1|\alpha] + [A|H_1|A] \qquad (1.117)$$

Then $[\alpha|H_1|\alpha]$ consists of the sum in (1.112) plus the mean of (1.113) taken over all the A's of the configuration. The latter mean is given in Table 1.5 as the mean interaction energy of pairs of electrons for the various kinds of shells.

TABLE 1.5. MEAN INTERACTION ENERGY OF PAIRS OF ELECTRONS

Pair	*Mean interaction energy*			
s^2	F^0			
ss'	F^0	$-\ G^0/2$		
sp	F^0	$-\ G^1/6$		
sd	F^0	$-\ G^2/10$		
sf	F^0	$-\ G^3/14$		
p^2	F^0	$-\ 2F^2/25$		
pp'	F^0	$-\ G^0/6$	$-\ G^2/15$	
pd	F^0	$-\ G^1/15$	$-\ 3G^3/70$	
pf	F^0	$-\ 3G^2/70$	$-\ 2G^4/63$	
d^2	F^0	$-\ 2F^2/63$	$-\ 2F^4/63$	
dd'	F^0	$-\ G^0/10$	$-\ G^2/35$	$-\ G^4/35$
df	F^0	$-\ 3G^1/20$	$-\ 2G^3/105$	$-\ 5G^5/231$
f^2	F^0	$-\ 4F^2/195$	$-\ 2F^4/143$	$-\ 100F^6/5{,}577$
ff'	F^0	$-\ 2G^2/105$	$-\ G^4/77$	$-\ 50G^6/3{,}003$

A table of values needed to calculate $[A|H_1|A]$ is given by Slater [6b, vol. 1, pp. 327–330]. From this it is possible to use the diagonal-sum rule on the (M_S, M_L) classes to derive the results for energies of Russell-Saunders terms, relative to the configuration mean, that are given in Table 1.6.

To illustrate with a particular example, the ground state of NaI, for which $Z = N = 11$, has the configuration $1s^2 2s^2 2p^6 3s$. This gives rise to a ^2S term,

TABLE 1.6. RUSSELL-SAUNDERS TERM ENERGIES ABOVE CONFIGURATION MEAN‡

Configuration open shells	Term	Energies	
0	^1S	0	
nl	^2L	0	
$ns \cdot n'l$	^1L	$3G/2(2l+1)$	
	^3L	$-1/2(2l+1)$	
$(np)^2$	^1S	$12F^2/25$	
$(np)^4$	^1D	$3F^2/25$	
	^3P	$-3F^2/25$	
$np \cdot n'p$	^1S, ^3S	$10F^2/25$	$\begin{cases} +7G^0/6 + 35G^2/75 \\ -5G^0/6 - 25G^2/75 \end{cases}$
	^1P, ^3P	$-5F^2/25$	$\begin{cases} -5G^0/6 + 10G^2/75 \\ +7G^0/6 - 10G^2/75 \end{cases}$
	^1D, ^3D	$+1F^2/25$	$\begin{cases} +7G^0/6 + 8G^2/75 \\ -5G^0/6 + 2G^2/75 \end{cases}$
$np \cdot n'd$	^1P, ^3P	$+7F^2/25$	$\begin{cases} +2G^2/15 + 21G^2/70 \\ 0G^2/15 - 15G^2/70 \end{cases}$
	^1D, ^3D	$-7F^2/25$	$\begin{cases} -2G^2/15 + 24G^2/70 \\ +4G^2/15 - 18G^2/70 \end{cases}$
	^1F, ^3F	$+2F^2/25$	$\begin{cases} +7G^2/15 + 17G^2/490 \\ -5G^2/15 + 5G^2/490 \end{cases}$
$(np)^3$	^2P	$+6F^2/25$	
	^2D	$0F^2/25$	
	^4S	$-9F^2/25$	
$(np)^5$		Same as (np)	
$(nd)^2$	^1S	$140F^2/441 + 140F^4/441$	
$(nd)^8$	^3P	$77F^2/441 - 70F^4/441$	
	^1G	$50F^2/441 + 15F^4/441$	
	^1D	$-13F^2/441 + 50F^4/441$	
	^3F	$-58F^2/441 + 5F^4/441$	

‡ A fuller compilation is given by Slater [16b, vol. 2, appendix 21].

the energy of which, in the first approximation, is

$$\begin{aligned} E = 2I(s) &+ 2I(2s) + 6I(2p) + I(3s) \\ &+ F^0(1s,1s) + F^0(2s,2s) \\ &+ 4[F^0(1s,2s) - G^0(1s,2s)/2] \\ &+ 15[F^0(2p,2p) - 2F^2(2p,2p)/25] \\ &+ 12[F^0(1s,2p) - G^1(1s,2p)/6] \\ &+ 12[F^0(2s,2p) - G^1(2s,2p)/6] \\ &+ 2[F^0(1s,3s) - G^0(1s,3s)/2] \\ &+ 2[F^0(2s,3s) - G^0(2s,3s)/2] \\ &+ 6[F^0(3s,2p) - G^1(3s,2p)/6] \end{aligned} \qquad (1.118)$$

This contains $N = 11$ one-electron terms and

$$\tfrac{1}{2}N(N-1) = 55 \text{ two-electron terms}$$

The energy of an excited level $(nl\,^2L)$ is given by the same expression on replacing the $3s$ terms in (1.118) by appropriate terms for the nl states. The energy needed to ionize Na by removal of the $3s$ electron is the difference between (1.118) and the same expression with omission of the 11 terms that involve the $3s$ orbital.

19. Spin-orbit Interaction

The most important correction to be added to (1.106) is the term H_2; it represents the magnetic spin-orbit interaction,

$$H_2 = \frac{u_B{}^2}{a^3} \sum_i \frac{1}{r_i} \frac{\partial}{\partial r_i} \left(\frac{Z(r_i)}{r_i} \right) \mathbf{L}_i \cdot \mathbf{S}_i \qquad (1.119)$$

in which $\mathbf{L}_i$ and $\mathbf{S}_i$ are measured in units of $\hbar$, and r_i is measured in units of a. The coefficient

$$\frac{\mu_B{}^2}{a^3} = \tfrac{1}{2}\alpha^2 \frac{e^2}{2a}$$

and so its wave-number equivalent is

$$\tfrac{1}{2}\alpha^2 R = 2.671 \text{ cm}^{-1}$$

The term H_2 does not commute with $\mathbf{S}$ and $\mathbf{L}$ separately but does commute with $\mathbf{J} = \mathbf{S} + \mathbf{L}$, their vector sum. Therefore, the energy levels can be accurately characterized by J and M, where $\mathbf{J}^2 = J(J+1)$ and $M = M_S + M_L$, but no longer by accurate values of S and L. This behavior is described by saying that spin-orbit interaction acts to break down (S,L) or Russell-Saunders coupling.

In many cases H_2 is small enough that its matrix components connecting different (S,L) terms can be neglected; then labeling of levels by (S,L) values continues to be accurate. The effect of the part of H_2 that is diagonal with respect to a term is to split the $(2S+1)(2L+1)$ states into levels, each labeled by J, where the J values run from $(S-L) \leq J \leq (S+L)$. The shift in energy of the levels due to this part of H_2 is given by

$$E_2(S,L,J) = (\mu_B{}^2/a^3)\zeta(S,L) \cdot \tfrac{1}{2}[J(J+1) - L(L+1) - S(S+1)] \qquad (1.120)$$

If S or L equals zero, then J is equal to L or S and (1.120) vanishes. According to (1.120) the interval between the J level and the $(J-1)$ level is $J(\mu_B{}^2/a^3)\zeta(S,L)$; that is, it is proportional to the larger of the two J values. This is known as the *Landé interval rule.*

The nondiagonal matrix elements of H_2 connect levels of the same (J,M) but different (S,L). Consequently, the accurate ψ's are linear combinations of states having different (S,L) values. Thus the conventionally assigned labels have to be understood as indicating the (S,L) values of the principal part of the wave function [16*a*, chap. 11].

The $\zeta(S,L)$ coefficients in (1.120), which give the magnitude of the intervals between levels in Russel-Saunders terms, are expressible in terms of one-electron radial integrals,

$$\zeta_{nl} = \int_0^\infty \frac{1}{r} \frac{d}{dr} \left(\frac{Z(r)}{r} \right) R^2(nl)\, dr \qquad (1.121)$$

in which r is in atomic units. For hydrogenic wave functions (constant Z) the value is

$$\zeta_{nl} = \frac{Z^4}{n^3 l} (l + \tfrac{1}{2})(l+1) \qquad (1.122)$$

which may serve as a very rough guide to the magnitude of the ζ_{nl} in nonhydrogenic cases.

Figure 1.6 shows some observed doublet intervals in the alkali atoms in comparison with the hydrogenic values as calculated from (1.120) and (1.122). Figure 1.7 shows the observed values of ζ_{3p} in the Na, Mg, and Al isoelectronic sequences.

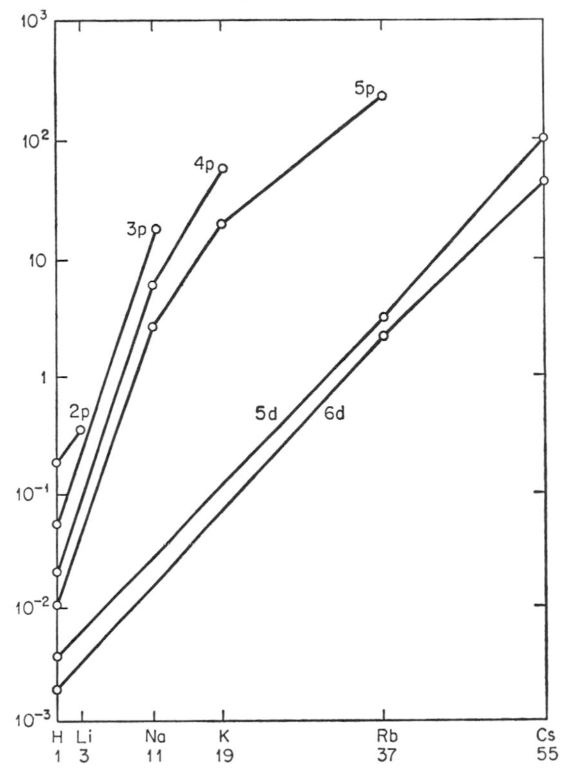

FIG. 1.6. Doublet intervals in hydrogen and alkali spectra. Ordinates are doublet intervals in cm^{-1} on a logarithmic scale and abscissas are the atomic numbers Z.

Some values of the $\zeta(S,L)$ in terms of the ζ_{nl} are given in Table 1.7. A characteristic feature is that the intervals in p^4 and p^5 are inverted relative to those in p^2 and p. A similar inversion occurs for d^8 and d^9 relative to d^2 and d, and also for f^{14-n} relative to f^n.

20. Hartree and Hartree-Fock Functions

The calculation of levels outlined in Secs. 18 and 19 depends on a suitable choice of the $V(r)$ of (1.107), which defines the radial wave functions $R(nl)$ and the $E_0(nl)$ of (1.111). Qualitatively the wave functions resemble those of Fig. 1.2 for hydrogen with regard to vanishing at $r \to 0$ and $r \to \infty$ and in having $n - l - 1$ other nodes. As shown in (1.51), the effect of Z is to increase the energy of binding, there called W, by a factor Z^2 and to contract the extent of the wave function on r, moving its maxima and its nodes to smaller values of r by a factor Z^{-1}.

If the mutual repulsion of the electrons were neglected altogether, the individual $E_0(nl)$ would be given by (1.51) and the $R(nl)$ by (1.52). But this would give a poor starting point for a perturbation calculation. The binding energies would be much too great and the wave functions much too contracted.

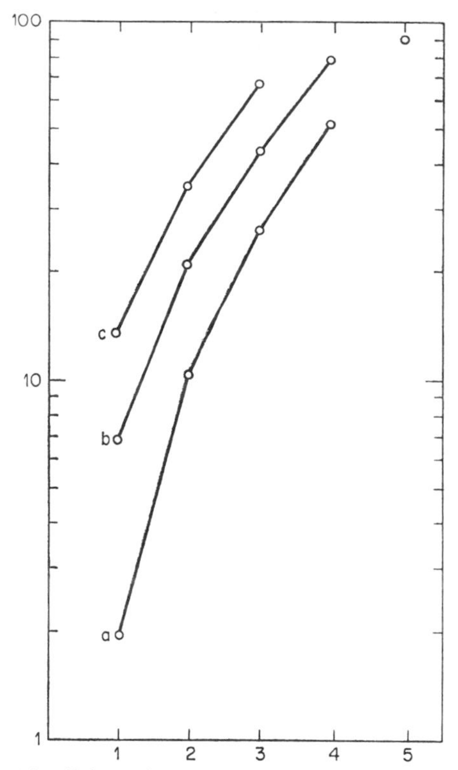

FIG. 1.7. Spin-orbit interaction parameters, ζ_{3p}, as inferred (*a*) from the ^2P interval in the Na isoelectronic sequence, (*b*) from the ^3P interval in the Mg isoelectronic sequence, and (*c*) from the ^2P interval in the Al isoelectronic sequence.

D. R. Hartree [17] in 1928 gave a systematic procedure for obtaining a good choice of $V(r)$ on which to base the calculations. Basically his idea is that, since each electron moves in the field of the nucleus and the average field of all the others, $V(r)$ is to be chosen in such a way that the $R(nl)$ determined by it through (1.49) leads to a radial-charge-density distribution whose electrostatic potential is equal to this same $V(r)$. He calls such a $V(r)$ a *self-consistent field*. The procedure he developed was equivalent to one in which the exchange integrals of (1.114) are neglected; this defect was remedied in 1930 by V. Fock, but the resulting equations were too complicated for extensive numerical handling until automatic digital computers became available. Slater showed how the Hartree and Hartree-Fock equations can be obtained by an application of the variation method of Sec. 9.

Using antisymmetric wave functions built of one-electron functions (1.110), the mean energy $(A|H|A)$ can be expressed in terms of integrals involving the unknown $R(nl)$, as in the special example of (1.118). We now vary the $R(nl)$, subject to the condition that they form an orthonormal set, to get a stationary value of $(A|H|A)$. This leads to a rather complicated set of simultaneous nonlinear integrodifferential equations for the $R(nl)$; they are known as the Hartree-Fock equations. Neglect of the terms arising from the exchange integrals leads to a simpler set, known as the Hartree equations.

TABLE 1.7. LANDÉ INTERVAL FACTORS, $\zeta(S,L)$

Configuration	*Interval factor*
p, p^5	$\zeta(^2P) = \pm\zeta_{np}$
p^2, p^4	$\zeta(^3P) = \pm\zeta_{np}$
p^3	$\zeta(^2P) = \zeta(^2D) = 0$
np, $n'p$	$\zeta(^3P) = \zeta(^3D) = \frac{1}{4}(\zeta_{np} + \zeta_{n'p'})$
p, d	$\zeta(^3P) = -\frac{1}{4}\zeta_p + \frac{3}{4}\zeta_d$
	$\zeta(^3D) = \frac{1}{12}\zeta_p + \frac{5}{12}\zeta_d$
	$\zeta(^3F) = \frac{1}{6}\zeta_p + \frac{1}{3}\zeta_d$
s, l	$\zeta(^3L) = \frac{1}{2}\zeta_{nl}$
p, f	$\zeta(^3D) = -\frac{1}{6}\zeta_p + \frac{2}{3}\zeta_f$
	$\zeta(^3F) = \frac{1}{24}\zeta_p + \frac{11}{24}\zeta_f$
	$\zeta(^3G) = \frac{1}{8}\zeta_p + \frac{3}{8}\zeta_f$
d, d^9	$\zeta(^2D) = \pm\zeta_d$
d^2, d^8	$\zeta(^3P) = \zeta(^3F) = \pm\frac{1}{2}\zeta_d$
d^3, d^7	$\zeta(^4P) = \zeta(^4F) = \pm\frac{1}{3}\zeta_d$
	$\zeta(^2H) = \pm\frac{1}{5}\zeta_d$
	$\zeta(^2G) = \pm\frac{3}{10}\zeta_d$
	$\zeta(^2F) = \pm\frac{1}{6}\zeta_d$
	Sum of two $\zeta(^2D) = \pm\frac{1}{3}\zeta_d$
	$\zeta(^2P) = \pm\frac{2}{3}\zeta_d$
d^4, d^6	$\zeta(^5D) = \pm\frac{1}{4}\zeta_d$
	$\zeta(^3H) = \pm\frac{1}{10}\zeta_d$
	$\zeta(^3G) = \pm\frac{3}{20}\zeta_d$
	Sum of two $\zeta(^3F) = \pm\frac{1}{12}\zeta_d$
	Sum of two $\zeta(^3P) = \pm\frac{1}{2}\zeta_d$
d^5	All zetas equal zero.
p^2s	$\zeta(^2P) = \frac{2}{3}\zeta_p$
	$\zeta(^4P) = \frac{1}{3}\zeta_d$
	$\zeta(^2D) = 0$
$nd \cdot n'd$	$\zeta(^3P) = \zeta(^3D) = \zeta(^3F) = \zeta(^3G)$
	$= \frac{1}{4}(\zeta_{nd} + \zeta_{n'd})$

In recent years a large amount of computing has been done to find self-consistent fields and radial functions for the elements. Herman and Skillman [18] have published a computer program and the results obtained with it for the ground configurations of all the elements in the neutral, $Z = N$, form. With the $V(r)$ and $R(nl)$ obtained in this way, the I, J, K of Sec. 18 and the ζ_{nl} of Sec. 19 can be computed to complete the task of obtaining theoretical first-order energies for the ground configurations. Some work has also been done on excited states and on states of ions as well as of neutral atoms.

A good idea of the way in which the $R(nl)$ contract with increasing Z of the normal configuration of neutral atoms is given in Fig. 1.8, based on the calculation of Herman and Skillman. Abscissas are values of r on a logarithmic scale; ordinates are values of Z on a logarithmic scale. The curves then show for each Z the value of r at which each $R^2(nl)$ has the greatest maximum radial charge density. From this we see that $R^2(1s)$ contracts in about the same way as it would if no other electrons were present. But $R^2(2p)$, for example, is always much more extended, because of the screening action of the $1s$ and $2s$ electrons which spend most of the time at distances closer

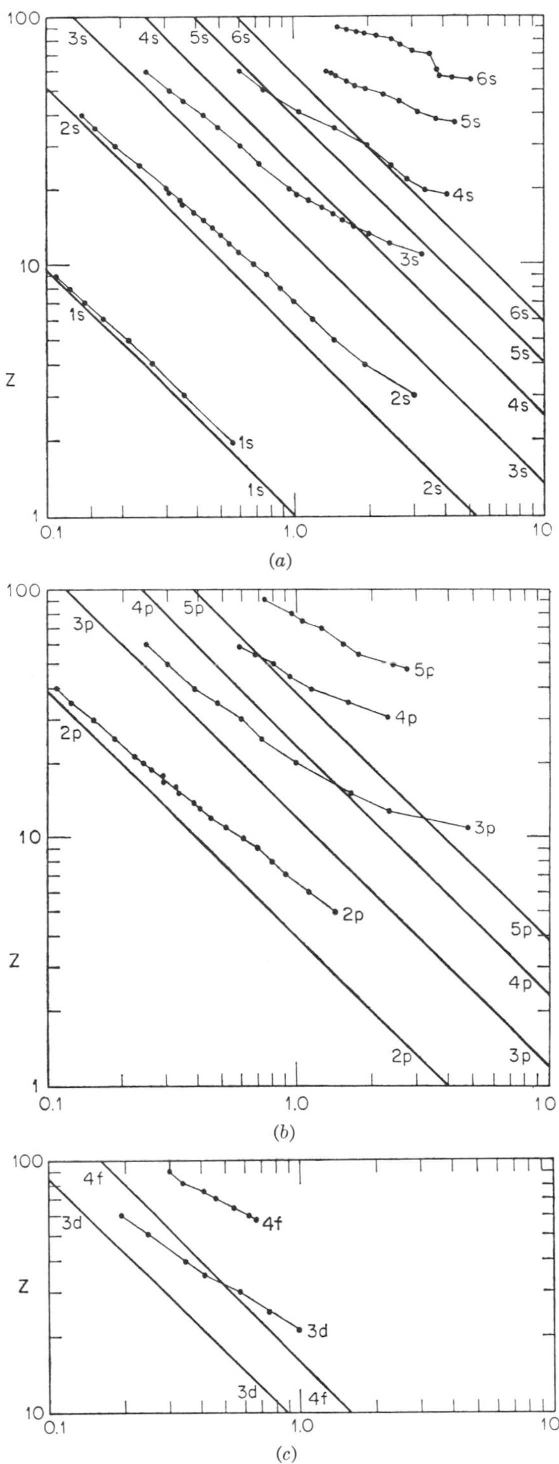

FIG. 1.8. Abscissas show the value of r in atomic units at which $R^2(nl)$ has its largest maximum on a logarithmic scale and ordinates are the atomic number Z. The straight lines give the hydrogenic values, and the plotted points are from the Hartree-Fock calculations of Herman and Skillman [18].

to the nucleus than the $2p$ electrons. By reading down from the point for $R^2(2p)$ for a given Z to the hydrogenic line for the $2p$ orbital, one can find the smaller screened value of Z for which the hydrogenic wave function would have its maximum at the same value of r. After $Z = 10$, the $2p$ shell is filled, and so the screening is 2 by the two $1s$ electrons and something less than 7 for the remaining $2s$ and $2p$ electrons—less because the $2p$ orbital under consideration does not lie at larger distances than the other $2s$ and $2p$ orbitals. The screening is much greater for the $n = 3$ orbitals and is greater for the $3d$ then for the $3p$, because the $3d$ orbital is less penetrating into the domain of the $n = 1$ and $n = 2$ shells. Similar trends are also found with the $n = 4$, 5, and 6 shells.

21. Series. Isoelectronic Sequences

As discussed in Sec. 13, the normal configurations are formed by assigning electrons successively to the most tightly bound state available to them in view of the states already occupied by other electrons.

The excited levels which combine with the levels of the ground level and with each other to give the observed spectral lines usually arise from configurations that are the same as the ground configuration except for the change of one (nl) value, usually that of the least tightly bound of the various electrons in the ground configuration. However, in some instances configurations occur in which a different nl value is excited or in which more than one electron has nl values different from those of the ground configuration.

Configurations can be arranged in *series,* which refers to a set of configurations in which the n of the excited electron increases through all integers from the lowest value to infinity. At this stage $n \to \infty$, the excited electron is free, and so the levels of the series of configuration extrapolate naturally into a configuration of the ion of one higher value of

$$Z_0 = (Z - N + 1)$$

If the series of configurations involves such a gradual loosening of the least tightly bound electron, the series extrapolates into the ground configuration of the next higher ion. Otherwise one may find a series of configurations that extrapolates to an excited configuration of the next higher ion.

An *isoelectronic* sequence is a series of ions having the same number N of electrons. As N increases, the energy values below the ionization limit of the successive members mainly becomes larger as Z_0^2; thus it is convenient to exhibit the trends by plotting the level energies relative to the ionization limit divided by Z_0^2 as ordinates against the value of Z_0. Figure 1.9 is such a plot for the potassium $(N = 19)$ isoelectronic sequence.

Here the underlying core is K II, which belongs to the argon $(N = 18)$ sequence, for which the ground configuration is $1s^22s^22p^63s^23p^6$. The binding energy of the $4s$ electron in the ground state of K I is 35,010 cm^{-1}, so that all the excited states up to the ground state (^1S) of K II are contained in the interval $-35{,}010$ up to 0. Analysis of the K II spectrum shows that the first excited configuration of K II $(1s^22s^22p^63s^23p^54s)$ gives rise to four levels

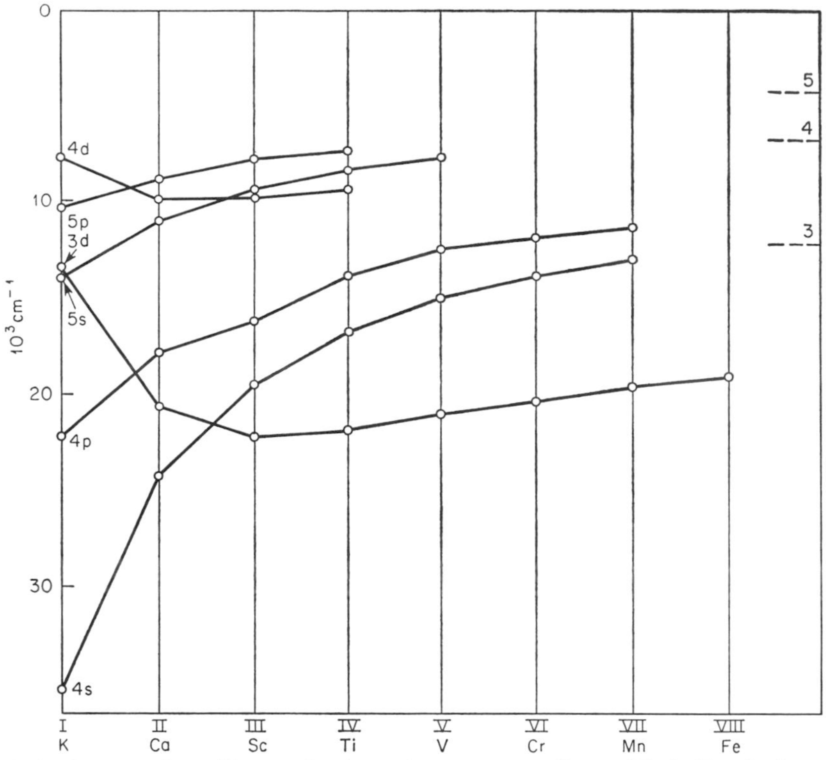

FIG. 1.9. Low levels in the potassium ($N = 19$) isoelectronic sequence; ordinates $W(nl)/Z_0^2$, abscissas Z_0. Hydrogenic values for $n = 3, 4, 5$ are shown at the extreme right.

from +162,507 to +166,461, nearly four times as great as the ionization energy of the $4s$ electron in K I.

As Fig. 1.9 shows, the $4s$ electron is much more tightly bound than the $3d$ electron in K I; $4s$ is more tightly bound than $3d$ also in Ca II, but here the difference is much reduced so that the opposite is true for Sc III and the higher members of the isoelectronic series. Further consideration of Fig. 1.9 shows that both the ns and np levels start, at $Z_0 = 1$, much more tightly bound than the nth hydrogenic value and gradually become relatively less tightly bound, moving to the hydrogenic values with increasing Z_0. But the behavior of the $3d$ and $4d$ levels is quite different. They start at $Z_0 = 1$ quite close to the hydrogenic value, corresponding to the fact that $R(3d)$ and $R(4d)$ are well outside the K II core. As Z_0 increases, the $R(3d)$ and $R(4d)$ contract more than the core contracts, and so they become at first relatively more tightly bound, then go through a minimum, and also start to approach the corresponding hydrogenic value.

It is often convenient to write the binding energy in the form

$$W(nl) = -\frac{RhcZ_0^2}{n^{*2}} \tag{1.123}$$

defining n^* by this relation, where $W(nl)$ is the observed value, so that n^* is generally nonintegral. Here n^* is called the effective quantum number, and

$$\Delta = n - n^* \tag{1.124}$$

is called the *quantum defect*. As the electrons are always more tightly bound than the corresponding hydrogenic value, Δ is always positive. Empirically it is found that $\Delta(n,l)$ remains nearly constant as n increases along a series. In this case the levels are said to obey a *Rydberg* series formula. More accurately the $\Delta(n,l)$ vary along a series in a manner given by the empirical *Ritz formula*

$$\Delta(n,l) = \Delta(\infty,l) - \delta n^{-2} \tag{1.125}$$

Another form of the Ritz formula regards Δ as a function of the term value $W(nl)/hc$, writing

$$\Delta(n,l) = \Delta(\infty,l) + \frac{\alpha W(nl)}{hc} \tag{1.126}$$

Study of trends in n^* or Δ along an isoelectronic sequence has proved to be an extremely valuable tool for making approximate empirical predictions of the location of spectral terms as an aid in the analysis of spectra.

22. Ionization Potentials

The ionization potential of an atom or ion is the amount of energy needed to remove one electron from it when it is in its normal state, leaving the next higher ion in its normal state. These are usually stated in electron volts even though most of them are determined spectroscopically and therefore given directly in cm^{-1}.

Table 1.8 gives the ionization potentials for the first six stages of ionization, where known, for the

TABLE 1.8. IONIZATION POTENTIALS
(In electron volts)

Z	Element	Stage of ionization, Z_0					
		I	II	III	IV	V	VI
1	H	13.595					
2	He	24.580	54.400				
3	Li	5.390	75.619	122.42			
4	Be	9.320	18.206	153.85	217.66		
5	B	8.296	25.149	37.920	259.30	340.13	
6	C	11.264	24.376	47.864	64.476	391.99	489.84
7	N	14.54	29.605	47.426	77.450	97.86	551.92
8	O	13.614	35.146	54.934	77.394	113.87	138.08
9	F	17.42	34.98	62.646	87.23	114.21	157.12
10	Ne	21.559	41.07	64 ± 1	97.16	126.4	157.91
11	Na	5.138	47.29	71.65	98.88	138.60	172.36
12	Mg	7.644	15.03	80.12	109.29	141.23	186.86
13	Ae	5.984	18.823	28.44	119.96	153.77	190.42
14	Si	8.149	16.34	33.46	45.13	166.73	205.11
15	P	10.55	19.65	30.156	51.354	65.01	220.41
16	S	10.357	23.4	35.0	47.29	72.5	88.03
17	Cl	13.01	23.80	39.90	53.5	67.80	96.7
18	Ar	15.755	27.62	40.90	59.79	75.0	91.3
19	K	4.339	31.81	46.	60.90	......	99.7
20	Ca	6.111	11.87	51.21	67.	84.39	
21	Sc	6.56	12.89	24.75	73.9	92.	111.1
22	Ti	6.83	13.63	28.14	43.24	99.8	120.
23	V	6.74	14.2	29.7	48.	65.2	128.9
24	Cr	6.763	16.49	30.95	49.6	73.	90.6
25	Mn	7.432	15.64	33.69	....	76.	
26	Fe	7.90	16.18	30.64			
27	Co	7.86	17.05	33.49			
28	Ni	7.633	18.15	36.16			
29	Cu	7.724	20.29	36.83			
30	Zn	9.391	17.96	39.70			
31	Ga	6.00	20.51	30.70	64.2		
32	Ge	7.88	15.93	34.21	45.7	93.4	
33	As	9.81	20.2	28.3	50.1	62.6	127.5
34	Se	9.75	21.5	32.0	42.9	73.1	81.7
35	Br	11.84	21.6	35.9			
36	Kr	13.99	24.56	36.9			
37	Rb	4.176	27.5	40			
38	Sr	5.692	11.027	......	57		
39	Y	6.5	12.4	20.5	....	77	
40	Zr	6.95	14.03	24.8	33.97	......	99
41	Nb	6.77	14				
42	Mo	7.06	16.15				
43	Tc	7.28	15.26				
44	Ru	7.364	16.76				
45	Rh	7.46	18.07				
46	Pd	8.33	19.42				
47	Ag	7.574	21.48				
48	Cd	8.991	16.904				
49	In	5.785	18.86				
50	Sn	7.342	14.628				
51	Sb	8.639	16.5				
52	Te	9.01	18.6 ± .4				
53	I	10.454	19.09				
54	Xe	12.127	21.2				

TABLE 1.8. IONIZATION POTENTIALS (*Continued*)

Z	Element	Stage of ionization, Z_0					
		I	II	III	IV	V	VI
55	Cs	3.893	25.1				
56	Ba	5.210	10.001				
57	La	5.61	11.43				
72	Hf	~7	14.9				
73	Ta	7.88	(16.2 ± .5)				
74	W	7.98	(17.7 ± .5)				
75	Re	7.87	(16.6 ± .5)				
76	Os	8.7	(17. ± 1.)				
77	Ir	9.					
78	Pt	9.0	18.56				
79	Au	9.22	20.5				
80	Hg	10.43	18.751				
81	Te	6.106	20.42				
82	Pb	7.415	15.028				
83	Bi	7.287	16.68				
84	Po	8.43	(19)				
85							
86	Rn	10.746	(20)				
87							
88	Ra	5.277	10.144				
89	Ac		12.1				

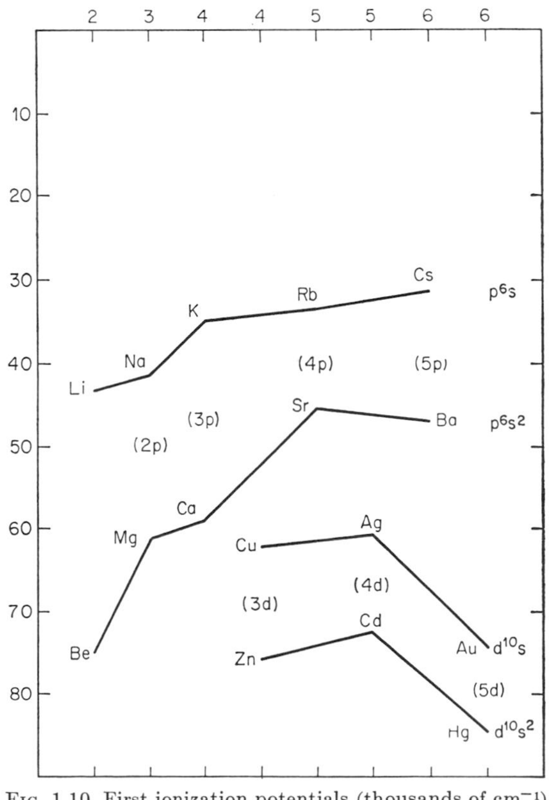

FIG. 1.10. First ionization potentials (thousands of cm^{-1}) of group I and group II elements.

first forty elements, followed by the values for the first two stages, where known, for the higher elements. Uncertain values are enclosed in parentheses.

The ionization potentials show systematic trends. That for an s^2 configuration is always much greater than that for the corresponding s configuration. Figure 1.10 shows the variation of the first ionization potentials of the groups I and II elements, indicating how one and two outer s electrons are much more tightly bound to a d^{10} shell than to a p^6 shell. Likewise the ionization potential increases rapidly during the filling of a p shell, except for a hesitation which reverses the general trend in going from p^3 to p^4. In contrast the ionization potentials remain much more nearly constant during the filling of a d shell.

All these trends are shown in Fig. 1.11. In part (a) of this figure are plotted the ionization potentials of the $2s,2p$, $3s,3p$, and $4s,4p$ elements, against the degree of filling of these shells. The close resemblance of the $4p$ shell to $2p$ and $3p$ is noteworthy in that this shell does not follow directly after $4s^2$, but instead the 10 elements between Ca(20) and Ga(31) intervene with the filling of the $3d$ shell.

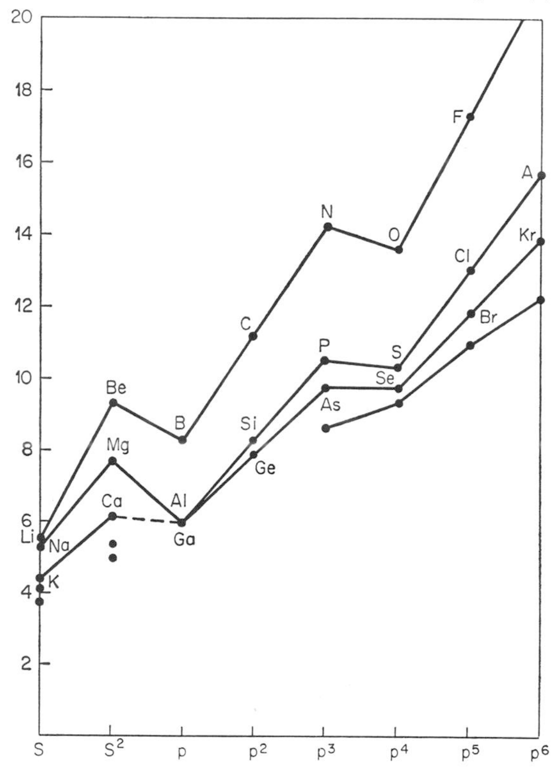

FIG. 1.11a. First ionization potentials (electron volts) for the $(2s,2p)$, $(3s,3p)$, and $(4s,4p)$ periods.

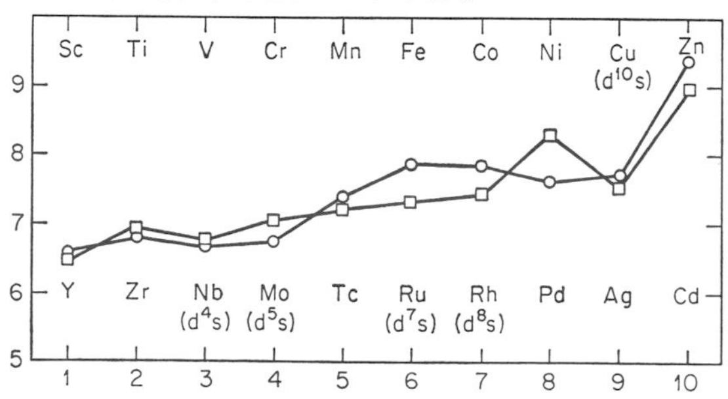

FIG. 1.11b. First ionization potentials (electron volts) for the $3d$ transition elements (circles) and the $4d$ transition elements (squares).

The knowledge is not available to make part (b) complete, yet enough is known to show the close general resemblance of the $3d$ and $4d$ shells. In these shells the d^ns^2 configuration is nearly of the same energy as the $d^{n+1}s$ configuration; so in some of these elements the normal state belongs to this latter configuration. This detail was ignored in preparing part (b), each element being plotted at an abscissa as if the normal configuration were d^ns^2 in all cases. Actually, since these two configurations are both even, there is a good deal of configuration interaction between them which often robs configuration assignments of precise meaning. Moreover, the totality of terms from each configuration extends over a considerable and overlapping range of energies so that it becomes more or less accidental which particular term happens to lie lowest.

References

1. *Expository Books on Quantum Mechanics:*

Introductory Textbooks:

a. Condon, E. U., and P. M. Morse: "Quantum Mechanics," McGraw-Hill, New York, 1929; paperback ed., 1964.
b. Pauling, L., and E. B. Wilson, Jr.: "Introduction to Quantum Mechanics," McGraw-Hill, New York, 1935.
c. Kemble, E. C.: "The Fundamental Principles of Quantum Mechanics," McGraw-Hill, New York, 1937.
d. Rojansky, V.: "Introductory Quantum Mechanics," Prentice-Hall, Englewood Cliffs, N.J., 1938.
e. Mott, N. F., and I. N. Sneddon: "Wave Mechanics and Its Applications," Oxford University Press, Fair Lawn, N.J., 1948.
f. Schiff, L. I.: "Quantum Mechanics," 2d ed., McGraw-Hill, New York, 1955.
g. Pauli, W.: Wellenmechanik, "Encyclopedia of Physics," vol. 24, pt. I, Springer, Berlin; reprinted by Edwards, Ann Arbor, Mich., 1950.
h. Bohm, D.: "Quantum Theory," Prentice-Hall, Englewood Cliffs, N.J., 1951.
i. Houston, W. V.: "Principles of Quantum Mechanics," McGraw-Hill, New York, 1951.
j. Kramers, H. A., "Quantum Mechanics," North Holland Publishing Co., Amsterdam, 1958.
k. Merzbacher, E.: "Quantum Mechanics," Wiley, New York, 1961.
l. Fong, P.: "Elementary Quantum Mechanics," Addison-Wesley, Reading, Mass., 1962.
m. Ikenberry, E.: "Quantum Mechanics for Mathematicians and Physicists," Oxford University Press, Fair Lawn, N.J., 1962.
n. Blokhintsev, D. I.: "Principles of Quantum Mechanics," Allyn and Bacon, Boston, 1964.

Advanced General Works:

o. Von Neumann, J.: "Mathematical Foundations of Quantum Mechanics," Princeton University Press, Princeton, N.J., 1955.
p. Landau, L. D., and E. M. Lifshitz: "Quantum

Mechanics: Non-relativistic Theory," Addison-Wesley, Reading, Mass., 1958.

q. Dirac, P. A. M.: "The Principles of Quantum Mechanics," 4th ed., Oxford University Press, Fair Lawn, N.J., 1958.

r. Messiah, A.: "Quantum Mechanics," 2 vols., North Holland Publishing Co., Amsterdam, 1961.

s. Kuruşnoğlu, B.: "Modern Quantum Theory," Freeman, San Francisco, 1962.

t. Mackey, G. W.: "Mathematical Foundations of Quantum Mechanics," Benjamin, New York, 1963.

2. *a*. Planck, M.: *Ann. Physik*, **4**: 553 (1901).

b. Einstein, A.: *Ann. Physik*, **17**: 132 (1905).

c. Condon, E. U.: Sixty Years of Quantum Physics, *Phys. Today*, **15**: 37 (1962).

3. *a*. Rose, M. E.: "Elementary Theory of Angular Momentum," Wiley, New York, 1957.

b. Edmonds, A. R.: "Angular Momentum in Quantum Mechanics," Princeton University Press, Princeton, N.J., 1957.

c. Feenberg, E., and G. E. Pake: "Notes on the Quantum Theory of Angular Momentum," Stanford University Press, Stanford, Calif., 1959.

4. *a*. Courant, R., and D. Hilbert: "Methods of Mathematical Physics," vol. 1, Interscience, New York, 1953.

b. Morse, P. M., and H. Feshbach: "Methods of Theoretical Physics," vol. 2, p. 1117, McGraw-Hill, New York, 1953.

5. *a*. Hasted, J. B.: "Physics of Atomic Collisions," Butterworth, Washington, D.C., 1964.

b. Mott, N., and H. Massey: "The Theory of Atomic Collisions," 3d ed., Oxford University Press, Fair Lawn, N.J., 1965.

6. *a*. Rutherford, E.: *Phil. Mag.*, **21**:(6): 669 (1911).

b. Darwin, C. G.: The Discovery of Atomic Number, in "Niels Bohr and the Development of Physics," McGraw-Hill, New York, 1955.

7. Bohr, N.: *Phil. Mag.*, **26** (6): 1, 476, 857 (1913); **27**, 506 (1914); **29**, 332 (1915); **30**, 394 (1915).

8. Uhlenbeck, G. E., and S. A. Goudsmit: *Nature*, **117**: 264 (1926); *Naturwiss.*, **13**: 953 (1925); *Physica*, **5**: 266 (1925).

9. Urey, H. C., F. G. Brickwedde, and G. M. Murphy: *Phys. Rev.*, **40**: 1 (1932).

10. Chadwick, J.: *Nature*, **129**: 312 (1932); *Proc. Roy. Soc. (London)*, **136**: 692 (1932).

11. Lozier, W. W., P. T. Smith, and Walker Bleakney: *Phys. Rev.*, **45**: 655 (1934). Sherr, R., L. G. Smith, and Walker Bleakney: *Phys. Rev.*, **54**: 388 (1938).

12. Mendéleeff, D.: *J. Russian Phys.-Chem. Soc.*, **1**: 60 (1869). Historical account in T. Moeller, "Inorganic Chemistry," chap. 4, Wiley, New York, 1952.

13. Hartree, D. R.: *Proc. Cambridge Phil. Soc.*, **24**: 89, (1926). Condon, E. U.: *Am. Phys. Teacher*, **2**: 63 (1934).

14. Kusch, P., and H. M. Foley: *Phys. Rev.*, **74**: 250 (1948). Karplus, R., and N. M. Kroll: *Phys. Rev.*, **77**: 536 (1950).

15. *a*. *National Bureau of Standards Technical News Bulletin*, October, 1963. *Physics Today*, **17**: 48 (February, 1964).

b. Cohen, E. R., and J. W. M. DuMond: *Revs. Mod. Phys.*, **37**: 537 (1965).

16. *a*. Condon, E. U., and G. H. Shortley: "The Theory of Atomic Spectra," Cambridge University Press, New York, 1935; paperback reprint, 1964.

b. Slater, J. C.: "Quantum Theory of Atomic Structure," 2 vols., McGraw-Hill, New York, 1960.

c. Griffith, J. S.: "The Theory of Transition-Metal Ions," Cambridge University Press, New York, 1961.

17. *a*. Hartree, D. R.: The Calculation of Atomic Structures, *Rept. Progr. Phys.*, **11**: 113 (1946–1947).

b. Freeman, A. J.: *Phys. Rev.*, **91**: 1410 (1953).

c. Hartree, D. R.: "The Calculation of Atomic Structures," Wiley, New York, 1957.

18. Herman, F., and S. Skillman: "Atomic Structure Calculations," Prentice-Hall, Englewood Cliffs, N.J., 1963.

Chapter 2

Atomic Spectra, Including Zeeman Effect and Stark Effect

By J. RAND McNALLY, Jr., Oak Ridge National Laboratory

1. Introduction

The status of our knowledge of the energy states of atoms up to the year 1951 has been discussed and summarized by Meggers [1].† Detailed information on atomic energy levels, ionization potentials, and original references may be found in the comprehensive National Bureau of Standards Tables of Charlotte E. Moore [2]. An excellent survey of atomic spectra is provided in the book of H. G. Kuhn [3].

The work of Edlén [4] has been in the direction of obtaining spectra of stripped atoms, i.e., atoms having a large number of the outer electrons removed. As an example, he has studied Sn XXIV, which has had 23 outer electrons removed. Certain brilliant lines appearing in the corona of the sun have been identified as forbidden transitions in the spectra of the highly ionized atoms of iron, nickel, and calcium.

A major contributing factor to understanding many of these spectra, especially such complex ones as the lanthanide and actinide rare-earth series, has been the use of the Zeeman effect, that is, the effect on the lines of placing the source in a magnetic field. The Zeeman effect at high magnetic resolution gives unique information regarding quantum numbers J_1 and J_2 of the transition levels as well as the Landé factors g_1 and g_2, which are indicative of the quantum nature of the levels.

Interpretation of spectra is based on the general theoretical principles outlined in Part 7, Chap. 1.

The main points of the terminology are summarized here:

1. A spectrum of an ion of atomic number Z and having N electrons is designated by giving the chemical symbol correlated with Z (Table 1.3) followed by a roman numeral giving the value of $Z_0 = Z - N + 1$. Thus Na IV designates the spectrum of sodium ($Z = 11$) from which three electrons have been removed so that $N = 8$.

2. The levels are rigorously characterized as odd or even, and strong transitions are associated with change of parity.

3. The levels are rigorously characterized by J, where $\mathbf{J}^2 = J(J + 1)$ in units of $\hbar$, $\mathbf{J}$ being the vector resultant of spin and orbital angular momentum of the electrons. The values of J are 0, 1, 2, . . . for N even and $\frac{1}{2}$, $\frac{3}{2}$, $\frac{5}{2}$, . . . for N odd.

4. To a good approximation in most cases the levels can be assigned to *configurations*, that is, to the levels arising from states in which the (nl) (principal and orbital angular momentum) quantum numbers of the N electrons have definite values. When configuration interaction occurs, such configuration assignments lose their exact validity, the wave function of the level now involving components from different configurations. Even so, in most cases one configuration is usually much more strongly represented than others. The parity of a configuration is that of the sum of the l values of the electrons; configuration interaction does not occur between configurations of opposite parity.

5. In most spectra studied thus far, the configurations of the excited states are obtained from that of the normal state by change of the (nl) of one or two of the loosely bound outer electrons. X-ray spectra (Part 7, Chap. 8) arise from the excitation of the tightly bound inner electrons.

6. In many spectra, particularly those of the low-Z elements, the magnetic spin-orbit interaction is small, so that the levels of a configuration are in close groups, called *Russell-Saunders terms*, which are characterized quite accurately by values of $\mathbf{S}^2 = S(S + 1)$, the vector resultant of the electron spins, and $\mathbf{L}^2 = L(L + 1)$, the vector resultant of orbital angular momenta. Such a group of levels is designated by a term symbol ${}^{(2S+1)}L$ in which the *multiplicity* $(2S + 1)$ is written as a left superscript to a capital letter designating L by the S, P, D, F . . . code (Chap. 1, Sec. 13) for $L = 0, 1, 2, 3, \ldots$.

The levels in such a term are distinguished by values of J, where J can take values differing by unity in the range $|L - S| \leq J \leq L + S$. The spin-orbit interaction gives rise to small energy differences between the levels of a term (Chap. 1, Sec. 19).

7. For elements of higher Z, the spin-orbit interaction is no longer small, and Russell-Saunders coupling goes over into *intermediate coupling* in which the levels are no longer accurately labeled by L and S values.

8. The systematic building up of the configurations of the normal states accounts for the main features of the periodic system of the elements (Chap. 1, Sec. 13) and of the trends in the structure of their spectra.

† Numbers in brackets refer to References at end of chapter.

2. Penetrating and Nonpenetrating Orbitals

An electron that is far away from the nucleus, compared with the other $N - 1$ in the ion, is acted on by a Coulomb field, which tends to

$$V(r) \to -\frac{Z_0 e^2}{r} \tag{2.1}$$

where $Z_0 = Z - N + 1$. At a lesser distance the potential energy decreases (becomes larger negatively) much more rapidly than this when the electron penetrates within distances occupied by the core of the other $N - 1$ electrons.

Thus there are excited levels whose wave function lies almost entirely in the range where (2.1) is valid; for these the term values are

$$T(nl) = -\frac{E}{hc} = \frac{RZ_0^2}{n^{*2}} \tag{2.2}$$

where n^* is nearly equal to an integer, as in hydrogenic spectra (Chap. 1, Sec. 21). These are called *nonpenetrating orbitals* because for them the electron stays mostly well outside the core. For other levels the opposite is true. The orbital penetrates into a region where the potential energy is much larger (negatively) than (2.1), and so the corresponding values of $T(nl)$ are greater; that is, the effective n^* in (2.2) is considerably less than n. These are called *penetrating orbitals*.

We may write

$$\Delta = n - n^* \tag{2.3}$$

and call Δ the *quantum defect* of a particular term. Then Δ is nearly zero for nonpenetrating orbitals but increases as the amount of penetration increases. The amount of penetration is greatest for s states and decreases with increasing l, so that f states are nonpenetrating for most of the elements of the periodic table.

These points are nicely illustrated in Fig. 2.1, which shows the main energy-level structure of the alkali metals, in comparison with hydrogenic values. Here, for example, one sees that the $n^*(6s)$ in $_{55}$Cs is less than 2, and so the corresponding quantum defect is more than 4. In contrast, the $n^*(4f)$ of Cs is exactly 4, as far as can be read from the diagram, indicating that $4f$ is a nonpenetrating orbital even for $Z = 55$. This same point is also indicated in Table 2.1, which shows the near constancy of the $4f$ term at its hydrogenic value for all the alkali metals. Table 2.2 gives the term values and quantum defects for the $n = 4$ levels in lithium ($Z = 3$), indicating clearly that $4s$ is more penetrating than $4p$ and that $4d$ and $4f$ are nonpenetrating.

Table 2.1 shows that even for nonpenetrating orbits there are small departures from the hydrogenic values. This is because the valence electron by its field induces an electric-dipole moment in the core so that the potential energy (2.1) (here $Z_0 = 1$) becomes modified to

$$V(r) = -\frac{e^2}{r}\left(1 + \frac{\alpha}{2r^3}\right) \tag{2.4}$$

Here α is the *polarizability* of the core [5], that is, the electric-dipole moment induced by unit applied electric field. The polarizability of the rare-gas cores can

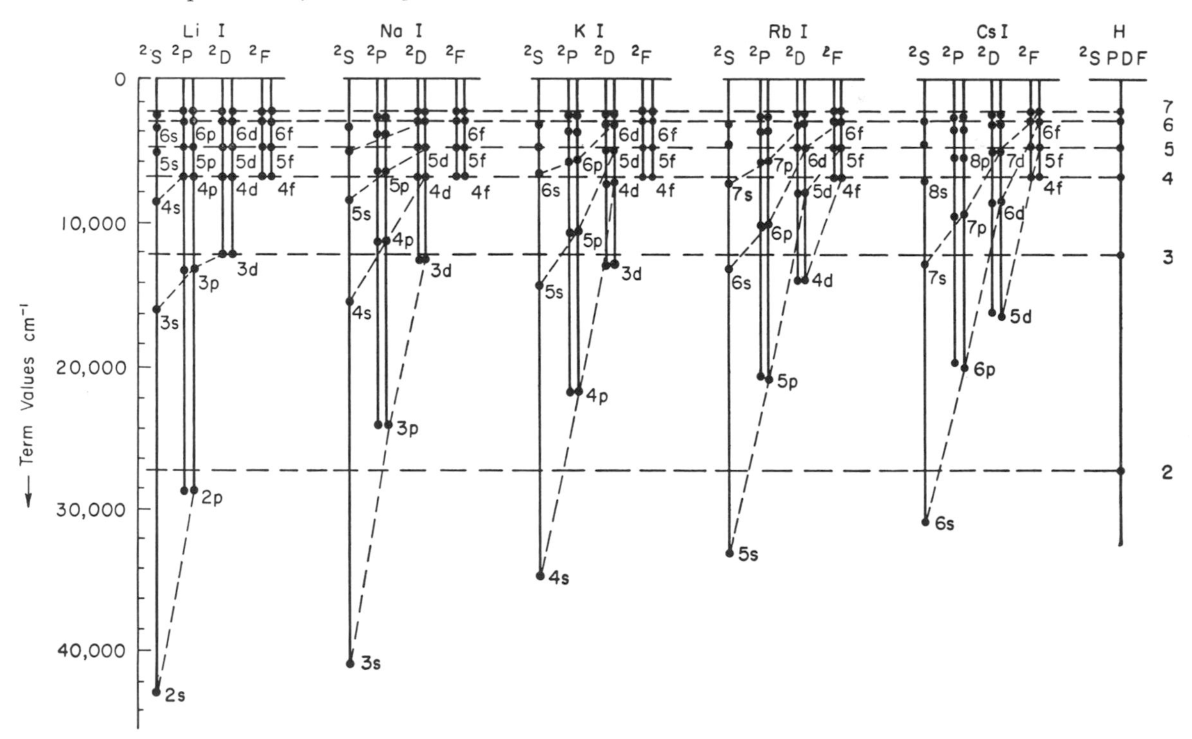

FIG. 2.1. Energy-level diagrams of the alkali metals lithium, sodium, potassium, rubidium, and caesium, and hydrogen. (*From: White, H. E.: "Introduction to Atomic Spectra," McGraw-Hill, New York,* 1934.)

TABLE 2.1. THE 4f TERMS IN THE ALKALI METALS

Element	$T(4f)$ cm^{-1}
H	6854.9
Li	6857.0
Na	6861.1
K	6882.1
Rb	6899.1
Cs	6935.2

TABLE 2.2. DEPENDENCE OF QUANTUM DEFECT ON l IN LITHIUM

Orbital	T, cm^{-1}	$\Delta(4l)$
4s	8475.13	0.402
4p	7017.64	0.046
4d	6863.80	0.001
4f	6857.0	0.000
(Hydrogen $n = 4$)	6854.9	0.000

be estimated from the index of refraction of the rare gases through the relation

$$\alpha = \frac{3}{4\pi N_0} \frac{n^2 - 1}{n^2 + 2} \tag{2.5}$$

where n is the index of refraction and N the number of atoms in unit volume (Chap. 6, Sec. 6). In this way these values of α are found for the rare-gas cores, measured in 10^{-24} cm^3:

	He	Ne	Ar	Kr	Xe
α	0.20	0.39	1.63	2.46	4.00

The trend with Z of the principal energy levels along the potassium isoelectronic sequence ($N = 19$) is shown in Fig. 1.9.

3. Two-electron Spectra

The helium ($N = 2$) isoelectronic sequence has a $^1S_0$ lowest level arising from the configuration $1s^2$, and most of the observed excited states are ^1L and ^3L levels of the $1s \cdot nl$ configurations. Here the spin-orbit interaction is quite small so that the singlet system hardly combines at all with the triplet system. Before the quantum-mechanical theory was developed it was thought for a time that there are two quite different kinds of helium atom; the singlet system was known as *parhelium*, and the triplet system was known as *orthohelium*.

The values of $T(1s^2,{}^1S)/Z_0{}^2$ are shown in Fig. 2.2 for the first seven members of the $N = 2$ isoelectronic sequence. This clearly shows a trend toward the hydrogenic value ($n = 1$) with increasing Z_0, analogous to the trends shown in Fig. 1.9 for the potassium isoelectronic sequence.

Figure 2.3 shows the principal low-lying energy levels of the two-electron spectra of the group II elements from beryllium to mercury. The ground state is in all cases $^1S_0$ from an ns^2 configuration.

Each excited configuration of the type ns, $n'l$ gives a ^1L and a ^3L. According to first-order theory

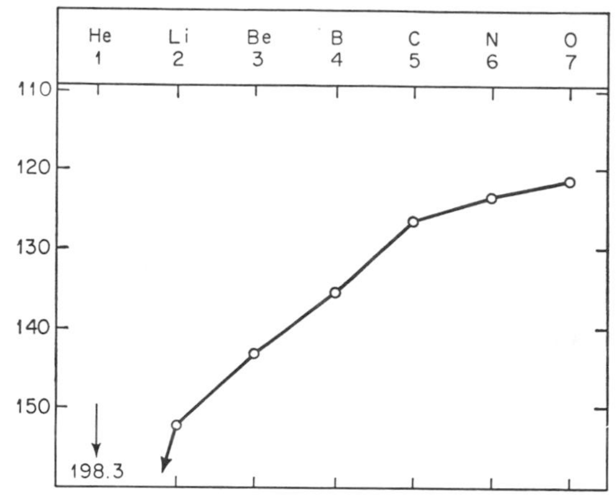

FIG. 2.2. Values of $T(1s^2,{}^1S)/Z_0{}^2$ for the first seven members of the helium ($N = 2$) sequence, showing the approach to the hydrogenic value at $n = 1$. T in thousands of cm^{-1}.

(Chap. 1, Sec. 18), the singlet should lie higher than the triplet by the amount of an exchange energy. This is usually observed to be the case, but in some of these spectra ^1D lies below the related ^3D, indicating a configuration interaction. In Fig. 2.3 the singlet and triplet energies are shown in columns appropriate to the l values, with the n' values labeled opposite each pair of terms.

Particularly noteworthy here is the large singlet-triplet interval for the $ns \cdot np$ configuration, the lowest of a particular p series, as compared with the higher-series members. Also evident is the tighter binding of an $ns \cdot np$ configuration when outside a d^{10} shell as compared with the binding when outside a p^6 shell. The f levels remain quite hydrogenic, all the way down the periodic table to Hg, where the $4f$ has been incorporated in a closed shell and the $5f$ has acquired unit quantum defect.

4. Compilations of Spectral Lines

For work in spectroscopy applied to spectrochemical analysis good compilations of spectral lines are essential. Some useful compilations are available but there is a great need for a thorough critical compilation of all the known material. For a compilation of levels, that of Moore [2] is the best available and gives detailed references to original sources. An older compilation of the same kind was made by Bacher and Goudsmit [6]. Another one is that of Brix and Kopfermann [7].

Dieke [8] has prepared a useful short list of important spectral lines. Longer lists are those of Harrison [9], Kayser and Ritschl [10], and Zaidel, Prokof'ev, and Raiskii [11]. A useful table for the vacuum ultraviolet has been prepared by Kelly [12]. For the complex spectra of the rare earths, tables have been published by Gatterer and Junkes [13].

Most tables of wavelengths provide only qualitative information about the relative intensity of spectral lines. An important list giving intensities on a quantitatively standardized basis has been prepared by Meggers, Corliss, and Scribner [14], from which

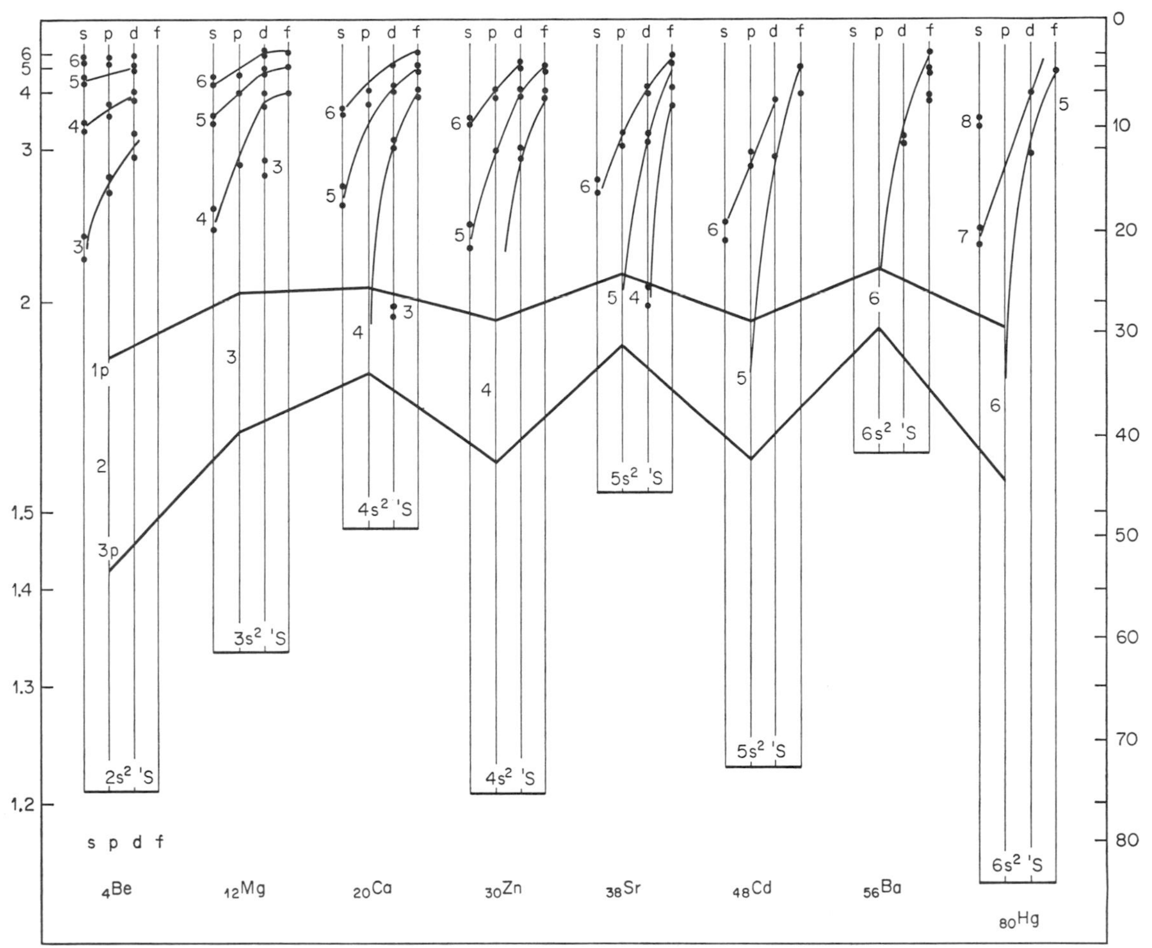

FIG. 2.3. Energy-level diagrams of the principal low-lying levels of the group II elements. Effective n values are shown on the left, and term values in thousands of cm^{-1} on the right.

Corliss and Bozman [15] have obtained measured transition probabilities for 25,000 lines.

Useful tables of multiplets have been published by Moore [16].

Of great usefulness for rapid interpretation of spectra are the Grotrian diagrams, of which an extensive compilation was published by W. Grotrian in 1928 [17]. These diagrams are somewhat outdated, but no comparable set is available. Figures 2.4 to 2.7 present modern diagrams for He I, Li II, Ne I, and Hg II. Dieke has also published some modern diagrams [8]. More of them are to be found in the book by Kuhn [3].

5. Zeeman Effect

In 1896 Zeeman [18] discovered a broadening effect in the yellow lines of sodium when the light source is placed in a strong magnetic field. Soon afterward, Lorentz predicted from classical electromagnetic theory that each line should split into three components with frequencies

$$\nu = \nu_0 - \frac{eH}{4\pi mc},\ \nu_0,\ \nu_0 + \frac{eH}{4\pi mc} \qquad (2.6)$$

where ν_0 is the zero-field frequency, H is the magnetic field strength in gauss, and e, m, and c are electron charge, electron mass, and the velocity of light. The positions and polarization of line components as predicted by Lorentz are shown in Fig. 2.8; p (or π) and s (or σ) refer to the parallel and normal polarizations observable when the light source is viewed perpendicular to the field direction. This splitting pattern is called a *normal*, or *Lorentz, triplet*. For directions of observation transverse to the magnetic field, the unshifted component is linearly polarized with its electric vector in the plane determined by the magnetic field and the direction of observation, the intensity falling to zero as $\sin^2\theta$ as the direction of observation approaches that along the magnetic field.

The shifted components show maximum intensity for observation along the magnetic field and drop to half this maximum for observation transverse to the magnetic field, the variation of their intensity with θ being $(1 + \cos^2\theta)/2$. For observation of light radiated along H, the high-frequency component shows left circular polarization, and the low-frequency component right circular polarization. For intermediate directions these become elliptic polarization until for

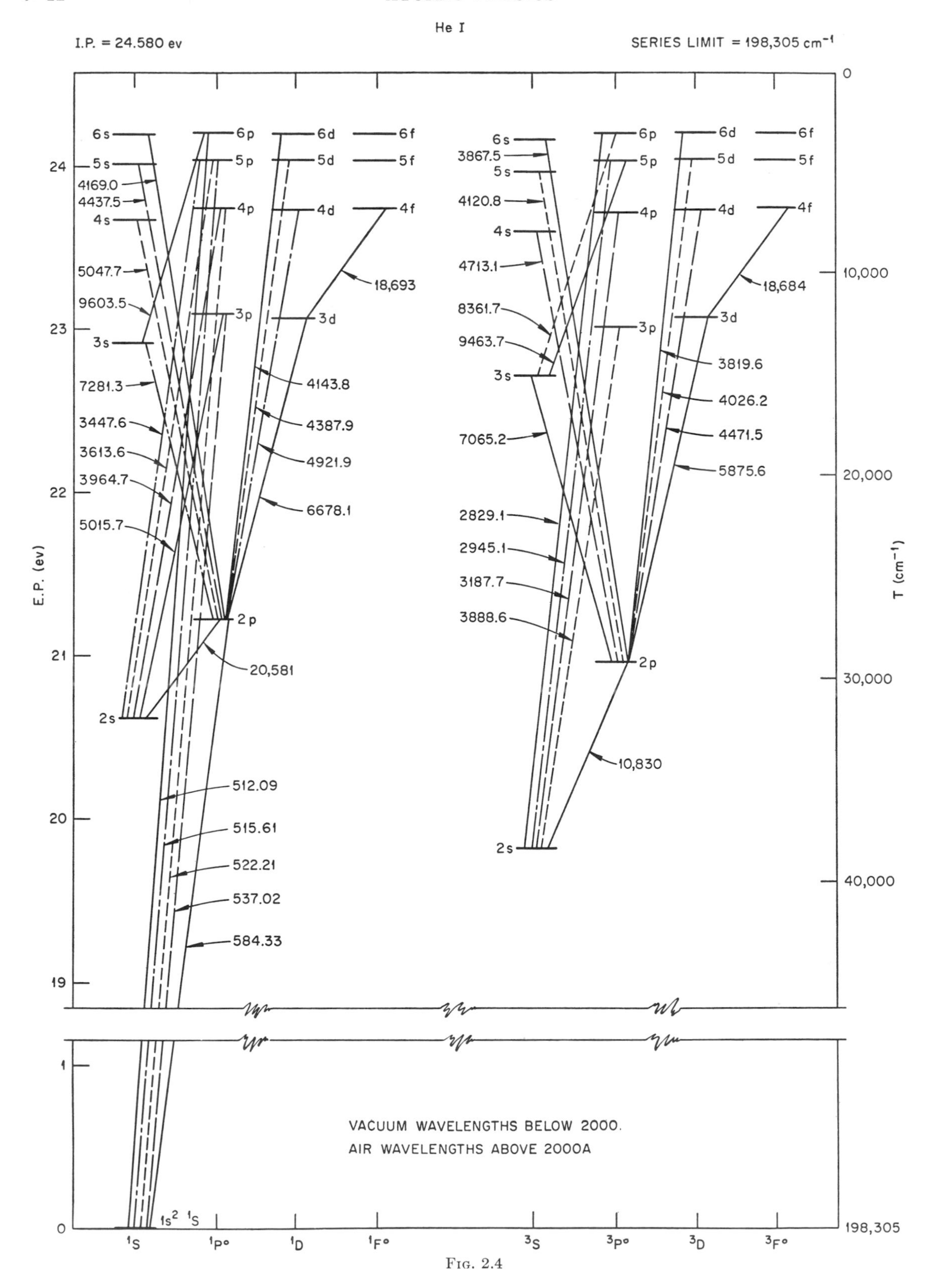

He I
I.P. = 24.580 ev
SERIES LIMIT = 198,305 cm⁻¹
E.P. (ev)
T (cm⁻¹)
6s
5s
4s
3s
2s
6p
5p
4p
3p
2p
6d
5d
4d
3d
6f
5f
4f
4169.0
4437.5
5047.7
9603.5
7281.3
3447.6
3613.6
3964.7
5015.7
18,693
4143.8
4387.9
4921.9
6678.1
20,581
512.09
515.61
522.21
537.02
584.33
3867.5
4120.8
4713.1
8361.7
9463.7
7065.2
2829.1
2945.1
3187.7
3888.6
18,684
3819.6
4026.2
4471.5
5875.6
10,830
1s² ¹S
¹S
¹P°
¹D
¹F°
³S
³P°
³D
³F°
0
10,000
20,000
30,000
40,000
198,305
24
23
22
21
20
19
1
0
VACUUM WAVELENGTHS BELOW 2000.
AIR WAVELENGTHS ABOVE 2000A

Fig. 2.4

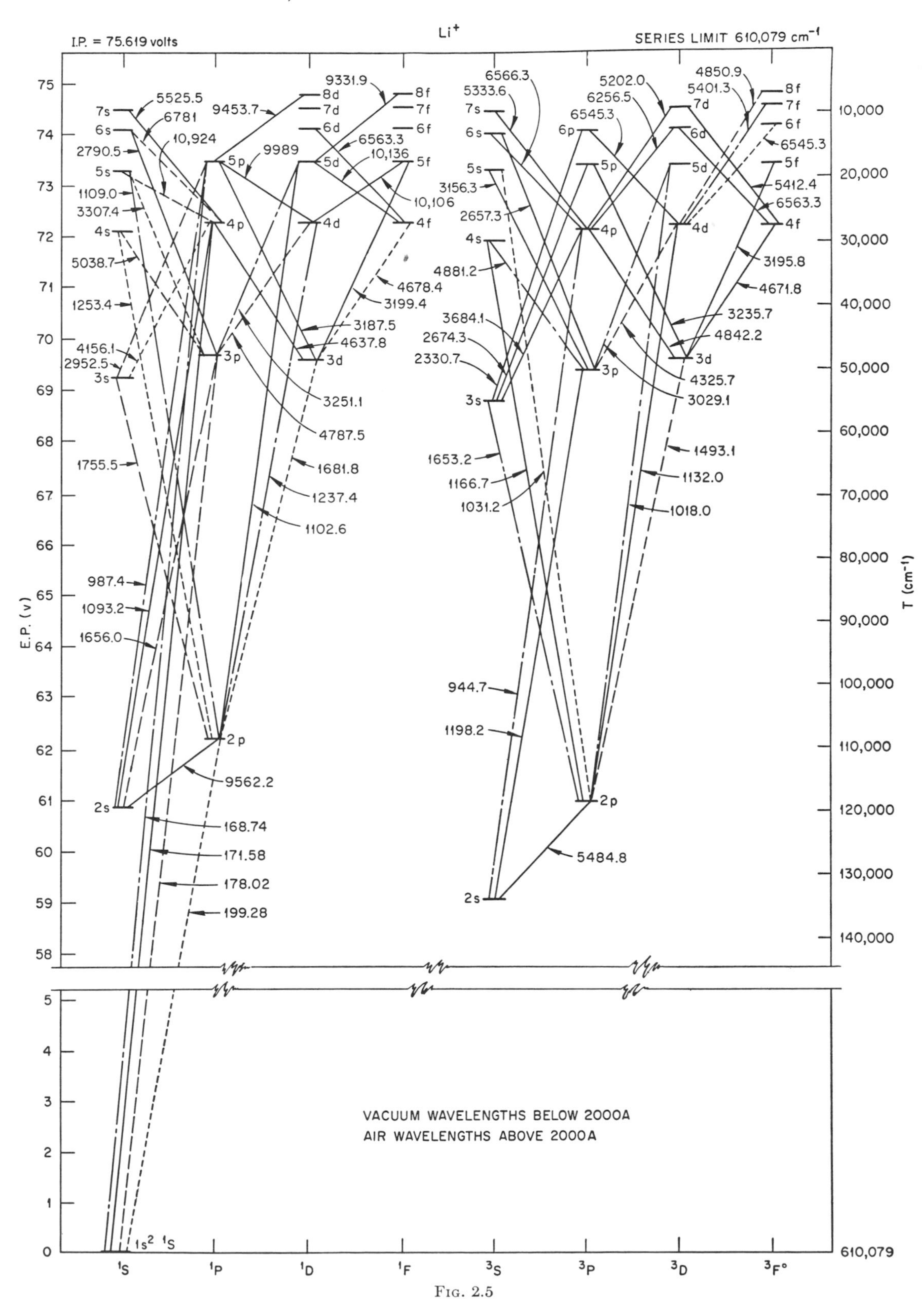
Li+
I.P. = 75.619 volts
SERIES LIMIT 610,079 cm-1
E.P. (v)
T (cm-1)
VACUUM WAVELENGTHS BELOW 2000A
AIR WAVELENGTHS ABOVE 2000A
1s2 1S
1S
1P
1D
1F
3S
3P
3D
3F°

Fig. 2.5

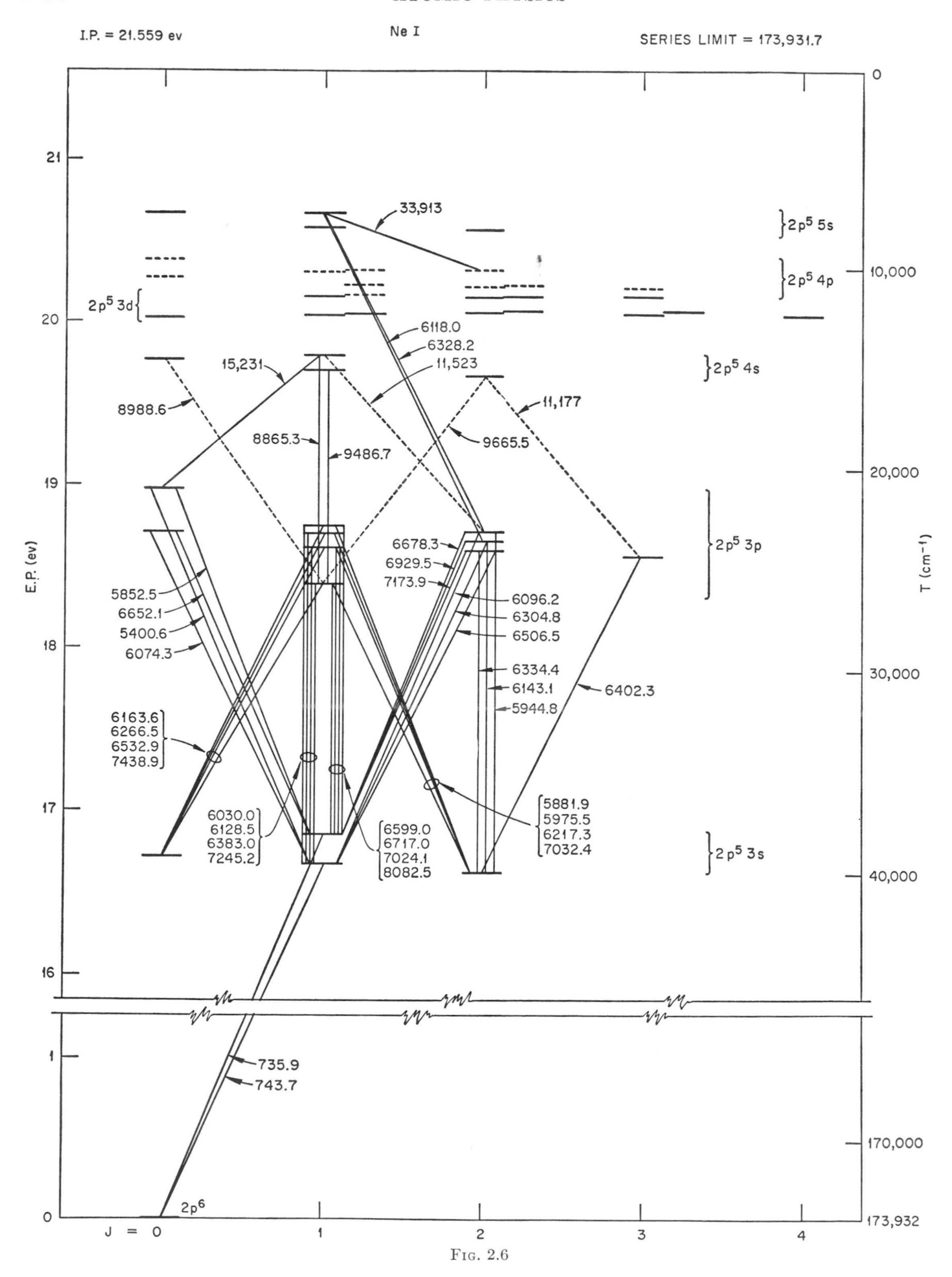
I.P. = 21.559 ev
Ne I
SERIES LIMIT = 173,931.7
2p5 5s
2p5 4p
2p5 3d
2p5 4s
2p5 3p
2p5 3s
2p6
33,913
6118.0
6328.2
11,523
15,231
8988.6
11,177
8865.3
9486.7
9665.5
6678.3
6929.5
7173.9
6096.2
6304.8
6506.5
6334.4
6143.1
5944.8
6402.3
5852.5
6652.1
5400.6
6074.3
6163.6
6266.5
6532.9
7438.9
6030.0
6128.5
6383.0
7245.2
6599.0
6717.0
7024.1
8082.5
5881.9
5975.5
6217.3
7032.4
735.9
743.7
E.P. (ev)
T (cm−1)
J = 0 1 2 3 4

Fig. 2.6

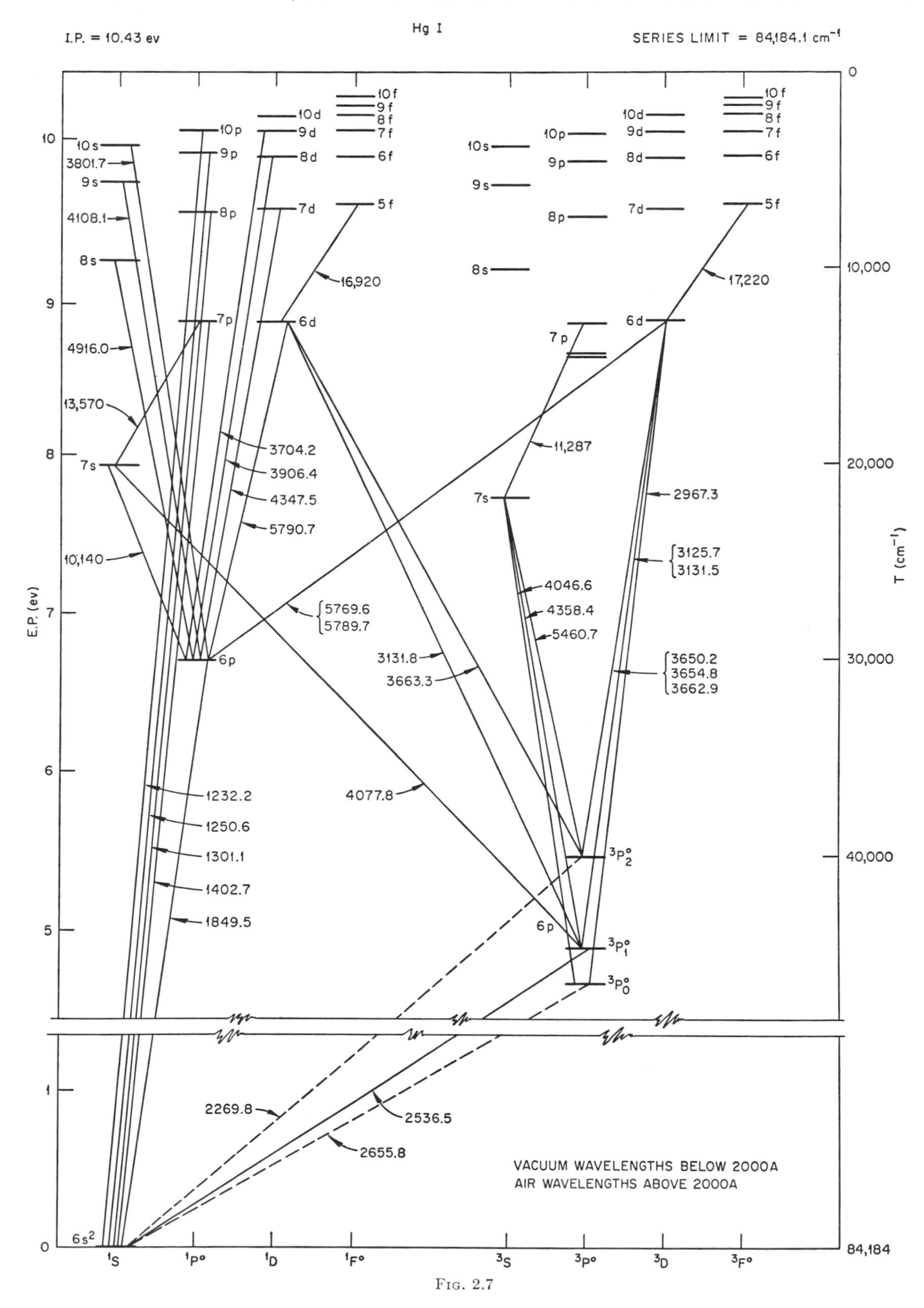

Hg I
I.P. = 10.43 ev
SERIES LIMIT = 84,184.1 cm⁻¹
E.P. (ev)
T (cm⁻¹)
3801.7
4108.1
4916.0
13,570
10,140
16,920
17,220
3704.2
3906.4
4347.5
5790.7
5769.6
5789.7
3131.8
3663.3
4077.8
11,287
2967.3
3125.7
3131.5
4046.6
4358.4
5460.7
3650.2
3654.8
3662.9
1232.2
1250.6
1301.1
1402.7
1849.5
2269.8
2536.5
2655.8
6s²
VACUUM WAVELENGTHS BELOW 2000A
AIR WAVELENGTHS ABOVE 2000A
84,184

FIG. 2.7

observation in transverse directions both components show linear polarization with their electric vector perpendicular to H. These are called s (normal) components, and the kind typified by the unshifted component is called p (parallel).

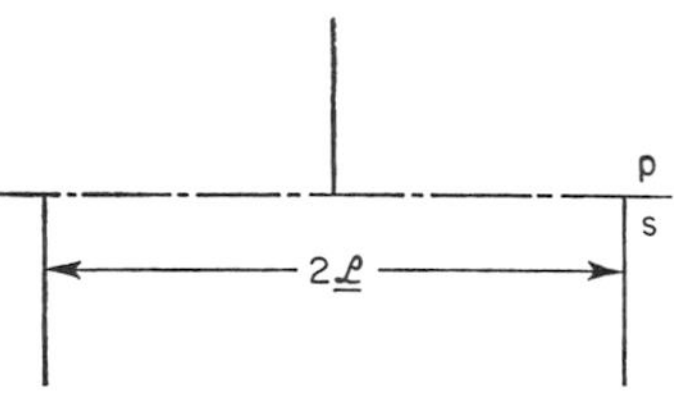

FIG. 2.8. The normal Zeeman effect: p (or π) refers to the parallel polarized components; s (or σ) refers to normal (senkrecht) polarized components. Separation of s components from p components is $4.669 \times 10^{-5}\, H$ cm^{-1} = £ (Lorentz unit).

Later investigation by Preston [19], using higher resolution, showed that most spectrum lines show more complex or "anomalous" patterns, which Runge correlated empirically with simple rational fractions of the normal-triplet separation. The Runge-law splitting for the D lines of sodium can be expressed fairly accurately as follows:

Wavelength	*Zeeman pattern*
5,896 A	$p(\pm 2/3$ £$)$, $n(\pm 4/3$ £$)$
5,890 A	$p(\pm 1/3$ £$)$, $n(\pm 3/3$ £, $\pm 5/3$ £$)$

Experimentally it is found that spectral lines arising from transitions between singlet terms show a Zeeman pattern in agreement with the normal triplet. Measurements of the splitting in known magnetic fields give an experimental determination of e/m which agrees with other determinations of this ratio for the electron. The agreement of this value with that from cathode-ray deflection experiments did much to establish the electronic theory of light emission long before the development of modern atomic theory.

For all other spectral lines, the observed patterns are more complicated than the normal triplet. For many years this *anomalous Zeeman effect* was a great puzzle that was only cleared up in 1925 by the electron-spin hypothesis [20]. The patterns in general can be described by supposing that the levels, characterized by J, become split into states characterized by different values of M, where M is the component of J along the magnetic field. Thus

$$W(J,M) = W_0 + gH \frac{e\hbar}{2mc} M \qquad -J \leq M \leq +J \tag{2.7}$$

so that each individual level becomes split into a close group of $2J + 1$ levels. Here g is a coefficient characteristic of each level that is known as the *Landé factor* [21].

The Zeeman pattern consists of lines arising from transitions between the level components of different M values, with M obeying the selection rule,

$$\Delta M = 0, \pm 1$$

so that the frequencies of the lines are

$$\nu = \nu_0 + \frac{eH}{4\pi mc} (g'M' - g''M'') \tag{2.8}$$

All the lines in the pattern for which $M'' = M' - 1$ show left circular polarization on observation along H, and those for which $M'' = M' + 1$ show right circular polarization. The lines for which $M'' = M'$ show parallel polarization in transverse observation.

Thus a general Zeeman pattern is an elaboration of the normal Zeeman triplet. If $g' = g'' = 1$, the pattern collapses into three groups of coincident lines forming the normal triplet. What is "anomalous" about the general case is therefore that the splittings of levels are different in magnitude from that given by classical Lorentz theory and are different in the initial and final states involved in forming the line.

The origin of this extra complication is now known to be that the magnetomechanical ratio for electron spin is twice that which applies to the orbital motion. The electron spin gives rise to a magnetic moment of $e\hbar/2mc$ (Bohr magneton) with an angular momentum of $\hbar/2$, whereas the same magnetic moment is produced by a full $\hbar$ of orbital angular momentum.

Therefore the interaction of the atom with the magnetic field is given to the first order by a term in the Hamiltonian

$$H_m = \frac{eH}{2mc} \hbar(L_z + 2S_z) \tag{2.9}$$

in which L_z is the resultant z component of orbital angular momentum and S_z is the resultant z component of spin angular momentum. Since

$$J_z = L_z + S_z$$

is a constant of the motion whose allowed values are M, it follows that the first-order alteration of the energy levels by the magnetic field is

$$\frac{e\hbar}{2mc} H(a\,J\,M|L_z + 2S_z|a\,J\,M) = g \frac{e\hbar}{2mc} H\,M$$

where

$$g = 1 + \frac{1}{M} (a\,J\,M|S_z|a\,J\,M) \tag{2.10}$$

Thus the anomaly whereby $g \neq 1$ arises entirely from the extra magnetic moment due to spin.

The extra structure arising in this way provides an experimental method of finding the initial and final J values involved in a particular spectral line. Moreover, in the Russell-Saunders case it is possible to give an explicit formula for g,

$$g = 1 + \frac{J(J+1) - L(L+1) + S(S+1)}{2J(J+1)} \tag{2.11}$$

so that measurement of the absolute intervals in a Zeeman pattern gives information helping to determine the initial and final LS values involved in production of a particular line. Similarly departures from Russell-Saunders coupling result in departures from the g values given by this formula. Some examples of the Zeeman patterns for transitions between doublet levels are given in Fig. 2.9.

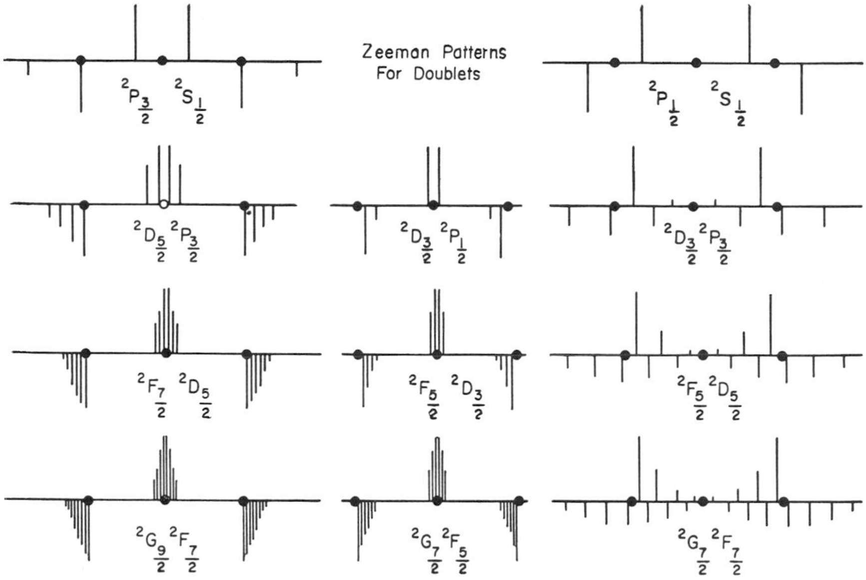

FIG. 2.9. Zeeman patterns of various doublet transitions. The dots represent the normal triplet separations. *(From: White, H. E.: "Introduction to Atomic Spectra," McGraw-Hill, New York,* 1935.)

In a more accurate approximation it is necessary to consider nondiagonal matrix elements of S_z in the magnetic term in the Hamiltonian. This matrix in diagonal in M; so M is an accurate quantum number at all field strengths, but it is not diagonal in J. Therefore strong magnetic fields produce an intermingling of states which tends to break down the coupling of **L** and **S** to give **J**. This gives rise to variations in the energy values which come from roots of a secular equation and which are therefore not linear in H. It also results in mixing the wave functions in such a way that they contain parts referring to more than one J. This results in an apparent breakdown of the J selection rule, by the appearance of Zeeman components at high magnetic fields which do not appear at weak fields. These more general variations of the Zeeman effect are known as the *Paschen-Back effect* after its discoverers [22].

The foregoing discussion neglects the effect of the magnetic field on the magnetic moment of the nucleus [23]. If the nuclear spin is **I** (with $\hbar$ as unit), the magnetic moment of the nucleus is commonly written

$$\mathbf{M} = g_N \frac{e\hbar}{2M_pc} \mathbf{I}$$

where M_p is the mass of the proton and $e\hbar/2M_pc$ is called the *nuclear magneton.* In terms of this natural unit, the values of g_N for nuclei are all of the order of unity, and therefore actual nuclear magnetic moments are of the order of 10^{-3} of those arising from the electrons in an atom.

The nuclear magnetic moment has a much greater effect on the Zeeman effect of hyperfine structure lines than might be supposed from the smallness of the nuclear magnetic moments. This is because the individual hyperfine levels are characterized by values of **F**, the vector resultant of **J** and **I**, which alters the way in which the electronic magnetic moment may orient along the magnetic field. The $g(F)$ value for a particular hyperfine level is related to the $g(J)$ value of the electronic level to which it belongs by the Goudsmit-Bacher formula,

$$g(F) = g(J) \frac{F(F+1) + J(J+1) - I(I+1)}{2F(F+1)}$$

The factor reduces to unity if $I = 0$, for then $F = J$.

Because the interval between hyperfine levels in the absence of a magnetic field is small, it is possible to produce Paschen-Back breakdowns on the coupling of **J** and **I** with relatively weak magnetic fields. This property is used by Rabi and associates in a variety of beautiful molecular beam methods for measuring the magnetic moments of nuclei [29].

Recently, a further correction to the magnetic moment of the spinning electron has been recognized [25]. This contributes to the spin magnetic moment in the form

$$\mu_s = \frac{(1.001145)S^*e\hbar}{2mc}$$

Theoretical g values can be revised to take into account this extra spin moment by the equation $g' = g + 0.00229(g - 1)$. This introduces a change of only 0.1 per cent in the theoretical g factor for the $^2S_{1/2}$ level (2.00229).

6. Intensity of Zeeman Components

The intensity of the ordinary Zeeman effect components follow the multiplet-type rules deduced by Burger, Dorgelo, and Ornstein [26] and known as the sum rules for intensities. The sum of the intensities of all transitions from (or to) any Zeeman level is equal to the sum of all transitions from (or to) any other level having the same n and l values. Formulas for the relative intensities of the Zeeman components are as follows (A and B are constants for any given Zeeman pattern):

$$J \to J \text{ type transition} \begin{cases} p \text{ components, } \Delta M = 0 \\ \qquad I = 4AM^2 \\ s \text{ components, } \Delta M = \pm 1 \end{cases}$$
$$I = A[J(J+1) - M(M \pm 1)]$$
$$J \to J+1 \text{ type transition} \begin{cases} p \text{ components, } \Delta M = 0 \\ \quad I = 4B[(J+1)^2 - M^2] \\ s \text{ components, } \Delta M = \pm 1 \end{cases}$$
$$I = B[J \pm M + 1][J \pm M + 2]$$

These formulas apply to components observed at right angles to the magnetic field. The s components can also be observed at this same intensity when viewed parallel to the field. This means that Zeeman

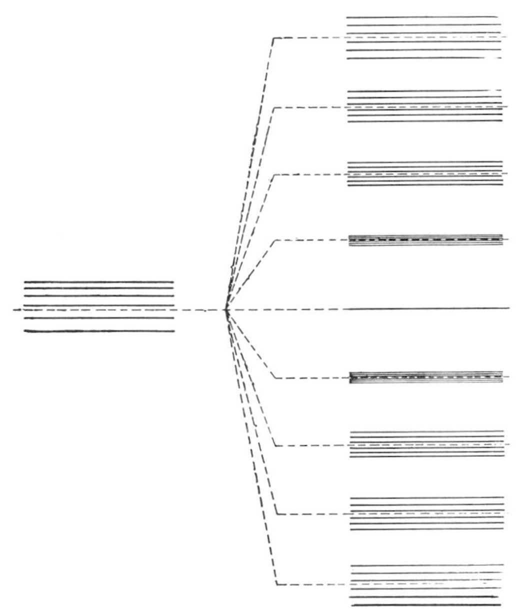

FIG. 2.10. Back-Goudsmit effect in praseodymium hyperfine level, $a^5I^0_4$. The nuclear spin of $I = \frac{5}{2}$ gives six hyperfine levels at zero field but 54 magnetic levels for strong fields. [*From: Rosen, N., G. R. Harrison, and J. R. McNally, Jr.: Phys. Rev.*, **60**: 722 (1941).]

exposures require approximately double the exposure time for the s polarization to obtain equivalent line densities as the p components. Typical Zeeman patterns of doublet transitions are shown in Fig. 2.9. Simple qualitative rules for the intensities are:

Transition Type	g Factors	Intensity Type
$J_1 = J_2$	$g_1 \neq g_2$	Symmetrical
$J_1 > J_2$	$g_1 > g_2$	s shade in
$J_1 > J_2$	$g_1 < g_2$	s shade out

The selection rules for weak field Zeeman effects were simply $\Delta M = 0, \pm 1$; however, for the strong field or Paschen-Back effect, M_j has no significance and the selection rules become $\Delta M_l = 0, \pm 1, \Delta M = 0$. In general, the Zeeman patterns at high fields approximate the normal triplet splitting, although several components may arise. The p and n components in the weak field case become p and n components in the strong field case or else fade out as forbidden lines (see Fig. 2.10).

The effect of a strong magnetic field on a hyperfine multiplet gives a somewhat analogous situation (see Fig. 2.11). Here, the energy change of the atom in the external magnetic field due to the magnetic moment of both nucleus and electrons is given by

$$-\Delta T = (Mg - M_I g_I)L + AM_I M \qquad \text{cm}^{-1}$$

where M_I and g_I refer to the nuclear magnetic quantum number and nuclear g factor, respectively. The structure is modified only slightly by the $M_I g_I L$ term, since g_I is of the order of 1/2,000 of the electronic g factor because of the nuclear mass occurring in the denominator of the expression for the nuclear magnetic moment.

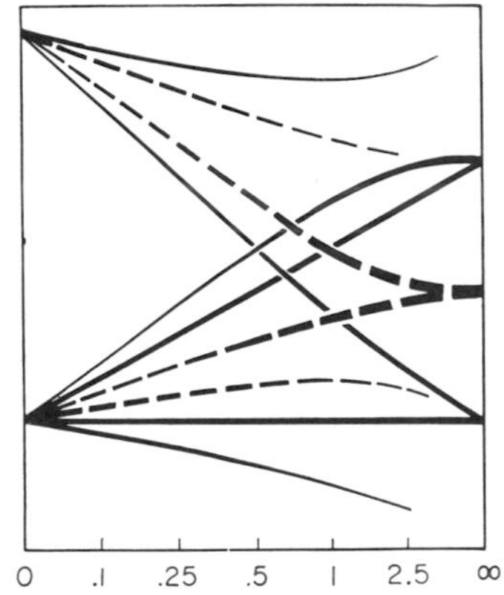

FIG. 2.11. Transition from the Zeeman effect (weak field) to the Paschen-Back effect (strong field) for the $^2S - {}^2P$ transition. (*From: Condon, E. U., and G. H. Shortley: "The Theory of Atomic Spectra."*)

7. Stark Effect

The effect of an electric field on spectrum lines was not discovered for some 16 years after Zeeman made his discovery of the effect of a magnetic field on spectrum lines. In 1913, Stark [27] studied the spectra emitted by atoms of hydrogen in a very strong electric field (100,000 volts/cm). Stark found that each line of the Balmer series of hydrogen was split into numerous components, that these components were field dependent as regards both position and intensity, and that the light was electrically polarized.

If the radiating atoms are viewed in a direction perpendicular to the field, some components are plane-polarized with the electric vector parallel to the field. These are referred to as π or p components in the literature. The remaining components are found to be plane-polarized with the electric vector perpendicular to the field. These are referred to as σ or s components. If the light source is observed along the direction of the electric field, only the σ or s components are observed and they are found to be unpolarized. The π or p Stark components are analogous to the p Zeeman components as regards polarization. However, the s Stark components differ from the s Zeeman components due to the noncoherent superposition of two components of opposite circular polarization, thus giving an unpolarized spectrum line.

The discovery of this electric field effect by Stark was made under fortunate circumstances. First, the magnitude of the effect in hydrogen is unusually large due to the large electric moment associated with a hydrogen-like atomic system. The second fortunate circumstance was the simultaneous development of the classical Bohr-Rutherford atom model. Epstein [28], Schwarzschild [29], and, later, Kramers [30] developed a highly satisfactory classical treatment of the hydrogen atom in an external electric field. Still later with the discovery of the spinning electron, Pauli, Wentzel, Epstein, Schroedinger, Waller, Schlapp, and Rojansky [31] investigated the problem from a quantum-mechanical viewpoint. It was found that minor second-order corrections to the classical model were necessary and a third-order term was also reported [32].

The energy of interaction between a hydrogen atom and the applied electric field may be represented by

$$\Delta T = AF + BF^2 + CF^3 + \cdots \quad \text{cm}^{-1} \qquad (2.12)$$

where ΔT is the correction to the term value of the atom, F is the electric field strength in volts per centimeter, and A, B, and C are constants referred to as the first-order, second-order, and third-order coefficients. The three terms in the equation are also referred to as the linear term, the quadratic term, and the cubic term, respectively. The theoretical expressions for these terms are

$$A = \frac{3h}{8\pi^2 mecZ(300)} nn_F \qquad (2.13)$$

$$B = \frac{h^5}{2^{10}\pi^6 m^3 e^6 cZ^3(300)^2} n^4(17n^2 - 3n_F{}^2 - 9m_l + 19) \qquad (2.14)$$

$$C = \frac{3h^9}{2^{15}\pi^{10} m^5 e^{11} cZ^5(300)^3} n^7 n_F(23n^2 - n_F{}^2 - 11m_l{}^2 + 39) \qquad (2.15)$$

where the coefficients of A, B, and C are 6.40×10^{-5} volts^{-1}, 5.19×10^{-16} cm volt^{-2}, and 1.52×10^{-25} cm^2 volt^{-3}, respectively, for hydrogen. In these expressions, n is the principal quantum number and is restricted to integral values 1, 2, 3, 4, . . . ; m_l is the component of the orbital quantum number l in the direction of the external field and is restricted to the integral values $-l$, . . . , $+l$; and n_F is called the electric quantum number and is restricted to integral values. This latter quantum number measures, in units of $\frac{3h^2n}{8\pi^2 Zem(300)}$, the average value of the electric moment of the atom in the direction of the external electric field. It is also frequently defined as $n_2 - n_1$ in terms of the so-called parabolic quantum numbers n_1 and n_2, which determine the classical range of the electron in the electric field parabolic coordinate system (see Fig. 2.12). If $n_2 = 0$, η_{min} is equal to η_{max} and the electron is restricted to move along the η paraboloid between the

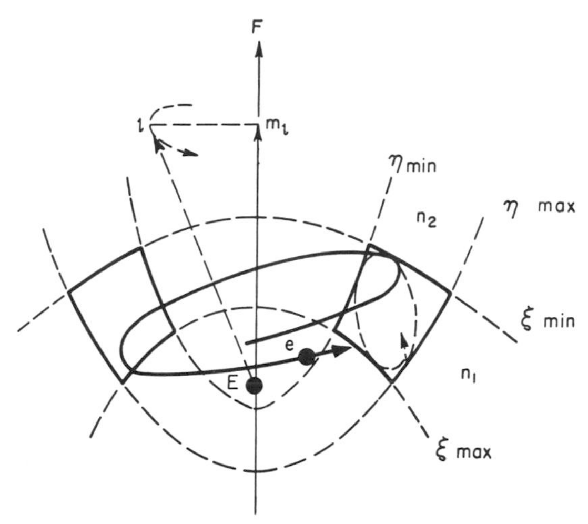

FIG. 2.12. Coordinate system for hydrogen-like atom in an electric field. (*From: White, H. E.: "Introduction to Atomic Spectra," McGraw-Hill, New York*, 1935.)

limits ξ_{min} and ξ_{max}. If n_1 and n_2 are both zero, the electron moves in a circular orbit with $m_l = \pm l$.

Weak Field Stark Effect in Hydrogen. The problem of the hydrogen atom in a weak electric field including spin-relativity corrections has been treated by Schlapp [33] and Rojansky [31]. A weak electric field is one for which the Stark splitting of levels is small compared to the fine-structure separation between levels. Electrostatic coupling occurs between the electric field and the electron orbital motion and only indirectly with the electron spin motion. Thus, the effect of the field is to induce a precession of the orbit or l^* vector about the field axis. Since the weak field is not able to disrupt the bond between the s^* and l^* vector momenta, the net result is that the resultant angular momentum vector j^* actually does the precessing. The splitting of hydrogenic levels in a weak electric field, to terms in F is [33]

$$\Delta T = -\frac{3hF}{8\pi^2 mecZ(300)} \frac{(l-j)n}{j(j+1)} \{[n^2 - (j+\tfrac{1}{2})^2]m_j{}^2\}^{1/2} \quad \text{cm}^{-1} \qquad (2.16)$$

This equation gives zero splitting for the highest j-valued term ($j = n - \frac{1}{2}$) associated with a given value of n. However, for $j = l + \frac{1}{2}$ we note an increase in the term value T, whereas for $j = |l - \frac{1}{2}|$, the ΔT is negative. The expected structures of the $n = 2$ levels of hydrogen for weak field and strong field are illustrated in Fig. 2.13. Until recently it had been thought that the $2\,^2S_{\frac{1}{2}}$ level and the $2\,^2P_{\frac{1}{2}}{}^0$ level fell together. The Lamb-Retherford [34] experiment showed a separation of these levels of 0.0353 cm^{-1}. In view of this, the weak field Stark effect in these hydrogen levels should be reinvestigated from a theoretical standpoint.

Strong Field Stark Effect in Hydrogen. In a strong field the coupling between l^* and s^* is broken down and l^* precesses rapidly about the field with s^* precessing more slowly about the time average

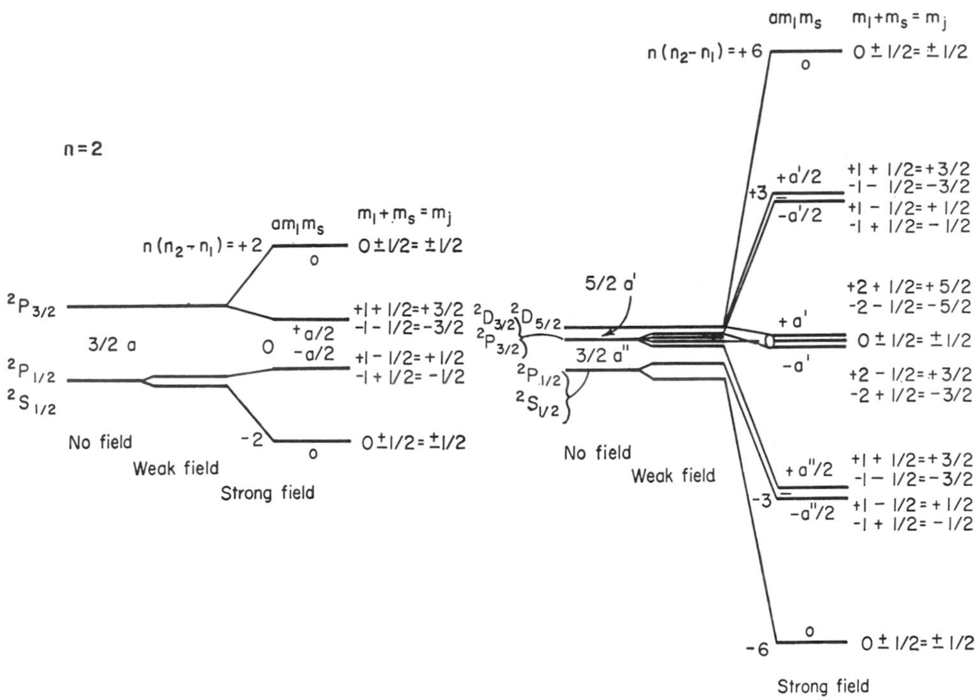

FIG. 2.13. Stark effect for $n = 2$ and $n = 3$ levels of hydrogen in weak and strong fields. (*From: White, H. E.: "Introduction to Atomic Spectra," McGraw-Hill, New York,* 1935.)

direction of l^*, which will be the field axis also. Thus, in the weak field case we refer to the quantum number m_j, whereas in the strong field case we use m_l and m_s. In the weak field case we obtain m_j values ranging from $\frac{1}{2}$ to j and in the strong field case the m_l values range from $+l$ to $-l$ and of m_s from $+s$ to $-s$. However, the energy states for $\pm m_l$ are degenerate and correspond to the same energy. The reason for this is that the energy contribution is determined by the orientation of the orbit in the field and not by the direction of electron motion in the orbit.

Another approach to the Stark effect picture is to consider the energy of (1) a permanent electric dipole in the electric field and (2) an electric dipole induced in the atom by the applied field. The former gives an energy $\mu(F/300)\cos\beta$, where μ is the electric moment of the atomic system and β the angle between the orbit major axis and the field. Using the expression for an electric moment of a hydrogen atom as $(\frac{3}{2})ae\epsilon$, where a is the semimajor axial distance and ϵ the eccentricity of the ellipse $(= \sqrt{n^2 - k^2}/n)$, we obtain for ΔE

$$\frac{3h^2nF}{8\pi^2meZ(300)}\sqrt{n^2 - k^2}\cos\beta$$

which corresponds to the linear effect

$$\Delta E = \frac{3h^2F}{8\pi^2meZ(300)}nn_F \tag{2.17}$$

The second-order Stark effect is essentially the energy contribution due to the induced electric moment. This amounts to $-\frac{1}{2}\alpha F^2$, where α is the polarizability of the atom; that is, $\alpha = -2Bhc$ from Eq. (2.12). The polarizabilities of numerous atoms and ions have been obtained in this way (see Table 2.3). Pauling [35] has made other calculations of the mole refractions by this same approach [see Eq. (2.5)].

TABLE 2.3. POLARIZABILITIES OF ATOMS AND IONS OBTAINED FROM STARK-EFFECT THEORY

Atom	*Polarizability,* $\alpha \times 10^{24}$
H	0.666
He^+	0.0416
Li^+	0.0292
Li^{++}	0.0082
Be^{++}	0.0079
B^{+3}	0.0030
C^{+4}	0.00134
Na^+	0.180
K^+	0.835
Rb^+	1.41
Cs^+	2.42

Stark Effect in Complex Atoms. The effect of an electric field on a penetrating electron in a nonhydrogen-like atom may be considered in terms of the time averaged electric moment of this system. Only when the electron is outside the electron core is there a sizable contribution to the average electric moment. Indeed, this contribution is practically negligible unless the electron is in a hydrogenic or nonpenetrating state. In this latter case, levels of the same n value have very closely the same energy and a first-order Stark effect will be observable; otherwise, only the second order or quadratic effect occurs. Studies on sodium [36] and potassium [37] resonance lines involving strongly penetrating orbits have shown no first-order effect and only small ($<0.1A$) second-order effects which correlate excellently with the expected F^2 dependence.

Sueoka and Sato [38] have studied the Stark effect of helium in fields up to 300 kv/cm and compared with theory. They find a perturbation of the 6^3H^0 term by the neighboring 7^3S, 7^3P^0, and 7^3D levels as the field gradient is increased. The pair of m_l levels approaching closest to each other and giving rise to resonance-type repulsion effects are as follows:

Repelling levels	m_l	Energy difference, K	Field strength, kv/cm
$6^3H - 7^3S$	0	6.6	177
$6^3H - 7^3P$	1	6.6	226
$6^3H - 7^3D$	2	11.0	301

Microwave and Radiofrequency Spectra. The study of microwave spectra has been considerably enhanced by application of the Stark effect. The Stark effect modulation of the microwave rotational transitions is a particularly useful technique in effectively increasing signal to noise ratios. Another use has been with molecules having no permanent electric-dipole moment (or in a Σ state for which the mean component of the electric moment in the field direction is zero) and hence no microwave spectrum. By the use of the Stark effect they will have a small dipole moment induced in them and transitions between the second-order Stark splittings can thereby be detected. This gives rise to the characteristic quadratic Stark effect and is observable in the radio-frequency region. For example, the splitting of the $J = 1$ state in HCl into $m = +1$ and $m = 0$ states is 0.0043 cm^{-1} (130 Mc) at 10,000 volts/cm. Using a field of only 265 volts/cm, Hughes [39] observed a splitting in CsF of only 0.001043 cm^{-1}. Molecules having a permanent electric-dipole moment give rise to the linear Stark effect in addition.

8. Intensity of Stark Components

Schroedinger [40] has calculated the intensities of the first four hydrogen line components, which are in quite good agreement with theory (see Fig. 2.14). However, the Stark effect produces many so-called "forbidden lines" or transitions involving other than the electric-dipole type of radiation characteristic of most line spectra. Long series of forbidden helium lines of the type $\Delta l = 0$ and $\Delta l = 2$ corresponding to transitions like $1s2s(^3S) - 1s\,ns(^3S)$ and $1s2s(^3S) - 1s\,nd(^3D)$, respectively, were obtained by Stark and others.

Another interesting feature of the Stark effect intensities is the coalescing of the series limit continuum with the highest-series members; i.e., the continuum extends to longer and longer wavelengths with increasing electric-field gradient. However, discrete spectrum lines appear superimposed on this continuum, but they are found to be relatively unsharp. This is somewhat analogous to the phe-

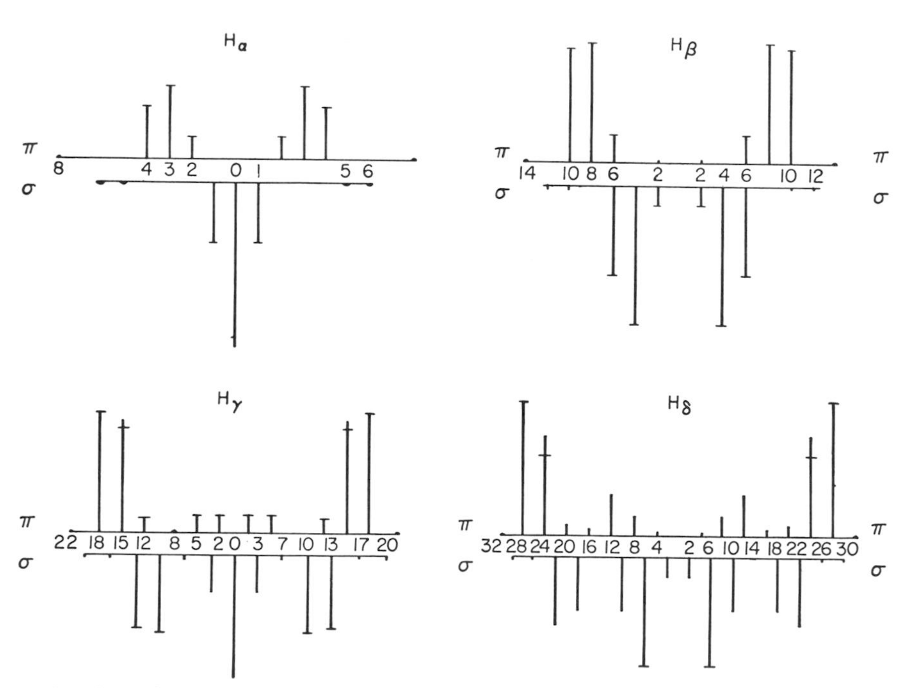

FIG. 2.14. Observed and predicted intensities in the first four Balmer lines of hydrogen. The theoretical relative intensities of the components are indicated by the line length; the cross bars indicate the measured values. (*From Condon, E. U., and G. H. Shortley: "The Theory of Atomic Spectra."*)

nomenon of autoionization broadening of copper lines involving transitions to levels above the series limit. This question of stability of the atomic system, where it has more than enough energy to self-ionize, has been compared to the analogy of a rolling ball in a chipped teacup. So long as the ball or electron does not approach the condition for leakage through the teacup or potential wall, it may be in a positive energy state. A comparison of the widely split red and violet components of H_2 showed the violet component to be sharper and stronger to much higher field strengths than the red component, despite the fact that the violet component involves a more positive energy state. This is aptly described by Condon and Shortley [41] as follows: "In the violet component's initial state the electron is cautious and conservative, avoiding the leaky barrier; in the red state the electron is reckless and wastrel, hurling itself against a less leaky barrier often enough to rob itself of the chance to shine!"

The Stark effect also produces a line broadening if the light source is in a field which changes rapidly during the spectrographic exposure. Such fields exist in high current density sources and give rise to what is called a collision-damping line breadth. Thus, a radiating atom may be perturbed by a closely approaching atom or ion such that the phase and amplitude of the radiation are changed. The half-intensity breadth of such collision damping is given by

$$\Delta\sigma = \frac{4Np\rho^2}{c\sqrt{\pi\mu RT}} \tag{2.18}$$

where $\Delta\sigma$ is the half-intensity breadth, N is Avogadro's number, p the pressure, ρ the collision dameter, μ the reduced mass of the colliding atom, R the gas constant, and T the absolute temperature. The pressure broadening observed for the 2,537 A line of mercury is about 0.46 cm^{-1} per atmosphere pressure of CO_2 or 0.28 cm^{-1} per atmosphere pressure of N_2. These pressure shifts are linear with pressure to at least 50 atm. Collision diameters calculated from Eq. (2.18) are about three times larger than those calculated from kinetic theory.

Collision damping incorporates the radiation wave train which is characteristic of the emitting atom prior to or subsequent to the collision itself. This interruption of the electromagnetic wave introduces a frequency broadening due to the introduction of additional Fourier components in the shortened wave trains. The radiation wave train characteristic of the period during the collision is affected even more seriously and introduces both a line broadening effect and an asymmetry or pressure shift effect. This is known also as the incipient Stark effect in that the close approach of a foreign atom or ion produces an induced electric moment or polarization of the radiating atom and the accompanying quadratic level shift. Usually this is manifested as a depression of the energy levels in question, with the less tightly bound levels being affected more seriously. However, pressure shifts are observed in both directions and are found to depend on both the particular energy levels involved and the type foreign atoms constituting the gas. As an example, the principal series lines of rubidium are displaced to the violet by helium and neon, but to the red by argon [42].

References

1. Meggers, W. F.: *J. Opt. Soc. Am.*, **41**: 143 (1951); **43**: 415 (1953).
2. Moore, C. E.: Atomic Energy Levels, *Natl. Bur. Std. (U.S.)*, *Circ.* 467, vol. 1, 1949; vol. 2, 1952; vol. 3, 1958.
3. Kuhn, H. G.: "Atomic Spectra," Academic, New York, 1962.
4. Edlén, B.: "Encyclopedia of Physics," vol. 27, pp. 80–201, Springer, Berlin, 1964.
5. Born, M., and W. Heisenberg: *Z. Physik.*, **23**: 388 (1924).
6. Bacher, R. E., and S. A. Goudsmit: "Atomic Energy States," McGraw-Hill, New York, 1932.
7. Brix, P., and H. Kopfermann: in "Landolt-Börnstein Tables," 6th ed., vol. 1, pp. 1–69, Springer, Berlin, 1951.
8. Dieke, G. H.: Atomic and Molecular Physics, "American Institute of Physics Handbook," sec. 7, McGraw-Hill, New York, 1963.
9. Harrison, G. R.: "M.I.T. Wavelength Tables," Wiley, New York, 1939.
10. Kayser, H., and R. Ritschl: "Tabelle der Hauptlinien der Linienspektren aller Element," Springer, Berlin, 1939.
11. Zaidel, A. N., V. K. Prokofiev, and S. M. Raiskii: "Tables of Spectral Lines," VEB Verlag Technik, Berlin, 1955.
12. Kelly, R. L.: Vacuum Ultraviolet Emission Lines, *Univ. Calif., Lawrence Radiation Lab. Rept.* UCRL 5612, 1959.
13. Gatterer, A., and J. Junkes: "Spektren der Seltenen Erden," Specola Vaticana, Vatican City, 1945. Junkes, J., and E. W. Salpeter: "Spectrum of Thorium from 9400 to 2000 Å," Specola Vaticana, Vatican City, 1964.
14. Meggers, W. F., C. H. Corliss, and B. F. Scribner: Tables of Spectral Line Intensities, *Natl. Bur. Std. (U.S.)*, *Monograph* 32, pts. 1 and 2, 1961.
15. Corliss, C. H., and W. R. Bozman: Experimental Transition Probabilities for Spectral Lines of Seventy Elements, *Natl. Bur. Std. (U.S.)*, *Monograph* 53, 1962.
16. Moore, C. E.: An Ultraviolet Multiplet Table, *Natl. Bur. Std. (U.S.)*, (*Circ.* 488), secs. 1–5, 1950–1961, and A Multiplet Table of Astrophysical Interest, *Natl. Bur. Std. (U.S.)*, *Tech. Note* 36, pts. 1 and 2, 1959.
17. Grotrian, W.: "Graphische Darstellung der Spektren von Atomen und Ionen mit Ein, Zwei und Drei Valenzelektronen," Springer, Berlin, 1928. Reprint: Edwards, Ann Arbor, Mich., 1946.
18. Zeeman, P.: *Phil. Mag.*, **43**: (5) 226; **44**: 55, 255 (1897). Lorentz, H. A.: *Ann. Physik*, **60**: 1519 (1897).
19. Preston, T.: *Phil. Mag.*, **45**: 325 (1898).
20. Uhlenbeck, G. E., and S. A. Goudsmit: *Naturwiss.*, **13**: 953 (1925); *Nature*, **107**: 264 (1926).
21. Landé, A.: *Z. Physik*, **5**: 234 (1921); **7**: 399 (1921); **15**: 189 (1923).
22. Paschen, F., and E. Back: *Ann. Physik*, **39**: 897 (1912); **40**: 960 (1913); *Physica*, **1**: 261 (1921).
23. Back, E., and S. A. Goudsmit: *Z. Physik*, **47**: 174 (1928).
24. Kellogg, J. B. M., and S. Millman: *Revs. Mod. Phys.*, **18**: 323 (1946).
25. Schwinger, J.: *Phys. Rev.*, **73**: 416 (1948). Karplus, R., and N. Kroll: *Phys. Rev.*, **77**: 536 (1950). McNally, J. R., Jr.: *Phys. Rev.*, **73**: 1130 (1948). Breit, G.: *Phys. Rev.*, **72**: 984 (1947).

26. Burger, H. C., and H. B. Dorgelo: *Z. Physik*, **23**: 258 (1924). Ornstein, L. S., and H. Burger: *Z. Physik*, **22**: 170; **24**: 41 (1924).
27. Stark, J.: *Sitzber. Deut. Akad. Wiss. Berlin*, **40**: 932 (1913).
28. Epstein, P. S.: *Ann. Physik*, **50**: 489 (1916); *Physik. Z.*, **17**: 148 (1916).
29. Schwarzschild, K.: *Sitzber. Deut. Akad. Wiss. Berlin*, **43**: 548 (1916).
30. Kramers, H. A., *Z. Physik*, **3**: 199 (1920).
31. Rojansky, V.: *Phys. Rev.*, **33**: 1 (1929).
32. Doi, S.: Results are quoted in a paper by Y. Ishida and S. Hiyama, *Sci. Papers Inst. Phys. Chem. Res. (Tokyo)*, **9**: 1 (1928).
33. Foster, J. S.: *Phys. Rev.*, **23**: 667 (1924). Foster found that the displacements are linear to at least 38,000 volts/cm.
34. Lamb, W. E., Jr., and R. D. Retherford: *Phys. Rev.*, **81**: 222 (1951).
35. Pauling, L.: *Proc. Roy. Soc. (London)*, **A114**: 181 (1927).
36. Ladenburg, R.: *Ann. Physik*, **78**: 675 (1925).
37. Grotrian, W., and G. Ramsauer: *Physik. Z.*, **28**: 846 (1927).
38. Sueoka, S., and M. Sato: *J. Phys. Soc. Japan*, **6**: 444 (1951).
39. Hughes, H. K.: *Phys. Rev.*, **72**: 614 (1947).
40. Schroedinger, E.: *Ann. Physik*, **80**: 457 (1926).
41. Condon, E. U., and G. H. Shortley, Jr.: "The Theory of Atomic Spectra," Cambridge University Press, New York, 1935.
42. For recent surveys of the extensive literature on collision broadening:
Griem, H. R.: "Plasma Spectroscopy," McGraw-Hill, New York, 1964.
Baranger, M.: Spectral Line Broadening in Plasmas, in D. R. Bates (ed.), "Atomic and Molecular Processes," chap. 13, Academic, New York, 1962.

Chapter 3

Atomic Line Strengths

By LAWRENCE ALLER, University of California

1. Atomic Radiation Processes

Classical theory, according to which all accelerated charges radiate, gives a suitable starting point for the discussion of atomic radiation processes. A system of moving charges will emit a radiation field that can be treated as though it originated from an electric-dipole, a magnetic-dipole, an electric-quadrupole, plus possibly high-order-multipole radiations that can be neglected in all practical applications [1].†

A dipole is defined as two equal and opposite charges q separated by a distance r. The dipole moment is

$$\mathbf{P} = q\mathbf{r} \tag{3.1}$$

If the dipole be set in oscillation with a frequency $\nu = \omega/2\pi$,

$$\mathbf{P} = \mathbf{P}_0 e^{i\omega t} = q\mathbf{r}_0 e^{i\omega t} \tag{3.2}$$

If the wavelength, λ, of the emitted radiation is very much larger than r_0, nearly pure dipole radiation will result. In fact, if r_0 approaches zero and q increases such that P_0 remains constant, pure electric-dipole radiation will be emitted.

Magnetic-dipole radiation is emitted when charges flow back and forth in a curved path, whereas quadrupole radiation is emitted by a system in which a charge $2q$ is placed at the origin and two charges $-q$, on opposite sides, vibrate back and forth in synchronism in such a way that the dipole moment is always zero. For optical transitions, electric-dipole emission, whenever it is allowed, is much more important than magnetic-dipole or electric-quadrupole radiation. In nuclear γ-ray transitions, however, the quadrupole component of the radiation can assume considerable importance.

Electromagnetic theory shows that if classical dipole oscillators of frequency ν_0 are subjected to a beam of natural monochromatic radiation of frequency ν, the dipoles will be set in forced oscillation with frequency ν. As a consequence the beam will suffer extinction and refraction. If $(\nu - \nu_0) \ll \nu_0$ the coefficients of extinction and refraction will be, respectively [2]

† Numbers in brackets refer to References at end of chapter.

$$k' = \frac{N\epsilon^2}{4\pi mc} \frac{\gamma}{(\nu - \nu_0)^2 + (\gamma/4\pi)^2} \tag{3.3}$$

$$n = 1 + \frac{N\epsilon^2}{4\pi m\nu} \frac{\nu_0 - \nu}{(\nu - \nu_0)^2 + (\gamma/4\pi)^2} \tag{3.4}$$

where the symbols ϵ, m, c have their usual meaning, N is the number of oscillators of frequency ν_0 per unit volume, and γ is the classical damping constant,

$$\gamma = \frac{8\pi^2\epsilon^2}{3mc^3}\nu^2 \tag{3.5}$$

Equation (3.4) is the expression for the anomalous dispersion in the neighborhood of a line.

Classical theory requires that the number of dispersion electrons of natural frequency, ν_0, per atom must be an integer. Weisskopf and Wigner [3] showed that Eq. (3.3) describes the absorption coefficient of an atom in the quantum theory, provided one replaces the classical damping constant γ by the quantum-mechanical damping constant Γ [see Eq. (3.11)] and assigns to the spectral line an *oscillator strength*, the Ladenburg f, which is no longer an integer.

The absorption coefficient per atom, α_ν, is

$$\alpha_\nu = \frac{\pi\epsilon^2}{mc} f \frac{\Gamma}{4\pi^2} \frac{1}{(\nu - \nu_0)^2 + (\Gamma/4\pi)^2} \tag{3.6}$$

Often the Einstein coefficients of emission and absorption are used. Suppose that from a level n atoms are able to escape to the level n' with the spontaneous emission of radiation of frequency $\nu(nn')$. If N_n atoms are constantly maintained in level n by collisions, recombinations, and cascade from higher levels, the number of quanta of frequency $\nu(nn')$ emitted per cubic centimeter per second will be $N_n A_{nn'}$, where $A_{nn'}$ is the Einstein coefficient of spontaneous emission. If radiation of frequency $\nu(nn')$ strikes atoms in level n, they will be triggered to emit further quanta of frequency $\nu(nn')$. The total number of downward transitions for an intensity $I_{\nu(nn')}$ will be $N_n A_{nn'} + N_n B_{nn'} I_{\nu(nn')}$, where $B_{nn'}$ is called the coefficient of induced emission or *negative absorption*. The latter designation is better since the induced quantum is ejected in the same direction as the quantum that does the triggering. Finally, if the atoms in a lower level n' are exposed to radiation of frequency $\nu(nn')$ of intensity $I_{\nu(nn')}$, the number of upward transitions per cubic centimeter per second will be $N_{n'}I_{\nu(nn')}B_{n'n}$. The Einstein A's and B's are invariant for any given line. They are independent of conditions of excitation, of the temperature, etc. By substituting for $I_{\nu(nn')}$ the Planckian function for intensity

$$I_\nu = \frac{2h\nu^3}{c^2}\frac{1}{e^{h\nu/kT} - 1}$$

the following relationships may be shown to hold between them (see, for example, [2, p. 169]):

$$\tilde{\omega}_n B_{nn'} = \tilde{\omega}_{n'} B_{n'n} \tag{3.7}$$

$$\frac{\tilde{\omega}_n}{\tilde{\omega}_{n'}} A_{nn'} = B_{n'n}\frac{2h\nu^3}{c^2} \tag{3.8}$$

Here $\tilde{\omega}_n$ is the statistical weight of level n. If its inner quantum number is J,

$$\tilde{\omega}_n = 2J + 1 \tag{3.9}$$

The Einstein coefficients are here defined in terms of intensity rather than in terms of radiation density, as is often done [1], because it is the intensity rather than the radiation density that is actually measured.

The quantum-mechanical damping constant for pure radiation damping $\Gamma_{nn'}$ may be expressed in terms of the Einstein coefficients.

$$\Gamma_{nn'} = \Gamma_n + \Gamma_{n'} \tag{3.10}$$

where

$$\Gamma_n = \sum_{n'} A_{nn'}(1 - e^{-h\nu/kT})^{-1} + \sum_{n''} A_{n''n}\frac{\tilde{\omega}_{n''}}{\tilde{\omega}_n}(e^{h\nu/kT} - 1)^{-1} \tag{3.11}$$

The first summation is taken over all levels n' lower than n; the second over all levels n'' above level n. If the intensity of the radiation is not large, it suffices to take

$$\Gamma_n = \sum_{n'} A_{nn'} \tag{3.12}$$

One may show (see, for example, [2, p. 175]) that the Einstein A value is related to the f value by

$$A_{nn'} = \frac{8\pi^2\epsilon^2\nu^2}{mc^3}\frac{\tilde{\omega}_{n'}}{\tilde{\omega}_n} f_{nn'} \tag{3.13}$$

The energy levels involved may be characterized by their configurations (denoted as γ) and the quantum numbers L, S, and J. Thus the transition between two levels may be indicated as

$$(\gamma LSJ) \rightarrow (\gamma' L'S'J')$$

If the transition is a "permitted" one, i.e., of the electric-dipole type, the parity of the configuration must change in accordance with the Laporte rule, L can go to L or $L \pm 1$, and J to J or $J \pm 1$ except that $J = 0$ to $J = 0$ is excluded. If the atom is in good LS coupling S cannot change.

In addition to f values and Einstein coefficients it is useful to introduce the "line strength," a quantity which is symmetrical in the upper and lower levels [1]. It is related to the f value by the expression:

$$f(\alpha'J';\alpha J) = \frac{8\pi^2 mc}{3h\epsilon^2}\frac{1}{\tilde{\omega}'\lambda} S(\alpha'J';\alpha J) \tag{3.14}$$

where the primes denote the lower level. The expressions for the strengths do not explicitly involve the wavelengths or frequencies of the emitted radiation, whereas the Einstein coefficients and f's do. This means that if we are concerned with multiplets or configuration arrays in which the lines fall at different wavelengths, we may obtain the relative strengths quickly from appropriate tables or formulas (at least for atoms in good LS coupling) whereas the relative f or A values require that the wavelength of each line be included.

The line strengths may be expressed in terms of the dipole moments for the transition involved between the individual Zeeman states. Thus

$$S(\alpha'J';\alpha J) = \sum_{M,M'} |<\alpha'J'M'|\mathbf{P}|\alpha JM>|^2 \tag{3.15}$$

where M' and M denote the magnetic quantum numbers for the lower and upper states, respectively. Hence the problem of determining S is essentially that of finding the wave functions for the states of a complex atom.

Like the f or A value, the strength of a line is an atomic constant. The *intensity*, however, depends on the physical excitation conditions. The relative, as well as the absolute, intensities of the lines in a given spectrum will vary in a more or less complicated fashion with change of excitation conditions. A knowledge of the relative intensities of the lines in a spectrum is of little help in obtaining empirical line strengths unless something is known about the mode of excitation.

In addition to the absorption oscillator strengths an emission oscillator strength is defined as follows:

$$f_{nn'}^e = -\frac{\tilde{\omega}_{n'}}{\tilde{\omega}_n} f_{n'n} \tag{3.16}$$

Although it is necessary to use the emission oscillator strengths in problems such as the application of sum rules, ordinarily it is better to employ conventional Ladenburg f values.

Some writers (e.g., Unsöld [4]) have introduced a pseudo f value:

$$f^*(\alpha'J';\alpha J) = f(\alpha'J';\alpha J)\frac{\tilde{\omega}_{J'}}{(2S + 1)(2L' + 1)}$$

but the use of such quantities should be avoided.

2. Formulas and Tables for Line Strengths

If the wavelength is expressed in angstrom units (10^{-8} cm) and S in atomic units ($a_0^2\epsilon^2$, where a_0 is the radius of the first Bohr orbit), Eq. (3.14) becomes

$$f(\alpha'J';\alpha J) = \frac{304}{\tilde{\omega}'\lambda} S(\alpha'J';\alpha J) \tag{3.17}$$

and the corresponding expression for the Einstein A is

$$A(\alpha J;\alpha' J') = \frac{2.02 \times 10^{18}}{\bar{\omega}\lambda^3} S(\alpha J;\alpha' J') \quad (3.18)$$

The strength S may be expressed as the product of three factors, namely,

$$S(\alpha J;\alpha' J') = \mathcal{S}(\mathfrak{M})\mathcal{S}(\mathfrak{L})\sigma(\alpha';\alpha)^2 \quad (3.19)$$

where $\mathcal{S}(\mathfrak{M})$ may be taken as the absolute strength of the whole multiplet, $\mathcal{S}(\mathfrak{L})$ is the strength of the line divided by the sum of the strengths of all the lines of that particular multiplet, and

$$\sigma(\alpha';\alpha) = \frac{1}{\sqrt{4l^2 - 1}} \int_0^\infty R(\alpha')R(\alpha)r^3\, dr \quad (3.20)$$

Here $R(\alpha')$ and $R(\alpha)$ are, respectively, the radial quantum wave functions for the lower and upper states of the transition. These wave functions are normalized and orthogonal in the usual ways, namely,

$$\int_0^\infty R^2(\alpha)r^2\, dr = \int R^2(\alpha')r^2\, dr = 1$$

$$\int_0^\infty R(\alpha)R(\alpha')r^2\, dr = 0$$

Here l is the greater of the two azimuthal quantum numbers involved in the wave functions $R(\alpha)$ and $R(\alpha')$.

The transition $\alpha LSJ - \alpha' L'S'J'$ gives rise to a single spectral line.

The transitions from all the levels of one term (common αLS) to all the levels of another term (common $\alpha' L'S'$) comprise a *multiplet.*

All the transitions from the totality of terms based on a single parent (polyad) to another polyad compose a *supermultiplet.*

Finally, all the transitions from one configuration to another configuration make up a *transition array.*

The relative strengths of the lines within a multiplet were computed first by formulas given by the old quantum theory and later by appropriate quantum-mechanical formulas. A convenient tabulation may be found in [2, pp. 601–603]. Goldberg [5] calculated extensive tables of relative multiplet strengths (see also [2, p. 604]). He also gives the factors for converting the relative strengths to S/σ^2. Rohrlich [6] has published formulas based on the Racah [7] methods for calculating line strengths. They apply to situations where Goldberg's tables are inadequate. Extensive tables of Racah coefficients necessary for calculations by Rohrlich's formulas are given, for example, by Griem [8].

Most atoms deviate from LS coupling although they are still far from jj coupling. Intercombination lines such as singlet-triplet or doublet-quartet combinations may attain considerable strength in this state of *intermediate coupling.* Line strengths may still be computed [9] but each element in each stage of ionization requires separate treatment because the parameter χ [see Eq. (3.35)] describing the departure from LS coupling is different for each of them. Thus far, relatively few calculations have been carried out. Garstang finds the deviations from LS coupling for the $3s$–$3p$ transition array to be small for O II (as one might have anticipated from the Landé g factors) and somewhat larger for the same transition array in Ne II. For some transitions Garstang found the effects of departure from LS coupling to be large and that intermediate coupling resulted in a considerable improvement in the line strengths. For example, Gottschalk [10] has calculated the relative strengths for the $3d^74s$–$3d^74p$ transition array in Fe I and finds fair agreement with the relative f values measured by R. B. King. In view of the astrophysical applications there is urgent need for further computations of this sort, especially for such metals as iron in higher stages of ionization where empirical f values are not available.

Exact f values may be calculated for hydrogenlike atoms where the radial wave functions can be stated as confluent hypergeometric functions and the integral (3.20) can be evaluated exactly. Green et al. [11] have calculated the necessary integrals; Infeld and Hull [12] provide recursion formulas that allow easy computation over a large range of (n,l). For hydrogen, we have [13]

$$f_{n'n} = \frac{64}{3\sqrt{3}\,\pi} \frac{1}{\bar{\omega}_{n'}} \frac{1}{[1/n'^2 - 1/n^2]^3} \left| \frac{1}{n^3} \frac{1}{n'^3} \right| g \quad (3.21)$$

Many years ago, Kramers derived a similar formula with the aid of the correspondence principle. The factor g is a correction required by the quantum theory. It was first introduced by Gaunt and accurate values have been calculated by Menzel and Pekeris [13].

In more complex atoms, the radial quantum wave functions are not accurately known; so approximate methods such as the perturbation theory or variation method must be used. In these techniques a minimum value of the energy, involving integrals of the form $\int \psi \frac{1}{r} \psi'\, d\tau$, is evaluated. On the other hand, in the computation of f values we are concerned with the dipole moment $\int \psi r \psi'\, d\tau$. Thus, for the energy the value of ψ near the origin is important, whereas the value of ψ at relatively great distances from the origin is important for the f value. An approximate ψ function that gives a good representation of the energy may yet give poor f values.

The negative hydrogen ion provides an extreme example in point. Bates and Massey calculated a wave function that represented the energy extremely well but gave a poor representation of the absorption curve of the H^- ion. S. Chandrasekhar [14] showed that if one supplemented the computation of the dipole moments by calculations of velocity and acceleration moments, much better estimates could be made of the absorption coefficient. That is, in the calculation of the transition probability for the Zeeman component $(\alpha JM - \alpha' J'M')$, namely,

$$A(\alpha JM;\alpha' J'M') = \frac{64\pi^4\nu^3}{3hc^3} |(\alpha' J'M'|\mathbf{P}|\alpha JM)|^2 \quad (3.22)$$

the radial quantum integral $\sigma(nl,n'l')$ is usually calculated as a dipole moment

$$\sigma(nl,n'l') = \frac{1}{\sqrt{4l_>^2 - 1}} \int_0^\infty R(nl;r)R(n'l';r)r^3\, dr \quad (3.23)$$

where $l_>$ is the greater of the two azimuthal quantum

numbers. Alternatively, following Chandrasekhar [14] we may evaluate

$$\sigma^2(nl,n'l') = \frac{1}{\sqrt{4l_>^2 - 1}} \left\{\frac{2}{W(nl) - W'(n'l')}\int_0^\infty \left[R(nl;r)\,\frac{dR(n'l';r)}{dr} + \begin{Bmatrix} l+2 \\ -(l-1)\end{Bmatrix} \frac{RR'}{r}\right] r^2\,dr\right\} \tag{3.24}$$

where $\begin{Bmatrix} l+2 \\ -(l-1)\end{Bmatrix}$ is $l+2$ if $l' = l+1$ and $-(l-1)$ if $l' = l-1$

or

$$\sigma^2(nl;n'l') = \frac{1}{\sqrt{4l_>^2 - 1}} \left\{\left[\frac{2}{W(nl) - W'(n'l')}\right]^2 \int_0^\infty R(n'l';r)\,\frac{dV}{dr}\,R(nl;r)r^2\,dr\right\} \tag{3.25}$$

If the values obtained from all three integrals agree, one may have reasonable confidence in the transition probabilities calculated.

With modern, high-speed electronic computers it is possible to calculate wave functions by Hartree's method of self-consistent fields not only for the ground level but for excited levels as well [15]. The calculations should include exchange and also configuration interaction if possible. Both numerical and analytical procedures have been developed by various investigators [16, 17]. The differential equations of the Hartree-Fock type are complicated to treat but wave functions have been obtained by both iterative and analytical methods for about 70 atoms and ions of the $2p^r$ and $3p^r$ types. Heavier atoms and ions, for example, Y II, Zr II, have not been studied extensively because of the complexities of the Hartree-Fock equations. It is important to recognize that in f-value calculations the effects of exchange must be included.

Subordinate transitions between the s, p, d, and f excited configurations in atoms such as C, N, O, S, etc., may be handled with the aid of the tables calculated by Bates and Damgaard [18] for the evaluation of the radial quantum integral σ defined by Eq. (3.20). They assume that for high levels the potential V varies as C/r, where C is the excess charge on the nucleus when the active electron is detached. For any given level one may assign an effective quantum number

$$n^* = \frac{C}{\sqrt{\epsilon}} \tag{3.26}$$

where ϵ is the energy necessary to remove the electron from the atom expressed in atomic units.† Then the wave equation can be solved to give an asymptotic solution which is deemed adequately valid for the outer part of the atom where the chief contribution to the f value originates. The value of σ may then be expressed in terms of C, n^*, and the azimuthal quantum number l of the jumping electron as parameters, namely,

$$\sigma(n^*_{l-1},l-1;n^*_l l;C) = \frac{1}{C}\,\mathfrak{F}(n^*_l,l)\mathfrak{I}(n^*_{l-1},n^*_l,l) \tag{3.27}$$

† That is, the energy is expressed in units of 13.60 ev, the ionization energy of hydrogen.

for the transition $(n^*,l-1) \rightarrow (n^*,l)$. $C = 1$ for a neutral atom, 2 for a singly ionized atom, etc. Bates and Damgaard tabulate the function $\mathfrak{F}(n^*_l,l)$ for $l = 1$ (s–p transitions), $l = 2$ (p–d transitions) and $l = 3$ (d–f transitions). For $\mathfrak{I}$ a double-entry table is required for each of the s–p, p–d, and d–f transitions. With the aid of a term table such as National Bureau of Standards circular 467 σ may be computed quickly for any given line and the total strength evaluated easily. Improvements in this method have been proposed by a number of workers.

The complexity of the computation of line strengths in heavy atoms is enhanced because of the failure of configuration assignments (configuration interaction) in some instances [1, chap. 15]. That is, a given level may partake not only of, say, a $3p^23d^3$ configuration but also of a $3p^23d^24d$ configuration.

These configuration interactions must be handled on an individual basis for each transition array. No general results are available, but on the basis of calculations by Biermann and Trefftz [19] and by Trees [20], Green [21], and others, Czyzak concluded that A values may be decreased by as much as 25 per cent. Kingsbury [22] has found that the effects of configuration interaction on f values for the $3s$–$3p$ transition array in O I are not negligible; the f values may differ from the Bates-Damgaard predictions by as much as 40 per cent. Various calculations indicate that by introducing this correction the positions of energy levels can be accounted for in a more satisfactory fashion than heretofore.

It appears that the usual spin-orbit, spin-spin, and electrostatic interactions as well as configuration interaction cannot account completely for the observed positions of energy levels and the deviations of the line intensities from the observed values. Layzer [23] and also Trees [20] have suggested a magnetic spin-free coupling between the electrons, the change in energy produced by such a coupling being proportional to $L(L+1)$. The effect is comparable in magnitude to the spin-orbit interaction. It affects only the term splitting and therefore may be grouped with the electrostatic rather than with the spin-orbit interaction. The numerical value of the new parameter varies smoothly from one configuration to another, its value being roughly one-quarter that of the spin-orbit parameter. Various calculations indicate that by introducing this correction the positions of energy levels can be accounted for in a more satisfactory fashion than heretofore.

Certain *sum rules* provide valuable checks on theoretical and experimental determinations of f values. The *Thomas-Kuhn f-sum rule* [24] states that the sum of the f values of all lines arising from a single atomic energy level shall equal the number of optical electrons, i.e., the number of electrons that participate in giving the optical line and continuous spectrum. Menzel's sum rule [25] for transition arrays says that the sum of the multiplet strengths over the transition array is equal to

$$\sum_{L'S'} S(\gamma nlLS;\gamma n'l'L'S') = \frac{\bar{\omega}}{2}[2(2l+1) - k][2(2l'+1) - k']l_>\sigma^2 \tag{3.28}$$

where γ denotes the parent configuration whose statistical weight as defined by Menzel is $\bar{\omega}$, $l_>$ is the larger of the angular momentum l values of the jumping electron, k and k' are the number of electrons in γ to which the jumping electron is equivalent, and σ is defined by Eq. (3.20). If configuration interaction is negligible, this sum is independent of the type of coupling.

As a third example consider resonance lines in a complex atom; the transition array is $nl^h - nl^{h-1}n''l''$. The jumping electron is one of h equivalent electrons in the lower configuration. Shortley's J-file sum rule [26] states that the sum of the strengths of all the multiplets ending in a term (LS) of nl^h is given by

$$h(2S+1)(2L+1)(l+1)(2l+3)\sigma^2 \tag{3.29}$$

if $l'' = l + 1$, and by

$$h(2S+1)(2L+1)l(2l-1)\sigma^2 \tag{3.30}$$

if $l'' = l - 1$. Examples of the application of these rules may be found in the literature. Rohrlich [27] has derived some useful sum rules.

3. Continuous Atomic Absorption Coefficients

In addition to their discrete line spectra, atoms absorb continuous spectra which correspond to the ionization of the electron from a bound level of the atom. If α_ν is the continuous absorption coefficient, one may write (see, for example, [2, p. 182])

$$\alpha_\nu = \frac{\pi\epsilon^2}{mc}\frac{df}{d\nu} \tag{3.31}$$

where $df/d\nu$ is the oscillator strength per unit frequency for the continuum. For hydrogen it is possible to obtain an exact expression for the continuous absorption coefficient. From Eq. (3.21) one may derive

$$\alpha_n(\nu) = \frac{32}{3\sqrt{3}}\frac{\pi^2\epsilon^6}{ch^3}\frac{R}{n^5\nu^3}g \tag{3.32}$$

where R is the Rydberg constant and g is the Gaunt factor, extensive tabulations of which have been published by Karzas and Latter [28]. At any frequency ν the absorption coefficient of hydrogen gas will depend on the number of levels which can contribute at that wavelength. Thus photoionizations from the $n = 1$ level can contribute to absorption only at wavelengths shorter than λ912, the Balmer continuum starts at λ3646, etc. Between two consecutive series limits the total atomic absorption coefficient falls off as ν^{-3}. It rises to a higher value as a series limit is crossed in the direction of increasing frequency. The relative importance of the Lyman, Balmer, and Paschen continua will depend on the number of atoms in the $n = 1$, 2, or 3 levels and therefore on the temperature.

For a more refined treatment of the hydrogen problems and for calculations involving hydrogenlike levels (such as highly excited states of atoms and ions of C, N, O, Ne, Si, S, Ar, etc.), Eq. (3.32) is inadequate and must be replaced by a more comprehensive formulation such as that given by Burgess [29]. The absorption coefficient for a level (n,l) and energy-dependent parameter k^2 is given by

$$\alpha_{nl}(k^2) = \left(\frac{4\pi\alpha a_0^2}{3}\right)\frac{n^2}{Z^2}\sum_{l''=l\pm1}\frac{\max(l'',l)}{2l+1}\Theta(n,l;k^2,l'') \tag{3.33}$$

where α = Bohr magneton, a_0 = Bohr radius, Z = atomic charge, and k^2 is defined by the Rydberg formula for the continuum:

$$\nu = Rc\left(\frac{Z^2}{n^2} + k^2\right) \tag{3.34}$$

Here $\Theta(n,l,k^2,l'')$ is proportional to the square of the radial quantum integral for the bound-free transitions.

In helium and more complex atoms the absorption coefficient does not fall off as ν^{-3}. Often it rises beyond the series limit, reaches a maximum, and then falls off. This reason for this dissimilarity with the hydrogen behavior lies in the fact that the electron in the hydrogen atom moves in a field of the same *effective* charge whether it is bound or not, whereas this is not true for other atoms. Consider nitrogen or oxygen as an example. The outer electrons are bound in a p shell and the screening of any one of these by the others is not complete. The effective charge $Z_{\text{eff}}e > e$. When one of these p electrons is detached from the atom, the screening by the remaining electrons is complete and $Z_{\text{eff}} = 1$.

Burgess and Seaton [30] have shown how one may generalize the Bates-Damgaard method to bound-free transitions. They use approximate free-state radial functions, which have exact asymptotic forms and derive a general formula at least as accurate as those obtained in the best alternative methods of calculation. They summarize results of their extensive numerical calculations in tables that permit a rapid calculation of transition integrals once the energy levels are known. Griem [8, Tables 5–8, pp. 534–536] tabulates the continuous absorptions, $\alpha_{me}(\nu)$, for various levels of H I, He I, C I, N I, and O I.

For the ground terms, one may use the Hartree-Fock wave functions with exchange and proper continuum wave functions for the free electrons. Much of the early theoretical work was done on the alkali metals, particularly by Hylleraas. These atoms have a simple structure—a single s electron outside a tightly closed shell. Their ionization potentials are small and they are easily vaporized; hence their continuous absorptions are easy to observe. Unfortunately, theoretical calculations have so far given no more than the correct order of magnitude. The wave functions show nodes and loops so arranged that the $\int R_b R_f r^3\, dr$ integral almost vanishes and a change of the order of 2 per cent in the wave function will lead to a 100 per cent error in the absorption coefficient. The situation is a little better for the alkaline earths where calculations have been made by Bates and Massey and by Louis Green and Nancy Weber; Biermann and Lübeck have calculated the continuous absorption coefficients of ionized magnesium and silicon [31].

4. Forbidden Lines

A forbidden transition is one that violates the Laporte parity rule which states that the change in the l value of the jumping electron is ± 1. Forbidden lines represent magnetic-dipole or electric-quadrupole transitions. If spin-spin and spin–other-orbit interactions can be neglected, the strengths S_m and S_q may be calculated for p^2 and p^4 configurations once and for all in terms of a parameter

$$\chi = \frac{\zeta_p}{5F_2} \tag{3.35}$$

which is essentially a measure of the departure from LS coupling. For pure LS coupling $\chi = 0$ and it increases without bound as jj coupling is approached. F_2 and ζ_p are parameters which depend on electrostatic and magnetic spin-orbit interactions, respectively.†

The Einstein coefficient for spontaneous emission is related to the strength, S_m, for magnetic-dipole radiation by [1, 32]

$$A_m(J,J') = 35{,}320\,\frac{(\tilde{\nu}/R)^3 S_m(J,J')}{2J+1} \quad \text{sec}^{-1} \tag{3.36}$$

where $\tilde{\nu}$ is the wave number of the line, R is that of Lyman limit, 109,737 cm^{-1}. The strength $S_m(J,J')$ is given by

$$S_m(J;J') = \sum_{M,M'} |(\alpha JM|\mathbf{M}|\alpha' J'M')|^2 \tag{3.37}$$

$$\mathbf{M} = -\frac{\epsilon}{2mc}(L + 2S) \tag{3.38}$$

In strict LS coupling, magnetic-dipole transitions can occur only between levels of the same term. Shortley [33] showed that

$$S_m(J, J+1) = \frac{(J - L + S + 1)(J + S + L + 2) \times (S + L - J)(J + S - L + 1)}{4(J+1)} \tag{3.39}$$

Transitions of the type of the green nebular lines λ4959, λ5007 of [O III] or the red [O I] lines observed in the auroral spectrum arise because of deviations from LS coupling. Their strength depends on χ^2.

Likewise, [1, 32] the electric-quadrupole transition probability is

$$A_q(J,J') = 2{,}648\left(\frac{\tilde{\nu}}{R}\right)^5 \frac{S_q(J,J')}{2J+1} \quad \text{sec}^{-1} \tag{3.40}$$

Here

$$S_q(\alpha J;\alpha' J') = S_q(\alpha' J';\alpha J) = \sum_{M,M'} |(\alpha JM|\mathfrak{M}|\alpha' J'M')|^2 \tag{3.41}$$

where $\mathfrak{M}$ is a dyad:

$$\mathfrak{M} = \sum_s \mathfrak{M}_s = \epsilon \sum_s (r_s r_s - \tfrac{1}{3} r_s^2 J) \tag{3.42}$$

with

$$J = ii + jj + kk \tag{3.43}$$

as the identity matrix. One can write

$$S_q(\alpha J;\alpha' J') = C_q(\alpha J;\alpha' J') S_q^2(nl,nl) \tag{3.44}$$

† For a definition of the F's and of ζ, see E. U. Condon and G. H. Shortley, "Theory of Atomic Spectra," pp. 177, 120, 195, Cambridge University Press, New York, 1951.

where

$$S_q(nl,nl) = \tfrac{2}{5}\epsilon \int_0^\infty r^4 R^2(nl)\, dr \tag{3.45}$$

Tables of S_m and C_q as a function of χ have been given by Shortley, Aller, Baker, and Menzel [34] for p^2, p^3, and p^4 configurations. The evaluation of S_q requires a knowledge of the radial wave function $R(nl)$.

Garstang [35] showed how Racah's methods could be applied to calculate the electric-quadrupole strengths $C_q(\alpha' J';\alpha J)$ for transitions involving configurations such as $d^n \to d^n$, $d^n \to d^{n-1}s$, $d^{n-1}s \to d^{n-2}s^2$. In the p^3 configurations of [N I], [O II], and [S II], the tables of Shortley and his colleagues are not valid. The theory must be refined to take into account spin–other-orbit and spin-spin interactions [36]. These normally second-order effects must be taken into account in refined treatments of many other lines, as illustrated, for example, in the calculations of Garstang [37] on [O III], [N II], etc.

Since the pioneer investigations by Condon, calculations of transition probabilities for forbidden lines have been carried out by many workers. Particular mention should be made of calculations by Pasternack [38] for many lines of astrophysical interest including transitions involving d electrons. B. Edlén, R. Garstang, Czyzak and Krueger, and D. Osterbrook have calculated A values for lines observed in the solar corona and other high excitation sources. Some of the most significant transitions for which forbidden-line data are needed involve the elements from scandium to nickel, in various stages of ionization. Iron is the most important of these elements; some work on [Fe VII] and [Fe VI] was done by Pasternack, but the most complete study is that by Garstang [39] who computed the line strengths for [Fe II], [Fe III], [Fe IV], and [Fe V]. Strengths for [Ni II], [Ni III], and many other elements are also available.

Among light elements, the lifetimes of the upper levels of prominent forbidden transitions range from about a half second for the green λ5577 [O I] auroral line to something like seven hours for the $^2D_{\frac{5}{2}}$ level in [O II]. The ^2D–^4S transition of [O II] is often prominent in the spectra of gaseous nebulae in our own and other galaxies.

Forbidden transitions are prominent in the aurora and light of the night sky [40], in the solar corona [2, p. 537], novae (exploding stars), certain peculiar stars such as η Carinae [41], gaseous nebulae, and the interstellar medium [42]. Probably the most important forbidden transition in celestial sources is the 21-cm "spin-flop" jump in ground-state hydrogen. This emission is observed in cold hydrogen gas in this and other galaxies; its measurement yields important information on the physical state and kinematics of the interstellar material.

5. The Atomic Line Absorption Coefficient

Under natural conditions of excitation spectral lines are broadened by [43]:

1. Doppler effect due to the random kinetic motions of the radiating atoms
2. Natural broadening due to the finite widths of atomic energy levels

If the radiating gas has an appreciable density, there also occurs:

3. Collisional broadening due to the perturbations produced upon the radiating atoms by other particles

Hyperfine structure causes an intrinsic broadening of some lines, e.g., certain lines of Mn. If a magnetic field is present (as occurs in nature in sunspots or magnetic stars), lines may also be broadened by the Zeeman effect.

We consider first the situation where the lines are broadened only by the Doppler effect and natural damping. The Doppler breadth of a line depends only on the kinetic temperature and mass of the atom. For a line broadened solely by the Doppler effect, we would have

$$\alpha_\lambda = \frac{\sqrt{\pi}\,\epsilon^2}{mc} f \frac{\lambda}{v} \exp\left[-\frac{c^2}{v^2}\left(\frac{\lambda - \lambda_0}{\lambda_0}\right)^2\right] \qquad (3.46)$$

where $v = \sqrt{2kT/M}$ is the most probable velocity of the atoms, λ_0 is the wavelength of the center of the line, and m is the mass of an electron. The Doppler effect is often a nuisance in laboratory studies of hyperfine structure. It becomes necessary to cool the source with liquid air.

The absorption coefficient of an atom at rest with respect to the observer is given by Eq. (3.6) so that at large distances from the line center:

$$\alpha_\nu = \frac{\pi\epsilon^2}{mc} f \frac{\Gamma}{4\pi^2} \frac{1}{(\nu - \nu_0)^2} \qquad (3.47)$$

If the gas is so attenuated that there are no collisions, Γ is simply the radiative damping constant [see Eq. (3.10)]. Since each atom radiates a profile whose form is given by Eq. (3.6), the final line shape is determined by the superposition of many such profiles according to a Maxwellian velocity distribution appropriate to the temperature concerned. The resultant expression for the absorption coefficient may be written as

$$\frac{\alpha_\nu}{\alpha_0} = H_0(u) + aH_1(u) + a^2H_2(u) + \cdots \qquad (3.48)$$

where

$$u = \frac{\nu - \nu_0}{\Delta\nu_0} \qquad (3.49)$$

$$a = \frac{\Gamma}{4\pi\,\Delta\nu_0} \qquad (3.50)$$

$$\alpha_0 = \frac{\sqrt{\pi}\,\epsilon^2}{mc} f \frac{1}{\Delta\nu_0} \qquad (3.51)$$

$$\Delta\nu_0 = \frac{\nu_0}{c}\sqrt{\frac{2kT}{M}} \qquad (3.52)$$

and H_0, H_1, H_2, etc., are functions whose numerical values have been tabulated by Harris [44]. See also the tabulation by Mugglestone and Finn [45].

When collisional broadening is important, the situation becomes much more complicated [8, 46, 47, 48]. For most spectral lines of elements other than hydrogen and helium, it is usually sufficiently accurate to employ Eq. (3.6) with

$$\Gamma = \Gamma_{\text{rad}} + \Gamma_{\text{coll}} \qquad (3.53)$$

where $\Gamma_{\text{coll}} = 2/T_0$. Here T_0 is the mean time between two collisions that are effective in broadening the line. In order to calculate T_0 by the gas kinetic theory it is necessary to define an effective target area $\pi\rho_0^2$ for a broadening encounter. This is possible as long as the force of interaction between the perturber and the emitting atom varies as r^{-4} (as in the quadratic Stark effect) or as r^{-6} (as for the van der Waals type of force between neutral particles). An exception must be made for hydrogen and certain lines of helium which are subject to the linear Stark effect. Here it is necessary to consider not only the contributions of the ions that broaden according to the statistical Holtsmark theory but also those of the electrons that broaden according to the "discrete encounter" or impact theory. An accurate quantitative theory has been given by Kolb and Griem, who have published extensive numerical tables so that for many hydrogen and certain helium lines the broadening can be calculated from the electron density, the temperature, and the known Stark pattern [8, chap. 4].

For atoms other than hydrogen and helium, collision-broadening target areas, $\pi\rho_0^2$, may be measured experimentally in the laboratory or calculated theoretically. One has to distinguish between broadening appropriate to the quadratic Stark effect and interactions of the van der Waals type (for a summary see [2, pp. 317–322]). A simple theory due to Lindholm [49] is often applied to calculate Γ_{coll} but the more detailed treatment by Kolb and Griem, by Baranger, and by van Regemorter shows that broadening of many lines must be handled on an individual basis.

The shape of the line absorption coefficient is of critical importance for the relation between the total intensity of a spectral line and the number of atoms acting to produce it. Consider a volume of metallic vapor placed in front of a hotter source which emits a continuous spectrum. A pipe containing a volatile metallic salt may be heated and placed in front of an incandescent lamp. An astronomical example is provided by the extended atmosphere of a supergiant star through which shines (at times near eclipse) the light of a much hotter companion star. A number of such eclipsing binary systems have been recognized and studied by astronomers. Under these simplified conditions, the radiation is simply extinguished according to the law:

$$I_\lambda = I_{0\lambda} \exp(-k_\lambda \bar{\rho} L) \qquad (3.54)$$

where L is the length of the column of vapor of mean density $\bar{\rho}$, and k_λ is the mass absorption coefficient at wavelength λ.

We define the intensity of an absorption line in terms of its equivalent width, W_λ, that is, the width in angstroms of a completely black line of rectangular profile that would subtract from the continuum the same amount of energy as does the real line. Unlike the line profile, it does not depend on the resolution of the spectrograph (unless, of course, the line is severely washed out by insufficient resolution).

The relation between the equivalent width of a line and the number of atoms acting to produce it is referred to as a *curve of growth*. When the number of atoms is small, W_λ is proportional to N. Soon, however, the intensity at the center of the line approaches zero and the shape of the line is determined by the width of the Doppler core of the absorption coefficient. Then W_λ increases slowly with N until substantial absorption takes place in the damping wings when W_λ increases as $\sqrt{Nf\Gamma}$ [2, chap. 8].

Figure 3.1 is calculated for the λ3933 line of ionized calcium for a kinetic temperature of 4000°K. We have plotted log W against log N where N is the number of atoms acting to produce the line (i.e., in the lower level of the transition involved) per square centimeter in the line of sight. In most practical examples N is not varied. Rather, one compares the equivalent widths of lines of different f values. We

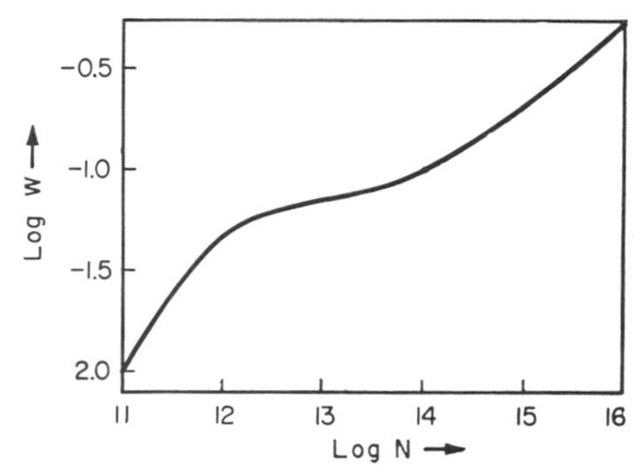

FIG. 3.1. Logarithm of equivalent width, log W, is plotted against log N where N is the number of atoms acting to produce the line for an atmosphere in which the gas kinetic temperature is 4000°K. It is assumed that pure extinction occurs in accordance with Eq. (3.54).

might expect to plot log W against log Nf but, actually, if lines of different spectral regions are employed, the proper procedure is to plot log $\frac{W}{\lambda}\frac{c}{v}$ (where v is the most probable kinetic velocity) against log $Nf\lambda$. Examples drawn from an astrophysical context are described in [2, chap. 8; 50].

6. Experimental Determination of f Values

Among methods proposed for the measurement of f values we may mention the following.

Anomalous Dispersion. In the neighborhood of a line, Eq. (3.4) shows that the index of refraction varies according to

$$n - 1 \sim \frac{Nf\lambda_0^3}{\lambda - \lambda_0} \tag{3.55}$$

Usually an interference technique is employed (involving, e.g., the Jamin interferometer) and $(n - 1)$ may be found from the curvature of the fringes. If the vapor density N along the line of sight is known, f may be computed at once. If N is not known, only relative f values can be found. This method has been employed particularly in the Soviet Union [51].

Faraday Effect. Measures of the rotation of the plane of polarization in a gas placed in a magnetic field also yield values of Nf.

The *Stark effect* has been employed to get upper and lower limits for the oscillator strengths for certain sodium lines. From experimentally measured splitting patterns and shifts of multiplet terms in electric fields it is possible to estimate the f values for jumps to nearby levels.

Absorption in Thin Layers. Some investigators, particularly R. B. and A. S. King and their collaborators, have determined relative f values for a number of metals by measuring the equivalent widths of absorption lines produced by thin layers of vapors. The equivalent width of a very weak line is related to the f value by [2, p. 302]

$$W_\lambda = hN\lambda_0^2 \frac{\pi\epsilon^2}{mc} f \tag{3.56}$$

The condition that the optical depth at the center of the line is always much less than 1 is exacting. Some experimenters (e.g., Fuchtbauer) have introduced foreign gases so that the lines are broadened by collisions and the flat portion of the curve of growth is eliminated. Others have attempted to allow for saturation effects in the curve of growth.

Emission from Optically Thin Layers. If the layer of emitting gas is thin,

$$I \sim N_n A_{nn'} h\nu_{nn'} \tag{3.57}$$

so that if the population of the excited level is given by Boltzmann's law for some temperature T, relative A's may be found from relative I's. Burger and Dorgelo applied this method to measure the relative strengths of lines in a multiplet. In analogy to the restrictions on absorption line intensities, the intensity I at the center of the line must be much less than I_0, the intensity of the radiation emitted by a black body at the temperature of the source. The method cannot be applied to get the relative strengths in a transition array unless Boltzmann's law is rather closely obeyed for some temperature. If the number of atoms per cubic centimeter and the length of the emitting column are known, absolute f values can be determined.

Astrophysical Methods. Experimentally, f values are usually difficult to obtain for many lines arising from high levels in a neutral atom or for lines of ionized atoms. Such lines are observed in the Fraunhofer spectra of the sun and stars. Consider the high-level lines of a metal such as iron in the sun. If the abundance of the element concerned is established from low-level lines whose f values are known and if the variation of temperature and density with depth in the sun are known, in principle it should be possible to get empirical f values from measured equivalent widths by means of the curve of growth. Actually, the damping constants must be known or estimated because the high-level lines are formed at great depths in the atmosphere and are more subject to "pressure" broadening. The use of a mean curve of growth (determined from low-level lines of known f value) is not valid for the interpretation of high-excitation lines; the effects imposed by the structure of the atmosphere on the line shapes and intensities must be taken into account. Since a stellar atmosphere is composed of many elements, line blending can cause trouble.

As another example, we mention Groth's determination [53] of log gf values for Fe II from line intensities measured in the spectrum of α Cygni. Attempts to use emission-line spectra of the solar chromosphere or bright-line stars have not been successful in f-value determinations.

Electric Arcs. Several different types of electric arc have been employed as interpretable sources of emission-line spectra. An electric arc has the advantage of simplicity and ease of operation, producing a bright spectrum over a period of time adequate for photographic exposures. A disadvantage is a tendency to instability; also, thermal diffusion may cause the incandescent-arc vapor to develop a chemical composition different from that of the arc material.

Copper Alloy Arc. In this technique a d-c arc is struck between two poles of copper alloys prepared with accurately known small amounts of other elements. If the vapor is assumed to have the same composition as the solid pole and if absolute f values for some elements are known, one may solve for the electron density and temperature; hence the absolute f values for lines of other elements may be found. C. W. Allen and A. W. Asaad determined absolute f values for certain lines in the spectra of Al, Si, Cr, Mn, Fe, Co, Ni, Cu, Ga, Ag, Sn, Pb, and Bi [54], while Corliss and Bozman [55] obtained f values for lines of 70 elements. More recently Corliss and Warner obtained extensive f values for Fe I and ions of the ion group. The method possibly gives good relative f values, but occasionally the absolute values are poor.

Stabilized Whirling-fluid Arc. At Kiel, W. Lochte-Holtgreven [56] developed an arc in which the discharge took place along the axis of a rapidly rotating tube through which water was flowing. The current was confined to a narrow region with the result that large currents and high temperatures could be obtained. Using direct currents up to 1,600 amperes, steady temperatures from 10,000°K to 50,000°K could be obtained in this way. From studies of the ionization equilibrium and the line broadening, the temperature and density may be fixed as a function of the distance from the axis of the arc. Modifications of the arc permit other liquids and gases to be used. Measurements of f values for carbon, nitrogen, and oxygen have been published, while extensive studies of line broadening have also been undertaken. Unfortunately this type of arc is subject to severe turbulence, causing the arc column to wander. Secondly, the concentration of atoms is affected by thermal diffusion.

To overcome these difficulties, Wiese [57] and others have used a wall-stabilized arc which is suitable for pure gases such as oxygen or argon.

Luminous Shock Tube [57]. When a shock wave is reflected from a fixed surface, much of its kinetic energy is turned into heat. Temperatures high enough to excite lines of hydrogen and other gases have been produced. Since conditions in the shock tube are amenable to theoretical analysis, ambient temperature and pressure can be obtained independently of measurements in the spectrum. Hence absolute f values can be obtained. E. B. Turner and L. Doherty, working in the laboratory of O. Laporte at the University of Michigan, studied line broadening in hydrogen and measured f values for neon. A. C. Kolb and his associates at the Naval Research Laboratory have demonstrated how very high temperatures may be obtained by shock-wave techniques. The shock-tube method promises to be of great importance for f values of astrophysical interest.

Since the gas temperature can be both determined and controlled easily, the gas is in a thermal-equilibrium state which is uniform along the line of sight, and the optical depth can be varied readily, shock excitation is better suited to spectroscopic work than is the glow or arc discharge. There are, however, difficulties: (1) The phenomenon is transient so that long exposures cannot be obtained, and (2) the theoretical temperatures and pressures do not agree with the predicted ones, and so independent measurements must be secured. Some of these problems have been discussed by Wilkerson and Charatis, who used a mixture of neon and chromium carbonyl and observed the emission lines of Cr I and Cr II behind the reflected shock. Studies with somewhat similar shock-tube equipment have been undertaken by Byard, Roll, and Slettebak at Ohio State University and by G. W. Wares and his associates at Air Force Cambridge Research Laboratories. Important advances have been made by Reeves and Parkinson and their associates at the Smithsonian Astrophysical Observatory.

In the technique developed by Kolb and his associates, a capacitor is discharged between electrodes in a T tube, creating a shock wave which is magnetically accelerated down the side arm of the shock tube. A gas in thermal equilibrium at a very high temperature may be studied by this technique.

Atomic Beam Method [59]. In this technique, which is due to Kopfermann and Wessell, one measures the weak absorption produced by a metallic vapor stream crossing the line of sight. The vapor is subsequently condensed on a surface whose rate of weight increase can be measured. From the rate of growth of the deposit and the speed of the metallic atoms, the atomic density in the line of sight may be found. R. B. King and his associates [60] have further developed this method to measure f values for the resonance lines of a number of elements of astrophysical importance.

Lifetimes of Atomic Levels. The lifetime of an excited level, αJ, is related to the sum of Einstein coefficients of spontaneous emission by

$$T = \frac{1}{\Sigma A(\alpha J;\alpha' J')} \tag{3.58}$$

where the summation is taken over *all* lower levels to which direct transitions may occur. One need not know the population of the level (αJ); only the rate of decay need be measured. In practice, decay rates may be measurable only for certain levels because of cascade effects [61, 62].

In an interesting variation of this procedure, developed by Bashkin and Meinel [63], a Van de Graaf accelerator is used to produce fast, singly charged, positive ions of carbon, nitrogen, and oxygen with energies between 0.5 and 5 Mev. These particles are then fired through a foil of carbon or beryllium where they are stripped of electrons. The lifetimes are determined by measuring the intensity as a function of the distance downstream.

Other techniques may be mentioned briefly. Prag, Fairchild, and Clark obtained f values from absorption measurements of O and N in their ground states [64]. Nitrogen atoms were produced in a flowing nitrogen afterglow; calibrated titration of this afterglow with NO was used to produce O atoms and to measure concentrations of both N and O. The measured f values are in good agreement with other determinations.

L. Huldt and A. Lagerquist measured intensities of lines of Cr and Mn in a source in which the oxides were dissociated at a known temperature [65]. The dissociation energies of MnO and CrO being known, it is possible in principle to derive absolute f values from the line intensities. Their f values appear to be much lower than those found by King's group and other investigators. The method probably cannot be considered reliable.

7. Tests and Applications of the Theory

Numerous experimental tests have given confidence in the principles of radiation theory as applied to spectral lines. For example, experiments involving subordinate lines show that Γ depends on the sum of the damping constants of both the upper and lower terms and not simply on that of the upper term. Measures of the absorption coefficient in the outer parts of the line ($\nu - \nu_0 \gg \Delta\nu_0$) in gases at high pressure verify general predictions of the theory of pressure broadening. The Lindholm theory appears to give promising results for line broadening but poor results for line displacements. Doubtless, improved theoretical treatments along the lines indicated by Kolb, Griem, and their co-workers will help.

R. Minkowski [67] gave a beautiful verification of the theory of line absorption utilizing the sodium D lines whose f values he found from the magnetorotation effects. The absorption in the far wings varies as $Nf\Gamma/(\lambda - \lambda_0)^2$ so that one may measure $Nf\Gamma$ from optically thick layers. At low densities, Γ approached the value appropriate for pure radiation damping.

Perhaps the simplest demonstration that the equivalent widths of strong lines vary as $\sqrt{Nf\Gamma}$ is provided by the telluric lines in the solar spectrum. These are lines that are produced by molecules in the earth's atmosphere and the number of them in the line of sight is proportional to the cosecant of the altitude of the sun for moderate zenith distance. Such studies of the infrared telluric lines have been carried out at the McMath-Hulbert Observatory [68].

Comparisons between experimental f values and theoretical predictions have been made for various elements. Both procedures are subject to difficulties. On the theoretical side are departures from LS coupling, failure of configuration assignment, some further required refinements in the theory to handle such things as correlation and polarization effects, and particularly the proper calculation of the radial quantum integral at great distances from the origin. There are practical choices, too. For example, should one use experimental or calculated energies in calculations of f values? Generally more consistent results are obtained if one employs calculated rather than observed energies [69], at least for one-electron spectra. Considerable progress appears to have been made since Garstang in 1955 commented on "the almost complete absence of any transition probabilities of absolutely certain accuracy for any complex atoms."

Experimental determinations of f values are affected by many sources of error, e.g., variations of temperature in a supposedly uniform source, deviations from thermal equilibrium, thermal diffusion in arcs, cascade phenomena in lifetime measurements, errors in vapor pressure in electric-furnace experiments, etc. Hence results by different groups often differ by intolerable amounts. The accuracy is gradually being improved, however, and it is hoped that reliable data will be available soon for the more important elements.

8. Compilations of Data

Because of the vast extent of the literature, we shall not present here a table or compilation of f values. Rather, we shall list some of the collections of data that have proved useful.

Glennon and Wiese [70] have presented a comprehensive bibliography on atomic transition probabilities. They list general references, conversion factors, review articles, etc., as well as giving detailed references to individual elements. They do not give any indications of the quality of the work.

Allen [71] has given a list of both permitted and forbidden atomic oscillator strengths for selected lines of many elements, with a concise bibliography.

In connection with a critical study of the quantitative chemical composition of the solar atmosphere, Goldberg, Müller, and Aller [72] list log gf values for a large number of lines of mostly neutral elements between $\lambda 3000$ and $\lambda 22084$ A. The compilation is selective in the sense that lines that are too strong or are blended in the solar spectrum are omitted. Also omitted are elements that were not of "astrophysical interest," i.e., were not represented in the solar spectrum because they were too rare or their lines fell outside the observable region.

A collection of translations of Russian articles on optical transition probabilities (1932–1962) has been published for the U.S. Department of Commerce and the National Science Foundation by the Israel Program for Scientific Translations [73]. It includes both theoretical and experimental papers.

Compilations of f values calculated by the Bates-Damgaard method for light elements in various stages of ionization are given by Griem [8] and by Aller and Jugaku [74]. These f values appear to be remarkably accurate for transitions of many ions.

References

1. Condon, E. U., and G. H. Shortley: "The Theory of Atomic Spectra," chap. 4, Cambridge University Press, New York, 1951.
2. Aller, L. H.: "Astrophysics I—The Atmospheres of the Sun and Stars," 2d ed., chap. 4, Ronald, New York, 1963.
3. Weisskopf, V., and E. Wigner: *Z. Physik*, **63**: 54 (1930); **65**: 18 (1930).
4. Unsöld, A.: *Z. Astrophys.*, **21**: 1, 1941.
5. Goldberg, L.: *Astrophys. J.*, **82**: 1, 1935; **84**: 11, 1936.
6. Rohrlich, F.: *Astrophys. J.*, **129**: 441.
7. Racah, G.: *Phys. Rev.*, **61**: 186; **62**: 438 (1942); **63**: 367 (1943); **76**: 1352 (1949).
8. Griem, H.: "Plasma Spectroscopy," McGraw-Hill,

New York, 1964. See Table 3-1, pp. 338–355, for Racah coefficients.
9. Ref. 1, chap. 11 (methods are indicated). For applications of the methods, see, for example, Garstang, R. H.: *Monthly Notices Roy. Astron. Soc.*, **114**: 118 (1954).
10. Gottschalk, W.: *Astrophys. J.*, **108**: 326 (1948).
11. Green, L. C., P. P. Rush, and C. D. Chandler: *Astrophys. J. Suppl.*, **3**: 37 (1957).
12. Infeld, L., and T. E. Hull: *Revs. Mod. Phys.*, **23**: 21 (1951).
13. Menzel, D. H., and C. L. Pekeris: *Monthly Notices Roy. Astron. Soc.*, **96**: 77 (1935).
14. Chandrasekhar, S.: *Astrophys. J.*, **102**: 223 (1945).
15. Hartree, D. R.: "Calculation of Atomic Structures," Wiley, New York, 1957.
16. For bibliographies of atomic wave functions, see Knox, R. S.: "Bibliography of Atomic Wave Functions in Solid-state Physics," Academic, New York, 1957. Additional wave functions have been calculated by:
Czyzak, S. J.: *Astrophys. J. Suppl.* 65, **7**: 53 (1962)
Clementi, E. J.: *Chem. Phys.*, **36**: 33 (1962); **38**: 2248 (1963); **39**: 175 (1963).
Hermann, F., and S. Skillman: "Atomic Structure Calculations," Prentice-Hall, Englewood Cliffs, N.J., 1963.
Douglas, A. S.: *Proc. Cambridge Phil. Soc.*, **52**: 687 (1956); **58,** 377 (1962).
Watson, R. E., and A. J. Freeman: *Phys. Rev.*, **123**: 521, 2027 (1961).
Routhann, C. C.: *Revs. Mod. Physics*, **32**: 179 (1960).
Piper, W. W.: *Phys. Rev.*, **123**: 1281 (1961).
See also Slater, J. C.: "Quantum Theory of Atomic Structure," vols. I and II, McGraw-Hill, New York, 1960.
17. For a concise summary of the techniques and procedures involved, see Czyzak, S.: in B. M. Middlehurst and L. H. Aller (eds.), "Stars and Stellar Systems, Gaseous Nebulae and Interstellar Matter," vol. 7, University of Chicago Press, Chicago, 1967.
Numerical procedures have been developed by Froese, C.: *Can. J. Phys.*, **41**: 50, 1895 (1963); and Mayers, D., and A. Hirsch: unpublished report to Aerospace Research Laboratory, 1963.
Analytic methods for handling the Hartree-Fock self-consistent-field equations have been given by, e.g.:
Boys, S. F.: *Proc. Roy. Soc. (London)*, **A201**: 128 (1950); **A217**: 136, 235 (1953).
Lowdin, P. O.: *Phys. Rev.*, **90**: 120 (1953); 1474, 1490, 1509 (1955); *Revs. Mod. Phys.*, **34**: 80 (1962); *J. Mod. Phys.*, **10**: 12 (1963).
Nesbet, R. K.: *Phys. Rev.*, **100**: 228 (1955); **109**: 1632 (1958); **119**: 658 (1960); **122**: 1497 (1961).
Nesbet, R. K., and R. E. Watson: *Phys. Rev.*, **110**: 1073 (1958).
18. Bates, D., and A. Damgaard: *Trans. Roy. Soc. (London)*, **A242**: 101 (1949). Applications of these tables have been made by a large number of workers.
19. Biermann, L., and E. Trefftz: *Z. Astrophys.*, **26**: 213, 240 (1949); **28**: 67 (1950); **29**: 287 (1951).
20. Trees, R. E.: *Phys. Rev.*, **85**: 382 (1952); *J. Res. Natl. Bur. Std.*, **53**: 35 (1954).
21. Green, L. C.: *Astrophys. J.*, **109**: 289 (1949).
22. Kingsbury, R. F.: *Phys. Rev.*, **99**: 1846 (1955).
23. Layzer, D.: Thesis, Harvard University, 1950.
24. *Naturwissenschaften*, **13**: 627 (1925); *Z. Physik*, **33**: 408 (1925).
25. Menzel, D. H.: *Astrophys. J.*, **105**: 126 (1947).
26. Shortley, G. H.: *Phys. Rev.*, **47**: 419 (1935). An illustrative example is given in ref. 2, p. 299.
27. Rohrlich, F.: *Astrophys. J.*, **129**: 449 (1959).
28. Karzas, W. J., and R. Latter: *Astrophys. J. Suppl.*, **6**: 167 (1961).
29. Burgess, A.: *Mem. Roy. Soc.*, **69**: pt. 1 (1964). See Eq. 1.
30. Burgess, A., and M. J. Seaton: *Monthly Notices Roy. Astron. Soc.*, **120**: 121 (1960).
31. Biermann, L., and K. Lübeck: *Z. Astrophys.*, **26**: 43 (1949).
32. Aller, L. H.: "Gaseous Nebulae," Chapman & Hall, London, chap. 5, 1956.
33. Shortley, G. H.: *Phys. Rev.*, **57**: 225 (1940).
34. Shortley, G. H., L. H. Aller, J. G. Baker, and D. H. Menzel: *Astrophys. J.*, **93**: 178 (1941).
35. Garstang, R. H.: *Proc. Cambridge Phil. Soc.*, **53**: 214 (1957); **54**: 383 (1958).
36. Aller, L. H., C. W. Ufford, and J. H. van Vleck: *Astrophys. J.*, **109**: 42 (1949).
37. Garstang, R. H.: *Monthly Notices Roy. Astron. Soc.*, **111**: 115 (1951).
38. Pasternack, S.: *Astrophys. J.*, **92**: 149 (1940).
39. Garstang, R. H.: *Monthly Notices Roy. Astron. Soc.*, Fe III, Fe V, **117**: 393 (1957); Fe IV, **118**: 572 (1958); Fe II, **124**: 32 (1962); Ni II, Ni III, **118**: 234 (1958). For other ions, see *J. Res. Natl. Bur. Std.*, **68A**: 61 (1964).
40. Chamberlain, J. W.: "Physics of Aurorae and Airglow," Academic, New York, 1961.
41. Thackeray, A. D.: *Monthly Notices Roy. Astron. Soc.*, **113**: 211 (1953). See also L. H. Aller, *Proc. Nat'l. Acad. Sci.*, **55**: 671 (1966).
42. Aller, L. H.: "Astrophysics-Nuclear Transformations, Stellar Interiors, and Nebulae," Ronald, New York, 1954.
43. See ref. 2, p. 111 (Doppler broadening); p. 172 (natural broadening); p. 310 (collisional broadening).
44. Harris, Daniel L., III: *Astrophys. J.*, **108**: 112 (1948).
45. Mugglestone, D., and G. D. Finn: *Monthly Notices Roy. Astron. Soc.*, **129**: 221 (1965).
46. van Regemorter, H.: in L. Goldberg, (ed.), "Annual Reviews of Astronomy and Astrophysics," Annual Reviews, Inc., Palo Alto, Calif., 1965.
47. Margenau, H.: *Revs. Mod. Phys.*, **31**: 569 (1959).
48. Traving, G.: *Mitt. Astron. Ges.*, Sonderheft Na1, Braun, Karlsruhe, Germany, 1960.
49. Lindholm, E.: *Ark. Mat. Astron. Fysik.*, **32A**: no. 17 (1946).
50. Aller, L. H.: "Abundance of the Elements," chap. 5, Interscience, New York, 1961.
51. Fillipov, A. N.: *Tr. Gos. Optich. Inst. Leningrad*, **8**: 8 (1932). See also ref. 73.
52. Paul, W.: *Z. Physik*, **124**: 121 (1944).
53. Groth, H. G.: *Z. Astrophys.*, **51**: 206 (1961).
54. Allen, C. W., and A. W. Asaad: *Monthly Notices Roy. Astron. Soc.*, **115**: 571 (1955); **117**: 36 (1957).
55. Corliss, C. H., and W. R. Bozman: *Natl. Bur. Std. (U.S.), Monograph* 53, 1962.
56. Lochte-Holtgreven, W.: *Rept., Progr. Phys.*, **21**: 312 (1958). Jurgens, G.: *Z. Physik*, **134**: 21 (1952); **138**: 613 (1954). Foster, E. W.: *Proc. Phys. Soc. London*, **A79**: 94 (1962). Richter, J.: *Z. Astrophys.*, **53**: 262 (1961).
57. Wiese, W. L., D. R. Paquette, and J. E. Solarski: *Phys. Rev.*, **129**: 1225 (1963).
58. Aller, L. H.: *Sky and Telescope*, **14**: 59 (1954). Charatis, G., and T. D. Wilkerson: *Phys. Fluids*, **2**: 578 (1959). McLean, E. A., C. E. Faneoff, A. C. Kolb, and H. R. Griem: *Phys. Fluids*, **3**: 843 (1959). Berg, H., F. Eckerle, R. W. Burris, and W. L. Wiese: *Astrophys. J.*, **139**: 751 (1964). Kolb, A. C., and H. Griem in: D. R. Bates (ed.), "Atomic and Molecular Processes," Academic, New York, 1962.
59. Kopfermann, H., and G. Wessell: *Z. Physik*, **130**: 100 (1951). Bell, G. D., and M. H. Davis, R. B. King, and P. M. Routley: *Astrophys. J.*, **127**: 775 (1958).
60. Lawrence, G. M., J. K. Link, and R. B. King: *Astrophys. J.*, **141**: 293 (1965).

61. Heron, S., R. W. P. McWhirter, and R. Roderick: *Proc. Roy. Soc. (London)*, **A234**: 565 (1956). Ziock, K.: *Z. Physik.*, **147**: 99 (1957).
62. Brannen, E., et al.: *Nature*, **175**: 810 (1955).
63. Bashkin, S., and A. B. Meinel: *Astrophys. J.*, **139**: 413 (1964).
64. Prag, A. B., C. E. Fairchild, and K. C. Clark: *Phys. Rev.*, (in press).
65. Huldt, L., and A. Lagerquist: *Arkiv Fysik*, **5**: 1, 2 (1952).
66. Hindmarsh, W. R.: *Monthly Notices Roy. Astron. Soc.*, **119**: 11 (1959).
67. Minkowski, R.: *Ann. Physik*, **66**: 206 (1921).
68. Goldberg, L.: G. P. Kuiper (ed.), "The Earth as a Planet," chap. 9, University of Chicago Press, Chicago, 1954.
69. Green, L. C., N. E. Weber, and E. Krawitz: *Astrophys. J.*, **113**: 690 (1951).
70. Glennon, B. M., and W. I. Wiese: *Natl. Bur. Std. (U.S.), Monograph* 50, 1962.
71. Allen, C. W.: "Astrophysical Quantities," pp. 65–76, University of London Athlone Press, London, 1963.
72. Goldberg, L. G., E. A. Müller, and L. H. Aller: *Astrophys. J. Suppl.* 45, **5**: 1–138 (1960).
73. Meroz, L. (ed.): Optical Transition Probabilities, U.S. Department of Commerce, Office of Technical Services. (Translations by A. Barouch, N. Kaner, Z. Lerman, and R. N. Sen.)
74. Aller, L. H., and J. Jugaku: *Astrophys. J. Suppl.* 38, **4**: 109 (1959).

Bibliography

Condon, E. U., and G. H. Shortley: "Theory of Atomic Spectra," secs. 1, 2, 4, Cambridge University Press, New York, 1951.

Aller, L. H.: "Astrophysics—The Atmospheres of the Sun and Stars," 2d ed., secs. 1, 2, 5, 6, 7, Ronald, New York, 1963.

Griem, H.: "Plasma Spectroscopy," secs. 1, 2, 3, 5, 7, McGraw-Hill, New York, 1964.

Aller, L. H.: "Gaseous Nebulae," chap. 5, sec. 4, Chapman & Hall, London, 1956.

Czyzak, S.: in B. M. Middlehurst and L. H. Aller (eds.), "Stars and Stellar Systems—Nebulae and the Interstellar Medium," vol. 7, University of Chicago Press, Chicago, 1967.

Chapter 4

Hyperfine Structure and Atomic Beam Methods

By NORMAN F. RAMSEY, Harvard University

1. Introduction

The gross features of atomic spectra correspond to transitions between energy levels of electrons moving in the Coulomb field of a positively charged nucleus of negligibly small dimensions. In this case, and for Russell-Saunders coupling, the most closely adjacent energy levels are usually those of atomic states which differ in the relative orientation of the orbital and spin angular momenta. The separation of this "fine structure" varies from less than one-tenth to several thousand wave numbers.

However, if this fine structure is experimentally examined more closely, it is often found that each line of the fine structure can in turn be resolved into further lines or "hyperfine structure" with a separation of the order of 1 cm^{-1}. Pauli first suggested that the hyperfine structure might be due to a magnetic interaction between the nucleus and the moving electrons of the atom, dependent upon the orientation of the nucleus.

Interactions between a nucleus and its atom or molecule have been studied by the methods of optical spectroscopy, molecular beams, nuclear paramagnetic resonance, microwave spectroscopy, and paramagnetic resonance (see also Part 9, Chap. 3). All of the measurements are consistent with these assumptions concerning atomic nuclei [1]:*

1. A nucleus has a charge Ze confined to a small region of the order of 10^{-12} cm diameter. Interaction of this charge with the electrons gives rise to the gross features of optical atomic spectra.

2. A nucleus whose mass number A is odd obeys Fermi-Dirac statistics (the sign of the wave function is reversed if two such identical nuclei are interchanged) and a nucleus whose mass number is even obeys Bose-Einstein statistics (wave function unaltered on interchange).

3. A nucleus possesses a spin angular momentum capable of being represented by a quantum-mechanical angular momentum vector **a** with all the properties [2] usually associated with such a vector. In hyperfine-structure work, it is usually convenient to use a dimensionless quantity **I** to measure the angular momentum in units of $\hbar$ where **I** is therefore defined by

$$\mathbf{a} = \hbar\mathbf{I} \tag{4.1}$$

The spin I of the nucleus is defined as the maximum possible component of **I** in any given direction.

* Numbers in brackets refer to References at end of chapter.

4. The nuclear spin I is half integral if the mass number A is odd and integral if the mass number is even.

5. A nucleus has a magnetic moment μ_I which can be represented as

$$\boldsymbol{\mu}_I = \gamma_I\hbar = g_I\mu_N\mathbf{I} \tag{4.2}$$

where γ_I and g_I are defined by the above equations and are called the nuclear gyromagnetic ratio and the nuclear g factor, respectively. μ_N is the nuclear magneton and is defined as $e\hbar/2Mc$, where M is the proton mass; μ_N has the numerical value of $(5.0505 \pm 0.00036) \times 10^{-24}$ erg gauss^{-1}. The scalar quantity which measures the magnitude of $\boldsymbol{\mu}_I$ and which is called the magnetic moment μ_I is

$$\mu_I = \gamma_I\hbar I \tag{4.3}$$

With this Eq. (4.2) can be written

$$\boldsymbol{\mu}_I = \left(\frac{\mu_I}{I}\right)\mathbf{I} \tag{4.4}$$

6. Many nuclei with spin equal to or greater than 1 possess an electric-quadrupole moment, i.e., have electric charge distributions which depart from spherical symmetry in a manner appropriate to an electric-quadrupole moment (a precise definition of an electric-quadrupole moment will be given subsequently).

7. A nucleus with a spin $\frac{3}{2}$ or greater may possess a magnetic-octupole moment (the definition of a magnetic-octupole moment will be given subsequently).

8. The nucleus is not infinitely heavy and of negligible size, but instead allowance must be made in some experiments for the finite mass and appreciable size of the nucleus.

9. The nucleus has a finite polarizability and can be polarized by a strong electric field.

2. Multipole Interactions

The general electrostatic interaction $\mathcal{H}_E$ between a charged nucleus and the charged electrons of the remainder of the atom is as follows if the finite extension of the nucleus is taken into account:

$$\mathcal{H}_E = \int_{\tau_e}\int_{\tau_n} \frac{\rho_e\rho_n\, d\tau_e\, d\tau_n}{r} \tag{4.5}$$

where ρ_e is the electron charge density, ρ_n is the nuclear charge density, and r is the distance between the electron volume element $d\tau_e$ and the nuclear volume element $d\tau_n$ where $d\tau_e$ is at the position $\mathbf{r}_e$ relative to the nuclear centroid and $d\tau_n$ is at $\mathbf{r}_n$. Then for $r_e > r_n$, $1/r$ can be expanded into the well-known power series in (r_n/r_e) [1, 3] so that

$$\mathcal{H}_E = \Sigma_k \mathcal{H}_{Ek} \tag{4.6}$$

where

$$\mathcal{H}_{Ek} = \int_{\tau_e} \int_{\tau_n} \frac{\rho_e \rho_n}{r_e} \left(\frac{r_n}{r_e}\right)^k P_k(\cos\theta_{en})\, d\tau_e\, d\tau_n \tag{4.7}$$

and $P_k(\cos\theta_{en})$ is the kth Legendre polynomial defined by

$$P_k = \frac{1}{2^k k!} \frac{d^k}{(d\cos\theta_{en})^k} (\cos^2\theta_{en} - 1)^k \tag{4.8}$$

with θ_{en} being the angle between $\mathbf{r}_e$ and $\mathbf{r}_n$. The $\mathcal{H}_k$ term of Eq. (4.6) which involves P_k is said to be the interaction energy from the multipole of order 2^k. Thus the first term of Eq. (4.6) corresponds to a monopole or simple charge, the second to the electric-dipole moment, the third to the electric-quadrupole moment, the fourth to the electric-octupole moment, etc.

A similar multipole expansion is possible with magnetic interactions [3] leading to magnetic-dipole terms, etc.

The number of possible multipole interactions is severely restricted by several general theorems that are derived from general assumptions, as in ref. 1. According to these theorems (1) no odd (k odd) nuclear electric-multipole moment can exist, (2) no even nuclear magnetic-multipole moment can exist, and (3) for a nucleus of spin I it is impossible to observe a nuclear multipole moment of order 2^k where k is greater than $2I$.

So far only the four lowest multipole moments have been observed in practice; and one of these, the electric monopole, is not of concern in hyperfine structure. Therefore the three orientation dependent interactions that are important in hyperfine-structure studies are the magnetic-dipole interaction $\mathcal{H}_{M1}$, the electric-quadrupole interaction $\mathcal{H}_{E2}$, and the magnetic-octupole interaction $\mathcal{H}_{M3}$. The relevant interaction Hamiltonian can be written

$$\mathcal{H} = \mathcal{H}_{M1} + \mathcal{H}_{E2} + \mathcal{H}_{M3} \tag{4.9}$$

Alternative and more convenient expressions for $\mathcal{H}_{M1}$, $\mathcal{H}_{E2}$, and $\mathcal{H}_{M3}$ are given below.

In the above discussion r_e has been assumed greater than r_n. Although this is approximately true, small corrections arise from the nonzero size of the nucleus and from the finite mass of the nucleus. Corrections for these lead to the isotope shift of atomic hyperfine structure and to the hyperfine-structure anomaly discussed further below.

3. Magnetic-dipole Interactions

The magnetic-dipole-interaction energy may be obtained either by a reexpression of the dipole term in the above formal multipole expansion or by direct physical interpretation. From the latter point of view the nuclear magnetic moment is given by Eq. (4.4) and the resultant magnetic field that acts upon it is the sum of the external magnetic field H_0 and the internal magnetic field H_J which arises from the atomic electrons. Therefore, from the usual expression for the interaction energy of a magnetic dipole,

$$\mathcal{H}_{M1} = -\left(\frac{\mu_I}{I}\right)\mathbf{I}\cdot(\mathbf{H}_J + \mathbf{H}_0) - \left(\frac{\mu_J}{J}\right)\mathbf{J}\cdot\mathbf{H}_0 \tag{4.10}$$

where the last term corresponds to the interaction of the electronic magnetic moment with the external magnetic field. As long as the atomic angular momentum quantum number J is a good quantum number [1], $\mathbf{H}_J$ can be taken as simply proportional to the electronic angular momentum $\mathbf{J}$ so that Eq. (4.10) can be rewritten as

$$\mathcal{H}_{M1} = ha\mathbf{I}\cdot\mathbf{J} - \left(\frac{\mu_J}{J}\right)\mathbf{J}\cdot\mathbf{H}_0 - \left(\frac{\mu_I}{I}\right)\mathbf{I}\cdot\mathbf{H}_0 \tag{4.11}$$

where the new parameter a contains the various proportionality constants. Approximate theoretical expressions for a are given in the literature [1, 3, 4]; therefore a can be related to the nuclear magnetic moment.

For any assumed values of a, μ_I, μ_J, J, and I the energy of the atom can be calculated as a function of the external field H_0 from Eq. (4.11). If this is done [3] the results are as shown in Fig. 4.1. At low

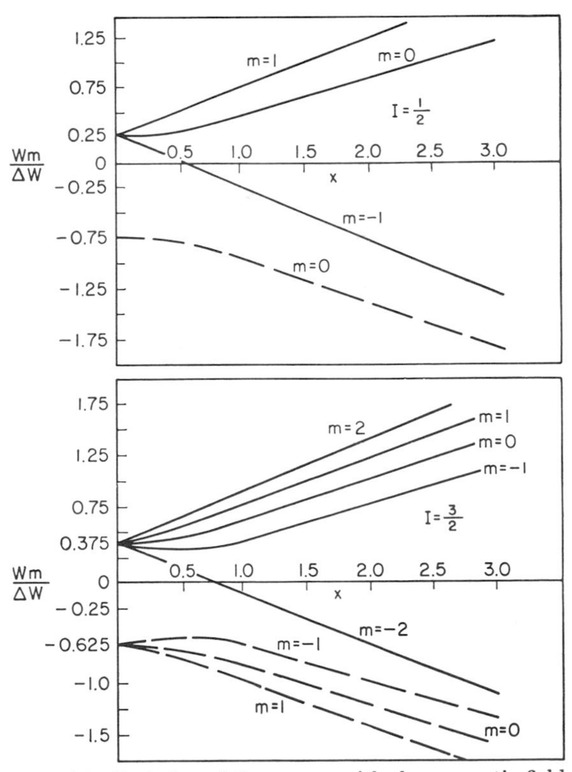

FIG. 4.1. Variation of the energy with the magnetic field for $J = \frac{1}{2}$. The nuclear magnetic moment is assumed positive. The dashed lines are the magnetic levels arising from the $F = I - \frac{1}{2}$ state. $x = (-\mu_J/J + \mu_I/I)H_0/\Delta W(R2)$.

magnetic fields I and J are tightly coupled to form a resultant angular momentum $\mathbf{F} = \mathbf{I} + \mathbf{J}$, whose quantum number F at low fields is a good quantum number. At high magnetic fields, such that $\mathbf{I}$ and $\mathbf{J}$ are fully decoupled from each other, the suitable quantum numbers are the magnetic quantum numbers m_I and m_J of I and J separately. The separation of the two F states at zero field is often called the hyperfine-structure separation $\Delta\nu$ and is related to the parameter a above by

$$\Delta\nu = \frac{\Delta W}{h} = \frac{a}{2}(2I + 1) \tag{4.12}$$

4. Electric-quadrupole Interaction

By suitable simplifications [1] one may express the electric-quadrupole-moment-interaction energy as follows, provided I and J are both good quantum numbers,

$$E_2 = hb\,\frac{\frac{3}{2}\mathbf{I}\cdot\mathbf{J}(2\mathbf{I}\cdot\mathbf{J} + 1) - \mathbf{I}^2\mathbf{J}^2}{2I(2I - 1)J(2J - 1)} \tag{4.13}$$

where

$$hb = eQ\int_{\tau_e} \frac{3z_e^2 - r_e^2}{r_e^5}(\rho_e)_{JJ}\,d\tau_e \tag{4.14}$$

and the symbol $(\ )_{JJ}$ indicates that ρ is to be taken for the electronic state with $m_J = J$ relative to the z axis. Q is the nuclear electric-quadrupole moment defined by

$$Q = \frac{1}{e}\int_{\tau_n}(3z_n^2 - r_n^2)(\rho_n)_{II}\,d\tau_n \tag{4.15}$$

5. Magnetic-octupole Interaction

By simplifications analogous to those in the quadrupole case [5], the electric-octupole-interaction energy may be expressed as follows, provided I and J are good quantum numbers,

$$\mathcal{H}_{M3} = hc\,\frac{5}{4}\,\frac{8(\mathbf{I}\cdot\mathbf{J})^3 + 16(\mathbf{I}\cdot\mathbf{J})^2 + (\frac{8}{5})\mathbf{I}\cdot\mathbf{J}[-3I(I+1)J(J+1) + I(I+1) + J(J+1) + 3] - 4I(I+1)J(J+1)}{I(I-1)(2I-1)J(J-1)(2J-1)} \tag{4.16}$$

where

$$hc = \Omega\int_{\tau_e}\frac{(5z_e^3 - 3z_e r_e^2)}{2r_e\tau}(\nabla\cdot\mathbf{m}_e)_{JJ}\,d\tau_e \tag{4.17}$$

and

$$\Omega = \int_{\tau_n}\tfrac{1}{2}(5z_n^3 - 3z_n r_n^2)(\nabla\cdot\mathbf{m}_n)_{II}\,d\tau_n \tag{4.18}$$

$\mathbf{m}_n$ is the nuclear magnetization conventionally defined so that

$$\mathbf{j}_n = c\nabla\times\mathbf{m}_n \tag{4.19}$$

where $\mathbf{j}_n$ is the nuclear current density that gives rise to the magnetic effects.

6. Optical Studies of Hyperfine Structure

The experimental techniques for the optical study of hyperfine structure are those of ordinary optical spectroscopy. However, because of the very close spacing ($\Delta\nu \sim 0.05\ \text{cm}^{-1}$) of many hyperfine-structure lines, highly refined techniques must be used. Carefully made Fabry-Perot etalons are often used to resolve the lines. Special light sources have been developed by Schüler and others which can be well cooled to reduce Doppler broadening and which are designed to reduce absorption broadening; use is often made of hollow cathode discharges in which the cathode is a hollow cylinder well cooled with liquid air. Light sources have also been designed so the emitting atoms are in directed atomic beams which can be viewed transversely to reduce Doppler effect. Fuller descriptions of the optical techniques employed and of the detailed methods of term analysis are given in books primarily devoted to optical hyperfine structure [4, 6].

Provided $J > I$, the nuclear spin can be directly determined from atomic hyperfine structure from the number of hyperfine components in the spectral line. The multiplicity of the electronic state in such a case is $2I + 1$ corresponding to the $2I + 1$ different possible orientation states of $\mathbf{I}$ relative to $\mathbf{J}$. Even when $I > J$, the value of I can be inferred (less reliably) from the relative spacing of the components of a term or from relative intensities of the components [4, 6]. The multiplicity of the lines in very strong magnetic fields also provides a means for determining the nuclear spin I, as can be seen from Fig. 4.1.

The magnitude and sign of the hyperfine-structure separation can be directly measured and from this the value of the interaction parameter a of Eq. (4.11) inferred. From the value of a an approximate value of the nuclear magnetic moment can be calculated, as discussed in Sec. 3.

If I and J are each greater than $\frac{1}{2}$, the separation of the lines may be in part due to a nuclear electric-quadrupole-moment interaction. Schüler and Schmidt first found in europium an optical hyperfine-structure spectrum which required for its interpretation the inclusion of a quadrupole interaction as in Eq. (4.15). Since then, many nuclear electric quadrupole moments have been observed both by optical hyperfine-structure methods and by other methods described in the present chapter and in Part 9, Chap. 3.

In interpreting the optical hyperfine spectrum of elements, one must take care to avoid confusion of the nuclear hyperfine structure with the isotope shift, since this shift is often of a magnitude comparable to the nuclear hyperfine structure. The isotope shift is of two kinds: (1) a pure mass effect corresponding to the fact that the nucleus with its noninfinite mass is not a motionless attracting center to the electrons, and (2) the volume and polarizability effects mentioned in Secs. 1 and 2. The first of these is important for light elements such as hydrogen and lithium, while the second is the important cause of isotope shift in heavy atoms. However, it should be noted that for light atoms with more than one electron the mass-dependent isotope shift is more complicated than a simple reduced mass calculation since it depends on how the electrons are coupled together in their motion [1]. The isotope shift in heavy nuclei depends on the distribution of the charge in the nucleus and upon the way the charge distribution may be distorted by the strong electric fields which arise when an orbital electron is very close to the nucleus; the study of the heavy isotope shift is therefore a promising source of information on nuclear charge distributions and polarizabilities.

7. Atomic Beam-deflection Experiments

A beam of paramagnetic atoms will be deflected by an inhomogeneous magnetic field since, for example, the north pole of the atomic magnet will be pushed north harder than the south pole is pushed south due to the inhomogeneity of the magnetic field. Alternatively, if the energy in the magnetic field is W

$$F = -\frac{\partial W}{\partial z} = -\frac{\partial W}{\partial H_0}\frac{\partial H_0}{\partial z} = \mu_{\text{eff}}\frac{\partial H_0}{\partial z} \tag{4.20}$$

where μ_{eff} is the effective component of the magnetic field along the direction z of the field gradient. Therefore

$$\mu_{\text{eff}} = -\frac{\partial W}{\partial H_0} \tag{4.21}$$

The dependence of μ_{eff} on H_0 can then be obtained by differentiation of curves as in Fig. 4.1 which represent the dependence of W on H_0. The result of such a differentiation is shown in Fig. 4.2.

The earliest beam-deflection studies of atomic states were those of Stern and Gerlach. In their experiments, however, the magnetic fields were so strong that the deflection depended merely on the atomic magnetic moment since large fields correspond to the extreme right limit of Fig. 4.2. These experiments, however, showed the reality of spatial quantization and the need of electron spin to account for the number and magnetic moments of the observed states.

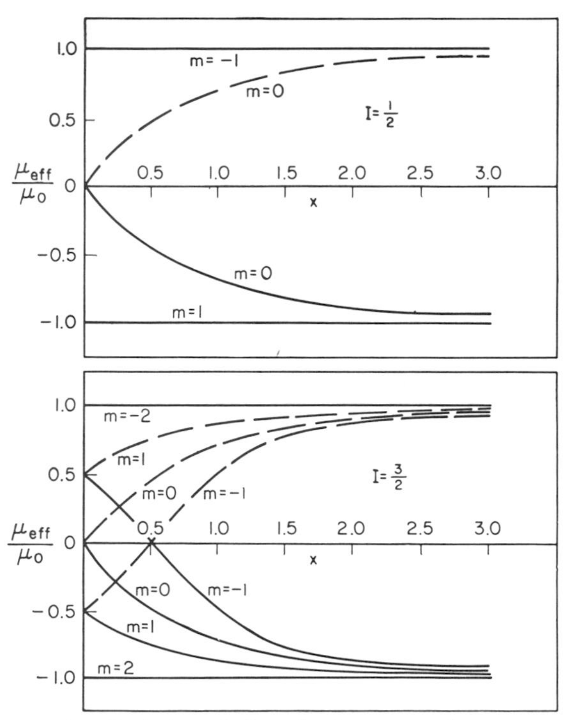

FIG. 4.2. Variation of the effective magnetic moment component μ_{eff} with magnetic field for $J = \frac{1}{2}$. The nuclear moment is assumed positive. The dashed lines are the moments of the magnetic levels arising from $F = I - \frac{1}{2}$ states. $x = (-\mu_J/J + \mu_I/I)H_0/\Delta W(R2)$.

Rabi pointed out that atomic hyperfine structure could be studied by measuring in this fashion the effective atomic magnetic moments at such weak magnetic fields that intermediate coupling in Fig. 4.2 applied. From the nature of the deflection pattern in a measured field H_0 the value of x could be determined and, from this and H_0, a could be calculated from the relation in the caption of Fig. 4.2 since μ_I makes a negligible contribution to x. In this way values of a were determined. A particularly effective atomic beam-deflection technique was the zero-moment method. The beam detector could be placed at a position corresponding to no deflection and the value of H_0 could then be varied. For certain values of H_0, corresponding, for example, to $x = 0.5$ in the lower part of Fig. 4.2, $\mu_{\text{eff}} = 0$ so that a large peak in the intensity of the undeflected beam occurs at such a value of H_0. From the value of H_0 at which the peak occurred and from the corresponding value of x inferred from Fig. 4.2, a could be calculated. From the number of zero-moment peaks the nuclear spin could be determined, as is shown by a comparison of the upper and lower parts of Fig. 4.2. For atomic states with $J > \frac{1}{2}$ like the metastable $2\,^2P_{\frac{3}{2}}$ states of In and Ga, quadrupole-interaction constants b could be determined from the zero-moment pattern. A detailed account of such experiments is included in ref. 3.

8. Atomic Beam Magnetic Resonance Experiments

By far the most effective techniques for the precision study of atomic hyperfine structure have been the atomic beam magnetic resonance methods [3, 7, 8]. A schematic diagram of an apparatus used for such experiments is shown in Fig. 4.3. The atoms are first deflected in an inhomogeneous magnetic field A and then refocused by an inhomogeneous magnetic field B. Between these two fields they are subjected to a uniform magnetic field C of arbitrary magnitude and a weak oscillatory magnetic field. When the oscillatory field is at a Bohr frequency,

$$\nu_{p,q} = \frac{W_p - W_q}{h} \tag{4.22}$$

for an allowed transition, the molecule may change its state in the C region with the result that it is no longer refocused by the B magnet. The occurrence of the resonance of Eq. (4.22) may therefore be detected by the reduction in the intensity of the refocused beam. In this way a radio-frequency spectrum of the atom may be obtained. The sharpness of the line is determined only by the length of time $\Delta\tau$ that the atom is in the magnetic field in accordance with the usual relation of Fourier analysis.

$$\Delta\nu \sim \frac{1}{\Delta\tau} \sim \frac{1}{(100/10^5)} = 10^3 \text{ cps} \qquad (4.23)$$
$$\sim 3 \times 10^{-8} \text{ cm}^{-1}$$

since a typical atom velocity is 10^5 cm/sec and a typical length of transition region is 100 cm. Ramsey has shown that increased sharpness of the resonance lines can be obtained by the use of two separated oscillatory fields [3].

When the transitions of Eq. (4.22) are between two different atomic states as in Fig. 4.3, the hyperfine-structure separation can be inferred with great accuracy. Likewise, if I and J each exceed $\frac{1}{2}$, quadrupole interaction, parameters b may also be inferred and, if I and J each exceed 1, octupole, interactions c may be determined as well. Nuclear spins can also be determined directly from the observed spectra.

The most accurate measurements of atomic hyperfine structure have been obtained with the atomic hydrogen maser [8]. In this device the atoms in the upper hyperfine state are focused by a six-pole focusing magnet onto the entrance aperture of a Teflon-coated bulb in which they are stored several seconds. Coherent maser oscillations are then established in the surrounding tuned cavities. With this device, hyperfine measurements have been made to accuracies of better than 10^{-12}.

Tables of the values of I, $\Delta\nu$, a, b, and c that have been precisely measured in this way are included in ref. [3]. One of the most interesting $\Delta\nu$'s that has been measured is for atomic hydrogen whose experimental value is $\Delta\nu = 1{,}420{,}405{,}751.800 \pm 0.028$ cps. The nuclear magnetic-moment dipole moments, electric-quadrupole moments, and magnetic-octupole moments that can be calculated from these results are given in Part 9, Chap. 3.

In some cases the atomic hyperfine structures of two different isotopes have been accurately measured in this way and the nuclear magnetic moments of the same two isotopes have also been accurately and directly measured by the techniques described in Part 9, Chap. 3. In such cases one would expect that the ratio of the first two measurements could be accurately calculated from the observed ratio of the last two. Although there is generally good agreement between the calculated and measured hyperfine-structure ratios for the two isotopes, the disagreement is sometimes as large as four parts in a thousand and is definitely beyond the experimental error. These discrepancies have been attributed to the effects of the distribution of the nuclear magnetism throughout the nucleus [3]. This data has consequently provided valuable information on nuclear structure. Tables listing the observed hyperfine-structure anomalies are included in the literature [3].

From Eq. (4.11) or Fig. 4.1 it is apparent that the value of the atomic magnetic moment can be accurately measured in this fashion. Such measurements have been made on a large number of atoms [3]. A particularly interesting result from such measurements has been the determination that the spin magnetic moment of the electron is not exactly 1 Bohr magneton but instead is 1.001165 ± 0.000012 Bohr magnetons. The significance of this discrepancy is discussed in Sec. 9.

9. Hydrogen Fine Structure. The Lamb Shift

One of the most important atomic beam resonance studies has been on atomic fine structure, rather than hyperfine structure. Lamb and Retherford studied the fine structure of the first excited ($n = 2$) state of atomic hydrogen with a specialized form of atomic beam apparatus. In this apparatus the hydrogen emerged from a heated tungsten oven which dissociated many of the hydrogen atoms to hydrogen molecules. This atomic beam was then cross bombarded by an electron beam of about 10 ev energy which excited a small fraction of the atoms to the $n = 2$ state. Those atoms which were excited to either a $2\,^2P_{3/2}$ state or a $2\,^2P_{1/2}$ state would

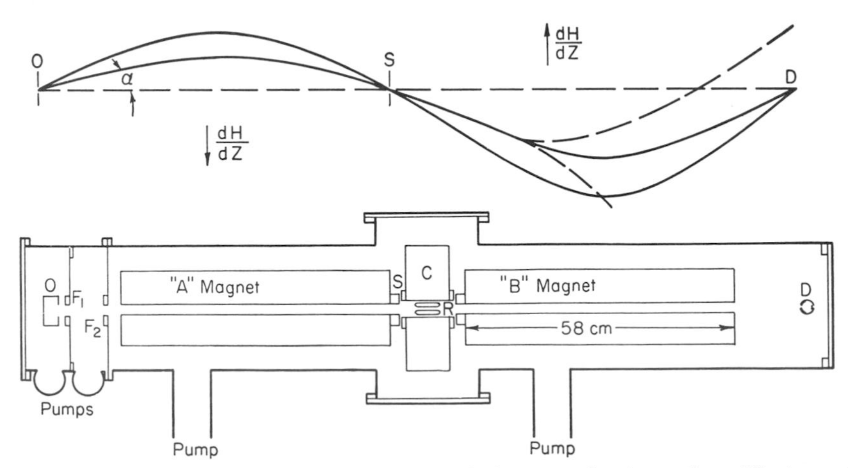

FIG. 4.3. Schematic diagram of a molecular-beam apparatus and of some molecular paths. The two solid curves in the upper part of the figure indicate the paths of two molecules having different moments and velocities and whose moments do not change during passage through the apparatus. The dotted curves indicate possible changes in path if the component of magnetic moment is changed. The motion in the z direction is greatly exaggerated. This apparatus has been used both with nonparamagnetic molecules and with paramagnetic atoms.

undergo an allowed transition to the 1 $^2S_{1/2}$ ground state almost immediately. However, those atoms which ended in the metastable 2 $^2S_{1/2}$ state remained there until the atoms struck the detector. The detector consisted of a tungsten target and an electron collector; an electron was emitted in an inelastic collision of the second kind when the metastable atom struck the detector. A microwave radiation field was then applied between the exciting and the detection region. When Eq. (4.22) was satisfied for an allowed transition to a 2 ^2P state, transitions would occur after which the atoms would almost immediately decay to the ground state, in which case they were no longer detectable. The occurrence of the resonance condition of Eq. (4.22) was therefore detected by the accompanying reduction in measured beam intensity.

In this way, the fine structures of atomic hydrogen and deuterium and their Zeeman effects have been studied with high precision. Similar results have been obtained with singly ionized helium. These experiments showed that the 2 $^2S_{1/2}$ state was not exactly degenerate with the 2 $^2P_{1/2}$ state, as predicted by the Dirac theory; instead for atomic hydrogen the 2 $^2S_{1/2}$ state was higher by 1057.77 $\pm$ 0.10 Mc, often called the Lamb shift. In similar experiments the separation of the 2 $^2P_{3/2}$ state and the 2 $^2P_{1/2}$ state was accurately measured and found to be 10,971.59 $\pm$ 0.20 Mc; from this result the fine-structure constant α was calculated to have the value $\alpha^{-1} = 137.0390 \pm 0.0012$.

The Lamb shift has been interpreted by Kramers, Bethe, Schwinger, Weisskopf, and others as resulting from changes in the electron self-energy which results from its interactions with the electromagnetic and electron-positron fields. Thus the potentiality of the electron virtually to emit and reabsorb photons contributes to its self-energy and the amount of this self-energy may be slightly different when the electron is in a 2 $^2P_{1/2}$ state and when it is in a 2 $^2S_{1/2}$ state and is thereby closer to the nucleus where it is subjected to stronger electrostatic fields. The theoretical magnitude for the Lamb shift has been calculated from this point of view to be 1,057.199 $\pm$ 0.13 Mc, in excellent agreement with experiment. Schwinger suggested that the same physical phenomena were responsible for the anomalous magnetic moment of the electron; a calculation on this basis leads to a theoretical value for the electron magnetic moment of 1.0011596 in excellent agreement with experimental result listed in Sec. 8. The observed atomic hydrogen hyperfine structure $\Delta\nu$ given in Sec. 8 has been similarly interpreted successfully.

References

1. Ramsey, N. F.: "Nuclear Moments," Wiley, New York, 1953.
2. Condon, E. U., and G. H. Shortley: "Theory of Atomic Spectra," Cambridge University Press, New York and London, 1935.
3. Ramsey, N. F.: "Molecular Beams," Oxford University Press, New York and London, 1955.
4. Kopfermann, H.: "Kernmomente," Akademische Verlagsgesellschaft, Leipzig, 1940.
5. Schwartz, C.: *Phys. Rev.*, **97**: 380 (1955).
6. Tolansky, S.: "High Resolution Spectroscopy," Methuen, London, 1948.
7. Kellogg, J. M. B., and S. Millman: *Revs. Mod. Phys.*, **18**: 323 (1946).
8. Kleppner, D. K., H. M. Goldenberg, and N. F. Ramsey: *Phys. Rev.*, **126**: 603 (1962); **138**: A972 (1965).

Chapter 5

Infrared Spectra of Molecules

By HARALD H. NIELSEN, The Ohio State University

1. Introduction

One of the most effective methods of arriving at information concerning the structure of molecules is that which the study of band spectra makes possible. The term *band* spectra, as distinguished from *line* spectra, is of historical origin and is a misleading designation. It stems from early investigations of molecular spectra under low resolving power in which the spectra, particularly in the visible ultraviolet, appeared as bands of continuous color. Later experiments carried out under much higher dispersive power have revealed that these bands may also be resolved into a pattern of discrete sharp lines.

Band spectra may, in a general way, be arranged into three different classifications according to what part of the spectrum they occupy, i.e., the visible ultraviolet (greater than 15,000 cm^{-1}), the near infrared (10,000 cm^{-1} to 500 cm^{-1}) or the far infrared and microwave region (500 cm^{-1} to 1 cm^{-1}).

The visible ultraviolet band spectra lie in much the same spectral region as do the lines in the spectra of atoms and are frequently referred to as electronic band spectra, their general positions in the spectrum being consistent with quantum transitions from one electronic energy state to another.

The frequency positions of band spectra in the near infrared can be identified with the oscillations of the atomic nuclei about their positions of equilibrium in the molecule. The forces which exist between atomic nuclei in a molecule are much the same as those existing between electrons or between electrons and the nuclei, but the masses of the nuclei are of the order of 2,000 times greater than the masses of the electrons. Thus the frequencies with which the particles will move are roughly proportional to the inverse square root of the masses. Therefore nuclear vibration frequencies should be about one-thirtieth as great as the electronic frequencies or in the neighborhood of 1,000 cm^{-1}. Thus the positions of bands in the near infrared spectrum of a molecule may be ascribed to quantum transitions from one nuclear vibration state to another.

The far infrared and microwave spectra can be identified with the pure rotation frequencies of the molecule. The molecules in an absorption cell may be considered to be in thermal equilibrium. The distribution of the molecules among the rotation energies is given by a Maxwell-Boltzmann law, so that the number of molecules dN having rotational frequencies between ν_r and $\nu_r + d\nu_r$ will be

$$dN(\nu_r) = \frac{4\pi^2 I \nu_r}{kT} N \exp\left(-E_r/kT\right) d\nu_r$$

where N is the total number of molecules present, E_r is the rotational energy, k is the Boltzmann constant, and T is the absolute temperature. The frequency for which $dN(\nu_r)/d\nu_r$ is a maximum is obtained by solving the equation $d[dN(\nu_r)/d\nu_r] = 0$ for ν. Moreover, since the intensity with which a frequency occurs depends upon the number of molecules, $dN(\nu_r)/d\nu_r$, which participate in the frequency, the frequency $\nu_{\max}$ for which $dN(\nu_r)/d\nu_r$ is a maximum will be the most intense line in the spectrum.

Setting $E_r = \frac{1}{2}I\omega^2 = 2\pi^2 I \nu_r^2$, where I is the moment of inertia and ν_r is the frequency of rotation of the molecule, $\nu_{\max} = (\frac{1}{2}\pi)(kT/I)^{1/2}$. For a diatomic molecule, $I = m_1 r_1^2 + m_2 r_2^2$, where $r_1 + r_2$ is the equilibrium distance of the two nuclei from each other, $r_1 + r_2$ will be about 1 A. The masses will be about 1.6×10^{-24} g. At room temperatures, $T \approx 300°K$, $\nu_{\max}$ is about 60 cm^{-1}. The lines in the far infrared may therefore be ascribed in general to quantum transitions from one pure rotational level to another. The lines have frequencies from a few cm^{-1} to about 500 cm^{-1} with the most intense lines near 50 cm^{-1} to 100 cm^{-1}.

The three periodic types of motion* of a molecule are to a fair approximation independent of each other, although the interactions between them cannot always be neglected. Thus for the energy of a molecule

$$E = E_e + E_v + E_r + E_i \tag{5.1}$$

where E_e, E_v, E_r, and E_i are, respectively, the electronic energy, the vibrational energy, the rotational energy, and the energy of interaction between the three types of motion.

According to the Bohr frequency condition the position of a line in the spectrum will be

$$\nu = \frac{E' - E''}{hc} \tag{5.2}$$

* A molecule taken as a unit also has translational motion. These are not periodic, however, and are of no spectroscopic interest.

If E_i is small enough* so that we may neglect it,

$$\nu = (hc)^{-1}[(E'_e - E''_e) + (E'_v - E''_v) + (E'_r - E''_r)] \quad (5.3)$$

where E' and E'' are the energies of the final and the initial quantum states, h is Planck's constant, and c is the velocity of light. Rules, known as selection rules, establish between what levels E' and E'' transitions may take place. Generally speaking, transitions will be from one state to another such that $E'_e \neq E''_e$, $E'_v \neq E''_v$, and $E'_r \neq E''_r$. In the visible ultraviolet spectrum associated with a given electron transition there is a series of vibration bands, each of which in turn has a band *structure* determined by the permitted rotational transitions. In the near infrared the transitions are between states for which $E'_e = E''_e$, whereas $E'_v \neq E''_v$ and $E'_r \neq E''_r$. A series of vibration frequencies may occur, each of which has a band structure depending upon the allowed rotational transitions between E'_r and E''_r. Finally, in the far infrared and the microwave regions, the spectra arise from transitions between levels where $E'_e = E''_e$, $E'_v = E''_v$, but where $E'_r \neq E''_r$.

Spectra may occur in emission or in absorption. Emission spectra arise when the molecule makes a transition from states of higher energies to states of lower energies. Conversely, absorption spectra occur when, by absorption of radiation, a molecule is raised from one quantum state to another of higher energy.

This chapter is concerned only with bands in the near infrared where $\Delta E_e = 0$ and the far infrared spectra where only $\Delta E_r \neq 0$.

2. The Energies of a Molecule

The energies of a molecule are obtained theoretically by the solution of the Schroedinger equation,

$$(H - E)\psi = 0 \quad (5.4)$$

appropriate to the molecule under consideration. H, in (5.4), is the Hamiltonian operator for the molecule, obtained from the classical Hamiltonian by replacing the linear momenta p_{x_i}, p_{y_i}, and p_{z_i}, conjugate to the coordinates x_i, y_i, and z_i of the electrons and the atomic nuclei, respectively, by the differential operators $i\hbar(\partial/\partial x_i)$, etc. The functions ψ are the wave functions or eigenfunctions and are solutions to the differential equation (5.4) which must be single-valued and must remain finite for all values of the coordinates x_i, y_i, etc. The eigenvalues of E are the energies of the molecule.

* The argument for assuming E_i to be small is somewhat as follows. One may regard the molecule as making a great many nuclear oscillations while it makes one complete rotation. The true moment of inertia must therefore be replaced by an effective moment of inertia averaged over the vibration of the nuclei. Since the deviations of the nuclei from their equilibrium positions are small compared to the equilibrium positions themselves, the effective moment of inertia is very little different from its equilibrium value I_e. Similarly the electrons go around in their orbits many times while the molecule executes one nuclear vibration. Therefore, the effective force field in which the nuclei move may be taken to be the true force field averaged over the motion of the electrons.

The operator H may be written:

$$H = H_e + H_v + H_r \quad (5.5)$$

in the approximation that the three types of motion are independent. Then the equation for ψ is satisfied by a product $\psi_e\psi_v\psi_r$, where ψ_e, ψ_v, and ψ_r are functions only of the electronic, vibrational, and rotational coordinates, respectively. Equation (5.4) then becomes separable in the coordinates of each type of motion so that

$$(H_e - E_e)\psi_e = 0 \qquad (H_v - E_v)\psi_v = 0 \qquad (H_r - E_r)\psi_r = 0 \quad (5.6)$$

So long as H_e is independent of the vibrational and the rotational coordinates, it may be solved by regarding the atomic nuclei as fixed. The energies E_e then become quantities which will depend upon the nuclear coordinates as parameters, the equilibrium values of these being those for which E_e is a minimum. For a diatomic molecule, E_e depends upon the internuclear distance r as a parameter, as is illustrated in Fig. 5.1. In general, E_e will depend upon many internuclear distances, s_{ij}, and $E_e(s_{ij})$ will define a surface.

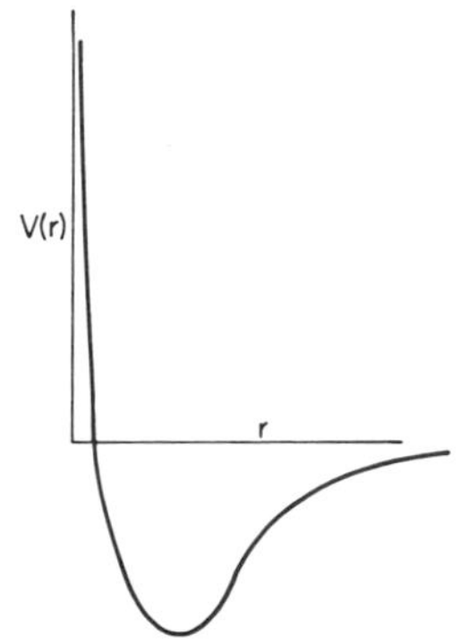

FIG. 5.1. The potential energy of a diatomic molecule as a function of the internuclear distance r.

3. The Vibrations of a Molecule

The electronic energies E_e depend upon the internuclear distances as parameters and the equilibrium values of these will be determined so that E_e is a minimum. $E_e(s_{ij})$ thus serves as a potential energy function, $V(s_{ij})$, for the oscillating atomic nuclei. The electronic energy problem of a molecule has, unfortunately, been solved in only the simplest of cases so that in general $E_e(s_{ij})$ must be regarded as unknown. It has been necessary in order to make progress, to replace the exact potential energy function $V(s_{ij})$ by one which simulates it, especially in the regions where the coordinates s_{ij} are close to the equilibrium values s_{ij}^0. The exact function $E_e(s_{ij})$ is replaced by an expansion in terms of s_{ij}

$$V(s_{ij}) = V_0 + \sum_{ij} \frac{\partial V}{\partial s_{ij}} \delta s_{ij} + \tfrac{1}{2} \sum_{ijkl} \frac{\partial^2 V}{\partial s_{ij}\, \partial s_{kl}} \delta s_{ij}\, \delta s_{kl} \quad (5.7)$$

where $\delta s_{ij} = s_{ij} - s_{ij}^0$ and the derivatives are evaluated at the equilibrium values. The first term is a constant which for our purpose may be set equal to zero. The second term will vanish also since when the nuclei are in equilibrium the quantities $(\partial V/\partial s_{ij})$ must, of course, be zero [see, for example, Fig. 5.1 where $\partial E(r)/\partial r$ is zero at $r = r_0$]. The constants

$$\left(\frac{\partial^2 V}{\partial s_{ij}\, \partial s_{kl}}\right)_{\substack{s_{ij}=s_{ij}^0 \\ s_{kl}=s_{kl}^0}}$$

may be regarded as generalized force constants designated as K_{ijkl}; so $V(s_{ij})$ becomes

$$V(s_{ij}) = \tfrac{1}{2} \sum_{ijkl} K_{ijkl}\, \delta s_{ij}\, \delta s_{kl} \tag{5.8}$$

The energy of vibration of a molecule may be written as the sum of its kinetic energy T and its potential energy V. Thus

$$E_v = T_v + V \tag{5.9}$$

where T_v will be

$$T_v = \tfrac{1}{2} \sum_i m_i(\dot{x}_i^2 + \dot{y}_i^2 + \dot{z}_i^2) \tag{5.10}$$

x_i, y_i, and z_i being the cartesian coordinates of the ith nucleus. It will be necessary to rewrite the relation for $V(s_{ij})$ in terms of the coordinates x_i, y_i, and z_i:

$$\begin{aligned} V(x_i,y_i,z_i) = \tfrac{1}{2} \sum_{ij} (k_{ij}^{(xx)}\, \delta x_i\, \delta x_j + k_{ij}^{(yy)}\, \delta y_i\, \delta y_j \\ + k_{ij}^{(zz)}\, \delta z_i\, \delta z_j + k_{ij}^{(xy)}\, \delta x_i\, \delta y_j + k_{ij}^{(zx)}\, \delta x_i\, \delta z_j \\ + k_{ij}^{(yz)}\, \delta y_i\, \delta z_j) \end{aligned} \tag{5.11}$$

where the constants $k_{ij}^{(\alpha\beta)}$ and the constants K_{ijkl} are related to each other in a way that is determined by the geometry of the model.

There are $3N$ of the coordinates x_i, y_i, z_i, where N represents the number of nuclei in the molecule. In studying the vibrational motion of the nuclei, one may require that the center of mass be fixed and that the angular momentum of the molecule be zero to the approximation that the molecule is a rigid rotator. This imposes six relationships between the coordinates which are generally taken to be

$$\sum_i m_i\, \delta\dot{\alpha}_i = 0 \quad \text{and} \quad \sum_i m_i(\alpha_i^0\, \delta\dot{\beta}_i - \beta_i^0\, \delta\dot{\alpha}_i) = 0 \tag{5.12}$$

There remain $3N - 6$ linearly independent coordinates which will describe the nuclear motions in the molecule.

Now introduce a set of coordinates, Q'_i, of which there will be $3N - 6$, and express the coordinates δx_i, δy_i, etc., as linear combinations of these, that is,

$$\begin{aligned} \delta x_1 &= a_{11}^{(x)}Q'_1 + a_{12}^{(x)}Q'_2 + \cdots + a_{1\,3N-6}^{(x)}Q'_{3N-6} \\ \delta x_2 &= a_{21}^{(x)}Q'_1 + a_{22}^{(x)}Q'_2 + \cdots + a_{2\,3N-6}^{(x)}Q'_{3N-6} \\ &\cdots \\ \delta \alpha_i &= a_{i1}^{(\alpha)}Q'_1 + a_{i2}^{(\alpha)}Q'_2 + \cdots + a_{i\,3N-6}^{(\alpha)}Q'_{3N-6} \end{aligned} \tag{5.13}$$

Further require the constants $a_{ij}^{(\alpha)}$ to be so chosen that when the relations (5.13) for $\delta\alpha_i$ are substituted into the relation (5.10) there will be no cross-product terms $\dot{Q}'_i\dot{Q}'_j$ in the kinetic energy and no cross-product terms in $Q'_iQ'_j$ in the potential energy function. It is convenient also to introduce a normalizing condition $\sum_{i\beta} a_{ij}^{(\alpha)}a_{ij}^{(\beta)} = \delta_{\alpha\beta}$ in which case

$$E_v = T_v + V = \tfrac{1}{2} \sum_{i=1}^{3N-6} \mu_i\, \dot{Q}'^2_i + \sum_{i=1}^{3N-6} \lambda'_i Q'^2_i$$

where the μ_i are reduced masses, which are related to the m_i, and where the λ'_i are generalized force constants which are related to $k_{ij}^{(\alpha\beta)}$. It is general practice to replace $(\mu_i)^{1/2}Q'_i$ by a new coordinate Q_i so that

$$E = \tfrac{1}{2} \sum_{i=1}^{3N-6} (\dot{Q}_i^2 + \lambda_i Q_i^2) \tag{5.14}$$

The procedure corresponds to the classical theory of normal modes of vibration in Part 2, Chap. 3, Sec. 6.

The $3N - 6$ equations of motion of Lagrange have the form $(d^2Q_i/dt^2) + \lambda_i Q_i = 0$ and therefore

$$Q_i = A_i \sin\,[(\lambda_i)^{1/2}t + \epsilon_i] \tag{5.15}$$

in which A_i is the amplitude and ϵ_i is a phase constant. Then

$$\begin{aligned} \delta x_i &= a_{i1}^{(x)}A_1 \sin\,[(\lambda_1)^{1/2}t + \epsilon_1] + a_{i2}^{(x)}A_2 \sin\,[(\lambda_2)^{1/2}t + \epsilon_2] \cdots \\ &\cdots \\ \delta \alpha_i &= a_{i1}^{(\alpha)}A_1 \sin\,[(\lambda_1)^{1/2}t + \epsilon_1] + a_{i2}^{(\alpha)}A_2 \sin\,[(\lambda_2)^{1/2}t + \epsilon_2] \cdots \end{aligned} \tag{5.16}$$

The coordinates, Q_i, chosen this way are known as *the normal coordinates.* When the molecule is oscillating in one of its modes, described, say, by Q_r (so that all the other $Q_i = 0$), then, from Eq. (5.16), all the particles move with the same frequency, namely, $\nu_i = (\frac{1}{2}\pi)(\lambda_i)^{1/2}$.

The vibrational Hamiltonian expressed in these coordinates is

$$H_v = \tfrac{1}{2} \sum_{i=1}^{3N-6} (P_i^2 + \lambda_i Q_i^2) \tag{5.17}$$

where the P_i are the linear momenta conjugate to Q_i. The transition to quantum mechanics is effected by replacing P_i by $i\hbar\partial/\partial Q_i$ where $\hbar$ is Planck's constant, h, divided by 2π. Replacing Q_i by $(\hbar/\lambda_i)^{1/2}q_i$ the Schroedinger equation for the oscillational motion of the nuclei in the molecule is

$$\left\{hc \sum_{i=1}^{3N-6} \nu_i \left[\left(\frac{p_i}{\hbar}\right)^2 + q_i^2\right] - E_{v_i}\right\} \psi = 0 \quad (5.18)$$

which is separable in the coordinates and leads to the value for E

$$E = hc \sum_{i=1}^{3N-6} (v_i + \tfrac{1}{2})\nu_i \quad (5.19)$$

where the frequency ν_i is measured in cm^{-1} (see Part 2, Chap. 3, Sec. 4).

The selection rule for the harmonic oscillator states that $\Delta v_i = 1$. For an aggregate of oscillators as in a molecule it states $\Delta v_i = 1$, $\Delta v_k = 0$ for $k \neq i$. This follows from the fact that the electric dipole moment is linear in the coordinates. Thus, from Eq. (5.2), the frequencies observed will be exactly the normal frequencies of the molecule.

One proceeds now to identify the fundamental frequencies ν_i with certain of the bands observed experimentally and, from this knowledge, to evaluate the constants K_{ijkl} in Eq. (5.8). If this can be accomplished successfully, so that the constants K_{ijkl} are all known, a useful approximation to the actual potential energy function $E_e(s_{ij})$ will have been achieved which will suffice for many chemical and physical purposes. Unfortunately, there are frequently more constants, K_{ijkl}, present in $V(s_{ij})$ than there are vibration frequencies so that a complete determination of the constants in the general potential energy relation cannot be carried out. It is then necessary to introduce simplifying assumptions which will provide new relationships between the constants K_{ijkl}. The assumption which has met with most success is the valence force concept which assumes the restoring forces to be of a valence bond type and a valence angle type. The substitution of isotopic atoms in molecules of the same species will, in some instances, provide useful information. In such cases the frequencies are altered, but the forces remain the same. Determination of the frequencies of the isotopic molecule will, therefore, give new relations which may be used to evaluate the K_{ijkl}.

As an example, consider the vibration of a symmetric nonlinear (XY_2) molecule. The model is shown in Fig. 5.2, the masses being m_1, m_2, and

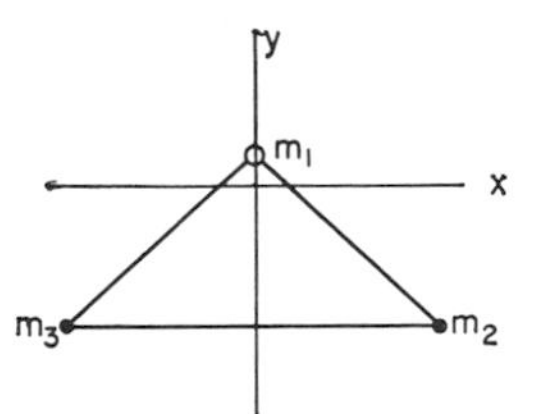

FIG. 5.2. The XY_2 molecular model.

$m_3(= m_2)$, and it is placed in the xy plane so that the center of mass is at the origin of the coordinate system and the y axis passes through the vertex of the triangle. The equilibrium values of the coordinates of the nuclei will therefore be $(0,y_1^0,0)$, $(x_2^0,y_2^0,0)$, and $(-x_2^0,y_2^0,0)$.

The kinetic energy T_v will be

$$T_v = \tfrac{1}{2} \sum_{i=1}^{3} m_i(\delta\dot{x}_i^2 + \delta\dot{y}_i^2 + \delta\dot{z}_i^2)$$

where δx_i, etc., are the displacement coordinates of the nuclei from their equilibrium positions. We shall assume, further, that the simple valence force law is valid here so that we may write for V

$$V = \frac{k}{2}(\delta s_{12}^2 + \delta s_{13}^2) + \frac{k_\alpha}{2} s_{12}^{02}\, \delta\alpha^2 \quad (5.20)$$

in which δs_{12}, etc., are the variations in the internuclear distances s_{12}, etc., and $\delta\alpha$ is the variation of the valence angle.

T_v (and V when written in terms of δx_i, etc.) will have nine coordinates. Next introduce the following six relations between these coordinates:

$$\sum_{i=1}^{3} m_i\, \delta\dot{x}_i = 0 \qquad \sum_{i=1}^{3} m_i\, \delta\dot{y}_i = 0 \qquad \sum_{i=1}^{3} m_i\, \delta\dot{z}_i = 0 \quad (5.21)$$

and

$$\begin{aligned} m_2(x_2^0\, \delta\dot{z}_2 + x_3^0\, \delta\dot{z})_3 &= 0 \\ m_1 y_1^0\, \delta\dot{z}_1 + m_2(y_2^0\, \delta\dot{z}_2 + y_3^0\, \delta\dot{z}_3) &= 0 \\ -m_1 y_1^0[\delta\dot{x}_1 - \tfrac{1}{2}(\delta\dot{x}_2 + \delta\dot{x}_3)] + m_2 x_2^0(\delta\dot{y}_2 - \delta\dot{y}_3^0) &= 0 \end{aligned} \quad (5.22)$$

This is the special form of (5.12). From the first two relations in (5.21), $\delta\dot{z}_1 = \delta\dot{z}_2 = \delta\dot{z}_3 = 0$; thus the motion of the nuclei will be entirely in the xy plane. There remain three coordinates to describe the vibrational motion.

Our work is simplified if we introduce three intermediate coordinates which reflect the symmetry of the model:

$$x = \delta x_1 - \tfrac{1}{2}(\delta x_2 + \delta x_3) \qquad y = \delta y_1 - \tfrac{1}{2}(\delta y_2 + \delta y_3)$$
$$q = \delta x_2 - \delta x_3 \quad (5.23)$$

The potential energy function must be invariant to a change of algebraic sign of x, but will not be so to y or to q. Thus V can contain x only to an even power, which precludes cross-product terms in V between x and q and x and y. Using (5.21) and (5.22) and the intermediate coordinates (5.23),

$$T_v = \frac{1}{2}\left(\frac{\mu I_{zz}}{I_{xx}}\right)\dot{x}^2 + \frac{\mu}{2}\dot{y}^2 + \frac{m_2}{2}\dot{q}^2 \quad (5.24)$$

and also

$$\begin{aligned} s_{12} &= \left(-\frac{2m_2 \sin^2\alpha + \mu\cos^2\alpha}{2m_2 \sin\alpha}\right)x \\ &\qquad + \cos\alpha y + \tfrac{1}{2}\sin\alpha q \\ s_{13} &= \left(\frac{2m_2 \sin^2\alpha + \mu\cos^2\alpha}{2m_2 \sin\alpha}\right)x \\ &\qquad + \cos\alpha y + \tfrac{1}{2}\sin\alpha q \\ s_{12}^0 &= \cos\alpha q - 2\sin\alpha y \end{aligned} \quad (5.25)$$

Substitution into the relation (5.20) for V yields

$$V = k\left(\frac{2m_2 \sin^2\alpha + \mu\cos^2\alpha}{2m_2\sin\alpha}\right)x^2 + (k\cos^2\alpha + 2k_\alpha\sin^2\alpha)y^2 + \left(\frac{k\sin^2\alpha}{4} + \frac{k_\alpha\cos^2\alpha}{2}\right)q^2 + (k - 2k_\alpha)\sin\alpha\cos\alpha yq \quad (5.26)$$

In the above $\mu = 2m_1m_2/(2m_2 + m_1)$ and I_{xx} and I_{zz} are two of the principal moments of inertia. Equations (5.24) and (5.25) are somewhat simplified by making the replacements $(\mu I_{zz}/I_{xx})^{1/2}x = Q_3$,

$$(\mu)^{1/2}y = Q'_2$$

and $(m_2/2)^{1/2}q = Q'_1$. Q_3 may be regarded as one of the normal coordinates. For Q'_1 and Q'_2, introduce the linear combinations $Q'_1 = aQ_2 + bQ_1$ and $Q'_2 = cQ'_2 + dQ'_1$. Replacing Q'_1 and Q'_2 by these linear relations and introducing the normalization conditions $a^2 + b^2 = 1$, $c^2 + d^2 = 1$ together with the orthogonality condition $ac + bd = 0$ and requiring further that the cross products in $\dot{Q}_1\dot{Q}_2$ in T_v and the cross product Q_1Q_2 in V must vanish,

$$\left.\begin{matrix} -a = d \\ c = b \end{matrix}\right\} = 2^{-1/2}\left[1 \mp \frac{A}{(A^2 + B)^{1/2}}\right]^{1/2}$$

in which

$$A = k(2m_2\cos^2\alpha - \mu\sin^2\alpha) + 2k_\alpha(2m_2\sin^2\alpha - \mu\cos^2\alpha)$$
$$B = 8m_2\mu(k - 2k_\alpha)^2\sin^2\alpha\cos^2\alpha$$

The following relations exist between the frequencies:

$$4\pi^2(\nu_1^2 + \nu_2^2) = (1 + 2m_2\cos^2\alpha/m_1)(k/2m_2) + (1 + 2m_2\sin^2\alpha/m_1)(k_\alpha/m_2)$$
$$16\pi^4\nu_1^2\nu_2^2 = (1 + 2m_2/m_1)(kk_\alpha/m_2^2)$$
$$4\pi^2\nu_3^2 = (1 + 2m_2\sin^2\alpha/m_1)(k/2m_2) \quad (5.27)$$

The vibration of the triatomic (XY_2) molecule is thus completely solved. The three relations (5.27) which exist between the frequencies contain three constants k, k_α, and α which are determined if the three fundamental frequencies can be identified in the spectrum. Moreover, the constants a, b, etc., can also be evaluated so that the form of the normal coordinates can be described. The form of these for the XY_2 model is illustrated in Fig. 5.3. If, for

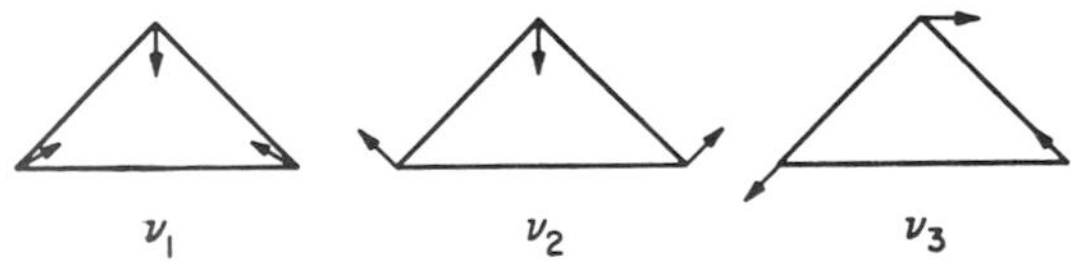

Fig. 5.3. The normal modes of vibration of the XY_2 model.

example, the potential energy function had been completely general so as to include also the cross-product terms $k_{23}\,\delta s_{12}\,\delta s_{13} + k_{1\alpha}s_{12}^0(\delta s_{12} + \delta s_{13})\,\delta\alpha$, there would be five constants to evaluate. Then it would have been necessary to make use of a molecule where, for example, isotopes of the Y atoms had been substituted for the ordinary Y atoms, to accomplish this, giving six frequencies to determine the five constants.

The oscillations described in Fig. 5.3 are essentially of two different types. One type, ν_1 and ν_2, involves oscillating electric moments parallel to the y axis; the other type, ν_3, involves an electric moment parallel to the x axis. All three involve electric moments different from zero and therefore give rise to infrared vibration bands in the spectrum.

In certain molecules the atoms are so arranged that a higher degree of symmetry than that exhibited by the XY_2 model exists. An example of such a molecule would be the regular pyramidal model. Another example would be the XYZ_3 type of molecule where the YZ_3 atoms form a regular pyramid and the X atom lies along the altitude of the pyramid extended beyond its vertex. Such molecules have axial symmetry and are called *axially symmetric molecules*. Figure 5.4 shows the vibrations of the XY_3 model.

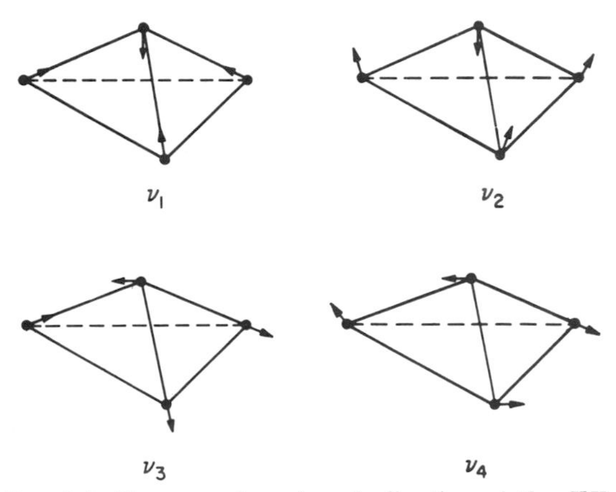

Fig. 5.4. The normal modes of vibration of the XY_3 model.

Frequencies ν_1 and ν_2 illustrate modes which induce electric moments directed along the axis of symmetry of the model. Such modes are called *parallel vibrations*. The modes ν_2 and ν_4, however, are such that they induce electric moments normal to the symmetry axis and such oscillations are called *perpendicular vibrations*. The nuclei in the latter are moving in force fields which are axially symmetric. The consequence of this is that a perpendicular vibration will have two components, one along the x axis and another along the y axis, with the same frequencies. Two coordinates Q_i, designated by Q_{i1} and Q_{i2}, will then be required to describe a single vibration frequency ν_i. Such vibrations are spoken of as being twofold degenerate.

The arrangement in a molecule of the atoms may, in some instances, be even more symmetric. An example is the regular tetrahedral XY_4 model of which such molecules as CH_4, SiH_4, etc., are specific cases. The vibrations of such a molecule are shown in Fig. 5.5. Frequency ν_1 may be visualized as a *breathing* frequency where the Y atoms are all moving exactly in phase along the XY distances. It is nondegenerate, induces no electric moment, and is, therefore, optically inactive in the infrared. Frequency ν_2 may be visualized as one where the X atom remains at rest and where the four Y atoms move in elliptic paths on the surface of a sphere of radius

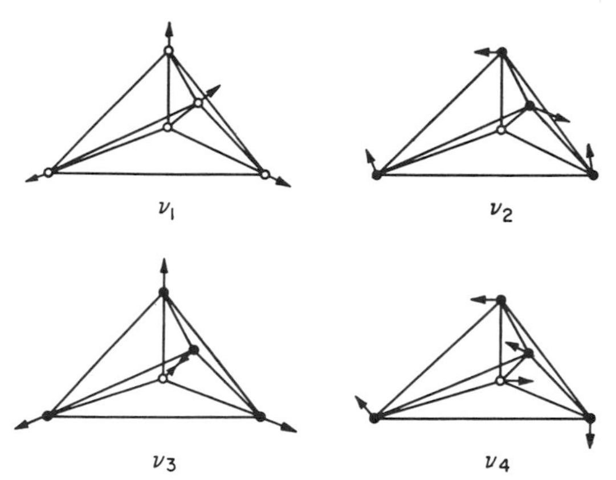

FIG. 5.5. The normal modes of vibration of the tetrahedral XY_4 model.

$(XY)_0$. This frequency is twofold degenerate. It induces zero electric moment and may therefore also be expected to be optically inactive in the infrared. Frequencies ν_3 and ν_4 are similar in character. The former may be thought of as a periodic motion of the Y atoms along the XY bonds, where one Y atom is exactly out of phase with the other three. The latter may be imagined to be a periodic deformation such that one XY_2 triangle is exactly out of phase with the remaining three. These latter two vibrations can each occur in three independent ways. ν_3 and ν_4 are, therefore, threefold degenerate and require each three coordinates Q_{i1}, Q_{i2}, and Q_{i3} to describe them.

The Schroedinger equation necessary to describe a vibration ν_i which is degenerate must therefore be written

$$\left\{hc \sum_{\sigma=1}^{g_i} \left[\nu_i \left(\frac{p_{i\sigma}}{\hbar}\right)^2 + q_{i\sigma}^2\right] - E_{v_i}\right\} \psi_i = 0 \qquad (5.28)$$

where g_i will be two or three as the vibration ν_i is twofold or threefold degenerate. The energy E_{v_i} will be

$$E_{v_i} = \sum_{\sigma=1}^{g_i} \left(v_{i\sigma} + \frac{g_i}{2}\right) hc\nu_i \qquad (5.29)$$

It is usually preferable in dealing with degenerate vibrations to replace the usual normal coordinates $q_{i\sigma}$ by their equivalents in polar coordinates. When this is done for a twofold degenerate oscillation ν_i, two new quantum numbers occur in place of v_{i1} and v_{i2} in (5.29), namely, v_i, which equals $v_{i1} + v_{i2}$, and l_i. The quantum number l_i does not occur in the energy of the doubly degenerate *harmonic* oscillator. It may be regarded as a quantum number which defines the angular momentum of vibration associated with the oscillation ν_i. The solution to the quantum-mechanical equation demands that l_i take only the values v_i, $v_i - 2$, . . . , 1 or 0. The actual angular momentum may be shown to be equal to $l_i\zeta_i\hbar$, where ζ_i depends upon the nature of the normal coordinate and may be thought of as a modulus of the angular momentum vector.

Similarly, if polar coordinates are substituted for the usual normal coordinates in a three dimensionally isotropic oscillator, three quantum numbers v_i, l_i, and m_i arise instead of the quantum numbers v_{i1}, v_{i2}, and v_{i3}. The total vibration quantum number $v_i = v_{i1} + v_{i2} + v_{i3}$; l_i is a quantum number of angular momentum which, as before, may take only the values v_i, $v_i - 2$, . . . , 1 or 0; m_i is a quantum number taking all integral values from l_i to $-l_i$ and may be regarded as the projection of l_i along an axis fixed in the molecule. Neither l_i nor m_i occurs in the energy of the three dimensionally isotropic *harmonic* oscillator. We have for the energies then of a two dimensionally isotropic oscillator and a three dimensionally isotropic oscillator, respectively,

$$E_{v_i} = (v_i + 1)hc\nu_i \qquad \text{and} \qquad E_{v_i} = (v_i + \tfrac{3}{2})hc\nu_i \qquad (5.30)$$

where v_i takes all integral values including zero. The vibrational selection rules for the molecule are not altered, but remain $\Delta v_i = 1$, $\Delta v_k = 0$ and $\Delta v_i = 0$, $\Delta v_k = 1$.

The linear molecule may be thought of as a special case of the axially symmetric molecule. Figure 5.6

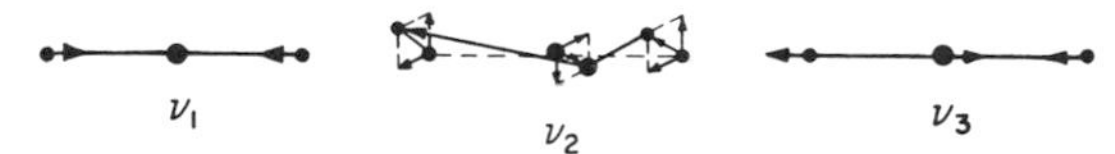

FIG. 5.6. The normal modes of vibration of the symmetric linear XY_2 model.

illustrates the normal vibrations of the symmetric XY_2 molecule. The linear molecule has three degrees of translational energy, but only *two* degrees of rotational motion. This type of molecule will, therefore, have $3N - 5$ degrees of vibrational freedom or in this specific case four frequencies. Only three of these are distinct, but one, ν_2, is twofold degenerate and must be counted twice. ν_2 is a perpendicular vibration and, as in the axially symmetric molecule, has associated with it an angular momentum $l_i\zeta_i\hbar$, where in the linear case, however, the ζ_i are always equal to 1. ν_2 induces an electric moment and is therefore optically active in the infrared. ν_3 induces an electric moment along the internuclear axis and should therefore give rise to a parallel infrared band. The frequency ν_1, on the other hand, induces no electric moment and should be optically inactive in the infrared.

Thus far the motion of the nuclei is assumed to be harmonic, and thus the equivalent expansion of the potential energy function includes only the quadratic terms. This assumption is too idealized and generally it is necessary to include also the anharmonic terms through the quartic terms. These may be expanded in terms of the normal coordinates and may be written

$$V_1 = hc \sum_{ijk} k_{ijk}q_iq_jq_k \qquad (5.31)$$

and

$$V_2 = hc \sum_{ijkl} k_{ijkl}q_iq_jq_kq_l \qquad (5.32)$$

respectively. The k_{ijk} and k_{ijkl} are constants.

4. The Rotational Energies of Molecules

Consider first the rotational motion of a molecule as that of a rigid rotator. The energies of a rigid rotator are obtained from the solution of the Schroedinger equation $(H_r - E_r)\Psi_r = 0$ where

$$H_r = \frac{1}{2}\left(\frac{P_x^2}{I_{xx}} + \frac{P_y^2}{I_{yy}} + \frac{P_z^2}{I_{zz}}\right) \tag{5.33}$$

P_α in (5.33) are the components of the total angular momentum directed along the α axis in the molecule and $I_{\alpha\alpha}$ are the principal moments of inertia. Classically the P_α are known to be

$$\begin{aligned} P_x &= \cos\phi p_\theta + \frac{\sin\phi}{\sin\theta}(p_\psi - \cos\theta p_\phi) \\ P_y &= \sin\theta p_\theta - \frac{\cos\phi}{\sin\theta}(p_\psi - \cos\theta p_\phi) \\ P_z &= p_\phi \end{aligned} \tag{5.34}$$

where θ, ϕ, and ψ are the Eulerian angles (see Part 2, Chap. 1, Sec. 3) relating the body fixed axes to the space fixed axes and where $p_\theta = \partial T/\partial\dot\theta$, etc., T being the kinetic energy of the rotator. Figure 5.7 shows the system of Eulerian angles referred to here. The translation to quantum mechanics is effected by replacing p_θ, p_ϕ, and p_ψ in (5.34) by the operators $i\hbar\,\partial/\partial\theta$, $i\hbar\,\partial/\partial\phi$ and $i\hbar\,\partial/\partial\psi$, respectively.

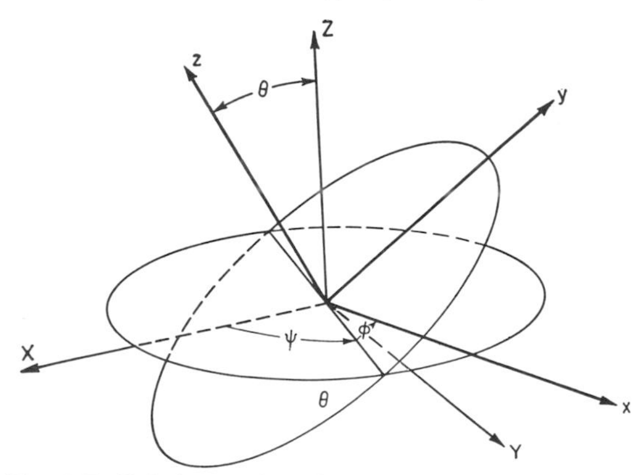

FIG. 5.7. Eulerian angles relating body fixed coordinates in the rotator to the space fixed coordinates.

Equation (5.33) may more advantageously be written

$$\begin{aligned} H_r = \frac{1}{2}\Bigg\{\frac{1}{2}\left(\frac{1}{I_{xx}} + \frac{1}{I_{yy}}\right)P^2 \\ + \left[\frac{1}{I_{zz}} - \frac{1}{2}\left(\frac{1}{I_{xx}} + \frac{1}{I_{yy}}\right)\right]P_z^2 \\ + \frac{1}{2}(P_x^2 - P_y^2)\left(\frac{1}{I_{xx}} - \frac{1}{I_{yy}}\right)\Bigg\} \end{aligned} \tag{5.35}$$

when one takes account of the fact that

$$P_x^2 + P_y^2 + P_z^2 = P^2$$

the total angular momentum. It is a general quantum-mechanical rule which also can be derived from a solution of Eq. (5.33) that $P^2 = J(J+1)\hbar^2$, where J takes all integral values including zero (see Part 2, Chap. 6, Sec. 5).

An important example arises when $I_{xx} = I_{yy}$: the last term in (5.35) vanishes. This is the case of the *symmetric rotator*, and a detailed study of Eq. (5.33) shows that here P_z may be replaced by $K\hbar$, where K is an integer such that $J \geq K \geq -J$. The energy of the symmetric rotator will therefore be

$$E_r = J(J+1)\frac{h^2}{8\pi^2 I_{xx}} + \frac{K^2h^2}{8\pi^2}\left(\frac{1}{I_{zz}} - \frac{1}{I_{xx}}\right) \tag{5.36}$$

While K may assume the $2J+1$ values from $+J$ to $-J$, the energy is independent of the algebraic sign of K. Hence, for a given value of J, there are only $J+1$ energy levels.

Many manifestations of the molecule as a rigid symmetric molecule are known, but only a few examples will be noted. One type is the XYZ_3 molecule where $I_{xx} > I_{zz}$, which is often referred to as the *prolate rotator*. Here $[(1/I_{zz}) - (1/I_{xx})]$ is positive. A second type is the flat pyramidal XY_3 model where $I_{zz} > I_{xx}$. This form of molecule is referred to as the *oblate rotator* and here $[(1/I_{zz}) - (1/I_{xx})$ is negative. An intermediate case is illustrated by the tetrahedrally symmetric XY_4 type of molecule where $I_{xx} = I_{zz}$. This is the *spherical rotator* where the last term in the energy vanishes. A fourth example of the rotator is the linear model where $I_{zz} = 0$.

When $I_{xx} \neq I_{yy}$, the rotator becomes asymmetric. As before, P^2 must be equal to $J(J+1)\hbar^2$, where J is integral. Here, however, P_z is no longer quantized (that is, the values of K are no longer integral) and there are $2J+1$ different values of P_z for a given value of J. Correspondingly there are $2J+1$ energy levels for a given value of J which are commonly identified by an index τ which takes all integral values from $+J$ to $-J$. It is customary to indicate the level of highest energy by a τ value equal to J and the level of lowest energy by a τ value equal to $-J$.

The energies of an asymmetric rotator lie, in general, somewhere intermediate between those of the two limiting types of symmetric rotator, the prolate and the oblate. It is advantageous in the case of the nearly symmetric prolate rotator to identify I_{xx} with the largest moment of inertia I_c and I_{zz} with the smallest moment of inertia I_a. One should, on the other hand, identify I_{xx} with the smallest moment of inertia I_a and I_{zz} with the largest moment of inertia I_c in the case of the nearly symmetric oblate rotator. I_{yy} should be identified, in both instances, with the intermediate moment of inertia I_b.

The quantity $[(1/I_{xx}) - (1/I_{yy})]$ will be small in both the above limiting examples and the energies may be approximated by the perturbation methods of quantum mechanics, using the term $\frac{1}{4}(P_x^2 - P_y^2)$ $[(1/I_{xx}) - (1/I_{yy})]$ as the perturbing term. This calculation leads in second order of approximation to

$$E_r = E_r^0 + E'_r \tag{5.37}$$

where

$$\begin{aligned} E_r^0 = J(J+1)\frac{h^2}{8\pi^2}\left[\frac{1}{2}\left(\frac{1}{I_{xx}} + \frac{1}{I_{yy}}\right)\right] \\ + \frac{K^2h^2}{8\pi^2}\left[\frac{1}{I_{zz}} - \frac{1}{2}\left(\frac{1}{I_{xx}} + \frac{1}{I_{yy}}\right)\right] \end{aligned}$$

and

$$E'_r = \left(\frac{h}{16\pi}\right)^2 \left[\frac{I_{xx}^{-1} - I_{yy}^{-1}}{I_{zz}^{-1} - \frac{1}{2}(I_{xx}^{-1} + I_{yy}^{-1})}\right]$$
$$\frac{1}{2}\left(\frac{1}{I_{xx}} - \frac{1}{I_{yy}}\right)$$
$$\left\{\frac{[(J+K)(J+K-1)(J-K+1)(J-K+2)}{K-1} - \left[\frac{(J-K)(J-K-1)(J+K+1)(J+K+2)}{K+1}\right]\right\}$$

When the values of K are equal to $+1$ or -1, the method fails and E'_r must, already in this approximation, be obtained by the method of the degenerate perturbation theory. In fourth order of approximation the levels $K = \pm 2$ split into two components, and so on. Wang gives the following expression for the splitting between the component levels for a given value of K

$$\Delta\nu = h\left(\frac{\frac{1}{2}[(1/I_{xx}) - (1/I_{yy})]}{\{(1/I_{zz}) - \frac{1}{2}[(1/I_{xx})(1/I_{yy})]\}}\right)^K \left\{\left[\frac{1}{I_{zz}} - \frac{1}{2}\left(\frac{1}{I_{xx}} + \frac{1}{I_{yy}}\right)\right](J+K)!^2\right\} [2^{3K}c\pi^2(J-K)!(K-1)!^2]^{-1} \quad (5.38)$$

No simple method for approximating to the levels is possible in the region intermediate between the two limiting cases. One may say, in a rough way, that in the region of $\tau = +J$ they will resemble the levels of the prolate rotator and in the region of $\tau = -J$ they will resemble the levels of the oblate rotator. In the region of $\tau = 0$ there is no resemblance discernible between the levels of the asymmetric rotator and those of either limiting case. Certainly the simplest method for arriving at the energies of the asymmetric rotator is by referring to King, Hainer, and Cross [9].* They write for the energy of the rotator

$$E_\tau^J(a,b,c) = \frac{a+c}{2} J(J+1) + \frac{a-c}{2} E_\tau^J(\varkappa) \quad (5.39)$$

where a, b, and c are, respectively, $(\hbar^2/2I_a)$, $(\hbar^2/2I_b)$, and $(\hbar^2/2I_c)$ and where $E_\tau^J(\varkappa)$ is a constant which depends upon J and K and upon $\varkappa$ which is a parameter of asymmetry defined as $\varkappa = (2b - a - c)/(a - c)$. $\varkappa$ varies from $\varkappa = 1$ for the oblate symmetric rotator $(b = a)$ to $\varkappa = -1$ for the prolate symmetric rotator $(b = c)$. King, Hainer, and Cross have prepared tables for the values of E_τ^J throughout the range of asymmetry $\varkappa = -1$ to $\varkappa = 0$, $\varkappa$ varying by intervals of 0.1. The tables become applicable to any degree of asymmetry, i.e., from $\varkappa = -1$ to $\varkappa = +1$ with the observation that $E_\tau^J(\varkappa) = -E_\tau^J(-\varkappa)$.

Figure 5.8 shows the levels of an asymmetric rotator for the value of $J = 3$, where I_a has been chosen equal to 1 and I_c equal to 2, plotted against a variable I_b as it ranges from $I_b = 1$ to $I_b = 2$. On the left (a) is shown the oblate limiting case where $I_a = I_b$ and on the right (b) is shown the prolate limiting case $I_b = I_c$. The lines joining (a) with (b) show the levels for the asymmetric rotator intermediate between the two limiting cases (a) and (b).

* Numbers in brackets refer to References at end of chapter.

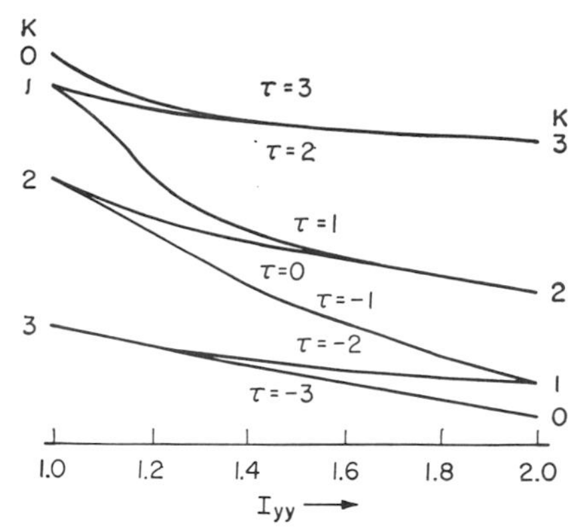

FIG. 5.8. Energy levels $J = 3$ of the asymmetric rotator for various values of asymmetry.

5. The Energy of Interaction, E_i

Experience with diatomic molecules shows that when E_v and E_r are known one may neglect E_i, the energy term arising from the interaction between vibration and rotation, to a good approximation, and proceed to a rather satisfactory interpretation of their spectra. This can be understood in the following manner: When the terms (5.31) and (5.32) in the potential energy V and higher-order terms in the kinetic energy arising from the nonrigidity of the rotating molecule are included in the Hamiltonian H for the molecule, the Schroedinger equation is no longer exactly separable in the coordinates. However, such terms may be regarded as small so that their effect upon the energy may be taken into account by the usual methods of the perturbation theory of quantum mechanics.

In the case of a polyatomic molecule—in particular, an axially symmetric molecule—the interaction energy E_i is anomalously large; if one were to neglect this term, only a rather qualitative interpretation of the band spectra would be obtained. It will, henceforth, be advantageous to write E_i as $E_i = E'_i + E''_i$, where E'_i is the large part of E_i and E''_i is the small part which, in a general way, corresponds to the small E_i in the energy of a diatomic molecule.

The explanation of the anomalously large E_i for many polyatomic molecules was first given by Teller and Tisza [14] and subsequently enlarged upon by Johnston and Dennison [8]. The explanation lies in the fact that treating a molecule even as a quasi-rigid rotator is not always a good approximation.

We have heretofore assumed that the molecule is a rigid rotator. To this approximation we do not differentiate between the angular momentum of the molecule and the angular momentum of the framework of the molecule. Such a differentiation must now be introduced for the symmetric molecule.

When the selection of the vibrational coordinates was made, three conditions were introduced which stated that to the approximation that the rotator was rigid, the angular momentum of the molecule should be zero. These are the second set of conditions stated in (5.12). Actually the complete angular momentum components will involve the instantaneous values of the coordinates x, y, and z

and not merely the equilibrium values, i.e., one should write for these, $\Sigma_i m_i(\alpha_i\, \delta\dot{\beta}_i - \beta_i\, \delta\dot{\alpha}_i)$ and not $\Sigma_i m_i(\alpha_i^0\, \delta\dot{\beta}_i - \beta_i^0\, \delta\dot{\alpha}_i)$. Since we have already set the latter equal to zero, we have left $\Sigma_i m_i(\delta\alpha_i\, \delta\dot{\beta}_i - \delta\beta_i\, \delta\dot{\alpha}_i)$ which is the angular momentum associated with the vibration of the nuclei. Expressed in terms of q_i and the conjugate momenta p_i, we have

$$p_\alpha = \sum_{ij} \zeta_{ij}^{(\alpha)} \left[\left(\frac{\nu_j}{\nu_i}\right)^{1/2} q_i p_j - \left(\frac{\nu_i}{\nu_j}\right)^{1/2} q_j p_i \right] \tag{5.40}$$

where the $\zeta_{ij}^{(\alpha)}$ here is the ζ referred to earlier and is often called the Coriolis coupling factor. Its magnitude depends in an involved manner on the nature of the normal coordinates.

When ν_i and ν_j are substantially different, the relation (5.40) contributes only to E''_i, which we shall neglect for the present. When $\nu_i = \nu_j$, however, the relation may give a large effect which we shall call E'_i.

Consider first the axially symmetric molecule. Certain of the oscillations in this type of molecule (the perpendicular oscillations) induce electric moments which have components along the x axis and the y axis. Each component has the same frequency (that is, $\nu_i = \nu_j$) and the vibration is twofold degenerate.

It has been suggested that it was desirable in such instances to replace q_i and p_i by their equivalents in the polar coordinates r_i and χ_i. When this is done in (5.41) the only component p_α which will occur is p_z which is equal to $\Sigma_i \zeta_i p_{\chi i}$. Quantum mechanically this says

$$p_z = \sum_i \zeta_i p_{\chi i} = -i\hbar \sum_i \zeta_i \frac{\partial}{\partial \chi_i} \tag{5.41}$$

The quantum-mechanical value for p_z, stated earlier, is $p_z = \Sigma_i l_i \zeta_i \hbar$.

The total angular momentum directed along the z axis, P_z, will no longer be just that of the framework of the rotator, but will be the vector sum of the angular momentum of the framework, which we shall designate by L and p_z, that is, $P_z = L + p_z$. The Hamiltonian for the rotator, i.e., the framework of the axially symmetric molecule, will, therefore, be

$$H_r = \frac{1}{2}\left(\frac{P_x^2 + P_y^2}{I_{xx}} + \frac{L^2}{I_{zz}}\right) \tag{5.42}$$

We have from an earlier section

$$P^2 = P_x^2 + P_y^2 + P_z^2 = J(J+1)\hbar^2$$

$P_z^2 = K^2\hbar^2$. This gives

$$P_x^2 + P_y^2 = P^2 - P_z^2 = [J(J+1) - K^2]\hbar^2$$

We have for L, $P_z - p_z = (K - \Sigma_i l_i \zeta_i)\hbar$. The energy of the rotator, including E'_i, will therefore be

$$E_r = [J(J+1) - K^2]\frac{h^2}{8\pi^2 I_{xx}} - \frac{L^2\hbar^2}{8\pi^2 I_{zz}} \tag{5.43}$$

A selection rule for the quantum number l_i is now also required. It has been shown to be $\Delta l_i = \Delta K = \pm 1$, that is, if K changes by $+1$, l_i must change by $+1$ also; if K changes by -1, l_i must change by -1. A transition where, for example, K changes by $+1$ and l_i by -1 is forbidden. The permitted transitions from the rotation levels $J = 3$ in the normal vibration state, where $v_i = l_i = 0$, to the rotation levels $J = 3$ in an excited perpendicular vibration state, where $v_i = 1$, $l_i = \pm 1$, are shown in Fig. 5.9.

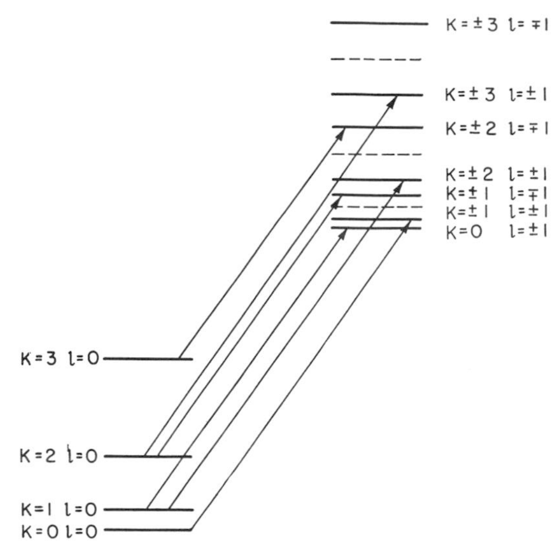

FIG. 5.9. Transitions from levels $J = 3$, $l_i = 0$ to $J = 3$, $l_i = \pm 1$ in a symmetric molecule.

It may be shown that $L = 0$, that is,

$$p_z = P_z = \Sigma_i l_i \zeta_i \hbar$$

in the case of the linear molecule. Because the ζ_i are all unity in the linear molecule E_r becomes

$$E_r = \left[J(J+1) - \left(\sum l_i\right)^2\right]\frac{h^2}{8\pi^2 I} \tag{5.44}$$

This emphasizes the fact that the linear molecule is simply a special case of the axially symmetric molecule.

In molecules of higher symmetry, where threefold degenerate vibrations may occur with components along the x, y, and z axes of the molecule p_x and p_y as well as p_z may be significant. Here the Hamiltonian for the rotator becomes

$$H_r = \frac{1}{2}\left[\frac{(P-p)^2}{I}\right] = \frac{1}{2}\frac{(P^2 - 2pP + p^2)}{I} \tag{5.45}$$

since $I_{xx} = I_{yy} = I_{zz}$ and

$$(P_x - p_x)^2 + (P_y - p_y)^2 + (P_z - p_z)^2 = (P - p)^2$$

with $p^2 = p_x^2 + p_y^2 + p_z^2$.

The term in p^2 in (5.45) may be included with the vibration energy since in a given vibration state it is a constant. P^2 is, of course, the square of the total angular momentum of the molecule. The term $-2pP$ is the term of interaction between vibration and rotation from which E'_i must be calculated. Writing

$$-pP = -\zeta_i l_i P \tag{5.46}$$

where l_i may take the values v_i, $v_i - 2$, . . . , 1 or 0. Now P is the vector sum of l and L, where L is the angular momentum of the framework of the molecule, that is, $P = l_i + L$. Then

$$L^2 = (P_i - l_i)^2 = P^2 + l_i^2 - 2l_iP$$

Equation (5.46) then becomes

$$-\zeta_i l_i P = -\tfrac{1}{2}\zeta_i(P^2 + l_i^2 - L^2) \qquad (5.47)$$

As usual $P^2 = J(J+1)\hbar^2$; l_i^2 is the square of the angular momentum associated with the vibration ν_i and is equal to $l_i(l_i+1)\hbar^2$; and L^2, the angular momentum of the framework, takes the value $L(L+1)\hbar^2$, L being equal to $J + l_i$, $J + l_i - 1$, . . . , $J - l_i$. Thus

$$-\zeta_i l_i P = -\tfrac{1}{2}\zeta_i[J(J+1) + l_i(l_i+1) - L(L+1)] \qquad (5.48)$$

Applied to the first excited state of a three dimensionally isotropic oscillator where $v_i = 1$ so that $|l_i| = 1$ and $L = J + 1$, J, $J - 1$, we obtain the three values for $\zeta_i l_i P$, $J\zeta_i\hbar^2$, $-\zeta_i\hbar^2$, and $-(J+1)\zeta_i\hbar^2$.

A selection rule also exists here. For the example just given it states that $\Delta J = +1$, 0, and -1 as the molecule makes a transition from the normal vibration state to the states characterized by $-\zeta_i l_i P$ equal to $J\zeta_i\hbar^2$, $-\zeta_i\hbar^2$, and $-(J+1)\zeta_i\hbar^2$, respectively. For vibration states of higher excitation the selection rules are more complicated. Moreover, the formula (5.46) for $-\zeta_i l_i P$ does not apply directly when more than one vibration state v_i is excited at the same time.

Finally, the smaller correction E''_i must be taken into account. It contributes in three ways. The first contribution is a set of terms that are vibrational in character and depend only upon the vibrational quantum numbers. These terms originate with the anharmonic nature of the vibration of the molecule. When they are added to the harmonic part of the vibration (5.29) and (5.30) one obtains for the vibration term values of a molecule

$$\frac{E(v_i)}{hc} = \sum_i \left(v_i + \frac{g_i}{2}\right)\nu_i + \sum_{ij} x_{ij}\left(v_i + \frac{g_i}{2}\right)\left(v_j + \frac{g_j}{2}\right) + \sum_{ij} g_{ij}l_il_j \qquad (5.49)$$

where the x_{ij} and the g_{ij} are involved functions of the constants k_{ijk} and k_{ijkl} and where g_i is the order of the degeneracy of the frequency. The last term in (5.49) is absent for molecules where no degenerate frequencies exist, since in that case l_i has no meaning. The relation (5.49) is not applicable to molecules in which threefold degenerate oscillations occur and where the anharmonic contributions to the vibration energy are considerably more complicated.

The coupling of the normal vibrations of a molecule by inclusion of anharmonic terms in the potential energy relaxes the selection rules so that one may write $\sum_i \Delta v_i = n$, where n is an integer; i.e., a single quantum number v_i may change by more than one unit or several quantum numbers v_i may change, at the same time, by more than one unit. This allows not only the fundamental vibrations to be present in the spectrum but also overtone and combination frequencies. It is a general rule, however, that the intensities with which bands occur decrease rapidly with n.

The anharmonic correction terms introduce many more constants into the vibration problem. They can be evaluated only if the spectrum is sufficiently rich in observed overtone and combination bands so that there can be set up as many independent relations involving them as there are constants to be determined. Frequently this is not possible. In many cases this has necessitated using as the harmonic frequencies ν_i simply the frequencies of the fundamental vibration bands as observed in the spectrum. This unfortunately reduces the effectiveness of using such equations as (5.27), which were derived for the triatomic molecules, to arrive at values for the constants of the molecule. This is especially true with respect to α, which depends critically upon the values of ν_i. The force constants k and k_α may probably be expected to be accurate to about 10 per cent.

The second contribution of E''_i is to the rotational constants. E''_i alters the moments of inertia $I_{\alpha\alpha}$ in (5.43) and replaces them by quantities $(I_{\alpha\alpha})_v$, which are moments of inertia effective for a given vibration state. These depend linearly upon the vibration numbers. The corrections to $I_{\alpha\alpha}$ come from three sources: the vibration in molecular dimensions due to the harmonic component of the vibration, the variation of molecular dimensions due to the anharmonic component of the vibration, and the portions of p_α neglected in the computations of E'_i. Replacing $h/8\pi^2 I_{xx}$ by B and $h/8\pi^2 I_{zz}$ by C and observing the fact that the corrections are small, we may write

$$B_v = B_e - \sum_i \alpha_i\left(v_i + \frac{g_i}{2}\right)$$

and

$$C_v = C_e - \sum_i \gamma_i\left(v_i + \frac{g_i}{2}\right)$$

B_e and C_e being the equilibrium values of B and C, where α_i and ν_i are constants. α_i and ν_i are sometimes written

$$\alpha_i = (\alpha_i)_{\text{harmonic}} + (\alpha_i)_{\text{anharmonic}} + (\alpha_i)_{\text{Coriolis}}$$

In the third instance, E''_i introduces a set of terms which may be looked upon as corrections to the B and C values due to the distortion of the molecule arising from centrifugal stretching. These may be written as follows

$$-[J(J+1) - K^2]^2 D_J - [J(J+1) - K^2]L^2 D_{JL} - L^4 D_L \qquad (5.50)$$

where the D's are constants. In linear molecules where $K = \Sigma_i l_i$, that is, $L = 0$, the centrifugal distortion terms reduce to $-[J(J+1) - (\Sigma_i l_i)^2]^2 D$. In molecules where no degenerate vibrations occur, the term $\Sigma_i \zeta_i l_i$ will vanish. This is tantamount to setting L everywhere equal to K in relations (5.43) and (5.44) for such molecules.

The term values for an axially symmetric polyatomic molecule will then be

$$\frac{E}{hc} = \frac{E_{vi}}{hc} + \frac{E_r}{hc} \qquad (5.51)$$

where E_{v_i}/hc has already been stated in (5.49) and where now

$$\frac{E_r}{hc} = [J(J+1) - K^2]B_v + L^2C_v - [J(J+1) - K^2]^2D_J - [J(J+1) - K^2]L^2D_{JL} - L^4D_L \quad (5.52)$$

6. The Selection Rules for the Rotator

The method for determining the selection rules in quantum mechanics consists in establishing the nonvanishing matrix components of the electric moment. The latter are obtained by evaluating the integrals $\int \bar{\psi}_{J,K} E \psi_{J',K'}\, d\tau$, where the $\psi_{J,K}$ and $\psi_{J',K'}$ are the wave functions of the rotational states characterized by the quantum numbers J, K, and J', K', respectively, where E is the classical expression for the electric moment, and where $d\tau$ is an element of volume in configuration space.

The electric moment E for a molecule will, in general, consist of a constant term, known as the permanent electric moment, and a variable part, the components of which will vary, respectively, with the frequencies of the oscillations which induce them. The permanent electric moment in the triatomic XY_2 model, for example, will lie along the y axis; in the pyramidal XY_3 model the permanent electric moment lies along the altitude of the pyramid. The permanent electric moment in all axially symmetric molecules lies, in fact, along the axis of symmetry. The linear XY_2 model and the tetrahedral XY_4 model, on the contrary have zero permanent electric moment. The induced electric moments for several models were described in discussing the nature of the normal vibrations.

One can infer from the classical method what the selection rules for the symmetric rotator are when the electric moment lies along the axis of symmetry. Let $E = E_0 + E_i \sin 2\pi\nu_i t$ in this case. Let the z axis in Fig. 5.7 be the symmetry axis of the rotator. The molecule will be spinning about this axis with a frequency $\nu_\phi = p_\phi/2\pi I_{zz}$. In the classical theory where

$$\int_0^{2\pi} p_\phi\, d_\phi = Kh$$

$\nu_\phi = Kh/4\pi^2 I_{zz}$ with the angular momentum $Kh/2\pi$ directed along the axis z. Choose the axis of total angular momentum to coincide with the axis z in which case $p_\theta = 0$. According to the classical theory we may then set ν_ψ equal to $Jh/4\pi^2 I_{xx}$ and

$$\cos\theta = \frac{K}{[J(J+1)]^{1/2}}$$

Consider now what are the components of E along the space fixed axes. These will evidently be $E_z = E\cos\theta$, $E_x = E\sin\theta\sin\psi$, and

$$E_y = -E\sin\theta\cos\psi$$

Replacing E by $E_0 + E_i \sin 2\pi\nu_i t$ and ψ by $2\pi\nu_\psi t$, where $\nu_\psi = Jh/4\pi^2 I_{xx}$, we have

$$E_x = (E_0\sin\theta)\sin 2\pi\nu_\psi t + \tfrac{1}{2}(E_i\sin\theta)[\cos 2\pi(\nu_i + \nu_\psi)t - \cos 2\pi(\nu_i - \nu_\psi)t]$$

$$E_y = (-E_0\sin\theta)\cos 2\pi\nu_\psi t + \tfrac{1}{2}(-E_i\sin\theta)[\sin 2\pi(\nu_i + \nu_\psi)t + \sin 2\pi(\nu_i - \nu_\psi)t]$$

and $E_z = (E_0\cos\theta) + (E_i\cos\theta)\sin 2\pi\nu_i t$. The relations for E_x, E_y, and E_z contain the frequencies ν_i, ν_ψ, and $\nu_i \pm \nu_\psi$ and these may be regarded as the frequencies permitted according to the classical theory.

The correspondence principle states that for large quantum numbers (that is, large values of J and K) the frequencies of the quantum theory go asymptotically over into the classical frequencies, that is, $(E' - E'')/hc = \nu_{\text{class}}$. Using the somewhat simplified relation for the term values,

$$\frac{E}{hc} = \sum_i \left(v_i + \frac{g_i}{2}\right)\nu_i + J(J+1)B + K^2(C - B)$$

we have

$$\sum_i \Delta v_i \nu_i + [J'(J'+1) - J''(J''+1)]B + (K'^2 - K''^2)(C - B) = \nu_i \quad (5.53)$$

and

$$\sum_i \Delta v_i \nu_i + [J'(J'+1) - J''(J''+1)]B + (K'^2 - K''^2)(C - B) = \nu_i \pm 2JB \quad (5.54)$$

Equating like terms on both sides of the equation, $\nu_i = \Sigma_i\, \Delta v_i \nu_i$, $(K'^2 - K''^2)(C - B) = 0$,

$$[J'(J'+1) - J''(J''+1)]B = 0$$

for E_z and $[J'(J'+1) - J''(J''+1)]B = \pm 2JB$ for E_x and E_y. The first term concerns the vibrational frequency only. The second term states that $K' - K'' = \Delta K = 0$ for oscillations of the electric moment along the axis of symmetry. We may neglect J with respect to J^2 for large values of J with respect to the last two relations. These reduce, therefore, to $[(J' + J'')(J' - J'')]B = 0$ and to $[(J' + J'')(J' - J'')]B = \pm 2JB$ in the two cases, respectively. We may, moreover, set $J' \approx J'' \approx J$ for sufficiently large values of J; therefore the selection rules for J are $J' - J'' = \Delta J = 0$ and

$$J' - J'' = \Delta J = \pm 1$$

When the electric moment lies along the z axis of the rotator, the projections along the space fixed axes are independent of the angle ϕ. This is not so when the electric moment lies in the xy plane. One may proceed to obtain relations for these projections in a manner similar to that described above for electric moments along the z axis and in this instance the classical frequencies which may occur are these, $\nu_i + \nu_\psi \pm \nu_\phi$ and $\nu_i - \nu_\psi \pm \nu_\phi$. One may, as before, establish the selection rules for J and K with the aid of the correspondence principle remembering that $\nu_\phi = Kh/4\pi^2 I_{zz}$. They are, this time, found to be $\Delta K = \pm 1$, $\Delta J = 0$, $\Delta J = \pm 1$.

There is no simple way of stating the selection rules for the asymmetric rotator except for J, which is still $\Delta J = 0, \pm 1$. When the asymmetric rotator simulates one of the limiting symmetric rotators, the important selection rules corresponding to ΔK in a symmetric rotator are those of the limiting symmetric rotator. Thus, for example, if the molecule is one which is close to the limiting prolate symmetric rotator (that is, $I_a \approx I_b$), then the selection rules which are important are those for the prolate symmetric rotator; on the other hand, if the asymmetric rotator is nearly oblate (that is, $I_c \approx I_b$), then the selection rules are essentially those of the oblate symmetric rotator. The rotator has characteristics of both limiting forms in the region of asymmetry intermediate between these two limiting cases. Correspondingly the selection rules allow transitions of both limiting types of symmetric rotator to take place. Certain other transitions which are characteristic of neither limiting symmetric rotator may, indeed, also take place, but since these are forbidden in both of the limiting extremities they may be regarded as of secondary significance. Here, as in the evaluation of the energies of the asymmetric rotator, it is advantageous to resort to the work of King, Hainer, and Cross where tables of the line strengths are given for rotators of varying degrees of asymmetry.

Closely associated with the problem of the selection rules is the question of the intensity distribution of the rotation lines in a band spectrum. The intensities of lines are proportional to the squares of the amplitudes of the electric moment projected along the space fixed axes (see Chap. 3). These quantities are proportional to $\cos^2 \theta$ and $\frac{1}{2} \sin^2 \theta$ in the example considered earlier where the electric moment lies along the axis of symmetry of an axially symmetric molecule. When $\cos^2 \theta$ and $\frac{1}{2} \sin^2 \theta$ are translated into quantum mechanics they become, in terms of J and K, $[K^2/J(J+1)]$ and $[(J^2 - K^2)/J(2J+1)]$, respectively.

7. The Interpretation of Band Spectra

We shall confine attention primarily to the interpretation of the band spectra of symmetric molecules. Denote values of B_v and C_v in the excited vibration state and in the normal vibration state by B' and C' and by B'' and C'', respectively. Consider first the type of band which arises when the electric moment lies along the symmetry axis of an axially symmetric molecule. The selection rules for the rotator in this case are $\Delta J = +1, 0, -1$ and $\Delta K = 0$.

In the pure rotation spectrum, $\Delta v_i = 0$. $B' = B''$ and $C' = C''$ in this instance and only $\Delta J = +1$ will give lines in absorption. Using the term values stated in Eq. (5.52) the positions of lines in the pure rotation spectrum of an axially symmetric molecule are

$$\nu_{J,K} = 2JB - 4J(J^2 - K^2)D_J - 2JK^2D_{JL} \quad (5.55)$$

where J takes the values $J = 1, 2, 3, \ldots$ and K is an integer less than or equal to J. There will then be a different set of lines for each value of K. The several sets of lines may be expected to be almost coincident, however, since the constants D are very small. The unresolved lines will, therefore, be separated by distances very nearly equal to $2B$.

The resolving power of far infrared spectrographs is scarcely great enough to separate a rotation line into its K components. Only in the case of the spectrum of NH_3 has such resolution been achieved and in that case only for lines of high J value. Such a partially resolved line is shown in Fig. 5.10. Extensive measurements on such lines have been made in the microwave region (see Part 6, Chap. 7). The resolving power is here sufficient to identify the several components of such a line. It is found that they do, indeed, fit into a formula of the form (5.55). Frequently, however, the frequency range available to the microwave spectroscopist does not allow him to investigate more than one set of J transitions.

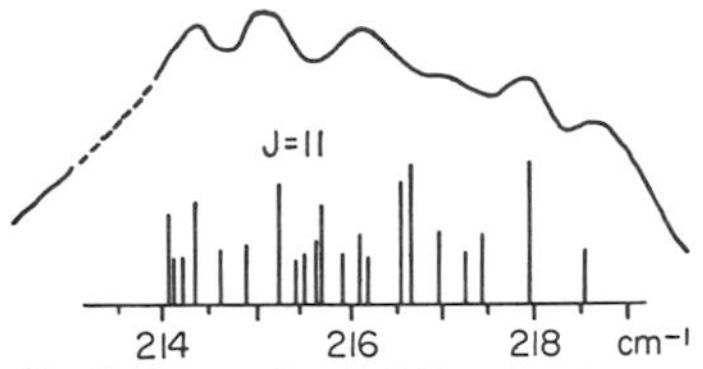

FIG. 5.10. Partially resolved J line in the pure rotation spectrum of NH_3.

When the electric moment is one induced by vibration, then $\Delta v_i \neq 0$. All the transitions $\Delta J = +1$, 0, and -1 may now give lines in absorption. The B and C values will no longer be entirely alike in the upper and lower states. The positions of lines in a *parallel* band are

$$\nu = \nu_i + J(J+1)(B' - B'') + K^2(C' - C'' + B'' - B') \quad \text{for } \Delta J = 0$$

and

$$\nu = \nu_i \pm J(B' + B'') + J^2(B' - B'') + K^2(C' - C'' + B'' - B') \mp 4J(J^2 - K^2)D_J \mp 2JK^2D_{JL} \quad \text{for } \Delta J = \pm 1 \quad (5.56)$$

where J takes the values $J = 1, 2, \ldots$ and K is an integer less than or equal to J. Such a set of lines form a parallel vibration-rotation band. The position of such a set of lines in the spectrum is predominantly established by the term ν_i which is the vibration frequency. Such bands lie in the near infrared.

The lines for which $\Delta J = 0$ are said to make up the Q branch of the band and the lines for which $\Delta J = -1$ and $+1$ are said to make up the P and R branches, respectively. As before, each value of K gives a separate set of lines forming what is called a sub-band. Figure 5.11 shows such a group of sub-bands schematically. $K = 0$ in the uppermost band in Fig. 5.11. Here $\cos \theta$, which equals $K/[J(J+1)]^{1/2}$, will evidently be zero. The central line will, as we have already noted, therefore be absent here. The next sub-band is for $K = 1$; $\cos \theta$ is now no longer zero and here a rather weak central line may occur. The first line on either side of the central line is absent because for such lines one of the participating levels would have to be one where $J = 0$. Since $K = 1$, however, the value of J cannot be zero. The next sub-band in order is for $K = 2$, etc. At the bottom, the composite band is shown as it might be observed with a spectrograph of ideal resolving power. The actual resolving power of an infrared spectrograph is, however, seldom sufficient to resolve a parallel band into its sub-band structure.

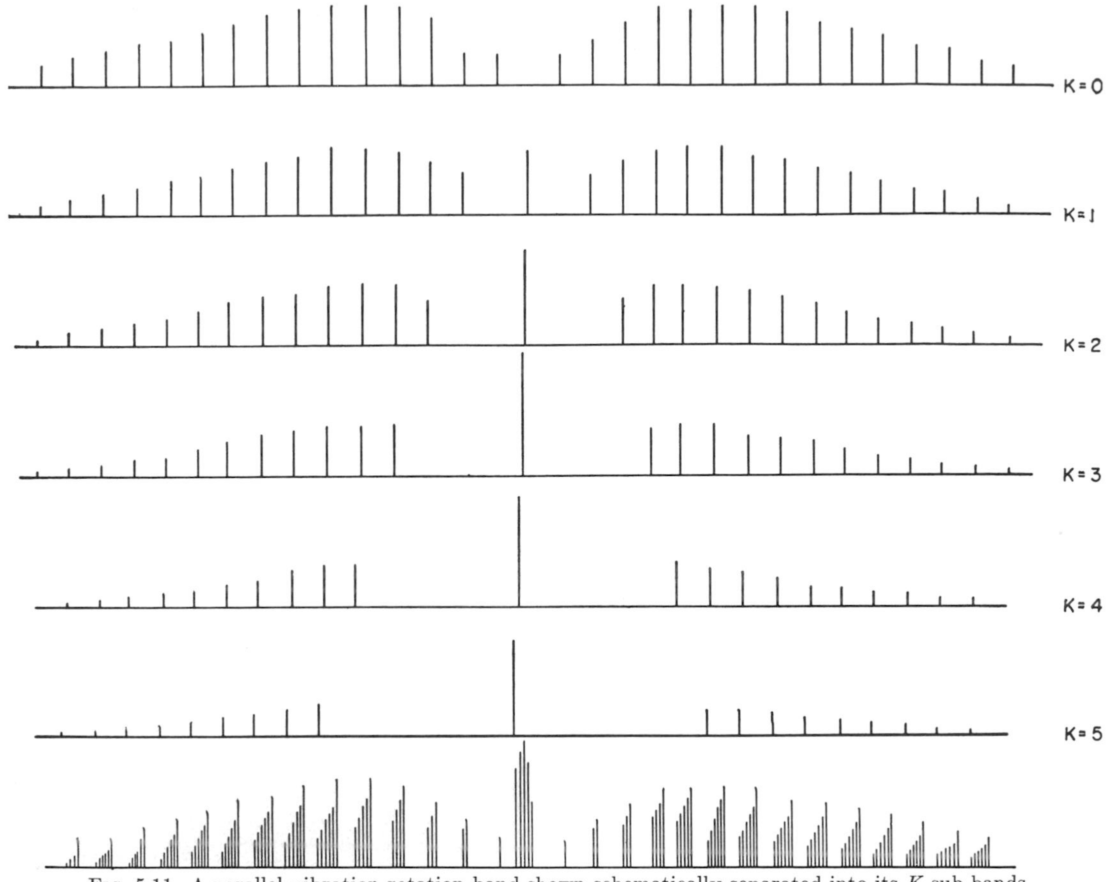

FIG. 5.11. A parallel vibration-rotation band shown schematically separated into its K sub-bands.

The linear polyatomic molecule affords an opportunity, however, to verify that such substructure really does exist. Suppose a parallel band to arise with transitions beginning in the normal vibration state. The value of $\Sigma_i l_i$, which is equal to K, is zero in both vibration states. The band that will occur will be due to transitions $\Delta J = +1$, 0, and -1 where K goes from $K = 0$ to $K = 0$. A parallel band will, in this case, consist of a single sub-band only with $K = 0$. Such a parallel band should be exactly like the sub-band shown at the top of Fig. 5.11. Figure 5.12 shows a parallel band of the kind just described from the infrared spectrum of CO_2 and it will be seen to confirm the structure predicted in the diagram.

It is practicable by thermal excitation to bring molecules into higher vibration states. Suppose by heating a gas the first excited state of one of the perpendicular vibrations of the molecule becomes populated. In this case $v_i = |l_i| = 1$. Assume further that a transition is observed from this state to another vibration state where $|l_i| = 1$. The value of $\Sigma_i l_i$ is equal to 1 and this is the only allowed K value in both states. The band that will occur will be due to transitions $\Delta J = +1$, 0, and -1 where now K goes from $K = 1$ to $K = 1$. A parallel band will in this instance again consist of a single sub-band only, this time K being equal to 1. Figure 5.13 shows such a band observed in the spectrum of C_2H_2: it does have the characteristics predicted in the diagram. A rather weak central line is present and the first line on either side of the center is absent.

The band ν_2 in the spectrum of the pyramidal molecule PH_3 is strongly perturbed by a neighboring band. The K substructure becomes anomalously large for that reason and may be partially resolved. The lower frequency branch (the P branch) of this band is shown in Fig. 5.14 and in it the K substructure is plainly visible.

Accurate values of the rotational constants B', C', B'', and C'' could be obtained for the initial and final states if it were possible to carry out a detailed analysis of the sub-bands which go to make up a parallel vibration-rotation band. The K structure can, however, seldom be resolved in practice. This becomes tantamount to considering that the terms in Eqs. (5.56) involving K can be neglected. The unresolved lines in the P and R branches will, therefore, on the average be grouped at distances close to $2B$ cm^{-1} from each other. Applying the combination relations, so successful in the spectroscopy of diatomic molecules, one may arrive here also at values for B' and B'' which are accurate to the approximation that the K structure may be neglected.

The structure of a *perpendicular* band arises from oscillations of the electric moment in the xy plane.

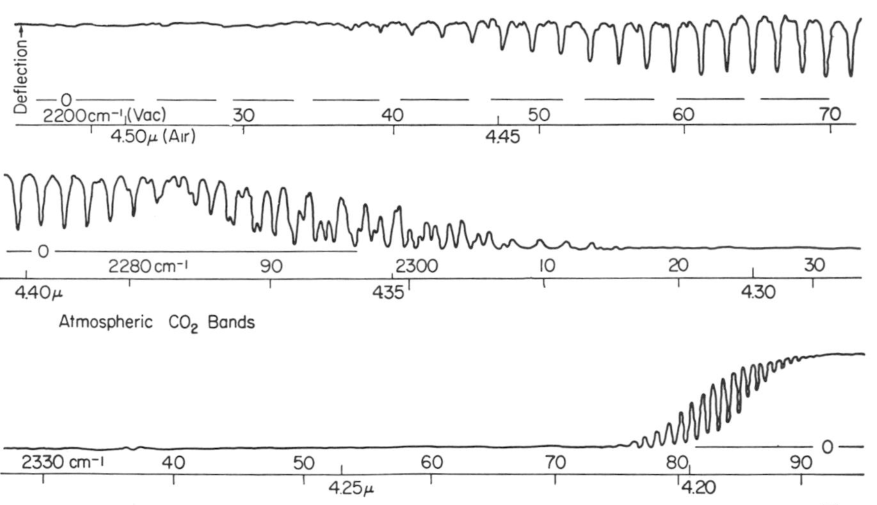

FIG. 5.12. A parallel vibration-rotation band in the infra-red spectrum of CO_2 where $K = \sum_i l_i = 0$.

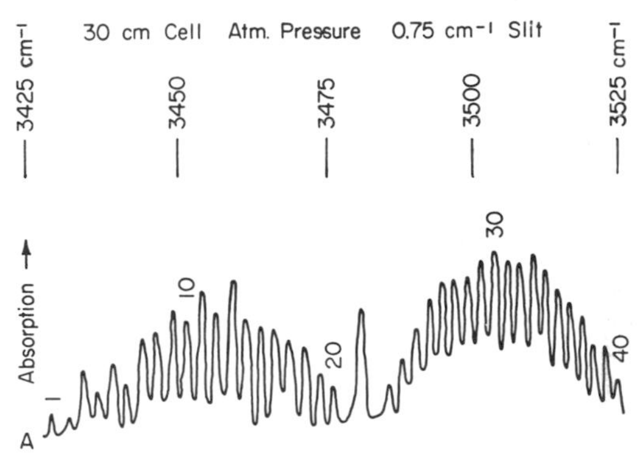

FIG. 5.13. A parallel vibration-rotation band in the infra-red spectrum of C_2H_2 where $K = \sum_i l_i = 1$.

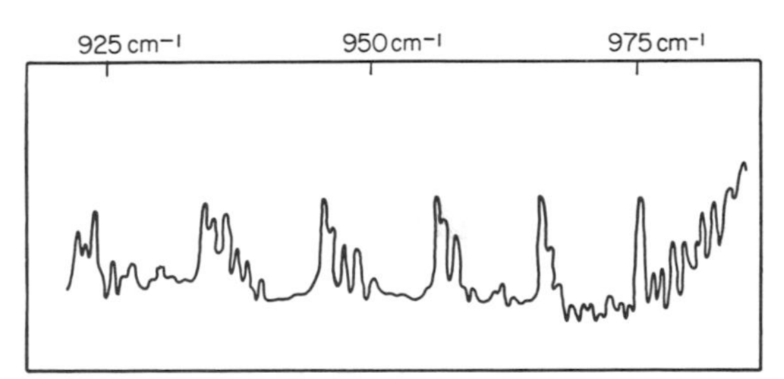

FIG. 5.14. The P branch of the parallel vibration-rotation band ν_2 in the spectrum of PH_3 showing the partially resolved K structure.

Here the selection rules are $\Delta J = +1$, 0, and -1; $\Delta K = \pm 1$ and further $\Delta l_i = \Delta K$. Using these selection rules and the term values (5.49) and (5.52), we arrive, first for $\Delta J = 0$

$$\nu = [\nu_i + (C' - B')] + J(J + 1)(B' - B'') + 2K[(1 + \zeta_i)C' - B'] + K^2(C' - B' + B'' - C'') \quad (5.57)$$

and where $\Delta J = \pm 1$,

$$\nu = [\nu_i + (C' - B')] \pm J(B' + B'') + J^2(B' - B'') + 2K[(1 - \zeta_i)C' - B'] + K^2(C' - B' + B'' - C'') \quad (5.58)$$

There will, as in the parallel band, also here be a separate set of lines for each value of K. The manner in which we have chosen the K values in the initial and final states requires K to take the values 0, ± 1, ± 2, etc., in the relations (5.57) and (5.58). Moreover, because J can never take values less than $K + 1$ in our notation, when K is zero or positive, J will take the values $K + 1$, $K + 2$, etc., in the R branch of the sub-band and the values $K + 2$, $K + 3$, etc., in the P branch of the sub-band. For the same reason, when K takes negative values, J takes the values $|K| + 1$, $|K| + 2$, etc., in the P branch of the sub-band and the values $|K| + 2$, $|K| + 3$, etc., in the R branch of the sub-band.

The sub-bands which constitute a perpendicular band are shown schematically in Fig. 5.15. The upper illustration is for $K = 0$ where all the lines corresponding to $J = 1, 2, 3$, etc., are present in the R branch, but where the first line in the P branch is for $J = 2$. The next diagram is for $K = -1$ and it is essentially a reflection of the diagram where $K = 0$. The next illustration is for $K = +1$, where the first line in the R branch is absent and the first two lines in the P branch are absent. Again, the next diagram in which $K = -2$ is essentially a reflection of the sub-band $K = +1$. The last illustration in Fig. 5.15 shows a composite picture of how the several sub-bands would appear if they were observed with a spectrograph of ideal resolving power. The diagrams shown in Fig. 5.15 are for a prolate molecule where $[(1 - \zeta_i)C' - B']$ is positive and large compared to B'.

A perpendicular band may vary widely in appearance primarily because the quantity $[(1 - \zeta_i)C' - B']$ may vary so extensively from one molecule to another. The quantity $[(1 - \zeta_i)C' - B']$ may, indeed, vary

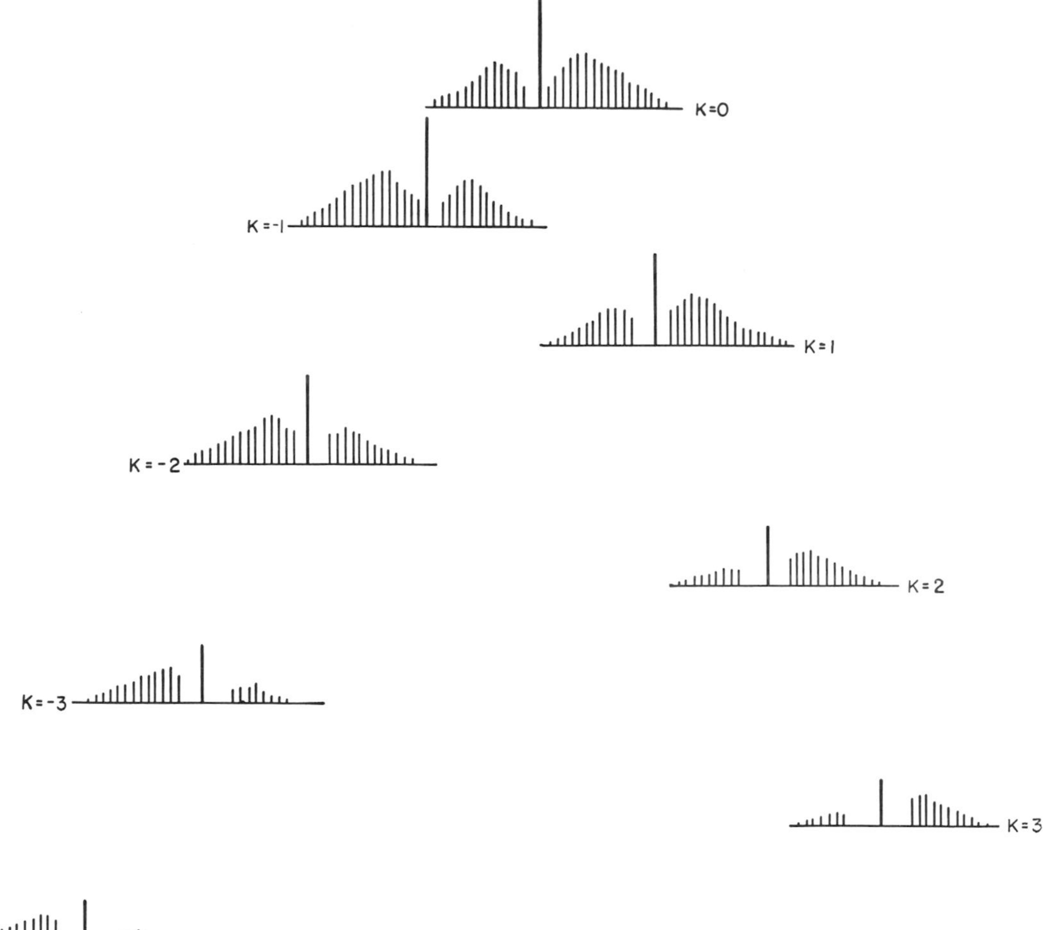

FIG. 5.15. A perpendicular vibration-rotation band shown schematically separated into its K sub-bands.

appreciably from one perpendicular band to another in the spectrum of the same molecule since ζ_i depends upon the nature of the vibration. Examples are known where the term $[(1 - \zeta_i)C' - B']$ is positive in one band, but negative in another band.

As in the parallel band, it is seldom, if ever, possible completely to resolve a perpendicular band into its substructure. The linear molecule again affords an opportunity to verify that a single sub-band really has the structure predicted by the theory. Consider that a linear molecule makes a transition from the normal vibration state where $\zeta_i l_i = 0$ to the first excited state of a perpendicular vibration where $\zeta_i l_i = 1$. Since $K = \zeta_i l_i$ in a linear molecule the resulting perpendicular band is a single sub-band where $K = 0$ goes to $K = 1$. Such a band is shown in Fig. 5.16 which is the perpendicular band ν_5 in the spectrum C_2H_2. It exhibits the characteristics predicted in the illustration of Fig. 5.15 where $K = 0$.

An analysis of a perpendicular band, in the event that it could be completely resolved, would lead to information concerning B', B'' $[(1 - \zeta_i)C' - B']$, and $(C' - B' + B'' - C'')$ for the states involved. The best one can hope for, in general, is probably a value for $[(1 - \zeta_i)C' - B']$ which is essentially one-half the frequency interval between two of the Q branches. This quantity is of only limited value since ζ_i is a quantity which depends upon the nature of the normal vibration and cannot conveniently be calculated. However, the sums of the ζ_i over all the perpendicular fundamental vibrations is a quantity which does not depend upon the potential field. This sum, as shown by Lord and Merrifield [11], can be computed quite in general for any axially symmetric molecule. We shall here state the values of this sum for four of the most important molecular models, namely, the XY_3 model, the XYZ_3 model, the $WXYZ_3$ model, and the $VWXYZ_3$ model. For

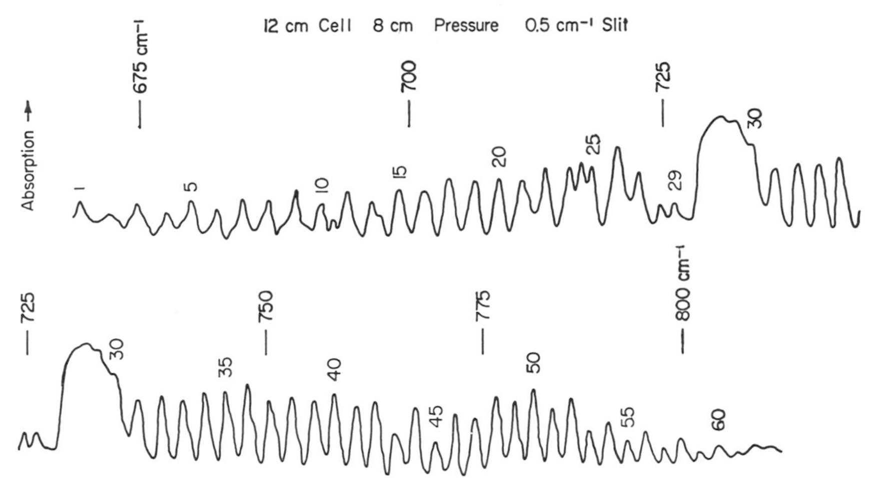

FIG. 5.16. The vibration-rotation band ν_5 in the infra-red spectrum of C_2H_2 illustrating the sub-band of a perpendicular band where $K = 0 \rightarrow K = 1$.

these molecular models $\Sigma_i\zeta_i$ is, respectively, equal to $-1 + (I_{zz}/2I_{xx})$, $(I_{zz}/2I_{xx})$, $1 + (I_{zz}/2I_{xx})$, and $2 + (I_{zz}/2I_{xx})$. When the intervals between the Q branches in the fundamental perpendicular bands are all known one may, with the aid of the values $\Sigma_i\zeta_i$, determine C' and B'.

It is possible, in principle, to evaluate not only B', B'', C', and C'', but also the equilibrium values B_e and C_e if enough bands, fundamentals and overtones, are known. This has been possible only in isolated cases so far. One may use instead of B_e and C_e the values B'' and C'', if one is willing to compromise, and from these derive estimates of the molecular dimensions. The values of the moments of inertia of a molecule alone will, however, seldom be sufficient to fix completely the size and the shape of the molecule.

8. Rotation-Vibration Resonance Interaction in Polyatomic Molecules

The energy relations (5.52) are adequate to account for many of the details observed in the infrared spectra of polyatomic molecules. Almost from the very beginning of infrared spectroscopy, however, certain observed details are not satisfactorily explained by these relations. This comes about in the following way. The constants $x_{ss'}$ and $g_{ss'}$, which occur in the vibration energy of the molecule computed to second order, contain terms having as denominators $2\omega_s - \omega_{s'}$ or $\omega_s + \omega_{s'} - \omega_{s''}$. Similarly, the coefficients α_i and γ_i multiplying the correction terms to the rotational constants B_e and C_e contain denominator $\omega_s - \omega_{s'}$. Often two frequencies, say $2\omega_s$ and $\omega_{s'}$ or ω_s and $\omega_{s'}$, are very nearly alike so that $2\omega_s \approx \omega_{s'}$ or $\omega_s \approx \omega_{s'}$. Then the perturbing terms become indefinitely large, the usual methods of perturbation theory fail, and the techniques of the degenerate perturbation theory must be used. In such cases resonance is said to exist. The above instances are of first-order importance, but second-order and even third-order resonances have been observed. Other effects to be considered in this section are those originating with l-type doubling.

l-type Doubling and l-type Resonance. l-type doubling (and resonance) exists only in axially symmetric molecules and in linear polyatomic molecules (which may be regarded as special instances of axially symmetric molecules). It can, moreover, to this approximation exist only for excited perpendicular vibration states where the quantum numbers $\sum_t l_t$ equal 1.

Let us first consider the case of l-type doubling in linear molecules. It was stated in Sec. 5 that for linear molecules $p_z = P_z = \sum_t l_t\hbar$. For simplicity, we assume that $v_t = 1$ and all other $v_{t'}$ are equal to zero. According to an earlier section, l_t may take the values $l_t = +1$ and $l_t = -1$ only. Moreover, the vibration-rotation energy is independent of the algebraic sign of l_t and is therefore degenerate in l_t. There exist, however, terms of interaction that couple states characterized by l_t with states characterized by $l_t \pm 2$. These are important when $l_t = 1$ or $l_t = -1$. The interaction terms that couple the levels where $l = \pm 1$ with levels where $l = \mp 1$ remove the degeneracy of these states. The interaction energies are found to be

$$\pm g_0(v_t + 1)J(J + 1) \qquad (5.59)$$

where g_0 is a constant equal to $-B_e(B_e/2\omega_t)\left\{1 + 4\sum_{s'}[\zeta_{ts'}^{(x)}]^2[\omega_t^2/(\omega_{s'}^2-\omega_t^2)]\right\}$ in which $\zeta_{s\neq s'}^{(x)}$ is the Coriolis coupling factor coupling ω_t with another frequency $\omega_{s'}$. Since the correction to the energy (5.59) is proportional to $J(J + 1)$, it may be added to the rotational energy (5.52) where, because of the linearity of the molecule, $L = 0$. The effect is that the two states have slightly different rotational constants. A selection rule may be derived which states that when $\Delta J = 0$ the transition is from the ground state to the upper component of the l-type doublet, but when $\Delta J = \pm 1$ the transition is from the ground state to the lower of the two l-type doublets. This may give rise to convergence in opposite senses of the lines in the Q branch and the lines in the P and R branches. This may be seen in Fig. 5.16.

In the case of the axially symmetric molecule, p_z and P_z are not necessarily equal. The level when $K = \pm 1$ with $l = \pm 1$ (which is the level where $p_z = P_z$) generally is split, the amount being as before, $\pm g_0(v_t + 2)J(J+1)$, but where q_0 is a considerably more complicated constant than that encountered in the linear example. In virtually all instances, as, for example, $K = \pm 1$, $l = \mp 1$, there is no splitting. The level that splits corresponds to the classical situation where the framework is at rest.

Consider also the case where $v_t = 2$ and where l_t takes the values $l_t = \pm 2$ and $l_t = 0$. Corresponding to these quantum numbers there are three levels, two of which ($l_t = \pm 2$) are degenerate. This time the l-type coupling terms couple the component level T_2, where $l_t = 2$, and the state T_0, where $l_t = 0$. The state T_0, in turn, is coupled to the component state T_{-2} with $l_t = -2$. The result is that one of the component levels T_2, where $l_t = 2$, remains unaffected by the perturbation while the other two levels corresponding to $l_t = 2$ and $l_t = 0$ have terms added to their energies so as to give

$$T'_2 = T_2 + \left\{ 4\left(\frac{B_e^2}{\omega_t}\right)^2 \frac{\left[1 + 4\sum_s (\zeta_{st}^{(x)})^2\omega_t^2/(\omega_s^2 - \omega_t^2)\right]^2}{g_{l_t l_t} - B_e} \right\} J^2(J+1)^2 \tag{5.60}$$

$$T'_0 = T_0 - \left\{ 4\left(\frac{B_e^2}{\omega_t}\right)^2 \frac{\left[1 + 4\sum_s (\zeta_{s,t}^{(x)})^2\omega_t^2/(\omega_s^2 - \omega_t^2)\right]^2}{g_{l_t l_t} - B} \right\} J^2(J+1)^2$$

$g_{l_t l_t}$ being the difference in the energies T_2 and T_0. In other words, two of the three sublevels have, in effect, their centrifugal distortion constants substantially altered while the third sublevel remains unaltered. The latter effect is referred to as rotational l-type resonance.

Fermi-type Resonance. As we have seen, the first-order potential function for a polyatomic molecule is made up of terms that are cubic in the normal coordinates, that is, $V_1 = hc\sum_s\sum_s\sum_s k_{ss's''}q_sq_{s'}q_{s''}$. Frequently the force fields are such that V_1 contains terms such as $k_{sss''}q_s^2q_{s''}$. Indeed the symmetries of linear and axially symmetric molecules require terms of the variety $k_{sss''}(q_{s1}^2 + q_{s2}^2)q_{s''}$ which may be written $k_{sss''}r_s^2q_{s''}$. These terms contribute to the interaction energy E_i amounts that are inversely proportional to the quantities $\omega_s + \omega_{s'} - \omega_{s''}$ and $2\omega_s - \omega_{s''}$. When these denominators accidentally become very small, the quantities $x_{ss''}$ become indefinitely large. In such cases one must use the degenerate perturbation theory to calculate the energy.

The energies cannot be expressed in closed form but must be calculated independently for each case. We shall here confine ourselves to Fermi-type resonance between the first overtone of an isotropic oscillation and a parallel vibration like $2\nu_2$ and ν_1 in a triatomic linear molecule such as CO_2. The perturbed energies may be written

$$\begin{aligned}\epsilon\pm = (2\nu_2 + \nu_1 \pm \Delta/2) + F(v_2 = 0, l_2 = 0, v_1 = 0) \\ + (\alpha/2\Delta)(\Delta \mp \Delta_0) + (\beta/2\Delta)(\Delta \pm \Delta_0) \\ \pm (\alpha^2 - \beta^2)(\Delta^2 - \Delta_0^2)/4\Delta^3\end{aligned} \tag{5.61}$$

where

$$F(v_2,l_2,v_1) = [J(J+1) - l_2^2]\,B(\sigma_2,l_1) - D[J(J+1) - l_2^2]^2$$

and where α and β are equal to

$$F(v_2 = 2, l_2 = 0, v_1 = 0) - F(v_2 = 0, l_2 = 0, v_1 = 0)$$

and $F(v_2 = l_2 = 0, v_1 = 1) - F(v_2 = l_2 = 0, v_1 = 0)$, respectively, and $\Delta_0 = 2\nu_2 - \nu_1$ and $\Delta^2 = 2k_{221}^2 + \Delta_0^2$. In the case of molecules like CO_2 the Fermi-type resonance is not always independent of l-type resonance. In the case we have just considered l-type resonance occurs and adds a term to the effective centrifugal distortion constant, D. The effective D values associated with $\epsilon+$ and $\epsilon-$ are known to be

$$D_{+\text{eff}} = D - (\alpha - \beta)^2\frac{\Delta^2 - \Delta_0^2}{4\Delta^3} - 16\frac{\Delta - \Delta_0}{\Delta_0 + \Delta - 8gu}\frac{g_0^2}{\Delta}$$

and

$$D_{-\text{eff}} = D + (\alpha - \beta)^2\frac{\Delta^2 - \Delta_0^2}{4\Delta^3} + 16\frac{\Delta + \Delta_0}{\Delta_0 - \Delta - 8gu}\frac{g_0^2}{\Delta} \tag{5.62}$$

Anharmonic resonances are known to occur in higher orders also. Second- and even third-order resonances have been observed in the spectra of H_2O and HCN, respectively. These effects are beyond the scope of this text.

Coriolis Resonance Interactions. We have already considered the contribution to the rotation-vibration energy of an axially symmetric molecule by the Coriolis operator

$$p_zP_z = \Sigma\zeta_{ij}^{(z)}[(\nu_i/\nu_j)^{1/2}q_ip_j - (\nu_j/\nu_i)^{1/2}q_jp_i](P_z/I_{zz}) \tag{5.63}$$

where $\nu_i = \nu_j$ because of an essential degeneracy. We have, moreover, made the observation that when ν_i and ν_j are widely different this operator contributes only slightly to the energy; i.e., it makes a contribution only to E''_i. We shall now consider the contribution to the energy by the operator (5.63) when ν_i and ν_j are accidentally nearly alike, i.e., where $\nu_i \approx \nu_j$. If we assume ν_i to be $\nu_j + \delta/2\pi c$, where δ is small and c is the velocity of light, the operator (5.63) may be written

$$\Sigma\zeta_{ij}^{(3)}[(q_ip_j - q_jp_i) - (\delta/4\pi c\nu_j)(q_ip_j + q_jp_i)](P_z/I_{zz}) \tag{5.64}$$

Evidently, since $\delta/2\pi c\nu_j$ is small, its importance may safely be relegated to inclusion as a part of E''_i. The first term of (5.64) is, however, of the order of magnitude of (5.40).

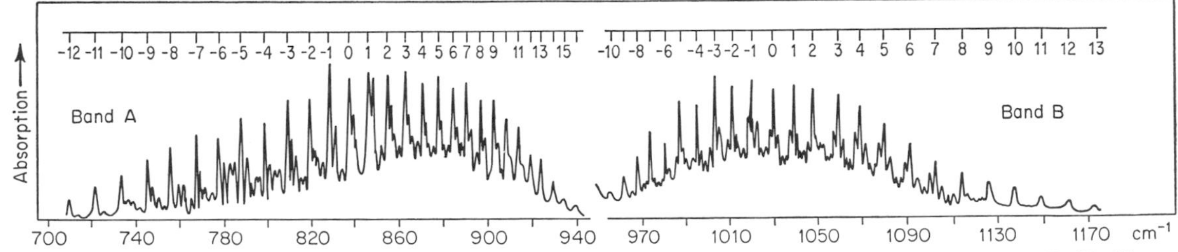

FIG. 5.17. Fundamental absorption bands of two degenerate vibrations in allene. [*From: C. H. Miller and H. W. Thompson, Proc. Roy. Soc. (London)*, **A200**: 1 (1949) *as adapted by J. de Heer, J. Chem. Phys.*, **20**(4): 637–641 (1952).]

To deal with this term, one must assume that ν_i and ν_j are accidentally degenerate and that the contribution to E_i by the operator (5.64) can be satisfactorily estimated by the methods of the degenerate perturbation theory. The correction cannot be computed in closed form for all levels in general, but each set of levels must be studied independently. We shall confine ourselves to studying the Coriolis interaction between the two fundamental frequencies ν_5 and ν_6 in the case of CH_2O. These two frequencies differ from each other by about 100 cm^{-1}; ν_5 is a deformation frequency in the plane of the molecule and has a magnitude of about 1,200 cm^{-1}, and ν_6 is a deformation frequency perpendicular to the plane of the molecule and has a value of about 1,100 cm^{-1}. These two frequencies have the right symmetry properties to perturb each other through Coriolis resonance. It will be assumed, for simplicity, that the CH_2O molecule has an asymmetry so small that it may be regarded as symmetric. As the energies of the two perturbed states, one obtains

$$\frac{E(\pm)}{hc} = \sum_{s''} \frac{\omega''_s}{2} + J(J+1)B_v + K^2(C_v - B_v) + (\nu_5 + \nu_6) \pm \left[\left(\frac{\Delta_0}{2}\right)^2 + K^2\left(\frac{\Delta_1}{2}\right)^2\right]^{1/2} \quad (5.65)$$

where $\Delta_0 = \nu_5 - \nu_6$ and $\Delta_1 = 2C_v\zeta^{(z)}_{5,6}(\nu_5 + \nu_6)/\sqrt{\nu_5\nu_6}$.

Subtracting the values of the energies in the ground state from $E(\pm)/hc$ and observing the appropriate selection rules (that is, $\Delta J = 0$, $\Delta K = \pm 1$) give the following values for the frequency positions of the Q branches of bands ν_5 and ν_6:

$$\nu(\pm) = \frac{\nu_5 + \nu_6}{2} - (C_v - B_v) \pm \left[\left(\frac{\Delta_0}{2}\right)^2 + K^2\left(\frac{\Delta}{2}\right)^2\right]^{1/2} \pm 2K(C_v - B_v) \quad (5.66)$$

K taking plus or minus integral values, the absolute values being always less than or equal to J.

It is of interest to note what Eq. (5.65) for $E(\pm)$ becomes in two limiting cases. Suppose $\Delta_0 \ll K\Delta_1$ so that Δ_0 may be neglected with respect to $K\Delta_1$. The term under the radical then becomes simply $K\Delta_1/2$ which, when added to the last term in Eq. (5.65), gives a relation equal to Eq. (5.44) plus the energy of an isotropic oscillator. If, however, $\Delta_0 \gg K\Delta_1$ so that the latter may be neglected with respect to Δ_0, the term under the square root becomes just $\Delta_0/2$. In this case the energy becomes that of two nondegenerate oscillators plus that of a rigid rotator, i.e., Eq. (5.33). Equation (5.66) for the line positions of the Q branches in the two bands shows that they obey a kind of convergence. The Q branches $\nu(+)$ may be seen to degrade toward the red while $\nu(-)$ degrades toward the violet. The convergence is not of a variety where a "head" is reached but is one where maximum and minimum spacings between the Q branches are reached in the two bands. This type of convergence is illustrated in Fig. 5.17.

9. The Raman Spectroscopy of Molecules

When light passes through a cell containing a gas or a liquid the radiation scattered by the medium may be viewed from the side. There will be observed, if the spectrum of this radiation is examined, primarily just the wavelengths of the incident light. This light is the *Rayleigh scattering*. Some much weaker lines will also be observed on both sides of the principal lines if these are greatly overexposed. The frequency displacements of these lines from the principal line are found to be independent of the incident light, but the pattern and the magnitude of the displacements are characteristic of the molecules in the cell. These much weaker lines are produced by *Raman scattering*.

This phenomenon has an explanation in the quantum mechanics. Its nature may, however, be understood in a general way from the following considerations. The incident radiation of frequency ν_0 subjects the molecules in the cell to an alternating electric field $E = E_0 \sin 2\pi\nu_0 t$. This electric field induces a dipole in the molecule which we shall designate as F. F will be proportional to E, the proportionality constant α being known as the polarizability, that is, $|F| = \alpha|E|$. The quantity α will depend upon the instantaneous positions of the nuclei in the molecule and will, therefore, to some extent also be influenced by its vibration and its rotation. It will, moreover, depend upon the orientation of the molecule with respect to the field. Thus for a nonrotating diatomic molecule with its internuclear axis aligned with the field we may represent α by the expansion

$$\alpha = \alpha_0 + \left(\frac{\partial \alpha}{\partial r}\right)_{r=r_0} \delta r + \cdots \quad (5.67)$$

where δr, for the molecule oscillating harmonically, will be $\delta r = \rho_0 \sin 2\pi\nu_v t$. Inserting the relation (5.67) for α into the expression for F,

$$F = \alpha_0 E_0 \sin 2\pi\nu_0 t + \rho_0 \left(\frac{\partial \alpha}{\partial r}\right)_{r=r_0} E_0 \sin 2\pi\nu_0 t \sin 2\pi\nu_v t$$

which is

$$F = \alpha_0 E_0 \sin 2\pi\nu_0 t + \frac{1}{2}\rho_0 \left(\frac{\partial\alpha}{\partial r}\right)_{r=r_0} E_0[\cos 2\pi(\nu_0 - \nu_v)t - \cos 2\pi(\nu_0 + \nu_v)t] \quad (5.68)$$

This must be multiplied by $\cos\psi$ if the internuclear axis of the molecule is oriented at an angle ψ with respect to the field. Designate $F\cos\psi$ by F'.

When the molecule is rotating, F' must be referred to the space fixed axes. We obtain for the components of F' directed along the space fixed X and Y axes the following:

$$F_x = F'\cos\psi \quad \text{and} \quad F_y = F'\sin\psi \quad (5.69)$$

Using in the relation (5.68) for F the expression (5.67) and replacing ψ in (5.69) by its equivalent $2\pi\nu_\psi t$, F_X and F_Y contain the frequencies ν_0, $\nu_0 \pm \nu_v$, $\nu_0 \pm 2\nu_\psi$, $\nu_0 + \nu_v \pm 2\nu_\psi$, and $\nu_0 - \nu_v \pm 2\nu_\psi$.

Quantum theoretically the Raman effect may be regarded as a collision process of photons with molecules. These collisions may be elastic, in which case the photon rebounds without altering its energy, or the collision may be inelastic, in which case the photon rebounds with an amount of energy ΔE greater or less than it had initially. This energy, ΔE, the photon will evidently have obtained from or given to the molecule with which it collides.

If the energy ΔE is added or taken away from the vibration-rotation energy of the molecule, then $(\Delta E/h)$ will be equal to the frequency displacements of the Raman lines from the incident frequency, ν_0 that is, $\nu_0 \pm 2\nu_\psi$, $\nu_0 \pm \nu_v$, $\nu_0 + \nu_v \pm 2\nu_\psi$, and $\nu_0 - \nu_v \pm 2\nu_\psi$. As in Sec. 6 we get $\nu_\psi = 2JB$ and equate the Raman frequencies to $(E' - E'')/hc$ where E/hc is a term value as used there. The selection rules governing Raman scattering are $\Delta v = 0$, $\Delta J = 0, \pm 2$; $\Delta v = \pm 1$, $\Delta J = 0, \pm 2$. When the anharmonicity of the vibrational motion is taken into account Δv may be greater than unity also. The transition $\Delta v = 0$, $\Delta J = 0$ is the principal scattered frequency and represents the Rayleigh scattering.

When the scattering is of such a nature that the energy ΔE is given to the molecules by the photons the scattered frequency will be on the low-frequency (long wavelength) side of the frequency ν_0. These lines are known as *Stokes* lines by analogy with the corresponding situation in fluorescence (Part 6, Chap. 7). When the scattering is such that ΔE is given to the photon by the molecule the scattered frequency will lie on the high-frequency (short wavelength) side of the frequency ν_0. These lines are known as *anti-Stokes* lines. The former are, in general, especially when $\Delta v \neq 0$, much more intense than the latter. For a molecule to surrender an amount of energy ΔE to a photon the molecule must be in an excited state. Since the vibration states are separated by an amount which is of the order of magnitude of 1,000 cm^{-1}, at ordinary temperatures the population of molecules in excited vibration states will be small.

The preceding argument may be extended to embrace polyatomic molecules. The selection rules here are considerably more complex and depend in an intimate manner upon the symmetry of the molecules. Consider, for example, the normal vibrations of the linear XY_2 molecular model, described in Sec. 3 and illustrated in Fig. 5.6. The polarization associated with the vibration ν_1 should, clearly, not be the same for positive values of the normal coordinates q_1 as for negative values. One should therefore, expect this vibration to produce a line in the Raman spectrum since the coefficient $(\partial\alpha/\partial q_1)_{q_1=0}$ may be expected to be different from zero. From considerations of symmetry the polarization associated with the frequencies ν_3 and ν_2, on the other hand, will have the same values for positive and negative values of the respective normal coordinates, q_3 and q_2. The polarization must in each of these cases, therefore, pass through a minimum as the atoms of the molecule pass through their equilibrium positions, i.e., the coefficients $(\partial\alpha/\partial q_3)_{q_3=0}$ and $(\partial\alpha/\partial q_2)_{q_2=0}$ must be zero. Such frequencies are Raman inactive to the approximation that expansions of the type (5.67) are valid.

It is interesting to note in this example how the infrared spectra and the Raman spectra complement each other. The frequency ν_1, which is Raman active, induces no electric dipole moment and is, therefore, optically inactive in the infrared. Similarly the two frequencies, ν_3 and ν_2, which are Raman inactive are the two frequencies which are optically active in the infrared.

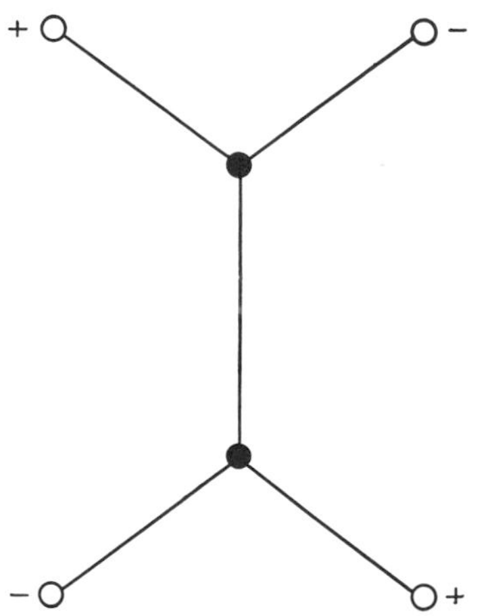

FIG. 5.18. The vibration ν_4 for a molecule of the X_2Y_4 variety.

There are, unfortunately, some vibrations in polyatomic molecules which are inactive in both the infrared spectrum and in the Raman spectrum. Such a frequency would be the frequency ν_4 in the planar X_2X_4 molecular model. This frequency is illustrated in Fig. 5.18. The electric moment generated by this vibration must be zero and it will, therefore, produce no band in the infrared spectrum. Moreover, when one regards the polarization of the molecule in this mode by a light wave, it will be the same for a positive as for a negative value of the normal coordinate. The polarizability must, therefore, pass through a minimum at $q_4 = 0$; so

$$\left(\frac{\partial\alpha}{\partial q_4}\right)_{q_4=0} = 0$$

and this frequency is also Raman inactive.

Rotational Raman transitions can be observed only for molecules in the gaseous phase while vibrational Raman transitions may be observed in the liquid and even in the solid state. Molecules in the gaseous phase scatter poorly and for this reason observations on the rotational Raman effect are difficult to make. They have been observed in only rather few cases. The common method is for Raman measurements to be made on molecules in the liquid state and in this field a wealth of data have been gathered.

The difference in the data revealed by the infrared spectroscopy of molecules and by the Raman spectroscopy of molecules is essentially one of the selection rules associated with the two phenomena.

References

1. Amat, G., and H. H. Nielsen: *J. Mol. Spect.*, **2**: 524 (1958).
2. Darling, B. T., and D. M. Dennison: *Phys. Rev.*, **57**: 128 (1940).
3. Dennison, D. M.: *Revs. Mod. Phys.*, **3**: 280 (1931).
4. Dennison, D. M.: *Revs. Mod Phys.*, **12**: 175–214 (1940).
5. Fermi, E.: *Z. Physik*, **71**: 250 (1930).
6. Hanson, H., and H. H. Nielsen: *J. Chem. Phys.*, **25**: 591 (1956).
7. Herzberg, G.: "Infra-red and Raman Spectra," Chap. 5, Van Nostrand, Princeton, N.J., 1945.
8. Johnston, M., and D. M. Dennison: *Phys. Rev.*, **48**: 869 (1935).
9. King, G. W., R. M. Hainer, and P. C. Cross: *J. Chem. Phys.*, **11**: 27 (1943).
10. King, G. W., R. M. Hainer, and P. C. Cross: *J. Chem. Phys.*, **12**: 210 (1944).
11. Lord, R., and R. E. Merrifield: *J. Chem. Phys.*, **20**: 1348 (1952).
12. Nielsen, H. H.: *Revs. Mod. Phys.*, **23**: 90–136 (1951).
13. Nielsen, H. H.: *J. Chem. Phys.*, **77**: 130 (1950).
14. Teller, E., and L. Tisza: *Z. Physik*, **73**: 791 (1933).
15. Nielsen, H. H.: "Encyclopedia of Physics," vol. XXXVII/1, p. 173, Springer, Berlin, 1959.
16. Nielsen, H. H.: M. Goldsmith, and G. Amat: *J. Chem. Phys.*, **24**: 1178 (1956).
17. Nielsen, H. H., M. Goldsmith, and G. Amat: *J. Chem. Phys.*, **27**: 838 (1957).
18. Nielsen, H. H., and G. Amat: *J. Chem. Phys.*, **27**: 845 (1957).
19. Nielsen, H. H., and G. Amat: *J. Chem. Phys.*, **29**: 665 (1958).
20. Nielsen, H. H., and G. Amat: *J. Chem. Phys.*, **34**: 339 (1961).
21. Nielsen, H. H., M. Grenier Besson, and G. Amat: *J. Chem. Phys.*, **36**: 3454 (1962).

Chapter 6

Microwave Spectroscopy

By WALTER GORDY, Duke University

1. Introduction

The microwave region lies in the electromagnetic spectrum between conventional radio which ranges approximately from a meter to miles in wavelength, and the infrared, which ranges approximately from a micron to a millimeter. Microwaves are thus millimeter or centimeter waves ranging in frequency from about 1 to 600 kilomegacycles (kMc)—30 cm to ½ mm in wavelength. The microwave region was the last great gap in the electromagnetic spectrum to be explored by the spectroscopist. Now the entire spectrum from the longest radio waves to the shortest known gamma rays is usable by the spectroscopist. Precise microwave electronic measurements have been extended down to the millimeter [1]* and into the submillimeter [2] region, thus overlapping modern infrared grating measurements.

The gap between infrared and radio waves was closed in 1923 by Nichols and Tear [3] in the sense that radiation was generated and detected. Not until 1935, when Cleeton and Williams [4] observed the inversion absorption of ammonia, were the first spectral measurements made in the microwave region. No further microwave spectral measurements were made until after the developments of microwave radar components during the Second World War. Earlier methods for generating and detecting microwaves were not practical for spectroscopy or for engineering purposes. In 1954 measurements of spectral lines by microwave electronic methods were extended [5*a*] down to wavelengths of 0.77 mm. During the same year infrared measurements of rotational spectra [6] were extended up to 0.99-mm wavelengths, thus effectively overlapping the infrared and microwave regions. Later work [5*b*] has extended the precise microwave measurements to 0.58 mm, then to 0.43 mm [5*c*] wavelength.

A particular region of the spectrum is usually characterized by the methods used in generating, transmitting, or detecting the radiation. Microwaves are most often generated by cavity resonators driven by velocity-modulated electron beams, transmitted by hollow waveguides, and detected by crystal rectifiers. The resistors, capacitors, and coils of conventional radio circuits are not effective at these high frequencies, nor are the usual radio tubes. Nevertheless the methods of microwave spectroscopy are basically similar to those of longer radio-wave spectroscopy in that both are electronic in nature as contrasted with the optical methods of infrared spectroscopy. In the heat sources used to produce infrared radiation the motions of the particles producing the radiation are essentially random. A very broad band of frequencies having no phase coherence is therefore produced. These frequencies must be sorted by an instrument of dispersion—prism or grating. When this is done, little energy is left in a chosen narrow band of frequencies. In contrast, radio waves, including microwaves, are produced by the orderly, synchronized motions of electrons moving under the control of a resonant circuit or cavity. To a good approximation, the electrons in the radio-frequency source oscillate with the same phase and frequency. The radiation produced is nearly monochromatic and phase-coherent, and the output frequency can be varied at the source so that no instrument of dispersion is required.

Because the band spread of the electronic source is narrow, it yields a much higher radiation density at a given frequency than can be obtained from an optical source. This superior source largely accounts for the exceptional resolving power possible in microwave spectroscopy—more than 10,000 times that obtainable in the infrared region.

A given region of the spectrum may also be characterized by the types of energy transformations which occur within its boundaries. Most microwave spectra so far investigated are pure rotational spectra of molecules in the gaseous phase and paramagnetic resonance spectra in solids. Transitions between hyperfine levels of atoms fall in the microwave region.

Energy differences which give rise to microwave frequencies are often revealed as fine or hyperfine structure in the spectral transitions occurring at optical frequencies. Hence many of the results obtained in microwave spectroscopy could be anticipated from optical spectroscopy. Furthermore much of the theory developed for interpretation of the finer details of optical spectra can be carried over to microwave spectroscopy, where its application is usually simpler. Advantages of the higher resolution and accuracy of microwave over optical spectroscopy in the investigation of the fine structure of matter are great. Most of the intervals observed directly in the microwave region are barely detectable as perturbations of optical spectra. The smaller per-

* Numbers in brackets refer to References at end of chapter.

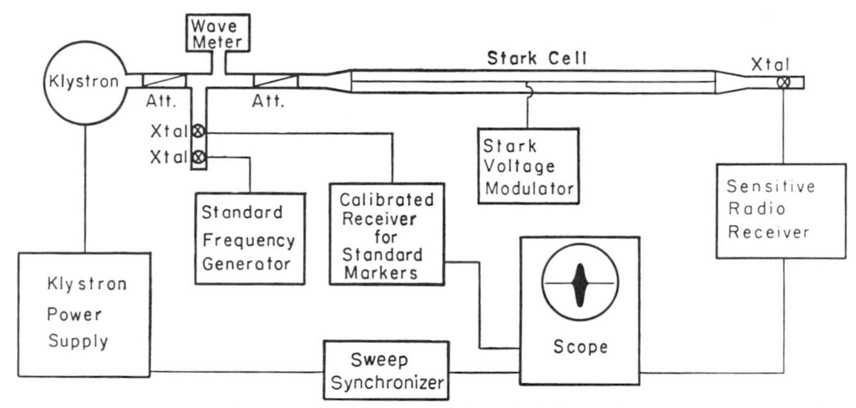

FIG. 6.1. Diagram of a common form of Stark-modulation microwave spectrograph.

turbations which give rise to well-resolved fine structure in microwave spectra cannot even be seen in the optical region.

2. The Microwave Spectroscope

In its simplest form the microwave spectroscope [7] consists of (1) a tunable source of radiation, (2) a frequency meter, (3) an absorption cell, (4) a detector of the radiation, (5) an amplifier, and (6) an indicator. An arrangement of these components is shown in Fig. 6.1.

The source is most often a reflex klystron, except in the high-frequency millimeter wave region where a crystal multiplier driven by a klystron is usually employed. Magnetrons yield higher power but are not often used because they do not give as pure a spectrum as do klystrons and are not as easily tuned.

Frequency measurements accurate to about 5 Mc can be made simply with a carefully measured cavity resonator. Accuracies to about 50 kc or better are commonly obtained with frequency markers produced by electronic multiplication of the frequencies from a crystal-controlled oscillator of low frequency (usually 5 or 10 Mc), which is monitored by comparison with the standard frequencies broadcast by Station WWV of the National Bureau of Standards.

Absorption cells are made in various ways. Most often the substance to be investigated is placed in a rectangular waveguide sealed at either end with mica windows. For Stark modulation a plane electrode is often inserted down the center of the guide, perpendicular to the electric vector of the microwave radiation. Teflon strips at either side of the guide make suitable supports for the electrode.

The silicon crystal rectifier is the detector most commonly used, although some use is made of germanium crystal rectifiers and also of thermal detectors. Some form of modulation is almost always used to aid detection. In one modulation method the source frequency is made to sweep periodically into and out of the absorption line so that an intensity modulation of the transmitted radiation is produced. More frequently the intensity modulation is produced through a modulation of the line itself by an electric or a magnetic field. In this manner the Stark or Zeeman components are pulled into and out of the frequency of the radiation source. The amplifier for the detected radiation is tuned to the frequency of the modulation. In addition to these modulations which are usually rapid (100 kc), the oscillator frequency is made to sweep slowly over the absorption line. The optimum rate at which the oscillator should sweep over the line is determined by the receiver response or band width. The slower the rate of sweep of the oscillator over the line, the narrower the receiver band width can be, and hence the greater is the sensitivity of the spectrometer. When a cathode-ray display is used, the oscillator must sweep over the line several times a second unless a special persistent screen is employed. Such rates of sweep require a receiver band width of a hundred or more cycles a second for response. On the other hand, if several seconds are spent in the sweep over the line, band widths of the order of a cycle a second can be used with an appreciable increase in sensitivity. For such a slow rate of sweep, automatic pen-and-ink recorders are employed to display the line.

3. Microwave Spectra of Free Atoms

The microwave region, 0.01 to 20 cm^{-1}, encompasses fine and hyperfine intervals of most atoms. Because of the low intensities of direct transitions between atomic fine or hyperfine levels and because of difficulties of getting large concentrations of free atoms or ions in an absorption cell, only a few atoms have been investigated at microwave frequencies. These include H, He^+, Na, Cs, N, O, and P.

Lamb-Retherford Experiment. By measurement at microwave frequencies with sensitive atomic beam techniques, Lamb and Retherford [8] proved that the 2 $^2S_{1/2}$ and 2 $^2P_{1/2}$ levels are not degenerate as was expected from Dirac theory but are separated by 1,062 Mc. This effect, which is commonly called the Lamb shift, is satisfactorily explained [9] as resulting from the interaction of the electron with the radiation field. A similar experiment has been performed by Lamb and Skinner [10] on ionized helium, in which the shift is larger, 14,000 Mc.

Atomic Hyperfine Intervals. The hyperfine transitions (Part 9, Chap. 3) result from interaction of the nuclear magnetic moment with the resultant magnetic field of the atomic electron. The vector **J** representing the total angular momentum of the electrons, is coupled to the nuclear spin vector **I** to form a resultant **F**. The total angular momentum quantum number F can take the values

$$F = J + I, J + I - 1 \cdots |J - I|$$

The hyperfine energy intervals arise from a change in the angle between **J** and **I** without a change in their magnitudes. First-order perturbation theory, using the interaction Hamiltonian $A\mathbf{J} \cdot \mathbf{I}$, shows that the quantized hyperfine energies are

$$E = \left(\frac{A}{2}\right) [F(F+1) - J(J+1) - I(I+1)] \qquad (6.1)$$

in which A is the coupling constant which depends upon the value of the nuclear magnetic moment and upon the average magnetic field at the nucleus, arising from the electronic state of the atom. Transitions which give rise to microwave absorption are $F \to F + 1$ with no change in J or I. The absorption frequencies are

$$\nu = \left(\frac{A}{h}\right)(F + 1) \qquad (6.2)$$

giving a series of equidistant lines to different values of F.

Both nuclear spins and nuclear magnetic moments can be evaluated from microwave atomic spectra of this type. Although much more accurate than optical methods, the method can not compete in accuracy and simplicity with the radio-frequency nuclear magnetic resonance method.

Microwave spectra are also useful for the evaluation of electric-quadrupole moments of nuclei. Nuclear-quadrupole [11] interactions in atoms are usually (but not always) weak compared with the magnetic-dipole interaction and can be treated as perturbations upon the magnetic hyperfine levels.

Significant astronomical information is being obtained with the atomic emission line of hydrogen observed in interstellar hydrogen [12].

4. Pure Rotational Spectra

The largest class of spectra in the microwave region is that associated with molecular rotation in gases. From these many important molecular and nuclear properties have been obtained.

The rotational energy of a molecule (when centrifugal distortions are neglected and no external field is applied) can be expressed in terms of its principal moments of inertia I_a, I_b, and I_c, and angular momenta P_a, P_b, and P_c, and is

$$E_{\text{rot}} = \frac{P_a^2}{2I_a} + \frac{P_b^2}{2I_b} + \frac{P_c^2}{2I_c} \qquad (6.3)$$

Here a, b, and c represent the principal axes of the momental ellipsoid, and conventionally $I_a \leq I_b \leq I_c$. Rotational motions are quantized (Part 2, Chap. 3 and Part 7, Chap. 5). The total angular momentum P can take only the values

$$P = [J(J+1)]^{1/2}\hbar \qquad (6.4)$$

where J is zero or an integer. Also, the component of P in a fixed direction in space is quantized according to the relation

$$P_Z = M\hbar \qquad (6.5)$$

where M can assume the $2J + 1$ values

$$M = J, J - 1 \cdots -J$$

When P has a fixed component along an internal axis, as in a symmetric-top rotor, this component takes the values

$$P_K = K\hbar \qquad (6.6)$$

where $K = 0, \pm 1, \pm 2, \ldots, \pm J$. With these quantum relations, the allowed rotational energies of a rigid linear or symmetric-top molecule can be readily obtained.

Linear Molecules. For a linear molecule, $I_a = 0$ and $I_b = I_c \equiv I$; thus (6.3) reduces to

$$E_{\text{rot}} = \frac{P^2}{2I} \qquad (6.7)$$

and substitution of (6.4) yields the characteristic energies

$$E_J = \frac{\hbar^2 J(J+1)}{2I} = hBJ(J+1) \qquad (6.8)$$

where $B = \hbar/(2I)$. If the molecule has a dipole moment, it can absorb radiation, with J increasing by integral steps. The selection rule for absorption is $J \to J + 1$, giving the absorption frequencies

$$\nu = 2B(J + 1) \qquad (6.9)$$

The pure rotational spectra of a linear molecule therefore consist, in this approximation, of equally spaced lines, the lowest frequency of which is $2B$. Measurement of a single rotational frequency gives B, and hence the moment of inertia. For low J values, (6.9) usually fits the observed spectrum very closely. For higher J transitions measured in the millimeter-wave region effects of centrifugal distortion are easily detectable, and (6.9) must be replaced by the formula

$$\nu = 2B(J + 1) - 4D_J(J + 1)^3 \qquad (6.10)$$

Here J represents the rotational quantum number of the lower state involved. Table 6.1 gives observed values of B and D_J for some linear molecules.

Table 6.1. Spectral Constants of Some Linear Molecules [7]

Molecule	B_0, Mc	D_J, Mc
$C^{12}O^{16}$	57897.75 ± 0.13	193 × 10^{-3}
$HC^{12}N^{14}$	44315.97 ± 0.10	99 × 10^{-3}
$N^{14}N^{14}O^{16}$	12561.66 ± 0.03	5.75 × 10^{-3}
$O^{16}C^{12}S^{32}$	6081.494 ± 0.010	1.27 × 10^{-3}
$O^{16}C^{12}Se^{78}$	4042.460 ± 0.005	0.83 × 10^{-3}
$Br^{81}C^{12}N^{14}$	4096.788 ± 0.007	0.81 × 10^{-3}
$I^{127}C^{12}N^{14}$	3225.578 ± 0.018	0.88 × 10^{-3}

The linear polyatomic molecule with n atoms has $n - 1$ independent interatomic distances, so that a single moment of inertia is insufficient to determine its structure. One must resort to isotopic substitution to obtain the complete structure. In this method it is assumed that the internuclear distances are the same for different isotopic combinations. Because of the difference in zero-point vibrations, this is not exactly true, but the error introduced by this effect is usually not greater than 0.1 per cent. Table 6.2 lists some distances obtained in this way.

TABLE 6.2. INTERNUCLEAR DISTANCES IN LINEAR MOLECULES [7]

Diatomic molecule	r_e, A	Diatomic molecule	r_e, A
CO	1.128227	ICl	2.323
FCl	1.62811	NaCl	2.3606
FBr	1.759	CsCl	2.9041
BrCl	2.138		

Linear XYZ molecule	XY distance, A	YZ distance, A
HCN	1.064	1.156
ClCN	1.629	1.163
BrCN	1.790	1.159
ICN	1.995	1.159
NNO	1.126	1.191
OCS	1.164	1.558
OCSe	1.159	1.709
TsCS	1.904	1.557

The B values are the average values for the lowest vibrational state. For any vibrational state

$$B_v = B_e - \sum_i \alpha_i \left(v_i + \frac{d_i}{2} \right) \tag{6.11}$$

where B_e is the equilibrium value. The α's are small constants, v_i represents the vibration quantum number of the ith mode, and d_i its degeneracy. Measurements of B for excited as well as ground vibrational states give the constants α, and hence B_e can be obtained.

In an excited degenerate bending vibration state, the interaction of vibration with rotation, gives rise to a doubling of all the rotational levels known as l-type doubling. The doublet separations are given by

$$\Delta E = qJ(J + 1) \tag{6.12}$$

where q is a constant of the order $2B^2/\omega_1$ in which ω is the frequency of the degenerate bending mode. The constant can be evaluated by measurement of the separation of the doublets. In many molecules q is so large that the doublet components for the higher rotational states are separated by microwave frequencies. Shulman and Townes [13] have observed in the microwave region direct transitions between l-type doublets of particular rotational states of HCN—with no change in J—and have made a very exacting test of (6.12).

Symmetric-top Molecules. Symmetric-top molecules have two equal principal moments of inertia. The prolate type has $I_a < I_b = I_c$ (momental ellipsoid elongated along the symmetry axis) and the oblate has $I_a = I_b < I_c$ (momental ellipsoid flattened along the symmetry axis). For the prolate top, a here represents the symmetry axis, and for the oblate top, c represents this axis. The rigid symmetric-top energy levels are

$$E_{JK} = \frac{\hbar^2 J(J+1)}{2I_b} + \frac{\hbar^2 K^2}{2}\left(\frac{1}{I_a} - \frac{1}{I_b}\right) = h[BJ(J+1) + (A - B)K^2] \tag{6.13}$$

where $A = h/8\pi^2 I_a$ and $B = h/8\pi^2 I_b$. The selection rule for absorption of radiation is $J \to J + 1$ with no change in K. The absorption frequencies are therefore the same as those for the linear molecule.

When effects of centrifugal distortion [14] are included, the levels are given by

$$E_{JK} = h[BJ(J+1) + (A - B)K^2 - D_J J^2(J+1)^2 - D_{JK} J(J+1)K^2 - D_K K^4] \tag{6.14}$$

where D_J, D_{JK}, and D_K represent distortion constants which are small compared with B or A, so that the absorption frequencies are

$$\nu = 2B(J+1) - 4D_J(J+1)^3 - 2D_{JK}(J+1)K^2 \tag{6.15}$$

where J represents the lower state and K is the same for both states ($\Delta K = 0$). Except for the last term, this is the same as for the nonrigid linear molecule. The last term causes a splitting of each line into $J + 1$ components. This splitting is small but usually resolvable in the microwave region. Table 6.3 gives some values of rotational constants as obtained from microwave spectroscopy.

In excited degenerate bending vibrational states, the rotational lines of symmetric rotors are split due to interaction of vibration and rotation. There are more components than for a linear molecule and the theory is more involved [15]. Parallel stretching vibrations simply change the effective moments of inertia and displace the rotational lines without splitting them. The equilibrium values B_e are given by a formula similar to that for the linear molecule (6.11).

TABLE 6.3. SPECTRAL CONSTANTS OF SOME SYMMETRIC-TOP MOLECULES [7]

Molecule	B_0, Mc	D_{JK}, Mc	D_J, Mc
$N^{14}F_3^{19}$	10,681.07 ± 0.45	-25.7×10^{-3}	14.2×10^{-3}
$P^{31}F_3^{19}$	7,820.01	-11.7×10^{-3}	7.5×10^{-3}
$P^{31}O^{16}F_3^{19}$	4,594.282 ± 0.009	1.25×10^{-3}	1.10×10^{-3}
$P^{31}S^{32}F_3^{19}$	2,657.663 ± 0.009	1.8×10^{-3}	0.30×10^{-3}
$C^{12}H_3F^{19}$	25,535.91 ± 0.05	443×10^{-3}	32.5×10^{-3}
$C^{12}H_3Cl^{35}$	13,292.95 ± 0.01	189×10^{-3}	26.4×10^{-3}
$C^{12}H_3Br^{81}$	9,531.845 ± 0.015	129×10^{-3}	10.7×10^{-3}
$C^{12}H_3I^{127}$	7,501.310 ± 0.007	99.4×10^{-3}	7.95×10^{-3}
$B^{10}H_3C^{12}O^{16}$	8,979.94 ±	390×10^{-3}	177×10^{-3}
$C^{12}H_3C^{12}C^{12}H$	8,545.87 ± 0.03	164×10^{-3}	3.12×10^{-3}
$C^{12}F_3^{19}C^{12}C^{12}H$	2,877.948	3×10^{-3}	0.24×10^{-3}
$Ge^{70}F_3^{19}Cl^{35}$	2,168.52	$< \lvert 10^{-3} \rvert$	0.6×10^{-3}

Only one principal moment of inertia of a symmetric-top molecule can be measured directly from pure rotational spectra, because of the selection rule $\Delta K = 0$. Hence one must use isotopic substitution to evaluate the structure of even the simplest molecule of this class. A large number of symmetric-top structures have been obtained in this way. Table 6.4 lists some of the simpler ones.

TABLE 6.4. STRUCTURES OF SOME SYMMETRIC-TOP MOLECULES [7]

X_3Y molecule	XY distance, A	XYX angle
NH_3	1.016	107°
PH_3	1.419	93.5°
AsH_3	1.523	92.0°
SbH_3	1.712	91.5°
NF_3	1.371	102°9′
PF_3	1.535	(100°)
AsF_3	1.712	(100°)
PCl_3	2.043	100°6′
$AsCl_3$	2.161	98°25′
$SbCl_3$	2.325	99.5°

X_3YZ molecule	XY distance, A	YZ distance, A	XYZ angle
H_3CF	1.109	1.385	110°0′
H_3CCl	1.103	1.782	110°20′
H_3CBr	1.101	1.938	110°48′
H_3CI	1.100	2.140	110°58′
H_3SiCl	1.50	2.048	110°57′
H_3SiBr	1.57	2.209	111°20′
H_3GeCl	1.52	2.147	111°4′
H_3GeBr	1.55	2.297	112°0′
F_3PO	1.52	1.45	102.5°
Cl_3PO	1.99	1.45	103.6°
F_3PS	1.53	1.87	100.3°
Cl_3PS	2.02	1.85	100.5°
O_3ReCl	1.761	2.230	108°20′

Asymmetric-top Molecules. If the three principal axes of inertia are unequal and none is zero, the problem of finding the allowed rotational energies is complicated, and no complete or general solution has been obtained. The energy levels can be expressed in closed algebraic formulas only for certain low J values, but useful approximations for higher J values for certain degrees and types of asymmetry have been obtained (Chap. 5). Hainer, Cross, and King [16] give selection rules, formulas, and extensive numerical tables for calculation of energy levels and intensities.

5. Inversion Spectra

Theoretically all nonplanar molecules have, because of the possibilities of inversion, two equivalent equilibrium configurations separated by a potential energy barrier. Quantum mechanically, the molecule can tunnel through the barrier from one of these configurations to the other—can "invert"—but cannot do so by any succession of simple rotations. In most molecules, however, the barrier is so high that the inversion time is extremely long, and effects on the spectrum without consequence. Only in NH_3 and ND_3 have inversion spectra been observed.

The inversion problem can be treated most simply by a consideration of the molecule as a one-dimensional oscillator of reduced mass μ oscillating in a potential field with a double minimum, as shown in Fig. 6.2. This treatment was first made by Hund [17] and extended by Dennison and Uhlenbeck [18]. The molecule resonates between the two potential minima, and the resonance energy splits the vibrational levels into doublets as indicated in Fig. 6.2. The magnitude of the splitting varies inversely with the area under the potential hill above the level of the vibrational state of the oscillator, and therefore increases with the vibrational quantum number. It can be expressed as [18]

$$\Delta E_{\text{inv}} = \frac{\Delta E_{\text{vib}}}{\pi A^2} \tag{6.16}$$

if A represents the area under the hill and ΔE_{vib} the separation of the vibrational levels.

In ammonia, the inversion barrier is sufficiently low that splitting of the ground vibrational level yields a microwave frequency. The wave functions of the two vibrational sublevels are of opposite symmetry so that dipole transitions between the sublevels are allowed. The observation of this inversion frequency at 23 kMc by Cleeton and Williams [4] in 1936 represented the first spectral measurement in the microwave region and gave confirmation to the inversion phenomenon predicted quantum mechanically but not classically.

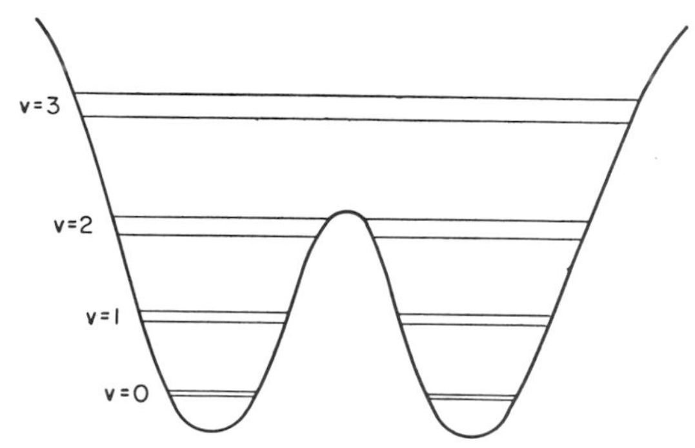

FIG. 6.2. Illustration of the inversion splitting of vibrational levels in a symmetric-top XY_3 molecule. The splitting is greatly exaggerated here.

The inversion frequency decreases rapidly with increase of the reduced mass of the oscillator. For ND_3 in the ground vibrational state it occurs [19] at about 2 kMc and at 125 kMc for ND_3 in the first excited vibrational state [20]. The slightly heavier molecules PH_3 and NF_3 have no microwave inversion spectrum.

Interaction of vibration and rotation makes the inversion splitting slightly different for different rotational states, giving a fine structure of the ammonia inversion transition [21, 22].

Nuclear quadrupole hyperfine structure, Stark and Zeeman splitting, radiation saturation, and certain other phenomena in microwave molecular spectra were demonstrated first in the NH_3 inversion spectrum.

6. Electronic Effects in Molecular Spectra

Transitions in Spin Multiplets. Molecules in nonsinglet sigma states have uncanceled electronic spin momenta which can interact with rotational motions of the molecule to give a first-order splitting of the rotational levels. In the case of O_2 (triplet sigma ground state) transitions between the resulting fine-structure components can absorb millimeter waves. Gaseous free radicals and ions and molecules in excited electronic states would show these effects.

The separation of the spin multiplets of the rotational levels varies slightly with rotational state, giving a fine structure to the microwave spectrum of O_2 which extends from about 4- to 6-mm wavelengths with the exception of a single component which falls at 2.5-mm wavelength. Beringer [23] measured the unresolved 5-mm wave absorption of O_2 at atmospheric pressure. Later, at Duke [24], the fine structure was resolved at low pressures with a magnetic modulation spectrograph. The O_2 absorption is due to magnetic-dipole interaction with the radiation. Oxygen has no permanent electric-dipole moment. The absorption is sufficiently strong to limit seriously the propagation of millimeter waves through the atmosphere.

The theory of the O_2 triplet has been worked out by Schlapp [25] and others. The formulas were found to fit approximately, but not within the experimental error of the microwave measurement. New calculations by Mizushima [26], taking into account centrifugal distortion effects, are in better agreement with observation.

Spin multiplets of rotational levels have $2S + 1$ components, where S is the resultant electronic spin quantum number. The total angular momentum **J** (excluding nuclear spins) of the molecule is the vector sum of **N**, the angular momentum of the molecule exclusive of spin, and **S**. Therefore, the quantum number J can take the values

$$J = N + S, N + S - 1, \ldots |N - S|$$

The electronic ground state of NO_2 is $^2\Sigma$, and therefore all the rotational levels are doublets. In this molecule, coupling of spin magnetic moment to molecular frame is too weak to give a microwave spectrum similar to that of O_2.

Molecules with Electronic Orbital Momentum. If the molecule has an electronic orbital momentum **L**, this will, in general, be coupled strongly to the molecular frame through the molecular electric field. Through spin-orbit interaction the electronic spin vector **S** will likewise be coupled to the molecular frame; when this coupling is strong only components of electronic momentum along the molecular fixed axis are defined. In a diatomic molecule this axis is the internuclear axis. The component of **L** along the internuclear axis is conventionally designated as Λ, the corresponding component of S as Σ, and the total electronic momentum along the axis $|\Lambda + \Sigma|$ as Ω. In this type of coupling, Hund's case (a), the total angular momentum **J** is the vector sum of **N** and Ω; the formula for the rotational energy of a diatomic molecule, if higher-order effects are neglected, is like that of a symmetric top (6.13)

$$E_{J,\Omega} = h[J(J + 1)B + (A - B)\Omega^2] \quad (6.17)$$

where $B = h/8\pi^2 I_B$ and $A = h/8\pi^2 I_A$, and $J = \Omega$, $\Omega + 1$, $\Omega + 2$, $\Omega + 3 \cdots$. The selection rules for rotational spectra are $\Delta\Omega = 0$ and $J = 1$. Hence, to a first approximation, the pure rotational lines are given by

$$\nu = 2B(J + 1) \quad (6.18)$$

which is exactly the same formula as that for linear and symmetric-top molecules in $^1\Sigma$ states. However, there is an important difference in the spectrum resulting from the fact that levels for $J < \Omega$ do not exist. The lowest transition is not a $0 \to 1$ as for molecules in $^1\Sigma$ states but a $J = \Omega \to J = \Omega + 1$. The outstanding example is NO, which has two electronic states, $^2\Pi_{1/2}$ and $^2\Pi_{3/2}$, significantly populated at room temperature. The lowest rotational frequency of the first is approximately $3B$ and has been measured [27a, b] at 150 kMc. The lowest rotational frequency of the $^2\Pi_{3/2}$ is approximately $5B$; it has been measured [27c] at 257 kMc. Another important difference in the rotational spectrum of a diatomic molecule having $\Lambda \neq 0$ from those with $\Lambda = 0$ is discussed below.

In the first approximation, levels for Λ and $-\Lambda$ have the same energy, and in this approximation all levels of molecules with $\Lambda \neq 0$ are doubly degenerate (Λ degeneracy). The interaction of L with the end-over-end rotation of the molecule causes a slight splitting of the levels for $+\Lambda$ and $-\Lambda$. This splitting is usually very small but increases rapidly with J. For light molecules, particularly those for which $S \neq 0$, it is frequently of the order of a microwave quantum or even larger. When $S \neq 0$ the Λ splitting is superimposed upon the splitting of the spin multiplet and causes a doubling of each spin component. Usually, but not always, splitting of the spin multiplet is much wider than the Λ splitting. For light molecules in high rotational states, Λ splitting may be comparable to, or even larger than, splitting of spin multiplets. Observation of direct transitions between Λ doublets in the OH radical at microwave frequencies has been observed by Sanders [28] et al.

For singlet Π states ($S = 0$) or for others in which Hund's case (b) applies, Λ splitting is small and can be expressed as

$$\Delta E = qJ(J + 1) \quad (6.19)$$

where q is approximately constant. Beringer [29] has observed this splitting as a barely resolved doublet in the paramagnetic resonance of the $J = \frac{3}{2}$ level of NO in the $^2\Pi_{3/2}$ state. In singlet Δ states the Λ splitting is smaller than in $^1\Pi$ states and is approximately

$$\Delta E = kJ^2(J + 1)^2 \quad (6.20)$$

in which the proportionality constant $k \ll q$.

The Λ doubling depends upon the electronic ground states and especially upon the nature of the internal coupling, whether it follows Hund's case (a) or (b), etc. Various cases have been treated by Kronig [30], Van Vleck [31], Mulliken and Christy [32], and others [33].

Magnetic Resonance. The splitting of Zeeman sublevels of paramagnetic substances is proportional to the magnetic field, and for paramagnetic gases (magnetic moment $\sim$ one Bohr magneton) only a few kilogauss is required to bring the absorption frequencies to the microwave region. This type of spectra, known as paramagnetic resonance or simply magnetic resonance, is observed in paramagnetic liquids and solids as well as in gases (Part 7, Chap. 4).

The paramagnetic resonance of most of the few stable paramagnetic gases has been studied by Beringer [29, 34] and coworkers at Yale.

Magnetic resonance spectra depend upon the couplings within the molecule and whether any of these are sufficiently weak to be broken down completely or partially by the applied field. These couplings are different for every molecule so far investigated [7, 29, 33, 34].

7. Nuclear Effects in Molecular Spectra

Nuclear Quadrupole Interactions. Nuclear electric quadrupole hyperfine structure in molecular spectra has yielded a number of nuclear spins and quadrupole moments. Much has been learned about chemical bonds from the nuclear quadrupole coupling constants. Whenever a molecule has a nucleus with a spin greater than $\frac{1}{2}$, its microwave spectrum is likely to have a resolvable nuclear quadrupole hyperfine structure.

The fundamental problem of nuclear quadrupole interactions was solved by Casimir [11]. Nuclear quadrupole hyperfine structure in microwave spectra was detected first by Good [22] in the ammonia inversion spectrum. The Hamiltonian which described the interaction is

$$H = \sum_i eQ_i \left\langle \left(\frac{\partial^2 V}{\partial Z^2} \right)_i \right\rangle_{\text{av}} \left[\frac{3(\mathbf{J} \cdot \mathbf{I}_i)^2 + \frac{3}{2}(\mathbf{J} \cdot \mathbf{I}_i) - \mathbf{J}^2\mathbf{I}_i^2}{2J(2J-1)I_i(2I_i-1)} \right] \tag{6.21}$$

in which e is the electronic charge, Q_i is the electric-quadrupole moment of the ith nucleus, $\left\langle \left(\frac{\partial^2 V}{\partial Z^2} \right)_i \right\rangle_{\text{av}}$ is the average electric field gradient at the ith nucleus in the space-fixed direction Z, which results from all the extranuclear charges of the molecule. $\mathbf{J}$ represents the total angular momentum of the molecule exclusive of nuclear spins and $\mathbf{I}_i$ represents the angular momentum of the ith nucleus. The expression in brackets is independent of the type of molecule considered and is also applicable to atoms.

For an atom or a molecule with a single nucleus having quadrupole coupling, the first-order quadrupole perturbation energies obtained with (6.21) can be expressed in convenient formulas. The numerator in brackets was evaluated by Casimir

$$E_Q = eQ \left\langle \frac{\partial^2 V}{\partial Z^2} \right\rangle_{\text{av}} \left[\frac{\frac{3}{4}C(C+1) - I(I+1)J(J+1)}{2J(2J-1)I(2I-1)} \right]$$

where

$$\begin{aligned} C &= F(F+1) - I(I+1) - J(J+1) \\ F &= J+I,\ J+I-1,\ \ldots,\ |J-I| \end{aligned} \tag{6.22}$$

In (6.22) the average field gradient $\left\langle \frac{\partial^2 V}{\partial Z^2} \right\rangle_{\text{av}}$ is peculiar to each individual atom or molecule. Since this quantity is evaluated along a fixed direction in space, its average value depends upon the motions of the charges from which it arises and therefore upon the electronic and molecular states. Classically,

$$\left\langle \frac{\partial^2 V}{\partial Z^2} \right\rangle_{\text{av}} = \left\langle \sum_k e \left[\frac{3\cos^2\theta_k - 1}{r_k^3} \right] \right\rangle_{\text{av}} \tag{6.23}$$

where r is the radius vector between the nucleus and the kth charge, and θ_0 the angle between this vector and the Z axis. The summation is taken over all the extranuclear charges, and the average is taken over the electronic state for atoms and over the rotational state as well as the electronic state for molecules. Small effects of the charges on other nuclei in the molecule and the effects of molecular vibration are neglected. The average is taken over the state which gives the maximum projection along the space-fixed axis, the state for which $M_J = J$. It is convenient to take the average over the rotational state and to express the field gradient with reference to the axis z, fixed in the molecule.

For diatomic and linear polyatomic molecules

$$\left\langle \frac{\partial^2 V}{\partial Z^2} \right\rangle_{\text{av}} = \frac{-J}{2J+3} q_{zz} \tag{6.24}$$

where $q_{zz} = \left\langle \frac{\partial^2 V}{\partial z^2} \right\rangle_{\text{av}}$ in which z is along the figure or bond axis. The quantity $\left\langle \frac{\partial^2 V}{\partial z^2} \right\rangle_{\text{av}}$ must now be averaged over only the electronic state of the molecule. The product eQq_{zz} is commonly designated as χ_{zz} and called the quadrupole coupling.

For symmetric-top molecules with the coupling atom on the symmetry axis

$$\left\langle \frac{\partial^2 V}{\partial Z^2} \right\rangle_{\text{av}} = \frac{J}{2J+3} \left[\frac{3K^2}{J(J+1)} - 1 \right] q_{zz} \tag{6.25}$$

where the molecular-fixed reference axis is along the symmetry axis.

A similarly compact expression for the asymmetric rotor cannot be given, but $\left\langle \frac{\partial^2 V}{\partial Z^2} \right\rangle_{\text{av}}$ has been expressed [35] in terms of the numerical intensity factors tabulated by Cross, Hainer, and King [36], and an alternate expression in terms of the reduced asymmetric-top energies $E(\varkappa)$, has been developed [37]. The latter form is

$$\begin{aligned} \left\langle \frac{\partial^2 V}{\partial Z^2} \right\rangle_{\text{av}} = \frac{J}{(2J+3)J(J+1)} & \left\{ q_{aa} \left[J(J+1) + E(\varkappa) - (\varkappa+1) \left(\frac{\partial E(\varkappa)}{\partial \varkappa} \right) \right] \right. \\ & + 2q_{bb} \left(\frac{E(\varkappa)}{\partial \varkappa} \right) \\ & \left. + q_{cc} \left[J(J+1) - E(\varkappa) + (\varkappa-1) \left(\frac{\partial E(\varkappa)}{\partial \varkappa} \right) \right] \right\} \end{aligned} \tag{6.26}$$

where q_{aa}, q_{bb}, and q_{cc} represent the field gradients,

$\partial^2 V/\partial a^2$, etc., with respect to the principal inertial axes. $E(\varkappa) = \dfrac{2E_{J_\tau} - (A + C)J(J+1)}{A - C}$ is the reduced energy for asymmetric rotors;

$$\varkappa = \frac{(2B - A - C)}{(A - C)}$$

is the asymmetric parameter; E_{J_τ} represents the entire rotational energy; and A, B, and C are the usual spectral constants.

The first-order quadrupole coupling energy for a linear molecule (single-coupling nucleus) is

$$E_Q = -eQq_{zz}Y(F) \qquad (6.27)$$

and the similar expression for the symmetric top is

$$E_Q = eQq_{zz}\left[\frac{3K^2}{J(J+1)} - 1\right]Y(F) \qquad (6.28)$$

For the asymmetric rotor (with single coupling nucleus)

$$E_Q = eQ\left\langle\frac{\partial^2 V}{\partial Z^2}\right\rangle_{\text{av}}\left[\frac{2J+3}{J}\right]Y(F) \qquad (6.29)$$

where $\left\langle\dfrac{\partial^2 V}{\partial Z^2}\right\rangle_{\text{av}}$ is given by (6.26)

The common factor

$$Y(F) = \frac{\frac{3}{4}C(C+1) - J(J+1)I(I+1)}{2(2J-1)(2J+3)I(2I-1)} \qquad (6.30)$$

is numerically tabulated [7] for $J = 1$ to 20.

To find the hyperfine spectrum, the E_Q values must be added to the energies of the unperturbed state using the selection rules, $\Delta F = 0, \pm 1$, and those governing the over-all transition.

Relative intensities of the hyperfine components are

$$\left.\begin{aligned} I_{\mp} &= \frac{1}{F}\cdot Q(F)\cdot Q(F-1) && \Delta F = \mp 1 \\ I_0 &= \frac{2F+1}{F(F+1)}\cdot P(F)\cdot Q(F) && \Delta F = 0 \\ I_{\pm} &= \frac{1}{F}\cdot P(F)\cdot P(F-1) && \Delta F = \pm 1 \end{aligned}\right\}\ \Delta J = \pm 1$$

$$\left.\begin{aligned} I_0 &= \frac{2F+1}{F(F+1)}\cdot R^2(F) && \Delta F = 0 \\ I_{\pm} &= \frac{1}{F}\cdot P(F)\cdot Q(F-1) && \Delta F = \pm 1 \end{aligned}\right\}\ \Delta J = 0$$

where
$$\begin{aligned} P(F) &= (F+J)(F+J+1) - I(I+1) \\ Q(F) &= I(I+1) - (F-J)(F-J+1) \\ R(F) &= F(F+1) + J(J+1) - I(I+1) \end{aligned}$$

in which J and F are the larger values involved in the transitions. These intensities are tabulated numerically for $\Delta J = 0, \pm 1$ for J up to 6 by White [38] and by Condon and Shortley [39] and for $\Delta J = \pm 1$ to $J = 20$ by Gordy, Smith, and Trambarulo [7].

These formulas give quadrupole perturbation energies to first order only. Second-order formulas for linear and symmetric-top molecules with a single coupling nucleus are given by Bardeen and Townes [40] and for an asymmetric rotor by Bragg [33]. The case of two nuclei with coupling in linear molecules is treated by Bardeen and Townes [41] and that of three symmetrically placed identical nuclei in a symmetric top by Bersohn [42] and by Mizushima and Ito [43].

The quantities obtained from the analysis of quadrupole hyperfine structure are coupling factors of the form eQq_{zz}. The coupling is a tensor, but the coordinates can always be chosen so that nondiagonal components vanish. Because the electronic charge at the nucleus is neglected (only the s electrons have significant densities at the nucleus, and this charge is spherically symmetric and makes no contribution to q), Laplace's equation $\nabla^2 V = 0$ can be applied to reduce the independent coupling factors to two. They are usually expressed as q_{zz} and $\epsilon = (q_{xx} - q_{yy})/q_{zz}$, where q_{zz} is the larger of the coupling factors and ϵ represents an asymmetry parameter. When the charge distribution about the nucleus is symmetric about the z axes, ϵ is zero and only one coupling constant eQq_{zz} is obtained for the particular nucleus. This is the case for linear and symmetric-top molecules. For an asymmetric rotor the coupling constants are experimentally obtained with reference to the principal inertial axes, which do not necessarily lie along bond axes.

The quantities q_{zz}, q_{yy}, and q_{xx} depend upon the electronic structure of the molecule. The potential V due to a single electron at a distance r from the nucleus is e/r, and the contribution to q_{zz} of this electron is

$$\frac{\partial^2}{\partial z^2}\frac{e}{r} = \frac{e(3\cos^2\theta - 1)}{r^3} \qquad (6.31)$$

where θ is the angle between z and r. The average contribution of an electron in a normalized orbital ψ_i is

$$(q_{zz})_i = e\int \psi^*_i\left(\frac{3\cos^2\theta - 1}{r^3}\right)\psi_i\,d\tau \qquad (6.32)$$

To obtain the total q_{zz}, contributions of all electrons must be added. Completely filled electronic shells and electrons in s orbitals can be neglected since they have spherically symmetric charge distributions around the nucleus (except for small distortion effects). A precise evaluation of the integral of (6.32) is extremely difficult and has been made for the simple hydrogen molecule only. In many cases, q for a molecule can be approximated from atomic orbitals if the time-averaged fractional charge in the different atomic orbitals is taken into account [7, 44]. Sometimes nuclear spins can be obtained from the hyperfine patterns simply by a count of the number of hyperfine components. Figure 6.3 shows some hyperfine patterns and illustrates a spin determination.

Nuclear Magnetic-dipole Interactions. With respect to nuclear magnetic interactions, molecules can be divided into two distinct classes: (1) those which have no unbalanced electronic moments and hence only very small molecular fields (of the order of a gauss) caused mainly by molecular rotations; (2) those which have unbalanced electronic magnetic moments.

The former group constitutes practically all stable molecules. In them the nuclear magnetic interaction is so small that its effects have been detected in only

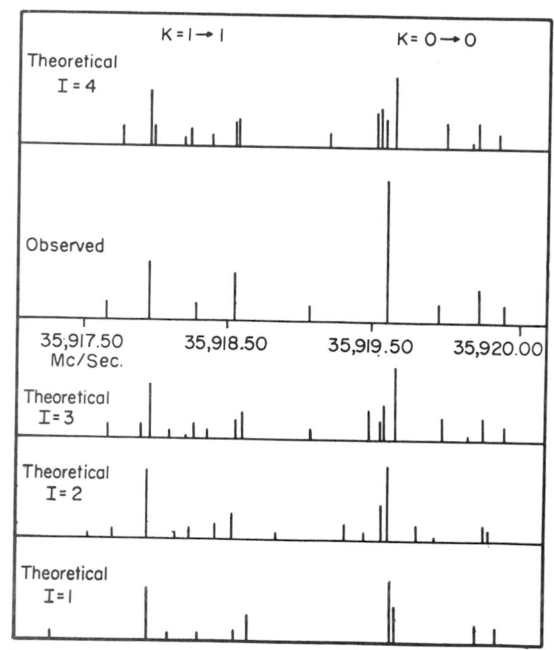

FIG. 6.3. Observed nuclear quadrupole hyperfine structure of the $J = 1 \rightarrow 2$ rotational transition of $B^{10}H_3CO$, with theoretical patterns for different integral spin values for the B^{10} nucleus. This comparison shows that the spin of B^{10} is 3. [*From: Gordy, W., H. Ring, and A. B. Burg: Phys. Rev.*, **78:** 512 (1950).]

a few instances. In the latter, the interactions are so large that a splitting of the order of a microwave quantum may be produced. The magnetic hyperfine structure depends on the direction as well as on the strength of the molecular field components. When there is an electronic spin magnetic moment only, it will not be coupled strongly to the molecular frame but will be coupled strongly to the nuclei, and the perturbation of the rotational levels will not be great. However, if there is an orbital electronic momentum, $\Lambda \neq 0$, the electronic moment and hence the nuclei spin vectors will be coupled strongly to the molecular frame, and the interactions of the nuclei with the molecular rotation will be large. Different quantitative treatments are therefore necessary for different types of paramagnetic molecules.

Nuclear Statistical Weights. If a molecule has identical nuclei symmetrically placed so that an indistinguishable configuration can be obtained by their exchange through rotation, an intensity alternation of the rotational lines results. Suppose that the spins of the identical nuclei are zero. Then, if the molecule is in an even electronic and an even vibrational state, odd rotational levels are missing; if the electronic function is odd and the vibrational function even, *even* rotational levels are missing. The latter condition applies to $O^{16}O^{16}$ or $O^{18}O^{18}$, for which only odd rotational levels occur. A rotation of 180° about the symmetry axis of $SO_2{}^{16}$ exchanges the identical O^{16} nuclei, and alternate rotational levels corresponding to rotation about this axis are missing. In this case the electronic ground state is even, and hence the *odd* rotational levels are missing.

When the nuclear spin I is greater than zero, the nuclear spin functions are linear combinations of the spin functions of the individual nuclei, and both odd and even spin functions exist. Either the Bose-Einstein or the Fermi-Dirac statistics can be satisfied regardless of the symmetry of $\psi_e\psi_v\psi_r$, and there are no missing levels. There is, however, an alternation in intensity of the levels because the odd and even nuclear spin functions are not of equal weight. For a homonuclear diatomic molecule the alternation in intensity is $I/(I + 1)$. This ratio also represents the intensity alternation for symmetrical linear molecules such as XYX and for certain levels of triangular YX_2 molecules. The intensity alternation for lines $K = 0$, 3, 6, . . . , to those with $K = 1, 2, 4, 5, \ldots$, for symmetric top X_3YZ molecules is $1 + \frac{3}{4}I(I + 1)$, where I is the spin of the symmetrically placed X atoms [7].

In microwave spectroscopy as in optical spectroscopy nuclear spins can be determined by intensity alternation in the spectral lines. However, most spins which could be obtained in this way have already been ascertained by other methods.

Isotopic Masses. Effects of isotopic masses on the microwave frequencies of molecules are so great that frequently a different microwave oscillator and different waveguide components are required to observe the same molecular transition with different isotopic molecules. The change depends upon the distance of the substituted atom from the molecular center of mass as well as upon the mass change. Occasionally an atom is so near the center of gravity that negligible shifts are produced [7].

Isotopic substitution is used to obtain additional parameters needed in determinations of molecular structures. Another important application is the measurement of relative isotopic masses. The most accurate mass determinations can be made with diatomic molecules. Here isotopic substitution may cause large line shifts which can be measured to an accuracy of five to six decimal places. The internuclear distances in diatomic molecules can be assumed constant to about the same degree, if equilibrium values r_e are obtained. Mass ratios have been obtained from microwave spectroscopy with linear, polyatomic, and symmetric-top molecules as well as with diatomic ones [7].

8. Stark and Zeeman Effects in Rotational Spectra

The spatial M degeneracy of a molecule can be lifted by the application of an electric (Stark effect) or a magnetic (Zeeman effect) field.

Stark Splitting. Electric fields ordinarily imposed produce splittings small compared with separations of internal energy levels so that the Stark splitting can be calculated by quantum mechanical perturbation theory. The Hamiltonian for an applied field $\boldsymbol{\mathcal{E}}$ and a molecular electric-dipole moment $\boldsymbol{\mu}$ is

$$H_{\mathcal{E}} = -\boldsymbol{\mu} \cdot \boldsymbol{\mathcal{E}} = -\mu\mathcal{E} \cos \theta \qquad (6.33)$$

The first-order Stark splitting $E_{\mathcal{E}}{}^{(1)}$ vanishes for a rigid linear rotor.

Second-order Stark effect in linear molecules is

given by

$$E_{J,M_J}{}^{(2)}_{J\neq 0} = -\frac{\mu^2\mathcal{E}^2}{2hB}\left[\frac{3M_J{}^2 - J(J+1)}{J(J+1)(2J-1)(2J+3)}\right] \tag{6.34}$$

For $J = 0$, the levels are displaced but not split,

$$E_{000} = -\frac{\mu^2\mathcal{E}^2}{6hB} \tag{6.35}$$

Selection rules for M_J are $\Delta M_J = 0, \pm 1$. In most microwave experiments, the field $\mathcal{E}$ is imposed parallel to the electric vector of the microwave radiation. In this case only the π components ($\Delta M_J = 0$) are obtained. The σ components ($\Delta M_J \pm 1$) are obtained when $\mathcal{E}$ is perpendicular to the microwave electric vector.

A symmetric-top rotor has a first-order Stark effect when $K \neq 0$. The first-order splitting is

$$E_{JKM_J}{}^{(1)} = \frac{\mu\mathcal{E}KM_J}{J(J+1)} \tag{6.36}$$

It is zero when $K = 0$ and decreases rapidly with increasing J.

Second-order Stark energies of a symmetric-top rotor are

$$E_{JKM_J}{}^{(2)}_{J\neq 0} = \frac{\mu^2\mathcal{E}^2}{2hB} \left\{\frac{[3K^2 - J(J+1)][3M_J{}^2 - J(J+1)]}{J^2(J+1)^2(2J-1)(2J+3)} - \frac{M_J{}^2K^2}{J^3(J+1)^3}\right\} \tag{6.37}$$

For $J = 0$, the level is displaced but not split. The displacement is given by (6.35).

Selection rules for M_J are those of a linear molecule, $\Delta M_J = 0$ for π components, and $\Delta M_J = \pm 1$ for σ components. These are applied with the selection rules for the over-all transition, which for pure rotational absorption lines are $\Delta J = 1$ and $\Delta K = 0$.

Stark effect of the asymmetric rotor is similar, but complexity of the asymmetric rotor wave functions makes evaluation difficult. The line-strength table of Cross, Hainer, and King [36] can be used for calculation of Stark splittings in asymmetric rotors [45].

Application of the Stark Effect. The Stark effect is used as an aid in identification of spectral transitions and in measurement of electric-dipole moments of molecules [46]. The absorption line is modulated at a low radio frequency (usually ~100 kc), and an AM receiver is tuned to the modulation frequency. The Stark modulation produces an intensity modulation of the microwave power by pulling the line component into and out of the frequency of the microwave radiation. The method avoids detection of power variations caused by reflections in the microwave components or other factors, and allows amplification of the signal at relatively high frequency with less trouble due to circuit noise.

Identification of the rotational transitions is made by counting the Stark components. This method is of principal advantage in asymmetric rotors where the identification cannot be made from ratios of line rotational frequencies as for linear and symmetric-top molecules. A first-order Stark effect of an asymmetric rotor has in general $2J + 1$, $\Delta M = 0$ components, and a second-order effect, $J + 1$, $\Delta M = 0$ components, where J is the smaller of the rotational quantum numbers involved.

Stark measurements of electric-dipole moments of molecules are usually more accurate than those with other methods. The advantage over the dielectric-constant method (Part 4, Chap. 7) is that its accuracy is not affected by impurities. To measure a dipole moment, the frequency splitting of a single Stark component is measured with a known field. Some dipole moments obtained with the Stark effect are given in Table 6.5.

TABLE 6.5. ELECTRIC-DIPOLE MOMENTS FROM THE STARK EFFECT IN MICROWAVE MOLECULAR SPECTRA [7]

Molecule	Moment (Debye units)	Molecule	Moment (Debye units)
Linear molecules			
FCl	0.88	HCN	3.00
FBr	1.29	ClCN	2.80
BrCl	0.57	N_2O	0.166
OCS	0.710	HCCCl	0.44
OCSe	0.754	HCCCN	3.6
Symmetric-top molecules			
NH_3	1.468	CH_3F	1.790
PH_3	0.55	CH_3Cl	1.869
AsH_3	0.22	CH_3I	1.647
SbH_3	0.116	SiH_3F	1.268
NF_3	0.234	SiH_3Cl	1.31
PF_3	1.025	SiH_3Br	1.31
AsF_3	2.815	GeH_3Cl	2.13
POF_3	1.77	BH_3CO	1.795
CHF_3	1.645		
Asymmetric-top molecules			
H_2O	1.94	CH_2O	2.39
SO_2	1.59	CH_2CO	1.42
O_3	0.53	CH_2F_2	1.93
SO_2F_2	0.228	CH_2Cl_2	1.62
		CH_2OCH_2	1.88

Zeeman Effect in Molecules without Hyperfine Structure. Practically all stable molecules have $^1\Sigma$ ground states and hence no primary magnetic-dipole moments. They do have small magnetic moments of the order of a nuclear magneton due to molecular rotation, interactions with higher electronic states, nuclear moments, etc. These secondary magnetic moments give rise to a Zeeman splitting of rotational lines usually resolvable with a field of about 10 kgauss. Nuclear magnetic moments cause splitting only when the nuclear spin axis is coupled to the molecular frame through some interaction such as the electric-quadrupole interaction.

The Zeeman interaction term is $\boldsymbol{\mu} \cdot \mathbf{H}$. In the absence of nuclear interactions the first-order perturbation energy is

$$E_H^{(1)} = -\frac{\mu_J H M_J}{[J(J+1)]^{1/2}}$$

or
$$E_{M_J}^{(1)} = -g_J \beta_I M_J H \tag{6.36a}$$

where
$$\mu_J = g_J \beta_I [J(J+1)]^{1/2} \tag{6.37a}$$

in which β_I is the nuclear magneton. The Zeeman splitting of the $J \to J'$ rotational lines are

$$\Delta\nu = [-g_{J'}\,\Delta M_J + (g_J - g_{J'})M_J]\beta_I \frac{H}{h} \tag{6.38}$$

Selection rules are $\Delta M = 0$ for H parallel to the electric vector of the radiation and $\Delta M = \pm 1$ for H perpendicular to this vector for electric-dipole radiation.

In many cases, particularly for linear molecules, μ_J is proportional to $[J(J+1)]^{1/2}$, and so g_J is then not dependent on J and $g_J = g_{J'}$. For nonlinear molecules g_J may be a complicated function of the rotational motions of the molecule. Table 6.6 illustrates some rotational g_J factors obtained from microwave spectroscopy.

TABLE 6.6. ROTATIONAL g_J FACTORS

Molecule	Transition	g_J (av. value)*
H_2O	5_{-1}–6_{-5}	0.586
HDO	5_0–5_1	0.439
NH_3	Inversion	+0.477
N_2O	$0 \to 1$	0.086
SO_2	$7_{2,6}$–$8_{1,7}$	0.084
OCS	$1 \to 2$	0.029

Values taken from C. K. Jen, *Physica*, **17**: 379 (1951).
* Sign uncertain except where specified.

Zeeman Effect in Molecules with Nuclear Couplings. When the molecule contains a nucleus with quadrupole coupling sufficiently strong that it is not broken down by the applied field **H**, then **J** and the nuclear spin **I** are coupled to form a resultant **F**. Writing $\mu_F = g_F\beta_I[F(F+1)]^{1/2}$,

$$E_H^{(1)} = -g_F \beta_I M_F H \tag{6.39}$$

where $g_F = g_J\alpha_J + g_I\alpha_I$,

$$\alpha_J = \frac{F(F+1) + J(J+1) - I(I+1)}{2F(F+1)} \tag{6.40}$$

$$\alpha_I = \frac{F(F+1) - J(J+1) + I(I+1)}{2F(F+1)} \tag{6.41}$$

$$F = J + I, J + I - 1, \ldots, |J - I|$$
$$M_F = F, F - 1, \ldots, -F$$

Selection rules for M_F are

$\Delta M_F = 0$ π components
$\Delta M_F = \pm 1$ σ components

When coupling is through an electric dipole, π components are observed when **H** is parallel to the electric vector of the radiation, and σ components are observed when **H** is perpendicular to it. Zeeman effect in hyperfine structure can be used to measure the nuclear g_I factor and hence the nuclear magnetic moment for certain radioactive nuclei for which the more accurate methods of nuclear resonance are inapplicable because of the large samples required [47].

The Paschen-Back effect in molecules is treated by Jen [48].

Relative Intensities of Stark and Zeeman Components. If P and Q are parameters independent of M, which depend upon the strength of the unsplit (zero field) rotational line, relative intensities of the Stark or Zeeman components of the line are

$$\left.\begin{aligned} I_{M\to M} &= PM^2 \\ I_{M\to M\pm1} &= (P/4)(J \mp M)(J \pm M + 1) \end{aligned}\right\} J \to J \tag{6.42}$$

$$\left.\begin{aligned} I_{M\to M} &= Q[(J+1)^2 - M^2] \\ I_{M\to M\pm1} &= (Q/4)(J \pm M + 1)(J \pm M + 2) \end{aligned}\right\} J \to J+1 \tag{6.43}$$

$$\left.\begin{aligned} I_{M\to M} &= Q(J^2 - M^2) \\ I_{M\to M\pm1} &= (Q/4)(J \mp M)(J \mp M - 1) \end{aligned}\right\} J \to J-1 \tag{6.44}$$

The same rules apply either to Stark or Zeeman effects of nuclear hyperfine structure (weak-field cases) if J is replaced by F and M by M_F.

9. Shapes and Intensities of Microwave Absorption Lines

Intensities. Microwave spectra are ordinarily observed in absorption. A microwave emission line has, however, been detected from an extended astronomical source [12]. In the optical region emission lines are commonly observed: The Einstein coefficient of spontaneous emission decreases as the cube of the frequency and in the centimeter wave region is smaller by the order of 10^{-12} than in the visible region (Part 7, Chap. 3). Although the Einstein coefficient of absorption is independent of frequency, even absorption spectra are much weaker at microwave than at optical frequencies (but not as much so as are emission lines). The lower intensity of microwave absorption arises partly from the smaller amount of energy per quantum at the lower frequency and partly from the fact that absorption is accompanied by induced emission which is coherent with the inducing radiation. At microwave frequencies this adds back to the radiation field almost as much energy as is taken out by absorption.

The absorption coefficient α is the fractional decrease in power loss in unit path of the radiation

$$\alpha_\nu = -\frac{1}{P}\frac{dP}{dx} \tag{6.45}$$

The "intensity" of a microwave line is frequently given in terms of the absorption coefficient at the frequency in the absorption line for which the absorption is maximum. Sometimes the integrated intensity, defined as

$$I_{\text{int}} = \int \frac{\alpha_\nu\,d\nu}{\nu^2} \tag{6.46}$$

is specified. Experimentally, the absorption coefficient can be obtained by a measurement of the change in power during its passage through the absorption cell. If α_c represents the coefficient of loss for the cell walls and α_ν that for the substance being investigated, then

$$\alpha_\nu = -\alpha_c - \frac{1}{l}\log\frac{P_r}{P_i} \qquad (6.47)$$

where l is effective cell length, P_i is the power input to the cell, and P_r is the power remaining at the cell output.

The microwave absorption coefficient is simply related to the Einstein coefficients [7]. For a transition $J \to J'$ it can be expressed as

$$\alpha_\nu = \frac{8\pi^3\nu^2 N_J}{ckT} |(J|\mu|J')|^2 S(\nu,\nu_0) \qquad (6.48)$$

where α_ν represents the power loss per centimeter path length, N_J is the number of absorbing particles per cubic centimeter in the lower state J, k is Boltzmann's constant, T is the absolute temperature, c is the velocity of light, ν is the frequency of absorption, $(J|\mu|J')$ represents the matrix elements of the dipole moment connecting the states J and J', and $S(\nu,\nu_0)$ is the line-shape function.

The population factor N_J depends upon the nature and structure of the absorbing particle. If J represents a rotational state of a molecule

$$N_J = \frac{FNg \exp(-E_J/kT)}{Q_r} \qquad (6.49)$$

in which N is the total number of molecules in unit volume, F is the fraction of these which are in the particular vibration state observed, g is the total degeneracy of the state J, E_J is the rotational energy, and Q_r is the rotational partition function (Part 5, Chap. 2). The dipole matrix elements also depend upon the type of molecules and upon the particular rotational state.

Line Shapes. Under most experimental conditions the shapes and widths of microwave absorption lines are determined almost completely by interactions of absorbing particles with their close neighbors. In gases at pressures lower than an atmosphere the significant interaction is through collisions. Collision broadening is the most important type of broadening of lines of gases. As the pressure of the gas is increased, the collision rate increases and the line is broadened. Hence collision broadening is often called pressure broadening. In solids, lattice vibrations and direct interactions of neighboring dipoles are responsible for the greater part of the broadening, although exchange interaction plays an important role in determining the shape and width of electronic paramagnetic resonance lines (Part 7, Chap. 4).

Since spontaneous emission is extremely small at microwave frequencies, the natural lifetime in a given state is very long, and hence the natural line width is small. Natural widths are much too small to be detected at microwave frequencies. Doppler broadening is insignificant in comparison to collision broadening except at very low pressures. The Doppler width is

$$2\Delta\nu = 71.5 \times 10^{-8}\nu_0 \left(\frac{T}{M}\right)^{1/2} \qquad (6.50)$$

in which M is the molecular weight, T the absolute temperature, and ν_0 the peak resonant frequency.

For collision-broadened lines at pressures ranging from tenths of millimeters to several centimeters of Hg, the observed shape is given by the Van Vleck-Weisskopf [49] shape function

$$S(\nu,\nu_0) = \frac{\nu}{\pi\nu_0}\left[\frac{\Delta\nu}{(\nu_0-\nu)^2+(\Delta\nu)^2} + \frac{\Delta\nu}{(\nu_0+\nu)^2+(\Delta\nu)^2}\right] \qquad (6.51)$$

In the expression, $\Delta\nu$ is the line breadth or half of the width measured between the half intensity points. If τ is defined as the mean time between collisions that end the lifetime in the state, then $\Delta\nu = 1/2\pi\tau$. For sharp lines where $\nu \approx \nu_0$, this reduces to the Lorentz shape function

$$S(\nu,\nu_0) = \frac{1}{\pi}\left[\frac{\Delta\nu}{(\nu_0-\nu)^2+(\Delta\nu)^2}\right] \qquad (6.52)$$

which is applicable to most microwave lines of gases at the pressures commonly employed.

At peak absorption the shape function equals $1/\pi\Delta\nu$. This value for S in (6.48) gives the peak absorption coefficient. Over the pressure range in which most gaseous absorption lines are measured, about 10^{-2} to 10^2 mm of Hg, line widths $\Delta\nu$ are found to be directly proportional to pressure. Under these conditions the peak absorption does not change with pressure.

10. Electronic Magnetic Resonance in Solids

Paramagnetic Resonance. This important class of spectra is created by imposition of a magnetic field upon a paramagnetic substance. The observed transitions occur between the Zeeman components of particular internal levels.

Electronic magnetic resonance was observed first by Zavoisky [50] in 1945. These resonances are not peculiar to the microwave region but can be observed, when not too broad, with smaller fields at lower radio frequencies. For most paramagnetic substances the resonances can be made to fall in the microwave region with fields of a few kilogauss, and it is in this region that most of these resonances have been observed. Because the strength of absorption increases rapidly with frequency (6.48), it is usually desirable to use microwave rather than lower frequencies. Also, the frequency used must be large compared with line width if the proper line shape is to be obtained.

If g represents the spectroscopic splitting factor, the separations of the Zeeman levels are given by

$$E_H = -g\beta MH \qquad (6.53)$$

where β is the Bohr magneton and M the magnetic quantum number, which measures in integral values of $h/2\pi$ the component of moment along $\mathbf{H}$, the applied field. The selection rule for absorption of radiation is $M \to M + 1$. Double jumps $M \to M + 2$ are sometimes observed, but with much lower intensity. The frequency of absorption is

$$\nu = \frac{g\beta H}{h}$$

or

$$\nu \text{ (cps)} = 1.400 \times 10^6 \, gH \text{ (gauss)} \qquad (6.54)$$

If the paramagnetic resonance were observed for a free atom with Russell-Saunders coupling, the g factor would be given by the Landé factor

$$g_J = 1 + \frac{J(J+1) + S(S+1) - L(L+1)}{2J(J+1)} \qquad (6.55)$$

which has still to be multiplied by $g_S/2$ where $g_S = 2.0023$ for the electron. This value is never observed in paramagnetic resonance of solids although in certain of the rare earth salts it is approximated. In iron group salts the experimentally observed g factor approximates much more nearly the free electronic spin value. Departures from the free-ion values are in the main caused by interactions of the internal electric field which partly quench the orbital angular momentum, leaving only the electron spin free to precess about the external field. These internal electric fields also affect the electric spin precession indirectly, through spin-orbit coupling. Hence paramagnetic resonance can be used to study internal crystalline fields. The nature and origin of magnetism in solids, nuclear moments, exchange interaction, and similar phenomena can also be investigated this way.

Because of various internal interactions in solids, the experimental g factor obtained from (6.54) does not correspond to a true gyromagnetic ratio. Kittel has suggested the "spectroscopic splitting factor" to designate the g observed in this type of measurement.

Although paramagnetic ions in solids are by no means isolated, it has been found possible to treat them as individual systems, perturbed by the fields of neighboring ions. One starts with the unperturbed levels of the corresponding free ions and calculates the changes which would be brought about by the crystalline field expected for the particular substance. If the symmetry of the crystalline field at the ion is unknown, various symmetries are assumed until one is obtained which predicts results in agreement with observation.

The symmetry of the field at the ion considered, is not necessarily the symmetry of the macroscopic crystal itself [51]. The former symmetry depends primarily upon the locations and types of ions or dipoles immediately surrounding the ion considered. There may be more than one paramagnetic ion in unit cell, with individual crystalline axes oriented differently, each having its own g value. It is often possible to predict the approximate crystalline field symmetry from X-ray data or other structural information.

It is usually possible to account for the microwave results by assumption of a predominantly cubic field, with a weaker component of lower symmetry superimposed. Frequently, the field of lower symmetry has axial symmetry, trigonal or tetragonal, but often it is rhombic.

Salts of the Iron Group. The various factors which constitute the internal interactions of the iron group salts are indicated in the order of decreasing strength by the Hamiltonian of Abragam and Pryce [52],

$$W = W_F + V + W_{LS} + W_{SS} + \beta \mathbf{H} \cdot (\mathbf{L} + 2\mathbf{S}) + W_I - g_I\beta_I \mathbf{H} \cdot \mathbf{I} \qquad (6.56)$$

In this expression W_F ($\sim 10^5$ cm^{-1}) represents the energy for the corresponding free ion; V ($\sim 10^4$ cm^{-1}), the electrostatic energy of the ion in the crystalline field; W_{LS} ($\sim 10^2$ cm^{-1}), the spin-orbit interaction; W_{SS} (~ 1 cm^{-1}), the electronic spin-spin interaction; $\beta\mathbf{H} \cdot (\mathbf{L} + 2\mathbf{S})$ (about a microwave quantum with customary H), the interaction of the electronic magnetic moment with the external field; W_I ($\sim 10^{-2}$ cm^{-1}), the interaction of nuclear magnetic and electric quadrupole moments with the internal field; and $g_I\beta_I\mathbf{H} \cdot \mathbf{I}$ ($\sim 10^{-3}$ cm^{-1} with customary H), interaction of the nuclear magnetic moment with H.

From the orders of magnitude given for an average iron group salt, one needs in the first approximation to consider only the lowest levels of $W0$ and V, since these will be the only ones significantly populated at room temperature ($kTh \approx 200$ cm^{-1}). Slight mixing or interaction of lower with higher levels has important consequences in higher-order perturbation effects of the ground levels.

The crystalline field interaction is in the nature of an internal Stark splitting of the orbital levels. The energy of a field-free ion is independent of the orientation of its electronic orbitals. In the strong crystalline fields the L orbital degeneracy is lifted with wide intervals between the levels, except for some degeneracies which may remain in fields of certain symmetries. The crystalline field in the iron group salts is usually sufficiently strong to break down spin-orbit coupling, leaving in the first approximation the $2S + 1$ spin orientation degeneracy of each orbital level. The spin components of the lowest (hence populated) Stark (orbital) level are the ones which give rise to microwave absorption. Even the spin orientation degeneracy is frequently lifted wholly or in part by indirect interaction of the crystalline field by way of residual spin-orbit coupling. When spin degeneracy is completely lifted by internal interactions with intervals larger than the microwave quantum, it is not possible to observe microwave paramagnetic resonance. An important theorem [53] shows that when the number of electrons of the ion is odd it is not possible for all spin degeneracies to be lifted by internal fields. There always remains a double spin degeneracy of each orbital level which can be lifted only by an externally imposed H. For odd electron ions it should then always be possible to observe microwave paramagnetic resonance, provided the resonance is not too broad for detection.

The magnitude of the splitting of the spin multiplets by the residual spin-orbit coupling depends on separation of orbital levels. The stronger the internal Stark interaction in comparison with spin-orbit interaction, the more effectively will spin-orbit coupling be broken down and the smaller will be the decomposition of the spin multiplets. Approximately,

$$\Delta_S \approx \frac{\lambda^2}{\Delta_L} \qquad (6.57)$$

in which Δ_S represents the separation of the spin components, Δ_L the separation of the orbital levels, and λ the spin-orbit coupling. The residual spin-orbit coupling produces marked effects on the line width also.

An important consequence of the residual spin-orbit coupling is anisotropy in the g factor. When the crystalline field has lower than cubic symmetry, the magnetic susceptibility, and hence the g factor, is different for different orientations of the crystal. Even a small field component of tetragonal or trigonal symmetry superimposed upon a strong cubic field can be detected by measurement of the g factor at different crystal orientations. When there is more than one ion in unit cell (with crystalline axes oriented differently), the analysis is more complicated.

An appropriate example of the iron group salts is copper sulfate, $CuSO_4 . 5H_2O$, which has two Cu^{++} ions in the unit cell. Each ion is in a strong cubic field, with an axially symmetric tetragonal component superimposed. The two crystalline axes have a relative orientation of 82°. Theoretical expressions [54] for the g factors parallel ($g_\parallel$) and perpendicular ($g_\perp$) to the tetragonal axis are

$$g_\parallel = 2 - \frac{8\lambda}{E_4 - E_2} \qquad g_\perp = 2 - \frac{2\lambda}{E_5 - E_2} \tag{6.58}$$

in which E_3 is the lowest orbital level, E_4 and E_5 are higher orbital levels, and λ is the spin-orbit coupling constant. Cu^{++}, with $^2D_{5/2}$ ground state, has only one unpaired electron. Hence, the spin degeneracy 2 is not lifted by the internal field (Kramers' degeneracy). The lowest orbital level is a "nonmagnetic" doublet. Microwave absorption results essentially from the flipping of the electron-spin vector in the magnetic field. Microwave measurement of $g_\perp = 2.06$ and $g_\parallel = 2.47$ shows [55] the splitting of the orbital levels to be

$$E_4 - E_3 = 14{,}000 \text{ cm}^{-1}$$

and $E_5 - E_3 = 2{,}800$ cm^{-1}. Numerous other cases have been investigated [3].

In general when the orbital motions are quenched and the g factor has axial symmetry, the Hamiltonian can be reduced [52] to

$$\begin{aligned} W = \beta[g_\parallel H_z S_z + g_\perp(H_x S_x + H_y S_y)] \\ + D[S_z^2 - \tfrac{1}{3}S(S+1)] + AS_z I_z + B(S_x I_x + S_y I_y) \\ + Q'[I_z^2 - \tfrac{1}{3}I(I+1)] - g_I\beta_I \mathbf{H}\cdot\mathbf{I} \end{aligned} \tag{6.59}$$

The first bracketed term represents the interaction of the field H with the electronic spin; the second bracketed term represents the initial splitting of the spin levels by internal fields; the terms in A and B represent the interaction of the nuclear magnetic moment with the electronic spin; the term in Q represents the nuclear quadrupole perturbations, and the last term represents the interaction of the nuclear magnetic moment with H. When the lowest orbital level is not single, S in this expression is treated as an effective spin obtained when $(2S + 1)$ is put equal to the total degeneracy of the level. This Hamiltonian has been used [56] to calculate the fine and hyperfine spectrum for this type of salt. This theory applies to the common case in which the field is predominately cubic but has a small axially symmetric component. Comprehensive reviews of the paramagnetic resonance of the iron group salts and other substances are given by Bleaney and Stevens [57] and by Bowers and Owen [58].

Rare Earth Salts. The unpaired electrons which give rise to paramagnetic resonance in rare earth salts are in subvalence shells and hence partly shielded from the crystalline field. Because the spin-orbit coupling in these elements is also large, $\lambda > \Delta_L$. Spin-orbit coupling is not then effectively broken down. As a starting approximation one might assume that **S** and **L** form a resultant **J** and that the external magnetic field decomposes, the $2J + 1$ multiplet. However, internal perturbations are so strong that the free-ion approximation is not adequate. The $2J + 1$ components are decomposed either wholly or partly by internal fields. If the $2J + 1$ levels are all separated by the internal interaction at intervals wide compared with the microwave quantum, microwave resonance cannot be observed. However, resonance has been found in a number of the rare earth salts. A theoretical treatment of the decomposition of the $2J + 1$ levels has been made [59]. Bleaney and others have measured several unknown nuclear spins from paramagnetic resonance in the rare earth salts (Table 6.5). Results for the rare earths are summarized by Bowers and Owen [58].

The fact that spin-orbit coupling is not broken down in the rare earths makes the spin-orbit relaxation time short and the lines very broad and difficult to detect except at low temperature. When $L = 0$, as in Gd^{+++}, this does not hold, of course, and the lines can be detected at room temperature.

Actinide Elements. A number of elements of the actinide group, including uranyl, peptunyl, and plutonyl compounds, have now been investigated with the paramagnetic resonance method. The results on these are summarized by Bleaney [60].

Irradiated Crystals. Paramagnetic resonances have been detected in LiF and KCl crystals irradiated with neutrons by Hutchinson [61] and in similar crystals irradiated with X rays by Schneider and England [62]. The absorption apparently arises from unpaired electrons in F centers in the crystals. These resonances are moderately sharp, ~75 gauss half width, and the g factors are very close to that of the free electron spin. Kip et al. [63] have observed F-center resonance in irradiated NaCl, KCl, NaBr, and KBr and have interpreted the line width as arising partly from unresolved hyperfine structure. Upon irradiation and observation of the samples at 93°K, Kanzig [64] found a resonance with a widely spaced hyperfine structure for LiF, NaCl, KCl, and KBr, which he showed is due to the halogen molecule ion F_2^-, Cl_2^-, or Br_2^-.

Organic Free Radicals. Resonances in these are generally found to be very sharp (half width ~1 gauss) with g factors very close to that, 2.0023, for the free electron spin. Such resonances were first detected in diphenyltrinitro phenyl hydroxyl [65]. When the radicals are diluted in nonparamagnetic solvents, hyperfine structure [66] is often observable.

Paramagnetic resonance has been observed [67]

in a large number of biologically significant substances, including amino acids and proteins, after irradiation with X rays. The resonances often exhibit a hyperfine structure arising from proton interaction which allows identification of the radicals produced by the irradiation.

Electron resonances, with hyperfine structure, have also been observed in X-irradiated plastics [68].

Alkali Metals in Solution. Resonances of free electrons in solutions of alkali metals in ammonia and certain other solvents have been found [69]. These are extremely sharp (width ~0.1 gauss), and the g factor almost exactly that of free electron spin.

Conduction Electrons in Metals. Spin resonance of conduction electrons in metallic sodium has been detected [70]. The observed g is 2.004 and the line width 78 gauss. Small particles of sodium (diameter 10^{-3} to 10^{-4} cm) dispersed by ultrasonic waves and suspended in paraffin wax were used.

Hyperfine Structure in Paramagnetic Resonance. The general theory of nuclear effects on paramagnetic resonance is complicated, although certain cases are simple. When the g factor has axial symmetry, and spacing of the magnetic hyperfine structure is small as compared with the over-all spacing of the Zeeman levels and when, in addition, nuclear quadrupole interaction is small compared with nuclear magnetic interactions, hyperfine structure of a single coupling nucleus with spin I to a good approximation consists of $2I + 1$ equally spaced, equally intense, components with displacement

$$\Delta_\nu = \frac{KM_I}{h} \tag{6.60}$$

$$M_I = I, I - 1, I - 2, \ldots, -I$$

and
$$K = \frac{1}{g}(A^2g_\parallel^2 \cos^2\theta + B^2g_\perp^2 \sin^2\theta)^{1/2}$$

in which A and B are constants in the Hamiltonian of (6.59), θ is the angle between H and the symmetry axes, and

$$g = (g_\parallel^2 \cos^2\theta + g_\perp^2 \sin^2\theta)^{1/2}$$

where $g_\parallel$ and $g_\perp$ are the values obtained with symmetry axes oriented parallel and perpendicular to H. The constants A and B contain as a factor the nuclear magnetic moment.

To a first approximation quadrupole interaction displaces all levels equally and hence does not affect spacing of the magnetic hyperfine components. In the higher-order approximation, quadrupole interactions produce observable effects useful in evaluation of nuclear moments [56]. The nuclear spin I can be obtained from the magnetic hyperfine structure by counting the $2I + 1$ components. A number of important spin determinations have been made in this manner. Approximate values of nuclear magnetic and nuclear quadrupole moments obtained from hyperfine structure in paramagnetic resonance are given in Table 6.5.

Line Breadths. When the paramagnetic ions are concentrated, the internal magnetic field produced by magnetic moments of the ions becomes an important factor in broadening paramagnetic resonance. If a Gaussian shape is assumed, the line half width produced by this "spin-spin" broadening process [71] is

$$\Delta H_{1/2} = 1.18[(\Delta H)_{av}^2]^{1/2} \tag{6.61}$$

in which

$$\langle\Delta H^2\rangle_{av} = \tfrac{3}{4}g^2\beta^2S(S+1)\sum_j (1 - 3\cos^2\theta_{ji})^2 r_{ij}^{-6}$$

Here it is assumed that the frequency is held constant and that the line contour is measured in terms of H. Since dipole-dipole interaction falls off rapidly, the summation is taken only over the close neighbors j of ion i. The angle θ_{ji} is the angle between H and the line joining ions j and i. Since spin-spin broadening arises from a spread in the internal magnetic field, it cannot be significantly reduced by cooling, though it can be made inconsequential by dilution of the magnetic ions.

In most paramagnetic experiments, the strongest broadening factor is the spin-lattice interaction. The spread in the energy levels is a function of the lifetime in the state. The spin-lattice relaxation time τ gives rise to a width $\Delta\nu \approx 1/2\pi\tau$. Since the spin-lattice relaxation time τ is short, the resonance is broad when spin-orbit coupling is incompletely broken down. Then the spin orientation energy can be converted readily to the lattice vibration energy. Quantitative calculation of spin-lattice relaxation is very difficult. This type of broadening can be reduced to insignificance by cooling to low temperatures. In rare earth salts, for which $\Delta_L < \lambda$, the spin-lattice relaxation is so short that cooling is usually necessary to make the lines sharp enough for detection. When $\Delta_L \gg \lambda$, as in many iron group salts, τ is sufficiently long that the resonances are easily detectable at room temperature.

Exchange interaction between like ions (those which have the same Zeeman energies) tends to *sharpen* the resonances and give them a peaked shape. This effect is called exchange narrowing [71]. Physically, it may be attributed to averaging out of slowly varying internal-field components (which broaden the line) by rapid exchange of electrons between ions. Only isotropic exchange narrows the lines. If the electron spins have different g factors or Zeeman energies on the different ions, the exchange will tend to average the two different Zeeman frequencies and thus will broaden the line.

Ferromagnetic Resonance. Magnetic resonance in ferromagnetic materials was observed first by Griffiths [72] and has been investigated in a number of substances. The resonances are entirely analogous to those in paramagnetic salts, but the resonance formula must be modified to take into account the demagnetization field [73] with the result that the ferromagnetic resonance frequency depends upon shape of the sample. For small spherical samples the resonance frequency is given by (6.54). For a plane specimen with both H and the microwave magnetic vector in the surface plane, the resonance frequency is given by

$$\nu = \frac{g\beta}{h}(BH)^{1/2} \tag{6.62}$$

where g is the spectroscopic splitting factor, close to the free spin value 2, β the Bohr magneton, H the applied field, and B the magnetic induction in the sample.

More generally, when H is along the z axis and the

microwave magnetic component along x, then

$$\nu = \frac{g\beta}{h}[\{H_z + (N_y - N_z)M\}\{H_z + (N_x - N_z)M_z\}]^{1/2} \tag{6.63}$$

M_z represents the magnetization along z, and N_x, N_y, and N_z represent the demagnetization factors along x, y, and z, respectively. This assumes uniform magnetization.

Results so far obtained agree well with the theory of Kittel. Because of effects of coupling of the angular momentum vector of the electron to the crystal lattice and of residual spin-orbit coupling, observed g factors do not correspond exactly to gyromagnetic ratios and are called spectroscopic splitting factors.

Exchange interaction affects measurably [74] both the ferromagnetic resonance frequency and line width at low temperatures where the skin depth becomes small. At room temperature where the sample is uniformly magnetized, effects of exchange interaction on resonant frequency are not significant.

Bibliography

Bleaney, B.: *Reports on Progress in Physics*, 1948.
Coles, D. K.: *Advances in Electronics*, **II**: 299 (1950).
Gordy, W.: *Revs. Mod. Phys.*, **20**: 668 (1948).
Gordy, W., W. V. Smith, and R. F. Trambarulo: "Microwave Spectroscopy," Wiley, New York, 1953.
Herzberg, G.: "Infrared and Raman Spectra of Polyatomic Molecules," Van Nostrand, Princeton, N.J., 1945.
Herzberg, G.: "Spectra of Diatomic Molecules," 2d ed., Van Nostrand, Princeton, N.J., 1950.
Maier, W.: *Ergeb. exakt. Naturwiss.*, **XXIV**: 275 (1951).
Strandberg, M. W. P.: "Microwave Spectroscopy," Methuen, London, 1954.
Townes, C. H., and A. L. Schawlow: "Microwave Spectroscopy," McGraw-Hill, New York, 1955.
Wilson, E. B., Jr.: *Annual Review of Physical Chemistry*, Annual Reviews, Inc., 1951.

References

1. King, W. C., and W. Gordy: *Phys. Rev.*, **90**: 319 (1953); **93**: 407 (1954).
2. Burrus, C. A., and W. Gordy: *Phys. Rev.*, **93**: 897 (1954); **101**: 599 (1956).
3. Nichols, E. F., and J. D. Tear: *Phys. Rev.*, **21**: 378 (1923); **21**: 587 (1923).
4. Cleeton, C. E., and N. H. Williams: *Phys. Rev.*, **45**: 234 (1934).
5a. Burrus, C. A., and W. Gordy: *Phys. Rev.*, **93**: 897 (1954). For experimental methods of the millimeter and submillimeter region, see W. C. King and W. Gordy: *Phys. Rev.*, **90**: 319 (1953).
5b. Cowan, M., and W. Gordy: *Phys. Rev.*, **104**: 551 (1956).
5c. Jones, G., and W. Gordy: *Phys. Rev.*, **135**: A295 (1964).
6. Genzel, L., and W. Eckhardt: *Z. Physik*, **139**: 592 (1954).
7. Gordy, W., W. V. Smith, and R. Trambarulo: "Microwave Spectroscopy," Wiley, New York, 1953.
8. Lamb, W. E., and R. C. Retherford: *Phys. Rev.*, **72**: 241 (1947). Retherford, R. C., and W. E. Lamb: *Phys. Rev.*, **75**: 1325 (1949).
9. Bethe, H. A.: *Phys. Rev.*, **72**: 339 (1947).
10. Lamb, W. E., and M. Skinner: *Phys. Rev.*, **78**: 539 (1950).
11. Casimir, H. B. G.: "On the Interaction between Atomic Nuclei and Electrons," Teyler's Tweede Genootschap, E. F. Bohn, Haarlem, 1936.
12. Ewen, H. I., and E. M. Purcell: *Phys. Rev.*, **83**: 881 (1951). Muller, C. A., and J. H. Oort: *Nature*, **168**: 357 (1952).
13. Shulman, R. G., and C. H. Townes: *Phys. Rev.*, **77**: 421 (1950).
14. Slawsky, Z. I., and D. M. Dennison: *J. Chem. Phys.*, **7**: 509 (1941).
15. Nielsen, H. H.: *Phys. Rev.*, **77**: 130 (1950).
16. Hainer, R. M., P. C. Cross, and G. W. King: *J. Chem. Phys.*, **11**: 27 (1943); **12**: 210 (1944); **17**: 826 (1949).
17. Hund, F.: *Z. Physik*, **43**: 805 (1927).
18. Dennison, D. M., and G. E. Uhlenbeck: *Phys. Rev.*, **41**: 313 (1932).
19. Lyons, H., L. J. Rueger, R. G. Nuckolls, and M. Kessler: *Phys. Rev.*, **81**: 630 (1951).
20. Loubser, J. H. N., and J. A. Klein: *Phys. Rev.*, **78**: 348A (1950).
21. Bleaney, B., and R. P. Penrose: *Nature*, **157**: 339 (1946).
22. Good, W. E.: *Phys. Rev.*, **70**: 213 (1946).
23. Beringer, R.: *Phys. Rev.*, **70**: 53 (1946).
24. Burkhalter, J. H., R. S. Anderson, W. V. Smith, and W. Gordy: *Phys. Rev.*, **77**: 152 (1950); **79**: 651 (1950). Anderson, R. S., C. M. Johnson, and W. Gordy: *Phys. Rev.*, **83**: 1061 (1951).
25. Schlapp, R.: *Phys. Rev.*, **51**: 342 (1937).
26. Mizushima, M., and R. M. Hill: *Phys. Rev.*, **93**: 745 (1954).
27a. Burrus, C. A., and W. Gordy: *Phys. Rev*, **92**: 1437. (1953).
27b. Gallagher, J. J., F. D. Bedard, and C. M. Johnson: *Phys. Rev.*, **93**: 729 (1954).
27c. Favero, P. G., A. Mirri and W. Gordy: *Phys. Rev.*, **114**: 1534 (1959).
28. Sanders, T. M., A. L. Schawlow, G. C. Dousanis, and C. H. Townes: *Phys. Rev.*, **89**: 1158 (1953).
29. Beringer, R.: *Ann. N. Y. Acad. Sci.*, **55**: 814 (1952).
30. De L. Kronig, R.: *Z. Physik*, **50**: 347 (1928).
31. Van Vleck, J. H.: *Phys. Rev.*, **33**: 467 (1929).
32. Mulliken, R. S., and A. Christy: *Phys. Rev.*, **38**: 87 (1931).
33. Herzberg, G.: "Spectra of Diatomic Molecules," 2d ed., Van Nostrand, Princeton, N.J., 1950.
34. Beringer, R., and J. G. Castle: *Phys. Rev.*, **75**: 1963 (1949). Castle, J. G., and R. Beringer: *Phys. Rev.*, **80**: 114 (1950).
35. Bragg, J. K.: *Phys. Rev.*, **74**: 533 (1948).
36. Cross, P. C., R. M. Hainer, and G. W. King: *J. Chem. Phys.*, **12**: 210 (1944).
37. Bragg, J. K., and S. Golden: *Phys. Rev.*, **75**: 735 (1949).
38. White, H. E.: "Introduction to Atomic Spectra," McGraw-Hill, New York, 1934.
39. Condon, E. U., and G. H. Shortley: "Theory of Atomic Spectra," Cambridge, New York and London, 1935.
40. Bardeen, J., and C. H. Townes: *Phys. Rev.*, **73**: 627, 1204 (1948).
41. Bardeen, J., and C. H. Townes: *Phys. Rev.*, **73**: 97 (1948).
42. Bersohn, R.: *J. Chem. Phys.*, **18**: 1124 (1950).
43. Mizushima, M., and T. Ito: *J. Chem. Phys.*, **19**: 739 (1951)
44. Townes, C. H., and B. P. Dailey: *J. Chem. Phys.*, **17**: 782 (1949).
45. Golden, S., and E. B. Wilson, Jr.: *J. Chem. Phys.*, **16**: 669 (1948).
46. Hughes, R. H., and E. B. Wilson, Jr.: *Phys. Rev.*, **71**: 562 (1947).
47. Gordy, W., O. R. Gilliam, and R. Livingston: *Phys. Rev.*, **76**: 443 (1949).
48. Jen, C. K.: *Phys. Rev.*, **76**: 1494 (1949).
49. Van Vleck, J. H., and V. F. Weisskopf: *Revs. Mod. Phys.*, **17**: 227 (1945).

Chapter 7

Electronic Structure of Molecules

By E. U. CONDON, University of Colorado

1. Energy Levels of Diatomic Molecules

The energy-level scheme of diatomic molecules is more complicated than for atoms because (1) the electrons move in a field lacking a center of symmetry and (2) the molecule as a whole can have quantized rotation-vibration energy levels. The large masses of the nuclei compared to that of the electrons is the basis of the Born-Oppenheimer [1]* approximation procedure by which the states of the electrons are worked out for the nuclei in fixed position. The energy levels and electronic wave functions will then depend parametrically on the assumed positions. The electronic energy-level function of the nuclear coordinates is then used as an effective potential energy function to determine the wave functions and energy levels of the motion of the heavy nuclei.

Because the electrons do not move in an effective central field, the quantum states of individual electrons cannot be assigned exact values of orbital angular momentum. However, in the nonrotating molecule the molecular axis is an axis of symmetry for the forces on the electrons, and so the resultant orbital angular momentum component along this axis is a constant of the motion and may be used as a quantum number for labeling molecular states. This is denoted by Λ, being the analogue of L in atomic spectra. In the nonrotating molecule the states with $+\Lambda$ and $-\Lambda$ have the same energy, giving rise to a twofold degeneracy except for the case of $\Lambda = 0$. In the rotating molecule gyroscopic reactions come into play that remove this degeneracy, producing a small splitting of the energy into two states which are appropriate linear combinations of the $+\Lambda$ and $-\Lambda$. This is known as Λ-type doubling. Because of the degeneracy in the sign of Λ it is customary to label states by the value of $|\Lambda|$, using a Greek letter code that is analogous to the code for L used in the atomic spectra:

$\|\Lambda\| = 0$	1	2	3	4	. . .
Σ	Π	Δ	Φ	Γ	. . .

The Russell-Saunders coupling approximation in atoms is that which starts by neglecting terms in the Hamiltonian representing spin-orbit couplings. This gives rise to the possibility of labeling each electronic state by a value of S, the resultant spin of all the electrons. A similar approximation in molecules leads to the possibility of labeling the electronic states by a definite S value: this determines the multiplicity $(2S + 1)$ and is attached to the Λ symbol as a left superscript. Thus $^3\Pi$ refers to an electronic state having $|\Lambda| = 1$ and $S = 1$.

In the nonrotating molecule the component of S along the molecule axis is also a constant of the motion. This quantum number, called Σ, can assume the values

$$\Sigma = S, S - 1, \ldots, -S$$

or $(2S + 1)$ different possibilities (Σ as quantum number is not to be confused with Σ meaning $\Lambda = 0$). The resultant (spin + orbital) component of electronic angular momentum along the figure axis is also a quantum number. It is customary to label the states with a quantum number Ω defined as

$$\Omega = |\,|\Lambda| + \Sigma| \tag{7.1}$$

and to write this as a subscript in the place occupied by J in the Russell-Saunders notation for an atomic term, thus $^3\Pi_2$ refers to a state with $\Omega = 2$.

The simplest kind of electronic state is $^1\Sigma$, with no resultant electronic spin or orbital angular momentum. For such a state there are no complications of its rotational motion due to gyroscopic actions of these angular momenta. The resultant angular momentum of the molecule as a whole is J. It consists entirely of the angular momentum of the rotation of the molecule about an axis normal to the figure axis.

The electronic energy $V_e(r)$ for a molecule AB as a function of internuclear distance for electronic states leading to stable molecule formation will be as in Fig. 7.1. At large distances, forces of interaction become negligible and the curve approaches an asymptotically constant value corresponding to the sum of the energies of the atoms A and B. The value of r for the minimum of $V_e(r)$ is written r_e and D_e is written for

$$D_e = V_e(\infty) - V_e(r_e) \tag{7.2}$$

the energy difference between the minimum and the asymptotic value.

Only in the case of a $^1\Sigma$ state is there a single $V_e(r)$ curve to represent the state. For $\Lambda = 0$ and S not zero, there is a close lying group of $(2S + 1)$ states all having nearly the same $V_e(r)$ curve, because the

* Numbers in brackets refer to References at end of chapter.

energy is nearly independent of the orientation of S relative to the figure axis. For $\Lambda \neq 0$, there will be a group of $2(2S + 1)$ states whose $V_e(r)$ curves lie quite close together.

These $V_e(r)$ curves usually have a relatively deep minimum so that to the first approximation the molecule behaves like a rigid rotator of moment of inertia $I_e = \mu r_e^2$ where $\mu^{-1} = \mu_A^{-1} + \mu_B^{-1}$ and so the rotational energy levels for a $^1\Sigma$ state are

$$hcB_eJ(J + 1) \text{ where } B_e = \frac{h}{8\pi^2 cI_e} \tag{7.3}$$

Near the minimum

$$V_e(r) = V_e(r_e) + \tfrac{1}{2}k_e(r - r_e)^2 + \cdots \tag{7.4}$$

so the vibrational motion will be like that of a harmonic oscillator whose levels are

$$hc\omega_e(v + \tfrac{1}{2}) \text{ where } hc\omega_e = (2\pi)^{-1}\sqrt{\frac{k_e}{\mu}} \tag{7.5}$$

in which v is the vibrational quantum number.

For other than $^1\Sigma$ states the rotation structure is complicated [2] by interaction between the electronic angular momentum and the angular momentum of molecular rotation.

More accurately, the vibrational levels are not equally spaced, because $V_e(r)$ is not accurately represented by the parabolic approximation. The interval between successive levels diminishes with increasing v. Also, when the molecule rotates with angular momentum J, there is an effective addition to $V_e(r)$ of the term $(\hbar^2/2\mu r^2)J(J + 1)$. This has the effect of swelling the molecule so that the rotational levels do not increase in spacing as rapidly as if the swelling did not occur. It is customary to represent the rotation vibration levels by

$$\begin{aligned} W(e,v,J) = V_e(r_e) + \omega_e hc[(v + \tfrac{1}{2}) - x_e(v + \tfrac{1}{2})^2 \\ + y_e(v + \tfrac{1}{2})^3 + \cdots] + B_{ev}hcJ(J + 1) \\ - D_{ev}hcJ^2(J + 1)^2 + \cdots \end{aligned} \tag{7.6}$$

in which B_{ev} depends on the vibrational quantum number.

$$B_{ev} = B_e - \alpha_e(v + \tfrac{1}{2}) + \gamma_e(v + \tfrac{1}{2})^2 + \cdots \tag{7.7}$$

In principle these expressions are infinite series, but in practice it is hardly necessary to go beyond the terms written.

The most complete calculations [3] relating the coefficients in the energy-level formula to the $V_e(r)$ curve are those of Dunham and of Sandeman. They based their work on a general power series for $V(r)$ in $\xi = (r - r_e)/r_e$.

Often it is found that the vibrational levels are given quite accurately by the closed expression

$$hc\omega_e[(v + \tfrac{1}{2}) - x_e(v + \tfrac{1}{2})^2] \tag{7.8}$$

This is rigorously the vibrational energy-level formula for the *Morse function* [4],

$$V_e(r) = V_e(r_e) + D_e\{1 - \exp[-a(r - r_e)]\}^2 \tag{7.9}$$

The relation of the constants in the energy level formula to the parameters D_e, a, and r_e is

$$\begin{aligned} \omega_e &= \frac{a}{2\pi}\left(\frac{2D_e}{\mu}\right)^{1/2} \qquad x_e = \frac{h\omega_e}{4D} \\ B_e &= \frac{h}{8\pi^2 I_e} \\ \alpha_e &= 6\left(\frac{x_e B_e^3}{\omega_e}\right)^{1/2} - \frac{6B_e^2}{\omega_e} \end{aligned} \tag{7.10}$$

The $V_e(r)$ curves are themselves independent of the values of μ_A and μ_B to a high order of approximation [5]. However, the rotation-vibration levels depend on μ in a definitely predicted way. The levels are different for different isotopic species of the same molecule AB, giving rise to the *isotope effect* in the band spectra of diatomic molecules. Study of the isotope effect has been of great importance because it gives direct experimental evidence for the appearance of $(v + \frac{1}{2})$ in the energy-level formula instead of the v of the older quantum theory and also because it was the means of discovery of the rarer natural isotopes of oxygen, O^{17} and O^{18}, carbon, C^{13}, and nitrogen, N^{15}.

2. Electronic Band Spectra of Diatomic Molecules

The emission and absorption spectra of diatomic molecules involve transitions between levels of the electronic rotation-vibration level scheme. A *band system* consists of the ensemble of all transitions between the same upper, e', and lower, e'', electronic levels, regarding the electronic fine-structure group as being one.

A *band* is the ensemble of transitions within a *system* corresponding to a particular pair of upper, v', and lower, v'', vibrational quantum numbers. It is customary to display data about a band system in a double-entry table in which v' labels the rows and v'' the columns. This is called the band system *array*. A group of bands having the same v' is called a v' *progression*, while those having the same v'' are called a v'' progression. Under most conditions of excitation for emission a number of v' progressions will be emitted but when I_2 is excited by absorption of the green Hg line the molecules are put into one particular vibration level and so emit a single v' progression. Likewise for a cold gas essentially all the molecules will be in the $v = 0$ state and so only the $v'' = 0$ progression appears in the absorption spectrum. A group of bands for which $(v' - v'')$ has a fixed value and therefore lying on a diagonal in the system array is called a band *sequence*.

The lines which make up a single band come from the ensemble of all possible changes in the rotational quantum number J and associated changes in other angular momentum quantum numbers involved in the fine structure. The band lines are severely limited by the *selection rules* on these quantum numbers.

The ensemble of lines resulting from a particular change in J is called a *branch* of the band. As J is always restricted by the rule $\Delta J = 0, \pm 1$, there can be at most three branches, which are known as the P, Q, and R branches

$$\begin{aligned} &P \text{ branch: } (J-1) \to J \text{ in emission} \\ &Q \text{ branch: } \quad J \to J \text{ (except } 0 \to 0) \\ &R \text{ branch: } (J+1) \to J \text{ in emission} \end{aligned} \tag{7.11}$$

In bands involving $\Lambda = 0$ in initial and final states the Q branch does not occur.

As to the electronic quantum numbers, the selection rule on S is that S does not change. As in atomic spectra this rule is only approximately valid and is violated for heavy molecules. The selection rule on Λ is that $\Delta\Lambda = 0, \pm 1$, which is like that for M_L in atoms. In cases in which Σ is a good quantum number the selection rule is $\Delta\Sigma = 0$ analogous to the rule for M_s in atoms. The selection rule on Ω is $\Delta\Omega = 0, \pm 1$ analogous to the rule for M_J in atoms. The relative intensity of the different lines in the branches for different fine-structure conditions is considered thoroughly by Herzberg [6].

The $BJ(J + 1)$ formula for the rotational energy gives for the P and R branches a single parabolic formula:

$$\nu = \nu_0 + (B' + B'')m + (B' - B'')m^2 \tag{7.12}$$

in which $m = -J$ for the P branch and $m = J + 1$ for the R branch, and ν_0, called the *band origin*, is written for the contribution from the change in electronic and vibrational energies. In most cases a large number of values of m are involved, giving of the order of 100 lines in the band. The interval between adjacent lines becomes very small near

$$m = -\frac{(B' + B'')}{2(B' - B'')} \tag{7.13}$$

giving rise to a dense piling up of lines which is the most conspicuous feature of a spectrogram, known as the *band head*. The frequency of the head is

$$\nu_h = \nu_0 - \frac{(B' + B'')^2}{4(B' - B'')} \tag{7.14}$$

so if $B' > B''$, the head is at lower frequency than the origin and the band is shaded toward higher frequencies, whereas if $B' < B''$ the head is at higher frequency than the origin and the shading is opposite. If the change in B is small, then the head may be so far from the origin that it is not observed.

The Q branch is represented by the formula

$$\nu = \nu_0 + (B' - B'')m + (B' - B'')m^2 \tag{7.15}$$

where $m = J$ so that $d\nu/m$ occurs at $m = -\frac{1}{2}$, that is, the Q branch head is always close to the band origin.

In carrying out the analysis of a band it is convenient to plot the lines on a *Fortrat diagram* in which observed frequencies are plotted as abscissas against values of m as ordinates. The assignment of m is at first arbitrary but is normalized to make the two branches lie on a single curve, with no line corresponding to $m = 0$.

In work of high accuracy it is not necessary to rely on the approximately valid quadratic $BJ(J + 1)$ formula for the rotation levels. Let $F'(J)$ and $F''(J)$ be the term values (cm^{-1}) of the rotational levels in the upper and lower states and $P(J)$ $Q(J)$ $R(J)$ be the observed frequencies for the lines in the respective branches, where J is written for J''. Then

$$\begin{aligned} P(J) &= \nu_0 + F'(J-1) - F''(J) \\ Q(J) &= \nu_0 + F'(J) - F''(J) \\ R(J) &= \nu_0 + F'(J+1) - F''(J) \end{aligned}$$

Therefore

$$F'(J + 1) - F'(J - 1) = R(J) - P(J)$$

and

$$F''(J + 1) - F''(J - 1) = R(J - 1) - P(J + 1)$$

Since $F'(0) = F''(0) = 0$, these relations allow one to build up the values of $F'(J)$ and $F''(J)$ for even values of J from observed line frequencies. They also allow determination of $F'(J)$ and $F''(J)$ for odd values from observations except for $F'(1)$ and $F''(1)$ which remain undetermined if only P and R branch data are available. If Q branch data are also available

$$\begin{aligned} F'(J) - F'(J-1) &= Q(J) - P(J) \\ F''(J) - F''(J-1) &= R(J-1) - Q(J) \end{aligned}$$

giving the missing intercombination of odd and even values of J.

In a complete analysis the rotational levels for a given v' or v'' must agree when determined in this way from the lines of each band involving a particular vibrational level.

3. Franck-Condon Principle

Although the angular quantum numbers obey simple, fairly rigorous selection rules in radiative transitions, there is no such exact rule concerning changes in the vibrational quantum number v. Nevertheless it is possible to make some statements [7] about the changes in v, which show how they are related to the change in the potential energy functions occurring in the electron transition.

Franck observed that the absorption spectrum of I_2 vapor consists of a v'' progression of bands converging to a limit at about $\lambda = 5000$ A, after which there is a continuous absorption of light. This corresponds to a series of transitions from the $v'' = 0$ level of the normal electronic state to high values of v' of an excited electronic state, even reaching to the unquantized radial motions at higher energy than the dissociation limit of this excited state. Similar behavior is shown by Cl_2 and Br_2, but in many other molecules absorption produces no dissociation but only transitions to low vibration levels of the excited electronic state. Franck saw that this behavior of I_2 could be explained on simple mechanical grounds if the two potential energy curves were related as in Fig. 7.1. Initially the molecule is in the lower electronic state, not vibrating. Absorption of light occurs by a mechanism which directly affects the electronic structure, substituting a new law of force, represented by the upper curve, for the old one, and doing it so rapidly that the heavy nuclei cannot move appreciably in the process. Thus the favored transition will be to a vibrational state in which the internuclear distance in the instant after the transition is the same as before, that is, vertically upward on the diagram. Later it was possible to substitute definite knowledge of the two potential energy curves as obtained from analysis of the bands and to see that the curves really do lie in the relative location required for this explanation.

Condon extended this simple picture to a theory of the relative intensities of the bands in a system. In emission the relative intensities of the different v' progressions is determined by the conditions of excitation. But the relative intensities of the bands in any one v' progression depend on the relative magnitudes of the various $v' \rightarrow v''$ transition probabilities. Extending Franck's idea, Condon supposed that (1) the electron jump occurs with equal likelihood at all phases of the initial vibratory motion and (2) the favored transition is one in which there is no instantaneous change of nuclear momentum or position in the transition. Referring to Fig. 7.2, this

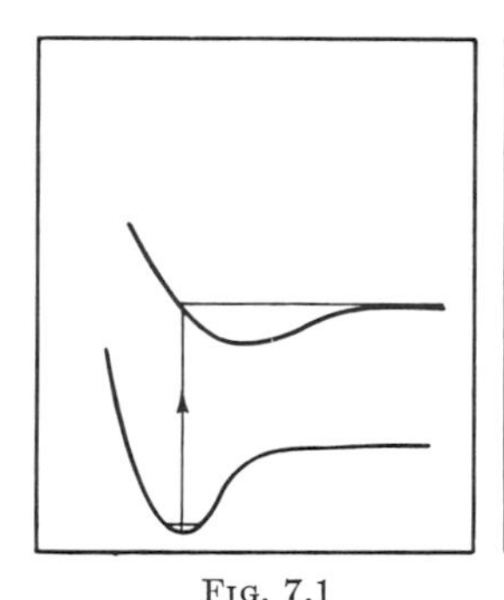

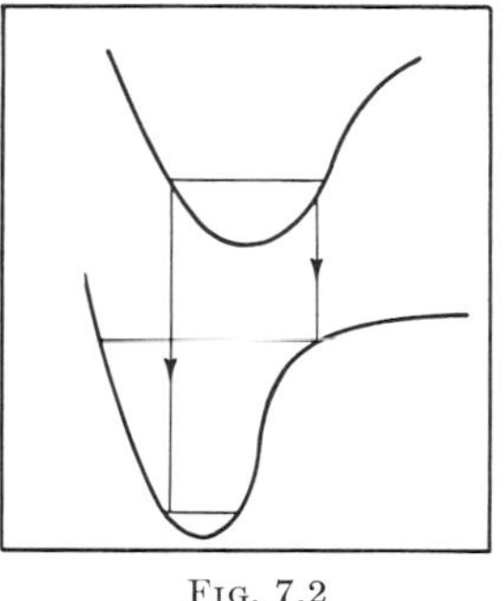

FIG. 7.1 FIG. 7.2

FIG. 7.1. Schematic relation of potential-energy curves involved in Franck's explanation of the photochemical dissociation of iodine molecules. Ordinates: electronic energy of molecule; abscissas: internuclear distance.
FIG. 7.2. Potential-energy curves illustrating Condon's extension of Franck's mechanism to cover the general case of prediction of most-favored vibrational transitions in a band system. Ordinates: electronic energy; abscissas: internuclear distance.

gives a geometrical construction by which the favored transitions can be found from a knowledge of the two potential energy curves. In particular if the two curves are similar and similarly located, then $\Delta v = 0$ is strongly favored, but if they are displaced, then the strong bands lie on an open and roughly parabolic curve on the (v',v'') array. This view has been tested on a large number of band systems and always found to agree with the facts on this semiquantitative basis.

Obviously the rules cannot be exact, for their strict application would lead to transitions to energies between the quantized allowed values of the final state. Likewise, for example, in I_2 there is quite appreciable probability of transition to a wide range of final vibrational energies and the simple classical picture gives no idea of what to expect about the extent of this range.

Applying quantum mechanics to the problem, let $R(e',v',r')/r$ be the radial wave function that describes the internuclear motion in the initial state and $R(e'',v'',r'')/r$ that for the final state. The transition probability is mainly governed by the square of the matrix component of electric-dipole moment P between initial and final states

$$|(e'v'J'|P|e''v''J'')|^2$$

In the calculation of this the integration over the electronic coordinates will give a function

$$A(e',e'') + B(e',e'')r + \cdots$$

that is approximately linear in r, so that the dominant factor for the (v,v) transition probability is

$$\left| \int_0^\infty (A + Br)R(e',v',r)R(e'',v'',r)\, dr \right|^2$$

The exact value of this integral depends on the relative value of B and A and on the exact nature of the radial wave functions. Qualitatively the integral will be small unless there is overlap of the two wave functions, that is, a range of values of r at which both have appreciable values. But the radial wave functions only have appreciable values inside the range of the classical motion and for a small range just outside. Also their largest values are near the ends of the range of the classical motion where the classical motion is slow and so, for observation at random phase, the nuclei are most likely to be found at this distance. Hence the transition probability will be large when the turning points of the classical motions are near the same value of r. Likewise R is a rapidly oscillating function at places where the velocity of the classical motion is great, so if such a rapidly oscillating part of $R(v'',r)$ overlaps a nonoscillatory part of $R(v',r)$ the integral will be small; that is the quantum-mechanical equivalent of the statement that no large changes in the vibrational momentum will directly accompany an electronic change of state.

Similar considerations apply if the final state corresponds to a continuum of nonquantized nuclear motions above the dissociation limit, or for transitions to electronic states which have a repulsive potential energy curve and therefore possess no discrete quantized states. Hydrogen, H_2, has a strong continuous emission spectrum in the far ultraviolet which is interpreted [8] as emitted in transitions from bound excited electronic states to the repulsive $1\,^3\Sigma$ state.

These ideas are applicable as well to the changes in nuclear motion which accompany an electronic transition caused by electron impact instead of light absorption. The dissociation energy of the normal state of molecular hydrogen is known to be 4.48 ev but there is no evidence of any dissociation when H_2 molecules are struck by electrons of this or slightly higher energy. If the electrons strike with energy of about 11 ev they are capable of causing a transition from the normal state to the repulsive $^3\Sigma$ state. The two nuclei find themselves near the normal distance $r_e = 0.74$ A and the extra 6 ev of energy is divided between the two H atoms in their normal state which are formed by dissociation of the H_2 molecule, so that each H atom is formed with 3 ev of translational kinetic energy.

4. Dissociation Energy

The energy needed to dissociate a molecule AB into the separated atoms, $A + B$, is the difference between the lowest vibrational-rotational energy level of the lowest electronic level, and the horizontal asymptote of the potential energy curve for that lowest level,

$$D_0 = V_e(\infty) - W(e,0,0)$$

It is less than D_e because of the zero-point vibrational energy which is $[W(e,0,0) - V_e(r_e)]$.

The thermodynamically defined ΔH of the reaction $A + B \rightarrow AB$ is a temperature-dependent quantity which involves the average thermal energies of AB and of the atoms A and B. However, ΔH approaches D_0 in the limit of low temperatures.

There is a quantity D_0 for each electronic state of the molecule. Each electronic state is characterized by the state of excitation in which the atoms A and B are left when the molecule dissociates from that electronic state, which is the level of the horizontal asymptote of the potential energy curve. Below $V_e(\infty)$ the vibrational levels are discrete, giving rise to discrete band structure. Above this limit, the radial motion is not quantized, giving a continuous spectrum.

If the transition probabilities are such that the absorption bands coming from the $v'' = 0$ level of the normal state are observed up to the limit of the quantized structure of an upper electronic state, then this limit, ν_l, may be observed spectroscopically. If also one knows ν_A, the excitation energy in which this dissociation leaves the atoms, then the dissociation energy of the normal electronic state is

$$D_0 = hc(\nu_l - \nu_A) \tag{7.16}$$

This method was applied by Franck [9] to the case of the visible absorption bands of the halogens. These gases appear colored because they absorb in the visible. In the case of F_2 the absorption is so far beyond ν_l that ν_l is not observed, but for Cl_2, Br_2, and I_2, the limiting ν_l is observed directly. The upper electronic state is one which dissociates into a normal halogen in the $^2P_{3/2}$ level and an excited halogen in the $^2P_{1/2}$ level and so ν_A is known from atomic spectra.

TABLE 7.1. DISSOCIATION ENERGIES OF HALOGENS

Gas	ν_l, cm^{-1}	A, cm^{-1}	D_0, ev	
			Spectrum	Thermochemical
Cl_2......	20,893	881	2.461	2.47
Br_2.....	19,575	3,685	1.960	2.00
I_2.......	20,037	7,593	1.534	1.50

The resulting values of D_0 shown in Table 7.1 for the normal state agree well with those obtained thermochemically; therefore the method is fully established for such a case.

Usually the transition probabilities are such that the vibrational levels are not observed all the way up to the *dissociation* limit. Birge and Sponer [10] introduced the method of extrapolating the vibrational levels to the place at which they show vanishing interval spacing, as a way of estimating ν_l. This is less certain than direct observation but sometimes gives useful results. It is convenient to make a graph of

$$\Delta\nu(v) = \frac{[W(e, v + 1, 0) - W(e, v, 0)]}{hc}$$

against v in connection with such an extrapolation. Sometimes it is more convenient [11] to plot $\Delta\nu(v)$ against $W(e,v,0)$ instead of v to carry out this extrapolation. In case the graph of $\Delta\nu(v)$ against v is linear, then $W(e,v,0)$ is quadratic in v, from which it may be inferred that the Morse curve represents the potential energy, and the value of D_0 can be obtained by fitting such a curve. In terms of its parameters,

$$D_0 = \frac{\omega_e}{4x_e}(1 - x_e)^2$$

Detailed application of spectroscopic data to the determination of dissociation energies has been rather uncertain for a number of the most important cases [12]. Important additional information is given by data on predissociation (Sec. 5).

Table 7.2 gives some of the important examples of D_0 or D_e in electron volts for the normal state.

5. Continuous and Diffuse Spectra. Predissociation

Continuous molecular absorption [14] arises if the potential energy curve of the upper electronic state lies out at sufficiently great internuclear distances relative to those characterizing the minimum of the normal state's potential energy curve. This is the case in gaseous F_2, for which only a continuous absorption spectrum is observed, with its intensity maximum at about 2750 A. Likewise continuous emission may occur in the same way. In particular the strong continuous emission of H_2 in the far ultraviolet is due [8] to transitions from stable upper electronic states of this molecule to the unstable $(1s\sigma)(1s\sigma)^*\ {}^3\Sigma$ state. In such transitions a considerable amount of the original excitation energy of the molecule is converted into kinetic energy of relative motion of the two H atoms produced. An intense far ultraviolet (600 to 1000 A) continuous spectrum in helium is similarly attributed to transitions from stable excited states of the He_2 molecule to the repulsive potential energy curve of the normal electronic state of He_2.

In the case of weakly bonded molecules, such as Hg_2, the potential energy curves are extremely shallow; so the absorption bands, both discrete and continuous, have a small range of frequencies and lie close to the absorption lines of the separated atoms. Such bands are also found in the absorption spectra of alkali metal vapors.

In addition to these truly continuous spectra, in which intensity is a smooth slowly varying function

TABLE 7.2. DISSOCIATION ENERGIES, EQUILIBRIUM LENGTHS, AND FORCE CONSTANTS OF CHEMICAL BONDS
[The following values are condensed from the critical review by T. L. Cottrell, "The Strength of Chemical Bonds," Academic Press Inc., New York, 1954. In polyatomic cases, $D(R_1,R_2)$ is the minimum energy difference for dissociation into $R_1 + R_2$.]

	Bond	D, kcal/mole	r_e, A	k, 10^5 dynes/cm
Hydrogen	HH	103.24	0.7417	5.733
Halogen acids	HF	134	0.917	9.655
	HCl	102.2	1.275	5.157
	HBr	86.5	1.414	4.116
	HI	70.5	1.604	3.141
Hydrides	LiH	~58	1.595	1.026
	BeH	~53	1.343	2.263
	BH	~70	1.233	3.03
	CH (diatomic)	80	1.130	4.484
	NH	~85	1.038	6.0
	OH	103	0.971	7.791
	NaH	47	1.887	0.781
	MgH	46	1.731	1.275
	AlH	67	1.646	1.620
	SiH (in SiH_4)	~76	1.456	2.77
	PH (in PH_3)	~77	1.42	3.2
	SH	85	1.34	4.20
	KH	43	2.244	0.561
	CaH	39	2.002	0.977
	RbH	39	2.367	0.515
	SrH	38	2.145	0.854
	CsH	42	2.494	0.467
	BaH	42	2.232	0.809
First row elements	Li_2	25	2.672	0.255
	LiF	137		
	BeO	124	1.331?	7.510?
	BeF	92?	1.361	5.767
	B_2	69?	1.589	3.583
	BC (in $B(CH_3)_3$)	(66)	1.56	
	BN	92?	1.281	8.328
	BO	~185	1.205	13.65
	BF	195?	1.262	8.045
	CC	137?	1.312	9.52
	CN	129?	1.172	16.29
	CO	224 ± 3?	1.128	19.02
	CF	~106		
	N_2	170.2	1.094	22.96
	NO	122?	1.151	15.94
	NF (in NF_3)	(56)	1.37	
	O_2	117.2	1.207	11.765
	OF (in F_2O)	(45.3)	1.41	5.27 or 4.26
Carbon bonds	CH	80	1.120	4.484
	(CH)H	~92		
	$(CH_2)H$	~87		
	$(CH_3)H$	101	1.094	5.394
	CH av	90.5	1.08	
	$(C_6H_5)H$	77.5		
	$(CCl_3)H$	90		
	C—C av	66.2	1.54	4.5
	C═C av	112.9	1.35	9.6
	C≡C av	150.3	1.21	15.6
	(CH)≡(CH)	166?	1.20	17.2
	(CH_2)═(CH_2)	125?	1.353	10.90
	(CH_3)—(CH_3)	83	1.54	4.57
	(CN)—(CN)	112	1.37	6.75 or 5.22
	(CH_3)—(CN)	103	1.460	5.2 or 4.94
	(CH_2)═(CO)	~80	1.30	9.8
	(CH_3)—(CO)	~17		
	(CF_3)—(CF_3)	124	1.52	5.45
	(CH_3)—(CF_3)	117	1.53	
	C—N av	(55.5)	(1.47)	
	C═N av	~112		
	C≡N av	(160.6)		(17.7)

TABLE 7.2. DISSOCIATION ENERGIES, EQUILIBRIUM LENGTHS, AND FORCE CONSTANTS OF CHEMICAL BONDS (*Continued*)

	Bond	D, kcal/mole	r_e, A	k, 10^5 dynes/cm
	(HC)≡N	164	1.153	18.583
	(CH_3)—(NH_2)	80	(1.47)	4.86 or 4.99
	(CO)═O	127	1.162	15.9
	C—O av	(77.1)		
	C═O aldehydes	(159)	1.22	
	C≡O ketones	(162)	1.22	
	(H_2C)═O	(149)	1.21	12.1
	(CH_3)—(OH)	90	(1.434)	5.77 or 4.99
	(C_2H_5)—(OH)	90		
	CF in CF_4	94?	1.36	9.14 or 5.4 or 4.32
	C—SI carborundum	64?		
	C—P	138?	1.562	7.830
	C—S diatomic	166?	1.534	8.488
	C═S (in CS_2)	121	1.554	(7.5)
	C—S av	59		
	C—Cl av	(73)	(1.76)	3.6
	C—Br av	(60)		
	C—I av	(53)	(2.139)	(2.3)
Oxides	(OH)—(OH)	48	(1.48)	(3.83)
	MgO	120?	1.749	3.484
	AlO	138?	1.618	5.660
	SiO	165	1.510	9.247
	SiO (in SiO_2)	(104)		
	PO	145?	1.449	9.409
	SO	119?	1.493	7.929
	S═O (in SO_2)	(128)	(1.43)	9.97
	S + O (in SO_3)	(113)	(1.43)	10.77 or 9.2
	ClO	63		4.0
	CaO	116?		2.84?
	ScO	138?		6.559
	TiO	160	1.620	7.184
	VO	127?	1.890	7.353
	CrO	97?		5.821
	MnO	92?		5.157
	FeO	92?		5.67
	NiO	<99		
	CuO	113		
	ZnO	<92		
	GaO	58?		4.516
	GeO	150?	1.651	7.525
	AsO	113		7.266
	AgO	32?		1.996
	BaO	129	1.940	3.786
	PbO	94?	1.922	4.557
Halogens	F_2	36	(1.435)	(4.45)
	Cl_2	57.07	1.988	3.286
	Br_2	45.46	2.284	2.458
	I_2	35.55	2.666	1.721
	FCl	60.5	1.628	4.56
	FBr	50.6	1.756	4.07
	FI	46?		3.64
	ClBr	52.1		(2.67)
	ClI	49.63	2.321	2.383
	BrI	41.90		2.064

of frequency, there are observed many cases of diffuse band spectra, that is, spectra in which some of the lines at least are quite wide, even under low-pressure discharge conditions where the possibility of line broadening by collisions is excluded. These are interpreted [14] as due to transitions between levels which are broadened by coupling of quasi-quantized states to the continuous energy levels of the molecule.

Clearly any energy level higher than the dissociation limit of the normal state is an allowed energy level for the molecule, for it can be realized by having the two dissociated atoms in their normal electronic states and moving relative to each other with any unquantized amount of kinetic energy. At considerably higher energies the continuum of allowed energies can be realized in additional ways, by having one or both of the atoms in various of their quantized excited states.

Thus it is generally true that the excited quantized levels of a molecule lie in the same range of energies as a possible continuum of levels of the molecule. Whenever a set of quantized levels overlaps in energy a range of continuous levels of the same system, it is possible that quantum-mechanical coupling between them will take place. This is produced [15] in the theory by the existence of nondiagonal matrix components of the Hamiltonian connecting states of the continuum with the quantized states of the same energy. Since the transitions so produced are between states of the same total energy, these are often called *radiationless transitions*. The same kind of coupling of quasi-quantized levels to a continuum occurs in the X-ray level scheme of atoms, giving rise to the *Auger effect* whereby an atom in an excited X-ray level may return to the normal state by ejecting a high-energy electron instead of radiating an X-ray quantum. For this reason all such transitions, even including those occurring in molecules, are sometimes called *Auger transitions*.

The coupling in question has the effect that if the system is in a discrete state, symbolized by the quantum number n, it has a probability per unit time γ of making transitions to the continuous state c of equal energy that is given to the first order by

$$\gamma = \frac{2\pi}{h} |(n|H|c)|^2$$

in which $(n|H|c)$ is the matrix component of the Hamiltonian connecting the quantized state and an appropriately normalized continuous state.

Such a coupling of the quantized state to the continuum reduces its mean life at least to γ^{-1}. This limits the length of wave train which can be emitted or absorbed in radiative transitions involving such a state. The shortened wave train manifests itself observationally as a broadened spectral line whose half width at half maximum in frequency units is equal to γ. The molecular spectroscopic effects associated with such transitions are known as *predissociation*.

A wide variety of special cases has been observed, all of which are reasonably interpreted along these lines. In the case of HgH, the $K(K + 1)\hbar^2/2\mu r^2$ term introduced by rotation into the effective potential energy function gives rise to a potential barrier separating the quasi-bound states above the dissociation limit from the dissociated continuum. Therefore these levels can be coupled to the continuum by barrier leakage [16] of the same kind as is used in interpreting radioactive decay of nuclei by alpha-particle emission. Similar barrier leakage effects are observed in the band spectra of AlH and AlD. In these cases interesting pressure effects are observed.

The coupling of the quantized level to the continuum also has as a consequence that two atoms colliding with the correct relative kinetic energy (that is, so the system is close in energy to a quasi-quantized or resonance level) have an unusually large probability of being united to form a stable molecule. Under low-pressure conditions, if the barrier leakage refers to the upper state, the higher rotation levels will dissociate by barrier leakage before radiating, and the high rotation lines of a band will be absent or very weak. As the pressure in the discharge is increased, more molecules are reformed into the high levels by the inverse of the dissociation process, so that the high rotation lines are considerably strengthened, although they remain diffuse.

Barrier leakage predissociation is limited to hydrides because of the strong dependence of barrier penetrability on the reduced mass of the molecule.

The more general case of predissociation is that in which there is coupling between the quasi-quantized levels of one electronic state with the continuous range of levels of another. In case the coupling becomes sufficiently great that γ is comparable with the rotation frequency of the molecule, that is, with the spacing of the lines in a band in frequency units, the diffuseness becomes so great as to wash out all observable rotational structure. This gives rise, in the absorption spectrum of S_2 vapor, to a progression of bands whose lower frequency members are quite sharp. Then abruptly, at a particular vibration level and higher ones of the upper electronic state, the bands become quite diffuse. This is because that vibration level lies above the dissociation limit of a different electronic state with which interaction takes place. Below the dissociation limit the matrix components of the Hamiltonian produce only small shifts in location of sharp levels, while above this limit the coupling is to a continuum, resulting in a great qualitative alteration of the structure of the level scheme.

Observation of such limits between sharp and diffuse bands thus serves to locate the dissociation limits of certain electronic states of molecules. In this way observations of predissociation are helpful in spectroscopic determination of molecular dissociation energies.

Because the effect is mainly determined by the interaction matrix component $(n|H|c)$, the phenomena of predissociation are restricted by electronic selection rules and the vibrational changes that can occur are governed by the Franck-Condon principle.

6. Hydrogen Molecule

Quantum mechanics provides a formulation of the problem of the chemical bond, in that the effective potential energy curves of diatomic molecules are given by the electronic energy levels as functions of the internuclear distance. For polyatomic molecules the potential energy curves become surfaces in a space whose coordinates are the relative positions of the nuclei. The mathematical problem is more complicated than that of determining the electronic levels of isolated atoms. The variety of cases encountered covers all the substances of interest in chemistry. Therefore in large part the theory of the chemical bond is based on qualitative and semi-quantitative reasoning derived from quantum-mechanical consideration of the electronic problem.

The only problem of this kind that has been dealt with exactly is H_2^+, the problem of one electron in the coulomb field of two nuclei [17]. It is important because it gives close agreement with experiment and illustrates many points of vital importance in all molecular structure problems. Experimentally the normal level of H_2^+ is known [18] as the limit of several series of electronic levels of H_2 in which one of the electrons is highly excited and therefore has, in the limit, no effect on the potential energy curve.

For one electron in the field of two protons a and b separated by distance R in atomic units the Hamiltonian is

$$H = -\frac{\hbar^2}{2m}\nabla^2 - \frac{e^2}{ar_a} - \frac{e^2}{ar_b} + \frac{e^2}{aR} \tag{7.17}$$

When $R \gg 1$, the system is twofold degenerate in its lowest state in that the electron can be in a 1*s* state around a or around b with wave functions

$$\begin{aligned} u_a &= (\pi)^{-1/2}e^{-r_a} \\ u_b &= (\pi)^{-1/2}e^{-r_b} \end{aligned} \tag{7.18}$$

For finite R these are not orthogonal, the overlap integral being

$$S = \int u_a u_b\, dv = e^{-R}\left(1 + R + \frac{R^2}{3}\right) \tag{7.19}$$

The symmetry of the problem is such that the Hamiltonian matrix component $H_{bb} = H_{aa}$. This is

$$H_{aa} = E_a + \frac{e^2}{aR} + C(R)\frac{e^2}{a} \tag{7.20}$$

in which $C(R)$ is called the Coulombic integral,

$$C(R) = \int r_b^{-1}u_a^2\, dv = -\frac{1}{R} - \frac{1+R}{R}e^{-2R} \tag{7.21}$$

This negative energy expresses the direct energy of interaction of an electron around one nucleus with the other proton. The nondiagonal matrix element is

$$H_{ab} = E_a + \frac{e^2}{aR}S + J\frac{e^2}{a} \tag{7.22}$$

in which J is the exchange integral

$$\begin{aligned} J &= -\int r_a^{-1}u_a u_b\, dv \\ &= -(1 + R)e^{-R} \end{aligned} \tag{7.23}$$

representing the coulomb interaction of either nucleus with the mixed charge density $u_a u_b$. $E_a = -\frac{1}{2}(e^2/a)$ is the unperturbed energy of the hydrogen atom in its normal state.

The roots of the secular equation in E

$$\begin{vmatrix} H_{aa} - E & H_{ab} - ES \\ H_{ab} - ES & H_{bb} - E \end{vmatrix} = 0$$

give the first-order perturbed energies as

$$(E_1, E_2) = \frac{e^2}{a}\left(-\frac{1}{2} + \frac{1}{R} + \frac{C \pm J}{1 \pm S}\right) \tag{7.24}$$

As C and J are both negative (interaction energy of an electron cloud with a proton), the lowest energy is E_1, corresponding to the upper sign. $E_1(R)$ gives a potential energy curve with a minimum because $(C + J)/(1 + S)$ is a little more negative than the l/R positive term representing the repulsion of the nuclei. It is this minimum which gives rise to chemical bonding to form stable H_2^+. $E_2(R)$ is everywhere repulsive, giving rise to no stable molecule formation in this state. In the range $2 < R < 3$, where $E_1(R)$ has its minimum, $J(R)$ is almost equal to $C(R)$; thus when the negative sign occurs the combined effect of the electronic energy terms is almost zero so that the full effect of the repulsion of the protons is active to prevent molecule formation.

The wave functions corresponding to these states in the zeroth approximation are

$$(U_1, U_2) = \frac{u_a \pm u_b}{(2 \pm 2S)^{1/2}} \tag{7.25}$$

so U_1 is symmetric with regard to interchange of a and b while U_2 is antisymmetric. The corresponding charge density is

$$(U_1^2, U_2^2) = (1 \pm S)^{-1}[\tfrac{1}{2}(u_a^2 + u_b^2) \pm u_a u_b] \tag{7.26}$$

In the range of R values for which $E_1(R)$ has its minimum the overlap charge density $u_a u_b$ has considerable values, in the region of space halfway between the two nuclei. For the stable state the electron has an extra probability of being localized here, whereas in the unstable state it has a reduced probability of being between the two nuclei. Thus the theory gives a picture of normal H_2^+ being held together by a one-electron chemical bond which derives its bond-forming ability from the effects of the exchange charge density $u_a u_b$. Here the exchange charge density is that arising from the possibility of a single electron being attached to either atom, rather than the possibility of its being in either of two quantum states in the same atom as in the exchange energies arising in the theory of atomic spectra.

More accurate calculations by Hylleraas and by Joffe using improved wave functions have given a theoretical $E_1(R)$ that is in complete agreement with experimental observations on this curve derived from the molecular spectrum of hydrogen.

In the case of the neutral H_2 molecule, the presence of two electrons makes the first-order [19] perturbation calculation more complicated and brings in important new features. Again there is a twofold degeneracy for large R in that electron 1 may be associated with nucleus a and 2 with b, or vice versa, giving the two starting states

$$u_a(1)\, u_b(2) \qquad \text{and} \qquad u_a(2)\, u_b(1)$$

But now arises a point where physical thinking supplements a formal perturbation theory approach. If the e^2/r_{12} repulsion of the two electrons were completely neglected in the starting approximation, then two more ionic starting states

$$u_a(1)\, u_a(2) \qquad \text{and} \qquad u_b(1)\, u_b(2)$$

corresponding to both electrons being attached to a or b, respectively, would have the same starting energy. However, the repulsion of the two electrons is so strong that these ionic states lie at high energy compared to the atomic states, and so they will not in fact make much contribution to the wave functions of the normal state of the molecule. Therefore in the method developed by Heitler and London they are omitted from the first approximation which is treated as derived from a linear combination of the two atomic states. Correction for the effects of the ionic states then can be brought in as a higher-order effect such as that of the inclusion of the effects of other higher quantum states of the interacting atoms.

It might appear that one could build up suitable two-electron wave functions for H_2 from the one-electron wave functions of H_2^+, getting a symmetric one

$$U_1(1)U_1(2)$$

to go with an antisymmetric spin factor to give the $^1\Sigma$ normal state, but this gives a large unwanted component of the ionic states and a poor approximation to the normal state wave function.

Using the two atomic starting functions, the overlap integral becomes S^2, the square of that for the one-electron functions (7.19). The energy levels are

$$(E_1,E_2) = \frac{H_{11} \pm H_{12}}{1 \pm S^2} \tag{7.27}$$

in which $H_{11} = H_{22}$ is the direct integral for the Hamiltonian

$$\begin{aligned} H_{11} &= \iint u_a(1)u_b(2)Hu_a(1)u_b(2)\, dv_1\, dv_2 \\ &= (2E_a + 2C + C_e)\frac{e^2}{a} \end{aligned} \tag{7.28}$$

where E_a is the atomic energy value and C is the Coulombic integral of the one-electron problem representing the interaction of each electron with the other nucleus than the one to which the wave function supposes it to be attached. It is given by (7.21), the same as in H_2^+. A new feature is the positive C_e representing the direct repulsive term for the Coulomb interaction of the two electrons, supposed attached to the two atoms,

$$\begin{aligned} C_e &= \iint r_{12}^{-1}u_a^2(1)u_b^2(2)\, dv_1\, dv_2 \\ &= \frac{1}{R} - e^{-2R}\left(\frac{1}{R} + \frac{11}{8} + \frac{3}{4}R + \frac{1}{16}R^2\right) \end{aligned} \tag{7.29}$$

The nondiagonal matrix element H_{12} also contains the same exchange integrals J as appeared in the H_2^+ problem and a new one J_e representing interaction of the electrons

$$H_{12} = (2E_aS^2 + 2JS + J_e)\frac{e^2}{a} \tag{7.30}$$

where

$$\begin{aligned} J_e &= \iint r_{12}^{-1}u_a(1)u_b(1)u_a(2)u_b(2)\, dv_1\, dv_2 \\ &= \tfrac{1}{5}\left\{-e^{-2R}(-\tfrac{25}{8} + \tfrac{23}{4}R + 3R^2 + \tfrac{1}{3}R^3)\right. \\ &\quad \left. + \frac{6}{R}[S^2(\gamma + \ln R) + S'^2Ei(-4R) - 2SS'Ei(-2R)\right\} \end{aligned}$$

in which $S' = e^R(1 - R + \frac{1}{3}R^2)$, $\gamma = 0.57722$, and $Ei(-x)$ is the exponential integral.

Again the situation is that the $E_1(R)$ curve gives rise to stable molecule formation and $E_2(R)$ gives a curve representing repulsive forces between the atoms at all distances. The wave function for $E_1(R)$ is symmetric in the two electrons and so must be associated with an antisymmetric spin factor giving a $^1\Sigma$ normal state. The other wave function is antisymmetric and so must be associated with a symmetric spin factor giving a $^3\Sigma$ unstable state. As in H_2^+ the bonding is made possible by the exchange charge density which gives an extra concentration of the electrons along the line joining the atoms. The bonding arises from the interaction of each electron with the nucleus to which it is not directly attached as in H_2^+, this action being doubled because there are two electrons but also reduced by the C_e and J_e terms representing the positive energy of interaction of the two electrons. The equations show that the bonding energy of H_2 is not simply double that of H_2^+ although it is roughly twice as great.

These calculations have been greatly improved [20] by various better choices of starting wave functions. Wang showed that a better result is obtained by using an effective Z on each nucleus which is not 1 and which varies with R to minimize the energy as an approximate way of allowing for change in the wave functions by mutual screening. Rosen considered that each atom will be polarized by the field of the other: this means the wave functions can be improved by introducing some $2p$ component in the atomic states. Coolidge and James made the most elaborate calculations for the normal state of H_2 and found exact agreement with experimental values.

Table 7.3 gives the experimental value of D_e (electron volts) and R_0 (in 10^{-8} cm) and the theoretical values calculated by various approximations.

TABLE 7.3. NORMAL D_e AND R_0 FOR H_2 MOLECULE

Experimental	D_e (ev)	R_0 (10^{-8} cm)
	4.72	0.740
Heitler-London	3.14	0.869
Screening (Wang)	3.76	0.743
Polarization (Rosen)	4.02	0.74
Ionic (Weinbaum)	4.10	0.749
Many-parameter (Coolidge-James)	4.27	0.740
	4.70	0.740

(The first many-parameter value is without inclusion of r_{12} terms in the molecular wave function, showing that about 10 per cent of D_e arises from such effects.)

The end result of this work is to give a strong conviction that quantum mechanics gives exact values but the labor of such calculations precludes their being made with high accuracy for many of the molecules of importance to chemistry.

7. Sketch of Chemical-bond Theory

Calculations like those made for H_2^+ and H_2 are too complicated to be feasible for other molecules, if the goal is exact description of electronic states of molecules. The energies of binding of atoms in molecules appear as small differences of much larger quantities, complicating the task of calculation.

Nevertheless a great deal of progress has been made [21] in understanding chemical valence forces by qualitative and semiquantitative extensions of the theory. The existence of stable molecules is due to their having lower energy states than the sum of the lowest states of the constituent atoms. Coulomb repulsion of nuclei always works against molecule formation. Molecules can be formed only if there are electronic levels which decrease in energy as the interatomic distances are diminished at a rate sufficient to overbalance the nuclear repulsion.

Repulsion of the nuclei is always reduced by the space charge of those electrons which are tightly bound to the nuclei, in inner closed shells. The closest distance of approach of nuclei in molecules is such that these inner shells remain well separated in space, and thus have little distorting influence on each other. This is the basis for the observed result that X-ray spectra arising from transitions between excited states of these inner shells are almost unaffected by the state of chemical combination of the atoms.

The essential thing therefore is the behavior of the outermost electrons of the atoms, when these come near each other. This is why these outermost electrons are called *valence electrons,* and why the chemical properties of an element are so closely correlated with the outer electronic structure as systematized in the periodic table (Chap. 1, Sec. 13). When two atoms, a and b, are brought closer together, their effective central fields V_a and V_b begin to overlap; thus $V_a + V_b$ lies lower than either V_a or V_b in the region near the bond axis or line of centers connecting the nucleus of a to that of b. Such a distortion of the effective field will result in a distortion of the atomic wave functions.

A wave function which in the separated atoms belonged entirely to a begins to reach out toward b, being drawn by the lower potential energy represented by V_b, and in a similar way the wave functions originally belonging entirely to b are drawn toward a by the action of V_a. Such individual electron states are lowered in energy if the wave function is one which has no node in the region between a and b (called the ab bond region), as was, for example, the case for the symmetrical molecular orbital wave function in the normal state of H_2^+. In this case the wave function is drawn strongly out into the ab bond region, and the energy of such an electron state becomes lower than it was in the separated atoms.

An electronic state that evolves in this way as the atoms come together is called a *bonding orbital.* Such a state might well be called *extroverted* because of the way it reaches out to widen its contacts with the other atom.

In contrast, some of the electronic states will be characterized by having a node in the bond region, as in the example of U_2 of the H_2^+ system (Sec. 6). In this case the node prevents the development of charge density in the bond region, and the electron tends therefore to be kept back on the atoms. Because it is thus withdrawn to its own atom it does not reach out into the region of lowered energy in the bond region; thus its energy is not much, if at all, lowered by the approach of the atoms a and b.

An electron in such a state therefore does little or nothing to lower the energy of the electronic system enough to counteract the nuclear repulsion and contribute to molecule formation. An electronic state that evolves in this way is called an *antibonding orbital.* It might well be called *introverted* because of the way its wave function avoids the bond region.

Therefore the first important point of the theory is the recognition of bonding and antibonding states for individual electrons on the basis of their behavior as the two atoms approach each other. The next point is to regard the molecule as built up by occupancy of these states by the available electrons, the lowest over-all state being arrived at by putting electrons into the lowest individual states that may be used consistently with the Pauli exclusion principle. In this respect the procedure is exactly like that which governs the finding of the normal electron configurations of isolated atoms. In atoms the order of tightness of binding of individual states, in the early part of the periodic table, is

$$1s,\ 2s,\ 2p,\ 3s,\ 3p,\ 3d,\ 4s\ \cdots$$

and the number of electrons is limited by the Pauli principle to 2 for s shells, 6 for p shells, and 10 for d shells.

In the H_2 molecule the lowest electronic states are $1s\sigma_g$ and $1s\sigma_u$. Here $1s$ is used to designate the states of the electron in the separated atoms, σ indicates that the combined state is one having zero orbital angular momentum about the bond axis, and g indicates it is the extroverted or "gerade" state, while u indicates the introverted or "ungerade" state. By the Pauli principle at most two electrons can go into a single orbital, and if two do go in, their spins must be opposite. Hence the lowest state for H_2 is that in which the electron configuration is $(1s\sigma_g)^2$, giving rise to a *covalent electron-pair bond,* in which two electrons with opposite spins fully occupy the same bonding orbital state.

Considering HeH, the details will be different because the molecule is no longer symmetric about the mid-plane of the internuclear line. The general classification into bonding and nonbonding states still obtains. However, only two of the three electrons can go into $1s\sigma_g$; thus the third must go into $1s\sigma_u$. The third electron therefore does not contribute to formation of a chemical bond and because of the greater Coulomb repulsion of the nuclei the other two that are in a bonding state are unable to overcome the enhanced nuclear repulsion. In fact, no stable state of this type exists for this molecule. Considering He_2 the lowest configuration is $(1s\sigma_g)^2(1s\sigma_u)^2$. Here the two antibonding electrons cancel out the two bonding electrons and the nuclear repulsion is even stronger, so that there is no stable molecule formation.

To go much farther along these lines requires a more detailed analysis of the orbital behavior of $2s$ and $2p$ atomic states with regard to molecule formation. This in turn is quite dependent on the relative Z's of the two atoms involved. The present discussion will be restricted to homonuclear molecules, A_2. The $2s$ states give $2s\sigma_g$ and $2s\sigma_u$ as before. The atomic states are spherically symmetric before distortion and much more spread out in space than the $1s$ states, much more weakly bound, and more polarizable by outside fields because of their nearness in energy to $2p$ states.

Detailed behavior of the $2p$ states is quite different from that of the s states, because their own wave functions are not spherically symmetric when the atoms are separated, but are of the form $P(2p) \cos \theta_x$, where θ_x is measured from the axis in space designated by the subscript x. Thus there are three orthogonal $2p$ functions, according as the factor is $\cos \theta_x$, $\cos \theta_y$, or $\cos \theta_z$. Schematically such a wave function looks like the sketch in Fig. 7.3, having two lobes in space around its axis, the wave function having opposite signs in the two lobes. Which end is called + is a matter of arbitrary normalization; here the end toward $+x$ is called +.

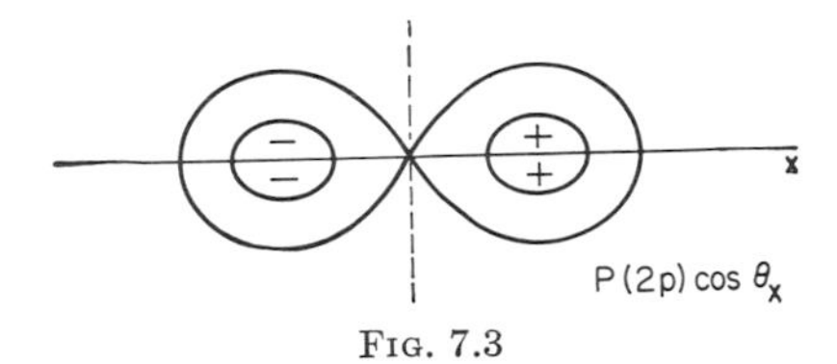

FIG. 7.3

When two such states come near each other along the x axis, with the sign convention just made, the picture becomes that in Fig. 7.4; therefore to get a bonding orbital with no node in the mid-plane, it is necessary to take

$$u_a(2p_x) - u_b(2p_x)$$

and the antibonding orbital will be the sum of these same two terms. The general appearance of the corresponding bonding and antibonding wave functions will be as in (b) and (c) in Fig. 7.4. These are labeled σ because they are symmetric around the bond axis

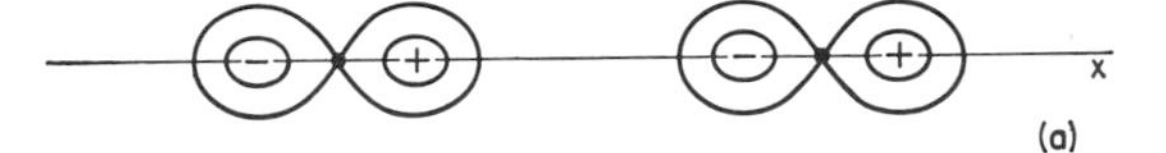

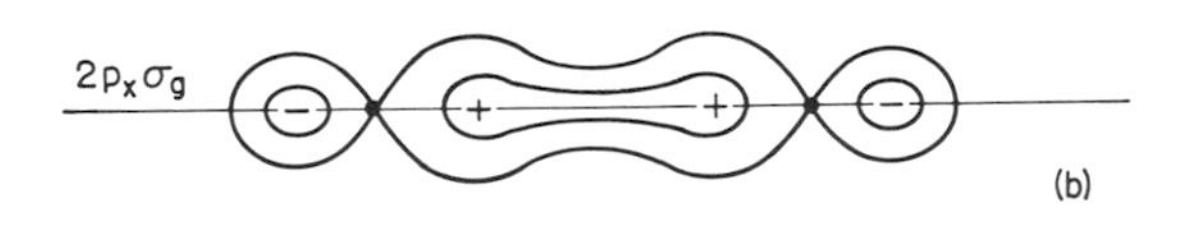

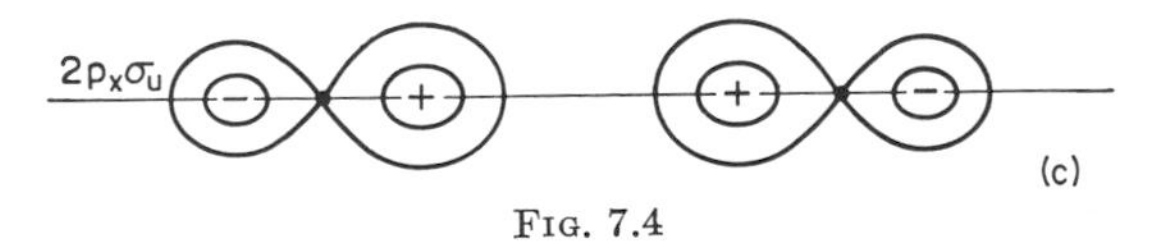

FIG. 7.4

and thus correspond to zero orbital anguler momentum about this axis. The sketches show the characteristic behavior that the extroverted state favors the electron's taking up a place in the bond region, while the introverted state is one in which the electron must avoid the bond region. The bonding of two boron atoms in B_2 proceeds by the formation of an electron pair bond in which two electrons with opposite spins fill the $(2p\sigma_g)$ state.

Consider now the approach toward each other along the x axis of the two states, $u_a(2p_y)$ and $u_b(2p_y)$, indicated in Fig. 7.5. These do not extend away transversely from their own axes of symmetry as far as they do lengthwise. Therefore their centers will have to come closer together than before for overlap to become appreciable, than in the case of the $u(2p_x)$ functions. When they do begin to overlap, it will be the sum of them, as normalized here, which gives rise to a bonding orbital and the difference which gives rise to an antibonding orbital. The corresponding distributions in space are shown in (b) and (c) of Fig. 7.5. In these figures the auxiliary diagrams at the right show the appearance as viewed end-on along the bond axis, serving to remind that the distribution is not symmetric about the bond axis. These are π states because of the $\cos \theta_y$ factor.

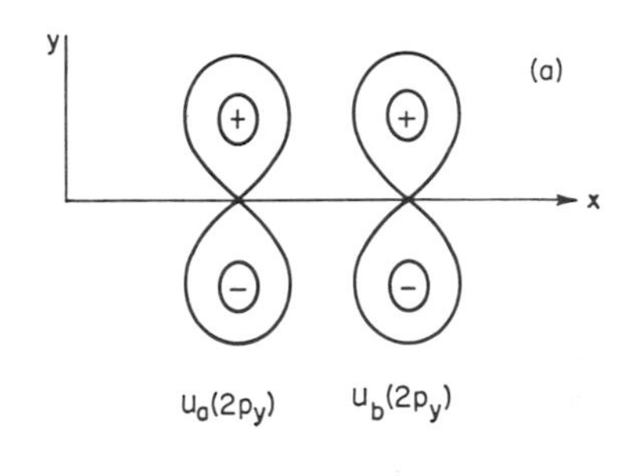

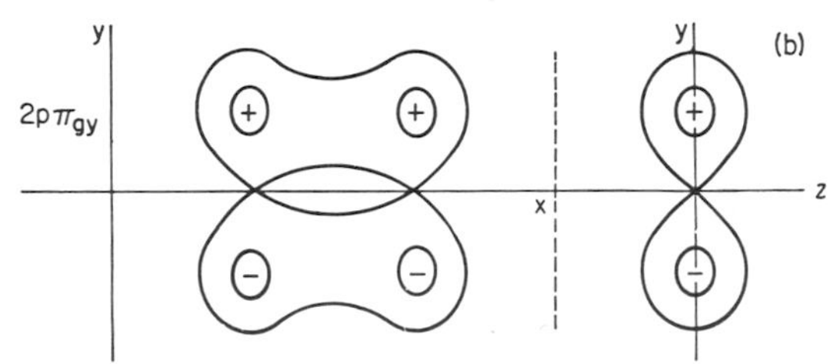

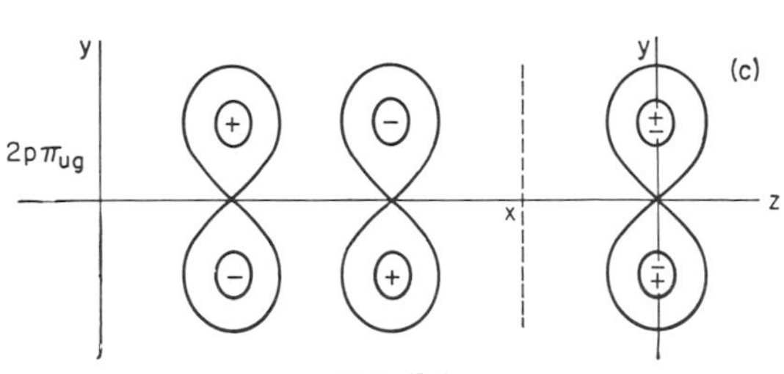

FIG. 7.5

Clearly the approach of $u_a(2p_z)$ and $u_b(2p_z)$ toward each other along the x axis will give rise to the same patterns for two more $2p\pi_g$ and $2p\pi_u$ states which are like those in Fig. 7.4, but rotated through $\pi/2$, so that they bear the same relation to the z axis as the others do to the y axis.

Thus with regard to behavior on bond formation, interaction of $2p$ electrons can give rise to three kinds of bonding states, $2p\sigma_g$, $2p\pi_{gy}$, and $2p\pi_{gz}$, and three kinds of antibonding states with u written in place of g.

The foregoing discussion paves the way for considering the homonuclear diatomic molecules of the $2p$ elements. The normal configuration of carbon is

$(1s)^2(2s)^2(2p)^2$; so when two carbon atoms come together in this configuration they could form a double electron-pair bond, with two electrons in the $2p\sigma_g$ state and two more in the $2p\pi_g$ state.

In traditional chemical valence notation each of these would be indicated by a — (although quantum mechanics shows that the bonds are of quite different character), and the molecule would be characterized as C=C. Likewise the normal configuration of nitrogen is $(1s)^2(2s)^2(2p)^3$, and so N_2 can form three electron-pair bonds putting the six $2p$ electrons in pairs into the three kinds of bonding states giving the molecular configuration

$$(2p\sigma_g)^2(2p\pi_{gy})^2(2p\pi_{gz})^2$$

This would be written as N≡N in traditional valence notation.

With oxygen, there are four $2p$ electrons in each atom. Six of them can go into making three covalent electron-pair bonds as in N_2, but the Pauli principle requires that the remaining two go into antibonding orbitals. Whether they go into the same one or two different ones cannot be said without more elaborate analysis. Evidently they go into different ones, permitting their spins to be parallel, for it is known that O_2 is one of the rare examples of a molecule with an even number of electrons that is paramagnetic.

Traditional valence theory would say that O_2 is double-bonded, writing it O=O. According to modern views the double bond is really the net result of three bonds and one antibond. Similarly in F_2 there are ten $2p$ electrons, forming three covalent bond pairs and two antibond pairs, leaving a resultant that may be called a single bond and written F—F.

Next may be considered the hydrides of the $2p$ elements, working back from HF. When hydrogen atoms in $1s$ states come up to a $2p$ orbital, the most favorable position for strong interaction is for its center to be along the axis of the corresponding $2p$ orbital. These can interact to give bonding and antibonding states, as indicated in Fig. 7.6. Here the notation sp indicates the sum, or bonding combination, and sp^* the difference or antibonding combination, as the notation g and u applies only to description of states of homonuclear molecules.

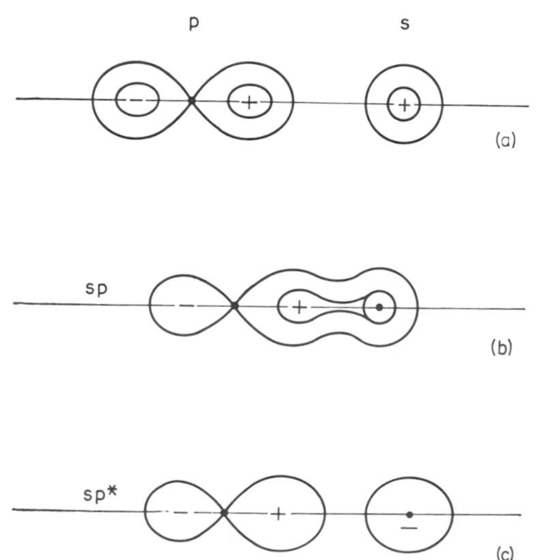

FIG. 7.6

In the normal state of HF, there are five $2p$ electrons in the valence shell of fluorine and one $1s$ electron in the valence shell of hydrogen. Four of the fluorine $2p$ electrons are paired internally. The state occupied by the unpaired one can combine with the $1s$ wave function of hydrogen to make the sp bonding orbital which is then occupied jointly by the pair of electrons to form an electron-pair bond, one of these electrons having been provided by the fluorine atom and the other by the hydrogen atom.

In the case of water, H_2O, two of the $2p$ electrons of the oxygen atom are paired internally in the oxygen atom. The other two occupy different orbitals which combine with $1s$ states of the two hydrogen atoms and jointly with the two electrons coming from the hydrogen bonds provide two electron-pair O—H bonds. The spatial distribution of the different $2p$ functions is such as to tend to make the two O—H bonds be at right angles to each other, making H_2O tend to be a rectangular molecule. Experimentally it is found that the angle between the two O—H bonds in water is 104°40′ instead of 90°, indicating that a strong distorting influence is at work.

By similar reasoning it is expected that NH_3 consists of three electron pair bonds of sp type, making angles of 90° with each other. Experimentally, however, it is again found that the angle is 109°, almost equal to the angle subtended at the center by pairs of corners of a regular tetrahedron.

Continuing in this way, one would be led to expect that carbon with two $2p$ electrons would combine with two hydrogen atoms to give CH_2 and that this should be a rectangular molecule as H_2O was predicted to be. Here is an even greater discrepancy with reality because the stable compound, methane, is not CH_2 but CH_4. Moreover, the methane molecule is known to possess tetrahedral symmetry; so that all four CH bonds are alike in it.

To understand quadrivalent carbon, one must recognize that the $2s^2$ shell in carbon is not very tightly bound, so that the levels in carbon atoms due to the configuration $2s2p^3$ do not lie very much higher than the normal state. In this configuration the carbon atom has four unpaired electrons which could combine with the electrons from four hydrogen atoms to form CH_4. However, this cannot be the whole story, because the four valence electrons are not alike. On this view there would be formed an ss bond between C and one H, and three mutually perpendicular bonds of sp type with the other H atoms. What is being looked for is the state of lowest energy, and if the energy is lowered enough by forming two additional bonds that this more than compensates for the energy required to excite the carbon atom from s^2p^2 to sp^3, then the quadrivalent state will be the normal one.

In molecules the effective field in which electrons move is far from being spherically symmetrical. Therefore identification of the electron states by the s,p,d,f labels has little if any validity. It may therefore be supposed that in methane the four orbitals, one $2s$ and three $2p$, no longer may be labeled with any degree of accuracy by continuing to call one of them an s electron and the others three p electrons. Instead it may be much more appropriate to replace these four wave functions by four appropriately chosen orthogonal linear combinations of these wave functions. Such atomic wave functions which are linear combinations of atomic wave functions connecting different orbital angular momentum values are called *hybrids*.

Writing $u(s)$ and $u(p)$ for s and p wave functions, respectively, supposed normalized, where $\mathbf{p}$ is a vector whose direction indicates the axis in space of the positive lobe of the p wave function, then an sp hybrid will have the form

$$U_1 = u(s) + c_1 u(\mathbf{p}_1)$$

The condition that two such hybrids be orthogonal functions then becomes

$$\int U_1 U_2 \, dv = 1 + c_1 c_2 \mathbf{p}_1 \cdot \mathbf{p}_2 = 0$$

If several of such hybrids are to be equivalent, that is, the same function but differently oriented in space, then their c's must all be equal. The orthogonality condition requires that $\mathbf{p}_1 \cdot \mathbf{p}_2$ be negative, that is, the two hybrid states must have their axes greater than 90° apart. Such hybridization therefore acts to open up the angle between sp bonds, as is observed in H_2O and NH_3 as well as in tetrahedral CH_4.

By making $\mathbf{p}_1 \cdot \mathbf{p}_2 = -\frac{1}{3}$, so that the angle between the hybrid functions is 109°28′, one can get four different orthogonal sp hybrids whose axes point toward the corners of a regular tetrahedron. This is the basis of the quadrivalence of carbon on which all organic chemistry is based; that the normal state of carbon as it exists in carbon compounds involves covalent bonds based on these tetrahedral sp hybrid wave functions.

Such a tetrahedral hybrid function will be denoted by t. The combination of s and p functions involved in it shifts the principal lobe of the function in the direction of its axis, p, and thus favors bond formation on one side (see Fig. 7.6). The sign on the small loop depends on the detailed variation of the s and p radial functions. The important thing is that the lobe on the $+\mathbf{p}$ side is made larger, improving its ability to interact with the field of the atom to which it is bonding. Finally, therefore, the kind of molecular orbital that is involved in the formation of a CH bond is obtained by combining the s function of the hydrogen atom with the t function of carbon in a bonding combination to form an st bonding orbital.

Similarly, in the formation of a single C—C bond, the electron-pair bond is based on bonding tt functions obtained by bringing together two tetrahedral hybrid functions together with their axes pointing toward each other along the bond axis. Such a bond has symmetry about the bond axis and therefore no restraints against free relative rotation of radicals joined in this way are brought into play, in accordance with a fundamental observation of organic chemistry. In crystalline diamond all the carbon atoms are linked together in a tetrahedral arrangement that extends indefinitely in all directions, each carbon atom being linked by a covalent tt bond to its four nearest neighbors.

These general ideas indicate how the electron-pair bond, formed by two electrons with opposite spins and sharing the same bonding orbital, gives a specific interpretation of the covalent bond of chemical valence theory.

Considering now C═C, the carbon-carbon double bond, this is built up by bringing up two carbon atoms toward each other in such a way that two pairs of tetrahedral orbitals overlap to form two different tt covalent bonds, which are somewhat bent. Such a double bond is no longer symmetrical about the internuclear line joining the two carbon atoms and accordingly in such unsaturated molecules there is no free rotation about the axis of a double bond. Similarly C≡C, the carbon-carbon triple bond, is formed by bringing the two tetrahedrally shaped carbon atoms together in such a way that there is overlapping of three pairs of tetrahedral hybrid wave functions. This automatically puts the fourth tetrahedral wave function of each carbon atom pointing away from the C≡C along the axis of the bond, accounting for the observed straight-line configuration of acetylene, HC≡CH.

The foregoing discussion will have indicated in a very sketchy way how the valence bond of chemistry finds its interpretation in the bonding action of a pair of electrons occupying the same bonding orbital with opposite spins in the bond region between two atoms. It will also have indicated how bond formation leads to considerable distortion of the atomic wave functions resulting in the formation of hybrid wave functions by the breakdown of orbital angular momentum assignments to the electronic states. Proceeding along these lines there can be established a reasonably definite one-to-one correspondence between the valence structural formulas of chemistry and the type of molecular orbital wave functions that have been described as being reasonably localized with their biggest portions mainly concentrated in the bond region between two atoms.

However, chemists had recognized, before the development of quantum-mechanical theory, that there were many cases in which traditional valence ideas required additional development. This was mainly in the description of the chemical properties of compounds containing conjugated double bonds. A good simple example is butadiene, which would be written

$$\begin{array}{c} H_2C{=}\underset{H}{C}{-}\underset{H}{C}{=}CH_2 \end{array}$$

Here the essential thing is the existence of two double bonds between carbon atoms, separated by a single bond. Such double bonds are said to be conjugated. Another example is the six-membered ring, C_6H_6, which is described by either of the formulas,

CH
HC CH
HC CH
CH

or

CH
HC CH
HC CH
CH

as originally proposed by Kekulé. Hundreds of other examples are known to organic chemists. The characteristic thing about all such compounds is that, contrary to what is implied by these classical formulas, there is no sharp distinction between the single and double carbon-carbon bonds in their properties. Rather they behave as if all of the carbon-carbon bonds were nearly alike in properties (especially in highly symmetrical cases like benzene) and as if the bond were stronger than a single bond and weaker than a double bond.

These properties find their explanation in a special feature of quantum-mechanical theory which has become known as *resonance* in chemical valence theory [22]. To each of the structural formulas of valence theory there will correspond a particular type of approximate wave function for the electrons in a molecule. In general the wave functions associated with particular structures will not be orthogonal, in the approximate forms usually used, nor will the Hamiltonian operator be represented by a diagonal matrix in terms of these approximate functions.

Therefore, by standard results of perturbation theory, a better approximation will be obtained by using as the improved wave function an appropriate linear combination of several wave functions corresponding to different classical structural formulas. When this is done, it cannot be said that the molecule is represented by any one of the classical formulas, but it can be said that it has properties related to several of them. Its properties are dependent on the properties of the linear combination; it is incorrect to imply that the molecule jumps back and forth between the condition of being in one or the other of the constituent wave functions used in arriving at the final composite result.

In particular if only two structures a and b, having wave functions U_a and U_b, are important in arriving at the resonant structure, and if the diagonal matrix components of the Hamiltonian with respect to these are H_{aa} and H_{bb}, then the perturbed energies arising from interaction of these two states are the roots of

$$\begin{vmatrix} H_{aa} - E & H_{ab} - ES \\ \bar{H}_{ab} - E\bar{S} & H_{bb} - E \end{vmatrix} = 0$$

in which H_{ab} is the nondiagonal matrix element of the Hamiltonian and S is the overlap integral for these two wave functions

$$H_{ab} = \int \bar{U}_a H U_b \, dv \qquad S = \int \bar{U}_a U_b \, dv$$

Thus the allowed values of E are given by

$$(E - H_{aa})(E - H_{bb}) = |H_{ab} - ES|^2$$

Therefore the lower of the two roots of this equation lies below the lower of the two unperturbed values H_{aa} and H_{bb}, this being a general quantum-mechanical result that two interacting levels repel each other. Thus the nondiagonal matrix element has the effect of making the lower level more stable than it would be in the absence of such interaction.

Similar results apply in case there are more than two such structures whose interaction in this way is appreciable [23]. This general behavior by which bonding is strengthened by interaction of several structures is known as *resonance* and the interacting states are called *resonating structures*. The extra energy of stabilization that is brought in in this way is called *resonance energy*. The importance of the resonance concept in chemistry has been emphasized by Pauling.

As with many theoretical interpretations that arise in the course of approximate treatments the exact amount of energy that is called resonance energy depends on the detailed assumptions that are made in setting up the first approximation to the wave functions.

8. Bond Energies, Lengths, and Force Constants

Another characteristic of the formation of electron-pair bonds between pairs of atoms is that the bonds are largely independent of each other in a molecule. For example, the CH bond in methane, CH_4, has pretty much the same length and energy as it has in ethane, C_2H_6, or in methyl chloride, CH_3Cl, or in methanol, CH_3OH.

This has its basic explanation in the fact that the molecular orbitals involved in the formation of an electron-pair bond are fairly well localized in space close to the bond region along the line joining the nuclei of the two atoms involved in the bond, and therefore the spaces occupied by different bonds do not overlap.

This makes it possible in many instances to regard molecular properties as the sum of properties associated with the bonds in a molecule, these properties remaining practically constant for the same bond in different molecules. Discussion of molecular properties in terms of approximate bond additivity relations is especially important because it draws attention to the exceptional cases in which additivity relations break down. These require special treatment on the basis of mesomerism or resonance between interacting structures.

Some important data on bonds have been given in Table 7.2 of Sec. 4. Table 7.4 gives an additional compilation of data for some of the important bonds involved in organic compounds arranged so as to emphasize the dependence on bond order.

Other properties which show additivity relations are diamagnetic susceptibility, refraction of visible light, and the dipole moment. The electric-dipole moment is especially interesting as it is a vectorial property.

The values shown for C—C for graphite and benzene show that the carbon-carbon bond in these substances has an intermediate character between that ascribed to C—C and to C=C in aliphatic compounds. This is a consequence of resonance between the two Kekulé forms, as discussed in Sec. 7. Another point of interest is the close correspondence between N_2 and CO, each of which have 10 valence electrons. Following the discussion of Sec. 7, it is natural to suppose that N_2 is triply bonded, N≡N, with the π bonds possibly being strengthened by hybridization of $2s$ and $2p$ orbitals in the nitrogen atoms. The fact that D and k for N_2 are so much greater than the values for N—N and N=N in compounds supports this conclusion.

TABLE 7.4. BOND ENERGIES (KCAL/MOLE) AND BOND LENGTHS (10^{-8} CM)

CH	98.6	1.06	NH	101.7	0.99	OH	117.2	0.96
		1.10			1.01			
						SH	94.5	1.33
CC	81.1	1.54	CN	77.9	1.47	CO	88.2	1.42
C—C*		1.42						
C=C	145	1.34	C=N	152.5	1.29	C=O	185	1.21
C≡C	190	1.21	C≡N	238	1.16	C≡O	254	1.13
						C—S	73	1.80
						C=S	140	
			N—N	60	1.47	O—O	66	
			N=N	152	1.20	O=O		
			N≡N	278	1.09			
HF	147.5	1.28	CF	118.3	1.38	O—F	65.6	
HCl	102.7	1.28	CCl	77.8	1.76	O—Cl	56.3	
HBr	87.3	1.42	CBr	65.3	1.94	N—F	86.8	
HI	71.4	1.62	CI	56.8	2.13	N—Cl	56.4	

* Bond length for graphite and benzene. C≡O is written for diatomic carbon monoxide, as discussed in text.

As for CO, a direct application of classical valence ideas leads to the picture C=O, as two is the maximum valence of oxygen. This interpretation involves the assumption that carbon is occasionally divalent, an idea that has been much discussed in chemistry. However, the large values for D and k for the diatomic CO molecule and the small value of r_e all point to the conclusion that the bonding is considerably stronger than the C=O bond in formaldehyde or in ketones. This would suggest that the structure is C≡O, analogous to N≡N.

Very early in the history of the electronic theory of valency, Langmuir proposed this model and suggested that it came about by the donation to the bond of the two electrons which are paired off in the oxygen atom [24]. Thus two of the bonds are ordinary covalent links involving electron pairs in which one electron is supplied by each of the bonded atoms, but the third bond is unusual in that both electrons involved in the bond come from the oxygen atom. This has been called *dative covalency*.

Continuing the discussion of N_2 and O_2 of Sec. 7, it is interesting to compare the properties of the normal states of these molecules and the normal states of the ions, N_2^+ and O_2^+:

	D_0, ev	ω_e, cm^{-1}	r_e, 10^{-8} cm
N_2^+	6.341 or 8.724	2,207.19	1.116
N_2	7.373 or 9.756	2,359.61	1.094
O_2^+	6.48	1,876.4	1.1227
O_2	5.080	1,580.36	1.207

In the case of N_2 the addition of an electron to N_2^+ completes the formation of the third electron-pair bond between the two nitrogen atoms. Accordingly the dissociation energy and the vibration frequency are increased and the bond shortened. In the case of O_2, addition of an electron to O_2^+ completes the formation of an antibonding pair. Accordingly the dissociation energy decreases, the vibration frequency decreases, and the bond lengthens.

9. Ionic Bonds and Dipole Moments

In the early nineteenth century Berzelius developed the idea of a chemical bond as originating in Coulomb forces of attraction between oppositely charged ions. This idea is applicable [25] in the alkali halides and to a lesser extent in alkaline earth oxides, but is clearly inapplicable for symmetrical homonuclear diatomic molecules such as O_2, N_2, and Cl_2.

On the extreme ionic view, NaCl is regarded as built by the coming together of Na^+ and Cl^-, one electron having gone over from the sodium to the chlorine, the bond deriving its stability from the energy of coulomb interaction, $-e^2/r$, of these two ions. In a better approximation it is recognized that the electric field of Na^+ will act to draw toward itself the electrons of the Cl^-. This polarizes the negative ion in a sense which tends to put the supposedly transferred electron back into the bond region between the two atoms as in a covalent bond.

The bond between Na and Cl is also describable in terms of electron-pair bond formation. The Cl atom has a p^5 configuration, leaving one unpaired p electron, and the Na atom has an s configuration with one unpaired s electron. As these two atoms come together, these two electrons will occupy, with opposite spins, the bonding orbital which evolves by combination of a chlorine-centered p wave function and a sodium-centered s wave function.

The degree to which the resulting bond can be described as ionic thus depends on the exact details of the distribution of the resulting wave function in relation to the two centers.

Experimentally this unsymmetrical distribution of charge manifests itself in the electric-dipole moments of molecules which can be inferred from the temperature dependence of the dielectric constant or by molecular beam methods [26]. The dipole moment μ is usually stated with

$$1 \text{ Debye (D)} = 10^{-18} \text{ esu-cm as unit}$$

Thus the dipole moment arising from two opposite unit electronic charges at a separation of 10^{-8} cm is 4.80 D.

Molecular beam measurements give for three alkali halides:

	KCl	KI	NaI
μ.	6.3	6.8	4.9
	13.4	15.5	13.9
Ratio.	0.47	0.44	0.35

The second line gives the value to be expected on a purely ionic model, using bond lengths from electron-diffraction work [27]. Thus the observed moments are less than half what is to be expected on a pure ionic model.

Similarly for the hydrogen halides [28]:

	HF	HCl	HBr	HI
μ.	1.91	1.03	0.78	0.38
	4.40	6.10	6.78	7.7
Ratio.	0.43	0.17	0.12	0.05

The second line is the pure ionic value based on the internuclear distances obtained from band spectra. The smallness of these values for all but HF indicates that these molecules are much more nearly covalent than ionic.

Such small values are associated with the one-sided conditions that exist in diatomic molecules. In ionic crystals, where each ion is more or less symmetrically surrounded by ions of the opposite sign, the distortions of the electron structure due to polarization are much smaller so that the pure ionic states represent a better approximation than in diatomic molecules.

Much effort has gone into attempts to treat the dipole moment of polyatomic molecules as the vector sum of individual bond moments associated with the structure. This is not as successful as in the case of other additive properties of bonds in molecules. For example, CH_4 and CCl_4 each have zero dipole moment because of their tetrahedral symmetry, even assuming there are nonvanishing $\mu(CH)$ and $\mu(CCl)$ associated with the individual bonds. From this it can be inferred that $\mu(CH_3)$ is equal and opposite to $\mu(CH)$, and also that $\mu(CCl_3)$ is equal and opposite to $\mu(CCl)$. Therefore if fixed bond moments were permanent attributes of particular bonds in different molecules, the dipole moment of $CHCl_3$ ought to equal that of CH_3Cl. However, the measured values are 0.95 D for $CHCl_3$ and 1.86 D for CH_3Cl.

Such discrepancies are probably related to internal induction effects whereby the over-all moment is not merely the vector sum of intrinsic bond moments, but is this vector sum modified by additional moments induced in polarizable parts of the molecule by the local electric fields set up by the permanent moments [29]. Similar variations of μ are observed in the *solvent effect* in which the dipole moment measured for a polar molecule in a nonpolar solution is found to depend on the solvent in which it is measured.

TABLE 7.5. BOND MOMENTS (*D* UNITS)
(The atom on the right is the negative end of the dipole)

H—N. . .	1.31	C—C. . .	0	C—I. . .	1.29	C—O. . .	0.86
H—O. . .	1.53	C═C. . .	0	C—Br. .	1.48	C═O. . .	2.4
		C≡C. . .	0	C—Cl. .	1.56	C≡N. .	3.6
		H—C. .	0.3	C—F. . .	1.51		

In spite of such complications it has been possible to set up a table of bond moments [30], Table 7.5, which can be used for rough calculations of dipole moments by summing the moment vectors if their relative spatial orientation is known, or to make inferences about molecular shape from experimental measurements of dipole moment. The sign of the bond moment as well as its magnitude is important, but there has been considerable uncertainty about what sign applies to the CH bond. It is usually considered that H is the positive end of the dipole but evidence strongly suggests that it is usually oriented with H at the negative end [31].

The value given in the table for C═O applies to this group in organic molecules such as acetone, for which $\mu = 2.75$ D. Diatomic carbon monoxide has a small moment of only 0.12 D. In acetone the large moment is in the direction to make oxygen negative and the smallness of the carbon monoxide moment indicates that this large moment is mainly neutralized by the negative charge that is put on the carbon atom in formation of a third bond by dative covalency.

Ordinarily the dipole moment originating in the bonds between a pair of atoms is regarded as primarily due to an unsymmetrical sharing of the pairs of electrons that are directly involved in bond formation. But Coulson [32] has pointed out that an additional contribution to the moment arises from other electrons. Usually the other electrons are symmetrically distributed about one or the other of the nuclei and therefore make no contribution to the bond moment. But if the bonding electrons have undergone hybridization in forming the bond, then the nonbonding electrons will also be forced into hybridized states which are also not symmetrical around either nucleus and thus also make a contribution to the bond moment.

References

1. Born, M., and J. R. Oppenheimer: *Ann. Physik,* **84:** 457 (1927). See also Part 8, Chap. 2, Sec. 1, and J. C. Slater, "Quantum Theory of Matter," p. 500, McGraw-Hill, New York, 1951.
2. Mulliken, R. S.: *Revs. Mod. Phys.,* **2:** 60 (1930); **3:** 89 (1931). Van Vleck, J. H.: *Revs. Mod. Phys.,* **23:** 213 (1951). Townes, C. H., and A. L. Schawlow: "Microwave Spectroscopy," Chap. 7, McGraw-Hill, New York, 1955. Herzberg, G.: "Spectra of Diatomic Molecules, Chap. 5, Van Nostrand, Princeton, N.J., 1950. Johnson, R. C.: "An Introduction to Molecular Spectra," Methuen, London, 1949.
3. Dunham, J. L.: *Phys. Rev.,* **41:** 721 (1932). Sandeman, I.: *Proc. Roy. Soc. Edinburgh,* **60:** 210 (1940).
4. Morse, P. M.: *Phys. Rev.,* **34:** 57 (1929). Pekeris, C. L.: *Phys. Rev.,* **45:** 98 (1934). ter Haar, D.: *Phys. Rev.,* **70:** 222 (1946).
5. Townes, C. H., and A. L. Schawlow: "Microwave Spectroscopy," p. 14, McGraw-Hill, New York, 1955. Kronig, R. de L.: *Physica,* **1:** 617 (1934). Dieke, G. H.: *Phys. Rev.,* **47:** 661 (1935). Van Vleck, J. H.: *J. Chem. Phys.,* **4:** 327 (1936). Watson, W. W.: *Phys. Rev.,* **49:** 70 (1936).
6. Herzberg, G.: "Spectra of Diatomic Molecules," Chap. 5, Van Nostrand, Princeton, N.J., 1950.
7. Franck, J.: *Trans. Faraday Soc.,* **21:** 536 (1925). *Z. physik. Chem.,* **120:** 144 (1926). Condon, E. U.: *Phys. Rev.,* **28:** 1182 (1926); **32:** 858 (1928); *Proc. Natl. Acad. Sci.,* **13:** 462 (1927); *Am. J. Phys.,* **15:** 365 (1947).
8. Winans, J. G., and E. C. G. Stueckelberg: *Proc. Natl. Acad. Sci.,* **14:** 867 (1928). Condon, E. U., and H. D. Smyth: *Proc. Natl. Acad. Sci.,* **14:** 871 (1928).

9. Franck, J.: *Trans. Faraday Soc.*, **21**: 536 (1925).
10. Birge, R. T., and H. Sponer: *Phys. Rev.*, **28**: 259 (1926).
11. Loomis, F. W., and R. E. Nusbaum: *Phys. Rev.*, **38**: 1447 (1931).
12. Herzberg, G.: "Spectra of Diatomic Molecules," p. 437, Van Nostrand, Princeton, N.J., 1950.
13. Cottrell, T. L.: "The Strength of Chemical Bonds," Academic Press Inc., New York, 1954.
14. Herzberg, G.: "Spectra of Diatomic Molecules," Chap. 7, Van Nostrand, Princeton, N.J., 1950.
15. Wentzel, G.: *Z. Physik*, **48**: 524 (1927); *Physik. Z.*, **29**: 321 (1928). The mathematics used here is nowadays familiar in its nuclear application in the Breit-Wigner formula for resonance capture of neutrons, *Phys. Rev.*, **49**: 519 (1936).
16. Villars, D. S., and E. U. Condon: *Phys. Rev.*, **35**: 1028 (1930).
17. Burrau, O.: *Kgl. Danske Vid. Selsk.*, **7**: 1 (1927). Hylleraas, E. A.: *Z. Physik*, **71**: 739 (1931). Steensholt, G.: *Z. Physik*, **100**: 547 (1936); *Norske Vid. Akad. Avhl.*, no. 4, 1936. Morse, P. M., and E. C. G. Stueckelberg: *Phys. Rev.*, **33**: 932 (1929). Jaffe, G.: *Z. Physik*, **87**: 535 (1934). Lennard-Jones, J. E.: *Trans. Faraday Soc.*, **24**: 668 (1929). Teller, E.: *Z. Physik*, **61**: 458 (1930). Finkelstein, B. N., and G. E. Horowitz: *Z. Physik*, **48**: 118 (1928). Dickinson, B. N.: *J. Chem. Phys.*, **1**: 317 (1933). Guillemin, V., and C. Zener: *Proc. Natl. Acad. Sci.*, **15**: 314 (1929). Pauling, L.: *Chem. Revs.*, **5**: 173 (1928).
18. Birge, R. T.: *Proc. Natl. Acad. Sci.*, **14**: 12 (1928). Richardson, O. W.: *Trans. Faraday Soc.*, **25**: 686 (1929).
19. Heitler, W., and F. London: *Z. Physik*, **44**: 455 (1927). Heitler, W.: *Z. Physik*, **46**: 47 (1927); **47**: 835 (1928); **51**: 805 (1929).
20. Sugiura, Y.: *Z. Physik*, **45**: 484 (1927). Wang, S. C.: *Phys. Rev.*, **31**: 579 (1928). Coolidge, A. S., and H. M. James: *J. Chem. Phys.*, **1**: 825 (1933). Coulson, C. A.: *Trans. Faraday Soc.*, **33**: 1473 (1937).
21. Pauling, Linus: "The Nature of the Chemical Bond," 3d ed., Cornell University Press, 1960. Coulson, C. A.: "Valence," Oxford University Press, New York and London, 1952. Mulliken, R. S.: *Chem. Revs.*, **9**: 347 (1931). Van Vleck, J. H., and A. Sherman: *Revs. Mod. Phys.*, **7**: 167 (1935). Rice, O. K.: "Electronic Structure and Chemical Binding," McGraw-Hill, New York, 1940.
22. Wheland, G. W.: "Resonance in Organic Chemistry," Wiley, New York, 1955.
23. Dewar, M. J. S.: "The Electronic Theory of Organic Chemistry," Oxford University Press, New York and London, 1949. Remick, A. E.: "Electronic Interpretations of Organic Chemistry," 2d ed., Wiley, New York, 1949.
24. Langmuir, I.: *J. Am. Chem. Soc.*, **41**: 1543 (1919). Sidgwick, N. V.: "Electronic Theory of Valency," p. 272, Oxford University Press, New York and London, 1927.
25. Kossel, W.: *Ann. Physik*, **49**: 229 (1916).
26. Smyth, C. P.: "Dielectric Behavior and Structure," McGraw-Hill, New York, 1955. Partington, J. R.: "An Advanced Treatise on Physical Chemistry," vol. 5, p. 287, Longmans, New York, 1954. Sutton, L. E.: Chap. 9, Braude, E. A., and F. C. Nachod: "Determination of Organic Structures by Physical Methods," Academic Press Inc., New York, 1955. Wesson, L. G.: "Tables of Electric Dipole Moments," Massachusetts Institute of Technology Press, 1948.
27. Maxwell, L. R., S. B. Hendricks, and V. M. Mosley: *Phys. Rev.*, **52**: 968 (1937).
28. Zahn, C. T.: *Phys. Rev.*, **27**: 455 (1926).
29. Smallwood, H. M., and K. F. Herzfeld: *J. Am. Chem. Soc.*, **52**: 1919 (1930). Hampson, G. C., and A. Weissberger: *J. Chem. Soc.*, p. 393, 1936.
30. Smyth, C. P.: *J. Phys. Chem.*, **41**: 209 (1937).
31. Gent, W. L. G.: *Quart. Revs. (London)*, **2**: 383 (1948).
32. Coulson, C. A.: *Trans. Faraday Soc.*, **38**: 433 (1942).

Chapter 8

X Rays

By E. U. CONDON, University of Colorado

1. Main Phenomena

X rays, or Roentgen rays, are electromagnetic radiations having wavelengths of a few angstrom units (10^{-8} cm), thus having quantum energies ranging from a few thousand up to several million electron volts. Common applications of X rays in medical practice involve radiations produced by X-ray tubes operating at 30 to 150 kv.

Modern X-ray tubes are high-vacuum tubes containing a tungsten filament as a thermionic source of electrons. This is maintained at high negative potential by a suitable rectified power supply. A beam of cathode rays from the filament strikes a massive metal target. The X rays are emitted in all directions from the target.

Spectroscopic analysis of X rays is carried out with X-ray spectrometers in which spectral resolution is accomplished by diffraction of the X rays from a crystal (Part 8, Chap. 1). The emission spectrum consists partly of a continuous spectrum and partly of a line spectrum characteristic of the atoms of which the target is made. The continuous spectrum is emitted by electrons which make transitions between two positive energy levels in the field of an atomic nucleus of the target. It therefore has a high-frequency limit, $h\nu = Ve$, where V is the potential difference applied to the tube. This high-frequency limit's relation to the applied potential is called the *Duane-Hunt law*.

The characteristic line spectrum is the result of excitation processes in the atoms of the target in which the electrons of the incident cathode ray beam knock out an electron from one of the inner normally closed electron shells of the atom. The line spectrum results when outer electrons make quantized radiative transitions to fill the hole in the inner shell.

The characteristic X-ray line spectrum of an atom cannot be observed in absorption because the final state of the line-emission act results in there being a hole in one of the outer electron shells, whereas in normal atoms these outer shells are filled so that there is no room for another electron to be put in them by an absorptive act. Instead, the absorption of X rays consists of continuous absorption bands corresponding to transitions in which an electron makes a transition from an inner closed shell to an outside free state.

The atomic cross section for absorption is highest at a frequency just sufficient to free the electron without much excess kinetic energy, giving rise to rather sharp absorption edges in the spectrum. The absorption spectrum may show structure just at these absorption edges, this being a manifestation of the structure of the loosely bound quasi-free states available to the electron as final states. This fine structure is studied as a way of learning about the electronic levels in solids.

When quantum energies considerably exceed the value corresponding to the absorption limit, the electrons are given considerable kinetic energy and so some of them, which are set free near enough to the surface of the solid, are able to escape from it. This is a natural extension into the X-ray region of the photoelectric effect as observed with visible and ultraviolet light (Part 8, Chap. 5).

In going through matter, X rays show refraction and dispersion which are describable in terms of a natural extension of the dispersion theory to very short waves.

Scattering of X rays may be of the coherent kind, which gives rise to the dispersive modifications of ordinary propagation, and also to the diffractive scattering which reveals the structural arrangement of electron density in the scattering material. It may also be of the incoherent kind, in which part of the quantum energy is given up to produce a change in the state of the scattering electron.

In case the quantum energy is large compared to the binding energy of the scattering electron, the electron will behave essentially as if free in the scattering process. The laws of conservation of energy and momentum then determine the distribution of the total energy between the scattered quantum and the recoil electron. This process is known as the Compton effect.

When the quantum energy $h\nu$ exceeds $2mc^2$, another absorptive process comes into play involving interaction in the strong electric fields close to an atomic nucleus. This process involves the disappearance of the quantum and the simultaneous creation or materialization of one electron and one positron. This process is known as *pair production*.

High-energy electromagnetic radiations may also interact directly with the nucleons of an atomic nucleus to produce various nuclear photodisintegrations (Part 9). Although there is no precise definition of the high-energy limit of the energy of quanta called X rays, this term is usually restricted to radiations of fewer than several million electron volts of energy, above which the radiation is referred to as γ radiation.

2. Emission: Continuous Spectrum

The continuous spectrum of radiation emitted when electrons strike the metal target of an X-ray tube is the result of transitions between the unquantized positive energy levels of the electrons in the fields of the nuclei of the target atoms. This radiation is often called *bremsstrahlung*.

Comparison of theory and experimental data that is obtained with an ordinary thick target X-ray tube is complicated by the fact that the radiation is then due to a continuous distribution of energies of the partially slowed down electrons, some of which also are traveling in directions widely different from that of the incident beam. The theory of the basic radiation process for energies in excess of about 100 kev requires relativistic quantum mechanics.

The actual thick target distribution of energy, $I(\nu)\,d\nu$, in the continuous spectrum is, to a good approximation, representable by an empirical formula due to Kulenkampff [1]*

$$I(\nu)\,d\nu = i[aZ(\nu_0 - \nu) + bZ^2]\,d\nu \qquad (8.1)$$

in which i is the current striking the target. The second term is usually small compared to the first and is often neglected. Above the Duane-Hunt limit, that is, $\nu > \nu_0$, the intensity is zero.

The total integrated intensity at all frequencies is

$$I = i(a'ZV^2 + b'Z^2V) \qquad (8.2)$$

in which $a' = a(e^2/h^2)/2$ and $b' = b(e/h)$. An approximate value for b'/a' is 16.3 volts; thus

$$I = ia'ZV(V + 16.3Z)$$

The total power input to the tube is Vi; therefore the efficiency of conversion of electric power input to X rays of all frequencies is

$$\epsilon = \frac{I}{Vi} = a'Z(V + 16.3Z) \qquad (8.3)$$

where V is in volts. Experimental measurements give $a' = (1.2 \pm 0.1)10^{-9}$.

The increase in $I(\nu)$ with decreasing ν is due to radiation from the partially slowed down electrons in the thick target. Studies made [2] with thin targets show that $I(\nu)$ is constant at least in the range down to $\nu_0/2$.

Earliest theoretical work [3] is based on the classical radiation of a pulse by an electron that is assumed to be uniformly decelerated from its initial velocity in going through the target material. The total rate of radiation in all directions for an acceleration a is $(2e^2a^2/3c^3)$ ergs/sec. If the electron's initial velocity is v_0 and it comes to rest in a distance s, the total energy radiated is $W = (e^2/3s)\beta_0^3$, where β_0 is the ratio of the initial velocity to the velocity of light. This will be in the form of a rectangular pulse of time duration $2s/v_0$ which on Fourier analysis will give a spectrum of frequencies extending to the order of $v_0/2s$. Therefore in order to get frequencies as high as the observed high-frequency limit, $mv_0^2/2h$, the distance s must be of the order h/mv_0, the de Broglie wavelength of the electron at its initial speed. Thus the radiation process must be occurring in individual atoms and not as a result of an average deceleration over a distance of the order of the mean depth of penetration of the X rays in the target.

Correspondence-principle treatments [4] indicate that $I(\nu)$ is independent of ν in accordance with observations. In terms of the basic unit, $\phi_0 = (8\pi/3)r_0^2$, where $r_0 = e^2/mc^2$, Kramers' result can be written as a cross section for radiation of energy in the range $d\nu$ at ν,

$$I(\nu)\,d\nu = 3^{-1/2}\phi_0\alpha Z^2mc^2d\left(\frac{\nu}{\nu_0}\right) \qquad \text{for } \nu < \nu_0 \qquad (8.4)$$

and $I = 0$ for $\nu > \nu_0$. Here $\alpha = e^2/\hbar c$.

Quantum-mechanical theory [5] has been put in a form suitable for comparison with experiment by Kirkpatrick and Wiedmann [6]

The main results are:

1. The over-all thin-target conversion efficiency of electron kinetic energy into continuous X rays by electrons of voltage V striking atoms Z is $2.9 \times 10^{-9}ZV$, giving $1.3 \times 10^{-9}ZV$ for the average thick-target conversion efficiency.
2. The radiation in $d\nu$ at ν that is emitted at an angle θ with the direction of motion of the incident electron is partially polarized.
3. The radiation in $d\nu$ at ν at angle θ is a function only of V/Z^2 and of ν/ν_0 and θ. The calculated theoretical values are given in Table II [6]. Let the x axis be the direction of the incident electron's motion and let the direction of observation of the radiation make an angle θ with the x axis in the xz plane. The radiation from one electron striking one atom Z in ergs per steradian in $d\nu$ (sec^{-1}) is given by

$$I_\theta = (I_x \sin^2\theta + I_z \cos^2\theta) + I_y \qquad (8.5)$$

where the first two terms (in parentheses) together represent the intensity that is polarized in the xz plane and I_y gives that polarized at right angles to this plane. Theory gives $I_y = I_z$.

Numerical values of I_x and I_y as a function of (Z^2/V) and of θ and of ν/ν_0 come out to be of the order of 10^{-50} in these units. If one expresses the frequency range as $d(\nu/\nu_0)$ and measures the radiated energy in units of $h\nu_0$, and the effective cross section in units of ϕ_0, then the quantity I in these units has the value $I/h\phi_0$ and $(h\phi_0)^{-1} = 2.270 \times 10^{50}$; so these units are appropriate ones in which to describe the effects. On integrating over all directions of emission of the radiation, the total radiation at ν in $d\nu$ becomes

$$W = \frac{8\pi}{3}(I_x + I_y + I_z) \qquad (8.6)$$

which is also tabulated by Kirkpatrick and Wiedmann. They also give numerical values for the theoretical effect of screening, using an atomic potential of the form $V = -(Ze/r)e^{-r/a}$.

3. Emission: Characteristic Line Spectrum

The system of atomic energy levels mainly involved in the emission of the characteristic X-ray line

* Numbers in brackets refer to References at end of chapter.

spectrum is that made up of levels in which one electron is missing from one of the inner normally closed electron shells. Some additional weak lines, known as satellite lines, require for their interpretation doubly excited states in which two electrons are missing from normally filled shells.

Study of X-ray spectra was of the greatest historical importance in development of atomic theory. The main facts were discovered in experimental researches of Moseley [7] made in 1913, the year after Laue's discovery of the diffraction of X rays by crystals and the same year as that in which Bohr [8] published the original quantum theory of the hydrogen atom. The applicability of Bohr's theory to the interpretation of Moseley's results was recognized at once and was a great triumph for the theory.

Very soon after, Webster [9] did an experiment which demonstrated the excitation of the $K\alpha$ line of Rh to start when the tube voltage was increased to 23.5 kv, which critical potential served to confirm the quantum view of spectral line excitation. The early work on interpreting X-ray spectra in terms of the Bohr atom model is mainly due to Kossel [10] and the application of the relativistic model to the interpretation of the doublet structure to Sommerfeld [11]. A good early systematic account of the X-ray energy levels is given by Bohr and Coster [12].

On the central field model of the atom (Chaps. 1 and 2) the structure of the energy levels resulting from one electron missing from closed shells is a doublet spectrum. The most tightly bound electrons are the two in the $1s$ shell; so the atom will be in the highest level of one-electron excitation when one electron is removed from this shell. Because the state involved is an s state, the level will be a $^2S_{1/2}$ level. In X-ray terminology this is called the K level.

Next in order of excitation energy will be the eight electrons in the $n = 2$ shells, $2s^2$ and $2p^6$. Removal of one $2s$ electron gives also a $^2S_{1/2}$ level which in X-ray terminology is called the L_{I} level. Removal of one $2p$ electron gives rise to two levels forming a $^2P_{1/2,3/2}$ term. Because the structure corresponds to one hole, the doublet is "inverted," that is, the $J = \frac{1}{2}$ level lies higher than the $J = \frac{3}{2}$ level. These are known as the $L_{\text{II}}(\frac{1}{2})$ and the $L_{\text{III}}(\frac{3}{2})$ levels, where the J values have been added for emphasis.

The difference in energy between L_{I} and the $(L_{\text{II}},L_{\text{III}})$ doublet is due to the fact that the $2s$ state penetrates more closely to the nucleus than does the $2p$ state and therefore is more strongly bound, as is also the case in the optical doublet spectra of alkalilike atoms and ions. In the X-ray literature this energy interval and corresponding ones in terms arising in other shells are called a *screening doublet.*

The difference in energy between L_{II} and L_{III} is due to the combination of relativistic and spin-orbit interactions which are given by the relativistic theory of hydrogenlike atoms. For this reason they are said to form a *relativistic doublet,* sometimes also called a *spin doublet.*

Similarly the $n = 3$ shells give rise to five levels:

M_{I}	M_{II}	M_{III}	M_{IV}	M_{V}
$^2S_{1/2}$	$^2P_{1/2}$	$^2P_{3/2}$	$^2D_{3/2}$	$^2D_{5/2}$

and correspondingly the $n = 4$ shell gives rise to seven levels designated N_{I} to N_{VII} in the same way, and so on.

Of course, a hole can exist only in shells that exist in atoms; therefore atoms of low Z have a relatively simple X-ray level scheme. As Z increases and more shells become filled, the X-ray level scheme becomes more fully developed. Before a particular shell is filled it is on the outside of the atom and determines the nature and structure of the optical spectrum. As Z increases, the shell is filled and goes into the inner structure of the atom where it contributes to the X-ray level structure. In this way there is a continuity between the development of the optical level scheme and the X-ray levels.

The selection rules for transitions between levels are the same as those that apply to electric dipole radiation in an optical one-electron spectrum:

$$\Delta l = \pm 1 \quad \text{and} \quad \Delta J = -1, 0, +1$$

except that 0 to 0 for J is forbidden. Quadrupole and magnetic-dipole radiation are relatively stronger in X-ray than in optical spectra [13]; therefore additional transitions which violate these rules are often observed.

In the literature of X-ray spectroscopy several different notations are to be found for the X-ray lines. These were developed before the level interpretation was fully worked out and so are not rationally related to the level notation. Lines originating on the K level are said to belong to the K series. Lines originating on the L levels are said to belong to the L series, which is really three different series, $^2S \to {}^2P$, $^2P \to {}^2S$, and $^2P \to {}^2D$ in optical terminology [12a].

The most commonly used notation for the X-ray lines is given in Table 8.1.

In the simple central field model of the atom each electron moves in an effective central field, $V(r)$, which is obtained by some procedure such as that of the Hartree self-consistent field. This can also be written as $Z(r)e^2/r$, in which $Z(r)$ is an effective nuclear charge at the distance r, and which varies from the full value Z near the nucleus down to 1 for neutral atoms, 2 for singly charged ions, and so on.

In the theoretical interpretation of the X-ray levels, it has proved to be fruitful to work with an even simpler model in which one uses the hydrogenic term values with an appropriately screened constant value of Z for each level. This procedure is not completely justified in terms of quantum mechanics, but it does give a useful basis for semiempirical correlation of the observed values of the levels.

The relativistic formula for the bound levels of an electron in the field of a nucleus of charge Ze may be written

$$\frac{W}{mc^2} = 1 - \frac{1}{\sqrt{1 + \kappa^2}} \tag{8.7}$$

in which W is the energy with sign reversed, as is appropriate for X-ray levels, and

$$\kappa = \frac{\alpha Z}{n - \epsilon} \quad \text{with } \epsilon = k - \sqrt{k^2 - \alpha^2 Z^2}$$

in which $k = j + \frac{1}{2}$ and $\alpha = e^2/\hbar c$.

The X-ray term values are given directly by the experimental values of the wavelengths of the absorp-

TABLE 8.1. NOMENCLATURE OF X-RAY LINES AND X-RAY LEVELS

Final level	Initial level								
	K series $K(\frac{1}{2})$	L series			M series				
		L_I	L_{II}	L_{III}	M_I	M_{II}	M_{III}	M_{IV}	M_V
$2s\ L_I\ \frac{1}{2}$									
$2p\ L_{II}\ \frac{1}{2}$	α_2								
$2p\ L_{III}\ \frac{3}{2}$	α_1								
$3s\ M_I\ \frac{1}{2}$				l					
$3p\ M_{II}\ \frac{1}{2}$	β_3	β_4		t					
$3p\ M_{III}\ \frac{3}{2}$	β_1	β_3		s					
$3d\ M_{IV}\ \frac{3}{2}$	β_{10}	β_1		α_2					
$3d\ M_V\ \frac{5}{2}$	β_9			α_1					
$4s\ N_I\ \frac{1}{2}$			γ_5	β_6					
$4p\ N_{II}\ \frac{1}{2}$	β_2	γ_2							
$4p\ N_{III}\ \frac{3}{2}$									
$4d\ N_{IV}\ \frac{3}{2}$			γ_1	β_{15}			γ_2		
$4d\ N_V\ \frac{5}{2}$				β_2			γ_1		
$4f\ N_{VI}\ \frac{5}{2}$			v					β_1	α_2
$4f\ N_{VII}\ \frac{7}{2}$									α_1
$5s\ O_I\ \frac{1}{2}$			γ_8	β_7					
$5p\ O_{II}\ \frac{1}{2}\ O_{III}\ \frac{3}{2}$			γ_4						
$5d\ O_{IV}\ \frac{3}{2}$			γ_6						
$5d\ O_V\ \frac{5}{2}$				β_5			ϵ		

tion edges (Sec. 4). If λ is the wavelength of an absorption edge (10^{-8} cm as unit), then

$$\frac{W}{mc^2} = \frac{0.024275}{\lambda}$$

The observed values of W/mc^2 for the K levels can be substituted in the relativistic formula with $n = 1$ and $k = 1$ and an effective value of Z calculated. When this is done, it is found that the effective values can be expressed as $(Z - \sigma_1)$, where σ_1 is a screening constant which increases gradually, having a value of about 3.0 at $Z = 30$ and about 5.0 at $Z = 90$. This smooth behavior depends essentially on the use of the relativistic formula. At high values of Z the relativistic increase in mass of the bound electron increases the binding energy of the levels by an amount which for uranium is about equal to the loosening effect of screening; so if a nonrelativistic calculation were made, it would give an apparent value $\sigma_1 = 0$ for large values of Z.

The relativistic formula for the (n,k) level may be developed in powers of αZ, giving

$$\frac{W}{mc^2} = \frac{\alpha^2(Z-\sigma_1)^2}{2n^2} + \left[\frac{\alpha^4(Z-\sigma_2)^2}{2n^4}\right]\left[\left(\frac{n}{k}-\frac{3}{4}\right) + \frac{\alpha^2(Z-\sigma_2)^2}{n^2}\left(\frac{n^3}{4k^3}+\frac{3n^2}{4k^2}-\frac{3n}{2k}+\frac{5}{8}\right) + \frac{\alpha^4(Z-\sigma_2)^4}{n^4} \cdot \left(\frac{n^5}{8k^5}+\frac{3n^4}{8k^4}+\frac{n^3}{8k^3}-\frac{15n^2}{8k^2}+\frac{15n}{8k}-\frac{35}{64}\right)\right] + \cdots$$

The formula as derived from exact theory would have a single effective value of Z. Sommerfeld and Wentzel [5] modify this heuristically by the introduction of two different screening constants, σ_1 and σ_2. The values of σ_2 can be chosen in such a way that the formula gives the observed relativity-spin doublet intervals such as $L_{II} - L_{III}$, $M_{II} - M_{III}$, etc. Having adopted suitable values of σ_2 in this way, one may subtract the calculated relativity-spin contribution from the observed term values. The reduced term values give a single value for the two levels of a relativity-spin doublet such as $L_{II}L_{III}$ which may be designated $(L_{II}L_{III})$. Then one can calculate the σ_1 values from the observed values of $(L_{II}L_{III})$ for $2p$, and so on. The curves of $\sigma_1(Z)$ for $(M_{II}M_{III})$ and $(M_{IV}M_V)$ show constant differences for σ_1, as indicated in Fig. 8.1. For L_I, M_I, N_I, etc., there is no actual doublet interval; thus Sommerfeld solved for a value of σ_2 which would make the L_I curve of $\sigma_1(Z)$ parallel to the curve of $\sigma_1(Z)$ for $(L_{II}L_{III})$. There is, of course, no theoretical basis for adopting this particular procedure.

When all this is done, it is found that the observations are well represented by the empirical σ_2 values in Table 8.2, which are quite accurately constant with respect to Z.

TABLE 8.2. RELATIVITY-SPIN SCREENING CONSTANT σ_2

	I ^2S	II,III ^2P	IV,V ^2D	VI,VII ^2F
$L(n = 2)$	2.0	3.50		
$M(n = 3)$	6.8	8.50	13.0	
$N(n = 4)$	14.0	17.0	24	34

With regard to relative intensities of X-ray lines, Webster's work [9] showed that the relative intensity of the K lines does not depend on excitation voltage, as is to be expected from their having a common initial level. Various measurements show that the intensity ratio of $K\alpha_1:K\alpha_2$ is 2:1 in accord with theory (proportionality to $2J + 1$ of the final levels). The ratio for the narrow $K\beta_1K\beta_3$ doublet in Mo was also found to be 2:1. With the L lines there are three initial levels involved; so the relative intensity depends on voltage and the problem of theoretical interpretation is partly a problem of the relative excitation functions of the three initial levels and partly one of relative transition probabilities [14]. There are outstanding discrepancies between theory and experiment.

4. Absorption

X-ray absorption involves the disappearance of X-ray quanta and the photoelectric ejection of one

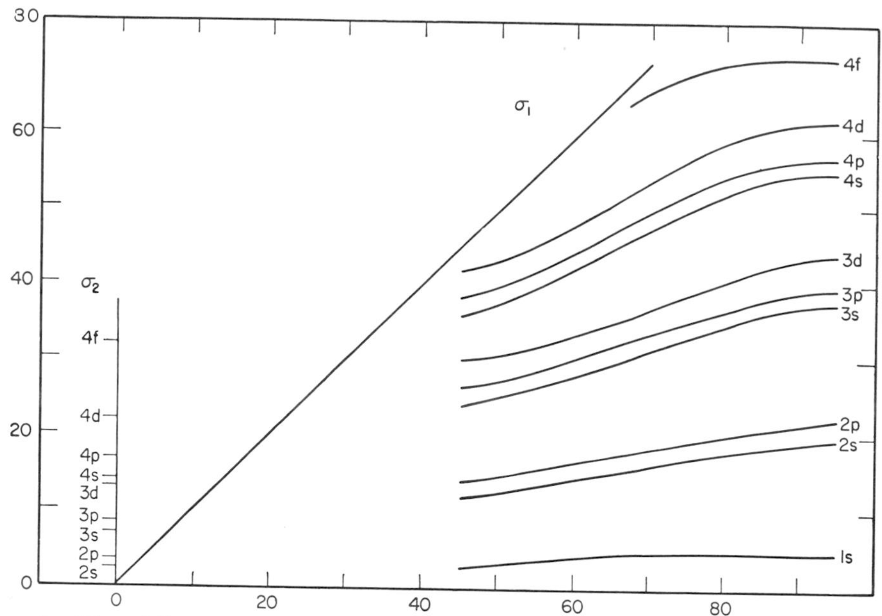

FIG. 8.1. Values of $\sigma_1(Z)$ and the constant values of σ_2 needed to represent observed values of the X-ray levels by the Sommerfeld-Wentzel application of the relativistic energy-level formula.

electron from the absorbing atom, leaving the atom in an excited ionized level. Starting at low frequencies in the ultraviolet, the quanta are able to eject only the outer loosely bound electrons. As the frequency is increased, the quantum suddenly becomes large enough to be able to eject electrons from an inner shell. This gives rise to an abrupt increase in the absorption coefficient. As the frequency is increased above such an *absorption edge*, the absorption coefficient decreases, approximately as ν^{-3}, until the frequency becomes great enough to allow electron ejection from the next deeper lying shell at which point there is another abrupt rise giving another absorption edge for this shell.

Thus the complete absorption spectrum consists of a series of sharp absorption edges, one for each X-ray level. The absorption edges were discovered by M. de Broglie [15]. Their location depends slightly on the state of chemical combination of the absorbing atoms. Also there is usually a fine structure on the low-frequency side of the absorption edge which depends on the physical-chemical state of the absorbing atoms.

Attenuation of a narrow X-ray beam is measured in terms of an absorption coefficient, μ (cm^{-1}) in $I(x) = I(0)e^{-\mu x}$, where $I(0)$ is the incident intensity and $I(x)$ the intensity after traversing the thickness x. In the experimental literature this is often expressed as a mass coefficient $\mu_m = (\mu/\rho)$, where ρ is the mass concentration (grams per cubic centimeter) of the material. In a mixture or compound in which $\rho(Z)$ is the mass concentration of the element Z, the total absorption coefficient is

$$\mu = \sum_Z \mu_m(Z)\rho(Z) \tag{8.8}$$

in which the mass absorption coefficients $\mu_m(Z)$ have the same values for all concentrations.

For many purposes it is more desirable to use an atomic absorption coefficient μ_a, defined by $\mu_a = (\mu/\rho)(A_Z/N_0)$ in which A_Z is the atomic weight and N_0 is Avogadro's number. In terms of the μ_a the over-all absorption coefficient of a mixture or compound is

$$\mu = \sum_Z \mu_a(Z)n(Z) \tag{8.9}$$

in which $n(Z)$ is the number of atoms per cubic centimeter of kind Z.

The entire attenuation is made up of a loss due to true absorption and a loss due to scattering. This means that the observed $\mu_a(Z)$ has to be regarded as the sum of a true photoelectric absorption $\tau_a(Z)$ and a scattering $\sigma_a(Z)$ (both are measured in square centimeters):

$$\mu_a(Z) = \tau_a(Z) + \sigma_a(Z) \tag{8.10}$$

In this section only the true absorption is discussed; scattering is considered in Secs. 9 and 10.

An idea of the range of values in these coefficients may be had from Table 8.3 (from notes by Prof. J. W. M. DuMond). For higher atomic numbers and longer wavelengths, the scattering becomes small compared to the total attenuation.

Early work on the dependence of μ_a on Z and λ is due to Siegbahn [16], Bragg and Pierce [17], and Hull and Rice [18]. It is found that $\tau_a(Z)$ is representable by an empirical formula of the type

$$\tau_a(Z) = CZ^m\lambda^n \tag{8.11}$$

where m is approximately 4, n approximately 3, and C is a coefficient which changes discontinuously on passing an absorption edge. The behavior at the K edge is indicated in Table 8.4, which gives the value of μ_a on the two sides of this edge for a number of

TABLE 8.3. TOTAL ABSORPTION AND SCATTERING CROSS SECTIONS (10^{-24} CM²)

λ	$\frac{\nu}{R}$	1 H μ_a	1 H σ_a	13 Al μ_a	13 Al σ_a	79 Au μ_a	79 Au σ_a
0.2	4,550	0.62	0.50	12	7	1,440	80
0.4	2,280	0.65	0.51	51.5	10	9,150	310
0.6	1,520	0.71	0.53	110	15	26,200	850
1.0	911	0.73	0.66	630	23		
1.4	650	0.78	0.67	1,660	33		

λ	$\frac{\nu}{R}$	6 C μ_a	78 Pt μ_a
0.05	18,200	2.39	111
11.9	76.6	33,900	316,000

TABLE 8.4. K EDGE DISCONTINUITY IN μ_a (10^{-24} CM²)

Z	λ	$\frac{\nu}{R}$	μ_a (low)	μ_a (high)
13 Al	7.95	114.5	12,540	161,000
26 Fe	1.739	523	5,000	42,600
42 Mo	0.618	1,462	1,990	14,000
56 Ba	0.331	2,750	1,230	6,400
74 W	0.178	5,120	825	3,425
82 Pb	0.142	6,420	722	2,660
92 U	0.1075	8,460	640	1,835

elements. An elaborate empirical analysis of the absorption coefficient as a function of Z and λ has been made by Jönsson [19].

Quantum-mechanical calculations of the atomic absorption coefficient for K electrons are reviewed by Heitler [20] and by Bethe and Ashkin [21]. These are for the absorption of light by a hydrogenic atom of charge Z, with the result multiplied by 2 because of the two K electrons in an atom. The calculations do not take account of screening. Writing $\nu_1 = Z^2me^4/4\pi\hbar^3$ for the ionization frequency of the hydrogenic level, and ν for the frequency of the X rays, then $h(\nu - \nu_1)$ is the kinetic energy of the ejected photoelectron whose speed in units of the velocity of light, β, is then $mc^2\beta^2/2 = h(\nu - \nu_1)$. It is convenient to define an auxiliary variable as

$$\xi = \frac{\alpha Z}{\beta} = \left(\frac{\nu_1}{\nu - \nu_1}\right)^{1/2}$$

This measures the importance of the Coulomb field of the nucleus in distorting the wave function of the ejected electron and ξ^2 is the ratio of the ionization energy to the kinetic energy of ejection.

The problem was treated by Kramers [22] by correspondence principle methods. He found

$$\tau_K = \frac{2g}{\sqrt{3}}\phi_0\alpha^{-2}\left(\frac{\nu_1}{\nu}\right)^3$$

where

$$\phi_0 = \frac{8\pi}{3}\left(\frac{e^2}{mc^2}\right)^2 = 0.665 \times 10^{-24}\text{ cm}^2$$

and g is an unevaluated numerical factor of the order of unity. The nonrelativistic calculation of Stobbe [23] gives this result

$$\tau_K = (2^7\pi)\phi_0\alpha^{-3}Z^{-2}\left(\frac{\nu_1}{\nu}\right)^4 \frac{\exp(-4\xi\cot^{-1}\xi)}{1 - \exp(-2\pi\xi)} \tag{8.12}$$

A simpler calculation which uses plane waves for the wave function of the ejected electron instead of the proper wave functions of the Coulomb field gives

$$\tau_K = 2^6\phi_0\alpha^{-3}Z^{-2}\left(\frac{\nu_1}{\nu}\right)^{7/2} \tag{8.13}$$

The relativistic calculation using relativistic wave functions for the bound state and for the ejected electron (but neglecting the nuclear Coulomb field on the ejected electron's wave function) was made by Sauter [24], who finds

$$\tau_K = (\tfrac{3}{2})\phi_0\alpha^4Z^5\left(\frac{mc^2}{h\nu}\right)^5 F(\gamma) \tag{8.14}$$

where $\gamma = 1 + h\nu/mc^2$ and

$$F(\gamma) = (\gamma^2 - 1)^{3/2}\left[\frac{4}{3} + \frac{\gamma(\gamma - 2)}{\gamma + 1} \cdot \left(1 - \frac{1}{2\gamma\sqrt{\gamma^2 - 1}}\log\frac{\gamma + \sqrt{\gamma^2 - 1}}{\gamma - \sqrt{\gamma^2 - 1}}\right)\right]$$

At extremely high energies $\gamma \gg 1$ and this approaches the simpler limiting form

$$\tau_K = (\tfrac{3}{2})\phi_0\alpha^4Z^5\frac{mc^2}{h\nu}\exp[-\pi\alpha Z + 2\alpha^2Z^2(1 - \ln\alpha Z)]$$

derived by Hall [25]. In this energy range the photo effect falls off much more slowly with increasing frequency than near the absorption edge. Calculations have been made [26] in which the effect of the Coulomb field of the nucleus on the ejected electron's wave functions is handled relativistically. Table 8.5 gives some indications of the magnitude of the effect of this improvement in accuracy. Analogous calculations have been made for the photoelectric absorption of X rays by electrons in the L shell [27].

TABLE 8.5. COMPARISON OF THEORETICAL VALUES OF τ_K IN UNITS OF $\phi_0\alpha^4Z^5$

z	$\gamma = 2$	$\gamma = 6$
Born approximation	10.4	0.45
13 Al	8.1	0.35
26 Fe	6.5	0.30
50 Sn	4.5	0.24
82 Pb	3.2	0.19

5. Angular Distribution of Photoelectrons

For low energy such that $\beta = v/c$ for the ejected electron is small, theory and experiment are in agreement in giving a distribution in solid angle of the direction of ejection of the photoelectrons as $\cos^2 \psi$, where ψ is the angle between the direction of ejection and the electric vector of the polarized X-ray beam [28]. At higher energies the momentum of the incident quantum results in a distortion of this angular distribution, causing it to lean forward toward the direction of motion of the incident quantum.

In the nonrelativistic Born approximation, the distribution in solid angle depends on the factor

$$\frac{\sin^2 \theta \cos^2 \phi}{(1 - \beta \cos \theta)^4} \doteq (1 + 4\beta \cos \theta) \sin^2 \theta \cos^2 \phi \qquad (8.15)$$

where θ is the angle between the direction of ejection of the electron and the direction of motion of the incident quantum and ϕ is the angle between the plane formed by these two directions and the plane of polarization of the incident quantum.

On a simple qualitative view one might suppose that the momentum of the incident quantum, $h\nu/c$, simply adds on to the momentum of each photoelectron. For $\beta = 0$, the angular distribution is symmetric around $\theta = \pi/2$; thus this view would simply result in a mean value of $h\nu/c$ for the average forward momentum of the ejected electrons. Calculation from the distribution given by theory gives $\frac{8}{5}(h\nu - h\nu_1)/c$ for this mean, where ν_1 is the critical frequency of the absorption edge. Therefore, for $h\nu \gg h\nu_1$, the forward momentum of the electrons is considerably more than that of the incident quantum, indicating that an electron in the atom is more likely to be ejected when moving in the direction of the incident quantum than when moving in the opposite direction.

6. Intensity Measurement

The intensity I of an X-ray beam in absolute units is an energy flux in ergs per square centimeter per sec, that is, the energy crossing unit area normal to the direction of propagation in unit time. The total exposure $E = It$, or in case of variable I, then $E = \int I\, dt$, is measured in ergs per square centimeter which have crossed a unit area of a plane normal to the direction of propagation.

The reciprocity law of blackening of a photographic plate, according to which the darkening D is a function of E, and not of the factors I and t separately, holds more accurately for X rays and γ rays than with visible or ultraviolet light [29]. The density of a plate, D, is defined as $D = \log (I_0/I)$, in which I_0 is the incident light intensity and I the transmitted light intensity. The characteristic curve relating D to E for photographic plates is quite different in shape for X rays and for ordinary light [30].

The most generally used method of measuring X-ray exposure is by means of the ionizing action of the X rays in the gas of an ionization chamber. The unit of exposure is the *roentgen*, defined as "The roentgen shall be that quantity of X or γ radiation such that the associated corpuscular emission per 0.001293 g of air produces in air ions carrying 1 esu of quantity of electricity of either sign."

Thus the roentgen is a measure of the ionizing effect of the X rays in air, rather than a direct measure of the energy flux. The same energy flux I of hard X rays gives a much lower value in roentgens than do soft X rays. It is a convenient measure for medical practice because animal and human tissue absorb in about the same way, as to frequency dependence, as does air, and the biological effects are related directly to the total amount of radiation absorbed in the tissues.

The relation [31] between exposure in energy units and the measure in roentgens for various wavelengths is:

λ, A	*Ergs per square centimeter for 1 roentgen*
0.1	3,200
0.2	1,900
0.3	800
0.4	400
0.5	250
0.6	150
0.7	125
0.8	100
0.9	80
1.0	75

In an ionization chamber most of the ionization is produced secondarily by the ionizing impacts of the photoelectrons and Compton recoil electrons that are ejected by primary actions of the X rays. A basic quantity of importance for conversion of the roentgen to absolute units is the average energy, ϵ, per ion pair formed in air as the over-all average of these processes. Experiments show that ϵ is independent of the energy of the incident quanta. Many determinations of ϵ have been made by various methods from which the best final result is believed to be $\epsilon = 33$ ev per ion pair in air, or 5.29×10^{-11} erg per ion pair in air. Since 1 roentgen corresponds to $e^{-1} = 2.08 \times 10^9$ ion pairs per cubic centimeter or 1.61×10^{12} ion pairs per gram of air, it follows that an exposure of 1 roentgen results in the absorption in air of 85 ergs per gram of air [32].* The corresponding energy in absorption in soft animal tissues for 1 roentgen is 95 ergs/g, the large value being caused by the high proportion of hydrogen which has about twice the mass absorption coefficient of other light elements.

7. Internal Conversion: Auger Effect

When an atom is in an excited X-ray level, having had one electron removed from an inner shell by whatever means, it may return to a state of lower energy by ejecting one of its own electrons as a process that is alternative to the ordinary quantized emission of radiation. This is called internal conversion or the Auger effect.

That some other process for returning to lower levels must be operative was revealed by experiments in which the *fluorescence yield* was measured and found to be less than unity. This is defined as the ratio of the total number of quanta emitted from a

* W. V. Mayneord and J. R. Clarkson give 83 ergs/g for this quantity.

given initial level to the number of atoms that have been excited to that state.

That atoms can get rid of the excitation energy by throwing off one or more electrons was discovered by Auger in cloud-chamber studies [33]. The excited atom already has one electron missing when in an excited X-ray state. Since another electron is ejected in an Auger process, this will leave the atom in a state in which two electrons are missing. The notation $W(K)$, $W(L_{\text{I}})$, etc., will be used for the energy of the ordinary X-ray levels, and $W(L_{\text{I}},L_{\text{I}})$, etc., for the energy of an atom in which two electrons are missing from the L shell. Roughly, $W(L,L) = 2W(L)$.

If the atom is initially in the K level, its energy is $W(K)$. After the Auger process the atom finds itself in the $W(L,L)$ level, one L electron having gone down to fill the hole in the K shell and the other L electron having been ejected with the kinetic energy $T = W(K) - W(L,L)$. Evidently it is also possible for the ejected electron to have the kinetic energy $T = W(K) - W(L,M)$ if the ejected electron comes from the M shell instead of the L shell.

With the atom left in a doubly ionized state such as $W(L,L)$ the atom may return to its normal state by single- or double-electron jumps with the emission of nondiagram or satellite lines. But alternatively another Auger process may occur in which a third electron is ejected from the M shell with kinetic energy $T = W(L,L) - W(M,M,M)$ in a process in which two M electrons go to fill the holes in the L shell and one M electron is ejected with the excess energy.

These various energetically possible radiationless transitions are sometimes called internal conversion. The name was originally suggested by the mental picture that regards the atom as first emitting a quantum of radiation in passing from the K to an L level, which is then absorbed on the way out of the same atom by an electron in the L or M shells. Actually, the effect is due to a coupling between two states of equal energy, namely, $W(K)$ and $W(L,L) + T$, but it is not correct to regard this coupling as arising solely in the way implied by thinking of a quantum as being emitted and then reabsorbed while in the same atom.

The first theoretical treatment is due to Wentzel [34]. He estimates that the Auger transition probability for the process in which an L electron is ejected is of the order $k = 10^{15}\ \text{sec}^{-1}$. The radiative transition probability from the K state varies roughly as Z^4 and is of the order of $10^{15}\ \text{sec}^{-1}$ for Z close to 30. Assuming that k does not depend on Z, this would give

$$\frac{Z^4}{30^4 + Z^4}$$

for the fluorescence yield, that is, the fraction of all transitions from the K level which are radiative. This is in good general agreement with the experimental results [35]. In heavy elements the relativistic effects are theoretically important according to calculations by Massey and Burhop [36]. Their calculations are carried out in detail for 79 Au for Auger transitions: $(L_{\text{I}}L_{\text{I}} \to K,T)$, $(L_{\text{I}}L_{\text{II}} \to K,T)$, and $(L_{\text{I}}L_{\text{III}} \to K,T)$, where T denotes the ejected electron. But the theoretical calculations have not been subjected to experimental test.

8. Pair Production

X-ray quanta for which $h\nu > 2mc^2$ may be absorbed in the field of an atomic nucleus with the creation or materialization of an electron-positron pair, as discovered by Anderson [37]. The quantum energy is partly converted into the rest energy of the two created particles, the excess going into their kinetic energy.

In the theoretical treatment [38] the positron is interpreted as being a hole in the normally filled negative energy states which are solutions of the Dirac relativistic wave equations for the electron. Thus the process is theoretically akin to X-ray absorption except that here the absorption is by one of the normally not observed electrons in negative energy states.

The simplest Born approximation works with plane wave functions and is valid only if the $\beta(=v/c)$ values of the positron and electron, β_+ and β_-, are each large compared with $2\pi\alpha Z$, and therefore it is valid only for light elements. Writing E_+ and E_- for the total energy (including mc^2) of positron and electron measured in mc^2 units, and p'_+/c and p_-/c for the momenta (where the p's are also in mc^2 units) and E for the energy of the incident quantum in the same units, the transition is one which carries an electron from an initial negative energy state $(-E_+,-p_+)$ to the final positive energy state $(+E_-,+p_-)$; so the energy balance requires $E = E_+ + E_-$.

The Born approximation gives for the cross section for which E_+ lies in dE_+ in the extreme relativistic limit in which both E_+ and E_- are large compared with unity,

$$\phi(E_+)\,dE_+ = 4\alpha Z^2 r_0^2\,dE_+\,E^{-3}\left(E^2 - \tfrac{4}{3}E_+E_-\right) \cdot \ln\left(\frac{2E_+E_-}{E} - \frac{1}{2}\right) \qquad (8.16)$$

in which $r_0 = e^2/mc^2$. Since the effect of the Coulomb field on the wave functions is neglected in this approximation, this is symmetric in E_+ and E_-. Actually the Coulomb field attracts the electron and repels the positron, giving rise to a distortion in the distribution of energy which tends to give the positrons more kinetic energy at the expense of that of the electrons.

Screening [39] of the Coulomb field becomes important if the energies of both particles are high, the criterion being that $4E_+E_-$ be large compared with $\alpha Z^{-1/3}$. The effect of screening is to modify the extreme relativistic limit for the pair production cross section to

$$\phi(E_+)\,dE_+ = 4\alpha Z^2 r_0^2\,dE_+ \cdot \left[E^{-3}\left(E^2 - \frac{4}{3}E_+E_-\right)\ln\frac{4\alpha Z^{-1/3}}{3} - \frac{E_+E}{9E^3}\right] \qquad (8.17)$$

The distribution in angle tends to be strongly in the forward direction for E large compared to unity. The average angle of emission of the electrons and positrons relative to the direction of the incident quantum is of the order E^{-1}, in radians.

Integration over E_+ from 1 to $(E - 1)$ gives the

following expressions for the total cross section for pair production in the extreme relativistic limit

$$\begin{aligned}\phi_{\text{pair}} &= \alpha Z^2 r_0^2 (\tfrac{28}{9} \ln 2E - \tfrac{218}{27}) \qquad \text{(no screening)} \\ &= \alpha Z^2 r_0^2 \left(\frac{28}{9} \ln \frac{4\alpha Z^{-1/3}}{3} - \frac{2}{27}\right) \qquad \text{(screening)}\end{aligned} \tag{8.18}$$

Figure 8.2 shows the theoretical values of this total cross section as functions of E, including some results

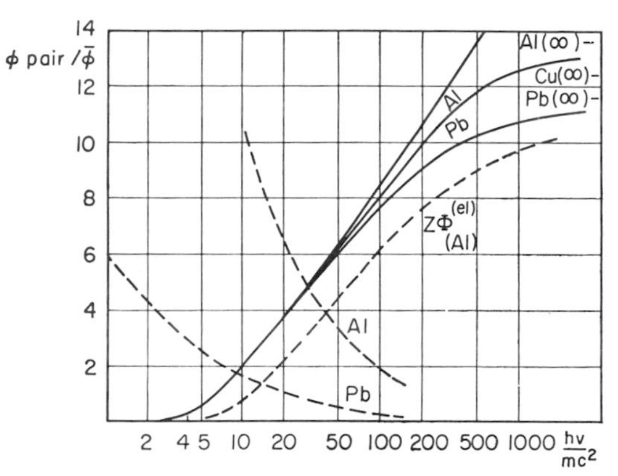

FIG. 8.2. Cross section for pair production in units of $\alpha Z^2 r_0^2$ as a function of incident quantum energy E in mc^2 units. The straight-line continuation for high energies corresponds to neglect of screening, while the actual curves for Al and Pb show its effect. On the right are the asymptotic high energy values for Al, Cu, and Pb. The descending dotted curves give the corresponding scattering cross section and the curve marked $Z\Phi_{el}$ is a rough estimate of the pair production in Al by direct action with its electrons. (*After Heitler* [38]).

of exact numerical calculations carried out at lower and intermediate energies. The following data are illustrative of the degree of validity of the Born approximation:

For 82 Pb, E	ϕ_{pair} in $\alpha Z^2 r_0^2$		$\frac{E_{+\text{av}} - 1}{E_{-\text{av}} - 1}$
	Born	Exact	
3.0	0.085	0.17	2.0
5.2	0.64	0.73	1.4

The last column gives an indication of the degree of asymmetry in energy distribution introduced by the Coulomb field, being the ratio of average positron kinetic energy to average electron kinetic energy, a quantity which the Born approximation gives as unity.

Besides the process considered, pair production can occur by absorption of the γ ray with an electron

$$h\nu + e \rightarrow e + e + e^+$$

The threshold value of E for this process is 4 instead of 2 owing to restrictions imposed by conservation of energy and momentum, and the cross section per electron at extremely high energies is, theoretically,

$$\phi_{\text{pair, electron}} = \alpha r_0^2 (\tfrac{28}{9} \ln 2E - a) \tag{8.19}$$

where a is estimated to be about 11.3.

There are not many experimental measurements. One by Blackett and Occhialini [37] gave $\phi_{\text{pair}} = 2.8$ barns for Pb, using the $E = 5.2$ γ rays emitted by Th C″, this value being in good agreement with the theory. Walker [40] measured the Z dependence of the pair-production cross section for γ rays of $E = 34.4$, finding the following relative values:

Z	3	13	29	50	82
ϕ	34	530	2,400	6,800	16,600
ϕ/Z^2	3.8	3.1	2.9	2.7	2.5

The slow decrease of ϕ/Z^2 with increasing Z is believed to be a manifestation of the contribution to the total cross section of the interaction with the electrons which tends to make the cross section vary as $Z(Z + b)$ instead of Z^2.

De Wire and Beach [41] measured the distribution in energy of the positrons produced by absorption of gamma rays of 280-Mev energy ($E = 548$) and confirmed the theoretical distribution to an experimental accuracy of 5 to 10 per cent.

9. Coherent Scattering

When a beam of X rays passes an electron, part of it is scattered into a spherical wave diverging from the location of the electron. When the scattering is due to a group of electrons, part of it is coherent, having a definite phase relation to the phase of the incident wave, and part is incoherent, being modified in frequency.

The coherent scattering is that which gives rise to the modified velocity of phase propagation, or refraction, in a medium containing many electrons, and also the diffraction effects that are used to study the periodic electron density distribution in crystals (Part 8, Chap. 1). The incoherent scattering of X rays is known also as the Compton effect.

The classical theory of scattering by free electrons is due to J. J. Thomson [42]. Under action of an electromagnetic wave of amplitude $E\mathbf{e}_0$, where $\mathbf{e}_0$ is its unit polarization vector, and of propagation vector $\mathbf{k}_0$, the forced amplitude of acceleration of a free electron is $a = eE/m$ and the resulting time-average rate of dipole radiation integrated over all directions is $\frac{1}{3}(e^2/c^3)a^2$ or $\sigma_0(cE^2/8\pi)$, where

$$\sigma_0 = \frac{8\pi}{3}\left(\frac{e^2}{mc^2}\right)^2 = 0.666 \text{ barns} \tag{8.20}$$

Here $cE^2/8\pi$ is the average energy crossing unit area normal to the direction of propagation in unit time; so σ_0 is the effective cross section for scattering in all directions, and is often called the Thomson scattering cross section.

As to polarization and distribution in angle, the forced amplitude is along $\mathbf{e}_0$. The radiation that is scattered outward in the direction $\mathbf{k}$ is polarized with its electric vector in the plane of $\mathbf{e}_0$ and $\mathbf{k}$. This defines the polarization vector $\mathbf{e}$ of the wave scattered in the direction $\mathbf{k}$. The intensity scattered in the direction $\mathbf{k}$ is proportional to $|\mathbf{e}\cdot\mathbf{e}_0|^2$; so the cross section for scattering into solid angle $d\Omega$ in any given direction is

$$\left(\frac{e^2}{mc^2}\right)^2 |\mathbf{e}\cdot\mathbf{e}_0|^2\, d\Omega \tag{8.21}$$

If the initial beam is unpolarized, then, for scattering through an angle θ, the component whose electric vector is in the plane of scattering is affected by the factor $\cos^2\theta$, and the component at right angles has a factor unity, so that the cross section for scattering an unpolarized incident beam into $d\Omega$ at θ is

$$\left(\frac{e^2}{mc^2}\right)^2 \frac{1}{2}(1+\cos^2\theta)\, d\Omega \tag{8.22}$$

Therefore, when an unpolarized beam is scattered through 90°, the resulting scattered beam is completely polarized with its electric vector normal to the plane of scattering. This was how Barkla [43] first demonstrated polarization of an X-ray beam scattered by carbon.

For the scattering of a group [44] of N free electrons, located at points whose position vectors are $\mathbf{r}_n$ one has to add the amplitudes of the scattered waves. For an incident wave whose propagation vector is $\mathbf{k}_0$ the phase of oscillation of the incident wave at the location of the nth electron will be $\omega_0 t - \mathbf{k}_0\cdot\mathbf{r}_n$. The scattered wave from this electron in the direction $\mathbf{k}$ at a large distance from the origin will have a further delay in phase of $2\pi R/\lambda - \mathbf{k}\cdot\mathbf{r}_n$ so that the phase of the nth scattered wave is

$$\phi - \delta_n = \left(\omega_0 t - \frac{2\pi R}{\lambda}\right) - (\mathbf{k}_0 - \mathbf{k})\cdot\mathbf{r}_n \tag{8.23}$$

The total wave amplitude will then be the real part of

$$E = E_0\left(\frac{r_0}{R}\right) P \sum_n \exp i(\phi - \delta_n) \tag{8.24}$$

in which $r_0 = e^2/mc^2$ and P is 1 or $\cos\theta$, depending on the plane of polarization of the incident wave. As in Part 6, Chap. 1, Sec. 4, the scattered intensity is the real part of $(c/8\pi)E^2$.

The separation into coherent and incoherent scattering involves averaging over statistical distributions of possible positions of each of the electrons. The average total scattering, which is the sum of the average coherent part and the average incoherent part, is given by calculating the mean-square value of the scattered amplitude $<E^2>_{\text{av}}$. The coherent part alone is given by calculating $(E)_{\text{av}}{}^2$, which is the scattered intensity associated with the average amplitude of scattering.

The simplest case is that in which there are no correlations between the motions of the electrons and $F_n(\mathbf{r}_n)\, dv_n$ is the probability of finding the nth electron in dv_n at $\mathbf{r}_n$. Then in calculating E_{av} one needs

$$\Phi_n = \iiint F_n(\mathbf{r}_n) \exp(-i\delta_n)\, dv_n \tag{8.25}$$

and obtains

$$I_{\text{coherent}} = I_0 r_0{}^2 P^2 \left(\sum_n |\Phi_n|^2 + \sum\sum_{n\neq m} \Phi_n\bar{\Phi}_m\right) \tag{8.26}$$

On the other hand the calculation for the total scattering gives

$$I_{\text{total}} = I_0 r_0{}^2 P^2 \left(N + \sum\sum_{n\neq m} \Phi_n\bar{\Phi}_m\right) \tag{8.27}$$

and therefore, by subtraction,

$$I_{\text{incoherent}} = I_0 r_0{}^2 P^2 \left(N - \sum_n |\Phi_n|^2\right) \tag{8.28}$$

In the general case there will be correlations between the motions of the electrons. For this case the electron probability distribution function is given by $F(\mathbf{r}_1, \mathbf{r}_2, \ldots, \mathbf{r}_N)\, dv_1\, dv_2\, dv_3 \cdots dv_N$ for the simultaneous probability that 1 is in dv_1, etc. The calculation of the average amplitude will proceed as before leading to the definition of the Φ_n with $F_n(\mathbf{r}_n)\, dv_n$ now defined as the result of integrating F over the coordinate space of all the electrons except the nth. The new feature arises in the calculation of the average squared amplitude, for this requires

$$\sum_{n,m} \iint \cdots \int F^2(\mathbf{r}_1, \mathbf{r}_2, \ldots, \mathbf{r}_N) \exp -i(\delta_n - \delta_m)\, dv_1\, dv_2 \cdots dv_N \tag{8.29}$$

the integral extending over the coordinates of all the electrons. Since the phase factor involves the electrons in all possible pairs, this leads to the definition of pairwise correlated electron density functions

$$G_{nm}(\mathbf{r}_n, \mathbf{r}_m) = \iint \cdots F^2(\mathbf{r}_1 \cdots \mathbf{r}_N)\, dv_1 \cdots dv_N \tag{8.30}$$

where the integration extends over all the dv's except dv_n and dv_m. This can be expressed in terms of an uncorrelated and a correlated part

$$G_{nm}(\mathbf{r}_n, \mathbf{r}_m) = F_n(\mathbf{r}_n)F_m(\mathbf{r}_m) + C_{nm}(\mathbf{r}_n, \mathbf{r}_m) \tag{8.31}$$

where the correlation term C_{nm} then gives rise to another term in the form factors

$$\Phi_{nm} = \iint C_{nm}(\mathbf{r}_n, \mathbf{r}_m) \exp[-i(\delta_n - \delta_m)]\, dv_n\, dv_m \tag{8.32}$$

and these give rise to an extra set of terms in the expression for the total scattering

$$I_{\text{total}} = I_0 r_0{}^2 P^2 \left[N + \sum\sum_{n\neq m} (\Phi_n\bar{\Phi}_m + \Phi_{nm})\right] \tag{8.33}$$

Since the coherent scattering is formally the same as before, the terms arising from correlations in the motion of the electrons are additional contributions to the incoherent scattering.

The total average amplitude involves only the sum $\Sigma_n F_n(\mathbf{r}_n)$, which can be written $F(\mathbf{r})\, dv$, a function which integrates to N over all space and gives the probable number of electrons in dv. When the electrons in question are those belonging to a single atom, this is called the electronic structure factor, or atomic scattering power f_0, defined by

$$f_0 = \iiint F(\mathbf{r}) e^{-i\delta}\, dv \tag{8.34}$$

If the atom has spherical symmetry, $F(\mathbf{r})$ is a function of r alone, where the origin is at the center of the atom. Writing $u(r) = 4\pi r^2 F(r)$ for the number of electrons between r and $r + dr$

$$f_0(s) = \int_0^\infty u(r) \frac{\sin sr}{sr}\, dr \tag{8.35}$$

in which $s = |\mathbf{k}_0 - \mathbf{k}| = (4\pi/\lambda) \sin \theta/2$. If $u(r)$ is known, $f_0(s)$ may be calculated, or conversely if $f_0(s)$ is measured experimentally, then $u(r)$ may be calculated, using the Fourier transform,

$$u(r) = \frac{2r}{\pi} \int_0^\infty s f_0(s) \sin sr\, ds \tag{8.36}$$

Calculations of electronic structure factors for various atoms based on Hartree or Thomas-Fermi density distributions of electrons have been made by James and Brindley [45]. Several representative values are given in Table 8.6.

TABLE 8.6. THEORETICAL ATOMIC STRUCTURE FACTORS, $f_0(s)$

Z	$\frac{s}{4\pi} = \lambda^{-1} \sin \frac{\theta}{2}$ (λ in angstroms)					
	0	0.2	0.4	0.6	0.8	1.0
1 H.	1	0.48	0.13	0.04	0.02	0.00
6 C.	6	3.0	1.9	1.6	1.3	1.0
13 Al.	13	8.95	6.6	4.5	3.1	2.3
29 Cu.	29	21.4	15.2	11.7	9.1	7.3
80 Hg.	80	66	50	41	33	28

10. Incoherent Scattering: Compton Effect

When relatively hard X rays are scattered by light elements and the radiation that is scattered through an angle θ is studied in an X-ray spectrometer, it is found [46] that, corresponding to each line in the incident radiation, there is, in the scattered radiation, an unmodified line at the same frequency and another line called the modified line that is shifted to longer wavelength given by

$$\lambda_2 = \lambda_1 + \lambda_0(1 - \cos \theta) \tag{8.37}$$

in which $\lambda_0 = 0.0243 \times 110^{-8}$ cm $= h/mc$. The process by which this modified scattered radiation is produced is known as the *Compton effect.*

The discovery of the Compton effect was of great importance in the development of quantum theory because it provided direct experimental evidence that light quanta carry momentum $h\nu/c$ in their direction of propagation as well as having energy $h\nu$. Compton's interpretation of the shift was based on the idea that the incident quantum $h\nu_1$ is scattered by an essentially free electron in the scatterer with conservation of energy and momentum. Letting E_1 and E_2 be the energy of the incident and scattered quanta, respectively, and E that of the electron, including the rest energy, and all measured in mc^2 units, then the conservation laws of momentum and energy give

$$\begin{aligned} E_1^2 + E_2^2 - 2E_1E_2 \cos \theta &= E^2 - 1 \\ E_1 &= E_2 + E - 1 \end{aligned} \tag{8.38}$$

since for quanta the momentum is $h\nu/c$ and for electrons c times the momentum is equal to $E^2 - 1$. The initial kinetic energy of the electron is neglected. Eliminating E between these

$$E_2^{-1} = E_1^{-1} + 1 - \cos \theta \tag{8.39}$$

This is the same as the experimentally observed law for the wavelength shift. Also the kinetic energy of the recoil electron, $E - 1$, becomes

$$E - 1 = E_1 \left[1 - \frac{1}{1 + E_1(1 - \cos \theta)}\right] \tag{8.40}$$

The angle of recoil ϕ of the electron is related to the angle of scatter θ of the quantum by

$$\cot \phi = (1 + E_1) \tan \frac{\theta}{2} \tag{8.41}$$

The existence of the recoil electrons was proved by the tracks they make in a cloud chamber [47]. The simultaneity of a scattered quantum and a recoil electron was shown by Bothe [48]. The kinetic energy of the recoil electrons was shown to agree with the theory by use of a magnetic spectrometer [49].

Polarization [50] of both the modified and the unmodified scattered radiation was shown to be complete for $\theta = \pi/2$.

For elements of low Z, the scattered energy is essentially all in the modified line. As Z is increased, some energy appears in the unmodified line [51], which for high Z may be more intense than the unmodified line. The unmodified scattering is that which is done by electrons that are so tightly bound that the recoil energy would be insufficient to set them free from the atom, so that the atom as a whole takes up the momentum change of the scattered quantum, which it can do without appreciable change of its energy because of its mass.

Accurate measurements [52] show that the modified line is shifted a little less than indicated by the theory for entirely free electrons, which is a consequence of their not being entirely free.

The modified line is experimentally found [53] to be broader than the line in the exciting radiation. This is due to the fact that the quasi-free electrons in the scatterer are not initially at rest, but have a distribution of momenta which manifests itself in a broadening of the Compton scattered radiation. Important information about the distribution of electronic momenta, as determined by application of Fermi statistics to the free electron gas in a metal, has been obtained this way.

The cross section for scattering an X-ray quantum into unit solid angle at angle θ by a free electron is given by the *Klein-Nishina formula* [54] based on use of the Dirac relativistic quantum mechanics of the electron. The scattering cross section for one electron initially at rest is

$$r_0^2\tfrac{1}{2}(1+\cos^2\theta)(1+E\text{ vers }\theta)^{-3}\cdot\left[1+\frac{E^2\text{ vers}^2\theta}{(1+\cos^2\theta)(1+E\text{ vers }\theta)}\right]$$

Here vers $\theta = 1 - \cos\theta$, and $E = h\nu/mc^2$. Although the formula is derived for strictly free electrons, it is a good approximation for bound electrons, provided the recoil energy of the scattering electron as calculated from the laws of conservation of energy and momentum is large compared with the binding energy of the electron.

In Fig. 8.3 is shown the angular dependence of the scattering for several representative values of E. At

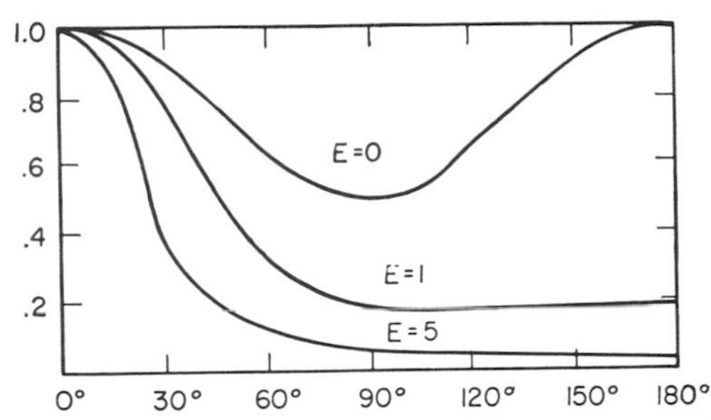

FIG. 8.3. The angular dependence of Compton scattering for several different values of the primary quantum energy in mc^2 units.

low energies, $E \doteq 1$, the angular dependence is given entirely by the factor $\frac{1}{2}(1+\cos^2\theta)$, which arises from the polarization and is symmetric around $\theta = \pi/2$. For larger energies, the scattering becomes more and more concentrated in the forward direction. This is mainly due to the factor $(1 + E\text{ vers }\theta)^{-3}$ which was first introduced by Breit [55]. The theory takes on a simpler form if expressed in coordinates moving relative to the laboratory coordinates with a vector velocity that is the mean of the initial and final velocities of the scattering electron. In such a system the effect of the scattering is to change the direction of the electron's momentum without altering its magnitude and therefore there is no change in energy of the quantum on scattering. The change in energy of the quantum as observed in the laboratory system is then due to the Doppler effect on its frequency when the same event is referred to the laboratory coordinates.

When integrated over all directions of scatter, the total cross section for Compton scatter is

$$\sigma = 2\pi r_0^2\left\{\frac{1+E}{E^2}\left[\frac{2(1+E)}{1+2E}-\frac{1}{E}\ln(1+2E)\right]+\frac{1}{2E}\ln(1+2E)-\frac{1+3E}{(1+2E)^2}\right\} \tag{8.42}$$

In the nonrelativistic limit $E \ll 1$, this approaches the classical Thomson cross section σ_0, defined in Sec. 9. More exactly,

$$\sigma = \sigma_0(1 - 2E + \tfrac{26}{5}E^2 + \cdots) \tag{8.43}$$

In the extreme relativistic case $E \gg 1$, the limiting form is

$$\sigma = \sigma_0\frac{3}{8E}(\ln 2E + \tfrac{1}{2}) \tag{8.44}$$

The dependence of total cross section for Compton scattering on primary energy of the X-ray quantum is shown in Table 8.7.

TABLE 8.7. CROSS SECTION FOR COMPTON SCATTERING AS FUNCTION OF PRIMARY ENERGY E IN mc^2 UNITS

E	$\frac{\sigma}{\sigma_0}$	E	$\frac{\sigma}{0} \times 10^2$
0.05	0.913	5	19.1
0.1	0.84	10	12.3
0.2	0.737	20	7.54
0.33	0.637	50	3.76
0.5	0.563	100	2.15
1	0.431	200	1.22
2	0.314	500	0.556
3	0.254	1,000	0.304

References

1. Kulenkampff, H.: *Ann. Physik*, **69**: 548 (1922). Siegbahn, M.: "Spektroscopie der Roentgenstrahlen," 2d ed., p. 454, Springer, Berlin, 1931.
2. Nicholas, W. W.: *J. Research Natl. Bur. Standards*, **2**: 837 (1929). Kulenkampff, H.: *Ann. Physik*, **87**: 579 (1928). Duane, W.: *Proc. Natl. Acad. Sci.*, **15**: 805 (1929).
3. Thomson, J. J.: *Phil. Mag.*, **45**: 172 (1898).
4. Kramers, H. A.: *Phil. Mag.*, **46**: 836 (1923). Wentzel, G.: *Z. Physik*, **27**: 257 (1924).
5. Heitler, W.: "Quantum Theory of Radiation," 3d ed., p. 242, Oxford University Press, New York and London, 1954.
6. Kirkpatrick, P., and L. Wiedmann: *Phys. Rev.*, **67**: 321 (1945).
7. Moseley, H. B. J.: *Phil. Mag.*, **26**: 1024 (1913); **27**: 703 (1914).
8. Bohr, N.: *Phil. Mag.*, **26**: 1, 426 (1913).
9. Webster, D. L.: *Phys. Rev.*, **7**: 559 (1916).
10. Kossel, W.: *Verhandl. der deut. physik. Ges.*, **18**: 339, 396 (1916); *Physik. Z.*, **18**: 240 (1917); *Z. Physik*, **1**: 119 (1920); **2**: 470 (1920).
11. Sommerfeld, A.: *Ann. Physik*, **51**: 1 (1916). Sommerfeld, A., and G. Wentzel: *Z. Physik*, **7**: 86 (1921). Wentzel, G.: *Z. Physik*, **16**: 51 (1922).
12. Bohr, N., and D. Coster: *Z. Physik*, **12**: 342 (1923).

12*a*. Bearden, J. A.: "X-ray Wavelengths," U.S. Atomic Energy Commission, Oak Ridge, Tenn., 1964.

13. Segré, E.: *Accad. nazl. Lincei Roma*, **14**: 501 (1931); **16**: 442 (1932).
14. Compton, A. H., and S. K. Allison: "X-rays in Theory and Experiment," pp. 637–654, Van Nostrand, Princeton, N.J., 1935.
15. De Broglie, M.: *Compt. rend.*, **158**: 1493 (1914); **163**: 87, 353 (1916). Wagner, E.: *Ann. Physik*, **46**: 868 (1915).
16. Siegbahn, M.: *Physik. Z.*, **15**: 753 (1914).
17. Bragg, W. H., and S. E. Pierce: *Phil. Mag.*, **28**: 626 (1914).
18. Hull, A. W., and M. Rice: *Phys. Rev.*, **8**: 326 (1916).
19. Jönsson, E.: Thesis, Uppsala, 1928. Compton, A. H., and S. K. Allison: "X-rays in Theory and Experiment," p. 537, Van Nostrand, Princeton, N.J., 1935.
20. Heitler, W.: "Quantum Theory of Radiation," 3d ed., pp. 204–211, Oxford University Press, New York and London, 1954.

21. Bethe, H., and J. Ashkin: "Experimental Nuclear Physics," vol. 1, p. 325, Wiley, New York, 1953.
22. Kramers, H. A.: *Phil. Mag.*, **46**: 836 (1923).
23. Stobbe, M.: *Ann. Physik*, **7**: 661 (1930).
24. Sauter, F.: *Ann. Physik*, **9**: 217 (1931); **11**: 454 (1931).
25. Hall, Harvey: *Revs. Mod. Phys.*, **8**: 358 (1936).
26. Hulme, H. R., J. McDougall, R. A. Buckingham, and R. H. Fowler: *Proc. Roy. Soc.* (*London*), **A149**: 131 (1935).
27. Compton, A. H., and S. K. Allison: "X-rays in Theory and Experiment," pp. 560–564, Van Nostrand, Princeton, N.J., 1935. Hall, Harvey: *Revs. Mod. Phys.*, **8**: 358 (1936); *Phys. Rev.*, **84**: 167 (1951).
28. Compton, A. H., and S. K. Allison: "X-rays in Theory and Experiment," pp. 564–582, Van Nostrand, Princeton, N.J., 1935.
29. Bell, G. E.: *British J. Radiology*, **9**: 587 (1936).
30. Sproull, W. T.: "X-rays in Practice," p. 173, McGraw-Hill, New York, 1946. May, A.: *J. Opt. Soc. Amer.*, **33**: 81 (1943). Silberstein, L., and A. P. H. Trivelli: *Phil. Mag.*, **9**: 787 (1930).
31. Mayneord, W. V.: *Nature*, **145**: 973 (1940); *British J. Radiology*, **13**: 235 (1940); **17**: 359 (1944).
32. Mayneord, W. V., and J. R. Clarkson: *British J. Radiology*, **17**: 151, 177 (1944).
33. Auger, P.: *Compt. rend.*, **180**: 65 (1925); **182**: 773, 1215 (1926); *J. phys.*, **6**: 205 (1925). Burhop, E. H. S.: "The Auger Effect and Other Radiationless Transitions," Cambridge University Press, New York and London, 1952.
34. Wentzel, G.: *Z. Physik*, **43**: 524 (1927).
35. Compton, A. H., and S. K. Allison: "X-rays in Theory and Experiment," p. 488, Van Nostrand, Princeton, N.J., 1935.
36. Massey, H. S. W., and E. H. S. Burhop: *Proc. Roy. Soc.* (*London*), **A153**: 661 (1936). Burhop, E. H. S.: *Proc. Roy. Soc.* (*London*), **A148**: 272 (1935). Pincherle, L.: *Nuovo cimento*, **12**: 81 (1935).
37. Anderson, C. D.: *Phys. Rev.*, **41**: 405 (1932); **44**: 406 (1935). Blackett, P. M. S., J. Chadwick, and G. P. S. Occhialini: *Proc. Roy. Soc.* (*London*), **A144**: 235 (1934).
38. Heitler, W.: "Quantum Theory of Radiation," 3d ed., p. 256, Oxford University Press, New York and London, 1954.
39. Bethe, H. A.: *Proc. Cambridge Phil. Soc.*, **30**: 524 (1934).
40. Walker, R. L.: *Phys. Rev.*, **76**: 1440 (1949).
41. De Wire, J. W., and L. A. Beach: *Phys. Rev.*, **83**: 476 (1951).
42. Thomson, J. J.: "Conduction of Electricity through Gases," 2d ed., p. 325, Cambridge University Press, New York and London, 1935.
43. Barkla, C. G.: *Proc. Roy. Soc.* (*London*), **A77**: 247 (1906). For later, more precise work, see A. H. Compton and C. F. Hagenow, *J. Opt. Soc. Amer.*, **8**: 487, 1924; H. Mark and L. Szilard, *Z. Physik*, **35**: 743, 1926.
44. Raman, C. V.: *Indian J. Phys.*, **3**: 357 (1928).
45. James, R. W., and G. W. Brindley: *Phil. Mag.*, **12**: 194 (1931). Compton, A. H., and S. K. Allison: "X-rays in Theory and Experiment," Appendix 4, p. 780, Van Nostrand, Princeton, N.J., 1935.
46. Compton, A. H.: *Phys. Rev.*, **21**: 715 (1923); **22**: 409 (1923).
47. Wilson, C. T. R.: *Proc. Roy. Soc.* (*London*), **104**: 1 (1923).
48. Bothe, W.: *Z. Physik*, **20**: 237 (1923).
49. Bless, A. A.: *Phys. Rev.*, **29**: 918 (1927).
50. Kallmann, H., and H. Mark: *Z. Physik*, **36**: 120 (1926).
51. Woo, Y. H.: *Phys. Rev.*, **27**: 119 (1926).
52. Ross, P. A., and P. Kirkpatrick: *Phys. Rev.*, **45**: 223 (1934); **46**: 668 (1934).
53. DuMond, J. W. M.: *Phys. Rev.*, **33**: 643 (1929); *Revs. Mod. Phys.*, **5**: 1 (1933). DuMond, J. W. M., and P. Kirkpatrick: *Phys. Rev.*, **52**: 419 (1937); **54**: 802 (1938).
54. Klein, O., and Y. Nishina: *Z. Physik*, **52**: 853 (1929).
55. Breit, G.: *Phys. Rev.*, **27**: 362 (1926).

Bibliography

Cauchois, Y., and H. Hulubei: "Longeurs d'onde des émissions X et des discontinuités d'absorption X," Hermann et Cie, Paris, 1947.

Compton, A. H., and S. K. Allison: "X-rays in Theory and Experiment," 2d ed., Van Nostrand, Princeton, N.J., 1935.

Heitler, W.: "The Quantum Theory of Radiation," 3d ed., Oxford University Press, New York and London, 1954.

Siegbahn, M.: "Spektroscopie der Roentgenstrahlen," 2d ed., Springer, Berlin, 1931.

Sproull, W. T.: "X-rays in Practice," McGraw-Hill, New York, 1946.

White, Gladys: "X-ray Attenuation Coefficients from 10 kev to 100 Mev," *Natl. Bur. Standards, Rept.* 1003, 1952.

Chapter 9

Mass Spectroscopy and Ionization Processes

By JOHN A. HIPPLE, Philips Laboratories

1. Introduction

J. J. Thomson with his parabola instrument pioneered in the mass analysis of ion beams. In his device a beam of ions having various masses and a range of energies passes transversely through uniform parallel electric and magnetic fields. After deflection by these fields, the ions pass onward and impinge on a fluorescent screen or photographic plate, forming a family of parabolas. Each parabola corresponds to a particular mass-to-charge ratio, the length of the parabola being dependent on the spread in energy of the incident ions.

Thomson showed an early awareness of the ultimate usefulness of mass analysis for the study of problems in physics and chemistry. The discovery of isotopes was one of the exciting events of this era. In this, mass analysis of ions played a key role. The parabola method remains an interesting means for simultaneously obtaining information on the types of ions present in an ion beam and their energy spread. However, most applications require a focusing action (analogous to that in optical spectroscopy) to maintain sufficient intensity as the resolution is increased. Thomson's arrangement had no focusing and therefore narrow collimation of the beam was necessary. Two instruments with focusing properties were soon introduced, one by Dempster and one by Aston. The analogy to optical spectroscopy now became evident to Aston although he carefully avoided calling Thomson's parabola instrument a spectrograph. In common usage today, the field of mass analysis is called mass spectroscopy regardless of the focusing properties of the instruments involved.

Dempster introduced direction focusing into mass spectroscopy, adopting the focusing scheme used earlier by Classen for the measurement of e/m of the electron. Ions diverging from the source are deflected by a uniform magnetic field being brought to an approximate focus after a deflection of 180°. A mass separation is effected if the ions leave the source and enter the magnetic analyzer with only a small distribution in energy. Dempster used electrical detection, adjusting his fields so the ion of the desired mass was focused on the exit slit.

In contrast, Aston's first instrument provided focusing for an initial energy distribution with no focusing for an initial divergence of the ions. Ions passed first through an electrostatic, deflecting field, traveling onward to a magnetic, deflecting field and thence to a photographic plate.

Very quickly it became apparent that the instruments of Aston and Dempster were applicable in different, although contiguous areas. Dempster's instrument with its simple magnetic analyzer and electrical detection was best suited for problems primarily concerned with intensity measurements such as those involved in studies of the ionization and dissociation of molecules by electron impact. The Dempster method and variations of it remain in wide use today for research and chemical analysis. In contrast, Aston applied his instrument to the measurement of atomic masses and the relative abundance of isotopes and for many years was the only scientist using mass spectroscopic methods for atomic mass measurements. With its sequential arrangement of electric and magnetic fields for deflecting the ions and with its photographic detection, it in turn was the prototype of the instruments later designed for this work. In the 1930s instruments were introduced which combined both velocity and direction focusing, the double-focusing instruments.

The contrast between the experimental methods just described began to disappear about 1950. High resolution became of increasing interest for chemical research and analysis; improved ion optical principles, for which the motivation for development in the past had been primarily the precise measurement of atomic masses, took on importance for studies involving the identification of the ion types composing a complex beam of ions and the measurement of their relative amounts. Also, direct photographic detection is here frequently the best available method. On the other hand, electrical detection has come into general use for atomic-mass measurements [1].*

Since the mid 1950s mass analyzers employing electric but no magnetic fields have gained a wide use [2]. They have offered attractive features for certain applications, such as small size and light weight. A capability to scan a range of masses at a high repetitive rate has provided valuable data in studies of fast reactions in chemistry. The analyzer employing electric-quadrupole fields developed by Paul and his coworkers [3] offers quite unique features: A desired type of ions can be filtered from a heterogeneous beam entering the relatively large aperture

* Numbers in brackets refer to References at end of chapter.

of the analyzer with a wide spread in velocity and direction of approach.

This chapter is devoted to the role of mass spectroscopy in ionization and dissociation studies and some related subjects. The earlier division in the application of mass-spectroscopic techniques emanating from the work of Aston and Dempster remains only of historical interest.

2. Ionization Studies. Experimental Background

For the study of ionization processes in gases many problems remained to be solved after adequate methods of mass analysis were devised. Until approximately 1930 much of the data was conflicting and not reproducible. The group at the University of Minnesota under Prof. J. T. Tate at this point made a major contribution in the design of an ion source with greatly improved control of the electron beam. Their electron impact studies of the ionization and dissociation of various diatomic molecules made with this ion source introduced the method which is today employed in many mass spectrometers using a controlled electron beam for gas analysis and ionization studies. In some of the early work at Minnesota ionization studies were made without a mass analysis although the kinetic energies of the fragment ions could be measured [4]. Bleakney [5], while at Minnesota, first used this source for mass spectroscopic work. Figure 9.1 illustrates the Minnesota source.

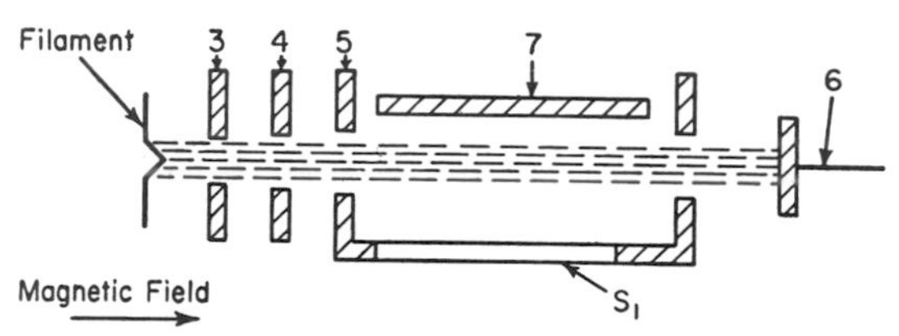

Fig. 9.1. Ion source for studies of ionization and dissociation by electron impact.

Electrons emitted from the filament and accelerated by appropriate voltages on the numbered electrodes are held in a tight beam by a uniform magnetic field. They are thus prevented from striking any electrode surfaces in the vicinity where the ions to be studied are formed, i.e., in the "ionization chamber" between electrode 7 and slit S_1. This is important in ensuring that the ionization can be attributed to the primary electrons from the filament and in minimizing distortion of the electric field through the accumulation of electric charges on insulating films, an effect otherwise difficult to control. The Minnesota source, with its tightly collimated electron beam, has another merit in producing an ion beam which emerges from slit S_1 with an energy spread much less than the full potential difference between electrodes 7 and 5 would produce in the absence of the collimating magnetic field. The commonly used mass analyzers require that the entering beam of ions be quite homogeneous in energy.

Nier was the first to use a magnetic deflection of less than 180° with electrical detection. The ion source is now far removed from the magnetic field that deflects the ion beam. To retain the desirable features of the Minnesota source, Nier introduced an additional magnet at the ion source to collimate the electron beam. The field of this magnet is considerably weaker than the main magnetic field, being usually no more than several hundred gauss. Nier's modification with his later refinements has served as the model for the many later sectored field mass spectrometers with ion sources employing an electron beam.

The pioneering studies of ionization processes with the mass spectrometer in the 1920s provided the groundwork for the more definitive studies of the 1930s which resulted largely from the availability of greatly improved experimental tools—the improved ion source, better vacuum techniques, and the introduction of vacuum tube amplifiers and controls. (One tends to forget the difficulties of the earlier experiments up to about 1935, when the only vacuum tube was the mass spectrometer tube itself!) The reproducible and still valid results of the studies of diatomic and polyatomic molecules in the 1930s set the stage for the mushrooming application of the mass spectrometer to analytical problems in research laboratories and industrial processes in the 1940s. Since then the mass spectrometer has become a conventional and routine tool for the chemist. Both physicists and chemists find it an invaluable means for the study of ionization processes in research on atomic and molecular phenomena.

3. Efficiency of Ionization by Electrons

An ionization efficiency curve is a plot of the intensity of ion current as a function of the energy of the electrons producing the ionization (in the case of ionization by electron impact). Ionization of atoms provides the simplest process for study. As the electron energy is gradually increased from zero in the ion source shown in Fig. 9.1, no ionization is detected until the electron energy attains a value equal to the ionization potential of the gas in the ionization chamber—the mass spectrometer associated with this ion source being set for measurement of the current corresponding to the mass-to-charge ratio appropriate for the particular atom. After this onset of ionization, the ion current increases with increasing electron energy, reaching a broad maximum at a region several times the ionization potential. In individual cases, the form of the ionization efficiency curve may depart in detail from this general picture. The introduction of the Minnesota source permitted for the first time the reproducible determination of ionization efficiency curves with the mass spectrometer. Study of the singly and multiply charged ions of mercury vapor by Bleakney [5] on his first instrument has been a favorite reference in this field.

Several features of the experimental results left much to be desired. The ionizing electrons were spread through an ill-defined range of energy, resulting in a smearing of the finer details of the ionization efficiency curve. One reason for this distribution in energy was the acceleration of the electrons in the ionization region by the electric field between electrodes 7 and 5 for sweeping the ions through S_1 (Fig. 9.1). The thermal distribution in energy of the electrons from the hot filament presents an even greater difficulty. These causes produce a long tailing of the ionization curve in the vicinity of the ionization potential even if the true curve is a linear function of the electron energy with a sharp break at onset.

However, the hot filament remains the most convenient source of electrons for studies with the mass spectrometer. The problem then is how to improve the curves obtained from apparatus using a hot filament. Operation at a lower temperature with an oxide-coated cathode is helpful. Uncertainty in the energy measurements remains approximately 0.1 volt at best. There being no objection-free method of determining the initial onset of ionization, the subjective nature of the procedures brings in differences of opinion. As the sensitivity of the ion detector is increased, the initial ion current will be detected at a lower and lower value of the applied voltage. An electron filter suggests itself. The first attempts were made by Lawrence [6] and Nottingham [7]. Nottingham studied the onset of ionization in mercury with electrons filtered by deflection through 180° in a magnetic field prior to entering the ionization chamber. In this work there was no mass analysis of the ions. In more recent years significant progress has been made in reducing the deleterious effect of the velocity distribution of electrons from a hot filament in mass-spectrometric studies. Before going into this, means for treating some other causes of the velocity spread will be mentioned.

The effect on the electron distribution of the filament heating current and the electric field between electrodes 7 and 5 in Fig. 9.1 can be eliminated by intermittent application of voltages to the appropriate terminals. For the filament, the heating current may be turned off in periodic fashion during those intervals when the square-wave voltage applied to an intermediate electrode (say electrode 3) permits the electrons to pass through to the ionization region. Similarly, a square-wave voltage applied between electrodes 7 and 5 is synchronized with the periodically interrupted electron beam in a manner such that no electric field exists between electrodes 7 and 5 when the electrons pass through the ionization region. Assuming other experimental precautions are taken to minimize the spread in electron energy (e.g., field penetration through the apertures in the ionization region because of voltages on adjacent electrodes), the thermal distribution of the electrons remains to smear the details of the ionization efficiency curve at the ionization potential.

Indirect methods for making more precise determinations of the onset of ionization have used modulation techniques on the electron beam and/or analytical treatment of the data, in contrast to the direct method in which the electron beam is filtered to permit only those within a very narrow velocity range (so-called "monoenergetic") to enter the ionization region. Both methods produce greatly improved data on critical potentials obtained by electron impact with the mass spectrometer and reveal structure on ionization efficiency curves heretofore evident indistinctly only in a few instances. The direct method applied to mass-spectrometric work has been so improved that the electron beam has a range of about 0.05 ev at half maximum of the distribution curve [8]. Figure 9.2 illustrates the filtering of the electrons in a series arrangement of two electrostatic analyzers for a study not involving mass analysis of ions; here the electrons scattered after making an atomic collision, as well as the incident electrons, are filtered for a selected energy. Several indirect methods are also currently interesting and produce results of comparable precision [9, 10]. They differ considerably from one another. However, for illustration here only the retarding-potential-difference method (RPD method) of Fox et al. will be described.

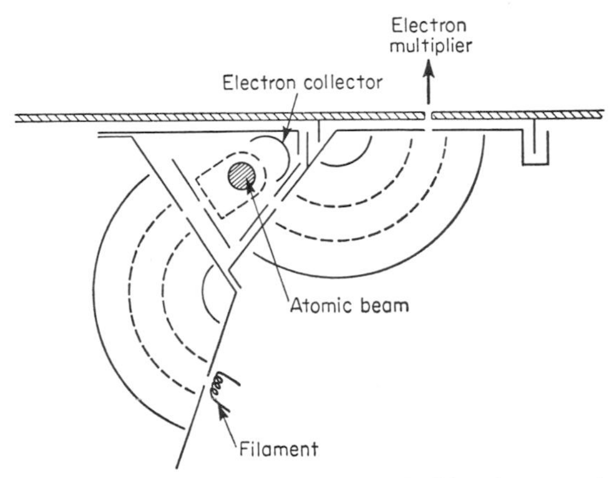

Fig. 9.2. Schematic diagram of double electrostatic analyzer. (*From ref.* 22.)

This method is illustrated in Fig. 9.3*a*. Electrons from the filament are accelerated into ionization chamber 3 by voltage V_1. Intermediate between the filament and the ionization chamber the electrons must pass through the aperture in electrode 2 at a voltage V_2 slightly negative relative to the filament. The slow electrons are thereby prevented from passing through electrode 2 and the energy distribution of the electrons reaching the ionization chamber is modified to that shown in Fig. 9.3*b*. The energy distribution now has a sharp cutoff on the low-energy side (actually an electrode system more complex than the single electrode 2 is required to achieve a sharp cutoff). In the ionization region 3 the electrons have the minimum energy eV_{2m} in Fig. 9.3*b*. If the voltage of 2 relative to the filament is now made more negative by the additional amount ΔV_R, the cutoff energy will be displaced and those electrons in the energy band $e\,\Delta V_R$ will also be prevented by electrode 2 from passing onward to the ionization chamber. The electrons within 3 now have a minimum energy eV_{2M}. The difference in the ion current produced by electrons accelerated through voltages V_{2m} and V_{2M} represents the number of ions created by electrons in the narrow band ΔV_R. A plot of this difference current against V_{2m} will show zero current at the ionization potential. For an ionization curve that is linear, the measured curve found in this manner will also be linear except for a region ΔV_R at onset. Experimentally ΔV_R is usually taken to be 0.1 volt and the average voltage $(V_{2m} + V_{2M})/2$ is plotted. For convenience of explanation no ion repeller electrode for accelerating the ions out of the

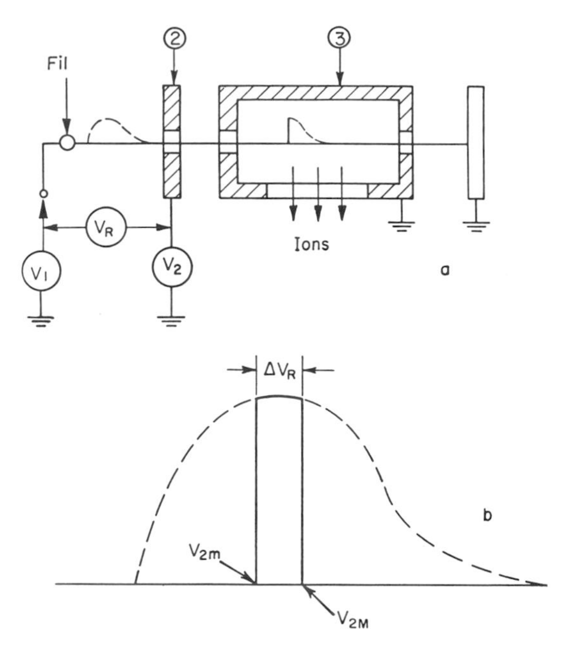

FIG. 9.3. (*a*) Method for reducing the effect of the velocity distribution of electrons in ionization studies. The slower electrons emitted from the filament are prevented from reaching the ionization chamber 3 by a suitable potential applied to 2. (*b*) By variation of the voltage V_A by the amount ΔV_R and recording of the corresponding difference in ion current, the ionization efficiency curve produced by electrons in the narrow band ΔV_R is obtained. (*From ref.* 10.)

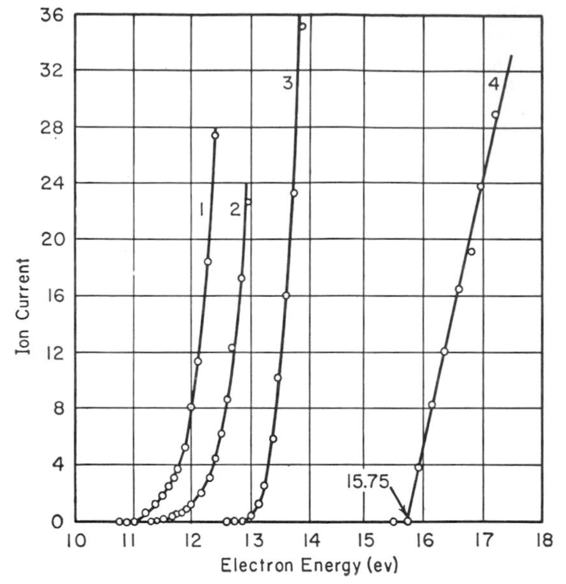

FIG. 9.4. Onset of ionization observed with A^{40} under the following conditions: 1, conventional method; 2, conventional method with 1/10 sensitivity for detection of ion current; 3, method of Fig. 9.3 with electric field in ionization region to eject ions (ion repeller field); 4, method of Fig. 9.3 plus pulsing of electron beam and ion repeller field in order that the repeller field does not disturb the electron beam. (*From ref.* 10.)

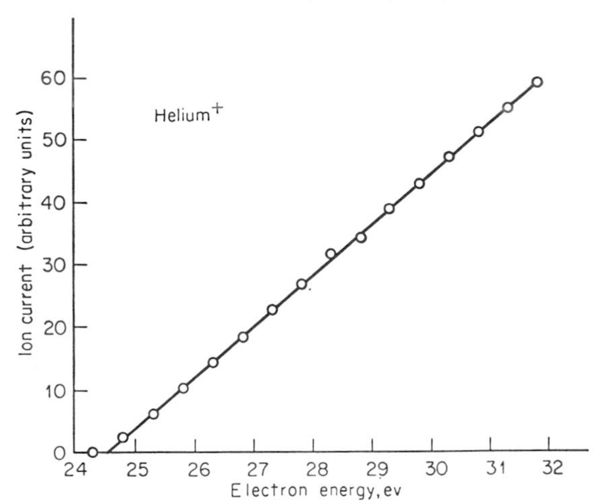

FIG. 9.5. Ionization efficiency curve for He^+. The onset has been made to agree with the spectroscopically determined ionization potential. (*From ref.* 14.)

ionization region is shown in Fig. 9.3. The smearing of the electron distribution by the repelling field can be eliminated by periodic pulsing. The improvement in the experimental curve for the ionization potential of argon is shown in Fig. 9.4.

The RPD method for improving the effective electron distribution can be effective only on the component of the electron velocity which lies along the direction of the retarding electric field. The velocity perpendicular to this direction is unaffected. The curves are in fact much sharper than the contribution from these perpendicular components permits. One can surmise that they are made ineffective by the particular geometry chosen for the electron gun.

Theoretical interest [11, 12, 13] in the problem and the development of improved techniques for measuring ionization potentials have induced many experimental studies concerned with the dependence of cross sections on electron energy above the onset of ionization. Theoretical studies suggest an nth-power variation with electron energy, where n is the degree of ionization. Results for singly charged He [14] with the RPD method are shown in Fig. 9.5. A linear relationship is evident. The fit of the data to a linear relationship as contrasted with $n = 1.1$, which was also theoretically predicted, is shown in Fig. 9.6. There have been other confirmations of the linear relationship for singly charged ions as well as the nth-power variations for multiple ionization. For multiple ionization, Fig. 9.7 shows the agreement with Xe for $n = 1$ to 6. Most of the recent studies confirm the nth-power dependence of the ionization efficiency near onset and the significance of the threshold energy determined by the extrapolation of the nth-root plots of electron energy. Competing ionization processes near the threshold can complicate the situation [16].

In some of the experiments on multiple ionization, detailed structure in the experimental curves near the onset of ionization did not emerge because of the width in the energy distribution of the electrons. Such structure can be attributed to excited electronic states

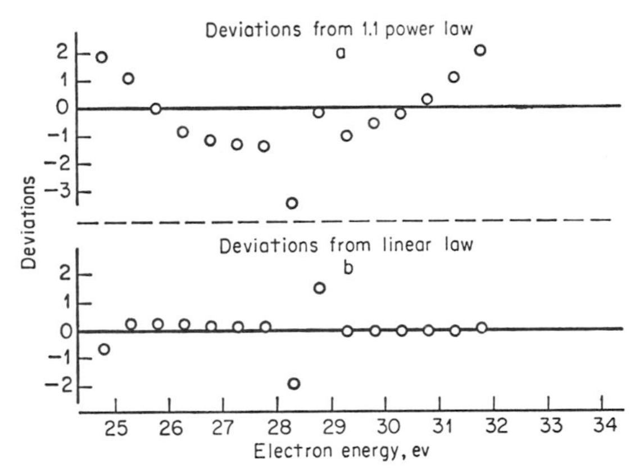

FIG. 9.6. (*a*) The deviation of the experimental points of Fig. 9.5 from the best fit by least squares to a 1.1 power curve. (*b*) The deviation of the same points from a linear (1.0) curve. (*From ref.* 14.)

of the ion close to the threshold for ionization of the unexcited ion, autoionization, or an Auger process. For more details of the work in this field reference should be made to the review of Kieffer and Dunn [17]. Figure 9.8 is illustrative [18]. The curve for the single ionization of Zn rises linearly for more than a volt. Then a pronounced hump appears, clearly indicating the occurrence of a different excitation. At 10.8 volts it is suggested that there occurs excitation of an inner (*d*) shell electron to Zn I^b states of the type identified by Beutler [19], which are above the ground state of Zn^+. By autoionization, radiationless transitions to the ground state of the ion occur with ejection of an electron. The suggested states do not clearly explain the structure observed. Recently Morrison [20] has reported a careful study of autoionization effects in the ionization-threshold curves for Xe and Kr. He cautions that competition between the cross section for direct ionization and that for autoionization can be interpreted as "spurious" structure and advises greater care than heretofore in interpreting ionization efficiency curves.

Recently developed techniques for the measurement of atomic and molecular excitation by electron impact are closely related to the subject of this chapter. Therefore one of these techniques will be described briefly. Baranger and Gerjuoy [21] postulated the existence of a compound state (He^-) to interpret excitation cross sections to the 2^3S level in helium and urged an examination of the elastic scattering of electrons with a more nearly monoenergetic distribution. With the experimental arrangement shown in Fig. 9.2, Schulz [22] has found a resonance at 19.3 ± 0.05 ev for electrons elastically scattered in helium, which is below the onset of the first excited state 2^3S (19.8 ev). Electrons leave the first electrostatic analyzer with a half width in energy spread of 0.06 ev and are accelerated into the collision chamber where they cross the beam of helium atoms. At an

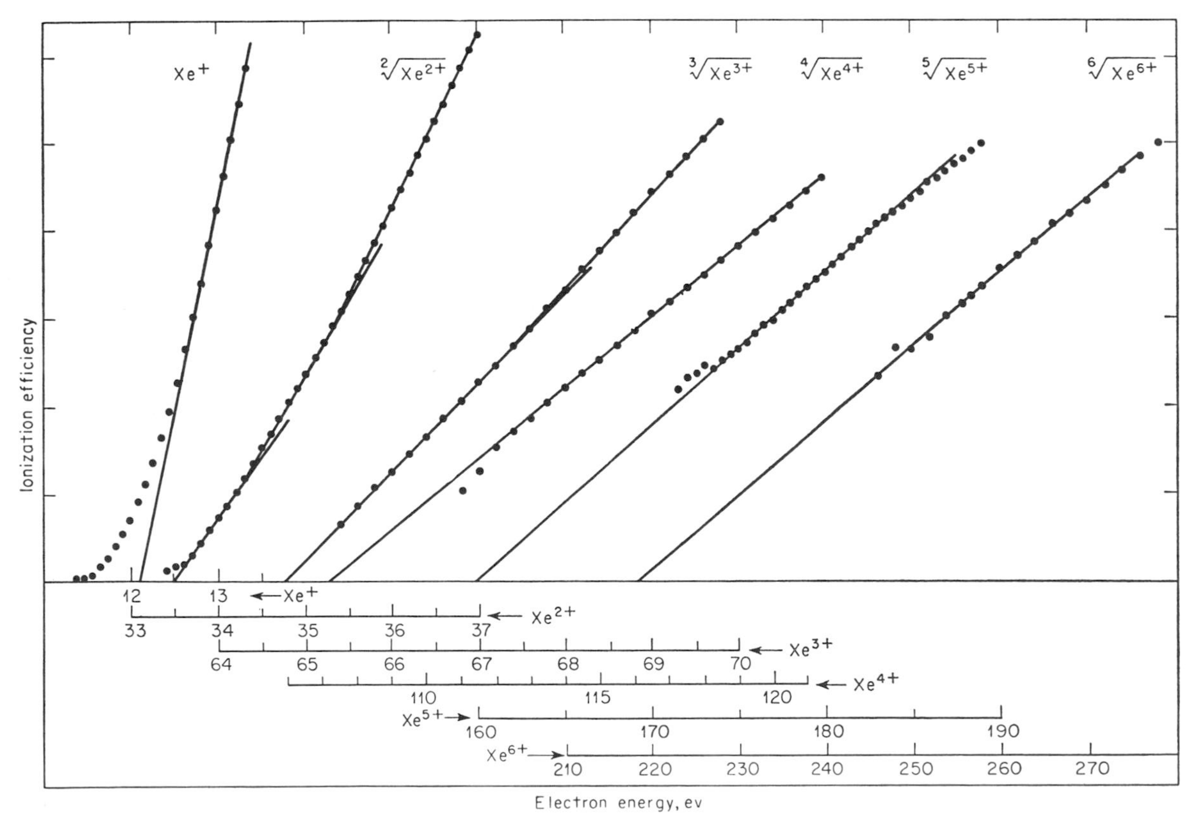

FIG. 9.7. Plots of the *n*th roots of the efficiency curves for *n*-fold ionization of Xe, for $n = 1$ to 6. (*From ref.* 15.)

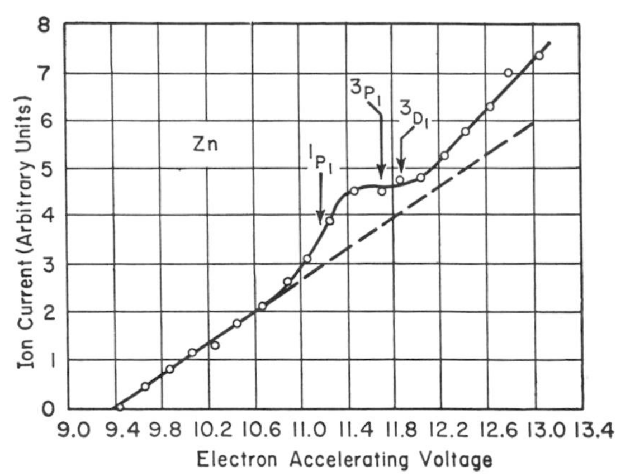

FIG. 9.8. Initial portion of the ionization efficiency curve for Zn obtained by the method of Fig. 9.3.

arbitrarily chosen angle of 72°, scattered electrons are decelerated into a second electrostatic analyzer. Both analyzers pass electrons of about 1.5 ev, and their relative potential is adjusted so that electrons that have not lost energy are transmitted through the second analyzer and reach the electron multiplier. The resonance for helium, as well as for neon, is shown in Fig. 9.9. Transmission experiments have confirmed the resonance in elastic cross sections [23]. Schulz applied the same apparatus to the study of the vibrational excitation of N_2 by electron impact and interpreted the resonances in terms of a temporary negative ion state, N_2^- [24].

Thus far the discussion has dealt with the form of the ionization efficiency curves for various processes, with no concern for the absolute magnitude of the ionization efficiency in particular processes. In fact, absolute collisional cross sections for atomic and molecular processes have been of great interest and have received much attention. The cross section Q is determined experimentally through the equation

$$I(E) = nLQI_B(E)$$

where $I(E)$ is the current signal resulting from an ionization process, $I_B(E)$ is the current of bombarding particles of energy E that induce the process, n is the number density of the target gas, and L is the length of the collision path for which $I(E)$ is collected. Because of the formidable difficulties of measuring absolute cross sections [17], usually cross sections are measured relatively. In a few instances careful experiments have attempted to measure absolute cross sections which are then used as reference standards by other experimenters.

For illustration here, the ionization of atomic hydrogen by electrons will be used [25]. For other types of bombarding and target particles (e.g., ion-impact studies), reference can be made to the review by Fite [26]. The experimental arrangement illustrated in Fig. 9.10 is particularly suitable when the target particles are unstable under the environmental conditions necessary for experiments with electron and ion beams. Hydrogen atoms produced by thermal dissociation in a tungsten furnace are collimated by an arrangement of slits to form a beam

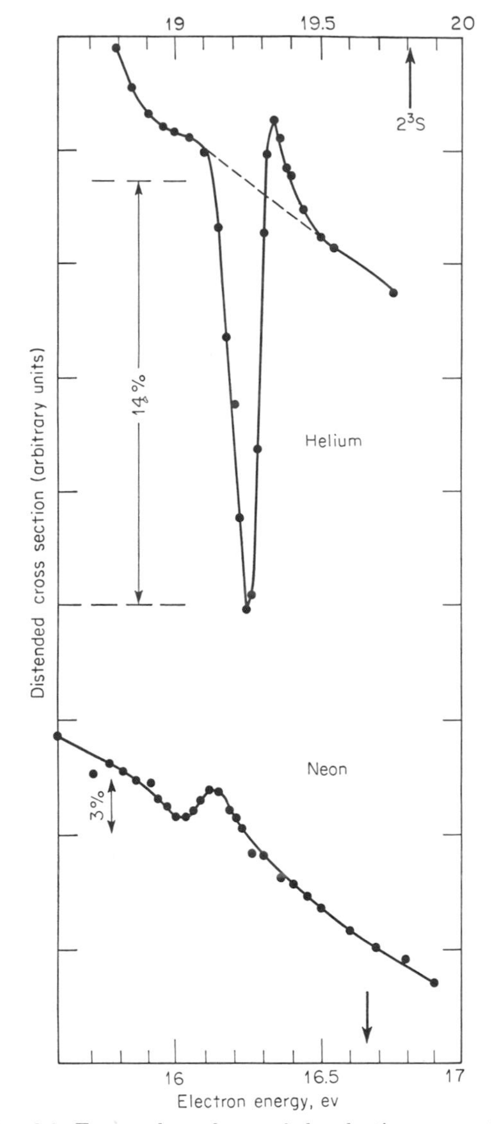

FIG. 9.9. Energy dependence of the elastic cross section at an angle of 72° in helium and neon. The energy scale on top is for the helium curve; on the bottom, for the neon curve. The dip in helium occurs at 19.30 ± 0.05 ev. (*From ref.* 22.)

which crosses perpendicularly the electron beam. The three regions divided by the two vacuum walls indicated in the figure are separately pumped to assure the lowest possible pressure in the ionization region. Nevertheless, the number of ions contributed by electron collisions in the gas producing this background pressure in the ionization region is sufficient to obscure the measurement of ions created in the beam emanating from the furnace. By interrupting the beam from the furnace at a frequency of 100 cps with a rotating toothed chopper wheel, the ions created from this beam in crossing the electron stream are selectively modulated in intensity; these ions contribute a current varying at 100 cps to the preamplifier at the exit of the mass spectrometer in contrast to the randomly fluctuating d-c signal contributed by the background pressure. The 100-cps signal is measured with a phase-sensitive amplifier.

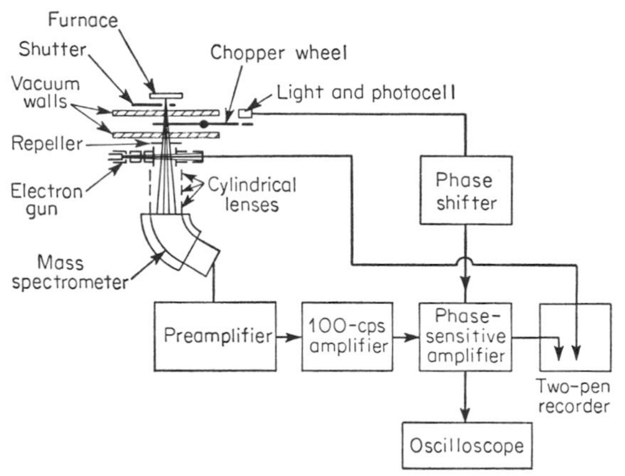

FIG. 9.10. Schematic diagram of ionization experiment of Fite and Brackman. (*From ref.* 25.)

With the method just described, the cross sections of atomic hydrogen for ionization by electrons have been measured relative to that of H_2 for a range of electron energy by varying the degree of dissociation in the neutral beam from the furnace. By using the measured ratios of the cross sections of H and H_2 for selected values of the electron energy and the values of Tate and Smith [27] for absolute cross sections of H_2, values for the absolute cross section for H were obtained. These data for atomic hydrogen, shown in Fig. 9.11, have unusual theoretical interest.

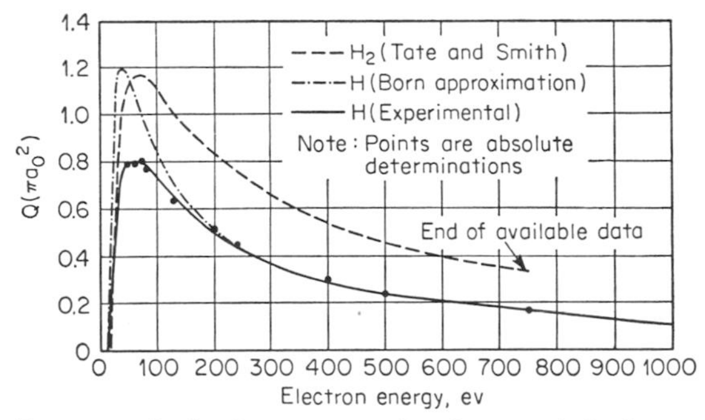

FIG. 9.11. Ionization cross section for atomic hydrogen measured with the experimental arrangement of Fig. 9.10 in comparison with that from Tate and Smith for molecular hydrogen and from the Born approximation. (*From ref.* 25.)

Figure 9.11 also provides a comparison with the results of Tate and Smith for ionization of H_2 and the theoretical predictions of the Born approximation for ionization of H. Above 250 ev the agreement between experiment and theory is excellent. Kieffer and Dunn [17] discuss this result and the later experiment of Boksenberg in connection with the effectiveness of collection of the ions after mass analysis.

4. Ionization and Dissociation of Molecules

Ionization of molecules without dissociation has a simplicity comparable with atomic ionization, for example, CO^+ produced from CO by electron impact. When molecular bonds are broken, the processes and their interpretation become more complex.

Consider that a molecule XY has the three electronic states indicated by the three potential energy curves of Fig. 9.12 [28]. Starting from the ground state I at the potential minimum, the most probable

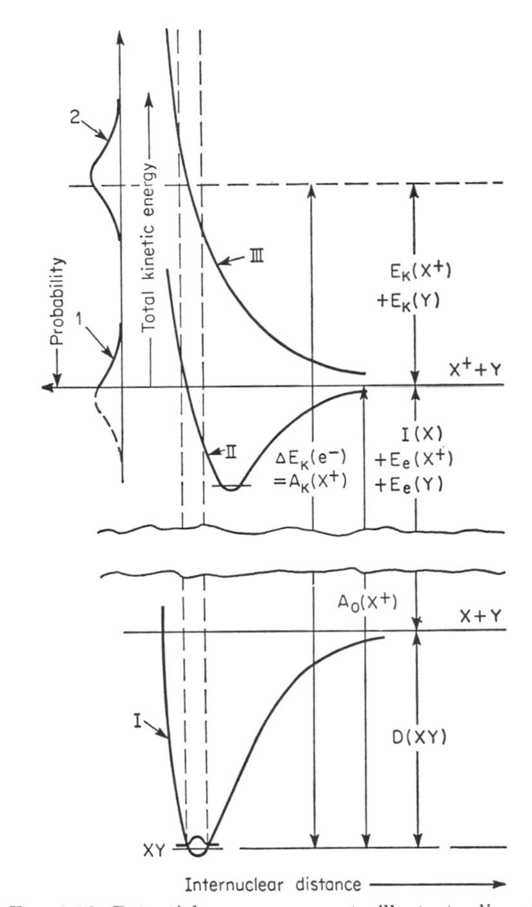

FIG. 9.12. Potential energy curves to illustrate dissociative ionization of the molecule XY yielding the fragments $X^+ + Y$. The two vertical dashed lines indicate the approximate limits for the Franck-Condon region for transitions from the ground state I of the molecule to the ionic states II and III. Transitions to state II produce either molecular (XY^+) or atomic (X^+) ions. Transitions to repulsive state III produce atomic ions (X^+) and the associated neutral fragments (Y) with considerable kinetic energy. Curves 1 and 2 indicate the predicted energy distributions for the dissociation products. These curves are drawn by reflecting the square of the eigenfunction of the lowest vibrational level of state I into the curves for states II and III. $D(XY)$ = dissociation energy of XY. $I(X)$ = ionization potential of X. $A_K(X^+)$ = appearance potential of X with kinetic energy. $A_0(X^+)$ = appearance potential of X with zero kinetic energy. $E_K(\ )$ is the kinetic energy and $E_0(\ )$ the excitational energy of the particle indicated within the parenthesis. (*From ref.* 28.)

transition on electron impact is vertically upward, according to the Franck-Condon principle—neither the relative positions nor the momenta of the nuclei change appreciably during the electronic transition. The approximate width of this Franck-Condon region is indicated by the two dashed lines for transitions from the $v'' = 0$ vibrational level of the electronic state I. The transition probability is proportional to $[\int \chi(v')\chi(v'')\, dr]^2$, where r is the internuclear distance and the vibrational eigenfunctions refer to the initial and final states of the transition. With the minimum of state II falling well outside the Franck-Condon region, the ion XY^+ has very little probability of being formed in its lowest vibrational level. In this case the vertical ionization potential for XY as measured in the mass spectrometer will differ from the adiabatic ionization potential corresponding to transition between the ground states of the two electronic states.

A portion of the Franck-Condon region crosses the curve for state II above the dissociation limit. This means that atomic ions having a range of kinetic energy very near to zero will be present. Atomic ions will first appear at the appearance potential given by

$$A_0(X^+) = D(XY) + I(X) + E_e(X^+) + E_e(Y) \qquad (9.1)$$

where $D(XY)$ is the dissociation energy of XY, $I(X)$ is the ionization potential of X, and E_e refers to the excitational energy of the quantity in parentheses immediately following.

The appearance potential $A_K(X^+)$ of ions of kinetic energy $E_K(X^+)$ will be

$$A_K(X^+) = D(XY) + I(X) + E_e(X^+) + E_e(Y) + E_K(X^+) + E_K(Y) \qquad (9.2)$$

The difficulty of definitive interpretation is apparent from these equations. From the measured appearance potential one wishes to infer information about the quantities on the right-hand side of the equation. Experiment can supply data on the kinetic energy of the fragments in certain cases. The amount of excitational energy E_e useful for Eq. (9.2) can usually be deduced only through arguments on consistency, with the plausibility varying in each instance. Here, however, the improved experimental techniques employing monoenergetic electron beams have contributed helpfully by revealing structure on the ionization-onset curves. For the atomic ionization potential $I(X)$ one refers to the results of optical spectroscopy. When confidence is established as to the values of all other terms, then the heat of dissociation $D(XY)$ is derived. This presentation has stressed the difficulties. Experience has shown that frequently the situation is really not so bad as it might seem. In a large number of cases, some ions do appear at the minimum electron energy energetically possible for negligible excitational and kinetic energy. Even though absolute confidence in the interpretation of dissociation processes must await the truly ingenious experiment which measures both the kinetic and excitational energy of all dissociation fragments, a carefully performed and properly interpreted study with the mass spectrometer can nevertheless contribute valuably to the solution of problems of molecular structure.

Referring again to Fig. 9.12, one sees that the Franck-Condon region crosses the curve for the repulsive state III. A transition to this state results in the dissociation of the molecule into fragments having considerable kinetic energy. This process in hydrogen was predicted theoretically [29] and confirmed experimentally [5]. This beautiful achievement stimulated an interest in measuring the kinetic energy of fragment ions. When the form of the curve for state III is known, the total energy of the fragments as a function of the energy of the incident electrons may be determined. The method parallels that for the explanation of the variation of intensity of absorption with wavelength in a transition from a stable lower state to a continuous upper state on absorption of a light quantum [30]. As in photodissociation, to a good approximation, the distribution in kinetic energy of the dissociation fragments for incident electrons of a specified energy can be determined by the probability that the nuclei will have a particular internuclear separation in the ground state, in other words by the square of the eigenfunction χ^2. In this approximation the eigenfunction of the upper state need not be known. The predicted distribution obtained by "reflecting" χ^2 for the ground state ($v'' = 0$) into the electronic state III is shown by the graph at the left of Fig. 9.12. The corresponding procedure for transitions to state II predicts that the maximum number of ions will be formed with very little kinetic energy. The predicted distribution is markedly affected by the form of the curves and positions of the Franck-Condon crossings relative to the dissociation asymptote.

The predicted distribution in itself provides information only on the kinetic energy of both fragments, ionic and neutral. Although past experiments have measured directly the kinetic energy of the ion only, the division of total kinetic energy between the two fragments is known from the conservation of momentum. The proportion of the total kinetic energy retained by the ion is

$$E_K(X^+) = \frac{m(Y)[E_K(X^+) + E_K(Y)]}{m(Y) + m(X^+)} \qquad (9.3)$$

where $m(X^+)$ is the mass of the ion and $m(Y)$ is the mass of the neutral fragment. Thus, the kinetic energy of the ion plotted against the total kinetic energy will be a straight line of slope $m(Y)/[m(Y) + m(X^+)]$. When the kinetic energy of the ions $E_K(X^+)$ is plotted against the appearance potential of ions of this kinetic energy $A_K(X^+)$, the intercept with the $A_K(X^+)$ axis will be the appearance potential of ions of zero kinetic energy. For the electronic state III the ions formed with zero kinetic energy will not be observed directly. However, a plot in the manner just described for the transitions in the Franck-Condon region will by extrapolation determine $A_0(X^+)$, the position of the dissociation asymptote of the repulsive curve. The plots of $E_K(X^+)$ vs. $A_K(X^+)$ for states II and III are given in Fig. 9.13. The intensity distributions from Fig. 9.12 are plotted along the curves [28]. These distributions indicate roughly the regions where the intensity of ion current is high enough for measurements to be made.

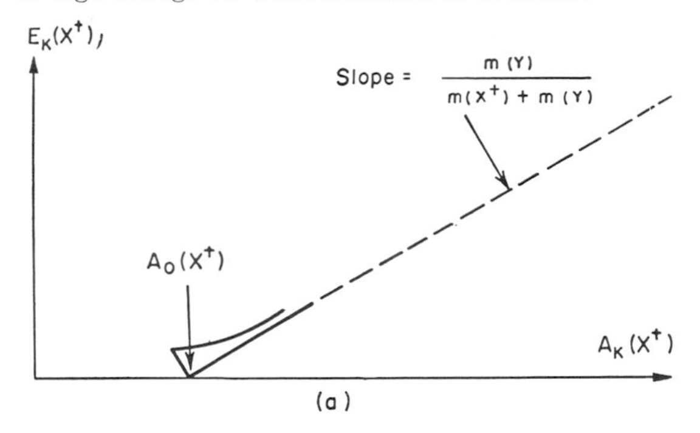

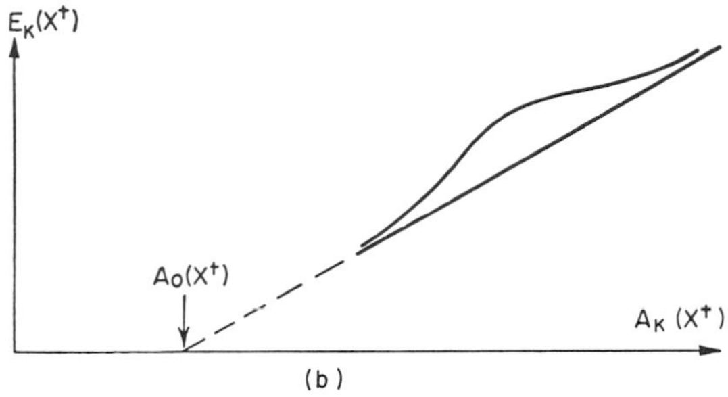

FIG. 9.13. Plot of the kinetic energy $E_K(X^+)$ as a function of the appearance potential $A_K(X^+)$ for the atomic ion X^+ formed by the processes described in Fig. 9.12. In the upper portion (*a*) of the figure the ion X^+ is formed by transitions to state II, and the intensity distribution is here replotted according to the prediction of curve 1 in Fig. 9.12. The dashed portion of the straight line indicates that the transitions are outside the Franck-Condon region. The lower plot (*b*) corresponds to transitions to state III in which no ions of zero kinetic energy will be detected although the appearance potential of such ions $A_0(X^+)$ may be obtained by extrapolation. The intensity distribution predicted in curve 2 of Fig. 9.12 is replotted in this figure. (*From ref.* 28.)

Where one of the fragments in the dissociation of a diatomic molecule is a negative ion, the other fragment may be a positive ion $(X^- + Y^+)$ or neutral $(X^- + Y)$. The former process is similar to the already discussed case of dissociation involving a positive ion and a neutral fragment in that ionization occurs when the electron energy is greater than $A_K(X^+)$. In contrast the capture process $(X^- + Y)$ is a resonant one requiring that the electrons have a particular, narrowly defined energy. If the electrons do not have this resonant energy, the process will not occur. The usual form of an ionization efficiency curve for negative ions is shown in Fig. 9.14. The peak at low voltage corresponds to the formation of O^- from CO through resonant capture of the impacting electron. The curve beginning at the higher electron energy corresponds to the simultaneous production of O^- and C^+ from CO. The resonant capture peak can be exceedingly sharp. In several intances it has been demonstrated that the entire measured width can be attributed to the spread in energy of the electron beam.

A suggestion of the relationship between the data on molecules obtained through mass spectroscopy and that obtained through thermochemistry or the study of molecular spectra can be obtained by considering

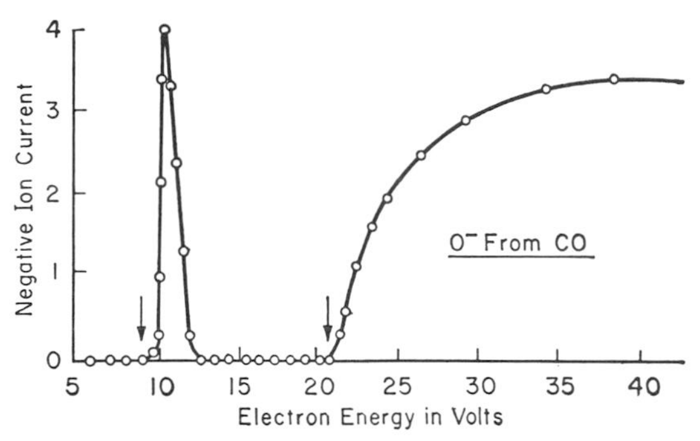

FIG. 9.14. Production of the negative ion O^- from CO by electron impact. [*From: Hagstrum, H. G., and J. T. Tate: Phys. Rev.*, **59**: 354 (1941).]

carbon monoxide. Referring to Fig. 9.15, one sees that the heat of dissociation of carbon monoxide $D(CO)$ may be determined if the heat of sublimation of carbon $L(C)$ is known, and vice versa, the energy required for the reaction $CO \rightarrow C_{solid} + \frac{1}{2}O_2$ being known from thermochemistry and that for $\frac{1}{2}O_2 \rightarrow O$ having been determined spectroscopically. For many years differing and sometimes very strong opinions existed on the correct values of $D(CO)$ and $L(C)$.

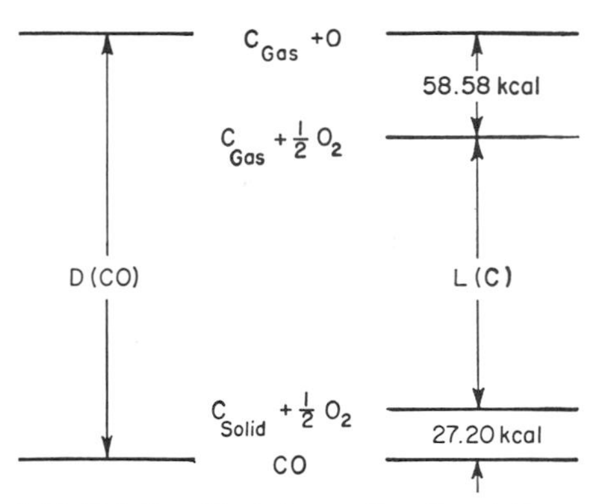

FIG. 9.15. Relationship between the dissociation energy carbon monoxide $D(CO)$ and the heat of sublimation of carbon $L(C)$.

Both chemists and physicists involved in the study of ionization processes now accept the value $D(CO) = 11.11$ ev and $L(C) = 170.4$ kcal/mole (1 ev = 23.06 kcal/mole), in sharp contrast to the situation in the early 1950s. In this, the definitive work of Branscomb and coworkers [31 and earlier publications] on the photodetachment of 0^- leading to a value for the electron affinity of oxygen $EA(O) = 1.465 \pm 0.005$ ev contributed importantly.

In experiments with Lozier-type tubes (one of which employed the RPD method on the electron beam) the dissociative ionization processes by electron impact in CO have been interpreted to conform with $D(C) = 11.11$ ev and $EA(O) = 1.465$ ev [32, 33].

These explanations have not been completely satisfying. In this connection the paper of Chantry and Schulz [34] is pertinent. They point out that the extrapolation method (illustrated in Fig. 9.13) as heretofore used was erroneous because the thermal motion of the target molecule was neglected. The random direction of the momentum of thermal motion of the target molecule contributes a spread in ion energies that must be considered even though the *energy* of thermal motion can justifiably be neglected in the energy equation. Taking this into account, they reinterpret data on the dissociative attachment of O_2 (measured by Schulz several years earlier) and obtain a result consistent with $EA(O) = 1.5$ ev, in contrast with the original interpretation by Schulz of the same data leading to $EA(O) = 2.0$ ev.

Another criticism has been directed at experiments on molecular dissociation that ignore anisotropies in angular distributions of dissociation products relative to the direction of the bombarding electron beam [35]. For example, mass spectrometers and Lozier tubes preferentially detect particles at right angles to the incident electrons. Dunn's theoretical studies predict anisotropies in a majority of processes. The results of many measurements of absolute cross section come into question. As an example of the error induced in measurement with the Lozier tube of the appearance potentials for ions of zero kinetic energy by extrapolation of measurements on ions formed with kinetic energy (similar to Fig. 9.13), Dunn cites processes where no ions are ejected at right angles to the electron beam. Fitting experimental data to a slope as given in Eq. (9.3) and Fig. 9.13 gives an erroneous value on extrapolation to the intercept. Anisotropic distribution of protons from the dissociative ionization of H_2 by electrons was measured experimentally many years ago by Sasaki and Nakao [36].

All the common diatomic molecules have been studied many times by the methods just described for CO. In addition an isotopic effect in dissociative ionization was predicted and observed [37]. Consider transitions from the ground state to state II in Fig. 9.12 by the impact of electrons of energy just insufficient to produce transitions to state III. Then all the atomic ions X^+ which are formed arise from transitions to state II above the dissociation limit. A study of Fig. 9.12 will make it evident that the proportion of X^+ ions relative to XY^+ ions will depend on the relative portion of the Franck-Condon region that crosses the potential energy curve for state II above the dissociation limit. This portion will be less for a narrower Franck-Condon region, the potential energy curves remaining unchanged. This condition will obtain for a molecule XY' where Y' is a heavier isotope of Y. The molecule XY' will have a lower zero-point energy than XY and its eigenfunction for the ground state will become more concentrated around the equilibrium value of the internuclear separation.

The simplest example of this isotopic effect and the first one to be studied was H_2 and D_2. If r_c is the internuclear separation corresponding to the dissociation limit of state II at the Franck-Condon crossing, then the ratio H^+/H_2^+ can be calculated from

$$\frac{H^+}{H_2^+} = \frac{\int_0^{r_c} \chi_0^2(r)\,dr}{\int_{r_c}^{\infty} \chi^2(r)\,dr} \tag{9.4}$$

This is illustrated in Fig. 9.16 in which the unshaded portion of the Franck-Condon region corresponds to the numerator and the shaded portion to the denominator [as previously, the region for small and large values of r is ignored because $\chi_0^2(r)$ is negligibly small]. More refined calculations of this type allow for the variation of transition probability with excess energy. The relative values for H_2, D_2, and T_2 calculated in this manner [39, 40] agree satisfactorily with experiment. This comparison is shown in Table 9.1.

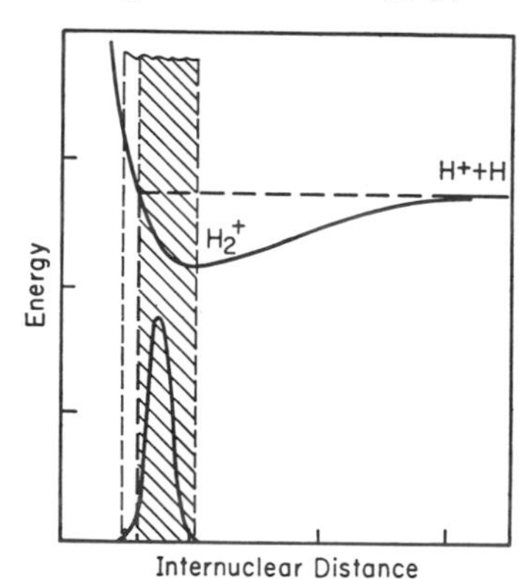

Fig. 9.16. The potential-energy curves governing the dissociative ionization of H_2 are similar to those of Fig. 9.12. Illustrated here are only the potential-energy curve corresponding to state II and the square of the eigenfunction of the lowest vibrational level of the ground state I, to illustrate an isotopic effect in dissociative ionization. Transitions producing H_2^+ occur in the shaded portion of the Franck-Condon region and those producing H^+ occur in the unshaded portion. For D_2 the Franck-Condon region is narrower than for H_2, resulting in the production of relatively less D^+.

Table 9.1. Comparison of Observed and Calculated Isotopic Effects in the Dissociative Ionization of Hydrogen

Ratio	Observed	Calculated
H^+/H_2^+	0.013	0.0159
D^+/D_2^+	0.0070	0.0072
T^+/T_2^+	0.0036	0.0038

After the study of methane with the modern experimental methods of mass spectroscopy showed that a distinctive and reproducible mass spectrum of the parent and fragment ions could be obtained with the relative peak heights being independent of pressure, the study of more complex molecules was stimulated. The application to the analysis of mixtures of hydrocarbons followed within a very few years under the urgency of this type of analysis during the Second World War. First emphasis was on measurement of appearance potentials and a correlation with thermochemical and spectroscopic data following the pattern of diatomic molecular studies. Qualitatively the mass spectrum of a molecule usually agreed with the prediction based on structure. There was no basis for a more quantitative prediction and in some instances a fragment ion appeared in abundance even though such a fragment could not arise from only the breaking of bonds in the original structure. The $C_2H_5^+$ ion in isobutane is such a rearrangement peak. The mass spectrum or "cracking pattern" of a substance is now better understood.

While the discussion of CO has presented the general problems that will also be encountered in the study of more complex molecules, additional factors are present. Some of these are illustrated by the studies of methane [40–42]. The following ions are observed in methane as primary products of electron impact: CH_4^+, CH_3^+, CH_2^+, CH^+, C^+, H_3^+, H_2^+, H^+, CH_2^-, CH^-, C^-, H^-. Of the total number of ions produced, CH_4^+ and CH_3^+ comprise 50 and 40 per cent, respectively, while the remainder are present only in trace amounts except for CH_2^+ (~4 per cent), CH^+ (~2 per cent), C^+ (~1 per cent), and H^+ (~3 per cent).

One would expect that experimental values for the simple ionization of methane

$$CH_4 + e \rightarrow CH_4^+ + 2e \qquad (9.5)$$

would agree fairly well. Five of the six values reported in the period 1948 to 1954 in the literature ranged from 13.0 to 13.1 volts with the estimates of error varying from 0.02 to 0.2 volt. For the process

$$CH_4 + e \rightarrow CH_3^+ + H + 2e \qquad (9.6)$$

the indication is that $A(CH_3^+) = 14.3$ to 14.4 volts with an error comparable to $I(CH_4)$. For the ions of lower mass the agreement gets progressively worse between the various workers and the cited errors become greater; this is because the ionization efficiency curve increases very slowly in the vicinity of the appearance potential. The thermal spread in energies of the electrons makes the selected point of initial onset of ionization particularly dependent on the sensitivity. Interpretation of the processes occurring in the formation of these lower mass ions becomes increasingly difficult. Accuracy of the data has deteriorated and an inference on the undetected atoms split from the parent with the detected ion tends toward surmise.

Equation (9.6) clearly shows that the bond dissociation energy $D(CH_3—H)$ emerges if $I(CH_3)$ is known and the excitational and kinetic energy (E) is negligible. $I(CH_3)$ has been measured in the mass spectrometer by introducing the free radial CH_3 directly into the ionization chamber through the decomposition of $Pb(CH_3)_4$, for instance [43–45]. If for this process

$$CH_3 + e \rightarrow CH_3^+ + 2e \qquad (9.7)$$

we have

$$I(CH_3) = 9.9 \pm 0.1 \text{ volt}$$

then $D(CH_3 - H) = A(CH_3^+) - I(CH_3) - E$

and taking $A(CH_3^+) = 14.3 \pm 0.1$ with the assumption that $E \approx 0$ then $D(CH_3—H) = 4.4$ volts. Stevenson [46] has calculated $D(CH_3—H)$ by six other methods using appearance potential measurements with thermochemical data. One of these he calls a "direct" method because it involves only $A(CH_3^+)$ in CH_3OH analogous to the example above of CH_4. The other five "indirect" methods involve in each instance two appearance potentials for the same ion from two substances, for example, $C_3H_7^+$ from C_3H_8 and iC_4H_{10}. Without going through the detailed steps, the result is a striking agreement for $D(CH_3—H)$ calculated by all seven methods with the average being 4.42 ± 0.04. That there can be any appreciable excess energy in the many reactions involved is unlikely.

A measurement of the kinetic energy of ions resulting from the dissociation of methane by electron impact, in particular the formation of C^+, permitted an interpretation of the data compatible with $L(C) = 170.4$ kcal/mole [47]. McDowell has reviewed the difficulties in deriving bond-dissociation energies from electron-impact processes with polyatomic molecules such as methane [48].

By use of essentially monoenergetic electrons, the ionization potential for the transition to the first excited electronic state of the CH_4^+ ion has been measured (19.42 ev) as well as that to the ground state [49]. More recent measurements with the RPD method show additional structure closer to the ground state of the CH_4^+ ion [50].

The absence of excess energy in the many processes used to determine $D(CH_3—H)$ and its presence in others indicate the importance of knowing this energy prior to analysis of the data. The result on $D(CH_3—H)$ was convincing without this prior knowledge only because of the consistency with seven methods. Stevenson has arrived at a useful generalization which seems to provide a necessary condition for a process to occur without excess energy. The process

$$R_1 - R_2 + e \rightarrow R_1^+ + R_2 + 2e \qquad (9.8)$$

will occur with no excess energy only if $I(R_1^+) < I(R_2)$. If $I(R_1) > I(R_2)$, then $I(R_1) + D(R_1 - R_2)$ will be less than the appearance potential $A(R_1^+)$.

This is to be expected in the theory of the mass spectra of polyatomic molecules of Eyring and his coworkers [51, 52], who applied absolute-reaction-rate theory. The density of electronic states of a polyatomic molecule is a justification for this statistical approach. They estimate that the average spacing of these levels for the propane ion with nineteen bonding electrons is about one millivolt (mv). After vertical ionization the molecule-ion does not in general decompose until it has undergone at least several vibrations during which time the excess energy above the ground state is distributed randomly by radiationless transitions. A reaction will occur when sufficient energy localizes in a reaction coordinate to permit it energetically. Successive fragments may dissociate in a similar fashion. Although immediate dissociation of a nonstatistical nature may occur, such occurrences are considered to be only a small fraction of the total. Metastable ions with measured lifetimes of about

10^{-6} sec which have been observed in the mass spectrometer [53] are a confirmation of the delayed dissociation inherent in the theory. The theory was applied to the calculation of the mass spectrum of propane. For comparison with the propane spectrum observed in the conventional mass spectrometer, the relative amount of unimolecular decomposition along possible reaction paths was calculated for a time of 10^{-5} sec, the approximate time for an ion to reach the ion collector after the ionizing collision by the electron with the parent molecule. This comparison is shown in Table 9.2. Qualitative arguments lend impressiveness to this agreement even though the parameters exceed the unknowns. Appearance potentials, ionization potentials, and thermal data are the basis for calculating activation energies. Isotope effects [54] and the temperature dependence in the spectra of polyatomic molecules can be reasonably explained. Theoretical aspects of the mass spectra resulting from the dissociative ionization of polyatomic molecules have also been studied by Lester [55].

TABLE 9.2. MASS SPECTRUM OF PROPANE CALCULATED BY EYRING ET AL. CONTRASTED WITH EXPERIMENTAL VALUES

	$C_3H_8^+$	$C_3H_7^+$	$C_3H_5^+$ $C_3H_3^+$ C_3H^+	$C_3H_6^+$ $C_3H_4^+$ $C_3H_2^+$	$C_2H_5^+$	$C_2H_4^+$	$C_2H_3^+$	$C_2H_2^+$
Calculated......	0.111	0.074	0.056	0.056	0.300	0.185	0.194	0.024
Experimental...	0.090	0.071	0.103	0.042	0.310	0.183	0.122	0.027

Positioned isotopes, both stable and radioactive, in reacting molecules have been a powerful tool in the study of chemical reactions. They have also been helpful in elucidating the processes of dissociative ionization. However, the dissociation probability is affected by the substitution of an isotope of different mass. The isotope effect in the relative production of H^+ and D^+ from H_2 and D_2 could be explained by the Franck-Condon principle. The substitution of D for H in hydrocarbons has more complex consequences. Two effects are noted. In the first place the probability of rupturing a C—D bond is found to be less than for a C—H bond. As in H_2 and D_2, this can be interpreted as a consequence of the Franck-Condon principle in the transition from the ground state to the electronic state resulting in dissociative ionization. The other effect is the increased probability of C—H rupture induced by the presence of D in the molecule. For the five compounds in Table 9.3, the weighting factors for the decreased probability of rupturing C—D and the enhanced probability of rupturing C—H are listed as deduced from experimental data [56].

An unusually large shift in ionization potential with the progressive substitution of deuterium atoms in methane was found by Lossing, Tickner, and Bryce [57]. The linear increase in ionization potential totals 0.18 volt from CH_4 to CD_4. This suggests that the potential minimum of the lowest state of the ion is not readily accessible to transitions from the ground state as governed by the Franck-Condon principle.

Inghram and Gomer [58] have made mass analyses of the positive ions desorbed from surfaces by high electrostatic fields. The beam of ions passed through a hole in a screen of the field emission microscope into a mass spectrometer equipped with an electron multiplier. The ions entering the mass spectrometer originated from an area of 900 A^2 at the edge of the 110 plane of the tungsten tip. The 10,000-volt ions had an energy spread of 20 volts. At electric fields greater than approximately 6 volts per angstrom autoionization (and fragmentation) occurs at 3 to 100 A from the tip with the increased spread in energy causing the peaks to broaden and decrease in height. For lower electric fields sharp peaks occur with ions originating within 5 A of the tip. For chemisorbed atoms or molecules, Inghram and Gomer suggest that the applied field permits one of the bonding electrons to tunnel, releasing the bond, and then the ion is formed by autoionization as described above. In almost all molecules studied, the parent peak predominates. For example, with acetone the parent peak is more than a thousand times greater than any fragment ion. Beckey with his co-workers [59] has applied the method to various ionization studies and analytical purposes.

TABLE 9.3. ISOTOPIC EFFECT IN THE RUPTURE OF CARBON-HYDROGEN BONDS IN DEUTEROMETHANES

Weighting factor	CH_3D	CH_2D_2	CHD_3	C_2H_5D*	CH_3CHD_2*
C—H rupture.	1.23	1.48	1.80	1.09	1.15
C—D rupture.	0.55	0.65	0.76	0.60	0.80

* In the last two columns only 2-carbon fragments are included.

5. Applications

The mass spectrometer is widely used as a necessary tool for analysis of unknown mixtures of gases or liquids. It has also been used for analysis in many instances in which the sample cannot be transported undisturbed to the mass spectrometer. An example is the study of free radicals and other reaction intermediates. Lossing has reviewed this work and lists 61 examples for which the ionization potentials have been measured (see "Mass Spectrometry," listed in the Bibliography at the end of this chapter). The example chosen here is related to the discussion on CO where the difficulty of determining with certainty the heat of sublimation of carbon $L(C)$ from electron-impact studies was indicated.

Two types of experiments had been made to determine $L(C)$—measurement of the rate of evaporation and measurement of the equilibrium vapor pressure. A difficulty with these experiments was identifying the molecular species which are evaporated and determining their relative amounts. The mass spectrometer is admirably suited for this. The first experiments measured the rate of evaporation as a function of temperature for a carbon filament placed very near to the electron beam. Some of the products of evaporation from the filament then would travel through the electron beam before striking the electrode structure [60, 61]. These experiments showed that carbon evaporates predominantly as C_3 (100) with the remainder mostly C_2 (13) and C_1 (35). Chupka and Inghram have measured the equilibrium vapor pressures of C_1, C_2, and C_3 above graphite by positioning a Knudsen cell adjacent to the ionization chamber of a mass spectrometer, the molecular species of carbon effusing from this cell being exposed to the ionizing electron beam before striking any obstacle. The relationship between the pressure inside the Knudsen cell and the intensity of the corresponding ion was measured by studying the ion current arising from a weighed amount of silver placed in the cell. To relate this to the carbon pressure, the relative ionization efficiencies of silver and carbon had to be estimated. The intensities of the carbon ions were measured in the temperature range 2150 to 2450°K. The mass-spectrometric studies show that the spectroscopic value of $L(C) = 170.4$ kcal/mole is correct. The mass spectrometer has been used by various workers to study the thermodynamic properties of many substances, both organic and inorganic. For the measurement of latent heats of vaporization and sublimation, the ability to measure the partial pressure of the substance of interest over a very wide pressure range in the presence of impurities is very attractive. Polyatomic species in the vapor phase from the evaporation of elements and compounds have been studied and their thermodynamic properties measured.

Research on the ionization and dissociation of molecules by electron impact has been the basis for the application of mass spectroscopy to chemical analysis. Instruments developed in this research with further refinements for the particular requirements of analysis produce mass spectra characteristic of each substance. The individual peaks in the spectrum of each substance vary linearly with pressure and are independently added to produce a combined spectrum characteristic of the mixture. Customarily the mixture is admitted to the mass spectrometer through a small leak designed for linear flow as a function of pressure. If the pressures of the n components of a gas mixture in a reservoir on the high-pressure side of this leak are $p_1, p_2, p_3, \ldots, p_n$ and peaks at selected masses in the mixture spectrum have intensities $A_1, A_2, A_3, \ldots, A_m$, then these peak heights are given in terms of the individual components of the mixture by the m equations

$$\Sigma \alpha_{ij} p_j = A_i \tag{9.9}$$

where the coefficients α_i are obtained experimentally by measuring the corresponding peaks when each component is admitted to the mass spectrometer separately. Equation (9.9) is solved for each p_j to give the composition of the unknown mixture.

Commercially available mass analyzers of very high resolution with direct photographic or electrical detectors are in wide use for both qualitative and quantitative analysis. Customarily, spark sources or electron-bombardment sources are employed. The high resolution permits, through packing fraction differences, the identification and intensity measurement of an additionally large number of ions that in mass analyzers of low resolution coincide at nominal whole numbers on the mass scale. In the determination of the molecular formula, precise mass measurement is a powerful means for ascertaining the identity of the atoms forming the parent molecule or its fragments from dissociative ionization. Isotope-abundance measurements are also helpful. After the molecular formula is known, other information can be used, frequently with great success, to deduce the structural formula, e.g., relative abundances of fragment ions, appearance potentials, metastable ions, and ionization efficiency curves. For quantitative analysis, the large amount of information presented by high-resolution instruments has greatly extended the usefulness of the mass-spectroscopic method for both organic and inorganic substances.

The role of mass-spectroscopic methods in research and applied fields of endeavor has become far-ranging, with a rapidly increasing number of publications appearing. Fortunately, excellent books of varying emphasis in their specific coverage are available (see Bibliography). A general impression one gets from the literature is that for approximately the first decade after the Second World War the instruments available for purchase were designed and therefore used predominantly for the analysis of samples of gases and vapors. Recently, the availability of instruments of additional types and enhanced versatility has extended the analytical applications and has permitted research scientists to undertake with success many studies formerly possible only with equipment designed and perfected by themselves. The well-attended conferences organized periodically by interested groups in the United States and Europe have been stimulating for the research of the participants. The proceedings of these conferences have often appeared later in book form.

References

1. Nier, A. O., and T. R. Roberts: *Phys. Rev.*, **81**: 507 (1951).
2. Wiley, W. C., and I. H. McLaren: *Rev. Sci. Instr.*, **26**: 1150 (1955).
3. Paul, W., H. P. Reinhard, and U. von Zahn: *Z. Physik*, **152**: 143 (1958).
4. Lozier, W. W.: *Phys. Rev.*, **36**: 1285 (1930).
5. Bleakney, W.: *Phys. Rev.*, **35**: 1180 (1930).
6. Lawrence, E. O.: *Phys. Rev.*, **28**: 947 (1926).
7. Nottingham, W. B.: *Phys. Rev.*, **55**: 203 (1939).
8. Marmet, P., and L. Kerwin: *Can. J. Phys.*, **38**: 787 (1960).
9. Morrison, J. D.: *J. Chem. Phys.*, **22**: 1219 (1954); **40**: 2488 (1964).
10. Fox, R. E., W. M. Hickam, T. Kjeldaas, Jr., and D. J. Grove: *Phys. Rev.*, **84**: 859 (1951).
11. Wigner, E.: *Phys. Rev.*, **73**: 1002 (1948).
12. Wannier, G. H.: *Phys. Rev.*, **100**: 1180 (1955).
13. Geltman, S.: *Phys. Rev.*, **102**: 171 (1956).
14. Hickam, W. M., R. E. Fox, and T. J. Kjeldaas, Jr.: *Phys. Rev.*, **96**: 63 (1954).
15. Dorman, F. H., J. D. Morrison, and A. J. C. Nicholson: *J. Chem. Phys.*, **31**: 1335 (1959).
16. Dorman, F. H., and J. D. Morrison: *J. Chem. Phys.*, **34**: 1407 (1961).
17. Dunn, G. H., and L. J. Kieffer: *Revs. Mod. Phys.*, **38**: 1 (1966).

18. Hickam, W. M.: *Phys. Rev.*, **95**: 703 (1954).
19. Beutler, H.: *Z. Physik*, **87**: 176 (1933).
20. Morrison, J. D.: *J. Chem. Phys.*, **40**: 2488 (1964).
21. Baranger, E., and E. Gerjuoy: *Proc. Phys. Soc. (London)*, **A72**: 326 (1958).
22. Schulz, G. J.: *Phys. Rev. Letters*, **10**: 104 (1963); *Phys. Rev.*, **136**: A650 (1964).
23. Simpson, J. A., and U. Fano: *Phys. Rev. Letters*, **11**: 158 (1963).
24. Schulz, G. J.: *Phys. Rev.*, **135**: A988 (1964).
25. Fite, W. P., and R. T. Brackman: *Phys. Rev.*, **112**: 1141 (1958).
26. Fite, W. L.: in D. R. Bates (ed.), "Atomic and Molecular Processes," pp. 421–492, Academic, New York, 1962.
27. Tate, J. T., and P. T. Smith: *Phys. Rev.*, **39**: 270 (1932).
28. Hagstrum, H. D.: *Revs. Mod. Phys.*, **23**: 185 (1951).
29. Condon, E. U.: *Phys. Rev.*, **35**: 658A (1930).
30. Herzberg, G.: "Molecular Spectra and Molecular Structure. I. Diatomic Molecules," p. 387, Van Nostrand, Princeton, N.J., 1950.
31. Branscomb, L. M., D. S. Burch, S. J. Smith, and S. Geltman: *Phys. Rev.*, **111**: 504 (1958).
32. Craggs, J. D., and B. A. Tozer: *Proc. Roy. Soc. (London)*, **A247**: 337 (1958); **A254**: 229 (1960).
33. Fineman, M. A., and A. W. Petrocelli: *J. Chem. Phys.*, **36**: 25 (1962).
34. Chantry, P. J., and G. J. Schulz: *Phys. Rev. Letters*, **12**: 449 (1964).
35. Dunn, G. H.: *Phys. Rev. Letters*, **8**: 62 (1962).
36. Sasaki, V. N., and T. Nakao: *Proc. Imp. Acad. Japan*, **17**: 75 (1941).
37. Bleakney, W., E. U. Condon, and L. G. Smith: *J. Phys. Chem.*, **41**: 197 (1937).
38. Stevenson, D. P.: *J. Chem. Phys.*, **15**: 409 (1947).
39. Schaeffer, O. A., and J. M. Hastings: *J. Chem. Phys.*, **18**: 1048 (1950).
40. Smith, L. G.: *Phys. Rev.*, **51**: 263 (1937).
41. McDowell, C. A., and J. W. Warren: *Trans. Faraday Soc.*, **48**: 1084 (1952).
42. Langer, A., J. A. Hipple, and D. P. Stevenson: *J. Chem. Phys.*, **22**: 1836 (1954).
43. Hipple, J. A., and D. P. Stevenson: *Phys. Rev.*, **63**: 121 (1943).
44. Waldron, J. D.: *Trans. Faraday Soc.*, **50**: 102 (1954).
45. Lossing, F. P., K. U. Ingold, and I. H. S. Henderson: "Applied Mass Spectrometry," Institute of Petroleum, London, 1954.
46. Stevenson, D. P.: *Discussions Faraday Soc.*, **10**: 35 (1951).
47. Morrison, J. D., and H. E. Stanton: *J. Chem. Phys.*, **28**: 9 (1958).
48. McDowell, C. A.: in C. A. McDowell (ed.), "Mass Spectrometry," pp. 549–564, McGraw-Hill, New York, 1963.
49. Frost, D. C., and C. A. McDowell: *Proc. Roy. Soc. (London)*, **A241**: 194 (1957).
50. Tsuda, S., C. E. Melton, and W. H. Hamill: *J. Chem. Phys.*, **41**: 689 (1964).
51. Wallenstein, M. B., A. L. Wahrhaftig, H. M. Rosenstock, and H. Eyring: "Symposium on Radiobiology," p. 70, Wiley, New York, 1952.
52. Rosenstock, H. M., M. B. Wallenstein, A. L. Wahrhaftig, and H. Eyring: *Proc. Nat. Acad. Sci.*, **38**: 667 (1952).
53. Hipple, J. A., R. E. Fox, and E. U. Condon: *Phys. Rev.*, **69**: 347 (1946).
54. Kropf, A., E. M. Eyring, A. L. Wahshaftig, and H. Eyring: *J. Chem. Phys.*, **32**: 149 (1960).
55. Lester, G. R.: in J. D. Waldron (ed.), "Advances in Mass Spectrometry," p. 287, Pergamon Press, New York, 1959.
56. Schissler, D. O., S. O. Thompson, and J. Turkevich: *Discussions Faraday Soc.*, **10**: 46 (1951).
57. Lossing, F. P., A. W. Tickner, and W. A. Bryce: *J. Chem. Phys.*, **19**: 1254 (1951).
58. Inghram, M. G., and R. Gomer: *J. Chem. Phys.*, **22**: 1279 (1954).
59. Beckey, H. D., and G. Wagner: *Z. Naturforsch.*, **20a**: 169 (1965).
60. Chupka, W. A., and M. G. Inghram: *J. Chem. Phys.*, **21**: 1313 (1953). Chupka, W. A., and M. G. Inghram: *J. Phys. Chem.*, **59**: 100 (1955).
61. Honig, R. E.: *J. Chem. Phys.*, **22**: 126 (1954).

Bibliography

"Advances in Mass Spectrometry," vol. 1, J. D. Waldron (ed.), Pergamon Press, New York, 1959; *ibid.* vol. 2, R. M. Elliott (ed.), 1962.

Aston, F. W.: "Mass Spectra and Isotopes," E. Arnold, London, 1942.

"Atomic and Molecular Processes," D. R. Bates (ed.), Academic, New York, 1962.

"Atomic Collision Processes," M. R. C. McDowell (ed.), Wiley, New York, 1964.

Beynon, J. H.: "Mass Spectrometry and Its Application to Organic Chemistry," Elsevier, Amsterdam, 1960.

Biemann, K.: "Mass Spectrometry," McGraw-Hill, New York, 1962.

Brunnée, C., and H. Voshage: "Massenspektrometrie," Karl Thiemig, Munich, 1964.

Craggs, J. D., and H. S. W. Massey: The Collisions of Electrons with Molecules, in S. Flugge (ed.), "Encyclopedia of Physics," vol. 37/1, pp. 314–415, Springer, Berlin, 1959.

Duckworth, H. E.: "Mass Spectroscopy," Cambridge University Press, London, 1958.

Ewald, H., and H. Hintenberger: "Methoden und Anwendungen der Massenspektroskopie," Verlag Chemie, GmbH, Weinhein, Germany, 1953.

Ewald, H.: Korpuskularoptik-Massenspektroskopische Apparate, in S. Flugge (ed.), "Encyclopedia of Physics," vol. 33, pp. 546–608, Springer, Berlin, 1956.

Field, F. H., and J. L. Franklin: "Electron Impact Phenomena and the Properties of Gaseous Ions," Academic, New York, 1957.

Hasted, J. B.: "Physics of Atomic Collisions," Butterworth, London, 1964.

Inghram, M. G., and R. J. Hayden: "A Handbook on Mass Spectroscopy," National Academy of Sciences—National Research Council, Washington, D.C., 1954.

Massey, H. S. W., and E. H. S. Burhop: "Electronic and Ionic Impact Phenomena," Oxford University Press, Fair Lawn, N.J., 1952.

"Mass Spectrometry," C. A. McDowell (ed.), McGraw-Hill, New York, 1963.

"Mass Spectrometry of Organic Ions," F. W. McLafferty (ed.), Academic, New York, 1963.

Chapter 10

Fundamental Constants of Atomic Physics

By E. RICHARD COHEN, North American Aviation Science Center *and*
JESSE W. M. DuMOND, California Institute of Technology

A review of the status of knowledge of the fundamental atomic constants as of January, 1963, has been prepared for the Committee on Fundamental Constants of the National Research Council. The resulting tables of numerical values have been published [1],† and recommended for general use. A complete description of all the experimental data upon which this adjustment is based has been published elsewhere [2, 3].

The 1963 adjustment replaces and supersedes all its predecessors. Previous complete adjustments of the fundamental constants were made by the present authors in 1955, 1952, 1950, and 1947, and the general methods and philosophy for performing such adjustments have been explained at length in earlier papers [4, 5]. These give the general approach, and so this chapter gives only a review of the present input data and the results.

1. Changes in Definitions of Units Since the 1955 Adjustment

1. The unified atomic-weight unit adopted by the International Union of Pure and Applied Physics and the International Union of Pure and Applied Chemistry in 1960 is equal to exactly one-twelfth the weight of the ^{16}C atom. The conversion factor from the unified scale to the old physical scale (on which ^{12}O = 16) is useful:

$$\frac{\text{Atomic weights on old physical scale } (^{16}\text{O} = 16)}{\text{Atomic weights on new unified scale } (^{12}\text{C} = 12)} = 1.000317917 \pm 0.000000017$$

2. Either the new thermodynamic scale of temperature or the new Celsius scale, both internationally adopted in 1954, is used exclusively in this adjustment. On the thermodynamic scale the fundamental fixed points are the absolute zero and the triple point of water, the latter defined as 273.16°K exactly. On the Celsius scale, the triple point of water is defined as 0.01°C, and the scale differs from the new thermodynamic scale throughout by the constant difference of 273.15°. This places the ice point at 0.0000 ± 0.0001°C and the steam point at 99.9964 ± 0.0036°C. Since the difference between the two is no longer exactly 100° the name "centigrade" is abandoned and "Celsius" is substituted.

† Numbers in brackets refer to References at end of chapter.

2. Auxiliary Constants

The experimental data are classified into two categories for practical reasons: (1) auxiliary constants (a name first bestowed by R. T. Birge in the 1930s) to denote those data that are believed to be known with such superior accuracy that their observational errors can be ignored and (2) stochastic data, that is to say, data subject to least-squares adjustment. These data are the results of physical measurements of more limited precision, considered, however, sufficiently accurate and reliable to qualify for use in such an adjustment.

Not *all* data by any means should be cast into the least-squares melting pot for the purposes of such adjustments. Indeed, the most time-consuming, difficult, and critical task of a reviewer preparing an adjustment of this sort is precisely the wise selection and rejection of the total budget of available data. In forming a weighted average or a least-squares adjustment, it is undesirable to retain some data of observational precision quite inferior to the rest, for the reason that the experimental physicist usually does not aim at avoiding systematic errors much smaller in magnitude than the measure of random error he wishes to achieve.

Table 10.1 lists the numerical values of the auxiliary constants adopted for the 1963 adjustment.

3. The Choice of Unknowns

Preliminary work on the adjustment of 1963 actually started early in 1960. The unknowns originally selected at that time as the objectives of the least-squares adjustment included Λ, the conversion factor between X-ray wavelengths expressed on Siegbahn's nominal X-unit scale and wavelengths expressed on the absolute-length scale in milliangstroms. As a result of an analysis of the compatibility of the input data, it was decided to reject all input data from X-ray sources as insufficiently reliable to qualify by comparison with other sources of information.

A great number of items of input data from many other sources besides those from the field of X-rays were rejected for various reasons. In no case, how-

TABLE 10.1. AUXILIARY CONSTANTS ADOPTED IN 1963 ADJUSTMENT

1. Velocity of light, c		299,792.5 km sec^{-1}
2. Electron moment in Bohr magnetons, μ_e/μ_0		1.001159615
3. Proton moment in Bohr magnetons, μ_p/μ_0		0.0015210325
4. Rydberg constant for infinite mass, R_∞		109,737.31 cm^{-1}
5. Atomic masses and ratios (on unified scale $^{12}C = 12$)		
Hydrogen	H	1.00782522
Hydrogen/proton	H/M_p	1.000544607
Deuterium	D	2.01410219
Deuterium/deuteron	D/M_d	1.000272448
6. 1 U.S. NBS coulomb = 0.1000012 ± 0.0000004 abs. emu		

Sources of information and references:

1. This coincides with the value adopted by the International Scientific Radio Union and accepted by the International Union of Geodesy and Geophysics in 1957. It is in close agreement with work of K. D. Froome [6] and of Bergstrand [7].

2. Computed from the theoretical formula $\mu_e/\mu_0 = 1 + \alpha/2\pi - 0.328\ (\alpha/\pi)^2 + \cdots$, using for α the value $\alpha^{-1} = 137.039$. D. T. Wilkinson and H. R. Crane have subsequently verified the value of $1 - \mu_e/\mu_0$ experimentally with high accuracy [8].

3. Measurement of E. B. D. Lambe [9].

4. Recalculated, taking Lamb shift into account [10].

5. Calculated from the 1962 least-squares adjustment of L. A. Koenig, J. H. E. Mattauch, and A. Wapstra [11], based on data from both nuclear reactions and mass spectroscopy.

6. This conversion factor corresponds to the most recent (1958) redetermination of the absolute ampere at the U.S. National Bureau of Standards by R. L. Driscoll and R. D. Cutkosky [12].

ever, was any datum rejected *solely* on the grounds of its being "outlying," that is to say, discordant from the consensus of similar items or discrepant from the consensus of the entire adjustment, as indicated by our variance analysis. In the case of every rejection, we believe we have either found good and sufficient reasons for suspecting the presence of systematic error rooted in the work itself or have omitted the item because its estimated precision of measurement was too low for it to exert appreciable weight in the adjustment.

4. The Stochastic Input Data in the 1963 Adjustment

The experimental input data retained for use in the 1963 adjustment are listed in Table 10.2 with references to our sources of information.

TABLE 10.2. STOCHASTIC INPUT DATA USED IN 1963 ADJUSTMENT

Fine-structure separation in deuterium [13]	10,971.59 ± 0.10 Mc sec^{-1}
Magnetic moment of the proton in nuclear magnetons [14]	2.792757 ± 0.000025
Magnetic moment of the proton in nuclear magnetons [15]	2.792770 ± 0.000070
Faraday constant by electrolysis of silver [16]	9,648.682 ± 0.066 coul mole^{-1}
Gyromagnetic ratio of the proton [17]	26,751.92 ± 0.08 sec^{-1} gauss^{-1}
Gyromagnetic ratio of the proton [18]	26,751.88 ± 0.08 sec^{-1} gauss^{-1}

These data, when appropriately combined with the auxiliary constants given in Table 10.1, permit formation of an overdetermined set of six equations in the three unknowns α, e, and N. The equations are therefore overdetermined so as to have 3 degrees of freedom. The adjustment of this set by least squares leads to the following list of "best" adjusted values for five important physical constants:

$$
\begin{aligned}
1/\alpha &= 137.0388 \pm 0.0006 \\
e &= (4.80298 \pm 0.00007) \times 10^{-10}\ \text{esu} \\
h &= (6.62559 \pm 0.00016) \times 10^{-27}\ \text{erg-sec} \\
m &= (9.10908 \pm 0.00013) \times 10^{-28}\ \text{g} \\
N &= (6{,}022.52 \pm 0.09) \times 10^{20}\ \text{mole}^{-1}\ (^{12}C = 12)
\end{aligned}
$$

X-ray data can be introduced into this adjustment only if a self-consistent system of X-ray wavelengths is used in which all wavelength values are defined with respect to a single reproducible X-ray standard. Bearden [19] has shown that inconsistencies of the order of as much as 20 ppm occur in existing precision measurements of X-ray wavelengths. He has proposed a definition of the X unit in terms of which $\lambda W K\alpha_1 = 208.5770$ and which implies

$$
\begin{aligned}
\lambda \text{Mo}K\alpha_1 &= 707.831 \\
\lambda \text{Cu}K\alpha_1 &= 1{,}537.370
\end{aligned}
$$

The conversion factor from X units to angstroms, $\Lambda = \lambda g/\lambda s$, is then [3]

$$\Lambda = 1.002080 \pm 0.000006$$

An evaluation of the experimental data that became available between 1963 and June, 1965, indicates that their inclusion in an analysis would not alter the "best" values given here by more than one-third of the standard deviations of these constants [3].

5. Assignment of Error Measures of the Output Values

The χ^2 for this adjustment indicated extremely good compatibility between the equations of the overdetermined set. Its very low value, $\chi^2 = 0.34$, in comparison with an expectation value of 3, indicates a value for R. T. Birge's ratio of errors by the criterion of external consistency to errors by internal consistency, $r_e/r_i = 0.33 \pm 0.14$.

It must be remembered, however, that this unusually low value results in part from the nonrandom character of our choice of input equations, a set which has indeed been radically *censored*, i.e., purged of "outlying" items. To achieve the *larger* and more realistic assignment of the output errors in the present case, the criterion of *internal* consistency has been chosen as the basis for calculating the estimated errors to be attached to all our output data.

In Tables 10.3 and 10.4 the digits in parentheses following each numerical value represent the standard deviation (computed by the criterion of internal consistency) in the final digits of the quoted value. (This is the only respect in which the values published here differ from those published by the National Research Council Committee [1], where the accuracy of the data was indicated by giving three-standard-deviation error limits.)

Table 10.5 lists the defined values and equivalents that constitute the basis in terms of which the entire

TABLE 10.3. GENERAL PHYSICAL CONSTANTS
LEAST-SQUARES ADJUSTED OUTPUT VALUES OF 1963

The digits in parentheses following each quoted value represent the standard deviation error in the final digits of the quoted value as computed on the criterion of internal consistency. The unified scale of atomic weights is used throughout (^{12}C = 12).

C = coulomb G = gauss Hz = hertz J = joule N = newton T = tesla u = unified nuclidic mass unit W = watt Wb = weber

Constant	Symbol	Value	Unit	
			mksa	cgs
Speed of light in vacuum	c	2.997925(1)	$\times 10^{8}$ m s^{-1}	$\times 10^{10}$ cm s^{-1}
Gravitational constant	G	6.670(5)*	10^{-11} N m^2 kg^{-2}	10^{-8} dynes cm^2 g^{-2}
Elementary charge	e	1.60210(2) 4.80298(7)	10^{-19} C	10^{-20} emu 10^{-10} esu
Avogadro constant	N_A	6.02252(9)	10^{26} kmole^{-1}	10^{23} mole^{-1}
Mass unit	u	1.66043(2)	10^{-27} kg	10^{-24} g
Electron rest mass	m_e	9.10908(13) 5.48597(3)	10^{-31} kg 10^{-4} u	10^{-28} g 10^{-4} u
Proton rest mass	m_p	1.67252(3) 1.00727663(8)	10^{-27} kg u	10^{-24} g u
Neutron rest mass	m_n	1.67482(3) 1.0086654(4)	10^{-27} kg u	10^{-24} g u
Faraday constant	F	9.64870(5) 2.89261(2)	10^{4} C mole^{-1}	10^{3} emu 10^{14} esu
Planck constant	h $h/2\pi$	6.62559(16) 1.054494(25)	10^{-34} J s 10^{-34} J s	10^{-27} erg s 10^{-27} erg s
Fine-structure constant, $2\pi e^2/hc$	α $1/\alpha$	7.29720(3) 137.0388(6)	10^{-3}	10^{-3}
Charge-mass ratio for electron	e/m_e	1.758796(6) 5.27274(2)	10^{11} C kg^{-1}	10^{7} emu 10^{17} esu
Quantum of magnetic flux	hc/e h/e	4.13556(4) 1.379474(13)	10^{-11} Wb	10^{-7} G cm^2 10^{-17} esu
Rydberg constant	R_∞	1.0973731(1)	10^{7} m^{-1}	10^{5} cm^{-1}
Bohr radius	a_0	5.29167(2)	10^{-11} m	10^{-9} cm
Compton wavelength of electron	h/m_ec $\lambda_C/2\pi$	2.42621(2) 3.86144(3)	10^{-12} m 10^{-13} m	10^{-10} cm 10^{-11} cm
Electron radius	$e^2/m_ec^2 = r_e$	2.81777(4)	10^{-15} m	10^{-13} cm
Thomson cross section	$8\pi r_e^2/3$	6.6516(2)	10^{-29} m^2	10^{-25} cm^2
Compton wavelength of proton	$\lambda_{C,p}$ $\lambda_{C,p}/2\pi$	1.321398(13) 2.10307(2)	10^{-15} m 10^{-16} m	10^{-13} cm 10^{-14} cm
Gyromagnetic ratio of proton	γ $\gamma/2\pi$	2.675192(7) 4.25770(1)	10^{8} rad s^{-1}T^{-1} 10^{7} Hz T^{-1}	10^{4} rad s^{-1}G^{-1} 10^{3} s^{-1}G^{-1}
(Uncorrected for diamagnetism H_2O)	γ' $\gamma'/2\pi$	2.675123(7) 4.25759(1)	10^{8} s^{-1} T^{-1} 10^{7} HzT^{-1}	10^{4} rad s^{-1}G^{-1} 10^{3} s^{-1}G^{-1}
Bohr magneton	μ_B	9.2732(2)	10^{-24} J T^{-1}	10^{-21} erg G^{-1}

TABLE 10.3 GENERAL PHYSICAL CONSTANTS (*Continued*)

Constant	Symbol	Value	Unit	
			mksa	cgs
Nuclear magneton	μ_N	5.05050(13)	10^{-27} J T^{-1}	10^{-21} s G^{-1}
Proton moment	μ_p μ_p/μ_N	1.41049(4) 2.79276(2)	10^{-26} J T^{-1}	10^{-23} erg G^{-1}
(Uncorrected for diamagnetism in H_2O sample)		2.79268(2)		
Gas constant	R_0	8.31434(35)	J deg^{-1} $mole^{-1}$	10^7 ergs deg^{-1} $mole^{-1}$
Boltzmann constant	k	1.38054(6)	10^{-23} J deg^{-1}	10^{-16} erg deg^{-1}
First radiation constant ($2\pi hc^2$)	c_1	3.74150(9)	10^{-16} W m^2	10^{-5} erg cm^2 s^{-1}
Second radiation constant (hc/k)	c_2	1.43879(6)	10^{-2} m deg	cm deg
Stefan-Boltzmann constant	σ	5.6697(10)	10^{-8} W m^{-2} deg^{-4}	10^{-5} erg cm^{-2} s^{-1} deg^{-4}

* The universal gravitational constant is not, and cannot be in our present state of knowledge, expressed in terms of other fundamental constants. The value given here is a direct determination by P. R. Heyl and P. Chrzanowski, *J. Res. Natl. Bur. Std.*, **29**: 1 (1942).

TABLE 10.4. ENERGY CONVERSION FACTORS

1 electron volt	= 1.60210(2) × 10^{-19} J
	= 1.60210(2) × 10^{-12} erg
	= 8,065.73(8) cm^{-1}
	= 2.41804(2) × 10^{14} s^{-1}
$V\lambda$	= 12,398.10(13) × 10^{-8} ev cm
1 ev per particle	= 11604.9(5)°K
	= 23061(1) cal_{th} $mole^{-1}$
	= 23045(1) cal_{IT} $mole^{-1}$
1 amu	= 931.478(5) Mev
Proton mass	= 938.256(5) Mev
Neutron mass	= 939.550(5) Mev
Electron mass	= 511,006(2) ev
Rydberg	= 2.17971(5) × 10^{-11} erg
	= 13.60535(13) ev
Gas constant, R_0	= 8.31434 × 10^7 erg $mole^{-1}$ deg^{-1}
	= 0.082053 liter atm $mole^{-1}$ deg^{-1}
	= 82.055 cm^3 atm $mole^{-1}$ deg^{-1}
	= 1.9872 cal_{th} $mole^{-1}$ deg^{-1}
	= 1.9858 cal_{IT} $mole^{-1}$ deg^{-1}
Standard volume of ideal gas, V_0	= 22,413.6 cm^3 $mole^{-1}$

system of physical measurements is defined and maintained. Most of these definitions now refer the unit quantity to some precisely reproducible natural phenomenon. The only arbitrarily defined fundamental unit is that of mass – the kilogram, defined as the mass of the platinum prototype preserved at the International Bureau of Weights and Measures at Sèvres, France. The "standard acceleration of free fall" (g_n), or "normal gravity," is of course a purely conventional number, having no particular relation to any terrestrial or other physical phenomenon, since the actual acceleration of gravity varies from this by ±0.3 per cent from pole to equator, and regional variations of several parts per million over the earth's surface are quite erratic.

TABLE 10.5. DEFINED VALUES AND EQUIVALENTS

Meter (m)	1,650,763.73 wavelengths of the unperturbed transition $2p_{10} - 5d_5$ in ^{86}Kr
Kilogram (kg)	Mass of the International kilogram
	Astronomical
Second (s)	1/31,556,925.9747 of the tropical year at 12^hET, 0 January, 1900 (yr = 365^d $5^h48^m45^s$.9747)
	Physical
	9,192,631,770 cycles of the hyperfine transition (4,0 → 3,0) of the ground state of ^{133}Cs unperturbed by external fields
Degree Kelvin (°K)	In the thermodynamic scale, 273.16°K = triple point of water T (°C) = T (°K) − 273.15 (freezing point of water, 0.0000 ± 0.0002°C)
Unified atomic mass unit (u)	1/12 the mass of an atom of ^{12}C nuclide
Standard acceleration of free fall (g_n)	9.80665 m s^{-2} 980.665 cm s^{-2}
Normal atmosphere (atm)	101,325 N m^{-2} 1.013250 dyne cm^{-2}
Thermochemical calorie (cal_{th})	4.184 J 4.184 × 10^7 ergs
International Steam Table calorie (cal_{IT})	4.1868 J 4.1868 × 10^7 ergs
Liter (l)	0.001000028 m^3 (recommended by CIPM 1950) 1,000.028 cm^3
Inch (in.)	0.0254 m 2.54 cm
Pound (avdp.) (lb)	0.45359237 kg 453.59237 kg

6. Computation of Propagated Errors When Two or More Output Data of the 1963 Adjustment Enter into Further Mathematical Formulas

The output values of any least-squares adjustment will, in general, have correlated error distributions, as explained elsewhere [3-5]. Therefore, when further numerical values are to be computed, using formulas (functions) that depend upon two or more such output values, care must be taken to use the generalized formula of error propagation to compute the error propagated into the function in question. This generalized formula involves not only the *variances* (squares of the standard deviations) of the individual values taken from our tables but also the *covariances* associated with any *pair* of these values. Since the covariances may have either positive or negative sign, failure to include such terms may result in either too small or too large an error estimate. The variances and covariances form a symmetric matrix; in Table 10.6 we give:

TABLE 10.6. VARIANCE MATRIX AND CORRELATION COEFFICIENT MATRIX OF 1963 ADJUSTMENT

Variances are given, in units of (ppm)², on and below the major diagonal. Correlation coefficients are given in italics above the diagonal. Since there are in fact only three independent variables in the adjustment, these matrices are degenerate and are only of rank 3.

	α	e	N	h	m	F
α	21	*0.96*	*-0.88*	*0.95*	*0.88*	*0*
e	62	199	*-0.93*	*0.99*	*0.97*	*-0.06*
N	-62	-204	240	*-0.93*	*-0.93*	*0.41*
h	104	336	-346	569	*0.98*	*-0.08*
m	62	212	-222	362	237	*-0.12*
F	0	-5	35	-10	-10	30

the variance-covariance matrix corresponding to the *relative errors* in parts per million (ppm) of the 1963 adjustment. The matrix is augmented to include the variances and covariances expressed as (ppm)² for six frequently used output quantities, the three primary "unknowns" α, e, and N and three others, h, m, and F, computed from these. The numbers on the major diagonal (variances) and the 15 covariances forming a triangle below it cover all the values needed. The upper triangle of 15 numbers in italics are the corresponding correlation coefficients.

The generalized formula of error propagation may be expressed either in terms of the variance matrix v_{ij} or in terms of the standard deviations σ_i and the correlation coefficients r_{ij}. We define v_{ij} as the mean value of the product $\epsilon_i\epsilon_j$ of relative deviations from the mean for two statistically correlated stochastic quantities, x_i and x_j. For $i = j$, v_{ii} is just σ_i^2, the mean-square deviation of a stochastic variable from its mean. The correlation coefficient is given by $r_{ij} = v_{ij}/\sigma_i\sigma_j$. It is clear that $v_{ij} = v_{ji}$ and $r_{ij} = r_{ji}$.

In order to compute the variance, σ_y^2, associated with a quantity y, which depends upon n stochastic, statistically correlated quantities x_i,

$$y = f(x_1,x_2, \ . \ . \ . \ ,x_n) \tag{10.1}$$

one must use the full variance matrix $V = (v_{ij})$.

The variance σ_y^2 of y is given by

$$\sigma_y^2 = \sum_{i=1}^{n} \sum_{j=1}^{n} \frac{\partial y}{\partial x_i} v_{ij} \frac{\partial y}{\partial x_j} \tag{10.2}$$

Formula (10.2) may also be written in terms of the correlation coefficients r_{ij}:

$$\sigma_y^2 = \sum_{i=1}^{n} \left(\frac{\partial y}{\partial x_i}\right)^2 \sigma_i^2 + 2 \sum_{i<j} r_{ij}\sigma_i\sigma_j \frac{\partial y}{\partial x_i}\frac{\partial y}{\partial x_j} \tag{10.2a}$$

The second summation in this equation represents the modification, caused by the presence of correlation, from the more familiar formula for the case of independent errors.

References

1. Announcement by Joint Committee of the Divisions of Physics and Chemistry of the National Research Council (A. G. McNish, Chairman), *Phys. Today*, **17**: 48 (1964); *J. Opt. Soc. Am.*, **54**: 281 (1964); *Natl. Bur. Std.* (*U.S.*), *Tech. News Bull.*, October, 1963.
2. Cohen, E. R., and J. W. M. DuMond: Present Status of Our Knowledge of the Numerical Values of the Fundamental Physical Constants, *Proc. Second Intern. Conf. Nuclidic Masses, Vienna*, Walter H. Johnson, Jr., (ed.), July, 1963.
3. Cohen, E. R., and J. W. M. DuMond: *Revs. Mod. Phys.*, **37**: 537 (1965).
4. Cohen, E. R., and J. W. M. DuMond: Fundamental Constants of Atomic Physics, in S. Flügge (ed.), "Encyclopedia of Physics," vol. XXXV, pp. 1-87, Springer, Berlin 1957.
5. Cohen, E. R., K. M. Crowe, and J. W. M. DuMond: "Fundamental Constants of Physics," Interscience, New York, 1957.
6. Froome, K. D.: *Proc. Roy. Soc.* (*London*), **A247**: 109 (1958).
7. Bergstrand, Eric: *Ann. Franc. Chronom.*, **II**: 97 (1957).
8. Wilkinson, T. D., and H. R. Crane: *Phys. Rev.*, **130**: 852 (1963).
9. Lambe, E. B. D.: Thesis, Princeton University, 1959.
10. Cohen, E. R.: *Phys. Rev.*, **88**: 353 (1952).
11. Koenig, L. A., J. H. E. Mattauch, and A. Wapstra: *Nucl. Phys.*, **31**: 18 (1962).
12. Driscoll, R. L., and R. D. Cutkosky: *J. Res. Natl. Bur. Std.*, **60**: 297 (1958).
13. Triebwasser, S., E. S. Dayhoff, and W. E. Lamb, Jr.: *Phys. Rev.*, **89**: 98 (1953).
14. Sommer, H., H. A. Thomas, and J. A. Hipple: *Phys. Rev.*, **82**: 697 (1951).
15. Collington, D. J., A. N. Dellis, J. H. Sanders, and K. C. Turberfield: *Phys. Rev.*, **99**: 1622 (1955). Sanders, J. H., and K. C. Turberfield: *Proc. Roy. Soc.* (*London*), **A272**: 79 (1962).
16. Craig, D. N., J. I. Hoffman, C. A. Law, and W. J. Hamer: *J. Res. Natl. Bur. Std.*, **64A**: 381 (1960). Craig, D. N., W. R. Shields, and V. H. Dibeler: *J. Am. Chem. Soc.*, **82**: 5033 (1960). Shields, W. R., E. L. Garner, and H. H. Dibeler: *J. Res. Natl. Bur. Std.*, **66A**: 1 (1962).
17. Bender, P. L., and R. L. Driscoll: *Trans. I.R.E.*, I-7, 176 (1958).
18. Vigoureux, P.: *Proc. Roy. Soc.* (*London*), **A270**: 72 (1962).
19. Bearden, J. A.: X-Ray Wavelengths, NYO-10586, U.S. Atomic Energy Commission, Division of Technical Information, Oak Ridge, Tenn., 1964.

Part 8 · The Solid State

Chapter I

Crystallography and X-ray Diffraction

By R. PEPINSKY *and* V. VAND, Pennsylvania State University

1. Classical Crystallography

Introduction. Most solid substances are crystalline, although large single crystals are of less common occurrence. Crystals can be defined on the macroscopic scale as homogeneous solids, in which some of the physical properties are a function of direction. This applies to the refractive index, elasticity, thermal, electric and magnetic properties, etc. Directional variation in cohesion gives rise to characteristic cleavages; and variation of velocity of growth, upon change of phase, often accounts for spectacular development of symmetry-related crystal faces. The crystal *habit* depends upon the relative shapes and areas of these faces. The habit may vary from specimen to specimen, since the velocity of growth is a function of many variables, such as temperature, concentration, pressure, and especially the presence of impurities. However, if one chooses any point as a center of the crystal and draws through it a set of normals to all the developed faces, this set of normals is usually a much more characteristic or *invariant* property of the crystal than its habit. A convenient procedure for examining this set of normals is to draw a sphere around the center and to mark the intersections of the normals with this sphere. These intersections can also be projected onto a plane in various ways. The most common methods, the stereographic and the gnomic projections, are also well known in cartography.

Examination of interfacial angles reveals: (1) the *constancy of angles* between corresponding faces for a given crystalline structure, discovered by Nicolaus Steno [1]† in 1669; (2) the *law of rational indices*, discovered by R. J. Haüy [2] in 1784; and (3) all crystals can be classified into a small number of *symmetry types*. All these properties find natural explanation in the atomicity of matter, as explained below. A discussion from the classical point of view can be found in the textbooks of Dana, Phillips, and others (see Bibliography).

Translational Symmetry. From the atomistic point of view, crystals represent atomic or molecular aggregates which repeat themselves indefinitely in space. The concept of *repetition*, whether translational or rotational, means an operation by which the system is brought into a state indistinguishable from the initial state, i.e., brought into an identity. In a crystal we can distinguish a *motif*, which can be an atom, an ion, or a cluster of atoms or ions, or one or more molecules, or even a part of a molecule. Such a molecule may be a very large one; even virus particles are said to crystallize. There is an infinite variety of motifs. Further, we can find in a crystal a definite *scheme of repetition* of the motif. There are two such schemes possible in one dimension, 17 in two, and 230 in three dimensions. These are called *line groups*, *plane groups*, and *space groups*, respectively. There are two basic kinds of repetition operations possible in a crystal: the *translation* operation and *rotations* or *rotation inversions*.

In three dimensions, there exist three *unit translations* which can be represented as three vectors not all in the same plane. If one of the unit translation vectors is multiplied by all the positive and negative integers, all the vectors so formed are said to form a *row of lattice points*. All the possible sums of the above set of vectors with another set generated from a different unit vector form a *net of lattice points*. All the possible sums of three sets generated from three unit translation vectors form a *lattice*. Any one row of lattice points can be described by one parameter, a, representing the separation of the neighboring points. A net has three parameters: distances a and b and angle γ. A lattice has six parameters: three distances a, b, and c, and three angles α, β, and γ (see Fig. 1.1). The atoms themselves should not be identified with the lattice, which is a geometrical concept only; atoms may be located at lattice points only as a special case.

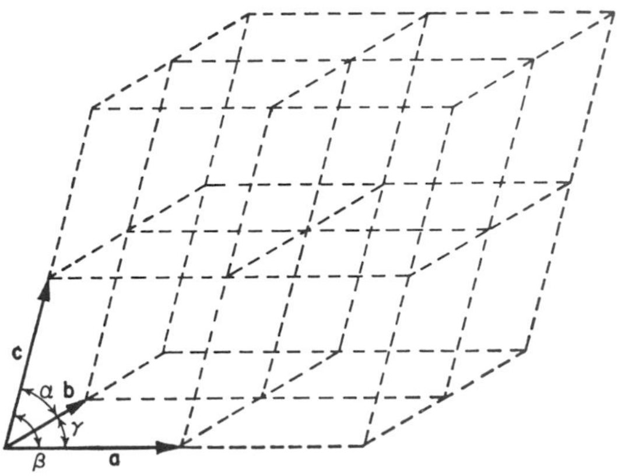

FIG. 1.1. Lattice parameters.

† Numbers in brackets refer to References at end of chapter.

In general there is an infinite number of ways in which the three unit translations or the vectors **a**, **b**, **c** can be chosen within the same lattice. Certain conventions limit the choice of the parameters. If no other symmetry elements are present, **a**, **b**, **c** are chosen so that they are the three shortest possible axes not in the same plane. They are then labeled so that $c < a < b$. (The convention $c > b > a$ is sometimes used, particularly in the field of long-chain compounds.) Then the positive senses of **a**, **b**, **c** are chosen in such a way that the angles α, β, γ are all obtuse if possible.

If symmetry elements are present, they overrule the above conventions. If a unique symmetry axis is present, **c** is taken along this axis, except in the monoclinic system where **b** is the unique axis. The introduction of symmetry is a more powerful simplification of description than the choice of a primitive cell. In space these simplifications lead to seven *crystal systems*, which in general can be recognized from the external appearance of crystals with developed faces. The seven systems are given in Table 1.1.

TABLE 1.1. SEVEN CRYSTAL SYSTEMS

Triclinic	$a \neq b \neq c$	$\alpha \neq \beta \neq \gamma$	
Monoclinic	$a \neq b \neq c$	$\alpha = \gamma = 90° \neq \beta$	
Orthorhombic	$a \neq b \neq c$	$\alpha = \beta = \gamma = 90°$	
Tetragonal	$a = b \neq c$	$\alpha = \beta = \gamma = 90°$	
Trigonal (or rhombohedral)	$a = b = c$	$\alpha = \beta = \gamma \neq 90°$	
Hexagonal	$a = b \neq c$	$\alpha = \beta = 90°$	$\gamma = 120°$
Cubic	$a = b = c$	$\alpha = \beta = \gamma = 90°$	

Bravais Lattices. The examination of all the possible lattices reveals that these 7 crystal systems can be further subdivided into 14 Bravais (or space) *lattices*, which are distinguishable by the presence or absence of *centering* of the unit cell. These were first introduced by Bravais [3]. The lattices can be conveniently described as *primitive* (no centering), symbol P; body-centered (German, *innerzentrierte*), symbol I; and face-centered. The latter can be either centered on all faces, symbol F, or only A, B, or C face-centered with symbols A, B, C. (In this nomenclature, the **a** and **b** axes define the C face, etc.) The space lattices cannot be recognized from the macroscopic properties of the crystal; an internal study of the atomic arrangement must be made.† The 14 Bravais lattices are given in Table 1.2.

TABLE 1.2. THE 14 BRAVAIS LATTICES

System	Centering P	I	F	C (or A or B)
Triclinic	P	—	—	—
Monoclinic	P	—	—	C
Orthorhombic	P	I	F	C
Tetragonal	P	I	—	—
Trigonal	P (conventionally called R)			
Hexagonal	P (conventionally called H, or sometimes C)			
Cubic	P	I	F	—

Miller Indices. It is useful to have a nomenclature describing all the possible sets of planes within a lattice which can form crystal faces (and also which can reflect X rays). The nomenclature is due to Miller [4]. It describes each set of planes by three integers h, k, l, positive or negative. Given Miller indices, the set of planes corresponding to them can be constructed as follows. Divide the **a** axis of the unit cell into h equal parts (that is, each of length a/h), the **b** axis into k parts, and the **c** axis into l parts. Choose a set of three dividing points so obtained which is nearest to the origin. Draw a plane through these (see Fig. 1.2). This gives one plane of the set. The next plane parallel to this will pass through the second nearest set of points from the origin, etc. A set of planes is thus obtained, having the interplanar spacing d equal to the distance of the first plane from the origin. Further parallel planes can thus be constructed by the use of the interplanar spacing. They will pass through the remaining dividing points. Furthermore, each such plane will eventually pass through lattice points of the crystal. Given any plane through three lattice points, and the set of lattice planes parallel to this,

† J. D. H. Donnay, *Bull. classe sci., Acad. roy. Belg.* [5], **23**: 749 (1937), for possible exceptions to this statement.

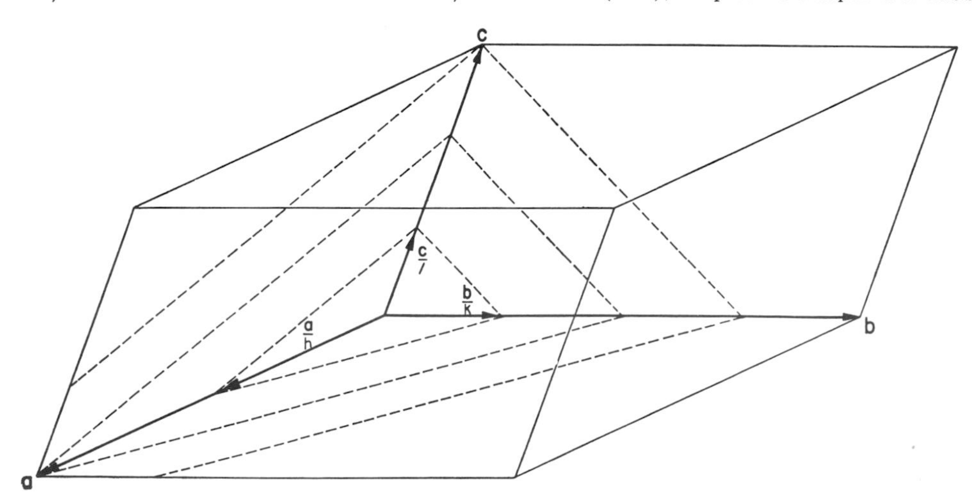

FIG. 1.2. Intercepts and Miller indices. $h, k, l = 2, 4, 3$.

the Miller indices of the set can be deduced by noting the intercepts a/h', b/k', c/l' within a given cell of a plane not passing through the origin of that cell. The Miller indices are the least set of integers proportional to $1/h'$, $1/k'$, $1/l'$.

Crystal faces arise during growth from the rate of deposition of the motifs within the crystal, as a function of direction. Faces will consequently always be parallel to lattice planes, and can be identified in terms of Miller indices. (Departures from this rule are very occasionally noted. Apparent "faces" may actually be due to alternating nonsystematic deposition on two different lattice planes. Such pseudo "faces" are said to be *vicinal*. Crystal growth via dislocations is not considered here; but it does not affect the above discussion significantly.) If the rules previously stated for selecting the vectors a, b, c are followed, the Miller indices of a face will generally be *small* integers. This is a statement of the empirical *law of rational indices*, first discovered by Haüy.

Point and Space Groups. There is only a small number of symmetry operations which do not violate the translational repetition of the crystal. As an example, crystals with a fivefold axis of rotation are not found in nature and are theoretically impossible, since such a symmetry is incompatible with any of the Bravais lattices. The only possible symmetry axes are 1-fold (identity operation), 2-fold, 3-fold, 4-fold, and 6-fold. Similar restrictions apply to other elements of symmetry. The elements of symmetry can be again divided into two kinds: one kind can be recognized from the development of the outside faces of the crystal; the other can be recognized only by methods which reveal the atomic arrangement within the crystal. The elements of the first kind are shown in Table 1.3.

TABLE 1.3. SYMMETRY ELEMENTS OF THE FIRST KIND

Element of symmetry	*Symbol*
1-fold rotation axis (identity operation)	1
2-fold rotation axis	2
3-fold rotation axis	3
4-fold rotation axis	4
6-fold rotation axis	6
Center of inversion	$\bar{1}$
2-fold rotary inversion axis (≡ mirror plane)	$\bar{2} = m$
3-fold rotary inversion axis	$\bar{3}$
4-fold rotary inversion axis	$\bar{4}$
6-fold rotary inversion axis	$\bar{6}$

By making all the possible combinations of the elements of the first kind which are consistent with the Bravais lattices, one arrives at 32 possible self-consistent combinations, which are called *point groups*. These are shown in Fig. 1.3 as equivalent points on stereographic projections: full points denote a point in the upper hemisphere; small circles denote correspondingly derived points in the lower hemisphere.

Each point group is denoted by the *Hermann-Mauguin* [5] symbol which enumerates the essential elements of symmetry present. As already stated, the point group to which a given crystal belongs can be determined from its macroscopic properties.

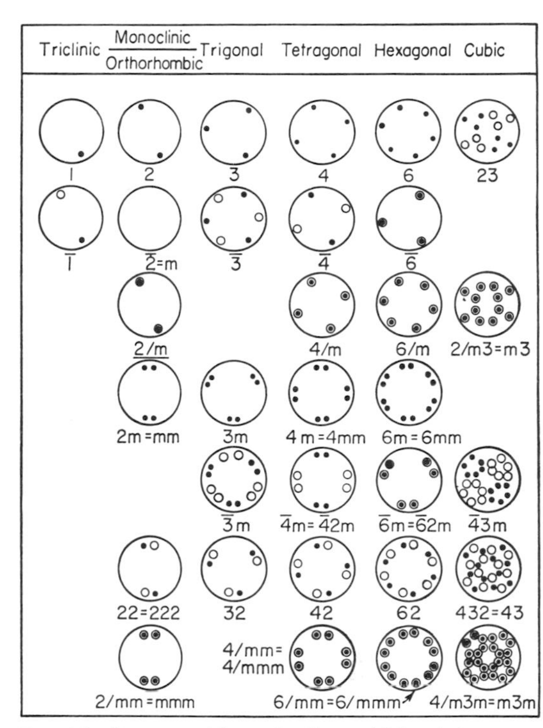

FIG. 1.3. Stereographic projections of 32 point groups.

The additional elements of symmetry not directly revealed by the macroscopic properties of the crystal are shown in Table 1.4.

TABLE 1.4. SYMMETRY ELEMENTS OF THE SECOND KIND

	Symbol
Axial glide plane (reflection plus translation of half axial vector)	a, b, or c
Diagonal glide plane (reflection plus translation of half of sum of two axial vectors)	n
"Diamond" glide plane (reflection plus translation of quarter of sum of two axial vectors)	d
2-fold screw axis (180° rotation plus translation of half unit vector along axis)	2_1
3-fold screw axis (120° rotation plus translation of $\frac{1}{3}$ or $\frac{2}{3}$ unit vector along axis)	3_1, 3_2
4-fold screw axis (90° rotation plus translation of $\frac{1}{4}$, $\frac{1}{2}$, or $\frac{3}{4}$ of unit vector along rotation axis)	4_1, 4_2, 4_3
6-fold screw axis (60° rotation plus translation of $\frac{1}{6}$, $\frac{1}{3}$, $\frac{1}{2}$, $\frac{2}{3}$, or $\frac{5}{6}$ of unit vector along rotation axis)	6_1, 6_2, 6_3, 6_4, 6_5

Again, making all the possible combinations of these and the previous group of elements of symmetry, one arrives in space at the 230 *space groups*, mentioned previously. The space groups are listed in Table 1.5. They were first derived independently by Federov in 1890 [6], Schoenflies in 1891 [7], and Barlow in 1895 [8]. They were also tabulated in much more detail in the *International Tables for X-ray Crystallography*, vol. I. These *Tables* are indispensable to any serious worker in crystallography, since the space group of a given crystal must be known before any further attempt is made at its structure determination.

TABLE 1.5†

(The eleven classes of distinct Laue symmetry are separated by double rulings)

System	Point groups Schfl.	Point groups H.-M.	Space groups						
Triclinic	C_1	1	$P1$						
	C_i	$\bar{1}$	$P\bar{1}$						
Monoclinic	$C_2^{(1-3)}$	2	$P2$	$P2_1$	$C2$				
	$C_s^{(1-4)}$	m	Pm	Pc	Cm	Cc			
	$C_{2h}^{(1-6)}$	$2/m$	$P2/m$	$P2_1/m$	$C2/m$	$P2/c$	$P2_1/c$	$C2/c$	
Orthorhombic	$D_2^{(1-9)}$	222	$P222$	$P222_1$	$P2_12_12$	$P2_12_12_1$	$C222_1$	$C222$	$F222$
			$I222$	$I2_12_12_1$					
	$C_{2v}^{(1-22)}$	$mm2$	$Pmm2$	$Pmc2_1$	$Pcc2$	$Pma2$	$Pca2_1$	$Pnc2$	$Pmn2_1$
			$Pba2$	$Pna2_1$	$Pnn2$	$Cmm2$	$Cmc2_1$	$Ccc2$	$Amm2$
			$Abm2$	$Ama2$	$Aba2$	$Fmm2$	$Fdd2$	$Imm2$	$Iba2$
			$Ima2$						
	$D_{2h}^{(1-28)}$	mmm	$Pmmm$	$Pnnn$	$Pccm$	$Pban$	$Pmma$	$Pnna$	$Pmna$
			$Pcca$	$Pbam$	$Pccn$	$Pbcm$	$Pnnm$	$Pmmn$	$Pbcn$
			$Pbca$	$Pnma$	$Cmcm$	$Cmca$	$Cmmm$	$Cccm$	$Cmma$
			$Ccca$	$Fmmm$	$Fddd$	$Immm$	$Ibam$	$Ibca$	$Imma$
Tetragonal	$C_4^{(1-6)}$	4	$P4$	$P4_1$	$P4_2$	$P4_3$	$I4$	$I4_1$	
	$S_4^{(1-2)}$	$\bar{4}$	$P\bar{4}$	$I\bar{4}$					
	$C_{4h}^{(1-6)}$	$4/m$	$P4/m$	$P4_2/m$	$P4/n$	$P4_2/n$	$I4/m$	$I4_1/a$	
	$D_4^{(1-10)}$	422	$P422$	$P42_12$	$P4_122$	$P4_12_12$	$P4_222$	$P4_22_12$	$P4_322$
			$P4_32_12$	$I422$	$I4_122$				
	$C_{4v}^{(1-12)}$	$4mm$	$P4mm$	$P4bm$	$P4_2cm$	$P4_2nm$	$P4cc$	$P4nc$	$P4_2mc$
			$P4_2bc$	$I4mm$	$I3cm$	$I4_1md$	$I4_1cd$		
	$D_{2d}^{(1-12)}$	$\bar{4}2m$	$P\bar{4}2m$	$P\bar{4}2c$	$P\bar{4}2_1m$	$P\bar{4}2_1c$	$P\bar{4}m2$	$P\bar{4}c2$	$P\bar{4}b2$
			$P\bar{4}n2$	$I\bar{4}m2$	$I\bar{4}c2$	$I\bar{4}2m$	$I\bar{4}2d$		
	$D_{4h}^{(1-20)}$	$4/mmm$	$P4/mmm$	$P4/mcc$	$P4/nbm$	$P4/nnc$	$P4/mbm$	$P4/mnc$	$P4/nmm$
			$P4/ncc$	$P4_2/mmc$	$P4_2/mcm$	$P4_2/nbc$	$P4_2/nnm$	$P4_2/mbc$	$P4_2/mnm$
			$P4_2/nmc$	$P4_2/ncm$	$I4/mmm$	$I4/mcm$	$I4_1/amd$	$I4_1/acd$	
Trigonal	$C_3^{(1-4)}$	3	$P3$	$P3_1$	$P3_2$	$R3$			
	$C_{3i}^{(1-2)}$	$\bar{3}$	$P\bar{3}$	$R\bar{3}$					
	$D_3^{(1-7)}$	32	$P312$	$P321$	$P3_112$	$P3_121$	$P3_312$	$P3_321$	$P32$
	$C_{3v}^{(1-6)}$	$3m$	$P3m1$	$P31m$	$P3c1$	$P31c$	$R3m$	$R3c$	
	$D_{3d}^{(1-6)}$	$\bar{3}m$	$P\bar{3}1m$	$P\bar{3}1c$	$P\bar{3}m1$	$P\bar{3}c1$	$R\bar{3}m$	$R\bar{3}c$	
Hexagonal	$C_6^{(1-6)}$	6	$P6$	$P6_1$	$P6_5$	$P6_2$	$P6_4$	$P6_3$	
	$C_{3h}^{(1)}$	$\bar{6}$	$P\bar{6}$						
	$C_{6h}^{(1-2)}$	$6/m$	$P6/m$	$P6_3/m$					
	$D_6^{(1-6)}$	622	$P622$	$P6_122$	$P6_522$	$P6_222$	$P6_422$	$P6_322$	
	$C_{6v}^{(1-4)}$	$6mm$	$P6mm$	$P6cc$	$P6_3cm$	$P6_3mc$			
	$D_{3h}^{(1-4)}$	$\bar{6}m2$	$P\bar{6}m2$	$P\bar{6}c2$	$P\bar{6}2m$	$P\bar{6}2c$			
	$D_{6h}^{(1-4)}$	$6/mmm$	$P6/mmm$	$P6/mcc$	$P6_3/mcm$	$P6_3mmc$			
Cubic	$T^{(1-5)}$	23	$P23$	$F23$	$I23$	$P2_13$	$I2_13$		
	$T_h^{(1-7)}$	$m3$	$Pm3$	$Pn3$	$Fm3$	$Fd3$	$Im3$	$Pa3$	$Ia3$
	$O^{(1-8)}$	432	$P432$	$P4_232$	$F432$	$F4_132$	$I432$	$P4_332$	$P4_132$
			$I4_132$						
	$T_d^{(1-6)}$	$\bar{4}3m$	$P\bar{4}3m$	$P\bar{4}3m$	$I\bar{4}3m$	$P\bar{4}3n$	$F\bar{4}3c$	$I\bar{4}3d$	
	$O_h^{(1-10)}$	$m3m$	$Pm3m$	$Pn3n$	$Pm3n$	$Pn3m$	$Fm3m$	$Fm3c$	$Fd3m$
			$Fd3c$	$Im3m$	$Ia3d$				

† Reproduced from J. M. Robertson, "Organic Molecules and Crystals," 44, Cornell University Press, Ithaca, N.Y., 1953.

2. X-ray Diffraction: Experimental

X-ray, Electron, and Neutron Diffraction. During the era of classical crystallography, which culminated in the derivation of all the space groups, practically no useful information was available concerning the *motifs*, or the details of the architecture of crystals on the atomic scale. Only axial ratios could be determined from the measurements of angles between the crystal faces, and dimensions of the unit cells were not known. A new epoch began with von Laue's discovery in 1912 of the diffraction of X rays by crystals. This established the nature of X rays, initiated the study of X-ray spectroscopy, and provided a first practical tool for exploration of crystals on the atomic scale.

Today, the tools of exploration are by no means limited to X-ray diffraction. Distances between neighboring atoms in a solid lie between 10^{-8} and 4×10^{-8} cm (10^{-8} cm = 1 A). Any radiation of wavelength of 2 A or less in general has resolving power high enough to distinguish individual atoms. Cu K_α radiation, with $\lambda = 1.54$ A, is readily obtainable in high intensity, and is most commonly used in crystal analyses. Need for higher or lower dispersion, higher or lower absorption, production or avoidance of fluorescence, etc., may dictate the choice of other wavelengths, however. In earlier researches, polychromatic ("white") radiation was frequently used; but monochromatic ("characteristic") radiation is now preferred.

Other types of radiation possessing the required wavelengths are in principle suitable for diffraction analyses. *Electron diffraction* represents a valuable tool, but comparatively high absorption of electrons restricts its application to very thin specimens or surface layers. *Neutron diffraction* is in some respects superior to X-ray diffraction for location of atom centers. It is limited in application chiefly by the relatively low primary-beam intensity presently available, necessitating the use of much larger crystals than are required for X-ray measurements. The detail revealed by these techniques differs according to the type of interaction of the radiation with the matter contained in the crystal. X rays interact with the electrons, the interaction with atomic nuclei being completely negligible. In addition, practically all the electrons behave as free electrons toward X rays of the wavelengths ordinarily used. X-ray scattering thus reveals the *electron density* within the crystal. Electrons interact with the electric fields within the atoms, so that they reveal primarily the distribution of the *electric potential* within the crystal. Neutrons interact primarily with the nuclei of the atoms, so that their diffraction reveals the positions of *atomic nuclei.* They also interact with the electrons with unpaired spins which are responsible for some of the magnetic properties of materials. Neutron diffraction thus reveals the distribution of such electrons, providing a tool for the study of magnetic phenomena on the atomic scale.

The theory of X-ray diffraction will be discussed below; conclusions will to a large extent be applicable to other types of diffraction as well.

Production and Measurement of X Rays. In X-ray crystal analysis, the X rays are produced by electron bombardment of an element, the characteristic radiation of which lies in the desired range of wavelengths for the material to be studied. The shorter the wavelength, the higher the resolution obtained of the atomic pattern in the crystal—in analogy with the optical microscope. Usually K_α characteristic radiation of the target element is employed, since this has a high intensity compared with the background of white radiation. The K_α line is a doublet, composed of K_{α_1} and K_{α_2} components. The doublet separation is slight, and it can be treated either as a single line, or corrections can be applied when splitting is significant. The bombarding electrons are usually produced from a hot cathode, and for copper radiation they are accelerated by a field of 30 to 45 kv. In older experiments, demountable gas X-ray tubes with cold cathodes were extensively used; their main advantages were low cost and the possibility of keeping the target clean and the radiation free of unwanted lines. The main problem of design is to arrange for dissipation of the heat generated by the electron bombardment of the focal spot of the target. This limits the X-ray output of the tube. The targets are generally water-cooled, and the brilliancy of the focus can be increased by making the focal spot as small as possible (microfocus tubes), or by moving the target so that a fresh cold area is always exposed to the electron beam (rotating anode tubes). It is also possible to excite the characteristic radiation by other means than by field-accelerated electrons: e.g., X rays of a shorter wavelength than the absorption edge of the target can be used (fluorescent X-ray tubes). Such a source has a very pure characteristic radiation, but is of a low brilliance. In principle, radioactivity could also be used for the excitation of a target.

The X rays, once generated, pass out of the tube through a low-absorption window. Beryllium is the most suitable material; but aluminum, Lindemann glass, mica, and other materials have been used. The X rays are then collimated by a slit system, so that only a narrow beam hits the crystal. The collimator may be a simple tube, or two orifices, or a system of Soller slits; or a suitably bent and shaped crystal may serve as a focusing collimator and a monochromator at the same time. The most suitable dimension of a crystal used for X-ray diffraction is such that the correction for absorption of X rays within it does not exceed perhaps 10 per cent, so that absorption can be neglected. Larger crystals can be used, but then the measurements must be corrected for absorption, which is laborious except for cylindrical crystals. The usual range of crystal sizes used is 0.1 to 0.5 mm in linear dimensions. Smaller crystals can be used, but the exposure times may then run into hundreds of hours—a practical inconvenience. The size of crystals used in electron diffraction may be several orders of magnitude smaller, and in neutron diffraction one order of magnitude larger. The crystal should be completely bathed in the primary beam.

The crystal then diffracts the X rays, and the scattered beams can be detected by a variety of devices. Among the first used were ionization chambers or photographic plates or films. The photographic method is the most used today, since it records a large amount of information rapidly and requires a minimum of monitoring; thus the equipment is simple to handle. Higher accuracy, however, can be attained by the use of Geiger-Müller, proportional, or scintillation counters. The latter two are faster and therefore have higher counting rates than Geiger-Müller counters, do not age as rapidly, and, when used with pulse-height discriminating circuits, can considerably suppress undesirable background. They require more complex associated equipment than do Geiger counters.

In electron diffraction, electrons are usually detected photographically, although other methods are possible. In neutron diffraction enriched boron trifluoride or scintillation counters are generally used.

It remains to discuss the arrangement of the cameras, i.e., the means by which the crystal can be oriented in the beam and the detector brought into a known position both in relation to the crystal and the primary X-ray beam.

General Conditions of Diffraction. Although X rays have a wavelength generally adequate for resolution of detail on the atomic scale, they are not suitable for direct microscopy, since no suitable lenses are available to form images of adequate quality, and single molecules are too transparent to X rays and are in general not maintainable in position and orientation. Images of atoms and molecules can be produced indirectly by arranging the molecules in a regular array in a crystal, and by constructing the magnified image of a single unit by a mathematical process, out of information contained in the diffraction pattern of the array. Otherwise, the analogy with microscope image formation is very close. X-ray diffraction theory is thus identical with that of light except for the numerical value of the wavelength. Analogous experiments can be made, when the scale of the object is changed in proportion [9–12].

If a monochromatic parallel beam of light falls on a plane grating, it is diffracted in the directions for which the phase difference of the scattered waves is exactly equal to $n\lambda$, since only in these directions do the wavelets reinforce each other (see Part 6, Chap. 3). Here n is the order of the spectrum, and is an integer. Similar conditions apply to a three-dimensional array of atoms, as found in a crystal.

In subsequent discussions, it is advantageous to abandon the classical crystallographic notation $\mathbf{a}$, $\mathbf{b}$, $\mathbf{c}$, α, β, γ for the axial vectors and interaxial angles, and adopt Zachariasen's notation (see Bibliography) $\mathbf{a}_1$, $\mathbf{a}_2$, $\mathbf{a}_3$, α_1, α_2, α_3, instead. Symbols in boldface are *vectors*. The Miller indices will be designated by the symbols h_1, h_2, h_3.

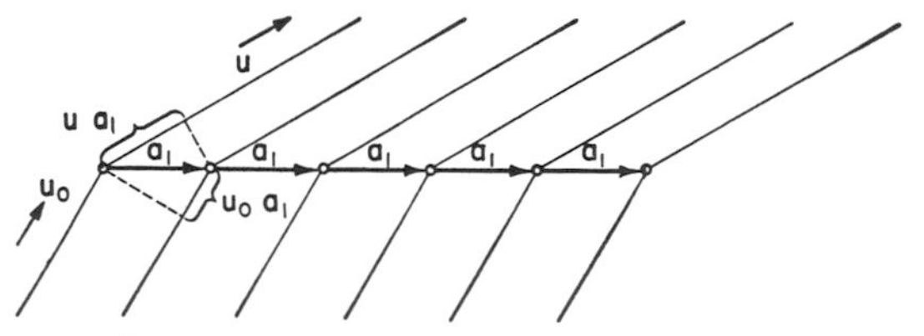

FIG. 1.4. a_1 component of Laue condition.

Let $\mathbf{u}_0$ and $\mathbf{u}$ be unit vectors in the directions of the incident and scattered beams, respectively. Then the three conditions for constructive interference (see Fig. 1.4) become

$$\begin{aligned} \mathbf{u}\cdot\mathbf{a}_1 - \mathbf{u}_0\cdot\mathbf{a}_1 = (\mathbf{u}-\mathbf{u}_0)\cdot\mathbf{a}_1 &= n_1\lambda \\ (\mathbf{u}-\mathbf{u}_0)\cdot\mathbf{a}_2 &= n_2\lambda \\ (\mathbf{u}-\mathbf{u}_0)\cdot\mathbf{a}_3 &= n_3\lambda \end{aligned} \tag{1.1}$$

These relations are known as the components of the *Laue equation*. They can readily be reduced to a single vector equation by the introduction of *reciprocal vectors*, as discussed in the next subsection. Each of these equations represents a condition for diffraction by a single row of lattice points. The triple set of integers, n_1, n_2, n_3, is the order of the spectrum of the integers, and is related to the Miller indices discussed previously.

A much simpler representation of the diffraction condition is known as Bragg's law (1912). This is based on the idea of reflection of X rays from lattice planes rather than on diffraction by lattice rows (see Fig. 1.5). Let AB be the primary X-ray beam of

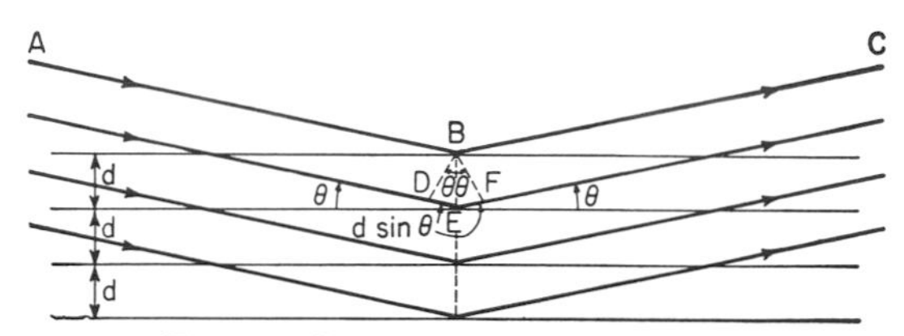

FIG. 1.5. Bragg's law: $2d \sin\theta = n\lambda$.

wavelength λ, incident on a crystal, the crystal planes of which are at an angle θ with the beam. A reflected beam BC will be formed by Huygens' principle. There will be reinforcement with a beam reflected from the next parallel lattice plane, at distance d, only when the path difference $DE + EF$ is equal to a whole number of wavelengths. This occurs when

$$n\lambda = 2d\sin\theta \tag{1.2}$$

which is the Bragg law.

When the Bragg law is fulfilled, the diffracted waves from all lattice points of the crystal are in phase. Just as in the plane grating the *shape* of the grooves does not influence the angles of the diffracted pattern but affects only the *intensities* of the individual diffracted beams; thus the presence of atoms or molecules modifies the intensities of the diffracted beams but not the Bragg angle θ. The diffraction by a crystal differs from that of a plane grating in that, for parallel monochromatic radiation, a crystal in a general position does not reflect X rays unless the Bragg condition is fulfilled. If the crystal is slowly rotated, diffracted beams flash out momentarily every time the crystal goes through the Bragg condition. On the other hand, if a crystal is placed in a beam of a white radiation (Laue experiment), there is always some wavelength λ for which the Bragg (or Laue) conditions are fulfilled, and there is always a number of diffracted beams present, all of different "colors." This is called the Laue pattern of a crystal. If such a crystal is rotated, the diffracted beams move, changing their "colors" as they do so.

According to the above picture, the angular spread of each reflection should be infinitely narrow. This is true for a perfect crystal of infinite extent and negligible absorption. Since ordinary crystals contain numerous imperfections, each reflection is spread over a narrow range of angles.

General Description of the Reciprocal Lattice. The interpretation of X-ray diffraction results is greatly simplified by means of the concept of reciprocal space and the reciprocal lattice. As will be seen, the relation between the direct and the reciprocal space description is equivalent to the relation between a function and its Fourier transform, or to the relation between contravariant and covariant vectors (see Part 1, Chaps. 10 and 11). It will be approached first in an elementary way.

In a diffraction experiment, one deals with *sets of planes*, a concept inconvenient for usual geometrical handling. A set of parallel equidistant planes is specified by only three parameters, so that it can be described by a point in space. Such a point, named a *reciprocal lattice point*, can be constructed as follows. Let the interplanar spacing be d. Construct a vector of length $1/d$, perpendicular to the planes. Choosing an origin, the vector, drawn from the origin, defines the reciprocal lattice point and completely specifies the set of planes. This vector is a *covariant vector*, as can easily be seen from its behavior when the coordinates are transformed. In particular, if the yardstick is increased in length, e.g., from centimeters to inches, then the numerical lengths of the ordinary vectors in the direct space decrease. This behavior is followed by all the *contravariant vectors*, such as distances between the atomic centers. However, the lengths of the covariant vectors increase numerically, so that they behave inversely, or reciprocally, compared with the distance vectors. Hence the nomenclature, as used in crystallography: the space of all the contravariant vectors is called *direct space;* the space of all the covariant vectors is *reciprocal space.*

If we take all the possible sets of crystallographic planes in a crystal, with all the possible Miller indices $h_1h_2h_3$, and construct all their reciprocal vectors as specified above, their end points again form a lattice. This is a most useful property, because it is much easier to construct this lattice, called the *reciprocal lattice*, than to construct all the sets of lattice planes by the construction given in the discussion of Sec. 1 on Miller indices. Each reciprocal lattice point can be labeled by three integers $h_1h_2h_3$, and the three unit vectors (100), (010), and (001) can easily be found. These unit vectors are denoted $\mathbf{b}_1$, $\mathbf{b}_2$, $\mathbf{b}_3$, and the angles between them β_1, β_2, β_3.

Owing to the importance of the concept of the reciprocal lattice, it is useful to practice the construction in two dimensions (see Fig. 1.6). In Fig. 1.6 the heavy lines represent one unit cell of the direct lattice with some of the lattice planes drawn. The normals to these are represented by thin lines. Open circles denote the corresponding reciprocal lattice points, with Miller indices also given.

Mathematical Relationships between Direct and Reciprocal Vectors. We can write down a mathematical definition of the reciprocal vector set as follows:

$$\mathbf{b}_1 = \frac{\mathbf{a}_2 \times \mathbf{a}_3}{\mathbf{a}_1 \cdot \mathbf{a}_2 \times \mathbf{a}_3} = \frac{\mathbf{a}_2 \times \mathbf{a}_3}{(\mathbf{a}_1\mathbf{a}_2\mathbf{a}_3)}$$

where $\mathbf{a}_1 \cdot \mathbf{a}_2 \times \mathbf{a}_3 = (\mathbf{a}_1\mathbf{a}_2\mathbf{a}_3) = V$, the volume of the cell defined by $\mathbf{a}_1$, $\mathbf{a}_2$, and $\mathbf{a}_3$.

$$\mathbf{b}_2 = \frac{\mathbf{a}_3 \times \mathbf{a}_1}{(\mathbf{a}_1\mathbf{a}_2\mathbf{a}_3)} \quad \text{and} \quad \mathbf{b}_3 = \frac{\mathbf{a}_1 \times \mathbf{a}_2}{(\mathbf{a}_1\mathbf{a}_2\mathbf{a}_3)} \tag{1.3}$$

It is immediately seen that

$$\mathbf{a}_i \cdot \mathbf{b}_j = \delta_{ij} \begin{cases} = 0 & \text{if } i \neq j \\ = 1 & \text{if } i = j \end{cases} \tag{1.4}$$

and furthermore one can show that

$$(\mathbf{a}_1\mathbf{a}_2\mathbf{a}_3)(\mathbf{b}_1\mathbf{b}_2\mathbf{b}_3) = 1 \tag{1.5}$$

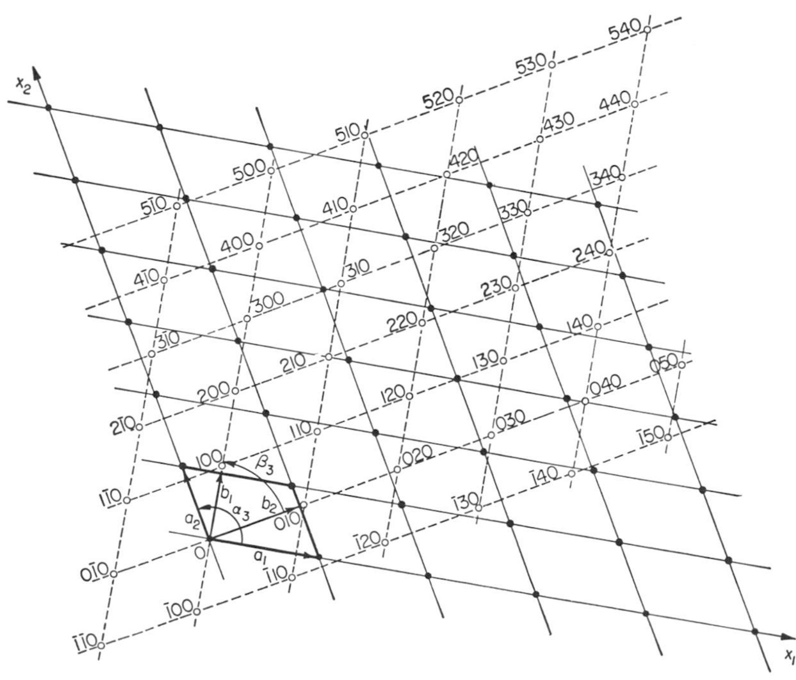

Fig. 1.6. Direct and reciprocal lattices, with Miller indices indicated. Direct lattice in solid lines; reciprocal lattice in dashed lines.

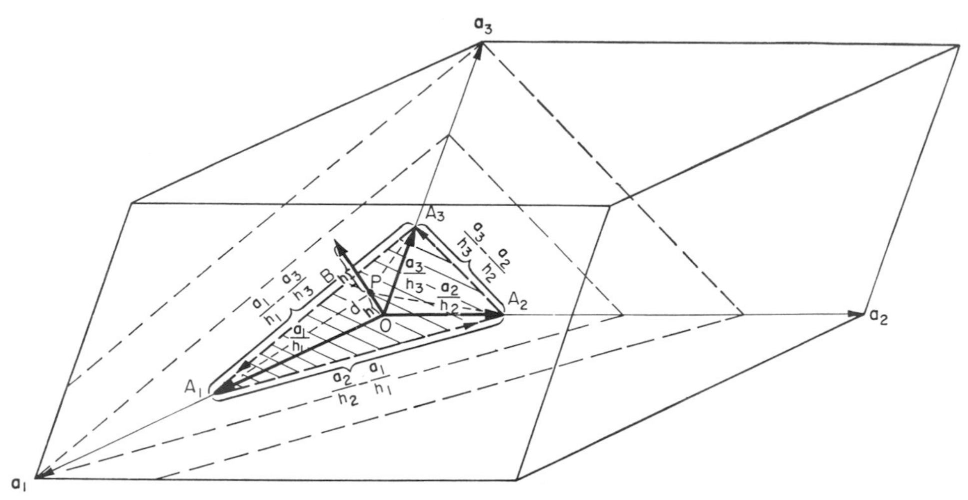

FIG. 1.7. Relation between $\mathbf{B}_h$, $\mathbf{d}_h$, and plane h_1,h_2,h_3. OP = perpendicular distance from O to plane h_1,h_2,h_3. P is in plane of h_1,h_2,h_3, and so are lines PA_1, PA_2, PA_3, and vectors $\mathbf{a}_2/h_2 - \mathbf{a}_1/h_1$, etc. Triangles OPA_1, OPA_2, and OPA_3 are all right triangles.

Furthermore [13],

$$\mathbf{a}_1 = \frac{\mathbf{b}_2 \times \mathbf{b}_3}{(\mathbf{b}_1\mathbf{b}_2\mathbf{b}_3)} \qquad \mathbf{a}_2 = \frac{\mathbf{b}_3 \times \mathbf{b}_1}{(\mathbf{b}_1\mathbf{b}_2\mathbf{b}_3)} \qquad \mathbf{a}_3 = \frac{\mathbf{b}_1 \times \mathbf{b}_2}{(\mathbf{b}_1\mathbf{b}_2\mathbf{b}_3)} \tag{1.6}$$

We now define a new vector

$$\mathbf{B}_h = h_1\mathbf{b}_1 + h_2\mathbf{b}_2 + h_3\mathbf{b}_3 \tag{1.7}$$

and examine its properties. Consider a plane with Miller indices h_1,h_2,h_3, as in Fig. 1.7. The intercepts of this plane are $\mathbf{a}_1/h_1$, $\mathbf{a}_2/h_2$, $\mathbf{a}_3/h_3$. From Fig. 1.7, it is apparent that the normal to plane h_1,h_2,h_3 is perpendicular to the vector sides of the intercepted triangle $\mathbf{a}_2/h_2 - \mathbf{a}_1/h_1$, $\mathbf{a}_3/h_3 - \mathbf{a}_2/h_2$, $\mathbf{a}_1/h_1 - \mathbf{a}_3/h_3$. The scalar product of $\mathbf{B}_h$ with each of these vector sides is zero. This is the condition that $\mathbf{B}_h$ is perpendicular to the plane h_1,h_2,h_3.

The perpendicular distance from the origin to the plane h_1,h_2,h_3 is the quantity we denote by $\mathbf{d}_h$, the interplanar spacing. The length of the vector $\mathbf{d}_h$ can be found by projecting $\mathbf{d}_h$ onto $\mathbf{B}_h/|\mathbf{B}_h|$. It is seen immediately that

$$|\mathbf{d}_h| = \mathbf{d}_h \cdot \frac{\mathbf{B}_h}{|\mathbf{B}_h|} = \frac{1}{|\mathbf{B}_h|} \tag{1.8}$$

The usefulness of $\mathbf{B}_h$ follows from these facts, just proved:

$\mathbf{B}_h$ is perpendicular to the plane h_1,h_2,h_3

$$|\mathbf{B}_h| = \frac{1}{|\mathbf{d}_h|}$$

i.e., $|\mathbf{B}_h|$ is the reciprocal of the interplanar spacing of the set of planes h_1,h_2,h_3.

Given a set of axial vectors $\mathbf{a}_1,\mathbf{a}_2,\mathbf{a}_3$, and Miller indices for various planes $h_1h_2h_3$, $h'_1h'_2h'_3$, $h''_1h''_2h''_3$, etc., one can compute $\mathbf{B}_h$, $\mathbf{B}_{h'}$, $\mathbf{B}_{h''}$, etc., by the use of Eqs. (1.3) and (1.7). Succinct vector expressions for $|\mathbf{d}_h|$, $|\mathbf{d}_{h'}|$, $|\mathbf{d}_{h''}|$, etc., then follow readily; and formulas for cell volumes, angles between planes h, h', h'', etc., directions of lattice lines (zones) formed by the intersection of two planes h and h', etc., all follow just as readily. Examples of such calculations are given in Zachariasen's text.

Vector Expression for the Laue Condition; the Ewald Construction. The three equations (1.1) can be reduced to a single expression by multiplying them, respectively, by $\mathbf{b}_1$, $\mathbf{b}_2$, and $\mathbf{b}_3$ and adding. One then obtains

$$[(\mathbf{u} - \mathbf{u}_0) \cdot \mathbf{a}_1]\mathbf{b}_1 + [(\mathbf{u} - \mathbf{u}_0) \cdot \mathbf{a}_2]\mathbf{b}_2 + [(\mathbf{u} - \mathbf{u}_0) \cdot \mathbf{a}_3]\mathbf{b}_3 = \lambda(n_1\mathbf{b}_1 + n_2\mathbf{b}_2 + n_3\mathbf{b}_3) = \lambda\mathbf{B}_n \tag{1.9}$$

The left side is equivalent to $\mathbf{u} - \mathbf{u}_0$, giving

$$\mathbf{u} - \mathbf{u}_0 = \lambda\mathbf{B}_n \tag{1.10}$$

which is the vector expression for the Laue condition. This contains the condition for specular reflection and Bragg's equation, as is seen most readily from Fig. 1.8. Equation (1.10) requires that $\mathbf{u} - \mathbf{u}_0$ be in the

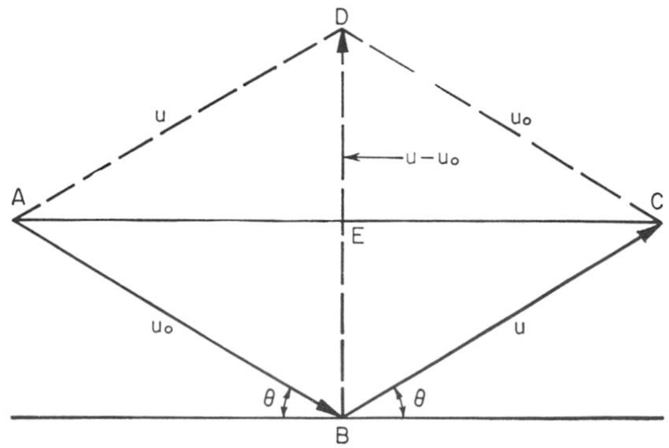

FIG. 1.8. Equivalence of $2 \sin \theta$ and $|\mathbf{u} - \mathbf{u}_0|$.

direction of $\mathbf{B}_n$, that is, perpendicular to the plane with indices $n = n_1,n_2,n_3$. Since $\mathbf{u}_0$ and $\mathbf{u}$ are unit vectors, the distance $BC = \sin \theta$, and $BD = 2 \sin \theta$. The scalar part of Eq. (1.10) requires that

$$2 \sin \theta = \frac{\lambda}{|d_n|}$$

or $2d_n \sin \theta = \lambda$. This is equivalent to Eq. (1.2), where $d_n = d/n$.

Once the reciprocal lattice is constructed, the geometric conditions for reflection of X rays can be found by the following construction.

Draw a sphere of radius $1/\lambda$, having the incoming X-ray beam in the direction $\mathbf{u}_0$, passing through its center, C. Where the primary beam emerges from the sphere, place the origin O of the crystal lattice and of its associated reciprocal lattice (see Fig. 1.9, representing the condition in which the equatorial plane of the sphere coincides with a reciprocal lattice plane. $\mathbf{u}$ must have its end point on the sphere, just as does $\mathbf{u}_0$.)

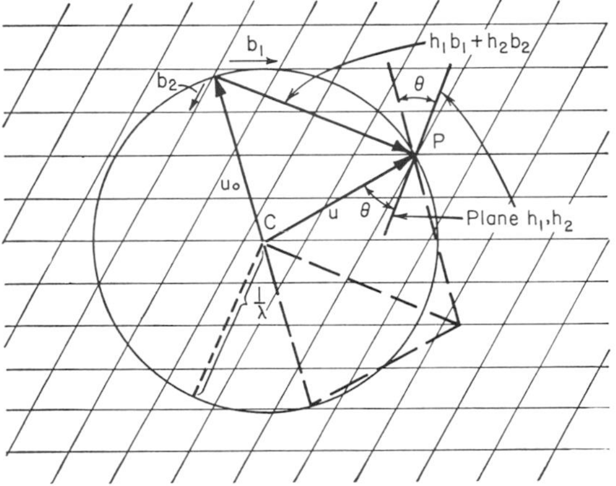

FIG. 1.9. Ewald's construction.

Turn the crystal along any axis passing through the origin O. The reciprocal lattice will turn with the crystal. Consider the situation when one of the reciprocal lattice points, say point P with indices $h_1h_2h_3$, crosses the surface of the sphere. Equation (1.10) specifies that a reflection can occur in the direction $\mathbf{u}$ only when the sphere is so oriented that its surface must intersect a reciprocal lattice point. The plane h_1,h_2 is perpendicular to $h_1\mathbf{b}_1 + h_2\mathbf{b}_2$.

The sphere is called the *sphere of reflection;* and whenever, upon rotation of the crystal, a reciprocal lattice point crosses its surface, a reflected beam flashes out of the crystal in the direction given by the construction. To determine what kind of pattern will be obtained on a film intercepting the diffracted beams, it is only necessary to imagine the film correctly oriented with respect to point C and to construct its intersections with the diffracted beams.

This useful construction is due to Ewald [14].

Single-crystal Cameras. The single-crystal diffraction photographs can thus easily be interpreted, whatever the geometry of the camera and the shape of the film (plane or cylindrical). One important case is that of the rotating crystal, the axis of rotation of which is perpendicular to the primary beam and concentric with the cylindrical film. If the crystal is aligned so that one of the crystallographic axes of the crystal (say the $\mathbf{a}_2$ axis) coincides with the rotation axis, all the reciprocal lattice planes of constant Miller index h_2 become perpendicular to the axis of rotation. In consequence, the sphere of reflection is intersected at its equator by reciprocal points h_10h_3, and at a number of smaller circles by reciprocal lattice points h_11h_3, h_22h_3, etc. After unfolding the concentric cylindrical film, a set of parallel lines of spots, called the *layer lines,* will appear, each line corresponding to one Miller index h_2. From their positions on the film, the length of the a_2 axis can easily be computed.

Often it is advantageous to oscillate the crystal over a small angular range, rather than to describe the whole rotation. This corresponds to the *oscillating crystal* method (see Fig. 1.10). Fewer diffraction spots will appear, which often makes their indexing in terms of Miller indices easier [15]. This method, however, lapsed into disuse after the introduction of the *moving film methods* into X-ray crystallography. Among these, the *Weissenberg method* [16] is most often used. In the Weissenberg camera a metal shield is placed between the crystal and the film, in such a way that only one layer line is allowed to reach the film. If the film holder is now mounted on a track so that it can move parallel to the axis of rotation of the crystal, and its motion is coupled with that of the rotation of the crystal, the spots, which would otherwise all fall along the same line, are spread out over the film in such a way that their indexing in terms of Miller indices becomes very easy. In fact, the film then represents one plane of the reciprocal lattice, distorted by a simple transformation of coordinates into a shape where the noncentral reciprocal lattice rows become characteristic festoons. From such photographs the unit vectors of the reciprocal lattice $\mathbf{b}_1$, $\mathbf{b}_2$, $\mathbf{b}_3$, and the angles between them, can be easily obtained.

There are several variants of moving film cameras, some of which use flat film. The most notable of these is the *Buerger precession camera* [17], which photographs on a film an undistorted image of the reciprocal lattice. Its only disadvantage is in the use of a flat film, which cannot record as much information as a cylindrical film. This disadvantage is in part obviated by use of radiation of a shorter wavelength; but this may be inconvenient for other reasons, particularly the lower absorption and hence reduced scattering efficiency of the shorter wavelengths. Also, it is sometimes impossible to obtain reflections of low Miller indices on the film; thus the data are not as complete for three-dimensional studies as those from the Weissenberg camera. However, the advantage of obtaining the reciprocal lattice directly renders this camera very suitable for determination of the reciprocal unit cell and the space group. The camera construction is also advantageous for low-temperature studies, etc.

In addition to these camera arrangements, direct-photon-counting methods may be employed for single-crystal studies. Indeed, single-crystal Geiger counter goniometers appear to offer great advantages over photographic techniques, in speed, sensitivity, and accuracy. These direct-counting goniometers are only now coming into general use. The arrange-

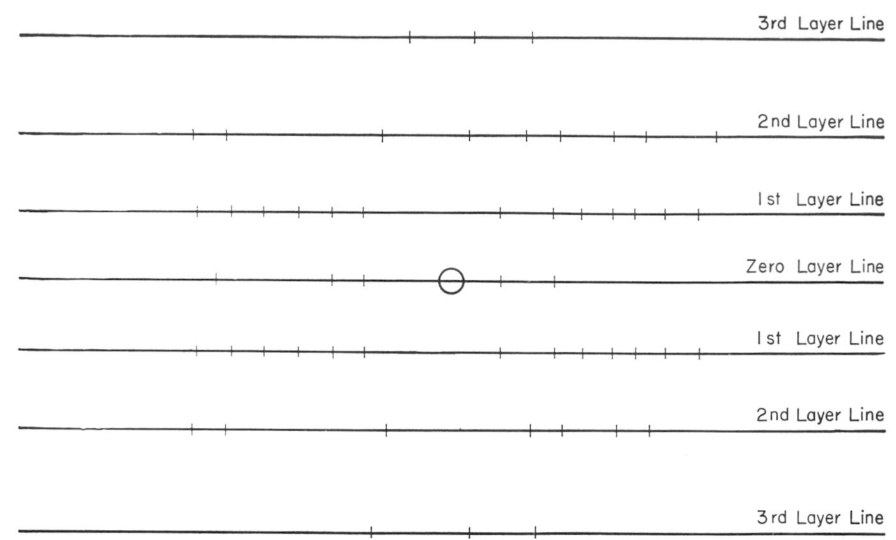

FIG. 1.10. Positions of reflections on typical rotation diagram. *C*-axis oscillation picture of aureomycin · HBr.

ments are more or less obvious, and will not be discussed in detail.

Powder Photographs. In the X-ray powder method the specimen consists of many crystals of diameter less than 0.05 mm, and the powder is exposed to a parallel beam of X rays. The X rays are diffracted by individual crystal grains of the powder in exactly the same way as discussed in the previous section, i.e., reflection occurs only when the specular and the Bragg conditions are simultaneously satisfied. With a few crystalline fragments these conditions would be simultaneously fulfilled only very rarely. However, when a very large number of crystalline grains is present, all oriented at random, there will always be a few grains oriented so that the reflection conditions are satisfied for any of the possible crystallographic planes. If a film is now placed so as to intercept the diffracted beams, a series of halos will be obtained, centered round the direct beam. In space these correspond to a series of cones, the vertex angles of each having a possible value of 2θ according to Bragg's law. This easily follows from the construction of the sphere of reflection, if each reciprocal lattice point is spread over a sphere of radius $1/d_h$. The Bragg angles can thus be directly measured by means of a photographic film or any other radiation detector. From these angles all the reflecting interplanar spacings can be computed, using Bragg's formula.

The spatial relations of the diffracted beams to the crystallographic axes are lost, however, so that individual planes cannot be indexed by mere inspection except for crystals in the very simple crystal systems (cubic, hexagonal, trigonal, or tetragonal).

If the flat film is placed at the opposite side of the specimen from the X-ray source, the photograph is called a *front reflection* pattern. When the film is placed at the *same* side as the source of radiation (it is then necessary to cut a hole in the film to admit the primary beam), a *back reflection* photograph is obtained. These photographs give accurate values of θ near 90°, and still more accurate values of the spacing d_h, since $\sin 2\theta$ is involved. If a cylindrical film is used and the specimen placed at its axis, then both regions may be covered on the same film.

The powder method is not as suitable for crystal structure determination as is the single-crystal method—except perhaps in special cases where the structure is very simple, so that only a limited number of data are sufficient for its solution. However, it is eminently suitable for a variety of practical purposes, such as identification of materials, qualitative and quantitative analysis, study of phase transitions, etc.

Many natural substances are microcrystalline, and give good powder photographs without any further treatment. Other substances, however, possess some kind of preferred orientation (especially cold-drawn metals). Such substances give rise to powder patterns in which the intensities of powder lines are distributed unevenly along the halos. In the extreme case, the powder pattern changes over into the diffraction pattern of a single crystal. From the intensity distribution along the halos, inferences can be made about the preferred orientation of the crystallites.

A large class of substances forms fibers, which are composed of crystallites well oriented in one direction (along the fiber axis), but are disoriented in the directions perpendicular to the fiber axis. These give characteristic *fiber photographs*, which closely resemble the rotation photographs of single crystals.

When the size of crystallites is so small that only a small number of planes reflects, which happens when their size is below about 10^{-4} mm, the X-ray reflections become noticeably broadened. The reflecting power is no longer concentrated at points of the reciprocal lattice, but becomes spread out over certain small volumes. The size of crystallites can thus be estimated from *line broadening* in the powder photograph.

Powder-diffraction techniques, including important methods utilizing bent-crystal monochromatized incident beams, and also direct photon-counting techniques, are described extensively in treatises by Guinier and by Klug and Alexander (see Bibliography).

3. Theory of X-ray Scattering

Scattering by an Electron. The interaction of radiation with matter will now be considered in more detail. For X rays interacting with a crystal, the electrons can be regarded as free at this stage of discussion. Consider a parallel monochromatic plane-polarized beam of electromagnetic radiation impinging on a free electron. The radiation is characterized by a field having an electric vector **E** and a magnetic vector **H**, which are perpendicular to each other and to the direction of the primary beam (see Part 6, Chap. 1). The magnetic forces are negligible compared with the electric forces, so that the acceleration of the electron will be

$$\ddot{\mathbf{x}} = \frac{e\mathbf{E}}{m} \tag{1.11}$$

where m is the mass of the electron and e its charge. Since **E** varies periodically, the electron acquires an oscillatory motion and in turn begins to radiate a field which, at the distance R from the electron, will have the components

$$|\mathbf{E}'| = |\mathbf{H}'| = \frac{e\ddot{x}\sin\phi}{c^2R} \tag{1.12}$$

where c is the velocity of light and ϕ is the angle between the scattered beam and the line along which the electron oscillates. Substituting for the acceleration,

$$|E'| = |H'| = \frac{e^2E\sin\phi}{mc^2R} \tag{1.13}$$

Since the intensity of radiation is given by Poynting's vector, after substitution,

$$I(\phi) = I_0K\sin^2\phi \tag{1.14}$$

where I_0 is the intensity of the primary beam and

$$K = \left(\frac{e^2}{mc^2R}\right)^2 \tag{1.15}$$

When the radiation of the primary beam is not polarized, which is the usual case for the beam emerging from the X-ray tube, it is necessary to average the above result for all values of angle φ, keeping the angle between the primary and the scattered beams, 2θ, constant. The integrations give

$$I(\theta) = I_0K\frac{1+\cos^2 2\theta}{2} \tag{1.16}$$

so the final expression for the intensity of radiation scattered by one electron illuminated by unpolarized primary radiation of intensity I_0 is

$$I(\theta) = I_0\left(\frac{e^2}{mc^2R}\right)^2\frac{1+\cos^2 2\theta}{2} \tag{1.17}$$

The term in parentheses on the right shows that the scattered wave decreases in intensity with the square of the distance from the electron, and the last term is called the *polarization factor*.

For most practical cases, the assumption of free electrons is valid with a sufficient degree of approximation. However, there are exceptions when strongly bonded electrons are considered to interact with radiation near their absorption edge. There then follows a more complex interaction, which in general results in a phase shift of the scattered wave and some modification of the intensity. It is possible to take advantage of this phenomenon in obtaining special information, such as the absolute configuration of atoms in an optically active crystal. Use of this fact was first made by Coster, Knoll, and Prins [18] in the determination of the absolute configuration of cubic ZnS; and it has been beautifully applied more recently by Bijvoet [19] in a determination of the absolute configuration of rubidium hydrogen tartrate.

General Scattering Formula. In the previous section, scattering by only one electron was considered. Let us consider a piece of matter of whatever shape, in which the spatial distribution of electron density is given as a function of a vector **r**, pointing from the origin to the element dV, the electron density being $\rho(\mathbf{r})$ at this point.

If this piece of matter is now exposed to a parallel beam of radiation of intensity I_0 and wavelength λ, each element dV scatters an amount which depends on the electron density within this element. In the final scattered beam, the electromagnetic *fields* of individual contributing scattered waves are strictly additive. Let us consider just two elements dV_1 and dV_2, one at the origin and the other at the end of vector **r**.

Let the amplitude of the electric vector of the primary beam be

$$E = E_0\exp\left[2\pi i\left(\nu t - \frac{x}{\lambda}\right)\right] \tag{1.18}$$

Then the amplitude of the electric vector of the phase-retarded scattered beam will be

$$E(\theta) = E_0C\rho\,dV\exp 2\pi i\left[\nu t - \frac{x}{\lambda} + (\mathbf{u}-\mathbf{u}_0)\cdot\frac{\mathbf{r}}{\lambda}\right] \tag{1.19}$$

where C is a proportionality constant. The resultant field will be given by integrating over the whole piece of matter:

$$E(\theta) = E_0C\exp 2\pi i\left(\nu t - \frac{x}{\lambda}\right)\iiint\rho\exp\left[2\pi i(\mathbf{u}-\mathbf{u}_0)\cdot\frac{\mathbf{r}}{\lambda}\right]dV \tag{1.20}$$

The retardation in phase of a wave scattered from dV_2 at **r**, with respect to the phase of the wave scattered from dV_1 at the origin, is found by considering the path difference $(\mathbf{u}-\mathbf{u}_0)\cdot\mathbf{r}$. The phase lag, in radians, is $(\mathbf{u}-\mathbf{u}_0)\cdot\mathbf{r}/\lambda$. Writing

$$\frac{\mathbf{u}-\mathbf{u}_0}{\lambda} = \mathbf{h} \tag{1.21}$$

which is Laue's equation with **h** in place of $\mathbf{B}_h$, for convenience, we have

$$I = |\mathbf{E}|^2 = I_0 C^2 |\iiint \rho \exp (2\pi i \mathbf{r} \cdot \mathbf{h})\, dV|^2 \quad (1.22)$$

or

$$I = I_0 C^2 |T(\mathbf{h})|^2 \quad (1.23)$$

The expression

$$T(\mathbf{h}) = \iiint \rho(\mathbf{r}) \exp (2\pi i \mathbf{r} \cdot \mathbf{h})\, d\ V \quad (1.24)$$

is called a *Fourier transform* of the electron distribution $\rho(\mathbf{r})$ (see Part 1, Chap. 3, Sec. 4). The theory of Fourier transforms is a well-known branch of mathematics; and the X-ray scattering process is a very practical example of that theory, of which more use will be made later. The best treatment of Fourier transform theory as applied to X-ray analysis is that of D. Sayre (see Bibliography).

The above expression for $I(\theta)$ is a general formula for scattering of a parallel beam of radiation of intensity I_0 and wavelength λ, in a direction 2θ, by any piece of matter characterized by the electron density distribution $\rho(\mathbf{r})$. Comparing the result with that of scattering by a single electron, the constant C has the following value for unpolarized radiation:

$$C^2 = \left(\frac{e^2}{mc^2R}\right)^2 \frac{1 + \cos^2 2\theta}{2} \quad (1.25)$$

Scattering by a Spherical Atom. One of the most important special cases is the scattering by an atom which has a spherically symmetric electron distribution.

$$\mathbf{r} \cdot \mathbf{h} = |\mathbf{r}| \frac{2 \sin \theta}{\lambda} \cos \phi_{\mathbf{r},\mathbf{h}} \quad (1.26)$$

let

$$|\mathbf{r}| \cos \phi_{\mathbf{r},\mathbf{h}} = z \quad (1.27)$$

and

$$2\pi|\mathbf{h}| = \mu = \frac{4\pi \sin \theta}{\lambda} \quad (1.28)$$

Then the transform of the electron density of the spherical atom is given by

$$T(\mathbf{h}) = \iiint_V \rho(\mathbf{r}) \exp (i\mu z)\, dV \quad (1.29)$$

where $\rho(\mathbf{r})$ is the radial distribution of the electron density. Since there is a center of symmetry at the origin, De Moivre's formula leads to

$$T(\mathbf{h}) = \iiint \rho(\mathbf{r}) \cos (\mu z)\, dV \quad (1.30)$$

Integrating first over a spherical shell of constant $|\mathbf{r}|$, and using cylindrical coordinates,

$$z = |\mathbf{r}| \cos \phi \qquad \mu = |\mathbf{r}| \sin \phi \quad (1.31)$$

we obtain

$$T(\mathbf{h}) = 4\pi \int \rho(\mathbf{r}) \frac{\sin \mu r}{\mu r} r^2\, dr \quad (1.32)$$

where μ is given in Eq. (1.28). This is the transform of the spherical atom.

Special Case. If $\rho(\mathbf{r}) = \rho_0 \exp(-\pi p^2 r^2)$, that is, the electron-density distribution is given by a Gaussian curve, where ρ_0 is the density at $r = 0$, the result is

$$T(\mathbf{h}) = \frac{\rho_0}{p^3} \exp\left(-\frac{\pi h^2}{p^2}\right) \quad (1.33)$$

The transform of a Gaussian function is again a Gaussian function. It is often of advantage to represent the true electron-density distribution as a sum of several Gaussian curves. The transform is then obtained as a sum of transforms of the individual terms.

These results can easily be generalized for the case where the electron distribution is ellipsoidal, since such a distribution is easily transformed into a spherical distribution by a linear transformation of coordinates. This case occurs when the atoms are subject to anisotropic thermal motion.

Application to a Finite Crystal. In order to apply the general scattering formula to a crystal, it is necessary to express the crystal in a mathematical form. In a crystal the electron density repeats itself periodically; so

$$\rho(\mathbf{r}) = \rho(\mathbf{R}) \quad (1.34)$$

for all vectors **r** given by

$$\mathbf{r} = \mathbf{u} + t_1\mathbf{a}_1 + t_2\mathbf{a}_2 + t_3\mathbf{a}_3 + \mathbf{R} \quad (1.35)$$

where **R** is called the *basis vector* for the unit cell, t_i are three integers enumerating the cells, $\mathbf{a}_i$ are the three cell-edge vectors, and **u** is a vector from the origin to the origin of the crystal.

Substituting in the expression for $T(\mathbf{h})$, we obtain for a crystal bounded by a parallelepiped around the origin:

$$T(\mathbf{h}) = \exp (2\pi i \mathbf{u} \cdot \mathbf{h}) L_1 L_2 L_3 \iiint \rho(\mathbf{R}) \exp (2\pi i \mathbf{R} \cdot \mathbf{h})\, dV \quad (1.36)$$

where the first factor is physically unimportant (and can be made equal to unity if we place the origin of the crystal so that $\mathbf{u} = 0$); L_i are given by

$$L_i = \frac{\sin \pi(2n_i + 1)\mathbf{a}_i \cdot \mathbf{h}}{\sin \pi \mathbf{a}_i \cdot \mathbf{h}} \quad (1.37)$$

and are called *Laue functions,* and the integral is over *one unit cell only,* and not over the whole crystal.

The Laue functions have important properties: for $\mathbf{a}_i \cdot \mathbf{h} = 0, 1, 2,$ etc., $L_i = 2n_i + 1$; if n_i, the number of the unit cells in the blocklike crystal, is large, the Laue functions drop rapidly to small values except for integral values of $\mathbf{a}_i \cdot \mathbf{h}$.

The result can also be viewed from a different aspect, using an important *convolution theorem* from Fourier transform theory (see Sayre's treatment).

The transform of a *convolution* of two functions is equal to the product of the transforms of the individual functions.

Conversely, the transform of a product of two functions is equal to a *convolution* of the transforms of the individual functions.

In the present case, the crystal can be regarded as a convolution of the contents of one unit cell with the crystal lattice,

$$\rho(\mathbf{r}) = \widehat{(\text{lattice})\, \rho(\mathbf{R})} \quad (1.38)$$

where the symbol $\frown$ indicates convolution. The transform of the point lattice is

$$T_1(\mathbf{h}) = L_1L_2L_3 \tag{1.39}$$

and the transform of the contents of one unit cell is

$$T_2(\mathbf{h}) = \int\int\int \rho(\mathbf{R}) \exp(2\pi i\mathbf{R}\cdot\mathbf{h})\, dV \tag{1.40}$$

The transform of the crystal is thus

$$T(\mathbf{h}) = T_1(\mathbf{h}) \times T_2(\mathbf{h}) \tag{1.41}$$

as in Eq. (1.36).

More general results can be obtained for a crystal of whatever boundary, by considering that such a crystal can be regarded as an infinite crystal multiplied by a shape function, the shape function having a value of unity within the boundary and zero outside the boundary. The transform of such a crystal is obtained by convoluting the transform of the infinite crystal with the transform of the shape function. For an infinite crystal the transform is zero except at the reciprocal lattice points given by $T_1(\mathbf{h}) \neq 0$. Convolution of such a reciprocal lattice with the shape function results in each reciprocal lattice point being spread out in the same way. These results are important when the diffraction by crystals of finite size is to be interpreted.

We shall limit ourselves, however, to crystals of a practically infinite extent. Then $T_1(\mathbf{h})$ can be regarded as zero except for points in space of vector $\mathbf{h}$ in which simultaneously L_1, L_2, $L_3 \neq 0$. This occurs when all three products $\mathbf{a}_1\cdot\mathbf{h}_1$, $\mathbf{a}_2\cdot\mathbf{h}_2$, $\mathbf{a}_3\cdot\mathbf{h}_3$ are simultaneously integers. These conditions are fulfilled only for values of $\mathbf{h}$ which lie on the reciprocal lattice.

Scattering by a Mosaic Crystal. The integral $T_2(\mathbf{h})$ of Eqs. (1.40) and (1.41), when sampled at *reciprocal lattice points only*, is usually denoted by $F(\mathbf{h})$ and is called the *structure factor*. For a large crystal

$$I = I_0 \cdot C^2 \cdot (\text{number of cells})^2 \cdot |F(\mathbf{h})|^2 \tag{1.42}$$

This formula would hold for a perfect crystal with negligible absorption. But practically all crystals are composed of small blocks at slight angular displacements one with respect to another, and therefore the formula is useless; for it would require the intensity to be proportional to the square of the volume of the crystal, whereas the incoherent radiation in fact adds up in intensity proportionally to the volume of the crystal. Furthermore, the quantity which is usually measured is the so-called *integrated reflection*, $E\omega/I_0$. The crystal is turned with angular velocity ω, and the total reflected energy E is recorded. For a very small crystal block of volume V with negligible absorption of X rays, the integrated reflection is independent of the shape of the crystal and is proportional to V. The following relation is then valid:

$$\frac{E\omega}{I_0} = \frac{N^2e^4|F|^2\lambda^3V}{m^2c^4}\,\frac{1+\cos^2 2\theta}{2\sin 2\theta} \tag{1.43}$$

Slightly modified formulas hold for powder photographs and for more complicated geometrical arrangements such as those used in various precession cameras. If a photographic film is used as a detector, it should be remembered that its sensitivity varies with the angle of the incidence of the radiation (Cox and Shaw factor).

In the above formula, I_0 is the intensity of the primary beam, N is the number of the unit cells per unit volume of the crystal, the factor e^2/mc^2 originates from the fact that the scattering is by the electrons, and the trigonometric factor is called the Lorentz-polarization factor, often denoted by $L.p.$

The expression holds only for very small crystals, bathed completely in the primary beam. If these conditions are not fulfilled, corrections must be applied: for absorption, and for modification of the intensity by the drainage of energy from the primary beam by the reflection itself. The latter occurs only for strong reflections, and is called the correction for *primary and secondary extinction*. Which correction is important depends on the size and disorientation of the assumed perfect blocks of the mosaic crystal.

For a powder photograph on a cylindrical film of radius r, the diffracted energy P per unit length of halo is given by

$$P = \frac{I_0N^2e^4\lambda^3V|F|^2p}{32\pi m^2c^4r}\,\frac{1+\cos^2 2\theta}{\sin^2\theta\cos\theta} \tag{1.44}$$

where I_0 is the energy of the primary beam per unit area per second, V is the volume of the powder in the beam, and p is a multiplicity factor, since lines of several Miller indices often overlap, due to the symmetry of the cell. The rest of the symbols have the same significance as in the formula for a single crystal. If there is a noticeable absorption in the specimen, the above expression is to be multiplied by the absorption factor. For substances other than cubic, it is experimentally difficult to produce powder specimens without noticeable *preferred orientation* of the crystalline grains. This introduces an error in the intensity measurements of powder lines. On the other hand, for really fine powders, errors due to primary and secondary extinction are as a rule negligible.

In spite of the many corrections involved, sufficient accuracy in the intensity measurements can be achieved for the purposes of crystal-structure determination. Photographic measurements can yield integrated intensities with an accuracy of between 10 and 20 per cent. Ionization chamber, Geiger-Müller counter, and other counting devices are capable of raising the accuracy by a factor of 10.

4. Fourier Transforms

Fourier Series and the Structure Factor. A crystal is a periodic function of electron density $\rho(\mathbf{R})$, and so can be represented by a triple Fourier series:

$$\rho(\mathbf{R}) = \sum\sum_{-\infty}^{\infty}\sum A(\mathbf{p}) \exp(2\pi i\mathbf{p}\cdot\mathbf{R}) \tag{1.45}$$

where $\mathbf{p}$ is a vector, the three components of which are integers, and $A(\mathbf{p})$ is a yet unspecified coefficient.

In order to determine the meaning of **p** and A, substitute in the equation for the structure factor:

$$F(\mathbf{h}) = \iiint \rho(\mathbf{R}) \exp(2\pi i \mathbf{R} \cdot \mathbf{h})\, dV \qquad (1.46)$$

We obtain

$$F(\mathbf{h}) = \sum_{-\infty}^{\infty}\sum\sum \iiint A(\mathbf{p}) \exp[2\pi i(\mathbf{p} + \mathbf{h}) \cdot \mathbf{R}]\, dV \qquad (1.47)$$

Noting that

$$\iiint \exp[2\pi i(\mathbf{p} + \mathbf{h}) \cdot \mathbf{R}]\, dV = \begin{cases} 0 & \text{if } \mathbf{p} = -\mathbf{h} \\ V & \text{if } \mathbf{p} = -\mathbf{h} \end{cases} \qquad (1.48)$$

we can identify

$$\mathbf{p} \text{ with } -\mathbf{h} \text{ and } F(\mathbf{h}) \text{ with } VA(\mathbf{h}) \qquad (1.49)$$

Substituting, we have the most important result:

$$\rho(\mathbf{R}) = \frac{1}{V}\sum_{-\infty}^{\infty}\sum\sum F(\mathbf{h}) \exp(-2\pi i \mathbf{h} \cdot \mathbf{R}) \qquad (1.50)$$

Thus if the structure factors $F(\mathbf{h})$ are known, it is possible to calculate the distribution of the electron density throughout the crystal. In Sec. 3 it was remarked that it is possible to measure the *magnitude* but not the *phase* of the structure factors $F(\mathbf{h})$ at the reciprocal lattice points of the crystal. This is due to the fact that our recording devices are sensitive to the intensity but not the phase of the radiation. One-half of the information is thus lost.

In practice, the *phase* of $F(\mathbf{h})$ must be inferred by various more or less indirect means. This constitutes the so-called *phase problem* of X-ray crystallography. Once the phases are determined, the electron density of the crystal can be calculated by means of the Fourier series with only two limitations: the first is that the measurements are of finite accuracy; the other is that it is possible to measure only a finite number of structure factors, so that the calculated series is not infinite, but *terminated.* This introduces so-called *termination errors.* Methods are available to correct for these. The termination errors are equivalent to viewing the structure through a microscope of a *finite aperture.* This introduces blurring of the image through diffraction ripples which cross the image.

Discussion of the Structure-factor Equation. The structure-factor equation

$$F(\mathbf{h}) = \iiint \rho(\mathbf{R}) \exp(2\pi i \mathbf{R} \cdot \mathbf{h})\, dV \qquad (1.51)$$

will now be discussed in more detail.

The crystals are built of atoms, which to a first approximation are spherically symmetrical. In this case, the electron density can be constructed by taking a crystal of *point atoms,* i.e., atoms in which the electron density is concentrated at a point corresponding to the atomic center, and then convoluting (or spreading out) the point atoms with their respective spherical electron density distribution functions. As the transform of a convolution is a product of the transforms of the individual functions, the structure factor is then a product, or a sum of products, of the transform of a *point atom crystal* with the transforms of the individual electron distributions. The latter are called the *atomic scattering factors,* and are denoted by a symbol f_j for the jth atom.

The great advantage of this treatment lies in the fact that the integral over the point atom crystal can be replaced by a sum; thus

$$F(\mathbf{h}) = \sum_{j}^{N} f_j \exp(2\pi i \mathbf{R}_j \cdot \mathbf{h}) \qquad (1.52)$$

where $\mathbf{R}_j$ are the coordinates of the jth atom ($j = 1, 2, \ldots, N$) and the summation is over the N atoms contained in the unit cell.

This formula is not limited to spherically symmetrical atoms; the atoms can be of any shape whatever, provided the atomic scattering factors f_j are the transforms of their electron densities.

The atomic scattering factors of atoms at rest have been tabulated for all the elements by a number of workers, notably by James and Brindley [20], Thomas [21], Fermi [22], and Pauling and Sherman [23], and the results are conveniently tabulated in the *Internationalle Tabellen* [24]. The values in the *Internationalle Tabellen,* however, are not considered sufficiently accurate for present-day work, especially for the lighter elements, and more accurate tables have been calculated by MacWeeny [25].

Since the atoms in the crystal are not at rest, even at absolute zero, but are subject to thermal motion, the scattering factors of atoms at rest cannot be used directly. The atoms oscillate so their centers on the time average are spread out into a Gaussian distribution function. The resulting electron density is a convolution of the electron density at rest with the Gaussian function for that particular atom; and since the transform of a Gaussian function is again a Gaussian function, the result is a product:

$$f_T = f_0 \exp\left(\frac{-B \sin^2\theta}{\lambda^2}\right) \qquad (1.53)$$

where B is the temperature factor for the atom. This is connected with the mean-square displacement of the atomic center at right angles to the reflecting plane, $\bar{\mu}^2$, by the relation

$$B = 8\pi^2\bar{\mu}^2 \qquad (1.54)$$

The X-ray data thus provide material for measurement of the mean atomic oscillation.

Recapitulation of the General Case. The most important results derived so far are the equation for the structure factor,

$$F(\mathbf{h}) = \sum_{j=1}^{N} f_j \exp(2\pi i \mathbf{R}_j \cdot \mathbf{h}) \qquad (1.55)$$

and the equation for the electron density

$$\rho(\mathbf{R}) = \frac{1}{V}\sum\sum\sum F(\mathbf{h}) \exp(-2\pi i \mathbf{h} \cdot \mathbf{R}) \qquad (1.56)$$

We can write

$$F(\mathbf{h}) = A(\mathbf{h}) + iB(\mathbf{h}) \qquad (1.57)$$

where

$$A(\mathbf{h}) = \sum_{j=1}^{N} f_j \cos 2\pi \mathbf{R}_j \cdot \mathbf{h}$$
$$B(\mathbf{h}) = \sum_{j=1}^{N} f_j \sin 2\pi \mathbf{R}_j \cdot \mathbf{h} \qquad (1.58)$$

The quantity observed is

$$|F(\mathbf{h})|^2 = A(\mathbf{h})^2 + B(\mathbf{h})^2 = F(\mathbf{h}) \times F(\mathbf{h})^* \qquad (1.59)$$

where F^* is conjugate to F,

$$F(\mathbf{h})^* = \sum_{j=1}^{N} f_j \exp(-2\pi i \mathbf{R}_j \cdot \mathbf{h}) \qquad (1.60)$$

and

$$F^* = A - iB \qquad (1.61)$$

We can also introduce a phase angle $\alpha(\mathbf{h})$ such that

$$A = |F| \cos \alpha, \qquad B = |F| \sin \alpha, \qquad \frac{B}{A} = \tan \alpha \qquad (1.62)$$

giving

$$|F| = A \cos \alpha + B \sin \alpha \qquad (1.63)$$

from which

$$|F| = \Sigma \mathbf{f}_j \cos (2\pi \mathbf{R}_j \cdot \mathbf{h} - \alpha) \qquad (1.64)$$

Since $\rho(\mathbf{R})$ is real (free-electron model),

$$\rho(\mathbf{R}) = \frac{1}{V} \sum \sum \sum [A(\mathbf{h}) \cos 2\pi \mathbf{R} \cdot \mathbf{h} + B(\mathbf{h}) \sin 2\pi \mathbf{R} \cdot \mathbf{h}] \qquad (1.65)$$

and a complementary equation is valid:

$$B(\mathbf{h}) \cos 2\pi \mathbf{R} \cdot \mathbf{h} - A(\mathbf{h}) \sin 2\pi \mathbf{R} \cdot \mathbf{h} = 0 \qquad (1.66)$$

Then

$$A(\mathbf{h}) = A(-\mathbf{h}) \qquad \text{and} \qquad B(\mathbf{h}) = -B(-\mathbf{h}) \qquad (1.67)$$

and thus

$$|F(\mathbf{h})| = |F(-\mathbf{h})| \qquad (1.68)$$

which is *Friedel's law,* a statement of the fact that, for the free-electron case, it is not possible, from measurements of scattered X-ray intensities, to determine whether the crystal has a center of symmetry or not. (The presence of certain translational symmetries may reveal that a symmetry center is present. See Sayre reference in Bibliography.) Also

$$\Sigma A(\mathbf{h}) \sin 2\pi \mathbf{R} \cdot \mathbf{h} = 0 \qquad \text{and} \qquad \Sigma B(\mathbf{h}) \cos 2\pi \mathbf{R} \cdot \mathbf{h} = 0 \qquad (1.69)$$

and

$$F(\mathbf{h}) = F(-\mathbf{h})^* \qquad (1.70)$$

Effects of Elements of Symmetry. Presence of elements of symmetry considerably simplifies the equations. One of the most important simplifications appears when there is actually a center of symmetry present. Choosing one of the centers as the origin of coordinates (if one center of symmetry exists, there are necessarily eight centers present in the unit cell), if there is an atom at $\mathbf{R}_j$, there is also necessarily an atom at $-\mathbf{R}_j$, and therefore $B(\mathbf{h}) = 0$; so

$$F(\mathbf{h}) = 2 \sum_{j=1}^{N/2} f_j \cos (2\pi \mathbf{R}_j \cdot \mathbf{h}) \qquad (1.71)$$

the summation extending over all the atoms per *asymmetric unit,* or half the unit cell—this being compensated for by the factor 2 in front of the summation.

Similarly, the Fourier series reduces to

$$\rho(\mathbf{R}) = \frac{1}{V} \sum_{-\infty}^{\infty} \sum_{-\infty}^{\infty} \sum_{-\infty}^{\infty} F(\mathbf{h}) \cos 2\pi \mathbf{R} \cdot \mathbf{h} \qquad (1.72)$$

The formulas are generally written in full as

$$F(h_1,h_2,h_3) = 2 \sum_{j=1}^{N/2} f_j \cos 2\pi (h_1 x_1^j + h_2 x_2^j + h_3 x_3^j) \qquad (1.73)$$

and

$$\rho(x_1,x_2,x_3) = \frac{1}{V} \sum_{-\infty}^{\infty} \sum_{-\infty}^{\infty} \sum_{-\infty}^{\infty} F(h_1 h_2 h_3) \cos 2\pi (h_1 x_1 + h_2 x_2 + h_3 x_3) \qquad (1.74)$$

Here, h_1, h_2, h_3 are the Miller indices of the planes concerned, x_1^j, x_2^j, x_3^j the coordinates of the jth atom, expressed in fractions of the cell edges a_1, a_2, a_3, and x_1, x_2, x_3 are similarly the fractional coordinates in the unit cell.

Presence of other elements of symmetry further reduces the formulas. Such reduced expressions are tabulated in the *Internationalle Tabellen,* and these should be consulted before calculations are undertaken.

The computation of the full three-dimensional series is rather laborious. It is used only when great accuracy is required. In preliminary stages of work, resort is made to a simpler *double* Fourier series, which gives the *projection* of the structure. For a centrosymmetric crystal, for example,

$$\sigma(x_1 x_2) = \frac{1}{A} \sum_{-\infty}^{\infty} \sum_{-\infty}^{\infty} F(h_1 h_2 0) \cos 2\pi (h_1 x_1 + h_2 x_2) \qquad (1.75)$$

which may be obtained from the three-dimensional formula by integrating over x_3. In this formula, A is the area of the cell face on which the electron density is projected. Similarly, the *one-dimensional* Fourier series is given by

$$l(x_1) = \frac{1}{d} \sum_{-\infty}^{\infty} F(h_1 00) \cos 2\pi h_1 x_1 \qquad (1.76)$$

which gives the projected electron density per unit length. Unfortunately this series is not of much use except for the very simplest structure, because of the overlap of the projected atomic peaks.

Computational Methods. In general, it is possible to solve a given problem in X-ray crystallography when the phases of the structure factors have been at least approximately determined, or when the atomic positions are at least roughly known. The condition is that the atomic positions must be known at the start of the computation within atomic diameters of the true positions. The computation then can proceed by successive iterations. Structure factors are computed from the assumed atomic positions. These will in general differ from the observed structure factors in magnitude. The next step is to take the *observed magnitudes* of the structure factors and to assign to them the *phases calculated* from a model. If a Fourier series is made using these, it may show improved positions of the peaks representing the atoms. The process is then repeated, the improved structure factors now being calculated using improved atomic coordinates, etc. The process is convergent in practice when the starting structure model is not far from the true model. The other condition is that there should be at least as many observed quantities as there are unknown parameters. This is usually fulfilled in practice; a structure with 10 atoms, or 30 unknown coordinates, may yield as many as 1,000 observed structure factors, so that the problem is usually overdetermined in this sense.

In following this procedure, repeated computations of the electron density and of the structure factors are required. Although both formulas are closely related, the practical computational requirements are different. In the structure-factor formula, the atomic coordinates x_j, y_j, z_j are to be entered with accuracy of at least one thousandth of the cell edge. On the other hand, the electron density can be calculated over a grid from perhaps 30ths to 120ths of the cell edge, so that only certain values of trigonometric functions enter repeatedly into the calculation.

For smaller structures, the structure factors are usually computed by hand. The operations can be speeded up by the use of suitable tables, however. The electron density is then computed by decomposition of the cosine of the sum into products of trigonometric functions, and rearrangement of these into one-dimensional Fourier series (preliminary and final). These are then computed by means of the method of Beevers and Lipson strips [26, 27], which are essentially sine and cosine tables printed on narrow strips of cardboard for all values of index h from 0 to 30. The one-dimensional summation is then reduced to a summation of rows of numbers. Related methods were devised by Robertson [28], and by Patterson and Tunell [29]; the various methods are approximately equally efficient, differing only in some detail.

The above methods are essentially digital in character. For larger structures, resort to IBM or Hollerith machines, or even to the more modern electronic digital computers such as the Manchester-Ferranti computer, is recommended. There exist several systems, among which are those described by Shaffer, Schomaker, and Pauling [30, 31], Cox and Jeffrey [32, 33], and Grems and Kasper [34] (see also R. Pepinsky [35]). The field is rapidly expanding.

In addition to digital methods, the field has proved to be very fruitful as regards design of various analogue methods and machines. These attempts have culminated in the design of the large electronic analogue computers X-RAC and S-FAC by Pepinsky [35], and the construction of these machines in his laboratory.

X-RAC adds electric potentials, generated as wave forms with proper frequencies, amplitudes, and phases, which can be set on panels of potentiometers and switches. The results are then presented two-dimensionally on the face of a cathode-ray oscilloscope. The machine can handle two-dimensional Fourier syntheses with h_1 from 0 to 20 and h_2 from -20 to $+20$. The electron density is produced in the form of a contour map, which is photographically recorded. The effects of phase changes are immediately observable in the oscilloscope pattern. S-FAC directly computes two-dimensional structure factors over the same range of indices for 20 atoms of 5 different scattering powers. Effects of shifts in atomic coordinates upon the structure-factor phases and amplitudes are again immediately observable. The two machines are interconnected in a manner which facilitates very rapid iterative procedures. The inputs of these machines are now being digitalized, to further increase their speed. The accuracy of X-RAC is high, and this machine has carried out more Fourier syntheses than all other machine methods together.

Refinement Methods. The basic iterative procedure discussed above was used for many years, and many structures have been solved by repeated application of Fourier series and structure-factor computation. The straightforward method, however, does not correct for the error introduced by termination of the series. Recently several improved methods have been introduced which increase the accuracy of the crystallographic method considerably.

One of these is the so-called *differential synthesis*, in which derivatives of the electron density are computed, and the positions of peaks determined from the zeros of the first derivatives. In the method of *least squares*, the nonlinear trigonometric functions are expanded into a series near the true solution, the higher terms are neglected, and the resulting linear equations solved by the standard least-squares procedure. In the method of *steepest descents*, the sum of the squares of discrepancies R, where

$$R = \Sigma(|F_0| - |F_c|)^2 \tag{1.77}$$

or a related function of residuals, is brought to a minimum by considering it as a function of the $3N$ atomic parameters in $3N$-dimensional space. The values of R lie on a surface, and the steepest descents method moves the point, representing the structure, downhill in the $3N$-dimensional space. Lastly, the *difference synthesis* uses a Fourier map in which the structure factors have been replaced by residuals $(F_0 - F_c)$. The atoms are to be shifted in the direction of the steepest *ascent* of this function. All these methods prove to be interrelated, each having some peculiar advantage. The differential synthesis is eminently suitable for use with large electronic computers, such as the Manchester-Ferranti computer. It gives the utmost accuracy in the very last stages of the refinement. The difference synthesis, on the other hand, has the advantage that it can be computed using Beevers-Lipson strips, just as an ordinary Fourier map, and that it gives the atomic positions corrected for the termination of the series; but it may converge slowly if there is an excessive overlap of the atoms. Difference syntheses are of course very readily and accurately computed on X-RAC.

5. The Phase Problem

Attempts at Solution in Reciprocal Space. If the phases of the structure factors were known, X-ray crystallography would be reduced to a routine computation of one Fourier series, which would then show a greatly magnified image of the structure, the resolution being limited by the finite "aperture" of the data. Since the phases are not known, at least at the outset of the investigation, the first step, after determination of the unit cell and the space group and the measurement of the intensities, is to determine these by some method. The magnitude of the problem can be visualized as follows.

If the crystal has no center of symmetry, the phases are continuously variable; and since any combination of these would produce a Fourier map, and the transform of that Fourier map would give exactly the absolute values of the Fourier coefficients which were put in, there is an infinite number of solutions, unless conditions are attached to the density function. If the crystal has a *center of symmetry*, the phase problem reduces to the determination of the *signs* of the structure factors, so that the number of solutions is now *finite*, but still very great; in fact, if n structure factors are observed, the number of possible combinations of signs is 2^n, which is a very great number, if n is of the order of 1,000.

The number of such combinations can be substantially reduced if the condition is introduced that the electron density must be everywhere *non-negative*. Some sign combinations can be rejected even before constructing 2^n electron density maps, since the non-negativity criterion imposes on the structure factors certain conditions, the *Harker-Kasper inequalities*. The problem becomes more feasible if the crystal contains *discrete atoms, all of the same shape*. Many crystals approximate quite closely to this condition, especially when composed of lighter atoms. Then $f_j = Z_j \cdot f$, where Z_j is the atomic number of the jth atom and f is a reduced atomic scattering curve common to all the atoms. Then the structure factor equation can be written

$$\frac{F(h_1h_2h_3)}{f(h_1h_2h_3)} = \sum Z_j \exp 2\pi i(h_1x_1{}^j + h_2x_2{}^j + h_3x_3{}^j) \tag{1.78}$$

where the coefficients on the right-hand side are known as the *sharpened structure factors*, denoted by $\hat{F}(h_1h_2h_3)$. It is now possible to set up polynomials, the roots of which give the required coordinates, as was first shown by Ott [36] and later by Avrami [37]. The method has been extended by Banerjee [38], Hughes [39], and Goedkoop [40, 41].

In order to demonstrate the Ott-Avrami method, let us take a one-dimensional problem.

Let $\exp 2\pi i x^j = \alpha^j$; then

$$\hat{F}(h) = \Sigma Z_j \alpha_h{}^j \tag{1.79}$$

If the cell contains N atoms, the α^j are roots of an Nth-degree equation, given by a determinant

$$\begin{vmatrix} 1 & \alpha & \alpha^2 & \cdots & \alpha^N \\ \hat{F}(p) & \hat{F}(p+1) & \hat{F}(p+2) & \cdots & \hat{F}(p+N) \\ \cdots & \cdots & \cdots & \cdots & \cdots \\ \hat{F}(p+N-1) & \hat{F}(p+N) & \hat{F}(p+N+1) & \cdots & \hat{F}(p+2N-1) \end{vmatrix} = 0 \tag{1.80}$$

This equation is difficult to solve, so that the method is at present largely of academic interest. This method requires knowledge of $2N$ structure factors, but this number can be reduced by a suitable choice of p and by making use of the relation $F(h) = F(-h)^*$. However, the analysis of the problem also leads to Harker-Kasper inequalities. In discussing inequalities, it is useful to introduce the *unitary structure factors* $U(h_1h_2h_3)$, defined as

$$U(h_1h_2h_3) = \frac{\hat{F}(h_1h_2h_3)}{F(000)} \tag{1.81}$$

The unitary structure factors have the property $|U| \leqq 1$ since no structure factor can be larger than $F(000)$ for structures with non-negative electron density. (In neutron diffraction, negative densities of scattering centers are encountered; and the results of this chapter do not hold when these negative scatterers are present. Except for the possible appearance of these, neutron data are otherwise *more* amenable to treatment by the present theory than are X-ray data.)

One of the simplest Harker-Kasper inequalities is

$$U(h_1h_2h_3)^2 = \tfrac{1}{2} + \tfrac{1}{2}U(2h_1,2h_2,2h_3) \tag{1.82}$$

for a centrosymmetric structure, from which the sign of $U(2h_1,2h_2,2h_3)$ can be determined if $U(h_1h_2h_3)$ is sufficiently large.

These inequalities were first derived from Schwarz's inequality by Harker and Kasper [42], and have been extended by Gillis [43], Grison [44], Karle and Hauptman [45], and others.

The inequalities of the above type are relations between a structure factor and its higher orders. This is not a necessary restriction, and more general relations exist. One of the most useful is Sayre's relation, discussed by Sayre [46], Zachariasen [47], and Cochran [48].

In its simplest form, Sayre's relation is: if $U(H)$, $U(K)$, and $U(H+K)$ are three large unitary structure factors (greater than about 1.5 times the root-mean-square value), the following relation between the signs (S) is probably, but not necessarily, correct:

$$S(H+K) = S(H) \times S(K) \tag{1.83}$$

Sayre's relation can be derived from the relation between the structure factors of the electron density and its square, using the convolution theorem.

The above relations are special cases of more general *probability distributions* of signs or phases of the structure factors, which can be obtained by letting the atoms roam uniformly through the unit cell. Such probability distributions were first derived by Hauptman and Karle [49]. However, caution must be used in the combining of probabilities. Vand and Pepinsky [50] have shown that the phase problem is not necessarily solved by derivation of probabilities of phases of individual structure factors; for errors can be introduced through the incorrect combination of these individual probabilities. This aspect of X-ray analysis is at present in a rapid state of evolution.

Attempts at Solution in Direct Space: the Patterson Function. The method first introduced by Patterson [51] in practice constitutes at present one of the most powerful phase-determining methods. Patterson considered a function

$$P(u_1,u_2,u_3) = V\iiint \rho(x_1,x_2,x_3)\, \rho(x_1+u_1,x_2+u_2,x_3+u_3)\,dx_1\,dx_2\,dx_3 \quad (1.84)$$

This is a convolution of the electron density with itself inverted in the origin. Hence the transform of P is a product of the structure factor and its complex conjugate, and therefore

$$P(u_1,u_2,u_3) = \frac{1}{V}\sum\sum\sum F(h_1h_2h_3)^2 \cos 2\pi(h_1u_1 + h_2u_2 + h_3u_3) \quad (1.85)$$

The Patterson function thus can be calculated in terms of *observable quantities*, since phases do not enter into the computation. In fact, it represents the complete presentation of the available experimental diffraction data in direct space. If the resulting convolution can be "unraveled" or *deconvoluted*, the phase problem is solved.

If the structure can be represented as composed of N point atoms per cell, the Patterson function is composed of N^2 points, some of which overlap: namely, there is a peak at the origin of weight N; and if the crystal has a center of symmetry, there are N so-called rotational peaks, each of weight 1, and $N \cdot (N/2 - 1)$ nonrotational peaks each of weight 2. The picture of the crystal is now a *vector map*, showing a superposition of all the interatomic vectors present in the crystal. In addition, for a centrosymmetric structure, the rotational peaks, being the vectors across the center of symmetry, represent the structure *on twice the scale.* If a Patterson map has sufficient resolution, the rotational peaks can be picked up and the image of the structure thus obtained. There is, however, an ambiguity due to the fact that as the unit cell of a Patterson map is unchanged in size, the enlarged images overlap so that it is necessary to disentangle the image from images arising from the neighboring cells. In other words, the rotational image of the structure determines the signs of the structure factors with $h_1h_2h_3$ *all even,* but does not convey any information about the signs of the other structure factors; in order to obtain information on these, the nonrotational vectors must be considered.

These considerations are modified by the presence of other elements of symmetry. For example, if a twofold axis of symmetry is present, say parallel to a_2, then each atom at $x_1x_2x_3$ is accompanied by an atom at $-x_1,x_2,-x_3$, so that there is a vector $2x_1,0,2x_3$ between each pair of atoms. If now a Patterson section $P(u_1,0,u_3)$ is constructed, all the peaks corresponding to these vectors will be found in this plane, and an image of the projection of the structure along the a_2 axis will be found, again on twice the scale. The three-dimensional summation is then greatly simplified, and the numerical work greatly reduced. The ambiguity of overlap of the images from neighboring cells remains.

This summation is the *Harker synthesis* [52]. It is very useful for structures possessing rotation axes, screw axes, mirror planes, and glide planes. A generalization of the Harker method has been developed by Buerger [53] and named the *implication theory*. It is, however, also subject to certain ambiguities, which Buerger has tabulated.

For a two-dimensional structure composed of point atoms, it is comparatively easy to construct the corresponding Patterson by carrying out the convolution graphically. The N points are plotted on a sheet of paper and numbered consecutively. A sheet of transparent paper is now taken with marked origin and axes. The origin is now superimposed on atom 1 and the structure copied; then the paper is translated until the origin is over point 2, the structure copied, and so on, until all the points of the structure are exhausted. The result gives the vector distribution, or *vector set.*

It is more difficult to *deconvolute* such a vector set, but general methods do exist. They have been studied by Wrinch [54], Buerger [55] (most extensively), MacLachlan and Harker [56], Clastre and Gay [57], and others.

It is of great theoretical interest that *it is in fact possible to deconvolute the Patterson function of point atoms, which means that the phase problem is generally soluble, the condition for solution being that the Patterson peaks be completely resolved.* This condition is fulfilled in practice only for structures with comparatively few atoms.

In order to obtain a rough estimate of this condition, let us assume that, in the usual structures, each atom occupies roughly 25 A^3 of space, so that the volume of the unit cell is of the order of $V = 25N\ A^3$, and the cell edge is $a = N^{1/3}$ A. The Patterson has in general $N(N-1)/2$ peaks, or roughly $N^2/2$ peaks. Each peak has thus, on the average, space available equal to $50/N\ A^3$, and the Patterson peaks are thus on the average separated by $4N^{-1/3}$ A. If radiation of wavelength λ is used, detail cannot be distinguished if peaks are separated by less than about $0.6\lambda/2 \sin \theta_{max}$. For $\theta_{max} = 90°$ and Cu K_α radiation with $\lambda = 1.54$ A, this distance is about 0.5 A. At these average distances, there would still be chance of overlapping of a great many peaks, so that this distance is to be increased by at least two or three times to reduce the number of overlapping peaks to a reasonable level. This leads to N of the order of 10 to 50 atoms per cell.

If the shape of the molecule is approximately known from the chemical evidence, Patterson maps can be helpful even if the separate vector peaks fail to resolve. The general appearance of the Patterson map may still convey information about the possible molecular arrangements.

The Possibility of Homometric Structures. Another question arises: whether the solution of the point atom problem is *unique* or not. There are, of course, certain freedoms of choice of origin; for a structure without a center of symmetry, the origin can be chosen quite arbitrarily, which amounts to a free choice of three phase angles. Similarly, if there is a center of symmetry, there are necessarily eight symmetry centers per cell and eight possible choices of origin. This point is fully discussed by Hauptman and Karle [58] for different space groups, and also by Lonsdale and Grenville-Wells [59]. Furthermore, for the free-electron model $|F(h)| = |F(-h)|$, so that the inability of registering phases introduces an ambiguity in *enantiomorphy,* i.e., a left-handed configuration scatters X rays the same as a right-handed configuration. However, in the region of wavelengths where there are deviations from the free-electron model, this limitation no longer holds and, in fact, left- and right-handed isomers have been successfully distinguished (see Coster, Knoll, and Prins, and Bijvoet, above).

Apart from the above ambiguities, there are certain special cases where two genuinely different structures in fact possess identical vector sets and therefore Patterson diagrams, and also diffraction patterns. These are the so-called Patterson *homometric structures* [60]. The simplest one-dimensional homometric case is a four-atom structure, in which there are special relations between coordinates so that the structure has in fact only one variable parameter. Similarly, the more complicated homometric structures have a number of constraints, so that for a structure in which there are no special conditions between coordinates, the probability of the structure being homometric is zero. In other words, if we imagine all the possible structures represented as points in a $3N$-dimensional space, the homometric structures lie on surfaces running through this space; and as these are infinitely thin, the probability of a random structure lying on them is zero. This does not mean that homometric structures do not occur in practice; in fact, the first example was found when the mineral bixbyite was investigated by Pauling and Shappell [61]. The space group of this mineral is Ia3, and the constraints on the positions are inherent in the space group.

Recently Pepinsky and Calderon [62] have shown ambiguities inherent in noncentrosymmetric structures. In practice, furthermore, some structures may be *near-homometric,* i.e., differing in their scattering only within certain limits. It is thus always important to confirm the results of X-ray crystal analysis from other independent evidence; above all, the correct structure should be physically and chemically reasonable.

Direct Phase-determining Methods. In Sec. 4 it has been shown that although in principle the phase problem is soluble by deconvolution of a Patterson map of point atoms this can be done only if the Patterson map is fully resolved. However, there are certain special cases, when the phases can be determined more directly. One of the cases is when the structure, composed of light atoms, contains in addition one or a small number of *heavy atoms.* In a Patterson map, the peaks corresponding to interatomic distances have weight corresponding to the product of scattering power of the two atoms concerned. Therefore the peaks between heavy atoms are usually much higher than peaks between a heavy and a light atom, and these in turn are much higher than peaks between the light atoms. If there is only one heavy atom at the origin, the heavy to light atom peaks give the structure directly, because the light to light atom peaks can be neglected. The extent to which the heavy-atom method can be applied to real structures depends on the ratio of the heavy-atom contribution to the structure factor to the average contribution of the rest of the atoms. The root-mean-square of the latter is proportional to $\sqrt{N}$, where N is the number of the light atoms per cell. Taking, for example, iodine and carbon, with atomic numbers 53 and 6, the contributions are equal for $N = 100$. This is about the upper range of practicability for this method.

A still more powerful method for phase determination is the method of *isomorphous replacement.* The phase relationship can then be determined from the differences between corresponding structure factors. It is only necessary to find two isomorphous crystals, identical in all respects except that in one of them one atom is replaced by another of different scattering power. It is of advantage when these atoms differ in scattering power as much as possible, and replacement series such as Cl, Br, and I or K or NH_4 and Rb are especially suitable. It is not necessary for the heavy-atom contribution to outweigh that of the rest of the molecule. Many complicated structures have been successfully solved by this technique. The method fails in general if the heavy atom is in a special position in a nonprimitive cell.

The neutron scattering of hydrogen and of deuterium differs not only in magnitude, but also in sign, hydrogen being a negative scatterer. Hydrogenated and deuterated compounds and their mixtures thus form suitable isomorphous pairs, as Pepinsky [63] has pointed out, in neutron studies. The location of negative scatterers in a structure comprised chiefly of positive scattering centers is also readily revealed in a Patterson map, since vectors between positive and negative centers will appear as negative peaks in the Patterson function.

The Fourier Transform Method. When the sterical configuration of the molecule within the unit cell is already known from chemical evidence, use can be made of the property of the structure factor that it is the transform of a single unit cell—which is a continuous function—*sampled* at the points of the reciprocal lattice.

The transform of the molecule is first constructed. Transforms of several molecules (if there are more than one per cell) are then added, account being taken of their different phases when the origins of the molecules are displaced from the origin of the unit cell; and the molecular orientations within the cell can then be found by rotating the reciprocal lattice with respect to the transform, until a position is found where the sampled transform gives magnitudes of calculated structure factors agreeing with the observed structure factors.

The procedure is especially simple if the molecule is *planar*, since the transform then becomes a cylindrical function (or, more exactly, a function with constant sections), with the cylinder axis perpendicular to the plane of the molecule. Since aromatic molecules are planar, crystals comprised of them are suitable for the Fourier transform treatment. The method has been discussed by Hettich [64], Knott [65], and others. Ingenious use of it has been made by Perutz and his coworkers [66] in studies of protein structures.

Recent Developments. Significant developments have appeared recently, the most notable being (1) advances in high-speed digital computing methods; (2) automatization of single-crystal diffraction data collection; (3) extension of the statistical approach to solution of the phase problem; (4) introduction of the Pepinsky-Okaya method for solution of noncentric structures, utilizing anomalous dispersion; and (5) introduction of the concept of "crystal designing," and its application. References to these can be found most easily in *Acta Crystallographica*, **10** (12) (1957), this issue of the journal consisting of abstracts of papers presented at the Montreal Congress of the International Union of Crystallography in June, 1957. Significant papers have appeared by the pioneer and leading expert in powder X-ray diffractometry, W. H. Parrish. The application of neutron diffraction in the study of magnetic structures, structures containing both light and heavy atoms, and in understanding crystal transitions has grown tremendously. Several new and important texts on crystallography, diffraction methods, and results have appeared. Significant among these is the treatise by Peiser, Rooksby, and Wilson on polycrystalline material diffraction (Institute of Physics, London, 1953); Bacon's volume on neutron diffraction (Oxford University Press, 1955); some chapters from Flügge's "Handbuch der Physik" (Springer); and several new collations of structural data [e.g., Landolt-Börnstein, Bd. I, Tl. 4 (Springer, 1955); and continuation volumes of *Structure Reports*]. The *Zeitschrift für Kristallographie* renewed publication in 1954 (vol. 106); the Russian journal *Kristallographia* began publication in 1956, and several important Russian crystallographic works are in process of translation. The great collation of chemical and physical crystallography of P. von Groth, "Chemisches Krystallographie," published between 1905 and 1919, is in process of long-needed revision and extension, by The Groth Institute.

References

1. Steno, N.: "Dissertationis Prodomus de Solido intra Solidum naturaliter contento," Florence, 1669.
2. Haüy, R. J.: "Essai d'une théorie sur la structure des crystaux," Paris, 1784.
3. Bravais, A.: *J. école polytech.* (*Paris*), **19**: 1 (1850).
4. Miller, W. H.: "Treatise on Crystallography," Cambridge, England, 1839.
5. "International Tables for X-ray Crystallography," Chap. 3.
6. Federov, R. S.: *Verhandl. Min. Gesell., St. Petersburg*, **27**: 448 (1891); *Z. Krist.*, **20**: 28 (1892); **32**: 22 (1903).
7. Schoenflies, A. M.: "Kristallsysteme und Kristallstruktur," Leipzig, 1891.
8. Barlow, W.: *Mining Mag.* (*London*), **11**: 119 (1895).
9. Bragg, W. L.: *Z. Krist.*, **70**: 491 (1929); *Nature*, **154**: 69 (1944).
10. Bunn, C. W.: "Chemical Crystallography," pp. 271–273; 348–350, Oxford University Press, New York and London, 1945; Pepinsky, R.: "Computing Methods and the Phase Problem in X-ray Crystal Analysis," pp. 106–118, The Pennsylvania State University, State College, Pa., 1952.
11. Hanson, A. W., H. Lipson, and C. A. Taylor: *Proc. Roy. Soc.* (*London*), **A218**: 371 (1953).
12. Buerger, M. J.: *Proc. Natl. Acad. Sci.*, **27**: 117 (1941); **36**: 330 (1950); *J. Appl. Phys.*, **21**: 909 (1950).
13. Gibbs, J. W., and E. B. Wilson: "Vector Analysis," Yale University Press, New Haven, 1901. See Part 1, Chap. 10, Sec. 3.
14. Ewald, P. P.: *Physik. Z.*, **14**: 465 (1913).
15. Bernal, J. D. H.: *Proc. Roy. Soc.* (*London*), **A113**: 117 (1926).
16. Weissenberg, K.: *Z. Physik*, **23**: 229 (1924).
17. Buerger, M. J.: "The Photography of the Reciprocal Lattice," ASXRED Monograph 1, Murray Printing Co., Cambridge, Mass., 1944.
18. Coster, D., K. S. Knoll, and J. Prins: *Z. Physik*, **63**: 345 (1930).
19. Bijvoet, J. M.: *Nature*, **173**: 88 (1954).
20. James, R. W., and G. W. Brindley: *Phil. Mag.*, **12**: 81 (1932).
21. Thomas, L. H.: *Proc. Cambridge Phil. Soc.*, **23**: 542 (1927).
22. Fermi, E.: *Z. Physik*, **48**: 73 (1928).
23. Pauling, L., and J. Sherman: *Z. Krist.*, **81**: 1 (1932).
24. "Internationalle Tabellen zur Bestimmung von Kristallstrukturen," vol. II, reprinted by J. W. Edwards, Ann Arbor, Mich., 1944.
25. MacWeeny, R.: *Acta Cryst.*, **4**: 513 (1951); **5**: 463 (1952).
26. Beevers, C. A., and H. Lipson: *Phil. Mag.*, **17**: 855 (1934); *Nature*, **17**: 825 (1936); *Proc. Phys. Soc.* (*London*), **48**: 772 (1936).
27. Beevers, C. A.: *Acta Cryst.*, **5**: 670 (1952).
28. Robertson, J. M.: *Phil. Mag.*, **21**: 176 (1936); *J. Sci. Instr.*, **25**: 28 (1948).
29. Patterson, A. L., and G. Tunell: *Am. Mineralologist*, **27**: 655 (1942).
30. Shaffer, P. A., V. Schomaker, and L. Pauling: *J. Chem. Phys.*, **14**: 648, 659 (1946).
31. Donohue, J., and V. Schomaker: *Acta Cryst.*, **2**: 344 (1949).
32. Cox, E. G., and G. A. Jeffrey: *Acta Cryst.*, **2**: 341 (1949).
33. Cox, E. G., L. Gross, and G A. Jeffrey: *Acta Cryst.*, **2**: 351 (1949).
34. Grems, M. D., and J. S. Kasper: *Acta Cryst.*, **2**: 347 (1949).
35. Pepinsky, R.: "Computing Methods and the Phase Problem in X-ray Crystal Analysis," X-ray and Crystal Analysis Laboratory, The Pennsylvania State University, State College, Pa., 1952.
36. Ott, H.: *Z. Krist.*, **66**: 136 (1927).
37. Avrami, M.: *Phys. Rev.*, **54**: 300 (1938).
38. Banerjee, K.: *Proc. Roy. Soc.* (*London*), **A141**: 188 (1933).
39. Hughes, E. W.: *Acta Cryst.*, **2**: 37 (1949).
40. Goedkoop, J. A.: "Theoretical Aspects of X-ray Crystal Structure Analysis," Thesis, Amsterdam, 1952.
41. Goedkoop, J. A.: "Computing Methods and the Phase Problem in X-ray Crystal Analysis," p. 61, R. Pepinsky (ed.), X-ray Crystal Analysis Laboratory, The Pennsylvania State University, State College, Pa., 1952.
42. Harker, D., and J. S. Kasper: *Acta Cryst.*, **1**: 70 (1948).
43. Gillis, J.: *Acta Cryst.*, **1**: 76, 174 (1948).

44. Grison, E.: *Acta Cryst.*, **4**: 489 (1951).
45. Karle, J., and H. Hauptman: *Acta Cryst.*, **3**: 181 (1950).
46. Sayre, D.: *Acta Cryst.*, **5**: 60 (1952).
47. Zachariasen, W. H.: *Acta Cryst.*, **5**: 68 (1952).
48. Cochran, W.: *Acta Cryst.*, **5**: 65 (1952).
49. Hauptman, H., and J. Karle: "Solution of the Phase Problem I. The Centrosymmetric Crystal," *ACA Monograph* no. 3, Wilmington, Del., 1953.
50. Vand, V., and R. Pepinsky: "The Statistical Approach to X-ray Structure Analysis," X-ray and Crystal Analysis Laboratory, The Pennsylvania State University, State College, Pa., 1953.
51. Patterson, A. L.: *Z. Krist.*, **90A**: 517 (1935).
52. Harker, D.: *J. Chem. Phys.*, **4**: 381 (1936).
53. Buerger, M. J.: *Phys. Rev.*, **73**: 927 (1948); *Acta Cryst.*, **1**: 259 (1948).
54. Wrinch, D. M.: *Phil. Mag.*, **27**: 98 (1939).
55. Buerger, M. J.: *Acta Cryst.*, **3**: 87 (1950); *Proc. Natl. Acad. Sci.*, **36**: 376, 738 (1950).
56. MacLachlan, D., and D. Harker: *Proc. Natl. Acad. Sci.*, **37**: 115 (1951).
57. Clastre, J., and R. Gay: *Compt. rend.*, **230**: 1876 (1950); *J. Phys. Radium*, **11**: 75 (1950).
58. Hauptman, H., and J. Karle: "Solution of the Phase Problem I. The Centrosymmetric Crystal," *ACA Monograph* No. 3, Wilmington, Del., 1953.
59. Lonsdale, K., and H. J. Grenville-Wells: *Acta Cryst.*, **7**: 490 (1954).
60. Patterson, A. L.: *Phys. Rev.*, **65**: 195 (1944).
61. Pauling, L., and M. D. Shappell: *Z. Krist.*, **75**: 128 (1930).
62. Pepinsky, R., and A. Calderon: "Computing Methods and the Phase Problem in X-ray Crystal Analysis," pp. 356–360, R. Pepinsky (ed.), X-ray Crystal Analysis Laboratory, The Pennsylvania State University, State College, Pa., 1952.
63. Pepinsky, R.: *Science*, **117**: 1 (1953).
64. Hettich, A.: *Z. Krist.*, **90A**: 473 (1935).
65. Knott, G.: *Proc. Phys. Soc. (London)*, **52**: 229 (1940).
66. Boyes-Watson, J., E. Davidson, and M. F. Perutz: *Proc. Roy. Soc. (London)*, **A191**: 83 (1947).

Bibliography

Barker, T. V.: "Graphical and Tabular Methods in Crystallography," T. Murby, London, 1922.
Barker, T. V.: "Systematic Crystallography," T. Murby, London, 1930.
Bragg, W. L. (ed.): "The Crystalline State," vols. I–III, Macmillan, New York, 1934–1953.
Buerger, M. J.: "X-ray Crystallography," Wiley, New York, 1942.
Compton, A. H., and S. K. Allison: "X-rays in Theory and Experiment," Van Nostrand, Princeton, N.J., 1935.
Dana, E. S.: "A Textbook of Mineralogy," 4th ed., revised and enlarged by W. E. Ford, Wiley, New York, 1932.
Ewald, P. P.: Die Erforschung des Aufbaues der Materie mit Röntgenstrahlen, "Handbuch der Physik," 2d ed., vol. 23, part 2, Springer, Berlin, 1933.
Halla, F., and H. Mark: "Röntgenographische Untersuchungen von Kristallen," Barth, Leipzig, 1937.
Hartshorne, N. H., and A. Stuart: "Crystals and the Polarizing Microscope," 2d ed., E. Arnold, London, 1950.
Henry, N. F., and K. Lonsdale (eds.): "International Tables for X-ray Crystallography," vol. I., Kynoch Press, Birmingham, England, 1952.
Lipson, H., and W. Cochran: "The Determination of Crystal Structures," vol. III of "The Crystalline State," Macmillan, New York, 1954.
Miers, Henry A.: "Mineralogy," 2d ed., revised by H. L. Bowman, Macmillan, London, 1929.
Pepinsky, R. (ed.): "Computing Methods and the Phase Problem in X-ray Crystal Analysis," The Pennsylvania State University, State College, Pa., 1952.
Phillips, F. C.: "An Introduction to Crystallography," Longmans, New York, 1946.
Robertson, J. N.: "Organic Molecules and Crystals," Cornell University Press, Ithaca, N.Y., 1953.
Tutton, A. E. H.: "Crystallography and Practical Crystal Measurement," Macmillan, London, 1922.
Von Laue, M.: "Röntgenstrahlinterferenzen," Akademische Verlagsgesellschaft m.b.H., Leipzig, 1941.
Von Laue, M.: "Materiewellen und ihre Interferenzen," Akademische Verlagsgesellschaft, Leipzig, 1948.
Zachariasen, W. H.: "Theory of X-ray Diffraction in Crystals," Wiley, New York, 1945.

Identification of Crystals by Classical Methods:

Donnay, J. D. H., and W. Nowacki: "Crystal Data," *Geol. Soc. Amer. Mem.* 60, New York, 1954.
Groth, P.: "Chemische Krystallographie," W. Engelmann, Leipzig, 1906–1919.
Palache, C., H. Berman, and C. Frondel: "Dana's System of Mineralogy," Wiley, New York, 1944–1951.
Porter, M. W., and R. C. Spiller: "The Barker Index of Crys als," Heffer, Cambridge, England, 1951.
Winchell, A. N.: "Elements of Optical Mineralogy," Wiley, New York, 1939.
Winchell, A. N.: "Microscopic Characters of Artificial Minerals," Wiley, New York, 1931.
Winchell, A. N.: "Optical Properties of Organic Compounds," 2d ed., Academic Press, New York, 1954.

Neutron Diffraction:

Hastings, J. M., and L. M. Corliss: "Neutron Diffraction," in Weissburger, A.: "Technique of Organic Chemistry," vol. I, part 3, pp. 2361–2398, Interscience, New York, 1954.

Powder Diffraction:

Guinier, A.: "X-ray Crystallographic Technology," Hilger and Watts, London, 1952.
Klug, H. P., and L. E. Alexander: "X-ray Diffraction Procedures for Polycrystalline and Amorphous Materials," Wiley, New York, 1954.

References on Fourier Transform Theory as Related to X-ray Diffraction:

Hettich, A.: *Z. Krist.*, **A90**: 473 (1935).
Sayre, D.: The Fourier Transform in X-ray Crystal Analysis, "Computing Methods and the Phase Problem in X-ray Crystal Analysis," R. Pepinsky (ed.), The Pennsylvania State University, State College, Pa., 1952.

Systematic Review of X-ray Structure Analyses:

Strukturbericht, vols. I–VII, Ergänzungsbände, Zeitschrift für Kristallographie, 1931–1939; reprinted by Edwards, Ann Arbor, Mich., 1943.
Wilson, A. J. C. (ed.): *Structure Reports*, vols. 10–13, Oosthoek, Utrecht, 1952–1954.
Wyckoff, R. W. G.: "Crystal Structures," vols. 1–3, Interscience, New York, 1948–1953.

Chapter 2

The Energy-band Theory of Solids

By HERBERT B. CALLEN, University of Pennsylvania

1. The Context of Energy-band Theory [1]†

It is customary to begin accounts of energy-band theory by a formal obeisance to the general and fundamental problem of which it is a part. If some 10^{23} nuclei and several times that many electrons are introduced into a container, the resultant properties are predictable, *in principle*, in terms of the solutions of the "simple" Schroedinger equation

$$H_{\text{total}}\Psi_{\text{total}} = E_{\text{total}}\Psi_{\text{total}}$$

where the Hamiltonian includes kinetic-energy terms, Coulomb-interaction terms, and certain less important and smaller spin-dependent terms. If some progress could be made by the disarmingly direct theoretical route of solution of this equation, the first result presumably would be the prediction of the crystal structure in which the system coalesces. Unfortunately even this initial step has proved to be intractable, primarily because of the very small energies that separate different crystal forms. Whereas the dominant energies in the problem are of the order of the Coulomb interaction of particles separated by distances of the order of atomic sizes (or $\simeq 10$ ev) the energy differences of different crystal structures are of the order of 10^{-3} ev per atom. Consequently the existing theory of solids simply accepts the crystalline structure of a particular material as an empirical fact and then attempts to account for other observable properties.

The model with which we begin, then, is one in which the nuclei are in given periodic positions and the electrons are free to accommodate to the nuclear configuration. In particular, the nuclei undergo small oscillations about their equilibrium positions, these oscillations constituting the phonon modes of the solid. The electrons also may be excited to higher energy states, thermally, optically, or by applied electric fields. Moreover, it is clear that the oscillations of the nuclei and the electronic response are not independent, and the first challenge to the theory is to dissociate these phenomena in such a way that they can be independently discussed and analyzed. The essential idea behind this dissociation is as follows: It may reasonably be expected that some of the electrons, at least, are tightly bound by individual nuclei, in atomic orbitals that are essentially identical with those of isolated atoms. These tightly bound orbitals have radii small compared with the internuclear distance in the crystal. On the other hand, the outer atomic orbitals, which overlap appreciably, are badly distorted by the nuclear vibrations. These orbitals consequently do not provide an adequate representation of the state of the "valence" electrons. The electrons in the inner-core states certainly follow the oscillations of the nuclei quasi-statically, whereas the valence electrons interact in a more complex way. Thus each nucleus and its tightly bound core electrons can be considered a coherent unit, and the crystal can be thought of as a system of *ions* and valence electrons. Their interaction is by no means trivial, and we shall make a few comments about it subsequently. But we shall first indicate the theoretical justification of the above qualitative remarks. This justification is provided by the Born-Oppenheimer "adiabatic approximation." For this purpose we return to the general Hamiltonian of the system, which we write in the more explicit form:

$$H_{\text{total}} = -\frac{\hbar^2}{2m}\sum_i \nabla_i^2 - \frac{\hbar^2}{2}\sum_n \frac{1}{M_n}\nabla_n^2 + \frac{1}{2}\sum_{i,j}{}' \frac{e^2}{r_{ij}} + \frac{1}{2}\sum_{n,p}{}' \frac{Z_n Z_p e^2}{r_{np}} + \sum_{i,n} \frac{Z_n e^2}{r_{ni}} \tag{2.1}$$

where i, j denote electrons and n, p denote nuclei.

The terms explicitly written in this Hamiltonian account only for the kinetic energy and Coulomb interactions of the particles. All other interactions, such as the spin-spin interaction and the direct magnetic interaction, are smaller by one or more powers of the fine-structure constant. These small interactions will be neglected in the following, although the electronic structure of certain semiconductors may be particularly sensitive to the spin-orbit coupling [2]. The standard procedure is to develop the band theory without such interactions and then to account for such interactions as are necessary by perturbation methods.

The Born-Oppenheimer approximation seeks a simplification of the general problem by writing the total wave function ψ_{total} as a product of a "nuclear wave function" ψ_{nuc} and an "electronic wave function" Ψ

$$H_{\text{total}}(\psi_{\text{nuc}}\Psi) = -\frac{\hbar}{i}\frac{\partial}{\partial t}(\psi_{\text{nuc}}\Psi) \tag{2.2}$$

† Numbers in brackets refer to References at end of chapter.

It is assumed that ψ_{nuc} depends only on nuclear coordinates and that the "electronic" wave function Ψ satisfies the Schroedinger equation of a group of electrons moving in the field of the *momentarily fixed* and arbitrarily situated nuclei:

$$\left(-\frac{\hbar^2}{2m}\sum_i \nabla_i^2 + \frac{1}{2}\sum_{i,j}{}' \frac{e^2}{r_{ij}} + \sum_{i,n}\frac{Z_n e^2}{r_{ni}}\right)\Psi = \epsilon\Psi \quad (2.3)$$

The electronic wave function Ψ is a function of the electronic coordinates $\mathbf{r}_i$, and depends parametrically on the coordinates of the nuclei $\mathbf{r}_n$. Similarly the eigenvalue of Eq. (2.3) has the physical significance of the energy of the electrons moving in the field of the fixed nuclei, and consequently itself depends upon the coordinates of the nuclei:

$$\epsilon = \epsilon(\cdot\cdot\cdot\ \mathbf{r}_n\ \cdot\cdot\cdot)$$

We insert (2.1) and (2.3) into (2.2), multiply by Ψ^*, and integrate over the electronic coordinates, obtaining

$$-\frac{\hbar^2}{2}\sum_n \frac{1}{M_n}\nabla_n^2\psi_{\text{nuc}} + \frac{1}{2}\sum_{n,p}{}'\frac{Z_nZ_pe^2}{r_{np}}\psi_{\text{nuc}} + \epsilon(\cdot\cdot\cdot\ \mathbf{r}_n\ \cdot\cdot\cdot)\psi_{\text{nuc}} - \hbar^2\sum_n\frac{1}{M_n}(\nabla_n\psi_{\text{nuc}})\cdot \int \Psi^*\nabla_n\Psi\,d\tau - \psi_{\text{nuc}}\frac{\hbar^2}{2}\sum_n\frac{1}{M_n}\int\Psi^*\nabla_n\Psi\,d\tau = -\frac{\hbar}{i}\frac{\partial\psi_{\text{nuc}}}{\partial t} \quad (2.4)$$

Consider the fourth term on the left of this equation. As Ψ is defined as independent of time, it can be taken as real and normalized, whence

$$(\nabla_n\psi_{\text{nuc}})\cdot\int\Psi^*\nabla_n\Psi\,d\tau = \tfrac{1}{2}(\nabla_n\psi_{\text{nuc}})\cdot\nabla_n\int\psi^2\,d\tau = 0 \quad (2.5)$$

With the fourth term on the left thus disposed of, we consider the remaining (fifth) term, within which the essential physical content lies. We consider first the electrons that are tightly bound in atomic orbitals on a particular nucleus. For these the atomic orbital quasi-statically follows the nuclear motion, and the term in question is $m/M \times$ (mean kinetic energy of electrons). That is, this term is of the order of m/M smaller than the third term $\epsilon(\cdot\cdot\cdot\ \mathbf{r}_n\ \cdot\cdot\cdot)\psi_{\text{nuc}}$, and is quite negligible. For the valence electrons, however, another approach is required. For them the fifth term above cannot simply be dismissed, for different orbital states lie close together in energy and the nuclear motion certainly causes transitions among different orbital valence states. However, we first consider a limiting case in which the valence electrons would be completely oblivious to the nuclear positions; then Ψ would be independent of $\mathbf{r}_n$, $\nabla_n^2\Psi$ would vanish, and the term in question would be zero. Thus, for the core electrons and for the valence electrons *if* they were insensitive to the nuclear motion, we would have

$$-\frac{\hbar^2}{2}\sum_n\frac{1}{M_n}\nabla_n^2\psi_{\text{nuc}} + \frac{1}{2}\sum_{n,p}{}'\frac{Z_nZ_pe^2}{r_{np}}\psi_{\text{nuc}} + \epsilon(\cdot\cdot\cdot\ \mathbf{r}_n\ \cdot\cdot\cdot)\psi_{\text{nuc}} = -\frac{\hbar}{i}\frac{\partial\psi_{\text{nuc}}}{\partial t} \quad (2.6)$$

Equation (2.6) has an appealing heuristic interpretation. As the heavy nuclei move in their sluggish vibrational motions the inner-core electrons readjust rapidly and quasi-statically, whereas the reaction of the valence electrons is (temporarily) neglected. At each particular position of the nuclei the electrons have a definite energy, and this energy is supplied by the work done on the electronic system by the nuclear motion. Consequently the electronic-energy eigenvalue appears as an effective potential for the nuclei. Equation (2.6) is then the Schroedinger equation for the nuclear motion, including both the direct nuclear-nuclear Coulomb potential and this effective potential.

The standard procedure in the theory of solids is first to solve Eq. (2.6) for the nuclear motion, and then to correct, by perturbation theory, for the invalid assumption that the valence electrons do not react to the nuclear motion. We describe briefly the general features of these two steps, preparatory to subsequent concentration on the purely electronic aspects of the problem.

In the Schroedinger equation (2.6) for the nuclear motion the effective potential has periodically spaced minima, corresponding to the observed crystal structure. Around these minima the effective potential is quadratic, and the motion of the nuclei is describable [3] in terms of normal modes of small-amplitude vibrations. For the very-long-wavelength modes the frequency of longitudinal vibrations is the ionic plasma frequency $(4\pi NZ^2e^2/M)^{1/2}$, where N is the number of ions per unit volume and Ze is their charge. This frequency results from the Coulomb repulsion of the charge concentrations induced by the ionic motion. However, one must then consider the response of the valence electrons to these vibrations. The long-wavelength ionic vibrations are sufficiently slow that the valence electrons completely screen the ionic charges, reducing the vibrational frequency from the plasma frequency to zero. Shorter-wavelength ionic modes are incompletely screened, and the result is that the frequency of the longitudinal vibrational modes increases from zero linearly with the wave vector. Furthermore, in this intermediate range of wave vector the non-quasi-static response of the electrons scatters the electrons among their states, leading to electrical resistivity [4].

We now focus attention on that restricted aspect of the problem which is generally denoted as the band theory. That is, we consider the electronic states (and particularly the valence-electron states) corresponding to the fixed periodic positions of the nuclei. If the nuclei are considered as fixed in their ideal periodic positions, Eq. (2.3) may be written in the form

$$H\Psi \equiv \left[-\frac{\hbar^2}{2m}\sum_i\nabla_i^2 + \frac{1}{2}\sum_{i,j}{}'\frac{e^2}{r_{ij}} + \sum_i V(\mathbf{r}_i)\right]\Psi = \epsilon\Psi \quad (2.7)$$

where $V(\mathbf{r}_i)$ is the potential energy of an electron in the field of the nuclei; it is a function which is invariant under the space group of the crystal and which therefore has, in particular, the periodicity of the lattice:

$$V(\mathbf{r}_i + \mathbf{R}) = V(\mathbf{r}_i) \quad (2.8)$$

Here **R** denotes any vector of the lattice—a sum of integral multiples of the basis vectors of the lattice. The band theory consists of the analysis of the form of the solutions of Eq. (2.7) with the periodicity condition (2.8).

2. Determinantal Wave Functions and the Hartree-Fock Equations

Exact solution of Eqs. (2.7) and (2.8) is impossible and so further approximations are made. The standard procedure in similar physical problems is to approximate the solution by the first few terms in an infinite series which would be an exact solution. If Ψ were a function of the coordinates of a single electron, such an infinite expansion would be made in terms of any complete orthonormal set of functions. But Ψ is a function of the coordinates of all N_e electrons; furthermore, by the Pauli principle, it is antisymmetric under the interchange of any two electrons. Such an antisymmetric many-electron function can be expanded in an infinite series of $N_e \times N_e$ determinants:

$$\Psi = \sum_{i,j,k,\ldots}' C_{i,j,k,\ldots} \begin{vmatrix} \psi_i(r_1) & \psi_j(r_1) & \psi_k(r_1) & \cdots \\ \psi_i(r_2) & \psi_j(r_2) & \psi_k(r_2) & \cdots \\ \cdot & \cdot & & \\ \cdot & \cdot & & \\ \cdot & \cdot & & \\ \psi_i(r_{N_e}) & \psi_j(r_{N_e}) & \psi_k(r_{N_e}) & \cdots \end{vmatrix} \tag{2.9}$$

The one-electron functions ψ_i which compose the determinants in this expansion can be the functions of any complete orthonormal set.

Stated differently, the set of all $N_e \times N_e$ determinants, formed by every combination of N_e one-electron functions selected from any complete orthonormal set, forms a complete orthonormal set over the manifold of antisymmetric N_e-electron functions.

Each of the one-electron functions ψ_i is the product of a spacelike function ϕ_i and a spin function $\zeta_\pm$. Throughout the remainder of this chapter an integral over a product of ψ's implies both a spatial integration and a summation over spin coordinates.

The number of determinants which can be handled feasibly in an attempt to approximate the true solution of Eqs. (2.7) and (2.8) has increased rapidly in recent years with the development of large-scale electronic digital computers. However, the classical development of the theory of energy bands utilizes the most extreme approximation conceivable; approximation to the true wave function by a single term of the series (2.9). That is, it is conventional to use a single $N_e \times N_e$ determinant to approximate the electronic wave function.

Although the single-determinant approximation sounds drastic indeed, the accuracy of the approximation can be enormously improved by a simple stratagem. In the expansion (2.9) the one-electron functions ψ_i may be chosen arbitrarily as any complete orthonormal set. In those calculations which employ a considerable number of determinants the one-electron functions are chosen for analytic convenience; sometimes as atomic orbitals [5] and sometimes as modified plane waves [6]. However, in employing a single determinant, we may pose the question as to which particular set of N_e orthonormal one-electron functions will make our approximation as accurate as possible. It would certainly seem plausible that a single determinant formed of the best possible N_e one-electron functions may provide as good a result as a series containing several determinants composed of arbitrarily chosen one-electron functions. The technique of selecting the best possible set of N_e one-electron functions to make up a single determinant is the Hartree-Fock method.

The N_e one-electron functions that will compose the determinantal wave function will be denoted as $\psi_1, \psi_2, \ldots, \psi_{N_e}$. The determinantal wave function is then

$$\Psi = \begin{vmatrix} \psi_1(\mathbf{r}_1)\psi_2(\mathbf{r}_1)\psi_3(\mathbf{r}_1) & \cdots & \psi_{N_e}(\mathbf{r}_1) \\ \psi_1(\mathbf{r}_2)\psi_2(\mathbf{r}_2) & \cdots & \\ \cdot & & \\ \cdot & & \\ \cdot & & \\ \psi_1(\mathbf{r}_{N_e}) & & \end{vmatrix} \tag{2.10}$$

$$= \sum_P (-1)^p P \prod_i \psi_i(\mathbf{r}_i) \tag{2.11}$$

In this latter way of writing the determinant the algebraic definition of a determinant as the sum of products is used. Each term is of the form $\psi_1\psi_2\psi_3 \cdots \psi_{N_e}$, but in each term the electronic coordinates $\mathbf{r}_1, \mathbf{r}_2, \ldots, \mathbf{r}_{N_e}$ are permuted differently among the functions ψ_i. There are $\underline{|N_e}$ possible permutation operators P which effect all possible permutations of the $\mathbf{r}_i$ among the ψ_i, and consequently there are $\underline{|N_e}$ terms in Eq. (2.11). In the sum of products, each product is assigned either a positive or a negative sign according to its parity p. A particular permutation has odd or even parity according to the oddness or evenness of the number of simple two-particle interchanges which are required to obtain the permutation, starting with the identity permutation.

The expectation value of the electronic energy is

$$<\epsilon> = \int \cdots \int \Psi^* H \Psi \, d\mathbf{r}_1 \, d\mathbf{r}_2 \cdots d\mathbf{r}_{N_e} \tag{2.12}$$

where Ψ is to be inserted from (2.11) and H from (2.7). Calculus of variations permits one to ascertain the particular set of orthonormal one-electron functions ψ_i which will make the integral (2.12) a minimum. That set will provide the best approximation to the true solution. The variational calculation indicates that the best one-electron functions satisfy the following set of coupled partial differential equations:

$$-\frac{\hbar^2}{2m}\nabla_1^2\psi_i(\mathbf{r}_1) + V(\mathbf{r}_1)\psi_i(\mathbf{r}_1) + \left[\sum_j{}' \int \psi^*_j(\mathbf{r}_2)\frac{e^2}{r_{12}}\psi_j(r_2)\,d\mathbf{r}_2\right]\psi_i(r_1) - \sum_j{}' \left[\int \psi^*_j(\mathbf{r}_2)\frac{e^2}{r_{12}}\psi_i(\mathbf{r}_2)\,dr_2\right]\psi_j(\mathbf{r}_1) = \epsilon_i\psi_i(\mathbf{r}_1) \tag{2.13}$$

These are the Hartree-Fock or Hartree-Fock-Slater equations, developed by Hartree [7], Fock [8], Dirac [9], and Slater [10] about 1930.

Up to a point, the intuitive content of the Hartree-Fock equations is apparent. In a determinantal eigenfunction no particular electron is assigned definitely to a particular one-electron function. The determinantal eigenfunction, according to Eq. (2.11), assigns each of the N_e electrons an equal probability of being found in each of the N_e one-electron functions. The electrons may be regarded, roughly, as exchanging their roles as occupants of the one-electron orbitals.

Denoting by $\mathbf{r}_1$ the position vector of the electron which momentarily occupies the one-electron function ψ_i, the first and second terms in Eq. (2.13) are the conventional Schroedinger terms corresponding to the kinetic energy and the potential energy in the field of the nuclei. The third term is the most naive possible way to account for the Coulomb interaction of the electrons; it is what we obtain if we assume that the electron in ψ_j is a sort of negative charge jelly, spread with density $\psi^*_j(\mathbf{r}_2)\psi_j(\mathbf{r}_2)$, and therefore producing a potential at the point $\mathbf{r}_1$ of magnitude

$$\int \psi^*_j(\mathbf{r}_2)\psi_j(\mathbf{r}_2) \frac{e}{r_{12}}\, d\mathbf{r}_2$$

Summing the potential over all electrons gives the third term in Eq. (2.13). The fourth term on the left we defer for a moment, to note that the right-hand member is of the conventional one-electron Schroedinger form. Thus, the entire equation (2.13), except perhaps for the fourth term, has intuitive relationship to the one-electron Schroedinger equation, of a single electron moving in the field of the nuclei and a negative charge jelly corresponding to the other electrons.

Rather than thinking of the Coulomb interaction as between electron 1 in ψ_i and a "charge jelly" $\psi^*_j(\mathbf{r}_2)\psi_j(\mathbf{r}_2)$, we may think of the interaction between two true point electrons at $\mathbf{r}_1$ and $\mathbf{r}_2$. If there were no correlation of the position of electron 1 within orbital ψ_i with the position of electron 2 within orbital ψ_j, then if electron 1 were at $\mathbf{r}_1$ the *average* interaction of the electrons would be $\int\psi^*_j(\mathbf{r}_2)\psi_j(\mathbf{r}_2) \frac{e^2}{r_{12}}\, d\mathbf{r}_2$. Again, summing over all electrons, we get the third term of Eq. (2.13). That is, the third term would represent the true average Coulomb interactions of the electrons *if there were no correlations of the positions of the various electrons within their individual orbits.*

A wave function which would imply no correlation of the electrons would be a simple product of one-electron functions, and indeed, if one uses the variation procedure to obtain the analogue of Eq. (2.13) for such a function, one obtains (2.13) except that the fourth term is then absent. As expected, the third term is then adequate to account for the Coulomb interaction of the electrons. This was the intuitive approach first followed by Hartree [7].

In a determinantal eigenfunction there is correlation of electrons, and the third term in Eq. (2.13) is not sufficient to account for the Coulomb interaction of the electrons. The fourth term appears as a correction term to account for the effects of electronic correlation. To summarize, the first term in Eq. (2.13) corresponds to the kinetic energy, the second to the potential energy in the field of the nuclei, and the third and fourth together to the Coulomb interaction of the electrons. The third term alone may be called the "uncorrelated Coulomb" term: it would represent the Coulomb interaction if the positions of the electrons were uncorrelated. The exchange term is the alteration in the Coulomb interaction due to the correlation of the electronic positions within their individual orbitals.

3. The Fermi Hole and the Exchange Term

To interpret the exchange term in more specific detail, the summations in the third and fourth terms of Eq. (2.13) can be augmented by the term for which $j = i$, the two new terms mutually canceling.

$$\begin{aligned} &-\frac{\hbar^2}{2m}\nabla_1{}^2\psi_i(\mathbf{r}_1) + V(\mathbf{r}_1)\psi_i(\mathbf{r}_1) \\ &+\left[\sum_j \int \psi^*_j(\mathbf{r}_2)\frac{e^2}{r_{12}}\psi_j(\mathbf{r}_2)\, d\mathbf{r}_2\right]\psi_i(\mathbf{r}_1) \\ &-\sum_j \left[\int \psi^*_j(\mathbf{r}_2)\frac{e^2}{r_{12}}\psi_i(\mathbf{r}_2)\, d\mathbf{r}_2\right]\psi_j(\mathbf{r}_1) \\ &= \epsilon_i\psi_i(\mathbf{r}_1) \qquad (2.14) \end{aligned}$$

As so written the uncorrelated Coulomb term includes the interaction of its electron in the ith orbit with its *own* smeared-out charge distribution. Part of the role of the exchange term is now to negate this spurious self-interaction. In addition the exchange term still must account for the effect of electronic correlation, and the manner in which this is accomplished is most conveniently seen in the form of Eq. (2.14).

The exchange term in Eq. (2.14) may be written in the form

$$-\left[\sum_j \int \frac{\psi^*_j(\mathbf{r}_2)\psi_i(\mathbf{r}_2)\psi_j(\mathbf{r}_1)}{\psi_i(\mathbf{r}_1)}\frac{e^2}{r_{12}}\, d\mathbf{r}_2\right]\psi_i(\mathbf{r}_1) \qquad (2.15)$$

the integration implying also a summation over spin coordinates. The Coulomb interaction of electron 1 (in the ith orbit) with a charge density ρ_i would contribute a term to the Schroedinger equation of the form

$$-\left[\int \rho_i(\mathbf{r}_2)\frac{e}{r_{12}}\, d\mathbf{r}_2\right]\psi_i(\mathbf{r}_1) \qquad (2.16)$$

This was the basis of the previous intuitive interpretation of the uncorrelated Coulomb term. The analogy may be stretched to interpret (2.15) as the interaction between electron 1 and a charge density ρ_i given by

$$\rho_i(\mathbf{r}_2,\mathbf{r}_1) = e\sum_j \sum_{\substack{\text{(spin}\\ \text{coordinates)}}} \frac{\psi^*_j(\mathbf{r}_2)\psi_i(\mathbf{r}_2)\psi_j(\mathbf{r}_1)}{\psi_i(\mathbf{r}_1)} \qquad (2.17)$$

The unique feature of this charge density is that it is a function of $\mathbf{r}_1$ as well as of $\mathbf{r}_2$. That is, electron 1 interacts with a charge which is not static, but which moves about in response to the position of electron 1 itself. This is precisely the correlation effect which we seek.

The total magnitude of the exchange charge is the integral of the exchange charge density over all space, or, by Eq. (2.17),

$$\int \rho_i(\mathbf{r}_2,\mathbf{r}_1)\, d\mathbf{r}_2 = +e \tag{2.18}$$

Thus the total magnitude of the exchange charge is equal to that of one positron. This effective positive charge which surrounds any electron really represents a region of electronic depletion, that is, a region surrounding the given electron in which there is less than the normal probability of finding other electrons. This region is called the *Fermi hole.*

Because of the implied summation over spin coordinates in Eqs. (2.15) and (2.17), the exchange interaction occurs only between electrons of like spin. That is, if the ith orbit is associated with spin "up," only those orbits ψ_j which also have spin "up" contribute to (2.17).

The magnitude of the exchange charge density at the position of electron 1 is

$$\rho_i(\mathbf{r}_1,\mathbf{r}_1) = e \sum_j \psi^*_j(\mathbf{r}_1)\psi_j(\mathbf{r}_1) \tag{2.19}$$

where spin j = spin i. Thus, the magnitude of the exchange charge density at the point $\mathbf{r}_1$ is equal to the total charge density at $\mathbf{r}_1$ of electrons with the same spin as that of electron 1.

Thus the exchange term in the Hartree-Fock equation represents the effect of Pauli electronic correlation on the Coulomb interaction. This correction is represented as the interaction of a given electron with an exchange charge density which surrounds and follows the given electron. This exchange charge density completely depletes the original charge density of electrons of the same spin at the position of the given electron, and has a total magnitude of one positive electronic charge.

The exchange term is difficult to handle analytically, despite its simple intuitive significance. A convenient approximation to the exchange term has been proposed by Slater [11]. He notes that each of the electrons in a solid is surrounded by an exchange charge of slightly different shape, but because of the common characteristics which we have derived above, these exchange charges are quite similar to each other. Slater therefore proposes that the exchange charges all be replaced by the common average. Multiplying $\rho_i(\mathbf{r}_2,\mathbf{r}_1)$ by the relative probability $\psi^*_i(\mathbf{r}_1)\psi_i(\mathbf{r}_1)/\Sigma_j\psi^*_j(\mathbf{r}_1)\psi_j(\mathbf{r}_1)$ that electron 1 be found in the orbit ψ_i, and summing over i gives the average charge $\rho(\mathbf{r}_2,\mathbf{r}_1)$

$$\rho(\mathbf{r}_2,\mathbf{r}_1) = \frac{\sum_i \rho_i(\mathbf{r}_2,\mathbf{r}_1)\psi^*_i(\mathbf{r}_1)\psi_i(\mathbf{r}_1)}{\sum_j \psi^*_j(\mathbf{r}_1)\psi_j(\mathbf{r}_1)} \tag{2.20}$$

$$= e\,\frac{\sum_{i,j} \psi^*_j(\mathbf{r}_2)\psi_i(\mathbf{r}_2)\psi_j(\mathbf{r}_1)\psi^*_i(\mathbf{r}_1)}{\sum_j \psi^*_j(\mathbf{r}_1)\psi_j(\mathbf{r}_1)} \tag{2.21}$$

where spin j = spin i.

With this averaged exchange charge density the Hartree-Fock equations become

$$\begin{aligned} &-\frac{\hbar^2}{2m}\nabla_1{}^2\psi_i(\mathbf{r}_1) + V(\mathbf{r}_1)\psi_i(\mathbf{r}_1) \\ &+ \left[\sum_j \int \psi^*_j(\mathbf{r}_2)\frac{e^2}{r_{12}}\psi_j(\mathbf{r}_2)\, d\mathbf{r}_2\right]\psi_i(\mathbf{r}_1) \\ &- \left\{\left[\sum_{j,k}\int \psi^*_j(\mathbf{r}_2)\psi_k(\mathbf{r}_2)\frac{e^2}{r_{12}}\psi_j(\mathbf{r}_1)\psi^*_k(\mathbf{r}_1)\, d\mathbf{r}_2\right. \right. \\ &\qquad \left. \Big/ \sum_j \psi^*_j(\mathbf{r}_1)\psi_j(\mathbf{r}_1)\right\}\psi_i(\mathbf{r}_1) = \epsilon_i\psi_i(\mathbf{r}_1) \end{aligned} \tag{2.22}$$

where spin j = spin k = spin i.

A further approximation, which yields an exceedingly simple expression for the exchange term, has also been suggested by Slater [11]. In this "free-electron approximation" the averaged exchange term in Eq. (2.22) is replaced by the value which it would have in a free-electron gas. This free-electron exchange term can be evaluated readily, and the modified Hartree-Fock equation finally becomes

$$\begin{aligned} &-\frac{\hbar^2}{2m}\nabla_1{}^2\psi_i(\mathbf{r}_1) + V(\mathbf{r}_1)\psi_i(\mathbf{r}_1) \\ &\qquad + \left[\sum_j \int \psi^*_j(\mathbf{r}_2)\frac{e^2}{r_{12}}\psi_j(\mathbf{r}_2)\, d\mathbf{r}_2\right]\psi_i(\mathbf{r}_1) \\ &- 3e^2\left[\frac{3}{4\pi}\sum_j \psi^*_j(\mathbf{r}_1)\psi_j(\mathbf{r}_1)\right]^{1/3}\psi_i(\mathbf{r}_1) = \epsilon_i\psi_i(\mathbf{r}_1) \end{aligned} \tag{2.23}$$

where spin j = spin i.

4. The Consequences of Symmetry

Each symmetry element of a crystal has implications with respect to the form of the solutions of the Hartree-Fock equations. A full discussion of the consequences of symmetry should be based on a knowledge of the complete space group of the crystal and of the symmetry of the Hamiltonian under spin inversion, under time reversal, and under any other symmetry operations which may be found. The translational group forms a subgroup of this complete group for any crystal, and it therefore is reasonable to begin the discussion by an analysis of the implications of translational symmetry alone.

Rewrite the exchange term in (2.14) in the form (2.15) obtaining

$$\begin{aligned} \tilde{H}\psi_i(\mathbf{r}_1) \equiv &\left\{-\frac{\hbar^2}{2m}\nabla_1{}^2 + V(\mathbf{r}_1)\right. \\ &\qquad + \sum_j \int \psi^*_j(\mathbf{r}_2)\frac{e^2}{r_{12}}\psi_j(\mathbf{r}_2)\, d\mathbf{r}_2 \\ &\left. - \sum_j \int \frac{\psi^*_j(\mathbf{r}_2)\psi_i(\mathbf{r}_2)\psi_j(\mathbf{r}_1)}{\psi_i(\mathbf{r}_1)}\frac{e^2}{r_{12}}\, d\mathbf{r}_2\right\}\psi_i(\mathbf{r}_1) = \epsilon_i\psi_i(\mathbf{r}_1) \end{aligned} \tag{2.24}$$

A stratagem to circumvent the difficulties due to the nonlinearity of this equation is now used. Suppose that the solutions of the Hartree-Fock equations are all known. These known functions are substituted *inside* the integrals in the Coulomb and exchange terms in Eq. (2.24). Then Eq. (2.24) becomes effectively a linear Schroedinger-like equation for ψ_i, with a definite effective Hamiltonian $\tilde{H}$. The wave function ψ_i satisfies this effective Schroedinger equation, and is certainly subject to any symmetry considerations based upon this equation.

The effective Hamiltonian $\tilde{H}$ in Eq. (2.24) is invariant under the translational group of the crystal. The potential $V(\mathbf{r}_1)$ is produced by the nuclei which themselves have translational symmetry; the "uncorrelated Coulomb" interaction is the interaction with the complete electronic charge density, which must have the symmetry of the crystal; and the exchange interaction is the interaction with a Fermi hole which travels with the given electron and which must have identical form if the electron is translated through a lattice period. Consequently the effective Hamiltonian is invariant under the translational group of the crystal.

The elements of the translational group commute with each other and with the effective Hamiltonian. Therefore the wave functions can be chosen as simultaneous eigenfunctions of all the translational operators and the Hamiltonian, and the eigenvalues of the translational operators have moduli unity. If T_1, T_2, T_3 are the operators corresponding to the three primitive translations of the lattice, then

$$\begin{aligned} T_1\psi_k &= \exp(i\delta_{k,1})\psi_k \\ T_2\psi_k &= \exp(i\delta_{k,2})\psi_k \\ T_3\psi_k &= \exp(i\delta_{k,3})\psi_k \end{aligned} \tag{2.25}$$

It follows that if a general translation of the lattice through the displacement $n\mathbf{a}_1 + m\mathbf{a}_2 + p\mathbf{a}_3$ is

$$T_{nmp} = T_1{}^n T_2{}^m T_3{}^p \tag{2.26}$$

that

$$T_{nmp}\psi_k = \exp[i(n\delta_{k,1} + m\delta_{k,2} + p\delta_{k,3})]\psi \tag{2.27}$$

Or, written in an alternative form,

$$\begin{aligned} \psi_k(\mathbf{r} &+ n\mathbf{a}_1 + m\mathbf{a}_2 + p\mathbf{a}_3) \\ &= \exp[i(n\delta_{k,1} + m\delta_{k,2} + p\delta_{k,3})]\psi_k(\mathbf{r}) \end{aligned} \tag{2.28}$$

Define a vector $\mathbf{k}$ with covariant components $\delta_{k,1}$, $\delta_{k,2}$, $\delta_{k,3}$:

$$\mathbf{k} \equiv \delta_{k,1}\mathbf{b}_1 + \delta_{k,2}\mathbf{b}_2 + \delta_{k,3}\mathbf{b}_3 \tag{2.29}$$

where $\mathbf{b}_1$,$\mathbf{b}_2$, and $\mathbf{b}_3$ is the reciprocal triplet to $\mathbf{a}_1$,$\mathbf{a}_2$,$\mathbf{a}_3$:

$$\mathbf{a}_i \cdot \mathbf{b}_j = \delta_{ij} \tag{2.30}$$

To each wave function ψ_k there is associated by Eqs. (2.25) a set of three parameters $\delta_{k,1}$, $\delta_{k,2}$, and $\delta_{k,3}$, and hence a particular vector $\mathbf{k}$. It is convenient to label ψ_k by this associated vector rather than by the scalar index k, so that we henceforth write $\psi_\mathbf{k}$. Then denoting the translational vector in Eq. (2.28) by $\mathbf{R}_{nmp}$,

$$\mathbf{R}_{nmp} \equiv n\mathbf{a}_1 + m\mathbf{a}_2 + p\mathbf{a}_3 \tag{2.31}$$

(2.28) becomes

$$\psi_\mathbf{k}(\mathbf{r} + \mathbf{R}_{nmp}) = \exp[i(\mathbf{R}_{nmp} \cdot \mathbf{k})\psi_\mathbf{k}(\mathbf{r}) \tag{2.32}$$

An equivalent statement is that if $\psi_\mathbf{k}(\mathbf{r})$ is written in the form

$$\psi_\mathbf{k}(\mathbf{r}) = \exp(i\mathbf{k} \cdot \mathbf{r})X_\mathbf{k}(\mathbf{r}) \tag{2.33}$$

then $X_\mathbf{k}(\mathbf{r})$ is a periodic function, such that

$$X_\mathbf{k}(\mathbf{r} + \mathbf{R}_{nmp}) = X_\mathbf{k}(\mathbf{r}) \tag{2.34}$$

This is the Bloch theorem; it is the basic theorem underlying the band theory of electronic structure.

It is convenient to consider that the macroscopic crystal is cut in the shape of a parallelepiped, with its edges parallel to the edges of a unit cell, and with N_1 unit translations parallel to a_1, N_2 parallel to a_2, and N_3 parallel to a_3. Simple periodic boundary conditions are adopted. These arbitrary assumptions are for analytic convenience, and do not have any significant effect on calculation of the electronic-energy spectrum. The periodic boundary conditions, together with Eq. (2.28), require that

$$N_1\delta_{k,1} = 2\pi \times (\text{integer}) \tag{2.35}$$

and similarly for $N_2\delta_{k,2}$ and $N_3\delta_{k,3}$. These restrictions limit the possible values of $\delta_{k,1}$, $\delta_{k,2}$, $\delta_{k,3}$, or of the vector $\mathbf{k}$. We must have, in fact,

$$\mathbf{k} = 2\pi\left(\frac{n_1}{N_1}\mathbf{b}_1 + \frac{n_2}{N_2}\mathbf{b}_2 + \frac{n_3}{N_3}\mathbf{b}_3\right) \tag{2.36}$$

where n_1, n_2, and n_3 are integers.

The relation above can be given a useful pictorial representation. Given a crystal with basis vectors $\mathbf{a}_1$,$\mathbf{a}_2$,$\mathbf{a}_3$ and macroscopic dimensions N_1a_1, N_2a_2, N_3a_3; associated with this crystal is a reciprocal lattice, with basis vectors $\mathbf{b}_1$,$\mathbf{b}_2$,$\mathbf{b}_3$. We now introduce a new lattice, called the *rationalized reciprocal lattice*, with basis vectors $2\pi\mathbf{b}_1$, $2\pi\mathbf{b}_2$, $2\pi\mathbf{b}_3$. The rationalized reciprocal lattice is simply a 2π-fold magnification of the reciprocal lattice. A two-dimensional representation of the rationalized reciprocal lattice is shown in Fig. 2.1. We subdivide the

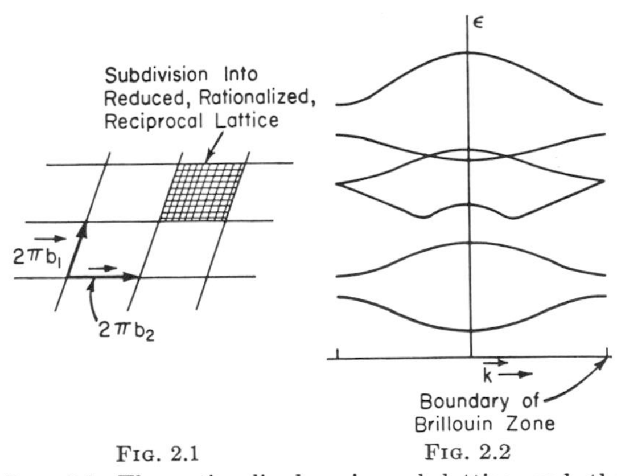

FIG. 2.1　　　FIG. 2.2

FIG. 2.1. The rationalized reciprocal lattice and the reduced, rationalized, reciprocal lattice.
FIG. 2.2. The dependence of the Hartree-Fock eigenvalue on electronic wave vector.

basis vector $2\pi\mathbf{b}_1$ into N_1 equal parts; similarly $2\pi\mathbf{b}_2$ into N_2 parts, and $2\pi\mathbf{b}_3$ into N_3 parts. The little vectors $2\pi\dfrac{\mathbf{b}_1}{N_1}$, $2\pi\dfrac{\mathbf{b}_2}{N_2}$, and $2\pi\dfrac{\mathbf{b}_3}{N_3}$ then serve as basis vectors for a much finer-scale lattice, called the *reduced, rationalized, reciprocal lattice.* The allowed values of $\mathbf{k}$, according to Eq. (2.36), are any of the vectors of the reduced, rationalized, reciprocal lattice.

The reduced, rationalized, reciprocal lattice subdivides a single unit cell of the rationalized reciprocal lattice into $N_1N_2N_3$ subcells; a number equal to the number of unit cells of the direct lattice contained in the original crystallite.

Given a particular solution of the Hartree-Fock equations, the association of a vector $\mathbf{k}$ with this solution is not unique. In fact, if $\mathbf{K}$ denotes any vector of the rationalized reciprocal lattice, then the vector $\mathbf{k} + \mathbf{K}$ is quite as appropriate as the vector $\mathbf{k}$ itself. For if the wave function is $\exp(i\mathbf{k} \cdot \mathbf{r})X_{\mathbf{k}}(\mathbf{r})$, we can write

$$\begin{aligned} \exp(i\mathbf{k} \cdot \mathbf{r})X_{\mathbf{k}}(\mathbf{r}) &= \exp[i(\mathbf{k} + \mathbf{K}) \cdot \mathbf{r}] \\ &\qquad [\exp(-i\mathbf{K} \cdot \mathbf{r})X_{\mathbf{k}}(\mathbf{r})] \qquad (2.37) \\ &= \exp[i(\mathbf{k} + \mathbf{K}) \cdot \mathbf{r}]X_{\mathbf{k}+\mathbf{K}}(\mathbf{r}) \end{aligned}$$

As $\exp(i\mathbf{K} \cdot \mathbf{R}) = 1$, the function $X_{\mathbf{k}+\mathbf{K}}(\mathbf{r})$ is periodic as well as $X_{\mathbf{k}}(\mathbf{r})$. Thus a wave function written in the Bloch form with wave vector $\mathbf{k}$ can also be written in the Bloch form with wave vector $\mathbf{k} + \mathbf{K}$.

For definiteness, a *fundamental region* or *zone* in reciprocal space is chosen. The $\mathbf{k}$ vector assigned to any wave function is understood to lie within this zone. Now it is immediately clear that a single unit cell of the rationalized reciprocal lattice is an acceptable choice for such a fundamental zone. For no two $\mathbf{k}$ vectors (that is, vectors of the reduced, rationalized, reciprocal lattice) lying within this unit cell can differ by a vector of the rationalized reciprocal lattice; this insures that the zone is not redundant. Furthermore, any $\mathbf{k}$ vector lying outside the zone differs from some $\mathbf{k}$ vector within it by a vector of the rationalized reciprocal lattice; this insures that the fundamental zone is complete.

Other choices of a fundamental zone can be made. To be an acceptable fundamental zone, a region of reciprocal space must satisfy the above two criteria of nonredundancy and completeness. Each such zone will have a volume of $(2\pi)^3/(\mathbf{a}_1 \cdot \mathbf{a}_2 \times \mathbf{a}_3)$, and will contain $N_1N_2N_3$ independent $\mathbf{k}$ vectors.

It is possible to choose a fundamental zone which is symmetric with respect to the origin (that is, it contains $-\mathbf{k}$ if it contains $\mathbf{k}$) and which is bounded by planes which satisfy the Laue condition

$$\mathbf{k} \cdot \mathbf{K} = \tfrac{1}{2}\mathbf{K} \cdot \mathbf{K} \qquad (2.38)$$

In analogy with X-ray diffraction, we may anticipate that the electronic wave function will undergo diffraction on such planes. The Laue planes in k-space consequently play a significant role in the band theory. The fundamental zone bounded by Laue planes is called the *Brillouin zone,* or more explicitly the *first Brillouin zone.*

Each wave function $\psi_{\mathbf{k}}$ is associated with a particular point $\mathbf{k}$ in the Brillouin zone; the corresponding eigenvalue $\epsilon(\mathbf{k})$ in the Hartree-Fock equation is similarly associated with the point $\mathbf{k}$. As $\mathbf{k}$ ranges over its $N_1N_2N_3$ allowed values within the zone, $\epsilon(\mathbf{k})$ takes on a set of scalar values. Because of the immensity of the number $N_1N_2N_3(\simeq 10^{23})$ the vector $\mathbf{k}$ can be considered as almost continuously variable, and $\epsilon(\mathbf{k})$ may be thought of as a scalar function of a continuous variable.

The scalar function $\epsilon(\mathbf{k})$ may be seen to be a continuous function of $\mathbf{k}$ as follows. Substitute

$$\psi_{\mathbf{k}}(\mathbf{r}) = \exp(i\mathbf{k} \cdot \mathbf{r})X_{\mathbf{k}}(\mathbf{r})$$

into the effective Schroedinger equation $\tilde{H}\psi_{\mathbf{k}} = \epsilon(\mathbf{k})\psi_{\mathbf{k}}$, to obtain a differential equation satisfied by $X_{\mathbf{k}}(\mathbf{r})$. This will have coefficients which are continuous functions of $\mathbf{k}$. In such a differential equation a small change in the coefficients leads to a small change in the eigenvalue $\epsilon(\mathbf{k})$, and this implies continuity of the function $\epsilon(\mathbf{k})$.

The differential equation for $X_{\mathbf{k}}(\mathbf{r})$ is identical to the differential equation for $\psi_{\mathbf{k}}(\mathbf{r})$, except that the operator ∇ is replaced by $(\nabla + i\mathbf{k})$. It therefore follows that

$$X_{-\mathbf{k}}(\mathbf{r}) = X^*_{\mathbf{k}}(\mathbf{r}) \qquad (2.39)$$

and

$$\epsilon(-\mathbf{k}) = \epsilon(\mathbf{k}) \qquad (2.40)$$

Thus the scalar function ϵ is invariant under inversion through the origin in k-space.

Associated with a particular $\mathbf{k}$ value there may be many different wave functions, each with a different periodic function $X_{\mathbf{k}}(\mathbf{r})$. Accordingly, there will be many different branches of the scalar function $\epsilon(\mathbf{k})$, each branch continuous within the Brillouin zone. If a line is passed through the center of the Brillouin zone, we may plot $\epsilon(\mathbf{k})$ against the scalar value of the $\mathbf{k}$ vectors lying on that line. Such a schematic representation is given in Fig. (2.2).

In Fig. (2.2) six branches of the $\epsilon(\mathbf{k})$ function are shown. The lowest two branches do not intersect with any others, and the corresponding wave functions are nondegenerate. From the point of view of group theory this is what we would expect if the only symmetry elements were the translational elements of the lattice. Except for "accidental degeneracies," intersection of different branches $\epsilon(\mathbf{k})$ should be associated with additional symmetry elements which generate a group having irreducible representations of higher dimensionality. The third, fourth, and fifth branches illustrate two possible types of intersection which may be generated by such additional symmetry elements; intersection at the boundaries and intersection within the volume of the Brillouin zone.

In the absence of particular additional symmetry elements the eigenvalue spectrum is nondegenerate, and each branch intercepts a separate, generally nonoverlapping interval on the energy axis. Separating these energy regions, within which the Hartree-Fock eigenvalues are quasi-densely distributed, are energy intervals in which no Hartree-Fock eigenvalues fall. The energy spectrum is thus composed of a series of "allowed bands" separated by "forbidden gaps."

For each $\mathbf{k}$ value within the first Brillouin zone there is a single wave function associated with each band. We may therefore label each wave function by the subscript $\mathbf{k}$ and by an index n which labels the band. In many cases, in which only the wave functions in a single band are important, the label n is omitted.

The Bloch form of the wave functions and the energy-band characteristic of the eigenvalue spectrum have been deduced on the basis of translational symmetry above. These are the key results of the band theory, and although other symmetry elements have consequences of great interest for particular crystals, they do not provide the same powerful and general results as does the translational symmetry.

As an illustration of the consequences of symmetry elements other than translational, suppose that a crystal is invariant under a reflection in the xz plane. Fig. (2.3) shows a two-dimensional yz section of the

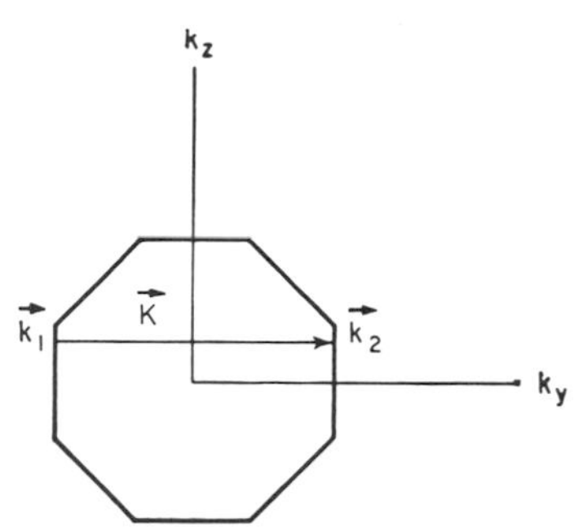

FIG. 2.3. Schematic two-dimensional section of Brillouin zone.

Brillouin zone. The two points $\mathbf{k}_1$ and $\mathbf{k}_2$ lie on opposite boundaries so that they differ by a vector of the rationalized reciprocal lattice.

$$\mathbf{k}_2 - \mathbf{k}_1 = \mathbf{K} \tag{2.41}$$

It is a matter of convention as to whether $\mathbf{k}_1$ or $\mathbf{k}_2$ is chosen to represent the wave vector of an electron; the two points are physically equivalent and

$$\epsilon(\mathbf{k}_1) = \epsilon(\mathbf{k}_2) \tag{2.42}$$

Similarly, $\mathbf{k}$ values to the immediate right of $\mathbf{k}_2$ (outside the Brillouin zone) are physically equivalent to $\mathbf{k}$ value to the immediate right of $\mathbf{k}_1$. Thus

$$\nabla_{\mathbf{k}}\epsilon(\mathbf{k}_1) = \nabla_{\mathbf{k}}\epsilon(\mathbf{k}_2) \tag{2.43}$$

where $\nabla_{\mathbf{k}}$ is the gradient operator in $\mathbf{k}$-space. Because of symmetry under reflection in the xz plane,

$$\epsilon(k_x,k_y,k_z) = \epsilon(k_x, -k_y,k_z) \tag{2.44}$$

and

$$\frac{\partial\epsilon(k_x,k_y,k_z)}{\partial k_y} = -\frac{\partial\epsilon(k_x, -k_y,k_z)}{\partial k_y} \tag{2.45}$$

Applying this equation to the points $\mathbf{k}_1$ and $\mathbf{k}_2$ and comparing with the k_y component of Eq. (2.43) finally gives

$$\frac{\partial\epsilon(\mathbf{k}_1)}{\partial k_y} = \frac{\partial\epsilon(\mathbf{k}_2)}{\partial k_y} = 0 \tag{2.46}$$

The vanishing of the normal derivative of ϵ at the Brillouin zone boundary thus requires more symmetry than merely that of the translational group.

5. Properties of Bloch Functions

The translational symmetry of a crystal requires that the one-electron wave functions be of the form $\psi_{\mathbf{k}}(\mathbf{r}) = \exp(i\mathbf{k}\cdot\mathbf{r})X_{\mathbf{k}}(\mathbf{r})$. If a wave-packet is constructed of the wave functions in a single band or branch (see Fig. 2.2) and in a narrow interval of $\mathbf{k}$, the wave-packet will travel through the crystal with a velocity $\mathbf{v}$. This velocity may be associated with the velocity of an electron in the crystal. The velocity of the wave-packet is the group velocity of its constituent waves, or

$$\mathbf{v} = \frac{\partial\omega}{\partial\mathbf{k}} = \nabla_{\mathbf{k}}\omega = \frac{1}{\hbar}\nabla_{\mathbf{k}}\epsilon(\mathbf{k}) \tag{2.47}$$

Because of the narrow range of $\mathbf{k}$ values in the packet we may say that "the velocity $\mathbf{v}(\mathbf{k})$ of an electron in the state $\psi_{\mathbf{k}}$ is $(1/\hbar)\nabla_{\mathbf{k}}\epsilon(\mathbf{k})$." The velocity is represented by the slope of the $\epsilon(\mathbf{k})$ curves in Fig. (2.2).

The vanishing of the normal derivative of ϵ at the Brillouin zone boundary, which has been demonstrated for sufficiently symmetric crystals, implies that the normal component of the electronic velocity vanishes at the zone boundary. The necessary condition that $\mathbf{k}$ be on the zone boundary is

$$2\mathbf{k}\cdot\mathbf{k} = \mathbf{k}\cdot\mathbf{K} \tag{2.48}$$

This is just the Laue condition for Bragg reflection of the waves $\mathbf{k}$ from the lattice planes associated with $\mathbf{K}$. The vanishing of the electronic velocity is associated with the Bragg reflection which prevents the propagation of the electron through the lattice.

If an electric field is applied to a crystal, an electron will undergo transitions from one eigenstate to another. Under the influence of small fields transitions will occur only within a single energy band, although the phenomenon of *Zener breakdown* [12] in dielectrics is associated with interband transitions induced by large fields ($\simeq 10^6$ volts/cm). As the $\mathbf{k}$ vector of an electron changes, the rate of increase in its energy is

$$[\nabla_{\mathbf{k}}\epsilon(\mathbf{k})]\cdot\frac{d\mathbf{k}}{dt} \tag{2.49}$$

This rate of increase in energy must be equal to the power delivered by the electric field $\mathbf{F}$, which is $e\mathbf{v}\cdot\mathbf{F}$. Utilizing Eq. (2.47) for $\mathbf{v}$,

$$[\nabla_{\mathbf{k}}\epsilon(\mathbf{k})]\cdot\frac{d\mathbf{k}}{dt} = \frac{e}{\hbar}\nabla_{\mathbf{k}}\epsilon(\mathbf{k})\cdot\mathbf{F} \tag{2.50}$$

whence it is at least plausible (and can be rigorously justified) that

$$\frac{d\mathbf{k}}{dt} = \frac{e}{\hbar}\mathbf{F} \tag{2.51}$$

The effect of an applied field is to move the electron with uniform velocity *in reciprocal space.*

To find the motion in *real space,* which results from the application of an electric field, take the time derivative of the equation

$$\mathbf{v} = \frac{1}{\hbar}\nabla_{\mathbf{k}}\epsilon(\mathbf{k}) \tag{2.52}$$

using the operator equation

$$\frac{d}{dt} = \left(\frac{d\mathbf{k}}{dt}\cdot\nabla_{\mathbf{k}}\right) \tag{2.53}$$

whence

$$\dot{\mathbf{v}} = \frac{1}{\hbar}\left(\frac{d\mathbf{k}}{dt}\cdot\nabla_{\mathbf{k}}\right)\nabla_{\mathbf{k}}\epsilon(\mathbf{k}) \tag{2.54}$$

or

$$\dot{\mathbf{v}} = \frac{e}{\hbar^2}(\mathbf{F}\cdot\nabla_{\mathbf{k}})\nabla_{\mathbf{k}}\epsilon(\mathbf{k}) \tag{2.55}$$

Introducing the diadic tensor operator,

$$(\nabla_{\mathbf{k}}\nabla_{\mathbf{k}}) \equiv \begin{Bmatrix} \dfrac{\partial^2}{\partial k_x\,\partial k_x} & \dfrac{\partial^2}{\partial k_x\,\partial k_y} & \dfrac{\partial^2}{\partial k_x\,\partial k_z} \\ \dfrac{\partial^2}{\partial k_y\,\partial k_x} & \dfrac{\partial^2}{\partial k_y\,\partial k_y} & \dfrac{\partial^2}{\partial k_y\,\partial k_z} \\ \dfrac{\partial^2}{\partial k_z\,\partial k_x} & \dfrac{\partial^2}{\partial k_z\,\partial k_y} & \dfrac{\partial^2}{\partial k_z\,\partial k_z} \end{Bmatrix} \tag{2.56}$$

the acceleration $\dot{\mathbf{v}}$ is

$$\dot{\mathbf{v}} = \frac{e}{\hbar^2}(\nabla_{\mathbf{k}}\nabla_{\mathbf{k}})\epsilon(\mathbf{k})\cdot\mathbf{F} \tag{2.57}$$

By analogy with Newton's law for free electrons,

$$\dot{\mathbf{v}} = \frac{e}{m}\mathbf{F} \tag{2.58}$$

it is conventional to define the coefficient of $\mathbf{F}$ in Eq. (2.57) above as the reciprocal "effective mass" tensor. That is, we write

$$\dot{\mathbf{v}} = \frac{1}{m^*}\cdot\mathbf{F} \tag{2.59}$$

where

$$\frac{1}{m^*} = \frac{1}{\hbar^2}(\nabla_{\mathbf{k}}\nabla_{\mathbf{k}})\epsilon(\mathbf{k}) \tag{2.60}$$

The effective mass tensor is a function of position in reciprocal space. At any point $\mathbf{k}$ the effective mass tensor has three orthogonal principal axes. A field applied in one of these directions will accelerate the electron in the same direction; this is not generally true of any other direction. The diagonal components of the effective mass tensor in a coordinate system coinciding with the principal axes are the "principal values of the effective mass."

6. Some Qualitative Comments

The foregoing description of the electronic structure of a crystal refers directly to the ground state. The effect of applied electric or magnetic fields, or of incident electromagnetic radiation, is to induce transitions between this ground state and particular excited states. In order to comprehend the phenomena associated with the application of external fields the excited states of the system must be studied. Fortunately, these are related to the ground state in a simple manner, and a description of the excited states requires only a minor extension of the concepts heretofore developed.

Consider a solid in which there are N_e electrons. The ground state is described by N_e one-electron functions which are the self-consistent solutions of the Hartree-Fock equations (2.24). Inserting these N_e functions into the integrands in the third and fourth terms of Eq. (2.24) to obtain the effective Hamiltonian $\tilde{H}$, the resultant Schroedinger-like equation has as solutions not only the N_e self-consistent wave functions, but in fact a complete orthonormal set ψ_k. The N_e functions which are relevant are the lowest N_e functions of this infinite set.

Consider now the set of N_e functions in which $\psi_j(j \le N_e)$ is replaced by $\psi_k(k > N_e)$, that is, the set $\psi_1, \psi_2, \ldots, \psi_{j-1}, \psi_{j+1}, \ldots, \psi_{N_e}, \psi_k$. We now ask whether this set also provides a self-consistent solution of the Hartree-Fock equations. Insert the new set of functions inside the integrands of Eq. (2.24), to obtain a new effective Hamiltonian $\tilde{H}'$. If this new Hamiltonian were identical with the old it would yield the same infinite set of solutions $\psi_1, \psi_2, \ldots$, and consequently the set $\psi_1, \psi_2, \ldots, \psi_{j-1}, \psi_{j+1}, \ldots, \psi_{N_e}, \psi_k$ would be included among the solutions. This set of functions would then be shown to be self-consistent. Therefore we ask whether the new Hamiltonian $\tilde{H}'$ might be equal to the old Hamiltonian $\tilde{H}$. Although the Hamiltonians are not rigorously and exactly equal, they differ by an exceedingly small and negligible quantity. Consequently the set of functions $\psi_1, \ldots, \psi_{j-1}, \psi_{j+1}, \ldots, \psi_{N_e}, \psi_k$ is, to a high precision, approximately self-consistent.

To see that the Hamiltonians $\tilde{H}$ and $\tilde{H}'$ are very nearly equal, consider the third term in Eq. (2.24). This term represents the potential energy of an electron moving in the Coulomb field of the charge distributions $e|\psi_1|^2 + e|\psi_2|^2 + e|\psi_3|^2 + \cdots + e|\psi_{N_e-1}|^2 + e|\psi_{N_e}|^2$. Each of the charge distributions $e|\psi_n|^2$ is periodic, because of the form of the Bloch function. That is, the charge distribution $e|\psi_n|^2$ assigns $1/N$ of an electron to each of the N unit cells in the crystal. This is true for each ψ_n, and the only difference between $e|\psi_k|^2$ and $e|\psi_j|^2$ is in the detailed distribution of the charge e/N within a unit cell. Thus, in changing the charge distribution from $e|\psi_1|^2 + \cdots + e|\psi_{N_e-1}|^2 + e|\psi_{N_e}|^2$ to $e|\psi_1|^2 + e|\psi_{j-1}|^2 + e|\psi_{j+1}|^2 + \cdots + e|\psi_{N_e}|^2 + e|\psi_k|^2$, we make no large-scale redistributions of charge, but merely alter the detailed manner in which one part in N_e of the charge in each unit cell is distributed. It follows that the fractional change in potential is smaller than $1/N_e$, where N_e is of the order of 10^{23}. Thus the third term in Eq. (2.24) is altered in a completely negligible way. The same reasoning applies to the fourth term; thus $\tilde{H} \simeq \tilde{H}'$, and the set of functions $\psi_1, \ldots, \psi_{j-1}, \psi_{j+1} \ldots, \psi_{N_e}, \psi_k$ is essentially a self-consistent solution of the Hartree-Fock equations. The determinantal function composed of this set of one-electron functions describes an excited state of the crystal. This state can be pictured as arising from the ground state through a transition of a single electron from the jth to the kth state.

Various two-electron excited states may be obtained by depopulating both ψ_j $(j \le N_e)$ and ψ_l $(l \le N_e)$ and populating both ψ_k $(k > N_e)$ and ψ_n $(n > N_e)$. Similarly we can describe three-electron excited states, etc.

The entire set of $N_e \times N_e$ determinants, formed from all possible choices of N_e one-electron functions from the complete orthonormal set ψ_k, constitutes a complete orthogonal set for the N_e-electron problem. If the predominant fraction of the N_e one-electron functions in a particular one of these determinants have indices less than $N_e + 1$, this determinant alone forms an approximately self-consistent solution of the Hartree-Fock equations and therefore describes an excited state of the system. If a given determinant has a large fraction of one-electron functions with indices greater than N_e, however, it will not alone give a good representation of an excited state. We then must use the completeness theorem which guarantees that a linear combination of such determinants can be used to represent these highly excited states.

The energy of a given state is not simply related to the Hartree-Fock eigenvalues ϵ_i of the one-electron functions ψ_i in the corresponding determinant. This is intuitively clear, because ϵ_i is the sum of the expectation values of the kinetic and potential energies of the ith electron, including the entire potential energy of interaction of the ith electron with all other electrons. The sum of the Hartree-Fock eigenvalues will therefore count the interactions of each pair of electrons twice. The total energy of a given state is the sum of the Hartree-Fock eigenvalues, minus the total Coulomb interaction energy.

If we consider the excited state in which ψ_j is depopulated and ψ_k is populated, we may easily find its energy relative to the ground state. Because its associated Hamiltonian $\tilde{H}'$ is approximately equal to the ground-state effective Hamiltonian $\tilde{H}$, all the Hartree-Fock eigenvalues ϵ_l $(l \neq j)$ are approximately equal to their values in the ground state. Similarly the total Coulomb interaction of all electrons is not appreciably changed. Therefore the difference in energy of the excited and ground states is just $\epsilon_k - \epsilon_j$. This is the content of Koopmans' [13] theorem, namely, that for all the low-lying excited states, which can be represented adequately by a single determinant, the energy relative to that of the ground state is the sum of the Hartree-Fock eigenvalues in the excited state minus the sum of those in the ground state.

This representation of the excited states of solids provides an insight into the classification of solids as metals, insulators, semiconductors, and semimetals. The remainder of this section is devoted to a discussion of this classification.

In the ground state of a solid there is no charge transport. This is so because the electronic states $\psi_{\mathbf{k}}$ and $\psi_{-\mathbf{k}}$ have the same energy, and if one is populated in the ground state the other is also. But the velocity in the state $\mathbf{k}$ is opposite to that in the state $-\mathbf{k}$, so that the net electronic current vanishes. Therefore the system must be elevated to an excited state if an electronic current is to flow.

If a sufficiently large electric field is applied to any crystal, the electrons can be excited into a state in which a net current flows. In diamond this occurs at about 10^8 volts/cm and is called dielectric breakdown. The distinction between a metal, such as copper, and an insulator, such as diamond, lies in the fact that an electronic current can be caused to flow in the metal by the application of an arbitrarily small electric field. There must exist in the metal excited states with arbitrarily small energies relative to the ground state. Or, in the set of one-electron functions, the Hartree-Fock eigenvalues ϵ_{N_e+1} must be extremely close to ϵ_{N_e} so that an electron can be excited easily from the highest filled state at $T = 0$ (that is, ψ_{N_e}) to the lowest unfilled state (ψ_{N_e+1}).

Consider a material composed of a single atom per unit cell, each atom carrying an odd number Z of electrons so that the total number of electrons is $N_e = ZN$. Each band in the crystal contains $2N$ states, where N is the number of unit cells. Because Z is odd, it follows that the ground state is one in which a number of low-lying bands are completely populated and one band is half populated. Above this lie completely empty bands. The situation is as shown in Fig. 2.4*a*, where the filled states are indicated by heavy shading. There exists empty states arbitrarily close to the filled states so that such a material is a metal. Although the monovalent metals Cu, Ag, and Au have a somewhat more complicated crystal structure than assumed in the above discussion, they nevertheless have half-filled bands quite analogous to the schematic representation of Fig. 2.4*a*.

In contrast consider a material composed of a single atom per unit cell, but with an even number Z of electrons per atom. We then arrive at a structure as indicated in Fig. 2.4*b*; that is, low-lying bands are entirely full, and higher-lying bands are entirely empty. The lowest excited state has an energy, relative to the ground state, equal to the width of the forbidden gap width separating the highest filled and lowest empty bands. This forbidden gap width is typically of the order of several electron volts. Such a material is an insulator. An example is diamond, which also has a more complicated crystal structure than described above, but for which the electronic states are nevertheless analogous to those in Fig. 2.4*b*.

It would appear on this basis that all materials in which the ground state contains a partially filled band would be metals, and all materials in which the ground state contains only completely filled and

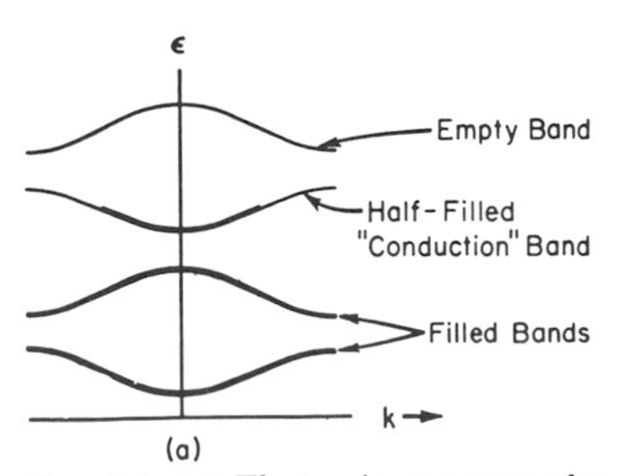

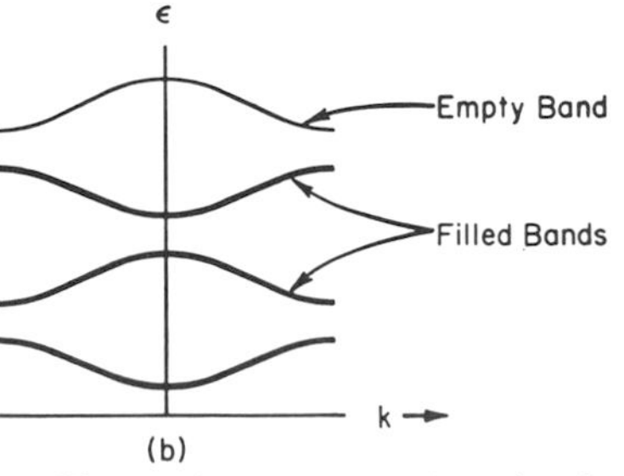

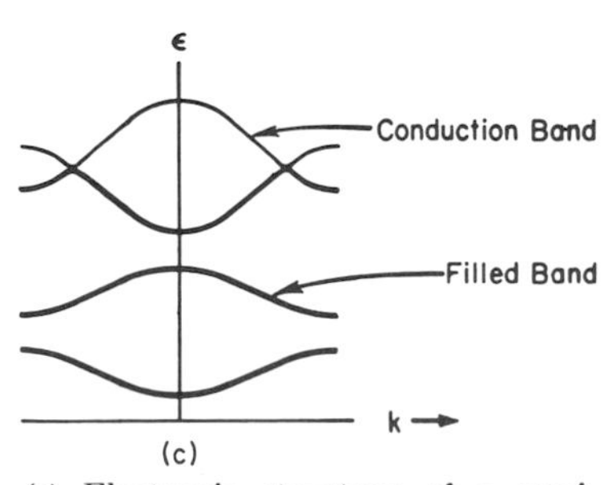

FIG. 2.4. (*a*) Electronic structure of a metal. (*b*) Electronic structure of an insulator or intrinsic semiconductor. (*c*) Electronic structure of a semimetal.

completely empty bands would be insulators. However, two special cases complicate this ideal situation and require special attention.

Consider a material in which there are an even number of electrons per atom, and for which an electronic structure such as that in Fig. 2.4*b* is expected. However, the bands may "accidentally" overlap, as indicated schematically in Fig. 2.4*c*. Such a material, which naïvely would be expected to be an insulator, is actually a conductor. However, the conductivity of such "semimetals" is much less than that of the true metals. This is so because the overlap of the bands is apt to occur only along certain directions in **k**-space, so that the number of excited states lying close to the ground state is less than that in a true metal. Examples of such semimetals are the divalent materials Mg and Zn.

The second special case of interest arises when the forbidden gap width separating the filled and empty bands in an insulator is very small. Such a material is an insulator at zero temperature, but at elevated temperatures it will be thermally excited into states which permit conductivity. The temperature at which this conductivity becomes appreciable is $kT \simeq E_g$, where E_g is the width of the forbidden gap. Such materials are called *intrinsic semiconductors*.

Finally, conductivity can be induced in a material which is nominally an insulator by the introduction of impurities. These impurities may either contribute electrons to a normally empty band or remove electrons from a normally filled band, thereby converting the insulator into an "impurity semiconductor."

7. Momentum Eigenfunctions

Two additional types of functions are closely related to the Bloch eigenfunctions and are very useful in the development of band theory. These are the momentum eigenfunctions [14] and the Wannier functions.† We shall now investigate the interrelationships among the three types of functions.

The Bloch eigenfunctions associated with the wave vector **k** and with some particular band are a product of the exponential factor $\exp(i\mathbf{k}\cdot\mathbf{r})$ and of a periodic function $X_\mathbf{k}(\mathbf{r})$. By virtue of its periodicity $X_\mathbf{k}(\mathbf{r})$ can be expanded in a Fourier series of exponentials $\exp(i\mathbf{K}\cdot\mathbf{r})$, where the **K** vectors assume all the values of the rationalized reciprocal lattice:

$$X_\mathbf{k}(\mathbf{r}) = \sum_\mathbf{K} \pi_\mathbf{k}(\mathbf{K}) \exp(i\mathbf{K}\cdot\mathbf{r}) \qquad (2.61)$$

The Fourier coefficients $\pi(\mathbf{K})$ are given by

$$\pi_\mathbf{k}(\mathbf{K}) = \frac{1}{v}\int X_\mathbf{k}(\mathbf{r}) \exp(-i\mathbf{K}\cdot\mathbf{r})\, dr \qquad (2.62)$$

where the integration extends over one unit cell, and where v is the volume of a unit cell. It follows from the above two equations that the Bloch function $\psi_\mathbf{k}(\mathbf{r}) = \exp(i\mathbf{k}\cdot\mathbf{r})X_\mathbf{k}(\mathbf{r})$ can be expanded in a Fourier series of exponentials $\exp[i(\mathbf{K}+\mathbf{k})\mathbf{r}]$:

$$\psi_\mathbf{k}(\mathbf{r}) = \sum_\mathbf{K} \pi(\mathbf{K}+\mathbf{k}) \exp[i(\mathbf{K}+\mathbf{k})\cdot\mathbf{r}] \qquad (2.63)$$

† This section and the following section draw heavily on the superb report by J. C. Slater (see ref. 14).

where

$$\pi(\mathbf{K}+\mathbf{k}) = \pi_\mathbf{k}(\mathbf{K}) = \frac{1}{v}\int \psi_\mathbf{k}(\mathbf{r}) \exp[-i(\mathbf{K}+\mathbf{k})\cdot\mathbf{r}]\, dr \qquad (2.64)$$

Accordingly, the wave function $\psi_\mathbf{k}(\mathbf{r})$ can be represented by the complete set of values of $\pi(\mathbf{K}+\mathbf{k})$ for each **K** in the reciprocal lattice. One value of π is associated thereby with each cell of the reciprocal lattice. In Fig. 2.5 the wave function $\psi_\mathbf{k}$ is represented by specification of π at each of the points indicated by small circles, and the wave function $\psi_{\mathbf{k}'}$ is represented by specification of π at each of the points indicated by crosses.

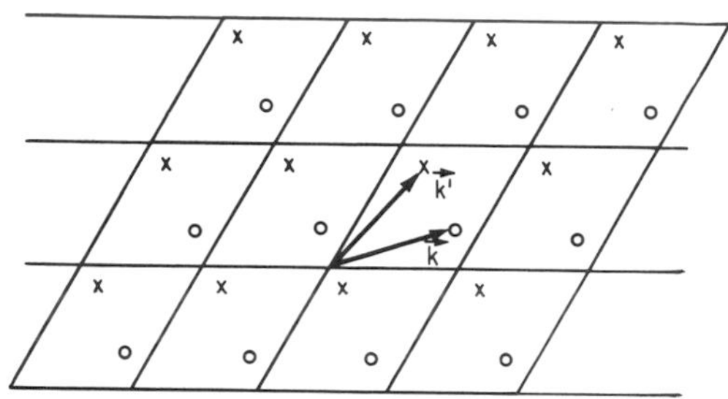

Fig. 2.5. Sets of points $(\mathbf{K}+\mathbf{k})$ and $(\mathbf{K}+\mathbf{k}')$ in reciprocal space.

By considering in turn each wave function $\psi_\mathbf{k}(\mathbf{r})$ in a given band, we obtain a succession of sets of values $\pi(\mathbf{K}+\mathbf{k})$. Taken together, these sets define a function $\pi(\mathbf{k})$ throughout reciprocal space. Actually, of course, $\pi(\mathbf{k})$ is defined only at the discrete points **k** of the reduced, rationalized, reciprocal lattice. But these points are so densely distributed (of the order of 10^{23} points in a single cell of the rationalized reciprocal lattice) that we can conveniently think of $\pi(\mathbf{k})$ as a continuous function in reciprocal space. Knowledge of $\pi(\mathbf{k})$ throughout reciprocal space (*not* simply within a single cell) is equivalent to the knowledge of all $\psi_\mathbf{k}(\mathbf{r})$ within a single band.

The physical significance of the function $\pi(\mathbf{k})$ is clear. The quantity $|\pi(\mathbf{K}+\mathbf{k})|^2$ is the probability that an electron in the state $\psi_\mathbf{k}(\mathbf{r})$ has the momentum $\hbar(\mathbf{K}+\mathbf{k})$. Thus $\pi(\mathbf{k})$ is the momentum eigenfunction.

There is one momentum eigenfunction for each band in the eigenvalue spectrum of the crystal. As to the orthogonality and normalization conditions obeyed by these momentum eigenfunctions, the Bloch functions $\psi_\mathbf{k}(\mathbf{r})$ are normalized within the periodicity region of the crystal. The allowed values of **k** were determined by applying periodic boundary conditions to a crystal in the shape of a parallelepiped with N_1 unit cells along the direction of $\mathbf{a}_1$, N_2 along $\mathbf{a}_2$, and N_3 along $\mathbf{a}_3$. It is in this region, containing $N_1N_2N_3 = N$ unit cells, that the $\psi_\mathbf{k}$ are normalized and orthogonal. Thus, writing the subscript n to label the bands with which $\psi_{\mathbf{k},n}$ is associated,

$$\int \psi^*_{\mathbf{k},n}\psi_{\mathbf{k}',n'}\, d\mathbf{r} = \delta_{\mathbf{k},\mathbf{k}'}\delta_{n,n'} \qquad (2.65)$$

where the integral extends over N unit cells. Various other normalizations are used; notably normalization within unit volume or normalization within a unit cell, and each has its particular advantages.

Inserting Eq. (2.63) into (2.65) gives the basic normalization and orthogonality relation for the momentum eigenfunctions:

$$\sum_{\mathbf{K}} \pi^*_n(\mathbf{K} + \mathbf{k})\pi_{n'}(\mathbf{K} + \mathbf{k}) = \frac{1}{Nv} \delta_{n,n'} \qquad (2.66)$$

Another form of the orthogonality relation can be obtained by integrating (2.66) over a unit cell of the rationalized reciprocal lattice. Combined with the summation over $\mathbf{K}$ on the left of (2.66) the integration over a unit cell becomes an integration over all reciprocal space. Recalling also that the volume of a unit cell of the rationalized reciprocal lattice is $(2\pi)^3/v$, we then obtain

$$\int_{\text{all }\mathbf{k}\text{-space}} \pi^*_n(\mathbf{k})\pi_{n'}(\mathbf{k})\, d\mathbf{k} = \frac{(2\pi)^3}{Nv^2} \delta_{n,n'} \qquad (2.67)$$

In either the sense of Eq. (2.66) or of (2.67) the momentum eigenfunctions of different bands thus are orthogonal.

The momentum eigenfunctions are the solutions of an eigenvalue equation in the momentum representation, just as the Bloch functions are solutions of an eigenvalue equation in the coordinate representation. This eigenvalue equation can be obtained directly by transforming the coordinate representation effective Schroedinger equation (2.24) to the momentum representation in the usual fashion. Assuming that the functions inside the integrals in Eq. (2.24) have been inserted explicitly, that equation can be written simply as

$$\tilde{H}\psi_{\mathbf{k}}(\mathbf{r}_1) = \left[-\frac{\hbar^2}{2m} \nabla_1{}^2 + \tilde{V}(\mathbf{r}_1) \right] \psi_{\mathbf{k}}(\mathbf{r}_1) = \epsilon(\mathbf{k})\psi_{\mathbf{k}}(\mathbf{r}_1) \qquad (2.68)$$

The momentum representation is then

$$\frac{\hbar^2 k^2}{2m} \pi(\mathbf{k}) + \sum_{\mathbf{K}} \tilde{v}(\mathbf{K})\pi(\mathbf{k} - \mathbf{K}) = \epsilon(\mathbf{k})\pi(\mathbf{k}) \qquad (2.69)$$

The quantities $\tilde{v}(\mathbf{K})$ in this equation are the Fourier components of the effective potential $\tilde{V}$. It is to be noted that Eq. (2.69) is a set of difference equations among the quantities $\pi(\mathbf{k} - \mathbf{K})$ for all $\mathbf{K}$. Solution of this set of equations yields the quantities $\pi(\mathbf{k} - \mathbf{K})$ for all $\mathbf{K}$, and thus specifies $\psi_{\mathbf{k}}$ by Eq. (2.63). In this way the set of difference equations (2.69) is completely equivalent to the differential equation (2.68).

8. The Wannier Function

The Fourier transform of the momentum eigenfunction is called the Wannier [15] function. We thus write

$$W(\mathbf{r}) = \frac{vN^{1/2}}{(2\pi)^3} \int \pi(\mathbf{k}) \exp(i\mathbf{k} \cdot \mathbf{r})\, d\mathbf{k} \qquad (2.70)$$

where the integral extends over all reciprocal space. The constant factor which appears before the integral above is chosen to yield a convenient normalization property for the Wannier function.

Here $\mathbf{k}$ is treated as a continuous variable despite the fact that $\mathbf{k}$ properly takes on only the values of the reduced, rationalized, reciprocal lattice. The discrete analogue of Eq. (2.70), which provides the more rigorous definition of the Wannier function, is

$$W(\mathbf{r}) = \frac{1}{N^{1/2}} \sum_{\mathbf{k}} \pi(\mathbf{k}) \exp(i\mathbf{k} \cdot \mathbf{r}) \qquad (2.71)$$

where the summation extends over all reciprocal space.

The Fourier transform (2.70) or (2.71) can be inverted to express the momentum eigenfunction in terms of the Wannier function:

$$\pi(\mathbf{k}) = \frac{1}{N^{1/2}v} \int W(\mathbf{r}) \exp(-i\mathbf{k} \cdot \mathbf{r})\, d\mathbf{r} \qquad (2.72)$$

where the integration extends over N unit cells.

With each band in the crystal there is associated a single momentum eigenfunction, and thence a single Wannier function. The Wannier function, like the momentum eigenfunction, should carry a subscript index denoting the band with which it is associated, but this index is frequently omitted. The Wannier function determines the momentum eigenfunction by Eq. (2.72) and the momentum eigenfunction determines the entire set of N Bloch functions in the band by Eq. (2.63). A simple relation enables one to obtain the N Bloch functions directly from the Wannier function without employing the momentum eigenfunction in an intermediate step.

The form of Eq. (2.71) provides an insight into the form of the Wannier function. The Wannier function is a wave-packet, built of the plane waves $\exp(i\mathbf{k} \cdot \mathbf{r})$, each with the amplitude $(1/N^{1/2})\pi(\mathbf{k})$. We expect such a wave-packet to have a small extension in $\mathbf{r}$-space if $\pi(\mathbf{k})$ has a large extension in $\mathbf{k}$-space, and vice versa. Referring back to Eqs. (2.63) and (2.64), we see that $\pi(\mathbf{k})$ is restricted to one cell of the reciprocal lattice if all the $\psi_{\mathbf{k}}$ are plane waves, but the $\pi(\mathbf{k})$ will necessarily be nonzero over large regions of $\mathbf{k}$-space if the $\psi_{\mathbf{k}}$ deviate greatly from plane waves. We may reasonably expect the latter to be the case, whence the Wannier functions may be expected to be a highly localized wave packet. This is indeed the case. The Wannier functions for the lowest lying bands have radii very nearly equal to the radii of the corresponding atomic orbitals, and consequently extend over less than a single unit cell. The Wannier functions of the upper or valence bands generally extend over several unit cells.

For many purposes it is convenient to define an entire set of Wannier functions to be associated with a single band. This set is obtained from the single Wannier function by translation with any of the N vectors $\mathbf{R}$ of the lattice. We thus have N Wannier functions defined by

$$W(\mathbf{r} - \mathbf{R}) = \frac{1}{N^{1/2}} \sum_{\mathbf{k}} \pi(\mathbf{k}) e^{i\mathbf{k} \cdot (\mathbf{r} - \mathbf{R})} \qquad (2.73)$$

The orthogonality and normalization properties of the Wannier functions can be related through

Eq. (2.4) to the orthogonality and normalization properties of the momentum eigenfunctions, as given in Eq. (2.66). Thus

$$\int W^*_n(\mathbf{r}-\mathbf{R})W_{n'}(\mathbf{r}-\mathbf{R}')\,d\mathbf{r} = \delta_{n,n'}\delta_{\mathbf{R},\mathbf{R}'} \tag{2.74}$$

where the integral extends over N unit cells. The Wannier functions are therefore normalized over N unit cells. Furthermore, a given Wannier function is orthogonal to any of the other $(N-1)$ Wannier functions associated with the same band as well as to any Wannier function associated with any other band.

We recall that the set of N Bloch orthonormal functions associated with each band, taken together for all bands, constitute a complete orthonormal set. Similarly, the set of N Wannier functions associated with each band, taken together for all bands, constitutes another complete orthonormal set.

The set of N Wannier functions associated with a given band is related to the set of N Bloch functions associated with that band by an orthogonal transformation. That is, any Wannier function can be expressed as a linear combination of the N Bloch functions, and inversely. This follows by comparison of Eqs. (2.63) and (2.73), whence

$$W(\mathbf{r}-\mathbf{R}) = \frac{1}{N^{1/2}}\sum_{\mathbf{k}} \psi_{\mathbf{k}}(\mathbf{r}) \exp(-i\mathbf{k}\cdot\mathbf{R}) \tag{2.75}$$

where the summation is over the N values of $\mathbf{k}$ within one unit cell of the rationalized reciprocal lattice. Inverting the above equation,

$$\psi_{\mathbf{k}}(\mathbf{r}) = \frac{1}{N^{1/2}}\sum_{\mathbf{R}} W(\mathbf{r}-\mathbf{R}) \exp(i\mathbf{k}\cdot\mathbf{R}) \tag{2.76}$$

A convenient integral analogue of Eq. (2.75), in which $\mathbf{k}$ is regarded as a continuous rather than a discrete variable, is

$$W(\mathbf{r}-\mathbf{R}) = \frac{vN^{1/2}}{(2\pi)^3}\int \psi_{\mathbf{k}}(\mathbf{r}) \exp(-i\mathbf{k}\cdot\mathbf{R})\,d\mathbf{k} \tag{2.77}$$

where the integral extends over one unit cell of the rationalized reciprocal lattice.

Although the Bloch and Wannier functions are thus related by simple linear transformations, each have unique and specific advantages. The central advantage of the Bloch functions is that they satisfy the Hartree-Fock equations directly, whereas the Wannier functions do not. The essential features of the Wannier functions which make them useful are their localization, orthogonality, and completeness, and the fact that a single Wannier function characterizes a complete band. An alternative statement of this latter fact is that all N Wannier functions of a given band are generated from any single one by simple translation operations.

Since the Wannier functions are not solutions of the Hartree-Fock equations, the effective Hamiltonian of Eq. (2.68) will have nonvanishing matrix elements between different Wannier functions. The values of these matrix elements are of considerable importance. Consider first the quantity $\tilde{H}W(\mathbf{r}-\mathbf{R})$:

$$\begin{aligned}\tilde{H}W(\mathbf{r}-\mathbf{R}) &= \frac{1}{N^{1/2}}\sum_{\mathbf{k}} \tilde{H}\psi_{\mathbf{k}}(\mathbf{r}) \exp(-i\mathbf{k}\cdot\mathbf{R}) \\ &= \frac{1}{N^{1/2}}\sum_{\mathbf{k}} \epsilon(\mathbf{k})\psi_{\mathbf{k}}(\mathbf{r}) \exp(-i\mathbf{k}\cdot\mathbf{R})\end{aligned} \tag{2.78}$$

Now, writing $\psi_{\mathbf{k}}(\mathbf{r})$ in terms of the Wannier functions once again,

$$\tilde{H}W(\mathbf{r}-\mathbf{R}) = \frac{1}{N}\sum_{\mathbf{R}'}\left\{\sum_{\mathbf{k}} \epsilon(\mathbf{k}) \exp[i\mathbf{k}\cdot(\mathbf{R}'-\mathbf{R})]\right\} W(\mathbf{r}-\mathbf{R}') \tag{2.79}$$

or
$$\tilde{H}W(\mathbf{r}-\mathbf{R}) = \sum_{\mathbf{R}'} \epsilon(\mathbf{R}'-\mathbf{R})W(\mathbf{r}-\mathbf{R}') \tag{2.80}$$

where $\epsilon(\mathbf{R})$ is the Fourier transform of $\epsilon(\mathbf{k})$. That is, $\epsilon(\mathbf{k})$ and $\epsilon(\mathbf{R})$ are related by the equations

$$\epsilon(\mathbf{R}) = \frac{1}{N}\sum_{\mathbf{k}} \epsilon(\mathbf{k}) \exp(i\mathbf{k}\cdot\mathbf{R}) \tag{2.81}$$

and inversely

$$\epsilon(\mathbf{k}) = \sum_{\mathbf{R}} \epsilon(\mathbf{R}) \exp(-i\mathbf{k}\cdot\mathbf{R}) \tag{2.82}$$

Equation (2.74) now permits us to find the desired matrix elements. Multiplying (2.80) by $W^*_{n'}(\mathbf{r}-\mathbf{R}')$ and integrating over the basic N unit cells gives

$$\int W^*_{n'}(\mathbf{r}-\mathbf{R}')\tilde{H}W_n(\mathbf{r}-\mathbf{R})\,d\mathbf{r} = \epsilon(\mathbf{R}'-\mathbf{R})\delta_{n,n'} \tag{2.83}$$

Thus, the matrix element of the effective Hamiltonian between Wannier functions in different bands vanishes, and the matrix element of the Hamiltonian between Wannier functions in the same band, centered a distance $\mathbf{R}'-\mathbf{R}$ apart, is just the corresponding Fourier component of the energy function.

As the internuclear distance is increased in a crystal, the solutions of the electronic eigenproblem obviously approach the isolated atomic orbitals. The energy bands become narrow and approach the discrete eigenvalues of the isolated atoms. As this happens $\epsilon(\mathbf{k})$ approaches a constant value independent of $\mathbf{k}$, and, by Eq. (2.81), $\epsilon(\mathbf{R})$ approaches zero for all $\mathbf{R}$ other than $\mathbf{R}=0$. The matrix element between different Wannier functions then vanishes, by Eq. (2.83). This means that the Wannier functions became solutions of the Hartree-Fock equations; they become identical to the atomic orbitals.

For any internuclear distance the Wannier functions maintain a close similarity to the atomic orbitals. For large internuclear distance they are identical. As the nuclei approach, the Wannier functions remain similar to the atomic orbitals in the regions very close to the nuclei, but their form is altered on the "fringes" so as to maintain orthogonality even as the functions overlap. The similarity to the atomic orbitals and the maintenance of orthogonality are the key features in the role of the Wannier functions in electronic band theory.

9. Perturbations of Periodicity

Perfectly periodic crystals are seldom the direct object of interest. Thermal vibrations of the nuclei insure that the perfectly periodic crystal is never realized. Further, the effect upon the crystal of an external electric field or of an incident radiation field imposes upon the crystal an additional potential which destroys the periodicity. The Wannier functions provide the formal instrument for the analysis of the electronic structure in the presence of such deviations from periodicity [15–17].

The manner in which the Wannier functions provide a solution of the perturbed problem may be visualized easily. According to Eq. (2.76) the electronic wave functions in a periodic potential are sums of Wannier functions centered at each lattice site. The coefficients of each of these Wannier functions have the same modulus, so that the electron has equal probability of being found in each of the Wannier orbitals. Now suppose that the perturbative potential varies so slowly that it is practically unchanged over the diameter of any single Wannier orbital. Then we would expect that each Wannier orbital will remain unaffected, but that the electron will no longer have equal probability of being found in each orbital. Rather, the electron should have large probability of being in the Wannier orbitals in the region where the perturbative potential is low, and small probability of being in those where the perturbative potential is high. Thus we expect that the true wave function can be expressed as a linear combination of Wannier orbitals, and the coefficients will depend upon the perturbative potential.

Now consider the wave function of a particle moving under the influence of the perturbative potential alone, without the underlying periodic potential. The wave function of such a particle is large where the perturbative potential is low, and small where the potential is high. This is just the qualitative behavior which we were lead to expect above for the coefficients of the Wannier functions in the expansion of the wave function in the perturbed crystal. The problem of finding those coefficients can be related to the solution of a hypothetical particle moving under the influence of the perturbative potential alone.

Consider first the periodic potential problem of Eq. (2.68), for which the Hamiltonian is

$$\tilde{H} = -\frac{\hbar^2}{2m}\nabla^2 + \hat{V}(\mathbf{r}) \qquad (2.84)$$

and the eigenfunctions are $\psi_{\mathbf{k}}(\mathbf{r})$, with the eigenvalues $\epsilon(\mathbf{k})$. Assume that the potential is altered by the additional potential $V_1(r)$. This perturbative potential is a complicated quantity, which must be evaluated by a self-consistent solution of the Hartree-Fock equations in the manner in which Eq. (2.68) itself emerged from Eq. (2.24). This vexatious problem has nowhere yet received a satisfactory treatment. We merely investigate the solutions of the perturbed Hamiltonian $\tilde{H} + V_1$, assuming it to be known. Then

$$(\tilde{H} + V_1)\phi = E\phi \qquad (2.85)$$

where $\phi(\mathbf{r})$ is a solution of the perturbed problem and E is the corresponding eigenvalue.

Expanding ϕ in the complete set of Wannier functions of all bands,

$$\phi = \sum_{n,\mathbf{R}} \phi_n(\mathbf{R}) W_n(\mathbf{r} - \mathbf{R}) \qquad (2.86)$$

The double sum here goes over all bands and over the N values of $\mathbf{R}$ in the fundamental region of the crystal. The coefficients $\phi_n(\mathbf{R})$ remain to be determined.

Substituting (2.86) into (2.85), multiplying by $W^*_{n'}(\mathbf{r} - \mathbf{R}')$, integrating over N cells, and employing Eq. (2.83) for the matrix elements of $\tilde{H}$,

$$\sum_{\mathbf{R}} \epsilon_{n'}(\mathbf{R}' - \mathbf{R})\phi_{n'}(\mathbf{R}) + \sum_{n,\mathbf{R}} \langle n',\mathbf{R}'|V_1|n,\mathbf{R}\rangle \phi_n(\mathbf{R}) = E\phi_{n'}(\mathbf{R}') \qquad (2.87)$$

This is the basic equation for the Wannier coefficients $\phi_n(\mathbf{R})$. Solution of this set of equations would yield the rigorous solution for the electronic states in the presence of any perturbation.

Because of the importance of the above equation for the Wannier coefficients it is of interest to survey the method which it provides for solution of the perturbed problem. We are given a problem with an effective Hamiltonian $\tilde{H} + V_1$, where $\tilde{H}$ refers to a periodic potential and V_1 describes a deviation from periodicity. Then we first solve the periodic problem $\tilde{H}\psi_{\mathbf{k}} = \epsilon(\mathbf{k})\psi_{\mathbf{k}}$, obtaining the Bloch functions and the eigenvalue spectrum $\epsilon(\mathbf{k})$. From the Bloch functions we then compute Wannier functions by Eq. (2.75). We also compute the matrix elements $\langle n',\mathbf{R}'|V_1|n,\mathbf{R}\rangle$ of the nonperiodic potential V_1 between pairs of Wannier functions. We also compute the Fourier transform $\epsilon(\mathbf{R})$ of $\epsilon(\mathbf{k})$ according to Eq. (2.81). With the matrix elements

$$\langle n',\mathbf{R}'|V_1|n,\mathbf{R}\rangle$$

and the quantities $\epsilon(\mathbf{R}' - \mathbf{R})$ thus known we undertake the solution of Eq. (2.87) for the $\phi_n(\mathbf{R})$. Finally, inserting these coefficients $\phi_n(\mathbf{R})$ into Eq. (2.86) we obtain the solution of the given problem.

Equation (2.87) provides the rigorous solution to the perturbed problem, but it does not exhibit clearly the intuitive features which the previous qualitative discussion suggested. Furthermore, Eq. (2.87) proves to be awkward to solve in real cases of interest. We shall now show that a simple approximation to Eq. (2.87) can be contrived, and that this approximation at once brings into sharp focus all the intuitive features which we have discussed, and that it also provides a more tractable and convenient analytic form of the problem. Unfortunately the approximations which we shall now exhibit are not frequently justifiable in practice, so that after all one is often forced into a brute force attack on Eq. (2.87).

There is a close relationship between the Wannier coefficients $\phi_n(\mathbf{R})$ and the wave function of a hypothetical particle moving under the influence of the perturbative potential $V_1(\mathbf{r})$ alone. To bring out this relationship, consider $\phi_n(\mathbf{R})$ to be a function of a continuous rather than a discrete variable, and convert the difference equation (2.87) into a differential equation. We assume that the nondiagonal matrix elements of V_1 all vanish, so that $<n',\mathbf{R}'|V_1|n,\mathbf{R}>$ is nonzero only if $n' = n$ and $\mathbf{R}' = \mathbf{R}$. It is this assumption which most frequently invalidates our approximation, so that it is well to inquire when it may reasonably be expected to be true. The Wannier functions are themselves orthogonal. Therefore if V_1 were a constant over the region of the two Wannier functions involved in the matrix element $<n'\mathbf{R}'|V_1|n,\mathbf{R}>$, the matrix element would vanish unless $n' = n$ and $\mathbf{R}' = \mathbf{R}$. Furthermore, because of the localization of the Wannier orbitals the matrix element will always vanish if $|\mathbf{R}' - \mathbf{R}|$ is large compared to the diameter of a Wannier orbital. Thus the nondiagonal matrix elements of V_1 will vanish if V_1 is so slowly varying that it is essentially constant over the diameter of a Wannier orbital. The approximation to be developed is restricted to very slowly varying perturbative potentials.

With our assumption of the vanishing of the nondiagonal matrix elements of V_1, Eq. (2.87) becomes

$$\sum_{\mathbf{R}} \epsilon(\mathbf{R}' - \mathbf{R})\phi(\mathbf{R}) + <\mathbf{R}'|V_1|\mathbf{R}'>\phi(\mathbf{R}') = E\phi(\mathbf{R}') \tag{2.88}$$

An "averaged perturbative potential" $\bar{V}(\mathbf{R}')$ is defined by

$$\bar{V}(\mathbf{R}') = <\mathbf{R}'|V_1|\mathbf{R}'> \tag{2.89}$$

Thus the value of the "averaged perturbative potential" at the point $\mathbf{R}'$ is the average value of $V_1(\mathbf{r})$ weighted with the probability distribution $|W(\mathbf{r} - \mathbf{R}')|^2$ of the Wannier wave function centered at $\mathbf{R}'$.

Considering $\mathbf{R}'$ and $\mathbf{R}$ as continuous variables, we can write

$$\phi(\mathbf{R}) = \exp\,[(\mathbf{R} - \mathbf{R}') \cdot \nabla]\phi(\mathbf{R}') \tag{2.90}$$

where $\exp\,[(\mathbf{R} - \mathbf{R}') \cdot \nabla]$ is a translation operator. Then Eq. (2.88) becomes

$$\sum_{\mathbf{R}} \epsilon(\mathbf{R}' - \mathbf{R}) \exp\,[(\mathbf{R} - \mathbf{R}') \cdot \nabla]\phi(\mathbf{R}') + \bar{V}(\mathbf{R}')\phi(\mathbf{R}') = E\phi(\mathbf{R}') \tag{2.91}$$

or

$$\left\{\sum_{\mathbf{R}} \epsilon(\mathbf{R}' - \mathbf{R}) \exp\,[-i(\mathbf{R}' - \mathbf{R}) \cdot \mathbf{k}_{op}]\right\} \phi(\mathbf{R}') + \bar{V}(\mathbf{R}')\phi(\mathbf{R}') = E\phi(\mathbf{R}') \tag{2.92}$$

where

$$\mathbf{k}_{op} \equiv -i\nabla \tag{2.93}$$

The expression in braces of Eq. (2.92) is a differential operator acting on $\phi(\mathbf{R}')$. By comparison with Eq. (2.82) this operator may be written simply as $\epsilon(\mathbf{k}_{op})$. That is, it may be obtained from $\epsilon(\mathbf{k})$ by replacing $\mathbf{k}$ by $\mathbf{k}_{op}$, as defined by Eq. (2.93). In the Schroedinger representation of quantum mechanics the momentum $\mathbf{p}$ is replaced by $-i\hbar\nabla$, and the wave vector $\mathbf{k}$, defined by $\hbar\mathbf{k} \equiv \mathbf{p}$, is therefore replaced by $-i\nabla$. Although we do not now have the identification of $\hbar\mathbf{k}$ as the momentum in a crystal, the association of the wave vector $\mathbf{k}$ with the operator $-i\nabla$, indicated in Eq. (2.93), is just that which is conventional in the Schroedinger quantum mechanics. Dropping the primes on $\mathbf{R}'$,

$$\epsilon(\mathbf{k}_{op})\phi(\mathbf{R}) + \bar{V}(\mathbf{R})\phi(\mathbf{R}) = E\phi(\mathbf{R}) \tag{2.94}$$

Equation (2.94) has the form of a conventional Schroedinger equation for the function $\phi(\mathbf{R})$. The energy $\epsilon(\mathbf{k})$, obtained from solution of the unperturbed problem, plays the role of the unperturbed Hamiltonian. The perturbed Hamiltonian is $\epsilon(\mathbf{k}) + \bar{V}(\mathbf{R})$. This Hamiltonian is converted into a Schroedinger operator by the operator association $\mathbf{k} \rightarrow -i\nabla$. The associated eigenvalue problem is Eq. (2.94).

Although Eq. (2.94) already provides an appealingly intuitive formulation of the problem, an even more drastic simplification can be made. Consider a dependence of ϵ on $\mathbf{k}$ such as depicted schematically in the lowest band in Fig. 2.2. Assume also that the electrons of interest have small energy and are therefore located near the minimum of $\epsilon(\mathbf{k})$. In this region $\epsilon(\mathbf{k})$ can be represented by a power series in $\mathbf{k}$, and over a sufficiently small range of $\mathbf{k}$ the power series may be terminated with the quadratic terms. Furthermore, the linear terms vanish, because $\epsilon(\mathbf{k})$ is minimum, and the zero-order term may be made zero by measuring the energy from the bottom of the band. Then $\epsilon(\mathbf{k})$ is represented simply by quadratic terms, which can be written with the aid of the effective mass tensor of Eq. (2.60) as

$$\epsilon(\mathbf{k}) = \frac{\hbar^2}{2}\mathbf{k} \cdot \frac{1}{m^*} \cdot \mathbf{k} \tag{2.95}$$

With this simplification Eq. (2.94) becomes

$$\frac{\hbar^2}{2}\mathbf{k}_{op} \cdot \frac{1}{m^*}\mathbf{k}_{op}\phi(\mathbf{R}) + \bar{V}(\mathbf{R})\phi(\mathbf{R}) = E\phi(R) \tag{2.96}$$

This is the basic equation of the "effective mass approximation" which is used very widely in solid state theory. Long before the justification of the equation was known the equation was introduced on intuitive grounds and widely used. In most such uses the effective mass has been taken as a spherical tensor, and the scalar value has been adjusted to fit experimental data.

As an illustration of the application of Eq. (2.96) the treatment of impurity states in germanium [18] is of interest. Suppose a Ge ion is replaced by an As ion in a germanium crystal. The As nucleus carries one more nuclear charge than the Ge nucleus which it replaces, and superimposes on the periodic potential the additional Coulomb potential of a single positive electronic charge. The wave functions of all the electrons adjust to this additional potential. The potential energy $V_1(\mathbf{r})$, which is to be inserted in Eq. (2.85), is the self-consistent potential of a single electron due to the additional nuclear charge and to the altered interaction of the electron with all other electrons. The perturbative potential energy induced by the additional nuclear charge alone would be $-e^2/r$. To account for the effect of the remaining electrons, Kittel and Mitchell simply insert the dielectric constant ϵ_0, taking $V_1(r)$ as $-e^2/\epsilon_0 r$. Furthermore, Eq. (2.96) is employed, even though the perturbative potential is not really slowly varying; and to this degree of approximation no distinction is made between $V_1(\mathbf{r})$ and the averaged perturbative potential $\bar{V}(\mathbf{R})$. Then choosing coordinate axes along the principal directions of the effective mass tensor, Eq. (2.96) becomes

$$\frac{\hbar^2}{2}\left[\frac{1}{m^*_1}k_1^2 + \frac{1}{m^*_2}k_2^2 + \frac{1}{m^*_3}k_3^2\right]\phi(\mathbf{R}) - \frac{e^2}{\epsilon_0 R}\phi(\mathbf{R}) = E\phi(\mathbf{R}) \quad (2.97)$$

Cyclotron resonance experiments in germanium indicate that $m^*_1 \simeq 1.58$ m and $m^*_2 = m^*_3 = 0.082$ m. The dielectric constant ϵ_0 is about 16. Thus all the parameters appearing in Eq. (2.97) are known, and the lowest eigenvalue can be found. Kittel and Mitchell use a variational method, with a trial function

$$\phi = \left[\frac{ab^2}{\pi r_0^3}\right]^{1/2} \exp\left\{-\frac{[a^2x^2 + b^2(y^2 + z^2)]^{1/2}}{r_0}\right\} \quad (2.98)$$

where

$$r_0 = \frac{\epsilon_0\hbar^2}{me^2} \quad (2.99)$$

Minimizing the expectation value of the energy with respect to the variational parameters a and b yields the binding energy

$$-E = 0.0090 \text{ ev} \quad (2.100)$$

The observed binding energy of an electron in the field of an As donor impurity is 0.0127 ev, so that the agreement is fair.

Returning now to the general aspects of the Wannier method for the perturbed periodic problems, we see from Eq. (2.86) that the solutions in the perturbed lattice are linear combinations of the Wannier functions, and hence, by Eq. (2.75), of the unperturbed Bloch functions. Now suppose that a localized impurity is introduced into the lattice, and that solution of the associated perturbed problem gives a localized function $\phi(\mathbf{R})$, describing an electron tightly bound to the impurity. Suppose that this solution has an eigenvalue lying in the "forbidden gap," just below or just above a particular energy band, and that the perturbed wave function is a linear combination of the Bloch functions associated with the band. Then it follows that the number of independent functions remaining in the band is reduced by one. That is, the discrete eigenvalues introduced into the forbidden gap in the perturbed problem are "split off" the unperturbed bands, and reduce the number of states remaining in the band. This situation is represented schematically in Fig. (2.6).

The splitting off of the impurity levels from the unperturbed bands has important consequences in semiconductor theory. Consider the introduction of gallium into germanium as a substitutional impurity. The atom of Ga has a nuclear charge which is one less than that of Ge, and carries one less electron. Because of the deficiency of nuclear charge the Ga acts as a repulsive center for electrons, or an attractive center for holes. The Wannier construction gives a level just above the filled band, as in Fig. (2.6). Germanium itself is an intrinsic semiconductor, with just sufficient electrons to fill the lower unperturbed band in Fig. 2.6, leaving none to go into the upper band. With one fewer electrons because of the substitution of one Ga atom for a Ge atom, once again the lower perturbed band is exactly filled. The important point is that because the perturbed band contains only $(N - 1)$ states the discrete level is left empty. At temperatures which are still insufficient to excite an electron from one band to another, an electron can now be excited from the filled band into the empty discrete impurity state, or "acceptor level." In this process a hole is created in the filled band, and electrical conductivity is permitted. Introduction of the Ga impurity converts the Ge into a "p-type impurity semiconductor."

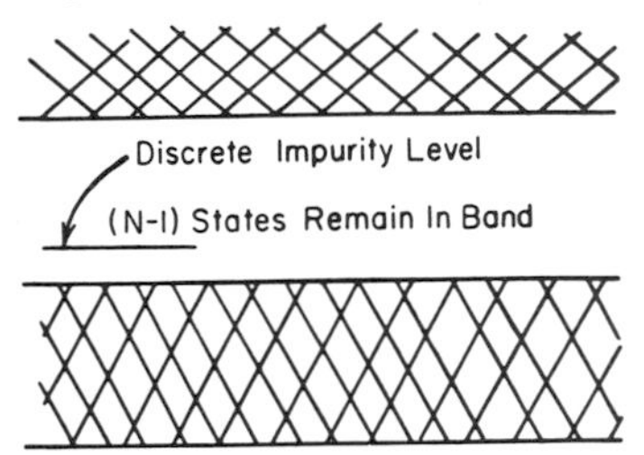

Fig. 2.6. Splitting off of impurity level from an allowed band.

10. Techniques of Calculation

The preceding nine sections have presented the "pattern" of the band theory of solids. The discussion has centered on the Bloch functions as solutions of the Hartree-Fock equations, on the properties of the Bloch functions, and on certain other functions (the momentum eigenfunctions and the Wannier functions) which are closely related to the Bloch functions. Because of the central role of the Bloch functions we consider the actual methods commonly employed in the numerical calculation of these functions [1].

The great difficulty inherent in the Hartree-Fock equations is that they are nonlinear coupled equations rather than linear differential equations of the Schroedinger type. In Sec. 4 we used a ruse to apply Schroedinger-type experience to study the consequences of translational symmetry of the Hartree-Fock equations. This same ruse is the key to the numerical solution of the equations. The method is that of the "self-consistent field."

The essential idea of the self-consistent field method is to approach the solution of the Hartree-Fock equations by an iterative process. That is, a set of crude solutions is first guessed, and the Hartree-Fock equations are used to improve this guess. By sufficiently many applications of this procedure we can, presumably, obtain solutions to any desired accuracy.

Consider the Hartree-Fock equations (2.13). As a crude approximation to the solution of these equations we exercise our ingenuity, physical intuition, and knowledge of the crystal symmetry to contrive a set of N_e orthonormal functions of the Bloch form, which we designate as $\psi_k^{(0)}$, $k = 1,2, \ldots ,N_e$. By inserting these functions into the integrands in the Hartree-Fock equations we obtain a Schroedinger-type equation, with an effective Hamiltonian which we designate by $\tilde{H}^{(0)}$. The Hartree-Fock equations thereby become

$$\tilde{H}^{(0)}\psi_k^{(1)} = \epsilon_k^{(1)}\psi_k^{(1)} \quad (2.101)$$

where the solutions are denoted by the superscript (1). These are the "first iterated functions," and they are supposedly better approximations to the true solutions than are the "zero iterates" $\psi_k^{(0)}$. Solution of the Schroedinger-like equation (2.101) actually yields a complete orthonormal set of functions $\psi_k^{(1)}$, of which only the lowest N_e are of direct interest here.

Inserting the first iterated functions into the integrands of the Hartree-Fock equations now provides a new effective Hamiltonian $\tilde{H}^{(1)}$, from which we obtain a set of "second iterated" functions by solving the Schroedinger-like equation

$$\tilde{H}^{(1)}\psi_k^{(2)} = \epsilon_k^{(2)}\psi_k^{(2)} \tag{2.102}$$

Again, we can recompute a new Hamiltonian $\tilde{H}^{(2)}$, and repeat the process. With n iterations we have the equation

$$\tilde{H}^{(n)}\psi_k^{(n+1)} = \epsilon_k^{(n+1)}\psi_k^{(n+1)} \tag{2.103}$$

The crucial test of the technique is a comparison of the nth and $(n + 1)$st iterates. If $\psi_k^{(n+1)}$ is approximately equal to $\psi_k^{(n)}$ (for all $k \leq N_e$), we clearly have obtained an approximate solution of the Hartree-Fock equations.

In actual practice the method of many successive iterations and corroboration of convergence by comparison of successive iterates is seldom used. Real calculations are most frequently made by carrying out only a single iteration. The set of functions $\psi_k^{(1)}$ is accepted more or less on faith as a reasonable approximation to the true solution. The functions $\psi_k^{(1)}$ almost always differ drastically from the zero iterates $\psi_k^{(0)}$, but one optimistically assumes (or hopes) that the second iterates $\psi_k^{(2)}$, if obtained, would be very nearly equal to the first iterates.

The method of the self-consistent field is but an outline of a method, rather than a fully explicit recipe. In particular we have stated merely that the Schroedinger-like equations $\tilde{H}^{(n-1)}\psi_k^{(n)} = \epsilon_k^{(n)}\psi_k^{(n)}$ "are solved" to yield the nth iterated set of functions $\psi_k^{(n)}$. A number of ingenious techniques have been devised to carry out this step in the procedure.

The earliest technique for obtaining approximate solutions of the Schroedinger-like equation (2.101) is the "tight binding approximation," introduced by Bloch [19] in a paper which essentially launched the modern theory of solids. The method can now be understood most easily by reference to Eq. (2.76), in which the electronic wave function is written in terms of the Wannier function. We recall also that if the nuclei are widely separated the Wannier function approaches a single free atomic orbital. The tight binding approximation consists in writing the Wannier function as a sum of several low-lying atomic orbitals, with undetermined coefficients. To determine the coefficients, the expression (2.76) is inserted into Eq. (2.101), and a secular equation is obtained. Solution of this secular equation by standard algebraic methods yields the unknown coefficients and the Hartree-Fock energy eigenvalue. For low-lying bands the Wannier function is well represented by a single atomic orbital, whereas an adequate representation of the Wannier functions of the higher bands generally requires several atomic orbitals. Of the many applications of the method we cite only the work of Fletcher and Wohlfarth [20] on nickel.

The bound-electron method is very unwieldy in practice. The formal difficulty arises from the fact that the atomic wave functions situated on neighboring atoms are not orthogonal; this gives rise to enormous numbers of overlap integrals in the nondiagonal elements of the secular determinant. Slater [14] has proposed a modification of the bound-electron method, in which the method is used as an interpolation scheme rather than as a true method of calculation. Slater merely treats the atomic orbitals as if they were orthogonal. He then carries out a tight binding calculation in a symbolic, formal fashion, without inserting the explicit form of the atomic orbitals. His results then depend upon certain matrix elements that have not been explicitly computed. Rather than insert the explicit form of the orbitals to compute these matrix elements, Slater treats the matrix elements as adjustable constants to be evaluated by comparison with empirical data, or with other more accurate but less complete calculations. In particular, he might employ empirical data on forbidden gap widths and principal effective mass values to evaluate the uncomputed matrix elements. The calculation then yields the full dependence of ϵ on $\mathbf{k}$. Slater's application of the method to various cubic structures, and in particular to diamond and germanium, yields results [14] which compare quite well with very much more pretentious calculations.

Another method upon which a great deal of early work is based was introduced by Wigner and Seitz [21] in 1933. This is the Wigner-Seitz cellular method. Because the wave functions are of the Bloch form they are completely determined if they are known within a single unit cell of the crystal. Thus the problem is to solve Eq. (2.101) within one unit cell. The Bloch form of the wave functions provides the boundary conditions

$$\psi(\mathbf{r}_s + \mathbf{R}) = \exp(i\mathbf{k}\cdot\mathbf{R})\psi(\mathbf{r}_s) \tag{2.104}$$

where $\mathbf{r}_s$ is a point on the boundary of the cell and where $\mathbf{r}_s + \mathbf{R}$ is on the opposite boundary.

Rather than choosing the crystalline unit cell as the fundamental region, we may clearly choose an alternate region of the same volume, provided that no two points within this volume differ by a vector of the lattice. We may choose a fundamental zone which bears precisely the same relationship to the real lattice as does the Brillouin zone to the reciprocal lattice. This zone is the "atomic polyhedron." The boundary condition (2.104) applies equally well to the atomic polyhedron.

The Schroedinger-like equation (2.101) is not generally separable in the nonorthogonal coordinates of the $\mathbf{a}_k$ vectors, and separability is almost our only hope in the solution of such partial differential equations. The essential Wigner-Seitz simplification consequently is to replace the atomic polyhedron by a spherical region of the same volume, and to adopt an approximation to the effective potential in Eq. (2.101) which is spherically symmetric. With these simplifications the Schroedinger-like equation becomes separable. The solution can be represented as a series of the form

$$\psi = \sum_{nlm} a_{nlm}R_nY_l^m \tag{2.105}$$

where the Y_l^m are spherical harmonics and the R_n are solutions of the radial equation. It is customary to take only a few terms of this series and to determine the coefficients by imposing the boundary conditions at a corresponding number of discrete points on the surface of the spherical region.

A marked improvement of the Wigner-Seitz method has been suggested by von der Lage and Bethe [22] in a paper which makes an explicit application to metallic sodium. These authors introduce certain linear combinations of the spherical harmonics, obtaining new angular functions which have the symmetry of the atomic polyhedra. For a cubic crystal the appropriate linear combinations are called the "Kubic Harmonics." By making an expansion similar to Eq. (2.105), in terms of the new harmonics instead of the spherical harmonics, it is possible to obtain equivalent results with many less terms in the truncated series. The shortcoming of the cellular method lies in the fact that it ignores the detailed structure of the solid, the entire calculation being limited to a single atomic polyhedron. Furthermore, the wave function is treated crudely (by approximate boundary conditions) between the nuclei, in the very region in which it is smoothest.

The smoothness of the wave function midway between nuclei is stressed in the augmented plane-wave method of Slater [23] and Safren [24]. The effective potential energy is presumed to be spherically symmetric inside a sphere surrounding each nucleus, and to be constant in the regions between the spheres. An augmented plane wave is then defined; it is a plane wave in the region between spheres and a sum of terms of the form (2.105) inside the spheres. The coefficients in this sum are chosen so that the function inside the sphere matches the external plane wave on the spherical surface. With the augmented plane waves thus computed, the wave function is expanded as a sum of augmented plane waves, and the coefficients are calculated by solution of the resultant secular equation. In this method the plane waves are taken as primary, and the spherical solutions are grafted on approximately. A shift in emphasis marks the Green-function method of Korringa [25] and of Kohn and Rostoker [26]. They choose the spherical solutions as primary and graft on the intermediate plane waves in an approximate manner. Both methods are quite successful in appropriate substances.

Another very successful method is the "orthogonalized plane wave" method of Herring [27], Phillips and Kleinman [28], and Cohen and Heine [29]. The Bloch functions can be written in terms of the Wannier function by Eq. (2.76), and for the low-lying bands the Wannier function can be represented adequately by a single free atomic orbital. This approximation is adopted for the low-lying bands, and we turn our attention to the comparatively difficult problem concerning the higher-lying bands.

In order to solve for the Bloch functions of electrons in the higher bands, we first make a formal expansion of the wave function in a series of functions called orthogonalized plane waves. Each orthogonalized plane wave is a true plane wave plus several free atomic orbital terms. These atomic orbital terms are chosen so as to make the orthogonalized plane wave orthogonal to all the Bloch functions of the low lying bands (which, we recall, are known approximately). Having thus set up the orthogonalized plane waves, the problem reduces to the evaluation of the coefficients in a truncated series expansion of the wave function in terms of these orthogonalized plane waves. This problem takes the usual form of a secular equation. The number of orthogonalized plane waves in the truncated series expansion determines the order of this secular equation. With modern computing machines, and with a well-developed computational technique for handling secular determinants, very accurate results can be obtained in practice.

11. The Fermi Surface

At zero temperature the electrons of a solid successively fill all energy states until the available number of electrons is exhausted. The maximum energy of the electrons is then the Fermi energy μ. The locus of states with $\epsilon(\mathbf{k}) = \mu$ determines a surface in reciprocal space, and this "Fermi surface" plays a dominant role in the theory of the electronic properties of the solid.

In a real crystal the various energy bands may be quite complex. Each band is represented by a scalar function $\epsilon(\mathbf{k})$ in reciprocal space, and these functions generally overlap. That is, the minimum energy in a higher band may well be below the maximum energy in a lower band. Consequently the states on the Fermi surface may belong to different bands in different regions of the surface, and the surface may be very complex in shape. Much contemporary research in the theory of solids is concerned with determining the shape of the Fermi surface for particular materials.

If the calculations of the previous section are carried out explicitly, the shape of the Fermi surface is determined. Either as a check on such calculations or as an alternative to them, there exist a number of experimental procedures which yield information on the Fermi surface. Because of the topical interest in this area of research we very briefly describe one particularly simple but representative type of experiment.

Perhaps the most direct measurement of the Fermi surface is made by *cyclotron resonance*. In the presence of a magnetic field the wave vector of every electron changes according to the law

$$\dot{\mathbf{k}} \equiv \frac{d}{dt}\mathbf{k} = \frac{e}{c\hbar}\mathbf{v} \times \mathbf{H} \tag{2.106}$$

Consequently each electron on the Fermi surface moves in $\mathbf{k}$-space in a direction perpendicular to the magnetic field. Furthermore we recall that the velocity vector $\mathbf{v}$ is perpendicular to the surface of constant energy [cf. Eq. (2.47)], so that the electron motion, being perpendicular to $\mathbf{v}$, is tangential to the Fermi surface. Therefore each electron on the Fermi surface describes an orbit that is the intersection of the Fermi surface with a plane perpendicular to $\mathbf{H}$.

For some shapes of Fermi surface the orbits are closed, or cyclic; for others, the orbits are open, or nonperiodic. We consider the special case of closed orbits.

The frequency ω_H of the electron traversing a closed "cyclotron" orbit on the Fermi surface can be written in the form

$$\omega_H = 2\pi \left[\oint \frac{dk}{\dot{k}}\right]^{-1} = \frac{2\pi eH}{c\hbar}\left[\oint \frac{dk}{v}\right]^{-1} \equiv \frac{eH}{m^*{}_H c} \tag{2.107}$$

where $m^*{}_H$, the "cyclotron mass," is defined by

$$m^*{}_H = \frac{\hbar}{2\pi}\oint \frac{dk}{v} = \frac{\hbar^2}{2\pi}\oint \frac{dk}{\nabla_k \epsilon} = \frac{\hbar^2}{2\pi}\left(\frac{dA}{d\epsilon}\right)_{\epsilon=\mu} \tag{2.108}$$

Here A denotes the area enclosed by the orbit. Thus a measurement of the period (or frequency) of traversal of the orbit determines the quantity $(dA/d\epsilon)_{\epsilon=\mu}$. Furthermore, the orientation of the plane of the orbit is variable, by rotation of the magnetic field H.

For semiconductors the cyclotron resonance can be observed by matching the frequency of an electromagnetic signal field to the cyclotron frequency and by observing the absorption of power. It is necessary, however, that the collision time τ of the electrons be short compared with the cyclotron period, or that $\omega_H\tau \ll 1$. If $\omega_H\tau \sim 1$, the resonance may be broadened beyond recognition. This restriction limits the method to very pure samples (small τ) or to very large magnetic fields (large ω_H).

For metals the method is further complicated by skin-depth problems. In fact, the skin depth of the electromagnetic field may be small compared with the size of the orbit, so that the field acts on the electron only for a small part of its cycle. When the magnetic field is parallel to the surface of this type of experiment is known as an Azbel-Kaner resonance [31]. Although its interpretation is considerably more complicated than that of simple cyclotron resonance, it yields similar information.

The characteristic of the Fermi surface given by cyclotron resonance is not as simple as one would like, but it is useful nevertheless. Other methods give independent characteristics. Thus the diamagnetic susceptibility shows periodic oscillations when plotted as a function of magnetic field (the de Haas–van Alphen effect), and the period of these oscillations yields the maximal or minimal cross sections of the Fermi surface normal to the applied magnetic field. As a result of such experiments and of band calculations, the detailed Fermi surfaces of many metals are now known with great accuracy.

References

General references in this field include:

1*a*. Callaway, J.: "Energy Band Theory," Academic, New York, 1963.
1*b*. Jones, H.: "The Theory of Brillouin Zones and Electronic States in Crystals," North Holland Publishing Company, Amsterdam, 1960.
1*c*. Reitz, J. R.: Methods of the One-electron Theory of Solids, in F. Seitz and D. Turnbull (eds.), "Solid State Physics," Academic, New York, vol. 1, 1955.
1*d*. Ham, F. S.: The Quantum Defect Method, in F. Seitz and D. Turnbull (eds.), "Solid State Physics," vol. 1, Academic, New York, 1955.
1*e*. Woodruff, T. O.: The Orthogonalized Plane Wave Method, in F. Seitz and D. Turnbull (eds.), "Solid State Physics," vol. 4, Academic, New York, 1957.
1*f*. Callaway, J.: Electron Energy Bands in Solids, in F. Seitz and D. Turnbull (eds.), "Solid State Physics," Academic, New York, 1958.
1*g*. Pincherle, L.: Band Structure Calculations in Solids, *Rept. Progr. Phys.*, **23**: (1961).

General background is also given in a number of books on the theory of solids, including the following:

1*h*. Ziman, J. M.: "Principles of the Theory of Solids," Cambridge University Press, New York, 1964.
1*i*. Anderson, P. W.: "Concepts in Solids," Benjamin, New York, 1963.
1*j*. Smith, R. A.: "The Wave Mechanics of Crystalline Solids," Chapman & Hall, London, 1961.
1*k*. Raimes, S.: "The Wave Mechanics of Electrons in Metals," Interscience, New York, 1961.
2. For an extensive discussion of the role of spin-orbit coupling, see Elliott, R. J.: *Phys. Rev.*, **96**: 266, 280 (1954).
3. Born, M., and K. Haung: "Dynamical Theory of Crystal Lattices," Oxford University Press, Fair Lawn, N.J., 1954.
4. Pines, D.: "Elementary Excitations in Solids," Benjamin, New York, 1963.
5. Fletcher, G. C., and E. P. Wohlfarth: *Phil. Mag.*, **42**: 106 (1951). Morita, A.: *Science Repts. Tohoku Imp. Univ.*, **33**: 92 (1949).
6. Herman, F.: *Phys. Rev.*, **93**: 1214 (1954).
7. Hartree, D. R.: *Proc. Cambridge Phil. Soc.*, **24**: 89, 111 (1928). Hartree, D. R., and W. Hartree: *Proc. Roy. Soc.* (*London*), **A150**: 9 (1935).
8. Fock, V.: *Z. Physik*, **61**: 126 (1930).
9. Dirac, P. A. M.: *Proc. Cambridge Phil. Soc.*, **26**: 376 (1930).
10. Slater, J. C.: *Phys. Rev.*, **35**: 210 (1930).
11. Slater, J. C.: *Phys. Rev.*, **81**: 385 (1951).
12. Zener, C.: *Proc. Roy. Soc.* (*London*), **A145**: 523 (1934).
13. Koopmans, T.: *Physica*, **1**: 104 (1933).
14. Slater, J. C.: "Electronic Structure of Solids: The Energy Band Method," *Tech. Rept.* 4, Solid State and Molecular Theory Group, Massachusetts Institute of Technology, July 15, 1953.
15. Wannier, G.: *Phys. Rev.*, **52**: 191 (1937).
16. Slater, J. C.: *Phys. Rev.*, **76**: 1592 (1949).
17. Slater, J. C.: "Electronic Structure of Solids II: The Perturbed Periodic Lattice," *Tech. Rept.* 5, Solid State and Molecular Theory Group, Massachusetts Institute of Technology, Dec. 15, 1953.
18. Kittel, C., and A. H. Mitchell: *Phys. Rev.*, **96**: 1488 (1954).
19. Bloch, F.: *Z. Physik*, **52**: 555 (1928).
20. Fletcher, G. C., and E. P. Wohlfarth: *Phil. Mag.*, **42**: 106 (1951). Fletcher, G. C.: *Proc. Phys. Soc.* (*London*), **A65**: 192 (1952).
21. Wigner, E., and F. Seitz: *Phys. Rev.*, **43**: 804 (1933); **46**: 509 (1934).
22. Von der Lage, F., and H. Bethe: *Phys. Rev.*, **71**: 612 (1947).
23. Slater, J. C.: *Phys. Rev.*, **92**: 603 (1953).
24. Slater, J. C., and M. M. Safren: *Phys. Rev.*, **92**: 1126 (1953).
25. Korringa, J.: *Physica*, **13**: 392 (1947).
26. Kohn, W., and N. Rostoker: *Phys. Rev.*, **94**: 1111 (1954).
27. Herring, C.: *Phys. Rev.*, **57**: 1169 (1940).
28. Phillips, J. C., and L. Kleinman: *Phys. Rev.*, **116**: 287 (1959).
29. Cohen, M. H., and V. Heine: *Phys. Rev.*, **122**: 1821 (1961).
30. Harrison, W. A.: *Phys. Rev.*, **118**: 1190 (1960); **126**: 497 (1962).
31. Azbel, M. I., and E. A. Kaner: *Soviet Phys.-JETP* (*English Transl.*), **3**: 772 (1956).

Chapter 3

Ionic Crystals

By R. W. GURNEY

1. The Perfect Ionic Lattice. The Cohesive Energy

In Born's theory of the cohesive energy of ionic crystals the constituent particles are assumed to be positively charged ions and negatively charged ions. The crystals may be divided into two classes according as the negative ions are molecular ions, as in calcite, or atomic ions, as in the metallic halides and oxides. For simplicity we shall discuss only the latter. It is assumed that these atomic ions are spherically symmetrical and that they interact with each other according to simple central-force laws. The ions adhere to form a crystal because, in each stable structure, the Coulomb attraction between unlike ions outweighs the Coulomb repulsion between like ions. To evaluate the strength of this over-all attraction, it is necessary to sum over all pairs of like and unlike ions.

All the alkali halides except CsCl, CsBr, and CsI crystallize in the sodium chloride structure. The positions of the nuclei in this structure are shown in Fig. 3.1. Each positive ion is surrounded by six

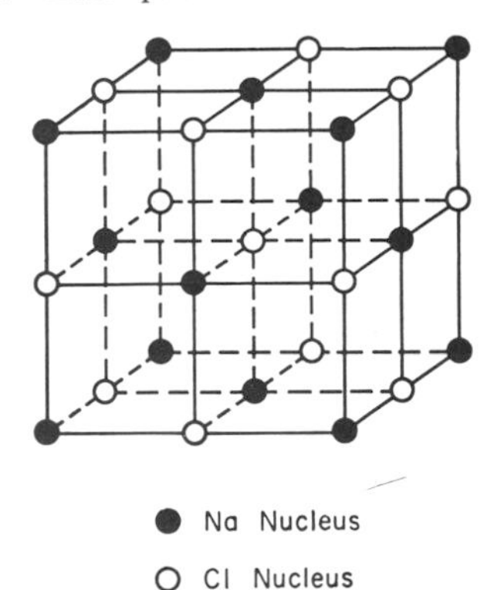

FIG. 3.1. The sodium chloride structure.

negatives, and, at the same time, each negative ion is surrounded by six positives. In any crystal let r be the shortest distance between the nuclei of ions of unlike sign. If the crystal is expanded or contracted uniformly in all directions, the distance between every pair of like ions, and between every pair of unlike ions, will change in proportion to r. With Coulomb forces the over-all electrostatic energy of the crystal will be proportional to $1/r$, whatever the form of the lattice. The constant of proportionality for each lattice structure is calculated by summing over all pairs of like and unlike ions. This numerical constant has been worked out for a number of structures [1],* and is known as the Madelung constant. For a uni-univalent crystal containing N ion pairs the Coulomb energy of the crystal is written

$$\frac{-MNe^2}{r} \tag{3.1}$$

where M is Madelung's constant. In what follows we shall discuss in detail only crystals having the sodium chloride structure, for which M has the value 1.7476.

Both in the oxygen negative ion $O^{=}$ and in the fluorine negative ion F^- the electronic shell has the same structure as a neutral neon atom; and in every alkali halide and oxide, the electronic shell of both the positive ion and the negative ion has the structure of a rare gas atom. When two rare gas atoms, or two ions having a similar electronic structure, are brought into contact with each other, there is, according to quantum mechanics, an intense repulsion, which increases exponentially, as the distance between the two nuclei is diminished. In the simplest form of the theory only this repulsion and the over-all Coulomb attraction are considered. The lattice spacing of the ions in the crystal is that distance at which the over-all Coulomb attraction is in equilibrium with the quantum-mechanical repulsion between adjacent ions: for we suppose that only the repulsion between nearest neighbors need be taken into account.

If the magnitude of the repulsion were known as a function of distance, the lattice constant of the crystal at the absolute zero of temperature could be calculated, since the Madelung constant is known. In practice, the reverse is done. The magnitude of the unknown repulsion is estimated from the observed lattice spacing and the observed compressibility of the crystal. The repulsive energy between one pair of unlike ions is assumed [2] to be of the form

$$w(r) = Ae^{-r/\rho} \tag{3.2}$$

where A and ρ are constants. In a crystal of the sodium chloride structure consisting of N ion pairs, since each ion has six nearest neighbors, the total repulsive energy is

$$6Nw(r)$$

* Numbers in brackets refer to References at end of chapter.

The resultant energy per ion pair of the crystal $U(r)$ will be

$$U(r) = -\frac{Me^2}{r} + 6w(r) \qquad (3.3)$$

The values of the unknown constants A and ρ in (3.2) may be determined from (3.1), the condition that the crystal is in equilibrium under zero pressure,

$$\frac{dU(r)}{dr} = 0 \qquad (3.4)$$

and (3.2), from the compressibility χ,

$$\frac{1}{\chi} = \frac{1}{18r}\frac{d^2U}{dr^2} \qquad (3.5)$$

From the observed lattice spacing and the observed compressibility the cohesive energy U may be calculated. This calculated value may be compared with the observed value derived from the Born-Haber cycle (see Sec. 2). Before making this comparison, we shall describe briefly a less simplified version of the theory. There are three additions to be made:

1. The existence of liquid argon and solid argon shows that between two argon atoms, in addition to the repulsion mentioned above, there is an attraction. The same remark applies to the atoms of all the rare gases. It was shown by London that such an attraction between the electronic shells of two like or two unlike atoms is to be expected from quantum-mechanical considerations. It is connected with a synchronization between the motions of the electrons in the two shells, and gives rise to an energy of the form $-C/r^6$, where

$$C = \frac{3(eh)^4}{2m^2} \sum_{k,k'} \frac{f_{0k} f_{0k'}}{(E_k - E_0)(E_{k'} - E_0)(E_k + E_{k'} - 2E_0)} \qquad (3.6)$$

Here E_0 is the energy of the ground state, E_k and $E_{k'}$ the energies of excited states, and f_{0k} and $f_{0k'}$ the oscillator strengths for the corresponding transitions. A similar van der Waals attraction exists between the electronic shells of two like or two unlike ions. In an aggregate of atoms or ions these van der Waals forces between pairs are additive; and in an ionic crystal they make a considerable contribution to the cohesive energy. In the alkali halides this amounts to 2 or 3 per cent, and in the silver halides to about 10 per cent of the total cohesive energy.

2. Instead of taking into account the quantum-mechanical repulsion between adjacent unlike ions only, we must include also the repulsion between nearest like ions. This is especially necessary in crystals where there is a great disparity between the sizes of the cation and the anion, as in lithium iodide.

3. The energy of the lattice vibrations at the absolute zero of temperature must be subtracted. This zero-point energy per ion pair is equal to

$$\tfrac{9}{4} h\nu_m \qquad (3.7)$$

where ν_m is the maximum frequency of the Debye frequency spectrum.

In this way a more accurate value may be obtained for the lattice energy of a crystal at the absolute zero of temperature—the work required to break up the crystal into its component ions and to leave them at rest in a vacuum.

2. The Born-Haber Cycle

The theoretical value of the lattice energy may be compared with the experimental value, which may be derived from thermochemical data by means of the Born-Haber cycle. This imaginary cycle (for the crystal of a metallic halide) is carried out by the following nine steps: (1) Starting with the crystal at room temperature, we abstract heat from it until it reaches the absolute zero. (2) We break up the crystal into free ions at rest in a vacuum. (3) From the negative ions we remove the supernumerary electrons, and (4) transfer them to the positive ions. (5) We allow the halogen atoms to combine to form diatomic molecules, still at rest in a vacuum, and then (6) bring this gas to room temperature at atmospheric pressure. (7) We bring the metallic vapor up to room temperature, and (8) condense the vapor to form solid metal. (9) We allow the metal and the halogen to combine chemically, thus recovering the original crystal at room temperature.

If the energy associated with each step of the cycle, except step 2, is known, an experimental value of the lattice energy may be obtained. At the time when theoretical lattice energies were first calculated, the energy of step 3—the electron affinity of the halogen atom—was not known for any of the halogens; the Born-Haber cycle was then used to deduce values for the electron affinities of all the halogens, by inserting for step 2 the calculated lattice energy. Later, when the electron affinities of halogen atoms were measured experimentally, the values were found to agree within 2 per cent with those which had been deduced. Column 3 of Table 3.1 gives the

TABLE 3.1. LATTICE ENERGIES IN KILOCALORIES [3] PER MOLE

Crystal	Experimental	Calculated	
NaCl	181.3 ± 3	183.1	183.5
RbBr	151.3	153.5	156.1
KI	153.8	150.8	152.4
CsI	141.5	139.1	142.5

theoretical values obtained by taking the value of ρ in (3.2) equal to 3.45×10^{-9} cm for each of the alkali halides. Column 4 gives values recalculated with $\rho = 3.0 \times 10^{-9}$ cm and with slightly different ionic radii.

3. Dielectric Constant

In alkali halide crystals the dielectric constant κ for static fields, or for radio frequencies, has a value considerably larger than the square of the refractive index n for red or near infrared light. For KCl the value of κ is 4.68 and for LiBr it is 12.1; the values of κ for all the other alkali halides lie between these extremes. On the other hand, if we write $n^2 = \kappa_0$, we find that the values of κ_0 for all the alkali halides lie between 1.74 and 3.80. A similar difference between κ and κ_0 is shown by all ionic crystals. For TlCl and TlBr, we have $\kappa \sim 30$ and $\kappa_0 \sim 5$; for MgO, CaO, and SrO, we have $\kappa \sim 10$ and $\kappa_0 \sim 3$.

The reason for this discrepancy is that in ionic crystals two distinct mechanisms contribute to the dielectric constant in static fields. In an applied field there is the polarization of each ion in the crystal, that is to say, a distortion of the electronic structure of the ion. In an alternating field of optical frequency the whole of the polarization arises in this way. In a static field, or in an alternating field of low frequency, each ion is slightly displaced from its normal position by the external field, the positive and negative ions being simultaneously displaced in opposite directions; to the polarization of the crystal this makes an additional contribution, which in many cases is about twice as large as that contributed by the other mechanism. No appreciable displacement of the ions can take place in an alternating field whose frequency exceeds the normal frequency of vibration of the ions, which is of the order 10^{12} to 10^{13} $\sec^{-1}$. For optical frequencies this contribution is absent; and in many crystals the value of n^2 is less than half that of κ.

4. Electronic Energy Levels

The way in which valence electrons give rise to bands of electronic levels in metal is suggestive in considering the bands of electronic levels in an ionic crystal. Consider, for example, the formation of a crystal of potassium chloride at the absolute zero of temperature from positive K^+ ions and negative Cl^- ions. Whereas a neutral chlorine atom possesses five $3p$ electrons, a chlorine negative ion has six $3p$ electrons. In a crystal of KCl, consisting of N negative Cl^- ions and N positive K^+ ions, these $3p$ levels of the Cl^- ions give rise to bands containing $6N$ levels which (at the absolute zero of temperature) will be completely filled by the $6N$ electrons. At the same time, whereas a neutral potassium atom has one $4s$ valence electron, a positive K^+ ion has a vacant $4s$ level. In the crystal the vacant levels of the N potassium ions combine to form a band of levels which (at the absolute zero of temperature) will be completely empty. Between the highest occupied level and the lowest vacant level in the crystal there is a zone of forbidden energy. In KCl the band of levels corresponding to the $3p$ levels of potassium will be completely filled. But since this band will lie well below the band corresponding to the $3p$ levels of chlorine, we shall not be concerned with it except in X-ray phenomena (see Part 8, Chap. 2).

In a crystal of Cu_2O the lowest empty band of levels will be provided by the vacant level of the Cu^+ ions; and the highest filled band will either be that provided by the occupied $3d$ levels of Cu^+ or that provided by the $2p$ levels of the $O^=$ ions, whichever band is the higher.

If some electrons are introduced into the lowest band of vacant levels in a crystal at room temperature, these electrons will describe a random Brownian motion. If a voltage is applied to the crystal, these electrons will drift in the field—a drift velocity will be superimposed on the random motion, and the crystal will show electronic conductivity. For this reason the lowest band of vacant levels is commonly called the *conduction band.* At room temperature few, if any, ionic crystals in the pure state show any electronic conductivity. This is because the conduction band is practically empty. The energy gap between the highest filled band and the conduction band is so wide (at least 2 electron volts) that the conduction band remains empty, in spite of the tendency for electrons to be raised by the thermal agitation. In certain crystals electrons can be raised to the conduction band by the absorption of light; during illumination such a crystal shows electronic conductivity (see Sec. 14). In most ionic crystals in the dark the electronic conductivity remains negligible at all temperatures up to the melting point; it is completely masked by the ionic conductivity (see Sec. 9).

5. Positive Holes

Suppose that, in the interior of the crystal of a metallic halide, we have removed an electron from a halide negative ion, leaving a neutral halogen atom. We can express this by saying that we have removed an electron from a filled band of levels, leaving in the band a vacancy for an electron. In a perfect crystal at room temperature this vacancy will be mobile. In fact, the vacancy (with which a positive charge is associated) will move with momentum, like an electron bearing a positive charge. Such a vacancy is usually known as a *positive hole.* As in the case of an electron, by interaction with the lattice vibrations of the crystal a positive hole will be deflected, with or without change of energy. It will describe a random Brownian motion. When a voltage is applied to a crystal where electrons have been removed from the highest filled band, the positive holes will drift in the applied field. This drift velocity being superimposed on the random motion, an electric current will flow during the short time that the holes remain mobile (see Sec. 14).

6. Excited Electronic States of a Crystal

A negative atomic ion, such as a halogen negative ion, in a vacuum is believed to have no excited states. This is because, when the electron is removed from the negative ion, the force of attraction falls rapidly to zero. On the other hand, when the negative ion is a constituent of an ionic crystal, the removal of the electron leaves behind a positive hole, and the mutual potential energy of the two charges tends to zero slowly as $1/r$. As a result, it can be shown [4] that, associated with the electron in an anion of the lattice, there will be an infinite series of excited states, leading up to a series limit beyond which there will be a continuum of states.

In greater detail the origin of these excited states of an ionic crystal may be described as follows: To produce simultaneously a free electron in the conduction band and a distant free positive hole in the filled band, a certain minimum quantum of energy will be required. It is not, however, necessary that the electron and the positive hole shall be free. There will exist a series of stationary states where the electron and the positive hole, caught in each other's field, will revolve around their common center of mass. An electron and a positive hole, coupled together in this way, are known as an *exciton*. An exciton may move as a whole through the crystal. For a given velocity of the exciton there will be a series of stationary states, leading up to a series limit.

Each of the excited states will be broadened by the presence of the lattice vibrations of the crystal. Even at the absolute zero of temperature there will be some broadening, owing to the zero-point vibrations of the lattice. If the lattice vibrations were absent, the optical absorption spectrum of an ionic crystal would be similar to that of a vapor of *neutral* atoms.

7. Lattice Imperfections. Schottky Defects

Hitherto in this chapter we have discussed the properties of a perfect ionic lattice in which every lattice point is occupied by the appropriate species of ion. In practice, the crystal may deviate from perfection in two different ways. In the first place the lattice itself may show the imperfections known as *Taylor dislocations* [5]. In the second place, even if the lattice itself is otherwise perfect, at any temperature a certain number of sites of the lattice, which should be occupied, will be vacant.

Consider then the process of removing one ion from the interior of a *perfect* crystal at temperature T and placing it on the surface of the crystal; to carry this out isothermally a certain amount of work will be required. In a crystal in thermodynamic equilibrium at any temperature a certain number of internal lattice points will be unoccupied. In a uni-univalent crystal, to preserve electrical neutrality, the number of vacant anion lattice points must be equal to the number of vacant cation lattice points. These pairs of vacant lattice points are usually known as *Schottky defects*. When the temperature of the crystal is raised, the number of vacant lattice points increases. The formation of additional vacancies will take place at the surface of the crystal, either by the process shown in Fig. 3.2 for an anion or by the corresponding process for a cation.

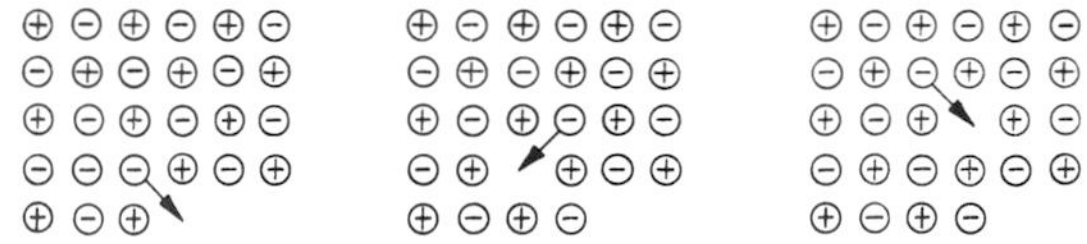

(a) (b) (c)

FIG. 3.2. The mechanism for the formation of a vacant anion lattice point, and its movement into the interior of the crystal.

The density of a crystal containing vacant lattice points is smaller than that of a perfect crystal. Accurate measurements of the density of sodium chloride, potassium chloride, and calcite have been made at room temperature. They show that the number of vacant lattice points must be smaller than one in 10^4, and may be much smaller.

Consider a uni-univalent crystal consisting of N pairs of ions and containing n vacant cation lattice points and n vacant anion lattice points. The number of sites in the crystal available to the N positive ions is $(N + n)$, and the number of recognizably different ways in which these N positive ions can be arranged over these $(N + n)$ sites is

$$\frac{(N+n)!}{N!n!} \tag{3.8}$$

At the same time, the number of sites available to the N negative ions is $(N + n)$, and the number of different ways in which these negative ions can be arranged is again given by (3.8). The number of different ways in which the N pairs of ions in the crystal can be arranged is therefore given by

$$W = \left[\frac{(N+n)!}{N!n!}\right]^2 \tag{3.9}$$

Now in a perfect crystal at temperature T let J be the change in the free energy when a positive ion is removed isothermally from a *given* lattice site and a negative ion is removed from a *given* distant lattice site, both ions then being built into a surface layer of the crystal. Then, if n is so small that the vacancies may be regarded as independent, the total change in the free energy of a crystal at temperature T when n pairs of vacant lattice points have been formed is

$$nJ - kT \ln W \tag{3.10}$$

where k is Boltzmann's constant. To allow for the decrease of J with temperature, as a first approximation we may take J to be of the form [6]

$$J = J_0 - cT \tag{3.11}$$

For equilibrium, the increment in (3.10), when additional pairs of vacancies are formed, must be zero. Using Stirling's approximation in (3.9), we find

$$\frac{n}{N+n} = \exp\left(-\frac{\frac{1}{2}J}{kT}\right) = C \exp -\frac{\frac{1}{2}J_0}{kT} \tag{3.12}$$

where the constant C is equal to $e^{c/2k}$. The number of vacant sites in a crystal will therefore increase exponentially with temperature.

Consider now the negative ions that happen to be adjacent to a vacant anion lattice site. At any moment any one of these ions may jump into the vacant site, as depicted in Fig. 3.2, with the result that the vacancy will have jumped in the opposite direction. The same will be true with regard to the positive ions that happen to be adjacent to a vacant cation lattice site. At any temperature each vacancy will therefore describe a random Brownian motion in the crystal.

8. Frenkel Defects

In certain crystals another type of lattice defect is believed to be present. An ion may leave its normal lattice site and wander away to a distant interstitial position, that is, to a position where it is squeezed in between ions that are in their normal lattice positions, but which are slightly displaced to make room for it. This kind of lattice defect, which is usually known as a

Frenkel defect, consists of an interstitial ion and a distant vacant lattice site.

Anions apparently do not form Frenkel defects. In a crystal let there be, available for the cations, N' interstitial sites all of which are equivalent. When n of the N cations of the crystal have moved to distant interstitial sites, they leave n cation lattice sites vacant. The number of recognizably different ways in which the remaining $(N - n)$ lattice cations can be distributed over the N cation lattice sites of the crystal is

$$\frac{N!}{n!(N - n)!} \tag{3.13}$$

At the same time, the number of different ways in which the n interstitial ions can be distributed over the N' available interstitial sites is

$$\frac{N'!}{n!(N' - n)!} \tag{3.14}$$

In a perfect crystal at temperature T let J be the work required to make a Frenkel defect isothermally. In the same way as before, we find that the number of Frenkel defects in a crystal in equilibrium at temperature T will be given by

$$\frac{n}{\sqrt{(NN')}} = C \exp\left(-\frac{\frac{1}{2}J_0}{kT}\right) \tag{3.15}$$

In any interstitial site an ion will have a set of vibrational energy levels; and at any equivalent interstitial site the ion will have an identical set of vibrational levels. The ion will be able to jump from one interstitial site to another. But in passing from one site to an adjacent site, the ion will have to squeeze its way through a region where its potential energy will be higher. (A lattice ion that jumps to an adjacent vacant lattice site, as in Fig. 3.2, will have to pass over a similar potential barrier.) Let the height of the potential barrier be denoted by U. For only a fraction $e^{-U/kT}$ of the time will the ion have sufficient energy to carry it over the barrier to the adjacent site. If the frequency of vibration of the ion is ν, the probability per unit time that the ion jumps to an adjacent site will be of the order

$$\nu e^{-U/kT} \tag{3.16}$$

Provided that U/kT is not too large, an interstitial ion will describe a lively Brownian motion in the crystal. When a voltage is applied to the crystal, the probability that the interstitial ion jumps in a direction having a component parallel to the field is greater than the probability of a jump having a component in the reverse direction. Consequently, in the applied field a drift velocity will be superimposed on the random Brownian motion; and if the crystal contains a sufficient number of interstitial ions and vacant lattice points, we expect the crystal to show ionic conductivity. Exactly the same remarks can be made for a crystal containing a sufficient number of Schottky defects. The variation of the mobility with temperature is discussed in Sec. 10.

9. Ionic Conductivity

Molten ionic crystals are among the best electrolytic conductors. Just below the melting point the conductivity of the solid crystal has a value of the order 10^{-4} ohm^{-1} cm^{-1}, and falls rapidly with decreasing temperature. By weighing the metal deposited at the cathode, the validity of Faraday's law over a wide range of temperature has been proved for several ionic crystals.

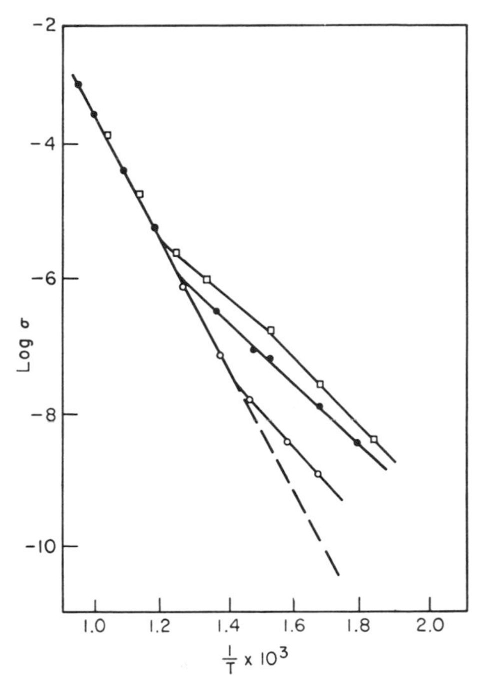

FIG. 3.3. Ionic conductivity of three specimens of sodium chloride. (*Smekal, A.: "Handbuch der Physik,"* 24/2, 881, Springer, Berlin, 1933.)

Some typical results are shown in Fig. 3.3. When the logarithm of the specific conductivity σ is plotted against the reciprocal of the absolute temperature, at high temperatures the experimental points lie on a straight line that is the same for all specimens of the crystal. At lower temperatures the experimental points break away from this straight line and lie on a curve which is different for each specimen. This individual behavior at low temperatures is ascribed to the presence of impurities.

When a small quantity of a divalent salt, such as $CdCl_2$ or $PbCl_2$, is intentionally added as an impurity to a crystal of AgCl or NaCl [7], the ionic conductivity may be increased 100-fold. The mechanism of this increase is discussed below.

10. Mobility of Lattice Defects

When discussing the Brownian motion of interstitial ions in Sec. 8, we denoted by U the height of the potential barrier between adjacent sites. When a voltage is applied to the crystal, the height of the barrier in a direction perpendicular to the field will be unchanged. If the direction of the applied field

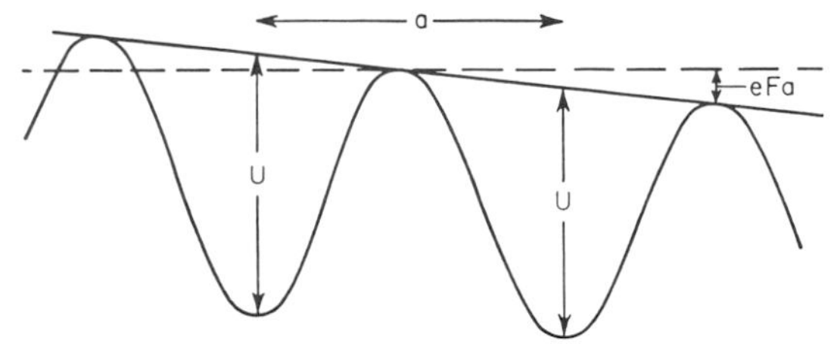

FIG. 3.4. The potential energy of an interstitial ion in an applied field of intensity F.

is parallel to the line joining adjacent interstitial sites, the situation in this direction will be as shown in Fig. 3.4. If a denotes the distance between adjacent interstitial sites, the height of the potential barrier in the direction of the field will be diminished from U to $U - \frac{1}{2}eFa$. In the reverse direction the height of the barrier will be increased to $U + \frac{1}{2}eFa$. The probability per unit time that the ions jump in the direction of the field will now be

$$\nu e^{-(U-\frac{1}{2}eFa)/kT}$$

while in the opposite direction it will be

$$\nu e^{-(U+\frac{1}{2}eFa)/kT}$$

Multiplying by a and subtracting, we find that the mean velocity of drift in the applied field will be given by

$$\nu a e^{-U/kT}\, 2 \sinh \frac{\frac{1}{2}eFa}{kT} \tag{3.17}$$

The value of U will decrease with rise of temperature, owing to the thermal expansion of the crystal. To allow for this, we may write

$$U = U_0 - bT$$

At the same time, since in practice eFa is very small compared with kT, we may expand the last factor in (3.17). For the ionic mobility in unit intensity of field, we then obtain

$$\frac{e\nu a^2}{kT} \exp\left(-\frac{U}{kT}\right) = \frac{Be\nu a^2}{kT} \exp\left(-\frac{U_0}{kT}\right) \tag{3.18}$$

where $B = e^{b/k}$.

The diagram of Fig. 3.4 and the discussion apply equally to the motion of interstitial ions and to the motion of vacant lattice points. In both cases the value of the specific conductivity of the crystal will be proportional to the product of the number of mobile lattice defects and their mobility. Since both these quantities, according to (3.18), (3.12), or (3.15), increase exponentially with temperature, the variation of the specific conductivity with temperature will be of the form

$$\sigma = \sigma_0 \exp\left[-\frac{(\frac{1}{2}J_0 + U_0)}{kT}\right] \tag{3.19}$$

where σ_0 depends only on quantities that are characteristic of the pure crystal. Thus the observed behavior of the ionic conductivity at high temperatures is accounted for.

Consider now a crystal of NaCl or AgCl containing a small quantity of $CdCl_2$ in solid solution. In these mixed crystals the Cd^{++} ions and the additional Cl^- ions occupy lattice sites. Thus each Cd^{++} ion must replace two Na^+ or two Ag^+ ions in the lattice; and the crystal must always contain at least as many vacant cation lattice points as there are Cd^{++} ions in the crystal. Since there is an effective negative charge associated with each cation vacancy in the crystal, the cation vacancy will be attracted by the divalent cation. If this attraction is weak, all the cation vacancies will describe the same free Brownian motion as in a pure crystal; and in an applied field they will give rise to an additional conductivity, whose variation with temperature will be of the form

$$\sigma' = \sigma'_0 \exp\left(-\frac{U_0}{kT}\right) \tag{3.20}$$

When the logarithm of σ' is plotted against $1/T$, the slope of the straight line will be only U_0/k, instead of $(\frac{1}{2}J_0 + U_0)/k$. At low temperatures the conductivity arising from Schottky defects, given by the broken line in Fig. 3.3, is small. Therefore, in a crystal containing a divalent cation impurity, however small in quantity, there will be a temperature below which the number of Schottky cation vacancies will be smaller than the number of additional free cation vacancies. Hence the gentler slope of the curves in Fig. 3.3 at low temperatures is accounted for, if the crystals contain small quantities of some divalent impurity. The amounts of impurity required to give the observed conductivities vary from one part in 10^5 to less than one part in 10^8.

On the other hand, if the attraction between cation vacancies and divalent cation is sufficient to bind a considerable fraction of the former, the number of free cation vacancies will increase with temperature, though less rapidly than the Schottky defects.

11. Crystals with Nonstoichiometric Composition

Many ionic crystals when heated in the vapor of one of their constituent elements acquire a small stoichiometric excess of this constituent. Thus when an alkali halide is heated in the vapor of the alkali metal, it acquires a stoichiometric excess of the metal, which may be greater than one part in 10^4; and when heated in the vapor of its halogen, it acquires a stoichiometric excess of halogen.

In each case the slight change in composition is accompanied by a change in the optical absorption spectrum of the crystal. Excess of alkali metal is accompanied by the presence of an additional absorption band whose maximum lies in or near the visible part of the spectrum, with the result that the crystal takes on a characteristic color (yellow for NaCl, violet for KCl). The centers responsible for this absorption are known as F centers (from the German *Farbcentren*) and the absorption band is known as the F band. As a typical example, the F band of KBr is shown in Fig. 3.5.

On the other hand, a stoichiometric excess of halogen in the crystal is accompanied by the presence of a similar absorption band in the ultraviolet region.

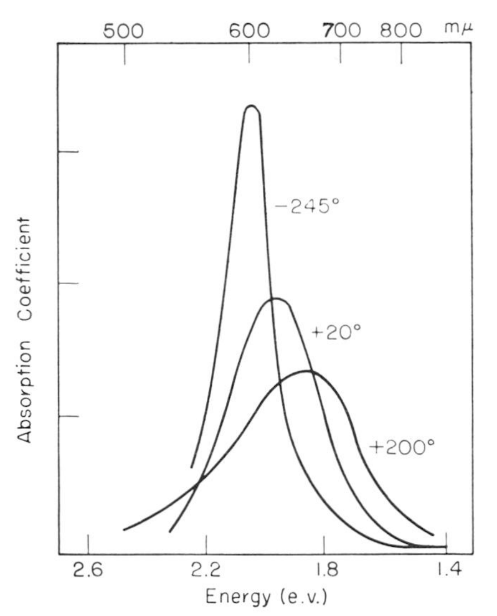

FIG. 3.5. The F band of potassium bromide at different temperatures. [*Pohl, R. W.: Proc. Phys. Soc.*, **49**: 3 (1937).]

The centers responsible for this absorption are called V centers, and the ultraviolet band is called the V band.

Returning to a crystal of correct stoichiometric proportions, irradiation of the crystal by X rays or by cathode rays causes the simultaneous appearance of both the F band mentioned above and of an absorption band in the ultraviolet similar to the V band. The irradiation by X rays or cathode rays produces in the crystal free electrons and free positive holes, both of which soon become trapped. The F band is associated with trapped electrons, and the V band with trapped positive holes. The form of the F band produced in any crystal by irradiation is identical with that characteristic of the same crystal when it contains a stoichiometric excess of metal; but the form of the V band depends on the conditions of excitation.

12. Trapped Electrons and Positive Holes

When a negative ion has been removed from the interior of a perfect crystal, leaving a vacant lattice point, there is a positive charge associated with the vacant lattice point; so if a free electron in the conduction band approaches this point, it will experience an attraction, as from a positively charged particle; and therefore, if it loses energy, the electron may be trapped in the field of the charge. For the trapped electron, there will be a series of stationary states, like those described in Sec. 6, leading up to a series limit, beyond which there will be a continuum of states. On the wave function of the trapped electron the periodic field of the lattice will impose periodic fluctuations, but it will not prevent the trapped electron from having, in each bound state, a quite definite wave function, similar to that of an electron in an atom in a vacuum; and this wave function may be spread over many atomic distances of the crystal. In the lowest state, the normal state, the wave function will (in a crystal having the NaCl structure) be to some extent concentrated on the six positive ions that are in contact with the vacant anion lattice point, and one can think of the electron as describing an "orbit" through these six positive ions.

In the same way, if a positive ion has been removed from the interior of a perfect crystal, a negative charge is associated with the vacant cation lattice point, as has already been mentioned in Sec. 10. If a free positive hole in the filled electronic band approaches this point, it will be attracted; and if it loses energy, it may be trapped in the field of the negative charge. For the trapped positive hole there will be a set of stationary states leading up to a series limit. In the lowest bound state, the normal state, the wave function will be largely concentrated on the six negative ions that are in contact with the vacant cation lattice point, that is to say, these six negative ions will share only five electrons.

13. The F Band and the V Band

In the absorption spectrum of any colored crystal the characteristic F band is ascribed to the absorption of light by electrons trapped at vacant anion lattice points, almost the whole of the band being due to the raising of the electron from the normal $1s$ level to the first excited p level.

The way in which these F centers come to be present in a crystal containing a stoichiometric excess of metal is as follows. When the crystal is exposed to the metal vapor, a metal atom adheres to the surface of the crystal. From this atom the valence electron escapes into the crystal, entering the conduction band. Now an anion from the crystal surface layer moves up to join the positive ion, as shown in Fig. 3.2; and by the process shown in Fig. 3.2 an internal vacant anion lattice point is then formed. If the electron has not been already trapped at one of the vacant anion lattice points that are already present in the crystal (Schottky defects), it can be trapped at this newly formed vacancy. In any case the number of trapped electrons will finally be equal to the number of metal atoms that have been incorporated into the crystal. For high densities of F centers the intensity of the F band absorption should be proportional to the density of the metallic vapor in which the metal has been heated. For KCl this has been shown to be the case.

On the other hand, in a crystal that has acquired a stoichiometric excess of halogen by exposure to halogen vapor the intensity of the V band was found to be proportional to the pressure of diatomic molecules in the vapor. This result, among others, has given rise to the idea [8] that the V band formed under these conditions is due to centers consisting of pairs of positive holes trapped at adjacent cation vacancies. Since in a halide crystal a positive hole means a neutral halogen atom, a pair of adjacent positive holes is equivalent to a halogen molecule in the lattice.

A crystal with a stoichiometric excess of either component and an equivalent number of vacant lattice sites has, of course, a density lower than that of a crystal of correct stoichiometric proportions.

For example, a crystal of calcium fluoride containing a large number of F centers has been found [9] to have a density more than 1 per cent below the normal value.

The formation of a weak F band by X-ray irradiation of a crystal of correct stoichiometric proportions is ascribed to the trapping of the photoelectrons by vacant anion lattice points already present in the crystal (Schottky defects). For many years it was thought that X irradiation of a nearly perfect crystal could produce only a relatively weak coloration, corresponding to about 10^{-16} F centers per cubic centimeter—this being the number of Schottky defects present in the crystal. More recently it has been discovered that prolonged irradiation with soft X rays causes a diminution in the density of the crystal itself [10]. The fact that the growth of a very intense F band is accompanied by a progressive diminution in the density of the crystal provides additional proof that the F center is an electron trapped at a vacant lattice point.

14. Photoconductivity

We return now to the discussion of a pure crystal, not containing F centers. If a voltage is applied to a crystal, which is illuminated with light of sufficiently short wavelengths (less than about 1300 A in NaCl), a photocurrent is observed [11] presumably due to the raising of electrons from the filled band to the empty conduction band. In this characteristic ultraviolet absorption band of the crystal the absorption coefficient has a very high value, of the order 10^5 cm^{-1}. Toward the visible region the absorption falls toward zero, either rapidly or comparatively slowly, so that the absorption band has a short or a long "tail" extending toward the visible. In a pure alkali halide crystal at room temperature absorption of light in this tail of the band does not give rise to any photoconductivity; the absorption evidently gives rise to excitons, which are not then dissociated to give free electrons by interaction with the lattice vibrations.

In pure silver halides, on the other hand, absorption of light in the long tail of the ultraviolet band is accompanied by photoconductivity, even at $-180°$C. If a slice of the crystal perpendicular to the applied field is illuminated by a flash of light, electrons liberated by the light will be drawn by the field into the unilluminated portion, where no positive holes have been created; and one might suppose that all the electrons would eventually reach the cathode, irrespective of the intensity of the field. In no ionic crystal, however, does the current behave in this way; for weak fields the current, instead of being independent of the voltage, is proportional to it. There is evidently a high probability of each electron being trapped somewhere in the crystal before reaching the cathode. Only in strong fields is there a high probability of the electrons reaching the electrode without having encountered a trap; under these conditions the current approaches a saturation value. An estimate of the number of trapping centers present in the crystal has shown that about 10^{12} traps per cubic centimeter would be sufficient to account for the phenomena observed in the silver and thallium halides. The nature of the trapping centers is uncertain. In cases where free positive holes are created, these get trapped equally readily, or probably more readily.

15. Crystals Containing F Centers

It has been mentioned that when an alkali halide crystal, to which a voltage is applied, is illuminated with light of sufficiently short wavelength this gives rise to photoconductivity, presumably through the raising of electrons into the conduction band. When a crystal containing F centers is illuminated at room temperature with light having a wavelength lying in the F band, photoconductivity is likewise observed, although the absorption only raises the electron to an excited state; the current is ascribed to the fact that the electrons are raised from the excited state to the conduction band, with high efficiency, by the thermal agitation.

In a crystal containing a stoichiometric excess of metal it was found that, for a given number of absorbed quanta, the current was smaller the greater the number of F centers per cubic centimeter; in fact, in crystals of KCl at $-100°$C the mean "range" of the photoelectrons was found to be inversely proportional to the number of F centers per cubic centimeter. This shows that in these crystals it is the F centers themselves that act as traps for the photoelectrons. Since an F center is an electron trapped at a vacant anion lattice point, this means that such a vacant lattice point can trap two electrons.

When the absorption spectrum of such a crystal is investigated, it is found, as is to be expected, that the removal of electrons from F centers causes a progressive disappearance of the F band. We have to distinguish two ranges of temperature. When a crystal is irradiated at a low temperature (below $-80°$C for KCl), the partial disappearance of the F band is accompanied by the appearance of a broad band extending to beyond 10,000 A in the infrared. This is called the F' band. When the quantum efficiency of absorbed light in converting the F band into the F' band was measured in a crystal at this temperature, it was found that one quantum of light of a wavelength lying within the F band destroys two F centers. An electron, ejected from one F center, leaves behind a vacant lattice point, and being captured by another F center, converts it into an F' center. Illumination of the crystal with light having a wavelength lying in the F' band gives rise to photoconductivity.

Suppose that, by absorption of a quantum of radiation, an electron is ejected into the conduction band from an F center or from an F' center. After the absorption of the quantum the neighboring ions will move into new positions. The adjacent negative ions, no longer repelled by the negative charge, will be displaced slightly inward, while the adjacent positive ions will be displaced slightly outward. These displacements will liberate a considerable amount of energy; but since they take place after the absorption of the quantum, they do not affect the frequency of the absorbed radiation. As a result, the amount of energy required to raise an electron *thermally* from such a center is considerably smaller than the quantum of energy required to raise the electron optically. The energy level of the electron in an F' center lies so near to the conduction band that the thermal energy is sufficient to raise the electron, even at room temperature. The same is true for electrons belonging to many impurity atoms contained in the crystal. In practice, even the purest crystals contain impurities of various kinds. In any crystal at room temperature there is a much greater number of electrons in the conduction band than would be characteristic of the pure crystal (see Sec. 16 and Chap. 4).

The excited state of an F center, to which the electron is raised by the absorption of light in the F band, likewise lies so near to the conduction band that, at room temperature, the thermal energy is sufficient to free the electron. To this fact we ascribe the observed photoconductivity. If this is the correct interpretation, we should expect that at sufficiently low temperatures the thermal energy would be insufficient and the photoconductivity should fall off rapidly. This is found to be the case. When NaCl containing F centers is illuminated with light in the F band and the temperature is lowered from $-150°C$ to $-225°C$, it is found that the photocurrent falls to less than one-thousandth of its value at $-150°C$. This behavior has been found to be consistent with an activation energy of 0.075 electron volt.

16. Dielectric Breakdown in Ionic Crystals

For weak and for moderately strong fields the ionic conductivity described in Sec. 9 obeys Ohm's law. But if the field is increased indefinitely, a certain critical intensity is reached at which the dielectric breaks down, and a very large current passes. For each crystal we have to distinguish two ranges of temperature. When the crystal is above a certain temperature T_0 characteristic of the crystal (usually in the neighborhood of 100°C) the breakdown takes place some seconds after the application of the field. In this case the breakdown is due to Joule heating, generated by the ionic conduction, which causes local melting. For temperatures below T_0, on the other hand, breakdown takes place in a much shorter time, of the order 10^{-8} sec, after the application of the field. The value of the critical field strength varies slowly with temperature. In KBr at room temperature, for example, the critical field strength is 8×10^5 volts/cm, and at $-185°$ it is 2.5×10^5 volts/cm.

In Sec. 15 it was pointed out that, even in the purest crystals, there will always be some electrons in the conduction band, released from impurity atoms. If some of these electrons, accelerated in the strong field, can gain enough energy to raise other electrons from the filled band, the effect will be cumulative, as in a spark discharge. Two quantitative theories of this process have been put forward; the relation between the two has been discussed by Fröhlich and Seitz [12].

17. Ionic Crystals in Photographic Emulsions

The emulsion that covers a photographic film consists of many small crystals of a silver halide, usually AgBr, embedded in gelatin. These crystals are commonly known as *grains*. Most of the grains have a diameter between 10^{-5} and 10^{-4} cm, and thus contain between 10^{11} and 10^{12} ion pairs. The absorption of light having a wavelength lying in the "tail" of the ultraviolet band releases photoelectrons in the crystal. Electrons are likewise released into the conduction band, when a fast nuclear or cosmic-ray particle passes through a grain. In both cases the release of a few electrons in a grain produces some permanent change, so that when the emulsion is subsequently exposed to a photographic developer the developer discriminates between exposed and unexposed grains, and attacks only the former.

The theory [13] not only makes use of the photoconductivity, but also of the ionic conductivity of the silver halides. At room temperature the ionic conductivity is many times greater than that of the alkali halides, but is suppressed at liquid-air temperatures. This enables the two parts of the dual process to be separated experimentally [14].

References

1. Seitz, F.: "Modern Theory of Solids," McGraw-Hill, New York, 1940.
2. Born, M., and J. E. Mayer: *Z. Physik*, **75**: 1 (1932).
3. Huggins, M. L.: *J. Chem. Phys.*, **5**: 143 (1937).
4. Mott, N. F., and R. W. Gurney: "Electronic Processes in Ionic Crystals," p. 83, Oxford University Press, New York and London, 1948.
5. Seitz, F.: *Phys. Rev.*, **80**: 239 (1950); *Revs. Mod. Phys.*, vol. 23, 1951.
6. Mott, N. F., and R. W. Gurney: Ref. 4, p. 136.
7. Etzel, H. W., and R. J. Maurer: *J. Chem. Phys.*, **18**: 1003 (1950).
8. Seitz, F.: *Phys. Rev.*, **79**: 529 (1950).
9. Mollwo, E.: *Nachr. Ges. Wiss. Göttingen, Math.-phys. Kl.*, **1**: 86 (1934).
10. Estermann, I., W. J. Leivo, and O. Stern: *Phys. Rev.*, **75**: 637 (1949).
11. Ferguson, J. N.: *Phys. Rev.*, **66**: 220 (1944).
12. Fröhlich, H., and F. Seitz: *Phys. Rev.*, **79**: 526 (1950). Seitz, F.: *Phys. Rev.*, **76**: 1376 (1949).
13. Gurney, R. W., and N. F. Mott: *Proc. Roy. Soc. (London)*, **A164**: 151 (1938).
14. Mees, C. E. K.: "Theory of the Photographic Image," Macmillan, New York, 1942.

Chapter 4

Flow of Electrons and Holes in Semiconductors (Physics of Transistor Effects)

By JOHN BARDEEN, University of Illinois

1. Introduction

Semiconductors have important applications in electronics as diodes, transistors, power rectifiers, etc. We shall be concerned in this chapter with a discussion of the basic physics of semiconductors required for an understanding of the operation of such devices. We shall be concerned particularly with current flow in junctions and in other situations in which the concentrations of electrons and holes depart from equilibrium values. The concepts with which we shall deal are similar to those familiar in gas discharges: mobilities of carriers and conduction in electric fields, diffusion, space charge, recombination and lifetime, and even avalanche formation and breakdown.

The rectifying property of metal-semiconductor contacts was first observed by Braun [1] and by Schuster [2] in 1874.† As pointed out by Lark-Horovitz [3], Braun showed great intuition when he described the process by saying that the more he studied these effects, the more apt seemed the analogy with gas discharges. Cat's-whisker detectors were used in the early days of radio. However, it was only after the discovery of the copper oxide rectifier by Grondahl [4] and subsequent studies made in the development of this and other area rectifiers that an adequate theory of rectification was given by Mott [5] and by Schottky [6]. Earlier, Wagner [7] had given the correct form of the rectifier equation and also the equations which describe current flow in semiconductors when both conduction and diffusion are important [8]. Research on semiconductors was stimulated during the Second World War by the development of the silicon detector for microwave applications [9]. It was during this period that it was observed by the group at Purdue University that a diode made with high-purity germanium will withstand a high voltage in the reverse direction. Subsequently, research on the properties of both germanium and silicon was carried out at a number of laboratories, including Purdue, Bell Telephone Laboratories, and the Radiation Laboratory of the Massachusetts Institute of Technology. A great deal was accomplished in obtaining materials of high purity and uniform properties and in understanding the semiconducting properties.

It was a fundamental research program on semiconductors initiated just after the war at the Bell Telephone Laboratories and under the general direction of W. Shockley that led to the invention of the transistor by Brattain and the author [10] and the recognition that it is necessary to consider flow of both holes and conduction electrons to understand the characteristics of germanium diodes. Much was learned about minority carrier flow in germanium in a beautiful series of experiments by Shockley, Pearson, and Haynes [11]. Research on semiconductors and transistors was aided greatly by the development by Teal and coworkers [12] of methods for producing large single crystals of germanium and later of silicon. Zone refining to produce materials of extremely high purity was initiated by W. G. Pfann [13].

In Sec. 2 we outline the physical ideas and basic equations required to describe current flow when both conduction electrons and holes are present and in Sec. 3 we apply them to carrier injection by point contacts, the Haynes-Shockley method for measuring drift mobility, and to the theory of *p-n* junctions and junction transistors. In Sec. 4 we discuss space-charge layers and metal-semiconductor contacts. We confine the discussion to an understanding of the operation of devices in physical terms. Little or no discussion is given of the characteristics as circuit elements.

2. Basic Equations

We present in this section an outline of the basic equations required to describe current flow in semiconductors. Included are discussions of impurity levels and charge balance, thermal generation and recombination, equations for current flow and continuity, and some solutions for special cases.

There are states with energies in the "forbidden" gap which correspond to electrons localized at impurities or other imperfections in the lattice. These latter are designated in various ways. *Donors* are neutral when occupied and are positively ionized on release of an electron to the conduction band or elsewhere. *Acceptors* are neutral when unoccupied, negative ions when occupied. Some impurities may have several states of ionization. Other levels are denoted as *traps*, and there may be no sharp distinction between these and donors and acceptors. In general, a level is called a trap if an appreciable time is required for equilibrium to be established between

† Numbers in brackets refer to References at end of chapter.

the traps and the conduction or valence bands. Donors usually are assumed to be in equilibrium with the conduction band and acceptors with the valence band.

The various concentrations (number per cubic centimeter) are denoted as follows: conduction electrons, n; holes, p; donors, N_D; ionized donors, N_D^+; acceptors, N_A; ionized acceptors, N_A^-; traps, N_T; ionized traps, N_T^+ or N_T^-. The equilibrium concentrations are determined from the Fermi-Dirac function, as described in Part 4, Chap. 6. Such equilibrium values will be denoted by the subscript zero.

The net charge density is

$$\rho = e(p - n + N_f) \tag{4.1}$$

where N_f is the net concentration of fixed charges:

$$N_f = N_{D^+} - N_{A^-} + N_{T^+} - N_{T^-} \tag{4.2}$$

In the bulk of semiconductors, where electrical neutrality pertains, one may take $\rho = 0$. This is true even when electric fields are present, because the unbalance of charge required to produce the field is usually negligible. Exceptions are the space-charge layers at metal-semiconductor contacts and at junctions between p- and n-type materials. In these regions, one uses Poisson's equation to relate the charge density with the electrostatic potential, ψ:

$$\nabla^2\psi = -\frac{4\pi\rho}{\kappa} \tag{4.3}$$

Here κ is the dielectric constant.

Thermal Generation and Recombination. One may represent the capture of a conduction electron, e, by a trap, T, by the reaction

$$e^- + T \rightleftarrows T^-$$

The reverse reaction is thermal release of the electron by the traps. The recombination rate is usually assumed proportional to the product of n and $N_T - N_T^-$ (the concentration of unoccupied traps), and the generation rate to N_T^-. The net rate at which electrons enter traps from the conduction band is then

$$r_c n(N_T - N_T^-) - g_c N_T^- \tag{4.4}$$

In thermal equilibrium, this vanishes, so that

$$r_c n_0(N_T - N_{T0}^-) = g_c N_{T0}^- \tag{4.5}$$

The subscript zero indicates equilibrium values. This gives a relation between the coefficients r and g. If the Fermi-Dirac functions are used for n_0 and N_{T0}^-, we find

$$g_c = r_c N \exp[-\beta(E_c - E_T)] = A_c \exp[-\beta(E_c - E_T)] \tag{4.6}$$

where $\beta = 1/kT$ and N_c is the factor $2(2\pi m_n kT/h^2)^{3/2}$, of the order of $10^{19}/\text{cm}^3$ at room temperature. An estimate for r is the product of the thermal velocity of the conduction electron ($\sim 10^7$ cm/sec) and the capture cross section ($\sim 10^{-15}$ cm^2). Thus the coefficient

$$A_c \sim 10^{11}\ \text{sec}^{-1} \tag{4.7}$$

Actual coefficients for different traps may vary by several orders of magnitude from this estimate.

There is a similar expression for capture and release of holes from traps. Capture of a hole corresponds to an electron from the trap dropping into the hole and release of a hole to thermal excitation of an electron from the valence band to the trap, forming a hole. The reaction may be expressed in the form

$$h^+ + T^- \leftrightarrows T$$

and the equilibrium rates are

$$r_v p_0 N_{T0}^- = g_v(N_T - N_{T0}^-) \tag{4.8}$$

This gives

$$g_v = A_v \exp[-\beta(E_T - E_v)] \tag{4.9}$$

Considering transfer of electrons to traps from both the valence and conduction bands, we have

$$\frac{dN_T^-}{dt} = r_c n(N_T - N_T^-) - g_c N_T^- + g_v(N_T - N_T^-) - r_v p N_T^- \tag{4.10}$$

Electrons may be thermally excited from the valence to the conduction bands. The reverse process, recombination of an electron and a hole, occurs at a rate proportional to the electron and hole concentrations:

$$\text{Recombination rate} = Cpn$$

The generation rate is equal to the equilibrium recombination rate

$$\text{Generation rate} = Cp_0 n_0$$

The net rate of recombination, when the concentrations are not in equilibrium, is then

$$C(pn - p_0 n_0) = C(pn - n_i^2) \tag{4.11}$$

where n_i is the intrinsic concentration.

We may write

$$n = n_0 + \Delta n \qquad p = p_0 + \Delta p \tag{4.12}$$

If it is assumed that the space charge of electrons in traps is negligible, electrical neutrality requires that $\Delta n = \Delta p$. If the change in concentration is small, the net recombination rate is

$$-\frac{d\,\Delta p}{dt} = c(n_0 + p_0)\,\Delta p = \frac{\Delta p}{\tau} \tag{4.13}$$

The last equality defines the lifetime τ of the added carriers. The lifetime of holes τ_p in n-type material is given by

$$\frac{1}{\tau_p} = Cn_0 \tag{4.14}$$

When the changes in concentration are large, C may depend on n and p.

Recombination in most cases occurs via traps, as illustrated in Fig. 4.1, rather than directly across the gap. The theory of this type of recombination was worked out independently by Shockley and Read [14] and by Hall [15]. They assume that $\Delta n = \Delta p$ and that the concentration of trapped electrons is determined by the requirement that under steady-state

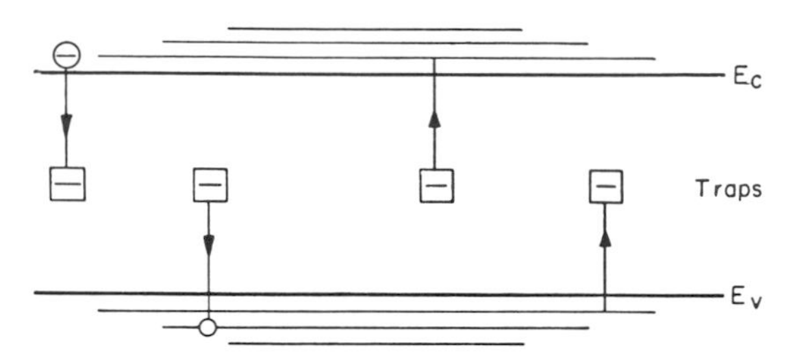

FIG. 4.1. Energy-band diagram illustrating recombination of electrons and holes via traps.

conditions electrons and holes enter the traps at equal rates, so that $dN_T^-/dt = 0$. They find that the net rate of recombination is equal to:

$$\frac{r_c r_v(np - n_i^2)}{r_c(n + n_1) + r_v(p + p_1)} \tag{4.15}$$

where

$$\begin{aligned} n_1 &= N_c \exp[-\beta(E_c - E_T)] \\ p_1 &= N_v \exp[-\beta(E_T - E_v)] \end{aligned} \tag{4.16}$$

It should be noted that the rate is proportional to Δp for Δp small and also for Δp very large.

Recombination traps often occur at the surface, so that a fraction of the minority carriers which hit the surface recombine. The rate of recombination may be expressed in terms of a *surface recombination velocity* [11], s, such that the net rate of recombination per square centimeter of surface per second is equal to $s\,\Delta p$. Values of s on carefully prepared germanium surfaces may be less than 100 cm²/sec. This means that the chance of a hole recombining when it hits the surface is less than 1 in 10^4. Values of s are estimated by measuring lifetime in filaments with varying surface to volume ratios.

Electrons and holes may also recombine directly, with emission of radiation. In a closely related process an electron bound in a shallow donor level may recombine with a hole bound in a neighboring acceptor, with the energy, a little less than the band gap, going into radiation. As we shall discuss later, coherent emission of light occurs by recombination of excess carriers injected in the forward direction in junction lasers. Recombination radiation occurs with greatest intensity in the so-called direct-gap semiconductors, such as GaAs, in which the minimum of the conduction band occurs at the same point in k-space as the maximum in the valence band.

In Ge and Si the gap is indirect; the intensity of recombination radiation is small but is nevertheless sufficiently large to be detected. The first observation of recombination radiation in Ge was made by Haynes and Briggs [16]. By applying the principles of detailed balancing, van Roosbroeck and Shockley [17] had earlier estimated the lifetime for radiative recombination from observed absorption coefficients. The estimated minority-carrier lifetime, of the order of 1 sec at room temperature, is much larger than that observed (usually less than 10^{-3} sec), indicating that recombination in Ge is most likely to occur via traps, without emission of radiation.

Current Flow and Continuity. When the concentrations vary in space, one must consider flow by diffusion as well as by conduction. The current densities for conduction electron and hole flow are then [8]

$$\begin{aligned} j_n &= ne\mu_p\mathcal{E} + eD_n \operatorname{grad} n \\ &= -ne\mu_n \operatorname{grad}\left(\psi - \frac{kT}{e}\log n\right) \end{aligned} \tag{4.17}$$

$$\begin{aligned} j_p &= pe\mu_p\mathcal{E} - eD_p \operatorname{grad} p \\ &= -ne\mu_p \operatorname{grad}\left(\psi + \frac{kT}{e}\log p\right) \end{aligned} \tag{4.18}$$

where $\mathcal{E} = -\operatorname{grad}\psi$ is the electric field intensity. As first pointed out by Einstein, the diffusion coefficient and mobility are related

$$D = \frac{\mu kT}{e} \tag{4.19}$$

The equations of continuity are

$$\frac{\partial n}{\partial t} = G_n - R_n + e^{-1}\operatorname{div} j_n \tag{4.20}$$

$$\frac{\partial p}{\partial t} = G_p - R_p - e^{-1}\operatorname{div} j_p \tag{4.21}$$

where G_n and R_p are the total rates of generation and of recombination for electrons and G_p and R_p are the corresponding quantities for holes. Equations (4.20) and (4.21) together with Poisson's equation (4.3) are a set of three partial differential equations in n, p, and ψ which determine the solution subject to appropriate initial and boundary conditions. If trapping needs to be considered, additional equations of the form (4.10) may be included, and trap concentrations, N_T^-, are additional independent variables. These equations are in general nonlinear; solutions have been obtained only for special cases [18].

Poisson's equation often may be replaced by the condition of electrical neutrality $\rho = 0$ or $\Delta n = \Delta p$. It should be emphasized that these latter conditions do *not* imply $\mathcal{E} = 0$, but that the unbalance of charge required to produce the existing field is negligible. It is often convenient to treat flow across space-charge layers at junctions separately from flow in the bulk, where one may assume $\rho = 0$. The properties of the junction then determine the boundary condition for bulk flow.

The boundary condition at a free surface usually is determined by surface recombination. More complicated conditions may be required to discuss flow when an inversion layer is present at the free surface (see Sec. 4).

Flow in Homogeneous Semiconductors. Some simplification of the equations results when it is assumed that the semiconductor is electrically neutral and homogeneous [19].† Traps are not considered explicitly, although they may indirectly affect the rates of generation and recombination of electrons and holes. Then, n_0 and p_0 are constants and since electrical neutrality requires that $\Delta n = \Delta p$,

$$\operatorname{grad} n = \operatorname{grad} p = \operatorname{grad}\Delta p \tag{4.22}$$

† Our treatment follows closely that of W. van Roosbroeck, ref. 19.

Adding (4.17) and (4.18), we find for the total current density

$$j = j_n + j_p = \sigma \mathcal{E} + e(D_n - D_p) \text{ grad } \Delta p \quad (4.23)$$

We may solve this equation for $\mathcal{E}$, substitute the result in (4.17) and (4.18), and find

$$j_p = \left(\frac{\sigma_p}{\sigma}\right) j - eD \text{ grad } \Delta p \quad (4.24)$$

$$j_n = \left(\frac{\sigma_n}{\sigma}\right) j + eD \text{ grad } \Delta p \quad (4.25)$$

in which D is an effective diffusion coefficient

$$D = \frac{\sigma_n D_p + \sigma_p D_n}{\sigma} \quad (4.26)$$

In n-type material, with $\sigma_n \gg \sigma_p$, $D \simeq D_p$; in p-type, $D \simeq D_n$. When (4.24) is substituted into the continuity Equation (4.20), there results

$$\frac{\partial \Delta p}{\partial t} = G - R - \frac{1}{e} \text{div} \frac{\sigma_n}{e} j + \text{div}(D \text{ grad } \Delta p) \quad (4.27)$$

Linear Equations for Δp Small. Linear equations which can be solved much more easily are obtained when the further assumption is made that the added carrier concentration is small. Thus, to discuss flow of added holes in n-type material, we assume $p_0 \ll n_0$ and $\Delta p \ll n_0$, and keep only the linear terms in Δp. Then

$$\text{div} \frac{\sigma_n}{\sigma} j = ev \cdot \text{grad } \Delta p \quad (4.28)$$

where v is the drift velocity in the field $\mathcal{E} = j/\sigma$:

$$v = \mu_p \mathcal{E} = \frac{\mu_p j}{\sigma} \quad (4.29)$$

We may express the net rate of recombination in terms of the lifetime of the added carriers.

$$R - G = \frac{\Delta p}{\tau_p} \quad (4.30)$$

With these substitutions, (4.27) becomes

$$\frac{\partial \Delta p}{\partial t} = -\frac{\Delta p}{\tau_p} - v \cdot \text{grad } \Delta p + D_p \nabla^2(\Delta p) \quad (4.31)$$

If the electric field is uniform, so that v is a constant, a solution can be obtained from any solution of the diffusion equation

$$\frac{\partial f}{\partial t} = D_p \nabla^2 f \quad (4.32)$$

If v is in the x direction, the corresponding solution of (4.31) is

$$\Delta p = e^{-t/\tau_p} f(x - vt, y, z) \quad (4.33)$$

Recombination gives the exponential decrease; drift and diffusion act independently.

A basic solution of the diffusion equation, which corresponds to $v = 0$ and the added carriers being localized at the origin at $t = 0$, is

$$\Delta p(r,t) = \frac{N_p}{(4\pi L_p)} \exp\left[-\frac{t}{\tau_p} - \frac{r^2}{4D_p t}\right] \quad (4.34)$$

where r is the distance from the origin. The mean net distance a particle diffuses during its lifetime is the diffusion length

$$L_p = \sqrt{D_p \tau_p} \quad (4.35)$$

Typical values for a well-prepared single crystal of germanium are $D_p \sim 40$ cm²/sec, $\tau_p \sim 5 \times 10^{-3}$ sec, $L_p \sim 0.5$ cm.

Similar solutions can be obtained for the near intrinsic case for which p_0 is not necessarily small compared with n_0, although we still suppose that Δp is small so that the linear approximation can be made. The only differences are that in place of (4.29), v is now

$$v = \frac{\mu_n(n_0 - p_0)}{\mu_n n_0 + \mu_p p_0} \mu_p \mathcal{E} \quad (4.36)$$

and that the lifetime τ now is given by

$$\tau^{-1} = C(p_0 + n_0) \quad (4.37)$$

Note that $v = \mu_p \mathcal{E}$ for n-type, $v = \mu_n \mathcal{E}$ for p-type, and $v = 0$ for intrinsic material ($p_0 = n_0$). In the latter case, there is no tendency for added carriers to move except by diffusion.

3. Examples of Flow

Carrier Injection by Point Contacts. An example of flow in which both conduction and diffusion play a role is injection of holes into n-type germanium by current flow from a small-area metal cat's-whisker contact [9, 20]. As illustrated schematically in Fig. 4.2, the diode is constructed from a small block of germanium with a large-area low-resistance contact on the base.

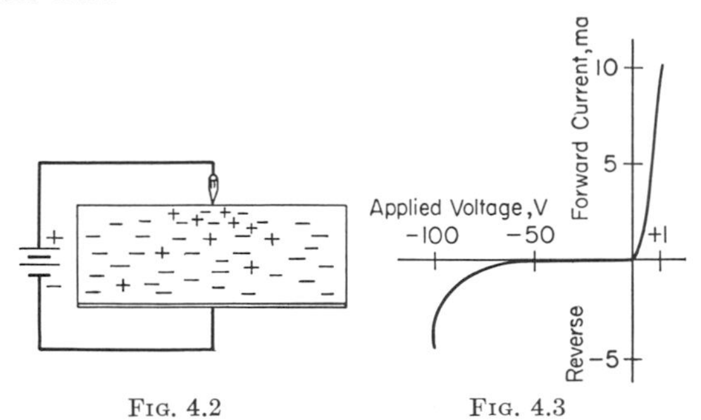

Fig. 4.2 Fig. 4.3

Fig. 4.2. Injection of holes from a metal point contact into n-type germanium.

Fig. 4-3. Typical characteristic of germanium point-contact diode.

It is an excellent rectifier, with a characteristic similar to that shown in Fig. 4.3. The direction of easy flow is that in which the metal is positive. A large part of the current consists of holes flowing into the block from the contact. The nature of the contact is such that there is a potential barrier which inhibits flow of conduction electrons between the germanium and metal. Instead, electrons are extracted from the valence band, and the holes resulting flow out. The space charge of these holes is balanced by an increased concentration of conduction electrons. With moderate forward currents, the increase in conductivity may be by more than a factor of 10 in the immediate vicinity of the contact [21].† Holes introduced in this way have a finite lifetime, since they tend to recombine with conduction electrons.

Nearly all the current may be carried by holes, in spite of the fact that there are nearly equal numbers of electrons and holes near the contact. Conduction and diffusion currents tend to cancel for electrons, while they add for holes. The tendency of electrons to flow away from the contact by diffusion may nearly balance the tendency to flow toward the contact under the influence of the electric field.

The direction of high resistance is that for which the metal is negative. The current must then consist of holes flowing toward the contact or electrons flowing away from the contact. Because there are very few holes normally present in the n-type germanium, the hole current is small. The potential barrier at the contact limits the flow of conduction electrons.

In the point-contact transistor [10], there are two contacts in close proximity, as shown in Fig. 4.4.

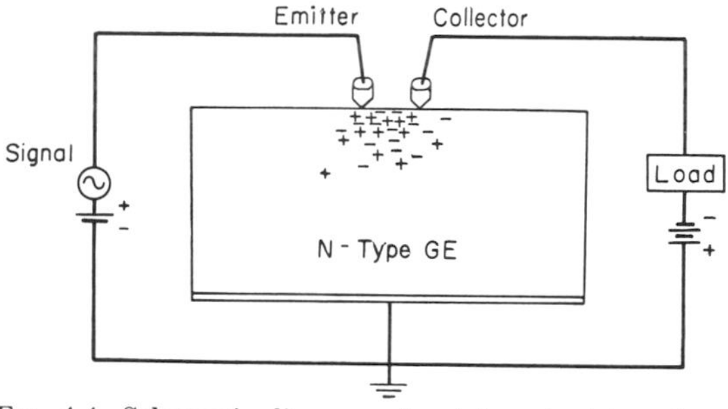

FIG. 4.4. Schematic diagram of point-contact transistor for grounded-base operation. Holes injected at the emitter contact flow to the collector and contribute to the collector current.

One, the emitter, is biased so as to inject holes into the block. The second, the collector, has a relatively large negative bias. When operated normally as an amplifier, the bias current in the collector circuit is several times the bias current in the emitter circuit. Holes injected at the emitter are drawn by the electric field in the block to the collector, so that a large part of the emitter current flows into the collector contact and through the load. Because of the low resistance of the emitter when biased in the forward direction, a small signal voltage applied between emitter and base will produce a relatively large change in emitter current. The load, matched to the collector, may have an impedance of the order of one hundred times that of the emitter. A signal current flowing through this high impedance produces a large voltage across it, giving a large voltage amplification of the input signal. It is found that with a properly "formed" collector contact there is some current amplification as well. Holes flowing into the contact modify the barrier in such a way as to allow more electrons to flow out. In a sense, then, the emitter current may be used to control the current between collector and base. The over-all power gain is of the order of a factor of 100, or 20 db.

† A discussion of the theory of diode characteristics, taking into account flow of both electrons and holes, is given in ref. 10. The theory is extended in ref. 21.

The Haynes-Shockley Experiment. A very beautiful and direct experiment which shows that holes are actually injected in n-type Ge and move as positive charges was made by Haynes and Shockley [22] not long after the initial transistor experiments. A small rod of germanium, 2 or 3 cm long and a few millimeters on a side, is cut from a single crystal. Large area low-resistance electrodes are soldered to the ends so that a field can be applied along the length. A point contact is used to inject a pulse of holes into the rod. These holes are swept down the rod by an applied field. They are detected by observing in an oscilloscope the current flowing to a formed collector contact placed further along the rod. A schematic diagram of the experiment is shown in Fig. 4.5.

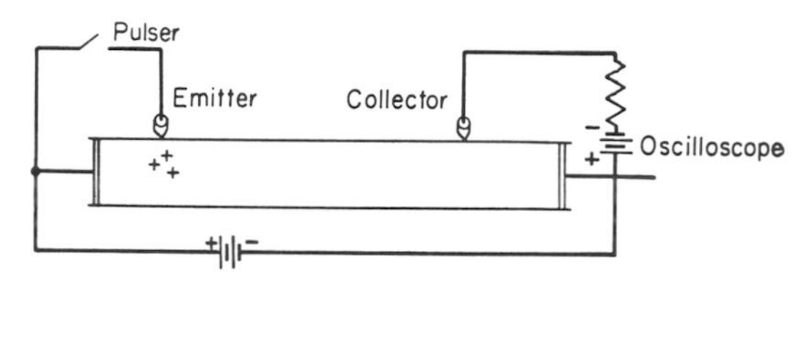

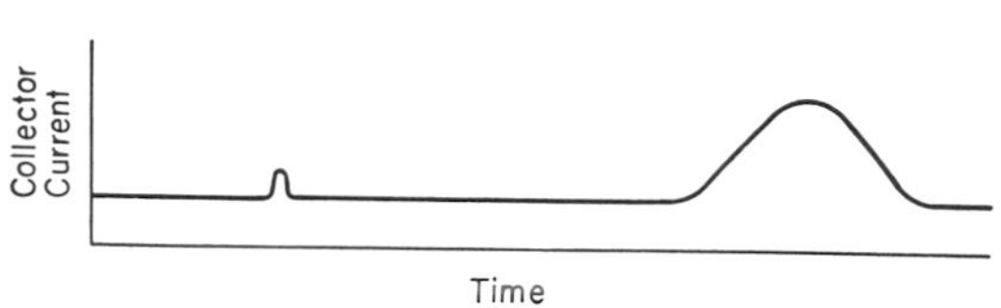

FIG. 4.5. Schematic diagram of Haynes-Shockley experiment for measurement of drift velocity, diffusion, and lifetime of injected carriers.

If the added hole concentration is small, the linear approximation of Eq. (4.31) may be used for analysis of the data. The pulse soon assumes a Gaussian shape given by

$$\Delta p = A \exp \left[-\frac{t}{\tau_p} - \frac{(x - vt)^2}{4D_p t} \right] \tag{4.38}$$

The center of the pulse moves with a velocity $v = \mu_p \mathcal{E}$. It spreads out in time as a result of diffusion and the integrated amplitude decreases as a result of recombination. The experiment yields values of τ_p and D_p as well as the drift mobility μ_p. Einstein's relation between μ_p and D_p has been confirmed [23].

Haynes and others have developed the experiment to make it a precise method for measuring mobility of minority carriers, and it has been widely used for this purpose. In one modification, a pulse of light is used to introduce the added carriers.

Theory of p-n Junctions. A p-n junction is the boundary between p- and n-type regions in the same semiconductor, as illustrated in Fig. 4.6. Best electrical characteristics are obtained if the junction is formed in a single crystal, in part of which there is an excess of acceptor impurities and in part of which there is an excess of donors. The theory of

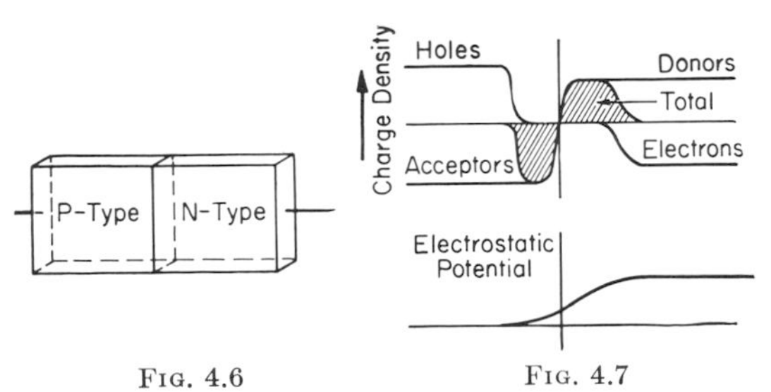

FIG. 4.6 FIG. 4.7

FIG. 4.6. A *p-n* junction, the boundary between *p-* and *n*-type regions of a single crystal.
FIG. 4.7. Variation of charge density and potential at a *p-n* junction in equilibrium.

current flow across such a junction was first worked out by Shockley [24], before any were actually made in the laboratory. Subsequent experiments have confirmed the theory.

A *p-n* junction makes an excellent rectifier. In order to understand how it operates, it is necessary to consider the separate flow of the conduction electrons and holes. The *direction of easy flow* is that for which the p side is positive, the n side negative. Holes can then flow across the junction from the p side to the n side and electrons from the n side to the p side. The space charge of the minority carriers injected in this way is balanced by an equal increase in concentration of majority carriers. The conductivity of the material in the vicinity of the junction is thereby enhanced. Current flow in the *high-resistance direction* must consist of electrons flowing from the p side to the n side. Since there are few minority carriers present on either side, the current is small. It consists of minority carriers thermally generated in the vicinity of the junction. It is interesting to note that the resistance is much higher than would be estimated from the resistivity of the material.

Let us first consider the equilibrium situation with no applied voltage. We then have $j_p = j_n = 0$. The solutions of Eqs. (4.17) and (4.18) with

$$\mathcal{E} = -\operatorname{grad}\psi$$

are then the Boltzmann distributions for electrons and holes:

$$p = p_0 e^{-e\psi(x)/kT} \qquad n = n_0 e^{e\psi(x)/kT} \tag{4.39}$$

If these values are substituted in Poisson's equation, there results a differential equation for ψ which can be solved to give the variation in potential across the junction. A schematic diagram of the variation of charge density and potential across the junction region is given in Fig. 4.7. We shall assume for simplicity that the impurity density is uniform on each side of the junction and, following Shockley, denote the equilibrium concentrations on the p side by p_p and n_p and on the n side by p_n, n_n. The potential also approaches a constant value, more positive on the n side than the p side. The difference is

$$\Delta\psi = \frac{kT}{e}\log\frac{n_n}{n_p} = \frac{kT}{e}\log\frac{p_p}{p_n} \tag{4.40}$$

It is this difference in potential which prevents the majority carriers from diffusing across the junction.

The electric double layer which gives the change in potential consists of uncompensated donors on the n side and uncompensated acceptors on the p side. The width of the region of appreciable charge density depends on the impurity concentration, but is generally of the order of 10^{-4} cm in germanium and silicon junctions.

When a voltage is applied, most of the drop in potential occurs across the double layer. In the forward direction, the potential barrier is decreased; in the reverse direction it is increased. There are corresponding changes in the space-charge regions. Because of the high resistance of the barrier, the current densities are usually so small that the electric field is negligible outside the space-charge region and the minority carriers move by diffusion alone. This greatly simplifies the analysis. This approximation is not valid for large forward currents for which the injected minority carrier density is large.

The diffusion equation for flow of holes in the n region is

$$\frac{\partial p}{\partial t} = \frac{p_n - p}{\partial \tau_p} + D_p\frac{\partial^2 p}{\partial x^2} \tag{4.41}$$

The solution for steady-state conditions is

$$p = p_n + p_1 \exp\left[-\frac{(x - x_R)}{L_p}\right] \tag{4.42}$$

where x_R is a point in the n region just to the right of the space-charge layer. Taking into account the change in height of the barrier by the applied voltage, V_a, the value of p_1 which gives the correct value of p at $x = x_R$ is

$$p_1 = p_n\left[\exp\left(\frac{eV_a}{kT}\right) - 1\right] \tag{4.43}$$

This gives

$$j_p = \left(\frac{eL_p p_n}{\tau_p}\right)\left[\exp\left(\frac{eV_a}{kT}\right) - 1\right] \tag{4.44}$$

There is a similar expression for the electron current density just to the left of the junction. If, as is consistent with (4.43), recombination in the space-charge region is neglected, the total density is the sum of these two:

$$j = j_p + j_n = I_0\left[\exp\left(\frac{eV_a}{kT}\right) - 1\right] \tag{4.45}$$

$$\text{where} \quad I_0 = \frac{eL_p p_n}{\tau_p} + \frac{eL_n n_p}{\tau_n} \tag{4.46}$$

Note that the largest current is transported by carriers from the side with highest conductivity. This expression is valid for the limiting case for which the resistance of the space-charge region proper is negligible compared with the total resistance of the junction. This is true when the diffusion lengths L_p and L_n are large compared with the width of the barrier.

A detailed quantitative check of Shockley's theory has been made at the Bell Telephone Laboratories [25]. The equilibrium concentrations were determined from the measured conductivities on each side of the junction. The diffusion lengths, and thus the lifetime of the minority carriers, were determined by a photoelectric method. As shown in Fig. 4.8, the observed current-voltage characteristic agrees with an equation of the form (4.45), with a value for I_0 close to the predicted one.

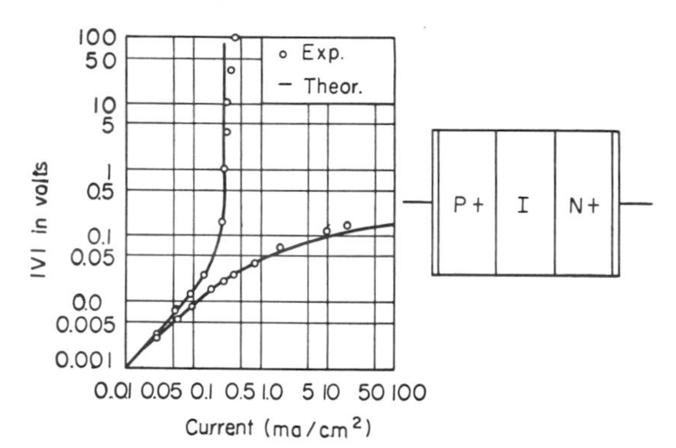

FIG. 4.8 FIG. 4.9

FIG. 4.8. Comparison between theoretical and observed current-voltage characteristics of a *p-n* junction (see ref. 25). [*From Phys. Rev.*, **81**: 637 (1951).]
FIG. 4.9. Schematic diagram of power diode in which high-resistance region separates heavily doped *n-* and *p*-type regions. (*After R. N. Hall* [30].)

It is often convenient to express the current in terms of an effective voltage, B_a, defined by

$$B_a = \frac{kT}{e}\left[\exp\left(\frac{eV_a}{kT}\right) - 1\right] \qquad (4.47)$$

This voltage approaches a constant value $-kT/e$ for V_a negative, increases exponentially for V_a positive, and is nearly equal to V_a when V_a is small. Equation (4.44) may be written

$$j_p = G_{p0}B_a \qquad (4.48)$$

where

$$G_{p0} = \frac{e\mu_p p_n}{L_p} \qquad (4.49)$$

is the limiting value of the conductance from hole flow for small applied voltages. The conductance is that of the minority carriers for a conductor of length L_p.

The theory can be extended readily so as to apply to alternating currents. As shown by Shockley [24], the small-signal complex admittance due to holes, for a direct bias voltage V_0, is

$$A_p = G_p + iS_p = (1 + i\omega\tau_p)^{1/2}G_{p0}\exp\left(\frac{eV_0}{kT}\right) \qquad (4.50)$$

in which ω is the angular frequency. There is an effective capacity associated with holes flowing in and out of the n region. The total admittance is the sum of A_p and the corresponding A_n. When V_0 is small or negative, it is necessary to include the susceptance of the space-charge layer, which acts as a capacitative shunt to flow of carriers across the barrier.

A new phenomenon occurs when a critical voltage is reached in the reverse direction. The reverse current increases very rapidly with little or no further increase in voltage. It was thought [26] originally that the breakdown is due to a mechanism once suggested by Zener, a direct pulling of electrons from the valence to the conduction band by the strong electric field in the junction region. For this reason, the effect has been called a Zener breakdown. However, present evidence indicates that the breakdown is nearly always due to avalanche formation [27]. When the field reaches a critical value, conduction electrons (or holes) are excited to sufficiently high energies to knock other electrons from the valence to the conduction bands, creating new conduction electrons and holes. These are in turn excited until the avalanche builds up. Evidence for avalanche formation is a large multiplication of electron-hole pairs created by light when the reverse voltage is close to the critical value.

A *p-n* junction makes an excellent photodiode [28]. Minority carriers created by adsorption of light within a diffusion length of the junction may diffuse to the junction and give a photocurrent. With favorable design, the conversion efficiency of light to electrical energy is fairly high. Theory indicates a possible efficiency of more than 20 per cent. A *solar battery* for conversion of sunlight designed and built by Chapin, Fuller, and Pearson [29] has an efficiency (1954) of about 8 per cent. It is a large-area *p-n* junction created close to the surface by diffusion of boron into silicon from the vapor phase.

Power Diodes. Hall [30] suggested a modified form of the diode for handling large amounts of power. As illustrated in Fig. 4.9, a narrow intrinsic or high-resistivity region separates junctions to highly conducting p material on the left to highly conducting n material on the right. The equilibrium energy-level diagram is also shown. When a voltage is applied, part of the drop occurs across the left junction and part across the right. In the forward directions, holes are injected from the left and electrons from the right, so that there is a large increase in carrier concentration of the central layer. Because of the conductivity ratios, few electrons transverse the left barrier and few holes the right. The current is determined almost entirely by recombination in the layer. Forward currents as high as several hundred amperes per square centimeter are obtained, with a rectification ratio as high as ten million.

Junction Lasers. Coherent emission of light has been observed in suitably prepared *p-n* junctions of GaAs [31–33] and of $GaAs_xP_{1-x}$ [34]. The former emits in the infrared and the latter in the red part of the spectrum. In these materials there is a direct gap between the valence and conduction bands such that both band edges occur at the same point in k-space. Carriers injected in the forward direction recombine by emission of light of a frequency such that $h\nu$ is about equal to the band gap. When the forward current is sufficiently large, recombination may occur by stimulated emission; under suitable conditions, coherent light is observed. A schematic energy-level diagram of a *p-n* junction with a forward bias is given in Fig. 4.10 illustrating both direct and indirect recombination via donor and acceptor levels.

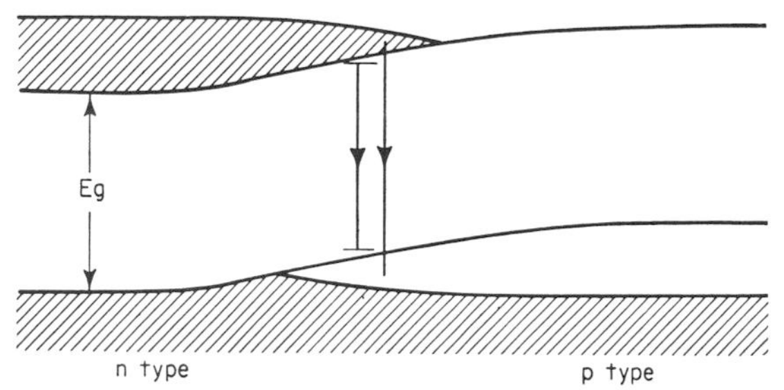

FIG. 4.10. Energy-level diagram for junction laser with forward bias. Electrons and holes in the inverted population may recombine directly or via shallow donor and acceptor levels.

Figure 4.11 is a diagram of a junction designed to give coherent emission of light. Opposite faces are polished as optical flats parallel to one another; the spacing determines the frequencies of possible resonant modes. When biased in the forward direction, coherent light is emitted from the junction regions in a direction normal to these faces.

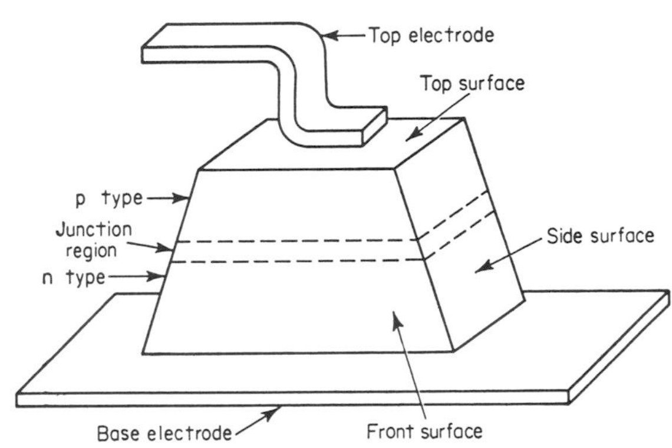

FIG. 4.11. Junction-laser geometry. Opposite faces are polished to be flat and parallel to form a Fabry-Perot interferometer. The spacing is typically less than 1 mm. (*After R. N. Hall* [31].)

To observe coherent light, it is necessary to apply sufficient voltage so that the separation between the quasi-Fermi levels for electrons and holes is greater than the energy of the radiation produced, or $eV > h\nu$. There is evidence that in suitable junctions nearly all the recombination occurs by radiation, so that conversion of d-c energy to light energy occurs with almost 100 per cent efficiency. However, it is difficult to get more than a small fraction of the light out of the crystal.

Esaki Tunnel Diode. In 1957 Esaki discovered that heavily doped *p-n* junctions with narrow space-charge layers exhibit a negative-resistance characteristic when a bias is applied in the forward direction [35, 36]. As illustrated in Fig. 4.12 the forward current increases linearly for small applied voltages, reaches a peak, and then decreases to a rather broad minimum after which there is another rapid increase. He correctly interpreted this unexpected result as a tunneling phenomenon.

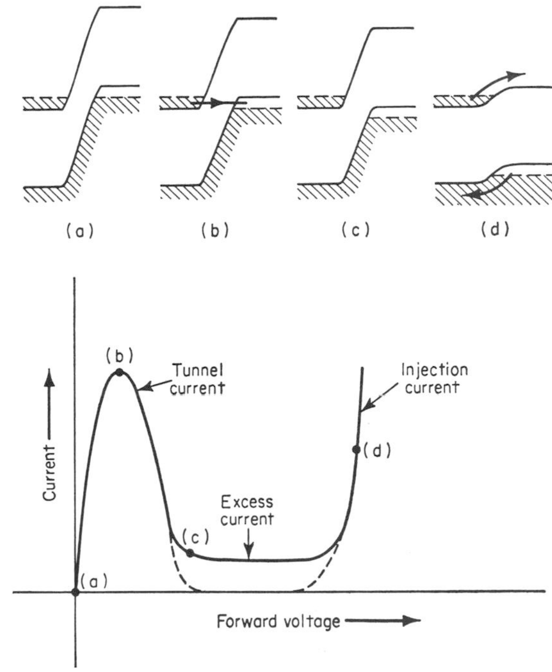

FIG. 4.12. Energy-level diagrams for various bias voltages and current-voltage characteristic for an Esaki tunnel diode. (*After R. N. Hall* [36].)

An energy-level diagram for an Esaki diode with no voltage applied is given in Fig. 4.12*a*. Both *p* and *n* regions are so heavily doped that the conductivity is in the metallic range, with the Fermi level above the bottom of the conduction band on the *n* side and below the top of the valence band on the *p* side of the junction. The occupied states below the Fermi levels are indicated by shading. The space-charge layer is so thin (less than $\sim 10^{-6}$ cm) that electrons can tunnel directly from the occupied states in the conduction band to unoccupied states in the valence band. When a small forward voltage is applied, raising the *n* side relative to the *p* side, a current flow proportional to the applied voltage occurs in this manner. However, when the voltage is so large as to raise the bottom of the conduction band to the Fermi level on the *p* side, as illustrated in Fig. 4.12*b*, this tunneling current starts to decrease and ceases when it is raised above the top of the valence band. As illustrated in Fig. 4.12*c*, there are then no unoccupied states of the same energy to which the electrons can tunnel. With still further increase, a voltage will be reached such that electrons can be injected in the usual way over the barrier into the *p* region, and correspondingly holes can flow into the *n* region (see Fig. 4.12*d*). This accounts for the second rapid rise in current with voltage, shown schematically in the lower diagram of Fig. 4.12.

The unusual current-voltage characteristics and the fact that they can operate at very high frequencies have given rise to a number of applications of Esaki diodes, particularly in computer circuits. Rise and fall times of the order of 10^{-10} sec have been observed in switching circuits employing such diodes.

Particle Detectors. There has been considerable use of *p-n* junctions as detectors for charged particles [37]. The junction is made close to the surface and is designed to have, when biased in the reverse direction, a broad space-charge layer with a uniform electric field. As the incoming particle loses energy by ionization, secondary electrons create electron-hole pairs. Those created in the space-charge layer are separated by the field. The time integral of the resulting current pulse is a good measure of the total energy loss of the incoming particle. If it is stopped in the space-charge layer, the integrated pulse is proportional to the energy. Since junctions can be made with small areas, one can get good spacial resolution with such detectors.

The Junction Transistor. As illustrated in Fig. 4.13, a *p-n-p* junction transistor consists of a narrow

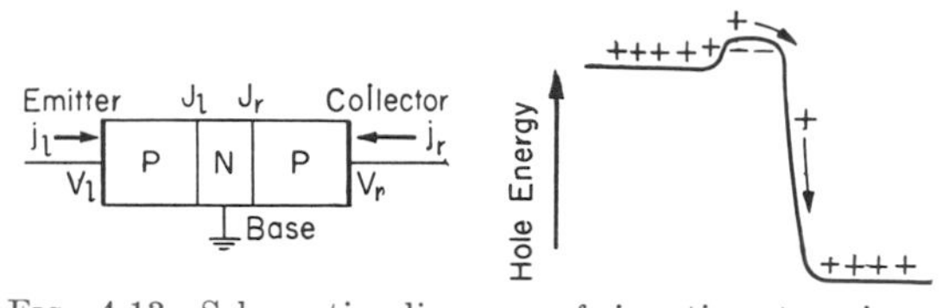

FIG. 4.13. Schematic diagram of junction transistor, showing energy-level diagram when biased for use as an amplifier.

n region sandwiched between two *p* regions of the same single crystal. Low-resistance contacts are made to each side and to the *n*-type base layer. When operated as a transistor, the left junction, biased positively with respect to the base, acts as an emitter and the right junction, biased negatively, acts as a collector. In an *n-p-n* transistor, a *p* region is sandwiched between *n* regions. These structures were suggested by Shockley [24] and the theory worked out long before they were made in the laboratory [38].

Transistors can be fabricated in various ways: by introduction of impurities during growth of the single crystal [39, 40], by fusion of metals or alloys containing donors or acceptors to the semiconductor [41], and by direct diffusion of impurities.

Minority carriers (holes in *p-n-p* structures) injected into the base layer at the emitter junction diffuse across the layer to the collector junction and into the collector circuit. The magnitude of this current is controlled by the voltage between emitter and base. There is a close analogy with a vacuum tube: the emitter acts like the cathode, the base layer like the grid which controls the flow of carriers, and the collector like the plate. In a favorable design, more than 98 per cent of the emitter current flows to the collector, only 2 per cent to the base.

When a number of approximations are made, calculation of direct current characteristics is quite simple. We shall assume that (1) minority carrier concentrations are small, so that the linear approximation can be made, (2) no lateral flow in the base layer, so that the flow is unidirectional, and (3) recombination in the base layer is negligible. Conventional directions for current flow and for bias voltages are as indicated in Fig. 4.13. The subscript l shall denote the left (emitter) junction and r the right (collector) junction. We shall use effective voltages as defined by (4.48) with B_l that for the left junction and B_r that for the right junction. The width of the base layer is W.

First consider the current flow due B_l with $B_r = 0$. The condition $B_r = 0$ implies that the hole concentration in the base layer near the right junction is equal to the equilibrium value, p_{b0}. The concentration in the layer near the left junction is

$$p_b = p_{b0} + \frac{e}{kT} B_l p_b \tag{4.51}$$

The current density from electrons diffusing across the base layer is

$$j_p = eD_p \frac{(p_l - p_b)}{W} = \frac{e\mu_p p_{b0}}{W} B_l = G_{plr} B_l \tag{4.52}$$

The expression is similar to (4.48) except that W replaces the diffusion length L_p. In good transistor design, $L_p \gg W$, so that hole flow is enhanced by a factor L_p/W. This current flows from the left junction through the right junction and contributes to j_r. The electron flow across the left junction due to B_l gives

$$j_{ln} = \frac{e\mu_n n_{l0}}{L_0} B_l = G_{nl} B_l \tag{4.53}$$

where n_{l0} is the equilibrium electron concentration in the p region on the left. There are corresponding expressions for currents due to B_r. The total currents flowing to the left and right junctions from both electron and hole flow are

$$j_l = G_{ll} B_l + G_{lr} B_r \tag{4.54}$$

$$j_r = G_{rl} + G_{rr} B_r \tag{4.55}$$

where $G_{ll} = G_{plr} + G_{nl} \qquad G_{rr} = G_{plr} + G_{nr}$ (4.56)

$$G_{lr} = G_{rl} = -G_{plr}$$

An important parameter of the transistor is the current multiplication factor α, defined by

$$\alpha = -\left(\frac{\partial j_r}{\partial j}\right)_{B_r=\text{const}} = \frac{-G_{rl}}{G_{ll}} \tag{4.57}$$

The low-level operation is thus characterized by three parameters, G_{ll}, G_{lr}, G_{rr}. When operated as a transistor, B_l is large and positive and $B_r = -kT/e$, the saturation value for negative applied voltages. It is desirable to make G_{nl} small compared with G_{plr} so that nearly all the flow across the left junction consists of holes and α is close to unity. This is accomplished by making W small (<0.01 cm) and the conductivity of the p region on the emitter side large compared with that of the base region, so that $n_{l0} \ll p_{b0}$.

According to the theory presented above, the emitter current is independent of collector voltage in the saturation region. This is not quite true. As pointed out by Early [42], the effective base width W decreases with increasing collector voltage because of the increase in width of the space-charge region of the collector junction. Another correction required is to include a base resistance resulting from lateral flow in the base layer. Rittner [43] has given equations appropriate for high-level operation. It is easy to take into account corrections for recombination in the base layer.

The theory has been extended to apply to alternating currents. We shall give a brief qualitative discussion of the frequency characteristics. The limiting factors are the transit time (or really dispersion of transit times) of minority carriers in the base layer and the time constant (RC) associated with charging the capacitance of the collector junction through the effective resistance of the base layer. Another factor is the emitter capacitance associated with added holes in the base layer. Thus to obtain operation at high frequencies it is desirable to make W, the collector junction capacitance and the base resistance, all as small as possible. The junction capacitance is inversely proportional to the width of the space-charge layer, and is smallest for high-resistivity materials with small donor and acceptor concentrations. However, use of high resistivity base material with small W increases the base resistance, so that a compromise must be made.

Continual progress has been made in increasing the frequency response of transistors, so that at present there are silicon transistors that operate beyond 10^9 cps and germanium transistors at even higher frequencies. This has been achieved by a number of innovations in design and technology. One of these is the use of a graded base layer, as first suggested by Early [44] in the p-n-i-p transistor illustrated in Fig. 4.14. A layer of substantially intrinsic (or high-

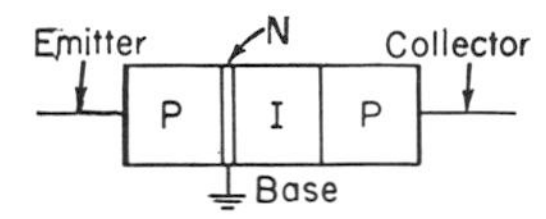

FIG. 4.14. Junction-transistor design for high-frequency operation. (*After J. M. Early* [39].)

resistivity n-type) material is sandwiched between the n-type base layer and the p-type collector region. The thin n region is made of relatively high conductivity, reducing the base-layer resistance. When a sufficiently large voltage is applied to the collector, the space-charge region extends through the intrinsic layer to the n region; thus the collector capacitance is small. Once the injected holes from the emitter diffuse through the thin n region, they are swept by the field to the collector junction so that the transit time is small.

The effect of a graded junction is obtained automatically if the donor impurities are introduced by diffusion from the emitter side. With controlled diffusion, there can be a large concentration gradient of donors with the highest concentration adjacent to the emitter. Acceptor impurities are then diffused in to make the emitter junction.

To reduce the series resistance to the collector, it is desirable to have high-conductivity material immediately adjacent to the junction. An abrupt change from high conductivity to low conductivity can be achieved by epitaxial growth. Starting from a single-crystal high-conductivity substrate, a low-conductivity layer is grown epitaxially. The appropriate diffusion steps can then be carried out in the epitaxial layer.

High-frequency performance requires transistors of extremely small dimensions. Fabrication of silicon transistors is carried out by a series of diffusion and etching steps, with use of masks and photoresistive methods to lay out the desired structures. A further important step is the use of an oxide grown on the surface to stabilize the surface, particularly where the junction is exposed. Use of oxide masking also permits selective diffusion; the oxide impedes diffusion of certain impurities such as boron and phosphorus. These techniques when combined with innovations in thin-film technology for deposition of resistors and capacitors form the basis for present-day microelectronics. References to some of the literature on this subject are given in the bibliography at the end of this chapter.

A planar design similar to that shown schematically in Fig. 4.15 is the one most frequently used in microelectronic circuits. This design is particularly suitable for mass production of transistors or, in combination with other elements, of monolithic integrated circuits.

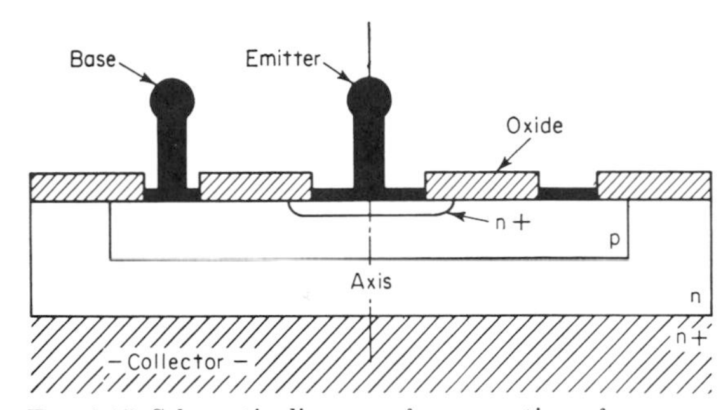

FIG. 4.15. Schematic diagram of cross section of an n-p-n planar transistor. Typically there is circular symmetry about the axis.

Another interesting device is the n-p-n (or p-n-p) "hook" collector, suggested by Shockley [38], and shown schematically in Fig. 4.16. The inner layers,

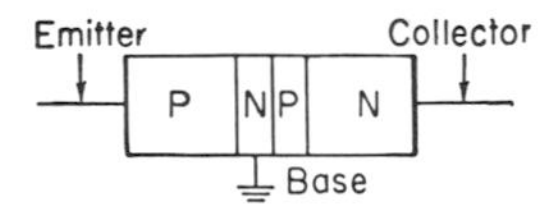

FIG. 4.16. Schematic diagram of p-n-p-n junction transistor with n-p-n "hook" collector. (*After Shockley et al.* [31].)

n_b and p_c, are both narrow, so that both the p-n-p structure on the left and the n-p-n structure on the right have characteristics like junction transistors. The base connection is to n_b. The device operates like a p-n-p transistor, but with a current multiplication factor α^* much greater than unity. Most of the collector bias voltage appears across J_2, which is in the reverse direction. Holes injected at the emitter junction, J_1, flow into the p layer "hook," biasing it more positively, and then across J_3. Neglecting recombination, the bias at which the layer "floats" is determined by the requirement that as many holes leave J_3 as enter across J_2.

The value of α^* may be estimated as follows. Let α be the current multiplication factor of the n-p-n structure with emitter junction at J_3. The ratio of electron to hole current flowing across J_3 is then $\alpha/(1 - \alpha)$. This ratio applies to the injected hole current coming from J_1. Thus

$$\alpha^* = 1 + \frac{\alpha}{1 - \alpha} = \frac{1}{1 - \alpha} \tag{4.58}$$

For example, if $\alpha = 0.98$, $\alpha^* = 50$.

Equivalent results [45] can be obtained by connecting *n-p-n* and *p-n-p* transistors as shown in Fig. 4.17.

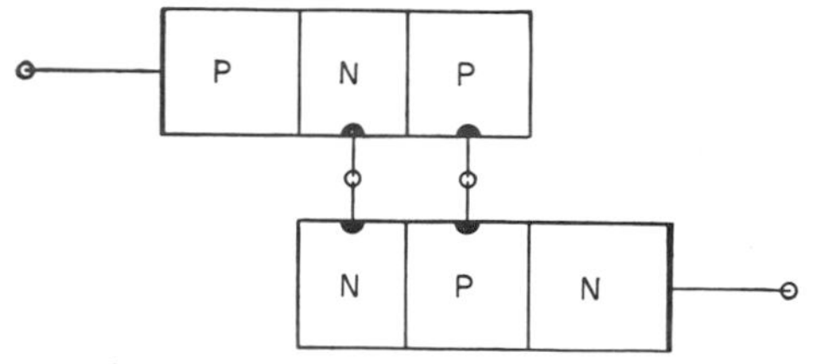

FIG. 4.17. Combination of *n.p.n.* and *p-n-p* transistors for current multiplication. (*After Ebers* [40].)

Without the base connection, Fig. 4.16 is the structure of a Shockley four-layer diode. With proper design, it has two stable states of operation, one with very high impedance and the other with very low impedance and high current flow. Switching from one state to the other can be accomplished with appropriate voltage pulses. When an additional control electrode is added to the base layer to facilitate switching, the device is called a silicon controlled rectifier (SCR).

4. Space-charge Layers and Metal-Semiconductor Contacts

In this section we discuss rectification at metal-semiconductor contacts and space-charge layers which may occur at the free surface of a semiconductor. When any two conductors in thermal equilibrium are brought into contact, an electric double layer is formed at the interface which adjusts the electrostatic potential of one relative to the other so that the Fermi levels are the same. In contacts between metals, the thickness of the double layer is of atomic dimensions [46]. In a metal-semiconductor contact, part of the double layer may consist of a space-charge layer extending to a depth of the order of 10^{-6} to 10^{-4} cm into the semiconductor, the charge in the layer being compensated by charges at the immediate interface. At a free surface, the charge in the layer is compensated by a surface charge.

As pointed out by Schottky [6], the space charge may consist of uncompensated donor or acceptor ions, as illustrated in Fig. 4.18 for an *n*-type semiconductor in contact with a metal. Except for appropriate changes in signs, the same theory can be used for both *n*- and *p*-type semiconductors. It is a little more convenient to talk about conduction electrons. Donor states which lie above the Fermi level on the diagram tend to be positively charged (unoccupied by electrons), those below, neutral. In the body of the semiconductor, there is a balance between conduction electrons and positive donors. In the space-charge region, where there are few electrons and nearly all of the donors are ionized, there is a net positive charge density. Charge densities are illustrated schematically in the lower part of

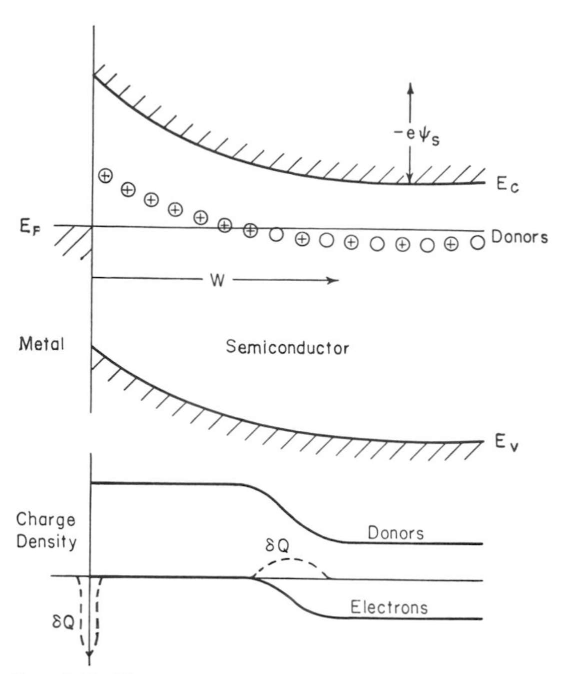

FIG. 4.18. Energy-level diagram and space charge at a metal-semiconductor contact.

Fig. 4.18. Schottky [47] denotes the region of complete ionization and few conduction electrons the *exhaustion* region. The transition region to the electrically neutral body of the semiconductor is called the *reserve* region. The contact is rectifying, current flowing most readily when the metal is positive.

Schottky Exhaustion-layer Theory. An approximate estimate of the width of the space-charge region can be obtained from the Schottky exhaustion-layer theory [6], in which the rather crude approximation is made that there is an abrupt transition from the exhaustion region to the electrically neutral region in the body of the semiconductor. In other words, it is assumed that the reserve region has negligible thickness [47]. Poisson's equation is used to determine the variation of the electrostatic potential ψ in the layer:

$$\frac{d^2\psi}{dx^2} = \frac{4\pi e}{\kappa} N_d \tag{4.59}$$

The total energy rise from the interior to the surface is, for a uniform donor concentration,

$$-e\psi_s = \frac{2\pi e^2}{\kappa} N_d w^2 \tag{4.60}$$

For a potential rise of the order of 0.5 volt and a dielectric constant of about 12, the following table gives the order of magnitude of the thickness l of the space-charge layer

N_d	10^{19}/cm³	10^{17}/cm³	10^{15}/cm²
w	10^{-6} cm	10^{-5} cm	10^{-4} cm
wN_d	10^{13}/cm²	10^{12}/cm²	10^{11}/cm²
w^3N_d	10	10^2	10^3

The value 10^{-4} cm is typical of copper oxide and germanium junctions. The number of charges per unit area, $N_d w$, given in the third row is equal to the number of compensating charges of opposite polarity at the interface. Note that this number is small compared with the number of surface atoms, which is $\sim 10^{15}/\text{cm}^2$. Also of interest is to estimate the number of charges in a cube of side w. There are only 10 for a donor concentration of 10^{19}; so the approximation of treating the discrete charges as a continuous distribution is a rather poor one when the density is as large as this.

When a voltage, V_a, is applied, most of the drop occurs across the barrier layer. The width increases in the reverse direction, decreases in the forward direction. The dependence of w on V_a is determined by adding V_a to ψ_s:

$$-e(V_a + \psi_s) = \frac{2\pi e^2}{\kappa} N_d w^2 \tag{4.61}$$

For our case, V_a is positive when the metal is positive and ψ_s is a negative number.

As shown by Schottky, the variation of w with V_a can be determined from the variation of the capacitance of the junction with V_a. The barrier acts like a parallel-plate capacitor with spacing w and dielectric constant κ, so that the capacity per unit area is

$$C = \frac{\kappa}{4\pi w} \tag{4.62}$$

Schottky plots $1/C^2$ as a function of V_a. According to (4.61), this should be a straight line with slope proportional to N_d. He shows that if N_d varies in depth, the plot is no longer straight, but the instantaneous slope is proportional to the density of donors at a depth w corresponding to the applied voltage.

In Fig. 4.18 is shown the change in charge, δQ, for a small increment in voltage. It can be seen that this charge is centered in the reserve region. The compensating charge is at the interface. Even though the reserve region is fairly wide, the parallel-plate formula should give a reasonably good estimate of the capacity.

The capacitance of the barrier is in parallel with the variable resistance resulting from flow of carriers across the barrier. Both are in series with the resistance corresponding to flow in the body of the semiconductor [48].

Other Types of Space-charge Layers. In the diagram of Fig. 4.18 it has been assumed that the valence band everywhere is well below the Fermi level, so that nearly all of the flow is by the majority carrier. This is the case considered by Mott and by Schottky, and it is the one which applies to Cu_2O (with holes the majority carrier). Other possibilities are illustrated in Fig. 4.19: (*a*) The energy bands may turn down at the surface, giving enhanced electron conductivity and an ohmic contact. (*b*) The barrier may raise the energy bands so high that at the interface the valence band is closer to the Fermi level than is the conduction band. This means that there is a change of conductivity type from n to p, even though there are no acceptor states in the bulk of the semiconductor. Both hole and electron flow must be considered in a barrier of this sort. A rough indication of the relative importance

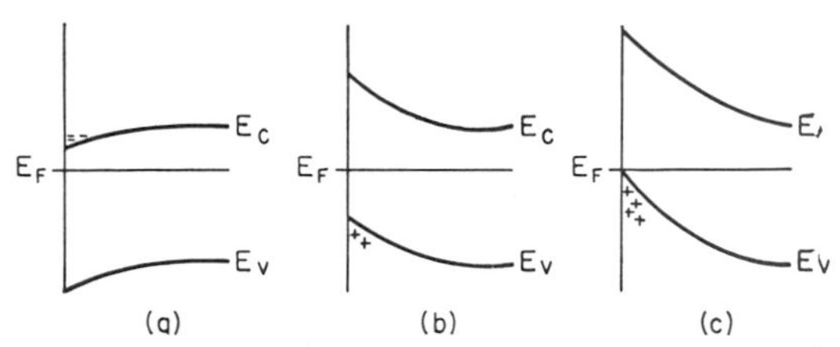

FIG. 4.19. Types of space-charge layers which may exist at a metal-semiconductor contact or at a free surface.

can be obtained by comparing the electron concentration where the barrier is a maximum with the minority hole concentration in the body. If $E_{cs} - E_F$ is greater than $E_F - E_V$, one may expect that most of the current flow across the junction is by holes. Here E_{cs} is the energy of the conduction-band edge at the surface. Holes are injected when the metal is positive. (*c*) In extreme cases, actually attained when an acceptor is a constituent of an alloy fused to the semiconductor, the valence band at the interface may be raised to or above the Fermi level. Nearly all the flow is then by the minority carrier. The reverse current consists of holes diffusing to the contact. In the forward direction, holes are injected. This case can equally well be considered as an extreme case of a *p-n* junction, in which the *p* region has very high conductivity so that nearly all the current is carried by holes. Cases analogous to (*a*), (*b*), and (*c*), of course, may also occur for a *p*-type semiconductor in contact with a metal.

Space-charge layers of the types illustrated in Figs. 4.18 and 4.19 may also occur at a free surface [49], although such an extreme case as Fig. 4.19*c* is unlikely. The formation of such a layer may be described as follows. Let E_{FS} be a position of the Fermi level such that surface itself is electrically neutral (no true surface charge) and let E_F be the position of the Fermi level in the interior, determined so that the body of the semiconductor is neutral. In equilibrium, the Fermi level must be the same everywhere, and so will be E_F at the surface too. If there were no barrier, there would be a net negative charge on the surface if $E_{FS} < E_F$ and a net positive charge if $E_{FS} > E_F$. Actually, a barrier is formed in such a way that the surface charge is balanced by the charge in the space-charge layer. The bands are raised if there is a negative surface charge, lowered if there is a positive surface charge. The surface charge may arise from adsorbed ions and from electrons in various types of surface states.

A number of physical properties are affected by the presence of such a double layer at the free surface of a semiconductor: (1) work function and contact potential [49, 50], (2) chemisorbtion and chemical reactions which involve electron transfer [51], (3) surface recombination [52], (4) change in contact potential with illumination [50], (5) surface conductance and change in surface conductance when an electric field applied normal to the surface [53] and (6) noise which originates at the surface. Surface conductance is increased by electron flow if the bands turn down (Fig. 4.19*a*), and it is increased by hole flow if they turn up to form an inversion layer (Fig. 4.19*b*). The conductance is a minimum at an intermediate position.

Brattain [50] has observed that the surface barriers of germanium and silicon change with the gaseous ambient surrounding the surface. Oxygen or ozone tends to make the bands at the surface of germanium move up, water vapor moves them down. Inversion layers have been observed at the surface of both n- and p-type germanium.

An inversion layer at the surface of the base layer in a junction transistor, for example, an n-type inversion layer in a n-p-n transistor, gives a direct path between emitter and collector which shorts out transistor action [54]. Such a path has been called a *channel*. Schrieffer [55] has estimated the reduction in mobility as a result of surface scattering when carriers are confined to a region near the surface by a barrier, as is the case for electrons in Fig. 4.19*a* and holes in Fig. 4.9*b*. Conductance in an inversion layer at the surface may also have a large effect on the characteristics of point-contact diodes and p-n junctions [56].

An interesting question is: What determines the barrier potential, ψ_s? No general answer can be given; ψ_s depends on the energy required to take an electron from the metal and place it in the conduction band of the semiconductor, and this in turn depends in part on the double layer at the immediate interface. In some cases, as suggested by Schottky [6, 47] the barrier is determined by the difference in work functions between metal and semiconductor. Characteristics of germanium point-contact diodes are nearly independent of the work function of the metal [57]. It has been suggested that the barrier layer in this case exists at the free surface and is only slightly modified by contact with the metal [49].

Current Flow by Majority Carrier. Different theories apply to the current-voltage characteristics of barrier layers of the sort illustrated in Fig. 4.18, depending on whether or not current is limited by scattering in the barrier itself. The diffusion theory of Mott [5] and Schottky [6], based on an equation similar to (4.17), leads to the current density:

$$I_{\text{diff}} = n_s e \mu_n |\mathcal{E}_s| [\exp (eV_a/kT) - 1] \quad (4.63)$$

where V_a is the applied voltage, taken positive in the forward direction, and n_s and $\mathcal{E}_s$ are the carrier concentration and electric field at the interface. It follows from Poisson's equation (4.59) that

$$|\mathcal{E}_s| = \frac{4\pi e N_d w}{\kappa} = \left[\frac{-8\pi e N_d(V_a + \psi_s)}{\kappa}\right]^{1/2} \quad (4.64)$$

The concentration n_s at the interface is maintained at the equilibrium value by transfer of electrons from the metal:

$$n_s = n_0 \exp (e\psi_s/kT) \quad (4.65)$$

where n_0 is the equilibrium concentration in the bulk. Some increase in current is obtained when the image force is considered, particularly if the barrier is not very wide. The reverse current (V_a large and negative) is just the conduction current at the interface in the field $\mathcal{E}_s$.

Bethe [9] suggested that the current may be limited by supply of electrons to the layer rather than by scattering in the layer. This assumption leads to the diode theory, and to a characteristic

$$I_{\text{diode}} = \left(\frac{n_s v}{4}\right) [\exp (eV_a/kT) - 1] \quad (4.66)$$

where v is the mean speed, $2(2kT/\pi m)^{1/2}$. The saturation current in the reverse direction is just the thermionic emission current from the metal to the conduction band of the semiconductor.

The theory which gives the smallest current is the one which should be applied. The criterion may be expressed in terms of the mean free path. If the barrier drops by more than kT in a distance equal to the mfp, the diode theory is applicable. The diffusion theory applies to copper oxide rectifiers, the diode theory to germanium and silicon.

The comparison between theory and experiment for metal-semiconductor contacts is not nearly as satisfactory as that for p-n junctions. One can often fit the d-c characteristic with an expression of the form of (4.63) or (4.66), but usually with e/kT replaced by a parameter of the order half of the theoretical value. Since the theoretical is approached in some cases, the theory is probably satisfactory, and the difficulty is that actual junctions are far from ideal. In particular, there may be a considerable variation of ψ_s over the interface (see ref. 9 for a discussion).

It is possible to inject majority carriers into a semiconductor or insulator if the junction is of the type shown in Fig. 4.19*a*. Since space charge is introduced, a large voltage is required to get an appreciable current to flow. For plane-parallel geometry, Mott and Gurney [58] obtain an expression for the space-charge limited current which may be written

$$I = 10^{-13} \frac{V^2 \mu \kappa}{L^3} \text{ amp/cm}^2 \quad (4.67)$$

where L is the separation between electrodes, μ is the mobility, and κ the dielectric constant. Such currents have been observed in Ge [59] and in CdS [60]. Shockley [61] has suggested making use of space-charge limited currents in semiconductor devices in exact analogy with vacuum tubes.

Another amplifying device in which the important flow is by majority carriers is the field effect or unipolar transistor, in which the number of carriers flowing in a thin layer is modulated by application of a transverse electric field. The general principle was known prior to the invention of the point-contact transistor, but did not prove practical with materials available at the time. In one form, a semiconductor in the form of a thin film is made one plate of a parallel-plate capacitor. If charge induced in the film by application of a voltage to the capacitor consists of mobile carriers, the conductance of the film changes. In experiments made to test this idea, it was found that a large part of the induced charge is trapped in surface states and so does not affect the conductance. This difficulty is avoided if the transverse field is applied across p-n junctions, as in the arrangement of Fig. 4.20 suggested by Shockley [62]. When a reverse bias voltage is applied across the junction, the width of the space-charge layer increases and the channel available for flow decreases. When sufficient voltage is applied, the channel can be "pinched off" with a large increase in resistance. A theoretical analysis by Shockley indicates a considerable power gain is possible when the device is used as an amplifier, and this was first verified in experiments of Pearson and of Dacey and Ross [63].

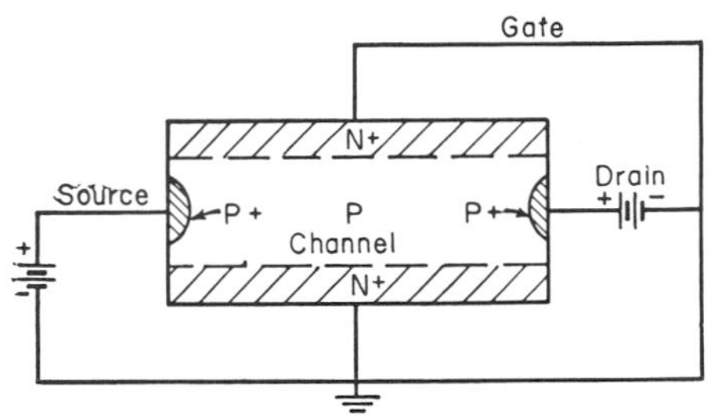

FIG. 4.20. Field effect or unipolar amplifier. (*After Shockley* [57].)

With the development of integrated circuits, there has been a revival of interest in field-effect devices in which the field is applied across a thin insulating layer rather than across a *p-n* junction. One version [64] makes use of thin-film technology; in another [65] the field is applied across the oxide layer on the surface of a semiconductor. The latter, called the metal-oxide-semiconductor or MOS transistor, is illustrated in Fig. 4.21. Silicon with a silicon-oxide

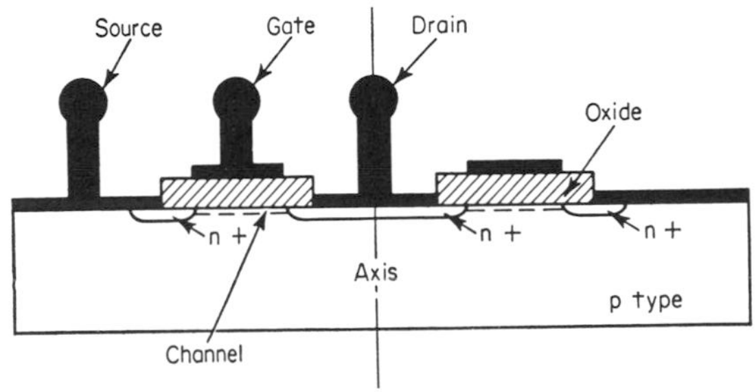

FIG. 4.21. Schematic cross section of an MOS (metal-oxide-semiconductor) transistor. In one form there is circular symmetry about the axis.

layer on the surface is most commonly used. The insulated gate has very high impedance. A positive voltage applied to the gate produces an *n*-type inversion layer at the surface of the *p*-type block. The diffused *n*-type layers contact this inversion layer at the source and the drain electrodes. Variation of voltage on the gate changes the conductance and thus the current flow between source and drain electrodes. As with the junction field-effect device, the inversion layer can be "pinched off" with a saturation of current as the drain voltage is increased.

Concluding Remarks. Many important topics such as trapping levels, magnetic effects, and noise have been omitted in this brief survey of a large and growing field of research. Perhaps in the long run one of the most important aspects from both a scientific and practical point of view is the production of nearly perfect crystals with controllable electrical properties. This makes it possible to design a material or structure with desired properties, and then to fabricate it. The transition from empirical art to engineering design of materials is also taking place in related fields: photoconductors, phosphors, magnetic materials, ferroelectrics, etc.

References

1. Braun, F.: *Ann. Phys. Pogg.*, **153**: 556 (1874).
2. Schuster, A.: *Phil. Mag.*, **48**: 251 (1874).
3. Lark-Horovitz, K.: article in *The Present State of Physics*, p. 66, American Association for the Advancement of Science, Washington, 1954.
4. Grondahl, L. O.: *Phys. Rev.*, **27**: 813 (1926).
5. Mott, N. F.: *Proc. Roy. Soc. (London)*, **171**: 27 (1939).
6. Schottky, W.: *Z. Physik*, **113**: 367 (1939).
7. Wagner, C.: *Physik. Z.*, **32**: 641 (1931).
8. Wagner, C.: *Z. physik. Chem.*, **B21**: 25 (1933). Wagner introduced the diffusion term in his theory of oxidation of metals. J. Frenkel, *Physik. Z. Sowjetunion*, **8**: 185 (1935), used equations identical with (4.17) and (4.18) to account for the change in contact potential with illumination (Dember effect) and the photomagnetoelectric effect in semiconductors. The latter is a voltage observed between the ends of a slab placed in a transverse magnetic field (parallel to the plane of the slab) when one face is illuminated. The effect, first reported by K. Kikoin and M. Noskov, *Physik. Z. Sowjetunion*, **5**: 586 (1934), in Cu_2O films, was rediscovered in germanium by P. Agrain and H. Bulliard, *Compt. rend.*, **236**: 595, 672 (1953), and by T. S. Moss, *Proc. Phys. Soc. (London)*, **B66**: 993 (1953). Electrons and holes created by absorption of light near the surface diffuse toward the interior. The paths are bent in opposite directions by the magnetic field (Hall effect) so that electrons tend to flow toward one end of the slab and holes toward the opposite, creating a voltage. The photomagnetoelectric effect is a good illustration of diffusive flow, B. Davydov, *J. Tech. Phys. (U.S.S.R.)*, **5**: 87 (1938), used equations equivalent to (4.20) and (4.21) to account for rectification at a junction between *p*- and *n*-type materials. Since he did not consider transfer of electrons and holes across the boundary, but only the current resulting from thermal generation at the boundary, he did not arrive at a satisfactory theory of rectification in a *p-n* junction in a single crystal. The diffusion equations were used by Mott, Schottky, and Davydov to explain rectification at a metal-semiconductor contact (see Sec. 4) and later by Shockley (1949) in his theory of *p-n* junctions, discussed in Sec. 3.
9. Torrey, H. C., and C. A. Whitmer: "Crystal Rectifiers," McGraw-Hill, New York, 1948.
10. Bardeen, J., and W. H. Brattain: *Phys. Rev.*, **74**: 230 (1948); **75**: 1208 (1949).
11. Shockley, W., G. L. Pearson, and J. R. Haynes: *Bell System Tech. J.*, **28**: 344 (1949). An excellent account is found in W. Shockley, "Electrons and Holes in Semiconductors," Van Nostrand, Princeton, N.J., 1950.
12. Teal, G. K., and J. B. Little: *Phys. Rev.*, **78**: 637 (1951). Teal, G. K., and E. Buehler: *Phys. Rev.*, **87**: 190 (1952).
13. Pfann, W. G., and K. M. Olsen: *Phys. Rev.*, **89**: 322 (1953).
14. Shockley, W., and W. T. Read, Jr.: *Phys. Rev.*, **87**: 835 (1952).
15. Hall, R. N.: *Phys. Rev.*, **87**: 387 (1952).
16. Haynes, J. R., and H. B. Briggs: *Phys. Rev.*, **86**: 647 (1952). Newman, R.: *Phys. Rev.*, **91**: 1313 (1953).
17. Van Roosbroeck, W., and W. Shockley: *Phys. Rev.*, **94**: 1558 (1954).
18. Van Roosbroeck, W.: *Bell System Tech. J.*, **29**: 560 (1950). Prim, R. C., III; *Bell System Tech. J.*, **30**: 1174 (1951). These papers deal specifically with the theory of flow.
19. Van Roosbroeck, W.: *Phys. Rev.*, **91**: 282 (1953).
20. Benzer, S.: *J. Appl. Phys.*, **20**: 804 (1949).

21. Banbury, R. C.: *Proc. Phys. Soc. (London)*, **B66**: 883 (1953).
22. Haynes, J. H., and W. Shockley: *Phys. Rev.*, **75**: 691 (1949); **81**: 835 (1951).
23. Haynes, J. H.: *Phys. Rev.*, **88**: 1368 (1952).
24. Shockley, W.: *Bell System Tech. J.*, **28**: 435 (1949).
25. Goucher, F. S., G. L. Pearson, M. Sparks, G. K. Teal, and W. Shockley: *Phys. Rev.*, **81**: 637 (1951). Hall, H. H., J. Bardeen, and G. L. Pearson: *Phys. Rev.*, **84**: 129 (1951).
26. McAfee, K. B., E. J. Ryder, W. Shockley, and M. Sparks: *Phys. Rev.*, **83**: 650 (1951).
27. McKay, K. G.: *Phys. Rev.*, **94**: 877 (1954). Wolff, P. A.: *Phys. Rev.*, **59**: 1415 (1954).
28. Shive, J. N.: *Bell Labs. Record*, August, 1950; *Proc IRE*, **40**: 1410 (1952).
29. Chapin, D. M., C. S. Fuller, and G. L. Pearson: *J. Appl. Phys.*, **25**: 676 (1954).
30. Hall, R. N.: *Proc. IRE*, **40**: 1512 (1952). Prim, R. C., III: *Bell System Tech. J.*, **32**: 665 (1954).
31. Hall, R. N., G. H. Fenner, J. D. Kingsley, T. J. Soltys, and R. O. Carlson: *Phys. Rev. Letters*, **9**: 366 (1962). Hall, R. N.: *Solid State Electron.*, **6**: 405 (1963).
32. Nathan, M. I., W. P. Dumke, G. Burns, F. H. Dill, and G. J. Lasher: *Appl. Phys. Letters*, **1**: 62 (1962).
33. Quist, T. M., R. H. Rediker, R. J. Keyes, W. E. Krag, B. Lax, A. L. McWhorter, and H. J. Ziegler: *Appl. Phys. Letters*, **1**: 91 (1962).
34. Holonyak, N., and S. F. Bevacqua: *Appl. Phys. Letters*, **1**: 82 (1962).
35. Esaki, L.: *Phys. Rev.*, **109**: 603 (1958).
36. Hall, R. N.: *IRE Trans. Electron Devices*, January, 1960, pp. 1–9.
37. Dearnaley, G., and D. C. Northrop: "Semiconductor Counters for Nuclear Radiations," Spon Press, London, 1964.
38. Shockley, W., M. Sparks, and G. K. Teal: *Phys. Rev.*, **83**: 151 (1951).
39. Teal, G. K., M. Sparks, and E. Buehler: *Phys. Rev.*, **81**: 637 (1951). Teal, G. K., and E. Buehler: *Phys. Rev.*, **87**: 190 (1952).
40. Hall, R. N.: *Phys. Rev.*, **88**: 139 (1952).
41. Hall, R. N., and W. Dunlap; *Phys. Rev.*, **80**: 467 (1950). Pearson, G. L., and B. Sawyer: *Proc. IRE*, **40**: 1348 (1952).
42. Early, J. M.: *Proc. IRE*, **40**: 1401 (1952); *Bell System Tech. J.*, **32**: 1271 (1954).
43. Rittner, E. S.: *Phys. Rev.*, **94**: 1161 (1954).
44. Early, J. M.: *Bell System Tech. J.*, **32**: 517 (1954).
45. Ebers, J. J.: *Proc. IRE*, **40**: 1361 (1952).
46. Fan, H. Y.: *Phys. Rev.*, **62**: 388 (1942).
47. More elaborate theories, in which this assumption is not made, have been given by W. Schottky and E. Spenke, *Wiss. Veröffemtl. Siemens-Werken*, **18**: 225 (1939), B. Davydov, and others.
48. Bardeen, J.: *Bell System Tech. J.*, **28**: 94 (1949) (a rather general derivation of this result).
49. Bardeen, J.: *Phys. Rev.*, **71**: 717 (1947).
50. Brattain, W. H., and J. Bardeen: *Bell System Tech. J.*, **32**: 1 (1953). Brattain, W. H., and C. G. B. Garrett: *Physica*, **20**: 885 (1954).
51. Agrain, P., and C. Dugas: *Z. Electrochem.*, **56**: 363 (1952). Weisz, P. B.: *J. Chem. Phys.*, **20**: 1483 (1952); **21**: 1531 (1953).
52. Stevenson, D. T., and R. J. Keyes: *Physica*, **20**: 1041 (1954).
53. Morrison, S. R.: *J. Phys. Chem.*, **57**: 860 (1953). Clarke, E. N.: *Phys. Rev.*, **91**: 756 (1953); **94**: 1420 (1954). Bardeen, J., and S. R. Morrison: *Physica*, **20**: 873 (1954). Low, G. G. E.: *Proc. Phys. Soc. (London)*, **B68**: 10 (1955).
54. Brown, W. L.: *Phys. Rev.*, **91**: 518 (1953). Kingston, R. H.: *Phys. Rev.*, **93**: 346 (1954).
55. Schrieffer, J. R.: *Phys. Rev.*, **97**: 641 (1955).
56. Brattain, W. H., and J. Bardeen: *Phys. Rev.*, **74**: 332 (1948). Holonyak, N.: Thesis, University of Illinois, 1954. Cutler, M., and H. M. Bath: *J. Appl. Phys.*, **25**: 1440 (1954).
57. Meyerhof, W. E.: *Phys. Rev.*, **71**: 727 (1947). Benzer, S.: ref. 20. Bocciarelli, C. V.: *Physica*, **20**: 1020 (1954).
58. Mott, N. F., and R. W. Gurney: "Electronic Processes in Ionic Crystals," p. 172, Oxford University Press, New York and London, 1940.
59. Dacey, G. C.: *Phys. Rev.*, **90**: 759 (1953).
60. Rose, A., and R. W. Smith: *Phys. Rev.*, **92**: 857 (1953).
61. Shockley, W.: *Proc. IRE*, **40**: 1289 (1952).
62. Shockley, W.: *Proc. IRE*, **40**: 1365 (1952).
63. Pearson, G. L.: *Phys. Rev.*, **90**: 336 (1953). Dacey, G. C., and I. M. Ross: *Proc. IRE*, **41**: 970 (1953).
64. Weimer, P. K.: *Proc. IRE*, **50**: 1462 (1962).
65. Sah, C. T.: *Trans IEEE*, **ED11**: 324 (1964).

Bibliography

Biondi, F. J. (ed.): "Transistor Technology," vols. 1–3, Van Nostrand, Princeton, N.J., 1958.
Coblenz, A., and H. L. Owens: "Transistors," McGraw-Hill, New York, 1955.
Dunlap, W. C., Jr.: "An Introduction to Semiconductors," Wiley, New York, 1957.
Gentry, F. E., F. W. Gutzwiller, N. Holonyak, Jr., and E. E. Von Zastrow: "Semiconductor Controlled Rectifiers," Prentice-Hall, Englewood Cliffs, N.J., 1964.
Henisch, H. K.: "Rectifying Semiconductor Contacts," Oxford University Press, Fair Lawn, N.J., 1957.
Jonscher, A. K.: "Principles of Semiconductor Device Operation," Wiley, New York, 1960.
Keonjian, E. (ed.): "Micropower Electronics," Pergamon Press, New York, 1964.
Keonjian, E. (ed.): "Microelectronics," McGraw-Hill, New York, 1963.
Kingston, R. H. (ed.): "Semiconductor Surface Physics," University of Pennsylvania Press, Philadelphia, 1957.
Madelung, O.: Halbleiter, in "Encyclopedia of Physics," vol. 20, pp. 1–245, Springer, Berlin, 1957.
Many, A., Y. Goldstein, and N. B. Grover: "Semiconductor Surfaces," North-Holland Pub. Co., Amsterdam, 1965.
Middlebrook, R. D.: "An Introduction to Junction Transistor Theory," Wiley, New York, 1957.
Moll, John L.: "Physics of Semiconductors," McGraw-Hill, New York, 1964.
Mott, N. F., and R. W. Gurney: "Electronic Processes in Ionic Crystals," Oxford University Press, Fair Lawn, N.J., 1940.
Nussbaum, A.: "Semiconductor Device Physics," Prentice-Hall, Englewood Cliffs, N.J., 1963.
Phillips, A. B.: "Transistor Engineering," McGraw-Hill, New York, 1962.
Runyan, W. R.: "Silicon Semiconductor Technology," McGraw-Hill, New York, 1965.
Schottky, W. (ed.): "Halbleiterprobleme," vols. 1–6, Friedr. Vieweg & Sohn, Brunswick, Germany, 1956–1961.
Sevin, L. J.: "Field-effect Transistors," McGraw-Hill, New York, 1965.
Shockley, W.: "Electrons and Holes in Semiconductors," Van Nostrand, Princeton, N.J., 1950.
Spenke, E.: "Elektronische Halbleiter," Springer, Berlin, 1955. (Transl.: "Electronic Semiconductors," McGraw-Hill, New York, 1958.)
Torrey, H. C., and C. A. Whitmer: "Crystal Rectifiers," McGraw-Hill, New York, 1948.
Valdes, L. B.: "The Physical Theory of Transistors," McGraw-Hill, New York, 1961.

Chapter 5

Photoelectric Effect

By R. J. MAURER, University of Illinois

1. General Considerations [1]†

The photoelectric effect was discovered by Hertz and Hallwachs in 1887–1888. The effect consists of the ejection of electrons from the surface of a solid when electromagnetic radiation is incident upon it. Although the photoelectric effect occurs at the surfaces of semiconductors and insulators, metallic surfaces have been the chief object of investigation.

The number of photoelectrons produced per unit time is proportional to the intensity of the incident electromagnetic radiation; so the photoelectric yield, which is the number of photoelectrons per incident quantum of radiation, is independent of intensity. The most important factors which determine the photoelectric yield are the nature of the metal, the state of contamination of its surface by adsorbed gas, and the frequency of the radiation. The state of polarization and the angle of incidence of the radiation may also be of considerable importance in particular if the surface exhibits specular reflection. Because of the difficulty in preparing gas-free surfaces there are almost no photoelectric data available which can be considered characteristic of clean metal surfaces.

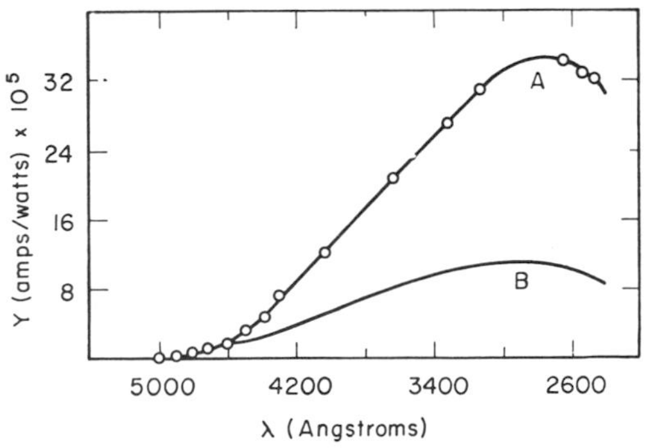

Fig. 5.1. The spectral distribution curve of barium. (*A*) Experimental; (*B*) theoretical. [*R. J. Maurer: Phys. Rev.*, **57**: 653 (1940).]

The typical dependence of the photoelectric yield on frequency is shown in Fig. 5.1 for a barium surface [2]. A characteristic feature of the spectral distribution curve is an apparent threshold frequency, a minimum frequency of radiation for which photoelectric emission is detectable. Since the photoelectric yield decreases rapidly as the frequency of the incident radiation is decreased but does not become zero, the apparent threshold frequency depends upon the sensitivity of the experimental apparatus. The apparent threshold also depends upon the temperature of the surface since the photoelectric yield increases with increasing temperature for frequencies near the apparent threshold. The photoelectric yield usually exhibits a maximum at a frequency somewhat less than twice the apparent threshold frequency. Because the apparent threshold for the majority of the metals lies in the vicinity of 4 ev, the maximum of the spectral distribution curve ordinarily occurs in a relatively inaccessible region of the ultraviolet. Of the pure metals, the alkalis and some of the alkaline earths possess apparent thresholds at sufficiently small frequencies to permit the maximum of the spectral distribution curve to be conveniently observed.

The photoelectric yield of pure metal surfaces at the maximum of the spectral distribution curve is of the order of 10^{-3} electrons per incident quantum. A few solids exhibit much larger yields. Caesium-antimony is an outstanding example with a maximum yield of approximately 10^{-1} electrons per incident quantum.

Specular surfaces usually exhibit a maximum yield for an angle of incidence of the radiation near 60°. The yield may be larger by a factor as much as 10 for radiation polarized with the electric vector in the plane of incidence than when the electric vector is perpendicular to the plane of incidence and therefore parallel to the surface.

The kinetic energies of the individual photoelectrons which are ejected by a fixed frequency and intensity of radiation are distributed over a range from zero to indefinitely large values. The form of the energy distribution function, $n(\epsilon)$, the relative number of electrons of energy ϵ, per unit energy range, is shown in Fig. 5.2. The relative number of fast electrons is

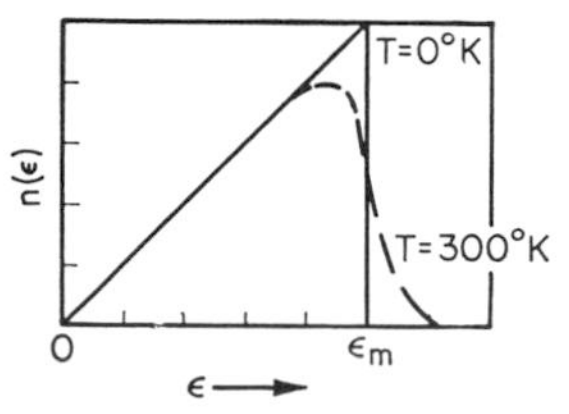

Fig. 5.2. The theoretical distribution in energy of photoelectrons from a metal, according to DuBridge.

† Numbers in brackets refer to References at end of chapter.

small and the original investigators concluded that a maximum kinetic energy of emission, ϵ_M, existed. As in the case of the photoelectric threshold, the apparent maximum emission energy is due to the finite sensitivity of experimental apparatus and the extremely rapid decrease in the number of fast electrons with increasing energy. A prime achievement of modern photoelectric research has been to give a definite and physical interpretation to the concepts of "threshold frequency" and "maximum kinetic energy of emission."

The original experiments on the distribution in energy of photoelectrons showed that the apparent maximum kinetic energy of emission was independent of the intensity but was a function of the frequency of the radiation. These results led Einstein to the hypothesis that photoelectric emission was a quantum effect in which the energy, $h\nu$, of a quantum of radiant energy was "absorbed" by an electron in the metal which thereby increased its kinetic energy by this amount. The observed distribution in energy of the photoelectrons was assumed to result from energy losses suffered by the electrons in escaping from the metal. Einstein suggested that the maximum kinetic energy of emission is given by

$$\epsilon_M = h\nu - \phi \tag{5.1}$$

where ϕ, the work function of the metal surface, is the minimum possible loss of kinetic energy by the escaping photoelectron. Despite the difficulties inherent in the concept of a maximum kinetic energy of emission, Millikan's experimental confirmation of this equation in 1916 offered powerful support to the quantum theory of radiation and provided an independent value of Planck's constant, h. It is to be noted that Eq. (5.1) implies a photoelectric threshold frequency $\nu_0 = \phi/h$ for $\epsilon_M = 0$. Further support for the quantum theory of the photoelectric effect was given by the experiments of Lawrence and Beams, who showed that the time lag between the incidence of radiation on a surface and the appearance of photoelectrons is less than 10^{-9} sec.

2. The Spectral Distribution Function

Within the framework of a general theory of the interaction of electromagnetic radiation with a solid, photoelectric emission appears as a by-product of the process of optical absorption. The theory of photoelectric emission begins with an assumed model of the solid which specifies the allowed states of the electrons in terms of their wave functions and energy levels; proceeds with a calculation of the optical transition probabilities connecting the initial and excited states of the electrons; and concludes with a calculation of the probability of an excited electron escaping through the surface. An excited electron which possesses sufficient kinetic energy to cross the potential energy barrier that exists at the surface of a solid may be reflected back into the solid by the barrier. In the case of a typical metal, although the radiation penetrates to a depth of approximately 10^{-5} cm below the surface, the photoelectrons come from a much thinner surface layer because electron-electron collisions limit the mean free path of an excited electron to a distance of the order of 10^{-7} cm [3]. In an insulator, electron-phonon collisions may limit the volume from which emission can be obtained.

In the Bloch approximation the allowed optical transitions of an electron moving in the periodic potential of a crystal are restricted to transitions between states of the same reduced wave number in different bands. This selection rule is illustrated in Fig. 5.3 for a one-dimensional metal by the vertical

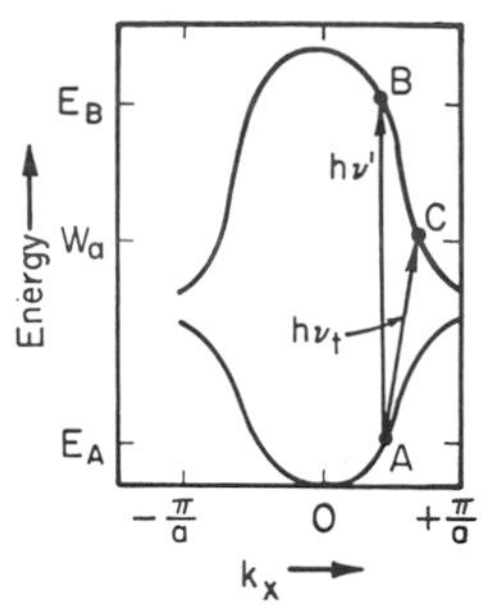

FIG. 5.3. Energy versus reduced wave-number vector for a one-dimensional metal with lattice constant a.

arrow connecting states A and B. If A represents a state lying at the surface of the Fermi distribution of electrons in a conduction band, then $E_B - E_A = h\nu'$ defines the threshold frequency for volume optical absorption by the conduction electrons. If W_a represents the height of the surface potential energy barrier, the energy difference, $W_a - E_A$, is the thermionic work function of the metal. In all cases where precise and comparable measurements have been made it has been found that the thermionic work function and the photoelectric work function, $h\nu_t$, of metals agree closely. It appears that optical transitions occurring between states such as A and C are the important ones for photoelectric emission near the threshold [4].

At the surface of a metal, the potential is not periodic and the optical transitions of an electron moving in the field of the surface potential energy barrier are not restricted by the volume optical selection rules. The observed photoelectric effect results from a *surface* optical absorption which is ignored as negligible in the conventional theory of the optical properties of metals [5]. Photoelectric measurements on the high-work-function metals, such as tungsten, have probably not been extended to sufficiently large frequencies to observe volume emission, while the alkali metals with their small work functions are optically transparent, the volume absorption being too small to contribute a detectable component to the observed surface emission.

The theory of surface photoelectric emission has been developed with the use of the Sommerfeld model of a metal (Fig. 5.4) which automatically excludes the possibility of volume absorption since the optical transition probabilities of an electron in a constant potential vanish [6]. At the surface of the metal, one may assume a discontinuous rise in potential energy of amount W_a or, more plausibly, a rapid but smoothly increasing potential energy which becomes the image

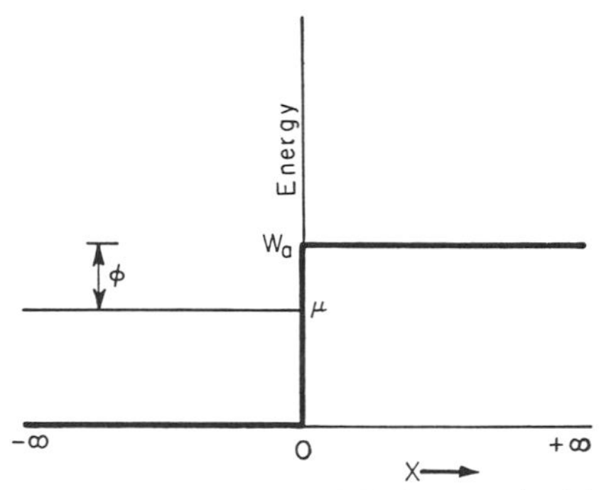

FIG. 5.4. The Sommerfeld model of a metal with a discontinuous potential step of magnitude W_a at the surface. The Fermi energy is μ and the work function is φ.

potential, $V = -(e^2/4x)$, at a distance of the order of 10^{-7} cm from the surface.

Inside the metal, the density of states, $D(E)$, is given by

$$D(E) = \frac{1}{2\pi^2}\left(\frac{2m}{\hbar^2}\right)^{3/2} E^{1/2} \tag{5.2}$$

where the zero of energy has been taken as the constant potential energy of an electron inside the metal. The population of the states by electrons is determined by the Fermi function

$$f(E) = \frac{1}{e^{E-\mu/kT} + 1} \tag{5.3}$$

where μ, the energy of the Fermi level, represents the kinetic energy of the most energetic electron in the metal at the absolute zero of temperature. Because of spin each state may be occupied by two electrons. The position of the Fermi level at 0°K can be calculated if the density of free electrons, n, is known.

$$\mu = \frac{h^2}{2m}\left(\frac{3}{8\pi} n\right)^{2/3} \tag{5.4}$$

The small temperature dependence of the Fermi energy may be neglected.

If the metal is assumed to extend indefinitely in the y, z and negative x directions with the surface at the plane $x = 0$, the initial unperturbed wave functions of the electrons are

$$\psi_k = \alpha_k[\exp(-ik_x x) + a_k \exp(ik_x x)] \exp(ik_y y + ik_z z) \qquad x < 0$$
$$\psi_k = \alpha_k b_k \exp(-px) \exp(ik_y y + ik_z z) \qquad x > 0 \tag{5.5}$$

where $b_k = 1 + a_k$, $pb_k = ik_x(1 - a_k)$, and p is a real constant. The wave-number vector $\mathbf{k}$ of an electron in the metal is related to its kinetic energy by the relation

$$k^2 = k_x^2 + k_y^2 + k_z^2 = \frac{8\pi^2 m}{h^2} E \tag{5.6}$$

The effect of the incident radiation upon the system can be calculated by first-order perturbation theory. The character of the radiation is defined by the form assigned to its vector potential $\mathbf{A}$, which appears as the perturbation in the Hamiltonian of the Schroedinger equation

$$-\frac{ihe}{2\pi mc}(\mathbf{A}, \text{grad } u) \tag{5.7}$$

Here u is the perturbed wave function of an electron. A plausible procedure for fixing the form of the vector potential inside and outside the metal is to use Maxwell's equations and the experimentally determined optical constants of the metal. A still simpler procedure is to ignore the optical constants and use the vector potential of a plane wave

$$\mathbf{A} = \mathbf{A}_0 \cos 2\pi\nu \left(t + \frac{x\cos\theta + y\sin\theta}{c}\right) \tag{5.8}$$

where θ is the angle of incidence. Reflection, refraction, and absorption are ignored by this simpler procedure so that the *absolute* yield obtained in this manner is not accurate.

Having obtained the perturbed wave functions for the region $x > 0$, the current density per electron, $\mathbf{j}_x$, is calculated

$$\mathbf{j}_x = \frac{eh}{4\pi mi}\left(u\frac{\partial u^*}{\partial x} - u^*\frac{\partial u}{\partial x}\right) \tag{5.9}$$

and summed over all initial states which after absorption of a quantum, $h\nu$, can contribute to the current. If this total photoelectric current density is divided by the rate at which quanta of radiation energy are incident upon the surface, the photoelectric yield results.

The theoretical spectral distribution curve for barium, which is shown in Fig. 5.1, was calculated by this procedure. A discontinuous rise in potential was assumed at the surface of the metal. The magnitude of the potential jump W_a was fixed by addition of the observed work function, 2.48 ev, to the calculated Fermi energy. It was assumed that barium contained 1.8 free electrons per atom. The simplest form for the vector potential, Eq. (5.8), was taken. The absolute yield as shown in Fig. 5.1 was fixed by a procedure which is described later.

The yield is given by the following integral, the integrand of which is made up of three easily interpretable terms.

$$Y_x = \frac{e^3\nu_a \sin^2\theta}{4\pi^2 m^2 c\cos\theta} \iiint \left(\frac{2}{8\pi^3}\right) \frac{dk_x\,dk_y\,dk_z}{1 + \exp(E-\mu)/kT} \cdot \frac{1}{\nu^4}\frac{k_x^2}{(k_x^2 + \bar{\mu}\nu)^{1/2}} \cdot \frac{4(k_x^2 + \bar{\mu}\nu)^{1/2}[k_x^2 + \bar{\mu}(\nu - \nu_a)]^{1/2}}{\{(k_x^2 + \bar{\mu}\nu)^{1/2} + [k_x^2 + \bar{\mu}(\nu - \nu_a)]^{1/2}\}^2} \tag{5.10}$$

where $W_a = h\nu_a$, $\bar{\mu} = 8\pi^2 m/h$, $E = hk^2/\bar{\mu}$, and μ is the Fermi energy. The integration is extended over all electrons in the metal whose wave-number vectors, k_x, normal to the surface are sufficient to enable them to escape after absorption of a quantum.

The first term in the integrand is the Fermi function; the exponential factor in it can be neglected at room temperature for $(\mu - E)$ greater than a few hundredths of an electron volt; the limits of integration being fixed in this case by taking μ as the energy of the most energetic electron in the metal. This is equivalent to treating the electron gas as if it were at 0°K. The temperature is therefore unimportant except very near the apparent threshold frequency.

The third term in the integrand is the transmission coefficient of a discontinuous potential step. For a simple continuous step, such as the image potential barrier, the transmission coefficient is almost unity for practically all electrons having sufficient energy to escape [7, 8]. The theory is improved, therefore, by discarding the third term of the integrand. This was done in calculating the theoretical curve of Fig. 5.1.

The second term in the integrand is a product of the probability that an electron in a state of energy E will be excited to a state of energy $E + h\nu$ and the probability that an electron, in the state of energy $E + h\nu$, will escape through the surface.

Only the component of the vector potential parallel to the plane of incidence gives rise to photoelectrons with the ideally plane surface which has been assumed. For this reason, the yield is zero for normal incidence of the radiation.

Mitchell has calculated absolute yields for the case of an image potential barrier and a rough surface which is composed of elements small compared to the wavelength of the radiation but large compared to the electron wavelength [6]. In addition, he has included the variation of the optical constants with wavelength for the case of sodium. With these assumptions the effect of the state of polarization and the angle of incidence of the radiation can be observed. The theory is only moderately successful in reproducing the experimentally observed spectral distribution curves. The calculated frequency dependence of the yield is not in very good agreement with experiment in the case of sodium and the absolute magnitude of the yield is too small by a factor of about 50 for both sodium and barium [2]. The extent to which the experimental data can be considered representative of the behavior of clean metal surfaces is, of course, always open to question.

Schiff and Thomas have given a quantum theory of reflection and refraction at a metallic surface and shown that the component of the electric vector which is perpendicular to the surface oscillates with large amplitude near the surface [9]. It must be concluded that the use of classical optical theory in photoelectric theory is of dubious validity. Makinson has given, however, a semiclassical treatment of photoelectric emission from a totally reflecting metal which approximates the procedure of Schiff and Thomas and compared his results with data from potassium [10]. The agreement is not very good and the results serve to emphasize the difficulties faced in attempting a realistic theory of the spectral distribution curve.

If the theory of the spectral distribution curve is restricted to a calculation of the relative yield for a narrow range of frequencies near the apparent threshold, Mitchell's equation (5.10) can be subjected to a number of approximations. As before, the transmission coefficient of the barrier can be omitted. The frequency dependence of the second term of the integrand can be ignored and this term taken proportional to k_x. The relative yield is then

$$Y(\nu,T) = \iiint \frac{k_x\,dk_x\,dk_y\,dk_z}{1 + \exp{(E - \mu)/kT}} \tag{5.11}$$

The range of integration is over all initial states of energy E which after absorption of a quantum possess sufficient kinetic energy associated with the component of the propagation vector k_x, normal to the surface, to escape. These are states for which k_x is greater than $[(8\pi^2m/h^2)(W_a - h\nu)]^{1/2}$. The relative yield of Eq. (5.11) is proportional to the number of electrons which strike unit area of surface per unit time and escape. In this approximation, the problem of photoemission is reduced to that of thermionic emission from a Sommerfeld metal with a nonreflecting barrier of magnitude $W_a - h\nu$ instead of W_a.

The theory, in this form, is DuBridge's modification of Fowler's theory which was published before Mitchell attempted a complete theory of the spectral distribution curve [11, 12]. The Fowler-DuBridge theory is extremely useful because it quantitatively accounts for the temperature dependence of the yield near the apparent threshold and provides an unambiguous and physically meaningful definition of a photoelectric threshold frequency. The yield Y, as given by Eq. (5.11), can be expressed as

$$Y = \alpha A T^2 \phi(x) \tag{5.12}$$

where $\phi(x)$ is the series

$$\phi(x) = \left[e^x - \frac{e^{2x}}{2^2} + \frac{e^{3x}}{3^2} - \cdots\right] \qquad x \le 0 \tag{5.13}$$

$$\phi(x) = \left[\frac{x^2}{2} + \frac{\pi^2}{6} - \left(e^{-x} - \frac{e^{-2x}}{2^2} + \frac{e^{-3x}}{3^2} - \cdots\right)\right] \qquad x \ge 0$$

The parameter x is

$$x = \frac{h\nu - (W_a - \mu)}{kT} \tag{5.14}$$

and Fowler *defined* the photoelectric threshold frequency, ν_0, by

$$h\nu_0 = W_a - \mu \tag{5.15}$$

so that $h\nu_0$ is, by definition, equal to the thermionic work function. The constant $A = 4\pi mk^2/h^3$ and is closely related to the universal constant of thermionic emission theory. The theoretically undetermined constant α is the fraction of electrons that arrive at unit area of the surface in unit time, absorb a quantum, and escape, when the incident radiation intensity is unity.

Extensive measurements by DuBridge, his coworkers, and others have amply verified that the above theory gives an adequate account of the temperature and frequency dependence of the relative yield near the threshold [12]. Theory and experiment are conveniently compared by using Eq. (5.12) in the form

$$\log \frac{Y}{T^2} = \log{(\alpha A)} + \log \phi(x) \tag{5.16}$$

and plotting $\log{(Y/T^2)}$ vs. x. Since $\log{(Y/T^2)}$ is a universal function of x, if the experimental data are plotted as log (Y/T^2) vs. $(h\nu/kT)$, the shift of the data parallel to the x axis necessary to bring it into coincidence with the theoretical curve determines $h\nu_0/kT$ and the threshold frequency ν_0. This is the presently accepted and universally used procedure for determining the photoelectric threshold. The threshold frequency ν_0 defined in this manner is the minimum frequency for emission if the temperature of the surface is reduced to 0°K without changing W_a or μ. At the absolute zero a true threshold frequency ν_0 exists and the yield rises parabolically for frequencies greater than the threshold frequency.

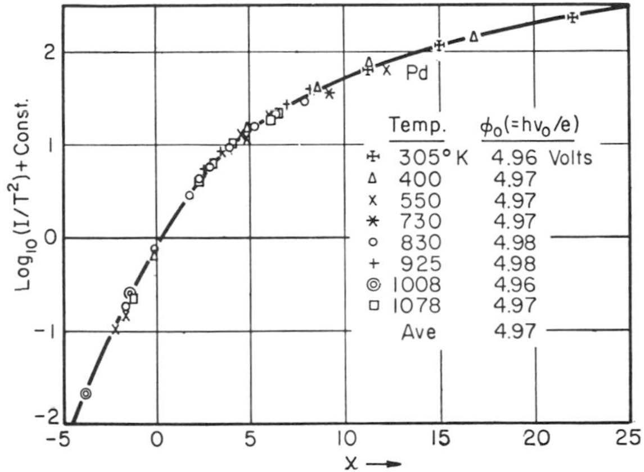

FIG. 5.5. The spectral distribution curve of palladium. The curve is Fowler's theory. [*L. A. DuBridge and W. W. Roehr: Phys. Rev.*, **39**: 99 (1932).]

The shift of the experimental data parallel to the log (Y/T^2) axis necessary to bring it into coincidence with the theoretical curve determines the constant α. This constant is ordinarily of the order of 10^{-33} cm²-sec/quantum. Figure 5.5 is a comparison of experimental spectral distribution curves of palladium with the Fowler theory [13]. The photoelectric threshold and the constant α of the barium surface whose behavior is illustrated in Figure 5.1 were obtained by comparison of the experimental spectral distribution curve with Fowler's theory. The theoretical spectral distribution curve at frequencies far from the threshold, calculated according to Mitchell's theory, was adjusted to agree with the Fowler curve at the threshold. Fowler's universal function log $\phi(x)$ is tabulated in Table 5.1.

TABLE 5.1. FOWLER'S FUNCTION

x	log $\phi(x)$	x	log $\phi(x)$
−8.0	−3.475	+2.0	+0.546
−6.0	−2.606	3.0	0.785
−5.0	−2.171	4.0	0.983
−4.0	−1.739	5.0	1.150
−3.0	−1.308	6.0	1.293
−2.5	−1.095	8.0	1.527
−2.0	−0.884	10.0	1.713
−1.5	−0.674	12.0	1.866
−1.0	−0.469	14.0	1.998
−0.5	−0.268	16.0	2.113
−0.2	−0.160	20.0	2.305
0.0	−0.085	25.0	2.497
+0.2	−0.015	30.0	2.655
0.4	−0.055	35.0	2.788
0.6	+0.125	40.0	2.904
1.0	0.249	50.0	3.097
1.5	0.400		

DuBridge also tested the Fowler theory by measuring the photocurrent as a function of the temperature for fixed values of the frequency [14]. In this case log (Y/T^2) is plotted against log T and superposition of the experimental and theoretical "isochromatic" curves determines the threshold frequency from data taken at a single frequency. This procedure eliminates the necessity of measuring the relative intensities of the incident radiation of various frequencies as must be done in obtaining a spectral distribution curve.

3. The Energy Distribution Function

The distribution in energy of photoelectrons from a Sommerfeld metal has been calculated by DuBridge [12, 15]. The assumptions used were those adopted by Fowler in his treatment of the spectral distribution function. The transmission coefficient of the surface barrier was assumed independent of the kinetic energy of the escaping electron; electron-electron and electron-phonon collisions were ignored since the photoelectrons are produced at the immediate surface; the probability of absorption of a quantum was assumed independent of the initial state of the electron; and the absorption of a quantum was assumed not to alter the direction of motion of the electron. The calculation is then that of the distribution in energy of thermionic electrons from a Sommerfeld metal with a surface barrier of magnitude $W_a - h\nu$ without the usual assumption that only electrons in the high-energy tail of the Fermi distribution possess sufficient energy to escape.

The number of electrons per unit volume inside the metal, with energy in dE, which come up to the surface per unit time is

$$v_x n(E)\, dE \tag{5.17}$$

where v_x is the component of electron velocity perpendicular to the surface and

$$n(E) = \frac{E^{1/2}}{e^{(E-\mu)/kT} + 1} 4\pi \left(\frac{2m}{h^2}\right)^{3/2} \tag{5.18}$$

is the distribution in energy of electrons in the metal.

Only the photoelectrons whose normal kinetic energy, $\frac{1}{2}mv_x^2$, is greater than the barrier height, $W_a - h\nu$, can escape. Since the kinetic energy associated with the component of velocity parallel to the surface remains unchanged as the electron passes through the surface, the total kinetic energy of a photoelectron outside the metal is

$$\epsilon = E - (W_a - h\nu) \tag{5.19}$$

The number of electrons leaving unit area of the surface in unit time with energy ϵ in the range $d\epsilon$ outside the metal is proportional to

$$n(\epsilon)\,d\epsilon = v_x\left\{1 - \left[\frac{W_a - h\nu}{\epsilon + (W_a - h\nu)}\right]^{1/2}\right\} \frac{(\epsilon + W_a - h\nu)^{1/2}\,d\epsilon}{\exp\,[(\epsilon - \epsilon_M)/kT] + 1} \quad (5.20)$$

The term $\{1 - [(W_a - h\nu)/(\epsilon + (W_a - h\nu))]^{1/2}\}$ is the probability that an electron of energy E inside the metal has its velocity vector directed so that the normal kinetic energy, $\frac{1}{2}mv_x^2$, is greater than or equal to $W_a - h\nu$. If $\frac{1}{2}mv_x^2$ is set equal to $W_a - h\nu$, an approximate form of Eq. (5.20) is obtained

$$n(\epsilon)\,d\epsilon = \frac{\epsilon\,d\epsilon}{\exp\,[(\epsilon - \epsilon_M)/kT] + 1} \quad (5.21)$$

where $\quad \epsilon_M = h\nu - (W_a - \mu) = h\nu - \varphi \quad (5.22)$

The energy ϵ_M is the maximum kinetic energy of emission of photoelectrons from a surface at 0°K. At higher temperatures, a maximum kinetic energy of emission does not exist. The energy distribution, according to DuBridge's equation (5.21), is shown in Fig. 5.2.

The distribution in energy of photoelectrons is usually investigated by observing the photocurrent as a function of applied potential in a spherical photocell, as illustrated in Fig. 5.6. The small photo-

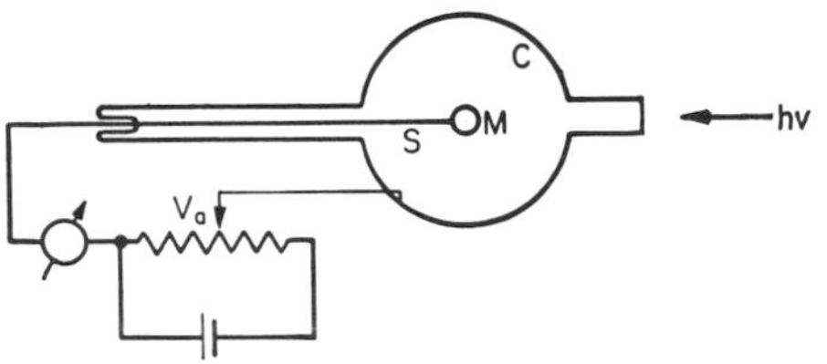

FIG. 5.6. Photocell for energy distribution measurements. M, emitter; S, support; C, collector; V_a, applied potential.

sensitive surface M is placed at the center of the spherical collector C, which is ordinarily a film of graphite or metal on the inner surface of the glass envelope. With this geometry, the velocity vectors of the photoelectrons are radially directed along the lines of force of the electric field between emitter and collector. Only the photoelectrons with kinetic energy $\epsilon > eV$ reach the collector C when the retarding potential difference V_a is applied between emitter and collector. Because of the contact difference of potential, $V_c = -(\varphi_c - \varphi_M)/e$, the potential difference between points just outside the surfaces of M and C is $V = V_a + V_c$.

According to DuBridge's theory the current-voltage curve $i(V_a)$ is given by the integral

$$i(V_a) = \int_{-e(V_a+V_c)}^{\infty} \frac{\epsilon\,d\epsilon}{\exp\,[(\epsilon - \epsilon_M)/kT] + 1} \quad (5.23)$$

The form of the current-voltage curve is shown in Fig. 5.7. For a surface at the absolute zero of temperature the exponential can be neglected and the upper limit of integration replaced by ϵ_M. The current-voltage curve is then a parabola. The observed photocurrent is zero for a retarding potential V_0 such that $-e(V_0 + V_c) = \epsilon_M$ or $V_0 = -(h\nu - \varphi_c)/e$ since $\epsilon_M = h\nu - \varphi_M$. φ_c is the work function of the collector. The current rises to a saturation value i_s

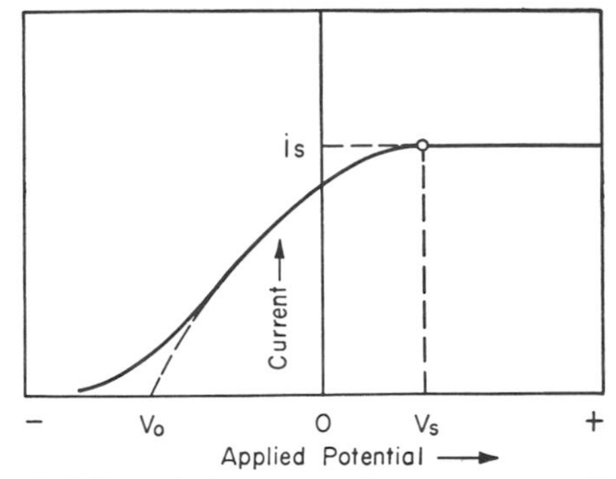

FIG. 5.7. Theoretical current-voltage curve according to DuBridge's theory. i_s, saturation photocurrent; V_s, saturation potential; V_0, stopping potential at 0°K.

for an applied potential $V_s = -V_c$. If V_s and V_0 are determined from the observed current-voltage curve, the work function of the emitter surface can be obtained

$$\varphi_M = h\nu - e(V_s - V_0) \quad (5.24)$$

For temperatures other than 0°K, the photocurrent decreases asymptotically to zero as the retarding potential is increased. The integral of Eq. (5.23) can be evaluated in terms of the series

$$i = \alpha AT^2\left(\frac{\pi^2}{6} - \frac{1}{2}(x^2 - x_M^2) + x\ln\,[1 + \exp\,(x - x_M)] - \left\{\exp\,(x - x_M) - \frac{\exp\,[2(x - x_M)]}{2^2} + \frac{\exp\,[3(x - x_M)]}{3^2} - \cdots\right\}\right) \quad \text{for } x \leq x_M \quad (5.25)$$

$$i = \alpha AT^2\left(x(x_M - x) + x\ln\,[1 + \exp\,(x - x_M)] + \left\{\exp\,[-(x - x_M)] - \frac{\exp\,[-2(x - x_M)]}{2^2} + \cdots\right\}\right) \quad \text{for } x \geq x_M$$

Here $x = \epsilon/kT$ and $x_M = \epsilon_M/kT$. For $x_M > 10$ and $x > x_M$, Eq. (5.25) may be approximated by

$$\log\frac{i}{xT^2} = \log\,(\alpha A) + \chi(x - x_M) \quad (5.26)$$

where $\chi(x - x_M)$ is the universal function of $(x - x_M)$

$$\chi(x - x_M) = \log\,\{\exp\,[-(x - x_M)] - (\tfrac{1}{4})\exp\,[-2(x - x_M)]\} \quad (5.27)$$

The experimental current-voltage curve can be plotted as $\log\,(i/xT^2)$ vs. x, where $x = \epsilon/kT$ and $\epsilon = -e(V_a - V_s)$. The shift parallel to the x axis necessary to bring the experimental curve into coincidence with the theoretical curve (5.27) determines x_M and ϵ_M. After ϵ_M is found in this manner, V_0 can be obtained from $\epsilon_M = -e(V_0 - V_s)$. In this manner, V_0 can be determined from current-voltage data obtained at any temperature. DuBridge and

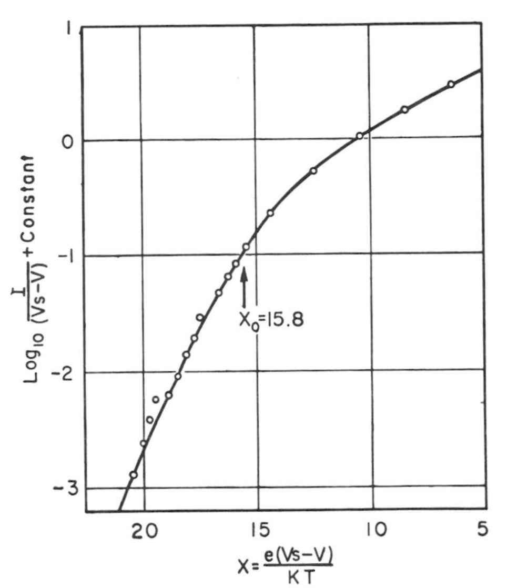

FIG. 5.8. The current-voltage curve of polycrystalline tungsten. The solid curve is DuBridge's theory. $T = 300°K$; $h\nu = 4.89$ ev; $x_0 = (h\nu - \varphi)/kT$;

$$V_0 = +0.21 \text{ volts}$$

$V_s = +0.62$ volts. [*L. Apker, E. Taft, and J. Dickey: Phys. Rev.*, **73**: 46 (1948).]

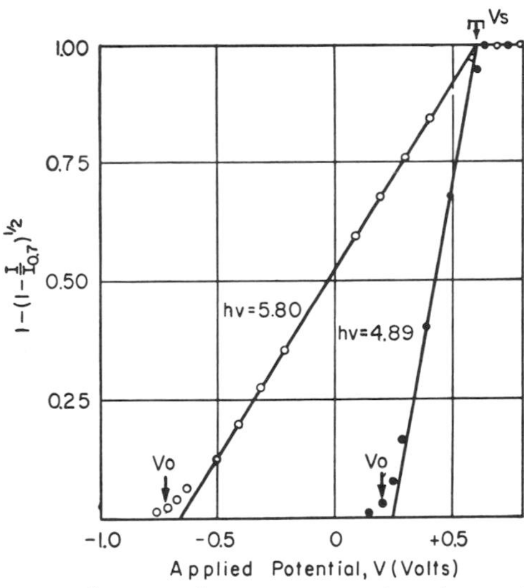

FIG. 5.9. Current-voltage curves of polycrystalline tungsten. $T = 300°K$; $h\nu = 4.89$ ev and 5.80 ev; $V_s = +0.62$ volts. [*L. Apker, E. Taft, and J. Dickey: Phys. Rev.*, **73**: 46 (1948).]

his coworkers have shown that the high-energy tails of experimental current-voltage curves are in excellent agreement with this theory [12]. Figure 5.8 shows a comparison of theory with data obtained from polycrystalline tungsten by Apker, Taft, and Dickey [16]. Figure 5.9 shows complete current-voltage curves for this surface. As predicted by DuBridge's theory, the current-voltage curve is parabolic except for the high-energy tail. The work function of this surface was determined as 4.48 ± 0.03 ev from the analysis of the energy distribution data. Spectral distribution curves and isochromatic curves analyzed by the Fowler-DuBridge procedure yielded identical work functions of 4.49 ± 0.02 ev.

Mitchell has extended his treatment of the surface photoelectric effect at an image potential barrier to include the energy distribution of the photoelectrons [6]. For electron energies near ϵ_M, his energy distribution function is the same as DuBridge's. Mitchell's theory predicts relatively fewer slow electrons than DuBridge's theory because the probability of absorption of a quantum depends upon the initial state of the electron.

Experimental energy distributions usually have for small energies the general form predicted by Mitchell's theory. Apker has shown, however, that energy distributions are, in general, untrustworthy at small energies because of distortion of the electric field in the photocells as a result of contact differences between the emitter and its support. Extreme care was exercised to remove this difficulty in obtaining the data shown in Figs. 5.8 and 5.9.

Berglund and Spicer [17, 18] have investigated photoemission from copper and silver excited by photons of energy considerably greater than the threshold energy. The band structure of the metal is of primary importance for this volume emission. Figure 5.10 shows the energy distribution of photoemitted electrons from a copper surface whose threshold has been reduced to 1.55 ev by a surface layer of cesium. The large, low-energy peak in the energy distribution that is observed with exciting radiation of photon energy $h\nu = 3.9$ ev is attributed to exci-

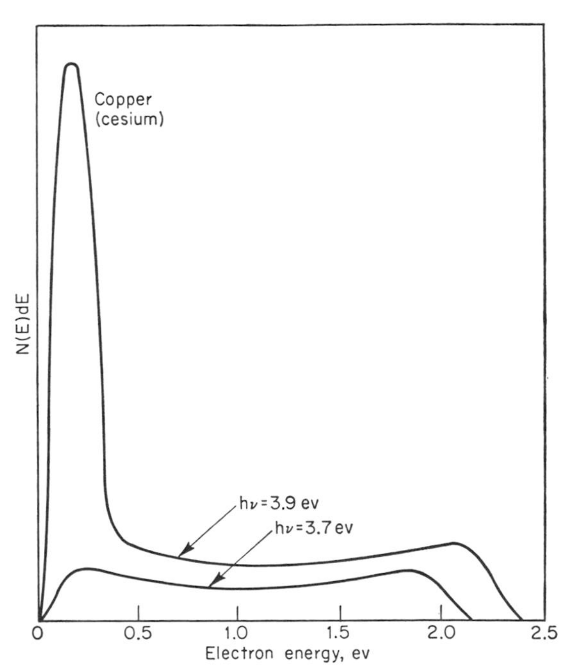

FIG. 5.10. Energy distribution of photoemitted electrons from copper. [*C. N. Berglund and W. E. Spicer: Phys. Rev.*, **136**: A1044 (1964).]

tation of electrons from the d band lying 2 ev below the Fermi level. The energy-distribution curve obtained with 3.7-ev photons shows only a trace of slow electrons excited from the d band. The photoemission is due to indirect optical transitions of the type illustrated by the arrow AC of Fig. 5.3. Direct transitions were observed for photon energies greater than 4.5 ev, but indirect transitions dominated the excitation process. The reason for this breakdown of the selection rules of the Bloch approximation is not understood [19].

Electron-electron scattering is of major importance in copper for photon energies greater than 6 ev, producing a large, low-energy peak in the energy distribution of the photoelectrons at about 0.5 ev. A low-energy peak in the energy distribution of the photoelectrons from silver, which appears for photon energies greater than 4 ev, can be attributed to a contribution to the photoemission from the Auger effect. For both silver and copper it was possible to deduce the density of states and the location of important symmetry points of the energy bands from the photoelectric data.

4. Semiconductors and Insulators

The distribution in energy of electrons in semiconductors and insulators is so different from that in metals that, for this reason alone, quite different photoelectric behavior is expected [20, 21]. Figure 5.11 compares the energy-level scheme for a typical

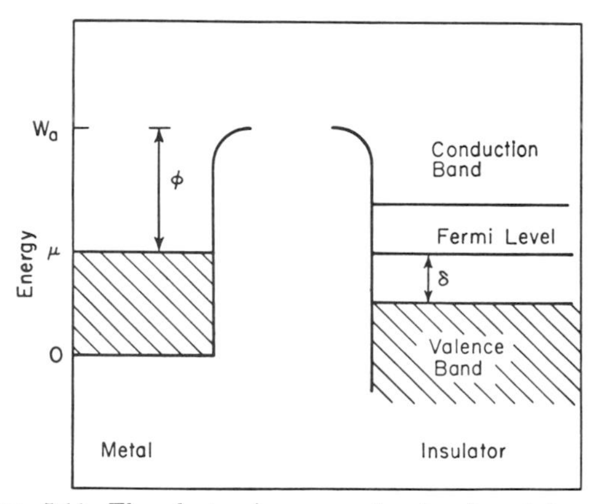

Fig. 5.11. The electronic energy levels of metals and insulators.

insulator and a metal of the same work function, $\varphi = W_a - \mu$. The Fermi level μ of the insulator lies within the forbidden energy region and approximately midway between the most energetic level of the filled valence band and the lowest level of the empty conduction band. The energy levels in the vicinity of the Fermi level μ in the case of a metal are filled with electrons and furnish the most energetic photoelectrons. In the insulator the electrons of largest energy have an energy δ less than μ. The distribution in energy of photoelectrons from an insulator may be expected to contain few fast electrons as compared with that of a metal of the same thermionic work function. The temperature dependence of the energy distribution will also be quite unlike that of a metal.

Figure 5.12 shows current-voltage curves for amorphous arsenic compared with a metal of the same

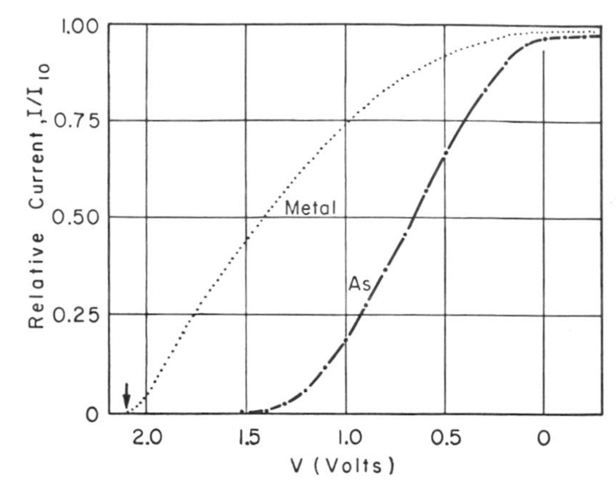

Fig. 5.12. The current-voltage curve of amorphous arsenic and of a metal with the same work function, 4.66 ev. $T = 300°K$; $h\nu = 6.71$ ev. The arrow marks V_0, the stopping potential for 0°K. [*E. Taft and L. Apker: Phys. Rev.*, **75**: 1181 (1949).]

work function $\varphi = 4.66$ ev, which satisfies DuBridge's theory [22]. As expected, the arsenic energy distribution contains few fast electrons and the current becomes unmeasurable at a retarding potential of -1.5 volts. The stopping potential, V_0, for a metal of the same work function was -2.05 ev in this experiment. The top of the valence band in arsenic appears to lie below the Fermi level by an amount $\delta = 0.5$ ev. Similar behavior is observed with tellurium, germanium, and boron surfaces [23]. There is evidence, however, that filled surface states which lie between the Fermi level and the top of the valence band may contribute some photoemission of fast electrons.

The absence of filled energy levels in the gap δ between the Fermi level and the top of the filled valence band affects the form of the spectral distribution curve. The photoelectric threshold energy $h\nu_0$ of an insulator or semiconductor is not equal to the thermionic work function $(W_a - \mu)$ but is given by $(W_a - \mu + \delta)$. Figure 5.13 shows spectral distribution curves of tellurium, germanium, and platinum surfaces of the same work function, $\varphi = 4.76$ ev [23]. The decrease in photoelectric response for quantum energies less than $h\nu = W_a - \mu$ is clearly visible; the photocurrents from the tellurium surface were not measurable for $h\nu < 5.0$ ev.

The spectral distributions of the quantum yields of atomically clean surfaces of silicon, germanium, gallium arsenide and antimonide, and indium arsenide and antimonide are approximately proportional to $(h\nu - h\nu_t)^3$ near ν_t, the photoelectric threshold frequency [24]. This emission is probably due to indirect optical transitions associated with the surface. A few tenths of an electron volt above the photoelectric threshold, volume optical absorption becomes more important than the surface absorption, and,

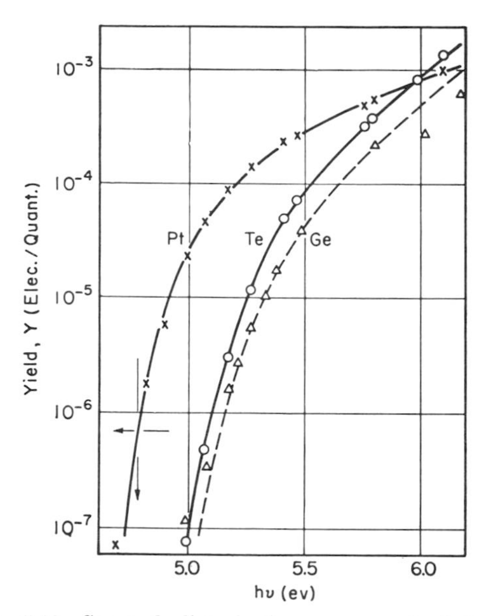

FIG. 5.13. Spectral distrubution curves of platinum, tellurium, and germanium. The solid curve through the platinum data is Fowler's theory. $T = 300°K$. The three surfaces have the same thermionic work function, 4.76 ev. [*L. Apker, E. Taft, and J. Dickey: Phys. Rev.*, **74**: 1462 (1948).]

in agreement with the theory of Kane, the yield increases linearly with $h\nu - h\nu'$, where ν' is the threshold for direct optical transitions [25, 26]. Both the spectral-distribution curves and the energy-distribution curves have yielded important information concerning energy-band structures. The band structure of lead telluride has been investigated in this manner [27]. Brust has made a detailed band-theoretical investigation of the photoelectric effect in silicon and obtained qualitative agreement with the results of Spicer [28, 29] and of Gobelli and Allen.

Allen and Gobelli [30] have compared the photoelectric thresholds and work functions of systematically doped p- and n-type silicon samples. As the Fermi level approaches the conduction or valence bands, the surface-state charge changes, with accompanying bending of the bands as they approach the surface. The resulting changes in the photoelectric threshold and work function make it possible to deduce the concentration of surface states and draw limited conclusions concerning their distribution in energy.

References

1. Hughes, A. L., and L. A. DuBridge: "Photoelectric Phenomena," McGraw-Hill, New York, 1932. Zworykin, V. K., and E. G. Ramberg: "Photoelectricity," Wiley, New York, 1949.
2. Maurer, R. J.: Photoelectric and Optical Properties of Sodium and Barium, *Phys. Rev.*, **57**: 653 (1940).
3. Sommerfeld, A., and H. Bethe: "Handbuch der Physik," Chap. 3, Springer, Berlin, 1933.
4. Fan, H. Y.: *Phys. Rev.*, **68**: 43 (1945).
5. Tamm, I., and Schubin, S.: *Z. Physik*, **68**: 97 (1931).
6. Mitchell, K.: Theory of the Surface Photoelectric Effect in Metals, *Proc. Roy. Soc.* (*London*), **A146**: 442 (1934); **A153**: 513 (1936).
7. L. Nordheim: *Proc. Roy. Soc.* (*London*), **A121**: 626 (1928).
8. A. G. Hill: Energy Distribution of Photoelectrons from Sodium, *Phys. Rev.*, **53**: 184 (1938).
9. Schiff, L., and L. Thomas: Quantum Theory of Metallic Reflection, *Phys. Rev.*, **47**: 860 (1935).
10. Makinson, R. E. B.: Metallic Reflexion and the Surface Photoelectric Effect, *Proc. Roy. Soc.* (*London*), **A162**: 367 (1937).
11. Fowler, R. H.: Analysis of Photoelectric Sensitivity Curves, *Phys. Rev.*, **38**: 45 (1931).
12. DuBridge, L. A.: "New Theories of the Photoelectric Effect," Hermann & Cie, Paris, 1935.
13. DuBridge, L. A., and W. W. Roehr: Photoelectric and Thermionic Properties of Palladium, *Phys. Rev.*, **39**: 99 (1932).
14. DuBridge, L. A.: A Test of Fowler's Theory of Photoelectric Emission, *Phys. Rev.*, **39**: 108 (1932).
15. DuBridge, L. A.: Theory of the Energy Distribution of Photoelectrons, *Phys. Rev.*, **43**: 727 (1933).
16. Apker, L., E. Taft, and J. Dickey: Energy Distribution of Photoelectrons from Polycrystalline Tungsten, *Phys. Rev.*, **73**: 46 (1948).
17. Berglund, C. N., and W. E. Spicer: Photoemission Studies of Copper and Silver: Theory, *Phys. Rev.*, **136**: A1030 (1964).
18. Berglund, C. N., and W. E. Spicer: Photoemission Studies of Copper and Silver: Experiment, *Phys. Rev.*, **136**: A1044 (1964).
19. Spicer, W. E.: Optical Transitions in Which Crystal Momentum Is Not Conserved, *Phys. Rev. Letters*, **11**: 243 (1963).
20. Fowler, R. H.: Statistical Mechanics, Cambridge University Press, New York and London, 1936.
21. Condon, E. U.: External Photoelectric Effect of Semiconductors, *Phys. Rev.*, **54**: 1089 (1938).
22. Taft, E., and L. Apker: Photoelectric Determination of the Fermi Level at Amorphous Arsenic Surfaces, *Phys. Rev.*, **75**: 1181 (1949).
23. Apker, L., E. Taft, and J. Dickey: Photoelectric Emission and Contact Potentials of Semiconductors, *Phys. Rev.*, **74**: 1462 (1948).
24. Gobelli, G. W., and F. G. Allen: Photoelectric Properties of Cleaved GaAs, GaSb, InAs, and InSb Surfaces; Comparison with Si and Ge, *Phys. Rev.*, **137**: A245 (1965).
25. Kane, E. O.: Theory of Photoelectric Emission from Semiconductors, *Phys. Rev.*, **127**: 131 (1962).
26. Gobelli, G. W., and F. G. Allen: Direct and Indirect Excitation Processes in Photoelectric Emission from Silicon, *Phys. Rev.*, **127**: 141 (1962).
27. Spicer, W. E., and G. J. Lapeyre: Photoemission Investigation of the Band Structure of PbTe, *Phys. Rev.*, **139**: A565 (1965).
28. Brust, D.: Band-theoretic Model for the Photoelectric Effect in Silicon, *Phys. Rev.*, **139**: A489 (1965).
29. Spicer, W. E., and R. E. Simon: Photoemissive Studies of the Band Structure of Silicon, *Phys. Rev. Letters*, **9**: 385 (1962).
30. Allen, F. G., and G. W. Gobelli: Work Function, Photoelectric Threshold, and Surface States of Atomically Clean Silicon, *Phys. Rev.*, **127**: 150 (1962).

Chapter 6

Thermionic Emission

By LLOYD P. SMITH, Stanford Research Institute

Introduction

Thermionic emission as it is understood today is that branch of physics having to do with the various phenomena connected with the ejection of electrons or positive ions from a solid when it is heated to a sufficiently high temperature. The study of this subject had its meager and floundering beginning over two hundred years ago when it was discovered that air in the neighborhood of hot solids could conduct electricity. It was gradually learned that the hot bodies emitted charged particles. A notable step forward was made by J. J. Thomson, who showed that the charged particles which were emitted by carbon at fairly high temperatures and in a partial vacuum were predominantly electrons. The important advances in the subject from this time on were closely tied to the degree of perfection of high-vacuum techniques. As it became possible to make measurements in better and better vacuums, effects of gas and surface contamination on the emission phenomena could be eliminated and the essential features of true thermionic emission could be investigated. The evolution of the field was greatly enhanced by the development of the quantum theory of the behavior of electrons in solids. At present the basic phenomena in electron and positive ion emission from essentially pure metals are well understood. Though significant advances have been made in the understanding of electron emission from composite materials such as the alkaline earths, the state of development is not as complete as for the pure metals.†

It is advantageous to divide the field into the following categories: the electron and ion emission from uniform pure metal crystals, the electron emission from polycrystalline metals, and the emission from metals with various adsorbed monolayers.

1. Uniform Pure Metal Crystals

Thermionic-emission Equation for Electrons. The current density j of electrons emitted from a uniform surface of a pure metal can be expressed in terms of the metal temperature T by the equation

$$j = A(1 - r)T^2 e^{-e\phi/kT} \qquad (6.1)$$

† For a good bibliography and critical treatment of electron emission from metals, see the article on thermionic emission by Conyers Herring, *Revs. Mod. Phys.*, **21**: 185 (1949).

This equation is the fundamental equation of thermionic emission and is usually known as the Richardson equation. The equation can be derived either from thermodynamical [1]† arguments or from the application of statistical mechanics [2] in connection with the quantum mechanics of electrons in metals.

Here A is a constant composed of a combination of fundamental physical constants

$$A = \frac{4\pi mk^2e}{h^3} = 120 \text{ amp/cm}^2 \cdot \text{deg}^2 \qquad (6.2)$$

where e is the absolute value of the electronic charge, k is Boltzmann's constant, and h is Planck's constant. r is a reflection coefficient for electrons crossing the potential barrier at the metal surface when the electric field just outside the metal surface is zero. It is possible that r could change somewhat with the temperature but it is thought that the quantity $1 - r$ is rather insensitive to temperature changes. For pure metals r is of the order 0.05. The reflection coefficient r will also depend slightly on the component of the electric field at the surface of the metal. This will be discussed in more detail under the section on the Schottky effect. ϕ is usually called the electronic work function and is defined so that ϕe is a characteristic amount of work required to remove an electron from the interior of the metal to a position just outside of the metal. In general ϕ is dependent to some extent on temperature and the normal component of electric field at the metal surface. Some experimentally determined values of ϕ for clean metals are given in Table 6.1. Since the emitted current density j depends on ϕ exponentially, small variations of ϕ caused by changes in temperature or electric field produce significant changes in j.

The current density given by Eq. (6.1) will not necessarily be that found by measuring the current to a plane electrode parallel to the surface of the hot metal because of the negative space charge produced between the two electrodes. Under some conditions this produces an electric field near the metal surface which retards the emitted electrons and causes the low-energy electrons to return to the surface. Care must be taken to make the potential difference such that no electrons are returned to the hot metal in order to obtain the current j given by the emission formula. If the potential difference is made too high, the field thus produced outside the metal surface

† Numbers in brackets refer to References at end of chapter.

changes the work function and the reflection coefficient as indicated above.

Statistical Derivation of the Richardson Equation. According to the modern quantum theory of electrons in metals [3], the outermost electrons in the atoms of which the metal is constituted are almost free to move through the metallic lattice. There exist certain discrete energy states which these electrons may occupy. These are represented schematically in Fig. 6.1. The number of such states in any energy interval is so large that the energy states are densely packed. The number of energy states dN which have energy between ϵ and $\epsilon + d\epsilon$ is

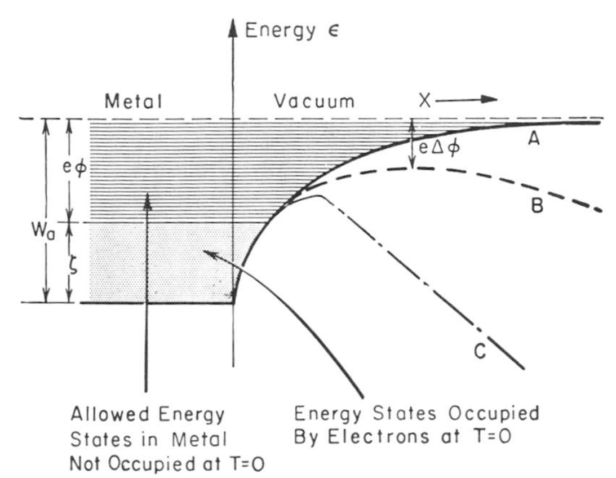

FIG. 6.1. Energy levels in a metal and potential energy of an electron near the metal surface.

$$dN = 2\pi(2m)^{3/2}h^{-3}\epsilon^{1/2}\,d\epsilon \tag{6.3}$$

The rather freely moving electrons which occupy these energy states are prevented from escaping from the metal by electrical forces at the surface of the metal. The potential energy of an electron due to the action of these forces is shown as a function of its distance away from the surface in Fig. 6.1 by curve A. It is assumed that there is no applied electric field at the surface nor fields due to space charge. From very near the surface to large distances away the potential energy of an electron is largely due to the image force $-e^2/2x^2$ and is essentially $W_a - (e^2/4x)$.

The Fermi distribution function [4] gives the number of electrons per unit volume in the metal and the manner in which they are distributed over the allowed energy states. This junction is

$$f = \frac{1}{1 + e^{(\epsilon-\zeta)/kT}} \tag{6.4}$$

where ϵ is the energy of a particular allowed state measured from the bottom of the conduction band (see Fig. 6.1). ζ is the energy below which all energy states are occupied by electrons at the absolute zero of temperature. ζ is consequently dependent on the number of electrons per unit volume in the conduction band at $T = 0$. Thus at $T = 0$, $W_a - \zeta = e\phi$. ζ is slightly dependent on temperature. The Fermi function f gives the probability that at temperature T the state of energy ϵ will be occupied. Therefore the probable number of electrons dn per unit volume occupying states between ϵ and $\epsilon + d\epsilon$ at temperature T is

$$dn = 2f(\epsilon)\,dN \tag{6.5}$$

The factor 2 comes from the fact that two electrons may occupy each energy state, one electron with spin in one direction and the other with spin in the opposite direction. Consequently at $T = 0$

$$n = 2\int_0^\zeta dN = \frac{8\sqrt{2}\pi m^{3/2}}{h^3}\int_0^\zeta \epsilon^{1/2}\,d\epsilon$$

$$= \frac{16\sqrt{2}\pi m^{3/2}}{3h^3}\zeta^{3/2} \tag{6.6}$$

Knowing the value of ζ at $T = 0$ is equivalent to knowing the number of electrons per unit volume in the metal. Since $\epsilon = \frac{1}{2}m(v_x^2 + v_y^2 + v_z^2)$, the number of energy states corresponding to velocity ranges dv_x at v_x and so on is

$$dN = \frac{m^3}{h^3}\,dv_x\,dv_y\,dv_z \tag{6.7}$$

Hence the probable number of electrons in this velocity range is

$$dn = 2f(\epsilon)\frac{m^3}{h^3}\,dv_x\,dv_y\,dv_z \tag{6.8}$$

With the help of this expression and Eq. (6.4) the electron current emitted per unit area of the metal is

$$j = \frac{2m^3}{h^3}e\int_{-\infty}^{+\infty}\int_{-\infty}^{+\infty}\int_{-\infty}^{+\infty} \frac{T(v_x)v_x\,dv_x\,dv_y\,dv_z}{\exp\left[\dfrac{\frac{1}{2}m(v_x^2 + v_y^2 + v_z^2) - \zeta}{kT}\right] + 1} \tag{6.9}$$

where $T(v_x)$ is the probability that an electron approaching the potential barrier at the metallic surface with velocity v_x will pass through or over the barrier. Evaluating the integrals over v_y and v_z, the current density is

$$j = \frac{4\pi m^2 ekT}{h^3}\int_0^\infty T(v_x)\log\left[1 + \exp\left(-\frac{\frac{1}{2}mv_x^2 - \zeta}{kT}\right)\right]v_x\,dv_x \tag{6.10}$$

It is necessary to discuss the nature of $T(v_x)$ and this requires a knowledge of the form of the potential energy barrier at the surface. Consider the case where the force acting on the electron is due to the image forces and an applied electric field E. The potential energy of an electron at a distance x from the metal surface can then be written as

$$W(x) = W_a - \frac{e^2}{4x} - eEx \qquad x > 0 \tag{6.11}$$

$W(x)$ is represented as curve B in Fig. 6.1. If one neglects the wave properties of an electron, then the only electrons contributing to the emitted current would be those which have sufficient velocity in the x direction to get over the barrier or those for which

$$\tfrac{1}{2}mv_x^2 \geq W_a - e\,\Delta\phi \tag{6.12}$$

From Eq. (6.11) $\Delta\phi = \sqrt{eE}$. For this model all electrons obeying (6.12) are transmitted and all others are reflected so that

$$T(v_x) = 1 \qquad \text{for } v_x \geq \sqrt{\frac{2(W_a - e\,\Delta\phi)}{m}} \tag{6.13a}$$

$$= 0 \qquad \text{for } v_x < \sqrt{\frac{2(W_a - e\,\Delta\phi)}{m}} \tag{6.13b}$$

Actually, however, even for condition (6.13a) there will be a small amount of reflection due to the fact that an electron with sufficient energy to be emitted will come too close to one or more atoms of the metal and be backscattered. To take account of this, we put

$$\begin{aligned} T(v_x) &= 1 - r \qquad \text{for condition (6.13}a\text{)} \\ &= 0 \qquad \text{for condition (6.13}b\text{)} \end{aligned}$$

Furthermore, it will be assumed that over the important range of v_x, r is independent of v_x. r is the reflection coefficient already discussed in connection with Eq. (6.1). The current density is

$$j = \frac{4\pi m^2 ekT}{h^3}(1 - r) \int_{v_{x0}}^{\infty} \log\left[1 + \exp\left(-\frac{\frac{1}{2}mv_x^2 - \zeta}{kT}\right)\right] v_x\,dv_x$$

where v_{x0} is the expression in (6.13b). For values of T and v_x of interest the second term in the logarithm is small compared to 1 and a sufficiently precise evaluation of the integral can be obtained by using the final term in the expansion of the log term. Using Eq. (6.12) and $e\phi = W_a - \zeta$, the current density is

$$j = A(1 - r)T^2 \exp\left(-\frac{e\phi - e\sqrt{eE}}{kT}\right) \tag{6.14}$$

where A has the value (6.2). This equation is the same as Eq. (6.1), when the applied electric field E is zero.

The Thermionic Work Function ϕ. The work functions of most metals have been determined experimentally from the thermionic-emission current or from a measurement of the frequency threshold for the photoelectric emission from the metal surface in question. Theoretical methods of calculating the work function of a metal are still not accurate enough to furnish reliable values. The customary way of determining the work function is to measure the emission current at a fixed temperature T for a range of applied fields E and extrapolate this curve to $E = 0$ in order to determine the emission current appropriate for zero field. This procedure is repeated for a range of temperatures and the current densities so measured divided by T^2 are plotted against $1/T$. The slope of this curve is the apparent work function. The analytical statement of this is

$$\phi^* = -\frac{k}{e}\frac{d(j/T^2)}{d(1/T)} \tag{6.15}$$

For a uniform surface Eq. (6.1) applies so that

$$\phi^* = \phi - T\frac{d\phi}{dT} - \frac{kT^2}{e}(1 - r)\frac{dr}{dT} \tag{6.16}$$

Thus the apparent work function as measured from the slope is equal to the true work function only when ϕ or r are independent of temperature. As already stated, r is small and insensitive to temperature changes for clean metals. The term $T(d\phi/dT)$ is not as certain. It is likely to be small for clean metals with a uniform surface but not negligible. From some attempts to measure the magnitude of this term for Ta, W, and Mo it appears likely that it is about 0.1 ev. It can be much larger than this for surfaces covered with monolayers or for surfaces of semiconductors.

Experiment has shown that the work function for a given metal single crystal depends on the particular crystal surface from which electron emission is taking place. Such a difference is to be expected because the arrangement of metal atoms on the surface differs with the different exposed crystallographic planes and this in turn would make the forces acting on an electron near the surface somewhat different. Table 6.1 shows the difference in the apparent work function of Cu, Ag, and W as a function of the exposed crystal surface. It will be seen that the differences become as high as 0.25 ev.

TABLE 6.1. WORK FUNCTION FOR SINGLE CRYSTAL SURFACES

Element	Index of crystal direction of normal to surface	Work function, ev
Cu	111	4.89
	100	5.64
Ag	111	4.75
	100	4.81
W	112	4.50
	310	4.30

Velocity Distribution of Emitted Electrons. It was demonstrated originally by Richardson that the electrons emitted from a hot metal possess a Maxwellian distribution of velocities. This is the case even though the velocity distribution law inside the metal is that of Fermi, i.e., (6.8), or

$$dn = \frac{2m^3}{h^3}\frac{dv_x\,dv_y\,dv_z}{\left[1 + \exp\frac{\frac{1}{2}m(v_x^2 + v_y^2 + v_z^2) - \zeta}{kT}\right]}$$

Let the velocities of the electrons outside the metal be v'_x, v'_y, v'_z. These will be related to the velocity components inside the metal by the relations

$$\tfrac{1}{2}mv'^2_x = \tfrac{1}{2}mv_x^2 - W_a \qquad v'_y = v_y \qquad v'_z = v_z$$

provided the transmission coefficient is such that

$$T(v_x) = 0 \qquad v_x < \sqrt{\frac{2W_a}{m}}$$

$$= 1 - r \qquad v_x \geq \sqrt{\frac{2W_a}{m}}$$

The number of electrons emitted per unit time whose velocities inside the metal lie in the range $v_x + dv_x$, etc., is

$$dn_0 = (1 - r)\frac{2m^3}{h^3}\frac{v_x\,dv_x\,dv_y\,dv_z}{1 + \exp\left[\dfrac{\frac{1}{2}m(v_x^2 + v_y^2 + v_z^2) - \zeta}{kT}\right]} \qquad v_x > \sqrt{\frac{2W_a}{m}}$$

The number of electrons emitted per unit time whose velocities outside the metal lie in the range $v'_x + dv'_x$, etc., is

$$dn' = (1 - r)\frac{2m^3}{h^3}\frac{v'_x\,dv'_x\,dv'_y\,dv'_z}{1 + \exp\left[\dfrac{\frac{1}{2}m(v'^2_x + v'^2_y + v'^2_z) + e\phi}{kT}\right]} \qquad v'_x > 0$$

Since $e^{e\phi/kT} \gg 1$, the distribution law for emitted electrons is

$$dn' = (1 - r)\frac{2m^3}{h^3}e^{-e\phi/kT}\exp\left[-\frac{\frac{1}{2}m(v'^2_x + v'^2_y + v'^2_z)}{kT}\right]v'_x\,dv'_x\,dv'_y\,dv'_z \tag{6.17}$$

This is the same distribution function that would have been obtained if the electrons inside the metal had been distributed in the velocity ranges according to Maxwell's law.

Field Effects. An electric field in the neighborhood of the surface of a clean metal which exhibits only one crystal face may exist most commonly because of a potential difference applied between the metal surface and a neighboring electrode or may exist because of the charge produced by electrons emitted into the space near the hot metal. Usually the field in the neighborhood of the surface during electron emission arises from both causes. For very low applied fields and appreciable electron emission the space-charge field in the neighborhood of the cathode is such as to repel many of the electrons emitted with the lower energies. For applied fields high enough to overcome the space-charge fields, all electrons emitted can be collected on a collecting electrode. At a constant emitter temperature the emission current will increase as the applied field increases in accordance with Eq. (6.14) if the applied field does not become too high (of the order 10^5 volts/cm). This increase in emission current is called the *Schottky effect* and is due to the lowering of the potential barrier. The validity of Eq. (6.14) is usually tested by plotting log j against $E^{1/2}$ at constant emitter temperature. This is a straight line and is called the *Schottky line.*

Except for small periodic deviations of the emission current from the Schottky line, the theory is very well substantiated by experiment. The small periodic deviations of the emission current from that predicted by the Schottky theory are also a field effect but come about through the interference effect of electron waves as the shape of the barrier is changed by means of the applied field. This effect is superposed on the lowering of the barrier due to the applied field, which is just the usual Schottky effect. The periodic deviation is an interference phenomenon [5] analogous to that which would occur in the case of transmissions of light through a medium with a continuously varying index of refraction. This theory leads to the conclusion that the reflection coefficient r in Eq. (6.14) depends on the field E. The actual dependence on E is complicated in form. The main feature of the dependence of r on E is that r is an oscillatory function of E whose amplitude and period increase as E increases.

A still different field effect sets in for really high fields of the order 10^5 volts/cm and greater. For such high fields the potential barrier looks more like curve C in Fig. 6.1, and the barrier thickness begins to approach the order of magnitude of a wavelength of the most energetic electrons in the metal. In this case some electrons tunnel through the barrier, and the dependence of the emitted current on temperature and field is entirely different from that portrayed by Eq. (6.14). Under these conditions the emission is referred to as "field emission" and is not properly thermionic emission.

For the case of small applied fields produced by applying a potential difference between a plane emitter and a plane collecting anode such that the space charge of the emitted electrons nevertheless produces a retarding field near the emitter surface, the electrostatic potential between the planes is depicted in curve I (Fig. 6.2). The minimum in the potential at a distance of d_m from the emitter surface is caused by the negative space charge produced by the emitted electrons. Since the emitted electrons have a Maxwellian distribution (see section above on velocity distribution of emitted electrons), not

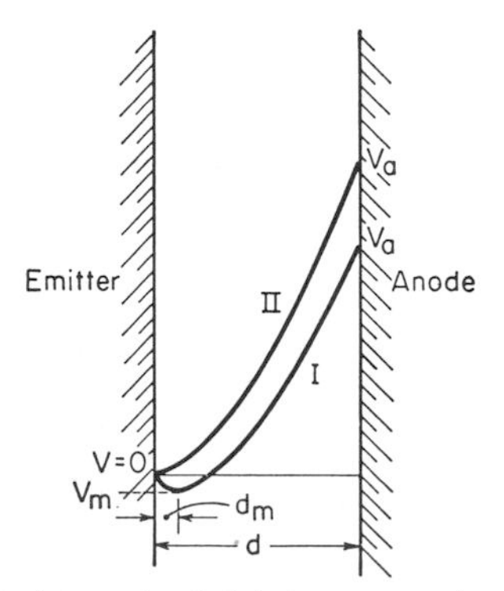

Fig. 6.2. Electric potential between an electron emitter and an anode.

all of the emitted electrons will have energy enough to overcome the potential minimum V_m and will return to the emitter. Consequently, the current density at the anode will be less than that given by the emission equation (6.1). Calling the emission current density given by Eq. (6.1) the saturation current and denoting it by j_s, Langmuir [6] has derived the relation between the actual current density j at the anode whose potential is V_a with respect to the emitter (emitter potential zero) and located a distance d from the emitter, as shown in Fig. 6.2

$$j = \frac{\sqrt{2e/m}}{9\pi} \frac{(V_a - V_m)^{3/2}}{(d - d_m)^2} \left[1 + \frac{2.658\sqrt{kT}}{\sqrt{e(V_d - V_m)}}\right] \tag{6.18a}$$

$$V_m = \frac{-kT}{e} \log \frac{j_s}{j} \tag{6.18b}$$

$$d_m = \frac{\xi(V_a - V_m)}{4(\pi/2kT)^{3/4} m^{1/4} \sqrt{ej}} \tag{6.18c}$$

Knowing the saturated emission current density j_s, the difference of potential between emitter and anode, and their distance d apart, the equations above determine j, the actual anode current. The function $\xi(V_a - V_m)$ is a slowly varying function of $V_a - V_m$ and is tabulated in ref. [6]. However

$$2 < \xi < 2.55 \qquad \text{for} \qquad 7 < \frac{j_s}{j} < \infty$$

so that for considerable range of currents ξ can be taken as 2.5. Since $d_m/d \ll 1$ in most cases, Eq. (6.18a) does not depend sensitively on d_m anyway. When d_m and V_m can be neglected compared with d and V_a and when

$$\frac{2.66\sqrt{kT}}{\sqrt{eV_d}} \ll 1$$

the anode current is given by

$$j = \frac{\sqrt{2e/m}}{9\pi} \frac{V_a^{3/2}}{d^2} \tag{6.19}$$

which is known as *Child's law*.

Expressions for the space-charge limited current for other geometrical arrangements of emitter and anode have been worked out [7].

Emission of Positive Ions. When metals are heated to sufficiently high temperatures, they emit positive ions. At the lower part of the temperature range the ions are usually impurities contained in the metal or its surroundings such as the tube wall, etc. However, mass spectrographic studies show that at high enough temperatures the metal emits positive ions of the metal itself. The ratio of the number of singly charged metal ions to the number of neutral atoms evaporated is small.

A thermodynamical [8] and statistical [9] theory of metallic ion emission shows that the emitted ion current density is related to the temperature as follows

$$J_+ = \frac{2\pi k e M}{h^3} (1 - r) T^2 \exp\left(-\frac{e\phi_+}{kT} - \frac{1}{k}\int_0^T \frac{dT}{T^2} \int_0^T C_{pm}\, dT\right) \tag{6.20}$$

where C_{pm} is the heat capacity, at constant pressure, of an ion in the condensed state, ϕ_+ is the positive ion work function, and r is a reflection coefficient. Except for a factor 2 multiplying the right-hand side and the additional term containing the specific heat of the ion in the solid, Eq. (6.20) would have the same form as Eq. (6.1) for electron emission. The specific heat term is missing in the electron emission case because electrons distributed among the energy states according to the Fermi law contribute only a negligible amount to the specific heat of the solid. The factor 2 comes about because of a difference in statistical weight.

The most precise and careful measurements of positive-ion emission have been made on molybdenum [10]. When necessary precautions are taken to eliminate ionization of low-ionization potential impurities, the measurements of ϕ_+ using Eq. (6.20) in the same way as for electron emission appear to be reliable. The constant $A_+ = 2\pi k^2 eM/h^3$ is also closely verified. A test for consistency in ϕ_+, if the emission of electrons, atoms, and ions takes place from one crystalline surface (it is difficult to guarantee this), is that

$$\phi_+ + \phi_- = U + V \tag{6.21}$$

where U is the heat of evaporation of a metal atom and V is the ionization potential of the metal atom, all quantities pertaining to the same temperature. The energy balance arises from a cycle in which an atom is removed from the surface, ionized at a distance, and the electron and ion separately are placed back on the surface. In the case of Mo

$$\phi_+ + \phi_- = 8.6 + 4.2 = 12.8 \text{ ev}$$
$$U + V = 5.75 + 7.15 = 12.9 \text{ ev}$$

balancing within experimental error. Because of the fact that several crystalline surfaces may be exposed during the measurements it could be that the experimentally determined quantities should not quite produce a closed energy cycle. The best experimentally determined values of ϕ_+ for Mo and W appear to be

	ϕ_+
Mo	8.6 ev
W	11.9 ev

It is to be expected that the metallic ions emitted from a hot metal should be emitted with a Maxwellian distribution of velocities just as electrons are. This has been verified [11] experimentally in the case of tungsten.

Other metals which are known to emit positive ions at temperatures below their melting points are Cr, Fe, Ni, Cu, Nb, Ru, Rh, and Ta.

2. Polycrystalline Metals

Field Effects Resulting from Polycrystalline Surfaces. In the section above on the thermionic work function ϕ it was stated that different exposed crystallographic planes of a crystal possessed somewhat different work functions. Consequently, a surface of polycrystalline metal will in general have a number of different crystallographic surfaces exposed so that the surface as a whole will exhibit areas or patches of different work functions. Thus at best the thermionic-emission current from such a surface would be a composite of emission currents characteristic of each patch. The situation is further complicated by the fact that there does not exist a situation where the electric field is zero over the entire surface. This situation arises from the fact that a local electric field always exists between neighboring patches with different work functions. This is illustrated in Fig. 6.3. The electric field is

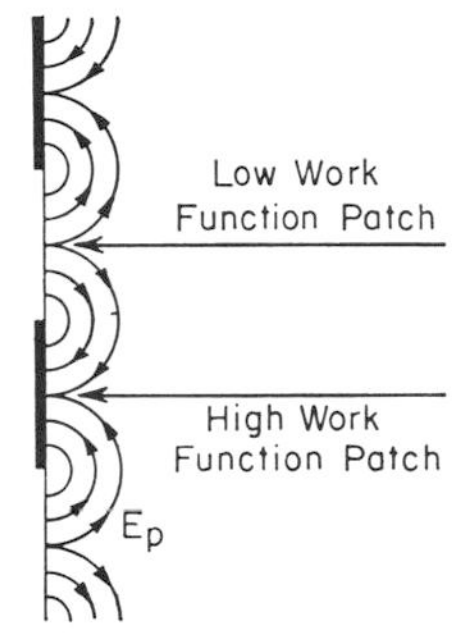

FIG. 6.3. Electric field in the neighborhood of a surface with patches of different work functions.

always directed so as to aid the electrons in getting away from the high work-function patch and to hinder the electrons from getting away from a low work-function patch.

The local field between two adjacent patches with work functions ϕ_a and ϕ_b will be the same as that produced by replacing the patches with conductors the difference in whose potentials is $\phi_a - \phi_b$. A surface with repetitive patches of different work functions will produce a local patch field which decreases exponentially with the distance away from the surface and will have reached $1/e$ of its value at the surface in a distance comparable with the mean diameter $\bar{D}$ of the patches. At distances large compared to $\bar{D}$ the potential assumes a value equal to the surface average of the potential of all the patches, i.e.,

$$\bar{\phi} = \sum_i f_i \phi_i \tag{6.22}$$

where f_i is the fraction of the area whose work function is ϕ_i.

In order to obtain some idea of the type of patch that contributes most of the electrons during thermionic emission, the potential energy of an electron as a function of distance away from the surface at the center of a low work-function patch will be compared to that of an electron leaving the surface at the center of a high work-function patch. Let the work function of the low and high work-function patches be ϕ_l and ϕ_h, respectively.

Referring to Fig. 6.1 and Eq. (6.11), the potential energies in the two cases are

$$W_l = \zeta + \bar{\phi}e - \frac{e^2}{4x} - e\int E_{pnl}\,dx \tag{6.23a}$$

$$W_h = \zeta + \bar{\phi}e - \frac{e^2}{4x} - e\int E_{pnh}\,dx \tag{6.23b}$$

where E_{pnl} and E_{pnh} are the normal components of the patch field at the centers of the low and high work-function patches, respectively. If the surface is covered with patches equal in size, then

$$E_{pnl} > 0 \qquad E_{pnh} < 0 \qquad \text{and} \qquad E_{pnh} = -E_{pnl}$$

Since

$$|E_{pnh}| \approx \frac{\phi_h - \phi_e}{\bar{D}} \qquad W_h \geq W_l$$

and for $x \approx \bar{D}$, W_h is likely to be greater than $\zeta + \bar{\phi}e$ for some values of x. The qualitative behavior of W_l and W_h as a function of x is shown in Fig. 6.4.

Thermionic Emission from a Patchy Surface. When there is no applied field present and space-charge fields can be neglected, Fig. 6.4 shows that it is

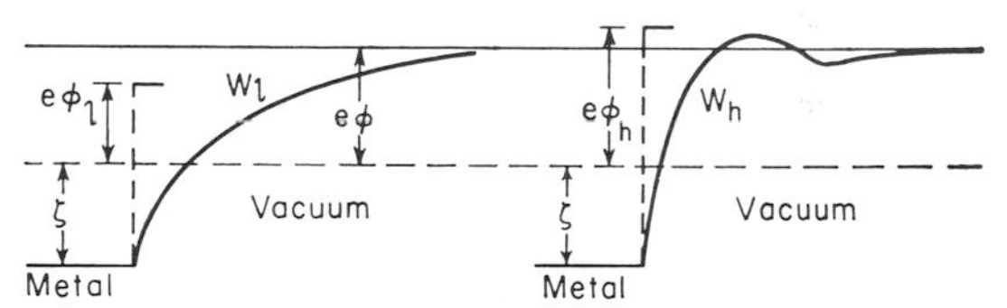

FIG. 6.4. Potential energy of an electron as a function of distance from the emitter surface for low and high work-function patches.

likely that most of the emission current will come from the low work-function patches because of the higher potential barrier in the case of the high work-function case. If, under these circumstances, a weak field E_a is applied so that $|E_a| \ll (\phi_h - \phi_l)/\bar{D}$, then this field will alter the low work-function barrier W_l in the same way as the normal barrier depicted in Fig. 6.1 was altered. Since the thermionically emitted electrons come over this barrier predominantly, the emission current will be given by a modification of Eq. (6.14) or

$$j = f_l A(1 - r_l)T^2 \exp\left(-\frac{\bar{\phi}e - e\sqrt{eE_a}}{kT}\right) \tag{6.24}$$

where f_l is the fraction of the surface area covered by low work-function patches. Thus for low applied fields the emission current will depend on the field like the normal Schottky effect. This is represented by the low field portion of Fig. 6.5.

An easier situation to treat occurs when the applied field is strong, that is, when $|E_a| \gg (\phi_h - \phi_l)/\bar{D}$. Then one may proceed as though there were no patch fields and the emission from each patch is independent of that of its neighbor. Hence the emission current from the ith patch is

$$j_i = f_i A(1 - r_i)T^2 \exp\left[-\frac{e(\phi_i - \sqrt{eE_a})}{kT}\right] \tag{6.25}$$

and the total current density is

$$j = AT^2 \left[\sum_i f_i(1 - r_i) \exp\left(-\frac{e\phi_i}{kT}\right) \right] \exp\left(+\frac{e\sqrt{eE_a}}{kT}\right) \quad (6.26)$$

For this high field case the dependence of j on E_a is again that of the normal Schottky effect and is illustrated by the high field portion of Fig. 6.5. The current is considerably higher through this region because the retarding field at the low work function has been neutralized and thus the low work-function areas are now contributing strongly and the effective surface area is increased over that of the low field region.

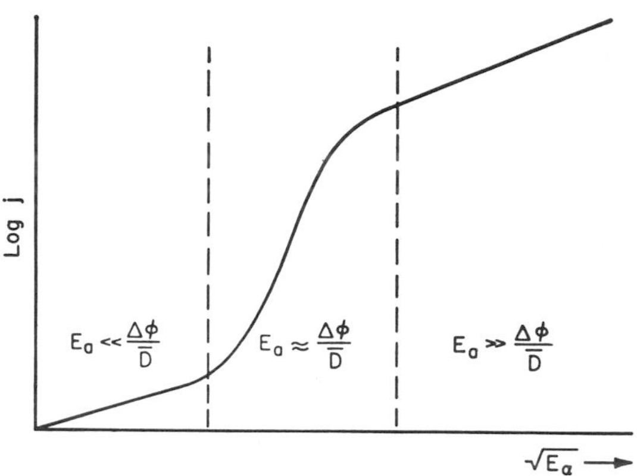

FIG. 6.5. Variation of emission current as a function of applied electric field for a surface with patches of different work functions.

In the intermediate field region where $E_a \approx (\phi_h - \phi_l)/\bar{D}$ the applied field is beginning to neutralize the patch field and by the time the patch field is neutralized the effective work function has been decreased approximately by $\delta\bar{\phi} = \phi - \phi_l$ in addition to the $\sqrt{E}$ effect. In this region the slope of the log j vs. $\sqrt{E}$ curve is considerably greater than for the other two regions. This is shown in the intermediate region of Fig. 6.5. The behavior in this region is sometimes referred to as the *anomalous Schottky effect*. It is noticeable in the emission from any patchy surface whether made up of clean polycrystalline surfaces or a surface with patches of foreign material. By carefully investigating the dependence of the emission current as a function of the applied field it is possible to obtain information about the size of the patches and the differences in their work functions.

3. Metals with Adsorbed Monolayers

General Facts. It has been realized since the experiments of Langmuir on thoriated tungsten wire that monolayers of adsorbed atoms on pure metal surfaces have a surprisingly large effect on the thermionic emission current at a given temperature. For a complete monolayer the zero-field-emission current obeys Eq. (6.1) but the values of r and ϕ are usually quite different from those for clean metals. Some interesting comparisons are shown in Table 6.2.

TABLE 6.2

Substance, polycrystalline base metal	$A(1 - r)$	ϕ, ev	Ionization potential, ev
Ba	....	2.49	5.19
Cs	....	1.81	3.87
Mg	....	3.60	7.61
Ni	30	4.6	7.61
Th	60.2	3.35	
W	60.2	4.52	8.1
Zr	3×10^3	4.50	6.92
Th on W	3.0	2.63	
Zr on W	5.0	3.15	
O on W	5×10^{11}	9.1	

When the ionization potential V_i of the adsorbed atoms is of the order of, or not too much greater than, the thermionic work function ϕ_b of the base metal, the effect of a monolayer of these atoms is to decrease the work function ϕ_c of the composite surface. For example, the work function ϕ_c of a surface composed of a monolayer of atoms which will remain adsorbed on tungsten at sufficiently high temperatures whose ionization potential V_i is less than about 7.5 ev will be less than the work function ϕ_W of pure tungsten. Thus $\phi_c < \phi_W$ for monolayers of the alkali metals, the alkaline earths, and such metal atoms as thorium and zirconium on tungsten. On the other hand the work function of tungsten is not appreciably reduced by an adsorbed monolayer of magnesium atoms ($V_i = 7.6$ ev).

Atoms possessing an appreciable electron affinity so that they can exist stably as negative ions can materially increase the work function of a metal when a monolayer is adsorbed on it. For example, a monolayer of oxygen or sulfur on tungsten increases the work function of tungsten by about a factor 2 (Table 6.2).

If θ is the fraction of the surface of the base metal covered by a monolayer of adsorbed atoms of a given kind, then the average work function of a partially coated surface can be found by applying formula (6.22). Consequently, if ϕ_b is the work function of clean base metal, ϕ_{ad} the work function when the metal has a completed monolayer, the average work function $\bar{\phi}_\theta$ of a partially coated surface is

$$\bar{\phi}_\theta = \theta\phi_{ad} + (1 - \theta)\phi_b \quad (6.27)$$

or

$$\bar{\phi}_\theta = \phi_b + (\phi_{ad} - \phi_b)\theta$$

$\bar{\phi}_\theta$ is shown as a function of θ in Fig. 6.6 for two cases; thorium on tungsten and oxygen on tungsten. The curve for thorium on tungsten shows that the work function is a minimum when the tungsten surface is completely covered with a monolayer of thorium. As more layers of thorium are added, the work function gradually increases until the work function of bulk thorium is attained. It is rather typical of some metal layers that the work function for a monolayer of one metal on another is lower than the work function of either metal alone.

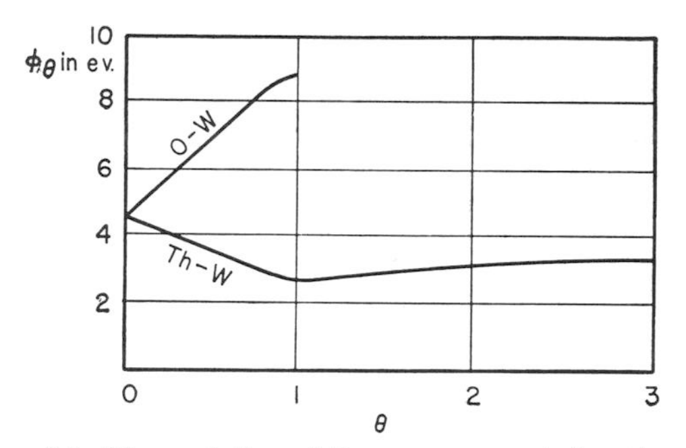

FIG. 6.6. The variation of the average work function of thorium on tungsten and oxygen on tungsten as a function of the fraction of tungsten surface covered.

Explanation of Modified Work Function. When an atom which has an ionization potential V_i is brought near a metal surface whose work function ϕ is greater than V_i, there is a tendency for the outer electron belonging to the atom to be captured by the metal since this situation represents a lower energy configuration. This leaves the atom ionized and when a monolayer of such ions covers the metal surface a charge double layer results and the field thus produced reduces the work function. This early explanation in this simple form does not quite suffice to explain why it is that some atoms whose ionization potential is somewhat greater than the work function of the base metal will nevertheless give rise to a reduced work function when deposited on a base metal. Such a situation exists when a monolayer of barium, $V_i = 5.19$ ev, is deposited on tungsten, $\phi_W = 4.52$ ev. In such cases the reduction in work function is not as great as for the case when $V_i < \phi_b$. It is very much as though the layer of adsorbed atoms was only partially ionized.

The phenomena described above can be understood with the help of Fig. 6.7*a* and 6.7*b*. In these figures the heavy curves denote the potential energy of an electron. At the left of the figure the potential energy of the surface of the metal is depicted together with the energy states in the metal filled with electrons similar to the situation in Fig. 6.1. At the right of Fig. 6.7, the potential energy of an outer electron in an isolated atom with ionization potential V_i is represented. Fig. 6.7*a* deals with the situation where V_i is a little less than the work function ϕ_b of the base metal. When the free atom is brought in the neighborhood of the metal surface, an interaction takes place and the discrete energy state occupied by the outermost electron in the free atom becomes broadened into a band of states of considerable width. In general the broadening is greater the less the ionization potential If none of the broadened states lies below the top of the Fermi level in the metal when the atom is on the surface, the atomic electron will leave the atom entirely and reside in the metal leaving the atom ionized, thereby establishing a strong double layer and producing an appreciable decrease in the work function. However, if part of the band of states lies below the Fermi level as actually depicted in Fig. 6.7*a* these states can be occupied by an electron part of the time by tunneling. Under these circumstances the adsorbed atoms will not be completely ionized

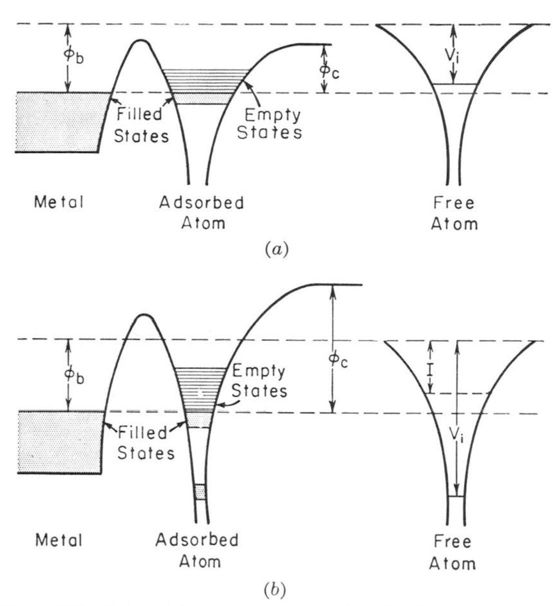

FIG. 6.7. Potential energy curves for single atoms and atoms adsorbed on the surface of a metal illustrating the mechanism of lowering or raising the work function.

and the resulting charge double layer will not be as strong and the work function of the monolayer ϕ_c will not be so much smaller than ϕ_b as in the first case. Thus it can readily be seen why the work function can be reduced even though $V_i > \phi_b$ because even here some of the states in the broadened band can lie above the Fermi level in the metal, in which case the layer of atoms will still be partially ionized. From experiments it has been found that when $V_i = 7.6$ ev (magnesium) and $\phi_b = 4.52$ (tungsten) no reduction in work function is found so that no energy states in the broadened band lie above the Fermi level. Figure 6.7*b* depicts the case where the adsorbed atom has an electron affinity as a free atom. Such atoms are oxygen, chlorine, etc., which can exist stably as negative ions. In oxygen the electron affinity I or work necessary to remove the extra electron is of the order of 3 ev. This situation is depicted at the right of Fig. 6.7*b*. The energy state occupied by the outermost electron in the neutral atom is shown by the solid line together with the corresponding ionization potential V_i. The dashed line denotes the energy state that can be occupied by an extra electron together with the corresponding electron affinity I. When this atom is brought to the surface, the deep lying energy state occupied by the outermost electron of the normal atom is only broadened a little and lies much lower than the Fermi level in the metal so these states remain filled, there being no exchange of electrons between atom and metal. The upper normally unfilled level represented by the dashed line at the right is greatly broadened and some of its states will almost certainly lie below the Fermi level. Since all these states are empty to begin with, some of them become occupied by electrons from the metal as shown. This again produces a charge double layer but in this case the adsorbed atoms are negative. This requires the electrons from the metal to do more work in escaping so that the work function ϕ_c is greater than ϕ_b as shown in Fig. 6.7*b*.

The mechanism described above by which the work function of a base metal is raised or lowered by the adsorption of monolayers of various atoms seems to be adequate to account for the changes in work function found experimentally.

References

1. Bridgman, P. W.: "The Thermodynamics of Electrical Phenomena in Metals," Chap. IV, Macmillan, New York, 1934.
2. Fowler, R. H., and E. A. Guggenheim: "Statistical Thermodynamics," p. 479, Cambridge University Press, New York and London, 1949.
3. Wilson, A. H.: "The Theory of Metals," Chap. II, Cambridge University Press, New York and London, 1953.
4. Reference 3, p. 330.
5. Guth, E., and C. J. Mullin: *Phys. Rev.*, **59**: 575 (1941).
6. Langmuir, Irving: *Phys. Rev.*, **21**: 419 (1923).
7. Langmuir, I., and K. T. Compton: *Revs. Mod. Phys.*, **3**: 245 (1931).
8. Smith, Lloyd P.: *Phys. Rev.*, **35**: 381 (1930).
9. Grover, Horace: *Phys. Rev.*, **52**: 982 (1937).
10. Wright, R. W.: *Phys. Rev.*, **60**: 465 (1941).
11. Mueller, G. J.: *Phys. Rev.*, **45**: 314 (1934).

Chapter 7

Glass

By S. D. STOOKEY *and* R. D. MAURER, Corning Glass Works

1. Structure of Glass

Considerations of atomic structure provide insight into the macroscopic behavior of glasses, even though quantitative connections between the two have rarely been achieved. A brief discussion of the atomic arrangement in oxide glasses is needed in later discussion [1].†

Silicon dioxide glass is the prime example because it not only illustrates many structural features but also is a major constituent of most commercial glasses. Figure 7.1 is a schematic two-dimensional diagram of the fused-silica structure. Each silicon is joined to four oxygen atoms in a tetrahedral form, with the oxygen atoms at the corners. Each tetrahedron is joined to four others by sharing these oxygen atoms at the corners. These tetrahedra form a three-dimensionally linked, disordered network—hence the term "network former" for SiO_2. The Si—O bond is a strong, covalent bond which makes this network rigid and causes the very high viscosity of the liquid from which the glass is formed.

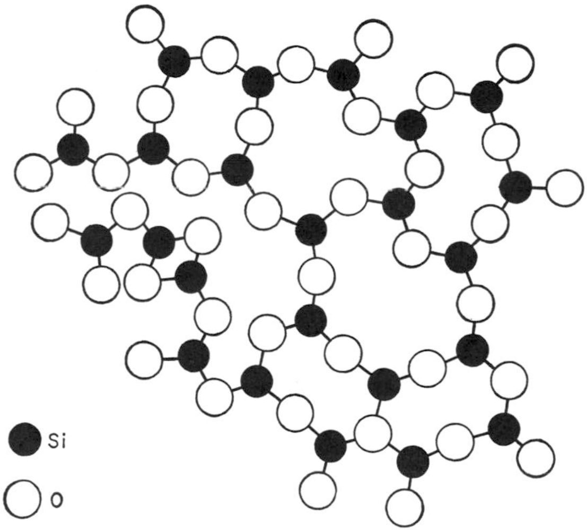

Fig. 7.1. Schematic two-dimensional diagram of silicon dioxide glass. The fourth oxygen of each tetrahedron is presumed outside the plane of the diagram.

† Numbers in brackets refer to References at end of chapter.

The effect of adding an alkali oxide, such as Na_2O, to silica glass is shown in Fig. 7.2. Each silicon atom remains at the center of a tetrahedron of oxygen atoms but all these oxygen atoms are no longer shared with adjacent tetrahedra. Some of the oxygens are a part of only one tetrahedron, with charge neutrality being maintained by a nearby alkali ion. The alkali-oxygen bond is called *ionic;* that is, there is little overlap of electron orbits with other ions. The alkali-oxygen bond is not as strong as the network bonds, and the alkali ion is relatively unrestricted in its movement. The alkali has thus broken the link between two tetrahedra, and the modified network is thus less rigid. Therefore, alkalis are sometimes called "network modifiers," or, more simply, "modifiers." Each sodium ion is matched by a nonbridging oxygen so that when there are two sodium ions to each silicon ion on an average ($Na_2O{\cdot}SiO_2$) the Na_2O content is large enough that some of the tetrahedra are quite unlinked to the network of linked tetrahedra. The viscosity of the liquid is markedly lower, and it becomes increasingly difficult to form glasses at higher alkali concentrations.

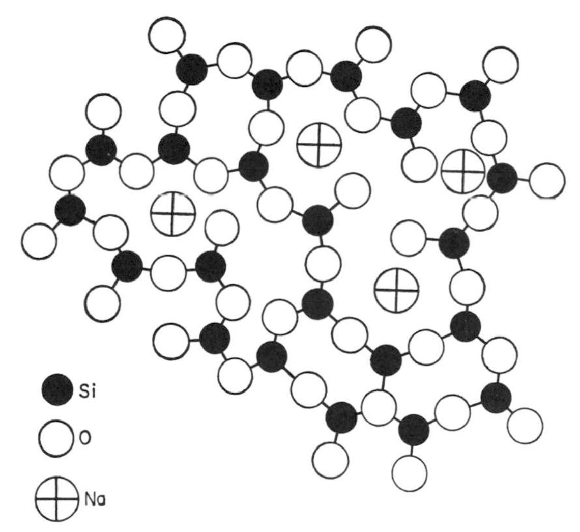

Fig. 7.2. Schematic two-dimensional diagram of sodium oxide–silicon dioxide glass. The fourth oxygen of each tetrahedron is again presumed outside the plane of the diagram.

Multivalent ions can be placed into the structure of silica glass in an analogous way. Each alkaline-earth ion, for example, breaks the link between two tetrahedra by requiring two nonbridging oxygen ions for charge neutrality. These multivalent ions are less mobile than the alkali ions.

Boric oxide, B_2O_3, forms a glass similar to SiO_2 except that the network is composed of linked planar triangles. However, as Na_2O is added an unusual change in the structure takes place: The average boron coordination gradually shifts from three to four. When all the boron is four-coordinated, the network consists of linked tetrahedra. Figure 7.3 shows values

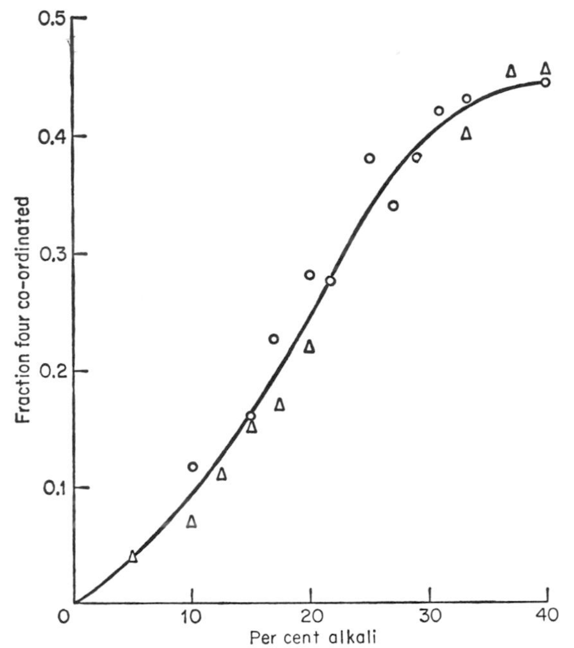

FIG. 7.3. The fraction of four-coordinated boron atoms vs. alkali oxide concentration in alkali oxide–boric oxide glasses. Δ – Li_2O; 0 – Na_2O (Bray [2]).

of the fraction of four-coordinated boron atoms obtained by studying the nuclear quadrupole resonance [2]. Metaphosphate glasses also are composed of tetrahedra but they are linked only at two corners to form chains. Ions alongside the chains complete the structure.

These simple glasses illustrate the major principles in understanding glass structure. Commercial glasses may contain as many as seven or eight oxides with consequent complication of the influence of each on the glass properties.

Between these atomic structures and macroscopic structures lies another region of heterogeneity. In a homogeneous glass, only "short-range order" is present. Starting from any particular atom, it is possible to predict the arrangement of nearest neighbors with a high degree of certainty. Proceeding farther, the certainty diminishes, and beyond about 10 A from the chosen atom, the atomic positions become completely random. However, many oxides form immiscible systems which can result in liquid-liquid separation on cooling from the melt. The structure then possesses a kind of compositional order which can extend to 100 A or so. Certain regions show a higher probability of a chemical composition extending for this distance than other regions. In addition to this type of compositional order, small crystals can form in the glass with dimensions of the order of 100 A. Some regions then possess long-range atomic order. Substances of both kinds possess many of the characteristics of homogeneous glasses, including transparency, glassy fracture, etc. Some materials like this are discussed below because they have important properties not present in homogeneous glasses.

2. Glass Types

The most common types of commercial glass may be classified into four groups in terms of chemical composition.

Silica Glass. But for its high viscosity and attendant difficulties in production, pure vitreous silica would be an ideal glass. Its linear expansion coefficient, $4 \times 10^{-7}/°C$, is among the lowest for all materials, and it possesses good chemical and weathering stability. However, extremely high temperatures are required to produce it; therefore large or complicated articles are costly. Slight contaminations affect its refractive index; thus, in its most common forms, its inhomogeneity makes it unsuitable for good optical systems. These difficulties have partly been overcome by (1) methods of producing SiO_2 glass, other than that of melting selected crystals of quartz, resulting in optical homogeneity comparable with other optical glasses and (2) the development of a glass containing at least 96 per cent SiO_2 whose properties are comparable in many respects to SiO_2 glass itself. This special product, sold under the trademark VYCOR, is made by leaching glass of more ordinary properties, after fabrication, to remove elements other than SiO_2. The article is then fired to reclose its structure.

Lime Glass. Window glasses, container glasses, and those employed in bulbs and tubing for incandescent and fluorescent lamps are usually in this classification. As a rule, they contain 70 to 85 per cent silica (by weight), 10 to 20 per cent alkali oxides (predominantly Na_2O), and 5 to 10 per cent CaO or a combination of CaO and MgO. Expansion coefficients are about $90 \times 10^{-7}/°C$.

Lead Glass. These contain a substantial content of PbO which frequently replaces part of the CaO in lime glasses. High density is accompanied by increased index of refraction and consequent brilliance. Use of lead also permits decrease in alkali content and low dielectric loss. Lead glasses have a long working range and low annealing temperature, without sacrifice of their stability against devitrification at high temperature or against weather at low temperature.

Borosilicate Glass. Fluxes such as alkali, lead, and lime added to the silica network always increase its expansion coefficient, but this effect can be minimized by the use of B_2O_3. Boric oxide is a network former in its own right but it lowers the viscosity of SiO_2. Thus, stable and workable glasses are produced by combining SiO_2 and B_2O_3 in weight ratios up to

5:1, with a limited amount of alkali. Some of these glasses retain enough of the low-expansion characteristic of SiO_2 glass to be classed as heat-resisting. Thus in making complicated laboratory apparatus, there is less risk of strain developing to the point of breakage during fabrication. Some borosilicates have very low dielectric-loss factors as well as relatively low dielectric constant.

3. Glass Melting and Forming

In the manufacture of glass, a premixed powdered batch, generally consisting of oxides or compounds that decompose to oxides at high temperature, is fed into a melting unit and is melted at 1400 to 1600°C until it becomes a clear, homogeneous viscous liquid. Molten glass is a universal solvent and can contain nearly every chemical element. After melting, the glass may be further homogenized by stirring.

During cooling from the melting temperature, viscosity increases rapidly and continuously with decreasing temperature until the rigid glassy state is reached. The viscous liquid is formed at appropriate viscosities by pressing, drawing, rolling, or blowing into molds and is frozen in shape; it is then annealed to remove temperature-induced stresses.

Physicochemical Reactions in Glass. Many of the colored and opaque glasses, as well as photosensitive and photochromic glasses, 96-per-cent-silica glass, and crystalline glass-ceramics, owe their behavior and properties to high-temperature chemical reactions occurring as the glass cools or is reheated. The molten glass is a solution of reactive oxides, which, depending on the composition, temperature, and viscosity, can undergo all the kinds of reaction characteristic of chemical solutions. Some of the more important of these are oxidation-reduction, precipitation of crystalline phases, and liquid-liquid phase separation.

The cooling of the glass often "freezes in" metastable high-temperature equilibrium states. With some glasses useful products are made by controlled reheating to alter the course of reaction in a favorable direction.

Oxidation-Reduction. Many elements can exist in more than one valence state. At the high temperatures employed in melting glass, higher-valence oxides of polyvalent elements generally evolve oxygen and become reduced to lower valences. This gas evolution, from oxides of antimony and arsenic, is often used to aid in "fining"—removing small bubbles from the molten glass.

When the glass is subsequently cooled or reheated in an intermediate temperature range, the polyvalent elements with the greatest affinity for oxygen can remove oxygen from other constituents of the glass.

Gold and copper ruby glasses and silver yellow glasses owe their colors to colloidal metal, thermally reduced by polyvalent ions such as those of tin, antimony, or selenium.

Crystallization and Liquid-Liquid Phase Separation. All glasses are in metastable equilibrium with respect to one or more crystal phases. Their high viscosity at all temperatures below crystal liquidi generally inhibits crystallization.

A number of compounds are slightly soluble in glass at the melting temperature but become less soluble as the glass cools and precipitate in small quantities as colloidal crystals. This results in colored or opaque glasses. Halides, sulfates, phosphates, heavy metal sulfides, and selenides fall in this category. The silicate glass network remains essentially unaffected by these trace precipitates, and only the optical properties are altered.

More far-reaching alterations in glass structure are found in certain borosilicate glasses. These glasses, homogeneous at the melting temperature, separate into two mutually insoluble glasses at lower temperature. The phase-separated droplets are generally smaller than 100 A.

In general, crystallization is an undesirable phenomenon in glass. In glass manufacture, crystals may grow when the hot glass is held for long times below the liquidus, in contact with refractory brick walls or with undissolved crystals of batch ingredients. This is called devitrification. In some cases, devitrification begins at glass-air interfaces.

Beginning in 1957, a family of useful materials called "glass-ceramics" has been developed, based on the discovery that controlled internal nucleation can initiate uniform fine-grained crystallization of glass. The properties of glass-ceramics depend on the composition and the unusually small size (0.1 to 1 μ in diameter) of the crystals. A large number of crystal compositions, including many oxides and silicates, can be precipitated in glass-ceramics. Properties of glass-ceramics include wide ranges of expansion coefficients and of dielectric constants, high mechanical strength, and resistance to high temperatures. Most are opaque, but some are as transparent as glass because their crystal diameters are only about 100 A.

The mechanism of the nucleation leading to such fine-grained crystallization must be very efficient. In some cases, formation of crystal nuclei is preceded by liquid-liquid phase separation at temperatures near the softening temperature of the glass. The low energy barrier to liquid-liquid nucleation, due to small interfacial energies between the liquid phases involved, means that metastable states of this type form very easily.

Glass-ceramics often contain titanium oxide, which causes liquid-liquid separation as the glass is cooled or reheated. As the glass is cooled into the softening range, titania-rich droplets a few angstroms in diameter are formed. Crystals nucleate readily as the fluid droplets grow with further heating.

A different type of heterogeneous nucleation is observed in photosensitive glasses and glass-ceramics nucleated by colloidal gold or silver crystals. A typical photosensitive glass contains in solution small concentrations of an easily reducible gold silicate, a polyvalent ion such as Sb^{+3} as a mild reducing agent, and a polyvalent optical sensitizer such as Ce^{+3}. Exposure of such a glass to ultraviolet light through a photographic negative and subsequent heating results in photoreduction of the gold silicate to colloidal gold by Ce^{+3}, with corresponding oxidation of the Ce^{+3} to Ce^{+4}:

$$Au^{+1} + Ce^{+3} \xrightarrow[\text{u.v.}]{h\nu} Ce^{+4} + Au^{\circ}$$

The polyvalent ions reduce most of the metal after the reaction is initiated by the ultraviolet light.

When the photographic three-dimensional pattern of metal crystals is precipitated in a glass that is a supersaturated solution of readily nucleated crystals such as sodium fluoride, further heating causes precipitation of NaF on the gold nuclei to form a white opaque photographic pattern.

In certain glasses, notably lithium silicates, the metal crystals can nucleate crystallization of a major constituent of the glass, lithium metasilicate (Li_2SiO_3). In this type of glass, the crystallized photographic pattern becomes soluble in dilute hydrofluoric acid so that the material is "chemically machineable."

If the crystallized glass is heated further without acid treatment, the metastable lithium metasilicate alters to lithium disilicate and quartz, and a high-strength glass-ceramic is formed.

Photochromic Glass. It has recently been found that copper-doped silver halides can be precipitated in glass, in such fine dispersion (crystals 40 to 100 A in diameter) that the glass is reversibly photochromic; i.e., it darkens when exposed to light and clears reversibly in the dark [3]. Figure 7.4 shows typical

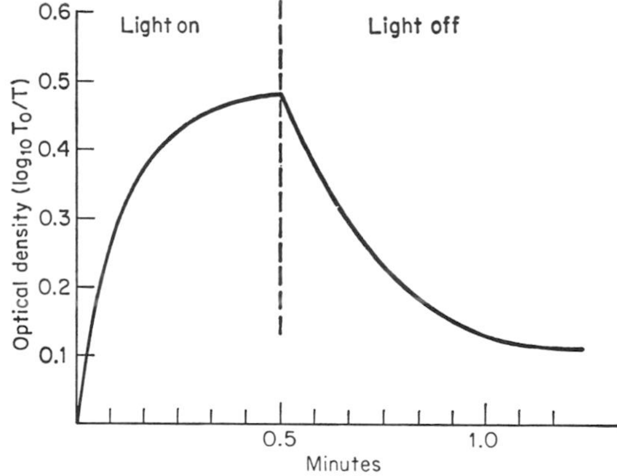

FIG. 7.4. Response of a typical photochromic glass 4 mm thick under an intensity of the order of 10^{15} photons (385 mμ) cm^{-2} sec^{-1}.

darkening and fading characteristics of a photochromic glass. This reversible behavior appears to be unique and is potentially useful in light-controlling, display, and memory applications.

Here again, crystallization is preceded by liquid-liquid separation. The silver halide dissolves in the molten glass, but on cooling the glass becomes supersaturated with silver halide at temperatures above its melting temperature (about 450°C). Droplets of the molten salt are formed by heating the glass at about 600°C; they crystallize on subsequent cooling.

4. The Transformation Range

Further cooling of the liquid beyond crystallization temperatures brings the appearance of a new and different kind of nonequilibrium behavior. Pioneer investigators interpreted it as a real change of state and called it "transformation."

In glass, the change in any property, such as density, with cooling is a combination of two effects: the ordinary temperature coefficient of a momentarily fixed system and changes in the system itself. The first is instantaneously reversible whereas the second, though reversible, is time-dependent. These latter changes require some statistical rearrangement toward a final state of equilibrium at each temperature. The relaxation time for adjustment toward equilibrium is related to the viscosity. Thus, for any given cooling rate the lag of the property with respect to temperature becomes greater with decreasing temperature and increasing viscosity. The lag is immeasurably small in the upper temperature ranges, but at some lower temperature it becomes measurable for that cooling rate, until finally it is effectively infinity. The result is a rather marked change in the temperature coefficient of the property.

The range of temperature within which the relaxation time passes through conventional values defines the transformation range of the glass. For ordinary glass articles the transformation range is in the neighborhood of 500°C; for fine-annealed optical glass it is lowered somewhat while for glass wool fibers it is higher because of the extremely fast cooling.

Behavior in the transformation range is illustrated by the density-temperature variation shown in Fig. 7.5. The high-temperature slope AB is the sum of

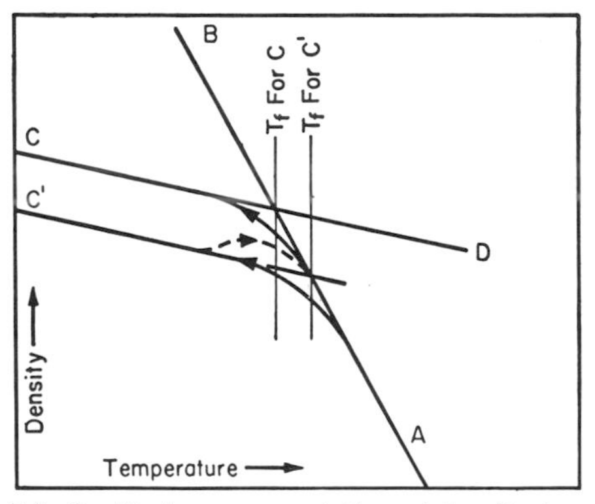

FIG. 7.5. Qualitative representation of density-temperature relation.

the ordinary thermal contraction of a system of fixed structure, represented by the slope of the line CD, and a compacting of the system as a result of structural adjustment, which proceeds at a rate depending on temperature. Progressively longer times are required at lower temperatures; thus the lag in adjustment becomes greater at any given cooling rate until finally only the slope CD remains. The glass is then in an arrested nonequilibrium state which, for any given composition, can be defined in terms of the "fictive temperature." This is that temperature at which, with infinitely long heat treatment followed by sudden chilling, the same room-temperature property would have been achieved. The fictive temperature corresponds to the point of intersection of AB with CD.

If the glass is cooled at a faster rate, it assumes the slope CD at a lower density level and reaches the point C'. Slow reheating of the C' glass causes the property to follow a course represented by the dashed line, approaching AB at a lower temperature than that at which it made its original departure.

An example of these effects is shown in Fig. 7.6 for refractive index [4], which is closely related to density (Part 6, Chap. 6).

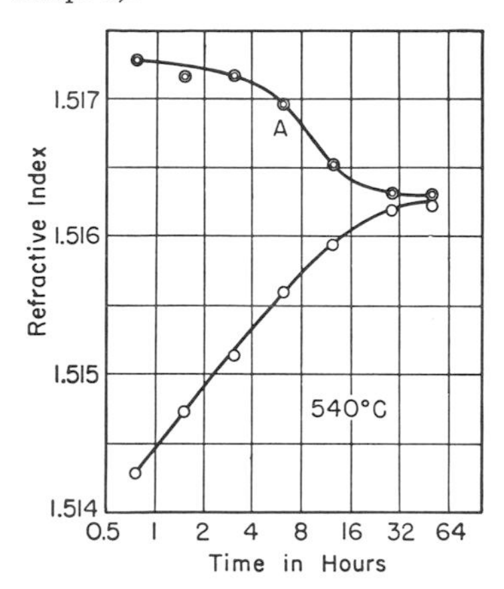

FIG. 7.6. Changes in refractive index with time in a borosilicate crown optical glass. The A glass was previously treated to $T_f < 540°C$.

Within and close to the transformation range of temperatures, relaxations are easily observed in the kinetic properties of glass-forming liquids as well as the thermodynamic properties mentioned above. This behavior is expected from the fact that all these phenomena have their origin in the sluggishness of molecular motion. Before discussing the appearance of nonequilibrium phenomena in viscosity, relevant information is provided about the high temperature, or equilibrium viscosity.

5. Viscosity

In the following discussion all viscosity curves are for the same soda-lime-silica glass.

The expression

$$\log \eta = A_2 + \frac{B_1}{T} \tag{7.1}$$

has been employed extensively in analyzing viscosities in simple liquids, including glasses. In the lower curve of Fig. 7.7, for $\log \eta$ versus $1/T$, the slope increases with decreasing temperature.

Fulcher [5], Robinson [6], and others have shown that the empirical relation

$$\log \eta = A_2 + \frac{B_2}{T - T_0} \tag{7.2}$$

accurately represents high-temperature viscosity data. The upper line of Fig. 7.7 demonstrates this, with $A_2 = 1.422$, $B_2 = 3612°C$, and $T_0 = 543.5°K$. Modified forms of Eq. (7.1) that make A_1 a function of fictive temperature and suggest that B_1 may equal B_2 are discussed by Lillie [7]. Under this concept, A_1, a function of structure, changes rapidly with temperature as long as equilibrium exists between

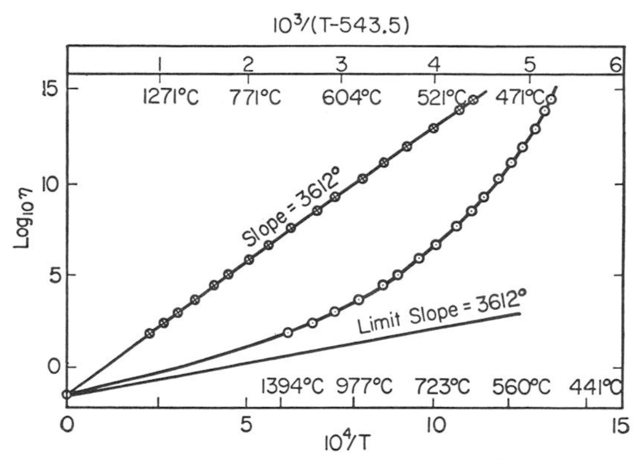

FIG. 7.7. Complete viscosity data for a soda-lime glass using $1/T$ and the Fulcher [5] hyperbola.

temperature and structure but can be arrested by either very rapid heating or cooling. Thus, well below the annealing range, A_1 and B_1 are fixed and Eq. (7.2) no longer applies because the glass is not then in structural equilibrium.

A material is brittle when its tensile strength is less than its elastic limit under the applied rate of loading. Thus, under sufficiently rapid application of load, even rather ductile materials show brittle fracture. Glass at ordinary temperatures is brittle under any loading rate, however slow, indicating that the relaxation time must be very long indeed.

A partial understanding of brittleness can be obtained from the Maxwell relaxation equation. If a glass is stressed sufficiently rapidly, it undergoes an elastic strain of $\xi = S/G$, with S the stress and G the shear modulus. The time rate of change of this strain equals the stress over the viscosity so that

$$\frac{dS}{dt} = \frac{GS}{\eta} \tag{7.3}$$

which gives a relaxation time η/G. Solids have shear moduli of the same order of magnitude; with a representative value of 3.5×10^{11} dynes/cm² for glass, a relaxation time of 1 min occurs at 2.1×10^{13} poises. At temperatures much lower, the relaxation time is too long to observe; at temperatures much higher it is so short that sophisticated techniques are needed to observe it. This viscosity is thus within the transformation range—the lower limit of applicability of (7.2).

Figure 7.8 shows how the viscosity for a typical glass follows the equilibrium curve until the relaxation time becomes great enough to introduce a lag. Starting near 10^{13} poises, the curve for 4°C/min cooling rate progressively departs from the equilibrium curve until, at about $10^{15.5}$ on the latter, the departure is a factor of 10.

If glass in equilibrium at T_0 is suddenly either heated or cooled, the instantaneous variation in viscosity follows the shallow slope $T_H T_L$, after which the equilibrium value for the new temperature is approached with time. The rate of approach is approximately

$$\frac{d\eta}{dt} = \frac{K_2(\eta_\infty - \eta)}{\eta_\infty} \tag{7.4}$$

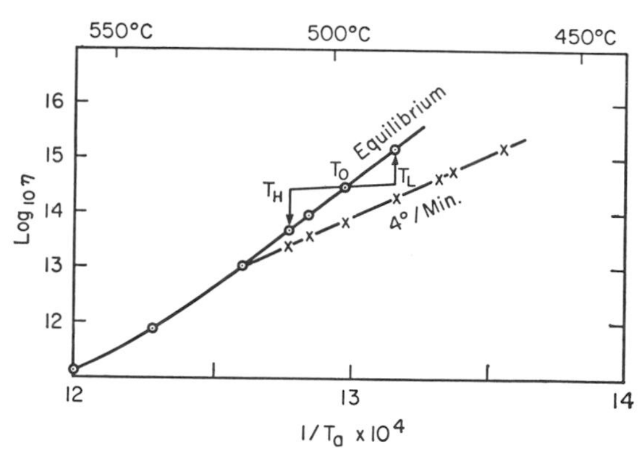

FIG. 7.8. Typical viscosity-temperature-time representation for a soda-lime glass.

where K_2 varies somewhat with temperature but is about 5.5×10^{11} poises/min for the glass represented in Fig. 7.8 and for $d\eta$ positive. Equation (7.4) gives for the relaxation time $\tau = \eta_\infty / K_2$.

Release of stress in glass can be interpreted and qualitatively explained, and practical annealing schedules inferred, on the basis of viscosity-time-temperature variations. For internal stress due to rapid cooling, Adams [8] found

$$\frac{-dS}{dt} = A_3 S^2 \tag{7.5}$$

where S is stress observed as birefringence. The term A_3 is the "annealing constant" which, however, was later found to be different for mechanically stressed annealed glass than for chilled glass [9]. The apparent variation with S^2 results from the time-variable viscosity, so that (7.3) holds where both S and η are functions of t and G is 6×10^{10} dynes/cm² for a soda-lime glass, and is independent of temperature in the range investigated. Variations in $1/\eta$ and S are such that (7.5) expresses stress-release rates; therefore it is widely employed in inferring optimum annealing schedules. However, A_3 is not a unique function of composition and temperature but is strongly dependent upon A_1, which represents fictive temperature.

6. Electrical Properties

In the application of glass as an insulator, properties other than electrical can assume major importance, among them thermal and mechanical endurance, resistance to weathering, and ease of fabrication. In low-voltage, low-frequency insulation at ordinary temperatures, the volume conductivity and dielectric properties are frequently less important than maintenance of high surface resistivity. For high voltages, particularly at high frequencies, low dielectric constant and low power factor are desirable. In capacitors, high dielectric constant, stable with temperature, and low power factor are needed. For applications such as radio-tube stems, involving elevated service temperatures and high d-c voltages, volume resistivity is important.

Table 7.1 indicates the range of properties available in commercial glasses. The superior qualities of

TABLE 7.1. ELECTRICAL PROPERTIES OF SOME COMMERCIAL GLASSES

Glass Corning code	Volume resistivity log of ohm-cm		Dielectric constant 1 Mc, 20°C	Power factor 1 Mc, 20°C	Loss factor 1 Mc, 20°C
	25°C	350°C			
0080 Soda-lime glass	12	5.1	7.2	0.9%	6.5%
0120 Lead glass	>17	8.0	6.6	0.16	1.1
7052 Borosilicate kovar sealing	17	7.4	5.1	0.26	1.3
7070 Low-loss borosilicate	>17	9.1	4.0	0.06	0.24
7740 Borosilicate	15	6.6	4.6	0.46	2.1
7900 96% SiO_2	>17	8.1	3.8	0.05	0.19
7911 96% SiO_2	>17	9.6	3.8	0.02	0.08
8870 High lead glass	>17	9.7	9.5	0.09	0.85
8871 Capacitor glass	>17	8.8	8.5	0.06	0.51
SiO_2 Fused quartz	>17	10.5	4.1	0.02	0.08

vitreous SiO_2 are evident, but this glass is not available in many of the desirable forms in which 7900 and 7911 (96 per cent silica) can be obtained. The borosilicates have rather low loss factors, the lead glasses high dielectric constants, and the lime glass, though least attractive electrically, is satisfactory for d-c or low frequencies at low temperature.

Polarization and Measurement of Electrical Properties. Glass is an electrolyte in which the principal charge carrier is the alkali ion, usually Na^+. On application of a steady field, time is required to attain equilibrium because of slow processes either within the glass or at the electrodes. When direct current flows through glass at elevated temperatures between metallic electrodes, electrolysis results in deposition of alkali metal at the cathode and evolution of oxygen at the anode. Both accumulations are partly reabsorbed on reversal of the potential, producing the effect of partial rectification. When direct current flows at low temperatures, the electrode reactions can be so slow that space charge (or polarization) results.

Polarization can be eliminated, for a true measure of d-c conductivity, by a liquid anode, e.g., sodium amalgam or molten $NaNO_3$. This method permitted Burt [10] to deposit metallic sodium inside a bulb for a photoelectric cell. The evacuated bulb is immersed in molten $NaNO_3$ as an anode, while the hot filament furnishes an electron current to the inside surface. The same process with K^+ is possible but more difficult.

Low-frequency dielectric properties are usually measured with a capacitor, or slab of dielectric between parallel electrodes (Part 4, Chap. 7). The current density is given by $J = (j\omega\epsilon' + \sigma)E$. If the real permittivity ϵ' and resistivity, $\rho = 1/\sigma$, are independent of frequency, the ratio of the real to imaginary impedance becomes increasingly larger as the frequency decreases. At low frequencies only the resistivity is measured. Actually the resistivity increases even faster at very low frequencies after passing through a region of constancy at conventional frequencies (Fig. 7.9). Glass often has these low-frequency relaxations, sometimes associated with polarization, which were formerly called "anomalous charging current." The plateau in Fig. 7.9 is commonly accepted as the d-c resistivity.

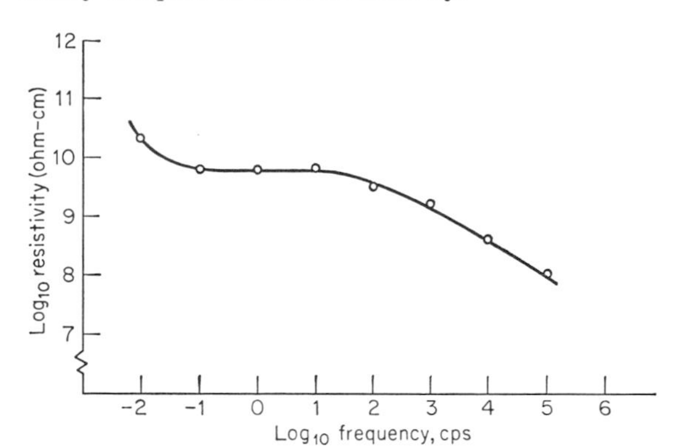

FIG. 7.9. Resistivity vs. frequency for an alumino-borosilicate glass.

Resistivity. Electric resistivity can be measured at much lower temperatures than viscosity and, since it is affected similarly by both temperature and thermal history, it may be an indicator of viscosities in the nonequilibrium state below the transformation range. Here, it is observed that [11–13]

$$\log \rho = A_\rho + \frac{B_\rho}{T} \tag{7.6}$$

similar to (7.1) for viscosity. As pointed out in that case, if A_ρ is taken as a function of fictive temperature, then B_ρ may be tentatively regarded as a constant equal to $d \log \rho / d\,(1/T)$ at $T = \infty$. Thus resistivity curves for temperatures below the transformation range, for the same glass with different thermal histories, would be parallel at levels determined by $A_\rho(T_f)$.

From (7.1) and (7.6)

$$\log \eta = C_1(\log \rho + C_2) \tag{7.7}$$

where $C_1 = \frac{B_1}{B_\rho}$ and $C_2 = \frac{A_1}{C_1} - A_\rho$

If C_1 and C_2 are independent of temperature, the ratio dA_1/dA_ρ with change of temperature (or fictive temperature) is the constant C_1.

Based on work later reported by Babcock [14], Littleton [15] has shown that (7.7), with C_1 and C_2 independent of temperature, is valid from the strain point to 1400°C, in four types of glass (Fig. 7.10).

Resistivity refers to migration of particular ions, while viscosity refers to shearing of the network itself. Some constituents (for example, Pb) play little part in conduction but substantially reduce the viscosity. Thus the activation energy for conductivity in an alkaline glass is generally less than that for viscosity, but the difference diminishes as other fluxes replace alkali. Littleton's data for C_1 and C_2 (partly recalculated) support this expectation.

Dielectric Loss. Qualitatively, the power factor tan δ parallels volume conductivity, varying in

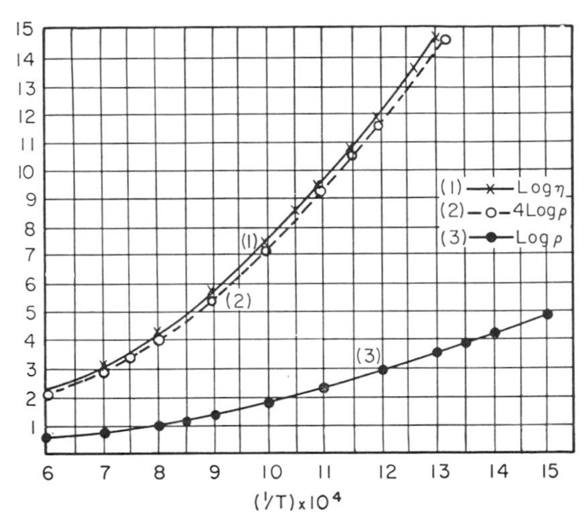

FIG. 7.10. Relationship between viscosity and electrical conductivity of a lime glass.

about the same way with composition, increasing with temperature, and decreasing with frequency. However, McDowell [16] and Strutt [17] show that variations in power factor persist even after correction has been made for the conductivity, probably due to relaxation effects. McDowell showed that

$$\tan \delta = Ae^{BT} \tag{7.8}$$

with $\tan \delta = 4.7 \times 10^{-3}$ at 0°C and $B = 7.7 \times 10^{-3}$ between 30 and 90°C for Corning borosilicate glass code 7740 having a room-temperature resistivity of 3.1×10^{14} ohm-cm. The slope B decreases with increasing frequency. Strutt's data on apparently the same glass are reproduced in Fig. 7.11 with McDowell's observations included for comparison.

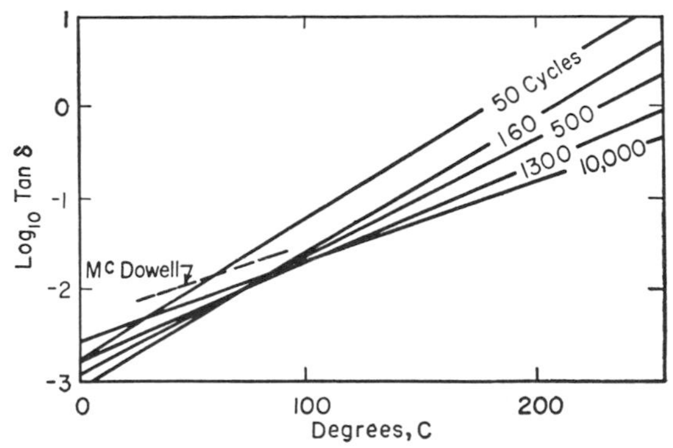

FIG. 7.11. Variation of power factor with temperature and frequency.

Glasses can be heated by dielectric loss at high frequencies. The loss per unit volume is proportional to the product of ϵ, tan δ, frequency f, and voltage gradient. Danzin [18] studied this effect for f between 3×10^5 and 10^7 cps.

Dielectric Constant. Morey [1e] gives extensive tables of dielectric properties and their variation with temperature, frequency, and composition. Glass dielectric constant k is seldom outside the range 3 to

13 and never reaches the extremely high values attainable in crystalline ceramics. It increases slightly up to about 100°C. On further heating the increase is pronounced, partly because of losses, since k decreases at room temperature with increasing frequency.

7. Thermal Properties

Specific Heat. Figure 7.12 shows the average specific heat c_m between 0°C and temperature T for representative glasses. The true specific heat c at T rises more sharply, being

$$c = \frac{T\,dc_m}{dT} + c_m$$

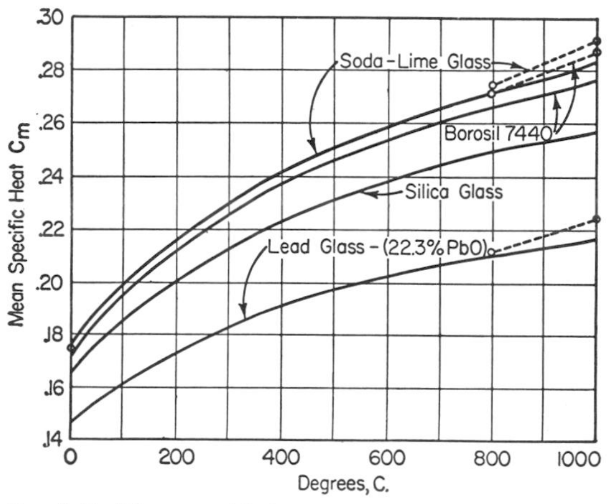

Fig. 7.12. Mean specific heat for representative glasses.

Thermal Conductivity. Glass differs from crystalline materials in becoming more conducting with increasing temperature [19] (Fig. 7.13). Among common glasses, SiO_2 glass has the highest conductivity, and the high-lead compositions have the lowest.

Kittel [20, 21] explains the increase in conductivity with temperature by making

$$K = \frac{c_p v \Lambda}{4}$$

where c_p is the heat capacity per unit volume, v the velocity of sound propagation, and Λ the mean free path of lattice phonons. Although c_p reaches extremely low values below 10°K (Fig. 7.14), the thermal conductivity levels off to about 0.00028 (cgs) and then drops sharply below 2.5°K. Resulting values for Λ are about 6,000 A at 2.4°K but only 20 A above about 40°K. Also, the thermal conductivity of a Thuringian glass is 8×10^{-5} at 1.3°K [22].

Curves such as in Fig. 7.13 continue their indicated course up to at least 300°C for clear glasses and probably to higher temperatures for those having appreciable infrared absorption. The effective conductivity rises as the transfer by radiation increases (Part 5, Chap. 5), Ginzburg [23] gives a value equivalent to 0.006 cgs for a clear glass at 1000 to 1300°C. This

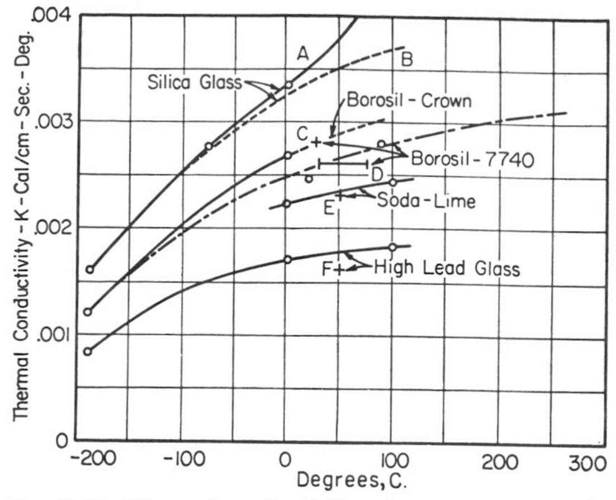

Fig. 7.13. Thermal conductivity of some representative glasses. At these temperatures radiation transfer is not significant. (*Sources for data:* SiO_2 *curve A, borosilicate crown and high lead* (69% PbO) *curves: Eucken* [100]; SiO_2 *curve B and soda lime curve: Kittel* [101]; 7740 *curve: Stevens* [176]; *point C: Nat. Bur. Standards (unpublished),* 1916; *bar D: Bridgman* [177]; *points E and F* (61% PbO): *Pittsburgh Plate Glass Co.)*

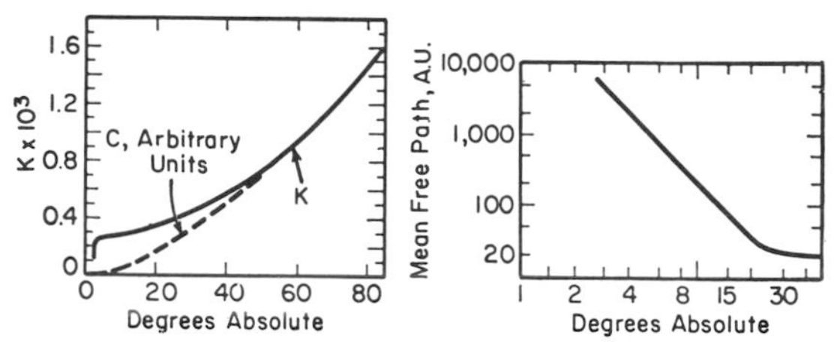

Fig. 7.14. Low-temperature thermal-conductivity relations in SiO_2 glass.

is about three times the room-temperature value. Rodnikova [24] reports a value of 0.034 cgs for colorless glass at 750°C and states that ordinary conduction persists to 300°C in colorless glass and to 800°C for colored glasses.

Heat transfer by radiation absorption and reemission is dominant at high temperatures [25–27]. For the thermal conductivity due to the thermal radiation

$$K'(T) = \frac{n^2 c}{3} \int_0^\infty A^{-1} \frac{\partial \rho}{\partial T}\, d\lambda \qquad (7.9)$$

where n is the refractive index, c the velocity of light, ρ the density of black-body radiation, and A the effective absorption coefficient of wavelengths of importance at temperature T. Kellett assumes an average $A = 0.3\ \text{cm}^{-1}$ at 1200°C and $5\ \text{cm}^{-1}$ at 500°C and finds radiation conductivity to be 0.117 and 0.0014 cgs units at the two temperatures, respectively. To this must be added the conventional conductivity, about 0.004 cgs (Part 5, Chap. 5, Sec. 10).

8. Strength of Glass

Changes of great practical importance are currently being made in the strength of commercial glass, which has been traditionally regarded as a fragile substance. Paradoxically, it has been known for many years that both the intrinsic tensile strength and the compressive strength of glass are extremely high ($>10^6$ psi). Its brittle nature, together with surface flaws, negate its intrinsic strength in service.

Brittleness. The brittleness of glass, even under slow loading, results from the absence of plastic yield. Preston [28] states requisites for a brittle material: It must fail in tension and not in shear; it must have a high modulus and high strength, to exclude materials that are rubbery or friable; and it must be capable of forked fracture.

Materials are compared, as to their brittleness, by breaking under impact. Williams [29] and Guyer [30] define the impact modulus:

$$S = \frac{18EW}{AL}$$

where E is Young's modulus, W is the work expended by the impacting ball, and A and L are the cross-section area and length of the sample. Preston [31] substitutes S^2 for S on dimensional grounds and finds better agreement with moduli of rupture obtained by other methods. Föppl [32, Ref. 1e, p. 348] compares various glasses in impact and finds borate flints the least and heavy silica flints the most brittle.

Flaws and Stress Concentration. Although the observed breaking stresses for sizable glass articles, having normal surface condition, is of the order of 10,000 psi, the strength of fine fibers and of larger samples with specially treated surfaces is much higher, approaching 10^6 psi. Thus, it is concluded that ordinary glass surfaces contain discontinuities that lead to high stress concentrations.

Griffith [33, 34] expressed the criterion of rupture, based on a two-dimensional crack, by

$$F = \left(\frac{2EW_s}{\pi L_c}\right)^{1/2} \qquad \text{or} \qquad F_c = \left(\frac{8EW_s}{\pi r_c}\right)^{1/2}$$

where F = observed stress across sample
F_c = concentrated stress at crack tip to produce rupture
E = Young's modulus
W_s = surface energy
L_c = one-half the length of two-dimensional elliptical crack
r_c = radius of crack tip

For artificially produced scratches, $FL_c^{1/2}$ has a mean value of 239 cgs, in fair agreement with that predicted from material constants. F_c was calculated to be about 2,000 kg/mm² (3×10^6 psi).

Strength of Large Samples. *Fabrication.* The strength of massive glass depends on the method of fabrication and the treatment in manufacture; the composition is of little importance. Holland [35] studied the strength of sheet-glass laths under various conditions of preparation and test. Table 7.2 shows the range of observed strength.

Composition. From Sosman's [36, p. 481] value of 5 kg/mm² for strength of SiO_2 glass and the data quoted by Morey [1e, pp. 333–338] it appears that composition is not an important variable. Heat treatment also has little effect except as it may incidentally govern the residual stresses or the surface condition.

TABLE 7.2. RANGE OF STRENGTH IN GLASSWARE

Ware or condition	*Tensile strength, kg/mm²*
Surface ground or sandblasted	2–4
Pressed	3–7
Blown	
Hot iron mold	4–7
Paste mold	7–11
Inside surface	14–35
Drawn tube or sheet	7–15
Polished	7–11
Fine fibers	
Unannealed	20–200
Annealed	7–30
Acid fortified, up to	175

Temperature. In tests between −190 and 520°C, Smekal [37] and Preston [38] found a minimum strength between 0 and 300°C. An 80 per cent increase in the strength of large fibers ($\approx$0.5 mm) is found upon cooling from room temperature to −190°C, while independent results show a 20 per cent increase at −40°C [39, 40]. The absence of moisture may fully account for the increase below 0°C. Jones [41] reports a substantially constant strength from room temperature up to some 40°C below the transformation range. At elevated temperatures, the strength depends markedly on experimental conditions.

Atmosphere. Baker [42] shows that glass gains about 20 per cent in strength when tested in dry air and 100 to 150 per cent *in vacuo*. Comparing the strength of lead, lime, and borosilicate glasses in water with that in air, Stockdale [43] found an immediate weakening of about 11 per cent, followed by a progressive strengthening. The increase undoubtedly resulted from attack by the water; it was most rapid for the lead glass and intermediate for the lime glass. Even the borosilicate attained double strength in 1,000 hr of immersion. The action of moisture is extremely rapid. A sample in vacuum under a load somewhat below its limit fails immediately on admission of room air.

Strengthening Methods. *Physical Tempering.* Until recently, the only practical and permanent way of strengthening glass has been physical tempering, i.e., chilling from near the softening point in order to produce compression in the surface layers. The effective tensile strength is the sum of the surface compression and the surface strength. Maximum practical strengths obtained by this method range up to about 30,000 psi, but the method is limited to simple shapes and relatively thick glass.

Acid Polishing. Very high strengths, up to 300,000 psi, have been obtained by "acid fortification," which removes the extreme surface layer containing scratches and flaws, by treatment in hydrofluoric and mineral acids [44]. The high strength is destroyed by the slightest mechanical touch but can be maintained for years if the surface is immediately lacquered or otherwise protected [45].

Chemical Strengthening. A number of methods have recently been developed for greatly and permanently enhancing the strength of glass and glass-ceramic articles by chemical methods. These methods alter the composition of the surface layer in such ways as to induce high compression [46, 47]. Practical strengths up to and exceeding 100,000 psi can be obtained.

One of the most effective methods consists of immersing a glass containing alkali-metal ions in a molten salt bath containing larger alkali-metal ions, at a temperature just below the strain temperature, until the alkali-metal ions in the surface layer about 10 μ deep have been replaced in a one-for-one exchange by the larger alkali-metal ions from the salt bath. This results in a swelling of the rigid surface layer and develops permanent compression whose magnitude depends on the concentration and the size differential of the exchanged ions. Very high strengths, of the order of 100,000 psi, can be obtained in this way, even in thin-walled and complex shapes.

One consequence of this increased strength is that light-weight glass sheets, thin enough to be flexible, can have strength equal to that of much thicker physically tempered glass.

Fatigue. Murgatroyd [48] found that time has no direct effect on strength in the absence of chemical attack. Previous stress history is important, including fatigue under continued load, as shown in Fig. 7.15. The strength in air appears to be comparable

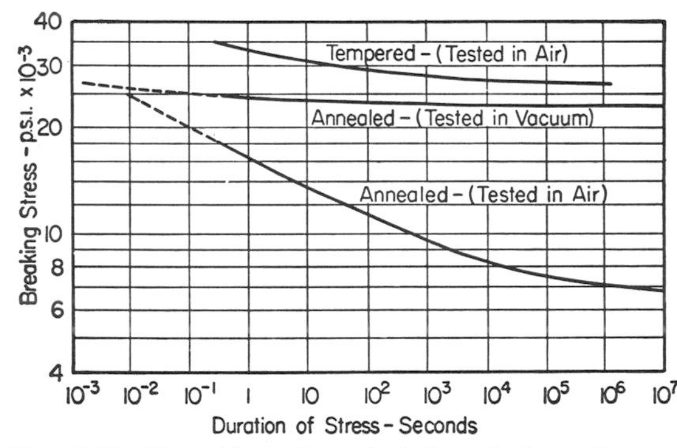

FIG. 7.15. Stress-time characteristics of glass at room temperature.

with that in vacuum for a duration of 0.01 sec, indicating that the action of atmospheric moisture requires a short but finite time. The fractional decrease in tempered glass is less than in annealed

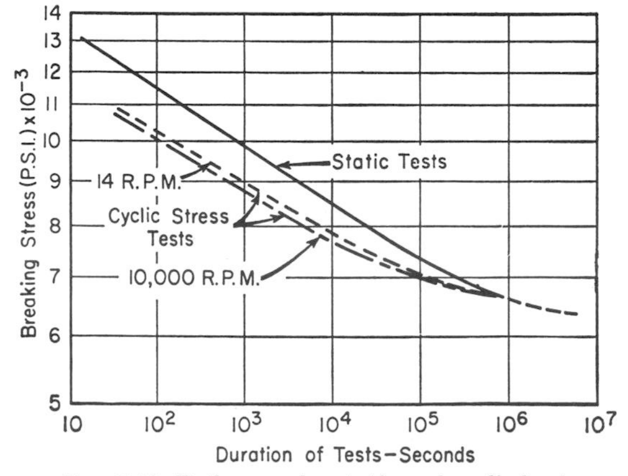

FIG. 7.16. Fatigue under static and cyclic load.

glass. Gurney [49] studied cyclic loading, with the results shown in Fig. 7.16. The frequency of cycling has little effect.

The effect of loading rate is shown by Smekal's [37] tests (Fig. 7.17) on fibers about 1.35 mm in diameter. After an interruption of loading at an intermediate stress, the samples break at lower values than for uninterrupted loading. This effect of previous stress history persists for a short time even if the load is completely removed.

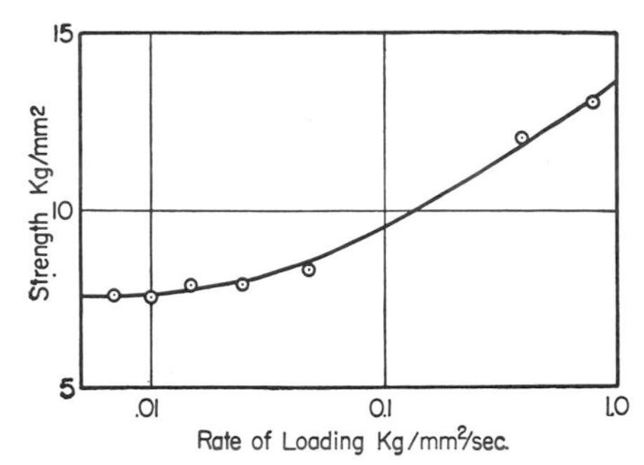

FIG. 7.17. Effect of loading rate on glass strength.

Bailey [50, 51] discusses stress concentrations due to large discontinuities and gives a survey of the effect of flaw intensity and distribution on glass strength. Greene [52] treats the flaw problem in such a way that conclusions regarding the flaw character and distribution may be drawn from a statistical analysis of strength data.

Fracture Velocity. Schardin [53] and Barstow [54] have studied photographically the spread of impact fractures in sheet glass. The velocity of crack propagation, under these violent conditions, is constant at about 1.5×10^3 m/sec. This is roughly one-third the velocity of sound.

Lundborg [55] observed electrically the rate of crack propagation from a score near the edge of sheet glass. He found that fracture was initiated at about 0.9 kg/mm² and that the velocity increased rapidly with increasing stress, to a constant 1.5×10^3 m/sec above 3 kg/mm².

Poncelet [56] gives a theoretical discussion of crack velocity and concludes that

$$V_r = \frac{V}{2} \exp \frac{-W_r}{kT} \tag{7.10}$$

where V_r = crack velocity
V = transverse wave velocity
W_r = energy required to rupture atomic bond, in excess of that furnished by externally applied stress

When the break is violent, W_r reaches zero and $V_r = V/2$. For higher energies, making W_r negative, branching occurs.

Yoffe [57] used Eq. (7.10) and $V = (G/\rho)^{1/2}$ to calculate the maximum velocity of crack propagation in lime glass and SiO_2 glass. The result was 1.6 and 1.9×10^3 m/sec, respectively, compared with 1.5 and 2.2 observed. Murgatroyd [58] shows that the rib and hackle marks frequently seen on a fracture surface can be used to trace the break to its origin and sometimes to determine the velocity of propagation.

Small Fibers. Figure 7.18, from Griffith's [33] early work, shows the increase in strength of fibers with decreasing diameter. The high strength results from small size, surface perfection, and thermal history.

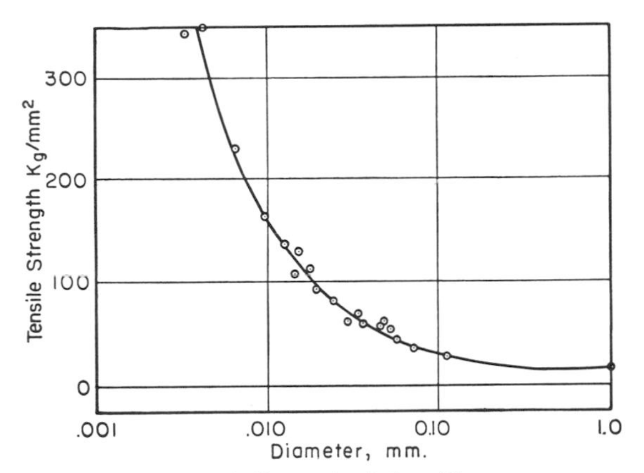

FIG. 7.18. Strength of glass fibers.

The small cross section precludes large flaws, especially those oriented normal to the longitudinal testing stress. In addition, the probability of a flaw occurring in the tested surface diminishes with decreasing surface area and shows that the breaking force per unit periphery is almost independent of diameter [59]. In support of this, strengths comparable to those of fine fibers are observed when small steel balls are pressed on the surface to fracture the glass, so as to involve only small areas [60]. Bikerman [61] reports that in threads composed of many small filaments the weakest of a batch of eight threads 1.7 cm long had the same strength as a single thread eight times as long.

Fine fibers are drawn rapidly from high temperatures, and their cooling rate is extreme. Further, the high rate of stretch requires a substantial tractive stress. Because of their high fictive temperature, their surface structure differs from that of massive glass. Fibers lose much of their strength on annealing, while an appreciable shortening also occurs; as freshly drawn, the flaws may have an abnormal geometry with low stress concentration.

9. Optical Properties

Optical Glass. Optical glass is made under the strictest possible control to insure that it is as free as possible from inhomogeneities, uniform in composition, well enough annealed to introduce negligible birefringence, highly transparent to some part of the spectrum (usually the visible), and meeting exacting specifications on refractive index and dispersion. Tolerances on these properties in general lie between 10^{-4} and 10^{-3} for index, about 0.3 for dispersion, and between 10^{-5} and 10^{-4} for index homogeneity.

Radiation Absorption. *Light.* Glasses are available with high transparencies for radiation between 1,800 A and about 5 μ. Addition of colorants results in selective absorption. A few of the colorants, notably colloidal cadmium sulfoselenides, have sharp cutoffs, but most absorption bands are diffuse.

High-energy Radiation. Glasses that are free of heavy elements are transparent to hard X rays, absorption increasing as X-ray wavelength increases. Glasses absorbing X rays but having good visible transmission are obtained by addition of such heavy elements as lead, barium, and tungsten. These are used as windows to permit observation of radioactive materials. X rays discolor most glasses by causing electron emission and retrapping. The discoloration is minimized by additives such as cerium oxide [62].

Glasses rich in CdO and B_2O_3 are used to absorb slow neutrons [63].

Radiation-induced Reactions in Glass. Very high fluxes of neutrons can cause nuclear reactions and change the atomic composition of certain glasses.

X rays, electron beams, and short-wave ultraviolet light can cause photoemission of electrons within glass. This can cause reactions ranging from the extreme of building up sufficient space charge to shatter the glass during discharge, to fluorescence, to "solarization" (coloration due to metastably trapped electrons or holes; subsequent heating erases the color and is accompanied by thermoluminescence), and to permanent photochemical reaction, as in the photosensitive glasses that develop photographs of colloidal gold, silver, or copper.

References

The authors have drawn heavily on the article by H. R. Lillie in the previous edition of this Handbook. Other general references are:

1*a*. Jones, G. O.: "Glass," Methuen, London, 1956.
1*b*. Stanworth, J. E.: "Physical Properties of Glass," Clarendon Press, Oxford, 1950.
1*c*. Condon, E. U.: *Am. J. Phys.*, **22**: 43–53, 132–142, 224–232, 310–317 (1954).
1*d*. Stevels, J. M.: The Structure and Physical Properties of Glass, "Encyclopedia of Physics," vol. 13, Springer, Berlin, 1962.
1*e*. Morey, G. W.: "The Properties of Glass," Reinhold, New York, 1954.
2. Bray, P. J., and J. G. O'Keefe: *J. Phys. Chem. Glasses*, **4**: 37 (1963).
3. Armistead, W. H., and S. D. Stookey: *Science*, **144**: 150 (1964).
4. Brandt, N. M.: *J. Am. Ceram. Soc.*, **34**: 332 (1951).
5. Fulcher, G. S.: *J. Am. Ceram. Soc.*, **8**: 339 (1925).
6. Robinson, H. A.: *J. Am. Ceram. Soc.*, **27**: 129 (1944).
7. Lillie, H. R.: *Proc. Intern. Comm. Glass*, 1955.
8. Adams, L. H., and E. D. Williamson: *J. Franklin Inst.*, **190**: 597, 835 (1920).
9. Lillie, H. R.: *J. Am. Ceram. Soc.*, **19**: 45 (1936).
10. Burt, R. C.: *J. Opt. Soc. Am.*, **11**: 87 (1925).
11. Gray, T., A. Gray, and J. J. Dobbie: *Proc. Roy. Soc. (London)*, **34**: 199 (1882); **36**: 488 (1884); **63**: 38 (1898); **67**: 197 (1900).
12. Fulda, M.: *Sprechsaal*, **62**: 653, 769, 789, 810, 831 (1929).
13. Robinson, D. M.: *Physics*, **2**: 52 (1932).
14. Babcock, C. L.: *J. Am. Ceram. Soc.*, **17**: 329 (1934).
15. Littleton, J. T.: *Ind. Eng. Chem.*, **25**: 748 (1933).
16. McDowell, L. S., and H. L. Begeman: *Phys. Rev.*, **33**: 55 (1929).
17. Strutt, M. J. O.: *Arch. Elektrotech.*, **25**: 715 (1931).
18. Danzin, A., and P. Meunier: *Compt. Rend.*, **228**: 391 (1949).
19. Eucken, A.: *Ann. Physik*, **34**: 185 (1911).

20. Kittel, C.: *Phys. Rev.*, **75**: 972 (1949).
21. Berman, R.: *Phys. Rev.*, **76**: 315 (1949).
22. Keesom, P. H.: *Physica*, **11**: 339 (1945).
23. Ginzburg, D. B.: *Steklo i Keram.*, **4**: 9 (1947).
24. Rodnikova, V. V.: *Steklo i Keram.*, **6**: 10 (1949); **8**: 9 (1951).
25. Preston, F. W.: *Bull. Am. Ceram. Soc.*, **15**: 409 (1936); *J. Soc. Glass Technol.*, **31**: 134 (1947).
26. Merren, W. J. R.: *J. Soc. Glass Technol.*, **35**: 230 (1951).
27. Kellett, B. S.: *J. Opt. Soc. Amer.*, **42**: 339 (1952); *J. Soc. Glass Technol.*, **36**: 115 (1952).
28. Preston, F. W.: *J. Am. Ceram. Soc.*, **15**: 176 (1932).
29. Williams, A. E.: *J. Am. Ceram. Soc.*, **6**: 980 (1923).
30. Guyer, E. M.: *J. Am. Ceram. Soc.*, **13**: 624 (1930).
31. Preston, F. W.: *J. Am. Ceram. Soc.*, **14**: 428 (1931).
32. Föppl, A.: *Sitzber.Math.-Phys.Kl. Akad. Wiss. München*, **3**: 505 (1911). See ref. 1e, p. 348.
33. Griffith, A. A.: *Trans. Roy. Soc. (London)*, **A221**: 162 (1920).
34. Griffith, A. A.: *Proc. First Intern. Congr. Appl. Mechs., Delft*, p. 55, 1924.
35. Holland, A. J., and W. E. S. Turner: *J. Soc. Glass Technol.*, **18**: 225 (1934).
36. Sosman, R. B.: "The Properties of Silica," p. 592, Reinhold, New York, 1927.
37. Smekal, A.: *Ergeb. Exakt. Naturw.*, **15**: 106 (1936).
38. Preston, F. W.: *J. Appl. Phys.*, **13**: 623 (1942).
39. Onnes, H. K., and C. Braak: *Koninkl. Ned. Akad. Wetenschap.*, **16**: 890 (1907–1908).
40. Holland, A. J.: *J. Soc. Glass Technol.*, **32**: 5 (1948).
41. Jones, G. O., and W. E. S. Turner: *J. Soc. Glass Technol.*, **26**: 35 (1942).
42. Baker, T. C., and F. W. Preston: *J. Appl. Phys.*, **17**: 179 (1946).
43. Stockdale, G. F., F. V. Tooley, and C. W. Ying: *J. Am. Ceram. Soc.*, **34**: 116 (1951).
44. Brodmann, C.: *Nachr. Kgl. Ges. Wiss. Göttingen*, **44**: (1894).
45. Phillips, C. J.: *Am. Scientist*, **53**: 20 (1965).
46. Nordberg, M. E., et al.: *J. Am. Ceram. Soc.*, **47**: 215 (1964).
47. Kistler, S. S.: *J. Am. Ceram. Soc.*, **45**: 59 (1962).
48. Murgatroyd, J. B.: *J. Soc. Glass Technol.*, **16**: 350 (1932).
49. Gurney, C., and S. Pearson: *Proc. Roy. Soc. (London)*, **A192**: 537 (1948).
50. Bailey, J.: *Glas-Ind.*, **16**: 137 (1935).
51. Bailey, J.: *Glas-Ind.*, **20**: 21, 59, 95, 143 (1939).
52. Greene, C. H.: *J. Am. Ceram. Soc.*, **39**: 66 (1956).
53. Schardin, H., and W. Struth: *Glastech. Ber.*, **16**: 219 (1938).
54. Barstow, F. E., and H. E. Edgerton: *J. Am. Ceram. Soc.*, **22**: 302 (1939).
55. Lundborg, N., and C. H. Johansson: *Arkiv Fysik*, **4**: 555 (1952).
56. Poncelet, E. F.: Fatigue in Brittle Solids, "Fracturing of Metals," American Society for Metals, 1948.
57. Yoffe, E. H.: *Phil. Mag.*, **42**: 739 (1951).
58. Murgatroyd, J. B.: *J. Soc. Glass Technol.*, **26**: 155 (1942).
59. Preston, F. W.: *J. Soc. Glass Technol.*, **10**: 234 (1926); **17**: 5 (1933).
60. Powell, H. E., and F. W. Preston: *J. Am. Ceram. Soc.*, **28**: 145 (1945).
61. Bikerman, J. J., and G. H. Passmore: *Glas-Ind.*, **29**: 144 (1948).
62. Stroud, J. S.: *J. Chem. Phys.*, **47**: 836 (1962).
63. Melnick, L. M., H. W. Safford, K. H. Sun, and A. Silverman: *J. Am. Ceram. Soc.*, **34**: 82 (1951).

Chapter 8

Phase Transformations in Solids

By R. SMOLUCHOWSKI, Princeton University

The field of phase transformations in solids is characterized by a huge amount of experimental (often only qualitatively significant) material and by a relative scarcity of satisfactory theories. This is particularly true of theories of the atomic mechanism and of kinetics of transformations. This chapter is concerned only with transformations which occur *within* the solid state. Thus melting, solidification, condensation, and evaporation are not included as specific topics. The equilibrium conditions are considered first and this is followed by a survey of kinetics of transformations in solids. Most of the illustrative material is taken from the fields of physics of pure elements, of alloys, and of simple ionic crystals which offer best chance for a theoretical understanding. Purely descriptive material is, as far as possible, omitted.

EQUILIBRIA

1. Classical Phase Transformations

The most fully developed theories of phase transformations are based on thermodynamics and, in particular, on statistical thermodynamics. An assembly of systems, for instance, of molecules or atoms of various kinds, can be physically and chemically either homogeneous or not homogeneous. A nonhomogeneous assembly can be divided into a number of homogeneous parts, called phases, each possessing its characteristic energy, temperature, and entropy. The statistical nature of these quantities indicates that a rigorous definition of a phase, and especially of a phase boundary, on an atomic scale is difficult if not impossible. This, as we shall see later, seriously limits our present understanding of the mechanism of many phase transformations in solids. In spite of this, thermodynamics provides a very convenient framework for description of these phenomena.

Gibbs, who was the first to give a rather complete analysis of transformations on this basis, introduced what is now called the Gibbs function

$$G = U - TS + PV \tag{8.1}$$

where the symbols have the usual significance (see Part 5, Chap. 1). An equilibrium between phase 1 and phase 2 at given T and P requires $G_1 - G_2 = 0$. Since the compressibility of solids is relatively small, one can assume $dV = 0$ and then the equilibrium condition can be expressed as $dF = 0$, where $F = U - TS$ is the (Helmholtz) free energy. It should be kept in mind, however, that for solids, the use of F instead of G in the equilibrium requirements is only an approximation. In general, whenever appreciable tensions and compressions occur, the thermodynamical analysis of the transformation or reaction becomes a delicate problem. An example here is the sign of the change of entropy associated with the vacancy mechanism of diffusion, where at the saddle point large local deformation can occur [1, 2].*

An estimate of the free energy of a solid as a function of temperature can be obtained in the following manner: From

$$dF = -P\,dV - S\,dT$$

at constant pressure we have

$$F = F_0 - \int_0^T S\,dT - P(V_T - V_0)$$

on the other hand

$$S = \int_0^T \frac{C_p}{T}\,dT \tag{8.2}$$

and thus

$$F = F_0 - P(V_T - V_0) - \int_0^T \left(\int_0^{T''} \frac{C_p}{T'}\,dT' \right) dT'' \tag{8.3}$$

where V_T and V_0 are molal volumes at pressure P and F_0 is free energy at absolute zero equal to the internal energy U_0 at that temperature. Formula (8.2) gives the absolute value of S by assuming that $S = 0$ at $T = 0$, which is the Nernst heat postulate or the third law of thermodynamics. For certain solids, this law appears not to be applicable because of orientation effects, ordering effects, isotopic composition, etc. Such cases are metastable states which in proximity of absolute zero have no chance to reach true equilibrium because of vanishing mobility of atoms. The calculation of the entropy requires addition of a small term to the value of S calculated from (8.2).

For solids, the second term in (8.3) is very small and thus in general

$$F = A - \int_0^T \left(\int_0^{T''} \frac{C_p}{T'}\,dT' \right) dT'' \tag{8.4}$$

* Numbers in brackets refer to References at end of chapter.

where A is a constant. This formula is useful in discussing the influence of the lattice specific heat and of the electronic specific heat on phase transformations, such as the alpha-gamma transformation in iron [3] and it permits an understanding of the influence of alloying elements on this transformation [4]. It can be used also to compare the free energy of solid phases if the specific heats are known experimentally or theoretically. The latter possibility is useful if the Debye specific heat theory is applicable.

At the transformation temperature, we have a change in the total energy

$$\Delta U = T\,\Delta S - P\,\Delta V$$

which gives for the change of enthalpy

$$\Delta H = \Delta U + P\,\Delta V = T\,\Delta S$$

This relation enables one to obtain the entropy of transformation ΔS from the experimentally known heat of transformation at constant pressure ΔH. If in turn, we know the enthalpies and entropies of both phases as a function of temperature we can, in principle, calculate the transformation temperature T. We have further

$$S = -\left(\frac{\partial G}{\partial T}\right)_P$$

which gives

$$G = H + T\left(\frac{\partial G}{\partial T}\right)_P$$

and

$$\Delta G = \Delta H + T\left(\frac{\partial\,\Delta G}{\partial T}\right)_P \tag{8.5}$$

the Gibbs-Helmholtz equation. The derivative in formula (8.5) vanishes at absolute zero and thus $\Delta G(T)$ can be calculated from measured $\Delta H(T)$.

Among the other thermodynamical formulas of importance for transformations, it is worth mentioning in particular

$$\left(\frac{\partial \alpha}{\partial P}\right)_T = -\left(\frac{\partial \beta}{\partial T}\right)_P$$

and the Maxwell relations:

$$\left(\frac{\partial S}{\partial V}\right)_T = \left(\frac{\partial P}{\partial T}\right)_V = \frac{\alpha}{\beta} \tag{8.6}$$

$$\left(\frac{\partial S}{\partial P}\right)_T = -\left(\frac{\partial V}{\partial T}\right)_P = -\alpha V \tag{8.7}$$

where

$$\alpha = V^{-1}\left(\frac{\partial V}{\partial T}\right)_P$$

is the cubic thermal expansion coefficient and

$$\beta = -V^{-1}\left(\frac{\partial V}{\partial P}\right)_T$$

is the isothermal compressibility. These formulas apply to a single phase only.

Differentiating (8.7), one obtains

$$\left(\frac{\partial C_p}{\partial P}\right)_T = -\alpha^2 TV - T\left(\frac{\partial \alpha}{\partial T}\right)_P V$$

in which for solids $(\partial\alpha/\partial T)_P$ is usually very small and thus approximately

$$\left(\frac{\partial C_p}{\partial V}\right)_T \simeq \left(\frac{\alpha^2 T}{\beta}\right) \tag{8.8}$$

here

$$C_p = T\left(\frac{\partial S}{\partial T}\right)_P = \left(\frac{\partial H}{\partial T}\right)_P$$

is the usual specific heat at constant pressure.

From relation (8.6) one can deduce that, at equilibrium between two phases 1 and 2,

$$\frac{dP}{dT} = \frac{S_1 - S_2}{V_1 - V_2} = \frac{H_1 - H_2}{T(V_1 - V_2)}$$

where the quantities S, V, and H refer to one mole of a single component. This so-called Clausius-Clapeyron relation describes the effect of pressure on transitions. It is useful for such transformations as melting, change of crystal structure, etc. For solids, the influence of temperature on the heat of transformation

$$\Delta H = H_\alpha - H_\beta$$

is given with sufficient accuracy by

$$\frac{d\,\Delta H}{dT} = \Delta C_p + \frac{\Delta H}{T} \tag{8.9}$$

If we have more than one component—say n_i moles of atomic species i and n_k moles of atomic species k—it may occur that a lower free energy is obtained by splitting the uniform phase of composition x into two phases of different chemical compositions x_1 and x_2 such that $x_1 < x < x_2$. The temperature at which $x_1 = x = x_2$ is the so-called critical temperature of mixing. There the first and the second derivatives of the chemical or thermodynamic potentials

$$\mu_i = \left(\frac{\partial U}{\partial n_i}\right)_{SVn_k} \quad \text{and} \quad \mu_k = \left(\frac{\partial U}{\partial n_k}\right)_{SVn_i}$$

with respect to the composition x are zero.

An important consequence of the existence of relationships among the various thermodynamic quantities is the so-called Gibbs phase rule:

$$C + 2 = F + P \tag{8.10}$$

in which C is the number of independent components, F is the number of degrees of freedom, and P is the number of phases. Degrees of freedom are independently variable parameters such as temperature, pressure, composition of phases, etc. In applying this rule to transition in solids containing more than one species of atoms or molecules, it should be kept in mind that if the two phases have the same chemical composition then C is equal to unity. Such conditions occur, for instance, at a congruent melting point or at the critical temperature of order-disorder transformations.

Classical illustrations of phase change in pure elements are provided by tin and sulfur. The measured heats of transformation ΔH and the corresponding differences in free energy ΔF as calculated for instance from Eq. (8.5) (in which for a solid we can put approximately $\Delta G = \Delta F$) are shown in Fig. 8.1. The transformation occurs at the temperature at which ΔF reverses its sign.

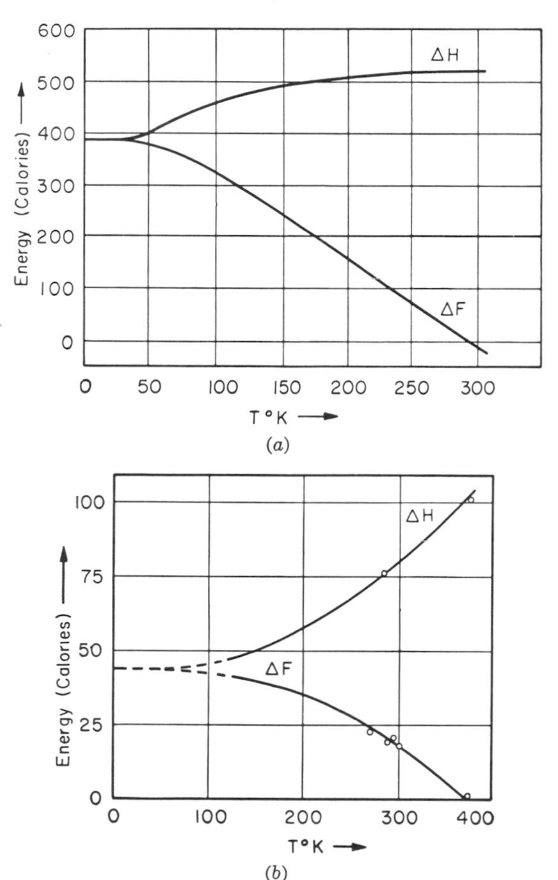

FIG. 8.1*a,b*. Heat of transformation ΔH and change of free energy ΔF in the transformation of (*a*) gray to white tin and (*b*) rhombic to monoclinic sulphur [3].

While the commonly encountered phase changes in solids are produced by changing temperature at constant pressure there are transformations which are produced by changing pressure. The classical examples of such phase changes in pure elements have been very extensively studied by Bridgman [5]. At high pressures, the densely packed crystal structures are favored over the more open structures (for instance, the transformation in potassium halides and in ribidium halides from NaCl lattice to CsCl lattice and in cesium from body-centered cubic to face-centered cubic at 23,000 kg/cm^2 as predicted theoretically by Bardeen [6]). An unusual type of transformation occurs in cesium at 45,000 kg/cm^2 in which a state of higher density is attained without a change of the face-centered cubic (i.e., close-packed) lattice. This is interpreted as a shift of valence electrons frcm the 6*s* zone to the empty 5*d* zone [7]. In compounds, for example, ice, several phases may be encountered as a function of pressure (near -30°C five phases of ice can exist at various pressures). In binary alloy systems, for example in Bi-Sn, compounds can be formed [8] under high pressure which do not exist at lower pressures. Finally, to proceed at a measurable rate, many transformations require a change both of temperature and of pressure. A classical example here is the transformation [9] of diamond to graphite which could be experimentally reversed at pressures around 50,000 atmospheres (atm). About a third of all solids tested by Bridgman showed a polymorphic transformation.

It should be stressed that while in the present state of the theory of solids [3] it is usually possible to get at least a qualitative estimate of how a transformation will behave under specified conditions it is, in general, not possible to predict by purely theoretical means thc crystal structurc or a phase change in a solid. With a few exceptions, such as the structure of alkali metals and of copper, the theoretical calculations have to use empirical data. However, some recent theoretical work on transition metals is very promising [10] in this respect. Also, the method of pseudo potentials [11] may lead to a better understanding of polymorphism, using first principles without empirical parameters.

Recently [12] it has been shown that UO_2 exhibits an antiferromagnetic transition near 30°K which is of the first order. The explanation [13] is based on the assumption that while the lowest state Γ_1 is nonmagnetic the next higher one Γ_5 can be split by the molecular field to such an extent that its lowest magnetic level drops below Γ_1. Depending upon the separation between Γ_5 and Γ_1, the intensity of the molecular field, etc., such systems can possess either no magnetic state or no stable magnetic state or can have antiferromagnetic transitions of the first or of second order.

2. Transformations of Higher Order

Ehrenfest suggested a classification of transitions according to which a transition is of the *n*th order when derivatives lower than the *n*th derivative of the function G are continuous at the transformation temperature while the *n*th derivative is discontinuous. These which can be treated in terms of the Gibbs formalism, and the majority of transformations in solids fall into this class, are of the first order, i.e., the first derivative of function G is discontinuous at the transformation temperature. Strictly speaking, a transition is a phase transformation only when it is of the first order but in accord with common usage the word phase will be used here in the broader sense.

The applicability of Ehrenfest's criterion turned out to be limited not only because of experimental difficulties in establishing whether a certain quantity is continuous or not, but also because of the existence of various intermediate kinds of anomalies [14]. Figure 8.2, taken from Mayer and Streeter [15], illustrates best the situation. The main difference between "first-order" and "anomalous first-order" transitions is that in the latter category each phase "anticipates" the change with approaching transformation temperature or transformation pressure. Instead of the "simple second-order" transition (as defined by Ehrenfest) one observes usually a

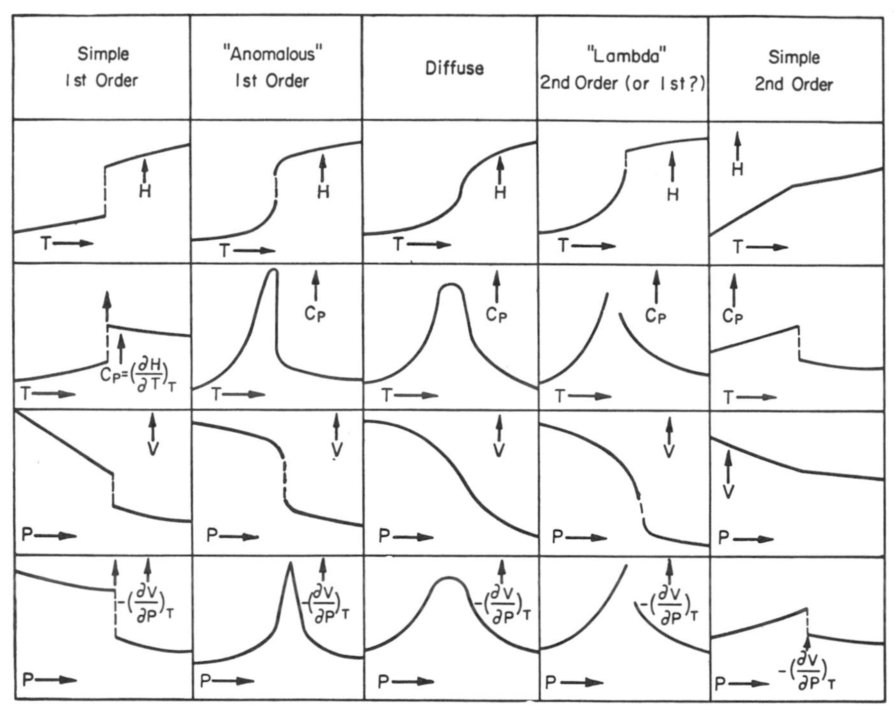

Fig. 8.2. Types of transitions.

"lambda" transformation in which near the transformation point both phases show a pronounced continuous change in the various thermodynamic quantities. This change is usually greater below than above the transition point. Finally, the "diffuse" transformation is spread over an appreciable range of temperatures and pressures. A clear-cut distinction between an anomalous first-order and a lambda transition on the basis of experimental equation of state, $U = U(T,V)$, or specific heats is usually impossible. The existence of a measurable heat of transformation is no rigorous criterion either because, as shown by Rutgers [16], the total "abnormal" heat capacity, if it occurs in a relatively narrow range of temperatures in second-order transitions, plays thermodynamically the same role as the latent heat in first-order transitions. On the other hand, if the two phases can actually coexist in equilibrium, that is, when the surface energy between the two phases is positive and thus a definite phase boundary is formed, then the transition is of the first order. This criterion, too, has its own experimental limitations, such as the minimum size of a crystallite which gives definite sharp X-ray diffraction lines, etc., and its usefulness depends strongly upon the rate of reaction, i.e., whether true equilibrium is experimentally attainable.

The continuity of the Gibbs function at a second-order transition implies a similarity to the critical point in gas-liquid equilibria. At the critical point, the fluctuations of density and other quantities are so great that a purely thermodynamical approach, which essentially assumes absence of fluctuations, breaks down. Only a statisticomechanical treatment can give rigorous results and the same seems to apply to second-order transitions. This analogy is the basis of Tisza's thermodynamic theory of transitions [17] of second kind in which, at the transition point, one of the compliance coefficients (i.e., specific heat, thermal-expansion coefficient, isothermal elastic coefficient, electric and magnetic permeabilities) has a singularity rather than a discontinuity. This result is similar to the exact solution of the two-dimensional Ising model discussed below. In view of the basic limitation of thermodynamic theories, Tisza's theory gives no picture of the mechanism of these transformations although in certain cases the general character of the change (i.e., change of order, displacement of atoms, etc.) can be deduced.

Since thermodynamics is unable to give an atomistic quantitative picture of second-order transitions, much effort has been directed toward obtaining a solution using purely statistical methods. These are based mostly on the so-called Ising model in which two kinds of atoms (or two spin orientations) are distributed in a regular one-, two-, or three-dimensional array. When a particular atom of one kind is substituted by an atom of the other kind (or one spin is reversed), then the energy changes by an amount which depends upon the kind of neighbors and upon their arrangement. This cooperative aspect of the phenomenon, which is characteristic of the second-order transitions, is implied by certain thermodynamic considerations [18, 19]. The solution of the statistical problem for the linear case is easily obtainable but leads to no transition point. The two-dimensional problem does lead to a transition point but the calculation is very difficult and for a long time only approximate solutions were known. An exact answer was finally obtained by Onsager [20] for a square lattice. Later, similar exact solutions were deduced for triangular and for two-dimensional hexagonal lattices [21, 22]. The three-dimensional case is as yet unsolved. Wannier [23] indicated that the approximate solutions (of the two-dimensional case) which are based on the concept of an "inner field" (such as "long-range order") lead to a discontinuity of the specific heat at the transition point, while Onsager's exact solution gives a logarithmic singularity which is more or less symmetrical around the transition point as shown in Fig. 8.3.

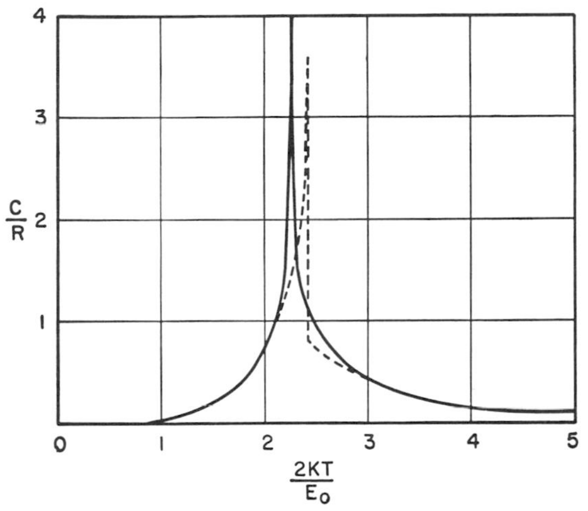

FIG. 8.3. Specific-heat anomaly at the transition point in a square lattice. Full curve, exact solution; dotted curve, "inner-field" solution [22].

This failure of long-range order theories in two dimensions casts doubt on their validity in three dimensions and, in particular, on the significance of the jump in the specific heat. On the other hand, there are few, if any, ordering phenomena where a three-dimensional Ising model is rigorously applicable. First, many of the transitions commonly classified as "second-order" are accompanied by an abrupt change of volume (and symmetry), and second, the notion of an interaction between neighbors, whether it includes first, second, or even more distant neighbors (as used by Yvon [24, 25]), has often little underlying physical significance. For instance, the metallic binding is primarily due to an interaction of *all* ionic cores (or nuclei) with *all* outer electrons rather than to a sum of interactions between neighbors. Such factors as atomic size, lattice strain, changes in band structure [26], etc., can hardly be expected to follow a pair-interaction formalism. This shows up, for instance, in the general lack of symmetry of binary solid-solution diagrams with respect to the two components and, in particular, in the difference of the critical temperature of $AuCu_3$ and Au_3Cu (see Table 8.3). Another phenomenon which does not seem to be suitable for description in terms of pairs is the ordering of atomic vacancies [27] as in pyrrhotite, Fe_7S_8. The effect of the difference in atomic size on the degree of order has been experimentally found [28] in Au-Ni alloys and a general discussion of such "asymmetric" binary alloys has been given by Lawson [29]. Thus, the qualitative and sometimes even semi-quantitative agreement between experiment and present approximate theories of order-disorder transformations is actually better than one might expect and it is doubtful whether it will be improved when an exact solution of the three-dimensional Ising model is obtained. The main theoretical problem yet to be solved here is not so much the mathematical as the physical picture of the various factors influencing ordering.

In fact, evidence [30–32] indicates that, at equilibrium, in the better known ordering alloys (AuCu, $AuCu_3$, etc.), the ordered phase is separated in the phase diagram from the disordered phase by a two-phase region (except at the critical temperature itself) and, thus, conforms to the classical phase rule and to the conditions for first-order transitions. The same seems to apply to the ferroelectric transition in $BaTiO_3$ [33, 34]. A rigorous application of the two-phase region criterion to CuZn is for various reasons as yet not possible. It is, however, rather certain that, with a few possible exceptions, order-disorder transitions exhibit, at the same time, characteristics both of first- and of second-order transformations. Whatever doubts one may have about the physical nature of long-range order and of pair-interaction it is so convenient that a summary of the main points of these theories is given below.

Transformations for which the Ising model and its variants seem to be quite appropriate are the condensation of a "lattice gas," exemplified by liquid helium, and the ferromagnetic Curie temperature.

3. Order-Disorder Theory

The simplest ordering systems are the simple cubic and the body-centered cubic lattices in which each atom has, respectively, six and eight nearest neighbors, i.e., the coordination number $z = 6$ or 8. Both these lattices are "open" lattices; i.e., the nearest neighbors are not nearest neighbors among themselves. If we have N atoms A and N atoms B then, in a perfectly ordered body-centered cubic lattice, each kind of atoms occupies a simple cubic lattice which we shall call α and β, respectively, and all zN pairs of nearest neighbors are of the AB type. In a partially ordered condition, there will be, using Guggenheim's [19] notation, $zN(1 - 2\xi)$ pairs of the AB type and the remaining $2zN\xi$ pairs will be equally divided among the AA and BB types. Finally, there will be Nr atoms A on α sites and $N(1 - r)$ atoms A on β sites and, *mutatis mutandis*, Nr atoms B on β sites and $N(1 - r)$ atoms B on α sites. The two parameters r and ξ describe the distribution of atoms. If the energy associated with AA, BB, and AB pairs is ϵ_{aa}, ϵ_{bb}, and ϵ_{ab}, respectively, and ϵ is defined by

$$\epsilon = \epsilon_{ab} - \frac{\epsilon_{aa} + \epsilon_{bb}}{2} \tag{8.11}$$

(for an ordering alloy $\epsilon < 0$), then the total configurational energy is

$$E = zN[(1 - 2\xi)\epsilon_{ab} + \xi(\epsilon_{aa} + \epsilon_{bb})] = zN(\epsilon_{ab} - 2\xi\epsilon) \tag{8.12}$$

and thus the equilibrium condition of the system is completely defined by the partition function

$$\Omega = \sum_r \sum_\xi g(N,r,\xi) \exp\left[-\frac{zN(\epsilon_{ab} - 2\xi\epsilon)}{kT}\right] \quad (8.13)$$

in which $g(N,r,\xi)$ is the number of various possible arrangements in our lattice of N atoms for a given r and ξ. The evaluation of this number and of the partition function can be done only approximately.

Zeroth Approximation (Bragg-Williams [35], Weiss [36]). If in (8.13) in the summation over r only the maximum term Ω_r is taken into account and if instead of ξ the quantity

$$\bar{\xi} = T \int_0^{1/T} \xi d\left(\frac{1}{T}\right)$$

is introduced, the combinational factor is simply the number of ways of rearranging, for a given r, the N atoms on lattices α and β. The configurational free energy is then, using Stirling's formula,

$$F_c = -kT \ln \Omega_r = 2NkT[r \ln r + (1 - r) \ln (1 - r)] + zN\epsilon_{ab} - 2zN\bar{\xi}\epsilon \quad (8.14)$$

The simplest assumption about the distribution of atoms is that all the atoms on the α lattice [that is, Nr atoms A and $N(1 - r)$ atoms B] are distributed at random and that the same applies to the β lattice. This is clearly a very rough assumption since it ignores the fact that there is a preference for AB pairs of neighbors, that is that $\epsilon < 0$. A consequence of this assumption is that the ratio of the probability for an A atom on an α site to have an A neighbor to the probability of having a B neighbor is equal to the ratio of the probability of a B atom on an α site to have an A neighbor to the probability of having a B neighbor. Thus

$$\frac{\bar{\xi}}{r - \bar{\xi}} = \frac{1 - r - \bar{\xi}}{\bar{\xi}}$$

or

$$\bar{\xi}^2 = (r - \bar{\xi})(1 - r - \bar{\xi}) \quad (8.15)$$

which is temperature-independent. Substituting $\bar{\xi}$ from (8.15) into (8.14) and equating the derivative with respect to r to zero, we obtain, for the dependence of r on T, the transcendental equation:

$$S^{-1} \ln \frac{1 + S}{1 - S} = \frac{2T_c}{T}$$

where $T_c = -(z\epsilon)/(2k)$ is the Curie point or critical temperature of order and $S = 2r - 1$ is the degree of order.

From (8.12) the change of the configurational energy between states of complete order and of complete randomness is $E_0 = -\frac{1}{2}Nz\epsilon$ and thus, assuming that $2N$ is one mole, we have $T_c = 2E_0/R$. The dependence of the configurational free energy on r and S, in this approximation, is shown in Fig. 8.4 for various temperatures. It is clear that at $T > T_c$ the lowest free energy corresponds to $S = 0$ (that is, $r = \frac{1}{2}$ and the two sublattices α and β are equivalent) but for $T < T_c$ the minimum free energy occurs at increasing degrees of order, reaching complete order $S = 1$ at $T = 0$. The dependence of order on temperature is shown by curve 1 in Fig. 8.5. It is clear that with increasing temperature, below T_c, the supplied heat not only increases the energy of thermal vibrations but also decreases the order. Thus there is an anomalous specific heat which disappears abruptly at T_c, as shown in Fig. 8.6. In this approxi-

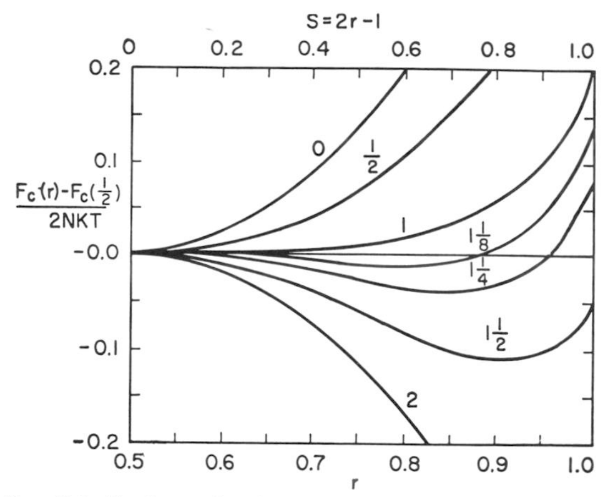

Fig. 8.4. Configurational free energy in an AB lattice as a function of r (or S) for various values of T_c/T in the zeroth approximation [18].

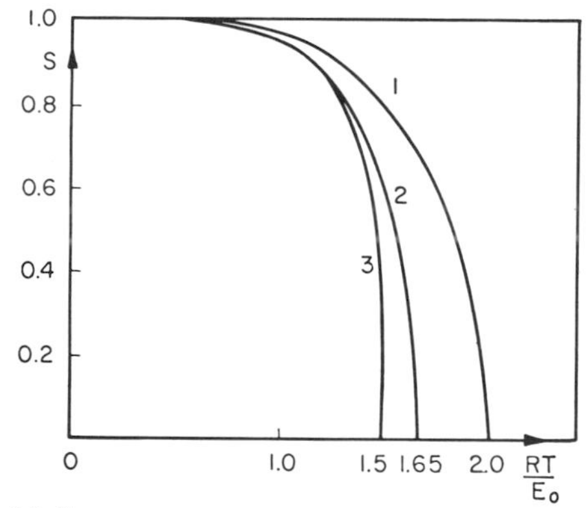

Fig. 8.5. Long-range order S in a simple cubic lattice as a function of temperature in (1) zeroth, (2) first, and (3) series expansion [41] approximations.

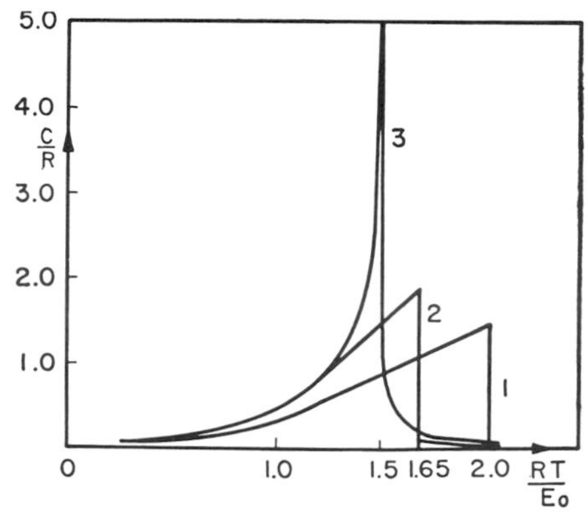

Fig. 8.6. Specific-heat anomaly in a simple cubic lattice in (1) zeroth, (2) first, and (3) series expansion [41] approximations.

mation, the calculated discontinuity of specific heat at the Curie temperature is $\frac{3}{2}k$.

First Approximation (**Quasi-chemical** [19], **Bethe** [37]). As mentioned before, in considering the distribution of atoms on the two sublattices α and β, in the zeroth approximation the preference for AB pairs is ignored. Since Eq. (8.15) has the form of an equation expressing the law of mass action in a gaseous equilibrium between AB, AA, and BB "molecules," all having the same "binding" energies, it is natural to take into account the energy required to change two AB pairs into an AA and a BB pair (that is, 2ϵ) by putting instead of (8.15),

$$\bar{\xi}^2 = (r - \bar{\xi})(1 - r - \bar{\xi}) \exp\left(\frac{2\epsilon}{kT}\right) \tag{8.16}$$

This is the basis of the so-called quasi-chemical approximation, which, as shown by Rushbrooke [38], is equivalent to the well-known Bethe's first approximation which was the first big improvement over the Bragg-Williams or zeroth approximation theory of ordering. Bethe's method consists of constructing a grand partition function for a central site and its "boundary," i.e., the nearest neighbors, and replacing the influence of all other atoms by one parameter and then eliminating the latter by means of the physically reasonable condition that there should be no difference whether the central site is an α site or a β site. In the second Bethe's approximation, which gives better results than the pair approximation of the quantum-chemical method, the central group consists of an atom and its nearest neighbors while the second nearest neighbors form the boundary.

In the quasi-chemical approximation, the configurational free energy is

$$\begin{aligned} F_c = {} & 2NkT[r \ln r + (1 - r) \ln (1 - r)] + Nz\epsilon_{ab} \\ & + NzkT \left\{ r \ln r \left[1 - (1 - r)^2 \frac{2\epsilon}{kT}\right] \right. \\ & \left. + (1 - r) \ln \left[1 - r^2 \frac{2\epsilon}{kT}\right] \right\} \end{aligned} \tag{8.17}$$

the dependence of long-range order on temperature is given by

$$\left(1 - \frac{2}{z}\right)^{T_c/T} = \frac{(1 + S)^{1-1/z}(1 - S)^{1/z} - (1 - S)^{1-1/z}(1 + S)^{1/z}}{2S} \tag{8.18}$$

where the Curie temperature is

$$T_c = -\frac{\epsilon}{k}\left(\ln \frac{z}{z - 2}\right)^{-1} = 4E_0 \left(Rz \ln \frac{z}{z - 2}\right)^{-1} \tag{8.19}$$

and the discontinuity of the specific heat at T_c is now

$$\Delta C = \frac{3}{2} k \frac{z - 2}{z - 1} \left(\frac{z}{2} \ln \frac{z}{z - 2}\right)^2 \tag{8.20}$$

By introducing the preference for AB pairs into the distribution of atoms on the two sublattices, account is taken of the local short-range order which in the zeroth approximation was neglected. This short-range order persists, of course, above the Curie temperature of the long-range order and its change with temperature increases somewhat the specific heat in that region, as can be seen from Fig. 8.6. (Appreciable short-range order has been observed several hundred degrees above critical temperatures: at 550°C for $AuCu_3$, at 770°C for AuCu, and at 300°C for AgAu.) If q denotes the proportion of AB pairs, then short-range order is defined by $\sigma = 2q - 1$ and for $T > T_c$ we have

$$\sigma = \frac{z^{T_c/T} - (z - 2)^{T_c/T}}{z^{T_c/T} + (z - 2)^{T_c/T}} \tag{8.21}$$

giving at T_c the value

$$\sigma_c = (z - 1)^{-1}$$

Bethe's method, the quasi-chemical method, and their variations have been applied to yield higher approximations by increasing the number of specifically considered atoms and configurations in "open" lattices and in "dense" lattices, such as face-centered cubic for AB and AB_3 compositions. These calculations are more complicated than the original theory [39–41].

Series Expansion Approximation. Several methods of calculating the partition function Ω are based on expanding it, or its logarithm, as a power series in $\beta = \exp(\epsilon/kT)$, in $\gamma = -\epsilon/kT$, or finally in $u = (1 - \beta)/(1 + \beta)$. The first expansion is valid at low temperatures ($\epsilon < 0$); the second [42] and the third are useful at high temperatures. The series converge rather slowly so that terms up to the 46th power in β (or 12th power in u) had to be included [43]. At the Curie temperature itself, neither one of the series is valid. Calculations have been made for simple cubic, body-centered cubic, and face-centered lattices. Of particular interest are the results of Wakefield [44] indicated in Figs. 8.5 and 8.6 for a simple cubic lattice. Here the specific-heat anomaly appears to have an asymmetric singularity rather than a discontinuity. This resembles Onsager's exact solution for a two-dimensional lattice (Fig. 8.2) and Tisza's quasi-thermodynamic conclusions. Bethe's approximation corresponds, as far as its accuracy is concerned, to a series development up to only the fourteenth power in β and up to only the second power in γ. Trefftz's [45] results are based on an extrapolation of reciprocal specific heats calculated from a series which near the critical temperature has still a satisfactory convergence. Domb and Sykes [46] used a series that contains both odd and even terms to represent high-temperature susceptibility near the Curie temperature. Table 8.1 allows a comparison of the various approximations for two- and three-dimensional lattices: β_c is the value of β at T_c, while the second column gives the Curie temperature T_c in terms of the total change of the configurational energy between states of complete order and complete randomness E_0.

A very powerful tool for evaluating theoretical expressions for various physical quantities in the neighborhood of the Curie point are the Padé approximants used by Baker [47]. Such quantities as magnetic susceptibility or specific heat may be given by a power series which either converges very slowly or

TABLE 8.1. CRITICAL TEMPERATURE RT_c/E_0 OF ISING MODELS IN VARIOUS APPROXIMATIONS

Approximation	Square ($Z = 4$)	Triangular ($Z = 6$)	Simple cubic ($Z = 6$)	Body-centered cubic ($Z = 8$)	Face-centered cubic ($Z = 12$)
Zeroth	2.00	2.00	2.00	2.00	2.00
First	1.44	1.65	1.65	1.74	1.77
Bethe second	.....	.....	1.58		
Kikuchi	1.21	1.30	1.53	1.74	1.67–1.54
Wakefield	.....	.....	1.50	1.67	
Trefftz	.....	.....	1.33	1.41	1.44
High-temp. susc	1.135	1.216	1.504	1.588	1.632
Exact	1.135	1.214			

TABLE 8.2. BEHAVIOR OF ISING LATTICE MODELS NEAR CURIE TEMPERATURE
(A is larger for $T < T_c$ then for $T > T_c$)

	Two-dimensional	Three-dimensional
Spontaneous magnetization	$(1 - T/T_c)^{1/8}$	$(1 - T/T_c)^{5/16}$
Magnetic susceptibility	$(1 - T/T_c)^{-7/4}$	$(1 - T/T_c)^{-5/4}$
Specific heat	$\ln\|1 - T/T_c\|$	$A \ln\|1 - T/T_c\|$

not at all even though the function itself is well behaved. If, however, this series is represented by a Padé approximant, which is a ratio of two polynomials of degree M and N with appropriately chosen coefficients, one can continue past nonphysical singularities, and the values of the function can be computed with very good approximation. Table 8.2 summarizes the results of various methods for the Ising model. For the Heisenberg model in three dimensions specific heat varies more rapidly than $(1 - T_c/T)^{-1}$ and the magnetic susceptibility varies as $(1 - T_c/T)^{-4/3}$.

Brout (see Bibliography at the end of this chapter) introduced a graph theoretical method which is based on topological characterization of various terms that arise when the logarithm of the partition function is expanded in powers of the interaction ϵ. One considers interactions in various clusters and attempts to sum over those which are responsible for a singularity.

An important result was obtained by Kac, Uhlenbeck, and Hemmer [48] who studied a one-dimensional system of hard-core molecules which are attracted to each other by an exponential potential. As the attractive interaction becomes weaker and weaker but at the same time longer and longer in range the equation of state approaches rigorously the van der Waals equation. At the same time one obtains the Maxwell "equal area rule" which determines the range of coexistence of two phases. The importance of this result lies in the fact that it shows how a one-dimensional condensation can occur. Similarly Langer [49] has shown that a so-called "spherical model," which is an approximation to the three-dimensional Ising model of a ferromagnet, can give a satisfactory description of condensation and supersaturation if a weak anisotropic field is added. The advantage of this model is that it uses finite range forces although it is not exactly soluble in the immediate neighborhood of the critical point.

For convenience, some of the ordering metallic systems are listed in Table 8.3. Only these systems are included for which (1) at the critical temperature the ordered phase transforms wholly into the disordered phase, and vice versa (thus congruent transitions only; which excludes such alloys as Ni_4Mo); (2) both the ordered and the disordered phase are equilibrium phases (this eliminates such alloys as AuNi), and (3) at least the symmetry of the crystalline structure of the two phases is known. A convenient rule of thumb is the relation $T_c = 0.6T_s$ in which T_s is the solidus point of the alloy. More detailed form of this relationship is discussed by Oriani [50].

4. Orientational Transitions

Certain crystals, such as ammonium halides, undergo a transformation with increasing temperature which has been interpreted by Pauling [51], as an onset of free rotation of the ammonium ion. Another explanation was suggested by Frenkel [52], who assumed a progressive decrease of orientation order of oscillation axes of these ions below the critical temperature and oscillation around randomly oriented axes above the critical temperature. Infrared and Raman spectroscopy [53] data and neutron-diffraction [54] measurements seem to confirm Frenkel's explanation. The same conclusion follows from specific heat measurements [55]. A good illustration of the difficulty in correlating and predicting the type of orientational transformation in the series HCl, HBr, and HI. In the first, the transition from the anisotropic low-temperature form to an isotropic high-temperature form appears to be of the first order, while in the other two a similar transition seems to be broken up into at least two transitions of second order [56]. Tisza [17] suggested that in the low-temperature form the dipoles of the molecules are all parallel (or antiparallel), in the intermediate phase (if it exists) the molecules are parallel but the dipoles not, while in the high-temperature phase molecules and their dipoles are oriented at random. The variation of the transformation temperatures, with changing halogen ion, appears to be in qualitative agreement with the variation of the dipole moments and of the van der Waals forces.

TABLE 8.3. ORDER-DISORDER TRANSFORMATIONS IN BINARY METAL SYSTEMS*

Alloy	Trans. temp., °C	Lattice type (disorder-order)	Alloy	Trans. temp., °C	Lattice type (disorder-order)
AgCd	440	B.c.c.–b.c.c.	CrPt	1000?	F.c.c.–f.c.c.?
Ag_3In	200	Hex.–hex.	Cu_4Pd	478	F.c.c.–tet.
Ag_3Pt	~800	F.c.c.–f.c.c.	Cu_3Pd_2	~600	F.c.c.–b.c.c.
AgPt	500	F.c.c.–complex	CuPd	600	F.c.c.–b.c.c.
$AgPt_3$	625	F.c.c.–f.c.c.	Cu_3Pt	650	F.c.c.–f.c.c.
AgZn	270	B.c.c.–b.c.c.	CuPt	812	F.c.c.–rhom.
$AlFe_3$	575	B.c.c.–b.c.c.	$CuPt_3$	640	F.c.c.–f.c.c.
$AlNi_3$	>1125	F.c.c.–f.c.c.	Cu_3Sb	432	Cub.–hex.
Au_3Cd	425	F.c.c.–f.c.c.	CuZn	460	B.c.c.–b.c.c.
AuCd	267	B.c.c.–orth.	$FeNi_3$	570	F.c.c.–f.c.c.
Au_3Cu	240	F.c.c.–f.c.c.	Fe_3Pd	780	F.c.c.–f.c.c.
AuCu	408	F.c.c.–(orth)–tet.	FePd	~750	F.c.c.–tet.
$AuCu_3$	391	F.c.c.–f.c.c.	$FePd_3$	760	F.c.c.–f.c.c.
Au_4Mn	420	F.c.c.–b.c.tet.	Fe_3Pt	~750	F.c.c.–f.c.c.
Au_3Mn	645	F.c.c.–tet.	FePt	1250–1500	F.c.c.–tet.
AuMn	615	B.c.c.–tet.	FePt	~700	F.c.c.–f.c.c.
Au_3Zn	270, 425	Complex?	FeV	1234	B.c.c.–complex
$AuZn_3$	225, 515	Complex?	$IrPt_3$	~900	F.c.c.–f.c.c.
$CdLi_3$	272	F.c.c.–f.c.c.	LiPb	~200	B.c.c.–b.c.c.
Cd_3Mg	85	Hex.–hex.	MnNi	~650	F.c.c.–tet.
CdMg	253	Hex.–hex.	$MnNi_3$	510	F.c.c.–f.c.c.
$CdMg_3$	153	Hex.–hex.	$MnPt_3$	~900	F.c.c.–f.c.c.
CoFe	750	B.c.c.–b.c.c.	Ni_3Pt	580	F.c.c.–f.c.c.
Co_2Ge	625	Hex.–hex.	NiPt	~600	F.c.c.–tet.
$CoPt_3$	700–800	F.c.c.–f.c.c.	Ni_3Si	~1050	F.c.c.–f.c.c.
CoPt	825	F.c.c.–tet.	Ni_3Sn	875	Hex.–hex.
CrFe	820	B.c.c.–complex	NiZn	~750	B.c.c.–tet.
Cr_3Pt	1000?	F.c.c.–f.c.c.?			

* Only true order-disorder transformations are included, i.e., in which at a stoichiometric composition and at a definite critical temperature an ordered lattice changes into a disordered lattice of the same composition. This eliminates such pseudo-ordering systems as AuNi or Ni_4W. Some of the data are tentative.

Table based on information collected by Drs. G. J. Dienes, C. E. Dixon, F. E. Jaumot, Jr., J. B. Newkirk, R. A. Oriani, and F. N. Rhines.

KINETICS OF TRANSFORMATIONS IN SOLIDS

While equilibria of phase transformations in solids are relatively well understood, their kinetics is, both experimentally and theoretically, a very difficult subject. The majority of these transformations can be described in terms of nucleation [57] and of subsequent growth of the nuclei. Others occur by the variously called shear or martensitic mechanism. The distinguishing characteristic is that in the transformations of the first category, the changes take place presumably atom by atom, while in the other, whole groups of atoms rearrange simultaneously. Change of positions of individual atoms implies a kinetics comparable to that of diffusion, while in the shear mechanism the transformation is exceedingly fast and proceeds, locally, with a speed comparable to that of sound [58]. This, also, implies that in the first category of transformations, neighbors of a given atom in a phase usually are not its neighbors after the transformation has taken place, while in the second category each atom retains its neighboring atoms at properly altered distances. In topological terminology, shear transformation occurs with a conformal mapping while nucleation and growth do not.

5. Nucleation and Growth

The success of Volmer's theory of *nucleation* [59] in condensation lead to early attempts to apply it to solids. Its basic idea is as follows: If a homogeneous phase is undercooled (or superheated), then another phase (or phase mixture) has a lower free energy. This drop of free energy is proportional to the volume of the transformed material. On the other hand, the presence of an inclusion of one phase in another phase necessitates an expenditure of energy to form the boundary layer. This increase of energy is proportional to the surface of the inclusion. If we assume, for simplicity, that we deal with spherical nuclei, then the difference between a term proportional to r^3 and another proportional to r^2 leads to a maximum energy which occurs at the so-called critical radius (Fig. 8.7):

$$r_c = \frac{2\sigma \nu_L}{\Delta f}$$

where σ is surface energy, ν_L is molecular volume of the liquid, and Δf is the drop of free energy per molecule going into the new phase. It is argued now that as a result of statistical fluctuations and collisions of atoms an "embryo" is formed and it grows (or disappears) by a random statistical process until it reaches the critical radius and becomes a nucleus. This original growth of embryos occurs with an increase of energy and, thus, according to classical thermodynamics, it would be impossible. Once the maximum energy is reached, a further growth is associated with a decrease of energy and it is thus thermodynamically allowed.

The nucleation rate is the rate at which embryos reach the critical size and, thus, it is proportional to

$$q \exp \left[\frac{-(\Delta\phi)_{\max}}{kT} \right] \tag{8.22}$$

where $(\Delta\phi)_{\max} = (4\pi/3)\sigma r_c^2$ is the increase of the thermodynamic potential between $r = 0$ and $r = r_c$ and q is a factor which depends on the supply of atoms from the unstable phase.

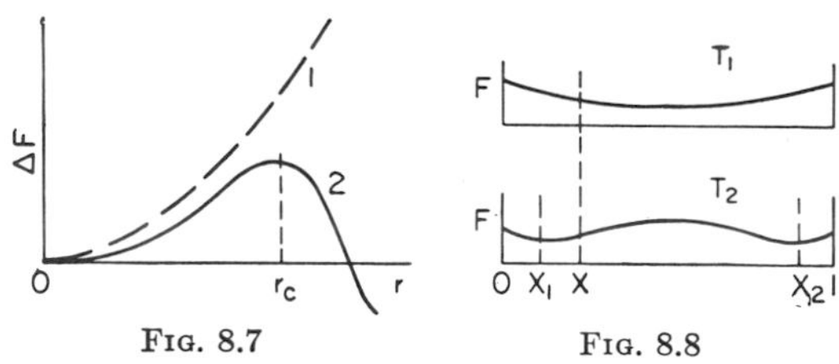

FIG. 8.7 FIG. 8.8

FIG. 8.7. Energy of formation of an embryo of an unstable (1) and stable (2) phase.
FIG. 8.8. Free energy of a simple system above (T_1) and below (T_2) the critical temperature of miscibility.

In this simple condensation process, the notion of a well-defined surface of an embryo and of a surface energy is very schematic and highly idealized. This is even more true of a phase transformation in a solid where such factors as differences in crystal structure, lattice matching, lattice strain, and concentration gradients complicate the picture considerably. Reiss [60] has shown how a more rigorous statisticomechanical theory of condensation avoids the troublesome notions of a surface and of a surface energy of an embryo or of a nucleus. No such improvement for solids has as yet been developed. In analogy to (8.22) Becker [61] suggested that the rate of nucleation in a solid is proportional to

$$\exp\left(-\frac{Q}{kT}\right) \exp\left(-\frac{A}{kT}\right) \tag{8.23}$$

where Q is the activation energy for diffusion in the solid and the energy A, in analogy to the condensation process, is given by $32\sigma^3/w^2$ for a cubic nucleus. Here σ and w are energies per unit surface and per unit volume, respectively, and they can be obtained from theoretical or experimental dependence of free energy F on composition x. In particular, if a solid solution of composition x at T_1 transforms at a lower temperature T_2 into two phases such that $x_1 < x < x_2$, as illustrated in Fig. 8.8, then in the nearest neighbor approximation

$$F(x) = \epsilon z x(1 - x) + kT[x \ln x + (1 - x) \ln (1 - x)]$$

$$w = \frac{d^2F}{dx^2}(x - x_1)(x_2 - x)$$

$$= \left[\frac{kT}{x_1(1 - x_1)} - 2\epsilon z\right](x - x_1)(x_2 - x) \tag{8.24}$$

and $$\sigma = \epsilon(x - x_2)^2$$

where $\epsilon > 0$ is given by (8.11) and is equal to $2RT_c/z$ with T_c being the critical temperature of mixing. A fair agreement with experimental data is obtained for an Au-Pt (30 at. per cent Pt) alloy.

With the advent of computers it became possible to examine the set of kinetic equations that control the formation of clusters. In particular, Turnbull [62] showed that the nucleation rate vanishes at zero time and that the rate of change of growth of clusters (bigger than two) goes through a maximum. A more quantitative expression for the pre-exponential factor in (8.23) was given by Turnbull and Fisher [63].

Another way to approach the problem of nucleation was used by Borelius [64], who considered explicitly the probability of the occurrence of various size fluctuations of composition: For every fluctuation richer $(+\Delta x)$ in one constituent there is a fluctuation poorer $(-\Delta x)$ in that constituent and thus the net change of free energy is

$$\Delta F = \frac{1}{2}[F(x + \Delta x) + F(x - \Delta x)] - F(x)$$

$$= \frac{1}{2}\frac{\partial^2 F}{\partial x^2}\overline{\Delta x^2} \tag{8.25}$$

which can be positive or negative. In the first case, the fluctuations will tend to disappear, in the second case, they will grow. Thus the condition $\partial^2F/\partial x^2 = 0$ (see Fig. 8.9) separates the area of low and of high

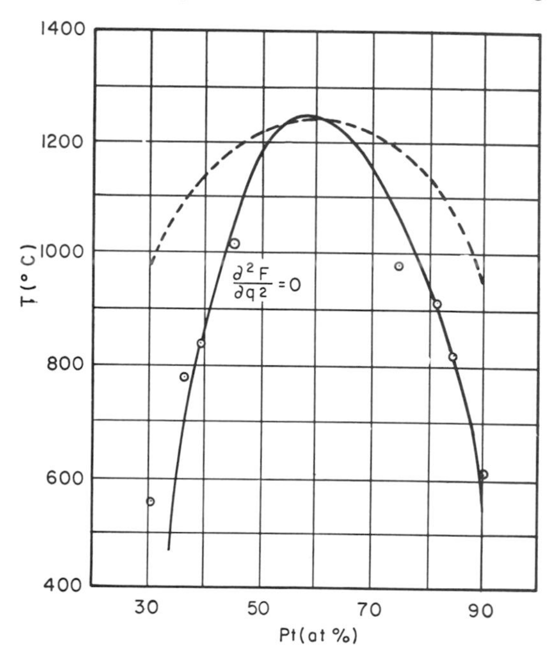

FIG. 8.9. Two-phase boundary line (dashed line) and inflection line (thick line) for the gold-platinum system. Open circles indicate experimental points [59].

nucleation rate. This seems to be in good agreement with experimental data for the Au-Pt and Al-Zn systems [65]. Furthermore, the rate of change of the nucleation rate with changing x or T will depend on the rate of change of $\partial^2F/\partial x^2$ and thus on $\partial^3F/\partial x^3$. This, too, seems to be the case [57].

The two compositions for which $\partial^2F/\partial x^2 = 0$ in Fig. 8.8 are called spinodal compositions x_s. Hillert [66] and Cahn [67] discussed in detail the behavior of alloys of intermediate compositions $x_s' < x < x_s''$

for which there is [see (8.25)] no energy barrier to nucleation and where the nucleus is actually a fluctuation of composition covering a large region in the lattice. One can assume that the local composition of the alloy deviates from the average value x_0 as $x - x_0 = A \cos \beta z$, where z is a space coordinate. This variation of composition introduces a gradient of free energy $\varkappa$ and a strain energy if there is a linear expansion η per unit change in composition. The change of free energy per unit volume is then given by

$$\Delta F = \frac{A^2}{4}\left(\frac{\partial^2 F}{\partial x^2} + 2\varkappa\beta^2 + \frac{2\eta^2 C}{1 - \nu}\right)$$

where C is the appropriate elastic modulus and ν is the Poisson ratio. If $\eta = 0$ there is a critical wavelength $2\pi/\beta_c$ such that all smaller fluctuations are unstable but all longer ones are stable with respect to a uniform X_0. Introducing an anisotropic elastic modulus C, one obtains in addition preferential orientation and shape of various composition fluctuations z, which account very nicely for the so-called satellite spots in X-ray diffraction patterns [68].

While Becker's theory does not consider in detail the various fluctuations, it does give an idea about the atomic mechanism, whereas Borelius' thermodynamic argument cannot give information about the actual mechanism, atomic configurations, etc. A compromise between these two points of view has been suggested by Hobstetter [69], who used the notions of an average free energy and an average internal energy per pair of neighboring atoms in an area of uniform concentration. There were several attempts [70, 71] to include, in the consideration of stability of embryos and of nuclei, the various complicating factors such as lattice strain, coherency, etc., mentioned above. Their general applicability is not yet certain.

Volmer's point of view according to which nucleation is an isothermal steady-state process is the most commonly used approach. Another kind of nucleation can occur when during a rapid cooling process the critical radius suddenly decreases. As a result of this change embryos which were unstable at a higher temperature now become stable nuclei and a new balance between the embryos and nuclei of various size is gradually established. This athermal nucleation [72] may in many cases account for the variation of nucleation rate with time.

In the above discussion nucleation is considered as an essentially random process, i.e., homogeneous nucleation, in which each atom in the lattice has, a priori, the same probability of becoming the center of a nuclear embryo. This is actually seldom the case and nucleation is often heterogeneous, that is, it is strongly influenced by the presence of impurities, inclusions, and lattice defects [73, 74] such as dislocations, twin boundaries, etc. It should be borne in mind that such imperfections may affect not only the rate of nucleation per se but also the local composition and the rate of diffusion which controls the rate of growth. Very few systematic data in that field are available although many transformations in metals are known to occur preferentially along grain boundaries. Important new information may be forthcoming from the study of the influence of neutron radiation on nucleation [75] in the beta-alpha transformation of tin.

Damask, Danielson, and Dienes [76] obtained a kinetic theory of homogeneous and heterogeneous nucleation and growth, assuming low solute concentration and rather low solute-solute binding energy. If it is assumed in addition that there is equilibrium among the subcritical particles and that there is no back reaction (dissolution) when a particle reaches a critical size p it is possible to obtain an analytical solution which agrees extremely well with an exact (computer) solution as shown in Fig. 8.10. If X_1^0

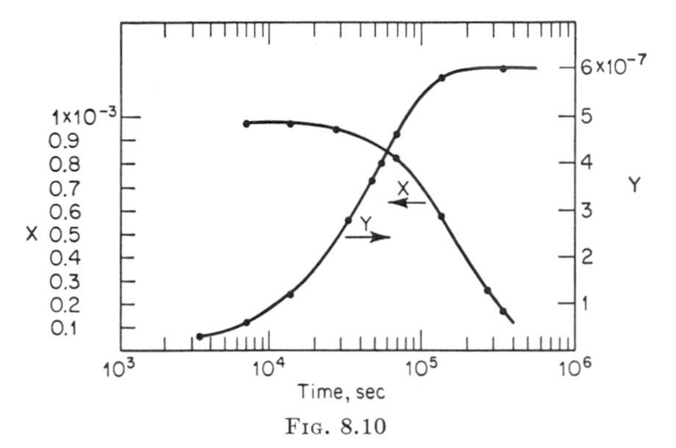

Fig. 8.10

is the initial concentration of single atoms, their concentration X_1 at time t, for large t, is given by

$$X_1 = X_1^0(2)^{2/p} \exp\left(-\frac{2}{p}\alpha t\right)$$

which permits obtaining the critical size p and α (which is a complicated function of various rate constants) from a plot of the experimentally determined function $X_1(t)$. One can also determine the distribution of precipitate particle size as shown in Fig. 8.11. Application of this theory to precipitation of

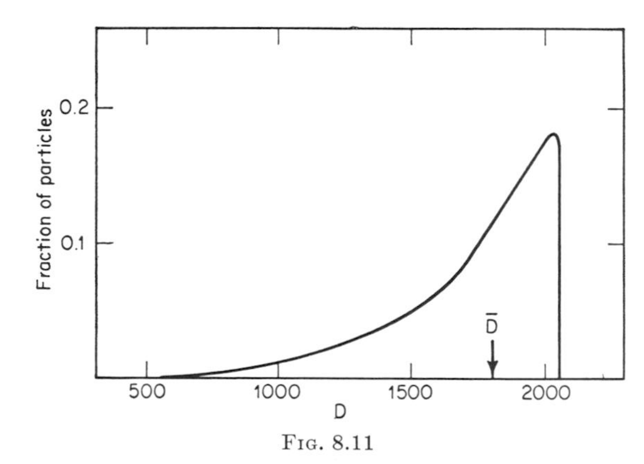

Fig. 8.11

carbon and of nitrogen in α-iron gave values for binding between solute atoms which are in very good agreement with other methods. It appeared, too, that the binding of a carbon atom to a cluster of carbon atoms is size-dependent and that the critical cluster size contained about seven atoms.

In the *growth* of nuclei, one has to distinguish between the rate of supply of new atoms from the matrix and the rate of two-dimensional nucleation (and growth) on the surface of the growing nucleus. The resulting rate of growth can be controlled by either one of these two processes. A big step toward explaining high rates of growth has been made by Frank [77], who pointed out that a screw dislocation which is perpendicular to the growing surface acts in effect as a continuous nucleus for depositing new layers of atoms. This idea found excellent experimental confirmation for crystals formed out of gaseous or liquid phases [78]. Presumably it is also applicable to processes in solids [79], though a direct experimental evidence is still lacking. The growth may occur by the formation of macroscopic rather than monoatomic new layers; this has been explained by Cabrera [80]. In general if the nucleus had no dislocations or other defects, then its growth would be slow and the nucleation of each new layer would occur by a random process [81]. In transformations in which a change of composition occurs the rate is usually controlled by the diffusion of atoms from the matrix to the nucleus, or vice versa. This process is not random but a steady "downhill" diffusion toward, or away from, the nucleus as schematically illustrated in Fig. 8.12.

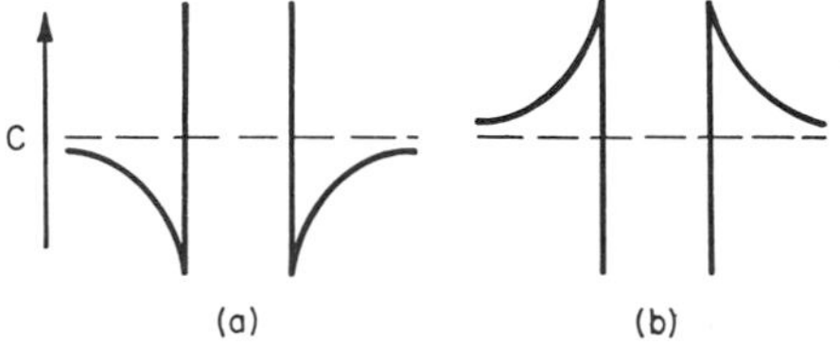

FIG. 8.12. Concentration gradient surrounding a particle of a new phase which has a (*a*) higher or (*b*) lower concentration than the matrix.

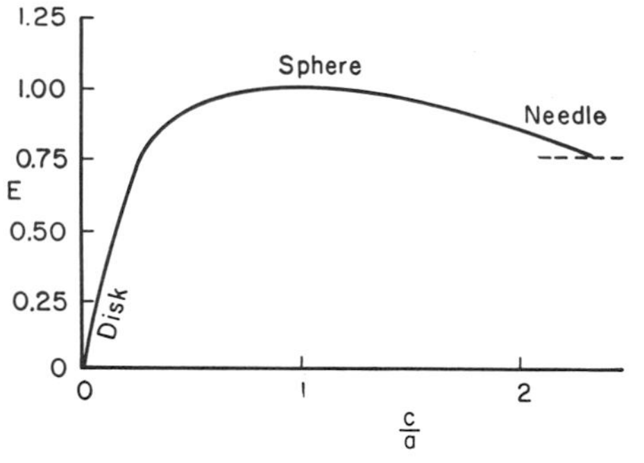

FIG. 8.13. Strain energy of an ellipsoidal particle embedded in an isotropic matrix. a = equatorial diameter, c = polar axis [78].

Many interesting aspects of the early stages of transformation in solids are related to the strain pattern and strain energy surrounding an inclusion. For instance, Nabarro [82] has shown that the strain energy is lowest if the inclusion, in an isotropic medium, is disk-shaped, as shown in Fig. 8.13. This argument may be extended to include the anisotropy of the crystalline lattice with the result that certain crystallographic planes are preferred for the formation of such disks [83, 84]. The choice of the particular plane depends upon the type of lattice and upon the relative importance of the lattice strains and the interaction parameter ϵ. Guinier [85] has shown that there is a close relationship between the differences in atomic sizes and the shape of the early cluster of atoms which will ultimately transform into the new phase, as illustrated in Table 8.4. Clustering of

TABLE 8.4. SHAPE OF ATOMIC CLUSTERS AND DIFFERENCE OF ATOMIC RADII

	ΔR, %	Shape
Al-Ag	0.7	Spheres
Al-Zn	−3	Spheres
Al-Cu	−11	Plates
Cu-Be	−20	Plates

vacancies into platelets in $NaCl$-$CaCl_2$ solid solutions has been shown by Miyake and Suzuki [86]. It is clear that since the nucleus and its early growth are so intimately connected with its surroundings, with which it is coherent, there usually exists a rigorously observed relationship between the orientations of the newly formed lattice and of the matrix. This and the above-mentioned preferred shape of the growing particle account for the existence of definite habit planes and of quite regular patterns (so-called Widmanstatten structures [87, 88]) formed by the precipitates. From this brief summary it appears that a nucleation and growth theory cannot be expected to account quantitatively for the reaction rates in solids until the strains and the lattice anisotropy, including a possible anisotropy of diffusion in a strained lattice, are properly taken into account.

The concept of nucleation has been applied to recrystallization by Turnbull [89] and to the formation of ferromagnetic domains by Dijkstra [90].

6. Shear Transformations

It is only recently that the shear transformation which was originally found in the formation of a metastable form of iron-carbon alloys (the so-called "martensite" and thus a "martensitic" reaction) has been found to occur in many other alloy systems and in pure metals. In spite of some unclear and controversial aspects of this phenomenon several theories have been proposed. In particular Cohen [91] suggested that the reaction is nucleated by a "strain embryo" and the transformation proceeds as a plastic shear wave. Assuming the strain embryo to have the characteristics of a screw dislocation, a fair agreement with the observed crystallographic orientation is obtained. A more detailed dislocation mechanism of nucleation and propagation of martensite has been proposed by Jaswon [92]. There are other theories in which this type of transformation is treated from the point of view of nucleation and growth [70, 93]. Both mechanisms lead to the establishment of a habit plane and of orientation relationships and it is not impossible that both kinds of theories will be found applicable in specific cases. In certain alloys a visible thin plate of transformed material appears instantaneously and then grows in thickness at a relatively slow rate [94].

A nice confirmation of the rapidity of the shear reactions is provided by the study of the lowering of the transformation temperature by rapid quenching. Recent measurements by Duwez [95] show that quenching rates up to 10^4°C/sec affect these temperatures very little: in Zr the drop is about 15°C, in Ti about 35°C (although in Ti-Mo alloys no drop was observed), in Tl less than 4°C, and in very pure iron about 250°C. The last one corresponds to a transformation from a close-packed (cubic) lattice to a body-centered cubic lattice, while the other ones are opposite, i.e., from a body-centered cubic to a close-packed (hexagonal). This may be the cause of the difference in the amount of drop of the transformation temperature. A more detailed discussion of these complicated transformations is beyond the scope of this survey and the reader is referred to special summaries of this field [96].

7. Rate of Ordering

The kinetics of ordering in a solid solution undercooled below the critical temperature is probably the least-known aspect of the order-disorder transformations, but since the experimental data are limited to a few particular alloys the total picture is fairly consistent. This is in contrast to other transformations in which the large diversity of available systems make any attempt at an all-embracing theory very difficult.

The process of ordering takes place presumably, like diffusion, by a vacancy mechanism and the very first stage is very likely an over-all increase of the degree of short-range order. This leads to formation of local areas which have such a high degree of short-range order that they can act as nuclei of long-range order domains. The "effective" order at a given stage of the process can be then described either in terms of an over-all short-range order [37, 97] or by a long-range order parameter. The interpretation of experimental data in the latter case is far from simple because of the many variables such as the number, size, and shape of domains which possess a definite long-range order and the condition of the remaining material in which no long-range order has yet developed. (This limitation applies, to a certain degree, also to the interpretation of equilibrium conditions.) The "driving force" for ordering will naturally depend on the degree of order already present and thus the dependence on time may be quite complicated.

In $AuCu_3$ ordering was believed to progress by a two-stage process [98, 99]. However, recent X-ray work by Edmunds and Hinde [100] seems to support the single-stage mechanism outlined above and indicates the basic importance of the short-range order. On the other hand, a magnetic and resistivity study of Ni_3Mn indicates the possibility of a multiple-step process [101]. A single-step process, at least in a temperature range around 300°C, has been observed in AuCu by X-ray, calorimetric, and resistivity methods. In particular, Dienes [102], using a pulse-annealing technique, found that the process obeys a third-order-rate law, that is, the rate is proportional to the third power of the fraction of disordered atoms. The physical significance of this fact is not clear, but it is probably due to the variation of the driving force with the degree of order which would be especially strong in an alloy such as AuCu in which the tetragonal ordered phase produces large strains in the cubic disordered phase. These strains are presumably responsible also for the fact that early areas of order in CoPt form thin platelets which are parallel to the dodekahedral planes of the random phase [84]. Studies of the influence of nuclear radiation on the rate of ordering and of disordering [103, 104] will undoubtedly throw additional light on the mechanism of these reactions.

Sato and Toth [105] made an extensive investigation of the formation of long-period superlattices. They were able to show experimentally that in noble-metal alloys the period is determined by the electron-atom ratio, and they gave a plausible explanation of this phenomenon in terms of an interaction between the Fermi surface and Brillouin zone boundaries. In transition-metal alloys the period seems to be independent of composition but a tetragonal distortion of the lattice is such as to keep the Brillouin zone boundary at the varying Fermi surface.

It should be stressed that conclusions concerning ordering reactions are often based on very rudimentary knowledge of the relation between order and the measured property. This applies especially to resistivity which can be affected by several factors such as the degree of long-range order, degree of short-range order, domain boundaries, and lattice strains. Significance of X-ray data on the other hand is limited by the considerable minimum size of particles required by this method.

8. Crystallographic Factors Affecting Transformation Rate

Buerger [106] pointed out that the actual crystallographic configuration affects greatly the rate of allotropic transformations (i.e., in which no composition changes occur). The main factor appears to be the number of bonds between atoms which have to be broken as expressed in coordination changes. On this basis, Buerger classified all such transformations as shown in Table 8.5. Type 1 is

TABLE 8.5. CRYSTALLOGRAPHIC CLASSIFICATION OF TRANSFORMATIONS

Change of		Rate of transformation
1. Secondary coordination (network)	a. Displacive	Rapid
	b. Reconstructive	Slow
2. Order	a. Rotational	Rapid
	b. Substitutional	Slow
3. First coordination	a. Dilatational	Rapid
	b. Reconstructive	Slow
4. Bond		Usually slow

quite common in nonmetallic systems: The structure can be either distorted without breaking the nearest neighbor pairs, case *a*, illustrated by the two forms of quartz, or it can be completely broken up and rearranged, case *b*, illustrated by the transformation of quartz to tridymite. Type 2 has been considered in some detail previously. Type 3 includes all changes of the first coordination number except when a definite change of type of binding is observed, these belong to type 4. The distinction is naturally often quite difficult. Case 3*a* is illustrated by the change of CsCl from its typical body-centered lattice to the NaCl-type lattice, a transformation which occurs by a rather simple deformation of the lattices. Presumably the transformations in iron, titanium, etc., belong to this category. Case 3*b* in which the structure has to be completely broken up before the new one is formed is illustrated by the change of aragonite to calcite. Type 4 is best illustrated by the diamond-graphite transformation and presumably by the $\alpha - \beta$ transformation in tin. Whenever the original structure has to be radically altered, the transformation is slow, while a transformation based on a simple deformation of the lattice is rapid.

Most systems to which the above classification applies are rather complicated open structures which are difficult for a quantitative theoretical treatment. In this category fall transformations in glasses [107] in silicates, etc. Only in certain ionic crystals such as silver halides good insight into the mechanism of transformation and even into the type of binding is possible: the transformations induced by temperature or by pressure seem to be understandable [108] in terms of the theories of Born and Mayer [109].

References

1. Dienes, J. G.: *Phys. Rev.*, **89**: 185 (1953); **93**: 265 (1954).
2. Huntington, H. B., G. A. Shirn, and E. S. Wajda: *Phys. Rev.*, **91**: 246 (1953).
3. Seitz, F.: "Modern Theory of Solids," McGraw-Hill, New York, 1950.
4. Smoluchowski, R.: *Metal Progr.*, **41**: 363 (1942). Smoluchowski, R., and J. S. Koehler: *Annual Rev. Phys. Chem.*, **2**: 187 (1951).
5. Bridgman, P. W.: "The Physics of High Pressure," G. Bell, London, 1949.
6. Bardeen, J.: *J. Chem. Phys.*, **6**: 372 (1938).
7. Sternheimer, R.: *Phys. Rev.*, **78**: 235 (1950).
8. Bridgman, P. W.: *Proc. Am. Acad. Arts Sci.*, **82**: 101 (1953).
9. Roth, W. A., and H. Wallasch: *Ber. deut. chem. Ges.*, **46**: 896 (1913).
10. Ganzhorn, K.: *Tech. Hochschule Stuttgart*, Thesis, 1952.
11. Harrison, W.: *Phys. Rev.*, **129**: 2503 (1963).
12. Frazer, B. C., G. Shirane, D. E. Cox, and C. E. Olsen: *Phys. Rev.* **140**: A 1448 (1965).
13. Blume, M.: *Phys. Rev.*, **141**: 517 (1966).
14. Bauer, E.: Changements de Phases, *Soc. chim. phys.*, p. 3, Paris, 1952.
15. Mayer, J. E., and S. F. Streeter: *J. Chem. Phys.*, **7**: 1019 (1939).
16. Rutgers, A. J., and A. Wouthuysen: *Physica*, **4**: 515 (1937).
17. Tisza, L.: "Phase Transformations in Solids," Wiley, New York, 1951.
18. Newell, G. F., and E. W. Montroll: *Revs. Mod. Phys.*, **25**: 352 (1953).
19. Guggenheim, E. A.: "Mixtures," Oxford University Press, New York and London, 1952.
20. Onsager, L.: *Phys. Rev.*, **65**: 117 (1944).
21. Newell, G. F.: *Phys. Rev.*, **79**: 876 (1950).
22. Temperley, H. N. V.: *Proc. Roy. Soc.* (*London*), **A202**: 202 (1950).
23. Wannier, G. H.: *Revs. Mod. Phys.*, **17**: 50 (1945).
24. Yvon, J.: *Cahiers phys.*, *no.* 28, 1, September, 1945.
25. Fournet, G.: *J. Phys. radium*, **13**: 14A (1952).
26. Slater, J. C.: *Phys. Rev.*, **84**: 179 (1951).
27. Neel, L.: *Revs. Mod. Phys.*, **25**: 58 (1953).
28. Warren, B. E., B. L. Averbach, and R. W. Roberts: *J. Appl. Phys.*, **22**: 1493 (1951).
29. Lawson, A. W.: *J. Chem. Phys.*, **15**: 831 (1947).
30. Borelius, G., L. E. Larsson, and H. Selberg: *Arkiv Fysik*, **2**: 161 (1950).
31. Rhines, F. N., and J. B. Newkirk: *Trans. ASM*, **45**: 1029 (1953).
32. Newkirk, J. B.: *J. Metals*, **5**: 823 (1953).
33. Merz, W. J.: *Bull. Am. Phys. Soc.*, Jan. 22, 1953.
34. Jaynes, E. T.: "Ferroelectricity," Princeton University Press, Princeton, N.J., 1953.
35. Bragg, W. L., and E. J. Williams: *Proc. Roy. Soc.* (*London*), **145**: 699 (1934); **151**: 540 (1935); **152**: 231 (1935).
36. Weiss, P.: *J. Phys.*, **6**: 661 (1907).
37. Bethe, H. A.: *Proc. Roy. Soc.* (*London*), **A150**: 552 (1935).
38. Rushbrooke, G. S.: *Proc. Roy. Soc.* (*London*), **A166**: 296 (1938).
39. Kikuchi, R.: *Phys. Rev.*, **81**: 988 (1951); *J. Chem. Phys.*, **19**: 1230 (9151).
40. Peierls, R.: *Proc. Roy. Soc.* (*London*), **A154**: 207 (1936).
41. Li, Y. Y.: *J. Chem. Phys.*, **17**: 447 (1949); *Phys. Rev.*, **76**: 972 (1949).
42. Kirkwood, J. G.: *J. Chem. Phys.*, **6**: 70 (1938).
43. Rushbrooke, G. S.: "Changements de Phases," p. 177, Société de Chimie Physique, Paris, 1952.
44. Wakefield, A. J.: *Proc. Cambridge Phil. Soc.*, **47**: 799 (1951).
45. Trefftz, E.: *Z. Physik*, **127**: 371 (1950).
46. Domb, C., and M. F. Sykes: *Proc. Roy. Soc.* (*London*), **A240**: 214 (1957).
47. Baker, G. A.: *Phys. Rev.*, **136**: A1376 (1965), and earlier papers.
48. Kac, M., P. C. Hemmer, and G. E. Uhlenbeck: *J. Math. Phys.*, **5**: 60 (1964), and two earlier papers.
49. Langer, J. S.: *Phys. Rev.*, **137**: A1531 (1965).
50. Oriani, R. A.: *Acta Met.*, **2**: 343 (1954).
51. Pauling, L.: *Phys. Rev.*, **36**: 430 (1930).
52. Frenkel, J.: *Acta Physicochem. U.S.S.R.*, **3**: 23 (1935).
53. Wagner, E. L., and D. F. Hornig: *J. Chem. Phys.*, **18**: 296, 305 (1950).
54. Levy, H. A., and S. W. Peterson: *Phys. Rev.*, **83**: 1270 (1951).
55. Lawson, A. W.: *Phys. Rev.*, **57**: 417 (1940).
56. Kruis, A., and R. Kaischew: *Z. physik. Chem.*, **41B**: 427 (1938).
57. Smoluchowski, R.: in "Phase Transformations in Solids," Wiley, 1951; *Ind. Eng. Chem.*, **44**: 1321 (1952).
58. Bunshah, R. F., and R. F. Mehl: *J. Metals*, **4**: 1042 (1952).
59. Volmer, M.: "Kinetik der Phasenbildung," Steinkopff, Dresden, 1939.
60. Reiss, H.: *J. Chem. Phys.*, **20**: 1216 (1952).
61. Becker, R.: *Ann. Physik*, **32**: 128 (1938).
62. Turnbull, D.: *Metals Techn.*, T.P. 2265 (1948).
63. Turnbull, D., and J. C. Fisher: *J. Chem. Phys.*, **17**: 71 (1949).
64. Borelius, G.: *Ann. Physik*, **28**: 507 (1937); **33**: 517 (1938); *Arkiv. Mat. Astron. Fysik*, **A32**: 1 (1945).
65. Wiktorin, C. G.: "Studies in Gold-Platinum Alloys," Ivar Hoeggstroms, Stockholm, 1947.
66. Hillert, M.: *Acta Met.*, **9**: 525 (1961).

67. Cahn, J. W.: *Acta Met.*, **10**: 907 (1962), and several earlier papers.
68. Van der Toorn, L. S.: *Acta Met.*, **8**: 715 (1960).
69. Hobstetter, J. N.: *Trans. AIME*, **180**: 121 (1949).
70. Fisher, J. C., J. H. Hollomon, and D. Turnbull: *Trans. AIME*, **185**: 691 (1949).
71. Sirota, N. N.: *Doklady Akad. Nauk S.S.S.R.*, **50**: 337, 343 (1945).
72. Fisher, J. C., J. H. Hollomon, and D. Turnbull: *J. Appl. Phys.*, **19**: 775 (1948).
73. Geisler, A. H.: *Trans. ASM*, **44A**: 269 (1952).
74. Turnbull, D.: *J. Chem. Phys.*, **18**: 198 (1950).
75. Fleeman, J.: *Bull. Am. Phys. Soc.*, March 26, 1953.
76. Damask, A. C., G. C. Danielson, and G. J. Dienes: *Acta Met.*, **13**: 973 (1965).
77. Frank, F. C.: *Discussions Faraday Soc.*, *no.* 5, p. 48, 1949.
78. Frank, F. C.: *Advances in Physics, London*, **1**: 1 (1952).
79. Bardeen, J., and C. Herring: in "Imperfections in Nearly Perfect Crystals," Wiley, New York, 1952.
80. Cabrera, N.: *J. Chem. Phys.*, **21**: 1111 (1953).
81. Stransky, I. N.: *Discussions Faraday Soc.*, *no.* 5, p. 13, 1949.
82. Nabarro, F. R. N.: *Proc. Phys. Soc. (London)*, **52**: 90 (1940).
83. Opinsky, A., and R. Smoluchowski: *Phys. Rev.*, **74**: 343 (1948). Smoluchowski, R.: *Physica*, **15**: 179 (1949).
84. Newkirk, J. B., R. Smoluchowski, A. H. Geisler, and D. L. Martin: *Acta Cryst.*, **4**: 507 (1951).
85. Guinier, A.: *Physica*, **16**: 148 (1949).
86. Miyake, S., and K. Suzuki: 3d Intern. Congr. Cryst., Paris, 1954.
87. Barrett, C. S.: "Structure of Metals," 2d ed., McGraw-Hill, New York, 1952.
88. Mehl, R. F., and C. S. Barrett: *Trans. AIME*, **93**: 78 (1931).
89. Burke, J. E., and D. Turnbull: in "Progress in Metal Physics," vol. 3, Interscience, New York, 1952.
90. Dijkstra, L. J.: *Trans. ASM*, **42A**: 271 (1950).
91. Cohen, M., E. S. Machlin, and V. G. Paranjpe: *Trans. ASM*, **42A**: 242 (1950).
92. Jaswon, M. A.: 3d Intern. Congr. Cryst., Paris, 1954.
93. Geisler, A. H.: *Acta Met.*, **1**: 260 (1953).
94. Chang, L. C., and T. A. Read: *Trans. AIME*, **189**: 47 (1951).
95. Duwez, P.: *J. Metals*, **3**: 765 (1951).
96. Barrett, C. S.: in "Progress in Metal Physics," vol. 3, Interscience, New York, 1952.
97. Cowley, J. M.: *Phys. Rev.*, **77**: 669 (1950).
98. Sykes, C., and H. Evans: *J. Inst. Metals*, **58**: 255 (1936).
99. Siegel, S.: *J. Chem. Phys.*, **8**: 860 (1940).
100. Edmunds, I. G., and R. M. Hinde: *Proc. Phys. Soc. (London)*, **A65**: 716 (1952).
101. Aronin, L. R.: *J. Appl. Phys.*, **23**: 642 (1952).
102. Dienes, J. G.: *J. Appl. Phys.*, **22**: 1020 (1951).
103. Stello, P. G.: A.E.C. Report NAA-SR-171.
104. Glick, H. L., F. C. Brooks, W. R. Witzig, and W. E. Johnson: *Phys. Rev.*, **87**: 1074 (1952).
105. Sato, H., and R. S. Toth: *Phys. Rev.*, **139**: A1581 (1965), and several earlier papers.
106. Buerger, M. J.: in "Phase Transformations in Solids," Wiley, New York, 1951.
107. Weyl, W. A.: in "Phase Transformations in Solids," Wiley, New York, 1951.
108. Huggins, M. L.: in "Phase Transformations in Solids," Wiley, New York, 1951.
109. Born, M., and J. E. Mayer: *Z. Physik*, **75**: 1 (1932). Mayer, J. E.: *J. Chem. Phys.*, **1**: 270, 327 (1933).

Bibliography

Smoluchowski, R., J. E. Mayer, and W. A. Weyl (eds.): "Phase Transformations in Solids," Wiley, New York, 1951.
"Changements de phases," Société de Chimie Physique, Paris, 1952.
"Reactions dans l'état solide," Centre National de la Recherche Scientifique, Paris, 1949.
Symposium on Nucleation, *Ind. Eng. Chem.*, **44**: 1269 (1952).
Seitz, F.: "The Modern Theory of Solids," McGraw-Hill, New York, 1940.
Mott, N. F., and H. Jones: "Theory of the Properties of Metals and Alloys," Oxford University Press, Fair Lawn, N.J., 1936.
Slater, J. C.: "Introduction to Chemical Physics," McGraw-Hill, New York, 1939.
Fowler, R. H., and E. A. Guggenheim: "Statistical Thermodynamics," Cambridge University Press, New York, 1949.
Cohn, G.: Reactions in the Solid State, *Chem. Revs.*, **42**: 527 (1948).
Christian, J. W.: "Theory of Transformations in Metals and Alloys," Pergamon Press, New York, 1964.
Fine, M.: "Introduction to Phase Transformations in Condensed Systems," Macmillan, New York, 1964.
Brout, R. H.: "Phase Transitions," Benjamin, New York, 1965.
Krivoglaz, M. A., and A. A. Smirnov: "The Theory of Order-Disorder in Alloys," Elsevier Publishing Company, Amsterdam, 1965.
Domb, C.: On the Theory of Cooperative Phenomena in Crystals, *Advan. Phys.*, **9**: 245 (1960).
Katsura, S.: Singularities in First Order Phase Transitions, *Advan. Phys.*, **12**: 391 (1963).

Chapter 9

Magnetic Resonance

By D. P. AMES, McDonnell Company

1. Introduction

Unpaired electrons in atomic or molecular species and atoms possessing unpaired nucleons (odd Z-odd A, even Z-odd A, and odd Z-even A) have magnetic dipole moments. These magnetic moments are associated with spin angular momentum. Magnetic resonance involves the collective macroscopic magnetic properties of these systems and implies correspondence of an external frequency with a natural frequency of the system.

When a system possessing unpaired electrons or nucleons is placed in a static magnetic field, a number of energy levels is differentiated by Zeeman splittings of the quantum states of the magnetic moment. Transitions of the magnetic moment between these levels are induced with an applied radiation at the resonant frequency. Such transitions are indicated by an energy absorption or emission. Detailed information regarding the environment of the atoms and electrons in the material are obtained from the intensities and the frequencies of resonance absorption and emission.

Magnetic resonance of assemblies of unpaired electrons is called either electron paramagnetic resonance (EPR) or electron spin resonance (ESR). The first experimental evidence for macroscopic magnetic resonance transitions between Zeeman levels of bulk $CuCl_2 \cdot 2H_2O$ was given by Zavoisky [1]. Cummerow and Halliday [2] employed 2.9 gigahertz (GHz) radiation to examine the magnetic resonance in bulk $MnSO_4 \cdot 4H_2O$. The first hyperfine structure EPR measurements were reported by Penrose [3] who doped magnesium ammonium nitrate hexahydrate with copper (II).

EPR spectroscopy has been used effectively for investigating transition (d electron) ions, inner transition (f electron) ions, stable or unstable mono- and bi-radicals, metal-organic compounds, ground and excited state triplet molecules, donors and acceptors in semiconductors, radiation-induced defects, metals and paramagnetic impurities in solids. The general theory and the applications of EPR spectroscopy have been described in detail in review articles and textbooks [4–27].

The first observation of bulk nuclear paramagnetism was made by Lasarev and Schubnikov [28] who measured the static magnetization of solid hydrogen. The first successful nuclear magnetic resonance experiments on bulk material were performed independently by Purcell, Torrey, and Pound [29] and by Bloch, Hansen, and Packard [30]. Purcell and his coworkers observed resonance absorption of protons in a hydrocarbon (paraffin) whereas Bloch and his colleagues found the proton resonance signal in water.

Nuclear magnetic-resonance spectroscopy (NMR) has been applied to many areas in physics and chemistry which include the elucidation of the intermolecular structure in solids, liquids, and gases, the determination of magnetic fields in single crystals, molecular adsorption, diffusion coefficients, and electric field effects. A number of review articles and books treat the general theory, experimental techniques and applications [31–51].

Useful literature surveys of the applications and advances of ESR and NMR have been given [52]. In addition, symposia to cover special topics and advances in NMR and EPR have been published [53–62]. A continuing series on magnetic resonance has been started [63].

2. Magnetic Properties of Nuclei and Electrons

If it is assumed that a charge, e, and mass, m, are distributed uniformly on a spherical shell which is spinning about a given axis, a magnetic field symmetrical about this axis will be generated by the rotating charge. The charge, e, rotating at $v/2\pi r$ revolutions $\sec^{-1}$ generates a current $ev/2\pi r$. Since the rotating charge behaves as a magnetic dipole whose moment, μ, is given by the product of the current i and the loop cross-sectional area, πr^2,

$$\mu = i\pi r^2 = \frac{evr}{2c} \text{ (emu)}$$

where c represents the velocity of light. The spinning mass generates an angular momentum, p, of magnitude mvr directed along the axis of rotation. Thus the magnetic moment is colinear and directly proportional to the angular momentum [64]

$$\vec{\mu} \propto \frac{e}{2mc} \vec{p} \tag{9.1}$$

The electron mass is used when unpaired electrons are involved, whereas when unpaired nucleons are present nuclidic masses are required.

The direction of the vectors is determined by the sign of the charge; however, this model does not explain nuclear moments (e.g., neutron). The nuclear model deficiencies suggest a complex structure involving orbital angular momentum and spin angular momentum coupling analogous to that well known for the electron [65].

The proportionality constant between $\vec{\mu}$ and $\vec{p}$ is designated g for the electron and g_N for nuclei. For free atoms the spectroscopic splitting factor, g, is identical with the Landé g value; however, this identity does not hold for all magnetic molecules as will be seen later. The nuclear g_N value or nuclear spectroscopic splitting factor has a characteristic value for each nuclide.

The angular momentum, p, of any particle or particle system can be expressed in units of $\hbar$ (Planck's constant divided by 2π). The spin is defined as the largest observable value of the time average of the angular momentum component in a specified direction ($<p>_{\max}/\hbar$ = spin). The electron spin is designated by S whereas the nuclear spin is termed I.

For the nucleus, the general expression for all permitted values of p is $p = m\hbar$, where m, the nuclear magnetic quantum number, has $(2I + 1)$ values I, $I - 1, \ldots 0, \ldots, -I + 1, -I$. The lowest lying energy level is characterized by an I which has a positive integral or half integral value. Then the nuclear magnetic moment is given by

$$\mu = g_N \frac{e\hbar}{2Mc} I = g_N \mu_N I \tag{9.2}$$

where μ_N, the nuclear magneton, has the value 5.05050×10^{-24} erg gaus^{-1} [66] for the proton. The term $g_N(e/2Mc)$ is often replaced by γ, the magnetogyric ratio; hence, $\mu = \gamma\hbar I$. The usual nuclear property tables [31, 32, 34–36, 40, 42, 50] list values of μ in nuclear magnetons and I in $\hbar$ units. The magnetic moment of the proton is 2.79277 ± 0.00007 nuclear magnetons [66, 67].

Nuclides having spins ≥ 1 do not always have a spherical nuclear charge. The charge asphericity is given by the nuclear electric quadrupole moment, Q, which is defined by the relation

$$eQ = \int \rho r^2 (3 \cos^2 \theta - 1)\, dv \tag{9.3}$$

where ρ is the charge density, dv is a volume element inside the nucleus located a distance r from the nucleus center and making an angle θ with the nuclear spin axis, e is the protonic charge, and the integration is performed for the state I. Positive eQ values are observed when the charge distribution is elongated along the spin axis (prolate spheroid) whereas a flattened charge distribution (oblate spheroid) along the spin axis is indicated by a negative eQ value. The Q values expressed in barns (10^{-24} cm^2) vary from 10.2 barns for Er^{167} to -1.7 barns for Ac^{227} [68–73].

The electron magnetic moment is related to the spin quantum number S by $\mu_e = -\gamma_e S\hbar$. Instead of using γ_e the electron magnetic moment is written in terms of the Bohr magneton ($\mu_B = e\hbar/2mc$) and the spectroscopic splitting factor, g_0,

$$\mu_e = -g_0 \mu_B S \tag{9.4}$$

where $\mu_B = 0.92732 \times 10^{-20}$ erg gauss^{-1} and $g_0 = 2.00232$ [66]. The g value is a measure of the contribution of the spin and orbital angular momentum of the electron to the total angular momentum and therefore depends on the electron environment. The spin S refers to the total spin of all unpaired electrons and has a multiplicity $2S + 1$ with values ranging from $-S$ to S. The Bohr magneton-proton moment ratio is 657.4481 [66]; hence the Bohr magneton-nuclear moment ratio is 1838.2403.

Static Magnetism. The magnetic induction B and the magnetic field H are related through classical electromagnetics by

$$B = H + 4\pi M \tag{9.5}$$

where M, the magnetization, is the macroscopic magnetic moment per unit volume. The static magnetic susceptibility, χ_0, is defined as the ratio of the magnetization to the magnetic field, H_0,

$$\chi_0 = \frac{M}{H_0} \tag{9.6}$$

When the susceptibility is positive, paramagnetism results whereas diamagnetism is indicated by a negative susceptibility.

When an unpaired electron is placed in a static magnetic field, the potential interaction energy, E_i, is given by

$$E_i = -\vec{\mu} \cdot \vec{H} \tag{9.7}$$

The total angular momentum, J, of the electron is related to its magnetic moment by

$$\mu = g\mu_B J \tag{9.8}$$

When the applied field direction lies along the z axis, then the resultant Zeeman energy levels are given by

$$E_i = -g\mu_B H_0 m_J \tag{9.9}$$

where m_J, the magnetic quantum number, has $2J + 1$ multiplicity ranging in value from $-J$ to J. Figure 9.1 shows the energy levels for a $J = 3/2$ system.

The probability P_i that an electron within the assembly has a potential energy E_i at a temperature T is given by the Boltzmann relation

$$P_i = \text{constant}\left(\exp - \frac{E_i}{kT}\right) = \frac{\exp(-g\mu_B H_0 m_J/kT)}{\sum_{m_J=-J}^{J} \exp(-g\mu_B H_0 m_J/kT)} \tag{9.10}$$

Boltzmann statistics are applicable since the magnetization is the property of similar, spatially distinguishable particles [74]. The energy when H_0 is

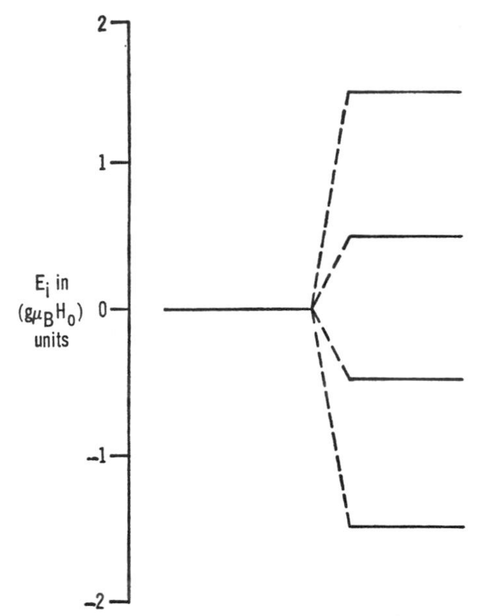

FIG. 9.1. Zeeman pattern for $J = 3/2$.

parallel to the z axis is

$$E = \frac{N \sum_{m_J=-J}^{J} g\mu_B H_0 m_J \exp(-g\mu_B H_0 m_J/kT)}{\sum_{m_J=-J}^{J} \exp(-g\mu_B H_0 m_J/kT)} \tag{9.11}$$

where N is the number of electrons per unit volume. For the high-temperature approximation $g\mu_B H_0/kT \ll 1$

$$M_z = \frac{\partial E}{\partial H_0} = NJ(J+1)\frac{g^2\mu_B{}^2 H_0}{3kT} \tag{9.12}$$

Hence the susceptibility, χ_0, is given by the Langevin-Brillouin relation

$$\chi_0 = \frac{NJ(J+1)g^2\mu_B{}^2}{3kT} \tag{9.13}$$

For conditions where $\frac{g\mu_B H_0}{kT} \approx 1$

then
$$M_z = Ng\mu_B J B_J\left(\frac{g\mu_B H_0}{kT}\right) \tag{9.14}$$

where
$$B_J\left(\frac{g\mu_B H_0}{kT}\right)$$

is the Brillouin function. [75, 76].

The volume susceptibility for unpaired electrons is of the order of 10^{-5}–10^{-6} cgs unit. For nonzero spin nuclei the static nuclear susceptibility is given by

$$\chi_0 = \frac{Ng_N{}^2\mu_N{}^2 I(I+1)}{3kT} \tag{9.15}$$

For protons in water at room temperature, $\chi_0 \approx 3 \times 10^{-10}$ cgs unit, which is considerably smaller than the diamagnetic susceptibility (-0.719×10^{-6} cgs unit) of water.

3. Classical Magnetic Resonance Description

When the spinning charged particle is placed in an external magnetic field, H_0, the magnetic moment experiences a torque, L, tending to align it in the direction of the field. The effect of the torque changes the angular momentum, p,

$$\frac{d\vec{p}}{dt} = \vec{\mu} \times \vec{H}$$

or

$$\frac{d\vec{\mu}}{dt} = \gamma(\vec{\mu} \times \vec{H}) \tag{9.16}$$

Since the magnetization M is defined by $N\langle\mu\rangle$ where $\langle\mu\rangle$ is the average moment,

$$\frac{dM}{dt} = \gamma(\vec{M} \times \vec{H}) \tag{9.17}$$

This equation (9.17) when written in terms of the vector components takes the form:

$$\begin{aligned} \frac{dM_x}{dt} &= \gamma(M_y H_z - M_z H_y) \\ \frac{dM_y}{dt} &= \gamma(M_z H_x - M_x H_z) \\ \frac{dM_z}{dt} &= \gamma(M_x H_y - M_y H_x) \end{aligned} \tag{9.18}$$

When the special case of a constant magnetic field in the z direction is considered, the following relations are obtained:

$$\begin{aligned} \frac{dM_x}{dt} &= \gamma M_y H_z \\ \frac{dM_y}{dt} &= -\gamma M_x H_z \\ \frac{dM_z}{dt} &= 0 \end{aligned} \tag{9.19}$$

The solutions of these equations are

$$\begin{aligned} M_x &= M_0 \sin\alpha \cos(\gamma H_0 t + \phi) \\ M_y &= -M_0 \sin\alpha \sin(\gamma H_0 t + \phi) \\ M_z &= M_0 \cos\alpha \end{aligned} \tag{9.20}$$

and represent the magnetization precession about H_z with an angular frequency $\omega = \gamma H_0$ as shown in Fig. 9.2. A special solution $M_z = \pm M_0$ independent of time exists when $\alpha = 0$ or π. The system has a characteristic frequency $\omega_0/2\pi$, termed the Larmor frequency, which equals 42.577 megahertz (MHz) for protons and 27.99 gigahertz (GHz) for electrons when each is placed in a 10-kilogauss (kG) field.

This system may be excited at the Larmor frequency, $\omega_0/2\pi$, with a small oscillating field, H_1, applied perpendicular to H_0. To first order, only the proper circular polarization component will synchronize with the precession frequency [77]; hence a

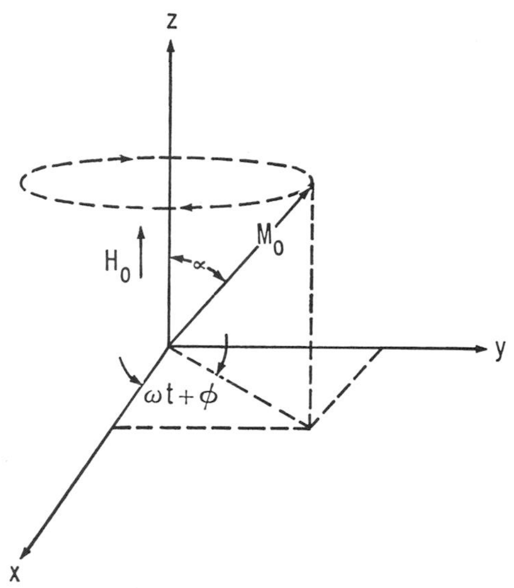

Fig. 9.2. Magnetization precession in a magnetic field.

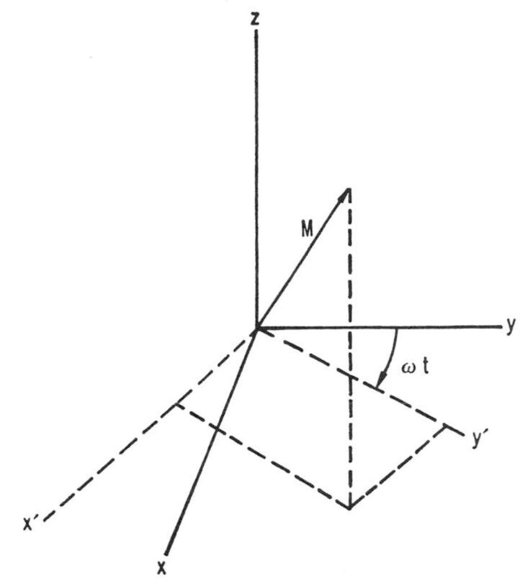

Fig. 9.3. Relation of rotating and laboratory coordinate systems.

linear oscillating field, composed of two counterrotating fields, is suitable. To describe this weak field addition, a coordinate system rotating in the xy plane at a frequency $\omega \approx \omega_0$ to hold H_1 constant is convenient [78]. Let $\partial M/\partial t$ represent the time variation of M in the rotating coordinate system. Then

$$\frac{d\vec{M}}{dt} = \frac{\partial \vec{M}}{\partial t} - (\vec{M} \times \vec{\omega}) \tag{9.21}$$

in the laboratory coordinates. If $\vec{\omega}$ is along the z axis (Fig. 9.3), then the components of M are

$$\begin{aligned} \frac{d\vec{M}_x}{dt} &= \frac{\partial \vec{M}_x}{\partial t} - \omega \vec{M}_y \\ \frac{d\vec{M}_y}{dt} &= \frac{\partial \vec{M}_y}{\partial t} + \omega \vec{M}_x \\ \frac{d\vec{M}_z}{dt} &= \frac{\partial \vec{M}_z}{\partial t} \end{aligned} \tag{9.22}$$

Thus the magnetization M' in the rotating system has the behavior

$$\frac{\partial \vec{M}'}{\partial t} = \gamma \left[\vec{M}' \times \left(\vec{H} + \frac{\vec{\omega}}{\gamma} \right) \right] \tag{9.23}$$

and the rotating system behaves as a stationary system with an effective magnetic field $[\vec{H} + (\vec{\omega}/\gamma)]$. No effective field exists when $\vec{\omega} = -\vec{\omega}_0 = -\gamma \vec{H}_0$. When H_1 is directed along the x' axis, the effective-magnetic-field components are $H_{x'} = H_1$ and

$$H_z = H_0 - \left(\frac{\omega}{\gamma}\right)$$

The effective magnetic field is directed at an angle

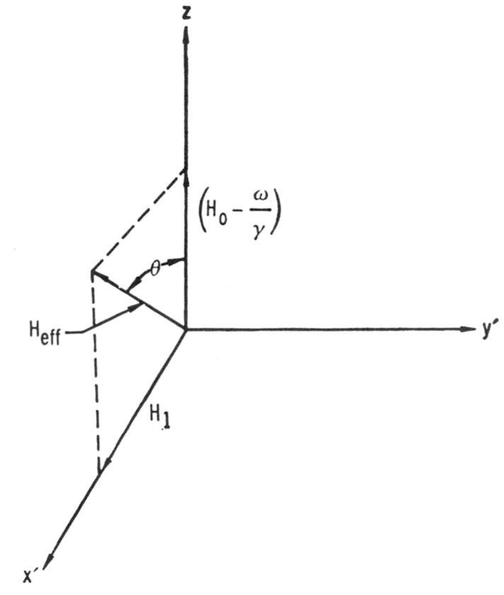

Fig. 9.4. Effective magnetic field in rotating frame.

θ to H_0 when the z axes are superimposed as shown in Fig. 9.4.

The explicit transformations to the rotating coordinate system for a positive γ are

$$\begin{aligned} H_x &= H_1 \cos \omega t \qquad H_y = -H_1 \sin \omega t \\ M_{x}' &= M_x' \cos \omega t + M_y' \sin \omega t \\ \text{and} \quad M_y' &= -M_x' \sin \omega t + M_y' \cos \omega t \end{aligned} \tag{9.24}$$

Then the time-independent equations for the rotating frame are

$$\begin{aligned} \frac{\partial M_x'}{\partial t} &= \gamma M_y' \left(H_z - \frac{\omega}{\gamma} \right) \\ \frac{\partial M_y'}{\partial t} &= -\gamma M_x' \left(H_z - \frac{\omega}{\gamma} \right) + \gamma H_1 M_z \\ \frac{\partial M_z'}{\partial t} &= -\gamma H_1 M_y' \end{aligned} \tag{9.25}$$

Now assume at $t = 0$ that M' is parallel to H_0 and that an oscillating field of amplitude H_1 is imposed on the system. At a later time t,

$$\begin{aligned}
M_x' &= 2M_0 \sin\theta \cos\theta \sin^2\left(\gamma H_{\text{eff}} \frac{t}{2}\right) \\
&= \frac{2M_0\gamma H_1(\omega_0 - \omega)}{(\omega_0 - \omega)^2 + \gamma^2 H_1^2} \sin^2\left(\gamma H_{\text{eff}} \frac{t}{2}\right) \\
M_y' &= M_0 \sin\theta \sin(\gamma H_{\text{eff}}\, t) \\
&= \frac{M_0 \gamma H_1}{[(\omega_0 - \omega)^2 + \gamma^2 H_1^2]^{1/2}} \sin(\gamma H_{\text{eff}}\, t) \\
M_z' &= M_0[\cos^2\theta + \sin^2\theta \cos(\gamma H_{\text{eff}}\, t)] \\
H_1^2 &= M_0\left[1 - \frac{2\gamma^2 H_1^2}{(\omega_0 - \omega)^2 + \gamma^2 H_1^2} \sin^2\left(\gamma H_{\text{eff}} \frac{t}{2}\right)\right]
\end{aligned} \tag{9.26}$$

where $\gamma H_{\text{eff}} = [(\omega_0 - \omega)^2 + \gamma^2 H_1^2]^{1/2}$ and $\omega_0 = \gamma H_0$. A relation between these quantities and the probability that a spin has made a transition between two Zeeman levels is required.

Let $P_{\pm 1/2}$ be the probability of a spin ½ system being in a state with the magnetic quantum number equal to $\pm 1/2$. Since $P_{1/2} - P_{-1/2} = M_z'/M_0$ and $P_{1/2} + P_{-1/2} = 1$,

$$\begin{aligned}
P_{-1/2} &= \frac{1 - (M_z'/M_0)}{2} \\
H_1^2 &= \frac{\gamma^2 H_1^2}{(\omega_0 - \omega)^2 + \gamma^2 H_1^2} \sin^2\left(\gamma H_{\text{eff}} \frac{t}{2}\right)
\end{aligned} \tag{9.27}$$

Now consider the special case $\omega = \gamma H_0 \equiv \omega_0$ then $P_{-1/2} = \sin^2[\gamma H_1(t/2)]$ since $H_{\text{eff}} = H_1$ (see Fig. 9.4). This result describes magnetic resonance and indicates that the magnetic moment oscillates between the z and $-z$ directions.

At equilibrium, the relative population of each Zeeman level is given by a Boltzmann distribution $N_j/N_i = \exp - \Delta E_{ij}/kT$ where N_j is the population of the upper level and N_i is the population of the lower level. For the nuclear case where $\Delta m = +1$,

$$\frac{N_j}{N_i} = \exp - \frac{g_N \mu_N H_0}{kT} \approx 1 - \frac{g_N \mu_N H_0}{kT} \tag{9.28}$$

For $I = 1/2$ system at room temperature in

$$H_0 = 10\ kG$$

$g_N \mu_N H_0/kT \approx 7 \times 10^{-6}$. In the analogous electron case $N_j/N_i = \exp(-g\mu_B H_0/kT) \approx 1 - g\mu_B H_0/kT$ and $g\mu_B H_0/kT \approx 5 \times 10^{-3}$ for a $S = 1/2$ system at room temperature in H_0 of 10 kG. Thus a net energy absorption will occur during the normal magnetic resonance experiment.

4. Quantum Mechanical Behavior [79–81]

A quantum mechanical description of magnetic resonance is required to explain coupled spin systems which are not handled by the classical theory. This analysis will be restricted to a nuclear spin of ½ and begins with the Schrödinger equation

$$i\hbar \frac{\partial \psi}{\partial t} = \mathcal{H}\psi \tag{9.29}$$

where ψ is the wavefunction and $\mathcal{H}$ is an operator representing the energy of the system. If $\mathcal{H}$ is time-independent, then

$$\psi(t) = u \exp\left(-\frac{iEt}{\hbar}\right) \tag{9.30}$$

where the wave functions u are time-independent, and $\mathcal{H}_0 u = Eu$ where E is the system energy and $\mathcal{H}_0$ represents the time-independent Hamiltonian.

The Hamiltonian of a nucleus of $I = 1/2$ placed in a magnetic field H is

$$\mathcal{H}_0 = -\gamma\hbar \vec{I} \cdot \vec{H} \tag{9.31}$$

and if $H_0 \| z$, then $\mathcal{H}_0 = -\gamma\hbar H_z I_z$. The energy, E_m, in a magnetic field is given by

$$E_m = -\gamma\hbar H_z m \tag{9.32}$$

where $m = \pm 1/2$. The two states correspond to spins parallel and antiparallel to H_0. The Bohr condition gives the natural frequency of the system

$$\frac{E_{-1/2} - E_{+1/2}}{\hbar} = \gamma H = \omega_0 \tag{9.33}$$

Application of an oscillating field requires the solution to the time dependent Schrödinger equation which can be treated as the sum of a time-independent Hamiltonian $\mathcal{H}_0$ and a time-dependent Hamiltonian $\mathcal{H}_1$. The time-independent solution is $\mathcal{H}_0 u_n = E_n u_n$. The wave function $\psi(t)$ can be expanded using time-dependent coefficients $a_n(t)$ in conjunction with the stationary wave functions, u_n,

$$\psi(t) = \sum_n a_n(t) u_n \exp\left(-\frac{iE_n t}{\hbar}\right) \tag{9.34}$$

Then the time-dependent Schrödinger equation becomes

$$\begin{aligned}
i\hbar \sum_n \frac{\partial a_n}{\partial t} u_n \exp\left(-\frac{iE_n t}{\hbar}\right) \\
= \sum_n a_n \mathcal{H}_1 u_n \exp\left(-\frac{iE_n t}{\hbar}\right)
\end{aligned} \tag{9.35}$$

Multiplying equation (9.35) by u_m and integrating gives

$$\frac{\partial a_m}{\partial t} = \frac{1}{i\hbar} \sum_n a_n \mathcal{H}_{mn} \exp(i\omega_{mn} t) \tag{9.36}$$

where $\quad \omega_{mn} = \dfrac{E_m - E_n}{\hbar}$

and $\quad \mathcal{H}_{mn} = (u_m^* | \mathcal{H}_1 | u_n)$

The time-dependent part of the Hamiltonian is given by

$$\begin{aligned}
\mathcal{H}_1 &= -\hbar\gamma H_1(I_x \cos\omega t - I_y \sin\omega t) \\
&= -\tfrac{1}{2}\hbar\gamma H_1[I_+ \exp i\omega t + I_- \exp(-i\omega t)]
\end{aligned} \tag{9.37}$$

where $I_\pm$, the raising and lowering operators, are $I_x \pm iI_y$. This equation corresponds to the oscillating field of magnitude H_1 rotating with frequency

ω in the xy plane.

Then
$$\frac{\partial a_{1/2}}{\partial t} = \tfrac{1}{2} i\gamma H_1 a_{-1/2} \exp\,[i(\omega_0 - \omega)t]$$
and
$$\frac{\partial a_{-1/2}}{\partial t} = \tfrac{1}{2} i\gamma H_1 a_{1/2} \exp\,[-i(\omega_0 - \omega)t] \tag{9.38}$$

Taking the second derivative, the differential equation involving $a_{-1/2}$ is obtained,

$$\frac{\partial^2 a_{-1/2}}{\partial t^2} + i(\omega_0 - \omega)\frac{\partial a_{-1/2}}{\partial t} + \frac{1}{4}\gamma^2 H_1{}^2 a_{-1/2} = 0$$

The solution of this equation is

$$a_{-1/2} = A \exp\,(iP_1 t) + B \exp\,(iP_2 t) \tag{9.39}$$

where P_1 and P_2 are determined from

$$-P^2 - (\omega_0 - \omega)P + \tfrac{1}{4}\gamma^2 H_1{}^2 = 0$$

yielding

$$P_1 = -\tfrac{1}{2}(\omega_0 - \omega) + \tfrac{1}{2}[(\omega_0 - \omega)^2 + \gamma^2 H_1{}^2]^{1/2}$$

and

$$P_2 = -\tfrac{1}{2}(\omega_0 - \omega) - \tfrac{1}{2}[(\omega_0 - \omega)^2 + \gamma^2 H_1{}^2]^{1/2}$$

For the conditions $t = 0$ and $a_{-1/2} = 0$, $B = -A$; hence

$$a_{-1/2} = A \exp\left[-\frac{i(\omega_0 - \omega)t}{2}\right] \sin\left(\gamma H_{\text{eff}} \frac{t}{2}\right) \tag{9.40}$$

where

$$\gamma H_{\text{eff}} = [(\omega_0 - \omega)^2 + \gamma^2 H_1{}^2]^{1/2}$$

The normalization condition

$$|a_{1/2}|^2 + |a_{-1/2}|^2 = 1$$

determines A. Since

$$|a_{1/2}|^2 = a_{1/2}{}^* a_{1/2} = \frac{|A|^2}{\gamma^2 H_1{}^2}\left[(\omega_0 - \omega)^2 + \gamma^2 H_1{}^2 - \gamma^2 H_1{}^2 \sin^2\left(\gamma H_{\text{eff}} \frac{t}{2}\right)\right]^{1/2}$$

then
$$A = \frac{\gamma H_1{}^2}{[(\omega_0 - \omega)^2 + \gamma^2 H_1{}^2]^{1/2}}$$
and

$$a_{-1/2} = \frac{\gamma H_1{}^2 \exp\left[-i(\omega_0 - \omega)\frac{t}{2}\right] \sin\left(\gamma H_{\text{eff}} \frac{t}{2}\right)}{[(\omega_0 - \omega)^2 + \gamma^2 H_1{}^2]^{1/2}} \tag{9.41}$$

Now the total probability, $P_{-1/2}$, of finding the nucleus in the $-\tfrac{1}{2}$ level is $|a_{-1/2}|^2 = a_{-1/2}{}^* a_{-1/2}$,

$$P_{-1/2} = \frac{\gamma^2 H_1{}^2 \sin^2\left(\gamma H_{\text{eff}} \frac{t}{2}\right)}{(\omega_0 - \omega)^2 + \gamma^2 H_1{}^2} \tag{9.42}$$

This equation is identical with the classical solution [Eq. (9.27)]. This development can be extended to higher-order spin systems.

The transition probability between two energy states m and n is given by $P_{mn} \propto |(\psi_m|I_x|\psi_n)|^2$ which is nonzero only when $m = n \pm 1$; thus the first-order selection rule for nuclear levels is $\Delta m = \pm 1$ [38, 79, 81, 82]. The first-order selection rule for many electronic Zeeman transitions is also $\Delta m = \pm 1$; however, other effects, e.g., crystal electric field, sufficiently mix the wave functions of the levels, thereby permitting $\Delta m > \pm 1$ transitions to occur.

5. Spin Lattices and Spin-Spin Relaxation

Both the classical and quantum mechanical solutions for magnetic resonance predict sharp absorption and emission lines. Actual lines are broadened by either homogeneous or inhomogeneous mechanisms. Homogeneous broadening mechanisms include:

1. Dipole-dipole interactions between like spins
2. Spin-lattice relaxation
3. Motional narrowing by local field fluctuations
4. Spin interactions with the exciting field
5. Diffusion of excitation through the spin system.

Examples of inhomogeneous broadening mechanisms are:

1. ESR hyperfine interactions
2. Anisotropic broadening
3. Dipolar and electric quadropolar coupling to unlike spins

The homogeneous broadening mechanisms (1) and (2) act as magnetization damping terms, and their concepts in terms of a nuclear spin system are presented below.

Spin-lattice Relaxation. Nuclear spins interact with their surroundings, but the magnitude of the interaction usually permits distinguishing between the temperature of the spin system and the lattice (bath) temperature. Absorption of energy at resonance tends to equalize the populations of the two levels involved, but heat flow from the spin system to the lattice via lattice vibrations (phonons) restores thermal equilibrium following resonance.

The Boltzmann distribution of a spin system in a magnetic field, H_0, is characterized by a spin temperature, T_s [83, 84] such that

$$\frac{N_n}{N_m} = \exp\left(-\frac{E_m - E_n}{kT_s}\right) \tag{9.43}$$

where N_n and N_m are the populations of the nth and mth nuclear Zeeman levels. The magnetization, M_0, at thermal equilibrium as given by Curie's law

$$M_0 = \frac{CH_0}{T_l} \tag{9.44}$$

where $C = N[\gamma^2\hbar^2 I(I + 1)/3k]$ defines the lattice temperature, T_l. When $T_s \neq T_l$ a semiequilibrium population distribution occurs where the spin system has internal equilibrium, but it is not in thermal equilibrium with the lattice; hence $M = CH_0/T_s$. A positive spin temperature occurs when M_z is parallel to H_0 and the level populations decrease with increasing energy. When M_z is antiparallel to H_0, the spin temperature is negative, indicating that the populations of the Zeeman levels increase with increasing energy. This situation is possible since antiparallel magnetization gives the state of greatest internal energy to the spin system. In many instances, the nuclear magnetization growth or decay rate is a single exponential and the time constant, T_1, is termed the spin-lattice relaxation time. Thus

$$\left(\frac{dM_z}{dt}\right) = \left[\frac{(M_0 - M_z)}{T_1}\right]$$

or in terms of the spin temperature

$$\frac{d}{dt}\left(\frac{1}{T_s}\right) = \frac{1}{T_1}\left(\frac{1}{T_l} - \frac{1}{T_s}\right). \tag{9.45}$$

The spin-lattice relaxation process can be described by an increase or decrease of a spin system temperature as thermal equilibrium is approached.

Another way of developing the spin-lattice relaxation involves the rate of change of the population difference between two Zeeman levels. Consider a nuclear spin of ½ in a magnetic field H_0. Let W_- represent the probability of $- \rightarrow +$ transition, W_+ is the probability of $+ \rightarrow -$ transition, and N_-, N_+ are the spin populations of the $m_I = -\frac{1}{2}$ and $m_I = \frac{1}{2}$ respectively; then according to the principle of detail balancing $N_+W_+ = N_-W_-$. The level populations N_+, N_- are given by the Boltzmann distribution

$$\frac{N_-}{N_+} = \exp -\frac{\Delta E}{kT} \tag{9.46}$$

Therefore,

$$\frac{W_+}{W_-} = \exp -\frac{\Delta E}{kT} \tag{9.47}$$

or the probability of emission exceeds the absorption probability. The population difference, n, is simply $(N_+ - N_-)$ and its rate of change is given by

$$\frac{dn}{dt} = 2N_-W_- - 2N_+W_+$$

$$= 2W_+\left[N_-\left(1 + \frac{\gamma\hbar H_0}{kT}\right) - N_+\right] \tag{9.48}$$

Then to first order

$$\frac{\gamma\hbar H_0}{kT} N_- \approx \frac{N_0}{2}\frac{\gamma\hbar H_0}{kT} \equiv n_0 \tag{9.49}$$

where n_0 is the population difference when thermal equilibrium between the lattice and spin system exists. Now

$$\frac{dn}{dt} = 2W_+(n_0 - n) \tag{9.50}$$

which has the solution

$$n_0 - n = (n_0 - n_i) \exp(-2W_+t) \tag{9.51}$$

where n_i is the initial value for n. The time constant $(2W)^{-1}$ is defined as the spin-lattice relaxation time, T_1. Thermal or longitudinal relaxation have been used interchangeably with spin-lattice relaxation.

The spin-lattice relaxation time can be written in terms of the magnetization by multiplying (9.51) by μ,

$$\mu\frac{dn}{dt} = 2W_+\mu\left[N_-\left(1 + \frac{\mu H_0}{kT}\right) - N_+\right] \tag{9.52}$$

Since $\mu(N_+ - N_-) = \mu n = M_z$

and $(N_+ + N_-)\dfrac{\mu^2 H_0}{kT} = M_0$

then the fundamental equation results,

$$\frac{dM_z}{dt} = 2W(M_0 - M_z) \tag{9.53}$$

where $T_1 = (2W)^{-1}$ and W is the transition probability for absorption.

Spin-Spin Relaxation. Each magnetic moment interacts with the local magnetic field,

$$\left[H_l \simeq \frac{\mu}{r^3}(3\cos^2\theta - 1)\right] \tag{9.54}$$

produced by neighboring nuclei. For hydrogen ($\mu = 1.41 \times 10^{-23}$ cm³ G) this local field is 14 gauss for a nearest neighbor distance of 1 Å. In the presence of an external magnetic field, H_z, the magnetic moment of nucleus f in a solid directed parallel to H_z will be constant if the time-dependent effects produced by precessing neighbor moments g are ignored. The mean-square local field at the f nucleus due to the remaining nuclei is

$$<\sum_g (H_{zg}^2)> = \sum_g \frac{<\mu_{zg}^2>(1 - 3\cos^2\theta_{fg})^2}{r^6_{fg}} \tag{9.55}$$

If all g nuclei are alike but differ from nucleus f,

$$<\mu_{zg}^2> = \tfrac{1}{3}I(I+1)\gamma^2\hbar^2$$

and
$$<(\Delta H)^2> = <(H - H_z)^2>$$
$$= \tfrac{1}{3}I(I+1)\gamma^2\hbar^2 \sum_g \frac{(1 - 3\cos^2\theta_{fg})^2}{r^6_{fg}} \tag{9.56}$$

when nuclei g and nucleus f have the same moment. Then Eq. (9.56) must be multiplied by 9/4 since a nucleus g precessing in the field, H_z, gives a resonant time-dependent field at nucleus f and vice versa, with the result that simultaneous spin flips, $\Delta m_f = \pm 1$ and $\Delta m_g = \mp 1$, occur. This moment reorientation (flip-flop) conserves total energy and limits the lifetime of the spin in an energy level, thereby contributing to the line broadening.

The precession frequency for $<H_l>$ from the Larmor equation is

$$\Delta\omega = \frac{\mu <H_l>}{I\hbar}$$

If the nuclear moments are precessing in phase at a time t, then the time $\Delta\omega^{-1}$ is required for the moments to get out of phase. These local field effects limit the phase memory lifetime of a spin state. Hence, the spin-spin or transverse relaxation time is defined as the characteristic time, T_2, required for the precessing spins to lose phase. Since local field variations give a range of absorption frequencies, a line width of the order of H_{local} is obtained. Usually the line shape function $g(\omega)$ is defined such that $T_2 = \pi g(\omega)_{\max}$.

6. Bloch Phenomenological Equations

Bloch [86, 87] introduced a set of phenomenological equations which describe the interaction of the macroscopic magnetization with a linear oscillating field. These classical equations are rigorous for many experimental conditions e.g., transient effects [88], spin echoes [89], and other time-dependent phenomena in both nuclear and electron magnetic resonance.

The rate of change of the components of nuclear magnetization in a static magnetic field, H_0, applied in the z direction and a linear oscillating field, H_1, applied perpendicular to H_0 are

$$\begin{aligned}\frac{dM_x}{dt} &= \gamma(M_yH_z - M_zH_y)\\ \frac{dM_y}{dt} &= \gamma(M_zH_x - M_xH_z)\\ \frac{dM_z}{dt} &= \gamma(M_xH_y - M_yH_x)\end{aligned} \tag{9.57}$$

The components of the magnetic field are

$$\begin{aligned}H_z &= H_0\\ H_x &= H_1 \cos \omega t\\ H_y &= H_1 \sin \omega t\end{aligned}$$

Hence

$$\begin{aligned}\frac{dM_x}{dt} &= \gamma(M_yH_0 + M_zH_1 \sin \omega t)\\ \frac{dM_y}{dt} &= \gamma(M_zH_1 \cos \omega t - M_xH_0)\\ \frac{dM_z}{dt} &= -\gamma(M_xH_1 \sin \omega t + M_yH_1 \cos \omega t)\end{aligned} \tag{9.58}$$

Bloch introduced two damping terms, the spin-lattice relaxation time T_1 and the spin-spin relaxation time T_2. The spin-spin relaxation causes the M_x and M_y components to decay whereas spin-lattice relaxation affects the M_z component. Thus the complete Bloch equations are

$$\begin{aligned}\frac{dM_x}{dt} &= \gamma(M_yH_0 + M_zH_1 \sin \omega t) - \frac{M_x}{T_2}\\ \frac{dM_y}{dt} &= \gamma(M_zH_1 \cos \omega t - M_xH_0) - \frac{M_y}{T_2}\\ \frac{dM_z}{dt} &= -\gamma(M_xH_1 \sin \omega t + M_yH_1 \cos \omega t) - \frac{M_z - M_0}{T_1}\end{aligned} \tag{9.59}$$

These equations take a simpler form when the coordinate reference system rotating with the applied field H_1 is employed. In this system u and v are defined as the components of M parallel (in phase) and perpendicular (out of phase) to the direction of H_1. Using the relations between the components

$$\begin{aligned}u &= M_x \cos \omega t - M_y \sin \omega t\\ v &= -M_x \sin \omega t - M_y \cos \omega t\end{aligned} \tag{9.60}$$

the Bloch equations in the rotating frame become

$$\begin{aligned}\frac{du}{dt} + \frac{u}{T_2} + (\omega_0 - \omega) &= 0\\ \frac{dv}{dt} + \frac{v}{T_2} - (\omega_0 - \omega)u + \gamma H_1 M_z &= 0\\ \frac{dM_z}{dt} + \frac{M_z - M_0}{T_1} - \gamma H_1 v &= 0\end{aligned} \tag{9.61}$$

The steady-state solutions for these equations are

$$\begin{aligned}u &= M_0 \frac{\gamma H_1 T_2^2(\omega_0 - \omega)}{1 + T_2^2(\omega_0 - \omega)^2 + \gamma^2 H_1^2 T_1 T_2}\\ v &= -M_0 \frac{\gamma H_1 T_2}{1 + T_2^2(\omega_0 - \omega)^2 + \gamma^2 H_1^2 T_1 T_2}\\ M_z &= M_0 \frac{1 + T_2^2(\omega_0 - \omega)^2}{1 + T_2^2(\omega_0 - \omega)^2 + \gamma^2 H_1^2 T_1 T_2}\\ M_x &= \tfrac{1}{2} M_0 \gamma T_2^2 \frac{T_2(\omega_0 - \omega)2H_1 \cos \omega t + 2H_1 \sin \omega t}{1 + T_2^2(\omega_0 - \omega)^2 + \gamma^2 H_1^2 T_1 T_2}\\ M_y &= \tfrac{1}{2} M_0 \gamma T_2^2 \frac{2H_1 \cos \omega t - T_2(\omega_0 - \omega)2H_1 \sin \omega t}{1 + T_2^2(\omega_0 - \omega)^2 + \gamma^2 H_1^2 T_1 T_2}\end{aligned} \tag{9.62}$$

Since both M_x and M_y contain a term in phase with $H_1(2H_1 \cos \omega t)$ and a term out of phase with $H_1(2H_1 \sin \omega t)$, Bloch suggested the use of complex susceptibility, $\chi = \chi' - i\chi''$.

Thus

$$\begin{aligned}M_x &= \chi'(2H_1 \cos \omega t) + \chi''(2H_1 \sin \omega t)\\ M_y &= \chi'(2H_1 \cos \omega t) - \chi''(2H_1 \sin \omega t)\end{aligned} \tag{9.63}$$

where

$$\begin{aligned}\chi' &= \tfrac{1}{2}\chi_0 T_2 \omega_0 \frac{T_2(\omega_0 - \omega)}{1 + T_2^2(\omega_0 - \omega)^2 + \gamma^2 H_1^2 T_1 T_2}\\ \chi'' &= \tfrac{1}{2}\chi_0 T_2 \omega_0 \frac{1}{1 + T_2^2(\omega_0 - \omega)^2 + \gamma^2 H_1^2 T_1 T_2}\end{aligned}$$

Figure 9.5 shows the dispersion χ' and absorption χ'' components as functions of the dimensionless product

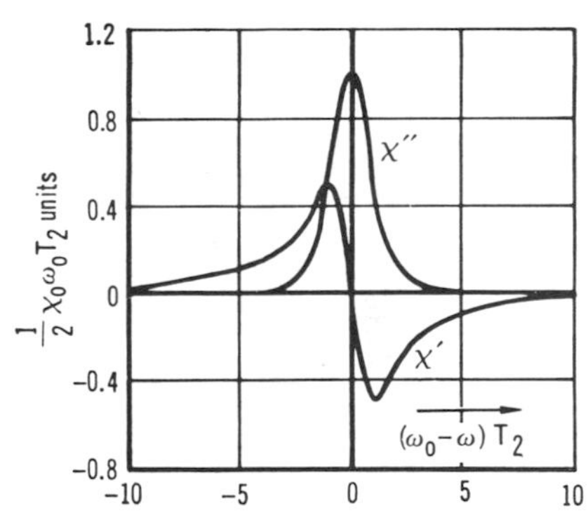

FIG. 9.5. Absorption (χ'') and dispersion (χ') Lorentzian line shapes.

$T_2(\omega_0 - \omega)$. If the spin temperature is equal to the lattice temperature, then χ'' can be written in terms of the line function $g(\omega)$ as

$$\chi'' = \tfrac{1}{4}\chi_0 \omega g(\omega) \tag{9.64}$$

Thus

$$g(\omega) = \frac{2T_2}{1 + T_2(\omega_0 - \omega)^2}$$

which is the Lorentzian line shape characteristic of a damped harmonic oscillator. Line shapes are not always approximated by a Lorentzian-type curve and hence limits the Bloch formalism. The spin-lattice relaxation theory [90] predicts a Lorentzian line shape but the spin-spin relaxation theory [91] indicates a Gaussian line shape.

Homogeneous broadening yields Lorentzian line shapes whereas Gaussian line shapes occur when inhomogeneous broadening is involved. Some magnetic resonance line shapes can be fitted by a combination of the Lorentzian and Gaussian functions.

The rate of energy absorption per unit volume, A, is given by

$$A = \frac{\omega}{2\pi} \int_0^{\frac{2\pi}{\omega}} 2H_1 \cos \omega t \left(\frac{dM_x}{dt}\right) dt$$

$$= \frac{\omega}{2\pi} \int_0^{\frac{2\pi}{\omega}} 4H_1{}^2\omega \cos \omega t[-\chi' \sin \omega t + \chi'' \cos \omega t]\, dt$$

$$= 2\omega H_1{}^2\chi'' \qquad (9.65)$$

This equation indicates that energy absorbed by the spin system from the oscillating field is proportional to the χ'' magnetization component.

The conditions under which the Bloch equations apply [87] are:

1. The effect of all other spins upon a given spin is equivalent to that of a "bath" in thermodynamic equilibrium.
2. The "bath" temperature remains constant.
3. γH_1, $T_1{}^{-1}$, $T_2{}^{-1} \ll \gamma H_0$ and γH_{local}.
4. The spin-bath interaction can be expanded in multipoles.
5. When two unpaired spins occur, $kT \gg \hbar\omega_0$, or for more than two unpaired spins, the quadrupole coupling, eqQ, $\ll \hbar\omega_0$.

The steady-state solution of the Bloch equations assumes that thermal equilibrium exists between the magnetization and the oscillating field. Thus the rate of passage through resonance influences the resulting line shape. The slow-passage condition $(d/dt)[(H_0 - H)/H_1] \ll (T_1T_2)^{-1/2}$ is required for the steady-state solution and the true line shape is observed when $dH/dt \ll \gamma(\Delta H)^2$, where ΔH is the line width at half maximum. The adiabatic rapid-passage condition, where $\gamma H_1 \gg (d/dt)[(H_0 - H)/H_1] \gg (T_1T_2)^{-1/2}$, may be employed to observe weak signals or long relaxation times. Other transient response effects which occur during resonance passage have been described in detail [89, 92–96].

When $\gamma^2H_1{}^2T_1T_2 \approx 1$, the resonance width increases and the resonance height decreases; at sufficiently high amplitudes of H_1 no resonance absorption is observed because the energy-level populations have been equalized or the resonance is saturated. The term $1 + \gamma^2H_1{}^2T_1T_2$ is called the saturation factor. The absorption mode χ'' saturates rapidly when $H_1 \geq (\gamma T_1T_2)^{-1/2}$, but the dispersion mode requires higher H_1 values for saturation [97].

7. Magnetic Resonance Spectrometers

The essential components of a magnetic resonance spectrometer are (1) a magnet to provide H_0, (2) a generator at the appropriate frequency to provide H_1, and (3) a detector to indicate the resonant condition.

Permanent magnets [98, 99], electromagnets [43], and superconducting solenoids [100] have been used in magnetic resonance spectrometers and the particular choice depends on the intended application. High-resolution NMR spectral measurements requiring extremely high homogeneity can be made with any of the above magnets; however, the electromagnet is preferred for ESR measurements where large field variations are desired. Significant signal-strength enhancement can be obtained with higher magnetic fields; therefore, the high field strengths which can be attained with superconducting solenoids may be used for low abundant nuclides or low spin concentrations. To traverse the resonance line, either frequency sweeping [101] or magnetic-field sweeping [43] as a function of time is required.

When signal-to-noise improvement is required, e.g., wide lines or low spin concentrations, a small sinusoidal modulation, ω_m, is applied to the magnetic field [102]; then

$$H = H_0 + H_m \cos \omega t$$

For wide NMR lines $\omega_m \approx 10$–10^3 hertz whereas 10^3–10^5 hertz is employed in EPR spectroscopy. In conjunction with magnetic-field modulation, a narrow-band amplifier and phase-sensitive detector are used. In this case, the first derivative, $d\chi'/dH_0$ or $d\chi''/dH_0$, is obtained.

NMR spectrometers can be divided into three major types: the double-coil type, the bridge type, and the marginal oscillator. The double-coil NMR spectrometer [30] shown in Fig. 9.6 is widely used.

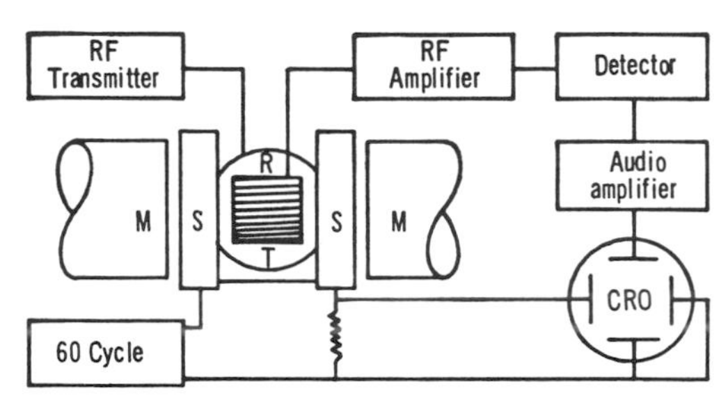

FIG. 9.6. Schematic diagram of the double-coil spectrometer used by Bloch, Hansen, and Packard [30]. M = magnet, S = sweep coils, T = transmitter coil, R = receiver coil.

In this spectrometer, the sample is placed inside the receiver coil with its axis perpendicular to both the transmitter coil and to the magnetic-field direction. Control of the radio-frequency leakage from the transmitter coil permits selection of either the absorption or dispersion magnetization component. Off resonance, no voltage is induced in the receiver coil, but at resonance the radiation emitted by the excited nuclei returning to lower energy levels induces a voltage E_r in the n turn receiver coil,

$$E_r = 4\pi An \frac{dM_y}{dt}$$

where A is the sample cross section. This voltage change is diode-detected and amplified for oscilloscope or recorder presentation.

In the bridge-type [29] NMR spectrometer a single inductance coil in a tuned LC circuit serves as both the transmitter and receiver coil. Off resonance, the radio-frequency voltage arriving at the amplifier is balanced out with an r-f bridge. Susceptibility component selection is made by unbalancing the bridge in the proper phase. This bridge unbalance must be significantly larger than the resonance signal for pure mode selection. At resonance the energy absorbed from the tuned circuit by the sample further unbalances the bridge and gives a voltage identical to that of the double-coil spectrometer. A double-bridge NMR spectrometer [103] providing both amplitude and phase balance minimizes fluctuations associated with large H_1 levels and gives better mode selection.

The sample coil of the marginal oscillator spectrometer [104–106] forms part of a tuned circuit which controls the oscillator frequency by regenerative feedback. In the marginal condition the feedback circuit time constant is sufficient to maintain nearly constant oscillation amplitude; however, at resonance, the long time constant does not permit the generator to follow the energy absorption. The great advantages of this spectrometer are pure absorption mode detection, high sensitivity, and the ease of frequency changes.

ESR spectrometers have operated in the frequency range 1 MHz [107, 108] to 158.5 GHz [109]. However, the usual commercial spectrometers operate at X band (8.2–12.4 GHz), K band (18–26.5 GHz), and K_a band (26.5–40 GHz) or Q band (33–50 GHz) because of increased sensitivity and higher resolution at microwave frequencies.

Two major types [12, 15, 16, 110] of ESR spectrometers, transmission and reflection, are used. In the transmission spectrometer, the sample is placed in the waveguide between the microwave generator (usually a reflex klystron) and the detector. At resonance, the microwave energy absorption is determined with a simple silicon crystal diode or bolometer.

A block diagram of a reflection-type ESR spectrometer is shown in Fig. 9.7. Microwaves pass from the stabilized klystron, through arm 1 of the magic tee to a high quality factor reflection cavity in which the sample is located. By mechanically and electronically tuning the frequency of the klystron, its operating mode can be matched to the resonant frequency of the reflection cavity. After being properly tuned to the cavity, the klystron is frequency- or phase-stabilized at the cavity resonant frequency. Off resonance the cavity will have a representative quality factor, Q, which is given by

$$Q = \frac{\omega[\text{total stored energy}]}{\text{power lost per cycle}} = \frac{\omega L}{R}$$

At resonance additional power is lost to the sample resulting in a decreased Q. This decrease, dQ, is given by

$$dQ = -4\pi\eta Q^2\chi''$$

where η is the filling factor which is the fraction of the coil volume occupied by the sample. Three general types of detectors [111–113] can be used: (1) bolometer, (2) crystal diode, and (3) superheterodyne, in the third arm of the magic tee to detect the resonance. Usually additional amplification is provided before the resonance is recorded.

The experimental methods for measuring relaxation processes are basically the same for both NMR [40, 44, 50, 56, 114, 115] and EPR [116–120]. These methods can be divided into saturation methods and pulse methods. The saturation methods can be divided into three techniques: the direct, the progressive or steady-state saturation, and the polarization reversal. In the direct method [44] H_1 is increased to produce saturation and then suddenly it is reduced to a small value. The observed signal recovery follows a single exponential $[1 - \exp(-t/T_1)]$ in many cases.

The progressive-saturation technique [44] depends on the saturation factor $(1 + \gamma^2 H_1{}^2 T_1 T_2)^{-1}$. At a low H_1 value T_2 is evaluated from the line width. The signal amplitude as a function of H_1 decreases

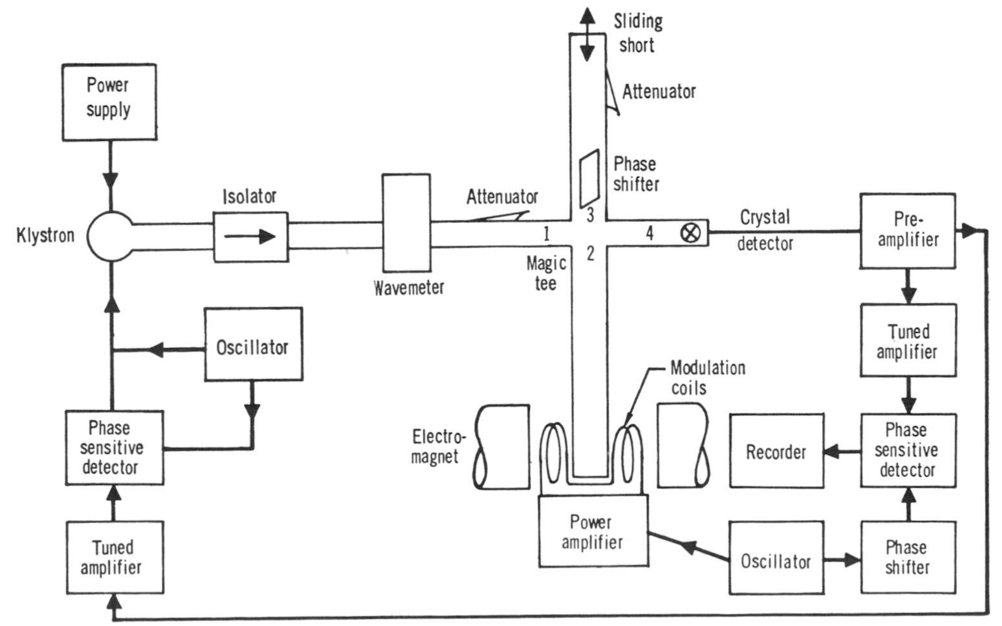

FIG. 9.7. Bloch diagram of a typical reflection cavity electron spin resonance spectrometer.

rapidly when $\gamma^2 H_1^2 T_1 T_2 = 1$. Then T_1 can be evaluated from T_2 and the value of H_1 at the sample.

The polarization-reversal method [114–121] employs an adiabatic fast passage through resonance plus a large H_1 field to reverse M_z from the z to $-z$ direction. At equilibrium the return signal $(-d\chi''/dH)$ will have the opposite sign and the same amplitude as the forward signal $(d\chi''/dH)$. The signal amplitude depends on the time above and below resonance and is proportional to the magnetization at each passage. Thus T_1 can be evaluated from the signal amplitudes at two different sweep frequencies or by using unequal time intervals above and below resonance.

Two pulse methods, often referred to as the incoherent and coherent techniques, are used for accurate relaxation-time measurements. Incoherent techniques usually refer to relaxation measurements which involve magnetization recovery from single high-power H_1 pulses. Most electron spin relaxation measurements are performed with the saturation recovery method [116, 117] because of experimental difficulties associated with a microwave spin echo spectrometer [119, 120].

The spin echo method [89] is an example of a coherent technique for relaxation measurements. In this method the magnetization is rotated away from the direction of static field with a timed series of pulses of amplitude and duration such that

$$\gamma H_1 t_\omega = \left(\frac{\pi}{2}\right) \text{ (90° pulse)}$$

[89] or a combination of a 90° pulse and a 180° pulse [122]. In the latter case a signal (the echo) will occur at time 2τ where τ is the time interval between pulses. The echo amplitude, A, is given by

$$A = \exp\left(-\frac{2\tau}{T_2} - k\frac{(2\tau)^3}{3}\right)$$

where $k = \frac{1}{4}\gamma^2 G^2 D$, with D being the diffusion coefficient and G being the magnetic-field gradient. T_1 can be obtained from a 180° − 90° sequence [122] or a series of 90° pulses spaced at variable time intervals [89, 123].

8. Characteristics of NMR Spectra

Solids. Nuclear resonances of solids are both homogeneously and inhomogeneously broadened. The major line broadening mechanisms are:

1. Dipole-dipole
2. Electric quadrupole
3. Electron-coupled spin-spin interactions

For $I = \frac{1}{2}$, the dipole-dipole coupling gives an H_z component at each nucleus of $H_0 \pm H_l$, where H_l is the local magnetic field [Eq. (9.66)] caused by neighboring nuclei. This magnetic [36, 124] interaction has the form

$$\mathcal{H}_d = \sum_{j>k} \frac{[\vec{\mu}_j \cdot \vec{\mu}_k - 3(\vec{\mu}_j \cdot \hat{r}_{jk})(\vec{\mu}_k \cdot \hat{r}_{jk})]}{r^3{}_{jk}} \quad (9.66)$$

and predicts that for a single crystal sample having isolated pairs of nuclei two equally intense resonance lines separated by $3\mu/r^3(3\cos^2\theta - 1)$ occur. This approach also applies to polycrystalline samples although averaging over all θ values causes appreciable resonance overlap. The prediction of line shapes for groups of three nuclei at the corners of an isosceles triangle [125] and a tetrahedral arrangement of nuclei [36, 47, 126, 127] has been experimentally verified.

When more complicated grouping of nuclei occur, the NMR line often is unresolved. Useful information can be obtained from this structureless line by the Van Vleck moment method [85] which expresses the internuclear distances in crystals in terms of the second and fourth moment. The general second moment for both like and unlike spins is

$$\begin{aligned} \langle(\Delta H)^2\rangle_{\text{ave}} = {} & \frac{3}{4}\gamma_i^4 \frac{I_i(I_i+1)}{N}\hbar^2 \sum_{i>j} (3\cos^2\theta_{ij} - 1)^2 r_{ij}^{-6} \\ & + \frac{1}{3}\frac{\gamma_i^2\gamma_j^2}{N} I_{j'}(I_{j'}+1)\hbar^2 \sum_{i,j'} (3\cos^2\theta_{ij'} - 1)^2 r_{ij}^{-6} \end{aligned} \quad (9.67)$$

Bond distances in crystals can be evaluated when one bond length of the crystal is unknown [36, 47, 127]. When an axial symmetric electric field gradient $\partial^2 V/\partial z^2$ due to nuclear electric quadrupolar coupling eQ, exists at the nucleus, the interaction energy is

$$\frac{eQ}{4I(2I-1)}\frac{\partial^2 V}{\partial z^2}\{3m^2 - I(I+1)\} \quad (9.68)$$

which must be added to the magnetic field nuclear moment energy [69, 73]. This electric field gradient introduces asymmetry in the nuclear Zeeman splittings, thereby allowing observation of $2I$ resonances instead of the single resonance characteristic of nuclei on which the quadrupole moment is relaxed.

The effective second moment in solids is reduced by motion of lattice points about an equilibrium position as well as molecular group vibration and rotation [20, 26, 46]. For free rotation a time average of all θ values [Eq. (9.67)] over all azimuthal angles is made. This averaging process increases the fourth moment but does not affect the second moment [128]. When the line width is much smaller than the rotation frequency the resonance tails become broadened beyond detection and an effective second-moment decrease results. Motional narrowing induced by increased temperature has been used to study phase transformations, tunneling, and hindered rotational barriers. In addition rapid rotation of solid samples at an axis inclined at $\cos^{-1} 1/\sqrt{3}$ to H_0 provides motional narrowing which permits resolution of nonequivalent site resonances [129–131]. The theory of motional narrowing in solids is directly applicable to the random Brownian motion of liquids [103].

Although rapid random motion averages the direct dipole-dipole coupling to zero, the resonance frequencies for identical nuclei located in chemically nonequivalent sites differ. Differences in the time average magnetic field at the nucleus are caused by

1. Interactions of the nucleus with the electrons of the same atom or molecular species
2. Nuclear interactions with other atoms or molecular species
3. Bulk susceptibility effects.

Thus these shielding mechanisms depend on the chemical environment of the nucleus, and hence resonance position differences between nonequivalent sites are called chemical shifts [132, 133]. The interaction Hamiltonian for a set of nuclei γ_i is given by

$$\mathcal{H}_0 = \sum_i \gamma_i H_i I_z(i) \tag{9.69}$$

Calculations of atomic shielding constants agree favorably with experimental resonance shifts [50, 134], but calculations of shielding constants for molecules require knowledge of the energies and wave functions of all excited electronic states which are not usually available [50]. Hence a relative chemical shift, δ, for molecules is used, where

$$\delta = \frac{H_S - H_R}{H_R} \times 10^6 \tag{9.70}$$

H_S is the magnetic field for the sample and H_R is the resonance field for a reference compound. Tables of these relative NMR chemical shifts for many nuclei have been presented [32, 35, 40–42, 50, 59, 63, 135, 136]. Chemical shifts as large as 10^4 parts per million have been measured for dimagnetic cobalt compounds but the usual range is 10–1,000 ppm.

The NMR spectrum of a molecule containing chemically nonequivalent groups will exhibit multiplet resonance structure caused by indirect spin-spin coupling transmitted between nuclei by the bonding electrons. This indirect coupling may be represented as an additional Hamiltonian involving scalar products of all pairs of nuclei,

$$\mathcal{H}_1 = \sum_{i<j} J_{ij} \vec{I}_i \cdot \vec{I}_j \tag{9.71}$$

where J_{ij} is the magnetic-field-independent spin-spin coupling constant. General procedures for the analysis of complex NMR spectra in terms of coupling constants and chemical shifts have been presented [40, 50, 63, 137–140].

Metals. Comparison of the Cu^{63} NMR resonances for Cu and CuCl shows the Cu resonance at a higher magnetic field [141, 142]. This shift is proportional to the field strength but the relative shift

$$\frac{H_{\text{metal}} - H_{\text{reference salt}}}{H_{\text{reference salt}}} = \frac{\Delta H}{H}$$

(the Knight shift) is independent of the field. Furthermore, Knight shifts do not depend on moments of different isotopes. They arise principally from the contact interaction between the nuclear magnetic moment and the spin magnetization field of the s conduction electrons. The shift is proportional to the spin magnetization density [143] or

$$\frac{\Delta H}{H} = \frac{8\pi}{3} \chi_v V_0 <|\psi(0)|^2>_F \tag{9.72}$$

where χ_v is the volume spin susceptibility

$$[\chi_v = 2\mu_B{}^2 N(E_F)$$

$N(E_F)$ is the density of states per unit energy interval at the Fermi surface], V_0 is the atomic volume and $<|\psi(0)|^2>_F$ is the probability density [47] at the nucleus for s conduction electrons of energies E_F averaged over the Fermi surface. Knight-shift measurements for pure solid and liquid metals [45], intermetallics [144], alloys (single-phase, multiple-phase, and ordered) [45] and superconductors [145] have provided useful information for characterization [45] of these materials.

Acoustic Wave—Spin Interaction [146–148]. One of the mechanisms responsible for spin-lattice relaxation of nuclear systems is the indirect or Raman interaction process which involves two high-frequency lattice vibrations. In this process one phonon is absorbed by the spin system and another phonon is emitted after de-exciting the spin system, The frequency difference between these phonons is $\pm\nu_0$ or $2\nu_0$ where ν_0 is the Larmor frequency. This inelastic phonon scattering process is highly probable since the frequencies, ν_L, of most of the thermal phonons are larger than the resonant frequency.

Application of a coherent acoustic phonon beam will modulate one of the internal interactions, thereby exchanging energy with the spin system, e.g., coupling 25.33 MHz acoustic waves to $KMnF_3$ causes $\Delta m = \pm 1$ transitions in the F^{19} NMR line [149].

9. Characteristics of EPR Spectra

EPR spectra are described by the spin Hamiltonian [4, 150–152] $\mathcal{H}_s$, which is essentially the system energy written in terms of the system parameters. The spin Hamiltonian can be divided into an electronic part $\mathcal{H}_e$ and a nuclear part $\mathcal{H}_n$. The electronic part $\mathcal{H}_e$ is given by

$$\mathcal{H}_e = \mu_B \vec{H} \cdot g \cdot \vec{S} + \vec{S} \cdot D \cdot \vec{S} \tag{9.73}$$

and the nuclear part $\mathcal{H}_n$ is given by

$$\mathcal{H}_n = \vec{S} \cdot T \cdot \vec{I} + \vec{I} \cdot P \cdot \vec{I} - g_N \mu_N \vec{H} \cdot \vec{I} \tag{9.74}$$

The first term of $\mathcal{H}_e$ describes the Zeeman splitting. The second term of $\mathcal{H}_e$ represents both the magnetic-dipole interaction between electrons and second order effects of spin orbit coupling. The first term of $\mathcal{H}_n$ is the hyperfine interaction and nuclear quadrupolar coupling is represented by the second term. The other terms of $\mathcal{H}_e$ and $\mathcal{H}_n$ represent the magnetic-field interactions with the electron and nucleus respectively.

For the case of a single electron with zero orbital angular momentum in a magnetic field,

$$\mathcal{H}_z = \mu_N g H S_z \tag{9.75}$$

with g being near g_0, the spin-only value. Quenching of orbital angular momentum is characteristic of most organic monoradicals [22–26, 153, 154]. When spin-orbit coupling is present, mixing of higher orbital states occurs, thereby causing an effective anisotropic magnetic moment. For this case the Zeeman interaction is written in terms of its components,

$$\mathcal{H}_z = \mu_B(g_x S_x H_x + g_y S_y H_y + g_z S_z H_z)$$

where g_x, g_y, and g_z are the principal values of the g tensor and S_x, S_y, and S_z refer to effective spin components. In the simple case of axial symmetry in a crystal only two g components, $g_\parallel = g_z$ (parallel component) and $g_\perp = g_x = g_y$ (perpendicular component) are required. The g value in a direction making an angle θ with the symmetry axis is

$$g = (g_\parallel^2 \cos^2 \theta + g_\perp^2 \sin^2 \theta)$$

Where spin-orbit coupling or crystal field effects are present, g values differing from g_0 are observed [13, 14, 17–19, 21–27].

Initial splittings of the effective spin levels of ions or biradicals in crystals often occur. The strong asymmetric electric field of the crystal interacting through the spin-orbit coupling is largely responsible for this zero field splitting. The second part of $\mathcal{H}_e$, the fine structure term, $\vec{S} \cdot D \cdot \vec{S}$, expresses this interaction. A twofold degeneracy always exists for a half-integral spin system in a crystal field (Kramer's rule), but for an integral spin system this level degeneracy may be resolved. For an axially symmetric crystal field

$$\vec{S} \cdot D \cdot \vec{S} = D[S_z^2 - \tfrac{1}{3}S(S+1)]$$

and for crystals with lower symmetry

$$\vec{S} \cdot D \cdot \vec{S} = D[S_z^2 - \tfrac{1}{3}S(S+1)] + E(S_x^2 - S_y^2)$$

Thus for $S = \tfrac{3}{2}$ the zero field splitting between $E_{\pm 3/2} - E_{\pm 1/2} = 2D$ for an axially symmetric crystal and $2(D^2 + 3E^2)^{1/2}$ for lower symmetry crystals; when $S = 1$ the axially symmetric zero field splitting, $E_{\pm 1} - E_0 = D$, whereas $E_{+1} - E_{-1} = 2E$ and

$$E_{-1} - E_0 = D - E$$

for positive D and E values when lower crystal symmetrics are involved. In the presence of both an applied magnetic field and a crystal field level crossing occurs and because of wave function mixing $\Delta M > \pm 1$ transitions are allowed.

The hyperfine term of $\mathcal{H}_n$ can be expressed as

$$\vec{S} \cdot T \cdot \vec{I} = AI_z S_z + B(I_x S_x + I_y S_y)$$

where A and B measure the hyperfine splitting parallel and perpendicular to the crystal axis. The effect of the hyperfine interaction is to split the electron resonance line into $2I + 1$ lines which to first order have the same separation. The resonance multiplicity serves to identify the coupling nucleus or nuclei, and information on the electron density in the molecule [26] is obtained from the separation.

The second term of $\mathcal{H}_n$ represents the nuclear electric quadrupole hyperfine interaction and to first order, it causes no change in the hyperfine resonance positions. However, it often is responsible for "forbidden" transitions which appear near the central resonance of the EPR spectrum. When small hyperfine interactions occur, the nuclear Zeeman term may give observable direct field effects by which the nuclear spin flips during an electron resonance [155].

Detailed procedural methods for deriving the spin Hamiltonian parameters from observed EPR spectral data have been developed [4, 5, 14–20, 25–27, 156], and the parameter values for various ions and biradicals have been listed [4–27].

10. Masers

Continuous high-gain microwave amplification by stimulated emission of radiation uses at least three unequally spaced Zeeman levels of a magnetically dilute paramagnetic ion $S \geq 1$ at 4.2°K [157]. In the three-level spin system the populations of the upper and lower energy levels are saturated by high-amplitude microwaves of the appropriate energy. When a weak microwave signal of energy equal to the separation between the upper and middle energy levels (or the middle and lower energy levels) is received by the hot spin system transitions from the higher to the lower energy level occur, thereby amplifying the triggering signal. Microwave signals of other energies do not affect the spin system. Since the separation between Zeeman levels is readily varied with the magnetic field, masers covering a wide frequency range are possible provided proper microwave or other saturating pumps are available. When the Zeeman level separation of the pumped levels approximates the zero field splitting, a decrease in the wave function mixing occurs which decreases the transition probability between the levels. Therefore, the frequency tunability of the maser crystal is limited. When the spin-lattice interaction between pump levels is strong (short T_1), these levels will not saturate and thus will limit the amplifier gain.

Two types of solid-state masers, the regenerative reflection cavity maser and the nonregenerative traveling-wave maser have been developed. In the cavity maser, the energy emitted by the maser crystal increases the microwave field in the cavity, thereby increasing the rate of emission from the crystal. Consequently, the maser gain bandwidth product ($G^{1/2}B$) is constant for the cavity maser. The energy emitted by a slow wave element of a traveling-wave maser increases the microwave field traveling through the structure and thus increases the rate of energy emission from succeeding active elements without reacting with the original element. Therefore, the gain of the traveling-wave maser is a function of its length and the $G^{1/2}B$ is not constant. The noise temperatures for masers are ~10–15°K and electronic gains of ~8 dB/in. are observed for traveling-wave masers, whereas $G^{1/2}B$ products of 40–250 MHz are obtained for cavity masers [158–162].

References

1. Zavoisky, E.: *J. Phys. U.S.S.R.*, **9**: 211, 245, 447 (1945).
2. Cummerow, R. L., and D. Halliday: *Phys. Rev.*, **70**: 433 (1946).
3. Penrose, R. P.: *Nature*, **163**: 992 (1949).

4. Bleaney, B., and K. W. H. Stevens: *Repts. Progr. Phys.*, **16**: 108 (1953).
5. Bowers, K. D., and J. Owen: *Repts. Progr. Phys.*, **18**: 304 (1955).
6. Wertz, J. E.: *Chem. Rev.*, **55**: 829 (1955).
7. Hutchison, C. A.: "Determination of Organic Structures by Physical Methods," Chap. VII, Academic, New York, 1955.
8. Ingram, D. J. E.: "Spectroscopy at Radio and Microwave Frequencies," Butterworth Scientific Publications, London, 1955.
9. Bagguley, D. M. S., and J. Owen: *Repts. Progr. Phys.*, **20**: 304 (1957).
10. Ingram, D. J. E.: "Free Radicals," Butterworth Scientific Publications, London, 1958.
11. Whiffen, D. H.: *Quart. Rev.*, **12**: 250 (1958).
12. Fraenkel, G. K.: "Physical Methods of Organic Chemistry," vol. 1, part IV, p. 2801, Interscience, New York, 1960.
13. Orton, J. W.: *Repts. Progr. Phys.*, **22**: 204 (1959).
14. Low, W.: "Paramagnetic Resonance in Solids," Supplement 2, Academic, New York, 1960.
15. Anderson, R. S.: "Methods of Experimental Physics: Molecular Physics," vol. 3, p. 441, Academic, New York, 1961.
16. Pake, G. E.: "Paramagnetic Resonance," W. A. Benjamin, New York, 1962.
17. Ludwig, G. W., and H. H. Woodbury: "Solid State Physics," vol. 13, p. 223, Academic, New York, 1962.
18. Yafet, Y.: "Solid State Physics," vol. 14, p. 2, Academic, New York, 1963.
19. Jarrett, H. J.: "Solid State Physics," vol. 14, p. 215, Academic, New York, 1963.
20. Slichter, C. P.: "Principles of Magnetic Resonance," Harper and Row, New York, 1963.
21. Low. W., ed: "Proceedings of First International Conference on Paramagnetic Resonance," vols. I and II, Academic, New York, 1963.
22. Carrington, A.: *Quart. Rev.*, (London) **17**: 67 (1963).
23. McDowell, C. A.: *Rev. Mod. Phys.*, **53**: 528 (1963).
24. Symons, M. C. R.: *Adv. Phys. Org. Chem.*, **1**: 284 Academic, New York, (1963).
25. Al'tshuler, S. A., and B. M. Kozyrev: "Electron Paramagnetic Resonance," Academic, New York, 1964. [See also Report FTD-TT-62-1086 AD 295 794, Air Force Systems Command (1963).]
26. O'Reilly, D. E., and J. H. Anderson: "Physics and Chemistry of the Organic Solid State," vol. 2, p. 121, Interscience, New York, 1965.
27. Low, W., and E. L. Offenbacher: "Solid State Physics," vol. 17, p. 135, Academic, New York, 1965.
28. Lasarev, B. E., and L. V. Schubnikov: *Physik. Z. U.S.S.R.*, **11**: 445 (1937).
29. Purcell, E. M., H. C. Torrey, and R. V. Pound: *Phys. Rev.*, **69**: 37 (1946).
30. Bloch, F., W. W. Hansen, and M. Packard: *Phys. Rev.*, **70**: 474 (1946).
31. Pake, G. E.: *Am. J. Phys.*, **18**: 438, 473 (1950).
32. Wertz, J. E.: *Chem. Rev.*, **55**: 829 (1955).
33. Grivet, P.: "La Resonance Paramagnetique Nucleaire," Centre National de la Recherche Scientifique, Paris, 1955.
34. Andrew, E. R.: "Nuclear Magnetic Resonance," Cambridge University Press, New York, 1955.
35. Gutowsky, H. S.: "Physical Methods in Chemical Analysis," vol. 3, p. 304, Academic, New York, 1956.
36. Pake, G. E.: "Solid State Physics," vol. 2, p. 1, Academic, New York, 1956.
37. Knight, W. D.: "Solid State Physics," vol. 2, p. 93, Academic, New York, 1956.
38. Saha, A. K., and T. P. Das: "Theory and Applications of Nuclear Induction," Saha Institute of Nuclear Physics, Calcutta, 1957.
39. Lösche, A.: "Kerninduktion" Verb. deutsch Verlag Wiss., E. Berlin, 1957.
40. Pople, J. A., W. G. Schneider, and H. J. Bernstein: "High Resolution Nuclear Magnetic Resonance," McGraw-Hill, New York, 1959.
41. Powles, J. G.: *Repts. Progr. Phys.*, **22**: 433 (1959).
42. Gutowsky, H. S.: "Techniques of Organic Chemistry: Physical Methods," vol. 1, part IV, p. 2663, Interscience, New York, 1960.
43. Varian Associates, "NMR and EPR Spectroscopy," Pergamon Press, New York, 1960.
44. Bloembergen, N.: "Nuclear Magnetic Relaxation," W. A. Benjamin, New York, 1961.
45. Rowland, T. J.: Nuclear Magnetic Resonance in Metals, "Progress in Material Science," vol. 9, no. 1, Pergamon Press, New York, 1961.
46. Abragam, A.: "The Principles of Nuclear Induction," Clarendon Press, Oxford, 1961.
47. Richards, R. E.: "Advances in Spectroscopy," vol. 2, p. 101, Interscience, New York, 1961.
48. Zimmerman, J. R.: "Methods of Experimental Physics: Molecular Physics," vol. 3, p. 359, Academic, New York, 1962.
49. Hebel, L. C.: "Solid State Physics," vol. 15, p. 409, Academic, New York, 1963.
50. Emsley, J. W., J. Feeney, and L. H. Sutcliffe: "High Resolution Nuclear Magnetic Resonance Spectroscopy," vols. I and II, Academic Press, New York, 1965.
51. Aleksandrov, I. V.: "The Theory of Nuclear Magnetic Resonance," Academic Press, New York, 1966.
52. *Annual Rev. Phys. Chem.*, **6–15** (1954–1964).
53. *Archives des Sciences (Genève)*, **9–15**: Special Number, Colloque Ampère (1956–1962).
54. Servant, R., and A. Charru, eds.: "Electronic Magnetic Resonance and Solid Dielectrics," 12th Colloque Ampère, North Holland, Amsterdam, 1964.
55. Van Gervan, L., ed.: "Nuclear Magnetic Resonance and Relaxation in Solids," 13th Colloque Ampère, North Holland, Amsterdam, 1965.
56. *Nuovo Cimento* **6** Supplement, 808 (1957).
57. *Discussions Faraday Soc.*, **19** (1955).
58. *Discussions Faraday Soc.*, **34** (1962).
59. *Ann. N. Y. Acad. Sci.*, **70**: 765 (1958).
60. Pesce, B., ed.: "Nuclear Magnetic Resonance in Chemistry," Academic, New York, 1965.
61. Ter Haar, D., ed.: "Fluctuation, Relaxation and Resonance in Magnetic Systems," Plenum, New York, 1962.
62. *Proc. Roy. Soc.* (London) **A283**: 433–480 (1965).
63. Waugh, J. S., ed.: "Advances in Magnetic Resonance," vol. 1, Academic, New York, 1965.
64. Feenberg, E., and G. E. Pake: "Notes on Quantum Theory of Angular Momentum," Stanford Press, Stanford, Calif., 1959.
65. Condon, E. U., and G. H. Shortley: "The Theory of Atomic Spectra," Cambridge University Press, New York, 1963.
66. Cohen, E. R., and J. W. DuMond: *Rev. Mod. Phys.*, **37**: 537 (1965).
67. Collington, D. J., A. N. Dellis, J. H. Sanders, and K. C. Turberfield: *Phys. Rev.*, **99**: 1622 (1955); J. H. Sanders and K. C. Turberfield: *Proc. Roy. Soc. (London)*, **A272**: 79 (1963); J. H. Sanders, K. F. Tittel, and J. F. Ward: *Proc. Roy. Soc. (London)*, **A272**: 103 (1963).
68. Blin-Stoyle, R. J.: "Theories of Nuclear Moments," Oxford University Press, London, 1957.
69. Cohen, M. H., and F. Reif: "Solid State Physics," vol. 5, p. 321, Academic, New York, 1957.

70. Townes, C. H.: "Handbuch der Physik" **38**: 1, 357 (1958).
71. Das, T. P., and E. L. Hahn: Nuclear Quadrupole Resonance, "Solid State Physics," Suppl. 1, Academic, New York, 1958.
72. O'Konski, C. T.: "Determination of Organic Structures by Physical Methods," vol. 2, p. 661, Academic, New York, 1962.
73. Jeffrey, G. A., and T. Sakurai: "Progress in Solid State Chemistry," vol. 1, p. 380 Macmillan, New York, 1964.
74. Fowler, R., and E. A. Guggenheim: "Statistical Thermodynamics," p. 77, Cambridge University Press, New York, 1956.
75. Fowler, R., and E. A. Guggenheim: "Statistical Thermodynamics," p. 629, Cambridge University Press, New York, 1956.
76. Pake, G. E.: "Paramagnetic Resonance," p. 8, W. A. Benjamin, New York, 1962.
77. Bloch, F., and A. Siegert: *Phys. Rev.*, **57**: 522 (1940).
78. Rabi, I. I., N. F. Ramsey, and J. Schwinger: *Rev. Mod. Phys.*, **26**: 167 (1954).
79. Schwinger, J.: *Phys. Rev.*, **51**: 648 (1937).
80. Rabi, I. I.: *Phys. Rev.*, **51**: 652 (1937).
81. Bloch, F., and I. I. Rabi: *Rev. Mod. Phys.*, **17**: 237 (1945).
82. Pauling, L., and E. B. Wilson: "Introduction to Quantum Mechanics," McGraw-Hill, New York, 1935.
83. Abragam, A., and W. G. Proctor: *Phys. Rev.*, **106**: 160 (1957).
84. Hebel, L. C., and C. P. Slichter: *Phys. Rev.*, **113**: 1504 (1959).
85. Van Vleck, J. H.: *Phys. Rev.*, **74**: 1168 (1948).
86. Bloch, F.: *Phys. Rev.*, **70**: 460 (1946).
87. Wangsness, R. K., and F. Bloch: *Phys. Rev.*, **89**: 728 (1953).
88. Torrey, H. C.: *Phys. Rev.*, **76**: 1049 (1949) and **104**: 563 (1956).
89. Hahn, E. L.: *Phys. Rev.*, **77**: 297 (1950) and **80**: 580 (1950).
90. Van Vleck, J. H.: *Phys. Rev.*, **57**: 426 (1940).
91. Bloembergen, N.: *Phys. Rev.*, **109**: 2209 (1958).
92. Jacobsohn, B. A., and R. K. Wangsness: *Phys. Rev.*, **73**: 942 (1948).
93. Gabillard, R.: *Compt. Rend.*, **232**: 1551 (1951) and **233**: 39 (1951); *Phys. Rev.*, **85**: 694 (1952).
94. Saltpeter, E. E.: *Proc. Phys. Soc.* (*London*), **A63**: 337 (1950).
95. Torrey, H. C.: *Phys. Rev.*, **76**: 1059 (1949).
96. Weger, M.: *Bell Syst. Tech. J.*, **39**: 1013 (1960).
97. Redfield, A. G.: *Phys. Rev.*, **98**: 1787 (1955).
98. Primas, H., and H. H. Gunthard: *Helv. Phys. Acta*, **30**: 315 (1957).
99. Leane, J. B., R. E. Richards, and T. P. Schaefer: *J. Sci. Instr.*, **36**: 230 (1959).
100. Nelson, F. A., and H. E. Weaver: *Science*, **146**: 223 (1964).
101. Baker, E. B., and L. W. Burd: *Rev. Sci. Instr.*, **34**: 238 (1963).
102. Glarum, S. H.: *Rev. Sci. Instr.*, **36**: 771 (1965).
103. Bloembergen, N., E. M. Purcell, and R. V. Pound: *Phys. Rev.*, **73**: 679 (1948).
104. Pound, R. V., and W. D. Knight: *Rev. Sci. Instr.*, **21**: 219 (1950).
105. Pound, R. V.: *Prog. Nuclear Phys.*, **2**: 21 (1952).
106. Robinson, F. N. H.: *J. Sci. Instr.*, **36**: 481, 484 (1959).
107. Codrington, R. S., J. D. Olds, and H. C. Torrey: *Phys. Rev.*, **95**: 607A (1954).
108. Garstens, M. A., L. S. Singer, and A. H. Ryan: *Phys. Rev.*, **96**: 53 (1954).
109. Sauzade, M.: *Compt. Rend.*, **258**: 4458 (1964).
110. Mock, J. B.: *Rev. Sci. Instr.*, **31**: 551 (1960).
111. Feher, G.: *Bell System Tech. J.*, **36**: 449 (1957).
112. Bruin, F.: "Electronics and Electron Physics," vol. 15, p. 327, Academic, New York, 1961.
113. Buckmaster, H. A., and J. C. Dering: *Can. J. Phys.*, **43**: 1088 (1965).
114. Drain, L. E.: *Proc. Phys. Soc.* (*London*), **62A**: 301 (1949).
115. Schwartz, J.: *Rev. Sci. Instr.*, **28**: 780 (1957).
116. Davis, C. F., M. W. P. Strandberg, and R. L. Kyhl: *Phys. Rev.*, **111**: 1268 (1958).
117. Bowers, K. D., and W. B. Mims: *Phys. Rev.*, **115**: 285 (1959).
118. Woonton, G. A.: "Electronics and Electron Physics," vol. 15, p. 163, Academic, New York, 1961.
119. Kaplan, D. E., M. E. Browne, and J. A. Cowen: *Rev. Sci. Instr.*, **32**: 1182 (1961).
120. Dyment, J. C.: *Can. J. Phys.*, **44**: 637 (1966).
121. Collins, S. A., R. L. Kyhl, and M. W. P. Strandberg: *Phys. Rev. Letters*, **2**: 88 (1959).
122. Carr, H. Y., and E. M. Purcell: *Phys. Rev.*, **94**: 630 (1954).
123. Luszczynski, K., and J. G. Powles: *J. Sci. Instr.*, **36**: 57 (1959).
124. Pake, G. E.: *J. Chem. Phys.*, **16**: 327 (1948).
125. Andrew, E. R., and N. Finch: *Proc. Roy. Soc.* (*London*), **B70**: 980 (1957).
126. Bersohn, R., and H. S. Gutowsky: *J. Chem. Phys.*, **22**: 651 (1954).
127. Deeley, C. M., and R. E. Richards: *J. Chem. Soc.*, 3697 (1954).
128. Kubo, R., and K. Tomita: *J. Phys. Soc. Japan*, **9**: 888 (1954).
129. Dreitlein, J., and H. Kessemeier: *Phys. Rev.*, **123**: 835 (1961).
130. Clough, S., and K. W. Gray: *Proc. Phys. Soc.* (*London*), **79**: 457 (1962); **80**: 1382 (1962).
131. Andrew, E. R., and V. T. Wynn: *Proc. Roy. Soc.* (*London*), **A291**, 257 (1966).
132. Proctor, W. G., and F. C. Yu: *Phys. Rev.*, **77**: 717 (1950).
133. Dickinson, W. C.: *Phys. Rev.*, **77**: 736 (1950).
134. Ramsey, N. F.: *Phys. Rev.*, **86**: 243 (1952).
135. Phillips, W. D.: "Determination of Organic Structures by Physical Methods," vol. 2, p. 401, Academic, New York, 1962.
136. Lauterbur, P. C.: "Determination of Organic Structures by Physical Methods," vol. 2, p. 465, Academic, New York, 1962.
137. Corio, P. L.: *Chem. Rev.*, **60**: 363 (1960).
138. Corio, P. L.: "Structure of High Resolution NMR Spectra," Academic, New York, 1967.
139. Swalen, J. D., and C. A. Reilly: *J. Chem. Phys.*, **37**: 21 (1962).
140. Castellano, S., and A. A. Bothner-By: *J. Chem. Phys.* **41**: 3863 (1964).
141. Knight, W. D.: *Phys. Rev.*, **76**: 1259 (1949).
142. Knight, W. D.: "Solid State Physics," vol. 2, p. 93, Academic, New York, 1956.
143. Townes, C. H., C. Herring, and W. D. Knight: *Phys. Rev.*, **77**: 852 (1950).
144. Schone, H. E., and W. D. Knight: *Acta Met.*, **11**: 179 (1963).
145. *Rev. Mod. Phys.*, **36**: I, 170–187 (1964).
146. Bolef, D. I.: in "Physical Acoustics" (W. P. Mason, ed.), vol. 4, Academic, New York, 1965.
147. Bolef, D. I., and R. K. Sundfors: *Proc. I.E.E.E.*, **53**: 1574 (1965).
148. Shutilov, V. V.: *Soviet Phys. Acoustics*, **8**: 303 (1963).
149. Dennison, A. B., L. W. James, J. D. Currin, W. H. Tanttilla, and R. J. Mahler: *Phys. Rev. Letters*, **12**: 244 (1964).

150. Abragam, A., and M. H. L. Pryce: *Proc. Roy. Soc. (London)*, **A205**: 135 (1951).
151. Koster, G., and H. Statz: *Phys. Rev.*, **115**: 1568 (1959).
152. Ray, T.: *Proc. Roy. Soc. (London)*, **A277**: 76 (1964).
153. McClelland, B. J.: *Chem. Rev.*, **64**: 301 (1964).
154. Morton, J. R.: *Chem. Rev.*, **64**: 453 (1964).
155. Weil, J. A., and J. H. Anderson: *J. Chem. Phys.*, **35**: 1410 (1961).
156. Swalen, J. D., and H. M. Gladney: *IBM J. Res. Develop.*, **8**: 515 (1964).
157. Bloembergen, N.: *Phys. Rev.*, **104**: 324 (1956).
158. Townes, C. H., ed.: "Quantum Electronics," Columbia, New York, 1960.
159. Singer, J. R., ed.: "Advances in Quantum Electronics," Columbia, New York, 1961.
160. Singer, J. R.: "Electronics and Electron Physics," vol. 15, p. 73, Academic, New York, 1961.
161. Grivet, P., and N. Bloembergen, eds.: "Quantum Electronics III," vols. I and II, Columbia, New York, 1964.
162. Geusic, J. E., and H. E. D. Scovil: *Repts. Progr. Phys.*, **27**: 241 (1964).

Part 9 · Nuclear Physics

Chapter 1

General Principles of Nuclear Structure*

By LEONARD EISENBUD, State University of New York, *and* G. T. GARVEY *and* EUGENE P. WIGNER
Princeton University

I. GENERAL FEATURES OF NUCLEI

1. Nuclear Composition

The concept of the atomic nucleus was first advanced by Rutherford (1911), who showed that the positive charge and all but a small fraction of the total mass of an atom are concentrated on a central nuclear core with radius of the order of 10^{-5} that of the atomic radius. It was soon recognized that the charge on the nucleus of an atom, Ze, is an integral multiple of the electronic charge ($-e$ = electronic charge) and that the integer Z specifies the position of the atom in the periodic table. J. J. Thomson showed (1913) that the mass of a nucleus is not determined by its charge; nuclei of the same charge but different mass are called isotopes. On a mass scale defined by assigning the value 16 to (the mass of the most abundant oxygen isotope the masses of all nuclei are close to integers. The integer closest to the mass of a nucleus on this scale is called the mass number, A, of the nucleus.

Modern theories of the nucleus stem from the discovery of the neutron—an uncharged particle of mass approximately that of the hydrogen atom (the mass number is unity)—by Chadwick (1932) and the suggestion by Heisenberg, shortly thereafter, that the elementary constituents of nuclei are neutrons and protons. A nucleus of charge number Z, mass number A, is composed of Z protons and $N = A - Z$ neutrons. *Isotopes* are nuclei of equal Z but unequal N. The atoms formed on isotopic nuclei have practically identical chemical properties. Nuclei of equal N but unequal Z are called *isotones*. The elementary nuclear constituents are often referred to indiscriminately as *nucleons*. The number of nucleons in a nucleus is the mass number. Nuclei composed of the same numbers of nucleons (equal A) are referred to as *isobars*. There is ample evidence for an approximate equivalence of the properties of the neutrons and protons within nuclei. As a consequence, the isobars play a role in the physics of nuclei comparable to the role of isotopes in atomic physics. For the designation of a particular nuclear species among isobars of given A the number $T_\zeta = \frac{1}{2}(N - Z)$, which measures excess of the number of neutrons over the number of protons in the nucleus, has considerable theoretical usefulness.

* Based in part on "Nuclear Structure," copyright 1958 by Princeton University Press.

Clearly any pair of the four numbers, A, Z, N, T_ζ, may be used to characterize the composition of a nucleus. The customary symbolic representation employs the numbers Z,A; the number Z is given by the chemical symbol of the element, while A appears as a superscript. Thus the nucleus with $Z = 79$, $A = 197$ is designated by Au^{197}. Sometimes the charge number is explicity given as a subscript: $Au_{79}{}^{197}$.

The ranges of Z, A for known nuclei are $Z = 0$ (neutron) to well over a hundred; A ranges from 1 (proton, neutron) to over 250.

The properties of the fundamental nuclear constituents are tabulated below:

Constituent	Mass	Charge	Spin	Magnetic moment in nuclear magnetons	Statistics
Neutron	1.00867	0	$\frac{1}{2}$	−1.9131	Fermi
Proton	1.00728	e	$\frac{1}{2}$	2.7928	Fermi

Masses are given in the $C^{12} = 12$ scale. Magnetic moments are expressed in nuclear magnetons ($e\hbar/2M_pc$, where M_p is the proton mass). The usual tables of atomic masses give the mass of the atom, i.e., the nuclear mass augmented by the mass of Z electrons or $548 \times 10^{-6}Z$.

The N, Z values for the known stable nuclei are presented in the Segré chart (see Fig. 2.1 of Sec. II). Isotopes are grouped along horizontal lines in the chart, isotones along vertical lines. We shall return to a discussion of the significance of the structure of this chart in Sec. II.

2. Nuclear Masses: Binding Energies

The energy required to decompose a nucleus into its constituent nucleons is called the *binding energy*, B, of the nucleus. The binding energy is related, through Einstein's mass–energy equation, to the difference between the mass of the nucleus and the sum of the masses of the constituent nucleons. If ΔM is this mass difference—the so-called mass defect of the nucleus—then $B = \Delta Mc^2$ (where c is the velocity of light). Thus, if $M(Z,N)$ is the mass of a neutral atom with Z protons and N neutrons, and M_H and M_n are the masses of the neutral hydrogen atom and of the neutron, the binding energy is given by

$$B(Z,N) = [M_H Z + M_n N - M(Z,N)]c^2$$

(the mass of the electrons in the neutral atom is taken into account through the use of M_H, the mass of the hydrogen atom, rather than the proton mass). It is customary to measure B in the unit Mev = millions of electron volts. The relation between mass on the C^{12} scale and this energy unit is 1 millimass unit = 0.931 Mev.

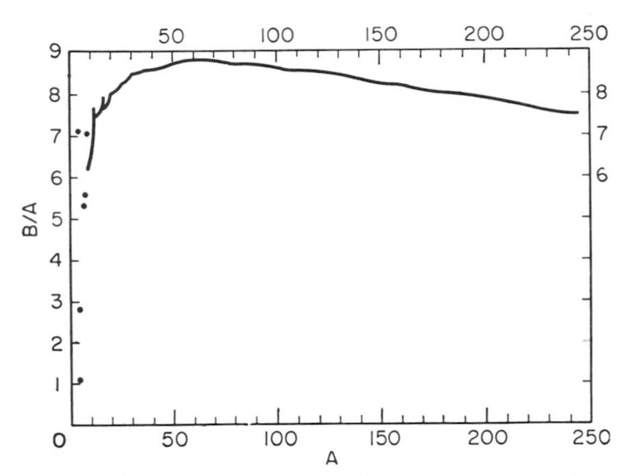

FIG. 1.1. Binding energies of stable nuclei. The binding energy, in Mev per nucleon, is plotted against the number A of nucleons.

In Fig. 1.1 the *binding energy per particle* B/A is plotted as a function of A for stable nuclei. The binding energy per particle is roughly constant for medium and heavy nuclei at about the value of 8 Mev per nucleon. This behavior contrasts sharply with that of the electron binding energy per electron in atoms, which increases (irregularly) with the number of electrons in the neutral atom. With respect to the binding energy the nucleus is similar to a solid or a liquid for which the heat needed for vaporization is proportional to the mass of the substance.

It is useful to consider, in addition to the binding energy B, or the *binding fraction* B/A, the separation energies of nucleons (neutrons, n, protons, p) or nuclear aggregates (deuterons, H_1^2, α particles, He_2^4, etc.) from a nucleus. The separation energy of a particle b (nucleon or aggregate) is defined as the energy required for the separation of a nucleus into two parts, one of which is the particle b. If the residual nucleus after the separation is designated by the index r, the separation energy is given by

$$S_b = [M_b + M_r - M(N,Z)]c^2$$

The separation energies characterize the relative stability of a nucleus with respect to transformations leading to the emission of a nucleon or nuclear aggregate.

The separation energies of neutrons S_n and of protons S_p are of the order of the binding fraction for stable nuclei. The separation energy S_α of an α particle (He_2^4) is, for intermediate A nuclei, of the order of 5 Mev. For the unstable α radioactive nuclei, the S_α is negative (see Sec. I, 3).

As already mentioned, the smooth variation of B/A with A suggests that the nucleus has properties similar to those of a liquid or solid. The fluctuations in S_b, however, indicate the inadequacy of such pictures. A close examination of the separation energies S_p and S_n of stable nuclei as functions of Z and N reveals discontinuities which are similar to those which occur in the electron separation energy (ionization potential) of stable atoms. Thus the separation energy of a neutron from Si^{28} is 17.2 Mev, that of a proton from N^{13} is 1.95 Mev. However, these are extreme cases, and on the whole, the fluctuations in the separation energies S_p and S_n are less marked than those in the ionization energies of atoms (see Secs. II, VIII). The largest separation energies S_p and S_n (20 Mev) are those for the α particle, He_2^4; as a result, He_2^4 has exceptional stability.

3. Types of Nuclear Instability. Spontaneous and Induced Transformations

Certain nuclei among those which occur naturally undergo spontaneous transformations. Thus the naturally radioactive substances with $Z > 82$ (Pb) exhibit transformations with emission of α particles, of electrons, and of electromagnetic radiation (γ rays). The processes are often referred to as α- , β- , and γ-decay processes, respectively. Not only the heavy natural nuclei undergo spontaneous transformation. For example, K^{40} also emits electrons; this nucleus also transforms by the absorption of an extranuclear electron—the K capture process (since a K electron of the atom is absorbed).

Unstable states of nuclei can also be formed in nuclear reactions. Nuclei can be formed which subsequently transform by emission of positive or negative electrons or by capture of an orbital electron. Nuclei may be excited in collision processes with subsequent emission of γ rays. It is also possible to form nuclear states (generally of very short lifetime—see below) which transform by emission of neutrons, protons, or α particles.

All the transformations mentioned above may be characterized by a transformation or decay probability— the probability per unit time of transition from the initial to the final state of the system. The reciprocal of this probability measures the *lifetime* of the initial state. The lifetime is defined as the time required for the population of the initial state to fall to e^{-1} (e is the base of natural logarithms) of its initial value. The emission probability, λ, is often expressed in terms of the level width instead of in units of reciprocal time; the width Γ, which may be considered as an uncertainty in the energy of the unstable state, is related to the average or mean decay time $\bar{\tau} = \lambda^{-1}$ through Heisenberg's relation, $\lambda = \Gamma/\hbar$. The relation between Γ (in electron volts) and λ (in reciprocal seconds) is

$$\Gamma = 0.65 \times 10^{-15}\ \lambda$$

The "half life" is the time required for the transformation of one-half of the original material, it is $\tau_{1/2} = \bar{\tau} \ln 2 = 0.693\bar{\tau}$. The general characteristics of the various types of nuclear instability are summarized below.

Natural α Radioactivity; Fission. The range of α-decay lifetimes is very broad. The lifetime is a sensitive function of the energy of the emitted α. The lifetime of Th^{232} is $\sim 10^{10}$ years for an α energy of 4 Mev; the corresponding numbers for Po^{212} are 3×10^{-7} sec, and 9 Mev.

The characteristics of the relation between energy and lifetime may be understood in terms of the concept of barrier penetration. Consider the potential energy as a function of the distance of the α particle from the residual nucleus which remains after transmutation. For large distances the potential energy is purely electrostatic. With decreasing r the potential energy increases as r^{-1}. When the distance is of the order of, or smaller than, the nuclear radius, the interaction is no longer purely electrostatic. Strictly speaking, the concept of the potential energy of the particle loses significance for such small distances, since the α particle does not maintain its identity and structure within the nucleus. We shall, however, consider the α particle as moving in some effective potential even inside the residual nucleus.

At some separation the attractive nuclear forces overcome the electrostatic forces, leading to an effective potential like that given in Fig. 1.2. In order to be released from the nucleus, the α particle must penetrate the electrostatic barrier, that is, it must pass through the region where the potential is higher than the energy of the α particle. On the basis of the theory of barrier penetration, the relation between the energy of the α particle and the lifetime of the unstable nucleus is readily understood (Sec. X, 5).

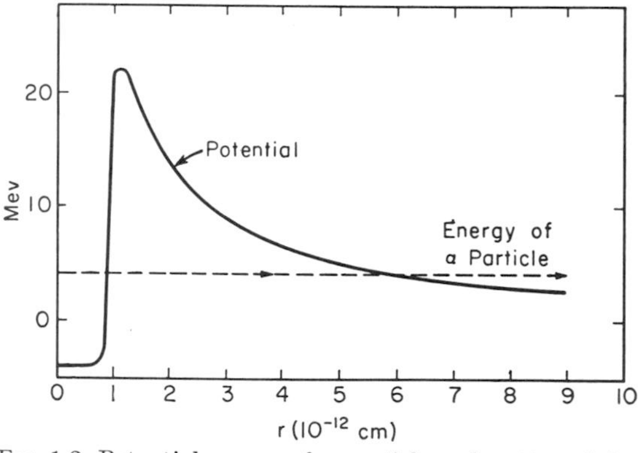

FIG. 1.2. Potential energy of α particle as function of the distance from the center of a nucleus (schematic). The potential energy is given in Mev; the energy of the α particle, as indicated, is a typical disintegration energy.

The height and breadth of the assumed electrostatic barrier depend critically on the value chosen for the nuclear radius, i.e., the radius at which the electrostatic forces are overcome by internal nuclear forces. This in turn strongly affects the theoretical results concerning the probability of barrier penetration. The comparison of theoretical predictions with observed lifetimes thus provides a means for determining the radii of α radioactive nuclei.

It may be of interest to note that the separation energy of an α particle is negative for many heavy elements which do not exhibit α activity. The reason is that the barrier is so effective at the low α emission energies available that the transformation probabilities are too small for the transformations to be observed.

The binding fraction curve indicates the possibility of another type of instability for the heavy nuclei. The masses of these nuclei are larger than the masses of the nuclei obtained by roughly splitting them in two. Thus "fission" of the heavy nuclei, i.e., splitting into two approximately equal parts, becomes possible with the release of considerable energy ($\sim$200 Mev). However, the barrier which impedes the process is of such height that the probability of transformation by splitting is very small. For the heaviest nuclei an excitation by the nucleus by a few Mev takes the system over the barrier and fission follows within a very small fraction of a second.

γ Radiation; Particle Emission. If a stable nucleus is excited but has insufficient energy for the emission of a particle, it will return to its normal state by emission of electromagnetic radiation. Lifetimes for such processes are extremely sensitive to spin changes between initial and final states (the energy dependence, which favors high energy changes, is less important). The lifetimes for γ emission range from 10^{-17} to 10^{-10} sec if the spin (J) change in the transition is small; that is, $\Delta J \leq 2$. For low-energy transitions with large spin changes, the lifetime can become abnormally long. For $\Delta J = 4$ and energy change $\sim$0.1 Mev, the lifetime can be several years. If the first excited level of a nucleus has a long life, the excited state of the nucleus is called an isomeric state.

In practice, the distinction between states which are unstable only with respect to the emission of γ rays and those which are unstable also with respect to particle emission is very important. Theoretically, the two processes take place side by side, each with its characteristic probability. For *light elements*, the particle emission, if it is at all possible energetically, is usually so much more probable than the emission of γ rays that the latter can be neglected altogether. This is also true for neutron emission in heavier elements, except when the neutron emission is only "barely possible," i.e., if the energy of the emitted neutrons is less than a few kilovolts. However, the emission of "low-energy" charged particles from *heavy nuclei* is so much impeded by the potential barrier that, as a rule, γ-ray emission is the faster process.

Just as the total decay probability is equal to the sum of the decay probabilities of the various processes (γ emission, emission of various particles) by which the nuclear state in question can decay, the total width of the state is the sum of the "partial widths" corresponding to the same processes (Sec. XI).

Examples of particle instability, other than those of natural radioactive elements, are the disintegrations of Be^8 into two α particles and the disintegrations of He^5 and Li^5 into an α particle and a neutron and proton, respectively. In addition, practically every nucleus has excited states with sufficient excitation energy to permit the emission of some of its constituents.

β Decay. Perhaps the most interesting forms of nuclear instability, both theoretically and experimentally, are those involving the emission of an electron or positron (a decay) or the capture of an orbital electron (K capture). Lifetimes of these processes range from around 10^{-2} sec to more than 10^{11} years. The emitted electrons do not carry the whole energy difference between initial and final states (as do α particles). Instead, they have a characteristic energy distribution (the Fermi distribution); the upper limit of this energy distribution is, to within the errors of observation, equal to the energy made available in the transition.

Since electrons are not fundamental nuclear constituents, they must be created in the β transformation. The process is analogous to the creation of photons in radiative transitions. The conservation of energy, momentum, and angular momentum in the β process and the distribution in energy of the emitted electrons can be understood only if it is assumed that, in addition to the electron (or positron), a second particle (called the neutrino and designated by the symbol ν) is emitted. To account for the charge, energy, and angular momentum conservation in β transitions, it must be further assumed that the neutrino is uncharged, has a mass less than a thousandth of the electron mass, and possesses an intrinsic spin angular momentum $\frac{1}{2}\hbar$. These assumptions, originally put forward by Pauli, enabled Fermi to explain the characteristic energy distribution of the β electrons (Sec. XII). The interaction of neutrinos with matter is so weak that they penetrate all matter practically unhindered. Direct evidence for the existence of the neutrino has been obtained only under great difficulties, by observing the reaction inverse to the beta decay.

In the β-decay processes, the number of nucleons, A, is the same in the initial and final nuclear states. The transition involves the transformation of a neutron (or proton) in the original nucleus into a proton (or neutron) in the final nucleus. Thus the β transitions may be indicated by

$$(A, T_\zeta) \rightarrow (A, T_\zeta \pm 1) + e^\pm + \nu$$

(positive or negative electron emission)

$$(A, T_\zeta) + e^- \rightarrow (A, T_\zeta + 1) + \nu$$

(capture of the orbital electron)

where the symbol (A,T_ζ) designates the initial nuclear species. Actually, in the first process, in case of electron emission, the other particle is an "antineutrino," the antiparticle of the neutrino. For our purposes, this fact has little significance. The energy available in the negative electron emission or in the capture process is equal to the difference in *atomic masses* of the initial and final nuclei. [For positron emission this mass difference must exceed $2mc^2$ (1.1×10^{-3} mass units) if the process is to be energetically possible.] Thus two neighboring isobars ($\Delta T_\zeta = 1$) cannot both be stable with respect to β decay. For the case of $A = 1$, for example, the neutron mass is greater than that of the hydrogen atom; the neutron decays with emission of negative electrons ($\tau_{1/2} = 13$ min). The theory of β decay shows that the lifetime depends upon the available energy E roughly as E^{-k}, where k is approximately 5. Thus the decay time for β disintegration depends on the decay energy much less sensitively than that of α disintegration so that low-energy β processes are not in general too strongly inhibited to be observed. However, the lifetime depends sensitively on spin and parity changes between initial and final nuclear states. For large spin change and low energy the lifetime may well become so long that the nucleus appears to be stable. Thus, one of the two $A = 50$ isobars must be unstable; yet neither exhibits observable β activity. The same applies for the pair Cd^{113}, In^{113}. An interesting case of an apparently stable isobar is Lu^{176}, which has a half life of about 4×10^{10} years. It was one of the odd-odd nuclei the radioactivity of which was difficult to discover. In^{115} and Re^{187} were also found to be radioactive with half lives of 6×10^{14} and about 10^{10} years, respectively.

In contrast to the scarcity of isobaric pairs with $\Delta T_\zeta = 1$, there are many examples of pairs of isobars with T_ζ differing by two units. The higher mass nucleus of such a pair may in principle decay to its partner by a double β process in which two electrons (positrons) are emitted (or by a double K capture). The double decay gives rise to two electrons and two antineutrinos, or to two positrons and two neutrinos. The electrons (or positrons) have a continuous distribution of energy up to the limit permitted by the mass difference between the isobars. The lifetimes for this process are extremely long—of the order of 10^{20} years for an available energy difference of 5 Mev. So far, this process has not been observed.

II. SYSTEMATICS OF STABLE NUCLEI. DETAILS OF BINDING-ENERGY SURFACES

Study of the stable nuclei reveals many interesting and remarkable regularities. For $A < 36$ the stable nuclei have almost equal numbers of protons and neutrons. The T_ζ values for all A in this region are ≤ 1. With increasing A the T_ζ values increase regularly. For heavy nuclei ($A \sim 200$) T_ζ is about 25. Figure 2.1, the so-called Segré chart, gives all nuclides occurring in nature. Most of these nuclides are stable. Some of them exhibit instability by α or β radioactivity; some of them are probably unstable but, owing to their long lifetime, their radioactivity has not been observed. All nuclei with $A > 209$ are radioactive.

There are many more stable nuclei with even A than with odd A. Moreover, the nuclei with even A almost invariably have even Z and even N. Indeed, only four odd Z, odd N stable nuclei are known, and all of these have $T_\zeta = 0$. The odd-odd nuclei are H_1^2, Li_3^6, B_5^{10}, N_7^{14}. (Two other cases, V_{23}^{50} and Ta_{73}^{180}, are believed to be β unstable with unobservably long lives.) The fact that even numbers of neutrons or protons lead to more stable structures can be seen from a counting of the numbers of stable isotopes for even or odd Z, or the number of stable isotones for even or odd N. For odd Z there are never more than two stable isotopes; for even Z the number ranges as high as 10. There are no stable isotopes for $Z = 43$ or 61, while there is only one even Z nucleus,

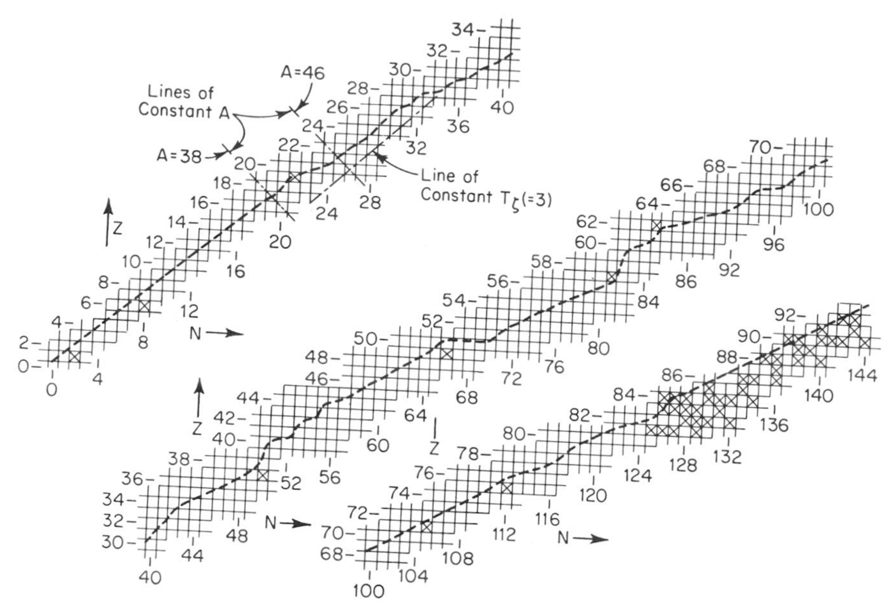

FIG. 2.1. Chart of natural nuclei. Each square corresponds to a charge number Z and neutron number N. The squares that correspond to stable nuclei are shaded; for those of natural radioactive nuclei the diagonals are drawn in.

Be_4^9, for which there is only one stable isotope. With very few exceptions, the number of stable isotones for odd N is one or zero. There are stable nuclei for all even N; for $N = 82$ the number of isotones is 6 (for $N = 50$, it is 5). A closer study of Fig. 2.1 permits expressing the enhancement of the stability of nuclei by the even character of Z and of N in several other ways.

For any given A the number of isobars is small. There is one stable isobar for each A between $A = 1$ and 36, except for $A = 5$ and 8 for which there is none. The number of isobars with odd A is generally one. In those cases where two are observed, it is believed that one of the isobars is β unstable (the ΔT_ζ for these isobaric pairs is 1). About 50 isobaric pairs are known for even A; the T_ζ values of the pair differ by 2. For $A = 96$, 124, 130, and 136 there are three isobars. In these cases also the difference in T_ζ between neighboring stable isobars is 2.

The even-odd relationships indicated above and the systematics of isobars can be explained by assuming that the binding energies of nuclei near the stable region can be represented not by one surface $B(N,Z)$ but by three close surfaces: one for the even-even, one for even-odd and odd-even, and one for odd-odd nuclei. Note that in the β-decay process the even-odd nuclei transform into odd-even nuclei, while odd-odd nuclei transform into even-even ones, or conversely. Existence of stable isobaric pairs of even-even nuclei and the absence of stable odd-odd nuclei show that the mass curves (as a function of T_ζ for constant A) are roughly as shown in Fig. 2.2. For the case illustrated all odd-odd nuclei are unstable and three even-even isobars will be stable. This picture also explains the fact that the T_ζ difference between stable isobars is almost always 2. From the energies of the β transitions from odd-odd to even-even nuclei, the displacement between the even-even and odd-odd mass curves may be measured; it is about 68 Mev/$A^{3/4}$. Reaction experiments locate the mass curve of the even-odd nuclei about midway between the even-even and odd-odd curves. The fact that the odd A nuclei exhibit only one stable isobar for each A indicates that the mass curve is the same for the odd N, even Z and even N, odd Z nuclei.

A number of fairly good semiempirical binding-energy or mass functions have been constructed which give a good approximation to the true masses over the wide range of both stable and unstable nuclei in terms of relatively few empirical constants. Perhaps the simplest such formula is that of Weizsäcker:

$$B(Z,N) = U_vA - U_cZ(Z-1)A^{-1/3} - U_sA^{2/3} - U_tT_\zeta^2/A \begin{cases} +\delta/A^{3/4} & \text{even-even} \\ +0 & \text{even-odd or odd-even} \\ -\delta/A^{3/4} & \text{odd-odd} \end{cases} \tag{2.1}$$

The first term provides a "volume" effect, i.e., a binding energy proportional to the number of nucleons. The second term expresses the fact that the binding energy is diminished by the electrostatic repulsion between the protons. There are $\frac{1}{2}Z(Z-1)$ interactions between the charges which are taken to be uniformly distributed over the volume of the nucleus. The $A^{-1/3}$ factor expresses the dependence of this energy contribution on the radius of the nucleus which is assumed to be proportional to $A^{1/3}$. The third term expresses the diminution of the volume-energy contribution due to surface effects. Terms similar to the first three occur also in the expression for the energy of a charged liquid drop. The fourth term takes into account the effects of symmetry properties of the nuclear states and how these are modified when the number of nucleons is kept constant and T_ζ, i.e., the neutron excess, is varied. The

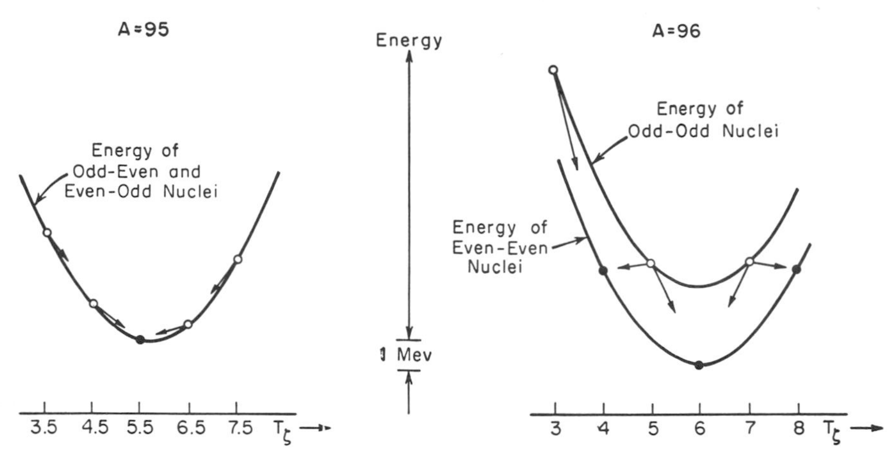

FIG. 2 2. The diagram on the left illustrates the energy (mass) of isobars with odd mass number A, as a function of the difference between neutron and proton numbers; $T_\zeta = \frac{1}{2}(N - Z)$. The diagram on the right applies for even A. In this case there are two lines: one for nuclei with even N and even Z, the other for nuclei with odd N and odd Z. The open circles represent unstable nuclei, the full circles stable nuclei. The energy of the latter is lower than that of either adjacent isobar. There is only one stable isobar in the case of odd A (in the diagram the isobar with $T_\zeta = 5\frac{1}{2}$); there is no stable nucleus with odd N and odd Z; there are, under the conditions of the diagram on the right, three stable isobars with even N and even Z.

last term also represents a symmetry effect. Constants which provide a fit with the binding-energy data are:

$$U_v = 14.0 \text{ Mev} \qquad U_c = 0.61 \text{ Mev}$$
$$U_s = 14.0 \text{ Mev} \qquad U_t = 84.2 \text{ Mev} \qquad (2.1a)$$
$$\delta \sim 34 \text{ Mev}$$

The existence of three binding-energy surfaces provides a qualitative understanding of the distribution of isobars. It also explains why the separation energy of a neutron $S_n = B(Z,N) - B(Z, N-1)$ is greater for even N than for odd N and why the separation energy of a proton $S_p = B(Z,N) - B(Z-1, N)$ is greater for even than for odd Z. For even N, the S_n contains the $\delta/A^{3/4}$ term with positive sign, for odd N with negative sign and a similar remark applies for the separation energy of the proton. However, as long as we compare S_n or S_p only for nuclei with even N and even Z (or, more generally, as long as we do not change the even or odd nature of Z and N), both S_n and S_p as calculated from (2.1) will be smooth functions of Z and N. Measurements of the separation energies show that this is true in general. However, there are a number of significant exceptions, which are connected with the so-called *magic numbers* 2, 8, 20, 50, 82, 126, and, to a lesser degree, some others. When Z passes these numbers, S_p drops by about 2 Mev. Thus, the separation energy of a proton S_p from Sn_{50} ranges from 8½ to 10½ Mev as A increases from 114 to 120. For Te_{52}, these numbers range from 8 to 9 Mev as A increases from 122 to 126. The increase of S_p with increasing A is due to the increasing number of neutrons which bind a constant number of protons with increasing strength and finds its expression in the symmetry term of (2.1). In spite of this increase, the S_p is higher for the "magic" Sn_{50} nuclei than for the nonmagic Te_{52}. The comparison of the In_{49} and Sb_{51} nuclei is less convincing: for the former, S_p ranges from 6 to 7½ Mev. For the latter, which has a single proton in excess of the magic 50, S_p ranges from 5½ to 7 Mev. Note the generally smaller value of S_p for odd than for even Z: the even-odd difference is larger than the magic-nonmagic difference. The situation at the other magic numbers is perhaps even more striking and a similar phenomenon is noted with respect to S_n when N passes through one of the magic numbers. The most marked discontinuity occurs at $N = 2$ and $Z = 2$, where both S_n and S_p drop from 20 Mev to about 7 Mev. These discontinuities are not reproduced in global formulas such as (2.1) which do not take into account properties of individual nucleon orbits. An explanation of the properties of the magic numbers is given in Sec. VIII by assuming that the nucleons are arranged in shells similar to the K, L, M, etc., shells of electrons in atoms.

The particular stability of the magic numbers manifests itself also in the Segré chart. The number of isotones for magic N, or isotopes for magic Z, is larger than the number of isotones or isotopes for adjoining even values of N and Z. The difference is even larger if comparison is made with adjoining odd values of N and Z, but this is a consequence of the $\delta/A^{3/4}$ terms in (2.1). The preferred stability of the magic N and Z numbers, as compared with adjoining even N and Z numbers, on the other hand, does not follow from (2.1) but only from the particular "magic" stability of these neutron and proton numbers.

III. PROPERTIES OF NUCLEAR STATES; GROUND STATES

Although much painstaking work has gone into investigation of excited states of nuclei, our knowledge and understanding of these states are still far from complete. The spectroscope can be employed to obtain quite rapidly a tremendous wealth of data on the excited states of atoms. Unfortunately, no instrument of corresponding power exists for nuclei.

Over the past several years a considerable effort has

been expended in the measurement of the properties of nuclear ground states for which special techniques are available. In addition to a definite energy, a stationary or quasi-stationary state also has a definite angular momentum characterized by the quantum number J [the square of the angular momentum is $J(J+1)\hbar^2$; the number J is referred to as the *spin*] and a parity. These properties are common to all stationary states of free systems. The spin, J, may be measured by a number of techniques. A few of the regularities of the spins of nuclei in their ground states follow. The spin of all even-even nuclei is zero. No exception to this rule is known. The even-odd and odd-even nuclei always have half integral spin while odd-odd nuclei have integer spin values. This strongly supports the assumption that nuclei are composed of neutrons and protons. The spins which follow from the earlier proton-electron composition theories are inconsistent with this rule. The spins of the stable nuclear ground states are in general rather small. The largest measured value for a nucleus with odd A is $\frac{9}{2}$.

Measurement of the electric quadrupole and magnetic dipole moments has greatly advanced our knowledge of nuclear structure. The magnetic 2^l poles vanish for even l (for free systems), while the electric 2^l poles vanish for odd l. Hence, the magnetic-dipole moment μ and the electric-quadrupole moment are the lowest nonzero moments. Apart from a few magnetic octupoles, which have been obtained by a careful analysis of hyperfine structure patterns, no higher moments have been measured. It follows from the general properties of the $J = 0$, $J = \frac{1}{2}$ states of a free system that μ vanishes for $J = 0$ while the quadrupole moment is zero for $J = 0$, or $\frac{1}{2}$.

The magnetic-dipole moment μ of a nucleus will affect the energy of a nuclear state in a uniform magnetic field. The energy change of the nucleus when its spin is aligned in the direction of the field H is $-\mu H$. A measurement of the energy change determines the dipole moment μ. For a single nucleon, moving in a central field, the magnetic dipole moment can be calculated if the spin and angular momentum of the nucleon are known. For a single particle (proton or neutron) of definite orbital angular momentum and definite total angular momentum—specified by the quantum numbers l and $j = l \pm \frac{1}{2}$—the magnetic moment in units of the nuclear magneton $e\hbar/2M_pc$ is given in Figs. 3.1*a* and *b*. The curves in these figures are known as the Schmidt lines. The points on Fig. 3.1 indicate the measured moments. The observed magnetic moments of nuclei with odd Z and even N lie between the proton Schmidt lines (except for H^3), while the nuclei with odd N even Z lie between the neutron Schmidt lines (except for He^3 and C^{13}). The magnetic moments of even A nuclei are generally zero, since, except for the few stable odd-odd nuclei, the spins of even A nuclei are zero.

The energy of a system, the charge distribution of which is axially symmetric, will depend on the gradient of the electric field parallel to the symmetry axis. The part of the energy which depends on the relative orientation of the field and system is

$$E = -\frac{1}{4}\left(\frac{\partial \mathcal{E}_z}{\partial z}\right)_0 eq \tag{3.1}$$

(terms in the higher field derivatives associated with higher-order poles have been dropped) where $\mathcal{E}$ is the electric field and q is the quadrupole moment of the axially symmetric distribution divided by e. The

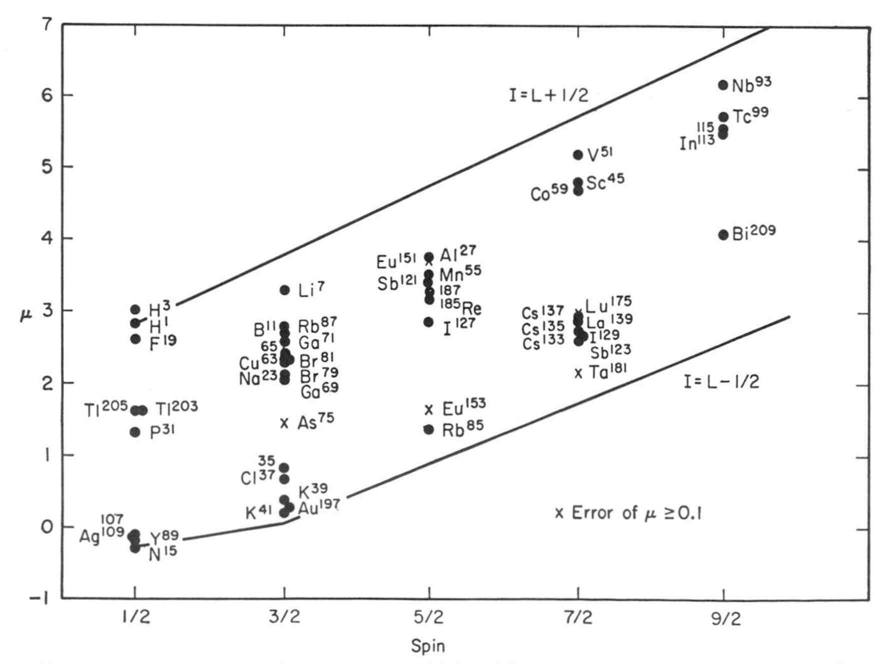

Fig. 3.1*a*. Magnetic moments of nuclei with even N and odd Z. The magnetic moments, in units of $e\hbar/2M_pc$, are plotted against the spin. The two full lines are the Schmidt lines; most magnetic moments are between these.

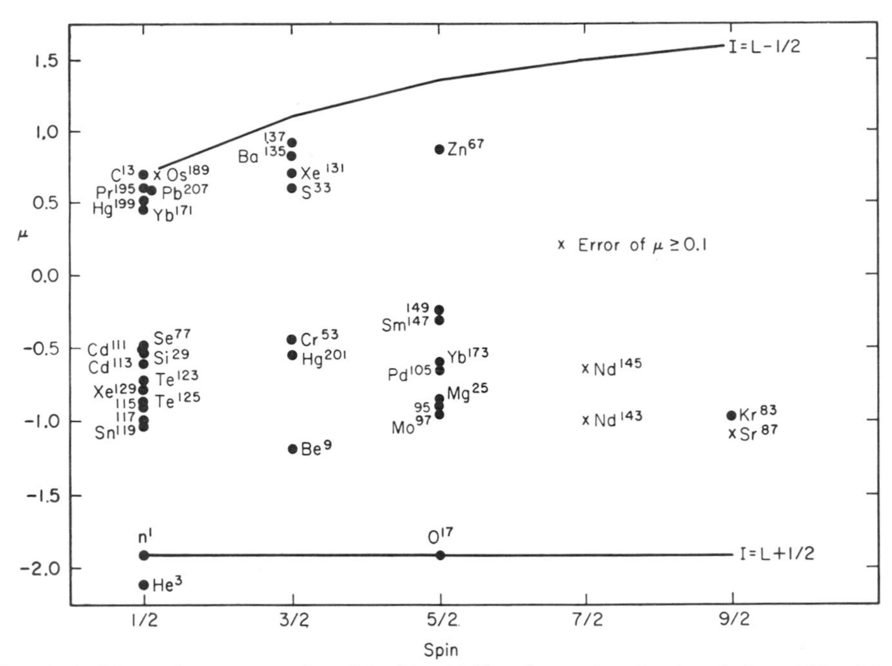

FIG. 3.1b. Magnetic moments of nuclei with odd N and even Z. See description of Fig. 3.1a.

zero indicates that the value of the gradient is to be taken at the center of the nucleus. In terms of the density distribution $\rho(\mathbf{r})$ of the protons

$$q = \int \rho(\mathbf{r})(3z^2 - r^2)\, dV \tag{3.2}$$

where the integration is carried out over the whole nucleus. The q for the nuclear state the spin of which is oriented along the z axis is called the nuclear quadrupole moment. Values for the observed quadrupole moments are given in Fig. 3.2. Except for nuclei with very large quadrupole moments, the distortion of the nucleus from spherical symmetry, as indicated by the quadrupole moment, is rather small. The "normal" quadrupole moments are equal in magnitude to the moments of uniformly charged ellipsoids of revolution with a difference of major and minor axes amounting to about 1 or 2 per cent of the nuclear

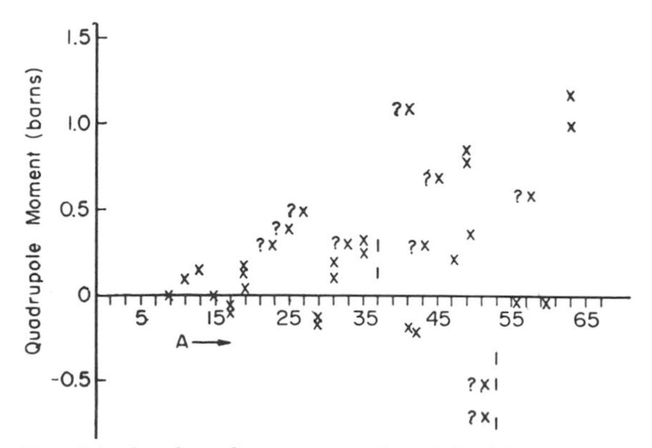

FIG. 3.2. Quadrupole moments of nuclei with odd mass numbers A, in barns (10^{-24} cm^2).

radius. For the anomalous quadrupole moments this difference rises to 10 and 15 per cent and even more.

1. The Size of the Nuclei

A variety of methods can be used to explore nuclear radii and are found to provide consistent results. The nuclear size can be estimated (1) from the scattering cross sections of nuclei for fast neutrons (neutron wavelength $\lambda \ll$ nuclear radius), (2) from the energy-lifetime relationship in α decay (Sec. I, 3), (3) from the difference of the binding energies of "mirror nuclei," (4) the energy levels of mesic atoms, and (5) from high-energy electron-diffraction experiments.

The first method—neutron scattering—depends on the assumption that the total cross section for high-energy neutrons is $2\pi R^2$, where R is the nuclear radius. In method (2), the theory of the penetration of the electrostatic barrier shows that the disintegration probability depends on the "radius" at which specifically nuclear forces overcome the electrostatic forces. The value of this radius which gives agreement with experiment is the nuclear radius. Methods (1) and (2) give only semiquantitative measures of the nuclear size. Method (3) requires a slightly more extended explanation. A pair of nuclei are called mirror nuclei if they have the same A and the T_ζ of one is the negative of the T_ζ of the other. The composition of the second nucleus is obtained by changing all nucleons of the first from protons to neutrons and from neutrons to protons. On the assumption that the only difference between the proton-proton and neutron-neutron interactions is given the electrostatic repulsion between the former, the kinetic energy and the potential of the nuclear forces will be the same in both nuclei. Their masses will differ only for two reasons: the greater number of neutrons in the positive T_ζ nucleus increases its mass because the neutron is heavier than the proton. On the other hand the electrostatic potential is greater in the negative T_ζ nucleus because this contains more protons. Except for the H^3-He^3 pair, this latter effect is larger than the former. It can be estimated on the assumption that the electric charge of the nucleus is uniformly spread over a sphere of radius R. (A small correction for exchange effects should be incorporated.) Hence, the last and most important contribution to the mass difference of mirror nuclei depends on the nuclear radius; it is inversely proportional thereto. Since the mass difference in question can be obtained experimentally (for instance, from the energy of the β decay of one into the other), one obtains a measure for the nuclear radius. The radii obtained in this way are closely approximated by

$$R = A^{1/3}r_0 \qquad r_0 = 1.37 \times 10^{-13}\text{ cm} \tag{3.3}$$

This gives for the root-mean-square distance of the nucleons from their center of mass

$$(\overline{r^2})^{1/2} = (\tfrac{3}{5})^{1/2}R = 1.05 \times 10^{-13}A^{1/3}\text{ cm}$$

Since R is proportional to the cube root of A, the densities of all nuclei are approximately the same. Lately, this method of obtaining nuclear radii could be extended to heavier nuclei, the mirrors of which cannot be produced. In this case a comparison is made between the binding energies of two isobars which have the same structure (the same wave function) except that a proton in one is replaced, in the other, by a neutron. (See also Sec. VI.)

When μ mesons are slowed down in a material, they may live long enough to be captured in orbits which are similar to the electronic orbits in atoms. The energy differences between these orbits, on which the fourth method of measuring nuclear radii is based, can be obtained by measuring the energy of the gamma rays which are emitted in the course of the transitions between these orbits. The "radius" of the innermost *electronic* orbit of an atom is $\hbar^2/Ze^2m$. This is, even for the heaviest nuclei (largest Z), about a hundred times larger than the nuclear radius. As a result, the electrostatic potential acting on the electron is very nearly that of a point charge and the energy values of the electrons are not affected appreciably by the finite size of the nucleus. The picture is markedly different, however, for the orbits of mesons. The formula $\hbar^2/Ze^2m$ again gives the "radius" of the lowest orbit but m is, in this case, the mass of the meson rather than that of the electron. Since the mass of the μ meson is about 207 times greater than the mass of the electron, the radii of the smallest mesonic orbits around heavy nuclei are quite comparable with the radii of these nuclei. As a result, the electrostatic potential, acting on a meson in its smallest orbit, differs greatly from the potential of a point charge; it is more nearly the potential of a reasonably large, uniformly charged sphere. The energy levels, therefore, show a considerable, or even very large, displacement from the positions which they would have if the electrostatic potential of the nucleus were that of a point charge. From the size of the displacement, the radius of the charge distribution can be calculated. This method yields the value $1.2 \times 10^{-13}A^{1/3}$ cm for the nuclear radius, if the nucleus is assumed to be spherical and to have a uniform charge distribution. The corresponding mean-square radius of the protons can be calculated to be

$$\overline{r^2} = \tfrac{3}{5}[1.2 \times 10^{-13}A^{1/3}]^2 = 0.85 \times 10^{-26}A^{2/3}\text{ cm}^2 \tag{3.4}$$

This is considerably less than the value given by the preceding methods. It should be noted, however, that this method is accurate only for heavy nuclei (large Z). For light nuclei, even the mesonic orbits lie outside the nucleus. As a result, the electrostatic potential at the orbit is so close to that of a point charge that the energy shift of the orbit is too small to be measured accurately.

The interpretation of the measurements of the energy levels of μ mesons presupposes that the interaction between nucleons and μ mesons is of purely electrostatic nature. If other forces also acted between μ mesons and nuclei, they would also change the mesonic orbits and energy levels, and the relation between energy levels and nuclear radii as developed above would require modification. There is, at present, no reason to believe that other than electromagnetic forces play a significant role in the μ-meson-nucleon interaction. The same is true of the interaction between electrons and nucleons. This makes it possible to investigate the charge distribution in nuclei by electron-scattering experiments. Naturally, as long as the wavelength of the electrons is very large as compared with the size of the nucleus, the scattering will closely resemble scattering by a point charge. However, if the wavelength of the electrons is comparable to the diameter of the nucleus, wavelets scattered from different parts of the nucleus will interfere—in some directions constructively, in others destructively. Under such conditions—which are obtained if the electron energy is of the order of 100 Mev—the scattering will differ considerably from that of a point charge, i.e., from Rutherford scattering. By analyzing the angular distribution of the scattered electrons one can, in principle, obtain not only the size of the nucleus (i.e., a number such as the mean-square distance of the protons from their center of mass) but the whole charge distribution. This is the basis of the fifth method of measuring nuclear radii.

The best high-energy electron-scattering experiments have been made by Hofstadter and his collaborators. Their interpretation, by Yennie, Schiff, and their collaborators, does not yet give the functional dependence of the proton density, as a function of the distance from the center. However, it does allow the determination of two constants characterizing the proton distribution, in contrast to the single constant given by all other measurements. These two constants can be interpreted as the mean-square distance of the protons from their center of mass and the diffuseness of the nuclear surface. This last quantity $2z$ is, crudely, the distance in which the proton density drops from three-fourths of its maximum value to one-fourth of this value.

A few values of these quantities follow (in units of 10^{-13} cm):

A	12	40	51	59	115	122	197	209
$[\overline{r^2}/A^{2/3}]^{1/2}$	1.05	1.02	0.97	0.98	0.92	0.93	0.91	0.93 $\times 10^{-13}$ cm
$2z$		1.1	1.0	1.1	1.0	1.1	1.1	1.2 $\times 10^{-13}$ cm

One notices that $\overline{r^2}$ increases somewhat less fast than $A^{2/3}$, i.e., that the average density increases somewhat with increasing size of the nucleus. This increase appears to be due to the decreasing role of the nuclear surface at which the density is, naturally, smaller than in the inside of the nucleus. The mean-square radius, as given by the electron-diffraction measurements, is consistent with the radii given by other methods.

The general agreement of the radii obtained by the different methods serves also to substantiate theories of barrier penetration and the theory of the equivalence of neutron-neutron and proton-proton interactions.

IV. SURVEY OF NUCLEAR REACTIONS

1. Cross Sections; Excitation Functions

In a broad class of nuclear experiments, studies are made of the collisions of neutrons (n), protons (p), deuterons (d), and α particles with various other nuclides. When the bombarded nucleus is transformed in any way in the collision, the process is called a nuclear reaction; otherwise one speaks of elastic scattering. If the species X is bombarded by a particle a, to form a nucleus Y and fragment b, the reaction is described by $a + X \rightarrow Y + b$, or more compactly by $X(a,b)Y$. The probabilities of various reactions or scattering processes may be characterized by *cross sections*, usually denoted by σ. If a totality of I particles crosses the target which has N target nuclei per unit area, the total number of reactions will be σIN. For a given pair of colliding nuclei, there are, in general, a number of possible reaction products. The total cross section for the process is the sum of the cross sections for the individual processes.

A convenient unit for nuclear cross sections is the *barn*, an area of 10^{-24} cm^2, which is approximately equal to the cross-sectional area of medium A nuclei for one of the most likely reactions.

One is often concerned with the probability of the production of a reaction product within an element of solid angle $d\Omega$ about a given direction; if $d\sigma$ is the cross section for this process, then $d\sigma/d\Omega$ is called the differential cross section (it is often designated simply as $d\sigma$). In a moving coordinate frame, with respect to which the center of mass of the collision system is at rest, if only two nuclei result from the reaction, they move off in opposite directions. Since the collision experiment (if the incident and target nuclei are not polarized) is characterized by axial symmetry about the direction of the beam, the differential cross section is a function only of the angle θ between the line of motion of the incident particle and the line of motion of the product particle in the center-of-mass coordinate frame. The differential cross section $d\sigma(\theta)$, as a function of θ, gives the angular distribution of the reaction products. The differential cross section is usually measured for the lighter of the two reaction products in the laboratory frame of reference. Figure 4.1 illustrates the connection between the velocity diagrams of a reaction in the laboratory frame of reference and in the center-of-mass coordinate system.

The difference between the internal energies of the reacting pairs a, X and the final products Y, b is called the Q of the reaction. If Q is positive (exothermic reaction), the reaction is energetically possible even in the limit of zero bombarding energy; if Q is negative (endothermic reaction), the reaction can occur only if the kinetic energy of the colliding particles in the center-of-mass coordinate system exceeds the energy difference between the Y, b and the X, a pairs. This energy difference is called the *threshold* of the reaction. The dependence of a cross section $\sigma(E)$ on the energy E is often referred to as the

Coordinate System	Time	Velocity Relations	Velocity Diagrams	Energy of Target Nucleus	Total Kinetic Energy
At Rest	Before Collision		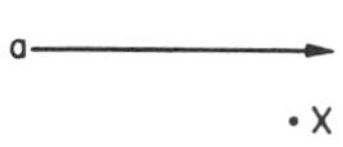 a, X	0	$\frac{1}{2}M_a v_i^2$
Center of Mass	Before Collision	$M_a v_i/(M_a+M_x)$ is Subtracted from Both Velocities	 a, X	$\frac{\frac{1}{2}M_x M_a^2 v_i^2}{(M_a+M_x)^2}$	$\frac{\frac{1}{2}M_a M_x v_i^2}{M_a+M_x}$
Center of Mass	After Collision	Only Direction of Velocities Changes	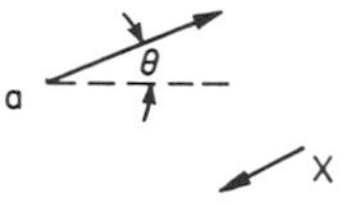 a, θ, X	$\frac{\frac{1}{2}M_x M_a^2 v_i^2}{(M_a+M_x)^2}$	$\frac{\frac{1}{2}M_a M_x v_i^2}{M_a+M_x}$
At Rest	After Collision	$M_a v_i/(M_a+M_x)$ is Added to Both Velocities	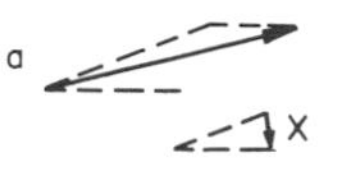 a, X	$\frac{2M_x M_a^2 v_i^2}{(M_a+M_x)^2}\sin^2\frac{1}{2}\vartheta$	$\frac{1}{2}M_a v_i^2$

FIG. 4.1. Reduction to the center-of-mass coordinate system in binary collisions.

excitation function. Energy in this connection means the kinetic energy of both particles in the center-of-mass coordinate system. This is smaller than the energy E_{lab}, of the single bombarding particle: one sees from Fig. 4.1 that

$$E = \frac{M_t}{M_t + M_b} E_{lab} \tag{4.1}$$

in which M_t is the mass of the target nucleus, at rest in the laboratory frame of reference and M_b the mass of the bombarding particle. In the relativistic region one has for the sum of the kinetic energies of the two particles in the center-of-mass coordinate system

$$E = [(M_t + M_b)^2c^4 + 2M_tc^2E_{lab}]^{1/2} - (M_t + M_b)c^2 \tag{4.1a}$$

2. Types of Reactions

Nuclear reactions can be classified from at least two points of view. The first classification refers to the complexity of the nuclei which react with each other and which emerge from the reaction. The second classification is based on the mechanism of the reaction, i.e., the picture that is most appropriate for its description.

In most reactions treated in the literature, at most one of the reacting nuclei and at most one of the reaction products are complex; the others are either protons or neutrons or, at any rate, nuclei with a mass number of 4 or below. Typical reactions of this nature are induced by bombarding targets with neutrons or protons and observing the outcoming neutrons or protons. If the outcoming particle is identical with the incident one, the process is called scattering; this is elastic if the target nucleus is left in its initial state, and inelastic if its state of excitation is changed. Reactions of this nature are often denoted by symbols such as (n,n) or (n,p), where the first letter refers to the incoming particle (neutron in these cases) and the second to the simple reaction product (neutron or proton in these instances). However, even if the incident, or emerging, particles are deuterons (d), tritons (t), He_3 or He_4 nuclei (α), or two neutrons $(2n)$, the reaction is considered simple. Typical simple reactions are

$$N^{14}(n,p)C^{14};\ C^{12}(d,p)C^{13};\ F^{19}(n,2n)F^{18};\ B^{10}(n,\alpha)Li^7$$

The first symbol gives the target nucleus; the last one gives the more complex reaction product.

In the second category of reactions under this classification are those in which a γ ray is either the bombarding particle or the reaction product. $Be^9(\gamma,n)2He^4$, $F^{19}(\gamma,p)O^{18}$, $In^{113}(n,\gamma)In^{114}$ are examples of such reactions.

In the last category of reactions under this classification, either the two colliding particles, or at least two of the reaction products, are complex. The fission reaction is the best known in this category; it results in two approximately equally heavy fragments of a very heavy nucleus, such as $U_{92}{}^{235}$ or $Th_{90}{}^{232}$. Fission can be most easily induced by neutron absorption; it can be induced also by γ rays (photofission) or by charged particles. The so-called heavy-ion reactions also fall into this category. $N^{14} + N^{14} \rightarrow N^{13} + N^{15}$, $Ho^{165} + B^{11} \rightarrow Hf^{172} + 4n$, and many similar ones are such reactions, as is also the scattering, elastic or inelastic, of two reasonably heavy nuclei.

Nuclear reactions can also be classified according to the reaction mechanism that provides the most appropriate description. We shall consider three such mechanisms: (1) processes in which the colliding partners temporarily fuse into a single "compound nucleus," (2) the direct processes in which the contact between partners is only peripheral, and (3) that of distant processes in which the colliding nuclei hardly come into direct contact.

3. Resonance Processes and Statistical Model

For many reactions the excitation function for the cross section, $\sigma(E)$, exhibits a number of more or less pronounced maxima. This behavior is particularly evident for low-energy neutron scattering on intermediate A nuclei and for neutron capture by heavy nuclei [(n,γ) reaction]. Often the cross section changes by as much as a factor of a thousand in passing through maxima and minima (see Fig. 10.1, Sec. X). The characteristic maxima in the excitation functions of these reactions can be understood by picturing the reaction as a *resonance process.* Indeed, the general theory of nuclear reactions, at least for energies of the incident particle which are a couple of Mev or less, may be usefully organized about the concept of resonance. (Sec. X).

In a general reaction $X(a,b)Y$, the nuclear system consisting of X and a (or equivalently of Y and b), is called the compound system of the reaction. Thus F^{20} is the compound nucleus for the $F^{19}(n,p)O^{19}$, $F^{19}(n,\alpha)N^{16}$, $O^{18}(d,\alpha)N^{16}$, etc., reactions. The compound nucleus will have, in general, a ground state in which the particle a is bound, low excited states which may decay only by gamma radiation, and excited states the energy of which is greater than the separation energy of a. These latter states may decay by gamma radiation, emission of a, and possibly by the emission of other particles. They are, of course, not stationary, but their lifetimes may be very long compared with the time of traversal of the particle a across the nucleus. They are called quasi-stationary states or "resonance levels." In such states the energy is not entirely sharply defined; the energy uncertainty Γ is related to the lifetime τ of the state by $\Gamma\tau = \hbar$.

If, in the bombardment of X by a, the energy of the system is close to the energy (average energy) of one of the quasi-stationary states of the compound system, the collision may be described as a two-step resonance process. The incident particle is first captured by X to form the quasi-stationary state: this state subsequently decays either by the reemission of the particle a, or by γ radiation, or if there is sufficient energy, by the emission of other particles. The probability of the capture of the incident particle varies in a characteristic manner with the difference between the energy E of the colliding systems and the energy E_λ of the resonance level. The total cross section for the process is proportional to the capture probability. This probability varies with energy roughly as $[(E - E_\lambda)^2 + (\Gamma/2)^2]^{-1}$, where Γ is the energy width of the resonance level. As the energy of the incident particle is varied over a range covering the resonance energy, the cross section goes through a characteristic resonance maximum.

The second part of the resonance reaction process, the disintegration of the intermediate state, is governed by probability laws. The emission of each of the possible products of the collision has a certain probability characteristic of the product and of the resonance level; the different emission processes compete with each other. The transition rate for each decay process may be expressed as an energy width; this width divided by $\hbar$ is the probability of the decay per second. The energy widths associated with the different modes of decay of the resonance levels are called the partial widths of the level. Thus if the partial widths for proton, neutron, and γ-ray emission from a particular level are Γ_p, Γ_n, Γ_r (and no other processes are energetically possible), the probability that the disintegration eventually yields a neutron is Γ_n/Γ where

$$\Gamma = \Gamma_p + \Gamma_n + \Gamma_r$$

is the total width of the compound state.

The essential quantities governing the behavior of cross sections when resonance phenomena are of importance are the level positions and level widths. *On the average,* the separation between adjacent resonance levels decreases with increasing energy of excitation. Similarly, the spacing of adjacent levels decreases with increasing A. The level widths increase on the average with the excitation energy; they decrease with increasing magnitude of the electrostatic barrier, and depend also on a number of other energy dependent factors (Sec. X). Separations between neighboring levels and level widths fluctuate markedly from level to level, but separations and widths averaged over several levels appear to be relatively smoothly varying functions of the excitation energy and of A.

The effect of the electrostatic barrier manifests itself in a reduction of the Γ for the process inhibited by the barrier. Thus the probability of an $X(n,p)Y$ process is decreased by the electrostatic barrier which inhibits the emission of the proton by decreasing Γ_p, and hence the probability Γ_p/Γ that the second process be the emission of the proton. The probability of the formation of the quasi-stationary state of the compound nucleus is not materially affected by the barrier in this case. In the $X(p,n)Y$ process, on the other hand, the approach of the proton to the target nucleus is inhibited by the barrier; in this case the cross section for the formation of the quasi-stationary state is diminished. The barrier penetration factor in the partial width decreases in importance as the energy of the bombarding particle increases. If the energy of the proton is sufficient to cross the barrier freely, the proton widths will be generally similar to neutron widths for the same energy.

The basic concept of the compound nucleus model is the intermediate state of relatively long lifetime. If the lifetime of any intermediate state that can be formed is so short that the corresponding width $\Gamma = \hbar/\tau$ exceeds the level spacing, the resonances begin to coalesce and the compound nucleus model ceases to be useful.

The picture of well-separated resonance levels is valid and useful for reactions of light nuclei up to several Mev and for reactions of heavy nuclei only to less than 1 Mev. Above these energies, the density of the levels and their widths increase to such an extent that the levels overlap. Even if this is not the case, the experimental energy resolution becomes insufficient to identify individual levels so that only average cross sections can be measured. Only statistical properties of the compound nucleus remain relevant under such conditions. The compound nucleus is pictured then as a hot liquid or solid from which various particles evaporate. From heavy nuclei, neutrons evaporate most easily even at high energies of excitation but proton emission is also common from lighter nuclei.

As a rule, if the compound nucleus has a high energy of excitation, the emission of a single nucleon does not leave the residual nucleus in its normal state. If its energy of excitation is below the separation energy of an additional nucleon, it emits electromagnetic radiation. If the residual nucleus' energy of excitation exceeds the separation energy of a neutron, this usually is emitted in preference to electromagnetic radiation. Hence $(n,2n)$ reactions are quite common, particularly if the separation energy of the neutron from the initial nucleus is low, as, for instance, in Be^9. Because of the Coulomb barrier, charged particles are less likely to appear as secondary reaction products from heavy compound nuclei, but in the case of light nuclei the (n,np) reaction is not uncommon.

Even the emission of the second particle may leave the residual nucleus in an excited state, and it may then again emit electromagnetic radiation, or a third particle. A very highly excited state of the compound nucleus can be obtained most readily by bombardment with a heavy ion, and the emission of as many as seven neutrons has been observed, for instance, in the reaction $Tb_{65}^{159} + N_7^{14} \rightarrow Hf_{72}^{166} + 7n$. The bombarding energy was in excess of 100 Mev, and the original energy of excitation of the compound nucleus around 125 Mev.

The basic assumption of the statistical model is that the behavior of the "compound nucleus" is independent of its method of creation and depends only on its constitution (number of protons and neutrons) and on its energy content. In its original form, it is also assumed that all energetically possible reaction products are formed with equal a priori probabilities, i.e., that the probabilities of the various forms of disintegration of the compound nucleus depend only on trivial factors such as the penetration probability of the emitted particle through the electrostatic potential barrier. This last assumption implies, in particular, that the properties of all resonance states of the compound nucleus are nearly alike and do not show individual preferences for the emission probabilities of a particular reaction product. This assumption, useful as it is in the description and interpretation of many reactions, is less well founded than the first assumption. Indeed, there are maxima in the energy dependence of cross sections which were first systematically observed by Barschall and which cannot be explained by variations of the penetration factors. These show that levels in certain energy intervals have individual properties. These are explained and characterized by the giant resonance model (Sec. X), a modification of the statistical model.

If one is interested only in the cross sections averaged over many resonances, the aforementioned maxima can be explained also by the optical model. This model assumes that the effect of the target nucleus on the incident particle can be well approximated by a potential which is, however, not real but has an imaginary part causing an "absorption" of the incident particle. Physically, this absorption corresponds to the fusion of the incident particle into the target, i.e., the formation of a compound nucleus wherein the state of the target nucleus is no longer its initial state. However, the imaginary part of the potential is small so that the incident particle's penetration into the target nucleus affects the state of the latter only slowly. Before the absorption, the motion of the incident nucleon can be described by an individual (one particle) wave function which is immediately diffracted but only slowly absorbed by the target. This picture of the optical model is very successful in describing the behavior of cross sections averaged over many individual resonances and is useful also in the region of overlapping resonances. The relation between the giant resonance and the optical models is somewhat subtle and will be dealt with in Sec. X.

4. Direct Processes

Even at low energies, collisions do not lead to the formation of a compound nucleus unless the collision energy is close to a resonance. Between resonances, the probability of reaction via the compound-nucleus mechanism is small even for a nearly head-on collision, i.e., even if the angular momentum of the colliding particles around their center of mass is quite low. Peripheral collisions, with relatively large angular momenta, in which only the outer regions of the colliding nuclei come into contact, can become important under such conditions. Usually, only one nucleon of each nucleus interacts with the other nucleus. Since no compound nucleus if formed as an intermediate stage of the reaction, one speaks of a direct process in this case.

The situation is similar if the energy of the incident nucleon is so high that the binding energies of the nucleons in the target nucleus become negligible in comparison. This then resembles a conglomerate of free nucleons, and the very fast incident nucleon interacts directly only with one of them. For this reason, in order to have at least approximately compound-nucleus behavior at high energy of excitation, this energy is best furnished as kinetic energy of several nucleons, i.e., in the form of a fast "heavy ion" such as the N^{14} in the $Tb_{65}{}^{159} + N_7{}^{14} \rightarrow Hf_{72}{}^{166} + 7n$ reaction mentioned previously. Direct-interaction models are most useful to describe reactions induced by deuterons. The deuteron, as a charged particle, is repelled by the target and this makes the formation of the compound nucleus rather improbable. However, the repulsion does not prevent the direct process equally effectively because the deuteron is a rather loose structure, with a low binding energy, in which relatively large neutron-proton separations are not improbable. As a result the neutron may enter the nucleus, and be absorbed by it, while the proton remains relatively far from it, not exposed to the electrostatic repulsion to the same extent as if both particles had to enter the nucleus. The resulting process is called "stripping" because the neutron is stripped off the proton by the collision. Similar processes play a role also in (d,p) reactions and in the inverse "pick-up" reactions (n,d), (p,d).

The stripping and pick-up processes are only examples of the more general class of direct processes. Common to the theory of all these processes is the idea that the effective interaction between target and incident particle takes place close to the surface of the nucleus and that it can be treated as a perturbation—a treatment which would yield incorrect results in the resonance region. The theory of the direct processes is considered in Sec. XI.

For heavy nuclei ($A > 90$) charged particle reactions are inhibited by the large electrostatic barrier if the incident energy is below about 5 Mev (the potential energy at the top of the barrier is of the order $0.4A^{2/3}$ Mev, that is, about 8 Mev for $A = 100$, 14 Mev for $A = 200$).

5. Distant Reactions

In distant reactions the reacting nuclei maintain their identity, at least essentially. Their paths are affected only by the electrostatic repulsion of the other nucleus, and the colliding nuclei influence each other only indirectly.

Coulomb excitation is a distant reaction. A charged particle of relatively low energy is prevented by the electrostatic barrier from reaching a nucleus with a large charge Ze. However, the passage of this nucleus nearby the heavy nucleus subjects the latter to a rapidly changing electromagnetic field. This field has a similar effect as a radiation field and causes transitions in the heavy nucleus. Thus, Coulomb excitation is a particular kind of photoeffect in which the passage of a charged particle provides the rapidly oscillating electromagnetic field.

The so-called transfer reactions provide another example of distant processes. In this case both colliding partners are reasonably heavy and carry rather high charges and, as a result of their strong repulsion, cannot come into direct contact. However, during the period in which they are in close proximity, one of the nucleons from one of the colliding nuclei may tunnel over to the other nucleus. Many of the heavy-ion ractions, such as $N^{14} + N^{14} \rightarrow N^{13} + N^{15}$, are of this nature; in this case a neutron tunnels over from one N^{14} to the other.

Nuclear reactions can be used to produce radioactive elements. They often lead also to excited states of nuclei, of longer or shorter lifetimes, the study of which is of great importance for the understanding of nuclear structure.

V. TWO-BODY SYSTEMS; INTERACTIONS BETWEEN NUCLEONS

1. Internucleon Forces

Our most detailed information on the nature of the interactions between nucleons is derived from a study of the behavior of two-body systems, in particular from study of properties of the deuteron and from analysis of neutron-proton and proton-proton scattering experiments. The two-body systems play a special role since only for them can the theoretical consequences of an assumed interaction between the nucleons be calculated with precision. The problem is to invent neutron-proton and proton-proton interactions in terms of which the observed properties of these two-particle systems can be derived theoretically. (We have no direct information on the neutron-neutron interaction, since two neutrons form no bound system and, as yet, it is impossible to perform neutron-neutron scattering experiments.)

The data which must be correlated by an assumed *n-p* interaction are very extensive. The neutron-proton system has a bound state, the normal state of the deuteron H^2, with a binding energy 2.23 Mev, spin 1, magnetic moment 0.8573 nuclear magnetons, quadrupole moment of 2.74×10^{-27} cm². Neutron scattering on protons and proton-proton scattering have been investigated over a broad range of energies, up to several hundred Mev. Over most of this range, not only the total scattering cross sections have been measured but also the angular distribution and the polarization of the scattered particles. The interactions used in present-day calculations are based on these measurements. Their complexity shows that they cannot be accepted as a unique, basic interaction, not even in the sense in which Coulomb's law is so accepted. Either they will prove a somewhat clumsy expression for a relation that is much simpler from a more fundamental point of view, or they will be derived as an elaborate consequence of some more basic theory. In fact, there is good reason to believe that the nuclear forces are due to a simple and fundamental interaction between nucleons with mesons, in particular the π meson, in a way similar to that in which the Coulomb interaction between electrically charged particles is due to a primitive interaction between charges and the photon (electromagnetic) field. Efforts to derive the nucleon-nucleon interaction from such a more fundamental interaction with mesons has had considerable but not complete success—hence the reliance on direct observations involving the nucleon-nucleon interaction itself.

The main qualitative features that the interaction must exhibit in order to account for the data on the behavior of two-particle systems have been known for a long time.

1. The interaction is of short range—1 to 2×10^{-13} cm—and, at this range, is much stronger than the electrostatic interaction. The short range is required by the observed isotropy of the neutron-proton angular distribution at energies up to 10 Mev, by the sensitive experiments on the scattering by ortho- and parahydrogen, as well as by other data.

2. The interaction is spin dependent, i.e., the force depends on the relative orientation of the spins of the two particles. This conclusion was forced by the behavior of the scattering cross section at low neutron energies ($\sim$1 ev) and is most clearly indicated by the large difference in the cross sections of ortho- and parahydrogen for thermal neutrons. These last cross sections also give a rather accurate value for the strength (effective depth of the potential) of the neutron-proton interaction for the singlet state (antiparallel spins). The binding energy of the deuteron provides a measure of this strength only for the triplet state (parallel spins). The cross sections for ortho- and parahydrogen also show that the neutron spin is $\frac{1}{2}$.

3. The neutron-proton interaction depends not only on the orientation of the proton and neutron spins with respect to each other, but also on their orientation with respect to the line passing through the two nucleons. An interaction of this character is required to explain the observed quadrupole moment of the deuteron; this moment indicates that the proton distribution in the ground state of the deuteron is not spherically symmetrical but is elongated in the direction of the total spin of the system. The simplest interaction which depends on the orientation of the spins $\boldsymbol{\sigma}_1$, $\boldsymbol{\sigma}_2$ of neutron and proton relative to the radius vector, $\mathbf{r}$, which specifies the separation of these particles, is

$$V_T = J_T(r)\left[\frac{3(\boldsymbol{\sigma}_1 \cdot \mathbf{r})(\boldsymbol{\sigma}_2 \cdot \mathbf{r})}{r^2} - (\boldsymbol{\sigma}_1 \cdot \boldsymbol{\sigma}_2)\right] = J_T(r)S_{12} \tag{5.1}$$

$J_T(r)$ specifies the radial dependence of the interaction. V_T is called the tensor interaction.

4. There is some evidence that the nuclear interaction is repulsive at very small separation of the nucleons. If this were not the case and the attraction at large distances continued down to short ones, a state of the A nucleon system in which all nucleons are within the range of each other would have a total binding energy proportional to the number of interacting pairs, i.e., proportional to $\frac{1}{2}A(A-1)$. This is not the case; the binding energy does not increase faster than the first power of A. It cannot be possible, therefore, to pack all the nucleons into the strongly attractive ranges of each other, as would be possible if the potential energy between two nucleons continued to be negative down to zero distance. This so-called saturation argument has played a great role in the early stages of the study of nuclear forces. It can be met in more sophisticated ways than by the assumption that the nuclear forces become repulsive at short distances (as do the forces between neutral atoms), but present thinking and evidence favor the assumption of a strong repulsive force, a "hard core," at short distances.

A particular potential which was originally proposed on the basis of the meson theory of nuclear forces and which gives a reasonably accurate description of two body systems is Levy's interaction. This is strongly repulsive at very short distances ($r < 0.55 \times 10^{-13}$ cm). Outside this range it consists of three parts. The first part is a so-called exchange interaction

$$V_1 = -0.55\frac{\hbar c}{r}e^{-r/a}\left[\mathbf{P} + \frac{1}{3}\boldsymbol{\sigma}_1 \cdot \boldsymbol{\sigma}_2(1 - \mathbf{P})\right] \tag{5.2}$$

where $a = 1.40 \times 10^{-13}$ cm is the Compton wavelength of the π meson. Since $0.55\hbar c = 75.4e^2$, at short distances this is indeed a much stronger interaction than the electrostatic one. $\mathbf{P}$ is the so-called Majorana exchange operator: it gives a factor 1 if the wave function is symmetric with respect to the interchange of the space coordinates of the interacting particles; it gives a factor -1 if the wave function is antisymmetric with respect to such an interchange. In the case of a two-body system, **P** is 1 if the orbital angular momentum is even (S, D, G, etc., states), $\mathbf{P} = -1$ if the orbital angular momentum is odd (P, F, etc., states).

The second part of the Levy's interaction is a tensor force

$$V_2 = -0.55 \frac{\hbar c}{r} e^{-r/a} \left(1 + \frac{3a}{r} + \frac{3a^2}{r^2}\right) S_{12}(2\mathbf{P} + 1) \tag{5.3}$$

where S_{12} is given by the bracket of (5.1). The last part of the interaction is given by an ordinary potential; it has neither exchange character nor does it depend on the orientation of the spin:

$$V_3 = -\frac{\hbar ca}{r^2} \left\{K_1\left(\frac{2r}{a}\right) + 0.05\left[K_1\left(\frac{r}{a}\right)\right]^2\right\} \tag{5.4}$$

The K_1 is the Hankel function, as defined, for example, in Watson's "Theory of Bessel Functions." The derivation, from meson theory, of this last part of the nuclear interaction has been questioned. From the point of view of the interpretation of the two-body systems, an attractive interaction, similar to (5.4), is undoubtedly needed. The whole two-body interaction is, therefore, according to Levy

$$\begin{aligned} &V = V_1 + V_2 + V_3 && \text{for } r > 0.38a \\ &V \text{ large and positive} && \text{for } r < 0.38a \end{aligned} \tag{5.5}$$

Levy's potential, which is given above, will not be used below for actual calculations. It is included here only to illustrate the attempts to obtain an expression for the nuclear interaction. Two of the constants in Levy's potential (the range $0.38a$ of the hard core and an over-all multiplicative constant) were so determined as to give the observed values for the binding energy of the deuteron and for the neutron-proton scattering at low energies. In addition to these, the potential reproduces the observed behavior of the neutron-proton and proton-proton scattering up to about 30 Mev and gives a fairly accurate value for the quadrupole moment of the deuteron. Actually, the majority of nuclear phenomena depend only on the interaction of particles with very low angular momenta about their common center of mass. In particular, unless the energy is well above 18 Mev, the two-body system's interaction is restricted to angular momenta 0 and 1. Hence, only the low angular momentum parts of the potentials can be tested by experiments below about 18 Mev on two-body systems.

Expressions proposed more recently for the two-nucleon interaction account more completely for the information provided by the experiments enumerated earlier. They are, however, quite complicated and contain many numerical constants. They proved useful for a variety of computations and unquestionably condensed an enormous amount of information. Nevertheless, since in their present form they are only a condensation of experimental data, they will not be given in detail. The most important qualitative difference between these potentials and that of Levy is the presence of a spin-orbit term, i.e., a spin dependence of the interaction that couples the spins of the interacting nucleons to their motion around their center of mass. Such a term is useful not only in explaining scattering experiments but also properties of complex nuclei. As a result of the spin-orbit interaction, the energy of the state in which the spin of a nucleon is parallel to its orbital angular momentum becomes different (lower) than the energy of the state in which spin and orbital angular momenta are antiparallel. This difference forms one of the bases of the most successful theory of complex nuclei, that of the j-j-coupling shell model.

The preceding discussion applies, principally, to the proton-neutron interaction. The proton-proton and the neutron-neutron systems differ from the proton-neutron system principally because of the exclusion principle. This principle demands that the wave function be antisymmetric with respect to the simultaneous interchange of space and spin coordinates, i.e., that the equation $\psi(r_1,\sigma_1,r_2,\sigma_2) = -\psi(r_2,\sigma_2,r_1,\sigma_1)$ be valid if both 1 and 2 refer to a proton, or if both refer to a neutron. All states of a two-body system can be characterized by an orbital angular momentum L and a total spin S which can assume the values 0 (singlet states) and 1 (triplet states). The spatial-coordinate part of the wave function is symmetric for even L and antisymmetric for odd L. The spin-dependent part of the wave function is symmetric for $S = 1$ and antisymmetric for $S = 0$. Since, according to the exclusion principle, the spin-dependent factor must be antisymmetric if the spatial factor is symmetric, and vice versa, only the states ${}^1S$, ${}^3P$, ${}^1D$, ${}^3F$, . . . are allowed for a two-proton system, and the same applies for a two-neutron system. It appears, then, that the proton-proton scattering data can be fitted at least very closely by assuming that the interaction between two protons is—except for the electrostatic part—the same as that between two neutrons *in the same state*. The same is then assumed for the interaction between two neutrons. Under this assumption the proton-neutron interaction determines also the proton-proton and the neutron-neutron interaction; in fact, in order to determine the latter interactions, the proton-neutron interaction must be known only for the ${}^1S$, ${}^3P$, ${}^1D$, etc., states. Naturally, the proton-proton interaction contains also the Coulomb term, e^2/r, in addition to the nuclear part which is the only one present in the proton-neutron and the neutron-neutron systems.

The assumption just made for the proton-proton and the neutron-neutron interactions is called the assumption of the charge independence of nuclear forces. Its consequences will be further pursued below. One of these consequences can be stated immediately: Since the proton-neutron system has only one bound state, and since this state is a ${}^3S$ state which is forbidden by the exclusion principle for both the proton-proton and the neutron-neutron systems, the latter systems have no bound states. This is a valid conclusion.

Whether the exact determination of the interaction between two nucleons would provide a firm foundation for the Hamiltonian of a system containing several nucleons has often been questioned. The presence of a third nucleon may modify the interaction between two nucleons, or, expressed differently, there may be three-particle interactions. Such interactions are indeed present in systems of three or more atoms, and the nucleon does not appear to be a very much simpler and more elementary system than an atom. However, the success of the assumption of only two-particle interactions in the shell model (Sec. VIII) appears to show that the effective interaction is, even in reasonably heavy nuclei, a sum of two-particle interactions. This would be unlikely if the basic interaction contained three-particle terms in a significant way.

It is by no means certain that the best way to express the interaction between two nucleons is the traditional way, by means of an interaction operator, such as (5.1) or the more elaborate one proposed by Hamada and Johnston. The experiments, in particular the scattering experiments, give characteristics of the wave function more directly than an interaction; the interaction operator is a derived quantity so chosen as to fit the information concerning the wave function that can be derived from the scattering experiments. The interaction operator is then used, conversely, to calculate the wave function, both for scattering—in which case it should merely reproduce the information that was put in—and also for other situations, in particular when a third particle is present. It would seem more natural to obtain some characteristic of the wave function and then use this also under conditions different from those for which it was obtained. Breit and Bouricius observed that the logarithmic derivative of the scattering wave function at a certain separation of the two interacting nucleons is nearly energy-independent and proposed to characterize the interaction by this logarithmic derivative. Unfortunately, the extension of such a condition from the two-particle to the three-particle case is not entirely unique and, in spite of several promising and interesting attempts, this novel method of describing the nuclear interaction is still in the exploratory stage.

The preceding review of the attempts to find a mathematical framework from which the properties of nuclei can be derived, at least in principle, is far from complete. It is interesting how rarely we shall have occasion to refer to these attempts later on and how little they have, on the whole, guided nuclear theory or experimentation. It is to be hoped that this situation will change when a more adequate and more firmly established description of the interaction between nuclei is found.

2. Saturation Properties and Internucleon Forces

If one assumes that the potential energy of a nuclear system is the sum of two particle interaction energies, then the fact that the binding energy per particle of nuclei is roughly constant—the saturation property—leads to important theoretical restrictions on the nature of the interactions between nucleons. The interaction cannot be attractive throughout and of the ordinary type, as this would not lead to saturation. If the interaction were of this nature, all particles would tend to be in the range of each other. Hence, the nuclear volume would be constant rather than proportional to A, and the binding energy proportional to A^2 rather than to A, as observed. The saturation property can be obtained by a mixture of ordinary and Majorana forces of the form

$$V = J(r)(\alpha + \beta \mathbf{P})$$

with $J(r)$ negative for all r only if $\beta \geq 4\alpha$, that is, if the Majorana force dominates. However, interactions which satisfy the saturation requirements predict high-energy n-p and p-p scattering distributions in disagreement with the scattering data. It appears, consequently, that it is necessary to assume either that $J(r)$ is not negative throughout (cf. the Levy potential which is positive for $r < 0.55 \times 10^{-13}$ cm) or that the interaction between nucleons within nuclear matter is different from the interaction between isolated pairs of nucleons, i.e., that many body forces (as distinguished from the usual two-body interactions) are of importance. It is believed at present that the former effect, the repulsive core, is more important.

VI. CHARGE INDEPENDENCE OF NUCLEAR FORCES: THE ISOTOPIC OR ISOBARIC SPIN QUANTUM NUMBER

1. Isotopic Spin Multiplets in Light Nuclei

The hypothesis of the charge independence of nuclear forces has proved increasingly fruitful. If nuclear interactions are in fact charge independent, i.e., the forces between nucleons are the same for neutrons and protons, then the states of a nuclear system may be characterized by an additional quantum number T, the so-called isotopic or isobaric spin quantum number.

The significance of T is analogous to that of the spin quantum number S for atomic states. The states of the atom are, according to Pauli's principle, wholly antisymmetric with respect to the exchange of space *and* spin coordinates of the electrons. As a consequence the *symmetry character* of an atomic state is, with respect to exchange of space coordinates only, completely defined by the quantum number S, the total spin angular momentum number of the state: S determines not only the total spin angular momentum but also the symmetry character with respect to the exchange of the spin variables of the electrons, and the symmetry character with respect to the exchange of space coordinates must be opposite thereto. If the atomic Hamiltonian is independent of the spin variables, and this is approximately true for light atoms, then S is a constant of motion and different states with different values of S are characterized by different *spatial* symmetries.

A similar argument can be applied to nuclear states. A new variable—the isotopic spin variable—may be introduced for nucleons. This variable plays a role similar to that of the electron-spin variable. It distinguishes between neutrons and protons just as the electron-spin variable is used to distinguish electrons with spins along or opposed to the z axis. With this additional variable, the states of a nuclear system must be antisymmetric in the exchange of space, ordinary spin, and isotopic spin coordinates of any two particles. The quantum number T bears the same realtion to the isotopic spin variable τ that S does to the electron-spin variable s. The symmetry character of a state with respect to exchanges of *space and ordinary spin* variables is uniquely defined if the state is characterized by a definite value of the total isotopic spin quantum number T. If the nuclear Hamiltonian is independent of the isotopic spin variable, i.e., if the forces are the same for protons and neutrons, then T is a constant of motion and has a well defined value for the stationary states of nuclei.

The isotopic spin quantum number T will regulate transitions between nuclear states in much the same way as do the angular momentum quantum numbers. Conversely, recent work on nuclear reactions, furnishing evidence that T is a constant of motion at least for the lighter nuclei, supports the assumption of the charge independence of nuclear forces.

In the atomic case there are $2S + 1$ independent states belonging to the same quantum number S with different values of the z component, S_z, of the spin angular momentum. S_z determines the difference between the number of electrons with spin up and the number with spin down. If the atomic Hamiltonian is spin independent, all these states have the same energy. Similarly, there are $2T + 1$ independent nuclear states associated with isotopic spin quantum number T belonging to different values of T_ζ ($T_\zeta = T, T - 1, \ldots, -T$). This latter variable measures the difference between the number of neutrons and protons in the nucleus and has the value $(N - Z)/2$ for a nucleus with N neutrons and Z protons. The different T_ζ associated with a particular T describe states of several isobaric nuclei; the dependence of all these states on the space and ordinary spin variables of the nucleons is the same. If the nuclear Hamiltonian is charge-independent, these states have the same energy. Hence, while the quantum number S stipulates the equality of the energy of those states of a given atom which differ in the component of the spin angular momentum in a given direction, the quantum number T postulates equality of the energy of states of several isobars, those united into a T multiplet. This is illustrated in Fig. 6.1*a*. Two states of a T multiplet differ only by neutrons being substituted in one of these states for some of the protons in the other state, or vice versa. Thus, naturally, the total angular momentum J and the parity are the same for all states of a T multiplet.

Figure 6.1*a* does not give an accurate picture of the energy values of isobars; it must be corrected for the difference between the interaction of two protons and that of two neutrons or of a proton-neutron pair. According to the hypothesis of charge independence, the latter two interactions are equal while the proton-proton interaction differs from them by the electrostatic repulsion of the protons. As a result of this interaction the energy is not independent of T_ζ; the members of the T multiplet with higher T_ζ (fewer protons) have lower energy. This is illustrated in Fig. 6.1*b*. Since the electrostatic potential within a nucleus can hardly be expected to depend critically on the nuclear state, the dependence on T_ζ is very nearly the same for all T multiplets. Hence the energy difference between the members of the two $T = 1$ multiplets is very nearly the same in Fig. 6.1*b*. It follows from this picture that the T value of the normal state of a nucleus which is stable against positron emission and K capture is equal to its T_ζ value, that is, equal to $\frac{1}{2}(N - Z)$ for that nucleus. Clearly, T cannot be smaller than the T_ζ of the stable isobar. If T were larger, the T multiplet would have a state with a larger T_ζ. The total energy of this state would then be smaller and the original nucleus would be able to decay into it by positron emission or K capture. It follows, for instance, that $T = 0$ for the normal states of H_1^2, He_2^4, Li_3^6, B_5^{10}, C_6^{12}, N_7^{14}, O_8^{16} and $T = \frac{1}{2}$ for the normal states of all odd A nuclei up to Cl_{17}^{35}. The T value of the normal state of Cl_{17}^{37} is, on the other hand, $\frac{3}{2}$; that of O_8^{18}, Ne_{10}^{22}, Mg_{12}^{26}, etc., is 1.

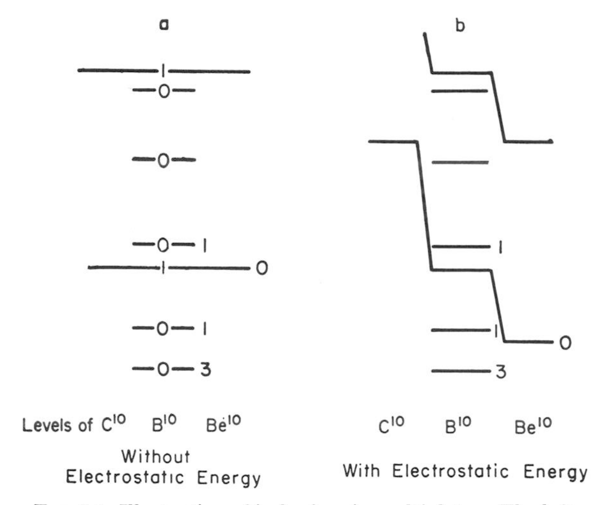

FIG. 6.1. Illustration of isobaric spin multiplets. The left side of both diagrams gives the levels of a nucleus with $T_\zeta = -1$ ($N - Z = -2$); the center gives the levels of the isobar with an equal number of neutrons and protons ($T_\zeta = 0$); and the right side, the levels of the isobar with $T_\zeta = 1$ ($N - Z = 2$). The diagram on the left would be valid if the electrostatic energy could be disregarded. In this case the binding energies of the levels of the $T_\zeta = \pm 1$ isobars would be equal and the $T_\zeta = 0$ isobar would have levels at the same binding energies. The three states with the same binding energies form an isobaric spin triplet. These are marked with a 1 in Fig. 6.1*a* at the center of the line representing the level. (The numbers at the right give the J value of the level, when known.) The nucleus with $T_\zeta = 0$ has further levels which are isobaric spin singlet levels ($T = 0$); these have a 0 at the center of the line representing the level. The diagram on the right illustrates the change in the binding energy introduced by the electrostatic forces. This diagram actually gives the low-lying levels of C^{10} (left side), of B^{10} (middle), and of Be^{10} (right side) of the diagram.

The normal states of mirror nuclei, such as Li^7 and Be^7, are the two members of the same isotopic spin doublet; their binding energies differ only as far as electrostatic effects are concerned. We have used this fact in Sec. III to calculate the difference between the electrostatic energies of mirror nuclei, and hence to estimate their size. The normal state of Be_4^{10} ($T = 1$) is stable against K capture or positron emission (it decays by electron emission). Hence $T = 1$ for this state. The other members of this T multiplet are the normal state of C_6^{10} ($T_\zeta = -1$) and the 1.74 Mev excited state of $B_5^{10}(T_\zeta = 0)$. The binding energies of these states differ from that of the normal state of Be_4^{10} only by the electrostatic energy and can easily be calculated therefrom. Many similar e· m-ples are known.

In particular, in addition to the $T = \frac{1}{2}$ and $T = 1$ multiplets, all the members of at least two $T = \frac{3}{2}$ multiplets have been observed. The first of these includes the normal states of C^9 and Li^9 and the highly excited states of B^9 and Be^9. Even more accurately known is the $T = \frac{3}{2}$ multiplet which contains the normal states of O^{13} and B^{13} and the states of C^{13} and N^{13} with 15.1-Mev excitation energies. That the last excitation energies are the same within 0.05 Mev is another example of the fact, already mentioned in connection with the two $T = 1$ states in Fig. 6.1, that the energy differences between corresponding members of isotopic spin multiplets depend mainly on Z and but little on the excitation energy.

If the nuclear interaction were truly charge-independent, the difference between the masses of the members of a T multiplet would be due to (1) the mass difference between neutron and proton plus electron and (2) to the difference in the electrostatic energy of these nuclei. The former quantity gives a term of $0.84T_\zeta$ millimass units in the mass formula. The second effect (2) can be calculated only in first approximation; this can be obtained easily only if one assumes that the charge distribution of the protons is uniform over the nuclear sphere. Both these approximations can be further supported, and they seem to be well confirmed also by the experimental data. The electrostatic energy then becomes $0.6Z(Z - 1)e^2/R$, where R is an effective nuclear radius, and the energy of the T_ζ member of an isobaric multiplet becomes, apart from a common constant (note that $Z = \frac{1}{2}A - T_\zeta$),

$$\frac{0.6e^2}{R}\left[\frac{1}{4}A(A-2) - (A-1)T_\zeta + T_\zeta^2\right] + 0.78T_\zeta \quad \text{Mev} \quad (6.1)$$

Essentially this same formula was used (Sec. III, 1) to obtain the nuclear radius from the measured mass difference between mirror nuclei, i.e., nuclei the normal states of which are members of a $T = \frac{1}{2}$ isotopic spin multiplet. The fact, noted previously, that the energy difference between members of various isotopic spin multiplets depend but little on the excitation energy shows that the effective nuclear radius R in (6.1) remains the same up to rather high excitation energies. The fact that the radii obtained by comparing (6.1) with experimental values agree with radii obtained by other methods is added evidence for the validity of (6.1).

2. *T* Multiplets in Heavier Nuclei

The mass difference between adjoining members of a T multiplet $[-(0.6e^2/R)(A - 2T_\zeta) + 0.78$ Mev, according to (6.1)] assumes a value of more than 10 Mev at $Z \sim 40$ and increases even further for heavier nuclei. On the other hand, the mass differences between the normal states of adjacent heavy nuclei are relatively small because of the balance between the symmetry and Coulomb energies along the valley of stability. Let us consider a stable nucleus such as Ge_{32}^{74}. The value of T_ζ for the nucleus is 5, and accordingly that is the T value of its normal state. The T_ζ of the adjoining isobar As_{31}^{74} is 4; this is also the T value of its normal state. The mass of this state is only slightly in excess of the mass of the normal state of Ge_{32}^{74}. However, the mass of the $T_\zeta = 4$ member of the isospin multiplet which includes the normal state of Ge_{32}^{74} ($T = 5$) lies some (10 − 0.78) Mev above the normal state of Ge^{74}. As the As^{74} normal state is 2.56 Mev above that of Ge^{74}, the $T = 5$, $T_\zeta = 4$ state occurs at approximately 6.6-Mev excitation energy in As^{74}. The $T = 5$ levels become more numerous in As^{74} at energies above 6.6 Mev as there is a $T = 5$ level in As^{74} corresponding to every low-lying level of Ge^{74}. However, the density of these levels is much less than the density of the $T = 4$ levels in the same energy interval.

The situation just discussed for As^{74} is characteristic of all medium-heavy and heavy nuclei. The total isotopic spin of the normal state and all the excited states up to about 10 or even 15 Mev is $\frac{1}{2}(N - Z)$. The first level with total isospin $\frac{1}{2}(N - Z) + 1$ appears at about 10 Mev or more, and from there on their level density increases with energy in much the same way as does the density of levels with $T = (N - Z)/2$ relative to the normal state. Thus it seems surprising that these levels can be found at all; one would suspect that they would dissolve entirely in the multitude of $T = \frac{1}{2}(N - Z)$ levels which lie nearby. This is not so, however, and the $T = \frac{1}{2}(N + Z) + 1$ levels can be seen either individually or in their effect in altering the character of the $T = \frac{1}{2}(N - Z)$ that lie nearby. The latter phenomenon, called giant resonance effect, will be discussed later (Sec. X). It suffices to note here that the states in question, currently referred to as analogue states, can be experimentally observed in a number of ways. However we shall discuss only the two most conventional techniques.

The first way of observing the "analogue states," i.e., states with isotopic spin $\frac{1}{2}(N - Z) \times 1$, is based on the fact that they are formed in some nuclear reactions with much greater probability than the neighboring states with isospin $\frac{1}{2}(N - Z)$. In the case of As^{74} just considered, this manifests itself in the energy distribution of neutrons from the direct reaction $Ge_{32}^{74}(p,n)As_{33}^{74}$. The neutron spectrum shows a narrow peak at an energy corresponding to a Q value equal to the Coulomb energy difference between Ge_{32}^{74} and As_{33}^{74}. The Q value for this reaction is clearly the energy difference between the neutron and proton in the same nuclear state, which is just a Coulonb energy difference. Alternatively one can observe the states with $\frac{1}{2}(N - Z) + 1$ as compound-nucleus resonance states which have a much greater probability of formation from, and disintegration into, normal states than the surrounding states with isotopic spin $\frac{1}{2}(N - Z)$. Hence, the proton elastic scattering cross section on Ge^{73} shows a sharp change where the energy of the $Ge^{73} + p$ compound nucleus is equal to the energy of the $T = 5$ state of As^{74}.

The reasons for the preferential formation of the "analogue states" [i.e., states with an isotopic spin larger than $\frac{1}{2}(N - Z)$] both as end products of a reaction and also as compound states are very similar. The wave functions of these states are the same as those of low-lying states of a nucleus—the isobaric nucleus with one less proton and one more neutron. They are relatively simple, therefore, and their formation does not require the extensive rearrangement in the structure of the target nucleus which the formation of a very highly excited $T = \frac{1}{2}(N - Z)$ state requires. This point can be made more explicit on the basis of the independent particle (shell) model but is probably true much more generally.

Comparison of the mass differences between adjoining isobars with (6.1) shows that R increases somewhat less fast than $A^{1/3}$. This is in accord with the decreasing tendency of $[r^2/A^{2/3}]^{1/2}$ noted at the end of Sec. III. The mass differences are now available over practically all the periodic table.

3. Selection Rules Following from Charge Independence

In addition to the consequences which one can derive from the concept of the T multiplet for the binding energies of the states of a T multiplet, a number of selection rules follow from the validity of the quantum number T. These are stated here without derivation.

Collision of two particles with isotopic spin T_1 and T_2 can give only compound states with T between $|T_1 - T_2|$ and $T_1 + T_2$. Conversely, disintegration of a compound state with a given T value always gives particles in such states that their isotopic spins T'_1 and T'_2 can form a triangle with T, that is, that $|T'_1 - T'_2| \leq T \leq T'_1 + T'_2$. Even if the reaction does not proceed via a compound state, the reaction is "forbidden" if the intervals $(|T_1 - T_2|, T_1 + T_2)$ and $(|T'_1 - T'_2|, T'_1 + T'_2)$ have no common element. Thus, the reaction $B_5^{10} + H_1^2$ should not lead to He^4 plus Be^8 in a $T = 1$ state. Electromagnetic radiation can change T only by 0 or ± 1 and $0 \to 0$ is forbidden for dipole transitions (naturally, electromagnetic radiation does not change T_ζ). Similarly, only $\Delta T = \pm 1$ is possible in allowed β transitions ($\Delta T_\zeta = \pm 1$ in this case). Even more specific is the rule for the so-called Fermi matrix element (see Sec. XII); this gives an allowed transition only from one member of an isotopic spin multiplet to another member of the *same* multiplet. Experimentally, all these rules seem to be valid in the great majority of cases.

All these rules can be derived from the invariance of the nuclear Hamiltonian with respect to interchanges of protons and neutrons. They are not precise because, even if the *nuclear* forces between all pairs of nucleons were the same, the electrostatic interaction would provide a perturbation as a result of which small violations of the preceding selection rules must be expected. However, in the vast majority of cases, these violations are very small. Thus the cross section of reactions which lead to the lower $T = 1$ state of B^{10} illustrated in Fig. 6.1 is very small if the isotopic spin of all other members in the reaction is 0. [It is less than 0.3×10^{-3} barn in the $C^{12}(d,\alpha)B^{10}$ reaction even if the energy is amply sufficient to produce the $T = 1$ state at 1.74 Mev.] There are innumerable similar examples. The situation is less favorable with respect to levels at higher energies where levels with different T values are quite close to each other, and even a relatively small interaction matrix element can cause a significant admixture of the wave function of one to the other. Thus the 9.17-Mev $T = 1$ state of N^{14} is excited with a considerable probability in the $B^{10} + Li^6 \to N^{14} + H^2$ reaction.

4. Validity of the Isotopic Spin Quantum Number

As was explained in the preceding section, the true nuclear Hamiltonian, since it includes also the electrostatic interaction between the protons, is surely not invariant with respect to interchanges of all the nucleons. The question remains whether the nuclear Hamiltonian is invariant under the interchange of the nucleons if one omits the Coulomb terms. In other words, are the specifically nuclear forces the same for protons and for neutrons? Second, one would like to know how accurately the true wave functions show the symmetry properties that they would show accurately if the invariance of the Hamiltonian under interchanges of all nucleons were accurate. This second question can also be formulated differently: How large admixtures of states with a different T are caused by the terms in the nuclear Hamiltonian that are not invariant with respect to interchanges of protons and neutrons? In particular, how large are these admixtures caused by the Coulomb terms?

The first question can be investigated on the basis of the theory of nuclear forces. This was done by Sugie and Fayazuddin with somewhat inconclusive results. The difference between the various nucleon-nucleon forces seems to be due, principally, to the difference between the masses of the charged $\pi^\pm$ and the uncharged π° mesons. (This difference, in its turn, is due to electromagnetic interaction.) According to Yukawa's theory, now generally accepted, the π mesons are the principal agents transmitting the nuclear forces, much as the electromagnetic field transmits the forces between electric charges. Since the proton-proton and the neutron-neutron forces are transmitted in first approximation only by the uncharged π° meson, whereas the charged mesons are principally responsible for the proton-neutron force, one expects the proton-proton and the neutron-neutron forces to be more nearly equal to each other than to the proton-neutron forces. However, even the latter difference is expected to be quite small—of the order of about 2 per cent.

One might think that scattering experiments would provide the most direct experimental test for the equality of proton-proton and proton-neutron interactions. Unfortunately, this is not the case because the effect of the electromagnetic and of the nuclear forces on the proton-proton scattering cannot be easily separated. The analysis, carried out principally by Noyes and by Breit, indicates that the scattering cross sections of the proton-proton and proton-neutron pairs would differ, in the absence of Coulomb forces, by a factor of about 2 in the singlet state; it would be

34×10^{-24} cm² for the *p-p* pair and 70×10^{-24} cm² for the *n-p* pair in the ^1S state. However, the scattering cross section is, in this case, a very sensitive function of the strength of the interaction: A change of not much more than 1 per cent in the strength of the interaction [in the 0.55 of (5.2)] causes such a change in the scattering cross section.

A similar conclusion can be arrived at by analyzing the very small deviations from (6.1), as found in the isobaric triplets Be^6-Li^6-He^6 and O^{14}-N^{14}-C^{14}. These also indicate a slight difference between the proton-proton, neutron-neutron, and proton-neutron forces. Even though neither of these comparisons is entirely conclusive, it seems probable that the proton-neutron interaction is larger by around 2 per cent than the proton-proton and the neutron-neutron interactions.

Even though the nuclear Hamiltonian is not invariant under all the interchanges of nucleons, the wave functions of low-lying stationary states exhibit to a very high approximation symmetry properties as if the invariance in question were accurate. The "isotopic spin purity" of low-lying states almost invariably exceeds 99 per cent. This conclusion was arrived at theoretically by assuming that the nuclear forces between all nucleons are the same and that all the deviation from isotopic spin purity is caused by the electromagnetic interaction. As we have just seen, this is not a safe assumption. The Coulomb interaction causes only little isotopic spin impurity, not because it is weak but because most of it can be taken into account as a constant potential which changes only the energy, not the wave functions. It has even been claimed that the isotopic spin purity increases with increasing A, that its effect on the wave functions is smallest when its effect on the energy is greatest. Whether or not this conclusion is fully valid, it seems certain that the isotopic impurity caused by the Coulomb term in the Hamiltonian causes very small admixtures only to the low-lying states—in most cases much less than 1 per cent.

VII. NUCLEAR MODELS. POWDER AND SHELL MODELS

1. General Remarks

Theoretical investigation of the properties of nuclear systems for $A > 2$ is made difficult by two features. First, we lack a clear conception of the nature of nuclear forces, although it becomes increasingly clear that they are rather complicated. Second, even if we assume forces of simple form but of short range, there appears to be no simple approximation procedure by means of which reasonably accurate solutions of the Schroedinger equation for many nucleon systems can be obtained. The situation for nuclei is in sharp contrast to that for atoms. For the latter systems the major interactions are well understood and substantiated; the nature of the forces is such that it is possible to justify, for the calculation of atomic wave functions, a rather simple, but yet very powerful, approximation procedure—the so-called shell model of the atom. As yet no nuclear model exists which has anything like the range of validity of the shell model for the atom.

Recently, it has become possible to calculate at least the properties of "nuclear matter" from an assumed interaction between pairs of nucleons. Such properties are, for instance, the energy density far from the nuclear surface or the probability for a given distance between two nucleons, both far from the surface. These properties are analogous to volume properties of a solid, or of a liquid, as contrasted with its surface properties. The significance of this progress, theoretical and practical, is very great. It should be recalled, nevertheless, that even in a nucleus as heavy as Pb, more than half of the nucleons are at the surface and also that the nucleons in the outer layers of the nucleus have more influence on its properties than the nucleons inside.

The concept of a "model" as we are using it here must be understood in a very broad sense. Any set of simplifying assumptions, physical or mathematical, by means of which the characteristics of nuclear systems may be computed to some degree of approximation may be called a model. So far, it has not been possible to justify by strict reasoning the simplifying assumptions of any of the proposed models. A model may be evaluated for the most part only by an investigation of the extent of its success, and failure, in describing the observed properties of nuclei. Even when a model is partially successful, it is often difficult to estimate to what extent its success reflects the accuracy of the model. The more or less extensively studied models are briefly reviewed in this and the following two sections.

2. Powder and Shell Models

Two broad classes of models have been proposed for nuclei. The powder models assume that the nuclear wave functions are very complicated and resemble the elementary chaos encountered in statistical mechanics. The shell models, on the contrary, liken the nuclear structure to a planetary system, or to the electrons in atoms, which move, at least in first approximation, independently of each other and are arranged in regular shells. Clearly, the two pictures are rather contradictory and their ranges of validity must be different. The shell or independent particle models are most nearly valid for the normal and low excited states of nuclei with closed shells, or with only a few particles missing from, or being present in addition to, closed shells. The powder model, on the other hand, can be expected to be valid for more highly excited states or, perhaps, if the nuclear constitution is far removed from that of a closed shell. In both cases, there are many close lying states of the nucleus and the interaction between these states creates conditions which resemble chaos.

The question naturally arises whether the powder model permits the calculation of any nuclear properties and, indeed, the powder model suffers from the paucity of conclusions which it permits. It has been, in its unadulterated form, useful principally in the theory of nuclear reactions which deals, naturally, with nuclear systems at rather high excitations. As far as stationary states are concerned, the conclusions from the powder model are much more limited than those of statistical mechanics dealing with molecular chaos.

We consider, first, the applications of the powder

model to regions of relatively high excitation. These applications bear the name statistical theory of spectra.

3. Powder Model at High Energy: Statistical Theory of Spectra

At an excitation energy of several Mev, there may be hundreds of levels within a few kev. If the energy region lies immediately above the neutron separation energy, so that the states within the energy region can decay with the emission of a very slow neutron, the states can be formed, conversely, by exposing a target containing the residual nucleus to a beam of very slow neutrons. The absorption and scattering cross sections then can be observed as functions of the energy. The technique of these measurements has been highly developed, and Fig. 10.1*a* shows one of the spectra obtained in this way. The maxima in the cross section correspond to the very long-lived, almost stationary states, also called resonance levels, of the compound nucleus. Both the positions and the widths of these levels can be determined.

The detailed information on the positions and widths of the levels is of little interest. Rather, the interesting questions concern (1) the density of the levels, (2) their average partial widths with respect to the processes that are energetically possible, (3) the probability that the actual separations between adjoining levels assume a certain value if the average separation is given (in other words, the statistics of the spacings between levels), and (4) the probability that the actual value of the width—or of a partial width—assumes a certain value, given the average total or partial width.

The first two questions—those concerning density and the average widths—are of considerable importance because they determine the cross sections averaged over an energy region containing many resonance levels. Several semiempirical formulas have been proposed to reproduce the density of the levels as a function of the mass number and of the excitation energy. Some of these apply for all the levels; others give the density of the levels with a definite total angular momentum J and definite parity. The most widely used formula, that of Weisskopf, applies to the density of all levels and gives for this

$$\rho(E) = \rho_0 \exp\,[2(aE)^{1/2}] \tag{7.1}$$

in which ρ_0 and a are empirical constant, depending on the mass number A. Unless a is very small, the density increases very rapidly with increasing excitation energy E, reaching values that correspond to an average spacing of a few electron volts at the excitation energy of 8 Mev, the separation energy of a neutron.

The average partial widths of the levels also play an important role in the theory of resonance reactions; they will be discussed in Sec. X. The statistics of the spacings and of the partial widths do not play a similarly important role in the calculation of the average cross sections. However, they are the subjects of mathematical analyses of considerable beauty and elegance. It is assumed that the distributions of these quantities can be obtained by replacing the Hamiltonian with a random symmetric real matrix. This reduces the problem to the determination of the properties of such a matrix. For the distribution of the partial widths of a definite process (e.g., neutron emission, leaving the residual nucleus in the ground state) the so-called Porter-Thomas distribution has been well confirmed experimentally:

$$p(\Gamma)\,d\Gamma = (2\pi\Gamma\bar{\Gamma})^{-1/2}\,e^{-\Gamma/2\bar{\Gamma}}\,d\Gamma \tag{7.2}$$

where Γ is the actual partial width, $\bar{\Gamma}$ the average thereof, and $P(\Gamma)\,d\Gamma$ the probability that the partial width, for an individual level, is between Γ and $\Gamma + d\Gamma$. Equation (7.2) is equivalent to a simple Gaussian distribution for $\Gamma^{1/2}$—a quantity that will be shown in Sec. X to have a simpler theoretical interpretation than Γ itself.

The expression for the distribution of the spacings is much more complicated and its derivation much more difficult. The probability vanishes for zero spacing, then increases to a maximum, and drops very fast beyond a spacing larger than twice the average.

4. Supermultiplet Theory

Naturally, the general conservation laws of quantum mechanics are valid for all stationary states no matter what the model from which they arise Each stationary state will have a total angular momentum J and a parity. The isobaric spin quantum number T will be a valid concept for not too highly excited states. However, more detailed results concerning the properties of low-lying stationary states can be derived only by assuming that the nuclear wave function is largely determined by the spin-independent forces, that is, by the ordinary and Majorana exchange forces. In terms of the Levy interaction, discussed in Sec. V, this would mean that the second term of (5.2), as well as (5.3), need be taken into account only in first approximation; the rest already determines the wave functions. This assumption, together with the powder model, leads to the so-called supermultiplet theory. Similar to charge independence, the spin independence of the forces does not determine the wave functions but only some of their properties. Furthermore, the conclusions resulting from the spin independence of the forces are of much more approximate character than the conclusions resulting from charge independence because the spin-dependent forces can modify the wave function much more severely than the Coulomb forces.

The assumption that the forces are spin-independent is made also in one of the modifications of the shell model: the L-S coupling model. The validity of this model is much more restricted than that of the other shell model; it is useful principally for very light nuclei. The supermultiplet theory seems to have much wider applicability, but most of its conclusions are of a global and somewhat unspecific nature.

Both the total spin S and the isotopic spin T are good quantum numbers in supermultiplet theory. The validity of the latter quantum number was discussed in the preceding section and found to be adequate; the validity of S as a good quantum number—which corresponds to the Russell-Saunders coupling in atomic theory—is more open to question. The validity of both quantum numbers S and T

implies that, in first approximation, $(2S + 1)(2T + 1)$ states of a set of isobars have the same energy: the S_z and T_ζ values of these states range independently from $-S$ to S and from $-T$ to T, respectively.

Rotations in ordinary and isotopic spin space are not the only symmetry operations: one can further interchange, *for instance,* the spin coordinates with the isotopic spin coordinates without affecting the Hamiltonian of the system. As a result of the last symmetry operation states with isotopic spin S and ordinary spin T will coincide, in the approximation considered, with states with isotopic spin T and ordinary spin S. The set of states the energies of which coincide in first approximation is much larger than in ordinary spectroscopy. It is called, therefore, a supermultiplet; it needs for its characterization three numbers, P, P', P''. The mathematical theory of the supermultiplets is much more involved than that of the multiplets of ordinary spectroscopy (or the mathematically identical theory of T multiplets of the preceding section). It is more similar to the SU theories of particle physics. The meaning of the numbers P, P', and P'' is as follows: P is the largest T_ζ of any state contained in the supermultiplet, P' is the largest S_z of any state with $T_\zeta = P$. The last quantum number P'' plays a subordinate role; it is the largest value of

$$\tfrac{1}{2}(\sigma_1\tau_1 + \sigma_2\tau_2 + \cdots + \sigma_A\tau_A) \tag{7.3}$$

for any state with $T = P$, $S_z = P'$. It follows that P, P', and P'' are all integers for even A, half integers for odd A.

The largest T multiplet contained in the supermultiplet is clearly the largest T, that is, is equal to P. The largest S value of any $T = P$ state is the largest S_z of any $T = P$ state, that is, it is equal to P'. Hence, the largest T, S combination is $T = P$, $S = P'$. Since the supermultiplet contains, along with a T, S combination, also an S, T combination and since P is the largest T contained the the supermultiplet, $P \geq P'$, and one can show further that

$$P \geq P' \geq P'' \tag{7.4}$$

While P and P' are always positive, P'' can be also negative. It may also be proved that

$$P + P' + P'' + \tfrac{1}{2}A = 2n \tag{7.4a}$$

is always an even number.

It is important for the understanding of ordinary spectroscopy that among all the spin functions of N spin variables, $S = \frac{1}{2}N, \frac{1}{2}N - 1, \frac{1}{2}N - 2, \ldots$, the last one, that is, either $S = 0$ or $S = \frac{1}{2}$, is the most nearly antisymmetric. (This is a concept which we shall not define rigorously.) As a result, and as a consequence of the antisymmetry of the total wave function, the coordinate function associated with $S = 0$ (in the case of even N) or $S = \frac{1}{2}$ (in the case of odd N) is most nearly symmetric. The situation is quite similar for supermultiplets. The most nearly antisymmetric functions of A variable pairs $\sigma_1, \tau_1; \sigma_2, \tau_2; \ldots; \sigma_A, \tau_A$ are those which have the smallest values of P, P', and P''. Antisymmetry is meant in this case with respect to the interchange of pairs of variables, for instance, the interchange of σ_2 with σ_3 *and* τ_2 with τ_3. Again, as a result of this and of the antisymmetry of the total wave function with respect to the interchanges of all coordinates (position, spin, and isobaric spin), the part of the wave function which depends only on the position coordinates is most nearly symmetric for the smallest possible values of P, P', and P''. As an example, for $A = 2$, the functions of $\sigma_1,\tau_1;\sigma_2,\tau_2$ are antisymmetric for $(P,P',P'') = (1,0,0)$ but symmetric for $(P,P',P'') = (1,1,1)$. Similarly for $A = 3$ and 4, the spin-isobaric spin functions are completely antisymmetric in the $(\frac{1}{2},\frac{1}{2},\frac{1}{2})$ and $(0,0,0)$ supermultiplets, respectively. (For higher A, there is no completely antisymmetric spin-isobaric spin function.) It follows that the position coordinate function is most nearly symmetric for the lowest possible values of P, P', P''; it is completely symmetric in this case for $A = 2,3,4$.

The more nearly symmetric the position coordinate function is, the lower is the kinetic energy and the more negative is the potential energy due to the spin- and isobaric-spin-independent ordinary and Majorana exchange forces. This is particularly clear for the Majorana interaction which has, as explained after (5.2), opposite sign for symmetric and antisymmetric position coordinate wave functions. Since the Majorana interaction is attractive for a symmetric wave function, it is repulsive for an antisymmetric one. The effect is similar, though less pronounced, for ordinary forces. These are (except for the repulsive core) attractive for both symmetric and antisymmetric wave functions. However, in the latter case, the wave function must vanish when the position coordinates coincide and remains relatively small within the range of interaction. As a result, the attractive short-range ordinary potential will have a much smaller value for an antisymmetric than for a symmetric wave function. A similar argument applies also for the kinetic energy. As a result, the quantum numbers P, P', and P'' of the lowest lying supermultiplets will be as low as in consistent with the T_ζ of the nucleus in question and with the rules (7.4) and (7.4a). Since the largest value of T_ζ in a supermultiplet is P, the smallest value which P can assume is T. The remaining quantum numbers P' and P'' will be as low as possible is view of (7.4) and (7.4a) and never larger than 1. More explicitly, they will be

(P,P',P'')

$= (\lvert T_\zeta\rvert,0,0)$	for even-even nuclei
$= (\lvert T_\zeta\rvert,\frac{1}{2},\pm\frac{1}{2})$	for even-odd or odd-even nuclei
$= (\lvert T_\zeta\rvert,1,0)$	for odd-odd nuclei, $T_\zeta > 0$
$= (1,0,0)$	for odd-odd nuclei, $T_\zeta = 0$

In the third case $P' = 1$ is required by (7.4a):

$$T_\zeta + 1 + \tfrac{1}{2}A = \tfrac{1}{2}(N - Z) + 1 + \tfrac{1}{2}(N + Z) = N + 1$$

is even in this case, but would be odd if the 1 were replaced by a 0. A similar remark applies to the last case.

Since P' is the largest value of S_z which is compatible with the value P for T_ζ, one infers at once $S = 0$ for the normal state of even-even nuclei and $S = \frac{1}{2}$ for the normal states of even-odd and odd-even nuclei. For odd-odd nuclei, one will infer $S = 1$. These consequences of the supermultiplet theory cannot be compared directly with experiment because S cannot be measured by itself; only the vector sum J of the spin angular momentum S and the orbital angular momentum L is directly observable as the "spin" of the nucleus. J is always 0 for even-even nuclei so that one has to postulate $L = 0$ for the normal states of all even-even nuclei. If one accepts the absence of spin-dependent forces, this is quite reasonable. However, the postulates concerning S stand in direct conflict with postulates of the j-j coupling model discussed later. According to present evidence, at least for medium heavy and heavy nuclei, the wave function furnished by the j-j coupling model comes closer to the actual wave function than any wave function compatible with the supermultiplet theory. However the supermultiplet theory and its quantum numbers P, P', and P'' constitute a good approximation for light nuclei. This statement means that the spin-dependent forces, and the electrostatic interaction, though large in magnitude, do not influence the wave function itself too radically. This is confirmed also by direct calculations which take the spin-dependent forces into account but which yield wave functions 90 per cent of which has the supermultiplet character described above.

The great weakness of the supermultiplet theory is its lack of specificity, i.e., that it postulates only some general characteristics for the wave function but does not give a definite expression therefor. As a result, those properties of the nuclei which do not follow from general symmetry arguments cannot be calculated on the basis of the supermultiplet theory alone. The two most important results of the theory which can be compared with experiment directly concern the magnitude of the allowed β decay matrix elements and the general trend of nuclear binding energies. This has been reviewed recently by Franzini and Radicati with satisfactory results, extending up to $A = 100$. With respect to β decay, supermultiplet theory provides a distinction between favored and unfavored transitions with the former having, for the same decay energy, much shorter lifetimes than the latter. They should occur only between states which belong to the same supermultiplet—a conclusion which seems to be confirmed by experimental evidence.

Concerning the general trend of nuclear binding energies, the supermultiplet theory leads to formulas for the binding energy similar to Weizsäcker's expression (2.1). In particular, the decreased binding energies of odd-odd nuclei, and of even-odd and odd-even nuclei, as compared with the binding energies of even-even nuclei [that is, the $\delta/A^{3/4}$ terms in (2.1)] result from the P', $P'' = 1,0$ and $P',P'' = \frac{1}{2}, \pm\frac{1}{2}$ in the symmetry symbols for the corresponding types of nuclei. One also obtains in this way an interpretation of the "symmetry term" $U_t T_\zeta^2/A$ and a connection between this term and the $\delta/A^{3/4}$ term which is, for not too heavy nuclei, well confirmed by the values of U_t and δ. The special role of the odd-odd nuclei with $T = 0$ deserves further comment. The only stable odd-odd nuclei (H^2, Li^6, B^{10}, N^{14}) belong to this type. Their supermultiplet (1,0,0) gives two S, T combinations: $S = 1$, $T = 0$ and $S = 0$, $T = 1$. The former gives the lowest state of the afore-mentioned odd-odd nuclei, which are therefore in the triplet state, the latter combination gives the normal states of the adjoining even-even nuclei (He^6, Be^6; Be^{10}, C^{10}; C^{14}, O^{14}) as well as excited states of the odd-odd nuclei. All these are singlets, in fact having $J = 0$. Neglecting all spin-dependent and electrostatic forces, their energies are therefore exactly equal. The spin-dependent forces will depress some of the triplet states, the electrostatic forces the states with larger T. For lower A, the former predominate; hence H^2, Li^6, B^{10}, and N^{14} are stable. However, as A and hence Z increase, the electrostatic forces begin to prevail and the normal state of F^{18} (which is a triplet state) lies higher than the normal state of O^{18}, which is a singlet but has less electrostatic energy.

The (1,0,0) supermultiplet is the only one which gives the normal states of two types of nuclei: those of odd-odd $T_\zeta = 0$ and of even-even $T_\zeta = \pm 1$ nuclei. This explanation of the stability of certain odd-odd nuclei is quite striking; it is, of course, common to all models in which the bulk of the interaction is due to spin (and isobaric spin) independent forces.

The nature of the model is such that it is not expected to be valid for very small or very large A. For low A, the statistical assumptions are incorrect. Nevertheless, the formulas which should be valid only for large A apply remarkably well also for light nuclei. At large A the model becomes grossly inaccurate, particularly in the regions in which the nucleus assumes an ellipsoidal rather than a spherical shape (Sec. IX).

VIII. NUCLEAR MODELS. INDEPENDENT PARTICLE MODELS

1. General Features of the Independent Particle or Shell Models

Nuclear theory was late in recognizing the usefulness of the independent particle model for nuclei. The reason for this is that the short range of the nuclear forces is hardly compatible with the independence of the motion of the nucleons. This assumes that the motion of each nucleon is determined by an average potential, due to the other nucleons. If the forces are of short range, one would expect that the nucleons suffer sudden collisions as a result of encounters with other nucleons. This means that the motion of the nucleons is not determined by the average field but is significantly influenced by the rapid variations of the forces, due to encounters. Indeed, the independent particle model is valid for nuclei in a much more subtle sense than for atoms (see Sec. 8). However, the consequences of the shell or independent particle model are hardly affected by the refinements.

The theory of the isobaric spin and the supermultiplet model do not provide definite wave functions for the nucleus; they only specify certain properties of such wave functions. On the contrary, the independent particle or shell models give a more or less explicit expression for the whole wave function whence expressions for all the properties of the nucleus in question can be derived.

As was pointed out in Sec. II, many of the properties of nuclei vary fairly smoothly with the numbers N, Z. Thus the binding energies of nuclei can be represented pretty well by three smooth surfaces for the even-even, even-odd, odd-odd species. Closer scrutiny reveals that there are strong discontinuities in some nuclear properties. At certain values of N, Z these discontinuities are particularly marked. These values, 2, 8, 20, 50, 82, 126, are called *magic numbers*. Nuclei with magic N or Z (or both) are called *magic nuclei*. The special stability of the magic nuclei manifests itself in various ways. The nuclei with magic Z or N have larger numbers of stable isotopes (or stable isotones) than their even Z or N neighbors; they are also anomalously abundant, indicating a greater stability associated with the magic numbers. The capture cross sections for neutrons drop sharply at the magic nuclei. All these features of magic nuclei are attributed to the binding-energy discontinuities at the magic numbers. The drop in binding energy may amount to as much as 2 or even 3 Mev at the passage of the magic number.

Fluctuations in the properties of nuclei in the neighborhood of the magic numbers can be understood in terms of a "shell" model of the nucleus similar to the highly successful shell model of atoms. On this basis the magic numbers are assumed to indicate the numbers of particles required to provide complete nuclear shells. As in the atomic case, a system consisting of complete shells will show a large separation energy (called ionization potential in the case of atoms). If the number of neutrons, or of protons, is one larger than can be accommodated in full shells, the last particle will have a small separation energy. According to this picture, magic nuclei play roles among nuclei similar to those of the noble gases among atoms.

The independent particle models for nuclei are similar to the independent particle or Hartree model for atomic electrons. Each nucleon is assumed to move, at least approximately, in an average field set up by the other particles or, at least, the calculations are carried out as if this were the case, i.e., as if the particles moved independently of each other—hence the alternative (and more descriptive) name independent particle model. The nuclear wave function is represented as a product of one-particle wave functions for the individual nucleons except that, in order to satisfy the exclusion principle, these products must be antisymmetrized, i.e., replaced by determinants. The average potential in which the individual nucleons are assumed to move is, or at least resembles, a spherical potential well. The states of the individual particles in such a field can be characterized by five quantum numbers. Three of these are the radial quantum number n (giving the number of zeros of the radial part of the wave function), the orbital angular momentum or azimuthal quantum number l, and the isotopic quantum number τ specifying the type of particle ($\tau = 1$ for neutron, $\tau = -1$ for proton). For the last two quantum numbers, one may take the projections of the orbital angular momentum and of the spin in a given direction, l_z and σ. Alternatively, one can use, as the last two quantum numbers, the total angular momentum j of the particle (which is equal to $l + \frac{1}{2}$ or $l - \frac{1}{2}$) and the projection j_z of this quantity in a given direction. The first set of quantum numbers is more adapted for the L-S coupling model, the latter set for the j-j model. In the first case, it is assumed that the total energy is given, in first approximation, by the n and l values of all the particles present; in the second case the j values of these particles are also given. The reason is that the j-j model assumes a large energy difference between the states $j = l + \frac{1}{2}$ and $j = l - \frac{1}{2}$ of the individual particles. Hence, in the first case the configuration is described by symbols of the type $(1s)^4(1p)^2$, in the second case by symbols of the type $(1s_{1/2})^4(1p_{3/2})^4$. The index $(\frac{3}{2})$ on the symbol giving the azimuthal quantum number (s for $l = 0$, p, d, f, g, h, i for $l = 1, 2, 3, 4, 5, 6$, respectively) gives the j value of the state.

It was stated before that the individual particle wave function is a determinant. This implies that a single determinant gives the whole wave function of the independent particle models. This is not correct. We have seen that, in the L-S model τ, l_z, and σ do not affect the energy substantially and that the same is true of τ and j_z in the j-j coupling model. As a result, the symbol of the configuration does not contain these quantum numbers. Hence, these quantum numbers can yet be given arbitrary values in the wave functions of the individual particles and the total wave function may be a sum of determinants formed of individual particle wave functions, the n, l of which are given by the symbol of the configuration but the τ, l_z, and σ of which may be different in each determinant. This applies for the L-S model; in the j-j model the n, l, and j values of the individual particle wave functions are specified by the configuration but τ and j_z can assume any value in each determinant. In order to obtain the coefficients of the various possible determinants of the wave function, one can use the requirements that the total angular momentum J, its projection J_z, the isobaric spin T, and T_ζ have given values. The L-S model assumes, in addition, the validity of the supermultiplet theory. These requirements usually determine the coefficients of the various possible determinants. If not, the corresponding model does not give a unique wave function and one has to resort to calculations based on further assumptions to obtain a wave function for the corresponding state of the nucleus.

2. The L-S Coupling Shell Model

In this model, the symbol of the configuration gives the radial and azimuthal quantum numbers of the individual particles. The lowest states of a particle in a spherical well are the four $1s$ states ($n = 1$, $l = 0$, $\tau = \pm 1$, $l_z = 0$, $\sigma = \pm 1$), the next twelve are $1p$ states ($n = 1$, $l = 1$, $\tau = \pm 1$, $l_z = \pm 1$ or 0, $\sigma = \pm 1$). The $2s$ and $1d$ states follow these. Note that the radial or principal quantum number is counted from 1 for all l. Hence the lowest p level is called $1p$, the lowest d level $1d$, rather than $2p$ and $3d$ as in atomic spectroscopy. The lowest configuration of a nucleus of mass A, for A between 4 and 16, is $(1s)^4(1p)^{A-4}$. Such a configuration gives rise in general to several states, with various J and T values. These states are further classified according to their symmetry properties P, P', P'' of the supermultiplet theory. If the forces are predominantly ordinary or Majorana forces, the states most nearly symmetric in the positional coordinates (lowest P, P', P'') will lead to the lowest energy values. This is assumed to be true in the simplest version of the shell model which we are considering. Hence, this model uses the assumptions of the supermultiplet theory and is in fact a specialization of that theory which assumes that the wave functions are sums of determinants, as described above. It follows, in particular, that the spin angular momentum S is a good quantum number, that it assumes the values 0 and $\frac{1}{2}$ for the low-lying states of even-even and even-odd or odd-even nuclei. The spin is 1 for the $T = 0$ states, and 0 for the $T = 1$ states of odd-odd nuclei with $T_\zeta = 0$. One has $S = 1$ or 0 for $T_\zeta \neq 0$ odd-odd nuclei.

Except for B^{10} the model gives the parity, the orbital angular momentum and the order of magnitude of the energy of the excited states of light nuclei. It also explains the difference between favored and unfavored β transitions (Sec. XII). This last point constitutes, along with the explanation of the general trend of the energy difference of isobars, the greatest success of the uniform and of the L-S coupling shell models. However, in order to obtain the spin (J value) of the normal states of nuclei if both L and S are different from zero, one has to obtain the relative positions of the levels which originate from the different orientations of L and S with respect to each other, that is, the levels with J values between $|L - S|$ and $L + S$. These levels all would have the same energy if the forces were entirely independent of the spin. The magnitude and sign of the energy differences between these levels depend on the nature and strength of the spin dependent forces. Very little is known on this last point and it appears that the type of spin-dependent forces, such as (5.3), which are suggested by the current meson theories do not lead to agreement with experiment. Considerable progress has been made, on the other hand, toward the explanation not only of the properties of the normal state but also of the position of the excited states by assuming other types of spin-dependent forces. The agreement obtained in this way is often quite striking. At the same time, the calculations with these spin-dependent forces also give an indication of the accuracy of the basic picture, i.e., of the wave functions obtained under the neglect of the spin-dependent forces. The results are quite favorable: even very strong spin-dependent forces yield wave functions in which the wave function of the supermultiplet theory has a probability of about 90%. However, this last type of calculation refers principally to nuclei in which a new shell has just been started and there are only a few particles in it. This is the case, for instance, for $A = 18$ for which the $1s$ and $1p$ shells are already completed with 4 and 12 particles, respectively, and there are only two particles outside these shells. There are indications that the situation is much less favorable toward the end of a shell, for example, for $A = 12$ with the configuration $(1s)^4(1p)^8$. The deviation from the picture of the L-S model appears to be quite strong in this case. The reason for this is not yet understood. A similar remark applies for $A = 14$.

3. Comparison of the L-S and j-j Shell Models

It was mentioned before that the L-S model loses validity around $A = 40$ and also that it appears to become increasingly inaccurate toward the ends of the shells. Apparently, the spin-dependent forces become increasingly important under these conditions. One is reminded that the situation is similar for atomic electrons: the Russell-Saunders coupling is most accurate for light elements and particularly so in the first part of each shell. However, the range of validity of Russell-Saunders coupling for atomic electrons far exceeds the validity of the L-S and supermultiplet models for nuclei. In fact, the opposite extreme, the the j-j model, is at present by far the most important nuclear model; it coordinates and explains a variety and wealth of phenomena which exceed the scope of any other known nuclear model.

The appellations L-S model (and supermultiplet model) express the assumptions that the concepts of orbital and spin angular momenta (and the concept of supermultiplets), have considerable accuracy. Expressed somewhat more mathematically, the orbital and spin angular momenta (or the triplet of quantum numbers P, P', P'') have with a large probability a single value. It also means that the forces which tend to destroy the validity of these quantum numbers are relatively unimportant and do not cause severe changes in energy. In a similar way, the term j-j coupling shell model expresses the assumption that the j values for individual nucleons are good quantum numbers, that the forces which couple the orbital angular momenta l of individual particles to their spins to form a resultant j are more effective than the forces which couple the individual l together to form a resultant orbital angular momentum L for the whole nucleus. The term j-j coupling should also mean that the forces which tend to destroy the validity of the individual quantum numbers j do not cause too large changes in energy and that the levels of the same configuration are grouped reasonably closely on the energy scale.

It appears that the L-S model, and even the much less specific supermultiplet model, are in severe conflict with the j-j model. This is, of course, not true if there is a single particle outside of closed shells, as in this case the individual l is already the L of the whole nucleus and this is, therefore, a good quantum number in both theories. Similarly, j is identical with the total J and is also a good quantum number. The same is true for closed shells or if only a single particle is missing from a closed shell. The L-S models and j-j models are identical also if the open shell is an s shell as the total angular momentum of each particle is necessarily $\frac{1}{2}$ if their orbital angular momenta vanish. However, even if we have only two particles in the shell, the wave functions of the two theories are very different. Thus the L-S model postulates, for the lowest state of Li^6, the wave function

$$C(x_1x_2 + y_1y_2 + z_1z_2)p(r_1)p(r_2)\delta_{\sigma_1,1}\delta_{\sigma_2,1}(\delta_{\tau_1,1}\delta_{\tau_2,-1} - \delta_{\tau_1,-1}\delta_{\tau_2,1}) \quad (8.1)$$

for the two particles outside the s^4 shell. This is a $^3S_1$ function (with $J_z = 1$) with $T = 0$ which belongs to the supermultiplet (1,0,0). $C = 6^{-1/2}$ is a normalization constant and $p(r)$ is the radial part of the wave function for p particles. The state with the same J, J_z, and T_ζ has, according to the j-j shell model, the wave function

$$
\begin{aligned}
20^{-1/2}[3^{1/2}P_{3/2}{}^{3/2}(\mathbf{r}_1,\sigma_1)P_{-1/2}{}^{3/2}(\mathbf{r}_2,\sigma_2) \\
-2P_{1/2}{}^{3/2}(\mathbf{r}_1,\sigma_1)P_{1/2}{}^{3/2}(\mathbf{r}_2,\sigma_2) \\
+3^{1/2}P_{-1/2}{}^{3/2}(\mathbf{r}_1,\sigma_1)P_{3/2}{}^{3/2}(\mathbf{r}_2,\sigma_2)] \\
(\delta_{\tau_1,1}\delta_{\tau_2,-1} - \delta_{\tau_1,-1}\delta_{\tau_2,1})
\end{aligned}
\tag{8.2}
$$

where

$$
\begin{aligned}
P_{3/2}{}^{3/2}(\mathbf{r},\sigma) &= 2^{-1/2}(x + iy)p(r)\delta_{\sigma,1} \\
P_{1/2}{}^{2/3}(\mathbf{r},\sigma) &= 6^{-1/2}(x + iy)p(r)\delta_{\sigma,-1} \\
&\quad -i(\tfrac{2}{3})^{1/2}zp(r)\delta_{\sigma,1} \qquad (8.2a) \\
P_{-1/2}{}^{3/2}(\mathbf{r},\sigma) &= -i(\tfrac{2}{3})^{1/2}zp(r)\delta_{\sigma,-1} \\
&\quad +6^{-1/2}(x - iy)p(r)\delta_{\sigma,1}
\end{aligned}
$$

The coefficients in these expressions are the so-called coefficients of the vector addition model (Clebsch-Gordan coefficients). The single particle functions (8.2*a*) have total angular momentum $j = \frac{3}{2}$; the lower index of the P is the j_z. In the function (8.2) of the *j-j* coupling model the quantum numbers which are good in both models, J, J_z, T, T_ζ, have the same values 1, 1, 0, 0 as in (8.1). However, the probabilities for the values 1 and 0 for S and L are not unity in (8.2); other values for these quantities ($S = 0$, $L = 1{,}2$) have finite probabilities. On the other hand,

$$j_1 = j_2 = \tfrac{3}{2}$$

for (8.2); that is, (8.2) represents a $(1p_{3/2})^2$ state while for the state (8.1) the configurations $(1p_{3/2})(1p_{1/2})$ and even $(1p_{1/2})^2$ have finite probabilities.

It is not known with certainty whether (8.1) or (8.2) is closer to the actual wave function of the normal state of Li^6. Since this is a light nucleus and has only two particles in the $1p$ shell, (8.1) is presumably more accurate. Both (8.1) and (8.2) give reasonable values for the magnetic moment (0.88 and 0.63); the experimental value of this quantity is 0.822. On the other hand, (8.2) gives a much too large value for the quadrupole moment: 7×10^{-27} cm². The quadrupole moment would vanish if (8.1) were accurate, but minor deviations from (8.1) suffice to explain the very small observed moment of about 8×10^{-28} cm². Even more convincingly in favor of the L-S model are the values for lifetimes of β active nuclei in this region. However, a similar comparison for heavier nuclei would almost surely favor the *j-j* wave function of the type (8.2), and this would be true particularly for nuclei in which each shell is almost filled.

Consequences of the L-S and *j-j* models differ in many cases less sharply than one would expect. Thus, for instance, according to the L-S model, the low-lying states of Li^6 should be ^3S, ^3D for $T = 0$ and ^1S, ^1D for $T = 1$. This gives levels with $J = 1, 1, 2, 3$ for $T = 0$ and $J = 0{,}2$ for $T = 1$. Actually, the first six energy levels of Li^6 have these quantum numbers. However, the *j-j* model, that is, the configuration $(p_{3/2})^2$, also gives $J = 1{,}3$ for $T = 0$ and $J = 0{,}2$ for $T = 1$. Furthermore, the $(p_{3/2})(p_{1/2})$ configuration gives $J = 1{,}2$ (for both $T = 0$ and $T = 1$) so that the levels $J = 1{,}2$, $T = 0$ find a natural interpretation in the *j-j* model also. On the other hand, the $T = 1$, $J = 1$, and $J = 2$ levels are missing; this is in conflict with the *j-j* model. A comparison of known states at the end of the p shell, for instance for C^{14}, would give a different result: the L-S model gives $J = 0{,}2$ for $T = 1$ (as for Li^6 or He^6), the *j-j* model only one low-lying state for $T = 1$, the $(p_{1/2})^2$ $J = 0$ state. Actually, $J = 0$ for the normal state of C^{14} and the first excited state seems to be at 6.1 Mev. The *j-j* model is more nearly in accord with the observations in this case.

4. The *j-j* Coupling Shell Model

As mentioned before, the configuration in this model specifies not only the radial (principal) and orbital angular momentum quantum numbers of the individual particles but also the relative orientation or orbital and spin angular momenta of each particle. It speaks, for instance, of $1d_{5/2}$ and $1d_{3/2}$ levels, rather than simply of a $1d$ level. In the former, spin and orbital angular momenta are parallel, in the latter antiparallel. The model assumes, furthermore, that the parallel orientation gives the lower energy and that the energy difference between parallel and antiparallel orientations is appreciable. It increases with increasing l. The order of the filling of the single particle levels is illustrated in Table 8.1. An important feature of the *j-j* model is the subdivision of each shell (except of the s shells) into subshells. Thus, the $1p$ shell is subdivided into a $1p_{3/2}$ and a $1p_{1/2}$ subshell; the energy differences between subshells and between shells are quite comparable.

TABLE 8.1

$1s_{1/2}(2)$	$1p_{3/2}(6)$	$1d_{5/2}(14)$	$1f_{7/2}(28)$	$2d_{5/2}(56)$	$2f_{7/2}(90)$	$1i_{11/2}(138)$
	$1p_{1/2}(8)$	$2s_{1/2}(16)$	$2p_{3/2}(32)$	$1g_{7/2}(64)$	$1h_{9/2}(100)$	$2g_{9/2}(148)$
		$1d_{3/2}(20)$	$1f_{5/2}(38)$	$3s_{1/2}(66)$	$3p_{3/2}(104)$	$2g_{7/2}(156)$
			$2p_{1/2}(40)$	$1h_{11/2}(78)$	$2f_{5/2}(110)$	$3d_{5/2}(162)$
			$1g_{9/2}(50)$	$2d_{3/2}(82)$	$3p_{1/2}(112)$	$3d_{3/2}(166)$
					$1i_{13/2}(126)$	$4s_{1/2}(168)$
						$1j_{15/2}(184)$

It is customary in the *j-j* model to give the configuration for protons and neutrons separately, and to separate them with a semicolon. Thus, the lowest configuration of Li^6 would be written as $(1s)^2 1p_{3/2}$; $(1s)^2 1p_{3/2}$. The lowest configuration for C^{12} would be $(1s)^2(1p_{3/2})^4$; $(1s)^2(1p_{3/2})^4$. Since j_z can assume only four values in a $p_{3/2}$ subshell, the normal state of C^{12} contains, both for protons and for neutrons, only closed shells. The normal state of B^{12} comes from the configuration $(1s)^2(1p_{3/2})^3$; $(1s)^2(1p_{3/2})^4(1p_{1/2})$. Since the exclusion principle prohibits more than four neutrons in the $1p_{3/2}$ subshell, one neutron must be in the next subshell. This is the $1p_{1/2}$ subshell. One often omits some of the closed shells and writes, for instance, for the lowest configuration of B^{12}, simply $(1p_{3/2})^3$; $(1p_{3/2})^4(1p_{1/2})$.

This notation gives not only the configuration in the sense in which this term was used before, but also the value of $T_\zeta = \frac{1}{2}(N - Z)$. The symbol of the configuration, as used before, can be obtained from the new symbol simply by contraction. Thus, we would have used in the preceding section the symbol $(1s)^4(1p_{3/2})^7(1p_{1/2})$, for the lowest configuration of B^{12}, and added $T_\zeta = 1$. Conversely, this symbol can be translated into the present notation if the configuration is the lowest for the nucleus in question. In such a case, there can be only one incomplete subshell for the protons and one for the neutrons. Hence, the joint symbol can be separated only in one way into a part before the semicolon and a part after the semicolon. This is not the case if the configuration is not the lowest: thus $(1s)^4(1p_{3/2})^7(1p_{1/2})$ for $T_\zeta = 0$ (C^{12}) can be separated into

$$(1s)^2(1p_{3/2})^4;\ (1s)^2(1p_{3/2})^3(1p_{1/2})$$

or

$$(1s)^2(1p_{3/2})^3(1p_{1/2});\ (1s)^2(1p_{3/2})^4$$

In this case, either the neutron configuration, or the proton configuration, contains two incomplete subshells. The reason for adopting a symbol for the configuration of a definite nucleus, rather than for a set of isobars as in the preceding section, is that the isobaric spin quantum number does not give additional clues for the wave function of those states which have the maximum T_ζ consistent with the configuration. The T of the wave functions with this T_ζ is automatically equal to this T_ζ: certainly no part of such a wave function can have a lower T than its T_ζ; higher T do not occur at all in the configuration. Hence, if dealing with the states of highest T_ζ of the configuration—this is the case, as we saw for B^{12} and is also the rule for heavier nuclei where T_ζ is relatively large—the isobaric spin quantum number is not useful for the determination of the wave function of low-lying levels. The table of successive subshells, i.e., energy levels of individual particles, follows. The lowest energy level is in the first column, and in order to obtain the levels in the order of their energy values, one has to read, next, the second column down, then proceed downward on the third column, and so on. The purpose of the columnar arrangement is to unite all those levels into a column the energy values of which are not sufficiently different to cause a pronounced discontinuity at the transition from one to the next. On the other hand, the energy difference between the last level in a column and the first level in the next one is so large that the nucleus is "magic" (particularly stable) when all the levels of a column are filled. The numbers in the parentheses after the symbol of each subshell indicate the number of neutrons which the subshell in question, and all the subshells of lower energy, can contain. The corresponding numbers for protons are, or course, the same. It follows that the parenthesized numbers at the end of the columns are the magic numbers. The number of neutrons, or of protons, which a subshell with angular momentum j can accommodate is $2j + 1$, corresponding to the $2j + 1$ values $-j, -j + 1, \ldots, j - 1, j$ which the quantum number j_z can assume. Hence, each parenthesized number differs by $2j + 1$ from the preceding one.

An important feature of the scheme is the large spin orbit splitting assumed for the $1g$, $1h$, and $1i$ levels; for these the $j = l + \frac{1}{2}$ and $j = l - \frac{1}{2}$ subshells are in different columns. The order for the levels in Table 8.1 is that obtained for a square well potential. For an oscillator potential ($V \sim r^2$), all the levels in a column, except the $1g_{9/2}$, $1h_{11/2}$, $1i_{13/2}$, $1j_{15/2}$, have the same energy; the energy of these levels is the same as that of all levels of the next column. Actually, the spin-orbit splitting plays a role also within the levels of a single column. However, the order of the levels within a column does not seem to be entirely fixed; it may change from nucleus to nucleus. Similar phenomena occur also for the electronic shells of atoms.

It is worth noting that all the subshells of one column have the same parity, except the $g_{9/2}$, $h_{11/2}$, $i_{13/2}$, which have opposite parity to the remaining states of the columns in which they appear.

5. Coupling Rules for the *j-j* Model

If we assume a particular order for the one-particle levels, the configuration of the normal state is uniquely given. However, in general, several states of the whole nucleus, with different J values, belong to each configuration. All these states have the same energy in first approximation so that, in order to find the characteristics, in particular the J value, of the normal state, one has to establish an ordering of the states arising from the same configuration. The rules which establish this order, or at least give the J value of the state of lowest energy, are called coupling rules.

Originally, these coupling rules were formulated on the basis of empirical evidence. It was postulated that an even number of protons always couples, for the normal state, to $J_p = 0$; likewise an even number of neutrons couples to $J_n = 0$. This already specifies the J of even-even nuclei to be $J = 0$. Similarly, it was postulated that an odd number of protons couple to the J value $J_p = j_p$ of the protons which are in an unfilled subshell—there should be only one such subshell. The same rule is postulated for neutrons if their number is odd. This gives for the J value of even-odd and odd-even nuclei uniquely the j value of the nucleon of which there is an odd number in the nucleus. It should be noted, however, that not infrequently $J = j - 1$ is 1 less than the j value of the shell being filled. This indicates that an even number of protons or neutrons does not invariably couple to $J = 0$, or, more probably, that an odd number of particles with angular momenta j may couple to $J = j - 1$ in the lowest state. The above rules do not give a unique answer to the question of the J value of the normal state of odd-odd nuclei; this could be anywhere between $|j_p - j_n|$ and $j_p + j_n$, where j_p and j_n are the j values of the unfilled proton and neutron subshells.

Attempts to provide a theoretical basis for the coupling rules necessarily involve assumptions concerning the nuclear interaction responsible for the splitting of the several levels due to a configuration. Assumptions which lead to the above coupling rules are: (1) The interaction is much stronger within the group of protons, and within the group of neutrons, than *between* protons and neutrons unless these are in the same subshell. (2) The interaction has extremely short range and is attractive. This last assumption makes it unnecessary to specify whether the interaction is of ordinary or Majorana exchange type: if the wave function is antisymmetric with respect to the interchange of the spatial coordinates of two particles, it must vanish at their coincidence. Then it will be so small within the assumed range of the forces that the interaction will vanish between particles for which the wave function is antisymmetric in the spatial coordinates; i.e., the interaction will vanish whenever the Majorana interaction is different from the ordinary interaction. Although these assumptions are not fully consonant with our knowledge of nuclear forces from other sources, they do lead to the empirical coupling scheme. Rule (1) is particularly difficult to understand: it leads to grossly inaccurate results if applied to light nuclei. For these, the postulate of the validity of the isobaric spin is in conflict with (1) and provides other coupling rules, leading to good agreement with experiment. Although, as has been pointed out before, for heavier nuclei the isobaric spin does not provide coupling rules, and is therefore not in conflict with (1), one does not see why postulate (1) should be valid for heavier nuclei if it is not valid for light ones.*

Prediction of the quantum numbers of the normal state remains difficult even assuming the above coupling rules. In the first place, the actual order of the one-particle levels (the subshells of Table 8.1) is uncertain. Furthermore, the configuration splitting may lead to a crossing of levels arising from close configurations. Indeed, in order to understand the values of the nuclear spins, and in particular the rather small spins which occur for nuclear ground states of even-odd nuclei, an explicit assumption of such crossing is required. Absence of spins of $\frac{11}{2}$ and $\frac{13}{2}$ can be explained by assuming that the interaction between the nucleons in a shell increases with increasing l and is proportional to the number of *pairs* of nucleons in the unfilled shell. Thus, for a set of 70 nucleons, the lowest configuration should have filled the $3s_{1/2}$ level. However, it may be that the configuration with 6 nucleons in the $1h_{11/2}$ level, and with the $3s_{1/2}$ level empty, leads to a lower state than the $(3s_{1/2})^2(1h_{11/2})^2$ configuration because of the increased interaction between the $h_{11/2}$ nucleons. If this is the case, the $3s_{1/2}$ shell will not be filled by 70 nucleons and the 71st nucleon can still be in a $3s_{1/2}$ state. As a result, a set of 71 nucleons will give a spin of $\frac{1}{2}$ rather than $\frac{11}{2}$.

Magnetic moments predicted by the model for the odd A nuclei are given by the Schmidt lines (see Sec. III). Magnetic moments of a number of nuclei do lie on these lines but, in general, the moments are not well described by the model (they are given about equally accurately by several other models). However, the magnetic moments may be used to provide a qualitative check of the assumptions of the model.

If the theory were exact so that the magnetic moments were given by the Schmidt lines, then from the spin J of an odd A nucleus, the j $(= J)$ *and* the l value of the last odd nucleon could be determined. We may use the magnetic moments to infer the l value by taking l to be that of the Schmidt line closer to the observed magnetic moment. This l value may then be compared with that which is to be expected on the assumed one-particle level structure, consistent, of course, with the given j. In all but a handful of cases, there is agreement with the theory. This result provides powerful support for the level grouping which has been assumed and gives confidence in the determination of the parities of nuclear ground states by means of the model.

6. Normal States and Low Excited States

The values of the magic numbers 50, 82, 126 provided the incentive and the first justification for the j-j model. However, evidence for the usefulness and validity of this model can now be found in almost every branch of nuclear physics. Perhaps the most impressive and convincing data which support this model derive from investigation of the normal states and low excited states of nuclei with mass numbers in excess of 70. Admittedly, the spin and parity assignments of these states are partially based on guesswork, influenced, no doubt, by the shell model which is to be proved. Nevertheless, the possibility of interpreting almost every low level of these nuclei on the basis of the j-j model remains a striking confirmation of the model.

Table 8.2 gives the parities and (for typographical reasons) twice the J value of the low states for odd-even and even-odd nuclei as far as they are known in the region of the table. The rows refer to Z or N, whichever is odd, the columns to the even one among Z and N. In the left side of the table the rows refer to N, the columns to Z; in the right side it is the other way around. If more than one symbol is given for a Z-N combination, the lowest refers to the normal state, the one above it to the first excited state, and so on. For the normal states of stable nuclei, J can be measured directly. The parity was obtained from the magnetic moment as explained before; "+" indicates even parity (even l of the odd proton), "−" indicates odd parity (odd l of that nucleon of which there is an odd number in the nucleus). The assignment of J and parity for the excited states and for unstable nuclei is based on the multipolarity of the γ rays emitted by the excited states (Sec. XII) and on the degree of forbiddenness of the β radiation of unstable nuclei (Sec. XII). Some of the assignments may be open to question, but the bulk of them can be expected to prove correct.

In the region between 41 and 49 the $g_{9/2}$ subshell should be filled, the parity and spin ideally should be $+\frac{9}{2}$ throughout this region. Actually, some of the normal states are $p_{1/2}$ states. Explanation of this in terms of pairing energy was given before. The frequent occurrence of $J = \frac{7}{2}$ with even parity is more difficult to understand from the point of view of the j-j model. Such a state occurs only for nuclei in which N or Z assume the values 43, 45, or 47. It does not occur for 41 or 49. Naturally the $(g_{9/2})$ and

* M. Redlich (personal communication) has exhibited a case in which the calculation yields, under the assumption of a wide separation of the $j = l + \frac{1}{2}$ and the $j = l - \frac{1}{2}$ levels, a surprisingly small interaction between the protons and the neutrons. He showed that the $J_p = 0$ level, arising from the $(1h_{9/2})^2$ configuration of the two protons, lies so much deeper than the other levels arising from the same configuration (J_p = 2, 4, 6, 8) that the interaction with a single $2g_{9/2}$ neutron does not introduce any appreciable admixture of the higher J_p states into the wave function of the $(1h_{9/2})^2$; $2g_{9/2}$ configuration. The generality of this result has not yet been established.

TABLE 8.2

	30	32	34	36	38	40	42	44	46	48	50	52	54	56	58	60	62	64	66	68	70	72	74	76	78	80	82
39	+9 −1	+9 −5 −1	+9							+9 −1	+9 −1	+9 −1	−1														
41	+9 −1		+5								−1 +9	−1 +9	−1 +9	−1 +9													
43		+7 −1	+7 −1	+7 −1	−1						−1 +9	−1 +9	−1 +9	+9	−5 +9												
45		−1 +7	−1 +7	−1 +7									+9 −1	+9 −1	+9 +7 −1	−1 +7											
47			+7 −1	−1 +7 +9	−1 +7 +9									+7	+7 −1	+7 −1	+9 +7 −1	+7 −1	+7 −1	−1 +7							
49			−1 +9	−1 +9	−1 +9	−1 +9	−1 +9								+9	−1 +9	−1 +9	−1 +9	−1 +9	−1 +9	−1 +9	−1 +9	+1 +9				
51				+7	+5	+5	+5											+5	+5	+7 +5	+7 +5	+5 +7	+5 +7	+7	+7		+7
53					+5	+3 +5	+5	+5	+5											+5	+5	+5	+3 +7 +5	+5 +7	+5 +7	+7	+7
55						+7 +5	+5															+1	+1	+1 +7 +5	+5 +7	+5 +7	+7

the $(g_{9/2})^9$ configurations do not give a $J = \frac{7}{2}$ level. However, three, five, or seven $g_{9/2}$ particles do. This suggests that the $J = \frac{7}{2}$ levels are other members of the $(g_{9/2})^n$ configuration, for $n = 3, 5, 7$, which, according to the coupling rules, should lie above the $J = \frac{9}{2}$ level. This is the case, as a rule, but not invariably. The most surprising and apparently well-established exception is the $J = \frac{5}{2}$ state of positive parity in ${}_{41}Se_{34}$.

The occurrence of low-lying states with $J = j - 1$ is not restricted to the $g_{9/2}$ subshell. It is a phenomenon which is rather common in all shells. Thus, the spins of Na^{23}, Ne^{21} are $\frac{3}{2}$ instead of $\frac{5}{2}$, those of Ti^{47}, Mn^{55} $\frac{5}{2}$ instead of $\frac{7}{2}$. The former two nuclei contain three $d_{5/2}$ particles, the latter ones five $f_{7/2}$ particles. The spin of V^{47}, with three $f_{7/2}$ particles, is $\frac{3}{2}$.

While the $p_{1/2}$ and $g_{9/2}$ states compete as long as the odd particle number is below 50, above 50 the competition is between the $g_{7/2}$ and $d_{5/2}$ states. This is quite in agreement with Table 8.1. It is striking, furthermore, how independent the low states are of the number of the particles of which there is an even number in the nucleus. Thus, if there are 45 protons in the nucleus, or 45 neutrons, the lowest state remains invariably a $J = \frac{7}{2}$ even state, the first excited state is a $J = \frac{1}{2}$ odd state. This is in full accord with the coupling rules and their explanation: an even number of neutrons (protons) couples to zero total angular momentum and the interaction of these neutrons (protons) with the protons (neutrons) is so weak that they do not affect the relative position of the states of the latter.

7. Magnetic and Quadrupole Moments

The reason for the deviations of the measured magnetic moments from the Schmidt lines (mentioned under Coupling Rules) has been the subject of much discussion. There is general agreement that the intrinsic moment of the nucleons (see Sec. I) will be affected by the proximity of other nucleons It appears quite likely, however, that this effect is quite small—of the order of a couple of tenths of nuclear magnetons. Since most of the moments lie quite far from the Schmidt lines, one has to assume that the wave function obtained by the simple coupling rules is inaccurate. This view is strengthened by the fact that the calculated magnetic moments agree quite well with the measured ones when one has a more reliable wave function. This is the case for light nuclei, the wave functions of which can be obtained on the basis of the isobaric spin concept. In fact, a study of the magnetic moments of these nuclei indicates that the changes in the intrinsic moment of nucleons tend to shift the magnetic moments beyond the Schmidt lines (cf. H^3, He^3, C^{13}) rather than pull them between these lines.

The modifications of the wave function to bring about agreement between observed and calculated moments do not seem to be large. It is worth recalling that the wave function of a state which is described, with 95 per cent probability, by the unperturbed state ψ_0, and contains a 5 per cent admixture of ψ_1, is

$$0.975\psi_0 + 0.225\psi_1$$

If a property, such as the magnetic moment, has a matrix element connecting ψ_0 with ψ_1, its coefficient will be $2 \times 0.975 \times 0.225 = 0.44$. This shows that a relatively small admixture can radically change some of the properties of stationary states. Let us denote the wave function of the shell model, as given by the coupling rules, by ψ_0; the operator of the magnetic moment by μ_z. Then, if we denote the actual wave function by $\alpha_0\psi_0 + \alpha_1\psi_1$, the total magnetic moment will be

$$(\alpha_0\psi_0 + \alpha_1\psi_1, \mu_z(\alpha_0\psi_0 + \alpha_1\psi_1)) \tag{8.3}$$

It appears that one can obtain reasonable agreement with the observed moment by omitting the $(\psi_1,\mu_z\psi_1)$ term of (8.3) as small. The Rayleigh-Schroedinger perturbation method gives a series expansion for ψ_1 but $(\psi_0,\mu_z\psi_1)$, and $(\psi_1,\mu_z\psi_0)$ are different from zero only for a few of the terms of the series. Their addition to the Schmidt value $(\psi_0,\mu_z\psi_0)$ greatly improves the agreement with the experimental values.

The agreement between observed magnetic moments and those calculated in the way described in the preceding paragraph may not be convincing because almost every modification of the wave function shifts the moments from the Schmidt lines to the region between these lines. However, a similar calculation of the quadrupole moments also greatly improves the agreement with the observed quadrupole moments. There is one exception to this rule: the very large quadrupole moments of nuclei in the rare earth region, amounting to several barns (10^{-24} cm^2), cannot be obtained in this way. These quadrupole moments are too large to be accounted for by a single particle with a reasonable wave function. The explanation of these quadrupole moments requires a more drastic modification of the wave function in which, then, the ψ_0 will play only a relatively small role.

8. Problems of the *j-j* Model

The general qualitative success of the shell model, at least for not too light nuclei, is most remarkable in view of the simplicity of the assumptions on which it is based. Several problems remain, however, which it is well to record.

The first stems from the fact that the independent particle model—the representation of the wave function by one or a few simple determinants—can hardly be expected to be accurate. This circumstance, together with the great success of the shell model, invites a reformulation of the independent particle model in which the determinants are replaced by some more general functions. Following the initiative provided by the work of Brueckner and Watson, several promising attempts have been made in this direction. These are formulated in rather abstract mathematical language so that it is not easy to describe them in terms of the wave function. It seems, nevertheless, that the wave function shows a characteristic departure from the wave function of the independent-particle model in those parts of the configuration space where two particles are very close together. Furthermore this departure is the same for all not too highly excited states so that it hardly affects the matrix elements of single-particle operators (such as are responsible for electromagnetic or beta transitions) between these states.

Second, the origin of the spin-orbit splitting, so characteristic for the model, is unclear. Meson theories do not yield a "vector force," i.e., an interaction of the form

$$[(\mathbf{p}_1 - \mathbf{p}_2) \times (\mathbf{r}_1 - \mathbf{r}_2) \cdot (\mathbf{s}_1 + \mathbf{s}_2)]V(r_{12}) \tag{8.4}$$

which would give the most simple and direct interpretation of the energy differences between the $j = l + \frac{1}{2}$ and the $j = l - \frac{1}{2}$ individual particle levels. Attempts to explain these energy differences as second-order effects of the tensor interaction have met only with partial success. The problem of a more detailed explanation of the coupling rules was mentioned before.

The preceding problems concern the foundations of the shell model from first principles. The problems to be enumerated below concern experimental results not accounted for by the model. The best known among these is the failure of the *j-j* model to distinguish between favored and unfavored β transitions. Comparison of the *L-S* and *j-j* models, carried out before, showed that the more detailed consideration of the interaction between the particles, which does lead to the explanation of the unfavored nature of most β transitions, so drastically modifies the wave function that this largely conforms again with the supermultiplet theory. It should be mentioned, however, that no favored β transitions exist for heavier nuclei (nor can any be expected on the basis of the supermultiplet theory) so that this problem applies only to relatively light nuclei ($A < 50$).

The problem of very large quadrupole moments is taken up again in the next section.

If the independent particle picture were accurate, the energy of a nuclear level would be determined, at least in first approximation, by the configuration from which it arises. That this is not the case can be seen most clearly by enumerating all the energy levels of the whole nucleus which arise from the lowest configuration in the middle of a shell. Consider, first, a small A; let us say $A = 8$. The lowest configuration is $(1s)^4(1p_{3/2})^4$; it gives the following levels: $J = 4, 2, 2, 0$ for $T = 0$, $J = 3, 2, 1$ for $T = 1$ and $J = 0$ for $T = 2$. Among the $T = 1$ states, the $J = 2$ is lowest. Its position is known from the energy difference between the normal states of Be^8 and Li^8: it lies 17.0 Mev above the normal state. The position of the $J = 0$, $T = 2$ state cannot be obtained from available information but can be estimated to lie about 12 Mev above the $T = 1$, $J = 2$ state. Thus the $(1s)^4(1p_{3/2})^4$ configuration is spread out over 29 Mev. It certainly does not form a closely spaced group of levels, as one would expect if the independent particle picture were valid. Naturally, $A = 8$ is a low mass number and one will not expect the *j-j* model to be very accurate. However, a similar situation prevails in general as long as neutrons and protons are in the same shell.

The *j-j* model underestimates the energy differences between isobars and the calculations which improve on this situation by considering the forces in more detail again modify the wave functions so drastically that they conform largely with the supermultiplet theory.

9. Semiempirical *j-j* Model

In a more recent modification of the *j-j* shell model, the attempt to explain the relative position of the levels by means of nuclear potentials similar to (5.5) is entirely abandoned. Instead a largely phenomenological approach is followed which is based on the two assumptions that the wave function is correctly given by the *j-j* model and that the interaction takes place between pairs of nuclei.

Let us consider, for instance, the $d_{5/2}$ shell. The $(d_{5/2})^2$ configuration gives six states with total angular momenta $J = 0, 1, 2, 3, 4, 5$. The $J = 0, 2, 4$ are isotopic spin triplets, the $J = 1, 3, 5$ isotopic spin singlets but this fact does not enter the following considerations. The next step is, ideally, the determination of the energy values of these states. This must be done experimentally. However, the six energy values of the two-particle system suffice to obtain the position of all energy levels of the rest of the $d_{5/2}$ shell, from F^{19} to Si^{28}. In order to obtain, for instance, the energy levels of the $(d_{5/2})^4$ configuration, one decomposes the wave function $\psi(1,2,3,4)$ of a level of this configuration, as given by the shell model, in the following way:

$$\psi(1,2,3,4) = \sum_{J=0}^{5} \sum_{m=-J}^{J} u_{Jm}(1,2) f_{Jm}(3,4) \quad (8.5)$$

The u_{Jm} are the wave functions of the two-nucleon problem mentioned above. Every antisymmetrized product $D_\mu^{5/2}(1)D_{\mu'}^{5/2}(2) - D_{\mu'}^{5/2}(1)D_\mu^{5/2}(2)$ can be expressed in terms of these u_{Jm}: the four-particle wave function (1,2,3,4) of the *j-j* model, as a function of the coordinates of the particles 1 and 2, is a linear combination of such antisymmetrized products. The $f_{Jm}(3.4)$ in (8.5) are simply the expansion coefficients of (1,2,3,4) in terms of the u_{Jm}; they depend, naturally, on the coordinates of the particles 3 and 4.

The probability w_{Jm} that the particles 1 and 2 are in the state u_{Jm} is

$$w_{Jm} = \int |f_{Jm}(3,4)|^2 \, d3 \, d4 \quad (8.5a)$$

where the integration includes summation over spin and isotopic spin variables. Hence, the interaction energy between particles 1 and 2 is, for the state $\psi(1,2,3,4)$,

$$\varepsilon_{12} = \sum_{J=0}^{5} \sum_{m} w_{Jm} E_J \quad (8.5b)$$

where E_J is the energy of the two-particle state with total angular momentum J, as obtained experimentally. The interaction energy between all other pairs is equally large so that the total energy in the state described by ψ becomes

$$E = \frac{4 \times 3}{2} \varepsilon_{12} \quad (8.5c)$$

The factor $\frac{1}{2} \times 4 \times 3$ is the number of pairs of $d_{5/2}$ particles.

If the position of the six states of the $(d_{5/2})^2$ configuration is not known, and the energy values of all levels of a configuration rarely are, the energy values of some of the levels of higher configurations must be used to determine the E_J. One has, at any rate, six constants available to fit all the levels of all the configurations $(d_{5/2})^n$, up to $(d_{5/2})^{12}$. The number of constants is $2j + 1$ if the angular momentum of the particles in the shell being filled is j.

If the partially filled proton and neutron shells are different, the preceding procedure has to be somewhat modified. Let us consider, for instance, the nuclei Zr, Nb, Mo, . . . with proton numbers $Z = 40, 41, 42$ . . . . The $1g_{9/2}$ shell is then the partially filled one. The neutron numbers which we shall consider range from 50 to 56; the partially filled neutron shell is $2d_{5/2}$. In order to calculate the proton-proton interaction, the possible wave functions Ψ of the system are expanded in the form

$$\Psi = \sum_{J_pm} \psi_{J_pm}(1,2) f_{J_pm}(\text{rest}) \quad (8.6)$$

where 1 and 2 refer to protons and f_{J_pm} depends on the coordinates of the remaining particles. The function ψ_{J_pm} is that linear combination of the products of single-particle wave functions $G_\mu^{9/2}$

$$\psi_{J_pm}(1,2) = \sum_{\mu\mu'} (\tfrac{9}{2}\mu\tfrac{9}{2}\mu';J_pm) G_\mu^{9/2}(1) G_{\mu'}^{9/2}(2) \quad (8.7)$$

for which the total angular momentum is J_p and its projection into the z axis is m. If the energies $E_{pp}(J_p)$ of the two proton states (8.7) are known, as they are from the spectrum of Mo^{92}, the matrix element of the interaction energy between protons 1 and 2 can be obtained as

$$\sum_{J_pm} E_{pp}(J_p) \int f^*_{J_pm} f'_{J_pm} \quad (8.8)$$

The integral is to be extended over the coordinates of the rest of the particles. The proton-proton energy's matrix element between Ψ and Ψ'

$$\Psi' = \sum_{J_pm} \psi_{J_pm}(1,2) f'_{J_pm}(\text{rest}) \quad (8.6a)$$

is (8.8), multiplied by the number of proton pairs, 1, 3, 6, . . . for $Z = 42, 43, 44$, ...

The matrix elements for the neutron-neutron interaction are calculated in the same way, except that the same Ψ which is expanded in (8.6) in terms of the ψ_{J_pm} functions of (8.7) is now expanded in terms of the neutron-pair functions

$$\phi_{J_nm} = \sum_{\mu\mu'} (\tfrac{5}{2}\mu\tfrac{5}{2}\mu';J_nm) D_\mu^{5/2}(1) D_{\mu'}^{5/2}(2) \quad (8.7a)$$

and the energy values $E_{nn}(J_n)$ for the analogue of (8.8) are obtained from the spectrum of Zr^{92} which has two neutrons in partially filled shells. Finally, to obtain the matrix elements of the proton-neutron interaction, one expands Ψ and Ψ' in terms of the functions of a proton-neutron pair

$$\chi_{Jm}(1,2) = \sum_{\mu\mu'} (\tfrac{9}{2}\mu,\tfrac{5}{2}\mu';Jm) G_\mu^{9/2}(1) D_{\mu'}^{5/2}(2) \quad (8.7b)$$

and obtains the matrix elements of the proton-neutron interaction energy E_{pn} from the spectrum of Nb^{92} which has a proton in the $1g_{9/2}$ shell and a neutron in the $2d_{5/2}$ shell. Again, in order to obtain the total matrix element, the analogue of (8.8) has to be multiplied by the number of proton-neutron pairs of the partially filled shells. Incidentally, the spectra of Mo^{92}, Tr^{92}, and Nb^{92} show that the variation of the $E_{np}(J)$ with J is about four times smaller than the variations of E_{pp} or E_{nn} with J_p and J_n, thus confirming the assumption made in Sec. V that the interactions between particles in different shells are less important for the determination of the wave function than the interactions of particles within the same shell. Hence, in first approximation, the protons couple to each other, and the same applies for the neutrons.

Having calculated the matrix of the interaction energy between the possible wave functions of the configurations considered, one can diagonalize it and obtain the energy values of all states of the nuclei with proton numbers between 40 and 50 and neutron numbers between 50 and 56, which originate from the $(1g_{9/2})^{Z-40}(2d_{5/2})^{N-50}$ configuration. The energy values obtained in this way agree, as a rule, with the observed values, within 0.1 Mev. There is some arbitrariness in the comparison of observed and calculated excitation energies because an observed level at a position where no such level would be expected can be attributed to another configuration. Conversely, if no level has been observed where one would be expected, this can be attributed to the difficulty of observing the level in question. There is no such arbitrariness in the comparison of the binding energies of the normal states with the calculated ones, and the comparison is, on the whole, satisfactory. The fluctuation of the separation energies from their average is about 1 Mev; the difference between calculated and observed values is about 0.1 Mev.

These calculations justify the assumptions (1) that the *j-j* model gives reasonably accurate wave functions in terms of single-particle wave functions ("orbitals"), (2) that the single-particle wave functions do not change substantially while the partially filled shell is being completed, and (3) that the total interaction energy is a sum of interactions between pairs of nucleons. The fact that the interaction between pairs has to be obtained experimentally (from the spectra of Mo^{92}, Zr^{92}, Nb^{92} in the example considered) indicates that the interaction operator is complicated; for the nucleons embedded in nuclear matter it also may well be different from the interaction operator describing the behavior of two isolated nucleons, discussed in Sec. V.

The calculations sketched also give the total wave functions of the stationary states in terms of the single-particle wave functions: $G_\mu{}^{9/2}$ and $D_{\mu'}{}^{5/2}$ in the example discussed. Hence they permit the calculation not only of the energy values but also of magnetic and quadrupole moments, transition probabilities, etc., at least as far as orders of magnitude are concerned. The comparison of these with the experimental values is less satisfactory than the comparison of the energy values. As was pointed out before, this is not surprising because small deviations of the assumed wave function from the correct one can influence significantly the expectation value of most quantities, except the energy. The energy does not depend sensitively on small deviations of the wave function from the actual one. There are a few cases in which the procedure just outlined gives inaccurate values for the energy also, but in these cases there is good a priori reason to assume significant mixing of the wave functions of different configurations.

IX. NUCLEAR MODELS. MANY-PARTICLE MODELS

1. The α-particle and Cluster Models

Correlations between motions of individual nucleons, which may exist in addition to those postulated by the exclusion principle, are entirely neglected in the independent particle model; they are described only in a global fashion in the supermultiplet model. The α-particle model is based on the assumption that these correlations are, in all nuclei, very similar to those in the α particle. It is suggested that a group of two neutrons and two protons forms a tight configuration in the nucleus and moves within the nucleus as a unit. On this picture the nucleus is considered as similar to a molecule, the individual units of which are α particles. It is not essential to presume that the α particles have a permanent identity. They may form, dissolve, and form again. If the time during which the α-particle structure is maintained is long compared with the intervening period, the properties of the nucleus may be calculated with good approximation on the basis of the molecular picture.

The binding energies of nuclei composed of an integral number of α particles are well accounted for by the α-particle model if, but only if, one disregards the increasing electrostatic energy. The binding energy of the α particles themselves accounts for a large part of the total binding energy. The remaining binding energy arises through the interactions between the α particles. It appears that the binding energy per α-particle bond in these nuclei is approximately constant (except for the case of Be^8 which does not form a bound state). This feature contributes strongly to the attractiveness of the model.

In general, however, the model has met only moderate success. It is difficult to reconcile the form of the model with the data on the α-α scattering and the absence of binding in the Be^8 system. The nuclei which are not composed of α-particle units are not easily incorporated into the model.

The α-particle model is a special case of the so-called cluster models. In its most extreme form the cluster model postulates a definite structure for every stationary state. Thus, some states of Be^9 are postulated to be composed of two α particles and a neutron, others of Li^7 and H^2, etc. The last statement means that the wave function of the corresponding state can be well approximated by the product of three functions, two of which are the wave functions of Li^7 and H^2 in their normal states, and the third describes the motion of these particles with respect to each other. In another excited state, the Li^7-H^2 pair may be replaced by a Li^6-H^3 pair or a triplet of $2He^4 + n$. In its original and less extreme form, the cluster model's wave functions of the stationary states are superpositions of the various wave functions which the extreme form of the cluster model attributes to the various states. Under this assumption, the Be^9 wave function would be a superposition of the cluster functions for 2α + neutron, for $Li^7 + H^2$, for $Li^6 + H^3$, etc. This is a reasonable assumption but does not lead to many very concrete conclusions.

2. The Collective Model; Evidence for the Nonspherical Form of the Nucleus

The very large quadrupole moments in the rare-earth ($Z = 57$ to 71) region suggest that the nuclei in this region are strongly nonspherical. For example, Lu^{175} with angular momentum $\frac{7}{2}$ in the ground state has a quadrupole moment of 5.6×10^{-24} cm². According to the shell model, such a nucleus should consist of a spherically symmetric core ($J_{core} = 0$) and a few extra protons. Since, because of $J_{core} = 0$, the quadrupole moment of the core is zero, all the quadrupole moments should be due to a few protons. For the quadrupole moment of a single proton one finds, using the angular distribution of a $g_{7/2}$ particle and a mean-square radius as discussed in Sec. III, a quadrupole moment of -0.2×10^{-24} cm². It is clear that the actual quadrupole moment, which is more than 25 times greater, cannot be due to these protons alone, but the core must also be deformed, i.e., cannot be in a $J_{core} = 0$ state. If one assumes an ellipsoidal shape of the nucleus with uniform charge (proton) distribution, and a density of nuclear matter as given in Sec. III, one arrives at major and minor axes of 7.6×10^{-13} cm and 6.3×10^{-13} cm in order to obtain the quadrupole moment in question.

Equally surprising, from the point of view of the *j-j* shell model, is the relatively large quadrupole moment of 0.8×10^{-24} cm² in In. Whereas the configuration of the Lu nucleus may be subject to some doubt, In has just one proton outside the magic shell of 50. Its four-times-too-large quadrupole moment can hardly be interpreted any other way but that the single proton outside the core deforms the core significantly so that other protons, in the closed shell, also contribute to the quadrupole moment. The large quadrupole moments of some even-Z–odd-N nuclei are almost equally convincing: According to the simple *j-j* model, the protons should couple to a $J_p = 0$ state which would have no quadrupole moment. Evidently, the neutrons deform the proton orbits in this case.

The large deviation from the spherical shape should not be too surprising since the energy of nuclear matter, inside the surface, does not depend on the form of the surface; the nucleons that are farther from the surface than the range of nuclear forces are in the same surroundings, no matter what the shape of the surface. The only obvious difference in the energy content of a spherical and a nonspherical nucleus is that more nucleons are close to the surface in the latter and are, hence, energetically in less favorable surroundings. However, the total surface is increased by the deformation only by about 1 per cent. With the value of the surface energy given in (2.1), this gives an increased surface energy of about 4 Mev. This increase in surface energy can be compensated by the lower energy that strongly aspherical orbits, with large values of the angular momentum, can assume in a potential due to the nonspherical potential of an ellipsoidal core.

The relatively easy deformability of nuclear matter stands in marked contrast to the spherical shape taken by the electronic cloud of a free atom. This spherical shape results because the attractive force on the electron emanates from a center that is essentially a point and thus gives the shape considerable rigidity.

There are two further indications which suggest a substantial revision of the shell model discussed previously. The first of these is the large transition probabilities for many electric-quadrupole transitions. These are, indeed, so large, in some cases even if there is only a single proton outside a closed shell, that they cannot be due to a single particle. Even in the case of the nuclei discussed at the end of Sec. VIII, for which the shell model gives rather accurate energies, the observed quadrupole transitions are about twice as fast as the calculated ones. The electric-quadrupole transition rates are even faster in the regions of large static quadrupole moments. These are particularly prevalent in the rare-earth region previously mentioned, with proton numbers in excess of 64 and neutron numbers between 90 and around 110, and for very heavy nuclei with proton numbers exceeding 88 and neutron numbers beyond 126.

The strength of the electric-dipole transitions also suggests that they are due to the concerted action of several protons, in fact to the rotation of the whole nonspherical nucleus.

The rotational spectra are the last, and most convincing, evidence of collective motion. They also are most prevalent in the regions of large quadrupole moments, but indications of them can be found almost throughout the periodic table. The rotational spectra consist of a succession of energy levels with J values 0, 2, 4, 6, . . . , or $K, K+1, K+2, \ldots$ with $K \neq 0$, the energy values of which are closely approximated by the formula

$$E_J = E_0 + \frac{J(J+1)}{2I} \tag{9.1}$$

with E_0 and I (the "moment of inertia") constant. In many cases, particularly if the spacing is low, of the order of 100 kev (that is, I is large), (9.1) is valid to within a few per cent up to $J = 8$ or 10. Rotational bands up to $J = 16$ have been observed. Their similarity with the rotational levels of diatomic molecules suggests most clearly that there exists some nonspherical, reasonably rigid structure that can rotate in space without undergoing much deformation.

3. The Collective Model; Mathematical Formulation

The easiest way to obtain a wave function representing a set of particles which form a nonspherical cloud is to assume a nonspherical potential, such as

$$V_0 = \frac{1}{2}\kappa^2 \left[\gamma(x^2 + y^2) + \frac{z^2}{\gamma^2}\right] \tag{9.2}$$

and to obtain the wave function of a set of noninteracting particles therein. In the case of the potential (9.2), this is a determinant formed of triple products of Hermite orthogonal functions

$$H_{n_x}\left(\frac{\gamma^{1/4}x}{r_0}\right) H_{n_y}\left(\frac{\gamma^{1/4}y}{r_0}\right) H_{n_z}\left(\frac{\gamma^{-1/2}z}{r_0}\right) \delta_{s\pm1}\delta_{\tau\pm1} \quad (9.3)$$

multiplied with suitable functions of the ordinary and isotopic spin coordinates s and τ; $r_0 = \hbar^{1/2}/M^{1/4}\kappa^{1/2}$ gives the mean-square distances $\bar{x}^2 = (n_x + \frac{1}{2})r_0^2/\gamma^{1/2}$, $\bar{y}^2 = (n_y + \frac{1}{2})r_0^2/\gamma^{1/2}$, $\bar{z}^2 = (n_z + \frac{1}{2})\gamma r_0$. By choosing $\gamma > 1$, a prolate, and by choosing $\gamma < 1$ an oblate, ellipsoid will result. The determinantal wave function obtained from the single-particle functions is called the intrinsic wave function Ψ_{intr}.

The energy of the intrinsic wave function, if the potential is given by (9.2), is smallest if the n_x, n_y, n_z are chosen as small as possible, consistent with the requirement that no two triplets n_x, n_y, n_z and n_x', n_y', n_z' can be the same unless the spin-isospin factors associated with them are different. However, the actual energy associated with the determinantal wave function Ψ_{intr} is not the sum of the energies of the particles in the potential (9.2). This would be

$$\hbar\kappa M^{-1/2}[\gamma^{1/2}(n_x + n_y + 1) + \gamma^{-1}(n_z + \tfrac{1}{2})] \quad (9.4)$$

the summation being extended over all triplets n_x, n_y, n_z that occur in the determinant representing Ψ_{intr}. The actual kinetic energy is one-half of (9.4), but the potential energy should be calculated not with the fictitious potential (9.2), which would give the other half of (9.4), but with the actual interaction between the nucleons. As was discussed in Sec. V, this is quite complicated. Hence this part of the calculation is usually carried out only schematically, by assuming only an ordinary interaction such as the V_3 of (5.4). Even this is often simplified by giving it a very short range but correspondingly greater magnitude within that range. The energy depends on γ, and γ then can be determined by postulating that the energy, as a function of γ, be a minimum.

The intrinsic wave function plays a role similar to that of the electronic wave function of diatomic molecules when the two nuclei are in fixed positions. Similar to that wave function, it can be characterized by the angular momentum around the symmetry axis of the potential. This is denoted by $K\hbar$, whereas the similar quantity in diatomic molecules is denoted by $\Lambda\hbar$. The intrinsic wave function is not the true wave function of the stationary states. This is already evident from the fact that there is a continuum of such wave functions, corresponding to the continuum of possible orientations of the preferred axis of the potential. All such directions give rise to a potential, similar to (9.2), and the corresponding Ψ_{intr} all have the same average energy.

The collective model's wave function for the stationary states is obtained by a superposition of the Ψ_{intr} which correspond to all possible directions θ, ϕ of the preferred axis of the potential generating the Ψ_{intr}. The wave functions of the state with angular momentum $J\hbar$ are

$$\Psi_m{}^J = \int\int D_{mK}{}^J(\phi,\theta)\Psi_{intr}(\theta,\phi) \sin\theta \, d\theta \, d\phi \quad (9.5)$$

where m is the component of the angular momentum J in the direction of the z axis; K is the angular momentum of Ψ_{intr} around the preferred axis; and $D_{mK}{}^J$, certain universal functions of the direction, $D_{m0}{}^J$ being the usual spherical harmonics. $\Psi_{intr}(\theta,\phi)$ is the intrinsic wave function with the preferred axis in the θ, ϕ direction. Actually, (9.5) is essentially identical with the expression for the total wave function of diatomic molecules: θ and ϕ give, in that case, the direction of the axis connecting the two nuclei; Ψ_{intr} is the electronic and vibrational wave function. The D functions are the same in both cases. Just as in the case of symmetric diatomic molecules, J can assume all values K, $K + 1$, $K + 2$, . . . , and only these, if $K \neq 0$; it can be 0, 2, 4, . . . , if $K = 0$ and Ψ_{intr} is symmetric with respect to reflection of the direction of the preferred axis. This is always assumed to be the case. Thus, one of the important characteristics of rotational spectra, the sequence of the J values of the rotational states, follows at once from the model's wave function.

In order to obtain the spectrum, the kinetic and potential energies associated with the wave functions (9.5) must be calculated. This is easiest in the case of very large or very small γ when two Ψ_{intr} are orthogonal as soon as their axes of orientation differ. In this case, just as in the case of diatomic molecules, the potential energy varies little with J; its variation is proportional to $J(J + 1)$. Most of the energy difference is due to a difference in the kinetic energies of the states (9.5) with different J; this also is proportional to $J(J + 1)$. Hence, in the extreme cases of $\gamma \gg 1$ and $1/\gamma \gg 1$, the simple picture of a rigid rotating ellipsoid is valid. The regions of large nuclear deformations come close to it. It might be observed, however, that the actual rotational energy always increases somewhat less fast than $J(J + 1)$ whereas, in the picture considered, the kinetic energy alone increases somewhat faster. It follows that the total neglect of the variation of the potential energy is never fully justified.

In the opposite case of $\gamma = 1$, that is, a spherical potential, let us consider the totality of the intrinsic wave functions which have the same energy in the original, fictitious potential (9.2). Let us assume that there are N such intrinsic wave functions. Then these are also the wave functions of the shell (independent particle) model which have, in the same potential (9.2) and again disregarding the interactions between nucleons, equal energies. They do not necessarily all belong to a single configuration in the usual sense because, in the particular potential (9.2) with $\gamma = 1$, several single-particle energy levels, such as $2s$ and $1d$, $2p$ and $1f$, coincide. However, if one wave function of a configuration is contained among the N intrinsic functions, all wave functions of the configuration are so contained.

If one now subjects the N intrinsic states to the operation (9.5), one will produce from them functions with definite J and m. Since the number of functions so produced will again be N, there will be, in general, several linear relations between the functions so produced. However, if only one function with a given J and m is produced [all other expressions (9.5) with the same J and m being either 0 or multiples of 1], this will be a proper wave function not only of the collective model but also of the shell model. In such a case, for $\gamma = 1$, the two models are identical.

If several different wave functions with the same J and m result from the N intrinsic functions, the most reasonable choice of the collective model would be to accept those for which the true energy of the producing intrinsic state (i.e., the energy with the actual interaction) is lowest. Such a function may not belong to a single configuration and may represent some configuration mixing from the point of view of the shell model. There are, however, indications that the wave functions obtained in this way are reasonable also from the point of view of the shell model. Hence, for $\gamma = 1$, the conflict between collective and shell models is less sharp than one might expect. It is important to realize, however, that the kinetic energies of all states, with any J and m, obtained from the N intrinsic states are the same; they all belong to configurations with the same kinetic energy. The energy differences between the members of a "rotational band," i.e., the differences between the energies associated with the functions (9.5) with different J, are due to differences of potential energy. In some cases, these still show the $J(J+1)$ dependence; this is not so in other cases. This is not in conflict with experiment, because (9.1) holds only in the regions of large deformations. As was mentioned before, the energy in the rotational bands outside the very strongly deformed regions increases less fast with J than if it were proportional to $J(J+1)$. In many cases, the order is even reversed and levels with higher J have lower energies than some with lower J. Of course, in such a case one does not speak of rotational bands.

The discussion of an intermediate γ, differing from 1 by only a few per cent, would be most important. Unfortunately, this is less simple than the two limiting cases. It seems clear that both kinetic and potential energies depend significantly on J. The former increases with J faster than $J(J+1)$. The latter also increases with J if $K = 0$, but less fast, so that the total energy increases less fast than $J(J+1)$. This is in agreement with experiment. For $K \neq 0$, the situation is less simple and the degree of agreement between calculated and observed spectrum is more difficult to ascertain.

4. The Collective Model; Nilsson's Modification

The choice of the intrinsic wave function which is then to be subjected to the transformation (9.5) was not fully specified in the preceding discussion. This is very important, however, because the functions $\Psi_m{}^J$ obtained from the different Ψ_{intr} are not orthogonal and, if all Ψ_{intr} were used, the number of states obtained would be much larger than that of the actual states. This became evident in the preceding discussion of the case $\gamma = 1$. There are some questions in this connection which have not been fully clarified.

At any rate, it is possible to form such linear combinations of the single-particle functions (9.3) which would have, for $\gamma = 1$, definite j and j_z values. The determinant of Ψ_{intr} can then be composed of such single-particle functions. This procedure is the original one of the collective model; since the wave functions obtained in this way show greater similarity to the wave functions of the shell model originating from a single configuration, one might expect that it is the valid procedure if the deformation is not too large. Alternatively, one can disregard all reference to j values and order the single-particle states in the potential (9.2) according to their energies. This is Nilsson's procedure; one would expect it to be valid for large deformations. Since the magnitude of the deformation itself depends on the Ψ_{intr}—it is supposed to make its energy content a minimum—the situation is somewhat complicated.

5. Further Consequences of the Collective Model

The wave function (9.5) permits one also to calculate, in addition to the energies, the rate of the electric-quadrupole transitions between members of a rotational band. The matrix element for these transitions is similar to that for the static quadrupole moment. However, since the transition rate is proportional to the square of the matrix element, it may become 100 times larger than that due to a single particle. Such very high transition rates have indeed been observed.

A number of selection rules also follow from the form (9.5) of the wave function. K seems to be a very good quantum number in strongly deformed nuclei; bands with different K may have levels with equal J in close proximity without interfering with each other. In γ and β decay, the transition probabilities between states in different bands, which differ in K by more than the multipolarity of the transition, should be greatly reduced below that normally observed for these types of transitions. In the case of Hf^{180} there is an electric dipole transition between an 8^- level and the 8^+ member of the ground-state rotational band ($K = 0$) which is 10^{15} times smaller than the standard estimate. The reason is, presumably, that the 8^- level is the first member of a $K = 8$ band so that the dipole transition is forbidden to a $K = 0$ member of the ground-state band. Similarly, in the β decay of Lu^{176} (7^-) to the 8^+ state of Hf^{176}, the decay is some 10^{12} times slower than what would be expected.

The early formulation of the collective model used the picture of irrotational flow of a liquid. In this form, the model could account for a number of regularities of the spectra of nuclei in the regions of strong deformation. However, the value of inertia as obtained from the spacing of the rotational levels differed from its value obtained from the static and radiative quadrupole moments by a factor as large as 5. The more modern picture, presented above, is largely free of this difficulty.

Excited states associated with the change of shape or vibration have not been observed as unambiguously as the rotational spectrum. The hydrodynamic model of quadrupole vibrations predicts in even-even nuclei a spectrum consisting of a one-quantum state with spin parity 2^+ at an energy E and, built upon this, three two-quantum states at energy $2E$ with spins 0, 2, and 4, all with positive parity. The second 2^+ state shows a preferred electric-quadrupole decay to the first 2^+ state rather than to the ground state. Away from the deformed region there is some evidence for such two-quantum states associated with a vibrational model, but the spacing is usually not correct nor are the two quantum levels always close to each other. Ni^{62} seems to be an example with vibrational states. However, the *j-j* coupling shell model with residual interactions can also account for the observed spectrum and the branching ratio observed in the γ decay. In favor of a collective model, however, it must be said that the observed quadrupole radiation between the first excited 2^+ states of even-even nuclei and their normal state is too swift to be accounted for by any simple shell-model theory.

The greatest success of the collective model is that it postulated the rotational states, and the large transition probabilities between these states both by spontaneous radiation and Coulomb excitation, even before these states were found experimentally. In addition, it has been applied lately to the phenomenon of fission.

X. NUCLEAR REACTIONS. I. CLOSE COLLISIONS

Nuclear reactions are collision phenomena and almost all methods developed for the treatment and understanding of collision phenomena have been used to interpret nuclear reactions. Many of these methods have, indeed, been developed for this purpose. Section V quotes some results for two-body problems—only these can be treated accurately with present-day techniques. If more than two particles are involved in the reaction, i.e., in all cases except proton-proton and proton-neutron scattering, one must be content either with approximate results, based on models similar to those used in Secs. VII, VIII, and IX or with general and accurate formulas which, however, contain several adjustable constants obtainable only experimentally or by using approximate calculations. We begin with an example for the second method.

1. The Collision Matrix

A collision between nuclei a and X may in general give rise to a variety of reaction products:

$$a + X \rightarrow \begin{cases} a + X & (s) \\ a + X^* & (t) \\ b + Y & (u) \\ c + Z & (w) \end{cases} \tag{10.1}$$

where X^* is an excited state of the nuclear system X. The first reaction is called elastic scattering, the second inelastic scattering, and the remainder exchange reactions. We consider only reactions in which two reaction products are formed. Each pair of products may be indicated by an index (s, t, etc.). The possibility of capture of a by X with subsequent γ emission is not considered in the analysis below; we take up such processes in connection with the considerations of resonance phenomena.

The nuclear system C consisting of all the nucleons in X and a (or Y and b, etc.) may be described in a many-dimensional configuration space. We distinguish two regions in this space: (1) the *internal region* within which the separations of all nucleons are of the order of nuclear dimensions; (2) the external region where at least the separations between some nucleons are greater than nuclear dimensions. For reasons which should become evident later, we imagine a many-dimensional surface S dividing the two regions; S is taken wholly in the external region, but everywhere near the internal region.

For a given energy E of the incident nuclei only a limited set of reaction products are energetically possible, and the energies available for the relative motion of any pair of products is definite. As a consequence, the wave function for the system at energy E will be different from zero in the external region only in certain "channels" of the configuration space which describe the system C as a pair of nuclei $a + X$, $b + Y$, etc. A particular state of motion of well-separated products t may be indicated by $R_t(\mathbf{r}_t)\chi_t$, where χ_t represents the state of the nuclei defined by the index t and $R_t(\mathbf{r}_t)$ describes the relative motion of the pair t of nuclei. The specifically nuclear forces have no influence on the function $R_t(\mathbf{r}_t)$. These functions will depend for a given E only on the presence or absence of the electrostatic field between the pair t, the relative angular momentum of the pair, and boundary conditions.

For simplicity, we assume that the spins of all nuclei involved in the reaction are zero. The removal of this assumption complicates the argument but adds nothing of importance in principle. We consider, to begin with, a collision in which the two nuclei of the pair s converge with relative orbital angular momentum quantum number l, that is, the incoming part of the wave function in the s channel is proportional to $r_s^{-1}P_l(\mathbf{\Omega}_s)I_{ls}(r_s)\chi_s$, where $r_s^{-1}P_l(\mathbf{\Omega}_s)I_{ls}(r_s)$ represents an incoming spherical wave of orbital angular momentum l. The direction of the internuclear line is $\mathbf{\Omega}_s$, its length r_s. The whole wave function for the system in the "open" channels—the channels which represent energetically possible reaction products—will consist, in the external region, of this incident wave together with emergent waves of the same angular momentum:

$$\begin{aligned}\Phi_l = {} & (M_s/\hbar)^{1/2}P_l(\mathbf{\Omega}_s)r_s^{-1}I_{ls}(r_s)\chi_s \\ & - \Sigma_t S_{st}{}^l(M_t/\hbar)^{1/2}P_l(\mathbf{\Omega}_t)r_t^{-1}E_{lt}(r_t)\chi_t\end{aligned} \tag{10.2}$$

where $P_l(\mathbf{\Omega}_t)r_t^{-1}E_{lt}(r_t)$ represents an emergent wave of orbital angular momentum l in the channel t. The I_{lt} and E_{lt} can be chosen as conjugate complex functions which satisfy the relation

$$I_{lt}E'_{lt} - E_{lt}I'_{lt} = I_{lt}I'^*_{lt} - I^*_{lt}I'_{lt} = 2i \tag{10.2a}$$

in which the primed quantities I'_{lt}, E'_{lt} denote the derivatives of I_{lt}, E_{lt} with respect to r_t, the internucleon separation. M_t is the reduced mass of the pair t and $(M_t/\hbar)^{1/2}I_{lt}$ and $(M_t/\hbar)^{1/2}E_{lt}$ are waves with unit inward and outward current per unit solid angle. The numbers $S_{st}{}^l$ describe the amplitudes of the emergent waves for the given incident wave; they are dependent, of course, on the nature of the collision, i.e., on the character of the colliding pair s, the product nuclei t, the angular momentum l, and also on the total energy E of the system. The quantities $S_{st}{}^l$ completely characterize the collision properties of the system. If only elastic scattering is possible, $S_{ss}{}^l = \exp 2i\delta_l$, where δ_l is the so-called phase shift. It is convenient to consider the set of quantities $S_{st}{}^l$ as an entity—the collision matrix for the system.

If no interaction were present between the particles of the pair s, the wave function for a system in which these particles stream toward each other in the Z direction and with unit flux per unit area would be simply $(M_s/\hbar k_s)^{1/2}\chi_s \exp ik_sr_s(\mathbf{e}_z \cdot \mathbf{\Omega}_s)$. In order to compare this expression with (10.2), one decomposes it into spherical waves; each such spherical wave represents a definite angular momentum

$$(M_s/\hbar k_s)^{1/2} \exp [ik_sr_s(\mathbf{e}_z \cdot \mathbf{\Omega}_s)]\chi_s = \sum_l \tfrac{1}{2}(2l+1)i^{l+1}(M_s/\hbar)^{1/2}P_l(\mathbf{\Omega}_s)k_s^{-1}r_s^{-1}(I_{ls} - E_{ls})\chi_s \tag{10.3}$$

This is a purely mathematical identity; $\mathbf{e}_z$ is the unit vector in the direction of motion of the incident particle in the pair s; the $P_l(\mathbf{\Omega}_s)$ are the well-known spherical harmonics, $P(\mathbf{e}_z) = 1$. The wave number k_s is the relative momentum of the colliding (or separating) pair s, divided by $\hbar$. Hence

$$(\hbar^2/2M_s)k_s^2 = E - \mathcal{E}_s$$

where $\mathcal{E}_s$ is the inner energy of the particles of the pair s. The k_t are defined in a similar way in terms of the inner energy $\mathcal{E}_t$ of the pairs t, the difference $\mathcal{E}_s - \mathcal{E}_t$ is the energy release (Q value) of the reaction $s \to t$.

If one superposes the wave functions (10.2) which take the interaction into account, with amplitudes $\frac{1}{2}(2l+1)i^{l+1}k_s^{-1}$, a wave function

$$\Phi = \sum_l \tfrac{1}{2}(2l+1)i^{l+1}k_s^{-1}\Phi_l \tag{10.2b}$$

is obtained, the incident waves in which are identical with those of (10.3). Φ gives, for the external region, the wave function for the collision of the particles s with unit flux, taking interaction into account. The difference between (10.2b) and (10.3) gives the products of the collision

$$\Phi' = \sum_l \sum_t (S_{st}^l - \delta_{st})\tfrac{1}{2}(2l+1)i^{l+1}(M_t/\hbar)^{1/2} P_l(\mathbf{\Omega}_t)k_s^{-1}r_t^{-1}E_{lt}(r_t)\chi_t \tag{10.2c}$$

The coefficient of χ_t describes the production of the pair t, the coefficient of χ_s the elastic scattering. If there is no interaction between the particles, the difference (10.2c) is zero; that is, the collision matrices S^l are all unit matrices in this case. The outgoing waves (10.2c) can be observed only for large separation r_t of the pairs, for large r_t the asymptotic form of E_t is

$$E_{lt}(r_t) = I_{lt}(r_t)^* \sim k_t^{-1/2}i^{-1} \exp (ik_tr_t) \tag{10.3a}$$

The reaction and scattering cross sections are all implicit in (10.2c). Since the integral of the square of $P_l(\mathbf{\Omega})$ over all directions is $4\pi/(2l+1)$, the total cross sections become

$$\sigma_{st} = \frac{\pi}{k_s^2}\sum_l (2l+1)|S_{st}^l|^2 \qquad \text{(reaction)} \tag{10.4}$$

$$\sigma_{ss} = \frac{\pi}{k_s^2}\sum_l (2l+1)|S_{ss}^l - 1|^2 \qquad \text{(scattering)} \tag{10.4a}$$

The last equation applies only if one of the pair s is uncharged, i.e., is a neutron. Otherwise, because of the long-range nature of the electrostatic interaction, the scattering extends to such high l that the last expression ceases to be useful and the formulas must be modified to take the electrostatic scattering into account more directly.

Certain properties of the collision matrix elements follow from general considerations. From the assumption of only two-particle breakup in the reaction, the number of pairs of incident particles per second is equal to the total number of emergent pairs per second. From this it follows that $\Sigma_t|S_{st}^l|^2 = 1$ and one can conclude also that

$$\Sigma_t S_{st}^l S_{s't}^{l*} = 0$$

for $s \neq s'$. An investigation of the properties of the system under time reversal shows further that $S_{st} = S_{ts}$: the collision matrix is unitary and symmetric. The symmetry relation leads to the so-called *reciprocity theorem*

$$k_\alpha^2\sigma_{\alpha\beta} = k_\beta^2\sigma_{\beta\alpha} \tag{10.4b}$$

which can be derived also from the thermodynamic principle of detailed balance.

Some properties, or more properly, some general limitations on the properties of collision processes may be derived from the results so far obtained, which do not involve the nature of the interactions between the colliding nuclei. For a particular pair of colliding systems upper limits on the reaction and scattering cross sections can be established. One can also obtain the general character of the expected angular distribution. For example, for incident neutrons, if the wavelength of the incident beam, $\lambda\!\!\!^- = \lambda/2\pi = 1/k$, is very much larger than the nuclear diameter, only the $l = 0$ wave (s wave) will be effective in the collision. The impact parameters for higher orbital angular momenta are large and the probability of effective collision is therefore small. In this case all the angular distributions are spherically symmetric. The maximum scattering cross section occurs for $S_{ss}^0 = -1$ and is $\sigma_{ss} = 4\pi/k_s^2$. The maximum total reaction cross section $\Sigma_{t\neq s}\sigma_{st}$ is π/k_s^2 (for $S_{ss}^0 = 0$). The associated scattering cross section is also π/k_s^2.

For the *total* cross section, (10.3) and (10.3a) give

$$\sum_t \sigma_{st} + \sigma_{ss} = \frac{\pi}{k_s^2}\sum_l (2l+1) \left\{\sum_t |S_{st}^l|^2 + |S_{ss}^l|^2 - S_{ss}^l - S_{ss}^{l*} + 1\right\} = \frac{\pi}{k_s^2}\sum_l (2l+1)\{2 - S_{ss}^l - S_{ss}^{l*}\} \tag{10.3b}$$

The sums in this equation exclude $t = s$ and the last line follows from $\Sigma|S_{st}^l|^2 = 1$ as extended over all t. The total cross section for the collision of the pair s depends only on the first power of the S_{ss}^l.

2. Qualitative Discussion of Resonance Phenomena

Under certain conditions the cross section σ_{st} for the reaction $a + X = Y + b$ rises, passes through a more or less sharp maximum, and falls to normal values as the energy of the incident particle a passes through certain energy intervals. The energy of the whole system (in the center of mass coordinate system) at the maximum cross section is referred to as the resonance energy E_λ of the reaction, the width in energy of the maximum in the cross-section function is called the width Γ_λ of the resonance level. The index λ specifies the resonance level. Examples for typical resonance behavior of the cross section are given in Fig. 10.1.

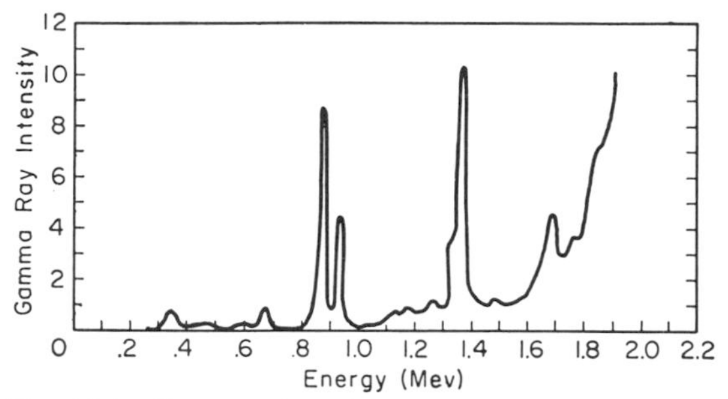

FIG. 10.1*a*. Cross section of the (p,γ) reaction in fluorine. The abscissa is the energy of the incident protons in Mev, the ordinate the γ-ray yield in arbitrary units. [*From E. J. Bernet, R. G. Herb, and D. B. Parkinson, Phys. Rev.*, **54**: 398 (1938), *one of the earliest observations of the resonance phenomenon.*]

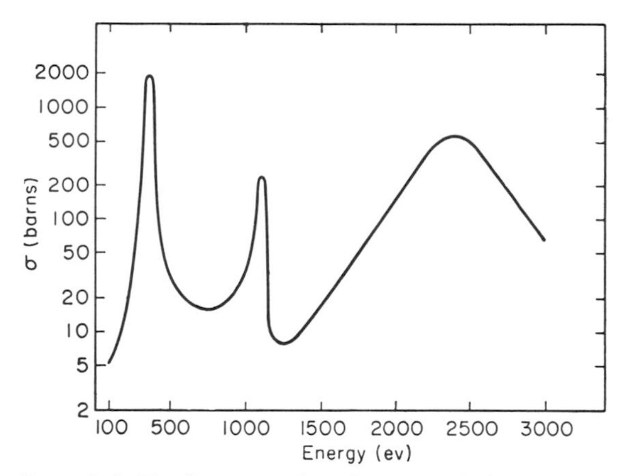

FIG. 10.1*b*. Total cross section of neutrons in Mn. Note the logarithmic scale for the cross section which is given in barns. [*From L. M. Bollinger, D. A. Dahlberg, R. R. Palmer, and G. E. Thomas, Phys. Rev.*, **100**: 126 (1955).]

This general behavior of the cross section may be understood on the basis of the picture given already in Sec. IV, 3. We assume that there exists a quasi-stationary state with average energy E_λ of the compound nuclear system C composed of the nucleons of the particles a and X. Since this state is quasi-stationary, its energy is not precisely defined; we denote the uncertainty in the energy of the state by Γ_λ. The reaction is presumed to proceed in two steps: (1) The incident particles make a transition to the quasi-stationary state of the compound nucleus. (2) This unstable state decays with the formation of the reaction products. The process is similar to that involved in the resonance scattering of radiation by atoms. On the basis of this picture, one may speak of the cross section σ_{sC} for the formation of the compound nucleus. This may be shown to be (we assume that only the $l = 0$ state is effective)

$$\sigma_{sC} = \frac{\pi}{k_s^2} \frac{\Gamma_{\lambda s}\Gamma_\lambda}{(E_\lambda - E)^2 + \frac{1}{4}\Gamma_\lambda^2} \tag{10.5}$$

where $\Gamma_{\lambda s}$ measures the probability of transition from the initial state of the incident s particles to the state λ of the compound nucleus. It is proportional to the square of a suitably formulated matrix element between these states. The behavior of the compound nucleus after formation depends on the properties of the state of the compound nucleus only. This compound nucleus breaks up with the formation of the pair t or the formation of other pairs. We write for the probability of the breakup of the compound nucleus into the t pair $\Gamma_{\lambda t}/\Gamma_\lambda$, where $\Gamma_{\lambda t}$ is called the partial width of the λ level for the emission of the pair t. Thus the cross section for the formation of the t particles becomes

$$\begin{aligned} \sigma_{st} &= \frac{\pi}{k_s^2} \frac{\Gamma_{\lambda s}\Gamma_\lambda}{(E_\lambda - E)^2 + \frac{1}{4}\Gamma_\lambda^2} \frac{\Gamma_{\lambda t}}{\Gamma_\lambda} \\ &= \frac{\pi}{k_s^2} \frac{\Gamma_{\lambda s}\Gamma_{\lambda t}}{(E_\lambda - E)^2 + \frac{1}{4}\Gamma_\lambda^2} \end{aligned} \tag{10.6a}$$

$\Gamma_{\lambda t}$ has a role similar to $\Gamma_{\lambda s}$: it is proportional to the square of a matrix element between the state of the compound nucleus and the final state of the t particles. Clearly we have $\Sigma_t(\Gamma_{\lambda t}/\Gamma_\lambda) = 1$ since the intermediate state of the compound nucleus must break up in some way. This formalism permits also description of the capture of a by X with subsequent emission of radiation. The compound system may radiate rather than break up by particle emission and we may assign a partial width for radiation $\Gamma_{\lambda r}$ to this process. $\Gamma_{\lambda r}/\hbar$ measures the probability per second that the excited state of the compound nucleus, formed in the collision, goes over to a lower state of the system by emission of radiation.

3. Derivation of the Resonance Formula

The preceding result is valid only to the extent to which the notion of a long-lived compound nuclear state is valid. The general condition for this is that the width Γ_λ be much smaller than the separation between adjacent levels.

It is difficult to judge the validity of the assumptions made above for actual nuclear processes. This can be done, however, on the basis of a mathematical derivation of (10.6*a*) using quantum-mechanical theory. We have seen that the cross sections of collision processes are given by the collision matrix. The elements of this matrix can be obtained from a knowledge of the stationary states of the system throughout the whole configuration space. Indeed what is required for the determination of the collision matrix elements are only the properties of the stationary state wave functions on the parts of the surface S (which separates internal and external regions of configuration space) where this surface is traversed by the open channels. If the wave function is known at this surface, the requirement of continuity across the surface completely determines the wave function in the open channels outside S and hence determines the collision matrix elements. The quantities which characterize the properties of the state at the surface of the channel t may be taken as (1) the coefficients v_t^l of $P_l(\Omega_t)r_t^{-1}\chi_t$ in the expansion of a wave function Φ on the surface S and (2) the similar quantities d_t^l in the expansion of the radial derivative $\partial\Phi/\partial r_t$ of the same wave function. These are related by a set of linear equations valid for all channels s, t.

$$M_s^{-1/2}v_s^l = \sum_t R_{st}^l M_t^{-1/2} d_t^l \tag{10.7}$$

M_s, M_t are again the reduced masses of the pairs s,t. The quantities R_{st}^l which enter this relation may be considered as elements of a matrix R^l. This matrix provides a generalization to reaction processes of the concept of the logarithmic derivative for potential scattering. Naturally, one has a different matrix R^l for each value of l.

It can be shown that the matrix elements R^l as functions of the energy E of the system can be expanded into a series of the form

$$R_{st}^l = \sum_\lambda \frac{\gamma_{\lambda s}\gamma_{\lambda t}}{E_\lambda - E} \tag{10.8}$$

where the $\gamma_{\lambda s}$, $\gamma_{\lambda t}$, E_λ are real, energy independent constants. They are different for different l. It follows from (10.8) that R^l is real and symmetric. The E_λ are the eigenvalues of a Hermitian boundary-value problem $HX_\lambda = E_\lambda X_\lambda$ for the internal region specified by the Hamiltonian operator H of the system together with the boundary condition that the normal derivative of X_λ vanish on S. In order to derive (10.8), one expands the actual wave function Φ of energy E and angular momentum $l\hbar$, inside the surface S, into a series in terms of the orthogonal functions X_λ

$$\Phi = \sum_\lambda c_\lambda X_\lambda \tag{10.9}$$

It is necessary to use, in (10.9), only X_λ with angular momentum $l\hbar$. The two expressions

$$(X_\lambda, H\Phi) = (X_\lambda, E\Phi) = Ec_\lambda$$

and $(HX_\lambda,\Phi) = (E_\lambda X_\lambda,\Phi) = E_\lambda c_\lambda$ are not equal because Φ does not satisfy the boundary condition on S. The difference is, in fact, an integral over the surface S:

$$(E_\lambda - E)c_\lambda = \frac{\hbar^2}{2M}\int (X_\lambda \operatorname{grad}_n \Phi - \Phi \operatorname{grad}_n X_\lambda)\, dS \tag{10.9a}$$

The last term vanishes because the normal derivative of X_λ on S is zero; the first term, however, is finite. In order to calculate it, we note that $\operatorname{grad}_n \Phi$ is, on S,

$$\operatorname{grad}_n \Phi = \sum_{l,s} \frac{M}{M_s} d_s^l \left(\frac{2l+1}{4\pi r_s^2}\right)^{1/2} P_l(\Omega_s)\chi_s \tag{10.9b}$$

Hence, if we define

$$\gamma_{\lambda t} = \left(\frac{\hbar}{2M_t}\right)^{1/2} \int X_\lambda \left(\frac{2l+1}{4\pi r_t^2}\right)^{1/2} P_l(\Omega_t)\chi_t\, dS \tag{10.9c}$$

we obtain from (10.9a)

$$\begin{aligned} c_\lambda &= \frac{\hbar^2}{2M}\frac{1}{E_\lambda - E}\int X_\lambda \sum_t \frac{M}{M_t} d_t \left(\frac{2l+1}{4\pi r_t^2}\right)^{1/2} P_l(\Omega_t)\chi_t\, dS \\ &= \sum_t \left(\frac{\hbar^2}{2M_t}\right)^{1/2} \frac{\gamma_{\lambda t}}{E_\lambda - E} d_t \end{aligned} \tag{10.9d}$$

Hence, we have for v_s

$$\begin{aligned} v_s &= \int \left(\frac{2l+1}{4\pi r_s^2}\right)^{1/2} P_l(\Omega_s)\chi_s\Phi\, dS \\ &= \sum_\lambda \int \left(\frac{2l+1}{4\pi r_s^2}\right)^{1/2} P_l(\Omega_s)\chi_s c_\lambda X_\lambda\, dS \\ &= \sum_\lambda \left(\frac{\hbar^2}{2M_t}\right)^{1/2} \frac{\gamma_{\lambda t}}{E_\lambda - E}\left(\frac{2M_s}{\hbar^2}\right)^{1/2} \gamma_{\lambda s} d_t \end{aligned} \tag{10.9e}$$

This is equivalent to (10.7) plus (10.8). Incidentally, the expression (10.9d) for c_λ shows that, for given d_t, all c_λ become small if E is removed from all resonances E_λ. Under these conditions, Φ hardly penetrates the internal region. This is in accord with the picture that the probability of the formation of the compound state is small far from resonances. It will be important to remember this conclusion also for the discussion of the direct reactions.

The matrix R^l is determined at all energies by the parameters E_λ, $\gamma_{\lambda s}$. The S^l matrix, and hence the cross sections, can be obtained from the R^l matrix by postulating the validity of (10.7) for the Φ^l of (10.2)

$$\begin{aligned} S^l &= (I_l - I'_l R^l)(E_l - E'_l R^l)^{-1} \\ &= (I_l - I'_l R)(I_l^* - I_l'^* R^l)^{-1} \end{aligned} \tag{10.10}$$

In this I_l and $E_l = I_l^*$ are diagonal matrices, the diagonal elements of which are the values of $I_{ls}(r_s)$ and $E_{ls}(r_s)$ for the r_s which correspond to the intersection of the channel s with the surface S. Similarly, the diagonal elements of the diagonal matrices I'_l and $E'_l = I_l'^*$ are the radial derivatives of these functions at the same r_s. One obtains by means of (10.4), (10.8), and (10.10) the same formula (10.6) for the reaction cross section which the more visualizable derivation at (10.6) gave, provided one can approximate R_l by a single term in the expansion (10.8). If this is the case, the E_λ play, apart from a minor correction discussed below, the roles of resonance energies. The partial widths are related to the $\gamma_{\lambda s}$ through

$$\Gamma_{\lambda s} = \frac{2\gamma_{\lambda s}^2}{|I_{ls}|^2} \tag{10.11}$$

For given $\gamma_{\lambda s}$, the partial width $\Gamma_{\lambda s}$, and hence also the probability of the reaction product s, decreases with increasing I_{ls}. If the incident particle of the pair s is a neutron, the I_{ls} are spherical waves of free particles. In particular, for $l = 0$

$$I_{0s}(r_s) = k_s^{-1/2} \exp(-ik_s r_s) \tag{10.12}$$

In this case $\Gamma_{\lambda s} = 2k_{\lambda s}\gamma_{\lambda s}^2$: the neutron width is proportional to the wave number $k_{\lambda s}$ and hence to the velocity of the neutron. From (10.6), the cross section for a neutron capture reaction is, at very low neutron energy, inversely proportional to the neutron velocity—a result which follows from the above considerations even if one does not approximate (10.8) by a single term, and has therefore general validity.

If both particles of the pair s are charged, $I_{ls}(r_s)$ is a Coulomb wave function, i.e., the wave function of a particle moving in an inverse square field of force. It is much more complicated than (10.12). It is customary in such a case to write

$$|I_{ls}|^2 = 1/k_s p \tag{10.13}$$

and call p the penetration factor. It expresses the fact that the density of the particles must be p^{-1} times higher at S for charged particles than for neutrons in order to give the same current at a very large distance. Accurate expressions and useful approximations were given for p in particular by Breit and his collaborators. A crude but in many cases useful approximation for $l = 0$ is

$$p = \exp\left[-\frac{ZZ'e^2}{\hbar}\left(\frac{2M_s}{E - \mathcal{E}_s}\right)^{1/2} + \frac{4}{\hbar}(2M_sZZ'e^2a_s)^{1/2}\right] \tag{10.13a}$$

In this, $ZZ'e^2$ is the product of the charges on the two nuclei of the pair s, their kinetic energy at large separation is $E - \mathcal{E}_s$, and a_s is the value of the radius r_s at which the channel s intersects the surface S. The formula is valid only if the first term in the bracket is considerably greater than the second. It

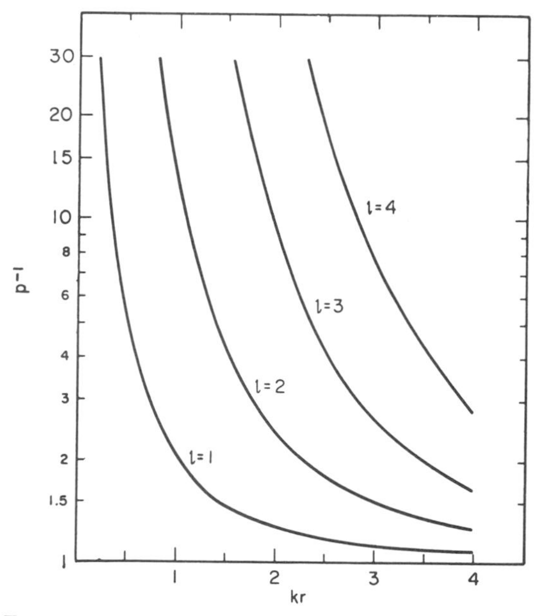

FIG. 10.2. Reciprocal penetration factors for neutrons. The reciprocal penetration factors are given for neutron with various angular momenta $l\hbar$ as function of the product kr of the wave number of the neutron and the radius of the nucleus. The penetration factor is 1 for $l = 0$.

gives a more quantitative measure for the decreased probability of charged particle reactions than the qualitative considerations of Sec. 6. For the case of neutrons, i.e., in the absence of electrostatic interaction, $p^{-1} = 1$ for $l = 0$ and is given, for higher l, as function of $ka_s = kr$ in Fig. 10.2.

The expression (10.11) for $\Gamma_{\lambda s}$ contains two factors: the first of these, $\gamma_{\lambda s}^2$, depends only on the properties of the internal region, while $|I_{\lambda s}|^{-2} = k_s p$ depends on the behavior of the wave function in the channels, that is, in the external region. For this reason, $\gamma_{\lambda s}^2$ is often called the reduced width of the level λ, for disintegration into the pair s, because the actual width $\Gamma_{\lambda s}$ would have this value for $p = 1$, $k_s = 1$. Division of the $\Gamma_{\lambda s}$ by $k_s p$ eliminates the factors characteristic for the external region.

4. Dependence of the Parameters on the Size of the Internal Region

Although the R matrix may always be characterized by parameters E_λ, $\gamma_{\lambda s}$, some care must be exercised in associating these quantities with the observed resonance energies and widths. The parameters E_λ, $\gamma_{\lambda s}$ depend on the position of the surface S surrounding the internal region while those for the energy of the maximum cross section and the level widths do not. Hence, the expressions for the latter involve, in addition to the E_λ, $\gamma_{\lambda s}$, other S dependent quantities (the I_{ls} and I'_{ls}) to cancel out the S dependence of the E_λ, $\gamma_{\lambda s}$. This is apparent, for instance, in (10.10), the left side of which is, naturally, independent of S, while both $\gamma_{\lambda s}$ and I_{ls} depend on S. Similarly, the energy for the maximum cross section, the E_λ of (10.6), is smaller than the E_λ of (10.8) by the real part of $\Sigma_t \gamma_{\lambda t}^2 I'_{lt}/I_{lt}$. In general, the replacement of (10.8) by a single term will be the more accurate the smaller the internal region is, consistent with the requirement that the specifically nuclear interaction be negligible for points of the configuration space which lie outside S. This suggests laying the surface S so that the channels intersect it close to the nuclear radius; the variables r_s, r_t to be inserted into I_l, I'_l in (10.10) or (10.11) then become equal to the sum of the radii of the two nuclei which constitute the pair of the channel. At the same time one can derive at least a crude estimate for the $\gamma_{\lambda s}$ which corresponds to this position of S if s corresponds to the emission of a single proton or neutron;

$$\gamma_{\lambda s}^2 \sim D/\pi K \tag{10.14}$$

in which D is the average level spacing (i.e., the average distance of successive E_λ) and K the mean wave number of the nucleons in the nucleus. The experimental value for $\gamma_{\lambda s}^2/D$ is about 10^{-13} cm; derivation of (10.14) will be given later. If the level spacing D and the $\gamma_{\lambda s}$ for processes other than neutron or proton emission are known, the above formulas suffice to obtain estimates for average reaction cross sections and similar quantities.

For instance, one obtains for the average cross section of the (n,γ) process

$$\bar{\sigma}_{n\gamma} \sim \frac{1{,}800\Gamma_r}{\Gamma_r E^{1/2} + 4.4 \times 10^{-4} DE} \tag{10.14a}$$

The orbital angular momentum of the neutron was assumed to be zero for (10.14a); the corresponding processes are the most important ones if the energy E of the neutrons, which is measured in electron volts for (10.14a), is below a few kev. Γ_r is the radiation width; it is usually of the order of 0.1 ev. Since (10.14) is only approximate, (10.14a) is also only an approximation to the average neutron absorption cross section. It is quoted here because it shows that the rather abstract formulas of the resonance theory can be reduced to numerical statements.

5. Radioactivity

Radioactivity is not properly a problem of collision theory. In spite of this, it can be treated by a slight modification of the preceding formalism.

A radioactive state contains only outgoing waves. At the same time, the amplitude of the wave function decreases in time at every point of space, reflecting the radioactive decay of the original nucleus. The wave function increases as a function of distance: the particles which are very far away from the source are very numerous because they originate at a time when the source was very much stronger than it is now. Such a wave function is not in accord with the principles of orthodox quantum mechanics; nevertheless, it gives a vivid picture of the process of radioactivity and is the one commonly used for the description thereof.

Let us assume that the wave emerging from the channel s is given by $\alpha_s(M_s/\hbar)^{1/2}P_l(\Omega_s)r_s^{-1}E_{ls}(r_s)$; it then carries a total current of $4\pi|\alpha_s|^2/(2l+1)$. The v_s on the left side of (10.7) then becomes just $\alpha_s(M_s/\hbar)^{1/2}I_{ls}^*$, the d_t becomes $\alpha_t(M_t/\hbar)^{1/2}I_{ls}'^*$ (we avoid the use of E_{ls} in order to avoid confusion with the energy E). Substituting again a single term for the sum in (10.8) and (10.7), one obtains

$$\alpha_s I^*_{ls} = \frac{\gamma_{\lambda s}}{E_\lambda - E} \sum_t \gamma_{\lambda t}\alpha_t I_{lt}'^* \qquad (10.15)$$

This shows that the $\alpha_s I_{ls}^*$ are proportional to the $\gamma_{\lambda s}$, that is, the current in the channel s is proportional to

$$|\alpha_s|^2 = \text{const}\,\frac{\gamma_{\lambda s}^2}{|I_{ls}|^2} = \text{const}\,\Gamma_{\lambda s} \qquad (10.16)$$

This confirms the interpretation of the $\Gamma_{\lambda s}$ given at (10.6). The constant of (10.16) corresponds to the fact that the source strength was left arbitrary above; it cannot be determined from (10.15) because these are linear and homogeneous in the α_s. However, the condition that (10.15) have a nonvanishing solution can be obtained by multiplying it with $\gamma_{\lambda s}I_{ls}'^*/I_{ls}^*$ and summing it over all channels s. The sum then drops out on both sides and one obtains for the energy of the radioactive state

$$E - E_\lambda = -\sum \gamma_{\lambda s}^2 I_{ls}'^*/I_{ls}^* \qquad (10.17)$$

This shows that the energy of the radioactive state is not real. Its imaginary part is, because of (10.2a),

$$\frac{E - E^*}{2i} = \sum_s \frac{\gamma_{\lambda s}^2}{|I_{ls}|^2} = -\frac{1}{2}\sum_s \Gamma_{\lambda s} = -\frac{1}{2}\Gamma_\lambda \qquad (10.18)$$

Hence, the amplitude of the radioactive state decays everywhere as $\exp(-\frac{1}{2}\Gamma_\lambda t/\hbar)$ and the decay constant of its intensity is $\Gamma_\lambda/\hbar$, as we had to expect. The real part of (10.17) is the energy shift mentioned in Sec. X, 4. This discussion of the radioactive decay confirms the qualitative picture of resonance levels developed in Sec. X, 2, as well as the interpretation of the p of Eq. (10.13) as a penetration factor. The complex energy E of the radioactive state and the partial widths $\Gamma_{\lambda s}$ are essentially observable quantities and hence independent of the size and shape of the internal region of Sec. 3.

6. The Optical Model

If the nuclear wave functions corresponded to complete chaos, as Bohr's original ideas and the powder models (Sec. VII) postulated, the $\gamma_{\lambda s}^2$ averaged over several resonances λ would not be expected to show maxima and minima as functions of the energy of the resonance. These models, which are also called strong coupling models, lead one to expect that the cross sections, after being averaged over several resonances, are smooth functions of the energy. The expression (10.14a), which was derived under the assumption (10.14) of the strong coupling model, indeed shows such a behavior. On the other hand, the success of the individual particle models leads one to expect that the effect of the target nucleus on the incident nucleon can be represented, at least approximately, by an average potential, or as a region in which the index of refraction of the incident nucleon's wave function is different from 1. This picture leads one to expect considerable fluctuations in the cross sections. These correspond to the interference maxima and minima in the scattering of light by a crystal ball the size of which is comparable to the wavelength of the incident light. Fluctuations in the cross section which correspond to this picture were indeed found and it is natural to describe these fluctuations by representing the effect of the target nucleus on the incident particle by a potential well.

The analogy between crystal ball and nucleus is incomplete because the amount of light that enters a nonabsorbing crystal ball is equal to the amount that leaves it. On the other hand, the particle which enters the nucleus may react with it so that a different particle may leave the nucleus and the intensity of the emerging wave which corresponds to the original particle alone is smaller than that of the incident beam. This circumstance will appear as an absorption in a theory which concerns itself only with the wave function of the incident particle. In order to account for this absorption, it must be assumed that the crystal ball is not entirely transparent, that it has the capacity to absorb, not only to refract. This leads to the concept of the optical model which can be described as a region with a complex refractive index, or its quantum-mechanical analogue, a region in which the potential is complex. The simplest form of the model assumes a potential

$$V = V_0(r)(1 + i\xi) \qquad (10.19)$$

where $V_0(r)$ has a fixed magnitude, $-V_0$, in the inside of the target nucleus and vanishes outside of it; ξ is a constant. This picture can be generalized by establishing a smooth transition between the value $-V_0$ and 0 of $V_0(r)$.

The optical model differs from standard theory in two regards. First, by replacing the target nucleus by a potential, it disregards the structure of this nucleus. As a result, it has no means for describing the changes in the target nucleus but gives an expression only for the wave function of the incident particle, that is, for elastic scattering. The other processes, including inelastic scattering, are not distinguished but lumped together as absorption, or rather formation of the compound nucleus with subsequent disintegration thereof. Second, by disregarding the intricate quantum-mechanical nature of the interaction, it fails to account for the resonance structure of scattering and absorption processes and gives only *average* values for these. However, these average values are the ones which are, as a rule, of principal interest and for these the model presents a very good picture. The value of ξ in (10.19) which gives best agreement for low energy processes is about 0.05, while the depth V_0 has the same magnitude which accounts for the binding energies of the stable orbits of the shell models. This is about 45 Mev for a radius of $1.25 \times 10^{-13}A^{1/3}$ cm. A ξ of about 0.3 would largely smooth out the maxima and minima of the average absorption and scattering and lead to the type of results as were expected on the basis of the strong coupling theories, that is, the original ideas of Bohr or the uniform models.

7. The Intermediate Coupling or Giant Resonance Model

The optical model provides a good description of the average cross sections and perhaps also of the actual cross sections in the energy region in which the resonances are so broad that they overlap and their individuality is lost. It remains desirable, however, to give a more detailed description of the nuclear reactions than this model provides, a description which accounts for the line structure of the cross sections and in which the average cross section is obtained as an average over the rapidly fluctuating line structure. It is also desirable to distinguish between the various nuclear reactions which the clouded crystal-ball model lumps together and follows only to the point of the "formation of a compound nucleus." Such a more detailed theory can also dispense with the concept of the "absorption" of the incident particle—a concept which has no direct foundation in quantum theory.

In order to develop a more complete theory, one first introduces a very crude optical model by replacing the target nucleus with an average potential, the real part of (10.19). This is a close approximation to the potential that would give the correct energy levels for the individual particles in the shell model. As was discussed previously, this potential gives a scattering cross section which shows, as a function of the energy, maxima and minima resulting from the diffraction of the incident beam. One may call the maxima single-particle resonances even though they differ in many respects from the resonances considered in Sec. 2. The positions of these maxima coincide with the maxima of the *average* cross section. This follows from the facts that the potential (10.19) reproduces the maxima and that the imaginary part has little influence on their positions. The model we are considering now is even less accurate than the optical model previously discussed because it does not account for the occurrence of nuclear reactions, whereas the optical model of the preceding section accounts for them, at least globally. However, the intermediate-coupling model also modifies the underlying crude optical model based on the real part $V_0(r)$ of (10.19) and does this not by making the potential complex but by replacing $V_0(r)$ by the true interaction, i.e., the sum of the potentials of the nucleons in the target nucleus.

Let us consider our crude optical model from the point of view of Sec. 3. Since the interaction between target nucleus and extra nucleon is given by a potential acting on the latter, the characteristic functions of the Hamiltonian are products of characteristic functions of the target nucleus with its own Hamiltonian and characteristic functions of the extra nucleon, moving in the potential $V_0(r)$. The former characteristic functions satisfy the boundary conditions for the internal functions X_λ automatically as long as the surface S extends far enough so that both the wave function of the target nucleus and its normal derivative vanish on S. The energy of the incident nucleon must be so chosen that its wave function—the other factor in X_λ—also satisfies the boundary condition. The spacing of energy values at which this happens is of the order of the spacing of single-particle levels with a definite l or j in the shell model, i.e., of the order of 10 Mev. On the other hand, the spacing of the energy levels of the target nucleus is quite narrow—usually less than 100 electron volts at an excitation energy of 8 or 10 Mev.

The internal functions X_λ of our crude optical model can be divided into two groups. Some of the X_λ (we shall denote them by X_n, and the corresponding energy values by F_n) are products of the wave function of the normal state of the target nucleus and a suitable characteristic function of the extra nucleon. Other X_λ (denoted by X_e) are products of wave functions of an excited state of the target nucleus and a suitable characteristic function of the extra nucleon. As was explained before, there are many more X_e per unit energy interval than X_n. However, the partial widths of the former, for a disintegration resulting in the normal state of the target nucleus, are 0 because of the factor $\chi_{\text{normal state}}$ in the integral of (10.9c). This situation is illustrated in Fig. 10.3*a*; it is a direct consequence of the absence of reaction processes in the crude optical model that we are considering.

Let us now introduce the difference between the actual interaction of the extra nucleon with the target nucleus and the $V_0(r)$. This results in new internal functions of the system which we denote by Y_λ:

$$Y_\lambda = \sum_e a_{\lambda e} X_e + \sum_n b_{\lambda n} X_n \qquad (10.20)$$

It is a simple consequence of perturbation theory (but can also be shown without its use) that the coefficients $a_{\lambda e}$ and $b_{\lambda n}$ will be large only for those

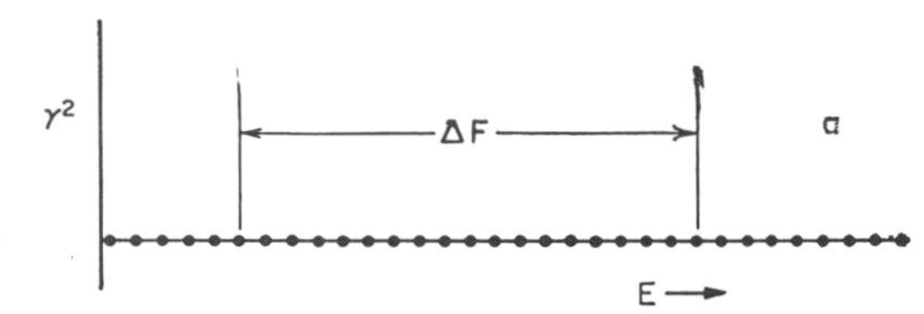

FIG. 10-3a. The reduced width for scattering in the pure single-particle model. The points along the energy axis represent schematically the energy levels in the single-particle picture. The reduced width, indicated by the lengths of the vertical lines, is zero for most levels; for a few of them it is very large. In the states corresponding to the latter case, the target nucleus is in the normal state, and the incident particle in a high orbit.

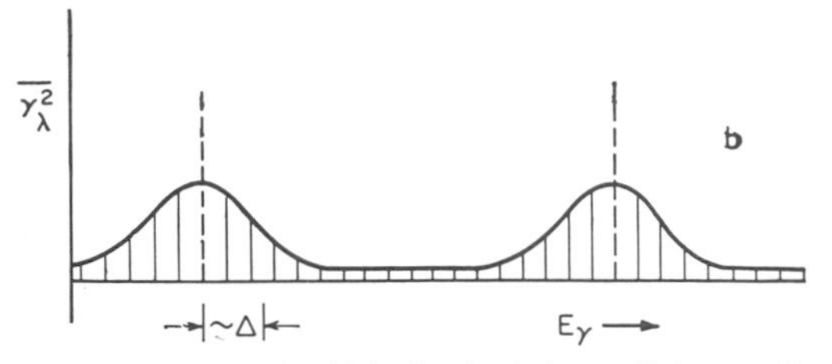

FIG. 10.3b. Reduced widths in the intermediate-coupling model (schematic). The very large widths of some of the levels of the independent-particle picture are spread out over several adjacent levels.

X_e and X_n the energy of which is close to the energy of Y_λ.

The range of energy, Δ, of the single-particle states X_e and X_n which are appreciably mixed into the state Y_λ furnishes a measure of the failure of the crude optical model. In the strong-coupling case, Δ is large compared with the spacing of the single-particle levels, i.e., is at least of the order of 10 Mev. Weak coupling, on the other hand, implies that, for every n, only one $a_{\lambda e}$ or one $b_{\lambda n}$ is appreciable. In this case Δ is small compared with the level spacing of the target nucleus. Neither of these two extremes is capable of providing a description of the actual situation.

The actual condition appears to be intermediate between the strong- and weak-coupling models: Δ is large compared with the separation of the levels of the target nucleus, i.e., large compared with the separation of all energy levels of the crude optical picture. It is small, on the other hand, compared with the separation of the energy levels F_n of the X_n. This is the basic assumption of the intermediate-coupling model. Hence, in (10.20), for every λ only one $b_{\lambda n}$ is appreciable; for most of them, those which lie roughly midway between energy levels of X_n, none is appreciable. The situation is illustrated in Fig. 10.3b. The reduced partial width of the level Y_λ, with respect to the formation of the target nucleus in its normal state, is due to the single term $b_{\lambda n}X_n$ which is appreciable in (10.20). Its magnitude η_λ^2 is given by

$$\eta_\lambda = b_{\lambda n}\gamma_n \tag{10.21}$$

where γ_n is the (enormous) reduced width of the state X_n.

It follows from the completeness of the states Y_λ that

$$\sum_\lambda \eta_\lambda^2 = \sum_\lambda |b_{\lambda n}|^2\gamma_n^2 = \gamma_n^2 \tag{10.22}$$

The sum has to be extended only over those states λ which have an appreciable contribution from X_n, that is, the group of states around the energy of X_n. It follows that the range of energy over which the levels have appreciable reduced widths for elastic scattering is given roughly by Δ. The energy dependence of the *average* of the reduced widths for scattering (averaged over several levels) as expected from this model is shown in Fig. 10.3b. The actual reduced widths, of course, fluctuate considerably about these averages.

The important quantity for the determination of the average value of the scattering elements of the collision matrix is the average value of the scattering reduced width, which is represented in the intermediate coupling model by Fig. 10.3b. Thus the average cross sections can be obtained from the model once the F_n are located and a choice of Δ is made. The F_n are located through the single-particle potential $V_0(r)$, while the quantity Δ may be shown to correspond to $V_0\xi$. With this identification of parameters the two models give the same average cross sections.

The order of magnitude of the γ_n^2 can be obtained on the basis of the assumption that the single-particle, wave function has, at the surface of the nucleus, about the same amplitude as inside the nucleus. This gives by means of the normalization condition of the wave function, $(M/\hbar^2)a\gamma_n^2 = 1$, where a is, as before, the nuclear radius. If the reduced width γ_n^2 is about evenly distributed among all the levels which lie between two levels F_n of the independent particle picture, the average reduced width will be, on account of (10.22), $\gamma_n^2D/\Delta F$. The spacing ΔF of the single-particle levels is about $\hbar^2K\pi/Ma$, where K is the wave number of the single particle. Hence, on this picture,

$$\eta_\lambda^2 \sim \frac{D}{\Delta F}\gamma_n^2 \sim \frac{D}{\hbar^2K\pi/Ma}\cdot\frac{\hbar^2}{Ma} = \frac{D}{\pi K} \tag{10.23}$$

This corresponds to the Bethe-Weisskopf estimate (10.14). In the giant resonance model the width γ_n^2 is not evenly distributed and the levels which are close to the broad levels F_n of the crude optical model have a substantially greater reduced width than that given by (10.23); the reduced widths of those levels far from the F_n are well below (10.23).

Rough estimates of $V_0(r)$ and of Δ may be made on the assumption of a definite law of interaction between nucleons. With simple two-body forces of the character discussed in Sec. V, 1, one obtains a $V_0(r)$ which is similar to that obtained empirically on the basis of the optical model. However, the calculation of Δ is quite intricate and has not yet been accomplished entirely satisfactorily. The value of Δ obtained by a naive straightforward calculation corresponds to a strong-coupling model. This *may* be an indication that the interactions between nucleons within heavy nuclei are rather different from those obtained in the study of two-nucleon systems.

The giant resonance picture has been applied not only to the states X_n in which all nucleons, except one, form the normal state of the nucleus with one less nucleon. It has been applied also to the states in which two nucleons are in excited states, and only A-2 form a stable nucleus. Whereas the cross section maxima caused by the former states are called giant resonances, the maxima caused by the two-particle excitation are called doorway states. The latter maxima are smaller and narrower than the giant resonances.

8. Giant Resonance Model and Analogue States

A picture very similar to the preceding one was postulated also for the analogue states discussed in Sec. VI, 2. In this case, the crude optical model is replaced by a model in which the nuclear forces are fully taken into account, but the electrostatic interaction

$$\sum_{i<k} \frac{e^2}{r_{ik}} \frac{(1-\tau_i)(1-\tau_k)}{4} \tag{10.24}$$

($\tau = 1$ for neutron, -1 for proton) is replaced by a kind of average

$$\sum_{i<k} \frac{e^2}{4r_{ik}} \left(1 + \frac{1}{3} \boldsymbol{\tau}_i \cdot \boldsymbol{\tau}_k\right) \tag{10.24a}$$

For the sum of (10.24) and the nuclear forces—if these are indeed charge-independent—the isotopic spin T is an accurate quantum number. Most levels belong to a T multiplet with the lowest possible value of T compatible with the neutron excess: $T = T_\zeta = \frac{1}{2}(N\text{-}Z)$. The wave functions of these levels correspond to the X_e of the preceding section. Among the many $T = T_\zeta$ levels, there are a few with an isotopic spin $T = T_\zeta + 1$. Their wave functions correspond to the X_n of the preceding section. Each of the latter levels is in the same T multiplet as one of the levels of the nucleus with one more neutron and one less proton.

If one now introduces the difference between (10.24) and (10.24a) as an addition to the Hamiltonian, i.e., if one goes over to the true Hamiltonian, an expansion similar to (10.20) applies for the characteristic functions of the true Hamiltonian. However, in contrast to the situation of the preceding section, the addition to the Hamiltonian is well known: It is the difference between (10.24) and (10.24a). All important quantities, such as the energy spread Δ of (10.20), can be explicitly calculated and seem to agree with experience. In particular, the levels that have acquired a significant share of the $T = T_\zeta + 1$ character are spread out over a range Δ of several kev—except in light nuclei, a range large compared with the spacing of the $T = T_\zeta$ levels but small compared with the spacing of the $T = T_\zeta + 1$ levels. The latter is of the order of 1 Mev. The energy spread Δ of the present case is so much smaller than that considered in the preceding section because the difference between (10.24) and its average (10.24a) is much smaller than the difference of the true nuclear forces and their average, the $V_0(r)$ of (10.10).

According to the picture projected, there is no single analogue state; rather, several states share the analogue character just as several Y_λ have a share of the X_n. In the present case, the sharing of the $T = T_\zeta + 1$ character manifests itself in a fine structure of the analogue state which appears to have the breadth 2Δ. The fine structure, which has been observed, actually means that there is not one broad analogue level but that there are several levels within the 2Δ width, all of which have acquired some share of the $T = T_\zeta + 1$ character.

XI. NUCLEAR REACTIONS. II. SURFACE REACTIONS

The considerations of the last section are valid under all circumstances; they are useful only if R can be simplified sufficiently so that S can be calculated by means of (10.10). That this is not always the case is indicated by our having omitted, in all applications, all but one of the infinitely many terms of (10.8). This is a useful approximation only if the Γ_λ are smaller than the spacing between successive E_λ. Furthermore, the decomposition of the total wave function into spherical waves, corresponding to different angular momenta $l\hbar$, gives a similar decomposition only for the total cross section. In the expression for the differential cross section, cross terms between waves of various angular momenta appear; the interference of the partial waves of (10.2) cancels only if the cross section is integrated over all directions. Hence the spherical wave decomposition will lead to particularly awkward expressions if many angular momenta contribute to the scattered wave. This is most likely to be the case in the forward and backward directions. Figure 11.1 illustrates this by giving the angular distribution of a wave with $S^0 = i$ (to give one-half of the maximum amount of spherically symmetrical scattering) and $S^3 = \exp(-i\pi/50)$. This last part of the scattering contributes hardly more than 1 per cent to the total cross section.

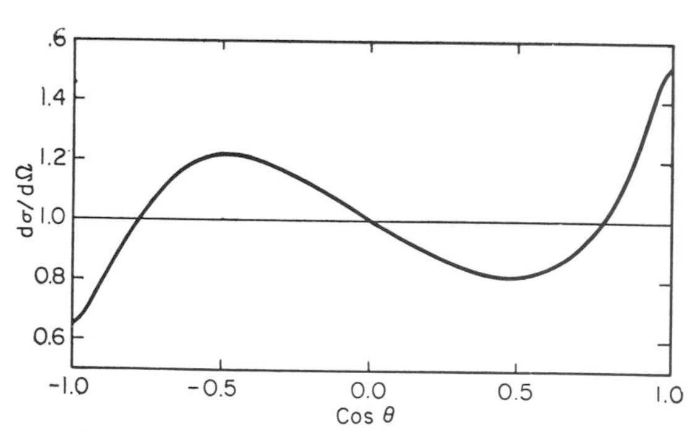

FIG. 11.1. Angular distribution of scattering (schematic). The figure illustrates the large effect of a small scattering amplitude with a high angular momentum. The non-spherically symmetric part (an f wave) contributes only 1 per cent to the total scattering.

It seems reasonable, under the conditions, to attempt applying Born's collision theory to obtain the angular distribution in the forward direction. This theory is always relatively easy to apply. It cannot be expected to reproduce the resonance character of the cross section, but it can be relied upon to give a good value for the average cross section as long as the partial cross sections are all small as compared with π/k_s^2. Butler first recognized that the complicated angular distribution of the reaction products points to such a situation for many (d,p) and (d,n) reactions. The following interpretation of the angular distribution of these reactions is due to him, although obtained by him on the basis of another type of mathematical analysis. As before, we disregard the spin of the deuteron as well as that of the emerging proton and neutron: even though they do affect the result of the calculation, they do not influence the underlying principles.

1. Angular Distribution in Stripping Reactions

Let us denote the deuteron wave function by $d(\mathbf{r}_p - \mathbf{r}_n)$, where $\mathbf{r}_p$ and $\mathbf{r}_n$ denote the position of the proton and neutron, respectively. The incident deuterons have the direction of $\mathbf{k}_i$, the wave function of the incident beam is then

$$d(\mathbf{r}_p - \mathbf{r}_n) \exp\left[\tfrac{1}{2}i\mathbf{k}_i \cdot (\mathbf{r}_p + \mathbf{r}_n)\right] \qquad (11.1)$$

We substitute a spherically symmetric potential field $V_p(r_p) + V_n(r_n)$ for the target nucleus. This approximation has no major influence either on the calculation or on the result. We calculate the probability of neutron capture into a state $u(\mathbf{r}_n)$ of angular momentum l, with the proton traveling in the direction $\mathbf{k}_e$. The principal interest centers at the relative probabilities of the various directions of $\mathbf{k}_e$, that is, the angular distribution of the protons. There is, naturally, a similar question concerning the angular distribution of the neutrons, resulting from the (d,n) process. Since this can be calculated in the same way as the angular distribution of the protons, we can confine attention to the latter question.

The absolute value of the wave number of the emerging proton is given by the energy Q of the reaction

$$(\hbar^2/4M)k_i^2 + Q = (\hbar^2/2M)k_e^2 \qquad (11.2)$$

in which the proton and deuteron masses are denoted by M and $2M$. According to Born's collision theory, the probability of the proton emerging within unit solid angle in the direction $\mathbf{k}_e$ is proportional to the absolute square of

$$\iint d\mathbf{r}_p\, d\mathbf{r}_n u(\mathbf{r}_n) e^{-i\mathbf{k}_e\cdot\mathbf{r}_p}[V_n(r_n) + V_p(r_p)] d(\mathbf{r}_p - \mathbf{r}_n) \exp\left[\tfrac{1}{2}i\mathbf{k}_i \cdot (\mathbf{r}_p + \mathbf{r}_n)\right] \qquad (11.3)$$

$d\mathbf{r}_p$ indicates integration over the components of $\mathbf{r}_p$ and $d\mathbf{r}_n$ has a similar significance. There are $2l + 1$ integrals of the form (11.3) if the angular momentum of the bound state u is l, corresponding to the possible orientations of this angular momentum. Hence, the total angular distribution is given by the sum of the absolute squares of $2l + 1$ integrals of the form (11.3).

It is possible to bring the two terms of (11.3) into somewhat simpler forms. Let us write, first

$$u(\mathbf{r}_n) = (2\pi)^{-3/2} \int U(\mathbf{k}) \exp(i\mathbf{k}\cdot\mathbf{r}_n)\, d\mathbf{k} \qquad (11.4)$$

Then

$$V(r_n)u(\mathbf{r}_n) = \left(\frac{\hbar^2}{2m}\Delta + E_n\right) u(\mathbf{r}_n) = (2\pi)^{-3/2}\int U(\mathbf{k})\left(E_n - \frac{\hbar^2\mathbf{k}^2}{2M}\right)\exp(i\mathbf{k}\cdot\mathbf{r}_n)\, d\mathbf{k}$$

where E_n is the binding energy of the neutron in the potential. It is a negative quantity. $U(\mathbf{k})$ is the wave function of the bound neutron state in momentum space; it has the same angular dependence as u. We can now write for the term of (11.3) which contains $V_n(r_n)$

$$\begin{aligned}(2\pi)^{-3/2} &\int d\mathbf{r}_p \int d\mathbf{r}_n \int d\mathbf{k}\, U(\mathbf{k})(E_n - \hbar^2\mathbf{k}^2/2M) d(\mathbf{r}_p - \mathbf{r}_n) \exp i(\mathbf{k} + \tfrac{1}{2}\mathbf{k}_i)\cdot(\mathbf{r}_n - \mathbf{r}_p) \exp i(\mathbf{k}_i - \mathbf{k}_e + \mathbf{k})\cdot\mathbf{r}_p \\ &= \iint d\mathbf{r}_p\, d\mathbf{k}\, U(\mathbf{k})(E_n - \hbar^2\mathbf{k}^2/2M) D(\mathbf{k} + \tfrac{1}{2}\mathbf{k}_i) \exp[i(\mathbf{k}_i - \mathbf{k}_e + \mathbf{k})\cdot\mathbf{r}_p] \\ &= (2\pi)^3 U(\mathbf{k}_e - \mathbf{k}_i)(E_n - \hbar^2(\mathbf{k}_e - \mathbf{k}_i)^2/2M) D(\mathbf{k}_e - \tfrac{1}{2}\mathbf{k}_i)\end{aligned} \qquad (11.5)$$

$D(\mathbf{k})$ is the deuteron wave function in momentum space; since the wave function is, in a very good approximation, spherically symmetric, $D(\mathbf{k}) = D(k)$ depends only on the absolute value of its argument.

As mentioned before, there are $2l + 1$ integrals similar to (11.3) corresponding to the possible orientations of the neutron orbit u. The U in the corresponding $2l + 1$ expressions (11.5) will contain the same dependence $U_r(|\mathbf{k}_e - \mathbf{k}_i|)$ on the absolute value of its argument $\mathbf{k}_e - \mathbf{k}_i$; the dependence on the direction of $\mathbf{k}_e - \mathbf{k}_i$ will be given by the $2l + 1$ spherical harmonics P_{lm} with $m = -l, -l + 1, \ldots, l - 1, l$. Omitting the $V_p(r_p)$ term in (11.3), the angular dependence of the emitted protons is given by

$$J(\mathbf{k}_e) = |U_r(|\mathbf{k}_e - \mathbf{k}_i|) D(|\mathbf{k}_e - \tfrac{1}{2}\mathbf{k}_i|)|^2 [(\hbar^2/2M)(\mathbf{k}_e - \mathbf{k}_i)^2 - E_n]^2 \qquad (11.6)$$

The $(2\pi)^6$ factor has been omitted, the sum of the squares of the $2l + 1$ spherical harmonics gives 1.

Since the deuteron wave function

$$d(\mathbf{r}_p - \mathbf{r}_n) = d(|\mathbf{r}_p - \mathbf{r}_n|)$$

is rather flat, its Fourier transform D drops rather rapidly with increasing

$$|\mathbf{k}_e - \tfrac{1}{2}\mathbf{k}_i| = (k_e^2 + \tfrac{1}{4}k_i^2 - k_e k_i \cos\theta)^{1/2} \qquad (11.7a)$$

where θ is the angle between incident deuterons and emitted protons. As a result, most of the protons are emitted at low θ, in the forward direction. The last factor of (11.6) tends to decrease this effect.

The kinetic energy of the neutrons in the bound orbit u is, in general, much higher than in the deuteron. As a result, U_r drops much less rapidly with increasing argument than D. On the other hand, the wave function in momentum space, $U(k)$, has a similar behavior to that of the coordinate wave function: it has an l-fold zero at $k = 0$. Hence, unless the neutron is captured into an s orbit ($l = 0$), the U_r will be very small for small

$$|\mathbf{k}_e - \mathbf{k}_i| = (k_e^2 + k_i^2 - 2k_e k_i \cos\theta)^{1/2} \qquad (11.7b)$$

i.e., for small θ. This depression at low θ is the more pronounced the larger is l. Naturally, for very large l, the whole cross section (11.6) remains very small: for low θ the first, for high θ the second factor becomes very small. For intermediate l, however, the maximum of $J(\mathbf{k}_e)$ wanders to higher and higher θ, decreasing at the same time in absolute magnitude. The angular distribution, calculated by Butler for various θ, taking the spins of the deuteron and the proton and neutron into account, is illustrated in Fig. 11.2. Comparison of the observed angular distributions with these curves permits a determination of the angular momentum of the orbit into which the neutron is captured. Similarly, the angular distribution of the neutrons from the (d,n) reaction permits the determination of the l value of the captured proton.

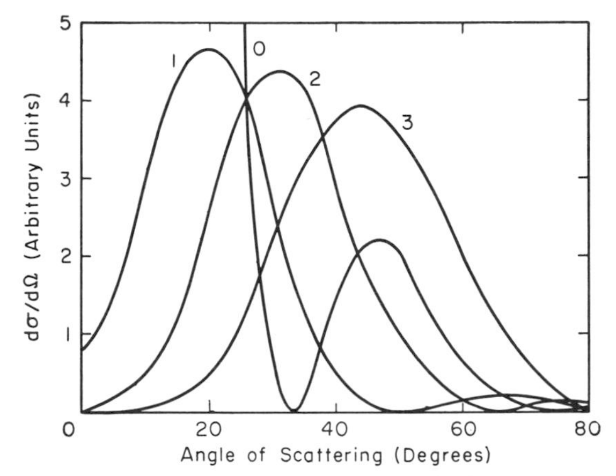

FIG. 11.2. Angular distribution of stripping reactions (Butler). The calculated angular distribution of the protons from a (d,p) reaction is given for various l values of the captured neutron.

The $V_p(r_p)$ term of (11.3) needs yet to be considered. The contribution of the $V_p(r_p)$ term corresponds to processes in which the proton is pushed away from the deutron by the nuclear potential and the neutron finds itself captured. The probability of such a process is considerably smaller than the one calculated above. The term with $V_p(r_p)$ can indeed be calculated to give

$$(2\pi)^{3/2} \int v_p(\mathbf{k}_i + \mathbf{k} - \mathbf{k}_e) U(\mathbf{k}) D(\mathbf{k} + \tfrac{1}{2}\mathbf{k}_i)\, d\mathbf{k} \quad (11.8)$$

in which v_p is the Fourier transform of V_p. Because of the behavior of D, as outlined above, the integrand is large only for $\mathbf{k} + \frac{1}{2}\mathbf{k}_i \approx 0$. Hence, one can write for (11.8) approximately

$$(2\pi)^{3/2} v_p(|\tfrac{1}{2}\mathbf{k}_i - \mathbf{k}_e|) U(\tfrac{1}{2}\mathbf{k}_i) \int D(\mathbf{k} + \tfrac{1}{2}\mathbf{k}_i)\, d\mathbf{k} = (2\pi)^3 v_p(|\tfrac{1}{2}\mathbf{k}_i - \mathbf{k}_e|) U(\tfrac{1}{2}\mathbf{k}_i) d(0) \quad (11.9)$$

This term does not show the characteristic behavior of (11.6). However, the $d(0)$ factor renders it considerably smaller than the latter and there is reason to believe that higher approximations further reduce its magnitude. Hence, (11.6) appears a reasonable expression for the angular distribution: the possibility to neglect the $V_p(r_p)$ term of (11.3) is hardly surprising since the neutron capture must be due, after all, principally to the interaction of the neutron, rather than that of the proton, with the nucleus.

Butler's method of determining the orbit into which a particle is captured often leads to a determination of the angular momentum and parity of the product nucleus. On this account, it is a particularly significant example to illustrate an approximate method for the description of nuclear collisions of a rather general kind.

2. Distorted-wave Born Approximation

The preceding calculation reproduces the characteristic features of the observed angular distributions. In order to obtain quantitative agreement, it was found necessary to restrict the integration over $\mathbf{r}_n$ in (11.3) to the outside of the nucleus so that only the tail of $u(\mathbf{r}_n)$, which extends beyond the nuclear radius as usually considered, plays a role. This shows that the reaction takes place on the surface of the nucleus and can be interpreted most naturally by assuming that the interaction between the nucleus and the incident deuteron distorts the wave function of the latter in such a way that its neutron does not enter the nucleus. This is reasonable because the stripping distribution is obtained at deuteron energies between resonance levels, and, as was observed after (10.9d) and (10.9e), the incident particle hardly penetrates into the target unless its energy is close to a resonance. Since the plane waves used in (11.3) for the calculation of the cross section penetrate the target nucleus with undiminished amplitude, much of the integral in (11.3) comes from a region in which, actually, no reaction takes place.

A greatly improved expression for the cross section can be obtained by an extension of the optical model, discussed in Sec. X, 6. The optical model gives an expression for the wave function of the incident particle, as affected both by the diffraction by the potential of the target nucleus and also by the absorption caused thereby. It therefore provides a much more accurate description of the initial state of the incident particle from which transitions can then take place. The transition rates are then calculated by the Born approximation but the final state is also considered to be given by a wave function of the optical model. The method of calculation, which usually requires extended use of computing machines, is called distorted-wave Born approximation. It has proved very useful not only for accounting for the angular distribution but also for the absolute value of the cross section. There is some arbitrariness in the choice of the optical potential (10.19), particularly as applied to composite incident particles. One way to obtain it is semiempirical, from the observed elastic cross section.

There are several modifications of the distorted-wave Born approximation, some of which have an even more solid theoretical foundation than the Born approximation itself. The principal assumption remains that the transition from the initial state (11.1) to the final state [in the foregoing example, $u(\mathbf{r}_n) \exp(-i\mathbf{k}_e \cdot \mathbf{r}_p)$] is direct. Mathematically, this amounts to the assumption of a wave function that is a superposition of the initial and final states only. The transition between them is due to the matrix element of the Hamiltonian between these states. The distorted-wave Born approximation differs from the original form by replacing the exponential factors which describe the relative motion of the colliding and of the separating nuclei by a more appropriate function, the solution of the wave equation with the optical potential.

It follows that the various modifications of the distorted-wave Born approximation are most accurate if the cross sections are small and most of the reaction takes place close to the surface of the target nucleus. In such a case the initial state's wave function is a reasonable approximation to the total wave function, and indirect processes—components of the wave function different from both initial and final states—do not become significant. These conditions are met for high angular momenta of the incident particle, when it only brushes the target nucleus, and if a similar situation prevails for the separating pair of nuclei. For low angular momentum, the approximation is reasonable between, but not close to, resonances. Under these conditions, the usefulness of the method is by no means restricted to the (d,p) and (d,n) reactions whose consideration initiated the development. It has been applied with equal success to (p,α), (He^3,d), (He^3,α), etc., processes as well as to heavy-ion reactions.

Comparing the theory of direct processes with that of resonance reactions, one finds that the latter is freer of approximations and that it accounts for the data with much greater accuracy. On the other hand, it leans very heavily on the experimental data themselves by its use of the positions of the resonance levels and their reduced partial widths which it uses as free parameters, given by experiment. The a priori information content of the theories of direct reactions probably exceeds that of the resonance theories.

3. Electric Excitation

In the collision of a charged particle with a nucleus, inelastic scattering and consequently nuclear excitation may occur even if specific nuclear forces do not come into play. The phenomenon may be described as follows. In the motion of the incident charged particle past the target nucleus, the latter is subjected to a rapidly varying electric field. This field may give rise to transitions from the ground state to excited states of the target nucleus. The process is thus a species of photoeffect, and is referred to as electric or Coulomb excitation. The cross section for electric excitation (about 10^{-28} or 10^{-29} cm^2) is small compared with direct nuclear transformations and is observable only when the latter are extremely improbable. This is the case, for example, in the collisions of protons or α particles of intermediate energy on heavy nuclei. In such cases the electrostatic repulsion between the colliding particles is so large that the classical distance of closest approach is large compared with the nuclear radius; the probability of direct nuclear interaction is, therefore, extremely small. When the direct nuclear transformation may be disregarded, the electric excitation is detected by observing the γ radiation emitted in the de-excitation of the target nucleus.

To calculate the cross section for electric excitation, we may neglect the nuclear forces between incident and target particles and write for the Hamiltonian of the system

$$H = T_r + H_A + Ze^2 \sum_p |\mathbf{r} - \mathbf{r}_p|^{-1} \qquad (11.10)$$

where T_r is the kinetic energy of relative motion, H_A is the Hamiltonian of the target nucleus, Z is the charge number of the incident particle, $\mathbf{r}$ is the vector from the center of mass of the target nucleus to the incident particle, and the $\mathbf{r}_p$ are the positional coordinates of the protons of the target with respect to its center of mass. The electrostatic energy term is

$$Ze^2 \sum_p |\mathbf{r} - \mathbf{r}_p|^{-1}$$
$$= \frac{ZZ'e^2}{r} + Ze \sum_{l=m}^{\infty} \sum_m \frac{4\pi}{2l+1} Q_{lm} r^{-l-1} P_{lm}(\theta,\phi) \qquad (11.11)$$

where Z' is the charge number of the target nucleus, θ, ϕ and θ_p, ϕ_p are the polar angles of $\mathbf{r}$ and $\mathbf{r}_p$, respectively, and

$$Q_{lm} = e \sum_p r_p{}^l P_{lm}{}^*(\theta_p,\phi_p) \qquad (11.11a)$$

The operators Q_{lm} are known as the electric multipole operators, l defines the multipole order of the operator (electric dipole for $l = 1$, electric quadrupole for $l = 2$, etc.). These operators occur also in the description of spontaneous electromagnetic transitions. The Hamiltonian H will lead to simple elastic Rutherford scattering if only the r^{-1} term in the potential energy is retained. The second term in the expression (11.11), designated below as V', may be considered as providing transitions between simple Rutherford scattering states with excitation of the target nucleus.

The probability of excitation may be obtained by means of first-order perturbation theory. The specifically nuclear contribution to the excitation probability occurs in the form of a matrix element of V' between the initial and final target states. In general, only the multipole operators of one particular order will contribute strongly to the matrix element (usually the lowest l value for which the matrix element of Q_{lm} is not zero) so that we may speak of electric-dipole excitations (E-1), electric-quadrupole excitations (E-2), etc. In the case of electric l-pole excitation, the only *nuclear* factor in the cross section is

$$B_l = \sum_{mM_iM_f} |(\psi_f, Q_{lm}\psi_i)|^2 \qquad (11.12)$$

where ψ_i and ψ_f are the initial and final nuclear states and M_i and M_f are the possible magnetic quantum numbers for these states. This same factor occurs in the rate of radiative transitions of polarity l.

The absolute value of the cross section has been calculated accurately for dipole and higher-pole transitions by considering the second term of the multipole expansion (11.11) to be a small perturbation. This is surely justified.

Experiments on the angular distribution of the γ radiation emitted in the process lead to information on the spin values of the nuclear states involved. Measurement of the excitation cross section as a function of incident energy is useful for the determination of the multipole order responsible for the process. A further measurement of the absolute probability of excitation makes possible the evaluation of B_l.

The different nuclear models imply rather different values for the quantities B_l. The single particle and collective models contrast rather strongly in this respect. The collective model provides relatively large values for B_2 (compared with those from a pure single-particle model) for transitions between the ground states and the low-lying "rotational levels" of heavy nuclei. The estimates of B_2 obtained by electric excitation lend support to the collective model.

XII. ELECTROMAGNETIC AND β RADIATION

1. Introduction

In so far as problems of nuclear structure are concerned, the study of radiative transitions and the study of β decay play similar roles. The relative ease of detection of the β particles and of the measurement of the slow transition rates stimulated the early study of the β-decay process. Analysis of the rate of β transitions has theoretical advantages also because the operators responsible for them are the simplest operators in the case of allowed transitions. Hence, inferences from the measured rates to the properties of nuclear states are most direct. Analysis of nuclear states by means of β decay suffers, however, from two features: (1) the slow transition rate for the process means that, for the most part, only the ground states of the parent nuclei are involved and (2), in the case of forbidden transitions, which form the majority of all transitions, matrix elements of several operators play a role. The radiative transitions connect states of a single nucleus. The operators responsible for the simplest transitions are more complex than the operator of allowed β decay. On the other hand, they are simpler than those responsible for forbidden β transitions, and the magnitudes of their matrix elements are easier to interpret. This is particularly true for electric, rather than magnetic, transitions.

2. Electromagnetic Radiation and Nuclear Structure

The study of electromagnetic transitions between the states of an atom or molecule offered the single most powerful means for the experimental investigation of the structure of these systems. Unfortunately, there exist no tools, comparable in resolving power with the grating spectrographs, with which to study the higher-energy radiations emitted by nuclei.

Some important differences between atoms and nuclei as radiating systems should be noted. In atomic transitions the radiation is almost invariably of dipole character. Between nuclear levels, transitions of multipolarity as high as four are not uncommon. The reason for this is that the ratio of the wavelength of the radiation typically emitted by nuclei to the nuclear size is smaller than the corresponding ratio for atoms by 2 orders of magnitude. A further difference is the change in the emitted radiation when the sources are placed in a magnetic field. The Zeeman effect readily observed in atoms is for all intents and purposes unobservable in nuclear transitions, as the energy splitting of the nuclear substates is usually small compared with the natural line width. Only in exceptional cases has this splitting been observed, by using the Mössbauer effect.

Although not occupying the unique position that it holds in atomic spectroscopy, the study of electromagnetic transitions in nuclei is an extremely important source of information with regard to nuclear structure. The data resulting from these studies are of two types. The angular distribution of γ rays following nuclear reaction and the angular correlations between successively emitted γ rays depend only on the "geometry" of the transitions and levels involved. That is, the angular variation of the radiation intensity depends only on the multipolarity of the transition and the spins of the levels involved. Measurement of the radiative capture cross sections or measurement of a radiative lifetime gives values for the absolute square of the matrix elements of the transition operators. These matrix elements are the off-diagonal elements of operators similar in structure to the magnetic-dipole or the electric-quadrupole operators considered in connection with the static moments of the normal states of nuclei.

The recent development of the solid-state ionization chamber and the subsequent availability of large lithium-drifted germanium detectors have greatly aided the experimentalist in studying these radiations. These detectors have a tenfold higher energy resolution than the earlier NaI scintillation counters. Nevertheless, NaI counters are still useful because of their larger size, particularly for the study of successively emitted radiations.

3. Multipolarity of Radiative Transitions

The theory of electromagnetic radiation is much older than that of β decay and the literature contains good comprehensive treatments of the subject. In fact, the theory of β decay was largely patterned on the quantum theory of radiation. The following treatment uses the decomposition of the electromagnetic field into spherical rather than plane waves. Such spherical waves have the advantage that they possess, in addition to a definite energy, also a definite total angular momentum, which will be called l, and even or odd parity. The total angular momentum is usually specified as the multipolarity of the radiation. The parity of electric radiation with angular momentum $l\hbar$ is $(-)^l$; the parity of magnetic radiation with the same angular momentum is $(-)^{l+1}$. The radiation field has, for a given l and parity, $2l + 1$ different states characterized by $2l + 1$ different values ($\mu = -l, -l + 1, \ldots, l - 1, l$) of the z component of total angular momentum. The names "electric" and "magnetic" derive from the classical distribution of charges and currents which give rise to these fields. Expansion of the field into spherical rather than plane waves is preferable if only one particle (in this case a γ quantum) is emitted, because energy, angular momentum, its z component μ, and parity completely specify the state. This is not the case if two particles are emitted, as they are in β decay. Use of spherical waves, in the case of the emission of two particles, would obscure the angular correlations between these.

The rate of radiative transitions is calculated by the theory of transition probabilities. Since the state of the emitted photon is completely determined by its energy and character, the probability of emission is given by

$$\lambda(f,i;l,\mu,E) = \frac{2\pi}{\hbar} |(\Psi_f, eQ(l,\mu,E)\Psi_i)|^2 \rho \qquad (12.1)$$

Ψ_f is the wave function of the final state and Ψ_i that of the initial state of the nucleus; $eQ(l,\mu,E)$ is the operator for the energy of the nucleus in an electromagnetic field of characteristics l, μ, E and frequency given by $h\nu = E_i - E_f$, so normalized that its energy is $h\nu$ in unit volume. If the components of the electromagnetic potential of this field are A_k, the expression for $eQ(l,\mu,E)$ would be $e\gamma_k A_k$ in Dirac's theory of a particle with normal magnetic moment, but nonrelativistic theory is used as a rule for the calculation of (12.1); ρ is the density of the states of the electromagnetic field in unit volume per unit energy interval. There is, of course, an expression similar to (12.1) for the magnetic transitions, with M replacing E.

The wavelengths of the γ quanta are much longer than the radius of the nucleus. One can, therefore, expand the operator Q in (12.1) as function of the radius and discard all but the first term. This gives

$$\lambda(f,i;l,\mu,E) = \frac{8\pi(l-1)!(l+1)!2^l}{[(2l+1)!]^2} k^{2l+1} |E_{l\mu} + E'_{l\mu}|^2 \qquad (12.2)$$

and a similar expression for magnetic transitions in which E is replaced by M. The wave number k of the photon is its energy divided by $\hbar c$. The matrix elements E and M are

$$E_{l\mu} = e \sum_{j=1}^{Z} (\Psi_f, r_j^l Y_{l\mu}(\theta_j,\phi_j)^* \Psi_1) \qquad (12.3a)$$

$$M_{l\mu} = \frac{e\hbar}{Mc} \frac{1}{l+1} \sum_{j=1}^{Z} \sum_{\alpha} \left(\Psi_f, \frac{\partial}{\partial x_{j\alpha}} [r_j^l Y_{l\mu}(\theta_j\phi_j)]^* \mathfrak{L}_{j\alpha} \Psi_i\right) \qquad (12.3b)$$

These expressions represent the interaction of the charge distribution and of the current due to the motion of the protons with the electromagnetic field. Hence, the summation is extended only over the coordinates of the protons. Alternately, a factor $\frac{1}{2}(1 - \tau_{j\zeta})$ could be introduced and the summation extended over all particles. M is the mass of the proton, r_j, θ_j, ϕ_j, its polar and x_j its rectangular coordinates; the summation over α is to be extended over the three coordinates x, y, z. The $\mathfrak{L}_j$ is the operator of the angular momentum

$$\mathfrak{L}_j = -i\mathbf{r} \times \mathbf{grad}_j \qquad (12.4)$$

the $Y_{l\mu}$ are normalized spherical harmonics. The interaction of the spin with the electromagnetic field is represented by the E' and M' terms. Only

$$M'_{l\mu} = \sum_{j=1}^{A} \sum_{\alpha} \mu_j \left(\Psi_f, \frac{\partial}{\partial x_{j\alpha}} [r_j^l Y_{l\mu}(\theta_j,\phi_j)]^* \sigma_{j\alpha} \Psi_i\right) \qquad (12.3c)$$

is important, $E'_{l\mu}$ is in general small as compared with $E_{l\mu}$. The summation here extends over all particles; μ_j is the magnetic moment of particle j (that is, $2.78e\hbar/2Mc$ for protons and $-1.91e\hbar/2Mc$ for neutrons). For $l = 1$, $\mu = 0$, a component of the dipole radiation, one obtains the expressions

$$E_{10} = e(\tfrac{3}{4}\pi)^{1/2} \sum_{j=1}^{Z} (\Psi_f, z_j \Psi_i) \qquad (12.4a)$$

$$M_{10} = \frac{ie\hbar}{2Mc} (\tfrac{3}{4}\pi)^{1/2} \sum_{j=1}^{Z} \left(\Psi_f, \left(y_i \frac{\partial}{\partial x_j} - x_j \frac{\partial}{\partial y_j}\right) \Psi_i\right) \qquad (12.4b)$$

$$M'_{10} = (\tfrac{3}{4}\pi)^{1/2} \sum_{j=1}^{A} \mu_j (\Psi_f, \sigma_{jz} \Psi_i) \qquad (12.4c)$$

The selection rules can be read off these expressions immediately, but they are independent of the assumption that the wavelength of the emitted light is much longer than the radius of the nucleus. If the parities of Ψ_i and of Ψ_f are the same, only even electric multipole and odd magnetic multipole radiations are possible. If the parities of Ψ_i and of Ψ_f are opposite, the polarity of the electric radiation is odd, that of the magnetic radiation even. In addition, the J values of Ψ_i and Ψ_f must form a vector triangle: $|J_i - J_f| \leq l \leq J_i + J_f$. Transition rates of the electric-multipole radiations decrease with increasing l and the same is true of transition rates due to magnetic multipole radiations. Hence, if the selection rules admit an electric l-pole radiation, transition rates due to higher electric multipoles can be, in general, disregarded, and the same is true of the magnetic multipoles. In general, the magnetic l pole should give a lower rate than the electric l pole; in the case of dipole radiation the two are often of the same order of magnitude. It is more usual, however, to consider magnetic l-pole and electric $(l+1)$-pole to be of the same order of magnitude; they also can occur in the same transition and often compete with each other. If J_i and J_f differ by several units, the multipole radiation will be of high order and the transition slow. This is the explanation of nuclear isomerism suggested by v. Weizsäcker (Sec. I).

The expressions for radiative transition rates are used, principally, to test the accuracy of the wave functions of the various models. They are given already in terms of the wave functions with two-valued spin; this accounts for the fact that they are, rather artificially, divided into two parts, one due to charges, the other due to spin. There is general agreement that the expressions for the electric transition rates are sufficiently accurate. The rate of the magnetic transitions, on the other hand, may be influenced by the currents which are not due to motion of the nucleons but to the mesons which transmit the nuclear interaction. These currents are held responsible also for the magnetic moment of the neutron and the anomalous magnitude of the magnetic moment of the proton. The use of the observed moments μ_j in (12.3c) therefore already partially accounts for these "exchange currents." It is to be expected, however, that the meson current associated with the free nucleons is modified by the proximity of other nucleons so that appreciable corrections to the transition rates may result. The magnetic transition rates will be more strongly influenced than the electric transition rates because the average velocity of the mesons is higher than that of the nucleons and the magnetic transition rates depend on the current distribution while the electric rates depend on the charge distribution. For this last reason also $E'_{l\mu}$ is in general much smaller than $E_{l\mu}$ and the electric transition rates are practically independent of the magnetic moments μ_j. The fact that the static magnetic moments of complex nuclei can be calculated, under the assumption of the additivity of the moments, with an accuracy of about 0.2 nuclear magnetons, from a sufficiently accurate wave function, indicates that the corrections to the transition rates may not be as large as might have been feared. This may be a less valid conclusion for the high multipole moments than for the low ones. The selection rules are unaffected by the exchange moments.

4. Weisskopf Units

The expressions for the radiative transition rates can be evaluated most easily in the independent particle models for states which differ in the quantum number of a single nucleon. Under somewhat schematic assumptions for the radial dependence of the wave function of this single nucleon, one obtains

$$\lambda(l,E) = \frac{18}{(l+3)^2} \frac{(l-1)!(l+1)!}{[(2l+1)!]^2} (2kR)^{2l} kc \qquad (12.5a)$$

$$\lambda(l,M) = \frac{180}{(l+3)^2} \frac{(l-1)!(l+1)!}{[(2l+1)!]^2} (2kR)^{2l} \frac{e^2}{Mc^2R} \frac{\hbar k}{MR} \qquad (12.5b)$$

R is the nuclear radius. These rates are called Weisskopf units and actual transition probabilities are conveniently expressed in terms of these units.

The expressions (12.5) are not the rates expected on the basis of the independent particle pictures; they are rates in terms of which the actual transition rates can be conveniently measured. Correction factors must be applied even if there is only a single particle outside of closed shells. These correction factors are less important for magnetic than for electric transitions because the neutron's and the proton's magnetic moments are of the same order of magnitude. They generate, therefore, magnetic fields of similar magnitude and (12.5*b*) gives the corresponding transition rate.

If there is a single proton outside of closed shells, this proton and the rest of the nucleus will move about their common center of mass. Since the rest of the nucleus, the "core" also has a positive charge, and since proton and core are always on opposite sides of the center of mass, the dipole moment of the system will be decreased. This will lead to a decrease of the transition rate by a factor $(1 - Z/A)^2 = (N/A)^2$. This factor can be obtained directly from (12.4*a*) but follows more easily from the remark that two bodies, moving about their common center of mass, and having the same charge-to-mass ratio, do not create any dipole field. If the extra particle outside the closed shells is a neutron, it is the core which creates the dipole field and the transition rate is given by (12.5*a*) multiplied by $(Z/A)^2$. For higher electric-multipole radiations, the correction factor for the proton is negligible and that for the neutron is Z^2/A^{2l}, a very small number.

If there are several particles outside of closed shells, the transition rate will vanish under the assumption of the independent particle model if more than one of the orbits of initial and final states are different: the operators in (12.3) change the state of only one particle. There is a similar selection rule in atomic spectroscopy which is, in general, quite well obeyed. Even if initial and final configurations differ in only one orbit, the calculated rate will be, as a rule, well below one Weisskopf unit because the wave functions will contain many terms. This is apparent already from (8.1) and (8.2) even though these expressions apply for the case of only two particles. Each term of Ψ_i is connected by the operators of (12.3) with only those terms of Ψ_f in which all factors, except one, are also factors of the term in Ψ_i. This, and the possibility of cancellation of the terms of which there still will be many, reduces the calculated transition rate below one Weisskopf unit unless there are phase relations between the contributions of the various terms.

5. Experimental Results

Experimental results bear out only partially the conclusions presented. A great deal of additional data on nuclear lifetimes has been accumulating over the past few years but the over-all picture is not much changed from that gleaned from earlier compilations of gamma-ray lifetimes. Electric- and magnetic-dipole transitions on the whole conform to the simple picture presented with transition probabilities ranging from 1 to 10^{-3} Weisskopf units. However, as has been mentioned earlier, E2 transitions, particularly those between the first excited 2^+ level and the normal level in even-even nuclei (0^+), are invariably many times faster than a transition due to a single particle can be. This is natural in the region where the collective model accounts for the large ground-state quadrupole moments and the spectra display rotational structure. However, even in nuclei which otherwise do not evidence collective behavior the E2 transition between some levels is faster than the single-particle estimate. Further, there seems to be little difference whether the "active" particle is a proton or a neutron. Thus the E2 transition from the 0.87-Mev state of O^{17} to the normal state proceeds at a rate of one Weisskopf unit. There is only one neutron outside the doubly closed shell of eight neutrons and eight protons so that the transition rate should be reduced below the single-particle estimate by $Z^2/A^4 \sim 10^{-3}$ so that the discrepancy between observation and the simple theory is flagrant. In Sm^{150} the rate is 55 times the Weisskopf estimate while in Sm^{152} the rate is increased to greater than 140 times the same single-particle estimate.

Similar collective behavior seems to apply to the E3 transition rates, many of which again are much faster than the single-particle estimates. These transitions, however, do not seem to be enhanced quite so much as the E2 rates. Also, the enhancement seems to be greatest near closed shells whereas the E2 transitions are strongest in the middle of a shell. The reason for this is, at least in part, that the E3 transitions normally involve transitions between different major shells whereas the quadrupole transitions occur principally between configurations in the same shell.

The magnetic transitions seem to show a slight increase in their transition rate relative to the single-particle estimate as the multipolarity increases. The implications of this are not understood. The detailed calculation of γ-ray lifetimes is beyond the scope of present-day nuclear theory.

6. Theory of β Decay

A qualitative description of the phenomenon of β decay is given in Sec. I. Here a somewhat more detailed account of the theory and of its experimental verification is given. Fermi's theory of the β process is patterned on the theory of radiation. In the radiative process we envisage a change of state of the charged radiating system accompanied by the emission of a photon; the β decay is pictured as a change of the state of a nucleon (or of a nuclear system), together with the emission of an electron and a neutrino. In the case of radiation, the form of the interaction between the charged particles and field is taken over from classical theory; the interaction strength is measured by the electronic charge. For the β decay, an interaction form had to be invented which is consistent with experimental observations.

Fermi's theory is formulated in terms of the operators of quantum field theories. The β decay consists either of the annihilation of a neutron and of a neutrino and the creation, instead, of a proton and of an electron, or of the inverse process. It is described, therefore, by two products of four field operators of the form

$$\psi_p{}^+\psi_n\psi_e{}^+\psi_\nu + \psi_\nu{}^+\psi_e\psi_n{}^+\psi_p \tag{12.6}$$

The indices p, n, e refer to protons, neutrons, electrons, and neutrinos, the ψ are annihilation, the ψ^+ creation operators. The latter are the Hermitian conjugate of the former.

The particles can be created and annihilated at any point of space and with any of the four values of Dirac's spin variable: all the $\psi = \psi(\mathbf{r},\zeta)$ depend on position and on the value of the spin variable. The total interaction is an integral of expressions of the form (12.6) over all space and could be an arbitrary linear combination of the 256 components which can be obtained by giving all four spin variables every possible value

$$H_{\text{interaction}} = \sum_{\zeta_p\zeta_n}\sum_{\zeta_e\zeta_n} C(\zeta_p,\zeta_n,\zeta_e,\zeta_\nu) \int \psi_p{}^+(\mathbf{r},\zeta_p)\psi_n(\mathbf{r},\zeta_n)\psi_e{}^+(\mathbf{r},\zeta_e)\psi_\nu(\mathbf{r},\zeta_\nu)\,d\mathbf{r} + \text{Hermitian adjoint} \tag{12.6a}$$

The "Hermitian adjoint" corresponds to the second term of (12.6). Certainly, (12.6a) is the simplest expression which accounts for the possibility of the β transformation at any point of space.

A condition on the coefficients C follows from the requirement of relativistic invariance of the interaction. There are sixteen products of the form $\psi_e{}^+(\mathbf{r},\zeta_e)\psi_\nu(\mathbf{r},\zeta_\nu)$ corresponding to the four values of ζ_e and ζ_ν each. From these sixteen products, one can form the following linear combinations: one scalar (S), the four components of a vector (V), the six components of an antisymmetric tensor (T), four components of an axial vector (A), and a pseudoscalar (P). The same linear combinations can be formed from the sixteen products $\psi_p{}^+(\mathbf{r},\zeta_p)\psi_n(\mathbf{r},\zeta_n)$. In order to obtain a scalar from these quantities, one can multiply the scalar, or the pseudoscalar, formed from the $\psi_e{}^+\psi_\nu$ products, with the scalar, or the pseudoscalar, formed from the $\psi_p{}^+\psi_n$ products. Similarly, one can multiply the vector components, or the components of the axial vector, formed from the $\psi_e{}^+\psi_\nu$ products, with the corresponding components of the vector, or of the axial vector, formed from the $\psi_p{}^+\psi_n$ products and add these with the proper signs. The result—the scalar product of two vectors of polar or axial nature—is invariant with respect to rotations or proper Lorentz transformations. Finally, one can obtain two invariants with respect to these transformations by combining the antisymmetric tensor components formed from the $\psi_e{}^+\psi_\nu$ products with the components of the antisymmetric tensor formed from the $\psi_p{}^+\psi_n$ products.

As long as it was believed that the interaction responsible for β decay is invariant also with respect to spatial reflections, it could be assumed that the scalar formed from the $\psi_e{}^+\psi_\nu$ products could be multiplied only with the scalar formed from the $\psi_p{}^+\psi_n$ products but not with the pseudoscalar. Similarly, it was assumed that the components of the vector formed from the $\psi_e{}^+\psi_\nu$ which are

$$v_k = \sum_{\zeta\zeta'} (\gamma_t\gamma_k)_{\zeta\zeta'}\psi_e{}^+(\zeta)\psi_\nu(\zeta') \tag{12.7a}$$

would have to be multiplied by the components of the vector

$$V_k = \sum_{\zeta\zeta'} (\gamma_t\gamma_k)_{\zeta\zeta'}\psi_p{}^+(\zeta)\psi_n(\zeta') \tag{12.7b}$$

(and not with the components of a pseudovector) to form an admissible expression for the interaction. The γ are Dirac's anticommuting matrices; γ_t is Hermitian, and the other γ, skew-Hermitian. The permissible interaction Hamiltonian (or Lagrangian) was then

$$H_{vV} = v_tV_t - v_xV_x - v_yV_y - v_zV_z \tag{12.7}$$

This could be multiplied with any real coupling constant C_{vV}. The operator of the axial vector interaction had a similar form, except that the $\gamma_t\gamma_k$ were replaced by $\gamma_t\gamma_k\gamma_5$, where $\gamma_5 = i\gamma_t\gamma_x\gamma_y\gamma_z$ anticommutes with the γ and hence commutes with the $\gamma_t\gamma_k$:

$$H_{aA} = a_tA_t - a_xA_x - a_yA_y - a_zA_z \tag{12.8}$$

Again, this could be multiplied with a real coupling constant which we shall call C_{aA}. Altogether, there are five linear combinations of the 256 products (12.6) which are invariant under rotations, Lorentz transformations, and spatial reflections. The real nature of the coupling constants C_{vV} and C_{aA} follows from invariance with respect to reflection of the time axis. Any linear combination of the five invariants was considered to be a possible interaction, and only these were considered permissible.

Actually, the preceding consideration implicitly assumes that the intrinsic parity of a proton-electron pair is the same as that of a neutron-neutrino pair. Were these parities opposite, the proper interactions would be

$$H_{vA} = v_t A_t - v_x A_x - v_y A_y - v_z A_z \qquad (12.9)$$

and

$$H_{aV} = a_t V_t - a_x V_x - a_y V_y - a_z V_z \qquad (12.10)$$

However, under the assumption of parity conservation, (12.7) and (12.8) were alternatives to (12.9) and (12.10), and either one pair or the other was admissible, but not both.

The assumption of parity conservation—an expression for the reflection symmetry of the interaction—was dramatically shattered by the experiment of Wu, Ambler, Hayward, Hoppes, and Hudson, following the theoretical suggestions of Lee and Yang. They showed that the electrons are emitted, by the decaying Co^{60} nucleus, preferentially in the direction of its spin. Since the spin is an axial vector and the direction of the electron's motion a polar vector, this contradicts reflection invariance; the mirror image of the process shows the electrons to be emitted preferentially in the direction opposite to that of the spin. The argument against the joint presence of all four kinds of interaction, (12.7), (12.8), (12.9), and (12.10), and in fact all 10 types mentioned earlier, has lost its foundation. If time inversion invariance is not valid either in the β-decay process, the constants C_{vV}, C_{vA}, etc., need not be real.

Actually, there seems to be good evidence, to which the aforementioned experiment of Wu and her collaborators contributed greatly, that the coefficients C_{vV} and C_{aV} are equal, the coefficients C_{vA} and C_{aA} are also equal, and all other coefficients are zero. Even the latter pair of coefficients, C_{vA} and C_{aA}, is nearly equal, or oppositely equal, to the former pair. They appear to be larger by about 17 per cent but this difference has been shown to have a secondary origin. Hence, the interaction of the nucleons with the electron-neutrino field seems to have, in principle, a very simple form

$$H_{\text{interaction}} = C \sum_k \sum_{\zeta\zeta'} \sum_{\eta\eta'} \psi_e{}^+(\zeta)(\gamma_t\gamma_k(1+\gamma_5))_{\zeta\zeta'}\psi_\nu(\zeta') \psi_p{}^+(\eta)(\gamma_t\gamma_k(1+\gamma_5))_{\eta\eta'}\psi_n(\eta')l_k + \text{Hermitian adjoint} \qquad (12.11)$$

where $l_t = 1$, $l_x = l_y = l_z = -1$. In practice, however, the coefficients of H_{aA} and H_{vA} are somewhat larger than those of H_{vV} and H_{aV}. The first term of (12.11) is responsible for the emission of an electron and absorption of a neutrino. The latter is equivalent to the emission of an antineutrino and manifests itself in practice in this way. The Hermitian adjoint is responsible for the emission of a neutrino and the absorption of an electron. The electron can be absorbed from one of the atomic orbitals, usually the innermost, K orbit. Alternatively, and more commonly, instead of the absorption of an electron, the emission of a positron takes place.

7. β Decay of Nuclei

The preceding section outlined the theory of β decay and ended with the expression for the interaction between nucleons and the electron-neutrino field. We now turn to the application of these results to the calculation of the decay rates and decay modes of nuclei.

The underlying principles of this calculation are the same as those used to calculate radiative transitions. The final states of the emitted particles are again known: They are states of a free particle, the neutrino, and another particle, the electron, under the influence of the Coulomb field of the nucleus. However, rather than using spherical waves to describe the states of these particles, it is more customary, and more useful, to characterize them by their momenta $\mathbf{p}$ and $\mathbf{q}$. The expression, which is the analogue of (12.1), for the transition probability from the initial state Ψ_i of the nucleus into its final state Ψ_f is

$$\lambda(f,i;\mathbf{p},\mathbf{q}) = \frac{2\pi}{\hbar} \sum_{Km} |(\Psi_f, C_K Q_{Km} \Psi_i)|^2 \qquad (12.12)$$

The C_K are the coupling constants which occur in (12.7), (12.8), etc.; K stands for vV, vA, etc.; m stands for t, x, y, z; and ρ is the density of the states of the electron-neutrino field. If one neglects the effect of the Coulomb field on the electron, the operators Q_{Km} become

$$Q_{Km} = \sum_{j=1}^{A} q_{Km}(j)\tau_{j\xi} \exp\left[-i(\mathbf{p}+\mathbf{q})\cdot\mathbf{r}_j\right] \qquad (12.12a)$$

The sum over j appears because each of the nucleons may, potentially, undergo a transformation. The exponential corresponds to the recoil of the transforming particle which is oppositely equal to the momentum of the electron-neutrino pair. The operator $\tau_{j\xi}$ is the first component of the isobaric spin operator of particle j; it transforms a proton into a neutron and conversely. The operators $q_{Km}(j)$ apply only to the spin coordinate ζ_j of particle j; they are essentially the matrices which form the component m of the quantity K from the products $\psi_p{}^+(\mathbf{r},\zeta_p)\psi_n(\mathbf{r},\zeta_n)$. They are γ matrices of Dirac or products of such matrices; their form is not essential for what follows.

The expressions (12.12) and (12.12a) imply that the nuclear wave functions Ψ_f and Ψ_i depend on four-valued spin variables, rather than the two-valued spin variables used hitherto. Hence, in order to evaluate these expressions, it would be necessary to convert the ordinary nuclear wave functions, with two-valued spin variables, to nuclear wave functions with four-valued spin variables. This is possible only with a relativistic theory of nuclear interaction. Since such a theory is lacking, calculation of the matrix elements in (12.12) has a certain amount of ambiguity, even if Ψ_i and Ψ_f are known as functions of the positional and two-valued spin coordinates. Nevertheless, (12.12) can be evaluated at least approximately because the velocity of the nucleons is, on the whole, small compared with that of light. As a result, the wave functions Ψ_i and Ψ_f are large only for two values of each spin variable ζ_j, and these two values can be identified with the two values of the ordinary spin variable.

Some of the matrices q_{Km} connect large components of Ψ_i with large components of Ψ_f (and small components with small components). The order of magnitude of these matrix elements is 1; they will be denoted by $q_{Km}{}^{(n)}$ (n for nonrelativistic). They can be obtained with a reasonable accuracy if the large components of Ψ_i and Ψ_f are known, that is, if these are known as a function of the two-valued spin variables. Other q_{Km} connect large components of Ψ_i with small components of Ψ_f, and vice versa. The order of magnitude of these is v/c, where v is the average velocity of the nucleons in the nucleus; they are denoted by $q_{Km}{}^{(r)}$ (r for relativistic).

8. Allowed and Forbidden Transitions

In order to obtain an estimate for the matrix element in (12.12), it is useful to expand the exponential

$$\exp - i(\mathbf{p} + \mathbf{q}) \cdot \mathbf{r}/\hbar = \sum i^{-l} \mathcal{E}^{(l)}$$

$$\mathcal{E}^{(l)} = \frac{1}{l!} \left[\frac{(\mathbf{p} + \mathbf{q}) \cdot \mathbf{r}}{\hbar} \right]^l \qquad (12.13)$$

Since the order of magnitude of $\mathbf{p}$ and $\mathbf{q}$ is mc, with m the electronic mass, and since only such r contribute significantly to the integral (12.12) which are not larger than the nuclear radius, every term $\mathcal{E}^{(l)}$ of (12.13) gives, in general, a smaller contribution than the preceding one by a factor of the order of magnitude 10^{-2}, unless the integral of the preceding $\mathcal{E}^{(l)}$ vanishes on account of a selection rule or some similar reason. It is customary, therefore, to replace the operators Q_{Km} by the operators of lowest l

$$M_{Km}{}^{(l)} = \sum_j [q_{Km}{}^{(n)}(j)\tau_{j\xi}\mathcal{E}^{(l)} + q_{Km}{}^{(r)}(j)\tau_{j\xi}\mathcal{E}^{(l-1)}] \qquad (12.14)$$

for which the integral (12.12) does not vanish. The two terms of (12.14) are of similar order of magnitude and their selection rules also have much in common so that if the matrix element of the second operator does not vanish, that of the first one is also finite. In particular, the $q^{(n)}$ do not change the parity of a state, while $\mathcal{E}^{(l)}$ changes it for odd l, leaves it unchanged for even l. Hence the first term may give a finite contribution only if the parities of Ψ_i and Ψ_f are the same and l is even, or if the parities of Ψ_i and Ψ_f are opposite and l is odd. This is true for all $\mathbf{p}$ and $\mathbf{q}$. The same is true of the second term of (12.14) also because the $q_{Km}{}^{(r)}$ are those Dirac matrices which do change the parity. For $l = 0$ there is no second term in (12.14). As the form of $\mathcal{E}^{(l)}$ indicates, successively higher l permit an increasing difference in the total angular momenta J of Ψ_i and Ψ_f. If the $l = 0$ term can give a finite contribution to the matrix element, the transition is called allowed; if $M_{Km}{}^{(l)}$ is the first term which gives a finite contribution, the transition is called lth forbidden.

We note that the $l = 0$ component arising from the H_{vV} and H_{aV} parts of the interaction is, essentially,

$$\sum_j \tau_{j\xi} = 2T_\xi \qquad (12.15)$$

and the $l = 0$ components of the H_{aA} and H_{vA} parts are

$$\sum_j \sigma_{jx}\tau_{j\xi} = 2Y_{x\xi} \quad \sum_j \sigma_{jy}\tau_{j\xi} = 2Y_{y\xi} \quad \sum_j \sigma_{jz}\tau_{j\xi} = 2Y_{z\xi} \qquad (12.15a)$$

The first of these, (12.15), is a scalar operator; its selection rule is $J_i = J_f$, with no change of parity. This is called the Fermi selection rule. If the isotopic spin quantum number is valid, T_ξ simply transforms one member of an isotopic spin multiplet into an adjoining member and (12.15) permits only transitions between "analogue states." This was mentioned in Sec. VI, 4.

The expressions (12.15a) are, on the other hand, components of a vector operator; they lead to the selection rule $J_i - J_f = \pm 1$ or 0, the transition from $J_i = 0$ to $J_f = 0$ being forbidden. This selection rule is named after Gamow and Teller; again it does not allow a change of parity. If the supermultiplet theory is valid, the operators (12.15a) would connect only members of the same supermultiplet. This follows from the fact that, if the nuclear Hamiltonian depends neither on the ordinary nor on the isotopic spin coordinates, it commutes with the operators (12.15a), and these transform characteristic functions of the nuclear part of the Hamiltonian into characteristic functions of the same Hamiltonian with the same energy. The corresponding states are those which form the members of a supermultiplet. The β transitions between them are called favored. In the regions in which the supermultiplet theory appears to be a good approximation, i.e., for very light elements, the favored transitions are about a hundred times larger than other transitions allowed by the Gamow-Teller selection rule.

9. Shape of the Spectrum

The expressions (12.15) and (12.15a) for the matrix elements of allowed transitions do not depend on the momentum of the emitted electron and neutrino. As a result, the shape of the allowed spectra is entirely determined by the factor ρ in (12.12). This depends in general, on the momentum and polarization of electron and neutrino. To obtain the energy spectrum of the electrons, one has to sum over all directions of polarization and all directions of momenta. The expression obtained is proportional to the volume in phase space per unit energy range available for electron and neutrino. This is

$$P(W_e) = pW_e qW_q \qquad (12.16)$$

where p and q are, as before, the momenta of electron and neutrino, W_e and W_q the corresponding energies, including rest mass. The sum of these is the energy of disintegration $W_0 = W_e + W_q$ so that (12.16) contains only one independent variable. It gives the spectral distribution of the electrons, that is, apart from a constant factor, the number of electrons the energy of which is W_e within unit energy interval.

If the rest mass of the neutrino is zero, $W_q = cq$ and (12.16) becomes

$$P(W_e) = \frac{pW_e(W_0 - W_e)^2}{c} \qquad (12.16a)$$

In this case the upper limit of the electron energy is W_0 and the spectral distribution (12.16a) behaves like $(W_0 - W_e)^2$ in the neighborhood of the upper limit. If the mass m of the neutrino is finite, the upper end of the spectrum is at $W_{\max} = W_0 - m_\nu c^2$ and the spectral distribution (12.16) is proportional to $(W_{\max} - W_e)^{1/2}$ in the neighborhood of the end point. Hence the electron spectrum depends, in the neighborhood of its end point, on the neutrino mass in a sensitive way. The lowest limit for the rest mass of the neutrino was obtained by careful investigation of the electron spectrum near the end point.

The effect of the electrostatic field of the nucleus on the wave funtion of the electron is not taken into-account in the preceding calculation. To take it into account, the plane wave exp $(i\mathbf{p} \cdot \mathbf{r})$ in (12.12a) should be replaced by a Coulomb wave function. This introduces a correction factor into the spectral distribution (12.16a) which is usually denoted by $F(Z_-,W_e)$ for electrons and by $F(Z_+,W_e)$ for positrons. These correction factors are not very important for allowed transitions and low Z unless the energy of disintegration is very small. They do not influence the behavior of the spectral distribution in the neighborhood of the end point.

The spectral distribution (12.16a) of the electrons, with the correction factor $F(Z_\pm,W_e)$ for the Coulomb field, was confirmed by very careful and accurate measurements. These form a strong support of Fermi's theory. The spectral distribution of the electrons or positrons contains a polynomial of lth or $(l - 1)$th degree of p and q for transitions forbidden to the lth degree. Such factors appear in the expression (12.13) for $\mathcal{E}^l$. In some cases therefore the degree of forbiddenness can be obtained from the electron spectrum directly. This is particularly true for certain second forbidden transitions. The influence of the Coulomb field gives most first forbidden transitions the allowed shape. The degree of forbiddenness manifests itself, as a rule, also in the absolute rate of the transition.

References

Since the first edition of this book, there has been a phenomenal increase in the amount of detailed and accurate information on nuclear physics, and the variety of the phenomena considered has also increased. Thus it is even less possible today for a single person, or for a small number of collaborators, such as the authors, to be intimately familiar with all parts of the subject. When rewriting this chapter we had to rely in several cases on reviews and summaries given by others and on reports on conferences which present several points of view and, we hope, a balanced picture. We hope that those whose work has been slighted will recognize our limitations of both space and knowledge. We also hope that the discussion will be useful in spite of its shortcomings, some of which we could not avoid.

I

The discovery of radioactivity is due to H. A. Becquerel [*Compt. rend.*, pp. 122, 420, 501, 559, 689, 762, 1086, (1896); pp. 123, 855, (1896)]. The exciting story of the determination of the properties of radioactive substances and their radiations is also described in Mme. P. Curie, "L'isotopie et les éléments isotopes," University of Paris Press, Paris, 1924; E. Rutherford, "Radioactive Substances and Their Radiations," Cambridge University Press, New York and London, 1913; and K. Fajans, *Jahrb. Rad.*, **14**: 314 (1917).

It is difficult to single out a few articles which were principally responsible for the development of concepts of nuclear structure. Four articles of major importance are E. Rutherford, *Phil. Mag.*, **21**: 669 (1911); H. Geiger and E. Marsden, *Phil. Mag.*, **25**: 610 (1913); J. Chadwick, *Proc. Roy. Soc.* (*London*), **A136**: 692 (1932); and W. Heisenberg, *Z. Physik*, **77**: 1 (1932).

The atmosphere in which Rutherford's model was discovered is recaptured by C. G. Darwin in the book "Niels Bohr and the Development of Physics," McGraw-Hill, New York, 1955.

The first measurements of nuclear masses are due to Aston and Dempster: A. J. Dempster, *Phys. Rev.*, **11**: 316 (1918), and later articles; F. W. Aston, *Phil. Mag.*, **38**: 709 (1919), and later articles.

The modern techniques of mass spectroscopy are due to Bainbridge, Mattauch, Nier, Duckworth, and many others. For a collection of nuclear masses, see J. H. E. Mattauch, W. Thiele, and A. H. Wapstra, *Nucl. Phys.* **67**: 1 (1965).

The relation between lifetime and decay energy of α emitters was given by H. Geiger and J. M. Nuttall, *Phil. Mag.*, **22**: 613 (1911).

The explanation of this relation marks the beginning of modern nuclear theory. It was given, simultaneously, by G. Gamow, *Z. Physik*, **51**: 204 (1928), and R. W. Gurney and E. U. Condon, *Nature*, **122**: 439 (1928); *Phys. Rev.*, **33**: 127 (1929).

Fission was discovered by Hahn and Strassmann: O. Hahn and F. Strassmann, *Nature*, **27**: 11 (1939).

The nature and properties of γ and β radiations were uncovered principally by Ellis and Meitner. Two very important articles in this connection are L. Meitner, *Z. Physik*, **34**: 807 (1925), and C. D. Ellis and W. A. Wooster, *Proc. Roy. Soc.* (*London*), **A117**: 109 (1927).

The original theory of β decay is due to E. Fermi, *Z. Physik*, **88**: 161 (1934). See also H. A. Bethe and R. F. Bacher, *Revs. Mod. Phys.*, **8**: 82 (1936), chap. VII.

The radioactivity of the neutron was first observed by A. H. Snell and L. C. Miller, *Phys. Rev.*, **74**: 1217 (1948). See also J. M. Robson, *Phys. Rev.*, **83**: 349 (1951).

The lowest limit for the mass of the neutrino was obtained by D. R. Hamilton, W. P. Alford, and L. Gross, *Phys. Rev.*, **92**: 1521 (1953).

Production of radioactive nuclei as a result of the interaction between neutrinos and stable nuclei was observed by F. Reines and C. L. Cowan, *Nature*, **178**: 446 (1956), and C. L. Cowan, F. Reines, F. B. Harrison, H. W. Kruse, and A. D. McGuire, *Science*, **124**: 103 (1956).

For an attempt to observe the process of the double β decay, see M. Awschalom, *Phys. Rev.*, **101**: 1041 (1956) and E. der Mateosian and M. Goldhaber, *Proc. Int. Conf. on High Energy Physics* (Dubna, 1965). See also R. C. Winter, *Phys. Rev.*, **99**: 88 (1955). The present lower limit on the lifetime of Ca^{48}, which is energetically unstable with respect to double β decay, is 10^{19} years.

The explanation of isomerism is due to C. F. V. Weizsäcker, *Naturwiss.*, **24**: 813 (1936).

II

Early work on the systematics of isotopes is principally due to Aston: F. W. Aston, "Mass Spectra and Isotopes," E. Arnold and Co., London, 1933. See also W. D. Harkins, *Phys. Rev.*, **19**: 136 (1922), and V. M. Goldschmidt, Geochemische Verteilunggesetze der Elemente, *Norske Videnshaps-Akad. Oslo* (1937).

The use of three binding-energy surfaces was proposed by G. Gamow. The first semiempirical mass formula was given by C. F. v. Weizsäcker, *Z. Physik*, **96**: 431 (1935). See also E. Feenberg, *Revs. Mod. Phys.*, **19**: 239 (1947).

The constants of the text are those of M. H. L. Pryce, *Proc. Phys. Soc. (London)*, **63**: 692 (1950).

The particular stability of the nucleon numbers which are now called "magic" was noticed by W. Elsasser, *J. Phys. Radium*, **5**: 389, 635 (1934).

Convincing evidence for the preferred nature of these numbers was presented by M. G. Mayer, *Phys. Rev.*, **75**: 1969 (1949).

III

First measurement of a nuclear magnetic moment, that of a proton, is due to O. Stern, I. Estermann, and R. Frisch, *Z. Physik*, **85**: 4 (1933); 132 (1933). See also I. I. Rabi, J. M. B. Kellogg, and J. R. Zacharias, *Phys. Rev.*, **46**: 157 (1934).

The article in which Schmidt's lines are defined is T. Schmidt, *Z. Physik*, **106**: 358 (1937).

The phenomenon of the quadrupole moment was discovered by H. Schüler and T. Schmidt, *Z. Physik*, **94**: 457 (1935).

The quadrupole moment of the deuteron was found by J. M. B. Kellogg, I. I. Rabi, N. F. Ramsey, and J. R. Zacharias, *Phys. Rev.*, **56**: 728 (1939); **57**: 677 (1940).

For the measurement of the magnetic-octopole moment, see V. Jaccarino, J. G. King, R. A. Satten, and H. H. Stroke, *Phys. Rev.*, **94**: 1798 (1954).

For measurement of nuclear moments in general, see N. F. Ramsey, "Molecular Beams," Oxford University Press, New York and London, 1956, and "Nuclear Moments," Wiley, New York, 1953.

The size of nuclei has been determined: (*a*) By scattering cross section of fast neutrons: R. Sherr, *Phys. Rev.*, **68**: 240 (1945), and E. Amaldi, D. Bocciarelli, C. Cacciaputo, and G. Trabacchi, *Nuovo Cimento*, **3**: 203 (1946). (*b*) By the lifetime-decay-energy relation: G. Gamow, "Constitution of Atomic Nuclei and Radioactivity," chap. II, Oxford University Press, New York and London, 1931. (*c*) By the comparison of the binding energies of mirror nuclei: E. P. Wigner and E. Feenberg, *Rept, Progr. Phys.*, **8**: 274 (1941). (*d*) The energy levels of μ mesons, bound to atomic nuclei, were measured by V. L. Fitch and J. Rainwater, *Phys. Rev.*, **92**: 789 (1953). This method of obtaining the size of nuclei was suggested by J. A. Wheeler, *Revs. Mod. Phys.*, **21**: 133 (1949). (*e*) For the high-energy electron-diffraction experiments, see R. Hofstadter, *Revs. Mod. Phys.*, **28**: 214 (1956). Our knowledge of nuclear sizes in the late 1950s was reviewed at the Stanford conference, *Revs. Mod. Phys.*, **30**: 412 (1958).

IV

The concepts of cross section and differential cross section originated in the kinetic theory of gases. See, for instance, L. Boltzmann, "Vorlesungen über Gastheorie," Johann Ambrosius Barth, Leipzig, 1898.

Much of the early work on nuclear theory was concerned with the effect of the electrostatic barrier on nuclear reactions. See, for instance, G. Gamow, "Structure of Atomic Nuclei and Nuclear Transformations," Oxford University Press, New York and London, 1937; G. Breit, M. Ostrovsky, and D. Johnson, *Phys. Rev.*, **49**: 22 (1936). Chapters IX and X of Gamow's book give an excellent summary of the ideas developed before 1937 by Gamow himself, by Bethe, Breit, Feenberg, Goldhaber, Landau Mott, Oppenheimer, Peierls, Teller, Wick, and their collaborators, and by many others. The book also reviews much of the experimental information available before 1937.

Experimental work provided the main stimulus for introducing the concept of resonance reactions. This was done simultaneously by N. Bohr, *Nature*, **137**: 344 (1936), and by G. Breit and E. Wigner, *Phys. Rev.*, **49**: 519 (1936). See also the review article, *Am. J. Phys.*, **23**: 371 (1955), which deals also with the limitations of the compound-nucleus model. The statistical model is due to V. F. Weisskopf and D. H. Ewing, *Phys. Rev.*, **57**: 472, 935 (1940). It is described by J. M. Blatt and V. F. Weisskopf in "Theoretical Nuclear Physics," chap. VIII, Wiley, New York, 1952. This book gives an admirable review of nuclear theory up to 1951.

The theory of stripping reactions is due to S. Butler, *Proc. Roy. Soc. (London)*, **A208**: 559 (1951). The subject is reviewed by R. M. Eisberg and N. Austern in the "Proceedings of the International Conference on Nuclear Structure in Kingston" (University of Toronto Press, Toronto, Canada, 1960). See also the articles by C. A. Levinson and by M. K. Banerjee in "Nuclear Spectroscopy," Part B, Academic Press, New York, 1960.

The optical model was used in early theoretical work on nuclear reactions but was abandoned when found to be inadequate in the resonance region. The revival and proper appreciation of the range of applicability are the work of H. Feshbach, C. E. Porter, and V. F. Weisskopf, *Phys. Rev.*, **96**: 448 (1954). This was stimulated by the observations of M. Walt and H. H. Barschall, *Phys. Rev.*, 1062 (1954). For the applications of the optical model to high-energy processes, see S. Fernbach, R. Serber, and T. B. Taylor, *Phys. Rev.*, **75**: 1352 (1949).

The phenomenon of Coulomb excitation was foreseen by V. F. Weisskopf, *Phys. Rev.*, **53**: 1018 (1938). It was found simultaneously, about 15 years later, by T. Huus and C. Zupancic, *Danske Mat. Fys. Medd.*, **28**(1): (1953), and C. McClelland and C. Goodman, *Phys. Rev.*, **91**: 760 (1953).

Heavy-ion reactions were reviewed at several conferences. See the "Proceedings of the Third Conference on Reactions between Complex Nuclei," University of California Press, Berkeley, Calif., 1963.

V

The approximate magnitude and range of the nuclear interaction were first determined by E. P. Wigner, *Phys. Rev.*, **43**: 252 (1933). See also E. Feenberg and J. Knipp, *Phys. Rev.*, **48**: 906 (1935).

The effect of the tensor forces was investigated by W. Rarita and J. Schwinger, *Phys. Rev.*, **59**: 436 (1941). L. Hulthén, *Arkiv Nat. Astron. Fysik*, **35A** (sec. 25): (1948) gave a general review of our knowledge concerning the wave function of the deuteron. A more up-to-date version may be found in Hulthén's article in vol. XXXIX of Flügge's "Handbuch der Physik," Springer-Verlag, Berlin, 1957.

The basic concepts of the meson theory of nuclear forces are due to H. Yukawa. For Levy's calculation and its evaluation, see M. M. Levy, *Phys. Rev.*, **88**: 725 (1952), and A. Klein, *Phys. Rev.*, **89**: 1158 (1953).

The significance of scattering experiments on ortho- and parahydrogen was first recognized by J. S. Schwinger and E. Teller, *Phys. Rev.*, **52**: 286 (1937).

The idea of the repulsive core was first used by R. Jastrow, *Phys. Rev.*, **81**: 165 (1951).

The concept of saturation and of exchange forces (a more sophisticated explanation of the fact that the binding energy does not increase with the *square* of the mass number A) is due to W. Heisenberg, *Z. Physik*, **77**: 1 (1932), and E. Majorana, *Z. Physik*, **82**: 137 (1933). The first of these articles already noted that the proton-proton and neutron-neutron interactions appear to be equal.

The most careful interpretation of the scattering experiments in terms of the modification of the wave function (called phase parameters by the authors) is due to Breit and his collaborators. The first in the series of articles is that of G. Breit, E. U. Condon, and R. D. Present, *Phys. Rev.*, **50**: 825 (1936); a recent and rather comprehensive article is by G. Breit, M. H. Hull, K. E. Lassila, K. D. Pyatt, and H. M. Ruppel, *Phys. Rev.*, **128**: 826 (1962). See also Breit's article in "Perspectives in Modern Physics," Interscience, New York, 1966. The articles proposing an interaction operator which, if inserted into the Schrödinger equation, reproduces these phase parameters are too numerous to cite. Probably the most accurate representation of the phase parameters can be obtained by a very complicated interaction operator, given by K. E. Lassila, M. H. Hull, H. M. Ruppel, F. A. McDonald, and G. Breit, *Phys. Rev.*, **126**: 881 (1962). The most widely used interaction operator is due to T. Hamada and I. D. Johnston, *Nucl. Phys.*, **34**: 382 (1962).

Calculations of the internucleon potential from first principles, and the comparison of the results with data obtained from scattering experiments, were reviewed by G. Breit. See his article in R. Hofstadter and L. I. Schiff (eds.), "Nucleon Structure," Stanford University Press, Stanford, Calif., 1964.

The boundary-condition model was conceived by G. Breit and W. G. Bouricius, *Phys. Rev.*, **74**: 1546 (1948). It was further developed by H. Feshbach and E. Lomon, *Phys. Rev.*, **102**: 891 (1956), and many others.

VI

Equality of the proton-proton and proton-neutron interactions, in contrast to the equality of proton-proton and neutron-neutron interactions already noted by Heisenberg, was noted by B. Cassen and E. U. Condon, *Phys. Rev.*, **50**: 846 (1936), and by G. Breit and E. Feenberg, *Phys. Rev.*, **50**: 850 (1936). The isobaric spin quantum number was introduced by E. P. Wigner, *Phys. Rev.*, **51**: 106 (1937).

Effect of the electrostatic interaction on the validity of the quantum number T was investigated by L. A. Radicati, *Proc. Phys. Soc. (London)*, **A66**: 139 (1953); **A67**: 39 (1954); and W. M. MacDonald, *Phys. Rev.*, **98**: 60 (1955); **100**: 51 (1955); **101**: 271 (1956) for light nuclei; for heavy nuclei, see A. M. Lane and J. M. Soper, *Nucl. Phys.*, **37**: 663 (1962). See also M. H. Macfarlane and J. B. French, *Revs. Mod. Phys.*, **32**: 567 (1960), and S. D. Bloom, *Nuovo Cimento*, **12**: 2417 (1964).

For experimental information pertaining to isobaric spin, see the series of article by Wilkinson and his collaborators, beginning with D. H. Wilkinson and G. A. Jones, *Phil. Mag.*, **44**: 542 (1953), to J. B. Woods and D. H. Wilkinson, *Nucl. Phys.*, **61**: 661 (1965). The "analogue states" of medium heavy nuclei, which are end products of nuclear reactions, were discovered by J. D. Anderson, C. Wong, and J. W. McClure, *Phys. Rev.*, **126**: 2170 (1962); **129**: 2718 (1963). In compound nuclei, they were found by the Tallahassee group. See, for example, J. D. Fox, C. F. Moore, and D. Robson, *Phys. Rev. Letters*, **12**: 198 (1964).

The role of the isobaric-spin concept in light nuclei was reviewed by E. P. Wigner, Proceedings of the Robert A. Welch Foundation Conferences (Houston, 1957), by W. M. MacDonald in "Nuclear Spectroscopy," Part B, F. Ajzenberg-Selove (ed.), Academic Press, New York, 1960. For the more recent developments, see the Proceedings of the 1966 Conference on Analogue States, Academic Press, New York and London, 1966.

VII

The method for calculating the properties of "nuclear matter" has been given by K. A. Brueckner and C. A. Levinson, *Phys. Rev.*, **97**: 1344 (1955). See also H. A. Bethe, *Phys. Rev.*, **103**: 1353 (1956), J. Goldstone, *Proc. Roy. Soc. (London)*, **A239**: 267 (1957), and N. M. Hugenholtz, *Physica*, **23**: 533 (1957).

The difficulties in using the nuclear potentials obtained from two-body experiments for the explanation of more complex systems become evident in the theory of H_1^3. See, for example, J. M. Blatt and L. M. Delves, *Phys. Rev. Letters*, **12**: 544 (1964).

The contrasting names "powder model" and "shell model" are due to M. H. L. Pryce. The most brilliant description of the powder model is due to N. Bohr, *Nature* **137**: 344 (1936). The statistical theory of spectra was reviewed by C. E. Porter in "Statistical Theory of Spectra," Academic Press, New York, 1965. The supermultiplet model was developed by E. P. Wigner, *Phys. Rev.*, **51**: 947 (1937), and W. H. Barkas, *Phys. Rev.*, **55**: 691 (1939). See also E. Feenberg and E. P. Wigner, *Rept. Progr. Phys.*, **8**: 274 (1941). The general trend of nuclear binding energies was reviewed from the point of view of the supermultiplet model by P. Franzini and L. A. Radicati, *Phys. Letters*, **6**: 322 (1963).

VIII

For early attempts to interpret nuclear structure on the basis of the independent-particle model, see G. Beck, *Z. Physik*, **61**: 615 (1930); G. Gamow, *Z. Physik*, **89**: 592 (1934); K. Guggenheimer, *J. Phys. Radium*, **5**: 475 (1934); W. Elsasser, *J. Phys. Radium*, **5**: 635 (1934); and F. Hund, *Phys. Zeits.*, **38**: 929 (1937).

The fact that the $1p$ shell is being completed in the region between $A = 4$ and $A = 16$ was recognized by J. H. Bartlett, *Nature*, **130**: 165 (1932).

Calculations on the L-S model were carried out by E. Feenberg and E. P. Wigner, *Phys. Rev.*, **51**: 95 (1937), and M. E. Rose and H. A. Bethe, *Phys. Rev.*, **51**: 205 (1937).

The j-j shell model was proposed by O. Haxel, J. H. D. Jensen, and H. E. Suess, *Z. Physik*, **128**: 295 (1950), and M. G. Mayer, *Phys. Rev.*, **75**: 1969 (1949); **78**: 16, 22 (1950); **79**: 1012 (1950).

Calculation of the fraction of the wave function that belongs to the lowest supermultiplet is contained in the article by A. R. Edmonds and B. H. Flowers, *Proc. Roy. Soc. (London)*, **A229**: 536 (1955). This is one of a series of articles by Edmonds, Elliot, Flowers, and Lane, on the transition from L-S to j-j coupling. See also M. G. Redlich, *Phys. Rev.*, **99**: 1427 (1955).

The problem of the transition between L-S and j-j models was first broached by D. R. Inglis, *Revs. Mod. Phys.*, **25**: 390 (1953), who first recognized the transitional nature of the states of light nuclei.

It is not possible to give an adequate review of the rapidly growing literature on the j-j model. Much of it is reviewed in the monographs: M. G. Mayer and J. H. D. Jensen, "Elementary Theory of Nuclear Shell Structures," Wiley, New York, 1955, and E. Feenberg, "Shell Theory of the Nucleus," Princeton University Press, Princeton, N.J., 1955.

Table 8.1 is based on several articles and reviews, including P. F. A. Klinkenberg, *Revs. Mod. Phys.*, **24**: 63 (1952), and N. Zeldes, *Nucl. Phys.*, **2**: 1 (1956).

The coupling rules for even-even, even-odd, and odd-even nuclei are contained in the two articles by Mayer that were mentioned last. The coupling rules for odd-odd nuclei were first formulated by L. W. Nordheim, *Phys. Rev.*, **78**: 294 (1950). The now-accepted ones are due to M. H. Brennan and A. M. Bernstein, *Phys. Rev.*, **120**: 927 (1960).

An explanation for these rules in terms of nuclear forces was proposed by C. Schwartz, *Phys. Rev.*, **94**: 95 (1954). See also A. de Shalit, *Phys. Rev.*, **91**: 1479 (1953), D. Kurath, *Phys. Rev.*, **91**: 1430 (1953), and A. de Shalit and J. D. Walecka, *Nucl. Phys.*, **22**: 184 (1961).

An appreciable fraction of the data contained in Table 8.2 was originally obtained by M. Goldhaber and R. D. Hill, *Revs. Mod. Phys.*, **24**: 179 (1952), and M. Goldhaber and A. W. Sunyar, *Phys. Rev.*, **83**: 906 (1951). The table itself is based on the data collection of Way, King, McGinnis, and van Lieshout, with a few additions.

The theories of the magnetic moments are reviewed by R. J. Blin-Stoyle, *Revs. Mod. Phys.*, **28**: 75 (1956).

Calculations on the effect of configuration interaction on magnetic and quadrupole moments are due to H. Horie and A. Arima, *Progr. Theoret. Phys.* (*Japan*), **11**: 509 (1954); *Phys. Rev.*, **99**: 778 (1955).

Section 9 is based on the work of I. Talmi and his collaborators, starting with S. Goldstein and I. Talmi, *Phys. Rev.*, **102**: 589 (1956), and I. Talmi and R. Theeberger, *Phys. Rev.*, **103**: 718 (1956). It is thoroughly reviewed in Amos de Shalit and I. Talmi, "Nuclear Shell Theory," Academic Press, New York, 1963. The second example of the text summarizes the article of K. H. Bhatt and J. B. Ball, *Nucl. Phys.*, **63**: 286 (1965).

A partial explanation of the spin-orbit coupling postulated by the *j-j* shell model was given by A. Feingold, *Phys. Rev.*, **101**: 258 (1956); **105**: 944 (1957). See also J. Keilson, *Phys. Rev.*, **82**: 759 (1951).

IX

The most significant papers on the α-particle model are by W. Wefelmeier, *Naturw.*, **25**: 525 (1937); J. A. Wheeler, *Phys., Rev.*, **52**: 1083 (1937); L. R. Hafstad and E. Teller, *Phys. Rev.*, **54**: 681 (1938); B. O. Gronblom and R. E. Marshak, *Phys. Rev.*, **55**: 229 (1939); D. M. Dennison, *Phys. Rev.*, **57**: 454 (1940); **96**: 378 (1954); H. Margenau, *Phys. Rev.*, **59**: 37 (1941); A. E. Glassgold and A. Galonsky, *Phys. Rev.*, **103**: 701 (1956); S. L. Kameny, *Phys. Rev.*, **103**: 358 (1956).

Observation of the easy deformability of the originally spherical nuclear matter was made by J. Rainwater, *Phys. Rev.*, **79**: 432 (1950). Starting from this observation, the collective model was developed in a series of articles by A. Bohr, Mottelson, and their collaborators, beginning with A. Bohr and B. R. Mottelson, *Physica*, **18**: 1066 (1952).

The rotational spectra were postulated by N. Bohr and F. Kalckar, *Danske Mat-fys. Medd.*, **14**(10): (1937), and H. A. Bethe, *Revs. Mod. Phys.*, **9**: 69 (1937). However, the modern theory of the rotational spectra begins with A. Bohr and B. R. Mottelson, *Kgl. Danske Videnskab. Selskab. Mat. Fys. Medd.*, **27**(16): (1953); *Phys. Rev.*, **90**: 717 (1953); and G. Alaga, K. Alder, A. Bohr, and B. R. Mottelson, *Kgl. Danske Videnskab. Selskab, Mat. Fys. Medd.*, **29**(9): (1955).

The existence of two regions in which the collective model has particular validity was first noted by G. Scharff-Goldhaber, *Phys. Rev.*, **103**: 837 (1956). See also G. Scharff-Goldhaber and J. Weneser, *Phys. Rev.*, **98**: 212 (1955), and K. Way, D. N. Kundu, C. L. McGinnis, and R. van Lieshout, *Annual Reviews of Nuclear Science* 6, Annual Reviews, Inc., Stanford, Calif., 1956.

K. W. Ford, *Phys. Rev.*, **95**: 1250 (1954) was the first to draw attention to the discrepancy between the magnitudes of nuclear deformations as determined from quadrupole moments and from rotational spectra. See also A. W. Sunyar, *Phys. Rev.*, **98**: 653 (1955).

Attempts to reconcile the two sets of data were initiated by D. R. Inglis, *Phys. Rev.*, **96**: 1059 (1954). See also A. Bohr and B. R. Mottelson, *Kgl. Danske Videnskab. Selskab, Mat. Fys. Medd.*, **30**(1): (1955).

The more modern theory, given in the text, was developed by R. E. Peierls and J. Yoccoz, *Proc. Phys. Soc.* (*London*), **A70**: 381 (1957). The foundation of the ideas was given by D. L. Hill and J. A. Wheeler, *Phys. Rev.*, **89**: 1102 (1953). The connection with the shell model, in case of small deformation, was elucidated principally by J. P. Elliott in several papers starting with *Proc. Roy. Soc.* (*London*), **A245**: 128 and 562 (1958). The work of Elliott covers, however, a much wider area than indicated above. Several points were clarified in the doctoral thesis of Y. Sharon (Princeton University, 1966). The model of S. G. Nilsson was given, in its initial form, in *Kgl. Danske Videnskab. Selskab., Mat. Fys. Medd.*, **29**(16): (1955). However, the number of important contributions to the subject is too large to list.

X

The best-known book on elementary scattering theory is N. F. Mott and H. S. W. Massey, "The Theory of Atomic Collisions," Oxford University Press, New York and London, 1949. M. L. Goldberger and K. M. Watson, "Collision Theory" (Wiley, New York, 1964), is an admirable book on the modern theory and its implications. Only a few of the most important contributions to the theory of nuclear collisions will be named. M. Born, *Z. Physik*, **37**: 803 (1926) is the originator of collision theory and of the "Born approximation." A generalization of this method is due to P. A. M. Dirac; see "The Principles of Quantum Mechanics," chap. VIII, Oxford University Press, New York and London, 1947. H. Faxen and J. Holtsmark, *Z. Physik*, **45**: 307 (1927), introduced the method of decomposing plane waves into spherical waves, with definite angular momenta. See also J. W. S. Rayleigh, "The Theory of Sound," vol. 2, pp. 334ff., Macmillan, London, 1926. An exact theory of scattering by an electric point charge was first given by W. Gordon, *Z. Physik*, **48**: 180 (1928).

The modern theory of proton-neutron scattering is due, in addition to Breit and his collaborators, principally to L. Landau and J. Smorodinsky, *J. Phys.* (U.S.S.R.), **8**: 154 (1944); J. S. Schwinger, *Phys. Rev.*, **72**: 742 (1947); and J. M. Blatt and J. D. Jackson, *Phys. Rev.*, **76**: 18 (1949). The variational principle was introduced by L. Hulthén, *Kgl. Fysiograf. Sallskap. Lund., Forh.*, **14**: 8, 21 (1944); *Arkiv. Mat. Astron. Fysik*, **35**(25): 1948; and by B. A. Lippmann and J. S. Schwinger, *Phys. Rev.*, **79**: 469 (1950).

The concept of the collision matrix is due to J. A. Wheeler, *Phys. Rev.*, **52**: 1107 (1937). See also P. L. Kapur and R. Peierls, *Proc. Roy. Soc.* (*London*), **A166**: 277 (1938), and W. Heisenberg, *Z. Physik*, **120**: 513, 673 (1943).

Extension of formulas (10.3) to take the electrostatic interaction into account is principally due to G. Breit and to his collaborators. The picture of the compound nucleus was given simultaneously by N. Bohr, *Nature*, **137**: 344 (1936); by E. P. Wigner and G. Breit, *Bull. Am. Phys. Soc.*, February, 1936; and G. Breit and E. P. Wigner, *Phys. Rev.*, **49**: 519 (1936). This last article gives the resonance formula.

The text derives the resonance formula by means of the so-called R-matrix theory [E. P. Wigner and L. Eisenbud, *Phys. Rev.*, **72**: 29 (1947)]. This leans heavily on earlier work, particularly on the article of P. L. Kapur and R. Peierls, mentioned above. The R-matrix theory is given only in its simplest form, for particles without spin and using the boundary condition for the internal functions which is adapted to slow neutron processes. For a more complete discussion, see A. M. Lane and R. G. Thomas, *Revs. Mod. Phys.*, **30**: 257 (1958), and also E. Vogt's article in "Nuclear Reactions," vol. 1, p. 215, North Holland Publishing Co., Amsterdam, 1959–1962. See also J. M. Blatt and V. F. Weisskopf, "Theoretical

Nuclear Physics," chaps. VIII and X, Wiley, New York, 1952, and R. G. Sachs, "Nuclear Theory," pp. 290–304 (Addison-Wesley, Reading, Mass., 1952). These two books give an excellent survey of the status of the theory of nuclear reactions in 1952.

The division of the configuration space into internal and external regions is an extraneous one, and there are many attempts, particularly by the Nordita School, to avoid it. See, for example, J. Humblet and L. Rosenfeld, *Nucl. Phys.*, **26**: 529 (1961), and C. V. Winter, *Nucl. Phys.*, **57**: 134 (1964). These theories make extended use of the radioactive states of Sec. X,5. A review of the most important papers on the subject, by A. M. Lane and D. Robson, will appear soon. It seems to us, however, that the short range of the nuclear forces and our resulting ability to calculate the wave function for the motion of nuclei as soon as their separation exceeds a rather modest distance provide important elements of information. They lead to estimates of the widely differing orders of magnitude of the partial widths (as compared with the more uniform magnitude of the reduced widths) and their dependence on the angular momentum and charge of the reacting nuclei. This renders it at least natural to subdivide the configuration space into two regions, external and internal, in the first of which we can calculate the wave function, and in the second of which, we cannot.

In addition, as the R-matrix theory shows, the wave function in the problematic, internal region does not show anomalies as a function of energy at thresholds. Since, on the other hand, the behavior of the wave function in the external region can be explicitly calculated, the discussion of the threshold anomalies can be carried out with relative ease. See, for example, E. P. Wigner, *Phys. Rev.*, **73**: 1002 (1948), and G. Breit, *Phys. Rev.*, **107**: 1612 (1957); also the article on Causality, R-matrix, and Collision Matrix in *Rend. Sc. Intern. di Fisica*, 1964, p. 40 (Academic Press, New York, 1964). Our knowledge of the behavior of the wave function down to small interparticle distances is more difficult to incorporate into theories which do not separate internal and external regions, in the latter of which the wave equation can be solved. It is for these reasons that we decided on using the R-matrix theory in this exposition.

For the effect of the electrostatic interaction, see G. Breit and J. S. McIntosh, *Phys. Rev.*, **106**: 1246 (1957).

For the first derivations of (10.14), see H. A. Bethe, *Revs. Mod. Phys.*, **9**: 69 (1937), and H. Feshbach, D. C. Peaslee, and V. F. Weisskopf, *Phys. Rev.*, **71**: 145 (1947).

Detailed development of the strong-coupling theory, in the form of the statistical theory, is due to V. F. Weisskopf and D. H. Ewing, *Phys. Rev.*, **57**: 472, 935 (1940). See also the book of Blatt and Weisskopf, cited above.

The experimental information on the average cross sections was obtained chiefly by Barschall and his collaborators. See M. Walt and H. H. Barschall, *Phys. Rev.*, **93**: 1062 (1954). See also R. K. Adair, *Revs. Mod. Phys.*, **22**: 249 (1950). He found the maxima in the average cross-section curve, which are often referred to as Barschall maxima. A maximum, predicted by the optical model, which was not known before, was found by R. Cote and L. M. Bollinger, *Phys. Rev.*, **98**: 1162 (1955). The optical model was proposed by H. Feshbach, C. Porter, and V. F. Weisskopf, *Phys. Rev.*, **96**: 448 (1954). See also R. K. Adair, *Phys. Rev.*, **94**: 737 (1954), and R. D. Woods and D. S. Saxon, *Phys. Rev.*, **95**: 377 (1954). For a more recent review, see P. E. Hodgson, Sec. 4b, in the C. R. du Congrès International de Physique Nucléaire, July, 1964 (Centre Nationale de la Recherche Scientifique, Paris, 1964).

The intermediate-coupling model was proposed simultaneously by J. M. Scott, *Phil. Mag.*, **45**: 1322 (1954), and E. P. Wigner, *Science*, **120**: 790 (1954). Its detailed development is, however, due to A. M. Lane, R. G. Thomas, and E. P. Wigner, *Phys. Rev.*, **98**: 693 (1955); E. Vogt, *Phys. Rev.*, **101**: 1792 (1956); and C. Bloch, *Nucl. Phys.*, **4**: 503 (1957).

For a discussion of the nature of the cross-section maxima in the crude optical model, see K. W. McVoy, *Phys. Letters*, **17**: 42 (1965). Doorway states were postulated by H. Feshbach, *Ann. Phys. (N.Y.)*, **5**: 357 (1958); **19**: 287 (1962). See also C. Shakin, *ibid.*, **22**: 373 (1963); B. Block and H. Feshbach, *ibid.*, **23**: 47 (1963).

For the application of the intermediate-coupling model to interpret the fine structure of analogue states, see E. P. Wigner, report on the meeting on Nuclear Spectroscopy with Direct Reactions, p. 244, Chicago, March, 1946 (ANL 6878) and C. Bloch, Sec. 3c, C. R. du Congrès International de Physique Nucléaire, previously cited; also C. Bloch and J. P. Schiffer, *Phys. Letters*, **12**: 22 (1964).

XI

For Born's collision theory, see notes to Sec. X. The theory of stripping reactions was given by S. T. Butler, *Proc. Roy. Soc. (London)*, **A208**: 559 (1951). See also S. T. Butler, "Nuclear Stripping Reaction," Wiley, New York, 1957. Similar results were obtained, independently, by F. Friedman and W. Tobocman (unpublished). The importance of (d,p) and (d,n) reactions was recognized by J. R. Oppenheimer and M. Phillips, *Phys. Rev.*, **48**: 500 (1935). The same results were obtained on the basis of Born's approximation by A. B. Bathia, Kun Huang, R. Huby, and H. C. Newns, *Phil. Mag.*, **43**: 485 (1952); and P. B. Daitch and J. A. French, *Phys., Rev.*, **87**: 900 (1952).

Conference reports are a good source of information on the theory of direct reactions and on the applications of the distorted-wave Born approximation. For earlier work on direct processes, see N. Austern and S. Yoshida, in "Proceedings of the International Conference on Nuclear Structure, August-September, 1960" (University of Toronto Press, Toronto, Canada, 1960), pp. 323 and 360. For more recent developments, see N. Austern's article in the report (ANL-6878) on the meeting Nuclear Spectroscopy with Direct Reactions, March, 1964, p. 1. The same report contains (p. 23) G. R. Satchler's review on the distorted-wave Born approximation. Much of the work on this subject is due to him and his collaborators. See also "Proceedings of the Conference on Direct Reactions, Padua, September, 1962" (Gordon and Breach, New York, 1963). This contains several articles also on heavy-ion reactions, some of which use pictures not reviewed in the text. See, for instance, the articles of G. Breit (p. 480) and of E. Almquist (p. 916).

For the early papers on Coulomb excitation, see the notes to Sec. IV. Accurate calculation of the dipole transitions was carried out by R. Huby and H. C. Newns, *Proc. Phys. Soc. (London)*, **A64**: 619 (1951), and C. T. Mullin and E. Guth, *Phys. Rev.*, **82**: 141 (1951).

The significance of the field was recognized and expressions for all multipole transitions were obtained by K. A. Ter Martirosyan, *J. Exptl.-Theoret. Phys.* (U.S.S.R.), **22**: 284 (1952); A. Bohr and B. R. Mottelson, *Kgl. Danske Videnskab. Selskab, Mat. Fys. Medd.*, **27**(3): (1953); K. Alder and A. Winther, *Phys. Rev.*, **91**: 1578 (1953); **96**: 237 (1954); and F. D. Benedict, P. B. Daitch, and G. Breit, *Phys. Rev.*, **101**: 171 (1956).

XII

Many topics covered in this section are dealt with in great detail in the 1955 edition of "Beta and Gamma Ray

Spectroscopy" and the more recent (1965) "Alpha, Beta and Gamma Ray Spectroscopy," both of which are edited by K. Siegbahn and published by North Holland Publishing Co., Amsterdam.

The quantum theory of radiation was founded by P. A. M. Dirac, *Proc. Roy. Soc.* (*London*), **A114**: 243, 710 (1927). However, the theory has been greatly generalized since. The best known comprehensive treatment is W. Heitler, "The Quantum Theory of Radiation," Oxford University Press, New York and London, 1954. The similarity between many details of Dirac's paper (e.g., choice of gauge) and the most modern treatments remains interesting. There are, of course, many excellent books on the relativistic theory of fields, including treatments of electromagnetic radiation. Steps leading to (12.2) are given in detail in Blatt and Weisskopf, "Theoretical Nuclear Physics," chap. XII, Wiley, New York, 1952.

The problem of exchange currents was raised by S. U. Condon and G. Breit, *Phys. Rev.*, **52**: 787 (1937). See also C. Moller and L. Rosenfeld, *Kgl. Danske Videnskab. Selskab, Mat. Fys. Medd.*, **20**(12): 1943. These articles are based on the field theory of nuclear forces. The phenomenological approach was proposed by R. G. Sachs and N. Austern, *Phys. Rev.*, **81**: 705 (1951). See also R. K. Osborn and L. L. Foldy, *Phys. Rev.*, **79**: 795 (1951); F. Villars, *Phys. Rev.*, **86**: 476 (1952); and R. G. Sachs, "Nuclear Theory," chap. 9, Addison Wesley, Cambridge, Mass., 1953, which gives a comprehensive account of the phenomenological point of view. For the absence of exchange currents in electric radiation, see A. S. F. Siegert, *Phys. Rev.*, **52**: 787 (1937).

The units (12.5) were established by V. F. Weisskopf, *Phys. Rev.*, **83**: 1073 (1951). See also J. M. Blatt and V. F. Weisskopf, "Theoretical Nuclear Physics," cited above.

Pair production in O^{16} was first treated by J. R. Oppenheimer, *Phys. Rev.*, **60**: 164 (1941). For a more complete discussion, R. H. Dalitz, *Proc. Roy. Soc.* (*London*), **A206**: 521 (1951).

Methods of detection of γ quanta are discussed, for instance, in "Beta and Gamma Ray Spectroscopy," edited by K. Siegbahn, chaps. IV–VII, North Holland Publishing Co., Amsterdam, 1955. The coefficient of internal conversion plays an important role in determination of the multipolarity of radiation. Early work of H. R. Hulme and of his collaborators is summarized in Gamow's book (see notes to Sec. IV). The most elaborate calculations are due to M. E. Rose. See Appendix IV of "Beta and Gamma Ray Spectroscopy," cited above. For the effect of the finite size of the nucleus, see L. A. Sliv, *J. Exptl.-Theoret. Phys.* (*U.S.S.R.*), **21**: 77 (1951); and L. A. Sliv and M. A. Listengarten, **22**: 29 (1952). F. K. McGowan and P. H. Stelson, *Phys. Rev.*, **103**: 133 (1956), and Appendix 5 by L. A. Sliv and I. M. Band of "Alpha, Beta and Gamma Ray Spectroscopy" cited above.

The angular correlation between successively emitted gamma rays was first pointed out by J. W. Dunworth, *Rev. Sci. Instr.*, **11**: 167 (1940). The first calculation of the directional correlation was carried out by D. R. Hamilton, *Phys. Rev.*, **58**: 122 (1940). More elaborate and general calculations have been carried out by many authors; for example, L. C. Biedenharn and M. E. Rose, *Rev. Mod. Phys.*, **25**: 729 (1953) and S. Devons and L. J. B. Goldfarb, "Handbuch der Physik," vol. XLII, S. Flügge (ed.), Springer-Verlag, Berlin, 1957. The first observation of a γ-γ correlation was by E. L. Brady and M. Deutsch, *Phys. Rev.*, **72**: 870 (1947).

The experimental data on γ-ray lifetimes have been reviewed recently by D. H. Wilkinson "Nuclear Spectroscopy, Part B" cited in Section VI, and by M. Goldhaber and A. W. Sunyar in Chap. XVIII of "Alpha, Beta and Gamma Ray Spectroscopy," cited above. A recent compilation of electric quadrupole transition probabilities by P. H. Stelson and L. Grodzins, *Nucl. Data*, **1**: 21 (1965) is most useful.

For the foundation of the theory of β decay see the references to Sec. I.3. For more recent reviews see E. J. Konopinski, "The Theory of Beta Radioactivity," Oxford University Press, New York, 1966, and C. S. Wu and S. A. Moszkowski, "Beta Decay," Interscience Publishers, New York, London and Sydney, 1966.

G. Gamow and E. Teller, *Phys. Rev.*, **49**: 895 (1936) were the first to suggest modifications of the form of the β decay interaction proposed by Fermi. The five invariant forms were given by H. A. Bethe and R. F. Bacher, *Rev. Mod. Phys.*, **8**: 32 (1936). This article summarizes the thinking on nuclear structure and reactions through 1935 in a most admirable fashion. H. Yukawa and S. Sakata, *Proc. Phys. Math. Soc.* (*Japan*), **17**: 467 (1935) extended Fermi's theory for the process of electron capture.

Beta transitions of various degrees of forbiddenness were classified by G. Uhlenbeck and E. J. Konopinski, *Phys. Rev.*, **60**: 308 (1941).

The shape of the allowed spectra was calculated by Fermi. Early experiments gave conflicting results for the shape; the discrepancies were cleared up principally by A. W. Taylor, *Phys. Rev.*, **56**: 125 (1939) and J. L. Lawson, *Phys. Rev.*, **56**: 131 (1939). For the accuracy with which the Beta spectrum agrees with theory see C. S. Wu's article in "Alpha, Beta and Gamma Ray Spectroscopy" and her article in *Rev. Mod. Phys.*, **36**: 618 (1964). Forbidden shapes were definitely identified first by L. M. Langer and H. C. Price, *Phys. Rev.*, **75**: 1109 and **76**: 641 (1949); P. R. Bell and J. M. Cassidy, *Phys. Rev.*, **76**: 183 (1949); C. L. Peacock and A. C. G. Mitchell, *Phys. Rev.*, **75**: 1273 (1949); C. S. Wu and L. Feldman, *Phys. Rev.*, **76**: 698 (1949).

The importance of the angular correlation between the directions of the emitted neutrons and beta particle was pointed out by D. R. Hamilton, *Phys. Rev.*, **71**: 456 (1947).

The inverse beta decay process was observed by F. Reines, C. L. Cowan, and coworkers in 1956. The current status of the pertinent experiments is summarized in an article by F. Reines in "Alpha, Beta and Gamma Ray Spectroscopy" cited above.

The modern theory of beta decay starts with the considerations of T. D. Lee and C. N. Yang, *Phys. Rev.*, **104**: 254 (1956) and T. D. Lee, C. N. Yang, and R. Oehme, *Phys. Rev.*, **106**: 340 (1957), that parity might not be concerned in the weak interactions. This was subsequently confirmed experimentally by measurements of the β decay from aligned Co^{60} nuclei by C. S. Wu, E. Ambler, R. W. Hayward, D. D. Hopper, and R. P. Hudson, *Phys. Rev.*, **105**: 1413 (1957).

The vector-axial vector theory of beta decay was first proposed by R. P. Feynman and M. Gell-Mann, *Phys. Rev.*, **109**: 193 (1958), E. Sudershan and R. Marshak, *Phys. Rev.*, **109**: 1860 (1958) and J. J. Sakurai, *Nuova Cimento*, **7**: 649 (1958). The experimental evidence that the beta decay interaction is vector-axial vector rather than scalar tensor is most convincingly furnished in the previously mentioned article of Wu et al., and was confirmed by an ingeneous experiment of M. Goldhaber, L. Grodzins, and A. W. Sunyar, *Phys. Rev.*, **109**: 1015 (1958).

The values for the coupling constants are reviewed in an article by O. Kofoed-Hansen in "Alpha, Beta and Gamma Ray Spectroscopy" cited above and in a recent letter by J. M. Freeman, J. G. Jenkin, G. Murray, and W. E. Burcham, *Phys. Rev. Letters*, **16**: 959 (1966) which contains the most accurately measured values for the vector coupling constant.

Chapter 2

Nuclear Masses

By WALTER H. JOHNSON, JR., *and* A. O. NIER,
University of Minnesota

Masses of atomic nuclei are less, by a measurable amount, than the sum of masses of the constituent protons and neutrons. Knowledge of nuclear masses is therefore useful in predicting, through the equivalence of mass and energy, the energy release in nuclear reactions and also in obtaining information regarding the nature of nuclear forces.

In practice, nuclear masses cannot be measured directly, and one deals instead with atomic masses. To obtain a nuclear mass, one must subtract from the atomic mass the sum of the masses of all the atomic electrons and the mass equivalent of the total electronic binding energy. This can be done precisely only where sufficient information concerning electron binding energies is available. Because the electron binding energy is small compared with the atomic binding energy, atomic masses may be employed for most calculations of nuclear reaction energies, β-decay energies, and nucleon binding energies.

Relative masses can be measured far more precisely than absolute masses. For this reason, all mass tables are based on a standard mass, and other masses are determined relative to the standard. A unified scale for atomic masses has been adopted by physicists and chemists in which the neutral C^{12} atom is assigned the mass of exactly 12 units. The symbol for masses in this mass scale is u.

1. Nuclear Transformations and Atomic Masses

Nuclear transformation energies can be employed through the equivalence of mass and energy ($E = Mc^2$) to verify directly measured atomic masses, to calculate radioactive atomic masses, and to obtain a number of stable atomic masses that have not been directly determined. Several examples are:

1. *β-decay.* The basic process involved in β^- decay is the conversion of a neutron to a proton and an electron. Any β^- decay can then be formed by adding a sufficient number of nucleons to the equation for this basic process. Consider the decay of Ne^{23} to Na^{23}:

$$\mathrm{Ne}^{23} \rightarrow \mathrm{Na}^{23} + \beta^- + E_{\beta^-} \tag{2.1}$$

where E_{β^-} is the energy released. Except for β decay in the light nuclei, this energy is the maximum energy of the emitted β^--particle energy spectrum. In the light nuclei, recoil energy of the daughter nucleus may not be negligible. From conservation of energy,

$$_{10}M_n{}^{23}c^2 = {}_{11}M_n{}^{23}c^2 + mc^2 + E_{\beta^-} \tag{2.2}$$

in which $_ZM_n{}^A$ is the mass of the nucleus having Z protons and A nucleons (that is, $A-Z$ neutrons) and m is the mass of the electron. One now adds to each side of this equation the energy equivalent of 10 electron masses. The left-hand side of the equation is then the energy equivalent of the atomic mass of Ne^{23}. Similarly, the right-hand side becomes the energy equivalent of the atomic mass of Na^{23}.

$$_{10}M_n{}^{23}c^2 + 10mc^2 = {}_{11}M_n{}^{23}c^2 + 11mc^2 + E_{\beta^-} \tag{2.3}$$

or

$$_{10}M_a{}^{23} = {}_{11}M_a{}^{23} + \frac{E_{\beta^-}}{c^2} \tag{2.4}$$

In these equations M_a is the atomic mass. This procedure neglects the nearly equal electron binding energies of Na^{23} and Ne^{23}. Thus β^- energies can be employed to determine atomic-mass differences, and, in the few cases where mass measurements of radioactive nuclei have been made, directly determined mass differences can be compared with β^--decay energies.

The energy released in the Ne^{23} decay is 4.39 ± 0.05 Mev [1].† Applying the conversion factor 1 u = 931.478 ± 0.015 Mev [2], this decay energy yields a Ne^{23}–Na^{23} mass difference of 0.004 71 ± 5 u.‡

2. *β^+ decay.* The decay of Ne^{19} to F^{19} is chosen.

$$_{10}\mathrm{Ne}^{19} \rightarrow {}_9\mathrm{F}^{19} + \beta^+ + E_{\beta^+} \tag{2.5}$$

From conservation of energy,

$$_{10}M_n{}^{19}c^2 = {}_9M_n{}^{19}c^2 + mc^2 + E_{\beta^+} \tag{2.6}$$

By adding $10mc^2$ to each side, one can convert to atomic masses:

$$_{10}M_n{}^{19}c^2 + 10mc^2 = {}_9M_n{}^{19}c^2 + 11mc^2 + E_{\beta^+} \tag{2.7}$$

or

$$_{10}M_a{}^{19}c^2 = {}_9M_a{}^{19}c^2 + 2mc^2 + E_{\beta^+} \tag{2.8}$$

† Numbers in brackets refer to References at end of chapter.

‡ Unless otherwise indicated, the error refers to the last place of the quoted figure.

Again the electronic binding energies may be neglected. In this case the energy equivalent of the atomic-mass difference is the energy released plus $2mc^2$. The β^+ energy for the Ne^{19} decay is 2.24 ± 0.01 Mev [3]. Adding $2mc^2$ (1.02 Mev) and converting the result to mass units yield a Ne^{19}–F^{19} mass difference of 0.003 50 ± 1 u.

3. *Nuclear reactions.* Reaction Q values may be employed to calculate the atomic-mass difference between the target and the residual nucleus. Electronic binding-energy differences may be neglected in most such reaction calculations. The reaction $O^{16}(d,p)O^{17}$ is chosen as an illustration of a reaction calculation.

$$O^{16} + d \rightarrow O^{17} + p + Q \tag{2.9}$$

From conservation of energy,

$$_8M_a{}^{16}c^2 + {}_1M_a{}^2c^2 = {}_8M_a{}^{17}c^2 + {}_1M_a{}^1c^2 + Q \tag{2.10}$$

or

$$_8M_a{}^{17} - {}_8M_a{}^{16} = {}_1M_a{}^2 - {}_1M_a{}^1 - \frac{Q}{c^2} \tag{2.11}$$

The Q value for this reaction is 1.920 ± 3 Mev [4]. By substituting the deuterium-minus-hydrogen mass difference of 1.006 277 u† into this equation, one obtains 1.004 216 ± 3 u.

Such comparisons furnish verification of both the experimental mass determinations and the nuclear energy measurements. In some instances it is difficult to assign reaction energies to particular isotopes of a multi-isotope element. In such cases the mass-spectroscopically determined mass differences between known isotopes can be compared with the unassigned reaction mass differences. Agreement suggests the proper assignment.

2. Atomic Masses from Mass Spectroscopy

Relative atomic-mass measurements are made with instruments which employ, in some manner, the interaction of an ion with electrostatic and magnetic fields A common type of instrument employs a combination of electrostatic and magnetic sector fields; it has the property that ions of different mass in the incident ion beam focus at different spatial positions on the instrument's focal plane, producing a mass spectrum. The electrostatic and magnetic fields are often combined to produce double focusing; that is, all ions of the same mass that enter the instrument within a small divergence angle and that have an energy different from the mean energy by a small amount will focus. If a single slit is placed at the focal point and is followed by a suitable ion detector (e.g., an electrometer-tube amplifier or electron multiplier) a mass spectrum may be obtained by plotting the collected current as a function of the electric or magnetic field in the instrument. Instruments using this detection system for mass measurements were first employed by Nier and his coworkers [5]. Oscilloscopic display of the variation of ion current with a continuous change in the electric or magnetic field provides a means for readily focusing mass spectrometers and also can be employed for a rapid and accurate mass measurement. The use of this technique in mass measurement, known as peak matching, was introduced by Smith [6]. If a photographic plate is placed in the focal plane, a mass spectrum appears on the plate, and by a suitable calibration procedure, distance along the plate may be correlated with mass. Many early measurements employed this technique.

Time-of-flight instruments have also been employed for mass measurement. The best example is the mass synchrometer designed by Lincoln Smith [7]. Mass measurements in the light-mass region reported by Smith are among the most accurate published. Excellent reviews describing mass spectroscopes and mass-spectroscopic techniques are given by Bainbridge [8], Ewald and Hintenberger [9], Duckworth [10], and McDowell [11].

Modern mass-spectroscopic measurements employ the "mass doublet" technique. One determines the difference in mass between ions of the same mass number having slightly different masses. Because there is only a small mass difference between the two kinds of ions, the instrument dispersion need not be known as accurately as would be the case if the masses differ widely. To employ the doublet method in calculating atomic masses, the mass of one of the doublet members must be accurately known. Hydrocarbon compounds

† Derived from mass values in Table 2.1.

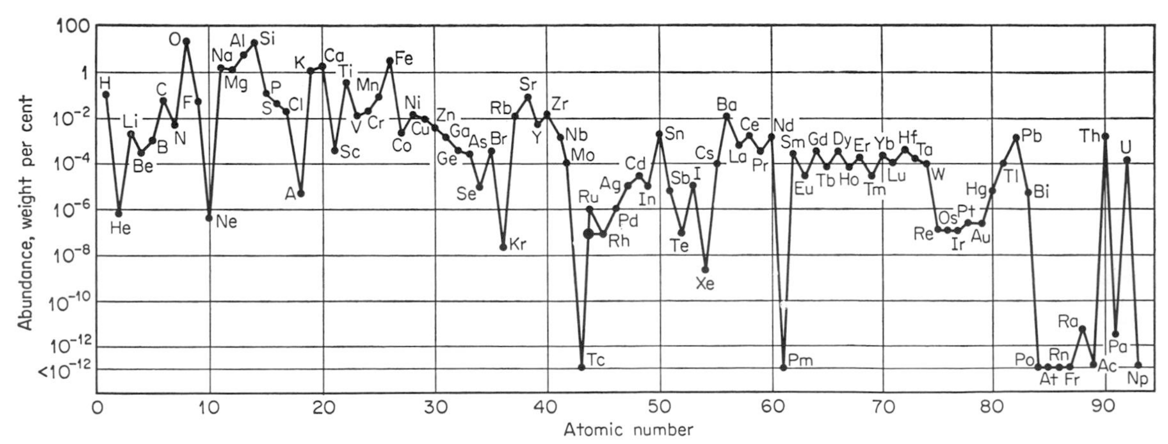

FIG. 2.1. Crustal abundances of elements of atomic numbers 1 to 93. (*From Brian Mason, "Principles of Geochemistry," Wiley, New York,* 1958.)

TABLE 2.1. MASS TABLE

In the column labeled $M(ZA)$ is listed the atomic mass of each indicated isotope. Listed opposite the name of the element in the same column is the atomic weight of the element. In the abundance column, the relative abundances of stable isotopes are listed as fractions which sum to 1.000 for each element. The decay path is indicated for radioactive isotopes in this column. Tabulated opposite the element name in the abundance column is the relative abundance of the element in terrestrial igneous rock in the scale that sets the number of silicon atoms equal to 100.

Z	A	$M(ZA)$ in u[a]	Error in μu	Abundance[b]
$_{0}n$	Neutron			
	1	1.008 665 20	0.10	β^-
$_{1}$H	Hydrogen	1.007 97[c]		
	1	1.007 825 19	0.08	0.999850
	2	2.014 102 22	0.12	149.2×10^{-6}
	3	3.016 049 71	0.21	β^-
$_{2}$He	Helium	4.002 6		7.6×10^{-6}
	3	3.016 029 73	0.21	137×10^{-8}
	4	4.002 603 12	0.42	0.99999863
	5	5.012 297	20	$n\alpha$
	6	6.018 892 7	4.3	β^-
$_{3}$Li	Lithium	6.939		
	5	5.012 538	40	$p\alpha$
	6	6.015 124 7	1.2	0.075632[d]
	7	7.016 003 9	1.2	0.924368[d]
	8	8.022 487 1	1.6	β^-
	9	9.026 802	22	β^-
$_{4}$Be	Beryllium	9.012 2		6.7×10^{-3}
	6	6.019 717	13	p
	7	7.016 928 9	1.2	EC
	8	8.005 307 9	0.9	2α
	9	9.012 185 5	1.0	1.00
	10	10.013 534 4	2.4	β^-
	11	11.021 666	16	β^-
$_{5}$B	Boron	10.811[e]		2.8×10^{-3}
	8	8.024 609 3	1.6	β^+
	9	9.013 332 2	1.4	p, 2α
	10	10.012 938 8	0.5	0.1961
	11	11.009 305 30	0.32	0.8039
	12	12.014 353 7	1.4	β^-
	13	13.017 780 0	4.3	β^-
$_{6}$C	Carbon	12.011 15[f]		0.27
	10	10.016 810	14	β^+
	11	11.011 431 7	1.2	β^+
	12	12.000 000 000	Standard	0.98893
	13	13.003 354 4	0.9	0.01107
	14	14.003 241 97	0.32	β^-
	15	15.010 599 5	0.9	β^-
	16	16.014 700	17	β^-
$_{7}$N	Nitrogen	14.006 7		0.033
	12	12.018 641	8	β^+
	13	13.005 738 4	1.2	β^+
	14	14.003 074 39	0.17	0.996337
	15	15.000 107 7	0.9	0.003663
	16	16.006 103 3	3.8	β^-
	17	17.008 450	16	β^-
$_{8}$O	Oxygen	15.999 4[g]		296
	14	14.008 597 09	0.45	β^+
	15	15.003 070 3	1.3	β^+
	16	15.994 915 02	0.28	0.99759
	17	16.999 132 9	1.0	0.000374
	18	17.999 160 02	0.36	0.002039
	19	19.003 577 9	3.1	β^-
	20	20.004 079	9	β^-

TABLE 2.1. MASS TABLE (*Continued*)

Z	A	$M(ZA)$ in u[a]	Error in μu	Abundance[b]	
$_{9}$F	Fluorine	18.998 4		0.32 to 0.48	
	16	16.011 706	13		β^+
	17	17.002 095 5	0.5		β^+
	18	18.000 936 6	0.8		β^+, EC
	19	18.998 404 6	0.8	1.000	
	20	19.999 987	5		β^-
	21	20.999 951	8		β^-
$_{10}$Ne	Neon	20.183			
	18	18.005 711	5		β^+
	19	19.001 880 9	1.7		β^+
	20	19.992 440 5	0.5	0.9092	
	21	20.993 848 6	1.6	0.00257	
	22	21.991 384 7	0.6	0.0882	
	23	22.994 472 9	3.6		β^-
	24	23.993 613	10		β^-
$_{11}$Na	Sodium	22.989 8			12.4
	20	20.008 880	320		β^+
	21	20.997 655	9		β^+
	22	21.994 436 6	2.9		β^+, EC
	23	22.989 770 7	2.0	1.000	
	24	23.990 962 3	3.5		β^-
	25	24.989 955	9		β^-
	26	25.991 740	320		β^-
$_{12}$Mg	Magnesium	24.312		8.76	
	22	21.999 850	90		2p
	23	22.994 125 0	3.1		β^+
	24	23.985 041 7	1.9	0.7870	
	25	24.985 839 0	2.0	0.1013	
	26	25.982 593 0	2.0	0.1117	
	27	26.984 344 7	4.0		β^-
	28	27.983 875	6		β^-
$_{13}$Al	Aluminum	26.981 5			30.5
	24	24.000 100	100		β^+
	25	24.990 412	7		β^+
	26	25.986 890 9	2.4		β^+, EC
	27	26.981 538 9	1.9	1.000	
	28	27.981 904 7	4.0		β^-
	29	28.980 442	7		β^-
	30	29.981 590	270		β^-
$_{14}$Si	Silicon	28.086[h]			100
	26	25.992 343	14		β^+
	27	26.986 702 8	2.8		β^+
	28	27.976 929 2	3.0	0.9221	
	29	28.976 495 8	4.0	0.0470	
	30	29.973 762 8	4.0	0.0309	
	31	30.975 349	6		β^-
	32	31.974 020	50		β^-
$_{15}$P	Phosphorus	30.973 8		0.38	
	28	27.991 780	300		β^+
	29	28.981 808	6		β^+
	30	29.978 317	8		β^+
	31	30.973 764 7	1.5	1.000	
	32	31.973 909 5	2.3		β^-
	33	32.971 728 2	3.7		β^-
	34	33.973 340	210		β^-
$_{16}$S	Sulfur	32.064[i]		0.16	
	30	29.984 873	29		β^+
	31	30.979 611	12		β^+
	32	31.972 073 7	0.9	0.950	
	33	32.971 461 9	3.0	0.00760	

Table 2.1. Mass Table (*Continued*)

Z	A	$M(ZA)$ in u[a]	Error in μu	Abundance[b]	
	34	33.967 864 6	2.9	0.0422	
	35	34.969 030 8	1.3		β^-
	36	35.967 090	9	0.000136	
	37	36.971 010	80		β^-
	38	37.971 230	160		β^-
$_{17}$Cl	Chlorine	35.453[j]		0.09	
	32	31.986 240	410		$\beta^+\alpha$
	33	32.977 440	13		β^+
	34	33.973 750	6		β^+
	35	34.968 851 1	1.3	0.75770[k]	
	36	35.968 308 9	4.4		β^-, EC
	37	36.965 898 5	1.1	0.24229[k]	
	38	37.968 005	9		β^-
	39	38.968 008	20		β^-
	40	39.970 400	500		β^-
$_{18}$Ar	Argon	39.948			
	35	34.975 254	18		β^+
	36	35.967 544 5	2.4	0.00337	
	37	36.966 772 2	1.3		EC
	38	37.962 727 8	2.7	0.00063	
	39	38.964 317	6		β^-
	40	39.962 384 2	0.8	0.99600	
	41	40.964 500	5		β^-
	42	41.963 048	43		β^-
$_{19}$K	Potassium	39.102		4.42	
	37	36.973 365	48		β^+
	38	37.969 097	11		β^+
	39	38.963 710 1	2.8	0.9310	
	40	39.963 999 8	1.3	0.0001181	
	41	40.961 832 3	3.8	0.0688	
	42	41.962 406	11		β^-
	43	42.960 730	12		β^-
	44	43.962 040	210		β^-
	45	44.960 680	210		β^-
$_{20}$Ca	Calcium	40.08		9.17	
	39	38.970 691	25		β^+
	40	39.962 588 9	3.5	0.9697	
	41	40.962 275	8		EC
	42	41.958 625 2	3.8	0.0064	
	43	42.958 779 6	4.2	0.00145	
	44	43.955 490 5	4.4	0.0206	
	45	44.956 189 5	3.8		β^-
	46	45.953 689	10	0.000033	
	47	46.954 538	6		β^-
	48	47.952 531	10	0.00185	
	49	48.955 675	12		β^-
$_{21}$Sc	Scandium	44.956			
	40	39.977 570	210		β^+
	41	40.969 247	10		β^+
	42	41.965 495	13		β^+
	43	42.961 165	9		β^+
	44	43.959 406	6		β^+, EC
	45	44.955 918 9	3.3	1.000	
	46	45.955 172 6	4.0		β^-
	47	46.952 412 9	3.4		β^-
	48	47.952 221	8		β^-
	49	48.950 026	6		β^-
	50	49.951 730	210		β^-
$_{22}$Ti	Titanium	47.90		0.92	
	43	42.968 500	160		β^+
	44	43.959 572	13		EC

TABLE 2.1. MASS TABLE (*Continued*)

Z	A	$M(ZA)$ in u[a]	Error in μu	Abundance[b]	
	45	44.958 129	5		β^+, EC
	46	45.952 631 6	2.4	0.0793	
	47	46.951 768 5	2.7	0.0728	
	48	47.947 950 3	2.1	0.7394	
	49	48.947 870 3	2.1	0.0551	
	50	49.944 785 9	3.5	0.0534	
	51	50.946 603	7		β^-
$_{23}$V	Vanadium	50.942		0.03	
	46	45.960 214	10		β^+
	47	46.954 899	9		β^+, EC
	48	47.952 258 7	3.6		β^+, EC
	49	48.948 522 5	4.8		EC
	50	49.947 163 8	3.5	0.0024	
	51	50.943 961 2	2.5	0.9976	
	52	51.944 780	5		β^-
	53	52.943 980	1,070		β^-
	54	53.946 720	1,070		β^-
$_{24}$Cr	Chromium	51.996[l]		0.039	
	48	47.953 760	210		EC
	49	48.951 271	12		β^+
	50	49.946 054 5	3.7	0.0435[m]	
	51	50.944 768 2	2.7		EC
	52	51.940 513 1	3.2	0.8376[m]	
	53	52.940 652 7	3.2	0.0951[m]	
	54	53.938 881 5	4.0	0.0238[m]	
	55	54.940 833	7		β^-
	56	55.940 640	160		β^-
$_{25}$Mn	Manganese	54.9380		0.18	
	50	49.954 215	29		β^+
	51	50.948 190	50		β^+
	52	51.945 568	6		β^+, EC
	53	52.941 295	7		EC
	54	53.940 362	6		EC
	55	54.938 050 3	3.5	1.000	
	56	55.938 910 2	4.6		β^-
	57	56.938 300	320		β^-
	58	57.940 260	1,070		β^-
$_{26}$Fe	Iron	55.847[n]		9.13	
	52	51.948 117	14		β^+, EC
	53	52.945 572	48		β^+
	54	53.939 617	5	0.0582	
	55	54.938 298 6	3.7		EC
	56	55.934 936 3	4.3	0.9166	
	57	56.935 397 8	4.5	0.0219	
	58	57.933 282	5	0.0033	
	59	58.934 877 8	4.6		β^-
	60	59.933 964	33		β^-
	61	60.936 520	1,070		β^-
$_{27}$Co	Cobalt	58.9332		4×10^{-3}	
	54	53.948 475	7		β^+
	55	54.942 013	11		β^+, EC
	56	55.939 847	8		β^+, EC
	57	56.936 296	5		EC
	58	57.935 761	6		β^+, EC
	59	58.933 189 3	3.8	1.000	
	60	59.933 813 4	4.8		β^-
	61	60.932 440	43		β^-
	62	61.933 946	43		β^-
	63	62.933 530	210		β^-
$_{28}$Ni	Nickel	58.71		0.014	
	56	55.942 116	16		EC

TABLE 2.1. MASS TABLE (*Continued*)

Z	A	$M(ZA)$ in u[a]	Error in μu	Abundance[b]	
	57	56.939 769	17		β^+, EC
	58	57.935 342	5	0.6788	
	59	58.934 342 3	4.3		EC
	60	59.930 787	5	0.2623	
	61	60.931 056	7	0.0119	
	62	61.928 342	5	0.0366	
	63	62.929 664	5		β^-
	64	63.927 958	6	0.0108	
	65	64.930 072	8		β^-
	66	65.929 085	33		β^-
$_{29}$Cu	Copper	63.54		0.011	
	58	57.944 541	8		β^+
	59	58.939 496	22		β^+
	60	59.937 362	9		β^+, EC
	61	60.933 457	7		β^+, EC
	62	61.932 566	11		β^+
	63	62.929 592	5	0.691739[c]	
	64	63.929 759	5		β^-, β^+, EC
	65	64.927 786	6	0.308261[c]	
	66	65.928 871	9		β^-
	67	66.927 759	13		β^-
	68	67.929 770	60		β^-
$_{30}$Zn	Zinc	65.37		0.02	
	61	60.939 250	210		β^+
	62	61.934 380	14		β^+, EC
	63	62.933 206	6		β^+, EC
	64	63.929 145	5	0.4889	
	65	64.929 234	6		β^+, EC
	66	65.926 052	6	0.2781	
	67	66.927 145	10	0.0411	
	68	67.924 857	6	0.1857	
	69	68.926 541	7		β^-
	70	69.925 334	6	0.0062	
	71	70.927 510	50		β^-
	72	71.926 843	10		β^-
$_{31}$Ga	Gallium	69.72		2.2×10^{-3}	
	64	63.936 737	33		β^+
	65	64.932 733	17		β^+, EC
	66	65.931 607	7		β^+, EC
	67	66.928 216	11		β^+, EC
	68	67.927 992	7		β^+, EC
	69	68.925 574 0	3.7	0.604	
	70	69.926 035	6		β^-
	71	70.924 706 0	4.6	0.396	
	72	71.926 372	7		β^-
	73	72.925 126	43		β^-
	74	73.927 190	50		β^-
$_{32}$Ge	Germanium	72.59		9.5×10^{-4}	
	65	64.939 600	1,070		β^+
	66	65.934 800	160		β^+, EC
	67	66.932 940	110		β^+, EC
	68	67.928 530	1,070		EC
	69	68.927 963 2	4.5		β^+, EC
	70	69.924 251 5	1.8	0.2052	
	71	70.924 956	6		EC
	72	71.922 081 8	1.7	0.2743	
	73	72.923 462 5	1.8	0.0776	
	74	73.921 180 6	1.7	0.3654	
	75	74.922 883	20		β^-
	76	75.921 405 2	2.0	0.0776	
	77	76.923 600	50		β^-

TABLE 2.1. MASS TABLE (*Continued*)

Z	A	$M(ZA)$ in u[a]	Error in μu	Abundance[b]
$_{33}$As	Arsenic	74.9216		6.7×10^{-4}
	69	68.932 150	320	β^+
	70	69.930 946	32	β^+, EC
	71	70.927 113	9	β^+, EC
	72	71.926 763	11	β^+, EC
	73	72.923 861	32	EC
	74	73.923 932 7	4.1	β^+, EC, β^-
	75	74.921 596 4	3.9	1.000
	76	75.922 397	12	β^-
	77	76.920 646	11	β^-
	78	77.921 900	210	β^-
	79	78.920 890	60	β^-
	80	79.922 970	210	β^-
$_{34}$Se	Selenium	78.96		
	71	70.931 840	320	β^+
	72	71.927 410	1,070	EC
	73	72.926 814	34	β^+
	74	73.922 476	5	0.0087
	75	74.922 524 9	4.3	EC
	76	75.919 207	7	0.0902
	77	76.919 911	5	0.0758
	78	77.917 313 7	2.6	0.2352
	79	78.918 494 3	4.7	β^-
	80	79.916 527 3	2.9	0.4982
	81	80.917 984	7	β^-
	82	81.916 707	7	0.0919
$_{35}$Br	Bromine	79.909[p]		2.0×10^{-4}
	74	73.929 780	1,070	β^+
	75	74.925 447	22	β^+, EC
	76	75.924 180	60	β^+, EC
	77	76.921 376	6	β^+, EC
	78	77.921 150	6	β^+
	79	78.918 329 1	3.3	0.506864[q]
	80	79.918 535 7	3.6	β^+, EC, β^-
	81	80.916 292	5	0.493136[q]
	82	81.916 802	5	β^-
	83	82.915 168	17	β^-
	84	83.916 550	50	β^-
	85	84.915 530	110	β^-
	86	85.918 200	500	β^-
$_{36}$Kr	Krypton	83.80		
	74	73.933 100	1,520	β^+
	75	74.930 920	1,070	?
	76	75.925 470	1,080	EC
	77	76.924 480	90	β^+, EC
	78	77.920 403	5	0.00354
	79	78.920 068	6	β^+, EC
	80	79.916 380	6	0.0227
	81	80.916 610	110	EC
	82	81.913 482	5	0.1156
	83	82.914 131 4	4.8	0.1155
	84	83.911 503 4	3.5	0.5690
	85	84.912 523	7	β^-
	86	85.910 615 9	4.2	0.1737
	87	86.913 365	10	β^-
	88	87.914 270	240	β^-
	89	88.916 600	500	β^-
	90	89.919 720	110	β^-
$_{37}$Rb	Rubidium	85.47		0.036
	80	79.921 900	600	β^+
	81	80.919 020	110	β^+, EC
	82	81.917 959	33	β^+
	83	82.914 730	1,070	EC

TABLE 2.1. MASS TABLE (*Continued*)

Z	A	$M(ZA)$ in u[a]	Error in μu	Abundance[b]	
	84	83.914 380 7	4.6		β^+, EC, β^-
	85	84.911 800	5	0.7215	
	86	85.911 193	7		β^-
	87	86.909 186 5	3.3	0.2785	
	88	87.911 270	100		β^-
	89	88.911 650	50		β^-
	90	89.914 820	110		β^-
	91	90.916 070	1,070		β^-
	92	91.919 140	1,080		β^-
$_{38}$Sr	Strontium	87.62		0.035	
	82	81.918 390	1,070		EC
	83	82.917 200	1,520		β^+, EC
	84	83.913 430 1	3.9	0.0056	
	85	84.912 989	33		EC
	86	85.909 285	5	0.0986	
	87	86.908 892 2	3.5	0.0702	
	88	87.905 641	6	0.8256	
	89	88.907 442	7		β^-
	90	89.907 747	9		β^-
	91	90.910 161	16		β^-
	92	91.910 980	80		β^-
	93	92.914 710	110		β^-
	94	93.915 380	240		β^-
$_{39}$Y	Yttrium	88.905		3.01×10^{-3}	
	84	83.920 190	110		β^+
	85	84.916 489	34		β^+
	86	85.914 946	18		β^+, EC
	87	86.910 740	210		β^+, EC
	88	87.909 528	8		β^+, EC
	89	88.905 871 9	4.8	1.000	
	90	89.907 163	8		β^-
	91	90.907 295	12		β^-
	92	91.908 926	22		β^-
	93	92.909 552	22		β^-
	94	93.911 680	210		β^-
	95	94.912 540	1,070		β^-
	96	95.915 690	1,070		β^-
$_{40}$Zr	Zirconium	91.22		0.026	
	86	85.916 230	1,070		EC
	87	86.914 490	220		β^+, EC
	88	87.910 060	1,070		EC
	89	88.908 914	6		β^+, EC
	90	89.904 699 6	4.1	0.5146	
	91	90.905 642	5	0.1123	
	92	91.905 030 9	3.5	0.1711	
	93	92.906 450	5		β^-
	94	93.906 313 4	3.7	0.1740	
	95	94.908 035	5		β^-
	96	95.908 286	5	0.0280	
	97	96.910 966	23		β^-
	98	97.911 960	1,520		?
$_{41}$Nb	Niobium	92.906		2.6×10^{-3}	
	89	88.913 080	100		β^+
	90	89.911 259	11		β^+
	91	90.906 860	70		EC
	92	91.907 211	10		EC
	93	92.906 382	5	1.000	
	94	93.907 303	15		β^-
	95	94.906 831 8	3.3		β^-
	96	95.908 056	27		β^-
	97	96.908 096	8		β^-
	98	97.910 350	1,070		β^-
	99	98.911 050	1,070		β^-
	100	99.914 020	1,070		β^-

TABLE 2.1. MASS TABLE (*Continued*)

Z	A	$M(ZA)$ in u[a]	Error in μu	Abundance[b]	
$_{42}$Mo	Molybdenum	95.94			
	90	89.913 940	110		β^+, EC
	91	90.911 650	60		β^+
	92	91.906 810 1	3.4	0.1584	
	93	92.906 830	14		EC
	94	93.905 090 1	2.9	0.0904	
	95	94.905 839 0	3.2	0.1572	
	96	95.904 673 8	2.6	0.1653	
	97	96.906 021 5	2.9	0.0946	
	98	97.905 408 8	2.8	0.2378	
	99	98.907 720	10		β^-
	100	99.907 474 7	3.7	0.0963	
	101	100.910 353	20		β^-
	102	101.910 250	1,520		β^-
$_{43}$Tc	Technetium				
	92	91.915 460	150		β^-
	93	92.910 251	20		β^+, EC
	94	93.909 663	7		β^-, EC
	95	94.907 620	23		EC
	96	95.907 830	50		EC
	97	96.906 340	1,070		EC
	98	97.907 110	210		β^-
	99	98.906 249	6		β^-
	100	99.907 840	60		β^-
	101	100.907 326	27		β^-
	102	101.909 180	1,070		β^-
	103	102.908 830	110		β^-
	104	103.911 710	110		β^-
	105	104.911 330	220		β^-
$_{44}$Ru	Ruthenium	101.07			
	95	94.909 801	40		β^+, EC
	96	95.907 598	6	0.0551	
	97	96.907 630	1,520		EC
	98	97.905 288 7	4.4	0.0187	
	99	98.905 935 5	4.3	0.1272	
	100	99.904 218	5	0.1262	
	101	100.905 576 8	3.3	0.1707	
	102	101.904 347 8	4.7	0.3161	
	103	102.906 306	21		β^-
	104	103.905 430	5	0.1858	
	105	104.907 679	17		β^-
	106	105.907 322	12		β^-
	107	106.910 130	320		β^-
	108	107.910 100	700		β^-
$_{45}$Rh	Rhodium	102.905			
	97	96.911 380	1,520		β^+
	98	97.909 800	320		β^+, EC
	99	98.908 190	22		β^+, EC
	100	99.908 126	22		β^+, EC
	101	100.906 178	19		EC
	102	101.906 842	9		β^+, EC, β^-
	103	102.905 511 0	4.8	1.000	
	104	103.906 659	7		β^-
	105	104.905 671	13		β^-
	106	105.907 279	12		β^-
	107	106.906 753	42		β^-
	108	107.908 700	600		β^-
	109	108.908 640	1,070		β^-
	110	109.911 100	500		β^-
$_{46}$Pd	Palladium	106.4			
	99	98.912 270	220		β^+
	100	99.908 770	1,070		EC

TABLE 2.1. MASS TABLE (*Continued*)

Z	A	$M(ZA)$ in u[a]	Error in μu	Abundance[b]	
	101	100.908 070	60		β^+, EC
	102	101.905 609	11	0.0096	
	103	102.906 107	22		EC
	104	103.904 011	11	0.1097	
	105	104.905 064	12	0.2223	
	106	105.903 479	6	0.2733	
	107	106.905 131 6	4.6		β^-
	108	107.903 891	8	0.2671	
	109	108.905 954	5		β^-
	110	109.905 164	14	0.1181	
	111	110.907 670	50		β^-
	112	111.907 386	33		β^-
$_{47}$Ag	Silver	107.870[r]			
	102	101.911 300	1,070		β^+, EC
	103	102.908 890	110		β^+
	104	103.908 596	16		β^+, EC
	105	104.906 460	1,070		EC
	106	105.906 661	9		β^+, EC
	107	106.905 094 0	4.5	0.51829[s]	
	108	107.905 949	8		β^+, EC, β^-
	109	108.904 756	5	0.48170[s]	
	110	109.906 095	7		β^-
	111	110.905 316	11		β^-
	112	111.907 064	25		β^-
	113	112.906 556	43		β^-
	114	113.908 300	430		β^-
	115	114.908 930	180		β^-
	116	115.911 310	1,070		β^-
$_{48}$Cd	Cadmium	112.40			
	104	103.909 880	1,070		β^+, EC
	105	104.909 470	1,520		β^+, EC
	106	105.906 462 6	4.1	0.01215	
	107	106.906 615	6		β^+, EC
	108	107.904 186 6	4.4	0.00875	
	109	108.904 928	7		EC
	110	109.903 011 8	3.8	0.1239	
	111	110.904 188 4	3.8	0.1275	
	112	111.902 762 5	3.2	0.2407	
	113	112.904 408 5	3.6	0.1226	
	114	113.903 360 3	3.0	0.2886	
	115	114.905 431	10		β^-
	116	115.904 761 8	3.3	0.0758	
	117	116.907 239	15		β^-
	118	117.906 970	1,160		β^-
	119	118.909 740	350		β^-
$_{49}$In	Indium	114.82			
	106	105.913 440	320		β^+
	107	106.910 360	160		β^+
	108	107.909 720	90		β^+, EC
	109	108.907 096	13		β^+, EC
	110	109.907 231	43		β^+, EC
	111	110.905 360	210		EC
	112	111.905 544	10		β^+, EC, β^-
	113	112.904 089	9	0.0428	
	114	113.904 905	9		β^+, EC, β^-
	115	114.903 871	8	0.9572	
	116	115.905 317	26		β^-
	117	116.904 534	10		β^-
	118	117.906 110	430		β^-
	119	118.905 990	130		β^-
	120	119.908 000	1,070		β^-
	121	120.908 090	1,070		β^-
	122	121.910 600	900		β^-
	123	122.910 570	1,070		β^-

TABLE 2.1. MASS TABLE (*Continued*)

Z	A	$M(ZA)$ in u[a]	Error in μu	Abundance[b]	
$_{50}$Sn	Tin	118.69		3.43×10^{-3}	
	111	110.908 060	220		β^+, EC
	112	111.904 835	10	0.0096	
	113	112.905 187	18		EC
	114	113.902 773	9	0.0066	
	115	114.903 346	7	0.0035	
	116	115.901 744 6	4.7	0.1430	
	117	116.902 958 1	3.4	0.0761	
	118	117.901 605 8	4.1	0.2403	
	119	118.903 313 3	3.3	0.0858	
	120	119.902 198 2	3.6	0.3285	
	121	120.904 227	6		β^-
	122	121.903 441 1	4.4	0.0472	
	123	122.905 738	11		β^-
	124	123.905 272	5	0.0594	
	125	124.907 746	13		β^-
	126	125.907 640	1,090		β^-
	127	126.910 260	1,070		β^-
	128	127.910 470	230		β^-
$_{51}$Sb	Antimony	121.75		8.3×10^{-5}	
	113	112.909 986	47		β^+, EC
	114	113.909 510	210		β^+
	115	114.906 599	23		β^+, EC
	116	115.906 630	50		β^+, EC
	117	116.904 912	32		β^+, EC
	118	117.905 574	8		β^+, EC
	119	118.903 935	22		EC
	120	119.905 081	8		β^+, EC
	121	120.903 816 1	2.8	0.5725	
	122	121.905 183	7		β^+, EC, β^-
	123	122.904 212 7	3.3	0.4275	
	124	123.905 973	6		β^-
	125	124.905 232	9		β^-
	126	125.907 320	160		β^-
	127	126.906 927	33		β^-
	128	127.909 070	160		β^-
	129	128.909 260	1,070		β^-
	130	129.912 040	1,070		β^-
$_{52}$Te	Tellurium	127.60			
	116	115.908 300	120		EC
	117	116.908 670	60		β^+, EC
	118	117.905 900	1,070		EC
	119	118.906 398	22		β^+, EC
	120	119.904 023	14	0.00089	
	121	120.905 199	48		EC
	122	121.903 066	6	0.0246	
	123	122.904 277	6	0.0087	
	124	123.902 842	6	0.0461	
	125	124.904 418	6	0.0699	
	126	125.903 322	5	0.1871	
	127	126.905 209	9		β^-
	128	127.904 476	6	0.3179	
	129	128.906 575	9		β^-
	130	129.906 238	6	0.3448	
	131	130.908 575	22		β^-
	132	131.908 523	18		β^-
$_{53}$I	Iodine	126.9044		2.4×10^{-5}	
	120	119.909 820	1,070		β^+
	121	120.907 730	70		β^+, EC
	122	121.907 511	43		β^+
	123	122.905 730	1,070		EC
	124	123.906 246	33		β^+, EC
	125	124.904 578	6		EC

TABLE 2.1. MASS TABLE (*Continued*)

Z	A	$M(ZA)$ in u[a]	Error in μu	Abundance[b]	
	126	125.905 631	7		β^+, EC, β^-
	127	126.904 469 8	4.3	1.000	
	128	127.905 838	9		EC, β^-
	129	128.904 987	7		β^-
	130	129.906 676	33		β^-
	131	130.906 127 1	4.3		β^-
	132	131.907 981	7		β^-
	133	132.907 750	70		β^-
	134	133.909 850	60		β^-
	135	134.910 020	1,080		β^-
	136	135.914 740	110		β^-
$_{54}$Xe	Xenon	131.30			
	123	122.908 730	1,080		β^+, EC
	124	123.906 120	150	0.00096	
	125	124.906 620	1,070		EC
	126	125.904 288	9	0.00090	
	127	126.905 220	380		EC
	128	127.903 540	6	0.01919	
	129	128.904 784	5	0.2644	
	130	129.903 509	6	0.0408	
	131	130.905 085 3	4.2	0.2118	
	132	131.904 161 0	4.7	0.2689	
	133	132.905 815	39		β^-
	134	133.905 397 1	4.9	0.1044	
	135	134.907 020	110		β^-
	136	135.907 221	6	0.0887	
	137	136.911 100	110		β^-
	138	137.913 810	1,100		β^-
	139	138.917 840	390		β^-
$_{55}$Cs	Cesium	132.905		5.3×10^{-4}	
	125	124.909 910	1,070		β^+, EC
	126	125.909 440	430		β^+, EC
	127	126.907 480	380		β^+, EC
	128	127.907 759	33		β^+, EC
	129	128.905 960	1,070		EC
	130	129.906 720	22		β^+, EC, β^-
	131	130.905 466	8		EC
	132	131.906 393	27		EC
	133	132.905 355	38	1.000	
	134	133.906 823	41		β^-
	135	134.905 770	110		β^-
	136	135.907 340	90		β^-
	137	136.906 770	80		β^-
	138	137.910 800	1,080		β^-
	139	138.912 900	330		β^-
	140	139.917 110	1,070		β^-
$_{56}$Ba	Barium	137.34		0.018	
	127	126.911 340	1,140		β^+
	128	127.908 510	1,070		EC
	129	128.908 590	1,070		β^+, EC
	130	129.906 245	23	0.00101	
	131	130.906 716	18		EC
	132	131.905 120	300	0.00097	
	133	132.905 879	39		EC
	134	133.904 612	41	0.0242	
	135	134.905 550	110	0.0659	
	136	135.904 300	80	0.0781	
	137	136.905 500	80	0.1132	
	138	137.905 000	60	0.7166	
	139	138.908 600	60		β^-
	140	139.910 565	23		β^-
	141	140.914 050	110		β^-
	142	141.916 350	120		β^-

TABLE 2.1. MASS TABLE (*Continued*)

Z	A	$M(ZA)$ in u[a]	Error in μu	Abundance[b]
$_{57}$La	Lanthanum	138.91		1.28×10^{-3}
	129	128.912 890	1,520	β^+, EC
	130	129.912 260	1,070	β^+, EC
	131	130.909 890	60	β^+, EC
	132	131.910 300	320	β^+
	133	132.908 240	220	β^+, EC
	134	133.908 660	70	β^+, EC
	135	134.906 890	1,080	EC
	136	135.907 380	110	β^+, EC
	137	136.906 040	1,080	EC
	138	137.906 910	60	0.00089
	139	138.906 140	50	0.99911
	140	139.909 438	20	β^-
	141	140.910 828	37	β^-
	142	141.913 980	60	β^-
	143	142.915 870	90	β^-
	144	143.919 600	1,070	β^-
$_{58}$Ce	Cerium	140.12		3.21×10^{-3}
	131	130.915 500	330	β^+
	132	131.911 590	1,120	β^+
	133	132.911 250	1,100	β^+, EC
	134	133.908 810	90	EC
	135	134.909 140	1,520	β^+, EC
	136	135.907 100	500	0.00193
	137	136.907 330	1,520	EC
	138	137.905 830	60	0.00250
	139	138.906 430	50	EC
	140	139.905 392	19	0.8848
	141	140.908 219	19	β^-
	142	141.909 140	50	0.1107
	143	142.912 327	19	β^-
	144	143.913 591	19	β^-
	145	144.917 270	1,070	β^-
	146	145.918 670	240	β^-
$_{59}$Pr	Praseodymium	140.907		3.89×10^{-4}
	137	136.910 360	1,520	β^+, EC
	138	137.910 460	120	β^+, EC
	139	138.908 580	120	β^+, EC
	140	139.909 007	27	β^+, EC
	141	140.907 596	18	1.000
	142	141.909 978	17	β^-
	143	142.910 781	16	β^-
	144	143.913 248	16	β^-
	145	144.914 476	19	β^-
	146	145.917 590	220	β^-
	147	146.918 800	1,070	β^-
	148	147.921 910	1,070	β^-
$_{60}$Nd	Neodymium	144.24		1.62×10^{-3}
	139	138.911 580	1,080	β^+, EC
	140	139.909 330	1,070	EC
	141	140.909 528	21	β^+, EC
	142	141.907 663	16	0.2711
	143	142.909 779	15	0.1217
	144	143.910 039	15	0.2385
	145	144.912 538	15	0.0830
	146	145.913 086	15	0.1722
	147	146.916 074	19	β^-
	148	147.916 869	15	0.0573
	149	148.920 122	18	β^-
	150	149.920 915	15	0.0562
	151	150.923 770	110	β^-
$_{61}$Pm	Promethium			
	142	141.912 820	320	β^+
	143	142.910 990	330	EC
	144	143.912 510	1,070	EC

TABLE 2.1. MASS TABLE (*Continued*)

Z	A	$M(ZA)$ in u[a]	Error in μu	Abundance[b]
	145	144.912 691	18	EC
	146	145.914 632	28	EC, β^-
	147	146.915 108	15	β^-
	148	147.917 421	26	β^-
	149	148.918 330	15	β^-
	150	149.920 960	70	β^-
	151	150.921 198	22	β^-
	152	151.923 510	1,070	β^-
	153	152.924 030	110	β^-
$_{62}$Sm	Samarium	150.35		4.19×10^{-4}
	143	142.914 550	90	β^+, EC
	144	143.911 989	15	0.0309
	145	144.913 394	18	EC
	146	145.912 992	23	α
	147	146.914 867	15	0.1497
	148	147.914 791	15	0.1124
	149	148.917 180	14	0.1383
	150	149.917 276	14	0.0744
	151	150.919 919	21	β^-
	152	151.919 756	15	0.2672
	153	152.922 102	17	β^-
	154	153.922 282	15	0.2271
	155	154.924 701	18	β^-
	156	155.925 569	30	β^-
$_{63}$Eu	Europium	151.96		6.8×10^{-5}
	145	144.916 390	60	EC
	146	145.917 138	37	β^+, EC
	147	146.916 800	330	EC, α
	148	147.918 110	60	EC
	149	148.918 000	1,070	EC
	150	149.919 689	24	EC
	151	150.919 838	21	0.4782
	152	151.921 749	15	β^+, EC
	153	152.921 242	18	0.5218
	154	153.923 053	20	β^-
	155	154.922 930	19	β^-
	156	155.924 802	25	β^-
	157	156.925 390	60	β^-
	158	157.927 940	220	β^-
	159	158.928 840	220	β^-
	160	159.931 000	500	β^-
$_{64}$Gd	Gadolinium	157.25		3.94×10^{-4}
	146	145.918 320	1,070	EC
	147	146.919 170	1,120	EC
	148	147.918 101	19	α
	149	148.919 300	160	EC, α
	150	149.918 605	24	α
	151	150.920 270	1,070	EC
	152	151.919 794	16	0.00200
	153	152.921 503	18	EC
	154	153.920 929	20	0.0215
	155	154.922 664	18	0.1473
	156	155.922 175	19	0.2047
	157	156.924 025	19	0.1568
	158	157.924 178	19	0.2487
	159	158.926 368	27	β^-
	160	159.927 115	20	0.2190
	161	160.929 720	80	β^-
	162	161 930 880	1,520	β^-
$_{65}$Tb	Terbium	158.924		5.6×10^{-5}
	148	147.924 130	320	β^+
	149	148.923 350	60	EC, α
	150	149.923 748	38	EC

TABLE 2.1. MASS TABLE (*Continued*)

Z	A	$M(ZA)$ in u[a]	Error in μu	Abundance[b]
	151	150.923 150	330	EC, α
	152	151.924 280	160	EC
	153	152.923 490	1,070	EC
	154	153.924 580	1,070	β^+, EC
	155	154.923 630	1,070	EC
	156	155.924 750	1,070	EC
	157	156.924 090	22	EC
	158	157.925 464	29	EC, β^-
	159	158.925 351	26	1.000
	160	159.927 146	25	β^-
	161	160.927 572	21	β^-
	162	161.929 810	1,070	β^-
	163	162.930 560	60	β^-
	164	163.933 280	1,070	β^-
$_{66}$Dy	Dysprosium	162.50		2.69 × 10^{-4}
	150	149.925 590	1,070	β^+, α
	151	150.926 250	1,120	α
	152	151.924 729	28	β^+, α
	153	152.925 740	160	EC, α
	154	153.924 350	60	α
	155	154.925 880	1,070	β^+, EC
	156	155.923 930	180	0.000524
	157	156.925 270	1,070	EC
	158	157.924 449	30	0.000902
	159	158.925 759	34	EC
	160	159.925 202	21	0.02294
	161	160.926 945	20	0.1888
	162	161.926 803	19	0.2553
	163	162.928 755	19	0.2497
	164	163.929 200	19	0.2818
	165	164.931 816	20	β^-
	166	165.932 807	30	β^-
$_{67}$Ho	Holmium	164.930		6.8 × 10^{-5}
	152	151.931 560	330	α
	153	152.930 270	60	α
	154	153.930 260	1,080	α
	158	157.928 790	31	β^+
	159	158.927 690	1,070	EC
	160	159.928 740	60	β^+, EC
	161	160.927 800	1,070	EC
	162	161.929 122	38	β^+
	163	162.928 766	22	EC
	164	163.930 390	41	EC, β^-
	165	164.930 421	21	1.000
	166	165.932 289	30	β^-
	167	166.933 130	110	β^-
	168	167.935 930	110	β^-
	169	168.936 860	110	β^-
	170	169.940 070	130	β^-
$_{68}$Er	Erbium	167.26		1.44 × 10^{-4}
	154	153.932 760	1,070	α
	161	160.929 950	1,080	β^+, EC
	162	161.928 740	90	0.00136
	163	162.930 065	23	EC
	164	163.929 287	43	0.0156
	165	164.930 819	22	EC
	166	165.930 307	29	0.3341
	167	166.932 060	29	0.2294
	168	167.932 383	32	0.2707
	169	168.934 610	34	β^-
	170	169.935 560	70	0.1488
	171	170.938 130	70	β^-
	172	171.939 330	80	β^-

TABLE 2.1. MASS TABLE (*Continued*)

Z	A	M(ZA) in u[a]	Error in μu	Abundance[b]
$_{69}$Tm	Thulium	168.934		1.15 × 10^{-5}
	161	160.933 730	1,080	EC
	162	161.933 990	140	EC
	163	162.932 502	40	EC
	164	163.933 541	48	β^+
	165	164.932 540	1,070	EC
	166	165.933 510	60	β^+, EC
	167	166.933 030	1,070	EC
	168	167.934 230	50	EC
	169	168.934 245	34	1.000
	170	169.936 060	60	β^-
	171	170.936 530	70	β^-
	172	171.938 380	80	β^-
	173	172.939 480	80	β^-
	174	173.941 970	120	β^-
	175	174.943 830	1,080	β^-
	176	175.947 190	130	β^-
$_{70}$Yb	Ytterbium	173.04		1.49 × 10^{-4}
	165	164.935 440	1,520	β^+
	166	165.933 850	110	EC
	167	166.935 130	1,070	EC
	168	167.934 160	160	0.00135
	169	168.935 530	1,070	EC
	170	169.935 020	60	0.0303
	171	170.936 430	70	0.1431
	172	171.936 360	70	0.2182
	173	172.938 060	70	0.1613
	174	173.938 740	60	0.3184
	175	174.941 140	60	β^-
	176	175.942 680	70	0.1273
	177	176.945 410	90	β^-
$_{71}$Lu	Lutetium	174.97		3.7 × 10^{-5}
	167	166.938 390	1,080	EC
	168	167.939 090	1,090	EC
	169	168.937 960	1,080	EC
	170	169.938 830	70	β^+, EC
	171	170.938 140	1,080	EC
	172	171.939 260	1,080	EC
	173	172.938 800	80	EC
	174	173.940 350	70	EC
	175	174.940 640	60	0.9741
	176	175.942 660	60	0.0259
	177	176.943 930	80	β^-
	178	177.946 300	90	β^-
	179	178.947 470	100	β^-
	180	179.950 370	150	β^-
$_{72}$Hf	Hafnium	178.49		3.0 × 10^{-4}
	174	173.940 360	70	0.0018
	175	174.941 610	1,080	EC
	176	175.941 570	60	0.0520
	177	176.943 400	80	0.1850
	178	177.943 880	80	0.2714
	179	178.946 030	90	0.1375
	180	179.946 820	100	0.3524
	181	180.949 105	42	β^-
	182	181.950 700	220	β^-
	183	182.953 830	220	β^-
$_{73}$Ta	Tantalum	180.948		1.2 × 10^{-4}
	177	176.944 650	80	EC
	178	177.945 930	130	β^+, EC
	179	178.946 160	90	EC

TABLE 2.1. MASS TABLE (*Continued*)

Z	A	$M(ZA)$ in u[a]	Error in μu	Abundance[b]
	180	179.947 544	48	0.000123
	181	180.948 007	42	0.999877
	182	181.950 167	42	β^-
	183	182.951 470	43	β^-
	184	183.953 980	50	β^-
	185	184.955 560	70	β^-
	186	185.958 410	330	β^-
$_{74}$W	Tungsten	183.85		8.2 to 380 $\times 10^{-5}$
	180	179.947 000	50	0.00135
	181	180.948 211	47	EC
	182	181.948 301	41	0.2641
	183	182.950 324	41	0.1440
	184	183.951 025	43	0.3064
	185	184.953 519	43	β^-
	186	185.954 440	45	0.2841
	187	186.957 244	45	β^-
	188	187.958 816	48	β^-
$_{75}$Re	Rhenium	186.2		5.4 $\times 10^{-8}$
	182	181.951 372	47	β^+, EC
	183	182.951 260	1,070	EC
	184	183.952 780	1,080	EC
	185	184.953 059	43	0.3707
	186	185.955 020	70	EC, β^-
	187	186.955 833	44	0.6293
	188	187.958 353	47	β^-
	189	188.959 370	90	β^-
	190	189.961 960	440	β^-
$_{76}$Os	Osmium	190.2		
	184	183.952 750	70	0.00018
	185	184.954 113	43	EC
	186	185.953 870	70	0.0159
	187	186.955 832	44	0.0164
	188	187.956 081	47	0.133
	189	188.958 300	90	0.161
	190	189.958 630	80	0.264
	191	190.960 970	60	β^-
	192	191.961 450	60	0.410
	193	192.964 227	35	β^-
	194	193.965 229	25	β^-
	195	194.968 000	500	β^-
$_{77}$Ir	Iridium	192.2		5 $\times 10^{-8}$
	186	185.957 990	80	β^+, EC
	187	186.957 560	1,070	EC
	188	187.959 122	49	β^+, EC
	189	188.958 910	1,080	EC
	190	189.960 830	180	EC
	191	190.960 640	60	0.373
	192	191.962 700	60	EC, β^-
	193	192.963 012	35	0.627
	194	193.965 125	25	β^-
	195	194.965 890	110	β^-
	196	195.968 250	1,070	β^-
	197	196.969 490	220	β^-
	198	197.972 620	320	β^-
$_{78}$Pt	Platinum	195.09		2.7 $\times 10^{-7}$
	188	187.959 670	70	EC
	189	188.960 610	1,520	EC
	190	189.959 950	70	0.000127
	191	190.961 450	1,080	EC
	192	191.961 150	60	0.0078
	193	192.963 060	31	EC
	194	193.962 725	23	0.329

TABLE 2.1. MASS TABLE (*Continued*)

Z	A	M(ZA) in u[a]	Error in μu	Abundance[b]
	195	194.964 813	18	0.338
	196	195.964 967	15	0.253
	197	196.967 347	13	β^-
	198	197.967 895	23	0.0721
	199	198.970 580	29	β^-
	200	199.971 430	1,080	β^-
	201	200.974 770	120	β^-
$_{79}$Au	Gold	196.967		2.6 × 10^{-7}
	190	189.964 710	1,080	β^+, EC
	191	190.963 550	1,520	EC
	192	191.964 620	80	β^+, EC
	193	192.964 240	1,070	EC
	194	193.965 418	28	β^+, EC
	195	194.965 051	19	EC
	196	195.966 555	14	EC, β^-
	197	196.966 541	10	1.000
	198	197.968 231	7	β^-
	199	198.968 773	13	β^-
	200	199.970 700	100	β^-
	201	200.971 920	110	β^-
	202	201.974 120	1,070	β^-
	203	202.975 130	1,070	β^-
$_{80}$Hg	Mercury	200.59		3.9 to 25 × 10^{-6}
	192	191.966 160	1,080	β^+, EC
	193	192.966 750	1,070	EC
	194	193.965 790	1,070	EC
	195	194.966 620	1,070	EC
	196	195.965 820	14	0.00146
	197	196.967 360	44	EC
	198	197.966 756	7	0.1002
	199	198.968 279	7	0.1684
	200	199.968 327	6	0.2313
	201	200.970 308	7	0.1322
	202	201.970 642	7	0.2980
	203	202.972 880	8	β^-
	204	203.973 495	7	0.0685
	205	204.976 210	110	β^-
	206	205.977 513	23	β^-
$_{81}$Tl	Thallium	204.37		1.5 to 15 × 10^{-5}
	194	193.971 570	1,520	EC
	195	194.969 840	1,090	EC
	196	195.970 760	160	EC
	197	196.969 720	170	EC
	198	197.970 470	90	FC
	199	198.969 460	320	EC
	200	199.970 962	8	β^+, EC
	201	200.970 750	60	EC
	202	201.971 950	25	EC
	203	202.972 353	8	0.2950
	204	203.973 865	8	EC, β^-
	205	204.974 442	8	0.7050
	206	205.976 104	8	β^-
	207	206.977 450	11	β^-
	208	207.982 013	9	β^-
	209	208.985 296	37	β^-
	210	209.990 054	29	β^-
$_{82}$Pb	Lead	207.19		8 × 10^{-4}
	196	195.973 800	1,090	EC
	197	196.974 090	1,090	?
	198	197.972 410	1,080	EC
	199	198.972 860	1,120	β^+, EC
	200	199.971 970	1,070	EC
	201	200.972 860	1,080	β^+, EC

TABLE 2.1. MASS TABLE (*Continued*)

Z	A	$M(ZA)$ in u[a]	Error in μu	Abundance[b]
	202	201.972 003	40	EC
	203	202.973 229	13	EC
	204	203.973 044	8	0.0148[t]
	205	204.974 480	9	EC
	206	205.974 468	7	0.236[t]
	207	206.975 903	7	0.226[t]
	208	207.976 650	7	0.523[t]
	209	208.981 082	11	β^-
	210	209.984 187	7	β^-
	211	210.988 742	22	β^-
	212	211.991 905	13	β^-
	213	212.996 290	1,070	β^-
	214	213.999 766	26	β^-
$_{83}$Bi	Bismuth	208.980		9.0×10^{-6}
	198	197.980 370	1,520	?
	199	198.978 440	1,090	EC, α
	200	199.978 940	1,520	EC
	201	200.977 370	1,520	EC, α
	202	201.977 880	1,070	EC
	203	202.976 650	60	β^+, EC
	204	203.977 810	1,070	EC
	205	204.977 382	13	β^+, EC
	206	205.978 389	28	β^+, EC
	207	206.978 438	8	EC
	208	207.979 731	9	EC
	209	208.980 394	8	1.000
	210	209.984 121	7	β^-
	211	210.987 300	11	β^-, α
	212	211.991 279	9	β^-, α
	213	212.994 317	19	β^-, α
	214	213.998 686	29	β^-, α
	215	215.001 830	100	β^-
	216	216.006 330	1,070	β^-
$_{84}$Po	Polonium			1.4×10^{-15}
	200	199.982 820	1,090	α
	201	200.983 020	1,090	EC, α
	202	201.981 130	1,080	EC, α
	203	202.981 470	1,120	EC, α
	204	203.980 460	1,070	EC, α
	205	204.981 200	1,080	EC, α
	206	205.980 324	41	EC, α
	207	206.981 558	11	β^+, EC, α
	208	207.981 243	12	EC, α
	209	208.982 426	13	EC, α
	210	209.982 876	7	α
	211	210.986 657	8	α
	212	211.988 866	7	α
	213	212.992 825	16	α
	214	213.995 201	7	α
	215	214.999 423	22	β^-, α
	216	216.001 922	13	α
	217	217.006 060	1,070	α
	218	218.008 930	26	β^-, α
$_{85}$At	Astatine			
	202	201.989 800	1,520	EC, α
	203	202.987 710	1,090	EC, α
	204	203.988 060	1,520	EC, α
	205	204.986 440	1,520	EC, α
	206	205.986 790	1,070	EC, α
	207	206.985 560	60	EC, α
	208	207.986 610	1,080	EC, α
	209	208.986 167	13	EC, α
	210	209.987 036	28	EC, α
	211	210.987 462	8	EC, α
	212	211.990 724	23	α

TABLE 2.1. MASS TABLE (*Continued*)

Z	A	$M(ZA)$ in u[a]	Error in μu	Abundance[b]
	213	212.993 070	210	α
	214	213.996 340	50	α
	215	214.998 663	24	α
	216	216.002 411	33	α
	217	217.004 648	19	α
	218	218.008 607	29	β^-, α
	219	219.011 290	90	β^-, α
	220	220.015 370	1,070	β^-, ?
$_{86}$Rn	Radon			
	204	203.992 300	1,090	α
	205	204.992 560	1,530	?
	206	205.990 580	1,080	EC, α
	207	206.990 760	1,120	EC, α
	208	207.989 790	1,070	EC, α
	209	208.990 420	1,080	EC, α
	210	209.989 540	42	EC, α
	211	210.990 566	11	EC, α
	212	211.990 707	13	α
	213	212.993 932	25	?
	214	213.995 380	1,070	?
	215	214.998 690	110	α
	216	216.000 272	13	α
	217	217.003 896	36	α
	218	218.005 603	13	α
	219	219.009 481	22	α
	220	220.011 401	14	α
	221	221.015 230	1,520	β^-, α
	222	222.017 531	26	α
$_{87}$Fr	Francium			
	206	205.999 840	1,520	α
	207	206.997 730	1,090	α
	208	207.997 950	1,520	α
	209	208.996 320	1,520	α
	210	209.996 570	1,070	α
	211	210.995 330	60	α
	212	211.996 230	1,080	EC, α
	213	212.996 184	17	?
	214	213.999 001	36	?
	215	215.000 356	33	?
	216	216.003 310	1,070	?
	217	217.004 750	300	α
	218	218.007 540	80	α
	219	219.009 257	32	α
	220	220.012 337	46	α
	221	221.014 183	19	α
	222	222.017 630	1,070	β^-, α
	223	223.019 736	22	β^-, α
	224	224.023 590	1,070	α
$_{88}$Ra	Radium			5.8×10^{-12}
	211	211.000 950	1,550	?
	212	211.999 950	1,070	?
	213	213.000 420	1,080	α
	214	213.999 990	50	?
	215	215.002 715	24	?
	216	216.003 490	35	?
	217	217.006 386	41	?
	218	218.007 170	1,520	?
	219	219.010 050	150	α
	220	220.011 029	17	α
	221	221.013 892	37	α
	222	222.015 376	17	
	223	223.018 501	22	α
	224	224.020 218	14	α
	225	225.023 528	22	β^-
	226	226.025 360	26	α

TABLE 2.1. MASS TABLE (*Continued*)

Z	A	M(ZA) in u[a]	Error in μu	Abundance[b]
	227	227.029 159	31	β^-
	228	228.031 139	21	β^-
$_{89}$Ac	Actinium			1.3×10^{-15}
	213	213.007 050	1,520	α
	214	214.007 100	1,080	α
	221	221.015 680	320	α
	222	222.017 760	90	α
	223	223.019 144	33	EC, α
	224	224.021 690	60	EC, α
	225	225.023 153	19	α
	226	226.026 160	110	EC, β^-
	227	227.027 753	22	β^-, α
	228	228.031 080	21	β^-
	229	229.032 800	1,070	β^-
	230	230.036 210	1,070	β^-
	231	231.038 550	110	β^-
$_{90}$Th	Thorium	232.038		5×10^{-4}
	223	223.020 920	190	α
	224	224.021 477	20	α
	225	225.023 927	39	EC, α
	226	226.024 901	20	α
	227	227.027 706	22	α
	228	228.028 750	14	α
	229	229.031 652	23	α
	230	230.033 087	25	α
	231	231.036 291	22	β^-
	232	232.038 124	21	α, SF
	233	233.041 469	23	β^-
	234	234.043 583	23	β^-
$_{91}$Pa	Protactinium			3.5×10^{-12}
	225	225.026 230	1,120	α
	226	226.027 810	110	α
	227	227.028 811	33	EC, α
	228	228.031 010	60	EC, α
	229	229.032 022	20	EC
	230	230.034 433	21	EC, β^-, α
	231	231.035 877	22	α
	232	232.038 612	26	β^-
	233	233.040 132	23	β^-
	234	234.043 298	30	β^-
	235	235.045 420	110	β^-
	236	236.049 230	110	β^-
	237	237.051 080	60	β^-
$_{92}$U	Uranium	238.03		1.6×10^{-4}
	228	228.031 387	23	EC, α
	229	229.033 481	40	EC, α
	230	230.033 937	21	α
	231	231.036 270	60	EC, α
	232	232.037 168	14	α, SF
	233	233.039 522	23	α
	234	234.040 904	25	0.000056[u]
	235	235.043 915	22	0.007205[u]
	236	236.045 637	21	α, SF
	237	237.048 608	23	β^-
	238	238.050 770	23	0.99274[u]
	239	239.054 300	23	β^-
	240	240.056 594	27	β^-
$_{93}$Np	Neptunium			
	230	230.037 680	1,080	α
	231	231.038 280	60	EC, α
	232	232.039 860	1,080	EC
	233	233.040 670	60	EC, α
	234	234.042 860	110	β^+, EC
	235	235.044 049	23	EC, α

TABLE 2.1. MASS TABLE (*Continued*)

Z	A	$M(ZA)$ in u[a]	Error in μu	Abundance[b]
	236	236.046 624	17	β^-
	237	237.048 056	23	α
	238	238.050 896	27	β^-
	239	239.052 924	22	β^-
	240	240.056 080	70	β^-
	241	241.058 200	110	β^-
	242	242.061 840	1,070	β^-
$_{94}$Pu	Plutonium			
	232	232.041 180	60	EC, α
	233	233.042 972	46	EC, α
	234	234.043 315	23	EC, α
	235	235.045 270	60	EC, α
	236	236.046 071	15	α, SF
	237	237.048 298	30	EC, α, SF
	238	238.049 511	25	α, SF
	239	239.052 146	22	α, SF
	240	240.053 882	21	α, SF
	241	241.056 737	22	β^-, α
	242	242.058 725	24	α, SF
	243	243.061 972	40	β^-
	244	244.064 100	800	α, SF
	245	245.067 830	1,070	β^-
	246	246.070 090	70	β^-
$_{95}$Am	Americium			
	236	236.049 160	1,070	?
	237	237.049 840	80	EC
	238	238.051 940	1,080	EC
	239	239.053 016	31	EC, α
	240	240.055 280	1,070	EC
	241	241.056 714	22	α
	242	242.059 502	27	EC, β^-
	243	243.061 367	23	α
	244	244.064 355	22	EC, β^-
	245	245.066 340	23	β^-
	246	246.069 660	60	β^-
	247	247.072 090	1,520	β^-
$_{96}$Cm	Curium			
	238	238.053 036	40	EC, α
	239	239.054 880	1,080	EC
	240	240.055 545	18	α, SF
	241	241.057 542	31	EC, α
	242	242.058 788	25	α, SF
	243	243.061 370	22	EC, α
	244	244.062 821	21	α
	245	245.065 371	23	α
	246	246.067 202	25	α
	247	247.070 280	1,070	α
	248	248.072 200	800	α
	249	249.075 810	110	β^-
$_{97}$Bk	Berkelium			
	241	241.060 100	1,070	?
	242	242.061 790	1,070	?
	243	243.062 965	33	EC, α
	244	244.065 170	1,070	EC, α
	245	245.066 272	31	EC, α
	246	246.068 770	1,070	EC
	247	247.070 260	60	α
	248	248.072 960	70	EC, β^-
	249	249.074 883	24	β^-, α, SF
	250	250.078 270	60	β^-
	251	251.080 810	1,520	β^-
$_{98}$Cf	Californium			
	243	243.065 310	1,070	?
	244	244.065 969	28	α

TABLE 2.1. MASS TABLE (*Continued*)

Z	A	M(ZA) in u[a]	Error in μu	Abundance[b]
	245	245.067 905	38	EC, α
	246	246.068 766	28	α
	247	247.071 070	1,070	EC
	248	248.072 262	39	α, SF
	249	249.074 749	24	α, SF
	250	250.076 384	26	α, SF
	251	251.079 260	1,070	α
	252	252.081 500	800	α, SF
	253	253.085 020	60	β^-
$_{99}$Es	Einsteinium			
	245	245.071 060	1,080	α
	246	246.072 430	1,080	α
	247	247.073 580	1,070	EC
	248	248.075 280	1,070	EC, α
	249	249.076 258	45	EC, α
	250	250.078 610	1,520	EC
	251	251.079 930	70	EC, α
	252	252.082 810	70	α
	253	253.084 730	25	α, SF
	254	254.087 900	60	α
$_{100}$Fm	Fermium			
	248	248.077 092	35	α
	249	249.079 140	320	α
	250	250.079 490	60	α
	251	251.081 190	1,080	EC, α
	252	252.082 562	44	α
	253	253.084 930	49	EC, α
	254	254.086 839	26	SF
	255	255.089 640	1,070	α
$_{101}$Md	Mendelevium			
	251	251.084 620	1,520	α
	252	252.086 120	1,520	α
	253	253.086 940	1,070	α
	254	254.089 470	1,860	α
	255	255.090 550	90	EC, α
$_{102}$No	Nobelium			
	253	253.091 340	340	α
	254	254.091 140	330	α
	255	255.092 730	1,080	α
$_{103}$Lw	Lawrencium			
	257	257.098 940	1,070	α

[a] See ref. 21 for atomic masses; see ref. 24 for atomic weights.
[b] See ref. 23 for isotopic abundances unless otherwise noted. See ref. 25 for atomic abundance data.
[c] Variable by 0.00001 owing to natural causes.
[d] Svec and Anderson, *Geochim. Cosmochim. Acta*, **29**: 633 (1965).
[e] Variable by 0.003 owing to natural causes.
[f] Variable by 0.00005 owing to natural causes.
[g] Variable by 0.0001 owing to natural variation.
[h] Variable by 0.001 owing to natural variations.
[i] Recommended range of 0.003.
[j] Recommended range of 0.001.
[k] Shields, Murphy, Garner, and Dibeler, *J. Am. Chem. Soc.*, **84**: 1519 (1961).
[l] Recommended uncertainty of 0.001.
[m] Flesch, Svec, and Staley, *Geochim. Cosmochim. Acta*, **20**: 300 (1960).
[n] Recommended uncertainty of 0.003.
[o] Shields, Murphy, and Garner, *J. Res. Natl. Bur. Std.*, **68A**: 589 (1964).
[p] Recommended uncertainty of 0.002.
[q] Catanzaro, Murphy, Garner, and Shields, *J. Res. Natl. Bur. Std.*, **68A**: 593 (1964).
[r] Recommended uncertainty of 0.003.
[s] Shields, Garner, and Dibeler, *J. Res. Natl. Bur. Std.*, **66A**: 1 (1962).
[t] Great Bear Lake galena.
[u] See ref. 24.

are used as sources of reference masses because of the relative ease with which comparison fragments of almost any mass number may be obtained. Thus one must have an accurate measurement of the mass of H^1. In certain cases hydrocarbon ions containing H^2 or C^{13} are also employed as reference masses. These secondary standard masses are readily measured by comparison with the C^{12} standard. A determination of the doublet $12H^1 - C^{12}$ was recently made by measurement of the doublet $C_{11}{}^{12}H_{22}{}^1 - C_{12}{}^{12}H_{10}{}^1$ by Benson and Johnson [12].

A series of doublets have been introduced by Duckworth [13] which measure the mass difference between two isotopes of the same element differing by approximately 2 units of mass. If X^A represents the isotope of element X with mass number A, the doublets have the form $X^AC1^{37} - X^{A+2}Cl^{35}$. These doublets have a small mass difference; thus, with knowledge of the $Cl^{37} - Cl^{35}$ mass difference, an accurate $X^{A+2} - X^A$ mass difference can be measured. Duckworth and coworkers have measured many mass differences in the medium and heavy elements to accuracies of about ± 4 μu with a large-sector-field instrument.

By combining measurements of the mass differences between neighboring isotopes with the more conventional hydrocarbon-isotope doublets, a least-squares adjustment can be made to determine the best mass. Measurements such as these have been made with the Minnesota sector-field instrument by Ries [14], Damerow [15], and Benson [16]. They yield masses of high precision for almost all the stable isotopes from $A = 70$ to 150. Errors quoted for these mass results are about ± 10 μu.

3. Atomic-mass Table

Atomic-mass tables can be constructed by several methods. In some cases spectroscopic doublet values have been employed alone in the construction of mass tables [17]. In other cases [18], nuclear-reaction Q values have been employed in chains to calculate the mass difference between the mass in question and the mass standard. Although mass tables of this sort are accurate for masses near the mass standard, they become less accurate as the chain becomes long and errors accumulate. The most complete and accurate mass tables, however, incorporate both groups of data. In a table such as this, the over-all trend of masses with increasing mass number depends rather heavily on mass doublet measurements, which prevents the possible cumulative errors made by employing Q values only. Local details of the trends in masses are given both by Q values and mass doublet measurements.

The mass table constructed by Mattauch, Wapstra, and coworkers is a good illustration of a very extensive table. This table represents a least-squares reduction of all available Q-value and mass doublet data. The first table by this group was constructed in 1956 [19]. This table was revised in 1960 [20] and 1964 [21] as significant new data became available. A computer program for the least-squares reduction has been developed in such a way that new input data can easily be incorporated [22]. A computer program of this sort is very time-consuming since the inversion of a very large matrix is involved. Mattauch and coworkers have chosen to reduce this problem to some extent by dividing the periodic table into a group of 10 smaller regions. Each of these regions then forms a separate least-squares problem.

The atomic masses found in Table 2.1 are taken from the 1964 least-squares reduction of data by Mattauch et al. [21]. The masses are given in the relative scale which assigns the mass of C^{12} to be exactly 12 units (symbol u). Included is a listing of radioactive masses specified by their decay path.

The relative abundances of the stable isotopes are listed in the last column. These data are obtained from a compilation by Nier and Way [23] which has been brought up to date by the inclusion of some new results. Listed opposite the element name is the atomic weight of the element taken from Cameron and Wichers [24]. Listed opposite the element name in the abundance column is the relative abundance of the element in terrestrial igneous rocks [25] in the scale that sets the number of silicon atoms equal to 100. Approximate crustal abundances are presented in Fig. 2.1.

References

1. Nuclear Data Sheets, NRC 59-4-23, Oak Ridge National Laboratory, Nuclear Data Group, Oak Ridge, Tenn.
2. Cohen, E. R., and J. W. M. DuMond: Present Status of Our Knowledge of the Numerical Values of Fundamental Physical Constants, in W. H. Johnson, Jr. (ed.), "Nuclidic Masses," p. 152, Springer, Berlin, 1964.
3. Nuclear Data Sheets, NRC 60-1-41 (see ref. 1).
4. Nuclear Data Sheets, NRC 58-1-19 (see ref. 1).
5. Nier, A. O., and T. R. Roberts: *Phys. Rev.*, **81**: 27 (1951).
6. Smith, L. G., and G. C. Damm: *Phys. Rev.*, **90**: 324 (1953).
7. Smith, L. G.: *Phys. Rev.*, **111**: 1606 (1958).
8. Bainbridge, K. T.: Charged Particle Dynamics and Optics, Relative Isotopic Abundances of the Elements, Atomic Masses, in E. Segrè (ed.), "Experimental Nuclear Physics," vol. 1, pp. 559–766, Wiley, New York, 1953.
9. Ewald, H., and H. Hintenberger: "Methoden und Anwendungen der Massenspektroskopie," Verlag Chemie GmbH, Weinheim, Germany, 1953.
10. Duckworth, H. E.: "Mass Spectrometry," Cambridge University Press, London, 1958.
11. McDowell, C. A.: "Mass Spectrometry," McGraw-Hill, New York, 1963.
12. Benson, J. L., and W. H. Johnson: *Phys. Rev. Letters*, **13**: 724 (1964).
13. Barber, R. C., R. L. Bishop, L. A. Cambey, H. E. Duckworth, J. D. Macdougall, W. McLatchie, J. H. Ormrod, and P. van Rookhuyzen: Recent Mass Values Obtained at McMaster University, "Nuclidic Masses," p. 393 (see ref. 2).
14. Ries, R. R., R. A. Damerow, and W. H. Johnson, Jr.: Atomic Masses from Gallium to Molybdenum, *Phys. Rev.*, **132**: 1662 (1963).
15. Damerow, R. A., R. R. Ries, and W. H. Johnson, Jr.: Atomic Masses from Ruthenium to Xenon, *Phys. Rev.*, **132**: 1673 (1963).
16. Benson, J. L., and W. H. Johnson: *Phys. Rev.*, **141**(2B): (1966).
17. Condon, E. U., and H. Odishaw: "Handbook of Physics," 1st ed., p. 9-58, McGraw-Hill, New York, 1958.

18. Li, C. W., W. Whaling, W. A. Fowler, and C. C. Lauritsen: *Phys. Rev.*, **83**: 512 (1959). Wapstra, A. H.: *Physica*, **21**: 367 (1955).
19. Mattauch, J. H. E., L. Waldmann, R. Bieri, and F. Everling: The Masses of the Light Nuclides, *Z. Naturforsch.*, **11a**: 525 (1956).
20. Everling, F., L. A. König, J. H. E. Mattauch, and A. H. Wapstra: Relative Nuclidic Masses, *Nucl. Phys.*, **18**: 529 (1960).
21. Mattauch, J. H. E.: private communication. Mattauch, J. H. E., E. Thiele, and A. H. Wapstra: *Nucl. Phys.*, **67**: 1 (1965).
22. Thiele, W.: Considerations about Programming Mass Adjustments, "Nuclidic Masses," p. 82 (see ref. 2).
23. Nuclear Data Sheets, Appendix 2, NRC 58-11-2 (see ref. 1).
24. Cameron, A. E., and E. Wichers: Report of the Commission on Atomic Weights, 1961 (unpublished).
25. Rankama, K., and T. G. Sahama: "Geochemistry," chap. 2, University of Chicago Press, Chicago, 1950.

Chapter 3

Nuclear Moments

By NORMAN F. RAMSEY, Harvard University

1. Introduction

Nuclear moments have been studied in a number of different ways. However, all of these measurements have so far been consistent with the nuclear properties listed in Part 7, Chap. 4, Sec. 1. These properties include the following:

1. A nucleus possesses a spin angular momentum ordinarily measured in units of $\hbar$ and designated by **I**. The spin I of the nucleus, defined as the maximum possible component of the **I**, is integral if the nuclear mass number A is even and half integral if A is odd.

2. A nucleus has a magnetic moment $\boldsymbol{\mu}_I$ which can be represented as

$$\boldsymbol{\mu}_I = \gamma_I \hbar \mathbf{I} = g_I \mu_N \mathbf{I} = \frac{\mu_I}{I} I \qquad (3.1)$$

where the scalar quantities γ_I, g_I, and μ_I are defined by the above equation and called the nuclear gyromagnetic ratio, the nuclear g factor, and the (scalar) nuclear magnetic moment and where μ_N is the nuclear magneton defined as $e\hbar/2Mc$.

3. Many nuclei with a spin equal to or greater than 1 possess an electric-quadrupole moment Q defined by

$$Q = \frac{1}{e} \int_{\tau_n} (3z_n^2 - r_n^2)(\rho_n)_{\text{II}}\, d\tau_n \qquad (3.2)$$

4. A nucleus with a spin of $\frac{3}{2}$ or greater may possess a magnetic-octupole moment Ω defined by

$$\Omega = \int_{\tau_n} \tfrac{1}{2}(5z_n^3 - 3z_n r_n^2)(\nabla \cdot \mathbf{m}_n)_{\text{II}}\, d\tau_n \qquad (3.3)$$

The methods for the measurement of nuclear moments of stable isotopes and radioactive isotopes of relatively long half life (at least several hours) are given in the present chapter. Nuclear spins and moments of excited nuclear states have also been determined by various nuclear-physics techniques such as the angular correlation of successive gamma rays and the angular distribution of the gamma rays emitted by nuclei that have been partially aligned by high fields at low temperatures [1].*

2. Optical Spectroscopy

First experimental evidence for nuclear spins, nuclear magnetic moments, and nuclear electric-quadrupole moments came from the study of optical hyperfine structure. Many of the nuclear moment measurements are still dependent upon such measurements. In general, however, these measurements are much less accurate than those by the other techniques described in this chapter. Results obtained from optical hyperfine structure measurements are included in the nuclear moment tables.

A different and important means of measuring nuclear spins has been to study the band spectra of homonuclear diatomic molecules, i.e., diatomic molecules whose two nuclei are identical, such as H_2. This measurement is possible because of the symmetry restrictions on the allowed complete wave functions with identical nuclei, i.e., the complete wave function must be symmetrical in the nuclear coordinates for nuclei of even mass number (which satisfy Bose-Einstein statistics) and the wave function is antisymmetrical for odd mass number (or Fermi-Dirac statistics).

A gas of homonuclear diatomic molecules essentially consists of molecules of two types: (1) molecules whose combined nuclear spin wave functions are symmetric in the two nuclei or so-called ortho-molecules, and (2) para-molecules whose nuclear spin wave functions are antisymmetric. There is little intercombination between the two, since a transition from one type to another can be induced only by a collision which acts quite differently on the spin of one nucleus than on the other so that the resultant spin state and the rotational state may simultaneously be changed. If n_o/n_p represents the ratio of the statistical weights of the ortho- and para-molecules [1],

$$\frac{n_o}{n_p} = \frac{I+1}{I} \qquad (3.4)$$

Consequently, the intensities of the alternate band spectra lines are populated in this ratio; so from a measurement of the alternate intensity ratio and from Eq. (3.4) I can be inferred. This technique is particularly effective at low values of I, as is apparent from Eq. (3.4).

3. Molecular Beam Experiments

Some of the most effective means for studying nuclear moments have been the different molecular beam methods. Application of these methods to the study of atomic hyperfine structure is described in Part 7, Chap. 4, Secs. 3 and 4. These experiments

* Numbers in brackets refer to References at end of chapter.

have been valuable sources of information on spins, nuclear magnetic-dipole moments, and electric-quadrupole moments; in addition the only experimental data so far on nuclear magnetic-octupole moments arises from study of atomic hyperfine structure by atomic beam methods.

In addition to the experiments previously described, much important data has been obtained by the application of the molecular beam magnetic resonance method to nonparamagnetic molecules and particularly to $^1\Sigma$ diatomic molecules. In such molecules there is no resultant electronic moment; thus the molecular interaction is dominantly with the external magnetic field whence the energy is

$$W = -\boldsymbol{\mu}_I \cdot \mathbf{H}_0 = -\frac{\mu_I}{I}\mathbf{I} \cdot \mathbf{H}_0 = -\frac{\mu_I}{I} H_0 m \quad (3.5)$$

The molecular beam magnetic resonance experiments with such molecules are dominantly similar to those described in Part 7, Chap. 4, Sec. 4; in fact, the apparatus shown in Fig. 4.1 of that section has been used both for atomic studies and for the study of $^1\Sigma$ diatomic molecules. From Eq. (4.1) of the above-mentioned section, from Eq. (3.5) above, and from the selection rule $\Delta m = \pm 1$, it is apparent that the resonance minimum in beam intensity will occur at

$$\nu_0 = \frac{W_m - W_{m+1}}{h} = \left(\frac{\mu_I}{hI}\right) H_0 = \frac{\gamma_I H_0}{2\pi} = \frac{g_I \mu_N H_0}{h} \quad (3.6)$$

where the various alternative expressions come from Eq. (3.1) of this chapter. By a measurement of ν_0 and H_0 in Eq. (3.6) it is apparent that μ_I, γ_I, or g_I can be obtained. Values of nuclear magnetic moments obtained in this way are included in Table 3.1.

In Eq. (3.5) above it is assumed that all the nuclear interaction is with the external field H_0. Actually, even in molecules, there are some interactions within the molecule itself. These interactions include the magnetic interaction of the two nuclear magnetic moments with each other, the interaction of a nuclear electric-quadrupole Q moment with the gradient q of the electric field from the rest of the molecules, the interaction of the nuclear magnetic moment with the magnetic field which results from the rotation of molecular charge distribution, etc. [1, 2]. Consequently nuclear quadrupole interactions eqQ can be determined from these measurements and in those cases where q can be calculated, the nuclear quadrupole moment Q can be determined. The values of Q so determined are included in the nuclear moments table. The quadrupole moment measured in this way that has had the most important theoretical consequences is the discovery by Kellogg, Rabi, Ramsey, and Zacharias [2] that the deuteron possesses a quadrupole moment. This discovery implies that there must be a tensor component [2] to the nuclear forces between a neutron and a proton. In addition to the determination of nuclear moments by molecular beam magnetic resonance experiments, many molecular properties have been determined as well [2].

Inhomogeneous electric fields have also been used for the deflection of the molecules and transitions have been produced by oscillatory electric fields in the presence of homogeneous electric fields. Such electric resonance experiments have been particularly valuable in the measurement of nuclear quadrupole interactions and in the study of molecular properties [2].

4. Nuclear Magnetic Resonance (NMR) Experiments

In the molecular beam magnetic resonance experiments described above, the occurrence of a resonance is detected by the effect of the transition upon the molecules. However, Purcell, Bloch, and their associates [1] in independent but closely related experiments have succeeded in detecting the occurrence of the resonance by the electromotive force induced by the reoriented nuclei at resonance. A typical apparatus for such an experiment is shown schematically in Fig. 3.1. Each of the two branches following A contains a resonant circuit, and one includes a coil surrounding the sample. The line AB contains an extra half wavelength so that a voltage node is produced at D if the bridge is perfectly balanced. When the magnetic field is swept through a nuclear resonance, the balance is disturbed and a signal is produced on the oscillograph and at M_3 following the indicated circuit, often called a phase-sensitive detector or lock-in amplifier. From the resonance frequencies and fields and from Eq. (3.6) the nuclear magnetic moment may be calculated.

One inherent feature of these experiments is that absorption of radiation is largely canceled by stimulated emission; only in so far as the lower energy levels are more abundantly occupied than the higher ones is there any net effect. Even if the nuclear spins are in complete thermal equilibrium, only a small fraction of the nuclei produce an uncanceled effect. For example, with hydrogen the fractional difference is just exp $(2\mu_I H_0/kT) \sim 10^{-5}$ at room temperature and 7,000 gauss. However, even with only this small fraction of the nuclei being effective, many resonances have been obtained. This method has been one of the most powerful for precision measurements of nuclear magnetic moments.

After a magnetic field is applied to a sample, a period of time is required before the above thermal equilibrium is achieved. This time is ordinarily measured by the so-called longitudinal or transverse relaxation time T_1 which is defined by

$$\dot{M}_z = \frac{M_z - M_0}{T_1}$$

where M_z is the z component of M, the resultant magnetization of the nuclear moments per unit volume, and M_0 is the equilibrium value of M_z under the influence of the external magnetic field, which is assumed to be applied along the z axis. For low values of the oscillatory field, the width of the nuclear resonance depends on inverse of the transverse relaxation time T_2 (the full width between half maximum points on a frequency scale is $1/\pi T_2$), where T_2 is defined by $T_2 = -M_x/\dot{M}_x = -M_y/\dot{M}_y$.

The nuclear paramagnetic resonance method (also called either nuclear induction or nuclear resonance

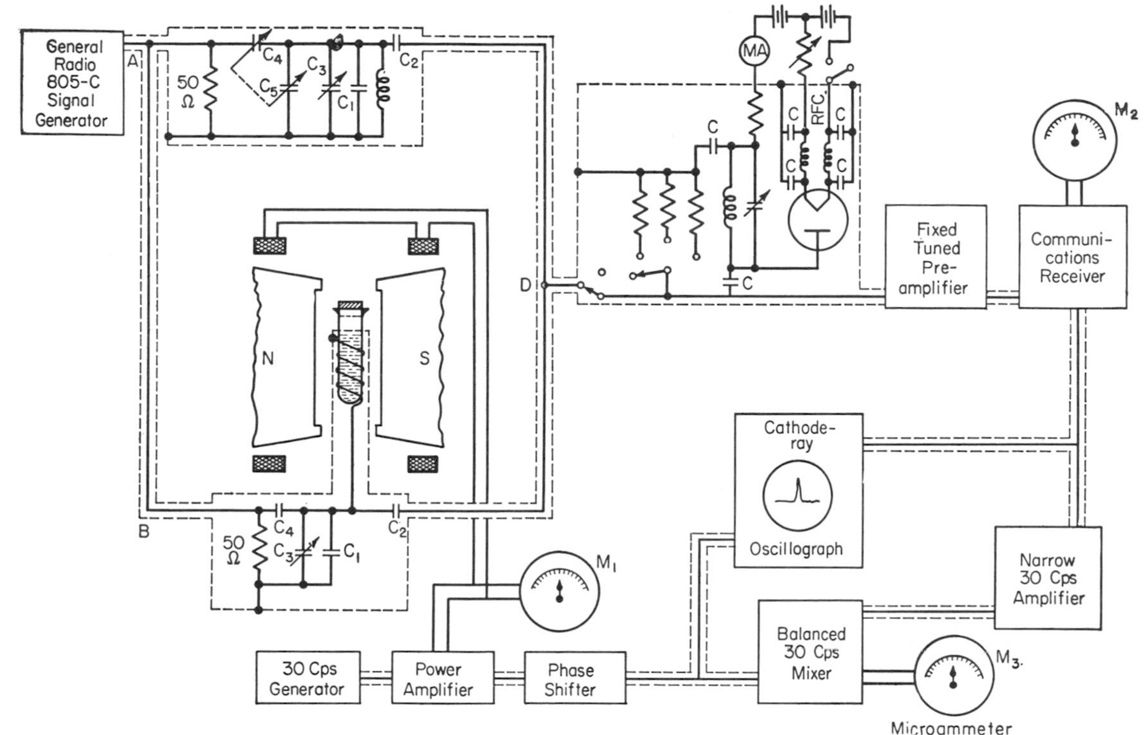

Fig. 3.1. Block diagram of a typical circuit for nuclear magnetic-resonance experiments. The sample and magnet are not shown in correct proportions.

absorption) has been particularly favorable for high precision measurements because the frequent collisions in liquids and high density gases tend to average out the effects of local molecular fields. When the collision frequency is large compared to the variations of the nuclear Larmor frequency, the local fields are partially averaged out and the width of the resonance becomes less than that to be expected from the local fields alone. This "collision narrowing" becomes more and more effective as the collision frequency increases.

Although the most extensive applications of the methods of nuclear paramagnetic resonance have been to nuclear magnetic-moment measurements, the measurement of the relaxation times T_1 and T_2 has provided nuclear quadrupole-moment data and much valuable information on molecular and crystalline structure. In crystals, resolved quadrupole structure can be observed [1].

Numerous modifications of the nuclear paramagnetic resonance techniques have been made, including the use of pulsed oscillatory fields to study transients. The "spin echo" technique of Hahn has been especially fruitful [1].

5. Microwave Spectroscopy and Electron Paramagnetic Resonance (EPR) Experiments

A very effective means of measuring nuclear spins, nuclear quadrupole interactions, and approximate values of nuclear magnetic moments is the study of the hyperfine structure of the microwave spectra of atoms and particularly of molecules at wavelengths of the order of 1 cm (Part 7, Chap. 6). In these experiments the nuclear effect is usually a hyperfine structure associated with some other energy transition in the molecule for which the Bohr frequency is in the microwave region. Frequently the other transition involves a charge in a molecular rotational state. Studies with linear molecules, such as ClCN, have been particularly effective in yielding important nuclear data.

Ordinarily the hyperfine structure is due dominantly to nuclear quadrupole interactions so that values of eqQ and the spin I can be inferred from the experiments. Townes pointed out that in polyatomic molecules approximate values of q can often be inferred from atomic configurations; in such cases then values of Q may be obtained from the results [1]. Values of the nuclear spin I can be inferred from the multiplicity of the hyperfine structure. In some cases Zeeman effects microwave spectra have been studied to obtain data on nuclear moments, although for magnetic-moment measurements this method is less accurate than other methods when the latter can be used at all.

Since the magnetic moments of paramagnetic molecules and ions are about a thousand times larger than nuclear magnetic moments, whereas electronic and nuclear angular momenta are both of the same order $\hbar$, the Larmor precession frequencies in paramagnetic materials are about a thousand times greater than the precession frequencies discussed in Sec. 4. Consequently, in magnetic fields of a few thousand oersted, the spectra corresponding to paramagnetic and ferromagnetic resonances are in the microwave region.

Ordinarily, measurements of paramagnetic resonance give data about the atom, ion, or crystal being studied. However, Penrose, Bleaney, and others [1] by using magnetically dilute crystals have observed a hyperfine structure in paramagnetic resonance which is attributed to the nuclear spin, magnetic moment, and electric-quadrupole moment. From such hyperfine structures in paramagnetic resonance spectra a number of spins and quadrupole interactions have been inferred.

6. Optical Pumping and Double Resonance Techniques

Brossel, Kastler, and others [11] have used optical radiation to alter the population of atomic states in order to study hyperfine structure by magnetic-resonance techniques. In these experiments [11, 12] the atom is excited from the ground state to a metastable state by the absorption of resonance radiation of a suitable frequency or polarization to increase selectively the population of certain magnetic substates. An oscillatory radio frequency is then applied to induce transitions between substates, altering thereby their relative populations and consequently changing the polarization, frequency, or intensity of the observed radiation.

7. Mössbauer Spectroscopy

The recoil-free emission and resonance absorption of γ rays (Mössbauer effect) can be used to study the hyperfine interactions between a nucleus and its environment [12, 13]. The lines of 100-kev γ rays may show emission and absorption widths of 10^{-8} ev. Shifts in energy levels due to nuclear moment interactions with the local magnetic and electric fields of the lattice may then be large compared with the radiation line width and may be observed as a structure in the resonance lines.

8. Results of Nuclear Moment Measurements

Magnetic and Electric Shielding. All measurements of nuclear magnetic moments are confused by the fact that the nucleus is in an atom or molecule at the time of the measurement. When the external magnetic field H_0 is applied, a diamagnetic circulation of the electrons in the atom or molecule is induced which in turn creates a small magnetic field $-\sigma H_0$ at the nucleus. The nucleus then precesses in the resultant of these two magnetic fields $(1 - \sigma)H_0$. Therefore, all nuclear magnetic-moment measurements must be corrected for shielding by the factor $1/(1 - \sigma)$ before the actual nuclear moment can be obtained. Theories of the shielding constant σ have been developed by Lamb and Ramsey [1]. The magnetic-moment values in the accompanying tables have all been corrected for magnetic shielding.

An analogous correction is necessary in nuclear electric-quadrupole-moment determinations as discussed by Sternheimer and others [1]. At first sight it might appear that all electrons forming closed shells in an atom would not contribute to the field gradient q because of their spherical symmetry so that the value of q could be calculated from the valence electron configurations alone. However, this is not the case since the inner core electrons are distorted by the nuclear quadrupole moment and thereby contribute to the quadrupole interaction, in other words they provide an electrical shielding or in some cases an antishielding. Sometimes this shielding or antishielding is very large.

Nuclear Moment Tables. Values for the nuclear spins, nuclear magnetic-dipole moments, nuclear electric-quadrupole moments, and nuclear magnetic-octupole moments that have been measured by any of the above techniques are included in Table 3.1. References to the original experiments which provided these data are not included since these references are given in other available tables [1, 3, 12].

Significance of Some Nuclear Moment Results. The proton magnetic moment is not exactly one nuclear magneton and the neutron magnetic moment is not exactly zero, as would be expected from the simplest considerations based on Dirac's theory. It is, however, always possible to add an additional magnetic moment to the one that naturally arises in Dirac's theory. Attempts have been made with partial success to account for the anomalous nucleon moments on the basis of the meson theory of nuclear forces, since the resultant magnetic moment is contributed to by the magnetic moment and currents of the mesons which have a finite probability of existence within the range of nucleon forces.

The existence of the quadrupole moment of the deuteron indicates that the ground state of the deuteron is not exactly spherically symmetrical, as had been expected on the basis of the theories used prior to the discovery of the deuteron quadrupole moment. This spherical asymmetry can be accounted for with a noncentral spin-dependent force (tensor force) which makes the ground state of the deuteron a mixture of ^3S and ^3D.

The magnetic moment of the proton plus that of the neutron is approximately, but not exactly, equal to the moment of the deuteron. The departure can be accounted for by the orbital contributions of the ^3D part of the state which arises from the tensor force.

The magnetic moments of $_1H^3$ and $_2He^3$ are not exactly equal to the moments of the proton and neutron, respectively, and even the tensor force does not account fully for the discrepancy. However, the discrepancy can be successfully attributed to the effects of meson exchange currents inside the complex nuclei.

As discussed in Part 7, Chap. 4, hyperfine structure ratios for pairs of isotopes do not agree exactly with the calculated values from the nuclear moment ratios of the two isotopes. The discrepancy can be attributed to the finite size of the nuclei and the distribution of the magnetism in the nuclei.

Isotopes whose constitutions differ by just two neutrons often have nearly equal magnetic moments. This empirical result lends support to the nuclear shell theories which predict it.

It has been pointed out by Schmidt [2] that almost all nuclear moments of odd mass number nuclei fall between two limits set by very simple considerations. These limits correspond to assuming that the entire

Table 3.1. Nuclear Moment Values

Explanation of Table

Nucleus — Chemical symbol with Z, A, and N numbers.
States, other than ground states, are designated by ^{m} following the A number.

$T_{1/2}$ or level (kev) — The half-life for radioactive nucleii or the energy of the level of the excited state.
For excited states either the half-life or the level energy may be given, depending on the type of experiment. For example, half-lives are used for atomic beam measurements and level energies for Mössbauer experiments since these are the quantities identifying the samples in these two types of experiment. m is minute, y is year, s second, ky thousand years, h hours, d day, My million years, Ty 10^{12} years, and Gy 10^{9} years.

I — Nuclear spin or angular momentum.
Values enclosed in brackets were not determined by spectroscopic or resonance measurements but were assumed in order to interpret data.

μ — Nuclear magnetic moment with diamagnetic correction in units of nuclear magnetons, $he/4\pi Mc = 5.0505 \times 10^{-22}$ erg/gauss.

Diam. corr. — The diamagnetic correction which was added to the last significant figure of the uncorrected magnetic-dipole moment to get the value quoted in the previous column.
For example, for Li^{6}, $\mu = \mu_{\text{uncorrected}} + \text{diam. corr.} = +0.82193 + 0.00008 = +0.82201$.

Q — Nuclear electric-quadrupole moment in units of barns or 10^{-24} cm^{2}.
Values marked by ^{s} are averages of the Q values obtained in different experiments.
Those marked by ^{r} have been calculated using Q^{s} and measured Q ratios.

Ω — Nuclear magnetic-octupole moment in units of nuclear magneton-barns.

Method — Indication of measurement technique in accordance with the following code:

A: Neutron, proton, and antiproton moments
B: Nuclear moments by paramagnetic resonance
C: Nuclear moments by microwave spectroscopy
D: Nuclear moments by quadrupole resonance
E: Nuclear moments by nuclear magnetic resonance
F: Nuclear moments by atomic and molecular beams
G: Nuclear moments by optical spectroscopy
H: Nuclear moments by optical double resonance and pumping techniques
I: Nuclear moments by Mössbauer spectroscopy

The listed values are all based on the tables of Fuller and Cohen [12] where experimental errors and reference to the original works can be found.

Nucleus	Nuclear moment values						Method
	$T_{1/2}$ or level (kev)	I	μ	Diam. corr.	Q	Ω	
${}_{0}n^{1}{}_{1}$	12 m	½	−1.9131				A
${}_{1}\bar{H}^{1}{}_{0}$ (antiproton)			−1.8				A
${}_{1}H^{1}{}_{0}$		½	+2.79278	8			A B F G H
${}_{1}H^{2}{}_{1}$		1	+0.85742	2	+0.0028		E F G H
${}_{1}H^{3}{}_{2}$	12 y	½	+2.9789	1			E F G H
${}_{2}He^{3}{}_{1}$		½	−2.1276	1			E F G
${}_{2}He^{3+}{}_{1}$		½					F
${}_{2}He^{4}{}_{2}$		0					G
${}_{2}He^{6}{}_{4}$	0.8 s	0^{a}					F
${}_{3}Li^{6}{}_{3}$		1	+0.82201	8	-0.0008^{r}		E F
${}_{3}Li^{7}{}_{4}$		3/2	+3.2564	3	-0.04^{s}		E F G H
${}_{3}Li^{8}{}_{5}$	0.8 s	[2]	+1.6532	2			E
${}_{4}Be^{9}{}_{5}$		3/2	−1.1776	2	±0.03		E F
${}_{5}B^{10}{}_{5}$		3	+1.8007	4	$+0.08^{r}$		C D E F
${}_{5}B^{11}{}_{6}$		3/2	+2.6885	5	$+0.04^{s}$		C D E F
${}_{6}C^{11}{}_{5}$	21 m	3/2	±1.03		±0.031	μ/Q neg	F
${}_{6}C^{12}{}_{6}$		0					G
${}_{6}C^{13}{}_{7}$		½	+0.7024	2			E F G
${}_{6}C^{14}{}_{8}$	5.6 ky	0					G
${}_{7}N^{13}{}_{6}$	10 m	½	±0.3223	1			F
${}_{7}N^{14}{}_{7}$		1	+0.4036	1	+0.01		C E F G H
${}_{7}N^{15}{}_{8}$		½	−0.2831	1			E F G H

TABLE 3.1. NUCLEAR MOMENT VALUES (*Continued*)

Nucleus	Nuclear moment values						Method
	$T_{1/2}$ or level (kev)	I	μ	Diam. corr.	Q	Ω	
$_{8}O^{15}_{7}$	2.1 m	$\frac{1}{2}$	± 0.7189	3			F
$_{8}O^{16}_{8}$		0					G
$_{8}O^{17}_{9}$		$\frac{5}{2}$	-1.8937	7	-0.026		B C E
$_{8}O^{18}_{10}$		0					C
$_{9}F^{19}_{10}$		$\frac{1}{2}$	$+2.6287$	12			B C E F G H
$_{9}F^{20}_{11}$	11 s	[2]	$+2.094$	1			E
$_{10}Ne^{19}_{9}$	18 s	a	-1.887	1			F
$_{10}Ne^{20}_{10}$		0					F G
$_{10}Ne^{21}_{11}$		$\frac{3}{2}$	-0.6618	4	$+0.09$		F G
$_{10}Ne^{22}_{12}$		0§					G
$_{11}Na^{21}_{10}$	23 s	$\frac{3}{2}$	$+2.386$	1			F
$_{11}Na^{22}_{11}$	2.6 y	3	$+1.746$	1			F
$_{11}Na^{23}_{12}$		$\frac{3}{2}$	$+2.2176$	14	$+0.11$		E F G H
$_{11}Na^{24}_{13}$	15 h	4	$+1.689$	1			F
$_{12}Mg^{24}_{12}$		0^a					G
$_{12}Mg^{25}_{13}$		$\frac{5}{2}$	-0.8553	6	$+0.22$		B E F G
$_{12}Mg^{26}_{14}$		0^a					G
$_{13}Al^{27}_{14}$		$\frac{5}{2}$	$+3.6414$	29	$+0.15$		F F G
$_{14}Si^{28}_{14}$		0^a					C E
$_{14}Si^{29}_{15}$		$\frac{1}{2}$	-0.5553	5			C E
$_{14}Si^{30}_{16}$		0^a					C
$_{15}P^{31}_{16}$		$\frac{1}{2}$	$+1.1317$	11			B C E F G
$_{15}P^{32}_{17}$	14 d	1	-0.2523	2			B
$_{16}S^{32}_{16}$		0					G
$_{16}S^{33}_{17}$		$\frac{3}{2}$	$+0.6436$	7	-0.055^s		C D E
$_{16}S^{34}_{18}$		0^a					C
$_{16}S^{35}_{19}$	87 d	$\frac{3}{2}$	$+1.00$ or -1.07		$+0.04^r$		C D
$_{16}S^{36}_{20}$		0^a					C
$_{17}Cl^{35}_{18}$		$\frac{3}{2}$	$+0.82183$	94	-0.079^s	-0.019	C D E F
$_{17}Cl^{36}_{19}$	0.3 My	2	$+1.285$	1	-0.017		C E
$_{17}Cl^{37}_{20}$		$\frac{3}{2}$	$+0.68411$	78	-0.062^r	-0.015	C D E F
$_{18}A^{36}_{18}$		0^a					G
$_{18}A^{37}_{19}$	34 d	$\frac{3}{2}$	± 1.0				G
$_{18}A^{38}_{20}$		0^a					G
$_{18}A^{40}_{22}$		0^a					G
$_{19}K^{39}_{20}$		$\frac{3}{2}$	$+0.3914$	5	$+0.09^s$		E F G H
$_{19}K^{40}_{21}$	1.3 Gy	4	-1.298	2	-0.07^r		F H
$_{19}K^{41}_{22}$		$\frac{3}{2}$	$+0.2148$	3	$\pm 0.11^s$		E F H
$_{19}K^{42}_{23}$	12 h	2	-1.140	2			F
$_{19}K^{43}_{24}$	22 h	$\frac{3}{2}$	± 0.163				F
$_{20}Ca^{40}_{20}$		0^a					G
$_{20}Ca^{41}_{21}$	110 ky	$\frac{7}{2}$	-1.595	2			E
$_{20}Ca^{43}_{23}$		$\frac{7}{2}$	-1.317	2			E F G
$_{21}Sc^{43}_{22}$	3.9 h	$\frac{7}{2}$	$+4.52$		≈ -0.05		F
$_{21}Sc^{44}_{23}$	3.9 h	2	$+2.56$		$+0.14^r$		F
$_{21}Sc^{44m}_{23}$	2.4 d	6	$+3.96$		-0.37^r		F
$_{21}Sc^{45}_{24}$		$\frac{7}{2}$	$+4.7564$	72	-0.22^s		E F G
$_{21}Sc^{46}_{25}$	84 d	4	$+3.03$		$+0.12^r$		F
$_{21}Sc^{47}_{26}$	3.4 d		$+5.31$		-0.22^r		F

TABLE 3.1. NUCLEAR MOMENT VALUES (*Continued*)

Nucleus	Nuclear moment values						Method
	$T_{1/2}$ or level (kev)	I	μ	Diam. corr.	Q	Ω	
${}_{22}Ti^{45}{}_{23}$	3.1 h	7/2					F
${}_{22}Ti^{47}{}_{25}$		5/2	−0.7884	13			B E
${}_{22}Ti^{49}{}_{27}$		7/2	−1.1040	18			B E
${}_{23}V^{50}{}_{27}$	10^{15} y	6	+3.3470	57			B E
${}_{23}V^{51}{}_{28}$		7/2	+5.148	9	+0.27		B E G
${}_{24}Cr^{49}{}_{25}$	42 m	5/2					F
${}_{24}Cr^{51}{}_{27}$	28 d	7/2					F
${}_{24}Cr^{53}{}_{29}$		3/2	−0.4744	9	−0.03		B E F H
${}_{25}Mn^{51}{}_{26}$	45 m	5/2					F
${}_{25}Mn^{52}{}_{27}$	5.7 d	6	±3.00	1			B
${}_{25}Mn^{53}{}_{28}$	140 y	7/2	±5.05	1			B
${}_{25}Mn^{54}{}_{29}$	290 d	3	±3.30	1			B
${}_{25}Mn^{55}{}_{30}$		5/2	+3.468	7	+0.4		B C E F G
${}_{25}Mn^{56}{}_{31}$	2.6 h	3	+3.240	6			F
${}_{26}Fe^{57}{}_{31}$		1/2	+0.0905	2			B F
${}_{26}Fe^{57m}{}_{31}$	14.4	[3/2]	−0.155		$+0.2^{b}$		I
${}_{26}Fe^{59}{}_{33}$	45 d	3/2					F
${}_{27}Co^{56}{}_{29}$	77 d	4	±3.85	1			B
${}_{27}Co^{57}{}_{30}$	270 d	7/2	±4.65	1			B
${}_{27}Co^{58}{}_{31}$	71 d	1,2	±4.06	1			B
${}_{27}Co^{59}{}_{32}$		7/2	+4.649	10	+0.4		B E F G
${}_{27}Co^{60}{}_{33}$	5.3 y	5	+3.81	1			B
${}_{28}Ni^{61}{}_{33}$		3/2	±0.75				B E G
${}_{28}Ni^{61m}{}_{33}$	71	[5/2]	+0.35				I
${}_{29}Cu^{60}{}_{31}$	24 m	2					F
${}_{29}Cu^{61}{}_{32}$	3.3 h	3/2	+2.16				F
${}_{29}Cu^{62}{}_{33}$	9.7 m	1					F
${}_{29}Cu^{63}{}_{34}$		3/2	+2.226	5	-0.18^{s}		B D E F G
${}_{29}Cu^{64}{}_{35}$	13 h	1	−0.216				F
${}_{29}Cu^{65}{}_{36}$		3/2	+2.385	6	-0.19^{r}		B D E F G
${}_{29}Cu^{66}{}_{37}$	5.2 m	1	±0.283	1			F
${}_{30}Zn^{64}{}_{34}$		0^{a}					G
${}_{30}Zn^{65}{}_{35}$	245 d	5/2	+0.769	2	-0.026^{r}		H
${}_{30}Zn^{66}{}_{36}$		0^{a}					G
${}_{30}Zn^{67}{}_{37}$		5/2	+0.8757	22	$+0.17^{s}$		E F G H
${}_{30}Zn^{68}{}_{38}$		0^{a}					G
${}_{31}Ga^{66}{}_{35}$	9.5 h	0^{a}					F
${}_{31}Ga^{67}{}_{36}$	78 h	3/2	+1.850	5	$+0.22^{r}$		F
${}_{31}Ga^{68}{}_{37}$	68 m	1	±0.0117		$+0.031^{r}$		F
${}_{31}Ga^{69}{}_{38}$		3/2	+2.016	5	$+1.19^{s}$	+0.14	D E F G
${}_{31}Ga^{70}{}_{39}$	21 m	1					F
${}_{31}Ga^{71}{}_{40}$		3/2	+2.562	7	$+0.12^{r}$	+0.18	D E F G
${}_{31}Ga^{72}{}_{41}$	14 h	3	−0.1322	4	$+0.59^{r}$		F
${}_{32}Ge^{70}{}_{38}$		0^{a}					C
${}_{32}Ge^{71}{}_{39}$	11 d	1/2	+0.546	1			F
${}_{32}Ge^{72}{}_{40}$		0^{a}					C
${}_{32}Ge^{73}{}_{41}$		9/2	−0.8792	24	−0.22		C E F
${}_{32}Ge^{74}{}_{42}$		0^{a}					C
${}_{32}Ge^{76}{}_{44}$		0^{a}					C
${}_{33}As^{75}{}_{42}$		3/2	+1.439	4	+0.29		B C D E F G
${}_{33}As^{76}{}_{43}$	26 h	2	−0.905	2	±7 *8*		B F

Table 3.1. Nuclear Moment Values (*Continued*)

Nucleus	Nuclear moment values						Method
	$T_{1/2}$ or level (kev)	I	μ	Diam. corr.	Q	Ω	
${}_{34}\mathrm{Se}^{74}{}_{40}$		0[a]					C
${}_{34}\mathrm{Se}^{75}{}_{41}$	120 d	5/2			+1.0[r]		C
${}_{34}\mathrm{Se}^{76}{}_{42}$		0[a]					C G
${}_{34}\mathrm{Se}^{77}{}_{43}$		1/2	+0.534	1			C E G
${}_{34}\mathrm{Se}^{78}{}_{44}$		0					C G
${}_{34}\mathrm{Se}^{79}{}_{45}$	60 ky	7/2	−1.02		+0.8[s]		C
${}_{34}\mathrm{Se}^{80}{}_{46}$		0					C G
${}_{34}\mathrm{Se}^{82}{}_{48}$		0[a]					C G
${}_{35}\mathrm{Br}^{76}{}_{41}$	17 h	1	±0.548	2	±0.25[r]	μ/Q neg.	F
${}_{35}\mathrm{Br}^{77}{}_{42}$	58 h	3/2					F
${}_{35}\mathrm{Br}^{79}{}_{44}$		3/2	+2.106	7	+0.31[s]		C D E F G
${}_{35}\mathrm{Br}^{80}{}_{45}$	18 m	1	±0.514	2	±0.18[r]	μ/Q pos.	F
${}_{35}\mathrm{Br}^{80m}{}_{45}$	4.5 h	5	+1.317	4	+0.71[r]		F
${}_{35}\mathrm{Br}^{81}{}_{46}$		3/2	+2.270	7	+0.26[r]		C D E F G
${}_{35}\mathrm{Br}^{82}{}_{47}$	36 h	5	±1.626	5	±0.70[r]		F
${}_{36}\mathrm{Kr}^{82}{}_{46}$		0[a]					G
${}_{36}\mathrm{Kr}^{83}{}_{47}$		9/2	−0.970	3	+0.23[s]	−0.18	E F G
${}_{36}\mathrm{Kr}^{84}{}_{48}$		0[a]					G
${}_{36}\mathrm{Kr}^{85}{}_{49}$	11 y	9/2	±1.005	3	+0.38[r]		F G
${}_{36}\mathrm{Kr}^{86}{}_{50}$		0[a]					G
${}_{37}\mathrm{Rb}^{81}{}_{44}$	4.7 h	3/2	+2.05				F
${}_{37}\mathrm{Rb}^{81m}{}_{44}$	32 m	9/2					F
${}_{37}\mathrm{Rb}^{82m}{}_{45}$	6.3 h	5	+1.5				F
${}_{37}\mathrm{Rb}^{83}{}_{46}$	83 d	5/2	+1.4				F
${}_{37}\mathrm{Rb}^{84}{}_{47}$	33 d	2	−1.32				F
${}_{37}\mathrm{Rb}^{85}{}_{48}$		5/2	+1.3527	45	+0.28[s]		E F G H
${}_{37}\mathrm{Rb}^{86}{}_{49}$	19 d	2	−1.691	6			F
${}_{37}\mathrm{Rb}^{87}{}_{50}$	47 Gy	3/2	+2.7506	92	+0.14[r]		E F G H
${}_{37}\mathrm{Rb}^{88}{}_{51}$	18 m	2					F
${}_{38}\mathrm{Sr}^{86}{}_{48}$		0[a]					G
${}_{38}\mathrm{Sr}^{87}{}_{49}$		9/2	−1.093	4	+0.36		E F G H
${}_{38}\mathrm{Sr}^{88}{}_{50}$		0[a]					G
${}_{39}\mathrm{Y}^{89}{}_{50}$		1/2	−0.1373	5			E F G
${}_{39}\mathrm{Y}^{90}{}_{51}$	64 h	2	−1.63	1			F
${}_{39}\mathrm{Y}^{91}{}_{52}$	58 d	1/2	±0.164	1			F
${}_{40}\mathrm{Zr}^{91}{}_{51}$		5/2	−1.303	5			E G
${}_{41}\mathrm{Nb}^{93}{}_{52}$		9/2	+6.167	24	−0.22		E G
${}_{42}\mathrm{Mo}^{92}{}_{50}$		0[a]					G
${}_{42}\mathrm{Mo}^{94}{}_{52}$		0[a]					G
${}_{42}\mathrm{Mo}^{95}{}_{53}$		5/2	−0.9135	36			B E G
${}_{42}\mathrm{Mo}^{96}{}_{54}$		0[a]					G
${}_{42}\mathrm{Mo}^{97}{}_{55}$		5/2	−0.9327	37	${}^{97}/_{95} > \pm 1$		B E G
${}_{42}\mathrm{Mo}^{98}{}_{56}$		0[a]					G
${}_{42}\mathrm{Mo}^{100}{}_{58}$		0[a]					G
${}_{43}\mathrm{Tc}^{99}{}_{56}$	210 ky	9/2	+5.68	2	+0.3		B E G
${}_{44}\mathrm{Ru}^{99}{}_{55}$		5/2	−0.63				B G
${}_{44}\mathrm{Ru}^{99m}{}_{55}$	90	3/2	−0.29				I
${}_{44}\mathrm{Ru}^{101}{}_{57}$		5/2	−0.69				B G
${}_{45}\mathrm{Rh}^{103}{}_{58}$		1/2	−0.0883	4			E G
${}_{46}\mathrm{Pd}^{105}{}_{59}$		5/2	−0.6015	26			E G

TABLE 3.1. NUCLEAR MOMENT VALUES (*Continued*)

Nucleus	Nuclear moment values						Method
	$T_{1/2}$ or level (kev)	I	μ	Diam. corr.	Q	Ω	
$_{47}Ag^{103}{}_{56}$	59 m	7/2					F
$_{47}Ag^{104}{}_{57}$	1.2 h	5	+4.0				F
$_{47}Ag^{104m}{}_{57}$	27 m	2	+3.7				F
$_{47}Ag^{105}{}_{58}$	40 d	1/2	±0.10[t]				F
$_{47}Ag^{106}{}_{59}$	24 m	1	+ large				F
$_{47}Ag^{106}{}_{59}$	8.3 d	6					F
$_{47}Ag^{107}{}_{60}$		1/2	−0.1135	5			E F G
$_{47}Ag^{108}{}_{61}$	2.4 m	1	+4.2				F
$_{47}Ag^{109}{}_{62}$		1/2	−0.1305	6			E F G
$_{47}Ag^{110m}{}_{63}$	253 d	6					F
$_{47}Ag^{111}{}_{64}$	7.5 d	1/2	−0.145	1			F
$_{47}Ag^{112}{}_{65}$	3.2 h	2	±0.054				F
$_{47}Ag^{113}{}_{66}$	5.3 h	1/2	±0.159	1			F
$_{48}Cd^{107}{}_{59}$	6.7 h	5/2	−0.616	3	+0.77[s]		H
$_{48}Cd^{109}{}_{61}$	470 d	5/2	−0.829	4	+0.78[r]		H
$_{48}Cd^{110}{}_{62}$		0[a]					G
$_{48}Cd^{111}{}_{63}$		1/2	−0.5950	28			E F G H
$_{48}Cd^{112}{}_{64}$		0[a]					G
$_{48}Cd^{113}{}_{65}$	>10^15 y	1/2	−0.6224	29.7			E F G H
$_{48}Cd^{113m}{}_{65}$	14 y	11/2	−1.086	5			H
$_{48}Cd^{114}{}_{66}$		0[a]					G
$_{48}Cd^{115}{}_{67}$	2.3 d	1/2	−0.647	3			H
$_{48}Cd^{115m}{}_{67}$	43 d	11/2	−1.044	5	−0.61[r]		H
$_{48}Cd^{116}{}_{68}$		0[a]					G
$_{49}In^{109}{}_{60}$	4.3 h	9/2	+5.53	3	+0.86[r]		F
$_{49}In^{110m}{}_{61}$	4.9 h	7	+10.4 or		−0.21[r] or		F
			−10.7		+0.22[r]		
$_{49}In^{111}{}_{62}$	2.8 d	9/2	+5.53	3	+0.85[r]		F
$_{49}In^{113}{}_{64}$		9/2	+5.523	27	+0.82[r]	+0.57	E F G
$_{49}In^{113m}{}_{64}$	1.7 h	1/2	−0.210	1			F
$_{49}In^{114m}{}_{65}$	50 d	5	+4.7				F
$_{49}In^{115}{}_{66}$	600 Ty	9/2	+5.534	27	+0.83[s]	+0.56	E F G
$_{49}In^{115m}{}_{66}$	4.5 h	1/2	−0.244	1			F
$_{49}In^{116m}{}_{67}$	54 m	5	+4.3				F
$_{49}In^{117}{}_{68}$	45 m	9/2					F
$_{49}In^{117m}{}_{68}$	1.9 h	1/2	±0.25				F
$_{50}Sn^{115}{}_{65}$		1/2	−0.918	5			E F G
$_{50}Sn^{116}{}_{66}$		0[a]					G
$_{50}Sn^{117}{}_{67}$		1/2	−1.000	5			E F G
$_{50}Sn^{118}{}_{68}$		0[a]					G
$_{50}Sn^{119}{}_{69}$		1/2	−1.046	5			E F G
$_{50}Sn^{119m}{}_{69}$	24	[3/2]	+0.71		−0.08		I
$_{50}Sn^{120}{}_{70}$		0[a]					G
$_{51}Sb^{121}{}_{70}$		5/2	+3.359	17	−0.29[c]		B C D E F G
$_{51}Sb^{122}{}_{71}$	2.8 d	2	−1.90	1	+0.69[r]		B F
$_{51}Sb^{123}{}_{72}$		7/2	+2.547	13	−0.37[r]		B C D E F G
$_{51}Sb^{124}{}_{73}$	60 d	3					F
$_{52}Te^{116}{}_{64}$	2.5 h	0[a]					F
$_{52}Te^{117}{}_{65}$	61 m	1/2					F
$_{52}Te^{119}{}_{67}$	16 h	1/2					F
$_{52}Te^{119m}{}_{67}$	4.5 d	11/2					F
$_{52}Te^{123}{}_{71}$	>50 Ty	1/2	−0.7359	39			E G
$_{52}Te^{125}{}_{73}$		1/2	−0.8871	47			E G
$_{52}Te^{125m}{}_{73}$	35.5	3/2	≈+0.7		±0.2		I
$_{52}Te^{126}{}_{74}$		0[a]					G
$_{52}Te^{128}{}_{76}$		0[a]					G
$_{52}Te^{130}{}_{78}$		0[a]					G

TABLE 3.1. NUCLEAR MOMENT VALUES (*Continued*)

Nucleus	Nuclear moment values						Method
	$T_{1/2}$ or level (kev)	I	μ	Diam. corr.	Q	Ω	
$_{53}I^{123}{}_{70}$	13 h	5/2					F
$_{53}I^{124}{}_{71}$	4.0 d	2					F
$_{53}I^{125}{}_{72}$	60 d	5/2	+3.0		−0.89[r]		C
$_{53}I^{126}{}_{73}$	13 d	2					F
$_{53}I^{127}{}_{74}$		5/2	+2.808	15	−0.79[d]	+0.18	C D E F G
$_{53}I^{128}{}_{75}$	25 m	1					F
$_{53}I^{129}{}_{76}$	16 My	7/2	+2.617	14	−0.55[r]		C D E
$_{53}I^{129m}{}_{76}$	26.8	[5/2]	+2.8		−0.68[r]		I
$_{53}I^{130}{}_{77}$	12 h	5					F
$_{53}I^{131}{}_{78}$	8.1 d	7/2	+2.74	2	−0.40[r]		C F
$_{53}I^{132}{}_{79}$	2.3 h	4	±3.08	2	±0.08	μ/Q neg.	F
$_{53}I^{133}{}_{80}$	21 h	7/2	+2.84	2	−0.26[r]		F
$_{53}I^{135}{}_{82}$	6.7 h	7/2					F
$_{54}Xe^{129}{}_{75}$		1/2	−0.7768	43			E F G
$_{54}Xe^{129m}{}_{75}$	40	[3/2]			±0.42[r]		
$_{54}Xe^{131}{}_{77}$		3/2	+0.6908	39	−0.12	+0.048	E F G
$_{54}Xe^{132}{}_{78}$		0[a]					G
$_{54}Xe^{134}{}_{80}$		0[a]					G
$_{54}Xe^{136}{}_{82}$		0[a]					G
$_{55}Cs^{127}{}_{72}$	6.2 h	1/2	+?1.43	1			F
$_{55}Cs^{129}{}_{74}$	31 h	1/2	+?1.479	8			F
$_{55}Cs^{130}{}_{75}$	30 m	1	+1.37 or −1.45	1			F
$_{55}Cs^{131}{}_{76}$	10 d	5/2	+3.54	2			F
$_{55}Cs^{132}{}_{77}$	6.2 d	2	+2.22	1			F
$_{55}Cs^{133}{}_{78}$		7/2	+2.579	15	−0.003		E F G H
$_{55}Cs^{134}{}_{79}$	2.2 y	4	+2.990	17			F
$_{55}Cs^{134m}{}_{79}$	3.1 h	8	+1.096	6			F
$_{55}Cs^{135}{}_{80}$	2 My	7/2	+2.729	16	+0.049		F H
$_{55}Cs^{137}{}_{82}$	30 y	7/2	+2.838	16	+0.050		F H
$_{56}Ba^{134}{}_{78}$		0[a]					G
$_{56}Ba^{135}{}_{79}$		3/2	+0.8372	49	+0.18[s]		D E F G H
$_{56}Ba^{136}{}_{80}$		0[a]					G
$_{56}Ba^{137}{}_{81}$		3/2	+0.9366	55	+0.28[r]		E F G H
$_{56}Ba^{138}{}_{82}$		0[a]					G
$_{57}La^{138}{}_{81}$	0.1 Ty	5	+3.707	22	±0.8[r]		E
$_{57}La^{139}{}_{82}$		7/2	+2.778	17	+0.22[s]		E F G
$_{57}La^{140}{}_{83}$	40 h	3					F
$_{58}Ce^{141}{}_{83}$	33 d	7/2	±0.9				B
$_{59}Pr^{141}{}_{82}$		5/2	+4.5		−0.06		B F G
$_{59}Pr^{142}{}_{83}$	19 h	2	±0.26		±0.03	μ/Q pos.	F
$_{59}Pr^{143}{}_{84}$	14 d	7/2					F
$_{60}Nd^{141}{}_{81}$	2.4 h	3/2					F
$_{60}Nd^{143}{}_{83}$		7/2	−1.1		−0.6		B F G
$_{60}Nd^{145}{}_{85}$		7/2	−0.71		−0.3		B F G
$_{60}Nd^{147}{}_{87}$	11 d	5/2	±0.59				B F
$_{60}Nd^{149}{}_{89}$	1.8 h	5/2					F
$_{61}Pm^{147}{}_{86}$	2.6 y	7/2	±3.2		±0.7	μ/Q pos.	B F G
$_{61}Pm^{148}{}_{87}$	5.4 d	1					F
$_{61}Pm^{149}{}_{88}$	54 h	7/2					F
$_{61}Pm^{151}{}_{90}$	28 h	5/2	±1.8		±1.9	μ/Q pos.	F

TABLE 3.1. NUCLEAR MOMENT VALUES (*Continued*)

Nucleus	Nuclear moment values						Method
	$T_{1/2}$ or level (kev)	I	μ	Diam. corr.	Q	Ω	
${}_{62}\text{Sm}^{147}{}_{85}$	0.1 Ty	7/2	−0.90		± <0.7		B F G
${}_{62}\text{Sm}^{149}{}_{87}$		7/2	−0.75		± <0.7		B F G
${}_{62}\text{Sm}^{149m}{}_{87}$	22	[5/2]	−0.62				I
${}_{62}\text{Sm}^{153}{}_{91}$	47 h	3/2	−0.03		+1.2		F
${}_{62}\text{Sm}^{155}{}_{93}$	24 m	3/2					F
${}_{63}\text{Eu}^{151}{}_{88}$		5/2	+3.464	24	$+0.95^{s}$		B F G I
${}_{63}\text{Eu}^{151m}{}_{88}$	21.7	7/2	+2.57	2	$+1.2^{r}$		I
${}_{63}\text{Eu}^{152}{}_{89}$	13 y	3	±1.924	13	$\pm 2.6^{r}$		B F G
${}_{63}\text{Eu}^{152m}{}_{89}$	9.3 h	0^{a}					F
${}_{63}\text{Eu}^{153}{}_{90}$		5/2	+1.530	11	$+2.42^{r}$		B F G
${}_{63}\text{Eu}^{154}{}_{91}$	16 y	3	±2.000	14			B
${}_{64}\text{Gd}^{153}{}_{89}$	225 d	3/2					F
${}_{64}\text{Gd}^{155}{}_{91}$		3/2	−0.27		+1.3		B E G
${}_{64}\text{Gd}^{157}{}_{93}$		3/2	−0.36		+1.5		B G
${}_{64}\text{Gd}^{159}{}_{95}$	18 h	3/2					F
${}_{65}\text{Tb}^{159}{}_{94}$		3/2	±1.7				B G
${}_{65}\text{Tb}^{160}{}_{95}$	73 d	3					F
${}_{65}\text{Tb}^{161}{}_{96}$	7.1 d	3/2					F
${}_{66}\text{Dy}^{161}{}_{95}$		5/2	±0.42		±1.1	μ/Q neg.	B F I
${}_{66}\text{Dy}^{161m}{}_{95}$	26	5/2	±0.5		±1		I
${}_{66}\text{Dy}^{161m}{}_{95}$	74	[3/2]	±1.5				I
${}_{66}\text{Dy}^{163}{}_{97}$		5/2	±0.58		±1.3	μ/Q pos.	B F
${}_{66}\text{Dy}^{165}{}_{99}$	2.3 h	7/2					F
${}_{66}\text{Dy}^{166}{}_{100}$	82 h	0^{a}					F
${}_{67}\text{Ho}^{161}{}_{94}$	2.5 h	7/2					F
${}_{67}\text{Ho}^{165}{}_{98}$		7/2	+4.1		+3.0		B F G
${}_{67}\text{Ho}^{166}{}_{99}$	27 h	0^{a}					F
${}_{68}\text{Er}^{165}{}_{97}$	10 h	5/2					F
${}_{68}\text{Er}^{166m}{}_{98}$	80	[2]	±0.62		−2		I
${}_{68}\text{Er}^{167}{}_{99}$		7/2	−0.56		+2.8		B F
${}_{68}\text{Er}^{169}{}_{101}$	9.4 d	1/2	+0.513	4			F
${}_{68}\text{Er}^{171}{}_{103}$	7.5 h	5/2	±0.70	1	±2.4	μ/Q neg.	F
${}_{69}\text{Tm}^{166}{}_{97}$	7.7 h	2	±0.05		±4.6	μ/Q pos.	F
${}_{69}\text{Tm}^{167}{}_{98}$	9.6 d	1/2					F
${}_{69}\text{Tm}^{169}{}_{100}$		1/2	−0.229	2			B F G
${}_{69}\text{Tm}^{169m}{}_{100}$	8.4	[3/2]	+0.52		−1.3		I
${}_{69}\text{Tm}^{170}{}_{101}$	127 d	1	±0.24		0.57	μ/Q pos.	F
${}_{69}\text{Tm}^{171}{}_{102}$	1.9 y	1/2	±0.23				F
${}_{70}\text{Yb}^{170m}{}_{100}$	84				negative		I
${}_{70}\text{Yb}^{171}{}_{101}$		1/2	+0.4930	40			B E G
${}_{70}\text{Yb}^{173}{}_{103}$		5/2	−0.678	5	+3.0		B E G
${}_{71}\text{Lu}^{175}{}_{104}$		7/2	+2.23	2	$+5.6^{s}$		E F G
${}_{71}\text{Lu}^{176}{}_{105}$	20 Gy	7	+3.18	3	$+8.0^{r}$		F G
${}_{71}\text{Lu}^{176m}{}_{105}$	3.7 h	1	+0.318	3	-2.3^{r}		F
${}_{71}\text{Lu}^{177}{}_{106}$	6.8 d	7/2	+2.24	2	$+5.4^{r}$		F
${}_{72}\text{Hf}^{177}{}_{105}$		7/2	+0.61		+3		G
${}_{72}\text{Hf}^{178}{}_{106}$		0^{a}					G
${}_{72}\text{Hf}^{179}{}_{107}$		9/2	−0.47		+3		G
${}_{72}\text{Hf}^{180}{}_{108}$		0^{a}					G
${}_{73}\text{Ta}^{181}{}_{108}$		7/2	+2.36	2	+4.2		E G
${}_{73}\text{Ta}^{183}{}_{110}$	5.0 d	7/2					F

TABLE 3.1. NUCLEAR MOMENT VALUES (*Continued*)

Nucleus	Nuclear moment values						Method
	$T_{1/2}$ or level (kev)	I	μ	Diam. corr.	Q	Ω	
$_{74}W^{182}{}_{108}$		0^a					G
$_{74}W^{182m}{}_{108}$	100	[2]			184m/182m ≈ 1		I
$_{74}W^{183}{}_{109}$		½	+0.117	1			E G
$_{74}W^{184}{}_{110}$		0^a					G
$_{74}W^{185}{}_{111}$	74 d	3/2					F
$_{74}W^{186}{}_{112}$		0^a					G
$_{74}W^{187}{}_{113}$	24 h	3/2					F
$_{75}Re^{185}{}_{110}$		5/2	+3.172	28	+2.6		C D E G
$_{75}Re^{186}{}_{111}$	90 h	1	±1.72				F
$_{75}Re^{187}{}_{112}$	60 Gy	5/2	+3.204	28	+2.6		C E G
$_{75}Re^{188}{}_{113}$	17 h	1	±1.76				F
$_{76}Os^{187}{}_{111}$		½	+0.067	1			G
$_{76}Os^{189}{}_{113}$		3/2	+0.6566	59	+0.8		E G
$_{77}Ir^{191}{}_{114}$		3/2	+0.18		+1.3		G
$_{77}Ir^{192}{}_{115}$	74 d	4					F
$_{77}Ir^{193}{}_{116}$		3/2	+0.18		+1.2		G
$_{77}Ir^{193m}{}_{116}$	73	[½?]	<2				I
$_{77}Ir^{194}{}_{117}$	19 h	1					F
$_{78}Pt^{194}{}_{116}$		0^a					G
$_{78}Pt^{195}{}_{117}$		½	+0.6060	56			E G
$_{78}Pt^{196}{}_{118}$		0^a					G
$_{79}Au^{190}{}_{111}$	40 m	1	±0.066	1			F
$_{79}Au^{191}{}_{112}$	3.0 h	3/2	±0.137	1			F
$_{79}Au^{192}{}_{113}$	4.1 h	1	±0.00785	7			F
$_{79}Au^{193}{}_{114}$	18 h	3/2	±0.139	1			F
$_{79}Au^{104}{}_{115}$	39 h	1	±0.074	1			F
$_{79}Au^{195}{}_{116}$	185 d	3/2	±0.1482	14			F
$_{79}Au^{196}{}_{117}$	6.2 d	2	+0.61 or −0.66				F
$_{79}Au^{196m}{}_{117}$	9.7 h	12					F
$_{79}Au^{197}{}_{118}$		3/2	+0.14486	137	+0.58	+0.0112	B F G
$_{79}Au^{197m}$	77	[½]	+0.37				I
$_{79}Au^{198}{}_{119}$	2.7 d	2	+0.58	1			F
$_{79}Au^{199}{}_{120}$	3.2 d	3/2	+0.266	3			F
$_{80}Hg^{193}{}_{113}$	6 h	½					H
$_{80}Hg^{193m}{}_{113}$	11 h	13/2	≈±1				G
$_{80}Hg^{195}{}_{115}$	9.5 h	½	+0.538	5			G H
$_{80}Hg^{195m}{}_{115}$	40 h	13/2	−1.049	10	+1.3[r]		G H
$_{80}Hg^{197}{}_{117}$	65 h	½	+0.524	5			E G H
$_{80}Hg^{197m}{}_{117}$	24 h	13/2	−1.032	10	+1.5[r]		G H
$_{80}Hg^{198}{}_{118}$		0^a					G
$_{80}Hg^{199}{}_{119}$		½	+0.5027	48			E F G H
$_{80}Hg^{200}{}_{120}$		0^a					G
$_{80}Hg^{201}{}_{121}$		3/2	−0.5567	54	+0.45[s]	−0.13	D E F G H
$_{80}Hg^{202}{}_{122}$		0^a					G
$_{80}Hg^{203}{}_{123}$	47 d	5/2	+0.84	1	± ≤13		G
$_{80}Hg^{204}{}_{124}$		0^a					G
$_{81}Tl^{195}{}_{114}$	1.2 h	½					F
$_{81}Tl^{197}{}_{116}$	2.7 h	½					F
$_{81}Tl^{198}{}_{117}$	5.3 h	2	± <0.002				F
$_{81}Tl^{198m}{}_{117}$	1.8 h	7					F

TABLE 3.1. NUCLEAR MOMENT VALUES (*Continued*)

Nucleus	Nuclear moment values						Method
	$T_{1/2}$ or level (kev)	I	μ	Diam. corr.	Q	Ω	
${}_{81}\mathrm{Tl}^{199}{}_{118}$	7.4 h	1/2	+1.59	2			F G
${}_{81}\mathrm{Tl}^{200}{}_{119}$	26 h	2	± ≤0.15				F G
${}_{81}\mathrm{Tl}^{201}{}_{120}$	72 h	1/2	+1.60	2			F G
${}_{81}\mathrm{Tl}^{202}{}_{121}$	12 d	2	± ≤0.15				F G
${}_{81}\mathrm{Tl}^{203}{}_{122}$		1/2	+1.6115	158			E F G
${}_{81}\mathrm{Tl}^{204}{}_{123}$	3.9 y	2	±0.089	1			F
${}_{81}\mathrm{Tl}^{205}{}_{124}$		1/2	+1.6274	160			E F G
${}_{82}\mathrm{Pb}^{206}{}_{124}$		0					G
${}_{82}\mathrm{Pb}^{207}{}_{125}$		1/2	+0.5895	59			E G
${}_{82}\mathrm{Pb}^{208}{}_{126}$		0[a]					G
${}_{83}\mathrm{Bi}^{199}{}_{116}$	25 m	9/2					F
${}_{83}\mathrm{Bi}^{200}{}_{117}$	35 m	7					F
${}_{83}\mathrm{Bi}^{201}{}_{118}$	1.8 h	9/2					F
${}_{83}\mathrm{Bi}^{202}{}_{119}$	1.6 h	5					F
${}_{83}\mathrm{Bi}^{203}{}_{120}$	12 h	9/2	+4.59	5	−0.62[r]		F
${}_{83}\mathrm{Bi}^{204}{}_{121}$	12 h	6	+4.25	4	−0.40[r]		F
${}_{83}\mathrm{Bi}^{205}{}_{122}$	15 d	9/2	+5.5				F
${}_{83}\mathrm{Bi}^{206}{}_{123}$	6.3 d	6	+4.56	5	−0.19[r]		F
${}_{83}\mathrm{Bi}^{209}{}_{126}$	>10^18 y	9/2	+4.080	41	−0.34[s]		D E F G
${}_{83}\mathrm{Bi}^{210}{}_{127}$	5 d	1	±0.0442	4	±0.13[r]	μ/Q neg.	F
${}_{84}\mathrm{Po}^{201}{}_{117}$	18 m	3/2					F
${}_{84}\mathrm{Po}^{202}{}_{118}$	51 m	0[a]					F
${}_{84}\mathrm{Po}^{203}{}_{119}$	42 m	5/2					F
${}_{84}\mathrm{Po}^{204}{}_{120}$	3.5 h	0[a]					F
${}_{84}\mathrm{Po}^{205}{}_{121}$	1.8 h	5/2	≈+0.26		+0.17		F
${}_{84}\mathrm{Po}^{206}{}_{122}$	8.8 d	0[a]					F
${}_{84}\mathrm{Po}^{207}{}_{123}$	6.0 h	5/2	≈+0.27		+0.28	+0.11	F
${}_{84}\mathrm{Po}^{209}{}_{125}$	103 y	1/2					G
${}_{84}\mathrm{Po}^{210}{}_{126}$	138 d	0[a]					F
${}_{85}\mathrm{At}^{211}{}_{126}$	7.2 h	9/2					F
${}_{89}\mathrm{Ac}^{227}{}_{138}$	22 y	3/2	+1.1		+1.7		G
${}_{90}\mathrm{Th}^{229}{}_{139}$	7.3 ky	5/2	+0.38		≈4.6		G
${}_{91}\mathrm{Pa}^{231}{}_{140}$	34 ky	3/2	±1.98	2			B G
${}_{91}\mathrm{Pa}^{233}{}_{142}$	27 d	3/2	+3.4		−3.0		F
${}_{92}\mathrm{U}^{233}{}_{141}$	0.2 My	5/2	+0.54		+3.5[s]		B G
${}_{92}\mathrm{U}^{235}{}_{143}$	0.7 Gy	7/2	−0.35		+4.1[r]		B G
${}_{93}\mathrm{Np}^{237}{}_{144}$	2.2 My	5/2	+5		negative		B G
${}_{93}\mathrm{Np}^{238}{}_{145}$	2.1 d	2					F
${}_{93}\mathrm{Np}^{239}{}_{146}$	2.3 d	5/2					B F
${}_{94}\mathrm{Pu}^{239}{}_{145}$	24 ky	1/2	+0.21				B F G
${}_{94}\mathrm{Pu}^{241}{}_{147}$	13 y	5/2	−0.73		+5.6		B G
${}_{95}\mathrm{Am}^{241}{}_{146}$	460 y	5/2	+1.4		+4.9		F G
${}_{95}\mathrm{Am}^{242}{}_{147}$	16 h	1	±0.33		±2.8	μ/Q neg.	F
${}_{95}\mathrm{Am}^{243}{}_{148}$	8 ky	5/2	+1.4		+4.9		G
${}_{96}\mathrm{Cm}^{242}{}_{146}$	160 d	0[a]					F

[r] From Q ratio and Q^s.
[s] Average of tabulated values.
[a] No hyperfine structure observed.
[b] Polarization correction included.
[c] Weighted average.
[d] Adopted value of 59S146.

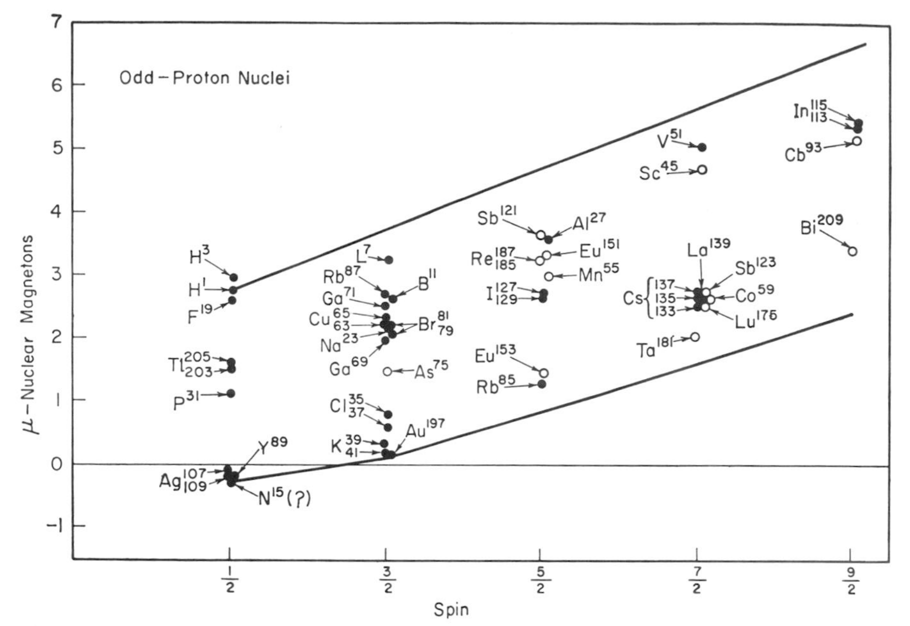

Fig. 3.2. Magnetic moments and Schmidt limits as a function of spin in nuclei of odd Z.

nuclear spin and magnetic moment arise from the one odd nucleon and that one limit corresponds to the nucleon's spin and orbital angular momentum being parallel and the other antiparallel. This is illustrated in Fig. 3.2 for odd proton nuclei; a similar figure applies to odd neutron nuclei.

The nuclear quadrupole moments vary with numbers of odd like nucleons in the nucleus in a fashion that fluctuates about zero in such a way that the quadrupole moments are low for numbers of such nucleons equal to the magic numbers—2, 8, 14, 20, 28, 50, 82, or 126—of nuclear shell theories.

The nuclear shell theories of Jensen, Mayer, and others have proved to be quite fruitful in predicting spins and magnetic moments of nuclei. However, the simple forms of such shell theories have predicted quadrupole moments that are much too small. Bohr and others have modified the shell models to allow for the distortion of the nuclear core. In such a way they improve not only the quadrupole predictions but also most other predictions of the theory.

The magnetic shielding varies from one compound to another and may even vary from one part to another of a single molecule. The study of such chemical shifts promises to be a useful means for the analysis of chemical compounds [1]. Relaxation time studies, direct and electron-coupled nuclear spin-spin interactions, quadrupole interactions, rotational magnetic moments, and similar measurements with nuclear moment techniques have proved to be valuable means for the study of molecular and solid-state properties [1].

References

1. Ramsey, N. F.: "Nuclear Moments," Wiley, New York, 1953.
2. Ramsey, N. F.: "Molecular Beams," Oxford University Press, Fair Lawn, N.J., 1955.
3. Walchli, H. E.: A Table of Nuclear Moment Data, *Report ORNL*-1469, Oak Ridge, Tenn.
4. Condon, E. U., and G. H. Shortley: "Theory of Atomic Spectra," Cambridge University Press, New York, 1935.
5. Foster, E. W.: Nuclear Effects in Atomic Spectra, *Repts. Progr. Phys.*, **14**: 288 (1951).
6. Hahn, E.: Spin-echo Method, *Phys. Rev.*, **80**: 580 (1950).
7. Kellogg, J. M. B., and S. Millman: *Revs. Mod. Phys.*, **18**: 323 (1946)
8. Kopfermann, H.: "Kernmomenta," Akademische Verlagsgesellschaft, m.b.H., Leipzig, 1940.
9. Pake, G. E.: Fundamentals of Nuclear Magnetic Resonance Absorption, *Am. J. Phys.*, **18**: 438, 473 (1950).
10. Schwartz, C.: *Phys. Rev.*, **97**: 380 (1955).
11. Brossel, J., and A. Kastler: *Compt. Rend.*, **229**: 1213 (1949); *J. Phys. Radium*, **11**: 255 (1950).
12. Fuller, G. H., and V. W. Cohen: Nuclear Moments, Oak Ridge National Laboratory, Oak Ridge, Tenn., 1965.
13. Mössbauer Effect Conference, *Revs. Mod. Phys.*, **36**(1, pt. 2): (1964).

Chapter 4

Alpha Particles and Alpha Radioactivity

By WILLIAM E. STEPHENS, University of Pennsylvania, *and*
THEODOR HURLIMANN, Federal Institute for Reactor Research (Switzerland)

1. Alpha Particles

Alpha particles played a central role in the early development of nuclear physics. Natural alpha radioactivity, discovered by Becquerel in 1896, was the first revelation of instability of nuclei. Examination of the chemical properties of the uranium and thorium series of radioactive transformations by Rutherford and Soddy in 1903 suggested the existence of isotopes. The interpretation of alpha-particle scattering by Rutherford in 1911 established the nuclear atom model. Natural alpha particles were used to cause the first observable nuclear transmutation, to produce the first artificial radioactivity, and in the discovery of the neutron and fission. The explanation of the alpha radioactivity energy–half-life relation by Gurney and Condon, and independently by Gamow, in 1928 was the first triumph for quantum theory applied to the nucleus and introduced the nonclassical concept of penetration of barriers. The stability of the alpha particle is important for the concept of short-range saturating forces in the nucleus.

The alpha particle is the nucleus of the helium atom, usually designated by ${}^4_2He$ while the doubly charged ion of helium, i.e., the helium nucleus, is often denoted by the Greek letter alpha (α). The mass of the helium-four atom on the mass scale in which ${}^{12}C$ has a mass of exactly 12 is 4.0026036 amu (atomic mass units, ${}^{12}C$ scale) [1]† with an uncertainty of 3 in the last place (see Part 9, Chap. 2). Using Avogadro's number [3] gives a mass of 6.6456×10^{-27} kg for the helium atom.

The charge on the alpha particle is two positive electronic charges. The charge-to-mass ratio (specific charge) of the alpha particle can be calculated by subtracting two electron masses from the helium mass and dividing into twice the faraday [3], giving 4.8225 coulombs/kg (see Part 7, Chap. 1). Measured values [4] are consistent with this when corrected for relativistic change of mass.

The size of the alpha particle depends on what aspect is important (see Part 9, Chap. 1, Sec. 3). As a moving particle it has associated with it a wave function whose wavelength is

$$\lambda\!\!\!^{-} = 2.26 \times 10^{-13} \frac{1 + 4/A}{E_0^{1/2}} \qquad \text{cm}$$

† Numbers in brackets refer to References at end of chapter.

where E_0 is the kinetic energy in Mev in the laboratory system and A is the atomic weight of the scattering nucleus. The helium nucleus needs a radius of about 2×10^{-13} cm [5] to allow theoretical calculations to fit the binding energies of light nuclei. The size calculated on this basis depends on the assumed nuclear potential. A radius of about 2.4×10^{-13} cm would be consistent with the observed Coulomb differences in light nuclei [6]. High-energy-electron scattering experiments [7] measure the extent of the charge distribution and are consistent with a Gaussian distribution $\rho = \exp(-r^2/a^2)$, where $a = 1.31$ fermis.

The alpha particle has zero spin as determined from band spectra [8] and from scattering experiments [9]. Its magnetic moment is presumably zero. Its statistics are Bose-Einstein.

The cosmic abundance of helium is about one-tenth that of hydrogen and several hundred times that of oxygen [10] (see Part 9, Chap. 12). This is thought to be related to the origin of the universe and stellar transmutation of hydrogen to helium in Bethe's nuclear energy cycles [11]. The relative terrestrial abundance of helium is much lower, presumably because of its loss during and since the formation of the planetary system [12].

Helium has two stable isotopes: ${}^4_2He$ and ${}^3_2He$. Helium three is found naturally in varying amounts. Atmospheric helium contains 1.3×10^{-6} of helium three [13], well helium about 10^{-7}, and spodumene (a lithium mineral) about 10^{-5}. These variations may well be due to cosmic-ray-induced transmutations ${}^{14}N + n \rightarrow 3{}^4He + {}^3H$ or ${}^6Li(n,\alpha){}^3H$ followed by ${}^3H \rightarrow {}^3He + \beta$. Helium enriched further in helium three has been produced from (1) decay of artificially produced tritium, (2) thermal diffusion isotope separation equipment, and (3) nuclear transmutations such as ${}^2D(dn){}^3He$ and ${}^6Li(p,\alpha){}^3He$. The mass of helium three is 3.0160299 ± 2 on the ${}^{12}C$ scale [1]. Helium three has a spin of $\frac{1}{2}$ [14] and a magnetic moment of -2.128 muclear magnetons [15]. (See Part 8, Chap. 3, Table 3.1.)

The total binding energies of the helium isotopes can be determined as ${}^3He$, 7.717 Mev; ${}^4He$, 28.295 Mev; ${}^5He$, 27.338 Mev; and, ${}^6He$, 29.259 Mev. This trend, especially the unstability of ${}^5He$, is the most direct evidence for the saturated character of the alpha particle. These binding energies also afford a test of hypotheses and theoretical calculations on nuclear forces. Theoretical calculations have been made also to predict excited states of the alpha particle and to explain details of nuclear transmutations involving ^{4}He. These calculations have been summarized by Bethe [16] and later by Trainor [17] with the conclusion that a bound excited state of the alpha particle remains an open question. Recent analysis of ^{4}He breakup reactions has suggested states in ^{4}He at 19.94 Mev excitation (narrow) and at 21.24 Mev excitation (broad) [18].

Sources of alpha particles include natural and artificially produced alpha radioactivity, artificially accelerated helium ions, and occasional cosmic rays near the top of the atmosphere. Much of the exploratory work in nuclear physics was accomplished with the natural alpha particles. They have the advantage of being monoenergetic or nearly so. The intensity available, however, is a stringent limitation on research. Sources of the order of 1 curie of alpha activity represent practical limits and are hazardous because of migration of the intense activity [19]. Such a source supplies approximately 10^{-11} amp or 10^7 alpha particles per second in a directed beam and 10^{-8} amp or 10^{10} alpha particles per second undirected. Their energies vary from about 4 to 9 Mev (see Table 4.10). Useful sources have been prepared of the following:

Element	E_α(Mev)	Half-life
^{210}Po	5.305†	138.4 days
^{212}Bi(ThC)	6.051‡	60.5 min
^{212}Po(ThC′)	8.785	3.0×10^{-7} sec
^{214}Po(RaC′)	7.687	1.6×10^{-4} sec
^{238}U	4.199‡	4.5×10^9 years
^{239}Pu	5.157‡	2.4×10^4 years
^{241}Am	5.484‡	458 years

† ^{210}Po has about 10^{-5} gamma ray per alpha.
‡ Complex spectrum. The E_α refers to the main group of alphas. (See Table 4.10.)

Radium is mixed with beryllium to provide sources of neutrons produced by the Be(α,n) reaction. Polonium-Be or plutonium-Be mixtures can be used when gamma rays are to be kept to a minimum.

Cockcroft and Walton in 1932 introduced the artificial acceleration of ions. Soon after, cyclotrons, statitrons, and more recently synchrocyclotrons and tandem accelerators have been used to produce high-velocity ions of hydrogen, deuterium, and helium (see Part 9, Chap. 9). Currents of the order of microamperes in directed beams of He$^+$ and He^{++} $(= \alpha)$ have been produced with energies up to hundreds of Mev. Artificially produced alpha particles are found as the result of transmutations. The ^{7}Li$(p,\alpha\alpha)$ reaction produces alphas recoiling from each other with over 17.34 Mev energy. Slow neutrons produce monoenergetic alphas in ^{6}Li$(n,\alpha)^3$He and ^{10}B$(n,\alpha)^7$Li. The yields of these secondary particles are usually very small, and their energy is fixed. Another sparse source of energetic alpha particles is the primary cosmic rays. An appreciable fraction ($\sim$20 per cent) of the primary cosmic rays is composed of nuclei heavier than protons. Alpha particles constitute a major fraction of these nuclei. Their energies extend up to billions of electron volts.

Detection of alpha particles is accomplished by the observation of the ionization produced in matter by the moving charge (see Sec. 2). This ionization can be observed either by scintillations produced by it in certain phosphors and organic crystals, by electrical means as in counters and ionization chambers, by the droplets condensed on the ions in its path in a cloud chamber, by the bubbles produced in an overheated liquid, or by the grains made developable in sensitive photographic emulsions (see Part 9, Chap. 10).

Zinc sulfide is one of the oldest and best phosphors for the detection of energetic alpha particles by scintillations. Table 4.1 gives the average energy lost by the alpha particle per scintillation photon produced in several sensitive substances. The ZnS-Ag activated phosphor discriminates against electrons both by having a greater energy loss per photon for electrons and by having a small effective thickness. This reduces background and allows the detection of alphas in the presence of electrons and gamma rays. Discrimination against protons is possible in some cases.

Table 4.1. Scintillation Materials for Alpha Detection[a]

Luminescent material	Wave-length emitted, A	Trans-parency thick-ness, mg/cm²	Emission time, μsec	Alpha energy loss per photon, ev/photon
ZnS-Ag	4,500	80	10	10
ZnS-Cu	5,200	200	15	9.6
CdS-Ag	7,600	Good	100	7.5
$CaWO_4$	4,300	100	6	160
NaI-Th[b]	4,100	Large	0.25	64[c]
Diphenyl	3,850	Large	1	320
anthracene[d]	4,400	Large	0.003	1,000

[a] Jordan and Bell, *Nucleonics*, **5**: 30 (1949).
[b] Deliquescent.
[c] Sensitive to surface conditions.
[d] W. H. Jordan, *Ann. Rev. Nucl. Sci.*, **1**: 207 (1952).

Ionization chambers measure the ionization produced by the passage of charged particles through gases. Pure argon yields an ionization that is very closely proportional to the energy lost by the alpha particle. Average energy per ion pair values are given in Table 4.3. The charge collected in an ion chamber has two components, the electronic part which is collected in times of the order of microseconds, because of the great mobility of the electrons (in the absence of electronegative gases), and the ionic part which takes milliseconds to collect. "Fast" amplifiers pass only the electronic contribution whose magnitude depends on the position of ionization. "Frisch grids" or cylindrical geometry reduces this dependence. "Slow" amplifiers respond to the total ionization but are hampered by relatively higher (low frequency) noise background, greater recombination, and more "pile-up" [20].

Semiconductor counters [21], a form of solid-state detector, are similar in concept to ionization chambers in semiconducting materials and offer important advantages in detection of nuclear radiations, particularly alpha particles. Table 4.3 shows that the average energy loss per ion pair for alpha particles in silicon and germanium is about 3 ev, compared with about 30 ev for gases. Hence, an alpha particle creates about ten times as many ion pairs in these solids, and the statistics are about three times better than for gas ionization detectors. In addition, the usual smaller distances involved in collection allow higher electric fields and faster collection times. These considerations allow improved resolution and decreased noise. Commercial surface-barrier detectors with a semiconductor barrier layer on the surface of silicon form particularly useful alpha counters. Typical values for resolution are 15 kev width for a 5.3-Mev alpha. Most of this width is statistics (11 kev) with smaller amounts due to thermal noise (7 kev) and amplifier noise (8 kev). A thin layer of gold, 20 to 50 $\mu g/cm^2$, may be used for surface contact and offers little absorption to protons or alphas. In addition to a fast pulse rise time of the order of 5 to 10 nanoseconds, no apparent dead time, and linearity of charge collected with particle energy, these counters offer 100 per cent detection efficiency, good stability, and insensitivity to gamma ray and neutron background.

A recent surge in development of spark chambers suggests that they will be increasingly useful in alpha detection, especially at higher energies [22].

Nuclear emulsions are of particular advantage in recording alpha particles since alpha-particle tracks can often be distinguished from those of other nuclear particles. The range of the track gives the alpha energy, and its path indicates its direction [23]. The charge of the particle producing the track can be determined by counting delta rays as a function of residual range or by determining the "thinning-down" length. The mass of the particle may be determined by grain counting the track as a function of residual range or by measuring small-angle scattering as a function of residual range (or grain counts).

The most precise measurements of alpha-particle energy are made by deflection in a magnetic field: $B\rho/c = (M/2e)V$, where B is the magnetic-field strength properly averaged over the ion path, ρ is the radius of curvature of the deflected path, $M/2e$ is the mass-charge ratio corrected for relativistic mass, that is, $M = M_0/(1 - V^2/c^2)^{1/2}$, and V is the alpha velocity. For natural alpha particles this can be approximated to

$$E_\alpha = \frac{B^2\rho^2(2e)^2}{2M_0c^2}\left[1 - \frac{B^2\rho^2(2e)^2}{2M_0{}^2c^4}\right]$$

A recent precision measurement [24] has been made of the ^{210}Po alphas, using modern NMR (nuclear magnetic resonance) magnetic determinations. This value may well serve as a primary standard. It agrees with several other recent absolute measurements but differs from older ones by a few kilo electron volts. The $B\rho$ value for ^{210}Po was determined to be 331,774 ± 20 gauss-cm corresponding to an energy of 5,304.8 ± 0.6 kev. An even more useful value is the product of the NMR proton frequency and the radius of curvature. For ^{210}Po this product is 1,412,561 kc-cm. Other alpha energies and $B\rho$ values consistent with this standard are as follows:

Emitter	$B\rho$, gauss-cm	E_α (kev)
^{212}Bi	354,346	6,050.6
^{211}Bi	370,720	6,622.2
^{214}Po	399,442	7,686.9
^{212}Po	427,060	8,785.4

2. Passage of Alpha Particles through Matter

When a fast charged particle, such as an alpha particle or proton, traverses matter, it interacts electromagnetically with the nearby atomic electrons, causing excitation or ionization of the atoms. At lower velocities, nuclear collisions and capture and loss of electrons become important. The energy lost by the original particle in an individual electronic interaction is relatively small, and its energy is gradually reduced by numerous such losses until it can no longer ionize. The distance it has traversed is called its range. Variations in the individual energy losses and statistical fluctuations in the number of ionizing events contribute to a straggling in energy loss per unit path length and also in the range. Excellent descriptions and summaries of theory and data concerning the passage of heavy particles through matter are given by Bethe and Ashkin [25], Allison and Warshaw [26], W. Whaling [27], U. Fano [28], Northcliffe [29], and in a recent report [30].

Primary ionization refers to the number of initial ionizations produced directly by the original particle. The total ionization is several times larger than the primary ionization and includes the secondary ionization produced by energetic primary electrons called delta rays. Mott [31] gives the number of delta rays initiated per unit length of path by a particle of charge z and velocity βc as $N = Kz^2/\beta^2$, where K is a constant best determined empirically for the particular technique used. Lea gives a table of the number of delta rays of energy greater than E produced by various-energy alpha particles in tissue [32]. Specific ionization denotes the number of ion pairs produced per unit length of path (reduced to 76 cm Hg pressure and 0°C) and varies with the charge of the initial particle, its velocity, and the matter through which it passes. Figure 4.1 shows a plot of specific ionization of a beam of alpha particles as a function of distance from the source and is called a Bragg curve.

The energy acquired by a free electron from a charged particle of mass M, charge ze, and velocity V passing a distance b away is

$$\epsilon = \frac{2z^2e^4}{mV^2b^2}$$

where m is the mass of the electron. Integrating over such energy losses, Bohr has shown on semiclassical grounds that the rate of energy loss is

$$-dE/dx = (4\pi z^2e^4n/mV^2) \ln (b_{max}/b_{min})$$

where n is the number of electrons per cubic centimeter and b_{max} and b_{min} are the maximum and minimum impact parameters. b_{max} is set by the limit that the time of interaction, roughly b/V, be longer than the

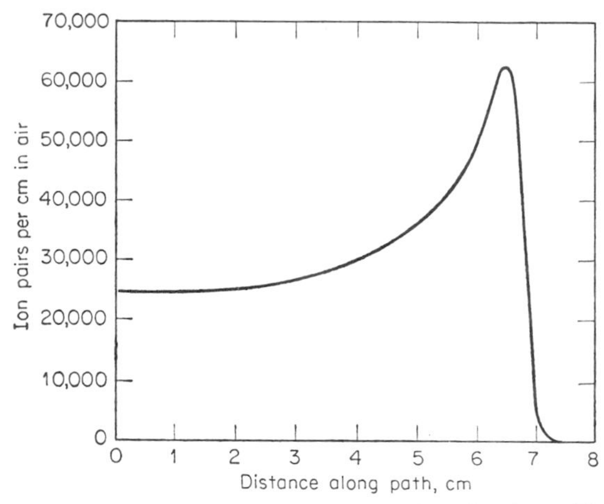

Fig. 4.1. Specific ionization in air (15°C and 760 mm Hg pressure) along a beam of Ra C′ alpha particles. A Bragg curve shows the number of ion pairs per centimeter per alpha particle as a function of distance in centimeters from the source.

period of orbital motion of the atomic electron involved. Consequently,

$$b_{\max} = V/\nu(1 - \beta^2)^{1/2}$$

where ν is an appropriate average rotational frequency for the atomic electrons in the absorbing material. $b_{\min}$ is limited by the maximum transfer of energy allowed by momentum considerations (that is, $2mV^2$), $b_{\min} \approx ze^2/mV^2$, or by uncertainty considerations of the electron position and momentum,

$$b_{\min} = \frac{\hbar(1 - \beta^2)^{1/2}}{mV}$$

whichever is larger.

An essentially similar relation is given by quantum-mechanical derivation by Bethe [33]:

$$-\frac{dE}{dx} = \frac{4\pi e^4 z^2 N}{mV^2}\left\{Z\left[\ln\frac{2mV^2}{I(1 - \beta^2)} - \beta^2\right] - C_k\right\}$$

where N is the number of atoms per cubic centimeter of atomic number Z and average ionization potential I. C_k is a correction term depending on the absorber and V and is unimportant at high velocities. This average ionization potential I cannot be calculated with sufficient accuracy for some purposes; therefore, it is determined empirically from data on stopping. Data summarized recently [30] are given in Table 4.2.

For low velocities this energy-loss formula breaks down because of neglect of capture and loss effects. The observed specific ionization drops to zero at zero velocity, as indicated in Fig. 4.1. For high velocities the ionization loss drops to a minimum at energies several times Mc^2 and then rises, owing to the relativistic terms in dE/dx (Fig. 4.2). At these high velocities, polarization [34] effects due to partial shielding by the absorbing electrons make small density corrections to dE/dx, and Cerenkov radiation [35] is produced as well as *bremsstrahlung*. These density-effect corrections are discussed in detail by Fano [28] and are included in the tables of Barkas and Berger [30].

Table 4.2. Adjusted Mean Excitation Energy [30]

Element	$I_{adj.}$ (ev)	Compounds	$I_{adj.}$ (ev)
		H_2	18.7
		H_2O	65.1
He	42	CO_2	85.9
		SiO_2	121
Li	38	AgCl	384
Be	60	AgBr	434
C	78	NaI	433
N	85 (76.8 for α)	LiI	473
O	89 (99 for α)	CH_4	44.6
Ne	131	Acetylene	63.6
Mg	156	Ethylene	54.6
Al	163	Stilbene	65.2
A	210 (184 for α)	Anthracene	67.0
Fe	273	Polyethylene	54.6
Ni	304		
Cu	314	Mixtures	
Kr	381		
Ag	487	Air	86.8 (84.0 for α)
Sn	516	Emulsion	320 (328 for Ilford G5)
Xe	555	Lucite	65.6
W	748	Mylar	72.6
Pt	787		
Au	797		
Pb	826		
Bi	835		
U	923		

Since $dE/dx = z^2 f(V)$, two particles of the same velocity ionize in proportion to the square of their ionic charge.

$$\frac{dE_1/dx}{dE_2/dx} = \frac{z_1^2}{z_2^2} \qquad \text{for } E_1 = \frac{M_1}{M_2}E_2$$

The energy loss is related to the specific ionization $\mathcal{I}$ by $-(dE/dx) = W'\mathcal{I}$, where W' is a differential average energy per ion pair. As a charged particle is slowed down in a gas, the relative importance of

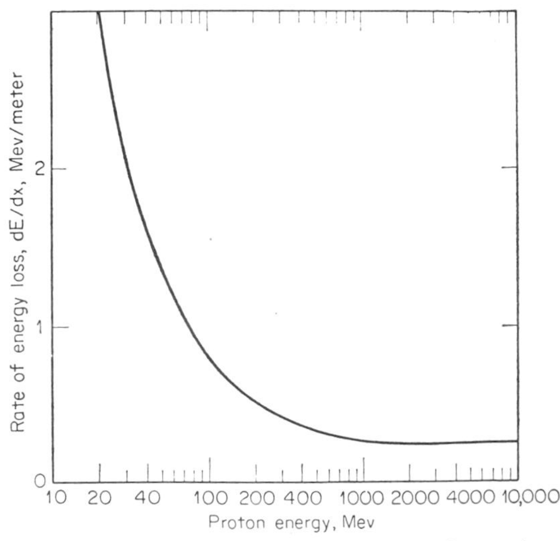

Fig. 4.2. Rate of energy loss for protons in air as a function of proton energy.

TABLE 4.3. AVERAGE ENERGY PER ION PAIR, W (EV)

Substance	Particle		
	Alpha particles	Protons	Electrons
Gases:			
He	31.7[c]; 42.7[e]; 29.6[f]; 46.0[g];	32.4[c]	31[a]; 32.5[c]; 42.3[j]
A	25.9[c]; 26.3[d]; 26.4[e]; 26.25[f]; 26.4[g];	27.7[c]	33[a]; 28.5[i]; 27.0[c]; 26.4[j]
H_2	37.0[c]; 36.3[e]; 37.0[g]; 35.96[h]	38.2[c]	37[a]; 38.0[c]; 36.3[j]
N_2	36.38[m]; 36.4[d]; 36.39[n]; 36.30[f]; 36.3[g]; 36.50[h]	36.3[c]	45[a]; 32.0[i]; 35.8[c]; 34.9[k]
Air	34.97[m]; 35.6[d]; 35.5[e]; 34.96[n]; 35.0[g]; 34.95[h]	36.0[b]	32.0[a]; 31.0[i]; 35.0[c]; 33.9[j]
O_2	32.2[c]; 32.9[d]; 32.5[e]; 32.17[f]; 32.2[g];	34.0[c]	28.8[i]; 32.2[c]; 30.8[j]
CO_2	34.2[d]; 34.5[e]; 33.5[f]; 34.3[g]; 34.3[h]		45[a]; 33.5[i]; 32.8[j]
CH_4	29.0[c]; 29.1[d]; 29.2[e]; 29.4[g]; 29.00[h]		28.5[i]; 30.2[c]; 27.3[k]
BF_3	36.0[g]; 35.3[h]		
Solids:			
Si	3.55[p]		3.55[l]
Ge	2.9[p]		

[a] J. F. Lehmann, *Proc. Roy. Soc. (London)*, **A115**: 624 (1927).
[b] L. H. Gray, *Proc. Cambridge Phil. Soc.*, **40**: 72 (1944).
[c] J. M. Valentine and S. C. Curran, *Phil. Mag.*, **43**: 964 (1952). Proton data normalized to air.
[d] J. Sharpe, *Proc. Phys. Soc. (London)*, **A65**: 859 (1952).
[e] W. P. Jesse and J. Sadauskis, *Phys. Rev.*, **90**: 1120 (1953).
[f] W. Haeberli, P. Huber, and E. Baldinger: *Helv. Phys. Acta*, **26**: 145 (1953).
[g] T. E. Brotner and J. S. Hurst, *Phys. Rev.*, **93**: 1237 (1954).
[h] C. Biber, P. Huber, and A. Müller, *Helv. Phys. Acta*, **28**: 503 (1955).
[i] S. C. Curran, *Phil. Mag.*, **40**: 36 (1949).
[j] R. L. Platzman, in Penetration of Charged Particles in Matter, *NAS-NRC Publ.* 752.
[k] W. P. Jesse, *Phys. Rev.*, **109**: 2002 (1958).
[l] L. Koch, J. Messier, and J. Velin, *I.R.E. Trans. Nucl. Sci.*, **N58**(1): 43 (1961).
[m] Z. Bay, P. E. Newman, and H. H. Seliger, *Radiation Res.*, **14**: 551 (1961).
[n] W. P. Jesse, *Radiation Res.*, **13**: 1 (1960).
[p] Semiconductor Nuclear Particle Detectors, *Natl. Acad. Sci. Natl. Res. Council Publ.* 871, 1961.

ionization, excitation, nuclear collisions, and charge exchange (capture and loss) in absorbing energy might be expected to change. Consequently, the average energy loss of the charged particle per ion pair produced might be expected to vary with its velocity. However, the total ionization produced is found to be approximately proportional to the total energy of the charged particle and the average energy per ion pair W is not very different for different particles [36]. Here $E_0 = W\mathfrak{g}_{\text{total}}$. Table 4.3 gives some representative values of W. The large discrepancies in the measured values of W for He are explained [37] by minute amounts of gaseous impurities. Measurements on W for mixtures of gases suggest that the ionization of the components cannot be considered independent [38]. Gurney [39] gives variations of up to 10 per cent in W as a function of velocity for alphas in various gases. Recent evidence suggests that W in argon is essentially constant for natural alphas [40] and that W in air varies by 15 per cent as the alpha energy is changed from 3.5 to 0.34 Mev. Even in argon slight differences in W are noted [41] for different-energy alpha particles and for different particles.

The energy-loss formula can be integrated to give the range of the charged particle.

$$R = \int_0^{E_0} \left(\frac{-dE}{dx}\right)^{-1} dE = z^{-2} \int_0^{E_0} f(V)\, dE = Mz^{-2}F(V)$$

This is properly called the CSDA range ("continuous slowing-down approximation" range). This sometimes underestimates the actual range because of the finite values of energy losses, but the difference is seldom appreciable. The depth of penetration or "projected range" can be effectively smaller (~1 per cent) than the mean path length in heavier elements because of small deflections of the particle track (see ref. 28, p. 48). These relations are useful mainly in interpolating and extrapolating empirical data. Since they neglect charge exchange and nuclear collisions, they are inaccurate for low velocities.

Because of straggling of energy loss, all particles of the same initial energy do not have the same range. This range straggling causes a Gaussian spread of ranges such that the number of tracks of range R can be written as

$$y = \frac{1}{2s} \exp \frac{\pi(R - \bar{R})^2}{4s^2}$$

where $\bar{R}$ is the mean range and s is a straggling parameter equal to the difference between the mean range and the integral-number extrapolated range

$$R_{extr} - \bar{R} = s$$

Figure 4.3 shows a differential and integral number range curve and the mean and extrapolated ranges; s

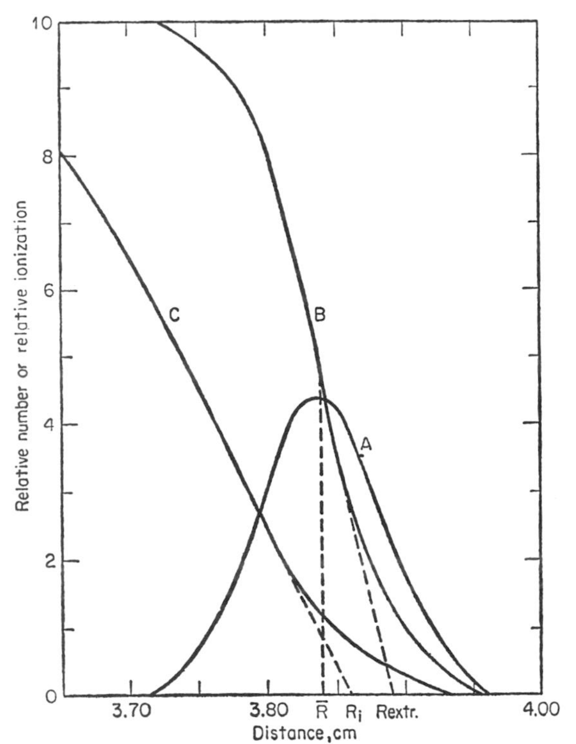

FIG. 4.3. Straggling curves. (A) Differential number-distance curve for polonium alpha particles, showing the distribution of ranges about the mean range due to straggling; (B) integral number-distance curve showing the extrapolated range, R_{extr}; (C) ionization-distance curve showing the ionization extrapolated range R_i.

can be calculated theoretically, yielding

$$s = \frac{(M)^{1/2}}{z^2} f(V)$$

Livingston and Bethe [42] have calculated this function and give the relative straggling in per cent as a function of mean range for protons and alphas. Their curve is reproduced in Fig. 4.4. If the range is determined by measuring the specific ionization as a function of distance and extrapolating this curve (see Fig. 4.3), then an ionization extrapolated range R_i is obtained. For the natural alpha particles the straggling in the ionization curve is about 0.5 per cent of the range. These statistical stragglings are not often the limiting experimental uncertainty. Geometric variations, thick-target effects, and lack of resolution in detectors often contribute to a spread or straggling in the range and have to be taken into account. Livingston and Bethe [42] give methods and corrections for doing this in a number of practical situations.

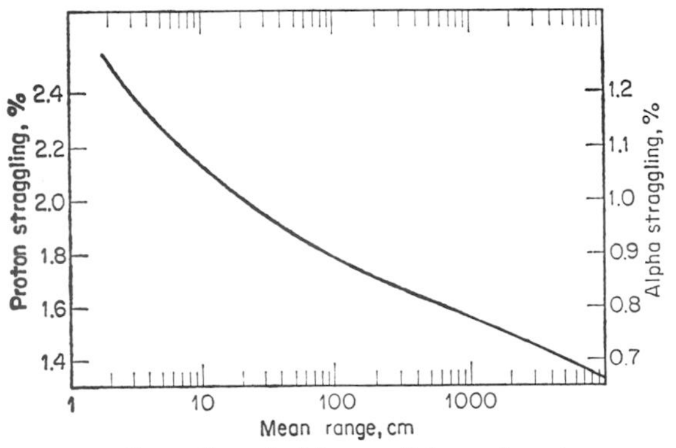

FIG. 4.4. Straggling of alpha particles and protons as a function of mean range. The straggling ($R_{extr} - R$) is given in per cent of mean range.

TABLE 4.4. RANGE AND ENERGY OF ALPHA PARTICLES AND PROTONS

	$\bar{R}$, cm	E_α, Mev	E_p, Mev
	0.20	0.271	
	0.50	0.994	0.357
$^{10}B(n,\alpha)$	0.725	1.472	
$^{14}N(n,p)$	1.00	1.995	0.585
$^{6}Li(n,\alpha)$	1.04	2.049	
	1.50	2.767	0.764
	2.00	3.425	0.919
	2.50	4.003	1.051
	3.00	4.502	1.179
^{234}U	3.258	4.772	
	3.50	4.987	1.296
^{210}Po	3.842	5.305	
$^{222}Em(Rn)$	4.051	5.486	1.406
$^{218}Po(Ra\ A)$	4.657	6.000	
$^{212}Bi(Th\ C)$	4.730	6.051	
	5.00		1.614
$^{216}Po(Th\ A)$	5.638	6.775	
$^{214}Po(Ra\ C')$	6.907	7.687	
$^{212}Po(Th\ C')$	8.570	8.785	
	9.724	9.492	
	11.580	10.543	

Range-energy curves for alpha particles can be determined from empirical data from a few tenths to 10 Mev, using the natural alpha particles and transmutation alphas. (Unless otherwise specified, ranges are mean ranges in dry air at 760 mm of Hg pressure at 15°C, called standard air.) Figure 4.5 gives such a curve. Table 4.4 gives some of the data from which Fig. 4.5 is constructed, including the energy and ranges of natural alpha emitters and transmutation particles.

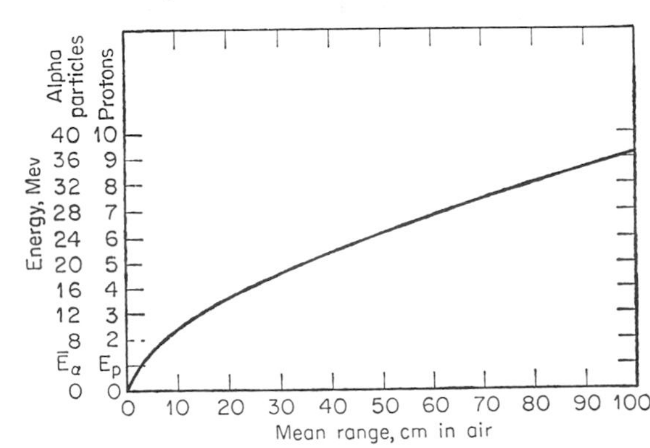

FIG. 4.5. Range-energy curves for alphas and protons. The energy in Mev is plotted against the mean range in centimeters.

Guided by the theoretical equations, relations can be set up between ranges of various particles. Proton ranges are related to alpha ranges by

$$\bar{R}_H(E) = 1.0072\bar{R}(3.971E) - 0.20 \qquad \text{cm}$$

where the range of a proton is given in terms of $M_H z_\alpha^2 / M_\alpha z_H^2$ times the range of an alpha of the same velocity with an empirical correction term. Similarly the range of a particle with charge z, mass M, and energy E can be given in terms of the range of an isotope of mass M_0 and energy (EM_0/M) by

$$R_{z,M}(E) = \frac{M}{M_0} R_{z,M_0} \left(\frac{EM_0}{M}\right)$$

For elements heavier than He, charge exchange becomes increasingly important, and these relations are no longer adequate. Empirical values have been determined for several gas ions by observing scattering recoils [29].

For some range corrections it is convenient to define the range exponent

$$n = \frac{2d \log R}{d \log E} \cong \frac{d \log R}{d \log V}$$

n varies [42] from about 1 to 4 but is approximately 3 for natural alpha particles. Geiger proposed such an approximate relation R (cm) ≈ 0.32 $[E$ (Mev)$]^{3/2}$ which is accurate to 10 per cent for natural alphas.

Integral stopping power is usually defined relative to air and is the ratio of the range in air to the range in the substance considered, $S = \bar{R}_{\text{air}}/\bar{R}$. Theoretically the differential atomic stopping power may be related to the stopping number B by $s_a = B/B_{\text{air}}$, where $B = Z \ln (2mV^2/I) - C_K$. This stopping power then varies with atomic number both because of Z effective and I. It also varies with the velocity of the particle stopped. Table 4.5 gives some integral atomic stopping powers relative to air. Since air consists mainly of diatomic molecules, it is conventional to specify its molecular stopping power as 2 in order to give its atomic stopping power as 1. Consequently, all the stopping powers in the table must be divided by 2. Livingston and Bethe also give curves of the stopping power of gold, carbon, and hydrogen as a function of particle range. The atomic stopping power is found to vary approximately as the square root of the atomic weight (Bragg's rule). The mass stopping power (that is, S/ρ) is then inversely proportional to the square root of the atomic weight.

Range-energy curves for alpha particles in air are given by Siri [43] (0 to 20 Mev), in aluminum (0.1 to 10^4 Mev), and in lead (0.4 to 10^4 Mev). Siri also gives proton ranges up to 10^4 Mev in hydrogen, helium, lithium, carbon, nitrogen, air, oxygen, aluminum, argon, neon, copper, and lead plus meson ranges up to 10^4 Mev in air, aluminum, and lead. Several of these curves are shown in Fig. 4.6. Range-energy curves for alpha particles and protons in nuclear emulsions are given by Bradner et al. [44]. Bethe and Ashkin [25] give the alpha ranges in air and several of the proton-range curves. More recent range-energy tables are due to Sternheimer [45] and Brandt [46] and are given in report form by Barkas [47] and Bichsel [48]. Stopping powers and range data are summarized in a recent report [30].

The importance of range in determining energy has decreased in practical experimental nuclear physics with the rise of high-resolution magnetic spectrometers and thick solid-state detectors. However, needed corrections for energy loss in sources required knowledge of the differential rate of energy loss. Consequently, for precision measurements, the emphasis has shifted from integral stopping powers to differential stopping cross sections per atom, $\epsilon = -\dfrac{1}{N}\dfrac{dE}{dx}$. Whaling has summarized the data establishing these

TABLE 4.5. ATOMIC STOPPING POWERS RELATIVE TO AIR

Vel. E_α E_H	1.0 2.07 0.52	1.5 4.66 1.17	1.75[a] 6.3 1.54	2.0 8.3 2.09	2.5 12.95 3.26	3 18.6 4.70	4 33.2 8.36	5.10^9 cm/sec 51.9 Mev 13.06 Mev
$\frac{1}{2}H_2$	0.26	0.224	(0.20)	0.209	0.200	0.194	0.186	0.181
He			0.35					
Li			0.50					
C	0.94	0.932		0.921	0.914	0.908	0.899	0.892
$\frac{1}{2}N_2$			0.99					
$\frac{1}{2}O_2$			1.07					
Ne			1.23					
Al	1.45	1.51	(1.50)	1.53	1.54	1.55	1.57	1.59
A			1.94					
Cu	[1.92][c]	2.41	(2.57)	2.62	2.73	2.80	2.89	2.95
Kr			2.92					
Mo			3.20					
Ag	[2.25][c]	3.08	(3.36)	3.43	3.64	3.76	3.93	4.04
Sn			3.59					
Xe			3.76					
Au	[2.42][c]	3.96	(4.50)	4.64	5.00	5.25	5.75	5.79
Pb			4.43					

[a] This column from Mano, *J. Phys. Radium*, **5**: 628 (1934).
[b] All other columns from Livingston and Bethe, *Revs. Mod. Phys.*, **9**: 272 (1937).
[c] Estimated.

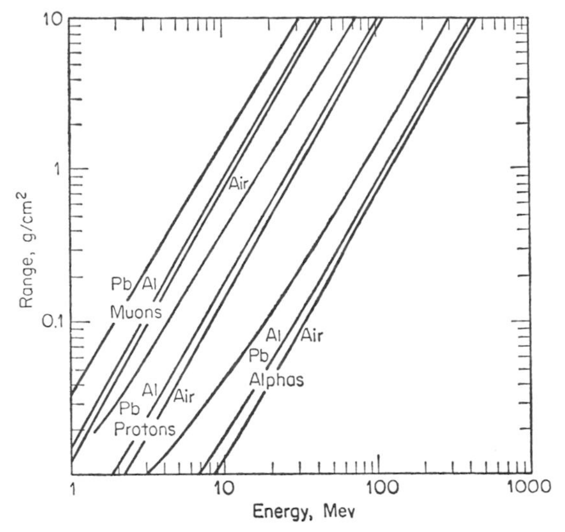

FIG. 4.6. Range-energy curves for alphas, protons, and μ mesons in air, aluminum, and lead. Ranges are given in grams per square centimeter and energies in Mev. (Density corrections included.)

values and has shown [27] that an empirical matching of the Bloch formula [2]

$$\epsilon = \frac{2\pi Z^2 e^4 m}{E m_e} Z \left[\ln \left(\frac{E}{Z} \right) + \text{const} \right]$$

to the experimental data gives a good approximation. Values of such constants for protons are given in Table 4.6. To use these curves for other particles, replace E_p by $E_d/2$ or $E_t/3$ for deuterons or tritons; replace E_p by $E_\alpha/4$ or $E_{H_e^3}/3$ and divide the resulting stopping cross section by 4 for alpha particles of energy greater than about 2 Mev, or He^{3++} of energy greater than 1.5 Mev. ϵ for other elements and compounds and other energy ranges are discussed in Whaling's article

TABLE 4.6. EMPIRICAL CONSTANTS IN STOPPING CROSS SECTION FORMULA FOR PROTONS [27]
$\epsilon = (a/E_p)[\ln (E_p/Z) + b] \times 10^{-15}$ ev cm²/atom for $E_p \geq c$ (Mev)

Element	a	b	c
He	0.479	4.68	1
Li	0.718	4.69	0.2
Be	0.960	5.04	2
C	1.44	5.14	2
N	1.68	5.08	2
O	1.92	5.12	1.5
Ne	2.40	5.14	4
Al	3.12	5.08	0.5
A	4.31	5.22	1
Ni	6.71	5.08	2
Cu	6.95	5.21	1
Kr	8.36	5.21	1.5
Ag	11.3	5.28	6
Sn	12.0	5.29	2
Xe	12.9	5.27	3
Au	18.9	5.22	6
Pb	19.6	5.28	4

[27] which also includes detailed range-energy tables for protons (0.04 to 10 Mev) and alphas (3 to 40 Mev) in many substances.

Charge exchange or capture and loss of electrons is an important aspect of the passage of slow alpha particles through matter. Table 4.7 gives mean free paths for capture of an electron by an alpha particle, λ_1, or loss of an electron by He^+, λ_2. It has been estimated that an alpha particle of 6.9-cm range spends

TABLE 4.7. MEAN FREE PATHS FOR CAPTURE AND LOSS OF ELECTRON BY ALPHA PARTICLES IN AIR AT NTP†

E_α, Mev	Velocity, cm/sec	Capture λ_1, mm	Loss λ_2, mm	λ_2/λ_1
6.75	1.81×10^9	2.2	0.011	0.005
4.45	1.46	0.52	0.0078	0.015
1.70	0.90	0.037	0.0050	0.13
0.65	0.56	(0.003)‡	(0.003)‡	(1.0)‡

† Rutherford, Chadwick, and Ellis, "Radiations from Radioactive Substances," Cambridge University Press, New York, 1930.
‡ Estimated.

about 0.5 cm as a singly charged ion. The loss of an electron by He^+ is essentially an ionization process and may be treated as such. Bohr [49] gives the cross section for such electron loss in light elements:

$$\sigma_l = \frac{4\pi\alpha_0^2}{z^2} (Z^2 + Z) \left(\frac{v_0}{v} \right)^2$$

where z is the alpha charge, Z is the charge of the absorber, $\alpha_0 = \hbar^2/me^2$, $v_0 = e^2/\hbar < v$. For large Z, $\sigma_l \approx \pi\alpha_0^2$, while for intermediate values

$$\sigma_l \approx \frac{\pi\alpha_0^2 Z^{2/3}}{z} \frac{v_0}{v}$$

The capture process is more complicated, but a rough approximation gives for the electron-capture cross section for $v > v_0$,

$$\sigma_c \sim 4\pi\alpha_0^2 z^5 Z^{1/3} \left(\frac{v_0}{v} \right)^6$$

These problems are of more importance for fission fragments, reaction recoil atoms, and heavy cosmic rays. Here the effective charge z^* is steadily changing as the particle slows down. Bohr gives $z^* \approx z^{1/3}(v/v_0)$, which seems a good approximation. Data on capture and loss are summarized by Northcliffe [29] who also gives stopping powers and range-energy curves for energetic heavy ions.

Nuclear collisions are important for low velocities of the incident particle. Bohr gives the cross section for nuclear collisions with transfer of energy between E_1 and E_2 as

$$\sigma_n = \frac{2\pi z^2 Z^2 e^4}{M V^2} \left(\frac{1}{E_1} - \frac{1}{E_2} \right)$$

The average number of such collisions in a distance t in matter containing N atoms per cubic centimeter would be $n = Nt\sigma_n$. These collisions are the cause of the branching or sudden changes of direction of the alpha particle observed near the end of occasional alpha tracks. Since N is proportional to z^2, the occurrence of branching is much more frequent in fission-fragment tracks. Also meson tracks, due to smaller M, show a characteristic wiggling due to branching. These nuclear collisions are the result of a Coulomb interaction between the alpha particle and the shielded nuclei. Consequently, there are more small-angle scatterings than branchings. The track will, therefore, have variable curvature in addition to recognizable abrupt changes of direction. See multiple scattering in Sec. 3.

3. Scattering of Alpha Particles

The scattering of natural alpha particles is the oldest method for exploring the nuclear field. The predominant interaction is the Coulomb force which gives rise to the Rutherford scattering law (see also Part 2, Chap. 6, Sec. 11, and Part 9, Chap. 1, Sec. 9). If a particle of charge ze, mass M, velocity V, and energy E_0 is scattered by a heavy nucleus of charge Ze, the fraction scattered through an angle between θ and $\theta + d\theta$ will be

$$\frac{n}{n_0} = Nt\sigma 2\pi \sin\theta\, d\theta$$

where N is the number of nuclei per cubic centimeter and t is the thickness of such matter traversed by the scattering particles. The number of particles scattered at the angle θ into unit solid angle per second is $n' = n_0 Nt\sigma$, where n_0 is the number of incident particles per second. The scattering cross section σ was derived classically by Rutherford as

$$\sigma_R = \frac{(zZe^2/2MV^2)^2}{(\sin\theta/2)^4}$$

Experimental verification of this relation was the basis of Rutherford's nuclear atom model [50]. The nuclear charge Z can be determined by accurate measurements of this Coulomb scattering [51]. Deviations from the Rutherford law occur in alpha scattering by light nuclei, owing to recoil of the scattering nuclei, and also occur when the alpha particle can penetrate to the region of the nuclear forces either at resonance or above the potential barrier. The latter deviations indicate nuclear dimensions and nuclear resonance levels.

For light nuclei of mass M_2, Darwin [52] gave the Coulomb scattering cross section as

$$\sigma_D = 4\left(\frac{zZe^2}{2MV^2}\right) \operatorname{cosec}^3\theta \times \frac{\{\cot\theta \pm [\operatorname{cosec}^2\theta - (M/M_2)^2]^{1/2}\}^2}{[\operatorname{cosec}^2\theta - (M/M_2)^2]^{1/2}}$$

where the plus sign is used if $M_2 > M$ and the minus sign if $M > M_2$. For $M \ll M_2$ this is approximately

$$\sigma = \sigma_R\left[1 - 2\left(\frac{M}{M_2}\right)^2 \sin^4\frac{\theta}{2}\right]$$

The kinetic energy associated with the relative motion in the center-of-mass system is

$$E = \frac{E_0 M_2}{M + M_2} = \frac{\mu V^2}{2}$$

where $\mu = MM_2/(M + M_2)$ is the reduced mass for the center-of-mass system of velocity $VM/(M_2 + M)$.

Inclusion of specific nuclear forces leads to more complex formulas. Bethe [53] gives a general treatment which is more conveniently approximated by a one-body theory of scattering. Mott and Massey [54] give the ratio of the differential cross section to the Rutherford one as

$$R = \left|1 + \left(\frac{i}{\alpha}\right)\sin^2\left(\frac{\theta}{2}\right)\exp\left(i\alpha \ln \sin^2\frac{\theta}{2}\right)\sum_l \exp 2i(\zeta_l - \zeta_0)(\exp 2iK_l - 1)(2l + 1)P_l(\cos\theta)\right|^2$$

where $\alpha = zZe^2/\hbar V$ for a scattering particle of charge z and velocity V scattered by nuclei of charge Z through an angle θ. The $P_l(\cos\theta)$ are Legendre polynomials of order l (see Part 1, Chap. 3, Sec. 12). The total phase shift ψ_l (see Part 2, Chap. 6, Sec. 11) of the lth partial wave may be considered as the sum $\psi_l = \eta_l + \delta_l$. η_l is the phase shift due to the potential scattering which is approximately that due to an infinite potential barrier, of a given radius, plus a Coulomb field. It arises partly before penetration of the nucleus and partly from shadow diffraction round the nucleus. δ_l is the phase-shift contribution from the neighboring resonance levels, where only elastic scattering and radiation are assumed:

$$\delta_l = \arctan\frac{\Gamma_t}{2(E_R - E)}$$

where Γ_t is the total width for the level at E_R (see Part 9, Chap. 1, Sec. 9,3).

The phase shift that would be produced by the pure Coulomb field in the absence of the nuclear potential is $\zeta_l = \arg\Gamma(l + 1 + i\alpha)$. The other phase shift K_l in the equation for R is defined by

$$K_l = \psi_l - \zeta_l = \eta_l + \delta_l - \zeta_l$$

However, K_l is different from the phase shift the particle would suffer by the nuclear potentials only, in the absence of the Coulomb forces. In general, the phases K_l cannot be calculated but are determined from experimental data. Since δ_l changes by $\pi/2$ in going through a resonance, K_l does the same since the phase shifts change gradually with energy. Hence an abrupt change in K_l indicates a resonance level; a more gradual change indicates a surmounting of the Coulomb barrier.

The most extensive analysis of alpha scattering has been done for alpha scattering in helium. The spinless nature of the two particles and the absence of other open channels for bombarding energies up to 35 Mev should permit a phase-shift analysis based entirely upon a knowledge of the differential cross section as a function of energy. The fact that the target nucleus and the incident particle are identical bosons allows only symmetric terms in the wave function and thus further simplifies the analysis by limiting the angular momentum l to even values. In the case of a collision of two identical particles of zero spin with only a Coulomb interaction, Mott [55] gives

$$\sigma = \left(\frac{z^2e^2}{MV^2}\right)^2 (\text{cosec}^4\,\theta + \sec^4\,\theta + 2\phi\,\text{cosec}^2\,\theta\,\sec^2\,\theta)4\cos\theta$$

$$\text{where} \quad \phi = \cos\,(\alpha \ln \tan^2\theta) \qquad \text{and} \qquad \alpha = \frac{z^2e^2}{\hbar V}$$

The corresponding classical relation is $\phi = 0$. (This term arises from the symmetry of the wave function for two particles; see Part 2, Chap. 6, Sec. 11.) At 45° the quantum-mechanical cross section is twice the classical prediction. This quantum result was checked by Chadwick [56] and Blackett and Champion [57], using slow alpha particles.

In recent years, the availability of intense alpha-particle beams from accelerators and the development of improved detection techniques have allowed precise experimental data to be taken for bombarding energies from 0.15 to 120 Mev [58]. These data have been analyzed to determine the empirical constants K_0, K_2, K_4, . . . of the general formula, including nuclear effects. The real parts of the phase shifts show a remarkably smooth and systematic behavior with the exception of some rapid fluctuations near 40 Mev, which are not fully understood at present [59]. The phase shifts all start from zero at zero energy since ^{8}Be has no bound states. As the energy increases, each partial-wave phase shift begins to deviate from zero in the positive direction, at an energy corresponding to an impact parameter of about 5 fermis. Resonances are indicated at center-of-mass energies of 0.094, 2.9, 11.7, and 26 Mev for the $l = 0$, 2, 4, and 6 partial waves, respectively. This behavior implies that the nuclear interaction must be attractive at large radii. Above the resonance, the $l = 0$ and $l = 2$ phase shifts decrease monotonically, eventually becoming negative at impact parameters of 1 to 2 fermis. The latter behavior indicates that the interaction has a repulsive core of radius 1 to 2 fermis. For the higher partial waves the centrifugal barrier prevents investigation of the central regions of the potential. The $l = 4$ phase shift has hardly begun to decline at 120 Mev, and the higher phase shifts are still rising. Thresholds for various nuclear reactions are reached soon after the bombarding energy exceeds 34.73 Mev. The imaginary parts of the phase shifts are already appreciable at 54 Mev. They show a smooth variation with angular momentum and energy. Their dependence indicates that most of the reactions occur when the two alpha particles are still far apart. In the theoretical interpretation of the scattering data various approximations have to be made which are valid only at low energies. In particular, the possibility of excitation of the alpha particles is not included. The first excited state in the ^{4}He nucleus is reported at 19.94 Mev, followed by the second one at 21.24 Mev [18, 60].

Alpha particles were scattered in hydrogen by Chadwick and Bieler [61], using natural alphas. For energies below about 3 Mev the scattering was the Rutherford type. Above this alpha energy the ratio of observed to Rutherford scattering increased to over 40 for $E_\alpha = 7.5$ Mev. This is consistent with a barrier of about 0.6 Mev, and the scattering ratio can be interpreted in terms of a single phase shift δ_0. The opposite process, the elastic scattering of protons in helium, was investigated more extensively, and these results are more reliable. Spin polarization may be measured in addition to the differential elastic-scattering cross section. The phase shifts are spin-orbit split [62]. An optical-model (see Part 9, Chap. 1), analysis including both surface and volume absorption fits well the measured cross sections of the elastic scattering and of the reaction $^4\text{He}(p,d)^3\text{He}$ [63]. From the proton-alpha phase shifts the corresponding neutron-alpha phase shifts can be obtained at the higher neutron energies by setting the Coulomb phase shifts equal to zero [64]. Such data agree well with measured polarization of helium scattered neutrons [65].

Alpha-particle scattering on light nuclei has been done with natural alpha particles by Riezler [66] and analyzed by Wenzel [67]. Increases in the ratio of observed to Rutherford scattering are found in Be and B and can be interpreted as surmounting the barrier with $l = 1$ alphas. Carbon shows resonance at $E_\alpha = 5$ Mev. Scattering from oxygen is less than Rutherford, also ascribed to overcoming the barrier. Neon exhibits Rutherford scattering up to 5.3 Mev, while aluminum shows deviations at 7 Mev. Since the distance of closest approach is

$$b = zZe^2\left(1 + \frac{M_2}{M}\right)E_\alpha$$

the alpha energy at which deviations from Rutherford scattering set in can be used to calculate a radius of the nucleus. Such radii are compatible with nuclear radii from other measurements but are not very accurate.

In recent years a large number of high-energy alpha scattering experiments have been carried out at various laboratories. Much impetus for these investigations was due to the success of Blair's [68] model in explaining the observed diffraction structure of the differential cross sections. Blair assumed a completely absorbing sphere of radius $R_v = 1.17A^{1/3}$ and $R_w = 1.40A^{1/3}$. Partial waves that would interact with the sphere in the classical limit are completely absorbed; otherwise they receive the phase-shift characteristic of the Coulomb potential. An extensive series of optical-model (see Part 9, Chap. 1), calculations by Igo [69] showed that the scattering process is determined essentially by the potential at the nuclear surface. He found that the surface potential for the interactions studied could be represented by (see Part 9, Chap. 8, Sec. 13)

$$V(r) + iW(r) = -1{,}100\exp\left(-\frac{r - 1.17A^{1/3}}{0.574}\right) - 45.7i\exp\left(-\frac{r - 1.40A^{1/3}}{0.578}\right)$$

where r is given in fermi units. Austern [70] has shown, on the basis of a WKB calculation, that Igo's suggestion is an oversimplification and that it is indeed possible for the nuclear interior to influence scattering. This conclusion is confirmed by experimental elastic-scattering data [71]. At backward angles the theoretical values do not fit the experimental curve.

For these single-scattering experiments to be uninfluenced by multiple scattering, Wentzel [72] has set up a criterion that the scattering angle θ always be greater than 10 times $\theta_{\min}$, where

$$\tan\frac{\theta_{\min}}{2} = \frac{Ze^2(\pi Nt)^{1/2}}{E_\alpha}$$

For thicker foils or smaller angles, multiple scattering occurs. Mott and Massey [73] give the mean-square deflection of a particle of mass M, charge ze, and velocity V passing through a thickness t of matter with N atoms per cubic centimeter of charge Ze and mass M_2 as

$$\overline{\alpha^2} = 2K \ln\frac{K}{2\theta_{\min}}$$

where $$K = \frac{4\pi Ntz^2Z^2e^4(M + M_2)^2}{M^2M_2^2V^4}$$

and $$\theta_{\min} = \frac{3.8Z^{4/3}e^2(M + M_2)}{MM_2V^2a_0}$$

for $zZe^2/\hbar V > 1$ (a_0 is the hydrogen orbit radius). Mayer [74] has measured the average angle of scattering of 1.5×10^9 cm/sec alpha particles passing through 1 cm of air equivalent foils. His results are in reasonable agreement with the calculated values:

	Mean angle of scattering, deg	
	Observed	Calculated
Gold	1.72	2.31
Platinum	1.85	2.28
Silver	2.00	1.72
Copper	1.36	1.26
Aluminum	0.93	0.81

4. Alpha-particle Radioactivity

Classically a particle cannot penetrate into a region where the potential energy is greater than the kinetic energy. Quantum mechanics allows such a penetration, a concept that was used by Gurney and Condon [75] and Gamow [76] to explain natural alpha-particle radioactivity. A particle of momentum p traveling in the positive x direction can be represented by an incident plane wave $\psi_i = A \exp i(kx - \omega t)$, where $k = p/\hbar$ (see Part 2, Chap. 6). Upon encountering a potential barrier of height U_0 and width b, this wave is partially reflected

$$\psi_r = B \exp i(-kx - \omega t)$$

and partially transmitted: $\psi_t = C \exp i(kx - \omega t)$. The transmission is $T = C^2/A^2$ and for a rectangular barrier is approximately

$$T \approx \exp\left\{-2b\left[\frac{2M(U_0 - E)}{\hbar^2}\right]^{1/2}\right\}$$

if E is of the order of U_0. For a barrier of arbitrary shape

$$T \approx \exp\left\{-2\int_a^b\left[2M\frac{U(x) - E}{\hbar^2}\right]^{1/2} dx\right\}$$

The cross section σ_c for an incident charged particle to penetrate the Coulomb barrier and reach the nucleus is related to the transmission T by

$$\sigma_c = (R + \lambda\!\!\!^-)^2\, T$$

where R is the effective radius of the nucleus and $\lambda\!\!\!^- = 1/k$ is the de Broglie wavelength of the incident particle. Shapiro [77] has calculated such cross sections for the formation of a compound nucleus (see Part 9, Chap. 8, Sec. 9) by charged-particle bombardment. Other values are listed as a function of alpha energy for several high atomic numbers in Table 4.8.

TABLE 4.8. CROSS SECTION σ_c FOR INCIDENT ALPHA PARTICLES TO PENETRATE COULOMB BARRIER†

$Z = 70$		$Z = 90$	
E_α, Mev	σ_c, cm^2	E_α, Mev	σ_c, cm^2
4.24	34×10^{-52}	4.98	44×10^{-57}
6.36	36×10^{-42}	7.47	35×10^{-45}
8.48	63×10^{-36}	9.96	42×10^{-38}
10.60	81×10^{-33}	12.45	46×10^{-34}
12.72	25×10^{-30}	14.94	42×10^{-31}
14.84	13×10^{-28}	17.43	46×10^{-29}
16.96	19.4×10^{-27}	19.92	12×10^{-27}
19.08	10.3×10^{-26}	22.41	12.3×10^{-26}
21.2	25.4	24.9	25.7
23.3	42	27.4	45
25.4	60	29.9	66
27.6	76	32.4	85
29.7	89	34.9	103
31.8	101	37.4	117
33.9	113	39.9	130
36.0	123	42.4	142
38.1	131	44.9	153

† Blatt and Weisskopf, "Theoretical Nuclear Physics," Wiley, New York, 1952.

Applying similar quantum-mechanical concepts to the problem of a Coulomb barrier enclosing an alpha particle, Gurney and Condon and also Gamow and Bethe [78] give a relation between the probability of escape of the alpha particle and its energy outside the nucleus in terms of the atomic number and radius of the nucleus. The number of alpha particles emitted per second is

$$n = \frac{2^{1/2}\pi^2\hbar^2[\exp(-2C)]}{M^{3/2}R^3(Zze^2/R - E)^{1/2}}$$

where z, M, and E are the alpha-particle charge, mass, and energy outside the nucleus and R and Z are the radius and charge of the nucleus.

$$C = \frac{2Zze^2}{\hbar V}\{\arccos(x)^{1/2} - [x(1 - x)]^{1/2}\}$$

where $x = E/B$ with the barrier height $B = Zze^2/R$ and V is the velocity of the alpha outside the nucleus. An approximation for $E < B$ gives

$$C = \frac{zZe^2}{\hbar V} - \frac{2e}{\hbar}(2ZzMR)^{1/2}$$

These relations are equivalent to an expression with the half-life for alpha radioactivity, τ.

$$\tau = \frac{\ln 2}{n} = \tau_0 \exp(2C)$$

where $\tau_0 = 3.3 \times 10^{-21}$ sec is the lifetime without barrier and $\exp(-2C)$ is the transmission coefficient of the Coulomb barrier. These formulas neglect any angular momentum of the alpha particle and assume that the alpha particle preserves its identity in the nucleus. Bethe has extended these calculations to a many-body model in which the probability of formation of the alpha particle out of neutrons and protons inside the nucleus is taken into account. Instead of being 10^{-21} sec, which is essentially the period of vibration of an alpha in the nuclear potential, τ_0 is much longer, because of the time of formation of the alpha particle, $\tau_0 \sim 10^{-15}$ sec. This model necessitates a lower barrier and hence a larger nuclear radius. More precise calculations have been made on the single-particle model by Preston [79]. Table 4.9 gives some nuclear radii calculated by Preston together with Bethe's one- and many-body values and some neutron-determined radii [80]. Nuclear radii calculated from alpha-decay theory depend on the detailed form of the model used; they are expected to differ from those obtained from other phenomena such as alpha-particle scattering and reaction cross sections (see Part 9, Chap. 1). Later calculations [81] use $R = 1.51A^{1/3}\,10^{-13}$ cm. The apparent nuclear radius changes abruptly when the closed shell of 126 neutrons is reached. Mang [82], using nuclear shell-model wave functions (see Part 9, Chap. 1), to calculate alpha-transition rates, has shown that the observed effect is not due to an actual change in nuclear size but is caused by the change in probability of formation of the alpha particle in the nuclear surface region.

The theoretical relation between half-life and alpha energy explains the empirical Geiger-Nuttall [83] law between the half-life (τ in seconds) and alpha range (R in centimeters) in air of natural alpha emitters

$$\log \tau = -57.5 \log R + A$$

where $A = 42.3$ for the uranium series, $A = 44.2$ for the thorium series, and $A = 46.3$ for the actinium series. Putting in the approximate range-energy relation for natural alphas (E in Mev),

$$\log \tau = -86.3 \log E + 28.5 + A$$

A more illustrative comparison between the experimental data is accomplished by plotting the log of the half-life τ against the total alpha-decay energy Q. For the even-even nuclei, Perlman, Ghiorso, and Seaborg [84] have shown that the experimental data fall on a series of smooth curves, one for each Z. For some purposes it is more convenient to plot the logarithm of the half-life against the reciprocal square root of the decay energy [85]. A family of straight lines is obtained as illustrated in Fig. 4.7. Each line represents a single element. Such plots serve also to show the forbiddenness of many even-odd, odd-even, and odd-odd nuclei by departures of experimental data from these semiempirical curves. Table 4.10 summarizes the alpha radioactivity data: half-life,

TABLE 4.9. HEAVY NUCLEUS RADII

Resultant nucleus			Barrier, Mev	Radius × 10^{-13} cm				
				Preston	Asaro	Bethe		Neutron
Z	A					One-body	Many-body	
94	238	Pu	....	....	9.32			
90	234	U X I	23.9	9.37	....	9.8	13.2	
88	226	Ra	....	....	9.35	...		
88	228	M Th 1	22.0	9.92	....	8.7	11.3	
86	222	Rn	....	....	9.35			
84	215	AcA	21.5	8.70	....	8.5	12.0	
84	218	Po	....	....	9.34			
83	...	Bi	....	....	....	...	....	7.9
82	211	Ac B	19.3	8.99	....	8.8	12.8	
82	212	Th B	19.5	9.12	....	8.9	12.7	
82	214	Ra B	23.8	8.27	9.23	8.2	11.5	
82	...	Pb	....	....	....	...	....	7.8
81	...	Ac C	24.3	7.90	....	7.3	10.6	
80	...	Hg	....	....	....	...	....	8.4

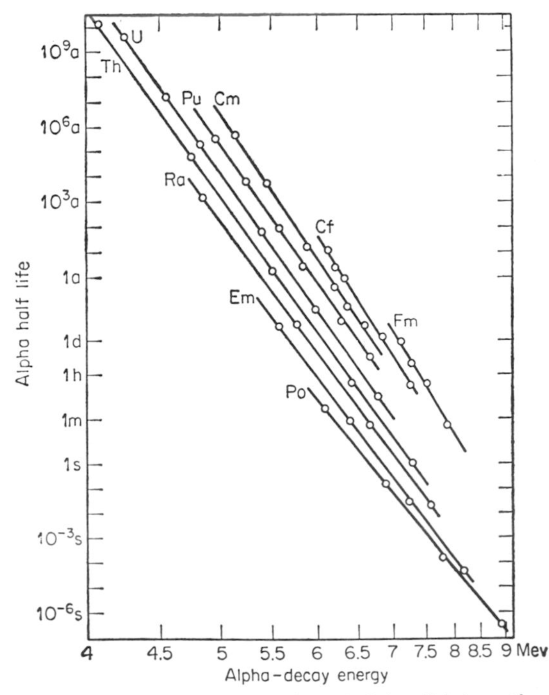

FIG. 4.7. Half-life τ as a function of alpha-disintegration energy Q for the even-even alpha-radioactive nuclei. To obtain a family of straight lines, the scales are linear in $\log \tau$ and $Q^{-1/2}$.

TABLE 4.10. ALPHA RADIOACTIVITY DATA
(a = years; d = days; h = hours; m = minutes; s = seconds; *long-range alpha group)
m = metastable; *h* = heavier; *l* = lighter

Element	Mass number	Half-life	Alpha energy, Mev	Abundance of group	Alpha branching ratio
$_{52}$Te	107	2.2 s	3.28		
	108	5.3 s	3.08		
$_{58}$Ce	142	>5 × 10^{16} a			
$_{60}$Nd	144	2.4 × 10^{15} a	1.83		
$_{61}$Pm	145	18 a	2.24		2.8 × 10^{-9}
$_{62}$Sm	146	5 × 10^{7} a	2.55		
	147	1.18 × 10^{11} a	2.24		
	148	>2 × 10^{14} a			
	149	>1 × 10^{15} a			
$_{63}$Eu	147	24 d	2.88		1.5 × 10^{-5}
$_{64}$Gd	148	1.3 × 10^{2} a	3.16		
	149	9.0 d	3.00		4.6 × 10^{-6}
	150	10^{5} a	2.70		0.3
	151	120 d	2.60		8 × 10^{-9}
	152	1.08 × 10^{14} a	2.14		
$_{65}$Tb	149	4.1 h	3.95		0.15
	149^m	4.3 m	3.99		2.5 × 10^{-4}
	151	17.2 h	3.42		4.8 × 10^{-6}
$_{66}$Dy	150	7.2 m	4.23		0.18
	151	18.0 m	4.06		0.059
	152	2.3 h	3.65		5 × 10^{-4}
	153	6.4 h	3.48		3.0 × 10^{-5}
	154	13 h	3.35		
$_{67}$Ho	151	35.6 s	4.51		0.20
	151^m	42 s	4.60		0.28
	152^h	52.3 s	4.45		0.19
	152^l	2.36 m	4.38		0.30
	153	9 m	3.92		3 × 10^{-3}
$_{68}$Er	152	10.7 s	4.80		0.90
	153	36 s	4.67		0.95
	154	4.5 m	4.15		
$_{69}$Tm	153	1.58 s	5.11		0.90
	154^h	2.98 s	5.04		0.85
	154^l	5 s	4.96		
$_{70}$Yb	154	0.39 s	5.33		0.98
	155	1.65 s	5.21		0.90
$_{72}$Hf	174	2.0 × 10^{15} a	2.50		
$_{74}$W	180	>9 × 10^{14} a			
$_{78}$Pt	190	6.9 × 10^{11} a	3.11		
$_{79}$Au	$183 \leq A \leq 187$	4.3 m	5.10		
$_{80}$Hg	$A < 185$	0.7 m	5.60		
	196	>1 × 10^{14} a			
$_{82}$Pb	204	1.4 × 10^{17} a	2.60		1
$_{83}$Bi	198	7 m	5.83		5 × 10^{-4}
	199	25 m	5.47		1 × 10^{-4}
	201	62 m	5.15		3 × 10^{-5}
	203	12 h	4.85		10^{-7}
	209	2 × 10^{17} a	3.00		
(R_aE)	210	5.01 d	4.941	0	1.3 × 10^{-6}
			4.686	0.4	
			4.649	0.6	
	210^m	2.6 × 10^{6} a	5.190	0	
			4.935	0.60	
			4.89	0.34	
			4.59	0.05	
			4.48	0.005	
(AcC)	211	2.16 m	6.617	0.83	0.9968
			6.273	0.17	
			5.941	3.7 × 10^{-5}	
(ThC)	212	60.5 m	6.090	0.271	0.337
			6.051	0.697	
			5.768	0.0178	

TABLE 4.10. ALPHA RADIOACTIVITY DATA (*Continued*)

Element	Mass number	Half-life	Alpha energy, Mev	Abundance of group	Alpha branching ratio
			5.626	0.00165	
			5.606	0.0119	
			5.483	1.4×10^{-4}	
			5.341	1×10^{-5}	
			5.298	1.1×10^{-6}	
			5.184	5×10^{-7}	
	213	47 m	5.86		0.02
(RaC)	214	19.7 m	5.507	0.392	2.1×10^{-4}
			5.443	0.539	
			5.263	0.058	
			5.179	0.0061	
			5.018	0.0021	
			4.936	0.0025	
$_{84}$Po	192	0.5 s	6.58		
	193	4 s	6.47		
	194	13 s	6.38		
	195	30 s	6.26		
	196	1.8 m	6.13		
	197	~4 m	6.040		
	198	~6 m	5.935		
	199	11.2 m	5.846		
	200	~8 m	5.770		
	201	17.5 m	5.671		
	202	44.5 m	5.575		0.02
	203	47 m	5.48		0.07
	204	3.6 h	5.370		0.0063
	205	1.5 h	5.20		7.4×10^{-4}
	206	9 d	5.224	1.00	0.05
	207	5.7 h	5.10		1.4×10^{-4}
	208	2.89 a	5.114	~1	~1
			4.21	weak	
	209	103 a	4.883	0.994	>0.99
			4.62	0.006	
			4.31		
			4.11		
	210	138.4 d	5.305	1.00	1
			4.518	1.07×10^{-3}	
(AcC′)	211	0.52 s	7.455	0.995	1
			6.90	0.005	
			6.59	$<7 \times 10^{-4}$	
	211^m	25 s	8.87	0.07	1
			8.30	0.0025	
			7.99	0.017	
			7.27	0.91	
(ThC′)	212	3.04×10^{-7} s	8.785	1.00	1.00
			9.503	3.5×10^{-5}*	
			10.433	2×10^{-5}*	
			10.554	18×10^{-5}*	
	212^m	45 s	11.65	0.97	1.00
			9.08	0.01	
			8.52	0.02	
	213	4.2×10^{-6} s	8.36	1.00	1.00
(RaC′)	214	1.64×10^{-4} s	7.687	1.00	1.00
			6.905	1×10^{-4}	
			8.284	4×10^{-7}*	
			8.945	4×10^{-7}*	
			9.072	2.2×10^{-5}*	
			9.320	4×10^{-7}*	
			9.496	1.4×10^{-6}*	
			9.664	4×10^{-7}*	
			9.786	1.1×10^{-6}*	
			9.912	4×10^{-7}*	
			10.081	1.7×10^{-6}*	
			10.153	4×10^{-7}*	

Table 4.10. Alpha Radioactivity Data (*Continued*)

Element	Mass number	Half-life	Alpha energy, Mev	Abundance of group	Alpha branching ratio
			10.333	1.1×10^{-6}*	
			10.513	2×10^{-7}*	
(AcA)	215	1.83×10^{-3} s	7.380	1.00	1
			6.950	3.4×10^{-4}	
			6.944	2.2×10^{-4}	
(ThA)	216	0.158 s	6.775	1.00	1
			5.982	2×10^{-5}	
	217	<10 s	6.54		~1
(RaA)	218	3.05 m	6.000	1.00	0.9998
			5.177	1.1×10^{-5}	
$_{85}$At	200	0.8 m	6.42	0.6	
			6.47	0.4	
	201	1.5 m	6.354	1.0	
	202	3 m	6.237	0.36	0.12
			6.139	0.64	
	203	7.4 m	6.092	1.0	0.14
	204	9.3 m	5.956	1.0	0.045
	205	26 m	5.905	1.0	0.18
	206	29 m	5.705		0.009
	207	1.8 h	5.756		0.1
	208	1.7 h	5.65	0.97	0.005
			5.53	0.03	
	209	5.3 h	5.648	1.0	0.05
	210	8.3 h	5.525	0.32	0.0017
			5.443	0.31	
			5.361	0.37	
	211	7.21 h	5.868	1.00	0.409
	212	0.305 s	7.60	0.2	
			7.66	0.8	
	212^m	0.120 s	7.82	0.8	
			7.88	0.2	
	213	short	9.2		
	214	2×10^{-6} s	8.78		
	215	10^{-4} s	8.00		
	216	3×10^{-4} s	7.79		
	217	0.018 s	7.066		
	218	~2 s	6.685	0.94	
			6.640	0.06	
	219	0.9 s	6.27		0.97
$_{86}$Em	204	3 m	6.28		<1
	206	6 m	6.25		0.65
	207	11 m	6.14		0.04
	208	23 m	6.141	1.0	0.20
	209	30 m	6.037		0.17
	210	2.7 h	6.042	1.0	0.96
	211	16 h	5.852	0.335	0.26
			5.785	0.645	
			5.619	0.02	
	212	23 m	6.270	1.0	
	213	19×10^{-3} s	8.13	1.0	
	215	10^{-6} s	8.6		
	216	4.5×10^{-5} s	8.04	1.0	
	217	5.4×10^{-4} s	7.741	1.0	
	218	0.030 s	7.127	0.998	
			6.529	0.002	
(An)	219	3.92 s	6.818	0.81	
			6.551	0.115	
			6.528	0.0012	
			6.424	0.075	
			6.310	5.4×10^{-4}	
			6.222	2.6×10^{-5}	
			6.157	1.74×10^{-4}	
			6.146	2.6×10^{-5}	
			6.101	3×10^{-5}	

TABLE 4.10. ALPHA RADIOACTIVITY DATA (*Continued*)

Element	Mass number	Half-life	Alpha energy, Mev	Abundance of group	Alpha branching ratio
			5.999	4.4×10^{-5}	
(Tn)	220	55.3 s	6.302	0.997	
			5.767	0.003	
	221	25 m	6.0		0.2
(Rn)	222	3.823 d	5.486	0.999	
		4.983	4.983	0.00078	
$_{87}$Fr	204	2 s			
	205	4 s	6.83	1.0	
	206	16 s	6.74	1.0	
	207	19 s	6.74	1.0	
	208	38 s	6.59	1.0	
	209	55 s	6.62	1.0	
	210	159 s	6.50	1.0	
	211	3 m	6.52	1.0	
	212	19.3 m	6.417	0.37	0.44
			6.393	0.39	
			6.348	0.24	
	213	34 s	6.77	1.0	0.995
	214	0.0039 s	8.55	1.0	
	215	$<10^{-3}$ s	9.6	1.0	
	217	short	8.3	1.0	
	218	5×10^{-3} s	7.85		
	219	0.02 s	7.30		1
	220	27.5 s	6.69		1
	221	4.8 m	6.340	0.84	
			6.125	0.16	
(AcK)	223	21 m	5.34		5×10^{-5}
$_{88}$Ra	212	18 s	6.90	1.0	
	213	2.7 m	6.74	0.5	
			6.61	0.5	
	214	2.6 s	7.17	1.0	
	215	1.6×10^{-3} s	8.7	1.0	
	216	<0.001 s	9.3	1.0	
	217		9.0	1.0	
	219	short	8.0		
	220	0.023 s	7.45	0.99	
			6.90	0.01	
	221	30 s	6.760	0.31	1
			6.671	0.20	
			6.612	0.38	
			6.589	0.08	
			6.579	0.03	
	222	38 s	6.557	0.968	1
			6.239	0.032	
			5.920	1×10^{-4}	
			5.77	3×10^{-4}	
			5.73	2×10^{-5}	
(AcX)	223	11.22 d	5.869	8.7×10^{-3}	1
			5.855	3.2×10^{-3}	
			5.745	0.091	
			5.714	0.537	
			5.605	0.26	
			5.537	0.091	
			5.500	8.6×10^{-3}	
			5.479	8×10^{-5}	
			5.431	0.024	
			5.364	0.0011	
			5.337	1.0×10^{-3}	
			5.285	1.3×10^{-3}	
			5.281	9.5×10^{-4}	
			5.257	4.3×10^{-4}	
			5.234	4.2×10^{-4}	
			5.210	5.4×10^{-5}	
			5.171	2.6×10^{-4}	

TABLE 4.10. ALPHA RADIOACTIVITY DATA (*Continued*)

Element	Mass number	Half-life	Alpha energy, Mev	Abundance of group	Alpha branching ratio
			5.150	2.1×10^{-4}	
			5.133	1.7×10^{-5}	
			5.110	6×10^{-6}	
			5.084	3×10^{-6}	
			5.054	2×10^{-6}	
			5.034	4×10^{-6}	
			5.024	6×10^{-6}	
			5.012	4×10^{-4}	
(ThX)	224	3.64 d	5.684	0.94	
			5.447	0.055	
			5.159	7.3×10^{-5}	
			5.049	7.2×10^{-5}	
			5.032	3.1×10^{-5}	
	226	1622 a	4.781	0.9445	
			4.598	0.0555	
			4.340	6.5×10^{-5}	
			4.191	1.0×10^{-5}	
			4.160	2.7×10^{-10}	
$_{89}$Ac	213	1 s	7.42	1.00	
	214	12 s	7.24	0.33	
			7.18	0.33	
			7.12	0.33	
	221	short	7.6		
	222	5.5 s	6.96		
	223	2.2 m	6.660	0.375	0.99
			6.646	0.421	
			6.564	0.133	
			6.52	0.038	
			6.47	0.032	
	224	2.9 h	6.17		0.1
	225	10.0 d	5.818	0.54	
			5.782	0.28	
			5.721	0.095	
			5.713	0.026	
			5.672	0.008	
			5.627	0.038	
			5.599	0.006	
			5.570	0.007	
			5.543	7×10^{-4}	
	227	21.6 a	4.949	0.49	0.012
			4.936	0.36	
			4.866	0.069	
			4.849	0.055	
			4.786	0.010	
			4.759	0.018	
			4.728	0.001	
			4.704	0.004	
			4.516	0.002	
$_{90}$Th	223	0.9 s	7.55		
	224	1.05 s	7.17	0.79	
			7.00	0.19	
			6.77	0.015	
			6.70	0.005	
	225	8.0 m	6.799	0.09	0.95
			6.745	0.07	
			6.701	0.02	
			6.651	0.03	
			6.627	0.03	
			6.502	0.14	
			6.479	0.43	
			6.442	0.15	
			6.346	0.02	
			6.313	0.02	
	226	30.9 m	6.336	0.79	
			6.227	0.19	
			6.101	0.017	

Table 4.10. Alpha Radioactivity Data (*Continued*)

Element	Mass number	Half-life	Alpha energy, Mev	Abundance of group	Alpha branching ratio
			6.035	0.006	
(Rd Ac)	227	18.17 d	6.041	0.23	
			6.012	0.028	
			5.981	0.24	
			5.963	0.035	
			5.919	0.009	
			5.870	0.030	
			5.810	0.010	
			5.798	0.003	
			5.766	0.003	
			5.760	0.021	
			5.717	0.050	
			5.713	0.087	
			5.704	0.040	
			5.697	0.015	
			5.672	0.019	
(Rd Th)	228	1.91 a	5.427	0.71	1
			5.344	0.28	
			5.214	0.004	
			5.179	0.002	
			5.143	3×10^{-4}	
	229	7340 a	5.048	0.067	1
			5.028	0.002	
			5.003	0.001	
			4.971	0.034	
			4.961	0.060	
			4.925	2.5×10^{-3}	
			4.894	0.107	
			4.837	0.582	
			4.806	0.114	
			4.788	0.010	
			4.751	0.015	
			4.678	0.004	
(Io)	230	7.52×10^{4} a	4.687	0.74	1
			4.620	0.26	
			4.481	0.002	
			4.438	3×10^{-4}	
			4.373	1×10^{-5}	
			4.278	8×10^{-8}	
			4.250	8×10^{-8}	
(Th)	232	1.39×10^{10} a	4.012	0.76	1
			3.954	0.24	
			3.829	0.002	
$_{91}$Pa	224	0.6 s			
	225	2 s			
	226	1.8 m	6.81		
	227	38.3 m	6.533	0.023	0.85
			6.522	0.003	
			6.467	0.50	
			6.425	0.115	
			6.417	0.148	
			6.403	0.093	
			6.378	0.026	
			6.358	0.078	
			6.338	0.007	
			6.328	0.004	
			6.301	0.008	
			6.145	0.025	
	228	22 h	6.121	0.10	0.02
			6.108	0.12	
			6.094	0.023	
			6.081	0.21	
			6.069	0.010	
			6.044	0.023	
			6.031	0.09	

TABLE 4.10. ALPHA RADIOACTIVITY DATA (*Continued*)

Element	Mass number	Half-life	Alpha energy, Mev	Abundance of group	Alpha branching ratio
			6.014	0.008	
			6.001	0.003	
			5.992	0.011	
			5.985	0.028	
			5.978	0.027	
			5.950	0.006	
			5.944	0.005	
			5.925	0.008	
			5.910	0.011	
			5.877	0.014	
			5.861	0.003	
			5.846	0.004	
			5.808	0.073	
			5.802	0.11	
			5.782	0.014	
			5.768	0.020	
			5.763	0.014	
			5.759	0.025	
			5.714	0.010	
	229	1.4 d	5.668	0.19	0.0025
			5.628	0.10	
			5.613	0.13	
			5.589	0.05	
			5.578	0.37	
			5.563	0.039	
			5.534	0.09	
			5.515	0.006	
			5.499	0.007	
			5.477	0.018	
			5.420	7×10^{-4}	
			5.411	15×10^{-4}	
			5.318	5×10^{-4}	
	230	17.7 d	5.14		3×10^{-5}
	231	34800 a	5.045	0.11	1
			5.018	0.025	
			5.016	0.20	
			4.999	0.254	
			4.972	0.014	
			4.962	0.004	
			4.938	0.228	
			4.921	0.030	
			4.887	2×10^{-5}	
			4.839	0.014	
			4.782	4×10^{-4}	
			4.724	0.084	
			4.700	0.01	
			4.668	0.015	
			4.630	10^{-4}	
			4.619	10^{-4}	
			4.586	1.5×10^{-4}	
			4.453	8×10^{-5}	
			4.495	3×10^{-5}	
$_{92}U$	227	1.3 m	6.8		
	228	9.3 m	6.68	0.70	0.95
			6.59	0.29	
			6.44	0.007	
			6.40	0.005	
	229	58 m	6.361	0.64	0.20
			6.333	0.20	
			6.298	0.11	
			6.261	0.01	
			6.224	0.03	
			6.186	0.01	
	230	20.8 d	5.891	0.672	1
			5.820	0.321	

TABLE 4.10. ALPHA RADIOACTIVITY DATA (*Continued*)

Element	Mass number	Half-life	Alpha energy, Mev	Abundance of group	Alpha branching ratio
			5.669	0.004	
			5.665	0.003	
	231	4.2 d	5.45		5.5×10^{-5}
	232	73.6 a	5.324	0.68	1
			5.267	0.32	
			5.241	0.0032	
			5.004	0.0001	
	233	1.62×10^5 a	4.821	0.83	1
			4.792	0.0048	
			4.778	0.146	
			4.751	0.003	
			4.724	0.015	
			4.695	8×10^{-4}	
			4.657	6×10^{-4}	
			4.625	1.5×10^{-4}	
			4.580	4×10^{-5}	
			4.505	3.3×10^{-4}	
			4.456	4×10^{-5}	
(UII)	234	2.48×10^5 a	4.772	0.72	1
			4.721	0.28	
			4.605	3.5×10^{-3}	
			4.27	3×10^{-7}	
			4.14	1.6×10^{-7}	
(AcU)	235	6.92×10^8 a	4.596	0.046	1
			4.555	0.037	
			4.501	0.012	
			4.444	0.006	
			4.414	0.04	
			4.395	0.57	
			4.365	0.18	
			4.322	0.034	
			4.265	0.006	
			4.215	0.057	
			4.156	0.005	
	236	239×10^7 a	4.503	0.73	
			4.455	0.27	
			4.343	0.005	
(UI)	238	4.51×10^9 a	4.199	0.77	
			4.152	0.23	
			4.042	0.0023	
$_{93}$Np	231	50 m	6.28		0.01
	233	35 m	5.53		10^{-7}
	235	410 d	5.095	0.038	1.6×10^{-5}
			5.015	0.836	
			4.925	0.118	
			4.864	0.008	
	237	2.14×10^6 a	4.877	0.0044	1
			4.875	0.0092	
			4.867	0.0024	
			4.821	0.0149	
			4.807	0.0156	
			4.792	0.5142	
			4.775	0.1938	
			4.770	0.1682	
			4.745	0.0002	
			4.716	0.0013	
			4.712	0.0029	
			4.703	0.0007	
			4.698	0.0018	
			4.668	0.016	
			4.663	0.0057	
			4.643	0.0462	
			4.603	0.0006	
			4.599	0.0008	
			4.585	0.0002	

TABLE 4.10. ALPHA RADIOACTIVITY DATA (*Continued*)

Element	Mass number	Half-life	Alpha energy, Mev	Abundance of group	Alpha branching ratio
			4.578	0.0005	
			4.519	0.0001	
			4.390	0.0002	
$_{94}$Pu	232	36 m	6.58		0.1
	233	20 m	6.30		10^{-3}
	234	9 h	6.203	0.68	0.06
			6.152	0.32	
			6.032	0.004	
	235	26 m	5.85		3×10^{-5}
	236	2.85 a	5.769	0.689	
			5.722	0.309	
			5.616	1.8×10^{-3}	
			5.454	2×10^{-5}	
	237	45.6 d	5.65	0.21	2.6×10^{-5}
			5.36	0.79	
	238	86.4 a	5.495	0.71	
			5.452	0.29	
			5.354	1.3×10^{-3}	
			5.204	4.3×10^{-5}	
			5.004	7×10^{-8}	
			4.699	1.2×10^{-6}	
	239	24360 a	5.157	0.733	1
			5.145	0.151	
			5.107	0.115	
			5.078	3.2×10^{-4}	
			5.066	9×10^{-6}	
			5.056	2.1×10^{-4}	
			5.031	5×10^{-5}	
			5.010	8×10^{-5}	
			5.001	6×10^{-6}	
			4.988	5×10^{-5}	
			4.963	3×10^{-5}	
			4.957	5×10^{-6}	
			4.937	3×10^{-5}	
			4.914	8×10^{-6}	
			4.873	7×10^{-6}	
			4.830	1.5×10^{-5}	
			4.801	6×10^{-6}	
			4.743	2.6×10^{-5}	
			4.695	4×10^{-6}	
			4.636	2×10^{-6}	
	240	6580 a	5.169	0.76	1
			5.125	0.24	
			5.024	9×10^{-4}	
			4.864	2×10^{-5}	
	241	13.3 a	4.895	0.75	2.3×10^{-5}
			4.850	0.25	
	242	3.73×10^5 a	4.905	0.74	
			4.861	0.26	
	244	7.6×10^7 a			1.0
$_{95}$Am	237	1.3 h	6.01		5×10^{-5}
	239	12 h	5.83		5×10^{-5}
			5.78	1.00	
	241	457.7 a	5.543	0.0025	
			5.510	0.0012	
			5.484	0.860	
			5.468	0.0004	
			5.442	0.127	
			5.416	0.0001	
			5.387	0.0133	
			5.320	1.5×10^{-4}	
			5.291	1×10^{-6}	
			5.277	5×10^{-6}	
			5 272	3×10^{-6}	

TABLE 4.10. ALPHA RADIOACTIVITY DATA (*Continued*)

Element	Mass number	Half-life	Alpha energy, Mev	Abundance of group	Alpha branching ratio
			5.242	24×10^{-6}	
			5.222	13×10^{-6}	
			5.192	6×10^{-6}	
			5.180	9×10^{-6}	
			5.176	3×10^{-6}	
			5.155	7×10^{-6}	
			5.137	3×10^{-6}	
			5.113	4×10^{-6}	
			5.099	7×10^{-6}	
			5.093	3×10^{-6}	
			5.086	3×10^{-6}	
	242	152 a	5.406	0.016	4.8×10^{-3}
			5.362	0.016	
			5.310	0.008	
			5.28	0.004	
			5.246	0.006	
			5.203	0.88	
			5.138	0.057	
			5.080	0.003	
			5.063	0.0025	
	243	7.95×10 a	5.347	0.0017	1.0
			5.316	0.0016	
			5.274	0.87	
			5.232	0.115	
			5.177	0.011	
$_{96}$Cm	238	2.5 h	6.52		0.025
	240	26.8 d	6.294	0.72	
			6.250	0.28	
			6.150	4×10^{-4}	
	241	35 d	5.942	0.70	
			5.932	0.17	
			5.886	0.13	
	242	162.5 d	6.115	0.737	
			6.072	0.263	
			5.972	3.0×10^{-4}	
			5.817	0.5×10^{-4}	
			5.611	3×10^{-7}	
			5.520	32×10^{-7}	
			5.190	6×10^{-7}	
			5.12	4×10^{-8}	
	243	32 a	6.066	0.010	1
			6.059	0.05	
			6.010	0.012	
			5.992	0.06	
			5.905	0.0015	
			5.877	0.007	
			5.785	0.73	
			5.741	0.115	
			5.685	0.016	
			5.681	0.0018	
			5.639	0.0015	
			5.613	0.0002	
			5.589	0.0005	
	244	18.11 a	5.802	0.767	
			5.760	0.233	
			5.663	1.7×10^{-4}	
			5.507	4×10^{5}	
			5.19	1×10^{-6}	
			4.950	1.5×10^{-6}	
			4.910	5×10^{-7}	
	245	9320 a	5.53		
			5.358	0.93	
			5.303	0.07	
	246	5480 a	5.384	0.79	
			5.340	0.21	

TABLE 4.10. ALPHA RADIOACTIVITY DATA (*Continued*)

Element	Mass number	Half-life	Alpha energy, Mev	Abundance of group	Alpha branching ratio
	247	1.64×10^7 a			
	248	4.7×10^5 a	5.076	0.82	0.89
			5.032	0.18	
$_{97}$Bk	243	4.5 h	6.72	0.30	1.5×10^{-3}
			6.55	0.53	
			6.20	0.17	
	244	4.5 h	6.67		6×10^{-5}
	245	4.98 d	6.37	0.33	10^{-3}
			6.17	0.41	
			5.89	0.26	
	247	1380 a	5.67	0.37	
			5.51	0.58	
			5.30	0.05	
	249	314 d	5.395	0.96	2.2×10^{-5}
			5.376		
			5.326		
			5.087	0.04	
$_{98}$Cf	244	25 m	7.17		
	245	44 m	7.11		0.3
	246	35.7 h	6.753	0.78	1.00
			6.711	0.22	
			6.615	0.003	
			6.468	0.0002	
	248	350 d	6.26	0.82	
			6.22	0.18	
	249	360 a	6.194	0.019	
			6.139	0.011	
			6.072	0.004	
			5.990	0.0008	
			5.941	0.033	
			5.898	0.030	
			5.842	0.012	
			5.806	0.84	
			5.778	0.005	
			5.749	0.044	
			5.687	0.004	
	250	13.2 a	6.044	0.83	1.00
			6.000	0.17	
			5.902	0.0032	
	251	800 a	5.844	0.50	
			5.667	0.50	
	252	2.65 a	6.132	0.84	0.97
			6.089	0.155	
			5.988	0.0028	
$_{99}$Es	245	75 s	7.65		
	246	7.3 m	7.35		
	248	25 m	6.87		3×10^{-3}
	249	2 h	6.76		1×10^{-3}
	251	1.5 d	6.48		5×10^{-3}
	252	140 d	6.64		
	253	20.0 d	6.633	0.90	1.0
			6.624	0.008	
			6.594	0.007	
			6.592	0.066	
			6.552	0.0075	
			6.541	0.0085	
			6.498	0.0026	
			6.480	0.0008	
			6.249	0.0004	
			6.209	0.0004	
			6.158	0.00015	
	254	270 d	6.479	0.0031	
			6.430	0.943	
			6.417	0.0122	
			6.385	0.0014	

TABLE 4.10. ALPHA RADIOACTIVITY DATA (*Continued*)

Element	Mass number	Half-life	Alpha energy, Mev	Abundance of group	Alpha branching ratio
			6.361	0.0243	
			6.349	0.0091	
			6.324	0.00055	
			6.279	0.0016	
			6.271	0.0023	
			6.260	0.00028	
			6.186	0.0020	
			6.138	0.00008	
			6.087	0.00024	
			6.062	0.00014	
$_{100}$Fm	248	0.6 m	7.8		
	249	150 s	7.9		
	250	30 m	7.43		
	251	7 h	6.89		0.01
	252	23 h	7.04		
	253	3 d	6.9		0.1
			6.7		
	254	3.24 h	7.192	0.85	
			7.150	0.14	
			7.053	0.009	
	255	20 h	7.122	0.0009	
			7.098	0.0010	
			7.076	0.0043	
			7.019	0.934	
			6.977	0.0011	
			6.960	0.053	
			6.887	0.0060	
			6.803	0.0012	
			6.58	4.5×10^{-4}	
	257	100 d			
$_{101}$Md	255	30 m	7.34		<1
	257	3 h	7.1		
$_{102}$No	253	15 s	8.8		
	254	3 s	8.3		
	255	15 s	8.2		
	256	8 s			
$_{103}$Lw	257	8 s	8.6		

References for Table 4.10

Nuclear Data Cards, National Research Council, Washington, D.C.

E. K. Hyde, I. Perlman, and G. T. Seaborg, "The Nuclear Properties of the Heavy Elements," Prentice-Hall, Englewood Cliffs, N.J., 1964.

Values of A. Rytz, *Helv. Phys. Acta*, **34**, 240 (1961) were used as energy standards.

energy of the alpha groups, abundance of group, and branching ratios.

Departures from the simple theory may be expected for several reasons, all of which lengthen the half-life: (1) Spin changes in the emission of alphas contribute to angular momentum of the alphas and increase the barrier and hence the half-life. The effect of angular momentum l is found by adding a potential $\hbar^2 l(l+1)/2\mu r^2$ to the Coulomb potential. An l of 3 changes τ only by a factor of 7 for these heavy alpha-radioactive nuclei. (2) Even-odd nuclei often show systems of "rotational levels" (see Part 9, Chap. 1). The hindrances for transitions to these increase rapidly and are only in part accounted for by the effect of angular momentum. (3) Departure of the nucleus from the spherical form may also produce hindrance. (4) The delay in assembling the components of the alpha particle is greater for nuclei having one or more unpaired nucleons, since the unpaired nucleon must be incorporated in the emitted alpha particle to leave the resultant nucleus in its ground state. The influence of this effect is shown conspicuously by Mang [86] in the alpha decay of ^{211}Po. Large discontinuities in the energy-mass number diagram occur near magic-number nuclei (see Part 9, Chap. 1). They are explained by energy changes due to the shell model and do not involve any anomaly in the alpha decay.

In many cases the alpha-particle spectrum is observed to be complex. The most common type of complexity consists of lower-energy groups of alpha particles of roughly equal intensity. This type of complexity is ascribed to transitions from the ground state of the initial nucleus to various excited states of the resultant nucleus. These excited states produce gamma rays or conversion electrons in coincidence with the lower-energy alphas. The lower-energy alphas are less probable because of decreased barrier penetration (except for spin changes or component assembly differences). These intensity differences are shown in Table 4.10. A second type of complexity is exhibited by the short-lived alpha emitters Th C′ and Ra C′. Their alpha spectrum contains very weak groups of higher-energy alphas and is attributed to various levels of the emitting nucleus. Here the alpha particle has more energy and should penetrate the barrier more easily than the ground-state alphas, but it must compete with the emission of gamma rays to the ground state. Such branching enables the calculation of gamma-ray half-lives. Lifetimes of 10^{-13} sec are found for the first levels of Ra C′ and Th C′. Branching occurs when several modes of decay of a nucleus have probabilities of comparable magnitudes so that they can be observed. Branching therefore designates the splitting that occurs in the radioactive series when the emission of alpha particles and beta rays are both possible.

The classical radioactive families of thorium, uranium, and actinium represent the $4n$, $4n + 2$, and $4n + 3$ series of natural radioactivities. Only alpha particles and beta rays are emitted in these series of activities; thus the series retains its even-odd characteristics, since its masses differ by 4. A new radioactive series, the $4n + 1$, called the neptunium series, has been created by the artificial production of ^{237}Np of half-life 2.14×10^6 years. Figure 4.8 shows this series. In addition, many new alpha-radioactive isotopes have been produced and detected in the heavy elements. Some are in the transuranic region, some connect with the natural series or make collateral series, and some result from neutron-deficient isotopes of heavy elements. These are included in Table 4.10. A few light alpha emitters are also included.

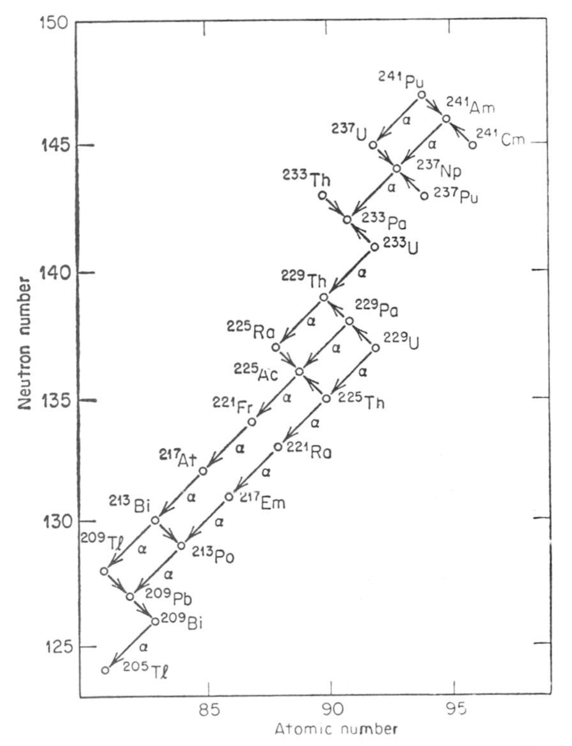

FIG. 4.8. The $4n + 1$ or neptunium radioactive series, showing the chain of radioactive disintegrations associated with ^{237}Np.

All isotopes with Z greater than 82 are radioactive; many of them emit alpha particles. The reason for this occurrence of alpha radioactivity is primarily one of nuclear energetics. The Coulomb repulsion increases with atomic number and reduces the binding of an alpha particle. The Coulomb barrier also increases with atomic number and retains the alpha particle except near variations in the binding-energy surface. The largest energy release in alpha emission occurs for the C′ members of the natural radioactive series. In these cases the resultant nucleus has a magic number, 82, of protons and almost a magic number, 126, of neutrons leading to increased stability. The medium elements (^{146}Sm, $_{52}$Te, etc.) showing alpha radioactivity are likewise situated near a drop in the binding-energy surface because of a magic number in neutrons or protons [84]. Alpha emission competes with beta emission (often K capture) and with fission. Heisenberg [87] and Turner [88] have discussed the competition of alpha emission with beta emission (or K capture) and with fission as it relates to the occurrence of heavy nuclei. The natural radioactive series owe their existence to the long-lived alpha-radioactive parents (^{238}U, ^{232}Th, and ^{235}U) which have not wholly decayed in the time ($\sim 3 \times 10^9$ years) since nuclei were formed.

References

1. Everling, F., L. A. Konig, J. H. E. Mattauch, and A. H. Wapstra: *Nucl. Phys.*, **18**: 529 (1960).
2. Bloch, F.: *Z. Physik*, **81**: 363 (1933).
3. National Bureau of Standards (U.S.): *Tech. News Bull.*, October, 1963.
4. Briggs, G. H.: *Proc. Roy. Soc.* (*London*), **A118**: 549 (1928). Rutherford and Robinson: *Phil. Mag.*, *Ser.* 6, **28**: 552 (1914).
5. Bethe and Bacher: *Revs. Mod. Phys.*, **8**: 145 (1936). Irving, J.: *Phil. Mag.*, **42**: 338 (1951).
6. Stephens, W. E.: *Phys. Rev.*, **57**: 938 (1940).
7. Herman, R., and R. Hofstadter: "High Energy Electron Scattering Tables," Stanford University Press, Stanford, Calif., 1960.
8. Mulliken, R. S.: *Trans. Faraday Soc.*, **25**: 634 (1929).
9. Blackett and Champion: *Proc. Roy. Soc.* (*London*), **A130**: 380 (1931).
10. Suess, H. E., and H. C. Urey: *Revs. Mod. Phys.*, **28**: 53 (1956). Waddington, C. J.: *Progr. Nucl. Phys.*, **8**: 5 (1960).
11. Bethe, H. A.: *Phys. Rev.*, **55**: 434 (1939). Alpher, Bethe, and Gamow: *Phys. Rev.*, **73**: 803 (1948).
12. Chandrasekhar, S.: *Revs. Mod. Phys.*, **18**: 94 (1946).
13. Aldrich and Nier: *Phys. Rev.*, **74**: 1590 (1948); **70**: 983 (1936).
14. Douglas and Nerzberg: *Phys. Rev.*, **76**: 1529 (1949).
15. Anderson, H. L.: *Phys. Rev.*, **76**: 1460 (1949).
16. Bethe, H. A., and Bacher: *Revs. Mod. Phys.*, **8**: 134 (1936).
17. Trainor, L. E. H.: *Phys. Rev.*, **85**: 962 (1952).
18. Parker, P. D., P. F. Donovan, J. V. Kane, and J. F. Mollenauer: *Phys. Rev. Letters*, **14**: 15 (1965).
19. Rutherford, Chadwick, and Ellis: "Radiations from Radioactive Substances," Cambridge University Press, New York, 1930. Erbacker, O.: *Z. Physik. Chem.*, **A156**: 142 (1931).

20. Rossi, B. R., and H. H. Staub: "Ionization Chambers and Counters," McGraw-Hill, New York, 1949. Wilkinson, D. H.: "Ionization Chambers and Counters," Cambridge University Press, New York, 1950. Staub, H. H.: in E. Segrè (ed.), "Detection Methods, Part I, Experimental Nuclear Physics," Wiley, New York, 1953.
21. Taylor, J. M.: "Semiconductor Particle Detectors," Butterworth, Washington, D.C., 1963. Dearnaley, G., and D. C. Northrop: "Semiconductor Counters for Nuclear Radiations," E. & F. N. Spon, Ltd., London, 1963. Miller, G. L., W. M. Gibson, and P. F. Donovan: *Ann. Rev. Nucl. Sci.*, **12**: 189 (1962).
22. See, for example, Fulbright, H. W., and D. Kohler: *Proc. Symp. Nucl. Instr.*, Harwell, 1961, Heywood & Co., London, 1962.
23. Rotblat, J.: Photographic Emulsion Technique, *Progr. Nucl. Phys.*, **1**: 37 (1950), Butterworth-Springer, Ltd., London. Powell, C. F., P. H. Fowler, and D. H. Perkins: "Study of Elementary Particles by Photographic Methods," Pergamon Press, New York, 1959.
24. Rytz, A.: *Helv. Phys. Acta*, **34**: 240 (1961).
25. Bethe, H. A., and J. Ashkin: in E. Segrè (ed.), "Experimental Nuclear Physics," pt. II, vol. 1, Wiley, New York, 1953.
26. Allison, S. K., and S. D. Warshaw: *Revs. Mod. Phys.*, **25**: 779 (1953).
27. Whaling, W.: "Encyclopedia of Physics," vol. 34, p. 193, Springer, Berlin, 1958.
28. Fano, U.: *Ann. Rev. Nucl. Sci.*, **13**: 1 (1963).
29. Northcliffe, L. C.: *Ann. Rev. Nucl. Sci.*, **13**: 67 (1963).
30. Penetration of Charged Particles in Matter, *Natl. Acad. Sci.–Natl. Res. Council Publ.* 1133, 1964.
31. Mott, N. F.: *Proc. Roy. Soc. (London)*, **A124**: 425 (1929).
32. Lea, D. E.: "Actions of Radiations on Living Cells," Cambridge University Press, New York, 1947.
33. Livingston and Bethe: *Revs. Mod. Phys.*, **9**: 260 (1937).
34. Fermi, E.: *Phys. Rev.*, **57**: 485 (1940). Uehling, E. A.: *Ann. Rev. Nucl. Sci.*, **4**: 315 (1954).
35. Cerenkov, P. A.: *Phys. Rev.*, **52**: 378 (1937).
36. Fano, U.: *Phys. Rev.*, **70**: 44 (1946).
37. Jesse, W. P., and J. Sadauskis: *Phys. Rev.*, **88**: 417 (1952).
38. Haeberli, W., P. Huber, and E. Baldinger: *Helv. Phys. Acta*, **26**: 145 (1953). Bortner, T. E., and J. L. Hurst: *Phys. Rev.*, **93**: 1237 (1954).
39. Gurney, R. W.: *Proc. Roy. Soc. (London)*, **A107**: 332 (1925).
40. Jesse, Forstat, and Sadauskis: *Phys. Rev.*, **77**: 782 (1950).
41. Rhodes, Franzen, and Stephens: *Phys. Rev.*, **87**: 141 (1952). Hanna, G. C.: *Phys. Rev.*, **80**: 530 (1950).
42. Livingston and Bethe: *Revs. Mod. Phys.*, **9**: 285 (1937).
43. Siri, W.: "Isotopic Tracers and Nuclear Radiations," McGraw-Hill, New York, 1949.
44. Bradner, Smith, Barkas, and Bishop, *Phys. Rev.*, **77**: 462 (1950).
45. Sternheimer, R. M.: *Phys. Rev.*, **115**: 137 (1959); Errata, **124**: 2051 (1961); **118**: 1045 (1960). Also, in L. C. Yuan and C. S. Wu (eds.), "Methods of Experimental Physics," vol. 5, pt. A, Academic, New York, 1955.
46. Brandt, W.: *Phys. Rev.*, **112**: 1624 (1958).
47. Barkas, W. H.: *Univ. Calif. Radiation Lab. Rept.* 10292, 1962.
48. Bichsel, H.: *Univ. So. Calif. Rept.* 2, Contract AT(04-3)-136, 1961.
49. Bohr, N.: *Kgl. Danske Videnskab. Selskab, Mat.-Fys. Medd.*, **18**: 8 (1948). Harvey, B. G.: *Ann. Rev. Nucl. Sci.*, **10**: 235 (1960).
50. Rutherford, E.: *Phil. Mag.*, **21**: 669 (1911).
51. Chadwick, J.: *Phil. Mag.*, **40**: 734 (1920).
52. Darwin, C. G.: *Phil. Mag.*, **27**: 499 (1914).
53. Bethe, H. A.: *Revs. Mod. Phys.*, **9**: 173 (1937).
54. Mott and Massey: "Theory of Atomic Collisions," Oxford University Press, Fair Lawn, N.J., 1949.
55. Mott, N. F.: *Proc. Roy. Soc. (London)*, **A126**: 259 (1929).
56. Chadwick, J.: *Proc. Roy. Soc. (London)*, **A128**: 114 (1930).
57. Blackett and Champion: *Proc. Roy. Soc. (London)*, **A130**: 380 (1931).
58. Darriulat, P., Igo, Pugh, and Holmgren: *Phys. Rev.*, **137**: B 315 (1965) and references therein.
59. Conzett, H. E., et al.: UCRL-11382, 1964.
60. Lefevre, H. W., R. R. Borchers, C. H. Poppe: *Phys. Rev.*, **128**: 1328 (1962).
61. Chadwick and Bieler: *Phil. Mag.*, **42**: 923 (1921).
62. Gammel, J. L., and R. M. Thaler: *Phys. Rev.*, **109**: 2041 (1958).
63. Bunch, S. M., et al.: *Nucl. Phys.*, **53**: 241 (1964).
64. Austin, S. M., et al.: *Phys. Rev.*, **126**: 1532 (1962).
65. May, T. H., et al.: *Nucl. Phys.*, **45**: 17 (1963).
66. Riezler, W.: *Proc. Roy. Soc. (London)*, **A134**: 154 (1932); *Ann. Physik*, **23**: 198 (1935).
67. Wenzel, P.: *Z. Physik. Chem.*, **90**: 754 (1934).
68. Blair, J. S.: *Phys. Rev.*, **108**: 827 (1957).
69. Igo, G.: *Phys. Rev. Letters*, **1**: 72 (1958).
70. Austern, N.: *Ann. Phys.*, **15**: 299 (1961).
71. Wilson, H. L., and M. B. Sampson: *Phys. Rev.*, **137**: B 305 (1965).
72. Wentzel, G.: *Ann. Physik*, **69**: 335 (1922).
73. Mott and Massey: "Theory of Atomic Collisions," Oxford University Press, Fair Lawn, N.J., 1949.
74. Mayer, F.: *Ann. Physik*, **41**: 931 (1913).
75. Gurney and Condon: *Phys. Rev.*, **33**: 127 (1929).
76. Gamow, G.: *Z. Physik*, **51**: 204 (1928).
77. Shapiro, M.: *Phys. Rev.*, **90**: 171 (1953).
78. Bethe, H. A.: *Revs. Mod. Phys.*, **9**: 161 (1937).
79. Preston, M. A.: *Phys. Rev.*, **69**: 535 (1946).
80. Barschall et al.: *Phys. Rev.*, **73**: 659 (1948).
81. Asaro, Reynolds, and Perlman: *Phys. Rev.*, **87**: 277 (1952). Asaro, F.: NCRL-2180, June, 1953.
82. Mang, H. J.: *Ann. Rev. Nucl. Sci.*, **14**: 1 (1964).
83. Geiger and Nuttall: *Phil. Mag.*, **23**: 439 (1912).
84. Perlman, Ghiorso, and Seaborg: *Phys. Rev.*, **77**: 26 (1950).
85. Gallagher, C. J., and J. O. Rasmussen: *J. Inorg. Nucl. Chem.*, **3**: 333 (1946).
86. Zeh, H. D., and H. J. Mang: *Nucl. Phys.*, **29**: 529 (1962).
87. Heisenberg, W.: Solvay Conference, 1933.
88. Turner, L.: *Revs. Mod. Phys.*, **17**: 292 (1945).

Bibliography

Allison, S. K., and S. O. Warshaw: *Revs. Mod. Phys.*, **25**: 779 (1953).
Bethe, R. A., et al.: Nuclear Physics, *Revs. Mod. Phys.*, **8**: 83 (1936); **9**: 69, 249 (1937).
Bethe, H. A., and J. Ashkin: in E. Segrè (ed.), "Experimental Nuclear Physics," vol. 1, Wiley, New York, 1953.
Blatt and Weisskopf: "Theoretical Nuclear Physics," Wiley, New York, 1952.
Bohr, N.: *Kgl. Danske Videnskab. Selskab, Mat.-Fys. Medd.*, **18**: 8 (1948).
Fano, U.: Penetration of Protons, Alpha Particles and Mesons, *Ann. Rev. Nucl. Sci.*, **13**: 1 (1963).
Hanna, G. C.: Alpha Radioactivity in E. Segrè (ed.), "Experimental Nuclear Physics," vol. 3, p. 55, Wiley, New York, 1959.
Hyde, Earl K., I. Perlman, and G. T. Seaborg: "Nuclear Properties of the Heavy Elements," vol. I: System-

atics of Nuclear Structure and Radioactivity; vol. II: Detailed Radioactivity Properties, Prentice-Hall, Englewood Cliffs, N.J., 1964.

Mang, H. J.: Alpha Radioactivity, *Ann. Rev. Nucl. Sci.*, **14**: 1 (1964).

Mott and Massey: "Theory of Atomic Collisions," Oxford University Press, Fair Lawn, N.J., 1949.

National Research Council: Studies in Penetration of Charged Particles in Matter, *Publ.* 1133, Washington, D.C., 1964.

Northcliffe, L. C.: Passage of Heavy Ions through Matter, *Ann. Rev. Nucl. Sci.*, **13**: 67 (1963).

Perlman, I.: Alpha Radioactivity, "Encyclopedia of Physics," vol. 42, p. 42, Springer, Berlin, 1957.

Preston, M. A.: α-Radioactivity, "Physics of the Nucleus," chap. 14, Addison-Wesley, Reading, Mass., 1962.

Rasmussen, J. O.: Alpha Radioactivity, in K. Siegbahn (ed.), "α, β, γ Spectroscopy," North Holland Publishing Company, Amsterdam, 1963.

Rutherford, Chadwick, and Ellis: "Radiations from Radioactive Substances," Cambridge University Press, New York, 1930.

Siri, W.: "Isotopic Tracers and Nuclear Radiations," McGraw-Hill, New York, 1949.

Sternheimer, R. M.: in L. C. Yuan and C. S. Wu (eds.), "Methods of Experimental Physics," vol. 5, pt. A, Academic, New York, 1955.

Strominger, D., J. M. Hollander, and G. T. Seaborg: Table of Isotopes, *Revs. Mod. Phys.*, **30**: 585 (1958).

Taylor, A. E.: *Rept. Progr. Phys.*, **15**: 1952.

Whaling, W.: The Energy Loss of Charged Particles in Matter, "Encyclopedia of Physics," vol. 34, p. 193, Springer, Berlin, 1958.

Chapter 5

Weak Interaction Processes

By J. NILSSON* *and* M. E. ROSE†, University of Virginia

1. Introduction

The first weak interaction process, observed almost seven decades ago, was identified some years later as a spontaneous decay of a naturally radioactive nucleus into an isobar with charge one unit higher with the emission of an electron. The widespread occurrence of both electron and positron decay of nuclei was revealed when artificial transmutation of nuclei became everyday events in the laboratory. The first major step toward an understanding of these processes was made by Fermi [1]‡ who exploited the suggestion of Pauli that β decay must be accompanied by a very light, neutral particle, the neutrino. Fermi's theory, which was remarkably close to current descriptions, provided a basis for assessing the strength of the coupling and was a first formulation of the formal structure of the interaction responsible for the decay. However, even without any detailed theory, the known fact that β-decay lifetimes were at least 10^{12} times longer than those pertaining to electromagnetic processes with comparable energy release is eloquent testimony of an extremely weak interaction, as compared with strong (e.g., nuclear) and electromagnetic interactions.

The period 1934 to 1957 was marked by two important developments. First, in nuclear β decay, including orbital capture, extensive measurements of spectral shapes and lifetimes made it clear that the Fermi theory, supplemented by spin-dependent (Gamow-Teller) interactions [2], gave an accurate representation of such data. Because of the intimate connection between the spectrum shape and the transition probabilities with relevant nuclear parameters (spins and parities) such studies were instrumental in deciphering many decay schemes and in providing information vital to description of nuclear structure based on a variety of nuclear models. Nevertheless, some ambiguity remained in the description of the β interaction itself in that, a priori, this interaction could contain five complex coupling constants (connected with S, V, T, A, and P couplings; see Sec. 2), and despite heroic efforts very few certain conclusions regarding them could be reached.

* Present address: Department of Mathematical Physics, Chalmers University of Technology, Gothenburg, Sweden.

† Work partially supported by U.S. Atomic Energy Commission Document ORO-2915-57.

‡ Numbers in brackets refer to References at end of chapter.

The second important development was the discovery of weak decays of other (elementary) particles. These, as well as some decay processes discovered more recently, are listed in Table 5.1. In addition to the decay modes listed for n which schematizes β decay in any nucleus, β^+ decay of complex nuclei involves the transformation

$$p \to n + e^+ + \nu_e \tag{5.1}$$

within the nucleus. This competes with the capture reaction

$$e^- + p \to n + \nu_e \tag{5.2}$$

Here and in Table 5.1 we have taken cognizance of the recent discovery of two kinds of neutrinos, ν_e, and ν_μ, associated with the electron and muon, respectively. Table 5.1 does not contain reactions induced by weak interactions. For example, the muon analogue of (5.2) is the well-known muon capture,

$$\mu^- + p \to n + \nu_\mu \tag{5.3}$$

which is a decay process of the muonic atom. In addition, neutrino-induced reactions (see Sec. 8) are not listed explicitly. An example is the inverse β decay [3]

$$\bar{\nu}_e + p \to n + e^+ \tag{5.4}$$

which is another variant of the decay process (5.1).

Only some of the decay processes listed in Table 5.1 were known in 1957. In particular, the 2π and 3π decays of K^+ attracted considerable attention since the odd parity of the pion implied that parity could not be conserved in this process [4]. This provocative idea of Yang and Lee [5] received dramatic confirmation in both nuclear β decay (asymmetry of e^- from Co^{60} [6]) and in μ decay (asymmetry of decay electrons [7]). That parity conservation and charge conjugation conservation were universally violated in weak interactions and, indeed, in a maximal way became firmly established. This seemed to complicate the theoretical description of weak interaction phenomena, but it soon developed that the breakdown of symmetry provided the basis for the decisive experiments whereby a complete determination of the basic interaction could be made. In place of the five invariant forms S, V, T, A, and P, weak interactions in the absence of coupling through meson fields (cf. μ decay) were simply described by the V-A combination. In nuclear decay the strong interactions had the effect of changing the effective coupling constant of the axial vector (A) interaction but the net weak interaction is

TABLE 5.1. DECAY PROCESSES AND PARTICLE PROPERTIES†

Particle[a]	Mass, Mev	Lifetime[b] sec	Typical decay modes	Energy release, Mev	Spin	Strangeness[c] S	Baryon number B	Isospin I
ν_e	<2(−4)	Stable			½			
ν_μ	<~4	Stable			½			
$e^\pm$	0.511006	Stable			½			
$\mu^\pm$	105.66	2.21(−6)	$e^\pm + \begin{Bmatrix} \nu_e + \bar{\nu}_\mu \\ \bar{\nu}_e + \nu_\mu \end{Bmatrix}$	105.2	½			
π^0	135.01	1.7(−16)	$\gamma + \gamma$[d]	135.0	0	0	0	1
$\pi^\pm$	139.60	2.55(−8)	$\mu^\pm + \nu_\mu/\bar{\nu}_\mu$	33.9	0	0	0	1
			$e^\pm + \nu_e/\bar{\nu}_e$	139.1				
			$\pi^0 + e^\pm + \nu_e/\bar{\nu}_e$	4.1				
K^+	493.9	1.22(−8)	$\mu^\pm + \nu_\mu/\bar{\nu}_\mu$	388.2	0	±1	0	½
			$\pi^\pm + \pi^0$	219.3				
			$\pi^\pm + \pi^+ + \pi^-$	75.1				
			$\pi^\pm + \pi^0 + \pi^0$	84.3				
			$\pi^0 + \mu^\pm + \nu_\mu/\bar{\nu}_\mu$	253.2				
			$\pi^0 + e^\pm + \nu_e/\bar{\nu}_e$	358.4				
$K_1{}^0$	497.8	1.00(−10)	$\pi^+ + \pi^-$	218.6	0		0	
			$\pi^0 + \pi^0$	227.8				
$K_2{}^0$	497.8	7 2(−8)	$\pi^+ + \pi^- + \pi^0$	83.6	0		0	
			$\pi + e + \nu$	357.7				
			$\pi + \mu + \nu$	252.6			0	
p	938.256	Stable			½	0	1	½
n	939.550	1.01(3)	$p + e^- + \bar{\nu}_e$	0.783	½	0	1	½
Λ	1115.40	2.5(−10)	$p + \pi^-$	37.6	½	−1	1	0
			$n + \pi^0$	40.9				
			$p + e^- + \bar{\nu}_e$	176.6				
			$p + \mu^- + \bar{\nu}_\mu$	71.5				
Σ^+	1189.41	0.81(−10)	$p + \pi^0$	116.2	½	−1	1	1
			$n + \pi^+$	110.3				
Σ^0	1192.3	Very short	$\Lambda + \gamma$[d]	76.1	½	−1	1	1
Σ^-	1197.08	1.6(−10)	$n + \pi^-$	116.9	½	−1	1	1
Ξ^0	1314.3	3.9(−10)	$\Lambda + \pi^0$	60.6	½	−2	1	½
Ξ^-	1320.8	1.4(−10)	$\Lambda + \pi^-$	63.4	½	−2	1	½
Ω^-	1675	0.7(−10)	$\Lambda + K^-$	66	3/2?	−3	1	0

† For further details, see A. H. Rosenfeld, A. Barbaro-Galtieri, W. H. Barkas, P. L. Bastien, J. Kirz, and M. Roos, *Revs. Mod. Phys.*, **36**: 977 (1964).

[a] In addition to the particles listed, the antiparticles for the two kinds of neutrinos and for all the baryons ($B = 1$) are known. They are designated by a bar (thus, $\bar{n} \equiv$ antineutron) except for the antiparticles explicitly listed. The quantum numbers, S, B, Q, I_3 for particles and antiparticles have opposite sign.

[b] The power of 10 is given by the number in parentheses.

[c] In terms of charge Q and the third component of isospin I_3 the following relations hold:

$$2Y = B + S = 2(Q - I_3)$$

where Y is the so-called hypercharge. For the antiparticle Y also changes sign.

[d] Electromagnetic process.

still basically *V-A*. (For experiments and results, see [S1].)

The above developments had a profound influence on the understanding of weak interaction processes. Nuclear β decay, which once stood as the single instance of this type of phenomenon, now becomes very closely linked with a more universal kind of process. We need no longer speculate as to the form of the basic interaction. Accepting it as the *V-A* interaction, we shall discuss the weak interactions from a more integrated point of view than was possible only a few years ago. For a better understanding of the events that occur in the laboratory, progress is made by exploring the more difficult question of the influence of the strong interactions in the weak processes; recent research has been concerned with just this problem.

In this chapter only the weak interactions of nonstrange particles will receive attention since they are better understood than those of the K mesons and hyperons with $S = 0$. The weak interaction processes of the $S = 0$ particles always involve leptons (ν, e, μ). Nuclear β decay is a prime example of weak interaction processes of nonstrange particles, and these constitute, from several points of view, a center of interest for the present discussion.

2. Formulation of the Theory of Weak Interactions

2.1. Empirical Conservation Laws. All experimental information about elementary particles stems from the study of either decay processes or scattering (production) processes. In the light of such data as lifetimes and scattering cross sections, these processes can be separated into three rather distinct groups called *strong processes, electromagnetic processes,* and *weak processes.* The characteristic lifetimes and cross sections are listed in Table 5.2. The underlying interactions basically responsible for these processes are similarly called *strong, electromagnetic,* and *weak interactions.* Apart from their difference in strength as reflected in cross sections and lifetimes, they also display very different symmetry properties, leading to different conservation laws and selection rules. However, there is strong evidence that the following conservation laws hold in the presence of all three types of interaction.

TABLE 5.2. COMPARISON OF TYPES OF INTERACTION

Type of process	Lifetime, sec	Cross section, cm^2
Strong	10^{-23}	10^{-26}
Electromagnetic	10^{-19}	10^{-30}
Weak	10^{-10}–10^{3}	$\lesssim 10^{-38}$

1. *Conservation of energy, momentum, and angular momentum.* As is well known, these conservation laws hold for any theory that is invariant under proper, inhomogeneous Lorentz transformations. The assumption of relativistic invariance being one of the cornerstones of modern physics, the validity of these selection rules is rarely questioned, and mostly they are employed to derive kinematical criteria to be used for the interpretation of processes as observed in experiments.

2. *Conservation of charge.* No process is known where electric charge is either created or destroyed.

3. *Conservation of baryon number.* If each baryon is assigned a baryonic number $B = +1$ and each antibaryon a baryonic number $B = -1$, then all known processes conserve the total baryon number; that is, the total baryon number of the final state equals the baryon number of the initial state. (This conservation law implies that the baryon with the lowest mass, namely, the proton, must be stable. The observation that the proton indeed is stable was one of the original motivations for requiring baryon-number conservation.)

4. *Conservation of lepton numbers.* To account for the observation that certain weak processes take place while others do not, in spite of the fact that they satisfy all the previously cited selection rules, two additional selection rules are introduced. They are expressed in terms of two new additive quantum numbers called the *muonic lepton number* L_μ and the *electronic lepton number* L_e. The particle μ^- and ν_μ are assigned $L_\mu = +1$, and the μ^+ and the $\bar{\nu}_\mu$ have $L_\mu = -1$; all other particles have $L_\mu = 0$. In the same way the e^- and ν_e are assigned $L_e = +1$ while the e^+ and the $\bar{\nu}_e$ have $L_e = -1$, and all other particles have $L_e = 0$. For a process to be allowed, these conservation laws require the total muonic and the total electronic lepton numbers to be conserved separately. Thus, for example, the following processes are allowed:

$$\mu^- \to e^- + \bar{\nu}_e + \nu_\mu$$
$$n \to p + e^- + \bar{\nu}_e$$
$$\bar{\nu}_\mu + p \to \mu^+ + n$$

whereas

$$\mu^+ \to e^+ + \gamma$$
$$\mu^+ \to e^+ + e^- + e^+$$
$$\mu^- + p \to e^- + p$$

are forbidden. In accordance with these selection rules, the first three processes have been observed to take place with a transition rate in good agreement with predictions while none of the others have been detected, although searched for extensively.

These conservation laws of the two lepton numbers were preceded by a single conservation law of the lepton number $L = L_\mu + L_e$, which is trivially fulfilled within the new scheme. The recent discovery of the existence of two different kinds of neutrinos, here denoted ν_e and ν_μ, prompted the introduction of two different lepton numbers. Thereby one could also account for the absence of the three last processes mentioned above. These processes all conserve the lepton number L and, therefore, they were allowed in the earlier scheme; they presented some embarrassment by their apparent absence.

In the development of a more detailed theory, these empirical selection rules are naturally incorporated from the beginning. As a consequence these rules will not be discussed explicitly in what follows although their full content is imposed upon the theory by requiring the S operator to respect them.

2.2. The Effective S Operator for Weak Processes. The S operator contains all information about possible scattering or decay processes (see Appendix A). It contains, therefore, many parts that are essentially irrelevant for the present discussion of weak processes. A trivial part has already been disposed of by introducing the T operator. We shall further restrict our attention to the aspects of the S operator that are directly related to the weak process under consideration. In doing so we are dealing with an effective S operator rather than with the true S operator. For the actual process the two operators are assumed to have identical matrix elements, but other matrix elements are, in general, not related at all.

The success of quantum electrodynamics in accounting for electromagnetic processes involving muons, electrons, and photons is well known. Here one is in the fortunate situation of having a complete dynamical theory. The field equations are solved within perturbation theory. Together with a specific prescription known as *renormalization* [S2] this yields results that agree extremely well with experiments. The applicability of perturbation theory is assumed to follow from the fact that the expansion parameter is so small. This expansion parameter is the fine-structure constant $\alpha \simeq 1/137$.

The corresponding expansion parameter for the weak interactions is less well defined, but the smallness of cross sections for weak processes makes it plausible here too that perturbation theory will work and, with equations of motion given, a full-fledged dynamical theory is attainable, following the pattern of quantum electrodynamics. However, dynamical questions for weak interactions are just as difficult to answer as in the case of strong interactions. Therefore, we shall aim at obtaining an effective S operator directly rather than try to develop a Lagrangian formalism [S2] for weak interactions.

In many weak processes that have been observed, four fermions participate. This observation has led to the assumption that the fundamental weak interaction always is *a four-fermion interaction.* This assumption seems to be violated in some cases, as, for example, in the decay

$$\pi^- \to \mu^- + \bar{\nu}_\mu$$

where only two fermions appear. It is known, however, that in the processes where the assumption of a basic four-fermion interaction does not seem to apply there is at least one meson present. For these processes there is good reason to believe that the transition from the initial to the final state generally is via an *intermediate state.*‡ For the process mentioned above a possible mechanism then is

$$\pi^- \to n + \bar{p} \to \mu^- + \bar{\nu}_\mu$$

The first step of the reaction is analogous to reactions well known in strong-interaction physics and is, in fact, caused by strong interactions. The intermediate state has a short lifetime and decays by a weak four-fermion interaction. A short lifetime for the intermediate state together with the uncertainty principle makes this process possible while the decay $\pi^- \to n + \bar{p}$ is forbidden because it cannot satisfy energy-momentum conservation. In the same way, other processes involving mesons can be reconciled with the assumption of a fundamental weak four-fermion interaction. Therefore, the basic process to be discussed is of the form

$$a \to b + c + \bar{d}$$

or equivalently $\quad a + \bar{b} \to c + \bar{d}$

or $\quad a + d \to b + c$

The equivalence of these three processes will be apparent when the appropriate effective S operator is obtained. The symbol a denotes a particle while $\bar{b}$ denotes an antiparticle, etc.

In the first of these three equivalent processes, it is seen that in the decay of particle a two particles and one antiparticle are created. What is observed experimentally is that in the interaction region particle a disappears while the three other particles are born. Whether the annihilation of particle a takes place at exactly the same point at which the other particles are created is impossible to determine, even in principle. All that can be said is that it takes place within a truly microscopic region. Furthermore, the possible location of such an interaction region in space and time is defined by the geometrical arrangements of the experiment and the length of time during which it is conducted. Thus the microscopic interaction region can be anywhere in a relatively large, macroscopic domain in space and time.

From the discussion following Eqs. (A.54) and (A.55) of Appendix A, the field operator $\psi(x)$ has the property of annihilating a particle or creating an antiparticle at the point x while $\bar{\psi}(x)$ creates a particle and annihilates an antiparticle. Therefore, the most general effective S operator describing the process can be expressed in terms of field operators. Allowing for a finite but microscopic interaction region, the location of which is undetermined to some extent, the most general form for this effective S operator is given by

$$S = \int d^4x_1 \cdots d^4x_4 \rho(x_1, \ldots, x_4) \bar{\psi}_b(x_1)\Gamma_i\psi_a(x_2)\bar{\psi}_c(x_3)\Gamma'_i\psi_d(x_4) \tag{5.5}$$

‡ The term "intermediate state" is here used in the sense of excluding a direct coupling. However, it cannot be denied that a direct $(\pi\mu\nu_\mu)$ vertex is possible.

where the function $\rho(x_1, \ldots, x_4)$ takes into account the structure of the interaction. It vanishes whenever any of the relative coordinates exceeds the size of the interaction region. For not too large momentum transfers no structure effects can be observed (see Sec. 8). It is then at least a good approximation to treat the interaction as a point interaction by putting

$$\rho(x_1, \ldots, x_4) = \delta^4(x_1 - x_2)\delta^4(x_1 - x_3)\delta^4(x_1 - x_4) \tag{5.6}$$

in which case (5.5) can be written

$$S = \int d^4x \bar{\psi}_b(x)\Gamma_i\psi_a(x)\bar{\psi}_c(x)\Gamma'_i\psi_d(x) \tag{5.7}$$

So far no restrictions on the quantities Γ_i and Γ'_i have been imposed except that they must be chosen so that invariance under proper Lorentz transformations obtains. To remove the prevailing arbitrariness we shall invoke the following principles:

1. There shall appear no derivatives in the expression for the effective S operator.‡
2. Each covariant of the S operator shall contain one charged field and one neutral field.
3. Only the positive chiral projections of the field operators appear in the coupling.

Because of the first restriction only γ matrices are available to construct the quantities Γ_i and Γ'_i. From the discussion of the Dirac equation in Appendix A it then follows that there are five types of couplings, referred to as S, V, T, A, and P coupling.

The second requirement is in itself no restriction because any ordering of the field operators in S can be rewritten in that form by means of the Fierz transformation [8]. However, requirements (2) and (3) together imply severe restrictions with respect to the type of coupling possible. Chiral projections of the field operators are obtained by means of the projection operators $\frac{1}{2}(1 \pm \gamma_5)$. To comply with requirement 3 we make the following replacements in (5.7):

$$\psi(x) \to \tfrac{1}{2}(1 + \gamma_5)\psi(x) \tag{5.8}$$

$$\bar{\psi}(x) \to \bar{\psi}(x)\tfrac{1}{2}(1 - \gamma_5) \tag{5.9}$$

The effect of these replacements in each possible covariant is given below:

$$\begin{aligned}
\Gamma_S &= \lambda_S I: & &\bar{\psi}\tfrac{1}{2}(1 - \gamma_5)\tfrac{1}{2}(1 + \gamma_5)\psi \equiv 0 \\
\Gamma_V &= \lambda_V\gamma_\mu: & &\bar{\psi}\tfrac{1}{2}(1 - \gamma_5)\gamma_\mu\tfrac{1}{2}(1 + \gamma_5)\psi \\
& & &= \tfrac{1}{2}\bar{\psi}\gamma_\mu(1 + \gamma_5)\psi \\
\Gamma_T &= \lambda_T\sigma_{\mu\nu}: & &\bar{\psi}\tfrac{1}{2}(1 - \gamma_5)\sigma_{\mu\nu}\tfrac{1}{2}(1 + \gamma_5)\psi \equiv 0 \\
\Gamma_A &= \lambda_A\gamma_5\gamma_\mu: & &\bar{\psi}\tfrac{1}{2}(1 - \gamma_5)\gamma_5\gamma_\mu\tfrac{1}{2}(1 + \gamma_5)\psi \\
& & &= -\tfrac{1}{2}\bar{\psi}\gamma_\mu(1 + \gamma_5)\psi \\
\Gamma_P &= \lambda_P\gamma_5: & &\bar{\psi}\tfrac{1}{2}(1 - \gamma_5)\gamma_5\tfrac{1}{2}(1 + \gamma_5)\psi \equiv 0
\end{aligned} \tag{5.10}$$

Thus only the vector and the axial vector covariants survive. Finally, the requirement that S be invariant under proper Lorentz transformations is equivalent to requiring it to transform as a scalar, a pseudoscalar, or as a mixture of a scalar and a pseudoscalar. To allow

‡ An early attempt to include derivative coupling led to predictions inconsistent with experimental data. Subsequent theories involving derivative coupling have led to the conclusion that such complications are unnecessary.

for this, Γ'_i is not necessarily identical with Γ_i. For example, with $\Gamma_V = \lambda_V\gamma_\mu$ one may have

$$\Gamma'_V = \lambda_V\gamma_\mu(1 + a\gamma_5)$$

However,

$$\lambda_V\gamma_\mu(1 + a\gamma_5)\tfrac{1}{2}(1 + \gamma_5) = \lambda_V(1 + a)\gamma_\mu\tfrac{1}{2}(1 + \gamma_5) \tag{5.11}$$

so that the effective freedom is only of the form

$$\Gamma'_V = \lambda'_V\gamma_\mu \qquad \lambda'_V = \lambda_V(1 + a) \tag{5.12}$$

and similarly for Γ'_A. Retaining this freedom, the most general S operator complying with principles 1, 2, and 3 is given by [10]

$$S = -\frac{iG}{\sqrt{2}} \int d^4x\bar{\psi}_b(x)\gamma^\mu(1 + \gamma_5)\psi_a(x)\bar{\psi}_c(x) \gamma_\mu(1 + \gamma_5)\psi_d(x) \tag{5.13}$$

where the notation

$$-\frac{iG}{\sqrt{2}} = \frac{1}{4}(\lambda_V\lambda'_V + \lambda_A\lambda'_A) \tag{5.14}$$

has been introduced in order to obtain the conventional expression for the effective S operator describing this process. In Eq. (5.13) it is now assumed that one of the particles a and b is charged and one is neutral and similarly for c and d so that requirement 2 is fulfilled.

The interaction appearing in (5.13) is referred to as the *V-A interaction* because historically the A part of this interaction was defined as the contraction of covariants with $i\gamma_\mu\gamma_5$ in each covariant. The minus sign then comes from the two factors i.

2.3. The Universal Fermi Interaction. Once the general expression of Eq. (5.13) is given for the effective S operator describing the process

$$a \to b + c + \bar{d}$$

it is a trivial matter to write the explicit expressions for physical processes of interest. Principle 2 above must be borne in mind separately, and one obtains

$\mu^- \to e^- + \bar{\nu}_e + \nu_\mu$:

$$S = -\frac{iG_1}{\sqrt{2}} \int d^4x\bar{\psi}_e(x)\gamma^\mu(1 + \gamma_5)\psi_{\nu_e}(x) \bar{\psi}_{\nu_\mu}(x)\gamma_\mu(1 + \gamma_5)\psi_\mu(x) \tag{5.15}$$

$\mu^+ \to e^+ + \nu_e + \bar{\nu}_\mu$:

$$S = -\frac{iG_1}{\sqrt{2}} \int d^4x\bar{\psi}_{\nu_e}(x)\gamma^\mu(1 + \gamma_5)\psi_e(x) \bar{\psi}_\mu(x)\gamma_\mu(1 + \gamma_5)\psi_{\nu_\mu}(x) \tag{5.16}$$

$\mu^- + p \to n + \nu_\mu$:

$$S = -\frac{iG_2}{\sqrt{2}} \int d^4x\bar{\psi}_n(x)\gamma^\mu(1 + \gamma_5)\psi_p(x) \bar{\psi}_{\nu_\mu}(x)\gamma_\mu(1 + \gamma_5)\psi_\mu(x) \tag{5.17}$$

$n \to p + e^- + \bar{\nu}_e$

$$S = -\frac{iG_3}{\sqrt{2}} \int d^4x\, \bar{\psi}_p(x)\gamma^\mu(1 + \gamma_5)\psi_n(x) \bar{\psi}_e(x)\gamma_\mu(1 + \gamma_5)\psi_{\nu e}(x) \tag{5.18}$$

There is no a priori reason why the various coupling constants G_i should be the same except for the two first-mentioned processes which are related by the so-called *CPT* transformation that will be discussed in Sec. 3. However, it is possible to interpret the experimental findings in a consistent fashion by assuming the existence of one common fundamental weak interaction coupling constant G, that is,

$$G_1 = G_2 = G_3 = G \tag{5.19}$$

This together with the unique V-A coupling for all weak processes is the basis for the assumption of a *universal Fermi interaction* (UFI). This concept is most elegantly dealt with in terms of weak interaction currents. To this end we introduce the following lepton and baryon vector-axial vector currents

$$j_\mu(x) \equiv (\bar{\mu}\nu_\mu) + (\bar{e}\nu_e) \tag{5.20}$$

$$\mathcal{J}_\mu(x) \equiv (\bar{n}p) \tag{5.21}$$

where the notation

$$(\bar{a}b) = \bar{\psi}_a(x)\gamma_\mu(1 + \gamma_5)\psi_b(x) \tag{5.22}$$

has been used. Together these two currents form the total nonstrange weak interaction current

$$J_\mu(x) = \mathcal{J}_\mu(x) + j_\mu(x) \tag{5.23}$$

The current $\times$ current-type effective S operator

$$S = -\frac{iG}{\sqrt{2}} \int d^4x\, J^\dagger_\mu(x)J^\mu(x) \tag{5.24}$$

now contains all the parts corresponding to Eqs. (5.15) to (5.18), provided (5.19) is true. By postulating (5.24) to be the full effective S operator for nonstrange weak interactions, one obtains some new and unique predictions. These are related to the diagonal terms in (5.24) corresponding to the self-coupling of the various currents. For example, there is a term

$$(\bar{e}\nu_e)(\bar{\nu}_e e)$$

corresponding to the process

$$e^- + \nu_e \to e^- + \nu_e$$

and (5.24) leads to a unique prediction of the cross section. It is still true, however, that the cross sections for weak processes are prohibitively small unless special circumstances such as coherence effects prevail; such is not the case for electron-neutrino scattering.

Most of the new predictions of (5.24) have not yet been observed. There is one notable exception: an observation of a parity-violating term in the interaction between nucleons arising from the $(\bar{p}n)$ $(\bar{n}p)$ contribution to the current $\times$ current interaction. This has been made evident by the observation of circular polarization of γ rays (the 482-kev transition in unpolarized Ta^{181}) by Boehm and Krankeleit [9]. In addition, one may perhaps expect to observe at some future date the electron-neutrino scattering as described by

$$e^- + \nu_e \to e^- + \nu_e$$

but the process

$$e^- + \nu_\mu \to e^- + \nu_\mu$$

should not be observed, according to the prescription provided by (5.24). In principle, muon-ν_μ scattering could occur (and no $\mu - \nu_e$ scattering) but this is experimentally much less feasible.

2.4. The Two-component Theory of the Neutrino [11]. The requirement that only the positive chiral projections of the field operators be present in the effective S operator describing weak interactions has important consequences for the helicity of zero-mass neutrinos.‡ To see this easily it is convenient to choose a somewhat different representation of the γ algebra than the one discussed in Appendix A. We take§

$$\gamma_0 = \begin{pmatrix} 0 & I \\ I & 0 \end{pmatrix} \qquad \gamma_k = \begin{pmatrix} 0 & \sigma_k \\ -\sigma_k & 0 \end{pmatrix} \qquad \gamma_5 = \begin{pmatrix} I & 0 \\ 0 & -I \end{pmatrix} \tag{5.25}$$

Introducing the two-component spinors φ_1 and φ_2 and

$$u(\mathbf{p}) = \begin{pmatrix} \varphi_1 \\ \varphi_2 \end{pmatrix} \tag{5.26}$$

the Dirac equation with $m = 0$ reads

$$\gamma_\mu p^\mu = \begin{pmatrix} 0 & E - \boldsymbol{\sigma} \cdot \mathbf{p} \\ E + \boldsymbol{\sigma} \cdot \mathbf{p} & 0 \end{pmatrix} \begin{pmatrix} \varphi_1 \\ \varphi_2 \end{pmatrix} = 0 \tag{5.27}$$

or

$$(E + \boldsymbol{\sigma} \cdot \mathbf{p})\varphi_1 = 0 \tag{5.28}$$

$$(E - \boldsymbol{\sigma} \cdot \mathbf{p})\varphi_2 = 0 \tag{5.29}$$

which means that φ_1 and φ_2 are decoupled in the absence of a mass term in the Dirac equation. From Eqs. (5.28) and (5.29) it immediately follows that

$$E^2 = |\mathbf{p}|^2 \qquad E = \pm|\mathbf{p}| \tag{5.30}$$

and furthermore for $E > 0$, that is, for $E = |\mathbf{p}|$, one obtains

$$\frac{\boldsymbol{\sigma} \cdot \mathbf{p}}{|\mathbf{p}|} \varphi_1 = -\varphi_1 \tag{5.31}$$

and

$$\frac{\boldsymbol{\sigma} \cdot \mathbf{p}}{|\mathbf{p}|} \varphi_2 = \varphi_2 \tag{5.32}$$

Similarly, for $E < 0$ so that $E = -|\mathbf{p}|$, one obtains

$$\frac{\boldsymbol{\sigma} \cdot \mathbf{p}}{|\mathbf{p}|} \varphi_1 = \varphi_1 \tag{5.33}$$

and

$$\frac{\boldsymbol{\sigma} \cdot \mathbf{p}}{|\mathbf{p}|} \varphi_2 = -\varphi_2 \tag{5.34}$$

The eigenvalue (expectation value, in general) of the operator $\boldsymbol{\sigma} \cdot \mathbf{p}/|\mathbf{p}|$ is called the helicity of the state. From the equations above it follows that:

1. Particles ($E > 0$) have negative helicity if they are described by the solution φ_1 and positive helicity if they are described by φ_2.

2. Antiparticles ($E < 0$) have positive helicity if described by φ_1 and negative helicity if described by φ_2.

‡ As is evident from Table 5.1 the assertion that $m_{\nu_e} = 0$ is, at least, an excellent approximation. While the same assumption is made for ν_μ, the present upper limit on the mass is undesirably large.

§ The choice of representation in Appendix A is motivated by the circumstance that for nonzero-mass Dirac particles this form of the γ matrices allows the spinors to be represented by large and small components in the nonrelativistic limit. Obviously, this property is irrelevant in the present context.

Acting with the chirality projection operators $\frac{1}{2}(1 \pm \gamma_5)$ on $u(\mathbf{p})$ one finds

$$\tfrac{1}{2}(1 + \gamma_5)u(\mathbf{p}) = \begin{pmatrix} I & 0 \\ 0 & 0 \end{pmatrix} \begin{pmatrix} \varphi_1 \\ \varphi_2 \end{pmatrix} = \begin{pmatrix} \varphi_1 \\ 0 \end{pmatrix} \tag{5.35}$$

and

$$\tfrac{1}{2}(1 - \gamma_5)u(\mathbf{p}) = \begin{pmatrix} 0 \\ \varphi_2 \end{pmatrix} \tag{5.36}$$

Thus, $\frac{1}{2}(1 + \gamma_5)$ projects out the solution φ_1 while $\frac{1}{2}(1 - \gamma_5)$ yields φ_2.

From these observations it is then concluded that the field operator $\frac{1}{2}(1 + \gamma_5)\psi_\nu$, which appears in the effective S operator, creates only antineutrinos described by the spinor wave function φ_1 and annihilates neutrinos described by the same φ_1. But this then implies that it creates only antineutrinos with helicity $+1$ and annihilates neutrinos of helicity -1. Physically this means that the antineutrinos have their spin along the momentum vector while the neutrinos have their spin antiparallel to the momentum. The neutrinos are then said to be left-handed and the antineutrinos are right-handed. In the same way it is seen that $\bar{\psi}_\nu \frac{1}{2}(1 - \gamma_5)$ creates left-handed neutrinos and annihilates right-handed antineutrinos.

Because neutrinos participate only in weak interactions, they can be created only in weak processes. But within the V-A theory only left-handed neutrinos and right-handed antineutrinos can appear in a weak process. If neutrinos of positive helicity or antineutrinos of negative helicity exist they do not interact with any known particle and there is no way of detecting their existence.

Also, the other particles involved in weak interactions are created and annihilated in a state corresponding to positive chirality. Because these particles all have nonvanishing mass, they will not remain in a state of definite chirality, and a mixing of the various solutions takes place. The larger the mass, the faster is this mixing process. Furthermore, there is no one-to-one correspondence between chirality and helicity for particles with mass. Therefore, particles with nonvanishing mass created in a weak process are in general not in an eigenstate of helicity unless other selection rules, such as angular-momentum conservation, require them to be so. This is, for example, the case with the muon in the decay

$$\pi^- \rightarrow \mu^- + \bar{\nu}_\mu$$

as will be discussed later.

The two-component neutrino is a necessary consequence of the V-A interaction, as has been demonstrated; the opposite is not true. The observation that a neutrino is of two-component nature can also be reconciled with other types of interaction.

3. Symmetry Operations

3.1. General Properties of Symmetry Operations. Symmetry operations form a very important part of elementary-particle theory. This is so because invariance under a specific symmetry operation leads to conclusions concerning observable quantities that can be subjected to experimental verification. Moreover, predictions concerning the symmetry properties of the observations can be made without recourse to the details of the theory.

We shall consider those symmetry operations for which there is a physical motivation; with one notable exception, they are represented by linear operators in the space of the state vectors. If the operator representing a certain transformation is denoted by Ω, then the transformation law for the state vectors is given by

$$|\alpha> \rightarrow |\alpha'> = \Omega|\alpha> \equiv |\Omega\alpha> \qquad (5.37a)$$

and

$$<\alpha| \rightarrow <\alpha'| = <\alpha|\Omega\dagger = <\Omega\alpha| \qquad (5.37b)$$

The operators acting on these state vectors obey the transformation law

$$\mathcal{O} \rightarrow \mathcal{O}' = \Omega\mathcal{O}\Omega^{-1} \qquad (5.38)$$

For discussions of invariance properties in physical processes the relevant operator to consider is the S operator. For a transformation to be a symmetry operation the following are required:

1. The S operator remains invariant under the transformation, which means

$$\Omega S \Omega^{-1} = S \qquad (5.39)$$

that is, S and Ω commute.

2. The S-matrix elements, from which all physical information can be derived, also remain invariant.

The second requirement primarily concerns the effects of the transformation on the state vectors. From the two conditions above it follows that

$$<f|S|i> = <f'|S|i'> \qquad (5.40)$$

where $|f'>$ and $|i'>$ are the state vectors obtained from $|f>$ and $|i>$, respectively, by means of the transformation. From the transformation properties of state vectors and Eq. (5.39) it is seen that

$$<f'|S|i'> = <f|\Omega\dagger S\Omega|i> = <f|\Omega\dagger\Omega S|i> \qquad (5.41)$$

To satisfy Eq. (5.40) it is then necessary that

$$\Omega\dagger\Omega = 1 \qquad (5.42)$$

This requires that Ω be unitary, that is, $\Omega\dagger = \Omega^{-1}$; we shall exclusively consider such symmetry operations.

We next consider some of the implications that follow from the existence of a physical symmetry operation. A unitary operator may or may not be Hermitian, but, if it is not, one can always define a Hermitian operator H related to it by

$$\Omega = \exp(iH) \qquad (5.43)$$

It is now assumed that the Hermitian operator H is related to or constitutes a symmetry operation. We consider the case in which the initial state is an eigenstate of this operator corresponding to the eigenvalue h, that is,

$$H|i> = h|i> \qquad (5.44)$$

Because H is related to a symmetry operation, it commutes with the S operator and therefore

$$H|f> = HS|i> = SH|i> = hS|i> = h|f> \qquad (5.45)$$

This means that the final state is also an eigenstate of H corresponding to the same eigenvalue as the initial state. If H corresponds to an observable, then h is a measurable quantum number which is conserved. It is thus seen how the existence of a symmetry operation implies the existence of a selection rule; under the influence of the interactions the initial state can transform only into a final state corresponding to the same quantum number h.

Returning to Eq. (5.40), we see that the matrix elements of the S operator remain the same whether evaluated in the original system where the state vectors are given by $|i>$ and $|f>$ or in the transformed system corresponding to $|i'>$ and $|f'>$. The same must then also be true for the observable quantities (cross sections, decay rates), as the entire effect of the interaction is contained in the square modulus of the S-matrix element. The existence of a symmetry operation therefore leads to relations between measurements with different experimental arrangements which, however, are related by the considered transformation. In many cases this implies the nonexistence of correlation functions that do not exhibit the same symmetry property. This will become more apparent in explicit examples, discussed later.

Invariance properties of the S operator may imply (1) the existence of selection rules and (2) the nonexistence of certain correlations.

A word of caution: Invariance of the S operator under given symmetry operations is neither sufficient nor necessary for the same invariance to appear in observed quantities. However, under most circumstances the symmetry properties of the S operator are reflected in corresponding symmetry properties of transition rates and cross sections.‡

3.2. Space Reflection. The space reflection transformation

$$x_k \rightarrow -x_k \qquad (5.46)$$

provides a basis for classification of states as even or odd under the parity operator. For a Dirac particle the wave function transformation [cf. Eq. (A.10)] is

$$\psi(x_0,\mathbf{x}) \rightarrow \psi'(x_0,-\mathbf{x}) = \gamma_0\psi(x_0,\mathbf{x}) \qquad (5.47)$$

By direct inspection it is immediately verified that the $\psi'(x_0,\mathbf{x})$ satisfies the Dirac equation (A.1) which makes the formalism invariant under the transformation.

In the second-quantized theory the space reflection is represented by a linear operator P which, according to (5.47), must have the following property:§

$$\psi'(x_0,\mathbf{x}) = P\psi(x_0,\mathbf{x})P^{-1} = \eta_P\gamma_0\psi(x_0,-\mathbf{x}) \qquad (5.48)$$

Inserting the expansion (A.54) on the right-hand side of (5.48) and observing the following properties for the spinors $u(\mathbf{p})$ and $v(\mathbf{p})$

$$\gamma_0 u(\mathbf{p}) = u(-\mathbf{p}) \qquad (5.49a)$$

$$\gamma_0 v(\mathbf{p}) = -v(-\mathbf{p}) \qquad (5.49b)$$

which follow directly from the defining equations (A.24) and (A.25), one obtains

$$P\psi(x_0,\mathbf{x})P^{-1} = \frac{1}{\sqrt{V}}\sum_{\mathbf{p},s}\sqrt{\frac{m}{E}}\,\{a_s(-\mathbf{p})u_s(\mathbf{p})e^{ipx} - b\dagger_s(-\mathbf{p})v_s(\mathbf{p})e^{-ipx}\} \qquad (5.50)$$

‡ The rather subtle relation between symmetry of the S operator and of observations is discussed in more detail in Secs. 3.8 and 3.9.

§ In Eq. (5.48) η_P is a phase factor. For the present we set $\eta_P = 1$; see, however, Eq. (5.101) et seq. The phase factor is unnecessary in the C and T transformations below.

from which it is concluded that

$$P a_s(\mathbf{p}) P^{-1} = a_s(-\mathbf{p}) \qquad (5.51a)$$
$$P b^\dagger_s(\mathbf{p}) P^{-1} = -b^\dagger_s(-\mathbf{p}) \qquad (5.51b)$$

Similarly it is found that

$$P a^\dagger_s(\mathbf{p}) P^{-1} = a^\dagger_s(-\mathbf{p}) \qquad (5.51c)$$
$$P b_s(\mathbf{p}) P^{-1} = -b_s(-\mathbf{p}) \qquad (5.51d)$$

Taking the vacuum state $|0>$ to be invariant under space reflection, it follows from the above that

$$\begin{aligned} P|\mathbf{p}_1 s_1, \mathbf{p}_2 s_2, \ldots, \mathbf{p}_m s_m; \mathbf{p}'_1 s'_1, \ldots, \mathbf{p}'_n s'_n> \\ = (-1)^n |-\mathbf{p}_1 s_1, \ldots, -\mathbf{p}_m s_m; -\mathbf{p}'_1 s_1, \ldots, -\mathbf{p}_n s_n> \end{aligned} \qquad (5.52)$$

where unprimed quantities refer to particles and primed to antiparticles. Equation (5.52) shows that the state obtained by space reflection differs from the original state in that the momentum vectors have the opposite sign; however, the spins are not affected by the transformation. Equation (5.52) also reveals that P is a unitary and Hermitian operator, i.e.,

$$P = P^\dagger = P^{-1} \qquad (5.53)$$

and consequently P defines a quantum number known as the *parity of the state*. Because

$$P^2 = 1 \qquad (5.54)$$

the eigenvalues are ± 1. Therefore, if the interaction is invariant under space reflection and the initial state is an eigenstate of P, the final state is also an eigenstate corresponding to the same eigenvalue, and a selection rule is operative. (The experimental consequences that follow if such a situation prevails are discussed in Secs. 3.8 and 3.9.) However, Eq. (5.52) implies that no state corresponding to particles with well-defined momenta can be an eigenstate of P unless all the particles are at rest. To form eigenstates of P with particles not at rest, one has to consider states of definite angular momenta rather than linear momenta (cf. Sec. 5).

3.3. Charge Conjugation. The *charge-conjugation operation*, more appropriately referred to as *particle-antiparticle conjugation*, relates a physical state involving particles and antiparticles to another physical state identical to the original one except that particles are replaced by antiparticles and vice versa. If the physical results of a theory are invariant under this replacement, the theory is said to be charge-conjugation invariant.

Under charge conjugation a state vector transforms according to the rule

$$\begin{aligned} |\mathbf{p}_1 s_1, \ldots, \mathbf{p}_m s_m; \mathbf{p}'_1 s'_1, \ldots, \mathbf{p}'_n s'_n> \\ \rightarrow |\mathbf{p}'_1 s'_1, \ldots, \mathbf{p}'_n s'_n; \mathbf{p}_1 s_1, \ldots, \mathbf{p}_m s_m> \end{aligned} \qquad (5.55)$$

where the original state contains m particles and n antiparticles while the transformed state contains n particles and m antiparticles. No change of momenta or spins is involved so that a particle of momentum and spin $(\mathbf{p},s)$ goes over into an antiparticle of the same momentum and spin $(\mathbf{p},s)$. The transformation (5.55) is represented by a linear operator C

$$\begin{aligned} |\mathbf{p}'_1 s'_1, \ldots, \mathbf{p}'_n s'_n; \mathbf{p}_1 s_1, \ldots, \mathbf{p}_m s_m> \\ = C|\mathbf{p}_1 s_1, \ldots, \mathbf{p}_m s_m; \mathbf{p}'_1 s'_1, \ldots, \mathbf{p}'_n s'_n> \end{aligned} \qquad (5.56)$$

From this definition of C it follows that

$$C^\dagger = C^{-1} = C \qquad (5.57)$$

which means that the operator C corresponds to a quantum number, often called the *charge parity*, which must be either $+1$ or -1, and the remarks made earlier with regard to parity are equally valid for the charge parity.

With the convention that the vacuum-state vector is invariant under charge conjugation, Eq. (5.56) implies the following transformation properties for creation and destruction operators:

$$C a^\dagger_s(\mathbf{p}) C^{-1} = b_s(\mathbf{p}) \qquad C a_s(\mathbf{p}) C^{-1} = b_s(\mathbf{p}) \qquad (5.58a)$$
$$C b^\dagger_s(\mathbf{p}) C^{-1} = a^\dagger_s(\mathbf{p}) \qquad C b_s(\mathbf{p}) C^{-1} = a_s(\mathbf{p}) \qquad (5.58b)$$

and thus

$$\begin{aligned} C\psi(x)C^{-1} = \frac{1}{\sqrt{V}} \sum_{\mathbf{p},s} \sqrt{\frac{m}{E}} \{ b_s(\mathbf{p}) u_s(\mathbf{p}) e^{-ipx} \\ + a^\dagger_s(\mathbf{p}) v_s(\mathbf{p}) e^{ipx} \} \end{aligned} \qquad (5.59)$$

This can be rewritten in a more convenient form by introducing a matrix $\mathcal{C}$ with the following properties:

$$\sum_\beta \mathcal{C}_{\alpha\beta} \bar{u}_s{}^\beta(\mathbf{p}) = v_s{}^\alpha(\mathbf{p}) \qquad (5.60a)$$

$$\sum_\beta \mathcal{C}_{\alpha\beta} \bar{v}_s{}^\beta(\mathbf{p}) = u_s{}^\alpha(\mathbf{p}) \qquad (5.60b)$$

where, for clarity, spinor indices are exhibited. In the more compact matrix notation these two equations read

$$\mathcal{C} \bar{u}_s{}^T(\mathbf{p}) = v_s(\mathbf{p}) \qquad (5.60c)$$
$$\mathcal{C} \bar{v}_s{}^T(\mathbf{p}) = u_s(\mathbf{p}) \qquad (5.60d)$$

where the upper index T means transpose in spinor space. Inserting this in Eq. (5.59), one obtains

$$C\psi(x)C^{-1} = \mathcal{C}\bar{\psi}^T(x) \qquad (5.61)$$

From the definition of the matrix $\mathcal{C}$ one can derive the following important properties for it by considering the properties of the spinors $u(\mathbf{p})$ and $v(\mathbf{p})$ as defined by the energy projection operators $\Delta \pm (p)$ of Appendix A:

$$\mathcal{C}^{-1}\gamma_\mu\mathcal{C} = -\gamma_\mu{}^T \qquad (5.62a)$$
$$\mathcal{C}^{-1}\gamma_5\mathcal{C} = \gamma_5{}^T \qquad (5.62b)$$
$$\mathcal{C}^\dagger = \mathcal{C}^{-1} \qquad (5.62c)$$
$$\mathcal{C}^T = -\mathcal{C} \qquad (5.62d)$$

The completeness of Dirac γ algebra requires that if there exists a matrix $\mathcal{C}$ with these properties it must be possible to express it in terms of γ matrices; in fact, with our choice of γ matrices, Eq. (A.5), it is found by inspection that

$$\mathcal{C} = \gamma_0\gamma_5 \qquad (5.63)$$

satisfies all the conditions put on $\mathcal{C}$.

Taking the Hermitian conjugate of Eq. (5.61) and exploiting the properties of the operator C and the matrix $\mathcal{C}$, $\bar{\psi}(x)$ transforms under charge conjugation according to the rule

$$C\bar{\psi}(x)C^{-1} = -\psi^T(x)\mathcal{C}^{-1} \qquad (5.64)$$

It is also of interest to consider the transformation properties of the electromagnetic current operator $j_\mu(x)$ for a Dirac field. By definition

$$j_\mu(x) = \tfrac{1}{2}\{\bar{\psi}(x)\gamma_\mu\psi(x) - \psi^T(x)\gamma_\mu{}^T\bar{\psi}^T(x)\} \quad (5.65)$$

Thus

$$\begin{aligned} Cj_\mu(x)C^{-1} &= \tfrac{1}{2}\{C\bar{\psi}(x)C^{-1}\gamma_\mu C\psi(x)C^{-1} \\ &\quad - C\psi^T(x)C^{-1}\gamma_\mu{}^T C\bar{\psi}^T(x)C^{-1}\} \\ &= -\tfrac{1}{2}\{\psi^T(x)\mathcal{C}^{-1}\gamma_\mu\mathcal{C}\bar{\psi}^T(x) - \bar{\psi}(x)\mathcal{C}\gamma_\mu{}^T\mathcal{C}^{-1}\psi(x)\} \\ &= \tfrac{1}{2}\{\psi^T(x)\gamma_\mu{}^T\bar{\psi}^T(x) - \bar{\psi}(x)\gamma_\mu\psi(x)\} = -j_\mu(x) \quad (5.66) \end{aligned}$$

where once more the properties of the matrix $\mathcal{C}$ have been exploited. Equation (5.66) spells out what one would expect intuitively; namely, the replacement of particles by antiparticles leads to a change in sign of the current operator.

Finally, a necessary although not sufficient condition on a state vector in order that it be an eigenvector of C is that it contain the same number of particles and antiparticles of each kind. In particular, charged states can never be eigenstates of C. On the other hand, multi-π^0 states are always eigenstates of C and so are multiphoton states because in these cases particles and antiparticles are identical.

3.4. The G Transformation. It was pointed out that very stringent selection rules may be obtained where a state vector is an eigenvector of a symmetry operator. Because no charged state can be an eigenstate of the charge-conjugation operator, no selection rules are obtained for such a state even if the interaction is invariant under C. As a remedy for this, one can generalize the concept of charge conjugation so as to include some charged states: those for which the total baryon number is zero.‡ Consider, for example, the π meson. From the very definition of the charge-conjugation operation it follows that

$$C|\pi^\pm\rangle = |\pi^\mp\rangle \quad (5.67a)$$
$$C|\pi^0\rangle = |\pi^0\rangle \quad (5.67b)$$

so that a charged state such as $|\pi^+\rangle$ goes over into the π-meson state of opposite charge. There is, however, another formalism in which the various charge states of the π meson are linked together: the isospin formalism [S3]. Combining the charge-conjugation operation with a suitably chosen transformation in isospace, one can, therefore, obtain an operation that leaves any π-meson state invariant, i.e., all three π-meson state vectors are eigenvectors of the corresponding operator.

π mesons are described by pseudoscalar fields. For uncharged particles the field can be chosen Hermitian while for charged particles complex fields are used. Denoting the π-meson fields $\varphi_0(x)$ if uncharged and $\varphi(x)$ if charged, one finds that $\varphi(x)$ creates π^+ mesons and destroys π^- mesons while $\varphi^*(x)$ creates π^- mesons and destroys π^+ mesons [S2]. The following linear combinations of these fields are real:

$$\begin{aligned} \varphi_1(x) &= \frac{1}{\sqrt{2}}[\varphi(x) + \varphi^*(x)] \\ \varphi_2(x) &= \frac{1}{i\sqrt{2}}[\varphi(x) - \varphi^*(x)] \quad (5.68) \\ \varphi_3(x) &= \varphi_0(x) \end{aligned}$$

They define the three components of an isovector $\boldsymbol{\phi}(x) \equiv (\varphi_1,\varphi_2,\varphi_3)$. Under a rotation in isospace represented by an operator Ω acting in the space of state vectors,‡ the isovector $\boldsymbol{\phi}(x)$ transforms in the following way:

$$\boldsymbol{\phi}(x) \rightarrow \boldsymbol{\phi}'(x) = \Omega\boldsymbol{\phi}(x)\Omega^{-1} = e^{i\epsilon_j I_j}\boldsymbol{\phi}(x) \quad (5.69)$$

where $\boldsymbol{\varepsilon} = (\epsilon_1,\epsilon_2,\epsilon_3)$ defines the rotation axis and the angle of rotation while the matrices I_j are conventional angular-momentum matrices in isospace. We use the representation

$$I_1 = \begin{pmatrix} 0 & 0 & 0 \\ 0 & 0 & -i \\ 0 & i & 0 \end{pmatrix} \quad I_2 = \begin{pmatrix} 0 & 0 & i \\ 0 & 0 & 0 \\ -i & 0 & 0 \end{pmatrix}$$
$$I_3 = \begin{pmatrix} 0 & -i & 0 \\ i & 0 & 0 \\ 0 & 0 & 0 \end{pmatrix} \quad (5.70)$$

If we choose a rotation around the second axis by 180° we obtain

$$\varphi_1 \rightarrow -\varphi_1 \qquad \varphi_2 \rightarrow \varphi_2 \qquad \varphi_3 \rightarrow -\varphi_3 \quad (5.71a)$$

as one would expect for a vector. In terms of φ_0, φ, and φ^*, Eq. (5.71a) reads

$$\varphi \rightarrow -\varphi^* \qquad \varphi^* \rightarrow -\varphi \qquad \varphi_0 \rightarrow -\varphi_0 \quad (5.71b)$$

This in turn implies the following transformation properties for the physical π-meson state vectors:

$$\Omega|\pi^{\pm,0}\rangle \equiv e^{-i\pi T_2}|\pi^{\pm,0}\rangle = -|\pi^{\mp,0}\rangle \quad (5.72)$$

Now, if the G operation is defined as

$$G = C\exp(-i\pi T_2) \quad (5.73)$$

all the π-meson state vectors are eigenvectors of G with the eigenvalue -1. As the G transformation, similar to the space reflection, gives rise to a multiplicative rather than additive quantum number, the eigenvalue is often referred to as the G parity. Thus, the π mesons have odd G parity, and a state with many π mesons has even or odd G parity, depending on whether the number of π mesons is even or odd. Clearly, the G transformation has the properties that were required of it at the beginning of this subsection.

The fact that isospin invariance holds only for strong interactions means that G parity is a useful concept only in strong processes. For systems of baryon number zero it yields absolute selection rules for fast (strong) decays [12]. However, later in the

‡ More exactly, the system under consideration and the system obtained by interchanging particles by antiparticles must belong to the same isomultiplet because the G transformation involves a charge conjugation which is compensated for by a rotation in isospace so as to obtain the original system again. This is possible only if the condition stated above is fulfilled.

‡ One can express the operator Ω in terms of the usual isospin operators T_j in the following way:

$$\Omega = \exp(-i,\epsilon_j T_j)$$

The operators T_j can be given in terms of creation and destruction operators. We shall never need these expressions, however, and omit them for that reason.

discussion involving nucleons, arguments based on the concept of G transformations will be used to cope with the complications due to strong interactions (cf. Sec. 5). For that purpose we need the transformation properties of the weak baryon current or rather its Hermitian conjugate

$$(\bar{p}n) = \frac{1}{\sqrt{2}} [\bar{\psi}_N(x)\gamma_\mu\tau_+\psi_N(x) + \bar{\psi}_N(x)\gamma_\mu\gamma_5\tau_+\psi_N(x)] \tag{5.74}$$

where the current has been separated into a vector and an axial-vector part. Furthermore we have written it in terms of the isospinor

$$\psi_N(x) = \begin{pmatrix} \psi_p(x) \\ \psi_n(x) \end{pmatrix} \tag{5.75}$$

introducing the matrices

$$\tau_\pm = \frac{1}{\sqrt{2}} (\tau_1 \pm i\tau_2);\ \tau_3 \tag{5.76}$$

where τ_1, τ_2, and τ_3 are the usual Pauli matrices (cf. Appendix A). In this way the isospin properties are properly displayed. Under rotations in isospace the operator $\psi_N(x)$ transforms in much the same way as the isovector $\boldsymbol{\phi}(x)$, or, more precisely, under an isorotation of 180° around the second axis one has

$$\psi_N(x) \to \psi'_N(x) = e^{-i\pi T_2}\psi_N(x)e^{i\pi T_2} = e^{(i/2)\pi\tau_2}\psi_N(x) = i\tau_2\psi_N(x) \tag{5.77a}$$

and similarly

$$\bar{\psi}_N(x) \to \bar{\psi}'_N(x) = e^{-i\pi T_2}\bar{\psi}_N(x)e^{i\pi T_2} = \bar{\psi}_N(x)e^{-(i/2)\pi\tau_2} = -i\bar{\psi}_N(x)\tau_2 \tag{5.77b}$$

or, equivalently,

$$\begin{array}{ll} \psi_p(x) \to \psi_n(x) & \psi_n(x) \to -\psi_p(x) \\ \bar{\psi}_p(x) \to \bar{\psi}_n(x) & \bar{\psi}_n(x) \to -\bar{\psi}_p(x) \end{array} \tag{5.78}$$

With these results and the previous results about the transformation properties of Dirac fields under charge conjugation we obtain

$$\begin{aligned} G\bar{\psi}_p(x)\gamma_\mu\psi_n(x)G^{-1} &= G\bar{\psi}_p(x)G^{-1}\gamma_\mu G\psi_n(x)G^{-1} \\ &= -C\bar{\psi}_n(x)C^{-1}\gamma_\mu C\psi_p(x)C^{-1} = \psi_n{}^T(x)\mathcal{C}^{-1}\gamma_\mu\mathcal{C}\bar{\psi}_p{}^T(x) \\ &= -\psi_n{}^T(x)\gamma_\mu{}^T\bar{\psi}_p{}^T(x) = \bar{\psi}_p(x)\gamma_\mu\psi_n(x) \end{aligned} \tag{5.79}$$

and similarly

$$G\bar{\psi}_p(x)\gamma_\mu\gamma_5\psi_n(x)G^{-1} = -\bar{\psi}_p(x)\gamma_\mu\gamma_5\psi_n(x) \tag{5.80}$$

that is, the vector current is even and the axial vector odd under a G transformation.

3.5. Time Reversal. Under the time reflection transformation

$$x_0 \to -x_0 \tag{5.81}$$

the Dirac equation (A.8) for the adjoint wave function $\bar{\psi}(x_0,\mathbf{x})$ is given by

$$i\frac{\partial}{\partial x^0}\bar{\psi}(-x_0,\mathbf{x})\gamma_0 + i\frac{\partial}{\partial x^k}\bar{\psi}(-x_0,\mathbf{x})\gamma_k - m\bar{\psi}(-x_0,\mathbf{x}) = 0 \tag{5.82}$$

Defining a matrix $\mathfrak{J}$ with the following properties

$$\mathfrak{J}^{-1}\gamma_k\mathfrak{J} = -\gamma_k{}^T \tag{5.83a}$$
$$\mathfrak{J}^{-1}\gamma_0\mathfrak{J} = \gamma_0{}^T \tag{5.83b}$$

we obtain from (5.82)

$$\left\{i\mathfrak{J}\gamma_0{}^T\frac{\partial}{\partial x^0} + i\mathfrak{J}\gamma_k{}^T\frac{\partial}{\partial x^k} - m\mathfrak{J}\right\}\bar{\psi}^T(-x_0,\mathbf{x}) = 0 \tag{5.84}$$

or

$$\left\{i\gamma_\mu\frac{\partial}{\partial x_\mu} - m\right\}\mathfrak{J}\bar{\psi}^T(-x_0,\mathbf{x}) = 0 \tag{5.85}$$

where use has been made of (5.83). Comparing (5.85) with the Dirac equation (A.1) we see that if the time-reflected wave function $\psi'(x)$ is defined by

$$\psi'(x_0,\mathbf{x}) = \mathfrak{J}\bar{\psi}^T(-x_0,\mathbf{x}) \tag{5.86}$$

it satisfies the original Dirac equation. Therefore, with this definition of the time-reflected wave function the theory is invariant under time reversal.

A matrix $\mathfrak{J}$ with the necessary properties to satisfy (5.83) is given by

$$\mathfrak{J} = \gamma_0\gamma_5\mathcal{C} = \gamma_2\gamma_5 \tag{5.87}$$

in the particular representation of Appendix A. From the properties derived earlier for $\mathcal{C}$ it is easily shown that the $\mathfrak{J}$ matrix satisfies the relations

$$\mathfrak{J}^\dagger = \mathfrak{J}^{-1} \tag{5.88a}$$
$$\mathfrak{J}^T = -\mathfrak{J} \tag{5.88b}$$

so that from (5.86) one obtains

$$\bar{\psi}'(x_0,\mathbf{x}) = \psi^T(-x_0,\mathbf{x})\mathfrak{J}^{-1} \tag{5.89}$$

For the second-quantized theory one could try to proceed in the same way as was done for space reflection, introducing a unitary operator acting on the space of state vectors. However, this cannot be done in a consistent fashion. For a consistent definition of time reversal in this case one must represent the operation by an antilinear operator, which is unitary, however. Operators of this kind are often referred to as *antiunitary operators*. The antilinear time-reversal operator is denoted by T. It is important to realize that operators and state vectors do not transform as in the case of linear operators given in Eqs. (5.37) and (5.38). The transformation laws in this case are

$$\Omega \to \Omega' = (T\Omega T^{-1})^\dagger \tag{5.90}$$
$$|\alpha\rangle \to |\alpha'\rangle = T|\alpha\rangle \equiv |T\alpha\rangle \tag{5.91a}$$

and

$$T\varkappa|\alpha\rangle = \varkappa^*T|\alpha\rangle \tag{5.91b}$$

where $\varkappa$ is a complex number. It is further noted that (5.90) implies that

$$(\Omega_1\Omega_2)' = \Omega'_2\Omega'_1 \tag{5.92}$$

Finally, to make the transformation laws consistent one must have

$$\langle\alpha|T^\dagger|\beta\rangle = \langle T\alpha|\beta\rangle^* \tag{5.93}$$

that is, transferring the operator from the ket to the bra vector involves an additional complex conjugation of the matrix element.

Taking the properties of the antilinear operator T into account, we now require the following transformation properties of the field operators under time reversal:‡

$$\psi(x_0,\mathbf{x}) \to \psi'(-x_0,\mathbf{x}) = (T\psi(-x_0,\mathbf{x})T^{-1})^\dagger = \bar{\psi}(x_0,\mathbf{x})\mathfrak{J} \quad (5.94a)$$

and

$$\bar{\psi}(x_0,\mathbf{x}) \to \bar{\psi}'(-x_0,\mathbf{x}) = (T\bar{\psi}(-x_0,\mathbf{x})T^{-1})^\dagger = \mathfrak{J}^{-1}\psi(x_0,\mathbf{x}) \quad (5.94b)$$

Inserting the Fourier expansions of the field operators in these expressions, we find

$$(Ta_s(\mathbf{p})T^{-1})^\dagger = \eta a_r{}^\dagger(-\mathbf{p}) \quad (5.95a)$$
$$(Tb_s(\mathbf{p})T^{-1})^\dagger = \eta' b_r{}^\dagger(-\mathbf{p}) \quad (5.95b)$$

where $r \neq s$ and η, η' are irrelevant phase factors. We conclude that, under time reflection, spin and momenta reverse sign, as one would expect from kinematic considerations.

Because of the complex conjugation in Eq. (5.93), the following relation holds:

$$\langle\alpha|\beta\rangle = \langle\alpha|T^{-1}T|\beta\rangle = \langle T\beta|T\alpha\rangle \quad (5.96)$$

from which it is seen that the sense of initial and final states is reversed under a time reflection, as one would expect.

3.6. Transformation Properties of the Dirac Covariants. We have obtained the transformation properties for the Dirac field operators $\psi(x)$ and $\bar{\psi}(x)$ under space reflection (P), charge conjugation (C), and time reflection (T). The results are collected in Table 5.3, where the transformation properties under the combined operation CPT are also included.

‡ Note that T implies complex conjugation for numbers and spinor quantities. Similarly the Hermitian conjugation of an operator O, denoted $O^\dagger$, means Hermitian conjugation in spinor space and in the space of state vectors, while it means complex conjugation for numbers.

From Table 5.3 and the properties of the matrices $\mathfrak{C}$ and $\mathfrak{J}$ one can now derive the transformation properties of the Dirac covariants in a straightforward fashion. The results are given in Table 5.4. To abbreviate the notation in the table we have introduced the following definitions:

$$x_{ab}(x) = \bar{\psi}_a(x)\Gamma_x\psi_b(x) \quad (5.97)$$

where $X = S, V, T, A, P$ (cf. Table A.1 in Appendix A).

3.7. The *CPT* Theorem. In Sec. 2 it was found that the most general S operator describing the processes

$$b \to a + c + \bar{d} \quad (5.98a)$$
$$\bar{b} \to \bar{a} + \bar{c} + d \quad (5.98b)$$

and which is invariant under proper Lorentz transformations is given by

$$S = -i \sum_{i=S,V,T,A,P} \int d^4x \{\bar{\psi}_a\Gamma_i\psi_b\bar{\psi}_c\Gamma_i(C_i + C'_i\gamma_5)\psi_d + \text{h.c.}\} \quad (5.99)$$

Noting that $\gamma_0\Gamma^\dagger_i\gamma_0 = \omega_i\Gamma_i$ with $\omega_{S,V,T,A} = +1$ and $\omega_P = -1$, one easily derives that the h.c. of Eq. (5.99) is given by

$$\text{h.c.} \equiv \bar{\psi}_b\Gamma_i\psi_a\bar{\psi}_d(C^*_i - C'^*_i\gamma_5)\Gamma_i\psi_c \quad (5.100)$$

From the transformation properties of the Dirac covariants under the combined transformation CPT (Table 5.4) it is immediately seen that the S operator (5.99) remains invariant. This result is just a special case of the CPT theorem which has been derived under much more general conditions than in the present context [13, S4].

3.8. Consequences of Separate *P*, *C*, and *T* Invariance for the *S* Operator. Expression (5.99) for the effective S operator was derived assuming invariance only under proper Lorentz transformations.

TABLE 5.3. TRANSFORMATION PROPERTIES OF THE DIRAC FIELD OPERATORS

Operator	Transformation			
	P	C	T	CPT
$\psi(x)$ $\bar{\psi}(x)$	$\psi'(x_0,\mathbf{x}) = \gamma_0\psi(x_0,-\mathbf{x})$ $\bar{\psi}'(x_0,\mathbf{x}) = \bar{\psi}(x_0,-\mathbf{x})\gamma_0$	$\psi'(x_0,\mathbf{x}) = \mathfrak{C}\bar{\psi}^T(x_0,\mathbf{x})$ $\bar{\psi}'(x_0,\mathbf{x}) = -\psi^T(x_0,\mathbf{x})\mathfrak{C}^{-1}$	$\psi'(x_0,\mathbf{x}) = \bar{\psi}(-x_0,\mathbf{x})\mathfrak{J}$ $\bar{\psi}'(x_0,\mathbf{x}) = \mathfrak{J}^{-1}\psi(-x_0,\mathbf{x})$	$\psi'(x_0,\mathbf{x}) = -(\gamma_5\psi(-x_0,-\mathbf{x}))^T$ $\bar{\psi}'(x_0,\mathbf{x}) = (\bar{\psi}(-x_0,-\mathbf{x})\gamma_5)^T$

TABLE 5.4. TRANSFORMATION PROPERTIES OF THE DIRAC COVARIANTS

Covariant	Transformation			
	P	C	T	CPT
$S_{ab}(x_0,\mathbf{x})$	$S'_{ab}(x_0,\mathbf{x}) = S_{ab}(x_0,-\mathbf{x})$	$S'_{ab}(x_0,\mathbf{x}) = S_{ba}(x_0,\mathbf{x})$	$S'_{ab}(x_0,\mathbf{x}) = S_{ba}(-x_0,\mathbf{x})$	$S'_{ab}(x_0,\mathbf{x}) = S_{ab}(-x_0,-\mathbf{x})$
$V_{ab}(x_0,\mathbf{x})$	$V'_{ab}(x_0,\mathbf{x}) = (-1)^{\delta_{\mu 0}+1}V_{ab}(x_0,-\mathbf{x})$	$V'_{ab}(x_0,\mathbf{x}) = -V_{ba}(x_0,\mathbf{x})$	$V'_{ab}(x_0,\mathbf{x}) = (-1)^{\delta_{\mu 0}+1}V_{ba}(-x_0,\mathbf{x})$	$V'_{ab}(x_0,\mathbf{x}) = -V_{ab}(-x_0,-\mathbf{x})$
$T_{ab}(x_0,\mathbf{x})$	$T'_{ab}(x_0,\mathbf{x}) = (-1)^{\delta_{\mu 0}+\delta_{\nu 0}}T_{ab}(x_0,-\mathbf{x})$	$T'_{ab}(x_0,\mathbf{x}) = -T_{ba}(x_0,\mathbf{x})$	$T'_{ab}(x_0,\mathbf{x}) = (-1)^{\delta_{\mu 0}+\delta_{\nu 0}+1}T_{ba}(-x_0,\mathbf{x})$	$T'_{ab}(x_0,\mathbf{x}) = T_{ab}(-x_0,-\mathbf{x})$
$A_{ab}(x_0,\mathbf{x})$	$A'_{ab}(x_0,\mathbf{x}) = (-1)^{\delta_{\mu 0}}A_{ab}(x_0,-\mathbf{x})$	$A'_{ab}(x_0,\mathbf{x}) = A_{ba}(x_0,\mathbf{x})$	$A'_{ab}(x_0,\mathbf{x}) = (-1)^{\delta_{\mu 0}+1}A_{ba}(-x_0,\mathbf{x})$	$A'_{ab}(x_0,\mathbf{x}) = -A_{ab}(-x_0,-\mathbf{x})$
$P_{ab}(x_0,\mathbf{x})$	$P'_{ab}(x_0,\mathbf{x}) = -P_{ab}(x_0,-\mathbf{x})$	$P'_{ab}(x_0,\mathbf{x}) = P_{ba}(x_0,\mathbf{x})$	$P'_{ab}(x_0,\mathbf{x}) = -P_{ba}(-x_0,\mathbf{x})$	$P'_{ab}(x_0,\mathbf{x}) = P_{ab}(-x_0,-\mathbf{x})$

If it is to be invariant under P, C, or T transformations separately, further restrictions are obtained.

Under space reflection we obtain from (5.99) and Table 5.4

$$PSP^{-1} = -i\eta_\pi \sum_i \int d^4x \{\bar{\psi}_a \Gamma_i \psi_b \bar{\psi}_c \Gamma_i (C_i - C'_i \gamma_5) \psi_d + \text{h.c.}\} \qquad (5.101)$$

where η_π is a product of four independent phase factors [cf. Eq. (5.48)]. One may choose $\eta_\pi = \pm 1$ without loss of generality. Then invariance under space reflection requires‡

$$C_i \neq 0 \qquad C'_i = 0 \qquad \text{or} \qquad C_i = 0 \qquad C'_i \neq 0 \qquad \text{all } i \qquad (5.102)$$

Similarly for charge conjugation we obtain

$$CSC^{-1} = -i \sum_i \int d^4x \{\bar{\psi}_b \Gamma_i \psi_a \bar{\psi}_d (C_i + C'_i \gamma_5) \Gamma_i \psi_c + \text{h.c.}\} \qquad (5.103)$$

The first term is to be compared with (5.100). We conclude that invariance under charge conjugation implies that

$$C^*_i = C_i \qquad C'^*_i = -C'_i \qquad (5.104)$$

This means that all the coupling constants C_i are real and the coupling constants C'_i are imaginary.§

Under time reversal, finally, the S operator (5.99) transforms in the following way:

$$(TST^{-1})^\dagger = -i \sum_i \int d^4x \{\bar{\psi}_b \Gamma_i \psi_a \bar{\psi}_d (C_i - C'_i \gamma_5) \Gamma_i \psi_c + \text{h.c.}\} \qquad (5.105)$$

which, compared with (5.100), yields

$$C^*_i = C_i \qquad C'^*_i = C'_i \qquad (5.106)$$

if the S operator is to be invariant under the transformation. Therefore, for time reversal to hold, all the coupling constants must be real.

From these observations we also conclude that, if S is invariant under T and C simultaneously, all C'_i must vanish. This, however, then implies P invariance, as one would expect in view of the CPT theorem. This is just an example reflecting a general property of the theory, namely, if any one of the three symmetries P, C, or T is proved to be broken, then by virtue of the CPT theorem, (at least) one more must be broken.

We finally note that the V-A theory corresponds to choosing

$$C_V = C'_V = C_A = C'_A = \frac{G}{\sqrt{2}} \qquad (5.107)$$

‡ The two alternative solutions in (5.102) imply symmetry in the coupling coefficients C and C'. This is actually the case for the parity-sensitive part of the observed rates, etc. If $m_c \neq 0$ and $m_d \neq 0$ the parity-nonsensitive part would not be symmetrical, but in all our applications either c or d is a neutrino and no practical consequences ensue.

§ An over-all phase angle in the coupling constants in (5.99) is undetectable experimentally. Strictly speaking, one deals with results regarding the relative phases.

in (5.99) and setting all other coupling constants equal to zero. From the results obtained it is then seen that the V-A theory is invariant under time reversal but not invariant under space reflection and charge conjugation separately. However, the time-reversal invariance together with the CPT theorem implies that the S operator is invariant under the combined operation CP. By inspection this is seen to be true.

3.9. Experimental Tests of P, C, and T Invariance for Weak Interactions. The problem of establishing the symmetry properties under P, C, T transformations of the weak interactions as described by (5.99) is intimately connected with the determination of the 10 complex coupling constants, C_i and C'_i, which are a priori possible. One approach to this question is to design experiments that serve to determine whether or not the relations (5.102), (5.104), and (5.106) properly describe the coupling constants as they appear in nature. For this purpose the detailed results of the theory for angular correlations, polarization effects, and so on are necessary. To obtain an insight without resorting to these details it is sufficient to ask what *kind* of correlations constitute a criterion for the absence of each symmetry. This approach is based on the fact that incoming and outgoing particles are represented by plane waves. These state vectors are eigenvectors of the spin and the momentum operators, and the only measurable quantities of a process are the corresponding eigenvalues of these operators. The expressions for transition rates pertaining to an actual experiment are functions only of these eigenvalues or, in some cases, expectation values as the observation corresponds to an average over an ensemble.‡ For the two discrete transformations P and T, which have classical analogues, we have previously derived the transformation properties of the state vectors corresponding to plane waves. The results are summarized in Table 5.5. In this table **p**

TABLE 5.5. TRANSFORMATION PROPERTIES OF LINEAR AND ANGULAR MOMENTA

Observable	Transformation	
	P	T
p	−**p**	−**p**
J	+**J**	−**J**

denotes a linear momentum and **J** an angular momentum (spin). It may be recalled that invariance under space reflection for the S operator would imply

$$\langle f|S|i\rangle = \langle Pf|S|Pi\rangle \qquad (5.108)$$

where $|Pf\rangle$ and $|Pi\rangle$ are the space-reflected final and initial state vectors. According to Table 5.5 these differ from the original state vectors only by the fact that momenta (but not angular momenta) have changed sign. Equation (5.108) implies that the S-matrix elements and, therefore, also the transition rates corresponding to these two experimental situ-

‡ For instance, β decay of polarized nuclei.

ations would be the same. Consider an experiment where one measures a correlation of the type $\mathbf{p} \cdot \mathbf{J}$: An example is the decay of polarized muons in which the polarization of the muons and the momentum of the outgoing electrons are determined (cf. Sec. 4). The transition rate is then a function of these two quantities, and it must be a scalar function under proper Lorentz transformations. Then in a specific reference system the most general expression is of the form

$$d\Lambda = f_1 + f_2 \langle \boldsymbol{\sigma}_\mu \rangle \cdot \mathbf{p}_e \qquad (5.109)$$

where f_1 and f_2 are scalar functions under Lorentz transformations including space reflections. However, the presence of the second term is not consistent with invariance of the S operator under space reflections since it changes sign under the transformation, in contrast to the P-invariance condition [Eq. (5.108)]. Therefore, it must be concluded that the presence of a nonvanishing term of the type $\langle \boldsymbol{\sigma} \rangle \cdot \mathbf{p}$ in an experimentally determined transition rate would demonstrate that the interaction is not invariant under P.

Similar conclusions regarding time-reversal invariance can be derived. The analogue of (5.108) is in this case

$$\langle f|S|i\rangle = \langle Tf|S|Ti\rangle^* \qquad (5.110)$$

where $|Tf\rangle$ and $|Ti\rangle$ correspond to states with a change in sign for both momenta and angular momenta. For time reversal there is another complication, however, which requires special attention. The simple argument in terms of vectors holds only to the extent that the "final"-state interactions are negligible. This is so because the time-reversal transformation also affects the boundary conditions of the problem: The incoming and outgoing states may in general participate in various scattering processes so that the incoming and outgoing waves are distorted. In the case of scattering due to the short-range strong forces, this simply introduces an extra phase shift for the plane waves. For the long-range electromagnetic forces the asymptotic wave functions are Coulomb wave functions rather than plane waves. We shall not give a detailed discussion of these effects (cf. ref. S5) but summarize by noting that correlations that would seem to violate time-reversal invariance (on the basis of the foregoing kinematic arguments) may appear in the case of final-state interactions. Reliable estimates of such corrections can be made in most cases of practical interest from the knowledge of the appropriate scattering data. However, this observation brings up an important point with regard to symmetry arguments. It is not sufficient that the basic interaction, that is, the S operator, displays the relevant symmetry properties in order that the observations should do so. It is also necessary that the boundary conditions must share this same symmetry, as is seen in the example above. Many similar situations occur in elementary-particle physics, and proper attention must always be paid to this point. However, the previous argument for space reflection in terms of vectors does not suffer from this complication.

In Table 5.6 most of the important experimental tests of the P, C, and T symmetry properties of the nonstrange-particle weak interactions have been summarized. Each of these tests involves the search for

TABLE 5.6. ACTUAL EXPERIMENTAL TESTS OF P, C, AND T SYMMETRY

Experiment	Correlation	Symmetry property under test	Asymmetry result	Ref.
1. Angular distribution of electrons from mu decay	$\langle \boldsymbol{\sigma}_\mu \rangle \cdot \mathbf{p}_e$	P, C	Present	7, 14
2. Electron polarization from mu decay	$\langle \boldsymbol{\sigma}_e \rangle \cdot \mathbf{p}_e$	P, C	Present	15
3. Angular distribution of electrons from polarized nuclei	$\langle \mathbf{J}_i \rangle \cdot \mathbf{p}_e$	P, C	Present	6, 16
4. Polarization of β particles	$\langle \boldsymbol{\sigma}_e \rangle \cdot \mathbf{p}_e$	P, C	Present	S6
5. Angular correlation of β particles and circularly polarized gamma rays	$\langle \mathbf{J}_f \rangle \cdot \mathbf{p}_\gamma$	P, C	Present	17
6. Neutrino helicity in orbital electron capture	$\boldsymbol{\sigma}_\nu \cdot \mathbf{p}_\nu$	P, C	Present	18
7. Muon helicity from π decay	$\langle \boldsymbol{\sigma}_\mu \rangle \cdot \mathbf{p}_\mu$	P, C	Present	7, 14
8. Electron-neutrino correlation from polarized nuclei	$\langle \mathbf{J}_i \rangle \cdot \mathbf{p}_e \times \mathbf{p}_\nu$	C, T	No effect	19

an anisotropy in the decay products which is expressed in terms of a scalar product of $\mathbf{p}$ and $\mathbf{J}$ vectors. Moreover, each anisotropy measurement provides an experimental criterion for two of the three symmetry properties‡ with no implication for the third. Each of the experimental tests listed in Table 5.6 showing a positive result involves a $(\mathbf{p} \cdot \mathbf{J})$ correlation and is a test for P and C breakdown. The last entry provides a test for time-reversal invariance, and the null effect observed indicates that this symmetry indeed applies to the weak interactions studied here. There is evidence that time-reversal invariance may not apply to strange-particle decays [20].

Experiments 1 and 3 were the first evidence for P, C breakdown. The existence of a polarization of the decaying muon is evidence for P, C breakdown in π decay (experiment 7). That is, experiment 1 is an analyzer for the longitudinal mu polarization produced in the latter type of process. The polarization of the electrons from mu decay was detected by transmission through magnetized iron of the *bremsstrahlung* radiation they emit. This provides a measure of the circular polarization of this radiation. A similar procedure is used in experiment 5 where one actually measures the $\mathbf{p}_e \cdot \mathbf{p}_\gamma$ correlation *with* a circular polarization analysis of the γ ray emitted from the final

‡ One way to see this is to recognize that each anisotropy term contains, as a factor, a bilinear combination of coupling coefficients and that the only possible combinations are real and imaginary parts of $B_1 = C_i C^*_j$, $B_2 = C'_i C^{*\prime}_j$, and $B_3 = C_i C'^*_j + C'_i C^*_j$. Apart from Re B_1 and Re B_2 each of these is connected with a test of two symmetries.

state of the β transition. The state of this residual nucleus is polarized along the direction $\mathbf{p}_e$ with a polarization proportional to $<\mathbf{J}_f>$. Similarly, in experiment 3, the initial nuclear state is prepared with a polarization proportional to $<\mathbf{J}_i>$. The neutrino helicity measurement, experiment 6, was decisive in distinguishing between a *V-A* beta interaction with $C_i = C'_i$ and an *S-T* interaction with $C_i = -C'_i$. Other experiments (3, 4, 5) in nuclear β decay, not involving an observation of neutrino helicity, could not make this distinction. Note that the experimental evidence is in excellent agreement with the two-component neutrino theory ($C_i = C'_i$) and hence no average of $\boldsymbol{\sigma}_\nu$ is involved.

Further discussion of certain key experiments listed in Table 5.6 will be given in following sections. It may be added that the observation of an anisotropy in the neutrons emitted from mu capture to excited states also provides evidence for *P*, *C* breakdown in that process. Anisotropy in the angular distribution of decay electrons from (negative) muons in bound orbits has also been observed.

4. Muon Decay

The muon normally decays into one electron and two neutrinos as noted in the Table 5.1:

$$\mu^- \rightarrow e^- + \bar{\nu}_e + \nu_\mu \qquad (5.111a)$$
$$\mu^+ \rightarrow e^+ + \nu_e + \bar{\nu}_\mu \qquad (5.111b)$$

Because all the particles participating in muon decay are leptons, no complications due to strong interactions are encountered. The electromagnetic corrections for these processes are well understood and present no difficulties. For these reasons the normal muon decay offers an essentially unique opportunity to study the weak interactions. Other weak processes that are experimentally observable are either too rare or include one or more strongly interacting particles; the weak-interaction aspect is then more or less masked by these strong effects.

On the other hand, in the final state only the electron can be detected in a practical way. The two neutrinos are uncharged, and they interact only weakly. Therefore, they escape any attempt of experimental detection. Their presence can be established only indirectly by invoking energy-momentum and angular-momentum conservation. This lack of knowledge about the final-state neutrinos strongly limits the variety of possible experiments. We shall follow the same approach as outlined in the previous sections and insist on a pure *V-A* theory to test its consistency with the experimental findings. In this way we are able to exhibit the extent to which the muon decay fits into the coherent picture of weak interactions. This is in contrast to an attempt to deduce a unique interaction responsible for the decay process. For reasons just given, the latter approach would not yield a unique result.

4.1. The Transition Probability. Because of the basic role of μ decay in weak interactions we shall present the formalism leading to observable results in some detail.

We shall consider process (5.111*a*). The results can be taken over for process (5.111*b*) with some trivial changes. Process (5.111*a*) can be represented by the graph of Fig. 5.1 where the notations for spins and momenta of the different particles have been introduced. The effective *S* operator for the process within the *V-A* theory has been given in Sec. 2 in the following form:

$$S = -\frac{iG}{\sqrt{2}} \int d^4x \bar{\psi}_e \gamma^\mu (1 + \gamma_5) \psi_{\nu_e} \bar{\psi}_{\nu_\mu} \gamma_\mu (1 + \gamma_5) \psi_\mu \qquad (5.112)$$

For computational reasons it is convenient to recast this expression by means of a Fierz transformation and write it in the form

$$S = -\frac{iG}{\sqrt{2}} \int d^4x \bar{\psi}_e \gamma^\mu (1 + \gamma_5) \psi_\mu \bar{\psi}_{\nu_\mu} \gamma_\mu (1 + \gamma_5) \psi_{\nu_e} \qquad (5.113)$$

By taking the matrix element of this operator between the initial and final state given in the graph one obtains, after some simplifications, the following expression for the corresponding *T*-matrix element as defined in Appendix A [Eq. (A.66)]:‡

$$T_{fi} \equiv < p's', qr, q'r'|T|ps = G\sqrt{8m_\mu m_e m_{\nu_\mu} m_{\nu_e}}\, M \qquad (5.114)$$

where

$$M = \bar{u}_{s'}(\mathbf{p}')\gamma^\mu(1 + \gamma_5)u_s(\mathbf{p})\bar{u}_{r'}(\mathbf{q}')\gamma_\mu(1 + \gamma_5)v_r(\mathbf{q}) \qquad (5.115)$$

By inserting this into Eq. (A.70) the following expression is obtained for the partial-decay rate of a muon with spin *s* and momentum $\mathbf{p}$ emitting an electron with spin s' and the momentum in the interval d^3p':

$$d\Lambda = \frac{G^2}{64\pi^5} \frac{m_\mu m_e m_{\nu_\mu} m_{\nu_e}}{E_\mu} \frac{d^3p'}{E_e} \int \frac{d^3q}{E_{\nu_e}} \frac{d^3q'}{E_{\nu_\mu}} \delta^4(Q-q-q') \sum_{r,r'=1}^{2} |M|^2 \qquad (5.116)$$

where $Q = p - p'$. In view of foregoing remarks we have integrated over neutrino momenta and summed over the spins r and r'. The latter operation gives

$$\sum_{r,r'=1}^{2} |M|^2 = -\mathfrak{M}^{\mu\nu}\,\mathrm{tr}\,[\gamma_\mu(1 + \gamma_5)\Delta_-(q)\gamma_\nu(1 + \gamma_5)\Delta_+(q')] \qquad (5.117)$$

with

$$\mathfrak{M}^{\mu\nu} = \bar{u}_{s'}(\mathbf{p}')\gamma^\mu(1 + \gamma_5)u_s(\mathbf{p})\bar{u}_s(\mathbf{p})\gamma^\nu(1 + \gamma_5)u_{s'}(\mathbf{p}') \qquad (5.118)$$

‡ The two neutrinos are treated as if they had small but nonvanishing masses. The usual normalization for spin ½ fields can then be used (see Appendix A), and the projection operators are the conventional ones. This makes the computation somewhat simpler. By setting the neutrino masses equal to zero in the final result, one then obtains exactly the same result as if they had been set equal to zero from the very beginning.

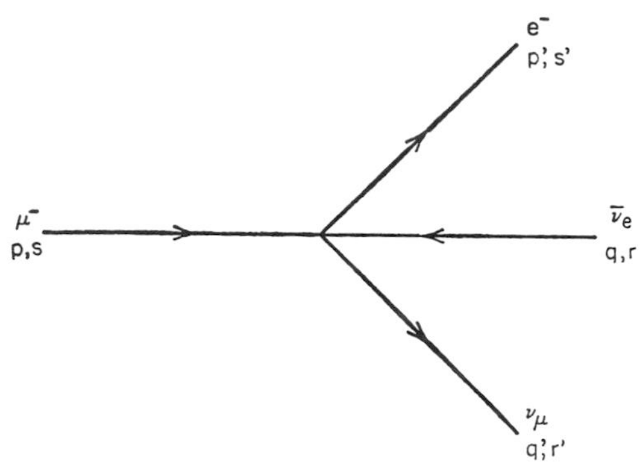

FIG. 5.1. Diagram describing muon decay.

Introducing the explicit expressions for the energy projection operators $\Delta_\pm$ the trace is easily evaluated with the following result:

$$d\Lambda = \frac{G^2}{32\pi^5}\frac{m_\mu m_e}{E_\mu}\frac{d^3p'}{E_e}\mathfrak{M}^{\mu\nu}\{g_{\mu\alpha}g_{\nu\beta} - g_{\mu\nu}g_{\alpha\beta} + g_{\mu\beta}g_{\nu\alpha} - i\epsilon_{\mu\alpha\nu\beta}\}\cdot\int\frac{d^3q}{E_{\nu_e}}\frac{d^3q'}{E_{\nu_\mu}}\delta^4(Q - q - q')q^\alpha q'^\beta \quad (5.119)$$

To evaluate the integral over the neutrino momenta, (5.119) is most conveniently rewritten in a manifestly covariant form. It is then easily shown that [S7]

$$\int\frac{d^3q}{E_{\nu_e}}\frac{d^3q'}{E_{\nu_\mu}}\delta^4(Q - q - q')q^\alpha q'^\beta = \frac{\pi}{6}\{Q^2g^{\alpha\beta} + 2Q^\alpha Q^\beta\} \quad (5.120)$$

which, inserted in (5.119), yields

$$d\Lambda = \frac{G^2}{48\pi^4}\frac{m_\mu m_e}{E_\mu}\frac{d^3p'}{E_e}\mathfrak{M}^{\mu\nu}\{Q_\mu Q_\nu - g_{\mu\nu}Q^2\} \quad (5.121)$$

From (5.121) we shall derive the various decay rates with which experimental data may be compared.

4.2. Polarization Effects. There are two different polarization effects that one can investigate in muon decay:

1. Polarization of the outgoing electron
2. Angular distribution of the outgoing electrons in the decay of polarized muons

These are experiments 2 and 1, respectively, in Table 5.6.

With regard to the polarization of the outgoing electron, we can apply the same argument that led to the two-component theory of the neutrino. Because of the phase-space factor the electrons are most likely to be emitted with large momenta, that is, $|\mathbf{p}'|^2 \gg m_e^2$. It is, therefore, an excellent approximation to neglect the electron mass. Within the *V-A* theory, the electrons are created in a state of positive chirality, and in the approximation of vanishing electron mass it maybe concluded that the emitted electrons are all left-handed, corresponding to helicity -1. In the same way positrons emitted in μ^+ decays should have the helicity $+1$.

Experimentally the helicity of the electron (positron) emitted in muon decay is determined by observing the polarization of the *bremsstrahlung*. The measurements [21] are not very accurate, however, but agree with the predictions of the *V-A* theory within the experimental errors, estimated to be about 20 per cent.

To obtain the prediction of the *V-A* theory regarding the angular distribution of electrons emitted from polarized muons, we return to Eq. (5.121). Before the calculations are carried out, one can easily be convinced that beams of polarized muons can be obtained.

The usual source of muons is π-meson beams from accelerators. Almost all π mesons decay into a muon and a neutrino, so that, for example,

$$\pi^- \to \mu^- + \bar{\nu}_\mu$$

In the rest system of the π meson the decay is described by Fig. 5.2. Conservation of momentum requires the two particles to emerge with equal but opposite momenta. The antineutrino has helicity $+1$ so that its spin is along its momentum vector. Because the spin of the π meson is zero, the muon must also come out fully polarized with helicity $+1$ in order to satisfy angular-momentum conservation.

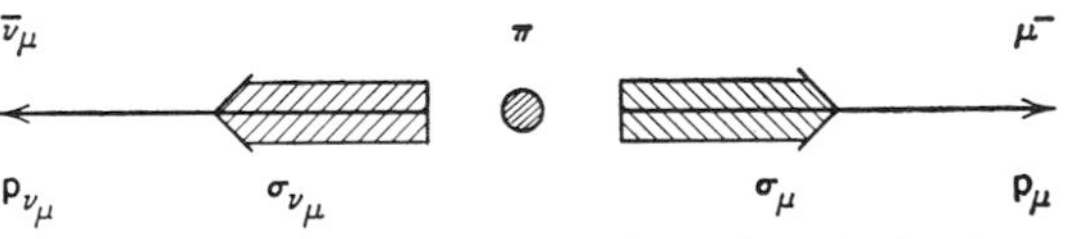

FIG. 5.2. Momentum and spin orientations in the decay of a charged π meson at rest, assuming a right-handed antineutrino as required by the *V-A* theory.

From Eq. (5.121) the following expression is obtained for the decay rate when the electron polarization is not measured:

$$d\Lambda = \frac{G^2}{48\pi^4}\frac{m_\mu m_e}{E_\mu E_e}d^3p'\sum_{s'=1}^{2}\mathfrak{M}^{\mu\nu}\{Q_\mu Q_\nu - g_{\mu\nu}Q^2\} \quad (5.122)$$

The fact that the muon is assumed to be polarized is most easily taken into account by means of the covariant spin projection operator [cf. Eq. (A.35*b*)]:

$$S_n = \tfrac{1}{2}(1 + w_\sigma\gamma^\sigma\gamma_5) \quad (5.123)$$

where in the muon rest frame $w = (0,\hat{\mathbf{n}})$, with $\hat{\mathbf{n}}$ a unit vector in the direction of the muon spin in its rest frame. We can then sum over s and obtain

$$\sum_{s'=1}^{2}\mathfrak{M}^{\mu\nu} = \sum_{s,s'=1}^{2}\bar{u}_{s'}(\mathbf{p}')\gamma^\mu(1+\gamma_5)S_nu_s(\mathbf{p})\bar{u}_s(\mathbf{p})\gamma^\nu(1+\gamma_5)u_{s'}(\mathbf{p}')$$
$$= \tfrac{1}{2}\,\mathrm{tr}[\gamma^\mu(1+\gamma_5)(1 + w_\sigma\gamma^\sigma\gamma_5)\Delta_+(p)\gamma^\nu(1+\gamma_5)\Delta_+(p')] \quad (5.124)$$

Evaluating this trace and neglecting the electron mass in the numerator of $\Delta_+(p')$, one obtains

$$\sum_{s'=1}^{2}\mathfrak{M}^{\mu\nu} = \frac{1}{m_\mu m_e}\{p_\sigma p'_\rho - m_\mu w_\sigma p'_\rho\}\{g^{\sigma\nu}g^{\rho\mu} - g^{\nu\mu}g^{\sigma\rho} + g^{\sigma\mu}g^{\nu\rho} - i\epsilon^{\sigma\nu\rho\mu}\} \quad (5.125)$$

Considering the decay in the muon rest frame and neglecting the electron mass, we now obtain from (5.122) and (5.125)

$$d\Lambda = \frac{G^2}{48\pi^4}\frac{d^3p'}{E_e} m_\mu \{3m_\mu E_e - 4E_e^2 + (m_\mu - 4E_e)\hat{\mathbf{n}}\cdot\mathbf{p}'\} \qquad (5.126)$$

or introducing the energy variable

$$x = \frac{2E_e}{m_\mu} \qquad (5.127)$$

which is referred to below as the "energy,"

$$d\Lambda = \frac{G^2 m_\mu^5}{384\pi^4} d\Omega_e\, dx\, x^2 \{3 - 2x + (1 - 2x)\hat{\mathbf{n}}\cdot\hat{\mathbf{p}}'\} \qquad (5.128)$$

If we define

$$N_+(x,\Omega) = \frac{G^2 m_\mu^5}{384\pi^4} x^2\{3 - 2x + (1 - 2x)\hat{\mathbf{n}}\cdot\hat{\mathbf{p}}'\}\Delta x\,\Delta\Omega \qquad (5.129)$$

then $N_+(x,\Omega)$ is the transition rate for the decay of a muon at rest with the spin along $\hat{\mathbf{n}}$ and with the emitted electron in the energy range $(x, x + \Delta x)$ and within the solid angle $(\Omega, \Omega + \Delta\Omega)$.

If we similarly define $N_-(x,\Omega)$ as the transition rate for the decay with the spin along $-\hat{\mathbf{n}}$, we obtain the expression for $N_-(x,\Omega)$ from $N_+(x,\Omega)$ by replacing $\hat{\mathbf{n}}$ by $-\hat{\mathbf{n}}$. Thus

$$N_-(x,\Omega) = \frac{G^2 m_\mu^5}{384\pi^4} x^2\{3 - 2x - (1 - 2x)\hat{\mathbf{n}}\cdot\mathbf{p}'\}\Delta x\,\Delta\Omega \qquad (5.130)$$

In general, only partially polarized muon beams can be obtained. With the quantization axis for the spin chosen along $\hat{\mathbf{n}}$ this means that in a beam of muons some have the spin along $\mathbf{n}$ but some have the spin pointing in the opposite direction. If we denote the probability for the spin of a decaying muon to point along $\pm\hat{\mathbf{n}}$ by $P_\pm$, the degree of polarization P in the beam is defined by

$$P = P_+ - P_- \qquad (5.131)$$

One can also express P as the expectation value for the spin of the muon if measured along the axis $\hat{\mathbf{n}}$. As $P_+ + P_- = 1$ we obtain from (5.131)

$$P_+ = \frac{1 + P}{2} \qquad P_- = \frac{1 - P}{2} \qquad (5.132)$$

The transition rate $N(x,\Omega)$ for the decay of a partially polarized muon (degree of polarization P) is now given by

$$N(x,\Omega) = P_+N_+(x,\Omega) + P_-N_-(x,\Omega) \qquad (5.133)$$

or

$$N(x,\Omega) = \frac{G^2 m_\mu^5}{384\pi^4} x^2\{3 - 2x + (1 - 2x)P\cos\Theta\}\Delta x\,\Delta\Omega \qquad (5.134)$$

where Θ is the angle between $\hat{\mathbf{n}}$ and $\mathbf{p}'$. From this equation it is seen that there is an anisotropic electron angular distribution provided the beam is at least partially polarized. Furthermore, this angular distribution is energy-dependent. In particular we note that for a fully polarized beam there are no electrons emitted in the forward direction with maximum energy ($x = 1$). In general, one can state that in a partially polarized beam high-energy electrons are more likely to emerge in the backward direction than in the forward direction, where forward and backward refers to the direction of polarization of the muons.

The angular correlation measured in most cases is not be be compared with (5.134) but rather with the angular correlation obtained after averaging over all electron energies. To make the comparison we rewrite (5.134) in the following way:

$$N(x,\Omega) = \frac{G^2 m_\mu^5}{64\pi^4} x^2\{1 - x + \tfrac{2}{3}\rho(\tfrac{4}{3}x - 1) + P\cos\Theta[\alpha(1 - x) + \tfrac{2}{3}\beta(\tfrac{4}{3}x - 1)]\}\Delta x\,\Delta\Omega \qquad (5.135)$$

with $\rho = \tfrac{3}{4} \qquad \alpha = -\tfrac{1}{3} \qquad \beta = -\tfrac{3}{4}$ (5.136)

The parameter ρ, known as the *Michel parameter* [22], characterizes the energy spectrum rather than the polarization effects (discussed, for that reason, in Sec. 4.3). Parameters α and β have been so chosen that only α appears in the averaged angular distribution. The most recent experimental [23] value for the parameter α is given by

$$\alpha_{\text{exp}} = -0.325 \pm 0.010 \qquad (5.137)$$

In a similar experiment [24] the parameter β has also been measured with the following result:

$$\beta_{\text{exp}} = (0.78 \pm 0.05)\cdot 3\alpha_{\text{exp}} \qquad (5.138)$$

Comparing these values with the theoretical predictions (5.136) we conclude that the experimental results are in excellent agreement with the V-A theory.

It is of interest to compare the experimental results with those calculated with arbitrary coupling constants C_i and C'_i (cf. Sec. 3.7). The calculated results for the most general case [25] are not of sufficient interest to repeat here, and we restrict attention to a V, A mixture. Then

$$-\alpha = \frac{2}{3}\frac{\mathrm{Re}(C_V C'^*_A + C_A C'^*_V)}{|C_V|^2 + |C'_V|^2 + |C_A|^2 + |C'_A|^2} \qquad (5.139)$$

and $\beta = -\tfrac{3}{4}$ again. While the measured value of α does not uniquely determine all the coupling constant ratios, it is important that the right-hand side of (5.139) has a maximum value of $\tfrac{1}{3}$ which strongly implies that the P, C violation is maximal. It is also to be emphasized that all asymmetries as observed for the other weak interaction processes are always maximal. This statement applies only to the nonstrange processes where, as stressed above, neutrinos are always involved.

4.3. The Electron Energy Spectrum and the Lifetime. To obtain the electron energy spectrum we integrate $N(x,\Omega)$ over all angles, with the following result:

$$d\Lambda = \frac{G^2 m_\mu^5}{16\pi^3} x^2\{1 - x + \tfrac{2}{3}\rho(\tfrac{4}{3}x - 1)\} \qquad (5.140)$$

As noted, the theoretical value of the Michel parameter is $\frac{3}{4}$. It has been determined experimentally [24]:

$$\rho_{\text{exp}} = 0.747 \pm 0.005 \tag{5.141}$$

In this result the electromagnetic corrections have been taken into account so that the experimental value in (5.141) is to be compared with the theoretical value $\frac{3}{4}$. A comparison between the experimental result and the *V-A* theory is given in Fig. 5.3. The agreement is quite striking.

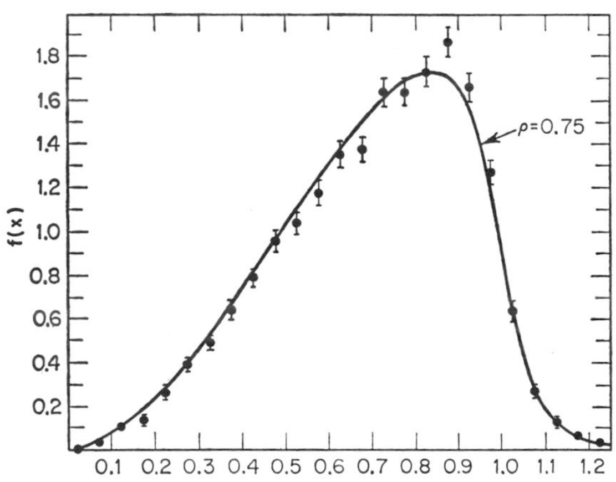

FIG. 5.3. The decay spectrum of electrons from muon decay. The solid curve corresponds to the theoretical prediction with the Michel parameter $\rho = \frac{3}{4}$ and with radiative corrections taken into account. [*Courtesy R. J. Plano, Phys. Rev.*, **119**: 1400 (1960).]

The experimental determination of the Michel parameter was one of the first tests of the *V-A* theory. The choice of other possible couplings in general give very different results, as summarized in Table 5.7. Obviously a suitable combination of *S*, *P*, and *T* could yield the experimental result, and μ decay by itself does not rule out such a possibility (cf. above). However, experimental results from other weak decays rule out such an alternative, and the conclusion is that all the results of μ decay are consistent with the *V-A* theory.

TABLE 5.7. MICHEL PARAMETER FOR VARIOUS COUPLINGS

Coupling	ρ
V-A	0.75
S	0
P	0
T	1.00

To obtain the lifetime we integrate (5.140) over all energies with the following result (note $0 \le x \le 1$ if the electron mass in neglected):

$$\frac{1}{\tau} = \Lambda = \frac{G^2 m_\mu^5}{192\pi^3} \tag{5.142}$$

From the experimental value of the lifetime [26]

$$\tau_{\text{exp}} = (2.200 \pm 0.002) \times 10^{-6} \text{ sec}$$

one can compute the coupling constant G. In cgs units one obtains

$$G = 1.431 \times 10^{-49} \text{ erg cm}^3$$

or, after electromagnetic corrections have been properly taken into account [25],

$$G = (1.434 \pm 0.001) \times 10^{-49} \text{ erg cm}^3$$

In natural units with $\hbar = c = 1$ one obtains

$$G = 1.02 \times \frac{10^{-5}}{m_p^2}$$

where m_p is the proton mass.

4.4. Rare Decay Modes of the Muon. Besides the dominant decay mode of the muon, others have also been established, although they all are extremely rare—so rare that essentially nothing is known about them except the branching ratios with respect to the normal decay.

The inner *bremsstrahlung* process

$$\mu^+ \to e^+ + \nu_e + \bar{\nu}_\mu + \gamma \tag{5.143}$$

has been observed, as has the internal conversion process [27]

$$\mu^+ \to e^+ + \nu_e + \bar{\nu}_\mu + e^+ + e^- \tag{5.144}$$

For experimental reasons, study has been limited to processes of type (5.143) for which the emitted proton has an energy larger than 10 Mev. For process (5.144) a corresponding limitation in the energy of the converted γ to $E_\gamma > 10$ Mev was necessary. The measured branching ratios are

$$R_{1\text{exp}} = \frac{\Lambda(\mu^+ \to e^+ + \nu_e + \bar{\nu}_\mu + \gamma)}{\Lambda(\mu^+ \to e^+ + \nu_e + \bar{\nu}_\mu)} = (1.4 \pm 0.4) \times 10^{-2}$$

$$R_{2\text{exp}} = \frac{\Lambda(\mu^+ \to e^+ + \nu_e + \bar{\nu}_\mu + e^+ + e^-)}{\Lambda(\mu^+ \to e^+ + \nu_e + \bar{\nu}_\mu)} = (2.2 \pm 1.5) \times 10^{-5}$$

These values are to be compared with the corresponding predictions [28] as derived from the *V-A* theory:

$$R_1 \simeq 1.3 \times 10^{-2}$$
$$R_2 \simeq 4 \times 10^{-5}$$

which are consistent with the experimental values.

There has been an extensive search for even more rare decay modes, namely,

$$\mu^\pm \to e^\pm + \gamma \tag{5.145}$$
$$\mu^\pm \to e^\pm + e^\mp + e^\pm \tag{5.146}$$

Experimentally neither one has ever been observed, and the upper limits for the branching ratios are presently given as [29, 30]

$$R_3 = \frac{\Lambda(\mu \to e + \gamma)}{\Lambda(\mu \to e + \nu + \bar{\nu})} \lesssim 10^{-8}$$

$$R_4 = \frac{\Lambda(\mu \to 3e)}{\Lambda(\mu \to e + \nu + \bar{\nu})} \lesssim 1.5 \times 10^{-7}$$

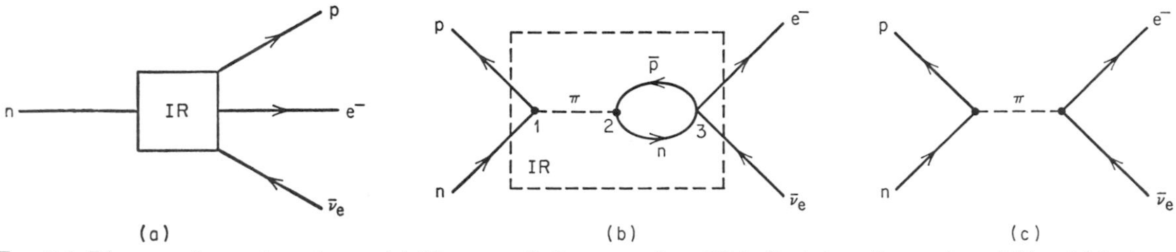

FIG. 5.4. Diagrams for neutron decay: (*a*) The over-all diagram, where IR is the interaction region within which strong interactions may occur. (*b*) An example of what may occur in the interaction region. Vertices 1 and 2 are, strictly speaking, not point interactions but contain processes involving exchanges of several strongly interacting particles. (*c*) This diagram summarizes a class of processes, of which (*b*) is one example, in which (p,n) is connected with $(e,\bar{\gamma}_e)$ by one pion exchange.

The fact that these processes are so rare did not receive a satisfactory explanation until the existence of two different kinds of neutrinos was established (see Sec. 8). With two kinds of neutrinos and correspondingly two lepton conservation laws these processes are absolutely forbidden [S8].

5. Nuclear Beta Decay

5.1. The Effective Interaction. As soon as nucleons are involved in any weak interaction process it is necessary to take into account the concomitant strong interactions. For example, the β^- decay of the neutron

$$n \rightarrow p + e^- + \bar{\nu}_e \tag{5.147a}$$

is represented by a diagram which is schematically described by Fig. 5.4*a*. What goes on in the interaction region (IR) may be very complicated, involving several strong interactions in addition to *one* weak interaction process. An example is shown in Fig. 5.4*b* where vertices 1 and 2 represent strong interaction and 3 represents a weak interaction. As a result, while the Hamiltonian‡ representing such decay processes is the same *V-A* interaction as in muon decay, the *S*-matrix element can have a radically different form, as will be evident from the following:

We consider a (symmetrical) process equivalent to (5.147*a*):

$$n + \nu_l \rightarrow p + l \tag{5.147b}$$

where l is a (negatively) charged lepton, e or μ, and ν_l is the corresponding kind of neutrino. The Hamiltonian is $H = H_\beta$ with

$$H_\beta = \frac{G}{\sqrt{2}} \int [J_\mu{}^+(x)j^\mu(x) + J_\mu(x)j^{\mu+}(x)]d^4x \tag{5.148}$$

and only the first term contributes to (5.147*b*). Also

$$\begin{aligned} J_\mu &= J_\mu{}^V + J_\mu{}^A \\ j_\mu &= j_\mu{}^V + j_\mu{}^A \end{aligned} \tag{5.148a}$$

with

$$\begin{aligned} j_\mu{}^V &= \bar{\psi}_l(x)\gamma_\mu\psi_v(x) \\ j_\mu{}^A &= \bar{\psi}_l(x)\gamma_\mu\gamma_5\psi_v(x) \end{aligned} \tag{5.148b}$$

‡ It will be convenient in this section to refer to the Hamiltonian H for the weak interaction. The connection with the S operator is

$$S = 1 - i\int_{-\infty}^{\infty} H\,dt$$

to the lowest order in H.

while $J_\mu{}^V$ and $J_\mu{}^A$ represent the vector and axial-vector parts of the nucleon current. The matrix element to be considered is

$$M = (pl|H_\beta|n\nu_l)$$

From translational invariance one can write

$$J_\mu{}^+(x)j^\mu(x) = e^{ipx}J_\mu{}^+(0)j^\mu(0)e^{-ipx}$$

where P is the total 4-momentum operator. Hence, neglecting small electromagnetic effects,

$$M = \frac{G}{\sqrt{2}}(2\pi)^4\delta(p + l - n - \nu) \\ <p|J_\mu{}^+(0)|n> \; <l|j^\mu(0)|\nu> \tag{5.149}$$

with p, l, n, and ν representing the corresponding 4-momenta.

The matrix element $\mathfrak{M} = <p|J_\mu{}^+(0)|n>$ is the factor that is affected by the strong interactions in a manner which is essentially unknown. However, the V and A parts of this matrix element must exhibit the same Lorentz transformation properties as are present in the interactions of bare particles. The most general form of $\mathfrak{M}$ is then [31]

$$\begin{aligned} \mathfrak{M} = \bar{u}(p)\{&F_1\gamma_\mu + F_2\sigma_{\mu\nu}q^\nu + F_3q_\mu + F_4\sigma_{\mu\nu}Q^\nu + F_5Q_\mu \\ &+ \gamma_5[G_1\gamma_\mu + G_2\sigma_{\mu\nu}q^\nu + G_3q_\mu + G_4\sigma_{\mu\nu}Q^\nu + G_5Q_\mu]\}u(n) \end{aligned} \tag{5.150}$$

where in place of p and n we have introduced $q = p - n$ and $Q = n + p$. The form factors F_n and G_n are invariant quantities and hence depend only on q^2, since $q^2 + Q^2 = 2(m_n{}^2 + m_p{}^2) \approx 4m^2$.

The 10 terms in (5.150) are not independent. In fact, four independent linear relations exist among them. Using the Dirac wave equation and the anti-commutation relations for the γ matrices (Appendix A), one finds

$$\begin{aligned} \bar{u}(p)[\sigma_{\mu\nu}q^\nu - iQ_\mu + 2im\gamma_\mu]u(n) &= 0 \\ \bar{u}(p)[\sigma_{\mu\nu}Q^\nu - iq_\mu]u(n) &= 0 \\ \bar{u}(p)[\gamma_5\sigma_{\mu\nu}q^\nu - i\gamma_5Q_\mu]u(n) &= 0 \\ \bar{u}(p)[\gamma_5\sigma_{\mu\nu}Q^\nu - i\gamma_5q_\mu - 2im\gamma_5\gamma_\mu]u(n) &= 0 \end{aligned}$$

Hence, only six independent terms remain. Therefore, in terms of a new set of form factors [31],

$$\begin{aligned} \mathfrak{M} = \bar{u}(p)[&f_1\gamma_\mu + if_2\sigma_{\mu\nu}q^\nu + if_3q_\mu \\ &+ (g_1\gamma_\mu + ig_2q_\mu + ig_3\sigma_{\mu\nu}q^\nu)\gamma_5]u(n) \end{aligned} \tag{5.151}$$

Under the well-justified assumption of time-reversal invariance of both the strong and the weak interactions, the form factors of f_n and g_n are real. The signs in (5.151) have been chosen to facilitate comparison with standard notation [S9].

The interpretation of the six terms in (5.151) is as follows: The f_1 and g_1 terms are the vector and axial-vector terms, respectively, with coefficients that are no longer necessarily equal. The f_2 term has the structure of a tensor interaction in the nuclear operator space. Its form is similar to the term that arises in the electromagnetic current from the anomalous magnetic moments of the neutron and proton. Hence, this term is referred to as *weak magnetism* [33] (discussed in Sec. 5.4). The g_2 term gives rise to the induced pseudoscalar interaction. Figure 5.4*c* gives the diagrammatic representation of the process primarily responsible for this interaction. Including the lepton matrix element, the corresponding term in M is

$$\begin{aligned} g_2\bar{u}_l[\gamma^\mu(1+\gamma_5)q_\mu]u_\nu\bar{u}(p)\gamma_5 u(n) \\ = -m_l g_2\bar{u}_l(1+\gamma_5)u_\nu\bar{u}(p)\gamma_5 u(n) \end{aligned} \quad (5.152)$$

where we have used Eq. (A.1) and $m_\nu = 0$. Because of the factor m_l this term makes a much larger contribution in mu capture (see Sec. 7) than in nuclear decay where it is completely negligible. In any case, the smaller momentum transfer in nuclear β transitions further minimizes the importance of all the terms in (5.151) which are explicitly proportional to q.

The terms with f_3 and g_3 are induced scalar and induced tensor interactions, respectively. With regard to these terms the transformation properties under the G parity are relevant [33]. From Sec. 3.4, it is seen that J^V is even under the G-parity transformation and J^A is odd. With the reasonable assumptions that (1) strong interactions conserve C and P separately and (2) that they conserve isospin, then only those weak interaction terms can be induced by the strong interactions that have the same G-parity transformation properties. It can be shown [33] that, under the assumptions made, the f_3 and g_3 terms in (5.151) have the wrong transformation properties and cannot be induced by the strong interactions. These so-called *second-class currents* should thus be absent,‡ so that four terms remain:

$$\mathfrak{M} = \bar{u}(p)[f_1\gamma_\mu + if_2\sigma_{\mu\nu}q^\nu + g_1\gamma_\mu\gamma_5 + ig_2\gamma_5 q_\mu]u(n) \quad (5.153)$$

For β decay only the first three terms need be considered. The weak magnetism term makes a small contribution at best but its role is of decisive importance for the interpretation of the vector interaction, as will be discussed in Sec. 5.4. For present purposes it is sufficient to set this term aside and consider an effective nuclear current

$$J_\mu = f_1\bar{\psi}_p\gamma_\mu(1+\lambda\gamma_5)\psi_n \quad (5.154)$$

‡ Because of the inevitable presence of electromagnetic interactions in the interaction region, one expects that the induced scalar and tensor terms will be present in the resulting weak interaction but that they will be small and extremely difficult to detect; in any case, it appears they produce no effects that cannot be attributed to other terms in the weak interactions involving the dressed particles (i.e., the particles subject to strong interactions.)

where the form factors have been evaluated at $q = 0$ and

$$\lambda = \frac{g_1(0)}{f_1(0)} \quad (5.155)$$

is the ratio of A to V effective coupling constants. The current (5.154) implies that the effective weak interaction density is

$$\mathcal{H}_{\text{eff}} = \frac{G_\beta}{\sqrt{2}}\bar{\psi}_p\gamma_\mu(1+\lambda\gamma_5)\psi_n\bar{\psi}_e\gamma^\mu(1+\gamma_5)\psi_\nu + \text{h.c.} \quad (5.156)$$

where $G_\beta = Gf_1(0)$. For complex nuclei the matrix element of $\mathcal{H}_{\text{eff}}$ is obtained from (5.156) by replacing $\bar{\psi}_p$ and ψ_n by $\bar{\psi}_f$ and ψ_i, where the latter are final- and initial-state nuclear wave functions, respectively.

5.2. Selection Rules: Forbiddenness Expansion. Nuclear β transitions are usefully classified according to angular-momentum and parity changes between the two nuclear states involved. To discuss them it is necessary to put $\mathcal{H}_{\text{eff}}$ in noncovariant form. Thus, for β^- decay,

$$\begin{aligned} \langle f|\mathcal{H}_{\text{eff}}^-|i\rangle = \frac{G_\beta}{\sqrt{2}}\langle f|\psi^\dagger_p\psi_n\psi^\dagger_e\varphi_\nu - \psi^\dagger_p\boldsymbol{\alpha}\psi_n\cdot\psi^\dagger_e\boldsymbol{\alpha}\varphi_\nu \\ -\lambda[\psi^\dagger_p\boldsymbol{\sigma}\psi_n\cdot\psi^\dagger_e\boldsymbol{\sigma}\varphi_\nu - \psi^\dagger_p\gamma_5\psi_n\psi^\dagger_e\varphi_\nu]|i\rangle \end{aligned} \quad (5.157)$$

where $\varphi_\nu = (1+\gamma_5)\psi_\nu = \gamma_5\varphi_\nu$. The matrix element in (5.157) includes the lepton space so that ψ_e and φ_ν are converted to the lepton wave function for which we use the same notation. For present purposes it is sufficient to use plane waves for the leptons, for example (disregarding irrelevant normalization factors), $\varphi_\nu = (1+\gamma_5)\exp(-i\mathbf{p}_\nu\cdot\mathbf{r})$, and

$$\begin{aligned} \psi_e^\dagger\omega\varphi_\nu &= u^\dagger_e\omega(1+\gamma_5)u_\nu e^{-i\mathbf{P}\cdot\mathbf{r}} \\ &\equiv L(\omega)e^{-i\mathbf{P}\cdot\mathbf{r}} \end{aligned} \quad (5.158)$$

where $\mathbf{P} = \mathbf{p}_e + \mathbf{p}_\nu = -\mathbf{P}_{\text{recoil}}$. We use the fact that $\mathbf{P}\cdot\mathbf{r} \ll 1$ for all nuclear dimensions. Considering the scalar‡ terms $\psi^\dagger_p\Omega\psi_n$, where $\Omega = 1$ or γ_5, one obtains after expanding the exponential in solid harmonics

$$\psi^\dagger_f\Omega\psi_i\psi^\dagger_e\omega\varphi_\nu \cong 4\pi L(\omega)\sum_{lm}\frac{i^{-l}\mathcal{Y}_l^{m*}(\mathbf{P})}{(2l+1)!!}\Psi^\dagger_f\Omega\mathcal{Y}_l^m(\mathbf{r})\Psi_i \quad (5.159)$$

Here $\mathcal{Y}_l^m(\mathbf{x}) = x^l\mathbf{Y}_l^m(\hat{x})$ is the solid harmonic. Similarly, for the vector terms with $\mathbf{\Omega} = \boldsymbol{\alpha}$ or $\boldsymbol{\sigma}$,

$$\Psi^\dagger_f\mathbf{\Omega}\psi_i\cdot\psi^\dagger_e\omega\varphi_\nu \cong 4\pi\mathbf{L}(\omega)\cdot\sum_{lm} i^{-l}\mathcal{Y}_e^{m*}(\mathbf{P})\Psi^\dagger_f\mathbf{\Omega}\mathcal{Y}_e^m(\mathbf{r})\Psi_i \quad (5.160)$$

In (5.160) it is convenient to introduce the irreducible tensors $T_{JL}(\mathbf{r},\mathbf{\Omega})$ (see Appendix B) according to

$$\mathbf{L}\cdot\mathbf{\Omega}\mathcal{Y}_l^m(\mathbf{r}) = \sum_{J\mu}(-)^\mu C(l|J;m,-\mu)L_\mu T_{Jl}^{m-\mu}(r,\Omega) \quad (5.161)$$

where

$$T_{Jl}^M(\mathbf{r},\mathbf{\Omega}) = \sum_m C(l|J;m,M-m)\mathcal{Y}_l^m(\mathbf{r})\Omega_{M-m}$$

Of course $J = l, l \pm 1$.

‡ Under three-dimensional rotations.

In (5.159) $\Omega\mathcal{Y}_l$ is a tensor of rank l and parity $\pi = (-)^l$ for $\Omega = 1$ and $\pi = (-)^{l+1}$ for $\Omega = \gamma_5$. In (5.161) the tensor rank is J and the parity is $\pi = (-)^l$ for $\mathbf{\Omega} = \boldsymbol{\sigma}$ and $(-)^{l+1}$ for $\mathbf{\Omega} = \boldsymbol{\alpha}$. The selection rules for transitions between nuclear states $J_i^{\pi_i} \to J_f^{\pi_f}$ is $\pi_i\pi_f\pi = 1$ and, for the matrix element of tensors of rank J, $\Delta(JJ_iJ_f)$, that is, J, J_i, and J_f must form a triangle. For the scalar term in (5.159) this result applies with $J = l$.

Since $\boldsymbol{\alpha}$ and γ_5 connect small to large components of the nuclear wave functions and hence introduce a factor of order $v_N/c \ll 1$, where v_N is the nucleon velocity, and with the term $\mathcal{Y}_l(\mathbf{r})$ there is associated a factor $(Pr)^l$, where Pr or $v_N/c \ll 1$, it follows that the most rapid transitions are obtained from $l = 0$ and $\Omega = 1$ in the V interaction and $\mathbf{\Omega} = \boldsymbol{\sigma}$ in the A interaction. These are the *allowed* transitions. The selection rules are:

$$\begin{array}{lll} \Delta(0J_iJ_f) & \pi_i = \pi_f & \textit{Fermi transitions} \\ \Delta(1J_iJ_f) & \pi_i = \pi_f & \textit{Gamow-Teller transitions} \end{array}$$

First forbidden transitions ($\pi_i = -\pi_f$) are obtained from the following tensors:

$$\begin{array}{ll} V\!: & \mathcal{Y}_1(\mathbf{r}),\ T_{10}(\mathbf{r},\boldsymbol{\alpha}) \\ A\!: & \gamma_5\mathcal{Y}_0(\mathbf{r}),\ T_{01}(\mathbf{r},\boldsymbol{\sigma}),\ T_{11}(\mathbf{r},\boldsymbol{\sigma}),\ T_{21}(\mathbf{r},\boldsymbol{\sigma}) \end{array}$$

Since there is only one second-rank tensor, transitions with $\Delta J = |J_i - J_f| = 2$, $\pi_i\pi_f = -1$ are referred to as *unique*. The selection rules and relevant tensors of nth-order forbidden transitions are summarized in Table 5.8. The second and third tensors in the row for $\Delta J = n$ are zero for $n = 0$. In addition to the tensors listed, the following tensors give matrix elements of nth order in the forbiddenness parameters but, because of ΔJ and parity selection rules, they constitute small corrections to transitions of forbiddenness order $n - 2$: $T_{n-1,n-1}(\mathbf{r},\boldsymbol{\alpha})$; $T_{n-2,n-1}(\mathbf{r},\boldsymbol{\alpha})$; $T_{n-1,n}(\mathbf{r},\boldsymbol{\sigma})$; $\gamma_5\mathcal{Y}_{n-1}(\mathbf{r})$. For $n = 1$ the only change is to add to Table 5.8 $\Delta J = n - 1 = 0$ in the selection rules and, as already noted, the last two tensors $T_{01}(\mathbf{r},\boldsymbol{\sigma})$ and $\gamma_5\mathcal{Y}_0$ become relevant. The first two tensors above vanish for $n = 1$.

TABLE 5.8. nth FORBIDDEN TRANSITIONS ($n \neq 1$)

ΔJ	$\pi_i\pi_f$	Relevant tensors
n	$(-)^n$	$\mathcal{Y}_n(\mathbf{M})$; $T_{n,\,n-1}(\mathbf{r},\boldsymbol{\alpha})$; $T_{nn}(\mathbf{r},\boldsymbol{\sigma})$
$n+1$	$(-)^n$	$T_{n+1,\,n}(\mathbf{r},\boldsymbol{\sigma})$

For allowed transitions, the case of greatest interest, one notes that for $\Delta J = 1$ only the A interaction contributes (pure Gamow-Teller transitions). For $J_i = J_f = 0$ only V contributes (pure Fermi transitions). In such cases only one nuclear matrix element enters in the results for transition rules; in a few cases these can be accurately evaluated without appeal to the details of nuclear models.

In writing the effective β interaction, for instance (5.156), we have omitted explicit reference to isospin operators. Evidently in β^- decay an $n \to p$ transition occurs; this is effected by a raising operator

$$t_+ = \tfrac{1}{2}(\tau_1 + i\tau_2)$$

acting on each nucleon in Ψ_i. In the Hermitian conjugate term this appears as $t_- = \tfrac{1}{2}(\tau_i - i\tau_2)$ acting on Ψ_f and thus effecting the $p \to n$ transition involved in β^+ decay. Here the isospin Pauli operators are such that τ_3 has eigenvalues $+1$ and -1 for proton and neutron isospin states. Taking these operators into account, the isospin selection rule in allowed transitions is $\Delta I = 0$ for Fermi transitions [S10]. Since the triangular rule $\Delta(1I_iI_f)$ holds, $I_i = I_f = 0$ is excluded. Gamow-Teller transitions are not quite as simple. For isobaric triads with mass number $4n + 2$ the transitions to the ground state of the daughter will be $|\Delta I| = 1$ but for ground-state transitions between mirror nuclei which are mixed Fermi and Gamow-Teller $\Delta I = 0(I_i = \tfrac{1}{2} \to I_f = \tfrac{1}{2})$. All these statements concerning isospin are relevant for light nuclei only.

5.3. Allowed Beta Transitions. Among other things we wish to discuss the manner in which P, C, and T symmetries are quantitatively investigated by means of allowed β transitions. The weak interaction Hamiltonian in (5.156) will be written as

$$\mathcal{H}'_{\text{eff}} = \frac{G_\beta}{\sqrt{2}}\{\bar{\psi}_p\gamma_\mu\psi_n\bar{\psi}_e\gamma^\mu(1 + \beta_V\gamma_5)\psi_\nu + \lambda\bar{\psi}_p\gamma_\mu\gamma_5\psi_n\bar{\psi}_e(1 + \beta_A\gamma_5)\psi_\nu\} + \text{h.c.} \quad (5.162)$$

with constants λ, $\beta_i = C'_i/C_i$ which, a priori, may be complex. Thus, for the present we ignore the two-component neutrino theory, which requires $\beta_V = \beta_A = 1$, and time-reversal invariance which requires β_V, β_A, and λ real. With the corresponding changes in (5.157) the time part of the V and the space part of the A couplings are considered without recoil effects; that is, p_eR and $p_\nu R \ll 1$, where R is the nuclear radius.

The transition rate for various types of experiments can then be obtained from a "master formula" [Eq. (5.163) below]. We shall consider only the following measurements:

1. Energy spectrum
2. Electron-neutrino correlation
3. Anisotropy of electrons from polarized nuclei
4. Electron polarization
5. Electron-neutrino correlation from polarized nuclei

The transition rate for the emission of $e^\mp$ with 4-momentum $(E_e,\mathbf{p}_e)$, of $\bar{\nu}_e$ (or ν_e) with 4-momentum $(E_\nu,\mathbf{p}_\nu)$, with the $e^\mp$ spin in the direction $\boldsymbol{\sigma}$ as measured in the $e^\mp$ rest system and with nuclear polarization $\mathbf{j}$ is (per $dE_e\,d\Omega_e\,d\Omega_\nu$) (cf. ref. 34)

$$\Lambda_\mp = \frac{G_\beta^2}{(2\pi)^5}2^{-1-\tau}F(\pm Z,E_e)S(E_e)|M_1|^2\xi \times \left\{1 + a_0\frac{\mathbf{p}_e\cdot\hat{\mathbf{p}}_\nu}{E_e} + \mathbf{j}\cdot\left[a_1\frac{\mathbf{p}_e}{E_e} + a_3\frac{\mathbf{p}_e\times\hat{\mathbf{p}}_\nu}{E_e}\right] + a_2\langle\boldsymbol{\sigma}\rangle\cdot\frac{\mathbf{p}_e}{E_e}\right\} \quad (5.163)$$

We put $m_e = 1$ as well as $\hbar = c = 1$ so that the rate is in units of $m_ec^2/\hbar$. In (5.163) $\tau = 0$ if the measurement sums over both spin states of the β particle and is otherwise 1. Also, here and in the following, the upper sign is for β^- decay, the lower for β^+ decay.

The Fermi nuclear matrix element M_1 is defined in (5.164) below, which also defines $\mathbf{M}_\sigma$, the Gamow-Teller matrix element. With

$$x = |\mathbf{M}_\sigma / M_1| \tag{5.163a}$$

we can write

$$\xi = 1 + |\beta_\nu|^2 + x^2|\lambda|^2(1 + |\beta_A|^2) \tag{5.163b}$$

The nuclear matrix elements, which are 3-vectors or scalars, are defined by

$$M_\Omega = \sum_{k=1}^{A} <f|\Omega(\mathbf{x}_k)t_\pm{}^k|i> \tag{5.164}$$

and for the vector operators, Ω is written as $\mathbf{\Omega}$ on the right-hand side. The Fermi function $F(\pm Z, E_e)$ contains the influence of the Coulomb field and is the sum of densities of $s_{1/2}$ and $p_{1/2}$ electrons at the nuclear surface. For a point nucleus this is [S11]

$$F(Z_1E_e) = 2(1 + \gamma)(2p_eR)^{2(\gamma-1)}e^{\pi y}\left|\frac{\Gamma(\gamma + iy)}{\Gamma(2\gamma + 1)}\right|^2 \tag{5.165}$$

where

$$\gamma = (1 - \alpha^2Z^2)^{1/2} \qquad y = \frac{\alpha Z E_e}{p_e} \qquad \alpha = \frac{e^2}{\hbar c} \approx \frac{1}{137}$$

Screening by the orbital electrons makes a modification of $F(\pm Z, E_e)$ which should be taken into account for precision results even for light nuclei [35]. Other than the density effect, we shall neglect the Coulomb field in what follows since the remaining changes produced by it are small and do not alter the sense of the present discussion [36]. The statistical factor $S(E_e)$ arises from the momentum space density:

$$\begin{aligned} S(E_e) &= p_e^2 \frac{dp_e}{dE_e} \int_0^{E_0-1} \delta(E_0 - E_e - E_\nu)q_\nu^2\, dq_\nu \\ &= p_eE_e(E_0 - E_e)^2 \end{aligned} \tag{5.166}$$

where E_0 is the energy release in the transition; that is, the maximum kinetic energy of $e^\mp$ is $E_0 - 1$.

The remaining coefficients $a_0 \cdots a_3$ are defined for a transition $J_i \rightarrow J_f$ between two nuclear states of the same parity. Dependence on isospin will be contained in the nuclear matrix elements only.

$$a_0\xi = 1 + |\beta_\nu|^2 - \tfrac{1}{3}\, x^2|\lambda|^2(1 + |\beta_A|^2) \tag{5.167a}$$

$$a_1\xi = 2\,\text{Re}\left[\mp \eta_{J_iJ_f}x^2|\lambda|^2 + \delta_{J_iJ_f}\left(\frac{J_i}{J_i+1}\right)^{1/2} x\lambda(\beta_A + \beta^*{}_V)\right] \tag{5.167b}$$

$$a_2\xi = \mp 2\,\text{Re}\,[\beta_V + x^2|\lambda|^2\beta_A] \tag{5.167c}$$

$$a_3\xi = -2\delta_{J_iJ_f}\left(\frac{J_i}{J_i+1}\right)^{1/2} x\,\text{Im}\,\lambda(1 + \beta^*{}_V\beta_A) \tag{5.167d}$$

$$\begin{aligned} \eta_{J_iJ_f} &= 1 && \text{for } J_f = J_i - 1 \\ &= (J_i + 1)^{-1} && \text{for } J_f = J_i \\ &= -\frac{J_i}{J_i+1} && \text{for } J_f = J_i + 1 \end{aligned}$$

a. Energy Spectrum. The distribution in E_e is obtained by integrating over $d\Omega_e$ and $d\Omega_\nu$. This removes the terms in a_n ($n = 0, 1, 2, 3$). The spectrum is then

$$\bar{\Lambda}_\mp(E_e) = \frac{1}{4\pi^3}G_\beta{}^2\xi|M_1|^2F(\mp Z, E_e)S(E_e) \tag{5.168}$$

The allowed shape predicted by Eq. (5.168) is tested by plotting

$$C_\mp(E_e) = \left[\frac{(\bar{\Lambda}_\mp)_{\text{obs}}}{F(\mp Z, E_e)p_eE_e}\right]^{1/2}$$

where $(\bar{\Lambda}_\mp)_{\text{obs}}$ is the observed counting rate as a function of E_e. A straight line (Kurie plot) checks (5.168), and the intercept on the E_e axis determines E_0 which can be compared to the energy release in $n - p$ or $p - n$ nuclear reactions. Experimental results on shapes are in complete agreement with expectations.

The reciprocal mean life $1/\tau$ is

$$\frac{1}{\tau} = \int_1^{E_0} \bar{\Lambda}_\mp(E_e)\, dE_e$$

Defining, [37],

$$f \equiv f(Z,E_e) = \int_1^{E_0} F(\mp Z,E_e)S(E_e)\, dE_e \tag{5.169}$$

it follows that

$$ft_{1/2} = \frac{4\pi^3 \log 2}{G_\beta{}^2\xi|M_1|^2}$$

where $t_{1/2}$ is the half-life, so that $\xi|M_1|^2ft_{1/2}$ should be a universal constant. For $\beta_A = \beta_V = 1$ this becomes

$$ft_{1/2}[|M_1|^2 + \lambda^2|M_\sigma|^2] = \frac{2\pi^3 \log 2}{G_\beta{}^2} \tag{5.170}$$

Therefore, for a group of transitions between nuclear states of similar structure, the $ft_{1/2}$ values should be very nearly the same. If the observed $ft_{1/2}$ values for allowed transitions are examined [38], it is seen that they fall into two fairly distinct groups. The first may be referred to as normal allowed and is characterized by log $ft_{1/2}$ in the range 4.5 to 6.0. The second group, comprising the so-called superallowed transitions, shows values of log $ft_{1/2}$ closely clustered around 3.5. An example is the neutron decay and other mirror transitions in light nuclei. The well-established charge symmetry of the nuclear forces would lead one to expect strong overlap of the initial- and final-state nuclear wave functions in these cases and hence favored transition probabilities in agreement with observations. Another important class of superallowed transitions are those between analogue states in mass $4n + 2$ triads, and again the favored position of these decays is understandable. While the $ft_{1/2}$ value alone is not an infallible criterion on which to base a forbiddenness classification, except for the superallowed transitions, this number together with a knowledge of the parity change usually leads to a unique decision. Thus if $\pi_i\pi_f = 1$ the ambiguity can be between only two orders of forbiddenness which differ by $\Delta n = 2$ ($n = 0$ and $n = 2$, say). Since the $ft_{1/2}$ values for normal allowed and second forbidden transitions differ by a factor of order 10^7, no confusion should arise between the two possibilities. Spectral shapes are also quite different, in general, for transitions differing by $\Delta n = 2$.

Transitions between analogue states in the $4n + 2$ nuclei are pure Fermi transitions. The $ft_{1/2}$ values listed for these positron emitters in Table 5.9 bear out, in a striking way, the validity of the foregoing remarks.

TABLE 5.9. $ft_{1/2}$ VALUES OF SOME ACCURATELY MEASURED ALLOWED FERMI TRANSITIONS‡

Transition	$t_{1/2}$(sec)	E_0(Mev)	$ft_{1/2}$	Ref.
$O^{14} \rightarrow N^{14*}$	71.36	2.324	3127	39, 40
$Al^{26*} \rightarrow Mg^{26}$	6.374	3.719	3086	40, S12, 41
$Cl^{34} \rightarrow S^{34}$	1.565	4.971	3140	40, S12, 41
$V^{46} \rightarrow Ti^{46}$	2.424	6.552	3138	42, 43
$Co^{54} \rightarrow Fe^{54}$	0.1937	7.740	3134	41, S12, 43

‡ Screening corrections made according to ref. 35. Finite nuclear size effect [44] and radiative corrections are included [25].

From these decays one can obtain the coupling constant G_β if the nuclear matrix element M_1 can be evaluated with reliability. This is indeed the case in this instance. In evaluating $|M_1|^2$ one notes that $I_i = I_f = 1$. For example,

$$M_1 = \left\langle f \left| \sum_{k=1}^{A} t^k \right| i \right\rangle = \langle fI_3 = 0|I_-|iI_3 = 1\rangle$$

$$= [(I + I_3)(I - I_3 + 1)]^{1/2} = \sqrt{2} \qquad (5.171)$$

provided that the nuclear forces are charge-independent (isospin is conserved). From the O^{14} decay results, or alternatively an average of all the decays listed in Table 5.9 [43], we obtain (in cgs units)

$$G_\beta = (1.4029 \pm 0.0022) \times 10^{-49} \text{ erg cm}^3 \qquad (5.172)$$

Comparing this with the coupling constant derived from μ decay (Sec. 4.3),

$$G_\mu = (1.4340 \pm 0.0011) \times 10^{-49} \text{ erg cm}^3,$$

wherein radiative corrections have been included [25], one is impressed with the support given to the hypothesis of a *universal Fermi interaction*. The discrepancy of 2.2 per cent has been discussed extensively in the literature [S9, S13]. It seems quite certain that the discrepancy is well within the uncertainty in the analysis of the radiative corrections.

For the neutron decay one has $|M_1|^2 = 1$, $|\mathbf{M}_\sigma|^2 = 3$, and $|\lambda|^2$ can be determined from the appropriate $ft_{1/2}$ value and the value of G_β from the O^{14} decay. The result [45] is $|\lambda| = 1.20 \pm 0.03$.

b. Electron-Neutrino Correlation. Integrating (5.163) over all angles except that between $\mathbf{p}_e$ and $\mathbf{p}_\nu$, one finds for the appropriate transition rate

$$\Lambda_0 = \tfrac{1}{2}\bar{\Lambda}_\mp(E_e)\left(1 + a_0 \frac{\mathbf{p}_e \cdot \hat{\mathbf{p}}_\nu}{E_e}\right) \qquad (5.173)$$

The quantity a_0 [cf (5.167*a*)] when plotted versus $y_0 = (1 + |\beta_v|^2)$ gives a straight line

$$a_0 = \tfrac{1}{3}(4y_0 - 1) \qquad (5.174)$$

for the V, A combination of interaction. For an ST combination the straight line would be $3a_0 = 1 - 4y_0$, with an obvious definition of y_0 ($V \rightarrow S$, $A \rightarrow T$). In such cases where reliable calculations of x are available a plot of a_0 versus y_0 can be made (Scott diagram [S1]). The data clearly favor the VA interaction. Obviously in the experiments one measures the recoil direction rather than $\mathbf{p}_\nu$.

c. Anisotropy of Electrons from Polarized Nuclei. The appropriate transition rate is obtained from (5.163) by integrating over $\hat{\mathbf{p}}_\nu$, and summing over spin directions:

$$\Lambda_1 = \tfrac{1}{2}\bar{\Lambda}_\mp(E_e)[1 + a_1\mathbf{j} \cdot \mathbf{p}_e/E_e] \qquad (5.175)$$

As is evident from (5.167*b*) and from the P, C, T transformation properties of the kinematic factor $\mathbf{j} \cdot \mathbf{p}_e$, the presence of the term in a_1 attests to a violation of P and C symmetry. This experiment was first carried out by Wu et al. [6] on polarized Co^{60} ($J_i = 5$, $J_f = 4$, $\pi_i = \pi_f = 1$). The result that there was indeed a large anisotropy with β^- preferentially emitted antiparallel to $\mathbf{j}$ (also β^+ from Co^{58} emitted preferentially parallel to $\mathbf{j}$) was the first evidence of P, C breakdown in nuclear β decay. Since the Co^{60} transition is pure Gamow-Teller, the result

$$a_1 = -2 \operatorname{Re} \frac{\beta_A}{1 + |\beta_A|^2} \qquad (5.176)$$

applies. Experimentally $a_1 = -1$, indicating that β_A is real. Moreover this corroborates the two-component neutrino theory since $\beta_A = 1$ is implied.

The relative sign of the V and A coupling coefficients, that is, the sign of λ, can be deduced from a measurement involving an interference between a Fermi and a Gamow-Teller interaction. In allowed transitions this implies an interference between M_1 and M_σ, which are matrix elements of tensors of rank 0 and 1. In general, it is easy to see that such interference terms between tensors of different rank (say L and L') can occur only if an observation of the orientation of the initial or final nuclear state is made. This is what is done when the initial state is polarized; in that case the following rules apply:

$$\Delta(J_iJ_fL) \qquad \Delta(J_iJ_fL') \qquad \Delta(LL'1)$$

and $L + L' + 1 =$ even. These rules are fulfilled in the present case where $L, L' = 0, 1$. In order to eliminate the uncertainties of nuclear structure the most appropriate case to consider is the neutron decay ($J_i = J_f = \tfrac{1}{2}$). The anisotropy of β^- from polarized neutrons is measured. In this case

$$a_1\xi = 2 \operatorname{Re}\left[-\frac{2}{3}x^2|\lambda|^2\beta_A + \frac{x\lambda}{\sqrt{3}}(\beta_A + \beta^*_V)\right] \qquad (5.177a)$$

or, with $\beta_A = \beta_V = 1$ and $x = \sqrt{3}$,

$$a_1 = 2 \operatorname{Re} \frac{(\lambda - |\lambda|^2)}{1 + 3|\lambda|^2} \qquad (5.177b)$$

The experimental value [46] of a_1 is -0.09 ± 0.03. Using $|\lambda| = 1.20$, (cf. Sec. 5.3*a*), this gives

$$\operatorname{Re} \lambda = 1.20 \pm 0.08$$

The fact that $\lambda > 0$ confirms the V-A interaction [cf. (5.157)]. Writing $\lambda = |\lambda|e^{i\varphi}$ gives

$$\cos \varphi = 1.00 \pm 0.03$$

which supports the results on time-reversal measurements (cf. Sec. 5.3*e*) below that all the coupling constants in (5.162) are real.

Another type of experiment in which Fermi and Gamow-Teller interference can occur (with $J_i = J_f$) is the observation of circularly polarized γ rays in coincidence with a preceding β transition. The existence of this effect, now firmly established [17], is evidence for P and C breakdown. However, the data are more useful for measuring x than for determining λ.

d. Electron Polarization. The only anisotropic term entering is the a_2 term in (5.163), and the appropriate rate is

$$\Lambda_2 = \tfrac{1}{2}\bar{\Lambda}_\mp(E_e)[1 + a_2 <\mathbf{\sigma}> \cdot \mathbf{p}_e/E_e] \quad (5.178)$$

The measurements [47,S14] agree with expectations in giving, quite accurately, polarizations in $\beta^\mp$ decay equal to

$$\mathcal{P}_\mp = \mp v_e/c \quad (5.179)$$

This implies

$$2 \text{ Re } [\beta_V + x^2|\lambda|^2\beta_A] = \xi \quad (5.180)$$

which is again possible only for $\beta_V = \beta_A = 1$. The observation of a polarization is again a testimony to P and C breakdown.

Another important experiment of somewhat similar character was performed by Goldhaber, Grodzins, and Sunyar [18] who succeeded in measuring the ν_e helicity. This was done using a Eu^{152} source. The initial state is 0^-, and K capture takes place to the 1^- excited state of Sm^{152}, followed by the emission of a γ ray to the 0^+ ground state. The recoil energy in the K capture is just sufficient to provide a Doppler shift of the γ ray so that resonance scattering of this γ ray in a Sm^{152} target is possible. Denoting the projections of angular momentum of electron, neutrino, and γ ray with respect to the γ-ray direction by m_e, m_ν and m_γ, we have

$$m_e = m_\nu + m_\gamma \qquad m_\gamma = \pm 1$$

since the over-all β-γ transition is $0 \to 0$. Since m_e and $m_\nu = \tfrac{1}{2}$ the following possibilities occur:

m_γ	m_e	m_ν	γ polarization
1	$\tfrac{1}{2}$	$-\tfrac{1}{2}$	right-handed
-1	$-\tfrac{1}{2}$	$\tfrac{1}{2}$	left-handed

Since $\mathbf{p}_\gamma$ and $\mathbf{p}_\nu$ are in opposite directions, a measurement of the γ-ray helicity, which is the same as the neutrino helicity, gives the desired information. This measurement was made by transmission of the γ ray through magnetized iron from which it is established that the neutrino is left-handed, in agreement with the V, A theory and the form of the two-component neutrino theory described here.

e. Tests of Time-reversal Symmetry. In the absence of external fields (Coulomb) any measurement indicating the presence of a term such as $\mathbf{q}_1 \cdot (\mathbf{q}_2 \times \mathbf{q}_3)$, where the $\mathbf{q}_i$ are either linear or angular momenta, would constitute evidence for a breakdown of time-reversal symmetry (cf. Sec. 3). The simplest experiment of this type, and the only one we shall mention in any detail here, is the electron-neutrino correlation from polarized nuclei. The neutron decay is again especially appealing because x is known and the source is easily polarized to a large degree. The relevant term in the transition rate now involves a_3 which for neutron decay becomes

$$a_3 = -\frac{2}{\xi} \text{ Im } (1 + \beta^*_V\beta_A) \quad (5.181)$$

In the experiment [19], one measures the difference in counting rates in two arrangements with the neutron polarization, electron, and recoil momenta all mutually perpendicular but with the recoil momentum reversed in sign. The result $a_3 = 0.04 \pm 0.07$, that is, a_3 essentially zero, indicates invariance under time reversal.

Similar conclusions were obtained by Ambler et al. [48], who measured the β-γ polarization from polarized Mn^{52}. The three vectors are $\mathbf{j}$, $\mathbf{p}_e$, and $\mathbf{k}$, the photon momentum. The results are consistent with time-reversal invariance, but uncertainty in the nuclear matrix element ratio x precluded an accurate measurement of the phase angle φ in the coupling constant ratio λ.

5.4. The Conserved Vector Current Hypothesis. The near equality of the coupling constants G_β and G_μ, discussed in Sec. 5.3, would seem to be a most gratifying justification of the assumption of a universal Fermi interaction. The small discrepancy is well within the present uncertainties of the radiative corrections, as discussed previously. However, this coincidence in the coupling constants poses a problem. The universal Fermi interaction requires the bare coupling constants for the various processes to be the same (cf. Sec. 2). In the case of muon decay the modifications of the coupling constant are negligible, as the muon only participates in weak and electromagnetic interactions. This means that $G_\mu \cong G$, where G is the universal weak coupling constant. In the case of β decay the situation is radically different, as the nucleons also participate in strong interactions. One would expect these to have a substantial influence on the effective coupling constant. The presence of strong interactions in this case has been taken care of by the introduction of form factors (cf. Sec. 5.1), and we have defined the effective vector coupling constant G_β by

$$G_\beta = Gf_1(0) \quad (5.182)$$

A priori, there is no reason why $f_1(0)$ should be equal to unity. However, within the experimental and theoretical uncertainties, $G_\beta = G_\mu = G$ and consequently $f_1(0) = 1$. This experimental result would be a very remarkable coincidence unless it is a consequence of some fundamental property of the interaction. In the search for a possible explanation along these lines it was noted [49] that an entirely analogous situation with regard to the charge, or more precisely the charge form factor, occurs in electromagnetic processes. As we have a clearer understanding of electromagnetic phenomena, it is instructive to consider the situation there first. In order to do so, an operational definition of the concept of charge is needed. One possible way of determining the charge of a particle, say an electron, is to use it as a target particle for the scattering of low-energy photons

(Compton scattering) [50]. In the rest system of the target particle and in the limit of zero momentum transfer the total cross section is given by the Thomson formula

$$\sigma = \frac{8\pi}{3}\left(\frac{e^2}{m_e c^2}\right)^2 \tag{5.183}$$

The quantity e in this formula is, by definition, the charge of the particle. A diagram describing the process in the case of photon scattering on electrons is given in Fig. 5.5.

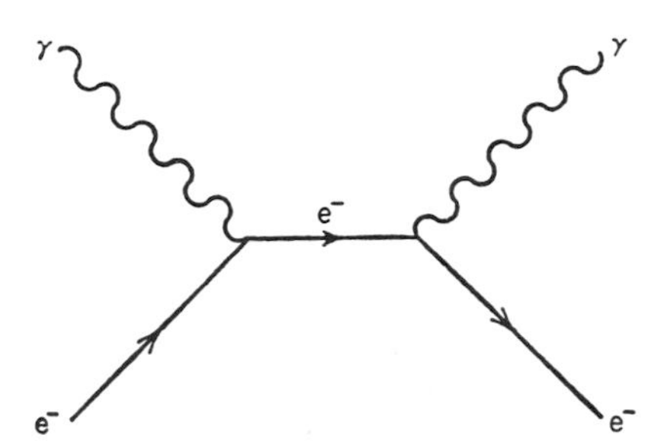

Fig. 5.5. Diagram describing electron-photon scattering.

Nucleons are subject to strong interactions with π mesons and other mesons, and virtual emission and absorption processes like

$$p \rightleftarrows p + \pi^0 \rightleftarrows n + \pi^+ \tag{5.184}$$

must occur. As a consequence the nucleons are surrounded by a cloud of mesons bound to the nucleon because of energy-momentum conservation (a cloud of virtual particles). The extension of the cloud is determined by the uncertainty relation and is of the order of the corresponding Compton wavelength (m_π^{-1}). This means that the most peripheral part of the cloud consists of virtual π mesons, as the π-meson mass is lower than the masses of other mesons. The fact that a nucleon is surrounded by a meson cloud leads to an extended charge structure of it, and one would expect that the Thomson formula would yield a different value for the proton charge than in the case of an electron, even if the charges were the same to start with. Just as in Sec. 5.1, one would obtain the modifications due to strong interactions by introducing form factors. In the case of Compton scattering there is a charge contribution to the cross section where the charge is multiplied by a charge form factor describing the charge distribution within the proton. In the limit of low-energy photons and zero momentum transfer, only this term in the cross section contributes, and one obtains the modified Thomson formula

$$\sigma = \frac{8\pi}{3}\left(\frac{e^2 f_Q{}^2(0)}{m_p c^2}\right)^2 \tag{5.185}$$

and the proton charge would be $ef_Q(0)$. Figure 5.6*a* is a diagram describing this process. The diagrams of Figs. 5.6*b* and *c* are special cases, indicating possible mechanisms for the scattering process. Diagram 5.6*c* is an example of new possible mechanisms that can occur because of the strong interactions. Because the charge of a particle is defined by means of low-energy (long wavelength compared with the meson cloud extension) photons, the actual distribution of charge is of no importance and only the total charge is significant. Furthermore, charge conservation requires the total charge of the proton with its cloud to be the same as that of a proton in the absence of virtual meson emission and absorption, i.e., in the absence of strong interactions. This is a restriction on the possible states in which the proton may virtually exist, and it has been explicitly recognized in Eq. (5.184). As the meson cloud extends only about 10^{-13} cm from the nucleon, the proton charge as defined by Thomson scattering must remain equal to e, also with strong interactions present. This can be reconciled with Eq. (5.185) only if $f(0) = 1$. This condition on the form factor is, therefore, an immediate consequence of charge conservation for the strong interactions.‡ But charge conservation merely reflects the fact that the total electromagnetic current $j_\mu(x)$ is conserved and satisfies the continuity equation

$$\frac{\partial j_\mu(x)}{\partial x_\mu} = 0 \tag{5.186}$$

For noninteracting nucleons the electromagnetic current is given by

$$j_\mu{}^N(x) = \frac{1}{2}\left[\bar{\psi}_N(x), \gamma_\mu \frac{1+\tau_3}{2}\psi_N(x)\right] \tag{5.187}$$

‡ This kind of argument must be used with care. It is not implied here that there is no charge renromalization in the theory (cf. [S2]); that is a purely electromagnetic effect, however, and it effects all charged particles in the same way. The arguments above are restricted to the short-range strong forces where they apply.

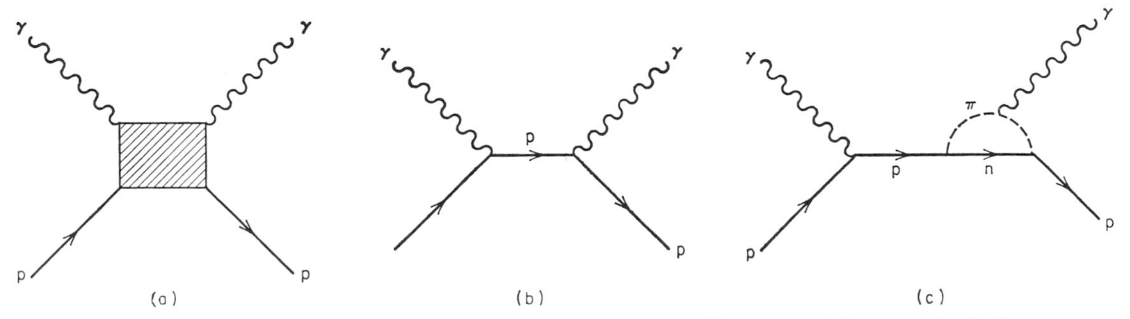

Fig. 5.6. Diagrams for scattering of photons on protons. (*a*) The over-all diagram. (*b*) An example of what may occur in the crosshatched region of diagram *a*. (*c*) An example of possible diagrams that occur as a consequence of the strong interactions in which the proton may participate.

By virtue of the free Dirac equation (A.1), that is, the Dirac equation without any source terms, this current is conserved. In the presence of a strong coupling‡ to the π mesons the field equations for the nucleons and the π mesons are coupled, and the Dirac equation describing the propagation of the nucleon field acquires a source term

$$\left(i\gamma^\mu \frac{\partial}{\partial x^\mu} - m\right) \psi_N(x) = g_{\pi N}\gamma_5(\boldsymbol{\tau} \cdot \boldsymbol{\phi}(x))\psi_N(x) \quad (5.188)$$

Because of the source term the nucleon current (5.187) is no longer conserved. However, it is readily shown that the field equations now require the total nucleon and π-meson current

$$j_\mu(x) = \left[\bar{\psi}_N(x), \gamma_\mu \frac{1+\tau_3}{2} \psi_N(x)\right] + i\left[\boldsymbol{\phi}(x) \times \frac{\partial \boldsymbol{\phi}(x)}{\partial x^\mu}\right]_3 \quad (5.189)$$

to be conserved. In (5.189) the conventional isospin notation has been used (cf. Sec. 3). If the coupling to other meson fields is taken into account in the same way, additional contributions to the electromagnetic-current operator must be included in order to get a conserved current. Therefore, the current satisfying Eq. (5.186) receives contributions from many sources, and it can be written

$$j_\mu^{EM}(x) = j_\mu^N(x) + j_\mu^\pi(x) + \cdots \quad (5.190)$$

It is now clear that, if the weak-interaction vector current satisfies a continuity equation analogous to (5.186), then the "weak charge" is unrenormalized by the strong interactions yielding $f_1(0)$ of Eq. (5.182) equal to unity. This is the conserved vector current (CVC) hypothesis put forward by Gershtein and Zeldovich [51] and independently by Feynman and Gell-Mann [49]. The latter actually went a step further, exploiting the close relationship between the weak vector current and the electromagnetic current (discussed below). We first note, however, that only the vector part of the weak current may be conserved. Already the fact that $\lambda \neq 1$ in nuclear β decay implies that the axial vector current is not conserved. Further and stronger evidence is obtained from π-meson decay (discussed in Sec. 6.1).

a. Formulation of the CVC Theory. The nucleon vector current in β decay bears a remarkable similarity to the corresponding part of the electromagnetic current. The latter can be written

$$\mathcal{J}_\mu^N(x) = \mathcal{J}_\mu^S(x) + \mathcal{J}_\mu^V(x) \quad (5.191)$$

where $\mathcal{J}_\mu^S(x)$ and $\mathcal{J}_\mu^V(x)$ are the isoscalar and the isovector parts, respectively. From Eq. (5.187) it follows that

$$\mathcal{J}_\mu^S(x) = \tfrac{1}{4}[\bar{\psi}_N, \gamma_\mu \psi_N] \quad (5.191a)$$

and

$$\mathcal{J}_\mu^V(x) = \tfrac{1}{4}[\bar{\psi}_N, \gamma_\mu \tau_3 \psi_N] \quad (5.191b)$$

On the other hand, the vector current in β decay is given by

$$\mathcal{J}_\mu(x) = \bar{\psi}_n(x)\gamma_\mu\psi_p(x) = \frac{1}{2\sqrt{2}}[\bar{\psi}_N, \gamma_\mu \tau_- \psi_N] \quad (5.192)$$

‡ For the present discussion we neglect electromagnetic and weak interactions, which in principle also give rise to source terms in the Dirac equation.

The current has been written in terms of a commutator to facilitate the comparison with the electromagnetic case. Comparing Eqs. (5.191*b*) and (5.192) it is seen that apart from a trivial numerical factor the two currents $\mathcal{J}_\mu^V(x)$ and $\mathcal{J}_\mu(x)$ are different spherical components of the same isovector current

$$\boldsymbol{\mathcal{J}}_\mu^V(x) = \tfrac{1}{4}[\bar{\psi}_N, \gamma_\mu \boldsymbol{\tau} \psi_N] \quad (5.193)$$

The nucleon contribution to the electromagnetic current is the third component of this current. This component is not conserved by itself, as was noted before. However, the isoscalar part given in Eq. (5.191*a*) is conserved as a consequence of baryon conservation. Therefore, to obtain a conserved current we must add to (5.191*b*) the contributions from π mesons and other mesons, and we conclude that the third component of the following isovector current is in fact conserved:

$$\boldsymbol{\mathcal{J}}_\mu^V(x) = \tfrac{1}{4}[\bar{\psi}_N, \gamma_\mu \boldsymbol{\tau} \psi_N] + i\boldsymbol{\phi}(x) \times \frac{\partial \boldsymbol{\phi}(x)}{\partial x^\mu} + \cdots \quad (5.194)$$

where the vector notation refers to isospace. Under the well-founded assumption of invariance of the strong interactions in isospace, there is no preferred direction in isospace, however, and we may perform rotations freely. An immediate consequence of this is that, if the third component is conserved with a particular choice of axis in isospace, then also the other components must be conserved, i.e.,

$$\frac{\partial \boldsymbol{\mathcal{J}}_\mu^V(x)}{\partial x_\mu} = 0 \quad (5.195)$$

and in particular the spherical component $\mathcal{J}_{\mu,-}^V$ is conserved. By postulating that the vector part of the weak-interaction current is proportional to this component of the isovector current, we are then assured that it is conserved. More precisely, we postulate that

$$\mathcal{J}_\mu(x) = \sqrt{2}\, \mathcal{J}_{\mu,-}^V(x) \quad (5.196)$$

With this relation one obtains the same nucleon current as before [Eq. (5.192)]. However, there are additional contributions from the π-meson part of $\boldsymbol{\mathcal{J}}_\mu^V(x)$. In terms of diagrams, this means that besides the conventional diagram for nuclear β^- decay given in Fig. 5.7*a* there are other diagrams of which Figs. 5.7*b* and *c* are examples.

The close relationship between the electromagnetic current and the vector current in β decay can be used to relate the matrix element of the isovector electromagnetic-current operator to the corresponding matrix element for the weak vector current in β decay. In the electromagnetic case the matrix element is given by [S15] (apart from some trivial factors)

$$\mathfrak{M}_V^{EM} = \bar{u}_p \left[f_Q(q^2)\gamma_\mu + i\frac{\mu'_p - \mu'_n}{2M} f_M(q^2)\sigma_{\mu\nu}q^\nu \right] u_p \quad (5.197)$$

Here $\mu'_p = 1.79$ and $\mu'_n = -1.91$ are the anomalous magnetic moments (in units of $e\hbar/2Mc$) of the proton and neutron, respectively. It has earlier been found that $f_Q(0) = 1$; similarly, with the usual definition of anomalous magnetic moments, one finds that also

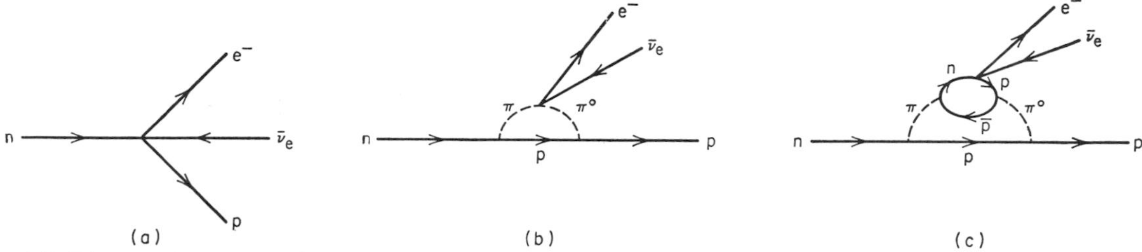

FIG. 5.7. Diagrams for neutron decay. (*a*) The diagram describing the process if the effects of strong interactions are neglected. (*b*) and (*c*) are examples of additional diagrams accounted for within the CVC theory.

$f_M(0) = 1$. The matrix element relevant for nuclear β decay has previously been given [cf. Eq. (5.151)]

$$\mathfrak{M}_V{}^\beta = \bar{u}_n[f_1(q^2)\gamma_\mu + if_2(q^2)\sigma_{\mu\nu}q^\nu]u_p \quad (5.198)$$

In the CVC theory, (5.197) and (5.198) are matrix elements of the same isovector current, and the form factors are, therefore, the same‡ so that

$$f_Q(q^2) = f_1(q^2) \quad (5.199a)$$

$$\frac{\mu'_p - \mu'_n}{2M} f_M(q^2) = f_2(q^2) \quad (5.199b)$$

and consequently

$$f_2(0) = \frac{\mu'_p - \mu'_n}{2M} \quad (5.200)$$

The term with $f_2(q^2)$ in the β-decay matrix element is the weak-interaction analogue of the contributions to the electromagnetic matrix element from the anomalous magnetic moments; hence the name *weak magnetism* is appropriate.

b. Tests of the CVC Theory. Perhaps the most clear-cut test of the *CVC* theory is the comparison of the shapes of the N^{12} and B^{12} β spectra [52]. Figure 5.8 shows the appropriate energy levels and transitions. The $I = 1$, 1^+ levels of the three members of the mass triad would be degenerate in the absence of electromagnetic effects. The weak magnetism term contributes, in addition, to the ordinary and dominant A interaction (since $\sigma_{12} \sim \sigma_3$, for example). The comparatively large values of E_0 in these β transitions make them favorable for observing a change in the spectrum as compared with the standard spectrum without CVC [cf. (5.168)].‡ The Ml γ transition $(1^+ \to 0^+)$ in C^{12} arises from the isovector part of the electromagnetic current and is of the same form as the second forbidden correction from the V interaction to the vector decay.§ Hence, the measured γ-ray width allows one to determine the transition magnetic moment $\mu(= 2.2)$.¶

The spectrum is given by [52]

$$\bar{\Lambda}_\mp(E_e)_{\text{CVC}} = \bar{\Lambda}_\mp(E_e)(1 \pm \tfrac{8}{3}aE_e)$$

where $\bar{\Lambda}_\mp(E_e)$ is given by (5.168) and

$$a = \frac{\mu}{\sqrt{2}\,M}\frac{\lambda}{M_\sigma}$$

when the theoretical uncertainties in the quantities involved are taken into account, the prediction is

‡ This can be seen by performing an isorotation.

‡ A number of other conditions that must be fulfilled to provide an adequate test of the CVC theory are uniquely satisfied by these transitions (see ref. S16).

§ This second forbidden correction arises from $T_{11}(\mathbf{r},\boldsymbol{\alpha}) \sim i\boldsymbol{\alpha} \times \mathbf{r}$.

¶ This transition moment is given by

$$\mu = \frac{1}{\sqrt{2}}(\mu_p - \mu_n)M_\sigma$$

and from the empirical value $M_\sigma \cong 0.8$ one obtains $\mu_p - \mu_n = 1 + \mu'_p - \mu'_n \cong 4.0$ which agrees fairly well with the expected value 4.7.

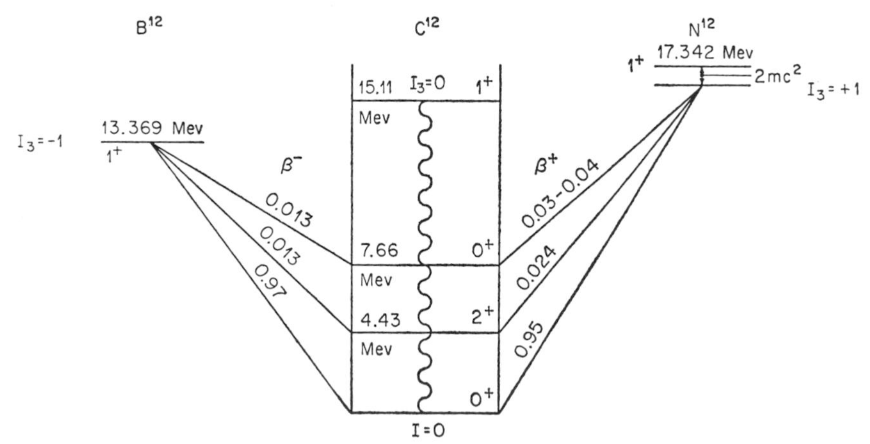

FIG. 5.8. Decay scheme for the mass 12 triad, showing the β and γ transitions from the three analogue 1^+ states to the ground state of $C.^{12}$

$a = (0.55 \pm 0.12)$ per cent per Mev. The experimental spectra show straight lines for $(\bar{\Lambda}_{\mp})_{CVC}/\bar{\Lambda}_{\mp}$, and the experimental slopes [53] which have the predicted sign in each case gives $a(B^{12}) = (0.55 \pm 0.10)$ per cent per Mev and $a(N^{12}) = (0.52 \pm 0.06)$ per cent per Mev, in excellent agreement with the CVC theory. A similar correction to the spectrum is expected in the theory without CVC (due to second-order corrections) but the a in this case is only 10 per cent of the observed values.

Further tests of the CVC theory can be obtained from the small anisotropy in the β-α angular correlation in the Li^8 and B^8 β decays but it is not possible at present to draw definite conclusions from the results. The same remark applies to the anisotropy in the β-γ circular-polarization correlation measurement in the Na^{24} and Al^{24} decays to Mg^{24} [54]. Additional evidence bearing on CVC is obtainable from the $\pi^+ \rightarrow \pi^0 + e^+ + \nu_e$ decay (see Sec. 6.2). All evidence either strongly favors or is at least consistent with the CVC theory.

6. Weak Decays of the π Meson

The two charged π mesons decay predominantly through the channels

$$\pi^+ \rightarrow \mu^+ + \nu_\mu \qquad (5.201a)$$
$$\pi^- \rightarrow \mu^- + \bar{\nu}_\mu \qquad (5.201b)$$

The analogous decays

$$\pi^+ \rightarrow e^+ + \nu_e \qquad (5.201c)$$
$$\pi^- \rightarrow e^- + \bar{\nu}_e \qquad (5.201d)$$

also occur but much less frequently, although the available phase space would seem to favor the electron mode. The apparent paradox is resolved within the V-A theory, as will be discussed in detail later. This result, in fact, lends very strong support to the theory, as any attempt to go beyond a V-A coupling in the weak-interaction part would predict essentially the same probability for the two modes, a result that is in violent disagreement with observations.

In the same way as in nuclear β decay, strong interactions are present in π-meson decay and must properly be taken into account. Following the pattern outlined in Sec. 5, we shall introduce form factors based on invariance considerations.

The charged π mesons can also decay through another, very rare channel:

$$\pi^+ \rightarrow \pi^0 + e^+ + \nu_e \qquad (5.201e)$$

This process is of particular interest in relation to the conserved-vector-current hypothesis discussed above. Within the CVC theory one obtains unique predictions for the process despite the fact that strong interactions are present. These predictions are in good agreement with observations, as will be explained in more detail in Sec. 6.2.

6.1. The $\pi \rightarrow \mu + \nu_\mu$ and $\pi \rightarrow e + \nu_e$ Decays. Process a is described by the diagram of Fig. 5.9. Within the current $\times$ current scheme of the V-A theory the effective S operator is given by

$$S = -\frac{iG}{\sqrt{2}} \int d^4x J_\mu{}^+(x) J^\mu(x) \qquad (5.202)$$

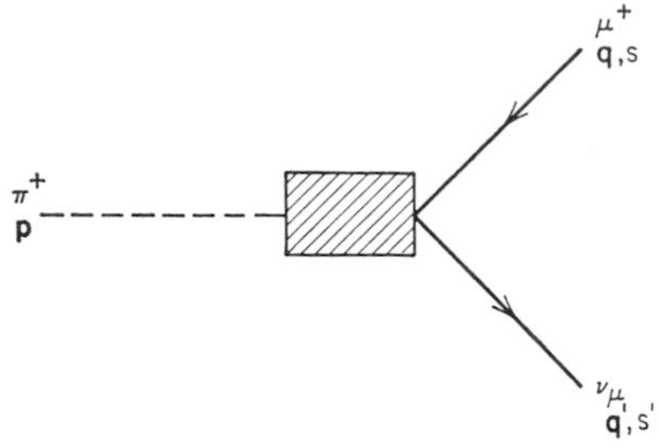

Fig. 5.9. Diagram for the process $\pi^+ \rightarrow \mu^+ + \nu_\mu$. The crosshatched region represents the effects of strong interactions.

and, therefore, the appropriate S-matrix element is

$$S_{fi} = -\frac{iG}{\sqrt{2}} <\mu,\nu_\mu| \int d^4x J_\mu{}^+(x) J^\mu(x) |\pi> \qquad (5.203)$$

As before, we shall neglect electromagnetic corrections, and Eq. (5.203) can be written

$$S_{fi} = -\frac{iG}{\sqrt{2}} \int d^4x <0|J_\mu{}^+(x)|\pi> <\mu,\nu_\mu|\bar{\psi}_{\nu_\mu}(x)\gamma^\mu(1+\gamma_5)\psi_\mu(x)|0> \qquad (5.204)$$

where only the relevant lepton part of the current $J^\mu(x)$ has been retained. The corresponding matrix element can be evaluated in standard fashion. Introducing the result and extracting the space-time dependence of the π-meson matrix element, one arrives at

$$S_{fi} = -\frac{iG}{\sqrt{2}} (2\pi)^4 \delta^4(p - q - q') \frac{1}{V} \sqrt{\frac{m_\mu m_{\nu_\mu}}{E_\mu E_{\nu_\mu}}} \cdot \bar{u}_{s'}(\mathrm{q}')\gamma^\mu(1+\gamma_5)v_s(\mathrm{q}) <0|J_\mu{}^+(0)|\pi> \qquad (5.205)$$

Absorbing some irrelevant normalization factors in the unknown π-mason matrix element, we can write the corresponding T-matrix element [see Eq. (A.66) of Appendix A]:

$$T_{fi} = -\frac{G}{\sqrt{2}} \sqrt{m_\mu m_{\nu_\mu}}\, \bar{u}_{s'}(\mathrm{q}')\gamma^\mu (1+\gamma_5)v_s(\mathrm{q}) <0|J_\mu{}^+(0)|\pi> \qquad (5.206)$$

From invariance under proper Lorentz transformations it is immediately concluded that

$$<0|J_\mu{}^+(0)|\pi> = f_\mu(p) \qquad (5.207)$$

where we have taken note of the fact that the π-meson matrix element can depend on only a single variable, namely, the π-meson momentum p. As it will transform as a vector, it can also be written

$$f_\mu(p) = p_\mu f(p^2) = p_\mu f(m_\pi{}^2) \equiv p_\mu f \qquad (5.208)$$

that is, a constant f times the 4-momentum of the π meson. Inserting this in (5.206), we obtain

$$T_{fi} = -\frac{Gf}{\sqrt{2}} \sqrt{m_\mu m_{\nu_\mu}}\, \bar{u}_{s'}(\mathrm{q}')p_\mu\gamma^\mu(1+\gamma_5)v_s(\mathrm{q}) \qquad (5.209)$$

By energy-momentum conservation it follows that $p = q + q'$. Inserting this and making use of the Dirac equation for the spinors we arrive at the final form

$$T_{fi} = \frac{Gf}{\sqrt{2}} \sqrt{m_\mu m_{\nu_\mu}}\, m_\mu \bar{u}_{s'}(\mathbf{q}')(1 - \gamma_5)v_5(\mathbf{q}) \qquad (5.210)$$

From the T-matrix element (5.210) the partial-decay rates can now be calculated easily by means of Eq. (A.70). As the neutrino cannot be observed, we sum over s' and integrate over $\mathbf{q}'$ and obtain, after some simplification, the following expression in the π-meson rest frame:

$$d\Lambda = \frac{G^2f^2}{64\pi^2}\frac{m_\mu{}^3}{m_\pi}\frac{d^3q}{E_\mu}\frac{1}{|\mathbf{q}|}\,\delta(m_\pi - \sqrt{\mathbf{q}^2 + m_\mu{}^2} - |\mathbf{q}|) \cdot v_s(\mathbf{q})(1 + \gamma_5)(\gamma_0 m_\pi + m_\mu)v_s(\mathbf{q}) \qquad (5.211)$$

As there are only two particles in the final state, energy-momentum conservation restricts the absolute value of the particles 3-momenta completely, as can be seen from the remaining δ function. We can perform the integration over $|\mathbf{q}|$ with the following result:

$$d\Lambda = \frac{G^2f^2}{16\pi}\frac{m_\mu{}^3}{m_\pi{}^2}\frac{d\Omega}{4\pi}(m_\pi - E_\mu)\bar{v}_s(\mathbf{q}) (1 + \gamma_5)(\gamma_0 m_\pi + m_\mu)v_s(\mathbf{q}) \qquad (5.212)$$

where now

$$|\mathbf{q}| = \frac{m_\pi}{2}\left[1 - \left(\frac{m_\mu}{m_\pi}\right)^2\right] \qquad (5.213)$$

From (5.212) one can easily derive the polarization of the outgoing muon in the usual way. However, we have already discussed this aspect of the problem when polarization effects in muon decay were examined and concluded that, because the π meson has spin zero, angular-momentum and momentum conservation require the muon to be in a helicity eigenstate with eigenvalue -1 for π^+ decay and eigenvalue $+1$ for π^- decay. This prediction is also borne out by experiments [55]. Therefore, we directly compute the total-decay rate of the π^+ meson into a muon and a neutrino from (5.212). This yields

$$\Lambda = \frac{G^2f^2}{8\pi}\frac{m_\mu{}^2}{m_\pi{}^2}(m_\pi - E_\mu)(m_\pi E_\mu - m_\mu{}^2) \qquad (5.214)$$

or

$$\Lambda = \frac{G^2f^2}{32\pi}\frac{m_\mu{}^2}{m_\pi{}^3}(m_\pi{}^2 - m_\mu{}^2)^2 \qquad (5.215)$$

This relation in itself is of rather little interest as it permits us only to determine the unknown number $f^2 = f^2(m_\pi{}^2)$. Inserting the experimental value for the decay rate [56]

$$\Lambda_{\text{exp}} = (0.392 \pm 0.005) \times 10^8 \text{ sec}^{-1} \qquad (5.216)$$

one obtains

$$f(m_\pi{}^2) \simeq 1.86\, m_\pi \qquad (5.217)$$

Of considerably more interest is the fact that the result (5.216) for the muon decay channel applies equally well to the electron mode [process (5.201c)] with a trivial change of the lepton mass. Thus, the V-A theory predicts a branching ratio

$$R = \frac{\Lambda(\pi^+ \to e^+ + \nu_e)}{\Lambda(\pi^+ \to \mu^+ + \nu_\mu)} = \left(\frac{m_e}{m_\mu}\right)^2\left(\frac{m_\pi{}^2 - m_e{}^2}{m_\pi{}^2 - m_\mu{}^2}\right)^2 \simeq 1.283 \times 10^{-4} \qquad (5.218)$$

which is independent of the unknown form factor. The theoretical value (5.218) is in excellent agreement with the observed branching ratio [57]

$$R_{\text{exp}} = (1.25 \pm 0.03) \times 10^{-4} \qquad (5.219)$$

As stated before, this branching ratio lends strong support to the V-A theory. The only alternative coupling would be an S-P mixture, in which case the branching ratio is of order unity.

The strong suppression of the electron mode within the V-A theory has a simple physical explanation. It was found previously that in $\pi^+ \to \mu^+ + \nu_\mu$ the muon is emitted with helicity -1, that is, with the spin antiparallel to the momentum. The same is true for the positron in the decay $\pi^+ \to e^+ + \nu_e$. But, to the extent that the positron rest mass can be neglected, the V-A theory implies that only positrons with positive helicity are created (compare the two-component theory of the neutrino). Thus, in this approximation, the positron mode is forbidden. The muon, on the other hand, has a large rest mass which cannot be neglected; for that reason the argument above fails in this case and there is no substantial suppression. This argument does not hold for couplings other than V-A, as in that case the neutrinos are not necessarily in a definite eigenstate of helicity and, therefore, neither the muons nor the positrons in the final state are forced to come out with the "wrong" helicity.

In passing, we note that the existence of the decays (5.201a) to (5.201d) rules out the possibility that the complete current $J_\mu(x)$ is conserved. Indications are that the vector part in fact is conserved, as has been discussed in connection with the conserved-vector-current hypothesis. Further evidence for that will be presented in the next subsection. If the axial-vector part were also conserved so that

$$\partial^\mu J_\mu(x) = 0 \qquad (5.220)$$

it would follow from (5.207) and (5.208) that

$$p^\mu p_\mu f(m_\pi{}^2) = m_\pi{}^2 f = 0 \qquad (5.221)$$

or

$$f = 0 \qquad (5.222)$$

and the processes (5.201a) to (5.201d) would then be absolutely forbidden.

6.2. The Decay $\pi^+ \to \pi^0 + e^+ + \nu_e$ and the Conserved-vector-current Hypothesis. To explain the remarkable agreement between the effective vector coupling constant in nuclear β decay and the fundamental weak-interaction coupling constant as it appears in μ decay, the hypothesis of a *conserved vector current* was introduced. The relevant quantity for the CVC theory is the isovector-current operator which was previously written in the following way:

$$\mathcal{J}_\mu(x) = \tfrac{1}{4}[\bar{\psi}_N(x), \gamma_\mu \boldsymbol{\tau} \psi_N(x)] + i\boldsymbol{\phi}(x) \times \frac{\partial \boldsymbol{\phi}(x)}{\partial x^\mu} + \cdots \qquad (5.223)$$

The third component of this isovector current is immediately identified as the isovector part of the electromagnetic current for nucleons and π mesons. This part is conserved by virtue of charge conservation. If invariance under rotations in isospace for the strong interactions is assumed, it follows that no direction in isospace is preferred and we can perform arbitrary rotations. This is an alternative way of arriving at the conclusion that the isovector current is conserved.

The conserved-vector-current hypothesis now asserts that the vector part of the weak-interaction current, except for the lepton contributions, is proportional to the component $\mathcal{J}_{\mu,-}(x)$ of the isovector current, or, more specifically,

$$\mathcal{J}_\mu{}^V(x) = \sqrt{2}\mathcal{J}_{\mu,-}(x) = \bar{\psi}_n(x)\gamma_\mu\psi_p(x) + \sqrt{2}\left\{\varphi_0(x)\frac{\partial\varphi^*(x)}{\partial x^\mu} - \varphi^*(x)\frac{\partial\varphi_0(x)}{\partial x^\mu}\right\} + \cdots \tag{5.224}$$

The presence of the second term in the conserved current has immediate physical implications which can be tested. With $\mathcal{J}_\mu{}^V(x)$ of Eq. (5.224) inserted in the effective S operator (5.13) there is a term

$$S = -iG\int d^4x \left\{\varphi_0(x)\frac{\partial\varphi^*(x)}{\partial x^\mu} - \varphi^*(x)\frac{\partial\varphi_0(x)}{\partial x^\mu}\right\} \bar{\psi}_{\nu_e}(x)\gamma^\mu\psi_e(x) \tag{5.225}$$

which accounts for the decay (see Fig. 5.10)

$$\pi^+ \to \pi^0 + e^+ + \nu_e$$

It is seen that in the π-meson current there is an extra factor $\sqrt{2}$ which strengthens the coupling to the lepton current.

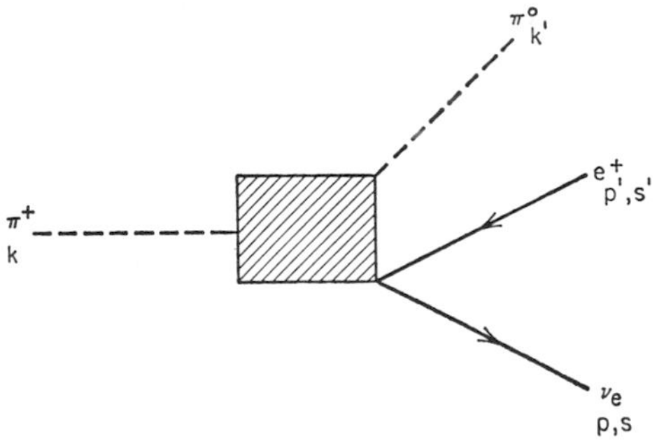

FIG. 5.10. Diagram describing the decay $\pi^+ \to \pi^0 + e^+ + \nu_e$.

To compute the $\pi^+ \to \pi^0 + e^+ + \nu_e$ decay we proceed in the same way as in the discussion of the normal decay modes of the charged π mesons. Neglecting electromagnetic effects, we can write the S-matrix element in the following way:

$$S_{fi} = -iG\int d^4x <\pi^0|j_\mu{}^\pi(x)|\pi^+> <e,\nu_e|\bar{\psi}_{\nu_e}(x)\gamma^\mu(1+\gamma_5)\psi_e(x)|0> \tag{5.226}$$

where $j_\mu{}^\pi(x) = \varphi_0(x)\dfrac{\partial\varphi^*(x)}{\partial x^\mu} - \varphi^*(x)\dfrac{\partial\varphi_0(x)}{\partial x^\mu}$

The lepton matrix element can be evaluated directly. From the unknown π-meson matrix element we extract some trivial normalization factors, and we then obtain the following expression for the T-matrix element:

$$T_{fi} = -2G\sqrt{m_e m_{\nu_e}} <\pi^0|\tilde{j}_\mu{}^\pi(0)|\pi^+> \bar{u}_s(\mathbf{p})\gamma^\mu(1+\gamma_5)v_{s'}(\mathbf{p}') \tag{5.227}$$

with

$$<\pi^0|j_\mu{}^\pi(0)|\pi^+> = \frac{1}{\sqrt{2\omega_0 V}}\frac{1}{\sqrt{2\omega_+ V}} <\pi^0|\tilde{j}_\mu{}^\pi(0)|\pi^+> \tag{5.228}$$

Here ω_0 and ω_+ denote the energy of the π^0 and π^+, respectively. In this way $<\pi^0|\tilde{j}_\mu{}^\pi(0)|\pi^+>$ has the dimension of a momentum.

From invariance considerations we know that the π-meson matrix element must transform as a vector. As it depends on at most two vectors, namely, k and k', it can generally be written

$$<\pi^0|\tilde{j}_\mu{}^\pi(0)|\pi^+> = q_\mu f_1(q^2) + q'_\mu f_2(q^2) \tag{5.229}$$

with $\quad q_\mu = k'_\mu - k_\mu \qquad q'_\mu = k'_\mu + k_\mu$

and where $f_1(q^2)$ and $f_2(q^2)$ are dimensionless form factors depending on the single independent invariant which has been chosen to be q^2. Invoking the hypothesis of a conserved vector current of which only $j_\mu{}^\pi(x)$ contributes to the matrix element considered here, we obtain

$$q^2 f_1(q^2) + q^\mu q'_\mu f_2(q^2) = 0 \tag{5.230}$$

However, a conserved-vector-current hypothesis is meaningful only to the extent that isospin invariance holds and, therefore, to the extent that mass differences within isomultiplets can be neglected. Neglecting the mass difference for the pions means that $q^\mu q'_\mu = 0$. Then we conclude from (5.230) that

$$f_1(q^2) = 0 \tag{5.231}$$

and we can write Eq. (5.229) as

$$<\pi^0|\tilde{j}_\mu{}^\pi(0)|\pi^+> = q'_\mu f(q^2) \tag{5.232}$$

where the index of the form factor has been dropped. The conserved-vector-current hypothesis further implies that the remaining form factor is the same as the π-meson electromagnetic form factor and furthermore with the normalization chosen here we have

$$f(0) = 1 \tag{5.233}$$

Somewhat loosely this last result is often said to imply that there are no renormalization effects for the vector coupling.

Inserting these results into (5.227) we now immediately obtain the following expression for the decay rate:

$$\Lambda = \frac{G^2}{(2\pi)^5}\int \frac{d^3k'\,d^3p\,d^3p'}{16\omega_0\omega_+E_eE_{\nu_e}}\delta^4(k-k'-p-p') (k'_\mu+k_\mu)(k'_\nu+k_\nu)\cdot \mathrm{Tr}[p_\lambda\gamma^\lambda\gamma^\mu(1+\gamma_5)(\gamma^\rho p'_\rho - m_e)\gamma^\nu (1+\gamma_5)]|f(q^2)|^2 \tag{5.234}$$

The available phase space is small as the mass difference between the two π mesons is small. For that reason we must retain the electro.. mass term in the trace. Furthermore, as we do not expect the form factor to vary much over the small range of the q^2 values involved in this process, we replace it by the value for $q^2 = 0$, that is,

$$f(q^2) \simeq 1 \tag{5.235}$$

The computation of the total-decay rate is now

straightforward and yields the following result [S7, p. 400]

$$\Lambda = \frac{G^2}{192\pi^3}\left(1 + \frac{m_{\pi^0}}{m_{\pi^+}}\right)^3 (m_{\pi^+} - m_{\pi^0})^5$$
$$\left\{\frac{2}{5}(2 - 9\epsilon - 8\epsilon^2)\sqrt{1-\epsilon} + 6\epsilon^2 \log \frac{1+\sqrt{1-\epsilon}}{\sqrt{\epsilon}}\right\} \quad (5.236)$$

with $$\epsilon = \frac{m_e^2}{(m_{\pi^+} - m_{\pi^0})^2} \quad (5.237)$$

Inserting the numerical values then gives the following branching ratio:

$$R = \frac{\Lambda(\pi^+ \to \pi^0 + e^+ + \nu_e)}{\Lambda(\pi^+ \to \text{anything})} = 1.02 \times 10^{-8} \quad (5.238)$$

The radiative corrections to this process have been evaluated [58] and amount to a few per cent. Despite the smallness of this branching ratio, it has been measured with the following result [59]:

$$R_{\text{exp}} = (0.97 \pm 0.02) \times 10^{-8}$$

This is in striking agreement with the prediction of the CVC theory. Without the CVC hypothesis there is no way of estimating the ratio, as it is not known how to handle the strong interactions. For that reason the branching ratio may attain any value. Consequently, the results obtained from the decay $\pi^+ \to \pi^0 + e^+ + \nu_e$ provide strong evidence in favor of the hypothesis.

7. Muon Capture

Muons from π decay have an initial kinetic energy in the π-rest system of $(m_\pi - m_\mu)^2/2m_\pi = 4.0$ Mev. Both $\mu^\pm$ slow down by ordinary ionizing collisions, the μ^+ eventually all decaying as described in Sec. 4. For the μ^- an alternative decay mode is available. At energies of a few kev, capture into highly excited atomic orbits occurs for all but the few μ^- that have decayed in flight. De-excitation to the $1s$ orbit follows by a cascade of Auger and X-ray transitions, and the $1s$ state is reached in a time of the order of 10^{-9} to 10^{-12} sec, so that virtually none of the μ^- decay before reaching the lowest muon orbit. Then, in addition to the decay into $e^- + \bar{\nu}_e + \nu_\mu$, the nuclear-capture process symbolized by

$$\mu^- + p \to n + \nu_\mu$$

can occur.

The branching ratio capture/decay strongly favors capture in all but the very lightest elements. This is evident from the experimental results; see Table 5.10. These data are in good agreement with the predictions of Primakoff [S17, 60] on the basis of which one expects a dependence on A and Z given by

$$\Lambda_{cap} = Z^n f(A,Z)$$

where n is about 4 for low Z and decreases slowly for larger Z, whereas $f(A,Z)$ is a slowly varying factor arising from the Pauli exclusion principle. For the decay process $\mu \to e + 2\nu$ inside a nucleus the decay rate (Λ_Z) is smaller than that of a free muon (Λ_0). The ratio Λ_Z/Λ_0 decreases very slowly and monotonically with increasing Z, primarily because the binding reduces the momentum space volume available to the decay products. For $Z < 25$, $\Lambda_Z/\Lambda_0 \approx 1$ and for Pb, $\Lambda_Z/\Lambda_0 \sim 0.6$. From Table 5.10 it is clear that $\Lambda_{cap} = \Lambda_Z$ for $Z \approx 9$.

Table 5.10. Capture Rates of μ^- in Various Elements [61]

Z	$\Lambda_{cap}(10^5\ \text{sec}^{-1})$
1	4.6×10^{-3} [62]
4	0.28
6	0.43
8	1.38
13	7.85
17	13.8
25	37.3
29	67.9
82	125

In addition to the total capture rate, data for capture to the ground state and various excited states of the daughter nucleus are available. In most cases, neutron emission follows capture to excited states. It is also feasible to measure the energy spectrum and angular distribution of these neutrons. Finally, it is possible to show that the capture rate in nuclei with nonzero spin exhibits a hyperfine effect, as will be discussed below. The motivation in many experiments is the verification of the universal Fermi interaction.‡ The evaluation of the data is not entirely straightforward since the theoretical capture rates depend on nuclear matrix elements and the limitations of present techniques and knowledge of nuclear structure stand in the way of precision comparisons. This difficulty is not present in the case of capture by protons but new difficulties arise there because the protons are found in molecular structure (cf. Sec. 7.3) such as p-μ-p, that is, muon molecular ion. Molecular wave functions are calculable in principle since all the forces are familiar but accuracy is obtained only after very laborious calculations. Nevertheless, it is possible to state in summary that all data are consistent with and, in some cases, in gratifying agreement with UFI.

It will be recognized immediately that μ capture involves non-negligible momentum transfer. In fact, the neutrino momentum is $q = -\Delta E + E_K$, where ΔE is the nuclear energy release in the inverse transition (β decay) and E_K is the muon energy in the K shell (including rest energy). For example, in the transition $O^{16} \to N^{16}$ (ground state) $q = 95.9$ Mev/c. Consequently, all the momentum-transfer-dependent terms in (5.153) are important. In particular, the g_2 term introduces a coupling constant not hitherto

‡ Other possible experiments include radiative muon capture, angular distribution of recoil nuclei [due to residual polarization (~15 per cent) of the μ^- in the $1s$ orbit], angular distribution, and circular polarization of gamma rays following capture to excited states. A large influence on the expected results arises in general from the induced pseudoscalar interaction and from the presence of weak magnetism (CVC; see Sec. 5). However, the data are sparse, and these effects will not be discussed further [S18].

considered—the induced pseudoscalar term. The coupling constant in the weak magnetism term is taken from the CVC theory. Therefore, μ-capture data are of particular interest for the determination of the induced pseudoscalar coupling constant g_2 (see Sec. 7.2).

7.1. The Effective Hamiltonian. The relevant matrix element for μ capture is

$$\mathfrak{M}_{cap} = \langle\Psi_f|H_{\text{eff}}|\Psi_i\rangle \qquad (5.239)$$

where initial and final nuclear states (i,f) are isobars with mass number A, proton number Z, $Z - 1$, respectively. The effective Hamiltonian is obtained in a manner similar to that leading to (5.156), with the following changes. First, there is the obvious fact that the Hermitian conjugate term describes the capture process. Second, as intimated, the momentum transfer terms in (5.153) are restored. The second-class current contributions will continue to be ignored so that weak magnetism and induced pseudoscalar terms will be added to the V and A interaction terms considered in nuclear β decay [cf. (5.156)]. Third, the form factors [cf. (5.153)] must be evaluated at the relevant 4-momentum transfer

$$q^2 = m_\mu^2(1 - m_\mu/M) = 0.9m_\mu^2$$

Fourth, a trivial change in notation will be made. Then, using (5.153), (5.158), and $H_{\text{eff}} = \psi_n^+\mathcal{H}_{\text{eff}}\psi_p$,

$$\mathcal{H}_{\text{eff}} = \frac{G}{\sqrt{2}}\{L(\beta\gamma^\lambda)[C\beta\gamma_\lambda - iD\beta\sigma_{\lambda\rho}q^\rho] + AL(\beta\gamma^\lambda\gamma_5)\beta\gamma_\lambda\gamma_5 + m_\mu BL(\beta\gamma_5)\beta\gamma_5\} \qquad (5.240)$$

Since the nucleons behave nonrelativistically, the nucleon operators are replaced by the appropriate limiting forms [63]. This gives

$$\mathcal{H}_{\text{eff}} = \frac{G}{\sqrt{2}}\tau_+\sum_{i=1}^{A}\tau^i\left\{1_iC\left[L(1) - \frac{\mathbf{q}\cdot\mathbf{L}(\boldsymbol{\sigma})}{2M}\right] + \boldsymbol{\sigma}_i\cdot\left[A\mathbf{L}(\boldsymbol{\sigma}) - \frac{iC}{2M}\mathbf{q}\times\mathbf{L}(\boldsymbol{\sigma}) + \frac{A}{2M}L(1)\mathbf{q} + \frac{C_p}{2M}L(1)\mathbf{q} - iC\frac{(\mu'_p - \mu'_n)}{2M}\mathbf{q}\times\mathbf{L}(\boldsymbol{\sigma})\right] + \text{velocity-dependent terms}\right\} \qquad (5.241)$$

The velocity-dependent terms‡ make small contributions [64] and are henceforth omitted. It is seen that the momentum-transfer terms are measured by $q/M \sim 0.1$. The CVC value for the weak magnetism term has been used. The facts that $\mathbf{L}(\boldsymbol{\alpha}) = -\mathbf{L}(\boldsymbol{\sigma})$, $L(\gamma_5) = L(1)$, and $L(\beta\gamma_5) = -L(\beta)$ have also been used. Finally, the approximation $L(\beta) = L(1)$, valid for a nonrelativistic muon, has been used; this enters only in the induced pseudoscalar terms

$$C_P = m_\mu g_2 = m_\mu B$$

‡ They are

$$\frac{C}{M}\mathbf{L}(\boldsymbol{\sigma})\cdot\mathbf{p}_i + \frac{A}{M}L(1)\boldsymbol{\sigma}_i\cdot\mathbf{p}_i$$

where $\mathbf{p}_i$ is the momentum operator for the ith nucleon.

Dispersion theoretic arguments give the following results [65]:

Vector coupling constant: $C \equiv g_V = 0.97g_V{}^\beta = 0.97$
Axial-vector coupling constant: $A \equiv g_A = 0.999g_A{}^\beta \cong g_A{}^\beta$

so that the form factors at $q^2 = 0.9m_\mu^2$ result in only a 3 per cent reduction in the vector coupling constant. By the same type of argument, Goldberger and Treiman [65] find

$$C_P/A \cong 7 \qquad (5.242)$$

based on the diagram in Fig. 5.4*c*. The momentum dependence of the V and A form factors in the relevant range was expected to be small, and the above results can presumably be trusted. Since the result (5.242) does not represent a small correction to a reliably determined constant, it is preferable to regard C_P/A as a parameter to be determined from experiment. The ratio of A and V coupling constants

$$g_A{}^\beta/g_V{}^\beta = -\lambda = -1.20$$

determines all other coupling parameters. Nevertheless, the experimental evidence will be used to verify these results as a test of UFI.

It is useful to introduce

$$G_V/G = g_V{}^\mu(1 + q/2M)$$
$$G_A/G = g_A{}^\mu - (g_V{}^\mu + g_M)q/2M$$
$$G_P/G = (C_P - g_A{}^\mu - g_V{}^\mu - g_M)q/2M$$

where $g_M = (\mu'_P - \mu'_n)g_V{}^\mu$, since the capture rates depend on bilinear combinations of these constants.

7.2. Hyperfine Effect. In the preceding discussion the interaction is $V - A$ if $\lambda > 0$ whereas it is $V + A$ if $\lambda < 0$. An elegant confirmation that the $V - A$ interaction is operative in the $(np)(\nu\mu)$ vertex, that is, the capture reaction, was obtained by Telegdi and coworkers on the basis of an analysis of Bernstein et al. [66].

The principle on which this analysis is based can be understood from the obvious fact that the interaction is spin-dependent and, in particular, depends on the relative orientation of the spins of the muon and capturing proton. In a capturing nucleus with spin $I \neq 0$, the muon in the ground state exists in either of one or the other of two hyperfine states with total angular momentum $F = I + \frac{1}{2}$ or $F = I - \frac{1}{2}$. Thus the system is a mixture of the two states with statistical weights $2F + 1$ or $2(I + 1)$ and $2I$, respectively. In addition, electromagnetic transitions (mainly Auger processes) from the higher $(F = I + \frac{1}{2})$ to the lower state take place at a rate which we designate by R.

For simplicity a free proton may be considered since this also represents with reasonable accuracy the situation in which a single (odd) proton outside a closed shell captures the muon. Then the capture rate from an arbitrary state F is [66] proportional to

$$\Lambda_F \sim G^2(a + b\langle\boldsymbol{\sigma}_p\cdot\boldsymbol{\sigma}_\mu\rangle)$$

where‡ $a = 1 + 3\lambda^2$ and $b = -2\lambda(1 + \lambda)$. The aver-

‡ Neglecting induced pseudoscalar contributions and the small reductions in coupling constants arising from the form factors.

age value of $\boldsymbol{\sigma}_p \cdot \boldsymbol{\sigma}_\mu$ is

$$\langle \boldsymbol{\sigma}_p \cdot \boldsymbol{\sigma}_\mu \rangle = \frac{1}{I(I+1)} [F(F+1) - I(I+1) - \tfrac{3}{4}] \\ [I(I+1) - L(L+1) - \tfrac{3}{4}]$$

where L is the orbital angular momentum of the (odd) proton. The first factor arises from the average projection of $\boldsymbol{\sigma}_\mu$ on the nuclear spin, $\boldsymbol{\sigma}_\mu \cdot \mathbf{I}$, and the second from the average projection of the proton spin on the nuclear spin, $\boldsymbol{\sigma}_p \cdot \mathbf{I}$. For the case of interest, $I = \frac{1}{2}$, $L = 0$, $F = 0$, 1. Then $\langle \boldsymbol{\sigma}_p \cdot \boldsymbol{\sigma}_\mu \rangle_{F=1} = 1$ and $\langle \boldsymbol{\sigma}_p \cdot \boldsymbol{\sigma}_\mu \rangle_{F=0} = -3$. Thus the capture rates are proportional to

$$\Lambda_1 \sim (1 - \lambda)^2$$
$$\Lambda_0 \sim (1 + 3\lambda)^2$$

with an over-all capture rate

$$\bar{\Lambda} = \tfrac{1}{4}(3\Lambda_1 + \Lambda_0) \sim 1 + 3\lambda^2$$

It is obvious that Λ_1 and Λ_0 are sensitive to the sign of λ, and for the value deduced from nuclear β decay one expects negligible capture from the $F = 1$ state. If, on the other hand, $\lambda \approx -1$, as the $V + A$ interaction would require, the decay rates Λ_1 and Λ_0 are almost equal and there would be only a very small hyperfine effect.

The last fact allows one to make a distinction between the two interactions. A $V + A$ interaction would give a mu capture rate (as measured by the appearance rate of neutrons and γ rays, or β activity from capture to the ground state of the daughter nucleus) which is independent of time. It would not matter whether the $F = 1 \rightarrow F = 0$ transitions took place. On the other hand, with $V - A$, Λ_1 and Λ_0 are very different. If R is neither much larger or much smaller than the capture rates, the time dependence of the capture products shows a typical rise-and-fall pattern, the rise due to the growth of the $F = 0$ population at the expense of the $F = 1$ population and then the fall due to the depletion of the $F = 0$ population as the capture process occurs.

The experiments [67] were performed with F^{19} (in a LiF target). Here, the odd proton should be in a $2s_{1/2}$ state, $I = \frac{1}{2}$ and $L = 0$. The rate R is of order 3×10^6 sec^{-1} $\sim \Lambda_{cap}$. The time-dependent part of the capture products represents only about 10 per cent ($\sim 1/Z$) of the total since all protons except the one outside the closed p shell have spins paired off and do not contribute to the hyperfine effect. Nevertheless, the data unmistakably show the time dependence appropriate to a $V - A$ interaction.

7.3. The Coupling Constants (UFI and the Induced Pseudoscalar Coupling). If UFI is accepted, the values of the coupling constants would be given by the results quoted in Sec. 7.1. For $q^2 = 0.9m_\mu{}^2$ this would give

$$G_V = 1.02G \qquad G_A = -1.40G \qquad G_P = -0.65G$$

This depends on the result (5.242) which, strictly speaking, applies only to capture in hydrogen. The two paramount problems are (1) to verify the UFI hypothesis and (2) to check the result (5.242). A priori, there is a greater confidence in the UFI hypothesis, because of its success elsewhere, than in the particular numerical value assigned to C_P on the basis of a less-than-compelling theoretical argument. Nevertheless, the two questions are more or less inextricably bound together. This is seen when we examine the various types of capture processes and determine which combinations of the coupling constants enter.

TABLE 5.11. SELECTION RULES IN μ^- CAPTURE

Operator	Relevant tensors	Parity change	Remarks
1 (Fermi)	$\mathcal{Y}_L(\mathbf{r})$	$(-)^L$	
$\boldsymbol{\sigma}$ (Gamow-Teller)	$T_{Ll}(\mathbf{r},\boldsymbol{\sigma})$	$(-)^l$	$\Delta(Ll1)$: for example, no $0 \rightarrow 0$ unless $l = 1$.
Induced pseudoscalar	$T_{L,L\pm1}(\mathbf{r},\boldsymbol{\sigma})$	$(-)^{L+1}$	
Weak magnetism	$T_{L,L\pm1}(\mathbf{r},\boldsymbol{\sigma})$	$(-)^{L+1}$	$L \neq 0$

The selection rules for the various operators that enter are summarized in Table 5.11. In all cases the angular-momentum transfer is L which makes a triangle with the initial and final nuclear spins, J_i and J_f. The first two entries (F and GT operators) are the same as in nuclear β decay. For either J_i or $J_f = 0$, $L = J_f$ or J_i, and the Fermi operator contributes only if the parity change is $(-)^{J_f}$ or $(-)^{J_i}$, respectively. For G-T either parity change is allowed but $\Delta(Ll1)$ applies. The induced pseudoscalar and weak magnetism terms lead to matrix elements that are related in exactly the same way as the longitudinal and transverse electromagnetic interactions, respectively. With these selection rules as a guide, one can deduce that for the following types of capture transitions the relevant coupling constant combinations are as given below:

Type	Transition	Coupling constant combinations
1	$0^+ \rightarrow 1^+, 2^-, 3^+ \ldots$	$2G_A{}^2 + (G_A - G_P)^2$
2	$\frac{1}{2}^+ \rightarrow \frac{1}{2}^+$	$G_V{}^2 + 2G_A{}^2 + (G_A - G_P)^2$
3	$0^+ \rightarrow 0^-$	$(G_A - G_P)^2$

The best examples of type 1 are $C^{12} \rightarrow B^{12}$, $Li^6 \rightarrow He^6$, $O^{16} \rightarrow N^{16}$, $Ca^{40} \rightarrow K^{40}$ in which the daughter nucleus is always in the ground state. These are all pure GT transitions. Moreover, for any reasonable value of G_P/G_A, the transition rate is insensitive to G_P: $G_A = G_P$ for $C_P = 21A$. Hence, the measurement of the rate allows a determination of G_A if the relevant nuclear matrix element can be evaluated. In the C^{12} capture the matrix element can be determined from the known $1^+ \rightarrow 0^+$ M1 γ transition and the valid assumption of isospin conservation [68]. The G_A thus obtained agrees with the expected value within the limits of experimental error [S19]. The Ca^{40} data also give agreement if in the analysis one uses experimental values for the outgoing neutrino energy (determined from known nuclear levels) instead of using values calculated with a nuclear model [69].

The V coupling constant can then be fixed from

data on transitions of type 2 of which the most extensively studied example is $He^3 \rightarrow H^3$. The nuclear matrix element is obtained from electron scattering data [70]. The measured capture rate is $1490 \pm 40\ sec^{-1}$ while the assumption of UFI gives $1470 \pm 55\ sec^{-1}$ [S18, 71]. In this comparison the result (5.242) is used.

The capture rates in O^{16} to the 0^- excited state of N^{16} (type 3) and to the 2^- ground state have been measured [72]; they provide particularly sensitive tests of the induced pseudoscalar coupling constant. The data for the capture to 0^- yield $C_P/A \approx 8$ but the ground-state capture indicates a considerably larger value of at least 16 for this ratio. It is to be noted that the nuclear wave function of the 2^- state is difficult to calculate with any degree of reliability. Additional information with regard to the induced pseudoscalar coupling is obtained from the radiative capture of muons in Ca^{40}. The basic process in this case is

$$\mu^- + p \rightarrow n + \nu + \gamma$$

A theoretical study of this process is given by Rood and Tolhoek [73]. Assuming the dominance of the one pion exchange term for the pseudoscalar coupling (cf. Fig. 5.1*c*) one obtains

$$\frac{C_P}{A} \sim \frac{1}{q^2 - m_\pi{}^2}$$

For normal capture $q^2 \leq 0$. In the case of radiative capture one can reach $q^2 \cong m_\mu{}^2$ and one expects the one pion exchange term to be even more dominant. As a consequence the radiative capture is particularly sensitive to the pseudoscalar coupling. Furthermore, the uncertainties in the nuclear dynamics can be essentially eliminated by comparison of the radiative capture rate to the total capture rate. The experimental results [74] give $C_P/A = 13.7 \pm 2.7$. Also this value is large compared with the theoretical estimate; so far there is no satisfactory explanation for the discrepancy.

An independent check on the consistency of the UFI theory for muon capture is obtained from the study of the capture process in liquid hydrogen, as mentioned previously. In this case the difficulties related to the nuclear dynamics are absent but, as was pointed out, the formation of various molecular states involving the muon complicates the interpretation of experiments.

Atomic capture of μ^- in hydrogen takes place in rather peripheral Bohr orbits $(n, l \sim 20)$. In less than 10^{-9} sec the muon cascades down to the 1*s* orbit of the muonic atom (μ^-p) where it can form a singlet or a triplet state. The energy difference between these two states is only a few tenths of an electron volt. The singlet state is the ground state so that eventually the muonic atoms are all in the singlet state. The singlet (μ^-p) atoms then form $(p\mu^-p)$ molecular ions. These ions are almost exclusively created in the ortho state [75], and the transition rate between the ortho and the para states is essentially negligible. Thus, the capture process takes place predominantly from the ortho state. One of the most crucial factors entering the capture rate is the probability of finding the muon at the position of either one of the protons. To estimate this probability, obviously the molecular wave functions must be known with rather high precision. The most accurate evaluation of the capture rate, assuming UFI [76], yields

$$\Lambda_{cap} = 480\ sec^{-1}$$

The agreement between the theoretical prediction and the experimental results is found to be quite satisfactory.

The general picture that emerges from the above discussion on muon capture is that the capture process fits into the UFI theory rather well. This conclusion has recently been questioned on the basis of some measurements of the emission asymmetry of neutrons in muon capture in Ca^{40} and S^{32} [S22]. The observed asymmetry is considerably larger than one would expect from the UFI theory, but uncertainties in the nuclear dynamics are once more present and could easily account for the discrepancy. It is fair to say that the muon-capture data are consistent with the UFI. It is also necessary to emphasize that the analysis of the data is beset with theoretical uncertainties which make precision comparisons impossible at the present time.

8. Weak Interactions at High Energies

All the processes hitherto discussed are basically low-energy processes involving small momentum transfers $(q^2 \ll m_p{}^2)$. Furthermore, in decay processes the relevant parameters such as energy and momentum transfer are not adjustable quantities and they must be taken as found in nature. Despite these limitations a great deal has been learned about the properties of weak interactions, and more can undoubtedly be expected to come from further studies of decay processes.

In the past few years a new development in weak-interaction physics has taken place, providing novel and more powerful tools for probing the very structure of the weak interactions. With the advent of recently constructed large accelerators sufficiently high-energy particle beams have become available for weak-scattering experiments. The greater flexibility of such experiments as compared with decay studies and the possibility of studying processes involving large momentum transfers should yield much richer information. There remain great experimental difficulties to be overcome before full advantage can be taken of these new facilities, but results obtained so far have already had a great impact on our understanding of weak interactions.

The reason for the circumstance that experimental attention had not been directed toward scattering processes earlier is obvious when one considers the relative order of magnitude for the cross sections of weak scattering processes as compared with strong and electromagnetic processes (see Sec. 2). In general the weak processes are drowned by the much more frequent strong or electromagnetic processes, and there is small chance to detect a weak event against that background. For weak decay processes the situation is totally different. For a particle stable against strong or electromagnetic decays there is no such competition; just by waiting long enough the weak decay will occur and it can be observed.

A characteristic feature of weak scattering processes is that the cross sections as calculated from the effective S operator [Eq. (5.25)] are proportional to the square of the energy in the center-of-mass system. Therefore, if sufficiently high energies can be reached the cross sections can be expected to attain values large enough to make the processes detectable; that is, in fact, just what has happened.

The foregoing remark about the rapid growth with energy of weak-scattering cross sections means that at sufficiently high energies the weak interactions *effectively* become "strong" and the cross sections may become arbitrarily large. This is not permissible, and it indicates that the simple description of weak interactions is too rough an approximation. This is not surprising; in fact, it was pointed out in Sec. 2 that one could expect only a point interaction to describe phenomena at low momentum transfers. For higher momentum transfers finer details of the interaction become important, and the actual structure of the function ρ in Eq. (5.5) must be taken into account. Experimental work is just about at the stage where an analysis of the structure function can begin. This program, although extremely difficult to carry out, should form the basis for a deeper understanding of the true nature of the weak forces. So far essentially nothing is known about these aspects of the problem, but various possible structure effects have been discussed already. Of these, the hypothesis of an intermediate vector boson as mediator of weak interactions has received considerable attention. However, before we discuss these matters in more detail we shall briefly consider the first important result derived from weak-scattering experiments, namely, the discovery that there are two different kinds of neutrinos.

8.1. The Existence of Two Kinds of Neutrinos. Throughout the present review of weak-interaction physics it has been explicitly recognized that the neutrino accompanying the electron in the lepton current is different from the one appearing together with the muon. This has been done despite the fact that there would be no actual differences for observed processes in the two neutrinos turned out to be the same. In connection with the rare decay modes of the muon it was pointed out that certain decay modes, which are expected to occur, never have been observed although they have been searched for extensively. A closer analysis of the situation shows that all these unobserved processes would be absolutely forbidden if the muon neutrino is different from the electron neutrino.

An obvious way to resolve this problem is to investigate whether the neutrinos emitted together with electrons can initiate the same kind of processes as the neutrinos that are emitted together with muons. It is a fortunate coincidence that neutrino-induced reactions provide the least difficulties of all weak-scattering processes from the experimental point of view. This is so because the neutrinos participate only in weak interactions, and a process induced by a neutrino, therefore, is necessarily weak. All other particles participate also in strong or/and electromagnetic interactions, and the detection problems are much more difficult. For this reason neutrino-induced reactions were the first processes of this kind to be investigated [78].

A neutrino beam is obtained from decaying π^+ mesons produced in a high-energy accelerator. As discussed in Sec. 6, the dominant decay mode for the π^+ meson is $\pi^+ \to \mu^+ + \nu_\mu$. A small admixture of electron neutrinos as well as neutrinos obtained from decays of K mesons present in the π-meson beam is unavoidable but can be kept under control and made small. If the muon neutrino were identical with the electron neutrino the following two processes induced by ν_μ neutrinos could occur in competition:

$$\nu_\mu + n \to p + \mu^- \qquad (a)$$
$$\nu_\mu + n \to p + e^- \qquad (b)$$

Process (*b*) is strictly forbidden if the two kinds of neutrinos are different. Several authors [79] have estimated the cross section for process (*b*) under the assumption that only one kind of neutrino exists. With reasonable assumptions about the neutrino energy spectrum for the neutrino beams at Brookhaven and CERN, the total cross section is of the order of 10^{-38} cm^2. The cross section for process (*a*) should be essentially the same or possibly somewhat smaller as a result of the smaller phase-space volume available in that case.

The results of experiments [78] so far overwhelmingly indicate that process (*a*) strongly dominates over (*b*). The few events that could possibly be interpreted in terms of process (*b*) can well be accounted for as a result of the small admixture of ν_e neutrinos in the beam. The presence of ν_e neutrinos can lead to a process

$$\nu_e + n \to p + e^- \qquad (c)$$

and experimentally there is no way of distinguishing a process (*c*) from one of type (*b*). Thus, this experiment carried out both at Brookhaven and at CERN clearly shows that the two neutrinos are different.

Although the ν_μ is different from ν_e they share many properties. Experiments clearly show that they both have a small mass even if the experimental upper limit is presently much less restrictive for the ν_μ than for the ν_e. They both have the same helicity. These are the properties most easily accessible for experimental investigations. The fact that they do not differ in these respects explains why they were considered to be identical until recently when new and more versatile experiments revealed their distinction.

8.2. The Intrinsic Structure of Weak Interactions. Since neutrino-induced reactions can be observed and thereby a wider momentum-transfer region is accessible for experiments, one can hope to learn about the intrinsic structure of weak interactions. In Sec. 2, leading to the expression (5.25) for the effective S operator, it was stated that for low momentum transfers the interaction appears to be a point four-fermion interaction. None of the experiments discussed in previous sections show a discrepancy requiring a modification of that assumption. However, after it was firmly established that the weak interactions are of the $V - A$ type, it seemed compelling to speculate that the fundamental interaction is not a pointlike four-fermion interaction but rather an interaction mediated by a vector particle. Within this framework one would then describe muon decay by the symbolic diagram of Fig. 5.11, where W denotes the intermediate vector boson. The vector boson would then be coupled to the muon current on one side and to the electron current on the other side.

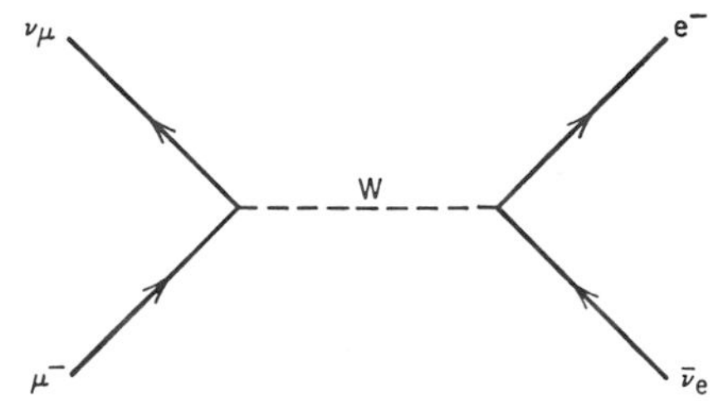

FIG. 5.11. A diagram describing muon decay if the weak interactions are mediated by an intermediate vector boson W.

The propagation of the boson corresponds to a particular structure of the function ρ in the general expression for the effective S operator. As indicated in Fig. 5.11, the absorption of the μ^- under emission of a ν_μ is described by a local current just as before. The same is true for the e^- and ν_e. However, now the currents are not directly coupled, but there appears a propagator function describing the exchange of a charged W particle. Thus,

$$S = ig^2 \int d^4x d^4x' j_\rho^{(e)}(x) \Delta^{\rho\sigma}(x - x') j_\sigma^{(\mu)+}(x') \quad (5.243)$$

now replaces the earlier point-interaction expression. The factor g^2 corresponds to a coupling constant g for the boson-lepton current vertex. The effect of the boson propagator function $\Delta_{\rho\sigma}(x - x')$ is most conveniently seen by considering its Fourier transform which is the quantity that appears directly in S-matrix elements. Then

$$\Delta_{\rho\sigma}(q) = \left(g_{\rho\sigma} - \frac{q_\rho q_\sigma}{m_W^2}\right) \frac{1}{q^2 - m_W^2} \quad (5.244)$$

where q is the momentum transfer carried by the boson. For $q^2 \ll m_W^2$ this structure function can be written

$$\Delta_{\rho\sigma}(q) \simeq -\frac{g_{\rho\sigma}}{m_W^2} \quad (5.245)$$

Therefore, in this approximation we obtain the usual four-fermion interaction S operator, provided

$$\frac{G}{\sqrt{2}} = \frac{g^2}{m_W^2} \quad (5.246)$$

This means that as long as we consider small momentum-transfer processes ($q^2 \ll m_W^2$) there would be no way to detect a possible intermediate boson. Furthermore, the mass of the intermediate boson is certainly larger than the K-meson mass, as one would otherwise have a rather fast (semiweak) decay of the K meson. This is sufficient to rule out observable effects on the usual decay processes considered so far [S8].

Besides its indirect effects on various weak processes the intermediate boson would also establish itself in various production processes—if it exists. A number of possible ways to produce W particles have been discussed in recent years [80]. Of particular interest is the process

$$\nu_\mu + \text{nucleus} \rightarrow W^+ + \mu^- + \text{nuclear products} \quad (d)$$

The nucleus does not directly participate in the process but acts by means of the Coulomb field. Its presence is necessary, however, in order that energy-momentum balance can be maintained.

Process (d) is a neutrino-induced reaction just as is process (a) discussed earlier. Therefore, provided the cross section for W production is sufficiently large, one would expect to observe it in the neutrino experiments referred to in Sec. 8.1. Accurate estimates of the cross section for process (d) require detailed knowledge of the boson mass and its electromagnetic properties, which are so far unknown; however, see (5.248) below. The predicted cross section decreases rapidly with increasing boson mass. For a boson mass of the same order as the nucleon mass the cross section is expected to be roughly 10^{-38} cm^2, that is, comparable to the cross section for process (a). These estimates are then based on presently available neutrino beams.

Before discussing the experimental results it is important to realize that if the intermediate boson exists one expects it to have a very short lifetime

$$\tau_W \lesssim 10^{-16} \text{ sec} \quad (5.247)$$

so that it can be detected only by its decay products. The estimate above is based on the decay channel

$$W^+ \rightarrow \mu^+ + \nu_\mu \quad (e)$$

which would presumably be one of the dominant decay modes. Furthermore, the coupling constant g is assumed to satisfy Eq. (5.246) in order to be consistent with all the observed weak processes. Thus a production of a W^+ particle will be established by observing $\mu^+\mu^-$ pairs or possibly an $e^+\mu^-$ pair. Other processes which would simulate a boson production give essentially negligible contributions.

So far no event has been observed that with certainty involves a W-particle production. This places an upper limit on the cross section and correspondingly a lower limit on the intermediate boson mass. The best value obtained so far indicates [81]

$$m_W \gtrsim 2 \text{ Bev} \quad (5.248)$$

If the intermediate boson exists, this means that the resulting structure effects for the weak decay processes discussed here are much finer-scaled than the present experimental accuracy. For all these processes the weak interactions are then very well represented by the local four-fermion interaction on which we have based our discussion in the foregoing sections.

APPENDIX A. MATHEMATICAL FORMALISM AND NOTATIONS

Throughout the discussion of the theory of weak interactions the characteristic properties of spin ½ particles play an important role. The basic formalism used for the description of such particles is given in all standard textbooks on relativistic quantum mechanics

[S2]. A brief account of the theory is given here to fix the notation.

The equation of motion for a free particle of spin ½ and mass m was first given by Dirac. It can be written in the following covariant form:

$$\left(i\gamma_\mu \frac{\partial}{\partial x_\mu} - m\right)\psi(x) = 0 \tag{A.1}$$

The wave function $\psi(x)$ has four components. In the Dirac equation (A.1) these four components are coupled through the four quantities γ which are 4×4 matrices satisfying the following anticommutation relations:

$$\gamma_\mu\gamma_\nu + \gamma_\nu\gamma_\mu = 2g_{\mu\nu}I \tag{A.2}$$

There Greek indices take on the values 0, 1, 2, 3, with 0 the time index: $x_0 = t$. Latin indices take on the values 1, 2, 3. Also, $g_{00} = 1$, $g_{kk} = -1$, and all other $g_{\mu\nu} = 0$. For any vector covariant, $A_\mu = g_{\mu\nu}A^\nu$ so that $A^0 = A_0$ and $A^k = -A_k$. Repeated indices are to be summed. Note also that in our units $c = \hbar = 1$.

From (A.2) we conclude that γ_0 can be chosen Hermitian while the γ_k's can be represented by anti-Hermitian matrices. Thus

$$\gamma^+_0 = \gamma_0 \qquad \gamma^+_k = -\gamma_k \tag{A.3}$$

or quite generally

$$\gamma^+_\mu = \gamma_0\gamma_\mu\gamma_0 \tag{A.4}$$

For most applications no explicit representations of the γ matrices are necessary. However, explicit representations are sometimes convenient. In particular, when $m \neq 0$ it is of interest to consider the non-relativistic limit of the theory; then the following representation turns out to be suitable:

$$\gamma_0 = \begin{pmatrix} I & 0 \\ 0 & -I \end{pmatrix} \qquad \gamma_k = \begin{pmatrix} 0 & \sigma_k \\ -\sigma_k & 0 \end{pmatrix} \tag{A.5}$$

where I is the 2×2 unit matrix and the matrices σ_k are the conventional Pauli matrices, that is,

$$\sigma_1 = \begin{pmatrix} 0 & 1 \\ 1 & 0 \end{pmatrix} \qquad \sigma_2 = \begin{pmatrix} 0 & -i \\ i & 0 \end{pmatrix} \qquad \sigma_3 = \begin{pmatrix} 1 & 0 \\ 0 & -1 \end{pmatrix} \tag{A.6}$$

If we define the adjoint wave function $\bar\psi(x)$ by

$$\bar\psi(x) = \psi^+(x)\gamma_0 \tag{A.7}$$

it can then easily be shown to satisfy the following equation of motion:

$$i\frac{\partial}{\partial x_\mu}\bar\psi(x)\gamma_\mu + m\bar\psi(x) = 0 \tag{A.8}$$

Under a proper Lorentz transformation

$$x_\mu \rightarrow x'_\mu = a_\mu{}^\nu x_\nu \tag{A.9}$$

the wave function $\psi(x)$ is transformed by means of a matrix Λ

$$\psi(x) \rightarrow \psi'(x') = \Lambda\psi(x) \tag{A.10}$$

For the Dirac equation to remain invariant under a Lorentz transformation the following relation must be fulfilled:

$$\Lambda^{-1}\gamma_\mu\Lambda = a_\mu{}^\nu\gamma_\nu \tag{A.11}$$

The matrix Λ can further be chosen to satisfy

$$\Lambda^{-1} = \gamma_0\Lambda^+\gamma_0 \tag{A.12}$$

For an infinitesimal transformation

$$a_{\mu\nu} = g_{\mu\nu} + \epsilon_{\mu\nu} \tag{A.13}$$

with $a_{\mu\nu} = g_{\nu\rho}a_\mu p$. The corresponding transformation matrix Λ is given by

$$\Lambda = 1 + \tfrac{1}{4}\epsilon_{\mu\nu}\gamma^\mu\gamma^\nu \tag{A.14}$$

and consequently for a finite transformation

$$\Lambda = \exp\left(\tfrac{1}{4}\epsilon_{\mu\nu}\gamma^\mu\gamma^\nu\right) \tag{A.15}$$

The transformation property for the adjoint spinor follows from Eqs. (A.7), (A.10), and (A.12):

$$\bar\psi(x) \rightarrow \bar\psi'(x') = \bar\psi(x)\Lambda^{-1} \tag{A.16}$$

From Eqs. (A.10) and (A.16) it is immediately seen that the bilinear form $\bar\psi(x)\psi(x)$ transforms as a scalar under Lorentz transformations, because

$$\bar\psi(x)\psi(x) \rightarrow \bar\psi'(x')\psi'(x') = \bar\psi(x)\Lambda^{-1}\Lambda\psi(x) = \bar\psi(x)\psi(x) \tag{A.17}$$

In the same way one can derive the transformation properties for the other bilinear expressions formed by means of γ matrices; these are listed in Table A.1.

TABLE A.1. DIRAC COVARIANTS

Covariant	Number of independent components	Transformation property	Notation
$\bar\psi(x)\psi(x)$	1	Scalar	S
$\bar\psi(x)\gamma_\mu\psi(x)$	4	Vector	V
$\bar\psi(x)\sigma_{\mu\nu}\psi(x)$	6	Second-rank tensor	T
$\bar\psi(x)\gamma_5\gamma_\mu\psi(x)$	4	Axial vector	A
$\bar\psi(x)\gamma_5\psi(x)$	1	Pseudoscalar	P

For convenience the following notations have been introduced:

$$\sigma_{\mu\nu} = \frac{i}{2}(\gamma_\mu\gamma_\nu - \gamma_\nu\gamma_\mu) \tag{A.18}$$

and

$$\gamma_5 = i\gamma_0\gamma_1\gamma_2\gamma_3 \tag{A.19}$$

The matrix γ_5 is thus Hermitian, and it anticommutes with all the other γ matrices:

$$\gamma_\mu\gamma_5 + \gamma_5\gamma_\mu = 0 \tag{A.20}$$

A complete set of solutions to the Dirac equation can be obtained in the form of plane waves. Equation (A.1) has two different solutions in the form of plane waves, namely,

$$\psi(x) = u(\mathbf{p})e^{-ipx} \tag{A.21}$$

and

$$\psi(x) = v(\mathbf{p})e^{ipx} \tag{A.22}$$

The Dirac equation requires the 4-vector p to satisfy the relativistic energy-momentum relation

$$p^2 - m^2 = 0 \tag{A.23}$$

and the momentum-dependent amplitudes must satisfy the equations

$$(\gamma_\mu p^\mu - m)u(\mathbf{p}) = 0 \tag{A.24}$$
$$(\gamma_\mu p^\mu + m)v(\mathbf{p}) = 0 \tag{A.25}$$

It is convenient to introduce the following energy projection operators acting in spinor space:

$$\Delta_\pm(p) = \frac{m \pm \gamma_\mu p^\mu}{2m} \tag{A.26}$$

By direct computation it is verified that these operators have the following properties:

$$\Delta_\pm(p)\Delta_\mp(p) = 0 \tag{A.27}$$
$$\Delta_+(p) + \Delta_-(p) = 1 \tag{A.28}$$
$$\Delta_\pm{}^2(p) = \Delta_\pm(p) \tag{A.29}$$

which shows that they are indeed projection operators. They decompose the four-dimensional spinor space into two two-dimensional subspaces. Equations (A.24) and (A.25) imply that the spinors $u(\mathbf{p})$ and $v(\mathbf{p})$ are eigenvectors of these operators:

$$\Delta_-(p)u(\mathbf{p}) = 0 \qquad \Delta_+(p)u(\mathbf{p}) = u(\mathbf{p}) \tag{A.30}$$
$$\Delta_-(p)v(\mathbf{p}) = v(\mathbf{p}) \qquad \Delta_+(p)v(\mathbf{p}) = 0 \tag{A.31}$$

From Eq. (A.4) it follows that

$$\gamma_0\Delta_\pm{}^+(p)\gamma_0 = \Delta_\pm(p) \tag{A.32}$$

and consequently we conclude from (A.27) that

$$\bar{u}(\mathbf{p})v(\mathbf{p}) = \bar{u}(\mathbf{p})\Delta_+(p)\Delta_-(p)v(\mathbf{p}) = 0 \tag{A.33}$$

Similarly

$$\bar{v}(\mathbf{p})u(\mathbf{p}) = 0 \tag{A.34}$$

To decompose the spinor space further we introduce the operators

$$\Sigma_\pm(\mathbf{p}) = \frac{1}{2}(1 \pm \sigma \cdot \hat{\mathbf{p}}) \tag{A.35a}$$

where $\hat{\mathbf{p}}$ is a unit vector along the 3-momentum $\mathbf{p}$ and $\boldsymbol{\sigma} = (\sigma_1, \sigma_2, \sigma_3)$ is defined by

$$\sigma_k = \sigma_{lm} = i\gamma_l\gamma_m \qquad k,\ l,\ m \text{ cyclic} \tag{A.36}$$

These operators $\Sigma_\pm(\mathbf{p})$ are special cases of the general spin projection operator

$$S_n = \tfrac{1}{2}(1 \pm w_\sigma\gamma^\sigma\gamma_5) \tag{A.35b}$$

Here w_σ is a 4-vector whose space part in the rest frame of a particle is a unit vector $\hat{\mathbf{n}}$ along the spin direction, it being recognized that the concept of spin in an arbitrary direction is not well defined except in the rest frame. The $\Sigma_\pm(\mathbf{p})$ are the special case of S_n in which $\hat{\mathbf{n}} = \pm\hat{\mathbf{p}}$. This case, where the spin is parallel or antiparallel to the momentum, corresponds to a constant of motion for a free particle. Therefore, it can be used in order to classify the solutions of the Dirac equation.

The operators $\Sigma_\pm(\mathbf{p})$ satisfy the following relations:

$$\Sigma_\pm(\mathbf{p})\Sigma_\mp(\mathbf{p}) = 0 \tag{A.37}$$
$$\Sigma_+(\mathbf{p}) + \Sigma_-(\mathbf{p}) = 1 \tag{A.38}$$
$$\Sigma_\pm{}^2(\mathbf{p}) = \Sigma_\pm(\mathbf{p}) \tag{A.39}$$

and further

$$[\Delta_\pm(\mathbf{p}), \Sigma_\pm(\mathbf{p})] = 0 \tag{A.40}$$

for all combinations of signs. This then means that the spinors $u(\mathbf{p})$ and $v(\mathbf{p})$ can be chosen to be simultaneous eigenvectors of $\Delta_\pm(\mathbf{p})$ and $\Sigma_\pm(\mathbf{p})$. We thus define four spinors $u_{1,2}(\mathbf{p})$ and $v_{1,2}(\mathbf{p})$ in the following way:

$$\Sigma_-(\mathbf{p})u_1(\mathbf{p}) = 0 \qquad \Sigma_+(\mathbf{p})u_1(\mathbf{p}) = u_1(\mathbf{p}) \tag{A.41}$$
$$\Sigma_-(\mathbf{p})u_2(\mathbf{p}) = u_2(\mathbf{p}) \qquad \Sigma_+(\mathbf{p})u_2(\mathbf{p}) = 0 \tag{A.42}$$
$$\Sigma_-(\mathbf{p})v_1(\mathbf{p}) = 0 \qquad \Sigma_+(\mathbf{p})v_1(\mathbf{p}) = v_1(\mathbf{p}) \tag{A.43}$$
$$\Sigma_-(\mathbf{p})v_2(\mathbf{p}) = v_2(\mathbf{p}) \qquad \Sigma_+(\mathbf{p})v_2(\mathbf{p}) = 0 \tag{A.44}$$

It is clearly implied that the indices $r = 1, 2$ refer to eigenvalues $+1$, -1, respectively, of the helicity operator $\boldsymbol{\sigma} \cdot \hat{\mathbf{p}}$. In the following we refer to r as the spin although what is meant is evidently spin projection.

As the operators $\Sigma_\pm(\mathbf{p})$ satisfy the relation

$$\gamma_0\Sigma_\pm{}^+(\mathbf{p})\gamma_0 = \Sigma_\pm(\mathbf{p}) \tag{A.45}$$

the following orthogonality relations hold:

$$u_1(\mathbf{p})u_2(\mathbf{p}) = 0 \qquad \bar{u}_2(\mathbf{p})u_1(\mathbf{p}) = 0 \tag{A.46}$$
$$v_1(\mathbf{p})v_2(\mathbf{p}) = 0 \qquad \bar{v}_2(\mathbf{p})v_1(\mathbf{p}) = 0 \tag{A.47}$$

To make these spinors $u_{1,2}(\mathbf{p})$ and $v_{1,2}(\mathbf{p})$ unique, we must impose a normalization condition for, say, $u_1(\mathbf{p})$ and a phase convention for the spinors $u_1(\mathbf{p})$ and $v_1(\mathbf{p})$. A convenient choice is

$$\bar{u}_1(\mathbf{p})u_1(\mathbf{p}) = 1 \tag{A.48}$$

and

$$\bar{v}_1(\mathbf{p})v_1(\mathbf{p}) = -1 \tag{A.49}$$

from which it can be shown to follow that

$$\bar{u}_2(\mathbf{p})u_2(\mathbf{p}) = -\bar{v}_2(\mathbf{p})v_2(\mathbf{p}) = 1 \tag{A.50}$$

The properties of the spinors $u_{1,2}(\mathbf{p})$ and $v_{1,2}(\mathbf{p})$ are collected in the following orthonormality and completeness relations:

$$\bar{u}_r(\mathbf{p})u_s(\mathbf{p}) = -\bar{v}_r(\mathbf{p})v_s(\mathbf{p}) = \delta_{rs} \tag{A.51}$$
$$\bar{u}_r(\mathbf{p})v_s(\mathbf{p}) = \bar{v}_r(\mathbf{p})u_s(\mathbf{p}) = 0 \tag{A.52}$$
$$\sum_{r=1,2} \{u_r{}^\alpha(\mathbf{p})\bar{u}_r{}^\beta(\mathbf{p}) - v_r{}^\alpha(\mathbf{p})\bar{v}_r{}^\beta(\mathbf{p})\} = \delta_{\alpha\beta} \tag{A.53}$$

where spinor indices α and β have been introduced in the last relation to emphasize that it is a matrix relation in spinor space.

The Dirac theory of spin ½ particles is a one-particle theory. A solution of the Dirac equation as given by Eq. (A.21) describes the propagation of a particle of spin ½ and mass m, while a solution of the type (A.22) describes the propagation of a corresponding antiparticle. However, a completely consistent particle-antiparticle theory can be given only within the framework of second quantization [S2], so that the appropriate symmetry properties of many-particle states are properly taken into account.

In the second-quantized version of the theory, $\psi(x)$ is a field operator acting in the space of state vectors. The state vectors are denoted $|\alpha>$, where α labels the corresponding state. A complete characterization of a state is given by naming the particles (and antiparticles) of the state and the appropriate quantum numbers of each particle (such as momentum, angular momentum). By imposing suitable commutation relations for the field operators the symmetry properties of many-particle states are taken care of. These commutation relations are most conveniently expressed in terms of the Fourier components of the field operators. The Fourier expansions for the operators $\psi(x)$ and $\bar{\psi}(x)$ are given by

$$\psi(x) = \frac{1}{\sqrt{V}} \sum_{\mathbf{p},r} \sqrt{\frac{m}{E}} \{a_r(\mathbf{p})u_r(\mathbf{p})e^{-ipx} + b\dagger_r(\mathbf{p})v_r(\mathbf{p})e^{ipx}\} \quad \text{(A.54)}$$

$$\bar{\psi}(x) = \frac{1}{\sqrt{V}} \sum_{\mathbf{p},r} \sqrt{\frac{m}{E}} \{a\dagger_r(\mathbf{p})\bar{u}_r(\mathbf{p})e^{ipx} + b_r(\mathbf{p})\bar{v}_r(\mathbf{p})e^{-ipx}\} \quad \text{(A.55)}$$

where V is the normalization volume. The spinor wave functions $u(\mathbf{p})$ and $v(\mathbf{p})$ are the same as before. The operators $a_r(\mathbf{p})$, $a\dagger_r(\mathbf{p})$, $b_r(\mathbf{p})$, and $b\dagger_r(\mathbf{p})$ satisfy the following anticommutation relations:

$$[a\dagger_r(\mathbf{p}),a_s(\mathbf{p}')]_+ = \delta_{rs}\delta_{\mathbf{pp}'} \quad \text{(A.56)}$$
$$[b\dagger_r(\mathbf{p}),b_s(\mathbf{p}')]_+ = \delta_{rs}\delta_{\mathbf{pp}'} \quad \text{(A.57)}$$

and all other anticommutators vanish. These operators act on the state vectors. It is consistent with these relations to interpret $a_r(\mathbf{p})$ as the annihilation operator of a particle of 3-momentum $\mathbf{p}$ described by the spinor wave function $u_r(\mathbf{p})$. In the same way, $b_r(\mathbf{p})$ acts as an annihilation operator for an antiparticle, and $a\dagger_r(\mathbf{p})$ and $b\dagger_r(\mathbf{p})$ as creation operators of a particle and an antiparticle, respectively. Thus, a state vector corresponding to a particle described by the spinor wave function $u_r(\mathbf{p})$ can be written

$$|\mathbf{p},r> = a\dagger_r(\mathbf{p})|0> \quad \text{(A.58)}$$

where $|0>$ denotes the vacuum (no particles) state vector. From Eq. (A.56) it follows that, acting on $|\mathbf{p},r>$ with another creation operator $a\dagger_r(\mathbf{p})$, one obtains

$$a\dagger_r(\mathbf{p})|\mathbf{p},r> = a\dagger_r(\mathbf{p})a\dagger_r(\mathbf{p})|0> = 0 \quad \text{(A.59)}$$

which means that no state vector containing two identical spin ½ particles in the same quantum state can be formed. The existence of such a state vector would obviously violate the Pauli principle.

In order to obtain a discrete momentum spectrum in the Fourier decomposition of the field operators a finite normalization volume V was used. For a plane wave resolution this volume can be taken to be a rectangular box with sides L_1, L_2, and L_3. In that case the spectrum of the momentum vectors in the summation is given by

$$p_i = n_i \frac{2\pi}{L_i} \qquad i = 1, 2, 3 \quad \text{(A.60)}$$

where n_i is an arbitrary integer. It follows that the total number Δn of momentum vectors $\mathbf{p}$ in the interval $(\mathbf{p}, \mathbf{p} + \Delta\mathbf{p})$ then is

$$\Delta n = \frac{V}{(2\pi)^3} \Delta\mathbf{p} \quad \text{(A.61)}$$

Therefore, if V is very large a summation over all momenta $\mathbf{p}$ of a slowly varying function $f(\mathbf{p})$ is replaced by an integral over phase space according to the prescription

$$\frac{1}{V} \sum_{\mathbf{p}} f(\mathbf{p}) \to \frac{1}{(2\pi)^3} \int d^3p f(\mathbf{p}) \quad \text{(A.62)}$$

An experiment with elementary particles consists in preparing a certain initial system and bringing the consituents of this initial state into interaction with each other. One then observes the properties of the resulting final state after the interaction has taken place. In the field-theoretic treatment [S2] of such a process the initial and final states are represented by state vectors $|i>$ and $|f>$. The physical process of the initial state being carried over into the final state under influence of the interaction is represented by an operator S acting in the space of the state vectors so that

$$|f> = S|i> \quad \text{(A.63)}$$

In general, an initial state may transform into many different final states. The probability $P(i \to f')$ for finding the system in a specific final state $|f'>$ is given by

$$P(i \to f') = |<f'|f>|^2 \quad \text{(A.64)}$$

provided the state vectors are properly normalized, which is assumed to be the case. From (A.63) and (A.64) one then obtains

$$P(i \to f') = |<f'|S|i>|^2 \quad \text{(A.65)}$$

Therefore, it is the modulus of the matrix elements of the S operator that determine physical quantities such as lifetimes and cross sections. If the S operator were known, then elementary-particle physics would be an exhausted subject. One could then predict the outcome of all possible experiments. This is not the situation. On the contrary, really very little is known about the S operator, and the research in elementary-particle physics aims at exploring its properties. In the restricted area of weak interactions, at least for presently attainable energies, the curtain of ignorance is lifted and a good account of observed phenomena can be obtained.

We next consider the explicit relation between the matrix elements of the S operator and measurable quantities such as various cross sections and decay rates. If the S operator is to describe everything that can happen to an initial state it must also account for the possibility that no interaction takes place and the initial state propagates undisturbed. This trivial part of the S operator is of no interest to us, and it is convenient to separate it out. This is done by defining a transition operator T through the relation

$$<f|S|i> = \delta_{fi} + i(2\pi)^4\delta^4(P_f - P_i)N<f|T|i> \quad \text{(A.66)}$$

where P_f and P_i denote the total 4-momenta of the final and the initial state, respectively, and N is a normalization factor which we shall choose to be

$$N = \prod_{\text{initial}} \frac{1}{\sqrt{2VE_i}} \prod_{\text{final}} \frac{1}{\sqrt{2VE_f}} \quad \text{(A.67)}$$

The symbols E_i and E_f stand for the energies of the particles in the initial and final states. V is the same normalization volume as before. With this choice of normalization factor N one assures relativistic invariance of the matrix elements of the T operator if all participating particles have spin zero. If particles with spin participate, the spin average of the T-matrix elements are invariant functions.

The transition probability δw is given by (A.65). Assuming that the initial state is different from the final state and employing the symbolic identity

$$[\delta^4(P_f - P_i)]^2 = \delta^4(P_f - P_i)\frac{Vt}{(2\pi)^4} \tag{A.68}$$

we obtain for the transition rate (transition probability per unit time)

$$d\Lambda = \frac{\delta w}{\delta t} = (2\pi)^4\delta^4(P_f - P_i)V \cdot |N|^2 \cdot |<f|T|i>|^2 \tag{A.69}$$

If the initial state contains two particles a and b the cross section σ for the process under consideration is defined as the total transition rate divided by the incoming flux of particles a striking the target particle b. If the relative velocity of the incoming particles with respect to the target is denoted v_{in}, the incoming flux is given by v_{in}/V, and for the cross section one obtains

$$\sigma = \frac{V^2}{v_{in}}(2\pi)^4 \sum_{\substack{\text{final} \\ \text{states}}} |N|^2\delta^4(P_f - P_i)|<f|T|i>|^2 \tag{A.70}$$

Assuming the final state to contain n particles and replacing the summation over final states by an integration over the momenta [cf. Eq. (A.62)] and a summation over the possible final spin directions, Eq. (A.70) can be written

$$\sigma = \frac{1}{4E_aE_bv_{in}}\frac{1}{(2\pi)^{3n-4}}\int\frac{d^3P_1}{2E_1}\cdots\int\frac{d^3P_n}{2E_n}\delta^4(P_f - P_i)\sum_{\substack{\text{final} \\ \text{spins}}}|<f|T|i>|^2 \tag{A.71}$$

If the initial-state particles carry spin and an unpolarized beam and target are used, the total cross section is obtained by taking the appropriate spin average. By restricting the summation over final states, various differential or partial cross sections are obtained.

For a particle a decaying into a final state containing n particles the total decay rate Λ is similarly given by

$$\Lambda = \frac{1}{(2\pi)^{3n-4}}\frac{1}{2E_a}\int\frac{d^3P_1}{2E_1}\cdots\int\frac{d^3P_n}{2E_n}\delta^4(P_f - P_i){\sum_{\substack{\text{final} \\ \text{spins}}}}'|<f|T|i>|^2 \tag{A.72}$$

where Σ' stands for summation over final spin directions and averaging over the initial spin directions. Various partial-decay rates are obtained by restricting the integrations over momenta or the summation over spin directions. The mean lifetime τ of the decaying particle is simply related to the total-decay rate, namely,

$$\tau = \Lambda^{-1} \tag{A.73}$$

APPENDIX B. IRREDUCIBLE TENSORS

An irreducible tensor of rank L is defined as a set of $2L + 1$ quantities which transforms under three-dimensional rotations according to [S10]

$$RT_L{}^MR^{-1} = \sum_{M'} D_{M'M}{}^L(\alpha\beta\gamma)T_L{}^{M'} \tag{B.1}$$

where R, the rotation operator, is defined in terms of the angular-momentum operators by

$$R = e^{i\mathbf{n}\cdot\mathbf{J}\theta} \tag{B.2}$$

Here $\mathbf{n}$ is a unit vector in the direction of the rotation axes and θ is the angle of rotation. In (B.1) the $D_{M'M}{}^L$ are elements of the rotation group representations, and α, β, γ are the Euler angles of rotation. For integer L, for example, explicit forms for the $D_{M'M}{}^L$ can be found by actually carrying out the rotation (B.2) on the spherical harmonics $Y_L{}^m$. If $T_{L_1}{}^{m_1}$ and $T_{L_2}{}^{m_2}$ are tensor components of the indicated ranks, then $T_{L_1}{}^{m_1}$ and $T_{L_2}{}^{m_2}$ can be combined to give tensors of rank L, where L_1, L_2, and L form a triangle. Thus

$$T_L{}^M = \sum_{m_2} C(L_1L_2L; M - m_2,m_2)T_{L_1}{}^{M-m_2}T_{L_2}{}^{m_2} \tag{B.3}$$

where $C(L_1L_2L;\ M - m_2,m_2)$ is a Clebsch-Gordan coefficient [S10]. From the unitary properties of these Clebsch-Gordan coefficients it follows that

$$T_{L_1}{}^{m_1}T_{L_2}{}^{m_2} = \sum_L C(L_1L_2L;m_1m_2)T_L{}^{m_1+m_2} \tag{B.4}$$

In β decay the tensors that occur arise from products of $Y_l{}^m(\hat{\mathbf{r}})$ and an operator Ω which is either a rotational scalar of a rotational vector; i.e., either $\Omega = T_0{}^0$ or $\Omega_m = T_1{}^m$, where the spherical components of $\boldsymbol{\Omega}$ are to be understood:

$$\Omega_{\pm1} = \pm\frac{1}{\sqrt{2}}(\Omega_x \pm i\Omega_y)$$

$$\Omega_0 = \Omega_z$$

Therefore the irreducible tensors constructed from these products according to (B.4) are written as $T_L{}^M(\mathbf{r},\boldsymbol{\Omega})$ to indicate the arguments on which the tensor depends. Moreover, the parity of $T_L{}^M(\mathbf{r},\boldsymbol{\Omega})$ is $\pi_\Omega\pi_l = (-)^l\pi_\Omega$, where π_Ω is the parity of the $\boldsymbol{\Omega}$ operator. Of course, $\pi_\Omega = +1$ for $\Omega = 1$, $\boldsymbol{\Omega} = \boldsymbol{\delta}$ and $\pi_\Omega = -1$ for $\Omega = \gamma_5$, $\boldsymbol{\Omega} = \boldsymbol{\alpha}$. The notation is made still more explicit by appending the index $L_1 = l$ to the irreducible tensor so that it reads $T_{Ll}{}^M(\mathbf{r},\boldsymbol{\Omega})$. The index L_2 is unnecessary because Ω or $\boldsymbol{\Omega}$ indicates whether this is 0 or 1. When $L_2 = 0$, $L_1 = L$, $m_1 = M$ and $T_L{}^M = Y_L{}^M(\hat{\mathbf{r}})\Omega$. In all the above, $\mathcal{Y}_l{}^m(\mathbf{r})$ can be read in place of $Y_l{}^m$.

A few special cases are of interest. In each case $L_2 = 1$ so that B occurs linearly.

$$T_{01}{}^0(\mathbf{A},\mathbf{B}) = -\frac{1}{\sqrt{4\pi}}\,\mathbf{A}\cdot\mathbf{B}$$

$$T_{10}{}^M(\mathbf{A},\mathbf{B}) = \frac{1}{\sqrt{4\pi}}\,B_M$$

$$T_{11}{}^M(\mathbf{A},\mathbf{B}) = i\sqrt{\frac{3}{8\pi}}\,(\mathbf{A}\times\mathbf{B})_M$$

References

1. Fermi, E.: *Z. Physik*, **88**: 161 (1934).
2. Gamow, G., and E. Teller: *Phys. Rev.*, **49**: 895 (1936).
3. Reines, F., and C. Cowan: *Phys. Rev.*, **111**: 273 (1959).
4. Dalitz, R. H.: *Phil. Mag.*, **44**: 1068 (1953); Proceedings of the Fifth Annual Rochester Conference, Interscience, New York, 1955.
5. Lee, T. D., and C. N. Yang: *Phys. Rev.*, **104**: 254 (1956); **106**: 1371 (1957).
6. Wu, C. S., E. Ambler, R. W. Hayward, D. D. Hoppes, and R. P. Hudson: *Phys. Rev.*, **105**: 1413 (1957).
7. Garwin, R. L., L. M. Lederman, and M. Weinrich: *Phys. Rev.*, **105**: 1415 (1957).
8. Fierz, M.: *Z. Physik*, **105**: 533 (1937).
9. Boehm, F., and E. Krankeleit: *Phys. Rev. Letters*, **14**: 312 (1965).
10. Sudarshan, E. C. G., and R. E. Marshak: Proceedings of the Padua-Venice Conference on Mesons and Newly Discovered Particles, 1957; *Phys. Rev.*, **109**: 1860 (1958). Feynman, R. P., and M. Gell-Mann: *Phys. Rev.*, **109**: 193 (1958).
11. Salam, A.: *Nuovo Cimento*, **5**: 299 (1957). Landau, L.: *Nucl. Phys.*, **3**: 127 (1957). Lee, T. D., and C. N. Yang: *Phys. Rev.*, **105**: 1671 (1957).
12. Pais, A., and R. Jost: *Phys. Rev.*, **87**: 871 (1952). Michel, L.: *Nuovo Cimento*, **10**: 319 (1953). Lee, T. D., and C. N. Yang: *Nuovo Cimento*, **3**: 749 (1956).
13. Schwinger, J.: *Phys. Rev.*, **82**: 664 (1951). Bell, J. S.: *Proc. Roy. Soc.* (*London*), **A231**: 479 (1955). Lüders, G.: *Ann. Phys.*, **2**: 1 (1957).
14. Plano, R. J.: *Phys. Rev.*, **119**: 1400 (1960).
15. Macq, P. C., K. M. Crowe, and R. P. Haddock: *Phys. Rev.*, **112**: 2061 (1958).
16. Burgy, M. T., V. E. Krohn, T. B. Novey, G. R. Ringo, and V. L. Telegdi: *Phys. Rev.*, **110**: 1214 (1958).
17. Boehm, F., and H. Wapstra: *Phys. Rev.*, **106**: 1364 (1957); **107**: 1202, 1462 (1957).
18. Goldhaber, M., L. Grodzins, and A. W. Sunyar: *Phys. Rev.*, **109**: 1015 (1958).
19. Burgy, M. T., V. E. Krohn, T. B. Novey, G. R. Ringo, and V. L. Telegdi: *Phys. Rev. Letters*, **1**: 324 (1958).
20. Christenson, J. H., J. W. Cronin, V. L. Fitch, and R. Turlay: *Phys. Rev. Letters*, **13**: 138 (1964).
21. Duclos, J., J. Heintze, A. DeRujula, and B. Soergel: *Phys. Rev. Letters*, **9**: 62 (1964). Bloom, S., L. A. Dick, L. Feuvrais, G. R. Henry, P. C. Macq, and M. Spighel: *Phys. Rev. Letters*, **8**: 87 (1964). Buhler, A., N. Cabibbo, M. Fidecaro, T. Massam, Th. Muller, M. Schneegans, and A. Zichichi: *Phys. Rev. Letters*, **7**: 368 (1963).
22. Michel, L.: *Proc. Phys. Soc.* (*London*), **A63**: 514 (1950).
23. Gurevich, I. I., L. A. Makariyna, B. A. Nikol'sky, B. V. Sokolov, L. V. Surkova, S. Kh. Khakumóv, V. D. Shestakov, Yu. P. Dobretsóv, and V. V. Akhmanov: *Phys. Rev. Letters*, **11**: 185 (1964).
24. Sachs, A. M.: *Bull. Am. Phys. Soc.*, **10**: 101 (1965). See also ref. 14.
25. Behrends, R. E., R. J. Finkelstein, and A. Sirlin: *Phys. Rev.*, **101**: 866 (1956). Kinoshita, T., and A. Sirlin: *Phys. Rev.*, **107**: 593 (1957); **113**: 1652 (1959). Berman, S. M.: *Phys. Rev.*, **112**: 267 (1958). Kinoshita, T., and A. Sirlin: *Phys. Rev. Letters*, **2**: 177 (1959). Berman, S. M., and A. Sirlin: *Ann. Phys.* (*N.Y.*), **20**: 20 (1962).
26. Meyer, S. L., E. W. Anderson, E. Bleser, L. M. Lederman, J. L. Rosen, J. Rothberg, and I. T. Wang: *Phys. Rev.*, **132**: 2693 (1963). Lundy, R. A.: *Phys. Rev.*, **125**: 1686 (1962).
27. Crittenden, R. R., and D. Walker: *Phys. Rev.*, **121**: 1823 (1961). Rey, C. A.: *Phys. Rev.*, **135**: B1215 (1964).
28. Fronsdal, C., and H. Überall: *Phys. Rev.*, **113**: 654 (1959). Eckstein, S. G., and R. H. Pratt: *Ann. Phys.* **8**: 297 (1959). Kinoshita, T., and A Sirlin: *Phys. Rev. Letters*, **2**: 177 (1959).
29. Parker, S., H. L. Anderson, and C. Rey: *Phys. Rev.*, **133**: B768 (1964).
30. Frankel, S., W. Frati, J. Halpern, L. Halloway, W. Wales, F. W. Betz, and O. Chamberlain: *Phys. Rev.*, **130**: 351 (1963). Babaev, A. I., M. Ya. Balats, V. S. Kaftanov, L. G. Landsberg, V. A. Lyubimov, and V. Obukhov: *Soviet Phys. JETP* (*English Transl.*), **16**: 1397 (1963).
31. Goldberger, M. L., and S. B. Treiman: *Phys. Rev.*, **111**: 354 (1958).
32. Gell-Mann, M.: *Phys. Rev.*, **111**: 362 (1958).
33. Weinberg, S.: *Phys. Rev.*, **112**: 1375 (1958).
34. Jackson, J. D., S. B. Treiman, and H. W. Wyld, Jr.: *Phys. Rev.*, **106**: 517 (1957).
35. Rose, M. E.: *Phys. Rev.*, **49**: 727 (1936). Durand, L., III; *Phys. Rev.*, **135**: B310 (1964).
36. Jackson, J. D., S. B. Treiman, and H. W. Wyld, Jr.: *Nucl. Phys.*, **4**: 206 (1957). Ebel, M. E., and G. Feldman: *Nucl. Phys.*, **4**: 213 (1957).
37. An extensive numerical tabulation of f is given by Feenberg, E., and G. Trigg: *Revs. Mod. Phys.*, **22**: 399 (1950).
38. Mayer, M. G., S. A. Moszkowski, and L. W. Nordheim: *Revs. Mod. Phys.*, **23**: 315 (1951). Nordheim, L. W.: *Revs. Mod. Phys.*, **23**: 322 (1951). See also Feingold, A. M.: *Revs. Mod. Phys.*, **23**: 10 (1951).
39. Bardin, R. K., C. A. Barnes, W. A. Fowler, and P. A. Seeger: *Phys. Rev.*, **127**: 583 (1962).
40. Durand, L., L. F. Landovitz, and R. B. Marr: *Phys. Rev.*, **130**: 1188 (1963).
41. Freeman, J. M., J. G. Montague, D. West, and R. E. White: *Phys. Rev. Letters*, **3**: 136 (1962).
42. Jänecke, J.: *Phys. Rev. Letters*, **6**: 69 (1963).
43. Freeman, J. M., R. E. White, J. H. Montague, G. Murray, and W. E. Burcham: *Phys. Rev. Letters*, **8**: 115 (1964).
44. Rose, M. E., and D. K. Holmes: *Phys. Rev.*, **83**: 190 (1951). See also reference S11.
45. Sosnovskii, A. N., P. E. Spivak, Iu. A. Prokof'ev, I. E. Kutikov, and Iu. P. Dobrynin: *Soviet Phys. JETP* (*English Transl.*) **8**: 739 (1959).
46. Burgy, M. T., V. E. Krohn, T. B. Novey, G. R. Ringo, and V. A. Telegdi: *Phys. Rev.*, **107**: 1731 (1957).
47. For example, Deutsch, M., B. Gittelman, R. W. Bauer, L. Grodzins, and A. W. Sunyar: *Phys. Rev.*, **107**: 1733 (1957).
48. Ambler, E., R. W. Hayward, D. D. Hoppes, and R. P. Hudson: *Phys. Rev.*, **110**: 787 (1958).

49. Feynman, R. P., and M. Gell-Mann: *Phys. Rev.*, **109**: 193 (1958).
50. Thirring, W.: *Phil. Mag.*, **41**: 1193 (1950).
51. Gershtein, S. S., and Ia. B. Zeldovich: *Zh. Eksperim. i Teor. Fiz.*, **29**: 698 (1955); *Soviet Phys. JETP, (English Transl.)* **2**: 576 (1955).
52. Gell-Mann, M.: *Phys. Rev.*, **111**: 362 (1958).
53. Lee, Y. K., L. W. Mo, and C. S. Wu: *Phys. Rev. Letters*, **10**: 253 (1963). See also ref. 54.
54. Wu, C. S.: *Revs. Mod. Phys.*, **36**: 618 (1964).
55. Bardon, M., P. Franzini, and J. Lee: *Phys. Rev. Letters*, **7**: 23 (1961). Backenstoss, G., B. D. Hyams, G. Knop, P. C. Marin, and U. Stierlin: *Phys. Rev. Letters*, **6**: 415 (1961). Franzini, P.: "Selected Topics on Elementary Particle Physics," Academic, New York, 1963.
56. Ashkin, J., T. Fazzini, G. Fidecaro, E. Goldschmidt-Clermont, N. H. Lipman, A. W. Merrison, and H. Paul: *Nuovo Cimento*, **16**: 490 (1960).
57. Di Capua, E., R. Garland, L. Pondrom, and A. Strelzoff: *Phys. Rev.*, **133**: B1333 (1964).
58. Da Prato, G., and G. Putzolu: *Nuovo Cimento*, **21**: 541 (1961). Chang, N. P.: *Phys. Rev.*, **131**: 1272 (1963).
59. Bartlett, D., S. Devons, S. L. Meyer, and J. L. Rosen: *Phys. Rev.*, **136**: B1452 (1964). Depommier, P., J. Heintze, C. Rubbia, and V. Soergel: *Phys. Rev. Letters*, **5**: 61 (1963). Dunajtsev, A. F., V. I. Petrukhin, Yu. D. Prokshkin, and V. I. Rykalin: Joint Institute for Nuclear Research Report, Dubna, 1964.
60. Primakoff, H.: *Revs. Mod. Phys.*, **31**: 802 (1959). Telegdi, V. L.: *Phys. Rev. Letters*, **8**: 327 (1962).
61. The data, with the exception of $Z = 1$, are from Sens, J. C., R. A. Swanson, V. L. Telegdi, and D. D. Yovanovitch: *Phys. Rev.*, **107**: 1464 (1957).
62. Hildebrand, R. H.: *Phys. Rev. Letters*, **8**: 34 (1962). See also ref. S18.
63. Morita, M., and A. Fujii: *Phys. Rev.*, **118**: 606 (1960).
64. Luyten, J. R., H. P. C. Rood, and H. A. Tolhoek: *Nucl. Phys.*, **41**: 236 (1963).
65. Goldberger, M. L., and S. B Treiman: *Phys. Rev.*, **111**: 355 (1958). Fujii, A., and H. Primakoff: *Nuovo Cimento*, **12**: 327 (1959). Primakoff, H.: *Revs. Mod. Phys.*, **31**: 802 (1959).
66. Bernstein, J., T. D. Lee, C. N. Yang, and H. Primakoff: *Phys. Rev.*, **111**: 313 (1958).
67. Culligan, G., J. F. Lathrop, V. L. Telegdi, and R. Winston: *Phys. Rev. Letters*, **7**: 458 (1961). See also Winston, R., and V. L. Telegdi: *Phys. Rev. Letters*, **7**: 104 (1961).
68. Foldy, L. L., and J. D. Walecka: Congres International de Physique Nucléaire, 1964, 5/C55.
69. Erickson, T., J. C. Sens, and H. P. C. Rood: *Nuovo Cimento*, **39**: 51 (1964).
70. Collard, H., R. Hofstadter, A. Johansson, R. Parks, M. Ryneveld, A. Walker, and M. R. Yearian: *Phys. Rev. Letters*, **11**: 132 (1963).
71. Auerbach, L. B., R. J. Esterling, R. E. Hill, D. A. Jenkins, J. T. Lach, and N. H. Lipman: *Phys. Rev.*, **138**: B127 (1965).
72. Cohen, R. C., S. Devons, and A. D. Kanaris: *Nucl. Phys.*, **57**: 255 (1964). Jenkins, D. A.: *Univ. Calif. Rept.* UCRL-11531, 1964.
73. Rood, H. P. C., and H. A. Tolhoek: *Phys. Rev. Letters*, **6**: 121 (1963).
74. Conversi, M., R. Diebold, and L. di Lella: *Phys. Rev.*, **136**: B1077 (1964).
75. Weinberg, S.: *Phys. Rev. Letters*, **4**: 575 (1960).
76. Wessel, W. R., and P. Phillipson: *Phys. Rev. Letters*, **13**: 23 (1964).
77. Rothberg, J. E., E. W., Anderson, E. J. Bleser, L. M. Lederman, S. L. Meyer, J. L. Rosen, and I. T. Wang: *Phys. Rev.*, **132**: 2664 (1963).
78. Danby, G., J. M. Gaillard, K. Goulianos, L. M. Lederman, N. Mistry, M. Schwartz, and J. Steinberger: *Phys. Rev. Letters*, **9**: 36 (1962). Bernardini, G., G. von Dardel, P. Egli et al.: Proceedings of the Sienna International Conference on Elementary Particles, vol. I, p. 571, 1963. Bingham, H. H., H. Burmeister, D. Cundy, et al.: Proceedings of the Sienna International Conference on Elementary Particles, vol. I, p. 555, 1963. For a theoretical survey of the neutrino experiment at CERN see Bell, J. S., J. Loevseth, and M. Veltman: Proceedings of the Sienna International Conference on Elementary Particles, vol. I, p. 584, 1963.
79. Lee, T. D., and C. N. Yang: *Phys. Rev. Letters*, **4**: 307 (1960). Yamaguchi, Y.: *Progr. Theoret. Phys. (Kyoto)*, **6**: 1117 (1960). Cabibbo, N., and R. Gatto: *Nuovo Cimento*, **15**: 304 (1960).
80. Lee, T. D., and C. N. Yang: *Phys. Rev. Letters*, **4**: 307 (1960). Lee, T. D., P. Markstein, and C. N. Yang: *Phys. Rev. Letters*, **7**: 429 (1961). Bell, J. S., and M. Veltman: *Phys. Rev. Letters*, **5**: 94 (1963). Wu, A. C. T., C. P. Yang, K. Fuchel, and S. Heller: *Phys. Rev. Letters*, **12**: 57 (1964).
81. Sunderland, J., R. Burns, G. Danby, K. Goulianos, E. Hyman, L. Lederman, W. Lee, N. Mistry, J. Rettberg, M. Schwartz, and J. Steinberger: *Bull. Am. Phys. Soc.*, **10**: 35 (1965).

Survey Articles and Books

S1. Rose, M. E.: Beta Radioactivity, in E. U. Condon and H. Odishaw (eds.), "Handbook of Physics," pt. 9, chap. 5, McGraw-Hill, New York, 1958. Other review articles which may be consulted are Jackson, J. D.: Weak Interactions, "Elementary Particle Physics and Field Theory," W. A. Benjamin, New York, 1963; Fronsdal, C. (ed)., "Lecture Notes on Weak Interactions," W. A. Benjamin, New York, 1962.
S2. Schweber, S. S.: "An Introduction to Relativistic Quantum Field Theory," Harper & Row, New York, 1961.
S3. Roman, P.: "Theory of Elementary Particles," North Holland Publishing Company, Amsterdam, 1960.
S4. Pauli, W.: in "Niels Bohr and the Development of Physics," McGraw-Hill, New York, 1955.
S5. Sakurai, J. J.: "Invariance Principles and Elementary Particles," chap. 4, Princeton University Press, Princeton, N.J., 1964.
S6. Lipkin, H. J. (ed.): "Proceedings of Rehovoth Conference on Nuclear Structure," pp. 376–403, North Holland Publishing Company, Amsterdam, 1958.
S7. Källen, G.: "Elementary Particle Physics," p. 191, Addison-Wesley, Reading, Mass., 1964.
S8. Nilsson, J.: On the Structure of Weak Interactions, *Trans. Chalmers Univ. Technol., Gothenberg*, 258, 1962.
S9. Blin-Stoyle, R. J.: in B. J. Verhaar (ed.), "Selected Topics in Nuclear Spectroscopy," p. 213, Wiley, New York, 1964.
S10. Rose, M. E.: "Elementary Theory of Angular Momentum," pp. 220–221, Wiley, New York, 1957.
S11. Tables for the Analysis of β Spectra, *Natl. Bur. Std. (U.S.), Appl. Math. Ser.* 13, 1952. See also Bhalla, C. P., and M. E. Rose: Table of Electronic Radial Functions at the Nuclear Surface and Tangents of Phase Shifts, *Oak Ridge Natl. Lab. Rept.* 3207, 1961.
S12. Freeman, J. M., R. E. White, J. H. Montague, G. Murray, and W. E. Burcham: *Proc. Intern. Conf. Nuclidic Masses, Vienna*, 1963.

S13. Blin-Stoyle, R. J.: ref. S9, p. 213.
S14. Alder, K.: in H. J. Lipkin (ed.), "Proceedings Rehovoth Conference on Nuclear Structure," North Holland Publishing Company, Amsterdam, 1958.
S15. Drell, S. D., and F. Zachariasen: "Electromagnetic Structure of Nucleons," Oxford University Press, London, 1961.
S16. Weidenmüller, H. A.: Nuclear Beta Decay, in B. J. Verhaar (ed.), "Selected Topics in Nuclear Spectroscopy," North Holland Publishing Company, Amsterdam, 1964.
S17. Primakoff, H.: "Proceedings of the Fifth Annual Rochester Conference in High-energy Physics," p. 174, Interscience, New York, 1955.
S18. Ericson, T. E. O.: Report at the International Conference on High Energy Physics, Dubna, 1964.
S19. Maier, E. J. R.: Thesis, Carnegie Institute of Technology, 1962.
S20. Hildebrand, R., and J. H. Doede: Proceedings of the 1962 International Conference on High Energy Physics, CERN, p. 418.
S21. Bertolini, E., A. Citron, G. Gialanella, S. Focardi, A. Murkhin, C. Rubbia, and S. Saporetti: Proceedings of the 1962 International Conference on High Energy Physics, CERN, p. 421.
S22. Evseev, V. S., V. S. Roganov, V. A. Chernogorova, M. M. Szymczak, and Chang Run-Hwa: Proceedings of the 1962 International Conference on High Energy Physics, CERN, p. 423.

Chapter 6

Nuclear Electromagnetic Radiation

By R. W. HAYWARD, National Bureau of Standards

1. Introduction

Much of our present knowledge of the structure of low-lying nuclear states has been obtained through the study of the interaction of the nucleus with the electromagnetic field. This interaction accounts for radiation processes in which one nuclear state undergoes a transition to another nuclear state by the emission or absorption of energy in the form of a quantum of the electromagnetic field. The electromagnetic interaction also causes shifts in the energy eigenvalue of a stationary state of a nucleus when the intrinsic static moments of the nucleus interact with static or quasi-static electric and magnetic fields. These static interactions are treated in Chap. 3.

When the nucleus is in an excited state but below the energy threshold for the emission of nucleons, the electromagnetic interaction is dominant in providing a mode of transition to a state of lower energy. Under certain favorable circumstances the beta-decay interaction may compete, however. In an electromagnetic transition, the internal electric charges, currents, and magnetic moments of the nuclear constituents are coupled to the electromagnetic field in two principal and independent ways. The first involves the direct interaction with a created field, a photon, which carries the energy, momentum, and angular momentum required by conservation laws and the properties of the initial and final nuclear states. The second involves the interaction with the electromagnetic field of one of the atomic electrons in the vicinity of the nucleus. Here the electron carries the energy, momentum, and angular momentum required, not only by the initial and final nuclear states, but also the initial and final electronic states. This process is called internal conversion.

The general implications of the empirical evidence for the structure of nuclear energy levels have been considered in Chap. 1. Each energy level may be characterized by certain stationary state properties, i.e., the eigenvalues of energy, angular momentum, and parity. The study of these properties by the electromagnetic interaction has been the most important tool in making experimental assignments of these eigenvalues. Similar information about these nuclear properties, depending on circumstances, can be obtained by the study of alpha decay discussed in Chap. 4, the study of beta decay discussed in Chap. 5, and by nuclear reactions discussed in Chap. 8. All this experimental information can be correlated with predictions from models of nuclear structure.

The study of electromagnetic transitions by the observation of gamma radiation and internal-conversion electrons has involved experimental procedures that are almost always rather indirect. In the past two decades, the obtainable precision has improved so that it is comparable to that obtained in the study of other physical phenomena. Here the quantum-mechanical features are so pronounced that even rather crude measurements give an unambiguous answer regarding the properties of the nuclear states involved. At the present development of the subject, there is a lack of precision in the theoretical description of emission and absorption of electromagnetic radiation by nuclei that is greater than the lack of experimental precision. This does not imply that the radiation process is not understood but rather that the description of the nucleus is not understood in detail and that most quantitative calculations are based upon simple approximate models of the nucleus.

2. Direct Nuclear Transitions

A complete description of the interaction process requires the quantum theory of radiation as well as the complete knowledge of the dynamical properties of nuclear matter. This would involve the solution of the many-body problem which in turn would require a knowledge of the nuclear forces. The details of such forces are not known. For this reason any theoretical prediction of properties must be made on a simplified model of the nucleus. In particular cases such as the deuteron, a two-nucleon system, the mathematical difficulties of the many-body problem are absent and it is possible to base the calculations on approximate features of the nuclear forces. In many calculations, however, the properties of the interaction process sometimes can be conveniently factored from the properties of the radiating nuclear system.

The quantum-mechanical transition probability for a process going from an initial nuclear state a to a final nuclear state b is given by [1, 2]†

$$\lambda = \frac{2\pi}{\hbar} |H_{ab}|^2 \frac{dn}{dE} \tag{6.1}$$

The quantity H_{ab} is the matrix element of this interaction between the states a and b. The factor dn/dE is the number of final states per unit energy interval.

† Numbers in brackets refer to References at end of chapter.

The interaction between a single nucleon of mass M and charge e and the electromagnetic field described by the 4-vector potential $(\mathbf{A}(\mathbf{r},t),\ i\Phi(\mathbf{r},t))$ is given by the quantum-mechanical Hamiltonian operator for the ith nucleon in the nonrelativistic approximation [1, 2]:

$$H_{int} = -\frac{e_i}{Mc}\mathbf{A}\cdot\mathbf{p}_i + i\frac{e_i\hbar}{2Mc}\nabla\cdot\mathbf{A} + \frac{e_i^2}{2Mc^2}\mathbf{A}^2 + e_i\Phi - g_i\frac{e\hbar}{2Mc}\boldsymbol{\sigma}_i\cdot\nabla\times\mathbf{A} \quad (6.2)$$

The quantity $g_i(e\hbar/2Mc)$ is the magnetic moment of the nucleon; g_i is a factor necessary to denote an anomalous moment; $\boldsymbol{\sigma}_i$ is the Pauli spin operator.

The matrix element H_{ab} is given by

$$H_{ab} = \int\int \psi^*_b(\mathbf{r}_n)H_{int}\psi_a(\mathbf{r}_n)\delta(\mathbf{r}_n - \mathbf{r})\,d\mathbf{r}_n\,d\mathbf{r} \quad (6.3)$$

The coordinates of the nucleon are denoted by $\mathbf{r}_n$ and those for the field by $\mathbf{r}$. The vector delta function specifies the locality of the interaction. Integration of the field coordinates over all space provides an effective evaluation of the field at the nuclear coordinates. To facilitate this calculation, a convenient expansion of the vector delta function [3] is given in terms of spherical harmonics involving the angular variables and a one-dimensional delta function involving the radial coordinate

$$\delta(\mathbf{r}_n - \mathbf{r}) = \sum_{l=0}^{\infty}\sum_{m=-l}^{l}\frac{\delta(r_n - r)}{r_n r}\hat{r}_n^l[i^lY_l^m(\vartheta_n,\varphi_n)]^*\,\hat{r}^l[i^lY_l^m(\vartheta\varphi)] \quad (6.4)$$

Here, we have chosen a spherical basis for the description of the nucleus and the electromagnetic field in order to achieve an irreducible multipole expansion.

We can write the transition Hamiltonian in terms of a generalized current operator, j_μ, a 4-vector which induces a transition from state a to state b of the nucleon as a scalar product with a 4-vector field, A_μ, which is implicitly a 4-vector operator that creates a photon from the vacuum state where none existed before. This Hamiltonian would be represented schematically after integration over the coordinates of the field by

$$H_{ab} = \sum_{\mu=1}^{4}\sum_{l=0}^{\infty}\sum_{m=-l}^{l} <b|j_\mu\hat{r}_n^l[i^lY_l^m\vartheta_n,\varphi_n)]^*|a>\ <A_\mu|\xi_\mu\hat{r}^li^lY_l^m(\vartheta,\varphi)|0> \quad (6.5)$$

We can make use of the unitary and orthogonality properties of vector coupling coefficients to put the operators in the form of irreducible tensor operators [4, 5] defined by

$$\mathbf{T}_{Jl1}{}^M(r,\boldsymbol{\xi}) = \sum_{m,\sigma} <lm1\sigma|JM>r^li^lY_l^m(\vartheta,\varphi)\xi_\sigma \quad (6.6)$$

and

$$T_{Jl0}{}^M(r,1) = \sum_m <lm00|JM>r^li^lY_l^m(\vartheta,\varphi) = \delta_{Jl}\delta_{Mm}i^lr^lY_l^m(\vartheta,\varphi) = \mathcal{Y}_J{}^M(\mathbf{r}) \quad (6.7)$$

The $\mathbf{T}_{Jl1}{}^M$ corresponds to the space part of the 4-vector operator and is proportional to a solid vector spherical harmonic. The $T_{Jl0}{}^M$ is the time part of the 4-vector operator and corresponds to a solid spherical harmonic. The formal transition Hamiltonian written in terms of these tensor operators is

$$H_{ab} = \sum_{J,l,M}[<b|\mathbf{T}_{Jl1}{}^{M+}(r,\mathbf{j})|a>\ <\mathbf{A}|\mathbf{T}_{Jl1}{}^M(r,\boldsymbol{\xi})|0> + <b|T_{Jl0}{}^{M+}(r,j_4)|a>\ <A_4|T_{Jl0}{}^M(r,\xi_4)|0>] \quad (6.8)$$

The Hamiltonian is a scalar quantity formed from the scalar product of two tensor operators, one in the space of the nucleons and one in the space of the electromagnetic field. It is a rotationally invariant quantity, and in this case where the initial (or final) state of the photon is a vacuum state, the Hamiltonian can be written, using the techniques of tensor algebra in quantum mechanics [4, 5], as a product of reduced matrix elements. Here the geometrical dependence, i.e., the M dependence, of the individual constituent spaces has been removed. Equation (6.8) becomes

$$H_{ab} = \sum_{J,l}\frac{(-1)^{j_a+j_b}}{[(2J+1)(2j_a+1)]^{1/2}}\{<b\|\mathbf{T}_{Jl1}(\hat{r},\mathbf{j})\|a>\ <A\|\mathbf{T}_{Jl1}(\hat{r},\boldsymbol{\xi})\|0> + <b\|T_{JJ0}(\hat{r},j_4)\|a>\ <\Phi\|T_{JJ0}(\hat{r}\xi_4)\|0>\} \quad (6.8a)$$

The reduced matrix elements are related to the ordinary matrix elements by means of the Wigner-Eckart theorem [4, 5].

The 4-vector potential, $A_\mu \equiv (\mathbf{A},i\Phi)$, obeys the wave equation in free space:

$$\left(\nabla^2 - \frac{1}{c^2}\frac{\partial^2}{\partial t^2}\right)A_\mu = 0 \quad (6.9)$$

subject to the Lorentz condition,

$$\sum_\mu\frac{\partial A_\mu}{\partial x_\mu} = 0 \quad (6.10)$$

The observable electric and magnetic fields are obtained from this vector potential by the relations

$$\mathbf{E} = -\nabla\Phi - \frac{1}{c}\frac{\partial}{\partial t}\mathbf{A} \quad (6.11a)$$

$$\mathbf{B} = \nabla\times\mathbf{A} \quad (6.11b)$$

Equations (6.11) do not uniquely define A_μ since one is free to make the gauge transformation

$$A'_\mu \rightarrow A_\mu + \frac{\partial}{\partial x_\mu}\Lambda \quad (6.12)$$

without changing the values of E or B, provided that Λ also satisfies the wave equation (6.9).

The energy density of the field is given by the relation [1]

$$w = \frac{1}{8\pi}(|\mathbf{E}|^2 + |\mathbf{B}|^2) \quad (6.13)$$

In a quantum-mechanical formulation the total energy of a photon is $\hbar\omega$, giving the relation [1, 2]

$$\int w\,d\mathbf{r} = \hbar\omega \quad (6.14)$$

A vector potential properly quantized that satisfies the above relations is given by [1, 2]

$$\mathbf{A}_\mu(\mathbf{r},t) = \sum_k \left(\frac{2\pi\hbar c}{kV}\right)^{1/2} \xi_\mu[a_{\mu k}e^{i(\mathbf{k}\cdot\mathbf{r}-\omega t)} + a_{\mu k}^{+}e^{-i(\mathbf{k}\cdot\mathbf{r}-\omega t)}] \tag{6.15}$$

where $\xi_\mu \equiv (\mathbf{e},i)$ is a unit 4-vector defining the polarization state of the photon, $a_{\mu k}^{+}$ is a creation operator and $a_{\mu k}$ is a destruction operator for a photon moving in the k direction. In this chapter we do not make explicit use of the second quantized hermitian operator form for $\mathbf{A}_\mu(\mathbf{r},t)$. Thus, $a_{\mu k}^{+}$ and $a_{\mu k}$ are to be considered as occupation numbers from plane wave photons moving in a particular k direction.

For the present we restrict the 4-vector potential to that involved in an emission process and to that which has a unique wave number k. Furthermore, we can describe the polarization vectors in a spherical basis [4, 5] referred to the direction of propagation, where

$$\xi_{\pm 1} = \mp \frac{1}{\sqrt{2}}(\mathbf{e}_1 \pm i\mathbf{e}_2) \tag{6.16a}$$

$$\xi_0 = \mathbf{e}_3 \tag{6.16b}$$

$$\xi_4 = i \tag{6.16c}$$

The vector $\xi_{\pm 1}$ describes the circular polarization along and opposed to the direction of propagation, the vector ξ_0 describes the longitudinal component, and ξ_4 describes the fourth component or "scalar" part of the potential. We write (6.15) again in simplified form as

$$\mathbf{A}_\sigma(\mathbf{r},t) = \left(\frac{2\pi\hbar c}{kV}\right)^{1/2} \xi_\sigma e^{i(\mathbf{k}\cdot\mathbf{r}-\omega t)} \tag{6.17a}$$

$$\Phi(\mathbf{r},t) = \left(\frac{2\pi\hbar c}{kV}\right)^{1/2} e^{i(\mathbf{k}\cdot\mathbf{r}-\omega t)} \tag{6.17b}$$

The polarization state, expressed in this spherical basis, can be the "spin" of the photon with the three quantized components σ. The exponential term, representing a plane wave, may be expanded into spherical waves by the Rayleigh expansion. Each of these spherical waves of angular momentum l with component m can be coupled to the intrinsic spin vector with component σ to a total angular momentum J with component M. The details are straightforward applications of angular-momentum coupling [4, 5].

The resultant expressions for the 4-vector potential can be put in the form [6]

$$\mathbf{A}(\mathbf{r},t) = \sum_{J,M,\sigma} [(2\pi)(2J+1)]^{1/2}(\sigma)^n \mathfrak{D}_{M\sigma}^{(J)} \left(\frac{2\pi\hbar c}{kV}\right)^{1/2} \mathbf{A}_{JM}(\mathbf{r},t) \tag{6.18a}$$

$$\Phi(\mathbf{r},t) = \sum_{J,M} [(2\pi)(2J+1)]^{1/2}\mathfrak{D}_{M0}^{(J)} \left(\frac{2\pi\hbar c}{kV}\right)^{1/2} \Phi_{JM}(\mathbf{r},t) \tag{6.18b}$$

The expansion is in terms of spherical waves, $\mathbf{A}_{JM}(\mathbf{r},t)$ and $\Phi_{JM}(\mathbf{r},t)$. The quantity $\mathfrak{D}_{M\sigma}^{(J)}$ is an angular function defined in ref. 4 that relates the polarization components σ of a plane wave traveling in a direction $\mathbf{k}$ related to the z axis by means of the Euler angles to the M components of a spherical wave quantized along the z axis. It should be emphasized that the polarization components are not observables for an individual spherical wave. The plane-wave representation is more convenient to use when directional and polarization correlations are to be considered. In the plane-wave representation the quantity dn/dE in expression (6.1) is given by

$$\frac{dn}{dE} = \frac{4\pi p^2\,dp}{(2\pi\hbar)^3\,dE}V = \frac{k^2V}{2\pi^2\hbar c} \tag{6.19}$$

This is used in contrast to the dn/dE in a spherical representation, often used where only transition probabilities are considered. The quantities $\mathbf{A}_{JM}(\mathbf{r},t)$ and $\Phi_{JM}(\mathbf{r},t)$ are listed in Table 6.1 for several different gauges. It should be noted that in the coupling of the spin to the orbital angular momentum l the vector spherical harmonics involve $J = l \pm 1$ and $J = l$. These two groupings have opposite parity, and it is customary to classify the fields involving $J = l \pm 1$ as *electric multipole fields* denoted by a superscript (e) and those fields involving $J = l$ as *magnetic multipole fields* denoted by a superscript (m) and to reserve these designations only for those components corresponding to $\sigma = \pm 1$, that is, transverse waves. The $\sigma = 0$ component is the longitudinal field and is designated by the superscript (l). The longitudinal component has the same parity as the electric fields. This separation into electric, magnetic, and longitudinal fields has been made in Table 6.1. The parameter n in expression (6.18a) is 0 for electric multipole fields and 1 for magnetic multipole fields. Note that the magnetic-multipole terms are the same in all gauges and obey the relations

$$\nabla \cdot \mathbf{A}_{JM}^{(m)} = 0 \tag{6.20a}$$

$$\Phi_{JM}^{(m)} \equiv 0 \tag{6.20b}$$

and furthermore in all gauges

$$\nabla \times \mathbf{A}_{JM}^{(e)} = k\mathbf{A}_{JM}^{(m)} \tag{6.21a}$$

$$\nabla \times \mathbf{A}_{JM}^{(m)} = k\mathbf{A}_{JM}^{(e)} \tag{6.21b}$$

In the long-wavelength limit when $kR \ll 1$, R being the radius of the radiating system, we may use the asymptotic form of the spherical Bessel function

$$j_l(kR) \xrightarrow[kR\to 0]{} \frac{(kR)^l}{(2l+1)!!} \tag{6.22}$$

where $(2l+1)!! = 1.3.5 \cdots (2l+1)$. In the long-wavelength limit the terms involving $j_{J+1}(kR)$ can usually be neglected in comparison with those involving $j_{J-1}(kR)$ since the former are of the order $(kR)^2$ smaller. These approximate forms are listed in Table 6.2. In both Tables 6.1 and 6.2, we use the notation $\mathcal{Y}_{JJ'M} = i^J\mathbf{Y}_{JJ'M}$ and $\mathcal{Y}_{JM} = i^J Y_{JM}$ where the $\mathbf{Y}_{JJ'M}$ are the vector spherical harmonics defined in ref. 4.

It is customary to treat radiation processes in field theory in the solenoidal gauge, as it is conceptually simpler, for both the longitudinal and scalar potentials are identically zero. The current operator takes a particularly simple form when higher-order terms are

TABLE 6.1. ELECTRIC AND MAGNETIC MULTIPOLE FIELDS IN THE J,M REPRESENTATION IN SEVERAL GAUGES

Lorentz gauge

$$\mathbf{A}_{JM}{}^{(e)} = \left(\frac{J+1}{2J+1}\right)^{1/2} j_{J-1}(kr)\mathbf{Y}_{JJ-1M} + \left(\frac{J}{2J+1}\right)^{1/2} j_{J+1}(kr)\mathbf{Y}_{JJ+1M}$$

$$\mathbf{A}_{JM}{}^{(l)} = \left(\frac{2J}{2J+1}\right)^{1/2} j_{J-1}(kr)\mathbf{Y}_{JJ-1M} - \left(\frac{2(J+1)}{2J+1}\right)^{1/2} j_{J+1}(kr)\mathbf{Y}_{JJ+1M}$$

$$\mathbf{A}_{JM}{}^{(m)} = -j_J(kr)\mathbf{Y}_{JJM}$$

$$\Phi_{JM} = (2)^{1/2} j_J(kr)\mathcal{Y}_{JM}$$

Solenoidal gauge

$$\mathbf{A}_{JM}{}^{(e)} = \left(\frac{J+1}{2J+1}\right)^{1/2} j_{J-1}(kr)\mathbf{Y}_{JJ-1M} + \left(\frac{J}{2J+1}\right)^{1/2} j_{J+1}(kr)\mathbf{Y}_{JJ+1M}$$

$$\mathbf{A}_{JM}{}^{(l)} = 0$$

$$\mathbf{A}_{JM}{}^{(m)} = -j_J(kr)\mathbf{Y}_{JJM}$$

$$\Phi_{JM} = 0$$

"Longitudinal" gauge

$$\mathbf{A}_{JM}{}^{(e)} = \left(\frac{J+1}{2J+1}\right)^{1/2} j_{J-1}(kr)\mathbf{Y}_{JJ-1M} + \left(\frac{J}{2J+1}\right)^{1/2} j_{J+1}(kr)\mathbf{Y}_{JJ+1M}$$

$$\mathbf{A}_{JM}{}^{(l)} = -\left(\frac{J+1}{2J+1}\right)^{1/2} j_{J-1}(kr)\mathbf{Y}_{JJ-1M} + \frac{J+1}{J}\left(\frac{J}{2J+1}\right)^{1/2} j_{J+1}(kr)\mathbf{Y}_{JJ+1M}$$

$$\mathbf{A}_{JM}{}^{(m)} = -j_J(kr)\mathbf{Y}_{JJM}$$

$$\Phi_{JM} = -\left(\frac{J+1}{J}\right)^{1/2} j_J(kr)\mathcal{Y}_{JM}$$

Gauge function

$$\Lambda = a\frac{i}{k} j_J(kr)\mathcal{Y}_{JM} \qquad \frac{1}{c}\frac{\partial}{\partial t}\Lambda = a j_J(kr)\mathcal{Y}_{JM}$$

$$\nabla\Lambda = -a\left(\frac{J}{2J+1}\right)^{1/2} j_{J-1}(kr)\mathbf{Y}_{JJ-1M} + a\left(\frac{J+1}{2J+1}\right)^{1/2} j_{J+1}(kr)\mathbf{Y}_{JJ+1M}$$

TABLE 6.2. ELECTRIC AND MAGNETIC MULTIPOLE FIELDS IN THE J,M REPRESENTATION IN SEVERAL GAUGES AND IN THE LONG-WAVELENGTH LIMIT

	Lorentz gauge	Solenoidal gauge	"Longitudinal" gauge
$\mathbf{A}_{JM}{}^{(e)}$	$\left(\frac{J+1}{2J+1}\right)^{1/2}\frac{(kr)^{J-1}}{(2J-1)!!}\mathbf{Y}_{JJ-1M}$	$\left(\frac{J+1}{2J+1}\right)^{1/2}\frac{(kr)^{J-1}}{(2J-1)!!}\mathbf{Y}_{JJ-1M}$	$\left(\frac{J+1}{2J+1}\right)^{1/2}\frac{(kr)^{J-1}}{(2J-1)!!}\mathbf{Y}_{JJ-1M}$
$\mathbf{A}_{JM}{}^{(l)}$	$\left(\frac{2J}{2J+1}\right)^{1/2}\frac{(kr)^{J-1}}{(2J-1)!!}\mathbf{Y}_{JJ-1M}$	0	$-\left(\frac{J+1}{2J+1}\right)^{1/2}\frac{(kr)^{J-1}}{(2J-1)!!}\mathbf{Y}_{JJ-1M}$
$\mathbf{A}_{JM}{}^{(m)}$	$-\frac{(kr)^J}{(2J+1)!!}\mathbf{Y}_{JJM}$	$-\frac{(kr)^J}{(2J+1)!!}\mathbf{Y}_{JJM}$	$-\frac{(kr)^J}{(2J+1)!!}\mathbf{Y}_{JJM}$
Φ_{JM}	$2^{1/2}\frac{(kr)^J}{(2J+1)!!}\mathcal{Y}_{JM}$	0	$-\left(\frac{J+1}{J}\right)^{1/2}\frac{(kr)^J}{(2J+1)!!}\mathcal{Y}_{JM}$

neglected. Equation (6.8) becomes

$$H_{ab} = \sum_{\sigma} \sum_{J=1}^{\infty} \sum_{M=-J}^{J} \left\{ -\frac{e_i}{Mc} <b|\mathbf{T}_{JJ-11}{}^{M+}(\hat{r}_i,\mathbf{p}_i)|a> <\mathbf{A}^{(e)}|\mathbf{T}_{JJ-11}{}^{M}(\hat{r},\xi)|0> - g_i \frac{ek}{2Mc} <b|\mathbf{T}_{JJ1}{}^{M+}(\hat{r}_i,\boldsymbol{\sigma}_i)|a> <\mathbf{A}^{(e)}|\mathbf{T}_{JJ1}(\hat{r},\xi)|0> - \frac{e_i}{Mc} <b|\mathbf{T}_{JJ1}{}^{M+}(\hat{r}_i,\mathbf{p}_i)|a> <\mathbf{A}^{(m)}|\mathbf{T}_{JJ1}(\hat{r},\xi)|0> - g_i \frac{ek}{2Mc} <b|\mathbf{T}_{JJ-11}{}^{M+}(\hat{r}_i,\boldsymbol{\sigma}_i)|a> <\mathbf{A}^{(m)}|\mathbf{T}_{JJ-11}(\hat{r},\xi)|0> \right\} \quad (6.23)$$

The terms involving T_{JJ0} can be dropped since the matrix elements are zero in this gauge. If Eq. (6.8*a*) is used, the transition Hamiltonian can be written in terms of reduced matrix elements where those for the photon have been explicitly evaluated:

$$H_{ab} = \sum_J \sum_\sigma (-1)^{j_a+j_b} \left[\frac{4\pi^2\hbar c}{kV}(2J+1)\frac{J+1}{J}\right]^{1/2} \mathfrak{D}_{M\sigma}{}^{(J)*} \frac{(k)^J}{(2J+1)!!} \times \{\mathfrak{M}(EJ) + \mathfrak{M}(E'J) + \mathfrak{M}(M'J) + \mathfrak{M}(MJ)\} \quad (6.24)$$

where the $\mathfrak{M}(EJ)$ are designated as the reduced matrix elements for electric-multipole emission and the $\mathfrak{M}(MJ)$ are those for magnetic-multipole emission.

Expressed in terms of reduced matrix elements of the tensor operators, the quantities $\mathfrak{M}$ are

$$\mathfrak{M}(EJ) = -\frac{e_i}{Mc}\frac{[J(2J+1)]^{1/2}}{k}\frac{<b\|\mathbf{T}_{JJ-11}(r_i,\mathbf{p}_i)\|a>}{(2j_a+1)^{1/2}} \quad (6.25a)$$

$$\mathfrak{M}(E'J) = -g_i\frac{e\hbar k}{2Mc}\left(\frac{J}{J+1}\right)^{1/2}\frac{<b\|\mathbf{T}_{JJ1}(r_i,\boldsymbol{\sigma}_i)\|a>}{(2j_a+1)^{1/2}} \quad (6.25b)$$

$$\mathfrak{M}(M'J) = -\frac{e_i}{Mc}\left(\frac{J}{J+1}\right)^{1/2}\frac{<b\|\mathbf{T}_{JJ1}(r_i,\mathbf{p}_i)\|a>}{(2j_a+1)^{1/2}} \quad (6.25c)$$

$$\mathfrak{M}(MJ) = -g_i\frac{e\hbar}{2Mc}[J(2J+1)]^{1/2}\frac{<b\|\mathbf{T}_{JJ-11}(r_i,\boldsymbol{\sigma}_i)\|a>}{(2j_a+1)^{1/2}} \quad (6.25d)$$

By neglecting factors of the order unity, the relative magnitudes of the two electric and the two magnetic reduced transition probabilities can be found by taking the ratio

$$\frac{\mathfrak{M}(E'J)}{\mathfrak{M}(EJ)} \approx \left(\frac{e\hbar k}{Mc}R^J\right)\left(\frac{e\hbar}{Mck}R^{J-2}\right)^{-1} = 0[(kR)^2] \quad (6.26a)$$

$$\frac{\mathfrak{M}(M'J)}{\mathfrak{M}(MJ)} \approx \left(\frac{e\hbar}{Mc}R^{J-1}\right)\left(\frac{e\hbar}{Mc}R^{J-1}\right)^{-1} = 0[1] \quad (6.26b)$$

For a transition of the order of 1 Mev energy occurring in a medium-weight nucleus, kR is of the order of magnitude 10^{-2} so that $\mathfrak{M}(E'J)$ can usually be neglected in comparison with $\mathfrak{M}(EJ)$. The $\mathfrak{M}(M'J)$ and $\mathfrak{M}(MJ)$ are of the same order of magnitude and both must be retained.

It is possible to calculate the transition Hamiltonian in another gauge, the "longitudinal gauge" where the contributions of $\mathbf{A}_{JM}{}^{(e)}$ plus $\mathbf{A}_{JM}{}^{(l)}$ are negligible and the major contribution to electric transitions comes from the scalar potential Φ_{JM}. The magnetic contributions are, of course, the same in any gauge. Here the portion of H_{ab} for electric transitions becomes

$$H_{ab} = \sum_J \left[\frac{4\pi^2\hbar c}{kV}(2J+1)\frac{J+1}{J}\right]^{1/2} \mathfrak{D}_{M0}{}^{(J)*} (-1)^{j_a+j_b}\frac{(k)^J}{(2J+1)!!}\mathfrak{M}'(EJ) \quad (6.27)$$

where

$$\mathfrak{M}'(EJ) = e_i\frac{<b\|T_{JJ0}(r_i,1)\|a>}{(2j_a+1)^{1/2}} \quad (6.28)$$

The transition probability, yet to be calculated, is independent of the gauge so that a relationship between the magnitudes of certain matrix elements exists:

$$<b\|T_{JJ0}(r_i,1)\|a> = -\frac{[J(2J+1)]^{1/2}}{Mck}<b\|\mathbf{T}_{JJ-11}(r_i,p_i)\|a> \quad (6.29)$$

This relationship constitutes Siegert's theorem that relates the current and charge matrix elements for electric-multipole radiation in the long-wavelength limit in a way that is totally independent of the details of the radiating system. The reduced matrix elements for the transition can be evaluated by conventional techniques of angular-momentum theory. They may be related to the matrix element of a single component of a tensor operator by the Wigner-Eckart theorem [4, 5]:

$$<j_bm_b|\mathbf{T}_{Jl1}{}^M|j_am_a> = (-1)^{j_a-m_a}<j_bm_bj_a-m_a|JM>\frac{<j_b\|\mathbf{T}_{Jl1}\|j_a>}{(2J+1)^{1/2}} \quad (6.30)$$

As a simple example, the z components of the matrix elements for electric- and magnetic-dipole transitions for an initial state of zero angular momentum are

$$[\mathfrak{M}(E1)]_0 = -\frac{e_i}{Mck}\left(\frac{3}{4\pi}\right)^{1/2}<b|(\mathbf{p}_i)_0|a> \quad (6.31a)$$

$$[\mathfrak{M}(E'1)]_0 = -ig_i\frac{ek}{4Mc}\left(\frac{3}{4\pi}\right)^{1/2}<b|i(\boldsymbol{\sigma}_i\times\mathbf{r}_i)_0|a> \quad (6.31b)$$

$$[\mathfrak{M}(M1)]_0 = -\frac{e_i}{2Mc}\left(\frac{3}{4\pi}\right)^{1/2}<b|i\mathbf{l}_i)_0|a> \quad (6.31c)$$

$$[\mathfrak{M}(M'1)]_0 = -g_i\frac{e}{2Mc}\left(\frac{3}{4\pi}\right)^{1/2}<b|(\boldsymbol{\sigma}_i)_0|a> \quad (6.31d)$$

and in the "longitudinal gauge," for example,

$$[\mathfrak{M}'(E1)]_0 = e_i\left(\frac{3}{4\pi}\right)^{1/2}<b|iz_i|a> \quad (6.32a)$$

$$[\mathfrak{M}'(E2)]_0 = e_i\left(\frac{5}{16\pi}\right)^{1/2}<b|(3z_i^2-r_i^2)|a>, \quad (6.32b)$$

In transitions of a given multipole order J, the parities of the electric and magnetic transitions are opposite and hence the transition probabilities for each type of radiation are to be separately considered. These transition probabilities are obtained by integrating expression (6.1) over all angles of the photon, giving [7, 8]

$$\lambda_\gamma(EJ) = \frac{8\pi(J+1)}{J[(2J+1)!!]^2}\left(\frac{\omega}{c}\right)^{2J+1}\frac{1}{\hbar}B(EJ) \tag{6.33a}$$

$$\lambda_\gamma(MJ) = \frac{8\pi(J+1)}{J[(2J+1)!!]^2}\left(\frac{\omega}{c}\right)^{2J+1}\frac{1}{\hbar}B(MJ) \tag{6.33b}$$

The quantities $B(EJ)$ and $B(MJ)$ are called the reduced transition probabilities for electric- and magnetic-multipole emission, respectively, and are given by

$$B(EJ) = |\mathfrak{M}'(EJ)|^2 = |\mathfrak{M}(EJ)|^2 \tag{6.34a}$$

$$B(MJ) = |\mathfrak{M}(MJ) + \mathfrak{M}(M'J)|^2 \tag{6.34b}$$

These reduced transition probabilities depend in detail on the radiating system and are parameters that are to be determined in an experiment or to be calculated when based on a model of the nucleus.

For given initial and final states, the transition proceeds usually by the lowest multipole order allowed by selection rules. For a given parity change, an estimate of possible mixtures can be obtained from the order of magnitude of the ratio of transition probabilities

$$\frac{\lambda_\gamma(EJ+1)}{\lambda_\gamma(MJ)} = 0\left[\left(\frac{Mc^2}{\hbar\omega}\right)^2 (kR)^4\right] \tag{6.35a}$$

$$\frac{\lambda_\gamma(MJ+1)}{\lambda_\gamma(EJ)} = 0\left[\left(\frac{\hbar\omega}{Mc^2}\right)^2\right] \tag{6.35b}$$

$$\frac{\lambda_\gamma(EJ+2)}{\lambda_\gamma(EJ)} = 0\,[(kR)^4] \tag{6.35c}$$

$$\frac{\lambda_\gamma(MJ+2)}{\lambda_\gamma(MJ)} = 0[(kR)^4] \tag{6.35d}$$

All ratios but the first are negligible, which indicates that $EJ + 1$ and MJ admixtures are possible in a transition. This fact is borne out by experiment. There are a few instances where the E1 transition probability is sufficiently small that E1-M2 mixtures have been observed, however.

Single-particle Transitions. The independent-particle model of the nucleus [9, 10] is a particularly simple model for theoretical calculation of the reduced transition probabilities, since the nuclear transitions may be considered to involve the transition of only one individual nucleon between states described by j-j coupling. The expressions can be written in center-of-mass coordinates that refer to this single particle.

It might appear that if a single neutron were responsible for the transition no electric-multipole moment would be present since there is no electric interaction with the electromagnetic field. However, the motion of the rest of the nucleus about the center of mass makes a contribution as though it had an effective charge, $-Ze/A^J$. This term is dependent on the multipole order, being largest in the electric-dipole case. For the above reasons the effective charge for an odd proton is $e(1 - Z/A^J)$. In a medium-weight nucleus the effective charge e_i is approximately $-(e/2)$ for an odd neutron and $e/2$ for an odd proton in an electric-dipole transition.

A nucleon in a central, velocity-independent potential is in an eigenstate

$$\psi_{jl}^{mj} = u_l(r)\Theta_{jl}^{mj} \tag{6.36}$$

where

$$\Theta_{jl}^{mj} = \sum_{m_l m_s} i^l Y_l^{m_l}\chi_s^{m_s} \langle l m_l s m_s | j m_j \rangle$$

where j, l, and m_j are the quantum numbers of the particular eigenstate formed by coupling $\mathbf{l}$ and $\mathbf{s}$ to a given $\mathbf{j}$. The corresponding state vector for a particular state a is written as

$$|a\rangle \equiv |j_a l_a m_a\rangle \tag{6.37}$$

The reduced electric transition probability for a pure configuration can be evaluated by using techniques of the quantum theory of angular momentum, giving [11, 12]

$$B(EJ) = \frac{e^2}{4\pi} S_E(j_a J j_b) \left| \int u^*_b(r_i) r_i^J u_a(r_i) r_i^2\, dr_i \right|^2 \tag{6.38}$$

The quantity $S_E(j_a J j_b)$ is a factor resulting from the angular-momentum parts of the matrix elements and is defined by

$$S_E(j_a J j_b) = (2j_b + 1)(2l_a + 1)(2J + 1)(2l_b + 1) \begin{pmatrix} l_b & l & l_a \\ 0 & 0 & 0 \end{pmatrix}^2 \begin{Bmatrix} l_b & j_b & \tfrac{1}{2} \\ j_a & l_a & J \end{Bmatrix}^2 \tag{6.39}$$

The last two terms are 3-j and 6-j symbols defined and tabulated in refs. 4 and 13, respectively.

The factor $S_E(j_a J j_b)$ is given in Table 6.3 for the special case where $J = |j_b - j_a|$.

TABLE 6.3 COEFFICIENTS $S_E(j_a J j_b)$
(For special case where $|j_a - j_b| = J$)

j_b \ j_a	$\frac{1}{2}$	$\frac{3}{2}$	$\frac{5}{2}$	$\frac{7}{2}$	$\frac{9}{2}$	$\frac{11}{2}$	$\frac{13}{2}$
$\frac{1}{2}$		1	1	1	1	1	1
$\frac{3}{2}$	2		$\frac{6}{5}$	$\frac{9}{7}$	$\frac{4}{3}$	$\frac{15}{11}$	$\frac{18}{13}$
$\frac{5}{2}$	3	$\frac{9}{5}$		$\frac{9}{7}$	$\frac{10}{7}$	$\frac{50}{33}$	$\frac{225}{143}$
$\frac{7}{2}$	4	$\frac{18}{7}$	$\frac{12}{7}$		$\frac{4}{3}$	$\frac{50}{33}$	$\frac{700}{429}$
$\frac{9}{9}$	5	$\frac{10}{3}$	$\frac{50}{21}$	$\frac{5}{3}$		$\frac{15}{11}$	$\frac{225}{143}$
$\frac{11}{2}$	6	$\frac{45}{11}$	$\frac{100}{33}$	$\frac{25}{11}$	$\frac{18}{11}$		$\frac{18}{13}$
$\frac{13}{2}$	7	$\frac{63}{13}$	$\frac{525}{143}$	$\frac{1225}{429}$	$\frac{315}{143}$	$\frac{21}{13}$	

The selection rules for the transition, resulting from the conservation of angular momentum and parity, are contained implicitly in the factor $S_E(j_a J j_b)$. The transition probability has a nonzero value only when

$$|j_a - j_b| \le J \le j_a + j_b \tag{6.40}$$

and when

$$l_a - l_b + J = 2n \qquad n \text{ an integer} \tag{6.41}$$

The latter restriction is the parity selection rule. The parity carried off by the radiation where $J = l$ is even (odd) must be that of the relative parity of the initial and final states, where $l_a - l_b$ is even (odd).

The radial integral depends on the form of the nuclear wave functions which result from a specific nuclear potential. If only order-of-magnitude estimates are desired, the radial wave functions can be assumed to be constant inside and zero outside the nuclear radius R. Then

$$|\int u^*_b(r_i) r_i^J u_a(r_i) r_i^2\, dr_i|^2 = \left(\frac{3}{3+J}\right)^2 R^{2J} \tag{6.42}$$

More refined calculations give similar functional dependences on J but modified by a factor $\mathcal{R}$ differing somewhat from unity.

Noting that $S_E(j_aJj_b)$ is of the order unity, a rough order-of-magnitude estimate for the reduced electric transition probability is

$$B(EJ) = \frac{e^2}{4\pi}\left(\frac{3}{3+J}\right)^2 R^{2J} \tag{6.43}$$

In a similar manner, the reduced transition probability for magnetic radiation can be calculated. Here the matrix elements must be evaluated in spin space as well as orbital space, and we get the simplest form when $J - 1 = l = |l_a - l_b|$, giving [11, 12]

$$B(MJ) = \frac{1}{4\pi}\left(\frac{e\hbar}{Mc}\right)^2 \left(g_i J - \frac{J}{J+1}\right)^2 S_M(j_aJj_b) \, |\int u^*_b(r_i) r_i^{J-1} u_a(r_i) r_i^2\, dr_i|^2 \tag{6.44}$$

Here the quantity $S_M(j_aJj_b)$ is of order unity but more complicated than (6.39), being defined by [14]

$$S_M(j_aJj_b) = \frac{3}{2}(2j_b+1)\frac{(2J+1)^2}{J}(2l_a+1)(2J-1)(2l_b+1)\begin{pmatrix} l_a & J-1 & l_b \\ 0 & 0 & 0\end{pmatrix}^2 \begin{Bmatrix} l_b & j_b & \frac{1}{2} \\ j_a & l_a & J-1\end{Bmatrix}^2 \begin{Bmatrix} l_a & l_b & J-1 \\ \frac{1}{2} & \frac{1}{2} & 1 \\ j_a & j_b & J\end{Bmatrix}^2 \tag{6.45}$$

The last term in the braces is the 9-j symbol defined in ref. 4. Again, S_M is a factor of order unity. The squared term containing the contribution of the orbital and intrinsic angular-momentum terms depends on whether the particle making the transition is an odd neutron or proton. The anomalous g factor is 2.792 for a free proton and -1.913 for a free neutron. A rough estimate of the magnitude of the squared term is about 10. The radial term becomes

$$|\int u^*_b(r_i) r_i^{J-1} u_a(r_i) r_i^2\, dr_i|^2 = \left(\frac{3}{2+J}\right)^2 R^{2J-2} \tag{6.46}$$

The reduced magnetic transition probability of order J is approximately

$$B(MJ) \approx \frac{10}{4\pi}\left(\frac{e\hbar}{Mc}\right)^2 \left(\frac{3}{2+J}\right)^2 R^{2J-2} \tag{6.47}$$

Putting these rough estimates for the reduced transition probabilities into Eqs. (6.33), we obtain

$$\lambda_\gamma(EJ) = 2\frac{J+1}{J[(2J+1)!!]^2}\left(\frac{3}{3+J}\right)^2 \frac{e^2}{\hbar c}(kR)^{2J}\omega \tag{6.48}$$

$$\lambda_\gamma(MJ) = 20\frac{J+1}{J[(2J+1)!!]^2}\left(\frac{3}{2+J}\right)^2 \frac{e^2}{\hbar c}\left(\frac{\hbar}{McR}\right)^2 (kR)^{2J}\omega \tag{6.49}$$

These transition probabilities are often referred to as the Weisskopf estimates [7] for single-particle transitions, and it is convenient to express any gamma-ray transition probability in units of these estimates.

These formulas are most sensitive to the radial distribution of the states involving the transforming nucleon and not necessarily the radial distribution of nuclear matter. The assumption that R is equal to $1.4A^{1/3}\, 10^{-13}$ cm gives reasonable predictions but may not be necessarily valid. A 10 per cent change in the nuclear radius would alter the transition probability by a factor of 2 for a M4 transition, for example.

The transition probability formulas have been derived on extremely simplified assumptions so that the currents within the nucleus may be treated nonrelativistically and so that all electromagnetic effects result from the motion of the transforming nucleon rather than partially resulting from mesonic exchange currents between the nucleons. The estimates for the transition probabilities $\lambda_\gamma(EJ)$ and $\lambda_\gamma(MJ)$, expressed as a function of the transition energy, are plotted in Figs. 6.1 and 6.2, respectively.

The half-life of the state for gamma-ray emission is related to the transition probability by

$$\tau_{1/2} = 0.693/\lambda_\gamma(J) \tag{6.50}$$

In detailed comparison with experiment, the exact expressions for $B(EJ)$ and $B(MJ)$ [(6.38) and (6.44)] are to be used. It should be remembered, however, that they have been calculated as transitions of a single nucleon in the independent-particle model assuming a j-j coupling scheme. In this model the low-lying states are given by the excitation of one nucleon to different states. In this model the total nuclear spin $\mathbf{J}_a = \sum_{i=1}^{A} \mathbf{j}_i$ which can be written as $\mathbf{J}_a = \mathbf{j}_a + \sum_{i=1}^{A-1} \mathbf{j}_i$ when the latter term represents the vector sum of the spins of all nucleons other than the single-particle spin j_a. In a transition from a state J_a to J_b, where $\mathbf{J}_b = \mathbf{j}_b + \sum_{i=1}^{A-1} \mathbf{j}_i$, the transition may be considered a single particle if the latter term $\sum_{i=1}^{A-1} \mathbf{j}_i = \mathbf{j}_{\text{eff}}$ is identical in each state J_a and J_b. However $\mathbf{j}_{\text{eff}}$ may be made up in a number of different ways from its constituents; i.e., different configurations give the same core spin. The matrix elements will

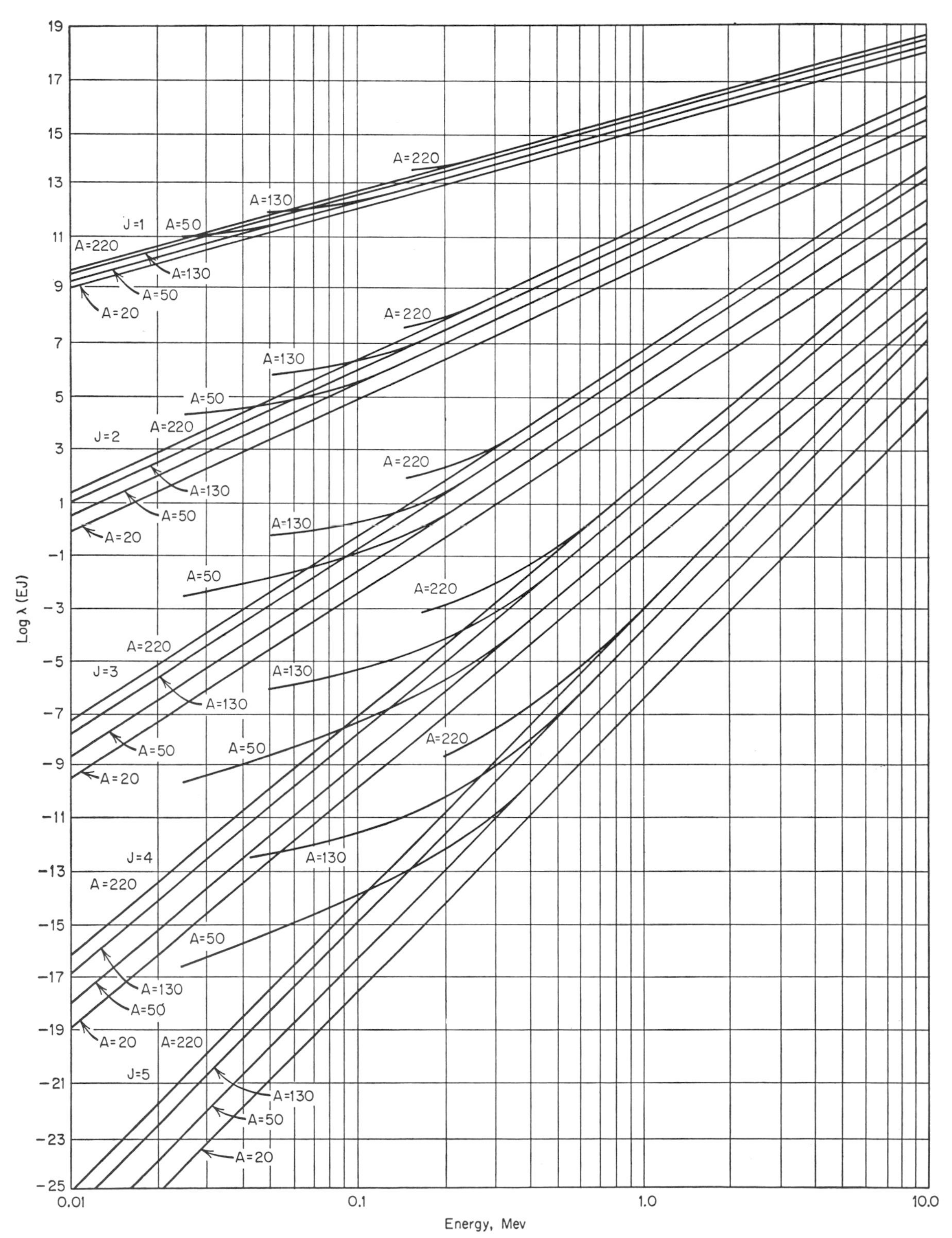

Fig. 6.1. The transition probability for electric transitions of a single odd proton in nuclei of mass number 20, 50, 130, and 220 for the first five multipole orders. The straight lines represent the gamma-ray transition probabilities, and the curved deviations from the straight lines represent the increase in the transition probability due to internal conversion.

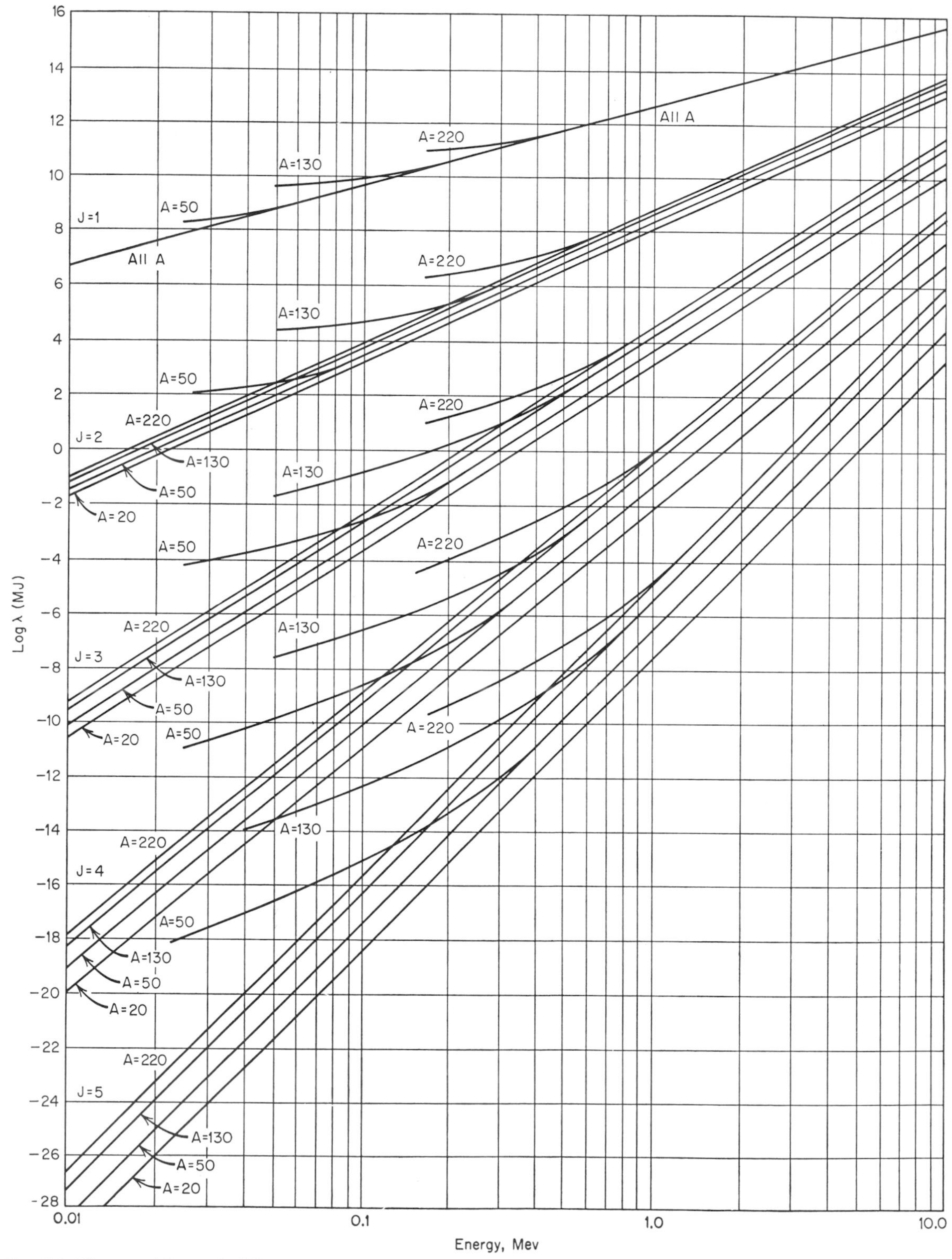

FIG. 6.2. The transition probability for magnetic transitions of a single odd proton in nuclei of mass number 20, 50, 130, and 220 for the first five multipole orders. The straight lines represent the gamma-ray transition probabilities, and the curved deviations from the straight lines represent the increase in the transition probability due to internal conversion.

be nonzero only to the degree of overlap of the configurations of the remaining nucleons in the actual initial and final states, for in this picture only the single particle is interacting with the electromagnetic field. When the so-called configuration mixing occurs the transition probability may be reduced by a factor even as large as 10^{-6}. These configurations may not involve all the nucleons but only those in unfilled shells because the angular momenta in filled shells uniquely couple to zero. Shell-model calculations of this type have been treated in detail; some of the procedures may be found in refs. 14 and 15. The evaluation of multiparticle matrix elements is rather involved.

On the other hand, many of the nucleons may be collectively and coherently coupled to the electromagnetic field in a transition and may actually enhance the transition probability over that for single-particle transitions. These effects can be calculated simply in the collective model of Bohr and Mottelson [16].

Collective Transitions. One of the assumptions of the independent-particle model is that the potential in which each particle moves is generated collectively by all the particles in the nucleus and that this potential can be considered to be static. However, this potential may have a variation associated with the collective motions of the system of particles, so bound together that they have the degrees of freedom of a hydrodynamical medium having characteristic vibrational and rotational motions. This generalization constitutes the collective model [16] which has been very successful in describing certain features of nuclei. The independent-particle and collective motions are not necessarily independent of one another but are coupled together, the amount of coupling depending on circumstances. In the region of closed shells when the coupling is weak and in the region of highly deformed nuclei where the coupling is strong, one expects to be able to distinguish between single-particle transitions which are associated with a change in the quantum state of the nucleon and collective transitions which involve vibrations and rotations of the nucleus as a whole, leaving the particle structure unaltered. There is an intermediate region of coupling that makes the separation of these two types of transitions more complicated theoretically. When single-particle effects are considered, the assumption is made that the coupling is so weak that the collective terms do not give any contribution. In general, each shell-model state has an associated rotational and vibrational band structure. The vibrational energy states are nearly equally spaced, being approximately 1 Mev for medium-weight nuclei. The rotational excitations, on the other hand, depend on the deformation and are lower than the vibrational excitations for highly deformed nuclei. The theoretical treatment of this collective behavior in terms of single-particle properties involves techniques similar to those used in the theory of superconductivity where certain short-range pairing forces and long-range distorting forces are introduced ad hoc. No discussion of this model can be given here. In the limit of extreme distortions, the rotational model of the nucleus is described by eigenfunctions of a rigid rotator with angular momentum **R** coupled to an intrinsic shell-model angular momentum **J** to give a resultant angular momentum **I**. A rather good semiquantitative picture of the collective motion results [16, 17]. This coupling scheme is shown in Fig. 6.3 for an axially symmetric nucleus. The energy eigenvalue is given by

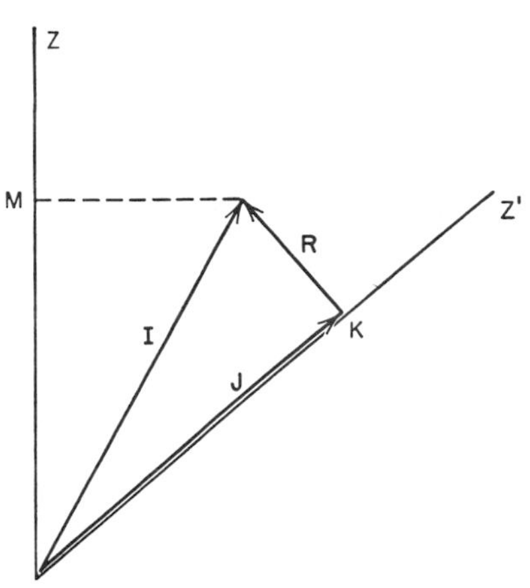

FIG. 6.3. The coupling scheme between the intrinsic and collective motions in the unified model. In the axially symmetric model, the nuclear spin **J** lies along the axis of symmetry of the nucleus z' with projection K. The collective angular momentum **R** is coupled to **J** to give a total angular momentum **I**. The component of **I** along the z axis is designated by M, and the component of **I** along the axis of symmetry is identical to K.

$$E_{\text{rot}} = \frac{\hbar^2}{2\mathcal{J}}[I(I+1) - K(K+1)] \qquad (6.51)$$

where $\mathcal{J}$ is the effective moment of inertia of the nucleus, K is equal to the angular momentum of the ground state, and I is the angular momentum of the excited state. I takes on certain restricted values, i.e., when

$$K = 0 \qquad I = 2, 4, 6$$
$$K \neq 0, \tfrac{1}{2} \qquad I = K+1, K+2, K+3$$

When $K = \frac{1}{2}$ there is an anomaly resulting in a more complicated expression for E_{rot}.

The eigenfunctions for the rotational states may be described in terms of an intrinsic shell-model wave function times a rotational wave function in the form of a $\mathfrak{D}$ function mentioned after expression (6.18). In a transition within a band, the intrinsic wave function is the same in both initial and final states. The reduced transition probability may be expressed in terms of the ground-state parameters.

The collective picture of the nucleus allows successive levels having a spin difference of one or two units of angular momentum and the same parity. Therefore, the transitions within a band are mainly of the electric-quadrupole or magnetic-dipole type. Calculation of the reduced matrix element is done in the same manner as in the single-particle transitions.

In general, the multipole operators are more complicated since the intrinsic and collective parts cannot be simultaneously diagonalized. For example, the electric-quadrupole operator has the following μ

component:

$$[\mathfrak{M}'(E2)_{op}]_\mu = -\sum_\nu er^2 \mathfrak{D}_{\mu\nu}^{(2)} Y_2^\nu \qquad (6.52)$$

resulting from a rotation of the intrinsic system's axis to the z axis. The magnetic-dipole operator has an additional term due to the collective angular momentum $\mathbf{R}$, that is

$$[\mathfrak{M}(M1)_{op}]_\mu = -\frac{e}{2Mc}\left(\frac{3}{4\pi}\right)^{1/2} \left\{\sum_\nu \mathfrak{D}_{\mu\nu}^{(1)}(ig_l l_\nu + g_\sigma \sigma_\nu) + g_R R_\mu\right\} \qquad (6.53)$$

The rotational eigenfunctions are not eigenfunctions of l_ν and σ_ν, but the z component in this basis leads to an expression for the magnetic moment given by

$$\frac{e\hbar}{2Mc}\left(\frac{3}{4\pi}\right)^{1/2}\left[g_R \frac{I(I+1) - K^2}{I+1} + g_K \frac{K^2}{I+1}\right] \qquad (6.54)$$

where g_R and g_K denote the gyromagnetic ratio of rotational and intrinsic motion, both depending on the details of the intrinsic structure. A rough estimate of g_R gives $g_R \approx Z/A$. Similar estimates occur from transitions involving vibrational states but will not be discussed. We get for the reduced electric-quadrupole transition probability between two states in a rotational band having spins I_a and I_b

$$B(E2) = \frac{5}{16\pi} e^2 Q_0^2 \langle I_a K 2 0 | I_b K\rangle^2 \qquad (6.55)$$

where Q_0 is the intrinsic quadrupole moment defined by the matrix element of (6.32*b*) between the intrinsic single-particle ground state

$$\langle a|\mathfrak{M}'(E2)_z|a\rangle = \left(\frac{5}{16\pi}\right)^{1/2} eQ_0 \qquad (6.56)$$

By similar methods we get the reduced magnetic-dipole transition probability

$$B(M1) = \frac{3}{4\pi}\left(\frac{e\hbar}{2Mc}\right)^2 (g_K - g_R)^2 K^2 \langle I_a K 1 0 | I_b K\rangle^2 \qquad (6.57)$$

If we consider decay transitions where we have $I + 1 \to I$, the above transition probabilities become

$$B(E2) = \frac{15}{16\pi} e^2 Q_0^2 K^2 \frac{(I+1-K)(I+1+K)}{I(I+1)(I+2)(2I+3)} \qquad (6.58)$$

$$B(M1) = \frac{3}{4\pi}\left(\frac{e\hbar}{2Mc}\right)^2 (g_K - g_R)^2 K^2 \frac{(I+1-K)(I+1+K)}{(I+1)(2I+3)} \qquad (6.59)$$

and for the case where $I + 2 \to I$ we have

$$B(E2) = \frac{15}{32\pi} e^2 Q_0^2 \frac{(I+1-K)(I+1+K)(I+2-K)(I+2+K)}{(I+1)(2I+3)(I+2)(2I+5)} \qquad (6.60)$$

In the special case when $K = \frac{1}{2}$ there is an anomalous sequence of states; the E2 transition probability is still given by (6.58) but the M1 transition probability depends on the detailed structure of the states.

Experimental Transition Probabilities. Experimental data for comparison with some of these theoretical predictions of the transition probability may be obtained from a number of sources. The most important are direct lifetime measurements. Transition probabilities may also be experimentally deduced from energy widths of levels obtained in nuclear reactions, from Coulomb excitation of levels by charged particles, from intensities of neutron-capture gamma rays, from intensities of long-range α-particle groups, etc. However, it is often necessary to have supporting information as to the correct assignment of the multipole order of the transition. This information is usually obtained from internal-conversion data and shell- or collective-model predictions.

Figures 6.4 and 6.5 compare some of the experimental information [18] with the calculated transition

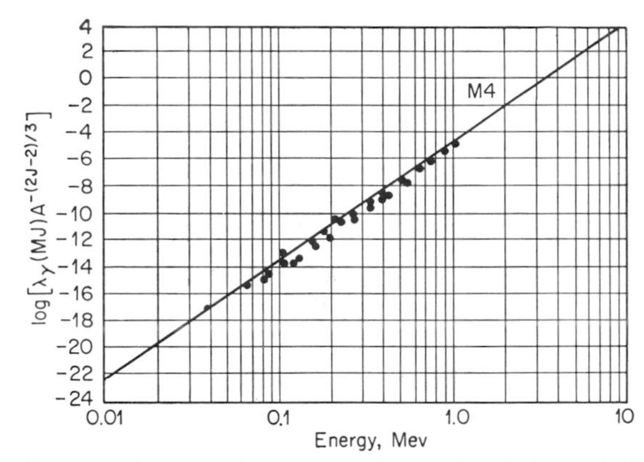

FIG. 6.4. A comparison of experimental and theoretical gamma-ray transition probabilities for a few M4 transitions. The experimental data have been corrected for internal-conversion and statistical effects.

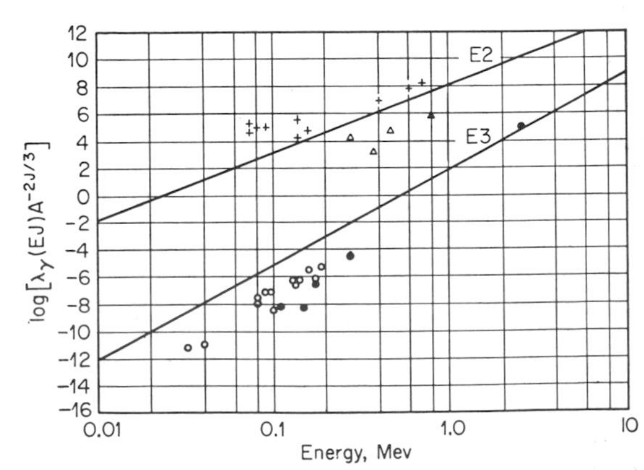

FIG. 6.5. A comparison of experimental and theoretical gamma-ray transition probabilities for a few E2 and E3 transitions. The experimental data have been corrected for internal conversion but statistical effects have been neglected.

probabilities. The mass effects on the transition probability have been removed by plotting the quantities log $[\lambda_\gamma(EJ)A^{-2J/3}]$ and log $[\lambda_\gamma(MJ)A^{-(2J-2)/3}]$ as a function of the transition energy.

All the observed transition probabilities are equal to, or smaller than, the transition probabilities theoretically expected on the basis of single-particle transitions, except for a large number of E2 transitions which have higher transition probabilities than expected on the basis of the single-particle model. The latter have been identified as collective rotational transitions. An additional piece of evidence that they are of a rotational nature comes from the comparison of the observed spectroscopic quadrupole moment Q and the intrinsic quadrupole moment Q_0 used in expressions (6.55) and (6.56). They are related by the expression

$$Q = Q_0 \frac{I(2I - 1)}{(I + 1)(2I + 3)}$$

When the nucleus has zero spin, a comparison may be made with a neighboring isotope of odd mass number.

The M4 transitions, for which a considerable amount of experimental data exists, are shown in Fig. 6.4. These transitions exhibit a markedly uniform tendency to agree with the calculated transition probability. The other magnetic transitions (M3, M2, and M1) scatter considerably more in the direction of lower transition probabilities. However, the multipolarity assignment is always clear-cut.

The largest deviations are to be found in the E3 and E2 transitions shown in Fig. 6.5. The E1 transition-probability data are rather limited for two reasons. The main reason is that there are very few low-lying nuclear states that would give rise to E1 radiation, a fact consistent with the single-particle or collective model. Also the lifetimes are usually shorter than what can be measured with present-day experimental techniques. However, a few E1 transition probabilities have been deduced from energy widths of levels observed in nuclear reactions occurring in light-mass nuclei. The E2 transitions are not unusual when a subclassification is made into single-particle and collective transitions.

The E3 transitions can be subdivided into two groups, those of the type consisting of pure single-particle transitions and those consisting of a transition of several particles. If one of the states involved in the transition is described by a configuration of several independent particles, the nuclear wave functions are then linear combinations of the products of the individual-particle wave functions.

The selection rules for the change in angular momenta apply to the nucleus as a whole, but according to the independent-particle picture only one particle at a time makes a transition so that the selection rules apply for any specific particle. If more than one particle simultaneously makes a transition, the independent-particle picture would no longer be valid.

The E3 transition probabilities provide a good example of the effects of multiple-particle transitions. The group that has substantially smaller transition probabilities has been identified as having one particle change its orbit, but the remaining particles change from a nonpaired configuration to a completely paired configuration, for example,

$$(g_{9/2})_{7/2}{}^5 \rightarrow (p_{1/2})_{1/2}{}^1(g_{9/2})_0{}^4$$

Such transitions would not be allowed according to the single-particle model.

In the practical case, however, the wave functions are not pure independent-particle wave functions but contain some impurities so that the selection rules would be expected to break down to some slight extent.

For single-particle transitions in the energy regions normally encountered, M1 radiation would usually predominate over E2 radiation by a factor of 10^3 or 10^4 when both are allowed by selection rules. However, the enhanced transition probabilities in the collective picture are such that E2 rotational transitions may strongly compete with M1 transitions. On the other hand, the mixing of the many possible configurations making up individual nuclear states reduces the over-all value of the transition matrix elements, often by considerable amounts, affecting the values for the $B(EJ)$ and $B(MJ)$. The calculation of these quantities provide good tests of the effects of different nuclear potentials giving rise to configuration mixing. Each nucleus is a separate problem and so the attenuation of the transition probabilities can be given only as order-of-magnitude ranges for a given multipolarity. Rough comparisons with experiment gives the attenuation ranges listed in Table 6.4.

TABLE 6.4. HINDRANCE FACTORS IN TRANSITION PROBABILITIES FOR ELECTRIC- AND MAGNETIC-MULTIPOLE RADIATION

Multipole order J	Hindrance factor	
	EJ	MJ
1	10^{-3}–10^{-7}	10^{-1}–10^{-3}
2	2×10^{2}–10^{-2}	1–10^{-3}
3	2–10^{-4}	2–10^{-3}
4	1–10^{-1}	2–10^{-1}
5	2–10^{-1}	No data

3. Interactions with Static Fields

When the nucleus is located in a static field, the source of which is outside the nuclear volume, it is possible to describe the interaction in terms of the multipole moments of the nucleus and the corresponding multipole-moment distribution of the field at the nuclear coordinates. The multipole-moment expansion of a scalar field is obtained by use of the translation operator [19] which translates the coordinate of the field to that of the nucleus. This translation operation is a three-dimensional Taylor expansion in spherical coordinates which can be written as

$$\Phi(\mathbf{r}_n) = T(\boldsymbol{\varrho})\Phi(\mathbf{r}) \tag{6.61}$$

where $\boldsymbol{\varrho} + \mathbf{r} = \mathbf{r}_n$

and where

$$T(\rho) = e^{\varrho\cdot\nabla} = \sum_{lm} \frac{4\pi}{(2l+1)!!} \mathcal{Y}_l^{m+}(\varrho)\, \mathcal{Y}_l^m(\nabla) \tag{6.62}$$

For this expansion to be valid the field $\Phi(\mathrm{r})$ must be a solution of Laplace's equation $\nabla^2\Phi(\mathrm{r}) = 0$. Expansions of vector fields in multipole moments proceed in a similar, but more complicated, manner and are treated in ref. 19.

At the present time, the only static moments of nuclei for which extensive experimental data exist are the electric-monopole moment, the magnetic-dipole moment, and the electric-quadrupole moment. The interaction Hamiltonian [Eq. (6.2)] can be expanded for static fields in a manner similar to the treatment in Sec. 2. Keeping only the terms corresponding to each of the above-mentioned moments and restricting to axially symmetric fields for the dipole and quadrupole terms, the interaction Hamiltonian becomes

$$\begin{aligned} H_a &= e_i \langle a|a\rangle \Phi(\rho) \\ &\quad - \frac{e_i}{2Mc} \langle a|il_z + g_i\sigma_z|a\rangle B_z(\rho) \\ &\quad + \frac{e_i}{4} \langle a|(3z^2 - r^2)|a\rangle \left[\frac{\partial^2}{\partial z^2}\Phi(r)\right]_{r=\rho} \end{aligned} \tag{6.63}$$

When the matrix elements are evaluated for a single nucleon in the j-j coupling scheme, the energy for state a is

$$E_a = E_c - \frac{e\hbar}{2Mc} g_j m_j B_z + \frac{eQ}{4} \frac{3m_j^2 - j_a(j_a+1)}{j_a(2j_a+1)} \tag{6.64}$$

E_c is the additional Coulomb energy of the nucleus resulting from the applied potential $\Phi(\rho)$. The Schmidt g factor, g_j, is given by [15]

$$g_j = \begin{cases} lg_l + g_\sigma & \text{for } j = l + \frac{1}{2} \\ \dfrac{j}{j+1}[(l+1)g_l - g_\sigma] & j = l - \frac{1}{2} \end{cases} \tag{6.65}$$

The quadrupole moment Q is the expectation value of the quadrupole operator $(3z^2 - r^2)$ when the nucleon is in the state j, $m_j = j$.

A level diagram of a hypothetical nucleus with a ground-state spin of $\frac{1}{2}$ and a first excited-state spin of $\frac{3}{2}$ is shown in Fig. 6.6. The perturbing static interactions are successively applied in order to demonstrate their effect on the level structure. The electric-monopole interactions of this form are usually generated by the presence of atomic electrons near the nuclear volume, particularly s electrons. This shift is often called an "isomer shift" since the effect can be observed in different chemical environments of the atom. Shifts of this type have been observed only recently, using the Mössbauer effect [20] discussed in Sec. 5. Similarly, isomer shifts have been observed in atomic spectra [21]. The magnetic-dipole and electric-quadrupole splittings are more familiar, since they have been observed for many years in the investigation of atomic energy levels. The fields giving rise to these interactions may result either from the electronic configuration of the atom or from an external source.

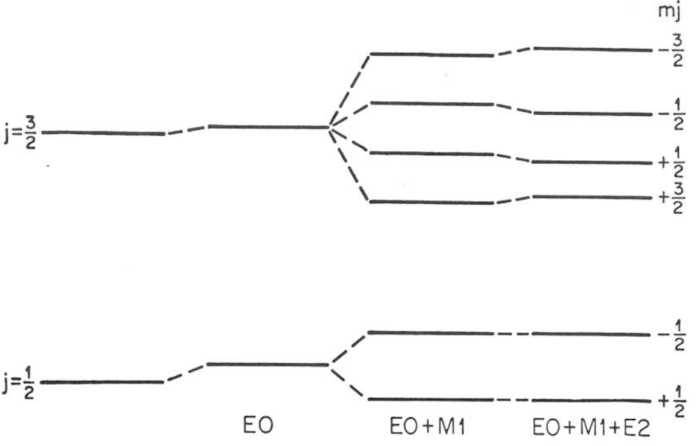

FIG. 6.6. An energy-level scheme for a hypothetical nucleus of ground-state spin $j = \frac{1}{2}$ and excited-state spin $j = \frac{3}{2}$. This nucleus is subsequently subjected to static perturbing interactions: (a) An electric-monopole field giving an isomer shift, (b) magnetic-dipole field giving a magnetic-dipole splitting, and (c) axially symmetric electric-quadrupole field giving an electric-quadrupole splitting or shift.

The splitting depends on the magnitude of the intrinsic magnetic-dipole and electric-quadrupole moments and the strength of the corresponding static fields. Precise energy measurements of electromagnetic transitions between different sublevels is of value in the determination of the intrinsic moments. In conventional magnetic-resonance experiments, direct observations are made between adjacent magnetic sublevels where the magnetic quantum number changes by one unit. These transitions are treated in Chap. 6. The gamma-ray transitions between two substates of different nuclear levels are of interest, particularly since recent high-resolution techniques involving the Mössbauer effect have allowed their study in detail. The individual transition probabilities for each of these transitions are related to the total integrated transition probability times the square of a Clebsch-Gordan coefficient $\langle j_a m_a JM|j_b m_b\rangle^2$.

The measurements of hyperfine splitting by nuclear magnetic resonance and by the Mössbauer effect complement one another. The splitting depends on two parameters—the intrinsic nuclear moments and the effective strength of the field—which cannot be separated experimentally, except in rare cases. However, if the magnetic moment of the ground state is known from, say, atomic-beam or microwave-resonance experiments, the Mössbauer hyperfine splitting will determine the magnetic moment of the excited state and the effective field B.

4. Other Phenomena Involving the Nuclear Electromagnetic Field

Internal Conversion. An alternative mode for a nucleus to undergo a transition from an excited state to one of lower energy is internal conversion. This process is competitive with that for gamma-ray emission and thus increases the total transition probability.

The general procedure for the calculation of the transition probability for the internal-conversion process is to consider a second-order process in which the nuclear current gives rise to a virtual photon which in

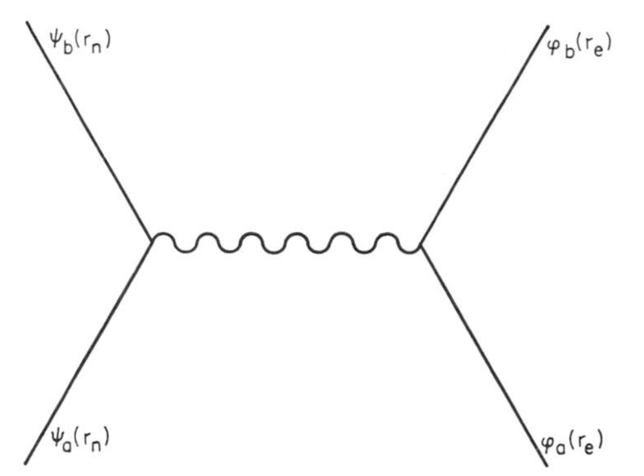

Fig. 6.7. A Feynman diagram illustrating the internal-conversion process.

turn interacts with the current of the atomic electron that is converted from a bound state to a free final state, as shown diagrammatically in Fig. 6.7. The interaction Hamiltonian is given by second-order perturbation theory as

$$H_{ab} = \sum_{\mu} \frac{1}{(2\pi)^3} \int d\mathbf{k}' \int d\mathbf{r}_n \int d\mathbf{r}_e \times \left\{ \frac{\psi^*_{k'}(\mathbf{r}_n) j_\mu^{(n)+} \psi_a(\mathbf{r}_n) \varphi^*_{-k'}(\mathbf{r}_e) j_\mu^{(e)} \varphi_a(\mathbf{r}_e)}{\hbar c(k - k' + i\eta)} \left(\frac{2\pi\hbar c}{k}\right) e^{i\mathbf{k}'\cdot(\mathbf{r}_n - \mathbf{r}_e)} + \frac{\psi_{k'}(\mathbf{r}_n) j_\mu^{(n)} \psi^*_a(\mathbf{r}_n) \varphi_{-k'}(\mathbf{r}_e) j_\mu^{(e)+} \varphi^*_a(\mathbf{r}_e)}{hc(k + k' - i\eta)} \left(\frac{2\pi\hbar c}{k}\right) e^{-i\mathbf{k}'\cdot(\mathbf{r}_n - \mathbf{r}_e)} \right\} \tag{6.66}$$

where the term

$$\left(\frac{2\pi\hbar c}{k}\right) e^{i\mathbf{k}'\cdot(\mathbf{r}_n - \mathbf{r}_e)}$$

represents the product of the emitted and absorbed virtual photon similar to (6.17) but with a normalization independent of volume since the normalization results from that of the nuclear and electronic wavefunctions.

Integrating over all virtual k' and passing to the limit $\eta \to 0$ gives the interaction Hamiltonian

$$H_{ab} = \int d\mathbf{r}_n \int d\mathbf{r}_e \psi^*_b(r_n) j_\mu^{(n)+} \psi_a(r_n) \varphi^*_b(r_e) j_\mu^{(e)} \varphi_a(r_e) \frac{e^{ik|\mathbf{r}_n - \mathbf{r}_e|}}{|\mathbf{r}_n - \mathbf{r}_e|} \tag{6.67}$$

A multipole decomposition is achieved by making use of the expansion for the Green's function of the operator $(\nabla^2 + k^2)$ [22]

$$\frac{e^{ik|\mathbf{r}_n - \mathbf{r}_e|}}{|\mathbf{r}_n - \mathbf{r}_e|} = 4\pi k \sum_{l=0}^{\infty} \sum_{m=-l}^{l} j_l(kr_<) h_l^{(+)}(kr_>) \times [i^l Y_l^m(\vartheta_n \varphi_n)]^* i^l Y_l^m(\vartheta_l \varphi_l) \tag{6.68}$$

where $j_l(kr)$ and $h_l^{(+)}(kr)$ are the spherical Bessel function and the Hankel function for an outgoing wave, respectively. The quantities $r_<$ and $r_>$ indicate that the radial coordinate is to be taken as less than and greater than, respectively. We introduce the tensor operators [(6.6) and (6.7)] and rewrite the Hamiltonian, taking into account the relative ranges of the nucleon and electron radial coordinates:

$$H_{ab} = 4\pi k \sum_{JlM} \int_0^\infty \psi^*_b(\mathbf{r}_n) j_l(kr_n) \mathbf{T}_{Jl1}{}^{M+}(\hat{r}_n, \mathbf{j}^{(n)}) \psi_a(\mathbf{r}_n)\, d\mathbf{r}_n \times \left\{ \int_0^\infty \varphi^*_b(\mathbf{r}_e) h_l^{(+)}(kr_e) \mathbf{T}_{Jl1}{}^M(\hat{r}_e, \mathbf{j}^{(e)}) \varphi_a(\mathbf{r}_e)\, d\mathbf{r}_e - \int_0^{r_n} \varphi^*_b(r_e) h_l^{(+)}(kr_e) \mathbf{T}_{Jl1}{}^M(\hat{r}_e, \mathbf{j}^{(e)}) \varphi_a(\mathbf{r}_e)\, d\mathbf{r}_e + \int_0^{r_n} \varphi^*_b(\mathbf{r}_e) j_l(kr_e) \mathbf{T}_{Jl1}(\hat{r}_e, \mathbf{j}^{(e)}) \varphi_a(\mathbf{r}_e) \frac{h_l^{(+)}(kr_n)}{j_l(kr_n)}\, d\mathbf{r}_e \right\} \tag{6.69}$$

The Hamiltonian is written as a product of a nuclear term and three electronic terms in order to illustrate the effects of nuclear structure in the internal-conversion process. If we consider the nucleus of finite extent R, the first term, which is usually the largest, corresponds to the case where the electron density extends to the origin. This term, taken by itself, is not valid since it includes contributions where the electron is within the nuclear volume and must be corrected by the, usually much smaller, second term. The neglect of the last two terms is called the "point nucleus approximation," although it is proper to consider the nucleus as of finite extent in so far as the wave functions of the electron are solutions in the field of an extended charge distribution. This dichotomy is usually referred to as a *static effect* of finite nuclear size. The last term takes into account the penetration of the electron inside the nuclear volume where $r_e < r_n$. The last two terms taken together are referred to as *dynamic effects* of nuclear size. These latter terms are usually neglected since the probability of finding the electron within the nuclear volume is small compared with that outside.

The operators of multipole order J can be separated into groups of opposite parity, corresponding to electric and magnetic transition operators, as is done in the gamma-ray emission case. The transition probability is given by expression (6.1) where the interaction Hamiltonian is that given by (6.69). The density of final states is that for the ejected electron

$$\frac{dn}{dE} = \frac{pE}{2\pi^2} \frac{V}{\hbar^3 c^2} \tag{6.70}$$

The transition probability λ_e for the internal conversion involves the details of nuclear structure in a way similar to that for gamma-ray emission if the so-called dynamic effects of nuclear size can be neglected. In this case, it is convenient to introduce a quantity called the internal-conversion coefficient defined as

$$\alpha(EJ) = \frac{\lambda_e(EJ)}{\lambda_\gamma(EJ)} \qquad \alpha(MJ) = \frac{\lambda_e(MJ)}{\lambda_\gamma(MJ)} \tag{6.71}$$

Each α is then approximately independent of the details of nuclear structure. As defined above, α is the sum of the conversion coefficients for electrons in a particular atomic shell, for example:

$$\alpha(EJ) = \alpha_k(EJ) + \alpha_L(EJ) + \alpha_M(EJ) \cdots \tag{6.72}$$

Each shell may have subshells, so that, for example,

$$\alpha_L(EJ) = \alpha_{LI}(EJ) + \alpha_{LII}(EJ) + \alpha_{LIII}(EJ) \tag{6.73}$$

It must be reemphasized that the total transition probability is the sum of the separate transition probabilities for internal conversion and gamma emission with a usually negligible correction due to interference effects between real and virtual processes. We may write the total transition probability as

$$\Lambda(EJ) = \lambda_\gamma(EJ) + \lambda_e(EJ) = [1 + \alpha(EJ)]\lambda_\gamma(EJ) \tag{6.74a}$$

$$\Lambda(MJ) = \lambda_\gamma(MJ) + \lambda_e(MJ) = [1 + \alpha(MJ)]\lambda_\gamma(MJ) \tag{6.74b}$$

Calculations of the electric and magnetic internal-conversion coefficients have been made by a number of authors employing various approximations in order to make the computational task easier, but in almost all cases the results have been disappointingly poor, even where the effects of finite nuclear size can be neglected and no nuclear-structure effects enter. The reason for this is that an accurate knowledge of the initial atomic energy state and the final Coulomb continuum is required and is meaningful only when a fully relativistic treatment is made. This information may be had only by extensive numerical calculation. A few sufficiently exact numerical calculations of K-shell conversion coefficients for isolated values of Z, energy, and multipole order have been carried out [23]. Rose and collaborators [24] have published an extensive compilation of the K-shell and three L-subshell internal-conversion coefficients for a wide range of energies, atomic numbers, and multipole orders for a nucleus with a finite charge distribution. Sliv and Band [25] have made a similar compilation, attempting to take into account finite-size effects by treating the nucleus as a uniform volume charge distribution and a surface distribution of transition currents, together with a Thomas-Fermi-Dirac distribution to account for atomic screening. The latter calculation when compared with that of Rose illustrates the trend of "dynamic" finite-size effects but should be used with caution as the model is rather idealized for computational purposes.

No simple analytical expression can be given to show the features of internal-conversion coefficients but a few qualitative observations can be made. The conversion coefficients are strongly decreasing functions of k for fixed values of Z and J, as can be seen from the asymptotic, $(kR)^{-(J+1)}$, dependence of the Hankel function. By the same argument the coefficients increase strongly with J for fixed k and Z since $kR \ll 1$. For fixed k and J they increase strongly with Z chiefly because of the density of electrons at the nucleus. The dependence is approximately Z^3 for $\alpha_k(EJ)$. Effects of nuclear size make a considerable contribution in $\alpha_k((\mathrm{M1})$ and $\alpha_{LI}(\mathrm{M1})$ coefficients where there is appreciable density of K and LI electrons near the origin. In general, the effect is small for the LII shell and negligible for all others. There is the possibility that the point-source nuclear matrix element may be zero (which also implies zero gamma-ray transition probability). The small transition probability due to finite-size internal-conversion effects then leads to a conversion coefficient equal to infinity.

A study of finite-size effects in internal conversion yields information not contained in the study of gamma radiation by itself. The feature was first pointed out by Church and Weneser, and some of the consequences are discussed in an excellent review paper by these authors [26].

The experimental observer can make use of the internal-conversion process in a number of ways. Measurement of the energies of the conversion electrons from the various shells provides a very accurate means of determining the transition energy. The measurement of the electron's energy in a magnetic spectrometer is usually an order of magnitude more accurate than the measurement of the energy of the gamma ray through its interaction with matter except, perhaps, when the gamma-ray energy is determined by interference effects in a crystal lattice, as in a diffraction spectrometer. Observation of a number of conversion lines with energy $T_i = \hbar\omega - E_i$, where E_i is the binding energy of the electron in subshell i, gives a number of independent determinations of the transition energy $\hbar\omega$. The conversion coefficient, say for K electrons, is equal to the ratio of the intensity of K conversion electrons to the intensity of gamma rays emitted from the source and provides information on the multipole order and parity of the transition. Often it is difficult or impossible to measure with accuracy the relative number of conversion electrons and gamma rays but comparatively simple to measure, at the same efficiency, the relative number of conversion electrons from the various shells such as N_K/N_L (sometimes called the K/L ratio), N_{LI}/N_{LII}, or N_{LI}/N_{LIII}. Sometimes one of the latter measurements is less ambiguous than a measurement of the K/L ratio [27].

Angular Correlations between Successive Radiations. Information can often be obtained about the nuclear states by observation of the angular correlation between successive gamma radiations, of the polarizations of the gamma rays, or of the angular distribution of gamma radiation from oriented nuclei. The derivation of the relevant expressions for these correlation effects involves considerable detail in the use of density matrix techniques of advanced quantum mechanics beyond the scope of this chapter. This section will be limited to a brief discussion of gamma-gamma angular correlations where no polarizations are detected. This correlation is of great practical importance because of the relative simplicity of the experimental techniques required. (Several comprehensive reviews [28, 29] provide more detail on the general subject of directional and polarization correlations.)

Figure 6.8 shows a state j_a that decays to a state j_b with the emission of radiation J_1. The state j_b in turn decays to j_c with the emission of radiation J_2. The radiations J_1 and J_2 are detected by counters 1 and 2 situated so as to define an angle ϑ between the successive radiations. If the lifetime of the intermediate state is sufficiently short, then "time coincidences" between the two detectors as a function of ϑ yield the angular correlation. The correlation function is given by [28, 29]

$$\begin{aligned} w(\vartheta) = 4\pi \sum_k \Bigg\{ \sum_{J_1J'_1} & F_k(J_1J'_1j_aj_b) \langle b\|\mathbf{T}_{J_1}\|a\rangle \\ & \langle b\|\mathbf{T}_{J'_1}\|a\rangle^* \times \sum_{J_2J'_2} F_k(J_2J'_2j_cj_b) \langle b\|\mathbf{T}_{J_2}\|c\rangle \\ & \langle b\|\mathbf{T}_{J'_2}\|c\rangle^* \Bigg\} P_k(\cos\vartheta) \end{aligned} \tag{6.75}$$

where

$$F_k(J_1J'_1j_aj_b) = (-1)^{j_a+j_b-1}[(2j_b+1)(2J_1+1)(2J'_1+1)(2k+1)]^{1/2}\begin{pmatrix}J_1 & J'_1 & k\\ 1 & -1 & 0\end{pmatrix}\begin{Bmatrix}j_b & j_b & k\\ J'_1 & J_1 & j_a\end{Bmatrix} \quad (6.76)$$

These coefficients satisfy

$$F_0(J_1J'_1j_aj_b) = \delta_{J_1J'_1}$$

As an example of the use of the angular correlation expression, we discuss the main gamma-gamma cascade in cadmium 111 following the electron capture occurring in indium 111. The decay scheme is shown in Fig. 6.8. The energy levels have been carefully

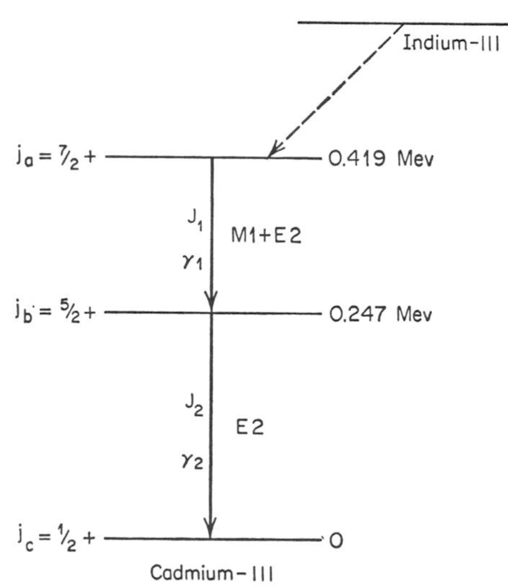

FIG. 6.8. The gamma-gamma cascade in cadmium 111 following the electron-capture transition in indium 111.

investigated, and spin assignments can be made without the help of directional correlation data. The measured internal-conversion coefficients indicate that γ_1 is primarily M1 and γ_2 is primarily E2. From arguments in Sec. 2 we conclude that there is a possibility of an appreciable admixture of E2 to M1 in γ_1 but a negligible admixture of M3 to E2 in γ_2. If we divide $w(\vartheta)$ in Eq. (6.75) by the reduced matrix elements and introduce the definition

$$\delta = \frac{\langle b\|T_2\|a\rangle}{\langle b\|T_1\|a\rangle} \quad (6.77)$$

a new correlation function is obtained for this particular case:

$$W(\vartheta) = 4\pi \sum_k \frac{F_k(11\ \tfrac{7}{2}\ \tfrac{5}{2}) + 2\delta F_k(12\ \tfrac{7}{2}\ \tfrac{5}{2}) + \delta^2 F_k(22\ \tfrac{7}{2}\ \tfrac{5}{2})}{1+\delta^2} \times F_k(22\ \tfrac{1}{2}\ \tfrac{5}{2})P_k(\cos\vartheta) \quad (6.78)$$

Evaluating the function F_k from Eq. (6.76) or from published tables [30] we get

$$W(\vartheta) = P_0(\cos\vartheta) - \frac{0.0714 + 0.7424\delta + 0.1734\delta^2}{1+\delta^2}P_2(\cos\vartheta) - \frac{0.0726\delta^2}{1+\delta^2}P_4(\cos\vartheta) \quad (6.79)$$

The experimentally determined angular correlations may be expressed as the anisotropy A defined by

$$A = \frac{W(\pi) - W(\pi/2)}{W(\pi/2)} = \frac{W(\pi)}{W(\pi/2)} - 1 \quad (6.80)$$

The experimental value, $A = -0.245 \pm 0.015$, is shown in Fig. 6.9 where the theoretical value for A is

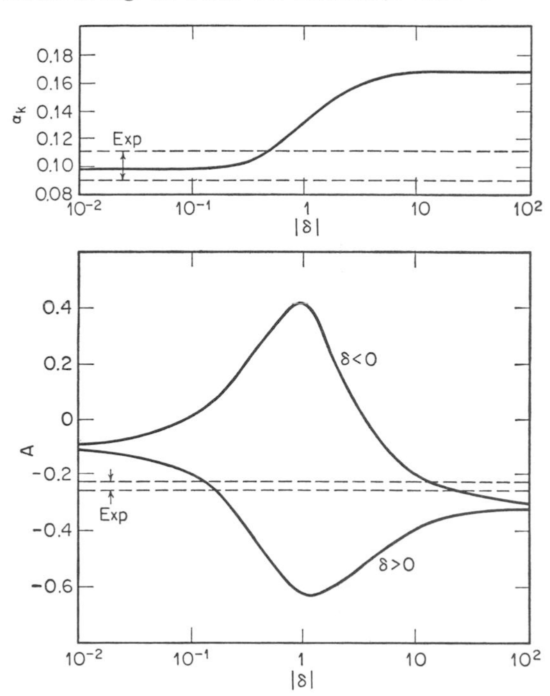

FIG. 6.9. (a) The K-electron conversion coefficient α_k, occurring in the transition between states a and b in cadmium 111 expressed as a function of δ, the E2-M1 mixing ratio. (b) The anisotropy A of the gamma-gamma angular correlation expressed as a function of δ for the upper transition. The combination of these two independent measurements gives a unique value for δ. (*After* [29].)

plotted vs. the mixing ratio $|\delta|$. The measured value of the anisotropy corresponds to two values of δ, one negative and one positive. Shown in the same figure is the value for the k conversion coefficient α_k as a function of $|\delta|$. The conversion coefficient for a mixed transition is given by

$$\alpha_k = \frac{\alpha_k(\mathrm{M1}) + \delta^2\alpha_k(\mathrm{E2})}{1+\delta^2} \quad (6.81)$$

The experimental value of α_k by itself allows a large range of δ but excludes one value determined by the angular correlation measurements. The resultant value is $\delta = 0.145 \pm 0.015$ corresponding to an intensity ratio $\delta^2 = 0.021$. The transition proceeds 2 per cent by E2 radiation and 98 per cent by M1 with a positive sign for the ratio of amplitudes.

The interpretation of angular-correlation data requires great care in consideration of effects which substantially alter the measured results. Small admixtures of multipoles can substantially alter the correlation function even to the amount of changing the sign of the coefficients of $P_k(\cos \vartheta)$. Extranuclear fields, particularly static magnetic-dipole and electric-quadrupole fields as discussed in Sec. 3, can act on the intermediate state j_b, causing precession of this state and thus attenuating the correlation. Competing cascades often mask the correlation function in a very complex spectrum, and sometimes very refined experimental techniques employing good energy and time resolutions are required.

Internal Pair Creation. If the energy separation of two nuclear levels is greater than $2mc^2$, another type of internal-conversion process is possible. Here the available energy is used to create a positron-electron pair with a kinetic energy $T = \hbar\omega - 2mc^2$. For not too small energies of the emerging particles the change of the wave functions due to the presence of the nucleus tends to be opposite for the positron and the electron so that the probability of this process is rather independent of Z.

As the gamma-ray energy is increased, the probability for internal pair creation becomes greater while the probability for internal conversion becomes less. Although internal pair creation is a rather infrequent process, it is of importance in low-Z elements where the probability of the internal-conversion process is small and the transition energies are high. The intensity relative to gamma-ray emission for a given multipole order and energy greater than $2mc^2$ is of the order of 10^{-3} to 10^{-4}.

Extensive calculations of internal-pair-creation coefficients have been made using the Born approximation [31]. It is not feasible to obtain an analytical formula for the total internal-pair-creation coefficient. However, by numerical integration, Rose [32] is able to present in graphic form the total internal-pair-formation coefficient as a function of transition energies for electric- and magnetic-multipole transitions for various energy divisions between the pair and for two angles separating the pair, 0 and 90°.

The process is of particular importance in several light even-even nuclei [33] where electromagnetic radiation between two levels of zero spin is completely forbidden in the first order.

The process has also allowed the multipolarity of a few transitions in light nuclei to be assigned experimentally [34]. This assignment may be made on the basis of the internal-pair-creation coefficient, the spectrum of the positrons, or the angular distribution of the positron-electron pair.

Auger Effect. When an atom is left in an excited state by the removal of one of the inner-shell electrons by the internal-conversion process mentioned above, a transition may occur in which the vacancy of the inner shell is filled by a transition of an electron from a less tightly bound state. The excess energy may then appear as an X-ray quantum of energy equal to the energy difference between the two states. This process may be successively repeated until all the excess energy of the atom has been radiated. Alternatively, the reorganization of the atom may occur without the emission ot electromagnetic radiation. In this case the energy is communicated to another electron of the same atom in a fashion somewhat analogous to the internal-conversion process. This process is called the Auger effect [35] in recognition of the work of Auger, who discovered it in 1925.

The probability of the ejection of an X-ray as opposed to the ejection of an Auger electron for each primary vacancy of a particular shell is called the fluorescent yield. Figure 6.10 [36] shows fluorescent yields of the K series; this yield is the probability that an atom with a vacancy in the K-shell will emit an X ray of the K series.

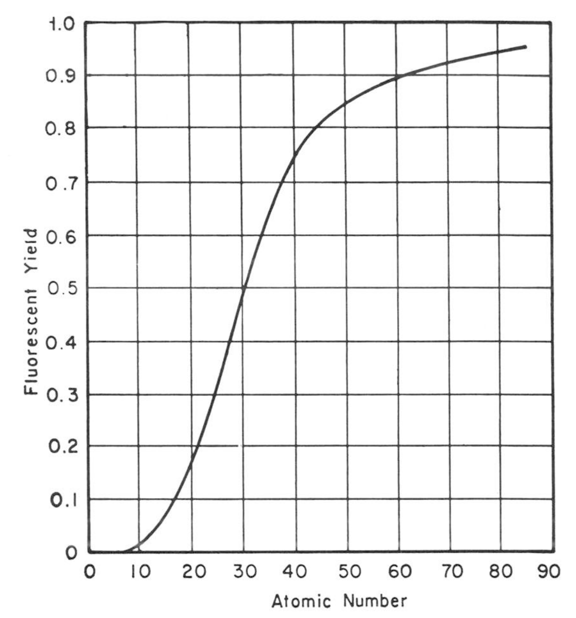

Fig. 6.10. The fluorescent yield in the K shell as a function of atomic number. The mean values of experimental measurements as compiled in ref. [36] have been used.

Although the Auger effect is strictly an atomic process, its probability is an important consideration when making an experimental determination involving an internal-conversion or orbital electron-capture processes. Examples are the determination of the K conversion coefficient, determination of K/L ratios, proof of electron capture in atoms where considerable internal conversion is also present, and the determination of the intensity ratio of K capture to positron emission.

Internal Bremsstrahlung. Accompanying beta decay and orbital electron capture is a weak continuous gamma-ray spectrum having an upper energy limit equal to the energy available for the transition. This gamma radiation is true *bremsstrahlung* emitted by the electron making a transition from the initial to the final quantum state.

In beta decay, theoretical calculations have been made [37] giving the spectral shape and intensity per disintegration as a function of the upper energy limit and type of beta interaction. In a typical beta emitter the number of photons per disintegration is of the order of 10^{-2}. The observation of the internal *bremsstrahlung* from beta emitters is complicated by the occurrence of the external *bremsstrahlung* due to the deflection of the beta particle by nuclei other than that in which the beta particle originated.

In orbital electron capture [38] the number of photons emitted per disintegration is somewhat smaller, being approximately 10^{-4} to 10^{-5}, depending on the transition energy. Since no external *bremsstrahlung* is present in the electron-capture process to mask the direct observation of the effect, the spectrum is probably easier to observe.

Also, since no beta spectrum is present and the emitted neutrino escapes detection, the only means of finding the energy separation of the initial and final nuclear states is through the determination of the upper energy limit W_0 of this continuous photon spectrum. This energy separation is given by $\Delta E = W_0 + E_k$, where E_k is the binding energy of the K-shell electron. When the energy separation is small, the transitions of L-shell electrons contribute appreciably to the photon spectrum.

Two representative spectra for inner *bremsstrahlung* from beta emission and electron capture are shown in Fig. 6.11.

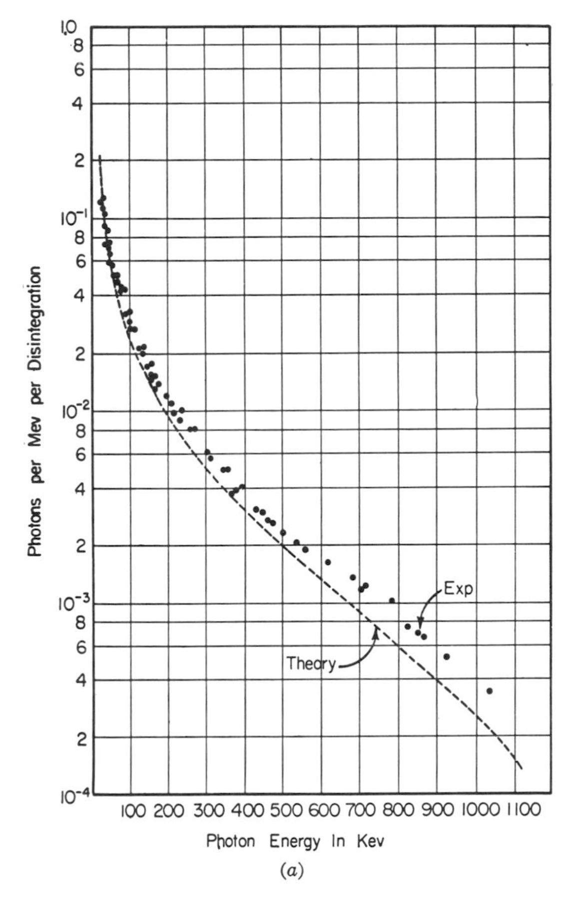

(*a*)

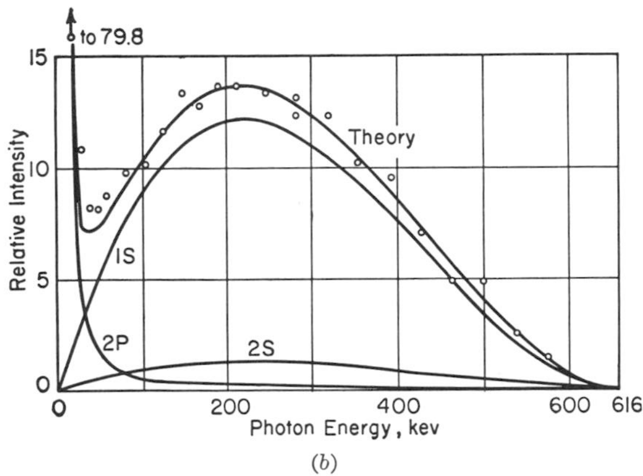

(*b*)

FIG. 6.11. Representative spectra of inner bremsstrahlung (*a*) resulting from the beta decay of phosphorus 32 [38] and (*b*) resulting from the orbital-electron capture in vanadium 49 [39].

5. Scattering and Absorption of Gamma Rays by Nuclei

Gamma rays interact with the material through which they are passing only by discrete elementary processes. These processes may be such that the photon is either absorbed outright or scattered from its original direction of propagation with or without a decrease in energy. The number of photons that experience an absorption or deflection is proportional to the differential thickness dx of the material and to I_0, the number of photons incident on the material. The number of photons undisturbed after traversal of a distance x is

$$I(x) = I_0 e^{-\mu x} \tag{6.82}$$

where μ is the absorption coefficient of the particular material. The customary units for x are grams per square centimeter. The units of μ are the reciprocal of x. However, from the fundamental standpoint it is desirable to express μ on a "per atom" basis $\mu = N\sigma/A$, where N is Avogadro's number, A is the atomic weight of the material expressed in grams per mole, and σ is the cross section per atom, representing the sum of the individual cross sections for each elementary process contributing to the absorption or scattering of the photon.

In nuclei, as in atoms, the elementary processes may be classified into a number of phenomenologically distinguishable types. There exists coherent scattering from the nuclei or atoms as a whole. When the energy of the gamma rays becomes so great that binding effects can be neglected, incoherent Compton scattering from the individual nucleons in the nucleus or electrons in the atom will occur. On the other hand, there are two discernible absorption processes in which the incident gamma ray disappears completely. The first of these, in which the nucleus or atom makes a transition to a real or virtual excited state, is designated resonant absorption. The second, in which a nucleon or electron is ejected directly, leaving the residual nucleus or atom in the low-lying state, is designated direct interaction. From an experimental point of view the resonant-absorption process may be difficult to distinguish from some of the other processes since the excited state may decay by emission of a gamma ray to the ground or a lower excited state, giving the appearance of an elastic or an inelastic scattering. Also, the excited state may decay by emission of a nucleon or electron, giving the appearance of a direct interaction.

Besides interacting with electrons and with nucleons, gamma rays can also interact with electric fields surrounding electrons or nucleons and meson fields surrounding nucleons. Gamma rays may be absorbed by electron-positron pair production or scattered by virtual electron-positron pair production. The latter is called Delbrück scattering. The meson field analogue corresponds to absorption and scattering by real and virtual meson production, respectively. These particular interactions do not become important until the gamma-ray energy is above the rest mass of the particles involved. The threshold for electron-positron pair production is $2mc^2$ or 1.02 Mev, and that for meson production about 140 Mev. For nuclear gamma rays in the range of 10^{-5} to 10 Mev, the latter meson production process plays no role.

The interaction of gamma rays with electrons and atoms are covered in Part 7, Chap. 8.

Direct Interaction with Nucleus. Coherent and Compton Scattering by the Nucleus. Gamma rays of wavelength long compared with the dimensions of the scattering nucleus set the nucleus as a whole in motion and are scattered coherently. The differential cross section for this process, in analogy with the Rayleigh scattering from electron systems, is [40]

$$\frac{d\sigma}{d\Omega} = \frac{1}{2}\left(\frac{e^2}{Mc^2}\right)^2 \frac{Z^4}{A^2}(1 + \cos^2\vartheta) \qquad \text{when } \lambda\!\!\!^- > R \tag{6.83}$$

where M is the proton mass.

As the energy of the gamma rays becomes so great that the wavelength becomes of the order of, or less than, nuclear dimensions, individual nucleons may act as more or less independent scattering centers so that the scattering becomes incoherent. It would be expected that, at about 100 Mev for heavy nuclei, the Compton scattering in the nonrelativistic limit by these nucleons would be

$$\frac{d\sigma}{d\Omega} = \frac{1}{2}\left(\frac{e^2}{Mc^2}\right)^2 Z(1 + \cos^2\vartheta) \qquad \lambda\!\!\!^- < R \tag{6.84}$$

The scattering process becomes incoherent, but the energy shift of the gamma ray is less than in the electronic Compton scattering case because the much greater nucleon mass takes up the momentum. Experimental detection of this scattering process could best be made by studies of the recoil nucleon in time coincidence with the scattered gamma ray.

These nuclear scattering cross sections are lower than the electronic Compton scattering cross section by a factor of 10^6 because of the dependence of the effect on the mass of the scatterer. The practical effect on the scattering of photons usually is negligible.

Nuclear Photoeffect and Resonance Absorption. The nucleus can absorb a photon in a transition to an excited state. These transitions are analogous to the bound transitions of atomic electrons. Analogy with the atomic electron processes suggests that the cross section should exhibit sharp line resonances at low energy, and as the energy is increased, these resonances should broaden and become more frequent. Finally, when the energy becomes above that of the nucleon binding energy the process will pass over into the nuclear photoeffect.

At the energy of a low-lying nuclear level the nuclear absorption and scattering are very pronounced. The expression for the cross section would be given by the Breit-Wigner single-level formula [1], provided that the level is isolated from other levels and has an intrinsic width Γ much less than the transition energy $\hbar\omega_0$. This formula is

$$\sigma = g\frac{\pi}{2k^2}\frac{\Gamma_\gamma\Gamma}{(\hbar\omega - \hbar\omega_0)^2 + \Gamma^2/4} \tag{6.85}$$

where k is the wave number and $\hbar\omega$ the energy of the incident gamma radiation. Γ is the total width of the excited state with spin j_b and is separated from the ground state with spin j_a by energy $\hbar\omega_0$. A statistical factor g takes into account the multiplicities of the ground and excited states. Here,

$$g = \frac{2j_b + 1}{2j_a + 1}$$

and $\Gamma_\gamma = \hbar\lambda_\gamma(EJ)$ or $\Gamma_\gamma = \hbar\lambda_\gamma(MJ)$. $\lambda_\gamma(EJ)$ and $\lambda_\gamma(MJ)$ are the transition probabilities discussed in Sec. 1, and $\Gamma = (1 + \alpha)\Gamma_\gamma$.

The total effective width of a nuclear energy level is usually much broader than the natural width because of the Doppler broadening by the thermal motion of the nucleons so that the observed peak cross section is reduced somewhat.

When the energy gets above the threshold for particle emission the level density is greater than the natural energy width of the levels so that a line absorption coalesces into a continuous absorption having a broad maximum in the neighborhood of 20 Mev. Causality considerations require that a certain amount of coherent scattering of photons be present also.

The general features of Eq. (6.85) have been observed for individual low-lying states where the various nuclear models provide information on Γ_γ and Γ. At higher excitations, the estimates of Γ_γ and Γ become more involved and properly belong to discussions of nuclear structure. Because of the strong nuclear interaction the problem becomes one involving the many-body collective approach and is at present the subject of vigorous experimental and theoretical investigation [41, 42].

The order of magnitude of the absorption cross section by the nucleus of 20-Mev photons is about 10^{-25} cm^2, and the cross section for elastic scattering about 100 times less. The nuclear absorption of photons may be of the order of 5 per cent of the electronic absorption at this energy, an amount not entirely negligible.

Resonance Absorption and the Mössbauer Effect. One of the features of nuclear electromagnetic transitions is the extreme sharpness of the emission lines. For example, a 14-kev transition occurring in iron 57 has a transition probability

$$\lambda(\text{M1}) = 10^{-7}\ \text{sec}^{-1}$$

This corresponds to a natural width $\Gamma = 1.4 \times 10^{-8}$ ev or a $\Gamma/E \approx 10^{-12}$, far sharper than any atomic absorption or emission line. The absorption cross section at resonance exhibits a rather large value

$\sigma = g2\pi\lambda^2\Gamma_\gamma/\Gamma = 2.2 \times 10^{-18}$ cm². This cross section is more than 200 times the photoelectric cross section, the next most important cross section. With this large cross section one might expect that observation of nuclear resonance absorption would be possible since it is similar to atomic resonance absorption such as that of absorption of sodium light by sodium atoms. This phenomenon in free nuclei is difficult to observe because of nuclear recoil effects and Doppler broadening at room temperature.

When a free nucleus decays to a state $\hbar\omega_0$ lower in energy, part of this energy is emitted as a photon of energy $\hbar\omega = \hbar\omega_0 - E_r$, while the nucleus recoils with kinetic energy E_r, where

$$E_r = \frac{(\hbar\omega)^2}{2Mc^2} \tag{6.86}$$

Likewise, when the nucleus absorbs a photon to make a transition to a state $\hbar\omega_0$ higher in energy it must absorb a photon of energy $\hbar\omega = \hbar\omega_0 + E_r$ to make up for the recoil energy of the absorbing nucleus. Furthermore, the Doppler width due to thermal motion of a free atom is

$$E_D = 2\hbar\omega\sqrt{\frac{kT}{Mc^2}}$$

There is no overlapping of emission and absorption energies, and no resonant scattering would occur for free atoms. Attempts to overcome this recoil energy shift by moving the source relative to the absorber, thus providing a compensating Doppler shift, have been successful. Placing a source on an arm in an ultracentrifuge [43] or making use of the recoil of the nucleus from a previous transition, where the energy of the recoil depends on its angular correlation with the transition [44], allows one to achieve resonance. A less selective method makes use of the Maxwellian distribution of velocities of the source in gaseous form at elevated temperatures [45].

A recent development making use of resonance absorption is that of the Mössbauer effect [46]. Mössbauer investigated the emission and absorption of gamma rays by nuclei bound in solids. An atom is bound in the lattice of a solid in a quantum-mechanical way that is characterized by vibrational energy levels. When the recoil energy is greater than the binding energy in this lattice site, the atom will be displaced and will behave as in the free-atom case discussed above. When the recoil energy is less than above displacement energy but larger than the lowest vibrational state, there is a finite probability that the atom will go to an excited lattice vibration state. When the recoil energy is less than the lowest vibrational state the atom remains in the ground state and the lattice takes up the recoil as a whole. The mass of this lattice is so great as to make the energy taken up by the recoiling lattice negligible and is sometimes called "recoil free." Even when E_r for a free atom is greater than the lattice vibration energy there is a finite nonzero probability that the atom remains in the ground state. This probability is called the recoil free fraction f and depends on the properties of the medium, the temperature, and the energy of the gamma ray, increasing with decreasing gamma-ray energy and temperature.

The existence of "recoil free" emitters and absorbers has allowed the use of the extremely sharp nuclear gamma rays for study of phenomena heretofore not possible because of the resolutions required.

The applications of the Mössbauer effect, discussed here, involve measurement of the shift in the energy eigenvalue of a level in a nucleus when placed in an external perturbing field. Either the emitter or the absorber is placed in this field which destroys the resonant condition and then resonance is reestablished by Doppler-shifting the radiation.

A very fundamental application is the measurement of the gravitational red shift done by Pound and Rebka [47]. The energy separation of the ground and first excited state of an iron-57 nucleus is ΔE at the earth's surface. In an emission or absorption process a gamma ray of energy $\hbar\omega = \Delta E$ is involved.

The energy separation of the ground and first excited states of a similar nucleus at height h above the earth's surface is $\Delta E' = \Delta E(1 + gh/c^2)$, being related to the work required to raise the nucleus a height h in a gravitational field. The difference in excitation energies of nuclei in the two positions reduces the chance for resonance absorption of one nucleus of the emission gamma ray by the other unless the energy of the emitted gamma ray is shifted by the amount $\delta E = \hbar\omega gh/c^2$. This can be achieved by moving either the emitter or absorber to give a Doppler shift [48].

$$\hbar\omega' = \hbar\omega\,\frac{1 + (\hat{\mathbf{r}}\cdot\mathbf{v})/c}{(1 - v^2/c^2)^{1/2}} \tag{6.87}$$

For very small velocities ($v/c \ll 1$) along the direction of propagation, this expression reduces to

$$\hbar\omega' = \hbar\omega(1 + v/c) \tag{6.88}$$

A velocity required to achieve resonance is $v = gh/c$. At the 22.6-m height separation in the Pound and Rebka experiment $v = 7.4 \times 10^{-5}$ cm/sec. The results obtained were consistent with the predicted shift.

Another experiment [49] has measured the transverse Doppler shift, using the Mössbauer effect. Equation (6.87) for $v \ll c$ and $\hat{\mathbf{r}}\cdot\mathbf{v} = 0$ becomes

$$\hbar\omega' \approx \hbar\omega\left(1 + \frac{1}{2}\frac{v^2}{c^2}\right) \tag{6.89}$$

where the source was placed on the axis of a rapidly rotating centrifuge and the absorber placed at a radius R from this axis. An angular velocity $\dot{\vartheta}$ of the rotor gives an energy shift

$$\delta(\hbar\omega) = \hbar\omega\,\frac{\dot{\vartheta}^2R^2}{2c^2} \tag{6.90}$$

For a rotor 10 cm in radius driven at 3.6×10^4 rpm the shift $\delta(\hbar\omega)/\hbar\omega$ is 7×10^{-13}, comparable to the line width of iron 57. This expression has been verified to a few per cent.

It is possible to look upon the transverse Doppler-shift experiment as a gravitational-type experiment, using the principle of equivalence. The centrifugal force can be interpreted as a gravitational force. The potential energy Φ at radius R is the negative of the work required to move a unit mass from radius R to the center

$$\Phi = \int_R^0 r\dot{\vartheta}^2 \, dr \qquad (6.91)$$

where $r\dot{\vartheta}^2$ is the centripetal acceleration.

6. Experimental Detection of Nuclear Gamma Rays

The determination of the intensity and the energy of nuclear gamma radiation is possible only by means of the study of the interaction of the gamma radiation with matter. This is to say that the direct physical observations of intensity and energy (or momentum) are usually made on the secondary electrons that are ejected in the interaction. From a knowledge of the mechanics and cross section for the interaction process, the desired information about the incident gamma radiation may be obtained from the measured quantities. Three elementary processes that have found extensive use as an experimental tool are the photoelectric absorption, Compton scattering, and pair production. In certain situations, use may be made of Rayleigh scattering, a coherent scattering process.

The instruments used for such measurements in modern research are generally of the same fundamental types as those employed in the measurement of beta-ray spectra. They include magnetic spectrometers of the lens and prismatic types [50], scintillation spectrometers [51, 52], proportional gas counters [53], and semiconductor or solid-state counters [52, 54].

The magnetic spectrometers make use of a converter of a heavy element such as tantalum, lead, or uranium in which the gamma radiation will interact and eject secondary electrons which are analyzed according to the radii of curvature in a known magnetic field. Depending on circumstances, the source of gamma radiation may be situated internally or externally to the instrument. Below 1 Mev the photoelectric process is dominant and the secondary electrons ejected from the converter will have discrete energies corresponding to that of the incident quantum less the binding energy in the converter of the particular electron that is ejected. In the intermediate energy region up to about 5 Mev, the Compton scattering process is dominant. Here the problem is complicated by the nonuniqueness of the energy of the secondary electrons. However, by accepting only those electrons scattered in the forward or near-forward direction a highly peaked distribution is obtained, corresponding to the energy of the incident photon less the energy of the back-scattered photon.

In the energy region higher than about 5 Mev, pair production is the most important contributing interaction. Here two secondary particles are produced, and in order to get a value of the incident quantum energy the momenta of both particles must be determined. Spectrometers of the flat 180° type have been employed [55] where the particles are detected in time coincidence with one another and a momentum selection is made so that the incident quantum energy is determined. Lens-type spectrometers have been used [56], attaining higher precision than the flat type. The requirement in this case is that the positron and electron have the same energy and be emitted with equal angles with respect to the forward direction.

The scintillation counter has come, in recent years, into very extensive use as an efficient detector of gamma radiation, particularly where the intensities involved are too small for the use of magnetic spectrometers. In a device such as this, one measures a voltage pulse output whose magnitude is proportional to the energy of the secondary electrons emitted in the interaction.

In the low-energy region where the photoelectric and Compton processes predominate, the interpretation of the pulse-height distribution is straightforward but becomes increasingly complex as the gamma-ray energy becomes greater than the pair-production threshold. There may be as many as four discrete peaks associated with a particular gamma-ray energy. Besides the peaks associated with a photoelectric absorption or a Compton scattering there may be two additional peaks corresponding to a pair production with the escape of one or two annihilation quanta from the crystal, giving lines 0.51 and 1.02 Mev lower in energy than the peak that arises from total absorption of the energy of the gamma ray. A distribution of pulse height always results from the Compton process since only a portion of the gamma-ray energy is transferred to an electron, leaving the remainder as a scattered photon, sometimes of energy almost as great as that of the incident photon. The scattered photon may be further absorbed or escape from the crystal. If the scintillation crystal is made large enough, total absorption of all the energy of the gamma ray becomes more probable, and the observed pulse-height spectrum tends to become a single line corresponding to the energy of the incident gamma ray.

The proportional counter is of particular value in the energy region below, say, 30 kev where the energy resolution of the scintillation counter leaves something to be desired. The absorption of the gamma ray is by the photoelectric process. There may be one or two discrete peaks, depending in the relative probability of the escape from the counter of the X ray emitted by the photoelectrically excited atom.

A new development, commencing about 1958, has substantially revolutionized the fields of charged-particle and gamma-ray spectroscopy. This is the semiconductor detector [52, 54]. The main advantage lies in the high resolutions possible because of the large number of carrier pairs produced. In solids, only about 3 ev is required to create the equivalent of one ion pair compared with about 30 ev in a gas ionization chamber. In scintillation counters the energy to create a photoelectron varies from about 200 ev for fast electrons in sodium iodide to something like 10^4 ev for alpha particles in an organic phosphor. If a detector is to measure accurately the amount of ionization produced by passage of a primary or secondary ionizing particle, the carriers must be collected without appreciable loss, and this will happen only if the collection time is short compared with the carrier lifetime. The carrier types, collection times, depth of the semiconductor junction, etc., are quantities that characterize the semiconductor detectors and will not be discussed here. There are many excellent references on the rapidly developing field.

The surface-barrier-junction and diffused-junction

type of detectors have limited sensitive depth and, therefore, are better suited to charged-particle spectroscopy. A means of increasing the barrier depth has been accomplished by drifting lithium ions into silicon or germanium. Lithium-drifted germanium, because of its high effective atomic number, has proved very effective in gamma-ray spectroscopy since counting volumes of 1 cm³ have been achieved. The resolution for gamma rays is of the order of a factor of 10 better than the sodium iodide scintillation detector. The rather low velocity of carriers in the semiconductor limits the ultimate practical size and hence the achievable efficiency of the solid-state detector for gamma radiation of moderate and high energy. An interesting comparison [57] of the spectra of the 1,173- and 1,333-kev gamma rays following the decay of cobalt 60 as observed with a lithium-drifted germanium detector and with a sodium iodide scintillation detector is shown in Fig. 6.12. The efficiency of the scintillation detector is substantially greater, being limited only by the economics of scintillator size, so that at the present time the two types of detectors complement one another.

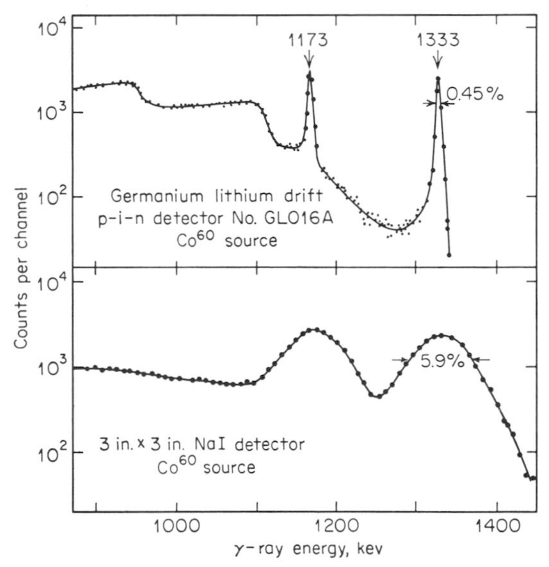

Fig. 6.12. A comparison of the pulse spectra from a lithium-drifted germanium solid-state detector and a sodium iodide scintillation detector [57].

The methods of voltage pulse-height analysis employed with scintillation, solid-state, and proportional counters are too numerous and diverse to be mentioned.

The coherent scattering processes allow the direct study of nuclear gamma rays by crystal diffraction. This has been accomplished [58] by utilizing a transmission-type curved-crystal focusing spectrometer in which the gamma-ray source in concentrated form is placed on a focal circle of a curved quartz crystal which acts as a monochromator of variable wavelength. The intensity of the diffracted gamma rays is recorded by a scintillation counter as a function of the source position which will delineate the profiles of spectral lines. Wavelengths corresponding to an energy as high as 1.3 Mev can be measured with an absolute precision of better than 1 part in 1,000. At lower energies the precision improves to about 1 part in 10,000.

The method of crystal diffraction may not always be used generally for practical reasons of intensity. It requires sources having an intensity of a thousand times that required for conventional magnetic spectrometers. Only very small solid angles of emergence from the source are accepted by the instrument so that the detection efficiency may be further limited in many cases.

For the detection of weak high-energy gamma rays in the presence of very intense low-energy gamma rays the nuclear photoeffect has been used with success [59]. The photoproton or photoneutron reactions employing nuclei with known proton or neutron binding energies provide a valuable tool for determining the energies and intensities of high-energy gamma rays. This method is free from the spurious effects due to the pile-up of less energetic gamma rays; however, the detection of the ejected nuclear particle may not always be unambiguous. The detection of the ejected nucleons may be made directly by using counters or photographic emulsions, or by means of observing induced radioactivity in the product nuclei.

Such detectors as Geiger counters, ionization chambers, electroscopes, etc., which have a complicated energy response to gamma radiation, are not extensively used as quantitative detectors of gamma radiation. They may be used, however, to make relative intensity comparisons in cases where the spectral distributions of the radiation are the same.

References

1. Heitler, W.: "The Quantum Theory of Radiation," 3d ed., Oxford University Press, London, 1954.
2. Schiff, L. I.: "Quantum Mechanics," 2d ed., McGraw-Hill, New York, 1955.
3. Biedenharn, L. C., and M. E. Rose: *Revs. Mod. Phys.*, **25**: 729 (1953).
4. Edmonds, A. R.: "Angular Momentum in Quantum Mechanics," Princeton University Press, Princeton, N.J., 1957.
5. Rose, M. E.: "Elementary Theory of Angular Momentum," Wiley, New York, 1957.
6. Rose, M. E.: "Multipole Fields," Wiley, New York, 1955.
7. Weisskopf, V. F.: *Phys. Rev.*, **83**: 1073 (1951).
8. Moszkowski, S. A.: *Phys. Rev.*, **83**: 1071 (1951).
9. Mayer, M. G.: *Phys. Rev.*, **78**: 16 (1950).
10. Haxel, O., J. H. D. Jensen, and H. Seuss: *Z. Physik*, **128**: 301 (1950).
11. Moszkowski, S. A.: *Phys. Rev.*, **89**: 474 (1953).
12. Stech, B.: *Z. Naturforsch.*, **7A**: 401 (1952); **9A**: 1 (1953).
13. Rotenberg, M., R. Bivins, N. Metropolis, and J. K. Wooten, Jr.: "The 3-*j* and 6-*j* Symbols," Technology Press, Cambridge, Mass., 1959.
14. Elliott, J. P.: The Shell Model, in P. M. Endt and M. Demeur (eds.), "Nuclear Reactions," vol. I, chap. 2, North Holland Publishing Company, Amsterdam, 1959.
15. De Shalit, A., and I. Talmi: "Nuclear Shell Theory," Academic, New York, 1963.
16. Bohr, A., and B. R. Mottelson: *Kgl. Danske Videnskab. Selskab., Mat.-Fys. Medd.*, **27**(16): (1953).
17. Kerman, A. K.: Nuclear Rotational Motion, in

P. M. Endt and M. Demeur (eds.), "Nuclear Reactions," vol. 1, chap. 10, North-Holland Publishing Company, Amsterdam, 1959.
18. Goldhaber, M., and A. W. Sunyar: Classification of Nuclear Transition Rates, in Kai Siegbahn (ed.), "Alpha, Beta, and Gamma-ray Spectroscopy," chap. 18, North Holland Publishing Company, Amsterdam, 1965.
19. Danos, M., and L. C. Maximon: *J. Math. Phys.*, **6**: 766 (1965).
20. Kistner, O. C., and A. W. Sunyar: *Phys. Rev. Letters*, **4**: 412 (1960).
21. Mellissinos, A. C., and S. P. Davis: *Phys. Rev.*, **115**: 130 (1959).
22. See, for example, Messiah, Albert: "Quantum Mechanics," Wiley, New York, 1961.
23. Reitz, J. R.: *Phys. Rev.*, **77**: 10 (1950). Gellman, H., B. A. Griffith, and J. P. Stanley: *Phys. Rev.*, **85**: 944 (1952).
24. Rose, M. E.: "Internal Conversion Coefficients," North Holland Publishing Company, Amsterdam, 1958.
25. Sliv, L. A., and I. M. Band: Tables of Internal Conversion Coefficients, in Kai Siegbahn (ed.), "Alpha Beta, and Gamma-ray Spectroscopy," North Holland Publishing Company, Amsterdam, 1965.
26. Church, E. L., and J. Weneser: *Ann. Rev. Nucl. Sci.*, **10**: 193 (1960).
27. Ewan, G. T., and R. L. Graham: Internal Conversion Studies at Very High Resolution, in Kai Siegbahn (ed.), "Alpha, Beta, and Gamma-ray Spectroscopy," chap. 18, North Holland Publishing Company, Amsterdam, 1965.
28. Devons, S., and L. J. B. Goldfarb: Angular Correlations, in S. Flügge (ed.), "Encyclopedia of Physics," vol. 42, Springer, Berlin, 1957.
29. Frauenfelder, H., and R. M. Steffen: Angular Correlations, in Kai Siegbahn (ed.), "Alpha, Beta, and Gamma-ray Spectroscopy," chap. 19, North Holland Publishing Company, Amsterdam, 1965.
30. Ferentz, M., and N. Rosenzweig: Table of *F* Coefficients, *Argonne Natl. Lab. Rept.*, ANL 5324 (unpublished), 1955.
31. Oppenheimer, J. R., and L. Nedelski: *Phys. Rev.*, **44**: 948 (1933). Horton, G. K.: *Proc. Phys. Soc. (London)*, **60**: 457 (1943).
32. Rose, M. E.: *Phys. Rev.*, **76**: 678 (1949).
33. See, for example, Wilson, R.: Internal Pair Formation, in Kai Siegbahn (ed.), "Alpha, Beta, and Gamma-ray Spectroscopy," chap. 25, North Holland Publishing Company, Amsterdam, 1965.
34. Bloom, S. D.: *Phys. Rev.*, **88**: 312 (1952).
35. Bergström, I., and C. Nordling: The Auger Effect, in Kai Siegbahn (ed.), "Alpha, Beta, and Gamma-ray Spectroscopy," North Holland Publishing Company, Amsterdam, 1965.
36. Broyles, C. D., D. A. Thomas, and S. K. Haynes: *Phys. Rev.*, **89**: 715 (1953).
37. Glauber, R. J., and P. C. Martin: *Phys. Rev.*, **104**: 158 (1956).
38. Liden, K., and N. Starfelt: *Phys. Rev.*, **97**: 419 (1955).
39. Hayward, R. W., and D. D. Hoppes: *Phys. Rev.*, **104**: 183 (1956).
40. Bethe, H. A., and J. Ashkin: in E. Segrè (ed.), "Experimental Nuclear Physics," vol. 1, Wiley, New York, 1953.
41. Hayward, E.: Photonuclear Reactions, in N. MacDonald (ed.), "Nuclear Structure and Electromagnetic Interactions," Oliver & Boyd, Edinburgh and London, 1965.
42. Brown, G. E.: "Unified Theory of Nuclear Models," North Holland Publishing Company, Amsterdam, 1964. This book contains a good survey of the many-body problem in nuclei.
43. Moon, P. B.: *Proc. Phys. Soc. (London)*, **64**: 76 (1951).
44. Goldhaber, M., L. Grodzins, and A. W. Sunyar: *Phys. Rev.*, **109**: 1015 (1958).
45. Metzger, F. R.: *Progr. Nucl. Phys.*, **7**: 53 (1959).
46. Mössbauer, R. L.: *Z. Physik*, **151**: 124 (1958); *Naturwissenschaften*, **45**: 538 (1958); *Z. Naturforsch.*, **14A**: 211 (1959).
47. Pound, R. V., and G. A. Rebka, Jr.: *Phys. Rev. Letters*, **4**: 337 (1960).
48. Hill, E. L.: Optics and Relativity, in E. U. Condon and Hugh Odishaw (eds.), "Handbook of Physics," pt. 6, chap. 8, McGraw-Hill, New York, 1958.
49. Hay, H. J., J. P. Schiffer, T. E. Cranshaw, and P. A. Egelstaff: *Phys. Rev. Letters*, **4**: 165 (1960).
50. Hayward, R. W.: *Advan. Electron.*, **5**: 97 (1953).
51. Jordan, W. H.: *Ann. Rev. Nucl. Sci.*, **1**: 207 (1952).
52. Price, W. J.: "Nuclear Radiation Detection," 2d ed., McGraw-Hill, New York, 1964.
53. Wilkinson, D. H.: "Ionization Chambers and Counters," Cambridge University Press, New York, 1950.
54. Dearnaley, G., and D. C. Northrop: "Semiconductor Counters for Nuclear Radiation," Wiley, New York, 1963.
55. Walker, R., and B. McDaniel: *Phys. Rev.*, **74**: 315 (1948).
56. Alburger, D. E.: *Rev. Sci. Instr.*, **23**: 671 (1952).
57. Tavendale, A. J., and G. T. Ewan: *Nucl. Inst. Methods*, **25**: 185 (1963).
58. Dumond, J. W. M.: *Rev. Sci. Instr.*, **18**: 626 (1947).
59. Wilson, R.: *Phys. Rev.*, **79**: 1004 (1950).

Chapter 7

Neutron Physics

By D. T. GOLDMAN *and* C. O. MUEHLHAUSE, National Bureau of Standards

1. Fundamental Properties

The belief that the *charge* of the neutron is identically zero is based on two fundamental points: (1) the highly substantial belief that charge is quantized, its absolute value being a multiple of the absolute value of the electron's charge, and (2) the overwhelming experimental evidence that the absolute value of the neutron's charge is very much less than that of an electron. Considering point 2 only, a very small upper limit to the charge on a neutron has been placed by experiments designed to observe the deflection of slow neutrons in a high electric field [1].† Utilizing a pair of perfect crystals of silicon, one as a beam collimator and the other to measure the electrostatic deflection produced by the electric field, Shull, Billman, and Wedgwood have been able to assign an upper limit for the neutron's charge of

$$-(2.0 \pm 3.8) \times 10^{-18}$$

that of the electron.

Ever since the existence of the neutron was demonstrated by Chadwick [2], the *mass* of the neutron has been recognized as being very nearly equal to that of the proton. The head-on collision of a fast neutron with a static proton results in transferring essentially all the neutron's kinetic energy to the proton. The precise means of determining the neutron's mass, however, is from a determination of the masses and Q values of several light-element reactions. For example,

$$_1H^3 + {_1H^1} \rightarrow {_2He^3} + {_0n^1} - 0.764 \text{ Mev} \quad (7.1)[3]$$

$$_1H^3 \rightarrow {_2He^3} + \beta^- + 0.0181 \text{ Mev} \quad (7.2)[4]$$

from which the following mass equation results:

$$e + {_1H^1} = {_0n^1} - 0.782 \text{ Mev} \tag{7.3}$$

which yields the value 1.008665 amu or 1.67482×10^{-24} g for the neutron mass.

Equation (7.3) implies an instability of the free neutron. It decays [5] with a mean life of about 12 min, a time long compared with mean absorption times in ordinary matter ($\sim 10^{-4}$ sec). The decay is to a proton by an allowed β^- transition, having an end point given by (7.3), i.e.,

$$_0n^1 \rightarrow {_1p^1} + \beta^- + 0.782 \tag{7.4}$$

† Numbers in brackets refer to References at end of chapter.

The mass and velocity of the neutron determine its de Broglie *wavelength* λ. The latter may be computed with high accuracy from $\lambda = \dfrac{3.956 \times 10^{-3}}{v(\text{cm/sec})}$ cm, or with fair accuracy from

$$4\pi\lambda\!\!\!^{-2} = \frac{2.60 \times 10^6}{E_n} \quad \text{barns} \tag{7.5}$$

The latter form is useful for the evaluation of resonant cross sections. Here $\lambda\!\!\!^-$ is $\lambda/2\pi$ in units of 10^{-12} cm (i.e., the cross section is in units of 10^{-24} cm² or barns), and E_n is the neutron energy in electron volts.

In so far as the neutron has mass, it is attracted to other matter (e.g., the earth) according to the classical long-range inverse-square law. The *acceleration due to gravity* has been measured with fair accuracy [6] and found to have the value 974 ± 5 cm/sec² (compared with the local value of 979.7 cm/sec²).

An overwhelming amount of evidence indicates that the neutron, like the proton, obeys *Fermi-Dirac statistics*. For example, all odd-particle nuclei have spins that are odd multiples $\neq 0$ of ½, and all even-particle nuclei have spins that are zero or multiples of 1. Further evidence is given by the fact that only two states are observed in the formation of a compound nucleus resulting from a slow neutron and an initial nucleus of total angular momentum I. This is to say that the compound angular momentum J is given by $I \pm \frac{1}{2}$, and therefore the *spin* of the neutron is ½. This value for the spin has been shown to be correct in a direct manner from the analysis of certain Stern-Gerlach-type magnetic scattering experiments [7].

As a stationary quantum-mechanical system the neutron is not expected to possess an *electric-dipole moment* nor any higher electric moment either, since the spin is only ½. Such a moment has been searched for but with negative results [8]. On the other hand, a *magnetic moment* parallel to its spin may and does exist. It is measured by comparison with the proton's magnetic moment. A beam of slow neutrons from a reactor is polarized and passed through a radio-frequency coil situated in a fixed magnetic field. The radio frequency necessary for depolarization is compared with the proton depolarization frequency in the same setup [9]. In this manner the value of the neutron's magnetic moment is found to be -1.9132 nuclear Bohr magnetons ($e\hbar/2Mc$). The negative sign indicates that the spin and moment are in opposite directions.

2. Interactions with Individual Nuclei

The probability or frequency of a nuclear interaction is measured in terms of the cross section; that is, the mean life τ_i for survival of a nucleus against an i-type interaction in a flux ϕ of radiation is simply

$$\tau_i = 1/(\phi\sigma_i) \tag{7.6}$$

where σ_i is the cross section for the i process between the nucleus and the radiation. The general theory and the various specific means for calculating cross sections and otherwise correlating experimentally observed data on reaction rates are outlined in Part 9, Chaps. 1 and 8. The reader is referred to these chapters for a fundamental discussion of the basic nuclear physics involved. This section will be restricted to certain of the more applied aspects of neutron nuclear physics; where appropriate, emphasis will be given to the portions of neutron scattering, reaction, and generation that are of particular significance to later portions of the chapter.

Certain nuclear reactions involving neutrons are particularly noteworthy and serve also to illustrate the variety of interactions possible. These are given below along with the approximate cross section in barns (b) (10^{-24} cm²) for the stated energy region of the incoming radiation.

Elastic Scattering:

$${}_0n^1 + {}_1H^1 \rightarrow {}_0n^1 + {}_1H^1 \qquad \sigma \simeq 20 \text{ barns} \qquad E_n \gtrsim 10 \text{ ev} \tag{7.7}$$ [10]

This reaction is of fundamental (see Sec. 5) and of practical importance. It is utilized in the shielding, moderation, and detection of fast neutrons. The detector may consist of some hydrogenous gas in an ionization or proportional chamber in which an electrical pulse is generated via the ionization induced by the proton recoils.

Inelastic Scattering:

$${}_0n^1 + {}_{26}Fe^{56} \rightarrow {}_0n^1 + {}_{26}Fe^{56*} - 0.85 \text{ Mev} \qquad \sigma \simeq 1 \text{ barn} \qquad E_n \simeq 1 \text{ Mev} \tag{7.8}$$ [11,12]

Following the inelastic event, the excited state of Fe^{56} decays by γ emission in a short time ($\sim 10^{-12}$ sec). In addition to providing a classic example for study, this reaction has practical application to neutron shielding. Iron or magnetite ore may be included as a component in high-density concrete shielding around a reactor. The relatively high cross section of iron is responsible for a rapid attenuation of the fast neutrons in the shield.

Double Neutron Emission:

$${}_0n^1 + {}_4Be^9 \rightarrow {}_4Be^8 + 2\,{}_0n^1 - 1.67 \text{ Mev} \qquad \sigma \sim 0.1 \text{ barn} \qquad E_n \gtrsim 1.8 \text{ Mev} \tag{7.9}$$ [13]

Following the emission of the two neutrons the Be^8 nucleus promptly breaks up into two alpha particles. Two alpha particles, like two protons or two neutrons, lack a bound state. The first excited state is virtual by about 100 kev and is of very short duration ($\sim 10^{-16}$ sec).

Fission:

$${}_0n^1(\text{slow}) + {}_{92}U^{235} \rightarrow X + Y + \nu\,{}_0n^1 + \sim 200 \text{ Mev} \qquad \sigma = 580 \text{ barns} \qquad E_n = 0.025 \text{ ev (thermal)} \tag{7.10}$$ [14,15]

$${}_0n^1(\text{fast}) + {}_{92}U^{238} \rightarrow X + Y + \nu\,{}_0n^1 + \sim 200 \text{ Mev} \qquad \sigma \simeq 0.6 \text{ barn} \qquad E_n \simeq 2 \text{ Mev} \tag{7.11}$$

These reactions are fundamental to the engineering of nuclear chain reactors [16]. X and Y represent fission products, and ν is the average number (~ 2.5) of neutrons emitted in fission. Scattering and capture compete with fission in each of the above. For example, in (7.10) U^{236} is formed about 18 per cent as often as fission. Also long-range alpha particles [17] (ternary fission) are emitted about once for every 500 binary fissions. The reader is referred to Part 9, Chap. 10, for a detailed discussion of the fission process.

Photodisintegration:

$${}_1H^2 + \gamma \rightarrow {}_1H^1 + {}_0n^1 - 2.23 \text{ Mev} \tag{7.12}$$ [18]

$${}_4Be^9 + \gamma \rightarrow {}_4Be^8 + {}_0n^1 - 1.67 \text{ Mev} \qquad \sigma \simeq 0.01 \text{ barn} \qquad E_n \gtrsim \text{threshold} \tag{7.13}$$ [19]

The first of these reactions, (7.12), is the one in which the photo effect was first observed. The second of these reactions, (7.13), is utilized at the National Bureau of Standards [20] to compose the primary standard of neutron source strength. The γ rays employed are those from radium in equilibrium with its daughter products.

Single Neutron Production:

$${}_3Li^7 + {}_1H^1 \rightarrow {}_4Be^7 + {}_0n^1 - 1.63 \text{ Mev} \tag{7.14}$$ [21]

$${}_1H^2 + {}_1H^2 \begin{cases} \xrightarrow{\sim 50\%} {}_2He^3 + {}_0n^1 + 3.27 \text{ Mev} \\ \xrightarrow[\sim 50\%]{} {}_1H^3 + {}_1H^1 + 4.03 \text{ Mev} \end{cases} \tag{7.15}$$ [22]

$${}_1H^3 + {}_1H^2 \rightarrow {}_2He^4 + {}_0n^1 + 17.60 \text{ Mev} \tag{7.16}$$ [23]

$${}_4Be^9 + {}_2He^4 \rightarrow {}_6C^{12} + {}_0n^1 + 5.75 \text{ Mev} \tag{7.17}$$ [2]

The above cross sections are $\simeq 0.1$ barn when the incoming charged particles have sufficient energy readily to overcome Coulomb repulsion. The first three along with (7.1) are commonly used with Van de Graaff, cyclotron, or linear accelerators to produce monoenergetic fast neutrons (for further use in study of fast neutron reactions). Reactions (7.15) and (7.16) constitute the so-called "fusion reactions"; (7.16) produces helium and 14-Mev neutrons, and (7.17) was the reaction from which neutrons were first observed [2].

Neutron Absorption:

$${}_0n^1(\text{slow}) + {}_{79}Au^{197} \rightarrow {}_{79}Au^{198} + \gamma + 6.50 \text{ Mev} \qquad \sigma = 98.8 \text{ barns} \quad E_n = \text{thermal}; \qquad \sigma \sim 4 \times 10^4 \text{ barns} \quad E_n = 4.91 \text{ ev (a resonance)} \tag{7.18}$$ [24]

$${}_0n^1(\text{slow}) + {}_7N^{14} \rightarrow {}_6C^{14} + {}_1H^1 + 0.63 \text{ Mev} \qquad \sigma = 1.80 \text{ barns} \qquad E_n = \text{thermal} \tag{7.19}$$ [25]

$${}_0n^1(\text{slow}) + {}_3Li^6 \rightarrow {}_1H^3 + {}_2He^4 + 4.79 \text{ Mev} \qquad \sigma = 950 \text{ barns} \qquad E_n = \text{thermal} \tag{7.20}$$ [26]

$$_0n^1(\text{slow}) + {_5B^{10}} \begin{cases} \xrightarrow{\sim 90\%} {_3Li^{7*}} + {_2He^4} + 2.31 \text{ Mev} \\ \xrightarrow{\sim 10\%} {_3Li^7} + {_2He^4} + 2.79 \text{ Mev} \end{cases} \qquad (7.21)^{[27]}$$

σ = 3,840 barns E_n = thermal

$$_0n^1(\text{slow}) + {_{48}Cd^{113}} \rightarrow {_{48}Cd^{114}} + \gamma + 9.05 \text{ Mev} \qquad (7.22)^{[28]}$$

σ = 20,000 barns E_n = thermal

$$_0n^1(\text{slow}) + {_2He^3} \rightarrow {_1H^3} + {_1H^1} + 0.764 \text{ Mev} \qquad (7.23)^{[29]}$$

σ = 5,330 barns E_n = thermal

The above reactions constitute a variety of neutron absorption processes. Reaction (7.18) is typical of many elements and discloses the ease with which slow neutrons (e.g., at thermal energies wherein the absorption cross section varies inversely as the velocity of the neutron) activate materials via the (n,γ) process [30]. In the example given, Au^{198} is a radioactive β^- and γ emitter with a half life of 2.7 days. Reactions (7.19) and (7.20) are utilized for the production of C^{14} (5,730 years) and H^3 (12.3 years) tracer elements, respectively. Reaction (7.21) is the one most commonly used for the detection of neutrons. BF_3 contained as a gas in a proportional counter gives rise to electrical pulses from the alpha and lithium charged-particle-induced ionization. The large resonance for reaction (7.22) at a neutron energy of 0.178 ev and its corresponding large value of the thermal absorption cross section enable one to use cadmium as a "black absorber" to remove low-energy neutrons ($E_n \lesssim 0.6$ ev) from a neutron beam. Reaction (7.23) is the reverse of (7.1). It is used for the detection of both thermal and fast neutrons. In a recent development fast neutrons are collimated through a thin volume of He^3 sandwiched by two silicon solid-state detectors. Coincidences between the proton and triton particles are required, and the two pulses are summed to give an output pulse proportional to neutron energy. Such a device constitutes a fast neutron spectrometer with an energy resolution of ~100 kev [31].

As discussed generally in Chap. 1 and elaborated upon in Chap. 8 of Part 9, the *optical model* of nuclear reactions constitutes a useful analytical representation of a large amount of cross-section data. The formalism is developed in some detail below in order to bring out the general features of elastic and inelastic neutron scattering and absorption processes. Recognizing, after Wigner [32], that the nuclear force is spin-dependent, the potential V is assumed to be central and to contain a spin-orbit term.

$$V = V(r) + V_{so}(r)\mathbf{L} \cdot \mathbf{S} \qquad (7.24)$$

The wave function ψ governing the interaction is found from the solution of the time-independent Schrödinger equation

$$-\frac{\hbar^2}{2\mu}\nabla^2\psi + V\psi = E\psi \qquad (7.25)$$

where μ is the reduced mass of the system and E is the energy of the colliding pair in their center of mass. $\mathbf{L}$ is the orbital angular-momentum operator and $\mathbf{S}$ is the spin angular-momentum operator. Using the vector spherical harmonics [33],

$$y_{ljm}(\hat{r},\sigma) = \sum_{m_l m_s} C(l,\tfrac{1}{2},j;m_l m_s m) Y_l^{m_l}(\hat{r}) \chi_{\frac{1}{2}}^{m_s}(\sigma) \qquad (7.26)$$

where C is the Clebsch-Gordan coefficient, $Y_l^{m_l}(\hat{r})$ is the spherical harmonic of order l, and $\chi_{\frac{1}{2}}^{m_s}(\sigma)$ is the spin wave function, the wave function ψ may be expanded as

$$\psi(\mathbf{r},\sigma) = \sum_{jlm} \frac{A_{ljm}}{r} u_{ljm}(r) y_{ljm}(\hat{r},\sigma) \qquad (7.27)$$

After substitution of (7.27) into (7.25) and noting that y_{ljm} is an eigenfunction of J^2, the square of the total spin operator, the following radial equation results:

$$\frac{d^2u_{lj}}{dx^2} + \left[1 - \frac{l(l+1)}{x^2} - \frac{V(x)}{E} - \frac{\gamma_{lj}V_{so}(x)}{E}\right] u_{lj}(x) = 0 \qquad (7.28)$$

where

$$x = kr \qquad k = \sqrt{\frac{2\mu E}{\hbar^2}}$$

$$\gamma_{jl} = \begin{cases} l/2 & \text{if } j = l + \frac{1}{2} \\ -\dfrac{l+1}{2} & \text{if } j = l - \frac{1}{2} \end{cases}$$

The potential in (7.28) may assume various forms such as the one given by (8.34) in Chap. 8, Part 9. For the analysis of scattering data it is necessary to solve (7.28) by numerical techniques. If the nucleus is deformed or if it is susceptible to vibration the separation of a single radial wave equation such as (7.28) is not valid. In the more general case, coupled radial wave equations result. This presentation is confined to the simple optical model. For a more general case the reader is referred to [34].

The regular asymptotic solution of (7.28) is

$$u_{lj}(kr) \underset{r\to\infty}{\sim} \sin\left(kr - \frac{l\pi}{2} + \delta_l^j\right) \qquad (7.29)$$

The total wave function far from the origin must have the form of a plane wave plus a spherical outgoing wave:

$$\psi(\mathbf{r},\sigma) \underset{r\to\infty}{\sim} e^{i\mathbf{k}\cdot\mathbf{r}}\chi_{\frac{1}{2}}^{m_s}(\sigma) + f^{m_s}(\hat{r},\sigma)\frac{e^{ikr}}{r} \qquad (7.30)$$

from which the differential *shape elastic* scattering cross section $\sigma_{se}(\theta)$ is given by

$$\sigma_{se}(\theta) = \sum_{m_s} |f^{m_s}(\hat{r},\sigma)|^2 \qquad (7.31)$$

$f^{m_s}(\hat{r},\sigma)$ is termed the *scattering amplitude*. It is a function of θ, the scattering angle. Two alternative procedures are available for calculating $f(\theta)$. One involves the substitution of (7.27) and

$$e^{i\mathbf{k}\cdot\mathbf{r}}\chi_{\frac{1}{2}}^{m_s}(\sigma) \underset{r\to\infty}{\sim} \sum_l \frac{(2l+1)i^l \sin(kr - l\pi/2)}{kr} P_l(\hat{k}\cdot\hat{r})\chi_{\frac{1}{2}}^{m_s}(\sigma) \qquad (7.32)$$

into (7.30) to evaluate $f^{m_s}(r,\sigma)$ directly. This expresses the shape elastic differential cross section as a sum over Legendre moments:

$$\sigma_{se}(\theta) = \frac{1}{8k^2}\sum_{\substack{lj\tilde{l}\tilde{j}\\L}} P_L(\cos\theta)(1-\eta_l^j)(1-\eta_{\tilde{l}}^{\tilde{j}})^* \cdot (2l+1)(2\tilde{l}+1)(2j+1)(2\tilde{j}+1) C^2(l\tilde{l}L;000)W^2(jl\tilde{j}\tilde{l};\tfrac{1}{2}L) \qquad (7.33)$$

where W is the Racah coefficient, and

$$\eta_l^j = e^{2i\delta_l^j} \tag{7.34}$$

The second method utilizes projection operators [35]

$$\Pi_l^+ = \frac{l + 1 + \boldsymbol{\sigma} \cdot \mathbf{L}/\hbar^2}{2l + 1} \qquad \Pi^- = \frac{l - \boldsymbol{\sigma} \cdot \mathbf{L}/\hbar^2}{2l + 1} \tag{7.35}$$

allowing (7.27) to be rewritten as

$$\psi(\mathbf{r},\sigma) \underset{r\to\infty}{\sim} \sum_l (2l + 1)i^l \left[A_l^+\Pi_l^+ \frac{u_l^+(kr)}{kr} + A_l^-\Pi_l^- \frac{u_l^-(kr)}{kr} \right] P_l(\hat{k} \cdot \hat{r})\chi_{1/2}^{m_s}(\sigma) \tag{7.36}$$

Using (7.30), (7.32), and (7.36) one finds

$$A_l^\pm = e^{i\delta_l^\pm} \tag{7.37}$$

$$f^{m_s}(\hat{r},\sigma) = \sum_l \left\{ \sum_j (j + \tfrac{1}{2})e^{i\delta_l^j} \sin \delta_l^j + [e^{i\delta_l^{l+1/2}} \sin \delta_l^{l+1/2} - e^{-i\delta_l^{l-1/2}} \sin \delta_l^{l-1/2}] \frac{\boldsymbol{\sigma} \cdot \mathbf{L}}{\hbar^2} \right\} P_l(\cos \theta)\chi_{1/2}^{m_s} \tag{7.38}$$

or, alternatively,

$$f^{m_s}(\theta) = \frac{1}{2k} \left[\frac{a(\theta)}{i} + \boldsymbol{\sigma} \cdot \mathbf{n}\, b(\theta) \right] \tag{7.39}$$

where $\mathbf{n}$ is the unit vector normal to the scattering plane, and

$$a(\theta) = \sum_{lj} (j + \tfrac{1}{2})(\eta_l^j - 1)P_l(\cos \theta) \tag{7.40}$$

$$b(\theta) = \sum_l \sin \theta \frac{dP_l(\cos \theta)}{d \cos \theta} (\eta_l^{l+1/2} - \eta_l^{l-1/2}) \tag{7.41}$$

$$\sigma_{se}(\theta) = \frac{1}{4k^2} \{|a(\theta)|^2 + |b(\theta)|^2\} \tag{7.42}$$

The potential scattering cross section which is synonymous with the angle-integrated shape elastic cross section can be calculated by integrating either (7.42) or (7.33). The result is

$$\sigma_{se} = \frac{\pi ƛ^2}{2} \sum_{lj} (2j + 1)|1 - \eta_l^j|^2 \tag{7.43}$$

The reaction σ_r and total σ_T cross sections are readily shown to be

$$\sigma_r = \frac{\pi ƛ^2}{2} \sum_{lj} (2j + 1)(1 - |\eta_l^j|^2) \tag{7.44}$$

$$\sigma_T = \pi ƛ^2 \sum_{lj} (2j + 1) \operatorname{Re} (1 - \eta_l^j) \tag{7.45}$$

By also making use of (7.39) the polarization $\mathbf{P}(\theta)$ can be shown to be

$$\mathbf{P}(\theta) = \frac{ƛ^2 \operatorname{Im} (ab^*)}{2\sigma_e(\theta)} \frac{\mathbf{k}_f \times \mathbf{k}_0}{|\mathbf{k}_f \times \mathbf{k}_0|} \tag{7.46}$$

where $\mathbf{k}_0$ and $\mathbf{k}_f$ are the initial and final wave vectors, and $\sigma_e(\theta)$ is the differential elastic cross section including *compound elastic scattering* to be discussed below.

In addition to the direct scattering the incoming neutron may amalgamate with the target nucleus to form a compound or intermediate state between the initial and some final state. The final state could be the initial state (compound elastic scattering), an excited state (inelastic scattering), the ground state (cascade γ emission from capture), or a nucleus with a different Z (charged-particle emission). Examples were given earlier [e.g., (7.18)].

A method for the calculation of compound elastic and inelastic cross sections was presented by Hauser and Feshbach [36]. It is based on the statistical assumption that the possible capture levels are sufficiently dense that the probability for the formation of the compound nucleus is proportional to the penetrability of the incident particle. This has been extended [37] to include the effect of well-separated nonoverlapping levels. The cross sections given below accounts only for the spin-orbit term in the potential.

$$\sigma(J_i,J_f;E,E';\theta) = \frac{ƛ^2}{4} \sum_{\substack{j_ij_fl_il_f \\ J}} (-)^{J_i-J_f} \times \frac{T_{l_i}{}^{j_i}(E)T_{l_f}{}^{j_f}(E')(2J + 1)^2}{2(2J_i + 1) \sum_p T_{l_p}{}^{j_p}(E'_p)} \sum_{L(\text{even})} P_L(\cos \theta)Z(l_ij_il_ij_i;\tfrac{1}{2}L)Z(l_fj_fl_fj_f;\tfrac{1}{2}L)W(Jj_iJj_i;J_iL)W(Jj_fJj_f;J_fL) \tag{7.47}$$

where

$W(a,b,c,d;e,f)$ = Racah coefficient

$Z(a,b,c,d;e,f) = \sqrt{(2a + 1)(2b + 1)(2c + 1)(2d + 1)}\, C(acf;\, 000)\, W(a,b,c,d;ef)$

E = initial energy of neutron

E' = final energy of final-state particle (neutron, proton)

J_i = total angular momentum of target nucleus

J_f = total angular momentum of final nucleus

$T_l{}^j(E)$ = $(1 - |\eta_l{}^j|^2)$, the nuclear penetrability. This quantity is calculated with the optical model described previously.

j_i = total angular-momentum quantum number of incident neutron and may assume only half-integral values.

J = total angular momentum of compound nucleus. If J_i is integral (half-integral), then J is half-integral (integral). All J values must be included up to values for which the contribution to the sum is negligible.

j_f = total angular momentum of the final-state particle (neutron or proton)

To perform compound-nucleus calculations, the parity and spin of the target nucleus and all energetically possible final states must be known.

If the parity of J_i and J_f are the same (different), the possible values of l_f are even (odd) and l_i is even and conversely when l_i is odd.

E'_p is a possible final value of the energy of the particle. The summation over (p) includes all final particle energies permitted by energy conservation. To each E'_p there corresponds a state of the final nucleus, $E(J_f)$, such that $E'_p + E_p(J_f) = E + E(J_i)$, where $E(J_i)$ is the initial energy of the target nucleus (zero for a target in the ground state).

An important characteristic of the compound nucleus angular distribution given by Eq. (7.47) is that it is symmetrical about the 90° scattering angle. This is a consequence of the fact that the formation and decay of the compound nucleus are independent.

The total cross section for the transition from the initial state (J_i,E) to the final state (J_f,E') is obtained by integrating (7.47) over all angles:

$$\sigma(J_i,J_f;E,E') = \pi\lambda\!\!\!^{-2} \sum_{l_i l_f j_i j_f} \frac{T_{l_i}{}^{j_i}(E)}{2(2J_i+1)} \sum_J \frac{T_{l_f}{}^{j_f}(E')(2J+1)}{\sum_p T_{l_p}{}^{j_p}(E'_p)} \tag{7.48}$$

When the incident energy of the neutron is below that of the first excited state in the target, only compound elastic scattering is energetically possible, and (7.48) reduces [since $E' = E'_p = E$, and

$$\sum_J (2J+1) = (2j_i+1)(2J_i+1)$$

the multiplicity of initial states of the system] to

$$\sigma(J_i,J_f;E,E) = \frac{\pi}{2}\lambda\!\!\!^{-2} \sum_{l_i j_i} (2j_i+1) T_{l_i}{}^{j_i}(E) \tag{7.49}$$

Equations (7.49) and (7.44) are identical, as they should be.

For the calculation of direct interactions such as (n,p) by the so-called *distorted-wave Born approximation*, the reader is referred to Part 9, Chap. 8, Sec. 11.

The compound nuclear formation discussed above was embodied in the cross-section expressions on a statistical basis. Its more evident manifestation is as resonance structure superimposed on the potential or shape elastic cross section vs. energy variation. The single resonance level function is given for slow neutrons by the Breit-Wigner dispersion formula [38]:

$$\sigma_s = \sigma_p + 4\pi\lambda\!\!\!^{-}{}_0{}^2 g \frac{\dfrac{\sqrt{\sigma_p/4\pi}}{\lambda\!\!\!^{-}{}_0{}^2}\Gamma_n(E-E_0) + \dfrac{\Gamma_n{}^2}{4}}{(E-E_0)^2 + \Gamma^2/4} \tag{7.50}$$

$$\sigma_a = 4\pi\lambda\!\!\!^{-}{}_0{}^2 g \frac{(\Gamma_n\Gamma_\gamma/4)\sqrt{E_0/E}}{(E-E_0)^2+\Gamma^2/4} \tag{7.51}$$

where σ_s and σ_a = scattering and absorption cross sections
E = neutron energy
E_0 = resonance energy
Γ_n = neutron width at resonance
Γ_γ = absorption width at resonance
Γ = total width at resonance
σ_p = potential scattering cross section, that is, σ_{se}
$\lambda\!\!\!^{-}{}_0$ = (neutron wavelength at E_0)/2π
$g = \dfrac{2J+1}{2(2I+1)}$ the statistical weight
$J = I \pm \frac{1}{2}$ for S-wave neutrons
I = initial nuclear spin

The important features of this expression are (1) the resonance character is governed by the denominator which has a minimum at $E = E_0$. At $|E - E_0| = \Gamma/2$ the cross section is approximately one-half its maximum value. (2) The total peak cross section σ_0 is given by $4\pi\lambda\!\!\!^{-}{}_0{}^2 g\Gamma_n/\Gamma$. (3) The absorption term, in addition to the resonance factor, contains a $1/v$ factor which becomes important at thermal energies (i.e., the $1/v$ law). (4) There are three scattering terms representing potential or shape elastic scattering, resonance or compound elastic scattering, and the interference between the two. For S waves the latter causes the scattering cross section to be low, that is, $\sigma_p(1-g)$, on the low-energy side of the level. In this region the phases of potential and resonance scattering are opposite.

The scattering features are best revealed by presenting (7.50) in amplitude form. Let R be the potential scattering amplitude ($R \sim$ nuclear radius). Then since $\sqrt{\sigma_p/4\pi} \simeq R$ as $\lambda\!\!\!^{-} \to \infty$,

$$\sqrt{\sigma_p/(4\pi)} \simeq R \qquad \text{as} \qquad \lambda\!\!\!^{-} \to \infty$$

$$\sigma_s = 4\pi(1-g)R^2 + 4\pi g \left| R + \frac{\lambda\!\!\!^{-}{}_0\Gamma_n/2}{E - E_0 + i\Gamma/2} \right|^2 \tag{7.52}$$

For the purposes of Sec. 4 it is important to note here that at thermal energies (7.52) reduces to

$$\sigma_{(th)s} \simeq 4\pi(1-g)R^2 + 4\pi g \left| R - \frac{\lambda\!\!\!^{-}{}_0\Gamma_n}{2E_0} \right|^2 \tag{7.53}$$

Additional levels would contribute additional terms to their respective amplitudes (i.e., for S waves to g_+ or g_- components, depending upon J).

The radiation widths in (7.52) are ~0.1 ev [i.e., typical compound nuclear lifetimes against capture are $\sim 10^{-14}$ sec, a time long compared with the time ($\sim 10^{-22}$ sec) required for a nucleon to transit a nuclear diameter]. This is to say that the levels are sharp. A consequence of this is that thermal motion of the target nuclei alters the effective cross section near resonance. This effect is commonly denoted as *Doppler broadening* and is presented below on the assumption that the target nuclei may be considered to be a free gas of mass M [39].

The effective cross section is given by

$$v\sigma_{\text{eff}}(v) = \int |\mathbf{v} - \mathbf{V}| \sigma(\mathbf{v} - \mathbf{V}) P(\mathbf{V})\, d\mathbf{V} \tag{7.54}$$

where $\mathbf{V}$ is the velocity of the target nucleus, $P(\mathbf{V})$ is the normalized distribution function of $\mathbf{V}$, and $\sigma(\mathbf{v} - \mathbf{V})$ is given by (7.50) and (7.51). For a perfect gas at temperature T

$$P(\mathbf{V})\, d\mathbf{V} = \left(\frac{M}{2\pi kT}\right)^{3/2} e^{-(MV^2/2kT)} 4\pi V^2\, dV \tag{7.55}$$

Equations (7.50) and (7.51) can be expressed as

$$\sigma_s = \sigma_p + \frac{\sigma_0}{1+x^2}\frac{\Gamma_n}{\Gamma} + \sqrt{\sigma_0\sigma_p g \frac{\Gamma_n}{\Gamma}}\,\frac{2x}{1+x^2} \tag{7.56}$$

$$\sigma_a = \frac{\sigma_0}{1+x^2}\sqrt{\frac{E_0}{E}}\,\frac{\Gamma_\gamma}{\Gamma} \tag{7.57}$$

where $x = (E - E_0)\Gamma/2$. It can then be shown that

$$\sigma_{s\,\mathrm{eff}} = \sigma_p + \sigma_0 \frac{\Gamma_n}{\Gamma} \psi(x,\theta) + \sqrt{\sigma_0 \sigma_p g \frac{\Gamma_n}{\Gamma}}\, \chi(x,\theta) \quad (7.58)$$

$$\sigma_{a\,\mathrm{eff}} = \sigma_0 \frac{\Gamma_\gamma}{\Gamma} \sqrt{\frac{E_0}{E}}\, \psi(x,\theta) \quad (7.59)$$

where
$$\psi(x,\theta) = \frac{\theta}{2\sqrt{\pi}} \int_{-\infty}^{\infty} \frac{e^{-(\theta^2/4)(x-y)^2}}{1+y^2}\, dy \quad (7.60)$$

$$\chi(x,\theta) = \frac{\theta}{2\sqrt{\pi}} \int_{-\infty}^{\infty} \frac{e^{-(\theta^2/4)(x-y)^2}}{1+y^2}\, 2y\, dy \quad (7.61)$$

$$\theta = \frac{\Gamma}{\Delta} \qquad \Delta = \sqrt{\frac{4mkTE_0}{m+M}} \quad (7.62)$$

Δ is the so-called *Doppler width.*

Certain useful properties of the broadening functions are apparent. For either $T = 0$, $m/M \to 0$, or $\theta \to \infty$ the unbroadened forms of the cross section result. Also the integrals around the resonance terms are simply

$$\int \sigma_0 \psi(x,\theta)\, dE = \frac{\pi}{2} \sigma_0 \Gamma \qquad \int \chi(x,\theta)\, dE = 0 \quad (7.63)$$

i.e., are independent of temperature. Thus, the experimental area under a thin target transmission curve yields $g\Gamma_n$. Other useful quantities to the experimentalist are the so-called *resonance integrals* which are quantities proportional to the absorption or scattering of a thin sample in a $1/E$ varying flux (e.g., such as pertains to the moderator of a nuclear reactor).

$$\int_{res} \sigma_a \frac{dE}{E} = \frac{\pi}{2} \sigma_0 \frac{\Gamma_\gamma}{E_0} \quad (7.64)$$

$$\int_{res} \sigma_s \frac{dE}{E} = \frac{\pi}{2} \sigma_0 \frac{\Gamma_n}{E_0} \quad (7.65)$$

As shown in Chap. 1, Part 9, a complete derivation of the nuclear cross section requires the inclusion of interference effects between levels of the compound nucleus. Wigner and Eisenbud [40] have expressed the cross section for going from channel c to channel c' as

$$\sigma_{cc'} = \pi \lambda\!\!\!^{-2} \sum_J \frac{2J+1}{(2I+1)(2s+1)} |\delta_{cc'} - U_{cc'}{}^J|^2 \quad (7.66)$$

where $U_{cc'}{}^J$ is denoted as the *collision matrix.* Vogt [41] has shown that a particular form of the collision matrix is

$$U_{cc'}{}^J = e^{i(\delta_c + \delta_{c'})} \left(\delta_{cc'} + i \sum_{\lambda\lambda'} \sqrt{\Gamma_{\lambda c} \Gamma_{\lambda' c'}}\, A_{\lambda\lambda'} \right) \quad (7.67)$$

where

$$A_{\lambda\lambda'}{}^{-1} = (E_\lambda - E)\delta_{\lambda\lambda'} - \frac{i}{2} \sum_c \sqrt{\Gamma_{\lambda c} \Gamma_{\lambda' c'}} \quad (7.68)$$

and δ_c is the potential scattering phase shift, i.e., determined from an optical-model potential. $\Gamma_{\lambda c}$ is the partial width for the decay of the level λ into channel c. Making use of (7.66) to (7.68) the *multi-level* formula for neutron scattering and capture [that is, (n,γ)] can be written

$$\sigma_{n,n} = 4\pi \lambda\!\!\!^{-2} \sum_J g_J \left| \sum_l e^{i\delta_l} \sin \delta_l + \tfrac{1}{2} \sum_{\lambda\lambda'} \sqrt{\Gamma_{\lambda n} \Gamma_{\lambda' n}}\, A_{\lambda\lambda'} \right|^2 \quad (7.69)$$

$$\sigma_{n,\gamma} = \pi \lambda\!\!\!^{-2} \sum_J g_J \sum_{\lambda\lambda'\lambda''} \Gamma_{\lambda\gamma} \sqrt{\Gamma_{\lambda' n} \Gamma_{\lambda'' n}}\, A_{\lambda\lambda'} A^*{}_{\lambda\lambda''} \quad (7.70)$$

In (7.70) it has been assumed there are very many capture channels and that $\sqrt{\Gamma_{\lambda\gamma}}$ fluctuates in sign. For other absorption processes, such as fission, there may be only very few exit channels, thus requiring explicit evaluation of the summation.

If there is no interference between levels, then A and A^{-1} are diagonal matrices, $A = [(E_\lambda - E) - i\Gamma_\lambda/2]^{-1}$, and (7.69) and (7.70) reduce to a simpler form:

$$\sigma_{n,n} = 4\pi \lambda\!\!\!^{-2} \sum_J g_J \left[\sum_l \sin^2 \delta_l + \frac{1}{2} \sum_l \Gamma_{\lambda n} \frac{(E - E_\lambda) \sin 2\delta_l - \Gamma(1 - \cos 2\delta_l)}{(E - E_\lambda)^2 + \Gamma^2/4} + \frac{1}{4} \frac{\Gamma_{\lambda n} \Gamma_{\lambda n}}{(E - E_\lambda)^2 + \Gamma^2/4} \right] \quad \text{for each resonance } \lambda \quad (7.71)$$

$$\sigma_{n,\gamma} = \pi \lambda\!\!\!^{-2} \sum_J g_J \frac{\Gamma_n \Gamma_\gamma}{(E - E_\lambda)^2 + \Gamma^2/4} \quad (7.72)$$

3. Transport of Neutrons in Matter

In this section consideration is given to the life history of a neutron from its production until it reaches thermal equilibrium with its surroundings. The process of slowing down and thermalization is important to a number of practical situations such as moderation in reactor media and in shielding.

The equation that governs the behavior of neutrons in a medium is the *Boltzmann transport equation.* This is given below for the reasonable conditions in which (1) the medium is static except for microscopic thermal motion, (2) the collision of neutrons with other neutrons may be neglected (i.e., the neutron density is much less than the density of the medium), and (3) free neutron decay may be neglected (i.e., wherein the mean time for capture is much less than the mean free decay time). Neutrons are described by the *angular flux vector* $f(\boldsymbol{\varrho},E,\boldsymbol{\Omega},t)\, d\boldsymbol{\varrho}\, dE\, d\boldsymbol{\Omega}\, dt$, the number of neutrons in a volume element $d\boldsymbol{\varrho}$ around $\boldsymbol{\varrho}$, the energy element dE around E, whose direction of motion lies within $d\boldsymbol{\Omega}$ around $\boldsymbol{\Omega}$, within the time element dt around t, and multiplied by the speed $v = \sqrt{2E/m}$. Then

$$\frac{1}{v} \frac{\partial f(\boldsymbol{\varrho},E,\boldsymbol{\Omega},t)}{\partial t} = \int f(\boldsymbol{\varrho},E',\boldsymbol{\Omega}',t)\, \Sigma_s(E' \to E, \boldsymbol{\Omega}' \to \boldsymbol{\Omega})\, dE'\, d\boldsymbol{\Omega}' - \boldsymbol{\Omega} \cdot \nabla f(\boldsymbol{\varrho},E,\boldsymbol{\Omega},t) - f(\boldsymbol{\varrho},E,\boldsymbol{\Omega},t) \left[\sum_s (E) + \sum_a (E) \right] + S(E,\boldsymbol{\varrho}) \quad (7.73)$$

where $\Sigma_s(E)$ and $\Sigma_a(E)$ are the macroscopic scattering and absorption cross sections, respectively. (That is, Σ is the product of the nuclear density and the cross section. It is the inverse mean free path for the process.) $\Sigma_s(E' \to E,\ \mathbf{\Omega}' \to \mathbf{\Omega})$ is the macroscopic differential in energy and angle cross section, and $S(E,\boldsymbol{\varrho})$ is a source term (e.g., from fission in a reactor core location).

The total density of neutrons, $n(\boldsymbol{\varrho},t)$, and the total flux $\phi(\boldsymbol{\varrho},t)$ are scalar quantities obtained by integration over E and $\mathbf{\Omega}$. The neutron *current* density $\mathbf{J}(\boldsymbol{\varrho},E,t)$ is obtained by integration only over $\mathbf{\Omega}$. That is,

$$n(\boldsymbol{\varrho},t) = \int \frac{1}{v} f(\boldsymbol{\varrho},E,\mathbf{\Omega},t)\, dE\, d\mathbf{\Omega} \qquad (7.74)$$

$$\phi(\boldsymbol{\varrho},t) = \int f(\boldsymbol{\varrho},E,\mathbf{\Omega},t)\, dE\, d\mathbf{\Omega} \qquad (7.75)$$

$$\mathbf{J}(\boldsymbol{\varrho},E,t) = \int f(\boldsymbol{\varrho},E,\mathbf{\Omega},t)\mathbf{\Omega}\, d\mathbf{\Omega} \qquad (7.76)$$

The solution of the Boltzmann equation is, in general, difficult, and simplifying techniques must be employed. The one most commonly used [42] is to expand each term in (7.73) into a suitable set of moments. Since the scattering cross section Σ_s can usually be described by a few Legendre moments, the Legendre expansion is used for each of the physical quantities. For example,

$$f(\boldsymbol{\varrho},E,\mathbf{\Omega},t) = \sum_{lm} \frac{2l+1}{4\pi} f_l{}^m(\boldsymbol{\varrho},E,t)P_l{}^m(\mathbf{\Omega}) \qquad (7.77)$$

The resulting equations are known as the spherical harmonic or Legendre functional form of the transport equation. If all moments beyond the first are set equal to zero, there results the following two-coupled equations:

$$\frac{1}{v}\frac{\partial\phi(\boldsymbol{\varrho},E,t)}{\partial t} + \nabla\cdot\mathbf{J}(\boldsymbol{\varrho},E,t) = \int\Sigma_s{}^0(E'\to E)\phi(\boldsymbol{\varrho},E',t)\,dE' + S(E) - [\Sigma_a(E) + \Sigma_s(E)]\,\phi(\boldsymbol{\varrho},E,t) \qquad (7.78)$$

$$\frac{1}{v}\frac{\partial\mathbf{J}(\boldsymbol{\varrho},E,t)}{\partial t} + \frac{1}{3}\nabla\phi(\boldsymbol{\varrho},E,t) = \int\Sigma_s{}^1(E'\to E)\mathbf{J}(\boldsymbol{\varrho},E',t)\,dE' - [\Sigma_s(E) + \Sigma_a(E)]\,\mathbf{J}(\boldsymbol{\varrho},E,t) \qquad (7.79)$$

where the source has been assumed to be isotropic, and $\phi(\boldsymbol{\varrho},E,t)$ is the flux of neutrons in the volume element $d\boldsymbol{\varrho}$ about $\boldsymbol{\varrho}$ with energies in dE about E.

Equation (7.79) can also be written

$$\frac{1}{v}\frac{\partial\mathbf{J}(\boldsymbol{\varrho},E,t)}{\partial t} + \frac{1}{3}\nabla\phi(\boldsymbol{\varrho},E,t) = \Sigma_{\text{tr}}(\boldsymbol{\varrho},E)\mathbf{J}(\boldsymbol{\varrho},E,t) + \int[\Sigma_s{}^1(E'\to E)\mathbf{J}(\boldsymbol{\varrho},E',t) - \Sigma_s{}^1(E\to E')\mathbf{J}(\boldsymbol{\varrho},E,t)]\,dE' \qquad (7.80)$$

where $\Sigma_{\text{tr}}(\boldsymbol{\varrho},E) = \Sigma_s(\boldsymbol{\varrho},E) + \Sigma_a(\boldsymbol{\varrho},E) - \Sigma^1(\boldsymbol{\varrho},E)$

If the integral in (7.80) is neglected, the energy-dependent diffusion approximation is obtained:

$$\mathbf{J}(\boldsymbol{\varrho},E,t) = -D(\boldsymbol{\varrho},E)\left[\nabla\phi(\boldsymbol{\varrho},E,t) + \frac{3}{v}\frac{\partial J(\boldsymbol{\varrho},E,t)}{\partial t}\right] \qquad (7.81)$$

where $D(\boldsymbol{\varrho},E) \equiv 1/[3\Sigma_{\text{tr}}(\boldsymbol{\varrho},E)]$ is the so-called *diffusion constant.*

Substituting (7.81) into (7.78), we obtain a second-order time-dependent equation for the flux which is called the "telegrapher's equation":

$$\frac{3D}{v^2}\frac{\partial^2\phi}{\partial t^2} + \frac{1}{v}(1 + 3D\,\Sigma_a)\frac{\partial\phi}{\partial t} = DV^2\phi - \Sigma_a\,\phi + S \qquad (7.82)$$

where the medium has been assumed to be homogeneous and the source independent of time. This equation has the general form of a wave equation, having second-order derivatives in both space and time, and its solution has a well-defined traveling wavefront. If we set $\frac{1}{v}\frac{\partial\mathbf{J}}{\partial t} = 0$ in (7.81), we obtain the time-dependent diffusion equation

$$\frac{1}{v}(1 + 3D\,\Sigma_a)\frac{\partial\phi}{\partial t} = D\nabla^2\phi - \Sigma_a\,\phi + S \qquad (7.83)$$

which corresponds to making the propagation velocity in the telegrapher's equation infinite. That is, a source at a given time is instantaneously propagated throughout the space but with an exponential attenuation.

A general method for the solution of neutron transport in a multiplying medium such as a reactor core is the so-called *multigroup diffusion* method [43]. It has great conceptual simplicity and at the same time effects reasonably practical solutions. The fastest neutrons are regarded as born at fission energies and slow down in the moderating medium by degradation through multiple collisions in successive energy groups until they reach thermal energies where they diffuse until captured by fissionable or other material. The presentation given below is limited to three groups though additional groups could readily be added. The groups are numbered from the top. Let f represent fast (i.e., virgin fission neutrons), i represent intermediate energy neutrons, and s represent slow or thermal energy neutrons. Then the conservation of neutrons requires

$$\begin{aligned} D_f\nabla^2\phi_f - \Sigma_f\,\phi_f + k\,\Sigma_s\,\phi_s &= 0 \\ D_i\nabla^2\phi_i - \Sigma_i\,\phi_i + \Sigma_f{}^s\,\phi_f &= 0 \\ D_s\nabla^2\phi_s - \Sigma_s\,\phi_s + \Sigma_i\,\phi_i &= 0 \end{aligned} \qquad (7.84)$$

Σ_f and Σ_i represent slowing down macroscopic cross sections from fast → intermediate and from intermediate → slow, and k is the number of neutrons produced per neutron absorbed by the medium. The eigen solution to (7.84) determines k, the multiplication (i.e., the *critical condition* of the reactor). The reactor, of course, contains regions without fuel (i.e., no fissionable material) so that these portions of the system lack the multiplying feature. At the boundary between two different media it is necessary to match fluxes and currents.

When a neutron (unit mass) of energy E_0 collides with a nucleus (mass A) which is initially at rest in the laboratory system, the cross section for energy lost by the neutron is

$$\sigma_l(E_0\to E) = \frac{(2l+1)\sigma(E_0)P_l(\mu)}{\alpha E_0} \qquad (7.85)$$

for $\qquad E_0 > E > (1-\alpha)E_0$

where $\alpha = 4A/(1 + A)^2$, $P_l(\mu)$ is the lth Legendre polynomial, μ is the cosine of the scattering angle in the laboratory system, and $\sigma(E_0)$ is the total (angular integrated scattering cross section). One notes that the integral in the transport therefore has a $1/E$ factor appearing in it, and it has been found convenient to define a new quantity, the "lethargy" u, such that

$$u = \log \frac{E_{up}}{E} \tag{7.86}$$

where E_{up} is usually set equal to 10 Mev, a convenient upper energy.

This simplifies the resulting transport (or diffusion) equation. If the resulting integrand is expanded as a Taylor series in powers of $u' - u$, the lethargy exchange, and only the first three terms in the expansion are retained, the steady-state-source free-diffusion equations take the form

$$\nabla \cdot \mathbf{J} - \Sigma_a(u)\phi(u) = -\frac{\partial q}{\partial u} \tag{7.87}$$

where

$$\lambda \frac{\partial q}{\partial u} + q(u) = \xi \, \Sigma_s \phi(u) \tag{7.88}$$

$$\begin{aligned} \xi &= \frac{1}{\sigma_0(u)} \int (u' - u)\sigma_0(u \to u') \, du' \\ &= 1 - \frac{(A-1)^2}{2A} \log \frac{A+1}{A-1} \end{aligned} \tag{7.89}$$

the "*slowing-down power*" of the material,

$$\lambda \xi = -\frac{1}{2\sigma_0(u)} \int \sigma_0(u \to u')(u' - u)^2 \, du' \tag{7.90}$$

and a similar set of two equations for the P_1 anisotropic part. For large A targets, $\lambda \to 0$ and (7.88) reduces to the Fermi approximation in which $q(E) = \xi \Sigma_s E \phi(E)$ can be interpreted as *the slowing-down density*, the number of neutrons slowed down past energy E per second per unit volume. When absorption takes place at energies $> E$, the above expression for q is reduced by the "resonance escape probability" factor p [44]:

$$p = e^{-1/\xi \int_E^{E_0} \frac{\Sigma_a}{\Sigma_s} du'} \tag{7.91}$$

Substituting $\mathbf{J} = -D\nabla\phi$ into (7.87), one obtains

$$\frac{\partial q(\mathbf{r},u)}{\partial u} = \frac{D}{\xi \Sigma_s} \nabla^2 q(\mathbf{r},u) - \frac{\Sigma_a q(\mathbf{r},u)}{\xi \Sigma_s} \tag{7.92}$$

Introducing the *Fermi age* $\tau(u)$ by

$$\tau(u) = \int_0^u \frac{D(u')}{\xi \Sigma_s(u')} \, du' \tag{7.93}$$

and noting that

$$D(u') = \frac{1}{3} \frac{\lambda_s}{1 - 2/3A} \qquad \text{for } A \gg 1 \tag{7.94}$$

where λ_s is $1/\Sigma_s$, the mean free scattering path, and $(2/3A)\sigma_s$ is the value of the P_1 component of the scattering cross section which is isotropic in the center-of-mass system, then one obtains

$$\tau(E) = \int_E^{E_0} \frac{\lambda_s^2(E')}{3\xi(1 - 2/3A)} \frac{dE'}{E'} \tag{7.95}$$

and (7.92) becomes

$$\frac{\partial q(\mathbf{r},\tau)}{\partial \tau} = \nabla^2 q(\mathbf{r},\tau) - \frac{\Sigma_a}{D} q(\mathbf{r},\tau) \tag{7.96}$$

With the substitution

$$q(\mathbf{r},\tau) = q(\mathbf{r},\tau)p(\tau) \tag{7.97}$$

where $p(\tau)$ is the resonance escape probability defined by (7.91), Eq. (7.96) reduces to [45]

$$\frac{\partial q(\mathbf{r},\tau)}{\partial \tau} = \nabla^2 q(\mathbf{r},\tau) \tag{7.98}$$

the *Fermi-age equation*.

The solution of (7.98) for a point source in an infinite homogeneous medium is given by

$$q(\mathbf{r},\tau) = \frac{Q}{(4\pi\tau)^{3/2}} e^{-(r^2/4\tau)} \tag{7.99}$$

where Q is the source strength at the origin and r is the distance from the origin. The "mean square distance" R_E^2 for neutrons of initial energy E_0 to degrade to energy E is given by the integral of (7.99) weighted by r^2.

$$\overline{R_E^2} = \frac{2}{\xi(1 - 2/3A)} \int_E^{E_0} \lambda_s^2(E') \frac{dE'}{E'} = 6\tau(E) \tag{7.100}$$

The diffusion of thermal neutrons is simplified by the fact that slowing down no longer takes place. Thermal neutrons diffuse through homogeneous matter for a mean time given by $1/(v\Sigma_a)$, which for $1/v$ absorbers is a characteristic value of the material. Regarding the thermal neutrons as a group,

$$D\nabla^2\phi - \Sigma_a\phi + S = 0 \tag{7.101}$$

where S is a source per unit volume.

The solution to (7.101) for a point source Q is given by

$$\phi = \frac{3Q}{4\pi\lambda_s} \frac{e^{-r/L}}{r} \tag{7.102}$$

where $L = \sqrt{D/\Sigma_a} \cong \sqrt{\lambda_s \lambda_s/3}$ is the so-called *diffusion length*. In graphite $L \cong 50$ cm and in heavy water $L \cong 100$ cm.

Equation (7.101) may be combined with (7.98), for example, wherein the latter constitutes the source term for the former. For the very extensive development of calculational techniques such as have been indicated above, the reader is referred to texts on reactor engineering and physics.

4. Interactions with Ordered Matter

The application of (7.5) discloses that the wavelength of thermal neutrons is $\simeq 1.8$ A, a value of the same order as interatomic distances. This makes possible the observation of diffraction effects in ordered matter, e.g., with crystals, using thermal neutrons.

That is, it is possible to observe *Bragg reflection* according to

$$n\lambda = 2d \sin \theta \qquad (7.103)$$

where n is the diffraction order, λ the neutron wavelength, d the spacing between crystal planes, and θ is the angle which the normal to the wavefront makes with the crystal planes. This basic relationship is the same as cited for X rays in Part 8, Chap. 1.

Except for magnetic effects to be discussed later, the neutron scattering is essentially via point scatterers (i.e., the nuclear radius $< 10^{-12}$ cm $= 10^{-4}$ A). As was shown in Sec. 2, the effect of the nuclear potential is to shift the phase of the neutron wave by an amount δ. Consider only S-wave scattering and let the value of r which makes the wave function ψ_0 zero be denoted by a'. The scattering amplitude a is then given by [see (7.73)]

$$a = \frac{1}{k} e^{i(\pi - ka')} \sin ka' \qquad (7.104)$$

that is, $a \rightarrow -a'$ as $E \rightarrow 0$.

The term a' is the *scattering length* and it is the distance by which the wave function is shifted because of the nuclear potential. A hard sphere of radius a' would also shift ψ_0 by this amount, as well as change the phase by π [46].

The scattering length can be related to a suitable average well depth U. Following Fermi, the scattering cross section σ_s may be written as

$$\sigma_s = \frac{\mu^2}{\pi \hbar^4} <M>^2 \qquad (7.105)$$

where M is the matrix element of the perturbation causing the interaction. It is a suitable product of the nuclear volume and the nuclear potential. That is,

$$<M> = \tfrac{4}{3}\pi R^3 U \qquad (7.106)$$

If now the scattering is considered to result from a point potential, and the delta function defined by

$$\int \delta(\mathbf{r})\, d\mathbf{r} = 1 \qquad (7.107)$$

is employed, it is readily shown that

$$U = \frac{2\pi \hbar^2 a'}{\mu} \delta(\mathbf{r}) \qquad (7.108)$$

The quantity U is termed the *Fermi pseudo-potential* and is the basis for the quantum-mechanical theory of localized impacts [47].

As was also disclosed in Sec. 2, the central force is spin-dependent and contains a term proportional to $\mathbf{I} \cdot \mathbf{S}$. The scattering amplitude a is therefore a certain mixture of the two components, a_+ and a_-, corresponding to $I + \frac{1}{2}$ and $I - \frac{1}{2}$, respectively. Using projection operators, it is readily shown that

$$a = \frac{I + 1 + 2I \cdot S}{2I + 1} a_+ + \frac{I - 2I \cdot S}{2I + 1} a_-$$
$$= g_+ a_+ + g_- a_- + \frac{a_+ - a_-}{2I + 1} 2\mathbf{I} \cdot \mathbf{S} \qquad (7.109)$$

where the g are the usual statistical weight factors, which for $s = \frac{1}{2}$ have the values

$$g_+ = \frac{I + 1}{2I + 1} \qquad g_- = \frac{I}{2I + 1} \qquad (7.110)$$

In diffraction and related interference phenomena the quantum-mechanical average value of the scattering amplitude determines coherent effects. That is,

$$a_{\text{coh}} = <a> = g_+ a_+ + g_- a_- \qquad (7.111)$$

since $\mathbf{I} \cdot \mathbf{S}$ averages to zero when the neutron and nuclear spins are uncorrelated.

The total scattering cross section σ_s, on the other hand, is determined by the average of the square of the amplitude. This can readily be shown to be

$$<a^2> = (g_+ a_+ + g_- a_-)^2 + g_+ g_- (a_+ - a_-)^2$$
$$\text{that is,} \quad g_+ g_- (a_+ - a_-)^2 = <a^2> - <a>^2 \qquad (7.112)$$

The latter constitutes the usual definition of the disordered or incoherent component of a. The total cross section is thus the sum of coherent and incoherent parts:

$$\sigma_s = \sigma_{\text{coh}} + \sigma_{\text{inc}} = 4\pi <a^2> \qquad (7.113)$$

It should be noted that the coherent scattering cross section is, in general, less than the total scattering cross section. If potential or hard sphere type of scattering predominates in the thermal region, then $a_+ = a_- \simeq R$ and $\sigma_{\text{inc}} = 0$. Inspection of (7.53), however, reveals that the presence of a near level in the compound nucleus will contribute a resonance amplitude, $a_0 = \lambda_0 \Gamma_n / 2E_0$, to the thermal region. This alone will contribute incoherence, for if $a_+ = a_-$ except for a resonance component a_0,

$$\sigma_{\text{inc}} = 4\pi g_+ g_- a_0^2 \qquad (7.114)$$

The above applies to a single isotope. The presence of additional isotopes is handled by treating their abundance factors p_i like other weighting factors. That is,

$$\frac{\sigma_s}{4\pi} = \sum_i p_i (g_{i+} a_{i+}^2 + g_{i-} a_{i-}^2) \qquad (7.115)$$

$$\frac{\sigma_{\text{coh}}}{4\pi} = \left| \sum_i (p_i g_{i+} a_{i+} + p_i g_{i-} a_{i-}) \right|^2 \qquad (7.116)$$

The case of two isotopes having essentially equal radii (and therefore essentially equal scattering amplitudes except for resonance) but where one isotope possesses a significant resonant amplitude a_0 will serve to illustrate the effect of multiple isotopes on the incoherent scattering cross section. Let g and p be associated with a_0, the resonant state. Then,

$$\frac{\sigma_s}{4\pi} = pg(R + a_0)^2 + p(1 - g)R^2 + (1 - p)R^2 \qquad (7.117)$$

$$\frac{\sigma_{\text{coh}}}{4\pi} = |pg(R + a_0) + p(1 - g)R + (1 - p)R|^2 \qquad (7.118)$$

and

$$\frac{\sigma_{\text{inc}}}{4\pi} = pg(1 - pg)a_0^2 \qquad (7.119)$$

Equation (7.119) points up the three nuclear factors contributing to incoherence: (1) spin, i.e., the g factor; (2) isotope, i.e., the p factor; and (3) resonance, i.e., the a_0 factor. It should be observed that, provided $a_0 \neq 0$ and $p \neq 1$, $\sigma_{inc} \neq 0$, even if $I = 0$.

Hard sphere (infinitely repulsive potential) or, more generally, the potential scattering effects a near 180° phase change of the neutron wave. By convention the coherent scattering length b, which is positive for most elements (that is, $\simeq R$ in the absence of resonances) is utilized to represent the coherent nuclear scattering employed for diffraction. That is,

$$\sigma_{coh} = 4\pi b^2 \tag{7.120}$$

where $b \simeq$ (R − resonant terms from virtual levels + resonant terms from bound levels). In a few cases, e.g., hydrogen, manganese, and titanium, strong virtual scattering resonances cause the coherent scattering length at thermal energies to be negative (that is $E_0 > 0$ and $|ga_0| > R$). In the case of vanadium the resonant terms approximately cancel the potential scattering term and $\sigma_{coh} \simeq 0$. In other cases such as cadmium, the phase shift may be significantly complex because of absorption.

If the coherent scattering length had been determined at energies above chemical binding (i.e., pure nuclear value) at least two corrections to this value, b_{free}, must be made for proper application to thermal neutron diffraction. One corrects for the reduced mass effect and the other for the effect of *temperature diffuse* scattering. The latter is termed the Debye-Waller factor [48].

$$b = b_{free}\left(\frac{M+m}{M}\right) e^{-W[(\sin\theta)/\lambda]^2} \tag{7.121}$$

where θ is the scattering angle, and W is a function of the lattice temperature T and the Debye temperature T_D.

$$W = \frac{6h^2}{Mk_BT_D}\left(\frac{1}{4} + \frac{T}{T_D}\int_0^{T_D/T}\frac{x\,dx}{e^x - 1}\right) \tag{7.122}$$

where k_B is the Boltzmann constant. The factor correcting for the reduced mass is due to the fact that for thermal elastic scattering the nucleus is bound (chemically) rather than free. The quantity b appearing in (7.121) is the appropriate scattering length for diffraction analysis.

In terms of wave vectors $\mathbf{k}$, elastic scattering may be written

$$\mathbf{k}_0 - \mathbf{k} = 2\hat{d}\,\frac{2\pi}{\lambda}\sin\theta \tag{7.123}$$

where $\hat{d}$ is a unit vector normal to the scattering planes, and $\mathbf{k}_0$ and $\mathbf{k}$ are the initial and final wave vectors, respectively. Equation (7.123) may now be written for the Bragg condition as

$$\mathbf{k}_0 - \mathbf{k} = 2\pi\frac{2d}{\lambda}\sin\theta\,\frac{\hat{d}}{d} = 2\pi n\boldsymbol{\tau} \tag{7.124}$$

where $\boldsymbol{\tau}$ is the *reciprocal lattice vector* appropriate to the direction $\mathbf{k}_0 - \mathbf{k}$.

The phase factor may also be expressed

$$\begin{aligned} e^{i(\mathbf{k}_0-\mathbf{k})\cdot\mathbf{r}} &= e^{2\pi i n\boldsymbol{\tau}\cdot\mathbf{r}} \\ &= e^{2\pi i(hx/a_0+ky/b_0+lz/c_0)} \end{aligned} \tag{7.125}$$

where (x,y,z) are the nuclear coordinates, (h,k,l) are the Miller indices, and (a_0,b_0,c_0) are the unit cell dimensions. The reader is referred to Part 8, Chap. 1.

The differential cross section for scattering in the direction from (h,k,l) by an assembly of N_0 atoms is thus

$$\frac{1}{N_0}\left|\sum_j a_j e^{2\pi i(hx/a_0+ky/b_0+lz/c_0)}\right|^2 \tag{7.126}$$

If the sum is taken only over the unit cell one obtains the so-called "structure factor" or "cell amplitude factor" F_{hkl}:

$$F_{hkl} = \sum_j a_j e^{2\pi i(hx_j+ky_j+lz_j)} \tag{7.127}$$

where the x_j, y_j, z_j are the coordinates of the jth atom in units of the cell dimensions.

Neutron crystallography has dealt mainly with the following two topics:

1. Determination of the atomic positions of light elements, especially hydrogen, in crystal structures
2. Determination of crystal structures involving atoms of neighboring atomic number. In both cases X-ray diffraction is of limited use. In topic 1 the X-ray amplitude is too small and in 2 the X-ray amplitudes are too nearly identical.

In 1 above, neutron diffraction has been applied mainly to the structural analysis of crystals containing hydrogen and to the crystal chemistry of the oxides and carbides of heavy elements [49, 50, 51, 52, 53]. The determination of hydrogen positions is vital not only to organic chemistry but also to the chemistry of metal hydrides, hydrated compounds, and hydroxides. The importance of neutron diffraction to the understanding of the hydrogen bond in crystals has been stressed by Pauling [54] and by Pimentel and McClellan [55]. Localization of hydrogen atoms in certain crystals has elucidated the mechanisms of some chemical reactions and has also accounted for changes in physical properties. Examples include the photochemical reaction in which o-nitrobenzaldehyde transforms into nitrobenzoic acid [56], and the mechanism of the ferroelectric transition in KH_2PO_4 [57]. Another example in which structure determinations by the techniques of neutron diffraction have effected greater understanding of the chemistry of the systems is the work on the rare-earth dicarbides and sesquicarbides. In these studies it was possible to relate the variations in the C-C bond length with the electronic state of the metal ions [52, 53].

In 2 above, neutron diffraction has been applied to the study of the order-disorder transitions in such alloys as FeCo and Ni_3Mn [58] for which the X-ray scattering amplitudes are nearly equal.

Finally, Peterson and Smith [61, 62, 63] in their study of the anomalous dispersion of neutrons in crystals such as the borates, while giving further substantiation to the general theory (i.e., including appreciable absorption in the thermal region), indicated the possibilities of this technique for the absolute determination of the structures of complex crystals. This comes about because highly absorbing nuclei such as cadmium possess phase shifts containing an imaginary component. That is, the phase angle δ of the scattered wave, instead of being either 0° or π, contains an imaginary term to effect attenuation via absorption. This condition permits a more direct and unambiguous analysis of the diffraction pattern, i.e., effects a solution to the so-called *phase problem of crystallography*.

One of the more unique applications of neutron diffraction is to the determination of *magnetic structures*. This is due to the fact that the neutron possesses a magnetic moment and therefore experiences magnetic as well as nuclear scattering. There exists no corresponding effect for X-rays.

Magnetic scattering will take place, for example, in a crystal containing ions of the transition or rare-earth elements. These ions contain a number of unpaired electrons which give rise to a resultant magnetic moment μ_I. The interaction of this atomic moment with the magnetic moment μ_N of the neutron causes the scattering. In the case of the transition elements the unpaired electrons are located in the incomplete 3*d* shell; in the case of the rare earths they are located in the incomplete 4*f* shell. Examples from the transition region are Mn^{++} with five unpaired electrons in the $^6S_{5/2}$ ground state, and Fe^{++} with four unpaired electrons in the $^5D_4$ ground state. The effective atomic magnetic moment can usually be computed from the ground-state term when spin only is involved or by assuming quenching of the orbital momenta.

In addition to the nuclear scattering amplitude a_n, there exists a magnetic amplitude a_m. The latter is given by [64]

$$a_m = p \sin \alpha \tag{7.128}$$

where α is the angle between the scattering and magnetization vectors, and

$$p = \frac{e^2 \gamma}{mc^2} If \tag{7.129}$$

in which $e^2/mc^2 = r_0$, the classical radius of the electron, $\cong 2.8 \times 10^{-13}$ cm, γ is the number of Bohr magnetons in a neutron's magnetic moment, I is the effective ionic spin momentum (i.e., in units of $\hbar$), and f is a form factor (i.e., some function of $\sin \theta/\lambda$).

It can be shown that the differential magnetic cross section is given by

$$\sigma_m(\theta) = p^2 q^2 \tag{7.130}$$

where $\mathbf{q}$ is the magnetic interaction vector defined by

$$\mathbf{q} = (\hat{e} \cdot \hat{\mathcal{H}})\hat{e} - \hat{\mathcal{H}} \tag{7.131}$$

$\hat{e}$ is a unit vector in the direction of momentum transfer (i.e., inward normal to planes for Bragg reflection), and $\hat{\mathcal{H}}$ is a unit vector in the direction of magnetization. Thus,

$$\hat{e} \cdot \hat{\mathcal{H}} = \cos \alpha \tag{7.132}$$

$$q^2 = 1 - (\hat{e} \cdot \hat{\mathcal{H}}) = \sin^2 \alpha \tag{7.133}$$

and

$$\mathbf{q} \perp \hat{e} \tag{7.134}$$

The total differential scattering cross section is then

$$\sigma_s(\theta) = (a_n + pq)^2 \tag{7.135}$$

which reduces to $a_n^2 + p^2q^2$ for unpolarized neutron beams. Thus

$$\frac{\sigma_s}{4\pi} = a_n^2 + \langle p^2q^2 \rangle \tag{7.136}$$

where $\langle p^2q^2 \rangle$ depends on the magnetic properties of the medium. Also, by analogy,

$$F^2 = F_n^2 + F_m^2 \sin^2 \alpha \tag{7.137}$$

where the F are the usual structure factors.

In paramagnetic crystals there is no preferential alignment of the magnetic moments which point at random in all directions; in this case the magnetic scattering is incoherent and contributes to the background of the diffraction pattern. From quantitative study of the background scattering it is possible to deduce the magnitude of the magnetic moment and the angular dependence of the magnetic scattering [65]. Such dependence, which does not exist in the case of the nuclear scattering, is due to the fact that the electrons that determine the magnetic moment are distributed over a volume having linear dimensions comparable with the wavelength of the neutrons, and it is somewhat similar to the angular dependence of the scattering of X-rays. Background measurements are by no means easy because the magnetic background has to be distinguished from other types of background scattering.

Ordered or partially ordered magnetic moments produce more or less sharp coherent diffraction peaks whose intensities and angular positions depend on the three-dimensional "magnetic structure" of a domain (a domain can be considered as the counterpart of a crystallite in ordinary nuclear scattering). In this case there is coherence between neutrons scattered by the magnetic atoms in the crystal. If the incident beam is unpolarized, however, there is no coherence between nuclear and magnetic scattering.

When the magnetic and nuclear unit cells are the same (ferromagnetic materials) each reflection results from the superposition of nuclear and magnetic scattering. For antiferromagnetic materials, on the other hand, the two unit cells are different, the magnetic cell being larger; in this case some of the magnetic peaks occur in new angular positions corresponding to the superlattice formed by the magnetic moments in the crystal structure. In order to clarify a magnetic structure one must be able to separate in some way the magnetic and nuclear contributions to the diffraction peaks. Three methods are available: (1) If the crystal structure is known (and, in general, this is the case), one can calculate the nuclear intensities, put them on an absolute scale, and deduce the calculated values from the observed ones; (2) since in the paramagnetic region there is no magnetic ordering, by making measurements above and below the Curie or Néel point, it is possible to measure the magnetic component of the reflection; (3) finally, by applying an external magnetic field one can change the value of α in (7.128). If the magnetic field is applied along the scattering vector ($\alpha = 0°$), and if the diffraction pattern is taken with and without the field, it is possible to measure the magnetic contribution (this method is restricted to materials that can be totally saturated by the fields available). Because of the angular dependence of the magnetic scattering, the magnetic contribution to the diffraction peaks is rather small at large scattering angles. It is rather difficult, then, to measure with reasonable accuracy the magnetic scattering from powder patterns. For greater precision single crystals have to be used whenever possible. Accurate measurements of small scattering amplitudes can be carried out using polarized neutrons; the production of a polarized beam and its use in magnetic scattering measurements have been discussed by Shull et al. [66] and by Nathans et al. [67].

Extensive studies have been carried out in the past 15 years on a variety of magnetic materials, such as metals and alloys, oxides, sulfides, halides, etc. [68]. Besides ferromagnetic and antiferromagnetic spin structures, also ferrimagnetic, spiral, and related spin configurations have been found. The description of magnetic structures requires an extension of symmetry operations to include antisymmetry operations, but so far little use has been made of this conception in the study of magnetic structures.

Experimentally the Bragg law is utilized in two ways, these being analogous to X-ray techniques. By using a crystal with planes of known spacing and selecting the proper diffraction angle, neutrons of a desired wavelength may be obtained (e.g., from a reactor). Thereafter these monochromatic neutrons are diffracted from a sample of unknown structure to determine the spacing and the intensities of the various planes in the specimen. An instrument operating on this principle is known as a *neutron diffractometer* [69].

The principal experimental methods of structure analysis are, in principle, the same with neutron diffraction as those employed in X-ray diffraction. They are the single-crystal method and the Debye-Scherrer or powder diffraction method. The reader is referred to more detailed accounts [70] of this subject for fuller comprehension.

The phenomenon of refraction may also be observed with neutrons. It results from the fact that the velocity of a neutron wave in a medium is dependent upon the nuclear properties of that medium. It is considered here even though the nuclei through which the neutron passes may be randomly arranged. The front of such a wave is coherent.

In a vacuum the wave number k_0 is given by

$$k_0 = \frac{\sqrt{2mE}}{\hbar} \tag{7.138}$$

and in a medium by

$$k = \frac{\sqrt{2m(E - V)}}{\hbar} \tag{7.139}$$

The index of refraction, n, is therefore

$$n = \sqrt{1 - \frac{V}{E}} \tag{7.140}$$

which can be greater or less than unity, depending on whether the average nuclear potential V is negative or positive.

According to (7.108),

$$\begin{aligned} V &= N\tfrac{4}{3}\pi R^3 U \\ &= \frac{2\pi N\hbar^2 a'}{\mu} \end{aligned} \tag{7.141}$$

where N is the number of nuclei per unit volume. Therefore

$$n^2 = 1 - \frac{Nb\lambda^2}{\pi} \tag{7.142}$$

or, since $n^2 \cong 1$,

$$n = 1 - \frac{Nb\lambda^2}{2\pi} \tag{7.143}$$

The condition for total reflection is given by Snell's law, namely, $n = \cos\theta_c$, where θ_c is the critical angle. This yields [71]

$$\theta_c = \lambda\sqrt{\frac{Nb}{\pi}} \tag{7.144}$$

Total reflection from mirror surfaces thus takes place when b is positive, i.e., for most materials where $b \simeq R$. The exceptions are those substances mentioned earlier, namely, the ones with strong virtual nuclear scattering levels such that the sign of b is reversed, for example, H, Mn, and Ti. Typical values for θ_c vary between 1 and 3 milliradians, i.e., $\sim 10'$ of arc.

It is also possible to make magnetic mirrors from ferromagnetic material such as cobalt. In this instance the magnetic contribution to the scattering cross section is larger than the nuclear cross section [72], so that the index of refraction can be made greater than unity for one spin state and less than unity for the other spin state. Such an arrangement allows for the production of completely polarized beams by total reflection of only one spin state from the magnetized surface.

In addition to elastic scattering, thermal neutrons readily undergo *inelastic scattering* with materials. That is, they exchange energy with, for example, a crystalline lattice. This phenomenon permits the determination of the fundamental frequency spectrum of the lattice and hence yields information on the interatomic forces. The conservation of energy and momentum requires that

$$\mathbf{k}_0 - \mathbf{k} = 2\pi\boldsymbol{\tau} + \mathbf{q} \tag{7.145}$$

$$|\mathbf{k}_0|^2 - |\mathbf{k}|^2 = \frac{2m}{\hbar}\,\omega(\mathbf{q}) \tag{7.146}$$

where $\mathbf{q}$ represents the creation of a phonon in the lattice, and $\omega(\mathbf{q})$ is the angular frequency spectrum of the lattice. Note that if $\mathbf{q} = 0$ the neutron is elastically scattered and Bragg's law results.

The differential cross section for the scattering of a beam of neutrons of initial energy $E_0 = \hbar^2 k_0^2/2m$ and final energy $E = h^2k^2/2m$ is given by the *Born approximation:*

$$\frac{d\sigma}{d\Omega} = \left(\frac{\mu}{2\pi\hbar^2}\right)^2 \frac{k}{k_0} \sum_{if} \rho(E_i) \left| \int \psi^*_f e^{-i\boldsymbol{\kappa}\cdot\mathbf{r}} U(\mathbf{r} - \mathbf{r}_j)\psi_i \, d\tau \right|^2 \tag{7.147}$$

where $\hbar\boldsymbol{\kappa} = \hbar\mathbf{k} - \hbar\mathbf{k}_0$, the momentum gained in the scattering, and $\mathbf{r}$ and $\mathbf{r}_j$ are the position vectors of the incident neutron and the jth scattering atom, respectively. The integral is taken over all space and summed over final states and averaged over initial states. The potential appearing in (7.147) is the Fermi pseudo-potential given by (7.108). That is,

$$U(\mathbf{r} - \mathbf{r}_j) = \frac{2\pi\hbar^2}{\mu} \sum_j a_j\, \delta(\mathbf{r} - \mathbf{r}_j) \qquad (7.148)$$

where a_j is the scattering length of the jth atom. Then

$$\frac{d\sigma}{d\Omega} = \frac{k}{k_0} \sum_{if} \rho(E_i) \int_{-(\hbar^2 k_0^2/2m)}^{\infty} \delta\left(\frac{E_i - E_f}{\hbar} - \omega\right) \frac{d\epsilon}{\hbar} \times \left| \int \psi^*_f \sum_j a_j e^{-i\boldsymbol{\kappa}\cdot\mathbf{r}_j} \psi_i \, d\tau \right|^2 \qquad (7.149)$$

where the integral is taken over all states of the target; and $\epsilon = \hbar^2 k^2/2m - \hbar^2 k_0^2/2m = \hbar\omega$ is the energy gained by the neutron.

By making use of the Fourier representation of the δ function [73], the differential cross section in angle and energy is obtained:

$$\frac{d^2\sigma}{d\Omega\, d\epsilon} = \frac{1}{2\pi} \frac{k}{\hbar k_0} \sum_{if} \rho(E_i) \int dt\, e^{-i\omega t} e^{i[(E_i - E_f)/\hbar]t} \times \sum_{jj'} \int d\tau\, \psi^*_i a^*_j e^{i\boldsymbol{\kappa}\cdot\mathbf{r}_j} \psi_f \int d\tau\, \psi^*_f a_{j'} e^{-i\mathbf{k}\cdot\mathbf{r}_{j'}} \psi_i \qquad (7.150)$$

which can also be written [74] using

$$e^{-i\boldsymbol{\kappa}\cdot\mathbf{r}_{j'}(t)} = e^{-iHt/\hbar} e^{-i\boldsymbol{\kappa}\cdot\mathbf{r}_{j'}} e^{iHt/\hbar}$$

where

$$e^{iHt/\hbar}\psi_i = e^{iE_i t/\hbar}\psi_i$$

$$\frac{d^2\sigma}{d\Omega\, d\epsilon} = \frac{1}{2\pi\hbar} \frac{k}{k_0} \int dt\, e^{-i\omega t} \sum_i \rho(E_i) \sum_{jj'} \int d\tau\, \psi^*_i \left\{\sum_f <a^*_j>_{fi} <a_{j'}>_{fi}\right\} e^{i\boldsymbol{\kappa}\cdot\mathbf{r}_j} e^{-i\boldsymbol{\kappa}\cdot\mathbf{r}_{j'}(t)} \psi_i \qquad (7.151)$$

where from the previous discussion on coherence and incoherence it may be appreciated that

$$\sum_f <a^*_j>_{fi} <a_{j'}>_{fi} = a_{\text{coh}}^2 + a_{\text{inc}}^2\, \delta_{jj'} \qquad (7.152)$$

Equation (7.151) may now be rewritten:

$$\frac{d^2\sigma}{d\Omega\, d\epsilon} = \frac{1}{8\pi^2\hbar} \frac{k}{k_0} \int dt\, e^{-i\omega t} \sum_{jj'} (\sigma_{\text{coh}} + \sigma_{\text{inc}}\delta_{jj'}) \times <e^{i\boldsymbol{\kappa}\cdot\mathbf{r}_j} e^{-i\boldsymbol{\kappa}\cdot\mathbf{r}_{j'}(t)}> \qquad (7.153)$$

where the average is taken over the initial states of the target system.

Considering $\mathbf{r}_j(t)$ to be steady-state position plus a time displacement, i.e.,

$$\mathbf{r}_j(t) = \mathbf{j} + \mathbf{u}_j(t) \qquad (7.154)$$

then

$$<e^{i\boldsymbol{\kappa}\cdot\mathbf{r}_j} e^{-i\boldsymbol{\kappa}\cdot\mathbf{r}_{j'}(t)}> = e^{-i\boldsymbol{\kappa}\cdot(\mathbf{j}'-\mathbf{j})} <e^{i\boldsymbol{\kappa}\cdot\mathbf{u}_j} e^{-i\boldsymbol{\kappa}\cdot\mathbf{u}_{j'}(t)}> \qquad (7.155)$$

Following Zemach and Glauber [74] the displacements may be resolved into a set of λ normal modes with amplitude vectors $\mathbf{C}_j{}^\lambda$ for each particle and mode:

$$\mathbf{u}_j(t) = \sum_\lambda \mathbf{C}_j{}^\lambda q_\lambda(t) \qquad (7.156)$$

where the $q_\lambda(t)$ have the properties of a one-dimensional harmonic oscillator.

$$q_\lambda(t) = i\sqrt{\frac{\hbar}{2M\omega_\lambda}}\,(a_\lambda e^{-i\omega_\lambda t} - a_\lambda{}^\dagger e^{i\omega_\lambda t}) \qquad (7.157)$$

where $a^\dagger_\lambda$ and a_λ are, respectively, the creation and annihilation operators for the normal mode. They satisfy the following relationships:

$$\begin{aligned} a_\lambda a^\dagger_\lambda &= n_\lambda + 1 \\ a^\dagger_\lambda a_\lambda &= n_\lambda \\ [a,a] &= [a^\dagger,a^\dagger] = 0 \end{aligned} \qquad (7.158)$$

where n_λ is the occupation number for the λ mode and $\Sigma_\lambda n_\lambda$ is the total number of particles in the system. Now

$$\begin{aligned} <e^{i\boldsymbol{\kappa}\cdot\mathbf{u}_j} e^{-i\boldsymbol{\kappa}\cdot\mathbf{u}_{j'}(t)}> &= <e^{i\boldsymbol{\kappa}\cdot(\mathbf{u}_j - \mathbf{u}_{j'}(t))} e^{-\frac{1}{2}[(\boldsymbol{\kappa}\cdot\mathbf{u}_j),\boldsymbol{\kappa}\cdot\mathbf{u}_{j'}(t)]}> \\ &= e^{-<(\boldsymbol{\kappa}\cdot\mathbf{u}_j)^2>} e^{<(\boldsymbol{\kappa}\cdot\mathbf{u}_j)(\boldsymbol{\kappa}\cdot\mathbf{u}_{j'}(t))>} \end{aligned} \qquad (7.159)$$

and:

$$<(\boldsymbol{\kappa}\cdot\mathbf{u}_j)(\boldsymbol{\kappa}\cdot\mathbf{u}_{j'}(t))> = \sum_\lambda <(\boldsymbol{\kappa}\cdot\mathbf{C}_j{}^\lambda)(\boldsymbol{\kappa}\cdot\mathbf{C}_{j'}{}^\lambda)> \times [(<n_\lambda> + 1)e^{i\omega_\lambda t} + <n_\lambda> e^{-i\omega_\lambda t}] \qquad (7.160)$$

where $<n_\lambda>$ is the average occupation number in the normal mode λ. For a target system in thermal equilibrium

$$<n_\lambda> = \frac{1}{e^{\hbar\omega_\lambda/k_B T} - 1} \qquad (7.161)$$

Replacing the summation over discrete states by an integral over the frequency distribution function $f(\omega)$, the incoherent cross section can be written as

$$\frac{d^2\sigma}{d\Omega\, d\epsilon} = \frac{\sigma_{\text{inc}}}{8\pi^2\hbar} \frac{k}{k_0} e^{-(\hbar^2 k^2/2M)g(0)} \int_{-\infty}^{\infty} e^{-i\omega t} e^{(\hbar^2\kappa^2/2M)g(t)}\, dt \qquad (7.162)$$

where

$$g(t) = \int_0^\infty d\omega \frac{f(\omega)}{\omega} [(<n> + 1)e^{i\omega t} + <n> e^{-i\omega t}] \qquad (7.163)$$

assuming isotropy in the target nuclei.

The coherent cross section can also be expressed as a Fourier transform but is considerably more complex in form. It is most important in the small energy transfer region and has in effect been discussed previously under crystallography.

In principle the normal mode frequency distribution $f(\omega)$ can be obtained by taking the inverse time transform of (7.162). In practice experimental limitations require that $g(0)$ and $g(t)$ be calculated from models and compared with experiment via (7.162) and (7.163). Simple forms of $f(\omega)$ are noted in passing:

$f(\omega) = \delta(\omega)$ — perfect gas

$$f(\omega) = \begin{cases} \dfrac{3\omega^2}{\omega_D{}^3} & \omega < \omega_D \\ 0 & \omega > \omega_D \end{cases}$$ — Debye approximation

$f(\omega) = \delta(\omega - \omega_E)$ — Einstein crystal

It is useful to display the differential cross section in terms of material properties. Returning to (7.153), which can be written as

$$\frac{d^2\sigma}{d\Omega\, d\epsilon} = <a^2> \frac{k}{k_0} S'_s(\boldsymbol{\kappa},\omega) + <a>^2 \frac{k}{k_0} S'_d(\boldsymbol{\kappa},\omega) \tag{7.164}$$

where $S'_s(\boldsymbol{\kappa},\omega)$ is the so-called "self part" and $S'_d(\boldsymbol{\kappa},\omega)$ is the so-called "different part":

$$S'_s(\boldsymbol{\kappa},\omega) = \frac{1}{2\pi\hbar} \int dt\, e^{-i\omega t} \sum_j <e^{i\boldsymbol{\kappa}\cdot\mathbf{r}_j} e^{-i\boldsymbol{\kappa}\cdot\mathbf{r}_j(t)}> \tag{7.165}$$

$$S'_d(\boldsymbol{\kappa},\omega) = \frac{1}{2\pi\hbar} \int dt\, e^{-i\omega t} \sum_{j\neq j'} <e^{i\boldsymbol{\kappa}\cdot\mathbf{r}_j} e^{-i\boldsymbol{\kappa}\cdot\mathbf{r}_{j'}(t)}> \tag{7.166}$$

The four-dimensional transform of (7.165) and (7.166) are, after Van Hove [75], given by

$$G_s(\mathbf{r},t) = \frac{1}{N}\left\langle \sum_{j=1}^{N} \int d\mathbf{r}'\, \delta(\mathbf{r} + \mathbf{r}_j - r')\delta(\mathbf{r}' - \mathbf{r}_j(t))\right\rangle \tag{7.167}$$

$$G_d(\mathbf{r},t) = \frac{1}{N}\left\langle \sum_{j\neq j'}^{N} \int d\mathbf{r}'\, \delta(\mathbf{r} + \mathbf{r}_j - \mathbf{r}')\delta(\mathbf{r}' - \mathbf{r}_{j'}(t))\right\rangle \tag{7.168}$$

and $\qquad G(\mathbf{r},t) = G_s(\mathbf{r},t) + G_d(\mathbf{r},t)$

is the so-called *space-time correlation function.* At $t = 0$

$$G(\mathbf{r},0) = \frac{1}{N}\left\langle \sum_j \delta(\mathbf{r} + \mathbf{r}_j - \mathbf{r}_j)\right\rangle + \frac{1}{N}\left\langle \sum_{j\neq j'} \delta(\mathbf{r} + \mathbf{r}_j - \mathbf{r}_{j'})\right\rangle = \underset{\text{Self}}{\delta(\mathbf{r})} + \underset{\text{Different}}{g(\mathbf{r})} \tag{7.169}$$

This states that the probability that a particle be found in position **r** is a delta function (self) and the probability that any other particle be found at **r** is given by a space correlation function (different). This latter quantity is customarily found from X-ray or neutron diffraction measurements.

Egelstaff [76, 77] following a suggestion of Vineyard reformulated the scattering in terms of the so-called "scattering law" $S(\alpha,\beta)$. For a system in thermal equilibrium

$$\frac{d^2\sigma(E_0 \to E)}{d\Omega\, d\epsilon} = \frac{E}{E_0} e^{-(E-E_0)/kT} \frac{d^2\sigma(E \to E_0)}{d\Omega\, d\epsilon} \tag{7.170}$$

The scattering law is defined:

$$S(\alpha,\beta) = kTe^{\beta/2}S'(\boldsymbol{\kappa},\omega) \tag{7.171}$$

where $$\beta = \frac{E - E_0}{k_BT} \qquad \alpha = \frac{\hbar^2k^2}{2Mk_BT} \tag{7.172}$$

and [78] $$\alpha = 2\int_0^\infty \beta S(\alpha,\beta) \sinh\frac{\beta}{2}\, d\beta \tag{7.173}$$

The self part of $S(\alpha,\beta)$ can be written as the Fourier transform of $G_s(\mathbf{r},t)$ which Vineyard [79] showed for the perfect-gas or diffusing-atom model of the target can be written as

$$G_s(\mathbf{r},t) = \frac{1}{[\pi w_1{}^2(t)]^{3/2}} e^{-r^2/w_1(t)} \tag{7.174}$$

where

$$\begin{aligned} w_1(t) &= \sqrt{\frac{2k_BT}{M}}\,|t| && \text{for a perfect gas} \\ &= \sqrt{\frac{2k_BT}{M\omega_E}}\sqrt{2 - 2\cos\omega_E t} && \text{for an Einstein crystal} \\ &= \sqrt{\frac{12k_BT}{M\omega_D{}^2}\left(1 - \frac{\sin\omega_D t}{\omega_D}\right)} && \text{for a Debye lattice} \\ &= 2\sqrt{D|t|} && \text{for a diffusing atom, } D \text{ being the coefficient of self-diffusion} \end{aligned} \tag{7.175}$$

One can thus write

$$S_s(\alpha,\beta) = \frac{\alpha}{2\pi\beta^2}\int_{-\infty}^{\infty} e^{-\alpha w(t')+i\beta t'}\, dt' \tag{7.176}$$

where $$w(t) = \frac{Mk_BT}{2\hbar^2} w_1{}^2(t) \qquad t' = \frac{tk_BT}{\hbar}$$

This is approximately

$$S_s(\alpha,\beta) = \frac{\alpha}{2\pi\beta^2}\int_{-\infty}^{\infty} e^{-\alpha w(t')+i\beta t'}[\ddot{w}(t') - \alpha\dot{w}^2(t')]\, dt' \tag{7.177}$$

from which

$$w(t') = 2\int_0^\infty \frac{p(\beta)}{\beta^2}\left[\cosh\frac{\beta}{2} - \cos\beta t'\right] d\beta \tag{7.178}$$

using

$$p(\beta) = \lim_{\alpha\to 0} \beta^2 \frac{S_s(\alpha,\beta)}{\alpha} = \int_{-\infty}^{\infty} \ddot{w}(t)e^{i\beta t'}\, dt' \tag{7.179}$$

Equation (7.178) is to be compared with $g(0) - g(t)$ appearing in (7.162) to determine the relationship between $f(\omega)$ and $p(\beta)$. When $\hbar\omega/2k_BT \ll 1$, this simplifies to

$$\frac{f(\omega)}{\sinh \beta/2} = \frac{2p(\beta)}{\beta} \tag{7.180}$$

which constitutes a direct prescription for obtaining the frequency distribution from the "scattering law" data.

It is interesting to compare other physical processes such as electromagnetic scattering with neutron scattering from chemical or solid-state systems. Following Lamb [80] the transition probability $W(E_n)$ for neutron resonance scattering is given by

$$W(E_n) = \frac{\Gamma}{2\pi}\sum_{if} \rho(E_i)\frac{|\int\psi^*{}_f e^{i\boldsymbol{\kappa}\cdot\mathbf{r}}\psi_i\, d\tau|^2}{[(E_n - E_R) - (E_f - E_i)]^2 + \Gamma^2/4} \tag{7.181}$$

where E_n is the neutron energy and E_R is the resonance energy. Using the techniques outlined above, (7.181) can be written

$$W(E_n) = \frac{\Gamma}{2\pi}\int_{-\infty}^{\infty} d\omega\, \frac{S'_s(\boldsymbol{\kappa},\omega)}{(E_n - E_R - \hbar\omega)^2 + \Gamma^2/4} \tag{7.182}$$

which for a perfect-gas model is readily shown to be

$$W(E_n) = \frac{2}{\pi\Gamma}\psi(\theta,x) \tag{7.183}$$

where
$$x = \frac{E_n - E_R - \hbar^2\kappa^2/2M}{\Gamma/2} \tag{7.184}$$

and θ is defined as in (7.62).

The term $\psi(\theta,x)$ is the Doppler-broadening function defined in Sec. 2. Thus in principle by measuring the precise shape of a resonance line it is possible to derive information on $G_s(r,t)$ [81, 82, 83, 84].

The absorption or emission of a γ ray of energy E_γ was shown by Singwi [85] to be similarly given by

$$\sigma_a(E_\gamma) = \sigma_0 \frac{\Gamma^2}{4} \sum_{if} \rho(E_i) \frac{|\int \psi^*{}_f e^{i\kappa\cdot\mathbf{r}}\psi_i \, d\tau|^2}{[(E_\gamma - E_R) - (E_f - E_i)]^2 + \Gamma^2/4} \tag{7.185}$$

where σ_0 is the peak absorption cross section (i.e., at E_R). Again using the methods outlined above, (7.185) can be expressed as

$$\sigma_a(\kappa,\omega) = \frac{\sigma_0\Gamma}{4\hbar} \int_{-\infty}^{\infty} e^{i(\kappa\mathbf{r}\cdot - \omega t) - (\Gamma/2\hbar)|t|} G_s(\mathbf{r},t) \, d\mathbf{r} \, dt \tag{7.186}$$

in terms of the "self part" of the space-time correlation function. When the target nuclei are tightly bound $G_s(\mathbf{r},t)$ is independent of time, and the cross section reproduces the natural line shape. This fact was first recognized by Mössbauer, and the effect is so named.

The scattering of electromagnetic waves by the electrons around the atom is a two-stage process that can be calculated by second-order perturbation theory [86].

$$W(\mathbf{k},\mathbf{k}_0) = \frac{2\pi}{\hbar} \sum_{if} \rho(E_i) \left| \sum_{\hbar q''} \frac{\langle \psi^*{}_f q' | \mathcal{H}'(k) | q'' \psi_{\text{int}} \rangle \langle \psi^*{}_{\text{int}} q'' | \mathcal{H}'(k_0) | q \psi_i \rangle}{E_q - E_{q'} + E_i - E_f \pm \hbar\omega_e} \right|^2 \times \delta(\pm\hbar\omega_e - E_i + E_f - E_q + E_{q'}) \tag{7.187}$$

where
$$H'(k) = \sum_j e^{i\mathbf{k}\cdot\mathbf{r}_j} \sum_{\nu}^{(j)} e^{i\mathbf{k}\cdot\mathbf{r}_\nu}\hat{e}\cdot\mathbf{p}_i \tag{7.188}$$

Σ_j is over the atoms and $\Sigma_\nu{}^{(j)}$ is over the electrons. $\hbar\omega_e$ represents the change in energy of the electron, $q = \hbar k_0$, and $q' = \hbar k$ are the initial and final momenta, respectively, of the photon. Once again,

$$W(\mathbf{k},\mathbf{k}_0) = \int e^{i(\kappa\cdot\mathbf{r} - \omega t)} G(\mathbf{r},t) \, d\mathbf{r} \, dt \, \langle M \rangle^2 \tag{7.189}$$

where $\langle M \rangle^2$ represents the matrix elements of that part of the Hamiltonian $H'(k)$ evaluated over the electronic configurations. The latter may be obtained from analysis or the scattering from a free atom. In practice the relationship between the photon wavelength and photon energy makes impractical the simultaneous determination of $G(\mathbf{r},t)$ as a function of both $\mathbf{r}$ and t. Only the neutron possesses both wavelength and energy comparable to the structure of chemical systems. This accounts for its power and subtlety as a probe of the constitution of ordinary matter.

5. Interactions with Fundamental Particles

Four basic low-energy neutron interactions are accessible to experimentation. They are the interactions between the neutron and the fundamental classical nuclear particles, namely, $(n - n)$, $(n - p)$, $(n - e)$, and $(\nu - n)$.

In $(n - n)$ interactions it is technically feasible to observe directly (n,n) scattering by using thermal neutron fluxes $\gtrsim 10^{15}/\text{cm}^2$ sec, such as are available in the very highest flux reactors [87]. To date the attempt has not been made for lack of adequate in-pile geometrical conditions. Measurements have been made [88, 89] utilizing reactions such as

$$_0n^1 + {}_1H^2 \rightarrow {}_1H^1 + 2{}_0n^1 - 2.23 \text{ Mev} \tag{7.190}$$ [88]

$$\pi^- + {}_1H^2 \rightarrow 2{}_0n^1 + \gamma \tag{7.191}$$ [89]

wherein the two neutrons may be emitted with small relative velocity. Under these conditions the details of the reaction are dependent upon a'_{nn}, the neutron-neutron scattering length. To date the measurements have not been sufficiently accurate to permit good quantitative comparison with $n - p$ and $p - p$ scattering.

For S-wave neutrons only the singlet-state scattering parameter, that is, $a_{(-)nn}$, is involved, since the two neutrons are identical particles. As was mentioned earlier, this state is not bound. A bound state has been searched for, but with negative results. This hypothetical configuration is termed the *dineutron*. Assuming its existence, one may search for reactions of the following type:

$$_{29}Cu^{65} + {}_0n^2 \rightarrow {}_{28}Ni^{66} + {}_1H^1 \tag{7.192}$$

in which the radioactive product $_{28}Ni^{66}$ could not be formed by any other process. No evidence of its formation by (7.192) was found [90]. Thus the lowest state for two neutrons is expected to be virtual, and the thermal neutron-neutron scattering length is expected to be negative. That is,

$$a'_{(-)nn} < 0 \qquad \text{and} \qquad \frac{\sigma_{nn}(\text{thermal})}{4\pi} = (a_{(-)nn})^2 \tag{7.193}$$

The $(n - p)$ interaction is fundamental to nuclear structure and forms the simplest multinucleon system, the deuteron. Such a configuration has as its two lowest states $J = \frac{1}{2} \pm \frac{1}{2}$, wherein $J = 1$ corresponds to the bound state and $J = 0$ to the virtual singlet state. Two scattering lengths again determine the thermal scattering.

$$a'_{\text{coh }np} = \tfrac{3}{4}a'_{(+)np} + \tfrac{1}{4}a'_{(-)np} < 0 \tag{7.194}$$

$$\frac{\sigma_{np}(\text{thermal})}{4\pi} = \frac{3}{4}(a'_{(+)np})^2 + \frac{1}{4}(a'_{(-)np})^2 \tag{7.195}$$

Equation (7.194) is measured by using the techniques of total reflection [91], and (7.195) by using the technique of transmission [10] (for immediate epithermal neutrons). The results of these measurements are

$$a'_{(+)np} = 0.538 \times 10^{-12} \text{ cm}$$
$$a'_{(-)np} = -2.37 \times 10^{-12} \text{ cm}$$

At neutron energies <10 Mev the scattering data may be used to evaluate two parameters of the nuclear potential [92]. These are the effective range r_0 and the effective well depth V_0. The triplet range is given by

$$r_{0+} = 2r_D\left(1 - \frac{r_D}{a'_{(+)np}}\right) \qquad (7.196)$$

where $r_D = \hbar/\sqrt{2\mu E_D}$, the so-called *radius of the deuteron*, and $E_D = 2.23$ Mev is the deuteron binding energy. Equation (7.196) yields the value 0.17×10^{-12} cm for r_{0+}.

The scattering cross section as a function of energy is given by [93]

$$\frac{\sigma_{np}(E)}{4\pi} = \lambda\!\!\!^{-2}\left\{\frac{3/4}{1 + [\lambda\!\!\!^{-}/a'_{(+)np} - r_{0+}/2\lambda\!\!\!^{-}]^2} + \frac{1/4}{1 + [\lambda\!\!\!^{-}/a'_{(-)np} - r_{0-}/2\lambda\!\!\!^{-}]^2}\right\} \qquad (7.197)$$

which reduces to (7.195) when $\lambda\!\!\!^{-} \to \infty$. Equation (7.197) is fitted to scattering data in the energy region below 10 Mev to yield $(0.24 \pm 0.5) \times 10^{-12}$ cm for r_{0-}. It is interesting to compare the latter value with the analogous one obtained from $p - p$ scattering, i.e., $(0.27 \pm 0.02) \times 10^{-12}$ cm. Additionally, the effective singlet well depths for $n - p$ and $p - p$ scattering are about equal, namely, $\simeq$18 Mev. The near equality of these values suggests that nuclear forces are *charge-independent*. It would be most valuable to make a similar comparison with $n - n$ scattering.

Radiative neutron-proton capture proceeds by formation of the singlet state, followed by a magnetic-dipole transition to ground (i.e., the triplet state). The photomagnetic cross section can be written in terms of r_{0+}, r_{0-} and the *magnetic exchange moment*. Since the latter cannot be precisely and independently evaluated σ_a (thermal) is used to evaluate [94] this quantity.

A weak attractive potential exists between the neutron and the electron. It may be represented by a square well of radius e^2/mc^2 ($\sim 2.8 \times 10^{-13}$ cm) and depth $\sim$5 kev. Such an interaction is to be expected from meson theory. It may be observed as an asymmetric component in thermal neutron scattering from a noble gas (i.e., closed shell) [95] or as an additional coherent amplitude in an appropriate experiment on total reflection of thermal neutrons from a bismuth-oxygen surface [96].

The neutrino interaction is termed a *weak interaction*. Its cross section with protons may be evaluated from the coupling constant obtained from (7.4) and is $\approx 10^{-44}$ cm^2. However, the neutron chain reactor gives rise to a very high antineutrino flux via β^- decay of fission products. These types of research have established (1) the existence of the inverse β^- reactions and (2) the nonidentity of neutrinos and antineutrinos.

Using a high-power reactor, type 1 has been demonstrated [97] via

$$\bar{\nu} + {}_1H^1 \to {}_0n^1 + \beta^+ - 1.8 \text{ Mev} \qquad (7.198)$$

and type 2 has been demonstrated by showing that the following reaction does not take place [98]; that is,

$$\bar{\nu} + {}_{17}Cl^{37} \not\to {}_{18}A^{37} + \beta^- \qquad (7.199)$$

This last result is in agreement with searches for double β^- decay which, with similar probability, have shown that the lifetime for the process is so long ($>10^{20}$ years) that neutrinos and antineutrinos must be different [99]. For a more complete discussion of the various possible modes of decay which could lead to double β^- emission the reader is referred to Rosen and Primakoff [100].

Acknowledgment

We would like to acknowledge the assistance of Drs. A. Santoro and M. Zocchi of the National Bureau of Standards in writing the section of this report dealing with the diffraction of neutrons.

References

1. Shull, C. G., K. Billman, and F. Wedgwood: (MIT) private communication.
2. Chadwick, J.: *Proc. Roy. Soc. (London)*, **A136**: 692 (1932).
3. Taschek, Jarvis, Hemmendinger, Everhart, and Gittings: *Phys. Rev.*, **75**: 1361 (1949). The Q values for this reaction and all others appearing in this chapter are taken from the compilation by Mattauch, J. H. E., W. Thiele, and A. H. Wapstra: *Nucl. Phys.*, **67**: 1 (1965). The values of the thermal cross sections were compiled by Goldman, D. T.: "The Chart of the Nuclides," 8th ed., 1965.
4. Alvarez, L. W., and R. Cornog: *Phys. Rev.*, **57**: 248 (1940).
5. Robson, J. M.: *Phys. Rev.*, **31**: 81, 297 (A) (1951).
6. Dabbs, J. W., J. A. Harvey, D. Paya, and H. Horstman: *Phys. Rev.*, **139**: B756 (1965).
7. Hamermesh, M., and E. Eisner: *Phys. Rev.*, **79**: 888 (1950).
8. Purcell, E. M., and N. F. Ramsey: *Phys. Rev.*, **78**: 807 (1950).
9. Corngold, N. R., V. F. Cohen, and N. F. Ramsey: *Bull. Am. Phys. Soc.*, **1**: 11 (1956).
10. Melkonian, E.: *Phys. Rev.*, **76**: 1744 (1949).
11. Cranberg, L., and J. S. Levin: *Phys. Rev.*, **103**: 343 (1956).
12. Muehlhause, C. O., S. E. Bloom, H. E. Wegner, and G. N. Glasoe: *Phys. Rev.*, **103**: 720 (1956).
13. Fünfer, E., and W. Z. Bothe: *Z. Physik*, **122**: 769 (1944).
14. Hahn, O., and F. Strassman: *Naturwiss.*, **27**: 11 (1939).
15. Szilard, L., and W. H. Zinn: *Phys. Rev.*, **55**: 799 (1939).
16. Fermi, E.: *Science*, Jan. 10, 1947.
17. Alvarez, L., as reported by Farwell, Segrè, and Wiegand: *Phys. Rev.*, **71**: 327 (1947).
18. Chadwick, J., and M. Goldhaber: *Proc. Phys. Soc. (London)*, **151**: 479 (1935).
19. Szilard, L., and T. A. Chalmers: *Nature*, **134**: 494 (1934).
20. Noyce, R. H., E. R. Mosberg, Jr., S. B. Garfinkel, and R. Caswell: *Reactor Sci. Technol.*, **17**: 313 (1963).
21. Du Bridge, L. A.: *Phys. Rev.*, **53**: 603A (1939).
22. Alvarez, L. W., and R. Cornog: *Phys. Rev.*, **56**: 613 (1939).
23. Bretscher, E., and A. P. French: *Phys. Rev.*, **75**: 1154 (1949).
24. McMillan, E. M., M. Kamen, and S. Ruben: *Phys. Rev.*, **52**: 375 (1937). Wood, R. E., H. H. Landon, and V. L. Sailor: *Phys. Rev.*, **98**: 639 (1955).
25. Ruben, S., and M. D. Kamen: *Phys. Rev.*, **59**: 349 (1941).
26. Taylor, H. J., and M. Goldhaber: *Nature*, **135**: 341 (1935).

27. Chadwick, J., and M. Goldhaber: *Nature*, **135**: 65 (1935).
28. Rainwater, J., W. Havens, C. Wu, and J. Dunning: *Phys. Rev.*, **71**: 65 (1947).
29. Hughes, D. J., and C. Eggler: *Phys. Rev.*, **73**: 809 (1948).
30. Amaldi, D'Agostino, Fermi, Pontecorvo, Rosetti, and Segrè: *Proc. Roy. Soc. (London)*, **149**: 522 (1935).
31. Landon, H. H., and I. G. Schroder: (NBS) private communication.
32. Wigner, E. P.: *Proc. Natl. Acad. Sci. U.S.*, **27**: 282 (1941).
33. Blatt, J. M., and V. F. Weisskopf: "Theoretical Nuclear Physics," p. 796 et seq., Wiley, New York, 1952.
34. Tamura, T.: *Revs. Mod. Phys.*, **37**: 679 (1965).
35. Lepore, J.: *Phys. Rev.*, **79**: 137 (1950).
36. Hauser, W., and H. Feshbach: *Phys. Rev.*, **87**: 366 (1952).
37. Moldauer, P. A., *Revs. Mod. Phys.*, **36**: 1079 (1964). Krieger, T. J.: *Ann. Phys.*, **31**: 88 (1965).
38. Breit, G., and E. P. Wigner: *Phys. Rev.*, **49**: 519 (1936).
39. See, for example, Dresner, L.: "Resonance Absorption in Nuclear Reactors," chap. 3, Pergamon Press, New York, 1960.
40. Wigner, E. P., and L. Eisenbud: *Phys. Rev.*, **72**: 29 (1947).
41. Vogt, E.: *Revs. Mod. Phys.*, **34**: 723 (1962).
42. Weinberg, A. M., and E. P. Wigner: "The Physical Theory of Neutron Chain Reactors," chap. 9, University of Chicago Press, Chicago, 1958.
43. Ehrlich, R., and H. Hurwitz, Jr.: *Nucleonics*, **12** (2): 23 (1954).
44. Bethe, H.: *Revs. Mod. Phys.*, **9**: 121 (1937).
45. Beckurts, K. H., and K. Wirtz: "Neutron Physics," chap. 8, Springer, Berlin, 1964.
46. Sachs, R. G.: "Nuclear Theory," chap. 4, Addison-Wesley, Reading, Mass., 1953.
47. Fermi, E.: *Ric., Sci.*, **7**: 13 (1936).
48. Waller, I.: *Z. Physik.*, **17**: 398 (1923).
49. Brown, G. M., and H. A. Levy: *Science*, **147**: 1038 (1965).
50. Loopstra, B. O.: *Acta Cryst.*, **17**: 651 (1964).
51. Willis, B. T. M.: *J. Phys. Radium*, **25**: 431 (1964).
52. Atoji, M.: *J., Chem. Phys.*, **35**: 1950 (1961).
53. Atoji, M., and D. E. Williams: *J. Chem. Phys.*, **35**: 1960 (1961).
54. Pauling, L.: "The Nature of the Chemical Bond," Cornell University Press, Ithaca, N.Y., 1960.
55. Pimentel, G. C., and A. L. McClellan: "The Hydrogen Bond," Freeman, San Francisco, 1960.
56. Copens, P.: *Acta Cryst.*, **17**: 573 (1964).
57. Bacon, G. E., and R. S. Pease: *Proc. Roy. Soc. (London)*, **A230**: 359 (1955).
58. Shull, C. G., and S. Siegel: *Phys. Rev.*, **75**: 1008 (1949).
59. Shull, C. G., and M. L. Wilkinsen: *Phys. Rev.*, **97**: 304 (1955).
60. Bacon, G. E., R. Street, and R. H. Tredgold: *Proc. Roy. Soc. (London)*, **A217**: 252 (1952).
61. Peterson, S. W., and H. G. Smith: *Phys. Rev. Letters*, **6, 7**: (1961).
62. Peterson, S. W., and H. G. Smith: *J. Phys. Soc. Japan*, **17** (Supp. B-II): 335 (1962).
63. Smith, H. G., and H. G. Peterson: *J. Phys. Radium*, **25**: 615 (1964).
64. Halperin, O., and M. H. Johnson: *Phys. Rev.*, **55**: 898 (1939).
65. Shull, C. G., W. A. Strauser, and E. O. Wollan: *Phys. Rev.*, **83**: 333 (1951). Ruderman, I. W.: *Phys. Rev.*, **76**: 1572 (1949).
66. Shull, C. G., E. O. Wollan, and W. C. Koehler: *Phys. Rev.*, **84**: 912 (1951).
67. Nathans, R., C. G. Shull, G. Shirane, and A. Andresen: *Phys. Chem. Solids*, **10**: 138 (1959).
68. Bacon, G. E.: "Neutron Diffraction," Clarendon Press, Oxford, 1926. Nathans, R., and S. J. Pickart: in G. T. Rado and H. Suhl, "Magnetism," vol. 3, Academic, New York, 1963.
69. Aloji, M.: ANL-6920, July, 1964. This report also contains an extensive bibliography on the subject.
70. See, for example, Klug, H. P., and L. E. Alexander: "X-ray Diffraction Procedures," Wiley, New York, 1962. Henry, N. F. M., H. Lipson, and W. A. Wooster: "The Interpretation of X-ray Diffraction Photographs," Macmillan, London, 1961. For bibliography on neutron diffractometers, see ref. 69.
71. Goldberger, M. L., and F. Seitz: *Phys. Rev.*, **71**: 294 (1947).
72. Hamermesh, M.: *Phys. Rev.*, **75**: 1766 (1949).
73. Wick, G. C.: *Phys. Rev.*, **94**: 1228 (1954).
74. Zemach, A. C., and R. J. Glauber: *Phys. Rev.*, **101**: 118 (1956).
75. Van Hove, L.: *Phys. Rev.*, **95**: 249 (1951).
76. Egelstaff, P. A.: *Nucl. Sci. Eng.*, **12**: 250 (1962).
77. Egelstaff, P. A., and P. Schofield: *Nucl. Sci. Eng.*, **12**: 260 (1962).
78. Placzek, G.: *Phys. Rev.*, **86**: 377 (1952).
79. Vineyard, G.: *Phys. Rev.*, **110**: 999 (1958).
80. Lamb, W. E.: *Phys. Rev.*, **55**: 190 (1939).
81. Jackson, H. E., L. M. Bollinger, and R. E. Coté: *Phys. Rev. Letters*, **6**: 187 (1961).
82. Michaudon, A., H. Derrien, C. Le Pipec, and P. Ribon: *Compt. Rend.*, **255**: 2086 (1962).
83. Bernabei, A.: BNL-860, 1964.
84. Nelkin, M. S., and D. E. Parks: *Phys. Rev.*, **119**: 1060 (1960).
85. Singwi, K. S.: in "Inelastic Scattering of Neutrons in Solids and Liquids," vol. 1, p. 1, IAEA, Vienna, 1963.
86. Heitler, W.: "The Quantum Theory of Radiation," 3d ed., p. 141, Oxford University Press, Fair Lawn, N.J., 1954, as presented by M. Bloom, BNL Conference on Inelastic Neutron Scattering, 1965.
87. Muehlhause, C. O.: ANL-6797, 1963.
88. Ilakovac, K., L. G. Kero, M. Petravic, and I. Slaus: *Phys. Rev.*, **124**: 1923 (1961).
89. Haddock, Salter, Zeller, Czirr, Nygren, and Maung: *Bull. Am. Phys. Soc.*, **9**: 443 (1964).
90. Cohen, B. L., and T. H. Handley: ORNL-1382, 1952.
91. Hamermesh, M.: *Phys. Rev.*, **77**: 140 (1950).
92. Blatt, J. M., and J. Jackson: *Phys. Rev.*, **76**: 18 (1949).
93. Bethe, H.: *Phys. Rev.*, **76**: 38 (1949).
94. Austern, N.: *Phys. Rev.*, **92**: 670 (1953).
95. Hamermesh, M., G. R. Ringo, and A. Wattenger: *Phys. Rev.*, **85**: 483 (1952).
96. Hughes, D. J., J. A. Harvey, M. D. Goldberg, and M. J. Stafne: *Phys. Rev.*, **90**: 497 (1953).
97. Cowan, Reines, Harrison, Kruse, and McGuire: *Science*, **124**: 103 (1956).
98. Davis, R.: *Bull. Am. Phys. Soc.*, **4**: 219 (1956). Alvarez, L. W.: UCRL-328, 1949.
99. Der Mateosian, E., and M. Goldhaber: private communication, BNL, 1965, re Ca^{48}. Takoaka, N.: *Shitsuryo Bunseki*, **12**: 195 (1965) re Te^{130}, and *Nucl. Sci. Abstr.*, 33351, 1965.
100. Rosen, S. P., and H. Primakoff: in K. Siegbahn (ed.), "Alpha-, Beta- and Gamma-Ray Spectroscopy," vol. 2, chap. 24(J), North Holland Publishing Company, Amsterdam, 1965.

Chapter 8

Nuclear Reactions

By BERNARD L. COHEN *and* DAVID HALLIDAY, University of Pittsburgh

1. Introduction

A nuclear reaction is the series of events that takes place when two nuclei collide and interact. In this sense it is analogous to chemical reactions which take place when atoms collide. One important difference arises from the fact that atoms are electrically neutral so that there are no strong forces keeping them apart; chemical reactions therefore occur commonly in nature. Nuclei, on the other hand, carry positive electric charges which result in strong electrical repulsion, so that nuclear reactions are very rare in our earthly environment. To overcome the Coulomb repulsion, it is usually necessary to accelerate nuclei to high energies by use of an accelerator. However, there are three outstanding exceptions to this rule:

1. A very special nucleus, the neutron, bears no electric charge so that neutron-induced reactions may occur at any energy, no matter how low. This feature makes possible the fission reactor. For most research purposes, however, it is desirable to have monoenergetic neutrons, and these can be produced only by nuclear reactions induced by energetic charged particles from accelerators.

2. At temperatures above 10^8 deg Kelvin, the lightest nuclei attain sufficient energy to overcome their relatively small Coulomb repulsion with an appreciable probability. Above 10^{10} deg this becomes true even for relatively heavy nuclei. These "thermonuclear reactions" are the source of energy in stars, where they are also responsible for the formation of all heavier nuclei from protons. There is hope that thermonuclear reactors can be used for producing energy on earth.

3. Particles emitted from nuclear radioactive decay often have sufficient energy to induce reactions. Alpha particles were used in the discovery of nuclear reactions by Rutherford in 1919 [1]† and are still commonly used for producing neutrons with nuclear reactions.

A typical reaction can be written

$$a + X \rightarrow Y + b$$

or in the conventional notation

$$X(a,b)Y$$

† Numbers in brackets refer to References at end of chapter.

Most work on nuclear reactions has been with relatively light bombarding particles as these need less energy to overcome Coulomb barriers. For the same reason, light particles are by far the most frequently emitted in reactions. The discussion to follow will therefore be slanted toward reactions in which a and b in the above expressions are nuclei of mass 1 to 4, but nearly all the discussion applies to reactions in which a and b are heavier ions. The discussion will be limited to energies below the meson production threshold and, to a large extent, to energies below 50 Mev.

If $a = b$, which implies $Y = X$, the reaction is called *scattering*. Further, if not only the identity but also the kinetic energy of the emerging particle (for an observer at the center of mass) is the same as for the projectile, there is *elastic scattering*. For *inelastic scattering*, the emerging particle energy for the center-of-mass observer is less than that carried in by the projectile. This means that Y has a certain amount of excitation energy above its ground state, which will be given off by further emission of nuclear particles or of gamma rays. If the later emitted particles are $c, d, \ldots$, the reaction is written $(a, bcd \ldots)$. If nothing but gamma rays are emitted, the reaction is called "capture" and it is written (a,γ).

In all nuclear reactions, certain properties of the system remain unchanged because of well-known conservation laws. They are total energy, linear momentum, angular momentum, electric charge, parity, statistics, and baryon number. In view of the rather well-established principle of charge independence of nuclear forces, isobaric spin is conserved to a rather good approximation in nuclear reactions.

2. Energetics

Assuming that X is at rest in the laboratory, conservation of energy gives

$$(E_a + m_a c^2) + m_X c^2 = (E_Y + m_Y c^2) + (E_b + m_b c^2) \tag{8.1}$$

The E's are kinetic energies measured with respect to the laboratory and the m's are rest masses. Rewriting and introducing a new symbol,

$$Q = E_Y + E_b - E_a \tag{8.2a}$$
$$Q = (m_X + m_a - m_Y - m_b)c^2 \tag{8.2b}$$

where c^2 can be conveniently written as (931.141 ±

0.010) Mev/amu (phys.) [2]. This *Q value* or *disintegration energy* may be positive or negative or, for elastic scattering, zero. It is a measure of the extent to which rest mass is converted into kinetic energy by the reaction.

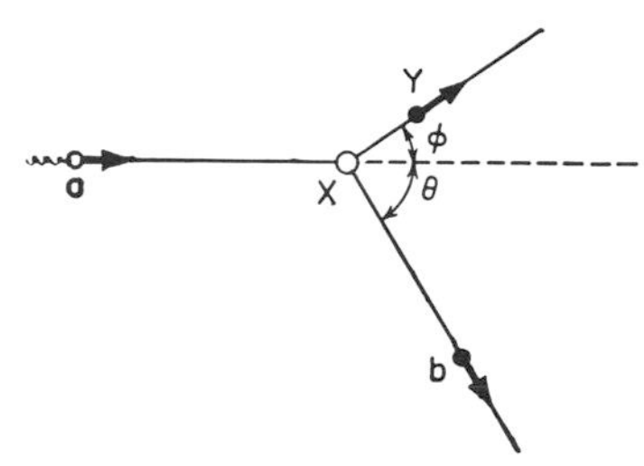

Fig. 8.1. A nuclear reaction as viewed in the laboratory frame of reference. The target X is taken to be initially at rest.

By using the law of conservation of linear momentum, the hard-to-measure E_Y can be eliminated, giving (see Fig. 8.1)

$$Q = E_b\left(1 + \frac{m_b}{m_Y}\right) - E_a\left(1 - \frac{m_a}{m_Y}\right) - (E_aE_bm_am_b)^{1/2}\cos\theta \quad (8.3)$$

where θ is the angle between the directions of motion of b and a. Although the very concept of a Q is relativistic, the speeds of a, b, and Y are often small enough so that the nonrelativistic relationship between kinetic energy and momentum can be used. This was done in deriving Eq. (8.3).

Mass numbers can usually be substituted for masses in Eq. (8.3). In fact, if this substitution makes a significant change in the calculated Q value, it usually means that Eq. (8.3) should be replaced by its relativistically correct equivalent. A relativistic treatment of nuclear reactions has been given by Morrison [3].

If Q is given, E_b can be computed from Eq. (8.3):

$$(A_Y + A_b)E_b^{1/2} = (A_aA_bE_a)^{1/2}\cos\theta \pm \{A_aA_bE_a\cos^2\theta + (A_Y + A_b)[A_YQ + (A_Y - A_a)E_a]\}^{1/2} \quad (8.4)$$

Here mass numbers (the A's) have been substituted for rest masses.

Figure 8.2 shows a plot for the reaction

$$H^3(p,n)He^3$$

for which $Q = -0.764 \pm 0.001$ Mev. (This is one of the most frequently used reactions for producing monoenergetic neutrons.) The figure shows that there is a minimum projectile energy E_{th} below which the reaction will not go. If a has this *threshold energy*, the emerging particle b is observed only in the forward direction ($\theta = 0$). Equation (8.4) shows that

$$E_{th} = -Q\,\frac{A_Y + A_b}{A_Y + A_b - A_a} \quad (8.5)$$

Thresholds occur only for reactions with negative Q's. These are called *endoergic* reactions. Because of the conservation of baryons, for any nuclear reaction,

$$A_a + A_X = A_Y + A_b \quad (8.6)$$

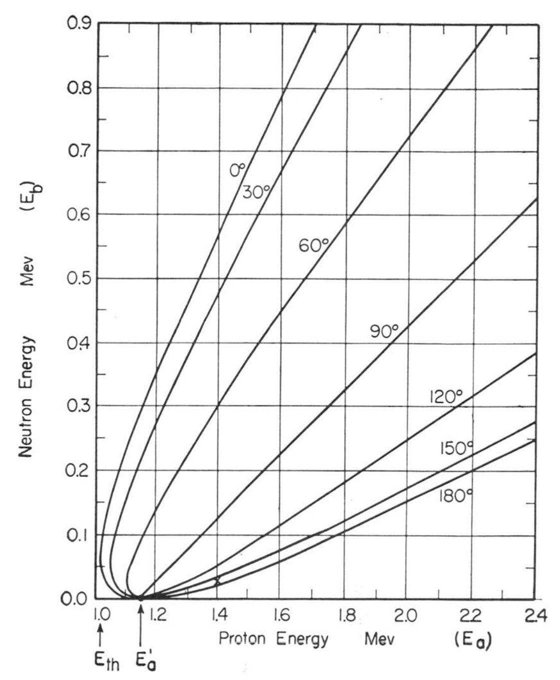

Fig. 8.2. Energy of neutrons from the reaction $H^3(p,n)He_3$ as a function of emission angle and bombarding energy.

where the symbols represent mass numbers. Equation (8.5) can be rewritten as

$$E_{th} = -Q\,\frac{A_X + A_a}{A_X} \quad (8.7)$$

The threshold energy for the reaction of Fig. 8.2 is 1.017 Mev.

Figure 8.2 shows that, for a given E_a and a given angle, E_b can sometimes be double-valued. This happens if E_a lies between E_{th} and E'_a, where

$$E'_a = -Q\,\frac{A_Y}{A_Y - A_a} \quad (8.8)$$

E'_a is 1.145 Mev for the reaction of Fig. 8.2.

For projectile energies slightly above threshold the paths of the emerging particle b lie in a cone around the axis of the incoming beam. The half angle of this cone, θ_m, is found from

$$A_aA_bE_a\cos^2\theta_m + (A_Y + A_b)[A_YQ + (A_Y - A_a)E_a] \quad (8.9)$$

as Eq. (8.4) shows. As the projectile energy is increased, the cone opens out until, at the value E'_a, it fills the forward hemisphere. For energies above E'_a it includes all space.

The Q value of the reaction may be found either from precise mass measurements [Eq. (8.2*b*)] or from energy measurements of the particle b [Eq. (8.3)]. Tables of reaction Q values are available in the literature [4], and new compilations appear periodically.

3. Center-of-mass Coordinates

Experimental observations such as angles, particle energies, etc., are measured with respect to a coordinate frame at rest in the laboratory. On the theoretical side, however, a certain simplicity results if the nuclear reaction is examined from the point of view of an observer who is at rest with respect to the center of mass of the interacting particles. In such a coordinate system, for example, the resultant linear momentum of the interacting particles is zero. For a two-particle system this requires that the particles, both before and after the collision, move in opposite directions along a straight line.

The relationship between the angle θ at which particle b is observed in the laboratory system and the corresponding angle θ^* in the center-of-mass system can be shown to be given by, in nonrelativistic approximation,

$$\tan \theta = \frac{\sin \theta^*}{\cos \theta^* + \gamma} \tag{8.10a}$$

where, identifying masses with mass numbers,

$$\gamma = \left[\frac{A_a A_b E_a}{A_Y(A_Y + A_b)Q + A_Y A_X E_a}\right]^{1/2} \tag{8.10b}$$

Figure 8.3 shows a plot of θ^* versus θ for various values of the parameter γ. For the special cases of $\gamma = 1$ and $\gamma = 0$, Eq. (8.10a) reduces to

$$\theta^* = 2\theta \qquad \text{and} \qquad \theta^* = \theta \tag{8.11}$$

respectively. The first case is that for nucleon-nucleon elastic scattering; the second case corresponds to an infinitely massive target nucleus. If the incident particle has the threshold energy given by Eq. (8.7), γ becomes infinite. This means that θ can have only the value zero, which agrees with the fact that the emergent particles come out in the forward direction at threshold.

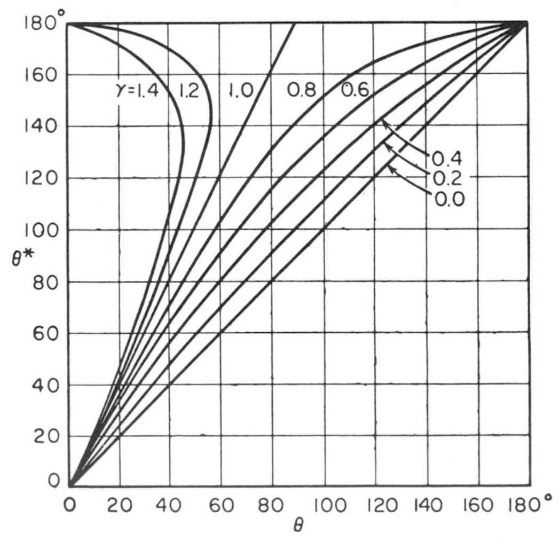

FIG. 8.3. Center-of-mass angle θ^* versus laboratory angle θ for various values of the parameter γ. This a plot of Eq. (8.10a).

For $\gamma > 1$, the maximum value for θ is less than $\pi/2$ (no back-scattering in the laboratory frame). For every allowed value of θ, there are two values of θ^*, corresponding to particles with two different energies. The condition of $\gamma > 1$ is realized if E_a is less than the E'_a of Eq. (8.8).

An additional quantity of interest with different values in the laboratory and the center-of-mass reference frames is the solid angle. Suppose that a laboratory worker is observing the emerging particles that lie between two cones whose elements make angles of θ and $\theta + d\theta$ with the projectile axis and whose common apex is at the target position. The solid angle is $d\Omega = 2\pi \sin \theta \, d\theta$. A center-of-mass observer would see these same particles filling the region extending from θ^* to $\theta^* + d\theta^*$, where θ and θ^* are related by Eqs. (8.10). The solid angle corresponding to $d\Omega$ is $d\Omega^* = 2\pi \sin \theta^* \, d\theta^*$. The relationship between $d\Omega$ and $d\Omega^*$ is

$$d\Omega^* = \frac{(1 + 2\gamma \cos \theta + \gamma^2)^{3/2}}{|1 + \gamma \cos \theta|} \, d\Omega \tag{8.12}$$

4. Cross Section

The intrinsic probability that a reaction will occur is measured by its *cross section.* If a collimated beam of projectiles strikes a target of thickness Δx at a rate R_0 particles per unit area per unit time, if n is the number of target nuclei per unit volume, and if R is the rate at which reactions occur, a cross section can be defined from

$$R = R_0 n\sigma \, \Delta x \tag{8.13}$$

where Δx is small enough that $n\sigma \, \Delta x \ll 1$. The geometrical interpretation of σ as an effective projected area is clear. Cross sections are usually measured in barns or millibarns (mbarns).

$$1 \text{ barn} = 10^3 \text{ mbarns} = 10^{-24} \text{ cm}^2$$

From the point of view of a single target nucleus, σ is the probability per unit time that a reaction will occur if the target is placed in a unit flux of incident particles.

All conceivable events that occur when a projectile strikes a nucleus have their separate cross sections. One speaks of the cross section for elastic scattering, for inelastic scattering, for absorption, for absorption followed by emission of, say, an alpha particle, etc.

A *total cross section* σ_t can be defined from Eq. (8.13) if R_0 is taken as the rate at which particles disappear from the collimated beam for any cause whatsoever. This attenuation of the incident beam by interposed absorbers is often described by a *linear absorption coefficient* μ_x or a *mass absorption coefficient* μ_m; these are related to σ_t by

$$\mu_x = \rho\mu_m = n\sigma_t$$

where ρ is the density of the target.

The relative probability that the reaction product will be emitted in various directions is measured by the *differential cross section* $d\sigma/d\Omega$, which is defined from

$$R_\Omega = R_0 n \frac{d\sigma}{d\Omega} \Delta x \tag{8.14}$$

where R_Ω is the rate per unit solid angle at which reaction products are emitted in the direction in question. Clearly

$$\sigma = \int \left(\frac{d\sigma}{d\Omega}\right) d\Omega = \int_0^\pi \int_0^{2\pi} \frac{d\sigma}{d\Omega}(\theta,\phi) \sin \theta \, d\phi \, d\theta \tag{8.15}$$

where the integration is carried out over all directions. Where polarization problems are not considered, the ϕ dependence is averaged out so that (8.15) becomes

$$\sigma = \int_0^\pi \frac{d\sigma}{d\Omega}(\theta) 2\pi \sin\theta \, d\theta \qquad (8.15')$$

5. Method of Partial Waves

Nuclear forces, excluding Coulomb forces, are known to have a rather well-defined range. This means that, if target nuclei are placed in a beam of projectiles, only those projectile-target pairs whose impact parameters are less than the range of the forces will be able to interact. Thus only part of the projectile beam is effective.

In passing from this classical description to a wave-mechanical description it is possible to keep the spirit of the above remarks. The incident wave can be written as a series of subwaves and their effects, in most cases, discussed separately. Again relying on the finite range of the nuclear forces, one can discard many of these subwaves (in many cases all but one) and thus find a simpler description of the nuclear event. The mathematical treatment is simplest if the incident particle is a neutron, so that case will be considered here. Analogous treatments for charged particles require the use of Coulomb wave functions; they may be found in the literature [5].

The wave function ψ for a pair of particles, a and X, that are approaching each other along the z axis, if the two particles do not interact with each other, from the viewpoint of a center-of-mass observer, is

$$\psi_0 = e^{i(kz-\omega t)} \qquad (8.16)$$

where

$$k = \frac{(2\mu E)^{1/2}}{\hbar} \qquad (8.17)$$

and

$$\omega = \frac{E}{\hbar}$$

E is the system energy in the center-of-mass system, and μ is the reduced mass of the pair of particles. Thus the two-particle system can be described in a one-particle symbolism [Eq. (8.16)].

In a purely formal way the space-varying part of Eq. (8.16) can be written as the sum of a series of terms that involve the angular-momentum quantum number l,

$$\psi_0 = \sum_{l=0}^{\infty} \frac{\mu_0^{(l)}(r)}{r} P_l(\cos\theta) \qquad (8.18)$$

The $P_l(\cos\theta)$ are Legendre polynomials of order l. For the limiting case of $kr \gg l$, the coefficients $\mu_0^{(l)}(r)$ are given by

$$\mu_0^{(l)}(r) = \frac{(2l+1)i^{l+1}}{2k} \left\{ \exp\left[-i\left(kr - \frac{\pi}{2}l\right)\right] - \exp\left[i\left(kr - \frac{\pi}{2}l\right)\right] \right\} \qquad (8.19)$$

Thus the plane wave of Eq. (8.16) is represented at large separations as a sum of spherical partial waves [Eq. (8.19)]. The first term in the braces represents a radially incoming spherical wave; the second term represents a radially outgoing spherical wave. The quantity l determines, in the usual way, the angular momentum to be associated with each partial wave. In classical terms, large l values are associated, for a given relative linear momentum, with large impact parameters. For $l = 0$ encounters (S-wave interactions), Eq. (8.18) reduces to

$$\psi_0 = (2ikr)^{-1}(e^{ikr} - e^{-ikr}) = \frac{\sin kr}{kr} \qquad (8.20)$$

If the energy is low enough it should be possible to ignore all but the $l = 0$ partial wave in Eq. (8.18). Encounters with $l = 1, 2, 3$, etc., correspond to maximum resolvable angular momenta of $\hbar$, $2\hbar$, $3\hbar$, etc. If a pair of particles with small relative energy is to have a large angular momentum, the pair must have a large impact parameter. If the energy is small enough, the required impact parameter can be bigger than the range of the nuclear forces so that no interaction occurs. If the nuclear forces did not have a finite range, the partial-wave series could not be limited to a few terms in this way. The condition for considering only $l = 0$ interactions is very easily satisfied for the case of nuclei bombarded with thermal neutrons.

6. Elastic Scattering and Reaction Cross Sections

The wave function ψ describing the actual situation in which the particles a and X interact with each other will now be considered. It again has the form

$$\psi = \sum \frac{\mu^{(l)}(r)}{r} P_l(\cos\theta) \qquad (8.21)$$

where, for $kr \gg l$,

$$\mu^{(l)}(r) = \frac{(2l+1)i^{l+1}}{2k} \left\{ \exp\left[i - \left(kr - \frac{\pi}{2}l\right)\right] - \eta_l \exp\left[i\left(kr - \frac{\pi}{2}l\right)\right] \right\} \qquad (8.22)$$

The term η_l is in general a complex number and can be written as

$$\eta_l = a_l \exp(i\delta_l) \qquad (8.23)$$

where a_l and δ_l are real. Only the outgoing subwaves of Eq. (8.19) are changed by the nuclear event, both in amplitude (a_l) and in phase (δ_l). If elastic scattering is the only process occurring, $a_l = 1$; if other reactions occur, and thus drain from the beam some particles that might otherwise have been elastically scattered, $a_l < 1$. The quantities η_l are determined by the scattering potential.

The elastically scattered wave is the difference between ψ_0 and ψ,

$$\psi_{el} = \psi - \psi_0 = \frac{1}{2kr} \sum_l (2l+1)i^{l+1}(1 - \eta_l) \exp\left[i\left(kr - \frac{\pi}{2}l\right)\right] P_l(\cos\theta) \qquad (8.24)$$

It necessarily contains only outgoing subwaves.

The differential cross section for elastic scattering is computed by first finding, by standard quantum-mechanical procedures, the rate per unit solid angle at which particles are elastically scattered in the direction θ; this quantity is then divided by the incident flux [$= v\psi\psi^* = v$; see Eq. (8.16)] to yield

$$\left(\frac{d\sigma}{d\Omega}\right)_{el} = \frac{1}{4k^2}\left|\sum_l (2l+1)(1-\eta_l)P_l(\cos\theta)\right|^2 \quad (8.25)$$

The elastic scattering cross section then follows from Eq. (8.15):

$$\sigma_{el} = \pi\bar{\lambda}^2 \sum_l (2l+1)|1-\eta_l|^2 \quad (8.26)$$

where we have used $\bar{\lambda} = k^{-1}$. We see that σ_{el} can be written as a sum of partial cross sections representing the separate scattering of the various subwaves; $(d\sigma/d\Omega)_{el}$ cannot be so written.

This development can throw light on reactions other than elastic scattering because of the fact that, if the number of elastically scattered particles is less than the number of incident particles, some reaction other than elastic scattering must have accounted for them. Weisskopf and his coworkers [6, 7] define a *reaction cross section* from

$$\sigma_t = \sigma_r + \sigma_{el} \quad (8.27)$$

Often only one reaction other than elastic scattering is at all likely; σ_r is then the cross section for that reaction.

The rate at which particles disappear from the incident beam by processes other than elastic scattering is found from Eq. (8.21) by computing the *net* inward flux through the surface of a large sphere surrounding the center of mass. Dividing this quantity by the incident flux ($= nv$) yields

$$\sigma_r = \frac{\pi}{k^2}\sum (2l+1)(1-a_l^2) \quad (8.28)$$

$$\sigma_r = \pi\bar{\lambda}^2 \sum_l (2l+1)(1-|\eta_l|^2) \quad (8.29)$$

Here σ_r is independent of the phase factor δ_l. The cross sections σ_r and σ_{el} are clearly closely related.

An especially simple model which is essentially valid at energies above a few Mev is to assume that the only hindrance to the penetration of an l wave into the nucleus to induce a nuclear reaction is the angular-momentum barrier. We may then replace $1 - |\eta_l|^2$ in (8.29) by T_l, the transmission through this barrier. That equation then becomes

$$\sigma_r = \pi\bar{\lambda}^2 \sum_l (2l+1)T_l \quad (8.30)$$

The penetration T_l may be readily calculated [7] in terms of the nuclear radius R, and the result obtained for (8.30) is, to a very good approximation for energies above about 2 Mev,

$$\sigma_r = \pi(R+\bar{\lambda})^2 \quad (8.31)$$

The physical significance of (8.31) is obvious. To a rather good approximation, this treatment also gives for the scattering cross section

$$\sigma_{el} = \pi(R+\bar{\lambda})^2 \quad (8.32)$$

For charged particles, one must also take into account the Coulomb barrier penetration in the calculation of T_l. The results for σ_r are given in tables [7, 8] for energies below or comparable to the Coulomb barrier height. For higher energies, the result is well approximated by

$$\sigma_r = \pi(R+\bar{\lambda})^2\left[1 - \frac{V(R+\bar{\lambda})}{E}\right] \quad (8.33)$$

where $V(R+\bar{\lambda})$ is the height of the Coulomb barrier at a distance $R + \bar{\lambda}$ from the center of the nucleus. This is the result that would be obtained classically for a grazing collision if the electrical deflection of the incident particle as it approaches the nucleus is taken into account.

7. Optical Model

While the above treatment is an interesting approximation, for best quantitative results, the η_l in (8.25), (8.26), and (8.29) should be obtained from experiment. Measurements of angular distributions in elastic scattering can be made with quite high precision. These could be compared with (8.25) to give the η_l, but the results would be only a list of numbers applicable for that particular energy and having no easily discernible physical significance. The approach therefore taken is to calculate the scattering from a complex potential ("optical") well and adjust the parameters of the well to obtain a fit to the measured angular distribution. The values of η_l are then a

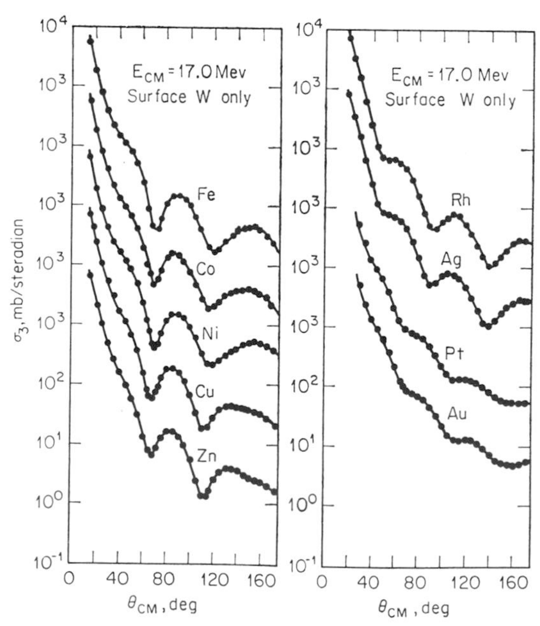

FIG. 8.4. Typical fits of optical-model calculations to elastic-scattering angular distributions. These data are for 17-Mev protons. [*From F. G. Perey, Phys. Rev.*, **131**: **745 (1963)**.]

byproduct of the calculation. An example of a fit of this type is shown in Fig. 8.4. The angular distribution contains a great deal of detailed structure, so that optical-model potentials with a great many parameters can be used with meaning. The most commonly used one [9] is

$$U(r) = V_c(r) - Vf(x) - iWf(x_W) - W' \exp[-(x')^2] - (V_{s0} + iW_{s0}) \left(\frac{\hbar}{M_\pi c}\right)^2 \frac{1}{r} \left| \frac{d}{dr} f(x) \right| l \cdot \sigma \quad (8.34)$$

where

$$f(x) = [1 + \exp(x)]^{-1} \qquad x = \frac{r - r_0 A^{1/3}}{a}$$

$$f(x_W) = [1 + \exp(x_W)]^{-1} \qquad x_W = \frac{r - r_W A^{1/3}}{a_W}$$

$$f(x') = [1 + \exp(x')]^{-1} \qquad x' = \frac{r - r'_0 A^{1/3}}{a'}$$

$$\left(\frac{\hbar}{M_\pi c}\right)^2 = 2.00 \text{ (fermi)}^2$$

The first term in (8.34) is the Coulomb potential, taken as

$$V_c(r) = \begin{cases} \dfrac{ZZ'e^2}{2R_c} = \left(3 - \dfrac{r^2}{r_0^2}\right) & \text{for } r \le R_c \\ \dfrac{ZZ'e^2}{r} & \text{for } r \ge R_c \end{cases} \quad (8.35)$$

where $R_c = r_c A^{1/3}$

The second term in (8.34) is the ordinary real potential, the next two terms are the imaginary potential which gives absorption (they are a volume and a surface absorption, respectively), and the remaining terms are the spin-orbit interaction, consisting of a real plus an imaginary term, modeled after the Thomas term used in atomic physics. The adjustable parameters in (8.34) are r_c, V, r_0, a, W, r_W, a_W, W', r'_0, a', V_{s0}, W_{s0}—a total of 12.

The usefulness of this "optical model" is that these parameters vary slowly and monotonically with energy and have values consistent with our knowledge about the size and shape of the nucleus. They also join smoothly onto the shell-model potential as the energy is lowered.

An enormous amount of information is now available on optical-model parameters for scattering of all particles of mass up to 4 from many nuclei throughout the periodic table with a wide range of bombarding energies [10]. The parameters vary sufficiently slowly that interpolations can be made with some confidence.

The principal reason for the great effort expended on this work is that the scattering wave functions are useful in analysis of direct reaction experiments with the distorted-wave Born approximation (see Sec. 11 below). Another application is that these calculations yield reaction cross sections [for example, from (8.29) for neutrons]. These do not exhibit the smooth behavior predicted by (8.31) but rather are characterized by "giant resonances" at energies where the size of the optical-model potential well is an integral number of half wavelengths of the incident particle [11]. Some of these effects for low-energy neutrons are shown in Fig. 8.5. At higher energies, these effects become less apparent because many l waves, each having resonances at different energies, contribute. Interesting effects do, however, appear [12].

The theory of the coefficient V of the second term in Eq. (8.34) is rather well developed for neutrons and protons.

It may be written

$$V = V_0 - \beta E + V' \left(\frac{N - Z}{A}\right)$$

The leading term, V_0, is about 50 Mev. The coefficient β in the second term is about $\frac{1}{4}$. This is the leading term in the velocity dependence of the nucleon-nucleus interaction; it is frequently represented by an "effective mass." The final term is the symmetry energy; V' is about ± 25 Mev, with the + sign applying for protons and the − sign applying for neutrons.

8. Models for Nuclear Reactions—The Compound Nucleus

The previous sections have considered reactions only in the most general sense, as particles removed from the incident beam. In order to proceed further, it is necessary to formulate more detailed models for what goes on in the reaction. The two types of models widely used are differentiated in accordance with the time over which the interaction takes place. If this time is of the order of the transit time of the incident particle across the target nucleus (a few times 10^{-22} sec), the reaction is referred to as "direct"; if it is appreciably longer than this, it is referred to as "compound nucleus." Various types of direct reactions have been differentiated. By virtue of their simplicity, they are rather easily analyzed in the light of our knowledge of nuclear structure. They will be discussed in some detail in Sec. 11. While these reactions are very interesting for research purposes, they represent a small minority of nuclear reactions. The great majority have the compound-nucleus mechanism.

A compound nucleus may be throught of as arising by virtue of a series of internucleon collisions. Classically speaking, the incident particle enters the nucleus and collides with one of its nucleons, thereby sharing with the struck particle the excess energy it brought in. Each of these particles then has sufficient energy to have collisions with other nucleons, thereby further distributing the available energy. Eventually, this energy is distributed among a relatively large number of nucleons in something of a statistical fashion; the nucleus in this stage is called a "compound nucleus." Eventually, in the course of statistical fluctuations, enough energy is concentrated on a single particle so that it can escape if its momentum and position are favorable.

From a quantum-mechanical shell-model point of view, an entering particle forms a nucleus in a highly excited virtual state. This state consists of a very complicated mixture of configurations which, relative to the target nucleus, may be viewed as (1) single-particle (i.e., target plus particle in a shell-model state), (2) two-particle–one-hole, (3) three-particle–two-hole, etc. configurations. The cross section for it to be captured into the state depends on the amplitude of configuration 1. The probability for a more complex reaction to take place depends on the amplitude of 2, as the system must go through this to reach more complex configurations [13]. States with a large amplitude of configuration 1 occur in the energy region of a shell-model state with the proper value of l; these are known [14] as "giant resonances." An example is shown in Fig. 8.5; for low-energy neutrons, $l = 0$ predominates so that the cross section is large in the region where the $3s_{1/2}$ and $4s_{1/2}$ shell-model states occur at small positive energies. These are near $A = 60$ and $A = 160$, respectively. (The explanation given previously in terms of the optical-model potential can be seen to be consistent with this, as the real part of the optical-model and the shell-model potentials are the same.) States with large amplitudes of configuration 2 are also expected to occur in groups, with spacings of a few hundred kev; they are known as "doorway states." A considerable amount of theoretical investigation has been devoted to them [13], although clear connections with experiment are not overly evident.

The wave function for the compound state we have been discussing may also be decomposed so as to have configurations consisting of various states plus a single particle, and the probability that it will decay with the emission of those particles depends on the amplitudes of those configurations. Thus, a compound nucleus can, in general, decay in many different ways.

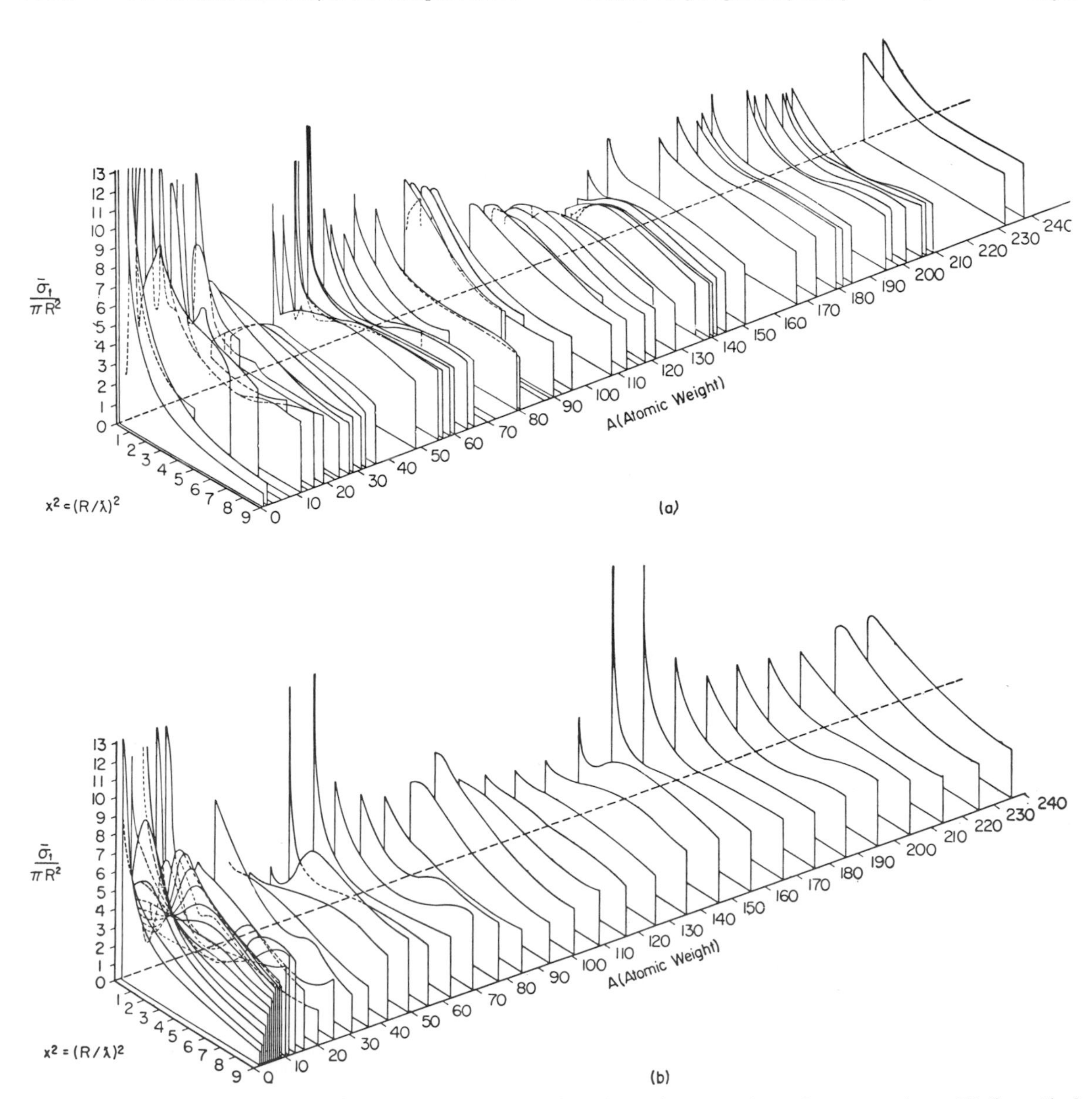

FIG. 8.5. (*b*) Smoothed experimental total neutron cross sections for various energies and mass numbers; (*b*) theoretical total reaction cross sections calculated from the optical model. [*Curves from H. Feshbach, C. E. Porter, and V. F. Weisskopf, Phys. Rev.*, **96**: 448 (1954).]

It is similarly clear that the same compound-nucleus state can be formed in many ways (by bombarding different targets with different particles) and that the mode of decay is independent of the mode of formation.

As an example, a compound state of Al^{27} can decay into

$Al^{26} + n$	$Na^{23} + \alpha$
$Mg^{26} + p$	$O^{16} + B^{11}$
$Al^{27} + \gamma$	$N^{14} + C^{13}$
$Mg^{25} + d$	etc.

In each case, the residual nucleus could be left in various states of excitation (in the last two cases, both emitted particles could be excited). The state of Al^{27} could have been formed by any of these combinations $Al^{26} + n$, $Mg^{26} + p$, etc., and the relative probability of the various modes of decay is independent of which of these ways it was formed.

9. Statistical Theory of Compound-nucleus Decay

It is clear that the decay of a highly excited compound nucleus is an extremely complex problem to work out in detail, so that statistical treatments are frequently employed. The assumption made is that all modes of decay are intrinsically equally likely, so that the probability of various modes is just proportional to the number of ways they can occur.

Consider a compound nucleus C that can decay into final nucleus B plus a light particle b, with a total energy release E_0; that is,

$$C \rightleftarrows B + b$$

This is a quantum-mechanical transition which can proceed in either direction; the relative probabilities for proceeding to the right, $P_{\rightarrow}$, and to the left, $P_{\leftarrow}$, are proportional to the densities of states on the two sides, or

$$\begin{aligned} P_{\rightarrow} &= |M|^2\omega_B(E_0 - E) \cdot \omega_b(E) \\ P_{\leftarrow} &= |M|^2\omega_c \end{aligned} \tag{8.36}$$

where $|M|^2$ is the transition matrix element, and ω is the density of states. But $P_{\leftarrow}$ may be expressed in terms of a cross section σ_b for particle b with energy E to form C by the inverse reaction. If the system is enclosed in a box of volume V, we then have

$$P_{\leftarrow} = \frac{\sigma_b v_b}{V} = \frac{2\sigma_b E^{1/2}}{V m_b^{1/2}} \tag{8.37}$$

The density of states of the light particle b is

$$\omega_b(E) = \frac{4\pi p^2\, dp \cdot V}{\hbar^3} \propto m_b^{3/2} E^{1/2}\, dE \cdot V \tag{8.38}$$

Equating the second of (8.36) with (8.37) and solving for $|M|^2$, and inserting this and (8.38) into the first of (8.36) give

$$P_{\rightarrow} \propto m_b \sigma_b E \omega_b(E_0 - E)\, dE \tag{8.39}$$

This is the expression for the energy distribution of emitted particles. For σ_b, one ordinarily uses the approximations discussed at the end of Sec. 6. Various expressions have been proposed for ω, the density of states in a nucleus as a function of its excitation energy ϵ. Probably the most widely used is that due to Weisskopf [7]

$$\omega(\epsilon) = C \exp\,[2(a\epsilon)^{1/2}] \tag{8.40}$$

The merits of this expression are as follows: In analogy with conventional thermodynamics, a nuclear entropy

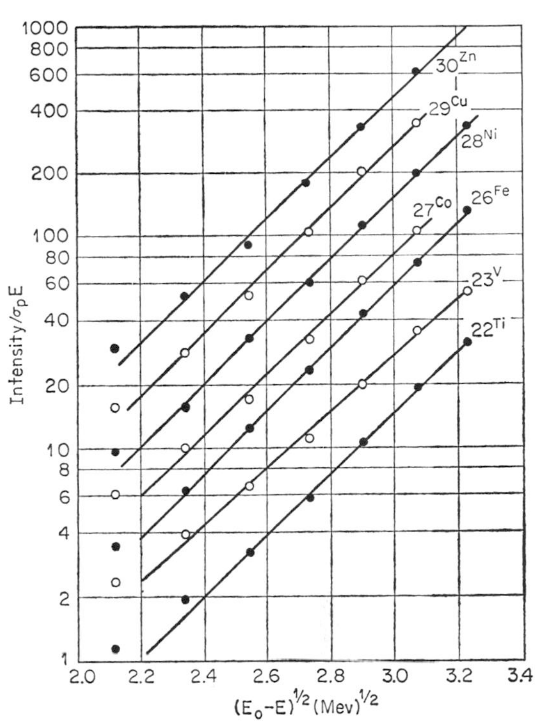

Fig. 8.6. Typical test of the use of the level density formula (8.40) in expression (8.39) for the energy distribution of particles emitted in a compound-nucleus nuclear reaction. Data are for inelastic scattering of 14.5-Mev protons from B. L. Cohen and A. G. Rubin, *Phys. Rev.*, **113**: 579 (1959). Intensities of neutrons emitted at 90° are divided by the energy and the proton reaction cross section as obtained from ref. 8 and plotted on a logarithmic scale vs. $(E_0 - E)^{1/2}$, where E_0 is the maximum proton energy which in this case is 14.5 Mev. According to (8.39) and (8.40) this should give straight lines. The lines through the data points correspond to a [in Eq. (8.40)] $\simeq 2.7$ Mev^{-1}

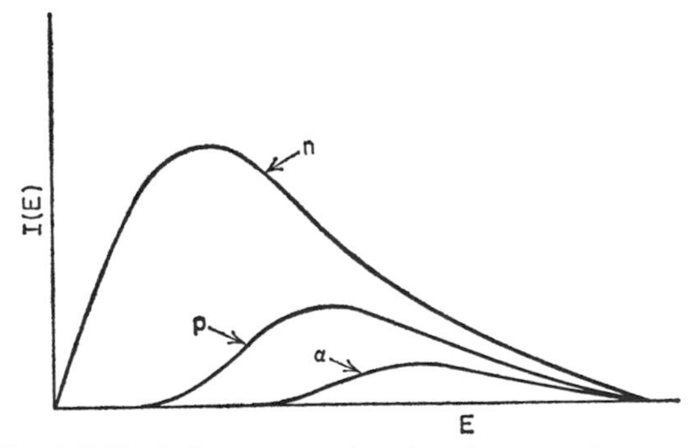

Fig. 8.7. Typical energy spectra of neutrons, protons, and alpha particles emitted from compound-nucleus reactions, as given by formula (8.42).

$S(\epsilon)$ and a nuclear temperature $T(\epsilon)$ may be defined as

$$S = \log \omega$$
$$\frac{1}{T} = \frac{\partial S}{\partial E} = \frac{\partial \log \omega}{\partial \epsilon} \tag{8.41}$$

whence, from (8.40)

$$\epsilon = aT^2$$

the well-known expression for a Fermi gas. Since a nucleus is a highly degenerate Fermi system, one expects this type of relationship.

Equation (8.39) gives the energy distribution of particles emitted from a nuclear reaction proceeding via a compound-nucleus interaction. Measurements of these energy distributions have been carried out by many groups, and they are analyzed to determine ω. Many groups have found good agreement with (8.40) and have determined the parameter a; an example is shown in Fig. 8.6. However, different experiments do not always give the same values. Some have interpreted experiments as casting doubt on the validity of (8.40). Where it is used, typical values of the parameter a are $\sim$3 Mev^{-1} for $A = 50$ ranging up to $\sim$20 Mev^{-1} for $A = 200$.

An interesting approximation for (8.39) is obtained by expanding $\log \omega(E_0 - \mathbf{E})$ about E_0 and using (8.41) as

$$\log \omega(E_0 - E)$$
$$= \log \omega(E_0) + \left[\frac{\partial \log \omega}{\partial(E_0 - E)}\right]_{E_0} E + \cdots$$
$$= \text{const} - \frac{E}{T(E_0)}$$

Inserting this into (8.39),

$$P_{\rightarrow} \propto \sigma_b E \exp(-E/T)\, dE \tag{8.42}$$

If b is a neutron, this is similar to the equation for a Maxwell distribution. It is shown schematically in Fig. 8.7. If b is a proton or an alpha particle, (8.42) is a Maxwell distribution multiplied by a Coulomb barrier penetration which cuts out the low-energy part, as shown by the other curves in Fig. 8.7. The relative probability for emission of neutrons, protons, and alpha particles is just proportional to the areas under these curves. It is qualitatively clear that neutron emission is generally predominant. This is because the density-of-states function ω weights the energy distribution of emitted particles heavily toward low energies, and low-energy charged-particle emission is strongly impeded by the Coulomb barrier.

Actually, this treatment is rather crude in that it ignores differences in E_0 for the emission of various particles due to differences in Q values. It is clearly more correct to define

$$F_b = m_b \int_0^{E_{0-b}} \sigma_b E\, \omega_b(E_{0-b} - E)\, dE \tag{8.43}$$

where E_{0-b} is the maximum energy available for emission of b, whence the probability of emission of any particular particle k is P_k, where

$$P_k = \frac{F_k}{F_n + F_p + F_\alpha} \tag{8.44}$$

Ordinarily, only neutrons, protons, and alpha particles need be considered, as other nuclei have much smaller values of E_0 and hence much smaller values of F_k. Plots of F_n, F_p, and F_α for particular choices of the parameter a are given in ref. 7.

Combining (8.44) with cross sections for formation of the compound nucleus (which is approximately equal to σ_r) from the latter part of Sec. 6 gives complete

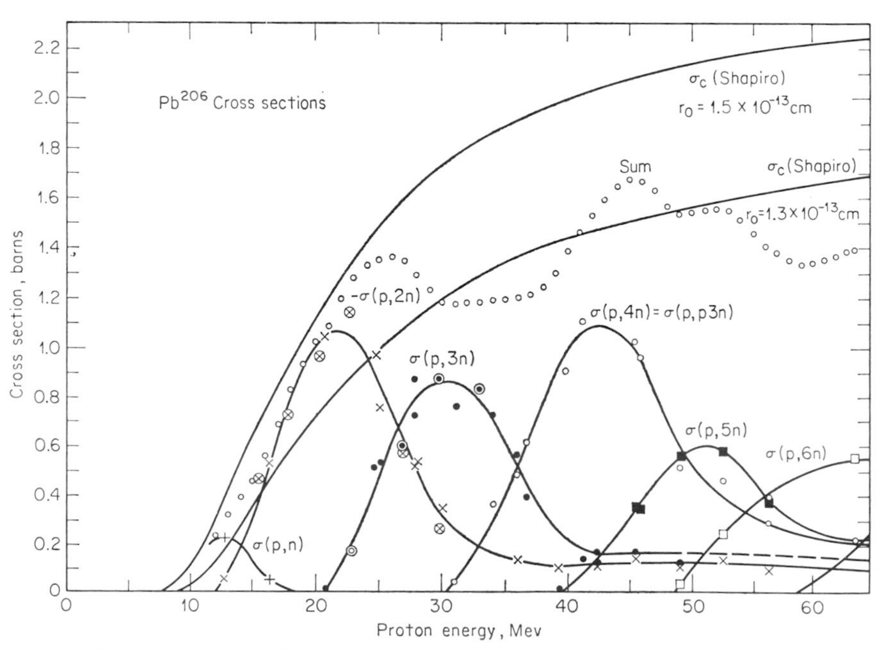

FIG. 8.8. Measured (p,xn) cross sections of Pb^{206} as a function of proton energy, from data of R. E. Bell and H. M. Skarsgard, *Can. J. Phys.*, **34**: 745 (1956). Curves labeled σ_c are total reaction cross sections for protons as obtained from ref. 8.

expressions for cross sections. For example, for an (n,p) cross section

$$\sigma(n,p) = \pi(R + \lambda\!\!\!^{-})^2 \frac{F_p}{F_n + F_p + F_\alpha}$$

If, after a particle is emitted from a nuclear reaction, there is still sufficient energy for emission of a second particle, a second particle is usually emitted, since particle emission is intrinsically a much more rapid process than its chief competitor, electromagnetic de-excitation (gamma-ray emission). Neutron emission usually becomes even more predominant in this case, since the available energy is lower. Thus, when, for example, 20-Mev protons are incident on a heavy nucleus, the predominant reaction is $(p,2n)$. A (p,pn) reaction would ordinarily have a much lower cross section, unless the energetics are much more favorable for proton emission than for neutron emission.

As the available energy is increased, emission of a third, then a fourth, etc., particle becomes likely. An example of this behavior is shown in Fig. 8.8. When a great number of particles are emitted, the process is referred to as "spallation."

10. Compound Nucleus—Resonances

The discussion since Sec. 8 has been based on the assumption that the probability for a target nucleus to absorb or scatter an incident particle varies smoothly with energy; this is valid in what is known as the "continuum region." Actually, the compound nucleus is a quantum-mechanical system so that it can be excited only into its discrete energy levels. If one considers the variation of the cross section for a nuclear reaction or scattering with bombarding energy, one therefore expects maxima at various energies E_R, corresponding to these levels. These maxima in the cross section are called "resonances."

The energy levels of the compound nucleus are not infinitely sharp since they can decay; indeed, they have a width Γ that is related to their lifetime before decay, τ, by the well-known relationship

$$\Gamma = h/\tau \tag{8.45}$$

The resonances in the cross section are of this width.

From (8.45), the width Γ is proportional to the probability per unit time for the state to decay. This is just equal to the sum of the probabilities per unit time for decay by various modes so that

$$\Gamma = \Gamma_1 + \Gamma_2 + \cdots \tag{8.46}$$

where Γ_i are the "partial" widths for decay by these modes. For example, Γ_1 might be neutron emission leaving the residual nucleus in the ground state—called Γ_n; Γ_2 might be decay by gamma-ray emission—called $\Gamma\gamma$; etc. As the available energy increases, many new modes of decay become possible, so that in accordance with (8.46) the total width Γ becomes larger.

The continuum region that we have been discussing is where the widths are larger than the distance between resonances; this causes the cross section to be a smooth function of energy. According to the last paragraph, this happens at higher energies. The continuum assumption is valid for bombarding energies above a few hundred kev in heavy nuclei, but in light nuclei, well-separated resonances occur up to 10 Mev or higher. Examples are shown in Figs. 8.9 and 8.10.

The cross section in the region of an isolated resonance is very well described by the Breit-Wigner formula

$$\sigma(a,b) = \pi\lambda\!\!\!^{-2}(2l + 1) \frac{\Gamma_a\Gamma_b}{(E - E_R)^2 + \Gamma^2/4} \tag{8.47}$$

where a and b are the incident and emitted particles

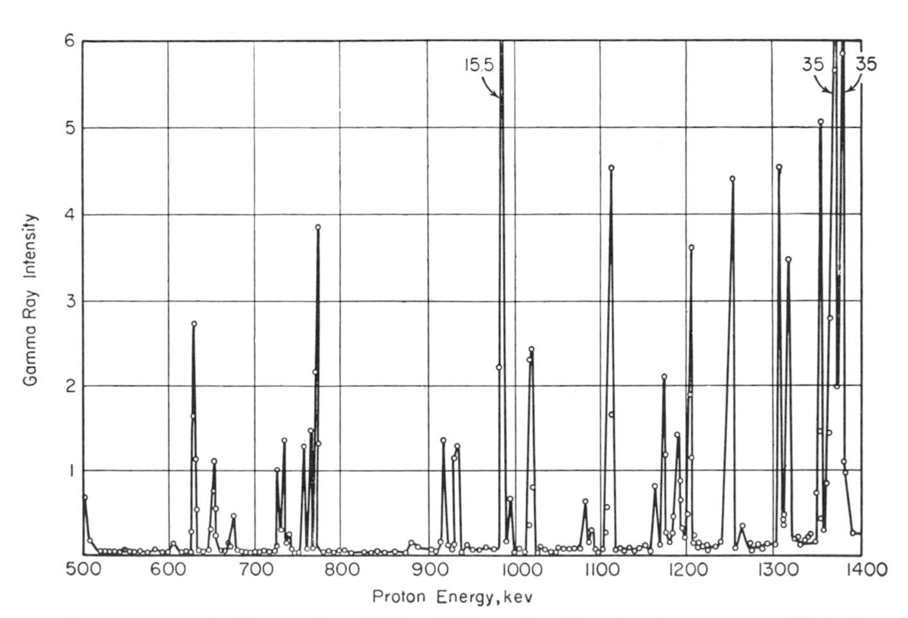

Fig. 8.9. Cross section vs. energy for the reaction $Al^{27}(p,\gamma)Si^{28}$ as a function of proton energy. [*Data from K. J. Brostrom, T. Huss, and R. Tange, Phys. Rev.,* **71**: 661 (1947).]

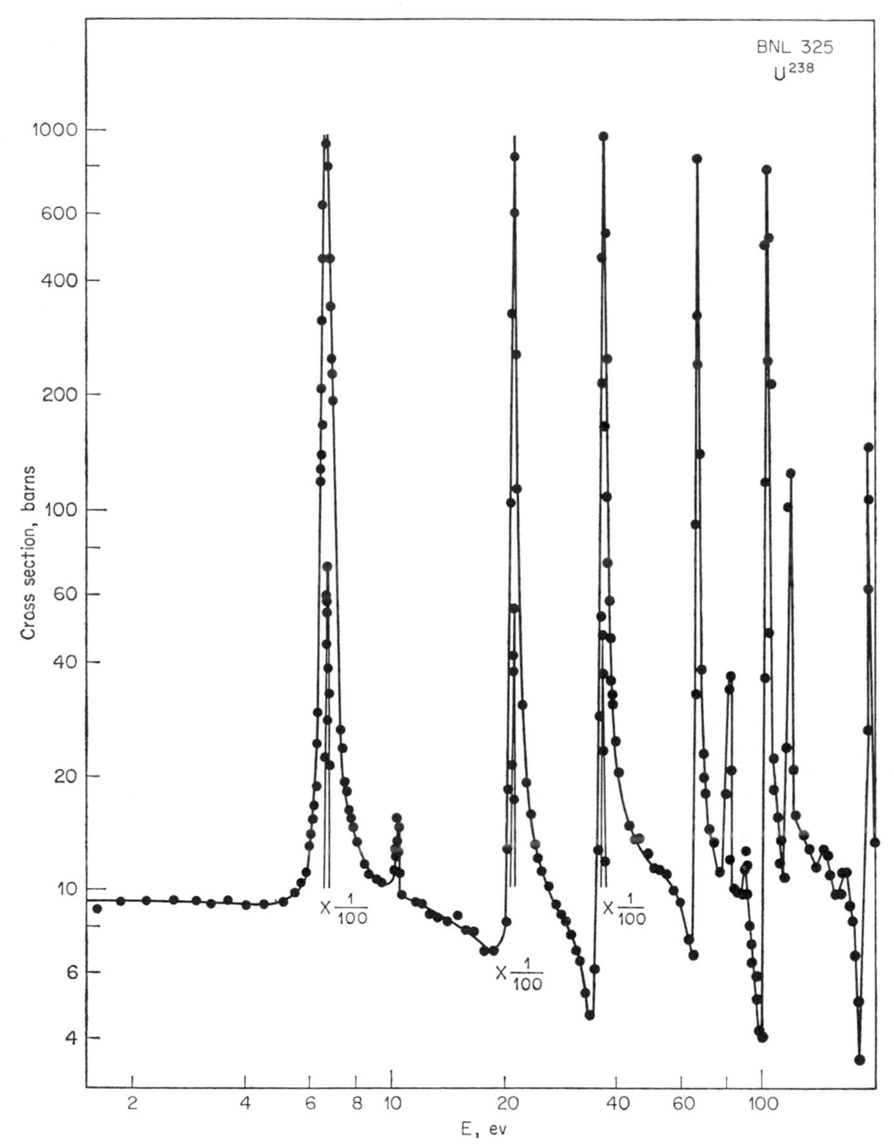

FIG. 8.10. Total neutron cross section vs. energy for U^{238}. [*Data from Brookhaven Natl. Lab. Rept.* 325, *a compilation of such data.*)

and spin degeneracy factors have been ignored. A more complete formula is

$$\sigma(a,b) = \pi \lambda\!\!\!^{-2} \frac{2I_f + 1}{(2I_i + 1)(2S + 1)} \sum_{ll'} \frac{\Gamma_{al}\Gamma_{bl'}}{(E - E_R)^2 + \Gamma^2/4} \tag{8.47'}$$

where S is the spin of a, and the sum extends over all values of l and l' consistent with conservation of angular momentum and parity. Derivation of the Breit-Wigner formula requires rather lengthy and intricate reasoning (satisfactory derivations [15] were not available until a decade after the formula was proposed) so that it will not be given here. (The simplest derivation is that of Feshbach, Peaslee, and Weisskopf given in ref. 6.) One may readily check that (8.47) gives a maximum at $E = E_R$ with width Γ, as expected.

For elastic scattering, account must be taken of potential scattering. The formula, ignoring spins, is then

$$\sigma(a,a) = (2l + 1)^2\pi\lambda\!\!\!^{-} \left| \frac{i\Gamma_a}{E - E_R + \frac{1}{2}i\Gamma} + A_{pot}{}^l \right|^2 + \text{nonresonant terms} \tag{8.48}$$

where A_{pot} is the amplitude for potential scattering. The expression (8.48) contains interference between the resonance and potential scattering, which can give a curve of various shapes. Particularly familiar is the case where $l = 0$, for which

$$A^0{}_{pot} = \exp\,(2iR/\lambda\!\!\!^{-}) - 1$$

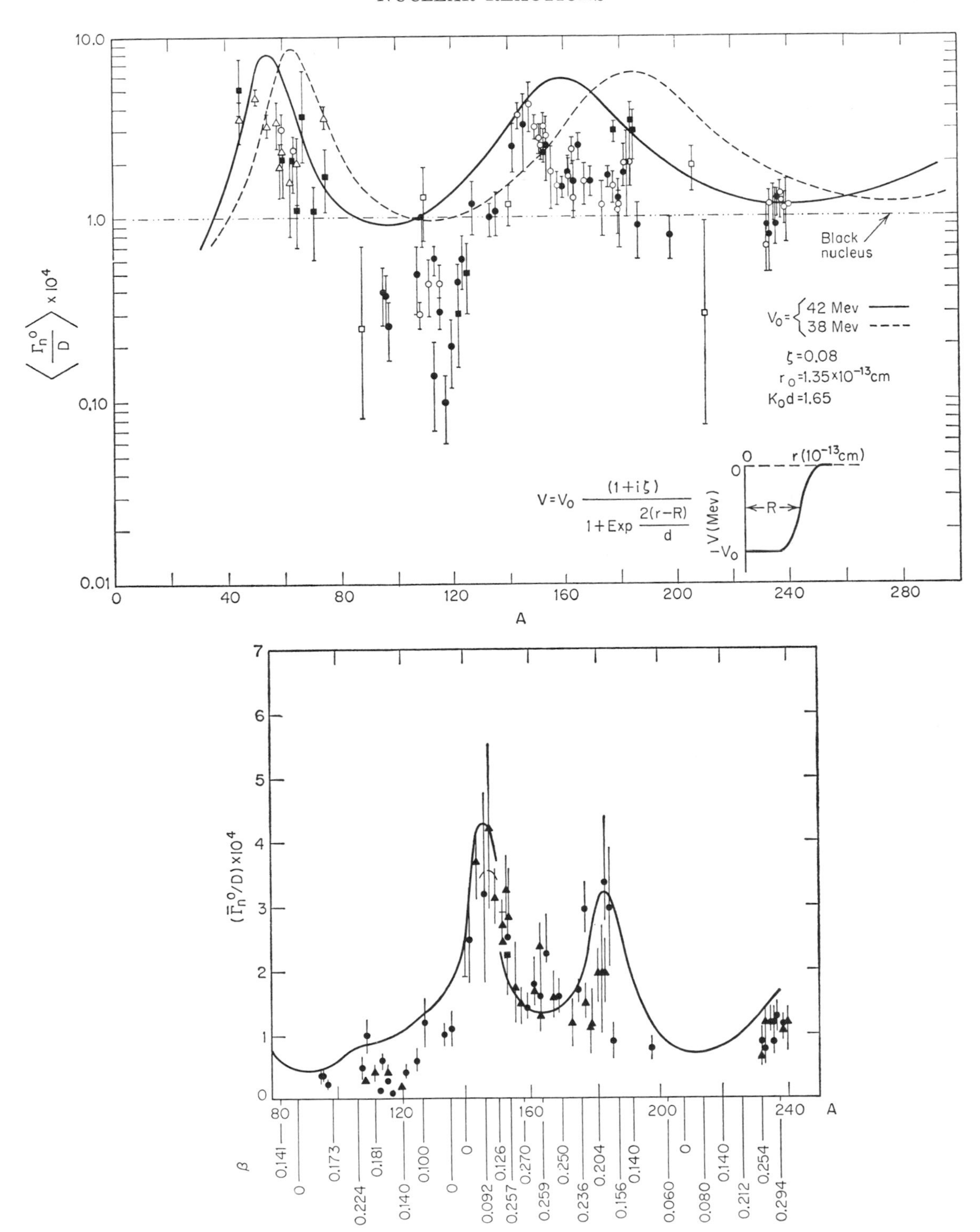

FIG. 8.11. Neutron strength function vs. mass number as obtained from neutron cross-section data. Upper figure shows optical-model fits, using the potential well shown at the lower right; it is from E. Vogt, *Brookhaven Natl. Lab. Rept.* 331. Lower figure is a more complete and detailed representation of the data in the region of deformed nuclei, and the curves show a fit as obtained for a deformed optical model by D. M. Chase, L. Wilets, and A. R. Edmonds, *Phys. Rev.*, **110**: 1091 (1958). The values of the deformation parameter β used are shown at the bottom.

and (8.48) gives a dip resulting from destructive interference on the low-energy side and constructive interference on the high-energy side of the resonance. An example is shown in Fig. 8.10.

Expressions for elastic-scattering differential cross sections as a function of angle of the emitted particle are available in the literature [16] for charged particles as well as neutrons. The shapes of the resonances at various angles are sufficiently characteristic that one can easily determine the l value of the entering particle from them.

Expressions for Γ include three factors:

1. A reflection factor, $4k/K$, where k and K are the wave numbers of the particle outside and inside the nucleus, respectively. A particle leaving the nucleus encounters a sharp change in potential at the surface which, from wave-optical considerations, frequently causes a reflection back into the nuclear interior, thus lengthening the lifetime and decreasing the width by this factor. This factor gives an explicit energy dependence to Γ as

$$\Gamma = \Gamma_0 E^{1/2} \tag{8.49}$$

2. A barrier-penetration factor v, which takes into account the fact that, if the outgoing particle must penetrate an angular momentum and/or Coulomb barrier, the probability that it will be reflected back into the interior is increased so that the width is reduced.

3. A factor γ^2 which is proportional to the square of the amplitude of the configuration "residual nucleus plus single particle" in the wave function of the compound state. If this configuration is the entire wave function, γ^2 is close to the "Wigner limit" $\hbar^2/MR^2$ (M is the mass of the particle). If this configuration is split up among many nuclear levels, the sum of the γ^2 for all these levels is close to the Wigner limit.

Perhaps the most widely studied resonances are those for low-energy neutrons bombarding heavy nuclei. The low energy requires $l = 0$ and restricts the modes of decay to elastic scattering and gamma-ray emission so that $\Gamma = \Gamma_n + \Gamma_\gamma$. Formula (8.47) then reduces to

$$\sigma(n,n) = \pi\lambda\!\!\!^{-2} \frac{\Gamma_n^2}{(E - E_R)^2 + \frac{1}{4}\Gamma^2} \tag{8.50}$$

$$\sigma(n,\gamma) = \pi\lambda\!\!\!^{-2} \frac{\Gamma_n\Gamma_\gamma}{(E - E_R)^2 + \frac{1}{4}\Gamma^2} \tag{8.51}$$

whence

$$\sigma_{\text{total}} = \sigma(n,n) + \sigma(n,\gamma)$$

$$= \pi\lambda\!\!\!^{-2} \frac{\Gamma_n\Gamma}{(E - E_R)^2 + \frac{1}{4}\Gamma^2} \tag{8.52}$$

Since the total cross sections are very easy to measure (by measuring transmission probabilities), vast amounts of data are available [17] for fitting to (8.52). These fits give Γ from the observed width of the peak, Γ_n from the maximum value of the cross section, and Γ_γ from the difference between them. By locating resonances over a broad energy range, one can determine D, the average spacing between resonances. The average value of the ratio Γ_{n0}/D, known as the "strength function," may then be determined [Γ_{n0} is $\Gamma/E^{1/2}$; cf. (8.49)]. The strength function is of considerable interest since it is proportional to the reaction cross section which can be compared with predictions from the optical model. A plot of experimentally determined strength functions and its comparison with an optical-model calculation [18] are shown in Fig. 8.11.

At neutron energies of the order of 1 ev or below, Γ_n becomes very small as a result of (8.49) so that (n,γ) reactions become predominant [gamma-ray emission is available with about 8 Mev energy release, so that Γ_γ is unaffected by (8.49) in this region]. From (8.49) and (8.51) with $\lambda\!\!\!^- = \hbar/(2ME)^{1/2}$, the cross section in this region is seen to be proportional to $1/v$. The absolute cross section in this region depends on Γ_n and E_R for the nearest resonance. The latter quantities vary over a wide range, so that cross sections for thermal neutrons (0.025 ev) vary from millibarns to megabarns. An example of an especially large one is that in Cd^{113}; a plot of its cross section vs. neutron energy, reproduced in Fig. 8.12, shows that this is due to a very strong resonance very close by.

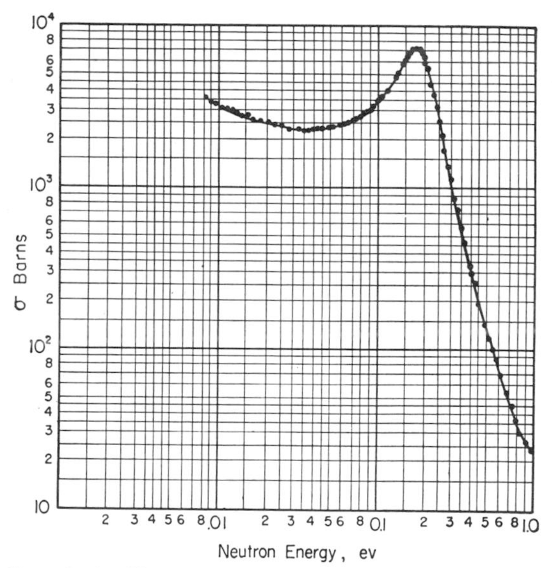

FIG. 8.12. The 0.176-ev resonance for the reaction $Cd^{113}(n,\gamma)Cd^{114}$. Data are from the report referred to in the legend for Fig. 8.10. Curve is fitted with Eq. (8.47), with Γ_n from (8.49).

In a very few special cases, reactions other than elastic scattering and (n,γ) can take place at thermal energies. For naturally occurring target materials, these are

$$Li^6(n,\alpha)H^3$$
$$B^{10}(n,\alpha)Li^7$$
$$N^{14}(n,p)C^{14}$$
$$U^{235}(n,\text{fission})$$

The first of these is extremely important as a source of H^3 (tritium) which is a key material in thermonuclear energy sources, both destructive and constructive. Both of the first two reactions are commonly used for detecting neutrons; especially common is the BF_3 (boron trifluoride, a gas) proportional counter which uses the second. The third reaction is induced frequently by cosmic-ray neutrons in the nitrogen of the atmosphere; it gives appreciable yields of the long-lived radioactive isotope C^{14} which is taken up by living organisms and can then be used for determining the age of objects made from them. The fourth reaction in the above list plays the key role in atomic energy; without it, there could be no nuclear reactors or atomic bombs.

11. Direct Reactions

Let us now depart from our discussion of compound-nucleus reactions and return to the other type of reaction mentioned in Sec. 8, direct reactions. These occur more frequently as the bombarding energy is raised. The ratio of direct to compound-nucleus reactions increases from about 5 per cent at 10 Mev to about 25 per cent at 30 Mev for incident neutrons or protons. At energies above about 50 Mev, the situation becomes confused because the lifetime of the compound nucleus is so short as to be comparable with the transit time of the incident particle, so that the distinction between the two processes as given in Sec. 8 breaks down. Furthermore, at these energies mean free paths become large and wavelengths become small compared with the nuclear radius, so that reactions are better described as single collisions between nucleons. We shall confine our discussion here to the lower-energy region of roughly 10 to 50 Mev.

To obtain detailed information from direct-reaction studies, it is necessary to formulate detailed models. The most frequently applicable model is a "transfer" process in which some nucleons are transferred between the target nucleus and the incident particle as it passes. Important examples of this are the deuteron "stripping" and "pick-up" reactions, (d,p) and (p,d), respectively, in which a neutron is transferred. Other widely studied reactions in which a single nucleon is transferred are (d,n), (d,t), (He_3,α), (C^{12},C^{11}), (N^{14},C^{13}), etc. Widely studied reactions in which two particles are transferred are (p,t), (d,α), (N^{14},C^{12}), etc. Three-particle transfers such as (p,α), (N^{14},C^{11}), etc.; four-particle transfers such as (d,Li^6), (α,Be^8), (O^{16},C^{12}), etc.; and five-particle transfers such as (d,Li^7) and (d,Be^7) have also been investigated. In heavy ion reactions transfer of much larger subunits have been observed.

In inelastic scattering the dominant mechanism is an interaction with the nuclear surface to set it into vibration or rotation. This leads to strong excitation of collective states, as can be seen in Fig. 8.13. On the other hand, noncollective states are also excited with appreciable strength in these reactions, so that less coherent combinations of nucleon-nucleon interactions must play a role. Inelastic scattering has been widely studied with protons, neutrons, deuterons, alpha particles, and several heavier nuclei as projectiles.

In (p,n) reactions, the dominant mechanism is a charge exchange with one of the nucleons in the nucleus; an example is shown in Fig. 8.14. In the process, a neutron in the nucleus is changed into a proton without changing its shell-model state. The states generated in this process are known as "isobaric analogue states" of the target nucleus; they will be discussed extensively below, and there will be further discussion of (p,n) direct reactions in this regard. Isobaric analogue states are also strongly excited in

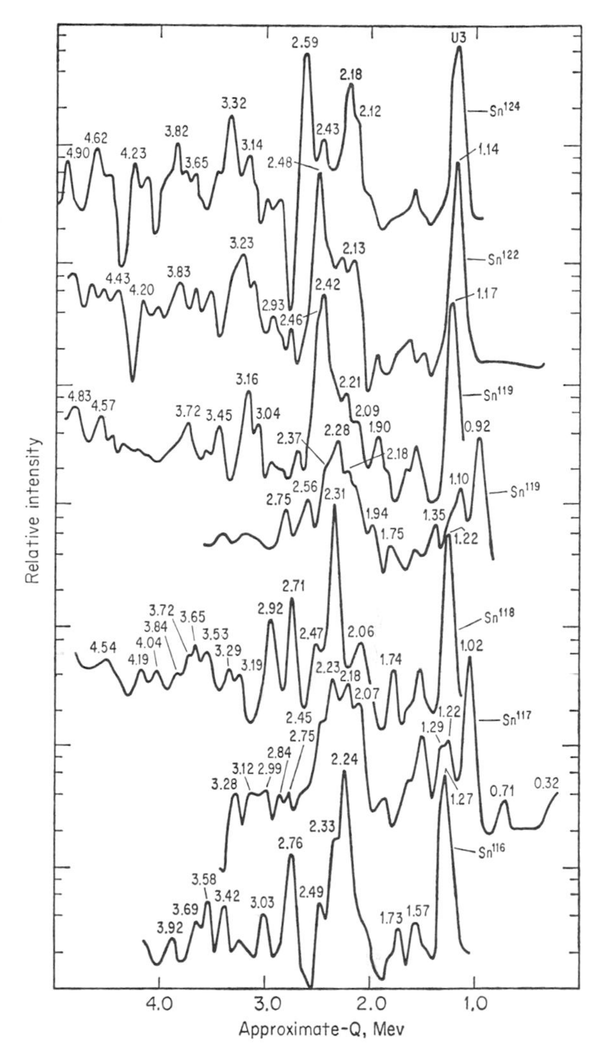

FIG. 8.13. Energy spectra of deuterons inelastically scattered from various isotopes of tin. Incident particles were 15-Mev deuterons, and emitted particles (at 60°) are of energy less than this by the amount shown on the abscissa. Figures are excitation energies of levels excited in Mev. Peaks near 1.5 Mev are due to elastic scattering from oxygen impurity. Data are from B. L. Cohen and R. E. Price, *Phys. Rev.*, **123**: 283 (1961). Note the strongly excited 2^+ and 3^- collective states at about 1.2 and 2.4 Mev, respectively.

(He_3,t) reactions. Other states in these reactions are excited rather weakly, presumably as a result of incoherent combinations of nucleon-nucleon collisions.

One process which had been expected and is often discussed, but for which there is no supporting experimental evidence in this energy region, is "knock-out," in which the entire reaction consists of a single collision in which the struck particle is knocked out of the nucleus. Such reactions are the dominant mechanism at very high energies.

Since the incident particle in a direct reaction brings in a great deal of momentum and the interactions are relatively simple, the particle emitted is very likely

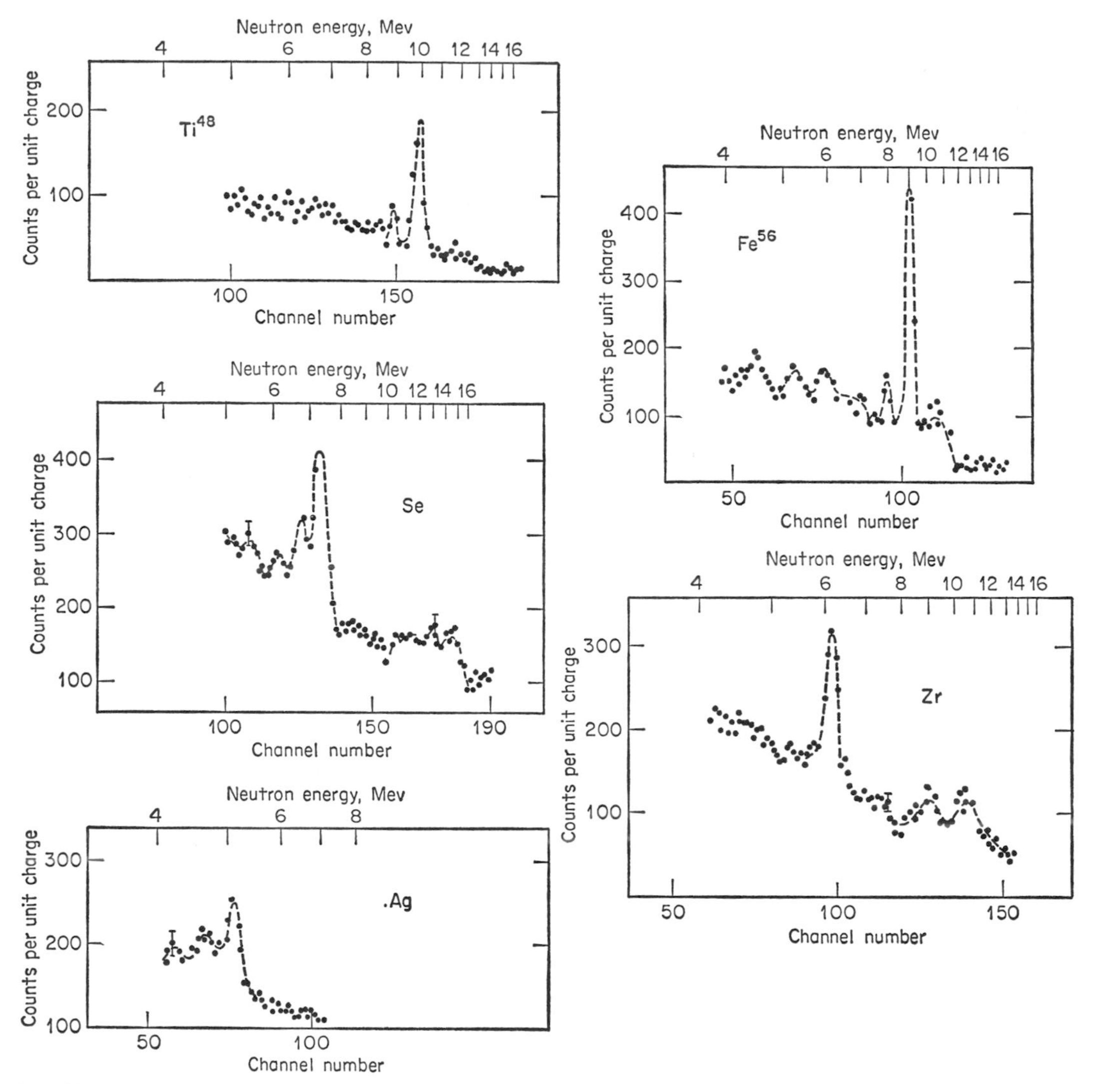

FIG. 8.14. Energy distribution of neutrons from (p,n) reactions induced by 18.0-Mev protons; emission angle 23°. Neutron energies are measured by their time of flight; this explains the nonlinear energy scale at the top. Data are from J. D. Anderson, C. Wong, and J. W. McClure, *Phys. Rev.*, **129**: 2718 (1963).

to be forward directed, so that the resulting angular distributions are peaked at small angles. This immediately distinguishes direct reactions from compound-nucleus reactions, as the latter have rather isotropic angular distributions. However, it frequently happens that angular-momentum conservation inhibits emission of products of direct reactions at small angles. Consider, for example, a situation in which the momenta of the incident and outgoing particles, P_i and P_c, are about equal. The momentum transferred to the nucleus, P_t, is then the vector difference between these, or

$$\begin{aligned} P_t^2 &= P_i^2 + P_o^2 - 2P_iP_o \cos\theta \\ &= 2P_i(1 - \cos\theta) \end{aligned}$$

The maximum orbital angular momentum that can be transferred is then P_t times the nuclear radius. Calling the transferred orbital angular momentum $\hbar l$, we thus have

$$\begin{aligned} l &\lesssim \frac{P_tR}{\hbar} \\ &\lesssim \frac{P_iR}{\hbar}[2(1 - \cos)]^{1/2} \\ &\lesssim \frac{P_i}{\hbar}R\theta \end{aligned} \tag{8.53}$$

Thus, for a given angular-momentum transfer l, there is classically a minimum angle θ given by (8.53) at which the reaction product can be emitted.

The initial and final nuclei are in definite quantum states with definite angular momenta, I_i and I_f, and parity. Moreover, a direct reaction does not generally flip spins, so that the difference between I_i and I_f must be made up of l plus the spin of any transferred particles. When the target nucleus is even-

even and therefore of 0^+ spin parity, the conservation laws usually limit l to a single value. For example, if no particles are transferred (as in inelastic scattering), or if the transferred particle has spin parity 0^+ [as in a (t,p) reaction], the conservation laws require

$$l = I_f \qquad \pi_f = (-1)^l \tag{8.54a}$$

(where π_f is the parity of the final nucleus) so that only a single l is involved. Moreover, only states with

$$\pi_f = (-1)^{I_f}$$

can be expected. These are the so-called "natural parity" states, 0^+, 1^-, 2^+, 3^-, ... Unnatural-parity states can be excited only to the extent that the interactions can flip a spin. This is typically two orders of magnitude less probable than a non-spin-flip interaction.

When the target nucleus for a single nucleon transfer reaction is even-even, the conservation laws require

$$l = I_f \pm \tfrac{1}{2} \qquad \pi_f = (-1)^l \tag{8.54b}$$

which again limits l to a single value, although here all states can be excited.

From (8.53) the angular distribution of the emitted particle should fall off at angles below θ_0 given by

$$\theta_0 \simeq \frac{\hbar l}{P_i R}$$

Combining this with the fact that angular distributions in direct reactions are on general principles forward directed, we expect the first maximum to occur at about the angle θ_0. We see that θ_0 increases with increasing l and is zero only for $l = 0$.

These arguments suggest that measurements of angular distribution could be rather easily analyzed to give values of l which can then be used to determine spins and parities of nuclear states and for other nuclear structure purposes. A great deal of attention has therefore been paid to angular distributions in direct interactions. The usual approach is to use the Born approximation; a simplified derivation using the plane-wave Born approximation (PWBA) follows:

In PWBA, the transition matrix element T_{if} is

$$T_{if} = \int \psi^*{}_f e^{-i\mathbf{k}_o \cdot \mathbf{r}_o} V \psi_i e^{i\mathbf{k}_i \cdot \mathbf{r}_i}\, d\tau \tag{8.55}$$

where ψ_i and ψ_f are wave functions for the initial and

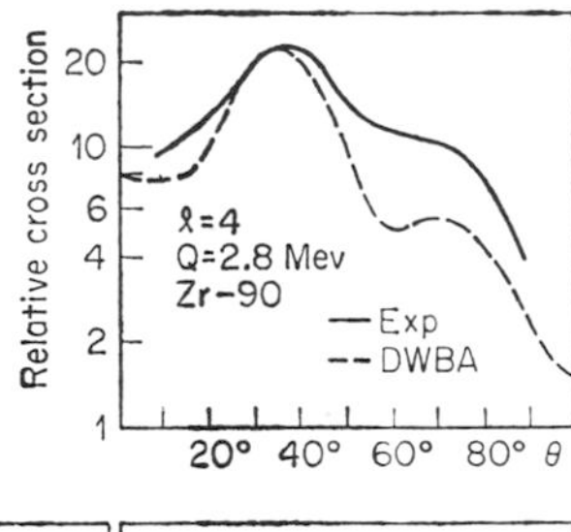

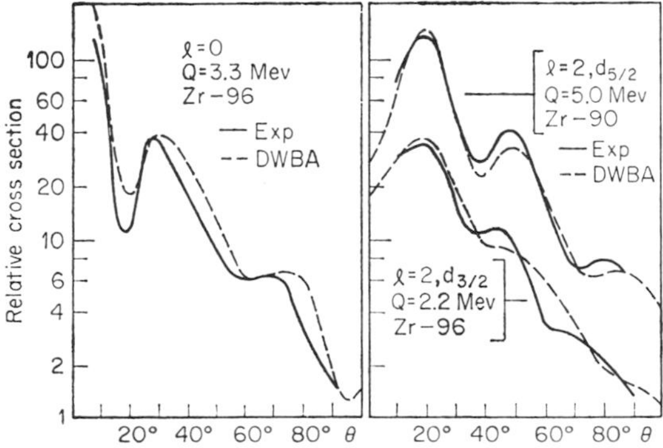

FIG. 8.15. Angular distributions of protons from (d,p) reactions on isotopes of Zr. Solid lines are curves through the data, and dashed lines are the results of distorted-wave Born-approximation calculations. [*Data from B. L. Cohen and O. Chubinsky, Phys. Rev.*, **131**: 2184 (1963).]

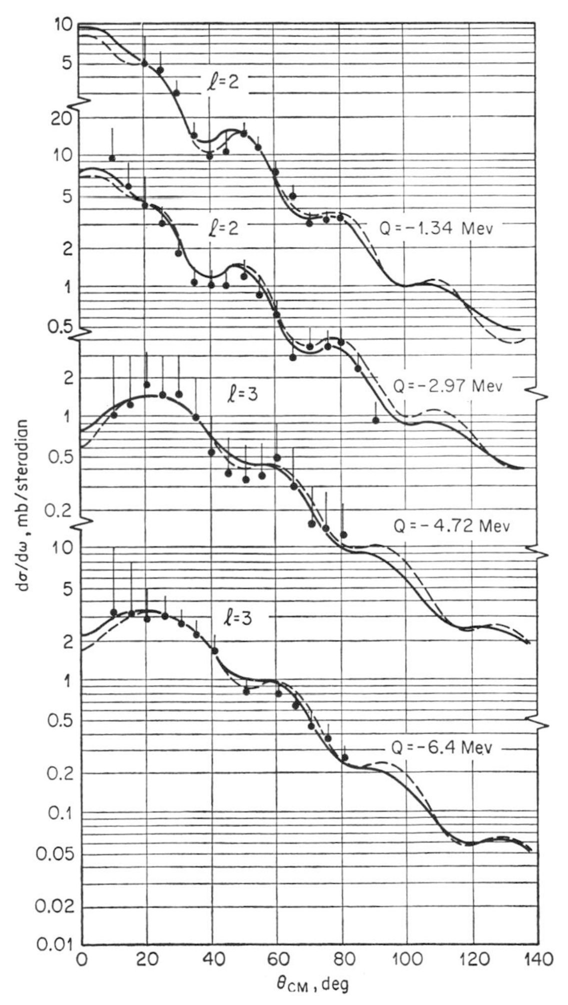

FIG. 8.16. Angular distributions of various proton groups from inelastic scattering on Fe^{54}. Incident energy was 40 Mev, and different curves are groups corresponding to the residual nucleus being left in states of excitation given by Q values shown. Curves are fits with distorted-wave Born approximation. Data are from M. Fricke and G. R. Satchler *Phys. Rev.*, **139**, B567 (1965).

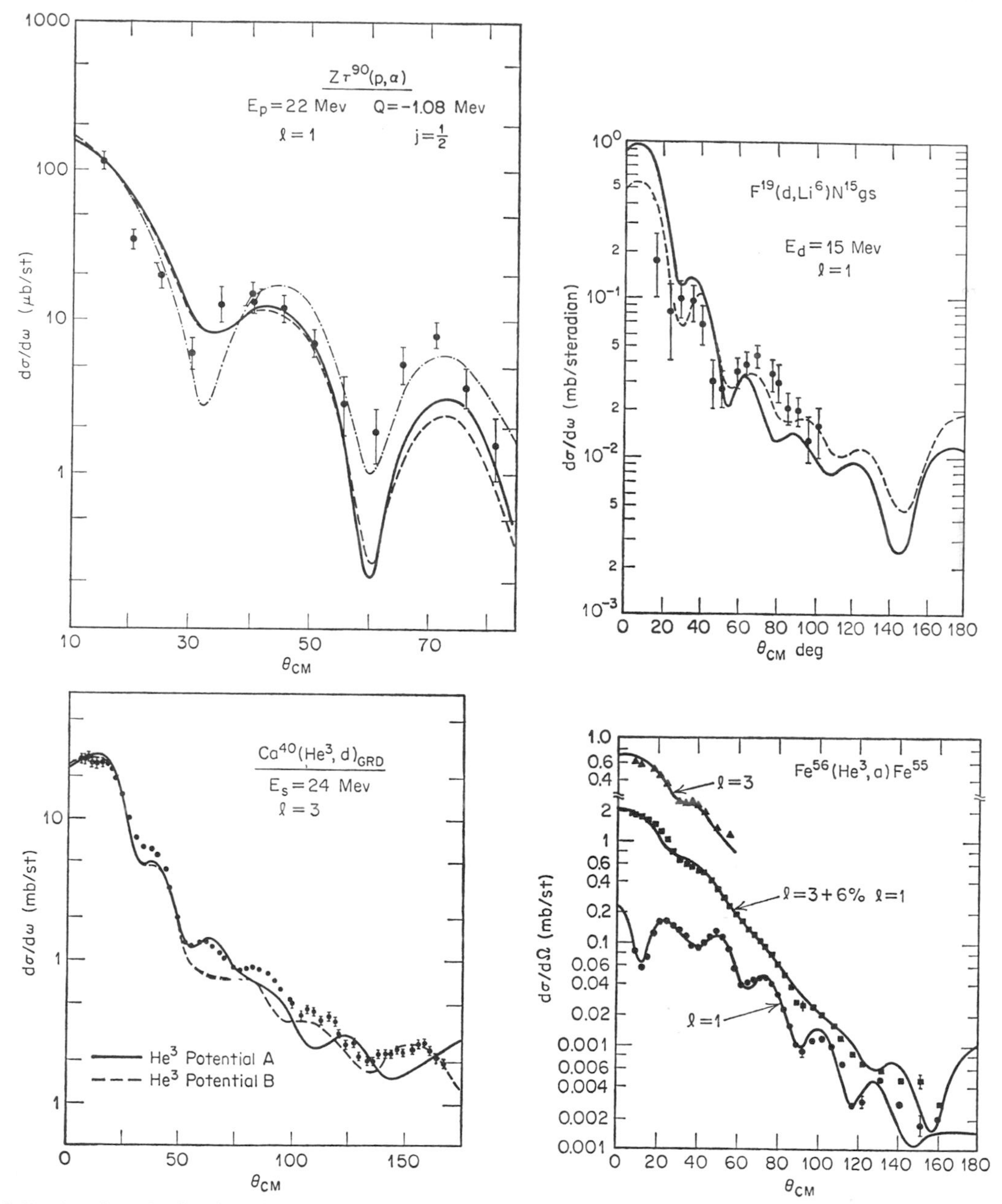

FIG. 8.17. Angular distributions from various types of direct interactions. Curves are fits to these by distorted-wave Born-approximation calculations. (*Figures from G. R. Satchler, Argonne Natl. Lab. Rept.* 6878, 1964.)

final nuclei, including internal wave functions of the incident and outgoing particles; $\mathbf{k}_i$ and $\mathbf{k}_o$ are the wave numbers of the incoming and outgoing particles; and $\mathbf{r}_i$ and $\mathbf{r}_o$ are their position vectors. For simplicity, we take a δ-function interaction acting only at the nuclear surface

$$V = V_0 \delta(\mathbf{r}_i - \mathbf{r}_o)\delta(r_i - R)$$

which gives

$$T_{if} = V_0 \int e^{i(\mathbf{q}\cdot\mathbf{R})} \psi^*_f \psi_i \, d\tau$$

where $\mathbf{q} = \mathbf{k}_i - \mathbf{k}_o = P_t/\hbar$ in our previous notation.

Using the expansion

$$e^{i(q\cdot R)} = \sum_{l=0}^{\infty} i^l \sqrt{4\pi(2l+1)}\, j_l(qR) Y_{l,0}(\theta)$$

and carrying out the integrations,

$$T_{if} = \text{const} \times j_l(qR) \tag{8.56}$$

Thus, the angular distribution is, in this approximation, the square of the spherical Bessel function $j_l(qR)$.

Great improvements have been made in this treatment by use of the "distorted-wave Born approximation," in which the plane waves of (8.55) are replaced by waves describing elastic scattering by the optical-model potential; i.e., the incident particle is being elastically scattered up to the instant of interaction, and the outgoing particle is being elastically scattered thereafter. The surface-interaction assumption is dropped, so that the integrals over r_i and r_o are extended from 0 to ∞ (although sometimes a cutoff is used to avoid the uncertainties of integration through the nuclear volume). In transfer reactions, the results are somewhat sensitive to the accuracy of the wave functions of the transferred particles; after several years of rather poor approximations, this problem seems to be well in hand. In some calculations, the "zero range approximation," i.e., use of a δ-function interaction, has been dropped, but this generally does not change things appreciably. Nonlocal potentials have been used, and their effects studied. The sensitivity to optical-model parameters (within the uncertainties allowed by fits to elastic scattering experiments) have been investigated. As a result, experimental work with direct-interaction angular distributions is by now a rather quantitative technique for studying nuclear structure [19]. DWBA fits to various direct-interaction angular distributions are shown in Figs. 8.15 to 8.17.

This is now probably the most active field of research in nuclear reactions, so that this problem will be dealt with in some detail in the next section, using (d,p) and (p,d) reactions on even-even target nuclei as an example.

12. Nuclear-structure Information from Stripping Reactions

In accordance with the Born approximation, the transition matrix element for a (p,d) reaction is

$$T_{if} = \int \psi^*_f \phi^*_d \chi^*_d V \psi_i \chi_p \, d\tau \qquad (8.57)$$

where ψ_i and ψ_f are the wave functions for the initial and final nuclei, ϕ_d is the internal wave function for the deuteron which is well known, and χ_p and χ_d are the wave functions for the proton and the center of mass of the deuteron as obtained from the optical model. We expand ψ_i as

$$\psi_i = \psi'_i \phi_n{}^l + \sum_{l' \neq l} \psi'_i(l') \phi_n{}^{l'}$$

where l, n are the quantum numbers in the shell model for the neutron which is to be picked up. In the zero-range approximation, (8.57) becomes

$$T_{if} = V_0 \int \psi^*_f \psi'_i \, d\tau_{nuc} \int \chi^*_d \phi^*_d \delta(r_p - r_n) \delta(r_p - r_d) \phi_n{}^l \chi_p \, d\tau_n \, d\tau_p \, d\tau_d \qquad (8.58)$$

The second integral determines the angular distribution. It is computed by using the DWBA as outlined in the preceding section; we call it (including the factor V_0)σ_0. The spectroscopic factor S is defined as

$$S = n\left[\int \psi^*_f \psi'_i \, d\tau\right]^2 \qquad (8.59)$$

where n is the number of neutrons in the shell-model state; it is intuitively clear that the cross section should include such a factor. Thus we have for the (p,d) differential cross section

$$\frac{d\sigma}{d\Omega}(p,d) = \sigma_0 S \qquad (8.60)$$

Application of the law of detailed balance then gives for the (d,p) differential cross section

$$\frac{d\sigma}{d\Omega}(d,p) = \frac{2I_f + 1}{2I_i + 1} \sigma_0 S \qquad (8.61)$$

An appreciation of the spectroscopic factor S, defined by (8.59), is perhaps best obtained by studying some examples, shown in Table 8.1. Simple and straightforward methods for obtaining S, once the wave function is known, are readily seen in Examples 1, 2, and 3 in the table.

In Example 1, one might ask why we have not included in the final state configuration $(g_{7/2})_2{}^2(g_{7/2})_2{}^2 d_{3/2}$. This wave function has two broken pairs more than $(g_{7/2})_0{}^2(g_{7/2})_0{}^2 d_{3/2}$, so that it is a "seniority-4" wave function, whereas the previous one is "seniority zero." We are assuming here that *seniority* is a good quantum number. If it were not, applications of stripping would be much more complicated and much less useful. It is easy to see how much more complicated Examples 2 and 3 would be.

In cases where the state entered by the neutron was not initially empty, the situation is somewhat more complex. Example 4 of Table 8.1 illustrates a case of this type for $(d_{5/2})_0{}^2 \rightarrow (d_{5/2})_{5/2}{}^3$. The three final configurations listed can all have the same seniority; seniority is not so simple when an odd number of identical particles is involved. In cases of this type, the coefficients a, b, and c are determined from geometric coupling rather than from nuclear dynamics, and they are known as coefficients of fractional parentage (cfp) of the configuration $(d_{5/2})_{5/2}{}^3$. Tabulations of cfp's are available, but for most cases of interest, the following formulas due to French [20] are sufficient:

$$\begin{aligned} S(n, n-1) &= n && \text{for } n\text{-even} \\ &= 1 - \frac{n-1}{2j+1} && \text{for } n\text{-odd} \end{aligned} \qquad (8.62)$$

They refer to the case where the configuration couples to total angular momentum j or 0, when the number of particles is odd or even, respectively. Thus in Example 4, $n = 3$ so that $S = 1 - 2/6 = 2/3$.

An interesting application of (8.62) is to calculate the cross sections for (d,p) and (p,d) reactions when the target nucleus is even-even, with configuration $(j)_0{}^m$, and the final nucleus is $(j)_j{}^{m\pm1}$. For (d,p) reactions, $n = m + 1$ is odd in (8.62), $I_i = 0$, $I_f = j$, whence application of (8.61) gives

$$\frac{d\sigma}{d\Omega}(d,p) = (2j+1)\sigma_0 \times \left(1 - \frac{m}{2J+1}\right)$$

$$= (2j+1)\sigma_0(1 - V_j{}^2)$$

where $V_j{}^2$ is the fraction by which the state j is full. For (p,d) reactions, $n = m$ is even so that (8.60) and

TABLE 8.1. EXAMPLES OF SPECTROSCOPIC FACTORS IN (d,p) REACTIONS

Example number	Initial		Final		l	S
	Configuration	$I - \pi$	Configuration	$I - \pi$		
1	$(g_{7/2})_0{}^4$	0^+	$(g_{7/2})_0{}^4 d_{3/2}$	$\frac{3}{2}^+$	2	1
2	$(g_{7/2})_0{}^4$	0^+	$a(g_{7/2})_0{}^4 d_{3/2}$ $+b(g_{7/2})_0{}^2(s_{1/2})_0{}^2 d_{3/2}$	$\frac{3}{2}^+$	2	a^2
3	$m(g_{7/2})_0{}^4$ $+n(g_{7/2})_0{}^2(s_{1/2})_0{}^2$	0^+	$a(g_{7/2})_0{}^4 d_{3/2}$ $+b(g_{7/2})_0{}^2(s_{1/2})_0{}^2 d_{3/2}$	$\frac{3}{2}^+$	2	$(am + bn)^2$
4	$(d_{5/2})_0{}^2$	0^+	$a(d_{5/2})_0{}^2 d_{5/2}$ $+b(d_{5/2})_2{}^2 d_{5/2}$ $+c(d_{5/2})_4{}^2 d_{5/2}$	$\frac{5}{2}^+$	2	$\frac{2}{3}$

(8.62) give simply

$$\frac{d\sigma}{d\Omega}(p,d) = \sigma_0 m$$
$$= (2j + 1)\sigma_0 V_j{}^2$$

Both of these results are intuitively appealing. For example, when the $d_{5/2}$ state has 0, 2, 4, and 6 particles, the cross sections for inserting another $d_{5/2}$ particle in a stripping reaction are $6\sigma_0$, $4\sigma_0$, $2\sigma_0$, and 0, respectively, just proportional to the number of vacancies in the $d_{5/2}$ state.

The validity of these results is not confined to the configuration j^n, but in pairing theory [21], it includes situations where many shell-model states are filling simultaneously. Thus measurements of cross sections for (d,p) and (p,d) reactions can be used to determine the fullness of states. Examples of results obtained in this way are shown in Fig. 8.18.

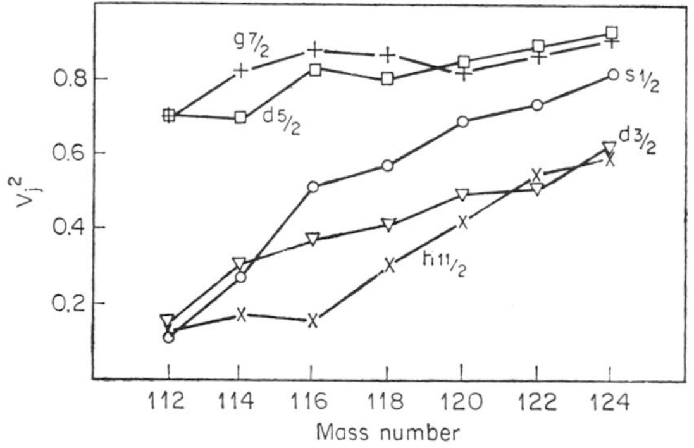

FIG. 8.18. Fullness ($V_j{}^2$) of various shell-model states in various isotopes of tin as determined from (d,p) and (d,t) reaction studies. [*Data from E. J. Schneid, A. Prakash, and B. L. Cohen, Phys. Rev.*, **156,** No. 4 (1967).]

Let us now return to the simpler situation where the neutron is inserted into an otherwise empty shell-model state, in order to study cases where configuration mixing is involved. For example, let us consider Example 2 of Table 8.1. The configuration mixture in the final state indicates that, in a pure shell model, there are two $\frac{3}{2}^+$ states in this energy region, namely, $(g_{7/2})_0{}^4 d_{3/2}$ and $(g_{7/2})_0{}^2(s_{1/2})_0{}^2 d_{3/2}$, and that these are mixed by residual interactions. Such a mixture leads to two mutually orthogonal states as

$$\psi_1 = a[(g_{7/2})_0{}^4 d_{3/2}] + b[(g_{7/2})_0{}^2(s_{1/2})_0{}^2 d_{3/2}]$$
$$\psi_2 = b[(g_{7/2})_0{}^4 d_{3/2}] - a[(g_{7/2})_0{}^2(s_{1/2})_0{}^2 d_{3/2}]$$

where $a^2 + b^2 = 1$. The spectroscopic factors for exciting these two states from a target nucleus $(g_{7/2})_0{}^4$ are a^2 (as in Table 8.1) and b^2, respectively, so that the sum of the S for the two states is unity. We see from this example that configuration mixing does not alter the sum of the S values. All the S may be in a single state, as in Example 1; in two states, as in Example 2; or in a very large number of states, but the sum is still unity. This may be shown to be true in general, as follows:

Let $\phi_0{}^{(j)}$ be a configuration consisting of the target nucleus plus a neutron in a given shell-model state j (for example, $d_{3/2}$ in Table 8.1). This may be considered as one of a complete set of mutually orthonormal basic functions, $\phi_k{}^{(j)}$, in the linear function space of the final nucleus with total angular momentum j.

Let $\psi_i{}^{(j)}$ be the various actual states of the final nucleus with spin j. Then these can be expanded as

$$\psi_i{}^{(j)} = \sum_k a_{ik}\phi_k{}^{(j)}$$

From (8.59) the spectroscopic factor $S_i{}^{(j)}$ for the state i is then just

$$S_i{}^{(j)} = a_{i0}{}^2$$

From quantum-mechanical completeness theorems, $\sum_i a_{i0}{}^2 = 1$ so that we have the sum rule

$$\sum_i S_i{}^{(j)} = 1 \tag{8.63}$$

It should be noted that this derivation is valid also in cases where there is configuration mixing in the target nucleus, as in Example 3 of Table 8.1. It is also valid if the proton configurations are different for the various states involved.

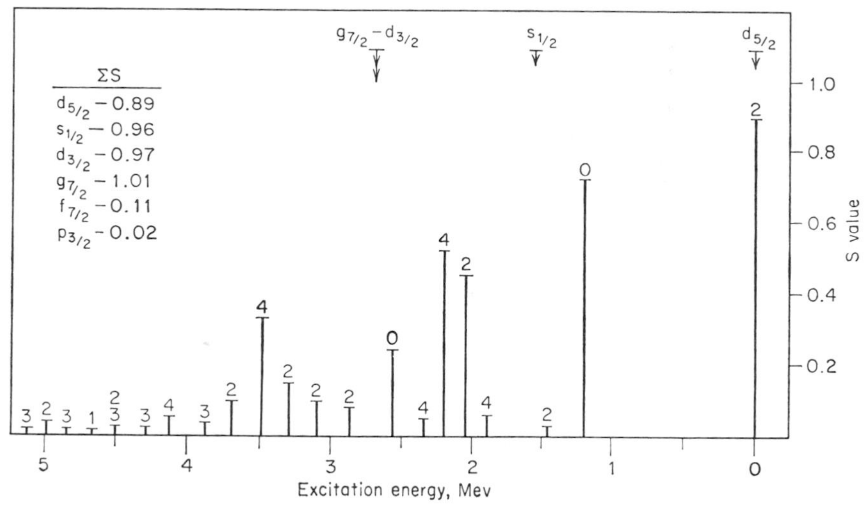

FIG. 8.19. Results from (d,p) reaction studies on Zr^{90}. Vertical lines show locations of levels in Zr^{91} excited; figures above them are l values as determined from angular distributions; and heights of lines are spectroscopic factors S. The sums of S values for various single-particle states are listed in the table at the left. The "center of gravity" for each single-particle state, as obtained from these data, is shown at the top of the figure. Data are from reference cited for Fig. 8.15.

An interesting application of the sum rule (8.63) is in determining locations of single-particle states. An experimental example is shown in Fig. 8.19. A (d,p) experiment [22] was done on Zr^{90} (some of the data are shown in Fig. 8.15) and the vertical lines represent the energies at which proton groups were found. The figures above the lines indicate the l values as obtained from the angular distributions, and the heights of the lines indicate the S values as found from the absolute cross sections. The ground state was known to be $\frac{5}{2}^+$, and so it is a $d_{5/2}$ state. Its S value just about fills the sum rule for $d_{5/2}$, so that one is tempted to conclude that all the other $l = 2$ states are $d_{3/2}$; on this assumption, the sum rule for $d_{3/2}$ is just about satisfied, which gives some confidence in the assumption. For $l = 0$, the states excited are $s_{1/2}$, and we see again that the sum rule for $s_{1/2}$ is just filled. Since Zr^{90} has 50 neutrons, the $g_{9/2}$ should be full, so that when a neutron is inserted with $l = 4$, it must enter a $g_{7/2}$ state. Note that the sum rule for $g_{7/2}$ is also approximately filled. The "center of gravity" of these states, weighting each with its S value, is the location of the single-particle states. They are shown by the arrows on the top of the figure.

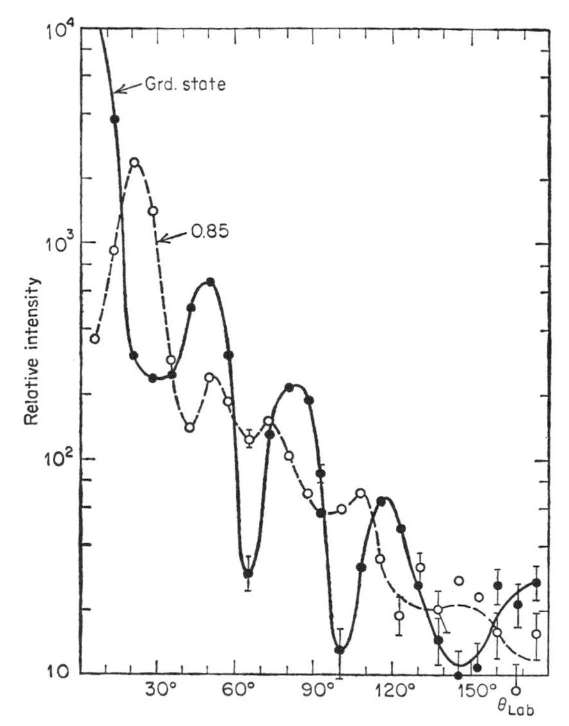

FIG. 8.20. Angular distributions of protons from the $Fe^{54}(t,p)Fe^{56}$ reaction in which the residual nucleus is left in its ground and 0.85-Mev excited states. These are for $l = 0$ and $l = 2$ angular-momentum transfer, respectively. [*Data from B. L. Cohen and R. Middleton, Phys. Rev.*, **146**: 748 (1966).]

There are many other interesting applications of stripping, including cases where the initial target is odd. There are also important nuclear-structure applications of two nucleon transfer reactions; for example, angular distributions from (t,p) reactions may be readily analyzed to give l which, from (8.54a), gives the spin and parity of the states excited. Some typical angular distributions and their l values are shown in Fig. 8.20. There is good reason to hope for useful results from more complex transfers. Inelastic scattering reactions have been very successfully used for locating collective vibrational 2^+ and 3^- states and for determining the amplitude of the vibrations. An example of this was shown in Fig. 8.16.

13. Excitation of Isobaric Analogue States

Now that we have discussed compound-nucleus and direct reactions, it is interesting to point out some areas of overlap between them. In this section and

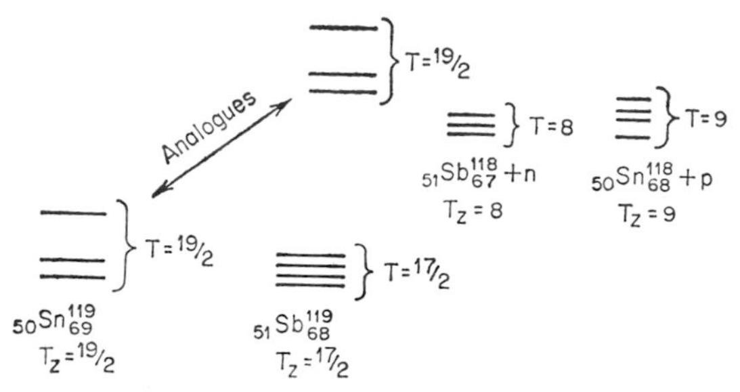

FIG. 8.21. States involved in analogue-state studies with the $Sn^{119}(p,p)$ reaction. See discussion in text.

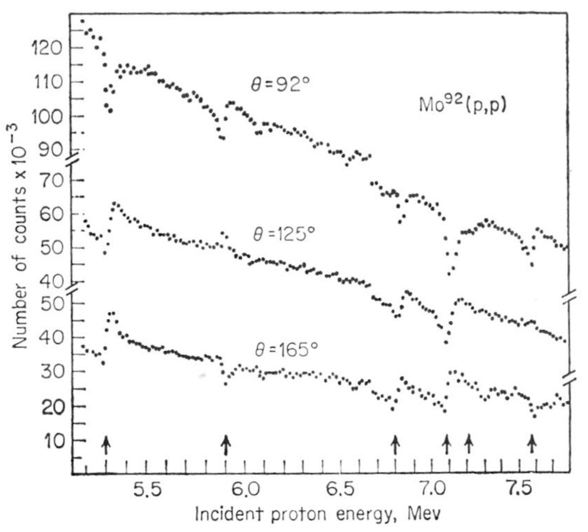

FIG. 8.22. Elastic scattering cross sections vs. bombarding energy in the $Mo^{92}(p,p)$ reaction. [*Data from C. F. Moore, P. Richard, C. E. Watson, D. Robson, and J. D. Fox, Phys. Rev.*, **141**: 1166 (1966).]

the next, we discuss two such subjects which are now receiving a great deal of attention.

If one operates on the wave function for a nuclear state with the T_- operator, one obtains the wave function for an isobaric analogue state. In the approximation that nuclear forces are charge-independent, the T_- operator commutes with the Hamiltonian so that the wave function thus generated is a true nuclear state; its energy differs from that of the original state

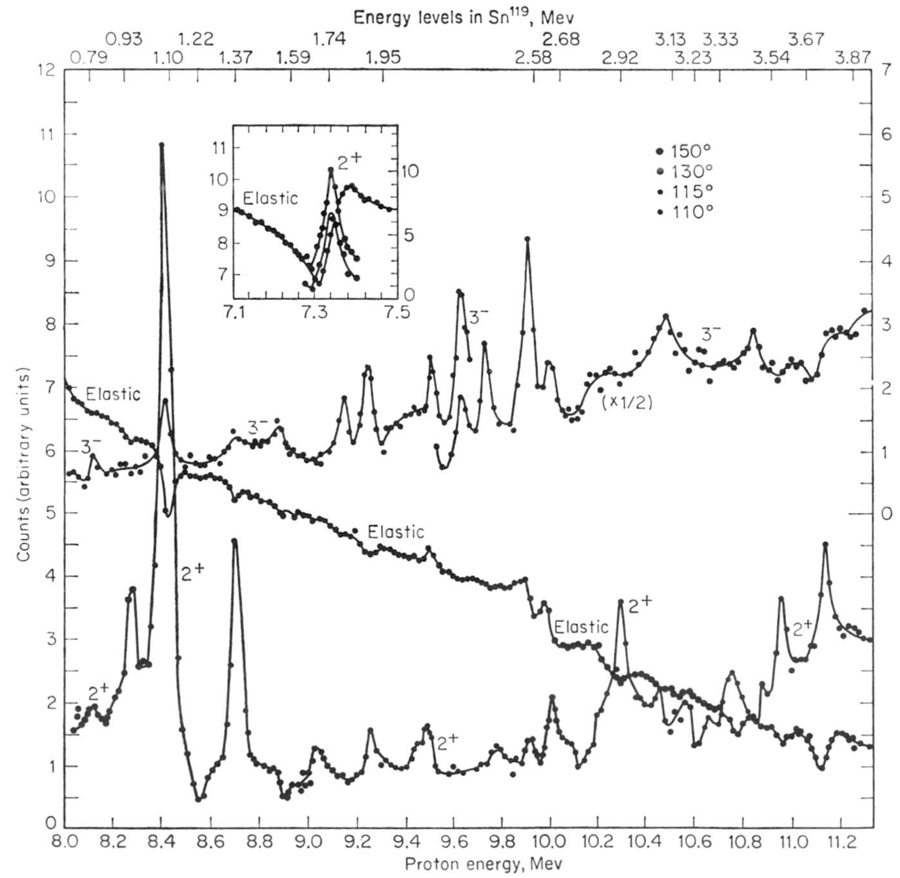

FIG. 8.23. Cross section vs. bombarding energy for (p,p) and (p,p') reactions on Sn^{118}. The data labeled 2^+ and 3^- are for reactions exciting the 1.2-Mev 2^+ and 2.5-Mev 3^- states, respectively. [*Data from D. L. Allen et al., Phys. Rev. Letters*, **17**: 57 (1965).]

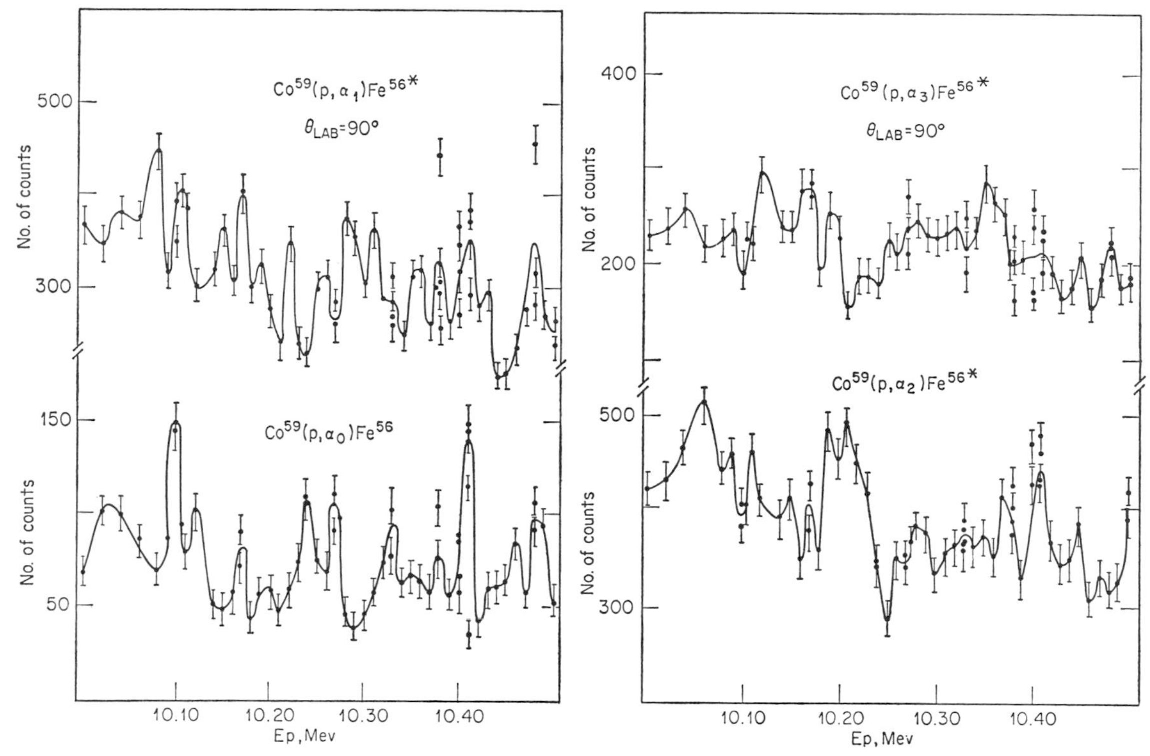

FIG. 8.24. Excitation curves for the ground state and first three excited states for the reaction $Co^{59}(p,\alpha)Fe^{56}$. Compound nucleus is Ni^{60} at about 20-Mev excitation energy. (*Data from G. Temmer, Argonne Natl. Lab. Rept.* 6878.)

by the Coulomb energy. The analogue states are highly excited states of the nucleus with the same atomic weight but with one higher atomic number. An example is illustrated in Fig. 8.21 for the low-lying states of Sn^{119} and their analogues in Sb^{119}. Note that, for the low-lying states of all heavy nuclei, T is its minimum value T_Z but for the analogue states $T = T_Z + 1$.

If one bombards Sn^{118} with protons in compound-nucleus reactions, one might expect to observe the analogue states as resonances. On the other hand, we recall that in high-excitation regions the widths of levels are greater than their spacing so that the cross section is smoothly varying with energy. These arguments are certainly valid for the $T = 17/2$ states, of which there are a great many in this energy region (excitation energy ~15 Mev). For the analogue states ($T = 19/2$), however, there are two special considerations. Firstly, the reason for the large widths of the $T = 17/2$ states is that they can decay very rapidly into $Sb^{118} + n$; for the analogue states, however, such a decay is forbidden because of the selection rule $\Delta T = \pm 1/2$ in nucleon emission. The analogue states (as well as the $T = 17/2$ states) can decay into $Sn^{118} + p$, but the widths for such decays are reduced by the Coulomb barrier penetration term (cf. Sec. 10). The second special consideration is that the reduced widths, γ^2, for exciting many of the analogue states are comparatively very large; their wave functions contain large amplitudes of the configuration Sn^{118} (Grd. st.) $+ p$. These reduced widths are essentially the same as the spectroscopic factor S for exciting by $Sn^{118}(d,p)$ the states of Sn^{119} of which these states are the analogues. The reason why low-lying states have large S values (and their analogues have large γ^2 values) is that configuration mixing is not so strong in the lower-energy region; there are not many levels of the same spin and parity with which they can mix, and also the energy range over which mixing occurs is smaller. Thus the ΣS sum-rule strength (cf. Sec. 12) is not divided among so many levels.

As a result of these special considerations, isobaric-analogue-state resonances stand out above the background of $T = T_Z$ states when measuring cross sections for proton-induced reactions as a function of energy. An example of such data is shown in Fig. 8.22. As discussed near the end of Sec. 10, the shapes of resonances can be readily analyzed to give the l value with which the proton is captured; this is the same as the l value for exciting the corresponding state in the (d,p) reaction. With rather elaborate analysis it is possible to extract the reduced width γ^2 from the proton elastic scattering data; this may be compared directly with S values determined from (d,p) reaction studies. The agreement is generally very good [23].

Up to this point, the analogue-state experiments give only the same nuclear-structure information that is more easily obtainable from (d,p) reactions. However, one can get further information from inelastic scattering experiments. Allen et al. [24] have studied inelastic scattering in which the Sn^{116} nucleus is left in the 2^+ or 3^- collective states; the data for this are shown in Fig. 8.23. The reduced width for emitting a proton to reach the 2^+ state is proportional to the amplitude of the configuration $Sn^{118}(2^+) + p$ in the compound state. This is very valuable nuclear-structure information which would be obtained, for example, by studying the (d,p) reaction from bombarding Sn^{118} in its 2^+ state; that experiment, of course, is impossible. By studies of inelastic scattering via analogue states one can therefore, in principle at least, experimentally determine the composition of the states of Sn^{119} as an expansion of the complete set of states [Sn^{118} (in various states) + neutron (in various single-particle states)].

Isobaric analogue states have also been studied with (p,n) reactions, as has been mentioned above (cf. Fig. 8.14). These reactions have been used for obtaining the energies of the analogue states and also to determine the charge-exchange part of the nucleon-nucleon interaction. This is usually taken to be

$$V_1(\tau_1 \cdot \tau_2)(1 + a\sigma_1 \cdot \sigma_2)$$

where τ and σ are the isobaric and ordinary spin operators, respectively. For even-even targets, the $\sigma_1 \cdot \sigma_2$ term is inoperative, so that a DWBA analysis of the differential cross sections for (p,n) reactions yields a value of V_1. Experiments on other nuclei determine a; it is approximately unity.

When the above term in the nucleon-nucleon interaction is summed over the particles in the target nucleus, one obtains a term in the nucleon-nucleus interaction or what we have been calling the optical-model potential. This term becomes

$$\frac{4V'}{A}(t \cdot T)$$

where t and T are the isobaric spin operators for the incident nucleon and the target nucleus, respectively. The coefficient V' here is identical with that discussed in Sec. 7.

14. Fluctuation in Cross Sections

In Sec. 10 we discussed the behavior of cross sections in the neighborhood of an isolated resonance and pointed out that at higher energies, where widths of resonances become larger than the distance between them, the cross section as a function of energy becomes smooth. This is indeed true for the total cross section but not for $d\sigma/d\Omega$, the differential cross section at some particular angle for a transition to a particular final state of the residual nucleus. An example of this is shown in Fig. 8.24.

In a region where several resonances are contributing to an (a,b) reaction, each contributes to the scattering amplitude a term

$$\frac{i\gamma_a\gamma_b}{(E - E_R) + \frac{1}{2}i\Gamma}$$

where γ_a^2 is proportional to the Γ_a used in Sec. 10. For a given set of spin orientations of a, b, the target nucleus, and the residual nucleus, $d\sigma/d\Omega$ is the square of the sum of these terms; i.e., they add coherently. For each contributing resonance, the phases of γ_a and γ_b are random in sign, so that the most probable value of the sum is zero. We might then expect $d\sigma/d\Omega$ frequently to be zero, which represents a rather large fluctuation from the average value. This is true no matter how much larger Γ is than D, the spacing between levels.

The contributions of $d\sigma/d\Omega$ from different spin couplings (channels), however, are incoherent so that if there are many of them the fluctuations are reduced by statistical averaging. The number of independent contributing channels, N, is

$$N = \tfrac{1}{2}(2S_a + 1)(2S_b + 1)(2I_i + 1)(2I_f + 1) \quad (8.64)$$

where the $\frac{1}{2}$ is introduced by the fact that one-half of the combinations are of the wrong parity to reach the final state. The degree of fluctuation is therefore expected to decrease with increasing N. From a detailed analysis one finds [25]

$$P_N(y) = \frac{N(Ny)^{N-1}}{(N-1)!}e^{-Ny} \quad (8.65)$$

where y is the ratio of the cross section at any energy to its average value. Plots of (8.65) are shown in the upper half of Fig. 8.25, and experimental tests are shown in Fig. 8.26.

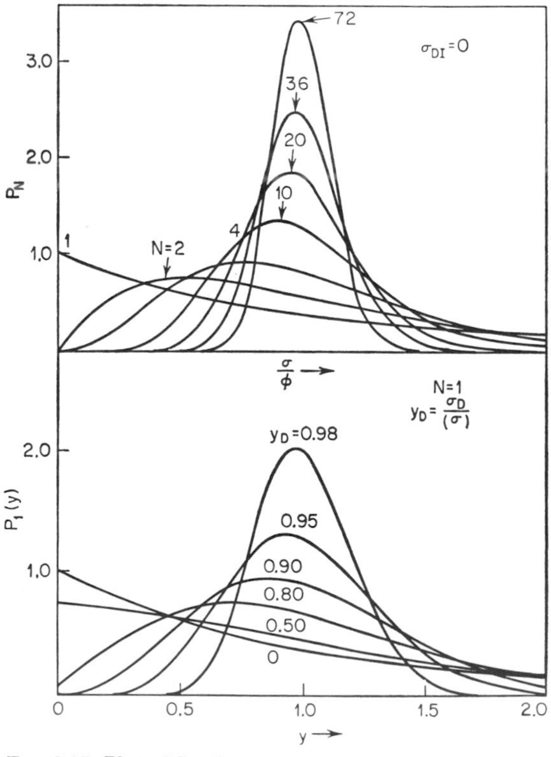

FIG. 8.25. Plot of Stephen's formula [Eq. (8.66)]. Upper curve is for $y_D = 0$ and various values of N [it is thus a plot of Eq. (8.65)] and lower curve is for $N = 1$ and various values of y_D.

In some cases, both direct and compound-nucleus interaction may be present. The amplitudes of these processes add coherently. The direct-interaction amplitude does not fluctuate, so that the presence of

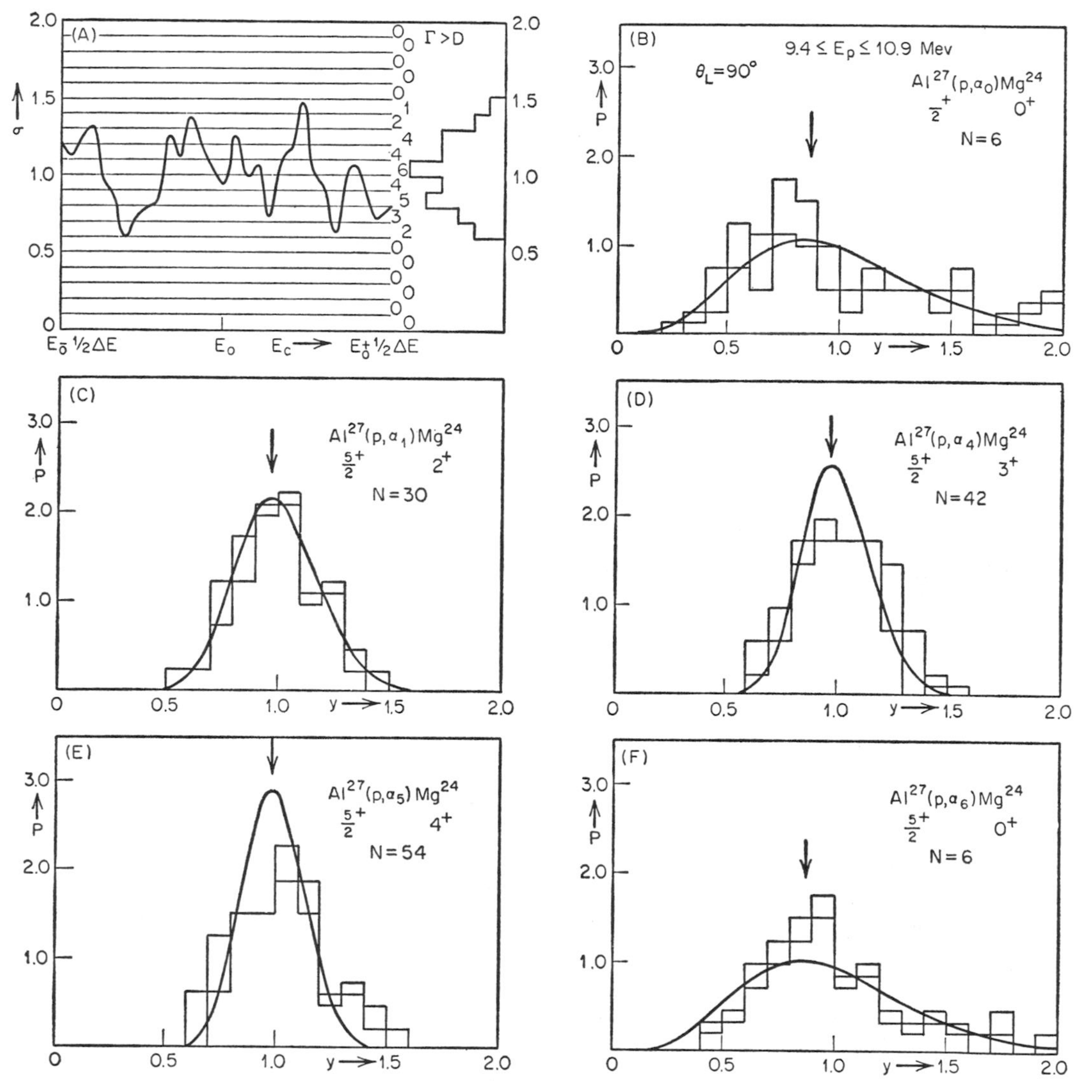

FIG. 8.26. Examples of fluctuation analyses, from reference cited for Fig. 8.24. (*A*) Illustration of fluctuation analysis by obtaining histogram of distribution of cross section σ about the average $[\sigma]$. Averaging interval ΔE. Vertical numbers indicate points per bin. $\Delta E \gg \Gamma \gg D$. (*B*) to (*F*) Probability distributions for $Al^{27}(p,\alpha)Mg^{24}$ at 90° (lab). Proton bombarding energy interval 9.40 to 10.90 Mev, corresponding to excitations in the compound nucleus Si^{28} between 20.6 and 22.1 Mev. Solid histograms for bins of 0.1; dotted histograms for bins of 0.2 in $y = \sigma/[\sigma]$. Solid curves are theoretical expressions for pure compound-nucleus formation and appropriate values of N. Numbers of bits are indicated on the figures and represent independent measurements. Small arrows indicate predicted peak positions. Theoretical and experimental distributions normalized to area unity. (*B*) Ground state of Mg^{24}, $I' = 0$, $N = 6$; (*C*) first excited state of Mg^{24} (1.368 Mev), $I' = 2$, $N = 30$; (*D*) fourth excited state of Mg^{24} (5.224 Mev), $I' = 3$, $N = 42$; (*E*) fifth excited state of Mg^{24} (6.005 Mev), $I' = 4$, $N = 54$; (*F*) sixth excited state of Mg^{24} (6.432 Mev), $I' = 0$, $N = 6$.

direct interaction reduces the fluctuations. For this case, the analysis gives [25]

$$P_N(y,y_D) = \frac{Ny^{N-1}}{1-y_D}\left[\exp\frac{-N(y+y_D)}{1-y_D}\right] \frac{J_{N-1}[2iN\sqrt{yy_D}/(1-y_D)]}{(i\sqrt{yy_D})^{N-1}} \tag{8.66}$$

where y_D is the fraction of the cross section that goes by direct interaction. A plot of (8.66) for $N = 1$ and various values of y_D is shown in the lower half of Fig. 8.25. A comparison with experimental data is shown in Fig. 8.27. We see that one can rather accurately determine the amount of direct interaction present; this is probably the best method of doing this, although the effort required is rather substantial.

A fluctuation analysis can also be used to determine the average value of Γ. This is best done by an autocorrelation analysis. We define

$$C(\epsilon) = \frac{\overline{(d\sigma/d\Omega)(E)\cdot(d\sigma/d\Omega)(E+\epsilon)}}{\overline{(d\sigma/d\Omega)(E)}\cdot\overline{(d\sigma/d\Omega)(E+\epsilon)}} - 1 \tag{8.67}$$

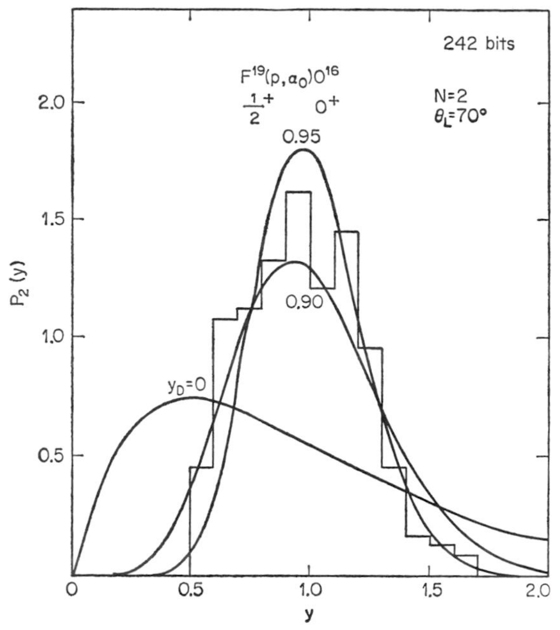

FIG. 8.27. Probability distribution for $F^{19}(p,\alpha)O^{16}$ at 70°. Proton bombarding energies correspond to excitation in the compound nucleus between 18.5 and 24.2 Mev. Curves are predictions from (8.66) for various values of y_D; it can be seen that the reaction is more than 90 per cent direct. Figure is from reference for Fig. 8.24.

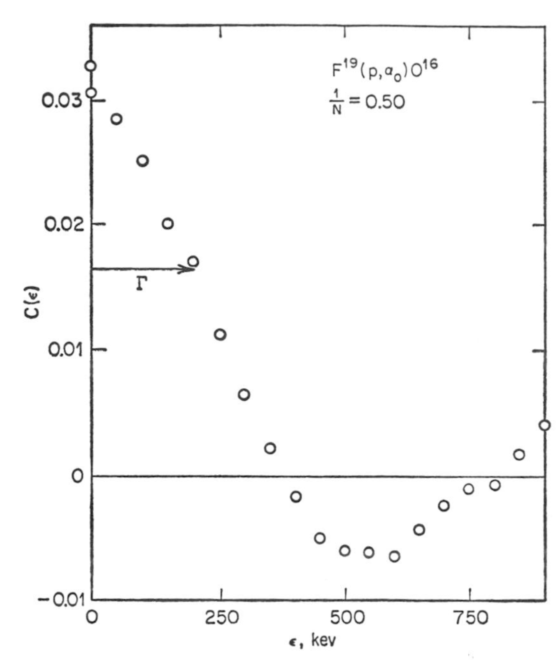

FIG. 8.28. Autocorrelation function for $F^{19}(p,\alpha_0)O^{16}$. Data are the same as in Fig. 8.27. See the discussion in the text for the explanation of Γ and zero intercept. The oscillation about the zero ordinate is a typical finite sample effect.

where ϵ is the energy difference between the two points where $d\sigma/d\Omega$ is measured. Plots of $C(\epsilon)$ can be readily made from experimental data, as shown in Fig. 8.28. It can be shown [25] that

$$C(\epsilon) = \frac{1 - y_D^2}{N} \frac{\Gamma^2}{\Gamma^2 + \epsilon^2} \qquad (8.68)$$

Thus, the half width of the curve in Fig. 8.28 gives the value of Γ. At about 20-Mev excitation energy, values of Γ are found to be about 200 kev for $A \sim 20$, decreasing to about 6 kev for $A \sim 60$. These correspond via (8.45) to compound-nucleus lifetimes of 3×10^{-21} and 1×10^{-19} sec, respectively. One should note in passing that a comparison of the plot in Fig. 8.28 with (8.68) gives another method of determining y_D; it is just

$$y_D^2 = 1 - NC\ (\epsilon = 0)$$

For the case shown it is 0.94, which compares well with a determination from Fig. 8.27 made with the same data.

A much simpler (but less accurate) alternative method of determining Γ is from counting the number of maxima per unit energy interval in the cross section vs. energy curve, ν. It has been shown that ν is given by [25]

$$\nu = 0.59 b_N/\Gamma \qquad (8.69)$$

where $b_1 = 1$, $b_2 = 0.78$, $b_3 = 0.75$, $b_5 = 0.70$, etc.

A considerable amount of work has been done on the cross correlations of cross sections at different angles. As the angular separation increases, the correlations fall off, reaching zero at an angle of about $\lambda\!\!\!^{-}/R$. Fluctuations are strongest near 0 and 180°. The expressions obtained are somewhat more complex and difficult to apply than for autocorrelations.

When cross sections are integrated over angles and/or over many levels of the final nucleus, the number of incoherent contributing terms becomes very large so that the fluctuations disappear. Thus, there are no fluctuations in total cross sections.

15. Electromagnetically Induced Reactions

When energetic gamma rays are incident on a nucleus, their energy can be absorbed, thereby initiating nuclear reactions. The absorption process is an electromagnetic interaction and is largely dominated by electric-dipole transitions. It is, of course, subject to the sum rules of electromagnetic transitions, the principal one being [26]

$$\int_0^\infty \sigma\, dE \simeq 20A \text{ Mev-mbarns} \qquad (8.70)$$

where A is the mass number. Various energy-weighted sum rules have also been developed [26].

A typical cross section vs. energy curve for gamma-ray absorption is shown in Fig. 8.29. It is dominated by the "giant resonance," a broad peak centered at about 15 Mev in heavy nuclei which essentially exhausts the sum rule (8.70). This is due to excitation of the 1^-, $T = 1$ collective state which corresponds classically to an oscillation of the protons relative to the neutrons. The nuclear-structure details of these states are now well understood as coherent configuration mixtures induced by particle-hole interactions

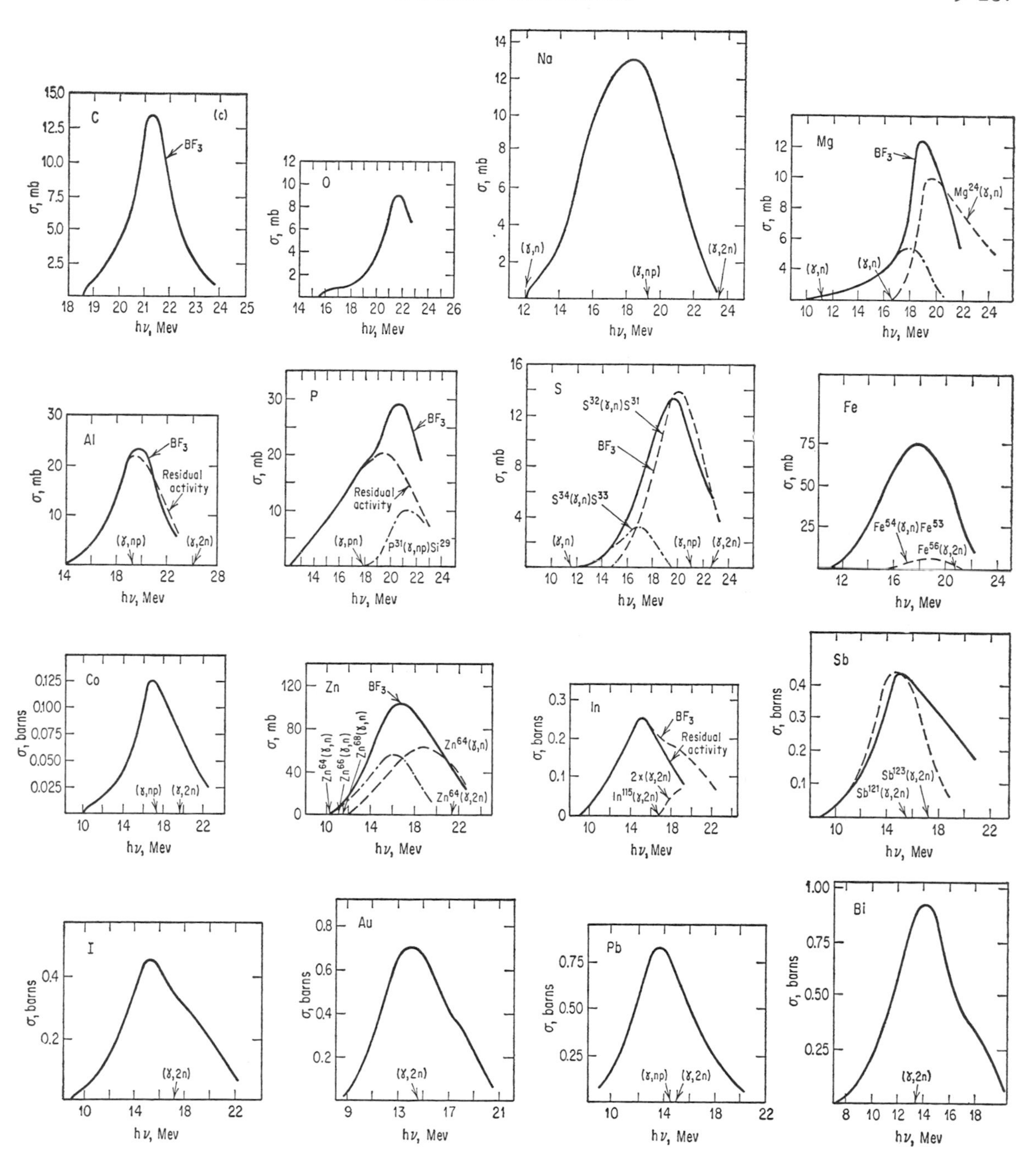

FIG. 8.29. Photonuclear giant resonances for various nuclei. Curves labeled BF_3 are total neutron yields, whereas curves labeled "residual activity" are yields for (γ,n) reactions as measured by the readioactivity produced. [*Data from R. Montalbetti, J. Goldemberg, and L. Katz, Phys. Rev.*, **91**: 659 (1953).]

[27], and the cross sections for exciting them can be calculated with reasonably good accuracy. The width and fine structure of the giant resonance have been the object of a great deal of theoretical and experimental investigation. The widths are plotted versus A in Fig. 8.30; they are especially narrow for closed-shell nuclei and are wide and split into two components for deformed nuclei. The correlation between widths and nuclear deformation is illustrated in Fig. 8.30.

About 90 per cent of the time, gamma-ray absorption results in formation of a compound nucleus. The subsequent decay of this compound nucleus is governed by the considerations discussed in Sec. 10, just as if it were formed in any other way. About 10 per cent of the time, nucleons are ejected directly in the absorp-

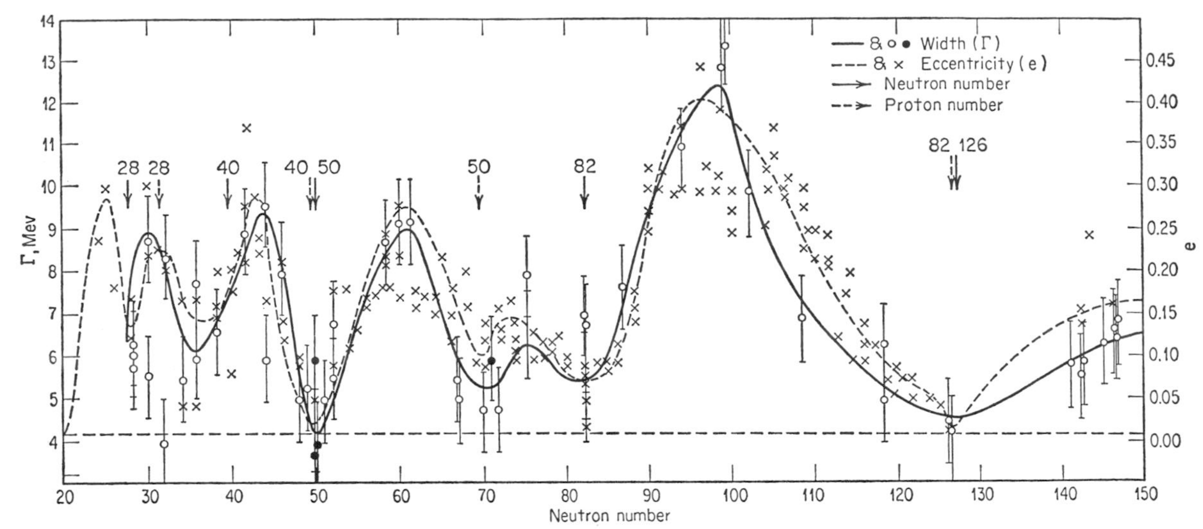

Fig. 8.30. Widths of photonuclear giant resonances plotted vs. neutron number (circles and solid lines). Nuclear eccentricity e (crosses and dashed line) as calculated from quadrupole moments is also shown. [*Data from K. Okamoto, Phys. Rev.*, **110**: 143 (1958).]

tion process. They come out with relatively high energy and predominantly at 90° to the incident-beam direction.

There has been a great amount of effort devoted to studies of these reactions because of the limited usefulness of betatrons for other research. In recent years, however, betatrons are being replaced by electron linear accelerators, so that much of this effort is being diverted to the large variety of activity for which these machines are well suited.

The simplest of these reactions is inelastic electron scattering by heavy nuclei. They predominantly excite collective states, and analysis of the angular distributions and absolute cross sections for inelastic electron scattering can be used to determine the multipolarity of the transitions—and hence the spin and parity of the state excited—and the amplitude of the collective vibrations.

Electron linear accelerators are also capable of producing interesting beams of positrons, and of monoenergetic gamma rays from the annihilation of positrons in flight. There is every reason to hope that studies of reactions produced by these will open up interesting new vistas.

Another very interesting electromagnetically induced reaction is Coulomb excitation—the excitation of nuclei by a passing charged particle due to their electromagnetic interaction. The transitions induced are most frequently electric quadrupole exciting the vibrational and rotational states. From the cross sections, the amplitude of the nuclear deformations involved can be deduced. When heavy ions are used in the bombardment, rather high members of rotational bands can be reached by successive E2 excitations. By careful comparison of the excitation of vibrational states with different projectiles, the static quadrupole moments of these states can be determined [28]. After more than a decade of research, Coulomb excitation continues to yield a steady stream of very interesting and pertinent results. It would merit much more thorough discussion than is devoted to it here were it not somewhat off the subject of nuclear reactions.

References

1. Rutherford, E.: *Phil. Mag.*, **37**: 581 (1919).
2. Cohen, E. R., J. W. DuMond, T. W. Layton, and J. S. Rollett: *Revs. Mod. Phys.*, **27**: 363 (1955).
3. Morrison, P.: in E. Segre (ed.), "Experimental Nuclear Physics," vol. 2, Wiley, New York, 1953.
4. Konig, L. A., J. H. E. Mattauch, and A. H. Wapstra: *Nucl. Phys.*, **28**: 1 (1961). Wapstra, A. H.: *Nucl. Phys.*, **28**: 29 (1961). Ries, R. R., R. A. Damerow, and W. H. Johnson: *Phys. Rev.*, **132**: 1662, 1673 (1963). Cohen, B. L., R. Patell, A. Prakash, and E. J. Schneid: *Phys. Rev.*, **135**: B383 (1964). Future compilations should appear periodically, edited by A. H. Wapstra.
5. See articles by Vogt, E., and H. E. Gove: in P. M. Endt and M. Demeur (eds.), "Nuclear Reactions," vol. 1, North Holland Publishing Company, Amsterdam, 1959, and by Richards, H. T.: in F. Ajzenberg-Selove (ed.), "Nuclear Spectroscopy," Pt. A, Academic, New York, 1960.
6. Feshbach, H., D. C. Peaslee, and V. F. Weisskopf: *Phys. Rev.*, **71**: 145 (1947).
7. Blatt, J. M. and V. F. Weisskopf: "Theoretical Nuclear Physics," Wiley, New York, 1952.
8. Shapiro, M.: *Phys. Rev.*, **90**: 171 (1953).
9. Bassel, R. H., R. M. Drisko, and G. R. Satchler: *Oak Ridge Natl. Lab. Rept.* 3240.
10. Perey, F. G.: *Phys. Rev.*, **131**: 745 (1963). Winner, D. R., and R. M. Drisko: *Univ. Pittsburgh Dept. Physics Rept.*, 1965.
11. Feshbach, H., C. E. Porter, and V. F. Weisskopf: *Phys. Rev.*, **96**: 448 (1954).
12. Peterson, J. M., A. Bratenahl, and J. P. Stoering: *Phys. Rev.*, **120**: 521 (1960).
13. Feshbach, H.: *Ann. Phys.*, **19**: 287 (1962). Lemmer, R. H., and C. M. Shakin: *Ann. Phys.*, **27**: 13 (1964).
14. Lane, A. M., R. G. Thomas, and E. P. Wigner: *Phys. Rev.*, **98**: 693 (1955).
15. Wigner, E. P., and L. Eisenbud: *Phys. Rev.*, **72**: 29 (1947).
16. Feshbach, H.: *Ann. Phys.*, **5**: 357 (1958). Lane,

A. M., and R. G. Thomas: *Revs. Mod. Phys.*, **30**: 257 (1958). Christy, R. F.: *Physica*, **22**: 1009 (1956). Wheeler, J. A.: *Phys. Rev.*, **59**: 19 (1941). See also ref. 5.

17. *Brookhaven Natl. Lab. Rept.* 325.
18. Vogt, E.: *Brookhaven Natl. Lab. Rept.* 331. Chase, D. M., L. Wilets, and A. R. Edmonds: *Phys. Rev.*, **110**: 1091 (1958).
19. Satchler, G. R.: in *Argonne Natl. Lab. Rept.* 6878, 1964.
20. French, J. B.: in F. Ajzenberg-Selove (ed.), "Nuclear Spectroscopy," Pt B, Academic, New York, 1960.
21. Kisslinger, L. S., and R. A. Sorenson: *Kgl. Danske Videnskab. Sekkab, Mat.-Fys.*, **32**: 9 (1960).
22. Cohen, B. L., and O. Chubinsky: *Phys. Rev.*, **131**: 2184 (1963).
23. Lee, L. L., A. Marinov, and J. P. Schiffer: *Phys. Rev. Letters*, **8**: 352 (1964).
24. Allen, D. L., G. A. Jones, G. C. Morrison, R. B. Taylor, and R. B. Weinberg: *Phys. Rev. Letters*, **17**: 56 (1965).
25. Stephen, R. O.: Thesis, Oxford University, 1964.
26. Levinger, J. S., and H. A. Bethe: *Phys. Rev.*, **78**: 115 (1950).
27. Brown, G. E., J. A. Evans, and D. J. Thouless: *Nucl. Phys.*, **45**: 164 (1963).
28. DeBoer, J., R. G. Stokstad, G. D. Symons, and A. Winther: *Phys. Rev. Letters*, **14**: 564 (1965).

Chapter 9

Acceleration of Charged Particles to High Energies

By JOHN P. BLEWETT, Brookhaven National Laboratory

1. Introduction

Particle accelerator art began in 1928 with a paper by Wideröe [1]* describing successful experiments on a two-stage linear accelerator. He gave also an account of an unsuccessful attempt to accelerate by the method later used in the betatron. Since 1928 new techniques have evolved in rapid succession; as each method approached its apparent upper limit, new ideas appeared that made higher energies attainable. As a result the energies of accelerated charged particles have increased roughly by a factor of 10 every six years. This rate is still being maintained, with no upper limit yet in view. A peak energy of 33 billion electron volts† was reached in 1960 at the Brookhaven National Laboratory. The only accelerator under construction in 1966 that will surpass this energy is the 70-Gev synchrotron at the Institute for High Energy Physics at Serpukhov (near Moscow) in the Soviet Union. This machine should be completed by about 1968. The progress toward higher energies is summarized in the graph of Fig. 9.1 in which energies of accelerated particles are plotted against the date of announcement of successful operation. The three dots to the right of the graph represent machines under construction in 1965.

All accelerators are limited by the fundamental fact that charged particles can be accelerated to high energies only by the application of electric fields. The methods vary only in the way in which the electric field is applied to the particles. In the early nuclear accelerators direct fields were used. Cockcroft and Walton used transformer-rectifier sets, while Van de Graaff generated high voltages in an electrostatic machine. Later a direct field was produced by electromagnetic induction in the betatron. The second stage in the art was reached with the repetitive application of the same field in the linear accelerator and the cyclotron. The third, present stage is characterized by the use of synchronous acceleration, so called because the particles automatically stay in synchronism with an alternating accelerating field. The modern linear accelerator, the synchrocyclotron, and the synchrotron all fall in this category; it is with devices of this type that the highest energies have been reached.

Two comprehensive reference works on accelerators are available: "Particle Accelerators" by Livingston and Blewett and "Cyclic Particle Accelerators" by Livingood. Also valuable collections of papers on high-energy accelerators are available in the proceedings of a series of five international accelerator conferences held between 1956 and 1965. These volumes are listed in the Bibliography at the end of this chapter.

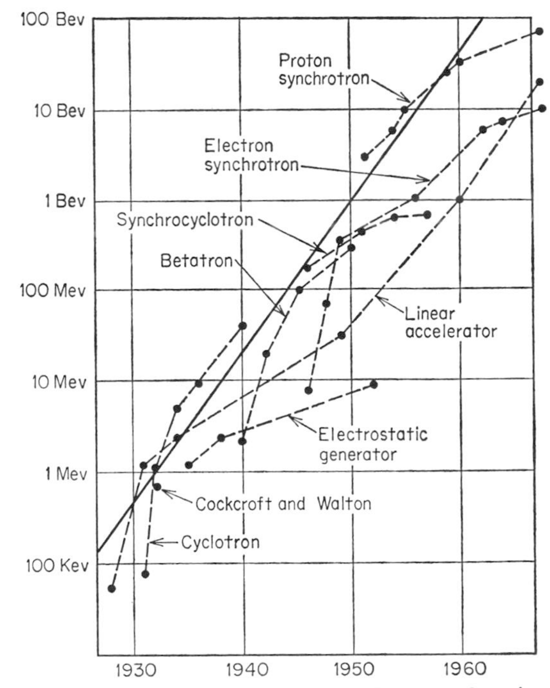

FIG. 9.1. Energy of accelerated particles plotted against the date of announcement of successful operation. The straight line represents a factor of 10 every six years.

2. Transformer-rectifier Systems

The first accelerator to find use in nuclear physics was the transformer-rectifier voltage multiplier of Cockcroft and Walton [2]. The circuit involved is

* Numbers in brackets refer to References at end of chapter.

† 10^9 electron volts = 1 Gev. The term Bev representing 10^9 electron volts is frequently used in the United States.

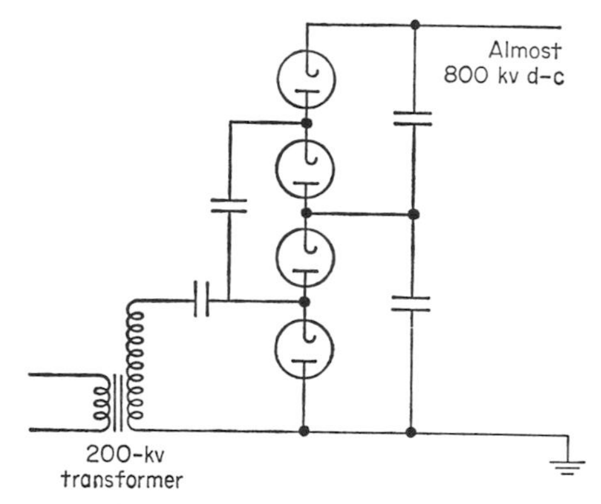

FIG. 9.2. Voltage-multiplier circuit of Cockcroft and Walton.

shown in Fig. 9.2. By using a 200-kv transformer and four stages of multiplication, the peak voltage achieved was 700 kv. Much higher voltages than this had been attained during the previous decade in surge generators but Cockcroft and Walton were successful for the first time in applying this voltage across an evacuated tube and accelerating protons from a gas-discharge ion source. With this device they made the first studies of artificial disintegration of lithium and several other elements.

Cockcroft-Walton generators are now commercially available from several sources. A pressurized high-frequency version of the Cockcroft-Walton developed by Radiation Dynamics, Inc., is available, under the trade name Dynamitron, at energies up to 3 million ev and delivering currents up to 10 ma.

Still another variation is the insulating-core transformer (ICT) developed by the High Voltage Engineering Corporation. The ICT is a multisecondary transformer in which the magnetic circuit is broken into insulated segments. The outputs of the secondaries wound on the individual segments are rectified and connected in series. The commercial units are pressurized and operated in a three-phase system. This device is available at voltages up to 4 million volts to deliver currents of the order of 10 ma.

3. The Van de Graaff Electrostatic Generator

The most useful high-voltage machine has proved to be the electrostatic generator (Fig. 9.3) developed in 1931 by R. J. Van de Graaff [3]. The corona discharge from a row of sharp metal needles is directed on a moving insulating belt. The charge on the belt is carried to the field-free region inside an insulated metal sphere where it is removed by corona discharge to another group of needles. As the charge on the sphere increases, its potential rises until breakdown takes place or until some other limitation is applied.

Van de Graaff generators have been operated reliably in air up to about 2 million volts. Above this voltage they become unreasonably large, and it is preferable to operate in a pressure tank at elevated pressures. A horizontal Van de Graaff generator for 4-Mev protons is shown in Fig. 9.4. In mixtures of nitrogen and carbon dioxide at pressures up to several hundred pounds per square inch satisfactory operation

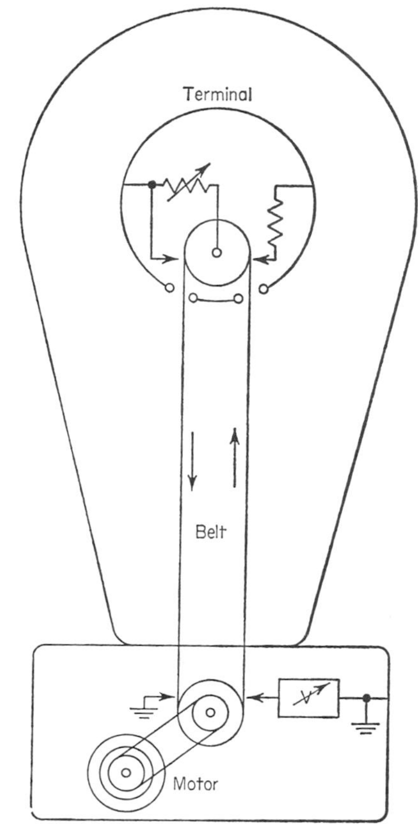

FIG. 9.3. Schematic diagram of generator enclosed in a grounded pressure tank. Circuits are shown for spraying charge on both the ascending and descending faces of the belt. *(From M. S. Livingston and J. P. Blewett, "Particle Accelerators," McGraw-Hill, New York,* 1962, p. 38.)

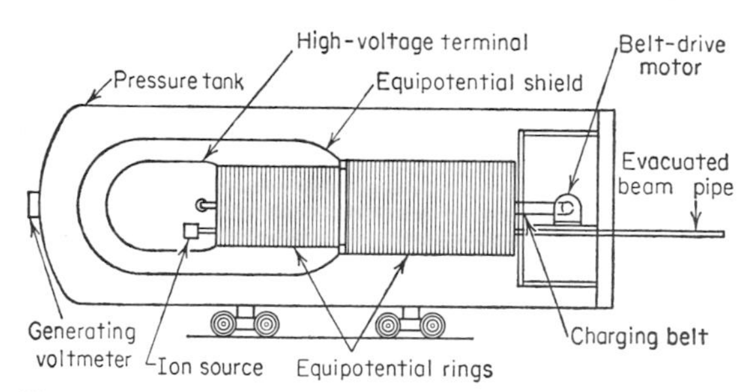

FIG. 9.4. Horizontal Van de Graaff generator for 4-Mev protons. *(From M. S. Livingston and J. P. Blewett, "Particle Accelerators," McGraw-Hill, New York,* 1962, p. 46.)

has been achieved at voltages as high as 10 million volts.

Although the Van de Graaff generator presents several unpleasant technical problems associated with operation of the ion sources and auxiliary equipment at high voltages, its constant, controllable voltage and the sharply focused beam that it emits have made it the most useful accelerator in its energy range for precise measurements of thresholds for nuclear events and nuclear energy levels. Voltage control systems operating from the measured energy of the accelerated beam make it possible to calibrate and control the beam energy to better than 1 part in 1,000.

The Van de Graaff accelerator has been used both for electrons and for nuclear particles. The highest-energy machines have generally used protons or other nuclear particles.

The latest application of the Van de Graaff generator is in the so-called tandem accelerator. The concept of the tandem generator has been discussed for many years [4] but the first practical applications have been made by the High Voltage Engineering Corporation. In the two-stage tandem shown in Fig. 9.5

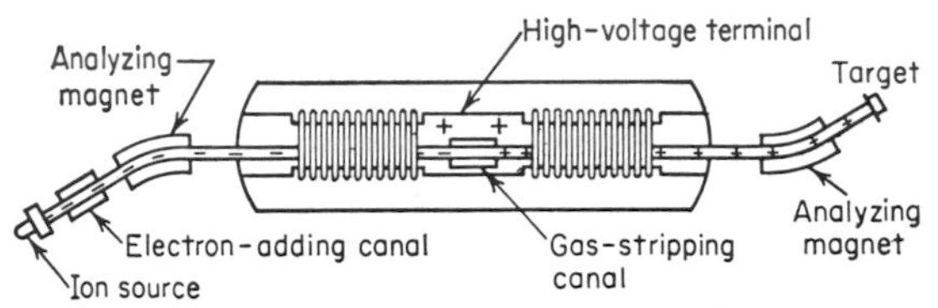

FIG. 9.5. Two-stage tandem electrostatic generator. *(From M. S. Livingston and J. P. Blewett "Particle Accelerators," McGraw-Hill, New York,* 1962, p. 70.)

negative ions are accelerated from an ion source at ground potential to the high-voltage terminal of a Van de Graaff generator where they are stripped in a jet of hydrogen gas and converted to positive ions. They are then accelerated again to ground potential where they arrive with twice the energy available when the Van de Graaff is used in the conventional fashion. Addition of still another two-stage tandem makes possible three and even four stages of acceleration. Use of these ideas makes available energies up to 30 Mev with the narrow energy spread and stable energy characteristic of the electrostatic generator.

4. The Betatron

Charged particles can be accelerated also in the electric field that circulates around a region in which a magnetic flux is changing. This accelerator can be described as a transformer in which the secondary winding is replaced by a beam of particles. The final particle energy, or the effective transformation ratio, is determined by the number of times the beam circulates around the magnetic flux. To make this device practical, it is necessary only to restrain the beam to a closed path around the flux and to maintain it in a stable orbit over some thousands of revolutions. Credit for the solution of the latter problem goes to D. W. Kerst, who built the first successful induction accelerator or "betatron" in 1940 [5].

The particles in the betatron are held in a circular path by a magnetic field which usually is generated by the same means that produce the accelerating flux. The relation between field and flux is easily derived as follows: The electric field E accelerating the particle is given in terms of the magnetic flux linked ϕ by

$$\oint E\,ds = \frac{d\phi}{dt} \tag{9.1}$$

If $\bar{B}$ is the average magnetic field inside a circular orbit of radius R,

$$2\pi RE = \pi R^2 \frac{d\bar{B}}{dt} \tag{9.2}$$

and the rate of change of momentum of the particle is given by

$$\frac{d}{dt}(mv) = eE = \frac{eR}{2}\frac{d\bar{B}}{dt} \tag{9.3}$$

But the momentum of a particle traveling in a circular path in a magnetic field B_0 is

$$mv = eRB_0 \tag{9.4}$$

whence

$$\frac{d}{dt}(mv) = er\frac{dB_0}{dt} \tag{9.5}$$

From (9.3) and (9.5) comes the fundamental betatron relation

$$\frac{d\bar{B}}{dt} = 2\frac{dB_0}{dt} \tag{9.6}$$

so that the rate of change of the average field inside the orbit must be twice the rate of change of the field at the orbit.

Stability of the betatron orbit is achieved by making the paraxial field B_z at the orbit decrease with radius according to the relation

$$B_z = B_0\left(\frac{r}{R}\right)^{-n} \tag{9.7}$$

where R is the radius of the "equilibrium orbit" and B_0 is the field at radius R.

To satisfy Maxwell's equations the radial field component B_r must have the form

$$B_r = -\frac{nz}{r}B_z + \text{higher-order terms in } \frac{z}{r} \tag{9.8}$$

The radial and axial motions of the particle follow from the force equations

$$\frac{d}{dt}(m\dot{r}) - \frac{mv^2}{r} = -evB_z \tag{9.9}$$

$$\frac{d}{dt}(m\dot{z}) = evB_r \tag{9.10}$$

Replacing r by $R + x$ in the first of these relations, and assuming that x and z are small compared with R and that m and v are sensibly constant during the period under consideration,

$$\ddot{x} + (1 - n)\omega_0^2 x = 0 \tag{9.11}$$

where $\omega_0 = v/R$ is the angular velocity of the particles in the equilibrium orbit. From (9.10) and (9.4),

$$\ddot{z} + n\omega_0^2 z = 0 \tag{9.12}$$

Equations (9.11) and (9.12) describe oscillations which are stable, provided only that

$$0 < n < 1 \tag{9.13}$$

These oscillations are known as *betatron oscillations.* Their period evidently is greater than the period of revolution of the particles in their orbits. Their amplitude is determined by initial conditions set up at the time of injection of the particles.

A more refined calculation shows that these oscillations are damped as the magnetic field rises; the oscillation amplitude is proportional to $B^{-1/2}$.

These principles are utilized in a structure which, in one of its simpler forms, is shown in Fig. 9.6. The same magnet structure is used to provide the accelerating flux across the short central gap and the guiding field in the wider gap which surrounds the flux gap. The gap lengths are chosen to satisfy (9.6), and the guide field gap is tapered to give an n value between 0 and 1.

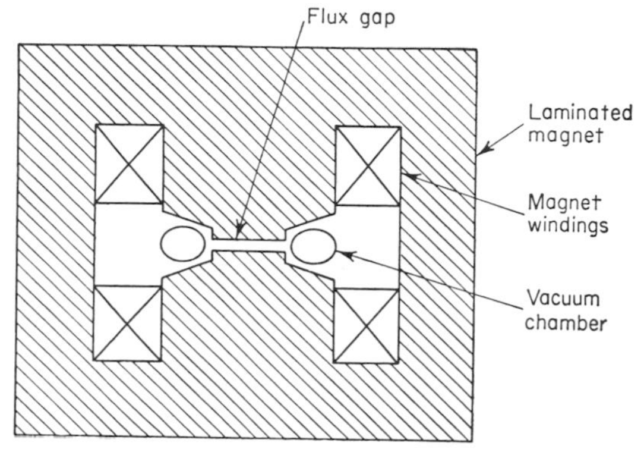

FIG. 9.6. The betatron in cross section.

Betatrons have been used thus far only with electrons. The electrons are injected from an electron gun displaced to one side of the equilibrium orbit. Injection is at some tens of kilovolts. Acceleration then takes place over a period of the order of 0.01 sec to energies which, thus far, have not exceeded 315 million ev.

The magnet structure for the betatron must be laminated because of the rapidly varying magnetic fields which it sustains. In the larger betatrons the structure weighs some hundreds of tons [6]. The magnet current is driven at a 60-cycle or similar rate, and the phase angle between magnet current and voltage is corrected by a shunt capacitor bank. Acceleration takes place over the quarter cycle during which the field rises from zero to peak in the appropriate direction.

At the end of the accelerating cycle the beam is displaced by perturbation of the magnetic field to strike a target. The finally useful radiation is the beam of X rays from the target. This beam has a half width which varies from 12° at 20 million volts to 2° at 100 million volts. Radiation intensities in the beam are of the order of 1,000 r/min for the 100-Mev machine.

The vacuum chamber is usually a ceramic "doughnut" coated internally with a conducting coating whose resistance is low enough that it can dissipate static charges but high enough that large eddy currents are not generated. The coating resistance is about 1,000 ohms per square.

An effective upper limit to betatron acceleration of electrons is set by the emission of electromagnetic radiation due to the continual radial acceleration of the electrons as they are restrained in their circular path. If the electron energy is higher than a few million volts so that the electron velocity is approximately the velocity of light, the rate of energy loss in electron volts per revolution is [7]

$$8.8 \times 10^{-32} W^4/R \qquad (9.14)$$

where W is the electron energy in electron volts and R is the orbit radius in meters. In a typical machine where W is 100 million volts and R is 1 m the electrons lose 8.8 ev per revolution. If electrons are to be accelerated to the billion-volt range, the orbit radius must be increased to very large values or extra accelerating mechanisms must be introduced to prevent complete loss of the beam as it loses energy and spirals in to the inner walls of the vacuum chamber. Fortunately, the energy loss is inversely proportional to the fourth power of the particle mass so that no such difficulties are experienced with protons or other nuclear particles until energies of the order of 10^{12} ev are reached.

5. Principles of Synchronous Accelerators

The possibilities of repetitive acceleration of particles by an alternating field have been explored by many investigators since Wideröe's first linear accelerator of 1928. Basically, this method consists in applying to the particles an alternating field whose phase is such that acceleration takes place and then shielding the particles from the field until they can again be exposed at a later accelerating phase. In the case of the linear accelerator, as indicated in Fig. 9.7, the particles cross a gap when the field is in the accelerating direction; they are then shielded in a "drift tube" whose length is such that the particles traverse it in one cycle of the alternating field. At the next gap between drift tubes the particles are again accelerated, and so forth. In other devices such as the cyclotron or the synchrotron the particles are returned to the original gap or gaps by a magnetic field so that they undergo acceleration many times and finally attain an energy many times the energy gained at each gap. The energy to be reached in such a device is evidently limited only by the ability to keep the particles in phase with the accelerating wave.

The synchronizing mechanism is simplest in the case of the linear accelerator where stable acceleration is possible on the rising part of the accelerating field wave, that is, between 0 and 90° in phase. If the particle arrives at an accelerating gap too soon, it will receive less energy than it should and so will take longer to arrive at the next gap where it will find itself closer to the correct phase. If it reaches a gap after the correct time, it will gain too much energy and so will reach the next gap sooner, or, again, closer to the desired phase. In the case of most circular accelerators this mechanism is overpowered by the increase in path length with increasing energy. In the circular accelerator a particle that receives too much energy travels on a larger orbit and so arrives later in phase at the accelerating gap for its next traversal. For this reason the stable or "equilibrium" phase in circular accelerators is on the falling side of the field wave or between 90 and 180°. In both accelerators the mechanism acting to restore the particle to the correct phase sets up phase oscillations that are stable and that gradually decay in amplitude as the particle is accelerated.

Synchronous accelerators are most easily understood by analyzing the rather complex electric-field patterns into traveling waves. The interactions of the particles and the various traveling-wave components average out to zero except in the case of the one component which travels with the velocity of the particles in the same direction as the particles are traveling. If s represents distance traveled by the particles, this component can be represented by

$$E_s = E \sin\left(\int \omega\, dt - \int \frac{\omega}{v}\, ds + \phi_0\right) \tag{9.15}$$

where ω ($= 2\pi \times$ frequency of accelerating signal) may be a function of time; v, the wave velocity, may be a function of time and position; ϕ_0 is the so-called "equilibrium phase"; and E is defined by

$$E = \frac{V}{s_0} \tag{9.16}$$

where V is the voltage across the accelerating gap and s_0 is the distance from one accelerating gap to the next. A choice of parameters can be made such that small errors in phase will not eventually cause the particles to fall out of step with the accelerating wave and so be lost.

If the particle is traveling on the "equilibrium orbit" at the equilibrium phase ϕ_0, ω and v must have been adjusted so that

$$\int \omega\, dt = \int \frac{\omega}{v}\, ds \tag{9.17}$$

Suppose now that the particle is not at the equilibrium phase but is at a phase $\phi_0 + \Delta\phi$. Associated with the phase error will be a momentum error Δp from the equilibrium momentum p_0 and a position error Δs from the position of the equilibrium particle. Since a phase change of 2π results in a position change Δs of one wavelength, it is evident, if the particle is traveling on a linear path, that in general

$$\Delta\phi = -\frac{\omega\, \Delta s}{v} \tag{9.18}$$

The momentum error Δp is given by

$$\begin{aligned}\frac{\Delta p}{p_0} &= \frac{1}{p_0}\left[\frac{m_0\dot{s}}{(1 - \dot{s}^2/c^2)^{1/2}} - \frac{m_0 v}{(1 - v^2/c^2)^{1/2}}\right] \\ &= \frac{\Delta\dot{s}}{v(1 - v^2/c^2)}\end{aligned} \tag{9.19}$$

Δp also appears in the equation of motion

$$\frac{d}{dt}(p_0 + \Delta p) = eE \sin(\phi_0 + \Delta\phi) \tag{9.20}$$

whence, for small $\Delta\phi$,

$$\Delta\dot{p} = eE\, \Delta\phi \cos\phi_0 \tag{9.21}$$

Equations (9.18), (9.19), and (9.21) can now be combined to give a second-order differential equation in one of the variables. Usually $\Delta\phi$ is the variable chosen, and the following equation is derived:

$$\frac{d}{dt}\left[\frac{p_0}{v(1 - v^2/c^2)}\frac{d}{dt}\left(\frac{v\, \Delta\phi}{\omega}\right)\right] + eE\, \Delta\phi \cos\phi_0 = 0 \tag{9.22}$$

This equation applies to the linear accelerator in which ω is kept constant. Rewriting in terms of $\beta(= v/c)$ and $\gamma[= (1 - v^2/c^2)^{-1/2}]$ we obtain

$$\frac{d}{dt}\left[\gamma^3 \frac{d}{dt}(\beta\, \Delta\phi)\right] + \frac{\omega e E}{m_0 c}\Delta\phi \cos\phi_0 = 0 \tag{9.23}$$

If the particle travels on a circular or spiral orbit in a magnetic field, as in the synchrotron or the synchro-cyclotron, the procedure for deriving the phase motion is analogous to that given above but differs slightly in detail. In this case a momentum error results not only in an azimuthal position error but also in a radius error Δr from the equilibrium radius r_0. In a circular machine the frequency of the accelerating signal is an integral multiple of the frequency of revolution; this integer will be represented by h. The phase error will have a magnitude h times the error $\Delta\theta$ in the azimuthal position of the particle

$$\Delta\phi = -h\, \Delta\theta \tag{9.24}$$

and

$$\begin{aligned}\Delta\dot{\phi} &= -h\, \Delta\dot{\theta} \\ &= -h\left(\frac{\Delta(r\dot{\theta}) - \dot{\theta}\, \Delta r}{r}\right)\end{aligned} \tag{9.25}$$

To the first approximation

$$\Delta\dot{\phi} = \frac{hv}{r_0}\left(\frac{\Delta r}{r_0} - \frac{\Delta\dot{s}}{v}\right) \tag{9.26}$$

Since $p = Ber$ and $B \propto r^{-n}$ for stability of betatron oscillation [cf. Eqs. (9.4) and (9.7) above]

$$\frac{\Delta p}{p_0} = (1 - n)\frac{\Delta r}{r_0} \tag{9.27}$$

Substituting in (9.26) for Δr from (9.27) and $\Delta\dot{s}$ from (9.19) we obtain

$$\Delta\dot{\phi} = \frac{hv}{r_0}\frac{\Delta p}{p}\left(\frac{1}{1 - n} - \frac{1}{\gamma^2}\right) \tag{9.28}$$

In a circular accelerator the equation of motion is

$$\frac{1}{r}\frac{d}{dt}(rp) = \dot{p} + \frac{\dot{r}p}{r} = eE \sin(\phi_0 + \Delta\phi) + eE' + e\dot{r}B_z$$

But $\dot{r}p/r = e\dot{r}B_z$; hence

$$\dot{p} = eE \sin(\phi_0 + \Delta\phi) + eE' \tag{9.29}$$

In this equation E' represents other accelerating or decelerating forces such as betatron acceleration by the changing flux inside the synchrotron orbit or energy loss forces due to electromagnetic radiation in electron synchrotrons. For the betatron acceleration term

$$E'_b = \frac{1}{2\pi r}\int \frac{\partial B}{\partial t} 2\pi r\, dr \tag{9.30}$$

The term describing radiation deceleration for electrons is

$$E'_r = 9.60 \times 10^{-10}\gamma^4/r^2 \qquad (9.31)$$

where E' is in volts per meter and r is in meters. The effective electric field E of Eq. (9.29) is given by

$$E = \frac{V}{2\pi r}$$

where V is the sum of the peak voltages applied across all accelerating gaps. We now rewrite (9.29) for the synchrotron case, neglecting, for this study, the radiation term

$$r\dot{p} = \frac{eV}{2\pi}\sin(\phi_0 + \Delta\phi) + \frac{\phi}{2\pi}\int \frac{\partial B}{\partial t} 2\pi r\,dr \qquad (9.32)$$

Hence $\dot{p}_0\,\Delta r + r_0\,\Delta\dot{p} = \frac{eV}{2\pi}\Delta\phi\cos\phi_0 + e\frac{\partial B}{\partial t} r_0\,\Delta r$

Since $\dot{p}_0 = er_0\,\partial B/\partial t$

$$\Delta\dot{p} = \frac{eV}{2\pi r_0}\Delta\phi\cos\phi_0 \qquad (9.33)$$

It is now possible to combine (9.28) and (9.33) to obtain the phase equation for the synchrotron:

$$\frac{d}{dt}\left[\frac{\gamma\,\Delta\phi}{1/(1-n) - 1/\gamma^2}\right] = \frac{heV\cos\phi_0}{2\pi r_0^2 m_0}\Delta\phi \qquad (9.34)$$

In the synchrocyclotron the magnetic field B is constant and radiation losses are negligible; hence E' in Eq. (9.29) is zero. r_0 is not constant as in the synchrotron. The harmonic number h is usually unity. Also n is very small and can be neglected in this treatment. Hence Eq. (9.28) becomes

$$\Delta\dot{\phi} = \frac{v}{r_0}\frac{\Delta p}{p_0}\left(1 - \frac{1}{\gamma^2}\right)$$
$$= \frac{eB}{m_0{}^2c}\frac{\beta}{\gamma^2}\Delta p \qquad (9.35)$$

Equation (9.32) takes the form

$$r\dot{p} = \frac{p\dot{p}}{Be} = \frac{eV}{\pi}\sin\phi \qquad (9.36)$$

where V is the dee-to-dee voltage. Hence

$$\frac{1}{m_0c}\frac{d}{dt}(p\,\Delta p) = \frac{d}{dt}(\beta\gamma\,\Delta p) = \frac{e^2BV}{\pi m_0 c}\Delta\phi\cos\phi_0 \qquad (9.37)$$

Combining Eqs. (9.35) and (9.37) gives the phase equation for the synchrocyclotron

$$\frac{d}{dt}(\gamma^3\,\Delta\dot{\phi}) = \frac{(e/m_0)^3VB^2}{\pi c^2}\Delta\phi\cos\phi_0 \qquad (9.38)$$

Equations (9.23), (9.34), and (9.38) describe stable, damped oscillations in phase, provided only that ϕ_0 lies in the appropriate range. Equation (9.23) indicates that ϕ_0 should lie between 0 and $\pi/2$ for the linear accelerator. In the magnetic accelerators, since $1/(1-n)$ and γ both are greater than unity, $\cos\phi_0$ must be negative for stability. Since $\sin\phi_0$ must be positive to permit acceleration, ϕ_0 must lie between $\pi/2$ and π.

The further implications of these equations will be discussed in the sections that follow.

6. The Linear Accelerator

Linear accelerators (often referred to as "linacs") have been used for acceleration of electrons, protons, and heavy ions. The accelerating systems in protons and heavy-ion machines are similar. Electron linear accelerators use a quite different structure as described below.

Operating accelerators of both types are described in detail in the "Linear Accelerator Issue" of the *Review of Scientific Instruments* and in Lloyd Smith's article in the "Encyclopedia of Physics" [8].

The linear accelerator is simple in concept and convenient for experimental use since the beam is easily extracted. Its chief drawback is its very high power consumption. A linear accelerator of length L meters, excited by a field whose free-space wavelength is λ meters, and accelerating particles to a final energy of T ev, requires a total power P given by (cf. "Particle Accelerators," p. 323)

$$P = \frac{CT^2\lambda^{1/2}}{L} \quad \text{watts} \qquad (9.39)$$

where C is about 3×10^{-8} for proton machines and about 10^{-7} for electron machines. For a 50-Mev proton linac 30 m long, excited at 200 Mc, this relation indicates a power consumption of 3 Mw. At such high power levels it is necessary to use linear accelerators in pulsed service at low duty cycle.

Equation (9.39) indicates a choice of frequency as high as possible. Electron linacs usually are operated at 3,000 Mc/sec. But in proton linacs the injection energy (500 to 750 kev) and velocity are relatively low. At so low a phase velocity the accelerating field pattern develops a strong radial variation which can be reduced only by lowering frequency. The highest tolerable frequency usable in a proton linac is about 200 Mc/sec.

The Proton Linear Accelerator. In 1965 the highest energy attained with a proton linear accelerator was 70 Mev. Several linear accelerators have been built for lower energies, some for use as injectors for large synchrotrons. At Brookhaven a design study has been made for a 500-Mev linac to replace the present 50-Mev injector of the 33-Gev AGS, and at Los Alamos a design has been evolved for an 800-Mev machine for use in meson studies.

All operating linacs take the form of cylindrical cavities excited at frequencies between 100 and 200 Mc/sec and operated in the TM_{010} waveguide mode. The drift tubes shown in Fig. 9.7 are supported from the wall by stems which make only minor perturbations in the field pattern. Each drift tube is carefully shaped to maintain a uniform average field along the accelerator. Since the field is kept uniform the gap voltages increase as the drift tubes become longer. To maintain synchronism the drift tube plus gap length is maintained at a value of $\beta\lambda$, where λ is the free-space wavelength of the exciting field. In the nonrelativistic range where existing proton linear

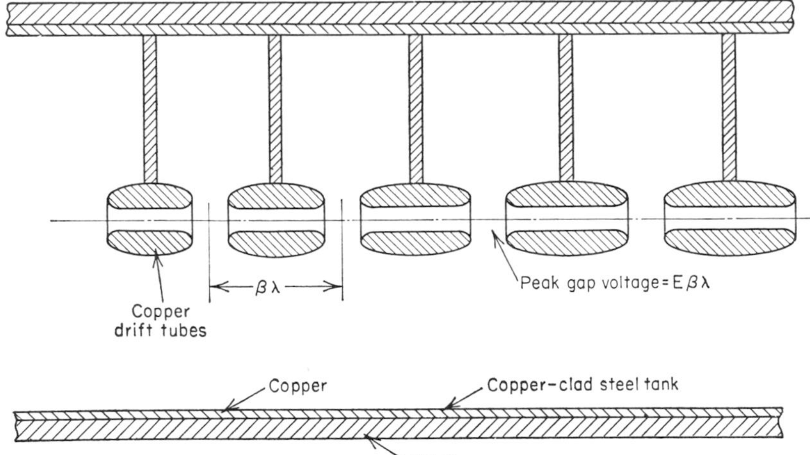

FIG. 9.7. The proton linear accelerator.

accelerators operate, it follows from Eq. (9.20) that β is proportional to the square root of distance along the proton path.

In the nonrelativistic range, Eq. (9.23) has the asymptotic solution

$$\Delta\phi = \Delta\phi_i(\beta_i/\beta)^{3/4} \cos [2(K\beta)^{1/2} - 2(K\beta_i)^{1/2}] \quad (9.40)$$

where the i indicates the initial value of the variable and

$$K = \frac{\omega m_0 c \cos \phi_0}{eE \sin^2 \phi_0} \quad (9.41)$$

K is existing proton linear accelerators has a value of about 1,000. The momentum error follows from (9.18) and (9.19). The relative error in kinetic energy ($\Delta T/T_0$) follows from the momentum error and is proportional to $(\beta)^{-5/4}$. The complete expression for the amplitude of the energy error is

$$(\Delta T/T_0)_{\max} = 2K^{-1/2} \cot \phi_0 \beta_i^{3/4} \beta^{-5/4} \Delta\phi_i \quad (9.42)$$

Radial Motions. The radial motions of the particles in a linear accelerator are influenced by two field components E_r and B_θ, both of which are derivable, using Maxwell's equations, from E_z. To a first approximation these components are given by

$$E_r = -\frac{r}{2}\frac{\partial E_z}{\partial z} \quad (9.43)$$

$$B_\theta = \frac{r}{2c^2}\frac{\partial E_z}{\partial t} \quad (9.44)$$

The form of these components can be derived from (9.15), which is the expression for E_z. The radial force equation, $m\ddot{r} = eE_r - evB_\theta$, becomes

$$m\ddot{r} = \frac{er\omega}{2v} E_z \left(1 - \frac{v^2}{c^2}\right) \cot \phi \quad (9.45)$$

The second term in the parentheses is negligible for proton linear accelerators. This term was the contribution of B_θ. The remaining term corresponds to a strong defocusing electric field. In modern proton linear accelerators this field is of the order of 1,000 volts/cm at a distance of 1 cm from the axis. Fields of this order are sufficient to blow up a beam of protons in a relatively short distance. Three procedures are available to circumvent this difficulty. One method requires that only the region very close to the peak of the accelerating wave be used, so that cot ϕ_0 is small. At this point weak electrostatic focusing can be made to compensate the reduced defocusing field. If this method is used, the phase acceptance angle is very small and only a small fraction of the injected particles will be accelerated. The second solution involves the use of grids across the upstream faces of the drift tubes. These modify Eq. (9.43) since charges are now present in the space under consideration. The focusing mechanism is easily seen from the sketches of Fig. 9.8.

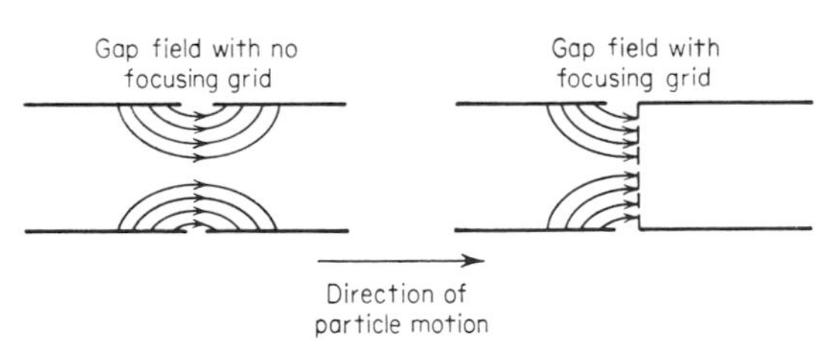

FIG. 9.8. Focusing by means of grids in the linear accelerator.

The defocusing field region to the right of the grid is removed and only focusing forces remain active. This method is quite effective in eliminating the defocusing force but it results in a considerable loss in intensity because of interception of charge by the grids. Both of these techniques have now been superseded by alternating-gradient focusing which is discussed in Sec. 11. Alternating-gradient focusing is achieved by the inclusion inside the structure of each drift tube of a quadrupole magnet. This method is entirely effective and results in negligible beam loss.

The Electron Linear Accelerator. In the electron linac the electron velocity is substantially that of light after a short perliminary acceleration to an energy of a few Mev. In this case it is possible to use the TM_{01} mode in a cylindrical waveguide for the accelerating field pattern. Since phase velocities in empty waveguides always are greater than the velocity of light it is necessary to load the waveguide periodically with irises, as shown in Fig. 9.9, to bring the phase velocity to the velocity of light.

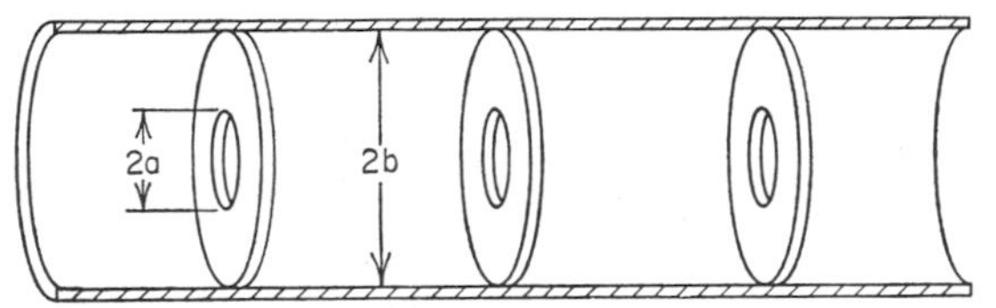

FIG. 9.9. Cutaway sketch of iris-loaded accelerating system for an electron linear accelerator. (*From M. S. Livingston and J. P. Blewett, "Particle Accelerators," McGraw-Hill, New York*, 1962, p. 324.)

In practice the phase velocity is set exactly equal to the velocity of light. This choice makes much simpler the analysis of the phase motion. Equation (9.18) can be replaced by

$$\phi = \omega\left(t - \frac{s}{c}\right) + \phi_0 \tag{9.46}$$

Equation (9.20) can be written simply

$$\frac{dp}{dt} = eE \sin \phi \tag{9.47}$$

The solution of these two equation is

$$\cos \phi = \cos \phi_i + \frac{\omega m_0 c}{eE} [\gamma_i(1 - \beta_i) - \gamma(1 - \beta)] \tag{9.48}$$

As energy increases and β approaches unity $\cos \phi$ tends to a fixed final value, provided that E is high enough that the cosine is not greater than unity.

Radial focusing is not a problem in the electron linac because the $1 - \beta^2$ term in the defocusing force [cf. (9.45)] reduces the defocusing effect to make it negligible.

In the absence of real phase stability, construction tolerances in the electron linac are very severe. Critical dimensions must be held to about 0.0002 in. in the 3,000-Mc waveguide structure.

In spite of these severe requirements electron linacs have been run at energies above 1 Gev and, in 1965, a 10- to 20-Gev electron linac was nearing completion at Stanford. Later, by addition of extra power sources this energy may be increased to about 40 Gev.

7. The Cyclotron and the Synchrocyclotron

The invention of the cyclotron was stimulated by Wideröe's paper in 1928, describing his linear accelerator. E. O. Lawrence investigated the possibility of using a magnetic field to return a beam of particles to an accelerating gap so that the same gap could be used for repetitive acceleration. Since the angular velocity of a particle of charged e an mass m in a magnetic field B is

$$\frac{v}{r} = \frac{eB}{m} \tag{9.49}$$

which is independent of the radius of the particle orbit, it appeared that repetitive acceleration could be achieved if an alternating electric field of constant frequency were applied across a radial gap. The first cyclotron was built by Lawrence and Livingston in 1931, and reports of acceleration to energies over a million volts were published in 1932 [9]. The configuration of the cyclotron is illustrated in Fig. 9.10. The ion beams are contained in a split cylindrical pill-box structure the two halves of which are referred to as "dees" for obvious reasons. These dees are enclosed in a vacuum chamber between the poles of an electromagnet capable of maintaining a field of the order of 18 kilogauss. No important change has been made in this design in the numerous cyclotrons that have been built since 1931.

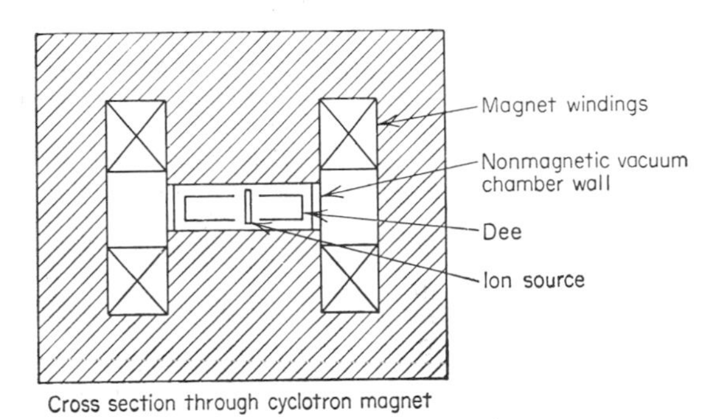

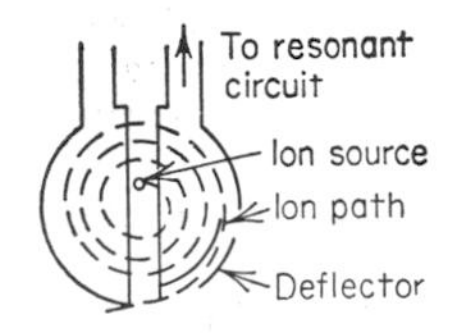

FIG. 9.10. The cyclotron.

Electric and Magnetic Focusing. As the particles in the cyclotron traverse their rather long spiral path in the horizontal plane they experience several vertical forces. The electric vertical forces are analogous to those already discussed in the linear accelerator. If the particles cross the gap when the field is rising, the focusing field they experience when they approach the gap is overcompensated by the stronger defocusing field that exists after the gap is passed; if the field is decreasing as they cross the gap, a net focusing force is experienced. If the gap is crossed at the peak field when the field is momentarily constant, a small focusing force is experienced because the particle has been accelerated and so passes through the defocusing field past the gap more rapidly than it traversed the focusing field as it approached the gap.

Space-charge forces are always present and contribute a small defocusing component because of the mutual repulsion of like charges.

To compensate for these various forces which generally add up to a defocusing total, a magnetic focusing force is built into the cyclotron by tapering the poles of the magnet slightly so that the magnetic field decreases slightly with radius. This introduces both radial and vertical focusing forces of exactly the type already discussed in connection with the betatron.

The rate of variation of field with radius is held to a minimum, both because as high a field as possible is required at the outer radius of the machine and because any gross change in magnetic field will destroy the resonance condition (9.49) and cause the particles to drift away from the accelerating phase of the electric field.

Operating Phase. If the operating frequency is chosen to correspond to the magnetic field at the center of the orbit, the angular velocity of the particles will decrease as their orbit radius increases into regions of lower field and they will fall behind the accelerating wave. Some improvement is made by choosing a compromise frequency corresponding to an intermediate region of field. The particles are started at or near the peak of the accelerating wave, a situation more or less automatically maintained by the fact that this is the point at which the field extracts the most ions from the ion source. For a time their angular velocity is higher than that of the field and their phase advances into the falling field region; then after the resonance field region is passed the phase drops back into the rising field region until it passes into the region of retarding fields. By this time it has reached the maximum energy attainable in this type of cyclotron.

Maximum Energy Attainable. The above considerations set a very effective upper limit to the energy attainable with a given field configuration. A further limitation is introduced in the energy range of the order of 20 million volts at which relativistic effects become appreciable, and the angular velocity of the particle experiences an additional deceleration. This also causes a continual phase retardation. Only one procedure is available in the conventional cyclotron to raise this upper limit. Since the phase error increases during each revolution by an amount depending only on its radial position, the upper limit can be increased by decreasing the number of revolutions required to reach a given radius. The upper limit is roughly proportional to the rate of increase of velocity and so to the square root of the accelerating peak voltage. The peak energy in millions of volts for deuterons, for example, is given very approximately by twice the square root of the dee voltage in kilovolts. For 100 kv on the dees, the maximum deuteron energy is thus about 20 million volts. Since the dee structure is massive and complicated, the radio-frequency power necessary to drive it to these voltages is of the order of 100 kw or more.

Ion Source. The ion source in the cyclotron is usually a semienclosed arc discharge located between the dees near the center of the chamber. The discharge is maintained in any one of several ways along the lines of force of the magnetic field. Probe electrodes graphically described by cyclotronists as "auspullers" are connected to one of the dees and extend almost to the hole in the side of the ion-source housing through which the ions emerge. The alternating field of the dee extracts an ion beam from the source, and the ions are accelerated in one revolution by an amount sufficient that their orbit radius expands to clear the ion-source structure on their second revolution.

Utilization of the Accelerated Beam. The accelerated beam can be used to bombard an internal target which is mounted at the maximum orbit radius, or it can be extracted by a high electrostatic field applied to an external deflecting electrode, as indicated in Fig. 9.10.

Sizes and Applications of Conventional Cyclotrons. With radio-frequency power supplies of the order of 200 kw the upper energy limit attainable is such that there is no advantage in building magnets having diameters greater than about 60 in. (For esoteric reasons the size of a cyclotron is usually quoted in terms of the diameter of the magnet pole in inches.) In a 60-in. cyclotron, deuterons, which are the particles most commonly used in cyclotrons, can be accelerated to about 20 million volts.

The beam obtainable from a cyclotron, if used internally, is copious, often of the order of 1 ma. It is not remarkably homogeneous in energy since there is an uncertainty of one or more in the number of accelerations each particle has undergone. The internal beam is more useful in the production of artificial radioactivities. In this field the cyclotron complements nuclear reactors since it can produce positron-emitting radioactive isotopes not found in the products of neutron bombardment.

Ten to twenty per cent of the cyclotron beam can be extracted and can be analyzed by auxiliary fields to permit precise studies of nuclear reaction thresholds or cross sections, or for use in scattering experiments. The extracted beam diverges rapidly but can be reconverged by focusing magnets and utilized at considerable distances from its source.

The Synchrocyclotron. It was evident for many years that cyclotron beams could be taken to higher energies if the frequency of the accelerating signal was varied as the particle was accelerated. That this process could be carried out stably and controllably was not appreciated, however, until about 1945 when the principles of synchronous accelerators outlined in Sec. 5 were first evolved with the invention of the cynchrotron. It was immediately evident [10] that the dee voltage of the cyclotron could be reduced materially and that the particles would travel in stable orbits whose radius depended only on the frequency of the accelerating signal. Equation (9.38) is applicable in this situation. The instantaneous angular frequency of the phase oscillation is

$$\frac{(e/m_0)^3 V B^2 \cos \phi_0}{\pi c^2 \gamma^3}$$

If V is, say, 10 kv and the particle energy is 300 Mev, the frequency of phase oscillation is about 1/1,000 of the rotation frequency of the ions. A complete solution of (9.38) shows that the oscillation is damped as energy increases so that the acceleration can be carried on to whatever limit is set by the greatest radius at which the magnetic field is usable. This upper limit is set by the fact that the field falls off rapidly with radius as the edge of the magnet pole is approached. The limit is not set in practice by the obvious limit at $n = 1$, where, from (9.11), the radial betatron type of oscillation becomes unstable, but rather by the passage of n through a value of 0.2 at which the radial betatron oscillation frequency is exactly twice the vertical betatron oscillation frequency. At this point a resonance takes place between the two oscillations and the vertical oscillation amplitude increases until the beam is lost. This limit is generally found about one magnet gap width in from the edge of the magnet pole.

The variation of frequency applied to the dees is accomplished mechanically. In the largest cyclotrons with magnet-pole diameters of the order of 200 in. and peak energies of the order of 700 Mev, the frequency changes during the acceleration cycle by a factor of almost 3. This swing is accomplished by a rotating or vibrating capacitor included in the resonant circuit connected to the dees.

With variable frequency acceleration the cyclotron is now referred to as a "synchrocyclotron." Aside from the variable frequency the synchrocyclotron differs from the conventional cyclotron in one other important feature. Whereas in the conventional cyclotron many bunches of ions are accelerated simultaneously, in the synchrocyclotron only the bunch that is synchronous with the accelerating frequency can be accelerated at one time. Consequently, the output of the synchrocyclotron consists of one short burst of particles at the end of each cycle of frequency variation, or roughly 100 times per second for most synchrocyclotrons. Its average intensity, accordingly, is much lower than that of the conventional cyclotron. This is not a very serious disadvantage, however, since the energy range around 200 to 500 Mev in which it is most useful is one in which the greatest interest is in single nuclear events. The synchrocyclotron has been most valuable to date in providing information about the production and properties of mesons.

The synchrocyclotron giving the highest energy is the 184-in machine at the University of California in which protons are accelerated to 720 Mev. At Dubna in the Soviet Union a 680-Mev synchrocyclotron has been in use since 1954. A 600-Mev synchrocyclotron was completed in 1958 at the CERN laboratory in Geneva, Switzerland.

Other Cyclotron Types. The latest cyclotron variation is the spiral-ridge or isochronous cyclotron using alternating-gradient focusing. Most new cyclotrons are of this type, which will be discussed in Sec. 11.

8. The Electron Synchrotron

The principles of synchronous acceleration discussed in Sec. 5 and already applied in the discussion of the linear accelerator and the synchrocyclotron were first presented [11] in 1945 and at that time were applied to an electron accelerator in which orbit radius in a magnetic field was to be held constant and the magnetic field and frequency of an accelerating signal were to be varied to keep in step with the momentum and frequency of revolution of the electrons as they were accelerated. The guiding magnet structure was similar to that in the betatron; in fact, as is evident from Fig. 9.11, the device differed in appearance from a betatron only in that the accelerating flux magnet structure was absent and was replaced at one point in the vacuum chamber by a radial insulated gap across which a radio-frequency electric field was applied.

The technical development of this device was materially aided by the restoration of a small flux

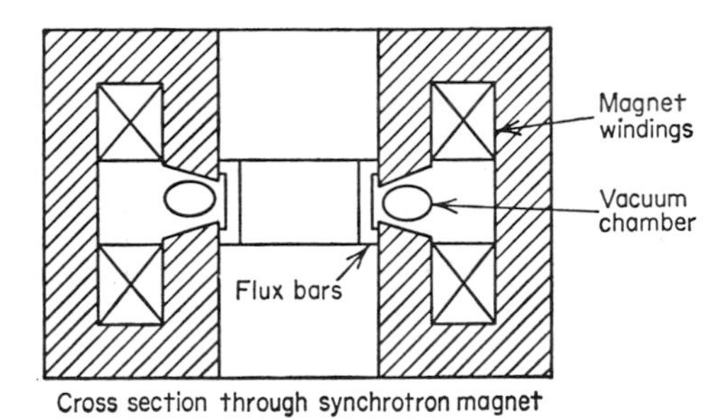

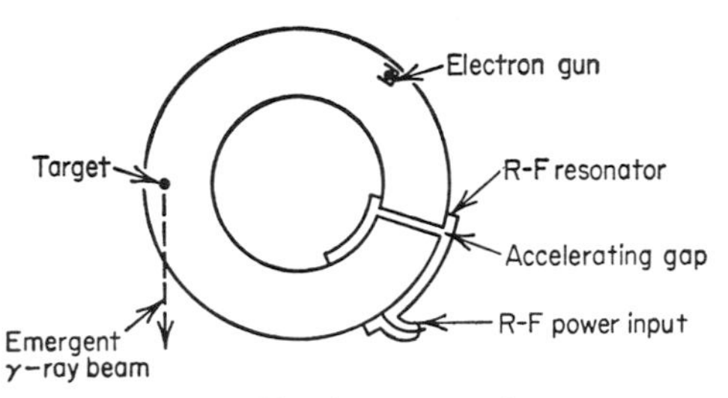

FIG. 9.11. The electron synchrotron.

structure which converted the machine to a betatron up to the energy range of 4 or 5 million volts. At this energy the velocity of an electron is within 1 per cent of the velocity of light and its frequency of revolution has become almost constant. It was thus possible to permit the flux structure to saturate at this energy and from that point on to accelerate with a constant frequency signal. The radio-frequency problems became negligible with frequency variation unnecessary. The final fraction of 1 per cent of variation in electron revolution frequency is automatically compensated during the early acceleration stages by a small variation in orbit radius so that initially, after the betatron stage, the device is invoking the synchrocyclotron mechanism to keep the particles in step with the accelerating wave.

The electron synchrotron presents two important advantages over the betatron. Its magnet structure consists only of a light ring-shaped guiding magnet and a few light "flux bars," a structure lighter and cheaper than that of a betatron by a factor of at least 3; and the radiation losses which require complex compensation in the betatron are automatically compensated in the synchrotron by phase shifts which change the stable phase to a value at which just enough extra energy is added to compensate for the radiation losses. This is easily seen from the fact that radiation energy losses tend to make the electron travel on an orbit of smaller radius in the guiding magnetic field and thus to arrive at an earlier time. Since, in the synchrotron as in other magnetic accelerators, the stable phase is on the falling side of the accelerating wave, this means that the earlier arrival at the accelerating gap results in addition of more energy.

Just as in the synchrocyclotron, the behavior of the electrons in the synchrotron is described by the betatron oscillation relations (9.11) and (9.12) and by the phase oscillation relation (9.34). The vertical motion consists only of an ordinary betatron oscillation. The radial motion includes a betatron oscillation whose frequency is about half the rotation frequency of the electrons and a phase or "synchrotron" oscillation whose frequency is about 1/1,000 of the rotation frequency.

Several electron synchrotrons have been built and operate in the energy range up to about 1 Gev. These machines are particularly useful in studies of photonuclear reactions. The electron beam is allowed to strike an internal target, and the resulting beam of high-energy gamma rays is employed outside of the machine.

9. The Proton Synchrotron

The only accelerator that has penetrated appreciably into the billion-volt range of energies is the proton synchrotron [12]. The Brookhaven proton synchrotron, nicknamed the Cosmotron, reached an energy of 3 Gev in 1952. In 1953 a proton synchrotron in Birmingham, England, reached its design energy of about 1 Gev. During 1954 the University of California Bevatron was operated successfully at 6 Gev. A 10-Gev machine has been built in the Soviet Union and was in operation in 1957. Other conventional proton synchrotrons are in operation at Saclay, France (3 Gev), Princeton (3 Gev), the Rutherford Laboratory in England (7 Gev), and the Argonne National Laboratory (12.5 Gev).

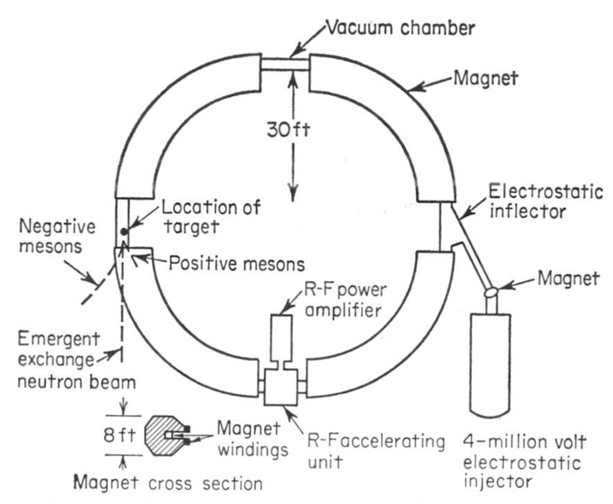

FIG. 9.12. The proton synchrotron (the Cosmotron).

The basic features of the proton synchrotron are shown in Fig. 9.12, which represents the Brookhaven Cosmotron. It differs from the electron synchrotron in two important respects. First, as was indicated in Sec. 4, radiation problems are negligible with protons whereas an electron synchrotron in the billion-volt range requires very high accelerating voltages to compensate for the radiation losses indicated by (9.13). Second, protons do not approach relativistic velocities until their energy is in the billion-volt range. If, as is the case in the Cosmotron, protons are injected at 3 or 4 million volts, at which energy their velocity is less than one-tenth of the velocity of light, the frequency of revolution of the protons in their orbit will vary during the acceleration cycle by a factor of more than 10. The frequency of the accelerating signal must vary over the same range and be accurately controlled at all times during the accelerating cycle. The only other problems encountered in the design of the multibillion-volt proton synchrotron are associated with its size. Using the highest magnetic fields easily attainable with iron magnets, the radius of the proton orbit in the Cosmotron is 30 ft.

The radial and vertical excursions of the protons from their equilibrium orbit consist, as in the electron synchrotron, of radial and vertical betatron oscillations derivable from the considerations outlined in Sec. 4 and radial synchrotron oscillations discussed in Sec. 5. The betatron oscillation amplitudes depend primarily on the initial conditions set up at injection. Injection in the Cosmotron, for example, is from a Van de Graaff electrostatic machine which delivers a beam so well focused that the initial betatron oscillation amplitudes are less than a couple of inches. The radial synchrotron oscillation amplitudes follow from Eqs. (9.27), (9.28), and (9.34). From (9.34) the initial frequency of synchrotron oscillation is about 2,000 cycles, since the initial frequency of revolution is about 300 kc, the n value is about 0.7, the accelerating voltage is about 2,000 volts peak, and ϕ_0 is 150°. This corresponds to radial excursions of maximum value about 4 in./radian of phase oscillation. Initially the acceptance range of the Cosmotron is about 2 radians in phase so that a total of about 8 in. in radius must be allowed for these oscillations. When allowances were being made for magnetic-field fringing, errors in radio frequency, gas scattering, thickness of vacuum chamber walls, etc., the final magnet-gap dimensions chosen for the Cosmotron were 9 in. vertically by 36 in. radially. The magnet required to supply a peak field of 14,000 gauss in this gap is about 8 by 8 ft in cross section and weighs about 2,000 tons.

At the peak field, the energy stored in the magnetic field in this magnet gap is about 13 million joules, and the magnetic field is taken from its remanent value to peak in 1 sec. The Cosmotron magnet supply is a motor-flywheel a-c generator-rectifier set capable of handling 20,000 kva.

Accurate timing of the magnet pulse is no longer possible at such power levels, and the magnetic field, rather than time, must be taken as the independent variable for control of the radio-frequency cycle. A signal proportional to the rate of change of magnetic field is derived from a pickup loop in the magnetic field; this signal is integrated electronically and supplied to a computer which yields a control signal for the radio-frequency oscillator. This oscillator has a saturable ferromagnetic core inductance as part of its resonant circuit, and the computer output supplies the saturation current at such a rate that the radio frequency has the appropriate value for the magnetic field at all times. The oscillator output is then amplified and supplied to the gap at which the proton acceleration takes place. The details of this and the other control procedures are presented in detail in the design studies in the literature [12].

The accelerated beam in the Cosmotron may be directed against an internal target or extracted from the machine [13] and brought to external target areas. By 1965 most of the Cosmotron's research program was carried on in three external beams.

The average intensity in the Cosmotron lies between 10^{11} and 10^{12} protons per pulse. The acceleration cycle is repeated every 5 sec. This intensity is quite adequate for the particle-physics studies for which the machine is intended.

Particles at energies below the peak energy available are easily and controllably obtainable by cutting off the accelerating signal at the magnetic field corresponding to the desired energy.

The Saclay and Rutherford Laboratory machines are similar in configuration to the Cosmotron. The Bevatron, the Dubna accelerator (known as the "synchrophasotron"), and the Argonne (ZGS) machine use H magnets (magnets with return legs on both sides of the vacuum chamber). The ZGS magnet has a uniform field rather than one that has a radial gradient. Focusing in the ZGS is by shaping of the ends of the magnet sectors.

10. Alternating-gradient Focusing

It is not possible, using electrostatic or magnetostatic or both fields in free space, to focus a beam of charged particles in all planes through the axis of the beam. This fact can be demonstrated as follows: Let E_x, E_y, B_x, and B_y be the components of the electrostatic and magnetostatic fields acting on a particle of charge e and mass m, traveling with velocity v in the z direction. The forces F_x and F_y acting on the particle are

$$F_x = eE_x - evB_y \tag{9.50}$$
$$F_y = eE_y + evB_x \tag{9.51}$$

whence

$$\frac{\partial F_x}{\partial x} + \frac{\partial F_y}{\partial y} = e \operatorname{div} \mathbf{E} - ev(\operatorname{curl} \mathbf{B})_z = 0 \tag{9.52}$$

Consequently, if F_x is a force tending to return the particle to the z axis, F_y must inevitably be a diverging force, and vice versa.

Fortunately this difficulty which applies to linear systems can be circumvented in circular systems in which the effect of the centrifugal-force term permits a weak focusing in both directions, provided the field configuration satisfies the type of condition already discussed for magnetic fields in connection with the betatron. This type of focusing is, however, weak, and the aperture required to contain the particle excursions in the proton synchrotron is so large that much higher energies than 10 billion volts are not attainable without unreasonable expenditures of money and materials.

This difficulty has been resolved by the strong focusing principle discovered in 1949 by Christofilos in Athens. His work was not published, and the principle was rediscovered and published in 1952 by Courant, Livingston, and Snyder [14]. These authors point out that successive application of strong focusing and defocusing fields can result in a very strong net focusing. This property is most easily seen through its optical analogue of a combination of two lenses having focal lengths f_1 and f_2 and separated by a distance d. The focal length F of the combination is given by

$$\frac{1}{F} = \frac{1}{f_1} + \frac{1}{f_2} - \frac{d}{f_1 f_2} \tag{9.53}$$

if $f_1 = -f_2$, then

$$F = \frac{f_1^2}{d} \tag{9.54}$$

which is positive regardless of the sign of f_1.

The simplest lens system of this type consists of a magnetic field having alternating transverse gradients. Suppose that a beam travels along the z axis with velocity v and that from $z = 0$ to $z = a$ we apply a magnetic field having the form $B_y = B'x$, where B' is a constant. Since curl $\mathbf{B}$ is zero, $B_x = B'y$. From $z = a$ to $z = 2a$ we apply a reversed field $B_y = -B'x$ and $B_x = -B'y$. If we trace through this field combination a particle having initially $\dot{x} = \dot{y} = 0$ and $x = y = 1$, we find that, at $z = 2a$,

$$\dot{x} = \dot{y} = K \sinh \frac{Ka}{v} \cos \frac{Ka}{v} - K \cosh \frac{Ka}{v} \sin \frac{Ka}{v} \tag{9.55}$$

where $K^2 = evB'/m$. If we expand in terms of the arguments of these functions,

$$\begin{aligned} \dot{x} = \dot{y} &= -\frac{2K}{3}\left(\frac{Ka}{v}\right)^3 + \text{higher-order terms} \\ &= -\frac{2e^2B'^2a^3}{3m^2v} + \cdots \end{aligned} \tag{9.56}$$

Evidently, so long as the higher-order terms in (9.56) are negligible, there will be focusing in both planes.

The technique works equally well with electric fields. To a first approximation it is necessary only to replace B' by E'/v, where the focusing fields have the form $E_x = \pm E'x$ and $E_y = \pm E'y$ and the above relations will still be valid.

In order that a succession of such lenses focus continually without increase in necessary aperture, the distance between successive image and object points must be less than twice the focal length of the separate lenses. This theorem also has its analogue in the case of a sequence of optical lenses. This limit is given by

$$\frac{Ka}{v} < 1.9 \tag{9.57}$$

or, if W is the energy of the particle,

$$B'va^2 < 7W \tag{9.58}$$

It is not essential that the gradients in the successive regions be exactly equal and opposite. The limits of variation permissible are discussed in detail in the paper by Courant, Livingston, and Snyder.

11. Application of Alternating-gradient Focusing to Accelerators

The alternating-gradient focusing technique described in Sec. 10 can serve to compensate the defocusing which has been so troublesome in the linear accelerator. Application of alternating-gradient focusing to the linear accelerator is discussed by Blewett [15]. In this case structures providing focusing electric fields or magnetic fields can be integrated into the accelerating structure. If a suitable choice of frequency is made, the required lens length can be identical with the drift-tube length and each drift tube can include lens elements that will supply the linear focusing gradients.

If a uniform magnetic field is superposed on the linear gradients of a succession of magnetic lenses, a focused beam can be made to travel in a circular path. If, now, all fields are made to increase with time, there is a suitable magnet structure for a synchrotron. The focusing can be made strong enough that, regardless of the orbit radius, apertures of the order of 4 by 4 in. are adequate to contain all the particle excursions from the equilibrium orbit. This results in a major reduction of the magnet cross section required. Figure 9.13 is a cross section of a magnet suitable for use in an alternating-gradient synchrotron. The pole shape of the next magnet sector, in which the gradient is reversed, is indicated in this figure by a dotted curve.

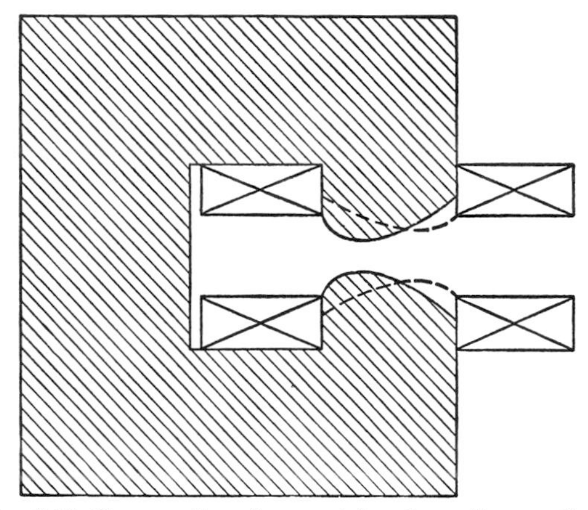

FIG. 9.13. Cross section of magnet for alternating-gradient synchrotron.

One danger is inherent in the designs of machines using alternating-gradient focusing. Because the focusing results from a combination of very strong focusing and defocusing forces, the particle beam is extremely sensitive to small misalignments of the focusing elements. The consequences of misalignment errors were first pointed out by Adams, Hine, and Lawson [16] and have been studied in detail by several design groups. The result of even the smallest misalignment is to produce a disastrous resonance if the number of betatron wavelengths around the machine is equal to an integer or close to an integral value. Machine design must be such as to avoid these and other dangerous "stop bands." In spite of this difficulty, however, it now appears that proton synchrotrons in the energy range up to and above 1,000 Gev are entirely within the realm of possibility.

A number of alternating-gradient synchrotrons are now in operation. The highest particle energy yet achieved has been reached in the 33-Gev AGS (alternating-gradient synchrotron) at Brookhaven. This machine reached its design energy in 1960. A similar machine is the 28-Gev CERN PS (proton synchrotron) built jointly by a group of Western European countries in Geneva, Switzerland, and brought into operation in 1959. In the Institute of Theoretical and Experimental Physics in Moscow (USSR), a 7-Gev alternating-gradient synchrotron is in operation. It was built as a model for a 70-Gev machine to be completed in Serpukhov (near Moscow) in about 1968.

Alternating-gradient synchrotrons are useful also for acceleration of electrons. A machine for about 1 Gev built at Cornell was the first electron accelerator of this type. The technique was extended to 6 Gev at Cambridge, Massachusetts, by a joint Harvard-MIT group whose accelerator came into operation in 1962. Three other machines have been based on the Cambridge design. A 7-Gev electron synchrotron in Hamburg, Germany, began operation in 1964. A similar machine for 6.5 Gev is under construction at Yerevan in the Soviet Union, and a 4-Gev accelerator is being built near Liverpool in England. Construction is also under way for a 10-Gev alternating-gradient synchrotron at Cornell University.

The Midwestern Universities Research Association (MURA) has explored fixed-field alternating-gradient (FFAG) accelerators very thoroughly. In these machines a field that increases rapidly with radius contains particles of all energies from injection to final energy. Alternating-gradient focusing is achieved by inclusion on the pole face of spiral ridges or by other pole-face modifications. Although FFAG accelerators give promise of quasi-continuous operation and high intensity, they are relatively very costly and hence proposals for construction of FFAG accelerators in the 10 to 15-Gev range have not received support. The FFAG principle has, however, been widely applied in the cyclotron field. A proposal for an FFAG cyclotron structure had been made in 1938 by L. H. Thomas [17] but was not given much attention. From the MURA studies it became evident that ridges on the pole face of the cyclotron could yield focusing forces and at the same time could keep the particles under acceleration in synchronism with a field of a constant frequency. FFAG cyclotrons with spiral or radial ridges are often called AVF (azimuthally varying field) cyclotrons. Many conventional cyclotrons and synchrocyclotrons have been converted or are being converted into AVF machines.

12. Meson Factories

The term "meson factory" is applied to an accelerator of high intensity providing protons in the energy range between 500 and 1,000 Mev, since such a machine would be peculiarly suited for studies of meson physics. In 1965 no meson factory was yet in construction in the United States but many design studies were completed or in progress.

At the Oak Ridge National Laboratory a study was completed for an 800-Mev spiral-ridge isochronous cyclotron which has now been superseded by a new Oak Ridge invention, the "separated orbit cyclotron" (SOC), in which protons travel a spiral path through a succession of small alternating-gradient magnets. Acceleration occurs as the protons pass through an array of radial cavities excited at a frequency of 50 Mc/sec. It is expected that this device will yield a continuous current of 1 to 5 ma at an energy of 800 Mev.

The SOC is under study also at the Canadian Chalk River Laboratory where it is considered to be promising for generation of intense beams of thermal neutrons. Currents of some tens of milliamperes would be required in this application.

At the University of California in Los Angeles an isochronous cyclotron in which negative ions are accelerated has been studied for use as a meson factory. At the periphery of the machine the negative ions would be stripped of their electrons by passing them through a foil or gas jet. As positive ions they would then be extracted by the cyclotron's magnetic field.

Seven-hundred-and-fifty to eight-hundred-Mev proton linacs with duty cycles of the order of 6 per cent and average currents of 1 ma have been proposed both at Yale and at Los Alamos. These linacs would use the conventional drift-tube structure (Fig. 9.7) excited at 200 Mc/sec up to an energy of about 200 Mev. At this point the structure would change to a variation of the iris-loaded structure (Fig. 9.9) excited at 800 Mc/sec.

The most advanced meson-factory project is a 500-Mev sector-focused machine now under detailed design at the E.T.H. in Zurich, Switzerland. This is to be a rather open structure into which 70-Mev protons are injected from a smaller sector-focused cyclotron.

13. Colliding Beams

Because relativistic particles gain mass as their energy is increased they become less effective in supplying energy for particle reactions. Most of the energy of a relativistic particle colliding with a particle at rest appears in the form of kinetic energy of the target particle. The maximum energy available for particle production can easily be shown to be

$$m_0c^2\left[\left(2+\frac{2W}{m_0c^2}\right)^{1/2}-2\right] \tag{9.59}$$

where W is the total energy of the particle. For example, to produce a proton-antiproton pair, an energy of about 6 Gev is required.

If, however, beams of equal energy could be made to collide, each beam need be only of 1 Gev to produce a proton-antiproton pair, since in this case all the kinetic energy of the particles is available for particle production. An appreciable number of collisions can take place between circulating beams of a few amperes. The best procedure appears to be to build up circulating particle beams in "storage rings" that intersect at one or more points. These are rings of alternating-gradient magnets whose field is held at a constant value. In 1965 no proton storage rings had been built but a design group at CERN was actively proposing construction of 28-Gev rings for use with the CERN PS. Electron-electron rings and electron-positron rings have been built and tested at lower energies, and it appears probable that a 3-Gev electron-positron ring will be built at Stanford for use with the new 2-mile linear accelerator.

14. Intensity Limits

In circular accelerators a space-charge limit is set when the charge density affects the focusing force to the extent that the number of betatron wavelengths around the ring becomes an integer. This effect has been analyzed in detail by Laslett [18] and others who show that an important part is played by image forces, both electrostatic images in the vacuum chamber wall and current images in the magnet poles. Both coherent and incoherent instabilities can develop, and some have been observed in the Cosmotron and in a 50-Mev electron FFAG model at MURA.

Space charge sets an upper limit to intensity dependent on aperture and injection energy since the space-charge limit first appears immediately after injection. The space-charge limit is roughly proportional to injection energy. In the Brookhaven AGS, for example, with 50-Mev injection the theoretical limit is about 3×10^{12} protons per pulse. It is proposed eventually to raise the injection energy to 500 Mev, in which case the limit will be about 4×10^{13} protons per pulse.

15. Conclusion

In 1965 the most important trend was toward higher proton energies. Design studies were in progress at Saclay, France, for a 60-Gev machine, at the Lawrence Radiation Laboratory for a 200-Gev alternating-gradient proton synchrotron, at CERN for a 300-Gev machine, and at Brookhaven for a 600 to 1,000-Gev accelerator. It is estimated that these machines will cost between $100 and $150 million per 100 Gev. These enormous expenditures are the subject of intense debate among high-energy physicists, accelerator designers, scientists in other disciplines, and representatives of the Government. The case for super-energy accelerators is presented in a collection of essays [19] by leading high-energy theorists. It seems probable that much further discussion will take place before these super-energy accelerators are built. In the minds of accelerator designers there is no doubt that they are entirely feasible.*

References

1. Wideröe, R.: On a New Principle for Production of High Potentials, *Arch. Elektrotech.*, **21**: 387–406 (1928).
2. Cockcroft, J. D., and E. T. S. Walton: Experiments with High Velocity Positive Ions, *Proc. Roy. Soc. (London)*, **A136**: 619–630 (1932).
3. Van de Graaff, R. J.: A 1,500,000 Volt Electrostatic Generator (A), *Phys. Rev.*, **38**: 1919 (1931). Tuve, M. A., L. R. Hafstad, and O. Dahl: High Voltage Techniques for Nuclear Physics Studies, *Phys. Rev.*, **48**: 315–37 (1935).

* *Note added in proof:* Several important developments have taken place since the above was written. (1) The Stanford two-mile linac has been brought into operation at an energy of about 18 Gev. (2) The CERN project for intersecting storage rings for colliding beam experiments has been approved and construction has been started. (3) The energy of the proposed new linac injector for the Brookhaven AGS has been dropped from 500 to 200 Mev. (4) Over a hundred sites in the United States have been proposed for a 200-Gev accelerator. A similarly large number have been submitted for a European 300-Gev machine. In Europe the list has been reduced by mutual agreement to twelve sites. In the United States all but six have been eliminated. A final choice among the six was made by the Atomic Energy Commission in December, 1966: Weston, Ill.

4. Bennett, W. H., and P. F. Darby: *Phys. Rev.*, **49**: 97, 422, 881 (1936). Alvarez, L. W.: *Rev. Sci. Instr.*, **22**: 705 (1951).
5. Kerst, D. W.: Acceleration of Electrons by Magnetic Induction, *Phys. Rev.*, **60**: 47–53 (1941).
6. Westendorp, W. F., and E. E. Charlton: A 100 Million Volt Induction Accelerator, *J. Appl. Phys.*, **16**: 581–593 (1945). Kerst, D. W.: Operation of a 300 MeV Betatron, *Phys. Rev.*, **78**: 297 (1950).
7. Blewett, J. P.: Radiation Losses in the Induction Electron Accelerator, *Phys. Rev.*, **69**: 87–95 (1946). Schwinger, J.: On the Classical Radiation of Accelerated Electrons, *Phys. Rev.*, **75**: 1912–1925 (1949).
8. *Rev. Sci. Instr.*, vol. 26, complete February issue, 1955. Smith, L.: Linear Accelerators, "Encyclopedia of Physics," vol. 44, pp. 341–389, Springer, Berlin, 1959.
9. Lawrence, E. O., and M. S. Livingston: The Production of High Speed Light Ions without the Use of High Voltages, *Phys. Rev.*, **40**: 19–35 (1932).
10. Bohm, D., and L. L. Foldy: Theory of the Synchro-cyclotron, *Phys. Rev.*, **72**: 649–661 (1947). Richardson, J. R., B. T. Wright, E. J. Lofgren, and B. Peters: Development of the Frequency Modulated Cyclotron, *Phys. Rev.*, **73**: 424–436 (1948).
11. McMillan, E. M.: The Synchrotron. A Proposed High Energy Particle Accelerator, *Phys. Rev.*, **68**: 143–144 (1945). Veksler, V.: A New Method of Acceleration of Relativistic Particles, *J. Phys. (U.S.S.R.)*, **9**: 153–158 (1945).
12. Livingston, M. S., J. P. Blewett, G. K. Green, and L. J. Haworth: Design Study for a Three-BeV Proton Accelerator, *Rev. Sci. Instr.*, **21**: 7–22 (1950). Brobeck, W. M.: Design Study for a Ten-BeV Magnetic Accelerator, *Rev. Sci. Instr.*, **19**: 545–551 (1948). Blewett, M. H., et al.: *Rev. Sci. Instr.*, **24**: 723–870 (1953).
13. Piccioni, O., D. Clark, R. Cool, G. Friedlander, and D. Kassner: *Rev. Sci. Instr.*, **26**: 232 (1955).
14. Courant, E. D., M. S. Livingston, and H. S. Snyder: The Strong-focusing Synchrotron—a New High Energy Accelerator, *Phys. Rev.*, **88**: 1190–1196 (1952).
15. Blewett, J. P.: Radial Focusing in the Linear Accelerator, *Phys. Rev.*, **88**: 1197–1199 (1952).
16. Adams, J. B., M. G. N. Hine, and J. D. Lawson: *Nature*, **171**: 926–927 (1953).
17. Thomas, L. H.: *Phys. Rev.*, **54**: 580 (1938).
18. Laslett, L. J.: On Intensity Limitations Imposed by Transverse Space-Charge Effects in Circular Particle Accelerators, *Proc.* 1963 *Summer Study on Storage Rings, Accelerators and Experimentation at Super-high Energies*, Brookhaven, 1963 (BNL 7534), p. 324, J. W. Bittner (ed.).
19. Nature of Matter, Purposes of High Energy Physics, L. C. L. Yuan (ed.), Brookhaven, 1965 (BNL 888).

Bibliography

Livingston, M. S., and J. P. Blewett: "Particle Accelerators," McGraw-Hill, New York, 1962.

Livingood, J. J.: "Principles of Cyclic Particle Accelerators," Van Nostrand, Princeton, N.J., 1961.

Proceedings of CERN Symposium on High Energy Accelerators and Pion Physics, CERN, 1956, E. Regenstreif (ed.).

Proceedings of International Conference on High Energy Accelerators and Instrumentation, CERN, 1959, L. Kowarski (ed.).

Proceedings of International Conference on High Energy Accelerators, Brookhaven, 1961, M. H. Blewett (ed.).

Proceedings of International Conference on High Energy Accelerators, Dubna, 1963, A. A. Kolomensky (chief ed.), Atomizdat, 1964.

Proceedings of International Conference on High Energy Accelerators, Frascati, 1965.

Chapter 10

Fission

By JOHN A. WHEELER, Princeton University, *and*
IVAN G. SCHRÖDER, National Bureau of Standards and Brookhaven National Laboratory

1. Introduction [1]†

The discovery of fission came only a few years after Chadwick's discovery of the neutron (1932) and Curie and Joliot's discovery of artificial radioactivity (1934). In 1934 Fermi and his collaborators used the neutron to produce artificial radioactivity in every element where it could be observed [2], according to the radioactive capture reaction

$$n + (Z,A) \rightarrow (Z, A + 1) + \text{one or more photons} \qquad (10.1)$$

In some cases the newly formed nucleus was stable; in other cases it was beta or electron radioactive according to the scheme

$$(Z, A + 1) \rightarrow (Z + 1, A + 1) + e^- + \text{neutrino} \qquad (10.2)$$

and sometimes the nuclear product of this first decay was itself unstable and underwent a second beta decay. But the products of neutron bombardment of Th ($Z = 90$) and U ($Z = 92$) fitted none of these patterns. They showed in each case at least four distinct half-lives. Before the number or the chemistry of the activities was thoroughly explored, the decay periods were attributed to transthoric or transuranic elements. Considerably later work proved that part of the activity is rightly to be explained in this way, in uranium, for example, by the process

$$\begin{aligned} {}_{92}U^{238} + n &\rightarrow {}_{92}U^{239}(23 \text{ min } \beta \text{ decay}) \\ &\rightarrow {}_{93}Np^{239}(2.3 \text{ min } \beta \text{ decay}) \rightarrow {}_{94}Pu^{239} \end{aligned} \qquad (10.3)$$

However, the more the bulk of the activity was investigated, the more puzzling became the chemical identification of the radioactive species. Finally, late in 1938 Hahn and Strassmann found themselves forced to the unexpected result [3] that barium and lanthanum are among the products of the neutron bombardment of uranium. It was necessary to conclude that the uranium nucleus divided into at least two nuclei of lighter weight.

A simple comparison of the mass of the original nucleus, M_0, with the masses of any two fragment nuclei, M_1 and M_2, of comparable size showed that the energy release in fission

$$\Delta E = c^2(M_0 - M_1 - M_2) \qquad (10.4)$$

ought to have the enormous order of magnitude of 200 Mev. Energies of the order of 100 Mev were observed early in 1939 for individual fragments ejected from a thin layer of neutron-irradiated uranium into an ionization chamber [4, 5]. It was clear in principle that nuclei much lighter than uranium should also give off very great amounts of energy on division into two fragments of comparable mass.

† Numbers in brackets refer to References at end of chapter.

2. The Probability of Fission

In the elementary analysis of nuclear binding energy, which disregards both closed-shell effects and deviations of the nuclear form from sphericity, the mass of a nucleus (Z,A) is given by the semiempirical mass formula [6]

$$\begin{aligned} M(Z,A) &= M_pZ + M_n(A - Z) \\ &\quad - A\left\{a_v - a_a\left[\frac{A/2 - Z}{A}\right]^2\right\} + a_sA^{2/3} + a_c\frac{Z^2}{A^{1/3}} + \Delta \end{aligned} \qquad (10.5)$$

A division into two equal parts corresponds, according to (10.5), to a release of mass energy equal to

$$\begin{aligned} \Delta M &= M(Z,A) - M(Z/2,A/2) - M(Z/2,A/2) \\ &= 4.63A^{2/3}\left(-1 + 0.0568\frac{Z^2}{A}\right) \quad \text{Mev} \end{aligned} \qquad (10.6)$$

This energy changes from negative to positive above the neighborhood of ${}_{40}Zr^{90}$, and heavier nuclei are, in principle, increasingly unstable as judged by release alone. The evidence for this instability, spontaneous fission, was first shown to exist in U^{238} [7].

Division of a heavy nucleus into three equal parts should release even a little more energy than bipartition, according to a simple modification of Eq. (10.6) for this type of process. Both the relative rareness of tripartition and the absence of observable spontaneous fission for nuclei, even those as heavy as lead or bismuth, show that high-energy release is not by itself sufficient for high fission probability. To explain the observed stability against fission of nuclei heavier than zirconium, one has to turn from the energy alteration when fission occurs to the probability that fission will occur.

The ideas of Bohr [8] found a natural expression in the newly discovered phenomenon of fission. The liquid-drop model of fission [9] compares the nucleus to a droplet of uniformly charged incompressible fluid endowed with a surface tension. When the charge is small, energy is always demanded to accomplish a change in the shape of the droplet against the tension of the surface. When, as will be seen, the square of the charge reaches a critical value, no external energy is needed to elongate the droplet. Once such a deformation starts, the electric forces of repulsion more and more take command of the situation. Along the narrowing neck of the fluid the surface tension has less and less purchase to resist the repulsion. The elongation accelerates, scission (Latin for cut) occurs, and even thereafter the fragments gain speed from the continuing electrostatic repulsion. When the charge is less than the critical value, the energy of the system no longer decreases monotonically as a function of elongation. Instead it rises slightly to a barrier summit (Fig. 10.1) and only then falls off for greater

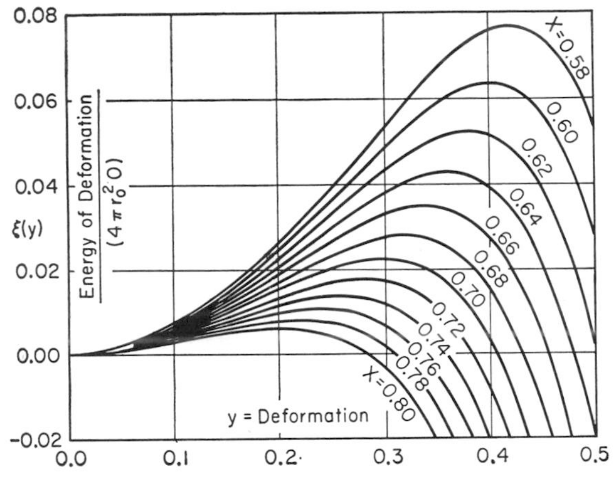

FIG. 10.1. Energy required for a deformation of a liquid drop from the spherical form of equilibrium. (Taken from Hill and Wheeler [46].) The quantity ξ represents the deformation energy divided by the surface energy of the undeformed drop. The quantity x is proportional to Z^2/A. The quantity y is a measure of deformation so defined that the shape for $y = 0.3$, for example, is identical with the saddle-point shape for $x = 1 - y = 0.7$ (Fig. 10.2).

deformations. The original spherical droplet is stable against small departures from sphericity. A finite amount of energy is required to surmount the barrier against fission, an energy E_f variously termed the "fission threshold" or "critical energy to initiate fission" or "transition-state energy" or "fission barrier." The droplet at the summit of this barrier has a shape (Fig. 10.2) (a "critical form of unstable equilibrium" or a "transition-state form" or a "fission form") in which electrostatic repulsions and surface tensions are in equilibrium—but in unstable equilibrium. If a mild disturbance slightly increases the diameter of the neck of the figure, the surface tension overcomes the electric repulsions and brings the mass of liquid back together. If a slight disturbance narrows the neck, then electrostatic forces win, the two halves of the figure start to separate, the neck grows

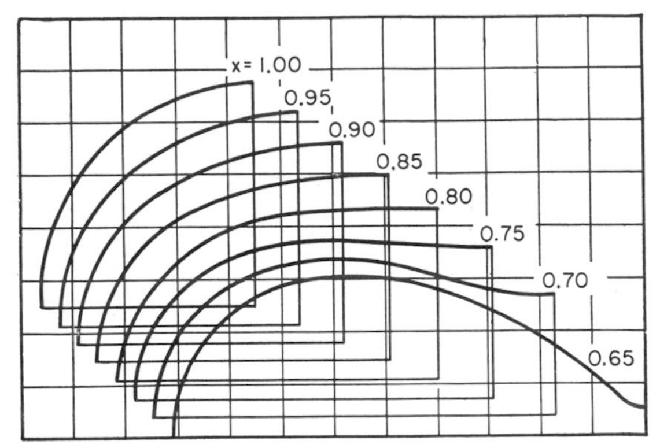

FIG. 10.2. Forms of unstable equilibrium for a uniformly charged liquid drop endowed with a (charge)^2/mass parameter x, as calculated by Frankel and Metropolis [12]. More recent calculations by Swiatecki [12] give slight changes in the curves for $x = 0.70$ and 0.65.

still thinner, the speed of deformation rises, and scission is guaranteed.

To determine the critical energy that must be supplied to an undisturbed spherical droplet of radius R_0 and surface tension σ to enable it to surmount the fission barrier one must solve, in principle, the problem of a distorted liquid drop in equilibrium. To do this, the droplet is considered to have an axis of symmetry. Its shape, when expressed in terms of spherical polar coordinates R and θ by a function $R = R(\theta)$, is specified by an infinite number of deformation parameters $\alpha_2, \alpha_3, \ldots$. Of these the first represents the magnitude of a spheroidlike deformation, the second the strength of a pearlike deformation, and so on, as illustrated, for example, in the formula

$$R(\theta) = R_0 \left[1 + \alpha_0 + \sum_{n=2}^{\infty} \alpha_n P_n(\cos\theta) \right] \tag{10.7}$$

Here the first parameter, α_0, is determined in terms of the other deformation parameters by the condition of volume normalization. The quantities α serve as coordinates in a space of infinitely many dimensions—a space that may be pictured as a plane if limited for simplicity to the two coordinates α_2 and α_4. The deformation energy of the system becomes a function of these coordinates, $V = V(\alpha_2, \ldots)$, and may be plotted in another dimension to give a contour diagram or topographical map of the deformation energy (Fig. 10.3). In this diagram the elevation of the saddle point measures the critical energy required for fission.

The energy for a deformed configuration is evaluated from a simple generalization of the semiempirical energy formula (10.5). The only terms considered to change are those that represent surface energy and electrostatic energy. The potential energy of deformation can then be expressed in terms of deformation parameters from the expression

$$\begin{aligned} V(\alpha) &= E_{\text{deformed}} - E_{\text{spherical}} \\ &= \sigma(\text{surface for configuration } \alpha - 4\pi \mathbf{R}_0^2) \\ &\quad + Z^2e^2 \left(\frac{1}{2} \int \frac{dv_1\, dv_2}{r_{12}} - \frac{3}{5}\frac{1}{\mathbf{R}_0} \right) \end{aligned} \tag{10.8}$$

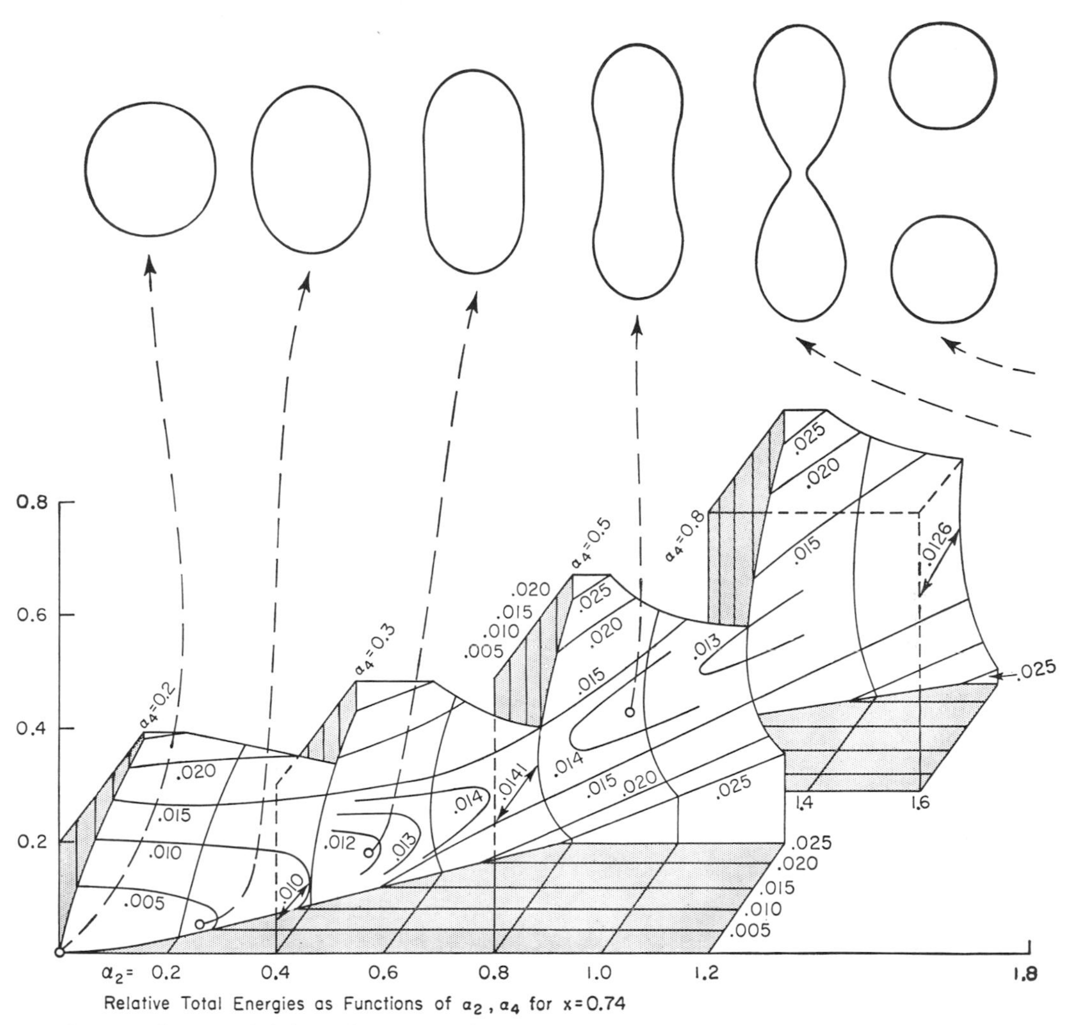

FIG. 10.3. Contour diagram of deformation energy (in dimensionless units $f = E/4\pi R^2\sigma$) for a droplet of fissionability parameter $x = 0.74$ as a function of parameters α_2 and α_4 that describe deformations of order 2 and 4, according to Frankel and Metropolis [12]. The origin corresponds to the undeformed sphere. Endowed with a small energy, the system will oscillate about this equilibrium point. With enough energy (f just in excess of 0.0141) the oscillating representative point of the system will ultimately cross the saddle point and proceed in the direction of fission.

For small symmetric deformations the terms in α_3 and α_4 are negligible in comparison with the terms in prolate deformation α_2 so that the potential energy of deformation, Eq. (10.8), becomes

$$V(\alpha) = \left(8\pi R_0^2\sigma - \frac{3}{5}\frac{Z^2e^2}{R_0}\right)\frac{\alpha_2^2}{5} \tag{10.9}$$

where the terms in parentheses represent twice the undisturbed surface energy and the Coulomb energy, respectively. The relative magnitude of these two quantities determines then, in this approximation, the equilibrium properties of the deformed liquid drop.

It is convenient to express the degree of instability by means of a dimensionless "fissionability parameter"

$$x = \frac{E_{\text{coulomb}}}{E_{\text{surface}}} = \left(\frac{3/5\,(e^2/r_0)}{8\pi r_0^2\sigma}\right)\frac{Z^2}{A} = \frac{Z^2/A}{(Z^2/A)_{\text{critical}}} \tag{10.10}$$

where $$\left(\frac{Z^2}{A}\right)_{\text{crit}} = 2\,\frac{4\pi r_0^2\sigma}{3e^2/5r_0} = 50.13 \tag{10.11}$$

a number uncertain by perhaps 5 units. This calculated value of Z^2/A for $x = 1$, that is, for instability in the normal spherical form, is about three times the Z^2/A value for purely energetic instability with respect to fission.

The energy that has to be imparted to the nucleus in order to reach this critical shape, i.e., to surmount the fission barrier, is given by an expression of the form

$$E_f = 4\pi R^2\sigma f(x) \tag{10.12}$$

where $f(x)$ is a decreasing function of the fissionability parameter. From the constants of the semiempirical mass formula (10.6) one finds

$$E_f = 4\pi r_0^2\sigma A^{2/3}f(x) = 17.8A^{2/3}f(x) \quad \text{Mev} \tag{10.13}$$

In order to calculate this energy, the threshold energy, it is necessary to maximize $V(\alpha)$ for a given choice of

TABLE 10.1. CRITICAL ENERGY FOR FISSION IN THE DIMENSIONLESS UNITS OF EQS. (10.10) TO (10.13)†

x	$f(x)$	$\frac{f(x)}{(1-x)^3}$	Values calculated for $A = 238$ assuming:			
			$(Z^2/A)_{\text{crit}} = 50.13$; $4\pi r_0^2\sigma = 17.8$ Mev		$(Z^2/A)_{\text{crit}} = 46$; $4\pi r_0^2\sigma = 15$ Mev	
			Z	E_f, Mev	Z	E_f, Mev
~1	$[\frac{98}{135}(1-x)^3 - \frac{11368}{34425}(1-x)^4 + \cdots]$	0.7259	109.2	0	104.7	0
0.90	0.0007059	0.7059	103.6	0.5	99.4	0.4
0.85	0.002400	0.7112	100.6	1.6	96.5	1.4
0.80	0.005834	0.7292	97.6	4.0	93.6	3.4
0.76	0.01040	0.7524	95.2	7.1	91.3	6.0
0.75	0.01187	0.7594	94.5	8.1	90.7	6.8
0.74	0.01348	0.7667	93.9	9.2	90.0	7.8
0.70	0.02159	0.7996	91.4	14.8	87.5	12.4
0.65	0.03627	0.8459	88.0	24.8	84.5	20.9
~0	$[0.260 - 0.215x]$	0.260	0	178.	0	150.

† Power series expansion calculated [9] by maximizing the $Y(\alpha)$ of Eq. (10.8). Numerical values of $f(x)$ from W. J. Swiatecki, *Phys. Rev.*, **104**: 993 (1956). Two choices for the nuclear constants are used. The first is based upon the latest values of the nuclear radius and the simple semiempirical mass constants [Eqs. (10.10) to (10.13)] and gives barrier heights for U^{238} and Th^{232} over twice as high as the observed values.

deformation parameters. This leads to a power series expansion for $f(x)$ (Table 10.1). In the original liquid-drop model these computations were carried out to the fourth order in α_2 and to the second order in α_4, yielding

$$E_f = 4\pi r_0^2 A^{2/3}[0.7259(1-x)^3 - 0.3302(1-x)^4] \tag{10.14}$$

These calculations show that fission can occur upon absorption of relatively small amounts of excitation energy.

Up to now threshold measurements have been performed using neutrons, photons, and deuterons (Table 10.2). Of these three forms of excitation, photons give, under ideal conditions, a direct measure of the fission threshold. They suffer, however, from the fact that sharply monochromatic sources of photons are not available. Consequently, the observed photofission thresholds represent an average over an appreciable range of energy. On the other hand, thresholds determined from neutron-induced fission are reduced by the binding energy of the neutron. As a consequence of this, as the binding energy of an even neutron is larger than that of an odd one, nuclei with an odd number of neutrons may be induced to fission by thermal neutrons. This fact led Bohr [10] to attribute the thermal neutron fissionability of uranium to the odd isotope 235. Typical examples of such odd-neutron isotopes are U^{233}, U^{235}, and Pu^{239}; their thresholds are in a region inaccessible to neutron-induced fission. They are accessible, nevertheless, experimentally by deuteron stripping reactions [11], as these permit the addition of a neutron to the above isotopes without adding its binding energy.

The values shown in Table 10.2 do not agree well with those calculated from the liquid-drop model, and though the measured values decrease with increasing Z^2/A they do not decrease as rapidly as the model would predict. This is not surprising as the simple liquid-drop model applies to nuclei having a spherical equilibrium form and not the deformed shape of actual fissionable nuclei. An accurate account of absolute barrier heights and their variation with Z and A cannot therefore be expected. Several attempts have been made [12], in the framework of the liquid-drop model, to improve these results by increasing not only the number of deformation parameters but also the range of variation of x. Nevertheless, the results have been poor. Lastly, it should be remembered that threshold values are experimentally hard to determine, as below the "classical" threshold the excited nucleus has, quantum-mechanically, a small probability of tunneling through the fission barrier [13].

Such tunneling can occur even in nuclei that have received no external excitation; this is the origin of spontaneous fission. It is the transparency of the barrier that determines the lifetime τ for this process. The primitive liquid-drop model evaluates this transparency as proportional to

$$\exp\left(-\frac{4\pi}{\hbar}\sqrt{2ME_f} \times b\right) \tag{10.15}$$

where M is the reduced mass of the separating fragments, E_f the fission threshold, and b the barrier thickness (more explicitly defined in ref. 9, eq. 28). Using the measured values of both the spontaneous-fission lifetimes and of fission thresholds, Eq. (10.15) leads to reasonable values of barrier thickness. Furthermore, it leads to a dependence of τ on Z^2/A which is not completely borne out by the measured values (Fig. 10.4). Therefore, one has only a rough guide to extrapolation for the half-lives of heavier nuclei so far unstudied. Such extrapolations demand the most drastic simplifications, including neglect of shell effects and unexpected trends in nuclear deformations. With these qualifications one gets a rough impression of the conceivable limits of stability of very heavy even-even

TABLE 10.2. FISSION THRESHOLD OBTAINED FROM NEUTRON IMPACT, PHOTOFISSION EXPERIMENTS, AND (*d,pf*) REACTIONS (UNCERTAINTIES ARE OF THE ORDER OF 0.2 MEV)

Neutron fission thresholds

Target nucleus	Compound nucleus	Neutron threshold E_n, Mev	Neutron binding B_n, Mev	Barrier $E_n + B_n$ E_f, Mev
$_{90}Th^{232}$	$_{90}Th^{233}$	1.4	5.1	6.5
$_{91}Pa^{231}$	$_{91}Pa^{232}$	0.7	5.5	6.2
$_{92}U^{238}$	$_{92}U^{239}$	1.3	4.8	6.1
$_{92}U^{234}$	$_{92}U^{235}$	0.5	5.3	5.8
$_{93}Np^{237}$	$_{93}Np^{238}$	0.5	5.4	519

Thresholds for photofission

Target nucleus (compound nucleus)	Observed photofission threshold, Mev
$_{90}Th^{232}$	5.9
$_{92}U^{233}$	5.4
$_{92}U^{235}$	5.7
$_{92}U^{238}$	5.8
$_{94}Pu^{239}$	5.5

Threshold for (*d,p*) fission [11]

Target nucleus	Observed (*d,pf*) threshold, Mev
$_{92}U^{233}$	−1.53
$_{92}U^{235}$	−0.62
$_{94}Pu^{239}$	−1.65

nuclei as affected by spontaneous fission and other decay processes (Fig. 10.5). For nuclei of odd mass number the rate of spontaneous fission is typically less than that of nearby even-even nuclei by a factor of the order of $10^{3\pm1}$ [15]. An explanation of this behavior is outside the scope of the liquid-drop model; one must look at the unified model for an interpretation (Sec. 5) of these results [16].

The probability of fission (fission cross section) depends on the nuclear species involved as well as on the excitation energy made available to it. Taking as a reference point the fission threshold, one can speak of energies below, near, and above this point and analyze the behavior of the fission cross section accordingly. As most fission thresholds are around 5 to 6 Mev, about this amount of energy is needed before one can observe fission. Up to this energy the compound nucleus decays through electromagnetic de-excitation. As the threshold is approached, fission becomes possible and its character is dominated by barrier penetration, as shown by low-energy photofission studies [17]. Close to the threshold energy the fission cross section rises rapidly until that energy

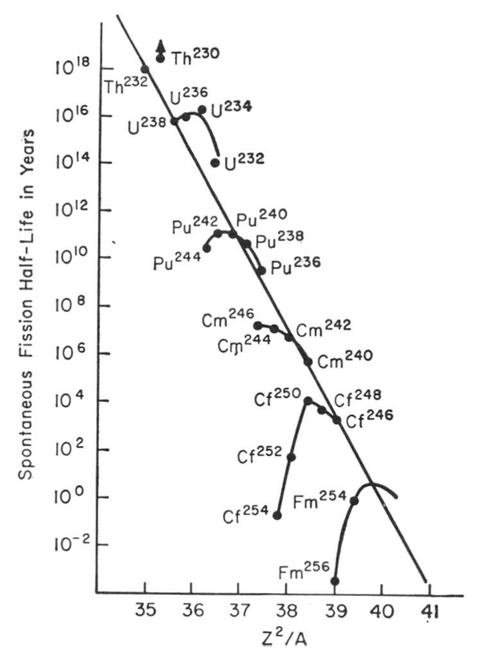

FIG. 10.4. The half-life with respect to spontaneous fission (actually 0.693/probability per year of spontaneous fission) shows a general trend with Z^2/A but cannot be expressed as a function of Z^2/A.

is reached at which neutron emission first becomes possible. Then the fission cross section drops (competition) only to rise again as the energy is further increased [18]. Beyond the threshold, further drops occur (Fig. 10.6) as additional channels for neutron reemission open out (energy of the compound nucleus enough to emit the neutron and leave the residual nucleus in one or another excited state). The neutron fission cross section in the Mev region is governed by the ratio of widths Γ_f/Γ_n for fission as compared with neutron emission [19]. A high value for this ratio is favored by (1) a low fission barrier (nuclei with large Z^2/A), (2) a high neutron binding (compound nucleus containing an even number of neutrons), and (3) a high angular momentum [20]. As the excitation energy becomes extremely high the compound nuclear model is no longer valid, the Serber model [21] has to be used, and Monte Carlo calculations performed [22]. For a description of the problems involved, see ref. 1.

The slow neutron fission cross section of heavy elements shows characteristics quite different from those described above (Fig. 10.7). As can be seen from this figure, in the thermal region the cross section changes approximately as $1/v$, to be followed by a "resonance region" which corresponds to individually excited states of the compound nucleus. The study of the properties of resonances has helped clarify many ideas on the fission process. Thus, for instance, the fact that fission widths show such large fluctuations from resonance to resonance (Table 10.3) has led to the conclusion [23] that fission is at most a few-channel process. Another confirmation of this result comes from the necessity of multilevel Breit-Wigner fits [24] to the resonance data due to interference effects

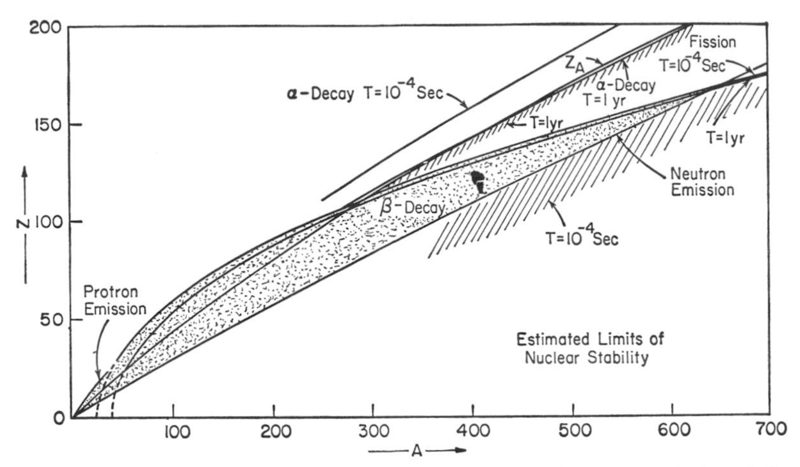

Fig. 10.5. Limits of nuclear stability estimated by Werner and Wheeler [14] by extrapolation of the semiempirical mass formula and by very rough and uncertain extrapolations of spontaneous fission rates. Z_A, charge for beta-stable nucleus of a given A. Lifetimes against beta decay grow shorter (apart from odd-even effects) as Z falls below Z_A, reaching roughly 10^{-4} sec at a neutron-proton ratio so extreme that spontaneous neutron emission occurs. Nuclei in the dotted region were estimated to live 10^{-4} sec or longer. If the effective nuclear surface tension itself is substantially reduced by an extreme neutron-proton ratio, the calculated limiting mass will be reduced from $A = 650$ to a substantially smaller figure.

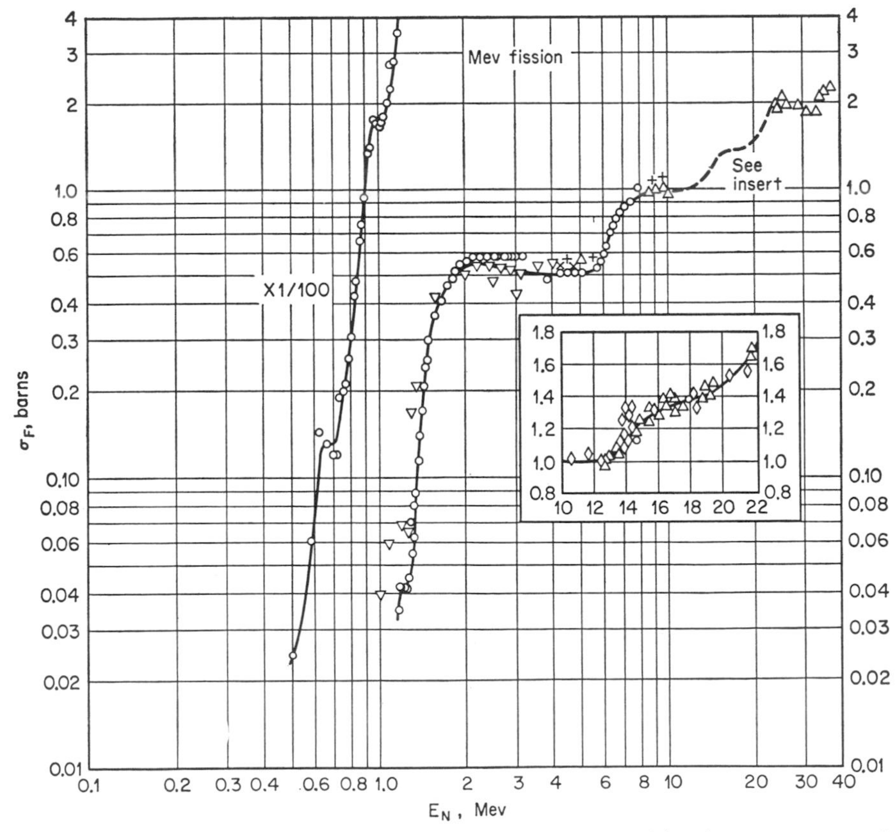

Fig. 10.6. Cross section for n-induced fission of U^{238} as affected by (*a*) competition from neutron reemission (kinks in "below-the-fission threshold" part of curve) and by (*b*) "second-try fission" (rise near $E_n = 6$ Mev; nucleus formed by reemission of neutron often left with enough energy to undergo fission itself), and (*c*) "third-try fission" and "multiple-try fission" (rises near 14 Mev and 20 Mev). *(From J. R. Stehn, M. D. Goldberg, R. Wiener-Chasman, S. F. Mughabghab, B. A. Magurno, and V. M. May, Neutron Cross Sections, Brookhaven Natl. Lab. Rept. BNL* 325, *2d ed., Suppl.* 2, *vol.* 3, 1965.)

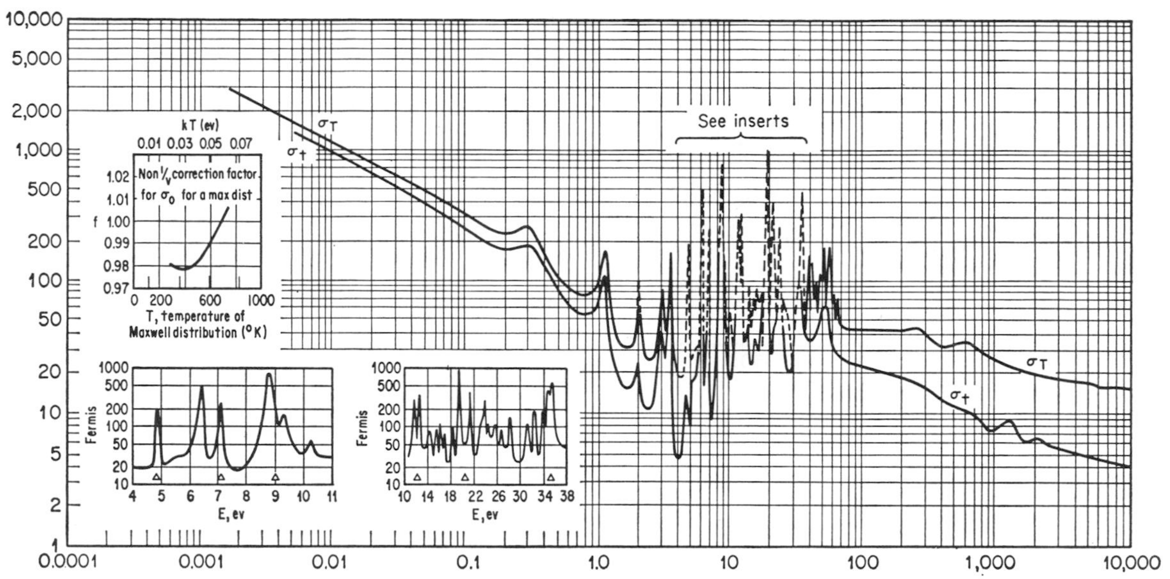

Fig. 10.7. Total (σ_T) and fission (σ_f) cross section of U^{235} in the low-resonance and thermal regions.

between different levels [25]. As will be seen in Sec. 5, these phenomena find a natural explanation in terms of the unified model of nuclear fission.

3. The Fission Process

One can think of fission as the splitting of a compound nucleus into two excited fragments (10^{-14} sec) of approximately equal mass and high kinetic energy. These highly excited fragments, because of their

Table 10.3. Energies and Widths of Resonance Levels of the Compound Nucleus U^{236} (Spin $I = 3^-$ or 4^-) Formed by Addition of a Slow Neutron to U^{235} ($I_0 = 7/2^-$)†

E_n, ev	Γ, mv	$\Gamma\gamma$, mv	Γ_f, mv	$2g\Gamma_n{}^0$, mv
−1.45	259	33	223	3.056
−0.02	97	34	63	0.00072
0.290	135	35	100	0.0055
1.135	157	42	115	0.0145
2.040	47	37	10	0.0054
2.84	180	40	160	0.005
3.15	140	47	90	0.0163
3.61	87	42	45	0.025
4.845	39	35	4	0.026
6.39	47	37	10	0.107
7.08	63	37	26	0.045
8.79	128	53	74	0.40

† J. R. Stehn, M. D. Goldberg, R. Wiener-Chasman, S. F. Mughabghab, B. A. Magurno, and V. M. May, Neutron Cross Sections, BNL 325, 2d ed., Suppl. 2, vol. III, Brookhaven National Laboratory, Upton, N.Y., 1965. Add B_n = 6.3 Mev to the neutron energy E_n to obtain the excitation energy E^*. Here Γ = full width at half maximum; $\Gamma\gamma = \Gamma_v$ = partial width for radiation; Γ_f partial width for fission; $\Gamma_n = [E_n(\text{ev})]^{1/2}\Gamma_n{}^0$ = partial width for neutron emission; and g = statistical factor $(2I + 1)/2(2I_0 + 1)$.

proton deficiency, promptly emit ($\sim 4 \times 10^{-14}$ sec) a number of neutrons, succeeded, once the excitation energy falls below the neutron binding energy, by photon emission ($\sim 10^{-14}$ sec). These processes leave a pair of charge-deficient fragments (the so-called fission products) which proceed to their stable configuration through a series of β decays or via excited states which are responsible for the emission of (delayed) photons and neutrons.

Depending on the amount of excitation energy made available to the fissioning nucleus and on the nature of the projectile furnishing it, the details of the above description will change but not its essential characteristics. In this section the experimental information available on these processes will be summarily examined.

The observed kinetic energy arises from the Coulomb repulsion between the fragments. A study of this energy should lead to a knowledge of an "effective separation distance" of the fragments or, more directly, to information on their deformation energy at the moment of scission. Nevertheless, the measured average kinetic energy of a number of nuclei shows that this quantity is not only insensitive to excitation energy but also to the mass number and angular momentum of the compound nucleus (Fig. 10.8). Only when one looks at the widths of the total kinetic energy distributions of individual nuclei (FWHM 20 to 30 Mev) does one find an indication that there exists a considerable variation in the "effective lengths" of the fission fragments at the time of scission [26]. These variations, as will be seen, are quite striking when one looks at fragment asymmetry.

Most of the excitation energy of the compound nucleus is transformed into excitation energy of the fragments; these subsequently emit a number of (prompt) neutrons and photons. Thus $\bar{\nu}$, the average number of neutrons emitted per fission, is strongly dependent on excitation energy (Fig. 10.9). Experimentally it is seen that prompt neutron emission is

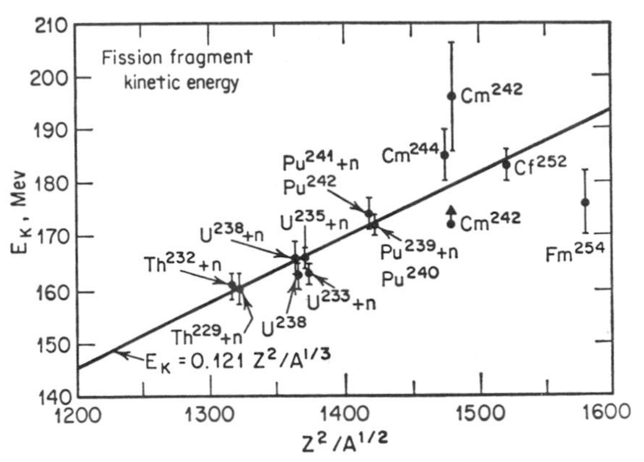

FIG. 10.8. Dependence of the average total kinetic energy $\bar{E}_k$ of the fragments (before neutron emission) on $Z^2/A^{1/3}$. [From J. Terrel, *Phys. Rev.*, **113**: 527 (1959).] For a detailed analysis of these results see ref. 26.

strongly peaked in the direction of motion of the fragments; this, together with their energy distribution, is consistent with the assumption that neutrons are evaporated isotropically from the moving fragments. These evaporated neutrons have an energy distribution that can be fitted with a Maxwellian distribution

$$N(E) \propto \sqrt{E} \exp(-E/T) \qquad (10.16)$$

This is one of a number of analytical expressions which approximately represent the observed distributions [27]. Furthermore, the average energy of these fission neutrons can be expressed as a function of $\bar{\nu}$ through the relation [27]

$$\bar{E} = 0.74 + 0.653\sqrt{1 + \bar{\nu}} \qquad (10.17)$$

Recent experiments on the angular distribution of fission neutrons [28] show systematic deviations from isotropy which can be explained by the assumption that a small number (~10 per cent) of neutrons are emitted isotropically from a source at rest with respect to the fragments. These would presumably arise from a nonadiabatic potential change at the instant of scission [29]. This mechanism may also be responsible for the observed nonlinear increase in $\bar{\nu}$ beyond the neutron threshold [30].

When the excitation energy of the fragments becomes comparable to the neutron binding energy, neutron emission is energetically forbidden, and the remaining excitation, 4 to 5 Mev, should be liberated in the form of prompt photons. This, however, is not the case, as measurements show that about twice this amount is used up in photon emission [31]. The discrepancy between the predicted and measured values finds a ready explanation in the assumption that fragments possess large angular momenta; this would hinder neutron emission, favoring that of photons [32].

Most of the information concerning the division of nuclear charge in fission has come from radiochemical sources; these have given rise to a number of empirical relationships [33] concerning the most probable charge division. Recent experiments [34] on the emission of prompt X rays as a function of mass in the spontaneous fission of Cf^{252} seem to point out that the equal-charge-displacement (ECD) hypothesis of Glendenin [35], which assumes that the fragments are equally displaced from the line of β stability, is closest to the experimental results.

Fission, whether spontaneous or induced, normally leads to fragments of unequal energy and, therefore, of unequal mass. The asymmetric character of the mass distribution is most pronounced in spontaneous fission and low-energy (near threshold) fission of heavy elements, becoming less pronounced and gradually giving way to symmetric fission as the excitation energy increases or the mass of the target nucleus decreases (Fig. 10.10).

Figure 10.10 shows that the relative behavior of the symmetric and asymmetric components in the mass distribution is strongly dependent on excitation energy. The emergence of symmetric mass distributions with increasing excitation energy has led some to the belief [36] that symmetric fission exists as a "mode" distinct from asymmetric fission and possibly related to a different saddle point. These ideas have recently received added support from the observation that the total kinetic energy release is greater for the asymmetric mode than for the symmetric one [37].

The dependence of symmetric fission on total kinetic energy has also been observed in the predominantly asymmetric fission of heavy elements at low excitation energy and in spontaneous fission [38]. Associated with a decrease in total kinetic energy for symmetric mass division is an above-average increase in the number of prompt neutrons emitted by these frag-

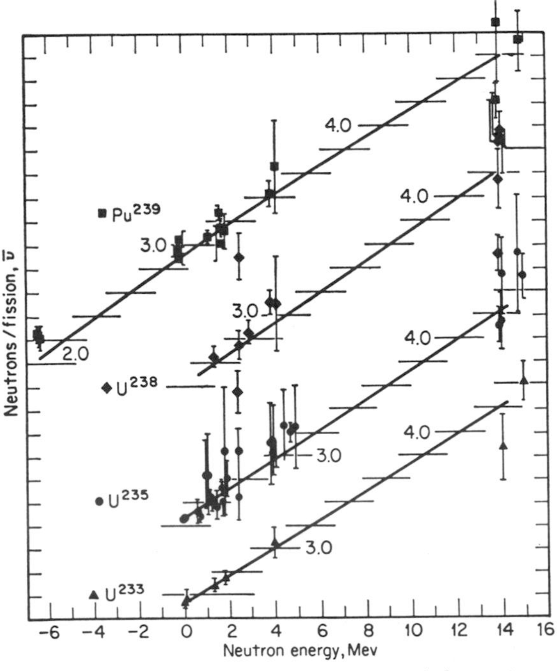

FIG. 10.9. Dependence of $\bar{\nu}$ on the energy of the neutron-inducing fission. (*From R. B. Leachman, Proc. U. N. Intern. Conf. Peaceful Uses At. Energy, 2nd, Geneva,* 1958, *vol.* 15, *p.* 229.)

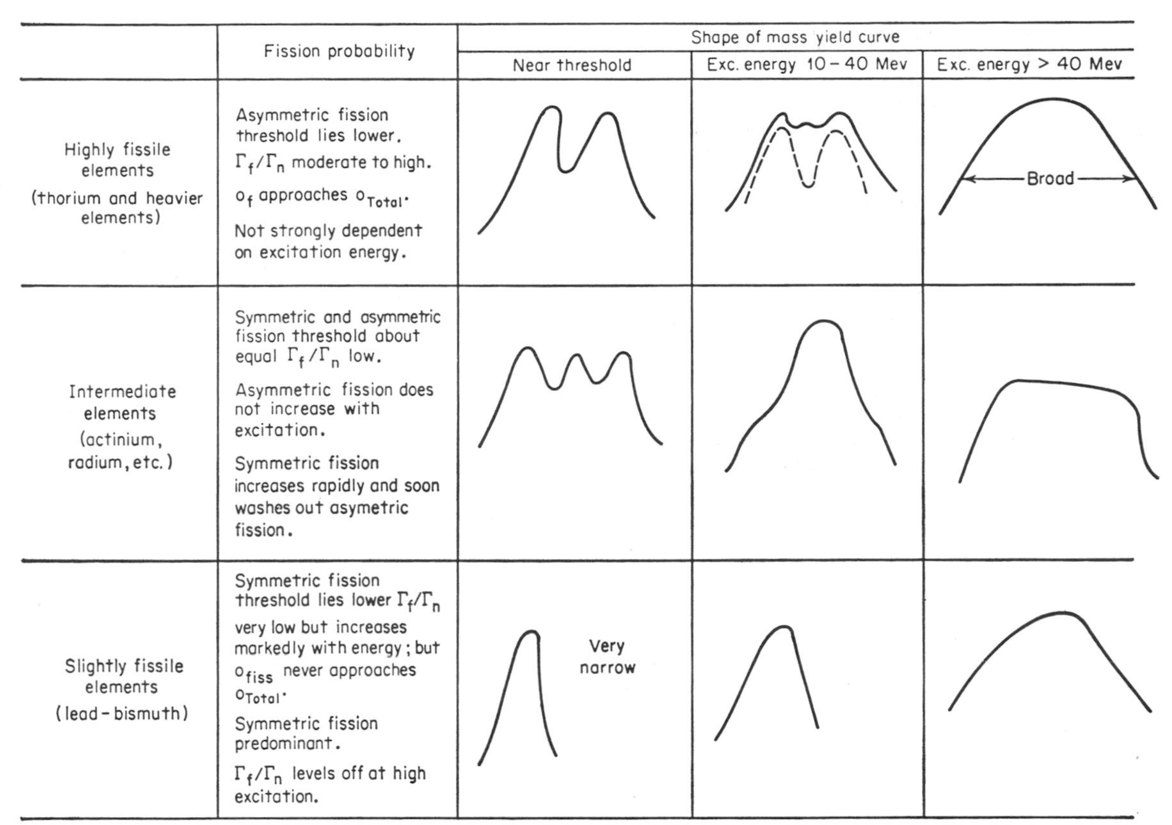

Fission probability
Shape of mass yield curve
Near threshold
Exc. energy 10 – 40 Mev
Exc. energy > 40 Mev
Highly fissile elements
(thorium and heavier elements)
Asymmetric fission threshold lies lower.
Γ_f/Γ_n moderate to high.
σ_f approaches σ_{Total}.
Not strongly dependent on excitation energy.
Broad
Intermediate elements
(actinium, radium, etc.)
Symmetric and asymmetric fission threshold about equal Γ_f/Γ_n low.
Asymmetric fission does not increase with excitation.
Symmetric fission increases rapidly and soon washes out asymetric fission.
Slightly fissile elements
(lead - bismuth)
Symmetric fission threshold lies lower Γ_f/Γ_n very low but increases markedly with energy; but σ_{fiss} never approaches σ_{Total}.
Symmetric fission predominant.
Γ_f/Γ_n levels off at high excitation.
Very narrow

FIG. 10.10. Influence of excitation energy and target atomic number on the mass distribution of fission fragments. (*After Hyde* [1].)

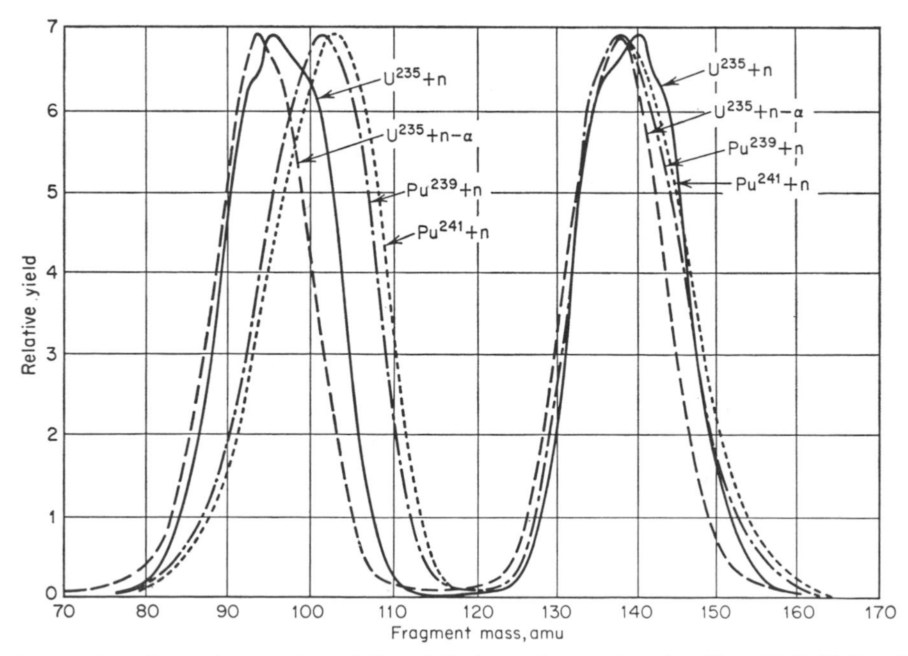

FIG. 10.11. Influence of magic numbers on the stability of the heavy-fragment peak. [*From F. J. Walter, H. W. Schmitt, and J. H. Neiler, International Conference on Nuclear Physics with Reactor Neutrons, Oct.* 15–17, 1963, p. 441, *ANL*-6797 (*unpublished*).]

ments. This phenomenon is accompanied by a below-average emission from fragments in the vicinity of the magic numbers $Z = 50$ and $N = 82$ [39]. This region (mass numbers 127 to 139) seems to play an important role in the low-energy fission of heavy elements, as seen from Fig. 10.11. This figure shows that the heavy-fragment peak remains stationary while the light-fragment peak shifts to conserve mass. This has led to the conclusion that fission is related to a common configuration which is initially asymmetric [40].

Aside from the change in the mass distribution, an increase in excitation energy brings about a change in the angular distribution of the fission fragments, from isotropy at low energies to extreme anisotropy at high energies. The angular distribution of fission fragments arises from the fraction of the incident angular momentum that is transformed into orbital angular momentum between the fragments, the remaining momentum becoming part of their individual spin. The first observation of such an effect was made in the photofission of even-even nuclei [41] when it was observed that the fragments were emitted in a direction perpendicular to the incident beam. These observations were soon followed by studies undertaken at intermediate energies with both neutrons and charged particles [42]. The results of these studies show that, contrary to photofission, the fragments are emitted backward and forward along the beam (Fig. 10.12); the anisotropy increases with projectile size, decreases with Z^2/A, is independent of target parity, and only changes slowly with bombarding energy. This surprising behavior found a ready qualitative explanation in the unified model of nuclear fission.

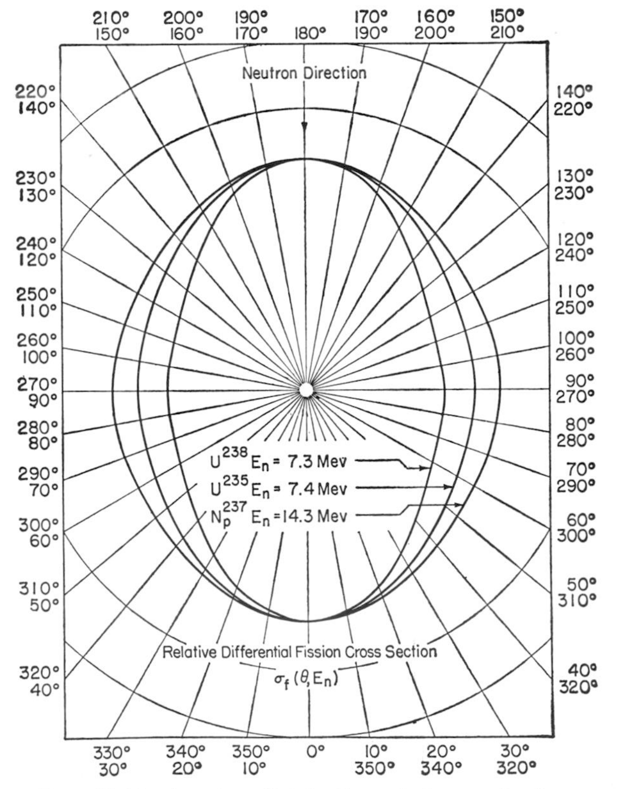

Fig. 10.12. Angular distribution of fragments from neutron-induced fission as observed by Brolley and Dickinson [42].

4. The Unified Model of Nuclear Structure

Of all nuclear phenomena, fission evidences most strikingly the existence of collective motions in which large numbers of particles move together in an organized way. A detailed account of collective motion is therefore indispensable to any quantitative analysis of fission. The first and simplest picture of collective motion is supplied by the liquid-drop model. However, that model is far from perfect. It envisages a nucleon as having a mean free path comparable to or less in order of magnitude than the average spacing between nucleons. In contrast, a variety of evidence indicates that the mean free path is comparable to nuclear dimensions. Therefore the actual situation is complicated by being intermediate between the two extreme ideal situations so thoroughly analyzed in other branches of fission: (1) a liquid, with short mean free path, and (2) an open system, like a solar system or an atom, where the mean free path is enormous compared with the other relevant dimensions. How it was learned that the ideal situation is intermediate between these two extremes and how an approximate way was found—through the "collective" or "unified" nuclear model—to account for both collective and individual particle aspects of nuclear behavior are essential ingredients in a modern account of fission phenomena.

Evidence that nucleons behave in a certain approximation as having a substantial mean free path, that is, evidence for individual-particle behavior, came in the beginning from two quite different sources: properties of nuclear ground states and energy distribution of inelastically scattered particles. From the study of nuclear ground states (see chapter on nuclear shell model) and especially the spins and magnetic moments of even-odd nuclei, it was recognized that (1) quantum numbers could be assigned to the last odd particle and (2) this particle normally governs the spin and magnetic moment of the entire nucleus—features that would be beyond explanation if the nucleonic mean free path were as short as assumed in the extreme liquid-drop model.

The substantial mean free path for nucleons in nuclear matter appears today not to be in contradiction with reasonable views about nucleon-nucleon forces and nuclear matter [43]. Regardless of the explanation of this mean free path, its magnitude gave further vitality to the shell or independent-particle model of nuclear structure, which views nucleons as behaving in some respects as approximately independent particles moving through a common nuclear potential [44].

The liquid-drop model and the independent-particle model have in common the feature that each describes reasonably some aspects of nuclear physics and fails to describe others, but there is one striking phenomenon that neither accounts for: the large nuclear quadrupole moment Q for nuclei with spin $I \geq 1$. For such nuclei the individual-particle picture predicts values of Q that are too small by factors of 10 or 20.

The liquid-drop model predicts a spherically symmetric distribution of charge and zero quadrupole moment. However, by combining the two pictures, one arrives at reasonable orders of magnitude for Q [45]. Consider those individual nucleons that are not paired off with other nucleons in closed shells. These particles move in orbits that create an unbalanced pressure against the nuclear wall. As a result, the surface of the liquid drop is deformed. This deformation causes a bulk transport of charge and thus—in qualitative agreement with observation—an asymmetry of the electric charge distribution larger in many cases by a factor of 10 to 20 than the asymmetry due to a single nucleon. Most heavy nuclei ($Z > 90$) have substantial deformations in the ground state. On this account one has to expect characteristic departures of individual nuclei from the simple predictions of the liquid-drop model of fission that starts with a spherical configuration.

The features of nuclear forces that make possible collective deformations for a nucleus in its ground state also lead to the existence of excited states describable in terms of collective oscillations and rotations. It is easier to define a region of rapid falloff of the potential for a nucleus than for an atom. Moreover, the location of this region depends, not upon the position of some massive central charge as in the atom, but upon the positions of the several nearby nucleons. These positions change with time. The nuclear boundary does not remain stationary. On the contrary, the typical particle feels an average potential, the limits of which change slowly with time. The motion of the particle adjusts itself nearly adiabatically to the configuration of what can loosely be termed the nuclear wall. Thus collective modes of motion arise. The forces of correlation in these motions are pictured more simply in terms of the couplings of the nucleons with each other. These motions include rotations of deformed nuclei as well as nuclear vibrations.

Nuclear rotation, as just described, illustrates the central feature of the collective or unified model [46]: an interaction between individual nucleons and the "nuclear wall," or a coupling between the particle and the system as a whole, by way of which energy and angular momentum are exchanged, energy levels are split, and nuclear reactions are brought about. Moreover, the coupling between nucleon and droplet in a nucleus is analogous to the coupling between electron motion and general vibration-plus-rotation in a molecule. Just as the motion of a nucleon is fast compared with the vibration or rotation of the nucleus as a whole, so in a molecule (Table 10.4) electrons move at high frequency in nearly independent orbits that adjust themselves adiabatically to the lower-frequency molecular vibrations.

In conclusion, those features of nuclear physics that depend principally on the properties of one nucleon—such as the spins and magnetic moments of certain nuclear ground states, many transition probabilities, and the elastic or nearly elastic scattering of merely "tasted" nucleons—can be described to a first approximation in terms of the independent-particle picture and more accurately by calculations that allow for perturbation of the individual nucleon orbit by other orbits and by wall deformations. In contrast, collective motions such as vibration and fission receive their first approximate description in the liquid-drop model. An improved analysis combines the two points of view and has the following features: (1) The equilibrium configuration of the surface is ordinarily no longer spherical but depends on the nucleonic state of the system; (2) the restoring force, or effective surface tension, associated with the vibrations also depends upon the nucleonic state; and (3) the partition

TABLE 10.4. ANALOGY BETWEEN THE COLLECTIVE NUCLEUS AND A MOLECULE

Property	Molecule	Nucleus
Individual particle state occupied by	Electron	Nucleon
Slowly varying parameters that affect energy of individual particle states	Internuclear separations; $r_{12}, r_{13}, \ldots$	Parameter α and other parameters that describe in more detail the configuration of the nuclear wall
Oscillation period in an example	8×10^{-15} sec in H_2	$\sim 5 \times 10^{-21}$ sec in U
Fundamental period of motion of most energetic particle in same example	$\sim 10^{-15}$ sec	$\sim 0.3 \times 10^{-21}$ sec
Vibrational potential energy described as function of these one or more parameters by a curve or surface	$V(r_{12}, r_{13}, \ldots) =$ sum of energies of individual electron states plus electrostatic interactions not otherwise taken into account	$V(\alpha_2, \alpha^2, \ldots) =$ sum of energies of individual nucleon states calculated for a deformed wall, plus other interactions not otherwise taken into account
Vibrational or rotational kinetic energy	Mainly localized in nuclei	Increment of kinetic energy of nucleons because wall is moving and nodes of individual-particle wave functions are undergoing displacement
Exchange of energy between individual particle excitation and general vibration of the system can take place at point of contact between one potential surface and an adjacent one, via a radiationless transition	Mechanism for excitation to be degraded into vibrational energy; important in polyatomic molecules, where the variation of two or more parameters ordinarily allows one to arrive at a point in configuration space where two successive potential surfaces make a cusplike contact	Mechanism for nucleonic motion to be degraded into collective motion, and conversely for energy of collective vibration to be imparted to an individual nucleon, as for example in a nucleonic evaporation process. To be distinguished from direct energy exchanges between a pair of nucleons—an independent mechanism for capture and evaporation. Both contribute to the absorption component of the complex nuclear potential

of the energy between nucleonic and collective excitation ordinarily changes slowly with time except as forbidden by a principle such as the law of conservation of angular momentum. A more and more comprehensive view of nuclear constitution has evidently evolved and continues to develop hand in hand with the experimental and theoretical study of many nuclear phenomena, including fission.

5. Fission and the Unified Nuclear Model

In the unified model of nuclear fission the fundamental assumption is made that as the excited nucleus approaches the saddle point it expends most of its energy in potential energy of deformation. The motion of this nucleus, thus "cooled," through the saddle point is sufficiently slow that the nucleonic states there are rather sharply defined. The spin and parity of these levels, which should resemble the levels of deformed nuclei near the ground-state configuration, have a pronounced effect on the mode of fission [47].

The anomalously long lifetimes for spontaneous fission of odd-A nuclei were the first to find a natural explanation in the framework of the unified model [16]. In even-even nuclei the nucleonic or intrinsic angular momentum Ω about the symmetry axis is zero for all deformations due to the pairing of spins; thus the saddle point is accessible directly to the nucleus as it distorts. This is not the case with odd-A nuclei where the intrinsic angular momentum Ω of the state of lowest energy changes from one value to another as the nucleus distorts. As the total angular momentum of the nucleus is conserved during fission, the saddle-point state that has the same spin as the ground state will generally lie higher ("specialization" energy) (Figs. 10.13 and 10.14). This contributes an added amount to the barrier height for odd-A nuclei relative to neighboring even-A nuclei that has been estimated to be of the order of ½ to 1 Mev. This extra barrier height is sufficient in order of magnitude to explain the observed $10^{3\pm1}$ slower spontaneous-fission rate of odd-A nuclei. In consequence of this effect, U^{238}, with a lower Z^2/A than U^{235} and, according to the liquid-drop model, a higher fission barrier, actually has a lower fission barrier.

A further initial success of the unified model was the elucidation of the fission anisotropies observed in the photofission of even-even nuclei. The lowest rotational band in even-even nuclei is characterized by $\Omega = 0$ or $K = 0$. Photofission proceeds through dipole absorption. Such absorption in the 0^+ ground state can readily excite the 1^- rotational level of one or another $K = 0$ band and thus raise the nucleus to a state with the angular-momentum vector parallel to the direction of the photon beam. In a nucleus so excited the axis of symmetry and therefore the direction of fission are perpendicular to the angular-momentum vector. Thus fission fragments come off preferentially perpendicular to the beam. As the energy of the photons increases, other intrinsic states become available. These tend to make the distribution more nearly isotropic (see last of ref. 42). The success of this explanation led to the application of these ideas to the angular distributions observed in particle-induced fission [49].

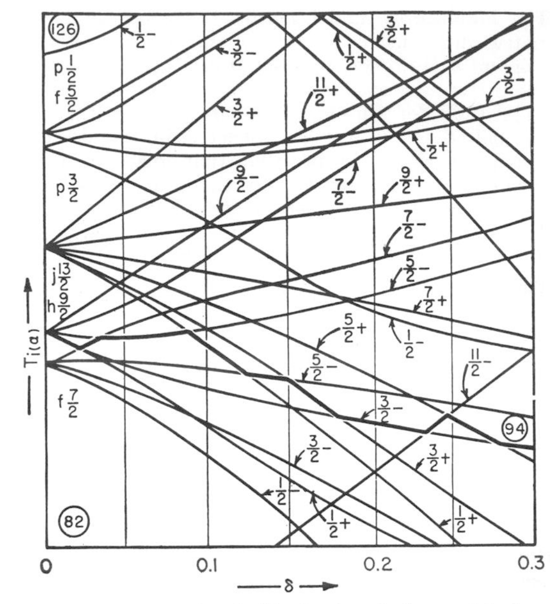

FIG. 10.13. Energy levels $T_i(\alpha)$ of a single particle in a deformed nuclear potential as a function of the deformation coordinate α or a related coordinate δ as calculated by Nilsson [48]. Only the small part of the whole diagram that is relevant for heavy nuclei is shown. The dark line indicates the point at which single-particle levels are filled by 94 protons.

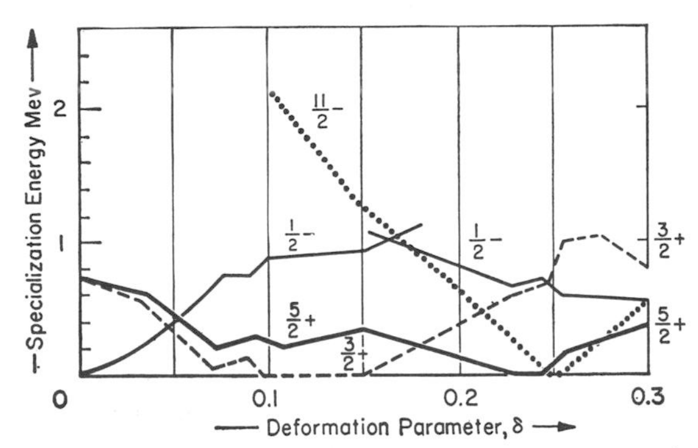

FIG. 10.14. Specialization energy as a function of Nilsson's deformation parameter δ as calculated [47] from the single-particle levels of Fig. 10.13. The specialization energy is defined for odd-A nuclei with axial symmetry. For any given distortion α and any given intrinsic or nucleonic angular momentum Ω about the symmetry axis (various curves in the diagram are labeled according to their Ω values) this quantity represents the difference between the deformation potential $V_\Omega(\alpha)$ for the given Ω value and the potential curve $V_{\Omega^*}(\alpha)$ that lies lowest at the given deformation. It gives a first estimate of the extra height of the fission barriers of odd-A nuclei compared with even-even nuclei.

Both the fission probability A_f and the corresponding fission width Γ_f are governed by the height of the fission barrier. The basic principles that underlie this dependence are shown in Fig. 10.15. The motion of a "representative point" in α space—the coordinates α represent the amplitudes of independent modes of collective deformation—describes the changes of the

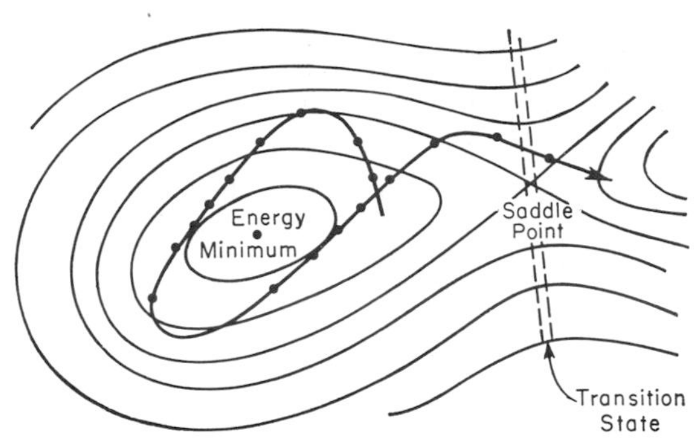

FIG. 10.15. The curves supply a topographic map of the deformation potential: coordinates such as α_2 and α_4 in the plane of the paper. The representative point of the system moving under the influence of this potential executes nearly random motion until almost as if by chance it crosses the saddle point leading to fission.

system with time. The figure shows deformation energy as a function of the position of the representative point. The motion of this point may be compared to the motion of a marble rolling freely on a curved surface. It moves about, passing from time to time near the energy minimum, until by chance it passes over the saddle point. Thereafter it will ordinarily never return; fission will occur.

The likelihood of barrier passage admits of statistical analysis. At the saddle point a thin slice is drawn in α space normal to the minimal path over the barrier. The accessible volume is phase space interior to this slice (a kind of surface), after multiplication with the normal velocity to give a "surface flux," is compared with the total accessible "volume" in phase space that lies to the left of the slice. The ratio has the dimensions of a reciprocal time and measures the probability per second of escape from the region of stability.

The quantum-mechanical result [9] is very close to this classical result. The "surface flux" in phase space is measured by the number Nf of transition-state energy levels or, more vividly, fission channels of the given spin Ω (about the symmetry axis) and parity which are accessible at the excitation reached by addition of the neutron. The "volume" is measured by the number of levels of the compound nucleus per unit energy interval, the reciprocal of the level spacing D. The average probability per second of fission $<A_f>$ or the corresponding average level width with respect to fission, $<\Gamma_f> = \hbar<A_f>$, is given by the fission-rate formula

$$<\Gamma_f>/D = N_f/2\pi \tag{10.18}$$

where $<\Gamma_f>$ should be understood to represent the average of a number of levels of the same spin and parity and nearly the same energy.

The effective channel number N_f does not increase as a function of energy in a series of sharply defined steps. It is not exactly zero below the energy of the first exit channel, nor unity just above it, nor 2 just above the energy of the second fission channel. Instead, barrier penetrability phenomena round the steps, as indicated in Fig. 10.16.

One can think of the width $N_{\text{step}\,k}(E)$ of a single fission channel as given by the probability of penetrat-

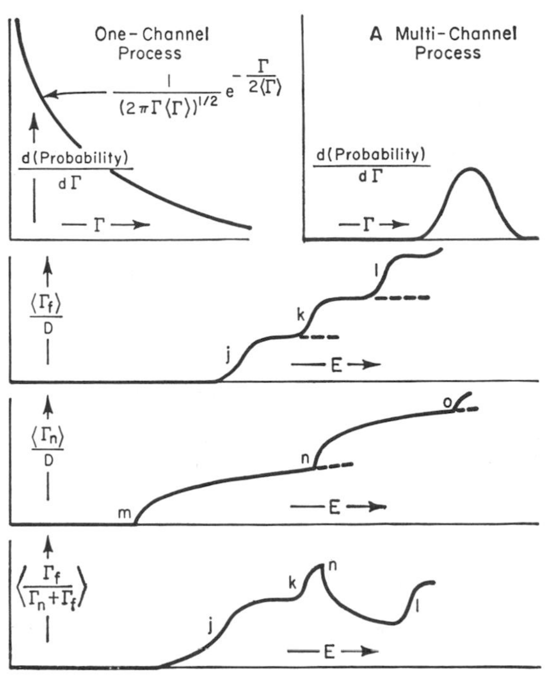

FIG. 10.16. Level widths. Top: their distribution in size for one-channel and many-channel processes. Middle: variation of average width with energy for fission and for neutron emission. Bottom: ratio that determines $\sigma(n,f)$ in the Mev range of energies. For greater details see refs. 1 and 50.

ing the fission barrier. The effective number of fission channels is then given by a "carpeted stair-case" function of energy

$$N_f(E) = \sum_k N_{\text{step}\,k}(E) \tag{10.19}$$

On the simplest assumption that the barrier resembles an inverted harmonic oscillator, the contribution $N_{\text{step}\,k}(E)$ from any one channel is described by

$$N_{\text{step}\,k}(E) = \frac{1}{1 + \exp\,[2\pi(E_k - E)\hbar\omega]} \tag{10.20}$$

Here E_k is the energy of the fission channel, and the characteristic energy $\hbar\omega$ is related to the width of the barrier against fission; it is probably of the order of a few hundred kev. [Equation (10.20) rises from nearly zero to nearly unity over a narrow range ($\sim$0.1 or 0.2 Mev).]

This steplike dependence upon energy of the fission cross section (Fig. 10.6) serves as a confirmation of the statistical analysis just outlined.

According to the unified model, the number of fission channels is determined by the number of states available to the fissioning nucleus at the saddle point. As the compound nucleus is relatively "cold" at this point, only a few channels should be available to it. This behavior should find a reflection in the distribution of fission widths. Recent experiments [51] seem to show that resonances may be separable into two groups having different fission widths.

Another question posed by this behavior of the fissioning system at the saddle point is whether the ratio of symmetric to asymmetric fission varies—for slow neutron-induced fission—from one resonance to another. The experiments so far performed [52] seem to point to a possible separation of this ratio into two groups, though no definite explanation for this behavior can be obtained from the data. It seems that the question of the dependence of fission asymmetry on the spin state of the fissioning nucleus will have to await more data and a determination of the spins of the various resonances.

Lastly, a few comments will be made on the non-adiabatic problem of viscosity. If one were to think of the nucleus as sliding over the particular potential curve it happened to be on when it went over the top of the barrier, one might well think of the kinetic energy of the separating fragments as being fairly uniquely determined—a conclusion very much in contrast to the observations. For this reason, there is a very strong experimental argument that the exchange of energy between the collective-motion and individual-particle excitation takes place rapidly. There must exist a great number of potential curves corresponding to different nucleonic or intrinsic excitations. What is the likelihood of passing across from one potential curve to another? The evidence suggests that this probability is great; that the exchange between the energy of collective-motion and individual-particle excitation is so rapid that only a small part of the energy available for collective motion really goes into collective motion and the bulk of it goes into individual-particle excitations. The experiments indicate that increase in the excitation of the compound nucleus makes very little change in the kinetic energy with which the two fragments separate (Fig. 10.8). The excess energy, as has been seen, goes into individual-particle excitation. The nucleus appears to be "sticky" in the sense that there is a rapid loss of collective energy into individual-particle excitations.

The possibility of a large viscosity suggests that the kinetic energy of the collective movement toward fission at the instant of actual split does not have the value of 10 to 20 Mev that one would expect from following the lowest potential energy curve. Instead, this collective energy should have a value much closer to a thermal energy of the order of 1 Mev. The 10 or 20 Mev will go into other forms of excitation of the slowly extending compound nucleus. Some will go into more complex modes of collective deformation of the nuclear surface. Therefore the shape of the nucleus at the moment of scission is expected to differ very much from one case to another. It seems reasonable to believe that this variation in form at the moment of breaking has great consequences for the partition of energy between (1) internal excitation of the individual fragment and (2) kinetic energy of separation. It appears natural to attribute to this variation in shape at the moment of scission most of the observed spread in kinetic energy of separation.

6. The Fission Chain Reaction

Granted an adequate number of kilograms of separated U^{235}, it is simple enough to start a self-sustaining chain reaction. Of the $2.5N$ neutrons produced in N acts of fission, $1.315N$ can be allowed to leak out of the mass or otherwise disappear. The remaining $1.185N$, caught in U^{235}, produce $0.185N$ nuclei of U^{236} and N fissions. Thereby $2.5N$ new neutrons are liberated and the chain reaction is continued.

With a smaller mass of fissionable material there will be a greater amount of leakage and only k fissions will occur in the second generation for each fission in the first generation, where k is a number less than unity. Then the total number of fissions will have the finite value

$$N + Nk + Nk^2 + \cdots = \frac{N}{1 - k} \tag{10.21}$$

Thus the chain reaction converges.

When k is a few per cent larger than unity, then the rate of reaction follows an exponential law with a simple exponent. The number of neutrons present increases in one generation by the factor

$$k = 1 + (k - 1) \doteq e^{k-1} \tag{10.22}$$

Let τ represent the reproduction time—the time between one generation and the next. Then in the time t there occur t/τ generations, in each of which the number of neutrons rises by the factor (10.21). Starting with n_0 neutrons one therefore has after the time t a number

$$n = n_0 e^{(k-1)t/\tau} \tag{10.23}$$

With even a few per cent excess reactivity—a value of k of 1.05, for example—the neutron population rises by many powers of 10 in a small fraction of a second. In such a period of time the delayed neutron emitters have not had an opportunity to give

Table 10.5. Thermal Cross Sections and Neutron Yields of the Principal Fissile Nuclear Species†
(All cross sections in units of 10^{-24} cm²)

Nucleus	σ_T	σ_f	$1 + \alpha$	η	ν
U^{233}	586 ± 2	524.5 ± 1.9	1.0926 ± 0.0027	2.292 ± 0.006	2.497 ± 0.008
U^{235}	693 ± 4	577.1 ± 0.9	1.175 ± 0.002	2.078 ± 0.005	2.426 ± 0.006
Pu^{239}	1,024 ± 10	740.6 ± 3.5	1.370 ± 0.006	2.116 ± 0.009	2.892 ± 0.011

† J. R. Stehn, M. D. Goldberg, R. Wiener-Chasman, S. F. Mughabghab, B. A. Magurno, and V. M. May, Neutron Cross Sections, BNL 325, 2d ed., Suppl. 2, vol. III, Brookhaven National Laboratory, Upton, N.Y., 1965.

off their 0.755 per cent contribution to the reproduction process. What counts is therefore not the total excess reactivity, but only that part of it which is *prompt*. Thus the exponent of (10.23) has to be revised to

$$(k - 1.00755)t/\tau \tag{10.24}$$

in the case of a rapidly multiplying system. For a slower rate of multiplication the situation is much more complicated—but also much safer! The chain reaction is of course still divergent, by however little k exceeds unity; but also the delayed neutrons still make the rate of multiplication much slower than one would expect from (10.23). Equation (10.24) is now inapplicable too. A fuller treatment of the multiplication rate under these conditions appears, for example, in the monograph of Glasstone and Edlund [53].

Natural uranium, with only one part is 139 of U^{235}, is incapable by itself of sustaining a chain reaction. No matter how large the mass employed, and no matter how great the consequent conservation of neutrons against leakage to the outside, the loss of neutrons by capture into U^{238} is too large. Too few neutrons are caught in U^{235} to give a factor of multiplication, k, as great as unity, contrary to what one might think from an inspection of the cross sections in Table 10.5. From them it follows that each thermal neutron caught in a nucleus of natural uranium produces a number of new neutrons equal to

$$\eta = \frac{\frac{1}{139}\,530 \times (2.5 \text{ neutrons per fission})}{\frac{1}{139}(530 + 61) + \frac{138}{139}\,2.8} = 1.33 \tag{10.25}$$

In the chain reaction the fission neutrons are produced fast, and must be used slow if the U^{235} is to have a chance at them as good as that implied by (10.25). In the process of moderation in solid uranium a large fraction of the 1.33 fast neutrons will be caught at resonance capture levels by the process

$$U^{238} + n \rightarrow U^{239} \xrightarrow[23\text{ min}]{} Np^{239} \xrightarrow[2.3\text{ days}]{} Pu^{239} \tag{10.26}$$

and much less than 1 will reach thermal energy (Table 10.6).

Dispersion of a sufficiently large mass of natural uranium is a still larger mass of a perfect moderator, like He, that absorbs no neutrons at all, or a practically perfect moderator, like D_2O, makes possible a chain reaction. The fast neutrons are brought to thermal energy by collisions with the abundant moderator nuclei, and run little chance of capture at intermediate energy in scarce uranium nuclei. Once at thermal energy, the neutrons diffuse around until they are caught in uranium. Thus it is possible in principle to realize a multiplication factor, k, arbitrarily close to the η of Eq. (10.25).

In practice the moderator is not perfect, and some neutrons are lost by leakage; the multiplication factor, k, is represented by the product of the following five factors:

1. ϵ, the factor of multiplication of neutrons while still in the fast condition—for example, by collision of a rare 10-Mev fission neutron with a U^{238} nucleus to produce fission. The value of ϵ for solid natural metal is of the order of 1.03.

2. p, the probability of escaping resonance absorption during moderation from high energy to thermal energy. It is unity for a perfect moderator, and 0.8 to 0.95 for various reasonable reactor designs.

3. l, the chance to escape leakage out of the syste

Table 10.6. General Character of Cross Sections of U^{235} and U^{238} in Their Dependence upon Energy

Neutron energy	U^{235}	U^{238}
Thermal	$\sigma_{rad} = 101 \times 10^{-24}$ cm^2 $\sigma_{fiss} = (577 \pm 1) \times 10^{-24}$ cm^2 Cross sections perhaps due to a single resonance close to thermal energy	$\sigma_{rad} = 2.7 \times 10^{-24}$ cm^2 $\sigma_{fiss} = 0$ Cross section due to one or more resonances separated from thermal energy by an amount of order of a few ev, large compared with the level width, $\Gamma_{total} \sim \Gamma_{rad}$, of the rough order of 10^{-1} ev (see Table 10.3)
Few ev	Numerous characteristic cross-section resonances for which Γ_{rad} will be expected to be nearly the same, but $\Gamma_n{}^1$ and Γ_f may vary from resonance to resonance as illustrated in the upper part of Fig. 10.16. In addition to such fluctuations $\Gamma_n{}^1$ will show a general increase in proportion to v	Characteristic resonances with a spacing greater than that for U^{235} because of lesser excitation. Similar fluctuations in $\Gamma_n{}^1$. This quantity of order of milli-ev and therefore smaller than Γ_{rad}. Cross section σ_{rad} at resonance less than $\pi\lambda^2$ by a factor of order $\Gamma_n{}^1/\Gamma_{rad}$
Many kev	Level spacing has decreased, and level width increased, so that levels now overlap and fission cross section varies relatively smoothly with energy. Beginning to approach a value of order of $\sigma_f \sim \pi R^2 \sim 2 \times 10^{-24}$ cm^2 Radiative capture cross section obtained by multiplication with Γ_r/Γ_f	$\Gamma_n{}^1$ becomes comparable with Γ_{rad}, and cross section at exact resonance $\sigma_{rad}(\text{res}) \sim \pi\lambda^2$. With further increase in energy, $\Gamma_n{}^1$ exceeds Γ_{rad}, and $\sigma_{rad}(\text{res}) \sim \pi\lambda^2 \frac{\Gamma_{rad}}{\Gamma_n{}^1}$ Ultimately level widths exceed level spacings and σ_{rad} becomes a nearly smooth function of energy
Region 1 Mev to 10 Mev	$\sigma_f \sim \pi R^2 \sim 2 \times 10^{-24}$ cm^2 $\sigma_{rad} \sim \pi R^2 \frac{\Gamma_{rad}}{\Gamma_{total}}$ Latter becomes very small, not that Γ_{rad} decreases, but $\Gamma_{total} = \Gamma_{rad} + \Gamma_n + \Gamma_f$ increases	Near 1 Mev the nuclear excitation becomes comparable to the fission barrier. Γ_f rapidly rises and overwhelms Γ_{rad}. The fission cross section approaches $\sigma_f \sim \pi R^2 \frac{\Gamma_f}{\Gamma_f + \Gamma_n} \sim 0.5 \times 10^{-24}$ cm^2 —only a fraction of geometrical because neutron emission is still more important than fission. When the energy of the bombarding neutron reaches 7 or 8 Mev, its re-evaporation with only 1 Mev leaves the residual nucleus with enough energy to have a second—and more hopeful—try at fission. Now σ_f approaches πR^2

—a function of the size of the system, that may range from 0.4 to a value arbitrarily close to unity, depending upon design.

4. f, the chance to be caught at thermal energy in uranium, relative to the chance to be caught in any material.

5. η, the effective multiplication *if* caught in uranium.

Thus, the multiplication factor is $k = \epsilon p l f \eta$.

For a more detailed analysis of these factors and their dependence upon design, reference is made to the treatise of Weinberg and Wigner [54].

References

1. For a survey of the phenomena associated with fission see: "Physics of Nuclear Fission," Atomnaya Energ., Suppl. 1, (transl., J. E. S. Bradley), Pergamon Press, New York, 1958. Halpern, I.: Nuclear Fission, in "Annual Review of Nuclear Science," vol. 9, Annual Reviews, Inc., Palo Alto, Calif., 1959. Huizenga, J. R., and R. Vandenbosh: Nuclear Fission, in P. M. Endt and P. B. Smith (eds.), "Nuclear Reactions," vol. 2, North Holland Publishing Company, Amsterdam, 1962. Wheeler, J. A.: Channel Analysis of Fission, in J. B. Marion and J. L. Fowler (eds.), "Fast Neutron Physics," Pt. II, Interscience, New York, 1963. Wilets, L.: "Theories of Nuclear Fission," Oxford University Press, London, 1964. Perfilov, N. A., and V. P. Eismont (eds.): "Physics of Nuclear Fission," Israel Program for Scientific Translations, Jerusalem, 1964. (AEC-tr-6205). Hyde, E. K.: Fission Phenomena, in E. K. Hyde, I. Perlman, and G. T. Seaborg, (eds.) "The Nuclear Properties of the Heavy Elements," vol. 3, Prentice-Hall, Englewood Cliffs, N.J., 1964.
2. Amaldi, E., O. D'Agostino, E. Fermi, B. Pontecorvo, F. Rasetti, and E. Segrè: *Proc. Roy. Soc. (London)*, **A146**: 483 (1934); **A149**: 522 (1935).
3. Hahn, O., and F. Strassmann: *Naturwiss.*, **27**: 11, 89 (1939).
4. Frisch, O. R.: *Nature*, **143**: 276 (1939).
5. For a review of the work undertaken during this period see: Turner, L. A.: *Revs. Mod. Phys.*, **12**: 1 (1940).
6. The constants used in the formulas in the text are those of Green, A. E. S.: *Phys. Rev.*, **95**: 1006 (1954). For a review of different formulations of the semi-empirical mass formula and comparison with experiment see Preston, M. A.: "Physics of the Nucleus," Addison-Wesley, Reading, Mass., 1962.
7. Petrzhak, K. A., and G. N. Flerov: *J. Exptl. Theoret. Phys.*, **101**: 1013 (1940).
8. Bohr, N.: *Nature*, **137**: 344, 351 (1936). Bohr, N., and F. Kalckar: *Kgl. Danske Videnskab. Selskab, Mat.-Fys. Medd,.* **14**: 10 (1937).
9. Bohr, N., and J. A. Wheeler: *Phys. Rev.*, **55**: 1124A (1939); **56**: 420 (1939).
10. Bohr, N.: *Phys. Rev.*, **55**: 418L (1939).
11. Stokes, R. H., J. A. Northrup, and K. Boyer: *Proc. U. N. Intern. Conf. Peaceful Uses At. Energy, 2nd Geneva*, 1958, vol. 15, p. 179. J. A., Northrup, R. H. Stokes, and K. Boyer: *Phys. Rev.*, **115**: 1277 (1959).
12. Present, R. D., F. Reines, and J. K. Knipp: *Phys. Rev.*, **70**: 557 (1946). Frankel, S., and N. Metropolis: *Phys. Rev.*, **72**: 914 (1947). U. L., Businaro, and S. Gallone: *Nuovo Cimento*, **1**: 629, 1277 (1955). Nosoff, V. G.: *Proc. U. N. Intern. Conf. Peaceful Uses At. Energy, New York*, 1956, vol. 2, p. 205. Swiatecki, W. J.: *Phys. Rev.* **101**: 651 (1956); **104**: 993 (1956); *Proc. U. N. Intern. Conf. Peaceful Uses At. Energy, 2nd Geneva*, 1958, vol. 15, p. 248. Cohen, S., and W. J. Swiatecki: *Ann. Phys.*, **67**: 164 (1962).
13. Bowman, C. D., M. S. Coops, G. F. Auchampaugh, and S. C. Fultz: *Phys. Rev.*, **137**: B326 (1965).
14. Werner, F. G., and J. A. Wheeler: *Phys. Rev.*, **109**: 126 (1958).
15. Hyde, E. K.: *op. cit.* [1], pp. 75–77.
16. Newton, J. O.: *Progr. Nucl. Phys.*, **4**: 234 (1955). Wheeler, J. A.: Nuclear Fission and Nuclear Stability, in W. Pauli (ed.), "Niels Bohr and the Development of Physics," Pergamon Press, New York, 1955. Johansson, S. A. E.: *Nucl. Phys.*, **12**: 449 (1959).
17. Katz, L., A. P. Baerg, and F. Brown: *Proc. U. N. Intern. Conf. Peaceful Uses At. Energy, 2nd Geneva*, 1958, vol. 15, p. 188.
18. Lamphere, R. W.: *Phys. Rev.*, **104**: 1654 (1956).
19. For a detailed study of the competition between fission and neutron emission, see Huizenga, J. R., and R. Vandenbosh: *op. cit.* [1].
20. Pik-Pichak, G. A.: *Soviet Phys. JETP (English Transl.)*, **7**: 238 (1958).
21. Serber, R.: *Phys. Rev.*, **72**: 1114 (1947).
22. Goldberger, M. L.: *Phys. Rev.*, **74**: 1269 (1948).
23. Porter, C. E., and R. G. Thomas: *Phys. Rev.*, **104**: 483 (1956).
24. Reich, C. W., and M. S. Moore: *Phys. Rev.*, **111**: 929 (1958). Vogt, E.: *Phys. Rev.*, **112**: 203 (1958).
25. Shore, F. J., and V. L. Sailor: *Phys. Rev.*, **112**: 191 (1958).
26. Milton, J. C. D., and J. S. Fraser: *Can. J. Phys.*, **40**: 1626 (1962).
27. Terrel, J.: *Phys. Rev.*, **113**: 527 (1959).
28. Bowman, H. R., S. G. Thompson, J. C. D. Milton, and W. J. Swiatecki: *Phys. Rev.*, **126**: 2120 (1962); *Phys. Rev.*, **129**: 2133 (1963).
29. Fuller, R. W.: *Phys. Rev.*, **126**: 684 (1962).
30. Mather, D. S., P. Fieldhouse, and A. Moat: *Phys.* Rev., **133**: B1403 (1964).
31. Maienschein, F. C., R. W. Peele, W. Zobel, and T. A. Love: *Proc. U. N. Intern. Conf. Peaceful Uses At. Energy, 2nd, Geneva*, 1958, vol. 15, p. 366. Rau, F. E. W.: *Ann. Physik*, **10**: 252 (1963).
32. Kapoor, S. S., and R. Ramanna: *Phys. Rev.*, **133**: B598 (1964). Hoffman, M. M.: *Phys. Rev.*, **133**: B714 (1964). Johansson, S. A. E.: *Nucl. Phys.*, **60**: 378 (1964).
33. Wahl, A. C., R. L. Ferguson, D. R. Nethaway, D. E. Troutner, and K. Wolfsberg: *Phys. Rev.*, **126**: 1112 (1962).
34. Glendenin, L. E., and J. P. Unik: *Phys. Rev.*, **140**: B1301 (1965). Kapoor, S. S., H. R. Bowman, and S. G. Thompson: *Phys. Rev.*, **140**: B1310 (1965).
35. Glendenin, L. E., C. D. Coryell, and R. R. Edwards: in C. D. Coryell and N. Sugarman (eds.), "Radiochemical Studies: The Fission Products," p. 489, McGraw-Hill, New York, 1951.
36. Turkevich, A., and J. B. Niday: *Phys. Rev.*, **84**: 52 (1951). Jensen, R. C., and A. W. Fairhall: *Phys. Rev.*, **109**: 942 (1958). Niday, J. B.: *Phys. Rev.*, **121**: 1471 (1961).
37. Britt, H. C., H. E. Wegner, and J. C. Gursky: *Phys. Rev.*, **129**: 2239 (1963).
38. Gibson, W. M., T. D. Thomas, and G. L. Miller: *Phys. Rev. Letters*, **7**: 65 (1961). Milton, J. C. D., and J. S. Fraser: *Can. J. Phys.*, **40**: 1626 (1962). Bowman, H. R., J. C. D. Milton, S. G. Thompson, and W. J. Swiatecki: *Phys. Rev.*, **129**: 2133 (1963).
39. Apalin, V. F., Yu. N. Gritsyuk, I. E. Kutikov, V. I. Lebedev, and L. A. Mikaelian: *Nucl. Phys.*, **71**: 553 (1965).
40. Whetstone, S. L., Jr.: *Phys. Rev.*, **114**: 581 (1959). Vladimirskii, V. V.: *Soviet Phys. JETP (English Transl.)*, **5**: 673 (1957).

41. Winhold, E. J., P. T. Demos, and I. Halpern: *Phys. Rev.*, **87**: 1139 (1952).
42. Brolley, J. E., Jr., and W. C. Dickinson: *Phys. Rev.*, **94**: 640 (1954). Henkel, R. L., and J. E. Brolley, Jr.: *Phys. Rev.*, **103**: 1292 (1956). Gindler, J. E., G. L. Bate, and J. R. Huizenga: *Phys. Rev.*, **136**: B1333 (1964). Leachman, R. B., and L. Blumberg: *Phys. Rev.*, **137**: B814 (1965). Viola, V. E., Jr., J. M. Alexander, and A. R. Trips: *Phys. Rev.*, **138**: B1434 (1965). Albertsson, E., and B. Forkman: *Nucl. Phys.*, **70**: 209 (1965).
43. For a summary and further development of the analysis by Brueckner and collaborators of the properties of nuclear matter, see Bethe, H. A., and J. Goldstone: *Proc. Roy. Soc.* (*London*), **A238**: 551 (1957); Gomes, L. C., J. D. Walecka, and V. F. Weisskopf: *Ann. Phys.*, **3**: 241 (1958).
44. For a summary of concepts and results of the shell model, see Mayer, M. G., and J. H. D. Jensen: "Elementary Theory of Nuclear Shell Structure," Wiley, New York, 1955, and Elliott, J. P., and A. M. Lane: The Nuclear Shell Model, in S. Flügge (ed.), "Encyclopedia of Physics," vol. 39, Springer, Berlin, 1957.
45. Rainwater, J.: *Phys. Rev.*, **79**: 432 (1950).
46. Bohr, A.: *Phys. Rev.*, **81**: 134 (1951); *Kgl. Danske Videnskab. Selskab, Mat.-Fys. Medd.*, **26**: 14 (1952). Hill, D. L., and J. A. Wheeler: *Phys. Rev.*, **89**: 1102 (1953). Bohr, A., and B. R. Mottelson: *Kgl. Danske Videnskab. Selskab, Mat.-Fys. Medd.*, **27**: 16 (1953). Ford, K. W.: *Phys. Rev.*, **90**: 29 (1953). Wilets, L., D. L. Hill, and K. W. Ford: *Phys. Rev.*, **91**: 1488 (1953). Bohr, A.: "Rotational States of Nuclei," Copenhagen, 1954.
47. Bohr, A.: *Proc. U. N. Intern. Conf. Peaceful Uses At. Energy, New York*, 1956, vol. 2, p. 151. Wheeler, J. A.: Nuclear Fission and Nuclear Stability, in W. Pauli (ed.), "Niels Bohr and the Development of Physics," Pergamon Press, New York, 1955.
48. Mottelson, B. R., and S. G. Nilsson: *Kgl. Danske Videnskab. Selskab, Mat.-Fys. Skrifter*, **1**: 8 (1959).
49. Halpern, I., and V. M. Strutinskii: *Proc. U. N. Intern. Conf. Peaceful Uses At. Energy, 2nd, Geneva*, 1958, vol. 5, p. 408.
50. Wheeler, J. A.: *Physica*, **22**: 1103 (1956).
51. Moore, M. S., O. D. Simpson, T. Watanabe, J. E. Russell, and R. W. Hockenbury: *Phys. Rev.*, **135**: B945 (1964). Michaudon, A., H. Derrien, P. Ribon, and M. Sanche: *Nucl. Phys.*, **69**: 545 (1965).
52. Cowan, G. A., A. Turkevich, C. I. Browne, and Los Alamos Radiochemistry Group: *Phys. Rev.*, **122**: 1286 (1961). Cowan, G. A., B. P. Bayhurst, and R. J. Prestwood: *Phys. Rev.*, **130**: 2380 (1963). Regier, R. B., W. H. Burgus, R. L. Tromp, and B. H. Sorensen: *Phys. Rev.*, **119**: 2017 (1960). Faler, K. T., and R. L. Tromp: *Phys. Rev.*, **131**: 1746 (1963). Mehta, G. K.: Ph.D. Thesis, Columbia University, 1963.
53. Glasstone, S., and M. C. Edlund: "The Elements of Nuclear Reactor Theory," Van Nostrand, Princeton, N.J., 1952.
54. Weinberg, A. M., and E. P. Wigner: "The Physical Theory of Neutron Chain Reactors," University of Chicago Press, Chicago, 1958.

Chapter 11

Cosmic Rays and Their Interactions

By YASH PAL, Tata Institute of Fundamental Research

Starting with investigations of the conductivity of gases in closed chambers, the subject of cosmic rays has, over the past four decades, come to have links with such diverse fields as nuclear physics, elementary-particle physics, geophysics, astronomy, astrophysics, solar physics, and magnetohydrodynamics. In cosmic-ray studies fundamental contributions have been made in each of these fields, for example: Most of the quasi-stable fundamental particles and their important properties were discovered; important characteristics of high-energy electromagnetic and nuclear collisions were established; insight into the solar surface phenomena and the sun-earth relationships was obtained; new information on the chemical composition of the sun and supernovae and on the matter and magnetic-field distribution in the galaxy was provided; and some important new aspects governing the basic stability of the galaxy were discovered.

In a subject as wide-ranging as this it is no longer convenient to follow the customary historical approach. Quite often discoveries made very recently lead to an understanding of important phenomena discovered much earlier. A large amount of work has properly become a classical part of other disciplines. Some of the new fields, such as sun-earth relationships and interplanetary phenomena, have become separate sciences in their own right and do not warrant a very detailed discussion in a chapter devoted to cosmic radiation in general. Therefore, subject to such restriction, the topics have been chosen to give a reasonably coherent picture of the areas with which cosmic-ray studies are currently concerned. To compensate partly for the omission of important original contributions, a short historical sketch of the development of cosmic-ray physics is included.

1. Brief History of Cosmic-ray Research†

At the beginning of the century it was discovered that an electroscope, even when enclosed in a hermetically sealed envelope, gradually loses its charge and that the conductivity of the surrounding air which is responsible for this leakage corresponds to a continuous production of ions at the rate of approximately 10 ions/cm^3-sec. The effect was attributed to rays from radioactive substances in the ground and contaminations in the chamber wall; yet no numerical agreement could be obtained on these assumptions. Why a thin lead shield reduces the leakage to $\sim$20 per cent, while additional lead shields induce no further reduction, also remained unexplained. In an attempt to escape from the effects of radioactivity at the ground, Gockel in 1910 and 1911 made three balloon ascents; his electroscopes indicated, however, a slight increase rather than the expected decrease in the leakage rate. Hess first proceeded seriously on the hypothesis that the observed effect is due to radiation entering the earth from outer space. In balloon ascents between 1911 and 1913 he not only demonstrated a real increase of ionization with altitude but also established that the radiation was equally strong day and night and was not reduced during the solar eclipse of 1912. He concluded that radiation is entering from outer space and that this radiation, if of electromagnetic nature, does not originate in the sun.

Cosmic-ray research during the decade following the First World War was characterized by two important discoveries. In 1922–1927 Millikan and his coworkers lowered electroscopes into the water of mountain lakes and measured the penetrating power of cosmic rays; they found that it greatly exceeded the penetrating power of the hardest known γ rays. In 1927–1929 Clay demonstrated that the intensity of cosmic radiation decreased with decreasing geographic latitude. Thus it appeared that the magnetic field of the earth influenced the primary radiation and that therefore at least part of it must be attributed to fast-moving electrically charged particles.

Until then cosmic-ray research had been carried out almost exclusively with electrometers and ionization chambers. In 1929 new methods were introduced. The corpuscular nature of cosmic rays was demonstrated by Skobeltzyn, who photographed particle tracks in a Wilson cloud chamber, and by Bothe and Kohlhörster, who used Geiger-Müller (G.M.) counters in coincidence. These two important techniques were combined by Blackett and Occhialini, who designed a cloud chamber whose expansion was initiated by the passage of cosmic-ray particles through G.M. counters placed on top of the chamber. New cloud chambers were constructed which were located between the pole faces of magnets in order to obtain from the measured curvature of the tracks both the particle's momentum and the sign of its charge.

These technical advances led to new and quite

† Part of this section is taken from the chapter on Cosmic Rays by B. Peters in the first edition of this Handbook.

unsuspected discoveries between 1929 and the beginning of the Second World War. With a cloud chamber in a magnetic field Anderson discovered the positron in 1932. Showers of cosmic-ray particles were discovered by Rossi, and it was found by Blackett that such showers may enter the cloud chamber from outside or be produced inside the chamber itself. The existence of the μ meson was established in 1936–1937 by the work of Neddermeyer and Anderson as well as that of Street and Stevenson. Nuclear disintegrations produced by cosmic rays in photographic emulsions were first observed in 1937 by Wambacher and Blau. In 1939 Auger and his collaborators showed that cosmic-ray showers produced in the atmosphere may contain more than 10^6 particles and spread over areas large enough to produce coincidences in counters separated by 300 m; they concluded that primary cosmic rays capable of producing such showers must have energies in excess of 10^{15} ev.

Important theoretical advances during this period influenced cosmic-ray research. Bethe and Bloch calculated the ionization produced by the traversal of charged particles through matter. The appearance of electronic showers in the atmosphere was explained by the cascade theories of Bhabha and Heitler and of Carlson and Oppenheimer. The theory of the latitude effect of cosmic rays was developed by Lemaître and Vallarta, who refined Störmer's original calculations on the motion of charged particles in the magnetic-dipole field of the earth.

After the conclusion of the Second World War, cosmic-ray research was accelerated still further. The development of nuclear emulsions capable of registering tracks of all charged particles (even those of relativistic velocities), as well as the development of improved stratosphere balloons, added new valuable techniques to cosmic-ray work.

In 1935 Yukawa had proposed the existence of a strongly interacting particle of mass 200 to 300 m_e to account for nuclear forces. It became increasingly clear that the Anderson-Neddermeyer particle, the muon, was too weakly interacting to be identified with the Yukawa particle. This was conclusively demonstrated in 1947 by Conversi, Pancini, and Piccioni by showing that a large fraction of negative muons decay after stopping in carbon, in spite of the fact that in their K orbit they spend part of their time inside the nucleus.

In 1947, using nuclear emulsions, Lattes, Occhialini, and Powell discovered the charged π meson, which was Yukawa-like in its properties and was the parent of the muon. Existence of the neutral π meson and its γ-ray decay mode was postulated by Sakata and Tanikawa in 1940 and discussed in greater detail by Oppenheimer in 1947. Experimental evidence for emergence of high-energy γ rays from nuclear interactions was first obtained by Bridge, Rossi, and Williams in 1947. Around the same time Rochester and Butler discovered the first heavy unstable particles in a cloud chamber. An intensive and systematic search led to the identification of a large fraction of the presently known quasi-stable particles.

In 1948 the primary cosmic radiation was shown to contain a mixture of atomic nuclei, whose chemical composition was later found to be richer in heavier nuclei when compared with the universal abundance of elements. High-energy interactions (up to primary energies $\sim 10^{13}$ ev) were studied with large stacks of nuclear emulsions and emulsion-metal sandwiches exposed at high altitude or with huge installations using cloud chambers and ionization calorimeters. Large arrays of different detectors were used by various groups to study extensive air showers, and the existence of primary particles of energy up to 10^{20} ev was demonstrated by Rossi and collaborators.

In 1957 cosmic-ray research entered a new era with the launching of the first sputnik. Groups of Van Allen and Vernov discovered intense belts of geomagnetically trapped radiation around the earth. Satellite technology made it possible to probe plasma clouds and magnetic fields in the interplanetary space and establish the existence of intimate sun-earth relationships.

Primary electrons were discovered in 1961 in balloon experiments; their existence in the galaxy and radio sources had already been inferred from the synchrotron nature of radioemission. Cosmic X rays and discrete X-ray sources were discovered in 1962, first in rocket experiments and later with balloons. Investigations of deep underground muon intensity up to depths of 3 km below ground led to the setting up of high-energy cosmic-ray neutrino experiments; the first cosmic-ray neutrino interactions were observed in 1965. Important theoretical advances based on the theory of solar wind, postulated in 1958, led to a systematic investigation of cosmic-ray modulation by solar influence.

The theories of the origin of cosmic rays have not crystallized thus far. A great many possible sources and acceleration processes have been identified. Their relative importance for generating different cosmic-ray components is still not clear. It also appears now that the highest-energy cosmic-ray particles may be of extragalactic origin.

2. Schematic Outline of Principal Cosmic-ray Phenomena in the Atmosphere

It may be helpful to describe briefly how the incident radiation is modified upon entering the atmosphere and what is the main sequence of processes by which the energy of individual primary particles is subdivided and dispersed.

The incident radiation is composed mainly of atomic nuclei consisting of ~ 90 per cent protons, ~ 10 per cent helium nuclei, and about 1 per cent heavier nuclei. These particles arrive more or less isotropically carrying kinetic energies ranging over nearly 14 orders of magnitude i.e., from a few Mev to 10^{20} ev. Their energy spectrum is very steep; above a few Bev the integral intensity falls by 5 orders of magnitude when the energy increases by 3 orders. Thus the flux above 10^{16} ev is about one particle per square meter per steradian per year.

Except for very low-energy particles, which stop by ionization loss, almost all the incident particles suffer nuclear collisions with air nuclei in the upper atmosphere.

Nuclear collisions at high energy are generally inelastic inasmuch as a significant fraction of the collision energy is used up in production of new particles most of which are unstable. The newly created particles, mostly charged and neutral π mesons (or pions), carry away one-third to one-half of the incident energy. The number of these particles increases with primary energy, though their total fractional energy remains nearly constant. The products of primary collisions, nucleons, pions, etc., if they have sufficient energy, produce secondary collisions leading to a further multiplication of particles in a cascade process. The development of a nuclear cascade in the atmosphere may be intercepted by the following processes:

Unstable particles decay directly or indirectly into lighter particles, muons, electrons, neutrinos, or photons which do not have any specific nuclear interaction with matter.

Charged particles suffer Coulomb energy losses, through ionization at low energy and by *bremsstrahlung* and creation of electron-positron pairs if their energy is high compared with their rest mass.

At least up to energies of 10^{14} ev, most of the secondary particles are charged and neutral pions, produced in the ratio of 2:1. In addition K particles, hyperons, and nucleon-antinucleon pairs are also produced, though their relative frequency remains small.

Neutral π mesons decay, within a very short time ($\sim 2 \times 10^{-16}$ sec), predominantly into two photons which by creation of electron-positron pairs initiate an electromagnetic cascade. The number of particles in the cascade continues to multiply until the energy of the electrons is degraded to such an extent that ionization loss dominates the energy loss by *bremsstrahlung* and until the photon energies have been reduced to the point where Compton scattering and photoelectric absorption dominate the pair-production process.

The charged π meson decays mostly into a muon and a neutrino with a lifetime of 2.55×10^{-8} sec. In the rarefied material of the atmosphere the decay of $\pi^{\pm}$ mesons is much more likely than a nuclear collision. Only if their lifetime is greatly dilated by relativistic effects (for energies $\gtrsim 100$ Bev) will nuclear interactions occur with appreciable probability. For a given energy this probability is larger in the dense lower regions of the atmosphere than in the rare upper regions.

Muons have no specific nuclear interaction with matter. At low energies ionization energy loss predominates. Above 100 Bev losses due to *bremsstrahlung*, pair production, Cerenkov radiation, and photonuclear reactions increase progressively (Sec 7.10). A muon decays into an electron (or positron) and two neutrinos with a lifetime of 2.2×10^{-6} sec. Because of time dilation, however, muons whose energy exceeds a few Bev have a very small probability of decaying before reaching the earth's surface, if they are traveling near the vertical direction. The decay probability increases at larger angles because of the increased path length available.

Thus there are four principal components of cosmic rays in the atmosphere:

1. A nuclear component containing particles capable of producing nuclear collisions. It is attenuated strongly in passing through the atmosphere. It consists mostly of nucleons; the fraction of pions in it increases with increasing depth in the atmosphere (Sec. 7.3).

2. An electron-photon component which owes its origin mainly to the production of neutral π mesons in nuclear collisions and multiplies by pair-production and *bremsstrahlung* processes. It reaches a broad maximum at about 13-km altitude; most of it dies out, however, before reaching the earth's surface.

3. A muon component with owes its origin mainly to the production of $\pi^{\pm}$ mesons in nuclear collisions. Muons do not multiply like electrons. They produce fast electrons, both by collisions with atomic electrons and decay. At sea level muons constitute about 80 per cent and electrons about 20 per cent of the charged particles incident on the earth's surface.

4. A neutrino component which arises from the decay of unstable particles in the atmosphere, mostly pions and muons. Neutrinos have only weak interaction with matter and are capable of passing through the earth without any sensible attenuation.

Figure 11.1 is a schematic diagram showing the principal generic relations between various types of secondary cosmic-ray particles.

3. Influence of the Earth's Magnetic Field

Charged cosmic-ray particles are affected by the earth's magnetic field long before they enter the atmosphere. To a first approximation the earth's magnetic field is represented by a dipole of magnetic moment $M = 8.1 \times 10^{25}$ gauss-cm^3, situated at the center of the earth with the magnetic axis cutting the earth's surface in the Northern Hemisphere at 78.2°N and 69°W (Finch and Leaton, 1957). A better representation is obtained if some of the quadrupole terms are taken into account by making the dipole slightly eccentric (Webber, 1958).

The magnetic field acts as a momentum analyzer. For each location on the earth and for each direction of incidence one can define a rigidity (ratio of momentum to charge, that is, pc/Ze) P_1, such that all particles of rigidity greater than P_1 can reach from infinity to the given point from the given direction. Then from Liouville's theorem, applicable if the cosmic-ray intensity is isotropic at infinity, these particles arrive with their full intensity. One can also define a rigidity P_c such that particles of rigidity lower than P_c cannot arrive from infinity at the given point in the given direction. Thus intensity for $P < P_c$ is zero. For P intermediate between P_1 and P_c, there are "forbidden" and "allowed" bands of rigidities, and the intensity would have an "average" value.

Obviously P_1 and P_c are functions of the geomagnetic location and also of the arrival direction. For a given direction of incidence, P_1 and P_c decrease with increasing latitude. This explains the so-called latitude effect, or the increase in intensity of cosmic radiation in going from the equator to the poles.

Threshold Rigidities for a Magnetic Dipole. Störmer (1955) has given a very useful relation defining the trajectory of a charged particle in a static magnetic field. In a plane containing the moving particle and the dipole axis, called the meridian plane, the motion is described by

$$2\gamma = R \cos \lambda \sin \xi + \frac{\cos^2 \lambda}{R} \tag{11.1}$$

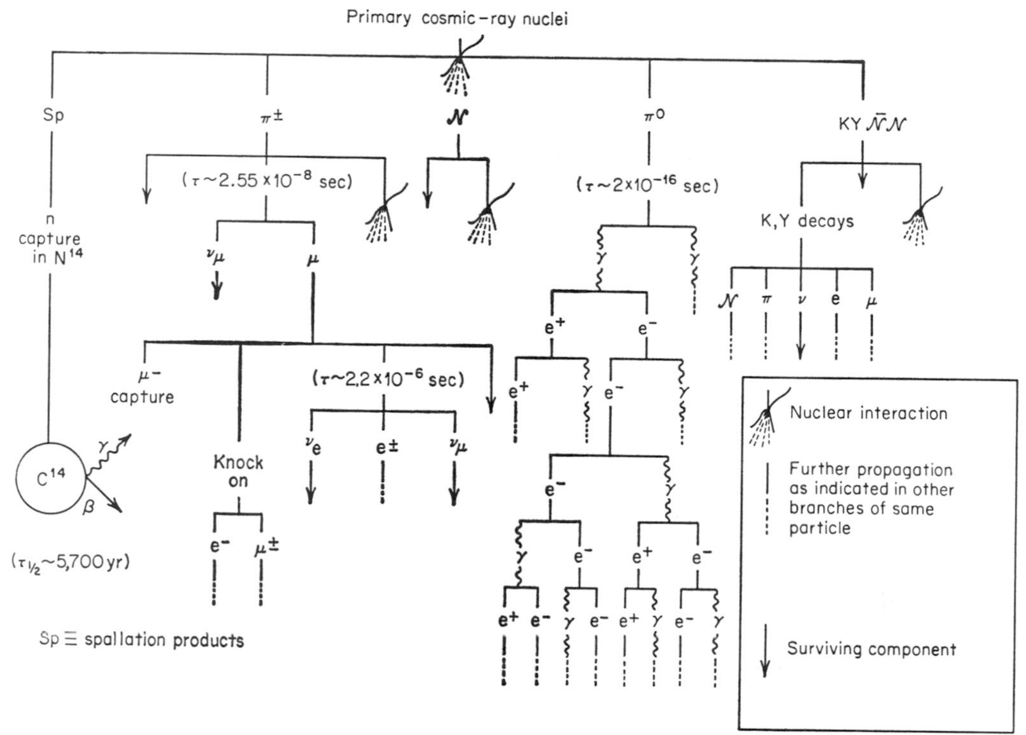

FIG. 11.1. Progeny of a high-energy cosmic-ray nucleon incident on top of the atmosphere.

where γ is a constant proportional to the impact parameter of the particle trajectory at great distance, λ is the geomagnetic latitude, and ξ is the angle between the velocity vector and the meridian plane and is taken positive if this plane is crossed from east to west. R is the radial distance measured in Störmer units where one Störmer unit is defined as $\sqrt{M/P}$ (thus the scale of distance is rigidity-dependent).

For a given λ and ξ, the minimum value of R is obtained by putting $\gamma = 1$; this is

$$R_m = \frac{\cos^2 \lambda}{1 + \sqrt{1 - \sin \xi \cos^3 \lambda}} \qquad (11.2)$$

The cutoff rigidity P_c is obtained by putting R_m equal to the radius of the earth (measured in Störmer units), which gives

$$P_c = \frac{M}{a^2} R_m{}^2$$

$$= 59.6 \frac{\cos^4 \lambda}{[1 + \sqrt{1 - \sin \theta \cos \phi \cos^3 \lambda}]^2} \quad \text{BV} \qquad (11.3)$$

In Eq. (11.3), a is the radius of the earth and $\sin \xi$ has been replaced by $\sin \theta \cos \phi$, where θ and ϕ are the conventional zenith and azimuthal angles of incidence.

Detailed calculations of Lemaître and Vallarta (1936) show that for particles of a given rigidity at a given geomagnetic latitude the sky may be divided into four regions:

1. The Störmer cone, within which particles of a given rigidity are forbidden.
2. The main cone, within which particles of a given rigidity are allowed.
3. The penumbra, which lies between the Störmer cone and the main cone and contains alternate bands of "forbidden" and "allowed" regions. This results from complicated particle trajectories intersecting the earth.
4. The shadow cone, which is a forbidden region near the horizon arising from simple particle trajectories intersecting the earth.

A schematic illustration of the various regions is shown in Fig. 11.2.

Experimental Observations and Nondipole Effects. General predictions of the geomagnetic theory are obviously fulfilled. Latitude effect exists. The intensity of penetrating particles arriving from the east is smaller than from the west, indicating that most of the primaries are positively charged (Neher, 1952). From a comparison of the energy spectrum of heavy nuclei ($Z \geq 6$) at $\lambda = 55°$ and $41°$ Kaplon et al. (1952) concluded that the cutoff is in rough agreement with theory if one assumes that the particles enter the atmosphere completely stripped of their electron shells.

However, in a neutron-monitor survey of sea-level intensity Rose et al. (1956) noticed that the minimum in cosmic-ray intensity was 9° north of the geomagnetic equator at 30° west longitude. Waddington (1956) found that helium nuclei had a rigidity cutoff lower than the theoretical value over North America and higher than the theoretical value over south England. Many such discrepancies were later established, including the observation by Rothwell and Quenby (1958) that the cosmic-ray intensity near the Cape Town magnetic anomaly was unusually high and positively correlated with the high value of local

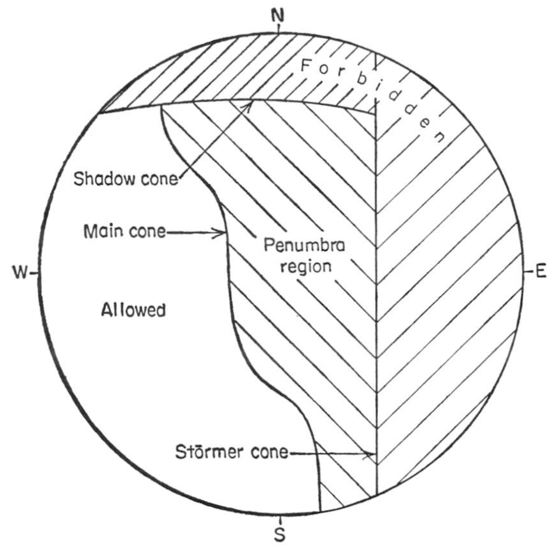

FIG. 11.2. Schematic diagram showing the division of the sky into the allowed, forbidden, and penumbra regions due to the effect of the earth's magnetic field on particles arriving from infinity. The Störmer cone, the main cone, and the shadow cone are also indicated (Sec. 3).

magnetic-dip angle. All these observations suggested that the earth's magnetic field cannot be approximated well by a dipole and that several higher-order harmonics in the magnetic potential need to be taken into account.

After the magnetic survey data are approximated by an expansion of the magnetic potential in multipoles, the calculations are made by a rigorous integration of a large number of individual trajectories. This has been done by Quenby and Webber (1959), Hultqvist (1958), Kellogg and Schwartz (1959), Wenk (1961), and Quenby and Wenk (1962). Some of these calculations are compared in Fig. 11.3 with an airplane neutron-monitor determination of the cosmic-ray equator by Katz, Meyer, and Simpson (1958). One sees the improvement in fitting the experimental data when successively higher-order harmonics in the magnetic potential are included. There is also an over-all better agreement between the measured and calculated cutoff rigidities at high latitudes, though one encounters occasionally a change in the effective value of the cutoff with time. It is believed that residual discrepancies are due to nongeocentric influences such as the possible existence of a westward flowing ring current at a distance of 6 to 8 earth radii, the effect of the steady interplanetary field, and the effect of scattering of particles from one orbit to another at the disturbed geomagnetic boundary and to the interaction of expanding solar corona with the earth's magnetic field (Sec. 8.4). A complete understanding of these phenomena has not yet been reached.

4. Primary Cosmic-ray Particles

Since the early experiments of Schein, Jesse, and Wollan (1941) in which it was demonstrated that most of the primary cosmic-ray particles do not multiply in thin layers of lead as electrons and γ rays do, it has been shown that the most dominant component of cosmic rays outside the atmosphere consists of protons. The presence of energetic multiply charged particles at residual pressures of less than 20 g/cm^2 was first observed by Freier, Lofgren, Ney, Oppen-

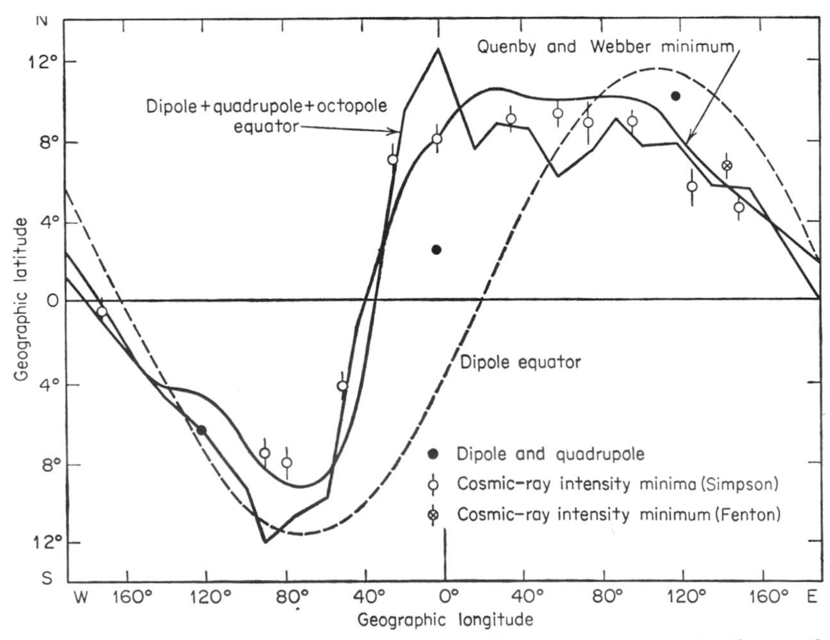

FIG. 11.3. Cosmic-ray equator as calculated according to dipole and higher-order approximations to the earth's magnetic field. The Quenby and Webber calculation includes terms up to $n = 6$ in the multipole expansion of the magnetic potential. The experimental points give the positions of observed cosmic-ray minima. [*From Webber* (1962).]

heimer, Bradt, and Peters (1948), who employed nuclear photographic emulsions. In recent times the multiply charged primary particles have also been studied by using electronic techniques, usually involving a combination of scintillation and Cerenkov counters. At high latitudes where nonrelativistic particles are not forbidden by the earth's magnetic field from reaching near the earth, the charge of the particles can be determined by a simultaneous measurement of their ionization and energy, or ionization and velocity. At low latitudes all the particles allowed by the geomagnetic field must be relativistic and hence a determination of ionization alone suffices. In this way these particles have been identified as atomic nuclei belonging to various chemical elements (Bradt and Peters, 1948, 1950a, b). Even before they enter the atmosphere the nuclei are completely stripped of orbital electrons; they do not therefore carry any high-energy electrons into the atmosphere as part of the primary radiation.

4.1. Relative Abundances of Nuclear Components. The chemical composition of the nuclear part of the primary cosmic radiation has been measured at several energies. It seems that the relative abundance of different elements is approximately the same at rigidities (pc/Ze), from 2 to at least a few hundred BV. The most accurate relative values are those obtained in experiments made from Texas (41°N) where the vertical cutoff rigidity is 4.5 BV; these are given in Table 11.1, where the universal chemical abundance of elements has also been given for comparison. The following features of cosmic-ray chemical composition should be noted:

1. Presence of light elements (L nuclei) Li, Be, B which, being important nuclear fuels, are almost entirely absent in the universal abundance column
2. Large relative overabundance of heavy elements in the cosmic radiation, which suggests a production region that is different in composition from the exteriors of stars (Sec. 11)
3. Carbon more abundant than oxygen, while in the universe oxygen is about three times as abundant as carbon

TABLE 11.1. CHEMICAL COMPOSITION OF PRIMARY COSMIC-RAY PARTICLES OF RIGIDITY GREATER THAN 4.5 BV

Element		Relative cosmic-ray abundance	Universal abundance[a]
Hydrogen		7,000	6,600
Helium		1,000[b]	1,000
$Z \geq 3$		100[c]	10.6
Li_3		5.3[d]	2.6×10^{-5}
Be_4		2.3	5.2×10^{-6}
B_5		7.4	6.2×10^{-6}
$3 \leq Z \leq 5$	L	16.0	3.7×10^{-5}
C_6		30.1	2.4
N_7		9.7	6.3×10^{-1}
O_8		19.4	6.6
F_9		2.4	4.2×10^{-4}
$6 \leq Z \leq 9$	M	61.6	9.6
$10 \leq Z \leq 15$		16.4[e]	7.6×10^{-1}
$16 \leq Z \leq 19$		0.13	1.9×10^{-1} (for $16 \leq Z \leq 28$)
$20 \leq Z \leq 28$		7.0	
$10 \leq Z \leq 28$	H	23.4	9.5×10^{-1}
$Z > 28$		10^{-3}[f]	1.6×10^{-5}

[a] After Cameron (1959).

[b] Hydrogen-to-helium ratio is about 7.0 from several measurements (for example, Freier and Waddington, 1964).

[c] Helium to $Z \geq 3$ ratio is close to that given by Waddington (1960).

[d] Charge composition up to fluorine and relative fraction of nuclei with $Z \geq 10$ is that given by O'Dell et al. (1962).

[e] Charge composition above $Z > 10$ is as given by Daniel and Durgaprasad (1962).

[f] Value for $Z > 28$ is from Fleisher et al. (1965).

Measurements made near the geomagnetic equator, which give the chemical composition above a rigidity of 16 BV per particle, give values consistent with those given in Table 11.1. The relative number of interactions produced by primary particles of different atomic numbers at about 1,000 Bev/nucleon in giant emulsion stacks flown at high altitude also indicate roughly the same chemical composition. There is not much definitive evidence about chemical composition at still higher energies, though it is likely that around 10^6 Bev most primaries are heavy nuclei and above 10^8 Bev they are mostly protons (see Sec. 8).

At low energies the galactic spectrum of particles is subject to strong solar modulation, though the relative abundance of nuclei of same charge to mass ratio is expected to remain independent of solar influence, irrespective of whether the degree of modulation depends on rigidity, velocity or a combination of these two parameters (Sec. 9). Recent experimental results (Comstock, Fan, and Simpson, 1966) show that the ratio M/He increases below 100 Mev/n and the ratio L/M probably goes through a broad maximum around 300 Mev/n. The first of these two observations may be due to an intrinsic difference in the shapes of low-energy spectra, while the second is most probably due to the opposing effects (as one goes to lower energies) of an increasing containment lifetime and a rapidly increasing rate of ionization energy loss in the interstellar medium (Sec. 4.3).

Recent low-energy measurements also show that the element fluorine is almost absent in the primary cosmic radiation (Simpson, 1966).

4.2. Isotopic Composition. Like chemical composition, perhaps more so, the isotopic composition of the nuclear component of primary cosmic radiation must also contain the imprint of its production and propagation history. Measurement of isotopic composition puts rather severe demands on the experimental techniques which have hitherto been used for this work. So far only the most abundant elements of the primary cosmic rays have been studied in this respect; they are hydrogen, helium, and carbon nuclei. Two different approaches have been followed: One is direct mass and charge determination of low-energy particles (Appa Rao, 1961), and the other uses the difference in interaction characteristics of nuclei of different isotopes of the same chemical element (Appa Rao et al., 1956). Thus far the results in this important field are sketchy but are likely to be improved in the near future. Present knowledge about isotopic composition of various elements may be summarized as follows:

Helium. About one-fifth of the helium nuclei in the rigidity range 1 to 1.5 BV are He^3 (Appa Rao,

1961; Aizu, 1963; Foster and Mulvey, 1963; Hildebrand et al., 1963; Dahanayake et al., 1964). He^3 is also present at higher energies, at a rigidity of 14.6 BV, though its relative abundance with respect to He^4 is not well known. (Balasubrahmanyan et al., 1963).

Hydrogen. Deuterons probably exist in the primary cosmic radiation and constitute 2 to 5 per cent of the proton flux at a rigidity of 0.7 BV and also at 14.6 BV (Webber and Ormes, 1963b; Hasegawa, Nakagawa, and Tamai, 1963b; Ganguli, Kameshwara Rao, and Swami, 1963).

Carbon. By measuring the interaction mean free path of carbon nuclei for events of the type carbon in-carbon out, Hasegawa et al. (1963a) conclude that one-half of the carbon nuclei of kinetic energy greater than 500 Mev/nucleon have a mass number 13. Alvial (1963) has made direct mass measurements on carbon nuclei in the energy range 200 to 680 Mev/nucleon and concludes again that two isotopes of carbon are involved and that the abundance ratio of heavier to lighter isotope is about 1:1. However, both these results should be treated with caution because of the difficulty of measurements.

4.3. Chemical and Isotopic Abundances. It was first pointed out by Bradt and Peters (1950b) that the relative abundance of heavy nuclei, particularly that of the light elements, Li, Be, and B, with respect to heavier nuclei could be fruitfully used to determine the total amount of matter traversed by cosmic rays between acceleration and arrival at earth. These nuclei, even if totally absent at the source, would be produced by fragmentation of heavier nuclei in collision with interstellar hydrogen. Taking into account the "fragmentation parameters," the observed relative abundance of L nuclei suggests that cosmic rays, at least in the energy range lying between ~2 Bev/n to about 10 Bev/n, have traversed ~2.5 g/cm² of interstellar hydrogen (Sec. 12.2). If an interstellar density of one proton per cubic centimeter is assumed, this figure corresponds to a lifetime of 2×10^6 years. However, the assumption about the density is unreliable, and it would be extremely useful to be able to determine the age of cosmic radiation independently. A method of doing this has been investigated by Durgaprasad (1963) following a suggestion by Peters (1961). The relative abundance Be/B depends on the age of cosmic rays, because one of the beryllium isotopes Be^{10}, which has a mean life of 2.4×10^6 years, would contribute, through its decay, to boron flux if the age is much greater than its mean life and to beryllium flux if it is shorter. Present experimental accuracy on the ratio Be/B is inadequate to settle this question.

The ratio of He^3 to all helium nuclei may also be used to estimate the amount of interstellar material traversed by cosmic rays. It is important to do this by independent methods because some difference in the observed relative abundances of different nuclei may be caused by a differential modulation effect in the planetary system (Sec. 9).

The inverted ratio for relative abundance of the elements C, N, O and the possible large abundance of C^{13} suggest that these elements may be synthesized in some nonequilibrium stage of stellar evolution. According to Hayakawa et al. (1960), for example, this may come about because of the rapid CNO cycle occurring in outer layers of supernovae which act as cosmic-ray sources (Sec. 12.3).

Large excess of nuclei with even charge number over those with odd charge number is against the hypothesis that a significant fraction of the observed medium nuclei could be spallation products of very heavy nuclei.

4.4. Nonnuclear Components. On any model of cosmic-ray production and propagation through space, one expects a small part of the primary flux to be due to electrons and γ rays. These may be produced either in the source region or by the interaction of the nuclear components with the interstellar gas. The presence of electrons of up to a few hundred Mev has been detected in the primary radiation by Earl (1961) who flew a cloud chamber in a balloon and by Meyer and Vogt (1961) by using a counter system. Agrinier et al. (1963, 1965) have observed primary electrons of energy greater than 4.5 Bev, using a spark-chamber arrangement. Positrons of energy up to 2 Bev have also been observed by De Shong, Hildebrand, and Meyer (1964), using spark chambers and a permanent magnet carried up by a high-flying balloon. They give a value $e^+/(e^+ + e^-) \simeq 0.3$ between 100 and 1,000 Mev. Recently Daniel and Stephens (1965) have measured the flux of electrons up to ~100 Bev, using an emulsion stack. Below ~50 Bev they have attempted to determine the e^+/e^- ratio, using geomagnetic theory (Sec. 3) and find evidence for the existence of a substantial flux of positrons. The flux of electrons and positrons has been found to be a few per cent of that of the nuclear component at energies up to 5 Bev. The experimental values above 1 Bev are given in Fig. 11.4.

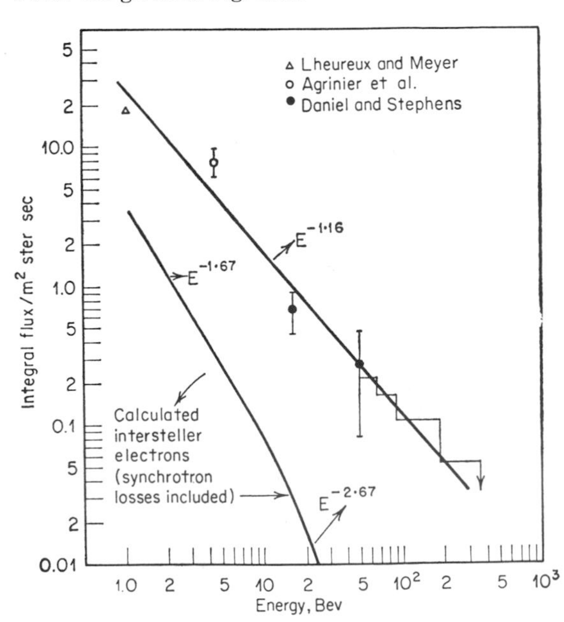

FIG. 11.4. The integral energy spectrum of primary cosmic electrons beyond 1 Bev. The calculated curve gives the flux of electrons from the decay of unstable particles produced in collisions of primary nucleons in traversing 2.5 g/cm² of interstellar hydrogen. [*After Daniel and Stephens* (1965).]

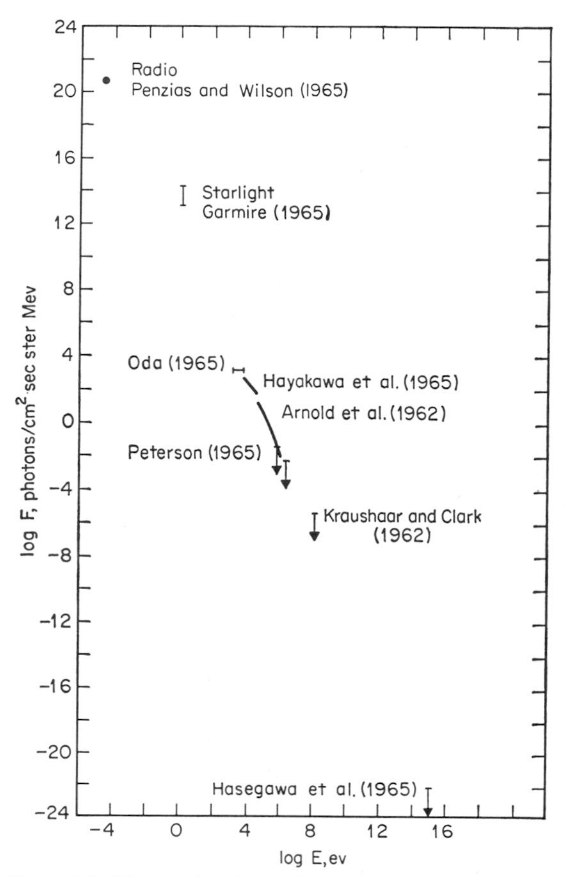

FIG. 11.5. Observed values of and upper limits on the differential flux of isotropic cosmic photons for a range of photon energies E. [*After Oda* (1965).]

It is now generally believed that the observed electron flux is of galactic origin. The existence of positrons in the primary flux would suggest that at least some of the electrons arise from collisions of cosmic-ray particles in the interstellar space, leading to production of positive and negative mesons which, through their decay, give rise to positrons and electrons. However, the observed flux of all electrons above 1 Bev is too large, by about a factor of 10, to be produced by cosmic-ray interactions if all nuclear components of cosmic rays have traversed only ~ 3 g/cm² of hydrogen before arrival on the earth, as suggested by the L/M ratio and by the ratio of He^3/He^4 at low energies.

Recently galactic electrons of a few Mev have been detected by Cline et al. (1964). Their flux is found to be such as would be obtained by simple extrapolation from high energies. The majority of these electrons can arise only from direct acceleration.

Gamma rays, X rays, and other neutral radiations are important because of their direction of arrival points to the source. For γ rays in the 50-Mev range and above, atmospheric contamination is a serious problem. For soft photons, below 30 Kev, the absorption effects are so serious that the measurements can be done only by rockets and satellites. The experimental results in this field have been accumulated only recently. They may be summarized thus:

1. Several discrete sources of X rays (1 to 10 Kev) have been discovered, all except one lying near the galactic plane.
2. Intensities of these X-ray sources are too high to be expected from normal astrophysical objects and, except for the Crab nebula (Taurus A), no visible or radio object is identified with these sources. (Intensity of X rays of more than 10 Kev from Taurus A is about 1 per cm²/sec.)
3. There is an isotropic background of X rays. (Intensity at 10 Kev is ~ 100 photons/cm²/sec/ster/Mev.)
4. No definite flux of γ rays of more than a few tens of Mev has been established, though valuable upper limits have been obtained for isotropic as well as discrete sources of γ rays of energies up to 10^{15} ev.

Experimental observations on isotropic flux of cosmic photons are summarized in Fig. 11.5 and on the fluxes from discrete sources in Fig. 11.6.

It is seen that most known forms of stable matter and radiations are present in the primary cosmic radiation. Existence of antimatter in general and antiprotons in particular has not so far been demonstrated; at low energies (E/nucleon $\ll 1$ Bev) antiprotons and antinuclei constitute less than 0.1 per cent of the nuclear flux, while for energies in the several-thousand-Bev range some rough arguments show that antiprotons, if any, are less than 10 per cent of the protons (Pal, 1963). The only other known stable particles that must be looked for are neutrinos. If γ rays and positrons exist there is no reason the neutrinos should not. During the next few years serious attempts are likely to be made to detect them.

The following significant developments have

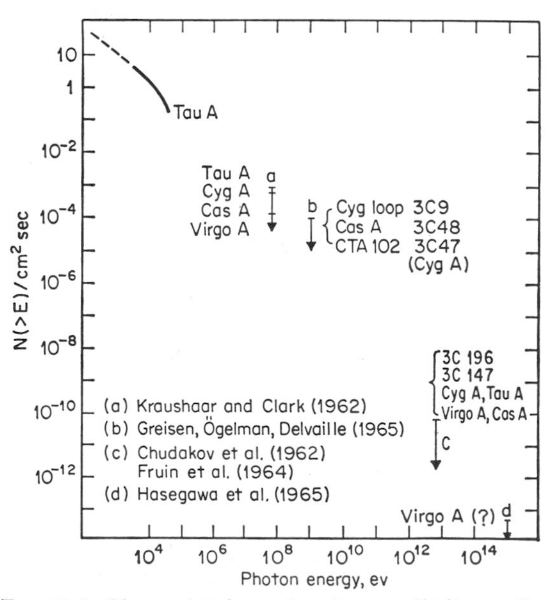

FIG. 11.6. Observed values of and upper limits on the integral flux of cosmic photons from some discrete sources in the sky. [*After Oda* (1965).]

taken place in our knowledge of the photon component of primary cosmic rays during 1965–1966:

1. The background radiation observed by Penzias and Wilson (1965) at ~7 cm wavelength (Fig. 11.5) is now generally believed to be part of a universal blackbody radiation at 3°K (Dicke et al., 1965). This has interesting cosmological consequences (see Sec. 12.7, note added in proof).

2. The strong X-ray source in Scorpio has been identified with a starlike object, situated at a distance of a few hundred light years, which has many of the morphological properties of an old nova (Sandage et al., 1966).

3. A definite flux of $(1.5 \pm 0.8) \times 10^{-4}$ per cm² per sec for γ rays of a few hundred Mev has been reported by Duthie et al. (1966) from the direction of the constellation Cygnus.

5. Primary Intensity and Energy Spectrum

Unless the flux measurements are carried out far outside the earth's atmosphere, it becomes difficult to discriminate against the secondary cosmic-ray particles that accompany the primary radiation at all latitudes. Therefore reliable measurements of flux of different components of cosmic radiation have become available only recently by making use of very high balloon flights, rockets, and satellites. Even in these experiments one has to contend with secondaries produced in the upper regions of the atmosphere, which may be recorded either when they are moving upward (splash albedo) or when, once having escaped from the earth's atmosphere, they are bent back toward the apparatus by the earth's magnetic field (return albedo). The albedo problem is more serious for the proton flux than for the flux of heavier nuclei because in interactions in the atmosphere mainly singly charged particles can be produced with sufficient velocities to travel up to the apparatus or to escape from the atmosphere. The splash albedo is eliminated by introducing in the measuring apparatus a direction-sensing device such as a Cerenkov counter or by observing the directional emanation of particles produced in nuclear interactions. The return albedo is absent for particles of rigidity higher than the local geomagnetic cutoff; such particles cannot be bent back toward the earth by the geomagnetic field.

5.1. Low-energy Part of the Spectrum. The flux of incident particles at low energies depends strongly on the intensity of solar activity. A measure of solar activity at any time is the sunspot number. For a measure of cosmic-ray activity one often uses large counting-rate neutron monitors which sample the intensity of nucleons of energy above ~500 Mev at a given location and reflect the intensity of several-Bev nucleons incident on top of the atmosphere (see Sec. 9). It is found that there is an inverse correlation between the neutron-monitor counting rate and sunspot number; this is shown in Fig. 11.7.

During the latest solar cycle primary cosmic-ray fluxes have been measured by using counter techniques and nuclear emulsions carried aloft by balloons, rockets, and satellites. In combining various measurements it is essential to ensure that they correspond to the same degree of cosmic-ray modulation as indicated by the counting rate in a continuously operating neutron monitor.

Integral Spectra in the Latitude Sensitive Region. Differential energy spectra measured to date are confined to rigidities of less than 1.5 BV for protons and less than 4.5 BV for helium nuclei. This is because so far no momentum-analyzing equipment of sufficient accuracy has been flown at high altitude. Above these energies the total flux measurements made at locations with different geomagnetic cutoff rigidities are used to obtain integral energy spectra of different primary components. Most of the existing measurements refer to three vertical cutoff rigidities: these are 1.3, 4.5, and 16.5 BV and correspond to the locations of Minneapolis, Texas, and some stations near the geomagnetic equator, respectively. Fluxes of hydrogen, helium, and S nuclei ($Z \geq 6$) have been measured in a large number of counter and emulsion experiments. Fluxes of L nuclei are obtained by using the L/S ratio which is measured only in some of the experiments. To reduce the measured flux

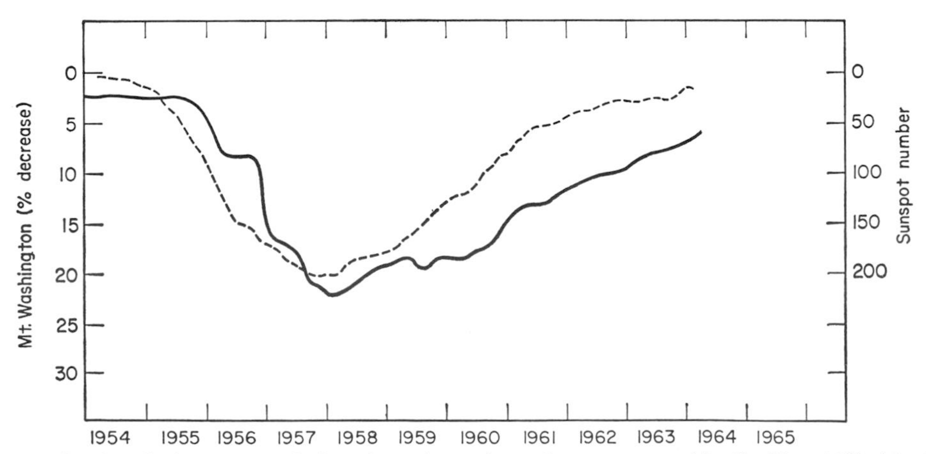

Fig. 11.7. Curves showing the inverse correlation of cosmic-ray intensity, as measured by the Mount Washington neutron-monitor counting rate (solid curve), with solar activity as indicated by the Zurich sunspot number (broken curve). [*After Webber* (1965).]

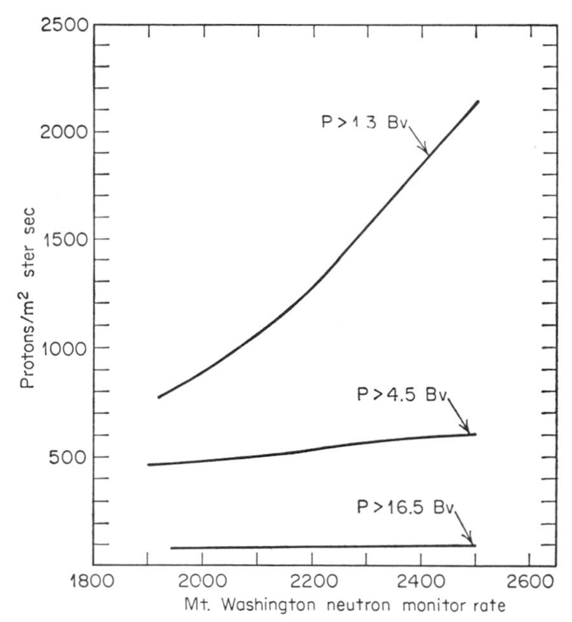

FIG. 11.8. Intensity of primary protons of rigidity greater than 1.3, 4.5, and 16.5 BV plotted against the Mount Washington counting rate. Increase of modulation with decreasing rigidity is obvious. [*After Webber* (1965).]

values to any given level of solar activity one normally uses regression curves of the type shown in Fig. 11.8 where the fluxes of protons of rigidity greater than 1.3, 4.5, and 16.5 BV measured over the latest solar cycle are plotted against the Mount Washington neutron-monitor counting rate. It is seen that the degree of solar modulation increases sharply with decreasing particle rigidity. Using curves of this type for all nuclear components, Webber (1965) has recently deduced the integral flux values, listed in Table 11.2, for the period of minimum solar activity

TABLE 11.2. INTEGRAL FLUX (M^{-2}/STER/SEC) OF VARIOUS COMPONENTS AT THE TIME OF MINIMUM OF SOLAR ACTIVITY

Rigidity, BV / Component	>1.3	>4.5	>16.5
Proton	2140 ± 40	610 ± 30	93 ± 5
Helium	328 ± 12	94 ± 4	17.5 ± 1.0
L	7.2 ± 1.0	2.1 ± 0.2	0.38 ± 0.08
$S = (M + H)$	28 ± 1.5	8.7 ± 0.3	1.35 ± 0.08
$R_{p\alpha}$†	6.5	6.5	5.3
$R_{\alpha S}$†	11.7	10.8	13.0

† $R_{p\alpha}$ and $R_{\alpha S}$ are the ratios of corresponding fluxes. (From a review by Webber, 1964.)

(Mount Washington counting rate = 2460). The integral fluxes of various nuclei are also plotted in Fig. 11.9. As was noticed in some of the earliest work in this field, the rigidity spectra of all nuclear components are nearly parallel above 1.3 BV. This general feature puts important restrictions on theories of origin, acceleration, and modulation of cosmic rays.

Differential Spectra at Low Energies. Differential spectrum measurements for protons have been limited to rigidities of less than 1.5 BV, and for helium and L nuclei to less than 4.5 BV. Some attempts have been made, using nuclear emulsions, to measure differential spectra of S nuclei at higher energies, either by making measurements on their breakup products or on high-energy knock-on electrons produced in collisions of these nuclei with atomic electrons. However, so far these measurements can be used only to confirm the validity of the integral spectra obtained by using geomagnetic theory.

Accurate measurements of energy spectra extending down to very low rigidities have become possible only recently with the availability of very high balloon flights and of satellite vehicles, in which the instrumental cutoff can be reduced to a very small value. Figure 11.10 gives a summary of the differential rigidity spectra of protons, helium nuclei, and S nuclei, in the regions in which each of them has been measured. Only the data obtained at a time when the Mount Washington neutron-monitor counting rate was close to 2,320 have been used for draw-

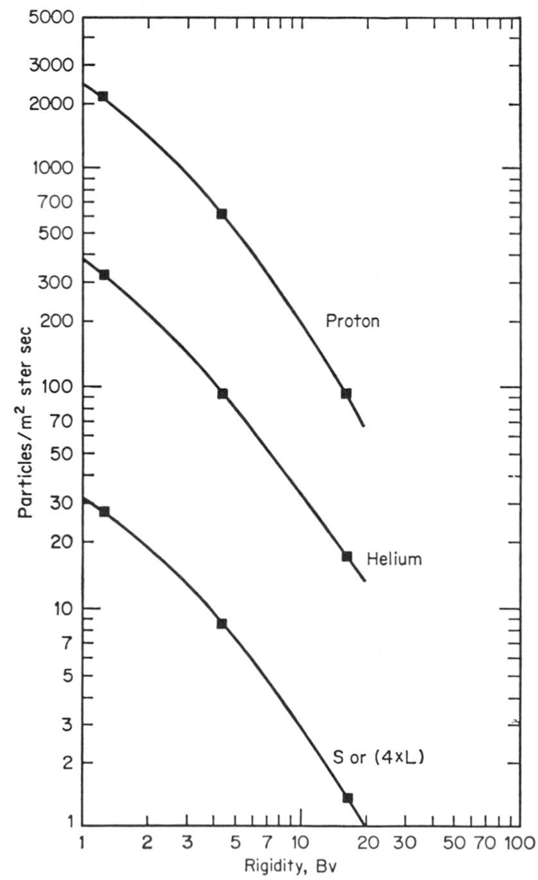

FIG. 11.9. Integral energy spectra of P, He, and S ($Z \geq 6$) nuclei at a time close to the solar minimum (Mount Washington counting rate ~ 2,320).

ing these curves; thus most of the measurements were made in 1963 and a few (for heavy nuclei) in 1954, and they all correspond to the state of the galactic radiation in the neighborhood of the earth, near the trough of the solar activity cycle, though not exactly at its minimum which corresponds to a Mount Washington rate of ~2,460. [Below 1.2 BV the helium spectrum measured by Balasubrahmanyan and McDonald (1964) on June 25, 1963, does not agree well with the curve adopted in Fig. 11.10, though at

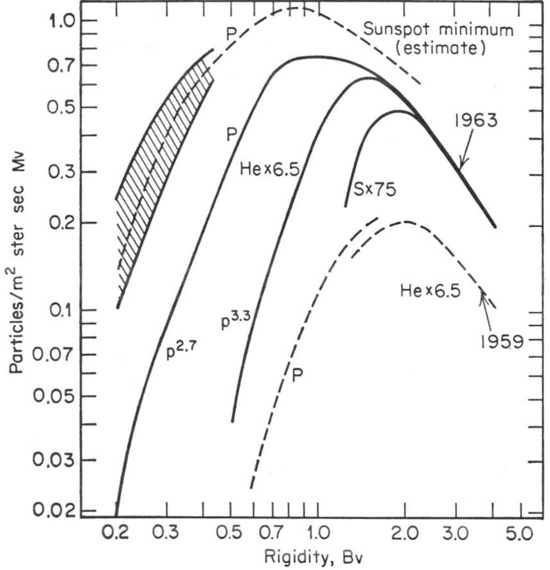

FIG. 11.10. Low-energy differential rigidity spectra of P, He, and S nuclei in 1963. He and P spectra for 1959 and Webber's (1965) estimate for the proton spectrum at the absolute sunspot minimum are also given for comparison.

low rigidities even this exhibits a splitting with respect to the proton spectrum.]

The following interesting features about these spectra are to be noted:

1. Every spectrum goes through a maximum, below which the flux drops sharply.

2. The position of the maximum shifts to higher rigidity with increasing mass; its position is at 0.9 BV for protons, 1.35 BV for helium nuclei, and 1.85 BV for S nuclei.

3. Various components have approximately parallel rigidity spectra above 2 BV.

4. Protons of energy as low as ~7 Mev, α particles of ~7 Mev/nucleon, and S nuclei of ~25 Mev/nucleon are present in the primary radiation (see below).

5. Low-energy spectra are extremely sensitive to the effect of solar modulation; this is already seen in Fig. 11.8 and again in Fig. 11.10 where proton and helium spectra for 1959 are given for comparison.

Solar activity seems to reduce the over-all intensity and to shift the maximum in the differential spectrum to higher rigidities. It is not certain that low-energy primary spectra are free from modulation effects even at the time of solar minimum; indeed the present ideas of modulation theory seem to indicate that they are not (Sec. 9).

Recently Fan et al. (1965) have shown the existence of protons and helium nuclei (during November–December, 1964) of very low-energy $7 \leq E \leq 20$ Mev/nucleon, whose spectrum falls rapidly with increasing energy. This very low-energy component of reversed slope is more easily observable at the time of low solar activity, though it is believed to be always present. It is not clear whether these nuclei are of solar origin or preferentially leak through the modulation barrier.

Comstock et al. (1966) have shown recently that 1964 and 1965 spectra of medium and heavy nuclei stay flat from a few hundred Mev/n down to 50 Mev/n. When corrected for solar modulation, these spectra would have negative slopes outside the solar system. As discussed in Sec. 12.2, this behavior cannot be understood in terms of the conventional approach to the problem of cosmic-ray propagation through interstellar space.

5.2. High-energy Part of the Spectrum. Protons above about 15 Bev and heavy nuclei above about 7 Bev per nucleon can reach the earth's atmosphere from the vertical direction at all latitudes. One may attempt to use the analyzing properties of the earth's magnetic field for measuring spectra up to ~59 Bev for protons and ~29 Bev per nucleon for heavier nuclei by using fluxes of these particles arriving near the eastern horizon. So far this method has not been successfully utilized.

For lack of any direct measurement on the primary energy, one allows the primary particle to interact and then one samples the energy going into a secondary component whose energy content can be measured. One such component is the electron-photon component which arises primarily from the decay of neutral pions which are believed to carry about 15 per cent of the primary nucleon energy (Sec. 6.7). In a material of high atomic weight γ rays produce a dense electromagnetic cascade close to their point of origin near the interaction, and the total energy in the cascade can, in principle, be estimated by studying its rate of growth or the position of its maximum density with respect to the origin. This principle was first utilized by Kaplon, Ritson, and Woodruff (1952), who used heavy-metal nuclear-emulsion sandwiches for the purpose. The method has been employed by Baradzei et al. (1962) who used ionization chambers to measure the energy released into the electron-photon component by primary interactions in a carbon target. Malhotra et al. (1965) used a large tungsten-emulsion assembly, exposed in a long balloon flight, to obtain energy spectrum of "nuclear cascades." Assuming that the primary energy is ~7 times the cascade energy (Sec. 6), one obtains several points on the primary energy spectrum up to energies of the order of 10^{13} to 10^{14} ev.

At energies beyond 10^{13} ev, the flux of primaries incident on a reasonable-sized detector flown in a balloon for a few hours is much too small for determining the energy spectrum. In this region the whole atmosphere is used as the developing medium in which the secondary particles interact over and over again; the resulting γ rays produce a large number of electrons and photons which spread over a large area and give rise to an "air shower" (Sec. 8). These showers can be observed by large detector arrays placed on the ground, near sea level, or at a mountain altitude. The size of an air shower or number of charged particles in it, most of which are electrons, can be related to the total primary energy in a fairly unique manner, particularly if the measurements are made close to the maximum of its development. However, even though the relation between the shower size and the primary energy is not very sensitive to the assumptions about the character of nuclear interactions, it is not easy to distinguish the showers produced by protons of energy E from those produced by heavy nuclei of atomic weight A and energy E/A per nucleon. To translate a given particle-energy spectrum into an energy spectrum for protons then involves an assumption about the chemical composition of the primary particles in the air-shower region about which there is little reliable information. On the other hand, up to energies of 10^{12} to 10^{13} ev, the relative number of high-energy interactions produced by protons, helium, and heavy nuclei in large nuclear emulsion stacks is consistent with the assumption that the chemical composition of the primary radiation has not undergone a drastic change in going up to these energies. It can be shown that, for a chemical composition like that observed for the latitude-sensitive part of the cosmic-ray spectrum, the total particle flux at any energy would be almost exactly twice the proton flux at that energy. [If α is the ratio of nuclei of charge Z to protons above a certain rigidity, then the ratio above the same total energy will be $\alpha(Z)^\gamma$, where γ is the power-law exponent of the integral energy spectrum.] Thus in drawing the particle-energy spectrum in Fig. 11.11, some of the points below 10^{14} ev are obtained by multiplying the measured or estimated proton fluxes by a factor of 2 and the data above 10^{14} ev are taken from air-shower experiments (Zatsepin et al., 1963; Greisen, 1960; Clark et al., 1963; Linsley, 1963).

Although there is some discrepancy in the absolute values of the flux in the air-shower region, the following features are fairly clear from this collection of data:

1. From 10^{10} to $\sim 10^{15}$ ev the particle-energy spectrum may be represented fairly well by the power-law expression:

$$N(>E) = (2.08 \times 10^4)(E/1 \text{ Bev})^{-1.67} \text{ per m}^2\text{/ster/sec} \quad (11.4)$$

The flux of protons is one-half, and of all nucleons (including those forming parts of heavy nuclei) about two-thirds (more nearly 0.675) of the flux of all nuclei above the same energy. The differential flux of all nucleons of 10^{10} ev to a few times 10^{14} ev is then

$$N(E) = (2.35 \times 10^4)(E/1 \text{ Bev})^{-2.67} \text{ per m}^2\text{/ster/sec/Bev} \quad (11.4a)$$

2. Beyond 5×10^{17} ev the slope of the spectrum is ~ 1.7.

3. From 6×10^{15} to 5×10^{17} Bev the slope of the spectrum is ~ 2.2.

Linsley and Scarsi (1962) have presented data suggesting that primary particles above 10^{17} ev have a pure composition and presumably contain only protons. If one accepts this, one can also construct a spectrum for primary protons at all energies up to 10^{19} ev, except for a gap between 10^{14} and 10^{17} ev (curve II in Fig. 11.11). Recent experimental evidence (McCusker, 1963) indicates that air showers of total energy $\geq 10^{15}$ ev have a predominantly multicored structure and are probably produced by heavy nuclei; in other words, primary protons are relatively scarce in the gap between 10^{14} to 10^{17} ev. This behavior can be explained by postulating two independent spectra: one having the normal chemical composition found at low energies, but subject to a rigidity cutoff at $\sim 10^{15}$ ev, and the other containing only protons (or antiprotons!) and not subject to a rigidity cutoff up to 10^{19} ev. The second spectrum may have a slope similar to the first one at low energies and about 1/15 of its proton intensity (also see Sec. 7.6). This type of explanation was first suggested by Peters (1961).

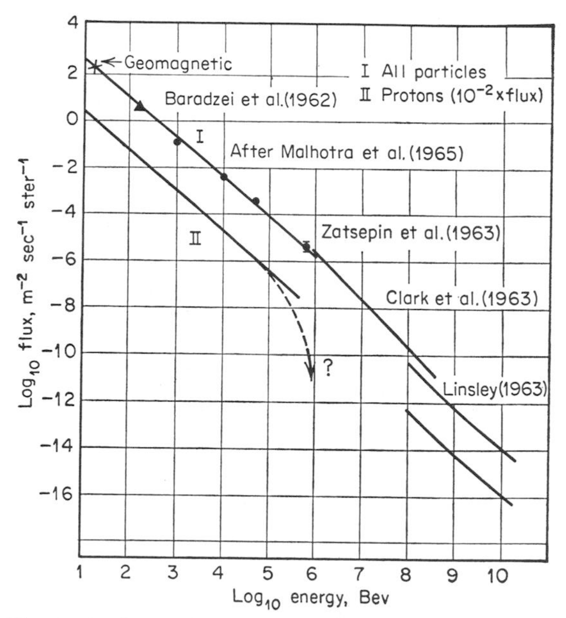

FIG. 11.11. Integral energy spectra of all nuclei (curve I) from 10 to 10^{10} Bev. The line between 10 and 10^6 Bev has a slope of -1.67. Curve II gives the integral spectrum of protons on the assumption that, at least up to 10^5 Bev, the chemical composition is the same as at low energies and that beyond 10^8 Bev there are only protons. The dotted branch on curve II indicates the possibility of a rigidity cutoff. Proton intensity between $\sim 10^6$ and 10^8 Bev is unknown, though it might be given by an upward linear extrapolation of the highest-energy spectrum.

6. High-energy Nuclear Interactions

High-energy primary cosmic-ray nuclei lose a small amount of energy by ionizing the gas of the upper atmosphere. Large transfers of energy occur only when the particles undergo nuclear collisions with nuclei of air molecules. In this process the nuclei are broken up and a significant fraction of the incident energy is used up in creating new particles. Therefore study of secondary cosmic radiation is closely linked up with study of ultrahigh-energy nuclear interactions. As a matter of fact, till about

1953, the field of high-energy physics and particle physics formed a part of cosmic-ray studies. Most of the presently known quasi-stable elementary particles, including the muon, the charged and neutral pions, charged and neutral K mesons, and various types of hyperons were first discovered in cosmic rays. The first evidence for production of a large number of mesons in a single high-energy nucleon-nucleon encounter was also obtained in a cosmic-ray-exposed nuclear-emulsion stack (Bradt, Kaplon, and Peters, 1950). With the advent of high-energy proton accelerators, extending at the time of writing to energies of ~30 Bev, the study of particle production and nuclear phenomena in this energy range has properly been taken over by the accelerator laboratories where more detailed and accurate investigations are possible by using intense and well-controlled particle beams. Modes of production and properties of all the previously discovered particles and a large number of new ones have been studied in detail.

6.1. Process of Star Production and Nuclear Spallation. Nuclear interactions lead to a sharing of the primary energy with a host of secondary particles, some of which are the fragments of the struck nucleus, while others are created in the collision process. At low energies, below the meson production threshold (~256 Mev), nucleon-nucleon collisions are elastic. When the target nucleon is bound in a nucleus, both the collision partners may share part or all of their energy with the nucleus which may remain in an excited state until, through a statistical process, a large part of the energy is concentrated on a few of the particles, resulting in their emission from the nucleus. This process is generally known as the evaporation process. In the evaporation process, protons, neutrons, α particles, and other nuclei acquire energies that may extend up to ~100 Mev. Rare isotopes like deuterons, tritons, and He^3 nuclei are also emitted. The evaporation process is a property of nuclear matter, and its character is not very sensitive to the incident primary energy. As the primary energy increases, in addition to the evaporated particles, some of the nuclear fragments emerge with significantly higher energies (extending up to several Bev) though the number of such particles is always small. Figure 11.12 gives the distribution of heavily ionizing prongs in "stars" produced by protons of different energies. This shows the relative insensitivity of nuclear excitation to primary energy, once it is significantly above the total binding energy of the nucleus. Thus it is seen that at high energies only a small part of the energy is expended in disintegrating the struck nucleus; the stable configurations of nuclear matter are expelled from the excitation region before being fragmented.

Charged nuclear fragments lose energy by ionization and come to rest in the atmosphere before suffering further collisions. Neutrons are slowed down by elastic and inelastic collisions, and the majority of them are absorbed by atmospheric nitrogen to produce C^{14}, which decays with a half-life of ~5,700 years. Carbon in the atmosphere is in continuous

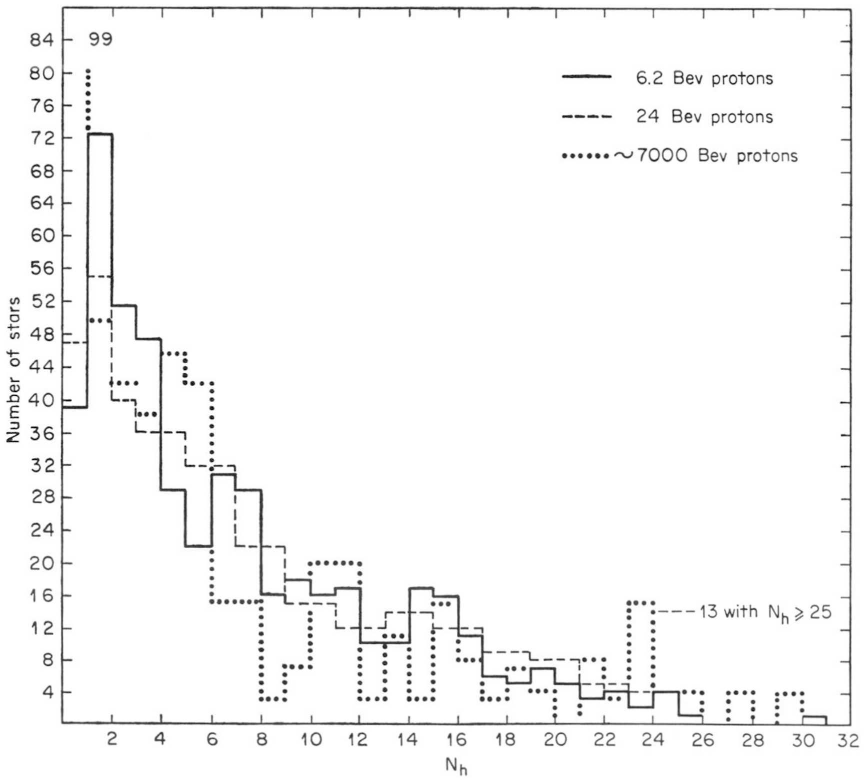

FIG. 11.12. The distribution of the number of heavily ionizing prongs, N_h, in stars produced by protons of various energies in emulsion. [*Malhotra* (1964).]

equilibrium with the biosphere. The relative amount of C^{14} is a dead organism, for example, wood, then provides a measure of the time it has been dead, provided one assumes that cosmic-ray activity was the same when it was alive as it is now. This is the basis for the C^{14} method for dating archaeological samples.

Some of the nuclides produced in the spallation of atmospheric nuclei of nitrogen, oxygen, and argon are also radioactive. These may also be used as tracers for the study of geophysical, meteorological and hydrological phenomena (Lal and Peters, 1962). Table 11.3 gives some of the cosmic-ray-produced

TABLE 11.3. ISOTOPES ($\tau_{1/2} > 1$ DAY) PRODUCED BY COSMIC RAYS IN THE ATMOSPHERE†

Isotope	Half-life	Main radiation	Main target nuclide(s)
He^3	Stable	. . . .	N, O
Be^{10}	2.7×10^6 years	β^- – 550 Kev	N, O
Al^{26}	7.4×10^5 years	β^+ – 1.17 Mev	Ar
Cl^{36}	3.1×10^5 years	β^- – 714 Kev	Ar
Kr^{81}	2.1×10^5 years	K – X ray	Kr
C^{14}	5,730 years	β^- – 156 Kev	N, O
Si^{32}	500 years	β^- – 100 Kev	Ar
A^{39}	270 years	β^- – 565 Kev	Ar
H^3	12.5 years	β^- – 18 Kev	N, O
Na^{22}	2.6 years	β^- – 540 Kev; γ – 1.3 Mev	Ar
S^{35}	87 days	β^- – 167 Kev	Ar
Be^7	53 days	γ – 480 Kev	N, O
A^{37}	35 days	K – X ray	Ar
P^{33}	25 days	β^- – 250 Kev	Ar
P^{32}	14.3 days	β^- – 1.7 Mev	Ar

† After Lal and Peters (1965).

isotopes which have so far been discovered. The record of cosmic-ray-produced activities in meteorites has been used to study the interrelation between the chronology of cosmic rays and meteorites. We shall discuss this in Sec. 12.2.

When the energy of the incident particle is above the π-meson production threshold, ~256 Mev for p-p collisions, some of the collision energy is carried away by the newly created mesons. The number of mesons produced in the interaction increases with energy. At an energy of 1 Bev, on an average, one pion is produced in each collision.

6.2. Meson Production in High-energy Collisions: Energy Measurement. Present-day accelerators produce proton beams of energies up to 30 Bev. From interactions of these protons, beams of secondary particles extending in energy, in the case of pions, to ~20 Bev are also available. By using these beams with known targets, which may be liquid hydrogen in a bubble chamber, the characteristics of particle production can be studied in detail.

For higher energies the only information comes from cosmic-ray experiments. These experiments suffer from several limitations. With increasing energy the cosmic-ray flux decreases very sharply (Sec. 5); for example, the flux of nucleons of energy greater than 300 Bev on top of a high mountain (~14,000 ft) is about 1.5×10^6 per m²/ster/year, as compared with a possible intensity of ~10^{12} protons per second for one of the present-day particle accelerators. Therefore the sizes of detectors required for cosmic-ray studies are rather large if one wants to investigate processes that occur with a probability ~1/100 of the total interaction probability. The second, and perhaps more important, limitation comes from the inherent difficulty of making measurements at such high energies. The momentum determination by magnetic curvature in cloud chambers is limited to about 30 Bev/c, while the identity of charged particles cannot be usually established above 1 Bev. (These are only practical limitations; in principle, these problems can be solved by using large magnetic volumes, spark chambers, and velocity-measuring gas counters, as pointed out is some recent experimental proposals.)

One of the earliest methods of estimating the primary energy was based on the assumption that in a high-energy collision between particles of equal mass the emission of particles in the C system (center-of-momentum system) should be symmetric with respect to the plane perpendicular to the collision axis. Then the 90° angle in the C system transforms into the laboratory angle $\theta_{1/2}$ defined as the angular opening of the cone which contains one-half of the emitted particles, provided the velocities of the colliding particles and the emitted particles in the C system are equal, a condition that is usually satisfied for ultrahigh-energy collisions. In this case, the Lorentz factor (Γ_c) of the C system with respect to the laboratory system is

$$\Gamma_c = \cot \theta_{1/2} \tag{11.5}$$

and the laboratory energy (E) of the incident particle in rest-mass units (MC^2) is

$$\Gamma = \frac{E}{MC^2} = 2\Gamma_c^2 - 1 = 2\cot^2 \theta_{1/2} - 1 \tag{11.6}$$

A different procedure for obtaining Γ_c from the angular distribution of the interaction was first suggested by Castagnoli et al. (1953) according to which

$$-\ln \Gamma_c = \frac{1}{n}\sum_{i=1}^{n} \ln \tan \theta_i + K \tag{11.7}$$

where n is the total number of secondary particles and the parameter K is zero if the velocities of the incident and secondary particles in the C system are equal ($\beta_c/\beta^* = 1$).

The assumption of symmetry in the C system, which is necessary for this method to be applicable, though certainly valid on a statistical basis for collisions of identical particles, may not hold in individual collisions. Furthermore, since the multiplicities are small ($n_s \sim 10$ at 1,000 Bev) and there can be strong fluctuations in the fraction of particles that are charged and hence observable in most instruments, the measure of confidence in an individual energy determination is quite small. Nevertheless, inasmuch as some of the characteristics of high-energy nuclear interactions are known to depend insensitively on the primary energy, this method has proved to be very valuable in getting an estimate of the primary energy, within a factor of 2 or so, when no other methods were possible.

Occasionally a multiply charged high-energy cosmic-ray nucleus breaks up into a large number of nuclear fragments on suffering a nuclear collision in an emulsion stack. All the fragments, which move in a narrow cone in the laboratory system, must have very nearly the same velocity, or energy per nucleon. This velocity may be determined by constructing a composite angular distribution of shower particles produced by the fragments in their subsequent interactions, because the requirement of fore-aft symmetry must necessarily be satisfied for this composite distribution (Lohrmann, Teucher, and Schein, 1961; Abraham et al., 1963). Alternatively one can measure the relative multiple Coulomb scattering between different fragments; this is possible up to very high energies because the local dislocations in the nuclear emulsion do not disturb the relative position of the tracks.

While working with magnet cloud chambers, occasionally the total longitudinal momentum of the secondary particles in an interaction, appropriately corrected for the missing neutral particles that are not observed, is taken as an estimate of the primary energy (Hansen and Fretter, 1960). A few years ago a more powerful variant of this method was introduced by Guseva et al. (1962). Below the cloud chamber in which the interactions are studied is placed an absorber several interaction mean free paths thick in which ionization chambers are interspersed through its depth. All the particles produced in the primary interaction interact over and over again in this absorber, and most of the energy dissipated in the process is sampled by the ionization chambers. The fraction of the energy leaking out of the absorber is estimated to be $\sim$20 per cent. It is estimated that by using this "ionization calorimeter" the primary energy can be measured to within $\sim$30 per cent. Ramana Murthy et al. (1963) used a similar instrument in which the ionization is sampled by liquid scintillators instead of ionization chambers.

6.3. Composition of Created Particles. Most of the created particles in high-energy interactions are pions; the rest are kaons, hyperons, and baryon-antibaryon pairs. The relative fraction of nonpions increases with incident energy, but only up to about 25 Bev, after which it appears to become constant. In cosmic-ray jets the fraction of nonpions among the charged secondaries is determined by measuring the ratio of γ rays to charged shower particles and assuming that all γ rays arise from the 2γ decay of neutral pions which are equal to one-half the number of charged pions (assumption of charge independence). In this way it has been shown that 20 per cent of the charged particles created in nuclear interactions of 20 to several thousand Bev are nonpions [Lal, Pal, and Raghavan, 1962; see Perkins (1960) for references to emulsion work]. The fraction of neutral particles that are nonpions may be obtained by measuring the relative number of interactions produced by neutral and charged secondary particles in the cores of high-energy primary interactions and again assuming charge independence. Perkins gives a value of 0.15 ± 0.05 for the ratio of neutral interacting particles for interactions of 10^3 to 10^5 Bev studied with nuclear emulsions. Lal et al. find 0.16 ± 0.06 for this ratio in a multiplate-cloud-chamber experiment at 20 to 150 Bev. The two values seem to be nearly the same. Since the cloud-chamber experiment would not detect nearly one-half of the neutral kaons that decay by the short-lived $K^\circ{}_1$ mode, the closeness of the emulsion and cloud-chamber values shows that an appreciable fraction of neutral nonpions are not kaons; most likely they are nucleon-antinucleon pairs.

Table 11.4 gives a summary of measurements on the fraction of nonpions produced in interactions of nucleons ranging in energy from $\sim$20 to $\sim 10^6$ Bev.

TABLE 11.4. COMPOSITION OF CREATED PARTICLES IN INTERACTIONS OF COSMIC-RAY NUCLEONS

Ratio of heavy created particles to all created particles	Primary energy, Bev	Method
0.20 ± 0.06	20–150	Multiplate chamber (interactions in carbon)[a]
0.22 ± 0.06	50–1,000	Direct observation in a magnetic chamber (interactions in carbon)[b]
0.20 ± 0.05	10^3–10^5	Emulsion[c]
0.18 ± 0.10	10^6	Emulsions (one event in Al)[d]

[a] Lal et al. (1962).
[b] Hansen and Fretter (1960).
[c] Perkins (1960).
[d] Ali et al. (1960).

The relative numbers of different types of particles produced in collisions of 10- to 30-Bev protons with light nuclei at two angles and momenta, as measured by Baker et al. (1961) at the Brookhaven AGS, are given in Figs. 11.13 and 11.14. From these figures it is evident that the composition of the produced particles at any angle depends rather strongly on their energy. The change with energy of the ratio K^+/K^-, for example, is shown explicitly in Fig. 11.15 taken from Dekkers et al. (1965), who have studied particle production at small angles in p-p collisions at 23.1 Bev energy. This figure suggests that positive K particles carry larger energies and tend to be emitted closer to the collision axis in the C system. This is indicated at cosmic-ray energies also. Stopping K mesons observed in balloon-exposed emulsion stacks, which must represent particles emitted fast in the backward direction in the C system, were found to show a K^+/K^- ratio of about 20 (Lal, Pal, and Peters, 1953). Kim (1963) has attempted to make mass measurements on particles emitted in the backward direction in the C system in high-energy interactions in the 1,000-Bev range. He finds that for charged particles the K/π ratio is $\sim$0.2 for the center-of-mass angle $\theta_c < 175°$ and $\sim$1 for $\theta_c \geq 175°$. He also finds that while there are very few baryons at $\theta_c < 175°$ the number of baryons is approximately equal to the number of pions in the extreme backward direction. The dependence of the composition on angle and energy of the emitted particles provides an important clue as to their production (Sec. 6.9). The over-all fraction of nonpions among the particles produced in proton interactions in the 20-Bev range is $\sim$20 per cent.

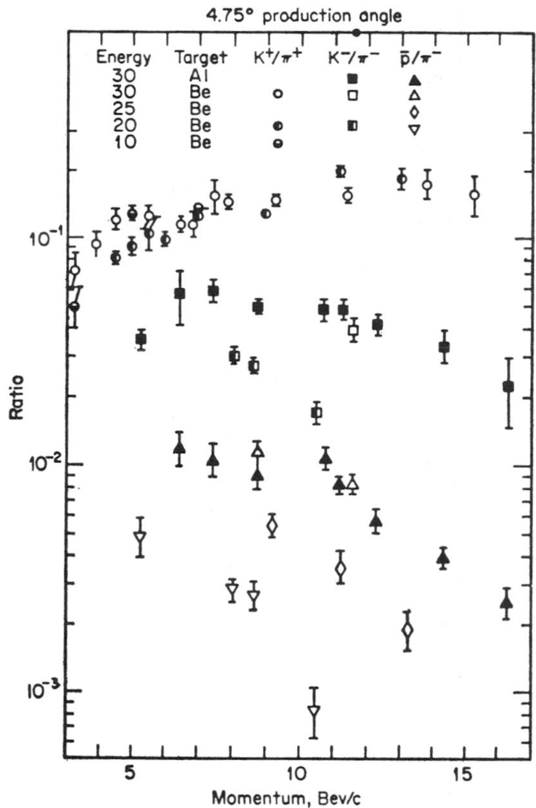

Fig. 11.13. Production ratios K^+/π^+, K^-/π^-, and $\bar{p}/\pi^-$ as functions of momentum for particles produced at 4.75° in proton collisions with light nuclei. [*Baker et al.* (1961).]

To summarize, the relative abundance of different kinds of particles produced in high-energy interactions does not seem to alter appreciably after an energy of ~20 Bev. Pions remain the most abundant component among secondary particles produced up to energies ~10^5 Bev. We shall see later that they also appear to carry the major fraction of the energy dissipated at all energies.

6.4. Multiplicity of Created Particles. If n_c is the number of charged particles produced in a shower, then, if 20 per cent of the charged particles are assumed to be nonpions, the total number of created particles would, on an average, be given by

$$n_t = (3/2 \times 0.8 + 2 \times 0.2)(n_c - 1) = 1.6(n_c - 1) \qquad (11.8)$$

The average multiplicity in the accelerator energy range is rather well measured. For cosmic-ray energies only a few measurements exist where the incident nucleon energy is high and measurable and where at the same time interactions have been collected without strong bias against low multiplicity events or against high charged multiplicity events. One should also restrict the measurements to collisions with light nuclei or, in emulsion, with a small number of evaporation prongs.

At 70, 100, 300, and 1,000 Bev there are cloud-chamber measurements (Lal et al., 1962; Hansen and Fretter, 1960; Guseva et al., 1962), interactions being produced in carbon or lithium hydride. Here incident energy is reliably measured only for the 300-Bev point, using an ionization calorimeter. Some bias against low multiplicity events may exist in all measurements, though the 100- and 1,000-Bev measurements, being made with thick carbon targets, may underestimate multiplicities consistently.

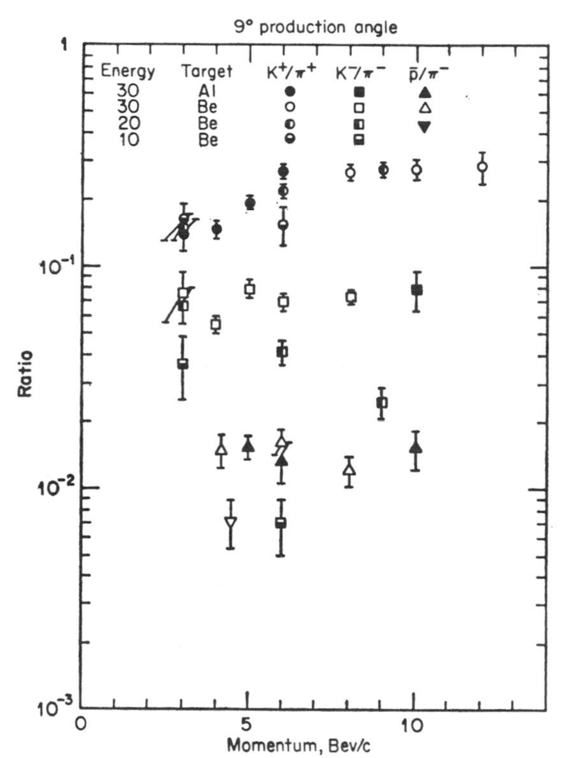

Fig. 11.14. Production ratios K^+/π^+, K^-/π^-, and $\bar{p}/\pi^-$ as functions of particle momentum for particles produced at 9° in proton collisions with light nuclei. [*Baker et al.* (1961).]

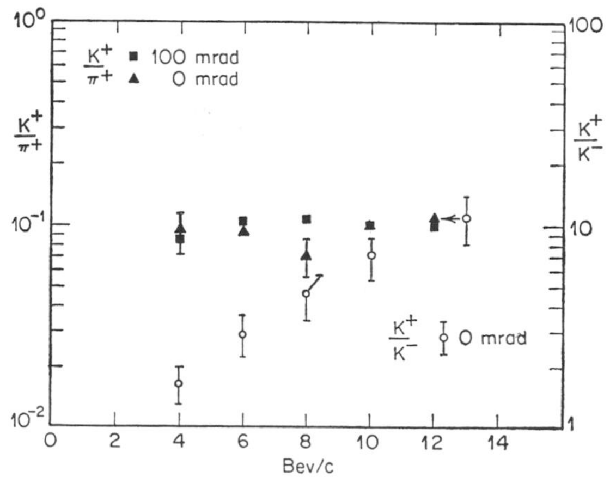

Fig. 11.15. Observed ratios K^+/π^+ at 0 and 100 milliradians and K^+/K^- at 0 milliradians in p-p collisions at 23.1 Bev/c. [*Dekkers et al.* (1965).]

At 250 Bev, 1.53 Tev, and 12.5 Tev (1 Tev = 10^{12} ev) fairly reliable, though statistically poor, measurements of the multiplicity of charged particles were obtained by Lohrmann, Teucher, and Schein (1961) and by Koshiba and Wolter (see Koshiba, 1963), who used fragmentation products of heavy muclei, for which the energy could be estimated by methods already mentioned. These measurements can also be in error sometimes because of the possible presence of deuterons and tritons among the fragmentation products and because some of the nucleons may have lost a considerable amount of their energy in the fragmentation process.

Perkins (1961), Malhotra (1963), and Malhotra et al. (1965) have given values for the charged multiplicity at an energy of about 10 Tev, the events in this case being selected by tracing back electromagnetic cascades recorded in emulsion detectors; the primary energy may be uncertain to within a factor of 2. Finally, there is a point at 500 Tev given by Perkins (1961) which is based on only six events obtained by tracing back electromagnetic cascades, and so its reliability is very much open to question.

Using most of the measurements discussed above, the total number of created particles n_t calculated according to relation (11.8) is plotted as a function of energy in Fig. 11.16. The line through the points indicated as $2.4 \times E^{1/4}$ gives the form of the multiplicity dependence expected in Fermi's theory of meson production (1950, 1951). We shall presently come to the justification for trying a fit of the type

$$n_t = A + BE^{1/2} \tag{11.9}$$

given by the other curve. It is found that $A \simeq 4.75$ and $B \simeq 1/4$ if E is in Bev.

6.5. Transverse Momentum Distribution. It is possible to measure the laboratory momenta and angles of many of the secondary particles if they are relatively slow or decay into γ rays that produce cascades. However, it is not usually possible to obtain a momentum distribution in the C system because accurate determination of primary energy is difficult. As a result, a large number of attempts are limited to the investigation of the distribution of tranverse momentum, p_T, which is a Lorentz invariant quantity. It is found that at all primary energies the average transverse momentum of the secondaries is ~400 Mev/c and, at least for pions, it rarely exceeds a value of ~1 Bev/c. The average values of p_T for neutral pions as measured at several primary energies are given in Table 11.5. The smallness of this quantity and its relative invariance with respect to primary energy are two of the noteworthy characteristics of all high-energy interactions. Very accurate

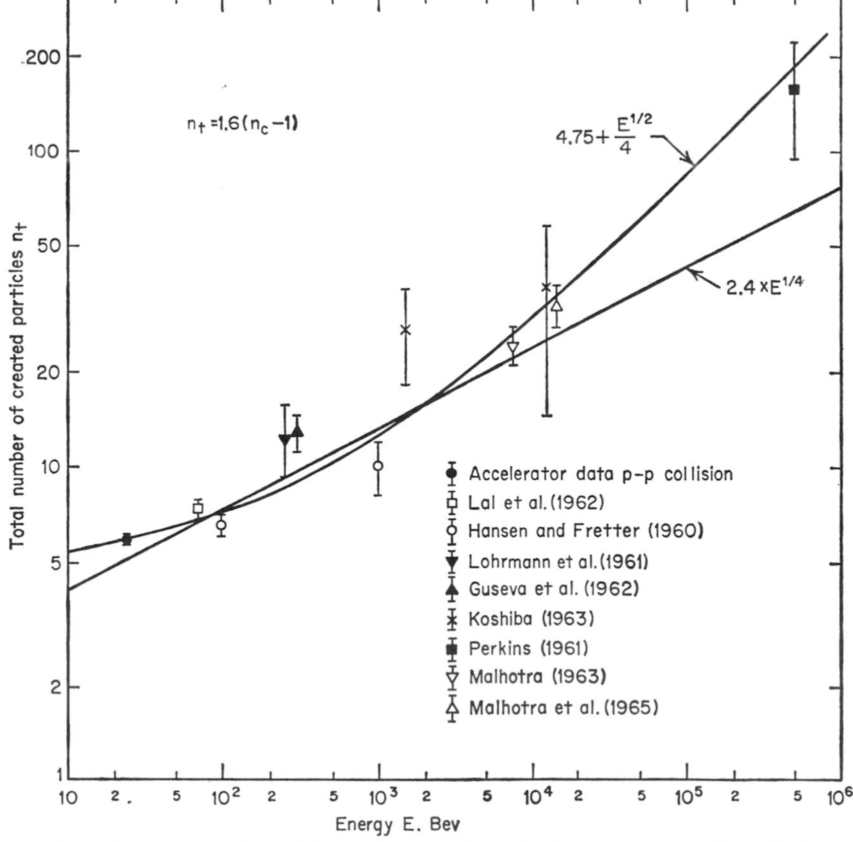

Fig. 11.16. The total number of created particles n_t as a function of primary energy. The relation $n_t = 1.6\ (n_c - 1)$ between n_t and the number of charged created particles $(n_c - 1)$ is obtained on the assumption that 20 per cent of these are particles of isotopic spin $\frac{1}{2}$. Two different analytical fits to the experimental data are given. (See text.)

TABLE 11.5. AVERAGE VALUES OF p_T FOR NEUTRAL PIONS

Primary energy, Bev	Average transverse momentum of π^0, p_T, in Bev/c	Reference
24	0.40	Pal and Rengarajan (1961)
20–150	0.39	Lal, Pal, and Raghavan (1962)
500	0.35	Akashi et al. (1962)
7×10^3†	0.39	Minakawa et al. (1959)
2×10^4†	0.46	Akashi et al. (1963)
3.5×10^4†	0.43	Akashi et al. (1963)
1.5×10^4†	0.51	Malhotra et al. (1965)

† This energy is seven times the total energy in the γ-ray component, which is measured in these experiments.

measurements with proton and pion beams of up to ~20 Bev show approximately similar values for p_T for charged and neutral pions. An experimental function

$$N(p_T)\,dp_T = p_T/p_0{}^2 \exp(-p_T/p_0)\,dp_T \qquad \text{with } \bar{p}_T = 2p_0 \qquad (11.10)$$

has been found to represent the experimental p_T distributions quite well. Aly, Kaplon, and Shen (1963) proposed an alternative form

$$N(p_T)dp_t = 2\alpha p_T \exp(-\alpha p_T{}^2)\,dp_T \qquad (11.11)$$

and claim that at accelerator energies this gives a better fit to the experiments than the exponential distribution.

In a study of 10-Bev π^- interactions Bigi et al. (1962) discovered that the average transverse momentum increases slowly with the mass of the emitted particle; the values obtained by them are plotted in Fig. 11.17. This mass dependence can be understood qualitatively if various particles are emitted from the decay of heavy intermediate states.

It is sometimes asserted that the value of the transverse momentum is independent of the angle of emission; however, the available experimental information does not seem to be adequate to prove this point.

6.6. Momentum Distribution. In order to determine the momentum spectra in the C system both the primary and the secondary momenta must be measured. This requirement limits the number of cosmic-ray experiments from which the C-system spectrum can be deduced. Among them are the experiments of Guseva et al. (1962) subsequently followed by Dobrotin et al. (1963) and by Lal, Raghavan, and collaborators (1962). The average value of the pion momentum in the C system is found to be approximately the same in all these experiments and is consistent with a value of ~450 Mev/c. Actually this is also quite close to the values found for 6-Bev p-p collisions, 9-Bev p-p collisions, and 24-Bev p-p collisions and not too different from the value obtained for $\bar{p}$-p annihilations. Thus, at least up to 1,000-Bev primary energy, not only the average transverse momentum is invariant but also the average total momentum. It is found that the relation $p = (4/\pi)p_T$ expected for an isotropic angular distribution is close to being valid. However the similarity

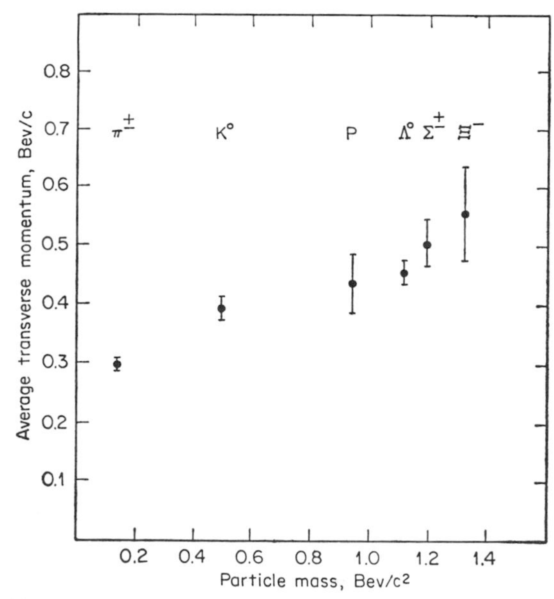

FIG. 11.17. Average transverse momentum of different particles produced in collisons of 10-Bev/c π^- with protons, as a function of mass. Increase of p_T with mass is apparent. [*From Bigi et al.* (1962).]

between accelerator energies and energies of hundreds of Bev or higher is only approximate and may be accidental. We shall discuss this presently.

Dobrotin et al. (1963) have carefully analyzed primary interactions of 250 ± 150 Bev, using a combination of magnet cloud chambers with an ionization calorimeter. The momentum spectrum of the secondary particles has been determined in two frames of reference, the rest frame of the target nucleon (laboratory system) and the rest frame of the incident nucleon (mirror system). By combining the two distributions which, when averaged for a large number of interactions, are expected to be the same except for measurement errors and statistical fluctuations, a laboratory momentum distribution has been obtained; this is given in Fig. 11.18. If the low-energy part of the momentum distribution is fitted by a Planck's function with a temperature $T = 0.65(m_\pi c^2/k)$ and transformed to the L system, the experimental points show a high-energy tail not indicated by the theoretical curve. An analysis of several high-energy interactions indicates that, while a lump of particles of comparatively low momenta can always be described in terms of characteristic temperature $\sim m_\pi c^2/k$, there exists a tail comprising a few particles whose extent is nearly proportional to the primary energy.

Cocconi, Koester, and Perkins (1961) have made an empirical fit to pion-production spectra for ~ 30-Bev incident proton energy, using two basic assumptions:

1. The total energy spectrum of pions valid above their median energy in the laboratory frame is given by

$$f(E)\,dE = \frac{n_\pi}{2}\frac{dE}{T}\exp\left(-\frac{E}{T}\right) \qquad (11.12)$$

where n_π is the total pion multiplicity and T is a characteristic energy.

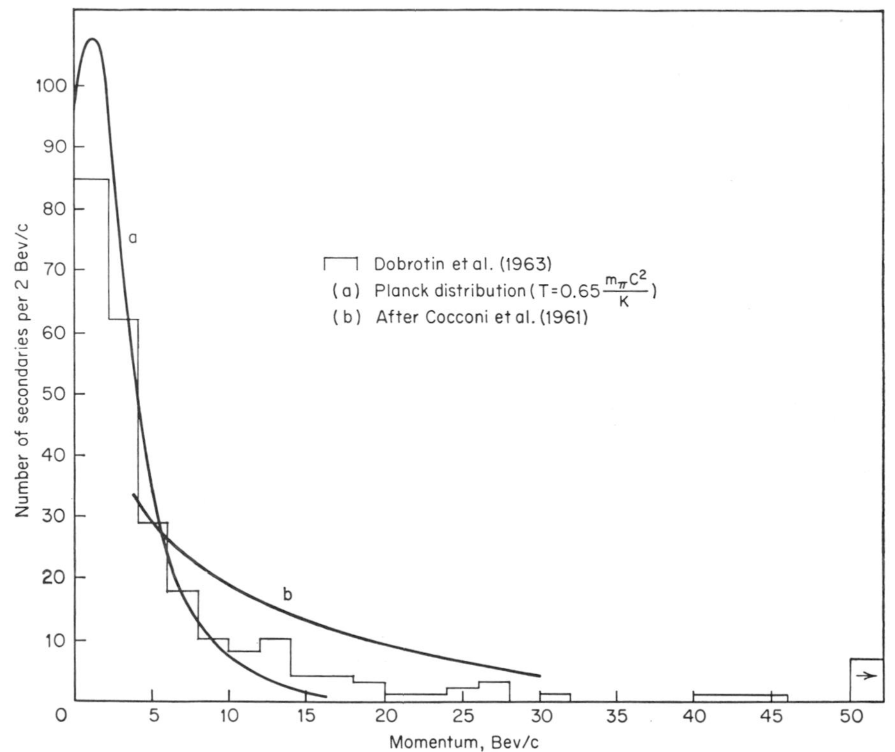

FIG. 11.18. Experimental momentum distribution of secondary particles in collisions of primary particles in the energy range 250 ± 150 Bev (most probable energy of 150 Bev) with lithium hydride. The Planck distribution fit, curve *a*, does not account for the high-energy tail while the energy spectrum *b* calculated from empirical formulas of Cocconi, Koester, and Perkins for a primary energy of 150 Bev is a very poor fit to the data.

2. The transverse momentum distribution is independent of the energy distribution and is

$$N(p_T)\,dp_T = \frac{p_T}{p_0^2}\,dp_T \exp\left(-\frac{p_T}{p_0}\right) \qquad (11.13)$$

Combining the two,

$$P(E,p_T)\,dE\,dp_T = \frac{n_\pi}{2}\frac{p_T}{p_0^2}\,dp_T\,\frac{dE}{T}\exp\left(-\frac{p_T}{p_0}-\frac{E}{T}\right) \qquad (11.14)$$

For small angles (θ) and high energy, $p_T = E\theta$, and the solid-angle element $d\Omega = 2\pi\theta\,d\theta$. Hence

$$\frac{d^2N(E,\theta)}{dE\,d\Omega} = \frac{n_\pi T}{4\pi p_0^2}\left(\frac{E}{T}\right)^2 \exp\left[-\frac{E}{T}\left(1+\frac{\theta T}{p_0}\right)\right] \qquad (11.15)$$

Cocconi et al. (1961) have shown that this expression gives a fair representation of the pion-production spectra at machine energies, though at C-system angles close to the collision axis some discrepancies are encountered (Dekkers et al., 1965; also Jordan, 1965).

In this approach, T gives the average energy of one-half the pions that are emitted in the forward hemisphere in the C system and hence carry most of the energy radiated in the form of pions. We therefore have the relation

$$T = \frac{2K_\pi E_0}{n_\pi} \approx 3.75 \qquad \text{at 30 Bev} \qquad (11.16)$$

where K_π is the fraction of the energy radiated into pions. If it is assumed that K_π is independent of primary energy (Sec. 6.7) and n_π increases as E^ρ, then the characteristic energy T would increase as $E_0^{1-\rho}$. Thus relation (11.12) can be used, on these assumptions, to predict the production spectrum of pions at high energies. The expected spectrum for $\rho = \frac{1}{4}$, which is commonly assumed, and a primary energy of 150 Bev is superimposed on the experimental spectrum of Dobrotin et al. (1963) in Fig. 11.18; 150 Bev is the most probable energy of the events used by these authors, though they all lie within the energy range 250 ± 150 Bev. It is noted that the spectrum of Cocconi et al., with the assumption $\rho = \frac{1}{4}$, gives more high-energy particles than are actually found. If this spectrum were to be used for calculating secondary cosmic-ray fluxes, the values obtained would be too high.

However, it may be possible to use a set of assumptions that would give an energy variation of the characteristic energy parameter T such that the experimental data, with their limited accuracy, may be well described in this manner. Although practically useful, such an empirical procedure is not equivalent to having a model of high-energy interactions, much less a theory.

6.7. Inelasticity. It is found that at all energies, both in nucleon-nucleon and in pion-nucleon collisions, the emerging nucleons are the most energetic particles in the C system. The nucleons taking part in the collision retain after the collision not only the bulk of their energy (in the C system) but also continue to move close to their original direction of motion; typical transverse momenta are of the order $p_T = Mc/2$. This property of nucleons in collisions is usually referred to as "persistence." If the fraction of the kinetic energy in the C system retained by either of the nucleons is given by η, the total energy of the forward and backward nucleons in the laboratory system is

$$\begin{pmatrix} E_f \\ E_b \end{pmatrix} = E_c(1 + \eta W_c)\left[1 \pm \beta_c\sqrt{1 - \frac{1 + p_T^2}{(1 + \eta W_c)^2}}\right] \tag{11.17}$$

Here $E_c = 1 + W_c$, where W_c is the kinetic energy of either of the nucleons in the C system before collision, and β_c is their velocity; all energies are measured in nucleon rest-mass units. The total primary energy in the laboratory system, E_0, is related to E_c as

$$E_0 = 2E_c^2 - 1 \tag{11.18}$$

For $W_c \gg 1$

$$\frac{E_f}{E_0} \simeq \eta \tag{11.19}$$

η is usually referred to as the elasticity of the collision. Sometimes the term inelasticity (K) is also used to denote the fraction of the primary energy given to secondary particles:

$$K = 1 - \eta \tag{11.20}$$

A large number of experiments have shown that, on an average, $\eta \simeq 0.55$, independent of primary energy up to several thousand Bev, though at any given energy is has a rather wide distribution. Some recent values of η are given in Table 11.6. That high-energy nucleon interactions are rather elastic was first inferred from the fact that the attenuation length of the nucleon component in the atmosphere is much longer than their interaction length. This question will be discussed in greater detail in Sec. 7.

6.8. Angular Distribution above 100 Bev and Fireballs. The ratio of average total momentum to average transverse momentum of shower particles in the C system suggests a fairly isotropic distribution for a large majority of the secondaries. This obviously cannot be true either for nucleons or for a small number of other created particles which seem to carry large longitudinal momenta but small transverse momenta.

A common method of studying the C-system angular distribution involves the plotting of the quantity $\ln \tan \theta$, where θ is the laboratory angle of emission. Since the transverse momentum distribution shows that most particles are relativistic in the C system, one may use, when transforming to the L system, the approximation

$$\beta_c/\beta^* = 1$$

TABLE 11.6. AVERAGE ELASTICITY IN COLLISION OF NUCLEONS

(η gives the average fraction of energy retained by nucleons. Sometimes only K_π, the fraction of energy radiated into pions, is measured; this is given in the last column.)

Primary energy, Bev	Target	η	K_π
10–24[a]	Protons or light nuclei	0.5–0.6	
24[b]	Nuclear emulsion	0.41 ± 0.05	
24[c]	C	....	0.43 ± 0.07
300[d]	Li H	....	~0.34 ± 0.05
70–700[e]	C	....	0.48 ± 0.06
≥1,000[f]	Nuclear emulsion	0.5 ± 0.15	
≥1,000[g]	Nuclear emulsion	0.5 ± 0.15	
≥1,000[h]	Nuclear emulsion	0.5	
~1,000[i]	Nuclear emulsion	0.5 ± 0.2	

[a] Lock (1963).
[b] Pal et al. (1963).
[c] Pal and Rengarajan (1961).
[d] Guseva et al. (1962).
[e] Azimov et al. (1963).
[f] Daniel et al. (1963).
[g] Hildebrand and Silberberg (1963).
[h] Huggett et al. (1963).
[i] Kim (1963).

which gives the relation

$$\Gamma_c \tan \theta = \frac{\sin \theta^*}{\cos \theta^* + 1} = \tan \frac{\theta^*}{2} \tag{11.21}$$

If the angular distribution in the C system is proportional to $-\cos^n\theta^*\, d\cos\theta^*$,

$$N(\theta^*)d\cos^*\theta = -\tfrac{1}{2}d(\cos^{n+1}\theta^*) = \frac{(-1)^n}{2}\, d\left(\frac{\tan\frac{\theta^*}{2} - \cot\frac{\theta^*}{2}}{\tan\frac{\theta^*}{2} + \cot\frac{\theta^*}{2}}\right)^{n+1} \tag{11.22}$$

Rewriting Eq. (11.21).

$$\tan \theta^*/2 = \exp(x + \ln \Gamma_c) \tag{11.23}$$

where $x \equiv \ln \tan \theta$, and substituting in Eq. (11.22),

$$N(x)\,dx = \frac{(-1)^n}{2}\, d[\tanh^{n+1}(x + \ln \Gamma_c)] \tag{11.24}$$

This distribution in general has two maxima which occur for

$$\tanh(x + \ln \Gamma_c) = \pm\sqrt{\frac{n}{n+2}} \tag{11.25}$$

and whose positions are

$$x_m = -\ln[\Gamma_c(\Gamma_F \pm \sqrt{\Gamma_F^2 - 1})] \tag{11.26}$$

where

$$\Gamma_F = \sqrt{1 + \frac{n}{2}} \tag{11.27}$$

For isotropic distribution ($n = 0$), the two maxima coalesce into a single maximum at $x_m = -\ln \Gamma_c$, and one obtains a quasi – Gaussian distribution with a root-mean-square dispersion $\sigma = 0.39$. A larger dispersion is an indication of anisotropy in the C system.

Log tan θ distribution is, therefore, frequently used to study the nature of angular distributions in the C system. At energies below 1,000 Bev, the observed distribution usually has a single maximum. Above 1,000 Bev one quite often finds a bimodal distribution. This has been interpreted by the Krakow group (Ciok et al., 1958), by Cocconi (1958), and by Niu (1958) as being due to the emission of two fireballs, moving in opposite directions with a Lorentz factor Γ_F in the C system, which subsequently decay isotropically by emission of several mesons. However, as is apparent from Eqs. (11.24) to (11.26), a single fireball decaying anisotropically can also lead to a bimodal ln tan θ distribution, where the two maxima are separated by $\Delta x = 2 \ln (\Gamma_F + \sqrt{\Gamma_F^2 - 1})$, where Γ_F depends on the degree of anisotropy in the C system through Eq. (11.27).

Another procedure due to Duller and Walker (1954), which is frequently used to study the angular distribution in the C system, involves plotting the quantity $\ln F/(1 - F)$ versus x ($\equiv \ln \tan \theta$), where F is the fraction of the particles emitted within a laboratory angle θ. For a C-system angular distribution $\cos^n \theta^* d \cos \theta^*$,

$$F(\theta^*) = \tfrac{1}{2}(1 - \cos \theta^{*n+1}) \qquad (11.28)$$

$$\frac{F}{1 - F} = \frac{1 - \cos^{n+1} \theta^*}{1 + \cos^{n+1} \theta^*} \qquad (11.29)$$

and on the assumption of symmetric emission, as for Eq. (11.22),

$$\frac{F}{1 - F} = \frac{1 + \tanh^{n+1} (x + \ln \Gamma_c)}{1 - \tanh^{n+1} (x + \ln \Gamma_c)} \qquad (11.30)$$

For $n = 0$ (isotropic case), this reduces to

$$\frac{F}{1 - F} = \exp [2(x + \ln \Gamma_c)]$$

or

$$\ln F/(1 - F) = 2(x + \ln \Gamma_c) \qquad (11.31)$$

Hence, for isotropic emission, the Duller-Walker plot would have a slope of 2 and an intercept equal to $-\ln \Gamma_c$ on the x axis.

The expected shapes for the Duller-Walker plots for several values of n are given in Fig. 11.19 [where $x + \ln \Gamma_c$ is plotted against $\ln F/(1 - F)$]. It is noted that, for higher values of n, the plot splits into two nearly straight lines which asymptotically (for $x + \ln \Gamma_c \gg \ln \sqrt{2n}$) have slopes of ~ 2 and are connected by a nearly horizontal curve. Some of the experimental angular distributions that are usually given as evidence for the production of two fireballs are also given in Fig. 11.19. For energies of $\sim$10,000 Bev a value of $n = 2$ to 3 is consistent with the observed distributions while below 1,000 Bev n is closer to zero. An increase in anisotropy with increasing energy would in general be expected from considerations of angular-momentum conservation; the fireball may be left with an increasing intrinsic angular momentum, resulting in a greater anisotropy in its decay.

Koba and Takagi (1958) have pointed out that, if the fireballs form real intermediate states in the collision process, in the event they possess high angular momenta, the emitted angular distribution should exhibit strong azimuthal asymmetry with preferential emission in the plane defined by the incident direction and the direction of emission of the fireball. No such asymmetries have so far been reported.

6.9. Meson Production at Energies up to 30 Bev: Importance of Resonance States. After the discovery of the first pion-nucleon resonance of mass 1,238 Mev, Lindenbaum and Sternheimer (1957) attempted to analyze inelastic nucleon-nucleon and pion-nucleon collisions at 2 to 3 Bev in terms of a model in which the colliding nucleons may be excited to their isobaric states and subsequently decay by emission of pions. This model proved to be very successful and was later extended to include higher pion-nucleon resonance states which were discovered subsequently. Since then a large number of resonance structures of nucleons and mesons have been found, and it has become apparent that production of these resonance states is an important intermediate step in the creation of new particles. Both in collisions of pions with nucleons and of nucleons with nucleons, two-body final states consisting of resonant states seem to play an important role. That in nucleon-nucleon collisions of up to $\sim$25 Bev, final states of nucleon isobars dominate has been indicated by the experiments of Damgaard and Hansen (1963) who conclude that about 90 per cent of pion production at 22 Bev may be explained in terms of excitation of two heavy objects of mass $\sim$1,400 to 2,600 Mev emitted with small transverse momenta and decaying by the emission, on an average, of two to three mesons each. Such a model accounts for the energy and angular distribution of secondary particles at all accelerator energies. In particular, it also accounts, in a natural way, for the large dominance of positive kaons over negative kaons at the high- and low-energy ends of their spectrum, because decay of nonstrange isobars into strange particles would ordinarily proceed by emission of kaons of positive strangeness only. Other recent experiments, particularly the one due to Dekkers et al. (1965), support this conclusion (Sec. 6.3 and Fig. 11.15).

6.10. Intermediate Resonant States in Ultra-high-energy Collisions. If isobars and resonant states in general play such a dominant role in particle production at accelerator energies, it may be suspected that they would also be important at higher energies. Actually isobar models for high-energy meson production were suggested long ago by Kraushaar and Marks (1954) and by Takagi (1952) and later investigated by Maor and Yekutieli (1960), among others. In these models, the collision of two nucleons results in the production of two massive isobars which disintegrate later by emission of mesons; this picture is supposed to hold up to the highest energies, the mass of the objects increasing with energy. It is quite obvious that isobar masses of 2 to 3 Bev are not adequate to account for the mean multiplicities observed at ultrahigh energies; also they would lead to a very high degree of anisotropy in angular distribution. If the mass of the isobar is allowed to increase with energy to account for the increased multiplicity,

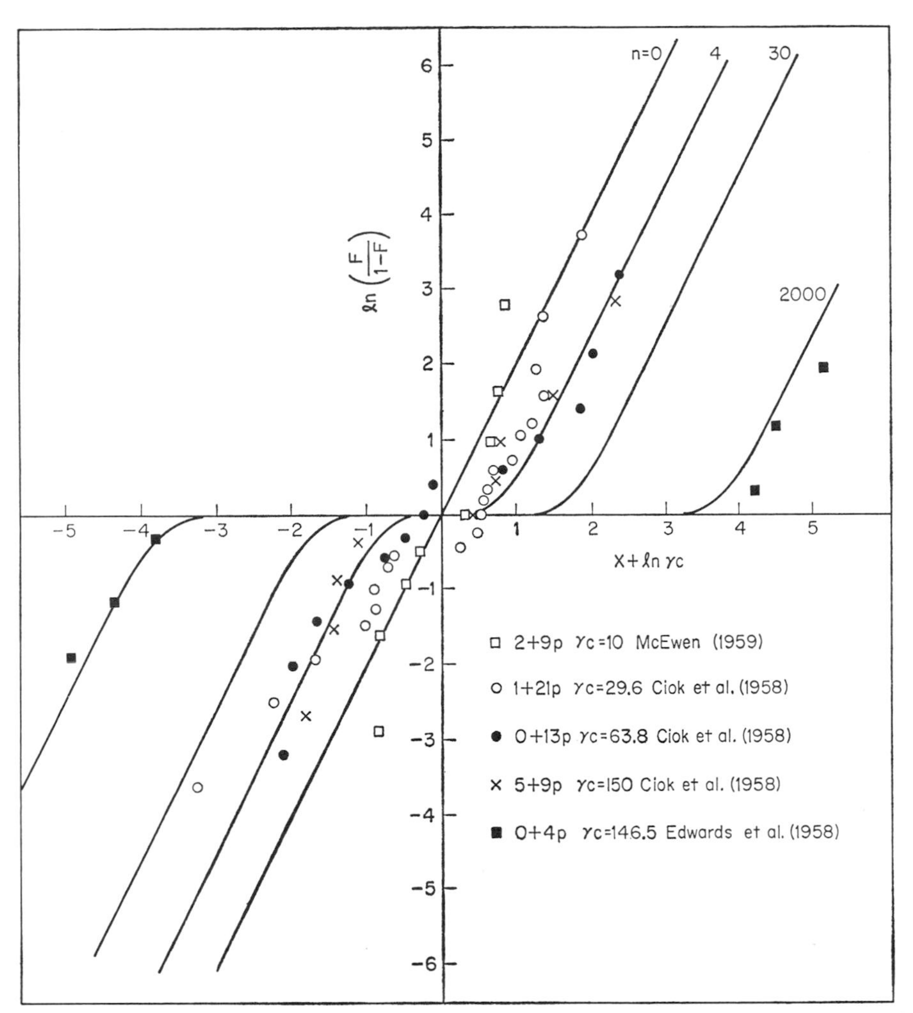

FIG. 11.19. Duller-Walker plots calculated for different values of the anisotropy index n and experimental angular distributions for some high-energy collisions. It is seen that anisotropic decay of a single fireball can reproduce the angular distributions of many jets which show a double maximum structure (see text).

it becomes difficult to reconcile the observed experimental feature that, while the C-system energy of the nucleon after the collision increases as the square root of the primary energy, the energy of most of the mesons remains approximately constant. Thus it seems that the fireballs cannot have the character of isobars in that the incident nucleons after collision do not belong to the fireball but, on an average, move faster than the fireball in the C system. However it appears that nucleons are not the only such particles; there is strong evidence that, at least in a considerable fraction of collisions, a small number of pions is generated with energies high in the rest system of the fireball and low in the rest system of one of the baryons. The particles in the tail of the energy spectrum shown in Fig. 11.18 would belong to this category. It is natural to expect strong final-state interactions between these mesons and the nucleon so that one may describe their creation as the result of de-excitation of a baryon isobar.

Production of isobars and resonant states in the cosmic-ray energy range has not yet been established by invariant mass analysis. What has been established is that there exist frames of reference (the fireball frame and the frames of surviving nucleons) in which several hadrons (strongly interacting particles) have relative energies of the order of a few hundred Mev, and it is known from accelerator physics that at these energies these particles form structures with definite quantum numbers and properties. Therefore it seems inescapable that the resonant states being discovered in large numbers in accelerator laboratories play a significant role in particle production, even at the highest cosmic-ray energies.

There is a growing impression that in all high-energy interactions the interaction regions break up into domains in each of which only the known physics of lower energies operates. There is some sort of limitation on the strength of strong interactions which manifests itself in a strong damping of high frequencies in the interaction. Thus whether it is a 6-Bev nucleon collision, a 6,000-Bev nucleon collision, or for that matter an antinucleon-nucleon annihilation at rest, the average energy of a pion in the C system is always of the order of a few hundred Mev. There is evidence that very-high-mass (2 to 3 Bev) isobars prefer to decay by multipion channels so that none of the pions gets too high a fraction of the available energy. Limitation on the value of transverse momentum and on the masses of isobars that can be excited in very-high-energy ($\sim 10^3$ Bev) nucleon-nucleon collisions may be another aspect of the same principle. Perhaps the most concise qualitative statement of this principle can be made by stating that all interactions are confined to low 4-momentum transfers.

That nucleon isobars of limited mass may be produced copiously in ultrahigh-energy interactions was suggested by Zatsepin (1962), Peters (1962), and Pal (1962), among others. A detailed investigation of this phenomenon and its relation to secondary cosmic rays was made by Pal and Peters (1964) [also see Peters (1963) and Pal (1963)]. One of the important experimental pieces of information suggesting isobar excitation was the following: The measured μ^+/μ^- ratio at sea level is nearly 5:4 and remains constant (or possibly increases) for muon energies above 100 Bev. This means that these muons are descendants of a subgroup of mesons (about three in each collision) which have a preferential share of the energy and charge of the incident nucleons and hence must be dynamically related to them. It was shown that if nucleon isobars are excited then their decay products and their progeny would dominate the steady flux of cosmic radiation (Sec. 7). A detailed calculation then indicated that the observed fluxes of all secondary particles would be reproduced if in seven-tenths of the collisions the forward nucleon emerges as an isobar of mass $\sim$2,300 Mev carrying 80 to 85 per cent of the incident energy and subsequently decaying by emitting $\sim$3 pions. The remaining energy goes into the production of a fireball (Sec. 7).

6.11. Phenomenological Model for Meson Production at High Energies. Following Pal and Peters (1964), and in the light of the discussion in previous sections, we may outline the salient features of a phenomenological model of meson production:

1. A fireball is created nearly at rest in the C system of nucleon-nucleon collisions and evaporates, giving n_F particles, with isotropic or moderately anisotropic angular distribution. The ratio of pions to nonpions among the fireball particles does not change with energy.

2. Each of the mesons, which are the dominant particles, receives a momentum whose average value is

$$\bar{P}_F = \frac{4}{\pi}\bar{P}_T \qquad \sim 450 \text{ Mev/c}$$

and whose maximum value is about 1 Bev/c. (There is a slight tendency for a very slow increase of $\bar{P}_T$ with primary energy above 10^4 Bev, but this increase, if real, is sufficiently slow so that one may assume that $\bar{P}_F$ is essentially independent of energy.)

3. The incident nucleons emerge in some excited state with a probability S and decay by emitting, on an average, n_B pions whose momentum in the baryon rest system is P_B. It will be assumed that the excited states are predominantly nonstrange; obviously their decay by nonpionic modes is not excluded. It does not have to be specified that some of or all the n_B pions may themselves form a resonant state of mesons.

4. If the character of the accessible baryon states is assumed to be independent of primary energy, beyond a certain threshold, the above assumptions, coupled with the indication that the average energy loss of nucleons is energy-independent (Sec. 6.7), imply that the mass of the fireball increases in proportion to the energy available in the C system. As a consequence, the primary nucleon expends a constant fraction, $1 - \eta'$, of its energy in fireball production, or pionization, and the number of fireball mesons, n_F, increases as the square root of the primary energy.

5. The multiplicity relation, for nucleon-nucleon collisions, then must be of the form given in Eq. (11.9), viz.,

$$\begin{aligned} n_t &= 2Sn_B + n_F \\ &= 2Sn_B + n_0E^{1/2} \end{aligned} \qquad (11.32)$$

It is clear that in this model fireballs are essentially a high-energy phenomenon; they become significant only above $\sim$100 Bev. The nature of the fireball remains essentially unspecified, because its detailed properties do not affect the steady component of the secondary cosmic radiation (Sec. 7). In particular, it is not clear whether the fireballs are themselves some sort of super-resonance states. It is also not clear what are most of the nonpions into which they decay. (It has been suggested that an appreciable fraction of them may be baryon-antibaryon pairs.)

A semiclassical description of high-energy interactions may be given: In a collision of two nucleons, the equivalent of one pion mass of each nucleon comes to rest in the C system, creating a fireball, which then has $\sim$15 per cent of the initial energy. The meson cloud of the emerging nucleon is disturbed; it is then reconstructed to form one of the baryon states which is "congruent" with it. This excited baryon then decays. When one adapts this description to a pion-nucleon collision, the equivalent of the fireball is now at rest in the C system of the incident pion and a pion in the nucleon cloud. At the same incident energy the fireball mass (and hence multiplicity) in a pion collision is $\sqrt{M_n/m_\pi}$ of that for a nucleon collision. This is very nearly true experimentally (Pal and Peters, 1964).

Figure 11.20 gives a schematic representation of the C-system momentum configuration of pions and nucleons produced in nucleon-nucleon collisions at three different energies.

7. Propagation of Cosmic Rays through the Atmosphere

The primary cosmic-ray energy spectrum is steep; as a consequence the secondary particles that carry

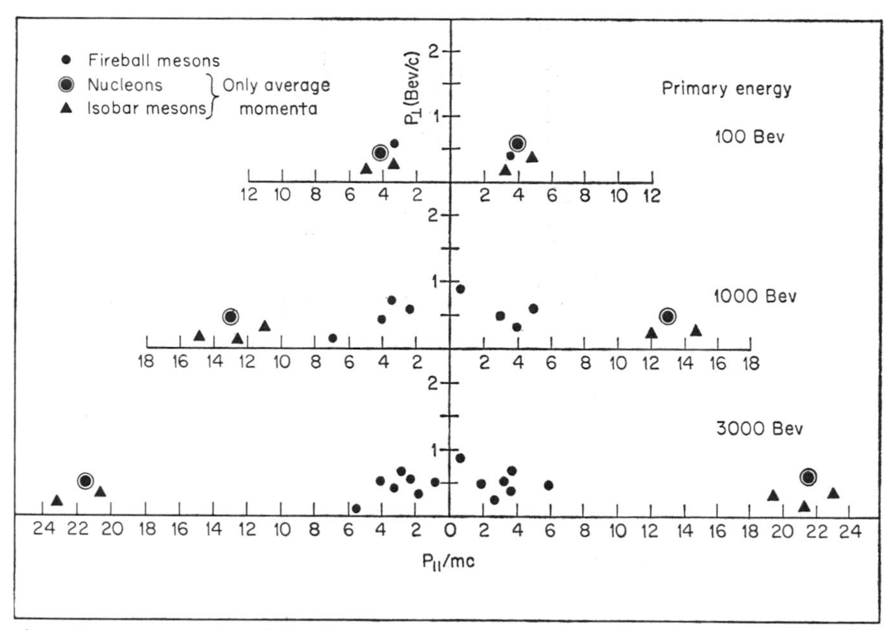

FIG. 11.20. A schematic representation of the C-system momentum configuration of mesons and nucleons after nucleon-nucleon collisions of various energies. For clarity, kaons and baryons produced in the interaction are not separately indicated. For the decay products of isobars only the average momenta are shown though their spread is so large that even at highest incident energy an isobar pion has a finite probability of having close to zero momentum. The longitudinal momentum is in units of mc where m is the rest mass of the particle concerned.

a large fraction of the collision energy dominate the secondary cosmic-ray flux. In fact, the production spectrum of particles that receive a fraction ϵ of the incident energy is proportional to $(\epsilon)^\gamma$, where γ is the index of the primary integral energy spectrum. In the isobar model of high-energy interactions, outlined in Sec. 6, the decay products of the forward emitted isobar form the most energetic particles and hence dominate the flux of secondary particles in cosmic rays at energies above 1 Bev or so. The ratio of pion production from the decay of the forward emitted baryon isobar to the production from evaporation of the fireball is plotted in Fig. 11.21 (Pal and Peters, 1964). It is seen that, for purely kinematical reasons, the mesons from the de-excitation of baryon isobars account for almost all particles above a few Bev, even if the probability of excitation is fairly small. The important consequence of this fact, from the point of view of calculating secondary fluxes, is that the effective production cross sections for all particles that contribute significantly to the secondary flux become homogeneous; the energy of the produced particle becomes proportional to the primary energy.

7.1. Energy Spectrum of Nucleons in the Atmosphere. The differential energy spectrum of primary cosmic-ray nucleons may be represented by

$$N(0,E)\,dE = S_0\,\frac{dE}{E^{\gamma+1}} \qquad (11.33)$$

If the total energy E is expressed in Bev and the flux is measured in nucleons per m²/ster/sec/Bev and γ is taken as 1.67, S_0 has a value of 23,500. This is the flux of all nucleons, including those coming as parts of heavy nuclei (Sec. 5.2).

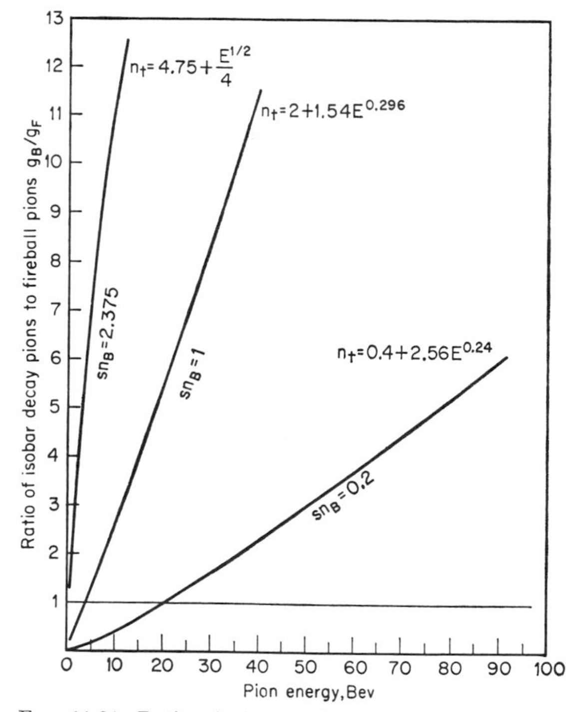

FIG. 11.21. Ratio of pion production (in cosmic rays) through isobar decay to that from fireball decay as a function of pion energy. The curves are drawn for various multiplicity relations consistent with experimental data. The first term in the expression for n_t gives the contribution of isobar pions and the second, that of the fireball pions. [*From Pal and Peters* (1964).]

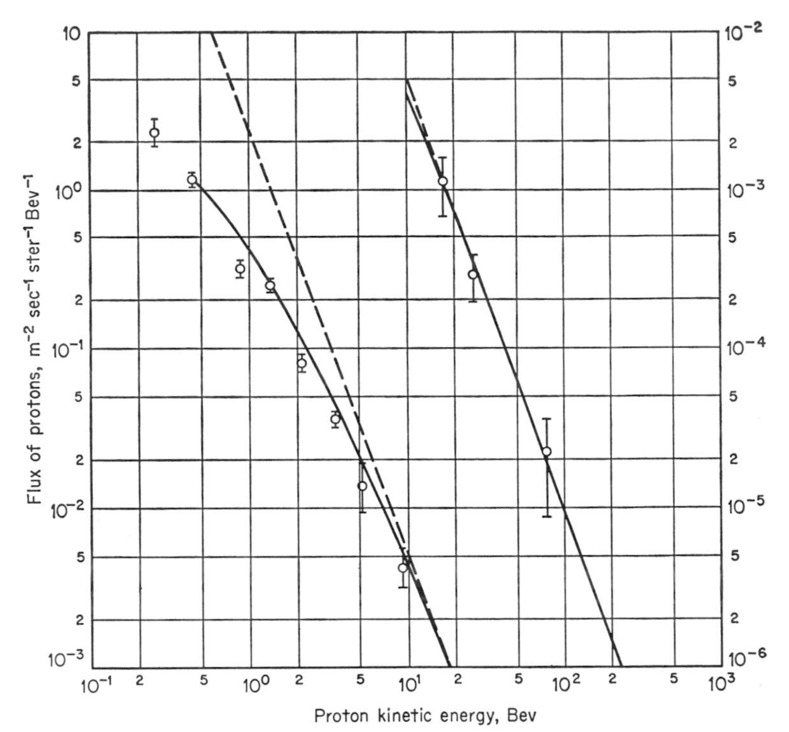

FIG. 11.22. The differential energy spectrum of protons at sea level. The dotted curve is a line of slope $\gamma + 1 = 2.67$. The full curve represents the calculated spectrum based on an attenuation length $\Lambda = 120$ g/cm² and taking into account the ionization correction, angular spread in collisions, and the fact that the primary spectrum is a power law only in total energy. Experimental points are from Brooke and Wolfendale (1964). [*Pal and Peters* (1964); *Rengarajan and Tandon* (1965).]

If λ is the interaction mean free path† and η the fraction of the energy retained by the nucleon in each collision, the flux of nucleons of energy E at a depth of x g/cm² is given by

$$N(x,E) = N(0,E) \exp\left(-\frac{x}{\lambda}\right) \sum_{j=0}^{\infty} \left(\frac{x}{\lambda}\right)^j \frac{\eta^{\gamma j}}{j!}$$
$$= N(0,E) \exp\left(-\frac{x}{\Lambda}\right) \qquad (11.34)$$

where $\Lambda = \lambda/(1 - \eta^\gamma)$ is the attenuation length of nucleons. It is easily shown that, if η has a distribution of values, η^γ in this expression has to be replaced by the average $\langle\eta^\gamma\rangle$. Physically $\langle\eta^\gamma\rangle$ represents the ratio of the production of nucleons per interaction to the incident flux of nucleons, at any energy E in dE.

At low energies, $E < 10$ Bev, the nucleon spectrum is altered because of a change in elasticity η and because of ionization loss suffered by protons. These corrections have been evaluated by Rengarajan and Tandon (1965), and are used for the calculated proton spectrum at sea level given in Fig. 11.22.

A comparison of the sea-level proton spectrum measured by Brooke and Wolfendale (1964) (also shown in Fig. 11.22) and the cosmic-ray primary spectrum serves to fix the value of the attenuation mean free path $\Lambda = 120 \pm 5$ g/cm². This value of Λ is consistent with a large number of measurements on the attenuation of the nucleon component in the atmosphere. The experimental data indicate that Λ is independent of energy though it is not possible to exclude a 10 per cent variation over one or two decades of energy; in interpreting experiments in which all nuclear active particles are measured, it must be remembered that the pion component in the atmosphere does not vary with depth in the same manner as the nucleon component (Sec. 7.3).

Assuming a value of the nucleon interaction length $\lambda = 75 \pm 5$ g/cm² (Sitte, 1961; Bozoki et al., 1962) and the above value of Λ, one obtains

$$\langle\eta^\gamma\rangle = 1 - \frac{\lambda}{\Lambda} = 0.37 \pm 0.06 \qquad (11.35)$$

almost independent of energy. This is the strongest cosmic-ray evidence indicating that in high-energy interactions the forward nucleons emerge with a large fraction $\eta \sim 0.55$ of the incident energy.

7.2. Neutron-to-Proton Ratio in the Atmosphere. If w is the probability of charge exchange for the persisting nucleon in an interaction with an air nucleus, then after j collisions the original charge composition of the nucleons δ_0 will be changed into

$$\delta_j = \left(\frac{p - n}{p + n}\right)_j = \delta_0(1 - 2w)^j \qquad (11.36)$$

† Because of the high elasticity of nucleon-nucleon collisions, we may assume that the interaction mean free path of bound nucleons is the same as that for protons.

Taking the Poisson fluctuations into account, as in the derivation of Eq. (11.34), one finds

$$\delta_x = \frac{N_p - N_n}{N} = \delta_0 \exp\left[-x\left(\frac{1}{\lambda} - \frac{1}{\Lambda}\right)\right] \sum_{j=0}^{\infty} \left(\frac{x\eta^\gamma}{\lambda}\right)^j \frac{(1-2w)^j}{j!}$$

$$= \delta_0 \exp\left(-\frac{2x}{\lambda} <\eta^\gamma w>\right) \qquad (11.37)$$

The < > sign in this expression signifies that in the case of fluctuating η and w an average of the quantity $\eta^\gamma w$ should be taken. Thus

$$N_p(x,E) = N(0,E) \exp\left(-\frac{x}{\Lambda}\right) \frac{1+\delta_x}{2}$$
$$N_n(x,E) = N(0,E) \exp\left(-\frac{x}{\Lambda}\right) \frac{1-\delta_x}{2} \qquad (11.38)$$

7.3. The Production Spectrum and Flux of Charged Pions. The pion flux at a depth x g/cm² in the atmosphere is obtained by the solution of the diffusion equation

$$\frac{dF_\pi}{dx} + F_\pi\left(\frac{1}{\lambda_\pi} + \frac{u}{x}\right) = P_\pi + P'_\pi \qquad (11.39)$$

where λ_π is the interaction length of pions,

$$u = \frac{h_0 m_\pi}{c\tau_\pi E} = \frac{\epsilon_\pi}{E} \qquad (11.40)$$

where ϵ_π is the constant characterizing the decay of pions in the atmosphere ($\epsilon_\pi = 128$ Bev if the scale height of the atmosphere $h_0 = 7$ km), P_π is the production of pions by nucleons, and P'_π is the production of pions by pions. As a consequence of the dominance of the homogeneous part of the pion-production cross section, the production spectrum of charged pions in nucleon collisions is given by

$$P_\pi(x,E)\,dx = <B> N(0,E) \exp\left(-\frac{x}{\Lambda}\right)\frac{dx}{\lambda} \qquad (11.41)$$

where, analogous to the quantity $<\eta^\gamma>$, $<B>$ represents the ratio of the production of charged pions per interaction to the incident flux of nucleons, at any energy E in dE. Analytically,

$$<B> \equiv \sum_i n_i \epsilon_i^\gamma \qquad (11.42)$$

where n_i is the number of charged mesons produced per interaction with a fractional energy ϵ_i. Since high-energy ($E \sim 100$ Bev) muons are produced mostly in the upper atmosphere where pion production by pions is unimportant, a comparison of the flux of muons of this energy range with the primary nucleon flux essentially determines $<B>$ without going into any model of nucleon-nucleon interactions (Sec. 7.7).

The source term P'_π is important in the lower regions of the atmosphere. Pal and Peters (1964) have used two extreme assumptions to specify P'_π:

1. Pion-nucleon collisions are completely elastic; the pions retain practically the whole of their energy and may be lost only through charge exchange.
2. Pion-nucleon collisions are completely inelastic in that the total energy in the C system of the incident pion and a target pion is shared equally by all particles (mostly pions). The multiplicity would have a square-root dependence on energy if one assumed that the average energy of pions in this system remains constant (Sec. 6.11).

With either of these assumptions it is possible to obtain an analytic solution for the pion flux at various depths in the atmosphere. The ratio of pion to nucleon flux so calculated is plotted in Fig. 11.23. The experimental points seem to be inconsistent with assumption 1, namely, that the collisions are extremely elastic. The experimental information on the pion fluxes in the atmosphere is still very poor for a more specific conclusion to be drawn from this type of analysis alone. It may, however, be noted that, if pion production by pions is not taken into account, it becomes impossible to explain the rather large flux of moderate-energy (~20 Bev) charged pions observed at 800 g/cm² and at sea level (Lal et al., 1963; Brooke et al., 1964).

7.4. Production Spectrum of γ Rays in the Atmosphere. Because of charge independence, the production spectrum of neutral pions must be related to the production spectrum of charged pions through

$$P_{\pi^\circ}(x,E) = \tfrac{1}{2}P_\pi(x,E) + \tfrac{1}{2}P'_\pi(x,E) \qquad (11.43)$$

At high altitudes P'_π can be neglected† in comparison with P_π. The production spectrum of γ rays arising from π° decay is then

$$P_\gamma(x,E) = \frac{P_{\pi^\circ}(x,E)}{\gamma+1} = \frac{1}{2}\frac{P_\pi(x,E)}{\gamma+1} \qquad (11.44)$$

assuming that the π° always decays into two γ rays which are emitted isotropically in its rest frame.

7.5. Passage of Photons and Electrons through Matter. The passage of energetic electrons and photons through matter leads to a rapid sharing of energy between a large number of electrons and photons of secondary origin. Quantitative understanding of this phenomenon is in terms of the quantum theory of radiation and the theory of electromagnetic interaction of charged particles, whose validity has been verified up to very high energies. The most important fundamental processes involved are the following:

1. Pair production by a photon in the Coulomb field of an atomic nucleus (or an electron)
2. Scattering of a photon by atomic electrons (Compton effect)
3. Radiation of photons by an electron scattered in the Coulomb field of an atomic nucleus or an electron (*bremsstrahlung*)
4. Collision of an electron with atomic electrons (ionization)

The first three of these processes lead to a multipli-

† On the model discussed in Sec. 6.11 and assumption (2) of Sec. 7.3, $P'_\pi \ll P_\pi$ at extremely high energies because an ultrahigh-energy pion collision results in the sharing of its energy by a large number of low-energy particles.

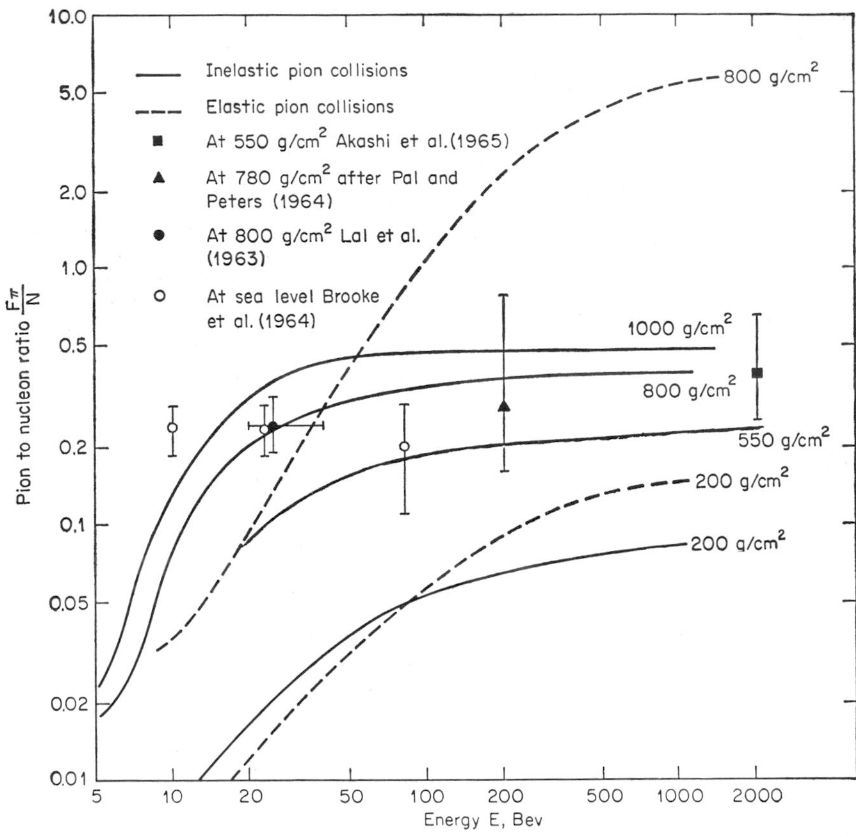

FIG. 11.23. The calculated ratio of charged pions to nucleons as a function of energy at various depths in the atmosphere with experimental points. [*After Pal and Peters* (1964).]

cation of the electron-photon component. Process 4 produces a continuous energy loss.

It is found that per gram per square centimeter of the medium both the fractional radiation energy loss by an electron and the probability of pair production by a γ ray are approximately proportional to a quantity

$$\frac{1}{X_0} = 4\frac{e^2}{\hbar c}\left(\frac{e^2}{mc^2}\right)^2 N\frac{Z(Z+1)}{A} \ln(183Z^{-1/3}) \quad \text{cm}^2/\text{g} \quad (11.45)$$

where N is Avogadro's number and Z, A are the atomic number and the mass number of the absorbing medium. (This applies for γ rays or electrons of energy $E \gg 137\ mc^2Z^{1/3}$.) Therefore it is convenient to introduce X_0 as a unit of length in the absorber. This is called the radiation length or the cascade unit and is equal to that thickness of the medium in which a high-energy electron loses by radiation, on an average, a fraction $(1 - 1/e)$ of its energy. The conversion length or mean free path for pair production by high-energy γ rays is more accurately equal to $9/7 X_0$. Usually one also defines a critical energy ϵ_c at which an electron loses equal amounts of energy per unit length by processes of ionization and radiation. The values of X_0 and ϵ_c for various materials are given in Table 11.7.

Figure 11.24 shows the rate (μ) at which a photon of energy E traversing one radiation length of air will produce an electron-positron pair or suffer Compton scattering. It is seen that, for $E \geq 10^8$ ev, the pair-production cross section dominates and with increasing energy approaches a constant value. Figure 11.25 gives the rate of fractional energy loss by electrons of various energies by collision and by radiation. These curves show that the ionization loss dominates for energies below 10^8 ev and radiation loss dominates above 10^8 ev. The radiation loss may occur in large finite steps and hence is subject to large fluctuations. The energy spectrum of the emitted photons has

TABLE 11.7

Material	Radiation length, X_0, g/cm²	Critical energy, ϵ_c, Mev
Air	37.7	84.2
Water	37.1	83.8
Carbon	44.6	102
Aluminum	24.4	49
Iron	13.9	24
Copper	12.9	22
Lead	5.83	7.8
Tungsten	6.3	8.6

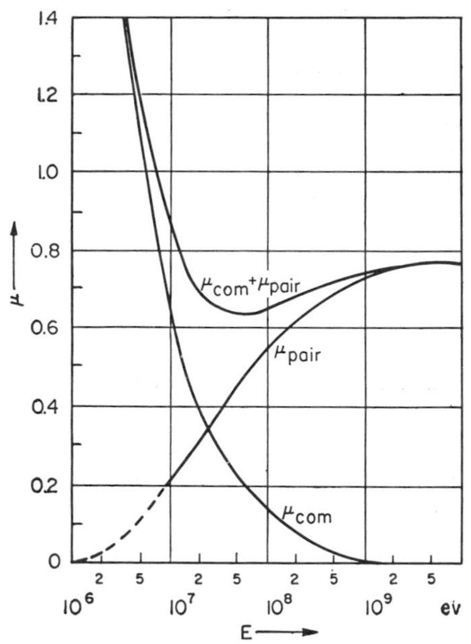

FIG. 11.24. The rate μ, per radiation length of pair production, of Compton scattering and of either process by a γ ray passing through air, as a function of its energy E. [*Rossi and Greisen* (1941).]

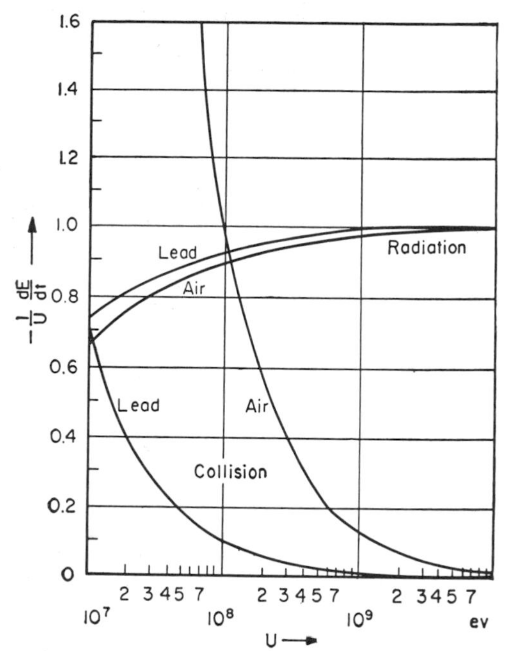

FIG. 11.25. The rate of fractional energy loss per radiation length by radiation and by collision processes, for an electron in air or lead. [*Rossi and Greisen* (1941).]

roughly a $1/E$ character and extends from zero to the total kinetic energy of the electron.

Direct pair production by electrons is a second-order process and hence relatively unimportant for the multiplication of electrons and photons.

Photon-Electron Cascades. High-energy γ rays are produced in the atmosphere mainly through π° decay. Within a thickness of the order of 50 g/cm² they are converted into electron-positron pairs. If the energy of the electrons is greater than ~100 Mev these would further multiply by emitting photons. The main qualitative features of the longitudinal development of a photon-electron cascade are illustrated by a simplified model due to Heitler (1948). Consider an incident particle (photon or electron) of energy E_0 and assume that either through pair creation or radiation it produces two particles of equal energy in a distance $X_0 \ln 2$. Let this process continue up to depth $t \ln 2$ (t being measured in units of X_0), where the total number of particles will become

$$N(t \ln 2) = 2^t \quad \text{or} \quad N(t) = e^t \tag{11.46}$$

and the energy of each particle will be

$$E = E_0 e^{-t} \tag{11.47}$$

The cascade multiplication will stop at an energy $E \sim \epsilon_c$; hence the maximum will occur at a distance

$$t_{\max} = \ln \frac{E_0}{\epsilon_c} \tag{11.48}$$

and the maximum number of particles will be

$$N_{\max} = \frac{E_0}{\epsilon_c} \tag{11.49}$$

Thus the buildup of the shower is nearly exponential, the maximum number of particles is proportional to E_0, and the position of the maximum is proportional to $\ln E_0$. After the maximum the shower is absorbed, primarily by collision and Compton losses. This absorption may also have an approximately exponential form.

Longitudinal solutions of the cascade problem are obtained by an iteration procedure (Bhabha and Heitler, 1937) or by setting up diffusion equations (Carlson and Oppenheimer, 1937) for the solution of which several approximations are made. In approximation A Compton scattering and collision loss of electrons are neglected, while in approximation B they are introduced as a constant energy loss independent of primary energy. The cross sections for pair production and radiation loss are replaced by their asymptotic forms which are simple homogeneous functions. Rossi (1952) has treated the Carlson-Oppenheimer method in some detail. It is shown that under approximation A, valid for high-energy shower particles, the diffusion equations yield two stationary solutions of the form

$$\begin{aligned} N(E,t) &= a_1 E^{-(S+1)} \exp [\lambda_1(S)t] \\ \gamma(E,t) &= a_1 b_1(S) E^{-(S+1)} \exp [\lambda_1(S)t] \\ \text{and} \quad N(E,t) &= a_2 E^{-(S+1)} \exp [\lambda_2(S)t] \\ \gamma(E,t) &= a_2 b_2(S) E^{-(S+1)} \exp [\lambda_2(S)t] \end{aligned} \tag{11.50}$$

Here N refers to the electron component and γ to the photon component; a_1 and a_2 are arbitrary constants depending on initial conditions. Expressions for λ_1, λ_2, b_1, and b_2 are given by Rossi (1952) who has also tabulated their values for a large number of S values. It can be seen that, in general, the propagation of an incident power-law spectrum of electrons and photons

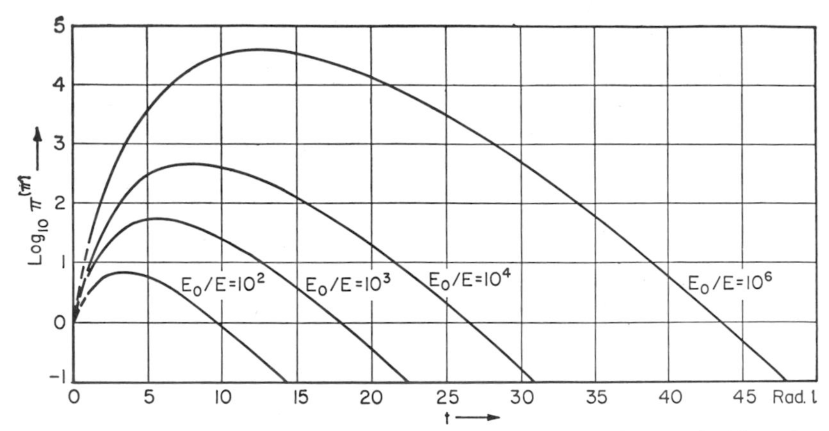

FIG. 11.26. Average number N of electrons with energy greater than E produced by an incident electron of energy E_0, as a function of the amount of material traversed (t in radiation lengths). Calculated with approximation A. [*Rossi and Greisen* (1941).]

is represented by a linear combination of the two stationary solutions given above. Propagation for a pure γ-ray source is given by this linear combination with $a_1 = -a_2$. Such a form is used for calculating the steady flux of high-energy γ rays and electrons because the γ-ray source spectrum in the atmosphere [Eqs. (11.44) and (11.41)] does have the required power-law shape.

It is found that λ_2 in Eqs. (11.50) is always negative and smaller than λ_1 and that for a sufficiently large value of t the integral spectrum of electrons in a shower initiated by an electron or a photon of energy E_0 is given by

$$N(t,>E) = F(S)\left(\frac{E_0}{E}\right)^S \exp\,[\lambda_1(S)t] \qquad (11.51)$$

for $E \ll E_0$. Here S is given by

$$t = -\frac{1}{d\lambda_1(S)/dS}\left[\ln\left(\frac{E_0}{E}\right) - \frac{n}{S}\right] \qquad (11.52)$$

where $n = 1$ for electron-initiated showers
$= \frac{1}{2}$ for photon-initiated showers

$F(S)$ is different for electron- and photon-initiated showers and varies slowly with t. Thus the depth dependence of the spectrum is dominated by the term $\exp\,[\lambda_1(S)t]$. λ_1 is found to be positive when $S < 1$, zero when $S = 1$, and negative when $S > 1$. Thus the shower builds up to a maximum at $S = 1$ and then decays. For this reason S is sometimes called the shower age, or the age parameter. Showers with $S < 1$ are called young and those with $S > 1$ are called old. When $S = 2$ the number of particles in the shower drops below 1.

The depth $t_{\max}$ is given by

$$t_{\max} = 1.01(\ln E_0/E - n) \qquad (11.53)$$

and the maximum number of electrons is

$$N_{\max}(E) = \frac{0.137}{(\ln E_0/E - 0.37n)^{\frac{1}{2}}}\frac{E_0}{E} \qquad (11.54)$$

Some of the results of calculations based on approximation A are represented in Fig. 11.26.

The total number of electrons in the shower as calculated under approximation B is larger than that under approximation A. Greisen (1956) has given the relation

$$N(t,E_0,>0) = \varphi(S)N(t,E_0,>\epsilon_c)_A \qquad (11.55)$$
$$\text{where} \quad \varphi(S) = 2.29[1 + 0.4(S - 1)]$$

Figure 11.27 gives the total number of electrons at various depths in showers initiated by γ rays of energies 10^{12} to 10^{18} ev, as calculated by Snyder (1949) with approximation B.

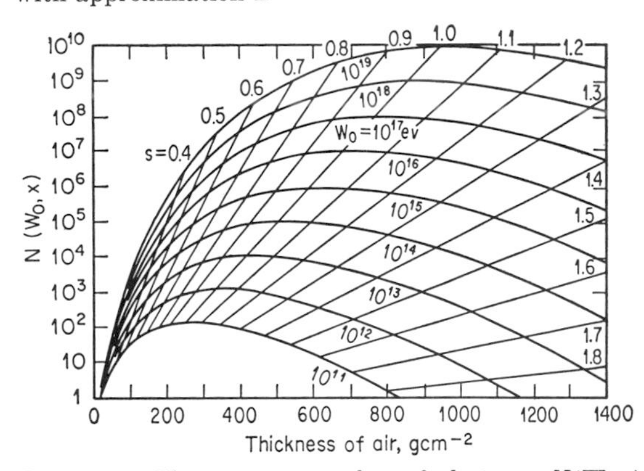

FIG. 11.27. The average number of electrons $N(W_0,x)$ at depth x in an air cascade initiated by a photon of energy W_0, under approximation B. Straight lines are the loci for various values of the age parameter S. [*Cocconi* (1961) *after calculation by Snyder* (1949).]

Greisen (1956) has given a convenient expression to calculate the total number of particles in a shower:

$$N(t,E_0,>0) \simeq \frac{0.31}{[\ln\,(E_0/\epsilon_c)]^{\frac{1}{2}}}\exp\,[t(1-\tfrac{3}{2}\ln S)] \qquad (11.56)$$

$$\text{where} \qquad S \simeq \frac{3t}{t + 2\ln\,(E_0/\epsilon_c)} \qquad (11.57)$$

Lateral Distribution of Cascades. The electrons and photons in a cascade can acquire an angular and lateral spread because of multiple Coulomb scattering of electrons. [The root-mean-square angle of scattering in a radiation length is

$$4\pi \frac{\hbar c}{e^2} \frac{mc^2}{E} = \frac{21}{E(\text{Mev})}$$

which is much larger than the intrinsic spread in the pair-creation or *bremsstrahlung* process which is $\sim mc^2/E = 0.51/E(\text{Mev})$.] The problem of the spread of air-shower particles has been analyzed by Molière (1946) for approximation *A*, by Nishimura and Kamata (1950, 1951, 1952) for approximation *B*, and by Eyges and Fernbach (1951) for both approximation *A* and *B*. Since the scattering angle is inversely proportional to *E*, only low-energy electrons are found at large distances from the shower axis. These electrons, which are close to the critical energy, are brought to rest by ionization and are continuously replenished by the multiplication of high-energy particles in the core. The lateral distance to which an electron of energy ϵ_c is scattered in traversing one radiation length is $21/\epsilon_c(\text{Mev}) \simeq \frac{1}{4}$ radiation lengths, which is ~ 9.5 g/cm² or about 79 m in air at sea level. This distance, r_1, is called the characteristic or scattering length. [Eyges (1948) has pointed out that particles at very large distances from the core are due to pair creation by photons radiated by electrons below the critical energy because, owing to their low cross section, such photons can travel large distances away from the axis before materializing.]

Figure 11.28 shows an example of a curve due to Molière (1946), and Fig. 11.29 gives the lateral distributions of Nishimura and Kamata for different shower ages, calculated under approximation *B*.

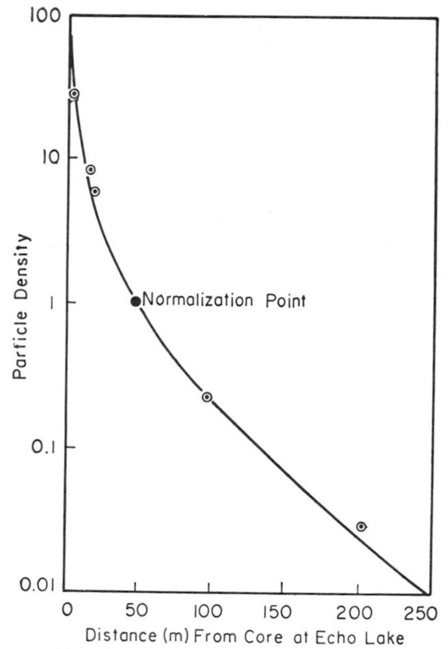

FIG. 11.28. Density of shower particles as a function of distance from the core at an atmospheric depth of 708 g/cm². The curve is according to Molière (1946) for showers initiated by a single photon. The experimental points are for extensive air showers. [*Cocconi, Cocconi-Tongiorgi and Greisen* (1949).]

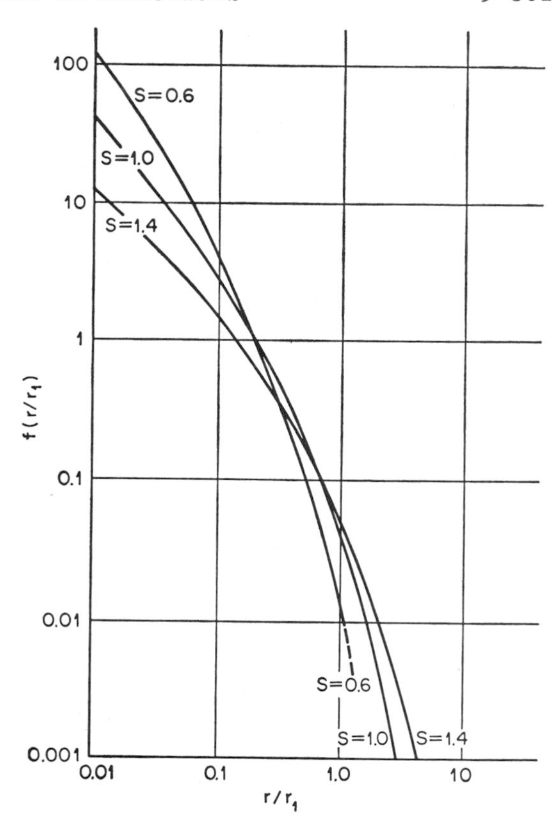

FIG. 11.29. Lateral distribution function $f(r/r_1)$ of electron density in a photon cascade in air for three different values of the age parameter *S* according to calculations of Nishimura and Kamata. [*After Galbraith* (1958).]

The density of particles $\Delta(r)$ at distance *r* is given by

$$\Delta(r) = \frac{Nf(r/r_1)}{r_1^2} \tag{11.58}$$

where *N* is the total number of particles in the shower and $f(r/r_1)$ is the function plotted in Fig. 11.29. Greisen (1956) has given an empirical fit to the Nishimura-Kamata function

$$f(r/r_1) = C(S)(r/r_1)^{S-2}(r/r_1 + 1)^{S-4.5} \tag{11.59}$$

which holds up to shower ages of 1.4. The values of the constants $C(S)$ are given in Table 11.8. In the calculations of Nishimura and Kamata *S* is defined by

$$S \approx \frac{3t}{t + 2 \ln (E_0/\epsilon_c) + 2 \ln r/r_1} \tag{11.60}$$

instead of by Eq. (11.57).

Curves given in Fig. 11.29 are for a uniform atmosphere and have to be modified slightly when used for the actual atmosphere (Greisen, 1956).

TABLE 11.8

S	0.5	0.75	1.0	1.25	1.50
$C(S)$	0.16	0.29	0.40	0.45	0.41

7.6. Flux of High-energy Electron-Photon Component. High-energy γ-ray and electron fluxes are measured by using sandwich emulsion stacks or ionization chambers. In nuclear emulsions the energies are estimated by measuring the opacity of the electromagnetic cascades or by counting the number of charged particles within a fixed radius around the axis at known distances from the point of origin. The theory of these measurements and the calibration techniques have been worked out by Kamata and Nishimura (1958), Pinkau (1958), and Duthie et al. (1961).

Figure 11.30 shows the experimental spectra at three depths in the atmosphere, viz., at 20, 220, and

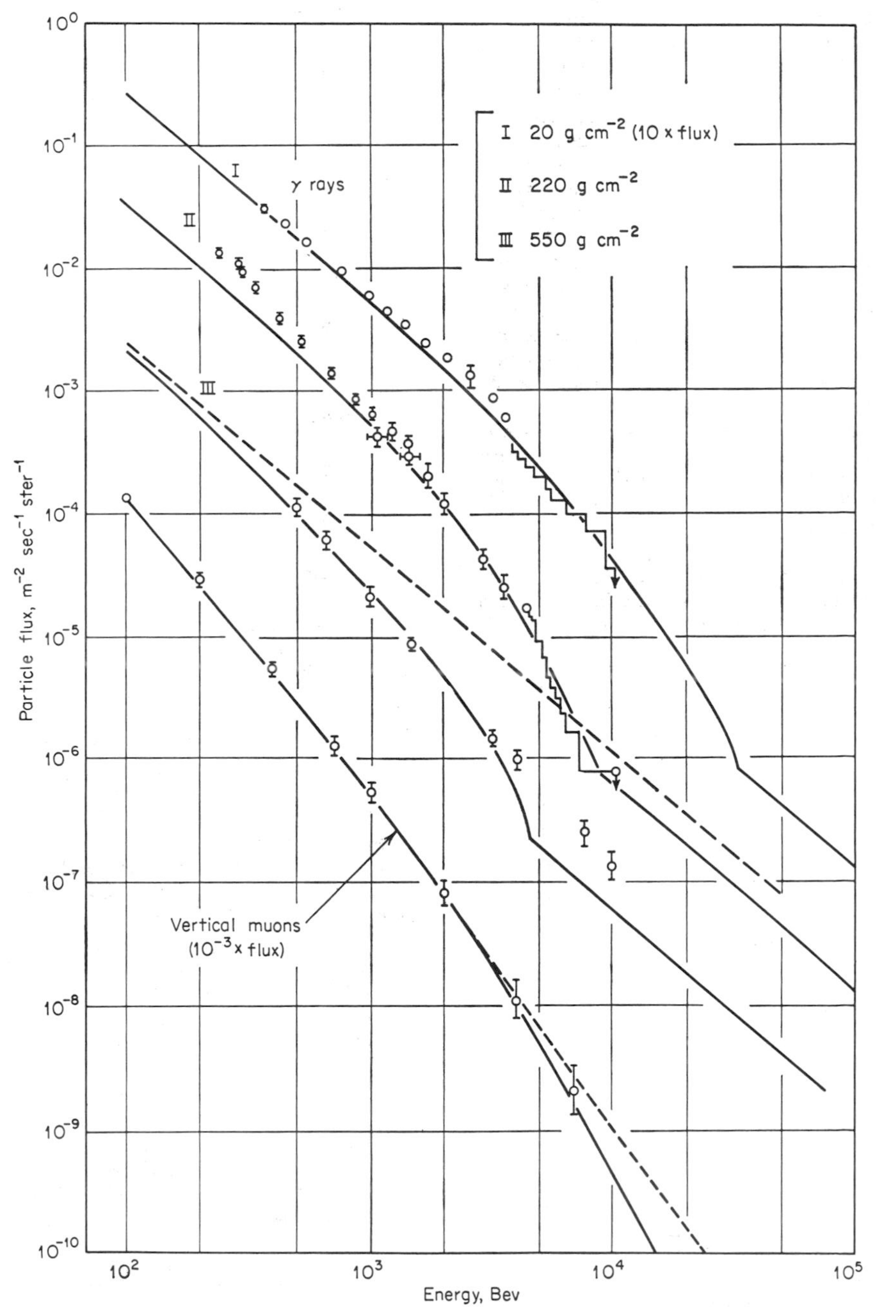

FIG. 11.30. Calculated integral energy spectra of γ rays at three depths and of vertical muons at sea level. The continuous curves correspond to a two-component primary spectrum (Secs. 5.2, 7.6, and 8.5), while dotted curves correspond to a continuous primary spectrum of slope $\gamma = 1.67$. The experimental points are: I, Malhotra et al. (1965); II, Bowler et al. (1962); III Akashi et al. (1963); and muons after Menon and Ramana Murthy (1966). The curves for comparison with experiments at vertical depths of 220 g cm^{-2} and 550 g cm^{-2} are calculated at 250 g cm^{-2} and 600 g cm^{-2} respectively.

550 g/cm², corresponding to balloon altitude, airplane altitude, and mountain altitude. The curves are calculated by using a two-component primary spectrum, one having the same chemical composition as at low energies and a rigidity cutoff around 2×10^{14} ev and the other consisting only of protons and having the same slope but $\sim\frac{1}{20}$ the nucleon intensity or $\sim\frac{1}{15}$ the proton intensity of the first. This form of the primary spectrum is consistent with other data on composition and characteristics of air showers (Secs. 5.2 and 8.5). The normalization for these calculations is obtained by simultaneous calculation of the high-energy muon spectrum (Sec. 7.7), experimental points for which are derived from intensity measurements at various depths below ground. A basic assumption in this is that charged and neutral π mesons produced in the ratio of 2:1 are the main sources of muons and γ rays, respectively. Fit to the muon spectrum is also given in Fig. 11.30. The following conclusions can be drawn from the data and theoretical calculations shown in Fig. 11.30:

1. Simultaneous fitting of muon and γ-ray spectra suggests that, even up to $\sim 10^{14}$ ev, pions constitute the most important component produced in high-energy interactions. [This was first shown by Duthie et al. (1962).]
2. The γ-ray component goes through a maximum in the atmosphere and below this the attenuation length of γ rays of a few hundred Bev is nearly the same as that of nucleons, that is, $\sim$120 g/cm²; however, this attenuation length is a function of energy and depth in the atmosphere.
3. Change in the shape of the γ-ray spectrum with depth is such as can be explained by a change in the shape of the primary spectrum in the manner discussed above and in Secs. 5.2 and 8.5.
4. No change with energy is required for the number and fractional energy of fast pions produced in high-energy interactions; that is, if these pions arise from isobar decay, the average mass of the isobars may remain nearly constant.

7.7. The Flux of Muons and Neutrinos. The decay of charged pions and other unstable particles ultimately gives rise to muons and neutrinos. The general expression of the spectrum for production of particles of type i from particles of type j decaying in the direction φ (as measured in their rest frame with respect to their direction of motion) is

$$P_i{}^j(x,E;\varphi) = \frac{m_j}{c\tau_j\rho(x)}\frac{b_{ij}}{E}F_j\left[x, \frac{E}{r_{ij}(\varphi)}\right] \tag{11.61}$$

where b_{ij} is the relevant branching ratio,

$$r_{ij}(\varphi) = r_{ij}(1 + \beta_{ij}\cos\varphi)$$

is the fractional energy received by particle i when it is emitted in the direction φ and β_{ij} its velocity, m_j and τ_j are the rest mass and lifetime of the parent particle, and $\rho(x)$ is the local density of the atmosphere. For isotropic decay,

$$\begin{aligned} P_i{}^j(x,E) &= \frac{m_j}{c\tau_j\rho(x)} b_{ij}\int_{-1}^{+1} F_j\left[x, \frac{E}{r_{ij}(\varphi)}\right]\frac{d\cos\varphi}{2E} \\ &= \frac{m_j}{2c\tau_j\rho(x)}\frac{b_{ij}}{r_{ij}\beta_{ij}}\int_{E/(1+\beta_{ij})r_{ij}}^{E/(1-\beta_{ij})r_{ij}} F_j[x,E']\frac{dE'}{E'^2} \end{aligned} \tag{11.62}$$

For muons arising from pion decay

$$r_{\mu\pi} = \tfrac{1}{2}[1 + (m_\mu/m_\pi)^2] \approx 0.79$$

and $\beta = 0.28$ while for neutrinos from this decay $r_{\nu\pi} = \frac{1}{2}[1 - (m_\mu/m_\pi)^2] \simeq 0.21$ and $\beta = 1$. For particles produced in three-body decays an additional integration over r_{ij} is necessary.

These expressions are valid for flux in all directions in the atmosphere, provided corresponding values of x, F, and ρ are taken. For the vertical direction [see Eq. (11.40)],

$$\frac{m_j}{c\tau_j\rho(x)} = \frac{h_0 m_j}{c\tau_j x} \equiv \frac{\epsilon_j}{x} \tag{11.63}$$

For neutrinos there is no attenuation after production, and the differential flux at x is given by

$$F_i{}^j(x,E) = \int_0^x P_i{}^j(y,E)\,dy \tag{11.64}$$

while for muons one must consider the probability of surviving decay and the energy loss by ionization. This has been considered by several authors, starting with the pioneering work of Sands (1950). The survival probability, defined as the probability that a muon of energy E survives decay in traversing from a depth z to x (arriving with energy $\{E - b(x - z)\}$, where b is the energy loss in traversing unit thickness) is given by

$$\omega(z,E;x) = \exp - \left[\int_z^x \frac{m_\mu}{c\tau_\mu}\frac{dy}{\rho(y)\{E - b(y - z)\}}\right] \tag{11.65}$$

From this it is easily shown that, for an exponential atmosphere with scale height h_0 (e.g., in a vertical direction), the probability that a muon of energy E_z produced at z survives to a depth x, arriving with energy $E = E_z - b(x - z)$, is

$$\omega(z,E_z;x) = \left\{\frac{z}{x}\left[1 - \frac{b(x - z)}{E_z}\right]\right\}^{\frac{\epsilon_\mu}{E_z + bz}} \tag{11.66}$$

$$\epsilon_\mu = \frac{h_0 m_\mu}{c\tau_\mu} \simeq 1.12 \text{ Bev} \tag{11.67}$$

Thus the muon flux is

$$\begin{aligned} F_\mu{}^j(x,E) = \int_0^x dz\, \omega[z, E \\ + b(x - z); x]P_\mu{}^j[z, E + b(x - z)] \end{aligned} \tag{11.68}$$

Detailed expressions for this have been derived by Pal and Peters (1964), and the calculated muon flux over an energy range extending from 0.5 to $\sim$500 Bev for vertically incident muons is given in Fig. 11.31. In the low-energy region the spectrum is corrected for the contribution of nonisobar pions. The experimental data are obtained from magnetic spectrograph measurements of Hayman and Wolfendale (1962) and Gardener et al. (1962). The calculated spectrum is normalized to the experimental data at 40 Bev; this fixes the value of $\langle B\rangle = 3.35 \times 10^{-2}$ if the interaction mean free path λ is chosen as 75 g/cm².

The integral energy spectrum of muons of very high energy is deduced from the depth intensity curve of cosmic-ray particles below ground (Sec. 7.10). This is plotted in Fig. 11.30. It is found that the 100-Bev

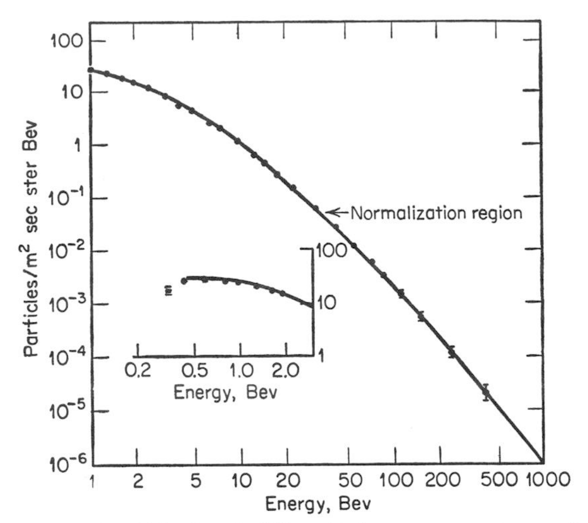

FIG. 11.31. Calculated differential energy spectrum of vertical muons at sea level up to 500 Bev with experimental points. [*After Pal and Peters* (1964).]

flux so obtained is about 40 per cent greater than the Hayman and Wolfendale magnetic spectrograph measurements and the normalization requires $<B> = 4.4 \times 10^{-2}$. The reason for the discrepancy is not clear.

The $<B>$ factor being assigned, it is possible to calculate muon and neutrino fluxes at any zenith angle, using the proper density distribution of the atmosphere at that angle. Such calculations for muons have been made by Jakeman (1956), Ashton et al. (1966), and Cowsik et al. (1966). The agreement between the measured fluxes and calculation indicates that up to several hundred Bev the contribution of muons arising from kaon decay is not large (<30 per cent); kaons have a shorter mean life than pions and their relative contribution at high energy in the vertical direction would be higher than at large zenith angles where competition between decay and interaction is not serious even for pions.

Thus by a comparison with experiments it has been determined that

$<\eta^\gamma> = 0.37$ (from nucleon flux)
$<B> = 3.35 \times 10^{-2}$ to 4.4×10^{-2} (from muon flux and incident nucleon flux)
$Sn_B = 2.37$ [from the multiplicity relation Eq. (11.9)] (11.69)

From this it can be deduced (Pal and Peters, 1964; Pal, 1963) that the secondary spectra could all be explained if, on an average, ~2.4 mesons (charged and neutral) are produced in a collision, each carrying 8 to 10 per cent of the initial energy, along with a nucleon carrying ~55 per cent of the initial energy, the rest of the 15 to 20 per cent of the energy being given to the fireball, which does not contribute significantly to the steady flux. The nearly constant value of the charge ratio of high-energy muons (Sec. 7.8) also suggests that approximately three mesons should be dynamically related with the nucleon, independent of its energy.

7.8. Charge Ratio of Cosmic-ray Muons. Figure 11.32 gives the observed charge ratio μ^+/μ^- for the vertically incident flux of muons at sea level. It is seen that this ratio is ~1.25 and does not reduce with increasing energy. This ratio can be calculated accurately in terms of the charge composition $\delta_0 = [(p - n)/(p + n)]_0$ of the incident nucleon flux and the charge composition

$$<\delta_\pi>_p = (\pi^+ - \pi^-)/(\pi^+ + \pi^-)$$

of the fast pions (which may be called isobar pions) for a proton collision with an air nucleus. From charge symmetry

$$<\delta_\pi>_n = - <\delta_\pi>_p$$

Since muons above a few Bev must arise predominantly from pions produced very high up in the atmosphere, their charge composition $<\delta_\pi>$, in the first approximation, is given by

$$\frac{\mu^+ - \mu^-}{\mu^+ + \mu^-} = \delta_0 <\delta_\pi> p \tag{11.70}$$

For $\mu^+/\mu^- = 1.25$ and $\delta_0 = 0.74$,

$$\frac{\pi^+}{\pi^-} = 1.35$$

being the average charge ratio of pions in the fast group of pions produced in proton interactions.

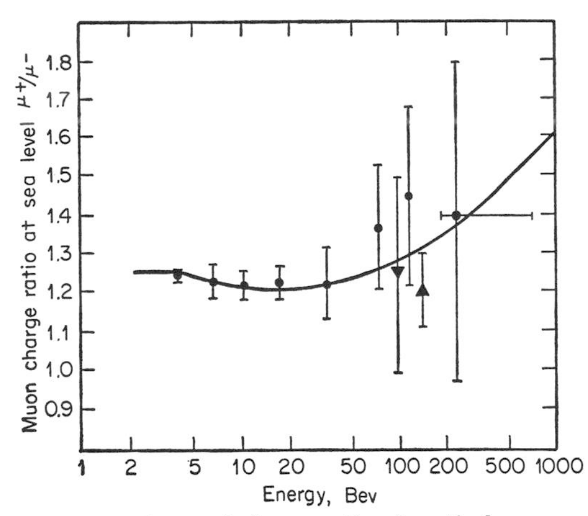

FIG. 11.32. Measured charge ratio of vertical muons at sea level with a calculation for a production ratio $K/\pi \sim$ 20 per cent for kaons to pions. [*After Pal and Peters* (1964).]

If one assumes the invariance of the properties of this group, one may expect the ratio μ^+/μ^- to remain constant with energy. However, the energy dependence of this ratio is modified by the following considerations:

1. Some muons are produced in the lower atmosphere where the charge composition of nucleons is more symmetric.

2. In the low region of the atmosphere, pion production by pions becomes important and contributes symmetrically to the positive and negative muons.

3. At high energies the relative contribution of

charged kaons becomes more important and, because of the large K^+/K^- ratio, tends to increase the μ^+/μ^- ratio (Sec. 6.3).

The theoretical curve in Fig. 11.32 takes all these facts into account and corresponds to a $(K/\pi)_{\text{total}}$ production ratio of ~20 per cent. From this fit one deduces that $\pi^+ - \pi^- = 0.35 \pm 0.15$ among the fast group of pions and that the upper limit on the production ratio $(K/\pi)_{\text{total}}$ is ~40 per cent for primary energies of several thousand Bev [see Pal and Peters (1964) and Pal (1963).]

7.9. Slow Neutrons in the Atmosphere. Low-energy neutrons in the atmosphere are produced mainly in disintegrations of atmospheric nuclei, except for those that arrive as parts of heavy nuclei or perhaps some that may be produced in interactions in the photosphere of the sun with sufficient energies ($E > 20$ Mev) to escape decay in their passage to the earth. The theory of diffusion of low-energy neutrons in the atmosphere was first developed by Bethe, Korff, and Placzek (1940). Recent calculations and a review of experimental data have been presented by Lingenfelter (1963) and Lingenfelter et al. (1965).

About 90 per cent of the neutrons arise from the evaporation process and have an energy spectrum peaked at about 1 Mev. The remaining 10 per cent are knock-on neutrons whose energy may extend up to ~1 Bev. Neutrons of energy in a few-Mev range are slowed down to thermal energies, acquiring a $1/E$ spectrum, by elastic collisions with air nuclei in distances (~50 to 100 g/cm²) small compared with the attenuation mean free path of the star-producing radiation. Most of them are finally captured in N^{14} or, near the top of the atmosphere, escape from the earth. Since the capture cross section for the reaction $N^{14} + n \rightarrow C^{14} + p$ has a $1/v$ dependence, the maximum in the energy spectrum of neutron shifts from thermal energies (0.025 ev) to about 0.1 ev.

Thus low-energy neutrons in the atmosphere have an approximately Gaussian spread about their origin point with a width less than ~100 g/cm². Hence their distribution in the atmosphere, as a function of latitude and altitude, is similar to the distribution of low-energy stars in which they are produced, except above 200 g/cm² where leakage into space is important. This has been experimentally demonstrated [see, for example, Simpson et al. (1951)].

Measurements of neutron density in the atmosphere are usually made by making use of the reaction $B^{10}(n,\alpha)Li^7$. $B^{10}F_3$ proportional counters, ionization chambers, and borated scintillators are used along with a counter using B^{11} instead of B^{10} to eliminate the charged-particle background. The counting rates yield a measurment of the density of neutrons because the cross section for neutron capture in B^{10} is inversely proportional to neutron velocity. To measure the flux of high-energy nonthermal neutrons the counters are surrounded by moderators of low-atomic-weight material. However, because of local production in the material of moderators, these experiments give only upper limits to the neutron fluxes.

Using results of various experimental measurements, Lingenfelter (1963) has summarized the data on the rate of neutron production at various altitudes and latitudes at the time of solar minimum. This is shown in Fig. 11.33. The production rate is normal-

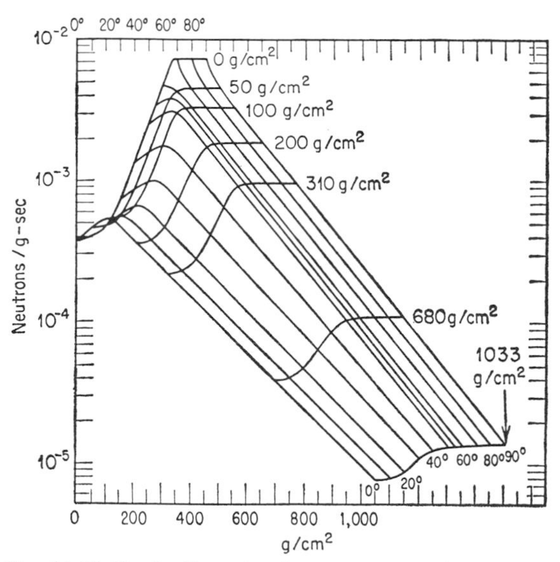

FIG. 11.33. Production rate of neutrons at various depths and latitudes in the atmosphere at the time of solar minimum, normalized to a total production rate of one neutron per square centimeter column of air per second at the geomagnetic pole (Lingenfelter, 1963).

ized to a total production rate of one neutron per square centimeter column of air per second at $\lambda = 90°$. Curves for different latitudes are shifted with respect to each other. The following features can be deduced from these curves:

1. Below 200 g/cm² all curves show nearly exponential absorption, the absorption mean free path varying from 212 g/cm² at 0° geomagnetic latitude (λ) to 163 g/cm² for $\lambda = 50°$. This is to be ascribed to the hardening of the primary spectrum (and hence greater penetrability) in going to lower latitudes.

2. The latitude effect increases with altitude, as expected.

3. Curves for low latitudes go through a maximum which shifts to higher altitudes with increasing latitude till for $\lambda > 60°$ the maximum disappears and the exponent of the curves near 0 g/cm² changes to correspond to the interaction length of nucleons (70 to 80 g/cm²). This latter effect is due to the fact that the majority of the primary nucleons at high latitudes do not have enough energy to produce more than one nuclear interaction.

Lingenfelter (1963) has also given a similar set of curves corresponding to the period of sunspot maximum. Such curves for the star-producing radiation were first given by Lal, Malhotra, and Peters (1958).

Since the energy dependence of the cross section for the reaction $B^{10}(n,\alpha)Li^7$ used for neutron density measurement is the same as that for the reaction $N^{14}(n,p)C^{14}$ responsible for absorption of neutrons, the counting rates of the $B^{10}F_3$ counters are proportional to the production rates of neutrons. The constant of proportionality is $k = \Sigma/A$, where Σ is the cross section in square centimeters per gram for the $N^{14}(n,p)C^{14}$ reaction at a neutron energy of 0.025 ev and A is the effective cross section of the counter in square centimeters at the same energy. Table 11.9 lists the production rates for various latitudes derived

TABLE 11.9. PRODUCTION RATES† FOR VARIOUS LATITUDES FOR PERIODS OF MINIMUM AND MAXIMUM SOLAR ACTIVITY

Geomagnetic latitude, deg	Neutrons produced per square centimeter of earth's surface per second	
	Solar minimum	Solar maximum
0	1.48	1.41
10	1.53	1.46
20	1.85	1.74
30	2.76	2.46
40	4.60	3.73
50	7.03	5.34
60	8.54	6.05
70–90	9.00	6.05
Global average	4.10	3.22

† Derived by Lingenfelter (1963).

by Lingenfelter (1963) for the periods of minimum and maximum solar activity.

Neutrons of energy greater than 0.66 ev can escape the gravitational field of the earth. Protons arising from their decay form an important component of the inner radiation belt (Sec. 10).

7.10. Cosmic Rays Underground. One meter of rock is roughly 3 nuclear-interaction mean free paths and about 12 radiation lengths. Therefore a few meters below ground the atmospheric nuclear and electromagnetic components are attenuated to very low intensities and practically all the particles encountered are muons, neutrinos, and their interaction products.

Measurements of the underground charged-particle intensity have recently been extended to depths of more than 9,000 hg/cm² (1 hg = 1 hectogram = 100 g). A depth-intensity curve for vertically incident particles based on a recent review of experimental data by Menon and Ramana Murthy (1966) is given in Fig. 11.34. (Here all depths have been converted to correspond to $<Z^2/A> = 5.5$, which is considered as representing the "standard" rock for this purpose.) Following Miyake (1963), this curve can be represented by the empirical relation

$$I(h) = h^{-\alpha} \frac{K}{h+H} \exp(-\beta h) \qquad \text{particles/m}^2\text{/sec/ster} \tag{11.71}$$

where h is the depth from the top of the atmosphere in hectograms per square centimeter of rock and the various constants are $\alpha = 1.53$, $K = 1.64 \times 10^{-2}$, $H = 400$, and $\beta = 6.5 \times 10^{-4}$.

In a solid medium the slowing-down time for muons is much shorter than their lifetime for decay. Therefore the attenuation of intensity is entirely due to collisions suffered in passing through matter. Muons do not possess any specific nuclear interaction, as has been confirmed by detailed laboratory experiments on their static and dynamic properties, including an accurate measurement of their gyromagnetic ratio. Thus electromagnetic interaction is responsible for all

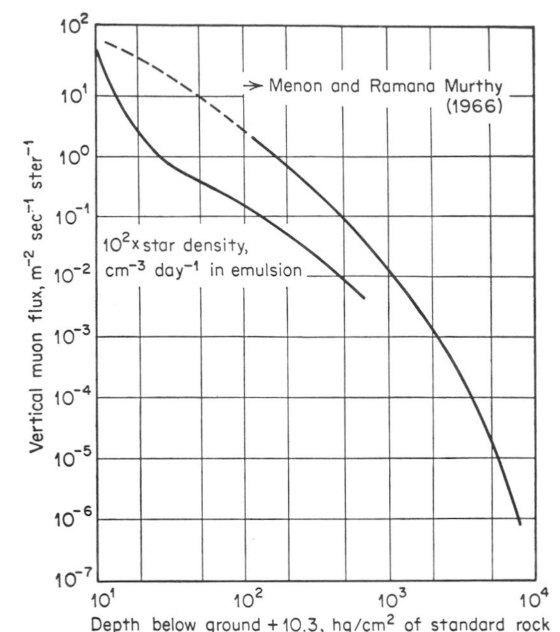

FIG. 11.34. Vertical intensity of muons as a function of depth below ground in hectograms per square centimeter of standard rock. The lower curve gives the star production rate in emulsion.

their collisions and energy transfer processes (weak interactions are relatively unimportant).

Rate of Energy Loss and Range-Energy Relation of Muons in Rock. The electromagnetic energy losses suffered by muons in rock can be divided into two classes as follows:

1. Losses that are nearly independent of muon energy. The losses due to ionization, excitation, and knock-on processes, called collision losses, are in this class. According to Barret et al. (1952), these are given by

$$-\left(\frac{dE}{dh}\right)_{\text{coll.}} = 1.88 + 0.0766 \ln \frac{E'_{\max}}{m_\mu c^2} \qquad \text{Mev cm}^2\text{/g} \tag{11.72}$$

where $E'_{\max} = \dfrac{E^2}{E + m_\mu^2 c^2/2m_e}$ is the maximum energy that may be transferred to an electron.

2. Losses that are proportional to the muon energy. Losses due to pair production, *bremsstrahlung*, and photo-nuclear interactions are in this class. The energy loss due to each of these processes, after Menon and Ramana Murthy (1966), is given below:

$$-\left(\frac{dE}{dh}\right)_{\text{pair}} = 1.58 \times 10^{-6} E \qquad \text{cm}^2\text{/g} \tag{11.73}$$

This is derived from the calculations of Mando and Ronchi (1952a, 1952b).

$$-\left(\frac{dE}{dh}\right)_{\text{brems}} = 1.77 \times 10^{-6} E \qquad \text{cm}^2\text{/g} \tag{11.74}$$

This is based on a calculation due to Bhabha (1938).

$$-\left(\frac{dE}{dh}\right)_{\text{nucl}} = 0.28 \times 10^{-6}E \qquad \text{cm}^2/\text{g} \quad (11.75)$$

which is a value obtained by Fowler and Wolfendale (1958). The contribution of the photonuclear process to the total energy loss is rather uncertain, except that Miyake et al. (1964) have put an upper limit of $0.9 \times 10^{-6}E$ cm²/g on $-(dE/dh)_{nucl}$ from their experimental measurements of muon intensity deep underground.

The relative importance of various energy losses at different muon energies can be directly inferred from Fig. 11.35. It is seen that for muons of energy

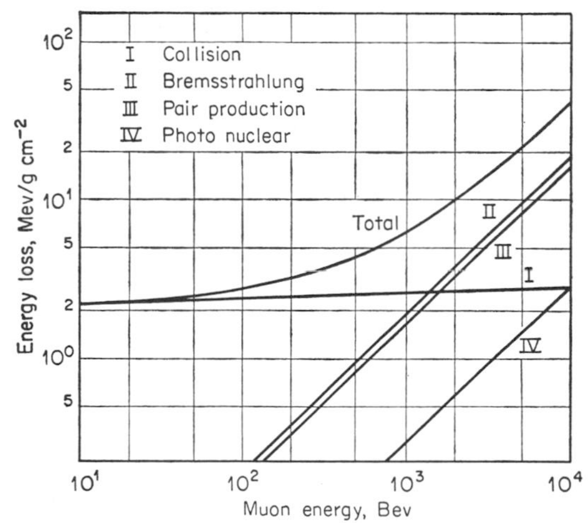

FIG. 11.35. Energy loss of muons per gram per square centimeter of standard rock through various processes as a function of their energy, based on equations given in the text.

$E \leq 800$ Bev the collision losses are dominant. At higher energies the losses proportional to muon energy become increasingly important and are dominant for $E \geq 1{,}500$ Bev.

Summing up all the losses belonging to classes 1 and 2, the average energy loss in standard rock ($\langle Z^2/A \rangle = 5.5$) is given by

$$-\left(\frac{dE}{dh}\right)_{\text{total}} = \left[\left(1.88 + 0.0766 \ln \frac{E'_{\max}}{m_\mu c^2}\right) \times 10^{-3} + 3.6 \times 10^{-6}E\right] \qquad \text{Bev cm}^2/\text{g} \quad (11.76)$$

where E, the muon energy, is expressed in Bev.

Barret et al. (1952) gave a approximate solution of Eq. (11.76) in the form of an average range-energy relation

$$E = \frac{a'}{b}\,[\exp{(bh)} - 1] \qquad (11.77)$$

where $a' = \left[1.88 + 0.0766 \ln \dfrac{E^2}{e m_\mu c^2 (E + eA)}\right] \times 10^{-3}$ Bev cm²/g

$A = 11.3$ Bev

and $b = 3.6 \times 10^{-6}$ cm²/g

At very high muon energies ($\geq 1{,}500$ Bev) the dominant energy losses are those of class 2, which occur in large steps so that the effect of fluctuations becomes very important when constructing an energy spectrum of incident muons from the observed depth-intensity curve. Corrections for this have been evaluated by several authors including Bollinger (1951), Zatsepin and Mikhalchi (1962), Nishimura (1963), Hayman et al. (1962), and Miyake et al. (1964). The energy spectrum of muons deduced by Menon and Ramana Murthy (1966) using the depth-intensity curve (Fig. 11.34) and the range-energy relation [Eq. (11.77)], and applying corrections for fluctuations is given in Fig. 11.30. The following observations can be made on the basis of a comparison of the energy spectrum so derived and several other cosmic-ray measurements.

1. The energy spectrum below 100 Bev is in good agreement with the magnetic spectrograph measurements (within ~40 per cent at 100 Bev), showing that the theoretical evaluation of collision losses is correct.

2. The assumption that γ rays and muons arise only from the decay of neutral and charged pions produced in the ratio 1:2 gives a simultaneous fitting of the muon and γ-ray spectra. Since kaons and pions contribute differently to muons and γ rays, this allows one to conclude that up to primary energies of $\sim 10^{13}$ ev (muon energy ~1,000 Bev) pion production dominates over kaon production. Such a conclusion cannot be extended to higher energies because of large errors on measured intensities of γ rays and muons. Similar arguments can also be used to put an upper limit of 1/100 to the ratio of direct pair production of muons to the production of charged pions of energy $\sim 10^4$ Bev.

3. A requirement of consistency between the sea-level muon spectrum derived from the spectrum of "bursts" produced by muons in ionization chambers and the one obtained by depth-intensity measurements enables one to conclude that the energy losses due to photonuclear or any new type of interactions are not very large. This may also be inferred from a comparison of the γ-ray production spectrum with the muon spectrum above a few thousand Bev deduced from underground measurements.

Secondary Components below Ground. Practically all secondary particles between depths of ~5 hg/cm² and ~8,000 hg/cm² arise from interactions of muons. Below this depth an increasing fraction of the observed charge particles of high energy would be due to interactions of neutrinos produced from the decay of unstable particles in the atmosphere.

The most important secondary component underground is the electron-photon component produced by collisions of muons with electrons and by pair-production and *bremsstrahlung* processes. As the cross section for these processes increases with muon energy, an increasing fraction of muons is accompanied by this soft component as one goes to great depths underground till at a depth of ~1,000 hg/cm² the fluxes of equilibrium electrons and muons become nearly equal.

Muons also produce nuclear interactions in which other nuclear particles, pions, nucleons, etc., can be produced. These interactions do not indicate a specific nuclear interaction for the muon but are produced by the high-frequency photon field surrounding a fast-moving charged particle. Following Williams and Weizsacker (Heitler, 1954) the cross section for this process can be calculated in terms of photonuclear cross sections measured in the laboratory. The equivalent photon spectrum associated with a muon of energy E is (for $k \ll E$) given by

$$f(k)\,dk = \frac{2}{\pi}\frac{e^2}{\hbar c}\ln\left(\frac{E}{k}\right)\frac{dk}{k} \tag{11.78}$$

The cross section for star production by a muon is then

$$\sigma_\mu = \int_{E_m}^{E} \sigma_\nu(k) f(k)\,dk \simeq \frac{\bar{\sigma}_\nu}{\pi}\left(\frac{e^2}{\hbar c}\right)\ln^2\left(\frac{E}{E_m}\right) \tag{11.79}$$

where E_m is the minimum photon energy for producing the star of the type under detection, $\sigma_\nu(k)$ is the photonuclear cross section at energy k, and $\bar{\sigma}_\nu$ is an average photonuclear cross section in the relevant energy region.

The rate of star production according to measurements of George and Evans (1955) and Cocconi and Cocconi-Tongiorgi (1951) at various depths is roughly represented by the curve given in Fig. 11.34 alongside the depth-intensity curve for charged particles. High-energy muon interactions have been analyzed by Higashi et al. (1962, 1963) using cloud chambers. When interpreted in terms of Heitler theory, all the measurements are consistent with the photo nuclear cross sections measured in the laboratory. In particular, Higashi et al. (1963) found the production cross section for showers to be $(2.58 \pm 0.3) \times 10^{-31}$ cm^2/nucleon for lead plates, which reduces to a photonuclear cross section of $(1.4 \pm 0.25) \times 10^{-28}$ cm^2/nucleon to be compared with a cross section of 1.4×10^{-28} cm^2/nucleon for real photons of 1 Bev, for producing one to three mesons, obtained in accelerator experiments. However, there have been some theoretical objections to the theory in its original form, and a complete understanding of the phenomena has not yet been reached (Kessler and Kessler, 1956; Kessler, 1960).

Cosmic-ray Neutrinos and Their Interactions. The flux of various types of neutrinos produced in the decay of unstable particles in the atmosphere can be calculated in the manner discussed in Sec. 7.7. Inasmuch as neutrinos are mostly produced in association with or in the decay of muons their energy spectrum is directly related to the measured muon energy spectrum and is subject to similar experimental uncertainties in addition to that due to a possible error in the K/π ratio assumed for the calculation. Neutrino fluxes have been calculated by Greisen (1960), Zatsepin and Kuzmin (1962), and in greater detail by Cowsik et al. (1963, 1966) and Osborne et al. (1965). The calculated vertical and horizontal differential spectra of neutrinos and antineutrinos of the muon type $(\nu_\mu + \bar{\nu}_\mu)$ can be approximated by power-law expressions:

$$(\nu_\mu + \bar{\nu}_\mu)_V = 450E^{-2.99} \quad \text{per m}^2\text{/sec/ster/Bev}$$
$$\text{and} \quad (\nu_\mu + \bar{\nu}_\mu)_H = 800E^{-2.89} \quad \text{per m}^2\text{/sec/ster/Bev} \tag{11.80}$$

for the energy range 2 to 100 Bev The uncertainty in this is not likely to be more than ~30 per cent. (Horizontal fluxes are larger because the decay of unstable particles is more likely in that direction.)

The known flux of high-energy cosmic-ray neutrinos may be used to study the nature of weak interactions at energies which are far higher than those possible with present accelerators. The feasibility of such experiments depends on being able to use a very large detector in a location with an exceptionally low background of normal cosmic rays because neutrino cross sections are very small (~10^{-38} cm^2/nucleon at 1 Bev). This may be achieved by setting up a large area detector at a sufficiently deep underground location. By increasing the thickness of the rock overburden, the background due to atmospheric muons can be reduced to an arbitrarily small value without reducing the neutrino flux, and the material in the surrounding rock can be used as the target, whose effective volume is essentially proportional to the range of the particles produced in neutrino interactions (Markov and Zhelezhnykh, 1961). The effective target volume is small for interactions of ν_e and $\bar{\nu}_e$ which produce electrons in their interactions but is large for ν_μ and $\bar{\nu}_\mu$ (because the resulting muons are penetrating particles) and increases linearly with neutrino energy. Thus if the incident differential flux of $\nu_\mu + \bar{\nu}_\mu$ is of the form $E^{-(\delta+1)}$ and the neutrino cross section varies as E^n up to an energy E_c then the counting rate will be proportional to

$$\int_1^{E_c} [E^{-(\delta+1)} \times E^n \times E]\,dE \sim \frac{E_c^{\,n-\delta+1} - 1}{n - \delta + 1} \sim \ln E_c \quad \text{for } (n - \delta + 1) = 0 \tag{11.81}$$

Experiments of Miyake et al. (1964) at a depth of 8,400 hg/cm^2 of rock demonstrated that suitable conditions for neutrino experiments exist at such depths and that charged particles arriving at large zenith angles would be exclusively those produced by neutrino interactions. Following this, two large experiments to detect cosmic-ray neutrinos have been set up; in both of them interactions of high-energy natural neutrinos have been observed (Achar et al., 1965; Reines et al., 1965). In order to eliminate the accidental counts due to radioactivity and instrumental noise, multifold coincidences are used. The flux of neutrino-induced muons is of the order of 0.6×10^{-12} to 0.9×10^{-12} particles/cm^2/sec/ster, and the total number of counts is still too small to draw any definite conclusions about the energy dependence of neutrino cross section.

These experiments open up a new branch of investigation which, like many other cosmic-ray studies, may be linked simultaneously with phenomena of very small and very large distances; it is likely that, besides providing valuable information about the form factor of weak interaction at small distances, such experiments may discover possible high-energy neutrino fluxes from cosmic sources, such as supernovae, quasi-stellar sources, and other active objects in the sky. When realized, this will be an extremely valuable new window to the external universe.

8. Extensive Air Showers

Nuclear-interaction mean free path in air (75 g/cm^2) being much smaller than the total thickness of the atmosphere (~1,030 g/cm²), a high-energy particle incident on top of the atmosphere can lose its energy in successive collisions as it propagates down. High-energy nuclear particles produced in each collision start their own nuclear cascades. About one-third of the radiated energy in each collision is given to neutral pions, which decay into γ rays, each of which initiates its own electromagnetic cascade (Sec. 7.5) which develops up to a maximum and then decays. For charged pions the decay process competes with nuclear interaction; the energy at which the decay probability is comparable to the interaction probability is ~10 Bev at sea level and varies inversely as the atmospheric density at higher altitudes. Thus the decay of charged pions results in muon (and neutrino) production, their mean height of production increasing with their energy. The muons come down to sea level after very little attenuation, owing to collision losses and decay. This process of energy transfer from the nuclear component in the core to the photon-electron component and the muon-neutrino component continues, with the consequent growth in the total number of particles in the family until, at a certain point, the electromagnetic component begins to die off because the energy transfer from the core to the electron component becomes inadequate to sustain its energy loss by ionization and the muon component reaches a saturation value, because high-energy ($E \gg 10$ Bev) muons cannot be produced in the lower atmosphere and low-energy muons ($E \sim 1$ Bev) are attenuated by decay and ionization loss. In traversal through the atmosphere the particles in the shower spread to great distances, the electron-photon component because of scattering and the muon component because of the angular spread acquired at production. Thus the arrival of a high-energy particle on top of the atmosphere leads to the production of a disk-shaped region of particles which grows as it propagates down with very nearly the velocity of light until, at a depth that increases with primary energy, its lateral extension and the number of particles attain their maximum values, depending on the initial energy. For a primary particle of 10 Bev the maximum number of particles in the family is the order of 10, and very few of them survive up to sea level. On the other hand, for a primary energy of 10^6 Bev, for example, the number of particles at sea level is $\sim 10^5$. This phenomenon then results in a nearly simultaneous incidence of particles at points widely distributed in space and is called an extensive air shower (EAS).

As already discussed (Sec. 5.2) the flux of primary cosmic-ray particles falls off steeply with increasing energy. For example, above 10^{16} ev one expects only ~1 particle/m²/ster/year. Therefore the practical problem of detecting and studying this high-energy radiation would have been insurmountable if the flux were not first intercepted by several interaction mean free paths of a tenuous material in the form of the terrestrial atmosphere which disperses the high energy concentrated in a single-incident particle over an area of several squares of kilometers, thus increasing the effective area of the detector by many orders of magnitude. The ultimate aim of air-shower experiments is to make enough relevant measurements on various shower components to determine (1) the character of interactions at high energies and (2) the nature, energy, and arrival directions of the initiating particles. The second of these objectives is important because only at air-shower energies may the charged particles retain some rough sense of their original directions in traversing through interstellar magnetic fields and thus may provide the only sample of matter from regions outside our galaxy.

8.1. General Features of Air Showers. Air showers are usually studied with large arrays of detectors connected in coincidence and spread over distances ranging up to many kilometers. While most of the detectors are unshielded charged-particle detectors, special instruments to detect muons and nuclear components are also used. About 90 per cent of the charged particles near the maximum of shower development are electrons, about 6 to 9 per cent are muons, and the rest, about 1 per cent, are strongly interacting particles. Therefore the over-all features of the air shower are dominated by the electromagnetic cascade. In spite of the fact that this cascade is a superposition of a large number of cascades initiated by individual γ rays generated along the length of the core of the air shower, its general appearance is not very different from that of a pure electromagnetic cascade with a suitably chosen age parameter S (Sec. 7.5).

Near the core of the shower the fraction of electrons may be as high as 98 per cent. Moving away from the core, the relative number of muons increases to become nearly equal to the number of electrons at a distance of a few hundred meters. Beyond this distance muons are the dominant charged particles. The lateral distribution of electrons is approximated by Nishimura-Kamata distributions given in Fig. 11.29 with the following age parameters S [see, for example, Zatsepin et al. (1963)]:

	Shower size (number of shower particles)	S
Sea level...........	$<10^8$	1.2–1.3
	$\geq 10^9$	~1
Mountain altitude..	$<10^7$	1.25
	$>10^7$	~1

The lateral distribution function steepens (i.e., becomes narrower) with size and altitude above sea level at a rate much slower than would be expected for a pure electromagnetic cascade. Also the absorption of the shower beyond the maximum occurs with a mean free path of ~200 g/cm² more or less independent of size. These slow variations are consequences of the fact that the electromagnetic component is generated by the nuclear component which penetrates deep into the atmosphere.

The energy spectrum of electrons is determined mainly by the electromagnetic processes and hence is similar to that expected for a pure cascade of the same age. (The energy spectrum near the core corresponds to that which would be obtained in pure electromagnetic showers of slightly greater age.)

8.2. Muon Component. Figures 11.36 and 11.37 give, for a shower size of 10^6, the schematic growth curves and the lateral distributions for various components at sea level. It is seen that, unlike electrons, the muon component reaches a near saturation value and does not attenuate significantly after that.

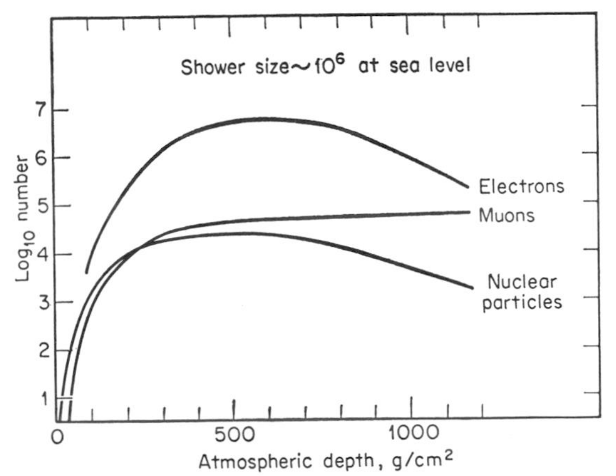

Fig. 11.36. Schematic growth curves for various components in an air shower of size $\sim 10^6$ at sea level.

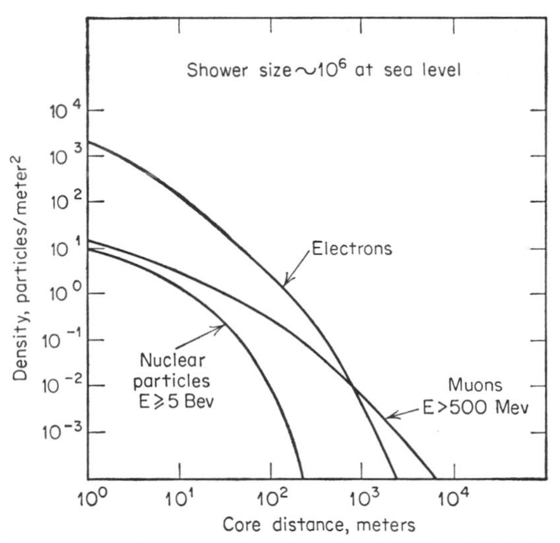

Fig. 11.37. Schematic representation of lateral distributions at sea level of principal components of an air shower of size $\sim 10^6$.

The lateral spread of ~4-Bev muons near sea level, for example, is ~300 m; this is much larger than a spread of ~50 m that would be produced by multiple scattering if these muons are mostly produced at a height of ~10 km. On the other hand, a lateral distribution calculated on the assumption of an exponential transverse momentum distribution [Eq. (11.11)] for parent pions at the point of production gives a fair representation of the experimentally observed spread of muons, if the average value of the transverse momentum is taken the same as measured at low energies. This does not yield any information about the character of interactions above 10^{12} ev, because the majority of the muons and nuclear particles observed in air showers are produced by primaries whose energy is lower than this.

The integral energy spectrum of muons is more or less independent of shower size (size N is essentially given by the total number of electrons in the shower) and is $\sim E^{-1.2}$ in the energy range of tens of Bev. Recent measurements, using deep underground muon detectors in association with air showers, show a slightly steeper spectrum $\sim E^{-1.5\pm0.1}$ for muons of 200 to 600 Bev in showers of size $N \simeq 10^5$ (Chatterjee et al., 1965a).

The number of muons in a shower as observed at a certain altitude is seen to increase as $\sim N^\beta$, where β is more than 0.5 and less than 1. There are two reasons for β being less than unity. First, probably less important, with increasing energy the average pion energy increases slightly with a consequent decrease in the decay probability. Another reason is that showers of size less than $\sim 10^8$ reach maximum above sea level and the position of the maximum moves down with increasing energy; thus the number of muons at the observation level may increase nearly proportionally to the primary energy while the size or number of electrons increases slightly faster. Experimentally it is found that $N_\mu \sim N^{0.65}$ for muons of energy greater than about 1 Bev (Matano et al., 1963). This means that the ratio of muons to electrons increases with shower size at the level of observation. If a shower of size N produced by a heavy nucleus of mass number A can be considered as a superposition of A showers of size N/A each, the ratio of the number of muons in a heavy-particle-produced shower to that in a proton-produced shower of the same size will be $A^{1-0.65}$. Thus showers with an abnormally large fraction of muons, or the so-called mu-rich showers, have sometimes been ascribed to heavy primaries.

Although some fluctuations in the number of muons for a fixed shower size must be due to the fluctuations in the level of the first collision in the atmosphere, a significant part of them may again be due to the complex composition of the incident primary radiation. Linsley and Scarsi (1962) have observed that for sizes larger than 10^8 these fluctuations become very small, indicating that the primary radiation becomes relatively pure, containing perhaps only protons.

Suga et al. (1963), Hasegawa et al. (1965), and Gawin et al. (1963) have been able to demonstrate the existence of a separate group of showers in the size range 10^5 to 10^6, containing an abnormally low number of muons. Although they may also be due to extreme fluctuations in the first interaction in the atmosphere, their separation from the main group suggests that they are produced by primary γ rays whose absolute intensity above an energy 2×10^{15} ev is $\sim 5 \times 10^{-10}$/m²/ster/sec. or $\sim 5 \times 10^{-4}$ times that of normal cosmic rays.

However, no such separate group of "low-mu" showers has been observed at energies greater than 10^{16} ev; since γ rays of this energy can be produced in collisions of protons of energy greater than 10^{18} ev with starlight, their absence can be used to put an upper limit of 0.01 ev/cm³ on the energy density of thermal photons in the intergalactic space, provided the intergalactic proton intensity above 10^{18} ev is assumed to be the same as that observed near earth (Hasegawa et al., 1965).

8.3. Nuclear Component. The nuclear component acts as the backbone of the air shower. It is found that high-energy nuclear particles are confined very close to the core, as expected from the smallness of the transverse momentum. The low-energy nuclear component is studied with neutron detectors. Danilova and Nikolsky (1963) find a relation $N_n \sim N^{0.7}$ for shower sizes 4×10^3 to 10^7 observed at an altitude of 3,330 m when the minimum energy of the nuclear particles is about 200 Mev. [At a depth of 800 g/cm^2, Chatterjee et al. (1963) find a dependence $N_n \sim N^{0.65 \pm 0.05}$ for nuclear particles of energy greater than 1 Bev within 50 m from the core for showers of size 2×10^5 to $\sim 10^7$.] Thus nuclear particles above 200 Mev are $\sim$1.2 per cent of N in showers of size 10^6 and $\sim$7 per cent of N for a size 4×10^3. The integral energy spectrum of the nuclear particles from 200 Mev to 3 Bev is $\sim E^{-0.45}$. From an analysis of over-all data it appears that variation of N_n with size cannot be represented well by a single power law. At a shower size of $\sim 10^5$, N_n varies very slowly with size while the variation becomes faster on both sides of $N \sim 10^5$. Thus above $N \sim 3 \times 10^5$ the number of nuclear particles increases almost linearly with size.

The lateral distribution of low-energy nuclear particles may be expressed as $\sim r^{-\alpha_n}$ for a large range of shower sizes. At 800 g/cm^2 for nuclear particles lying between 50 m of the core, the value of α_n increases gradually from 0.65 for a size of 5×10^4 to 2.3 for a size of 2×10^7 (Sreekantan, 1963). A schematic representation of the lateral distribution for a shower of $N \sim 10^6$ at sea level is given in Fig. 11.37.

The existence of moderate-energy (a few tens of Bev) nuclear particles, some of them neutral, has also been established in air showers. In view of relation (11.17), it can be shown that these particles cannot be the recoil nucleons from high-energy interactions, because such nucleons emerge with barely relativistic energies. Therefore it was conjectured that a significant fraction of nuclear particles of several Bev may be nucleon-antinucleon pairs created in high-energy interactions. The presence of such particles has recently been demonstrated by detecting energetic nuclear particles arriving with large delays with respect to the air-shower front; such particles cannot be pions because at the observed energies they cannot be substantially delayed (Damgaard et al., 1965; Chatterjee et al., 1965b; Pal and Tandon, 1965a). Such experiments have also attempted to detect the production of hypothetical particles of mass much greater than the nucleon, so far without any success.

8.4. Energy Dissipation in Air Showers. Several calculations have been made to relate the shower size at a given depth with the primary energy. It has been found that the conversion factor from size to primary energy is more reliably estimated for showers at their maximum of development, where this factor is also independent of size. Calculations of Nikolsky (1962) and Greisen (1960) indicate a relation $E = 1.9 \times 10^9 (N_{\max})$ ev. After a detailed investigation Zatsepin, Nikolsky, and Khristiansen (1963) have given a breakdown of energy carried by various components and the energy expended in ionization for showers of size 3.5×10^5 observed at Pamir (altitude $\sim$4 km) where they are near their maximum of development. This is given in Table 11.10. Thus it is seen that the bulk of the energy is expended in ionization in the atmosphere.

TABLE 11.10. ENERGY IN UNITS OF 10^{14} EV FOR A SHOWER SIZE = 3.5×10^5 AT $\sim$4 KM

Ionization above	Electrons and photons	Nuclear particles	Muons	Neutrinos	Total
(3.4 ± 1)	$(0.75 \pm .16)$	$0.3 \begin{smallmatrix} +0.3 \\ -0.2 \end{smallmatrix}$	$0.9 \begin{smallmatrix} +0.9 \\ -0.2 \end{smallmatrix}$	~ 0.65	$6 \begin{smallmatrix} +1.8 \\ -1.1 \end{smallmatrix}$

8.5. Core Structure of Air Showers. It is found that the density distribution of particles very close to cores of air showers has no unique structure. The Sydney group has shown that, for a fixed shower size below 10^6, the maximal central density Δ_c of the soft component varies over a broad range, the ratio of the largest to the smallest density being

$$\frac{\Delta_c(\max)}{\Delta_c(\min)} \approx 60 \sim \frac{\text{atomic weight of Fe}}{\text{atomic weight of H}}$$

$\Delta_c(\max)$ and $\Delta_c(\min)$ are found to increase linearly with size (McCusker, 1963). Both the electron component and the high-energy nuclear component are structured into cores whose number varies from one to many. The single-cored showers have a central particle of highest energy and around it the energy and density of particles fall radially. The multicored showers contain a number of high-energy nuclear particles of comparable energy separated from each other. For a given shower size the energy of the most energetic nuclear particle in a single-cored shower considerably exceeds that in the multicored showers. All these features tend to support the conclusion that the multicored showers are produced by heavy primaries, the separation between the cores being due to the relative transverse momenta among its nucleons after a few collisions.

A very significant observation is that the fraction of multicored showers increases very fast beyond a size of $\sim 10^6$ (at sea level). Actually it is found that at sizes near 10^5 to 10^6 air showers exhibit an anomalous behavior with respect to several properties: Size spectrum becomes steeper, lateral distribution of various components changes, the energy spectrum of the electron component becomes softer, dependence of N_n on N shows a kink, and multiple cores become dominant. All these features may be understood in terms of a model of the primary cosmic-ray spectrum originally proposed by Peters (1959) and later discussed in various forms by Zatsepin et al. (1963), Chatterjee et al. (1965a), and McCusker (1963). A version of this model, which is also consistent with the energy per nucleon spectrum as exhibited in the γ-ray spectra at various altitudes (Sec. 7.6), would involve a rigidity cutoff at about 2×10^{14} ev above which the relative contribution of heavy particles to air showers increases, the average mass of the primary increasing with the shower size. A second component of cosmic rays, perhaps of extragalactic origin, is then needed to account for particles of total energy $\gtrsim 10^{16}$ ev which, according to Linsley and Scarsi (1962), may consist only of protons. Although plausible, this picture of the primary energy spectrum and composition is not yet definitely established.

9. Time Variations; Solar Influences

9.1. Variation of Cosmic-ray Intensity in Time. Systematic measurements of cosmic-ray intensity have been carried out for the past 40 years. Until recently continuous measurements were possible only with ground-based equipment. Thus at least a part of the observed changes in intensity were due to changes in meteorological conditions. Charged-particle detectors at sea level used in early measurements were sensitive mainly to the low-energy muons and the equilibrium component of electrons. Duperier (1949) showed that the change in intensity of the muon component on the ground is represented by a relation of the type

$$\delta I = \beta_1 \, \delta B + \beta_2 \, \delta H + \beta_3 \, \delta T$$

Here δI is the percentage deviation from the average counting rate on the ground, δB the deviation from the average atmospheric pressure, δH the deviation from the average in the height of the 100-mbarn pressure level, and δT the variation from the average of some suitable mean atmospheric temperature. The dependence on the atmospheric pressure is connected with the increased absorption of the muons, while dependence on H and T arises from the change in survival probability of muons due to the change in the density distribution of the atmosphere. Under sufficient amounts of shielding material (≥ 400 g/cm^2) the atmospheric-intensity variation can be represented by the following constants:

$$\beta_1 = -0.13\% \text{ per mbarn}$$
$$\beta_2 = -4.8\% \text{ per km}$$
$$\beta_3 = +0.04\% \text{ per C}^\circ$$

The charged-particle detectors on the ground respond to muons of a few hundred Mev which were produced with sufficient energy to penetrate the atmosphere (≥ 2 Bev) and, therefore, reflect intensity variations in the primary component of 10 to 15 times that energy. Over 10 years ago Simpson, Fonger, and Treiman (1953) introduced neutron detectors as indicators of the intensity of the local nuclear active component which is correlated with the primary cosmic-ray intensity. Of the various meteorological variables only the pressure changes affect the neutron-monitor counting rate (the pressure coefficient at sea level is ~ 0.72 per cent per millibarn). Since the intensity of low-energy nucleons on the ground is not linked with the primary intensity through any unstable particle, a detailed knowledge of atmospheric density distribution is not required to make a correction for the meteorological effects.

Use of large neutron monitors and, over recent years, of direct intensity measurements of primary cosmic-ray intensity of various components has helped to establish four general phenomenological types of variation. They are the 11-year variation which is related to the 11-year cycle of solar activity, the 27-day recurrence which is connected with the rotation of the sun around its axis, the diurnal (and maybe a semidiurnal) variation connected with the rotation of the earth in the extended solar corona, and a Forbush-type variation connected with short-lived solar activity on the sun. All these variations are thus ascribed to the solar influence on the galactic intensity. As a general rule, except for the diurnal variation, the intensity on earth is inversely correlated with the intensity of solar activity.

Various authors have looked for variations with the period of one sidereal day; these would be related to a possible anisotropy in the galactic radiation. No sidereal variations have so far been established (amplitude of anisotropy for total cosmic rays ≤ 0.02 per cent, for particles of energy greater than 10^{15} ev ≤ 0.3 per cent). To some extent this is surprising because, as pointed out by Compton and Getting (1935), even if the radiation were completely isotropic throughout the universe the rotation of our galaxy should result in an increased flux incident on the parts of the earth that face the direction of motion. [The expected value of this variation has been calculated by Vallarta, Graef, and Kusaka (1939) as 0.17 per cent near the equator.] This may mean that the majority of cosmic-ray particles (which are of low energy, about a few Bev) are embedded in the local interstellar magnetic fields which share the motion of the galaxy.

Eleven-year Cycle. Extensive measurements over the past 20 years have established the existence of the 11-year cycle for long-term cosmic-ray variations. This is clearly seen in Fig. 11.7 where the change in the Mount Washington neutron-monitor rate and the Zurich sunspot number have been plotted for the latest solar cycle. The counting rate of an ion chamber operated at 10 g/cm^2 over Minneapolis between 1957 and 1964 shows a similar behavior. The neutron monitor is sensitive to a mean primary energy of ~ 15 Bev while the ion chamber is sensitive to a mean energy of 3 Bev. The amplitude of variation is ~ 20 per cent for the neutron monitor and ~ 45 per cent for the ion chamber, reflecting the larger percentage change in the lower-energy component.

The cosmic-ray intensity lags behind the change in the sunspot number. This "hysteresis" lag seems to be of the order of 12 months. The existence of "hysteresis" shows up clearly in Fig. 11.38 (Webber, 1965), where the neutron-monitor and the ion-chamber data are plotted directly against the Zurich sunspot number. The hysteresis gap provides a rough time constant for buildup and decay of the electromagnetic and plasma conditions in the interplanetary space which modulate the galactic cosmic-ray intensity.

The important features of the 11-year variation have recently been discussed by Webber (1965), Freier and Waddington (1965), and several other workers. Figures 11.8 and 11.10 show how the spectra of different components change with the degree of modulation. Important aspects of observations in this respect may be summarized as follows:

1. The cosmic-ray intensity is inversely correlated

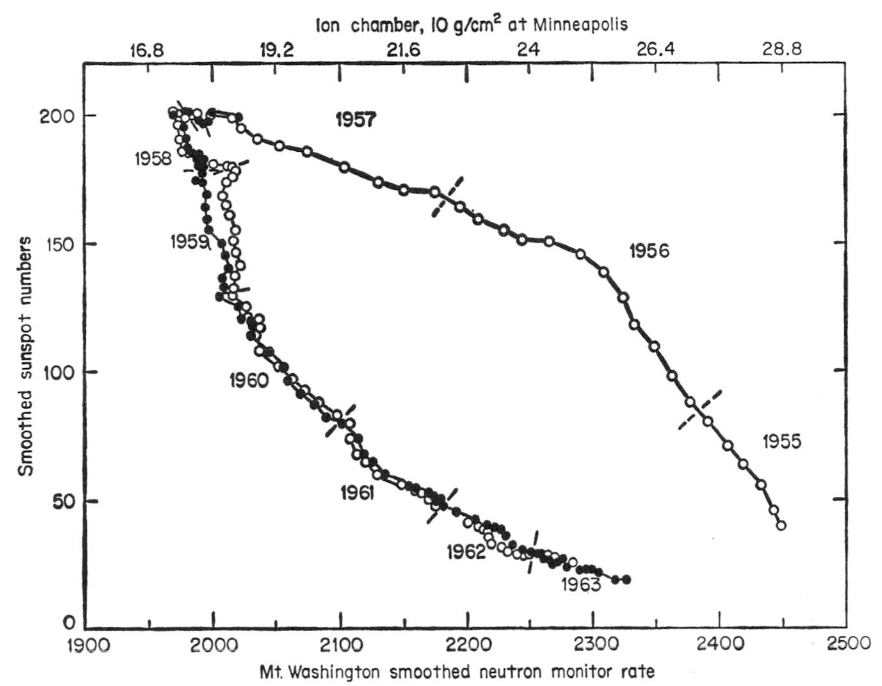

Fig. 11.38. Variation, over the latest solar cycle, of the Mount Washington neutron-monitor counting rate (open circles) and the Minneapolis ion-chamber counting rate (black circles) with Zurich sunspot number. The figure shows the phase lag between solar activity and the cosmic-ray intensity. [*From Webber* (1965).]

with the solar activity; changes in the cosmic-ray intensity have a phase lag of the order of a year with respect to the solar cycle.

2. The shape of the observed spectrum depends only on the total intensity as indicated by large neutron monitors.

3. Solar modulation is very large at low momenta and small at high momenta ($\sim$30 BV).

4. Subject to some uncertainty, the change in the spectra of protons and helium nuclei suggests that the modulation is better described in terms of energy or velocity rather than rigidity (Webber, 1965).

5. As the modulation decreases the maximum in the intensity of various components shifts toward lower energies (for example, the proton spectrum showed a maximum at $\sim$1.5 BV at the time of solar maximum in 1958 and at $\sim$0.9 BV at the time of solar minimum in 1964).

Twenty-seven-day Recurrence. The 27-day recurrence of maxima and minima in cosmic-ray intensity has now been firmly established. This recurrence which follows the same period as fluctuations in the geomagnetic field is obviously connected with the rotation of the sun; the configuration of the active regions on the sun with respect to the earth sun line will repeat with this period. Simpson (1954) demonstrated that the variation is not connected with the changes in geomagnetic cutoff rigidities by observing changes in neutron-monitor rates which could not be produced by admittance or exclusion of only low-energy particles. Figure 11.39 shows clearly the existence of this recurrence period. The intensity of variation depends, to some extent, on the phase of the solar cycle when the observations are made. Typically the amplitude of this variation is $\sim$1 to 2 per cent at sea level and about 10 per cent at the top of the atmosphere.

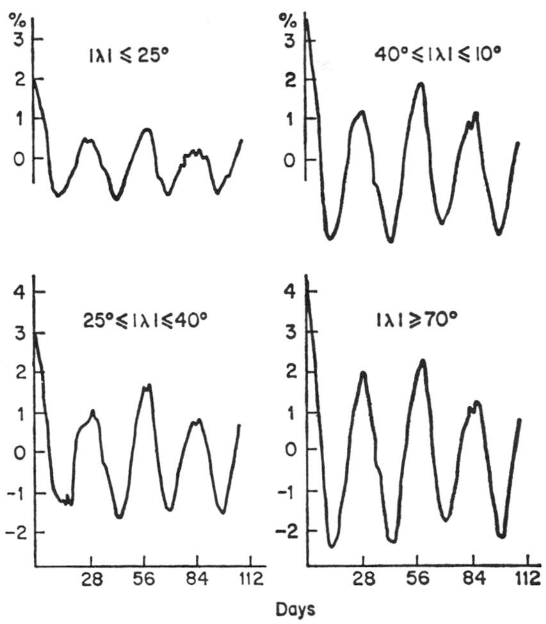

Fig. 11.39. The 27-day variation of cosmic-ray intensity at various geomagnetic latitudes according to the analysis of Dorman and Shatashvili (1961), using neutron-monitor data during July–December, 1957, from 38 stations.

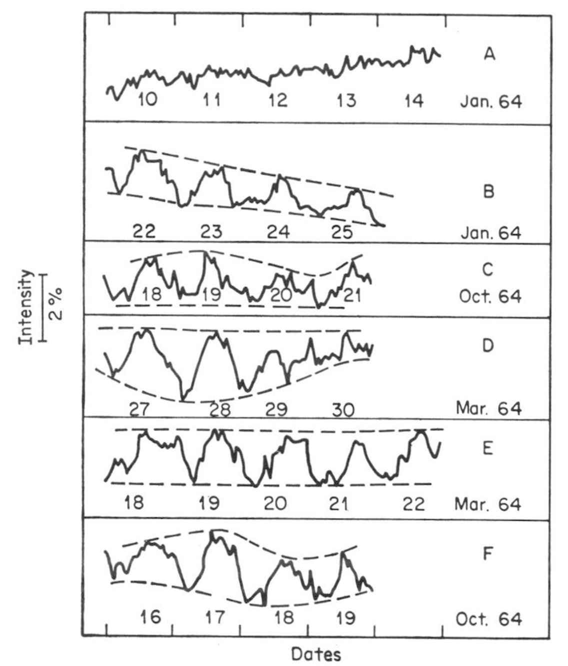

FIG. 11.40. Some sample patterns of diurnal variation in the year 1964. The curves show the deviations in the averaged rate for neutron monitors at Sulphur Mountain, Climax, and Calgary in North America. [*From Kane* (1965).]

Diurnal Variations. Figure 11.40 shows an analysis of diurnal variations as observed by superneutron monitors (Kane, 1965). It is seen that the maximum occurs at ~1800 hr solar time and the minimum at ~0600 hr. Harmonic analysis also yields a semidiurnal component in the variation. The amplitude of diurnal variation is typically ~0.6 to 0.75 per cent.

9.2. Solar Flares. During periods of sunspot maximum, events of intense activity and short duration occur in the solar corona. These are accompanied by enhanced emission in optical, ultraviolet, X-ray, and radio wavelengths. The flares can usually be observed and localized on the solar surface and are found to arise among groups of sunspots.

The solar flares are usually related with a host of geophysical phenomena. Some of the solar flares are followed within 1 to 10 hr by radio fade-outs in the polar regions and a sharp increase in the absorption of cosmic radio noise as measured by riometers (radio ionospheric opacity meters). These so-called polar-cap absorption events (PCA) are believed to be due to intense ionization at a height of 60 to 80 km in the D layer of the ionosphere, caused by a large flux of low-energy solar protons produced in these flares. The emission of these particles has been directly confirmed by observations at high altitude (see Sec. 11). It is shown that the spectrum of observed particles is very steep, though their energies may extend up to several tens of Bev. The spectrum may be deduced from the intensity of the PCA, direct particle-energy measurements with rockets and balloons, and the relative response of instruments sensitive to particles above different energy thresholds operating at different geomagnetic latitudes. The flares that accelerate particles that arrive at the earth are of giant size and are usually classed as 3 or 4 in terms of their optical intensity. Production of energetic particles is also accompanied by a large solar radio burst. It is found that wide-band radio emission, known as type IV burst, has a high probability (~85 per cent) of being accompanied by emission of energetic protons.

It is found that solar particle events are almost never observed following a flare that may have occurred on a nonvisible part of the sun; the majority of them are related to flares occurring on the western limb of the visible solar disk.

Within a day or two of the coronal activity, magnetic storms and auroral activity are observed at the earth. The horizontal component of the magnetic field suddenly increases by a few gamma to a few hundred gamma (1 gamma = 10^{-5} gauss). This sudden commencement of the magnetic storm is generally accompanied by a sudden decrease in the cosmic-ray intensity (Forbush decrease). A few hours later the magnetic field swings to negative values which may be as much as 500 gamma below normal. Then comes the recovery phase lasting a few days during which the cosmic-ray intensity and the magnetic field return to their normal values, unless there has been another flare in between. A typical flare sequence is exhibited in Fig. 11.41.

Theories of Cosmic-ray Modulation. Present theories of solar modulation of cosmic rays started with a proposal by Morrison (1956) that occasionally the sun emits plasma clouds which, while they expand, carry tangled magnetic fields that are essentially the irregular field of the solar corona. The expanding corona is empty of cosmic rays which then have to diffuse into it by scattering on the tangled magnetic fields; hence the intensity in the region covered by the plasma must be less than the intensity in the galactic space. This then would be an explanation of the main cosmic-ray variations which are all reductions in intensity.

9.3. Solar Wind. Now it is believed that the sun is continuously emitting neutral proton-electron gas which moves out radially with velocities of the order of 500 km/sec at least to a distance well beyond the orbit of the earth. There it breaks up into irregularities, and the gas escapes into the interstellar medium. The idea of a "solar wind" dates back to 1896 when Birkeland suggested that auroral phenomena are caused by low-energy electrically charged particles emitted by the sun being sucked into the earth's magnetic field. This was later worked out by Störmer (1907) and then used by Chapman and Ferraro (1931) to formulate a theory of magnetic storms. An important development came in 1950 when Alfvén showed that ionized gas clouds in their motion out from the sun must carry frozen magnetic fields characteristic of the part of the sun where they originate. Bierman (1951) showed that, contrary to general belief, radiation pressure was not adequate to push the comet tails away from the sun, but that a steady wind of solar plasma moving with a velocity of ~300 km/sec was. This was also used for understanding the phenomenon of zodiacal lights and the heating of the upper atmosphere (Chapman, 1957, 1959). The theory of solar wind and its intimate

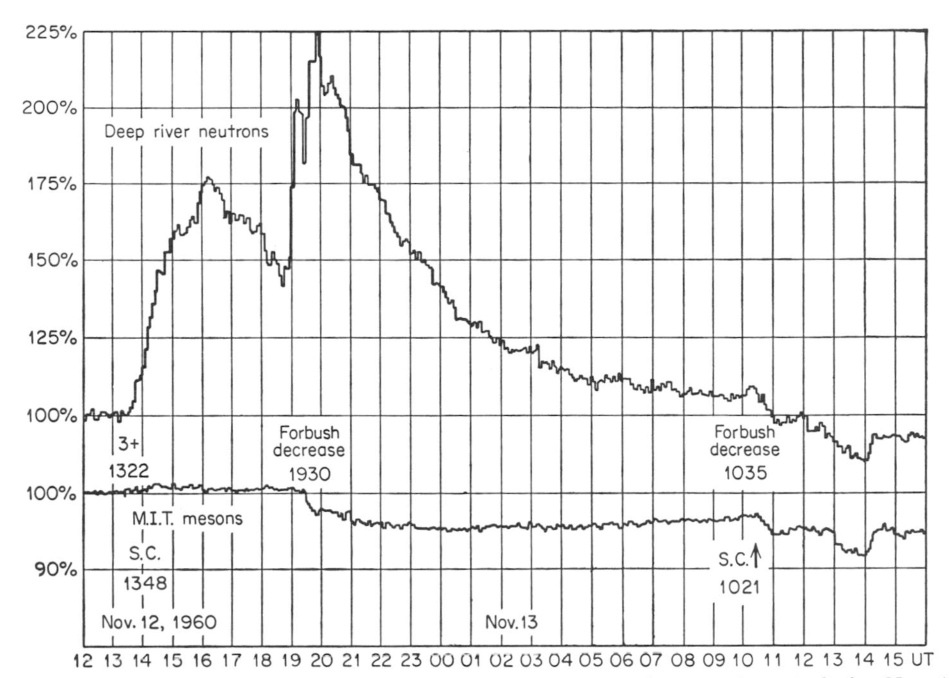

FIG. 11.41. Deep River neutron-monitor counting rate and M.I.T. meson-monitor counting rate during Nov. 12–13, 1960. A 3^+ flare was recorded at 1322 hr on Nov. 12. This was followed by a large increase in the neutron counting rate and a small increase in the meson counting rate, indicating the arrival of low-energy solar particles. (The double structure in the neutron peak is due to the complicated interplanetary conditions existing at that time because of previous flares on Nov. 10 and 11.) The sudden commencement of a magnetic storm at 1021 and the Forbush decrease at 1035 on Nov. 13 are connected with the Nov. 12 3^+ flare and are probably caused by the arrival of the enhanced solar plasma. The other sudden commencement and Forbush decrease indicated in the figure are due to previous flares. [*After Steljes, Carmichael, and McCracken* (1961).]

connection with solar modulation has been developed in some detail by Parker (1957, 1963). The existence of a steady solar wind has recently been established by direct measurements with plasma probes carried by various space probes (Gringauz et al., 1960, 1963; Bridge et al., 1960, 1963; Bonetti et al., 1963; Bridge et al., 1964).

It was found that a solar wind is always present outside the magnetosphere (see Sec. 9.4) and has a velocity of $\sim$300 km/sec with a total flux of a few times 10^8 particles/cm^2/sec. An increase in the wind velocity and flux was also noticed following solar activity.

9.4. Interaction of Solar Wind with Magnetosphere. Interaction of moving plasma with the earth's magnetic field was first considered theoretically by Chapman and Ferraro (1931, 1932) and Ferraro (1952, 1960). This led to the prediction of a geomagnetic cavity, meaning that in the solar direction the geomagnetic field will be terminated by the pressure of the solar wind and the advance of plasma across this boundary will be inhibited. The existence of the magnetopause and the rough shape of the magnetosphere have recently been established by space probes carrying plasma detectors and magnetometers. The experimental data have recently been reviewed by Ness (1965).

It is found that in the subsolar direction (12 noon), the regular geomagnetic field terminates at approximately 10 Re (Re = earth radius). Enclosing the magnetosphere is a turbulent boundary layer which is separated from the interplanetary region by a magnetohydrodynamic shock wave situated approximately at 13.4 Re. (The shock wave is produced because the velocity of the plasma is "supersonic" with respect to the velocity of magnetohydrodynamic waves in the plasma.) On the midnight side of the earth the geomagnetic field is stretched into a tail which is observed trailing away from the sun out to 40 Re and probably extends out to the moon (Fig. 11.42).

In the interplanetary space the magnetized plasma contains a regular field of approximately 5 gamma, and this field is directed along an Archimedean spiral arising from the rotation of the sun. Both the intensity of this field and its direction ($\sim$45° with respect to the earth sun line) were theoretically predicted by Parker's model of expanding solar corona, taking a field of 2 gauss at the photospheric surface. The field strength shows a strong recurrence tendency with a period of 27 days, the rotation period of the equatorial region of the sun, thus supporting the conclusion that the interplanetary field is of solar origin. It is further found that this field is structured into four sectors, with the field direction changing from one sector to a neighboring sector.

9.5. Parker's Theory of Solar Modulation. The earth is embedded in the continuously expanding solar corona. Originally the plasma cloud is empty of cosmic rays which diffuse into it by scatter-

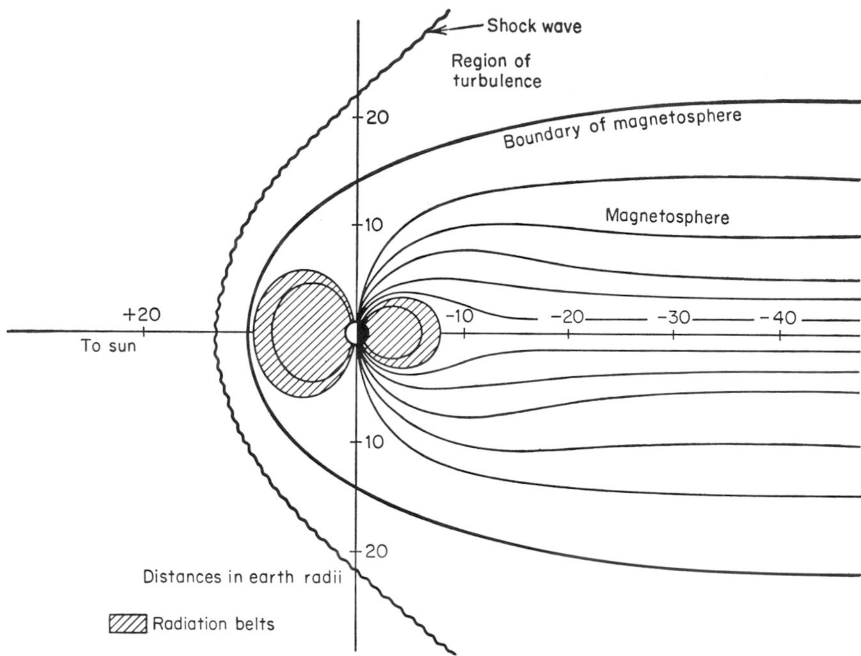

FIG. 11.42. The shape of the magnetosphere, showing the magnetopause, the turbulent region, and the shock wave induced by interaction of solar plasma with the geomagnetic field. The region of radiation belts is also indicated. [*After Ness* (1965).]

ing off magnetic irregularities embedded in the cloud. The diffusion constant for isotropic diffusion is $K = \frac{1}{3}W\lambda$ (Morrison, 1956), where W is the cosmic-ray particle velocity and λ is the scattering mean free path. Inward diffusion is opposed by the outward radial velocity (v) of the plasma. Then, qualitatively, for a uniform v, λ, the cosmic-ray density at a radial distance r from the sun is

$$n(r) = n_0 \exp\left(-\frac{3v}{W}\frac{R-r}{\lambda}\right) \qquad (11.82)$$

where R is the distance to the boundary of the solar wind.

Although the assumptions, such as, isotropy of diffusion and independence of v with r, are not sound (see below), the main features of large-scale cosmic-ray modulation can be understood in terms of Eq. (11.82). This predicts a gradient of cosmic-ray density in the interplanetary space; the 11-year and the 27-day variations are to be assigned to a change in the parameter vR/λ, which involves properties of coronal emission. It is reasonable that with increased activity v and R will increase while λ will decrease (because trapped magnetic irregularities will be stronger); hence the inverse correlation of cosmic-ray intensity with solar activity.

Dependence of modulation on particle energy is governed by the parameter $W\lambda$. This is smaller for low-energy particles, thus reducing the transparency of the plasma [Eq. (11.82)]. λ depends on the rigidity of the particle. If λ is very small, as at low energy, the modulation becomes purely velocity-dependent. This would be a qualitative explanation of the splitting of p and He spectrum at low energies (Sec. 5.1). At high energies the modulation would obviously be rigidity-dependent.

Consistent values of some of the parameters entering Eq. (11.82) are $\lambda = 10^{11}$ to 10^{12} cm, the diffusion coefficient for nearly relativistic particles $K = \frac{1}{3}\lambda W = 10^{21}$ to 10^{22} cm, and the distance to which the wind extends $R \sim 0$ (10) a.u. (Parker, 1965). Parker has also shown that typically the density of cosmic rays in the plasma achieves an equilibrium in 10^5 to 10^7 sec. This may perhaps account for the phase lag between the solar cycle and 11-year cosmic-ray intensity variation.

The existence of the diffusion process and its parameters are better studied in observations of arrival times of solar particles, following flares. Consider, for example, a hypothetical case (Parker, 1965) in which the diffusion coefficient is independent of position and the net particle flow is radial. Then, neglecting the convection velocity of the wind, the change in density at distance r is

$$\begin{aligned}\frac{dn}{dt} &= \nabla \cdot (K\,\nabla n)\\ &= \frac{K}{r^2}\frac{\partial}{\partial r}\left(r^2\frac{dn}{dr}\right)\end{aligned} \qquad (11.83)$$

Solution of this for a burst of n_0 particles per steradian at $t = 0$ is

$$n(r,t) = \frac{n_0}{2\pi^{1/2}(Kt)^{3/2}}\exp\left(-\frac{r^2}{4Kt}\right) \qquad (11.84)$$

The maximum occurs at $t = r^2/6K$, which at $r = 1$ a.u. is 1 to 10 hr for typical values $K = 10^{21}$ to 10^{22} cm^2/sec. The density distribution at later times decays as $t^{-3/2}$.

These are found to be approximately the characteristic times associated with SPR (solar particle radiation) buildup and decay.

Forbush Decrease. Some of the sharp Forbush decreases may be explained by the following mechanism. At the time of a flare a very fast plasma cloud is shot out which, overtaking the normal solar wind, creates a blast wave ahead of it (Fig. 11.43). This

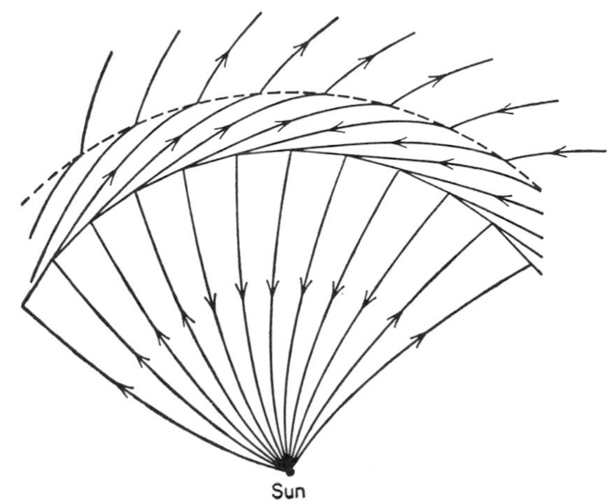

FIG. 11.43. A schematic representation of the magnetic lines of force and the magnetic bottleneck produced when a blast wave of a fast plasma from the sun overtakes an already existing slow plasma.

creates an advance front of compressed magnetic field which is like a partially opaque screen for cosmic rays entering behind the wave.† The onset of the Forbush decrease is then connected with this bottleneck sweeping past the vicinity of the earth.

Part of the Forbush decrease arises from increased compression of the geomagnetic field as a result of the impact of faster plasma. The two effects are in phase although their relative contribution is not yet known.

The Diurnal Variation. Parker (1964) shows that long-term daily variation of cosmic-ray intensity is not a real variation but that it arises from a net streaming of cosmic-ray particles in the Archimedean spiral (Fig. 11.43) direction. This streaming leads to a net azimuthal transport velocity $U_\phi = r\Omega$, where Ω is the angular velocity of solar rotation. The observed long-term diurnal variation is then caused by the Compton-Getting effect, the total fractional variation being equal to $6U_\phi/c \simeq 0.8 \times 10^{-2}$, that is, an amplitude of 0.4 per cent. This amplitude would be larger for instruments with a fixed energy threshold, because of increase and decrease of the effective particle energy when looking upstream and downstream, respectively. The predicted direction of the maximum is obviously $\sim$3 hr after noon (i.e., at 1500 hr), to be compared with 1800 hr as the experimentally observed time of maximum.

Several refinements of the solar-wind modulation theory are being introduced. The concept of isotropic diffusion is necessarily artificial and wrong; there must be more diffusion along the large-scale magnetic fields than across them. One obvious pointer to the anisotropy of the diffusion process is that flare particles are almost never observed for solar bursts occurring on the invisible parts of the photosphere. The whole subject is in a state of rapid development, and most of the parameters of the theory, which are after all connected with the magnetohydrodynamical nature of events occurring at the sun, should soon be directly supplied by experiment. For example, it should soon be possible to measure the actual cosmic-ray density gradient over a range of about 2 astronomical units. As it is, the general picture of the earth being embedded in the solar corona and the "weather" changes around it being directly connected with solar events has been amply verified.

† Similar overtaking of slow wind by fast plasma can also occur when, because of solar rotation, a cool coronal region is followed by a hotter coronal region. Sarabhai (1963a, b) has discussed the consequences of this in some detail in relation to the character of geomagnetic activity and cosmic-ray reduction.

10. Radiation Belts

Geiger counters carried aloft the United States satellites Explorer I and Explorer III early in 1958 (Van Allen, 1958; Van Allen et al., 1958) proved the existence of a high intensity of energetic charged particles trapped in the geomagnetic field of the earth. These findings were soon confirmed by Vernov et al. (1959) who employed a Na I scintillation crystal and a ZnS fluorescent screen in an apparatus carried by sputnik III. Subsequent experiments with satellites have provided a large amount of information on the nature of this trapped radiation.

After the early investigations of motion of charged particles in the earth's magnetic field (Poincaré, 1896; Birkeland, 1908, 1913; Störmer, 1907, 1955), it was clear that there exists a region that is completely forbidden to particles arriving from infinity. The particles in this region would then be trapped and cannot escape to infinity in the absence of an external perturbation. Analyzing the motion of trapped particles along the magnetic lines of the inhomogeneous geomagnetic field, Alfvén (1950) showed that the main component of the motion is that of oscillation between two mirror points, one in the Northern Hemisphere and the other in the Southern Hemisphere. The period of oscillation varies from a fraction of a second to a few seconds, depending on the velocity of the particle. Superimposed on this oscillation is a longitudinal drift which is westward for positively charged particles and eastward for negative particles. This gives rise to a net westward-flowing ring current. The formation of such a ring current due to the motion of charged particles was proposed by Singer (1957). The theory of the trapping of charged particles in the earth's field has been put to an experimental test by artificially injecting particles, using high-altitude nuclear explosions (Christofilos, 1959; O'Brien et al., 1962; Durney et al., 1962).

The naturally occurring radiation belt consists of two regions, an inner zone stretching from 1.2 to 2 earth radii and an outer zone lying between 3 and 4 earth radii (Fig. 11.44). The intermediate region known as the "slot" is not well defined for low-energy particles. The existing data on the flux, composition, energy spectra, spatial distribution, time variation,

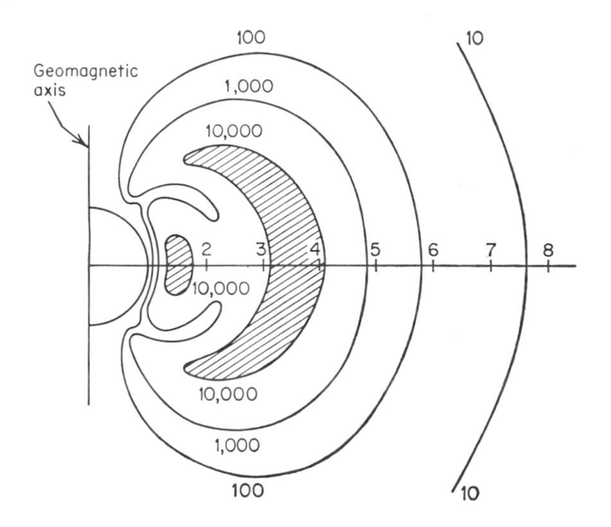

FIG. 11.44. The original diagram of the intensity structure of the trapped radiation around the earth. Contours of constant intensity are labeled with numbers that are the true counting rates of a Geiger counter. [*After Van Allen and Frank* (1959).]

and origin of trapped particles have been reviewed by O'Brien (1962), Hess (1960), Elliot (1963), and Van Allen (1963). The important information is summarized below.

The Inner Belt. The positive particles in the inner belt consist mainly of protons with deuterons and tritons constituting about 1 per cent and α particles having an upper limit of 0.1 ± 0.05 per cent (Freden and White, 1960; Heckman and Armstrong, 1962). The differential energy spectrum of protons is $J(E)\, dE = kE^{-1.8}\, dE$ for $75 < E < 700$ Mev. The peak in intensity of protons of $E > 40$ Mev occurs at a distance $L \simeq 1.5$ (measured in earth radii) and has an intensity of $\sim 2 \times 10^4$/cm²/sec.

Intensity of electrons in the inner belt is $\sim 3 \times 10^7$/cm²/sec for $E > 40$ Kev and $\sim 5 \times 10^6$/cm²/sec for $E > 500$ Kev. The energy spectrum is not known precisely. The electrons in this belt are perhaps an extension of the outer belt.

The estimated lifetime of protons of ~ 50 Mev in the inner belt is ~ 3 years. Time variation of their intensity is very small.

The Outer Belt. The outer belt begins at $\sim 10{,}000$ km and extends up to $\sim 70{,}000$ km from the earth. The location changes with the degree of solar activity.

Most of the particles in this belt are electrons. The average flux in the equatorial plane for $2 \leq L \leq L_{\max}$ is $\sim 10^7$/cm²/sec for $E > 40$ Kev. The fluxes of high-energy electrons ($E > 1.5$ Mev) vary from 10^3 to 10^6 particles/cm²/sec.

Large variations in the flux and energy spectrum of particles in this belt occur in association with solar activity and geomagnetic storms. Within a few hours or a day from the onset of a geomagnetic storm the content of particles is greatly depleted. After a day or so the intensity starts increasing and returns to its original value within a week or so. Balloon experiments on the *bremsstrahlung* X rays generated by energetic electrons entering the atmosphere at high latitudes suggest that at least a part of them are due to the trapped particles being dumped into the atmosphere at the time of the magnetic disturbance.

Sources of Trapped Particles. It is believed that most of the protons in the inner belt arise from the decay of the low-energy albedo neutrons (Singer, 1958; Kellogg, 1959; Vernov et al., 1959). These neutrons are produced in the atmosphere by steady cosmic rays and their intensity may be greatly enhanced in the polar regions at the time of solar particle events. This theory cannot, however, account for protons of energy $E < 50$ Mev nor can it explain the intensity and spectrum of electrons, which form the main component of the outer belt. It is felt that these electrons arise from the interaction of solar plasma and the geomagnetic field, resulting in local acceleration (Dessler, 1958; Pizzella et al., 1962). Recently Fan et al. (1964) have found evidence for acceleration of electrons above 30 Kev beyond the magnetosphere boundary but it is not certain whether this provides an adequate contribution for the electrons in the outer belt. Thus, although some possible mechanisms have been identified, the precise roles of solar injection, local acceleration, and trapping of albedo particles have not been determined.

11. Composition of Solar Particle Radiation (SPR)

During the maximum of the latest solar cycle, energy spectra and particle composition for several solar particle events were analyzed. The size of these events varied greatly. In the highest events the flux of particles above 20 Mev was several thousand times that of the normal cosmic rays. On the other hand, small events may have intensities that are barely detectable. Besides protons, which are the most dominant component, helium nuclei and other multiply charged nuclei have also been detected (see, for example, McDonald, 1963; Roederer, 1964; Webber, 1964; Bryant et al., 1962; Biswas, Freier, and Stein, 1962; Biswas, Fichtel, Guss, and Waddington, 1963). Recently Biswas and Fichtel (1965) have summarized the available information. The main features are the following:

1. Both proton and multiply charged nuclei spectra can be represented as exponentials in the rigidity scale, in the range 200 to 900 mv,

$$\frac{dJ}{dR} \propto \exp\left(-\frac{R}{R_0}\right) \tag{11.85}$$

where the characteristic rigidity R_0 is the same for He and heavier nuclei and almost the same for protons. Values of R_0 lie between 100 and 170 mv for different events (see Fig. 11.45).

2. The relative intensity of helium and of various heavier nuclei is the same within errors for all the events while the relative intensity of hydrogen and helium nuclei varies from event to event.

3. The relative composition of multiply charged nuclei ($Z > 2$) corresponds very closely to the relative abundance of these elements at the solar surface and is different from their relative abundance in galactic

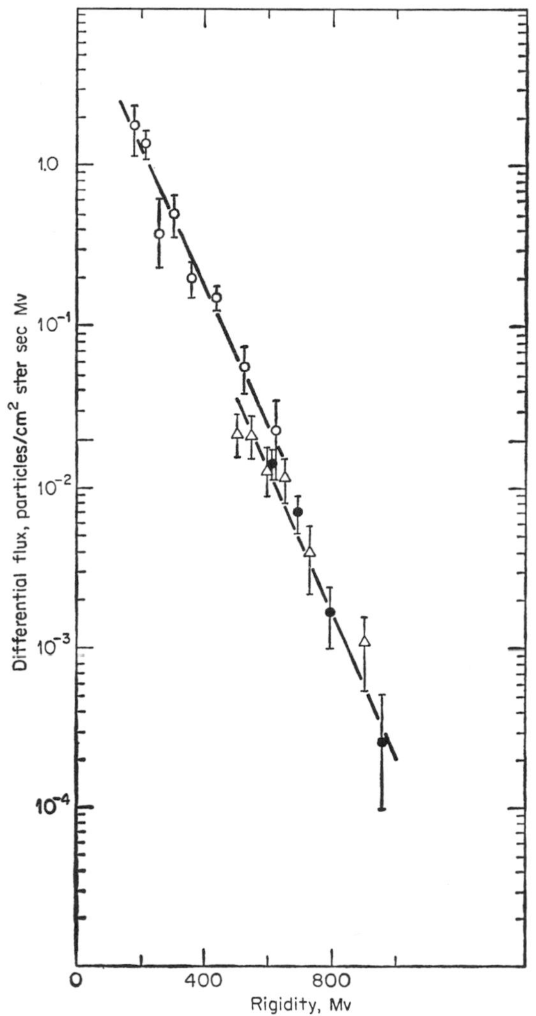

FIG. 11.45. Differential rigidity spectrum of solar protons, helium, and medium nuclei at 1951 UT, Nov. 16, 1960. Open circles: protons; triangles: He nuclei; black circles: 60 times the intensity of medium nuclei. [*Biswas et al.*, 1963).]

cosmic radiation. In particular, the light elements Li, Be, B are almost absent in solar particle radiation, while in the normal cosmic rays their abundance is about one-fourth that of medium nuclei (see Table 11.11).

4. Detectable fluxes ($\sim 5 \times 10^{-3}$ of protons) of deuterons have been found in two SPR events (McDonald et al., 1965). While H^3 and He^3 nuclei have also been detected by some authors (Fireman, DeFelice, and Tilles, 1961; Tilles, DeFelice, and Fireman, 1963; Schaeffer and Zähringer, 1962) their presence is still under some dispute (Lal, Rajagopalan, and Venkatavardhan, 1963).

The fact that in most events relative abundances of multiply charged nuclei follow closely those of the solar photosphere indicates that there is relatively small differentiation in the process of acceleration and escape. This then reinforces the argument that galactic cosmic rays, which have a very different chemical composition, particularly when extrapolated back to the source, cannot arise from normal stars like the sun. Furthermore, this observation allows one to measure, in the solar photosphere, the relative abundance of elements, such as He and Ne, that cannot be easily detected by other means.

TABLE 11.11†

Element	SPR	Sun	Universal abundance	Galactic cosmic rays	Galactic SPR
$_2$He	107 ± 14	.....	152	51.5	0.48
$_3$Li	....	0.001	4.00×10^{-6}	0.27	
$_4$Be–$_5$B	0.02	0.001	1.75×10^{-6}	0.5	25
$_6$C	0.59 ± 0.07	0.6	0.37	1.55	2.6
$_7$N	0.19 ± 0.04	0.1	0.096	0.5	2.6
$_8$O	1.0	1.0	1.0	1.0	1.0
$_9$F	0.03	0.001	6.4×10^{-5}	0.12	4.0
$_{10}$Ne	0.13 ± 0.02	.....	0.032		
$_{11}$Na	....	0.002	1.75×10^{-3}		
$_{12}$Mg	0.043 ± .011	0.027	0.036	0.85 (Ne–Si)	4.1 (Ne–Si)
$_{13}$Al	....	0.002	3.79×10^{-3}		
$_{14}$Si	0.033 ± 0.011	0.035	0.040		
$_{15}$P–$_{21}$Sc	0.057 ± 0.017	0.032	0.024	0.37 (P–Ni)	4.8 (P–Ni)
$_{22}$Ti–$_{28}$Ni	0.02	0.006	5.24×10^{-3}		

† SPR abundances are from Biswas and Fichtel (1965). Solar abundances are from Goldberg, Muller, and Aller (1960) and Aller (1961). Universal abundances are those given by Cameron (1959).

12. Origin of Cosmic Rays

Various attempts to understand the origin of cosmic radiation have led to a realization that this phenomenon is by no means a minor extraneous occurrence in the scheme of things but plays an essential role in the dynamics of the galaxy and perhaps the universe. Some of these aspects will be discussed here.

12.1. Isotropy of Cosmic Rays and Their Energy Density. It is found that cosmic rays of up to 10^{14} ev are isotropic to at least within 1 part in 10^3. An obvious way of ensuring this would be to assume that the cosmic rays fill the whole universe with uniform density. However one is immediately faced with a serious problem. One finds that the kinetic-energy density of the nuclear component of cosmic rays in out neighborhood is ~ 1 ev/cm^3, which is of the same order as the energy density of starlight.

The energy density is easily calculated from the relation

$$\epsilon \simeq \frac{4\pi \times 2.35}{c} \left[\int_1^\infty \frac{dE}{E^{2.67}} \times E - \int_1^\infty \frac{dE}{E^{2.67}} \right] \qquad (11.86)$$

$$\simeq 10^{-9} \text{ Bev/cm}^3$$

This energy density over the whole universe would require that the total energy in cosmic rays be many orders of magnitude greater than the total energy in thermal photons. Actually the situation is even worse than it appears at first sight. If the high-energy cosmic-ray gas has existed since the beginning of the universe, it must have been subject to adiabatic cooling due to expansion of the universe, thus requiring a large fraction of the initial energy of the universe to have been in the form of cosmic rays.

This situation is relieved a good deal if one assumes that cosmic rays are mainly galactic in origin and are confined within its volume by large-scale magnetic fields in the galaxy. If the storage factor is taken roughly as the inverse of the degree of anisotropy, the power going into cosmic rays becomes only $\sim 10^{-4}$ of the power in starlight.

An estimate of the lower limit on the magnetic field strength may be obtained from the requirement that out in the halo of the galaxy, (see Fig. 11.46), where it

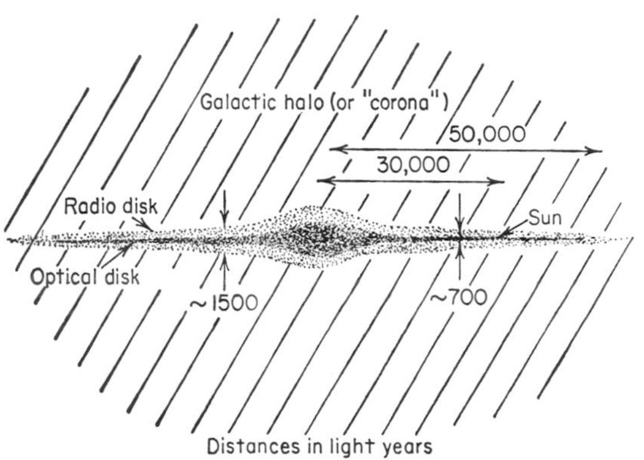

FIG. 11.46. A schematic representation of the shape of the galaxy and its halo.

is not well anchored in thermal plasma, the field be able to support the pressure of the cosmic-ray gas. Thus setting $B^2/8\pi \simeq \epsilon/3$, where

$$\epsilon = 1 \text{ ev/cm}^3 = 1.6 \times 10^{-12} \text{ dyne/cm}^2$$

one gets B 3.5×10^{-6} gauss (Davis, 1963).

Existence of magnetic fields of the order of 1 gamma was originally indicated by the observation that the light from many red stars is partially plane-polarized. The reddening is produced by scattering from dust clouds in the way. In the presence of magnetic fields the dust grains may be partially aligned, in which case the component of light whose electric vector is parallel to the longer axis of the grain would be preferentially weakened (Davis and Greenstein, 1951).

Chandrasekhar and Fermi (1953) pointed out that the stresses associated with the existence of a magnetic field of the order of 1 gamma would explain why, because of its rotation, the galaxy does not become a much thinner disk than it is. Their general argument would also highlight the fact that the pressure of the cosmic-ray gas ($\sim 10^{-12}$ dyne/cm^2) would not be insignificant in discussions of general galactic structure.

Further information about the magnetic field structure is provided by radio astronomy. General radio emission from the galaxy has the character of synchrotron radiation produced by high-energy electrons circulating in magnetic fields. Actually this observation led to the discovery of the radio halo of the galaxy, indicating that magnetically the galaxy is almost spherical (Fig. 11.46). This is of great importance in extending the energy range up to which cosmic-ray particles may be contained in the galaxy. Assuming a reasonable source strength and a uniform spacial distribution for cosmic-ray electrons, it is estimated that the field is ~ 2 gamma in the spiral arms and ~ 0.3 gamma in the halo. Magnetic fields of this order are also indicated by measurements of Faraday rotation of the plane of polarisation of radio waves from extragalactic radio sources and an upper limit of 0.5 gamma is obtained from the null result of an attempt to measure the Zeeman splitting of the 21-cm line of neutral hydrogen.

The interstellar space is also filled with plasma in which the magnetic fields are presumably embedded. There is most probably a generic relationship between the plasma and the magnetic fields. We know that wherever there are turbulent plasma, on the surface of the sun for example, magnetic fields are generated which are then frozen in the plasma and move with it with all their inhomogeneities. Ultimately there should arise an equipartition of energy between the plasma and magnetic fields. This must hold for the interstellar plasma also. Thus the kinetic-energy density of the plasma is also of the order of ~ 1 ev/cm^3. That the cosmic-ray energy density is also of the same order may not be accidental but may indicate a slow process in which the energy of the clouds is imparted to a few preferred particles in a statistical manner. Existence of such a process was first shown by Fermi (1949). In principle this process can provide an effective mechanism for damping the energy of plasma clouds; a fraction of high-energy cosmic-ray particles, after soaking up some of their energy, may leak out of the galaxy.

12.2. Residence Time and Prehistory. The light nuclei Li, Be B, being important nuclear fuels, have a very low cosmic abundance ($\sim 10^{-5}$ of heavier nuclei). However it is found that their abundance in primary cosmic rays is about one-quarter that of heavier nuclei. If one assumes that all the light nuclei are produced by fragmentation of heavier nuclei in collisions with interstellar hydrogen, then, feeding in the measured cross sections for fragmentation, one can estimate the total amount of matter through which they have passed between the time of acceleration and their arrival at earth (Bradt and Peters, 1950a, 1950b). For heavy particles of energy $E \leq 30$ Bev/nucleon this is found to be ~ 2.5 g/cm^2 (Badhwar et al., 1962).

Recent measurements (Anand et al., 1966; Balasubrahmanyan et al., 1965; Comstock et al., 1966) indicate that the ratio of light to medium nuclei goes through a maximum at around 300 Mev/n. It is difficult to understand this behavior of the L/M ratio or the low-energy spectral shape of multiply charged nuclei (Sec. 5.1) if the amount of matter between the source and the observer is approximated by a matter-slab of a definite thickness. Most probably one needs a model of cosmic-ray propagation (like the steady-state equilibrium model) in which the distribution of possible path lengths is very broad; because of increasing ionizing energy loss at low energies, very long path lengths are not realized.

Gas density in out neighborhood is of the order of ~ 1 atom/cm^3. Then 2.5 g/cm^2 of hydrogen corresponds to a residence time of

$$\tau = \frac{(2.5 \times N)}{c} = 2 \times 10^6 \text{ years}$$

On the other hand, residence time of starlight would be about 1,000 years, thus giving a storage factor of 10^3 to 10^4 for cosmic rays. This shows directly that the power of the cosmic-ray source need be only $\sim 10^{-4}$ that of light (also see Sec. 4.3).

However, this value of the residence time is not very reliable because there is no a priori knowledge about the fraction of the time spent in regions of different density, namely the source, the galactic disk and the galactic halo. A method suggested for a direct measurement of the residence time depends on determining the ratio Be/B in the cosmic-ray flux; this ratio is expected to be ~ 0.36 if life time is much shorter than the time dilated life time of Be^{10} ($\tau \sim 4 \times 10^6$ years) and ~ 0.51 if it is much longer (Daniel and Durgaprasad, 1966). The experimental accuracies achieved thus far are insufficient to provide a definite answer.

If all space is filled with a radiation field with an energy density ~ 0.5 ev/cm^3, cosmic-ray electrons in the energy range of a few hundred Bev would suffer serious energy degradation by Compton collisions in time scales of the order of ten million years. Detailed arguments of this type enable one to put an upper limit of $\sim 2 \times 10^7$ years on the residence time of high-energy electrons, if the recently discovered microwave background radiation (Dicke et al., 1965) is definitely proved to be a universal blackbody radiation at 3°K.

It has been experimentally demonstrated, using C^{14} dating techniques on artifacts of known antiquity, that the long-term average cosmic-ray intensity has been nearly constant over about 10 mean lives of C^{14}, that is, for $\sim 60{,}000$ years. Similar investigations carried out with radioactive isotopes produced in meteoritic materials have shown that the average intensity has not substantially changed during the last 10^9 years. Both these results refer to cosmic rays of relatively low energy ($E \ll 100$ Bev) and so far it has not been possible to get any information about the prehistory of the very high-energy ($E \gg 100$ Bev) component. This may perhaps be possible by investigating the activity levels of long-lived radioisotopes produced by muon interactions deep underground.

12.3. Possible Cosmic-ray Sources. It has been known for a long time that the sun can accelerate particles, and sometimes it has been seriously suggested that most cosmic rays may be of solar origin and their intensity may be built up over long periods by storage around the planetary system (Alfvén, 1950). However, now one can give several obvious arguments against such a possibility. Solar-particle spectra have been found to be very steep, the diffusion times are not very large, solar-particle events are associated mostly with visible flares (indicating high anisotropy in diffusion), and the chemical composition of solar rays is different from that of galactic cosmic rays (Sec. 11). Actually one can go further and say that even if all the sunlike stars in the galaxy were to produce radiations like the sun, they would not account for the observed galactic radiation. If there is no subsequent acceleration in the interstellar space, the energy density of the radiation would be too small by several orders of magnitude, and the spectrum would be far too steep. If this were patched up by some acceleration process subsequent to emission, the chemical composition would still remain a problem. As it is, the relative abundance of heavy elements in solar rays is less than that observed for galactic cosmic radiation (Sec. 11); further acceleration and traversal through additional amount of matter would make this discrepancy even wider. A few years ago ordinary stars could not be ruled out as injectors, because there was the possibility that during acceleration a fractionation may occur favoring the emission of heavier elements. This has now been excluded.

Therefore for the supply of initial "ions" to be accelerated one has to look to sources that are comparatively richer in heavy elements and, in order to dominate injection of material from ordinary stars, extraordinarily active. Such sources are novae and supernovae (Ginzburg, 1953, 1958; Hayakawa et al, 1958), red giant stars, and perhaps the magnetic stars of Babcock which are relatively newly condensed from interstellar gas and hence may be richer in heavy elements (Davis, 1963).

Particles from these source may be injected into interstellar space after a short primary acceleration when they may gain energy by a slow statistical process (Fermi 1949, 1954), or they may be accelerated to almost their full energy in the source region itself, through a combination of inductive, statistical, and shock processes (Colgate and Johnson, 1960).

12.4. Acceleration Mechanisms. It is clear at the start that the acceleration process cannot be "sudden" in the sense that a large amount of energy is concentrated on a chosen favorite particle in a single explosion. It is inconceivable, for example, that an iron nucleus could be so accelerated to 10^{12} ev without breaking or for that matter that a 10^{18} ev proton is produced by suddenly concentrating on a single proton equivalent of the rest mass energy of 10^9 protons. Acceleration theories involving large electrostatic forces are also not possible, because the gases in stellar environments, as in the interstellar space, are highly conducting.

Thus the only possible processes are electromagnetic or, rather, electrohydrodynamic. There may be more of these processes than we know of at present. One finds that, whenever there is turbulent plasma on a large scale, magnetic fields build up and particles are accelerated. This certainly happens on the sun, probably also happens in the outer radiation belt where it seems difficult to explain the behavior of electrons without bringing in some acceleration mechanism, and perhaps also happens around Jupiter whose polarized radio emission in the decimeter range indicates the presence of relativistic electrons. Two of the well-understood processes are the inductive process, and the statistical process of Fermi (1949).

An inductive process was first suggested by Swann (1933) as a likely mechanism for accelerating particles around sunspots. A rapidly changing magnetic field induces an emf that would accelerate particles in the manner of a betatron. The efficacy of this process in plasma conditions is somewhat in doubt because of high conductivity and because of the difficulty of selective operation on a few particles at a time to give them high energies. However, there may be special conditions under which some particles are accelerated.

In 1949 Fermi suggested that acceleration of cosmic rays may occur by collisions of particles with randomly moving magnetic clouds in interstellar space (Sec.

12.1). Since collisions with approaching fields are more frequent than overtaking collisions, there would be a slow gain in energy. It is obvious that the total energy gain would depend on the time a particle had spent in the galaxy, as well as on the time rate of gain of energy. Thus if

$$\frac{dE}{dt} = \bar{\alpha}(E)E \tag{11.87}$$

and the number of particles of total energy E decreases with a mean life $\tau(E)$,

$$\frac{dN(E)}{dt} = -\frac{N(E)}{\tau(E)} \tag{11.88}$$

Then the integral spectrum would have a slope

$$\gamma(E) = -\frac{E[dN(E)/dE]}{N(E)} = \frac{1}{\bar{\alpha}(E)\tau(E)} \tag{11.89}$$

If $\bar{\alpha}(E)$ and $\tau(E)$ have a weak dependence on E, the spectrum is a power law. The energy gain being proportional to the total energy, this process discriminates strongly against electrons if electrons and protons are injected at comparable kinetic energies. Recent values of gas-cloud velocities and fields do not seem to yield a value of the exponent in agreement with experiment. The mechanism would not be effective for heavy nuclei unless they are injected with high enough energies, because ionization loss dominates over energy gain. In fact the Fermi mechanism cannot operate for any particle unless there is a minimum starting energy; this may either be provided by a nonstatistical process or more likely from a source like a supernova.

It is now clear that the whole of the cosmic-ray acceleration need not occur in the interstellar space. The actual acceleration may be a hierarchical process starting, for example, with an explosion in a supernova, followed by a statistical and inductive process in its envelope, and later by the statistical process in the galactic space. The relative importance of these in different stages and some of the other related problems have been discussed recently by Hayakawa et al. (1964).

12.5. Energetics of Supernovae as Sources. Most of our present ideas about supernova explosions and their capacity for generating high-energy particles come from the study of the Crab nebula (Taurus A) which was formed by a supernova seen and recorded by Chinese astronomers in A.D. 1054. It is found that light from the Crab nebula is strongly polarized, indicating magnetic fields of the order 10^{-4} to 10^{-3} gauss. In addition, it is found to be a strong source of highly polarized radio waves and X rays. Radio and light spectrum has the character of synchrotron radiation emitted by electrons whose energies extend to $\sim 10^{12}$ ev. It is argued that a system that can accelerate electrons should be even more efficient in accelerating nuclei. However, Hayakawa et al. (1964) have wondered whether the rather high injection energies needed near the center of the star in the explosion stage would not lead to nuclear interactions and fragmentation of a large number of heavy nuclei, thus partly invalidating the very arguments that suggested a preferred role for supernovae as cosmic-ray sources. So far there is no direct evidence for the presence of high-energy nucleons in the Crab. This may be provided by observation of $\sim$100-Mev γ rays or of high-energy neutrinos which can only arise from interactions of nuclear particles.

Total cosmic-ray energy in the galaxy is (3×10^{68} cm^3 $\times$ 1 ev/cm^3) $\simeq 5 \times 10^{56}$ ergs. If lifetime is taken as 2×10^6 years, the power of the cosmic source is $\sim 10^{43}$ ergs/sec. This power may be reduced by about a factor of 100 (to $\sim 10^{41}$ ergs/sec) if it is assumed that most of the cosmic-ray nuclei are stored in the halo region where the gas density is $\sim 3 \times 10^{-2}$ p/cm^3, thus increasing the cosmic-ray lifetime to 10^8 years (Sec. 12.2). The power of the Crab nebula for generating relativistic electrons is estimated as 10^{36} to 10^{39} ergs/sec; hence that for generating protons may be 10^{39} to 10^{42} ergs/sec (Ginzburg, 1958). The lifetime of a supernova is $\sim$3,000 years while there is one such occurrence every 300 years. Thus the total cosmic-ray power of all supernovae may be 10^{40} to 10^{43} ergs/sec. This seems to be adequate when compared with the observed power of 10^{41} to 10^{43} ergs/sec. However it is found that the total cosmic-ray energy release required of a supernova comes uncomfortably close to its total energy release in all forms of radiation. On the other hand, all these arguments have only a qualitative significance, particularly because of the increasing evidence in recent years suggesting that the Crab is not a typical supernova.

12.6. Cosmic Rays from Outside the Galaxy. All radio stars and radio galaxies must have relativistic electrons and hence be capable of producing cosmic rays. Following this argument, Fujimoto, Hasegawa, and Taketani (1964) have estimated that there may be an extragalactic component of cosmic radiation, which may have an intensity of $\sim 10^{-3}$ to 10^{-4} times that of the galactic component. Since the hierarchy of acceleration processes is likely to be the same in all turbulent plasma, the shape of the extragalactic energy spectrum may be similar to that of the galactic spectrum. However, while the galactic spectrum may be limited at the high-energy end essentially by the size of the accelerating and/or containment volumes, to about 10^{15} ev, the extragalactic spectrum may extend to much higher energies.

This general picture of primary cosmic-ray spectrum was first suggested by Peters (1959). He showed that the more or less continuous shape for the number spectrum of air showers up to energies of the order of 10^{17} ev does not exclude the possibility that galactic radiation is subject to a rigidity cutoff at $\sim 10^{15}$ ev for protons, because for the same rigidity a heavy nucleus of mass number A has $A/2$ times the total energy of a proton. Beyond 10^{17} ev a flatter extragalactic spectrum may come in.

This proposal is now supported by Bray et al. (1964) who think that a large majority of showers of size $>10^6$ are produced by heavy nuclei. It has also been shown (Pal and Tandon, 1965b) that the muon spectrum up to 10^{13} ev and γ-ray spectra at various altitudes up to 10^{13} ev can all be understood in terms of a rigidity steepening starting around 2×10^{14} ev, if in addition to the main spectrum there exists another one, with an intensity of about 5 per cent, having the same slope as the first one ($\gamma \sim 1.7$) and continuing up to much higher energies. This second spectrum, which may or may not extend down into energies below $\sim 10^{12}$ ev, could then have an extragalactic origin and may have a different chemical composition; Linsley and Scarsi (1962) think it may contain only protons (see Secs. 5.2 and 8.5).

12.7. Role of Electrons, γ Rays, X Rays, and Neutrinos. Electrons and X rays have recently been detected in the primary cosmic radiation. Definite fluxes of γ rays or neutrinos have so far not been detected. These components are important because they presumably arise predominantly as secondary radiation due to interaction of nuclear components in the envelopes of sources in the interstellar and intergalactic regions and hence enable us to probe the spacial energy density of radiation, magnetic fields, and different forms of matter. Even null results are important for putting experimental upper limits on various unknown quantities. For example:

1. An upper limit of 10^{-2} photons per cm^2 per sec on the flux of 0.51 Mev γ rays (which would be produced in e^+-e^- annihilation) enabled Arnold et al. (1962) to show that less than 10^{-6} of the matter created in the steady-state continuous creation hypothesis can be in the form of matter-antimatter.

2. The fact that γ rays of ~ 100 Mev are less than 10^{-3} of the nuclear component enabled Kraushaar and Clark (1962) to rule out substantial nucleon-antinucleon annihilations in the interstellar gas, thus eliminating some versions of the continuous creation theory of matter.

3. A limit of $\sim 2 \times 10^{-4}$ compared with nuclear flux, on γ rays of $\gtrsim 10^{16}$ ev, enabled Hasegawa et al. (1965) to put an upper limit of 0.01 ev/cm^3 on the thermal photon energy density in the intergalactic space.†

4. A limit of $\sim 1/3$ on the ratio of γ rays to nucleons of $> 10^{17}$ ev was used by Cowsik et al. (1964) to put an upper limit of 2 ev on the depth of a possible Fermi sea of neutrinos in the universe. This suggested that, if the universe is an oscillating type, in the contraction phase its size must reduce at least to one-millionth of its present size.†

Fluxes of detected electrons, at least above a few Bev (Daniel and Stephens, 1965) can be shown to be at least an order of magnitude greater than those expected from production in interstellar nuclear collisions. The fluxes estimated in the halo of the galaxy from the power of the radio emission are consistent with the observed fluxes on earth. Therefore one is forced to conclude that directly accelerated electrons must form an important part of this flux, in which case the ratio of positrons to electrons would be expected to be negligible small, at least at high energies. The present experimental situation on this is not clear, though positrons seem to constitute an important fraction of the electron flux up to ~ 50 Bev.

Intensity of X-ray flux from localized sources as also the background isotropic flux from the galaxy remains unexplained thus far (Hayakawa and Matsuoka, 1964; Oda, 1965). It is particularly difficult to account for the background radiation and the radiation from sources, which are quiescent in the optical and radio wavelengths. A hypothesis of neutron stars has been suggested to provide sources that are hot enough (10^7 K°) so that the peak emission lies in the X-ray region. However Bahcall and Wolf (1965) have recently shown that neutron stars would cool off by neutrino emission in a few years.

The coming years should see new experimental advances in all these fields, particularly the new field of neutrino astronomy. Detection of high-energy neutrinos from some of the very active objects of the universe (like quasi-stellar radio sources) would provide extremely valuable information about the processes occurring in them and in general about high-energy aspects of cosmic-ray origin. It may also become possible to detect the existence of galaxies made of antimatter by observations of high-energy antiprotons (Pal and Peters, 1964; Pal, 1964) or on high-energy positrons.

† The existence of the 3°K radiation, if confirmed, would have several interesting consequences: Griesen (1966) has pointed out that, because of photonuclear reactions, there would be an end to the cosmic-ray proton spectrum at $\sim 10^{20}$ ev. Jelly (1966) and Gould and Schreder (1966) have shown that, because of $\gamma + \gamma \rightarrow e^+ + e^-$ reactions, space would become virtually opaque to very high-energy γ rays. For example, γ rays of energy $\geqq 10^{15}$ ev would suffer very serious attenuation even while arriving from the center of the galaxy. Therefore nonobservation of such γ rays may have no relevance to the energy density of very low-energy neutrinos in the universe. (However, the depth of a possible Fermi sea of neutrinos in the universe can still be limited to values of the order of 1 ev by putting limits on the attenuation of the nucleon flux at the high-energy end of the cosmic ray spectrum.)

References

Abraham, F., J. Kidd, M. Koshiba, R. Levi-setti, C. H. Tsao, W. Wolter, C. L. Deney, R. L. Fricken, and R. W. Huggett: Bristol Conference on Ultra-high Energy Physics, 1963.

Achar, C. V., M. G. K. Menon, V. S. Narasimham, P. V. Ramana Murthy, B. V. Sreekantan, H. Hinotani, S. Miyake, D. R. Creed, J. L. Osborne, J. B. M. Pattison, and A. W. Wolfendale: *Phys. Letters*, **18**: 196 (1965).

Agrinier, B., Y. Koechlin, B. Parlier, G. Boella, G. Degli Antoni, C. Dilworth, L. Scarsi, and G. Sironi: *Proc. Intern. Conf. Cosmic Rays, Jaipur*, vol. 3, p. 167, 1963.

Agrinier, B., Y. Koechlin, B. Parlier, J. Vasseur, C. J. Bland, G. Boella, G. Degli Antoni, C. Dilworth, L. Scarsi, and G. Sironi: *Proc. Intern. Conf. Cosmic Rays, London*, vol. 1, p. 331, 1965.

Aizu, H.: *Proc. Intern. Conf. Cosmic Rays, Jaipur*, vol. 3, p. 90, 1963.

Akashi, M., K. Shimiju, Z. Watanabe, T. Ogata, N. Ogita, A. Misaki, I. Mito, S. Oyama, S. Tokunaga, M. Tamura, Y. Fujimoto, S. Hasegawa, J. Nishimura, K. Niu, and K. Yokoi: *J. Phys. Soc. Japan*, **17** (Suppl. A-III): 427 (1962).

Akashi, M., K. Shimizu, Z. Watanabe, J. Nishimura, K. Niu, N. Ogita, Y. Tsuneoka, T. Taira, T. Ogata, A. Misaki, I. Mito, Y. Oyama, S. Tokunaga, A. Nishio, S. Dake, Y. Yokoi, Y. Fujimoto, T. Suzuki, C. M. C. Lattes, C. Q. Orsini, I. G. Pacca, M. T. Cruz, E. Okuno, and S. Hasegawa: *Proc. Intern. Conf. Cosmic Rays, Jaipur*, vol. 5, p. 326, 1963.

Akashi, M., Z. Watanabe, J. Nishimura, K. Niu, T. Taira, N. Ogita, K. Ogata, Y. Tsuneoka, A. Misaki, I. Mito, K. Nishikawa, Y. Oyama, A. Nishio, I. Ota, S. Dake, K. Yokoi, Y. Fujimoto, S. Hasegawa, A. Osawa, T. Shibata, T. Suzuki, C. M. G. Lattes, C. Q. Orsini, I. G. Pacca, M. T. Cruz, E. Okuno, T.

Borello, M. Kawabata, S. Hasegawa, J. Nishimura, and A. M. Endler: *Proc. Intern. Conf. Cosmic Rays., London*, vol. 2, p. 878, 1965.
Alfvén, H.: "Cosmical Electrodynamics," Oxford University Press, Fair Lawn, N.J., 1950.
Ali, H., J. Duthie, A. Kaddoura, D. H. Perkins, and P. H. Fowler: *Proc. Intern. Conf. High Energy Physics*, Rochester, 1960, p. 829.
Aller, L. H.: "The Abundance of Elements," Interscience, New York, 1961.
Alvial, G.: *Proc. Intern. Conf. Cosmic Rays, Jaipur*, vol. 3, p. 116, 1963.
Aly, H. H., M. F. Kaplon, and M. L. Shen: *Proc. Intern. Conf. Cosmic Rays, Jaipur*, vol. 5, p. 85, 1963.
Anand, K. C., S. Biswas, P. J. Lavakare, S. Ramadurai, N. Sreenivasan, V. S. Bhatia, V. S. Chohan, and S. D. Pabbi: *J. Geophys. Res.*, **71**: 4687 (1966).
Appa Rao, M. V. K., R. R. Daniel, and K. A. Neelakantan: *Proc. Indian Acad. Sci.*, **43**: 181 (1956).
Appa Rao, M. V. K.: *Phys. Rev.*, **123**: 295 (1961).
Arnold, J. R., A. E. Metzger, E. C. Anderson, and M. A. Van Dills: *J. Geophys. Res.*, **67**: 4878 (1962).
Ashton, F., Y. Kamiya, P. K. Mackeown, J. L. Osborne, J. B. M. Pattison, P. V. Ramana Murthy, and A. W. Wolfendale: *Proc. Phys. Soc. (London)*, **87**: 79 (1966).
Azimov, S. A., A. M. Abdullaev, V. M. Myalkovsky, and T. S. Yuldashbaev: *Proc. Intern. Conf. Cosmic Rays, Jaipur*, vol. 5, p. 69, 1963.
Badhwar, G. D., R. R. Daniel, and B. Vijayalakshmi: *Progr. Theoret. Phys.*, **28**: 607 (1962).
Bahcall, J. N., and R. A. Wolf: *Phys. Rev.*, **140**: B 1445, 1452 (1965).
Baker, W. F., Q. L. Cool, E. W. Jenkins, T. F. Kycia, S. J. Lindenbaum, W. A. Love, D. Lüers, J. A. Niederer, S. Ozaki, A. L. Read, J. J. Russel, and L. C. L. Yuan: *Phys. Rev. Letters*, **7**: 101 (1961).
Balasubrahmanyan, V. K., S. V. Damle, G. S. Gokhale, M. G. K. Menon, and S. K. Roy: *Proc. Intern. Conf. Cosmic Rays, Jaipur*, vol. 3, p. 110, 1963.
Balasubrahmanyan, V. K., D. E. Hagge, G. H. Ludwig, and F. B. McDonald: *NASA Goddard Space Flight Center, Preprint* X-661-65-480, 1965.
Balasubrahmanyan, V. K., and F. B. McDonald: *J. Geophys. Res.*, **69**: 3289 (1964).
Baradzei, L. T., V. I. Rubtsov, Y. A. Smorodin, M. V. Solovyov, and B. V. Tolkachev: *J. Phys. Soc. Japan*, **17** (Suppl. A-III): 433 (1962).
Barret, P. H., L. M. Bollinger, G. Cocconi, Y. Eisenberg, and K. Greisen: *Revs. Mod. Phys.*, **24**: 133 (1952).
Bethe, H. A., S. A. Korff, and G. Placzek: *Phys. Rev.*, **57**: 573 (1940).
Bhabha, H. J., and W. Heitler: *Proc. Roy. Soc. (London)*, **A159**: 432 (1937).
Bhabha, H. J.: *Proc. Roy. Soc. (London)*, **A164**: 257 (1938).
Bierman, L.: *Z. Astrophys.*, **29**: 274 (1951).
Bigi, A., S. Brandt, R. Carrara, W. A. Cooper, G. R. Macleod, Aurelia de Marco, Ch. Peyrou, R. Sosnowski, and A. Wroblewski: *Proc. High Energy Conf. CERN*, p. 247, 1962.
Birkeland, K.: "On the Cause of Magnetic Storms and the Origin of Terrestrial Magnetism," H. Aschehong and Co., Christiania, Norway, First Section, 1908, Second Section, 1913.
Biswas, S., and C. E. Fichtel: *Space Sci. Revs.*, **4**: 709 (1965).
Biswas, S., C. E. Fichtel, D. E. Guss, and C. J. Waddington: *J. Geophys. Res.*, **68**: 3109 (1963).
Biswas, S., P. S. Freier, and W. Stein: *J. Geophys. Res.*, **67**: 13 (1962).
Bollinger, L. M.: Ph.D. Thesis, Cornell University, 1951.
Bonetti, A., H. S. Bridge, A. J. Lazarus, E. F. Lyon, B. Rossi, and F. Sherb: "Space Research," III, p. 540, North Holland Publishing Company, Amsterdam, 1963.
Bowler, M., J. Duthie, P. Fowler, A. Kaddoura, D. Perkins, K. Pinkau, and W. Wolter: *J. Phys. Soc. Japan*, **17** (Suppl. A-III): 424 (1962).
Bozoki, G., E. Fenyves, and L. Janossy: *Nucl. Phys.*, **33**: 236 (1962).
Bradt, H. L., M. F. Kaplon, and B. Peters: *Helv. Phys. Acta*, **23**: 24 (1950).
Bradt, H. L., and B. Peters: *Phys. Rev.*, **74**: 1828 (1948).
Bradt, H. L., and B. Peters: *Phys. Rev.*, **77**: 54 (1950a).
Bradt, H. L., and B. Peters: *Phys. Rev.*, **80**: 943 (1950b).
Bridge, H. S., C. Dilworth, B. Rossi, F. Sherb, and E. F. Lyon: *J. Geophys. Res.*, **65**: 3053 (1960).
Bridge, H. S., A. Egidi, A. J. Lazarus, E. Lyon, and L. Jacobsen: "Space Research," IV, North Holland Publishing Company, Amsterdam, 1964.
Bridge, H. S., A. J. Lazarus, E. F. Lyon, B. Rossi, and F. Scherb: "Space Research," III, p. 1113, North Holland Publishing Company, Amsterdam, 1963.
Brooke, G., M. A. Meyer, and A. W. Wolfendale: *Proc. Phys. Soc. (London)*, **83**: 871 (1964).
Brooke, G., and A. W. Wolfendale: *Proc. Phys. Soc. (London)*, **83**: 843 (1964).
Bryant, D. A., T. L. Cline, U. D. Desai, and F. B. McDonald: *J. Geophys. Res.*, **67**: 4983 (1962).
Cameron, A. G. W.: *J. Astrophys.*, **129**: 676 (1959).
Carlson, J. F., and J. R. Oppenheimer: *Phys. Rev.*, **51**: 220 (1937).
Castagnoli, C., G. Cortini, C. Franzinetti, A. Manfredini, and D. Moreno: *Nuovo Cimento*, **10**: 1539 (1953).
Chandrasekhar, S., and E. Fermi: *Astrophys. J.*, **118**: 113 (1953).
Chapman, S.: *Smithsonian Contrib. Astrophys.*, **2**: 1 (1957).
Chapman, S.: *Proc. Roy. Soc. (London)*, **A253**: 462 (1959).
Chapman, S., and V. C. A. Ferraro, *Terr. Mgn. Atm. Elec.*, **35**: 77, 171 (1931).
Chapman, S., and V. C. A. Ferraro: *Terr. Mgn. Atm. Elec.*, **37**: 147, 421 (1932).
Chatterjee, B. K., S. Lal, T. Matano, G. T. Murthy, S. Naranan, K. Sivaprasad, B. V. Sreekantan, M. V. Srinivasa Rao, and P. R. Vishwanath: *Proc. Intern. Conf. Cosmic Rays, London*, vol. p. 2, 627, 1965a.
Chatterjee, B. K., G. T. Murthy, S. Naranan, B. V. Sreekantan, and M. V. Srinivasa Rao: *Proc. Intern. Conf. on Cosmic Rays, Jaipur*, vol. 4, p. 227, 1963.
Chatterjee, B. K., G. T. Murthy, S. Naranan, B. V. Sreekantan, M. V., Srinivasa Rao, and S. C. Tonwar: *Proc. Intern. Conf. Cosmic Rays, London*, vol. 2, p. 805, 1965b.
Christofilos, N. C.: *J. Geophys. Res.*, **64**: 869 (1959).
Chudakov, A. E., V. L. Dadykin, V. I. Zatsepin, and N. W. Nesterova: *J. Phys. Soc. Japan*, **17** (Suppl. A-III): 106 (1962).
Ciok, P., T. Coghen, J. Gierula, R. Holynski, A. Jurak, M. Miesowicz, T. Saniewska, and J. Pernegr: *Nuovo Cimento*, **10**: 741 (1958).
Clark, G., H. Bradt, M. La Pointe, V. Domingo, I. Escobar, K. Murakami, K. Suga, Y. Toyoda, and J. Hersil: *Proc. Intern. Conf. Cosmic Rays, Jaipur*, vol. 4, p. 65, 1963.
Cline, T. L., G. H. Ludwig, and F. B. McDonald: *Phys. Rev. Letters*, **13**: 786 (1946).
Cocconi, G.: *Phys. Rev.*, **111**: 1699 (1958).
Cocconi, G.: "Encyclopedia of Physics," vol. 46/1, p. 215, Springer, Berlin, 1961.
Cocconi, G., and V. Cocconi-Tongiorgi: *Phys. Rev.*, **84**: 29 (1951).
Cocconi, G., V. Cocconi-Tongiorgi, and K. Greisen: *Phys. Rev.*, **75**: 1063, **76**: 1020 (1949).

Cocconi, G., A. Koester, and D. H. Perkins: *Lawrence Radiation Lab.*, *UCID*-1444, 1961.

Colgate, S. A., and M. H. Johnson: *Phys. Rev. Letters*, **5**: 235 (1960).

Compton, A. H., and I. A. Getting: *Phys. Rev.*, **47**: 817 (1935).

Comstock, S. W., C. Y. Fan, and J. A. Simpson: private communication (1966).

Cowsik, R., Yash Pal, T. N. Rengarajan, and S. N. Tandon: *Proc. Intern. Conf. Cosmic Rays, Jaipur*, vol. 6, p. 211, 1963.

Cowsik, R., Yash Pal, and S. N. Tandon: *Phys. Rev. Letters*, **13**: 265 (1964).

Cowsik, R., Yash Pal, and S. N. Tandon: *Proc. Indian Acad. Sci.*, **63**: 217 (1966).

Dahanayake, C., M. F. Kaplon, and P. J. Lavakare: *J. Geophys. Res.*, **69**: 3651 (1964).

Damgaard, G., P. Grieder, K. H. Hansen, C. Iversen, E. Lohse, B. Peters, A. Klovning, T. Rengarajan, and A. Lillethun: *Proc. Intern. Conf. Cosmic Rays, London*, vol. 2, 808, 1965.

Damgaard, G., and K. H. Hansen: *Proc. Sienna Intern. Conf. Elementary Particles*, vol. 1, p. 643, 1963.

Daniel, R. R., and N. Durgaprasad: *Nuovo Cimento, Suppl.* 23, 82 (1962).

Daniel, R. R., N. Durgaprasad, P. K. Malhotra, and B. Vijayalakshmi: *Proc. Intern. Conf. Cosmic Rays, Jaipur*, vol. 5, p. 9, 1963.

Daniel, R. R., and S. A. Stephens: *Phys. Rev. Letters*, **15**: 769 (1965).

Danilova, T. V., and S. I. Nikolsky: *Proc. Intern. Conf. Cosmic Rays, Jaipur*, vol. 4, p. 221, 1963.

Davis, L., Jr.: "Space Science," p. 486, Wiley, New York, 1963.

Davis, L., Jr., and J. L. Greenstein: *Astrophys. J.*, **114**: 206 (1951).

Dekkers, D., J. A. Giebel, R. Mermod, G. Weber, T. R. Willits, K. Winter, B. Jordon, M. Vivargent, N. M. King, and E. J. N. Wilson: *Phys. Rev.*, **137**: 962 (1965).

De Shong, J. A., R. H. Hildebrand, and P. Meyer: *Phys. Rev. Letters*, **12**: 3 (1964).

Dessler, A. J.: *Phys. Rev. Letters*, **1**: 68 (1958).

Dicke, R., P. J. E. Peebles, P. G. Roll, and D. J. Wilkinson: *Astrophys. J.*, **142**: 414 (1965).

Dobrotin, N. A., N. G. Zelevinskaya, K. A. Kotelnikov, V. M. Maximenko, V. S. Puchkov, S. A. Salavatinsky, and I. N. Fetisov: *Proc. Intern. Conf. Cosmic Rays, Jaipur*, vol. 5, p. 79, 1963.

Dorman, L. I., and L. Kh. Shatashvili: 'Results of IGY' (Moscow) Cosmic Rays No. 4, 1961.

Duller, N. M., and W. D. Walker: *Phys. Rev.*, **93**: 215 (1954).

Duperier, A.: *Proc. Phys. Soc. (London)*, **A62**: 684 (1949).

Durgaprasad, N.: *Proc. Intern. Conf. Cosmic Rays, Jaipur*, vol. 3, p. 17, 1963.

Durney, A. C., H. Elliot, R. J. Hynds, and J. J. Quenby: *Nature*, **195**: 1245 (1962).

Duthie, J. G., R. Cobb, and J. Stewart: *Phys. Rev. Letters*, **17**: 263 (1966).

Duthie, J. G., C. M. Fisher, P. H. Fowler, A. Kaddoura, D. H. Perkins, K. Pinkau, and W. Wolters: *Phil. Mag.*, **6**: 113 (1961).

Duthie, J. G., P. H. Fowler, A. Kaddoura, D. H. Perkins, and K. Pinkau: *Nuovo Cimento*, **24**: 122 (1962).

Earl, J. A.: *Phys. Rev. Letters*, **6**: 125 (1961).

Edwards, B., J. Losty, D. H. Perkins, K. Pinkau, and J. Reynolds: *Phil. Mag.*, **3**: 237 (1958).

Elliot, H.: *Report on Progress in Physics*, **26**: 145 (1963).

Eyges, L.: *Phys. Rev.*, **74**: 1801 (1948).

Eyges, L., and S. Fernbach: *Phys. Rev.*, **82**, 23, 287 (1951).

Fan, C. Y., G. Gloeckler, and J. A. Simpson: *Phys. Rev. Letters*, **13**: 149 (1964).

Fan, C. Y., G. Gloeckler, and J. A. Simpson: *Proc. Intern. Conf. Cosmic Rays, London*, vol. 1, p. 380, 1965.

Fermi, E.: *Phys. Rev.*, **75**: 1169 (1949).

Fermi, E.: *Phys. Rev.*, **81**: 681 (1951).

Fermi, E.: *Progr. Theoret. Phys.*, **5**: 570 (1950).

Fermi, E.: *Astrophys. J.*, **119**: 1 (1954).

Ferraro, V. C. A.: *J. Geophys. Res.*, **57**: 15 (1952).

Ferraro, V. C. A.: *J. Geophys. Res.*, **65**: 3951 (1960).

Finch, H. F., and B. R. Leaton: *Monthly Notices Roy. Astron. Soc. (Geophys. Suppl.)*, **7** (6): 314 (1957).

Fireman, E. L., J. DeFelice, and D. Tilles: *Phys. Rev.*, **123**: 1935 (1961).

Fleisher, R. L., P. B. Price, and R. M. Walker: private communication, 1965.

Foster, F., and J. H. Mulvey: *Nuovo Cimento*, **27**: 93 (1963).

Fowler, G. N., and A. W. Wolfendale: "Encyclopedia of Physics," vol. 46/1, p. 272, Springer, Berlin, 1961.

Fowler, G. N., and A. W. Wolfendale: *Prog. in Cosmic Rays and Elementary Particle Physics*, **4**: 107 (1958).

Freden, S. C., and R. S. White: *J. Geophys. Res.*, **65**: 1377 (1960).

Freier, P., E. J. Lofgren, E. P. Ney, F. Oppenheimer, H. L. Bradt, and B. Peters: *Phys. Rev.*, **74**: 213 (1948).

Freier, P., and C. J. Waddington: *Phys. Rev. Letters*, **13**: 108 (1964).

Freier, P. S., and C. J. Waddington: *Space Sci. Revs.*, **4**: 313 (1965).

Fruin, J. H., J. V. Jelly, C. O. Long, N. A. Porter, and T. C. Weekes: *Phys. Letters*, **10**: 176 (1964).

Fujimoto, Y., H. Hasegawa, and M. Taketani: *Progr. Theoret. Phys., Suppl.*, no. 30, p. 32, 1964.

Galbraith, W.: "Extensive Air Showers," Butterworth, London, 1958.

Ganguli, S. N., N. Kameshwara Rao, and M. S. Swami: *Proc. Intern. Conf. Cosmic Rays, Jaipur*, vol. 3, p. 65, 1963.

Gardener, M., D. G. Jones, F. E. Taylor, and A. W. Wolfendale: *Proc. Phys. Soc. (London)*, **80**: 697 (1962).

Garmire, G.: *Proc. Intern. Conf. on Cosmic Rays, London*, vol. 1, p. 315 (1965).

Gawin, J., J. Hibner, and A. Zawadzki: *Proc. Intern. Conf. Cosmic Rays, Jaipur*, vol. 4, p. 180, 1963.

George, E. P., and J. Evans: *Proc. Phys. Soc. (London)*, **A68**: 829 (1955).

Ginzburg, V. L.: *Usp. Fiz. Nauk SSSR*, **51**: 343 (1953).

Ginzburg, V. L.: *Progr. Elem. Particle Cosmic Ray Phys.*, **4**: 339 (1958).

Goldberg, L., E. A. Muller, and L. H. Aller: *Astrophys. J. Suppl.*, **45** (5): 1 (1960).

Gould, R. J., and G. Schereder: *Phys. Rev. Letters*, **16**: 252 (1966).

Greisen, K.: *Progr. Cosmic Ray Phys.*, **3**: 3 (1956).

Greisen, K.: *Ann. Rev. Nucl. Sci.*, **10**: 63 (1960).

Greisen, K.: *Phys. Rev. Letters*, **16**: 748 (1966).

Greisen, K., H. Ogelman, and J. Delville, after M. Oda: *Proc. Intern. Conf. on Cosmic Rays, London*, vol. 1, p. 69 (1965).

Gringauz, K. I., V. V. Bezrukikh, V. D. Ozerov, and R. E. Rybchinskii: *Dokl. Akad. Nauk SSSR*, **131**: 1301 (1960).

Gringauz, K. I., V. V. Bezrukikh, S. M. Balandina, V. D. Ozerov, and R. E. Rybchinskii: "Space Research," III, p. 602, North Holland Publishing Company, Amsterdam, 1963.

Guseva, V. V., N. A. Dobrotin, N. G. Zelevinskaya, K. A. Kotelnikov, A. M. Lebedev, and S. A. Slavotinsky: *J. Phys. Soc. Japan*, **17** (Suppl. *A*-III): 375 (1962).

Hansen, L. F., and W. B. Fretter: *Phys. Rev.*, **118**: 812 (1960).

Hasegawa H., H. Aizu, and K. Ito: *Proc. Intern. Conf. Cosmic Rays, Jaipur*, vol. 3, p. 83, 1963a.
Hasegawa, H., K. Murakami, S. Shibata, K. Suga, Y. Toyoda, V. Domingo, I. Escobar, K. Kamata, H. Bradt, G. Clark, and M. Lal Pointe: *Proc. Intern. Conf. Cosmic Rays, London*, vol. 1, p. 708, 1965.
Hasegawa, H., S. Nakagawa, and E. Tamai: *Proc. Intern. Conf. Cosmic Rays, Jaipur*, vol. 3, p. 86, 1963.
Hayakawa, S., C. Hayashi, and M. Nishida: *Progr. Theoret. Phys., Suppl.*, no. 16, p. 169, 1960.
Hayakawa, S., K. Ito, and Y. Terashima: *Progr. Theoret. Phys. Suppl.*, no. 6, p. 1, 1958.
Hayakawa, S., and M. Matsuoka: *Progr. Theoret. Phys. Suppl.*, no. 30, p. 204, 1964.
Hayakawa, S., M. Matsuoka, and K. Yamashita: *Proc. Intern. Conf. on Cosmic Rays, London*, vol. 1, p. 119 (1965).
Hayakawa, S., J. Nishimura, H. Obayaski, and H. Sato: *Progr. Theoret. Phys. Suppl.*, no. 30, p. 86, 1964.
Hayman, P. J., N. S. Palmer, and A. W. Wolfendale: *Proc. Phys. Soc. (London)*, **80**: 800 (1962).
Hayman, P. J., and A. W. Wolfendale: *Proc. Phys. Soc. (London)*, **80**: 710 (1962).
Heckman, H. H., and A. H. Armstrong: *J. Geophys. Res.*, **67**: 1255 (1962).
Heitler, W.: "The Quantum Theory of Radiation," Oxford University Press, London, 1956.
Hess, W. N.: *J. Geophys. Res.*, **65**: 3107 (1960).
Higashi, S., T. Kitamura, Y. Mishima, S. Mitani, S. Miyamoto, T. Oshio, H. Shibata, K. Watanabe, and Y. Watase: *J. Phys. Soc. Japan*, **17** (Suppl. A-III): 362 (1962).
Higashi, S., T. Kitamura, Y. Mishima, S. Miyamoto, H. Shibata, and Y. Watase: *Proc. Intern. Conf. Cosmic Rays, Jaipur*, vol. 6, p. 53, 1963.
Hildebrand, B., F. W. O'Dell, M. M. Shapiro, R. Silberberg, and B. Stiller: *Proc. Intern. Conf. Cosmic Rays, Jaipur*, vol. 3, p. 101, 1963.
Hildebrand, B., and R. Silberberg: *Proc. Intern. Conf. Cosmic Rays, Jaipur*, vol. 5, p. 20, 1963.
Huggett, R. W., K. Mori, C. O. Kim, and R. Levi Setti: *Proc. Intern. Conf. Cosmic Rays, Jaipur*, vol. 5, p. 3, 1963.
Hultqvist, B.: *Arkiv Geofysik*, **3**: 63 (1958).
Jakeman, D.: *Can. J. Phys.*, **34**: 432 (1956).
Jelley, J. V.: *Phys. Rev. Letters*, **16**: 479 (1966).
Jordan, B.: Thesis, *CERN* 65-14, 1965.
Kamata, K., and J. Nishimura: *Progr. Theoret. Phys. Suppl.*, no. 6, p. 93, 1958.
Kane, R. P.: *Proc. Symp. Cosmic Rays, Elem. Particle Phys. Astrophys.*, Bombay, 1965.
Kaplon, M. F., B. Peters, H. L. Reynolds, and D. M. Ritson: *Phys. Rev.*, **85**: 295 (1952).
Kaplon, M. F., D. M. Ritson, and E. Woodruff: *Phys. Rev.*, **85**: 933 (1952).
Katz, L., P. Meyer, and J. A. Simpson: *Nuovo Cimento*, **8** (Suppl. II): 277 (1958).
Kellogg, P. J.: *Nuovo Cimento*, **11**: 48 (1959).
Kellogg, P. J., and M. Schwartz: *Nuovo Cimento*, **13**: 761 (1959).
Kessler, P.: *Nuovo Cimento*, **17**: 809 (1960).
Kessler, D., and P. Kessler: *Nuovo Cimento*, **4**: 601 (1956).
Kim, C. O.: *Proc. Intern. Conf. Cosmic Rays, Jaipur*, vol. 5, p. 382, 1963.
Koba, Z., and S. Takagi: *Nuovo Cimento*, **10**: 755 (1958).
Koshiba, M.: *Proc. Intern. Conf. Cosmic Rays, Jaipur*, vol. 5, p. 293, 1963.
Kraushaar, W., and G. W. Clark: *Phys. Rev. Letters*, **8**: 106, (1962).
Kraushaar, W. L., and L. J. Marks: *Phys. Rev.*, **93**: 326 (1954).
Lal, D., P. K. Malhotra, and B. Peters: *J. Atmospheric Terrest. Phys.*, **12**: 306 (1958).
Lal, D., Yash Pal, and B. Peters: *Proc. Indian Acad. Sci.*, **38**: 398 (1953).
Lal, S., Yash Pal, and R. Raghavan: *Nuclear Phys.*, **31**: 415 (1962).
Lal, D., and B. Peters: *Progr. Elem. Particle Cosmic Ray Phys.*, **6**: 3 (1962).
Lal, D., and B. Peters: "Encyclopedia of Physics," Springer, Berlin 1965.
Lal, D., G. Rajagopalan, and V. S. Venkatavardhan: *Proc. Intern. Conf. Cosmic Rays, Jaipur*, vol. 1, p. 99, 1963.
Lal, S., R. Raghavan, T. N. Rengaswami, B. V. Sreekantan, and A. Subramanian: *Proc. Intern. Conf. Cosmic Rays, Jaipur*, vol. 5, p. 260, 1963.
Lemaître and M. S. Vallarta: *Phys. Rev.*, **50**: 493 (1936).
L'Heureux, J., and P. Weyer: *Phys. Rev. Letters*, **15**: 93 (1965).
Lindenbaum, S. J., and R. M. Sternheimer: *Phys. Rev.*, **105**: 1874 (1957); **106**: 1107 (1957).
Lindgren, S.: *Tellus*, **14**: 44 (1962).
Lingenfelter, R. E.: *J. Geophys. Res.*, **68**: 5633 (1963).
Lingenfelter, R. E., E. J. Flamm, E. H. Canfield, and S. Kellman: *J. Geophys. Res.*, **70**: 4077, 4087 (1965).
Linsley, J.: *Proc. Intern. Conf. Cosmic Rays, Jaipur*, vol. 4, p. 77, 1963.
Linsley, J., and L. Scarsi: *Phys. Rev. Letters*, **9**: 123 (1962).
Lock, W. O.: *Proc. Intern. Conf. Cosmic Rays, Jaipur*, vol. 5, p. 105, 1963.
Lohrmann, E., M. W. Teucher, and M. Schein: *Phys. Rev.*, **122**: 672 (1961).
Malhotra, P. K.: *Proc. Intern. Conf. Cosmic Rays, Jaipur*, vol. 5, p. 40, 1963.
Malhotra, P. K.: Ph.D. Thesis, University of Bombay, 1964.
Malhotra, P. K., P. G. Shukla, S. A. Stephens, B. Vijayalakshmi, J. Boult, M. G. Bowler, P. H. Fowler, H. L. Hockforth, J. Keereetaveep, V. W. Mayes, and S. N. Tovey: *Nuovo Cimento*, **40**: 385, 404, (1965).
Mando, M., and L. Ronchi: *Nuovo Cimento*, **9**: 105 (1952a).
Mando, M., and L. Ronchi: *Nuovo Cimento*, **9**: 517 (1952b).
Maor, V., and G. Yekutieli: *Nuovo Cimento*, **17**: 45 (1960).
Markov, M. A., and I. M. Zhelezhnykh: *Nucl. Phys.*, **27**: 385 (1961).
Matano, T., I. Miura, M. Nagano, M. Oda, S. Shibata, Y. Tanaka, G. Tanahashi and H. Hasegawa: *Proc. Intern. Conf. Cosmic Rays, Jaipur*, vol. 4, p. 129, 1963.
McCusker, C. B. A.: *Proc. Intern. Conf. Cosmic Rays, Jaipur*, vol. 4, p. 35, 1963.
McDonald, F. B.: *NASA Tech. Rept.*-169, 1963.
McDonald, F. B., V. K. Balasubrahmanyan, K. A. Brunstein, D. E. Hagge, G. H. Ludwig, and R. A. Palmeira: *Trans. Am. Geophys. Union*, **46**: 124 (1965).
McEwen, J. G.: *Phys. Rev.*, **115**: 1712 (1959).
Menon, M. G. K., and P. V. Ramana Murthy: *Progr. Elem. Particle Cosmic Ray Phys.*, **9**: (1966).
Meyer, P., and R. Vogt: *Phys. Rev. Letters*, **6**: 193 (1961).
Minakawa, O., Y. Nishimura, M. Tsuzuki, H. Yamanouchi, H. Aizu, H. Hasegawa, Y. Ishii, S. Tokunaga, Y. Fujimoto, S. Hasegawa, J. Nishimura, K. Niu, K. Nishikawa, K. Imaeda, and M. Kazuno: *Nuovo Cimento, Suppl.*, **11**: 125 (1959).
Miyake, S.: *J. Phys. Soc. Japan*, **18**: 1093 (1963).
Miyake, S., V. S. Narasimham, and P. V. Ramana Murthy: *Nuovo Cimento*, **32**: 1505, 1524 (1964).
Molière, G.: in W. Heisenberg (ed.), "Cosmic Radiation," chap. 3, Dover, New York, 1946.
Morrison, P.: *Phys. Rev.*, **101**: 1397 (1956).

Neher, H. V.: *Progr. Cosmic Ray Phys.*, **1**: 243 (1952).
Ness, N. F.: *Proc. Intern. Conf. Cosmic Rays, London*, vol. 1, p. 14, 1965.
Nikolsky, S. J.: *Proc. Fifth Interamerican Seminar Cosmic Rays, Lapaz*, vol. 2, 1962.
Nishimura, J.: *Proc. Intern. Conf. Cosmic Rays, Jaipur*, vol. 6, p. 224, 1963.
Nishimura, J., and K. Kamata: *Progr. Theoret. Phys.*, **5**: 899 (1950); **6**: 628 (1951); **7**: 185 (1952).
Niu, K.: *Nuovo Cimento*, **10**: 994 (1958).
O'Brien, B. J.: *Space Sci. Revs.*, **1**: 415 (1962).
O'Brien, B. J., C. D. Laughlin, and J. A. Van Allen: *Nature*, **195**: 939 (1962).
Oda, M.: *Proc. Intern. Conf. Cosmic Rays, London*, vol. 1, p. 68, 1965.
O'Dell, F. W., M. M. Shapiro, and B. Stiller: *J. Phys. Soc. Japan*, **17** (Suppl. A-III): 23 (1962).
Osborne, J. L., S. S. Said, and A. W. Wolfendale: *Proc. Phys. Soc. (London)*, **86**: 93 (1965).
Pal, Y.: Seminars at Summer School on Theoretical Physics, Bangalore, 1962 (unpublished).
Pal, Y.: *Proc. Intern. Conf. Cosmic Rays, Jaipur*, vol. 5, p. 445, 1963.
Pal, Y.: *Proc. Conf. Interaction Between Cosmic Rays and High Energy Phys.*, Cleveland, III-1, 1964.
Pal, Y., and B. Peters: *Kgl. Danske Videnskab. Selskab, Mat.-Fys. Medd.*, **33** (15): (1964).
Pal, Y., A. K. Ray, and T. N. Rengarajan: *Nuovo Cimento*, **28**: 1177 (1963).
Pal, Y., and T. N. Rengarajan: *Phys. Rev.*, **124**: 1575 (1961).
Pal, Y., and S. N. Tandon: *Proc. Intern. Conf. Cosmic Rays, London*, 1965.
Pal, Y., and S. N. Tandon: *Proc. Intern. Conf. on Cosmic Rays, London*, vol. 2, p. 890 (1965b).
Pal, Y. and S. N. Tandon: *Proc. Intern. Conf. on Cosmic Rays, London*, vol. 2, p. 1111 (1965c).
Parker, E. N.: "Interplanetary Dynamical Processes," Wiley, New York, 1963.
Parker, E. N.: *Phys. Rev.*, **107**: 830 (1957).
Parker, E. N.: *Planetary Space Sci.*, **12**: 735 (1964).
Parker, E. N.: *Proc. Intern. Conf. Cosmic Rays, London*, vol. 1, p. 26, 1965a.
Penzias, A. S., and R. W. Wilson: *Astrophys. J.*, **142**: 419 (1965).
Perkins, D. H.: *Progr. Elem. Particle Cosmic Ray Phys.*, **5**: 257 (1960).
Perkins, D. H.: *Intern. Conf. on Theoretical Aspects of Very High Energy Phenomena, CERN*, p. 99 (1961).
Peters, B.: *Proc. Intern. Conf. Cosmic Rays, Moscow*, vol. 3, p. 157, 1959.
Peters, B.: *Nuovo Cimento*, **22**: 800 (1961).
Peters, B.: *Proc. Intern. Conf. High Energy Phys., CERN*, p. 623, 1962.
Peters, B.: *Proc. Intern. Conf. Cosmic Rays, Jaipur*, vol. 5, p. 423, 1963.
Peterson, L. E.: *COSPAR Symposium*, Buenos Aires (1965).
Pinkau, K.: Ph.D. Thesis, Bristol University, 1958.
Pizzella, G., C. E. McIlwain, and J. A. Van Allen: *J. Geophys. Res.*, **67**: 1235 (1962).
Poincaré, H.: *Compt. Rend.*, **123**: 930 (1896).
Quenby, J. J., and W. R. Webber: *Phil. Mag.*, **4**: 90 (1959).
Quenby, J. J., and G. J. Wenk: *Phil. Mag.*, **7**: 1457 (1962).
Ramana Murthy, P. V., B. V. Sreekantan, S. D. Verma, and A. Subramanian: *Nuclear Instruments and Methods*, **23**: 245 (1963).
Reines, F., M. F. Crouch, T. L. Jenkins, W. R. Kropp, H. S. Gurr, G. R. Smith, J. P. F. Sellschop, and B. Meyer: *Phys. Rev. Letters*, **15**: 429 (1965).
Rengarajan, T. N., and S. N. Tandon: private communication, 1965.
Roederer, J. G.: *Space Sci. Rev.*, **3**: 847 (1964).
Rose, D. C., K. B. Fenton, J. Katzman, and J. A. Simpson: *Can. J. Phys.*, **34**: 968 (1956).
Rossi, B.: "High Energy Particles," Prentice-Hall, Englewood Cliffs, N.J., 1952.
Rossi, B., and K. Greisen: *Revs. Mod. Phys.*, **13**: 240 (1941).
Rothwell, P., and J. J. Quenby: *Nuovo Cimento, Suppl.* II, p. 249, 1958.
Sandage, A. R., P. Osmer, R. Giacconi, P. Gorenstein, H. Gursky, J. Waters, H. Bradt, G. Garmire, B. V. Sreekantan, M. Oda, K. Osawa and J. Jugaku: private communication (1966).
Sands, M.: *Phys. Rev.*, **77**: 180 (1950).
Sarabhai, V. A.: *J. Geophys. Res.*, **68**: 1555 (1963a).
Sarabhai, V. A.: *Proc. Intern. Conf. Cosmic Rays, Jaipur*, vol. 2, p. 117, 1963b.
Schaeffer, O. A., and J. Zähringer: *Phys. Rev. Letters*, **8**: 389 (1962).
Schein, M., W. P. Jesse, and E. O. Wollan: *Phys. Rev.*, **59**: 615 (1941).
Simpson, J. A.: *Phys. Rev.*, **94**: 426 (1954).
Simpson, J. A.: private communication (1966).
Simpson, J. A., H. W. Baldwin, and R. B. Uretz: *Phys. Rev.*, **84**: 332 (1951).
Simpson, J. A., W. Fonger, and S. B. Treiman: *Phys. Rev.*, **90**: 934 (1953).
Singer, S. F.: *Trans. Am. Geophys. Union*, **38**: 175 (1957).
Singer, S. F.: *Phys. Rev. Letters*, **1**: 171, 181 (1958).
Sitte, K.: "Encyclopedia of Physics," vol. 46/1, p. 174, Springer, Berlin, 1961.
Snyder, H. S.: *Phys. Rev.*, **76**: 1563 (1949).
Sreekantan, B. V.: *Proc. Intern. Conf. Cosmic Rays, Jaipur*, vol. 4, p. 143, 1963.
Steljes, J. F., H. Carmichael, and K. G. McCracken: *J. Geophys. Res.*, **66**: 1363 (1961).
Störmer, C.: *Arch. Sci. Phys. Naturelles (Geneva)*, **24**: 317 (1907).
Störmer, C.: "The Polar Aurora," Clarendon Press, Oxford, 1955.
Suga, K., I. Escobar, K. Murakami, V. Domingo, Y. Toyoda, G. Clark, and M. La Pointe: *Proc. Intern. Conf. Cosmic Rays, Jaipur*, vol. 4, p. 9, 1963.
Swann, W. F. G.: *Phys. Rev.*, **43**: 217 (1933).
Takagi, S.: *Progr. Theoret. Phys.*, **7**: 123 (1952).
Tilles, D., J. DeFelice, and E. L. Fireman: *Icarus*, **2**: 258 (1963).
Vallarta, M. S., C. Graef, and S. Kusaka: *Phys. Rev.*, **55**: 1 (1939).
Van Allen, J. A.: Special Joint Meeting of National Academy of Sciences and American Physical Society, Washington, May 1, 1958. (Verbatim transcript of lecture published in IGY Satellite Report 13, January, 1961, IGY World Data Centre A.)
Van Allen, J. A.: "Space Science," p. 226, Wiley New York, 1963.
Van Allen, J. A., and L. A. Frank: *Nature*, **183**: 430 (1959).
Van Allen, J. A., G. H. Ludwig, E. C. Ray, and C. E. McIlwain; *Jet Propulsion*, **28**: 588 (1958).
Vernov, S. N., A. Ye Chudakov, P. V. Vakulov, and Yu. I. Logachev: *Dokl. Akad. Nauk SSSR*, **125**: 304 (1959).
Waddington, C. J.: *Nuovo Cimento*, **3**: 930 (1956).
Waddington, C. J.: *Nuovo Cimento*, **14**: 1205 (1959).
Webber, W. R.: *Nuovo Cimento, Suppl.* II, **8**: 532 (1958).
Webber, W. R.: *Progr. Elem. Particles Cosmic Ray Phys.*, **6**: 77 (1962).
Webber, W. R.: *AAS-NASA Symp. Physics of Solar Flare*, 419, 1964.
Webber, W. R.: *Univ. Minnesota, School of Phys. and Astron., Tech. Rept.* CR-76; also in "Encyclopedia of Physics," vol. 46/2, Springer, Berlin, 1965.

Webber, W. R., and F. B. McDonald: *J. Geophys. Res.*, **69**: 3097 (1964).

Webber, W. R., and J. Ormes: *Proc. Intern. Conf. Cosmic Rays, Jaipur*, vol. 3, p. 3 1963a.

Webber, W. R., and J. Ormes: *Proc. Intern. Conf. Cosmic Rays, Jaipur*, vol. 3, p. 69, 1963b.

Wenk, G.: Thesis, Imperial College, London, 1961.

Zatsepin, G. T.: *J. Phys. Soc. Japan*, **17** (Suppl. A-III): 495 (1962).

Zatsepin, G. T., and V. A. Kuzmin: *Soviet Phys.* JETP (*English Transl.*), **14**: 1294 (1962).

Zatsepin, G. T., and E. D. Mikhalchi: *J. Phys. Soc. Japan*, **17** (Suppl. A-III): 356 (1962).

Zatsepin, G. T., S. I. Nikolsky, and G. B. Khristiansen: *Proc. Intern. Conf. Cosmic Rays, Jaipur*, vol. 4, p. 100, 1963.

Bibliography

Galbraith, W.: "Extensive Air Showers," Butterworth, London, 1958.

Ginzburg, V. L., and S. I. Syrovatskii: "Origin of Cosmic Rays," Pergamon Press, New York, 1964.

Janossy, L.: "Cosmic Rays," Clarendon Press, Oxford, 1948.

Le Galley, Donald P. (ed.): "Space Science," Wiley, New York 1963.

Peters, B. (ed): "Cosmic Rays, Solar Particles and Space Research," Proceedings International School of Physics ≪Enrico Fermi≫, Course XIX, Academic, New York, 1963.

Proceedings, International Conference on Cosmic Rays, Moscow, vols. 1–4, 1959.

Proceedings, International Conference on Cosmic Rays and Earth Storms, Kyoto, *J. Proc. Phys. Soc. Japan*, **17**, A-II, A-III (1962).

Proceedings, International Conference on Cosmic Rays, Jaipur, vols. 1–6, 1963.

Proceedings, International Conference on Cosmic Rays, London, vols. 1, 2, 1965.

Ramakrishnan, A.: "Elementary Particles and Cosmic Rays," Pergamon Press, New York, 1962.

Rossi, B.: "High Energy Particles," Prentice-Hall, Englewood Cliffs, N.J., 1952.

Sandstrom, A. E.: "Cosmic Ray Physics," North Holland Publishing Company, Amsterdam, 1965.

Wilson, J. G. (ed.): "Progress in Cosmic Rays," vols. 1, 2, 3, North Holland Publishing Company, Amsterdam, 1952, 1954, and 1956.

Wilson, J. G., and S. A. Wouthuysen (eds.): "Progress in Elementary Particle and Cosmic Rays," vols. 4, 5, 6, and 7, North Holland Publishing Company, Amsterdam, 1958, 1960, 1962, 1963.

Wolfendale, A. W.: "Cosmic Rays," George Newnes Ltd., London, 1963.

Chapter 12

Physics of Strongly Interacting Particles

By R. D. HILL, University of Illinois

1. Introduction

The particle-physics era of the mid-twentieth century began in 1937 with the discovery of the mu meson. It seems very likely that this discovery was sparked by theoretical interest in an entirely different type of meson. Experimental work, performed between 1937 and 1947, showed that the mu meson was a weakly interacting particle and was not the strongly interacting meson predicted by theory. In 1947, a strongly interacting particle, the pi meson, was discovered. Then followed the production of π mesons by accelerators and the rapid elucidation of the π-meson characteristics. In the meantime, cosmic-ray physics had led to the discovery of newer particles, the K mesons and hyperons (known as strange particles). There followed once more the production of these particles by a second round of higher-energy machines and the further detailed investigation of the properties of these strange particles. This phase has continued up to the present time (1964) and has led to a new aspect of interacting particles, the so-called resonant states or resonance particles.

In this chapter, the subject of particle physics will be discussed largely in the order of the historical development, i.e., under the following headings: (1) Pi-meson Physics, (2) Antiparticles, (3) Strange Particles, (4) Resonant States, (5) Classification of Particles. Thus it is intended to cover only the field of particles known as strongly interacting particles. This field sometimes ties in with electromagnetic and weak interactions, involving electrons, μ mesons, and neutrinos. Except in so far as to indicate the nature of the decays of the strongly interacting particles, these other particles will be discussed elsewhere.

The subject of particle physics has bloomed so rapidly since the first edition of this chapter was written that it is, unfortunately, impossible to give a complete coverage of this field within the limitation of this space. There follows, therefore, only the most condensed review of the subject, and much significant work and many detailed references have regrettably been omitted.

PI-MESON PHYSICS

2. Discovery of π Mesons

It is clear that nucleons (neutrons and protons) are not simple Dirac particles; otherwise they would have normal magnetic moments (that is, $e\hbar/2Mc = 1$ or 0 nuclear magnetons). Instead their moments are $\mu_p = +2.79$ n.m. and $\mu_n = -1.91$ n.m. The anomalous moments (that is, $+1.79$ n.m. for the proton and -1.91 n.m. for the neutron) are attributed mainly to the meson field with which the nucleons interact. The particles associated with this quantized field were predicted by Yukawa [1]† in 1935. According to Yukawa, the meson field is responsible for the strong interactions between nucleons, and from the then known range of this nuclear interaction, Yukawa predicted a meson mass of approximately 200 electron masses. These strongly interacting particles were discovered as a product of high-energy cosmic rays by Lattes, Occhialini, and Powell [2] in 1947.

3. Properties of π Mesons

Charge States. There are three charges of π mesons: π^+, π°, and π^-. These charge states are customarily referred to as members of the isotopic spin triplet with $I = 1$. This conception was first developed for π mesons by Kemmer and grew out of the ideas of isotopic spin for nuclei. For example, the isotopic spins of the proton and the neutron, the simplest nuclei, are $+\frac{1}{2}$ and $-\frac{1}{2}$, respectively. These are the third or z components of the isotopic spin, $I = \frac{1}{2}$, of the nucleon. Likewise, $+1$, 0, and -1 corresponding to π^+, π°, and π^-, respectively, are the z components of the isotopic spin, $I = 1$, of the pion.

Masses. The masses of π^+ and π^- (being particle and antiparticle of one another and therefore having identical mass) are (139.60 ± 0.05) Mev [3]. The mass of the π° is (135.01 ± 0.05) Mev [3]. The mass difference is attributed to electromagnetic-field interactions.

Pion Production. The production of π mesons may take place in any reaction in which there is enough energy available in the center-of-mass system to supply the mass of the pion, provided that energy, momentum, charge, parity, and isotopic spin are conserved. An important feature of π-meson production is that a single pion may be produced. Several examples of pion production, including production by photons, are given in the following

† Numbers in brackets refer to References at end of chapter.

reactions:

$$\begin{aligned} p + p &\rightarrow \pi^+ + n + p \\ p + p &\rightarrow \pi^\circ + p + p \\ p + n &\rightarrow \pi^- + p + p \\ \gamma + p &\rightarrow \pi^+ + n \\ \gamma + n &\rightarrow \pi^- + p \end{aligned} \tag{12.1}$$

Decays. Pi mesons are unstable: π^+ and π^- *in vacuo* have the same mean lives of $(2.551 \pm 0.026) \times 10^{-8}$ sec; π° has a mean life of $(1.8 \pm 0.3) \times 10^{-16}$ sec. In the free state, $\pi^\pm$ decay almost exclusively by the following reactions:

$$\begin{aligned} \pi^+ &\rightarrow \mu^+ + \nu_\mu \\ \pi^- &\rightarrow \mu^- + \bar{\nu}_\mu \end{aligned} \tag{12.2}$$

where ν_μ and $\bar{\nu}_\mu$ are neutrino and antineutrino, respectively, associated with mu mesons. (The existence of mu-meson-associated neutrinos was inferred from the neutrino-interaction experiments by Danby et al. [4].) In materials, π^- mesons generally are captured by nuclei before the pions have a chance to decay. Several rare alternative decays of the charged π^+ meson have been observed:

$$\begin{aligned} &\pi^+ \rightarrow e^+ + \nu_e && (1.24 \times 10^{-4} \text{ of normal decay}) \\ &\pi^+ \rightarrow \mu^+ + \nu_e + \gamma && (1.24 \times 10^{-4} \text{ of normal decay}) \\ &\pi^+ \rightarrow e^+ + \nu_e + \pi^\circ && (1.5 \times 10^{-8} \text{ of normal decay}) \end{aligned} \tag{12.3}$$

The decay modes of the π° are

$$\begin{aligned} &\pi^\circ \rightarrow \gamma + \gamma && (98.8\%) \\ &\pi^\circ \rightarrow \gamma + e^+ + e^- && (1.2\%) \\ &\pi^\circ \rightarrow e^+ + e^- + e^+ + e^- && (0.003\%) \end{aligned} \tag{12.4}$$

Spin. The spin of the π° has been inferred to be even (presumably zero) from the fact that only two γ-ray photons, and never three, are emitted in reaction (12.4). The spin of the π^+ has been determined from a comparison of the following two reactions:

$$p + p \rightarrow \pi^+ + d \tag{12.5}$$

$$\pi^+ + d \rightarrow p + p \tag{12.6}$$

If these reactions are observed at the same energy in their center-of-mass system, then by the principle of detailed balancing

$$\sigma_\theta\,(12.5) = \sigma_\theta\,(12.6)\,\frac{(2S_\pi + 1)(2S_d + 1)}{(2S_p + 1)(2S_p + 1)}\,\frac{q^2}{p^2} \tag{12.7}$$

where σ_θ (12.5) and σ_θ (12.6) are the differential cross sections for the reactions at a particular angle θ; S_π, S_d, S_p are the spins of the pion, deuteron, and proton, and q and p are the momenta of the π^+ and proton in the center-of-mass system, respectively. Equation (12.7), integrated over all angles to obtain total cross sections, becomes

$$\sigma_{\text{total}}\,(12.5) = 2\sigma_{\text{total}}\,(12.6)\,\frac{3(2S_\pi + 1)}{4}\,\frac{q^2}{p^2} \tag{12.8}$$

where the factor 2 multiplying σ (12.6) arises because no distinction can be made experimentally between the two protons from reaction (12.6). Reaction (12.5) was investigated [5] at a laboratory proton energy of 341 Mev. The total cross section observed was $(1.8 \pm 0.6) \times 10^{-28}$ cm^2. At this proton energy, in the center-of-mass system the π^+ meson is produced with 21.4 Mev. If the π meson has spin 1, the total cross section of reaction (12.6) evaluated by (12.8) should be $(1.0 \pm 0.3) \times 10^{-27}$ cm^2; if the π meson has spin zero, σ_{total} (12.6) should be $(3.0 \pm 1.0) \times 10^{-27}$ cm^2. According to experiments by Clark et al. [6] the observed σ_{total} (12.6) at a π^+ energy of 23 Mev is $(4.5 \pm 0.8) \times 10^{-27}$ cm^2, and according to Durbin et al. [7] the observed σ_{total} (12.6) at a pion energy of 25 Mev is $(3.1 \pm 0.3) \times 10^{-27}$ cm^2. The results are therefore consistent with a π^+ spin of zero.

Parity. The parity of the pion is negative. The parity of the π^- has been measured by the following experiments. Slow π^- mesons are absorbed by nuclei and in particular by the nuclei of hydrogen according to the following reactions first observed by Panofsky et al. [8].

$$\pi^- + p \rightarrow \pi^\circ + n \tag{12.9}$$

$$\pi^- + p \rightarrow \gamma + n \tag{12.10}$$

The ratio of these two reactions (that is, σ (12.9)/σ (12.10), the so-called Panofsky ratio) has been the subject of a large number of independent determinations [9]. Probably the latest value of 1.53 ± 0.02 is the best estimate of this ratio. By analogy with the hydrogen reactions, π^- mesons might be expected at first sight to cause the following reactions in deuterium:

$$\pi^- + d \rightarrow \pi^\circ + n + n \tag{12.11}$$

$$\pi^- + d \rightarrow \gamma + n + n \tag{12.12}$$

$$\pi^- + d \rightarrow n + n \tag{12.13}$$

However, reaction (12.11) with the production of π° is not observed. The ratio of the reaction (12.12) to (12.13) has been experimentally determined [10] as 2.36 ± 0.36.

In order to interpret what these two ratios mean, it is usually assumed that the π^- mesons are absorbed from s states of the mesic atoms formed in the capture of the π^- by the atoms of hydrogen and deuterium. (The arguments supporting this assumption are mainly theoretical [11].) Another assumption is that the proton and neutron have the same parity. (This is a natural assumption because the p and n belong to the same isotopic spin doublet, and isotopic spin and space inversion are separately conserved in the reactions.) From the deuterium reactions (12.12) and (12.13), in order for the final states to be consistent with Pauli's principle, the two neutrons can exist only in ${}^1S_0$; ${}^3P_{0,1,2}$; ${}^1D_2$ or ${}^3F_{2,3,4}$, etc., states. In order to conserve angular momentum, which in the initial state is $J = 1$ arising from the deuteron spin, the final state chosen must be ${}^3P_1$. Since this is an odd-parity state and since the parity of the deuteron is even (mixture of ${}^3S_1$ and ${}^3D_1$), if parity is conserved in the fast reaction (12.13), then the π^- must possess an intrinsic parity that is negative. This conclusion is consistent with the suppression of reaction (12.11), for in this case, since the π° also has negative parity, it would need to be emitted in a p-orbital state relative to the center of mass of the two neutrons. However, as the Q of this reaction is only $\sim$1 Mev, there is insufficient energy in this reaction of π^- capture at rest either for the π° to be emitted in a p state or for the neutrons to be emitted in a p state.

All that one can conclude from reaction (12.13) is that the $(\pi^- p)$ system has a negative parity relative to n. From reaction (12.9) one can similarly conclude that the $(\pi^\circ n)$ system is the same as that of $(\pi^- p)$. However, the π° can be isolated from the nucleon because of its decay into two photons, i.e.,

$$\pi^\circ \rightarrow \gamma + \gamma \tag{12.14}$$

Yang [12] has discussed the symmetry and conservation laws as applied to π° decay. For a spin-zero π° meson, the form of the wave functions of the two outgoing photons depends on whether the π° is a pseudoscalar or scalar field particle (i.e., whether the particle wave function has odd or even properties with regard to the parity operation). It was inferred by Yang that, if the π° meson were a pseudoscalar, the two photons would be perpendicularly plane-polarized. In the case of the π° being a scalar, the two photons would be polarized in the same plane. Kroll and Wada [13] have shown that this polarization persists even when Dalitz pairs [such as the weak branchings in reaction (12.4)] are emitted in π° decay. If ϕ is the angle between the two planes containing the separate pairs resulting from double Dalitz decay of the π°, the distribution functions $E(\phi)$ of these angles for scalar and pseudoscalar particles are

$$E_S(\phi) = 1 + \alpha(x_1,y_1;x_2,y_2) \cos 2\phi \tag{12.15}$$
$$E_{PS}(\phi) = 1 - \alpha(x_1,y_1;x_2,y_2) \cos 2\phi \tag{12.16}$$

where $\alpha(x_1,y_1;x_2,y_2)$ is a function of four observables: x_1, y_1 for one pair and x_2, y_2 for the second pair. (These observables can be considered as the angle between and energy division within each pair.) The parity dependence of the particle is therefore entirely in the observed sign of α. This experiment was performed by Plano et al. [14] who measured approximately 100 events of the type

$$\pi^\circ \rightarrow e^+ + e^- + e^+ + e^- \tag{12.17}$$

The π° mesons were produced by charge-exchange scattering of π^- mesons in hydrogen [i.e., reaction (12.9)]. Of the order of 10 million slow π^- mesons stopping in a hydrogen bubble chamber were required to produce the 100 events of type (12.17). (The fraction of π° decays that produce double Dalitz pairs is approximately 1/30,000 of the abundant 2γ decay.) On the basis of 64 accurately measured events, Plano et al. observed the distribution (12.16) and found $\alpha_{exp} = 0.75 \pm 0.42$, whereas the expected theoretical distribution in α for a pseudoscalar meson was $\alpha_{th} = 0.48$. The observed weighted frequency distribution of the angle between the planes of polarization is shown in Fig. 12.1.

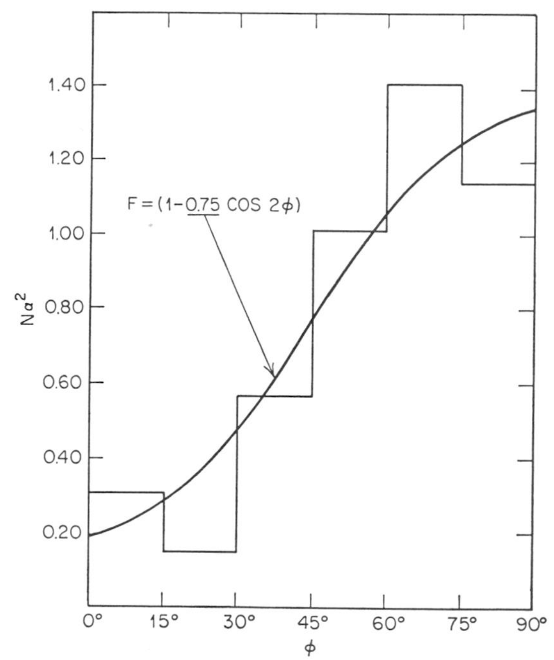

FIG. 12.1. Histogram of weighted frequency distribution, $\alpha^2 N$, of the angle ϕ between the planes of polarization of pairs from π° decay. The curve shown is for the distribution function $E_{PS}(\phi)$, Eq. (12.15) for $\alpha = 0.75$. (*After Plano et al.* [14].)

4. Nature of π-meson Interaction with Nucleons

The earliest experiments [15] on the scattering of $\pi^\pm$ mesons by hydrogen showed that the total scattering cross sections were a few millibarns at very low π-meson momenta and then increased rapidly as q^4 (where q is the meson momentum in the CM system of the meson and proton). From the fourth-power dependence of the total cross section on momentum it could be tentatively concluded that the pion-nucleon interaction was of a p-wave character (i.e., the π meson and nucleon interacted in an $l = 1$ orbital angular-momentum state). These results are consistent with the π meson being a pseudoscalar particle, with having a definite range of interaction, and with interacting very strongly with the nucleon in one angular-momentum channel [16].

The basic scattering reactions of charged π mesons on protons are the following:

$$\pi^+ + p \rightarrow \pi^+ + p \tag{12.18}$$
$$\pi^- + p \rightarrow \pi^- + p \tag{12.19}$$
$$\pi^- + p \rightarrow \pi^\circ + n \tag{12.20}$$

Reactions (12.18) and (12.19) are purely elastic scattering; (12.20) is described as charge-exchange scattering. The first accurate measurements up to approximately 200-Mev π meson (laboratory energy) indicated that the differential cross sections of reactions (12.18) to (12.20) were in the approximate ratio of 9:1:2. These results can readily be understood in terms of scattering in which the pion-proton interaction occurs in a total isotopic spin $\frac{3}{2}$ state. Thus, the six possible states of the above πp configurations have the isotopic spin assignments given in Table 12.1. If only the isotopic spin $\frac{3}{2}$ is the strongly interacting state, the cross sections of the reactions are given by

TABLE 12.1. TOTAL ISOTOPIC SPIN ASSIGNMENTS FOR πp CONFIGURATIONS

Particle configuration	*Total isotopic spin function*†
$\pi^+ p$	$I(\frac{3}{2},\frac{3}{2})$
$\pi^\circ p$	$\sqrt{\frac{2}{3}}\, I(\frac{3}{2},\frac{1}{2}) - \sqrt{\frac{1}{3}}\, I(\frac{1}{2},\frac{1}{2})$
$\pi^- p$	$\sqrt{\frac{1}{3}}\, I(\frac{3}{2},-\frac{1}{2}) - \sqrt{\frac{2}{3}}\, I(\frac{1}{2},-\frac{1}{2})$
$\pi^+ n$	$\sqrt{\frac{1}{3}}\, I(\frac{3}{2},\frac{1}{2}) + \sqrt{\frac{2}{3}}\, I(\frac{1}{2},\frac{1}{2})$
$\pi^\circ n$	$\sqrt{\frac{2}{3}}\, I(\frac{3}{2},-\frac{1}{2}) + \sqrt{\frac{1}{3}}\, I(\frac{1}{2},-\frac{1}{2})$
$\pi^- n$	$I(\frac{3}{2},-\frac{3}{2})$

† Note that the eigenfunction $I(i,i_3)$ specifies the eigenvalues $i(i+1)$ of the total isotopic spin operator I^2 and i_3 of the third-component operator I_3.

the following matrix elements with the same proportionality constant:

$$\sigma\ (12.18) \sim |(I(\tfrac{3}{2},\tfrac{3}{2})|H_3|I(\tfrac{3}{2},\tfrac{3}{2}))|^2 \qquad (12.21)$$

$$\sigma\ (12.19) \sim |(\sqrt{\tfrac{1}{3}}\, I(\tfrac{3}{2},-\tfrac{1}{2})|H_3|\ \sqrt{\tfrac{1}{3}}\, I(\tfrac{3}{2},-\tfrac{1}{2}))|^2 \qquad (12.22)$$

$$\sigma\ (12.20) \sim |(\sqrt{\tfrac{2}{3}}\, I(\tfrac{3}{2},-\tfrac{1}{2})|H_3|\ \sqrt{\tfrac{1}{3}}\, I(\tfrac{3}{2},-\tfrac{1}{2}))|^2 \qquad (12.23)$$

where H_3 is the Hamiltonian for the πp system in the $I = \frac{3}{2}$ state. The three cross sections are therefore in the ratio [writing $M_3 = I(\frac{3}{2},i_3)|H_3|I(\frac{3}{2},i_3)$]

$$\sigma\ (12.18):\sigma\ (12.19):\sigma\ (12.20) = |M_3|^2 : |\tfrac{1}{3}M_3|^2 : \text{or } |\ \sqrt{2}/3\ M_3|^2 \qquad (12.24)$$

TABLE 12.2. CHANNELS, PHASE SHIFTS, AND SCATTERING AMPLITUDES OF LOW-ENERGY πp SCATTERING

Channel			Phase shift	Amplitude
l	I	J		
0	½	½	δ_1	A_1
0	3/2	½	δ_3	A_3
1	½	½	δ_{11}	A_{11}
1	3/2	½	δ_{31}	A_{31}
1	½	3/2	δ_{13}	A_{13}
1	3/2	3/2	δ_{33}	A_{33}

that is,

$$\sigma_{el}(\pi^+p)\ :\sigma_{el}(\pi^-p)\ :\sigma_{\text{c.e.}}(\pi^-p) = 9 : 1 : 2$$

In general, for π-meson scattering of less than approximately 200 Mev, there are six phase shifts required to describe the scattering by s and p waves and in the $I = \frac{1}{2}$ and $\frac{3}{2}$ isotopic spin channels. The usual nomenclature [in terms of phase shifts δ and scattering amplitudes A, related by $A = (e^{2i\delta} - 1)/2i$, in units of $1/k$, where k is the π meson CM momentum divided by $\hbar$] are given in Table 12.2.

The three differential cross sections corresponding

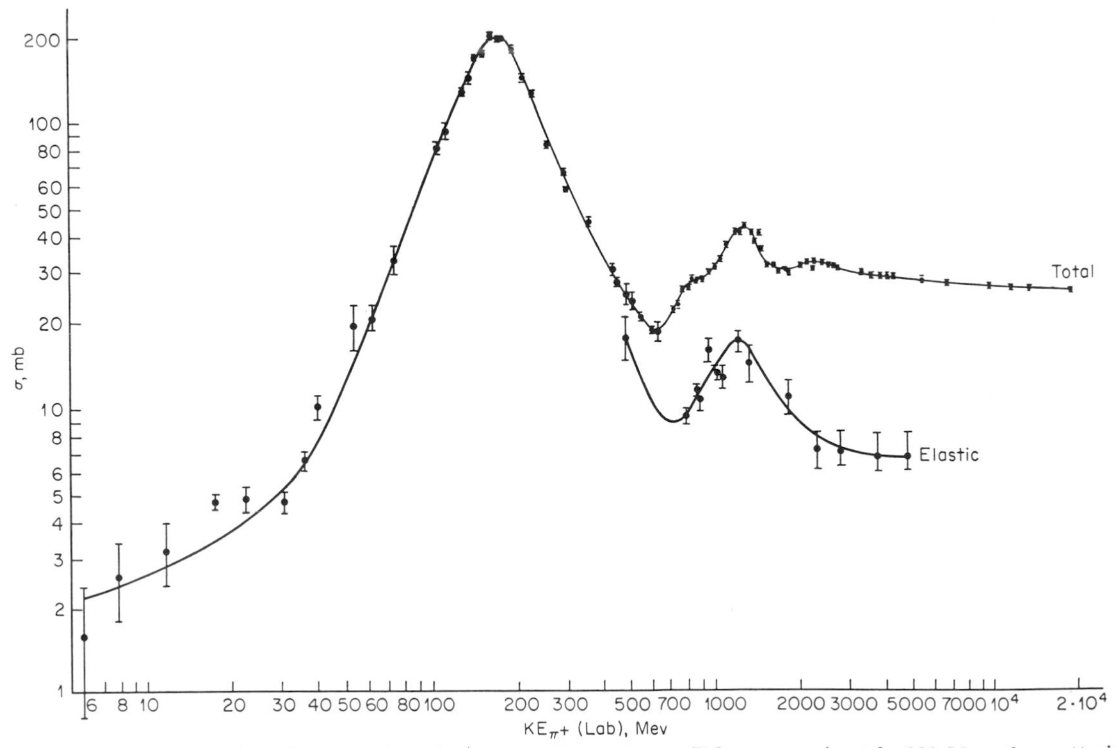

FIG. 12.2. Total cross sections for scattering of π^+ mesons by protons. Below approximately 200 Mev, the scattering is entirely elastic. The higher values of the total cross section above 200 Mev are attributable to inelastic π-meson production. (*After G. Källén, "Elementary Particle Physics," Addison-Wesley, Reading, Mass.,* 1964.)

to reactions (12.18) to (12.20) are

$$\pi^+p \text{ elastic:} \quad \frac{d\sigma}{d\Omega} = \frac{1}{k^2}[|\alpha_3 + (2\alpha_{33} + \alpha_{31})\cos\theta|^2 + |(\alpha_{33} - \alpha_{31})\sin\theta|^2] \quad (12.25)$$

$$\pi^-p \text{ elastic:} \quad \frac{d\sigma}{d\Omega} = \frac{1}{9k^2}[|(\alpha_3 + 2\alpha_1) + (2\alpha_{33} + \alpha_{31} + 4\alpha_{13} + 2\alpha_{11})\cos\theta|^2 + |(\alpha_{33} - \alpha_{31} + 2\alpha_{13} - 2\alpha_{11})\sin\theta|^2] \quad (12.26)$$

$$\pi^-p \text{ ch. ex.:} \quad \frac{d\sigma}{d\Omega} = \frac{2}{9k^2}[|(\alpha_3 - \alpha_1) + (2\alpha_{33} + \alpha_{31} - 2\alpha_{13} - \alpha_{11})\cos\theta|^2 + |(\alpha_{33} - \alpha_{31} - \alpha_{13} + \alpha_{11})\sin\theta|^2] \quad (12.27)$$

The so-called Fermi phase shifts are characterized by a large value of δ_{33} (going through 90° near 200-Mev incident π energy) and a value of δ_{31} approximately equal to zero. Below 100 Mev, the experiments are reasonably well represented by

$$\delta_1 = 9.2°q \qquad \delta_3 = -6.3°q \qquad \delta_{33} = 13.5°q^3$$
$$\text{and} \qquad \delta_{11} \approx \delta_{31} \approx \delta_{13} \approx 0$$

q being the pion momentum in the CM system of the π and nucleon in units of $m_\pi c$.

Curves of $\sigma_{\text{total}}(\pi^+p)$ and $\sigma_{\text{total}}(\pi^-p)$, based on a compilation of experimental data [17], are shown in Figs. 12.2 and 12.3. In the neighborhood of 200-Mev pion energy, the scattering is dominated by the $I = \frac{3}{2}$, $J = \frac{3}{2}$ state. The important amplitude is α_{33}, and the differential cross section may be obtained to a good approximation by setting the other phase shifts equal to zero. Thus, for example,

$$\pi^+p \text{ elastic:} \quad \frac{d\sigma}{d\Omega} = \frac{1}{k^2}(3\cos^2\theta + 1)\sin^2\delta_{33} \quad (12.28)$$

and the total cross section may then be obtained by integration, giving

$$\pi^+p \text{ elastic:} \quad \sigma_{el}(\pi^+p) = \frac{8\pi}{k^2}\sin^2\delta_{33} \quad (12.29)$$

An accurate analysis of the experimental data on $\sigma_{el}(\pi^+p)$ in the neighborhood of the peak at 190 Mev pion energy has been made by Lindenbaum and Yuan [18]. They found that the $\sigma_{el}(\pi^+p)$ cross section is at least equal to $8\pi/k^2$ at 190 Mev, so that if contributions from other states are small, it is clear that δ_{33} must be close to 90° at 190 Mev; i.e., there is a resonance in the $I = \frac{3}{2}$, $J = \frac{3}{2}$ state of the π^+p system at this energy. The existence of a resonance had been proposed much earlier by Brueckner [19] in order to account for the intensely strong pion-nucleon interaction. A one-level Breit-Wigner resonance curve, fitted to the experimental data available in 1954, is shown in Fig. 12.4. In the center-of-mass system this resonance is located at 159-Mev π-meson energy, and the full width at half maximum is approximately 125 Mev.

The manner in which the $\frac{3}{2}$, $\frac{3}{2}$ scattering phase shift δ_{33} varies with energy is rather well described by the Chew-Low theory of pion-nucleon interaction

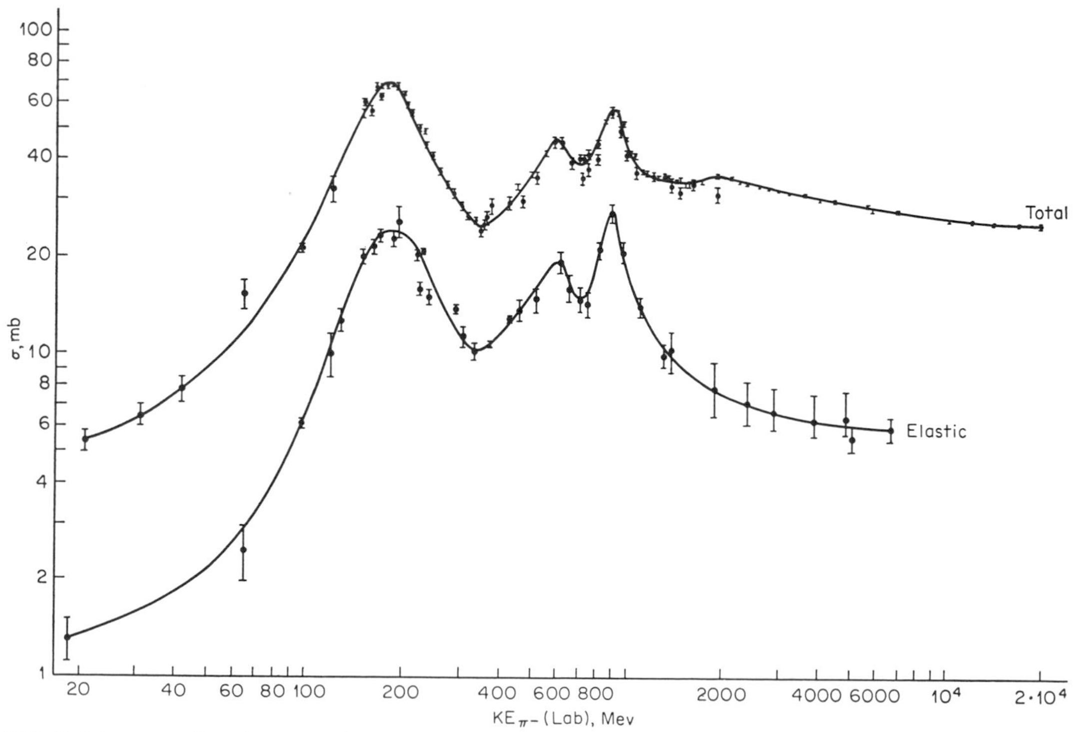

FIG. 12.3. Total cross sections for scattering of π^- mesons by protons. The difference between σ_{total} and σ_{elastic} at energies below approximately 200 Mev is due to $\sigma_{\text{charge-exchange}}$. Above 200 Mev there is also a contribution from inelastic pion production to σ_{total}. (*After G. Källén, "Elementary Particle Physics," Addison-Wesley, Reading, Mass.,* 1964.)

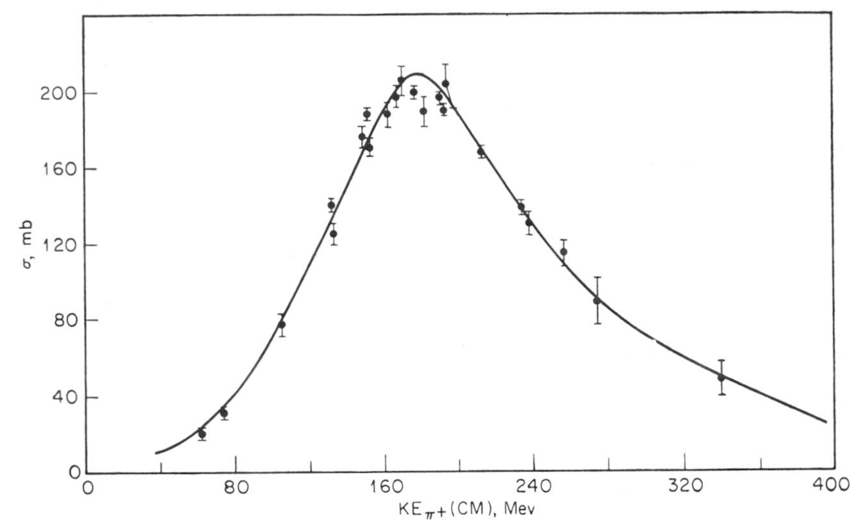

FIG. 12.4. Breit-Wigner resonance curve fitted to experimental π^+p σ_{elastic} data. (*After Gell-Mann and Watson* [16].)

[20]. This theory might be outlined briefly as follows: If pions are the main seat of nuclear forces, then it is necessary to account for two main features: the spin dependency (as, for example, in the bound triplet state of the deuteron) and the charge independency (as evidenced in the equality of the singlet *n-p*, *p-p*, and *n-n* forces). Thus the pion-nucleon interaction Hamiltonian must involve the spin of the nucleon and must be charge symmetric. An interaction Hamiltonian density which satisfies these conditions and which is linear in the meson field $\boldsymbol{\phi}$ is

$$H = F\rho(\mathbf{r})\boldsymbol{\sigma} \cdot \nabla(\mathbf{L} \cdot \boldsymbol{\phi}) \tag{12.30}$$

where F is a coupling constant between the nucleon and the pion field and is a measure of the strength of the interaction, $\rho(\mathbf{r})$ is a spherically symmetrical source density representing the nucleon, $\boldsymbol{\sigma}$ is the nucleon spin vector, and $\nabla(\mathbf{I} \cdot \boldsymbol{\phi})$ is the gradient of the scalar product of the isotopic spin vector of the nucleon with the meson field vector $\boldsymbol{\phi}$ in charge (i.e., isotopic spin) space, that is,

$$\mathbf{I} \cdot \boldsymbol{\phi} = I_1\phi_1 + I_2\phi_2 + I_3\phi_3$$

Such a Hamiltonian leads directly to a two-body nucleon force which involves $\boldsymbol{\sigma}_1 \cdot \boldsymbol{\sigma}_2$ of the two nucleons. Also the meson field has to be a pseudoscalar in order that H be scalar. Equation (12.30) is a nonrelativistic form of the interaction and is described by Chew and Low as a "static" model. Such a theory is valid only at low energies, and Chew and Low introduced a cutoff energy, $\omega_{\max}$, the maximum total energy of the pion, which is approximately the value of the nucleon rest-mass energy.

By using the interaction Hamiltonian (12.30), the scattering of π mesons by nucleons was solved at low energies. Chew and Low obtained the following effective-range formula for the phase shift δ_{33}:

$$\frac{q^3 \cot \delta_{33}}{\omega} = \frac{3}{4f^2}\left(1 - \frac{\omega}{\omega_0}\right) \tag{12.31}$$

where q is the center-of-mass pion momentum, ω is the total energy of the pion in the CM system, ω_0 is the value of ω at the center of the $\frac{3}{2}$, $\frac{3}{2}$ resonance, and f^2 is a coupling parameter equal to $F^2/4\pi$. The value of f^2 can be determined from a Chew-Low plot of the scattering data in the neighborhood of the $\frac{3}{2}$, $\frac{3}{2}$ peak. Such a plot is shown in Fig. 12.5. Probably the best value of the coupling constant as determined in this manner is f^2 equals 0.081 ± 0.003 [21]. (Note that f^2 is given in dimensionless units when q is expressed in momentum units of $m_\pi c$ and ω is expressed in energy units of $m_\pi c^2$.)

5. Photopion Production

Low-energy photoproduction of pions from nucleons generally confirms the conclusions already reached on the nature of the pion and of its interaction with nucleons. Above the thresholds, which are 145 Mev for π° and 150 Mev for π^+, photons produce the following reactions on protons:

$$\gamma + p \rightarrow \pi^+ + n \tag{12.32}$$

$$\gamma + p \rightarrow \pi^\circ + p \tag{12.33}$$

Experimental total cross sections [23] for the production of π^+ and π° mesons from hydrogen are shown as a function of photon energy in Figs. 12.6 and 12.7.

Although only observable indirectly through irradiation of deuterium, the photopion production from neutrons presumably follows reactions in which a proton (p) in deuterium is a spectator:

$$\gamma + n + (p) \rightarrow \pi^- + p + (p) \tag{12.34}$$

$$\gamma + n + (p) \rightarrow \pi^\circ + n + (p) \tag{12.35}$$

The determination of the π^-/π^+ ratio in the low-energy photoproduction from deuterium is an important constant in the over-all check of pion physics. Beneventano et al. [24] obtained a value of $\sigma(\gamma\pi^-)/\sigma(\gamma\pi^+)$ equal to 1.87 ± 0.13 for the pion production from deuterium when the production cross sections

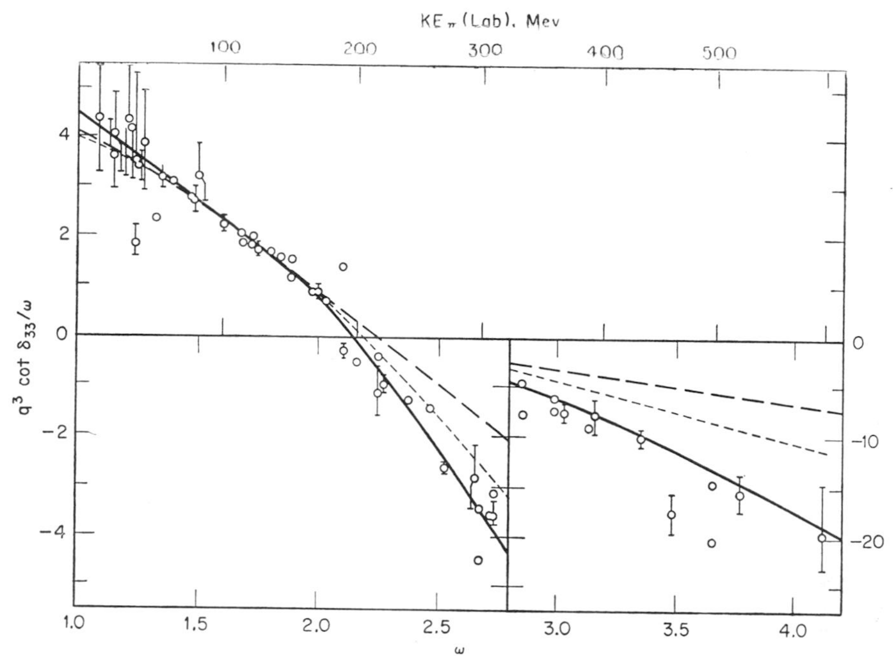

FIG. 12.5. Chew-Low plot: $q^3 \cot \delta_{33}/\omega$ as a function of ω. The solid curve is a three-parameter fit of McKinley [21]. The short-dash curve is a polynomial fit of Anderson [22]. The long-dash curve is a low-energy two-parameter fit [21]. (*After McKinley* [21].)

were extrapolated to threshold energy. [The latest observations of the π^-/π^+ ratio from deuterium by the University of Illinois group (T. S. Yoon, Thesis, May, 1964) yield a value equal to 1.34 ± 0.06. This value is consistent with a somewhat lower value of the coupling constant, viz., $f^2 = 0.076 \pm 0.003$.]

The broad features of the near-threshold production of photopions may be summarized as follows: Just above threshold the charged pions are produced in an s state. This is indicated experimentally by an isotropic angular distribution of π^+ mesons from reaction (12.32) and a dependence of the total π^+ cross section on the first power of the outgoing pion momentum. This photopion production process is consistent with electric-dipole absorption by a proton. The π° production cross section does not have an E1 s-wave part. This is apparent from the small value of σ just above threshold in the curve of Fig. 12.7. At photon energies somewhat above 200 Mev, the pions are produced in a p state. This is indicated by a $2 + 3 \sin^2 \theta$ angular distribution of π° mesons from reaction (12.33) and by an anisotropic

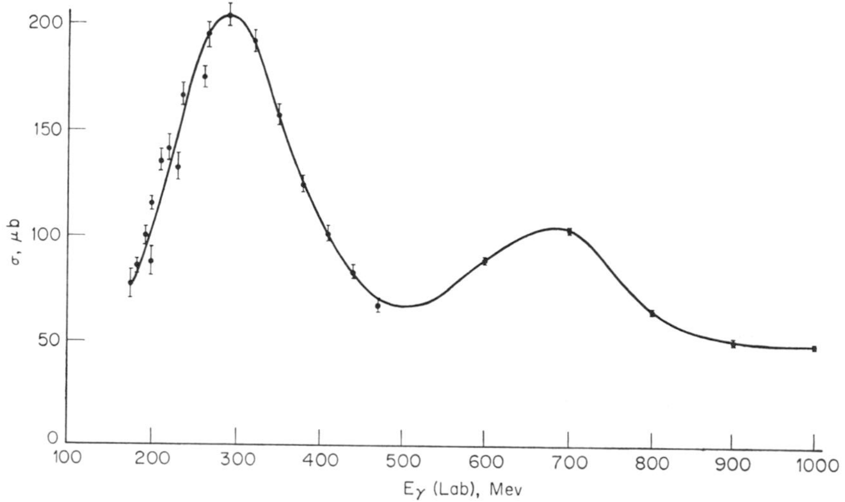

FIG. 12.6. Total cross section for production of π^+ mesons by photons on hydrogen. (*After G. Källén, "Elementary Particle Physics," Addison-Wesley, Reading, Mass.*, 1964.)

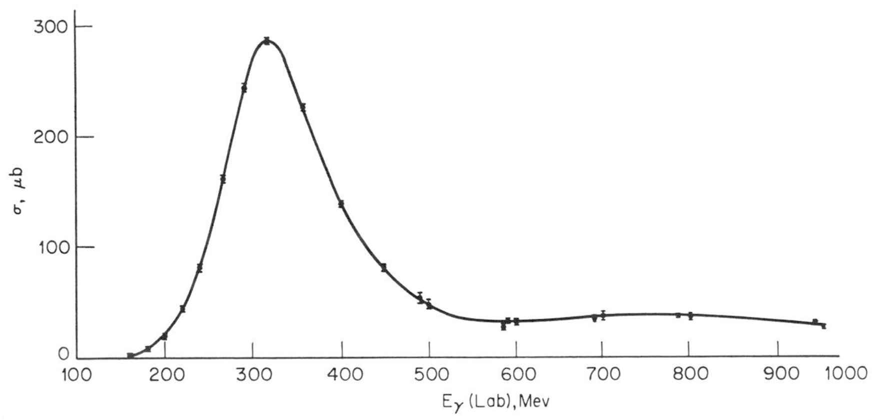

FIG. 12.7. Total cross section for production of π° mesons by photons on hydrogen. *(After G. Källén, "Elementary Particle Physics," Addison-Wesley, Reading, Mass.,* 1964.)

distribution in the π^+ production of (12.32). The dependence of the total cross sections on the cube of the pion momenta in this region is also indicative of a p-wave interaction. The occurrence of the photopion maxima at approximately 330 Mev is also consistent with the $\frac{3}{2}$, $\frac{3}{2}$ resonance (p wave) peak in the π-nucleon scattering observations. (The photopion maximum of $\sigma_{\gamma\pi}k_\gamma^2$ occurs at 121 Mev in the CM system; the pion-nucleon scattering maximum of $\sigma_{\pi p}k_\pi^2$ occurs at 127 Mev in the CM system, where k_π and k_γ are the pion and photon momenta in their CM systems.) The photopion production in the vicinity of the $\frac{3}{2}$, $\frac{3}{2}$ peak is consistent with magnetic-dipole absorption of the photon, the nucleon and photopion being left in a final $p_{3/2}$ state.

A number of theoretical analyses of photopion production have been given [25]. The Chew-Low theory indicates a close connection between photo-meson production and π-meson scattering. Koester and Mills [26] have determined the total cross section $\sigma(\gamma\pi^\circ)$ for the production reaction (12.33) and compared their observations with the values predicted by the Chew-Low theory from π-meson scattering data, viz.,

$$\sigma(\gamma\pi^\circ) = \left(\frac{\mu_p - \mu_n}{2}\right)^2 \frac{\sigma_0}{4\pi f^2\beta} \qquad (12.36)$$

where σ_0 is a total cross section for elastic scattering of π° mesons on protons, μ_p and μ_n are the magnetic moments of the proton and neutron, respectively, and β is the velocity of the π° meson in the center of mass of the π°-nucleon system. The value of σ_0 is not directly observable because of the unavailability of π° mesons as an incident particle, but it can be straightforwardly calculated from the known phase shifts obtained from the π^+ and π^- meson scattering on protons. The agreement between the observed and calculated values of $\sigma(\gamma\pi^\circ)$, which are shown in Fig. 12.8, is excellent. It should be pointed out that the Chew-Low theory of low-energy pion interactions appears to be in good agreement with observations where p waves are predominantly involved. Complications appear to set in at higher energies where other states are involved (e.g., see Fig. 12.6). Further developments of a relativistic pion theory are required [27].

6. Other π-meson Topics

There still remain a number of topics concerning production and interaction of pions, especially with atomic nuclei other than hydrogen and deuterium, which will not be discussed in this chapter. The very briefest reference only to some of the literature associated with these topics will be made. Some of these include the interaction of π mesons at rest with light and heavy nuclei [28], interaction and scattering for high-energy π mesons with nuclei [29], mesic X rays [30], multiple pion production at high energies [31], and photopion production from nuclei [32]. For more extensive references, especially to the earlier phases of

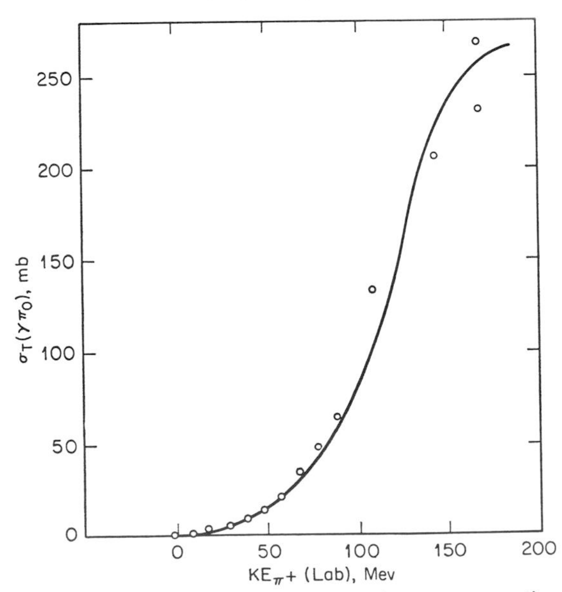

FIG. 12.8. Photopion (π_0) production cross section $\sigma_T(\gamma\pi_0)$ as a function of photon energy. The experimental values of $\sigma_T(\gamma\pi_0)$ are represented as points and are from ref. 26. The theoretical curve for scattering of π° mesons has been computed from π^+ scattering on hydrogen. Thus the abscissa is labeled kinetic energy of π^+. Photon energies can be obtained from the abscissa by adding 150 Mev to π^+ kinetic energies. *(After Koester and Mills* [26].)

pion physics, the chapter on this subject in the first edition of "Handbook of Physics" should be consulted.

ANTIPARTICLES

7. Discovery and Production

The particles and antiparticles to be discussed in this section are fermions of spin $\frac{1}{2}$. (K and π-meson particles and antiparticles will be considered later in connection with the nature of neutral K mesons.) The conception of antiparticles began with Dirac who observed that solutions of a relativistic wave equation also existed for "antiparticles" of opposite charge and magnetic moment to the "ordinary" particles. Particles and antiparticles may be said to exist because the laws of physics are invariant under the operations of charge conjugation (C, charge reversal). The discovery of the positron by Anderson in 1932 gave confirmation to Dirac's theory of the antielectron. The later discovery of the positive and negative mu mesons also gave added support to the Dirac theory. However, the unequivocal discovery of the antiproton had to wait until 1955 when Chamberlain, Segrè, Wiegand, and Ypsilantis [33] were able to detect $\bar{p}$ at the Bevatron, the University of California 6-Bev proton accelerator. The detection of antineutrons was made shortly afterward by Cork, Lambertson, Piccioni, and Wenzel [34].

Nucleon-antinucleon pairs may be produced in a number of reactions, some of which are the following:

$$p + p \rightarrow p + p + p + \bar{p} \qquad (12.37)$$

$$p + p \rightarrow p + p + n + \bar{n} \qquad (12.38)$$

$$\pi^- + p \rightarrow n + p + \bar{p} \qquad (12.39)$$

The Q values for reactions (12.37) and (12.38) are -1.88 Bev in the center-of-mass system. In the laboratory system, for target protons at rest, the threshold energies of the protons in reactions (12.37) and (12.38) are approximately 5.6 Bev and of the pion in reaction (12.39) is approximately 4.0 Bev. For target nucleons possessing Fermi energies of 22 Mev in nuclei, the threshold proton energy in reaction (12.37) is reduced to 4.6 Bev.

8. Annihilation

Probably the most interesting property of an antiparticle is its ability to annihilate itself and its particle partner. The property of a positron, for example, to produce photons from an annihilation process with a negative electron has been recognized since 1932. The interesting new feature of an antinucleon and nucleon annihilation, however, is its high probability of producing mesons, especially π mesons. This characteristic feature arises because of the much stronger coupling of the nucleon and antinucleon with the pion field than with the electromagnetic field. According to Brown and Peshkin [37] the cross section for two-photon annihilation of a nucleon-antinucleon pair is equal to

$$\sigma_{e-m} \simeq \pi \left(\frac{e^2}{Mc^2}\right)^2 \left(\frac{c}{v}\right) X \qquad (12.40)$$

where Mc^2 is the nucleon rest energy, v is the relative nucleon-antinucleon velocity, and X is a parameter depending on the anomalous nucleon magnetic moment. For antiprotons of 30 Mev, the value of σ_{e-m} can be estimated to be approximately 10^{-29} cm^2, whereas the observed cross section for annihilations producing mesons is approximately 2×10^{-25} cm^2 [38]. This ratio of the mesic to the electromagnetic annihilation is approximately what would be expected from the strengths of the pion-nucleon coupling constant, $F^2/\hbar c \approx 15$, and the electromagnetic coupling constant, $e^2/\hbar c \approx \frac{1}{137}$.

For purely mesic annihilation, there are certain selection rules, based on conservation of angular momentum, parity, charge conjugation, isotopic spin, and charge, which determine the charges and the number of pions emitted. In general, however, no experimental check has been made to observe whether the complicated predictions of the pion yield from particular initial angular-momentum states of the nucleon-antinucleon system are valid. Instead, the number and momentum distributions have been compared with statistical theories [39]. The theoretical distributions of pion multiplicities, N_π, depend rather sensitively on the interaction volume Ω assumed. Experiments on the annihilation of antiprotons in hydrogen bubble chambers appear to bear out the statistical-theory values, shown in Table 12.3, rather

TABLE 12.3. DISTRIBUTION OF PION MULTIPLICITIES FROM p-$\bar{p}$ ANNIHILATION (AT REST) ACCORDING TO THE FERMI STATISTICAL THEORY

N_π	Ω† = 1	Ω = 10	Ω = 15
2	6.4%	0.1%	0.0%
3	63.7%	5.6%	2.3%
4	24.5%	21.7%	13.4%
5	5.0%	44.0%	40.6%
6	0.3%	23.7%	33.1%
7	0.0%	5.1%	10.6%
$\langle N_\pi \rangle$	3.3	5.0	5.4

† Ω is the interaction volume in units of $(4\pi/3)(\hbar/m_\pi c)^3$.

well except that the interaction volume is required to be considerably larger than was originally suggested by Fermi. The values of the average numbers of charged pions at a number of different antiproton energies are given in Fig. 12.9. It would appear that the volume Ω required to fit the data is somewhere between 5 and 10.

The observed momentum distribution of emitted charged pions also fits rather well with that predicted by the statistical model, as shown in Fig. 12.10. Koba and Takeda [42] have discussed a possible mechanism that results in an interaction volume much larger than that corresponding to a radius of the pion Compton wavelength ($\hbar/m_\pi c \simeq 1.41$ fermis). They postulated that the bare nucleon and antinucleon destroy one another in a very short time and in a volume of the order of $\Omega \simeq 1$. Because of the rapid annihilation time, however, they postulate that the pion clouds, surrounding the bare nucleons, are merely released. This feature of the annihilation, according to Koba and Takeda, considerably enlarges the vol-

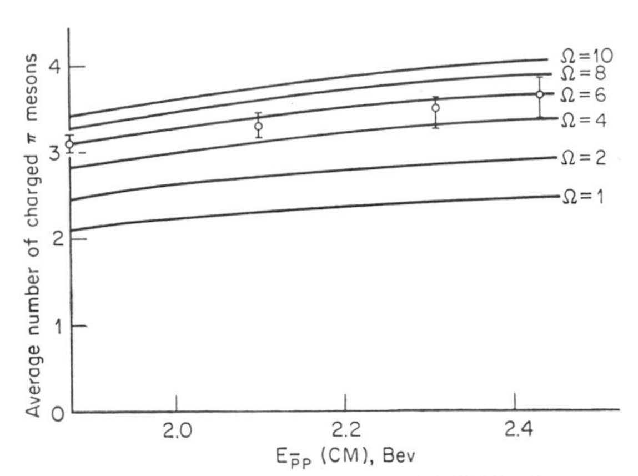

FIG. 12.9. Average charged pion multiplicities from p-p annihilation at different $\bar{p}$ CM energies. The curves are for statistical-model predictions for different interaction volumes. (*After Lynch* [40].)

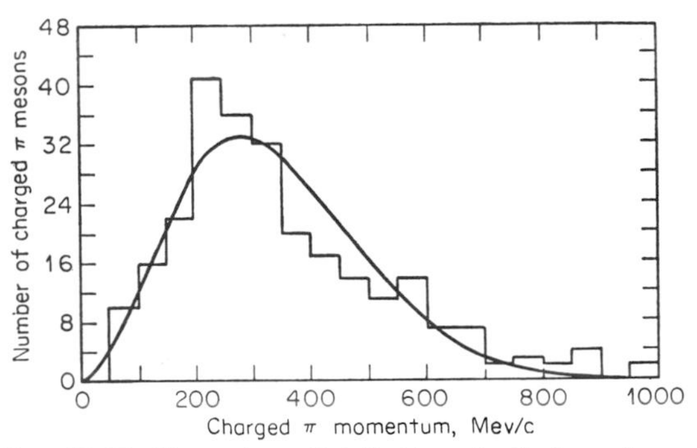

FIG. 12.10. Momentum distribution of all charged particles emitted in $\bar{p}$-p annihilations. The curve is the predicted momentum distribution according to the statistical theory for an interaction volume of $10(4\pi/3)(\hbar/m_\pi c)^3$. (*After Horwitz et al.* [41].)

ume in which the pions are statistically thermalized before emission. On the assumption that there are 1.3 mesons in each of the nucleon clouds and that the bare nucleon has a radius of $2\hbar/3m_\pi c$, they calculate that the average multiplicity should be approximately 5.

Because it indicates that the K-meson coupling with the nucleon is also strong, another interesting aspect of the nucleon-antinucleon annihilation process is the production of K mesons. Armenteros et al. [43] and Chadwick et al. [43] have studied the production of K-particle–anti-K-particle pairs. According to Armenteros et al., the following rates occur for antiprotons annihilating essentially at rest in liquid hydrogen:

$$\begin{aligned} &\bar{p} + p \to \pi^+ + \pi^- && (0.395\% \text{ of all annihilations})\\ &\bar{p} + p \to K^+ + K^- && (0.131\% \text{ of all annihilations})\\ &\bar{p} + p \to K^\circ + \bar{K}^\circ && (0.056\% \text{ of all annihilations}) \end{aligned} \tag{12.41}$$

The total intensity of charged K-meson production as a fraction of charged π-meson production from $\bar{p}$-p annihilations at rest is approximately 9 per cent [40].

At higher antiproton energies, many other fermion-antifermion pairs can also be produced from $\bar{p}$-p annihilations. Thus it has been found that 3 to 4 Bev/c antiprotons in liquid hydrogen produce the following relative intensities of hyperon-antihyperon pairs [44]:

$$\begin{aligned} &\bar{p} + p \to \bar{\Lambda} + \Lambda\ [\sigma \approx (80 \pm 25) \times 10^{-30}\,\text{cm}^2]\\ &\left.\begin{aligned}&\bar{p} + p \to \bar{\Lambda} + \Sigma\\ &\bar{p} + p \to \Lambda + \bar{\Sigma}\end{aligned}\right\}[\sigma \approx (90 \pm 25) \times 10^{-30}\,\text{cm}^2]\\ &\left.\begin{aligned}&\bar{p} + p \to \bar{\Lambda} + K + n\\ &\bar{p} + p \to \Lambda + K + \bar{n}\end{aligned}\right\}[\sigma \approx (25 \pm 15) \times 10^{-30}\,\text{cm}^2] \end{aligned} \tag{12.42}$$

In the early phase of antiproton investigations, one of the most surprising characteristics observed was the very large scattering cross sections exhibited by antiprotons. Summary curves of the measured $\bar{p}$-p cross sections have been given by a number of experimental groups [45]. One of these curves is shown in Fig. 12.11. It will be seen that the total $\bar{p}$-p cross

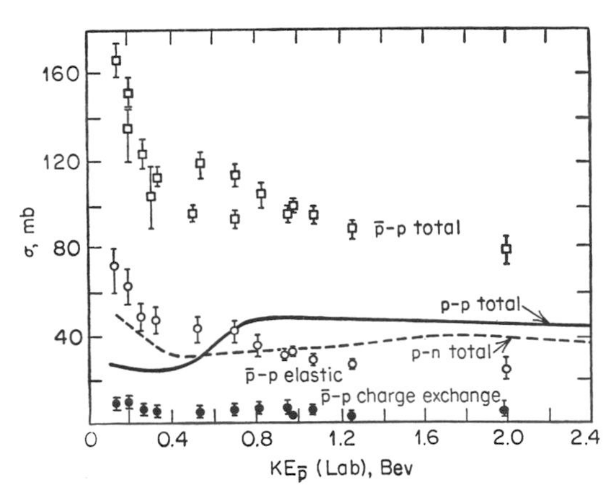

FIG. 12.11. Observed $\bar{p}$ scattering cross sections on hydrogen as function of $\bar{p}$ energy. The total $\bar{p}$-p cross section may be compared with the p-p cross section (full line) and p-n cross section (dotted line). (*After Wenzel* [45].)

section below 1 Bev is more than twice the total p-p cross section. According to Chew et al. [46], this is attributable to a strong attraction between the cores of the bare $\bar{p}$ and p, in contradistinction to the well-known strong repulsion between the cores of ordinary protons.

At the highest $\bar{p}$ kinetic energies (~20 Bev) thus far available, it appears that the total $\bar{p}$-p cross section is still approximately 20 per cent higher than the corresponding p-p cross section. On the basis of Pomeranchuk's theorem [47] that particle and antiparticle cross sections tend to the same value at high enough energies, one might have expected, since the p-p cross section has reached a constant value at 15 Bev, that the $\bar{p}$-p cross section would also have equaled this same constant value. However, this does not appear to be the case even at the highest energies yet reached.

STRANGE PARTICLES

9. *K* Mesons and Hyperons: Historical

Entirely new types of particles were uncovered in 1947. Rochester and Butler [48] observed certain new types of forked or V-shaped tracks in cloud-chamber photographs of cosmic rays. It gradually became clear that there were two groups of V particles [49]: those with a mass of approximately 500 Mev and others with a mass rather larger than the nucleon mass of approximately 1,000 Mev. These observations were supported by nuclear emulsion data on cosmic-ray particles [50]. By 1953, at an International Conference on Elementary Particles at Bagnères, a decision was taken to name those particles of mass greater than π mesons, but smaller than protons, K particles, while those particles whose masses were greater than protons were to be termed hyperons [51]. With the commencement of operation of the 3-Bev Brookhaven Cosmotron in 1953, the confirmation of the existence of K particles and hyperons was rapidly accomplished [52].

10. Summary of K Mesons and Hyperons

The elucidation of the number of strange particles, their masses and their decay properties, will not be retraced. The number of these known particles has been increased by only one since about 1957. This was in 1964 with the addition of the heaviest-known long-lived hyperon, the Ω^-. According to certain organizations of particles, the groups of long-lived hyperons and K particles are now regarded as complete. All these particles are unstable, yet they possess relatively long lives ($\gtrsim 10^{-10}$ sec) on a time scale which measures reaction times of strongly interacting particles ($\sim 10^{-23}$ sec); in fact, in the latest review [53] on elementary particles, the K mesons and hyperons are classified with the stable particles, i.e., with the nucleons and relatively long-lived π and μ mesons.

As shown in Table 12.4, there are essentially four K mesons, i.e., two charged mesons, K^+ and K^-, which are regarded as particle and antiparticle of one another, and two uncharged mesons, K° and $\bar{K}^\circ$, also particle and antiparticle of one another. [Because of a new property of "strangeness" (which will be later discussed), the neutral particles are not identical with each other.] The mass of the charged $K^\pm$ mesons is 493.8 Mev, and their mean lifetime against seven known types of decay is 2.55×10^{-8} sec. The mass of the K° is 498.0 Mev. The mean lifetime of the K° meson has dual values: K°_1 of 0.92×10^{-10} sec, K°_2 of 5.62×10^{-8} sec, and again there are multiple types of decay. The K mesons, like the π mesons, are bosons; i.e., they have zero spins.

As shown in Table 12.5, there are seven hyperons. The lowest-mass hyperon, the Λ°, is a neutral particle of 1,115.40 Mev and 2.62×10^{-10} sec mean life. The next massive hyperon is the Σ which has three charge states: Σ^+, Σ°, and Σ^-. Each Σ hyperon has a slightly different mass, attributable to the differences in the interactions of the virtual charge states of the particles with the electromagnetic field. The mean lifetimes of the Σ^+ and Σ^- are 0.788×10^{-10} sec and 1.58×10^{-10} sec, respectively, and are internally consistent with phase-space considerations for the decays, according to Table 12.5. The decay of Σ°, however, is somewhat anomalous since it has an extremely short lifetime (as yet undetermined but which is certainly less than 10^{-14} sec). Like the π°, the Σ° is stable under strong interactions but can decay electromagnetically. An approximate estimate of its lifetime according to its decay in Table 12.5 is a few times 10^{-20} sec. It should be pointed out that, unlike the π°, the Σ° is a fermion—as are all other hyperons as well. There are two Ξ hyperons: Ξ° of 1,314.3 Mev and mean life of 3.06×10^{-10} sec, and Ξ^- of 1,320.8 Mev and 1.74×10^{-10} sec mean life. Their decays are given in Table 12.5. The heaviest hyperon is the recently discovered Ω^- of 1,675-Mev mass and of mean lifetime 1.3×10^{-10} sec [54]. Of the four cases of Ω^- observed in 1964, two decays were into $\Xi^\circ + \pi^-$ and two were into $\Lambda^\circ + K^-$.

11. Production of Strange Particles

The production of K mesons and hyperons clearly requires higher energies than are necessary for the production of π mesons. In cosmic-ray and high-energy accelerator experiments it was early found that

TABLE 12.4. PROPERTIES OF K MESONS

Particle	Spin	Mass (Mev)	Mean life, (sec)	Decay (fraction)
K^+, K^- $(K^- = \overline{K^+})$	0	493.8	1.229×10^{-8}	$\mu^\pm + \nu(0.631)$ $\pi^\pm + \pi^\circ(0.215)$ $\pi^\pm + \pi^+ + \pi^-(0.055)$ $\pi^\pm + \pi^\circ + \pi^\circ(0.017)$ $\pi^\circ + \mu^\pm + \nu(0.034)$ $\pi^\circ + e^\pm + \nu(0.048)$ $\pi^\pm + \pi^\mp + e^\pm + \nu(4.3 \times 10^{-5})$
K°, $\bar{K}^\circ$ $[K^\circ = (K^\circ_1 + iK^\circ_2)/\sqrt{2}]$ $[\bar{K}^\circ = (K^\circ_1 - iK^\circ_2)/\sqrt{2}]$	0	498.0	$0.92 \times 10^{-10}(K^\circ_1)$ $5.62 \times 10^{-8}(K^\circ_2)$	$(0.50K^\circ_1, 0.50K^\circ_2)$ $\pi^+ + \pi^-(0.694)$ $\pi^\circ + \pi^\circ(0.306)$ $\pi^\circ + \pi^\circ + \pi^\circ(0.271)$ $\pi^+ + \pi^- + \pi^\circ(0.127)$ $\pi^+ + \pi^-(\sim 2 \times 10^{-3})$ $\pi^\pm + \mu^\mp + \nu(0.266)$ $\pi^\pm + e^\mp + \nu(0.366)$

TABLE 12.5. PROPERTIES OF HYPERONS

Particle	Spin	Mass (Mev)	Mean life (sec)	Decay (fraction)
Λ°	½	1,115.40	2.62×10^{-10}	$p + \pi^-(0.677)$ $n + \pi^\circ(0.316)$ $p + e^- + \nu(0.88 \times 10^{-3})$
Σ^+	½	1,189.41	0.80×10^{-10}	$p + \pi^\circ(0.54)$ $n + \pi^+(0.46)$ $n + \pi^+ + \gamma(\sim 4 \times 10^{-5})$ $\Lambda^\circ + e^+ + \nu(\sim 2 \times 10^{-5})$ $p + \gamma(\sim 3 \times 10^{-3})$
Σ°	½	1,192.3	$<10^{-14}$	$\Lambda^\circ + \gamma(1.00)$
Σ^-	½	1,197.08	1.60×10^{-10}	$n + \pi^-(\sim 1.00)$ $n + \pi^- + \gamma(\sim 10^{-5})$ $n + \mu^- + \nu(6.6 \times 10^{-4})$ $n + e^- + \nu(1.4 \times 10^{-3})$ $\Lambda^\circ + e^- + \nu(7.5 \times 10^{-5})$
Ξ°	½	1,314.3	3.06×10^{-10}	$\Lambda^\circ + \pi^\circ(\sim 1.00)$
Ξ^-	½	1,320.8	1.74×10^{-10}	$\Lambda^\circ + \pi^-(\sim 1.00)$ $\Lambda^\circ + e^- + \nu(3.0 \times 10^{-3})$
Ω^-	3/2(?)	1,675	1.3×10^{-10}	$\Xi^{\bar{\circ}} + \pi^{\underline{\circ}}(0.5?)$ $\Lambda^\circ + K^-(0.5?)$

the cross sections for the production of these particles were significantly large (that is, ~ 0.1 to 0.01×10^{-27} cm^2 or 0.1 to 0.01 mbarn). Cross sections of this order of magnitude indicate strong interactions. (It has also been pointed out in the previous section that K mesons are produced almost as intensely as pions in the annihilation of high-energy antiprotons.) However, herein lay a dilemma. The puzzling question was why do the K mesons and hyperons decay so slowly? (This can be appreciated, for example, by considering the production and decay of the Λ°. Back in 1952 there was no reason not to have considered that the Λ° could have been produced in a straightforward reaction at high energies, such as $\pi^- + p \rightarrow \Lambda^\circ + \pi^\circ$. Then one would also have expected that the Λ° could be virtually equivalent to $\Lambda^\circ \rightarrow \pi^- + p + (\overline{\pi^\circ})$, and since $\overline{\pi^\circ} \equiv \pi^\circ$, that is, $\Lambda^\circ \rightarrow \pi^- + p + (\pi^\circ)$. Since the interaction between p and π° is so strong, the π° could be absorbed by the proton and one should therefore have expected that the Λ° would interact strongly with the π^- and p. (That is, the decay into these particles would have been exoergic and fast, taking $\sim 10^{-23}$ sec. However, the observed mean lifetime of the Λ° is $\sim 10^{-10}$ sec) The way out of the dilemma was seen by Pais [55] who put forward the principle of associated production of K mesons and hyperons in strong interactions, and the separate slow decay of these particles in weak interactions. Experimental evidence has confirmed this principle which states that K mesons and hyperons shall be made in pairs in strong interactions. Typical observed reactions are

$$p + p \rightarrow \Sigma^+ + p + K^\circ \tag{12.43}$$

$$\pi^- + p \rightarrow \Lambda^\circ + K^\circ \tag{12.44}$$

[Reactions (12.41) and (12.42), in which strange-particle–antiparticle pairs are produced, are also examples of associated production.]

The slow decays of K mesons and hyperons suggest that there is some characteristic property of the particles that is not readily changed between the initial and final states of the systems. Although at first a certain amount of consideration was given to properties such as angular momentum (large changes of which can hold back electromagnetic transitions and beta decays), the evidence is against any large changes of angular momenta in strange-particle decays. Gell-Mann and Nishijima [56] independently suggested that a new property, "strangeness," was not conserved in the decays of K particles and hyperons. In the Gell-Mann–Nishijima scheme, all strongly interacting particles are assigned quantum numbers shown in Table 12.6. The following quantum numbers are included: I is the value of the isotopic spin, and the number of charge states is given by $2I + 1$. B is the baryon number. B is $+1$ for nucleons and hyperons and -1 for antinucleons and antibaryons. S is an assigned strangeness number. The S values of Table 12.6 have been assigned thus

TABLE 12.6. QUANTUM NUMBERS OF STRONGLY INTERACTING PARTICLES

Particle (Q) +1 0 −1	Isotopic spin (I)	Third comp. of I (I_3)	Strangeness (S)	Baryon (B)	Hypercharge (Y)
π^+, π°, π^-	1	+1 0 −1	0	0	0
K^+, K°	½	+½ −½	+1	0	+1
$\overline{K^\circ}$, K^-	½	+½ −½	−1	0	−1
p, n	½	+½ −½	0	+1	+1
$\bar{n}$, $\bar{p}$	½	+½ −½	0	−1	−1
Λ°	0	0	−1	+1	0
$\overline{\Lambda^\circ}$	0	0	+1	−1	0
Σ^+, Σ°, Σ^-	1	+1 0 −1	−1	+1	0
$\overline{\Sigma^-}$, $\overline{\Sigma^\circ}$, $\overline{\Sigma^+}$	1	+1 0 −1	+1	−1	0
Ξ°, Ξ^-	½	+½ −½	−2	+1	−1
$\overline{\Xi^-}$, $\overline{\Xi^\circ}$,	½	+½ −½	+2	−1	+1
Ω^-	0	0	−3	+1	−2
$\overline{\Omega^-}$	0	0	+3	−1	+2

far in a purely phenomenological way, that is, so as to be consistent with the principle of associated production. However, S is related to other quantum numbers by the empirical relationship

$$Q = I_3 + (B/2) + (S/2) \tag{12.45}$$

In strong and electromagnetic interactions, the total values of Q, S, and B are conserved. In weak interactions, S is not conserved but Q and B are. Another useful form of strangeness specification is hypercharge, Y, defined by the relationship

$$Y = S + B \tag{12.46}$$

The usefulness of the strangeness concept can be illustrated by certain experimental examples: (1) Associated production follows immediately from the application of the conservation laws; for example, $\pi^- + p \rightarrow \Sigma^- + K^+$ conserves Q, B, and S. (2) In the absorption of strange particles, as, for example, in the reaction $K^- + p \rightarrow \Lambda^\circ + \pi^\circ$, strangeness is conserved through the production of a hyperon of the same strangeness as the K meson. (3) In certain scattering processes, because of the conservation laws, only elastic and charge-exchange scattering may occur, as, for example, in the reactions $K^+ + p \rightarrow K^+ + p$ and $K^+ + n \rightarrow K^\circ + p$. (4) Hypernuclei may be formed from bound states of Λ° with nuclei. These systems, such as ${}_\Lambda H^4 = \Lambda^\circ + p + 2n$, endure for approximately the mean lifetime of the Λ° because the Λ° cannot convert into another particle except by a slow strangeness change.

A cornerstone of the Gell-Mann scheme is the K-meson doublet, that is, K^+, K° or $\bar{K}^\circ$, K^-. There are three charge states, but instead of considering them as parts of a triplet ($I = 1$), they were assigned to two doublets. One of the prevailing arguments supporting this assumption was the lack of experimental evidence for the reaction $n + n \rightarrow \Lambda^\circ + \Lambda^\circ$. This reaction is now interpreted as being forbidden because it does not satisfy the principle of conservation of strangeness. Initially, however, it was realized that the K° and $\bar{K}^\circ$ would have to be of opposite strangeness; otherwise the virtual processes $n \rightarrow K^\circ + \Lambda^\circ$ and $K^\circ + n \rightarrow \Lambda^\circ$ would occur and in so doing would bring about the reaction $n + n \rightarrow \Lambda^\circ + \Lambda^\circ$. The assignments of strangeness $+1$ to the K° and -1 to the $\bar{K}^\circ$ mean that the two neutral K particles, unlike the neutral π mesons, are now distinct particles. In certain strong interactions, such as, for example, $\pi^- + p \rightarrow \Lambda^\circ + K^\circ$, only one K° particle may be formed; but in decay, where strangeness is not conserved, the K° particle is a mixture of K° and $\bar{K}^\circ$. When Gell-Mann and Pais [57] first pointed out the possibility of two different K° decay modes in 1955, the principle of CP invariance was believed to be strictly valid. (Although this principle was seriously questioned in 1964, the extent of its possible invalidity will not affect the broad issues of the Gell-Mann–Pais argument. In fact, two K° decay modes persist even if the principle of CP invariance fails [58].)

If then, the phases of the state functions $|K^\circ\rangle$ and $|\bar{K}^\circ\rangle$, representing the neutral K particle and antiparticle, respectively, are chosen so that

$$\mathbf{CP}|K^\circ\rangle = |\bar{K}^\circ\rangle \tag{12.47}$$

the eigenstates corresponding to $\mathbf{CP} = +1$ and -1 are

$$\mathbf{CP} = +1: \quad |K^\circ_1\rangle = (|K^\circ\rangle + |\bar{K}^\circ\rangle)/\sqrt{2} \tag{12.48}$$

$$\mathbf{CP} = -1: \quad |K^\circ_2\rangle = (|K^\circ\rangle - |\bar{K}^\circ\rangle)/\sqrt{2} \tag{12.49}$$

Now, since both K° and $\bar{K}^\circ$ can decay via the slow reactions $K^\circ \rightarrow \pi^+ + \pi^-$ and $\bar{K}^\circ \rightarrow \pi^+ + \pi^-$, and since $|(\pi^+ + \pi^-)\rangle$ is an eigenstate only of $\mathbf{CP} = +1$, only K°_1 can decay into two π mesons if CP is conserved. Thus the decay of the K°_1 would be an "allowed" decay and would have a relatively short lifetime. The K°_2, however, could decay by other three-body processes, such as $K^\circ_2 \rightarrow \pi^+ + \pi^- + \pi^\circ$, etc., which are allowed by CP invariance but are somewhat slower. It has already been seen from Table 12.4 that these properties of the K° meson have indeed been experimentally confirmed. The K°_1 and K°_2 particles also possess slightly different masses because of their different couplings with the decay channels. A recent determination of the mass difference is 0.6×10^{-5} ev. [For a summary of the determinations to date (1964) see Fujii et al. [59].] Evidence has also recently been obtained for a weak 2π decay of the K°_2 meson [60]. K° mesons, which were produced at a distant target, were allowed to decay until the K°_1 component had disappeared. Decays of the K°_2 were then observed with spark chambers and scintillation counters. In a small fraction (2×10^{-3}) of the decays of K°_2, there were events having only 2π decay products of K° decays. The implication of these important experiments, which are still being widely discussed in 1964, is that CP invariance breaks down in certain instances.

12. Spins of Strange Particles

The spin of the K meson is undoubtedly zero. This conclusion has been reached by a number of consistent and indirect arguments, among which are the following: (1) The spectrum of the unlike pion (that is, π^- in the 3π decay of the K^+ meson) agrees closely with that theoretically predicted for a spinless K particle. [This was the original argument used by Dalitz [61] to draw conclusions of the spin and parity of the τ meson, i.e., the $K^\pm$ meson that was observed to decay into three pions. The τ decay does not involve a change of parity, whereas the 2π decay (or θ mode) does involve a parity change. It is well known that these arguments and their analysis by Lee and Yang [62] led to the discovery of the breakdown of parity conservation in weak interactions.] (2) The observation by Glaser and coworkers [63] of the neutral decay of the K°_1 into two π° mesons is consistent only with an even spin of the neutral K meson. [Two identical bosons (π°) must be in an even-parity state and therefore in a state of even orbital angular momentum. Since the π° mesons have zero spins, the initial state (that is, K°) must have even, or zero, spin.] (3) The longitudinal polarization of the μ^+ meson arising from the decay of K^+ into μ^+ and ν has been measured [64]. It is essentially 100 per cent and is in the same direction as the μ^+ from the decay of the π^+, that is, reaction (12.2). This implies that the K^+ is also, like the π meson, a spinless particle.

Except for the Ω^-, the spins of the hyperons are very probably $\frac{1}{2}$. The spin of the Ω^- is expected to be $\frac{3}{2}$. By 1964, however, there were only four known cases of Ω^- decay, and this sample is too small for a spin determination to have been made. The spins of Λ° and charged Σ hyperons have been inferred from angular distributions of their decay products. The results are all consistent with a value of $\frac{1}{2}$. The angular distribution measurements are based on the assumption either that the hyperons are made in an s state, as, for example, in the K^- capture at rest on protons such as $K^- + p \rightarrow \Sigma^\pm + \pi^\mp$, or that the hyperons are emitted in the forward or backward direction with respect to the incident production particle, as, for example, in the reaction $\pi^- + p \rightarrow \Lambda^\circ + K^\circ$. Choice of an axis of quantization along the direction of the emitted $\Sigma^\pm$ or Λ° then ensures that the magnetic substate of the hyperon is the same as that of the proton, that is, $m_p = \pm\frac{1}{2}$. This condition in turn restricts the angle (with respect to the quantization axis) of the emitted pion from the $\Sigma^\pm$ or Λ° decay to have a definite distribution dependent on the spin of the $\Sigma^\pm$ or Λ°. All the observed distributions have been consistent with isotropy, which is consistent with a hyperon spin of $\frac{1}{2}$ [65].

Determinations of the magnetic moments of Λ° and Σ^+ have also been made. The magnetic moment of Λ°, like the neutron, is negative and equal to approximately $-(0.77 \pm 0.27)$ n.m. [66]. The value of the magnetic moment of Σ^+ is $+(4.3 \pm 0.6)$ n.m. [67].

13. Parities of Strange Particles

Although parity is not conserved in decays of both K mesons and hyperons, nevertheless these particles must be considered to decay from states of definite parity. It is normally assumed, and there is strong experimental support for it, that parity is conserved in strong interactions. Because conservation of strangeness is also involved in fast reactions between K mesons and hyperons, only relative parities of Λ and K, or of Σ and K, systems can be measured with respect to nucleons. Conventionally, the Λ is often defined as having positive parity, as for the nucleon, and then the parity of the K meson is described as positive or negative in that sense relative to the Λ°.

The capture in helium of K^- mesons at rest and the production of ${}_\Lambda He^4$ hyperfragment is a reaction suggested by Dalitz [68] as a method for determining the parity of the K meson. This reaction

$$He^4 + K^- \rightarrow {}_\Lambda He^4 + \pi^- \tag{12.50}$$

has in fact been observed [69], and if it were certain that the spin of ${}_\Lambda He^4$ produced directly in this reaction is zero, then the conclusion would be unequivocally that the parity of the K^- is negative. Dalitz [68] has given strong theoretical arguments which favor a spin zero for the ${}_\Lambda He^4$ hyperfragment. Block et al. [70] have also found in a number of independent experiments that the spin of ${}_\Lambda H^4$ is zero, and one should then expect on symmetry arguments that the spin of ${}_\Lambda He^4$ is zero. The spin of ${}_\Lambda H^4$ can be determined from the angular distribution of the π^- mesons emitted in the decay reaction of ${}_\Lambda H^4$ in flight:

$${}_\Lambda H^4 \rightarrow He^4 + \pi^- \tag{12.51}$$

Block et al. have found the decay of ${}_\Lambda H^4$ to be isotropic. This requires the parity of the K^- to be negative—again if one can be certain that the ${}_\Lambda H^4$ in decay is the same as the ${}_\Lambda H^4$ produced directly in the reaction

$$He^4 + K^- \rightarrow {}_\Lambda H^4 + \pi^\circ \tag{12.52}$$

Moravscik [71] has analyzed experimental data on the reaction

$$\gamma + p \rightarrow \Lambda^\circ + K^+ \tag{12.53}$$

and from the differential cross section has tentatively concluded that the K^+ meson is a pseudoscalar particle. The evidence, therefore, from all these data is that the K meson is a pseudoscalar boson, like the π meson.

Two independent experiments have recently determined that the parity of the Σ° relative to Λ° is positive [72]. Negative K mesons stopped in liquid hydrogen were used to produce Σ° in a fraction of the events according to the reaction

$$K^- + p \rightarrow \Sigma^\circ + \pi^\circ \tag{12.54}$$

The most common decay of the Σ° is to a Λ° and γ ray, but in approximately 1 in 200 cases the Σ° decay proceeds via a Dalitz pair:

$$\Sigma^\circ \rightarrow \Lambda^\circ + e^+ + e^- \tag{12.55}$$

It has been shown theoretically [73] that the invariant mass distribution of the electron pairs from reaction (12.55) depends on the nature of the virtual γ-ray transition of the Σ° decay and therefore on the parity change between Σ° and Λ°. The observed distribution was found [72] to be characteristic of a magnetic-dipole transition, thus indicating that the Σ° and Λ° have the same parities. As Dalitz [74] points out, the conclusions given above on the relative K and Σ (or Λ) parities are consistent with other experimental evidence [75] on the polarization of Σ^+ produced in the reaction of ~150 Mev K^- mesons on protons.

14. Interactions of Strange Particles with Nucleons

The scattering of K^+ mesons by protons has been found to show almost a constant nuclear scattering cross section of approximately 11 mbarns over a range of K^+ laboratory energies from 20 to 450 Mev [76]. The scattering reaction

$$K^+ + p \rightarrow K^+ + p \tag{12.56}$$

in this range is found to be purely elastic and results largely from a repulsive K^+-nuclear potential except for a small part that appears to be attractive at its outer fringe [77]. The cross sections can be described quite well in terms of s-wave scattering for which the phase shifts vary linearly with momentum of the K meson from $-7°$ at 20 Mev to $-36°$ at 300 Mev. Zero-range approximation analysis yields a scattering length of $-(0.3 \pm 0.1)$ fermi.

By contrast with K^+ scattering, however, K^--nucleon scattering is found to have a large inelastic contribution and to vary strongly with K^--meson energy. For K^- energies less than approximately 300 Mev, the reactions that occur are

Elastic:	$K^- + p \rightarrow K^- + p$	(12.57)
Charge-exchange:	$K^- + p \rightarrow \bar{K}^\circ + n$	(12.58)
Inelastic:	$K^- + p \rightarrow \Sigma^- + \pi^+$	(12.59)
	$\rightarrow \Sigma^+ + \pi^-$	(12.60)
	$\rightarrow \Sigma^\circ + \pi^\circ$	(12.61)
	$\rightarrow \Lambda^\circ + \pi^\circ$	(12.62)

These reactions have been the subject of extensive experimental analyses [78] since the production of adequately intense beams of K mesons first designed at Berkeley in 1955 [79]. Both elastic and inelastic scattering are characterized by rapidly increasing cross sections as the K^- energy approaches zero. The inelastic reactions (12.59) to (12.62) are all strongly exoergic and, in particular, the reactions leading to $\Sigma^\pm$ exhibit an energy variation that is probably steeper than $1/v$ (v being the laboratory velocity of the incident K^- meson). Some experimental results of the $\Sigma^\pm$ production cross sections are shown in Fig. 12.12. The angular distributions

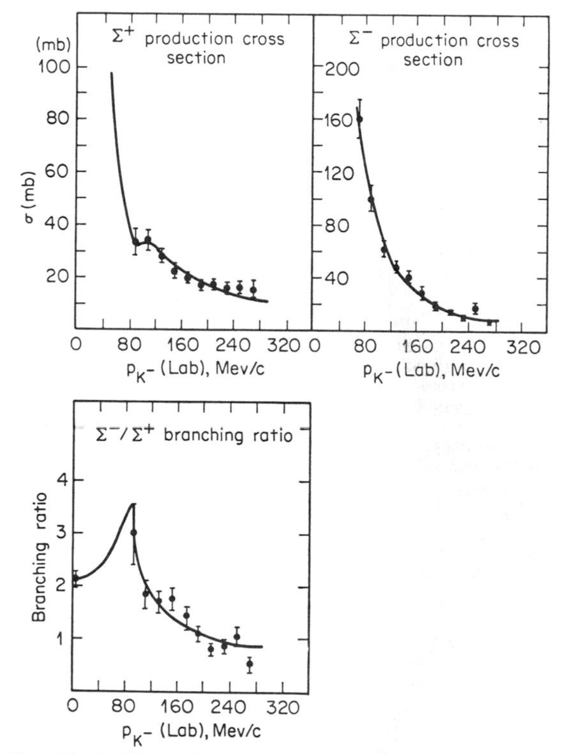

Fig. 12.12. $\Sigma^\pm$-production cross sections from capture of low-energy K^- mesons on protons. The Σ^+ cross section shows a kink that reflects the onset of the charge-exchange channel (12.57). This effect shows up even more strongly in the Σ^-/Σ^+ ratio at the $\bar{K}^\circ$ production threshold at approximately 95 Mev/c. The curves are for a particular solution of phase shifts according to the Dalitz-Tuan analysis [80]. (*After Kim* [78].)

of both the elastically scattered K^- and of the charged hyperons from the inelastic reactions are isotropic, certainly below an energy of 100 Mev. The low-energy scattering of K^- mesons on nucleons is therefore, like K^+ mesons, consistent with a strong s-wave interaction. The phenomenological theory of low-energy K^- scattering has also been extensively studied [80]. The scattering, following Jackson, Ravenhall, and Wyld, is usually described in terms of a zero-range s-wave interaction requiring complex phase shifts in the total isotopic spin channels $I = 0$ and $I = 1$. (Complex phase shifts are introduced in order to take into account the strong absorptive part of the K^--meson scattering.) The 1965 data of Kim [78] appear to give the best values of the scattering in terms of the following Dalitz-Tuan type of solution. The following complex, zero-range, scattering lengths, which are also fitted by the curves to the $\Sigma^\pm$ and Σ^-/Σ^+ observations in Fig. 12.12, are

$$I = 0 \text{ channel: } a_0 + ib_0 = (-1.674 \pm 0.038) + i(0.722 \pm 0.040) \tag{12.63}$$

$$I = 1 \text{ channel: } a_1 + ib_1 = (-0.003 \pm 0.058) + i(0.688 \pm 0.033) \tag{12.64}$$

Thus there is strong absorption in both the $I = 0$ and $I = 1$ channels, but there is very little real part of the scattering in the $I = 1$ state. The value of the phase angle ϕ between the absorption amplitudes in the $I = 0$ and $I = 1$ channels was found by Kim to be $-53.8°$ in the neighborhood of K^- energies from ~10 to 50 Mev. This conforms with the value of $\phi = -104°$ found by Watson et al. [81] at a K^- energy of 155 Mev. It was first pointed out by

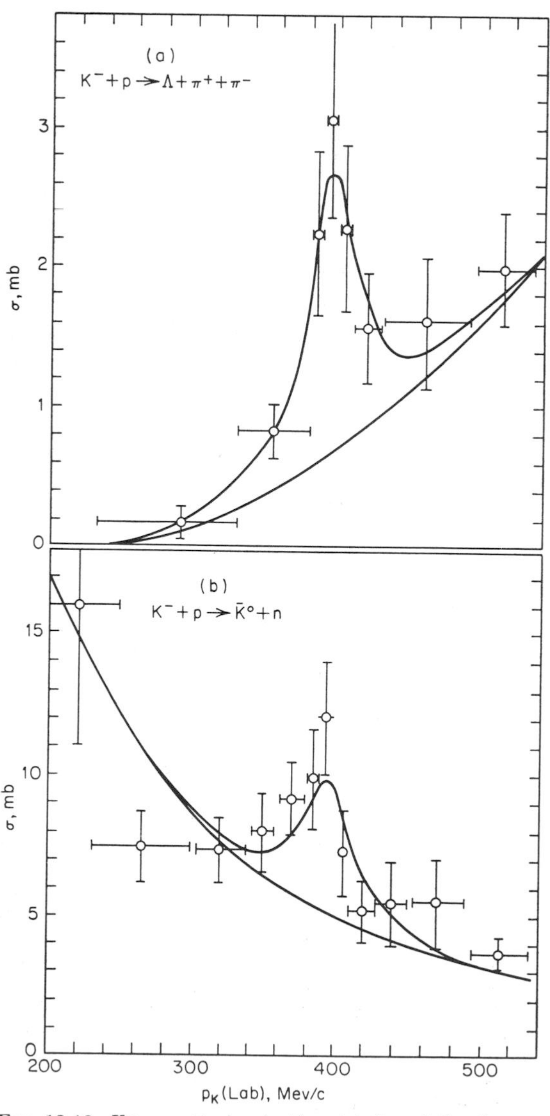

Fig. 12.13. K^--p scattering in the vicinity of the d-wave resonance at 135-Mev K^- energy. (*After Ferro-Luzzi, Tripp, and Watson* [81].)

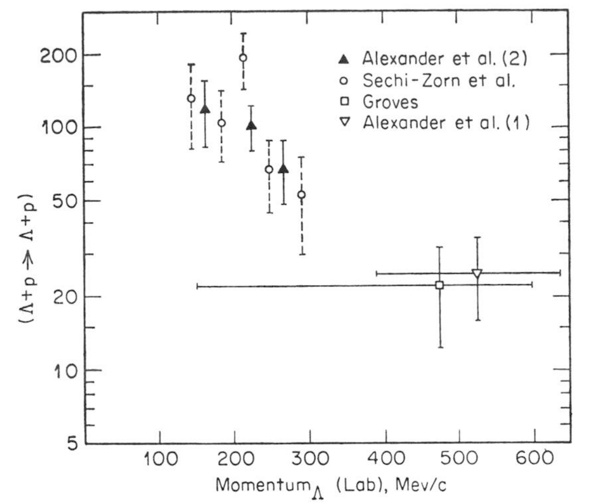

FIG. 12.14. Λ°-p total scattering cross sections as function of Λ° energy. The data points are from various groups cited in ref. 85. (*After Alexander et al.* [85].)

Akiba and Capps [82] that ϕ should probably vary smoothly in the energy range between $\sim$10 and 150 Mev because the Σ^-/Σ^+ ratio (to which ϕ can be directly related) is observed to change smoothly in this range.

At a K^- laboratory kinetic energy of approximately 135 Mev (momentum 400 Mev/c), a very sharp K^--p scattering resonance has been found in both the elastic and inelastic cross sections [81]. Some of the experimental results demonstrating this resonance, which is attributed to a d-wave interaction in an $I = 0$ channel, are shown in Fig. 12.13. It will be noticed that a pronounced peak occurs in the $\Lambda^\circ\pi^+\pi^-$ production, whereas none is found to occur in $\Lambda^\circ\pi^\circ$ production, which occurs only in the $I = 1$ channel. [Λ° production from reaction (12.62) is still present at 135 Mev. In fact, the nonresonant Λ° production is as intense as the resonant production at the peak of the $I = 0$ resonance.]

No further resonances in the initial scattering state of K meson and nucleon appear to have yet been observed. K^+-scattering experiments on protons and neutrons have been performed at a large number of energies between 100 and 1,000 Mev, and the total cross sections in both cases (that is, K^+-p and K^+-n) appear to be reasonably constant and equal to approximately 17 mbarns [83]. Elastic K^+-p-scattering cross sections have also been measured at 6.8, 9.8, 12.8, and 14.8 Bev/c, K^+ laboratory momenta, and the cross sections found to be approximately constant and equal to 3.4 mbarns. At the same momenta, the total K^+-p-scattering cross sections are likewise approximately constant and equal to 18.4 mbarns [84]. The K^--p-elastic and total cross sections have been measured at 7.2 and 9.0 Bev/c, and they also have been found to be equal at the two momenta and are 4 mbarns (elastic) and 24.2 mbarns (total). Bearing in mind Pomeranchuk's theorem [47] again, there seems to be a slight excess of the K^--p cross sections over the K^+-p cross sections at these energies [84].

Some data on the scattering of low-energy hyperons by protons have recently become available. The data on Λ°-p total scattering cross sections are summarized in Fig. 12.14 [85]. Some preliminary data are also available on $\Sigma^\pm$-proton scattering [86]. Stannard found $\sigma_{\text{elastic}}(\Sigma^+\text{-}p) \simeq (38_{-14}{}^{+18})$ mbarns and $\sigma_{\text{elastic}}(\Sigma^-\text{-}p) \simeq (10_{-4}{}^{+6})$ mbarns for $\Sigma^\pm$ energies in the range 100 to 700 Mev.

RESONANT STATES

15. Discovery of New Resonances

The existence of a strong resonant interaction between a pion of approximately 190 Mev and a nucleon has already been noted. Other peaks in pion-nucleon scattering, as shown in Figs. 12.2 and 12.3, and in photopion production, as shown in Fig. 12.6, were also subsequently observed. The first suggestion of the possibility of strange-particle resonances appears to have been made by Gell-Mann [87] who predicted that all baryons would interact strongly with pions. This was the so-called hypothesis of global symmetry. A second suggestion was made by Dalitz and Tuan [88], who pointed out that if certain solutions of the strong K^--nucleon interaction existed then there should also exist strong resonant state interactions in the pion-hyperon system. It has just been seen in the preceding section that such a K^--p scattering solution exists and the $\Sigma\pi$ resonance at a total energy of 1,405 Mev is very probably attributable to this mechanism.

In 1960, however, the interest in interparticle resonances took a different turn with the introduction of a new experimental technique by Alston et al. [89]. These workers studied the energy spectra of the π mesons emitted in the following reaction, resulting from the interaction of 1,150-Mev/c K^- mesons in a liquid H_2 bubble chamber:

$$K^- + p \rightarrow \Lambda^\circ + \pi^+ + \pi^- \qquad (12.65)$$

The kinetic energies of the pions, measured in the CM frame of the K^--p system, are shown in Fig. 12.15. It is observed that a group of pions stands out above the solid lines, which are the distributions on the basis of a three-body phase space. Furthermore, the elliptical area, within which points of definite π^+ and π^- kinetic energies are expected to fall uniformly if the Λ°, π^+, and π^- are emitted completely independently, is populated rather in bands which are associated with the peaks of the histograms. These groups of essentially monoenergetic pions were interpreted by Alston et al. as indicative of the existence of two-particle final states according to the reactions

$$\left.\begin{array}{l} K^- + p \rightarrow Y^{*+}{}_1 + \pi^- \\ \qquad\quad \rightarrow Y^{*-}{}_1 + \pi^+ \end{array}\right\} \rightarrow \Lambda^\circ + \pi^+ + \pi^- \qquad (12.66)$$

where the $Y^*{}_1$ represents an intermediate resonant state of the Λ and π, which must be an isotopic triplet (hence suffix 1) since it may appear in the three-charge states: $\Lambda^\circ\pi^+$, $\Lambda^\circ\pi^-$, and $\Lambda^\circ\pi^\circ$. From the observation that the π^+ and π^- groups had an energy of 280 Mev, and applying conservation of energy and momentum to the two-body decay reaction, Alston et al. found that the $Y^*{}_1$ "particle" had a mass of 1,385 Mev with a "decay" width of 60 Mev. The mean lifetime of the $Y^*{}_1$, as inferred from the reso-

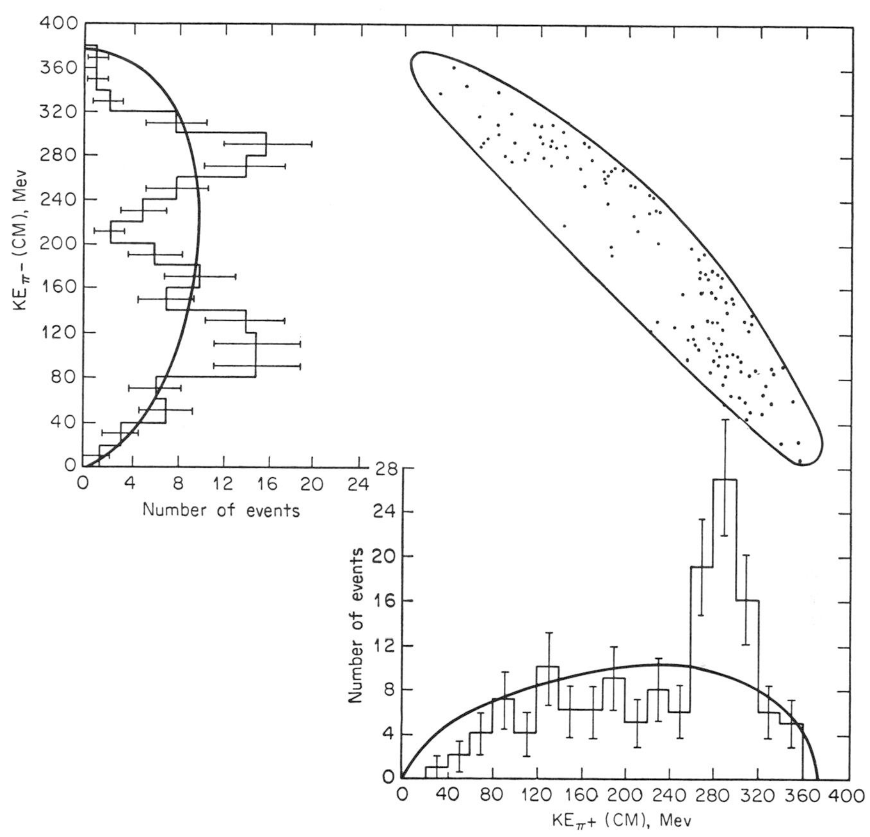

FIG. 12.15. Energy distributions of pions from reaction $K^-p \rightarrow \Lambda\ \pi\pi$. The histograms are of the numbers of pions as a function of pion kinetic energies in the CM system. The solid-line curves are the expected energy distributions on the basis of three-body phase space. The oval region in the upper right corner is a Dalitz plot of the individual events contained in the histograms. (*After Alston et al.* [89].)

nance width, is approximately 10^{-23} sec. It need hardly be pointed out that this lifetime is at most only a few times longer than the transit time of a high-energy pion across the Λ° diameter. Clearly the Y^*_1 and similar particles are more highly evanescent than the strange particles already referred to in the previous section and which, as already noted, are sometimes classified with the stable particles. It was immediately perceived that the pion-nucleon resonance, which has a width of 100 Mev and occurs at a CM π energy of 159 Mev, belongs to the same category of states as the Y^*_1. In fact, production experiments have been performed in the bubble chamber, just as in the case of Y^*_1, and they have shown that the πN pair form an intermediate state, now normally referred to as $N^*_{3/2}$ (1,236). [The suffix $\frac{3}{2}$ again refers to the isotopic spin of the state. Since there are several $N^*_{3/2}$ particles, in order to be more specific, the mass (1,236 Mev) is often attached.] Since the strong resonant interaction between the π and N has been amply demonstrated, it is presumed that all the short-lived intermediate particles owe their existence to some resonance system of particles. The practice has been, therefore, to refer to all these states as resonance particles even though the resonance character has not been shown. In fact, for most resonance particles, no accessible initial scattering state exists, and information on the state must come from a study of the strong interactions in the final state.

Although the experimental evidence for numerous resonance particles now known will not be discussed, reference will be made briefly to the first nonstrange multipion resonance particle discovered [90]. This experiment was performed by Maglic et al. using 1.61 Bev/c antiprotons incident on protons in a liquid-hydrogen bubble chamber. Among other reactions, the following events were selected:

$$\bar{p} + p \rightarrow \pi^+ + \pi^- + \pi^+ + \pi^- + \pi^\circ \quad (12.67)$$

A search for a three-pion resonant state was motivated by theoretical discussions of Nambu [91] and of Chew [92], who suggested that a virtual heavy meson in the nucleon could help explain the observed electromagnetic form factors of the neutron and proton in terms of virtual mesonic states. If, in the $\bar{p}$-p annihilation process (12.67), a three-pion state were temporarily produced, the total energy of these three pions in the CM frame of the $\bar{p}$-p system should have a very definite value, even after the breakup of the three-particle state. Maglic et al. therefore looked for an invariant mass value of M_3 given by

$$M_3 = [(E_1 + E_2 + E_3)^2 - (\mathbf{P}_1 + \mathbf{P}_2 + \mathbf{P}_3)^2]^{1/2} \quad (12.68)$$

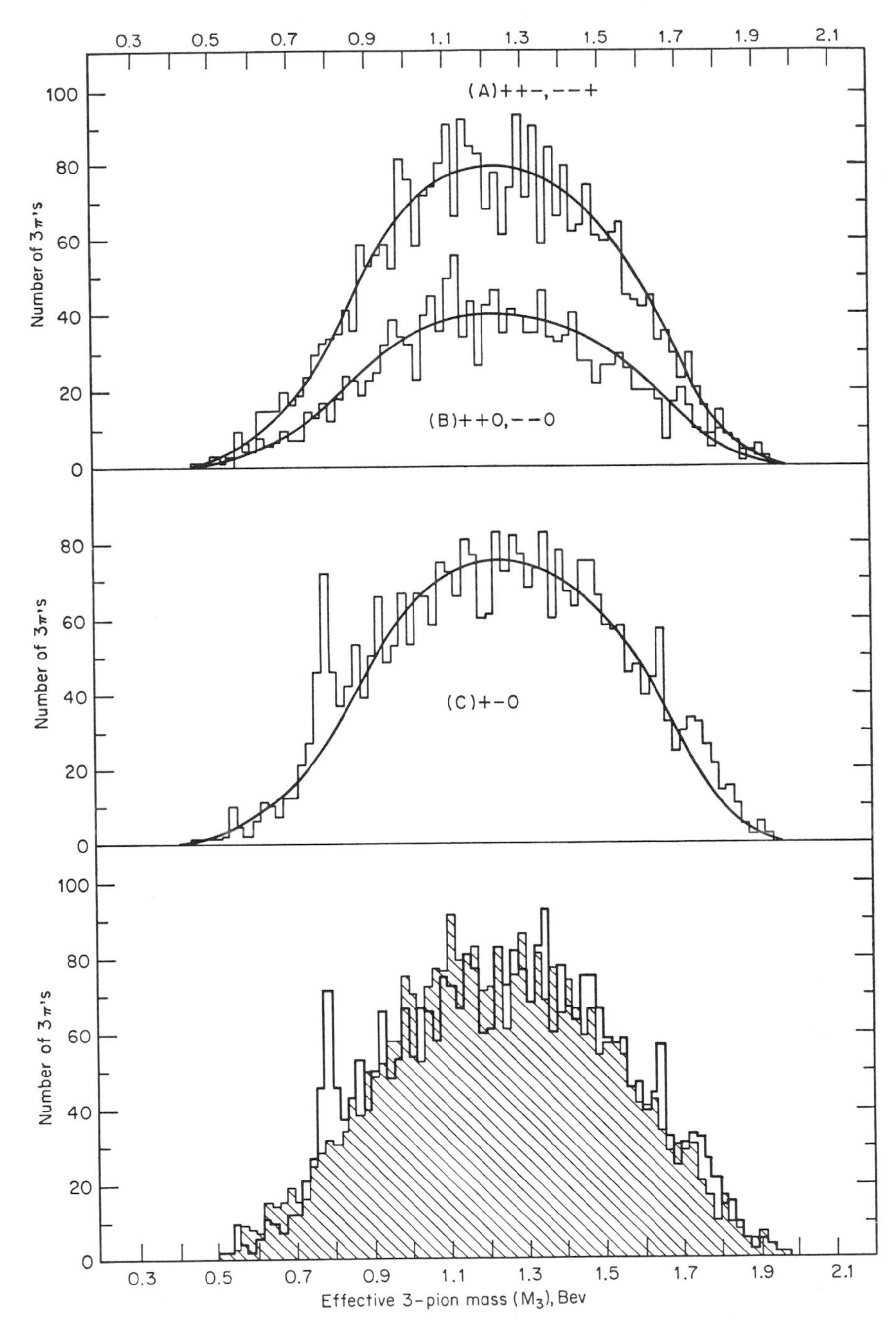

FIG. 12.16. Histograms of 3π invariant mass from $\bar{p}p \rightarrow 5\pi$ reactions. The histograms in the upper two plots are for $A(\pi^+\pi^+\pi^-$ and $\pi^-\pi^-\pi^+)$ and $B(\pi^+\pi^+\pi^\circ$ and $\pi^-\pi^-\pi^\circ)$ combinations. The center histogram is for $\pi^+\pi^-\pi^\circ$ combinations only. This shows a peak at an effective mass M_3 of approximately 0.8 Bev, which gives evidence for the ω° resonance. The lowest histogram is for an over-all combination of the three histograms A, B, and C. (*After Maglic* [90].)

where E_1, E_2, and E_3 are the CM total energies of three separate pions selected and $\mathbf{P}_1$, $\mathbf{P}_2$, and $\mathbf{P}_3$ are their momenta. Only in one combination of three pions did they find evidence of an invariant mass peak; that was when one of the π^+ and one of the π^- mesons was combined with a π°. The histograms observed by Maglic et al. are given in Fig. 12.16. The mass of the resonance particle corresponding to the peak above the continuous background of pions constituting five-body phase space is 783 Mev. The resonance particle, which was called ω°, has a width of approximately 10 Mev. Production and decay reactions were considered to be

$$\bar{p} + p \rightarrow \pi^+ + \pi^- + \omega^\circ \tag{12.69}$$

$$\omega^\circ \rightarrow \pi^+ + \pi^- + \pi^\circ \tag{12.70}$$

Although the ω° has a similar decay to $K^\circ{}_2$ it is clearly not a strange particle and has quantum number $S = 0$. This is strikingly seen in the completely different magnitudes of their lifetimes: $\sim 6 \times 10^{-23}$ sec for ω°; 5.6×10^{-8} sec for $K^\circ{}_2$; the ω° decay is not held back by any strangeness change.

16. List of Particles (1964)

Before referring to a table of particles and their properties, it will be useful to summarize the quantum numbers that are used to characterize the strongly interacting particles. These are

B, baryon number (+1 for all particles of mass greater than or equal to the proton, −1 for antibaryons, zero for all particles less than the mass of the proton).
S, strangeness number.
Y, hypercharge number (not to be confused with Y or Y^* for hyperon states). $Y = S + B$.
I, isospin. Charge Q is related to the third component of isospin. $Q = I_3 + (Y/2)$.
J, spin (or angular momentum). l is the orbital angular momentum of the state.
P, parity.
C, charge-conjugation parity. For neutral mesons C_n has the value ± 1.
G, G parity. For a state with $B = 0$ and $Y = 0$ (e.g., pions, ω particle, etc.), G is a good quantum number. For a neutral meson, $G = C(-1)^I$. For a pair of mesons, $G = (-1)^{l+I}$.

For purposes of classification, all strongly interacting particles, no matter whether stable, relatively long-lived (10^{-8} to 10^{-10} sec), or relatively unstable ($\sim 10^{-23}$ sec), are considered on an equal basis. In fact, if one inspects the Feynman diagrams for the decays of all these particles, one finds that they

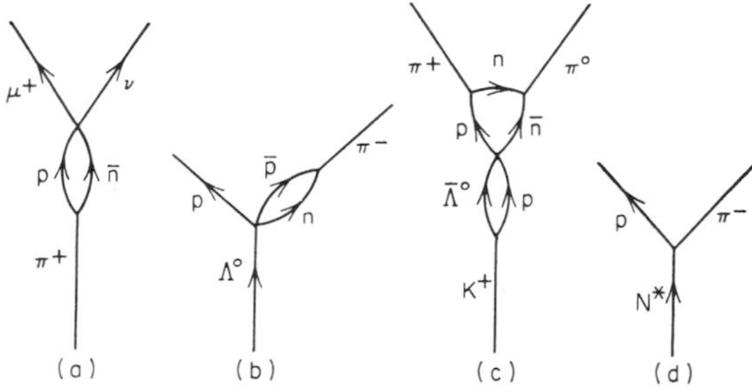

FIG. 12.17. Feynman diagrams for the decays of (*a*) π^+, (*b*) Λ°, (*c*) K^+, and (*d*) N^*.

possess ingredients of the same or similar strong interactions. In Fig. 12.17, for example, there are illustrations of four decays: (*a*) A π^+ meson decays through the strong-interaction vertex of a p and $\bar{n}$, followed by a four-fermion weak-interaction vertex which is responsible for holding back the decay of the π^+ into μ^+ and ν. (*b*) A Λ° hyperon decays first through a four-fermion weak-interaction vertex into a virtual $\bar{p}$,n pair and p. Strong interaction between $\bar{p}$ and n gives rise to a real π^- (together with the p). (*c*) A K^+ meson decays through a strong-interaction vertex into an antihyperon and nucleon, and then through a weak-interaction vertex (with a change of strangeness) into a final π°,π^+ pair. (*d*) A N^* (1,236) resonance particle, which has the shortest observed lifetime of the four examples, may decay directly through a strong-interaction vertex into p and π^-.

According to the viewpoint presented by Chew, Gell-Mann, and Rosenfeld [93] a reappraisal of the entire elementary-particle concept is necessary. Any quantum state of a strongly interacting particle, e.g., a proton, a hyperon, a resonance particle, etc., consists for a part of the time of both real and virtual states (or channels) that connect with the particles. Chew, Gell-Mann, and Rosenfeld conclude that, if the properties of a particle can be calculated (even only in principle) by treating it as a composite, the particle should not be regarded as elementary. According to the "bootstrap" hypothesis of Chew and Frautschi [94], the strongly interacting particles are all composite structures and they owe their existence to the same dynamical forces through which they mutually interact. Some qualitative success has been achieved in calculating the properties of some of the basic particles from an analysis of the channels with which they interact.

CLASSIFICATION OF PARTICLES

17. SU(3) Supermultiplets

The most successful organization of the strongly interacting particles to date (1964) has been according to the SU(3) classification scheme. This system, which is also known as the eightfold way, was put forward independently in 1961 by Ne'eman [95] and Gell-Mann [96]. The underlying idea of this scheme is that the strong interactions of certain families of particles (sometimes called supermultiplets) are largely invariant with respect to rotations of isotopic spin (I) and hypercharge (Y). In particular, the rotations involved are the unitary transformations belonging to the mathematical group SU(3). This group of transformations can be built up from the infinitesimal transformations generated by the operators of a Lie algebra of rank 2. The SU(3) multiplets are generated by successive operations on any state within the multiplet with the eight operations of the Lie algebra. The states of the multiplet are represented by points on a two-dimensional plot of the eigenvalues of I_3 and Y. All members of a particular supermultiplet have the same (ordinary) spin and parity. The known supermultiplets, as presently (1964) constituted, are shown in Fig. 12.18*a* to *h*. In these plots there are multiplets corresponding to 1, 8, and 10 members. These are all possible multiplets of the SU(3) group. The next highest multiplet is a family of 27 members. It is possible that some of the resonance particles listed in Tables 12.7 and 12.8 and not represented in the plots of Fig. 12.18 may be

TABLE 12.7. BOSONS AND COMBINED BOSON STATES

Particle	Mass (Mev)	Lifetime or energy width	Y	I	JP	C_n	G	Decay modes
π°	135.01	1.80×10^{-16}(sec)	0	1	0−	+	−	$\gamma\gamma$ (100 %)
$\pi^\pm$	139.60	2.55×10^{-8} (sec)	0	1	0−	~	~	$\mu\nu$ (~100 %)
$\eta(3\pi)$	548.7	<10 (Mev)	0	0	0−	+	+	$\pi^+\pi^-\pi^\circ$ (27.4 %) $\pi^\circ\pi^\circ\pi^\circ$ (31.8 %) $\gamma\gamma$ (35.3 %) $\pi^+\pi^-\gamma$ (5.5 %)
$\rho(2\pi)$	763	106 (Mev)	0	1	1−	−	+	$\pi\pi$ (100 %)
$\omega(3\pi)$	782.8	9.4(Mev)	0	0	1−	−	−	$\pi^+\pi^-\pi^\circ$ (86 %) $\pi^\circ\gamma$ (11 %) $\pi^+\pi^-\gamma$ (3.2 %)
$\eta 2\pi(5\pi)$	959	<12 (Mev)	0	0	(0−)	(+)	(+)	$\eta 2\pi$ (large)
$\phi(K\bar{K})$	1,019.5	3.1(Mev)	0	0	1−	−	−	K_1K_2 (41 %) K^+K^- (59 %)
A1$(\rho\pi)$	1,090	125 (Mev)	0	1	0−	+	−	$\rho\pi$ (~100 %)
$B(\omega\pi)$	1,215	122 (Mev)	0	1	(1+)	−	+	$\omega\pi$ (~100 %)
$f_0(\pi\pi)$	1,253	100 (Mev)	0	0	2+	+	+	$\pi\pi$ (large) 4π (8 %)
$A2(\rho\pi)$	1,310	80 (Mev)	0	1	2+	+	−	$\rho\pi$ (70 %) $\bar{K}K$ (~30 %)
$KK\pi$	1,410	60 (Mev)	0	≤1	(0−)	+	+	K^*K (large) $K\bar{K}\pi$ (small)
$K^\pm$	493.8	1.23×10^{-8}(sec)	+1	½	0−	~	~	$\mu\nu$ (63 %) $\pi^\pm\pi^\circ$ (21.5 %) $\pi^\pm\pi^-\pi^+$ (5.5 %)
K°	498.0	(See Table 12.4)	+1	½	0−	~	~	(See Table 12.4)
$K^*(K\pi)$	891	50 (Mev)	+1	½	1−	~	~	$K\pi$ (~100 %)
$K\pi\pi$	1,215	60 (Mev)	+1	≤3/2	1+	~	~	$K\rho$ (large)

members of a higher multiplet. To the baryon multiplets there should also be added similar antibaryon multiplets.

A characteristic of the two-dimensional plots is that the points along a horizontal line (of constant strangeness) represent members of an isotopic multiplet of the subgroup SU(2). Note that sometimes (e.g., in the octets) two states can be represented at one location. This degeneracy is characteristic of SU(3) (and can be much greater in the higher member multiplets). The degeneracy can be resolved only by consideration of the actual particles, in the cases of Fig. 12.18*a*, *b*, *e*, and *f* by recognizing that the eigenvalue of I has two values, $I = 1$ for the triplet members and $I = 0$ for the singlet member of the supermultiplet.

An outstanding achievement of the SU(3) scheme was the prediction of the existence of the Ω^- hyperon. Although the spin and parity of this particle have not yet been measured, it seems very likely that they will be $\frac{3}{2}+$ and fit in with the other members of the baryon decuplet (Fig. 12.18*g*). A mass of approximately 1,675 Mev was predicted for the Ω^-, and a strangeness of -3 (that is, $Y = -2$) is consistent with the observed Ω^- production and decay [54].

It is clear that the SU(3) system arose out of an earlier attempt to classify the particles, known as the Sakata model [97]. Initially Sakata proposed that all the elementary particles could be formed by certain combinations of three basic particles, usually assumed to be n, p, and Λ°. (Previously to this, Yang and Fermi [98] had proposed that pions could be regarded as strongly bound states of nucleons and antinucleons. In order to introduce strangeness into bound states producing K mesons, it is clearly necessary to have a third strange baryon as in the Sakata scheme.) In the Sakata model it is possible, by using the three basic fields with the same quantum numbers as n, p, and Λ, to build up any desired set of the three quantum numbers normally used: baryon number (B), isotopic-spin third component (I_3), and hypercharge (Y). In fact, families of supermultiplets in this model, similar to those represented in Fig. 12.18 and ascribed to SU(3), were originally discussed on the basis of the Sakata model by Ikeda, Ogawa, and Ohnuki [99] in 1959. It is now known, however, that the Sakata model cannot be correct because it does not allow the formation of a $\frac{1}{2}+$ baryon octet, nor does it explain the production of K°_1, K°_2 pairs from $\bar{p}$-p annihilation [100].

In an isotopic-spin multiplet (for example, $\Sigma^+ \Sigma^\circ \Sigma^-$) the small differences of mass between the members of a multiplet are attributed to differences of charge and to the fact that the electromagnetic-field interaction with the particles violates the principle of conservation of isotopic spin. In general, these differences of mass due to nonconservation of isotopic spin are small. In a supermultiplet of the SU(3) group, the mass variations between members may be large, though not excessively so compared with the masses of the particles themselves. These mass differences are attributed to violations of the eight operations of SU(3) in strong interactions. Until now there has

TABLE 12.8. STRONGLY INTERACTING FERMIONS AND COMBINED FERMION-BOSON STATES

Particle	Mass (Mev)	Lifetime or energy width	Y	I	JP	Decay modes
$N\{p$	938.2	Stable	+1	½	½+	Stable
$N\{n$	939.6	"Stable"	+1	½	½+	"Stable"
$N^*_{3/2}(1,236)$	1,236	125 (Mev)	+1	3/2	3/2+	πN (100 %)
$N^*_{1/2}(1,480)$	~1,480	~240 (Mev)	+1	½	½+	πN (~50 %)
$N^*_{1/2}(1,518)$	1,518	125 (Mev)	+1	½	3/2−	πN (~80 %) $\pi\pi N$
$N^*_{1/2}(1,688)$	1,688	100 (Mev)	+1	½	5/2+	πN (~80 %) $\pi\pi N$
$N^*_{3/2}(1,924)$	1,924	170 (Mev)	+1	3/2	7/2+	πN (34 %) $K\Sigma$
$N^*_{1/2}(2,190)$	2,190	~200 (Mev)	+1	½	7/2−	πN (~30 %) $K\Lambda$
$N^*_{3/2}(2,360)$	2,360	~200 (Mev)	+1	3/2	11/2+	πN (~10 %)
$N^*_{1/2}(2,700)$	2,700	~100 (Mev)	+1	½	11/2−	ηN (large) πN (~6 %)
$Y\{\Lambda$	1,115.4	2.62 × 10⁻¹⁰(sec)	0	0	½+	(See Table 12.5)
$Y\{\Sigma$	(Table 12.5)	(Table 12.5)	0	1	½+	(See Table 12.5)
$Y^*_1(1,382)$	1,382.1	53 (Mev)	0	1	3/2+	$\pi\Lambda$ (94 %) $\pi\Sigma$ (6 %)
$Y^*_0(1,405)$	1,405	50 (Mev)	0	0	½−	$\pi\Sigma$ (~100 %)
$Y^*_0(1,519)$	1,519	16 (Mev)	0	0	3/2−	$\pi\Sigma$ (55 %) $\bar{K}N$ (29 %) $\pi\pi\Lambda$ (16 %)
$Y^*_1(1,660)$	1,660	44 (Mev)	0	1	(3/2)	$\bar{K}N$ (~16 %) $\pi\Sigma$ (~32 %) $\pi\Lambda$ (~6 %) $\pi\pi\Sigma$ (~33 %) $\pi\pi\Lambda$ (~23 %)
$Y^*_1(1,765)$	1,765	60 (Mev)	0	1	5/2−	$\bar{K}N$ (~60 %)
$Y^*_0(1,815)$	1,815	70 (Mev)	0	0	5/2+	$\bar{K}N$ (~80 %)
Ξ	(Table 12.5)	(See Table 12.5)	−1	½	½+	(See Table 12.5)
$\Xi^*(1,529)$	1,529.1	7.5(Mev)	−1	½	3/2+	$\pi\Xi$ (~100 %)
$\Xi^*(1,810)$	1,810	~70 (Mev)	−1	½	3/2 −	$\pi\Xi^*$ (~45 %) $K\Lambda$ (~45 %)
Ω^-	1,675	1.3 × 10⁻¹⁰(sec)	−2	0	(3/2+)	$\pi\Xi$ (2 events '64) $K\Lambda$ (2 events '64)

been a great deal of discussion on this point, i.e., concerning the nature of the symmetry-breaking interactions and whether calculations of the mass separations based on various approaches are valid [101]. Although it is generally conceded that the foundations of the derivations are obscure, the mass-splitting formulas of Gell-Mann and Okubo [102] are satisfactorily close and have been quite successful in predicting masses of certain unknown member states.

For the octet mesons, the relationship

$$3m_0^2 + m_1^2 = 4m_{1/2}^2 \qquad (12.71)$$

is good to a few per cent. For example, for the members of the 0 − octet of Fig. 12.18*a*, m_0 ($\eta \approx 548.7$), m_1 ($\pi \approx 138$), $m_{1/2}$ ($K \approx 496$), and (13.70) is satisfied to approximately 7 per cent. Another interesting case is the 1 − octet of Fig. 12.18*b*. Neither the mass of ϕ° nor ω° fits with the expected mass of m_0 according to formula (12.70). The expected value of m_0 can be obtained from other members of the octet: m_1 $(= \rho) = 763$; $m_{1/2}$ $(= K^*) = 891$. These values yield, from (12.70), $m_0 = 930$. However, as first pointed out by Gell-Mann [102] and later analyzed by Sakurai [103], the ω° and ϕ° masses may arise from a symmetry-breaking mixing of the $I_3 = 0$ states in the octet (Fig. 12.18*b*) and the singlet (Fig. 12.18*c*). Real ω° and ϕ° are approximately the following mixtures of the octet and singlet states:

$$\text{real } \omega = \sqrt{1/3}\,\phi - \sqrt{2/3}\,\omega$$

$$\text{real } \phi = \sqrt{2/3}\,\phi + \sqrt{1/3}\,\omega$$

Therefore the masses of the ω° (783 Mev) and ϕ° (1,020 Mev) are approximately equally distant from the expected 930-Mev mass. Without really knowing which particle belongs in the SU(3) octet and which in the SU(3) singlet, the ϕ° has arbitrarily been placed in the octet.

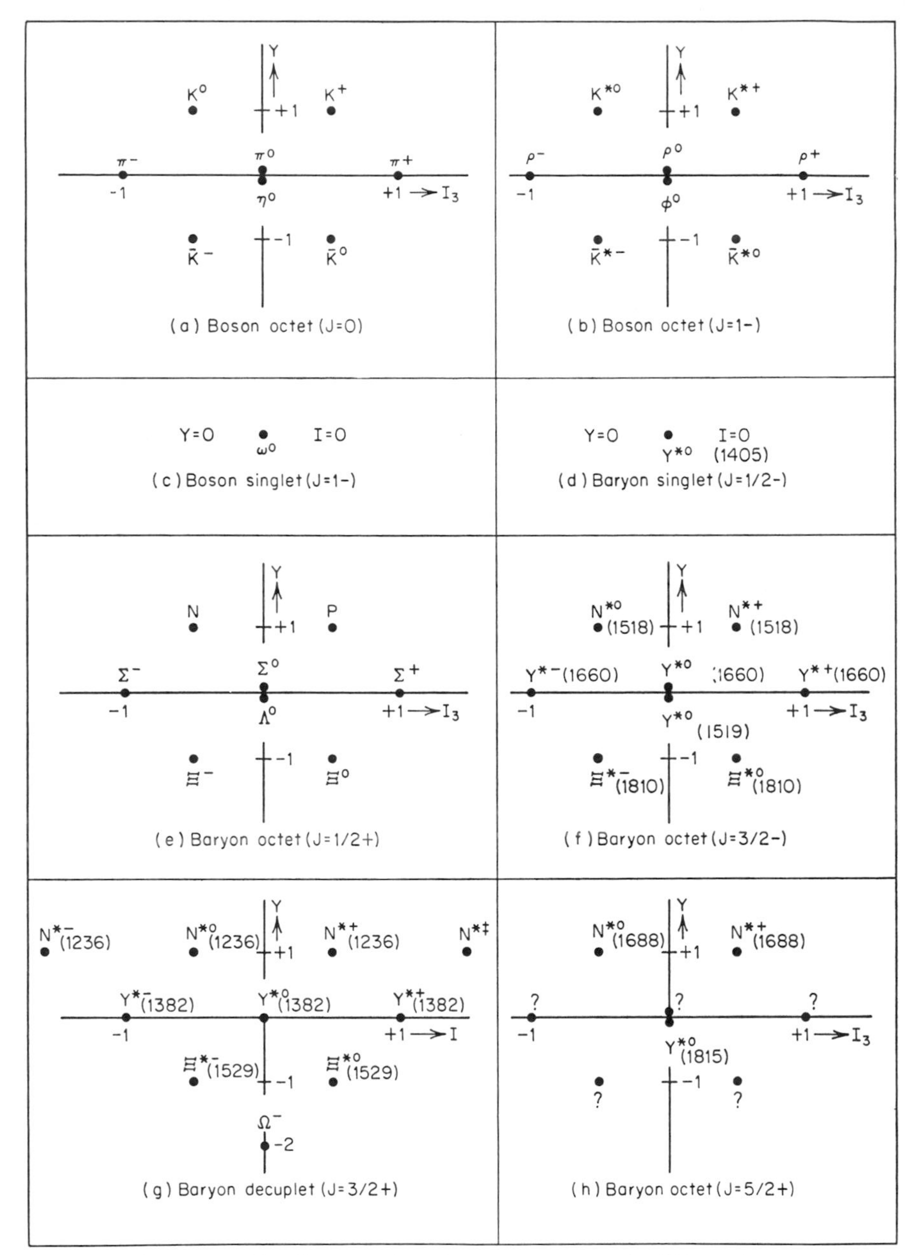

FIG. 12.18. Classification of particles according to the SU(3) scheme.

For the octet baryons of Fig. 12.18*e* the relationship

$$(3M_\Lambda + M_\Sigma) = 2(M_N + M_\Xi) \qquad (12.72)$$

is found to hold to a precision of better than 0.5 per cent. For the $\frac{3}{2}-$ baryon octet of Fig. 12.18*f* replacing: M_Λ by $Y^* = 1{,}519$, M_Σ by $Y^* = 1{,}660$, M_N by $N^* = 1{,}518$, and M_Ξ by $\Xi^* = 1{,}810$, the same mass formula is found to fit to approximately 7 per cent. (The mass of Ξ^* appears to be out of line. A value of approximately 1,600 would be more appropriate.) The formulas (12.72) and (12.71) are special cases of the general Okubo formula

$$m(I,Y) = m_0[1 + aY + b\{I(I+1) - (\tfrac{1}{4})Y^2\}] \qquad (12.73)$$

where a and b are constants for a particular supermultiplet, and I and Y are the isotopic-spin and hypercharge quantum numbers, respectively. This formula can be applied to the baryon decuplet of Fig. 12.18*g*. In this case, $I = (Y/2) + 1$, and $m(I,Y)$ is a linear function of Y. Before the Ω^- had been discovered, the values of a and b could be found in order to fit the $Y = +1$, 0, and -1 rows of the multiplet. Thus the mass of the $Y = -2$ member of the decuplet could be predicted. Its predicted value was 1,676 Mev, and the best value of the Ω^- mass thus far measured is 1,675 Mev.

In a recent development [104] attempts have been made to combine ordinary spin with isotopic spin and hypercharge into yet a higher symmetry. The new group, which is described as SU(6) and which contains SU(2) and SU(3) as subgroups, may contain the 0 − and 1 − meson octets as members of a single large 35-member multiplet, and may also contain the ½ + octet and 3/2 + decuplet baryons as members of a single large 56-member multiplet.

In conclusion, one may perhaps be permitted to take stock of the progress that has been made in particle physics since the original chapter in this "Handbook of Physics" was written. This period has seen a tremendous amount of new data amassed (and covered only in the very briefest fashion here). There have also been considerable advances in the phenomenological and empirical description of particles. However, one nevertheless feels that the basic theoretical understanding of where the many particles fit into our broad comprehension of matter and forces is still awaiting solution.

References

1. Yukawa, H.: *Proc. Phys.-Math. Soc. Japan*, **17**: 48 (1935).
2. Lattes, C. M. G., G. P. S. Occhialini, and C. F. Powell: *Nature*, **160**: 453 (1947).
3. Rosenfeld, A. H., A. Barbaro-Galtieri, W. H. Barkas, P. L. Bastien, J. Kirz, and M. Roos: *Revs. Mod. Phys.*, **36**: 977 (1964).
4. Danby, G., J.-M. Gaillard, K. Goulianos, L. M. Lederman, N. Mistry, M. Schwartz, and J. Steinberger: *Phys. Rev. Letters*, **9**: 36 (1962).
5. Cartwright, W. F., C. Richman, M. N. Whitehead, and H. A. Wilcox: *Phys. Rev.*, **91**: 677 (1953).
6. Clark, D. L., A. Roberts, and R. Wilson: *Phys. Rev.*, **83**: 649 (1951); **85**: 523 (1952).
7. Durbin, R., H. Loar, and J. Steinberger: *Phys. Rev.*, **83**: 646 (1951); **84**: 581 (1951).
8. Panofsky, W. K. H., R. L. Aamodt, and J. Hadley: *Phys. Rev.*, **81**: 565 (1951).
9. Cocconi, V. T., T. Fazzini, G. Fidecaro, M. Legros, N. H. Lipman, and A. M. Morrison: *Nuovo Cimento*, **22**: 494 (1961). Cassels, J. M., G. Fidecaro, A. M. Wetherall, and J. R. Wormald: *Proc. Phys. Soc. (London)*, **A70**: 405 (1957).
10. Kuehner, J. A., A. W. Merrison, and S. Tornabene: *Proc. Phys. Soc. (London)*, **A73**: 551 (1958). Panofsky, W. K. H., R. L. Aamodt, and J. Hadley: *Phys. Rev.*, **81**: 565 (1951).
11. Day, T. B., G. A. Snow, and J. Sucher: *Phys. Rev. Letters*, **3**: 61 (1959). Leon, M., and H. A. Bethe: *Phys. Rev.*, **127**: 636 (1962).
12. Yang, C. N. *Phys. Rev.*, **77**: 242 (1950).
13. Kroll, N. M., and W. Wada: *Phys. Rev.*, **98**: 1355 (1955).
14. Plano, R., A. Prodell, N. Samios, M. Schwartz, and J. Steinberger: *Phys. Rev. Letters*, **3**: 525 (1959).
15. Anderson, H. L., E. Fermi, R. Martin, and D. E. Nagle: *Phys. Rev.*, **91**: 155 (1953). Leonard, S. L., and D. H. Stork: *Phys. Rev.*, **93**: 568 (1954). Bodansky, D., A. M. Sachs, and J. Steinberger: *Phys. Rev.*, **93**: 1367 (1954). Ashkin, J., J. P. Blaser, F. Feiner, J. Borman, and M. O. Stern: *Phys. Rev.*, **93**: 1129 (1954).
16. Gell-Mann, M., and K. M. Watson: *Ann. Rev. Nucl. Sci.*, **4**: 219 (1954).
17. Barashenkov, V. S., and V. M. Maltsev: *Fortschr. Physik*, **9**: 549 (1961).
18. Lindenbaum, S. J., and L. C. L. Yuan: *Phys. Rev.*, **111**: 1380 (1958).
19. Brueckner, K. A.: *Phys. Rev.*, **86**: 106 (1952).
20. Chew, G. F., and F. E. Low: *Phys. Rev.*, **101**: 1570, 1579 (1956).
21. Hamilton, J., and W. S. Woolcock: *Revs. Mod. Phys.*, **35**: 737 (1963). McKinley, J. M.: *Revs. Mod. Phys.*, **35**: 788 (1963).
22. Anderson, H. L.: *Proc. Conf. High Energy Phys.*, Rochester, sec. I, p. 20, 1956.
23. Goldschmidt-Clermont, Y., L. S. Osborne, and M. Scott: *Phys. Rev.*, **97**: 188 (1955). Oakley, D. C., and R. L. Walker: *Phys. Rev.*, **97**: 1283 (1955). Mills, F. E., and L. J. Koester, Jr.: *Phys. Rev.*, **98**: 210 (1955). Bernardini, G., and E. L. Goldwasser: *Phys. Rev.*, **94**: 729 (1954). Jenkins, T. L., D. Luckey, T. R. Palfrey, and R. R. Wilson: *Phys. Rev.*, **95**: 179 (1954). Leiss, J. E., C. S. Robinson, and S. Penner: *Phys. Rev.*, **98**: 201 (1955). Walker, R. L., J. G. Teasdale, V. Z. Peterson, and J. I. Vette: *Phys. Rev.*, **99**: 210 (1955).
24. Beneventano, M., G. Bernardini, D. Carlson-Lee, and G. Stoppini: *Nuovo Cimento*, **4**: 323 (1956).
25. Watson, K. M.: *Phys. Rev.*, **85**: 852 (1952). Gell-Mann, M., and K. M. Watson: *Ann. Rev. Nucl. Sci.*, **4**: 219 (1954). Watson, K. M., J. C. Keck, A. V. Tollestrup, and R. L. Walker: *Phys. Rev.*, **101**: 1159 (1956). Moravcsik, M. J.: *Phys. Rev.*, **104**: 1451 (1956); **105**: 267 (1957). Chew, G. F., and F. E. Low: *Phys. Rev.*, **101**: 1579 (1956).
26. Koester, L. J., Jr., and F. E. Mills: *Phys. Rev.*, **105**: 1900 (1957). Oakley, D. C., and R. L. Walker: *Phys. Rev.*, **97**: 1283 (1955). McDonald, W. S., V. Z. Peterson, and D. R. Corson: *Phys. Rev.*, **107**: 577 (1957).
27. Lindenbaum, S. J.: *Ann. Rev. Nucl. Sci.*, **7**: 317 (1957).
28. Fermi, E., and E. Teller: *Phys. Rev.*, **72**: 399 (1947). Demeur, M., A. Huleux, and G. Vanderhaeghe: *Nuovo Cimento*, **4**: 509 (1956). Ozaki, S., R. Weinstein, G. Glass, E. Loh, L. Neimala, and A. Wattenberg: *Phys. Rev. Letters*, **4**: 533 (1960).
29. Bernardini, G., E. T. Booth, and L. Lederman: *Phys. Rev.*, **83**: 1075 (1951). Devlin, T. J., B. J. Moyer, and V. Perez-Mendez: *Phys. Rev.*, **125**: 690 (1962). Longo, M. J., B. J. Moyer: *Phys. Rev. Letters*, **9**: 466 (1962).
30. Fry, W. F.: *Nuovo Cimento*, **10**: 490 (1953). Stearns, M., and M. B. Stearns: *Phys. Rev.*, **107**: 1709 (1957). Stearns, M. B., M. Stearns, and L. Leipuner: *Phys. Rev.*, **108**: 445 (1957). Condo, G. T., R. D. Hill, and A. D. Martin: *Phys. Rev.*, **133**: A1280 (1964).
31. Heisenberg, W.: *Z. Physik*, **126**: 569 (1949). Heitler, W., and L. Janossy: *Helv. Phys. Acta*, **23**: 417 (1950). Fermi, E.: *Progr. Theoret. Phys. (Kyoto)*, **5**: 570 (1950); *Phys. Rev.*, **81**: 683 (1951). Heisenberg, W.: *Naturwiss.*, **39**: 69 (1952). Blokhintsev, D. I.: Symposium on Pion Physics CERN (Geneva), p. 155, 1956. Barashenkov, V. S.: *Nucl. Phys.*, **15**: 486 (1960). Barbaro-Galtieri, A., A. Manfredini, B. Quassiati, C. Castagnoli, A. Gainotti, and I. Ortalli: *Nuovo Cimento*, **21**: 469 (1961). Dobrotin, N. A., V. V. Guseva, K. A. Kotelnikov, A. M. Lebedev, S. V. Ryabikov, S. A. Slavatinsky, and N. G. Zelevinskaya: *Nucl. Phys.*, **35**: 152 (1962). Lim, Y. K.: *Nuovo Cimento*, **28**: 1228 (1963).

32. Imhof, W., E. A. Knapp, H. Easterday, and V. Perez-Mendez: *Phys. Rev.*, **108**: 1040 (1957). Waters, J. R.: *Phys. Rev.*, **113**: 1133 (1959). Odian, A., G. Stoppini, and T. Yamagata: *Phys. Rev.*, **120**: 1468 (1960).
33. Chamberlain, O., E. Segrè, C. Wiegand, and T. Ypsilantis: *Phys. Rev.*, **100**: 947 (1955).
34. Cork, B., G. R. Lambertson, O. Piccioni, and W. A. Wenzel: *Phys. Rev.*, **104**: 1193 (1956).
35. Segre, E.: *Ann. Rev. Nucl. Sci.*, **8**: 127 (1958).
36. Baker, W. F., R. L. Cool, E. W. Jenkins, T. F. Kycia, S. J. Lindenbaum, W. A. Love, D. Lüers, J. A. Niederer, S. Ozaki, A. L. Read, J. J. Russell, and L. C. L. Yuan: *Phys. Rev. Letters*, **7**: 101 (1961).
37. Brown, L. M., and M. Peshkin: *Phys. Rev.*, **103**: 751 (1956).
38. Loken, J. G., and M. Derrick: *Phys. Letters*, **3**: 334 (1963).
39. Fermi, E.: *Progr. Theoret. Phys. (Kyoto)*, **5**: 570 (1950). Lepore, J. V., and M. Neuman: *Phys. Rev.*, **98**: 1484 (1955).
40. Lynch, G. R.: *Revs. Mod. Phys.*, **33**: 395 (1961).
41. Horwitz, N., D. Miller, J. Murray, and R. Tripp: *Phys. Rev.*, **115**: 472 (1959). Agnew, L. E., Jr., T. Elioff, W. B. Fowler, R. L. Lander, W. M. Powell, E. Segrè, H. M. Steiner, H. S. White, C. Wiegand, and T. Ypsilantis: *Phys. Rev.*, **118**: 1371 (1960).
42. Koba, Z., and G. Takeda: *Progr. Theoret. Phys. (Kyoto)*, **19**: 269 (1958).
43. Armenteros, R., L. Montanet, D. R. O. Morrison, S. Nilsson, A. Shapira, J. Vandermeulen, Ch. d'Andlau, A. Astier, J. Ballam, C. Ghesquière, B. P. Gregory, D. Rahm, P. Rivet, and F. Solmitz: *Proc. Intern. Conf. High Energy Phys. (CERN)*, p. 351, 1962. Chadwick, G. B., W. T. Davies, M. Derrick, J. H. Mulvey, D. Radojicic, C. A. Wilkinson, M. Cresti, S. Limentani, A. Loria, and R. Santangelo: *Proc. Aix-en-Provence Intern. Conf.*, p. 269, 1961.
44. Armenteros, R., E. Fett, B. French, L. Montanet, V. Nikitin, M. Szeptycka, Ch. Peyron, R. Böck, A. Shapira, J. Badier, L. Blaskovicz, B. Equer, B. Gregory, F. Muller, S. J. Goldsack, D. H. Miller, C. C. Butler, B. Tallini, J. Kinson, L. Riddiford, A. Leveque, J. Meyer, A. Verglas, and S. Zylberach: *Intern. Conf. High Energy Phys. (CERN)*, p. 236, 1962. Baltay, C., E. C. Fowler, J. Sandweiss, J. R. Sanford, H. D. Taft, B. B. Culwick, W. B. Fowler, J. K. Kopp, R. I. Louttit, R. P. Shutt, A. M. Thorndike, and M. S. Webster: *Intern. Conf. High Energy Phys. (CERN)*, p. 233, 1962.
45. Cork, B., G. R. Lambertson, O. Piccioni, and W. Wenzel: *Phys. Rev.*, **107**: 248 (1957). Coombes, C. A., B. Cork, W. Galbraith, G. R. Lambertson, and W. Wenzel: *Phys. Rev.*, **112**: 1303 (1958). Armenteros, R., C. A. Coombes, B. Cork, G. R. Lambertson, and W. A. Wenzel: *Phys. Rev.*, **119**: 2068 (1960). Wenzel, W. A.: *Proc. Intern. Conf. High Energy Phys.*, Rochester, p. 151, 1960. Lindenbaum, S. J., W. A. Love, J. A. Niederer, S. Ozaki, J. J. Russell, and L. C. L. Yuan: *Phys. Rev. Letters*, **7**: 185 (1961). Cocconi, G.: *Intern. Conf. High Energy Phys. (CERN)*, p. 883, 1962.
46. Ball, J. S., and G. F. Chew: *Phys. Rev.*, **109**: 1385 (1958). Ball, J. S., and J. R. Fulco: *UCRL*-8365, July 30, 1958.
47. Pomeranchuk, I.: *J. Exptl. Theoret. Phys. (U.S.S.R.)*, **34**: 725 (1958), *Soviet Physics JETP (English Transl.)*, **34**: 499 (1958).
48. Rochester, G. D., and C. C. Butler: *Nature*, **160**: 855 (1947).
49. Seriff, A. J., R. B. Leighton, C. Hsiao, E. W. Cowan, and C. D. Anderson: *Phys. Rev.*, **78**: 290 (1950). Armenteros, R., K. H. Barker, C. C. Butler, A. Cachon: *Phil. Mag.*, **43**: 1113 (1951). Leighton, R. B., S. D. Wanlass, and W. L. Alford: *Phys. Rev.*, **83**: 843 (1951). Thompson, R. W., A. V. Buskirk, L. R. Etter, C. J. Karzmark, and R. H. Rediker: *Phys. Rev.*, **90**: 329 (1953). Rochester, G. D., and C. C. Butler: *Repts. Progr. Phys.*, **16**: 364 (1953).
50. Brown, R., U. Camerini, P. H. Fowler, H. Muirhead, C. F. Powell, and D. M. Ritson: *Nature*, **163**: 82 (1949). O'Ceallaigh, C.: *Phil. Mag.*, **42**: 1032 (1951). Menon, M. G. K., and C. O'Ceallaigh: *Proc. Roy. Soc. (London)*, **A221**: 292 (1954). Bonetti, A., R. Levi Setti, M. Panetti, and G. Tomasini: *Nuovo Cimento*, **10**: 1736 (1953). Amaldi, E., G. Baroni, C. Castagnoli, G. Cortini, and A. Manfredini: *Nuovo Cimento*, **10**: 937 (1953).
51. Thompson, R. W.: *Science*, **120**: 585 (1954).
52. Fowler, W. B., R. P. Shutt, A. M. Thorndike, and W. L. Whittemore: *Phys. Rev.*, **93**: 861 (1954); **98**: 121 (1955). Hornbostel, J., and E. O. Salant: *Phys. Rev.*, **93**: 902 (1954). Hill, R. D., E. O. Salant, and M. Widgoff: *Phys. Rev.*, **94**: 1794 (1954); **95**: 1699 (1954).
53. Rosenfeld, A. H., A. Barbaro-Galtieri, W. H. Barkas, P. L. Bastien, J. Kirz, and M. Roos: *Revs. Mod. Phys.*, **36**: 977 (1964).
54. Barnes, V. E., P. L. Connolly, D. J. Crennell, B. B. Culwick, W. C. Delaney, W. B. Fowler, P. E. Hagerty, E. L. Hart, N. Horwitz, P. V. C. Hough, J. E. Jensen, J. K. Kopp, K. W. Lai, J. Leitner, J. L. Lloyd, G. W. London, T. W. Morris, Y. Oren, R. B. Palmer, A. G. Prodell, D. Radojičič, D. C. Rahm, C. R. Richardson, N. P. Samios, J. R. Sanford, R. P. Shutt, J. R. Smith, D. L. Stonehill, R. C. Strand, A. M. Thorndike, M. S. Webster, W. J. Willis, and S. S. Yamamoto: *Phys. Rev. Letters*, **12**: 204 (1964). Abrams, G. S., R. A. Burnstein, G. R. Charlton, T. B. Day, B. Kekoe, B. Sechi-Zorn, G. A. Snow, M. C. Whatley, G. Wolsky, G. B. Yodh, and R. G. Glasser: *Phys. Rev. Letters*, **13**: 670 (1964).
55. A. Pais: *Phys. Rev.*, **86**: 663 (1952).
56. Gell-Mann, M., and A. Pais: *Proc. Glasgow Conf. Nuclear and Meson Phys.*, p. 342, 1954. Nishijima, K.: *Progr. Theoret. Phys. (Kyoto)*, **12**: 107 (1954).
57. Gell-Mann, M., and A. Pais: *Phys. Rev.*, **97**: 1387 (1955).
58. Lee, T. D., R. Oehme, and C. N. Yang: *Phys. Rev.*, **106**: 340 (1957).
59. Fujii, T., J. V. Jovanovich, F. Turkot, and G. T. Zorn: *Phys. Rev. Letters*, **13**: 253 (1964).
60. Christenson, J. H., J. W. Cronin, V. L. Fitch, and R. Turley: *Phys. Rev. Letters*, **13**: 138 (1964). Abashian, A., R. J. Abrams, D. W. Carpenter, G. P. Fisher, B. M. K. Nefkens, and J. H. Smith: *Phys. Rev. Letters*, **13**: 243 (1964).
61. Dalitz, R. H.: *Phil. Mag.*, **44**: 1068 (1953); *Phys. Rev.*, **94**: 1046 (1954); *Phys. Rev.*, **99**: 915 (1955).
62. Lee, T. D., and C. N. Yang: *Phys. Rev.*, **104**: 254 (1956).
63. Brown, J. L., H. C. Bryant, R. A. Burnstein, R. W. Hartung, D. A. Glaser, J. A. Kadyk, D. Sinclair, G. H. Trilling, J. O. Vander Velde, and J. D. van Putten: *Phys. Rev. Letters*, **3**: 51 (1959).
64. Coombes, C. A., B. Cork, W. Galbraith, G. R. Lambertson, and W. A. Wenzel: *Phys. Rev.*, **108**: 1348 (1957).
65. Eisler, F., R. Plano, A. Prodell, N. Samios, M. Schwartz, J. Steinberger, P. Bassi, V. Borelli, G. Puppi, H. Tanaka, P. Waloschek, V. Zoboli, M. Conversi, P. Franzini, I. Manelli, R. Santangelo, V. Silverstrini, G. L. Brown, D. A. Glaser, and C. Graves: *Nuovo Cimento*, **7**: 222 (1958). Alvarez,

L.: *Ninth Intern. Conf. High Energy Phys.*, Kiev, pts. 1–5, p. 471, 1959.
66. Anderson, J. A., and F. S. Crawford: *Phys. Rev. Letters*, **13**: 246 (1964). Cool, R. L., E. W. Jenkins, T. F. Kycia, D. A. Hill, L. Marshall, and R. A. Schluter: *Phys. Rev.*, **127**: 2223 (1962). Kernan, W., T. B. Novey, S. D. Warshaw, and A. Wattenberg: *Phys. Rev.*, **129**: 870 (1963).
67. McInturff, A. D., and C. E. Roos: *Phys. Rev. Letters*, **13**: 246 (1964).
68. Dalitz, R. H.: *Proc. Intern. Conf. High Energy Phys.*, Rochester, sec. V, p. 40, 1956. *Proc. Conf. High Energy Phys.*, Kiev, p. 587, 1959.
69. Block, M. M., E. B. Brucker, I. S. Hughes, T. Kikuchi, C. Meltzar, F. Anderson, A. Pevsner, E. M. Harth, J. Leitner, and H. O. Cohn: *Phys. Rev. Letters*, **3**: 291 (1959).
70. Block, M. M., L. Lendinara, and L. Monari: *Proc. Conf. High Energy Phys. at CERN*, Geneva, p. 371, 1962.
71. Moravscik, M. J.: *Phys. Rev. Letters*, **2**: 352 (1959).
72. Courant, H., H. Filthuth, P. Franzini, R. G. Glasser, A. Minguzzi-Ranzi, A. Segar, W. Willis, R. A. Burnstein, T. B. Day, B. Kehoe, A. J. Herz, M. Sakitt, B. Sechi-Zorn, N. Seeman, and G. A. Snow: *Phys. Rev. Letters*, **10**: 409 (1963). Alff, C., U. Nauenberg, M. Nussbaum, J. Ratau, J. Schultz, J. Steinberger, L. Kirsch, R. Plano, D. Berley, and A. Prodell: *Bull. Am. Phys. Soc.*, **8**: 514 (1963).
73. Feinberg, G.: *Phys. Rev.*, **109**: 1019 (1958). Feldman, G., and T. Fulton: *Nucl. Phys.*, **8**: 106 (1958). Dalitz, R. H.: *Proc. Aix-en-Provence Conf. Elem. Particles*, 1961.
74. Dalitz, R. H.: *Ann. Rev. Nucl. Sci.*, **13**: 339 (1963).
75. Watson, M. B., M. Ferro-Luzzi, and R. D. Tripp: *Phys. Rev.*, **131**: 2248 (1963).
76. Kycia, T. F., L. T. Kerth, and R. G. Baender: *Phys. Rev.*, **118**: 553 (1960). Cook, V., D. Keefe, L. T. Kerth, P. G. Murphy, W. A. Wenzel, and T. F. Zipf: *Phys. Rev. Letters*, **7**: 182 (1961). Goldhaber, S., W. Chinowsky, G. Goldhaber, W. Lee, T. O'Halloran, T. F. Stubbs, G. M. Pjerrou, D. H. Stork, and H. K. Ticho: *Phys. Rev. Letters*, **9**: 135 (1962).
77. Ravenhall, D. G.: *Phys. Rev. Letters*, **9**: 504 (1962).
78. Davis, D. H., R. D. Hill, B. D. Jones, B. Sanjeevaiah, J. Zakrzewski, and J. P. Lagnaux: *Phys. Rev. Letters*, **6**: 132 (1961). Humphrey, W. E., and R. R. Ross: *Phys. Rev.*, **127**: 1305 (1962). Sakitt, M.: *Univ. Maryland, Tech. Rept.* 410, September, 1964. Kim, J. K.: *Phys. Rev. Letters*, **14**: 29 (1965).
79. Murray, J. J.: UCRL-3492, 1957 (unpublished).
80. Jackson, J. D., D. G. Ravenhall, and H. W. Wyld, Jr.: *Nuovo Cimento*, **9**: 834 (1958). Dalitz, R. H.: *Intern. Conf. High Energy Phys., CERN*, Geneva, p. 187, 1958. Dalitz, R. H., and S. F. Tuan: *Ann. Phys.*, **8**: 100 (1959); **10**: 307 (1960). Jackson, J. D., and H. W. Wyld, Jr.: *Phys. Rev.*, **13**: 84 (1959); *Phys. Rev. Letters*, **2**: 355 (1959). Ross, M. H., and G. L. Shaw: *Phys. Rev.*, **115**: 1773 (1959).
81. Ferro-Luzzi, M., R. D. Tripp, and M. B. Watson: *Phys. Rev. Letters*, **8**: 28 (1962). Watson, M. B., M. Ferro-Luzzi, and R. D. Tripp: *Phys. Rev.*, **131**: 2248 (1963).
82. Akiba, T., and R. H. Capps: *Phys. Rev. Letters*, **8**: 457 (1962).
83. Kycia, T. F., L. T. Kerth, R. Baender: *Phys. Rev.*, **118**: 553 (1960). Burrowes, H. C., D. O. Caldwell, D. H. Frisch, D. A. Hill, D. M. Ritson, and R. A. Schluter: *Phys. Rev. Letters*, **2**: 117 (1959). Cook, V., D. Keefe, L. T. Kerth, P. G. Murphy, W. Wenzel, and T. F. Zipf: *Phys. Rev. Letters*, **7**: 182 (1961).
84. Foley, K. J., S. J. Lindenbaum, W. A. Love, S. Ozaki, J. J. Russell, and L. C. L. Yuan: *Phys. Rev. Letters*, **11**: 503 (1963).
85. Alexander, G., U. Karshon, A. Shapira, G. Yekutieli, R. Englemann, H. Filthuth, A. Fridman, A. Minguzzi-Ranzi: *Phys. Rev. Letters*, **13**: 484 (1964). Sechi-Zorn, B., R. A. Burnstein, T. B. Day, B. Kehoe, and G. A. Snow: *Phys. Rev. Letters*, **13**: 282 (1964). Alexander, G., J. A. Anderson, F. S. Crawford, Jr., W. Laskar, and L. J. Lloyd: *Phys. Rev. Letters*, **7**: 348 (1961). Groves, T. H.: *Phys. Rev.*, **129**: 1372 (1963).
86. Stannard, F. R.: *Phys. Rev.*, **121**: 1513 (1961). Fisk, H. E., and D. J. Prowse: *Proc. Rutherford Jubilee Intern. Conf.*, Manchester, p. 185, 1961.
87. Gell-Mann, M.: *Phys. Rev.*, **106**: 1296 (1957).
88. Dalitz, R. H., and S. F. Tuan: *Phys. Rev. Letters*, **2**: 425 (1959).
89. Alston M., L. W. Alvarez, P. Eberhard, M. L. Good, W. Graziano, H. K. Ticho, and S. G. Wojcicki: *Phys. Rev. Letters*, **5**: 520 (1960).
90. Maglic, B. C., L. W. Alvarez, A. H. Rosenfeld, and M. L. Stevenson: *Phys. Rev. Letters*, **7**: 178 (1961).
91. Nambu, Y.: *Phys. Rev.*, **106**: 1366 (1957).
92. Chew, G. F.: *Phys. Rev. Letters*, **4**: 142 (1960).
93. Chew, G. F., M. Gell-Mann, and A. Rosenfeld: *Sci. American*, **210**(2): 74 (1964).
94. Chew, G. F., and S. C. Frautschi: *Phys. Rev. Letters*, **7**: 394 (1961); **8**: 41 (1962).
95. Ne'eman, Y.: *Nucl. Phys.*, **26**: 222 (1961).
96. Gell-Mann, M.: *Calif. Inst. Tech. Rept.* CTSL-20, 1961; *Phys. Rev.*, **125**: 1067 (1962).
97. Sakata, S.: *Progr. Theoret. Phys. (Kyoto)*, **16**: 686 (1956).
98. Yang, C. N., and E. Fermi: *Phys. Rev.*, **76**: 1739 (1949).
99. Ikeda, M., S. Ogawa, and Y. Ohnuki: *Progr. Theoret. Phys. (Kyoto)*, **22**: 715 (1959); **23**: 1073 (1960).
100. Armenteros, R., L. Montanet, D. R. O. Morrison, S. Nilsson, A. Shapira, J. Vandermeulen, C. d'Andlau, J. Ballam, C. Ghesquiere, B. P. Gregory, D. Rahm, P. Rivet, and F. Solmitz: *Proc. Intern. Conf. High Energy Phys. CERN*, Geneva, p. 351, 1962. Levinson, C. A., H. J. Lipkin, S. Meshkov, A. Salam, and R. Munir, *Phys. Rev. Letters*, **1**: 125 (1962). Salam, A.: *Proc. Phys. Soc. (London)*, **80**: 13 (1962).
101. Freund, P. G. O., and Y. Nambu: *Phys. Rev. Letters*, **12**: 714 (1964); **13**: 221 (1964). Oakes, R. J., and C. N. Yang: *Phys. Rev. Letters*, **11**: 174 (1963). Colemann, S., and S. H. Glashow: *Phys. Rev.*, **134**: B671 (1964).
102. Gell-Mann, M.: *Phys. Rev.*, **125**: 1067 (1962). Okubo, S.: *Progr. Theoret. Phys. (Kyoto)*, **27**: 949 (1962).
103. Sakurai, J. J.: *Phys. Rev. Letters*, **9**: 472 (1962); *Phys. Rev.*, **132**: 434 (1963).
104. Gürsey, F., and L. A. Radicati: *Phys. Rev. Letters*, **13**: 173 (1964). Sakita, B.: *Phys. Rev.* **136, B**1756 (1964).

Bibliography

Bellamy, E. H.: *Progr. Nucl. Phys.*, **8**: 237 (1960).
Burhop, E. H. S., D. H. Davis, and J. Zakrzewski: *Progr. Nucl. Phys.*, **9**: 155 (1964).
Chew, G., M. Gell-Mann, and A. Rosenfeld: *Sci. American*, **210**(2): 74 (1964).
Cutkowsky, R. E.: *Ann. Rev. Nucl. Sci.*, **14**: 175 (1964).
Dalitz, R. H.: "Strange Particles and Strong Interactions," Oxford University Press, Fair Lawn, N.J., 1962.

Gell-Mann, M., and K. M. Watson: *Ann. Rev. Nucl. Sci.*, **4**: 219 (1954).

Hill, R. D.: *Sci. American*, **208**(1): 38 (1963).

Jackson, J. D.: "Physics of Elementary Particles," Princeton University Press, Princeton, N.J., 1958.

Källén, G.: "Elementary Particle Physics," Addison-Wesley, Reading, Mass., 1964.

Lock, W. O.: "High Energy Nuclear Physics," Methuen, London, 1960.

Marshak, R. E.: "Meson Physics," Dover, New York, 1958.

Morpurgo, G.: *Ann. Rev. Nucl. Sci.*, **11**: 41 (1961).

Perkins, D. H.: *Progr. Elem. Particle Cosmic Ray Phys.*, **5**: 257 (1960).

Powell, C. F., P. H. Fowler, and D. H. Perkins: "The Study of Elementary Particles by the Photographic Method," Pergamon Press, New York, 1959.

Schweber, S. S., H. A. Bethe, and F. de Hoffmann: "Mesons and Fields," Harper & Row, New York, 1955.

Thorndike, A. M.: "Mesons: A Summary of Experimental Facts," McGraw-Hill, New York, 1952.

Williams, W. S. C.: "An Introduction to Elementary Particles," Academic Press, New York, 1961.

Appendix: Units and Conversion Factors

The following tables summarize the physical units commonly used in physics. A statement of magnitude of a physical quantity implies an accepted method of comparison of like magnitudes, leading to the statement that one of them is a certain multiple of another one which serves as a *unit*.

Wherever possible the estimated precision of the multiple ought to be stated. This is conveniently done, for example, by stating that a certain length is 3.703 ± 0.004 miles. Preferably this means that 0.004 is the *standard deviation* (page 1–171) of the mean of several measurements that were made. However some authors mean that 0.004 is 0.6745 times the standard deviation, which is the probable error in case the residuals are distributed according to the Gauss error curve (page 1–179).

A *conversion factor* is the numerical value of one unit in terms of another. It is convenient to label such factors with a name in the form of a ratio of the units being compared. Thus the conversion factor associated with the relation 1 mile = 5,280 feet can be written 5280 ft/mi. Then the name can be treated algebraically to give the correct unit name of the result. Thus

$$3.70 \text{ mi} \times 5{,}280 \text{ ft/mi} = 19{,}536 \text{ ft}$$

the unit name of the product being given by cancelling out the mi in numerator and denominator of the unit names of the two factors.

Derived physical units are customarily defined in such a way as to give simple form to some physical law of basic importance. Thus in general, Newton's second law of motion is $F = kma$, but if unit force is defined as that force which gives unit acceleration to unit mass, this choice makes $k = 1$, giving a simpler form. Having done this, the law of gravitation takes the form $F = Gmm'/r^2$, where G is a constant which has to be found experimentally (page 2–57). Alternatively one could define unit force by the gravitation law, assigning to G an arbitrary value, but at the price of having $k \neq 1$, if units of mass, length, and time are chosen arbitrarily. Having chosen units of length and time arbitrarily, one could determine unit mass and unit force in such a way as to give assigned values to k and G. This is done in celestial mechanics (page 2–60) when the astronomical unit (mean earth-sun distance) is used as unit length, the sidereal year is used as unit time, and the sun's mass as unit mass. This choice implies $k = 1$ and $G = 4\pi^2$.

A multiplicity of units occurs in the literature from different ways of handling the geometric factors 2π and 4π. Angles can be measured in cycles (a cycle is a complete rotation about an axis) or in radians (a radian is the angle subtended by an arc of length equal to the radius of the circle on which it is measured) and the relation between these two units is 2π radian/cycle. No confusion need arise if angular velocities are always labelled as cycle/sec or radian/sec respectively and the factor 2π radian/cycle applied as required by the algebraic handling of unit names.

Similarly 4π arises in connection with alternative natural choices for measuring solid angle. Solid angle is proportional to the area subtended on a sphere of unit radius. When the solid angle is set equal to this area it is said to be measured in *steradians*. The total solid angle of *all* directions around a point is then 4π steradians. One can define this "all" as a unit of solid angle, the appropriate conversion factor being 4π steradian/all.

Dimensional formulas for derived quantities show the relation of the derived unit to the fundamental units in terms of which it is defined. They are useful as giving a concise way of stating the rule for changing the magnitude of such a derived unit when a change in the fundamental units is made. For example, the statement that LT^{-1} is the dimensional formula for velocity implies

$$1 \text{ mi/hr} = \frac{5{,}280 \text{ ft/mi}}{3{,}600 \text{ sec/hr}} = \frac{5{,}280}{3{,}600} \text{ ft/sec}$$

A standard set of prefixes to names of units designates multiples and submultiples of any unit by powers of 10 as follows:

tera	10^{12}	pico	10^{-12}
giga	10^{9}	nano	10^{-9}
mega	10^{6}	micro	10^{-6}
kilo	10^{3}	milli	10^{-3}
hecto	10^{2}	centi	10^{-2}
deka	10	deci	10^{-1}

MECHANICAL QUANTITIES

L. Length

The fundamental arbitrary unit is the meter (m), formerly defined as the distance between fiducial marks on a bar of platinum-iridium kept at the International Bureau of Weights and Measures at Sèvres, France.

The meter is now defined by international agreement in 1960 as 1,650,763.73 wavelengths in vacuum of the radiation corresponding to the transition between the $2p_{10}$ and $5d_5$ levels (orange-red) of the Kr^{86} atom. The inch becomes 41,929.399 wavelengths of the same line.

Other metric units are

Angstrom unit	(A)	$= 10^{-10}$ m
Micron	(μ)	$= 10^{-6}$ m
Millimeter	(mm)	$= 10^{-3}$ m
Centimeter	(cm)	$= 10^{-2}$ m
Kilometer	(km)	$= 10^{3}$ m

In the United States of America the common units are defined legally by the relation

39.37 inches = 1 m

This gives the following metric equivalents

Mil	(− 10^{-3} in)	= 25.400051	μ
Inch	(in)	= 2.5400051	cm
Foot	(ft = 12 in)	= 30.4800612	cm
Yard	(yd = 3 ft)	= 91.4401836	cm
Mile	(mi = 5,280 ft)	= 1.60935	km
Nautical mile	(= 6,080.2 ft)	= 1.8532	km

Astronomical Lengths

Earth's equatorial radius	= 6,378.388	km
Solar radius	= (6.960 ± 0.001) × 10^{10}	cm
Astronomical unit (AU) (mean earth-sun distance) (page 2–60)	= (1.4960 ± 0.0003) × 10^{13}	cm
Parsec (distance at which 1 AU subtends 1″ of arc)	= 206,265 AU	
	= (3.857 ± 0.0006) × 10^{18}	cm
Radius of observable universe (approximately)	= 10^9 parsec	
	= 3 × 10^{27}	cm

Atomic Lengths

Fermi	= 10^{-13}	cm
Electron radius, $r_0 = e^2/mc^2$	= (2.81785 ± 0.0004) × 10^{-13}	cm
X-ray unit (Seegbahn)	= 1.00203 × 10^{-11}	cm
Compton wavelength, = h/mc	= (2.42626 ± 0.0002) × 10^{-10}	cm
Bohr radius, $a_0 = \hbar^2/me^2$	= (5.29172 ± 0.00002) × 10^{-9}	cm

L^{-1}. Number Per Unit Length

This quantity appears in connection with the wave number as the number of complete wave cycles in unit length, or alternatively as radians per unit length, these two units differing by the factor 2π radian/cycle.

It also occurs in connection with absorption coefficients, in the most simple case as the α in a factor $e^{-\alpha x}$. Sometimes the product αx is said to be measured in *nepers*, in which case α is in nepers/L and x is in L where L stands for any length unit.

Alternatively the exponential factor may be expressed to a different base. Thus $e^{-\alpha x} = 10^{-M\alpha x}$ where $M = \log_{10} e = 0.4342944819$ and the quantity $M\alpha x$ is said to be expressed in *bels*, and likewise $10M\alpha x$ is the same quantity expressed in *decibels*. Thus one may describe the attenuation as α nepers/L or as $10M\alpha$ decibels/L, where L is any length unit.

In astronomical work light intensities are compared on an exponential *magnitude* scale, where $I(m) = I(0)\ 10^{-0.4m}$ and m is in magnitudes, while $I(m)$ is the light intensity going with the mth magnitude. Absorption of light in space gives rise to a more rapid reduction of light intensity with distance than the inverse square law, which may be expressed as magnitudes/kiloparsec

$$1 \text{ mag/kps} = 1.296 \times 10^{-22} \text{ bels/cm} = 2.984 \times 10^{-22} \text{ nepers/cm}$$

In atomic spectroscopy the natural unit of wave number is the Rydberg, R, defined (page 7–24) as the number of cycles per unit length of the limit of the Lyman series in "hydrogen," the theoretical nonrelativistic hydrogen whose nucleus has infinite mass (page 7–154)

$$1 \text{ R} = 109{,}737.311 \pm 0.012 \text{ cm}^{-1}$$

L^2. Area

Unit area is defined as the area of a square whose side is of unit length. A square whose side is L length units has an area of L^2 such area units

$$1 \text{ m}^2 = 10^4 \text{ cm}^2$$

The *circular mil* is defined as the area of a circle whose diameter is one mil, and the circular inch is the area of a circle whose diameter is one inch:

1 circular mil	= 0.78540 mil²	= 5.0671 × 10^{-6}	cm²
1 circular inch	= 0.78540 in²	= 5.0671	cm²

Common Units

1 in²		= 6.4516259	cm²
1 ft²	(= 144 in²)	= 929.0341296	cm²
1 yd²	(= 9 ft²)	= 8,361.30717	cm²
1 acre	(= 4840 yd²)	= 4,046.773	m²
1 hectare	(= 2.471044 acre)	= 10^4	m²
1 mi²	(= 640 acre)	= 2.5899987	km²

Atomic Units

Area occurs in atomic physics as a measure of the effective cross section for collision processes (page 7–18)

1 millibarn	= 10^{-27}	cm²
1 barn (page 9–13)	= 10^{-24}	cm²
$(8\pi/3)r_0^2$ (page 7–134)	= 6.65205 ± 0.00018 × 10^{-25}	cm²
Area first Bohr orbit: πa_0^2	= 8.797 × 10^{-17}	cm²

L^3. Volume

Unit volume is defined as that of a cube whose edge is of unit length. A cube whose edge is L length units has a volume of L^3 such volume units

$$1 \text{ m}^3 = 10^6 \text{ cm}^3$$

Common Units

1 in.^{3}	=	16.3870253 cm^3
1 fluid ounce = 480 minims	=	29.5737 cm^3
1 pint (liq) = 16 fl oz	=	473.179 cm^3
1 quart " = 2 pints	=	976.358 cm^3
1 U.S. gallon = 4 quarts	=	3,785.432 cm^3
1 British Imperial and Canadian gallon = 1.20094 U.S. gallon	=	4,546.1 cm^3
1 ft^3	=	28,316.77 cm^3
1 yd^3 = 202.0 U.S. gallon	=	0.76455945 m^3
1 acre-foot (volume of rectangular solid having base area of one acre and height of 1 ft)	=	1,233.46 m^3
Volume of sun	=	1.4147×10^{27} m^3

T. Time

The fundamental unit is the second (sec) defined as 1/31,556,925.9747 of the tropical year 1900. Time measured in these units is called ephemeris time; it is rigorously related to the orbital motion of the earth around the sun, but is free from the perturbations that the earth undergoes from the action of the moon and the other planets. The tropical year is the mean interval between successive March crossings of the sun over the terrestrial equator; it contains approximately 365.2422 calendar (mean solar) days.

1 calendar day = 24 calendar hours = 1,440 calendar minutes = 86,400 calendar seconds precisely.

The (Gregorian) calendar year contains 365 or 366 days, on the average 365.2425 precisely. It departs progressively from the tropical year at the rate of a day in about 3,000 years.

Mean solar time is rigorously related to the rotation of the earth and is not strictly uniform, since the earth's rate of rotation varies slightly by unpredictable amounts. Mean solar time of the meridian of Greenwich is required for navigation and surveying; it is called Greenwich mean time. All clocks in ordinary use show minutes and seconds of Greenwich mean time, the hours being adjusted to make noon by the clock occur near the middle of the daylight period.

The difference between ephemeris time and Greenwich mean time reached extreme values of −8 sec and +35 sec between 1820 and 1964.

Atomic time is obtained by mechanically integrating a suitable transition frequency of a substance such as cesium or hydrogen, the epoch (time of day) being an arbitrary constant of integration. Standard frequency emissions from radio stations are (1964) controlled by cesium standards. It seems likely that the second may soon be officially redefined in terms of some transition frequency under specified standard conditions.

The nominal frequency of cesium standards is 9,192,631,770 cycles per second.

Sidereal second = 0.9972696s
Atomic unit (1st Bohr period/2π) = 2.4190×10^{-17}s

T^{-1}. Angular Velocity. Time Constant

Unit angular velocity may be defined either as 1 radian/T or 1 cycle/T where T is any time unit. These two units differ by the factor 2π radian/cycle.

1″ of arc/day = 5.6113×10^{-11} rad/s
Earth orbital = 1.991×10^{-7} rad/s
Earth rotation = 7.292×10^{-5} rad/s
Hertz (Hz) = 1 cycle/sec

Quantities of the nature T^{-1} also appear as the rate constants in exponential decay factors like the k in e^{-kt} (compare L^{-1} above). Here k is expressed in nepers/T where T stands for the time unit used for t.

In radioactivity work, the decay rate is often expressed by giving the half life, $T_{1/2}$ which is the value of t for which $e^{-kt} = \frac{1}{2}$:

$$T_{1/2} = k^{-1} \ln 2 = 0.693k^{-1}$$

T^{-2}. Angular Acceleration

As with angular velocity, unit angular acceleration may be defined either as that acceleration in which the angular velocity increases at the rate of 1 radian/sec in 1 sec (or correspondingly any other unit of time) or as that acceleration in which the angular velocity increases at the rate of 1 cycle/sec in one second. The two units differ by the factor 2π radian/cycle.

LT^{-1}. Velocity (page 2–3)

Unit velocity is defined as that velocity in which unit distance is traversed in unit time

$$1 \text{ m/sec} = 100 \text{ cm/sec}$$

Common Units

1 ft/min		= 0.5080 cm/sec
1 ft/sec		= 30.4801 cm/sec
1 mi/hr	(= 88 ft/min)	= 44.7041 cm/sec
1 knot	(nautical mi/hr)	= 51.48 cm/sec
1 mi/min	(= 88 ft/sec)	= 28.622 m/sec

Astronomical Speeds

1 AU/yr	=	4.741 km/sec
Speed of point on earth's equator due to diurnal rotation	=	0.46512 km/sec
Mean orbital speed of earth (2π AU/sid yr)	=	29.785 km/sec
Speed of sun relative to mean speed of nearby stars	=	19.4 km/sec

Atomic Units (pages 7–24 and 7–156)

Velocity of light, c = 2.997930×10^{10} cm/sec
Speed of electron in first Bohr orbit,
$e^2/\hbar$ = 2.187667×10^8 cm/sec
= $7.29729 \pm 0.00003 \times 10^{-3}c$
$\hbar c/e^2$ = 137.0373 ± 0.0006

LT^{-2}. Acceleration (page 2–3)

Unit acceleration is that acceleration in which there is unit increase of velocity in unit time

$$1 \text{ m/sec}^2 = 100 \text{ cm/sec}^2 = 100 \text{ gal}$$

Common Units

1 (mi/hr)/min = 1 (mi/min)/hr	= 0.74507	cm/sec^2
1 km/hr-sec	= 27.778	cm/sec^2
1 ft/sec^2	= 30.4801	cm/sec^2
1 mi/hr-sec	= 44.7401	cm/sec^2

Gravity

Adopted "normal" acceleration of gravity (page 2–57) is

g = 32.174 ft/sec^2	= 980.665	cm/sec^2
Gravity at sun's surface	= 2.740 × 10^4	cm/sec^2
Gravity due to sun at 1 AU	= 0.5930	cm/sec^2
Centripetal acceleration of point on earth's equator due to diurnal rotation	= 3.392	cm/sec^2

Atomic Unit (page 7–24)

Centripetal acceleration of electron in first Bohr orbit in "hydrogen":

$$me^6/\hbar^4 = 9.044128 \times 10^{24} \text{ cm/sec}^2$$

L^2T^{-1}. Diffusivity (page 5–54). **Kinematic Viscosity** (pages 3–13 and 3–27). **Thermal Diffusivity** (page 5–61)

Diffusivity is the factor multiplying the concentration gradient to give the net transport of material across unit area in unit time. The unit of quantity cancels out as it appears both in the flow rate and in the measure of the concentration.

1 cm^2/day = (86,400)$^{-1}$ cm^2/sec	= 1.1574 × 10^{-5}	cm^2/sec
1 m^2/hr	= 1.7919 × 10^{-3}	cm^2/sec
1 in.^{2}/sec	= 6.451	cm^2/sec

For kinematic viscosity

1 stokes	= 1 cm^2/sec
1 myriastokes = 1 m^2/sec	= 10^4 cm^2/sec

Thermal diffusivity (page 5–61) is $k/c\rho$ where k is the heat conductivity, c the specific heat and ρ the density.

ILLUSTRATIVE VALUES

Diffusivity

$O_2 - N_2$ at standard conditions	= 0.181	cm^2/sec
H_2 — air at standard conditions	= 0.611	cm^2/sec
para H_2 — ortho H_2 at standard conditions	= 1.285	cm^2/sec
NaCl, 25°C, infinite dilution aq.	= 1.612 × 10^{-5}	cm^2/sec
HCl, 25°C, infinite dilution aq.	= 3.337 × 10^{-5}	cm^2/sec
Hg liq. — Hg liq., 23°C	= 1.70 × 10^{-5}	cm^2/sec
Cu sol. — Cu sol., 700°C	= 4.06 × 10^{-12}	cm^2/sec
Ag sol. — Ag sol. 666°C	= 2.45 × 10^{-11}	cm^2/sec

Kinematic Viscosity

Air, standard conditions	= 0.136	cm^2/sec
Hydrogen, standard conditions	= 0.929	cm^2/sec
Water liq. 20°C	= 0.0101	cm^2/sec

Thermal Diffusivity

Water, liq. 20°C	= 1.43 × 10^{-3}	cm^2/sec
Copper, sol. 20°C	= 1.12	cm^2/sec

L^2T^{-2}. Energy Per Unit Mass (or Per Mole)

Energy is measured in ML^2T^{-2}, hence energy content, or energy change in any form that is associated with unit mass, is measured in L^2T^{-2}:

$$1 \text{ erg/g} = 1 \text{ (cm/sec)}^2$$
$$1 \text{ joule/kg} = 1 \text{ (m/sec)}^2 = 10^4 \text{ (cm/sec)}^2$$

The rest energy of any body per unit mass is equal to c^2 where c is the velocity of light. The energy per unit mass of any body moving with speed v is (page 2–18)

$$c^2/\sqrt{1 - \beta^2} \text{ with } \beta = v/c$$

At low speeds this becomes ($v \ll c$)

$$c^2 + \tfrac{1}{2}v^2 + \cdots$$

the second term corresponding to the nonrelativistic kinetic energy.

Energy of formation of atoms from hydrogen and neutrons manifests itself in the mass defect by which the mass of the atom is less than the sum of the separate masses of these constituents (Part 9, Chap. 2).

Energy of formation of chemical compounds from elements, and energies of transition or change of state are often expressed in calories/g, joules/g, calories/mole, or joules/mole, where 1 mole = Mg, in which M is the molecular weight of the substance.

1 calorie/g	= 4.1840	joule/g
1 Btu/lb	= 2.325	joule/g
1 electron-volt/molecule	= 9.65177 × 10^4	joule/mole

L^3T^{-1}. Volume Flow Rate

Flow of liquids in a conduit is often expressed as the number of volume units passing a section of the conduit in unit time

1 U.S. gallon/min	= 63.0905	cm^3/sec
1 ft^3/min	= 471.946	cm^3/sec
1 ft^3/sec	= 28,316.77	cm^3/sec
1 yd^3/min	= 12.47	liter/sec

***M*. Mass** (page 2–11)

Fundamental unit is the kilogram (kg) defined as the mass of an adopted platinum standard kept at the International Bureau of Weights and Measures, Sèvres, France

1 microgram (μg)	= 10^{-6} gram (g)	= 10^{-9} kg
1 milligram (mg)	= 10^{-3} g	= 10^{-6} kg
1 gram (g)		= 10^{-3} kg
1 millier (metric ton)	= 10^6 g	= 10^3 kg

Common Units

Legal definition of the avoirdupois pound (lb) in United States of America is based on the relations

$$1 \text{ lb} = 453.5924277 \text{ g}$$
$$1 \text{ kg} = 2.2046223 \text{ lb}$$

giving for the related avoirdupois units

1 grain	$(= \frac{1}{7000}$ lb)	= 64.798918	mg
1 carat		= 0.2000	g
1 dram	$(= \frac{1}{256}$ lb)	= 1.7718	g
1 ounce	$(= \frac{1}{16}$ lb)	= 28.3495	g
1 ton	(= 2000 lb)	= 907.18486	kg
1 long ton	(= 2240 lb)	= 1016.0470	kg

The troy and apothecary's weights are based on the grain of the same value as in the avoirdupois system

1 grain	$(= \frac{1}{5670}$ lb)	= 64.798918	mg
1 scruple (ap)	(= 20 grains)	= 1.2959784	g
1 pennyweight (tr)	(24 grains)	= 1.555174	g
1 dram	(= 60 grains)	= 3.8879351	g
1 ounce	(= 480 ")	= 31.103481	g
1 pound	(= 12 ounces)	= 373.24177	g

Astronomical Masses

Earth (page 2–57)	= (5.977 ± 0.004)	$\times 10^{27}$ g
Sun (= 333,432 earths)	= 1.993	$\times 10^{33}$ g
Galaxy (= 1.6×10^{11} suns)	= 3.2	$\times 10^{44}$ g
Observable universe, more than	2	$\times 10^{54}$ g

Atomic Masses (page 7–24 and 7–156)

Electron, $m = (9.1083 \pm 0.0003) \times 10^{-28}$ g

Atomic mass unit (amu) is defined as $\frac{1}{12}$ of the mass of one atom of the abundant C^{12} isotope of carbon

$$N^{-1} = (1.65979 \pm 0.00004) \times 10^{-24} \text{ g/amu}$$

Avogadro number, N, is defined as the number of amu in 1 g

$$N = (6.02486 \pm 0.00016) \times 10^{23} \text{ amu/g}$$

Mass of the electron expressed in amu is

$$Nm = (5.4875 \pm 0.0004) \times 10^{-4} \text{ amu}$$

For complete table of atomic masses see Part 9, Chap. 2. Some basic values:

Entity	amu	10^{-24} g
Proton	1.007593 ± 0.000003	1.67239 ± 0.00004
Hydrogen	1.008142 ± 0.000003	1.67330 ± 0.00004
Neutron	1.008982 ± 0.000003	1.67470 ± 0.00004
Deuterium	2.014735 ± 0.000006	3.34404 ± 0.00008

The chemical scale of atomic masses uses as its unit $\frac{1}{16}$ the average mass of the stable isotopes of oxygen in their naturally occurring abundances. This definition is not unique as the relative abundances vary slightly with the source of the oxygen.

M^{-1}. **Molal Quantity**

One mole of a substance whose molecular weight is M is defined as M unit masses of that material. Unless another mass unit is explicitly stated it is understood that one mole means one gram-mole.

One mole of any substance contains a fixed number of molecules, this number being known as Avogadro's number. The value of Avogadro's number depends on whether the molecular weights are expressed on the physical or chemical scale, and also on the unit of mass that is chosen.

In terms of the physical scale of molecular weights several values of the Avogadro number are:

$$N = (6.02486 \pm 0.00016) \times 10^{23} \text{ amu/g}$$
$$= (2.7328 \pm 0.0007) \times 10^{26} \text{ amu/lb}$$
$$= (5.4657 \pm 0.0014) \times 10^{29} \text{ amu/ton}$$

ML^2. **Moment of Inertia** (page 2–23)

Unit moment of inertia is that associated with unit mass at unit distance from the axis of rotation

1 kg-m²	= 10^7	g-cm²
1 amu-A²	= 1.65979×10^{-40}	g-cm²
1 lb-in²	= 2926.4	g-cm²
1 lb-ft²	= 42.140	g-m²

ML^{-3}. **Density and Concentration**
$M^{-1}L^3$. **Specific Volume**

Density is defined as the ratio of mass to the volume occupied by that mass. Specific volume is the reciprocal of density and is the ratio of volume to the mass contained in that volume:

$$1 \text{ kg/m}^3 = 10^{-3} \text{ g/cm}^3$$
$$1 \text{ m}^3/\text{kg} = 10^3 \text{ cm}^3/\text{g}$$

Molal density and molal volume give respectively the number of moles contained in unit volume, and the volume occupied by one mole of substance. Referring to gram-moles, if ρ g/cm³ is the density and M is the molecular weight, then ρ/M mole/cm³ is the molal density and M/ρ cm³/mole is the molal volume

$$1 \text{ solar mass/parsec}^3 = 6.768 \times 10^{-23} \text{ g/cm}^3$$

1 g/ml	= 0.999973	g/cm³
1 lb/ft³	= 0.016018	g/cm³
1 lb/in³	= 27.680	g/cm³
1 lb/U.S. gallon	= 0.119826	g/cm³
1 ml/g	= 1.000027	cm³/g
1 U.S. gallon/lb	= 8.34543	cm³/g
1 ft³/lb	= 62.4298	cm³/g
1 in³/lb	= 0.036127	cm³/g

Molal volume of ideal gas at 273.16°K and normal atmosphere is

$$V_0 = (22.4146 \pm 0.0006) \times 10^3 \text{ cm}^3/\text{mole}$$
$$= (22.4140 \pm 0.0006) \text{ liter/mole}$$

Some illustrative values (at 20°C):

Substance	Density g/cm³	M	Molal volume cm³/mole
Benzene, C_6H_6	0.8784	78.11	88.82
Water, H_2O	1.00	18.02	18.02
Aluminum	2.70	26.97	9.99
Copper	8.933	63.57	7.116
Iron	7.86	55.84	7.10
Platinum	21.37	195.23	9.135

MT^{-1}. Mass Flow Rate

Unit mass flow rate is that in which unit mass of material passes a given place in unit time.

MT^{-2}. Surface Tension. Surface Energy (page 5–92)

Unit surface tension is that in which unit force acts across unit length in the surface. It is usually expressed in dyne/cm = 10^{-3} newton/m.

Some illustrative values

Substance	Temperature, °C	Surface tension in air, dyne/cm
Water	10	74.22
	20	72.75
	30	71.18
Mercury	15	487.
Sulphur	445	38.97
Carbon tetrachloride	20	26.95

MT^{-3}. Power Per Unit Area

For measuring intensity of a beam of radiant energy the unit corresponds to unit energy flowing across unit area in unit time

$$1 \text{ watt/m}^2 = 10^3 \text{ erg/cm}^2\text{-sec}$$

Solar constant is the total radiant energy flux per unit area per unit time due to sun at distance of earth at a place outside the earth's atmosphere

$$\begin{aligned}\text{Solar constant} &= 1.94 \text{ cal/cm}^2\text{-min.}\\ &= 1.35 \times 10^6 \text{ erg/cm}^2\text{-sec}\\ &= 0.135 \text{ watt/cm}^2\\ &= 1.35 \text{ kw/m}^2\end{aligned}$$

In solar energy work the total time integral of radiant power received per unit area is often expressed in a unit called the langley, where 1 langley = 1 cal/cm², and therefore the solar constant is also expressible as 1.94 langley/min.

MLT^{-1}. Momentum. Impulse (page 2–11)

Unit momentum is that associated with a body of unit mass moving with unit velocity:

$$\begin{aligned}1 \text{ kg-m/sec} &= 10^5 \text{ g-cm/sec}\\ 1 \text{ lb-ft/sec} &= 1.38255 \times 10^4 \text{ g-cm/sec}\end{aligned}$$

Unit impulse is that associated with unit force acting for unit time. By Newton's second law of motion the change of momentum of a body is equal to the impulse acting on it.

Atomic Units (page 7–24)

Relativistic unit associated with electron,

$$mc = 27.3046 \times 10^{-18} \text{ g-cm/sec}$$

Momentum of electron in first Bohr orbit,

$$me^2/\hbar = 1.9926 \times 10^{-19} \text{ g-cm/sec}$$

ML^2T^{-1}. Angular Momentum (page 2–11)

Unit angular momentum is that associated with a body of unit moment of inertia having an angular velocity of one radian per unit time. For a particle at position, **r**, having linear momentum, **p**, the angular momentum about the origin is **r × p**:

$$\begin{aligned}1 \text{ kg-m}^2\text{/sec} &= 10^7 \text{ g-cm}^2\text{/sec}\\ 1 \text{ lb-ft}^2\text{/sec} &= 4.21402 \times 10^5 \text{ g-cm}^2\text{/sec}\end{aligned}$$

Astronomical Values

Earth, diurnal rotation = 5.910×10^{40} g-cm²/sec
Earth-moon system = 3.442×10^{41} g-cm²/sec
Earth's orbital motion around sun = 4.231×10^{46} g-cm²/sec

Atomic Unit

Planck's constant divided by 2π is the natural quantum unit of angular momentum (pages 7–4, 7–24, and 7–155),

$$\hbar = (1.05443 \pm 0.00004) \times 10^{-27} \text{ g-cm}^2\text{/sec}$$

MLT^{-2}. Force (page 2–11)

Unit force (absolute) is that which gives unit acceleration to unit mass. One *dyne* is the force which gives an acceleration of 1 cm/sec² to 1 g. One *newton* gives an acceleration of 1 m/sec² to 1 kg

$$1 \text{ newton} = 10^5 \text{ dyne}$$

One *poundal* gives an acceleration of 1 ft/sec² to 1 lb

$$1 \text{ poundal} = 1.3825 \times 10^4 \text{ dyne}$$

Unit force (gravitational) is the force of gravitational attraction on unit mass at the locality in question and is therefore expressed in mass units. The conversion factor to absolute force units is the *local* value of g, the acceleration of gravity. Thus a mass of M grams is attracted to the earth by a force of M grams weight which is equal to Mg dynes. Similarly a mass of M pounds is attracted to the earth by a force of M pounds weight which is equal to Mg poundals. Here g is the local acceleration of gravity in cm/sec² and ft/sec² respectively.

Atomic Unit (page 7–24)

The natural atomic unit is the force of attraction between an electron and a proton when the distance between them is equal to a_0, the first Bohr radius. Its value is 8.2378×10^{-3} dyne.

ML^2T^{-2}. **Energy or Work. Torque**

Unit energy or work (page 2–13) is that exchanged when unit force acts through unit displacement in the direction of the force. The energy exchanged when force **F** acts through displacement **s** is **F · s**.

When force **F** acts on a body at a point whose position vector is **r**, the *torque* (page 2–34) is **G** = **r** × **F**. Thus torque is expressed as a force multiplied by a distance although its physical significance is quite different from that of work.

One *erg* is the work done when one dyne acts through one cm. The corresponding torque unit is always called one dyne-cm, never one erg. One *joule* is the work done when one newton acts through one meter

1 joule = 10^7 erg

One foot-poundal is the work done when a force of one poundal acts through a distance of one foot

1 ft/poundal = 421,402 erg

Energy or torque may also be expressed in gravitational units by using the corresponding values for the force in gravitational units. The conversion factor involves the local value of g, but, using g = 980.665 cm/sec² for illustrative purposes,

1 g-cm = 980.665 erg
1 ft-lb = 1.355821 joule

Other units in common use are

1 watt-hour = 3600 joule
1 kilowatt-hr = 3.6×10^6 joule

Using the normal atmosphere as unit stress

1 cm³-atm = 0.101325 joule
1 ft³-atm = 2869.4 joule

Many definitions of the calorie have been given but the one preferred here is the thermochemical calorie (page 5–4) defined as

1 calorie = 4.1840 joule

Sometimes this is more explicitly referred to as the gram-calorie to distinguish from the kilogram-calorie or large calorie, written also as Calorie:

1 Calorie = 4184.0 joule

The equivalent British units are the British Thermal Unit and the Therm

1 Btu = 1.0548 kilojoule
1 Therm = 10^5 Btu = 105.48 megajoule

Atomic Units (page 7–24)

Mass-energy equivalence of one electron

mc^2 = 8.18674×10^{-7} erg

Mass-energy equivalence of one atomic mass unit

$N^{-1}c^2$ = 1.4915×10^{-3} erg

Electrostatic potential energy of interaction of electron and proton when separated by one Bohr radius

= 4.35918×10^{-11} erg

One electron-volt is the work done when one electronic charge is moved through a potential difference of one volt

$e \cdot 10^8/c = (1.60199 \pm 0.00016) \times 10^{-12}$ erg

ML^2T^{-3}. **Power** (page 2–12)

Unit power is that rate of doing work in which unit energy is exchanged in unit time

1 watt	= 1 joule/sec	= 10^7	erg/sec
1 horse power	= 550 ft-lb/sec	= 745.70	watt
Force de cheval		= 735.5	watt
1 ft-lb/minute		= 2.2597×10^{-2}	watt

Astronomical Values

Rate of radiation of a star of absolute bolometric magnitude zero ($M_{bol} = 0$) = 2.72×10^{28} watt
Solar luminosity = 3.86×10^{26} watt

$ML^{-1}T^{-1}$. **Viscosity** (pages 3–12 and 3–26). **Fluidity.** $M^{-1}LT$

Unit viscosity is that of a material which requires unit shear stress to maintain unit shear velocity gradient. One *poise* measures the viscosity of that material in which 1 dyne/cm² maintains a shear velocity gradient of 1 (cm/sec)/cm or 1 sec⁻¹. One *dekapoise* is defined as that viscosity in which 1 newton/m² maintains the same shear velocity gradient

1 dekapoise = 10 poise

A viscosity such that 1 lb/in² is required to maintain a shear rate of 1 (ft/sec)/ft or 1 sec⁻¹ is equal to 6.8947×10^4 poise.

Fluidity is defined as the reciprocal of viscosity. A liquid which has a viscosity of η poise is said to have a fluidity of η^{-1} rhe.

For illustrative values, see Table 1.2, page 3–13.

$ML^{-1}T^{-2} = (MLT^{-2})/L^2$. **Pressure and Stress** (page 3–4). **Elastic Moduli** (page 3–10)

Pressure and stress are defined as unit force acting across unit area of the surface across which the force is exerted

1 pascal = 1 newton/m² = 10 dyne/cm²

Gravitational units of stress result from the use of gravitational units of force. The conversion factor is the local value of g, the acceleration of gravity, but g = 980.665 cm/sec² is used here for illustrative purposes.

Relations between units:

1 barye	= 1 microbar	= 1	dyne/cm²
1 millibar	= 10^{-3} bar	= 10^3	dyne/cm²
1 bar		= 10^6	dyne/cm²
1 kg/mm²		= 9.80665×10^7	dyne/cm²
1 psi(lb/in²)		= 6.8947×10^4	dyne/cm²

The *normal atmosphere* (atm) is defined as the pressure exerted by a vertical column of 76 cm of mercury of density 13.5951 g/cm³ at a place where g = 980.665 cm/sec²:

$$\begin{aligned} 1 \text{ atm} &= 1.013246 \times 10^6 \text{ dyne/cm}^2 \\ &= 1{,}033.22 \text{ g/cm}^2 \\ &= 14.696 \text{ psi} \\ &= 1.0581 \text{ ton/ft}^2 \\ &= 29.291 \text{ inches of Hg at } 32°\text{F} \\ &= 33.899 \text{ ft of } H_2O \text{ at } 39.1°\text{F} \end{aligned}$$

Vacuum Technique (page 5–73)

$$1 \text{ micron} = 10^{-3} \text{ mm of Hg} = 1.3332 \text{ dyne/cm}^2$$
$$1 \text{ mm Hg} = 1 \text{ Torr} = 1333.22 \text{ dyne/cm}^2$$

ML^3T^{-2}. Energy Times Length

This is the nature of the factor which multiplies the wave-number (cycles/cm) of a radiation to give the quantum energy in ergs

$$hc = (1.98564 \pm 0.00032) \times 10^{-16} \text{ erg-sec}$$

The constant, c_1, in the Planck radiation law (pages 6–14 and 7–157) is a quantity of the same kind:

$$c_1 = 8\pi hc = (4.9918 \pm 0.0002) \times 10^{-15} \text{ erg-sec}$$

$M^{-1}L^3T^{-2} = MLT^{-2} \times L^2/M^2$ Gravitational Constant

The gravitational constant (pages 2–31 and 2–57), is the force of gravitational attraction between two unit masses, spherical in shape, whose centers are separated by unit distance

$$G = (6.670 \pm 0.005) \times 10^{-8} \text{ dyne-cm}^2/\text{g}^2$$

G is the force of attraction in dynes between two one-gram masses separated by a distance of one cm. Also

$$G = (6.670 \pm 0.005) \times 10^{-11} \text{ newton-m}^2/\text{kg}^2$$

The acceleration of gravity, g, at distance r from a mass M is given by $g = GM/r^2$.

In a system in which unit length is one astronomical unit, unit time is the sidereal year, and unit mass is the sun's mass, the value of G is $4\pi^2$.

HEAT QUANTITIES

Q. Quantity of Heat

Quantity of heat, Q, may be measured in terms of a calorie, defined as the heat needed to raise the temperature of 1 g of water 1°C at a specified temperature. The large calorie, or Calorie, is similarly defined with respect to 1 kg of water.

By the first law of thermodynamics (page 5–4), heat is recognized as a form of energy. The conversion factor between calories and joules, or between any other heat and energy units, is known as the mechanical equivalent of heat, J joule/calorie.

The modern tendency is to divorce the definition of the calorie from properties of water by assigning an adopted value, J = 4.1840 joule/calorie. This is called the thermochemical calorie (page 5–4). The corresponding definition of the British thermal unit, originally defined as being the heat needed to raise the temperature of 1 lb of water 1°F becomes

$$1 \text{ Btu} = 251.996 \text{ calorie} = 1{,}054.8 \text{ joule}$$

Θ. Temperature (pages 5–7 and 5–39)

Originally the Celsius degree (°C) was defined as one one hundredth of the temperature interval on the absolute thermodynamic scale between the freezing point and the boiling point of water under one normal atmosphere of pressure. In 1954 it was decided to base the definition of the degree on the adopted value

$$T_0 = 273.16°\text{K}$$

for the triple point of water.

With the size of the degree so defined, temperatures stated as degrees Kelvin (°K) are measured from absolute zero, and those stated as degrees Celsius (°C) are measured from the triple point of water.

The Fahrenheit scale uses degrees whose size is $\frac{5}{9}$ the size of the Celsius degree and assigns the value 32 °F to the freezing point of water. Temperatures expressed in Fahrenheit degrees but measured from absolute zero are said to be expressed in degrees Rankine (°R).

In fundamental physics, temperature always is expressed as absolute temperature multiplied by a coefficient, k, known as the *Boltzmann constant*, so that the product kT is an energy quantity appearing as a parameter in the statistical distribution of atoms and molecules among the allowed energy levels (pages 5–14 and 5–21). Using deg to signify the Celsius degree (page 5–157)

$$k = (1.38044 \pm 0.00007) \times 10^{-16} \text{ erg/deg}$$

Closely related to k is the *molal gas constant* R defined as the product of the Avogadro number and the Boltzmann constant (pages 5–7 and 7–156)

$$R = Nk = (8.31696 \pm 0.00034) \text{ joule/deg-mole}$$

Θ^{-1}. Expansion Coefficient

Linear expansion coefficient is the fractional increase in length per degree increase in temperature. Volume expansion coefficient is the analogous fractional increase in volume, and for isotropic solids this is three times the linear coefficient. The expansion coefficient per °F is 0 that per °C.

For ideal gases the volume coefficient of expansion at absolute temperature T is T^{-1}.

For liquids the convenient unit is thousandths per °C. Some illustrative values at 20°C are

Substance	*Volume coefficient, thousandths per °C*
Mercury	0.18176
Water	0.207
Acetone	1.487
Ether	1.656

For solids the convenient unit is millionths per °C. Some illustrative values are

Substance	*Linear coefficient, millionths per °C*
Diamond, at 40°C	1.18
Aluminum, at 20°C	25.5
NaCl, at 40°C	40.4

$QM^{-1}\Theta^{-1}$. Specific Heat, Molal Heat Capacity (page 5–5). Specific Entropy, Molal Entropy (pages 5–6 and 5–116)

Specific heat, c, is the ratio of heat added, Q to the temperature rise produced, T, under some specified process such as at constant volume (c_v) or constant pressure (c_p) and for unit mass of the material. Molal heat capacity is Mc, where M is the molecular weight. Units are energy per unit mass (or mole) per degree

$$1 \text{ Btu/lb °F} = 1 \text{ cal/g°C}$$

Of fundamental theoretical significance for the molal heat capacity is the molal gas constant (page 7–156)

$$R = Nk = (8.31696 \pm 0.00034) \text{ joule/deg-mole}$$

Illustrative values of specific heat at constant pressure (c_p):

Substance	Temperature, °K	c_p, cal/g °C
Gases		
He	93	1.25
H_2	288	3.389
O_2	288	0.2178
N_2	288	0.2477
NH_3	288	0.5232
CO_2	288	0.1989
Liquids		
H_2O	288	1.0000
Hg	293	0.03325
C_6H_6	278	0.389
Solids		
Al	293	0.214
Diamond	273	0.1044
Cu	273	0.0910
Pb	273	0.0297

Entropy increase, dS, involved in the reversible addition of heat δQ to a system at temperature T is $dS = \delta Q/T$ (pages 5–6 and 5–116). Hence the entropy change of unit mass (or unit molal quantity) is expressed in the same units as heat capacity. Fundamental theory (page 5–22) shows that k is the natural unit of entropy per molecule and therefore R is the natural unit of entropy per mole.

$Q\Theta^{-1}L^{-1}T^{-1} = (QL^{-1}T^{-1})/(\Theta L^{-1})$. Thermal Conductivity

Thermal conductivity (page 5–61) is defined as the coefficient which multiplies the temperature gradient to give the rate of heat transfer by conduction expressed in heat energy crossing unit area in unit time.

For conversion factors and illustrative values, see pages 5–61 and 5–62.

ELECTROMAGNETIC QUANTITIES

Q and QT^{-1}. Charge and Current

Electric charge and electric current, that is, time rate of passage of charge in a linear conductor, are the most basic electrical quantities, in terms of which all others are defined (pages 4–3 and 4–13).

Electrostatic units define a measure of quantity of electric charge in relation to force and length units and an arbitrarily adopted value of ϵ_0 in Eq. (1.1) (page 4–3). Unit charge is that which, when at unit distance from another unit charge, repels it with a force of $4\pi\epsilon_0$ units of force. Unit current is then defined as that which passes unit charge in unit time. Thus the size of the unit varies as $\epsilon_0^{1/2}$ and the general dimensional relation is

$$Q_e = \epsilon_0^{1/2} M^{1/2} L^{3/2} T^{-1}$$

The cgs unit of charge is called the *esu* or the *statcoulomb*. It is the unit defined by using dynes of force and cm of distance and adopting $4\pi\epsilon_0 = 1$. The corresponding unit of current is called the *statampere*, defined as the flow of one statcoulomb per second.

Electromagnetic units define electric current in relation to force and length units and an arbitrarily chosen μ_0 in Eq. (1.65) (page 4–12). Suppose two rigid conducting circuits are so related that the double integral there (dimensionless) equals S, then unit current is defined so that when unit current flows in each circuit, the magnetic force of interaction between them is $4\pi/\mu_0 S$.

Thus the size of unit current depends on μ_0 as $\mu_0^{-1/2}$, and the general dimensional relation for the current so defined is

$$Q_m T^{-1} = \mu_0^{-1/2} M^{1/2} L^{1/2} T^{-1}$$

giving for the magnetic definition of charge

$$Q_m = \mu_0^{-1/2} M^{1/2} L^{1/2}$$

The cgs electromagnetic unit of current is called the *abampere*, and is the unit defined on using dynes of force and choosing $\mu_0/4\pi = 1$ in the general definition. The corresponding unit of charge, called the *abcoulomb*, is defined as that which passes when one abampere flows for one second.

It follows from the dimensional formulas that the ratio of the electrostatic and electromagnetic measures of the same charge has the form

$$Q_e/Q_m = (\epsilon_0\mu_0)^{1/2} L T^{-1}$$

and therefore

$$1 \text{ abcoulomb} = c \text{ statcoulomb}$$

where c is a conversion factor measured in cm/sec. This is found experimentally to have the value

$$c = 2.997930 \times 10^{10} \text{ cm/sec}$$

the same as the velocity of light (page 7–156).

Using any units of M, L, and T, and any choice of ϵ_0 and μ_0 one can define a unit which may be denoted as the Statcoulomb and another which may be called the Abcoulomb; the relation between these is

$$1 \text{ Abcoulomb} = c\,(\epsilon_0\mu_0)^{1/2} \text{ Statcoulomb}$$

where c stands for this same velocity expressed in the LT^{-1} units being used.

Meter-kilogram-second (*mks*) units of charge and current are based on the use of the meter, kilogram and second as units of length, mass and time, and also on a particular choice of ϵ_0 and μ_0 that is designed to make the units of charge and current have the same magnitude on either definition, and to agree with the already well-established coulomb and ampere units of practical electrical usage.

This is accomplished by the choices

$$\frac{4\pi}{\mu_0} = 10^{-7} \text{ and } 4\pi\epsilon_0 = \frac{10^7}{c^2}$$

where c is the velocity of light expressed in m/sec. The coulomb, by these definitions, becomes exactly $\frac{1}{10}$ abcoulomb, giving:

$$1 \text{ coulomb} = \tfrac{1}{10} \text{ abcoulomb} = 2.997930 \times 10^9 \text{ statcoulomb}$$

Atomic Unit

Electric charge seems always to occur in nature in integral multiples of the *electronic charge* (pages 7–25 and 7–156):

$$\begin{aligned} e &= (4.80286 \pm 0.00009) \times 10^{-10} \text{ esu} \\ &= (1.60206 \pm 0.00003) \times 10^{-19} \text{ coulomb} \end{aligned}$$

Electric current has a natural atomic unit corresponding to the passage of one electron per atomic unit of time (page 7–24):

$$1.98558 \times 10^7 \text{ esu/sec} = 6.62319 \text{ milliampere}$$

The *Faraday* is the charge associated with one gram mole of any singly charged ion (pages 4–147, 7–25, and 7–157)

$$\begin{aligned} \text{F} = \text{Ne} &= (2.89366 \pm 0.00003) \times 10^{14} \text{ esu/mole} \\ &= 96521.9 \pm 1.1 \text{ coulomb/mole} \end{aligned}$$

The Faraday may also be regarded as the charge-to-mass ratio of a particle having one electronic charge and a mass of one atomic mass unit.

Charge and Current Densities and Their Moments

A wide variety of auxiliary quantities relating to linear, surface and volume charge density, having dimensions QL^{-1}, QL^{-2}, QL^{-3}, and also of current density, $QT^{-1}L^{-2}$, and their moments, occur in the analysis of electromagnetic phenomena. Units for these can be built up easily from appropriate charge and current units in combination with length units.

Atomic unit of charge density is one electronic charge per cubic Bohr radius

$$e/a^3 = 3.24126 \times 10^{15} \text{ esu/cm}^3$$

Unit *dipole moment* (page 4–105) is defined as the strength of a dipole consisting of unit + and − charges separated by unit distance

$$\begin{aligned} 1 \text{ debye} &= 10^{-18} \quad \text{esu-cm} \\ ea &= 2.54155 \text{ debyes} \end{aligned}$$

Unit *quadrupole moment* (pages 4–17 and 4–18) is of the form QL^2 and the atomic unit is

$$ea^2 = 1.34491 \times 10^{-26} \text{ esu-cm}^2$$

$Q^{-1}ML^2T^{-2}$. Electromotive Force (page 4–10). Electric Potential (page 4–4)

Electromotive force refers to the work done in moving unit electric charge around a closed circuit. Electric potential is defined as the work done against the electrostatic forces of an electric field in moving unit charge from a reference position to the final position. Both are thus expressed in units of energy per unit charge

cgs esu: 1 statvolt = 1 erg/esu
cgs emu: 1 abvolt = 1 erg/abcoulomb
mks 1 volt = 1 joule/coulomb
1 statvolt = 299.7930 volts = 2.997930 $\times$ 10^{10} abvolts

Atomic Unit (page 7–25)

Potential at a distance of one Bohr radius from one electronic charge is

$$9.07623 \times 10^{-2} \text{ statvolts} = 27.20989 \text{ volts}$$

Potential at one electron radius, e^2/mc^2, from one electron is equal to the rest energy, mc^2 per electronic charge, equals 0.510984×10^6 volts.

$Q^{-1}MLT^{-2}$. Electric Vector, or Electric Field Strength

This is defined (page 4–4) as the electric force per unit charge acting on a small test charge in the field. The electrostatic measure of the field at distance r from a charge q in electrostatic units is $q/4\pi\epsilon_0 r^2$

cgs esu: 1 statvolt/cm = 1 dyne/esu
cgs emu: 1 abvolt/cm = 1 dyne/abcoulomb
mks: 1 volt/m = 1 newton/coulomb
1 statvolt/cm = 29979.30 volt/m = 2.997930 $\times$ 10^{10} abvolt/cm

Atomic Unit (page 7–25)

The electric field at a distance of one Bohr radius from one electronic charge is

$$1.71518 \times 10^7 \text{ statvolt/cm} = 5.19199 \times 10^{11} \text{ volt/m}$$

$QL^{-2} = (QL)/L^3$. Dielectric Polarization. Dielectric Displacement

Dielectric polarization (pages 4–5 and 4–105) is defined as the vector resultant electric dipole moment contained in unit volume

$$\begin{aligned} 1 \text{ coulomb/m}^2 &= 10^{-5} \text{ abcoulomb/cm}^2 \\ &= 2.997930 \times 10^5 \text{ esu/cm}^2 \end{aligned}$$

Dielectric displacement (page 4–20) usually denoted by **D**, is also a field quantity of the type QL^{-2}. If a plane conductor has a charge of σ units of charge per unit area, this is numerically equal to the magnitude of **D** just outside the conductor and moreover **D** is normal to the conducting surface.

A direct interpretation of ϵ_0 may be given: start with an adopted unit of electric charge, which leads to a unit of electric field, **E**, as that field in which unit force is exerted on unit charge. This unit field can be produced by equal and opposite charges per unit area on the plates of a parallel plate condenser, and ϵ_0 is the numerical value of the requisite surface charge density. To get a field of 1 dyne/esu, a charge density of $(4\pi)^{-1}$ esu/cm² is needed on the plates, since $4\pi\epsilon_0 = 1$ in the cgs esu system. To get a field of 1 newton/coulomb (= 1 volt/meter) a charge density of $10^7/4\pi c^2$ coulomb/m² is needed on the plates. Hence the value of ϵ_0 in the mks system may be said to be expressed in farad/m, since the farad is the same as the coulomb/volt.

$Q^2M^{-2}L^{-2}T^2$. **Capacitance** (page 4–20)

Capacitance of an isolated conductor is the ratio of the charge on it to the potential to which it is raised by the presence of that charge. The formulas (2.29) to (2.33) on (page 4–21) are written in the special form applicable to cgs esu. To give them more general form replace ϵ by $4\pi\epsilon\epsilon_0$.

The units of capacitance on the three systems are

cgs esu: 1 statfarad (= 1 cm) = 1 statcoulomb/statvolt
cgs emu: 1 abfarad = 1 abcoulomb/abvolt
mks: 1 farad = 1 coulomb/volt

These are related in this way:

$$1 \text{ abfarad} = 10^9 \text{ farad}$$
$$1 \text{ farad} = (2.997930)^2 \times 10^{11} \text{ cm}$$

$Q^{-2}ML^2T^{-1}$. **Resistance** (page 4–9)

The resistance of a conductor obeying Ohm's law is the potential difference between two contact electrodes needed to maintain unit current

cgs esu: 1 statohm = 1 statvolt/statampere
cgs emu: 1 abohm = 1 abvolt/abampere
mks: 1 ohm = 1 volt/ampere

These are related as

$$1 \text{ statohm} = c^2/10^9 \text{ ohms} = c^2 \text{ abohms}$$

with

$$c = 2.997930 \times 10^{10} \text{ cm/sec}$$

$Q^2M^{-1}L^{-2}T$. **Conductance** (page 4–9)

The conductance of a conductor obeying Ohm's law is the current flowing through it when unit potential difference is maintained between two contact electrodes, and is therefore the reciprocal of resistance: a conductor having a resistance R units has a conductance R^{-1} of the corresponding conductance unit

cgs esu: 1 statmho = 1 statampere/statvolt
cgs emu: 1 abmho = 1 adampere/abvolt
mks: 1 mho = 1 ampere/volt

These are related as

$$c^2 \text{ statmho} = 10^9 \text{ mho} = 1 \text{ abmho}$$

$Q^{-2}ML^3T^{-1}$. **Resistivity** (pages 4–9, 4–72, and 4–149)

Resistivity, ρ, of an ohmic conducting material is a specific property of the material that is equal to the resistance of a right-cylindrical conductor, R, multiplied by its cross-sectional area, S, and divided by its length, s. The cgs esu and cgs emu are almost never used. The mks unit is 1 ohm-m = 100 ohm-cm. Most of the values of resistivity to be found in the literature are in ohm-cm, which is the unit applying in the definition if the resistance R is in ohms, the area S in cm² and the length s in cm.

Conductivity is the specific property of a conducting material that is reciprocal to resistivity, the unit corresponding to ohm-cm being the mho/cm. Thus a material having a resistivity of ρ ohm-cm has a conductivity of ρ^{-1} mho/cm.

Mass resistivity is the name given to the product $\rho_m = \rho d$, where ρ is the resistivity and d is the density. It can be used in the formula, $R = \rho_m s/\lambda$ where s is the length of a conductor and λ is its mass per unit length. The ordinary unit is ohm-g/cm² but for some practical work it is convenient to use

$$1 \text{ ohm-lb/mi}^2 = 1.7513 \times 10^{-8} \text{ ohm-g/cm}^2$$

as the unit which gives the resistance in ohms when multiplied by the length in miles and the linear density in lb/mi.

Illustrative values for resistivity:

Substance	Temperature, °C	Ohm-cm
Silver	20	1.59×10^{-6}
Copper	20	1.7241×10^{-6}
Aluminum	20	2.824×10^{-6}
Lead	20	$22. \times 10^{-6}$
Mercury	20	95.783×10^{-6}
NaCl (5 g in 100 cm³ H_2O)	18	14.94
HCl same conc.	18	2.54
$CuSO_4$ same conc.	18	52.9
Glass (page 8-93)	20	about 10^{14}

$Q^{-1}MT^{-1} = (MLT^{-2})/(QLT^{-1})$. **Magnetic Induction** (pages 4–12 and 4–15)

Magnetic induction, **B**, is also called magnetic flux density. Unit is derived from Eq. (1.70) giving **F** as force on unit length of a linear conductor carrying current **J** in a field **B**

$$\mathbf{F} = \mathbf{J} \times \mathbf{B}$$

or the relation for the force **F** on a moving charge (emu) e moving with velocity **v**

$$\mathbf{F} = e\mathbf{v} \times \mathbf{B}$$

The cgs unit is 1 *gauss*, used with **J** in abamperes to give **F** in dyne/cm, or with e in abcoulombs and

$\mathbf{v}$ in cm/sec to give $\mathbf{F}$ in dynes. In relation to flux units

1 gauss = 1 line/cm^2 = 1 maxwell/cm^2

The mks unit is called 1 *tesla,* used with $\mathbf{J}$ in amperes to give $\mathbf{F}$ in newton/m, or with e in coulombs and $\mathbf{v}$ in m/sec to give $\mathbf{F}$ in newtons

1 tesla = 10^4 gauss
= 1 volt/sec/m^2

$Q^{-1}ML^2T^{-1}$. **Magnetic Flux** (page 4–15)

Magnetic flux, Φ, defined by Eq. (1.85) is the surface integral of $\mathbf{B}$.

The cgs unit is called the *line* or the *maxwell*

1 line = 1 maxwell = 1 gauss-cm^2

The mks unit is called the *weber*

1 weber = 1 tesla-m^2 = 10^8 maxwell

These units are also named in relation to the law relating induced electromotive force to rate of change of flux.

1 maxwell = 1 abvolt-sec
1 weber = 1 volt-sec

$Q^{-1}MLT^{-1}$. **Magnetic Vector Potential** (page 4–13)

The magnetic vector potential, $\mathbf{A}$, is related to $\mathbf{B}$ by $\mathbf{B}$ = curl $\mathbf{A}$ Eq. (1.72) so its unit is any unit of $\mathbf{B}$ combined with the length unit involved in calculating the curl

1 weber/m = 10^6 gauss-cm

$Q^{-2}ML^2$. **Inductance** (page 4–16)

Inductance is the ratio of the flux through a circuit to the current flowing in a circuit which gives rise to the flux. If the circuits are separate it is called mutual inductance, if the same it is called self inductance.

The cgs unit is called the *cm* or *abhenry*

1 abhenry = 1 maxwell/abampere

The mks unit is called the *henry*

1 henry = 1 weber/ampere = 10^9 abhenry

As with magnetic flux, these units can be named in a way that indicates the induced electromotive force, $L(di/dt)$ due to changing current

1 abhenry = 1 abvolt-sec/abampere = 1 abohm-sec
1 henry = 1 volt-sec/ampere = 1 ohm-sec

QL^2T^{-1}. **Magnetic Moment** (page 4–14)

The magnetic moment of a current J flowing in a small plane circuit of vector area $\mathbf{A}$ is normal to the circuit in the direction related to the positive sense of current by the right-hand rule. Unit magnetic moment is that due to unit current in a circuit of unit area

1 ampere-m^2 = 10^3 abampere-cm^2

These units are also named in relation to the energy, $-\mathbf{m}\cdot\mathbf{B}$, of the magnetic moment $\mathbf{m}$ in a field $\mathbf{B}$

1 abampere-cm^2 = 1 erg/gauss
1 ampere-m^2 = 1 joule/(weber/m^2)

The magnetic dipole moment of the earth's permanent magnetization is approximately 10^{23} ampere m^2.

Atomic Units (pages 4–132, 7–25 and 7–156).

The Bohr magneton, defined as

$$\mu_0 = e\hbar/2mc = (0.92837 \pm 0.00002) \times 10^{-20} \text{ erg/gauss}$$

is equal to the magnetic moment of the electron except for small corrections arising from quantum electrodynamics.

The nuclear magneton, μ_n, is defined in the same way but using m_p, the proton mass, in place of m, the electron mass

$$\mu_n = e\hbar/2m_pc = \mu_0/1836.12 = (0.505038 \pm 0.000018) \times 10^{-23} \text{ erg/gauss}$$

Proton magnetic moment

$$\mu = (2.79275 \pm 0.00002)\mu_n = (1.41044 \pm 0.00004) \times 10^{-23} \text{ erg/gauss}$$

$QL^{-1}T^{-1}$. **Magnetization. Magnetic Moment Per Unit Volume**

Intensity of magnetization, $\mathbf{M}$, is defined Eq. (1.49) as the vector resultant magnetic moment in unit volume

cgs: 1 abampere-cm^2/cm^3 = 1 abampere/cm
mks: 1 ampere-m^2/m^3 = 1 ampere/m
1 ampere/m = 10^{-3} abampere/cm

$\mathbf{M}$ defined in this way is measured in the same units as $\mathbf{H}$, magnetic field.

$QL^{-1}T^{-1}$. **Magnetic Field**

The magnetic field vector, $\mathbf{H}$, is also expressed as current per unit length, in the same units as magnetization, $\mathbf{M}$.

Unit magnetic field is that produced inside a long solenoid in which a unit number of current-turns per unit of axial length is flowing. For this reason it is customary to write the units as 1 abampere-turn/cm and 1 ampere-turn/m respectively.

The *oersted* is another unit of $\mathbf{H}$ defined by the relations

1 oersted = $1/4\pi$ abamp/cm = $10^3/4\pi$ amp/m

Since in the cgs system, $\mu_0 = 4\pi$, in vacuum a magnetic field of 1 oersted is associated with a flux density of 1 gauss. (In older literature, *oersted* has a quite different meaning as a unit of reluctance.)

A direct interpretation of μ_0 may be given, paralleling that given for ϵ_0 under QL^{-2}, Dielectric Displacement: Starting with an adopted unit of electric current, this leads to a definition of unit flux density as the case in which a linear conductor, normal to the flux density, carrying unit current, experiences unit force per unit length. This unit flux density can be produced in a long solenoid in which the number of current-turns per unit length is μ_0^{-1}. Thus to get a field of 1 gauss, the current must be $(4\pi)^{-1}$ abampere-turn/cm, since here $\mu_0 = 4\pi$. To get a field of 1 tesla, since here $\mu_0 = 4\pi \cdot 10^{-7}$, the solenoid must have $10^7/4\pi$ ampere-turn/m.

In the relation $\mathbf{B} = \mu_0\mathbf{H}$, since the mks unit **B** can be called 1 volt-sec/m², and the unit **H** is 1 ampere/m, an appropriate name for the permeability, μ_0, is $\mu_0 = 4\pi \cdot 10^{-7}$ henry/m.

QT^{-1}. Magnetomotive Force. Magnetic Potential

Magnetomotive force is defined as the line integral of **H** between two points. Its units are therefore 1 abampere (cgs) and 1 ampere (mks). The magnetomotive force associated with a closed path is equal to the free current that passes through a surface bounded by the closed path. With **H** in oersteds, the related unit is

$$1 \text{ gilbert} = 1 \text{ oersted-cm} = (4\pi)^{-1} \text{ abampere}$$

The same units apply to the multiple-valued scalar potential, Ω, in $\mathbf{H} = -\operatorname{grad}\Omega$.

$Q^2M^{-1}L^{-2}$. Reluctance

Reluctance in a magnetic circuit is defined as the ratio of the magnetomotive force around the circuit to the magnetic flux in the circuit.

cgs: 1 abampere/turn/maxwell = 4π gilbert/maxwell
mks: 1 ampere-turn/weber = 10^{-9} abamp-turn/maxwell

(In the older literature, a reluctance of 1 gilbert/maxwell was called 1 oersted.)

Index